VAN NOSTRAND'S SCIENTIFIC ENCYCLOPEDIA

Sixth Edition

VAN NOSTRAND'S SCIENTIFIC ENCYCLOPEDIA

Sixth Edition

VOLUME I

Animal Life
Biosciences
Chemistry
Earth and Atmospheric Sciences
Energy Sources and Power Techonology
Mathematics and Information Sciences
Materials and Engineering Sciences
Medicine, Anatomy, and Physiology
Physics
Plant Sciences
Space and Planetary Sciences

DOUGLAS M. CONSIDINE, P.E.
Editor

GLENN D. CONSIDINE
Managing Editor

VNR VAN NOSTRAND REINHOLD COMPANY
NEW YORK CINCINNATI TORONTO LONDON MELBOURNE

Library of Congress Catalog Card Number: 82–4936
ISBN: 0–442–25161–0
ISBN: 0-442-25168-8 (Volume I)
ISBN: 0-442-25166-1 (Volume II)
ISBN: 0-442-25164-5 (2 volume set)

Manufactured in the United States of America

Published by Van Nostrand Reinhold Company Inc.
135 West 50th Street, New York, N.Y. 10020

Van Nostrand Reinhold Publishing
1410 Birchmount Road
Scarborough, Ontario M1P 2E7, Canada

Van Nostrand Reinhold Australia Pty. Ltd.
17 Queen Street
Mitcham, Victoria 3132, Australia

Van Nostrand Reinhold Company Limited
Molly Millars Lane
Wokingham, Berkshire, England

15 14 13 12 11 10 9 8 7 6 5 4 3 2 1

Library of Congress Cataloging in Publication Data

Main entry under title:

Van Nostrand's scientific encyclopedia.

1. Science—Dictionaries. 2. Engineering—Dictionaries. I. Considine, Douglas Maxwell.
Q121.V3 1982 503'.21 82–4936
ISBN 0–442–25161–0 AACR2
ISBN 0-442-25168-8 (Volume I)
ISBN 0-442-25166-1 (Volume II)
ISBN 0-442-25164-5 (2 volume set)

REPRESENTATIVE TOPICAL COVERAGE

ANIMAL LIFE

Amphibians	Coelenterates	Mammals	Protozoa
Annelida	Echinoderms	Mesozoa	Reptiles
Arthropods	Fishes	Mollusks	Rotifers
Birds	Insects	Paleontology	Zoology

BIOSCIENCES

Amino Acids	Biophysics	Genetics	Proteins
Bacteriology	Cytology	Hormones	Recombinant DNA
Biochemistry	Enzymes	Microbiology	Viruses
Biology	Fermentation	Molecular Biology	Vitamins

CHEMISTRY

Acids and Bases	Corrosion	Inorganic Chemistry	Oxidation-Reduction
Catalysts	Crystals	Ions	Photochemistry
Chemical Elements	Electrochemistry	Macromolecular Science	Physical Chemistry
Colloid Systems	Free Radicals	Organic Chemistry	Solutions and Salts

EARTH AND ATMOSPHERIC SCIENCES

Climatology	Geodynamics	Hydrology	Tectonics
Ecology	Geology	Meteorology	Seismology
Geochemistry	Geophysics	Oceanography	Volcanology

ENERGY SOURCES AND POWER TECHNOLOGY

Batteries	Electric Power	Nuclear Energy	Steam Generation
Biomass and Wastes	Geothermal Energy	Ocean Energy Resources	Tidal Energy
Coal	Hydroelectric Power	Petroleum	Turbines
Combustion	Natural Gas	Solar Energy	Wind Power

MATHEMATICS AND INFORMATION SCIENCES

Automatic Control	Computing	Measurements	Statistics
Communications	Data Processing	Navigation and Guidance	Units and Standards

MATERIALS AND ENGINEERING SCIENCES

Chemical Engineering	Laser Technology	Mining	Process Engineering
Civil Engineering	Mechanical Engineering	Microelectronics	Structural Engineering
Glass and Ceramics	Metallurgy	Plastics and Fibers	Transportation

MEDICINE, ANATOMY, AND PHYSIOLOGY

Brain and Nervous System	Genetic Disorders	Ophthalmology
Cancer and Oncology	Gerontology	Otorhinolaryngology/Dental
Cardiovascular System	Hematology	Parasitology
Chemotherapy	Immunology	Pharmacology
Dermatology	Infectious Diseases	Reproductive System
Diagnostics	Kidney and Urinary Tract	Respiratory System
Digestive System	Mental Illness	Rheumatology
Endocrine System	Muscular System	Skeletal System

PHYSICS

Atoms and Molecules	Gravitation	Optics	Subatomic Particles
Electricity	Magnetism	Radiation	Surfaces
Electronics	Mechanics	Solid State	Theoretical Physics
Fluid State	Motion	Sound	Waves

PLANT SCIENCES

Agriculture	Diseases and Pests	Growth Modifiers	Seeds and Germ Plasm
Algae	Fruits	Nutritional Values	Trees
Botany	Fungi	Plant Breeding	Yeasts and Molds

SPACE AND PLANETARY SCIENCES

Astrochemistry	Astronautics	Astrophysics	Probes and Satellites
Astrodynamics	Astronomy	Cosmology	Solar System

Preface

To the tens of thousands of regular users of this encyclopedia, it will be of interest to compare this Sixth Edition with the prior edition. Since the last edition was published in 1976, there has been unusual progress in a number of scientific fields. To accommodate the reporting of these achievements, as well as to increase the overall scope of scientific coverage, the Sixth Edition has been expanded to accommodate about 30% more pages, representing an increase of over 700 pages. Nearly 100 entirely new topics have been added—with a coverage of over 7300 separate editorial entries. The policy of providing generous cross references, which make the encyclopedia virtually self-indexing, has been extended—with a total of nearly 9600 such entry referrals, up by over 1500 cross references from the last edition. The Sixth Edition contains over 2550 diagrams, graphs, and photographs as well as nearly 600 tables—also representing significant increases from the prior edition. Introduced for the first time for the Fifth Edition, the policy of including lists of references with the many hundreds of major entries has been continued and expanded—with several thousand references listed throughout the encyclopedia. Thus, in addition to providing comprehensive descriptions of numerous topics, the encyclopedia also can serve as a bridge to further information for those readers who are seeking the ultimate of detail.

From the list of acknowledgments which follows this preface, it will be noted that over 200 authors and advisors contributed to the content of this encyclopedia, not to mention the activities of the staff. As in the past, an effort was made to reflect worldwide scientific achievement in the book.

For the first time, this Sixth Edition includes both metric and English units for the convenience of readers worldwide.

As with past editions, much emphasis has been placed on scientific fundamentals and the logical ordering and classification of information. It is believed that many users of the encyclopedia not only seek detailed data on numerous subjects, but also expect to find well-organized overviews so that any subsequent researching of periodicals and specialized shelf literature can be pursued in the most workmanlike and time-saving manner. Obviously, an encyclopedia of this type cannot serve the same purpose as a news medium. Science, too, has its own noise level. This is particularly evident from the hundreds of prematurely announced and exaggerated claims one frequently finds in the general communications media. Consequently, the authors and editors of this encyclopedia must carefully sift through the vast scientific data bank and sort out the trivia from real progress. Even though culling of the unproven may make a book a bit less interesting and exciting to read, there is no room for rumor and the untried and very little room indeed for the controversial in a permanent scientific reference such as this. For example, the broad economic and social aspects of science, in its growing interface with society in general, are best left to competent coverage by the weekly and monthly scientific journals of repute. Consequently, although the temptations are often great, the authors and editors of this book avoid editorializing.

Although the editorial approach varies somewhat from one kind of entry to the next, the well-founded practice of commencing with the general and working toward the specifics is followed. This is particularly true in connection with hundreds of the broader, longer entries on topics of wide scope. For many of the shorter, highly specific entries, it is assumed that the reader who will seek information on such terms already possesses a reasonable topical background, but in many of these cases, readers who are less well informed are given the titles of other related entries in the encyclopedia.

A very abridged list of topics, which have been markedly expanded for the Sixth Edition and including some of the new topics, is given here:

Aquaculture; arterial and venous disorders; automotive electronics.
Battery; biomass and wastes as energy sources; brain and nervous system.
Cancer research; cardiac arrhythmias; coal conversion processes; communications.
Data processing; dietary requirements and trends; digital computer.
Earthquakes, seismology, and plate tectonics; earth resources satellites and geologic remote sensors; echocardiography; electron beam lithography.
Foodborne diseases; flavorings; fuel cells.
Galaxies; genes; genetics.
Halley's comet; heart and circulatory system; helicopters and V/STOL craft; holography; hormones; human factors engineering.
Information theory; infrared astronomy; input/output devices (computing system); insecticide and pesticide technology; insulation (thermal); irrigation; ischemic heart disease.
Josephson tunnel-junction; Jupiter.
Kidney and urinary tract.
Lasers; leavening agents; lightning; lipids; liquid crystals; lithium (for thermonuclear reactors); liver.
Magnetohydrodynamic generators; mathematics in transition; Mars; microelectronics; microstructure fabrication.

Natural gas; nematodes; Neptune; neutron; nondestructive testing; nuclear reactor.
Oceanography; optical fibers.
Particles (subatomic); photography and imagery; plant breeding; pollution (air); poultry; precipitation and hydrometeors.
Quarks; quasars.
Radioactivity; radioactivity and other dating techniques; radio and radar astronomy; radio communications; rare-earth elements and metals; recombinant DNA.
Satellites (communications); Saturn; signal processing devices; solar energy; space shuttle; Sun (The); superconducting electric generators; superconductors; sweeteners.
Telecommunications; telephony; transistors.
Ultrasonics; ultrasonography; ultraviolet astronomy; units and standards; Uranus.
Vegetable oils (edible); Venus; Viking mission to Mars; viruses; vitamins; voice recognition and synthesis; volcanology; Voyager mission to Jupiter and Saturn.
Water; water pollution; weather observations and forecasting; wind power.
X-ray analysis; x-ray astronomy; x-ray scanner (CAT).
Yeasts and molds.
Zinc; zinc in biological systems.

Although nearly all topics covered in the Fifth Edition have been expanded and strengthened in this edition, particular attention was given to the so-called natural sciences, including birds, fishes, insects, mammals, plants, and trees; and on a much greater scope in coverage of the medical sciences.

DOUGLAS M. CONSIDINE
Editor

Acknowledgments

Hundreds of scientists, engineers, and technologists worldwide have made this Sixth Edition of the Van Nostrand Scientific Encyclopedia a reality. The first edition, published in 1938, set a high standard for a broad-based scientific treatise compressed into one convenient volume.

Authorities from the many disciplines of science have assisted effectively and in numerous ways in the compilation of this work—ranging from fundamental information and counsel for the editorial staff to the preparation of comprehensive manuscripts on complex topics. Unfortunately, it is impractical to acknowledge individually the efforts of all persons who helped in some way to produce this volume. In particular, the major efforts of the following persons and organizations are gratefully acknowledged.

Further; recognition is made of the work provided by the authors and editors of earlier editions and for the sources of fundamental information, diagrams, and tables in certain selected and specialized fields from which summary excerpts are presented here. Notably these sources included: "The Encyclopedia of Biochemistry," Roger J. Williams and Edwin M. Lansford, Jr., Editors; "The Encyclopedia of Geochemistry and Environmental Sciences," Rhodes W. Fairbridge, Editor; "The Encyclopedia of Physics," Robert M. Besancon, Editor; "Grzimek's Animal Life Encyclopedia," Bernhard Grzimek, Editor-in-Chief; and the "Dictionary of Astronomy, Space and Atmospheric Phenomena," by David F. Tver—all books published by the Van Nostrand Reinhold Company. Appreciation is also extended to Omnibix U.S.A. for the nonexclusive use of "Spectrum of Various Energy Quantities," and the polar version of the periodic table, as well as most of the entries on the medical sciences. The editorial staff is also grateful to The Babcock and Wilcox Company for the use of material from the 39th edition of "Steam—Its Generation and Uses."

Acknowledgment is also made of several contributors who prepared and/or coordinated many entries in various topical areas: Dr. Thomas J. Harrison, IBM Corporation (computing and data processing); Peter Kraght, Certified Consulting Meteorologist (meteorology); Dr. Peter Pesch, Case Western Reserve University (several entries on astronomy); Elmer Rowley, formerly of Union College (mineralogy); W. G. Shequen, Bausch & Lomb-ARL (x-ray and associated technology); Dr. Karl A. Gschneidner, Jr. and B. Evans, Rare-Earth Information Center, Iowa State University (rare-earth elements); Dr. R. C. Vickery, Hudson Laboratories (several entries on physical chemistry); Dr. A. C. Vickery, University of South Florida (immunology and several entries on the biosciences); and Dr. A. L. Vincent, University of South Florida (parasitology).

Adlhart, O. J.
Engelhard Minerals and Chemicals Corporation
Iselin, New Jersey

American Forestry Association (The)
Washington, D.C.

Ames Research Center
National Aeronautics and Space Administration
Moffett Field, California

Baldwin, M. S.
Westinghouse Electric Corporation
East Pittsburgh, Pennsylvania

Bakos, J.
J. H. Fletcher & Company
Huntington, West Virginia

Bane, D.
Jet Propulsion Laboratory
Pasadena, California

Barr, R. Q.
Climax Molybdenum Company
Greenwich, Connecticut

Barrett, W. T.
Foote Mineral Company
Exton, Pennsylvania

Battelle Memorial Institute
Columbus, Ohio

Bell Laboratories
Murray Hill, New Jersey

Berman, A. I.
Rensselaer Polytechnic Institute
Troy, New York

Berring, H. (deceased)
Sangamo-Weston, Inc.
Newark, New Jersey

Bituminous Coal Research, Inc.
Monroeville, Pennsylvania

Blank, J.
GTE Laboratories Incorporated
Waltham, Massachusetts

Bloor, W. S.
Leeds and Northrup Company
North Wales, Pennsylvania

Bolton, R. S.
Ministry of Works
Wellington North, New Zealand

Bounds, C. O.
St. Joe Minerals Corporation
Monaca, Pennsylvania

Bouissières, G.
Institute of Nuclear Physics
University of Paris
Orsay, France

Brodersen, R. W.
College of Engineering
Dept. of Electrical Engineering and Computer Sciences
University of California
Berkeley, California

Brown, P. M.
Foote Mineral Company
Exton, Pennsylvania

Brown, V. M.
National Petroleum Council
Washington, D.C.

Browne, N. W.
Davy McKee (Oil & Chemicals) Ltd.
London, England

Bureau Internationale de l'Heure
Paris, France

Busker, L. H.
Beloit Corporation
Beloit, Wisconsin

Caianiello, E. R.
Instituto di Fisica Teorica
Università di Napoli
Naples, Italy

Canadian Petroleum Association
Calgary, Alberta, Canada

Carapella, S. C., Jr.
ASARCO Inc.
South Plainfield, New Jersey

Carlson, F. G.
The Oil Shale Corporation
Golden, Colorado

Carrigy, M. S.
Alberta Oil Sands Technology and Research Authority
Calgary, Alberta, Canada

Centre National de la Recherche Scientifique
Solar Energy Laboratory
Font Romeau, France

Cherry, R. H.
Consultant
Huntingdon Valley, Pennsylvania

Chiaviello, A.
Satellite Communications
Denver, Colorado

Comsat General Corporation
Washington, D.C.

Cook, P. H.
The Dow Chemical Company
Freeport, Texas

Cooper, G. R.
School of Electrical Engineering
Purdue University
West Lafayette, Indiana

Corrigan, D. A.
Handy & Harman
Fairfield, Connecticut

Coscia, A. T.
American Cyanamid Company
Stamford, Connecticut

Cronin, J. H.
Westinghouse Electric Corporation
East Pittsburgh, Pennsylvania

Crossman, A. B.
Brown & Root, Inc.
Houston, Texas

Cullen, V.
Woods Hole Oceanographic Institution
Woods Hole, Massachusetts

Dailey, W. W.
Mine Safety Appliances Company
Pittsburgh, Pennsylvania

DeCraene, D. F.
Chemetals Corporation
Baltimore, Maryland

Degenhard, W. E.
Carl Zeiss, Inc.
New York, N.Y.

Dennen, W. F.
University of Kentucky
Lexington, Kentucky

Desai, S. C.
Davy McKee Iron & Steel Division
Stockton-on-Tees, England

Dexter, D. L.
University of Rochester
Rochester, New York

Dickie, B.
Ministry of Mines and Minerals
Edmonton, Alberta, Canada

Dietz, E. D.
Owens-Illinois
Toledo, Ohio

Dietz, W.
Wacker Chemie, GMBH
Munich, West Germany

Dietl, J.
Wacker Chemie, GMBH
Munich, West Germany

Dilling, M. L. and W. L.
The Dow Chemical Company
Midland, Michigan

Doran, R. K.
Denver Research Institute
University of Denver
Denver, Colorado

Douglas, R. G.
State University of New York
Stony Brook, New York

Downing, R. C.
E. I. DuPont DeNemours & Co., Inc.
Wilmington, Delaware

Draeger, E. A.
McNally Pittsburg Mfg. Corp.
Pittsburg, Kansas

Dressler, H.
Koppers Company, Inc.
Monroeville, Pennsylvania

Duby, P. F.
Columbia University
New York, N.Y.

Dugger, G. L.
Applied Physics Laboratory
The Johns Hopkins University
Laurel, Maryland

Electric Power Research Institute
Palo Alto, California

Evans, B.
Rare-Earth Information Center
Iowa State University
Ames, Iowa

Evans, D. M.
Colorado School of Mines
Golden, Colorado

Faran, J. J., Jr.
GenRad, Inc.
Concord, Massachusetts

Fenninger, H.
Wacker Chemie, GMBH
Munich, West Germany

Fiber Controls, Inc.
Gastonia, North Carolina

File, J.
Plasma Physics Laboratory
Princeton University
Princeton, New Jersey

Fletcher, R.
J. H. Fletcher & Co.
Huntington, West Virginia

Flindt, F. C.
Monsanto Textiles Company
Decatur, Alabama

FMC Corporation
San Jose, California

Fredrick, L. W.
Leander McCormick Observatory
University of Virginia
Charlottesville, Virginia

Garman, J. A.
Great Lakes Chemical Corporation
West Lafayette, Indiana

Gilmour, I.
Polaroid Corporation
Cambridge, Massachusetts

Goddard Space Flight Center
Greenbelt, Maryland

Golden, J.
Environmental Research Laboratories
National Oceanic and Atmospheric Administration
Boulder, Colorado

Granger, L. R.
Military Sealift Command
Department of the Navy
Washington, D.C.

Greene, N. M.
Yale University School of Medicine
New Haven, Connecticut

Gregory, D. L.
Boeing Aerospace Company
Seattle, Washington

Groh, E. A.
Consulting Engineer
Portland, Oregon

Gschneidner, K. A., Jr.
Rare-Earth Information Center
Iowa State University
Ames, Iowa

Guggenheim, E. A.
The University
Reading, England

Halbmeyer, P. J.
Shell Internationale Petroleum Maatschappij B.V.
The Hague, Netherlands

Hall, W. S.
Amchem Products, Inc.
Ambler, Pennsylvania

Hamilton, R. C.
Cornell University
Ithaca, New York

Hansen, R. S.
Iowa State University
Ames, Iowa

Hanson, A. O.
University of Illinois
Urbana, Illinois

Harrison, T. J.
IBM Corporation
Boca Raton, Florida

Hartman, S. C.
Harvard University
Cambridge, Massachusetts

Haubert, J. L.
Westinghouse Electric Corporation
Pittsburgh, Pennsylvania

Havemann, W.
Carl Zeiss, Inc.
New York, N.Y.

Heinemeyer, B. W.
The Dow Chemical Company
Freeport, Texas

Herdeg, P. M.
Case Western Reserve University
Cleveland, Ohio

Hewson, E. W.
Oregon State University
Corvallis, Oregon

Hildebrandt, A. F.
Energy Laboratory
University of Houston
Houston, Texas

Hilton, P. J.
Department of Mathematics and Statistics
Case Western Reserve University
Cleveland, Ohio

Hluchan, S. E.
Pfizer Inc.
Wallingford, Connecticut

Hodge, D. R.
Alexandria, Virginia

Hodges, J. E.
American Petroleum Institute
Washington, D.C.

Hodges, R. E.
University of Iowa
Iowa City, Iowa

Hoogendorn, J. C.
South African Coal, Oil and Gas Corp., Ltd.
Sasolburg, Republic of South Africa

Hoover, L.
American Geological Institute
Washington, D.C.

Hopkins, H. S.
Olin Corporation
New Haven, Connecticut

Horvick, E. W.
Zinc Institute Inc.
New York, N.Y.

Humphreys, G. C.
Davy McKee (Oil & Chemicals) Ltd.
London, England

Institute of Gas Technology
Chicago, Illinois

Jacques, R. B.
Black Mesa Pipeline, Inc.
Flagstaff, Arizona

Jones, J. K.
Petrocarbon Developments Ltd.,
Manchester, England

Jones, T. O.
TRW Inc.
Cleveland, Ohio

Jet Propulsion Laboratory
California Institute of Technology
Pasadena, California

Kendall, Sir Maurice
International Statistical Institute
London, England

Keyes, R. W.
IBM Corporation
Thomas J. Watson Research Center
Yorktown Heights, New York

Kraght, P. E.
Consulting Meteorologist
Mabank, Texas

Kunasz, I. A.
Foote Mineral Company
Exton, Pennsylvania

Kupper, W.
Mettler Instrument Corporation
Hightstown, New Jersey

Lando, J. B.
Department of Macromolecular Science
Case Western Reserve University
Cleveland, Ohio

Lawrence, R. F.
Westinghouse Electric Corporation
East Pittsburgh, Pennsylvania

Lawrence, W.
Ethyl Corporation
Baton Rouge, Louisiana

Lebarbier, C.
Electricité de France
Paris, France

Lee, J. M.
Pullman Kellogg
Houston, Texas

Libby, L.
Simmons Refining Company
Chicago, Illinois

Lindal, B.
Virkir Consulting Group Ltd.
Reykjavik, Iceland

Lomartire, J.
Monsanto Textiles Company
New York, N.Y.

Madsen, E. W.
The Superior Electric Company
Bristol, Connecticut

Mamzic, C. L.
Moore Products Company
Spring House, Pennsylvania

Marshall Space Flight Center
National Aeronautics and Space Administration
Huntsville, Alabama

Massey, S.
The Babcock & Wilcox Company
Barberton, Ohio

Masson, J. R.
Davy McKee (Oil & Chemicals) Ltd.
London, England

McCown, W. R.
Westinghouse Electric Corporation
Pittsburgh, Pennsylvania

McGrath, J. J.
Mcdonnell Aircraft Company
St. Louis, Missouri

McIlhenny, W. F.
The Dow Chemical Company
Midland, Michigan

Meyer, R. A.
AT&T Long Lines
Bedminster, New Jersey

Moore, L. D.
PPG Industries, Inc.
Pittsburgh, Pennsylvania

Moran, M. K.
M & T Chemicals Inc.
Rahway, New Jersey

Moran, P. E.
Universal Oil Products Company
Des Plaines, Illinois

Morgan, W. L.
Communications Center of Clarksburg
Clarksburg, Maryland

Morris, R. A.
Society of Automotive Engineers, Inc.
Warrendale, Pennsylvania

Nojiima, S.
Japan Gasoline Company, Ltd.
Tokyo, Japan

Northeastern Forest Experiment Station
U.S. Department of Agriculture
Darby, Pennsylvania

Oak Ridge National Laboratory
Oak Ridge, Tennessee

Pasachoff, J. M.
Hopkins Observatory
Williams College
Williamstown, Massachusetts

Patchett, J. E.
Norton Research Corp. (Canada) Ltd.
Niagara Falls, Ontario, Canada

Pesch, P.
Warner and Swasey Observatory
Case Western Reserve University
East Cleveland, Ohio

Pfaender, L. V.
Owens-Illinois
Toledo, Ohio

Pierce, A. K.
Kitt Peak National Observatory
Tucson, Arizona

Plass, W. T.
Forest Service, U.S. Department of Agriculture
Princeton, West Virginia

Priddy, D. B.
The Dow Chemical Company
Midland, Michigan

Reincke, R. D.
Caterpillar Tractor Company
Peoria, Illinois

Rich, R. P.
Eastman Chemical Products, Inc.
Kingsport, Tennessee

Riddick, J. A.
Baton Rouge, Louisiana

Riley, J. C.
Metrologist and Consulting Engineer
Portland, Oregon

Rogers, T. H.
Elastomers Consultant
Clearwater, Florida

Romovacek, G. R.
Koppers Company, Inc.
Monroeville, Pennsylvania

Rowley, E. B.
Union College (formerly)
Schenectady, New York

Ross, D. M.
Propellants Consultant
Lancaster, California

Rudolph, P. F. H.
Lurgi Mineralotechnik, GMBH
Frankfurt (Main), West Germany

Sanderson, R. T.
Arizona State University
Flagstaff, Arizona

Sansonetti, S. J.
Reynolds Metals Company
Richmond, Virginia

Sargent, W. L. W.
Royal Greenwich Observatory
Sussex, England

Schappel, J. W.
Avtex Fibers Inc.
Front Royal, Virginia

Schiller, W. R.
Wacker Chemie, GMBH
Munich, West Germany

Schmidt-Nielsen, K.
Duke University
Durham, North Carolina

Schussler, M.
Fansteel
North Chicago, Illinois

Schweiker, G. C.
The PQ Corporatin
Lafayette Hill, Pennsylvania

Serson, P. H.
Division of Geomagnetism
Energy, Mines and Resources Canada
Ottawa, Canada

Shequen, W. G.
Bausch & Lomb-ARL
Sunland, California

Shock, N. W.
Baltimore City Hospital
Baltimore, Maryland

Shore, S. N.
Warner and Swasey Observatory
Case Western Reserve University
East Cleveland, Ohio

Shuman, E. C.
Consulting Engineer
State College, Pennsylvania

Simmons, L. E.
Simmons Refining Company
Chicago, Illinois

Slater, L. E.
The Food and Climate Forum
Aspen Institute for Humanistic Studies
Boulder, Colorado

Sleeman, D. C.
Davy McKee (Oil & Chemicals) Ltd.
London, England

Small, L. F.
Oregon State University
Corvallis, Oregon

Sperry, E.
Beckman Instruments, Inc.
Cedar Grove, New Jersey

Steffenson, M. R.
Parr Instrument Company
Moline, Illinois

Stephenson, C. B.
Warner and Swasey Observatory
Case Western Reserve University
East Cleveland, Ohio

Sterba, M. J. (deceased)
Universal Oil Products Company
Des Plaines, Illinois

Szepan, H. F.
Ingersoll-Rand (IMPCO Div.)
Nashua, New Hampshire

Taylor, C. H.
Woodall-Duckham Ltd.
Crawley, Sussex, England

Taylor, N. J. P.
BIF, a unit of General Signal Ltd.
West Warwick, Rhode Island

Terry, D. G.
Ingersoll-Rand (IMPCO Div.)
Nashua, New Hampshire

Tree, W.
Cornell University
Ithaca, New York

Troeger, W. A.
Weston Instruments Div., Sangamo-Weston, Inc.
Newark, New Jersey

Urry, L. F.
Battery Products Division
Union Carbide Corporation
Cleveland, Ohio

Van Den Berg, G. V.
Shell Internationale Petroleum Maatschappij B.V.
The Hague, Netherlands

Vermeulen, L. W.
Rio Algom Mines Ltd.
Toronto, Ontario, Canada

Vanderlugt, A.
Harris Corporation
Melbourne, Florida

Vant-Hull, L. L.
Energy Laboratory
University of Houston
Houston, Texas

Vickery, A. C.
College of Medicine
University of South Florida
Tampa, Florida

Vickery, R. C.
Hudson Laboratories
Hudson, Florida

Vincent, A. L.
Department of Comprehensive Medicine
University of South Florida
Tampa, Florida

Walker, J. R.
DeVlieg Machine Company
Birmingham, Michigan

Walton, J. D., Jr.
Georgia Institute of Technology
Atlanta, Georgia

Walsh, K. A.
Brush Wellman Inc.
Elmore, Ohio

Welsh, J. Y.
Chemetals Corporation
Baltimore, Maryland

White, R. M.
College of Engineering
Dept. of Electrical Engineering and Computer Sciences
University of California
Berkeley, California

Williams, E.
Cobalt Information Centre
London, England

Woodcock, G. R.
Boeing Aerospace Company
Seattle, Washington

Ziomek, J. F.
TRW, Inc.
Farmington Hills, Michigan

VAN NOSTRAND'S SCIENTIFIC ENCYCLOPEDIA

Sixth Edition

A

AA. An Hawaiian term introduced into geological nomenclature by C. E. Dutton, in 1883, and signifying the jagged, scoriaceous, blocky and exceedingly rough surface of some basic lava flows. Pronounced *ah-ah*.

AARD-VARK (*Mammalia, Tubulidentata*). African animals of peculiar form and ancient lineage, including an Ethiopian and a South African species. All are anteaters, feeding exclusively on ants and termites, nocturnal in habit, with acute hearing. The southern species has been called the ant bear. The aard-vark is the only living representative of its order. The animal's spine, curved from neck to tail in a near-half circle, gives it a truly prehistoric appearance.

AARD-WOLF. Hyena.

ABACA. The sclerenchyma bundles from the sheathing leaf bases of *Musa textilis*, a plant closely resembling the edible banana plant. These bundles are stripped by hand, after which they are cleaned by drawing over a rough knife. The fiber bundles are now whitish and lustrous, and from six to twelve feet (1.8–3.6 meters) long. Being coarse, extremely strong and capable of resisting tension, they are much used in the manufacture of ropes and cables. Since the fibers swell only slightly when wet, they are particularly suited for rope which will be used in water. Waste manila fibers from rope manufacture and other sources are used in the making of a very tough grade of paper, known as manila paper. The fibers may be obtained from both wild and cultivated plants, the latter yielding a product of better grade. The cultivated plants, propagated by seeds, by cuttings of the thick *rhizomes* or by suckers, are ready for harvest at the end of three years, after which a crop may be expected approximately every three years.

ABACUS. Calculator (Abacus).

ABALONE (*Mollusca, Gasteropoda; Haliotis*). Marine species, mostly of the Pacific and Indian Oceans. The single broad shallow shell has a richly colored iridescent inner surface and is an important source of mother-of-pearl and blister pearls for costume jewelry. The flesh is palatable. See **Mollusks.**

ABALONE POISONING. Foodborne Diseases.

ABBE CONDENSER. A compound lens used for directing light through the object of a compound microscope. All the light enters the object at an angle with the axis of the microscope. See also **Microscope.**

ABBE NUMBER. The reciprocal of the dispersive power of a material. It is also called the ν-number.

ABBE REFRACTOMETER. Refractometers.

ABBE SINE CONDITION. The relationship

$$ny \sin\theta = n'y' \sin\theta',$$

where n,n' are refractive indices, y,y' are distances from optical axis, and θ,θ' are angles light rays make with the optical axis. A failure of an optical surface to satisfy the sine condition is a measure of the coma of the surface.

ABDOMEN. The abdomen is the posterior division of the body in many arthropods. It is the *posterior* portion of the trunk in vertebrates. In the vertebrates this region of the body contains most of the alimentary tract, the excretory system, and the reproductive organs. It contains part of the coelom and in mammals is separated from the thorax by the diaphragm.

The abdominal cavity of the human body is subdivided into thc abdomen proper and the pelvic cavity.

The walls of the abdominal cavity are lined with a smooth membrane called the peritoneum, which also provides partial or complete covering for the organs within the cavity.

The abdomen proper is bounded above by the diaphragm; below it is continuous with the pelvic cavity; posteriorly it is bounded by the spinal column, and the back muscles; and on each side by muscles and the lower portion of the ribs. In front, the abdominal wall is made up of layers of fascia and muscles. The abdomen is divided into nine regions whose boundaries may be indicated by lines drawn on the surface. The mid-section above the navel between the angle of the ribs is known as the epigastric region; that portion around the navel, as the umbilical; below the navel and above the pubic bone, as the hypogastric region. It is further divided into right and left upper quadrants on each side above the navel, and right and left lower quadrants on each side below the navel. The lumbar region extends on either side of the navel posteriorly and laterally.

The principal organs of the abdominal cavity are the stomach, duodenum, jejunum, ileum, and colon or large intestine, the liver, gall bladder and biliary system, the spleen, pancreas and their blood and lymphatic vessels, lymph glands, and nerves, the kidneys and ureters.

The pelvic portion of the abdomen contains the sigmoid colon and rectrum, a portion of the small intestine, the bladder, in the male the prostate gland and seminal vesicles, in the female the uterus. Fallopian tubes and ovaries.

ABEL EQUATION. A mass point moves along a smooth curve in a vertical plane and under the influence of gravity alone. Given the time, t, required for the particle to fall from a point, x, to the lowest point on the curve as a function of x, what is the equation of the curve? The problem leads to a Volterra integral equation of the first kind

$$f(x) = \int_0^x \frac{\phi(t)\,dt}{\sqrt{2g(x-t)}}$$

where g is the acceleration of gravity. The solution is

$$\phi(x) = \frac{\sqrt{2g}}{\pi}\int_0^x \frac{f'(t)\,dt}{\sqrt{x-t}}$$

and the equation of the curve is

$$y = \int_0^x \sqrt{|\phi^2(t) - 1|}\,dt$$

A closely related problem is that of the brachistochrone, where the path is required for a minimum time of descent. Such matters were of considerable interest to many seventeenth and eighteenth century mathematicians; the one described here was solved by the Norwegian, N. H. Abel (1802–1829). See also **Brachistochrone.**

A more general case of the Abel equation is

$$f(x) = \int_0^x (x-y)^{-\alpha}\phi(y)\,dy$$

where $f(x)$ is continuously differentiable for $x \geq 0$ and $0 < \alpha < 1$. The solution is

$$\phi(y) = \frac{\sin \alpha\pi}{\pi}\left[\int_0^y (y-x)^{\alpha-1} f'(x)\,dx + f(0)y^{\alpha-1}\right]$$

A first-order differential equation

$$y' = f_0(x) + f_1(x)y + f_2(x)y^2 + f_3(x)y^3$$

is also known as an Abel equation. When the $f_i(x)$ are given explicitly, the equation can often be converted into one of simpler type and solved in terms of elementary functions. In the general case the solution involves elliptic functions.

ABELIAN GROUP. A commutative group, namely such that $AB = BA$ where A, B are any two elements contained in it.

ABERRATION OF LIGHT. The apparent change of position of an object, due to the speed of motion of the observer. Care must be taken not to confuse this effect with that of parallax.

If a telescope, assumed to be stationary, is pointed at a source of light, the light that enters the object glass centrally and in the direction of the optic axis will pass through the telescope along that axis and emerge through the center of the eyepiece. If the telescope is in motion relative to the source, in any direction other than parallel to the optic axis, the light that enters centrally will emerge off the center of the eyepiece. If this light is to emerge centrally, the telescope must be tilted forward in the plane containing the direction of motion of the instrument and the source. The amount of tilt will depend on the direction of the source and the ratio of the speed of the telescope to the speed of light.

This aberrational effect was first announced by Bradley in 1726. He noticed that stars had apparent periodic motions with a period of one sidereal year, and that the character of the apparent motion depended upon the celestial latitude of the star. He correctly interpreted the effect as due to the motion of the earth about the sun. Statistical discussions of the observations of a large number of stars have shown that the maximum value of this aberration due to the earth's orbital motion is 20".47. This is known as the "aberration angle" or the "constant of aberration," and is given by

$$\kappa = \frac{2\pi a \operatorname{cosec} 1''}{cT(1-e^2)^{1/2}}$$

where a is the mean radius of the earth's orbit, c is the velocity of light, T is the length of the year in seconds, and e is the eccentricity of the orbit. An aberrational effect of about 0".3, at maximum, is observed, due to the rotation of the earth on its axis, and is given by

$$k = \frac{2\pi\rho \cos\phi \operatorname{cosec} 1''}{ct}$$

where ρ is the radius of the earth, ϕ is the latitude of the place, and t is the length of the day in seconds.

In 1871, Airy made a series of observations for determination of the aberration constant, using a telescope filled with water. Because the value of the index of refraction of water is about $1\frac{1}{3}$, Airy expected that the value of the aberration would be 27".3 when using the water-filled tube. He found, however, that the value was 20".5 no matter what substance was placed in the telescope. The result of this so-called "Airy's Experiment" caused much discussion, but was eventually explained on the basis of the Michelson-Morley experiment and the theory of relativity.

All observations, in which the positions of the stars are involved, must be corrected for aberration of light if the results are to be accurate to within 20". Both the motion of the earth about the sun and the rotation of the earth must be considered. The magnitude of the correction depends upon the celestial coordinates of the star, the position of the observer on the earth, and the date and time of observation.

ABERRATION (Optical). The failure of an optical system to form an image of a point as a point, of a straight line as a straight line, and of an angle as an equal angle. See also **Astigmatism; Chromatic Aberration; Coma (Optics); Curvature of Field (Optics); Spherical Aberration.**

ABIES. Fir Trees.

ABLATION (Geomorphology). Essentially, the wasting away of rocks; the separation of rock material and formation of residual deposits, as caused by wind action or the washing away of loose and soluble materials.

ABLATION (Glaciology). The combined processes (sublimation, melting, evaporation) by which snow or ice is removed from the surface of a glacier or snowfield. In this sense, the opposite of alimentation. Ablation also refers to the amount of snow or ice removed by the aforementioned processes (the opposite of accumulation). The term may be applied to reduction of the entire snow-ice mass, and may also include losses by wind action and by calving (the breaking off of ice masses). Air temperature is the dominant factor in controlling ablation. During the ablation season, an ablation rate of about two millimeters/hour is typical of most glaciers. An ablatograph is an instrument that measures the distance through which the surface of snow, ice, or firn changes, as caused by ablation, during a specific period.

ABLATION (Meteorite). The direct vaporization of molten surface layers of meteorites and tektites during flight.

ABLATION (Spacecraft). In the interest of cooling space vehicles upon re-entry into the earth's atmosphere, ablation is used to control the temperature of strongly heated surfaces, such as parts of combustion chambers or nose cones. The process usually consists of the use of surface layers of materials which by their fusion, followed often by evaporation, absorb heat.

The heat of ablation is a measure of the effective heat capacity of an ablating material. Numerically, this is the heating rate input divided by the mass loss rate which results from ablation. In the most general case, heat of ablation is given by

$$(q_c + q_r - \sigma\epsilon T_w^4)/\mathring{m}$$

where q_c is the convective heat transfer in the absence of ablation; q_r is the radiative heat transfer from hot gases to ablation material; $\sigma\epsilon T_w^4$ is the rate of heat rejection by radiation from external surface of ablation material; and $\mathring{m}$ is rate at which gaseous ablation products are injected into the boundary layer.

Heat of ablation is sometimes evaluated neglecting the heat rejected by radiation and as a result unrealistically high heats of ablation are obtained.

If $q_r < \sigma\epsilon T_w^4$, for moderate values of stream enthalpy h_s, the heat of ablation is given by

$$H_v + \eta(h_s - h_w)$$

where H_v is the heat required to cause a unit weight of mass to be injected into boundary layer; η is the blocking factor, with numerical value from about 0.2 to 0.6 depending on material and type of flow; and h_w is the enthalpy at wall temperature.

ABLATIVE MATERIALS. Insulation (Thermal).

ABNEY EFFECT. A shift in hue which is the result of a variation in purity and, therefore, in saturation. The Abney effect may be represented by chromaticity loci, of specified luminance, with the hue and brightness constant, when purity and, therefore, saturation are varied. It is a relationship, of psychophysical nature, between psychophysical specifications and color sensation attributes.

ABNEY MOUNT. Relfection Grating.

ABOMASUM. Digestive System (Ruminants).

ABO SYSTEM. Blood.

ABRAHAM'S TREE. Clouds and Cloud Formation.

ABRASION. All metallic and nonmetallic surfaces, no matter how smooth, consist of minute serrations and ridges which induce a cutting or tearing action when two surfaces in contact move with respect to each other. This wearing of the surfaces is termed abrasion. Undesirable abrasion may occur in bearings and other machine elements, but abrasion is also adapted to surface finishing and machining, where the material is too hard to be cut by other means, or where precision is a primary requisite.

ABRASION pH. A term originated by Stevens and Carron in 1948, "to designate the pH values obtained by grinding minerals in water." Abrasion pH measurements are useful in the field identification of minerals. The pH values range from 1 for ferric sulfate minerals, such as coquimbite, konelite, and rhomboclase, to 12 for calcium-sodium carbonates, such as gaylussite, pirssonite, and shortite. The recommended technique for determination abrasion pH is to grind, in a nonreactive mortar, a small amount of the mineral in a few drops of water for about one minute. Usually, a pH test paper is used. Values obtained in this manner are given in the left-hand column of the accompanying table. Another method, proposed by Keller et al. in 1963 involves the grinding of 10 grams of crushed mineral in 100 milliliters of water and noting the pH of the resulting slurry electronically. Values obtained in this manner are given in the right-hand column of the accompanying table. See also terms listed under **Mineralogy.**

ABRASION pH VALUES OF REPRESENTATIVE MINERALS

Mineral	pH by Stevens-Carron Method	pH by Keller et al. Method*
Coquimbite	1	
Melanterite	2	
Alum	3	
Glauconite	5	5.5*
Kaolinite	5, 6, 7	5.5*
Anhydrite	6	
Barite	6	
Gypsum	6	
Quartz	6, 7	6.5
Muscovite	7, 8	8.0
Calcite	8	8.4
Biotite	8, 9	8.5
Microcline	8, 9	8.0 9.0*
Labradorite		8.0 9.2*
Albite	9, 10	
Dolomite	9, 10	8.5
Hornblende	10	8.9
Leucite	10	
Diopside	10, 11	9.9
Olivine	10, 11	9.6*
Magnesite	10, 11	

* indicates more recent values published in literature.

References

Stevens, R. E. and M. K. Carron: Simple Field Test for Distinguishing Minerals by Abrasion pH, *American Mineralogist*, **33**, 31–49, 1948.

Keller, W. D., Balgord, W. D., and A. L. Reesman: Dissolved Products of Artifically Pulverized Silicate Minerals and Rocks, *Jrnl. Sediment. Petrol.* **33**(1), 191–204, 1963.

ABS (Acrylonitrile-Butadiene-Styrene) RESINS. Thermoplastic resins which are produced by grafting styrene and acrylonitrile onto a diene-rubber backbone. The usually preferred substrate is polybutadiene because of its low glass-transition temperature (just above −80°C). Where ABS resin is prepared by suspension or mass polymerization methods, stereospecific diene rubber made by solution polymerization is the preferred diene. Otherwise, the diene used normally is a high-gel or cross-linked latex made by a "hot-emulsion" process.

ABS resins possess an attractive balance of impact resistance, hardness, tensile strength, and elastic modulus properties. The temperature range is wide, from −40 to +107°C (−40 to +225°F). Other advantages include chemical resistance, high gloss, and nonstaining properties. The dimension stability of ABS is good and creep resistance is excellent. The resins show low water absorption or volume change at varying humidities. Commercially available ABS is in the form of custom color-matched compounded pellets, or granular resin for compounding or alloying with other plastics. A representative alloying ingredient is polyvinyl chloride (PVC). Almost all standard thermoplastic converting processes can be used with ABS plastics. Injection molded parts include telephone sets, refrigerator parts, plumbing fixtures and fittings, radio, television, and appliance housings, and auto parts. ABS can be extruded into sheet, pipe, and various cross sections. Thermoforming of large surface areas and deep draws from sheet stock is possible. Examples of parts which involve extrusion and subsequent thermoforming include lawn-mower housings, refrigerator liners, pipe and conduit, vehicle bodies, snowmobile shrouds, and camper bodies. The various thermoforming techniques applicable to ABS include plug and air assist, vacuum snapback, vacuum-plug forming, and drape forming.

The compatibility of ABS with other plastics makes them useful as impact modifiers and as processing additives with many other polymers to achieve a variety of final product specifications. Substitution of α-methyl styrene for styrene increases heat-distortion temperatures; or of methacrylonitrile for acrylonitrile improves barrier properties to gases such as carbon dioxide in connection with carbonated beverage containers. Over 75 grades of ABS are commercially available, including self-extinguishing, electroplating, antistatic-expandable, glass-reinforced, high-heat, cold-forming, and low-gloss sheet grades.

Three polymerization steps are involved in ABS manufacture. The process is shown in block diagram format in the accompanying illustration.

STEP 1—Polybutadiene rubber is formulated by feeding butadiene, water, an emulsifier, and catalyst into a glass-lined reactor. This is an exothermic reaction. About 80% conversion is achieved in a period of about 50 hours. The residual butadiene monomer is recovered by steam-stripping and recycled.

STEP 2—Polybutadiene rubber is further polymerized, but in the presence of styrene and acrylonitrile monomers. This is done in low-pressure reactors under a nitrogen atmosphere. In this operation, the monomers are grafted onto the rubber backbone through the residual unsaturation remaining from the first step.

STEP 3—In a separate step, styrene-acrylonitrile (SAN) resin is prepared by emulsion, suspension, or mass polymerization by free-radical techniques. The operation is carried out in stainless-steel reac-

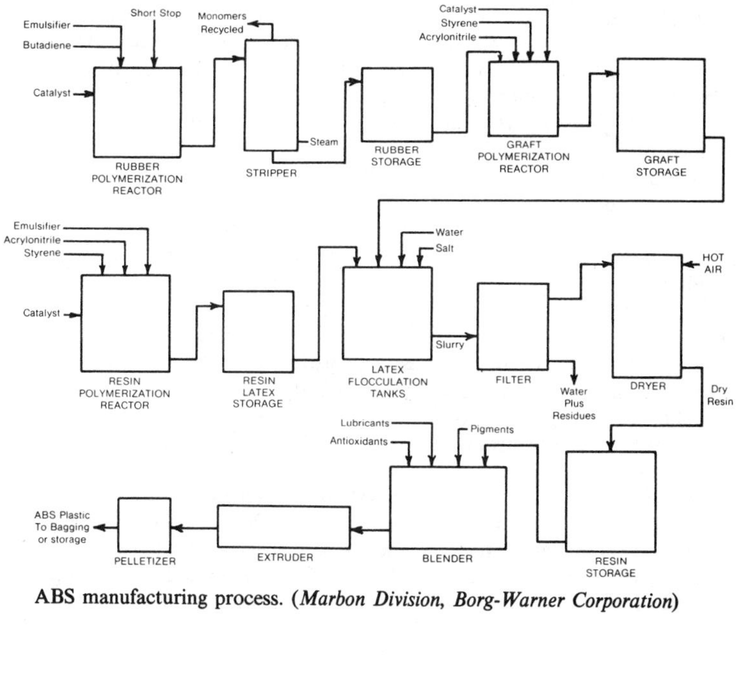

ABS manufacturing process. (*Marbon Division, Borg-Warner Corporation*)

tors operated at about 75°C (167°F) and 5 psig for about 7 hours. The final chemical operation is the blending of the ABS graft phase with the SAN resin, plus adding various antioxidants, lubricants, stabilizers, and pigments. Final operations involve preparation of a slurry of fine resin particles (via chemical flocculation), filtering, and drying in a standard fluid-bed dryer at 121–132°C (250–270°F) inlet air temperature.

Assistance of the Marbon Division, Borg-Warner Corporation in preparation of this entry is appreciated.

ABSAROKITE. A geologic term proposed by Iddings in 1805 for a porphyritic basalt containing phenocrysts of olivine and augite in a ground mass of smaller labradorite crystals. Type locality, Absaroka Range, Wyoming.

ABSCESS. A localized collection of pus within a cavity. An abscess may occur in many organs of the body. Abscesses can present a range of symptoms, depending upon location and cause, varying from severe, acute forms to milder, chronic forms. The presence of an abscess is always considered a serious matter by the physician because without immediate treatment, very serious consequences may occur.

Lung abscesses are among the more serious types. With the advent of antibiotics, the occurrence and severity of lung abscesses decreased markedly. Over half of the lung abscesses seen originate from a necrotizing suppurative bronchopneumonia resulting from the aspiration of mixed bacteria from the mouth and throat. Lung abscesses are sometimes associated with peridontal disease. In most cases, mixed anaerobic bacteria, such as *Fusobacterium nucleatum, Bacteroides melaninogenicus,* and anaerobic or microaerophilic streptococci and *Peptostreptoccus* predominate. *B. Fragilis* also may be present. Abscess formation is uncommon in pneumococcal pneumonia. Lung abscesses also may result from tumors or foreign bodies that cause bronchial obstruction. Other possible causes include pulmonary tuberculosis, fungal infection, and actinomycosis. Cough is present in nearly all patients. Copious foul sputum may result from drainage of the abscess into the bronchial tree. Chest pain and fever are common symptoms. X-rays may be required to confirm the presence of a lung abscess. Sputum examination is critical to diagnosis. Bronchoscopy is usually reserved for more difficult cases. Intravenous penicillin followed by oral penicillin V may be indicated. For penicillin-allergic patients, clindamycin may be used. Adequate drainage is also an important element of therapy. Where empyema is a complication of lung abscess, external drainage is required. Surgery is required in only a minimum of cases of lung abscesses, but was frequently required prior to antibiotics.

Intra-abdominal abscesses usually contain multiple bacterial species. Anaerobic bacteria are present in 60–70% of cases because of the proximity of the peritoneum to the bowel. The abscesses range from a small, acute disorder to a chronic process that causes intermittent fever, weight loss, and anemia. These "smouldering" abscesses sometimes result from a prior infection that was not fully eradicated by antibiotic therapy, allowing a pocket of infection to persist and slowly develop. Physicians, when attempting to diagnose a fever of undetermined origin, will usually ask the patient if there has been relatively recent abdominal surgery. The principal treatment for subphrenic abscesses is surgical drainage. Multiple antibiotic therapy is also used to reach a spectrum of possible causative organisms.

Mediastinitis is an inflammation of the wall dividing the two pleural cavities; a common complication is mediastinal abscess. Sometimes the abscess opens and empties its contents into the trachea; either the patient will cough up large amounts of pus, or he may suffocate. Mediastinitis may occur as a result of perforation of the esophagus. This can happen when a sharp foreign body becomes lodged in the esophagus, during attempts to remove it, or during examination of the organ for other reasons. Mediastinitis also can result from a bullet or stab wound.

Abscess of the breast may occur within the first month after childbirth. It is caused by infections entering through a "cracked nipple." The unfortunate consequences of a breast abscess are that the infant is deprived of breast milk, plus the fact that the mother has a long period of discomfort and pain. Treatment is instituted as quickly as possible in order to avoid a prolonged convalescent period, as well as the possibility of the destruction of a large amount of breast tissue.

Anal infections may cause anal fissure, hemorrhoids, abscess, and fistula, and is usually the result of invasion of the numerous tiny glands or crypts, which abound in the tissues adjacent to the anus. If the infection spreads through the wall of the anus, an abscess may occur in the tissues around the anus, and this may burst through the skin around the anus or back into the rectum. In either case, the abscess cavity has two openings, the original site of entry of the infection and the point where it bursts through. Fistula is the term by which such a condition is designated.

In *abscess of the external ear,* there is pain and tenderness over the affected area. The auricle may enlarge to two or three times the normal size. If proper care is not given, the ear may be permanently distorted in shape. Antibiotics and sulfonamide drugs may be used effectively. Surgical treatment may be required, but only after careful examination by a specialist.

Periapical abscesses occur at the apical (apex) region of a tooth as the result of death of the pulp tissue. Periodontal abscesses occur in the tissues closely surrounding a tooth, such as gingiva, bone, or the periodontal membrane. When an abscess breaks through a limiting membrane, working through surrounding bone to external soft tissue, a gum boil may result.

ABSCISSA. Coordinate Systems.

ABSCISSION. This term is applied to the process whereby leaves, leaflets, fruits, or other plant parts become detached from the plant. Leaf abscission is a characteristic phenomenon of many species of woody dicots and is especially conspicuous during the autumn period of leaf fall. The onset of abscission seems to be regulated by plant hormones. Three main stages can be distinguished in the usual process of leaf abscission. The first is the formation of an abscission layer which is typically a transverse zone of parenchymatous cells located at the base of the petiole. The cells of this layer may become differentiated weeks or even months before abscission actually occurs. The second step is the abscission process proper which occurs as a result of a dissolution of the middle lamellae of the cells of the abscission layer. This results in the leaf remaining attached to the stem only by the vascular elements which are soon broken by the pressure of wind or the pull of gravity and the leaf falls from the plant. In the final stage of the process the exposed cells of the leaf scar are rendered impervious to water by lignification and suberization of the walls.

Subsequently other layers of corky cells develop beneath the outer layer. These layers eventually become a part of the periderm of the stem. The broken xylem elements of the leaf scar become plugged with gums or tyloses and the phloem elements become compressed and sealed off.

In some kinds of plants an abscission layer is only imperfectly formed and in many others, especially herbaceous species, no abscission layer develops at the base of the petiole. In a few herbaceous species, of which coleus, begonia, and fuchsia are examples, an abscission layer develops. In the majority of herbaceous species, however, and in some woody species, there is no true abscission process. In such herbaceous plants most or all of the leaves are retained until the death of the plant. In the woody plants falling in this category (example: shingle oak, *Quercus imbricaria*) the leaves are shed only by mechanical disruption from the plant. Abscission of the fruits of apple and doubtless of many other species occurs in much the same manner as abscission of leaves. The abscission of apple fruits can be artifically retarded by spraying with certain growth regulators.

Various plant hormones and plant growth regulators can be of help to the fruit producer in terms of controlling the timing of abscission. See also **Gibberellic Acid and Gibberellin Plant Growth Hormones;** and **Plant Growth Modification and Regulation.** See also related entries under **Tree.**

ABSCISSION. Plant Growth Modification and Regulation.

ABSINTHE. Artemisia.

ABSOLUTE MAGNITUDE (Stellar). **Stellar Magnitude.**

ABSOLUTE SPACE-TIME. A fundamental concept underlying Newtonian mechanics is that there exists a preferred reference system to which all measurements should be referred. This is known as absolute space-time. The assumption of such a system is replaced in relativistic mechanics by the principle of equivalence. See **Equivalence Principle; Relativity and Relativity Theory.**

ABSOLUTE TENSOR (Tensor Field). Tensor (tensor field) of weight zero. Often called tensor (tensor field) when context admits no confusion. See also **Tensor Field.**

ABSOLUTE ZERO. Conceptually that temperature where there is no molecular motion, no heat. On the Celsius scale, absolute zero is −273.15°C; on the Fahrenheit scale, −459.67°F; and zero degrees Kelvin (0K). The concept of absolute zero stems from thermodynamic postulations.

Heat and temperature were poorly understood prior to Carnot's analysis of heat engines in 1824. The Carnot cycle became the conceptual foundation for the definition of temperature. This led to the somewhat later work of Lord Kelvin, who proposed the Kelvin scale based upon a consideration of the second law of thermodynamics. This leads to a temperature at which all the thermal motion of the atoms stops. By using this as the zero point or absolute zero and another reference point to determine the size of the degrees, a scale can be defined. The Comité Consultative of the International Committee of Weights and Measures selected 273.16K as the value for the triple point for water. This set the ice-point at 273.15K.

From the standpoint of thermodynamics, the thermal efficiency E of an engine is equal to the work W derived from the engine divided by the heat supplied to the engine, $Q2$. If $Q1$ is the heat exhausted from the engine,

$$E = (W/Q2) = (Q2 - Q1)/Q2 = 1 - (Q1/Q2)$$

where W, $Q1$, and $Q2$ are all in the same units. A Carnot engine is a theoretical one in which all the heat is supplied at a single high temperature and the heat output is rejected at a single temperature. The cycle consists of two adiabatics and two isothermals. Here the ratio $Q1/Q2$ must depend only on the two temperatures and on nothing else. The Kelvin temperatures are then defined by the relation

$$\frac{Q1}{Q2} = \frac{T1}{T2}$$

where $Q1/Q2$ is the ratio of the heats rejected and absorbed, and $T1/T2$ is the ratio of the Kelvin temperatures of the reservoir and the source. If one starts with a given size for the degree, then the equation completely defines a thermodynamic temperature scale.

A series of Carnot engines can be postulated so that the first engine absorbs heat Q from a source, does work W, and rejected a smaller amount of heat at a lower temperature. The second engine absorbs all the heat rejected by the first one, does work and rejects a still smaller amount of heat which is absorbed by a third engine, and so on. The temperature at which each successive engine rejects its heat becomes smaller and smaller, and in the limit this becomes zero so that an engine is reached which rejects no heat at a temperature which is absolute zero. A reservoir at absolute zero cannot have heat rejected to it by a Carnot engine operating between a higher temperature reservoir and the one at absolute zero. This can be used as the definition of absolute zero. Absolute zero is then such a temperature that a reservoir at that temperature cannot have heat rejected to it by a Carnot engine which uses a heat source at some higher temperature.

ABSORBANCE. By combining the laws of Bouguer and Beer, the absorbance

$$A = -\log T = \log \frac{I_0}{I} = abc$$

where T is the transmittance, I_0 and I are the intensities of light incident and transmitted by a sample of thickness b, concentration c (if the sample is in solution) and absorptivity a. It is assumed that all necessary corrections have been made in a reported value of A, hence terms such as absorbancy, absorptance, and absorptancy should now not be used.

Absorbancy is the common logarithm of the reciprocal of the transmittancy. The quantity is sometimes referred to as the *absorbancy index.*

Absorptancy. If T_s is the transmittancy of a dissolved solute, then the absorptancy may be defined as $1 - T_s$. The term is no longer used in precise absorptimetry.

ABSORBER. In general, a medium, substance or functional part that takes up matter or energy. In radiation and particle physics, an absorber is a body of material introduced between a source of radiation and a detector to (1) determine the energy or nature of the radiation; (2) to shield the detector from the radiation; or (3) to transmit selectively one or more components of the radiation, so that the radiation undergoes a change in its energy spectrum. Such an absorber may function through a combination of processes of true absorption, scattering and slowing-down.

ABSORPTIMETRY. A method of instrumental analysis, frequently chemical, in which the absorption (or absence thereof) of selected electromagnetic radiation is a qualitative (and often quantitative) indication of the chemical composition of other characteristics of the material under observation. The type of radiation utilized in various absorption-type instruments ranges from radio and microwaves through infrared, visible, and ultraviolet radiation to x-rays and gamma rays. See also **Analysis (Chemical);** and **Spectro Instruments.**

ABSORPTION COEFFICIENT. 1. For the absorption of one substance or phase in another, as in the absorption of a gas in a liquid, the absorption coefficient is the volume of gas dissolved by a specified volume of solvent; thus a widely used coefficient is the quantity α in the expression $\alpha = V_0/V_p$, where V_0 is the volume of gas reduced to standard conditions, V is the volume of liquid and p is the partial pressure of the gas.

2. In the case of sound, the absorption coefficient (which is also called the acoustical absorptivity) is defined as the fraction of the incident sound energy absorbed by a surface or medium, the surface being considered part of an infinite area.

3. In the most general use of the term absorption coefficient, applied to electromagnetic radiation and atomic and sub-atomic particles, it is a measure of the rate of decrease in intensity of a beam of photons or particles in its passage through a particular substance. One complication in the statement of the absorption coefficient arises from the cause of the decrease in intensity. When light, x-rays, or other electromagnetic radiation enters a body of matter, it experiences in general two types of attenuation. Part of it is subjected to scattering, being reflected in all directions, while another portion is absorbed by being converted into other forms of energy. The scattered radiation may still be effective in the same ways as the original, but the absorbed portion ceases to exist as radiation or is re-emitted as secondary radiation. Strictly, therefore, we have to distinguish the true absorption coefficient from the scattering coefficient; but for practical purposes it is sometimes convenient to add them together as the total attenuation or extinction coefficient.

If appropriate corrections are made for scattering and related effects, the ratio I/I_0 is given by the laws of Bouguer and Beer. Here, I_0 is the intensity or radiant power of the light incident on the sample and I is the intensity of the transmitted light. This ratio $I/I_0 = T$ is known as the transmittance. See also **Spectrochemical Analysis (Visible).**

ABSORPTION CURVE. The graphical relationship between thickness of absorbing material or concentration of dissolved substance and intensity of transmitted radiation.

ABSORPTION DISCONTINUITY. A discontinuity appearing in the absorption coefficient of a substance for a particular type of radiation when expressed as a function of the energy (or frequency or wavelength) of this radiation. An absorption discontinuity is often associated with anomalies in other variables such as the refractive index. See **Anomalous Dispersion.**

ABSORPTION DYNAMOMETER. Dynamometer.

ABSORPTION EDGE. The wavelength corresponding to an abrupt discontinuity in the intensity of an absorption spectrum, notably an x-ray absorption spectrum, which gives the appearance of a sharp edge in the photograph of such a spectrum.

ABSORPTION (Energy). The process whereby the total number of particles emerging from a body of matter is reduced relative to the number entering as a result of interaction of the particles with the body. Also, the process whereby the kinetic energy of a particle is reduced while traversing a body of matter. This loss of kinetic energy or radiation is also referred to as moderation, slowing, or stopping. See also **Blackbody.** The absorption of mechanical energy by dynamometers, which convert the mechanical energy to heat or electricity, has led to the use of the term "absorption dynamometer" to distinguish these machines. See also **Dynamometer.** In acoustics, absorption is the process whereby some or all of the energy of sound waves is transferred to a substance on which they are incident or which they traverse.

ABSORPTION LAWS. Spectrochemical Analysis (Visible).

ABSORPTION (Physiology). The process by which materials enter the living substance of which the organism is composed. Materials including food and oxygen are taken into special organs by ingestion and respiration, but they must pass through the cell wall to become an integral part of the organism by absorption. The basic physical forces involved are those of osmosis and diffusion.

ABSORPTION (Process). Absorption is commonly used in the process industries for separating materials, notably a specific gas from a mixture of gases; and in the production of solutions such as hydrochloric and sulfuric acids. Absorption operations are very important to many air pollution abatement systems where it is desired to remove a noxious gas, such as sulfur dioxide or hydrogen sulfide, from an effluent gas prior to releasing the material to the atmosphere. The absorption medium is a liquid in which (1) the gas to be removed, i.e., absorbed is soluble in the liquid, or (2) a chemical reaction takes place between the gas and the absorbing liquid. In some instances a chemical reagent is added to the absorbing liquid to increase the ability of the solvent to absorb.

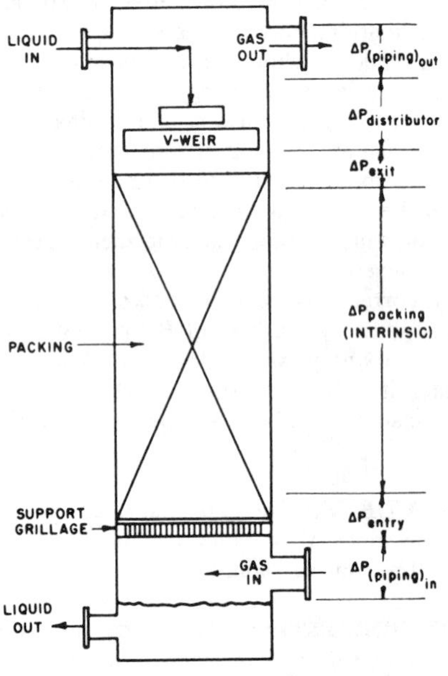

Fig. 1. Section of representative packed absorption tower.

Wherever possible, it is desired to select an absorbing liquid that can be regenerated and thus recycled and used over and over. An example of absorption with chemical reaction is the absorption of carbon dioxide from a flue gas with aqueous sodium hydroxide. In this reaction, sodium carbonate is formed. This reaction is irreversible. However, continued absorption of the carbon dioxide with the sodium carbonate solution results in the formation of sodium acid carbonate. The latter can be decomposed upon heating to carbon dioxide, water, and sodium carbonate and thus the sodium carbonate can be recycled.

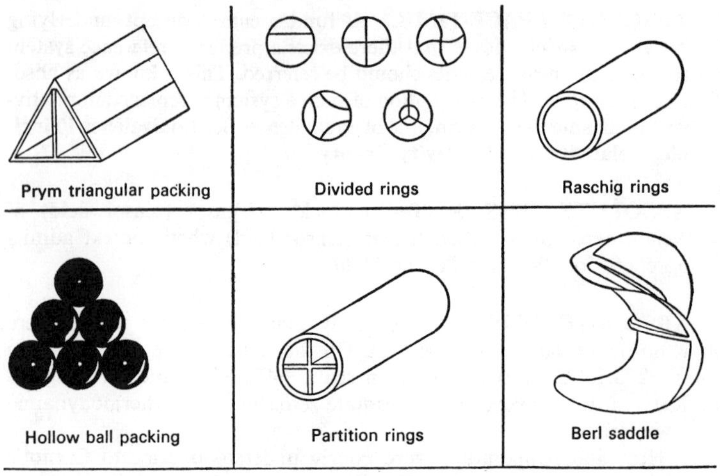

Fig. 2. Types of packing used in absorption towers.

Types of equipment used for absorption include (1) a packed tower filled with packing material, absorbent liquid flowing down through the packing (designed to provide a maximum of contact surface), and gas flowing upward in a countercurrent fashion; (2) a spray tower in which the absorbing liquid is sprayed into essentially an empty tower with the gas flowing upward; (3) a tray tower containing bubble caps, sieve trays, or valve trays; (4) a falling-film absorber or wetted-wall column; and (5) stirred vessels. Packed towers are the most commonly used.

A representative packed-type absorption tower is shown in Fig. 1. In addition to absorption efficiency, a primary concern of the tower designer is that of minimizing the pressure drop through the tower. The principal elements of pressure drop are shown at the right of the diagram. Important to efficiency of absorption and pressure drop is the type of packing used. As shown by Fig. 2, over the years numerous types of packings (mostly ceramic) have been developed to meet a wide variety of operating parameters. A major objective is that of providing as much contact surface as is possible with a minimum of pressure drop. Where corrosion conditions permit, metal packing sometimes can be used. Of the packing designs illustrated, the berl saddles range in size from $\frac{1}{4}$ inch (6 millimeters) up to 2 inches (5 centimeters); raschig rings range from $\frac{1}{4}$ inch (6 millimeters) up to 4 inches (10 centimeters); lessing rings range from 1 inch (2.5 centimeters) up to 2 inches (5 centimeters); partition and spiral rings range from 3 inches (7.5 centimeters) up to 6 inches (15 centimeters) when grouped as shown.

In operation, the absorbing liquid is pumped into the top of the column where it is distributed by means of a weir to provide uniform distribution of the liquid over the underlying packing. Gas enters at the base of the tower and flows upward (countercurrent with the liquid) and out the top of the tower. The liquid may or may not be recycled without regeneration, depending upon the strength of the absorbent versus the quantity of material (concentration) in the gas to be removed. In a continuous operation, of course, a point is reached where fresh absorbing liquid must be added.

It is interesting to note that over 100,000 of the $\frac{1}{4}$-inch (6-millimeter) size packing shapes will be contained in each cubic foot (0.02832 cubic meter) of tower space if dense packing is desired.

In the purification of natural gas, the gas is fed into the bottom of an absorption tower where the gas is contacted countercurrently by a lean absorption oil. Hydrochloric acid is produced by absorbing gaseous hydrogen chloride in water, usually in a spray-type tower. Unreacted ammonia in the manufacture of hydrogen cyanide is absorbed in dilute sulfuric acid. In the production of nitric acid, ammonia is catalytically oxidized and the gaseous products are absorbed in water. The ethanolamines are widely used in scrubbing gases for removal of acid compounds. Hydrocarbon gases containing hydrogen sulfide can be scrubbed with monoethanolamine, which combines with it by salt formation and effectively removes it from the gas stream. In plants synthesizing ammonia, hydrogen and carbon dioxide are formed. The hydrogen can be obtained by countercurrently scrubbing the gas mixture in a packed or tray column with monoethanolamine which absorbs the carbon dioxide. The latter can be recovered by heating the monoethanolamine. In a nonliquid system, sulfur dioxide can be absorbed by dry cupric oxide on activated alumina, thus avoiding the disadvantages of a wet process. Sulfuric acid is produced by absorbing sulfur trioxide in weak acid or water.

ABSORPTION SPECTROMETRY. Spectro Instruments.

ABSORPTION SPECTRUM. The spectrum of radiation which has been filtered through a material medium. When white light traverses a transparent medium, a certain portion of it is absorbed, the amount varying, in general, progressively with the frequency, of which the absorption coefficient is a function. Analysis of the transmitted light may, however, reveal that certain frequency ranges are absorbed to a degree out of all proportion to the adjacent regions; that is, with a distinct selectivity. These abnormally absorbed frequencies constitute, collectively, the "absorption spectrum" of the medium, and appear as dark lines or bands in the otherwise continous spectrum of the transmitted light. The phenomenon is not confined to the visible range, but may be found to extend throughout the spectrum from the far infrared to the extreme ultraviolet and into the x-ray region.

A study of such spectra shows that the lines or bands therein accurately coincide in frequency with certain lines or bands of the emission spectra of the same substances. This was formerly attributed to resonance of electronic vibrations, but is now more satisfactorily explained by quantum theory on the assumption that those quanta of the incident radiation which are absorbed are able to excite atoms or molecules of the medium to some (but not all) of the energy levels involved in the production of the complete emission spectrum.

A very familiar example is the spectrum of sunlight, which is crossed by innumerable dark lines—the Fraunhofer lines—from which so much has been learned about the constitution of the sun, stars, and other astronomical objects.

A noteworthy characteristic of selective absorption is found in the existence of certain anomalies in the refractive index in the neighborhood of absorption frequencies; discussed under **Dispersion (Radiation).** See also **Emission Spectrum; Fraunhofer Lines.**

References

Bauman, Robert P.: "Absorption Spectroscopy," Wiley, New York, 1962.

Harrison, George R., Lord, Richard C., and John R. Loofbourow: "Practical Spectroscopy," Prentice-Hall, Englewood Cliffs, N.J., 1948.

Evans, Ralph M.: "An Introduction to Color," Wiley, New York, 1948.

Jenkins, Francis A. and Harvey E. White: "Fundamentals of Physical Optics," 3rd edition, McGraw-Hill, New York, 1957.

Stearns, E. I.: "Practice of Absorption Spectrophotometry," Wiley, New York, 1969.

Harrick, N. J.: "Internal Reflection Spectroscopy," Wiley, New York, 1967.

ABSORPTIVITY. Solar Energy.

ABSORPTIVITY (Optical). If A is the absorbance of a solution b cm. in thickness and at a concentration c, the absorptivity is $a = A/bc$. See also **Beer's Law.**

ABUNDANCE (Chemical Elements). Chemical Elements.

ABUNDANCE RATIO. The proportions of the various isotopes making up a particular specimen of an element. See **Chemical Elements.**

ABYSSAL HILLS. Small hills up to several hundred feet in height that occupy the ocean floor. These may be nearly isolated or may occupy virtually the whole floor. See **Abyssal Plain.**

ABYSSAL PELAGIC ZONE. The lowest layer of water in the oceans. It is not penetrated by sunlight.

ABYSSAL PLAIN. An area on the ocean floor having a flat bottom and a very slight slope of less than 1 part in 1000. It is believed that these very flat surfaces arise from the continued deposition of mud and silt from turbidity currents. See **Ocean.** Seismographic studies support that these surfaces consist of such deposits. Mid-ocean canyons may be found on these abyssal plains; these are flat-bottomed depressions in the plains, varying from one to several miles in width and varying in depth up to several hundred feet. These, too, are believed to be the product of certain turbidity currents.

ABYSSAL ROCKS. Proposed by Brögger as a general term for deep-seated igneous rocks, or those which have crystallized from magmas far below the surface of the earth, very slowly and under great pressure. Granite is a typical abyssal rock. The term plutonic is synonymous.

ABYSSAL ZONE. The region of the ocean beyond the point of penetration of light, including the ocean floor in the deep areas. According to various investigators who have descended into the ocean depths, no light penetrates beyond about 1,500 feet (450 meters), and penetration may be much less if the water is murky with suspended particles. The water is always extremely cold in the abyssal zone and the pressure is very great. Still many forms of animal life are to be found at these great depths, feeding upon the organic matter which drifts down from the upper waters.

Abyssal animals fall into two groups: *scavengers,* living on the shower of organic matter, and *predators,* which prey upon the scavengers or upon each other. The most abundant deep-sea animals are the sea cucumbers, snails, crustaceans, tunicates, cephalopods, and fish. The predaceous fish have large mouths filled with long, sharp teeth, and stomachs capable of great stretching; they are actually known to swallow fish larger than themselves. Many of the fish have a lure with a light on the end which attracts prey. Others have rows of light-producing organs on the sides of their bodies. Other animals with light-producing organs are coelenterates, echinoderms, annelids, crustaceans, and cephalopods. It is believed that the lights not only help these animals in finding food, but also in finding each other during the reproductive season. See also **Ecology; Ocean.**

ACACIA GUM. Gums and Mucilages.

ACACIA TREES. Of the family *Leguminosae* (pea family), the genus *Acacia* represents a large number of mostly evergreen trees and shrubs, particularly abundant in Africa and Australia. The trees like warmth and full sun. The small flowers are aggregated into ball-like or elongate clusters, which are quite conspicuous. The leaves are rather diverse in shape; quite commonly they are dissected into compound pinnate forms; in other instances, especially in the Australian species, they are reduced even to a point where only the flattened petiole, called a phyllode, remains. This petiole grows with the edges vertical, a fact which some observers consider a protective adaptation against too intense sunlight on the surface. Some species, particularly those growing in Africa and tropical Asia, yield products of commercial value. Gum arabic is obtained from the *Acacia senegal.* A brown or black dye called clutch is obtained from *A. catechu.* Some acacias are used for timber. Shittinwood referred to in the scriptures, "And thou shalt make staves of shittinwood and overlay them with gold," (Exodus 26:26–37), is considered by authorities as wood from *Acacia seyal* (then referred to as the shittah tree).

Certain tropical American species are of particular interest because of the curious pairs of thorns, which are united at their base. These thorns are often hollowed out and used as nests by species of stinging ants. The leaves of some species, notably *Mimosa pudica,* are sensitive to the touch. The mimosa tree or silver wattle, native to Australia, is the *Acacia dealbata.* The leaves are fernlike and of a silver-green coloration. They attain a height of about 50 feet (15 meters) within 20 years, prefer full sun, and can be severely damaged by prolonged frosts. The tree has been introduced into warm regions of other parts of the world and has done well. The so-called catclaw acacia *(A.*

greggii) has done well in southwestern United States. One specimen, selected by The American Forestry Association for its "Social Register of Big Trees," is located at Red Rock, New Mexico. The circumference at 4-1/2 feet (1.4 meters) above the base is 6 feet, 5 inches (1.8 meters, 13 centimeters); the height is 49 feet (15 meters); and the spread is 46 feet (14 meters).

A. baileyana, also a native of Australia, is known as the Cootamundra wattle or Bailey's mimosa. It attains a height of 20 feet (6 meters) or more, has long, narrow, waxy evergreen leaves of a silver-green color. The contour of the tree is often weeping.

The *Robinia pseudoacacia*, also referred to as black locust, common acacia, or false acacia, is found in the eastern United States. The tree is highly tolerant of dryness and industrial environments. This tree may attain a height of from 60 to 80 feet (18 to 24 meters), with a trunk diameter up to 4 feet (1.2 meters). It is a highly favored tree for gardens, often described as graceful and decorative.

ACANTHOCEPHALA (Thorny-Headed Worms). Worms, slender and hollow (pseudocoelom) with recurved hooks on invaginate proboscis, no digestive tracts, and adults parasitic in intestine of vertebrates with larva in intermediate arthropod host. They are usually regarded as a class of roundworm (*Nemathelminthes*), but ranking as a separate phylum is now favored.

ACANTHUS. Genus of the family *Acanthaceae* (acanthus family). This is a relatively small genus of Mediterranean plants grown mainly for ornamental purposes. The flowers are white or various shades of red. The term *ancanthus* also is used in architecture with reference to an ornamental design patterned after the leaves of the acanthus.

ACARICIDE. A substance, natural or synthetic, used to destroy or control infestations of the animals making up (*Arachnida, Acarina*), mainly mites and ticks, some forms of which are very injurious to both plants and livestock, including poultry. There are numerous substances that are effective both as acaricides and insecticides; others of a more narrow spectrum are strictly acaricides. See also **Insecticides;** and **Insecticide and Pesticide Technology.**

ACARINA. The order of *Arachnida* which includes the mites and ticks.

ACCELERATED FLIGHT (Airplane). When the velocity of an airplane along its flight path contains elements of acceleration, the structure receives increments of inertial or dynamic loading that may prove to be far more severe upon the structure than the loading imposed by the static weight of the airplane and its contents. Consequently, accelerated flight has been the subject of extensive analytical and experimental investigation. Acceleration of rectilinear velocity below the speed of sound, as by increasing the thrust of the power plant in straight level flight, is of small import, since radial accelerations resulting from curvilinear flight at constant speed are so large as to be the critical influence. Cases of curved flight paths capable of accelerations of several *g* (acceleration due to gravity, i.e., 32.2 feet/second^2) are quick pull-ups (or "zooms") from high-speed rectilinear flight, spins, steeply banked turns, and loops. The magnitude of the effect of accelerated flight is well illustrated by considering the centrifugal force on an airplane following a curved flight path in the vertical plane. With a constant tangential speed as low as 120 mph (193 kph), the airplane experiences a radial acceleration of 4 *g* (4 times the acceleration of gravity) even though the radius of curvature be about 240 feet (72 meters). See **Load Factor.**

An acceleration also is caused when the airplane encounters a gust.

ACCELERATION. The rate of change of the velocity with respect to the time is called acceleration. It is expressed mathematically by $d\mathbf{v}/dt$, the vector derivative of the velocity, $\mathbf{v}$ with respect to the time, t. If the motion is in a straight line whose position is clearly understood, it is convenient to treat the velocity v and the acceleration dv/dt as scalars with appropriate algebraic signs; otherwise they must be treated by vector methods.

Acceleration may be rectilinear or curvilinear, depending upon whether the path of motion is a straight line or a curved line. A body which moves along a curved path has acceleration components at every point. One component is in the direction of the tangent to the curve and is equal to the rate of change of the speed at the point. For uniform circular motion this component is zero. The second component is normal to the tangent and is equal to the square of the tangential speed divided by the radius of curvature at the point. This normal component, which is directed toward the center of curvature, also equals the square of the angular velocity multiplied by the radius of curvature. The acceleration due to gravity is equal to an increase in the velocity of about 32.2 feet/second/second at the earth's surface and is of prime importance since it is the ratio of the weight to the mass of a body. For examples of acceleration in both curved and linear motion, see **Kinematics.** See also **Angular Velocity and Angular Acceleration.**

ACCELERATION (Due to Gravity). The universal character of the gravitational force for point masses or spherical bodies can be expressed by the equation:

$$F = \frac{GM_1M_2}{R^2}, \qquad (1)$$

where

M_1, M_2 = masses of two bodies

R = distance between two bodies

G = a constant = 6.670×10^{-8} dyne cm^2 gm^{-2}

The constant G is independent of all properties of the particular bodies involved.

VARIATION OF ACCELERATION DUE TO GRAVITY ON EARTH WITH LATITUDE (At Sea Level)

Latitude	g centimeters/(second)^2	g feet/(second)^2
0°	977.989	32.0862
10°	978.147	32.0916
20°	978.600	32.1062
30°	979.295	32.1290
40°	980.147	32.1570
50°	981.053	32.1867
60°	981.905	32.2147
70°	982.600	32.2375
80°	983.053	32.2523
90°	983.210	32.2575

The weight of a body of mass M on the earth is the force with which it is attracted to the center of the earth. On the surface of the earth, the weight is given by:

$$W = Mg, \qquad (2)$$

where the acceleration due to gravity is obtained from Equation (1):

$$g = \frac{GM_E}{R_E^2}$$

$$= 980.665 \text{ cm/(second)}^2$$

$$= 32.174 \text{ feet/(second)}^2 \ (9.81 \text{ meters/(second)}^2) \qquad (3)$$

Variation of the acceleration due to gravity at sea level for different latitudes on earth is given in the accompanying table.

Gravitational force at the surface of the sun, moon, and planets is given in specific entries on these solar system bodies. See also **Gravitation.**

ACCELERATION MEASUREMENT. Acceleration cannot be measured directly, but must be computed by measuring the force exerted by restraints that are placed on a mass to hold its position fixed in an accelerating body. The relationship between restraint and

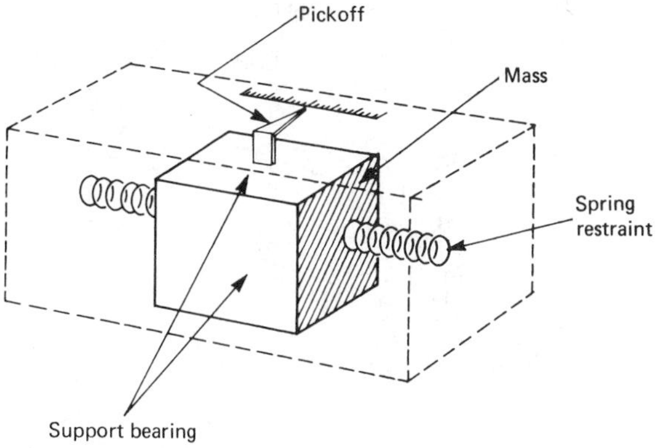

Fig. 1. Accelerometer based on a spring-restrained mass.

acceleration is defined by Newton's second law, namely, force = mass × acceleration.

A very simple form of accelerometer is the spring-restrained mass as shown in Fig. 1. The principal components are a constant seismic mass, a calibrated restraint, a support bearing, and a pickoff. Displacement of the mass must occur before the pickoff will work. Thus, a transient error is introduced during the time the mass is acquiring displacement. The mathematical relationships that pertain to the classical spring-mass device are shown in Fig. 2.

In order to reduce the transient error, developments have been undertaken to increase the stiffness of the system. Because of increased stiffness, low displacements result, requiring much more sensitive pickoffs and also pickoffs that operate over a smaller range. The useful output then becomes the restraint upon the mass. A basic servoed accelerometer is shown in Fig. 3; and the relevant block diagram and mathematical relationships are given in Fig. 4. The pickoff simply detects small changes in mass position and commands a proportional change in a current which is sent through the force generator. This current is used as a direct measure of restraining force. The accuracy of this current measurement depends upon maintaining the magnetic field around the forcer coil constant. Any forces of an extraneous nature acting on the mass will cause an erroneous reading of acceleration. Hence, elimination of such forces often demands sophisticated designs.

Types of Accelerometers. A convenient classification of accelerometers, based upon basic operating principles used, is:

(1) *Dynamic-Acceleration (Vibration) Instruments.* Included in this class are crystal types of accelerometers that do not respond to static acceleration. Devices of this type are used mainly in the control of vibration-test systems and in the monitoring of localized vibration response on an item under test.

(2) *Nonservoed Static and Dynamic Instruments.* Included are strain-gage devices, and spring-mass units with either potentiometric or inductive pickoffs. The devices are used in instrumentation and control systems. They have an accuracy of from 1 to 5%, depending mainly on the ambient temperature.

(3) *Servoed Static and Dynamic Instruments.* Included are the active devices that essentially measure the force required to accelerate the proof mass with the case rather than measure the deflection of the mass at the end of a spring. Servoed devices of this type are usually accurate within ±0.1%, although in some designs and under adverse ambient conditions the accuracy may be much poorer.

(4) *Gyroscopic Instruments.* These devices are used for navigation and guidance applications where the acceleration signal is integrated twice to obtain distance traveled. Accuracies obtainable are ±0.001 to 0.01%. This higher accuracy is attainable because inertia can be increased by spinning the proof mass and making use of gyroscopic-precession principles.

Velocity meters and vibrating-string accelerometers would fall into a special class of accelerometers.

Strain-Gage Accelerometer. In this device, the strain in a material whose stress is used to provide the rebalance force of an accelerometer mass is measured. Accuracy depends on linearity of stress and strain in the material. In a strain-gage accelerometer, the moving mass causes the resistance to increase in two resistive elements and to decrease in two resistive elements. The elements are placed in a bridge arrangement to provide an output that is proportional to acceleration. See Fig. 5.

In another configuration, a single-crystal silicon beam is used to obtain linear stress versus strain. Two piezoresistive strain elements are diffused into both sides of the beam to provide the basic sensing elements. Interconnection of the four elements into an electric bridge permits measurement of beam stress as a function of acceleration input. Locating two elements on one side of the beam and two elements on the opposite side provides a bridge wherein the resistance of two elements increases while the resistance of the other two elements de-

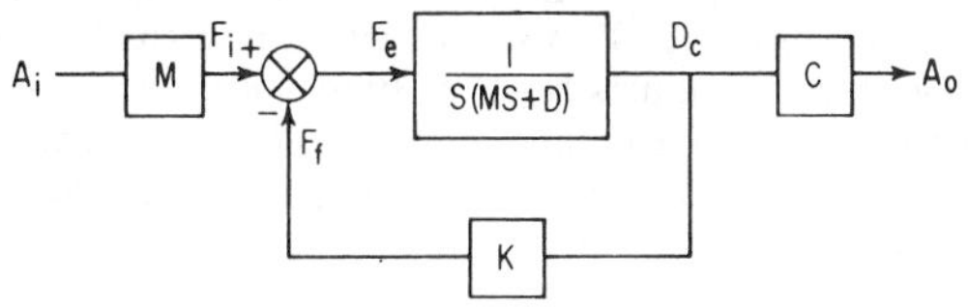

K = spring constant
C = readout device characteristic, e.g., volts/inch for a potentiometric pickoff excited by a fixed voltage

Transfer function is

$$\frac{A_O(s)}{A_i(s)} = \frac{MC/K}{(M/K)S^2 + (D/K)S + 1}$$

or

$$\frac{A_O(s)}{A_i(s)} = \frac{C/\omega_n^2}{(S^2/\omega_n^2) + 2\zeta(S/\omega_n) + 1}$$

where ω_n = the undamped natural frequency
ζ = the damping ratio D/D_C

Fig. 2. Technical description of classical spring-mass accelerometer.

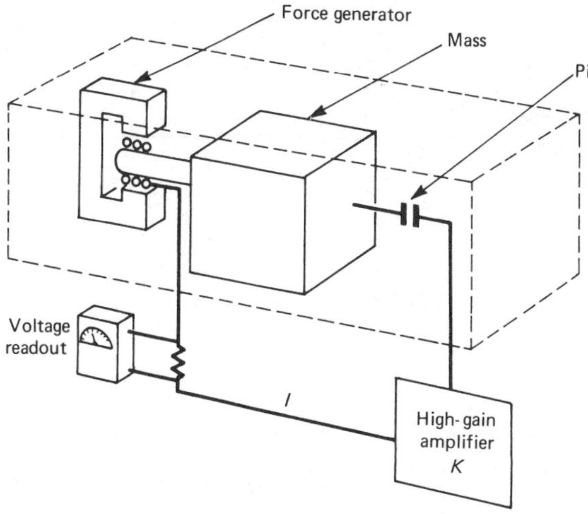

Fig. 3. Simple servoed accelerometer.

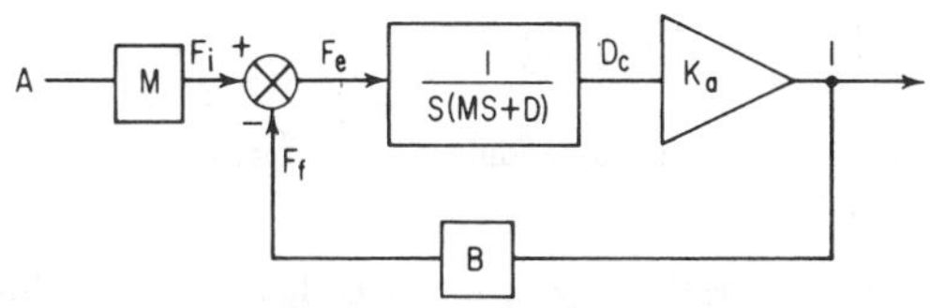

Transfer function is

$$\frac{I(s)}{A(s)} = \frac{M/B}{(M/K_aB)S^2 + (D/K_aB)S + 1}$$

or

$$\frac{I(s)}{A(s)} = \frac{M/B}{(S^2/\omega_n^2) + 2\zeta(S/\omega_n) + 1}$$

Fig. 4. Technical description of servoed accelerometer.

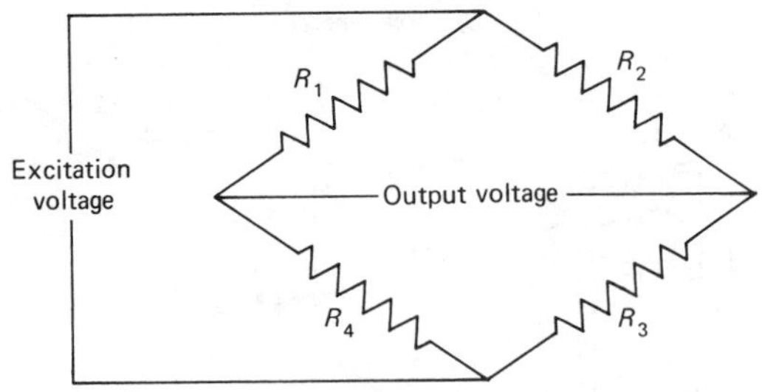

Fig. 5. Bridge of strain-gage accelerometer.

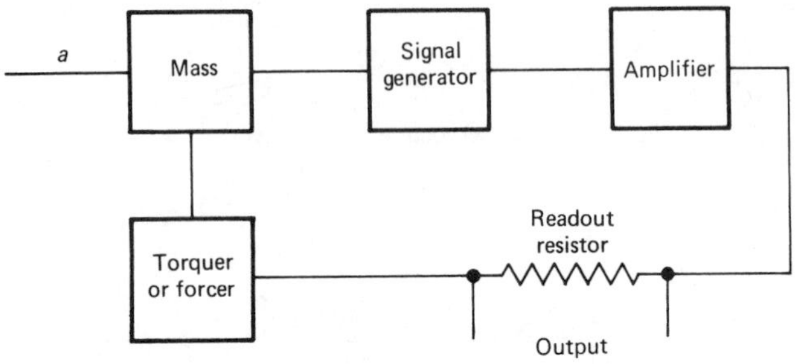

Fig. 6. Force-balance hinged-pendulum accelerometer. a = response to an acceleration.

creases when acceleration is applied. Accelerometers of this type are used for control and stabilization.

Force-Balance Hinged-Pendulum Accelerometer. The force-balance unit includes a torquer or forcer which performs the function of restraining the mass near its null position. The device is used with external electronics to provide the appropriate gain or amplification. A block diagram of the unit is given in Fig. 6. In another force-balance configuration, there is no external amplifier. The small pendulous mass is supported on a pair of flexural pivots. The pivots deflect freely in one axis, but are highly resistant to displacement in the other axis, thus reducing cross-coupling errors. Torque rebalance of the pendulum is accomplished by two permanent-magnet torquing elements operating in a push-pull manner. A moving-coil pickoff element is used. The device is filled with fluid to provide viscous damping of the pendulous element and to provide resistance to high vibration and shock. Accelerometers of this type are used for inertial navigation and are capable of sensing acceleration of less than micro-g's.

Pendulous Accelerometer. The effort to reduce friction has led to pendulous accelerometers which use rotational bearings. By floating the pendulum, the bearing loads also have been reduced. One advantage of the pendulous accelerometer is its cross-axis suspension. This is usually achieved by floating a pendulous cylinder in a viscous heavy liquid. The cylinder is pivoted on its axis and is pendulous about it. Because the cylinder is floated, cross-axis accelerations do not displace it. Some damping is also provided by the flotation liquid. Operating principles of the device are shown in Fig. 7. Acceleration along the sensitive axis causes deflection of the pendulum. The pickoff senses this, after which the signal is amplified and applied to a torquer in a direction to reduce the deflection. Response is given by:

$$A = \frac{IK_{T_c}}{P}$$

where

P = pendulosity = ML;
A = acceleration;
K_{T_c} = torque-generator transfer function.

The pendulous accelerometer allows some cross-axis sensitivity when the pendulum is deflected. The cross-axis sensitivity is proportional to the sine of the pickoff angle θ. The higher the loop gain on the sensitive axis, the smaller will be θ. As θ is kept near zero, so is the cross-axis sensitivity. The cross-coupling error is proportional

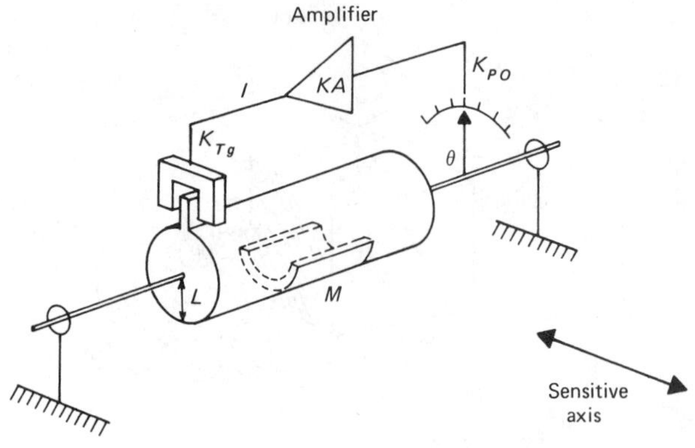

Fig. 7. Operating principle of pendulous accelerometer.

to the product of the magnitude of the cross acceleration and sensitive-axis acceleration. Thus, this is a g^2 type of error.

Pendulous-Gyro Integrating Accelerometer. In this device the measurable force developed by a processing gyroscope is used. This produces a very broad range of measurable restraints. The overall device bucks the torque of an accelerating pendulum against the reaction torque of a precessing gyro. A pendulous-gyro accelerometer is essentially a single-axis gyro with an unbalanced gimbal mounted on a single-axis platform controlled by the gyro. An elementary gyro is shown in Fig. 8. Once the wheel is spinning, a torque Wr exists about

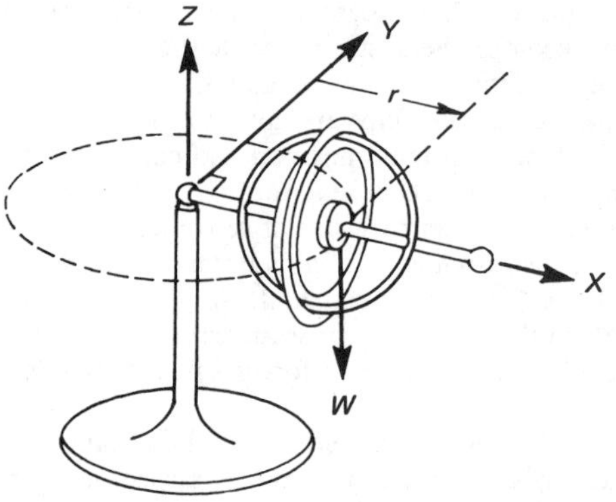

Fig. 8. Pendulous gyro accelerometer in elementary form.

the Y-axis because of the gyro weight. The Y-axis torque is initially resisted by only the gyro Y-axis moment of inertia. Thus, the gyro starts to fall and an angular velocity starts to build up around the Y-axis. But the angular velocity ω_y causes a gyroscopic reaction torque $H\omega_y$ around the Z-axis. Thus, the gyro starts to turn around the Z-axis. The angular rate ω_z then builds up until the gyroscopic reaction torque $H\omega_z$ exactly opposes the unbalance torque Wr, at which time the turning rate about the Y-axis is decreased to zero. This is a pendulous gyro with the pendulosity, $P = Wr/g = mr$. An equality can be written between the gyroscopic torque and the pendulous torque: $H\omega = mgr = Pg$. The turning (or precession) rate of the gyro about the Z-axis is:

$$\omega = \frac{P}{H} g$$

A pendulous 2-axis gyro is shown in Fig. 9. The spin motor is mounted on the inner gimbal with an off-balance mass or pendulum along the spin axis. The inner gimbal is free to turn about an axis perpendicular to the spin axis and is mounted in an outer gimbal. The outer gimbal is free to turn about an axis perpendicular to the spin axis and the inner-gimbal axis. If the base is accelerated along the input axis, the inertial reaction of the pendulum mass creates a torque about the inner-gimbal axis. Then, the gyro will precess around the outer-gimbal axis, that is, the input axis. The gyroscopic torque then balances the pendulous torque. The practical problems of this device are gimbal friction and the inertias, particularly around the outer-gimbal axis. These two factors cause the inner gimbal to turn about its axis and, therefore, the desired orthogonal condition deteriorates. A solution to these problems is to mount a single-axis pendulous gyro on a single-axis platform or turntable, as shown in Fig. 10. The turntable takes the place of the outer gimbal of the two-axis gyro. The output of the pendulous gyro is used to energize a servomotor which drives the turntable. The servomotor overcomes the friction of the turntable and drives it at a rate that causes the gyroscopic torque to equal the pendulous torque except during periods of transient accelerations. Since the gyro is an integrating gyro, the device will

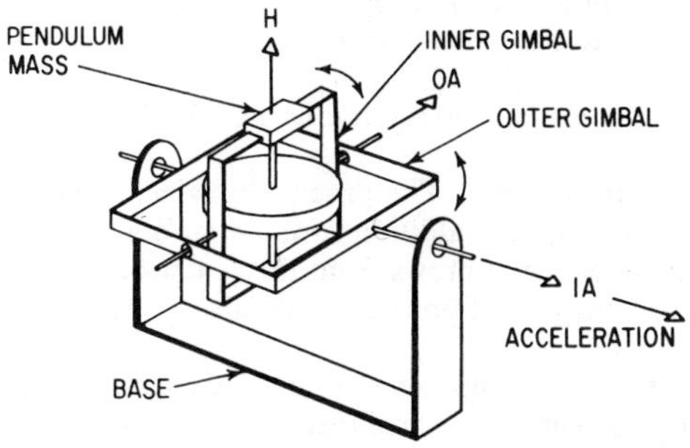

Fig. 9. Pendulous two-axis gyro.

integrate the input accelerations during transient conditions until the turntable has time to catch up. The output of the accelerometer can be either analog or digital. Technical characteristics of the pendulous-gyro integrating accelerometer are given in Fig. 11.

Viscous-Drag Integrating Accelerometer. When a body is moved through a newtonian fluid, a force proportional to velocity gradient is developed. If a floated gimbal is surrounded by a turning motor, the gimbal will experience a torque which is proportional to the relative speed between the rotor and gimbal, as shown in Fig. 12. The gimbal pickoff signal is amplified and drives the motor-rotor combination to establish the relative velocity between the rotor and the pendulous gimbal. The shear torque rebalances the acceleration torque and is proportional to the speed of the rotor. Thus, rotor position is proportional to velocity attained. An accelerometer of this type is particularly subject to temperature-dependent errors.

Double-Integrating Accelerometer. If a motor is mounted inside a pendulous gimbal with the spin axis *OA*, the device is called a double-integrating accelerometer or distance meter.

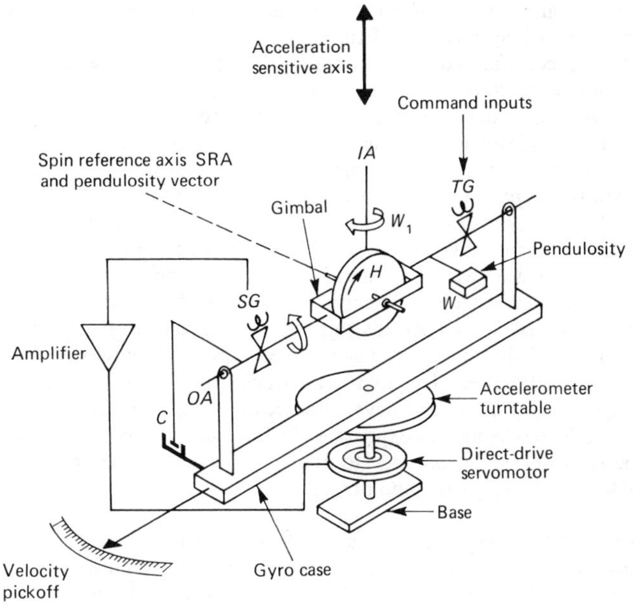

Fig. 10. Pendulous-gyro integrating accelerometer.

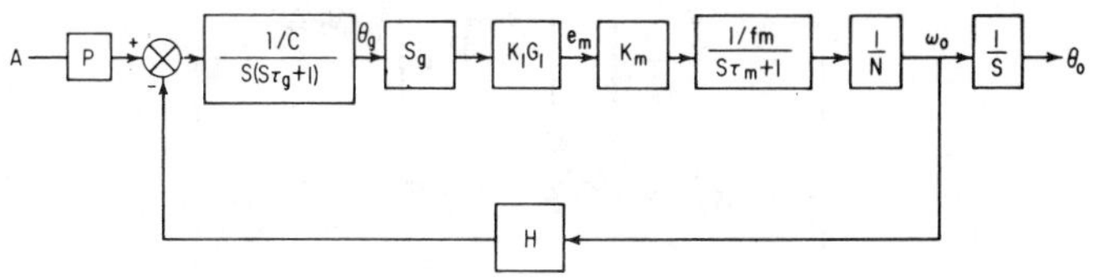

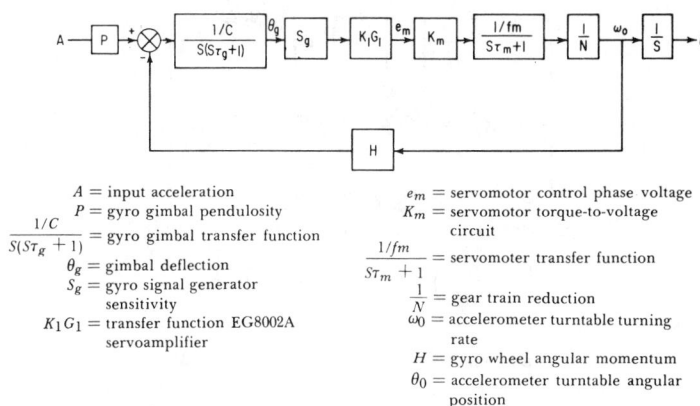

A = input acceleration
P = gyro gimbal pendulosity
$\frac{1/C}{S(S\tau_g+1)}$ = gyro gimbal transfer function
θ_g = gimbal deflection
S_g = gyro signal generator sensitivity
K_1G_1 = transfer function EG8002A servoamplifier
e_m = servomotor control phase voltage
K_m = servomotor torque-to-voltage circuit
$\frac{1/fm}{S\tau_m+1}$ = servomoter transfer function
$\frac{1}{N}$ = gear train reduction
ω_0 = accelerometer turntable turning rate
H = gyro wheel angular momentum
θ_0 = accelerometer turntable angular position

Fig. 11. Technical description of pendulous-gyro integrating accelerometer.

Vibrating-String Accelerometer. With reference to the device shown in Fig. 13, the strings are caused to vibrate. Their frequencies of vibration are measured. Acceleration along the sensitive axis causes an increase in tension and consequently in the frequency on one string, while producing a decrease in tension and lowering of frequency on the other string. The readout is nonlinear and the device is sensitive to ambient temperature.

Magnetic-Field-Supported Accelerometer. Servocontrolled electrostatic or magnetic fields can provide precise frictionless support and restraint for an accelerometer seismic mass. Pulse-rebalance techniques permit a digital readout when desired. Field supports are inherently nonlinear, but can be linearized by push-pull schemes. The dynamic range of field-supported devices is about 10^6, but the range can be

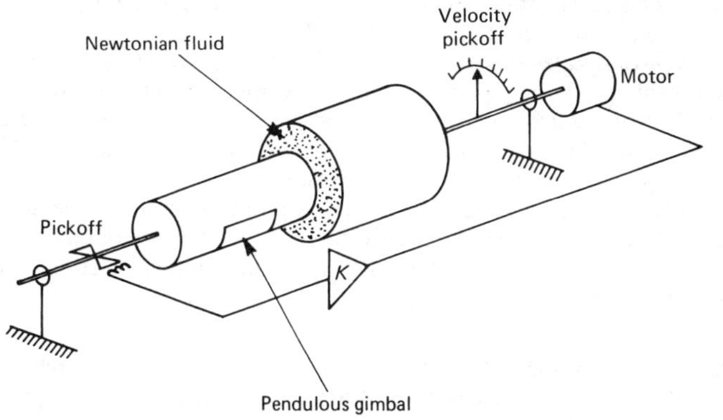

Fig. 12. Viscous drag integrating accelerometer.

shifted to nearly any portion of the acceleration spectrum through appropriate switching electronics. An accelerometer of this type is the electrically-suspended instrument which uses a seismic mass which is a conductor that is held floating, either in a vacuum or in gas under pressure, by an electrostatic-field attraction. The position of the seismic mass is sensed, and changes are used to command acceleration of the mass back toward its null position. The accelerating force is generated by changing the voltage gradient between the mass and electrodes. The force is proportional to the voltage gradient squared and is always a force of attraction. Thus, on each axis, fields must

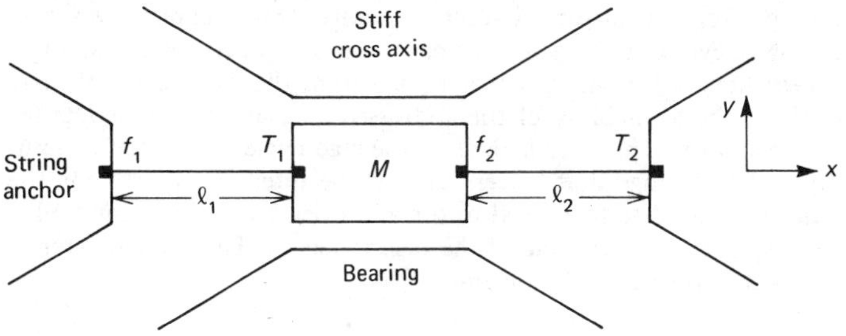

Fig. 13. Vibrating-string accelerometer.

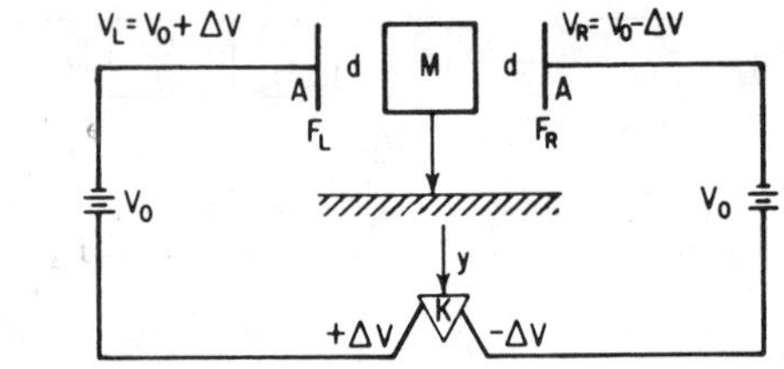

At the left force: At the right force: Net force:

$$F_L = \frac{kk_0}{2} A \left(\frac{V_L^2}{d}\right) \qquad F_R = \frac{kk_0}{2} A \left(\frac{V_R^2}{d}\right) \qquad F_N = F_L - F_R$$

but

$$V_L = V_0 + \Delta V \quad \text{and} \quad V_R = V_0 - \Delta V$$

thus

$$F_N = \frac{kk_0}{2} A[(V_0^2 + 2V_0\Delta V + \Delta V^2) - (V_0^2 - 2V_0\Delta V) + \Delta V^2)] = 2kk_0A\,\frac{V_0\Delta V}{d^2}$$

but

$$\Delta V = Ky \quad \text{and} \quad F_N = \frac{2kk_0 A V_0 K}{d^2}\, y = Ma$$

F_N now is proportional to y, as a linear spring would be. In the process a bias voltage V_0 had to be added. Linearity is destroyed if $\Delta V > V_0$. Note, too, that stiffness is proportional to V_0. The g spectrum of response is readily tailored or changed by switching the magnitude of V_0, K, and M.

Fig. 14. Technical description of electrostatic accelerometer.

be established in balanced opposition to one another. As one is increased, the other is decreased. The net force on the seismic mass is thus proportional to voltage change, and the readout of restraint voltage versus acceleration is linear. A technical description of the electrostatic accelerometer is given in Fig. 14.

High-G Accelerometers. When devices for surviving conditions of many hundreds of g's are required, piezoelectric and piezoresistive accelerometers are used. In the piezoelectric device, the seismic mass is cemented to a piezoelectric crystal, such as barium titanate. The latter, in turn, is cemented to the case. Acceleration stress causes the crystal to generate a voltage that is proportional to acceleration. The device cannot measure steady acceleration. The use of piezoresistive elements was included in the earlier description of strain-gage accelerometer.

ACCELERATION TIME (Computer System). Sometimes referred to as "Start Time," acceleration time is the brief time span required for a magnetic tape transport or any type of mechanical device to attain operating speed. Several milliseconds may be required to start a tape in motion and attain a speed at which data can be written or read.

ACCELERATOR (Particle). Particles (Subatomic).

ACCELERATOR (Rubber). Rubber (Natural).

ACCESS TIME (Computer System). The time required for a computer to locate data or an instruction in storage and transfer it to an arithmetic unit where the required computations are performed. Also, the time required to transfer information which has been operated on from the arithmetic unit to storage. On an immediate access storage device, such as a magnetic core or semiconductor storage, access time is the interval that occurs from the start of the storage cycle to the availability of the addressed data at the output register of the storage. This corresponds to the read cycle of the storage unit. On a disk storage unit, access time is the interval required for the read-write head to be moved to the addressed track (seek time), plus the required rotation time of the disk (latency). The average latency is one-half the disk-rotation time.

ACCIDENT PREVENTION SYSTEMS. Automotive Electronics.

ACCOMMODATION COEFFICIENT. A quantity defined by the equation

$$a = \frac{T_3 - T_1}{T_2 - T_1}$$

where T_1 is the temperature of gas molecules striking a surface which is at temperature T_2, and T_3 is the temperature of the gas molecules as they leave the surface, a is the accommodation coefficient. It is, therefore, a measure of the extent to which the gas molecules leaving the surface are in thermal equilibrium with it.

ACCOMMODATION (Ocular). The mechanism whereby the equatorial diameter of the lens of the eye may be decreased and its thickness increased to focus clearly on the retina the image of a near object.

ACCRETION (Clouds). Precipitation and Hydrometeors.

ACCRETION (Geology). The process by which crystals and other solid bodies grow by the addition of material onto their surfaces. A concretion is a body that grows from the center outward in a regular manner by successive additions of material.

ACCUMULATOR (Computer System). Generally, a register and associated equipment in the arithmetic unit of a computer in which arithmetical and logical operations are performed. The term also applies to a unit in a digital computer where numbers are totaled, i.e., accumulated. Often the accumulator stores one operand and upon receipt of any second operand, the accumulator forms and stores the result of performing the indicated operation on both the first and second operands. Commonly, an accumulator will work in conjunction with an accumulator extension, the latter being used in multiplication, division, and some shifting operations. In a division operation, for example, the quotient may appear in the accumulator—with the remainder appearing in the extension. Where a machine may use several accumulators, odd- and even-numbered accumulators may be paired for accomplishing multiplication and division operations.

Recent designs generally provide for multiple general-purpose registers which provide the function normally associated with an accumulator, in addition (typically) to other functions. See also list of related terms in entry on **Data Processing.**

ACCUMULATOR (Hydraulic). The hydraulic accumulator is a hydraulic device consisting of a cyclinder and piston which is actuated by weight, springs, or compressed fluid. On the opposite side of the piston a fluid such as water, oil, air, etc., is stored and, consequently, is available to do work when the pressure on it is reduced. This work is obtained by virtue of the fact that while the fluid is discharged rapidly, giving large hydraulic power for short periods of time, it may be refilled by a comparatively small and low-powered pump working a much longer period of time.

One successful type of accumulator consists of a closed cylinder containing a rubber bag filled with air or gas. Fluid is pumped into the cylinder under pressure, compressing the air or gas in the bag. When the release valve is opened the air or gas in the bag expands to eject the fluid from the cylinder.

Another type of hydraulic accumulator is the pumped storage plant, now being looked on with considerable favor by electric power systems for the economic carrying of variable loads. As employed in conjunction with steam generating stations, steam turbine-driven centrifugal pumps raise water from a lower to an upper pool with off-peak power. During the peak-load periods this water is released to the lower pool through a hydraulic turbo-generator as rapidly as is needed to give the required power. The hydraulic storage of power of this nature is essentially a high head development, low head equipment, and hydraulic losses being too expensive. In favorable locations the overall efficiency of conversion and storage may not need to be greater than 50% in order to justify the project.

ACCURACY. In terms of instruments and scientific measuring systems, accuracy may be defined as the conformity of an indicated value to an accepted standard value, or true value. Accuracy is usually measured in terms of *inaccuracy* and expressed as *accuracy.* As a

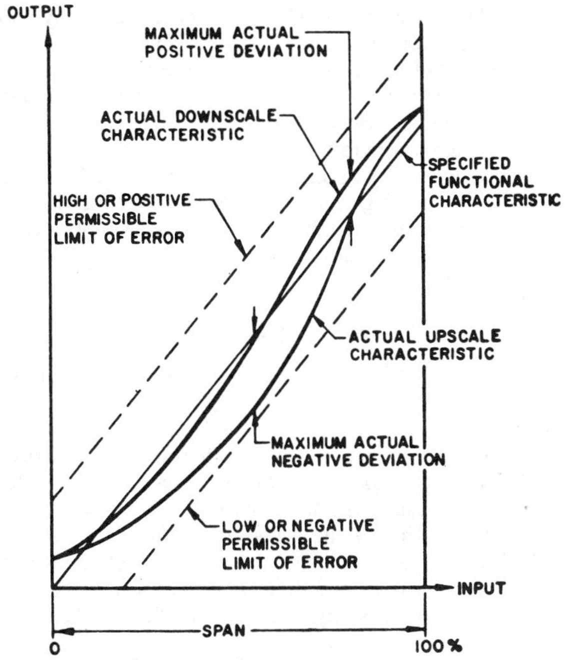

Fundamental relationships pertaining to reference accuracy.

performance specification, accuracy should be assumed to mean *reference accuracy* unless otherwise stated.

Reference accuracy may be defined as a number or quantity which defines the limit that errors will not exceed when the device is used under reference operating conditions. See accompanying diagram. Reference accuracy includes the combined conformity, hysteresis, and repeatability errors. The units being used must be stated explicitly. It is preferred that a + and − sign precede the number or quantity. The absence of a sign infers both a plus and a minus sign. Reference accuracy can be expressed in a number of forms, of which the five following examples are typical:

(1) Reference accuracy expressed in terms of the *measured variable.* Typical expression: The reference accuracy is ± 1°F (± 0.6°C).

(2) Reference accuracy expressed in percent of *span.* Typical expression: The reference accuracy is ± 1/2% of span. This percentage is calculated using units, such as deg. F or C, psi, and so on.

(3) Reference accuracy expressed in percent of the *upper range-value.* Typical expression: The reference accuracy is ± 1/2% of upper range-value. This percentage is also calculated, using units as previously given.

(4) Reference accuracy expressed in percent of *scale length.* Typical expression: The reference accuracy is ± 1/2% of scale length.

(5) Reference accuracy expressed in percent of *actual output reading.* Typical expression: The reference accuracy is ± 1% of actual output reading. Again, this percentage is calculated, using scale units such as previously given.

The foregoing definitions and explanations are from SAMA Standards, Scientific Apparatus Makers Association, New York.

ACERACEAE. Maple Tree.

ACETALDEHYDE. CH_3CHO, formula weight 44.05, colorless, odorous liquid, mp − 123.5°C, bp 20.2°C, sp gr 0.783. Also known as *ethanal,* acetaldehyde is miscible with H_2O, alcohol, or ether in all proportions. Because of its versatile chemical reactivity, acetaldehyde is widely used as a commencing material in organic syntheses, including the production of resins, dyestuffs, and explosives. The compound also is used as a reducing agent, preservative, and as a medium for silvering mirrors. In resin manufacture, paraldehyde $(CH_3CHO)_3$ sometimes is preferred because of its higher boiling and flash points.

In tonnage production, acetaldehyde may be manufactured by (1) the direct oxidation of ethylene, requiring a catalytic solution of copper chloride plus small quantities of palladium chloride, (2) the oxidation of ethyl alcohol with sodium dichromate, and (3) the dry distillation of calcium acetate with calcium formate.

Acetaldehyde reacts with many chemicals in a marked manner, (1) with ammonio-silver nitrate ("Tollen's solution"), to form metallic silver, either as a black precipitate or as an adherent mirror film on glass, (2) with alkaline cupric solution ("Fehling's solution") to form cuprous oxide, red to yellow precipitate, (3) with rosaniline (fuchsine, magenta), which has been decolorized by sulfurous acid ("Schiff's solution"), the pink color of rosaniline is restored, (4) with NaOH, upon warming, a yellow to brown resin of unpleasant odor separates (this reaction is given by aldehydes immediately following acetaldehyde in the series, but not by formaldehyde, furfuraldehyde or benzaldehyde), (5) with anhydrous ammonia, to form aldehyde-ammonia $CH_3 \cdot CHOH \cdot NH_2$, white solid, mp 97°C, bp 111°C, with decomposition, (6) with concentrated H_2SO_4, heat is evolved, and with rise of temperature, paraldehyde $(C_2H_4O)_3$ or

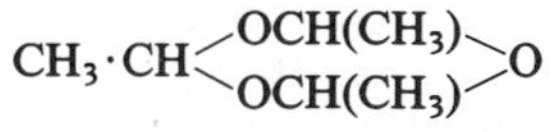

colorless liquid bp 124°C, slightly soluble in H_2O, is formed, (7) with acids, below 0°C, forms metaldehyde $(C_2H_4O)_x$ white solid, sublimes at about 115°C without melting but with partial conversion to acetaldehyde, (8) with dilute HCl or dilute NaOH, aldol $CH_3 \cdot CHOH \cdot CH_2 \cdot CHO$ slowly forms, (9) with phosphorus pentachloride, forms ethylidene chloride $CH_3 \cdot CHCl_2$, colorless liquid, bp 58°C, (10) with ethyl alcohol and dry hydrogen chloride, forms acetal, 1,1-diethyoxyethane $CH_3 \cdot CH(OC_2H_5)_2$, colorless liquid, bp 104°C, (11) with hydrocyanic acid, forms acetaldehyde cyanhydrin $CH_3 \cdot CHOH \cdot CN$, readily converted into alphahydroxypropionic acid $CH_3 \cdot CHOH \cdot COOH$, (12) with sodium hydrogen sulfite, forms acetaldehyde sodium bisulfite $CH_3 \cdot CHOH \cdot SO_3Na$, white solid, from which acetaldehyde is readily recoverable by treatment with sodium carbonate solution, (13) with hydroxylamine hydrochloride forms acetaldoxime $CH_3 \cdot CH{:}NOH$, white solid, mp 47°C, (14) with phenylhydrazine, forms acetaldehyde phenylhydrazone $CH_3 \cdot CH{:}N \cdot NH \cdot C_6H_5$, white solid, mp 98°C, (15) with magnesium methyl iodide in anhydrous ether ("Grignard's solution"), yields, after reaction with water, isopropyl alcohol $(CH_3)_2CHOH$, a secondary alcohol, (16) with semicarbazide, forms acetaldehyde semicarbazone $CH_3 \cdot CH{:}N \cdot NH \cdot CO \cdot NH_2$, white solid, mp 162°C, (17) with chlorine, forms trichloroacetaldehyde ("chloral") $CCL_3 \cdot CHO$, (18) with H_2S, forms thioacetaldehyde $CH_3 \cdot CHS$ or $(CH_3 \cdot CHS)_3$. Acetaldehyde stands chemically between ethyl alcohol on one hand—to which it can be reduced—and acetic acid on the other hand—to which it can be oxidized. These reactions of acetaldehyde, coupled with its ready formation from acetylene by mercuric sulfate solution as a catalyzer, open up a vast field of organic chemistry with acetaldehyde as raw material: acetaldehyde hydrogenated to ethyl alcohol; oxygenated to acetic acid, thence to acetone, acetic anhydride, vinyl acetate, vinly alcohol. Acetaldehyde is also formed by the regulated oxidation of ethyl alcohol by such a reagent as sodium dichromate in H_2SO_4 (chromic sulfate also produced). Reactions (1), (3), (14), and (16) above are most commonly used in the detection of acetaldehyde. See also **Aldehydes.**

ACETAL GROUP. An organic compound of the general formula RCH(OR′)(OR″) is termed an *acetal* and is formed by the reaction of an aldehyde with an alcohol, usually in the presence of small amounts of acids or appropriate inorganic salts. Acetals are stable toward alkali, are volatile, insoluble in H_2O, and generally are similar structurally to ethers. Unlike ethers, acetals are hydrolyzed by acids into their respective aldehydes. $H(R)CO + (HO \cdot C_2H_5)_2 \rightarrow H(R)COC_2H_5)_2 + H_2O$. Representative acetals include: $CH_2(OCH_3)_2$, methylene dimethyl ether, bp 42°C; $CH_3CH(OCH_3)_2$, ethylidene dimethyl ether, bp 64°C; and $CH_3CH(OC_2H_5)_2$, ethylidene diethyl ether, bp 104°C.

ACETAL RESINS. A very useful family of thermoplastic resins, obtainable both as homopolymers and copolymers, produced mainly from formaldehyde or a formaldehyde derivative. Acetals have the highest fatigue endurance of commercial thermoplastics. In their manufacture, a variety of ionic initiators, such as tertiary amines and quaternary ammonium salts, are used to effect polymerization of

formaldehyde. Chain transfer, as shown by the following reactions, controls the molecular weight of the resulting resins:
Step 1, initiation of new chain:

$$—CH_2OCH_2O^- + H_2O \rightarrow —CH_2OCH_2OH + OH^-$$

Step 2, reaction of growing chain with H_2O, releasing hydroxyl ion:

$$OH^- + CH_2O \rightarrow HOCH_2O^-$$

Step 3, end-capping of high-molecular-weight polyoxymethylene glycol to provide stable commercial resin:

$$HOCH_2O^- + nCH_2O \rightarrow HOCH_2O(CH_2O)_{n-1}CH_2O^-$$

Starting ingredients may be formaldehyde or the cyclic trimer trioxane, $CH_2OCH_2OCH_2O$. Both form polymers of similar properties. Boron trifluoride or other Lewis acids are used to promote polymerization where trioxane is the raw material.

Acetals provide excellent resistance to most organic compounds except when exposed for long periods at elevated temperatures. The resins have limited resistance to strong acids and oxidizing agents. The copolymers and some of the homopolymers are resistant to the action of weak bases. Normally, where resistance to burning, weathering, and radiation are required, acetals are not specified. The resins are used for cams, gears, bearings, springs, sprockets, and other mechanical parts, as well as for electrical parts, housings, and hardware.

ACETATE FIBERS. Cellulose acetate fiber, or acetate, is a chemical derivative of the naturally occurring polymer, cellulose. Two types of acetate fibers are produced: (1) Fibers made from partially hydrolyzed cellulose triacetate, called *secondary acetate* (or, more simply, *acetate*); and (2) fibers that are fully acetylated cellulose, called *triacetate fiber.*

Since the early 1970s, the total amount of acetate fiber produced per year has remained quite steady, but the distribution in its various markets has shifted. Acetate yarn has retained a strong position in the textile market despite the competitive pressures of newer synthetic textile yarns. The market strength for acetate derives from its pleasing and attractive aesthetic properties and its competitive pricing. In 1979,

PHYSICAL PROPERTIES OF ACETATE YARN

Property	Secondary Acetate	Triacetate
Tenacity, g/denier:		
Conditioned	1.2–1.4	1.1–1.3
Wet	0.8–1.0	0.8–1.0
Breaking elongation, %:		
Conditioned	25–45	26–35
Wet	35–50	30–40
Density, g/cm³	1.32	1.30
Percent moisture regain (65% relative humidity at 22°C)	6.3–6.5	3.2

a total of 690 million pounds (312 million kilograms) were produced, of which 317 million pounds (144 million kilograms) comprised filament yarn. This consisting of about 85% diacetate yarn and 15% triacetate yarn. Seven million pounds (3.1 million kilograms) of diacetate staple were produced; and 366 million pounds (166 million kilograms) of diacetate tow were produced. The tow product is used primarily for cigarette filters and represents a considerable volume increase over a decade ago.

Chemistry. Highly purified cellulose wood pulps (greater than 95% α-cellulose) is the basic raw material for manufacture of cellulose acetate. The natural polymer, cellulose, in wood pulp has a degree of polymerization of 500 to 1000, the basic repeating unit of which is cellobiose:

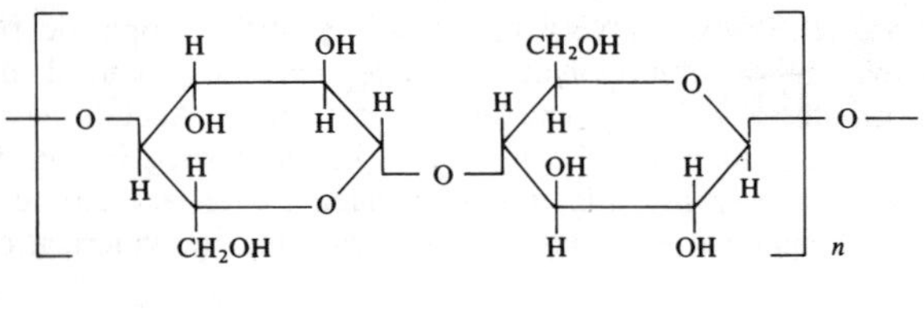

Cellobiose contains two anhydroglucose units connected by a B-1,4-glucoside linkage. Each anhydroglucose unit contains two secondary alcohols in the 2 and 3 positions and one primary alcohol in the 6 position. Esterification of the three hydroxyl groups produces cellulose triacetate, but in practice, the commercial triacetate has a DS of 2.80–2.95;* and secondary acetate has a DS of 2.35–2.40. Since cellulose and acetic acid do not react directly to any appreciable extent, the ester is prepared by reacting cellulose with acetic anhydride in a solvent of glacial acetic acid, using sulfuric acid as a catalyst. The chemical reaction for converting cellulose to the triacetate is:

$$C_6H_7O_2(HO)_3 + (CH_3CO_2)_2O \rightarrow C_6H_7O_2(OCOCH_3)_3 + 3\ CH_3COOH$$

where $C_6H_7O_2(OH)_3$ stands for one anhydroglucose unit in the cellulose molecule.

The solution of triacetate is converted to secondary acetate through the addition of aqueous acetic acid which results in hydrolysis of about 20% of the ester groups. Addition of water precipitates the secondary acetate and halts the deesterification. This procedure results in a random distribution of the acetyl and hydroxyl groups on the polymer which, in turn, enhances the solubility of the secondary acetate in acetone. The latter is commonly used as solvent for spinning. The triacetate is produced by precipitating the product without hydrolysis.

Production Processes. Acetic anhydride is produced by many of the acetate producers in the United States inasmuch as cellulose acetate is the major end use for this chemical. The anhydride is the product of reaction with ketene and acetic acid. Ketene is made by the catalytic pyrolysis of either acetic acid or acetone.

(1) Wood pulp (high alpha type) in roll or sheet form is shredded to permit rapid penetration of the pulp by the reactants.

(2) The shredded pulp is treated with glacial acetic acid which may contain part or all of the sulfuric acid catalyst. This "pretreatment" swells the cellulose fibers and increases the accessibility of the acetylating agent. Ratios of acetic acid to cellulose may be 1 : 1 to 3 : 1 and the treatment time may vary from one-half to several hours, depending upon temperature.

(3) The activated pulp is then treated with the acetylation solution, a mixture of acetic anhydride and acetic acid in about 2 : 3 ratio and containing the remainder of the sulfuric acid catalyst (5–20% based on cellulose). Since the reaction is exothermic and it is necessary to keep the reaction temperature low (< 50°C), precooling of the acetylation mixture and/or external cooling of the reaction vessel are employed. Reaction times of 5 to 10 hours, under controlled temperature, result in the pulp mass dissolving in the acetylating mixture. At this point, the cellulose is completely acetylated in the triacetate form. If triacetate is the desired product, aqueous acetic acid is applied to destroy the anhydride and the product is precipitated, washed, and dried.

(4) In producing the secondary acetate, the excess anhydride is reacted with aqueous acetic acid under careful temperature control. When all the anhydride is reacted, excess water is added to the extent of 10–30% of the mixture. The acetate is allowed to hydrolyze until a DS of 2.35–2.40 is obtained. Modern practice permits hydrolysis to occur in 4 to 8 hours at temperatures of 70–80°C. Sulfur esters are also hydrolyzed and sodium or magnesium acetate may be introduced with the hydrolysis water to neutralize the acid produced. Additional water precipitates the secondary acetate and control is exercised to obtain an open "flake" rather than a pelletized material. The open flake makes more efficient the removal of acetic acid in subsequent washing. The flake is dried and conveyed to storage bins.

Cellulose acetate is converted into fibers by solvating the polymer in an acetone solution which is then spun by a process called *dry spinning.* In the spinning process, the polymer solution is metered through a spinneret into a column containing warm air which evaporates the acetone. The solidified filaments are oiled and package collected at the base of the spinning tube.

(5) Preparation of the polymer spinning solution or "dope" begins

* The DS number is the average number of hydroxyl groups esterified per anhydroglucose unit. DS = degree of substitution.

with blending of flake batches to ensure uniformity of chemical composition, particularly acetyl content, which is important for uniform dyeing properties of the spun yarn. Either batch or continuous mixers are used for dissolving the blended flake in acetone-water solution. The final composition of the "dope" is about 26% acetate, 72% acetone, and 2% water. The water serves as a cosolvent with the acetone, leading to a marked reduction in the viscosity of the solution, which is approximately 900–1000 P (poises). A delustrant, such as titanium dioxide (TiO_2) may be added to the dope in concentrations of 1–2%, based on the acetate content. Colored pigments may also be added to the dope to produce mass-dyed fibers.

(6) The viscous "dope" solution is filtered several times to eliminate insoluble fiber and debris. Plate-and-frame as well as continuous filtration may be used. After filtration, the dope is transported to storage tanks where it is permitted to deaerate just prior to spinning.

(7) The deaerated dope is pumped to the spinning machine headers which distribute the dope to the individual spinning positions. Each machine has a hundred or more spinning positions where a single yarn of acetate yarn is produced. A metering pump provides accurate control of the quantity of dope pumped through a small line filter and heat exchanger which raises the temperature to 50–70°C just prior to reaching the spinneret. Spinnerets are made of stainless steel and incorporate 14–100 or more holes. The holes are 0.0015–0.0025 inch (0.04–0.06 millimeter) in diameter, having a depth slightly less than their diameter.

(8) The nascent acetate filaments leaving the jet are carried through the vertical spinning tube which may be 10–20 feet (3–6 meters) long and from 6 to 12 inches (15 to 30 centimeters) in diameter. Hot air moving concurrent or countercurrent to the path of the filaments evaporates most of the acetone and solidifies the liquid stream into the acetate filament.

(9) The acetone-laden air removed from the spinning tubes is stripped of acetone as it passes through activated carbon absorbers. The absorbers are steam-stripped and the acetone recovered by distillation. An acetone recovery efficiency of 96% overall is commonly experienced in the industry.

(10) As the yarn leaves the spinning tube, it is passed over an oiling device which applies 2–3% of lubricant of the weight of the yarn. The lubricant serves to reduce friction in subsequent textile processing and to diminish static electrification. From the oiler, the yarn passes over a rotating *godet* or wheel which supplies the force to transport the yarn from the jet to the take-up mechanism which packages the product.

Spinning speeds for acetate yarn range from 492 to 2297 feet (150 to 700 meters) per minute. Since acetone must be removed from the yarn in the spinning tube, the spinning speed depends upon the denier of the yarn, air velocity and temperature, dope temperature, and composition and spinning tube length.

Spinning of cellulose triacetate yarns follows the procedures used for the secondary acetate except that a different solvent is used. This is generally a 90 : 10 mixture of methylene chloride and methanol.

The manufacture of acetate staple and tow follow the same scheme as employed for continuous filament yarns, except that the yarns are combined as they leave the spinning tube to form a tow which is mechanically crimped and cut into staple lengths. Acetate and triacetate are packaged in bales of about 400 pounds (181 kilograms).

Properties. Commercial acetate and triacetate yarns are produced in a range of 45–900 denier,* the largest portion being in the range of 55–150. The filament count for these yarns ranges from 14 to 100 and the denier per filament ranges from 2.5 to 9. The yarns are produced in both bright and dull lusters, the latter type containing 1–2% TiO_2. Color-pigmented or "solution dyed" acetate yarns are also produced. The cross-sectional shape of acetate fibers is typically crenulated and symmetrically circular. The crenulated cross sections result from the formation of a rigid skin during the initial filament formation process when the fiber is largely fluid. As the acetone volatilizes through the skin, the fiber solidifies and shrinks, which results in skin folds or crenulations. Shaped fiber cross sections, such as Y shapes, are also produced and provide the capability for increasing the bulk of yarns and fabrics.

Thermal Properties. Secondary acetate, having a random substitution of the hydroxyl groups, shows very little crystallinity and cannot be induced to crystallize or heat set. On the other hand, triacetate is a stereospecific molecule capable of crystallizing when heated above 205°C, the glass transition temperature, for a short period. Hence, triacetate offers heat-setting advantages in fabric finishing processes. Triacetate fabrics, heated at 240°C for 30 seconds, in a state of tension, exhibit good retention of the flat geometry and resist wrinkling at lower temperatures. Pleats and creases applied to triacetate garments, which are then heated above the glass transition temperature, will be retained over the normal period of usage. The melting points for secondary acetate and triacetate are 260° and 300°C, respectively. Secondary acetate softens in the range of 205–230°C, which necessitates a moderate ironing temperature. Triacetate does not soften to the sticking temperature of 232°C and hence can accept ironing temperatures usually used for cotton.

Physical Properties. Key physical properties are given in the accompanying table.

Chemical Properties and Dyeing. Acetate and triacetate fibers are readily dyed with disperse and azoic dyestuffs. Acid dyes may be used for print dyeing the triacetate fibers. Acetate and triacetate fibers, being organic esters, are susceptible to hydrolysis in strong alkaline solutions. Weakly basic or acid solutions have little effect on these fibers. Severe degradation occurs through hydrolysis when these fibers are subjected to strong mineral acids. Both the secondary acetate and triacetate show strong resistance to hypochlorite and peroxide bleaches. The fibers are not affected by normal dry-cleaning solvents. See also **Fibers.**

Joseph W. Schappel, Avtex Fibers Inc., Front Royal, Virginia.

ACETIC ACID. CH_3COOH, formula weight 60.05, colorless, acrid liquid, mp 16.7°C, bp 118.1°C, sp gr 1.049. Also known as ethanoic acid or vinegar acid, this compound is miscible with H_2O, alcohol, and ether in all proportions. Acetic acid is available commercially in several concentrations. The CH_3COOH content of glacial acetic is approximately 99.7% with H_2O the principal impurity. Reagent acetic acid generally contains 36% CH_3COOH by weight. Standard commercial aqueous solutions are 28, 56, 70, 80, 85, and 90% CH_3COOH. Acetic acid is the active ingredient in vinegar in which the content ranges from 4 to 5% CH_3COOH. Acetic acid is classified as a weak, monobasic acid. The three hydrogen atoms linked to one of the two carbon atoms are not replaceable by metals.

In addition to the large quantities of vinegar produced, acetic acid in its more concentrated forms is an important high-tonnage industrial chemical, both as a reactive raw and intermediate material for various organic syntheses and as an excellent solvent. Acetic acid is required in the production of several synthetic resins and fibers, pharmaceuticals, photographic chemicals, flavorants, and bleaching and etching compounds.

Early commercial sources of acetic acid included (1) the combined action of *Bacterium aceti* and air on ethyl alcohol in an oxidation-fermentation process: $C_2H_5OH + O_2 \rightarrow CH_3COOH + H_2O$, the same reaction which occurs when weakly alcoholic beverages, such as beer or wine, are exposed to air for a prolonged period and which turn sour because of the formation of acetic acid; and (2) the destructive distillation of wood. A number of natural vinegars still are made by fermentation and marketed as the natural product, but diluted commercially and synthetically produced acetic acid is a much more economic route to follow. The wood distillation route was phased out because of shortages of raw materials and the much more attractive economy of synthetic processes.

The most important synthetic processes are (1) the oxidation of acetaldehyde, and (2) the direct synthesis from methyl alcohol and carbon monoxide. The latter reaction must proceed under very high pressure (approximately 650 atmospheres) and at about 250°C. The reaction takes place in the liquid phase and dissolved cobaltous iodide is the catalyst. $CH_3OH + CO \rightarrow CH_3COOH$ and $CH_3OCH_3 + H_2O + 2CO \rightarrow 2CH_3COOH$. The crude acid produced first is separated from the catalyst and then dehydrated and purified in an azeo-

* Denier is a measure of the fineness of a yarn and is the weight in grams of 9000 meters of yarn.

tropic distillation column. The final product is approximately 99.8% pure CH_3COOH.

Acetic acid solution reacts with alkalis to form acetates, e.g., sodium acetate, calcium acetate; similarly, with some oxides, e.g., lead acetate; with carbonates, e.g., sodium acetate, calcium acetate, magnesium acetate; with some sulfides, e.g., zinc acetate, manganese acetate. Ferric acetate solution, upon boiling, yields red precipitate of basic ferric acetate. Acetic acid solution attacks many metals, liberating hydrogen and forming acetate, e.g., magnesium, zinc, iron. Acetic acid is an important organic substance, with alcohols forming esters (acetates); with phosphorus trichloride forming acetyl chloride $CH_3 \cdot CO \cdot Cl$, which is an important reagent for transfer of the acetyl (CH_3CO—) group; forming acetic anhydride, also an acetyl reagent; forming acetone and calcium carbonate when passed over lime and a catalyzer (barium carbonate) or when calcium acetate is heated; forming methane (and sodium carbonate) when sodium acetate is heated with NaOH; forming mono-, di-, tri-chloroacetic (or bromoacetic) acids by reaction with chlorine (or bromine) from which hydroxy- and amino-, aldehydic-, dibasic acids, respectively, may be made; forming acetamide when ammonium acetate is distilled. Acetic acid dissolves sulfur and phosphorus, is an important solvent for organic substances, and causes painful wounds when it comes in contact with the skin. Normal acetates are soluble, basic acetates insoluble. The latter are important in their compounds with lead, copper ("verdigris").

A large number of acetic acid esters are important industrially, including methyl, ethyl, propyl, butyl, amyl, and cetyl acetates; glycol mono- and diacetate; glyceryl mono-, di-, and triacetate; glucose pentacetate; and cellulose tri-, tetra-, and pentacetate.

Acetates may be detected by formation of foul-smelling cacodyl (poisonous) on heating with dry arsenic trioxide. Other tests for acetate are the lanthanum nitrate test in which a blue or bluish-brown ring forms when a drop of 2.5% $La(NO)_3)_3$ solution, a drop of 0.01 *N* iodine solution, and a drop of 0.1% NH_4OH solution are added to a drop of a neutral acetate solution; the ferric chloride test in which a reddish color is produced by the addition of 1 *N* ferric chloride solution to a neutral solution of acetate; and the ethyl acetate test in which ethyl alcohol and H_2SO_4 are added to the acetate solution and warmed to form the odorous compound.

ACETIC ACID AND ACETATES (Foods). Antimicrobial Agents (Foods).

ACETIC ANHYDRIDE. Acetate Fibers.

ACETOACETIC ESTER CONDENSATION. A class of reactions occasioned by the dehydrating power of metallic sodium or sodium ethoxide on the ethyl esters of monobasic aliphatic acids and a few other esters. It is best known in the formation of acetoacetic ester:

$$2CH_3 \cdot COOC_2H_5 + 2CH_3 \cdot COOC_2H_5 + 2Na \rightarrow 2CH_3 \cdot C(ONa){:}CH \cdot COOC_2H_5 + 2C_2H_5OH + H_2$$

The actual course of the reaction is complex. By the action of acids the sodium may be eliminated from the first product of the reaction and the free ester obtained. This may exist in the tautometic enol and keto forms ($CH_3 \cdot COH{:}CH \cdot COOC_2H_5$ and $CH_3 \cdot CO \cdot CH_2 \cdot COOC_2H_5$).

On boiling ester with acids or alkalies it will split in two ways, the circumstances determining the nature of the main product. Thus, if moderately strong acid or weak alkali is employed, acetone is formed with very little acetic acid (ketone splitting). In the presence of strong alkalies however, very little acetone and much acetic acid result (acid splitting). Derivatives of acetoacetic ester may be decomposed in the same fashion, and this fact is responsible for the great utility of this condensation in organic synthesis. This is also due to the reactivity of the $\cdot CH_2 \cdot$ group, which reacts readily with various groups, notably halogen compounds. Usually the sodium salt of the ester is used, and the condensation is followed by decarboxylation with dilute alkali, or deacylation with concentrated alkali.

$$CH_3 \cdot CO \cdot CHNa \cdot COOC_2H_5 + RI \rightarrow CH_3 \cdot CO \cdot CHR \cdot COOC_2H_5 + NaI$$

$$CH_3 \cdot CO \cdot CHR \cdot COOC_2H_5 \xrightarrow[\text{Dilute alkali}]{H_2O} CH_3 \cdot CO \cdot CH_2R + C_2H_5OH + CO_2$$

$$CH_3 \cdot CO \cdot CHR \cdot COOC_2H_5 \xrightarrow[\text{Concentrated alkali}]{2H_2O} HOOC \cdot CH_2 \cdot R + C_2H_5OH + CH_3COOH$$

ACETONE. $CH_3 \cdot CO \cdot CH_3$, formula weight 58.08, colorless, odorous liquid ketone, mp −94.6°C, bp 56.5°C, sp gr 0.792. Also known as dimethyl ketone or propanone, this compound is miscible in all proportions with H_2O, alcohol, or ether. Acetone is a very important solvent and is widely used in the manufacture of plastics and lacquers. For storage purposes, acetylene may be dissolved in acetone. A high-tonnage chemical, acetone is the starting ingredient or intermediate for numerous organic syntheses. Closely related, industrially important compounds are diacetone alcohol (DAA) $CH_3 \cdot CO \cdot CH_2 \cdot COH(CH_3)_2$ which is used as a solvent for cellulose acetate and nitrocellulose, as well as for various resins and gums, and as a thinner for lacquers and inking materials. Sometimes DAA is mixed with castor oil for use as a hydraulic brake fluid for which its physical properties are well suited, mp −54°C, bp 166°C, sp gr 0.938. A product known as synthetic methyl acetone is prepared by mixing acetone (50%), methyl acetate (30%), and methyl alcohol (20%) and is used widely for coagulating latex and in paint removers and lacquers.

In older industrial processes, acetone is prepared (1) by passing the vapors of acetic acid over heated lime. In a first step, calcium acetate is produced, followed by a breakdown of the acetate into acetone and calcium carbonate:

$$CH_3 \cdot CO \cdot O \cdot Ca \cdot OOC \cdot CH_3 \rightarrow CH_3 \cdot CO \cdot CH_3 + CaCO_3;$$

and (2) by fermentation of starches, such as maize, which produce acetone along with butyl alcohol. Modern industrial processes include (3) the use of cumene as a chargestock, in which cumene first is oxidized to cumene hydroperoxide (CHP), this followed by the decomposition of CHP into acetone and phenol; and (4) by the direct oxidation of propylene, using air and catalysts. The catalyst solution consists of copper chloride and small amounts of palladium chloride. The reaction: $CH_3CH = CH_2 + 1/2O_2 \rightarrow CH_3COCH_3$. During the reaction, the palladium chloride is reduced to elemental palladium and HCl. Reoxidation is effected by cupric chloride. The cuprous chloride resulting is reoxidized during the catalyst regeneration cycle. The process is carried out under moderate pressure at about 100°C.

Acetone reacts with many chemicals in a marked manner: (1) with phosphorus pentachloride, yields acetone chloride $(CH_3)_2CCl_2$, (2) with hydrogen chloride dry, yields both mesityl oxide $CH_3COCH{:}C(CH_3)_2$, liquid, bp 132°C, and phorone $(CH_3)_2C{:}CHCOCH{:}C(CH_3)_2$, yellow solid, mp 28°C, (3) with concentrated H_2SO_4, yields mesitylene $C_6H_3(CH_3)_3$ (1,3,5), (4) with NH_3, yields acetone amines, e.g., diacetoneamine $C_6H_{12}ONH$, (5) with HCN, yields acetone cyanhydrin $(CH_3)_2CHOH \cdot CN$, readily converted into alpha-hydroxy acid $(CH_3)_2CHOH \cdot COOH$, (6) with sodium hydrogen sulfite, forms acetonesodiumbisulfite $(CH_3)_2COH \cdot SO_3Na$, white solid, from which acetone is readily recoverable by treatment with sodium carbonate solution, (7) with hydroxylamine hydrochloride, forms acetoxime $(CH_3)_2C{:}NOH$, solid, mp 60°C, (8) with phenylhydrazine, yields acetonephenyl-hydrazone $(CH_3)_2C{:}NNHC_6H_5 \cdot H_2O$, solid, mp 16°C, anhydrous compound, mp 42°C, (9) with semicarbazide, forms acetonesemicarbazone $(CH_3)C{:}NNHCONH_2$, solid, mp 189°C, (10) with magnesium methyl iodide in anhydrous ether ("Grignard's solution"), yields, after reaction with H_2O, trimethylcarbinol $(CH_3)_3COH$, a tertiary alcohol, (11) with ethyl thioalcohol and hydrogen chloride dry, yields mercaptol $(CH_3)_2C(SC_2H_5)_2$, (12) with hypochlorite, hypobromite, or hypoiodite solution, yields chloroform $CHCl_3$, bromoform $CHBr_3$ or iodoform CHI_3, respectively, (13) with most reducing agents, forms isopropyl alcohol $(CH_3)_2CHOH$, a secondary alcohol, but with sodium amalgam forms pinacone $(CH_3)_2COH \cdot COH(CH_3)_2$, (14) with sodium dichromate and H_2SO_4, forms acetic acid CH_3COOH plus CO_2. When acetone vapor is passed through a tube at a dull red heat, ketene $CH_2{:}CO$ and methane CH_4 are formed.

ACETONITRILE. Acrylonitrile.

ACETOSULFAM. **Sweeteners.**

ACETYLCHOLINE. **Brain and Nervous System; Choline and Cholinesterase.**

ACETYL CHLORIDE. **Chlorinated Organics.**

ACETYL COENZYME. **Carbohydrates.**

ACETYLENE. CH⋮CH, formula weight 26.04, mp −81.5°C, bp −84°C, sp gr 0.905 (air = 1.000). Sometimes referred to as *ethyne*, *ethine*, or *gaseous carbon* (92.3% of compound is C), acetylene is moderately soluble in H_2O or alcohol, and exceptionally soluble in acetone (300 volumes of acetylene in 1 volume of acetone at 12 atmospheres pressure). The gas burns when ignited in air with a luminous sooty flame, requiring a specially devised burner for illumination purposes. An explosive mixture is formed with air over a wide range (about 3 to 80% acetylene), but safe handling is improved when the gas is dissolved in acetone. The heating value is 1455 Btu/ft^3 (8.9 Cal/m^3).

Although acetylene still is used in a number of organic syntheses on an industrial scale, its use on a high-tonnage basis has diminished because of the lower cost of other starting materials, such as ethylene and propylene. Acetylene has been widely used in the production of halogen derivatives, acrylonitrile, acetaldehyde, and vinyl chloride. Within recent years, producers of acrylonitrile switched to propylene as a starting material.

Commercially, acetylene is produced from the pyrolysis of naphtha in a two-stage cracking process. Both acetylene and ethylene are end-products. The ratio of the two products can be changed by varying the naphtha feed rate. Acetylene also has been produced by a submerged-flame process from crude oil. In essence, gasification of the crude oil occurs by means of the flame which is supported by oxygen beneath the surface of the oil. Combustion and cracking of the oil take place at the boundaries of the flame. The composition of the cracked gas includes about 6.3% acetylene and 6.7% ethylene. Thus, further separation and purification are required. Several years ago when procedures were developed for the safe handling of acetylene on a large scale, J. W. Reppe worked out a series of reactions that later became known as "Reppe chemistry." These reactions were particularly important to the manufacture of many high polymers and other synthetic products. Reppe and his associates were able to effect synthesis of chemicals that had been commercially unavailable. An example is the synthesis of cyclooctatetraene by heating a solution of acetylene under pressure in tetrahydrofuran in the presence of a nickel cyanide catalyst: $C_2H_2 \rightarrow C_8H_8$. In another reaction, acrylic acid was produced from CO and H_2O in the presence of a nickel catalyst: $C_2H_2 + CO + H_2O \rightarrow CH_2{:}CH \cdot COOH$. These two reactions are representative of a much larger number of reactions, both those that are straight-chain only, and those involving ring closure.

Acetylene reacts (1) with chlorine, to form acetylene tetrachloride $C_2H_2Cl_4$ or $CHCl_2 \cdot CHCl_2$ or acetylene dichloride $C_2H_2Cl_2$ or $CHCl{:}CHCl$, (2) with bromine, to form acetylene tetrabromide $C_2H_2Br_4$ or $CHBr_2 \cdot CHBr_2$ or acetylene dibromide $C_2H_2Br_2$ or $CHBr{:}CHBr$, (3) with hydrogen chloride (bromide, iodide), to form ethylene monochloride $CH_2{:}CHCl$ (monobromide, monoiodide), and 1,1-dichloroethane, ethylidene chloride $CH_3 \cdot CHCl_2$ (dibromide, diiodide), (4) with H_2O in the presence of a catalyzer, e.g., mercuric sulfate, to form acetaldehyde $CH_3 \cdot CHO$, (5) with hydrogen, in the presence of a catalyzer, e.g., finely divided nickel heated, to form ethylene C_2H_4 or ethane C_2H_6, (6) with metals, such as copper or nickel, when moist, also lead or zinc, when moist and unpurified. Tin is not attacked. Sodium yields, upon heating, the compounds C_2HNa and C_2Na_2. (7) With ammonio-cuprous (or silver) salt solution, to form cuprous (or silver) acetylide C_2Cu_2, dark red precipitate, explosive when dry, and yielding acetylene upon treatment with acid, (8) with mercuric chloride solution, to form trichloromercuric acetaldehyde $C(HgCl)_3 \cdot CHO$, precipitate, which yields with HCl acetaldehyde plus mercuric chloride.

ACETYLENE SERIES. A series of unsaturated hydrocarbons having the general formula C_nH_{2n-2}, and containing a triple bond between two carbon atoms. The series is named after the simplest compound of the series, acetylene HC⋮CH. In more modern terminology, this series of compounds is termed the *alkynes*. See also **Alkynes.**

ACETYLSALICYCLIC ACID. $C_6H_4(COOH)CO_2CH_3$, formula weight, 180.06, melting point, 133.5°C, colorless, crystalline, slightly soluble in water, soluble in alcohol and ether, commonly known as aspirin, also called orthoacetoxybenzoic acid. The substance is commonly used as a relief for mild forms of pain, including headache, joint, and muscle pain. The drug tends to reduce fever. Aspirin and other forms of salicylates have been used in large doses in acute rheumatic fever, but must be administered with extreme care in such cases by a physician. Commercially available aspirin commonly is mixed with other pain relievers, so-called buffering agents, and other ingredients for the treatment of common colds.

ACHENE. A single-seeded, indehiscent fruit, in which the seed is free from the ovary wall except at the point of attachment. An example is the sunflower "seed."

ACHERNAR (α Eridani). Ranking tenth in apparent brightness among the stars, Achernar has a true brightness value of 200 as compared with unity for the sun. Achernar is a blue-white, spectral B type star and is one of the end stars in the constellation Eridanus, located south of the ecliptic and in the viewing vicinity of the Magellanic Clouds. Estimated distance from the earth is 65 light years. See also **Constellations;** and **Star.**

ACHILLES TENDON. In humans, the prominent tendon at the back of the ankle, extending from the muscle of the calf to the heel. Technically, it is the tendon which attaches the gastrocnemius and soleus muscles to the calcaneum or heel bone. The name derived from human anatomy is used in relation to other vertebrates.

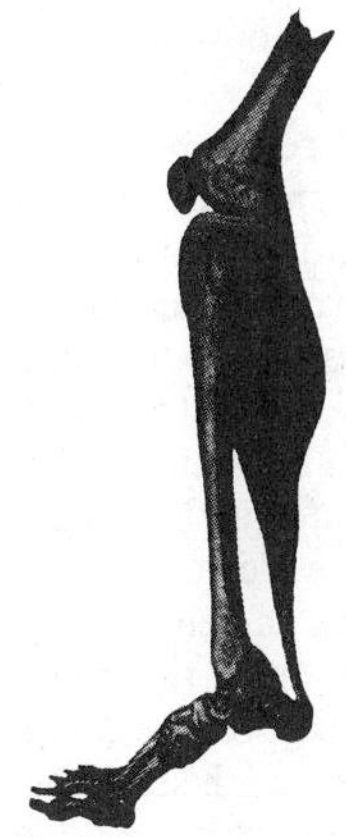

Achilles tendon.

ACHLORHYDRIA. The cessation of acid production by the stomach. The condition is relatively common among people of about 50 years of age and older, affecting 15 to 20% of the population in this age group. A well-balanced diet of easily digestible foods minimizes the discomforting effects of complete absence of hydrochloric acid in the stomach. The condition does not preclude full digestion of fats and proteins, the latter being attacked by intestinal and pancreatic enzymes. In rare cases, where diarrhea may result from achlorhydria, dilute hydrochloric acid may be administered by mouth.

Commonly, achlorhydria may not be accompanied by other diseases, but in some cases there is a connection. For example, achlorhydria is an abnormality which sometimes occurs with severe iron deficiency. Histalog-fast achlorhydria, resulting from intrinsic factor deficiency in gastric juice, may be an indication of pernicious anemia. Hyperplastic polyps are often found in association with achlorhydria.

ACHOLIA. Absence or lack of secretion of bile.

ACHONDRITES. A form of stony meteorites without chondri, and having textures similar to those of some terrestrial rocks.

ACHONDROPLASIA. Pituitary Gland.

ACHROMAT. A compound lens corrected so as to have the same focal length for two or more different wavelengths. Commonly, the F- and C-lines are the chosen wavelengths. See **Fraunhofer Lines.**

Achromats are used in optical microscopes for routine work, although they show some field curvature.

ACHROMATIC. Free from hue. Transmitting light without showing its constituent colors, or separating it into them. An achromatic color sometimes is referred to as gray.

Achromatic Combination. If reversed crown and flint prisms are made of such angle that the angles of dispersion between any two different wavelengths of light are alike but reversed in direction, then these two colors will not be separated and all colors lying between them will be separated little, if any, from each other. By using three kinds of glass, it is possible to bring three colors together. When the dispersions balance, the angles of deviation will, in general, not balance. This same principle is used in making achromatic lenses. Achromatic prisms have a maximum of deviation and a minimum of dispersion, whereas an Amici prism disperses the light with a minimum of deviation.

Achromatic Locus. Chromaticities that may be acceptable reference standards under circumstances of common occurrence are represented in a chromaticity diagram by points in a region which may be called the "achromatic locus." Any point within the achromatic locus, chosen as a reference point, may be called an "achromatic point." Such points have also been called "white points." However, the term "white point" is best used to specify the intersection of the various achromatic loci obtained under different conditions of adaptation.

Achromatic Stimulus. A visual stimulus that is capable of exciting a color sensation of no hue. In practice, an arbitrarily chosen chromaticity, such as that of the prevailing illumination.

ACID-BASE REGULATION (Blood). The hydrogen ion concentration of the blood is maintained at a constant level of pH 7.4 by a complex system of physico-chemical processes, involving, among others, neutralization, buffering, and excretion by the lungs and kidneys. Most physiological activities, and especially muscular exercise, are accompanied by the production of acid, to neutralize which, a substantial alkali reserve, mainly in the form of bicarbonate, is maintained in the plasma, and so long as the ratio of carbon dioxide to bicarbonate remains constant, the hydrogen ion concentration of the blood does not alter. Any non-volatile acid, such as lactic or phosphoric, entering the blood reacts with the bicarbonate of the alkali reserve to form carbon dioxide, which is volatile, and which combines with hemoglobin by which it is transported to the lungs and eliminated by the processes of respiration. It will also be evident from this that no acid stronger than carbon dioxide can exist in the blood. The foregoing neutralizing and buffering effects of bicarbonate and hemoglobin are short-term effects; to insure final elimination of excess acid or alkali, certain vital reactions come into play. The rate and depth of respiration are governed by the level of carbon dioxide in the blood, through the action of the respiratory center in the brain; by this means the pulmonary ventilation rate is continually adjusted to secure adequate elimination of carbon dioxide. In the kidneys two mechanisms operate; ammonia is formed, whereby acidic substances in process of excretion are neutralized, setting free basic ions such as sodium to return to the blood to help maintain the alkali reserve. Where there is a tendency to the development of increased acidity in the blood, the kidneys are able selectively to re-absorb sodium bicarbonate from the urine being excreted, and to release into it acid sodium phosphate; where there is a tendency to alkalemia, alkaline sodium phosphate is excreted, the hydrogen ions thus liberated being re-absorbed to restore the diminishing hydrogen ion concentration. See also **Acidosis; Alkalosis; Blood; Potassium** and **Sodium (In Biological Systems).**

ACIDIC SOLVENT. A solvent which is strongly protogenic, i.e., which has a strong tendency to donate protons and little tendency to accept them. Liquid hydrogen chloride and hydrogen fluoride are acidic solvents, and in them even such normally strong acids as nitric acid do not exhibit acidic properties, since there are no molecules which can accept protons; but, on the contrary, behave to some extent as bases by accepting protons yielded by the dissociation of the HCl or the H_2F_2. See **Acids and Bases.**

ACIDIMETRY. An analytical method for determining the quantity of acid in a given sample by titration against a standard solution of a base, or, more broadly, a method of analysis by titration where the end point is recognized by a change in pH (hydrogen ion concentration). See also **Analysis (Chemical); pH (Hydrogen Ion Concentration; Titration (Potentiometric);** and **Titration (Thermometric).**

ACID INTOXICATION. Acidosis.

ACIDITY. The amount of acid present, expressed for a solution either as the molecular concentration of acid, in terms of normality, molality, etc., or the ionic concentration (hydrogen ions or protons) in terms of pH (the logarithm of the reciprocal of the hydrogen ion concentration). The acidity of a base is the number of molecules of monoatomic acid which one molecule of the base can neutralize. See **Acids and Bases.**

ACIDITY (Corrosion). Corrosion.

ACID NUMBER. A term used in the analysis of fats or waxes to designate the number of milligrams of potassium hydroxide required to neutralize the free fatty acids in 1 gram of substance. The determination is performed by titrating an alcoholic solution of the wax or fat with tenth or half-normal alkali, using phenolphthalein as indicator.

ACIDOSIS. A condition of excess acidity (or depletion of alkali) in the body, in which acids are absorbed or formed in excess of their elimination, thus increasing the hydrogen ion concentration of the blood, exceeding the normal limit of 7.4. The acidity-alkalinity ratio in body tissue normally is delicately controlled by several mechanisms, notably the regulation of carbon dioxide-oxygen transfer in the lungs, the presence of buffer compounds in the blood, and the numerous sensing areas that are a part of the central nervous system. Normally, acidic materials are produced in excess in the body, this excess being neutralized by the presence of free alkaline elements, such as sodium occurring in plasma. The combination of sodium with excess acids produces carbon dioxide which is exhaled. Acidosis may result from: (1) severe exercise, leading to increased carbon dioxide content of the blood, (2) sleep, especially under narcosis, where the elimination of carbon dioxide is depressed, (3) heart failure, where there is diminished ventilation of carbon dioxide through the lungs, (4) diabetes and starvation, in which organic acids, such as β-hydroxybutyric and acetoacetic acids, accumulate, (5) kidney failure, in which the damaged kidneys cannot excrete acid radicals, and (6) severe diarrhea, in which there is loss of alkaline substances. Nausea, vomiting, and weakness sometimes may accompany acidosis.

See also **Acid-Base Regulation (Blood); Blood; Kidney and Urinary Tract;** and **Potassium and Sodium (In Biological Systems).**

ACID RAIN. Pollution (Air).

ACIDS AND BASES. The conventional definition of an acid is that it is an electrolyte that furnishes protons, i.e., hydrogen ions, H^+. An acid is sour to the taste and usually quite corrosive. A base is an electrolyte that furnishes hydroxyl ions, OH^-. A base is bitter to the taste and also usually quite corrosive. These definitions were formulated in terms of water solutions and, consequently, do not embrace situations where some ionizing medium other than water may be involved. In the definition of Lowry and Brønsted, an acid is a proton donor and a base is a proton acceptor. This concept is described later.

Acidification is the operation of creating an excess of hydrogen ions, normally involving the addition of an acid to a neutral or alkaline solution until a pH below 7 is achieved, thus indicating an excess of hydrogen ions. In *neutralization*, a balance between hydrogen and hydroxyl ions is effected. An acid solution may be neutralized by the addition of a base; and vice versa. The products of neutralization are a salt and water.

Some of the inorganic acids, such as hydrochloric acid, HCl, nitric acid, HNO_3, and sulfuric acid, H_2SO_4, are very high tonnage products and are considered as very important chemical raw materials. The most common inorganic bases (or alkalis) include sodium hydroxide, NaOH, and potassium hydroxide, KOH, and also are high tonnage materials, particularly NaOH.

Several classes of organic substances are classified as acids, notably the carboxylic acids, the amino acids, and the nucleic acids. These and the previously mentioned materials are described elsewhere in this volume.

In terms of the definition that an acid is a proton donor and a base is a proton acceptor, hydrochloric acid, water, and ammonia (NH_3) are acids in the reactions:

$$HCl \rightleftarrows H^+ + Cl^-$$

$$H_2O \rightleftarrows H^+ + OH^-$$

$$NH_3 \rightleftarrows H^+ + NH_2^-$$

Note that this definition is different in at least two major respects from the conventional definition of an acid as a substance dissociating to give H^+ in water. The Lowry-Brønsted definition states that for every acid there be a "conjugate" base, and vice versa. Thus, in the examples cited above, Cl^-, OH^-, and NH_2^- are the conjugate bases of HCl, H_2O, and NH_3. Furthermore, since the equations given above should more properly be written:

$$HCl + H_2O \rightleftarrows H_3O^+ + Cl^-$$

$$H_2O + H_2O \rightleftarrows H_3O^+ + OH^-$$

$$NH_3 + H_2O \rightleftarrows H_3O^+ + NH_2^-$$

It can be seen that every acid-base reaction involving transfer of a proton will involve two conjugate acid-base pairs, e.g., in the last equation NH_3 and H_3O^+ are the acids and NH_2^- and H_2O the respective conjugate bases. On the other hand, in the reaction:

$$NH_3 + H_2O \rightleftarrows NH_4^+ + OH^-$$

H_2O and NH_4^+ are the acids and NH_3 and OH^- the bases. In other reactions, e.g.,

Base$_1$	*Acid*$_2$		*Acid*$_1$	*Base*$_2$
$C_2H_3O_2^-$	$+ H_2O$	$\rightleftarrows$	$HC_2H_3O_2$	$+ OH^-$
HCO_3^-	$+ HCO_3^-$	$\rightleftarrows$	H_2CO_3	$+ CO_3^{-2}$
$N_2H_5^+$	$+ N_2H_5^+$	$\rightleftarrows$	$N_2H_6^{+2}$	$+ N_2H_4$
H_2O	$+ Cr(H_2)_6^{+3}$	$\rightleftarrows$	N_3O^+	$+ Cr(H_2O)_5OH^{2+}$

the conjugate acids and bases are as indicated. The theory is not limited to the aqueous solution; for example, the following reactions can be considered in exactly the same light:

Base$_1$	*Acid*$_2$		*Acid*$_1$	*Base*$_2$
NH_3	$+ HCl$	$\rightleftarrows$	NH_4^+	$+ Cl^-$
CH_3CO_2H	$+ HF$	$\rightleftarrows$	$CH_3CO_2H_2^+$	$+ F^-$
HF	$+ HClO_4$	$\rightleftarrows$	H_2F^+	$+ ClO_4^-$
$(CH_3)_2O$	$+ HI$	$\rightleftarrows$	$(CH_3)_2OH^+$	$+ I^-$
C_6H_6	$+ HSO_3F$	$\rightleftarrows$	$C_6H_7^+$	$+ SO_3F^-$

Acids may be classified according to their charge or lack of it. Thus, in the reactions cited above, there are "molecular" acids and bases, such as HCl, H_2CO_3, $HClO_4$, etc., and N_2H_4, $(CH_3)_2O$, C_6H_6, etc., and also cationic acids and bases, such as H_3O^+, $N_2H_5^+$, $N_2H_6^{2+}$, NH_4^+, $(CH_3)_2OH^+$, etc., as well as anionic acids and bases, such as HCO_3^-, Cl^-, NH_2^-, NH_3^{-2}, etc. In a more general definition, Lewis calls a base any substance with a free pair of electrons that it is capable of sharing with an electron pair acceptor, which is called an acid. For example, in the reaction:

$$(C_2H_5)_2O\!: + BF_3 \rightarrow (C_2H_5)_2O\!:\!BF_3$$

the ethyl ether molecule is called a base, the boron trifluoride, an acid. The complex is called a *Lewis salt*, or *addition compound*.

Acids are classified as monobasic, dibasic, tribasic, polybasic, etc., according to the number (one, two, three, several, etc.) of hydrogen atoms, replaceable by bases, contained in a molecule. They are further classified as (1) organic when the molecule contains carbon; (1a) carboxylic, when the proton is from a —COOH group; (2) normal, if they are derived from phosphorus or arsenic, and contain three hydroxyl groups; (3) ortho, meta, or para, according to the location of the carboxyl group in relation to another substituent in a cyclic compound; or (4) ortho, meta, or pyro, according to their composition.

Superacids. Although mentioned in the literature as early as 1927, *superacids* were not investigated aggressively until the 1970s. Prior to the concept of superacids, scientists generally regarded the familiar mineral acids (HF, HNO_3, H_2SO_4, etc.) as the strongest acids attainable. Relatively recently, acidities up to 10^{12} times that of H_2SO_4 have been produced.

In very highly concentrated acid solutions, the commonly used measurement of pH is not applicable. See also **pH (Hydrogen Ion Concentration)**. Rather, the acidity must be related to the degree of transformation of a base with its conjugate acid. In the *Hammett acidity function*, developed by Hammett and Deyrup in 1932,

$$H_0 = pK_{BH^+} - \log \frac{BH^+}{B}$$

where $pKBH^+$ is the dissociation constant of the conjugate acid (BH^+), and BH^+/B is the ionization ratio, measurable by spectroscopic means (UV or NMR). In the Hammett acidity function, acidity is a logarithmic scale wherein H_2SO_4 (100%) has an H_0 of -11.9; and HF, an H_0 of -11.0.

As pointed out by Olah, et al. (1979), "The acidity of a sulfuric acid solution can be increased by the addition of solutes that behave as acids in the system: $HA + H_2SO_4 \rightleftarrows H_3SO_4^+ + A^-$. These solutes increase the concentration of the highly acidic $H_3SO_4^+$ cation just as the addition of an acid to water increases the concentration of the oxonium ion, H_3O^+. Fuming sulfuric acid (oleum) contains a series of such acids, the polysulfuric acids, the simplest of which is disulfuric acid, $H_2S_2O_7$, which ionizes as a moderately strong acid in sulfuric acid: $H_2S_2O_7 + H_2SO_4 \rightleftarrows H_3SO_4^+ + HS_2O_7^-$. Higher polysulfuric acids, such as $H_2S_3O_{10}$ and $H_2S_4O_{13}$, also behave as acids and appear somewhat stronger than $H_2S_2O_7$."

Hull and Conant in 1927 showed that weak organic bases (ketones and aldehydes) will form salts with perchloric acid in nonaqueous solvents. This results from the ability of perchloric acid in nonaqueous systems to protonate these weak bases. These early investigators called such a system a superacid. Some authorities believe that any protic acid that is stronger than sulfuric acid (100%) should be typed as a superacid. Based upon this criterion, fluorosulfuric acid and trifluoromethanesulfonic acid, among others, are so classified. Acidic oxides (silica and silica-alumina) have been used as solid acid catalysts for many years. Within the last few years, solid acid systems of considerably greater strength have been developed and can be classified as *solid superacids*.

Superacids have found a number of practical uses. Fluoroantimonic acid, sometimes called *Magic Acid*, is particularly effective in preparing stable, long-lived carbocations. Such substances are too reactive to exist as stable species in less acidic solvents. These acids permit the protonation of very weak bases. For example, superacids, such as Magic Acid, can protonate saturated hydrocarbons (alkanes) and thus can play an important role in the chemical transformation of hydrocarbons, including the processes of isomerization and alkylation. See also **Alkylation;** and **Isomerization.** Superacids also can play key roles in polymerization and in various organic syntheses involving dienone-phenol rearrangement, reduction, carbonylation, oxidation, among others. Superacids also play a role in inorganic chemistry, notably in the case of halogen cations and the cations of nonmetallic elements, such as sulfur, selenium, and tellurium.

An excellent summary of superacids, their nature and utility, is given in "Superacids" by G. A. Olah, G. K. Surya Prakash, and J. Sommer in *Science*, **206**, 13–20 (1979).

ACID SULFATE PROCESS. Pulp (Wood) Production and Processing.

ACIDULANTS AND ALKALIZERS (Foods). Well over 50 chemical additives are commonly used in food processing or as ingredients of final food products, essentially to control the pH (hydrogen ion concentration) of the process and/or product. An excess of hydrogen ions, as contributed by acid substances, produces a sour taste, whereas an excess of hydroxyl ions, as contributed by alkaline substances, creates a bitter taste. Soft drinks and instant fruit drinks, for example, owe their tart flavor to acidic substances, such as citric acid. Certain candies, chewing gums, jellies, jams, and salad dressings are among the many other products where a certain degree of tartness contributes to the overall taste and appeal.

Taste is only one of several qualities of a process or product which is affected by an excess of either of these ions. Some raw materials are naturally too acidic, others too alkaline—so that neutralizers must be added to adjust the pH within an acceptable range. In the dairy industry, for example, the acid in sour cream must be adjusted by the addition of alkaline compounds in order that satisfactory butter can be churned. Quite often, the pH may be difficult to adjust or to maintain after adjustment. Stability of pH can be accomplished by the addition of buffering agents which, within limits, effectively maintain the desired pH even when additional acid or alkali is added. For example, orange-flavored instant breakfast drink has just enough "bite" from the addition of potassium citrate (a buffering agent) to regulate the tart flavor imparted by another ingredient, citric acid. In some instances, the presence of acids or alkalies assist mechanical processing operations in food preparation. Acids, for example, make it easier to peel fruits and tubers. Alkaline solutions are widely used in dehairing animal carcasses.

The pH values of various food substances cover a wide range. Plant tissues and fluids (about 5.2); animal tissues and fluids (about 7.0 to 7.5); lemon juice (2.0 to 2.2); acid fruits (3.0 to 4.5); fruit jellies (3.0 to 3.5).

Acidulants commonly used in food processing include: Acetic acid (glacial), citric acid, fumaric acid, glucono delta-lactone, hydrochloric acid, lactic acid, malic acid, phosphoric acid, potassium acid tartrate, sodium bisulfate, sulfuric acid, and tartaric acid. Alkalies commonly used include: Ammonium bicarbonate, ammonium hydroxide, calcium carbonate, calcium oxide, magnesium carbonate, magnesium hydroxide, magnesium oxide, potassium bicarbonate, potassium carbonate, potassium hydroxide, sodium bicarbonate, sodium carbonate, sodium hydroxide, and sodium sesquicarbonate. Among the buffers and neutralizing agents favored are: Adipic acid, aluminum ammonium sulfate, ammonium phosphate (di- or monobasic), calcium citrate, calcium gluconate, sodium acid pyrophosphate, sodium phosphate (di-, mono-, and tri-basic), sodium pyrophosphate, and succinic acid.

See also **Buffer (Chemical);** and **pH (Hydrogen Ion Concentration).**

ACINOUS GLAND. Gland.

ACLINIC LINE (or Dip Equator; Magnetic Equator). The line through those points on the earth's surface at which the magnetic inclination is zero. The aclinic line is a particular case of an isoclinic line.

ACMITE-AEGERINE. Acmite is a comparatively rare rockmaking mineral, usually found in nephelite syenites or other nephelite or leucite-bearing rocks, as phonolites. Chemically, it is a soda-iron silicate, and its name refers to its sharply pointed monoclinic crystals. Bluntly terminated crystals form the variety aegerine, named for Aegir, the Icelandic sea god.

Acmite has a hardness of 6 to 6.5, specific gravity 3.5, vitreous, color brown to greenish-black (aegerine); red-brown to dark green and black (acmite). Acmite is synonymous with aegerine, but usually restricted to the long slender crystalled variety of brown color.

The original acmite locality is in Greenland. Norway, U.S.S.R., Kenya, India, and Mt. St. Hilaire, Quebec, Canada furnish fine specimens. United States localities are Magnet Cove, Arkansas, and Libby, Montana, where a variety carrying vanadium occurs.

ACNAR CELLS. Diabetes Mellitus.

ACNE VULGARIS. A chronic disorder of the pilosebaceous units and generally confined to the face, chest, and back. The primary lesions are horny plugs (blackheads), which later develop into pink papules, pustules, or nodules. Nodules are tender, acute, and localized collections of pus deep in the dermis. Large pustular lesions may develop and may break down adjacent tissue to form lakes of pus, sinuses, and characteristically pitted scars. Studies have shown that heredity is a major predisposing factor in the disease. It has been described by some authorities as polygenic (summed effects of many genes) and thus difficult to clearly delineate in terms of heredity.

Prior to puberty, the small pilosebaceous units are dormant. At puberty, these units enlarge and produce sebum by action of the sebaceous gland which converts circulating testosterone to 5α-dihydrostestosterone, which is the tissue androgen. See **Androgens.** Acne may occur when an excess of sebum is produced; or when the pilosebaceous follicular openings are too small to allow increased sebum flow; or when both conditions are present. The duration of acne vulgaris is highly variable and can persist well into the third decade of life. But, normally the disease peaks during the teen years and early twenties.

After years of research, the ideal agent for use in treating acne vulgaris remains to be found. It is mandatory that the skin be cleansed thoroughly one or more times daily so that the population of microflora (*Corynebacterium acne* and *Staphylococcus*) can be kept to a minimum. Ordinary soap and water are effective. Greasy preparations should be avoided. An antiseborrheic shampoo is frequently suggested to maintain a thoroughly clean area over and around the face, head, shoulders, and back. The hair should be worn off the face. The lesions should not be manipulated (picking; squeezing). *Moderate* exposure to the sun may be beneficial, but this has not been proved. Excessive exposure predisposes skin cancer. The function of diet has been overemphasized, but may be a factor in some persons with acne. Experimentation with various kinds of foods may help to ascertain whether or not these aggravate the condition in specific individuals. Studies have not shown that traditional taboos (chocolate, coffee, cola drinks, fatty foods, nuts, ice cream, and sweets) play any role in this disease.

In severe cases, broad-spectrum systemic antibiotics are usually very effective in treating inflammatory acne. Tetracycline is the usual drug of choice.

ACNODE. Singular Point of a Curve.

ACOELA. An order of free-living flatworms in which the alimentary tract is without a cavity.

ACOELOMATA. Animals without a coelom. The term is applied especially to the flatworms, nemertine worms, and roundworms; these animals have attained the mesoderm in which the body cavity develops but it remains a more or less continuous mass with small spaces if any.

ACOUSTIC BOOM (Mysterious). Brontide.

ACOUSTIC COMPENSATOR. A device for adjusting acoustical path lengths for matching purposes in binaural listening.

ACOUSTIC FLOWMETER. Flow Measurement.

ACOUSTIC MODE. A type of thermal vibration of a crystal lattice which, in the limit of long wavelengths, is equivalent to an acoustic wave traveling with nearly constant velocity as if through an elastic continuum. At high frequencies, approaching the Debye frequency, the phase velocity of the acoustic modes tends to decrease, owing to dispersion. See also **Crystal.**

ACOUSTIC NOISE. Noise Generator.

ACOUSTIC PRINCIPLE OF SIMILARITY. For any acoustic system involving diffraction phenomena, it is possible to construct a new system on a different scale which will perform in similar fashion, provided that the wavelength of the sound is altered in the same ratio as the linear dimensions of the original system.

ACOUSTIC RECIPROCITY THEOREM. Reciprocity Theorem (Acoustical).

ACOUSTIC RESISTANCE. Resistance.

ACOUSTICS. In its broad interpretation, the term *acoustics* refers to the science of sound—generation, transmission, reception. The subject perhaps can be described best by considering the major categories of acoustics: (1) Physical acoustics; (2) architectural acoustics; (3) psychological acoustics; (4) physiological acoustics; and (5) electro-acoustics.

Physical acoustics deals with the properties and behavior of longitudinal waves of "infinitesimal" amplitude in solid, liquid, or gaseous media. These waves are propagated at the velocity of sound, or phase velocity, which is independent of frequency in a nondissipating free medium. In such a case, the shape of a complex wave remains unchanged during its propagation, although its amplitude may change. When the velocity of sound, or phase velocity, becomes dependent on frequency, the shape of a complex wave changes during propagation and dispersion is said to occur. In such cases groups of waves comprising a limited range of frequencies travel at a velocity called the group velocity, different from the phase velocity. It is the group velocity which carries the energy of such complex waves.

Acoustic waves are dispersive (1) in a *free* medium in which viscosity, heat conduction, and molecular, thermal, or chemical relaxation cause an increase in phase velocity with frequency, (2) in a *confined* medium in a capillary tube in which viscosity causes a decrease in phase velocity with frequency, (3) in a *confined* medium in *nondissipative* tubes of increasing cross section, where the rate of change of cross-sectional area differs from the conical (i.e., different from proportionality to the square of the distance along the tube)—examples of such tubes being the exponential and catenoidal horns, in which the phase velocity increases with decreasing frequency, (4) in nondissipative cylindrical tubes with *flexible* walls, and (5) in waves of *finite* amplitude, where the higher-frequency components have a higher phase velocity than the lower-frequency components, a transfer of energy occurring from the lower-frequency components to the higher-frequency components.

In physical acoustics, waves are reflected, refracted, diffracted, and absorbed. They exhibit all the properties of wave motion, such as reinforcement and destructive interference. They are accompanied by pressure and particle-velocity fluctuations detectable by the ear or by instruments capable of measuring the frequency instantaneous values, and mean intensity of these fluctuations.

Geometrical acoustics is a special case of physical acoustics in which diffraction and interference are disregarded. Energies of direct and reflected waves are considered to add irrespective of relative phase, a condition applicable to incoherent (i.e., uncorrelated) waves.

Architectural acoustics deals with the problems of distribution of beneficial sounds within buildings and with the exclusion or reduction of undesirable sounds. Here it is shown that mass and limpness of barriers such as partitions are most significant in providing high sound transmission loss. It is also shown that sound transmission loss tests on relatively small partitions (of the order of 66 × 80 inches) often give significantly higher transmission loss values than the full size partitions (of the order of 100 × 180 inches) usually used in practice. This points up the essential need to rely on full-scale tests, rather than on smaller-scale tests, for sound insulation data. Moreover, the application of sound absorbing materials, such as acoustical tiles or fibrous sound-absorbing blankets, to walls or ceilings, have a minor effect on transmission loss through partitions, vertical or horizontal. This applies equally well to such remedial measures as blowing rock wool between studs in walls or between joists in floors or ceilings. The only significant improvement in sound transmission loss results from massive structures or from discontinuous constructions such as floating floors, heavy flexibly suspended ceilings, staggered studs or multiple leaves in walls.

In auditoriums, reflective ceilings and reflective walls, combined with convex irregularities of random design, provide for reinforcement and diffuseness of sound found so beneficial for speech and music. Reflecting surfaces, giving short time-delay reflections (about 20 milliseconds or less), are particularly desirable in concert halls. Delays of 65 milliseconds or more may result in echoes and speech unintelligibility.

Psychological acoustics deals with the emotional and mental reactions of persons and animals to various sounds. Here questions arise as to which sounds are acceptable to most people under various living conditions and which are not. For this purpose, various noise criteria have been developed, related to the so-called speech interference level (SIL). By definition, SIL is the average of the sound pressure levels in decibels (L_p, see below) in three octave frequency bands 600 to 1200 Hz, 1200 to 2400 Hz, and 2400 to 4800 Hz. The loudness level in phons (L_N, see below) of a broad-band noise (no outstanding pure tones) should be not over 22 phons (at the most, not over 30 phons) greater than the SIL, in decibels, of the background noise. Two noise-control criteria, so-called NC and NCA, are designed to fulfill these conditions for various sound fields ranging from radio broadcasting studios, through bedrooms, offices, restaurants, sports arenas, and factories.

Physiological acoustics deals with hearing and its impairment, the voice mechanism, and the physical effects in general of sounds on living bodies. See also **Musical Sound.**

The frequency range of sound is divided into three somewhat overlapping regions, namely, an *audio-frequency* band ranging from approximately 20 to 20000 Hz flanked by an *infrasonic* region below 30 Hz and an *ultrasonic* region above 15000 Hz. Human ears do not respond in general to frequencies outside the audio band, although small animals such as cats and bats do hear in the lower ultrasonic region. At one time called supersonics, the term ultrasonics is now accepted to distinguish this area of high-frequency sound propagation from the cases of supersonic aircraft, supersonic fluid flow, and shock waves in fluids, which have to do with speeds higher than the speeds of sound.

The *strength* of a sound field is measured by its mean square pressure expressed as sound pressure level (L_p) in decibels. Decibels are logarithmic units defining the range of sound pressure levels (L_p) between the minimum audible value at 1000 Hz (4 dB*—the threshold of hearing) for the average pair of good young (high school age) ears and the maximum audible value of L_p at which effects other than hearing (such as tickling in the ears—the threshold of *feeling*) begin to appear. This upper limit shows up at about 120 dB at 1000 Hz.

Higher values of L_p (e.g., 130 dB) begin to cause pain in the average ear, and values of 160 dB may well cause instantaneous physical damage (perforation) to the tympanic membrane. The minimum audible sound pressure, p_0, at 1000 Hz is internationally accepted as 0.0002 microbar rms (i.e., 0.0002 dyne/cm^2, rms), and the sound pressure level at any other rms value of sound pressure, p, irrespective of frequency, is given by $L_p = 20 \log_{10} (p/p_0)$ dB. The bel (seldom used) is simply equal to 10 decibels. The bel appears first used in connection with power loss in telephone lines and is named in honor of Alexander Graham Bell.

Other reference pressures, p_0, may be used in special applications, instead of 0.0002 microbar, so it is essential to specify the reference pressure when quoting values of L_p.

The *loudness* of a sound field is judged by the ear in the audio frequency range. Loudness judgments by groups of observers have established a *loudness level* scale. The loudness level (L_N) in phons is arbitrarily taken equal to the sound pressure level L_p in dB at the reference frequency of 1000 Hz over the range from the threshold of hearing to the threshold of feeling. Jury judgment of equality in loudness between test tones at different frequencies (f) and 1000 Hz reference tones of known sound pressure level (L_p) have established *equal loudness* contours (contours of constant L_N) in the L_p-f plane.

These contours show in general a marked decrease in ear sensitivity to sounds at frequencies below about 200 Hz, and this decrease is much more pronounced in the lower loudness levels. For example, at 50 Hz the 4-phon contour has an L_p of about 43 dB, the 80-phon contour about 93 dB. At higher frequencies, the ear shows some 8-dB increase in sensitivity in the region around 3500 Hz, then a loss in sensitivity beyond about 6000 Hz. These characteristics of hearing are significant in the design of lecture and music halls, noise-control devices, and high-fidelity audio equipment.

Also based on jury judgments, a scale of *loudness*, *N*, (in sones) has been established for sounds (for pure tones and for broad band noise). On this scale, a given percentage change in sone value denotes

* 0 dB at 1000 Hz is defined in older work as the threshold of hearing.

an equal percentage change in the subjective loudness of the sound. The scale provides single numbers for judging the relative loudnesses of different acoustical environments, for evaluating the percentage reduction in noise due to various noise control measures, and for setting limits on permissible noise in factories, from motor vehicles, etc.

Loudness N is related to loudness level L_N in the range 40 to 100 phons by the equation $\log_{10} N = 0.03L_N - 1.2$. A loudness of 1 sone corresponds to a loudness level of 40 phons and is typical of the low-level background noise in a quiet home.

Various methods are available for estimating loudness of complex sounds from their sound pressure levels in octave, half-octave, or third-octave bands. For traffic noises, readings on a standard sound level meter using the A-scale (which incorporates a frequency-weighting network approximating the variation of ear sensitivity with frequency to tones of 40-dB sound pressure level) appear to correlate reasonably well with jury judgments of vehicle loudness.

The *noisiness* of a broad-band noise is more related to the annoyance it causes than to its loudness. Thus, corresponding to the scale of sones created to measure loudness, a scale of *noys* has been developed as a measure of the noisiness of jet aircraft noise in particular. Noys give more importance to the high-frequency bands of noise and less importance to the low-frequency bands than do sones. Also, corresponding to the scale of loudness levels in phons, there has been established a scale of *perceived noise levels* in PN dB. Rules have been established for converting sound pressure level measured in octave bands, half-octave bands and third-octave bands into noys and then into PN dB. Although originally developed as a means for the assessment of the "noisiness" of jet aircraft flying over inhabited communities, the concept of noisiness is being applied to traffic and other broad-band noises. See also **Hearing and the Ear.**

Electroacoustics. This field is concerned with the principles (transduction process) and devices (transducers) by which electrical energy may be converted into acoustic energy and vice versa. Consider the familiar electrodynamic transducer. A periodic electric current passing through a coil interacts with a steady radial magnetic flux causing the coil to vibrate. The coil in turn drives a diaphragm which radiates sound waves from one side. (The other side is usually enclosed to avoid cancellation of the acoustic output.) The entire process is *reversible* since sound waves striking the diaphragm set up a periodic variation in air pressure adjacent to the diaphragm causing it to vibrate. As the moving coil cuts the magnetic flux, an emf is generated which causes a current to flow when a load is connected to the coil terminals.

Many, but not all, types of transducer are similarly reversible. A reversible transducer may be made to perform sending and receiving functions successively in such a manner that an absolute sensitivity may be determined (*reciprocity* calibration).

The electrodynamic transducer may further be classified as *passive* since all of the energy appearing in the acoustic load is derived from the electrical input energy, and *linear* in the sense that there is a substantially linear relationship between the input and output variables (electric current and acoustic pressure in the present case).

Irreversible Transducers. These depend on a variety of special effects of which the best known is (a) the variation of surface contact electrical resistance with pressure (carbon microphone). Other effects are (b) the variation of bulk resistance with elastic strain (piezoresistance), (c) variation of transistor parameters with strain, (d) cooling effect of periodic air movement (hot wire microphone), (e) pressure wave generated by an electrical spark, (f) dependence of air pressure on level of corona discharge (ionophone). See also **Microphone.**

Reversible Transducers. An important class of reversible transducer depends on relative movement of suitable components linked by an electric or magnetic field traversing a gap. Examples are (a) the electrodynamic transducer already described; (b) electrostatic depending on the relative movement of charged condenser plates; (c) magnetic or variable reluctance depending on relative movement of magnetic poles in a magnetic circuit linked with a fixed coil.

Other *reversible transducers* are dependent on dimensional changes connected with the state of magnetic or electric polarization of certain crystalline materials (piezomagnetism and piezoelectricity). Since strain may be longitudinal or shear and since both strain and polarization are directional quantities, many possible relationships between strain and polarization exist. The behavior of an X-cut quartz disk may serve as an illustration. When such a disk is axially compressed, electric charges appear on the plane surfaces. Conversely, if a potential difference is established between the two surfaces, contraction or expansion occurs depending on the direction of the electric field. Other important single-crystal piezoelectric materials are ammonium dihydrogen phosphate (ADP) and Rochelle salt. During the past decade, polycrystalline ceramic materials based on barium titanate and lead zirconate titanate have replaced single-crystal materials in many applications. These materials are ferroelectric and, when prepolarized, exhibit piezoelectric behavior.

To date, only polycrystalline piezomagnetic materials (often termed magnetostrictive) have been found useful. Some are metals such as nickel and permendur. Others are ferrite ceramics [basic composition: $(NiO)(Fe_2O_3)$] which have such a high electrical resistivity that eddy current losses are negligible making lamination unnecessary.

Electromechanical Coupling. Transducer performance is closely connected with the tightness of coupling between mechanical and electrical aspects. Consider a piezoelectric disk which is compressed by putting in *mechanical* energy W_m. The appearance of surface charges shows that *electrical energy* W_e is stored in the self capacitance and is available when an external circuit is connected to suitable electrodes. The ratio W_e/W_m (electromechanical coupling coefficient) sets a limit to the efficiency for a given bandwidth (frequency range). The coefficient may reach 70% for lead zirconate titanate.

Transducer Design. Impedance matching is of primary importance in electroacoustics. It may be likened to the choice of gear ratio and wheel size in automobile design. Impedance matching is generally closely related to transducer parameters such as beam width of projected or received sound and frequency response, as well as efficiency. The many available matching techniques include (a) Resonance, (b) horn systems (acoustic transformers), (c) lever systems (mechanical transformers). In the direct radiator electrodynamic loudspeaker, the diaphragm is made large enough to interact with the acoustic medium (air) and yet small enough in relation to the sound wavelength (at low frequencies, at least) to ensure uniform projection of sound over a wide angle. In the condenser loudspeaker, a large transducer area compensates for the weakness of electrostatic forces. In the underwater sonar project, slabs of piezoelectric ceramic may be sandwiched between metal plates to form a resonant device which radiates a narrow beam of sound with high efficiency over a narrow frequency range.

During the 1970s, much progress was made toward refining an acoustic (voice) interface between people and computers. This topic is explored in the entry on **Telephony;** and **Voice Recognition,** among others. In the development of systems and components for various kinds of voice communications, it is necessary in the performance of various tests to isolate a chamber (anechoic chamber) as much as possible from ambient radiation, including sound and other electromagnetic radiation. A chamber of this type is shown in the accompanying illustration.

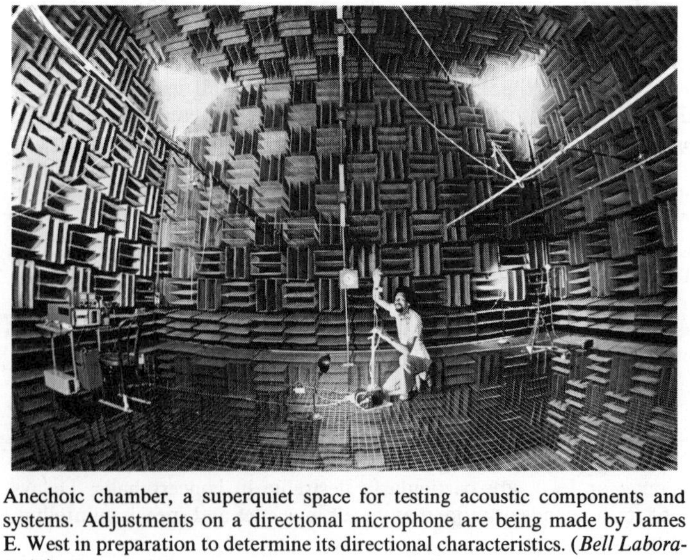

Anechoic chamber, a superquiet space for testing acoustic components and systems. Adjustments on a directional microphone are being made by James E. West in preparation to determine its directional characteristics. (*Bell Laboratories*)

ACOUSTICAL AND SOUND-RELATED TOPICS DESCRIBED IN THIS ENCYCLOPEDIA

Acoustic Compensator	Divergence Loss (Sound)	Octave
Acoustic Principle of Similarity	Drumskin Action	Pythagorean Scale
Acoustic Scintillation	Earphone Coupler	Radio Communications
Aeolian Tones	Echo	Reciprocity Theorem (Acoustics)
Anechoic Room	Fidelity (Communications)	Reciprocity Theorem (Electro-acoustical)
Antiresonance	Harmonic	Semitone (Half-step)
Anti-Side-Tone (Telephone)	Harmonic Analysis	Sonar
Articulation (Communications)	Harmonic Synthesizer	Speech Clipping
Audibility	Hearing and the Ear	Stereo Broadcasting
Audio Frequence Peak Limiter	High Fidelity	Telephony
Audiogram	Horn (Electromagnetic)	Thermophone
Automatic Volume Control	Hydrophone	Tone Control
Communications Satellites	Loudness Level	Transonic
Consonance	Loudspeaker	Ultrasonics
Debye-Sears Effect	Microphone	Voice Recognition
Dissonance	Musical Sound	
Distortion (Acoustic)	Noise	

Some of the other entries relating to acoustics in this encyclopedia are listed in the accompanying table.

References

Auld, B. A.: "Acoustic Fields and Waves in Solids," Wiley, New York, 1973.
Close, P. D.: "Sound Control and Thermal Insulation of Buildings," Van Nostrand Reinhold, New York, 1966.
Eargle, J.: "Sound Recording," Van Nostrand Reinhold, New York, 1976.
Goldstein, M. E.: "Aeroacoustics," McGraw-Hill, New York, 19767.
Izenour, G. C.: "Theater Design," McGraw-Hill, New York, 1977.
Lamb, H.: "The Dynamical Theory of Sound," Dover, New York, 1973.
Morse, P. M., and K. U. Ingard: "Theoretical Acoustics," McGraw-Hill, New York, 1966.
Newman, C.: "Sound in the Movies," Van Nostrand Reinhold, New York, 1979.
Pain, H. J.: "Physics of Vibrations and Waves," Wiley, New York, 1968.
Staff: "Sonochemical Engineering," American Institute of Chemical Engineers, New York, 1971.
Staff: "ISA Handbook of Control Valves," 2nd Edition (Chapter on Valve Noise), Instrument Society of America, Research Triangle Park, North Carolina, 1976.
Stephens, R. W. B., Editor: "Underwater Acoustics," Wiley, New York, 1971.
Tolstoy, I., and C. S. Clay: "Ocean Acoustics," McGraw-Hill, New York, 1966.
Wainstein, L. A., and V. D. Zubakov: "Extraction of Signals from Noise," Dover, New York, 1973.
Yerges, L. F.: "Sound, Noise and Vibration Control," 2nd Edition, Van Nostrand Reinhold, New York, 1978.

ACOUSTIC SCINTILLATION. Irregular fluctuations in the received intensity of sounds propagated through the atmosphere from a source of uniform output. These variations are produced by the nonhomogenous structure of the atmosphere along the path of sound. Turbulence and its concomitant variations in temperature and moisture are the chief causes of the inhomogeneities that lead to sonic refraction, diffraction, and scattering responsible for acoustic scintillation.

ACOUSTICS (Music). Musical Sound.

ACOUSTIC SPECTRUM ANALYSIS. Spectrum Analysis.

ACOUSTICS (Underwater). Sonar.

ACOUSTIC-WAVE DEVICES. Signal Processing Devices (CTD and SAW).

ACROMEGALY. Pituitary Gland.

ACRYLAMIDE POLYMERS. Using acid or base catalysis, acrylamide is derived from acrylonitrile by a hydration reaction. Although catalytic processes for the direct hydration of acrylonitrile over solid surfaces of metals, metallic salts, and oxides have been reported, commercial processes for its production use sulfuric acid. The acrylamide-sulfate salt initially formed may be isolated. When aqueous solutions of this product are neutralized with bases, such as ammonia, sodium hydroxide, and lime, acrylamide is yielded. Acrylamide is a white crystalline solid which melts at 84.5°C.

Several techniques may be used to polymerize acrylamide under controlled conditions. Radiation, photopolymerization, and ultrasonic methods may be used. Standard free-radical initiation of acrylamide in aqueous solution is most frequently used. The catalysts used include azo catalysts and many inorganic redox couples. Ionizing radiation in the solid state also may effect polymerization of acrylamide. Polyacrylamide may be "grown" to very high molecular size, with molecular weights about 10 million frequently obtained.

Acrylamide copolymerizes with several polar vinyl monomers. The vinyl polymer of acrylamide is a white solid having a high glass transition temperature (165°C) and softening temperature in excess of 200°C. The material is soluble in water, but not in most organic solvents. Polyacrylamides undergo the typical reactions of simple aliphatic amides.

Industrially, polyacrylamides are usually used in aqueous solution. To date, fabrication of bulk polymers into fibers, films, sheets, and molded products has not been achieved. The most important uses for aqueous polyacrylamide solutions are in flocculation, in which suspended matter is removed or concentrated in aqueous system.

ACRYLATES AND METHACRYLATES. A wide range of plastic materials dates back to the pioneering work of Redtenbacher before 1850 who prepared acrylic acid by oxidizing acrolein

$$CH_2{=}CHCHO \xrightarrow{O} CH_2{=}CHCOOH$$

At a considerably later date, Frankland prepared ethyl methacrylate and methacrylic acid from ethyl α-hydroxyisobutyrate and phosphorus trichloride. Tollen prepared acrylate esters from 2,3-dibromopropionate esters and zinc. Otto Rohm, in 1901, described the structures of the liquid condensation products (including dimers and trimers) obtained from the action of sodium alkoxides on methyl and ethyl acrylate. Shortly after World War I, Rohm introduced a new acrylate synthesis, noting that an acrylate is formed in good yield from heating ethylene cyanohydrin and sulfuric acid and alcohol. A major incentive for the development of a clear, tough plastic acrylate was for use in the manufacture of safety glass.

Ethyl methacrylate went into commercial production in 1933. The synthesis proceeded in the following steps:

(1) Acetone and hydrogen cyanide, generated from sodium cyanide and acid, gave acetone cyanohydrin

$$HCN + CH_3COCH_3 \rightarrow (CH_3)_2C(OH)CN$$

(2) The acetone cyanohydrin was converted to ethyl α-hydroxyisobutyrate by reaction with ethyl alcohol and dilute sulfuric acid

$$(CH_3)_2C(OH)CN + C_2H_5OH \xrightarrow{H_2SO_4} (CH_3)_2C(OH)COOC_2H_5$$

(3) The hydroxy ester was dehydrated with phosphorus pentoxide to produce ethyl methacrylate

$$(CH_3)_2C(OH)COOC_2H_5 \xrightarrow{P_2O_5} CH_2{=}C(CH_3)COOC_2H_5$$

In 1936, the methyl ester of methacrylic acid was introduced and used to produce an "organic glass" by cast polymerization. Methyl methacrylate was made initially through methyl α-hydroxyisobutyrate by the same process previously indicated for the ethyl ester. Over the years, numerous process changes have taken place and costs lowered, making these plastics available on a very high tonnage basis for thousands of uses. For example, the hydrogen cyanide required is now produced catalytically from natural gas, ammonia, and air.

As with most synthetic plastic materials, they commence with the monomers. Any of the common processes, including bulk, solution, emulsion, or suspension systems may be used in the free-radical polymerization or copolymerization of acrylic monomers. The molecular weight and physical properties of the products may be varied over a wide range by proper selection of acrylic monomer and monomer mixes, type of process, and process conditions.

In bulk polymerization, no solvents are employed and the monomer acts as the solvent and continuous phase in which the process is carried out. Commercial bulk processes for acrylic polymers are used mainly in the production of sheets, rods and tubes. Bulk processes are also used on a much smaller scale in the preparation of dentures and novelty items and in the preservation of biological specimens. Acrylic castings are produced by pouring monomers or partially polymerized sirups into suitably designed molds and completing the polymerization. Acrylic bulk polymers consist essentially of poly(methyl methacrylate) or copolymers with methyl methacrylate as the major component. Free radical initiators soluble in the monomer, such as benzoyl peroxide, are the catalysts for the polymerization. Aromatic tertiary amines, such as dimethylaniline, may be used as accelerators in conjunction with the peroxide to permit curing at room temperature. However, colorless products cannot be obtained with amine accelerators because of the formation of red or yellow colors. As the polymerization proceeds, a considerable reduction in volume occurs which must be taken into consideration in the design of molds. At 25°C, the shrinkage of methyl methacrylate in the formation of the homopolymer is 21%.

Solutions of acrylic polymers and copolymers find wide use as thermoplastic coatings and impregnating fluids, adhesives, laminating materials, and cements. Solutions of interpolymers convertible to thermosetting compositions can also be prepared by inclusion of monomers bearing reactive functional groups which are capable of further reaction with appropriate crosslinking agents to give three-dimensional polymer networks. These polymer systems may be used in automotive coatings and appliance enamels, and as binders for paper, textiles, and glass or nonwoven fabrics. Despite the relatively low molecular weight of the polymers obtained in solution, such products are often the most appropriate for the foregoing uses. Solution polymerization of acrylic esters is usually carried out in large stainless steel, nickel, or glass-lined cylindrical kettles, designed to withstand at least 50 psig. The usual reaction mixture is a 40–60% solution of the monomers in solvent. Acrylic polymers are soluble in aromatic hydrocarbons and chlorohydrocarbons.

Acrylic emulsion polymers and copolymers have found wide acceptance in many fields, including sizes, finishes and binders for textiles, coatings and impregnants for paper and leather, thermoplastic and thermosetting protective coatings, floor finishing materials, adhesives, high-impact plastics, elastomers for gaskets, and impregnants for asphalt and concrete.

Advantages of emulsion polymerization are rapidity and production of high-molecular-weight polymers in a system of relatively low viscosity. Difficulties in agitation, heat transfer, and transfer of materials are minimized. The handling of hazardous solvents is eliminated. The two principal variations in technique used for emulsion polymerization are the redox and the reflux methods.

Suspension polymerization also is used. When acrylic monomers or their mixtures with other monomers are polymerized while suspended (usually in aqueous system), the polymeric product is obtained in the form of small beads, sometimes called pearls or granules. Bead polymers are the basis of the production of molding powders and denture materials. Polymers derived from acrylic or methacrylic acid furnish exchange resins of the carboxylic acid type. Solutions in organic solvents furnish lacquers, coatings and cements, while water-soluble hydrolysates are used as thickeners, adhesives, and sizes.

The basic difference between suspension and emulsion processes lies in the site of the polymerization, since initiators insoluble in water are used in the suspension process. Suspensions are produced by vigorous and continuous agitation of the monomer and solvent phases. The size of the drop will be determined by the rate of agitation, the interfacial tension, and the presence of impurities and minor constituents of the recipe. If agitation is stopped, the droplets coalesce into a monomer layer. The water serves as a dispersion medium and heat-transfer agent to remove the heat of polymerization. The process and resulting product can be influenced by the addition of colloidal suspending agents, thickeners, and salts.

Product Groupings. The principal acrylic plastics are cast sheet, molding powder, and high-impact molding powder. The cast acrylic sheet is formable, transparent, stable, and strong. Representative uses include architectural panels, aircraft glazing, skylights, lighted outdoor signs, models, product prototypes, and novelties. Molding powders are used in the mass production of numerous intricate shapes, such as automotive lights, lighting fixture lenses, and instrument dials and control panels for autos, aircraft, and appliances. The high-impact acrylic molding powder yields a somewhat less transparent product, but possesses unusual toughness for such applications as toys, business machine components, blow-molded bottles, and outboard motor shrouds. The various acrylic resins find numerous uses as previously mentioned, with varied and wide use in coatings. Acrylic latexes are composed mainly of monomers of the acrylic family, such as methyl methacrylate, butyl methacrylate, methyl acrylate, and 2-ethyl hexylacrylate. Additional monomers, such as styrene or acrylonitrile, can be polymerized with acrylic monomers. Acrylic latexes vary considerably in their properties, mainly affected by the monomers used, the particle size, and the surfactant system of the latex. Generally, acrylic latexes are cured by loss of water only, do not yellow, possess good exterior durability, are tough, and usually have good abrasion resistance. The acrylic polymers are reasonably costly and some latexes do not have very good color compatibility. Acrylic latex paints can be used for concrete floors, interior flat and semigloss finishes, and exterior surfaces.

ACRYLIC ACID. $CH_2{:}CH{\cdot}COOH$, formula weight 72.06, colorless liquid monocarboxylic acid, mp 12°C, bp 141°C, sp gr 1.062. Also called propenoic acid, this compound is miscible in all proportions with H_2O or alcohol. The acid forms esters and metallic salts and forms addition products. The compound is of particular interest because of the large number of synthetic plastics and resins which are made as the result of polymerizing various acrylic derivatives, notably the esters of acrylic acid. The anhydrous monomer, glacial acrylic acid, contains less than 2% H_2O. It yields esters when reacted with alcohols, including ethyl acrylate and methyl acrylate. See also **ABS (Acrylonitrile-Butadiene-Styrene) Resins; Acrylic Fibers;** and **Acrylonitrile.**

ACRYLIC AND MODACRYLIC FIBERS. The U.S. Federal Trade Commission defines an acrylic fiber as one "in which the fiber-forming substance is any long-chain synthetic polymer composed of at least 85% by weight of acrylonitrile units ($-CH_2-\underset{\displaystyle CN}{\underset{|}{CH}}-$)." Further, when a fiber is "composed of less than 85%, but at least 35% by weight of acrylonitrile units, it is properly known as a modacrylic fiber."

Acrylonitrile polymers were reported in the German patent literature as early as the 1920s, but because of the instability of the fibers at the melting point and the lack of appropriate solvents, conversion into synthetic fibers by either melt spinning or solution spinning was not practical at that time. H. Rein of I. G. Farbenindustrie, in 1938, described fibers obtained from polymer dissolved in aqueous solutions of quaternary ammonium compounds, such as benzylpyridinium chloride, or of metal salts, such as lithium bromide, sodium thiocyanate, or aluminum perchlorate. In the early 1940s, Dupont, after studying the suitability of many solvents, chose dimethyl formamide and commenced development of "Fiber A," which became *Orlon*®, the first acrylic fiber to be commercially manufactured (1949) primarily for industrial and apparel use. Monsanto (1952) chose a different solvent, and *Acrilan*® acrylic fiber was introduced for broad use in textiles and home furnishings.

Subsequently, other U.S. firms entered into acrylic fiber production, including Dow (*Zefran*®, 1958) and American Cyanamid (*Creslan*®, 1959). *Zefran*® acrylic fiber is now produced by Badische Corp. Modacrylic fibers are produced in the U.S. by Tennessee Eastman (*Verel*®), and, Monsanto (*SEF*®), and Badische (*Zefran*®) type M-281.

Acrylic fiber production also commenced in various other locations in the world as well. By 1979, some 62 plants were producing acrylic and modacrylic fibers worldwide generating about 4500 million pounds (2041 million kilograms) of these fibers. Approximately one-sixth of this production came from plants in the United States. See accompanying table.

ACRYLIC AND MODACRYLIC FIBERS—SHIPMENTS IN UNITED STATES[1]

	Shipments		
Reported End-Use	Pounds (Millions)	Kilograms (Millions)	Percent of Total U.S. Use
Sweaters	71	32.2	13.1
Craft yarns	87	39.5	16.0
Half hose	75	34.0	13.8
Pile	66	29.9	12.1
Single knit	85}	38.6}	15.6}
Double knit	13}	5.9}	2.4}
Broadwovens	6	2.7	1.1
Blankets	48	21.8	8.8
Drapery	12	5.4	2.2
Upholstered furniture	8	3.6	1.5
Other home furnishings	11	5.0	2.0
Carpet face yarn	50	22.7	9.2
Industrial end uses	12	5.4	2.2
Total Use (United States)	544	246.7	100.0
Exported from the United States	230	104.3	
Grand Total	774	351.0	

[1] For 1979.

The basic raw material for acrylic fibers is acrylonitrile, a colorless liquid that, through the process of bulk, emulsion, solution, or suspension polymerization, is converted into large linear molecules ("polyacrylonitrile") of the type:

$$n(\mathrm{CH_2{=}CHCN}) \rightarrow \left(\begin{array}{c} -\mathrm{CH_2CH}- \\ | \\ \mathrm{CN} \end{array} \right)_n$$

The value of n is normally in the range 600–2000 for commercial fibers. This is equivalent to weight-average molecular weights of 100,000 to 150,000 or number-average molecular weights of 35,000 to 50,000.

Because of difficulties with dyeing and other limiting characteristics, fibers are seldom made from 100% polyacrylonitrile, but are usually copolymerized with one or more different monomers.

Commercially important comonomers include methylacrylate, methylmethacrylate, vinylacetate, and vinylbenzene. Terpolymers and tetrapolymers based on these and other comonomers, plus other compounds, particularly those containing dye-receptive groups, also are commercially important.

There are, however, some 100% polyacrylonitrile fibers available, such as Badische's *Zefran*® A-405, *Zefran*® A-507, and Bayer's *Dralon*® T-100. The A-405 is a producer-colored product for use in apparel knit goods, while the others are uncolored fibers for industrial uses, such as filter cloths. Homopolymer fibers are more mechanically and thermally stable and more chemical-resistant than the copolymer types.

Sites for attaching dyestuffs may be introduced either by polymerizable additives or by the generation of residual end groups through the selection of suitable radical-generating catalyst systems.

Fiber properties can be strongly influenced during the manufacturing step as a result of changing: (1) the inherent fiber structure; (2) the kind and amount of modifier used; (3) the spinning method used; and (4) the degree of stretching during fiber formation. A schematic representation of an acrylic (or modacrylic) fiber production process is shown in Fig. 1.

Acrylic fibers are produced by wet- or dry-spinning methods. Melt spinning is not commercially feasible because of polymer degradation for temperatures below or at the melting point. Both wet and dry spinning require that the polymer be dissolved in a suitable solvent to form a viscous solution that is forced through a spinneret. The spinneret is a metal plate or disk perforated with holes typically ranging from 0.002 to 0.010 inch (0.05 to 0.25 millimeter) in diameter. Spinnerets for dry spinning have 300 to 900 holes while those for wet spinning have 10,000 to 60,000 (or more) holes. About 60% of U.S. acrylic fiber production is by the wet-spinning technique.

Upon extrusion of the fiber from the spinneret hole, solvent is removed from the plastic mass, thereby regenerating the acrylic polymer in filamentary form. If solvent is removed by hot gases, the system is called dry spinning. If the solvent is leached out by another liquid, the process is called wet spinning. In either case, the extruded fibers must be drawn (extended to impart satisfactory strength and elongation). Then, the fibrous strand is crimped to make processing on textile-staple equipment possible and to give products made from acrylic staple or tow better aesthetic and performance properties.

Acrylic fibers usually are manufactured in the form of staple and tow. Staple length varies to suit the type of yarn spinning system used to convert the fiber into spun yarn. In the United States, lengths range from $1\frac{1}{8}$ inches (2.85 centimeters) for the cotton system to 6 inches (15.2 centimeters) for the worsted system. Variable cuts are also produced for special uses, such as blankets, pile, and fleece. For certain special applications, even shorter fiber lengths, e.g., $\frac{3}{4}$ inch (1.9 centimeters) have been provided. Acrylic fibers are also sold as tow for processing on stretch-breaking equipment (*Seydel, Turbo, Tematex*) and straight-cutting units (*Pacific Converter*). A small amount of continuous-filament yarn is produced in Japan.

Fiber fineness, reported by denier (d = weight in grams of 9000 meters of fiber) or tex (weight of 1000 meters of fiber), varies commercially from the apparel range, 1–6 d, up to the carpet range, 8–20 d. Microfibers (less than 0.75 d per filament) also have been produced.

Modacrylic fiber processes are similar to the acrylic manufacturing methods.

In general, all acrylics are resistant to ordinary chemicals, but are degraded readily by hot, concentrated alkalies. The acrylics, as a class, are very resistant to light, insects, and microbiological attack. Moth and carpet bettle larvae, for example, have virtually no effect on acrylic fibers. The fibers are not weakened by mold or mildew. Modacrylics have lower acrylonitrile content (<85%) and more comonomer. As a result certain physical properties are altered. In particular, behavior toward heat and solvents differ from their acrylic counterparts. Lower softening temperatures and solubility in a few more solvents are often noted in modacrylic fibers although these effects are found to varying degrees in fibers from different producers. For example, see Fig. 2.

Cross sections of acrylic and modacrylic fibers can vary, but in general, dry spun fibers are dog-bone (Fig. 3(b)), ribbon (Fig. 3(d)), or peanut-shaped (Fig. 3(e)). Modified cross sections can, however, be obtained from either spinning process. Different cross sections can influence the luster, sparkle, and other properties of a fabric. Fibers may be produced in bright form or delustered with titanium dioxide (TiO_2).

Fabrics made from acrylic fibers have a warm, soft touch, launder readily, dry rapidly, and are dimensionally stable. A low specific gravity coupled with a high bulk procedure results in thick, bulky fabrics of low weight. The inherent characteristics of acrylic fibers, especially resilience, coloration potential, hand, and blending potential with other fibers make them the fiber of choice for certain end uses, such as sweaters, craft yarns, pile fabrics, and blankets. Significant use is also evident in circular and flat knit goods, hosiery, carpets, and broadwov-

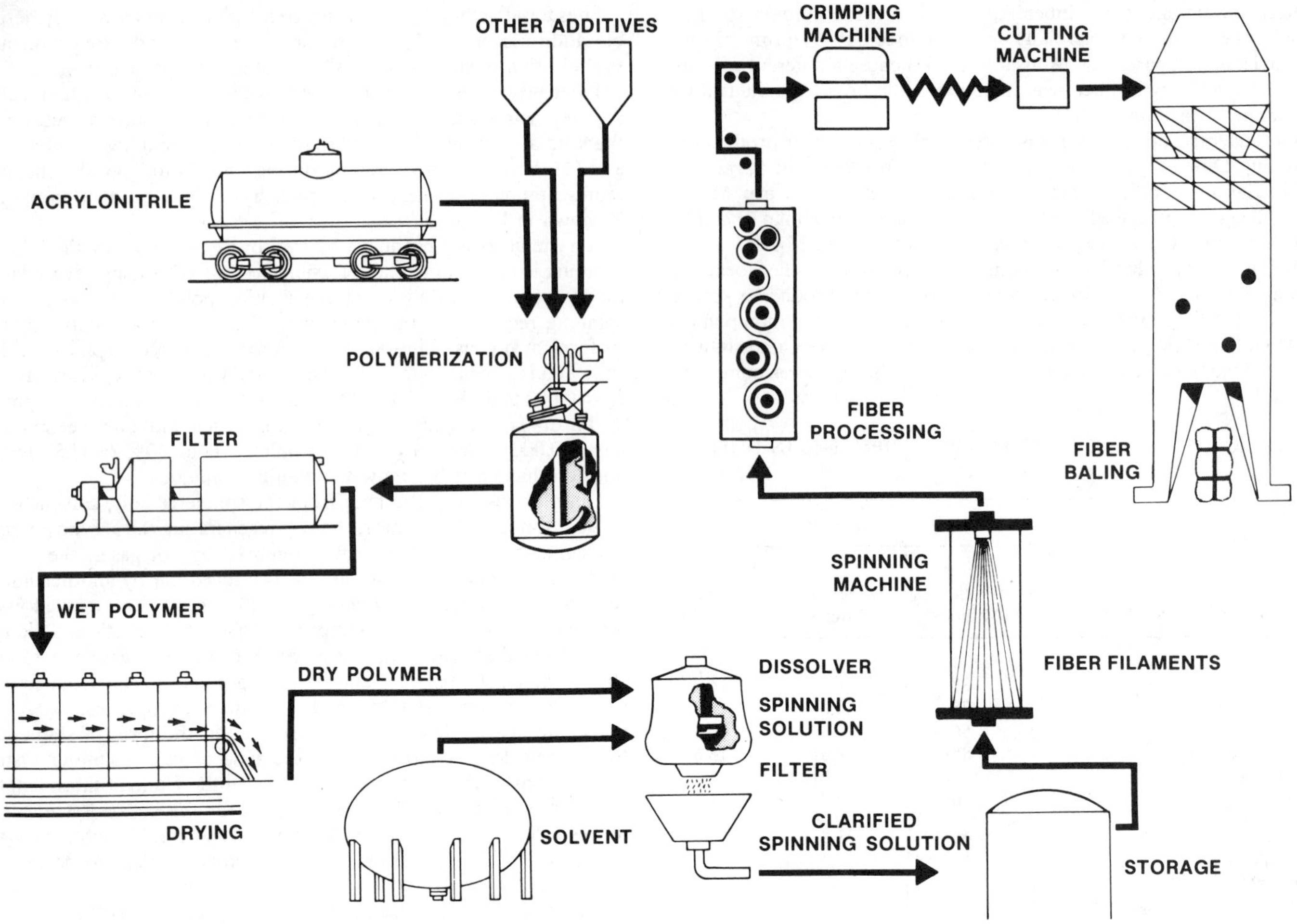

Fig. 1. Flow chart for production of a typical acrylic fiber. (*Monsanto*)

ens. Use of acrylic fibers in fabrics for upholstered furniture and draperies is also increasing.

Of all the synthetic fibers, acrylics, and modacrylics have aesthetics closest to wool in many applications. In contrast to wool, most acrylics are supplied with a good white that does not generally need bleaching. Where light hues require a better white base, a sodium chlorite bleach and addition of the proper optical brightener will allow virtually any clear bright color to be obtained on bright acrylics. Semidull acrylics readily accept the muted shades characteristic of wool. The different lusters are frequently blended to achieve the desired effect.

Like wool, acrylics are colored on a broad variety of conventional textile dyeing equipment to optimize the excellent aesthetic qualities of the fibers. For regular-dyeing acrylics, disperse or cationic (basic) dyes are the dyestuffs of choice. Dyes may be applied to staple (stock dyeing) or the staple may be spun into yarn for skein or package dyeing. The dyeing of fabrics or semifinished garments (piece dyeing) is also practiced to get the desired bulk level. The dyeing of acrylics can be done under pressure or at atmospheric conditions, depending on the equipment available. Acrylic fabrics and garments are often printed on standard textile equipment. These and other coloration techniques allow textile designers to utilize acrylics in a wide range of styles and applications in apparel, home furnishings, or industrial areas.

To enhance performance or styling, acrylics are often blended with polyester, nylon, cotton, rayon, or wool. Dyeing techniques are available to color even three-way and four-way blends.

Fiber Modifications. Certain properties can be built into or excluded from basic forms of the fiber. Some properties that are manipulated include crimp, residual shrinkage, elongation to break, strength, hand, and dyeability. Of special significance are the variants listed below that have become so important to particular end uses.

High-Bulk Fibers. It is well known that for every high polymer there is a narow temperature region in which the polymer characteristically changes from a more or less highly resilient state to a viscoelastic state. If acrylic fiber is heated to 140°F (60°C) or above, depending upon the individual fiber, stretched 10 to 20%, and then cooled under tension, the fiber is put into a metastable state. In this form, the fiber producer calls it a "hi-bulk" fiber. It is stable at ordinary temperatures and humidities. This modified fiber may be processed either alone or, preferably, in blends with regular acrylic or other fibers. When the hi-bulk fiber in the blended yarn is relaxed in a high-temperature dyebath or by steaming, it shrinks to its original length and thus shortens the entire yarn. This causes the nonshrinking fibers to pucker. The unusual softness, lightness, pleasing appearance, and high cover thus obtained are important to certain styles of garments.

The fiber producer may use this principle to produce regular, lo-bulk, and hi-bulk variants. The fiber producer may also use it to produce bulky yarns and fabrics by stretch-breaking or other processes as mentioned earlier.

Bicomponent Acrylic Fibers. Introduced in 1959 (by DuPont), these fibers are increasing in commercial importance. In one common type of bicomponent acrylic, a homopolymer and a copolymer are brought together as separate entities, the two sides being joined together along the entire length of the fiber. When such a fiber shrinks, it develops a helical twist because the two sides shrink by a different amount. Skin-core, biconstituent, and other types of two-element fibers also are known. When the two elements belong to different fiber classes, the product must be designated a biconstituent fiber, according to a Federal Trade Commission regulation. Bicomponent or biconstituent fibers have improved bulk, cover, and resilience, generally because of built-in, permanent crimp (Fig. 4). The newer bicomponent acrylic fibers, produced by either wet or dry spinning, have cross sections ranging from bean to popcorn to worm shapes, or round. These fiber types are particularly suitable for hand craft yarns and sweaters.

One very special and important type of bicomponent acrylic fiber uses selected polymer pairs that respond differently to a combination

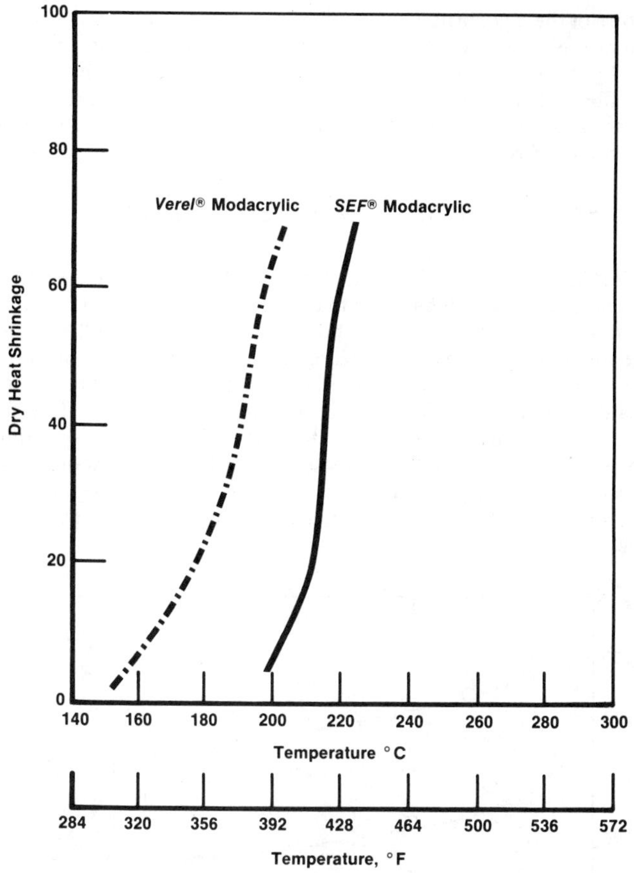

Fig. 2. Fiber dry heat shrinkage comparison. Note that the conventional modacrylic undergoes severe shrinkage from 140°C upward, while Type S-06 *SEF*® modacrylic does not show appreciable shrinkage at temperatures below 190°C. (*Monsanto*)

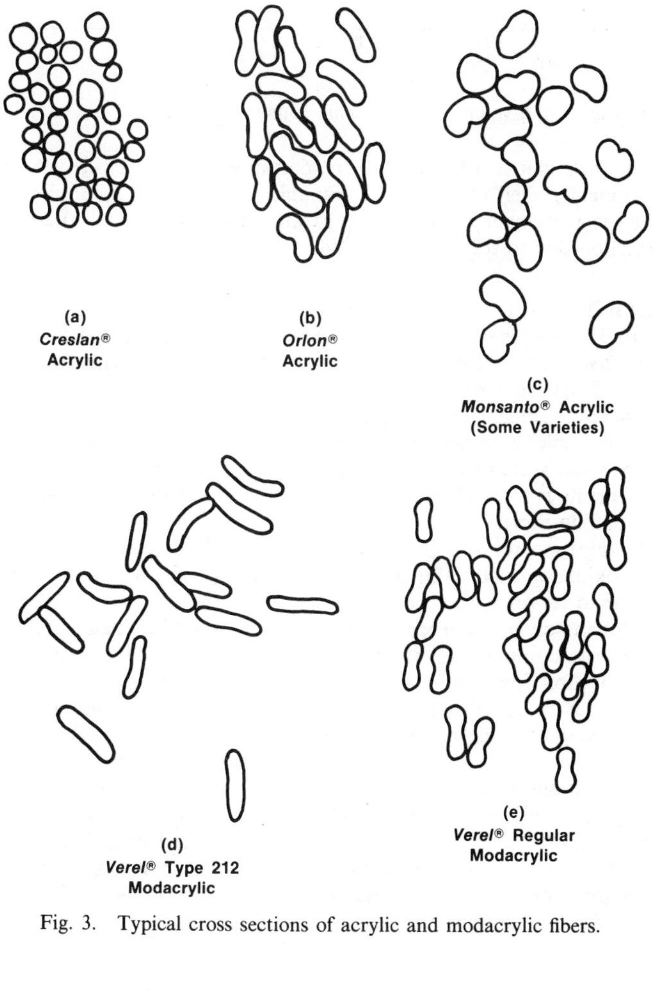

Fig. 3. Typical cross sections of acrylic and modacrylic fibers.

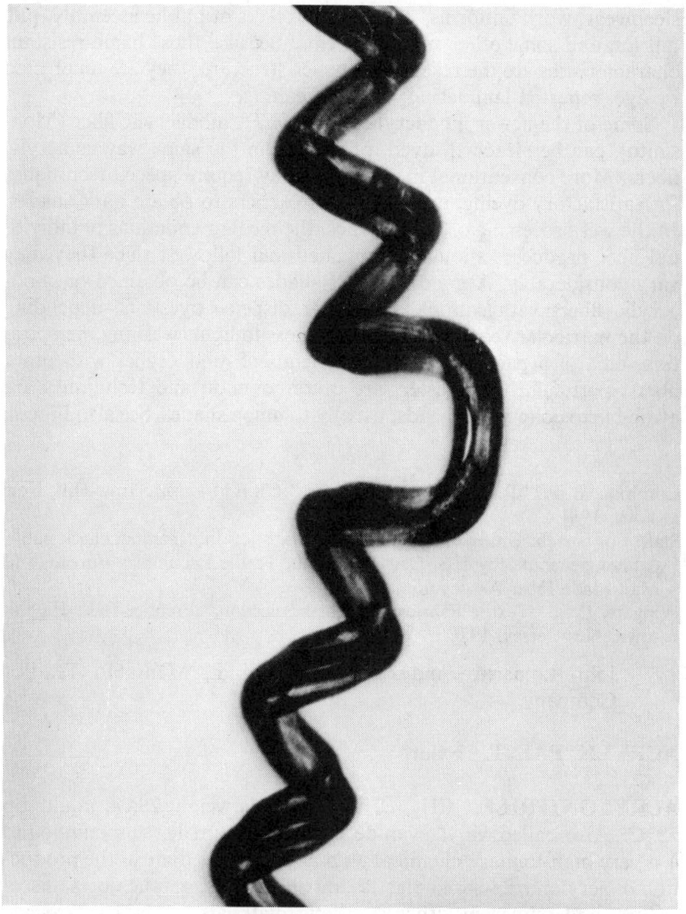

Fig. 4. Representation of *Monsanto*® bicomponent acrylic carpet fiber after crimp development. Note the helix direction reversal. Reversals of this type add greatly to softness, bulkiness, and resilience. The mechanism of crimp development is similar to that causing curvature in a bimetallic thermostat. Due to large recovery forces, helical bicomponent crimp is much more energetic in resisting and recovering from deformation than the planar crimp mechanically induced in monocomponent fiber. (*Monsanto*)

of heat and moisture. This approach leads to bulkier garments with greatly improved shape restorability after wearing in comparison with apparel made from other acrylic fiber types.

Producer-Colored Fibers. Several producers of acrylic fibers also offer colored fibers. The coloration is imparted to the fiber during the spinning step and is of value for several practical reasons. By using producer-colored fiber, fabrics can be produced that require no dyeing. If the producer-colored fiber is not delustered, then fabrics with enhanced luster can be obtained. Excellent color continuity is also assured by this technique.

Acid-Dyeability Fibers. Acrylic fibers which will accept acid dyes and resist cationic dyes are made and can be used for cross-dyed effects. In cross dyeing, cationic and acid dyes are combined in a single bath set at the proper pH and containing suitable auxiliaries. A fabric prepared from regular and acid-dyeable acrylic fibers in heather or pattern designs is treated in this dye bath to achieve two color effects simultaneously. Color and white effects are also possible in this way.

Modacrylics. Many of the statements made above for acrylics apply equally well to modacrylics, but there are some special differences that should be noted. Modacrylic fibers have lower heat stability and greater solubility in some solvents, as noted earlier and shown in Fig. 2. Although the definition of modacrylic does not specify the other raw materials to be used besides acrylonitrile, in actual fact the acrylonitrile is usually combined with one or more halogenated monomers to produce so-called inherently flame-resistant fibers. When these products are converted into properly constructed fabrics, the fabrics are then able to comply with specific flammability test criteria. As a result, modacrylics have found their main usage in those areas where flammability is of particular concern; for example, children's sleepwear, work uniforms, draperies in places of public assembly, pile fabrics, and some other specialty items. Because these flame-resistant characteristics are the result of polymer structure, they are unaffected by age, repeated laundering, or dry cleaning.

Some of the newer modacrylics, e.g., *SEF*® modacrylic fiber (Monsanto), can be bleached, dyed, or printed in the same way as acrylic fibers. More conventional modacrylics may require special techniques for satisfactory dyeing, such as use of carriers to obtain dark shades. In the wet processing of modacrylics, the recommendations of individual fiber producers should be obtained and followed since they may vary considerably. A good range of shades can be obtained on modacrylic fibers with cationic (basic) or disperse dyestuffs, depending on the particular requirements for fastness to light, washing, perspiration, etc., of a particular end use. Blends of modacrylics with other fibers, particularly polyester, are quite common and techniques are available to color such blends, usually to union shades. See also **Fibers.**

References

Corbman, B. P.: "Textiles: Fiber to Fabric," 6th Edition, McGraw-Hill, New York, 1981.

Staff: For current United States and world statistical information, check publications generated by U.S. Census Bureau, Textile Economics Bureau, and Man-Made Fiber Association.

Wingate, I. B.: "Textile Fabrics and Their Selection," Prentice-Hall, Hightstown, New Jersey, 1976.

John Lomartire and Frederic C. Flindt, Monsanto Textiles Company.

ACRYLIC PAINT. Paint.

ACRYLONITRILE. $CH_2{:}CHCN$, formula weight 29.04, liquid, bp 78°C. Also called vinyl cyanide or propene nitrile, this compound is a very high-tonnage chemical used as an intermediate in the production of acrylonitrile-based plastics, nitrile rubbers, acrylic fibers, insecticides, and numerous other synthetic materials. Several large-scale processes for the manufacture of acrylonitrile have appeared since the early 1960's. The majority of these processes use propylene, NH_3, and air as raw materials in what may be termed an ammonoxidation or oxyamination reaction:

$$CH_3CH{:}CH_2 + NH_3 + 1\text{-}1/2O_2 \rightarrow CH_2{:}CHCN + 3H_2O.$$

In one process, the starting ingredients are mixed with steam, preheated, and fed to the reactor. There are two main by-products, acetonitrile ($CH_3 \cdot CN$) and HCN, with accompanying formation of small quantities of acrolein, acetone, and acetaldehyde. The acrylonitrile is separated from the other materials in a series of fractionation and absorption operations. A number of catalysts have been used, including phosphorus, molybdenum, bismuth, antimony, tin, and cobalt.

ACRYLONITRILE-BUTADIENE RUBBER (ABR). Elastometers.

ACRYLONITRILE-BUTADIENE-STYRENE RESINS. ABS (Acrylonitrile-Butadiene-Styrene) Resins.

ACTH. The adrenocorticotropic hormone of the anterior lobe of the pituitary gland, which specifically stimulates the adrenal cortex to secrete cortisone, and hence has effects identical with those of cortisone. ACTH differs in its chemistry, absorption, and metabolism from the other adrenal steroids. Chemically, it is a water-soluble polypeptide having a molecular weight of about 3000. Its complete amino acid sequence has been determined. It produces its peripheral physiological effects by causing discharge of the adrenocortical steroids into the circulation. ACTH has been extracted from pituitary glands. In purified form, ACTH is useful in treating some forms of arthritis, lupus erythematosus, and severe skin disorders. The action of ACTH injections parallels the result of large quantities of naturally formed cortisone if they were released naturally. See also **Adrenal Glands; Brain and Nervous System; Hormones; Pituitary Gland;** and **Steroids.**

ACTIN. One of the two proteins that make up the myofibrils of striated muscles. The other protein is myosin. The combination of these two proteins is sometimes spoken of as actinomyosin. The banded nature of the myofibrils is due to the fact that both proteins are present where the bands are dark and only one or the other is present in the light bands. Since these bands lie side by side in the different myofibrils that go to make up a muscle fiber, the entire muscle fiber shows a banded or striated appearance. See also **Contractility and Contractile Proteins.**

ACTINIC KERATOSIS. Dermatitis and Dermatosis.

ACTINIDE CONTRACTION. An effect analogous to the Lanthanide contraction, which has been found in certain elements of the Actinide series. Those elements from thorium (atomic number 90) to curium (atomic number 96) exhibit a decreasing molecular volume in certain compounds, such as those which the actinide tetrafluorides form with alkali metal fluorides, plotted in the accompanying diagram. The effect here is due to the decreasing crystal radius of the tetrapositive actinide ions as the atomic number increases. Note that in the Actinides the tetravalent ions are compared instead of the trivalent ones as in the case of the Lanthanides, in which the trivalent state is by far the most common.

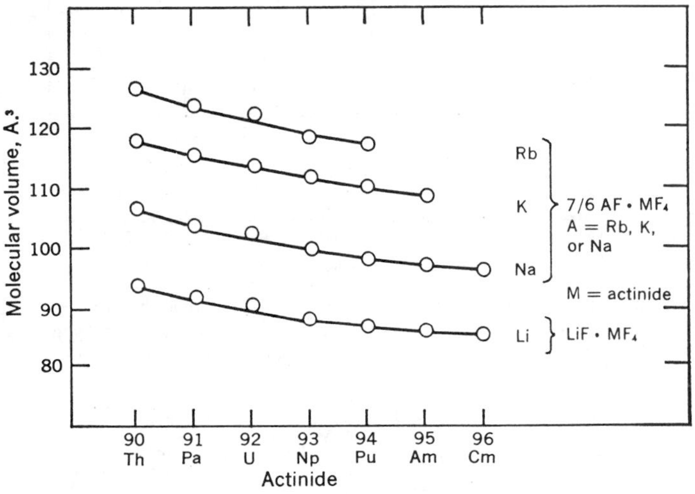

Plot of molecular volume versus atomic number of the tetravalent Actinides.

The behavior is attributed to the entrance of added electrons into an (inner) *f* shell (4*f* for the Lanthanides, 5*f* for the Actinides) so that the increment they produce in atomic volume is less than the reduction due to the greater nuclear charge.

ACTINIDE SERIES. The chemical elements with atomic numbers 90 to 103, inclusively, commencing with 90 (thorium) and through 103 (lawrencium) frequently are termed, collectively, the Actinide Series. The term derives from actinium (at. no. 89) which is considered the anchor element of the series, also appearing in group 3a of the periodic table. Members of the series are listed in the accompanying table. Some authorities place actinium in the series per se. This series of elements is somewhat analogous to the Lanthanide Series. See also **Lanthanide Series.**

Justification for the grouping is found in the higher elements of (III) oxidation states similar to actinium, and (IV) oxidation states similar to thorium. Certain similarities also exist between the atomic spectra and magnetic properties in the Actinide and Lanthanide Series. Note that actinium ($Z = 89$) and thorium ($Z = 90$) differ in electronic configuration from their immediate predecessor in atomic number, radium ($Z = 88$) in having, respectively, 1 and 2 electrons in their 6*d* subshells. The next element, protactinium ($Z = 91$) is the first to have an electron of the 5*f* subshell. Note also that the configurations of the last seven elements as given in the table are enclosed in parentheses to indicate that they are predicted, rather than determined, configurations. Two major methods have been used in making these determinations for the first eight elements: emission spectroscopy for actinium, thorium, uranium and americium; and atomic-beam experiments for protactinium, neptunium, plutonium and curium.

ELECTRONIC CONFIGURATIONS FOR NEUTRAL ATOMS OF THE ACTINIDE ELEMENTS

Element	Atomic Number (Z)	Electronic Configuration
Actinium	89	$6d7s^2$
Thorium	90	$6d^27s^2$
Protactinium	91	$5f^26d7s^2$
Uranium	92	$5f^36d7s^2$
Neptunium	93	$5f^46d7s^2$
Plutonium	94	$5f^67s^2$
Americium	95	$5f^77s^2$
Curium	96	$5f^76d7s^2$
Berkelium	97	$(5f^86d7s^2$ or $5f^97s^2)$
Californium	98	$(5f^{10}7s^2)$
Einsteinium	99	$(5f^{11}7s^2)$
Fermium	100	$(5f^{12}7s^2)$
Mendelevium	101	$(5f^{13}7s^2)$
Nobelium	102	$(5f^{14}7s^2)$
Lawrencium	103	$(5f^{14}6d7s^2)$

While in many respects the electronic configurations and chemical properties of the Actinide elements are similar to those of the Lanthanide series, the $4f$, $5d$ and $6s$ subshells of the latter corresponding to the $5f$, $6d$ and $7s$ of the latter, there are, however, significant differences. Cerium in the Lanthanide Series, unlike its analog thorium in the Actinide Series, has an electron in its $4f$ subshell. Moreover, for the first few members of each series, the $5f$ and $6d$ electrons are less energetically bound to the atomic nucleus than the $4f$ and $5d$ ones, so that the first few Actinide elements (except actinium) have in general higher oxidation states (lose electrons more readily) than the corresponding Lanthanides. Thus uranium, neptunium, plutonium, and americium have all four of the oxidation states 3, 4, 5 and 6. Later in the series, the Actinides correspond more closely to the Lanthanides in this respect.

In their electronic configurations the Actinide elements all have their innermost 86 electrons arranged in the configuration of radon, and their additional electrons as shown in the accompanying table.

ACTINIUM. Chemical element symbol Ac, at. no. 89, at. wt. 227 (mass number of the most stable isotope), periodic table group 3b, classed in the periodic system as a higher homologue of lanthanum. The electronic configuration for actinium is

$$1s^22s^22p^63s^23p^63d^{10}4s^24p^64d^{10}4f^{14}5s^25p^65d^{10}6s^26p^66d^17s^2.$$

The ionic radius (Ac^{+3}) is 1.11Å.

Presently, 24 isotopes of actinium, with mass numbers ranging from 207 to 230, have been identified. All are radioactive. One year after the discovery of polonium and radium by the Curies, A. Debierne found an unidentified radioactive substance in the residue after treatment of pitchblende. Debierne named the new material *actinium* after the Greek word for ray. F. Giesel, independently in 1902, also found a radioactive material in the rare-earth extracts of pitchblende. He named this material *emanium.* In 1904, Debierne and Giesel compared the results of their experimentation and established the identical behavior of the two substances. Until formulation of the law of radioactive displacement by Fajans and Soddy about ten years later, however, actinium definitely could not be classed in the periodic system as a higher homologue of lanthanum.

The isotope discovered by Debierne and also noted by Giesel was ^{227}Ac which has a half-life of 21.7 years. The isotope results from the decay of ^{235}U (AcU-*actinouranium*) and is present in natural uranium to the extent of approximately 0.715%. The proportion of Ac/U in uranium ores is estimated to be approximately 2.10^{-10} at radioactive equilibrium. O. Hahn established the existence of a second isotope of actinium in nature, ^{228}Ac, in 1908. This isotope is a product of thorium decay and logically also is referred to as *meso*-thorium, with a half-life of 6.13 hours. The proportion of mesothorium to thorium ($MsTh_2$/Th) in thorium ores is about 5.10^{-14}. The other isotopes of actinium were found experimentally as the result of bombarding thorium targets. The half-life of 10 days of ^{225}Ac is the longest of the artificially-produced isotopes. Although occurring in nature as a member of the neptunium family, ^{225}Ac is present in extremely small quantities and thus is very difficult to detect.

^{227}Ac can be extracted from uranium ores where present to the extent of 0.2 mg/ton of uranium and it is the only isotope that is obtainable on a macroscopic scale and that is reasonably stable. Because of the difficulties of separating ^{227}Ac from uranium ores, in which it accompanies the rare earths and with which it is very similar chemically, fractional crystallization or precipitation of relevant compounds no longer is practiced. Easier separations of actinium from lanthanum may be effected through the use of ion-exchange methods. A cationic resin and elution, mainly with a solution of ammonium citrate or ammonium-α-hydroxyisobutyrate, are used. To avoid the problems attendant with the treatment of ores, ^{227}Ac now is generally obtained on a gram-scale by the transmutation of radium by neutron irradiation in the core of a nuclear reactor. Formation of actinium occurs by the following process:

$$^{226}Ra(n, \gamma)^{227}Ra \xrightarrow{\beta^-} {}^{227}Ac$$

In connection with this method, the cross section for the capture of thermal neutrons by radium is 23 barns (23×10^{-24} cm^2). Thus, prolonged radiation must be avoided because the accumulation of actinium is limited by the reaction ($\sigma = 500$ barns):

$$^{227}Ac(n, \gamma)^{228}Ac(MsTh_2) \longrightarrow {}^{228}Th(RdTh)$$

In 1947, F. Hagemann produced 1 mg actinium by this process and, for the first time, isolated a pure compound of the element. It has been found that when 25 g of $RaCO_3$ (radium carbonate) are irradiated at a flux of 2.6×10^{14} ncm $^{-2}s^{-1}$ for a period of 13 days, approximately 108 mg of ^{227}Ac (8 Ci) and 13 mg of ^{228}Th (11 Ci) will be yielded. In an intensive research program by the Centre d'Etude de l'Energie Nucléaire Belge, Union Minière, carried out in 1970–1971, more than 10 g of actinium were produced. The process is difficult for at least two reasons: (1) the irradiated products are highly radioactive, and (2) radon gas, resulting from the disintegration of radium, is evolved. The methods followed in Belgium for the separation of ^{226}Ra, ^{227}Ac, and ^{228}Th involved the precipitation of $Ra(NO_3)_2$ (radium nitrate) from concentrated HNO_3, after which followed the elimination of thorium by adsorption on a mineral ion exchanger (zirconium phosphate) which withstand high levels of radiation without decomposition.

Metallic actinium cannot be obtained by electrolytic means because it is too electropositive. It has been prepared on a milligram-scale through the reduction of actinium fluoride in a vacuum with lithium vapor at about 350°C. The metal is silvery white, faintly emits a blue-tinted light which is visible in darkness because of its radioactivity. The metal takes the form of a face-centered cubic lattice and has a melting point of 1050 ± 50°C. By extrapolation, it is estimated that the metal boils at about 3300°C. An amalgam of metallic actinium may be prepared by electrolysis on a mercury cathode, or by the action of a lithium amalgam on an actinium citrate solution (pH = 1.7 to 6.8).

In chemical behavior, actinium acts even more basic than lanthanum (the most basic element of the lanthanide series). The mineral salts of actinium are extracted with difficulty from their aqueous solutions by means of an organic solvent. Thus, they generally are extracted as chelates with trifluoroacetone or diethylhexylphosphoric acid. The water-insoluble salts of actinium follow those of lanthanum, namely, the carbonate, fluoride, fluosilicate, oxalate, phosphate, double sulfate of potassium. With exception of the black sulfide, all actinium compounds are white and form colorless solutions. The crystalline compounds are isomorphic.

In addition to its close resemblance to lanthanum, actinium also is analgous to curium ($Z = 96$) and lawrencium ($Z = 103$), both of the group of trivalent transuranium elements. This analogy led G. T. Seaborg to postulate the actinide theory, wherein actinium begins a new series of rare earths which are characterized by the filling of the $5f$ inner electron shell, just as the filling of the $4f$ electron shell characterizes the Lanthanide series of elements. However, the first elements of the Actinide series differ markedly from those of actinium. Notably, there is a multiplicity of valences for which there is no equivalent among the lanthanides. See **Chemical Elements** for other properties of actinium.

Mainly, actinium has been of interest from a scientific standpoint. However ^{227}Ac has been proposed as a source of heat in space vehicles. It is interesting to note that the heat produced from the absorption of the radiation emitted by 1 g of actinium, when in equilibrium with its daughters, is 12,500 cal/hour.

References

Bouissières, G.: "Actinium dans le nouveau traité de chimie minérale de P. Pascal," pp. 7 and 1413–1446, Masson, Paris, 1960.
Centre d'Étude de l'Énergie Nucléaire Belge, Union Minière: Programme Actinium, Annual Report, 1969, Brussels.
Kirby, H. W.: "The Analytical Chemistry of Actinium," Mound Laboratory, Miamisburg, Ohio, 1966.
Bagnall, K. W.: "Chemistry of the Rare Radioelements," pp. 150–165, Academic, New York, 1957.
Hageman, F.: The Chemistry of Actinium, in G. T. Seaborg and J. J. Katz (editors), "The Actinide Elements," National Nuclear Energy Series, IV–14A, p. 14, McGraw-Hill, New York, 1954.
Katz, J. J. and G. T. Seaborg: "The Chemistry of the Actinide Elements," pp. 5–15, Methuen, London, 1957.

ACTINIUM SERIES. **Radioactivity.**

ACTINOLITE. The term for a calcium-iron-magnesium amphibole, the formula being $Ca_2(Mg,Fe)_5Si_8O_{22}(OH)_2$ but the amount of iron varies considerably. It occurs as bladed crystals or in fibrous or granular masses. Its hardness is 5–6, sp gr 3–3.2, color green to grayish green, transparent to opaque, luster vitreous to silky or waxy. Iron in the ferrous state is believed to be the cause of its green color. Actinolite derives its name from the frequent radiated groups of crystals. Essentially it is an iron-rich tremolite; the division between the two minerals is quite arbitrary, with color the macroscopic definitive factor—white for tremolite, green for actinolite. Actinolite is found in schists, often with serpentine, and in igneous rocks, probably as the result of the alteration of pyroxene. The schists of the Swiss Alps carry actinolite. It is also found in Austria, Saxony, Norway, Japan, and Canada in the provinces of Quebec and Ontario. In the United States actinolite occurs in Massachusetts, Pennsylvania, Maryland, and as a zinc-manganese bearing variety in New Jersey. See also **Amphibole; Tremolite;** and **Uralite.**

ACTINOMETER. The general name for any instrument used to measure the intensity of radiant energy. In earlier usage, the term was often restricted to the measurement of photochemically active radiation, but is now used more generally.

In meteorological and astronomical applications, actinometers may be classified according to the type of radiation they measure: (1) The *pyrheliometer* measures the intensity of direct solar radiation. It consists of a radiation sensing element enclosed in a casing that has a small aperture through which the direct solar rays enter, and a recorder unit. The amount of radiant energy absorbed is determined from the temperature rise of the sensing element. (2) The *pyranometer* measures the combined intensity of direct solar radiation and diffuse sky radiation, i.e., radiation reaching the earth's surface after having been scattered from the direct solar beam by molecules or suspensoids in the atmosphere. It consists of a recorder and a radiation sensing element mounted so that it views the entire sky. (3) The *pyrgeometer* measures the effective terrestrial radiation, i.e., the difference between the total outgoing infrared radiation of the earth's surface and the downcoming atmospheric radiation. It consists of four manganin strips, two blackened and two polished. The blackened strips are allowed to radiate to the atmosphere, whereas the polished strips are shielded. The electrical power required to equalize the temperature of the four strips is taken as a measure of the outgoing radiation.

ACTINOMYCIN D. **Antibiotic.**

ACTINOMYCOSIS. An infectious disease caused by actinomycetes, which are gram-positive bacteria with a characteristic filamentous, branching shape. Because of their unusual shape, the actinomycetes were once thought to be fungi. There are two genera of *Actinomycetaceae, Nocardia* and *Actinomyces.* The majority of cases of human actinomycosis are caused by one species, *A. israelii.* This is an anaerobe which normally resides in gingivodental crevices and tonsilar crypts. In cattle, the species, *A. bovis,* causes a condition known as *lumpy jaw,* and rarely if ever causes human disease.

In the past, a majority of cases of actinomycosis occurred in men, but with the growing use of contraceptive intrauterine devices, pelvic actinomycosis has gained some prominence. Types of actinomycosis are classified on the basis of location—*cervofacial* actinomycosis, frequently resulting from orofacial trauma, dental manipulation or gingival infection with poor dental hygiene, is found in about 50% of cases. *Pulmonary* actinomycosis usually is caused by aspiration of organisms from the oropharynx. Symptoms include cough, chest wall discomfort, fever, and loss of weight. This form occurs in about 20% of cases. *Gastrointestinal* actinomycosis frequently occurs at a point in the intestinal tract where the mucosa has been damaged, allowing ingested substances to invade the mucosa. This sometimes happens after appendicitis, usually with ileocecal involvement. Differential diagnosis of gastrointestinal actinomycosis requires consideration of carcinoma, Crohn's disease, and tuberculosis, which may produce similar symptoms. *Pelvic* actinomycosis is believed to originate from the upward spread of organisms that reach the perineum from the oropharynx: via the intestinal tract. Although not normally resident in the female genital tract, *A. israelii* may occur. Clinical observations of women using intrauterine devices indicate that this residence of the microorganisms is relatively common. Treatment of actinomycosis is by administration of antibiotics, usually penicillin. Tetracycline is the drug of choice for penicillin-alergic patients. Antibiotic therapy is usually required for several weeks. Prognosis in most cases is excellent.

ACTINON. The name of the isotope of randon (emanation), which occurs in the naturally-occurring actinium series (see **Chemical Elements; Radioactivity**), being produced by alpha-decay of actinium X, which is itself a radium isotope. Actinon has an atomic number of 86, a mass number of 219, and a half-life of 3.92 seconds, emitting an alpha particle to form polonium-215 (Actinium A).

ACTION. The action of a dynamical system is the space integral of the total momentum of the system. Specifically, if $\mathbf{r}_j$ is the position vector of the jth particle of the system, $\dot{\mathbf{r}}_j$ is its time rate of change, and m_j is the mass, the action for the path going from P_1 to P_2 is

$$\int_{P_1}^{P_2} \sum m_j \dot{\mathbf{r}}_j \cdot d\mathbf{r}_j$$

where the integral is taken along the actual path from P_1 to P_2. The integral can be shown to reduce to the form

$$2\int_{t_1}^{t_2} E_K\, dt$$

where E_K is the total kinetic energy of the system and t_1 and t_2 are the times at which the system is in positions P_1 and P_2 respectively. See **Least Action (Principle of).**

ACTION AND REACTION. **Newton's Laws of Dynamics.**

ACTION CURRENT. The local flow of current into the depolarized region of the cell membrane during the generation of the action potential. Since the flow of current from adjacent regions is an outward current—namely, a depolarizing current—under normal conditions it serves to depolarize the cell membrane in such regions beyond the threshold. By repetition of this process in successive regions, a self-propagated depolarization of the membrane sweeps along the nerve fiber. The nerve impulse is thus a consequence of the local action currents.

ACTION POTENTIAL. A characteristic variation in the membrane potential (i.e., potential across cell membrane) of excitable cells when the cell is stimulated. The potential falls rapidly in time toward zero, "overshoots," making the inside positive for a brief interval of time, and finally returns to the resting state. In many nerve and muscle cells, the duration of the action potential is found to be fractions of a millisecond to several milliseconds (see figure). The action potential is triggered by a depolarizing current when the membrane potential falls to the threshold level. See **Action Current.**

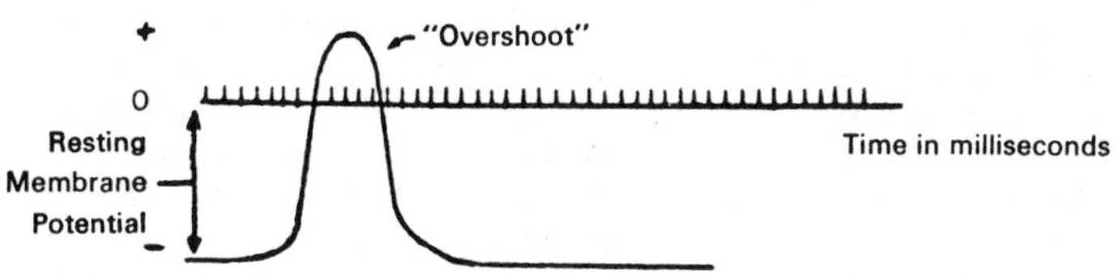

Action potential.

The action polarization is of particular interest in the passage of an impulse along a nerve fiber. At rest, the interior of a nerve fiber is negative to the exterior, with a potential difference of 50 to 100 millivolts; stimulation of the nerve sets up a negative wave of 70 to 100 millivolts, lasting one millisecond, followed by a brief refractory period during which further stimulation is ineffective. The record of such a change, as shown by a sensitive galvanometer, exhibits a characteristic sharp spike. Somewhat similar changes accompany muscular contraction. The study of action potentials has much practical importance e.g., electrocardiography, electroencephalography, and electromyography. See also **Brain and Nervous System.**

ACTION SPECTRUM. A graph of the amount of biological response produced by incident light as a function of wavelength. Thus, one might construct a dose-response curve for each wavelength of light used in the inactivation of an enzyme by ultraviolet light. The relative efficiency with which the inactivation is produced plotted versus wavelength of the incident light is the action spectrum for the inactivation of that enzyme.

ACTIVATED CARBON. Adsorption Operations; Decolorizing Agents.

ACTIVATED SILICA SOLS. Water Pollution.

ACTIVATED SLUDGE PROCESS. Water Pollution.

ACTIVATION. 1. The transformation of any material into a more reactive form, or into a form in which it functions more effectively, as in the regeneration of a metallic or inorganic catalyst, the transformation of an enzyme from inactive form to active form, and the treatment of various forms of finely-divided silica or carbon to render them more adsorbent.

2. The transfer of a sufficient quantity of energy to an atomic or molecular system to raise it to an excited state in which it can participate in a process not possible when the system is in its ground state.

3. In nuclear physics, the process of inducing radioactivity through neutron bombardment or by other types of radiation.

4. In electron-tube technology, the process by which the cathode is treated in order that maximum emission may occur.

ACTIVATION ENERGY. 1. The excess energy over the ground state which must be acquired by an atomic or molecular system in order that a particular process may occur. Examples are the energy needed by the molecule to take part in a chemical reaction, by an electron to reach the conduction band in a semiconductor, and by a lattice defect to move to a neighboring site.

In the first example cited, the rate of an elementary chemical reaction can usually be expressed as a product of a function of the concentrations of the participants and of a rate constant. This latter can be written as $A \exp(-E_a/kT)$, where k is the Boltzmann constant; T, the absolute temperature; A, a frequency factor that varies slowly with the activation energy E_a, which appears in the exponential. It is the minimum height of the potential barrier that must be crossed when one follows the reaction coordinate from the reactants to the products. The figure illustrates four typical situations. It is to be noted that it is only in case (*a*) that E_a is equal to the energy of the reaction.

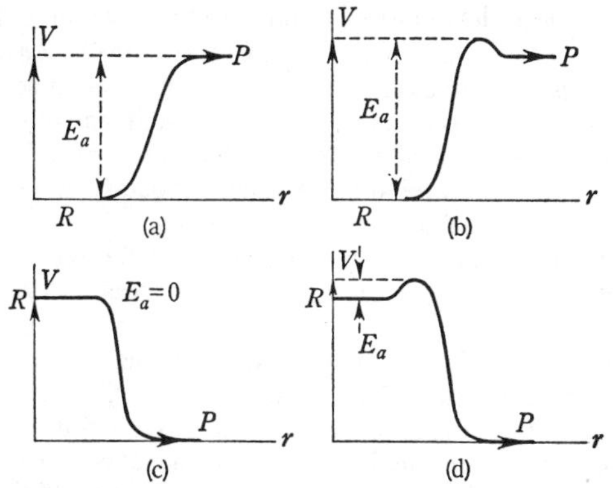

Activation energy. Variation of the energy along the reaction coordinate for two endothermic (a) and (b) and two exothermic (c) and (d) reactions. E_a is the activation energy.

2. If a liquid is regarded as an imperfect solid, the yielding to an applied shear stress takes place at a rate that depends on the frequency with which molecules leave their positions in the imperfect crystal lattice. The variation of this frequency with temperature is described by the energy required for an interchange of a molecule between the lattice and the free volume in the liquid. If this activation energy is linearly dependent on temperature at constant pressure, the slope of the $\log_e \eta : 1/kT$ plot gives the activation energy for liquid flow at low temperatures, where η is viscosity; k, the Boltzmann constant; and T, the absolute temperature.

ACTIVATOR. A substance that renders a material or a system reactive; commonly, a catalyst. 2. A special use of this term occurs in the flotation process, where an activator assists the action of the collector. 3. An impurity atom, present in a solid, that makes possible the effects of luminescence, or markedly increases their efficiency. Examples are copper in zinc sulfide, and thallium in potassium chloride.

ACTIVATOR (Enzyme). Enzyme.

ACTIVATOR (Flotation). Classifying (Process).

ACTIVE ACETATE. Coenzymes; Vitamin.

ACTIVE CENTER. Atoms which by their position on a surface, such as at the apex of a peak, at a step on the surface or a kink in a step, or on the edge or corner of a crystal, share with neighboring atoms an abnormally small portion of their electrostatic field and, therefore, have a large residual field available for catalytic activity or for adsorption.

ACTIVE DEPOSIT. The name given to the radioactive material that is deposited on the surface of any substance placed in the neighborhood of a preparation containing any of the naturally occurring radioactive chains (uranium, thorium, or actinium chains). This deposit results from deposition of the nongaseous products of the gaseous radon nuclides that have escaped from the parent substance. An active deposit can be concentrated on a negatively charged metal wire or surface placed in closed vessels containing the radon. See also **Radioactivity.**

ACTIVE MASS. Mass per unit volume, usually expressed in moles per liter (a concentration factor).

ACTIVE TRANSPORT (Cell). Cell (Biology).

ACTIVITY COEFFICIENT. A fractional number which when multiplied by the molar concentration of a substance in solution yields the chemical activity. This term provides an approximation of how much interaction exists between molecules at higher concentrations. Activity coefficients and activities are most commonly obtained from measurements of vapor-pressure lowering, freezing-point depression, boiling-point elevation, solubility, and electromotive force. In certain cases, activity coefficients can be estimated theoretically. As commonly used, activity is a relative quantity having unit value in some chosen standard state. Thus, the standard state of unit activity for water, a_w, in aqueous solutions of potassium chloride is pure liquid water at one atmosphere pressure and the given temperature. The standard state for the activity of a solute like potassium chloride is often so defined as to make the ratio of the activity to the concentration of solute approach unity as the concentration decreases to zero.

ACTIVITY (Radioactivity). The activity of a quantity of radioactive nuclide is defined by the ICRU as $\Delta N/\Delta t$, where N is the number of nuclear transformations that occur in this quantity in time Δt. The symbol Δ preceding the letters N and t denotes that these letters represent quantities that can be deduced only from multiple measurements that involve averaging procedures. The special unit of activity is the curie, defined as exactly 3.7×10^{10} transformations per second. See **Radioactivity.**

ACTIVITY SERIES. Also referred to as the *electromotive series* or the *displacement series,* this is an arrangement of the metals (other elements can be included) in the order of their tendency to react with water and acids, so that each metal displaces from solution those below it in the series and is displaced by those above it. See accompanying table. Since the electrode potential of a metal in equilibrium with a solution of its ions cannot be measured directly, the values in the activity series are, in each case, the difference between the electrode potential of the given metal (or element) in equilibrium with a solution of its ions, and that of hydrogen in equilibrium with a solution of its ions. Thus in the table, it will be noted that hydrogen has a value of 0.000. In experimental procedure, the hydrogen electrode is used as the standard with which the electrode potentials of other substances are compared. The theory of displacement plays a major role in electrochemistry and corrosion engineering. See also **Corrosion;** and **Electrochemistry.**

STANDARD ELECTRODE POTENTIALS (25°C)

Reaction		Volts
$Li^+ + e^-$	$\rightleftharpoons Li$	−3.045
$K^+ + e^-$	$\rightleftharpoons K$	−2.924
$Ba^{2+} + 2e^-$	$\rightleftharpoons Ba$	−2.90
$Ca^{2+} + 2e^-$	$\rightleftharpoons Ca$	−2.76
$Na^+ + e^-$	$\rightleftharpoons Na$	−2.711
$Mg^{2+} + 2e^-$	$\rightleftharpoons Mg$	−2.375
$Al^{3+} + 3e^-$	$\rightleftharpoons Al$	−1.706
$2H_2O + 2e^-$	$\rightleftharpoons H_2 + 2OH^-$	−0.828
$Zn^{2+} + 2e^-$	$\rightleftharpoons Zn$	−0.763
$Cr^{3+} + 3e^-$	$\rightleftharpoons Cr$	−0.744
$Fe^{2+} + 2e^-$	$\rightleftharpoons Fe$	−0.41
$Cd^{2+} + 2e^-$	$\rightleftharpoons Cd$	−0.403
$Ni^{2+} + 2e^-$	$\rightleftharpoons Ni$	−0.23
$Sn^{2+} + 2e^-$	$\rightleftharpoons Sn$	−0.136
$Pb^{2+} + 2e^-$	$\rightleftharpoons Pb$	−0.127
$2H^+ + 2e^-$	$\rightleftharpoons H^2$	0.000
$Cu^{2+} + 2e^-$	$\rightleftharpoons Cu$	+0.34
$I_2 + 2e^-$	$\rightleftharpoons 2I^-$	+0.535
$Fe^{3+} + e^-$	$\rightleftharpoons Fe^{2+}$	+0.77
$Ag^+ + e^-$	$\rightleftharpoons Ag$	+0.799
$Hg^{2+} + 2e^-$	$\rightleftharpoons Hg$	+0.851
$Br_2 + 2e^-$	$\rightleftharpoons 2Br$	+1.065
$O_2 + 4H^+ + 4e^-$	$\rightleftharpoons 2H_2O$	+1.229
$Cr_2O_7^{2-} + 14H^+ + 6e^-$	$\rightleftharpoons 2Cr^{3+} + 7H_2O$	+1.33
$Cl_2(gas) + 2e^-$	$\rightleftharpoons 2Cl$	+1.358
$Au^{3+} + 3e^-$	$\rightleftharpoons Au$	+1.42
$MnO_4^- + 8H^+ + 5e^-$	$\rightleftharpoons Mn^{2+} + 4H_2O$	+1.491
$F_2 + 2e^-$	$\rightleftharpoons 2F^-$	+2.85

ACTUATING ERROR SIGNAL. Signal (Instrument).

ACTUATOR (Control System). An actuator is that portion of a final controlling element in a control system that furnishes the power to change and/or to maintain the valve plug position in response to a signal received from the automatic controller. In some applications, the actuator may position elements other than a valve, such as a louver, damper, or pump speed governor. Actuators may be grouped into several categories: (1) mechanical, (2) pneumatic, (3) hydraulic, and (4) electric. The handwheel on a valve is the simplest form of mechanical actuator. A simple spring-return diaphragm actuator is an example of a pneumatic actuator. Oil-operated cylinders are hydraulic actuators. Motor-operated and solenoid actuators fall into the electrical category. Combination forms include electrohydraulic and electropneumatic actuation. Actuators generally mount directly on the valve body. See **Final Controlling Element; Valve (Control).**

ACUTE (Medical). Having a rapid onset, severe symptoms, and a relatively short duration; not chronic.

ACUTE MYOCARDIAL INFARCTION. Ischemic Heart Disease.

ACUTE NEPHRITIC SYNDROME. Kidney and Urinary Tract.

AC VOLTAGE REGULATOR. Power Sources and Supplies.

ACYCLIC HYDROCARBONS. Organic Chemistry.

ACYL. An organic radical of the general formula, RCO—. These radicals are also called acid radicals, because they are often produced from organic acids by loss of a hydroxyl group. Typical acyl radicals are acetyl, CH_3CO—, benzyl C_6H_5CO—, etc.

ACYLATION. A reaction or process whereby an acyl radical, such as acetyl, benzoyl, etc., is introduced into an organic compound. Reagents often used for acylation are the acid anhydride, acid chloride, or the acid of the particular acyl radical to be introduced into the compound.

ADAMANTINE COMPOUND. A compound having in its crystal structure an arrangement of atoms essentially that of diamond, in which every atom is linked to its four neighbors mainly by covalent bonds. An example is zinc sulfide, but it is to be noted that the eight electrons involved in forming the four bonds are not provided equally by the zinc and sulfur atoms, the sulfur yielding its six valence electrons, and the zinc, two. This is the structure of typical semiconductors, e.g., silicon and germanium.

ADAMS-STOKES ATTACK. Arrhythmias (Cardiac).

ADAPTATION (Ecology). The process of modification of the living organism to adjust it to the conditions of its environment. Also, an inherited character that enables the organism to meet certain environmental conditions.

All living things are adapted for a mode of life characteristic of their kind, under equally characteristic environmental conditions. They receive from previous generations a heritage that fits them for this mode of life, and all characters in the hereditary complex that are of definite use are adaptive. Wings, for example, are essential flight adaptation, and fins or other similar appendages are commonly found as adaptations for swimming.

Regardless of its adaptive heritage, however, each individual encounters some fluctuations in its environment to which it must adjust itself. The resulting changes in its body are adaptive, no less than its inherited structures. They are the acquired characters of biological literature, and have also been called individual adaptations. Human beings commonly experience two fine examples of this kind of adaptation in the calluses formed by the skin in response to friction, and the deposition of pigment, or tanning, as a protection against excessive ultraviolet light. A less evident result of exposure to ultraviolet light is a protective thickening of the epidermis, probably as important as the accompanying increase in pigmentation. It is interesting to note that the dark-skinned natives of central Africa, for example, do not sunburn as easily as persons with light-colored skin.

Although of different species, it is interesting to note that animals and plants from different parts of the earth often appear to be related as the result of certain adaptations to a given regional environment.

The resemblance of cacti which are prevalent in the deserts of the New World, but unknown in the deserts of Africa, nevertheless look very much like the euphorbias found in Africa. Botanically they are unalike. The cava found in the Argentine pampas is related to the guinea pig, although its resemblance to the jack rabbit of similar grassland environment in North America is indeed striking. Adaptation of this type is referred to as *convergent adaptation.*

These adptations have long been recognized, bringing about a series of generalizations which usually apply with few exceptions. For example, Gloger's rule states that cold dry climates encourage light colorations in animals whereas warm and moist climates encourage darker colors. Allen's rule states that parts protruding from the body tend to be shorter in colder climates (long noses are not found among the natives of cold central Asia, for example). Bergmann's rule states that individuals tend to be smaller in warmer climates.

ADAPTIVE CONTROL. A closed-loop control system in which system performance is monitored in relation to what might be termed an index of performance. In adaptive control, a method must be provided for the system to change parameters automatically within the closed loop. Auxiliary control variables are measured and evaluated with the objective of modifying the principal control functions so that selected control system performance criteria or indexes can be better realized. Adaptive control is similar to optimal control, in that various parameters are changed as a function of time. Adaptive control differs from optimal control in that parameters in the model of the process are to be evaluated *on-line.* Thus, adaptive control *combines control with the solution* of the identification problem. The resulting control may or may not be optimal. In particular, adaptive control is useful in cases where the process dynamics are not fully defined or change with time. Candidates for application of adaptive control include aerospace applications, such as rockets and aircraft where there is a change of control response because of changes brought about by fuel consumption, load, altitude, and other factors subject to alteration. Adaptive control also can be applied to advantage in connection with chemical and nuclear reactions which are affected by rapid temperature changes, alteration of catalyst activity, and similar factors.

Professional thinking in terms of adaptive control of processes and machines has been somewhat revised, particularly during the late 1970s and early 1980s as the result of introduction of microprocessors and minicomputers to the hardware of automatic control. It is expected that the technology will not settle down for another 5 to 10 years. Therefore, the descriptions and diagrams used here represent the more conventional approach to adaptive control and probably are more effective in conveying the fundamental principles involved than attempts to portray the more recent technology. Several papers pertaining to future configurations of adaptive control are included in the list of references at end of this entry.

Fundamental to the consideration of adaptive control are: (1) sufficient betterment in control performance must be obtainable to justify the additional computation required, often extensive, (2) the state variables or the model parameters that are to be evaluated must be observable directly or indirectly, (3) where parameters are not directly computable from measured signals, there must be sufficient variation on inputs to permit the estimation of these parameters by their influence, and (4) the dynamics of the system must be such that the parameters can be evaluated in a time that is reasonable based upon their rate of change. The fundamental operation of adaptive control is shown in Fig. 1.

The basic concepts can be demonstrated by use of a model-adaptive system, such as shown in Fig. 2. In the model-adaptive system, the controlled variable c is compared with the model output c_d. The difference in the two signals is a measure of performance or performance error. Control parameters are adjusted as a function of the performance error to bring the controlled variable to the desired value. One advantage of this type of system is the independent operation of the adaptive and main control loops. Operation of the main control loop can be maintained in event of a failure in the adaptive loop. The complexity of adaptive systems is a major disadvantage in many practical applications. Because of the feedback, which is a part of all closed-loop systems, there is a problem of stability against oscilla-

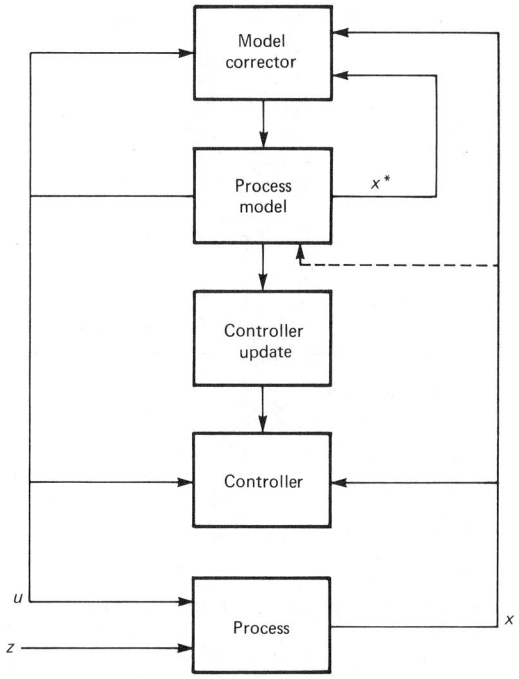

Fig. 1. Hierarchical control structure. (x, state factor; x^*, new state factor; u, control factor; z, z-transform)

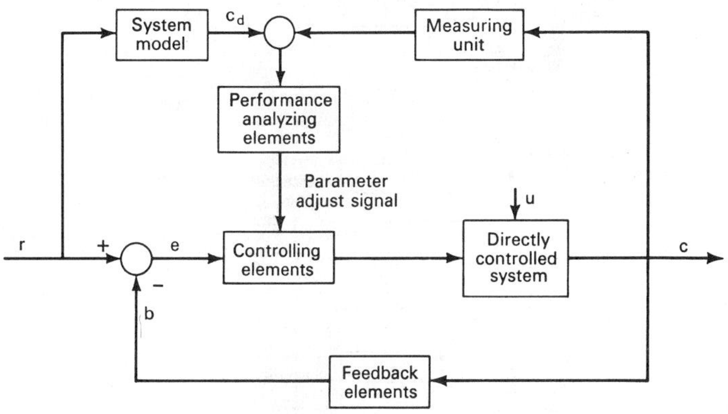

Fig. 2. Model adaptive system.

tion. Various forms of Nyquist criteria may be applied in the design of closed-loop systems to ensure satisfactory performance from the standpoint of stability.

The adaptive concept as applied to a 2-axis milling machine is shown in Fig. 3. The cutter velocity signal ω and torque signal τ are used in a simple multiplier to obtain a signal proportional to the power delivered to the workpiece by the cutter. From a previously determined relationship between cutter power and workpiece feed rate (axis velocity), a model signal that is useful in determining maximum cutter deflection is obtained. This signal is used to modify position-loop command (axis velocity) for optimum conditions of stock removal. The adaptive control loop for a milling machine is shown in Fig. 4.

Adaptive control as applied to the heat balance problem of a semi-batch reactor is shown in Fig. 5. Reactant gas A is added to liquid B to make liquid product C, also with the production of liquid by-product D. For maximum yield of C, it is desirable to operate the reactor near the batch freeze point. The setpoint of the temperature control loop must be adjusted to follow the freeze point. The freeze point increases with changing chemical composition during the batch. As shown, an analog computer is used to compute the incoming flow and thus determines the freeze point of the batch system to be used as the setpoint of the temperature controller. A different approach to the problem is shown in Fig. 6. This system does not depend upon external data. Rather, the adaptive control signal which controls the primary loop gain is obtained from analysis of the error signal. The signal is separated into high and low components, multiplied by a

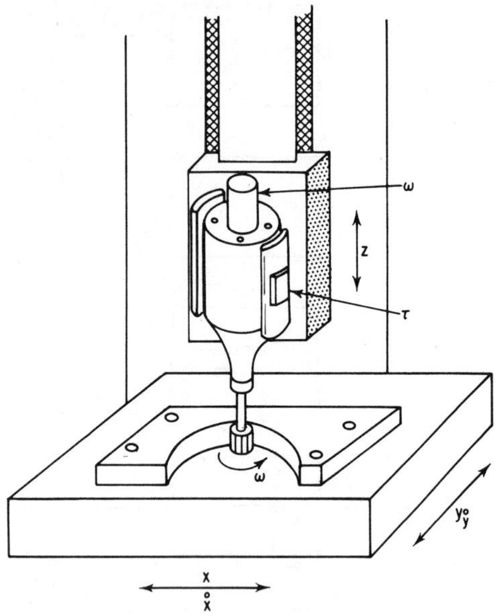

Fig. 3. Variables involved in adaptive control system for a 2-axis milling machine.

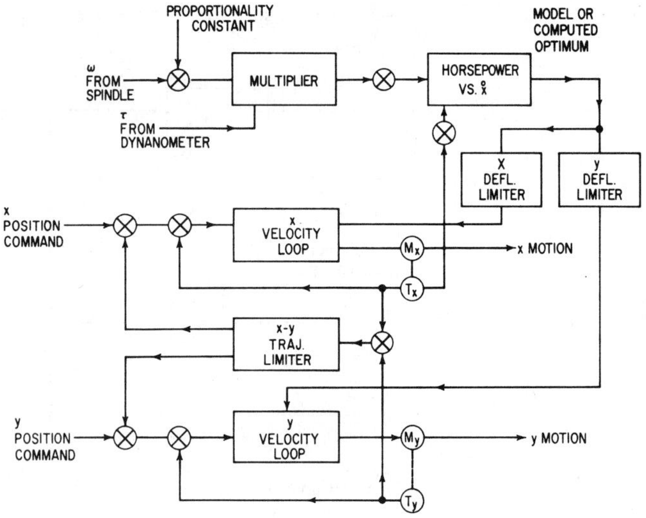

Fig. 4. Adaptive control system for milling machine.

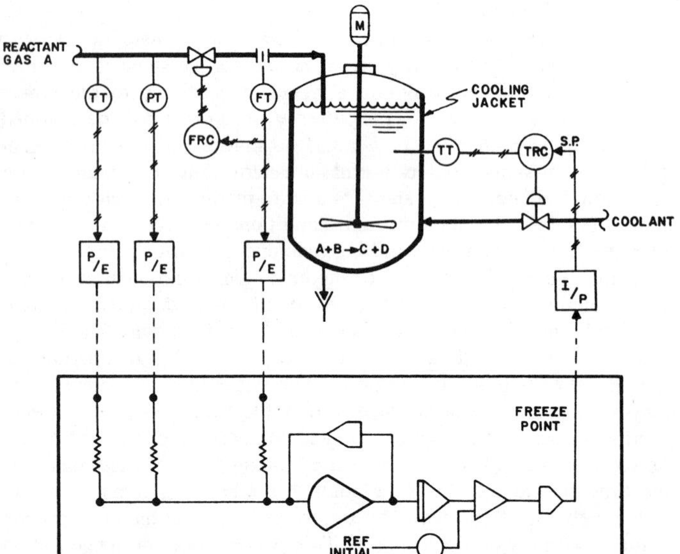

Fig. 5. An adaptive control approach to heat-balance problem of a semibatch reactor.

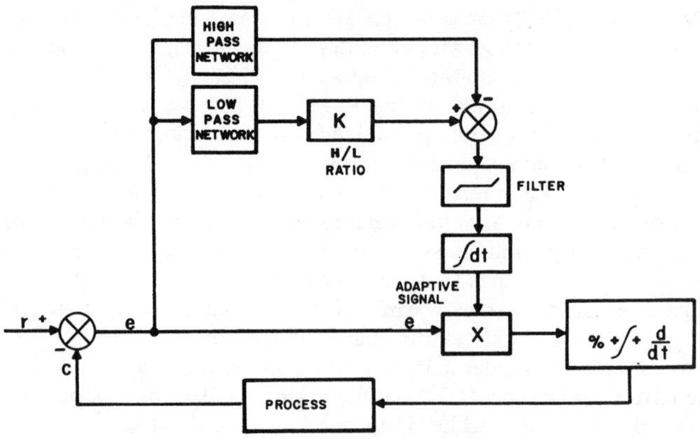

Fig. 6. Adaptive control approach to heat-balance problem based upon analysis of error signal.

ratio factor, combined, and then passed through a filtering and integrating network. A system of this type may be designed for universal purposes and thus the performance characteristics and reliability may not be expected to compete with the programmed controller, inasmuch as the latter is specifically designed for the particular process.

References

Andreiev, N.: "A Process Controller that Adapts to Signal and Process Conditions," *Control Engineering*, **24**, 12, 38–40 (1977).

Considine, D. M. (editor): "Process Instruments and Controls Handbook," 2nd edition, (Principles of Automatic Control, G. A. Hall, Jr. and Richard H. Kennedy), McGraw-Hill, New York, 1974.

Ibid: (Applied Automatic Control Theory, Stephen P. Higgins, Jr. and Joe M. Nelson).

Khandheria, J., and J. P. Shunta: "Adaptive Sampling Increases Sampling Rate As Process Deviations Increase," *Control Engineering*, **24**, 2, 33–35 (1977).

Uram, R.: "Computers Use Error-adaptive Control in Combined Cycle Power Plant," *Control Engineering*, **25**, 7, 37–39 (1978).

ADCOCK ANTENNA. Antenna.

ADDAX (*Mammalia, Artiodactyla*). A screw-horned antelope, *Addax nasomaculatus*, of northern Africa and Arabia.

ADDER (Computer System). A digital circuit which provides the sum of two or more input numbers as an output. A one-bit binary adder is illustrated in the accompanying figure. In this diagram, *A* and *B* are the input bits and *C* and $\overline{C}$ are the carry and no carry bits from the previous position. There are both serial and parallel adders. In a serial adder, only one adder position is required and the bits to be added are sequentially gated to the input. See also **Gate (Computer System).** The carry or no carry from the prior position

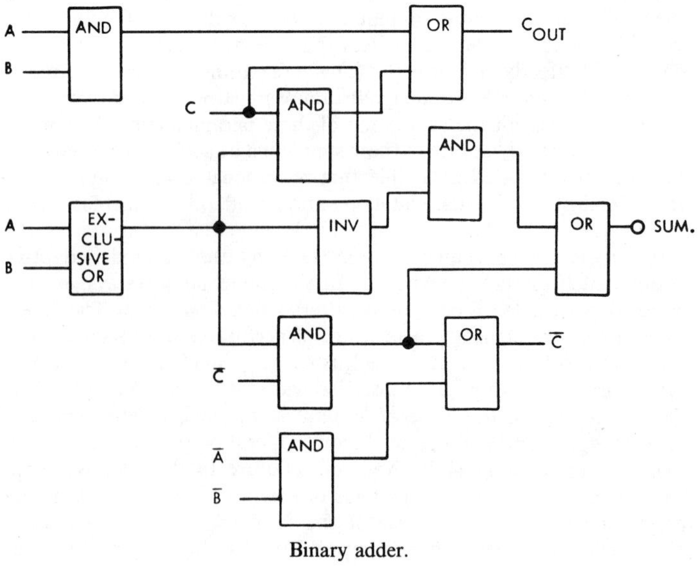

Binary adder.

is remembered and provided as an input along with the bits from the next position. In a parallel adder, all the bits are added simultaneously with the carry or no carry from the lower-order position propagated to the higher position. In a parallel adder, there may be a delay due to the carry propagation time. See also **Half-Adder.**

An adder may perform subtraction as well as the addition of two numbers. Generally, this is effected by complementing one of the numbers and then adding the two factors. The following is an example of a two's-complement binary-subtraction operation.

(a)		0 1 1 0	+6	(True)
	(+)	1 0 1 0	−6	(Complement)
		1 0 0 0 0	0	(True)
(b)		0 1 0 1	+5	(True)
	(+)	1 0 1 0	−6	(Complement)
		1 1 1 1	−1	(Complement)
(c)		1 1 1 1	(Complement) = −0 0 0 1 (True)	

The two's complement of a binary number is obtained by replacing all **1**'s with **0**'s, all **0**'s with **1**'s, and adding **1** to the units position. In (a) above, **6** is subtracted from **6**, and the result is all **0**'s; the carry implies that the answer is in true form. In (b), **6** is subtracted from **5**, and the result is all **1**'s with no carry. The no carry indicates the result is in complement form and that the result must be recomplemented as shown in (c)

Thomas J. Harrison, International Business Machines Corporation, Boca Raton, Florida.

ADDERS. Snakes.

ADDICTION (Drug). Drug Addiction.

ADDISON'S DISEASE. A disease caused by malfunction of the adrenal glands. It is characterized by a bronze color of the skin, prostration, anemia, disturbance of electrolyte metabolism, and diarrhea. Many cases are believed to result from an autoimmune disorder, i.e., an immune reaction by the body against some stimulus naturally present within itself. Sometimes it is caused by involvement of the glands by tuberculosis. Adrenal insufficiency may also result from insufficient pituitary ACTH. The course of the disease follows a definite pattern, characterized by the multiplicity of disturbances that occur in the chemistry of the body.

With the advent of modern hormonal treatment, over 70% of the patients survive. The usual treatment is the administration of cortisol or cortisone, usually supplemented by a sodium-retaining hormone, to maintain the salt level in the body. The existence of purified hormones simplifies the treatment, and each symptom can be alleviated with a corresponding hormone. Thus, there is available in pure form, for example, one of the hormones responsible for maintaining the water and salt balance in the body. Patients suffering a crisis caused by adrenal insufficiency must receive large quantities of salt and sugar solution intravenously and large doses of cortisone. Meals must be at frequent intervals and consist of abundant amounts of starches.

ADDITION. A fundamental operation for combining mathematical terms. The symbol + is used to denote the operation. To perform the operation of addition of two quantities x and y, it is necessary to identify like terms. If x and y are polynomials in z, then like terms are the terms in x and y containing z to the same power. For complex numbers, $x = a + ib$, $y = c + id$, $x + y = (a + c) + i(b + d)$. For fractions to have like terms, one must use common denominators.

$$\frac{a}{b} + \frac{c}{d} = \frac{ad}{bd} + \frac{cb}{bd} = \frac{ad + cb}{bd}.$$

Frequently, common factors can be cancelled in the numerator and denominator, e.g.,

$$\frac{1}{3} + \frac{1}{6} = \frac{6 + 3}{18} = \frac{9}{18} = \frac{1}{2}$$

Addition consists of finding the algebraic sum of like terms. Eventually, one must use the same process as in arithmetic, determine the numerical value of an addition by the addition tables for the base in use.

Donald R. Hodge, Alexandria, Virginia.

ADDITION COMPOUND. Compound (Chemical).

ADDITIVE COLOR PROCESS. A system of color photography in which the color synthesis is obtained by the addition of colors one to another in the form of light rather than as colorants. This color addition may take place (1) by the simultaneous projection of two or more (usually three) color images onto a screen, (2) by the projection of the color images in rapid succession onto a screen or (3) by viewing minutely divided juxtaposed color images.

In the case of a three-color process, three color records are made from the subject recording, in terms of silver densities, the relative amounts of red, green and blue present in various areas of the subject.

When the additive synthesis is to be made by simultaneous projection, positives are made from the color separation negatives and projected with a triple lantern onto a screen through red, green and blue filters. The registered color images give all colors of the subject due to simple color addition, red plus green making yellow, red plus blue appearing magenta, etc.

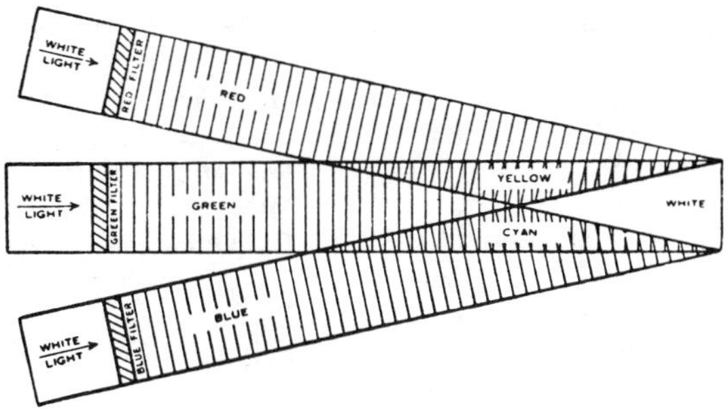

Mechanism of color addition.

When the additive synthesis is made by successive viewing, the same three color images must be flashed onto the screen in such rapid succession that the individual red, green and blue images are not apparent. Simple color addition is again obtained but this time use is made of the persistence of vision to "mix" the colors. See accompanying diagram.

The third type of additive synthesis makes use of the fact that small dots of different colors, when viewed from such a distance that they are no longer individually visible, form a single color by simple color addition. The three color images in this type of process are generally side-by-side in the space normally occupied by a single image. The red record image will be composed of a number of red dots or markings of differing density which in their sum total will compose the red record image. Alongside the red markings will be green and blue markings, without any overlapping. When viewed at such a distance that the colored markings are at, or below, the limit of visual resolution, the color sensation from any given area will be the integrated color of the markings comprising the area—an additive color mixture.

ADDITIVE (Fuel). Petroleum.

ADDITIVES (Food). The use of food additives is centuries old. One of the earliest and still most widely practiced example is the addition of common salt to foods, either by a food preparer or processor, or by the consumer at the table. Normally, the food producing professionals and the lay public do not view salt as a food additive, probably because salt is regarded as a substance widely found in nature, and particularly in seawater. Nevertheless within the last 10 to 20 years, there has been a growing awareness that salt is truly an additive and to certain controlled diets can have an adverse role as well as the numerous positive roles that it plays in food processing and preservation.

The term *food additive* is most frequently associated with the more complex substances, the presence or function of which the average consumer seldom notes unless the package labels are read. Particularly where additives contribute to the pleasure of the consumer, they are commonly passed over unnoticed. By far, the majority of consumers prefer breads that do not stale readily, salad dressings that do not separate into layers after a period of shelf life, dairy products that do not become rancid after a few days, meats that retain their color and wholesome appearance and taste after relatively long periods of refrigeration. These food qualities, of course, are made possible because of the many advances over the years in food additive technology.

Additives also contribute to the ease of processing and marketing foods and, even though indirectly, have played a positive role in terms of food transportation, distribution, and inventorying costs. Conversely, some additives in the past have been found to be damaging to health. Some of the colorants formerly used, for example, have been banned because of their carcinogenic qualities when tested in laboratory animals. Considering the vast quantities of additives used, the majority used to date have not been shown to be harmful.

There are tens of categories of food chemicals, most of which are described in various entries throughout this encyclopedia. Consult the following entries:

Acidulants and Alkalizers (Foods)
Anticaking Agents
Antimicrobial Agents (Foods)
Antioxidant
Bleaching Agents
Bodying and Bulking Agents (Foods)
Buffer (Chemical)
Carrier (Food Additive)
Clarifying Agents
Coating Agents
Colorants (Foods)
  Annatto Food Colors
  Anthocyanins
  Betalaines
  Carotenoids
Enzyme Preparations
Firming Agents (Foods)
Flavorings
  Flavor Enhancers and Potentiators
Humectants and Moisture-Retaining Agents
Leavening Agents
Masticatory Substances
Oxidation and Oxidizing Agents
Pectins
Polymeric Food Additives
Sweeteners

Dispersing, emulsifying, gelling, thickening, and stabilizing agents are covered under **Colloid Systems.** Dough conditioners are described under **Oxidation and Oxidizing Agents.** Nonnutritive sweeteners are included in entry on **Sweeteners.**

References

Cole, M. S., and J. F. Wintermantel, Co-Chairpersons: "Food Additives and Ingredients," *Institute of Food Technologists Symposium*, Saint Louis, Missouri, 1979.

Davis, R. B.: "Regulatory Aspects of Migration of Indirect Additives to Food," *Food Technology*, **33**, 4, 55–60 (1979).

FAO-UN: "Additives Evaluated for Their Safety-in-Use in Foods," Codex Alimentarius Commission, World Health Organization (WHO) and Food and Agriculture Organization (FAO), United Nations, Rome (revised periodically).

Fazio, T.: "Extraction Testing Methods for Evaluation of Food Packing Materials," *Food Technology*, **33**, 4, 61–62 (1979).

Gilbert, S. G.: "Modeling the Migration of Indirect Additives to Food," *Food Technology*, **33**, 4, 63–65 (1979).

Grice, H. C.: "Food Additive Evaluation and Regulation: The Canadian Approach," *Food Product Development*, **11**, 4, 28–32 (1977).

Kermode, G. L.: "Food Additives," *Sci. Amer.*, **226**, 5, 15–21 (1972).

Kohl, P.: "Food and Additives," General Foods Corporation, White Plains, New York (revised periodically).

Koros, W. J., and H. B. Hopfenberg: "Scientific Aspects of Migration of Indirect Additives from Plastics to Food," *Food Technology*, **33**, 4, 55–60 (1979).

Lowenstein, M.: "What Does 'Natural' Mean?" *Dairy Record* (May 1977).

Roberts, H. R.: "Regulatory Aspects of the Food Additive Risk/Benefit Problem," *Food Technology*, **32**, 8, 59–61 (1978).

Staff: "Food Chemicals Codex," National Academy of Sciences, Washington, D.C. (2nd Edition, 1972, but with updated supplements issued periodically).

ADDRESS (Computer System). An identification, representated by a name, label, or number, for a digital computer register, device, or location in storage. Addresses are also a part of an instruction word along with commands, tags, and other symbols. The part of an instruction which specifies an operand for the instruction may be an address.

Absolute address or *specific address* indicates the exact physical storage location where the referenced operand is to be found or stored in the actual machine code address numbering system.

Direct address or *first-level address* indicates the location where the referenced operand is to be found or stored with no reference to an index register.

Indirect address or *second-level address* in a computer instruction indicates a location where the address of the referenced operand is to be found. In some computers, the machine address indicated can in itself be indirect. Such multiple levels of addressing are terminated either by prior control, or by a termination symbol.

Machine address is an absolute, direct, unindexed address expressed as such, or resulting after indexing and other processing has been completed.

Symbolic address is a label, alphabetic or alphameric, used to specify a storage location in the context of a particular program. Sometimes programs may be written using symbolic addresses in some convenient code, which then are translated into absolute addresses by an assembly program.

Base address permits derivation of an absolute address from a relative address.

Effective address is derived from applying specific indexing or indirect addressing rules to a specified address.

Four-plus-one address incorporates four operand addresses and a control address.

Immediate address incorporates the value of the operand in the address portion instead of the address of the operand.

N-level address is a multilevel address in which N levels of addressing are specified.

One-level address directly indicates the location of an instruction.

One-plus-one address contains two address portions. One address may indicate the operand required in the operation. The other may indicate the following instruction to be executed.

Relative address is the numerical difference between a desired address and a known reference address.

Three-plus-one address incorporates an operation code, three operand address parts, and a control address.

Zero-level address enables immediate use of the operand.

For related terms in this volume, see list of entries under **Data Processing.**

Thomas J. Harrison, International Business Machines Corporation, Boca Raton, Florida.

ADDRESS REGISTER. A collection of single-bit logical storage elements for the temporary storage of an address in a computer. Typically, a computer contains a number of address registers, each with a specific purpose. An adjective modifier, such as "storage address register" is often used to indicate the specific use of the register. The *storage address register* contains the address of the data currently being accessed in storage. See also **Storage (Computer).** This register and its associated decoding logic provide the signals to the storage-selection circuits while storage is being cycled.

An *instruction address register* contains the address of the instruction which is to be read from storage. When the instruction is called for, the contents of this register typically will be transferred to the storage address register.

A *data address register* contains the address of the data to be obtained from storage, or the address at which the data are to be stored. When data are being read from or written to storage, the contents of this register is transferred to the storage address register. Two data address registers may be required in machines which provide the function of moving data from one storage location to another storage location with a single instruction. In this case, one register contains the address of the present location of the data; the other contains the address of the location to which the data are to be transferred.

ADENINE. A prominent member of the family of naturally occurring purines with the accompanying formula. Adenine occurs not only in ribonucleic acids (RNA), and deoxyribonucleic acids (DNA), but in nucleosides, such as adenosine, and nucleotides, such as adenylic acid, which may be linked with enzymatic functions quite apart from nucleic acids. Adenine, in the form of its ribonucleotide, is produced in mammals and fowls endogenously from smaller molecules and no nutritional essentiality is ascribed to it. In the nucleosides, nucleotides, and nucleic acids, the attachment or the sugar moiety is at position 9.

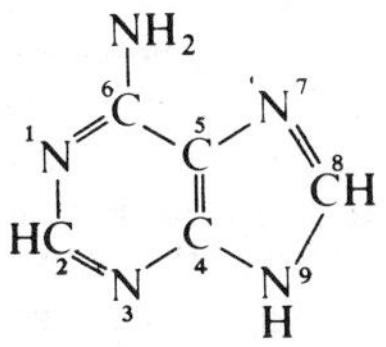

Structure of adenine.

The purines and pyrimidines absorb ultraviolet light readily, with absorption peaks at characteristic frequencies. This has aided in their identification and quantitative determination.

ADENINE ARABINOSIDE. **Antiviral Drugs.**

ADENOMA. A benign tumor consisting of an encapsulated overgrowth of epithelial cells of a glandular structure. Adenomata may occur in the endocrine glands, the gastro-intestinal tract, the respiratory system, the breast, and wherever glandular epithelium occurs. A malignant adenoma is referred to as a *adenocarcinoma.*

ADENOSINE. An important nucleoside. The upper portion of the accompanying formula represents the adenine moiety, and the lower portion of the pentose, D-ribose.

Structure of adenosine.

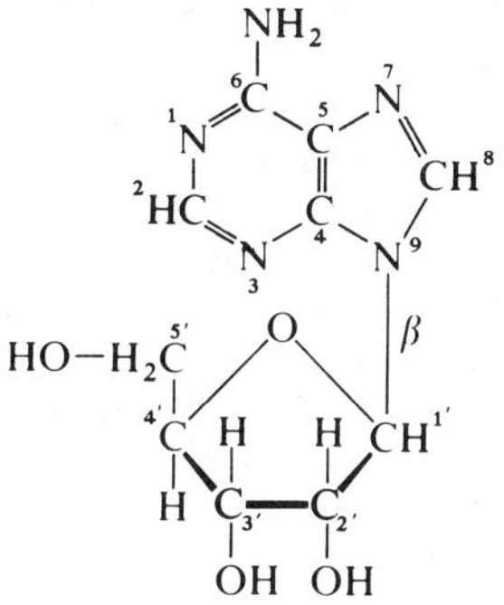

ADENOSINE DI- AND TRIPHOSPHATE. **Carbohydrates; Phosphorylation (Oxidative); Phosphorylation (Photosynthetic).**

ADENOSINE PHOSPHATES. The adenosine phosphates include *adenylic acid* (adenosine monophosphate, AMP) in which adenosine is esterified with phosphoric acid at the 5′-position; *adenosine diphosphate* (ADP) in which esterification at the same position is with pyrophosphoric acid,

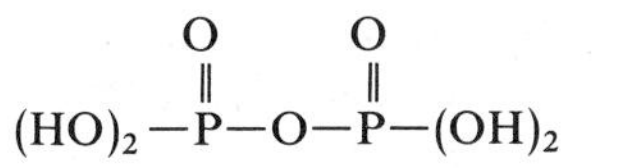

and *adenosine triphosphate* (ATP) in which three phosphate residues

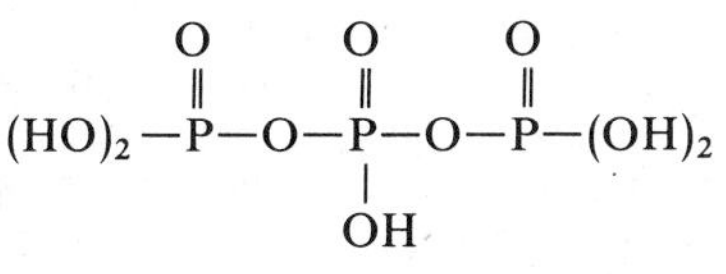

are attached at the 5′-position. Adenosine-3′-phosphate is an isomer of adenylic acid, and adenosine-2′,3′-phosphate is esterified in two positions with the same molecules of phosphoric acid and contains the radical

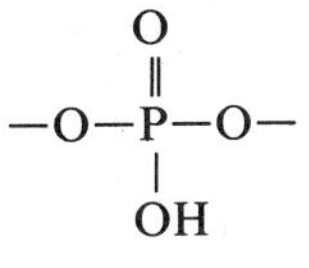

ADENOSIS. Any disease of a gland, or glands.

ADENOVIRUSES. **Virus.**

ADHARA (ε Canis Majoris). Ranking twenty-second in apparent brightness among the stars, Adhara has a true brightness value of 8,000 as compared with unity for the sun. Adhara is a blue-white, spectral type B star and is located in the constellation Canis Majoris. Estimated distance from the earth is 600 light years. See also **Constellations.**

ADHERENT SURFACE. **Conversion Coatings.**

ADHESION (Physics). The terms adhesion and cohesion designate intermolecular forces holding matter together. The tendency of matter to hold itself together or to cling to other matter is one of its most characteristic properties. Adhesion and cohesion are merely different aspects of the same phenomenon, which is apparently of the nature of an intermolecular attraction. One speaks of cohesion as an interaction between adjacent parts of the same body and as acting throughout the interior of its substance, while adhesion refers to a similar interaction between the closely contiguous surfaces of adjacent bodies.

In the case of solids, the experimental study of adhesion and cohesion, in which two solids are brought into intimate contact, can depend on the nature of adsorbed gases. For clean surfaces it is necessary to carry out the experiments in "ultra high vacuum" at pressures better than 10^{-10} torr.

There is reason to believe that as two neutral molecules or atoms approach each other, their mutual potential energy reaches a minimum value at a certain equilibrium distance; so that work would be necessary either to push them closer or to pull them farther apart, because of forces which are probably electrical. See **Least Energy Principle.** The distribution of molecules, ions, or atoms in a solid is determined by this type of equilibrium, and the regular spacing of crystal structure and the architecture of the molecule itself are dependent upon it. Any force tending to diminish the equilibrium distance meets with the rapidly increasing reaction of compressive elasticity, while any force tending to increase it is opposed by cohesion, which increases at first and then rapidly diminishes toward zero as the point of fracture is reached.

The behavior of bodies which are aggregates of crystals or of fibers is complicated by the friction and the adhesion of the adjacent particles, so that the ultimate strength of a material is not a safe measure of its true cohesion. A filament of spun quartz may be much stronger when freshly drawn than later when crystallization replaces its initial cohesion by the adhesion between separate crystals; and yarn is not nearly so strong as the cotton or wool fiber composing it.

Adhesion increases with closenss of contact. This explains why one must bear down with a pencil to make a mark on paper, why fine dust adheres more firmly than coarse sand, and why a liquid or a gum usually sticks to a solid better than another solid does.

Cohesion in liquids is usually less, and in gases it is always much less, than in solids. Aside from the pressure in liquids due to external causes, there is presumably a very great internal or intrinsic pressure, due to intermolecular attraction, but not capable of direct measurement by means at our disposal. The clearest evidences of its existence are the work required for thermal expansion and the phenomenon of surface tension.

ADHESION (Work of). The work of adhesion W_{AB} between two liquids A and B is the increase in free surface energy on separating 1 cm^2 of interface AB

$$W_{AB} = \gamma_A + \gamma_B - \gamma_{AB}$$

where γ_A and γ_B are the surface tensions of A and B respectively against their vapors, and γ_{AB} is the interfacial tension. For a solid-liquid interface the work of adhesion W_{SL} is defined as the work required to separate 1 cm^2 of interface in a vacuum to give a naked solid surface

$$W_{SL} = \gamma_S + \gamma_L - \gamma_{SL}$$

where γ_S and γ_L are the surface tensions measured in a vacuum.

It may be shown that

$$W_{SL} = \gamma_S - \gamma_{SVO} + \gamma_L(1 + \cos \theta_E)$$

where γ_{SV_0} is the surface tension of the solid covered by an absorbed film of liquid in equilibrium with the vapor, and θ_E the equilibrium contact angle.

ADHESIVES. Materials or compositions which enable two surfaces to become united are known as adhesives. Years ago the terms *adhesive* and *glue* were considered synonymous, but the general term glue now implies a sticky substance, whereas many adhesives are not sticky. Adhesives may be broadly classified into two main groups; organic and inorganic adhesives with the organic materials being subdivided into those of animal origin, vegetable origin, and synthetic origin. A more useful classification is one based on the chemical nature of the adhesive. Such a chemical classification comprises (1) protein or protein derivatives, (2) starch, cellulose, or gums and their derivatives, (3) thermoplastic synthetic resins, (4) thermosetting synthetic resins, (5) natural resins and bitumens, (6) natural and synthetic rubbers, and (7) inorganic adhesives.

Adhesives may also be classified according to the purpose for which they are used as (a) for bonding rigid surfaces such as wood, glass, porcelain, rigid plastics, or metal and (b) for bonding flexible surfaces such as paper, textiles, leather, flexible plastics, thin metallic sheets, and the like. In the latter category it is necessary that the adhesive have as high a degree of flexibility as the less flexible of the bonded surfaces.

The protein and protein derivative adhesives include those made from casein, zein, soybean proteins, and other proteins and the very important groups of glues made from hides, bones, etc., fish glues from fish offal, and those from blood albumen.

The group of adhesives made from materials such as starch and the vegetable gums are also known as vegetable adhesives or glues. They comprise the adhesives made from starch and processed starch, the dextrins, and the water soluble gums such as gum arabic, ghatti, tragacanth, Indian gum and the like. These are water soluble or water suspendable materials. The starch and dextrin products are used for the bonding of paper, wood, and textiles; the gum products are used as adhesives for paper, postage stamps and other stamps, and for miscellaneous items.

Cellulose adhesives are principally cellulose derivatives such as methylcellulose, ethylcellulose, cellulose acetate and nitrate, and sodium carboxymethyl cellulose. These products are used as components in adhesive compositions such as those used for the bonding of leather, cloth, paper, and many other materials.

Thermoplastic synthetic resin adhesives comprise a variety of polymerized materials such as polyvinyl acetate, polyvinyl butyral, polyvinyl alcohol, and other polyvinyl resins; polystyrene resins; acrylic and methacrylic acid ester resins; cyanoacrylates; and various other synthetic resins such as polyisobutylene, polyamides, coumarone-indene products, and silicones. Such thermoplastic resins usually have permanent solubility and fusibility so that they creep under stress and soften when heated. They are used for the manufacture of tapes, safety glass, and shoe cements and for the bonding of foils, metals, woods, rubber, paper, and many other materials.

Thermosetting synthetic resin adhesives comprise a variety of phenol-aldehyde, urea-aldehyde, melamine-aldehyde, and other condensation-polymerization materials like the furane and polyurethane resins. Thermosetting adhesives are characterized by being converted to insoluble and infusible materials by means of either heat or catalytic action. Adhesive compositions containing phenol-, resorcinol-, urea-, melamine-formaldehyde, phenolfurfuraldehyde, and the like are used for the bonding of wood, textiles, paper, plastics, rubbers, and many other materials.

The adhesives of the natural resin and bitumen group consist of those made from asphalts, shellac, rosin and its esters, and similar materials. They are used for the bonding of various materials including minerals, linoleum, and the like.

There are a number of different types of rubber adhesives. Some are simply solutions of rubber latex, rubber, or synthetic and modified rubbers in a solvent, and others are compositions consisting of one of the aforementioned and casein or a synthetic resin. Such compositions are very widely used from the bonding of flexible materials like paper, textiles, leather and rubber to rigid materials like metals and plastics.

The inorganic adhesives are a very important group. They comprise: sodium silicate principally used for corrugated paper and other paper products, laminating metal foils, plywood bonding, and in the production of building and insulating boards; plaster of paris for ceramic and similar products; magnesium oxychloride for ceramics; litharge-glycerin for joints in plumbing fixtures; and Portland cement for bonding mineral aggregates. See also **Silicates (Soluble).**

While it has been convenient to put the various types of materials from which adhesives are made into groups, many adhesive compositions are made of substances or materials from a number of groups.

In this connection it should be noted that the term *cement* is often used as a synonym for adhesive. A cement is a particular kind of adhesive which consists of any material that can be prepared in a plastic form which hardens to bond together various solid surfaces.

ADIABATIC. Occurring without gain or loss of heat by the system involved: for example, adiabatic expansion. A completely adiabatic process is unrealizable, but it can be closely approximated in practice by providing good thermal insulation and by carrying out the process rapidly so that little time is allowed for heat flow in or out of the system. If a thermally isolated system moves through a series of equilibrium states, i.e., undergoes an adiabatic and reversible process, the locus of the points representing these states on a graph is called an *adiabatic* or *adiabat.* The present tendency is to avoid this usage of adiabatic to mean isentropic. See **Adiabatic Process.**

ADIABATIC ATMOSPHERE. Atmosphere (Earth).

ADIABATIC FLAME TEMPERATURE. Combustion.

ADIABATIC LAPSE RATE. Atmosphere (Earth).

ADIABATIC PROCESS. Any thermodynamic process, reversible or irreversible, which takes place in a system without the exchange of heat with the surroundings. When the process is also reversible, it is called *isentropic,* because then the entropy of the system remains constant at every step of the process. (In older usage, isentropic processes were called simply adiabatic, or quasistatic adiabatic; the distinction between adiabatic and isentropic processes was not always sharply drawn.)

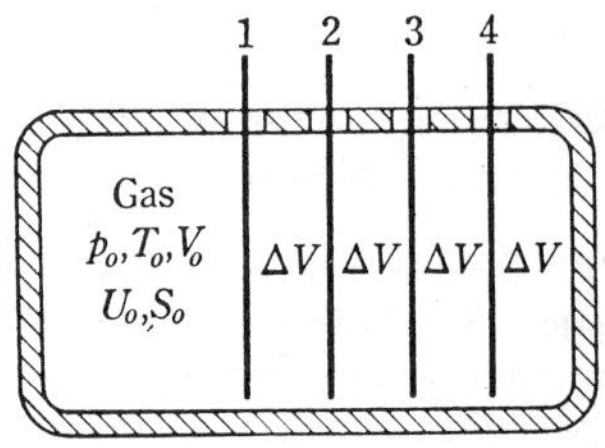

Fig. 1. Successive adiabatic expansions of gas by withdrawing slides.

When a closed system undergoes an adiabatic process without performing work (*unresisted expansion*), its internal energy remains constant whenever the system is allowed to reach thermal equilibrium. Such a process is necessarily irreversible. At each successive state of equilibrium, the entropy of the system S_i, has a higher value than the initial entropy, S_0. Example: When a gas at pressure p_0, temperature T_0, occupying a volume V_0 (see Fig. 1) is allowed to expand progressively into volumes $V_1 = V_0 + \Delta V$, etc., by withdrawing slides 1, 2, etc., one after another, it undergoes such a process if it is enclosed in an adiabatic container. After each withdrawal of a slide, the irreversibility of the process causes the system to depart from equilibrium; equilibrium sets in after a sufficiently long waiting period. At each successive state of equilibrium $U_1 = U_2 = \cdots = U_0$, but $S_0 < S_1 < S_2$, etc.

When an open system in steady flow undergoes an adiabatic process without performing external work, the enthalpy of the system regains its initial value at each equilibrium state, and the entropy increases as before. Example: Successive, *slow* expansions through porous plugs P_1, $P_2 \cdots$ (Fig. 2), when we have

$$H_1 = H_2 = \cdots = H_0$$

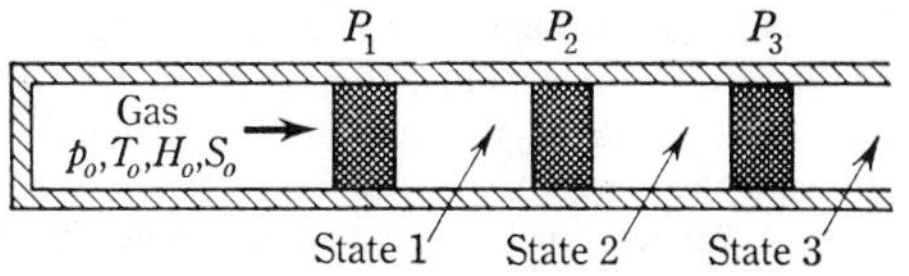

Fig. 2. Successive, slow adiabatic expansions of gas through porous plugs.

but

$$S_0 < S_1 < S_2, \text{ etc.}$$

This process is also necessarily irreversible.

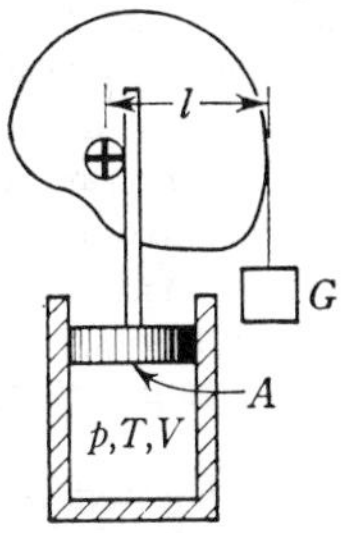

Fig. 3. Isentropic compression (or expansion) in cylinder.

A closed system cannot perform an isentropic process without performing work. Example (Fig. 3): A quantity of gas enclosed by an ideal, frictionless, adiabatic piston in an adiabatic cylinder is maintained at a pressure p by a suitable ideal mechanism, so that $Gl = pA$ (A being the area of piston). When the weight G is increased (or decreased) by an infinitesimal amount dG, the gas will undergo an isentropic compression (or expansion). In this case,

$$S = \text{constant}, \qquad dS = 0$$

at any stage of the process, but

$$U \neq \text{constant}, \qquad H \neq \text{constant}$$

During an isentropic process of a closed system between states 1 and 2, the change in internal energy equals *minus* the work done between the two states, or

$$U_2 - U_1 = -W_{12}$$

work is done "at the expense" of the internal energy.

ADIABATIC WALL. A perfect heat insulator. Since in a rigorous development of the principles of thermodynamics it is necessary to introduce the concept of an adiabatic wall before the concept of heat, it is convenient to adopt the following alternative definition. If two closed systems are placed in contact through an adiabatic wall, their states can be varied independently of one another. Any state of one system can co-exist with any state of the other system through such a wall; the systems are not coupled in any way. See **Diathermal Wall.** The number of independent properties of the combined system is equal to the sum of the number of independent properties of the component systems.

ADIABATIC WALL TEMPERATURE. Supersonic Aerodynamics.

ADIATHERMAL. Diathermal Wall.

ADIPOSE FIN. Fishes.

ADJOINT SPACE. Linear Space (Vector Space).

ADJUTANT BIRD. Storks.

ADJUVANT CHEMOTHERAPY. Cancer and Oncology.

ADJUVANT (Immune System). Immune System and Immunology.

ADOBE. An extremely fine-grained wind-blown clay particularly characteristic of the arid and semi-arid southwestern United States, Mexico, and South America. Used by the southwestern Indians and Mexicans for huts and buildings from prehistoric times.

ADOLESCENCE. The period of youth, extending from the beginning of puberty to adulthood. For human beings, this period usually ranges between 12 and 20 years. Physical developments in adolescence of the American male are shown in the accompanying chart.

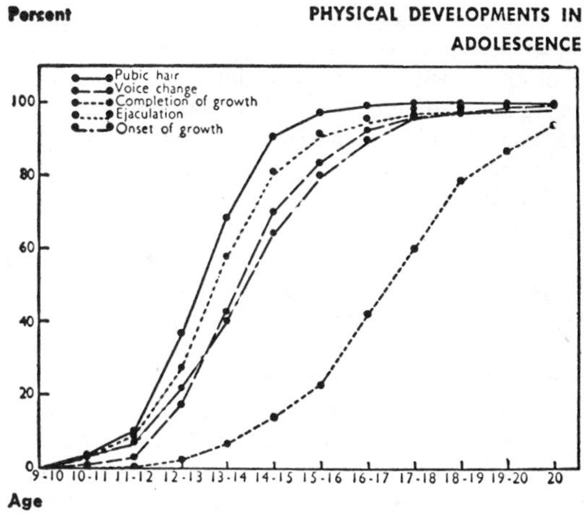

This chart portrays the percent of American males undergoing the indicated changes in physical development at the various ages during adolescence.

Growth during the first 10 years of life proceeds at a fairly uniform rate. The child's desires usually keep pace with his ability to meet his needs. However, during the adolescent or teenage period, there is disharmony between physiological development, growth, and emotional maturation. The teen-ager becomes a blend of maturity and immaturity in body, mind, and emotion.

Puberty, which refers to the sexual maturation of the individual, is only part of adolescence. Puberty and adolescence begin in the girl at 11 to 15 years of age. Boys lag about a year behind. The average age for puberty in girls in the United States is around 13.5 years and in boys, 14.5 years. Adolescence in most American youth terminates at about 19 years of age.

The first half of adolescence is characterized by physical growth and change; the latter half brings more intellectual and emotional changes.

Adolescent growth varies widely between individuals. Failure to realize this may cause unnecessary anxiety to both parents and child. Growth rate is influenced by the state of health and nutrition. Not only is there variability in the growth of different individuals, but growth within the single individual is not harmonious. Certain parts of the body grow, while other parts lag behind. The child may grow tall without putting on weight, or he may show more sexual maturity than maturity of the digestive organs. This organ imbalance leads to organ instability. Thus, laziness and awkwardness may result. In the average boy, the period of greatest height increase is from about 12 to 14.5 years, although they continue to grow until about the age of 18. They gain most of their weight between the ages of 13 and 16. Girls grow tallest from about 10.5 to 14 years; maximum weight development is from about ages 11.5 to 14.5. As much as 6 inches (15 centimeters) in height and 25 pounds (11.3 kilograms) in weight may be gained within a single year.

Puberty in the boy is marked by growth of the sexual organs and the beginning of their function. Secondary sexual changes occur. His voice deepens; his shoulders broaden; his muscles harden; his legs lengthen; his hands and feet grow disproportionately large.

Arrival of the first menstrual period in a girl may be shocking and frightening if improperly prepared for the event. Menstrual irregularity and pain are common the first year or so, usually without significance. The broadening of hips, development of breasts, and appearance of pubic hair usually parallel the beginning of menstruation, but are subject to wide variation. Both sexes are preoccupied with their bodies and unable to take their growth for granted.

Obesity is often a problem in this period and may be a factor in delayed puberty. Whether the condition is the result of poor eating habits, of emotional instability, or of endocrine dysfunction, it merits careful investigation.

To ignore the emotional factors in adolescence is to ignore the most crucial and important aspect of the adolescent years. The physical changes can be handled relatively easily by the young people. But, many other problems arise as a result of, and in addition to bodily changes. No longer children, teen-agers still have many childish wishes and habits. Although not adults, yet they have many adult emotions and adult problems. They long for the pleasures of childhood, yet rebel against the restrictions imposed on children. They long for the pleasures of adulthood and rebel against the responsibilities. Since teen-agers are often confused about their position in the family during this period, they often lose former intimate contact with their parents. They tend to mask their insecurity with cockiness or rationalize their isolation with arrogance.

Normally, adolescence should be a period in which the main emotional ties with the parents begin to give way to wider interests and relationships. Usually, this is accomplished with a good deal of rebellion against all forms of authority, particularly parental authority. Teen-agers may begin to think that their parents do not know what is going on in the modern world. The more strenuous the parents are in efforts to keep their children from becoming emotionally mature, the more apt the rebellion is to be pronounced and painful.

Teen-agers may attempt to show their new status by merely discarding their former good manners or by the excessive use of slang to bewilder their elders. They may be careless in dress and hygiene in order to show this revolt. A phase quite marked in adolescence, the wish to reform the whole social and economic system, is based not only on revolt against the established forms, but also on a certain amount of idealism.

Teen-agers have less fear of showing their aggression than do younger children. Normally, the aggression finds its outlets in hard-fought games and sports, in arguing, and often careless, dangerous activities.

Apparently paradoxical with this rebellion against authority is the conformity to the group so paramount in the teen-ager's thinking. Adolescents may set up their own standards of morals and values and exert the pressure of those standards on other teen-agers. Psychiatrists call this "adolescent peer-culture." Boys and girls desiring the approval of their friends must follow the fashion of their set in matters of morals, dress, speech, and so on—at least, many believe so. This encourages social participation, group loyalty, and individual achievement and responsibility. But, it also brings about problems. At times, particularly in the smaller groups, cliques may dominate either the school or the church, electing their own candidates to school offices and setting the social pace to the exclusion of those teen-agers not accepted. Those excluded from the social life of the reigning set attempt to make life for themselves with other unaccepted adolescents. This rejected group may react with defiant behavior, rowdyism, truancy, and so on, which isolate them even more from their other schoolmates. The individual who cannot make the grade with either group may be either delinquent, or bookish and withdrawn, depending upon his inner needs.

The emotional life of adolescence is full of paradoxes. Utterly self-centered, teen-agers are nevertheless capable of much self-sacrificing devotion. They may turn from their gangs to complete solitude. Idealism of the loftiest sort is present side by side with the most selfish materialism. Sensitive, teen-agers can be cruelly inconsiderate of others. Sensual indulgence may alternate with asceticism.

Physical maturity is reached when the body has its final height and has assumed adult proportions. The secondary sexual characteristics are fully developed, and the sexual functions have been established. While physical maturity is easy to assess, emotional maturity is much more difficult. People who become angry at trivial things, who are easily hurt, or who run away from reality, are not emotionally mature adults. To accept reality is difficult and is the major characteristic of maturity. To know one's limitations, to be able to work with them contentedly, and to suit one's life to one's capacities, is to be emotionally mature. To search for thrills, to refuse to compromise, is adolescent.

Each teen-ager, as an individual person, represents a unique situation and, consequently, patterns of behavior range widely.

ADRENAL GLANDS. Part of the endocrine system, the adrenal glands are two small bodies located at the upper end of each kidney. The right adrenal gland is somewhat triangular in shape; the left gland is more semilunar. These glands range in size, but on the average weigh from 5 to 9 grams. Surrounding each gland is a thin capsule. There are two parts—the *cortex* (external tissue) and the *medulla* (chromophil tissue). These parts differ both in origin and function. In proportion to total body weight, the adrenals are much larger at birth and weigh about 8 grams. In adults, proportionately, the adrenals are 1/20th the size they are in infants. Most of the shrinkage in size and weight occurs during the first year of life. The adrenals are well supplied with blood and nerves.

While the adrenal glands perform a number of functions, the two most important are: (1) Control of the body's adjustment to an upright posture; and (2) accommodation of the body to intermittent rather than constant intake of food. The adrenals also participate importantly in the regulation of electrolyte and water balance; the activity of lymphoid tissue and the number of eosinophils circulating in the blood (see **Blood**); the response to stress situations as may be encountered in infection, anesthesia, surgery, and volume loss; and the secretion of hormones. These hormones influence immune reactivity, blood cell formation, cerebral function, protein synthesis, and numerous other body processes. See also **Endocrine System;** and **Hormones.**

Several serious problems occur when there is imbalance or dysfunction of the adrenal glands, as, for example, in hyperfunction of the adrenal cortex, which precipitates Cushing's syndrome, of which there are several sub-forms; in hypofunction of the adrenal cortex; in adrenal neoplasms; and in pheochromocytoma.

Adrenal Cortical Hormones. The hormones elaborated by the adrenal cortex are steroidal derivatives of cyclopentanoperhydrophenanthrene related to the sex hormones. The structural formulas of the important members of this group are shown in Fig. 1. With the exception of *aldosterone*, the compounds may be considered derivatives of *corticoterone*, the first of the series to be identified and named. The C_{21} steroids derived from the adrenal cortex and their metabolities are designated collectively as *corticosteroids.* They belong to two principal groups: (1) those processing an O or OH substituent at C_{11} (*corticosterone*) and an OH group at C_{17} (*cortisone* and *cortisol*) exert their chief action on organic metabolism and are designated as *glucocorticoids*; (2) those lacking the oxygenated group at C_{17} (*desoxycorticosterone* and *aldosterone*) act primarily on electrolyte and water metabolism and are designated as *mineralocorticoids.* In humans, the chief glucocorticoid is *cortisol.* The chief mineralocorticoid is *aldosterone.*

The *glucocorticoids* are concerned in organic metabolism and in the organism's response to stress. They accelerate the rate of catabolism (destructive metabolism) and inhibit the rate of anabolism (constructive metabolism) of protein. They also reduce the utilization of carbohydrate and increase the rate of gluconeogenesis (formation of glucose) from protein. They also exert a lipogenic as well as lipolytic action, potentiating the release of fatty acids from adipose tissue. In addition to these effects on the organic metabolism of the basic foodstuffs, the glucocorticoids affect the body's allergic, immune, inflammatory, antibody, anamnestic, and general responses of the organism to environmental disturbances. It is these reactions which are the basis for the wide use of the corticosteroids therapeutically. See also **Immune System and Immunology.**

Aldosterone exerts its main action in controlling the water and electrolyte metabolism. Its presence is essential for the reabsorption of sodium by the renal tube, and it is the loss of salt and water which is responsible for the acute manifestations of adrenocortical insufficiency. The action of aldosterone is not limited to the kidney, but is manifested on the cells generally, this hormone affecting the distribution of sodium, potassium, water, and hydrogen ions between the cellular and extracellular fluids independently of its action on the kidney.

The differentiation in action of the glucocorticoids and the mineralocorticoids is not an absolute one. Aldosterone is about 500 times as effective as cortisol in its salt and water retaining activity, but is one-third as effective in its capacity to restore liver glycogen in the adrenalectomized animal. Cortisol in large doses, on the other hand, exerts a water and salt retaining action. Corticosterone is less active than cortisol as a glucocorticoid, but exerts a more pronounced mineralocorticoid action than does the latter. See also **Steroids.**

In addition to the aforementioned corticosteroidal hormones, the adrenal glands produce several oxysteroids and small amounts of testosterone and other androgens, estrogens, progesterone, and their metabolites.

Adrenal Medulla Hormones. Adrenaline (*epinephrine*) and its immediate biological precursor *noradrenaline* (*norepinephrine,* levarternol) are the principal hormones of the adult adrenal medulla. See Fig. 2. Some of the physiological effects produced by adrenaline are: contraction of the dilator muscle of the pupil of the eye (*mydriasis*); relaxation of the smooth muscle of the bronchi; constriction of most small blood vessels; dilation of some blood vessels, notably those in skeletal muscle; increase in heart rate and force of ventricular contraction; relaxation of the smooth muscle of the intestinal tract; and either contraction or relaxation, or both, of uterine smooth muscle. Electrical stimulation of appropriate sympathetic (*adrenergic*) nerves can produce all the aforementioned effects with exception of vasodilation in skeletal muscle.

Noradrenaline, when administered, produces the same general effects as adrenaline, but is less potent. Isoproternol, a synthetic analogue of noradrenaline, is more potent than adrenaline in relaxing some smooth muscle, producing vasodilation and increasing the rate and force of cardiac contraction.

Malfunctions of the Adrenals. Progressive destruction of the adrenal cortex, as found in *Addison's disease*, gives rise to symptoms resulting from deficiencies of the cortical hormones previously mentioned. In the Waterhouse-Friderichsen syndrome, destruction of part or the whole of one or both glands by hemorrhage, as may occur in the course of meningitis due to meningococci, leads to sudden collapse and death unless very prompt treatment by replacement of the absent hormones is available. See also **Addison's Disease.** In *Cushing's syndrome* (adrenal cortical excess), a rare condition, there is obesity of the abdomen, face, and buttocks, but not of the limbs. The skin about the face and hands is redder than normal. Hair grows profusely, and women may grow excessive hair about the face. Bones become brittle and suffer a considerable loss of mineral components. Sexual functions may fall to a low level. The adrenal cortex may be overly stimulated by an excess secretion of ACTH (*adrenocorticotropic hormone*) in the pituitary gland. See also **Pituitary Gland.** Also, excessive production of ACTH may be caused by malignant tumors, among which lung, thymus, pancreas, and kidney are the most common. Similar symptoms also may be produced by the excessive pharmacologic (*iatrogenic*) use of steroids. Because of the multifaceted causes of Cushing's syndrome, differential diagnosis is important.

References

Axelrod, L.: "Glucocorticoid Therapy," *Medicine* (*Baltimore*), **55**, 39 (1976).

Besser, G. M., and W. J. Jeffcoate: "Endocrine and Metabolic Disease: Adrenal Diseases," *Br. Med. J.*, **1**, 448 (1976).

Eddy, R. L., et al.: "Cushing's Syndrome," *Am. J. Med.*, **55**, 621 (1973).

Fauci, A. S., Dale, D. C., and J. E. Balow: "Glucocorticosteroid Therapy," *Ann. Intern. Med.*, **84**, 304 (1976).

Liddle, G. W., and K. L. Melmon: "The Adrenals," in "Textbook of Endocrinology," 5th edition (R. H. Williams, editor), Saunders, Philadelphia, 1974.

Rees, L. H., et al.: "Alcohol-Induced Pseudo-Cushing's Syndrome," *Lancet*, **1**, 726 (1977).

Sachar, E. J.: "Hormonal Changes in Stress and Mental Illness," *Hosp. Pract.*, **10**, 7, 49 (1975).

Tyrrell, J. B., et al.: "Cushing's Disease," *N. Engl. J. Med.*, **298**, 753 (1978).

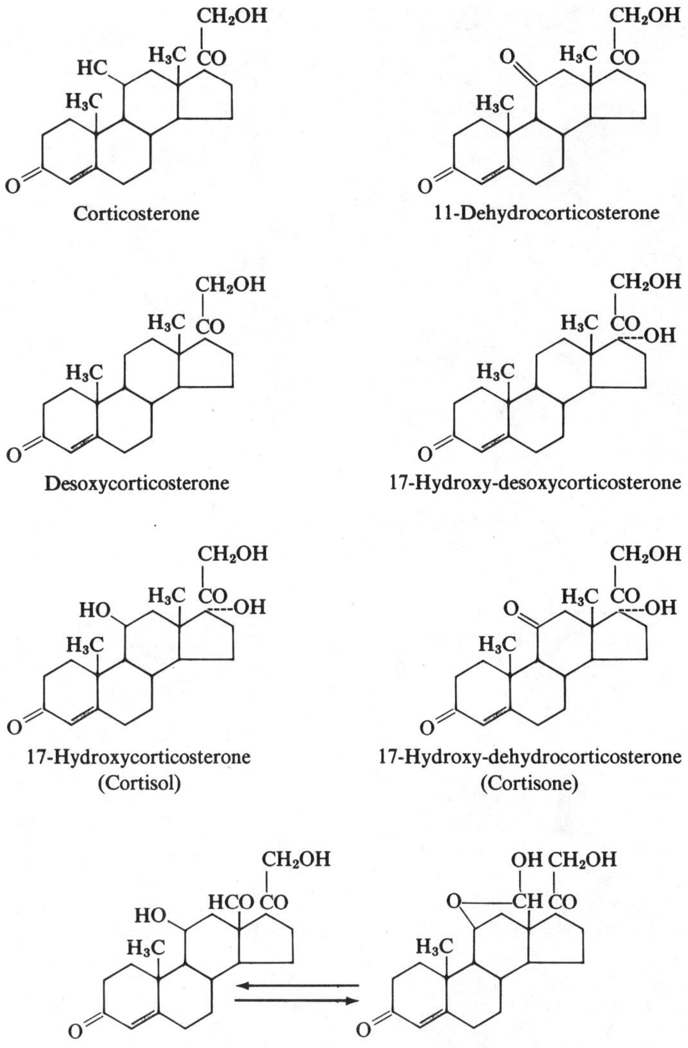

Aldosterone

Fig. 1. Adrenal cortical hormones.

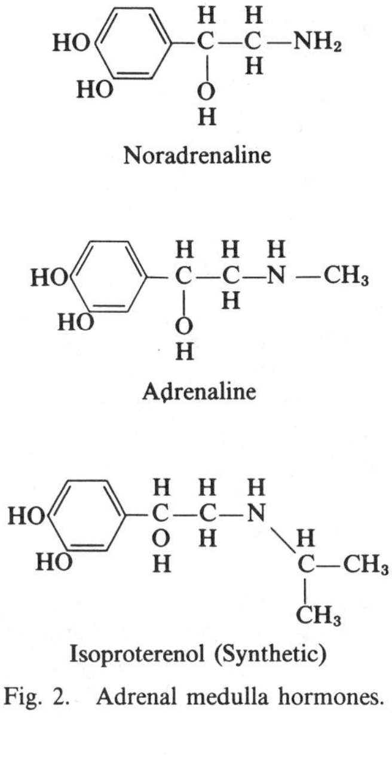

Fig. 2. Adrenal medulla hormones.

ADP. 1. Automatic data processing. 2. Adenosine diphosphate.

ADP CRYSTAL. Piezoelectric Effect.

ADRENAL CORTEX. Adrenal Glands.

ADRENALINE. Alkaloids; Brain and Nervous System.

ADRENOCORTICOTROPIC HORMONE (ACTH). Adrenal Glands; Brain and Nervous System; Pituitary Gland.

ADSORPTION. Adsorption is a type of adhesion which takes place at the surface of a solid or a liquid in contact with another medium, resulting in an accumulation or increased concentration of molecules from that medium in the immediate vicinity of the surface. For example, if freshly heated charcoal is placed in an enclosure with ordinary air, a condensation of certain gases occurs upon it, resulting in a reduction of pressure; or if it is placed in a solution of unrefined sugar, some of the impurities are likewise adsorbed, and thus removed from the solution. Charcoal, when activated (i.e., freed from adsorbed matter by heating), is especially effective in adsorption, probably because of the great surface area presented by its porous structure. Its

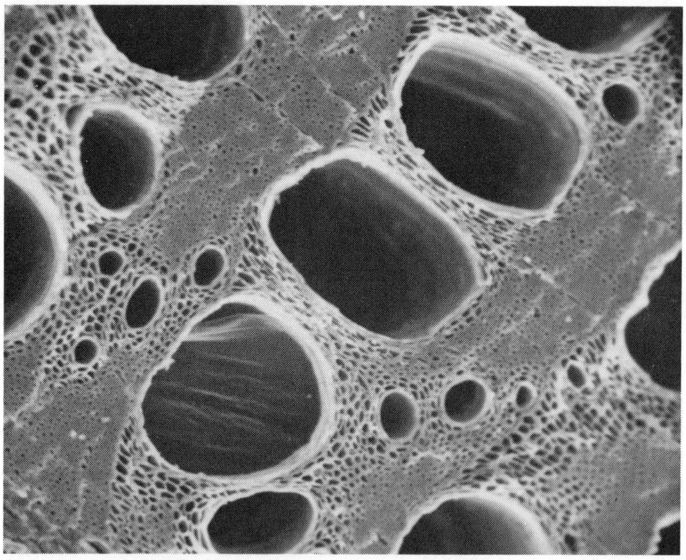

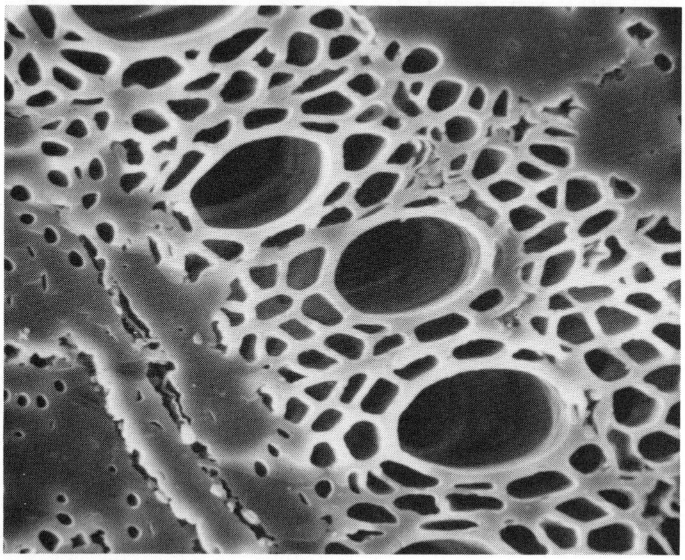

Scanning electron micrographs of charcoal magnified 100× (top) and 500× (bottom). (*Polaroid Type 105 Land Film*)

use in gas masks is dependent upon this fact. Penicillin is recovered in one state of the process by adsorption on activated carbon.

When colloidal hydroxides, notably aluminum hydroxide, are precipitated in a solution of acidic dyes, that is, those containing the groups —OH or —COOH, the dye adheres to the precipitate, yielding what is termed a lake. The "adsorption" of dirt on one's hands results from the unequal distribution of the dirt between the skin of the hands and the air or solid with which the skin comes in contact. Water is frequently ineffective in removing the dirt. The efficacy of soap in accomplishing its removal is due to the unequal distribution of dirt between skin and soap "solution," this time favoring the soap and leaving the hands clean.

At a given fixed temperature, there is a definite relation between the number of molecules adsorbed upon a surface and the pressure (if a gas) or the concentration (if a solution), which may be represented by an equation, or graphically by a curve called the *adsorption isotherm.*

The Freundlich or classical adsorption isotherm is of the form:

$$\frac{x}{m} = kp^{1/n}$$

in which x is the mass of gas adsorbed, m is the mass of adsorbent, p is the gas pressure, and k and n are constants for the temperature and system. In certain systems, it is necessary to express this relationship as:

$$\frac{x}{m} = k(h\gamma)^{1/n}$$

where h is the relationship of the partial pressure of the vapor to its saturation value, and γ is the surface tension. Numerous isotherm equations have been proposed in the chemical literature in the last fifty years. The Langmuir adsorption isotherm is of the form:

$$\frac{x}{m} = \frac{k_1 k_2 p}{1 + k_1 p}$$

The Brunauer, Emmett and Teller equation is more general than those of Freundlich or Langmuir; for among other limitations, those two equations apply only to the adsorption of gases. Even in those cases, the degree of adsorption depends upon five factors: (1) the composition of the adsorbing material, (2) the condition of the surface of the adsorbing material, (3) the material to be adsorbed, (4) the temperature, and (5) the pressure (if a gas). A notable case in point is carbon. Of the finely divided varieties of carbon there are important sugar charcoal, bone black or animal black, blood charcoal, wood charcoal, coconut-shell charcoal, activated carbon. The temperature of preparation of adsorbent charcoal is an important factor, high temperatures being deleterious, and the removal (or non-removal) of gases by passing steam over the heated carbon, which operation increases the adsorptive power. Bone black is used for removing the coloring matter from raw sugar solutions. Fusel oil is removed from whiskey and poison gases from air by adsorption with the proper form of carbon. By cooling carbon in a vacuum to the temperature of liquid air, the concentration of residual gas is greatly decreased. Dewar (1906) found that 5 grams of charcoal (presumably coconut-shell charcoal) at the temperature of liquid air reduced the pressure of air in a 1-liter container from 1.7 to 0.00005 millimeters. This is now a standard method for the production of a high vacuum, as in the neon lighting tube. See accompanying diagram.

Besides carbon, other important adsorbents in use are infusorial or diatomaceous earth, fuller's earth, clay, activated silica (silica gel), and activated alumina. All surfaces that behave indifferently towards non-electrolytes have the ability to adsorb electrolytes.

In comparing the adsorption properties of silica gel and activated carbon it may be noted that the latter is nonpolar and since it has no affinity for water will adsorb organic compounds in preference to water. Silica gel, on the other hand, is polar. It retains water and may thus reject an organic compound. It can also discriminate more selectively than activated carbon and consequently can be used for the fractionation of organic solvents.

Adsorption may be classified into types in various ways. On the basis of the adsorbate (or the substance which is adsorbed), adsorption may be *polar,* when the material adsorbed consists of positive or negative ions, so that the adsorbed film has an overall electrical charge. The term polar adsorption is also applied to adsorption chiefly attributable to attraction between polar groups of adsorbate and adsorbent. *Specific* adsorption is the preferential adsorption of one substance over another, or the quantity of adsorbate held per unit area of adsorbent.

On the basis of the process involved, adsorption may be classified as *chemical adsorption* (or chemisorption) where forces of chemical or valence nature between adsorbate and adsorbent are involved; and *van der Waals adsorption,* involving chiefly van der Waals forces. The difference is usually indicated experimentally by the greater heat of adsorption and more specific nature of the chemical process.

Other types of adsorption are *oriented adsorption,* in which the adsorbed molecules or other entities are directionally arranged on the surface of the adsorbent; and *negative adsorption,* a phenomenon exhibited by certain solutions, in which the concentration of solute is less on the surface than in the body of the solution.

Adsorption plays an important role in the process of dyeing, and in contact catalytic processes such as the conversion of sulfur dioxide to trioxide, and of nitrogen plus hydrogen to ammonia. In the case of insoluble organic acids (containing —COOH group) and substances containing hydroxyl (—OH) groups on the surface of water, the film is oriented so that the —COOH or —OH groups are attracted into the surface of the water, while their hydrocarbon ends project away from the surface of the water showing no tendency to dissolve (Langmuir).

The heat of adsorption, or wetting in this case, of starch by water is 29 calories per gram of dry starch. The heat of adsorption of various vapors and adsorbents has been measured. Since increase of temperature reduces adsorption, the adsorption process is accompanied by the evolution of heat. It appears that the heat liberated for a given volume of liquid filling the capillary spaces of a given adsorbent is practically constant. The heat of adsorption of hydrogen is, on nickel, palladium, platinum, copper, 11,700, 18,000, 13,800, 9500 calories respectively per gram mol (2 grams) of hydrogen; and of carbon monoxide on platinum 35,000 calories per gram mol (28 grams) of carbon monoxide; and of ethylene on copper 9500 calories per gram mol (28 grams) of ethylene.

Occlusion is a type of adsorption, or perhaps more properly absorption, exhibited by metals or other solids toward gases, in which the gas is apparently incorporated in the crystal structure of the solid. Palladium thus occludes extraordinary quantities of hydrogen, with the simultaneous liberation of much heat.

Adsorption is extremely important in biological reactions. Many of the constituents of plant and animal cells are colloidal in nature, and materials of various sorts are adsorbed on the surfaces of these colloids. For example, proteins frequently depend upon hydration or the adsorption of water for their activity, and the cellulose walls of plant cells often adsorb much water. See also **Chromatography.**

Adsorption viewed from the standpoint of a chemical engineering unit operation is described under **Adsorption Operations.**

ADSORPTION CHROMATOGRAPHY. Chromatography.

ADSORPTION INDICATORS. Dyestuffs or other chemicals which are used to detect the end point of a precipitation titration. These substances are dissolved in the solution to be titrated and lend color to it. The end point of the titration is signaled by disappearance of the color from the solution, or a change of color in the solution, attributable to adsorption of the indicator by the precipitate. See also **Indicator (Chemical).**

ADSORPTION OPERATIONS. A common requirement in the process industries is that of separating materials and numerous operations, such as filtering, evaporating, distilling, and screening, are used, depending upon the nature of the materials to be separated. Adsorption is frequently used.

In the gas-solids system, the simplest form of contactor is a static bed in which particulate solids are held in fixed positions, one particle resting upon another with no relative motion among particles. Gas is made to flow over, impinge upon, or flow through the voids among the particles. Contacting is confined to the interface between the gas and solid phases. Fixed beds usually require cyclic operation because it is difficult to add or remove solids for replenishment of adsorption effectiveness while a unit is operating. Fixed beds also tend to channel, resulting in a portion of the gas flowing through the bed without contacting the adsorbent solid.

Solid adsorbents used industrially include active alumina, alumina impregnated with calcium chloride, activated bauxite, fuller's earth, silica gel, shell-base carbon, wood-base carbon, coal-base carbon, petroleum-base carbon, and anhydrous calcium sulfate. See also accompanying table.

To avoid some of the problems of fixed adsorption beds, fluid beds are frequently used. In a moving-bed (fluid bed) system, adsorbent solid particles move countercurrently by gravity through a rising pressurized-gas stream. As the particles adsorb the desired materials from the gas stream, they become heavier and ultimately gravitate to the bottom of the vessel where they are removed and regenerated externally from the fluid bed per se. The operation is continuous. Moving-bed systems also are used for heat transfer, the mixing of solids with gases, drying solids and gases, and implementing chemical reactions as in a fluid catalytic cracker.

Adsorption operations also are used to remove impurities from liquids. In purifying the products of organic chemical syntheses, for example, decolorizing carbons and clays often are used as contact adsorbents. They are stirred directly into a liquid-phase mixture or solution and subsequently removed by filtration. In some cases, as in contact filtration, the adsorbents are incorporated in the filtering medium. Contact filtration is widely used for removing colored and carbon-forming materials from lubricating oils. In another configuration, a liquid may be allowed to percolate by gravity or under pressure through a bed of solid adsorbents. Packed columns or towers filled with adsorgents also are used, particularly for drying gases.

ADULT. A full-grown animal. In species having a pronounced metamorphosis, the last stage is known as the adult or, with insects, as the imago. Sexual maturity is characteristic of most adults, although some animals are without sexual functions, and in some species, reproduction may take place at an earlier stage of development.

Adulthood in the normal human implies emotional as well as physical maturity and, although followed by continuing learning and intellectual development, commences at an age of 19 to 21, depending upon specific individuals. Adulthood follows adolescence. See also **Adolescence.**

ADVECTION FOG. Fog and Fog Clearing.

ADVECTION (Meteorology). Atmosphere (Earth).

ADVENTITIOUS BUDS. Buds which appear elsewhere than in the leaf axils or above them. They may appear anywhere in the internode, or on roots or even on leaves, and develop either naturally or as a result of injury. The dense bunches of buds, which frequently appear on burls or in witches'-broom, may be adventitious. The buds which appear at the tops of thistle roots, particularly when the natural top of the plant is cut off, are adventitious. So also are the buds which develop on the leaves of the begonia and bryophyllum. The practice of pollarding, or cutting off the branches of a tree in such a way as to leave only the main trunk or perhaps the stumps of a few large branches, results in the development of dense groups of

PHYSICAL PROPERTIES OF REPRESENTATIVE ADSORBENTS

Adsorbent Substance	Internal Porosity (%)	External Void Fraction (%)	Average Pore Diameter (10^{-10} meter)	Surface Area (square meters/gram)	Adsorptive Capacity (gram/gram of dry solid)
Activated alumina	25	49	34	250	0.14[a]
Activated bauxite	35	40	~50	—	0.04–0.2[b]
Fuller's earth	~54	40	—	130–250	—
Silica gel	~70	30–40	25–30	~320	1.0[c]
Shell-base carbon	~50	~37	20	800–1100	45
Wood-base carbon	55–75	~40	20–40	625–14,000	6–9[d]
Coal-base carbon	65–75	45–70	20–38	500–1200	~0.4[e]
Petroleum-base carbon	70–85	26–34	18–22	800–1100	0.6–0.7[f]
Anhydrous calcium sulfate	38	~45	—	—	0.1[g]

[a] Water at 60% relative humidity.
[b] Water; test condition not specified.
[c] Water at 100% relative humidity.
[d] Phenol value.
[e] Benzene at 20°C; 7.5 millimeters partial pressure.
[f] Test conditions not specified.
[g] Water; test conditions not specified.

adventitious buds. These grow into adventitious branches which in certain willows may be long and supple and so useful in manufacturing wicker furniture. The habit of forming adventitious buds on roots is of material value, since in consequence it is possible to propagate many plants by means of root cuttings.

Adventitious roots also exist. They may appear from the stem where they arise in the pericycle, or from other tissues of the plant. The roots which appear on slips or stem cuttings are adventitious. See also **Bud;** and **Budding.**

AEGERINE. Acmite.

AEOLIAN TONES. The tones produced by a gas stream striking a stretched wire in a direction normal to the length of the wire.

AEOLIGHT. A glow discharge lamp employing a cold cathode, and a mixture of permanent gases in which the intensity of illumination varies with the applied signal voltage. The aeolight lamp is used to produce a modulated light for motion-picture sound recording.

AERATION. A process of contacting a liquid with air, often for the purpose of releasing other dissolved gases, or for increasing the quantity of oxygen dissolved in the liquid. Aeration is commonly used to remove obnoxious odors or disagreeable tastes from raw water. The principle of aeration is also used in the treatment of sewage by a method known as the activated sludge process. The sewage is allowed to flow into an aeration tank where it is mixed with a predetermined volume of sludge. Compressed air is introduced which agitates the mixture and furnishes oxygen which is necessary for certain biological changes which take place. Sewage may also be aerated by mechanically actuated paddles which rotate the liquid and constantly bring a fresh surface in contact with the atmosphere.

Aeration is of importance in the fermentation industries. In the manufacture of baker's yeast, penicillin, and other antibiotics, an adequate air supply is required for optimum yields in certain submerged fermentation processes.

Aeration can be accomplished by allowing the liquid to fall in a thin film or sprayed in the form of droplets in air at atmospheric pressure; or the air, under pressure, may be bubbled into the liquid as by means of a sparger, or other device that creates thousands of small bubbles, thus providing maximum contact area between the air and the liquid.

AERENCHYMA. Spongy tissue occurring chiefly in the stems of many aquatic or marsh plants. Such porous tissue gives great buoyancy to the stem and so helps keep the leaves up in the air and permits gas diffusion within the plant. The term aerenchyma is frequently applied to any loose porous tissue found in plants, as for example that occurring in lenticels.

AERIAL PHOTOGRAPHY. Photography and Imagery.

AERIAL ROOT. Root (Plant).

AEROBE. An organism that utilizes atmospheric oxygen in its metabolic processes; i.e., the so-called aerobic bacteria, or aerobes, which use oxygen. See also **Anaerobe.**

AEROBIC OXIDATION. Carbohydrates; Glycolysis.

AEROBIC RESPIRATION. Respiration (Plants).

AERODYNAMIC HEATING. Supersonic Aerodynamics.

AERODYNAMICS. Fluid mechanics is the study of reactions between fluids and the solids immersed in them. Aerodynamics is a phase of the mechanics of fluids, its study being limited to the reactions caused by the relative motion between the fluid (air) and the solid. Sometimes this strict definition is broadened so that the term aerodynamics may also include the reactions of gases other than air. Aerodynamics encompasses the flow of gases in conduits, the effects of winds on static structures (buildings, chimneys, bridges), and the effect on moving bodies on the atmosphere through which they move. Thus, the air resistance of automotive vehicles, trains, the flight of airplanes, and the sway of suspension bridges are examples of aerodynamic action.

Whether an object is standing still and the air around it is in motion or whether an object is moving in essentially still air, provided the velocity of the object in the one case is the same as the velocity of the air in the other case, the aerodynamic forces prevailing are the same. This concept of inversion is advantageous in determining aerodynamic forces created by air streams upon stationary airfoils and other objects as they may be tested in the moving air streams of a wind tunnel.

Air has a viscosity of 0.00019 poise under standard conditions, compared with 0.01 poise for water and 1 poise for light oil. Thus, there are fields of aerodynamic study in which the air can be considered frictionless, especially in subsonic aerodynamics—with the exception of the conditions existing in a very thin layer immediately adjacent to the solid object where there is relative motion between object and the air. However, at speeds greater than that of sound, friction effects become important. In aerodynamic expressions, the absolute viscosity, μ, usually is associated with mass density, ρ, as a ratio. This is called the *kinematic viscosity*, ν, which has a magnitude of 0.000157 ft^2 (0.0000146m^2) per second in standard sea level atmosphere.

At subsonic air speeds, it is customary to consider the air incompressible, and to rely on Bernoulli's theorem of the interchangeability of static head and velocity head. This permits the employment of the principle of continuity for flow, represented by the equation

$$AV = \text{constant}$$

which means, symbolically, that the product of the cross-sectional area of flow and the velocity is a constant at all points in the path taken by a finite quantity of the fluid. If the fluid were compressible, this continuity would be

$$\rho AV = \text{constant}$$

A streamtube in air may be visualized as an imaginary conduit through which the condition of continuity of flow of the incompressible fluid takes place. Although the flow is steady, the tube is not necessarily of uniform size, but where its sectional area decreases, we would anticipate an increase of velocity; furthermore, a consideration of Bernoulli's theorem would indicate that these diminished sections of the streamtube are regions of lower static pressure than elsewhere. Although a streamtube is ordinarily imagined to have a circular cross section, it will be convenient here to think of a streamtube having a rectangular cross section. If we have enough of these streamtubes, all of the same width, but whose heights may vary as velocity varies, we might imagine them stacked, one on another, as shown in Fig. 1, and be thus encompassing the whole flow of air through a given

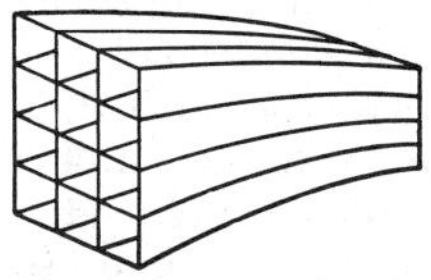

Fig. 1. Stacked streamtubes.

region. It will be convenient to think of the edges of these streamtubes as equivalent to the streamline pattern that is so useful in the study of the effect of airflows past solid objects. It follows that where the streamlines approach each other, there is a region of lower pressure and higher velocity; conversely when they diverge. Thus, converging streamtubes (and streamlines) are like nozzles, whereas diverging streamtubes are like diffusers. The reader, upon viewing a streamline pattern, may then judge fluid pressure by the density of streamlines. This is readily done, qualitatively, if not quantitatively.

It is now in order to state the nature of elementary streamline patterns and explain how a complex streamline pattern may be considered the result of the addition of two potential forces, each capable of producing a different character of streamline in the same region. A simple potential cylinder rotating in still air will drag with it a circulatory pattern of flow, the streamlines of which could be represented by a series of concentric circles. The circulation stream pattern is of great importance, because an airfoil is a shape that automatically induces a circulating component of airflow, which, when superimposed upon a rectilinear stream, provides the typical streamline pattern around an airfoil. The lifting streamline pattern of the typical airfoil could be induced in the absence of the airfoil by imposing some specific circulatory pattern of stream flow upon a rectilinear field, and this fact has been put to useful service in theoretical mathematical consideration of the aerodynamics of airfoils.

In Fig. 2a, a rectilinear flow is passing a static cylinder. At some distance from the cylinder, the flow is unaffected by the presence of the cylinder, and the streamtubes remain of constant size throughout, but it is obvious that between this undisturbed region and the cylinder, the intermediate tubes must be smaller in order that all can be accommodated in the space immediately above and below the cylinder. The action of the individual streamtube approaching a point abreast of the center line of the cylinder is, therefore, that of a nozzle which

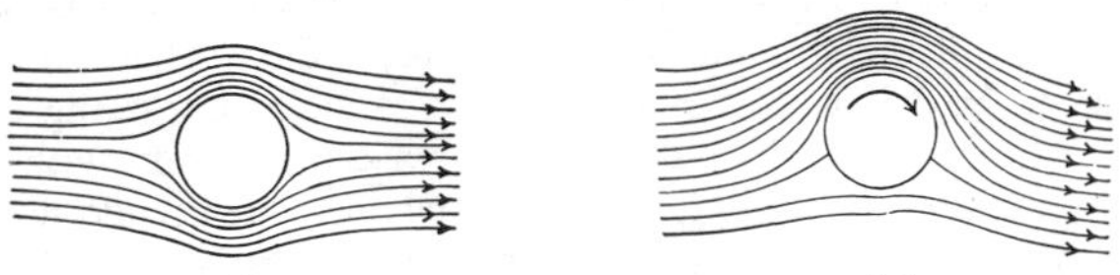

Fig. 2. Comparison of air passing over (a) cylinder at rest and over (b) rotating cylinder.

has increased the air speed and decreased its pressure. Hence, above and below the cylinder, static pressure is less than in the free, unaffected stream. If the cylinder were in rotation, tending to set up its concentric field of streamlines, the resulting streamline pattern would be like that shown in Fig. 2b, where, on one side of the cylinder, there is a reduction of static pressure due to crowding together of the streamlines, and on the other side, there may be no change of pressure, or even an increase. Briefly, then, the presence of a circulatory component of flow in a rectilinear field is to produce a net transverse force (in this case, upwards) on the object, since it will be subjected to unbalanced static pressure of the fluid against its surfaces. This is the force which causes a spinning tennis ball to assume a sharply curved trajectory, and which causes the lift to exist upon an airplane wing.

The influence of viscosity of air in aerodynamics is confined chiefly to the action of air in the boundary layer. Within this boundary layer (which, although extremely thin, constitutes the atmosphere immediately adjacent to the solid) the frictional qualities of air are of importance in determining the airflow. The thin boundary layer might be imagined subdivided into a great many lamina of air parallel to the surface and extending out to the limits of the boundary layer. A nonturbulent boundary layer is considered, although, as will be developed later, the boundary layer may become turbulent under certain conditions except for the lamina immediately adjacent to the stationary surface. As Fig. 3 displays, the laminar velocity, u, increases above

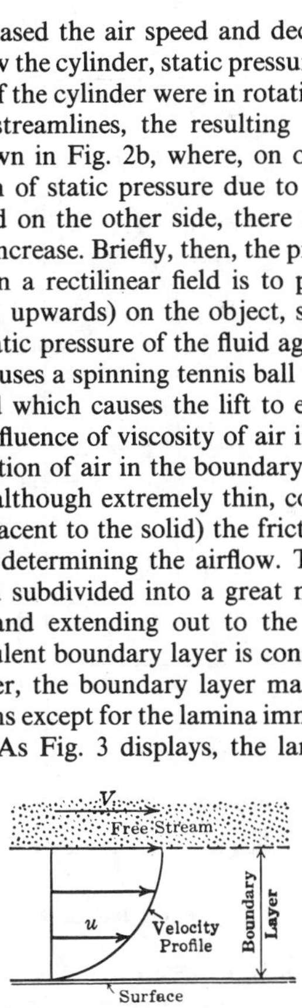

Fig. 3. Velocity profile.

the surface until, at the edge of the boundary layer, it equals the free stream velocity. The smaller the viscosity of the fluid, the thinner the boundary layer will be. Consequently, it is extremely thin for air, but what goes on in this boundary layer is of utmost importance in aerodynamics. The first layer of air sticks or adheres to the surface, and lamina above it successively slide on each other, exerting drags that are proportional to the viscosity. The rate of change of the velocity between adjacent lamina is a measure of the unit shearing force between them, and is the unit skin friction when the lamina considered is at the one in motion nearest the stationary surface. This occurs at practically zero boundary layer thickness. A curve joining the tips of velocity vectors plotted for the different lamina in the boundary layer is called a velocity profile. If u is the variable velocity, increasing from 0 to V across the boundary layer, then du/dy is the rate of change of velocity (also the tangent to the velocity profile) and, through the application of one of the Newtonian principles, the unit shear stress,

$$\tau = \mu\left(\frac{du}{dy}\right)_{y=0}$$

where y is measured in the vertical direction.

The proximity of streamlines to a stationary surface is now seen to produce a reaction resulting from the static pressure p existing in the flow nearest the surface, and a skin friction τ, which is, of course, tangential to the surface. The combination of these two is a force oblique to the surface. The foregoing statement premises an adherence of the streamlines to the surface, for, if they have separated from the surface, the region between will be filled with vortices of random motion, and the surface pressure cannot be definitely predicted from the character of the streamline flow. It is apparent that aerodynamic reaction is a function of μ, ρ, V, and the extent of the surface area on which the reaction takes place. Usually some one dimension l is taken to describe the size of the surface. In ft-lb sec units, the aerodynamic force F is in pounds, μ is in poises, ρ is in slugs per square foot, and l is in feet. To discover the relationship existing between the reaction and its controlling factors, it is customary to set up the dimensional relationship:

$$|F| = |\rho^a \mu^b V^c l^d|$$

and employ the methods of dimensional analysis to discover the value of a, b, c, and d. It is found that

$$|F| = \left|\left(\frac{\nu}{Vl}\right)^b \rho V^2 l^2\right| \quad \text{dimensionally}$$

and

$$F = K\left(\frac{\nu}{Vl}\right)^b \rho V^2 l^2 \quad \text{numerically}$$

in which

ν = kinematic viscosity

K = a dimensionless constant

Vl/ν = the Reynolds number (also dimensionless)

The product, $2K(\nu/Vl)^b$, is frequently given as a dimensionless coefficient C. The exponent b is of such magnitude that the Reynolds number has a minor (but not negligible) effect on C. The coefficient is affected primarily by the attitude of the body in the airstream. If it is completely symmetrical, as is a sphere, this variation is nonexistent, and C is a true constant except for the above-mentioned minor effect of the Reynolds number variation. However, the nonsymmetrical shape of an airfoil causes large and typical variations of C, depending on the attitude with which the airfoil is presented to the airstream. The force F would then be $C(\rho V^2/2)S$, or CqS, in which q is dynamic pressure, and S is some significant surface area, in square feet, of the object.

Flow within a streamtube may be laminar on slightly turbulent. It was stated that, in laminar flow, drag per unit area is $\mu(\partial u/\partial y)_{y=0}$, but turbulent flow produces higher skin friction because the velocity profile is fuller close to the surface due to turbulent energy interchange between lamina. Although at the beginning of contact between airstream and surface, true laminar flow existed, it has been found that a certain critical value of Reynolds number, determined chiefly by the linear dimension of contact passed over, the boundary layer suddenly becomes turbulent, and thickens. In spite of the higher skin friction, the drag on bluff bodies may actually become less with turbulence, because of an action within the turbulent boundary layer that resists separation of the streamlines from the surface. Viscous drag gradually slows down the fluid in a laminar boundary layer, and if contact with the surface is maintained over sufficient length, the innermost parts of the boundary layer will be brought to rest. This becomes the separation point. If a negative pressure gradient exists beyond the separation point, as it often does, there will be a reverse flow toward the separation point, all of which will produce a breaking away of the streamlines from the surface, leaving the surface between the two separation points, on either side of body, in contact with air of random vortices and low pressure. See Figures 4 and 5. It

Fig. 4. Low-pressure wake.

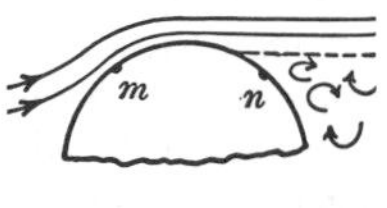

can be seen that because of the vortices, the pressure at n has failed to build up to the pressure at m (as it would have had the fluid

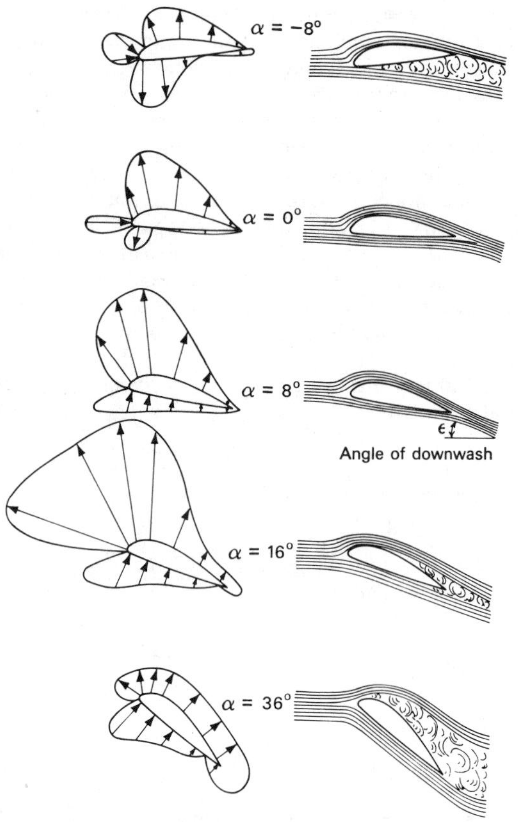

Fig. 5. Distribution of pressure on a cambered wing at different angles of attack but at the same airspeed as obtained in a wind tunnel. The arrows are force vectors, their length representing the relative intensities of the pressures at the points at which they are located. The end points (arrow-heads) are connected to form an envelope. To obtain the relative intensity at any point, erect an arrow normal to the airfoil surface, and its length is then determined by its intersection with the envelope drawn through the end points of the other force vectors. Sketches to the right show, approximately, the manner of air flow at the different angles of attack.

been frictionless and without premature separation point), resulting in the creation of a downstream dragging force. This is characteristic of any body having a turbulent wake. Early turbulence in the boundary layer is desirable, so that the greater average momentum of boundary layer air can carry it against skin friction farther around the bluff surface, and so delay the separation point and narrow the turbulent wake. Thus, the coefficient of drag for a sphere [C of $Cq(\pi d^2/4)$] is about 0.5 for laminar boundary layer and 0.1 for turbulent, with the transition occurring at a Reynolds number of approximately 300,000.

Airfoil. An airfoil is any body whose shape causes it to receive a useful reaction from an air stream moving relative to it. This definition is broad and could include many shapes not ordinarily considered airfoils. The term is usually associated with a body of the shape shown in Fig. 6, a profile or chordwise section of the wing. The dimension

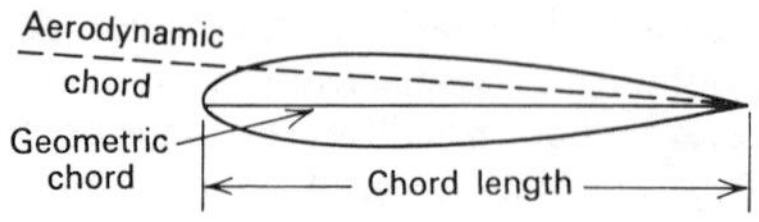

Fig. 6. Airfoil profile.

perpendicular to this section is called the span when referred to the wing. Many different airfoil shapes have been used, proposed for use, or tested. Some of them have flat lower surfaces, and others have convex or concave surfaces.

The characteristic lifting airfoil profile has a maximum thickness of 6 to 18% of the chord at 20 to 40% of the chord aft of the leading edge. Early experiments were made with flat-plate airfoils, and later with thin curved-plate airfoils, but neither possessed as large a ratio of lift to drag at a particular angle of attack as the double-surface cambered airfoil shown, nor as low resistance or drag. Also, an airfoil of finite thickness provides space for the foundation structure of a lightweight wing, thus removing the structural elements from the drag of the air stream, and thereby reducing the drag.

The value of the airfoil shape to aviation resides in the magnitude and direction of the resultant air reaction on it when employed at attitudes below the stall (about 15° angle of attack). The component of the air reaction normal (called the lift) to the free air stream is several times the magnitude of the parallel component (called the *drag*). The normal component, or lift, may be expressed in equation form as

$$L = C_L q S$$

The variation of C_L with geometric angle of attack is seen in Fig. 7. The slope of the straight portion is a per radian. Theory indicates a value of 2π for a, for infinite aspect ratio, and tests agree well with the theory. The lift coefficient increases uniformly with attack until the stall or burble point is reached, where it breaks and decreases rapidly with further increase of angle of attack, because the high attack angle creates a bluff body and the stagnation point advances forward on the airfoil. Normal controlled flight is not likely beyond the stall or burble point. The moment coefficient, C_{M_a}, about the aerodynamic center of the airfoil is shown in Fig. 7. It will be noted

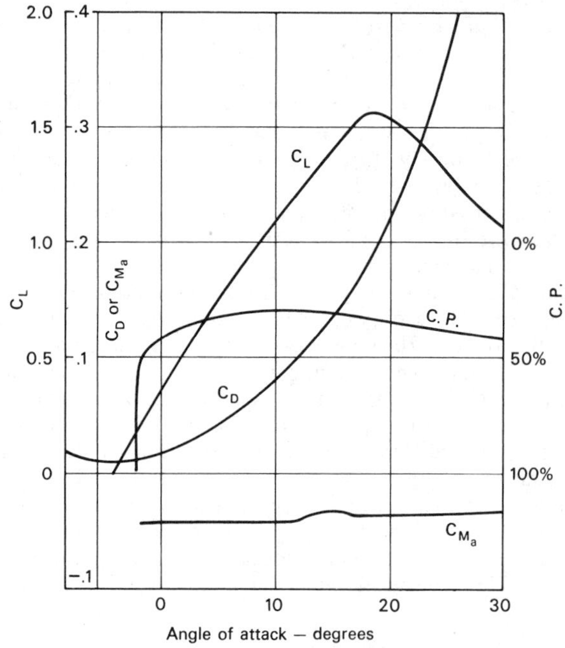

Fig. 7. Common aerodynamic characteristics of an airfoil.

that it has practically a constant value. When listed, the value of C_{M_a} at $C_L = 0$ is usually given.

The full aerodynamic reaction upon an airfoil is a force of which the lift is the component normal to the free stream velocity. The other component is measured in the direction of the air velocity and is called drag. Drag coefficient has a characteristic parabolic variation, as Fig. 7 shows. The drag equation is analogous to that for the lift:

$$D = C_D q S$$

If the resultant of all air reactions on an airfoil is consolidated into a single imaginary force, it must act at the center of pressure of the airfoil. The figure shows that the location of this center of pressure of an airfoil varies with its attitude. The qualities of an airfoil that are sought for when employed in aviation are:

(1) High maximum C_L, in order to give low landing speed for a given size wing and weight.

(2) Low minimum C_D, so that the high speed, which occurs at small angles of attack, may be the greatest possible.

(3) High ratio of C_L to C_D, so that an efficient, economical airplane will result.

(4) A shape well suited to the construction of a strong, but lightweight, wing, at minimum cost.

Downwash. From an airfoil of finite span in flight, there is a multitude of vortices trailing, forming a vortex sheet. As a result of circulatory

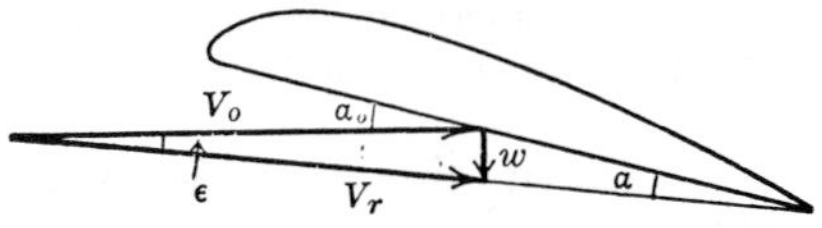

Fig. 8. Downwash.

flow about each vortex filament a net downwash flow of air is produced, having an important effect upon the air reactions. If the airfoil is untwisted, and elliptical in plan form (or tapered in approximation of an ellipse), the vortices bound to the airfoil will produce a circulation of approximately elliptical strength distribution over the span. This produces a uniform downwash of air in the region of the wing. The combined effect of the forward motion of the airfoil and the downwash will be better understood upon reference to Fig. 8.

The forward speed V_0 and the vortex-induced downwash w produce a resultant air velocity of V_r (substantially equal to V_0 in magnitude) which meets the airfoil at angle of attack α. The angle ϵ is called the downwash angle. Without downwash, the lift produced would be that corresponding to α_0 at velocity V_0, but downwash reduces the angle of attack to α thus decreasing the lift. To lift the same this airfoil would require α with downwash to be larger than α_0 without downwash. The effect of downwash is now seen to be deleterious since, for the same V and lift, it increases the needed α and produces extra drag. Furthermore this drag (induced drag) is not associated with aerodynamic cleanliness but with aspect ratio. Disposing the area of the wing as a slender wing of large wing span (i.e., high aspect ratio) is the way to reduce this form of drag.

The effect of downwash of the wing has to be accounted for in the disposition of the empennage when disposed in the rear of the wing. In front of the wing, there is an upwash.

High-Lift Devices. The lifting ability of an airfoil is improved by increasing its camber (over limited range), and also by delaying the separation of the air flow from the lifting surface as the angle of attack is increased. Most airfoils employed as airplane wings are of low camber because their drag coefficients at low attack angles (high speed) are more favorable than those of the thick, highly cambered airfoils.

It is known that the maximum lift coefficient of a highly cambered airfoil is greater than that of a flat, thin one. Safety in landing an airplane demands low landing speed, which is obtainable by using wings with high maximum lift coefficients. On the other hand, high maximum speed is only to be obtained with the thin, low-drag type airfoil. Many manufacturers of aircraft have adopted the flap as a means for securing low landing speeds on aircraft fitted with inherently high-speed wings. The flap, as illustrated in Fig. 9, is akin to a flexible

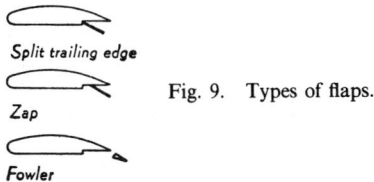

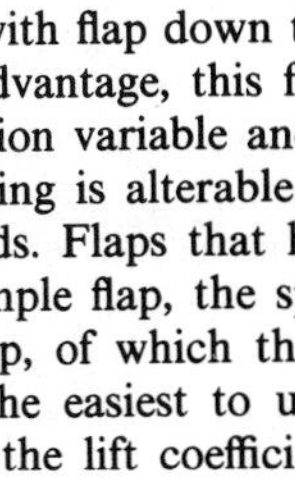

Fig. 9. Types of flaps.

trailing edge, and by depressing the flap, the effective camber of the airfoil may be increased. The drag is also increased, giving a steeper gliding angle with flap down than without the use of the flap. Apart from any disadvantage, this factor may actually be beneficial, since with flap position variable and under control of the pilot, the angle of normal gliding is alterable, permitting steep approaches to small obstructed fields. Flaps that have received a great deal of attention include the simple flap, the split flap, zap flap, and the fowler flap. The simple flap, of which the normal aileron, elevator and rudder are types, is the easiest to use, requiring only rotation downward for increasing the lift coefficient. The split flap is extensively used. The operation is simply that of a rotation of a panel on the underside of the trailing edge about its forward edge. In the zap flap, the rotation is accompanied by a backward movement of the pivoting point. In the fowler flap, the motion is not one of simple rotation alone, but is combined with a rearward movement of the flap area as well. The effect is one of increase of airfoil area as well as change of camber.

Increasing lift by delaying separation in the boundary layer is an

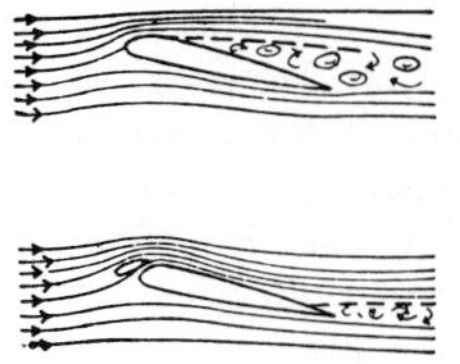

Fig. 10. Air flow at high angle of attack with and without slot.

independent action. Both flaps and partial boundary layer control with the aid of a leading edge slot (see Fig. 10) could be added to the same airfoil—and have been on some airplane wings. The effect of flaps and slots on lift and drag is illustrated in Fig. 11. Delay of airflow separation by sucking off the incipient vortices around the

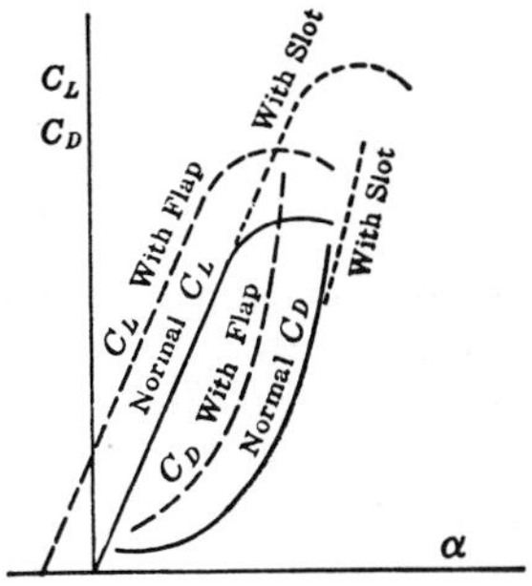

Fig. 11. Effect of flaps and slots on lift and drag.

stagnation point produced amazing lift coefficients but has not been exploited on account of the compressor energy needed, although, with the introduction of the jet engine, the interest in boundary layer control has revived. Also, very small high-speed jets to impart energy to the boundary layer, "blowing off" the air (as opposed to sucking off), and rotating cylindrical nose sections, have been tested with varying results.

Boundary layer control by leading edge slots is definitely valuable for lifting airfoils, and lacks the structural and mechanical complications attending the other boundary layer control proposals. Hence slots and flaps have been the only high-lift additions commercially employed to date. In the light of probable future aeronautic development, the other possibilities should not be discounted. Some attempts have also been made to intensify the action of the flap by using multiple-hinged flaps, which have been applied experimentally to a "vertical lift" airplane.

Drag. An object subjected to an air stream is acted upon by a resultant air pressure. Drag is the component of air reaction which is parallel to the air stream. Its origin is profile impact of molecules of air against the face of the object, the skin friction of the molecules of air as they slide along the object, the vortices and eddying air currents set up in an otherwise undisturbed air stream by the presence of the object, and the induction effects of downwash, if any. Drag is that quantity which imposes limitations upon the top speed of vehicles. As it is proportional to the square of the velocity, its magnitude mounts rapidly as velocities are increased.

There are two kinds of drag—drag on surfaces which obtain a useful reaction from the air stream as well as a drag, and drag upon surfaces where the only reaction is the drag. A wing has both drag and lift. Drag on the wing is the price paid for lift, and is so accepted. A strut, or wire, or wheel, creates no lift, and the drag is wholly undesirable. A drag of this type is called parasite drag. Careful streamlining and reduction of parts exposed to the air stream are ways of reducing parasite drag. The drag of a wing may be divided into profile drag and induced drag. Induced drag depends upon the lift coefficient and the aspect ratio of a wing. The induced drag coefficient may be expressed as

$$C_{D_i} = C_L^2/\pi(AR)$$

where C_{D_i} is the induced drag coefficient, C_L the lift coefficient, and AR the aspect ratio of the wing.

During a test, profile and induced drag are not separable. Parasite drag depends upon the surface roughness and shape of the object.

Profile drag of airfoils is the result of skin friction and turbulent wake. It is greatly influenced by the Reynolds number (scale effect) and by initial turbulence in the air stream. Although some authors have proceeded on the basis that this is the sole influence on profile drag coefficient, others have found that the profile drag coefficient has a small variation with coefficient of lift. However, the generally accepted theories of origin of aerodynamic forces imply that scale effects modify profile and parasitic drags only and are absent in lift and induced drag. Experience confirms the validity of theory except near the burble, or stall, attitude of the airfoil, where viscosity has an effect on lift.

Experiments in the wind tunnel show that factors affecting reaction of air on airfoils are:

(1) The relative velocity of the air and airfoil.
(2) Ratio of span to chord of the surface area.
(3) Density of the air.
(4) The angle of inclination of the airfoil to the air stream.

All this may be stated somewhat as follows:

$$F \sim R^n \alpha \rho S V^2$$

F is wind reaction, α the angle of attack of the wing, ρ the mass density of the air, S the surface area, V the air velocity, and R the Reynolds number.

If a proper constant K be inserted, the similarity can be made into an equality. K may also be made to include the effect of the Reynolds number, angle of attack, and air density. The equation is then simplified to:

$$F = KSV^2$$

This reaction is neither perpendicular nor parallel to the wind stream. It is convenient to divide it into its components of lift and drag. Letting K_L and K_D be the corresponding coefficients.

$$\text{Lift } L = K_L S V^2$$

$$\text{Drag } D = K_D S V^2$$

There will be a different value of K for each angle of attack and each R, although the effect of the latter is usually minor.

The defect of this equation is that K is not dimensionless, as a true coefficient should be; also, the formula is not flexible with respect to air density. If it were rewritten:

$$R = C\rho S V^2$$

C would be dimensionless. The equation is usually written thus for drag:

$$D = \tfrac{1}{2} C_D \rho S V^2$$

because $\rho V^2/2$ is the pressure necessary to give air of mass density ρ at a velocity of V. This is called the *dynamic pressure.* The drag is also $C_D q S$, where q stands for dynamic pressure. The drag coefficient C_D varies with angle of attack, but it is dimensionless. Consistent units have to be used. ρ is the ratio of the weight of the air, or fluid, in pounds per cubic foot, to the acceleration due to gravity or 32.2 ft/sec (9.8 m/sec); S is the wing area, projected upon the plane of chords, in square feet, and V is the velocity in feet per second. C_D is the nondimensional drag coefficient. The subscript D is used when the coefficient refers to the *average* coefficient for the *entire* wing, and the lower case letter d is used when the drag coefficient refers to the airfoil of unit width. The wind-tunnel test is used to establish the law of variation, and the results are plotted as a drag curve, with the coefficient of drag as the ordinate, and either coefficient of lift or angle of attack as the abscissa.

A few typical coefficients of drag are:

(1) Flat plate normal to air stream. Average value 1.28 often used, although subject to variation from Æ (aspect ratio) and R.N. (Reynolds number).

Flat plate parallel to air stream. C_D = skin friction coefficient as there is no turbulent wake. Varies, 0.002–0.008 depending on initial turbulence and R.N.

(2) Sphere (S = projected area), about 4×10^5 critical R.N. Laminar boundary, 0.45; turbulent, 0.1.

(3) Cylinder, axis transverse to air stream (S = projected area), about 4×10^5 critical R.N. Laminar boundary, 0.7; turbulent, 0.3.

(4) Thin wing, infinite Æ, R.N. 3.5×10^6. $C_{D_p} = 0.01$.

(5) Airplane. $C_D = C_{D_0} + C_L^2/C_{D_0}$ varies greatly, as it depends upon a large number of design and operational factors.

The drag of an object is subject to other factors also, such as compressibility of the fluid medium in which the object moves, and surface roughness.

Under supersonic conditions, the drag coefficient may be approximated by the following formula,

$$C_d = \frac{4\alpha^2}{\sqrt{M^2 - 1}}$$

where α is the angle of attack in radians, and M is the Mach number for the fluid speed considered.

Theory of Aerodynamic Compressibility. The "incompressible fluid" theory of classical hydrodynamics has proved useful for the estimation of aerodynamic parameters, and when applied to problems of low-speed flight has yielded sufficiently accurate results. It has been found, however, that the flow pattern about a body moving through the air at high speeds is affected to a large degree by changes in density resulting from compression or expansion of the fluid. Consequently, aerodynamic coefficients based on an incompressible-flow theory are in considerable error when applied to airplanes moving at high-speeds. An understanding of compressible flows is, therefore, of the utmost importance to the designer of high-speed aircraft.

In the study of compressibility phenomena as applied to aerodynamic problems, airplane speeds are classified according to their relation to the speed of sound in air. The speed of sound is taken as a reference velocity because it is a function of fluid elasticity. As applied to compressible flows this means that the amount of pressure necessary to cause a given change in density in any fluid is proportional to the speed of sound in the fluid. Since the pressure is proportional to the square of the velocity, the velocity which a body may attain before appreciable density changes occur is also proportional to the velocity of sound in the fluid. It is apparent, therefore, that the flow pattern about a body will be altered by density changes to a degree dependent upon the ratio of the velocity of the body to the velocity of sound. This ratio is known as the *Mach number* and is taken as an index of the effects of compressibility on the flow pattern. A curve showing the variation of the speed of sound with altitude is presented in Fig. 12. The local velocity of sound is directly proportional to the square root of the absolute temperature. Therefore, the variation of the speed of sound with altitude is essentially due to the variation in the temperature of the air.

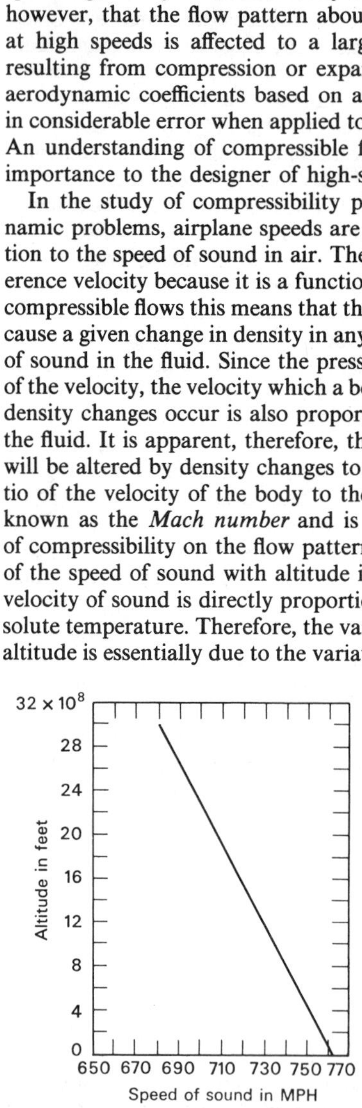

Fig. 12. Speed of sound as a function of altitude. Sea-level temperature, 60°F.

In order to show quantitatively the changes in air density associated with increasing Mach number the following table has been prepared. This table is based on an adiabatic flow through a converging nozzle. M is the stream Mach number and ρ/ρ_s is the ratio of the corresponding air density to the density at static conditions.

M	ρ/ρ_s
0.2	0.981
0.4	0.925
0.6	0.841
0.8	0.742
1.0	0.635

The preceding discussion fully accounts for the effects of compressibility only at Mach numbers less than one, or at subsonic speeds.

As air speeds approach and attain the velocity of sound, radical changes occur in the flow pattern which do not result entirely from changes in air density. The flow pattern in a perfect incompressible fluid is instantaneously influenced at all points by pressure changes occurring at any point in the flow field. A consideration of the theory of elasticity as applied to fluids, however, indicates that the effects of small pressure changes in a real fluid are transmitted throughout the fluid in the form of waves which travel at the speed of sound. It may be seen, then, that the effects of a pressure change which occurs behind the critical point at which the speed of sound has been reached cannot influence the flow field ahead of the point.

Since at the critical point the forward motion of the pressure waves is completely arrested by an air stream velocity equal to the velocity of wave propagation, a wave front is formed at the critical point. This wave front constitutes a sharp discontinuity in the flow with which is associated large increases in pressure, density, and temperature and a decrease in velocity. Such a wave front with its attendant discontinuities is known as a shock wave. The flow field about a body traveling at or near sonic velocities will be radically different from that at low speeds.

Once the velocity has increased beyond the sonic range and attained a sufficiently high supersonic value, the shock wave will be forced downstream and the flow at the original point of shock will be comparatively smooth. However, the flow pattern at supersonic speeds will bear little resemblance to that which obtains at low speeds.

High-speed flight is greatly complicated by the compressibility phenomena which have been described. The following paragraphs give an outline of the more important considerations in the aerodynamics of high-speed flight.

The change in aerodynamic characteristics which results from fluid compressibility is greatest when shock waves form on some part of the airplane such as an airfoil or cowling. The forward velocity of the airplane which corresponds to the formation of shock waves is called the critical speed. The critical airplane speed is always less than the velocity of sound since the local velocity over the airfoil surfaces and other components is in excess of the forward speed of the airplane. The critical Mach number is the ratio of the critical airplane speed to the speed of sound.

The airfoil section usually predominates as the factor which controls the effects of compressibility on the characteristics of high-speed airplanes. The variation in pressure distribution around an airfoil as the Mach number is increased is shown in Fig. 13. Both flight and

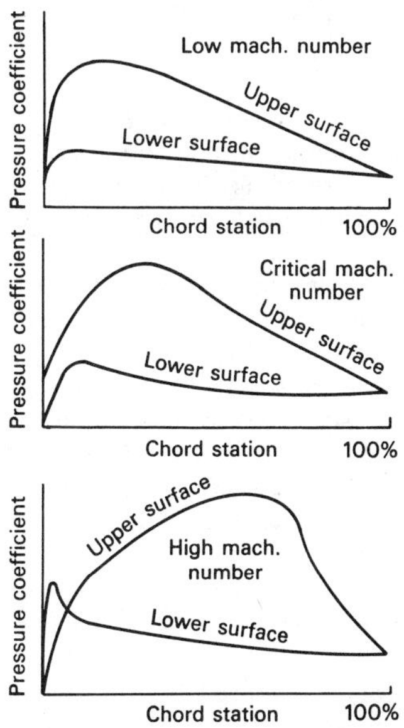

Fig. 13. Pressure distribution of an airfoil section at three different Mach numbers.

wind-tunnel tests have shown that this change in pressure distribution affects the following important airfoil parameters:

(1) The drag coefficient, C_D.
(2) The slope of the lift curve $dC_l/d\alpha$.
(3) The pitching-moment coefficient, C_M.
(4) The maximum lift coefficient, $C_{l\max}$.

The drag coefficient suffers most, its value increasing tremendously as the critical Mach number is reached. The large increase in drag results not only from energy loss in the shock wave but also from the large positive pressure gradient existing across the shock. Such a pressure gradient causes boundary-layer separation which results in a wide, turbulent wake with its attendant form-drag. A curve is presented in Fig. 14 which shows qualitatively the variation of airfoil-

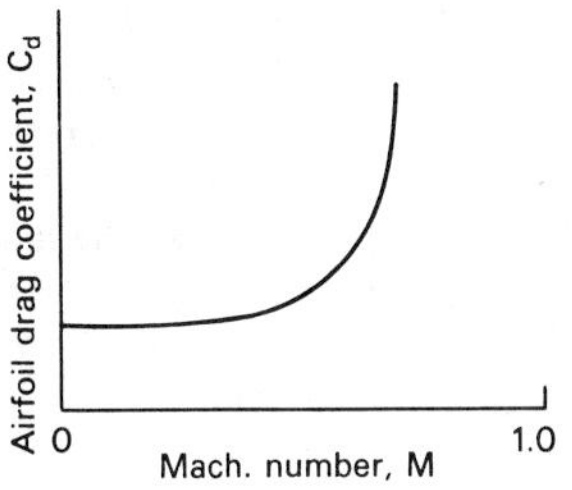

Fig. 14. Variation of airfoil drag coefficient with Mach number.

drag coefficient with Mach number. In terms of airplane performance, the increase in drag which occurs at the critical speed indicates that a tremendous amount of power would be required for an airplane to fly through the sonic range of speeds and reach speeds above the speed of sound.

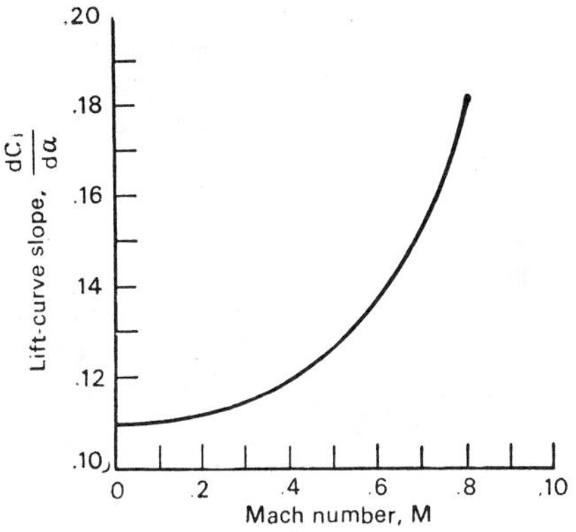

Fig. 15. Change of lift-curve slope with increasing Mach number.

A loss in maximum lift coefficient occurs as the critical speed is approached. However, airplanes do not operate at high lift coefficients when traveling at high speeds. The changes in lift-curve slope and pitching-moment coefficient which occur with increasing Mach numbers are shown in Figs. 15 and 16. The variation of lift-curve slope

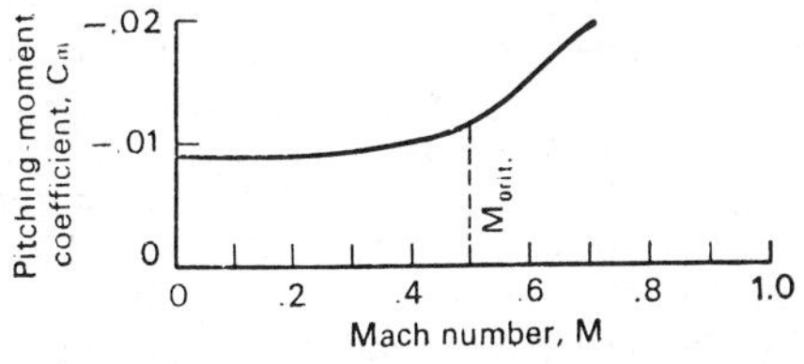

Fig. 16. Variation of pitching-moment coefficient with increasing Mach number.

with Mach number was calculated from Glauert's approximately correct theoretical relation which states that the change in lift-curve slope is proportional to $1/\sqrt{1-M^2}$, where M is the free-stream Mach number.

The changes in pitching-moment coefficient and lift-curve slope indicate a considerable variation in the external forces acting on the lifting surfaces of airplanes traveling at high speeds. Unless the changes in C_M and $dC_l/d\alpha$ are taken into consideration in the design of high-speed airplanes, complete loss of control and stability may result at high speeds. The large turbulent wake discussed in connection with the drag of bodies at shock speeds also has an adverse effect on the control and stability. The angle of downwash and thus the trim of the airplane will change instantly when the wake widens and the plane may be thrown completely out of control. The large turbulent wake may also cause serious control-surface buffeting.

Knowledge of compressible flows is steadily increasing. The most fruitful methods of obtaining information about compressibility phenomena have been found in the field of wind-tunnel testing and through careful instrumentation of rocket-propelled test vehicles for which the measured data are telemetered to the ground.

Theory of Aerodynamic Circulation. A mass of air in rotary motion is said to be in circulatory flow if its velocities at various radii from the center of rotation are of the proper magnitude to induce radial equilibrium of the circulating mass. Consideration of the requirements for equilibrium consists in balancing centrifugal forces against static pressures derived from Bernoulli's theorem and results in the specification that velocity of a particle must be inversely proportional to its radius from the center of rotation. Note that this is not like the motion of portions of a wheel, whose velocities are proportional to their radii. A simple case of circulation could be visualized as the flow pattern (concentric circles) induced by a rough cylinder rotating rapidly in still air.

If air is in circulation and no object such as a cylinder occupies the central core, the velocity at the center of rotation reaches a theoretical value of infinity. This is impossible and the center of circulatory flow must be occupied by a small core of air in simple rotary motion. Air in this condition is described as a free vortex. Exceptionally high velocities may exist at the edge of the rotary core, and in vortices of high circulatory strength such as tornadoes the atmospheric energy may reach the destructive proportions.

The strength of aerodynamic circulation (usually called the *circulation* and designated by symbol Γ) is the line integral of the tangential component of the velocity along any closed line encircling the vortex center, or some object.

$$\Gamma = \oint v\,dl$$

It has been proved that Γ has the same value for any closed path through the flow around the object (such as an airfoil, cylinder, etc.) and this fact is of value in various phases of aerodynamic analysis. A simple case of a flow pattern of concentric circles illustrates the constancy of Γ. The velocity at radius r is K/r and

$$\Gamma = \oint \frac{K}{r}\,dl = \oint \frac{K}{r}\,r\,d\theta = 2\pi K$$

Therefore Γ is not a function of r; its magnitude is constant throughout the flow pattern.

The generally accepted demonstration of the origin of lift on an airfoil employs a vortex circulation imposed on a rectilinear velocity field to produce the typical stream flow pattern, after which the aerodynamic forces are analyzed as originating from the impulse required to alter the momentum of the air stream and the static pressure. Fortunately, it is possible to analyze lift on the premise of frictionless flow. Viscosity effects do not enter until drag is sought. The dependence of lift of an airfoil of infinite span upon circulation Γ, free stream velocity V, and mass density ρ was proved independently by Kutta and Joukowski early in this century. Later Prandtl and others extended the "circulation theory" to cover the lift of finite wings. Joukowski's proof, considerably abbreviated, follows. The inherent tendency of an airfoil to create circulation is replaced by a bound vortex. A section of airfoil of unit span length is taken. The two flows are shown covering the same region in Fig. 17. Of course this figure does not represent the actual stream flow pattern since the latter would follow the vectorial combination of these two flows. Consider the region enclosed by the imaginary cylindrical surface at radius r from the vortex center. Let r be large enough to have circulatory velocity v small compared to rectilinear velocity V and integrate for vectorial change in momentum in this cylinder and vertical component of static pressure acting against it.

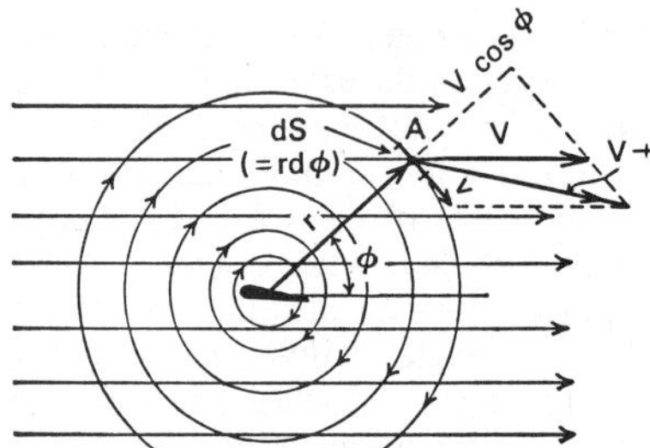

Fig. 17. Circulatory superimposed on rectilinear motion.

The velocity of circulation, v, is $\Gamma/(2\pi r)$ and its combination with V gives the direction and magnitude of the stream line at point A. The mass of air leaving the cylindrical region at A is $\rho ds(V\cos\phi)$ and its vertical component of velocity is $v\cos\phi$. The net vertical momentum in the cylindrical mass of air, due to inflow and outflow, is

$$\oint (\rho r\,d\phi)(V\cos\phi)\left(\frac{\Gamma}{2\pi r}\cos\phi\right)$$

which simplifies to

$$\frac{\rho\Gamma V}{2\pi}\oint \cos\phi\,d\phi$$

whence, by integration, the vertical impulse force is $(\rho\Gamma V)/2$. Since the change of momentum is downward the impulse force is upward, i.e., it is the lift. An integration of the horizontal component of momentum yields zero net momentum.

Proceeding next to evaluate the static pressure, note that Bernoulli's theorem applied at point A is covered by the following statement. Let p be the static pressure at A, and p_0 the free stream static pressure; then

$$p + \frac{\rho}{2}(V \dotplus v)^2 = p_0 + \frac{\rho}{2}V^2$$

After expanding $(V \dotplus v)^2$ and dropping all v^2 terms (since A was chosen to make v small relative to V),

$$p = p_0 - \rho V v \sin\phi$$

Since the second term on the right-hand side of this equation represents the variation from free stream pressure at point A, the lift (if any) will be the line integral of its vertical component; i.e.,

$$\text{vertical pressure component} = \oint (\rho V v\sin\phi)(\sin\phi)(r\,d\phi)$$

$$= \rho\frac{V\Gamma}{2\pi}\oint \sin^2\phi\,d\phi$$

$$= \rho\frac{V\Gamma}{2}$$

The horizontal integral yields zero net pressure. The total lift per unit of span is the sum of that originating from momentum and that originating from pressure. From the preceding demonstration it is seen that each contributes equally to the total lift, which becomes $\rho V\Gamma$.

Theory and experiment show that Γ is always sufficient to cause the divided flow over the upper and lower surfaces to reunite at the trailing edge without reverse flow.

If the aerodynamic chord of the wing is parallel to the free stream velocity V, the lift is zero and obviously Γ is zero. As the angle of attack is increased Γ increases nearly proportionately. The strength of the circulation for an airfoil of chord c at aerodynamic angle of attack α_a is $\frac{1}{2}aVc\alpha_a$, where a is the proportionating factor. Since lift, L, $= \rho V\Gamma b$ (b = span), it also equals

$$a\alpha_a\frac{\rho V^2}{2}(bc)$$

Call $a\alpha_a$ the coefficient of lift, bc the area S, and $(\rho V^2)/2$ the dynamic pressure q; then the lift equation becomes

$$L = C_L qS$$

If the wing has a finite span (i.e., definite tips), the bound vortex, assumed for the purpose of accounting for lift, is found to extend from the tips downstream as a free vortex. The effect of these tip vortices is to modify the air reaction since a downwash, w, is produced at the lifting line (see Fig. 18). The effect of downwash is to change

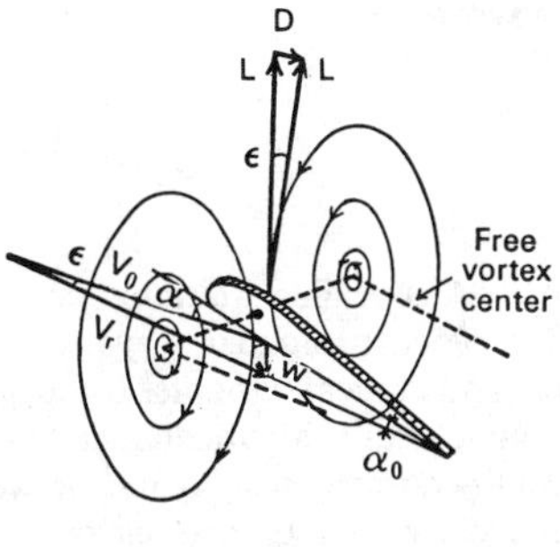

Fig. 18. Perspective of a section of airfoil subject to downwash from the free vortices at the tips.

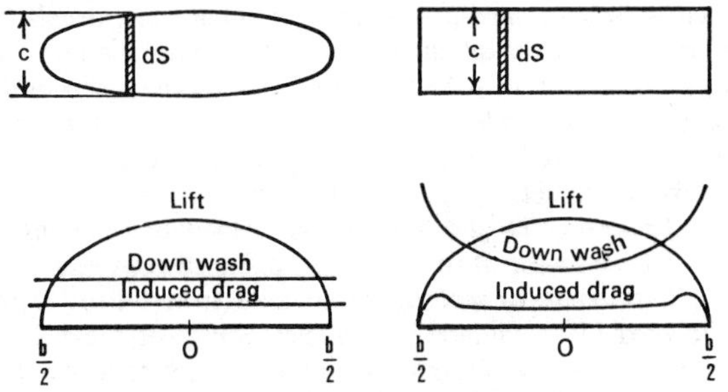

Fig. 19. Spanwise variations.

the lift direction from L to L' since the relative wind, V_r, is now inclined. The angle ϵ is always small, so L' may be considered of the same magnitude as L. However, L still remains the significant lift since the free stream velocity is V_0. Consequently, a drag D_i is introduced, although the original assumption of zero air viscosity remains in order.

This *induced drag* is a feature of the motion of ideal, frictionless air over a wing of finite span. The longer the span, the less the average effect of downwash (since in vortex motion $v \sim r^{-1}$) and the less the induced drag, but it will not disappear unless the span extends to infinity. By application of hydrodynamic theory beyond the scope of this article, aerodynamicists have shown that:

(1) Minimum induced drag for a given lift and span will be had if downwash is of constant magnitude over the span.

(2) Constant downwash will result from a bound vortex circulation strength of elliptical spanwise character.

(3) Elliptical spanwise Γ occurs on an untwisted airfoil of elliptical planform.

(4) With elliptical Γ a simple relation connections induced drag and aspect ratio.

$$D_i = L^2/\pi AR$$

where AR represents the aspect ratio.

This relation holds for the coefficient, too.

$$C_{D_i} = C_L^2/\pi AR$$

As

$$\frac{D_i}{L} = \frac{C_{Di}}{C_L} = \frac{w}{V_0} \text{ (nearly)}$$

$$\alpha = \alpha_0 + C_L/\pi AR$$

It appears that a finite wing possesses an induced drag of C_{D_i} not possessed by an infinite wing in ideal flow, also an induced angle of attack, requiring the angle of attack α of the finite span to be more than α_0 of an infinite span for the same lift per unit of span. The differences of induced drags of two wings of the same airfoil profile at the same lift coefficient but with different planforms follow naturally from the above:

$$C_{Di1} - C_{Di2} = \frac{C_L^2}{\pi}\left(\frac{1}{AR_1} - \frac{1}{AR_2}\right)$$

It is commonly assumed that the difference between the total drag coefficients of two wings of the same airfoil profile but different aspect ratios is the difference in induced drag. Likewise, the difference in necessary angle of attack to the free stream velocity for the same C_L is:

$$\alpha_1 - \alpha_2 = \frac{C_L}{\pi}\left(\frac{1}{AR_1} - \frac{1}{AR_2}\right)$$

While these induced effects are premised on elliptical planform for the wing, they hold very well for rectilinear wings with elliptical tips and those with moderate taper ratio.

The lateral or spanwise distribution of lift on a wing depends on the planform and the distribution of downwash. Two cases are illustrated—an elliptical wing and a rectangular one. (See Fig. 19.)

(1) The effect of downwash is to reduce the effective angle of attack and hence the lift coefficient. The lift of any lateral element dS is proportional to the product of C_L and c; hence the lift distribution of an elliptical wing is elliptical, while its induced drag, mirroring the downwash, is constant.

(2) The downwash strength increases near the tips of a rectangular wing. The lift coefficient accordingly declines and lift falls off at the tip even though chord is maintained. Induced drag is strong at the tips.

See also **Supersonic Aerodynamics.**

References

Abramson, H. N.: "An Introduction to the Dynamics of Airplanes," Dover, New York, 1973.

Abbott, I. H., and A. E. Von Doenhoff: "Theory of Wing Sections," Dover, New York, 1973.

Culick, F. E. C.: "The Origins of the First Powered, Man-Carrying Airplane," *Sci. Amer.*, **241**, 1, 86–100 (1979).

Kerr, R. A.: "East Coast Mystery Booms," *Science*, **203**, 256 (1979).

Hooven, F. J.: "The Wright Brothers' Flight-Control System," *Sci. Amer.*, **239**, 5, 166–185 (1978).

Meredith, D. L.: "Brain-Powered Flight," *Technology Review* (*MIT*), **80**, 3, 28–29 (1978).

Olsen, J. H., and A. Goldburg: "Aircraft Wake Turbulence and Its Detection," Plenum, New York, 1970.

Prandtl, L., and O. G. Tietjens: "Fundamentals of Hydro- and Aero-Mechanics," Dover, New York, 1973.

Reay, D. A.: "The History of Man-Powered Flight," Pergamon, New York, 1977.

Robinson, A. L.: "Human-Powered Flight—Californians and Kremer Prize," *Science*, **197**, 1171 (1977).

Sechler, E. E., and L. G. Dunn: "Airplane Structural Analysis and Design," Dover, New York, 1973.

Van Sickle, N. D.: "Modern Airmanship," 4th Edition, Van Nostrand Reinhold, New York, 1971.

Walker, J.: "Boomerangs," *Sci. Amer.*, **240**, 3, 162–172; and **240**, 4, 180–186 (1979).

AERODYNAMIC TRAIL. **Precipitation and Hydrometeors.**

AEROELASTICITY. The study of both the static and dynamic effects of aerodynamic forces on elastic bodies. The swaying of bridges, trees, smokestacks, and buildings are examples of the interplay of aerodynamic forces, inertia forces, and elastic properties of the structures. The flutter of flags, aircraft wings, and sails are more examples. For aircraft design, aeroelasticity is a most important study.

AEROGEL. A colloidal solution of a gaseous phase in a solid phase, obtained usually be replacement of the liquid in the dispersed phase by air or gas. Contrast with **Aerosol.**

AEROGENERATOR. **Wind Power.**

AEROLITE. A general term for meteorites that are richer in the basic silicates than in nickel and iron.

AEROLOGY. **Meteorology.**

AERONOMY. **Meteorology.**

AERO-OTITIS MEDIA. **Hearing and the Ear.**

AEROSOL. A colloidal system in which a gas, frequently air, is the continuous medium, and particles of solids or liquid are dispersed in it. Aerosol thus is a common term used in connection with air pollution control. Studies of the particle size distribution of atmospheric aerosols have shown a multimodal character, usually with a bimodal mass, volume, or surface area distribution and frequently trimodal surface area distribution near sources of fresh combustion aerosols. The coarse mode (2 micrometers and greater) is formed by relatively large particles generated mechanically or by evaporation of liquid from droplets containing dissolved substances. The nuclei mode (0.03 micrometer and smaller) is formed by condensation of vapors from high-temperature processes, or by gaseous reaction products. The intermediate or accumulation mode (from 0.1 to 1.0 micrometer) is formed by coagulation of nuclei. Study of the behavior of the particles in each mode has led to the belief that the particles tend to form a stable aerosol having a size distribution ranging from about 0.1 to 1.0 micrometer. The larger, settleable particles (in excess of 1.0 micrometer in size) fall out, whereas the very fine particles (smaller than 0.1 micrometer) tend to agglomerate to form larger particles which remain suspended. The nuclei mode tends to be highly transient and is concentration limited by coagulation with both other nuclei and also particles in the accumulation mode. It further appears that additional growth of particle size from the accumulation mode to the coarse mode is limited to 5% or less (by mass). Thus, the particulate content of a source emission and the ambient air can be viewed as composed of two portions, i.e., settleable and suspended.

Both settleable and suspended atmospheric particulates have deleterious effects upon the environment. The settleable particles can affect health if assimilated and also can cause adverse effects on materials, crops, and vegetation. Further, such particles settle out in streams and upon land where soluble substances, sometimes including hazardous materials, are dissolved out of the particles and thus become pollutants of soils and surface and ground waters. Suspended atmospheric particulate matter has undesirable effects on visibility and, if continuous and of sufficient concentration, possible modifying effects on the climate. Importantly, it is particles within a size range from 2 to 5 micrometers and smaller that are considered most harmful to health because particles of this size tend to penetrate the body's defense mechanisms and reach most deeply into the lungs.

The term aerosol is also applied to a form of packaging in which a gas under pressure, or a liquefied gas which has a pressure greater than atmospheric pressure at ordinary temperatures, is used to spray a liquid. The result of the spraying process is to produce a mist of small liquid droplets in air, although not necessarily a stable colloidal system. Numerous products, such as paints, clear plastic solutions, fire-extinguishing compounds, insecticides, and waxes and cleaners, are packaged in this fashion for convenience. Food products, such as topping and whipped cream, also are packaged in aerosol cans.

For a number of years, chlorofluorocarbons were the most popular source of pressure for these cans. Because of concern in recent years over the reactions of chlorofluorocarbons in the upper atmosphere of the earth that appear to be leading to a deterioration of the ozone layer, some countries have banned their use in aerosol cans. Manufacturers have turned to other gases or to conveniently operated hand pumps. See **Ozone.**

See also **Colloid System;** and **Pollution (Air).**

AEROSTATICS. The science of gases at rest (mechanical equilibrium).

AESTIVATION. Summer dormancy, the antithesis of the more familiar hibernation. As an example, the African lungfish is known to burrow down into the mud when the water begins to get low in the rivers and lakes where it lives. Here, even though the mud dries into a hard cake, the lungfish survives in a state of aestivation. Some have even been cut out along with the surrounding dirt and shipped abroad. When the mud again becomes moist with water, the lungfish becomes active and resumes normal activity.

AFFINE GEOMETRY. Geometry.

AFFINE TENSORS AND FREE VECTORS. A quantity which behaves like a tensor under a linear (affine) coordinate transformation, but not under a general coordinate transformation is called an affine tensor. From an affine tensor it is possible to construct a *free vector*, that is, a vector not related to a given point (non-localized vector).

AFLATOXICOSIS. Foodborne Diseases.

AFLATOXIN. Yeasts and Molds.

AFRICAN BUSH-PIG. Suines.

AFRICAN TRYPANOSOMIASIS. Also called *sleeping sickness*, this is an infection with the flagellate blood protozoan *Trypanosoma brucei.* There are two epidemiological and clinical variants, Rhodesian and Gambian, caused by two morphologically identical races of *T. brucei.* African trypanosomiasis is confined to tropical Africa, between 15°N and 20°S. latitude. It is carried by various species of tsetse flies (*Glossina*). The Gambian form of West and Central-Africa is without a known animal reservoir. It is a milder, endemic disease associated with tsetse flies breeding near streams. Rhodesian sleeping sickness of East and Central Africa is clinically more severe; it is carried by tsetse flies breeding in savannas and brushlands, where it is a zoonosis of hoofed game animals. In the blood, cerebrospinal fluid, lymph nodes, and other organs, both trypanosomes appear as motile, pleomorphic organisms, 15–30 μm long, with a flagellum and undulating membrane. The disease is characterized by fever, insomnia, lymph node enlargement, anemia, local edema and rash. After about 6 months there may be invasion of the central nervous system, producing somnolence, confusion, wasting, and coma. Untreated Rhodesian disease usually causes death within 1 year. Diagnosis depends on finding the trypanosomes in blood or in aspirates of lymph nodes or spinal fluid. In early infections suramin is used for the Rhodesian form and pentamidine for the Gambian form. Melarsoprol is effective in the later stages. Control is by destruction of tsetse habitats, the use of insecticides (DDT or dieldrin), personal prophylatic measures, and relocation of populations at risk.

Albert L. Vincent, Assistant Professor, Department of Comprehensive Medicine, University of South Florida, Tampa Florida.

AFRICAN VIOLET. Asexual Reproduction.

AFTERBIRTH. The membranes and placenta expelled from the uterus a short time after the birth of the child. See **Embryo.**

AFTERBURNER. Airplane; Catalytic Converter.

AFTERGLOW. Atmospheric Optical Phenomena.

AFTER-IMAGE. The image "seen" after a portion of the retina has been fatigued by continued fixed stimuli. For instance, if a person stares fixedly at a black cross-mark on a sheet of white paper for a minute or two and then suddenly looks at a blank wall, he will "see" a white cross-mark on the wall. If the image he stares at is colored, he will usually see an image of the complementary color on the wall. A green cross on the paper, for instance, will result in the appearance of a red cross on the wall. This is because the green-sensitive cones of the retina have become fatigued but the red-sensitive cones have not. The red-sensitive cones pick up the red light reflected from the wall, but the green-sensitive cones do not pick up the green light in a proper proportion.

AFTERSHOCK. Earthquakes, Seismology, and Plate Tectonics.

AGAMA. A large South African lizard of the genus *Agama.*

AGAR-AGAR. Now more commonly called "agar." A gelatine-like substance which is prepared from various species of red algae growing in Asiatic waters. The prepared product appears in the form of cakes, coarse granules, long shreds, or in thin sheets. It is used extensively alone or in combination with various nutritive substances, as a medium for culturing bacteria and various fungi. See also **Gums and Mucilages.**

AGARICS (*Agaricaceae*; *Fungi*). This family of fungi is probably better known than any other, since it contains most of the plants commonly described by the names toadstool and mushroom, which popularly and mistakenly denote poisonous and edible fungi, respectively.

The Agarics are mostly fleshy fungi of that very definite structure, the familiar parasollike toadstool. This is composed of convex pileus or cap, usually supported on an evident stalk. The underside of the cap shows a series of radiating plates, or gills, which are formed in agaric fungi only, and so serve to separate them from all others. The two sides of the gills are covered with the microscopic spore-bearing bodies called basidia, which are club-shaped or cylindrical cells bearing spores, generally four each. Agarics vary in size from delicate species with a cap a millimeter or so in diameter, supported by a slender threadlike stalk, to massive forms twelve inches in diameter: the larger species form millions of spores. See accompanying figure.

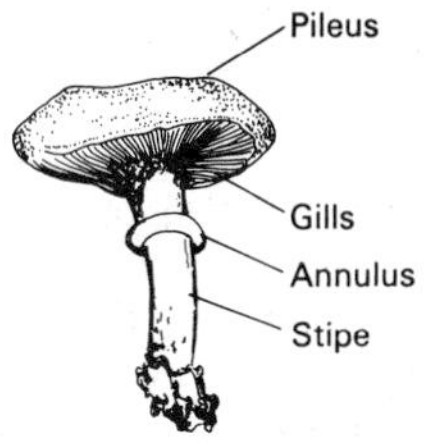

Sporophore of a mushroom (*Agaricus campestris*).

The spores float in the air for considerable distances, and finally come to rest on some solid substance. Should this be favorable for germination, the spore puts out a slender tube, which elongates rapidly and penetrates the substratum, from which it absorbs substances necessary for its continued growth. Gradually this threadlike body, known as the mycelium, spreads through extensive masses of substratum, branching frequently as it does so.

Finally there is accumulated in the mycelium a supply of reserve food sufficient for fruiting: then, if atmospheric conditions, such as moisture and temperature, be suitable, the familiar toadstool appears, it being only the reproductive stage of the fungus. Its rate of growth is often phenomenal, as is also the force it may exert in its growth. Seemingly delicate bodies not only break open the hard-packed surface of the ground, but also may push aside pebbles of considerable weight. Not infrequently whole rings of toadstools appear in a field, springing to maturity in a single night—these are the familiar fairy rings, resulting from the growing outward from a common source of the unseen mycelium and not from the dancing of fairy forms.

When the young fruit-body first comes up it is completely enclosed in a membranous skin known as the velum. As enlargement continues this skin is broken. Often traces of the velum remain in the form of flakes on the upper surface of the cap, and as a ring or annulus around the stalk. Attempts have been made to find in these characteristics a means for separating the edible from the poisonous species. However, no reliable distinction is found here. Actually, unless one is absolutely sure of the identity of a given species, the only safe rule is complete abstinence. Classification is based on the color of the spore-masses, and also the determination of the way in which the gills are attached to the stalk, the color changes shown as a cut or broken surface dries, and various other means.

A few species of Agarics, notably *Agaricus* (*Psalliota*) *campestris*, are extensively cultivated, and justly esteemed as food. Other species are violently poisonous, particularly species of the genus *Amanita*. Yet certain Siberian people use *Amanita muscaria*, a poisonous species, to produce a form of intoxication. Another common mushroom, the Inky-cap, a species of *Coprinus*, has been used as a writing fluid, the substance of the toadstool breaking down into a fluid mass containing vast numbers of black spores. See **Basidiomycetes.**

Mushrooms are cultivated extensively in certain regions, usually in caves or specially constructed buildings. "Spawn" is prepared by inoculating grain, commonly rye, with spores or mycelium of the desired variety, and allowing the fungus to grow for a time. The spawn is spread as "seed" upon beds of compost where the fungus develops the characteristic mycelium described above. The compost is prepared from horse manure, including the bedding straw, from ground corn cobs, hay, or other organic materials. After the mycelium is established, the bed is covered by a thin layer of carefully selected soil. Usually thousands of mushrooms break through the soil at one time, and are harvested at the proper stage of growth. The entire growth takes place in the dark under controlled temperature and humidity.

AGATE. Agate is a variety of chalcedony, whose variegated colors are distributed in regular bands or zones, in clouds or in dendritic forms, as in moss agate.

The banding is often very delicate with parallel lines of different colors, sometimes straight, sometimes undulating or concentric. The parallel bands represent the edges of successive layers of deposition from solution in cavities in rocks which generally conform to the shape of the enclosing cavity.

As agate is an impure variety of quartz it has the same physical properties as that mineral. It is named from the river Achates in Sicily where it has been known from the time of Theophrastus.

Agate is found in many localities; India, Brazil, Uruguay, and Germany are notable for fine specimens.

Onyx is a variety of agate in which the parallel bands are perfectly straight and can be used for the cutting of cameos. Sardonyx has layers of dark reddish-brown carnelian alternating with light and dark colored layers of onyx.

See also **Chalcedony; Quartz;** and terms listed under **Mineralogy.**

AGAVE (*Amaryllidaceae*). A large genus of plants, particularly abundant in Mexico, in which the thick rigid leaves form a basal rosette from the center of which rises the tall flower stalk. Because of the time required to store sufficient food reserves for flowering, certain species, notably *Agave americana*, are called century plants from the belief that they flower but once a century; actually flowering may occur in from 5 to 50 or more years. Once started, the flower stalk develops very rapidly, requiring immense quantities of sap. In Mexico, the flower stalk is cut off early in its formation and the stump scooped out to form a cup into which quantities of sweet sap exude. This sap is collected and fermented to form pulque, a strong drink with an unpleasant odor. Distilled pulque gives a more potent drink, mescal. From the leaves of several species, particularly *Agave sisalana* and *A. fourcroydes*, are obtained fibers. These fibers occur as sclerenchyma sheathes surrounding the vascular bundles in the leaves. To obtain the fibers, the leaves are cut off and the spiny tip and margin removed. Machines then heat and scrape the leaves and wash them until the clean fibers are obtained. These are then dried either in the sun or by artificial heat, and are ready for export under the name of sisal or henequen, according to the species from which they were obtained. Many species of *Agave* are cultivated for their ornamental value. See **Century Plant.**

AGE DETERMINATION. **Pine Trees; Radioactivity and Other Dating Techniques.**

AGE (Geology). **Geologic Time Scale.**

AGENT ORANGE. Common name for a 50–50% mixture of the herbicides 2,4,5-T and 2,4-D, once widely used by the military as a defoliant. The mixture contains dioxin as a contaminant. See also **Dioxin;** and **Herbicides.**

AGE (Universe). **Cosmology.**

AGGLOMERATE. A term proposed by Sir Charles Lyell in 1831 for coarsely graded volcanic ejectamenta similar in appearance to ordinary conglomerates or breccias. An extremely thick and widespread accumulation of so-called agglomerates occurs on the borders of the Yellowstone Park. These deposits, however, include numerous beds of water-laid pebbles, gravels and sands, the latter containing fossil plants of early tertiary age.

Because of the various interpretations of the term, it should be defined in context.

AGGLOMERATION. This term connotes a gathering together of smaller pieces or particles into larger size units. This is a very important operation in the process industries and takes a number of forms. Specific advantages of agglomeration include increasing the bulk density of a material, reducing storage-space needs, improving the handling qualities of bulk materials, improving heat-transfer properties, improving control over solubility, reducing material loss and lessening of pollution, particularly of dust, converting waste materials into a more useful form, and reducing labor costs because of resulting improved handling efficiency.

The principal means used for agglomerating materials include (1) compaction, (2) extrusion, (3) agitation, and (4) fusion.

Tableting is an excellent example of compaction. In this operation, loose material, such as a powder, is compressed between two opposing surfaces, or compacted in a die or cavity. Some tableting machines use the action of two opposing plungers which operate within a cavity. Resulting tablets may range from $\frac{1}{8}$ to 4 inches in diameter. Uniformity and dimensional precision are outstanding. Numerous pharmaceutical products are formed in this manner, as well as some metallic powders and industrial catalysts.

Pellet mills exemplify the use of extrusion. In some designs the charge material is forced out of cylindrical or other shaped holes located on the periphery of a cylinder within which rollers and spreaders force the bulk materials through the openings. A knife cuts the extruded pellets to length as they are forced through the dies.

The *rolling drum* is the simplest form of aggregation using agitation. Aggregates are formed by the collision and adherence of the bulk particles in the presence of a liquid binder or wetting agent to produce what essentially is a "snow ball" effect. As the operation continues, the spheroids become larger. The strength and hardness of the enlarged particles is determined by the binder and wetting agent used. The operation is followed by screening, with recycling of the fines.

The *sintering process* utilizes fusion as a means of size-enlargement. This process, used mainly for ores and minerals and some powdered metals, employs heated air which is passed through a loose bed of finely ground material. The particles partially fuse together without the assistance of a binder. Sintering frequently is accompanied by the volatilization of impurities and the removal of undesired moisture.

The *spray-type* agglomerator utilizes several principles. Loosely bound clusters or aggregates are formed by the collision and coherence of the fine particles and a liquid binder in a turbulent stream. The mixing vessel consists of a vertical tank, around the lower periphery of which are mounted spray nozzles for introduction of the liquid. A suction fan draws air through the bottom of the tank and creates an updraft within the mixing vessel. Materials spiral downward through the mixing chamber, where they meet the updraft and are held in suspension near the portion of the vessel where the liquids are injected. The liquids are introduced in a fine mist. Individual droplets gather the solid particles until the resulting agglomerate overcomes the force of the updraft and falls to the bottom of the vessel as finished product.

AGGLOMERATION (Cloud Physics). Precipitation and Hydrometeors.

AGGLUTINATION. 1. The gathering of particles. 2. The clumping together of bacteria or cells, resulting often from their reaction with the corresponding immune or modified serum.

AGGLUTININ. One of a class of substances found in blood to which certain foreign substances or organisms have been added or admixed. As the name indicates, agglutinins have the characteristic property of causing agglutination, especially of the foreign substances or organisms responsible for their formation.

AGGRADATION. In geology, the building up of the surface of the earth by deposition, as of sediment by a river in its valley. More specifically, the upbuilding caused by a stream so as to establish and maintain a uniform grade or slope. The term also is used sometimes as a synonym for accretion, as in the case of development of a beach. See also **Accretion; Alluvial Fan.**

AGGREGATE. The solid conglomerate of inert particles which are cemented together to form concrete are called aggregate. A well-graded mixture of fine and coarse aggregates is used to obtain a workable, dense mix. The aggregate may be classed as fine or coarse depending upon the size of the individual particles. The specifications for the concrete on any project will give the limiting sizes which will distinguish between the two classifications. Fine aggregate generally consists of sand or stone screenings while crushed stone, gravel, slag or cinders are used for the coarse aggregate. The aggregates should be strong, clean, durable, chemically inert, free of organic matter, and reasonably free from flat and elongated particles since the strength of the concrete is dependent upon the quality of the aggregates as well as the matrix of cementing material. See also **Concrete.**

AGGRESSIN. A product of bacterial metabolism which impairs the defensive mechanisms of the blood of the host.

AGING. Gerontology.

AGONIC LINE. Isogonic Line.

AGOUTI. Rodentia.

AGRANULOCYTOSIS. This is a potentially serious syndrome in which the white cells may be greatly decreased or almost absent from circulation. Because the granulocytes are important in protecting the body against infection, an individual deprived of these defensive forces for long may have an overwhelming invasion of the bloodstream and organs with dangerous disease-producing organisms. See also **Blood.** Important symptoms include general weakness, prostration, headache, shaking chills, and progressive ulcerative throat lesions. Diagnosis must be confirmed by studies of the bone marrow.

The syndrome of agranulocytosis may result from an overwhelming infection, releasing toxins specifically destructive to the bone marrow and lymphatic systems. It may develop secondary to allergic sensitization to drugs with antigenic properties, such as aminopyrine, cimetidine, potassium-sparing diuretics, procainamide, propanolol, and sulfapyridine. Industrial toxins, particularly benzol, may damage the marrow similarly.

AGREEMENT (Coefficient of). This coefficient relates to the situation where m observers each provide paired comparisons for n objects. A coefficient of agreement between the verdicts of the m observers is given by

$$u = \frac{8\Sigma}{m(m-1)(n-1)n} - 1$$

where Σ is the sum of the number of agreements between pairs of judges.

The coefficient may vary from $-1/(m-1)$, if m is even, or $-1/m$, if m is odd, up to $+1$ if there is complete agreement.

AGRICULTURAL MONITORING. Earth Resources Satellites and Geologic Remote Sensors.

AGROSTOLOGY. The science of grasses, including classification, management, and utilization. See **Grasses.**

AGULHAS CURRENT (also called Agulhas Stream). A generally southwestward-flowing ocean current of the Indian Ocean; one of the swiftest of ocean currents.

Throughout the year, part of the south equatorial current turns south along the east coast of Africa and feeds the strong Agulhas current. To the south of latitude 30°S, the Agulhas current is a well-defined and narrow current that extends less than 100 kilometers (62 miles) from the coast. To the south of South Africa, the greatest volume of its waters bends sharply to the south and then toward the east, thus returning to the Indian Ocean by joining the flow from South Africa toward Australia across the southern part of that ocean. However, a small portion of the Agulhas current water appears to round the Cape of Good Hope from the Indian Ocean and continue into the Atlantic Ocean.

AILANTHUS. Mahogany Trees.

AILERON. The aileron is one of the three aerodynamic surfaces of an airplane which are variable in position at the will of the pilot, the purpose of which is to provide the required degree of maneuverability and control of the aircraft about its longitudinal axis. The aileron is that surface which causes the necessary forces to be produced to induce rotation of the aircraft.

This motion is known as *roll*, which is employed to correct other rolls produced unintentionally, as by gusts—when not used to accomplish such maneuvers as banks or sideslips.

The conventional type of aileron is a flap inserted in the trailing edge of the wing, usually at the tip. This flap is rotatable around its forward axis, upward and downward, and in effect changes the camber of the airfoil. The result is a change of pressure on that portion of the wing over which the aileron extends. The change in pressure is greater on the side where the aileron is rotated downward and less on the side where it is rotated upward. The difference in pressure causes the airplane to roll. When the airplane has reached the required angle of bank, the ailerons are returned to neutral. The ailerons are then used in the opposite direction to bring the airplane to its level or normal position, when again the ailerons are returned to neutral. The principal defect of the flap-type aileron is that it becomes relatively ineffective when, for safety's sake, it is needed to be most effective, i.e., when the airplane approaches the stall, or is stalled. (This stall situation refers to the aerodynamic stall of the wing, not to the stall of the engine.) With adequate control of roll during a stall, many serious accidents involving tailspin can be avoided. This defect can be minimized by careful design, and avoided by proper handling of the aircraft. The flap-type ailerons produce some adverse yawing moments, in addition to the desirable rolling moments. This yawing moment is counteracted by use of the rudders.

The aerodynamic efficiency, simplicity, and reliability of the flap-type aileron are better than for other types, such as wing-tip ailerons and spoiler ailerons. See also **Aerodynamics.**

AIMLESS DRAINAGE. A type of drainage or stream pattern that occurs in low swampy lands; particularly characteristic of glaciated regions of low relief. Essentially, drainage without a well-developed system.

AIR. In addition to being the principal substance of the earth's atmosphere, air is a major industrial medium and chemical raw material.

TABLE 1. COMPOSITION OF AIR

Constituent	Percent by Weight	Percent by Volume
Oxygen (O_2)	23.15	20.95
Ozone (O_3)	1.7×10^{-6}	0.00005
Nitrogen (N_2)	75.54	78.08
Carbon dioxide (CO_2)	0.05	0.03
Argon (Ar)	1.26	0.93
Neon (Ne)	0.0012	0.0018
Krypton (Kr)	0.0003	0.0001
Helium (He)	0.00007	0.0005
Xenon (Xe)	5.6×10^{-5}	0.000008
Hydrogen (H_2)	0.000004	0.00005
Methane (CH_4)	trace	trace
Nitrous oxide (N_2O)	trace	trace

TABLE 2. WATER CONTENT OF SATURATED AIR

Temperature (°F)	Temperature (°C)	Water Content (Pounds in 1 Pound of Air, or Kilograms in 1 Kilogram of Air)
40	4.44	0.00520
45	7.22	0.00632
50	10	0.00765
55	12.8	0.00920
60	15.6	0.01105
65	18.3	0.01322
70	21.1	0.01578
75	23.9	0.01877
80	26.7	0.02226
85	29.4	0.02634
90	32.2	0.03108
95	35.0	0.03662
100	37.8	0.04305
105	40.6	0.05052

The average composition of dry air at sea level, disregarding unusual concentrations of certain pollutants, is given in Table 1. The amount of water vapor in the air varies seasonally and geographically and is a factor of large importance where air in stoichiometric quantities is required for reaction processes, or where water vapor must be removed in air-conditioning and compressed-air systems. The water content of air for varying conditions of temperature and pressure is shown in Table 2. The water content of saturated air at various temperatures is shown in the accompanying figure. See also **Oxygen; Nitrogen;** and **Pollution (Air).**

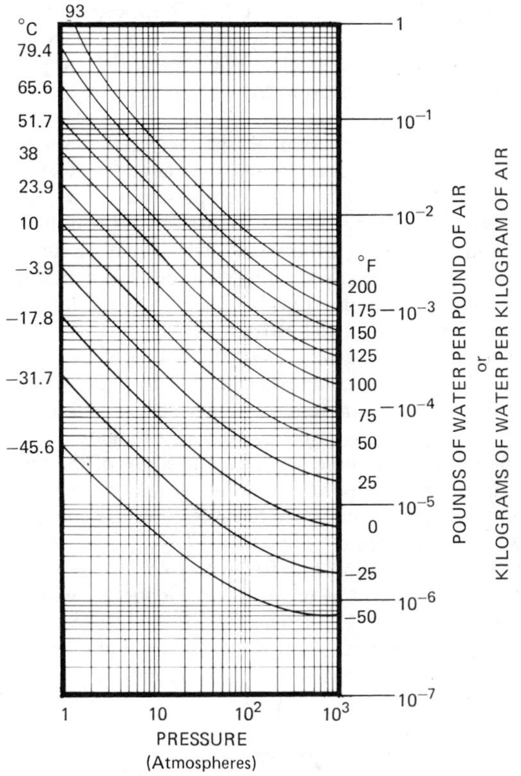

Water content of saturated air at various temperatures and pressures.

AIR ATOMIZER. Burner.

AIR BLADDER. Fishes.

AIRBORNE RADAR (Weather). Weather Observations and Forecasting.

AIR (Combustion). Burner; Combustion.

AIR COMPRESSION. The compression of air by mechanical means, and the raising of it to some desired pressure above that of the atmosphere, is effected, usually, by an approximate adiabatic change of state.

If the ideal compression were possible, it would be represented by the following equation showing the relation between pressure and volume:

$$PV^{1.4} = \text{a constant}$$

A compression of this nature may heat the air to temperatures which would interfere with reliable action of an air compressor and introduce lubrication difficulties, were there no provision for cooling the cylinder walls. Therefore, in compressors we find the cylinders to be externally finned or water-jacketed so that sufficient cooling is secured to keep the temperatures from becoming excessive. The extraction of heat from the cycle in this way modifies the conditions of compression from the ideal to some change more nearly represented by

$$PV^n = C$$

in which n usually lies between 1.35 and 1.4. The ratio of the temperature before and after compression is expressed by the following equation, the temperatures being degrees Fahrenheit absolute.

$$\frac{T_2}{T_1} = \left[\frac{V_1}{V_2}\right]^{n-1}$$

In compression to high pressures, the temperature rise may be too great to permit the compression to be carried to completion in one cylinder, even though it is cooled as mentioned above. In high-pressure compressors, the compression is carried out in stages, with a partial increase of the pressure in each stage, and cooling of the air between the stages. Two- and three-stage compression is very common where pressures of 300–1000 lbs. per sq. in. (20–68 atmospheres) are needed.

The volume of clearance air should be made as small as possible in order to improve the volumetric efficiency of the compressor, since the clearance air must expand to the suction pressure before the cylinder can begin to be charged.

The mechanical construction of air compressors varies with the amount of compression required and the character of service.

Air Compressors. Mechanical air compressors can be classified into two major groups (1) *reciprocating compressors* of the piston-and-cylinder type; and (2) *rotating compressors,* which may be further divided into a number of kinds. Since the reciprocating compressors were the earliest type to be developed, at any rate for higher pressors, they are discussed first in this entry.

Reciprocating Compressors. A common type of air compressor is the piston and cylinder compressor in which a reciprocating piston positively displaces the air from a cylinder during its discharge stroke. Compressors for charging tanks of air used to inflate pneumatic tires at the numerous automotive service stations are of the reciprocating type. Being of small capacity they are generally single acting and air cooled (by exterior fins) since those features are common in small compressors. Larger compressors, unless for extremely high pressure, are usually double acting and frequently cooled by water jackets. One or two cylinder arrangements are conventional, with a tendency to secure large capacity by increased bore and stroke rather than by multiple cylinders. Pistons are reciprocated by a crankshaft and connecting rod mechanism commonly deriving motion from the driving source by belt. Valves are springloaded to open upon slight differential pressures.

The compressor cycle is shown in Fig. 1. The discharge stroke which begins at A builds up the pressure to B where it exceeds the receiver pressure sufficiently to open a discharge valve. Discharge then takes place at a constant control pressure from B to C. The volume C is the clearance volume of the compressor. The air in the clearance volume must expand to D during the suction stroke before the inlet valve will open. Thus the volume of air drawn in per stroke is only that from D to A. Obviously the compressor should have as small a clearance as possible in order to obtain good volumetric efficiency, especially at high discharge pressures. Small compressors may be operated with high compression ratios (8–12) if desired because cooling is more effective in small cylinders and mechanical strength is readily provided. Large volumes compressed to ratios exceeding 4 will need a multi-stage compressor to permit cooling between stages and to lessen the structural loads on the large first stage cylinders.

Rotating compressors may be classified into four types ranging from the (1) centrifugal and (2) axial compressors that are often used for high pressure service to (3) blowers and (4) fans used for low pressure—large volume operation. Since, however, the term compressor is so often restricted to higher pressure service, blowers and fans are treated in separate entries in this book, leaving this one to deal with centrifugal and axial flow compressors.

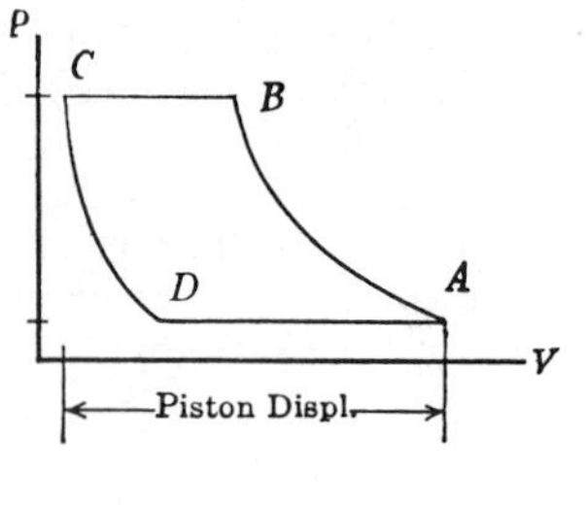

Fig. 1. Reciprocating compressor cycle.

Centrifugal Compressors. A rotating impeller mounted in a casing and revolved at high speed will cause a fluid which is continuously admitted near the center of rotation to experience an outward flow and a pressure rise due to centrifugal action.

Assume that an impeller with radial blades of depth $r_2 - r_1$ is revolving at a speed of ω radians per minute. This is illustrated in Fig. 2. Consider that a compressible fluid (a gas) is admitted at the center and flows into the impeller radially. Relative to the impeller blades it has an outward radial flow, finally emerging with some absolute velocity v_2 which is partially diffused into pressure. In addition a pressure gradient must exist to balance the sum of all the incremental $mr\omega^2$ inertia forces arising from the inward acceleration $r\omega^2$ given each particle of the fluid. The interrelation of r, ω, P, can readily be developed by considering the power required as (1) that necessary for the thermodynamics of compression, and (2) that which would account for the action of the impeller in effecting certain momentum changes on the fluid.

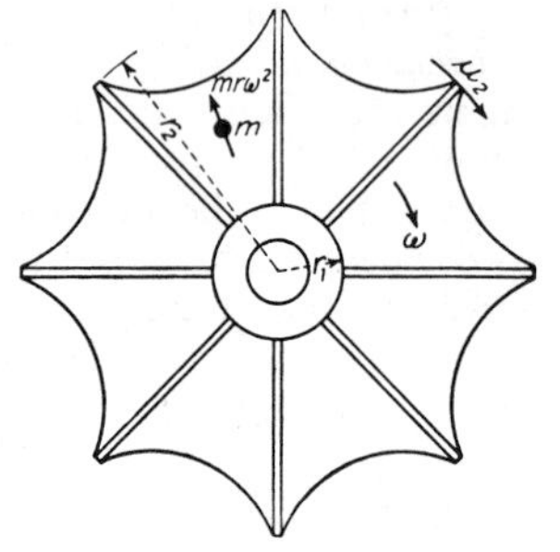

Fig. 2. Impeller for centrifugal compressor.

Without detailing the algebraic procedure, it may be stated that the ability of an impeller to raise the fluid pressure is expressed by the ratio of discharge to inlet pressure, i.e., P_2/P_1

$$\frac{P_2}{P_1} = \left[1 + \frac{\eta z u_2^2}{gRT_1}\right]^{1/z}$$

$g = 32.2$

$R = 53.4$ for air

$z =$ a gas coefficient, about 0.286 for air

$T_1 =$ energy coefficient. This would be 1.0 if the flow through the impeller were non-turbulent and frictionless. Typically, $\eta = .75$ to $.85$.

$u_2 =$ impeller rim velocity, ft per sec.

For pressure ratios higher than can be obtained from the action defined above, several impellers may be mounted on the same shaft and enclosed in a compound casing with passages arranged to lead the output from one impeller to the "eye" of the next. This multistaging principle is used to produce pressures above the capability of a single impeller compressor. Multi-stage compression may have cooling between the stages so the overall compression may be more isothermal than adiabatic. If compression were isothermal

$$\frac{P_2'}{P_1} = e^{\eta u_2 / gRT}$$

Axial-flow Compressors. Figure 3 illustrates the compressing action in an axial-flow multistage compressor. Air flows over a set of stationary airfoils (arranged circumferentially as fixed blading) with an angle of attack α. As the airfoil aspect ratio is small, the downwash created is considerable. Downwash turns the air stream through angle ϵ, which results in a flow channel of increasing cross section. The diffusion thus effected slows down the air velocity to V_2 and increases pressure. The moving airfoils (rotor blades) receive this air stream at V_3 as a result of the vectorial combination of air stream velocity V_2 and blade speed U. Relative to the moving airfoils the downwash again causes a diffusion of velocity and another increase of pressure. The relative velocity V_4 leaving the moving blades produces an absolute final velocity of V_5 because of blade motion. Suitable combinations of U and α enable V_5 to duplicate V_1 and permit a similar following stage. The blade height in the following stage is decreased because of the smaller specific volume of the compressed air. Increasing α will increase the pressure rise without much decrease in volume. The axial-

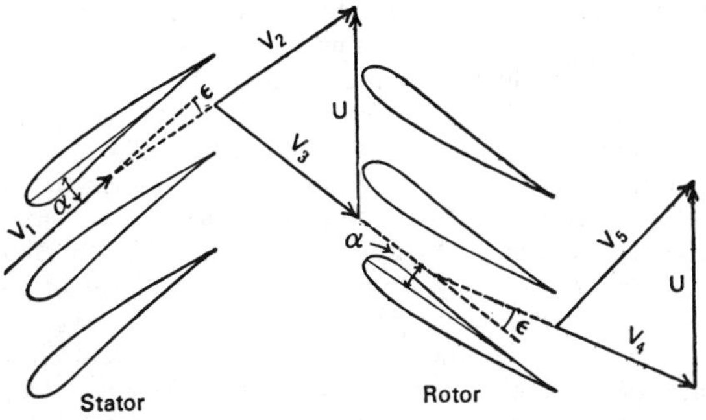

Fig. 3. Velocity relations in axial-flow compressing blading.

flow compressor should be designed for conservative α's for two reasons. First, an operation near the stalling angle would be hazardous as small variations of α might occur which could burble the airfoils and cause an unstable, rough, or even hazardous condition to exist. Secondly, axial-flow compressors may be employed under conditions where utmost efficiency is imperative (as in gas turbine power units) and α should create the optimum favorable balance between good downwash and minimum turbulent airfoil wake. Axial-flow compressors have been built with energy efficiences as high as 90%. They may be operated effectively at high speeds, i.e., 5000–50,000 rpm.

The compression of air and other gases may also be secured by the employment of steam jets if an admixture of vapor in the compressed gas is not undesirable. High-pressure steam is blown through nozzles which create a high-velocity jet. The gas to be compressed is led into the regions about the nozzle discharge where it is entrained in the steam jet. The mixture then travels into a diffuser for compression and attendant velocity reduction. Although the compression thus achieved is of limited magnitude, staging the compression in a series of nozzles, with intermediate coolers for partial condensation of vapor, allows moderate compression ratios to be achieved.

AIR COMPRESSOR (Gas Turbine). Gas and Expansion Turbines.

AIR CONDITIONING SYSTEM. Heap Pump; Solar Energy.

AIRCRAFT FUELS. Petroleum.

AIRCRAFT ICING. Aircraft flying in airspace that is also occupied by liquid water drops and droplets at temperatures below 0°C experience ice accumulation on the leading edges and surfaces of their structures. Water can remain in liquid form to temperatures near −40°C although most clouds tend to be composed of ice crystals in the temperature range of −20 to −30°C or lower. The airspace in which aircraft icing is most commonly encountered lies between the earth's surface and 20,000 feet (6000 meters).

There are several factors involved in aircraft icing. The amount of supercooled water in the space swept out by the moving plane determines in major part the intensity of icing, i.e., the rate of accumulation. The speed of the plane is another factor because at high speeds the plane sweeps out more space in less time. Drop and droplet size has a modifying effect on accumulation. Very small droplets tend to follow the airflow around the plane's surfaces and do not impact to freeze. Large drops like rain tend to abandon the airstream, impact on the plane and freeze. A cloud with large numbers of very small droplets therefore may be much less of a hazard than airspace containing many fewer drops of falling rain.

The temperature of the airspace determines in large part the type of icing. When drops and droplets freeze instantly the water material does not have time to assume a crystalline structure and the ice formation tends to be opaque, often granular with occluded air, usually less dense than clear ice, and frequently an irregular surface. This is *rime ice.* On the other hand, when liquid droplets and drops do have time to crystallize, the ice formation tends to be more or less clear, usually near the denisty of ordinary ice and tends to assume the shape of the external aircraft surfaces over which it spreads as it freezes. This is *clear ice.* Temperatures just less than freezing are conducive to clear icing and temperatures well below 0°C are associated with rime icing. See Figs. 1 and 2. Aircraft icing is usually a mix of the

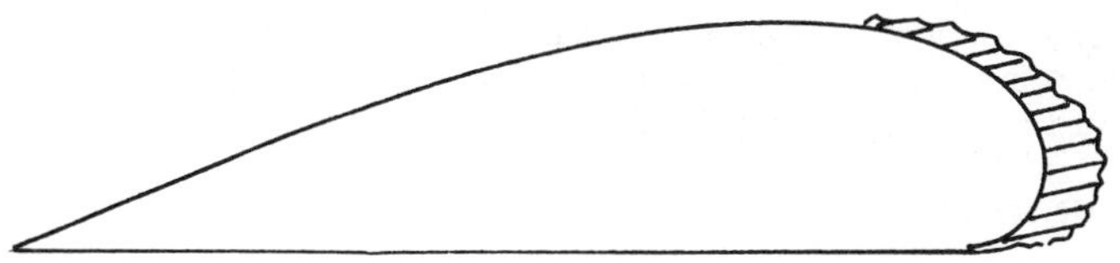

Fig. 1. Rime ice accumulation on the leading edge of a wing tends to be irregular, rough, and a spoiler of the air-flow, but not as heavy as clear ice.

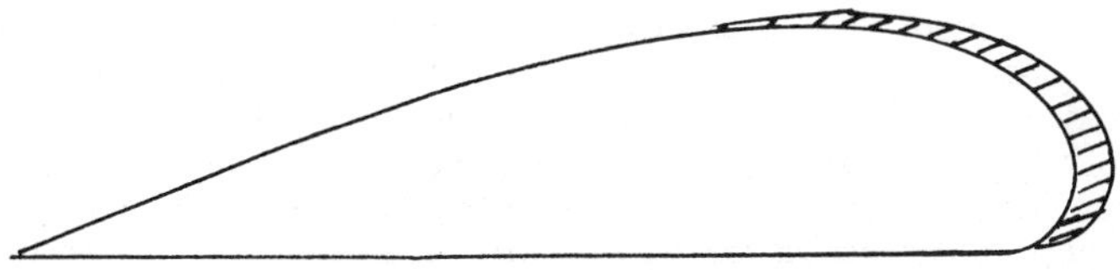

Fig. 2. Clear ice accumulation on the leading edge of a wing tends to be smooth and symmetrical, but heavy.

two types with the proportion of the two related to the temperature.

Clouds most prone to contain an icing hazard are those of cumulo form including billowing stratocumulus. Clouds least prone to contain icing or to contain icing of lesser intensity are stratus. Rain falling from a warm layer of air into a cold lower layer whose temperature is less than 0°C is the most intense of all icing conditions. Meteorological arrangements that produce freezing rain or supercooled rain are predominantly along the cold side of warm fronts and secondarily behind cold fronts.

Aircraft icing at one time in aviation history was a very serious problem, usually in winter, to all aircraft. It continues to be a serious problem for many smaller aircraft. The current generation of transport and military aircraft, however, are more capable of nullifying the presence of icing conditions than past generations. Among other factors are the heated leading edges and surfaces built into the aircraft structure. Warm leading edges tend to evaporate the water material. Also the speed of jet-engined aircraft causes a thermodynamic warm-up of the air immediately above the leading surfaces and edges. Also jet aircraft tend to be operated well above 20,000 feet (6000 meters) where clouds are predominantly composed of ice crystals.

Smaller propeller-driven aircraft, even those equipped with anti-icing devices, tend to be subjected to icing hazards in airspace where supercooled clouds and rain are present.

For references, see entries on **Climate**; and **Meteorology.**

Peter E. Kraght, Certified Consulting Meteorologist, Mabank, Texas.

AIRCRAFT INSTRUMENTS. Airspeed Indicator; Air Traffic Control; Altimetry; Artificial Horizon (Aircraft); Automatic Pilot; Compass (Navigation); Flight Data Recorder; Turn-and-Bank Indicator.

AIR-CUSHION VEHICLE. Ground-Effect Machine.

AIRFLOW (Aircraft). Aerodynamics; Helicopters and V/STOL Craft; Supersonic Aerodynamics.

AIRFOIL. Aerodynamics; Airplane; Helicopters and V/STOL Craft; Supersonic Aerodynamics.

AIR GAP. This term is commonly used in connection with various magnetic circuits and denotes a gap left in the magnetic material. In the construction of various chokes and transformers used in communications circuits a short gap is usually left in the core material to prevent saturation of the magnetic material by the d.c. which often flows in the one or more windings wound on the core material. See **Electromagnet.**

In rotating electrical machinery the rotating part of the magnetic circuit must, of course, be separated by a gap. In these machines this gap is kept as small as consistent with adequate mechanical clearance. In most instances the gap introduces no desirable electrical or magnetic characteristics and necessitates the application of additional electrical energy to overcome its reluctance. The term air gap is sometimes used to denote any spark gap, comprising two electrodes separated by a fluid, commonly air.

AIR HEATER. Boiler.

AIR LIFT PUMP. An air lift is a water pumping method whereby water may be raised from a well through the medium of compressed air. The drop pipe in the well is supplied at the bottom with compressed air from a small air pipe, and the effect of mixture of air and water at the bottom of the drop pipe is to bring water to the surface. This is accomplished either by the water acting as pistons, trapping intermediate layers of air, the expansion of which drives the water pistons to the surface, or it may be accomplished by the mingling of air and water, forming a mixture which is sufficiently lighter than the undisturbed water in the well so that the mixture rises above the surrounding water. In order for this rise to reach the surface the discharge pipe must be submerged in the water of the well an amount varying from 100–300% of the actual lift. The pumping efficiency of the system is very low, but it is very suitable for handling gritty or corrosive waters.

AIR LOCK. An air lock is an air-tight compartment in which the air pressure may be regulated to any desired intensity. When men are required to work in regions where the air pressure is above that of the atmosphere, an air lock must be provided to permit passage of the workmen from the open atmosphere to the pressure region. Thus, in the case of caissons, where workmen must labor under a high enough pressure to equalize the hydrostatic pressure existing at the bottom of the caisson, or in tunnels where flooding is avoided by forcing compressed air into it at sufficiently high pressures to hold the water back, the air lock is a feature essential to the maintenance of pressure during the admission of workmen. Separate air locks are generally provided for materials and spoil removal. In construction work, the air lock is a chamber of sufficient size to hold the number of men that must be accommodated in it at one time. It is provided with well-braced doors having sealing-type edges and tightening locks. The chamber is equipped with valves for admitting and releasing air and with safety devices to prevent excessive pressures endangering the lives of the occupants. The air lock must have two air-tight doors, one leading to the atmosphere, the other leading to the pressure region. These doors open inward so that the pressure in the air lock tends to tighten them against the frame. To enter a pressure region, a caisson for example, the workmen enter the air lock, after which the door leading to the atmosphere is tightly fastened. Compressed air is then slowly admitted until the pressure in the air lock equals that in the caisson, after which the connecting door may be opened without trouble or loss of air from the working chamber. After the workmen enter the caisson, the air lock door is tightly closed, after which the air lock may be opened from the outside without affecting conditions within the caisson.

AIR MASS. A very large bulk of air, having properties (temperature, humidity, thermal structure) that vary only slightly, or vary linearly from point to point within the parcel. Air masses range from about 500–5000 miles (805–8045 kilometers) in lateral dimensions and from several thousand feet to several miles deep. They develop over large, relatively homogeneous, geographical areas where air is stagnant for a sufficient period to acquire the characteristics of that region. These regions are either continental or maritime and are known as air-mass source regions. After an air mass begins to move from its source region, it acquires modifying features characteristic of the surface over which it travels. Modification continues until the air mass loses its identity in the general atmospheric circulation.

A number of systems have been proposed for the classification of air masses, but the Bergeron classification has been the most widely accepted. In this system, air masses are designated first according to the thermal properties of their source regions: tropical (T); polar (P); and less frequently, arctic or antarctic (A). For characterizing the moisture distribution, air masses are distinguished as to continental (c) and maritime (m) source regions. Further classification according to whether the air is cold (k) or warm (w) relative to the surface over which it is moving indicates the low-level stability conditions of the air, the type of modification from below, and is also related to the weather occurring within the air mass. This outline of classification yields the following identifiers for air masses: cTk, cTw, mTk, mTw, cPk, cPw, mPk, mPw, cAk, cAw, mAk, mAw.

Other classification systems introduce further distinction between stable (s) and unstable (u) conditions in upper levels. Some include equatorial (E), monsoon (M), or superior (S) air. Others omit the arctic (A) type and classify all air masses on the basis of polar and tropical air, separated by the polar front.

The major air classification types may be defined as follows:

Tropical air is developed over low latitudes. Maritime tropical air (mT), the principal type, is produced over the tropical and subtropical seas. It is very warm and humid, and is frequently carried poleward on the western flanks of the subtropical highs. Continental tropical air (cT) is produced over subtropical arid regions, and is hot and very dry.

Polar air is developed over high latitudes, especially within the subpolar highs. Continental polar air (cP) has low surface temperature, low moisture content, and, especially in its source regions, great stability in the lower layers. It is shallow in comparison with arctic air. Maritime polar air (mP) initially possesses similar properties to those of continental polar air; but in passing over warmer water, it becomes unstable with a higher moisture content.

Arctic air is developed mostly in winter over arctic surfaces of ice and snow. It is cold aloft and extends to great heights, but the surface temperatures are often higher than those of polar air. For two or three months in summer, arctic air masses are shallow and rapidly lose their characteristics as they move southward.

Continental air is developed over a large land area, and has the basic continental characteristic of relatively low moisture content.

Maritime air is developed over an extensive water surface, and has the basic maritime quality of high moisture content in at least its lower levels.

Equatorial air, according to some classifications, is the air of the doldrums, or the equatorial trough, to be distinguished somewhat vaguely from the tropical air of the trade-wind zones. Tropical air "becomes" equatorial air when the former enters the equatorial zone and stagnates. There is no significant distinction between the physical properties of these two types of air in the lower troposphere.

Superior air is an exceptionally dry mass of air usually found aloft but occasionally reaching the earth's surface during extreme subsidence processes (that is, the descending motion of air in the atmosphere, usually with the implication that the condition extends over a rather broad area). It is often found above tropical maritime air. See other entries listed under **Meteorology.**

For references see entries on **Climate;** and **Meteorology.**

Peter E. Kraght, Certified Consulting Meteorologist, Mabank, Texas.

AIR-MASS SHOWER. Precipitation and Hydrometeors.

AIR-OCEAN INTERFACE. Atmosphere-Ocean Interface.

AIR PARCEL. Atmosphere (Earth); Winds and Air Movement.

AIRPLANE. An aerodynamically designed, winged, powered, heavier-than-air craft intended for sustained flight in the earth's atmosphere. A glider is an engineless airplane. The airplane is part of the total family of aircraft, which includes other heavier-than-air forms, such as helicopters and autogiros, and lighter-than-air forms, such as balloons and airships (blimps, dirigibles, zeppelins).

Various configurations of airplanes serve two major ones: (1) As vehicles for transporting people and goods; and (2) as military weapons systems. Other important uses, although relatively minor in terms of the number of aircraft used, include: (a) As carriers of airborne instruments for collecting geophysical, geodetic, and meteorological information (mapping, surveying, mineral and natural resource exploration, weather observation); (b) as searching tools—for lost persons and craft at sea, in jungles, mountainous terrain, and desert—and for determining the presence of dangerous conditions (forest fires, rising flood waters, hurricane centers, and movement); (c) as airborne inspection and command posts for the supervision and management of geographically sprawling projects (ranching operations, construction projects, such as tunnels, highways, pipelines, commercial fishing operations)—however, these operations are frequently assigned to helicopters; (d) as "dusters" of crops for spreading insecticides and treating chemicals required by certain types of agricultural products; (e) as forest fire fighters, for carrying and dropping water and chemicals; and (f) as subsystems for controlling automotive vehicular traffic and for seeking out and helping in the apprehension of crime suspects—roles also often assigned to helicopters.

Military aircraft fall into three major classifications: (1) For bombing, strafing, and other offensive operations; (2) for fighting enemy aircraft in an effort to attain and maintain control over the air space of a given region—and to intercept enemy bombers and missiles; and (3) for conducting reconnaissance and command operations, including determination of enemy maneuvers and positions, as well as directing land and sea operations from an airborne observation platform. Most other military aircraft are used mainly for transporting people and equipment for numerous purposes.

There are, of course, still other specialized applications for commercial, military, and civil aircraft, not the least of which, in terms of small planes, is their use for recreation and the sport of flying.

With the perfection and refinement of helicopters a few decades ago, some of the former more specialized uses of aircraft have been assumed by the helicopter, with its superior maneuverability and hovering capability as, for example, in rescue missions.

While some valiant efforts were made in the 1800s and even before to obtain some form of sustained flight in a heavier-than-air craft, it is essentially universally accepted that this achievement was first successfully claimed by the Wright brothers of Dayton, Ohio. On December 17, 1903, the Wright brothers made several flights in their wood-and-muslin craft at Kill Devil Hills beach (North Carolina). The first successful flight had a duration of 12 seconds and a distance of 120 feet (36.6 meters). A later flight on the same day was of 59 seconds duration and covered a distance of 852 feet (259.7 meters). Less than two years later (October 5, 1905), in their third major design, the Wright brothers covered a distance of $24\frac{1}{5}$ miles (38.9 kilometers) in a time span of 38 minutes, 3 seconds, at a location called Huffman Prairie, near Dayton.

Although apparently very close to mastering powered flight, Prof. S. P. Langley of the Smithsonian Institution suffered two crashes in his *Aerodrome*, the last crash occurring only nine days prior to the successful flights of the Wright brothers. In the 1890s, the German engineer, Otto Lilienthal, had gained extensive experience with both monoplane and biplane gliders and was at work on motorizing one of his gliders. Unfortunately, he crashed in one of his gliders in 1896.

Propulsion

Following the success of the Wright brothers in developing the first true airplane power plant, engine development progressed rapidly. The Gnome "monosoupape" air-cooled engine of World War I fame was developed in France in 1909. This engine was unique in that the crankcase and cylinders rotated with the propeller about the stationary crankshaft. Good cylinder cooling was obtained by this arrangement, but engine lubrication was a difficult problem. Later in the war, this engine was producing 125 horsepower* at 900 revolutions per minute. Major developments by the British and American designers of that period were in liquid-cooled engines. Two famous engines of this type were the 8-cylinder LX 105, which produced 90 horsepower, and the 12-cylinder Liberty engine, which developed up to 400 horsepower. The latter engine was used in a number of different Allied combat airplanes during the latter part of the war and for several years thereafter. The 8-cylinder liquid-cooled Hispano-Suiza engine also powered many World War I airplanes and was the first to incorporate cast aluminum-alloy cylinder blocks with steel liners.

Advancements in aircraft propulsion diminished considerably for a number of years following the close of World War I. However, in the 1930s, the development of the controllable-pitch propeller was a tremendous advance and opened many new possibilities for airplanes of greater weight and speed. The next milestone was the perfection of the gas-turbine (jet) engine into a useful airplane power plant. By the end of World War II, reciprocating engines were producing as much as 3500 horsepower and weighed less than one pound per horsepower. This was the peak of development for that type of engine. However, it could no longer compete with the gas-turbine engine which offered far more power output for each pound of installed weight.

* One horsepower = 1.014 metric horsepower.

Theory of Propulsion. A dynamic flying machine must be set in motion by a force called thrust, provided either by a propeller, the forward component of the lifting force of rotating helicopter wings, the thrust of gas-turbine engine exhaust, or the thrust of rocket engine exhaust. Gliders and soaring machines are initially set in motion by the thrust provided by a launching or towing device. Continued flight is maintained only by descending into rising air currents.

With an airplane, the thrust force must be equal to total airplane drag to maintain level flight at any given speed or altitude. Since airplane drag increases with speed, greater thrust is required for any increase in speed. Thrust force must also be greater than total drag when the airplane is accelerating, as at take-off, or when the airplane is climbing. Generally, an airplane has reached its maximum speed or altitude when all available thrust is being utilized. However, in some high-altitude jet airplanes, the amount of thrust available to meet take-off acceleration and high-altitude climbing requirements is somewhat greater than that required to obtain maximum permissible speeds at lower altitudes because of airframe structural limitations.

In the case of rocket-powered missiles, the thrust is utilized in a somewhat different manner. A large thrust force is used for a short, initial period to gain a very high speed. After the rocket motor has depleted its fuel, the missile continues on its trajectory by means of the energy imparted to it by the high initial thrust. Once outside the earth's atmosphere, space missiles and satellites will continue in motion indefinitely with no friction forces to absorb their energy, until pulled back again by the earth's gravitational field, or "captured" by some other large body in space.

Propellers. The function of a propeller is the conversion of engine shaft torque (turning force) into thrust. In the early days of propeller design, it was thought that thrust was obtained by the rear face of the propeller blade pushing against the air. Later, it was learned that propeller blades should be considered as small rotating airplane wings with the shape of the airfoil being of primary importance. Therefore, the upper surface of the airfoil, or the front rather than the back of the propeller blade, should be given primary design consideration.

Propellers may be installed in two different ways. In the "tractor" installation, the propeller is mounted forward of the wing. The tractor propeller "pulls" the airplane through the air and the thrust bearings in the engine are designed for thrust loads in this direction. Nearly all propeller-driven airplanes are designed for tractor-type installations because of the unrestricted air inflow to the propeller. As the terminology implies, "pusher-type" propellers are installed aft of the wing and push the airplane forward. The first Wright airplane was of the pusher type. The advantage of this arrangement in multiengined airplanes is that the absence of engine nacelles forward of the wing permits unrestricted airflow over the wing.

Two theories are recognized on the operation of a propeller in converting engine shaft torque to thrust. The momentum theory deals with the energy change to the air mass acted upon by the propeller. The other and most commonly used concept in determining propeller performance is the blade element theory. In this theory (known as the Drzewiecki theory), each propeller blade is considered as being composed of an infinite number of airfoils (called blade elements) joined end to end, forming a shape similar to a twisted airplane wing. Because the propeller is rotating about the engine shaft centerline, each blade element will be rotating in a different arc. The greater the distance of each blade element from the shaft centerline, the larger is the arc it circumscribes and thus the greater the distance it must travel for each revolution. Each blade element, therefore, is moving at a different velocity which is greatest at the blade tip. If each blade element is to operate at maximum L/D (lift/drag), the angle of attack α for each element must diminish at the distance from the hub increases. This gradual decrease in blade angle gives the propeller blade its twisted appearance.

Since the circumference of a circle is $2\pi r$, each blade element rotates through a circular path of $2\pi r$ during each revolution, with r the distance from the center of rotation, expressed in feet or meters. If the propeller is turning at n revolutions per second, the linear velocity of each blade element in its individual plane of rotation must be $2\pi rn$ feet (meters) per second. In flight, the propeller is moving forward at the same speed as the airplane, as well as rotating about its own axis. This forward velocity must also be expressed in feet (meters) per second to keep all terms of the equation in common dimensions. The total velocity of each blade element is the resultant of the two component velocities, the forward velocity V and the rotational velocity $2\pi rn$, or $2\pi rnV$. This is shown graphically in Fig. 1 for each blade element.

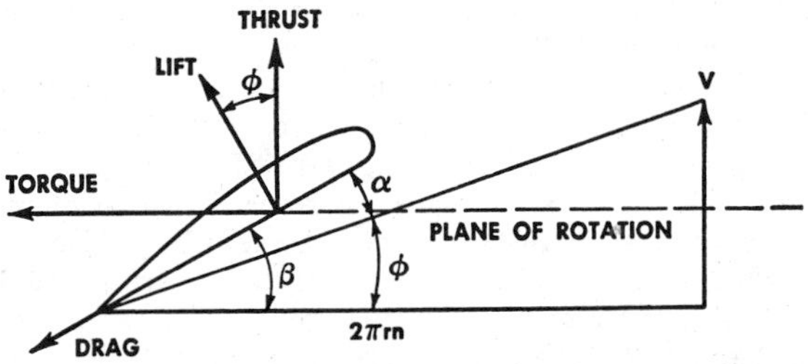

Fig. 1. Vector diagram of propeller blade element.

As shown on the diagram, the angle of attack α is equal to the fixed blade setting angle β, minus the angle ϕ which is determined by airplane velocity V and propeller rotational velocity $2\pi rn$ (or πdn since diameter = $2r$).

The flow of air about the propeller blade creates lift and drag forces in the same manner as airflow about an airplane wing. The resultant of these forces can be divided into two components, the thrust component acting in the direction of the propeller axis and the torque component. The torque component acting in the plane of rotation resists the operation of the propeller and is the component that must be overcome by the engine shaft torque. Similar to the airplane wing, the magnitude of the thrust and torque components depends upon angle of attack and velocity.

The efficiency of a propeller in converting engine power to thrust is obviously of tremendous importance to airplane performance. The efficiency of a propeller during cruising flight determines the range of an airplane equally as much as the efficiency of the engine itself in converting fuel energy to shaft power. Propeller efficiency is expressed as the ratio of thrust power delivered to engine power required to turn the propeller. Thrust power is thrust force (pounds) multiplied by airplane velocity V (feet per second), or TV (foot-pound/second). The power required to turn the propeller may also be expressed in foot-pound/second by multiplying engine shaft horsepower by 550 since one horsepower is 550 foot-pounds/second.* The airspeed at which maximum efficiency occurs is called the design airspeed for the propeller.

Employing fixed-pitch propellers in high-performance airplane design presents numerous problems. If a wing airfoil section is chosen for a transport airplane that will result in maximum L/D at high cruising speeds, a propeller of relatively high angle must also be chosen to yield good efficiency at high cruising speed. However, this propeller will give poor high-speed performance as well as inferior performance at take-off and climb. A high-speed propeller chosen for a high-speed airplane will have reduced efficiency during cruising and give less than optimum range for the airplane, in addition to reduced climb and take-off performance.

The advent of the controllable-pitch propeller offered a solution to these difficulties. Provided with a means for varying pitch angle, the pilot could select the optimum angles for take-off, climb, efficient cruising, or high-speed flight. It is notable to observe that modern propeller airplane performance essentially commenced with the development of the controllable-pitch propeller. Fixed-pitch propellers are used only on small, light airplanes where cost and lightness of installation are major considerations. The speed range of such airplanes is limited by power output of the engine as well as the propeller.

The velocity of the blade elements has little effect on efficiency until velocities near the speed of sound are approached. When the velocity of the blade elements close to the tip approaches the speed of sound, the effect of compressibility increases power input requirements at no increase in thrust, thereby reducing efficiency. Since compressibility of an airfoil is affected by airfoil thickness ratio, propeller tips are built as thin as possible. Speed of sound is proportional to temperature and, normally, tip speed is limited to 800 to 1,000 feet per second (244 to 305 meters/second). To prevent tip speeds from exceeding these limitations, high-powered engines, requiring propellers of large diameter, are equipped with reduction gears. This arrangement permits lower propeller rpm than the engine rpm required for efficient engine power output.

In addition to providing efficient thrust during cruising conditions, the propeller must be able to convert total engine power to thrust for take-off. This is necessary to provide rapid acceleration to flying speed and keep take-off distances within reasonable limits.

Some important propeller terms include:

(*Fixed-pitch Propeller*)—A propeller with blade angle fixed.

(*Adjustable-pitch Propeller*)—A propeller on which blade angles can be adjusted on the ground with propellers not turning. This is especially convenient for a take-off on a hot day from a high-altitude airport.

(*Controllable-pitch Propeller*)—A propeller on which the blade angle may be changed in flight by controls in the cockpit.

(*Constant-speed Propeller*)—A controllable-pitch propeller with a speed governor which maintains selected rpm constant by automatically changing blade angle regardless of airspeed or engine power. Most propellers of this type also provide for full "feathering" of the blades in flight to prevent "autorotation" or "windmilling" of the propeller, thus reducing drag when the engine is dead.

(*Feathering*)—Mechanically increasing pitch angle until the blade is turned approximately to the direction of flight. Feathering is used most frequently on multiengine airplanes.

(*Reverse Thrust*)—Thrust in the direction opposite to flight. Nearly all multiengine airplanes utilize constant-speed, full-feathering, reversible-pitch propellers. The pitch change mechanism is designed to permit the pilot to select negative blade angles immediately after landing, thus creating reverse thrust for rapid airplane deceleration. This feature saves brakes on heavy airplanes and permits safe landings on icy runways where wheel braking is quite ineffective.

(*Dual Rotation*)—Two propellers on the same engine shaft and operating in opposite direction. One propeller shaft rotates inside the other while both are driven by reduction gears in the engine nose case.

(*Windmilling Drag*)—Perhaps the most undesirable action in the use of propellers as a thrust-producing device is the drag that results from propeller windmilling. If engine power is lost in flight, the airflow over the propeller blades will cause the propellers to rotate and, in turn, rotate the engine. When the blade angle is great enough to be commensurate with the flying speed, propeller efficiency will remain high and the drag will not be appreciably greater than engine load. However, if a failure occurs in the pitch-change mechanism and the blade angle is reduced, excessively high rotational speeds may result. The low blade angle and the resultant negative angle of attack of the blade airfoil element cause a tremendous reduction in propeller efficiency. This means the load required to turn the dead engine is supplied by an inefficient propeller, and the resulting windmilling drag can become dangerously high. In such instances, it is necessary to reduce airspeed or altitude to reduce propeller rotational speed. However, unless actual failure occurs to the pitch-change mechanism, the propeller speed governor will maintain rpm as it was before engine failure, and the feathering system will feather the propeller to stop windmilling rotation altogether. See also **Aerodynamics;** and **Helicopters and V/STOL Craft.**

Jet Propulsion. The jet or rocket engine moves the airplane or missile forward by reacting to the momentum of a mass of air or gaseous matter accelerating out the rear (nozzle) of the engine. The

* One metric horsepower = 542.5 foot-pounds/second.

recoil of a rifle and the reactive thrust of a fire hose are typical examples of this principle. The thrust force is not the result of the jet gases pushing against the air behind, but instead is the reaction to the momentum of the escaping gases. This fact is demonstrated by rocket motors, which produce full thrust in outer space beyond the atmosphere.

Power Measurement. Jet thrust force is measured in pounds as is the thrust force of a propeller. The magnitude of jet gross thrust force is determined by the mass flow of air or gaseous matter multiplied by the velocity with which it escapes from the nozzle. This may be expressed by

$$F = MV_J \quad (1)$$

where F = thrust force, pounds
M = mass flow rate = W/g
V_J = jet velocity, feet per second
W = weight flow, pounds/second
g = gravitational acceleration + 32.2 feet/(second)^2 [9.8 meters/(second)^2]

Because aerodynamic equations require thrust to be expressed as a force, to equate thrust requirements against drag and acceleration forces, there usually is no reason to express jet engine output in terms of horsepower. But, where the equivalent is desired, jet engine "power" can be determined by multiplying thrust force times flying speed in feet per second and dividing by the horsepower constant.

$$\text{Horsepower} = \frac{(\text{Jet Thrust}) \times (\text{True Airspeed}) \times 1.467}{550} \quad (2)$$

Therefore, when flying speed is 550 feet per second (375 miles (603 kilometers) per hour), the jet thrust and jet "power" are the same. At speeds below this value, jet "power" is less than jet thrust in pounds. In the case of the turboprop engine, where both propeller thrust and jet thrust are utilized, it is sometimes desirable to convert jet thrust into horsepower. By adding this value to the horsepower rating of the engine, the total propulsive output of the engine can be expressed as "equivalent shaft horsepower" (eshp). Jet horsepower must be multiplied by propeller efficiency to convert to shaft horsepower. Assuming a propeller efficiency of 80%,

$$\text{eshp} = \frac{(\text{Jet Thrust}) \times (\text{True Airspeed}) \times 1.467 \times 0.80}{550} + \text{Shaft Horsepower} \quad (3)$$

Propulsive Efficiency. The efficiency with which a propulsive system is utilized in propelling the airplane or missile is called propulsive efficiency. A thorough understanding and investigation of propulsive efficiency are necessary for successful design when comparing one propulsion system with another. This is especially true when comparing propeller propulsion with jet propulsion.

Propulsive efficiency may be expressed by

$$\text{Efficiency} = 2V_a/(V_J + V_a) \quad (4)$$

where V_a = airplane velocity, feet per second
V_J = jet velocity (or propeller slipstream velocity), feet per second

In the case of propeller-driven airplanes, propulsive efficiency very closely approximates propeller efficiency because the propeller moves a large mass of air at relatively slow speeds. During cruising conditions, the air behind the propeller is moving at approximately the same speed as the airplane; hence, V_a and V_J are approximately equal. In contrast, the thrust-producing gases escaping from the jet engine nozzle are extremely high in velocity, relatively smaller in mass flow, and cannot be varied extensively in flight.

Because it is difficult to design a pure jet engine that would have satisfactory fuel economy and light weight without high jet velocity, jet-powered airplanes are generally designed for high-speed flight to obtain maximum propulsive efficiency.

High-altitude operation must be considered in propulsive efficiency analysis. Because of compressibility effects on the propeller airfoil, propeller efficiency decreases with both airspeed and altitude. This factor limits propellers to approximately 500 to 600 miles (805 to 965 kilometers) per hour true airspeed and 40,000 to 45,000 feet (12,192 to 13,716 meters) altitude. In contrast, since jet thrust is accomplished at high jet velocities, the higher the airplane speed, the greater is the propulsive efficiency. A comparison of propulsive efficiency for the various propulsion methods is given in Fig. 2.

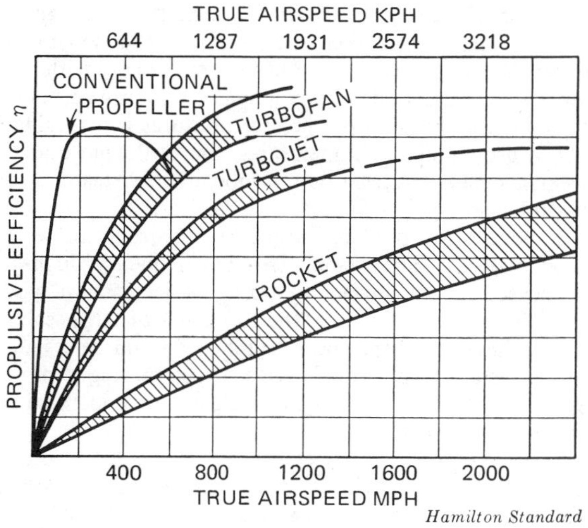

Fig. 2. Comparison of propulsive efficiency.

Turbojet Engines. The principle of jet propulsion was recognized early in aviation development as possibly the best method for obtaining greater speeds. The practical application of the principle proved difficult, however, because of the efficiency of early engine designs. Early developments centered around the turbojet and turboprop engines, patterned to some extent upon developments in stationary gas-turbine and steam-turbine engines. In 1930, Frank A. Whittle was granted an English patent for a turbojet engine design, but it was not until 1937 that intensive work was commenced on his design in England. The Germans initiated turbojet development in 1936 along with other jet engine and rocket motor programs. This work was carried on in an intensified manner, and the first turbojet-powered airplane flight took place in Germany in 1939. The airplane was a Heinkel HE178 powered by a Heinkel HE53B engine. Meanwhile, the Whittle engine program progressed rapidly in England. In 1941, the fourth Whittle engine, Model W1 delivering 855 pounds of thrust, propelled a Gloster E-28 airplane to 339 miles per hour at 20,000 feet. In this same year, the General Electric Company agreed to develop turbojet engines for the U.S. Army Air Forces from plans of the Whittle W2-B engine. The first American flight took place in 1942 when a Bell Aircraft P-59, powered by two General Electric IA engines of 1,300 pounds of thrust each, flew successfully. By 1944, production was started on the 4,000-pound thrust J-33 engine for the P-80 airplane. During 1944, the first axial-flow compressor turbojet engine was successfully tested by General Electric.

The three basic parts of the turbojet engine are: (1) Air compressor; (2) combustion chambers; and (3) a turbine wheel(s) for driving the compressor. The compressor in the forward part of the engine draws in large quantities of air, compresses it to high pressure, and forces it through the combustion chambers. Fuel is forced through spray nozzles into the combustion chambers where it is mixed with part of the air. This mixture is then burned in a continuous combustion process and produces a very high temperature (4000°F; ~2205°C or higher), which heats the entire air mass to 1600 to 2000°F (871 to 1903°C) or higher. The hot gases then pass through the turbine wheel(s) directly behind the combustion chambers. Part of the energy in the hot gases is used to rotate the turbine wheel, which in turn drives the compressor by means of a direct shaft. The hot gases then expand and blast out of the tailpipe nozzle at high velocity. The reaction to the momentum of the escaping gases is the thrust force which propels the airplane.

The efficiency with which the three basic parts of the turbojet engine operate, particularly the turbine wheel(s) and compressor, is very important to thrust output. Since approximately two-thirds of the energy available in the hot gases is absorbed by the turbine wheel to drive the compressor, the need for high efficiency is obvious. Early models failed because of inefficient turbine and compressor design, coupled with inferior materials of construction. Since thermodynamic laws reveal that the energy available in hot gases increases with both temperature and pressure, design development constantly strives for compressors that will provide even greater pressures and turbine wheels that will withstand higher temperatures. One of the most outstanding improvements made has been that of more effective materials for turbine-wheel blades.

Special nozzles are used to reduce the disturbing noise intensity of the jet blast during take-off. This low-frequency, high-intensity noise is caused by the turbulent mixing of jet exhaust gases with the atmosphere. Nozzles reduce the formation of large-scale eddies, lowering the intensity of the lower "rumbling" frequencies and changing the character of the sound. Also, by permitting a more rapid mixing of the jet gases with surrounding air, they increase the rate of sound-energy dissipation.

Thrust reversal to assist wheel braking during the landing roll is accomplished by stopping the aft flow of the jet and directing it partially forward in the direction of airplane movement. To prevent the hot gases from entering the compressor inlet, of the same or adjacent engines, the reverser directs the jet blast outward at a suitable angle from the nacelle. Because of this angle, and losses due to turning the path of the gases, thrust available in the reverse direction is about 40% of forward thrust for the same engine rpm.

The magnitude of thrust output of a turbojet engine is determined by the mass airflow through the engine and the velocity of the mass leaving the jet nozzle. This is expressed by previously given equation (1). That equation represents the thrust output when the engine and airplane are at rest. When the airplane and engine are in motion, air is being rammed into the engine compressor inlet, tending to retard the forward motion. This retarding force is called *ram drag* and its magnitude is determined by the mass air flow through the engine and the airplane velocity:

$$\text{Ram Drag} = MV_a \quad (5)$$

The net thrust for the engine in motion is, therefore, the *gross* thrust output minus the ram drag:

$$F = MV_J - MV_a \quad (6)$$

or

$$F = M(V_J - V_a) \quad (7)$$

Thus, the net thrust is equal to the product of mass airflow and the change in velocity through the engine. Actually, the mass flow of gases from the nozzle is increased by the amount of fuel added for combustion. However, since the airflow is about seventy times greater than fuel flow, the mass of the added fuel is neglected in basic equations.

Augmentation. This term applies to means for increasing thrust during take-off in addition to that delivered by the basic engine. *Afterburning* implies that fuel is added to the hot gases in the tailpipe after they have passed through the turbine wheel. The burning of this additional fuel greatly increases the gas temperature and correspondingly the jet velocity, resulting in as much as 50% increase in thrust. Because the hot gases pass through the tailpipe at high velocity, it is necessary to introduce the fuel through flameholders in the tailpipe similar in some respects to the flameholders used in the combustion burners. The increase in gas temperature from afterburning causes the gases to expand, which tends to increase tailpipe pressure and turbine wheel temperatures. *Water injection* can produce about 25% additional thrust. A desirable feature of this method is the small increase in engine weight over the basic engine as compared with considerable weight increase for afterburning. Water injection equipment causes little change in normal engine performance, whereas afterburning causes a slight increase in fuel flow during normal (afterburner off) operation because of the blocking effect of the flameholders in the tailpipe. Water is introduced into the compressor or is forced through nozzles into the combustion chamber, or both. When introduced into the compressor, the water cools the air by evaporation, permitting the compressor to deliver more mass flow of air at maximum rpm. To offset the additional airflow, a signal is sent to the fuel control, increasing fuel flow slightly to heat the additional air. Thrust is increased by the increased mass airflow, plus the mass of the added water and fuel. When water is introduced into the combustion chambers (or diffuser section), the resulting increase in compressor airflow is less than by direct compressor injection. However, large amounts of water injected into the combustion chambers, plus the slight increase in compressor airflow, result in a substantial gain in thrust. In the combustion-chamber injection method, alcohol may be mixed with the water to assist in heating the added mass rather than increasing fuel flow. The injection water must be very pure to prevent mineral deposits in the machinery.

Turbojet engines can be classified by the type of compressor used. The *centrifugal compressor* was used in early engines and in some engines today. In the centrifugal compressor, air enters the rotating impeller at the center through the inlet guide vanes. Rotating at high speed, the impeller compresses the air by centrifugal action. The air passes from the rim of the impeller through a diffuser, which retards the tendency of the air to rotate with the impeller, into the combustion chambers. The *axial-flow compressor* is used on large turbojet and turboprop engines. An advantage is the small diameter of the compressor. Many stages can be added together without increasing the diameter of the engine. See also **Gas and Expansion Turbines.**

Turbofan Engines. These engines were developed to provide greater propulsive efficiency than obtainable with conventional turbojets. This is accomplished by increasing the total compressor airflow and reducing the net velocity of the jet. The additional airflow does not pass through the combustion chambers and turbine wheels, but is directed aft by way of a fan outside the engine cases. In one engine design, about 25% of the air from the forward compressor is ducted outside the engine case and is mixed with the hot gases before passing through the nozzle. This type of engine is called "bypass" by the British. Bypass ratios permit from 30 to 60% or more of the propulsive force to be produced by the fan. In comparison, turboprop engine designs generally permit an energy distribution of 90% for the propeller and 10% for the jet nozzle. A basic difference between the turbofan and the turboprop is that the airflow through the turbofan is controlled by design so that the air velocity relative to the fan blades is unaffected by airspeed of the airplane. This eliminates the efficiency loss at high airspeeds which limits airspeed capability of the turboprop engine.

Turboprop Engines. Turboprop engine development began in 1926 when the English scientist Griffith proposed an axial-flow compressor and turbine engine for driving a propeller. After wind tunnel tests, development was discontinued because of the economic depression and scientific reluctance to accept the engine as having any practical significance. Following the successful development of the turbojet engine, interest was again directed toward the turboprop engine and successful flights in England and the United States were made with turboprop installations in 1945. For several years following these flights, the majority of effort was devoted to improving turbojet design in the United States and turboprop development lagged. More emphasis was placed on turboprop development in England and France during this period and the turboprop-powered English Viscount 630 was placed in commercial passenger service in 1950.

Basically, the turboprop engine is a turbojet with most of the heat energy converted to shaft power by the turbine wheels, leaving very little for jet reaction. The basic parts of the engine compressor, combustion chambers, and turbine wheels are the same in both engines, except that larger or a greater number of turbine wheels are used. In most turboprop engines, the compressor rotor shaft extends forward and drives the propeller shaft through a system of reduction gears.

A significant difference exists between turboprop and turbojet engine controls. In the turbojet engine, fuel is metered by the control as required in proportion to airflow to maintain constant rpm, or is metered in the proper amount to effect changes from one rpm to another. The control must be designed to meter fuel within precise limitations to prevent overheating of the turbine wheels, compressor surge, or burner blowout. The control must also correct for varying conditions of altitude, air temperature, and airspeed for a given power lever position.

All of the control requirements of the turbojet are present in the turboprop engine with additional requirement imposed by the propeller. By means of the blade pitch-angle control, the speed and power absorption of the propeller can be varied independently of the normal engine control. In addition, the rather large inertia (resistance to rapid rpm change) of the propeller and direct-connected engine rotor further complicates the control requirements. To provide the pilot with a simple single-lever control, as used for the turbojet, the propeller control of the turboprop is integrated with the engine fuel control. With most turboprop controls, the engine is operated within a narrow range in the high-rpm region for all flight conditions. This operation minimizes the effect of propeller inertia on power changes.

Ramjet and Pulsejet Engines. The ramjet or athodyd (aerodynamic-thermodynamic-duct) was invented by Lorin in France in 1913. The pulsejet, or resonant-type jet propulsion motor, was invented in 1910, also in France. The only extensive use of the pulsejet engine was made by the Germans during World War II when it was used to propel the V-1 "buzz bomb." Intensive development of the ramjet engine principle was begun in the United States following World War II for use with long-range, high-altitude guided missiles.

Pulsejet Engines. Because of its inefficiency and principle of operation, the pulsejet engine cannot compete with the turbojet or ramjet engines and finds no practical applications. With reference to Fig 3,

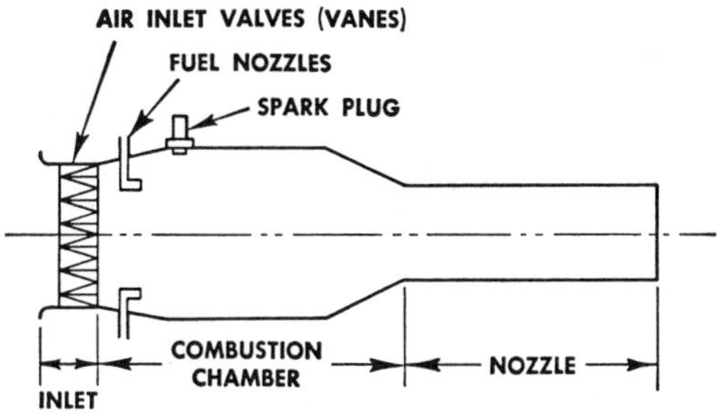

Fig. 3. Pulsejet engine.

the pulsejet engine consists simply of a combustion chamber with an air intake, incorporating check valves at the forward end, and a nozzle for expelling the gases at the aft end. Forward velocity is required in order to ram air into the combustion chamber to start the combustion process. With air pressure in the combustion chamber, fuel is added and the mixture is ignited by a spark plug. The resulting pressure in the chamber closes the check valves in the air inlet and forces the hot gases out of the nozzle. The velocity energy of the gases passing out of the nozzle reduces the pressure in the combustion chamber below the atmospheric level, causing the air inlet valves to open and admit a fresh charge of air. The combustion process is then repeated and, after a few cycles, is self-sustaining and requires no further spark ignition. For the engine to operate properly, it is necessary that the vibratory resonance of the air inlet valves be "tuned" with the impulses of the gases and sound waves escaping from the nozzle. This is accomplished by shortening or lengthening the tailpipe until resonance is established for the desired flight velocity. The V-1 engine used a grill of thin metal reeds, similar to reeds of a harmonica, for the air-inlet valves. The resonant frequency of explosions for the V-1 engine was approximately 200 to 300 Hz, creating the buzzing sound for which the missile was named.

Ramjet Engines. Mechanically, the ramjet engine represents the simplest air-breathing jet engine, since it contains no moving parts. The principal parts are shown in Fig. 4. As with the pulsejet engine, forward velocity is required for ramjet operation to induce air into the combustion chamber, or "ram" the air in, as the name implies.

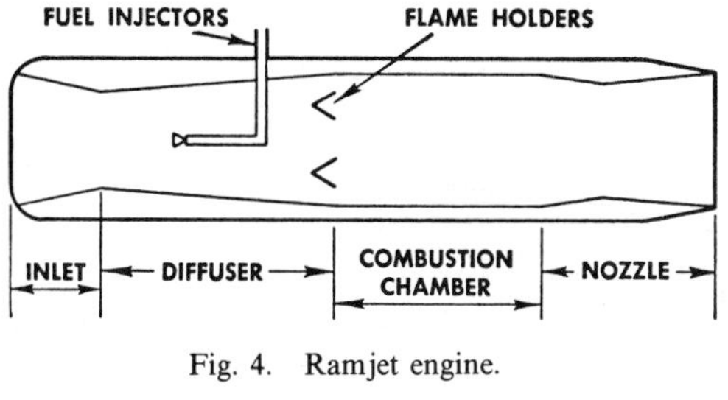

Fig. 4. Ramjet engine.

The forward velocity of the engine rams air into the inlet and thence into the diffuser section where the air velocity is slowed down by the divergent walls of the diffuser. In the combustion chamber, heat is added by the combustion of kerosine or other jet engine fuels which tends to expand the gases and increase their pressure. Because the entering ram air and the diffuser prevent expansion of the hot gases in a forward direction, the gases accelerate out the nozzle at a higher velocity than the velocity of the entering air. The reaction to the rearward velocity of the gases produces forward thrust.

The diffuser section is important in the functional operation of the engine. The outward taper must be carefully designed to permit proper slowing of the ram air with a minimum of losses. The shape of the air inlet is equally important, as is the design of the nozzle. In ramjet engines, combustion is continuous after having been started by spark ignition or other methods. Because of the rather high velocity in the combustion chamber (300 to 500 feet (91 to 152 meters) per second), flameholders are required, as in the combustion chambers of turbojet engines.

For long-range missile applications, provided the missile remains at an altitude where there is sufficient oxygen, the ramjet engine has the advantages of light weight and no requirement for oxidizing chemicals (also keeping the total weight down). These advantages also spell out low initial cost.

Aircraft Types

High-Performance Aircraft. Descriptions and specifications of modern aircraft are well beyond the scope of this encyclopedia. Several of the references listed at the end of this entry contain such information. Two fighter aircraft are described and illustrated to convey an appreciation for what is involved in the design and construction of aircraft of very high performance.

The F/A-18 Hornet Strike Fighter. The first flight of the F-18A *Hornet*, shown in Fig. 5, occurred in November 1978. This is a single-

Fig. 5. The F/A-18 *Hornet* strike fighter. (*McDonnell Aircraft Company; McDonnell Douglas Corporation*)

seat, twin-turbofan aircraft for fighter and attack missions. The power plant consists of two General Electric F404-GE-400 low bypass turbofan engines, each in the 16,000-pound thrust class, with a thrust-to-weight ratio of 8 to 1. The aircraft is 56 feet (17.1 meters) long, 15.3 feet (4.7 meters) high, with a wingspan of 37.5 feet (11.4 meters), and a wing area of 400 square feet (37.2 square meters). Speed of the aircraft is Mach 1.8+ (Mach 1+ at intermediate power). There is a crew of one except two in the trainer version. Range for combat is 400+ miles (644 kilometers) for combat missions and 2000 miles (3218 kilometers) or more ferry range. Combat ceiling exceeds 50,000 feet (15,240 meters). Hornet pilots have a full 360-degree visibility. The *Hornet* carries 11,000 pounds (4990 kilograms) of fuel internal, and 16,000 pounds (7258 kilograms) with external tanks. Take-off weight is approximately 35,000 pounds (15,876 kilograms). There are up to 19,000 pounds (8618 kilograms) of armament on line stations—two wing-tip for Sidewinder heat-seeking missiles; two outboard wing for air-to-ground ordnance; two inbroad wing for Sparrow radar-guided missiles, air-to-ground, or fuel tanks; two nacelle fuselage for

Sparrow missiles or sensor pods; one centerline for weapons, sensor pods, or tank. Internal 20-millimeter cannon is mounted in the nose.

The F-18 and A-18 are identical aircraft in all respects, including software. When delivered to a fighter squadron, the *Hornet* designated an F-18 is fitted with fuselage-mounted Sparrow launchers and three store pylons. When delivered to an attack squadron, the *Hornet* designated an A-18 is fitted with fuselage-mounted pod adapters, forward-looking infrared and laser spot tracker/strike camera pods and five store pylons. At the squadron level, an F-18 can be reconfigured to an A-18 in less than 1 hour and vice versa.

Fig. 6. The F-15 *Eagle* fighter. (*McDonnell Aircraft Company; McDonnell Douglas Corporation*)

The F-15 Eagle. The first flight of the F-15 *Eagle*, shown in Fig. 6, occurred in late July 1972. In the F-15A and F-15C, there is one pilot; two pilots in the F-15B and F-15D. The powerplant consists of two Pratt & Whitney F100-PW-100 turbofan engines, each developing ~25,000 pounds (11,340 kilograms) of thrust. The aircraft is 64 feet (19.4 meters) long; 18.5 feet (5.6 meters) high; with a wingspan of 43 feet (13 meters). Speed of the aircraft is Mach 2.5+. Maximum duration of flight is 10.5 hours with refueling; 5.25 hours unrefueled. The *Eagle* carries four *AIM-9* Sidewinders; four AIM-7 Advanced Sparrows; 940 rounds of 20-millimeter ammunition for the General Electric M-16A1 six-barrel gun. There are five weapons stations capable of carrying up to 12,000 pounds (5443 kilograms) of munitions

Fig. 7. The F-15 *Eagle* fighter showing complement of armament. (*McDonnell Aircraft Company; McDonnell Douglas Corporation*)

or additional electronic countermeasures gear in addition to air-to-air weaponry. See Fig. 7.

Historical Aircraft. *Early History.* The following paragraphs are included here for those readers who may be interested in the early days of aircraft. Difficult to comprehend for all but the older readers, the early days of flying were indeed thrilling times and the phrase "air minded" was very popular in that era preceding the late 1930s. There was much competition in terms of aircraft performance and pilots. Unquestionably, the major meets and well-funded competitions, such as the Gordon Bennett Cup, the Schneider Cup, the Pulitzer Trophy, the Thompson Trophy, and the Bendix Trophy greatly catalyzed the development of superior airplanes. The progress in airplane design also was grossly assisted during the two major World Wars. It is sometimes difficult to realize, in perspective, that modern massive movements of people and cargo by air all have occurred within the life span of many people still living. Progress made during the early period of 1906 to 1914 is shown in Fig. 8. The speed records of the major competitions held during the period 1909 to 1939 are plotted in Fig. 9. The Reims (France) Meet of 1909 was the first extensive meet in aircraft history. The Gordon Bennett meets, commenced in 1909, were the first airplane speed races. The Schneider Cup meets were designed to promote the advancement of seaplanes and flying-

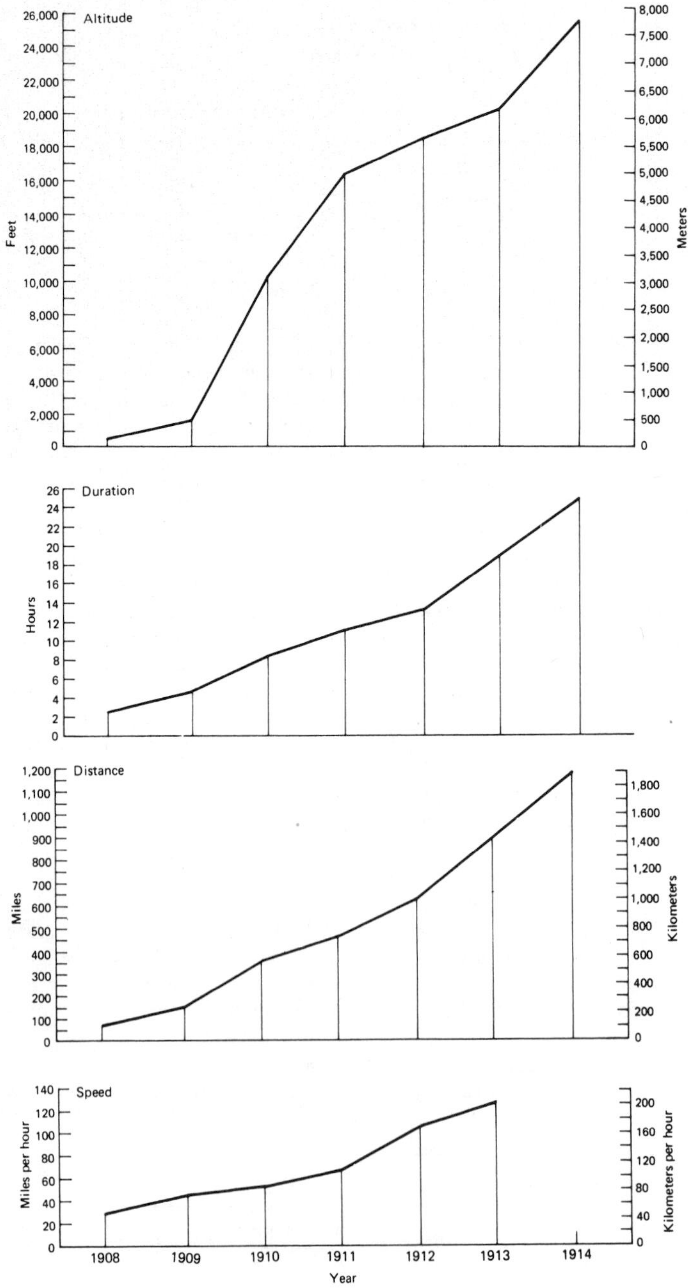

Fig. 8. Records set by early airplanes during period (1908–1914).

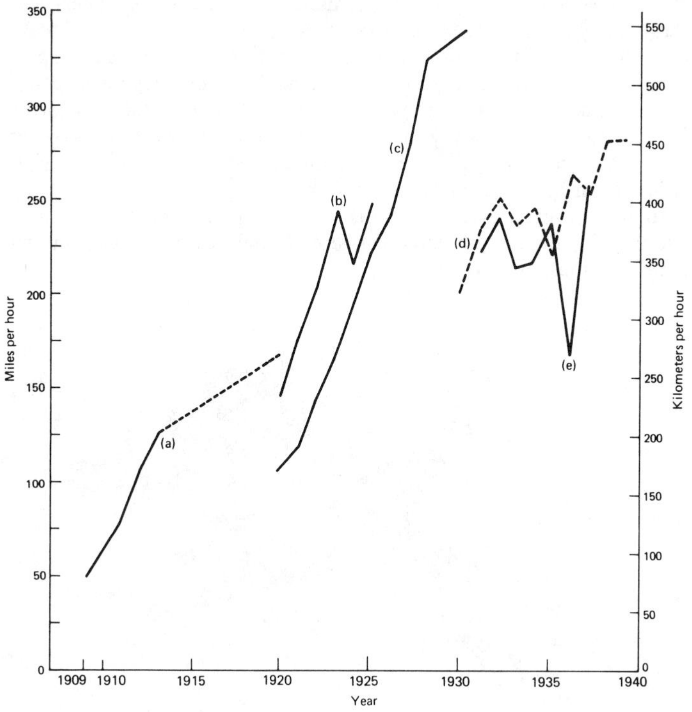

Fig. 9. Speed records of the major trophy races held during period (1909–1939): (a) The Gordon Bennett Cup; (b) The Schneider Cup; (c) The Pulitzer Trophy; (d) The Thompson Trophy; (e) The Bendix Trophy.

boats and were confined to that category of aircraft. A primary incentive of the Pulitzer Trophy races was that of promoting the design of more effective fighter aircraft for the military. These meets were abandoned when U.S. General Mitchell was court-martialled. The Thompson Trophy races were sponsored by Cleveland industrialist, Charles E. Thompson, and were geared to speed. The Bendix Trophy races were of a different nature, namely, cross-country flights from Burbank, California to Cleveland, Ohio, and thus the speeds represented an average over long straight distances. These races highlighted the role of women as competent pilots, the 1938 race having been won by Jacqueline Cochran.

There was essentially no aircraft industry at the time of World War I in the United States, with just a little over 50 planes produced in 1916. The war, however, hastened design and production activities even though few American planes were used during the war. American pilots mainly flew French aircraft. The French introduced nearly 20 new planes during the period between 1914 and 1917. The fastest of these was the Spad S XIII, with a speed of 130 miles (209 kilometers) per hour and a ceiling of 22,300 feet (6797 meters). It had a 200-horsepower Hispano-Suiza engine and a wingspan of just a little over 26 feet (7.9 meters). The British also introduced nearly 20 planes during the same period. The S.E.5a was the fastest plane of that period, reaching a top speed of 138 miles (222 kilometers) per hour. It was powered by a 200-horsepower Wolseley Viper engine, had a wingspan of nearly 27 feet (8.2 meters), a range of about 340 miles (547 kilometers), and a ceiling of 19,500 feet (5944 meters).

Although Italy only introduced seven aircraft during the 1915–1917 period, it was the first country to employ aircraft in warfare. The fastest of the Italian planes was the Ansaldo S.V.A-5, with a speed of 143 miles (230 kilometers) per hour, powered by a 220-horsepower S.P.A. 6A engine, a wingspan of 32 feet (9.8 meters), and a ceiling of 22,000 feet (6706 meters). The fastest plane produced by the Germans was the Hansa-Brandenburg D1, with a speed of 116 miles (187 kilometers) per hour, a wingspan of a little over 26 feet (7.9 meters), powered by a 160-horsepower Austro-Daimler engine. It had a ceiling of 16,400 feet (4999 meters). Probably the most famous of the German World War I planes were those in the Albatros series, most of which were powered by Mercedes engines. The first airplane to have a machine gun synchronized with the propeller was the Fokker E-III, introduced in 1915. Other famous names in production in Germany during the war period included the Etrich A-11 Taube, introduced in 1910 and extensively used for pilot training; the Fokker E-III and D-VII; the LVG C-1, the Rum per C-1; the Lloyd C-II (Austrian); the Pfalz D-III; the Hannover CL-IIIa; the Gotah G-VIII; the Friedrichshafen G-III (a heavy bomber). The Zeppelin R-VI (not to be confused with a dirigible) was also introduced during that period.

Airplanes of World War II. A representative few of the famous airplanes of World War II are described briefly in the accompanying table. Airplane production during the period just prior and carrying through the war years numbered in the tens of thousands, notably in the United States, Great Britain, Germany, Russia, and Japan. It is interesting to note that concentration on production of conventional propeller airplanes essentially delayed final development and refinement of jet aircraft during that period. The only true jet aircraft to see action during World War II were the British-built Gloster Meteor

REPRESENTATIVE MILITARY AIRPLANES USED IN WORLD WAR II.

		Wing Spread		Weight Empty		Cruising Speed		Range		Ceiling		
Designation (Nationality) (Date Introduced)(Type)	Horsepower	Feet	Meters	Pounds	Kilograms	Miles/Hour	Kilometers/Hour	Miles	Kilometers	Feet	Meters	Number in Crew
Vultee BT-13A Valiant (USA) (1940) (trainer)	450	42	13	3,375	1,530	182	293	725	1,167	21,650	6,600	2
Northrop P-61 Black Widow (USA) (1943) (night fighter)	4,000	66	20	22,000	9,980	366*	589*	2,500	4,025	33,100	10,090	3
Republic P-47 Thunderbolt (USA) (1943) (fighter)	2,300	41	12	10,000	4,535	428*	689*	475	765	42,000	12,800	1
North American P-51D Mustang (USA) (1942) (fighter)	1,490	37	11	7,125	3,230	362	583	950	1,530	41,900	12,770	1
Grumman F6F Hellcat (USA) (1942) (carrier-based fighter)	2,000	42	13	9,040	4,100	268	431	1,090	1,754	37,300	11,370	1
Martin B-26G Marauder (USA) (1942) (light bomber)	4,000	71	22	23,800	10,800	216	348	1,100	1,770	19,800	6,050	7
Boeing B-17F Flying Fortress (USA) (1941) (bomber)	4,800	104	32	32,250	14,630	210	338	2,400	3,860	36,600	11,155	10
Boeing B-29A Superfortress (USA) (1943) (bomber)	8,800	141	43	71,360	32,370	230	370	3,250	5,230	31,850	9,710	10
Douglas C-53 Sky Trooper (USA) (1941) (transport)	2,400	96	29	18,200	8,255	230*	370*	1,350	2,170	24,000	7,315	4
Dewontine D-520 (France) (1939) (fighter)	910	34	10	4,610	2,090	249	400	620	998	36,000	11,000	1
Morane-Saulnier M.S. 406 (France) (1938) (fighter)	860	35	11	4,190	1,900	249	400	497	800	30,840	9,400	1
Poetz 63.11 (France) (1938) (reconnaissance)	1,400	53	16	6,910	3,135	230	370	930	1,500	27,900	8,500	3
DeHavilland Tiger Moth II (Gt. Britain) (1932) (bomber-transport)	130	29	9	1,115	505	93	150	302	486	13,600	4,145	2
Supermarine Spitfire Mk 1A (Gt. Britain) (1938) (fighter)	1,030	36	11	4,810	2,180	315	507	575	925	34,000	10,360	1
Hawker Hurricane Mk 1 (Gt. Britain) (1937) (fighter)	1,030	40	12	4,670	2,120	324*	521*	460	740	34,200	10,425	1
De Havilland B-IV Mosquito (Gt. Britain) (1941) (light bomber)	2,920	54	17	13,400	6,080	265	425	2,040	3,283	34,000	10,360	2
Vickers Wellington III (Gt. Britain) (1938) (medium bomber)	3,000	86	26	18,555	8,417	255*	410*	2,200	3,540	19,000	5,800	6
Handley Page Halifax Mk III (Gt. Britain) (1940) (heavy bomber)	6,460	104	32	38,240	17,346	282*	454*	1,030	1,658	24,000	7,315	7
Avro Lancaster (Gt. Britain) (1942) (heavy bomber)	5,840	102	31	36,900	16,740	210	338	1,660	2,671	24,500	7,470	7
Mikoyan-Garevich MIG-1 (USSR) (1940) (fighter)	1,200	34	10	5,720	2,595	280	450	453	730	39,370	12,000	1
Lavochkin LaGG-3 (USSR) (1941) (fighter)	1,100	32	10	5,775	2,620	276	446	398	640	29,530	9,000	1
Ilyushin Il-2 Shturmovik (USSR) (1941) (attack bomber)	1,600	47	15	9,260	4,200	199	320	466	750	24,600	7,500	2
Messerschmitt Bf-109E (Germany) (1938) (fighter)	1,100	33	10	4,430	2,010	298	480	410	660	34,450	10,500	1
Junkers Ju-87B-1 Stuka (Germany) (1938) (dive bomber)	1,200	45	14	6,085	2,760	174	280	370	595	26,250	8,000	2
Junkers Ju-88A-1 (Germany) (1939) (bomber)	2,400	60	18	16,975	7,700	217	350	1,056	1,700	26,250	8,000	4
Heinkel He-111-H (Germany) (1936) (medium bomber)	2,700	74	23	19,135	8,680	211	340	1,280	2,060	27,900	8,500	5
Macchi MC-202 Folgore (Italy) (1941) (fighter)	1,075	35	11	5,180	2,350	370*	595*	475	765	37,730	11,500	1
Savoia-Marchetti SM 7911 (Italy) (1936) (torpedo bomber)	3,000	70	21	16,755	7,600	255	410	1,243	2,000	23,000	7,000	4
Mitsubishi A6M3 Zero (Japan) (1939) (carrier-based fighter)	1,100	36	11	3,985	1,805	230	370	1,479	2,380	36,250	11,050	1
Nakajima Ki-84 (Japan) (1943) (fighter)	1,900	36	11	5,865	2,660	277	445	1,053	1,695	34,450	10,500	1
Mitsubishi Ki-46-II (Japan) (1941) (reconnaissance)	1,050	48	15	7,195	3,665	249	400	1,305	2,100	35,530	10,830	2
Yokosuka MXY7 Ohka (Japan) (1944) (kamikaze)	jet	14	4	1,200	545	277*	445*	81	130	—	—	1

* Maximum speed.

III and the German-built Me-262 jet fighters. The first mission for the Meteors, of which 280 were produced, was set for mid-April 1945 for the purpose of intercepting the German jet fighters. The encounter did not occur, however, and the war was ended without the Meteors seeing action. The meteors were equipped with two jet engines, each capable of 2000 pounds (907 kilograms) of thrust. They had a speed of 493 miles (793 kilometers) per hour, a range of 1340 miles (2156 kilometers) and a ceiling of 44,000 feet (13,411 meters)—with a crew of one. The German Messerschmitt Me-262A Sturmvogel was commissioned in mid-1942, but did not see action until mid-1944. Nearly 1,500 of these aircraft were produced, but only about 400 saw action during the war. The Sturmvogels were equipped with two jet engines, each capable of 1984 pounds (900 kilograms) of thrust. They had a speed of 541 miles (871 kilometers) per hour, a range of 650 miles (1046 kilometers), and a ceiling of a little over 37,500 feet (11,430 meters)—with a crew of one. Also of interest was the Messerschmitt Me-163B Komet. It was the first and, in fact, only rocket-propulsion fighter of the war. It was first commissioned at the end of July 1944. The craft had extremely limited performance parameters—flying time, about 10 minutes; altitude, about 6.2 miles (10 kilometers) (in 2.5 minutes). About 350 were produced and it is reported that more of these aircraft exploded upon landing than were knocked out by Allied fighters. The single power plant of the Komet was capable of producing about 3,750 pounds (1,701 kilograms) of thrust, with a maximum flying speed of just under 600 miles (965 kilometers) per hour. It carried a crew of one. The Japanese Yokosuka MXY7 Ohka 22, also known as the kamikaze or suicide aircraft, was equipped with a single jet engine in the tail. The little craft with a wing spread of under 14 feet (4.3 meters) and length of about 23 feet (7.0 meters), carried a 1,300-pound (590 kilograms) charge of dynamite. In operation, the Ohka was carried to near its target by a G4M2e, at which time the Ohka was piloted for a short distance to the target, with no escape for the pilot just prior to impact.

The war faced a challenge in refurbishing the fleets of not only the various airlines in the United States which had been utilized to the full during the war, but also of demands for aircraft by much of the rest of the world. Jet engine and jet technology, in general, still was not quite ready for application in large commercial planes and required further testing and refining in military aircraft, for which funding was available. Thus, during the late 1940s and early 1950s, a number of interesting and well-designed and performing propeller aircraft were introduced.

The first jet passenger service was offered by BOAC in early May 1952. The plane used was the DeHavilland DH 106 Comet. A couple of accidents in 1954 precipitated early retirement for these airplanes. However, in 1958, a revised version, the Comet 4, was introduced. About 75 of these airplanes were built. The early Comet 1 was powered by 4 jet engines, each with a thrust of 4,450 pounds (2019 kilograms). The wing spread was 115 feet (35 meters), the cruising speed was 490 miles (788 kilometers) per hour, with a ceiling of 35,000 feet (10,668 meters). The airplane could accommodate up to 44 passengers with a crew of 4.

The Boeing 707–121 made its first flight in mid-July 1954. However, it was not used for nonstop flights across the Atlantic until late 1958. This airplane generally is considered as one of the most successful passenger aircraft of all times. The early airplane was powered by 4 jet engines, each producing a thrust of 13,000 pounds (5897 kilograms). The wing spread was 130 feet (39 meters), cruising speed 585 miles (941 kilometers) per hour, a range of 3,750 miles (6034 kilometers), and ceiling of 36,000 feet (10,973 meters). At maximum loading, it could accommodate 189 passengers with a crew of 4.

The Sud-Aviation SE-210 Caravelle made its first flight in late May 1955. It commenced commercial passenger service in Europe in 1959. Nearly 300 aircraft of the SE-210 were produced and, of course, there have been subsequent model changes. The SE-210 was powered by 2 jet engines, each with 10,500 pounds (4763 kilograms) of thrust, wing spread was nearly 113 feet (34 meters), cruising speed 456 miles (734 kilometers) per hour, a range of 1150 miles (1850 kilometers), and ceiling of 30,000 feet (9144 meters). Maximum capacity was 80 passengers with a crew of 3.

The 1960s, 1970s, and continuing into the 1980s has been an era of emphasis on economy of performance in terms of passenger- and cargo-carrying aircraft. Unquestionably, there will be major breakthroughs in the future, but the past two or three decades have largely been concerned with safety and economy and greater carrying capacity. The supersonic aircraft for passenger traffic, notably the Concorde, in retrospect, was possibly introduced at an unfortunate time—when there was greater stress on economy than speed—and during a period of great sensitivity to environmental matters.

In recent years, great emphasis has been placed upon the use of stronger, lighter-weight materials—both in military and passenger aircraft. See Figs. 10 and 11.

Fig. 10. Airframe of the F/A-18 strike fighter is a balance of conventional materials and graphite epoxy providing an optimum lightweight structure. The wing skins, trailing edge flaps, stabilators, vertical tails and rudders, speedbrake, and many access doors are made of graphite composite material comprising about 9% of the aircraft structural weight and resulting in appreciable weight savings, plus increased aircraft performance. Other materials used (as percent of structural weight): aluminum, 50.4%; steel, 16.0%; titanium, 13.2%; and other materials, 11.4%. (*McDonnell Aircraft Company*; *McDonnell Douglas Corporation*)

Experimental Aircraft. Airplanes in this category tend to create a great deal of news for a short period during the conceptual, design, and testing phases, and then frequently disappear into oblivion. A few examples are worthy of note: (1) The Northrop XB-35 Flying Wing. This aircraft, without wing or tail in the conventional sense, was first flown in 1947. The prototypes were equipped with four propellers; later two of the three prototypes were fitted out with jet engines. The airplane was designed as a bomber, capable of transporting 4400 pounds (1996 kilograms) of bombs and with an armor of 20 machine guns. After thorough testing, technical problems cancelled production runs. (2) The Bell X-1 was designed to be dropped by a Flying Superfortress B-29. In October 1947, the pilot of this craft was the first to break the "sound barrier," flying at a speed over Mach 1. The X1-A flew at a speed of Mach 2.42 in 1953 and reached an altitude

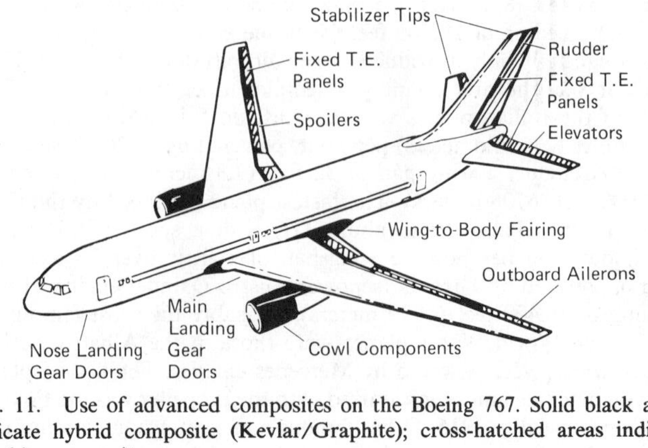

Fig. 11. Use of advanced composites on the Boeing 767. Solid black areas indicate hybrid composite (Kevlar/Graphite); cross-hatched areas indicate graphite composite.

of approximately 98,400 feet (29,992 meters) in 1954. (3) The Bell X-5 was designed around the concept of the German Messerschmitt P. 1101, which was taken at the end of World War II. The X-5 was designed so that the wings of the craft could be changed while in flight from an angle of 20 degrees to one of 60 degrees. It is interesting to note that subsequent variable-wing aircraft produced in the United States were based upon the experience gained with the X-5. (4) Possibly the best known of the experimental models is the North American X-15 which reached an altitude of 354,108 feet (107,932 meters) and a speed of 4105 miles (6605 kilometers) per hour in different flights. Three X-15s were built. They provided much needed information for the later space ventures. The planes were fitted with a rocket engine and taken to altitude by a B52-B bomber. The testing period extended from 1959 to 1962. (5) The North American XB-70A Valkyrie supersonic bomber intended to replace the B-52 did not become operational, but two prototypes were built. Currently aircraft experimentalists are working on the design of a space shuttle for "routinely" carrying passengers and equipment from earth to space stations and back. See **Space Shuttle.**

Appropriate and well deserved is mention here of the achievements of a group of Californians (Paul MacCready, Jr., Bryan Allen, et al.) who won the Kremer Prize of £50,000, administered by the Royal Aeronautical Society, for successfully piloting a human-powered machine over a figure-eight course around two pylons half a mile apart and clearing a 10-foot (3 meter) high obstacle at the start and finish. The flight was achieved on August 23, 1977 at an airport near Shafter, California. A professional bicyclist, Allen pedaled to generate about $\frac{1}{3}$-horsepower to lift and fly the *Gossamer Condor*, with a wingspan of 96 feet (29.3 meters) and weighing only 70 pounds (31.8 kilograms). On June 12, 1979, Allen "pedaled" the craft across the English Channel in 2 hours and 55 minutes over a distance of about 25 miles (40 kilometers). An official of the Royal Aeronautical Society called the flight "a tremendous achievement."

Passenger Aircraft. As with military aircraft, detailed descriptions and specifications of the several widely used passenger aircraft are beyond the scope of this encyclopedia. Refer to Fig. 12 through 16 which portray two of these craft and facilities used in their manufacture.

See also **Air Traffic Control; Automatic Pilot;** and **Helicopters and V/STOL Craft.**

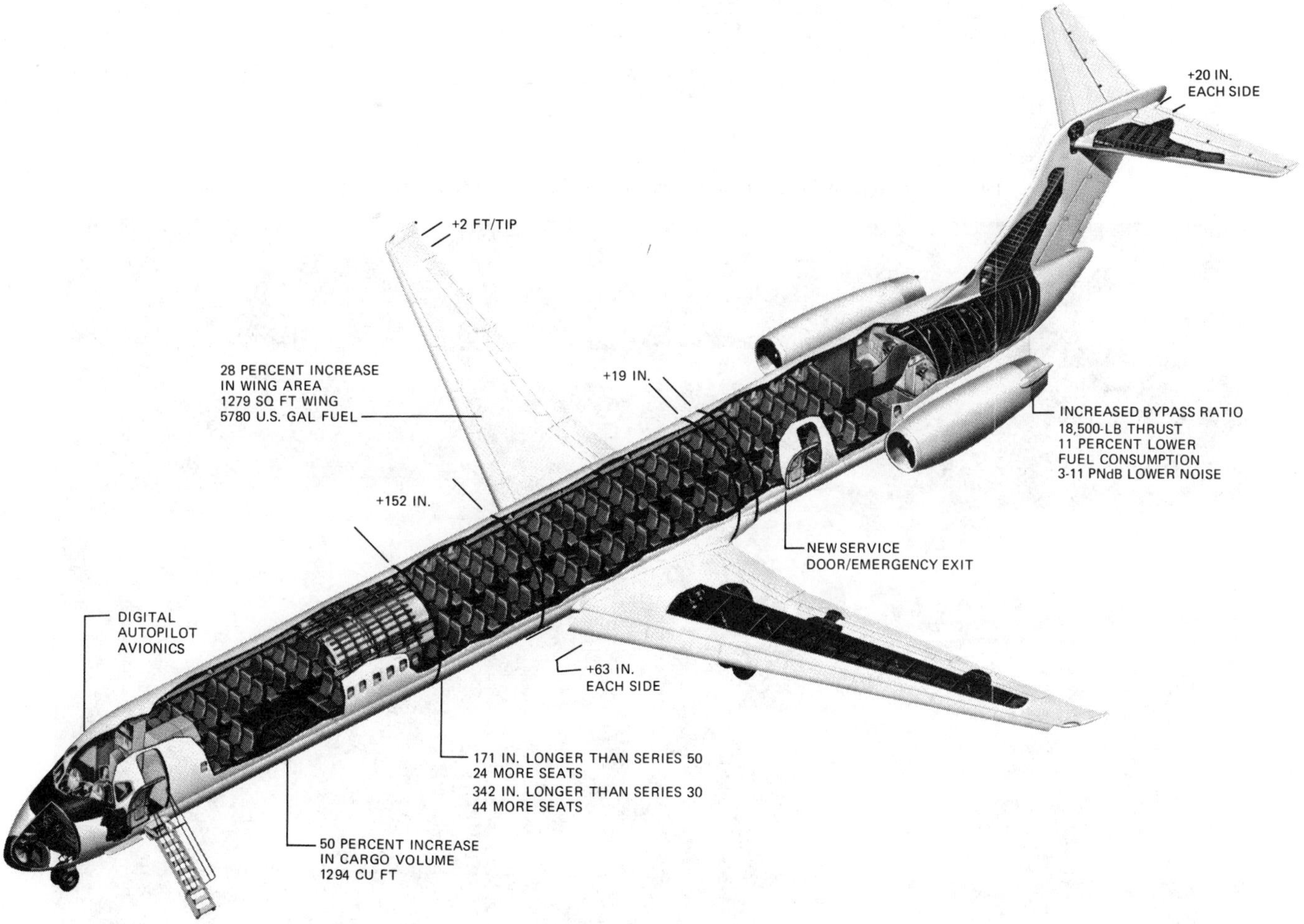

Fig. 12. Sectional view of the DC-9 Super 80. The final decision to produce this aircraft was made in late 1977 and the first deliveries were made in 1980. There are three interior versions—167 passenger capacity (charter service); 155 passengers (single class service); and 137 passengers (mixed-class service). The statistics given in this diagram compare the Super 80 with earlier models. The craft is powered by two Pratt & Whitney JT8D-209 engines. Maximum takeoff weight is 140,000 pounds (63,504 kilograms); maximum zero fuel weight is 118,000 pounds (53,525 kilograms); fuel capacity is 38,719 pounds (17,563 kilograms). The aircraft is designed for long-range cruising at 31,000/35,000 feet (9449/10,668 meters). (*McDonnell Douglas Corporation*)

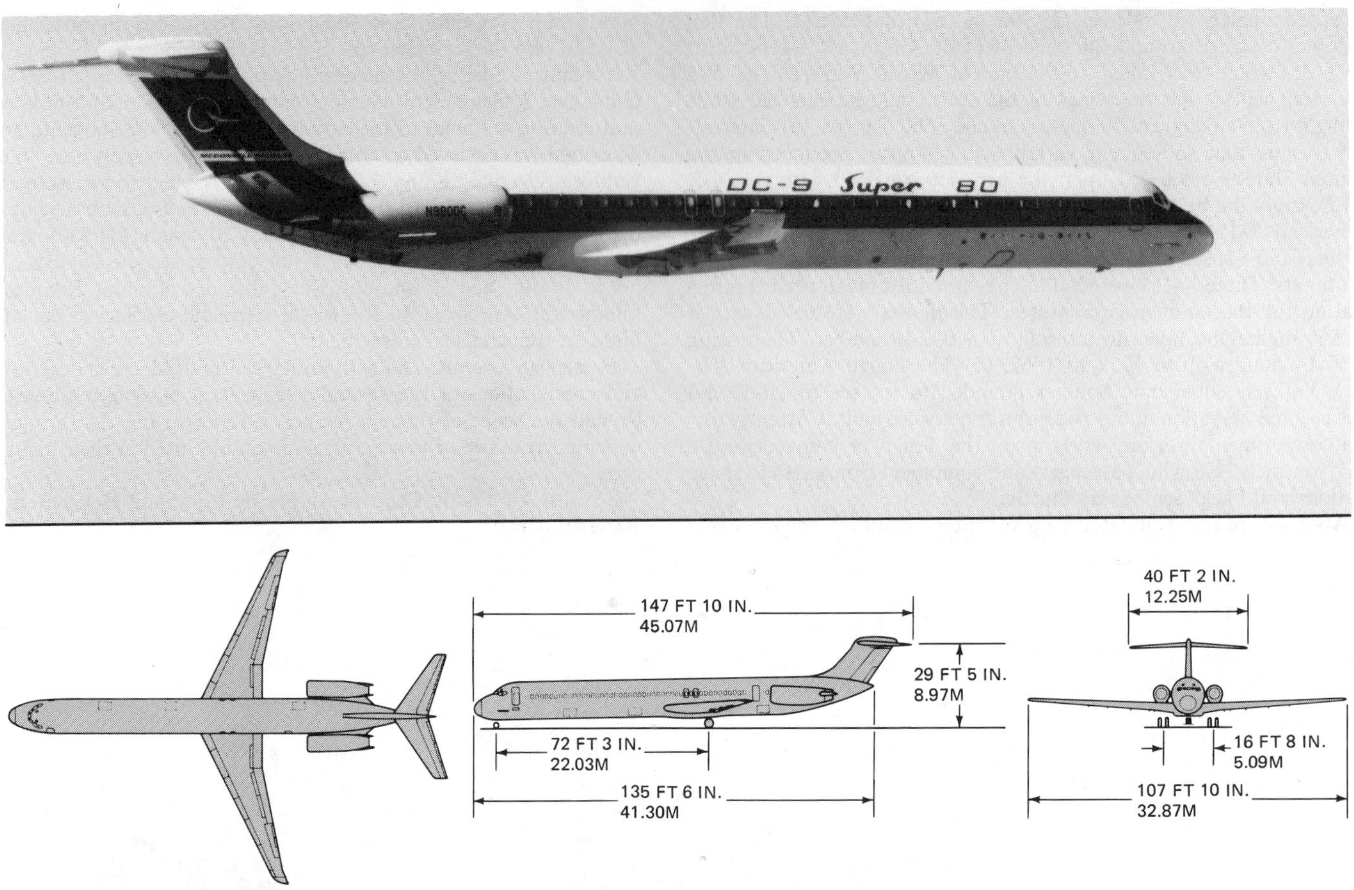

Fig. 13. DC-9 Super 80 in flight. Principal dimensions shown below. (*McDonnell Douglas Corporation*)

Fig. 14. Manufacturing bay with several DC-9s in late stage of production. (*McDonnell Douglas Corporation*)

Fig. 15. The DC-10, a workhorse of modern aviation, has flown considerably more than 2 billion miles (3.2+ billion kilometers) since its first flight in August 1970. The aircraft is powered by 3 engines, ranging in thrust from 40,000 to 53,000 pounds (18,144–24,041 kilograms). The length is approximately 182 feet (55.4 meters), wing span is over 164 feet (50 meters) on some models; the height of all models is 57.7 feet (17.5 meters). Maximum gross weights, depending upon model, range from 440,000 to 572,000 pounds (199,581 to 259,455 kilograms). Maximum passengers carried is up to 380 persons, depending upon interior arrangement. Again, depending upon model, the range for a typical mixed class is from 6950 to 9950 kilometers (4318 to 6183 miles). Operating altitude is 42,000 feet (12,802 meters). Cruise speed is 945 kilometers (510 knots or 587 miles) per hour. In terms of time, Alitalia's Rome–Rio de Janeiro nonstop flight requires about 11.2 hours. The briefest scheduled flight between Fort Lauderdale and Miami, Florida requires 22 minutes. The average DC-10 flight duration is 3.4 hours. The average daily flight utilization for a DC-10 is 9 hours. Forty-six airlines operate DC-10s. (*McDonnell Douglas Corporation*)

Fig. 16. Several DC-10s in fairly late stage of production. (*McDonnell Douglas Corporation*)

References

NOTE: See also references listed at end of entry on **Aerodynamics.**

Angelucci, E.: "Airplanes: From the Dawn of Flight to the Present Day," McGraw-Hill, New York, 1973.

Beardmore, P., et al.: "Fiber-Reinforced Composites: Engineered Structural Materials," *Science*, **208**, 833–840 (1980).

Green, W.: "Warplanes of the Third Reich," Macdonald & Co., London, 1970.

Kear, B. H., and E. R. Thompson: "Aircraft Gas Turbine Materials and Processes," *Science*, **208**, 847–856 (1980).

Kerr, R. A.: "East Coast Mystery Booms," *Science*, **203**, 256 (1979).

Long, M.: "The Flight of the Gossamer Condor," *National Geographic*, **153**, 1, 130–140 (1978).

Munson, K.: "Aircraft of World War II," Ian Allan Ltd., London, 1962.

Robinson, A. L.: "Human-Powered Flight—Californians and Kremer Prize," *Science*, **197**, 1171 (1977).

Staff: "Jane's All the World's Aircraft," Sampson Low, Merston & Co., Ltd., London (published annually).

Staff: "The Aerospace Year Book," Book Inc., Washington, D.C. (published annually).

Treager, I. W.: "Aircraft Gas Turbine Engine Technology," McGraw-Hill, New York, 1970.

AIR PLOT. A plot showing the movements of an airplane relative to the air is known, in navigation, as an air plot. It is similar in every respect to the old-fashioned method for obtaining the dead reckoning (DR) position of a sea-borne ship, disregarding the effects of ocean currents.

The air plot is made up of a series of distance vectors, added in the usual manner, the direction of each vector being a heading of the plane and the length being proportional to the distance moved through the air along the heading (distance = air speed × time on heading). The summation of the vectors gives a no wind (NW) position of the plane at any time. To obtain the DR position of the plane it is only necessary to add the total wind vector, whose direction is that of the motion of the air and whose length is the total movement (wind speed × total time since departure) of the air, to the NW position. Such a method avoids the necessity of constructing the velocity-vector diagram to obtain course and speed along course for each heading, as is necessary in the standard method for obtaining a DR position in air navigation.

In cases where the plane is frequently changing headings, the air-plot method saves from 30–60% of the time involved in obtaining the DR position by standard methods. Furthermore, in the frequent cases where the predicted wind for high altitude is very inaccurate, the air plot provides a method for determining the wind.

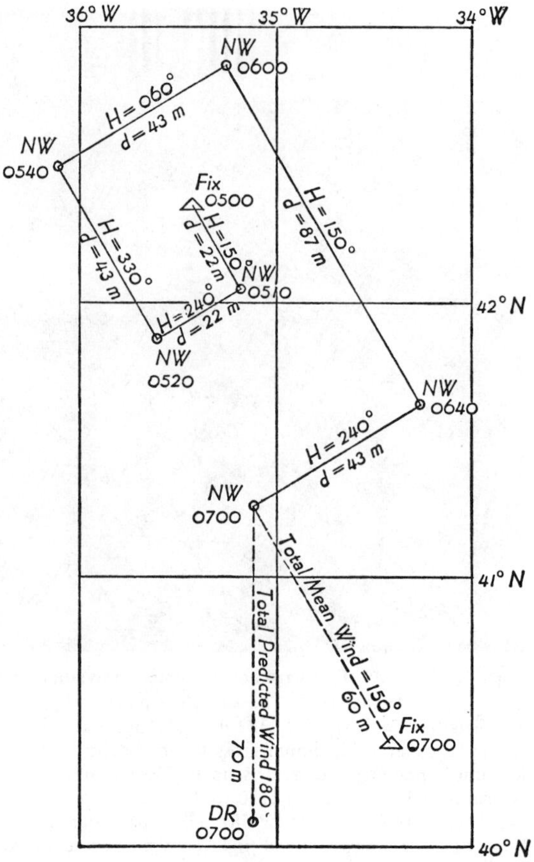

Scale figure for air plot.

The solution of the following problem indicates that advantages and value of the air plot. On a day when the predicted winds are uncertain, a plane is to search an area centered about a point in latitude L = 42°22′ N and longitude Lo = 35°25′ W. The plane, flying with true air speed TAS = 130 knots at altitude 10,000 feet (3,000 meters) is over the point at 0500 and heads 150°. The altitude and air speed are maintained, but the plane alters headings at the given times as follows: 0510 heading 240°, 0520 heading 330°, 0540 heading 060°, 0600 heading 150°, 0640 heading 240°. The air plot, properly labeled, is shown in the figure from which the NW position at 0700 is found to be L = 41°15′ N and Lo = 35°07′ W.

The predicted wind at the altitude of flight is from 000° and speed 35 knots. From the 0700 NW position a vector in direction 180° and of length 70 miles (112 kilometers) is drawn. This is shown as a dotted line in the figure and represents the total movement of the plane due to the predicted wind in the two hours from 0500. This gives a dead-reckoning position in latitude 40°05′ N and longitude 35°07′ W.

A fix is obtained at 0700 in latitude 40°23′ N and longitude 34°27′ W. If a vector is drawn from the NW position to the fix, it will represent the actual movement of the air in the interval from 0500 to 0700. From the diagram this is found to be in the direction 150° and 60 miles (96 kilometers) long. Since this represents 2 hours total motion of the air, we have a mean wind from 330°, speed 30 knots.

To obtain the DR position, using the standard methods of dead reckoning, would have required the plotting of at least four velocity-vector diagrams to obtain the course and ground speed on each heading. Each of these would have taken at least twice as much time as is required to plot an air vector and, after the courses and ground distances had been obtained, the geographic-motion vector diagram would have had to be plotted. Then to obtain the mean wind, the assumed air motion would have had to be subtracted from the DR position to obtain the NW position, which has to be used with the fix to obtain the actual total wind motion. See also **Course; Dead Reckoning; Heading;** and **Navigation.**

AIR POLLUTION. Aerosol; Coal; Natural Gas; Petroleum; Pollution (Air).

AIRPORT (Fog Clearance). Fog and Fog Clearning.

AIRPORT VISIBILITY. Visibility.

AIR PREHEATER. There are many devices of which the purpose is to heat air for some specific usage. However, in speaking of air preheaters, what is ordinarily meant is the heater employed for raising the temperature of air used for combustion of a fuel. This may occur in some industrial process such as preliminary heating of the air supplied to blast furnaces (see accompanying photo), but the most frequent use of air preheaters today is in connection with steam boilers. This type of air preheater is a heating surface installed between the boiler flue gas outlet and the stack. In arrangement the heating surface is composed either of tubes with flue gas inside and the air to be heated outside, or of rectangular plates spaced about one-half inch apart, leaving alternate gas and air passages. Its use is chiefly justified on economic grounds.

Two principles are employed for heat transfer in air preheaters. The recuperative principle implies transfer of heat through a separating partition, such as the walls of a tube, by continuously recuperating the cool side with conduction of heat from the hot side. Regenerative heaters are those which alternately heat and cool the same mass, regenerating it thermally by passing hot spent gas over its surface. Regenerative heaters are frequently used with blast furnaces, and are composed of two heating chambers in which are piled checkerworks of brick having sufficient heat storage capacity for the purpose. The burned gas leaving the furnace passes through one chamber, heating up the checkerwork, while in the other chamber the heated bricks are being cooled by air passing to the combustion region. When the air-heating chamber is thermally exhausted, valves shift the flow of hot gas through that chamber, and air is drawn through the hot one.

Three silo-like stoves at left preheat immense quantities of air for blast furnace in production of iron. The stoves are filled with refractory checkerwork and gas-fired in cylces so that one stove is always ready to go 'on wind' to supply the furnace with air at a temperature of approximately 1000°F (538°C).

AIR QUALITY SURVEILLANCE. **Earth Resources Satellites and Geologic Remote Sensors.**

AIRSHIP. **Dirigibles and Airships.**

AIRSPEED INDICATOR. Many air-data flight instruments operate from static or "pitot" pressure, or both. These pressures are usually obtained from a pitot-static tube of the type shown in Fig. 1. On the smooth side of the pitot tube, the pressure differs little from that of undisturbed air, and slots placed there provide a source of static pressure. Tubes usually are electrically heated to melt off any ice that might form and which would partly or fully seal off the openings and give erroneous readings in various air-data instruments. See Fig. 2. The difference between the pitot pressure and static pressure is a measure of *indicated airspeed.* Airspeed indicators are calibrated in knots (nautical mph) under standard pressure (29.92 inches of mercury) and temperature (59°F). A relationship between pressure difference and airspeed in knots is given in the accompanying table. See also **Pitot Tube.**

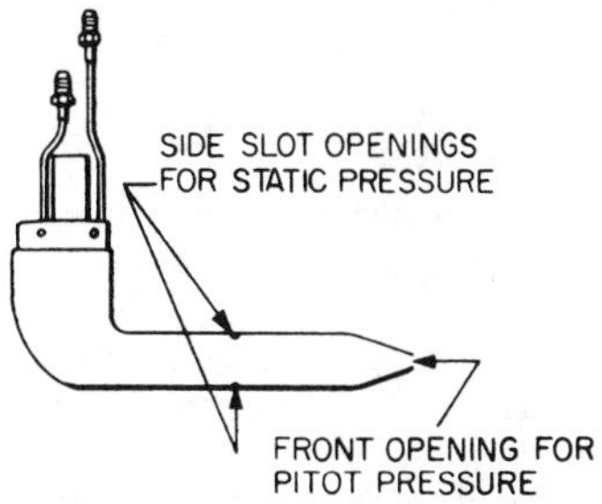

Fig. 1. Pitot tube of type used in aircraft.

True-Airspeed Indicator. This instrument is primarily used for navi-

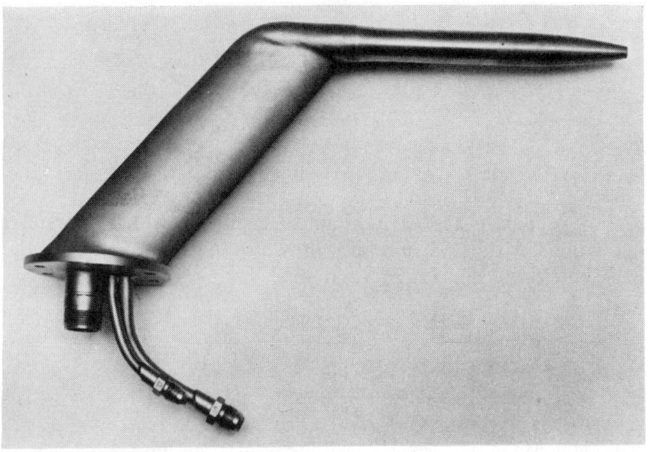

Fig. 2. Electrically heated pitot-static tube. (*Rosemount Engineering*)

AIRSPEED VERSUS DIFFERENTIAL PRESSURE AT STANDARD TEMPERATURE AND PRESSURE

Indicated Airspeed, knots	Differential Pressure, in. Hg	Indicated Airspeed, knots	Differential Pressure, in. Hg
0	0.000000	400	8.3973
10	0.004795	450	10.8837
20	0.019185	500	13.7967
30	0.043179	550	17.1859
40	0.076793	600	21.1088
50	0.1200	650	25.6316
60	0.1729	700	30.8159
70	0.2355	750	36.6276
80	0.3079	800	43.0092
90	0.3901	850	49.9241
100	0.4821	900	57.3481
125	0.7557	950	65.2644
150	1.0924	1,000	73.6613
175	1.4938	1,050	82.5302
200	1.9616	1,100	91.8649
225	2.4977	1,150	101.6606
250	3.1045	1,200	111.9136
275	3.7845	1,250	122.6210
300	4.5407	1,300	133.7805
350	6.2949	1,320	138.3706

SOURCE: Battelle Memorial Institute.

gation. The measurement is a function not only of difference in pitot and static pressure, but also of temperature. In effect, a true-airspeed indicator (Fig. 3) is a computer operating on a rather complex relationship of two pressures and air temperature to give the correct indication. The relationship is:

$$V = 38.94 \frac{M\sqrt{T_1}}{\sqrt{1 + 0.2KM^2}}$$

where V = true airspeed, knots; T_1 = indicated air temperature, Kelvin; K = recovery factor of temperature probe (usually 1.00 or slightly less); and M = Mach number.

Another approach is to use the speed of an air turbine as a measure of true airspeed, the intake pressure to the turbine being static pressure and the output being balanced to pitot pressure.

Machmeter. The ratio of true airspeed to the velocity of sound is called *Mach number.* The safe top speed of fast aircraft is usually expressed in terms of Mach number and is indicated on a Machmeter.

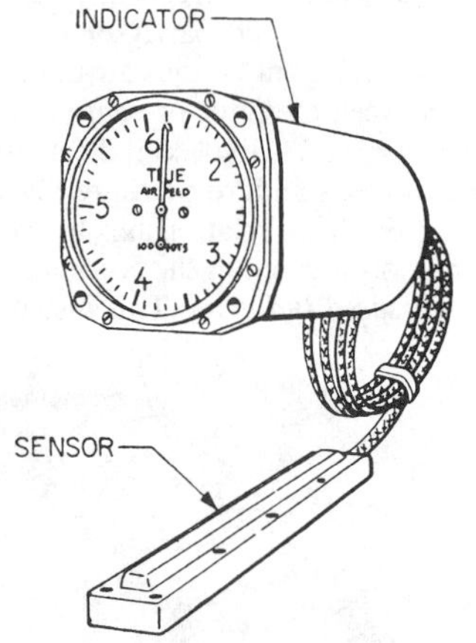

Fig. 3. True-airspeed indicator.

See Fig. 4. This is also a mechanical computer which operates on the relationship of static and pitot pressures to give the proper indication. The relationship for Mach number is:

$$\frac{dP}{Ps} = (1 + 0.2M)^{23.5} - 1 \quad \text{for } M < 1$$

$$\frac{dP}{Ps} = \frac{166.9215M^7}{(7M^2 - 1)^{2.5}} - 1 \quad \text{for } M > 1$$

where M = Mach number; dP = differential (pitot less static) pressure; and Ps = static pressure. The true airspeed and Mach number are combined in Fig. 4.

Fig. 4. Combined true-airspeed and Mach number computer indicator. (*Bendix Corporation*)

AIR TRAFFIC CONTROL. The use of a voice radio link for position reporting by the pilot to the air-traffic controller is a basic source of air-traffic-control position data. Pilots, of course, also use information from other navigation aids, such as VOR, DME, etc. See also **Navigation.** Two major and independent methods of checking on aircraft location are used by ground controllers: (1) Primary radar, which operates on reflection by the aircraft of the pulse signals which the radar transmits; and (2) secondary radar, which operates on replies from pulsed radio beacons (on the aircraft), verified by the secondary-radar pulses.

Primary radars include: (1) Relatively short range (airport-surveillance radars or ASRs); and (2) longer-range systems, known as airport-route surveillance radars (ARSRs). Both systems are pulse transmissions and rotating antennas. Both systems have moving-target indicators which compare the rf phase of successive returns in order to determine whether the target is moving or stationary. Stationary targets are canceled from display, using either a delay-line analog technique to make the comparison, or a digital-storage system. In some designs of either type of system, the pulse-repetition frequency may be "jittered" or staggered in order to avoid blind speeds where the aircraft has moved by precisely one or more integral wavelengths between successive pulses.

Secondary radars are commonly co-located with the primary radar and use a line-array antenna fastened on top of the primary-radar antenna structure. Frequency of operation is L-band (1030 MHz). Because one-way transmission elicits a reply from aircraft transponders, secondary radars pose a special problem. The difference in antenna gain from main lobe to side lobe is not as effective in suppressing false side-lobe replies as the primary-radar system, which uses the antenna gains on both transmission and reception to enhance the difference between main-lobe and side-lobe signals. Thus, the secondary-radar system transmits a signal from an omnidirectional (or moderately directional) antenna which has slightly more gain than any of the side-lobe signals.

The beacon in the aircraft receives the omnidirectional transmission and the signal from the high-gain directional antenna in a standard time sequence. If the pulse from the omnidirectional antenna is not smaller than the other pulses from the high-gain antenna, the beacon assumes that the reception is from a side lobe and does not reply. Reply frequency from the transponder is 1090 MHz, thus there is no problem with ground echo from the transmitted signals.

In a precision-approach radar, there are primary radars which use two very narrow beams that scan a relatively narrow sector aligned with the approach course to a particular runway. There is a beam that is broad in vertical dimension and narrow in horizontal dimension which scans at a relatively high rate (several sweeps per second) in the horizontal dimension. The controller, watching the cathode-ray-tube display, can tell the pilot whether he is right or left of the true approach course. There is another beam that is narrow in the vertical dimension and broad in the horizontal dimension and which scans

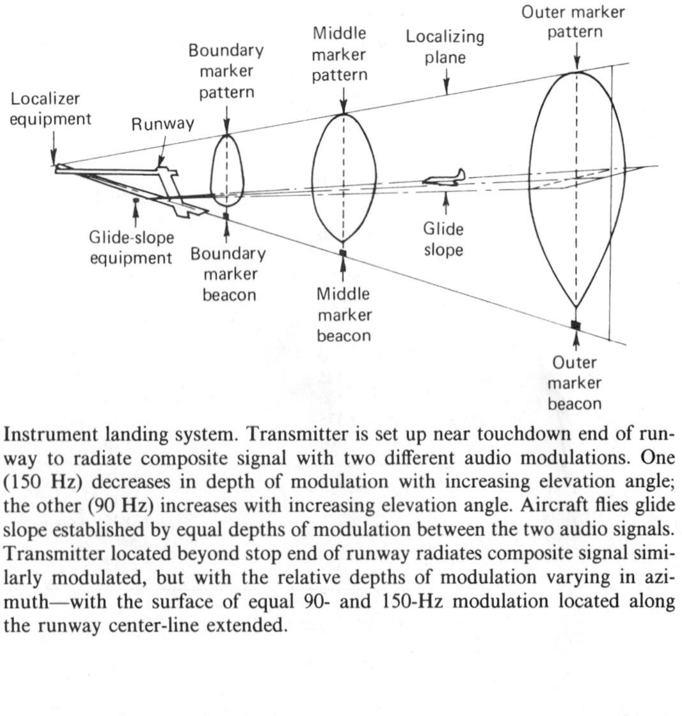

Instrument landing system. Transmitter is set up near touchdown end of runway to radiate composite signal with two different audio modulations. One (150 Hz) decreases in depth of modulation with increasing elevation angle; the other (90 Hz) increases with increasing elevation angle. Aircraft flies glide slope established by equal depths of modulation between the two audio signals. Transmitter located beyond stop end of runway radiates composite signal similarly modulated, but with the relative depths of modulation varying in azimuth—with the surface of equal 90- and 150-Hz modulation located along the runway center-line extended.

vertically. An expert controller watching this display can advise the pilot whether the aircraft is above or below the correct approach path. A low-approach system (ILS) is shown and briefly described in the accompanying diagram.

An automatic altitude reporting system can eliminate many of the manual steps required in air-traffic control. This consists of an air-data computer that corrects static-pressure errors and provides synchro-driven altitude information to the pilot's altimeter and altitude in digital form to the aircraft transponder in high-performance aircraft, or a direct readout to the pilot and digital altitude data to the aircraft transponder in lower-performance aircraft. The digital information then is transmitted to the ground interrogator and presented on a radarscope in alphanumeric form.

AITKEN'S METHOD OF INTERPOLATION. Given the arguments, a, b, c, d and the corresponding values of the function u_a, u_b, u_c, and u_d, we desire the value of u_x, $a < x < d$. We use the simple method provided by A. C. Aitken. This method works for unequal intervals for the argument as well as equal intervals, for inverse interpolation as well as for direct interpolation, and it provides a way of determining the number of significant figures in u_x.

A table is constructed as follows to find u_x:

u_a				$a-x$
u_b	$u_x(a, b)$			$b-x$
u_c	$u_x(a, c)$	$u_x(a, b, c)$		$c-x$
u_d	$u_x(a, d)$	$u_x(a, b, d)$	$u_x(a, b, c, d)$	$d-x$

We calculate the entries in the table (see **Determinant**):

$$u_x(a, b) = \frac{\begin{vmatrix} u_a & a-x \\ u_b & b-x \end{vmatrix}}{b-a}$$

$$u_x(a, c) = \frac{\begin{vmatrix} u_a & a-x \\ u_b & c-x \end{vmatrix}}{c-a}$$

$$u_x(a, d) = \frac{\begin{vmatrix} u_a & a-x \\ u_d & d-x \end{vmatrix}}{d-a}$$

$$u_x(a, b, c) = \frac{\begin{vmatrix} u_x(a, b) & b-x \\ u_x(a, c) & c-x \end{vmatrix}}{c-b}$$

$$u_x(a, b, d) = \frac{\begin{vmatrix} u_x(a, b) & b-x \\ u_x(a, d) & d-x \end{vmatrix}}{d-b}$$

$$u_x(a, b, c, d) = \frac{\begin{vmatrix} u_x(a, b, c) & c-x \\ u_x(a, b, d) & d-x \end{vmatrix}}{d-c}$$

$$= u_x$$

We may stop the calculations whenever the particular column $u_x(a, b, \ldots)$ shows an agreement to the number of significant figures desired. For example, if $u_x(a, b, c)$ and $u_x(a, b, d)$ agree to 4 significant figures, then u_x would be that result to 4 significant figures. Aitken's method corresponds to polynomial interpolation, and in the illustration cited with 4 points to a polynomial of the third degree is being used. The points (a, u_a), (b, u_b), (c, u_c) and (d, u_d) may be arranged in any order without affecting the value of u_x.

Aitken has also described a quadratic method of interpolation which is particularly useful when the arguments are equally spaced. If x lies between b and c and we write $\theta = (x - b)/(c - b)$ for the fraction of the interval between successive arguments and $\phi = 1 - \theta$, then to find the 4-point interpolate $u_x(a, b, c, d)$ we first form the two linear interpolates $u_x(b, c)$, $u_x(a, d)$ in the manner described above; we then have immediately:

$$u_x(a, b, c, d) = \tfrac{1}{2}[(2 + \theta\phi)u_x(b, c) - \theta\phi u_x(a, d)]$$

See also **Interpolation;** and terms listed under **Mathematics.**

AKREMITE. Explosive.

AKUND. Silk Cotton Trees.

ALABANDITE. Manganese sulfide, MnS. Associated with pyrite, sphalerite, and galena in metallic sulfide vein deposits.

ALABASTER. A fine-grained variety of the mineral gypsum, formerly much used for vases and statuary. It is usually white in color or may be of other light, pleasing tints.

The word alabaster is derived from the Greek name for this substance. See also **Gypsum.**

ALANINE. Amino Acids.

ALASKA CEDAR. Cedar Trees.

ALASKA POLLACK. Codfishes.

ALBACORE. Tunas.

ALBATROSS. Petrels and Albatrosses.

ALBEDO (Astronomy). A term used to indicate the reflecting power of an object. Technically defined, albedo (often referred to as the Bond albedo) is the ratio of the radiation reflected from an object to the total amount incident upon it. Thus,

$$A = pq$$

where p is the ratio of the brightness at the phase angle of zero to the brightness of a perfectly diffusing disk under the same conditions, and where q is a factor representing the phase law. For example, the albedo of the moon is 0.068, which means that the moon reflects that fraction of the sunlight which is incident upon it. The value of the albedo of a planet is a measure of the quantity of atmosphere that surrounds the object. The higher the albedo, the thicker the atmospheric layer. In the case of objects without atmosphere, as in the case of the moon, the albedo, combined with the color of the reflected light, may be used to make estimates of the character of the material making up the surface of the object.

ALBEDO (Physics). In nuclear physics, the albedo is the ratio of the neutron current density out of a (non-source) medium to the neutron current density into it. In cosmic ray physics, albedo refers to those energetic, charged secondary cosmic rays that move generally upward. Because some of these particles go far into the upper atmosphere and return to the earth's surface (due to the effect of the earth's magnetic field), measurements of primary cosmic radiation must be corrected for the contribution of the albedo particles.

ALBEDO (Polar Region). Polar Research.

ALBERTITE. An oxygenated hydrocarbon which differs from asphaltum slightly in that it is not completely soluble in turpentine, nor can it be perfectly fused. Specific gravity, 1.097, pitchy luster, dark brown to black color. Occurs in veins from 1 to 16 feet wide in the Albert Shale of Albert County, New Brunswick.

ALBINISM. Absence of pigmentation. The condition has been noted in occasional individuals of many species which are normally pigmented, including man. In some cases the term is applied to a partial lack of pigment, as in the white form of certain normally yellow butterflies; in this form the black markings characteristic of the species are fully developed. In contrast, the total lack of pigment in albino birds and mammals is shown by the pink eyes. In these organs pigment is functionally important but the albino fails to develop it, hence the color of the blood is seen through the tissues.

Albinism in man is known to be inherited as a recessive to normal pigmentation. The term is also applied to normally green plants which fail to develop chlorophyll. As in man, albinism is inherited as a simple recessive. Albino plants die as soon as food is exhausted from the seed. Partial albinism in plants is known as variegation.

ALBITE. Feldspar.

ALBUMIN. An albumin is a member of a class of proteins which is widely distributed in animal and vegetable tissues. Albumins are soluble in water and in dilute salt solutions, and are coagulable by heat.

Albumin is of great importance in animal physiology; in man it constitutes about 50% of the plasma proteins (blood) and is responsible to a great extent for the maintenance of osmotic equilibrium in the blood. The high molecular weight (68,000) of the albumin molecule prevents its excretion in the urine; the appearance of albumin may indicate kidney damage. See **Kidney and Urinary Tract.**

ALBUMINURIA. Kidney and Urinary Tract.

ALCOHOL. A term commonly used to designate ethyl alcohol or ethanol. See **Ethyl Alcohol.**

ALCOHOLATE. Replacement of the hydroxyl group of an alcohol by a metal, particularly a metal that forms a strong base, results in formation of an alcoholate. An example is sodium ethylate, C_2H_5ONa.

ALCOHOLIC CARDIOMYOPATHY. Heart and Circulatory System (Human).

ALCOHOLIC CIRRHOSIS. Liver.

ALCOHOLISM. This is a chronic illness, usually requiring from 10 to 15 years to develop into a major syndrome where serious physical and/or mental manifestations are presented. Although scientists have been struggling for years to define moderate drinking, social drinking, heavy drinking, and, in the extreme case, alcoholism, it is well established that alcohol, as a systemic poison and agent of degeneration, does not recognize such artificial lines of demarcation. Tolerance of alcohol varies widely with individuals. Often, when considering alcoholism, too much stress is given to statistics and too little emphasis is given to the acute organic disorientations which can and do occur as the result of the inhibiting role played by alcohol in body chemistry. While much remains to be learned, some genetic connections already have been made (Goodwin et al., 1973, 1974, and 1977; Partanen et al., 1966; Fenna et al., 1971; Bennion and Li, 1976).

Alcohol affects body chemistry in many ways. Williams (1974) describes alcohol as "the most active drug known to affect the human body." Drinking in excess of the individual's tolerance level (which may decrease with age) precipitates the degradation of the central nervous system, including cerebellar degeneration, central pontine myelinolysis, Marchiafava-Bignami disease, and the familiar psychiatric disorders, such as depression, anxiety, antisocial behavior, and intensified aggessiveness. Cerebral atrophy and ventricular enlargement are described by Fox et al. (1976). Permanent impairment of memory (Korsakoff's psychosis) is described in the entries on **Amnesia;** and **Memory.**

Similarly, gastrointestinal functions can be adversely affected by alcohol. The resulting disorders include esophagitis, gastritis, and ulcer. There may be illeal changes, with decreased absorption of folic acid, as well as pancreatitis, and fatty degeneration of liver. Cirrhosis is found in 10% of alcoholics. See **Liver.** There are also cardivascular effects from excessive alcohol intake (Mitchell and Cohen, 1970). Orlando et al. (1976) report how angina can be worsened; Klatsky et al. (1977) report on the aggravation of hypertension; Wu et al. (1976), Asokan et al. (1972), and Demakis et al. (1974) report on cardiomegaly in alcoholics; Orlando et al. (1976) report on the effect of ethanol on angina pectoris. Fisher and Abrams (1977) report on life-threatening ventricular tachyarrhythmias in delirium tremens.

The metabolic and hepatic effects of alcohol are reported by Isselbacher (1977). These include, in part, hypoglycemia (decreased gluconeogenesis), decreased albumin and transferrin synthesis, accompanied by increased lipoprotein synthesis, an increased level of lipids in the serum and of liver triglycerides in the liver (fatty liver), hyperuricemia, and increased aldehyde levels in blood and tissue, as well as decreased serum magnesium and phosphate. Gordon et al. (1976) report on the effect of ethanol ingestion on sex-hormone metabolism in normal men (decreased plasma testosterone; impotence). Musculoskeletal effects of alcohol ingestion include myopathy and osteoporosis.

Physicians have a growing concern over the interference of alcohol with other drugs, mainly because of the comparatively easy availability of drugs and the fact that increasing numbers of alcohol-prone persons have strong tendencies to take drugs. Alcohol, for example, increases the potency of benzodiazepines, antipsychotic drugs, and chloral hydrate, among others. Alcohol decreases the effectiveness of anticoagulants and tricyclic antidepressants. Alcohol produces effects similar to mild antacid abuse when taken with antidiabetic sulfonylureas.

Numerous socioscientific programs have been developed over the years for the treatment of alcoholism. Detailed descriptions of such programs are beyond the scope of this encyclopedia. Within recent years, physicians and psychiatrists have directed increasing attention to the physiological and psychological problems associated with withdrawal. As pointed out by Sellers and Kalant (1976), the abrupt discontinuation of ethanol is followed by compensatory neuronal excitability and catecholamine release, which produce muscle tension and tremor, hyperacuity of sensory modalities, hyperreflexia, overalertness, anxiety, sleeplessness, and a reduction of seizure threshold. In the management of mild withdrawal, thiamine is administered intramuscularly and a drug for sedation, such as chlordiazepoxide, may be used. Severe withdrawal (delirium tremens) infrequently occurs in persons under 30 years of age. Symptoms of severe withdrawal in older persons usually appear within 48 to 60 hours after drinking has ceased. These may include severe tremulousness, seizure (in about 10% of cases), hallucinations (auditory and visual or both), tachycardia, a confused mental state, diaphoresis, and agitation. Physicians treat delirium tremens as an emergency situation because approximately 15% of cases are fatal. A hospital environment is required for effective treatment.

Hospital personnel and law enforcement officers must always be aware of the fact that hypoglycemia (as resulting from an overdosage of insulin) can produce seizures that mimic those of extreme intoxication.

It is estimated that well over 100 various drugs have been used over the past few decades to assist in alcoholic withdrawal; the drugs of choice are the benzodiazepines. Heavy smokers usually require a higher dosage than nonsmokers.

While consistent drinkers face numerous risks as previously outlined, the pregnant woman faces a double risk. There is a growing and convincing body of evidence that alcohol consumption is connected with dangers to the offspring, such as microcephaly and multiple congenital anomalies (Ouellette et al., 1977).

References

Asokan, S. K., Frank, M. J., and A. J. Witham: "Cardiomyopathy Without Cardiomegaly in Alcoholics," *Am. Heart J.*, **84**, 13 (1972).

Bennion, L. J., and T.-K. Li: "Alcohol Metabolism in American Indians and Whites; Lack of Racial Differences in Metabolic Rate and Liver Alcohol Dehydrogenase," *N. Engl. J. Med.*, **294**, 9 (1976).

Berry, R. E., Jr.: "Estimating the Economic Costs of Alcohol Abuse," *N. Engl. J. Med.*, **295**, 620 (1976).

Demakis, J. G., et al.: "The Natural Course of Alcoholic Cardiomyopathy," *Ann. Intern. Med.*, **80**, 293 (1974).

Fenna, D., et al.: "Ethanol Metabolism in Various Racial Groups," *Can. Med. Assoc. J.*, **105**, 472 (1971).

Fisher, J., and J. Abrams: "Life-Threatening Ventricular Tachyarrhythmias in Delirium Tremens," *Arch. Intern. Med.*, **137**, 1238 (1977).

Fox, J. H., et al.: "Cerebral Ventricular Enlargement: Chronic Alcoholics Examined by Computerized Tomography," *J. Amer. Med. Assn.*, **236**, 365 (1976).

Gordon, G. G., et al.: "Effect of Alcohol (Ethanol) Administration on Sex-Hormone Metabolism in Normal Men," *N. Engl. J. Med.*, **295**, 793 (1976).

Goodwin, D. W., et al.: "Alcohol Problems in Adoptees Raised Apart from Alcoholic Biological Parents," *Arch. Gen. Psychiatry*, **28**, 238 (1973).

Goodwin, D. W., et al.: "Drinking Problems in Adopted and Nonadopted Sons of Alcholics," *Arch. Gen. Psychiatry*, **31**, 164 (1974).

Goodwin, D. W., et al.: "Alcoholism and Depression in Adopted-Out Daughters of Alcoholics," *Arch. Gen. Psychiatry*, **34**, 751 (1977).

Klatsky, A. L., et al.: "Alcohol Consumption and Blood Pressure: Kaiser-Permanente Multiphasic Health Examination Data," *N. Engl. J. Med.*, **296**, 1194 (1977).

Mitchell, J. H., and L. S. Cohen: "Alcohol and the Heart," *Mod. Concepts Cardiovasc. Dis.*, **39**, 109 (1970).

Orlando, J., et al.: "Effect of Ethanol on Angina Pectoris," *Ann. Intern. Med.*, **84**, 652 (1976).

Ouellette, E. M., et al.: "Adverse Effects on Offspring of Maternal Alcohol Abuse During Pregnancy," *N. Engl. J. Med.*, **197**, 528 (1977).

Partanen, J., Bruun, K., and T. Markkanen: "Inheritance of Drinking Behavior: A Study on Intelligence, Personality, and Use of Alcohol of Adult Twins," *Finnish Foundation for Alcoholic Studies*, Helsinki, 1966.

Sellers, E. M., and H. Kalant: "Alcoholic Intoxication and Withdrawal," *N. Engl. J. Med.*, **294**, 757 (1976).

Williams, K. H.: "Medical Consequences of Alcoholism," *Ann. Intern. Med.*, **81**, 265 (1974).
Wu, C. F., Sudhakar, M., and G. Jaferi: "Preclinical Cardiomyopathy in Chronic Alcoholics: A Sex Difference," *Am. Heart J.*, **91**, 281 (1976).

ALCOHOL (Pancreatitis). Pancreas.

ALCOHOLS. The alcohols may be regarded as hydrocarbon derivatives in which the hydroxyl group (OH) replaces hydrogen on a saturated carbon. Alcohols are classified as *primary, secondary,* or *tertiary,* according to the number of hydrogen atoms that are bonded to the carbon atom with the hydroxyl substituent. Alcohols also may be regarded as alkyl derivatives of water. Thus, alcohols with a small hydrocarbon group tend to be more like water in properties than a hydrocarbon of the same number. Alcohols with a large hydrocarbon group are found to have physical properties similar to a hydrocarbon of the same structure. Some comparisons are given in Table 1. Structures are summarized by:

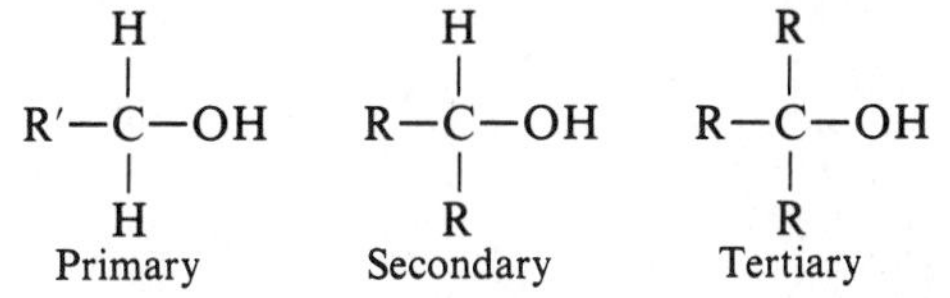

Primary Secondary Tertiary

where $R' = H$, alkyl, aryl; $R =$ alkyl, aryl.

In addition to the basic classification as primary, secondary, or tertiary, alcohols may be further grouped according to other structural features. *Aromatic alcohols* contain an aryl group attached to the carbon having the hydroxyl function; *aliphatic alcohols* contain only aliphatic groups. The prefix *iso* usually indicates branching of the carbon chain.

Alcohols containing two hydroxyl groups are called *dihydric alcohols* or *glycols.* Ethylene glycol, $HOCH_2CH_2OH$, trimethylene glycol, $HOCH_2CH_2CH_2CH_2OH$, and 1,4-butanediol are examples of industrially important glycols. Glycerol, $HOCH_2CHOHCH_2OH$, has three hydroxyl groups per molecule and is a *trihydric* alcohol. Physical properties of alcohols containing more than one hydroxyl group can be estimated by considering the number of carbons for each hydroxyl group as in the case of simple alcohols.

Reactions of Alcohols. Alcohols undergo a large number of reactions. However, these reactions may be grouped into a few general types. Reactions of alcohols may involve the O—H or C—O bonds. Ester formation and salt formation are examples of the former class, while conversion to halides is an example of the latter type.

—O—H Bond Cleavage

$$(1)\ ROH + CH_3COOH \overset{H^+}{\rightleftharpoons} CH_3COOR + H_2O$$
$$(2)\ ROH + K \rightarrow RO^-K^+ + \tfrac{1}{2}H_2$$

C—O Bond Cleavage

$$(3)\ RCH_2OH + HCl \xrightarrow[170°C]{ZnCl_2} RCH_2Cl + H_2O$$

Many industrially important substitution reactions of alcohols are conducted in the vapor phase over a catalyst. Only primary alcohols give satisfactory yields of product under these conditions.

$$(4)\ CH_3OH + H_2S \xrightarrow{K_2WO_4} CH_3SH + H_2O$$
$$(5)\ RCH_2OH + (CH_3)_2NH \xrightarrow{Al_2O_3} RCH_2N(CH_3)_2 + H_2O$$

Production of Alcohols. Lower alcohols (amyl and below) are prepared by (a) hydrogenation of carbon monoxide (yields methanol), (b) olefin hydration (yields ethanol, isopropanol, secondary and tertiary butanol), (c) hydrolysis of alkyl chlorides, (d) direct oxidation, and (e) the OXO process.

$$(6)\ C{=}O + 2H_2 \rightarrow CH_3OH$$
$$(7)\ CH_3CH{=}CH_2 + H_2O \xrightarrow{H^+} CH_3CHOHCH_3$$
$$(8)\ C_5H_{11}Cl + H_2O \rightarrow \underset{\text{(mixture of isomers)}}{C_5H_{11}OH}$$

Most higher alcohols (hexanol and higher) and primary alcohols of three carbons or more are synthesized by one of four general processes, or derived from a structurally related natural product. See also **Organic Chemistry.**

The OXO Process. An olefin may be hydroformylated to a mixture of aldehydes. The aldehydes are readily converted to alcohols by hydrogenation. Many olefins from ethylene to dodecenes are used in the OXO reaction. OXO alcohols are typically a mixture of linear and methyl branched primary alcohols.

$$(9)\ RCH{=}CH_2 + CO + H_2 \rightarrow [RCH_2CH_2CHO + RCH(CHO)CH_3$$

Aldol Condensation. Aldehydes may also be dimerized by an aldol condensation reaction to give a branched unsaturated aldehyde. This may be converted to a branched alcohol by hydrogenation.

$$(10)\ 2CH_3CH_2CH_2CHO \rightarrow CH_3CH_2CH_2CH{=}\underset{C_2H_5}{\underset{|}{C}}CHO \xrightarrow{H_2} CH_3(CH_2)_3\underset{C_2H_5}{\underset{|}{C}}HCH_2OH$$

Alcohols from an aldol reaction may be linear if acetaldehyde is a reactant, but usually aldol alcohols are branched primary alcohols. An aldol condensation sometimes is done with an OXO reaction. The combined process is called the ALDOX process.

Oxidation of Hydrocarbons. Using air, the oxidation of hydrocarbons generally results in a mixture of oxygenated compounds and is not a useful synthesis of alcohols except under special circumstances. Cyclohexanol may be prepared by air oxidation of cyclohexane inasmuch as only one isomer can result.

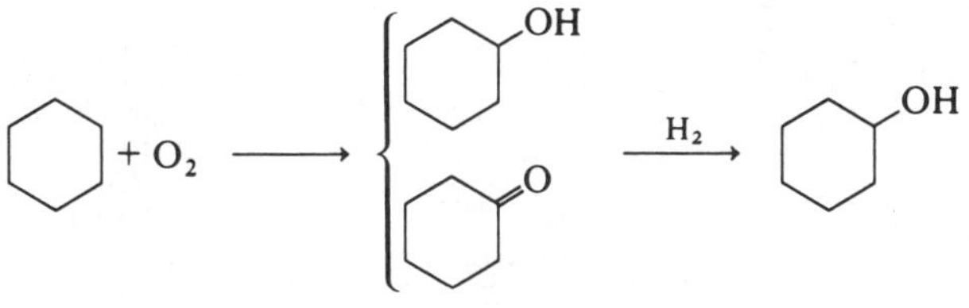

The yield of alcohol from normal paraffin oxidation may be improved to a commercially useful level by oxidizing in the presence of boric acid.

$$(12)\ 3RH + \tfrac{3}{2}O_2 + H_3BO_3 \rightarrow (RO)_3B + 3H_2O$$
$$(13)\ (RO)_3B + 3H_2O \rightarrow H_3BO_3 + 3ROH$$

A borate ester is formed which is more stable to further oxidation than the free alcohol. This is easily hydrolyzed to recover the alcohol. These alcohols which are predominately secondary are used in surfactant manufacture.

TABLE 1. COMPARISON OF PHYSICAL PROPERTIES OF ALCOHOLS AND HYDROCARBONS

Alcohol	Hydrocarbon	Formula	Properties
Methanol		CH_3OH	Liquid, water soluble, bp, 65°C
	Methane	$CH_3{-}H$	Gas, water insoluble
Ethanol		CH_3CH_2OH	Liquid, water soluble, bp, 78.5°C
	Ethane	$CH_3CH_2{-}H$	Gas, water insoluble
Tetradecanol		$CH_3(CH_2)_{12}CH_2OH$	Liquid, water insoluble, bp, 263.2°C
	Tetradecane	$CH_3(CH_2)_{12}CH_2{-}H$	Liquid, water insoluble, bp, 253.5°C

TABLE 2. PHYSICAL PROPERTIES OF SOME ALCOHOLS

Alcohol*	Formula	Formula Weight	Sp gr	Mp, °C	Bp, °C
Allyl	$CH_2{:}CH{\cdot}CH_2OH$	58.08	0.854	−129	96.6
Amyl (*n*-)	$CH_3(CH_2)_3CH_2OH$	88.15	0.817	−78.5	137.9
Amyl (*s*-, *n*-)	$C_2H_5CH_2CH(OH)CH_3$	88.15	0.810	....	119.5
Amyl (pri-, iso-)	$(CH_3)_2CHCH_2CH_2OH$	88.15	0.813	−117.2	132.0
Amyl (*s*-, iso-)	$(CH_3)_2CHCH(OH)CH_3$	88.15	0.819	....	113.4
Amyl (*t*-)	$(CH_3)_2C(OH)C_2H_5$	88.15	0.809	−11.9	102
Amyl, active	$C_2H_5CH(CH_3)CH_2OH$	88.15	0.816	....	128
Arabitol	$CH_2OH(CHOH)_3CH_2OH$	152.14	....	103	
Benzhydrol	$(C_6H_5)_2CHOH$	184.24	....	68	299
Benzyl	$C_6H_5CH_2OH$	108.13	1.043	−15.3	204.7
Borneol	$C_{10}H_{17}OH$	154.25	1.011	209	213
Butyl (*n*-)	$C_2H_5CH_2CH_2OH$	74.12	0.810	−79.9	117
Butyl (*s*-)	$C_2H_5CH(OH)CH_3$	74.12	0.808	−114.7	99.5
Butyl (iso-)	$(CH_3)_2CHCH_2OH$	74.12	0.805	−108	107–108
Butyl (*t*-)	$(CH_3)_3COH$	74.12	0.779	25.5	82.9
Ceryl	$C_{25}H_{51}CH_2OH$	382.72	....	80	
Cetyl	$CH_3(CH_2)_{14}CH_2OH$	242.43	0.818	49–50	189.5
Cholesterol (*n*-)	$C_{27}H_{45}OH$	386.66	1.067	148	< 360
Cholesterol (iso-)	$C_{27}H_{45}OH$	386.66	....	138	
Cinnamyl	$C_6H_5CH{:}CHCH_2OH$	134.18	1.040	33	254
Crotonyl	$CH_3CH{:}CHCH_2OH$	72.11	....	....	118
Cycloheptanol	$CH_2(CH_2)_5CHOH$	114.19	0.957	....	185
Cyclohexanol	$CH_2(CH_2)_4CHOH$	100.16	0.962	24	162
Decyl (*n*-)	$CH_3(CH_2)_8CH_2OH$	158.28	0.830	7	232.9
Diacetone	$(CH_3)_2C(OH){\cdot}CH_2COCH_3$	116.16	0.931	−47	167.9
Diglycerol	$[(HO)_2C_3H_5]_2O$	166.17	....	....	220–230
Dulcitol	$CH_2OH(CHOH)_4CH_2OH$	182.17	1.466	189	290–295
Eicosyl	$C_{19}H_{39}{\cdot}CH_2OH$	299.57	....	68	
Ergosterol	$C_{27}H_{41}OH$	382.63	....	160	
Erythritol	$CH_2OH(CHOH)_2CH_2OH$	122.12	1.451	126	329–331
Ethyl	CH_3CH_2OH	46.07	0.789	−112	78.4
Fenchyl	$C_{10}H_{17}OH$	154.24	0.935	35	201
Fluroenoyl	$(C_6H_4)_2CHOH$	182.22	....	116	156
Furfuryl	$C_4H_3O{\cdot}CH_2OH$	98.10	1.129	−14.6	169.5
Geraniol	$C_{10}H_{17}OH$	154.25	0.883	7–15	229
Glycerol	$CH_2OH{\cdot}CHOH{\cdot}CH_2OH$	92.09	1.260	17.9	290
Glycol (ethylene)	$CH_2OH{\cdot}CH_2OH$	62.07	1.113	−15.6	197.4
Heptyl (*n*-)	$CH_3(CH_2)_5CH_2OH$	116.20	0.824	34.6	175
Hexamethylene glycol	$HO(CH_2)_6OH$	118.17	....	42	250
Hexyl (*n*-)	$CH_3(CH_2)_4CH_2OH$	102.17	0.820	−51.6	157.2
Lauryl	$CH_3(CH_2)_{10}CH_2OH$	186.33	0.831	24	255–259
Mannitol	$CH_2OH(CHOH)_4CH_2OH$	182.17	1.489	166	290–295
Methyl	CH_3OH	32.04	0.729	−97.8	64.7
Methylphenyl carbinol	$(CH_3)(C_6H_5)CHOH$	122.17	1.003	....	205
Myricyl	$C_{31}H_{63}OH(?)$	452.82	0.777	88	
Myristyl	$CH_3(CH_2)_{12}CH_2OH$	214.38	0.824	38	167
Nitrobenzyl (*m*-)	$NO_2{\cdot}C_6H_4CH_2OH$	153.13	....	27	175–180
Nonyl (*n*-)	$C_8H_{17}CH_2OH$	144.26	0.828	−5	215
Octadecyl	$C_{17}H_{35}CH_2OH$	270.50	....	59	
Octyl (*n*-)	$CH_3(CH_2)_6CH_2OH$	130.22	0.827	−16	194–195
Octyl (*s*-, *n*-)	$CH_3(CH_2)_5CH(OH)CH_3$	130.22	0.822	38.6	179–180
1,4-Pentadiol	$HOCH_2(CH_2)_3CH_2OH$	104.15	0.994	....	239.4
Phenylethyl	$C_6H_5CH_2CH_2OH$	122.16	1.023	....	219–221
Phenylpropyl (*n*)	$C_6H_5(CH_2)_3OH$	136.19	1.008	< −18	235–237
Phytol	$C_{20}H_{37}CH_2OH$	308.55	0.850	....	145
Pinacol	$[(CH_3)_2C{\cdot}OH]_2$	118.17	0.967	43	171–172
Propargyl	$CH_3{:}C{\cdot}CH_2OH$	58.08	0.922	−17	115
Propyl (*n*-)	$CH_3CH_2CH_2OH$	60.09	0.804	−127	97–98
Propyl (iso-)	$(CH_3)_2CHOH$	60.09	0.789	−85.8	82.5
Propylene glycol	$CH_3CH(OH)CH_2OH$	76.09	1.040	....	188–189
Sorbitol	$[CH_2OH(CHOH)_2]_2$	182.17	1.47	110–112	
Terpineol	$C_{10}H_{17}OH$	154.25	0.933	35	220
Tetrahydrofurfuryl	$C_4H_4O{\cdot}CH_2OH$	102.13	1.050	....	177.8
Triethylene glycol	$({\cdot}CH_2OCH_2CH_2OH)_2$	150.17	1.125	−5	290
Trimethyl carbinol	$(CH_3)_3COH$	74.12	....	25.5	83
Trimethylene glycol	$HOCH_2CH_2CH_2OH$	76.09	1.060	....	214
Triphenyl carbinol	$(C_6H_5)_3COH$	260.32	1.188	162	< 360
Vanillic	$CH_3O(OH)C_6H_3CH_2OH$	154.16	....	115	d
Vinyl	$CH_2{:}CHOH$	44.06	....	....	
Vinyl (poly-)	$(CH_2{:}CHOH)_x$	$(44.06)_x$	1.3	220d	

* d, decomposes; *m*, meta; *n*, normal; pri-, primary; *s*, secondary; *t*, tertiary.

Synthesis from Alkylaluminums. Fundamental work on organoaluminum chemistry by Prof. Karl Ziegler and co-workers at the Max Planck Institute provided the basis for a commercial synthesis of even-carbon-numbered straight chain primary alcohols. These alcohols are identical with products derived from naturally occurring fats. In this process, ethylene is reacted with aluminum triethyl to form a higher alkylaluminum which then is oxidized and hydrolyzed to give the corresponding alcohols.

$$(14)\quad (C_2H_5)_3Al \xrightarrow{CH^2{=}CH^2} [CH_3(CH_2CH_2)_nCH_2]_3Al \xrightarrow[(2)\ H_2O]{(1)\ O_2} 3CH_3(CH_2CH_2)_nCH_2OH + Al(OH)_3$$

Commercialization of this route to higher alcohols is the most significant development in this area in recent years.

Synthesis from Natural Products. Many alcohols are prepared by reduction of the corresponding methyl esters which are derived from animal or vegetable fats. These alcohols are straight chain even-carbon-numbered compounds. Tallow and coconut oil are two major raw materials for higher alcohol manufacture.

$$(15)\ \underset{\text{Triglyceride}}{(RCOO)_3C_3H_5} + 3CH_3OH \rightarrow 3RCOOCH_3 + CH_2OHCHOHCH_2OH$$

$$(16)\ RCOOCH_3 + 2H_2 \xrightarrow{\text{Catalyst}} RCH_2OH + CH_3OH$$

Because of high prices, unreliability of crude oil supplies, and long-range outlook for petroleum shortages, considerable interest in alcohol production from wood, grains, sugar, and other renewable sources was rekindled during the late 1970s and is continuing into the 1980s. As of 1981, only a few new large ethyl alcohol plants designed essentially for production of alcohol to be used in "gasohol" fuels went on stream. This trend is expected to continue. Earlier interest in relatively small conversion units for use by individual farmers or small groups of farmers to provide fuel for tractors and other agricultural equipment has waned considerably during the last few years simply because of the unfavorable economics of production on a small scale. This also applies to making methyl alcohol from wood on a small scale. So long as natural gas supplies remain fairly reliable with relatively optimistic long-term forecasts, this source of raw material for alcohols will continue to factor heavily in the economics of alcohol production. Of course, the economics can be seriously balanced in favor of renewable sources if the governments of various countries offer subsidies for production, including taxation favorable to use of renewable sources. See also **Biomass and Wastes as Energy Sources; Ethyl Alcohol;** and **Methanol.** The properties of several alcohols of varying commercial importance are listed in Table 2 on facing page.

ALCOHOLYSIS. If a triglyceride oil is heated with a polyol, such as glycerol or pentaerythritol, mixed partial esters are produced in a reaction known as *alcoholysis.*

ALCYONARIA. An order of the class *Anthozoa* (*Actinozoa*) in the phylum *Coelenterata.* The animals of this order are marine and because of the hard deposits formed in and around their bodies they are often known as corals. Among them are the sea-pen, the sea-fan, organ-pipe coral, and precious coral.

Alcyonaria are usually found in colonies. The individuals are polyps connected together by living structures and by the hard skeletal structures. They differ from the true corals in having only eight tentacles, pinnately branched. See also **Coelenterata.**

ALDEBARAN (α Tauri). A star whose name is derived from an Arabic phrase indicating that the star is the "leader of the followers," i.e., the leader of the asterism known as the Hyades, which follow the Pleiades in their nightly journey across the sky. Astrologically, Aldebaran was a fortunate star, portending riches and honor. This star was one of the four royal stars of the Persians about 3000 B.C.

Aldebaran is one of the smaller stars, whose diameter has been measured with the stellar interferometer. The diameter is found to be about 6.4×10^7 kilometers, or 46 times the diameter of our sun. Ranking fourteenth in apparent brightness among the stars. Aldebaran has a true brightness value of 100 as compared with unity for the sun. Estimated distance from the earth is 53 light years. Aldebaran is classified as a star of orange color and of spectral type K. See also **Constellations;** and **Star.**

ALDEHYDES. The homologous series of aldehydes (like ketones) has the formula $C_nH_{2n}O$. The removal of two hydrogen atoms from an alcohol yields an aldehyde. Thus, two hydrogens taken away from ethyl alcohol $CH_3 \cdot C(H_2)OH$ yields acetaldehyde CH_3CH_2OH; and two hydrogens removed from propyl alcohol $C_2H_5 \cdot C(H_2)OH$ yields propaldehyde $C_2H_5 \cdot CHO$. The trivial names of aldehydes derive from the fatty acid which an aldehyde will yield upon oxidation. Thus, formaldehyde is named from formic acid, the latter being the oxidation product of formaldehyde. Similarly, acetaldehyde is oxidized to acetic acid. Or, the aldehyde may be named after the alcohol from which it may be derived. Thus, formaldehyde which may be derived from methyl alcohol may be named methaldehyde; or acetaldehyde may be named ethaldehyde since it may be derived from ethyl alcohol. In still another system, the aldehyde may take its name from the parent hydrocarbon from which it theoretically may be derived. Thus, propa*nal* (not to be confused with propa*nol*) may signify propaldehyde (as a derivative of propane).

Essentially aldehydes exhibit the following properties: (1) with exception of the gaseous formaldehyde, all aldehydes up to C_{11} are neutral, mobile, volatile liquids. Aldehydes above C_{11} are solids under usual ambient conditions; (2) formaldehyde and the liquid aldehydes have an unpleasant, pungent, irritating odor, (3) although the low-carbon aldehydes are soluble in H_2O, the solubility decreases with formula weight, and (4) the high-carbon aldehydes are essentially insoluble in H_2O, but are soluble in alcohol or ether.

The presence of the double bond (carbonyl group C:O) markedly determines the chemical behavior of the aldehydes. The hydrogen atom connected directly to the carbonyl group is not easily displaced. The chemical properties of the aldehydes may be summarized by: (1) they react with alcohols, with elimination of H_2O, to form *acetals;* (2) they combine readily with HCN to form *cyanohydrins,* (3) they react with hydroxylamine to yield *aldoximes;* (4) they react with hydrazine to form *hydrazones;* (5) they can be oxidized into *fatty acids* which contain the same number of carbons as in the initial aldehyde; (5) they can be reduced readily to form *primary alcohols.* When benzaldehyde is reduced with sodium amalgam and H_2O, benzyl alcohol $C_6H_5 \cdot CH_2 \cdot OH$ is obtained. The latter compound also may be obtained by treating benzaldehyde with a solution of cold KOH in which benzyl alcohol and potassium benzoate are produced. The latter reaction is known as Cannizzaro's reaction.

In the industrial production of higher alcohols (above butyls), aldehydes play the role of an intermediate in a complete process that involves aldol condensation and hydrogenation. In the OXO process, olefins are catalytically converted into aldehydes that contain one more carbon than the olefin in the feedstock. Aldehydes also serve as starting materials in the synthesis of several amino acids. See also **Acetaldehyde; Aldol Condensation; Benzaldehyde;** and **Furfuraldehyde.**

ALDEHYDIC CARBOXYLIC ACIDS. **Carboxylic Acids.**

ALDERFLY (*Insecta, Neuroptera*). The alderflies belong to the order *Neuroptera,* suborder *Megaloptera.* These are the most primitive members of the order, and the most primitive of all holometabolous insects. There are only two members of the *Megaloptera* in Europe, both of which are Alderflies (family *Sialidae*). These are small, with a wingspan of only 3 centimeters (1 inch). *Sialis lutaria* is commonly found in central Europe in the spring or summer, sitting on shore plants by still or slowly flowing waters. The second species, *Sialis fuliginosa,* is usually rather more dusky in color and differs from the first in the wing venation and the structure of the hind end of the body. The Smoky Alderfly (*Sialis infumata*) of the U.S., with brownish-black wings, is similar in size and habits to the European species. At rest, the alderflies lay their wings together over the body like a roof, as do almost all the *Neuroptera.* They are hunters, both as adults and as larvae. The larvae, however, are aquatic.

ALDER TREES. Members of the family *Betulaceae* (birch family), these trees are of the genus *Alnus.* They are deciduous trees or shrubs and are considered very hardy. Important species include:

Arizona alder	*Alnus oblongifolia*
Caucasian alder	*A. subcordata*
European alder	*A. glutinosa*
Gray alder	*A. incana*
Green alder	*A. viridis*
Hazel alder	*A. serrulata*
Italian alder	*A. cordata*
Red alder	*A. rubra*
Sitka alder	*A. sinuata*
Smooth or speckled alder	*A. rugosa*
Thinleaf alder	*A. tenuifolia*
White alder	*A. rhombifolia*

The red alder is found in forests on the west coast of the United States and Canada, from California northward to Alaska. The average height is from 40 to 60 feet (12 to 18 meters). However, as will be noted from the accompanying table of record alder trees, the red alder can reach heights of close to 100 feet (30 meters). The white alder occurs on the Pacific slopes and can attain a height of over 100 feet (30 meters), although the average is from 50 to 70 feet (15 to 21 meters). Staminate flowers are contained in catkins of about 5 inches (13 centimeters) in length. The cones range from $\frac{1}{2}$ to $\frac{7}{8}$ inch (1.2 to 2.2 centimeters) in length. The leaves are ovate and of a light yellow-green color. The tree ranges from northern Idaho westward to southeastern Oregon, with preference for the eastern slopes of the Cascade Mountains. The tree also is found in the Coastal Range and Sierra Nevadas down into southern California.

Alder wood is durable and well suited for applications in wet and damp areas, as may be required for sashes and bridge construction. In areas where forest fires have cleared the land, alders are fast-growing and come up abundantly. Alders have nodules which form on the root and which help to enrich the soil. However, if hemlock and Douglas fir are present, these latter trees can soon overtake the alder in height, smothering out all except the very strongest trees. Alder lumber is usually shipped from west coast forests. In the green condition, red alder has a moisture content of 98% and weighs 46 pounds per cubic foot (737 kilograms per cubic meter). When air-dried to 12% moisture content, the weight is 28 pounds per cubic foot (448.5 kilograms per cubic meter), with 1,000 board-feet (2.36 cubic meters) of nominal sizes weighing about 2330 pounds (1057 kilograms). Crushing strength when compression is applied parallel to the grain is 2960 psi (20.4 MPa) for the green wood; 5820 psi (40.1 MPa) for the dry wood. Tensile strength when tension is applied perpendicular to the green wood is 390 psi (2.7 MPa) for the green wood, 420 psi (2.9 MPa) for the air-dried wood.

Champion Alder trees in the United States are listed in the accompanying table.

ALDOL CONDENSATION. A reaction between aldehydes or aldehydes and ketones which occurs without the elimination of any secondary product, and yields β-hydroxycarbonyl compounds. It is distinguished from polymerization by the facts that it occurs between aldehydes and ketones and is not generally reversible. In its simplest form it may be represented by the condensation of two molecules of acetaldehyde to aldol:

$$CH_3 \cdot CHO + CH_3 \cdot CHO \rightarrow CH_3 \cdot CHOH \cdot CH_2 \cdot CHO$$

Weak alkalies and acids are employed to effect the condensation.

ALDOSES. **Carbohydrates.**

ALDOSTERONE. **Adrenal Glands; Chloride; Hormones; Hypertension (High Blood Pressure); Steroids.**

ALDOXIMES. **Hydroxylamine.**

ALEURITES. **Euphorbiaceae.**

ALEURONE GRAINS. Protein reserves found in the seeds of several different kinds of plants. In many plants there is a special aleurone layer of definite thickness found in the endosperm. In corn (maize), the layer is a single cell in thickness and may contain a colored pigment; in oats, it is two cells thick. In the castor oil plant, the aleurone grains are not restricted to a single layer, but are distributed rather generally in the endosperm, and have a complex structure. See accompanying figure.

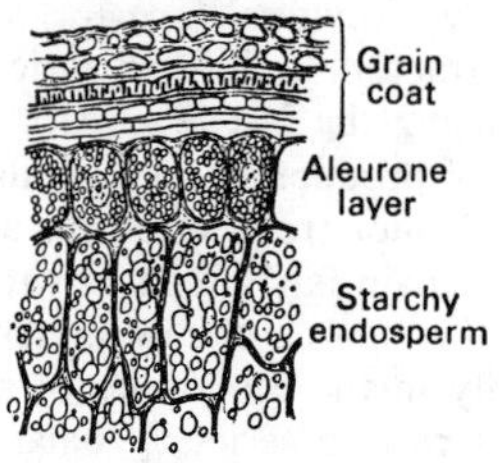

The aleurone layer in wheat. Part of a section of a wheat grain showing aleurone layer with cells filled with granules of protein.

ALEUTIAN LOW. **Atmosphere (Earth).**

ALEWIFE (*Osteichthyes*). A common fish (*Pomolobus pseudoharengus*) of the Atlantic coast. It is related to the herring and shad. It enters the streams to spawn and is found in lakes of New York. The alewife is generally considered as an unsatisfactory food fish. Normally found in marine waters, Nova Scotia to Florida. Maximum length about 11 inches (28 centimeters).

ALEXANDRITE. A variety of chrysoberyl, originally found in the schists of the Ural Mountains. It absorbs yellow and blue light rays to such an extent that it appears emerald green by daylight but columbine-red by artificial light. It is used as a gem, and was named in honor of Czar Alexander II of Russia. See also **Chrysoberyl.**

ALFALFA. Of the family *Leguminosae* (pea family), alfalfa (*Medicago sativa*), is a plant native to southwestern Asia. It is an important

RECORD ALDER TREES IN THE UNITED STATES[1]

Specimen	Circumference[2] (inches)	Circumference[2] (centimeters)	Height (feet)	Height (meters)	Spread (feet)	Spread (meters)	Location
Arizona alder (1973) (*Alnus oblongifolia*)	100	254	90	27	40	12	Arizona
European alder (1974) (*A. glutinosa*)	88	224	68	20.4	44	13.2	Illinois
Hazel alder (1965) (*A. serrulata*)	17	43	40	12	22	6.6	Ohio
Red alder (1975) (*A. rubra*)	240	610	75	22.5	69	20.7	Oregon
Seaside alder (1975) (*A. maritima*)	13	33	28	8.4	11	3.3	Virginia
Sitka alder (1967) (*A. sinuata*)	21	53	25–30	7.5–9	—	—	Oregon
Speckled alder (1972) (*A. rugosa*)	32	81	57	17.1	28	8.4	Michigan
Thinleaf alder (1976) (*A. tenuifolia*)	52	132	68	20.4	25	7.5	New Mexico
Thinleaf alder (1976) (*A. tenuifolia*)	60	152	57	17.1	37	11.1	New Mexico
White alder (1977) (*A. rhombifolia*)	174	442	99	29.7	45	13.5	Oregon

[1] From the "Social Register of Big Trees," The American Forestry Association (by permission).
[2] At 4.5 feet (1.4 meters) from ground level.

perennial forage plant, extensively cultivated since the Roman civilization. The plant has a very deeply penetrating root system, reaching down 25 feet (7.5 meters) or more and thus is well suited for growing in dry lands where resistance to drought is important. The fragrant, purplish flowers are a good source of honey. The plants is cultivated principally for hay. See also **Leguminosae.**

ALFÉN WAVE. A transverse wave in a hydromagnetic field in which the driving force is the tension introduced by the magnetic field along the lines of force. The dynamics of such waves are analogous to those in a vibrating string, the phase speed being given by

$$C^2 = \mu H^2/4\pi\rho \tag{1}$$

where μ is the permeability, H the magnitude of the magnetic field, and ρ the fluid density. Dissipative effects due to fluid viscosity and electrical resistance may also be present.

For proton-electron plasmas, a simpler expression than that of Equation (1) may be obtained in terms of the magnetic field strength, B, and the number of charged particles, n, per unit volume. This expression gives the phase velocity as

$$C = \frac{(1.7 \times 10^6)B}{n^{1/2}}$$

The Alfén wave behaves like a conventional shock wave, such as a supersonic sound wave, in having a high energy front resulting from an excess of the plasma velocity over the Alfén velocity. However, it differs in that it is produced by the interaction of waves instead of particles, since the latter, in, for example, a plasma in the upper atmosphere or in space, are too widely spaced to have an appreciable number of collisions.

ALFISOLS. **Soil.**

ALFORD ANTENNA. **Antenna.**

ALGAE. Thallophytes characterized by possessing chlorophyll, and so capable of elaborating their food by photosynthesis. Often the green pigment is completely concealed by other pigments, so that the plant is brown, red, or even black.

Algae are found in almost every habitat. In the oceans vast numbers of minute species float suspended in the upper levels of the water, while the shores are covered with many and varied forms from high tide level to depths of 30 feet (9 meters) or more. In fresh water they are equally abundant, but due to their smaller size are seldom so conspicuous as the marine forms; they occur in running water, in ponds, and in stagnant, often putrid, water. Many species are found only in hot springs. They are found on the surface of the ground, on the bark of trees, on rocks, and even underground to a depth of several feet. A few species have found a favorable habitat within the bodies of higher plants and animals. In fact, wherever they find support and can obtain the necessary materials for growth, there algae may be found.

Algal plants offer a wonderful diversity of forms. In size they range from unicellular microscopic plants to structures having dimensions comparable to the larger land plants. The plankton forms, those free-floating plants often so abundant in both fresh and salt water, are nearly all unicellular; other free-floating forms, usually found near shore or in small fresh water ponds and streams, are multicellular organisms of various shapes, filamentous forms being especially common. Finally, attached marine forms often attain massive dimensions. The common kelp, or devil's apron, of the colder coastal waters of North America may grow to a length of 30 feet (9 meters) or more, and to a width of 2 or 3 feet (0.6 or 0.9 meters), while related species found in the Pacific Ocean far exceed them in size, reaching lengths of 100 feet (30 meters) and more.

Not only do algae vary greatly in size, but they also show almost every conceivable shape. Unicellular types are often adorned with a complex but beautifully symmetrical series of arms, or bristles, which may be of service in keeping them floating in the water. The filamentous forms may be simple or very much branched; often they are delicate plants of rare beauty. Other algae grow in flat sheets or membranes, either spreading over the substratum or rising gracefully in the water. The larger forms are of coarser habit, varying from irregular tumorous plants to long slender cords and broad flat fronds. Plants of the genus *Sargassum*, one of the brown algae found in warm regions, have an appearance very similar to that of flowering plants. Each plant has a slender branching stalk, often 2 feet (0.6 meters) or more in length. From this stalk flat lateral branches arise, which look very much like leaves, except for their brown color. Other short lateral branches end in small sub-spherical balls easily mistaken for fruits. Actually these structures are hollow bladders which help to keep the plant floating. Other branches, the real reproductive parts of the plant, are short cylindrical objects which might be mistaken for buds. At first the plant grows on rocks and other solid objects. It is easily broken loose, however, and floats about in the ocean currents. Thus, these plants are frequently washed up on northern beaches.

There are several systems of classification of algae, varying in details but all using the various pigments found in their cells as a basis for separation. Obviously such a classification is very artificial, but in the algae it seems to agree quite closely with natural systems based on such other criteria as the structure of the thallus or plant body, as the substances formed by the cells and stored in them, and especially as the reproductive processes which are found in the different groups.

The algae are now commonly separated into seven divisions or phyla, although in an older system these groups had the status of Class. The four larger phyla are the *Cyanophyta* or blue-green algae. *Chlorophyta* or green algae, *Phaeophyta* or brown algae, and the *Rhodophyta* or red algae. The smaller phyla are the Euglenophyta or Flagellates, the Chrysophyta, including the yellow-green and golden-brown algae and the diatoms, and the Pyrrophyta or cryptomonads.

The blue-green algae are characterized by having within the cell, in addition to chlorophyll, a bluish pigment, phycocyanin. These pigments are not localized in a definite pigment-bearing body or plastid, but are frequently diffused in the outer zone of protoplasm. Surrounding the protoplasm is a definite cell wall of cellulose (see **Carbohydrates**). In most blue-green algae the outer part of the wall is modified and becomes a soft slimy substance which often forms a layer of considerable thickness. This slime substance may be of great value as an insulation against heat and desiccation, thus enabling blue-green algae to live in what seem to be most unfavorable environments. In the central portion of the protoplast are found many chromatin bodies, which, however, are not organized into a definite nucleus. The structure of the cell of these algae, with its absence of plastids and any definite nucleus, seems to indicate a relatively primitive organism, and is suggestive of the structure found in bacteria. See Fig. 1.

Many blue-green algae are single-celled organisms. The individual cells are often separate or they may be held together by the gelatinous outer wall in aggregates sometimes of considerable size. Representatives of this group are frequently observed in temporary puddles formed by a summer shower or in quiet shallow ponds in which the water often becomes very warm. Some of them are of considerable economic importance, when they appear in water supply reservoirs.

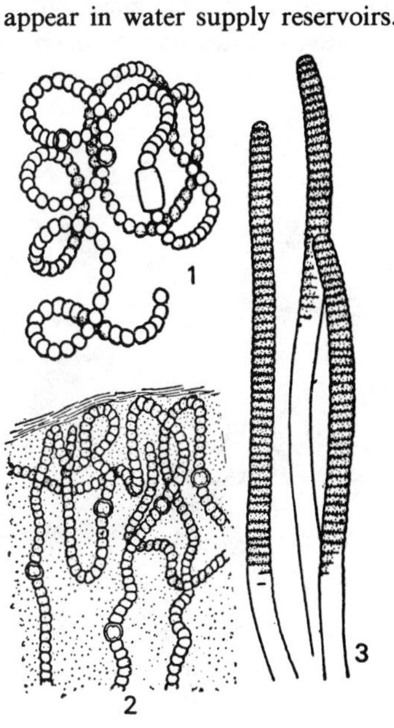

Fig. 1. Three species of *Myxophyceae*: (1) filaments of a species of *Anabaena*; (2) filaments of *nostoc* encased in the characteristic jelly; (3) *Oscillatoria*. (*After Smith and Fremy*)

In these they sometimes occur in numbers so great as to color the water and to be only too obvious to the most casual observer. Such occurrences are frequently described as "water-blooms." Due to the products formed by the metabolism of the cells and liberated into the water, such "blooms" are real problems, for not only do these substances give to the water a distinctly unpleasant oily fishy taste and color, but several cases are recorded in which drinking of such water has been quickly fatal to livestock.

Other blue-green algae occur as simple filaments of cells. Single filaments may occur among other algae or they may exist in extensive masses covering considerable areas with a soft felt-like layer. In some genera many filaments are held together in a common gelatinous sheath. Many blue-green genera show what is called false branching. The rapid division of cells in the middle of a filament causes them to grow out laterally. Sometimes the filaments grow out as a single branch, as in *Tolypothrix*, or in pairs, as in *Scytonema*. True branching is found in a few genera.

In all the non-filamentous species the only method of reproduction is that of cell division. In the filamentous forms continued division may produce a filament of indefinite length. However, it eventually breaks up. This fragmentation may be due to animals feeding on cells of the filament, or to the death of certain cells. In some forms it is due to the development of cells which do not adhere tenaciously to the cells adjoining them. Specialized cells known as heterocysts are largely responsible for the last condition. They are large transparent cells which develop from ordinary cells. See Fig. 2.

Blue-green algae, particularly abundant in regions having a warm climate, develop elsewhere in great abundance, during the warmer seasons of the year. Besides being a source of trouble in water reservoirs, blue-green algae sometimes form unsightly stains by growing on the walls of stone buildings, especially if the latter be constantly wet. Some of the blue-green algae have the ability to fix nitrogen (see **Bacteria**). It has been suggested that this may be important to such crops as rice.

The green algae, or Chlorophyta, are found in both salt and fresh water, where they often form conspicuous masses. They are characterized by a bright green color. The reserve food stored by these algae is starch, which is usually found around certain bodies known as pyrenoids, located in the chloroplasts.

The green algae are of considerable interest because of the possibility that from them the higher plants may have arisen, and because of the diversity of forms which are found within in the group.

The protoplasts of the green algae are with very few exceptions enclosed within a rigid wall composed of two layers, an inner made up wholly or largley of cellulose and an outer layer of pectin (see **Carbohydrates**). Within the protoplast of the cell is located one or more conspicuous chloroplasts containing pigments approximately like those occurring in the plastids of higher plants. The chloroplasts of any single genus are usually very constant in appearance but in the different genera remarkable diversity of size and shape exists. The primitive form seems to be the massive cup-shaped type such as occurs in many lower Chlorophyta. The nucleus of all green algae is a definitely organized body possessing a nuclear membrane, one or more nucleoli, nuclear sap (karolymph), and chromosomes. Many unicellular forms have one or more cilia which persist throughout their existence. The reproductive cells of most green algae have cilia. See **Cilium.**

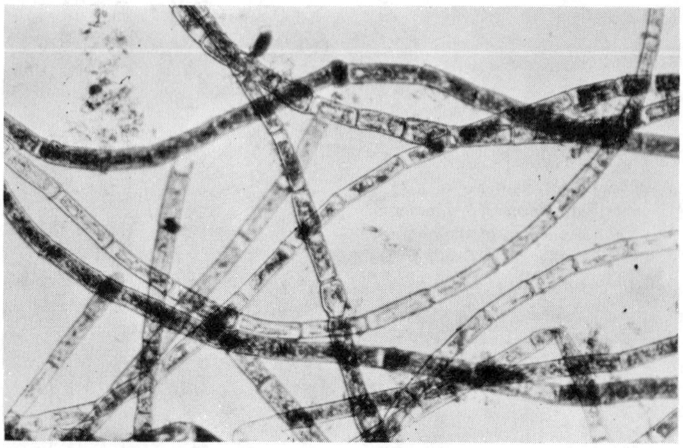

Fig. 2. *Oedogonium.* Filamentous cells in fresh water. (*Photomicrograph by B. J. Ford; copyright*)

Reproduction in green algae takes place in various ways. One is a strictly vegetative process in which the colony, or filament, of cells is broken up by various external agents, after which each fragment becomes a new colony, or filament.

Asexual reproduction is by means of cell division (nuclear division followed by division of the cytoplasm) or by zoöspore formation. These zoöspores are generally formed from the protoplast of any vegetative cell, and may appear singly or in numbers by division of a single protoplast. They are expelled from the cell in a manner as yet unknown and are frequently enclosed in delicate vesicle at the time of expulsion. The zoöspores are naked bodies, having no cell wall, and possessing apical cilia. *Vaucheria* is an exception, for its zoöspores are covered with cilia. After periods of motility varying from a few minutes to many hours, zoöspores become quiescent, withdraw their cilia, secrete a cell wall and develop to new colonies or organisms like the parent form.

Finally, many green algae possess a sexual reproduction which is often very complicated. In sexual reproduction there are formed two sets of reproductive bodies known as gametes, which fuse in pairs to form a zygote. From the zygote a new plant develops.

The great diversity of form found in the green algae makes them interesting plants to examine for lines of evolution which have produced the many forms existent today. Several different lines have been found. One of these includes many of the forms which remain motile throughout their existence. A relatively simple unicellular organism is the starting point for such a line. *Chlamydomonas* is of this type. The plant is a small spherical or oval cell enclosed in a definite cell wall and having two apical cilia. Within the protoplast there is a single massive chloroplast which has a shape like that of a soft rubber ball pushed deeply in on one side. Within the hollow of the plastid the single distinct nucleus is located. At the apex of the cell near the origin of the two cilia there is a minute red body called an eye-spot. This is a light-sensitive organ which when stimulated causes the organism to move toward or away from the light source. In the apical end there is a pair of contractile vacuoles. *Chlamydomonas* reproduces asexually by means of zoöspores. These are formed by divisions of the protoplast to form 2, 4 or 8 daughter protoplasts, contained within the wall of the original cell. This wall softens and liberates these naked cells. At once they become small replicas of the parent cell. They soon grow to the size of the original cell and again form a new group of zoöspores. In a suitable environment this is a very rapid method of reproduction. *Chlamydomonas* also reproduces sexually.

In sexual reproduction the protoplast of the cell divides to form motile bodies called gametes, which are quite like zoöspores, but smaller. On liberation from the parent cell wall, gametes from different cells unite in pairs, and form zygotes which develop into zygospores. A zygospore is a resistant spore with a thick wall which enables the organism to survive periods of adverse conditions. When favorable conditions return, the contents of the zygospore divide to form zoöspores, which behave as do similar spores from motile cells.

The primitive character of *Chlamydomonas* is found in its contractile vacuoles, its eye-spot, its single massive plastid, and its cilia. From Flagellates it differs only in having a definite cell wall. Comparing other motile green algae with *Chlamydomonas* makes it possible to discover an interesting series of species of increasing complexity. First in this series is *Gonium sociale*, with colonies of four cells, and *Gonium pectorale*, with sixteen cells, held together loosely in a gelatinous matrix. All the cells of a colony are alike and any cell may form either zoöspores or gametes. All gametes are alike, but fusion occurs between gametes from different colonies. Next in the line of increasing complexity is *Pandorina*, in which a spherical colony is formed. In this genus the gametes are slightly different in size and behavior, some being small and active, while others, slightly larger, are more sluggish. This is an indication of a differentiation of sex.

Still further advance is shown by *Eudorina*, in which a colony is composed usually of 32 cells located in the peripheral portion of the gelatinous matrix. Each cell of the colony is like a *Chlamydomonas* cell and each is capable of reproducing asexually. But in sexual reproduction a very obvious difference in sexes is apparent. Some colonies are definitely female, the cells enlarging slightly and functioning as eggs. In other colonies, each cell divides to form 64 minute biciliate sperm. Fusion between an egg and a sperm produces an oöspore which gives rise to a new colony. In *Pandorina* then a very obvious distinction between sexes has appeared, but the vegetative cells remain alike. In *Pleodorina* each colony is composed of small, purely vegetative cells and larger reproductive cells. The smaller vegetative cells are formed in the anterior end of the colony. In *Volvox* we find the highest degree of differentiation exhibited in this line of motile algae. In this plant the number of cells in a colony is very great, in some species there being as many as 25,000. Of these cells only a few are reproductive, while thousands remain vegetative. Many *Volvox* colonies are so large as to be readily visible to the unaided human eye. In these the many cells form a single layer embedded in the gelatinous matrix. The center of the colony is either water or a thin gelatinous substance. The cells of the colony have the same structure as *Chlamydomonas* cells. They are joined together by fine protoplasmic strands. The beating of the cilia causes the large colony to roll rapidly about in the water, making it a fascinating object to watch. In asexual reproduction certain cells of the colony enlarge and move to the central region. There they lose their cilia, after which they divide rapidly to form new colonies which remain for some time within the parent. Often a single colony will contain a dozen or more of these small colonies. The latter are liberated by the disintegration of the parent colony. In sexual reproduction, cells in the posterior region of the colony differentiate. Some lose their cilia and become very large; these are eggs. Other cells, either in the same or different colonies, divide many times to form large numbers of minute biciliate sperm. These swim to the eggs. A single sperm enters an egg, its nucleus fusing with that of the egg. As a result a zygote is formed. This secretes around itself a thick wall and becomes an oöspore, capable of enduring protracted periods of unfavorable conditions. With the return of favorable conditions the thick wall of the oöspore breaks, the protoplast emerges and divides rapidly, forming a new colony. *Volvox* represents the climax reached in this line of evolution. There is not only a very great increase in the number of cells forming a colony, but also a distinct separation of vegetative and reproductive cells. The reproductive cells are of two kinds, eggs and sperm. But every cell of the plant retains the primitive character of the individual cell.

It is possible to build up other evolutionary series of green algae in which the vegetative cells are non-motile. *Chlamydomonas* often assumes a non-motile condition; the cells become embedded in a copious gelatinous matrix and lose their cilia. This condition is known as the palmella stage. *Tetraspora* is an alga which has the appearance of the palmelloid stage of *Chlamydomonas*. See Fig. 3. Another genus, *Palmella*, normally exists as a shapeless colony of cells held together in a gelatinous matrix. Cilia are lacking, but may be developed by any cell in the colony. A ciliated cell escapes and swims about freely for a time, then settles down and divides to form a new colony. Asexual zoöspores are formed in *Palmella* and also isogametes, that is, gametes of equal size. *Palmella* shows the beginning of a non-motile habit, with a restriction of the motile stages to the reproductive cells. In *Geminella* the amorphous habit of the colony is lost; divisions take place in such a way that the resulting cells tend to exist in a single series, the individual cells being held together only by the gelatinous

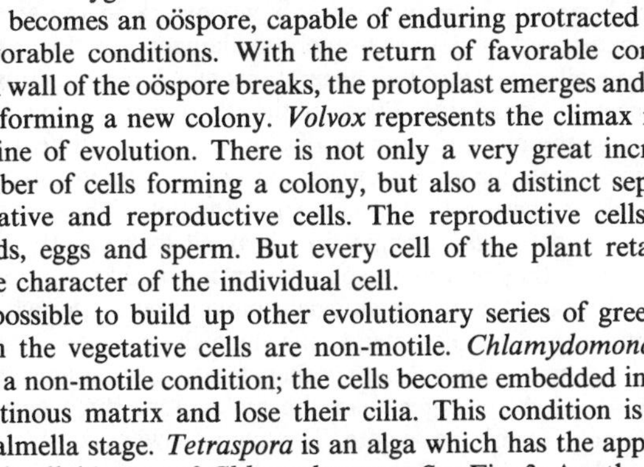

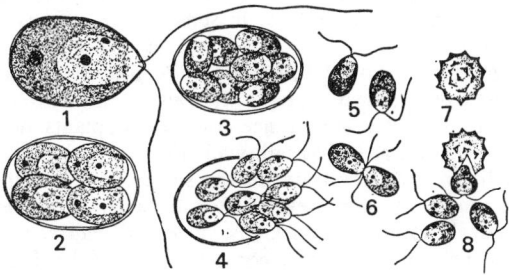

Fig. 3. *Chlamydomonas*: (1) mature cell; (2) four young cells formed in asexual reproduction; (3) eight gametes formed by the division of one cell; (4) gametes escaping from old cell wall; (4) free-swimming gametes; (6) two gametes conjugating; (7) resting zygote; (8) four cells formed from the zygote.

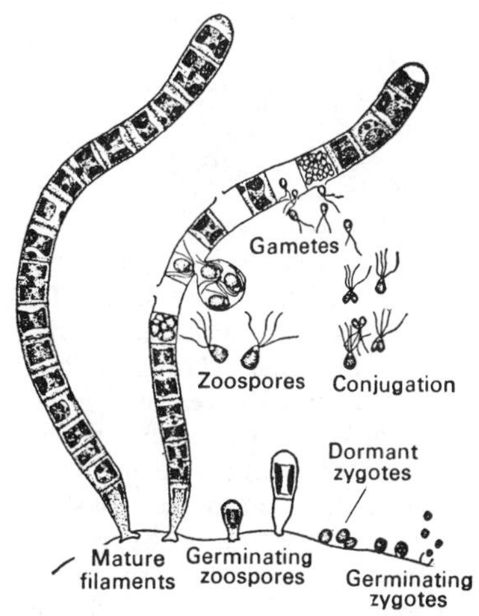

Fig. 4. *Ulothrix*. Stages in development.

matrix around them. In *Ulothrix* further advance is made. See Fig. 4. In this plant the protoplast of the cell divides within the wall of the cell. But it does not escape therefrom; instead cross-walls are formed between daughter protoplasts which remain permanently joined. Repeated divisions in a single direction result in the formation of a long unbranched filament of cells. Asexual reproduction in *Ulothrix* is by zoöspores. These are very similar to the cells of *Chlamydomonas*, but each has 4 cilia. From each zoöspore new filaments are formed directly. Sexual reproduction is by biciliate gametes which are formed in the usual manner. Two gametes fuse to form a zygote, which develops a thick wall, becoming a zygospore, and on germinating produces zoöspores. In some species of *Ulothrix* the gametes are alike, while in others slight differences in size of the gametes produced from different cells indicate the beginning of sex differentiation.

Other genera of algae, related to *Ulothrix*, show greater differentiation in the vegetative cells. Branching occurs in many; certain cells produce the zoöspores or gametes. Other genera, notably the marine *Ulva*, the Sea Lettuce, have cell divisions in two planes, so that extensive membranes are formed, often two meters or more in length.

The highest stage of development in this series is found in *Coleochaete*. This alga appears as small disks epiphytic on other aquatic plants. It consists of branching filaments which in some species grow out to form a flat shield-like thallus. Asexual reproduction is by biciliate zoöspores, which are formed singly from the protoplast of any cell. In its sexual reproduction, *Coleochaete* is especially interesting, exhibiting an advanced degree of differentiation of sex cells called oögamy. The eggs are formed in special cells at the tips of certain filaments. By continued growth, the vegetative cells form a layer enclosing the oögonium. From the oögonium a projection called the trichogyne grows out; in some species it remains short, but in one at least it becomes long and slender. The biciliate sperm are formed singly in special cells known as antheridia. A single sperm enters the egg through the trichogyne or papilla, through a soft place which forms in the wall. Following fertilization, or the union of egg and sperm, the fertilized egg enlarges greatly and secretes around itself a thick wall, in which condition it remains very resistant to external changes. On germination its contents divide to produce 16 or 32 cells, each of which produces a zoöspore. These give rise to new plants. In *Coleochaete* we have a highly developed end product to an advancing degree of differentiation of cells and development of sex.

Many other equally interesting lines of evolution can be found in green algae. One leads to *Vaucheria*, a branching filamentous plant in which cross walls are not formed, nuclear divisions occurring until an extensive multinucleate filament is formed. Many marine relatives of *Vaucheria*, especially abundant in tropical seas, have elaborate bodies, often thickly encrusted with lime. In the genus *Caulerpa* the thallus of some species appears to be differentiated into leaves, stems and roots; however, no differentiation of tissue occurs, the whole structure being essentially filamentous.

Another group of algae, found only in fresh water, is the Conjugales. This includes the frequently observed *Spirogyra*, a genus having spiral chloroplasts. See Fig. 5. Sexual reproduction in *Spirogyra*, as in all the Conjugales, is by a process called conjugation. When the cells of two filaments are to conjugate they come into contact one with

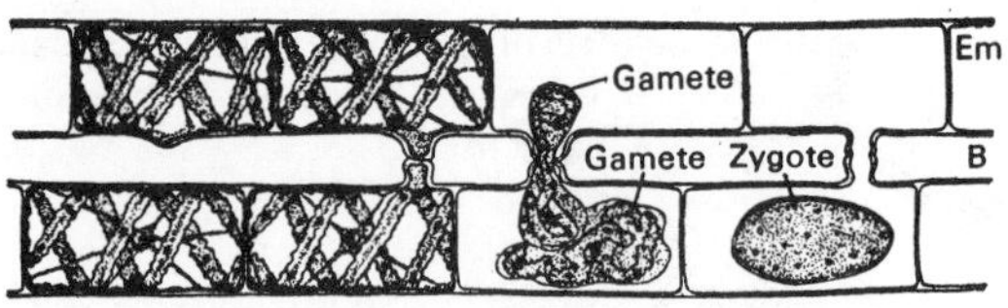

Fig. 5. *Spirogyra.* Stages in conjugation.

another. From the cells of one filament projections grow outward to meet similar projections from the cells of the other filament. These projections meet, and the walls at their tips are absorbed, leaving a tube connecting two cells of opposite filaments. Through this tube the protoplast of one cell moves into the opposite cell, where fusion occurs. Following this the zygote thus formed surrounds itself with a thick wall. This zygospore is extremely resistant to unfavorable external conditions, such as drought. After a prolonged rest period, if environmental conditions are favorable, the wall of the zygospore breaks, and a tube grows out from the protoplast within. By divisions, this tube becomes a new filament. Observation of conjugating filaments of *Spirogyra* reveal two facts: first, that when one cell of a filament conjugates with a cell of another filament, commonly all the cells of the first filament are conjugating with those of the other; second, that movement is largely from one filament to the other, so that at the end of the process one filament is composed of empty cell walls and the other filled with zygospores. This may be conceived as a sexual condition, the empty cells having been male and the cells in which the zygospores formed, female. Many related algae in the Conjugales show no indication of sexuality, the protoplasts uniting in the tube joining the two cells. No ciliated cells of any sort occur in this order.

Related to *Spirogyra* is a family containing many species whose cells are very symmetrical and beautiful objects. This is the Desmidiaceae, a family whose members commonly occur as single cells composed of two symmetrical halves. In many species the halves are distinctly indicated by a deep constriction which leaves only a slender isthmus connecting them. Conjugation in the Desmids is essentially like that in the other Conjugales.

Except for their botanical interest and for the fact that they provide food for many animals, the green algae are of little significance to man. Recently, however, investigations have been conducted to determine the feasibility of raising green algae as human food. One of the most promising of these algae is Chlorella, and some of its possibilities in food production are described under **Chlorella.**

The brown algae, or Phaeophyta, are distinguished from other algae by the presence of the brown pigment fucoxanthin, which masks the chlorophyll present and which imparts to them a brown color. Nearly all are marine plants. In this group there are no very simple primitive forms, comparable to *Chlamydomonas* of the greens. The simplest forms, such as *Ectocarpus*, are branched filamentous plants.

Other brown algae are very large, with tough bodies of many often complex shapes. They are most numerous on rocky coasts, where they grow attached to the rocks and are often exposed to the severest pounding of the tides. Brown algae are found in all regions, but are especially abundant in the cold temperature and arctic waters. One genus, *Sargassum*, occurs in great abundance floating in the Atlantic Ocean, forming the Sargasso Sea, through which Columbus passed so slowly on his voyage of discovery. Probably these plants are carried by ocean currents from the shores where they grow to the Sargasso Sea, where they float endlessly. The brown algae include the largest plants of this division of the plant kingdom, many of the kelps growing 30–40 feet (9 to 12 meters) long, and some of the plant forms of the Pacific Ocean attaining lengths of a hundred feet and more.

All brown algae are multicellular plants. Each cell contains a single distinct nucleus and several chloroplasts which contain chlorophyll, carotene, xanthophyll and fucoxanthin. The presence of the latter hides the other pigments. Photosynthesis in brown algae, as in other plants, results in the formation of sugar. This sugar however is changed into a compound, laminarin, instead of starch, which is never found in this group of algae.

The Phaeophyta are divided into several groups. These show interesting differences in their life histories. In most of them there is a very definite alternation of generations. Plants of the sexual generation form gametes; the asexual generation forms zoöspores. The motile cells of the brown algae are quite distinct from those of the green algae, having 2 unequal cilia which are attached laterally and extend in opposite directions.

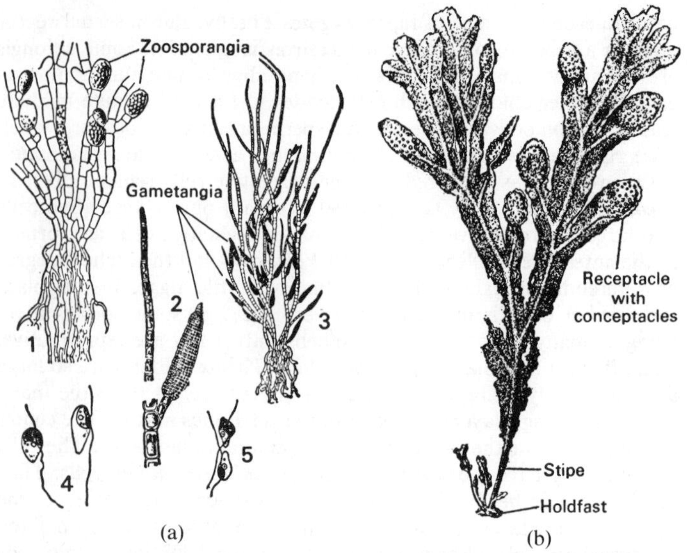

Fig. 6. (a) *Ectocarpus.* Stages in development: (1) sporophyte plant; (2,3) gametophyte plants; (4) two zoöspores; (5) conjugation of gametes to form zygotes. (*Holman and Robbins, "Textbook of General Botany," Wiley*). (b) *Fucus.* Mature plant. No bladders are shown. (*"Morphologie und Biologie der Algen," Gustav Fischer, Jena*)

One of the orders of this class is the Ectocarpales, of which *Ectocarpus* is a common form. See Fig. 6. It occurs as soft brown tufts growing on other larger algae or on stones or woodwork in the water. Each plant is composed of slender branching filaments. In this genus there is a definite alternation of two generations which are indistinguishable in the vegetative condition. Plants of the sexual generation are haploid, that is, have the reduced number of chromosomes and bear gametangia. These are elongated structures composed of many small, cubical cells. Each of these cells forms a single gamete. The plants of this generation and the gametangia they bear all look exactly alike. The gametes which they produce also look very much alike. But in behavior they are different. Some are sluggish, moving but little, while others swim actively and are attracted to the sluggish ones from other plants. The active gametes are males, the others females; sometimes the female gametes are slightly larger than the males. One male gamete fuses with a female, forming a zygote. The zygote always forms an asexual plant which is diploid. There are two kinds of asexual plants, which look exactly alike, and also like a sexual plant. Each forms zoösporangia. One form of zoöspórangium, called a plurilocular zoösporangium, is composed of many small cubical cells, each of which forms a single zoöspore. These zoöspores are liberated, swim about for a time, sink to the bottom and settle against any solid substratum and give rise to new asexual plants of the same type as that producing them. The other type of zoösporangium, often found on the same plant as the first, is composed of a single, usually much-enlarged, cell. The protoplast of this cell divides many times and gives rise to several zoöspores which are haploid. Therefore reduction division takes place in the sporangia, which produce these zoöspores. These zoöspores produce sexual plants. In some species of *Ectocarpus* the alternation of generations is not as regular as that described.

Another order of brown algae is the Cutleriales, of which the genus *Cutleria* is a well-known example. In *Cutleria* the two generations have a very different appearance, the sexual plants being much branched, and several inches tall, while the asexual plants are small lobed thalli growing prostrate on the substratum. So different are these two generations that they are often mistaken for different plants. The sexual plants are of two kinds, male and female, which differ very little. The sex organs are borne on the surface of the thallus. The male gametangia are elongate structures borne on branching filaments; the female gametangia are stouter and composed of few cells. The male gametes are minute and biciliate, the female, also biciliate, are many times larger. Many male gametes swim to a single female, one fuses with it, producing a zygote. The zygote develops into an asexual plant. The zoöspores, formed from small sac-like zoösporangia, which develop in large numbers on the upper surface of the thallus, are biciliate and very similar to gametes. They form sexual plants. In this order there is a very distinct difference in the two gametes, male and female, and a striking difference between sexual and asexual plants.

A third order is the Laminariales, which includes the largest algae known, commonly known as kelps. Some of them have a very striking appearance. *Postelsia*, for example, has a plant body composed of an erect stiff stalk often several inches in diameter. From its base many thick root-like outgrowths spread out and fix the plant firmly to the substratum. At the top of the stalk, long spreading branches are found. When the plant is seen growing in water it has much the appearance of a palm tree, whence the common name, sea palm. Another plant in this order is *Chorda filum*. Its thallus is a tough cord-like object 3–8 feet (0.9 to 2.4 meters) long and about $\frac{1}{4}$ inch (0.6 centimeters) in diameter. It looks very much like a coarse round leather shoe-string. A very common genus is *Laminaria*, which has many species of various forms, mostly large. Some of them consist of long cord-like stalks which bear at their upper end a broad flat expansion often 6–12 feet (1.8 to 3.6 meters) long and 8–15 inches (20.3 to 38.1 centimeters) broad. Colloquially these are known as devil's aprons. These large plants are the asexual generation and so are diploid. The zoösporangia are formed in immense numbers on the surface of the thallus. Each zoösporangium is a cylindrical object which produces many small biciliate zoöspores. These zoöspores swim down to the sea bed, where they develop into haploid or sexual plants. The latter are minute, usually consisting of a few cells which form a branching filament. Some of the plants are male, others female. In the female plant, any cell may become a sexual cell; often the plant is only a single cell. This sexual cell is an oögonium and forms a large non-motile egg which remains in the parent plant. Any cell of the male plant may become sexual, producing minute biciliate sperms which swim to the egg and fuse with it, forming a zygote. The latter at once develops into an asexual plant. In this order, there is also a distinct alternation of generations, but the sexual generation contains the small plants, the asexual usually very large plants. The sexual cells are distinctly different: the large non-motile egg and the small biciliate sperm.

A fourth order of brown algae is the Fucales. Members of this order are tough, much-branched plants which are particularly abundant between the tide levels on rocky shores in regions where the water is cold. *Fucus*, the common rockweed or bladder wrack, is a common and well-known plant. In this, as in all members of this order, there is no asexual reproduction. Therefore no distinct alternation of generations can occur. A *Fucus* plant consists of a tough dichotomously branched frond, which is attached to the rock on which it grows by a disk-shaped holdfast. In many species hollow bladders develop along the frond and serve to bring the plant into an erect position at high tide. At the tips of the branches of the thallus the reproductive bodies, receptacles, are formed. In some species these tips are swollen to form hollow bladders, in others they are flat and little differentiated from the rest of the thallus. The reproductive cells are formed in spherical cavities which are connected with the surface by small pores. Each cavity is called a conceptacle. Numerous branching filaments rise from the lower part of the conceptacle wall. Branches of these filaments bear the sexual organs. In some species the two sexes are borne in the same receptacle, in others they occur on different plants. The male sex organs or antheridia are oval sacs. The protoplast of each sac divides to form 64 cells, each of which becomes a laterally biciliate sperm. When mature these antheridia are extruded through the ostiole or opening of the conceptacle into the water. There the wall of the antheridium bursts, liberating the sperms. Each oögonium consists of a single cell. Its protoplast divides to form eight eggs. These also are extruded from the conceptacle, while still within the wall of the oögonium, and freed by the bursting of the same. Both types of conceptacle secrete a gelatinous matrix which surrounds the oögonia or antheridia and aids in bringing them to the surface of the receptacle. This matrix is squeezed out of the conceptacles by the partial drying of the frond at low tide. Each egg is a very large non-motile cell. Thousands of sperms are attracted to each egg and swim about it, causing it to revolve rapidly. Finally one sperm gains entrance to the egg and fertilizes it. The other sperms immediately swim away. The fertilized egg settles to the bottom, attaches itself to the substratum and at once starts to develop into a new plant. In the Fucales there is no alternation of generations. The gametes of the Fucales are very distinct, one being a large non-motile egg, the other a very small swimming sperm. One is tempted to arrange the various orders of brown algae in a series showing the way in which each may have evolved. However, such evolutionary relationships are purely speculative and not supported by any real evidence. It is impossible to trace the ancestry of the brown algae back to any simple ancestor, since no simple forms of brown algae are known.

The economic importance of the brown algae, while slight, is much greater than that of the green algae. Large quantities of these plants are gathered and used for fertilizer, wherever agriculture is carried on near the coast. From the ash produced by burning the larger forms, the kelps and Fucales, iodine and also potassium are obtained. In the Orient and in some of the north Atlantic islands, some of the brown algae are used as food, both for human beings and for livestock.

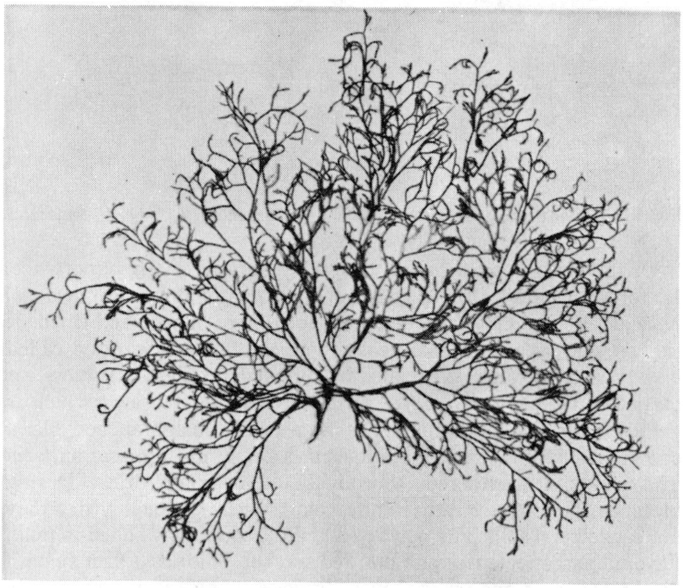

Fig. 7. Red algae. (*A. M. Winchester*)

The red algae or Rhodophyta form a very large group of plants, nearly all marine, of small to medium size. See Fig. 7. They are particularly abundant in warm coastal waters and are often plants of great beauty and extremely delicate habit. The red color to which they owe their name is caused by phycoerythrin, a red pigment which is present with the common chlorophyll-carotene group of pigments. These pigments are present in definite bodies or plastids, and not diffused through the protoplasts, as in the blue-green algae.

The forms of red algae are numerous. In many species the thallus is an extremely delicate filament. In other species the thallus is a tough, branched body 6–15 inches (15.2 to 38.1 centimeters) long. Others are flat membranes which may be a single cell in thickness or may be many cells thick. Some are thickly covered with a calcareous deposit, so that they are hard and stony, resembling corals. No motile reproductive cells are produced by members of this group. In sexual reproduction there is always a large female cell which is fertilized by a small male cell. This sexual reproduction is a rather complicated process. The antheridia are single-celled bodies; in some species the whole cell is liberated, in others the protoplast of the antheridium is freed. In either case the male cell floats in the water, carried only by the currents. The female reproductive organ is known as the procarp. In simpler forms this consists of a swollen basal portion called a carpogonium and a long slender portion called a trichogyne. Chance brings the male cell to the surface of the trichogyne, against which it sticks. The wall of the trichogyne is dissolved, allowing the nucleus of the male cell to enter the trichogyne. This nucleus passes down the trichogyne and enters the carpogonium, where it fuses with the female nucleus. From the fertilized carpogonium asexual spores called carpospores are formed, usually at the ends of branches which grow out from the carpogonium or from cells which are formed from those surrounding the carpogonium. Into these carpospores, nuclei from the carpogonium pass. The carpospores of simpler red algae at once produce sexual plants. In most red algae, however, they produce asexual plants which may be identical with the sexual plants in appearance. These asexual plants bear reproductive cells called tetraspores, because 4 of them are borne in a single sporangium. Each tetraspore gives rise to a sexual plant. So in the majority of red algae there is a distinct alternation of generations.

Fig. 8. *Chondrus crispus.* Frond. (*Photomicrograph by B. J. Ford; copyright*)

Of the many species of red algae very few are of any importance. In northern waters of both coasts of the Atlantic, dulse, *Rhodymenia palmata,* is found. It is gathered, cleaned more or less, and dried. It is then sold as a food or a relish. Species of *Porphyra,* often called laver, are also eaten, especially by Oriental people. Irish moss, or carrageen, which is *Chondrus crispus,* is another red alga which is gathered for food. See Fig. 8. It is a small much-branched plant, commonly dark red in color and with a beautiful iridescent surface. The plants are gathered, thoroughly cleaned and dried. Drying bleaches them to a creamy white color. When thoroughly dry they are bagged and sold. The powdered plant is commonly boiled in milk, flavored and sweetened, and allowed to cool. It forms a firm smooth gel known as blanc-mange. From species of red algae growing in the Pacific Ocean, agar is obtained.

The origin of the red algae and their relationships with other algae is a matter of considerable speculation. A few trace them from the blue-green algae, finding in some of the more primitive red algae "connecting links" which support this view. There are no ciliated reproductive cells in either of the two groups. But their distinct well-developed nucleus, their plastids, and their complex reproductive process set the red algae off very clearly. In view of these facts, it is perhaps more logical to derive the red algae from the green, using a form like *Coleochaete.*

ALGAE PROTEIN. Protein.

ALGA (Lichen). Lichen.

ALGEBRA. 1. A generalization of arithmetic involving the study of relations between numbers represented by symbols and obtained by the operations of addition, subtraction, multiplication, division, raising to a power, and extracting a root. If only the first four operations are performed a finite number of times the resulting numbers, functions, or equations are rational; more precisely, rational integral or polynomial and rational fractional. When fractional powers and extraction of roots are also considered the quantities are irrational. A common problem in algebra is that of finding the value of an unknown satisfying an equation written in terms of the operations just stated. Other kinds of numbers, functions, or equations are not algebraic but transcendental.

2. An algebra over a field is defined as follows:

Let K be any field (in physical practice, usualy a complex field) whose elements we shall call scalars. Let the set γ_{ijk}, with $i, j, k = 1, 2, \dots, n$, be any n^3 elements of K. Then the set of all ordered n-tuples $(x_1, x_2, \dots, x_n)$ of elements of K is called an algebra over K if addition, multiplication and scalar multiplication are defined thus:

$$(x_1, x_2, \dots, x_n) + (y_1, y_2, \dots, y_n) = (x_1 + y_1, x_2 + y_2, \dots, x_n + y_n)$$

$$(x_1, x_2, \dots, x_n)(y_1, y_2, \dots, y_n) = (\gamma_{ij1}x_iy_j, \gamma_{ij2}x_iy_j, \dots, \gamma_{ijn}x_iy_j)$$

$$\lambda(x_1, x_2, \dots, x_n) = (\lambda x_1, \lambda x_2, \dots, \lambda x_n)$$

where λ is a scalar and we have used the summation convention. Setting

$$l_1 = (1, 0, \dots, 0),\ l_2 = (0, 1, \dots, 0), \dots, l_n = (0, 0, \dots, 1)$$

we have $(x_1, x_2, \dots, x_n) = x_il_i$, and $l_il_j = \gamma_{ijk}l_k$. The l_i form a basis of the algebra and the γ_{ijk} are its structure constants.

3. An algebra of subsets of a set X is a class of subsets of X which contains the complement of each of its members and the union of any two of its members (or the intersection of any two of its members). An algebra of subsets ia a Boolean algebra relative to the operations of union and intersection. (For definitions of union and intersection as used here, see **Boolean algebra.**)

4. The algebra of a group is that of polynomials, with coefficients in a field K, of the elements $a_1, a_2, \dots, a_n$ of a finite group G, where if

$$x = k_1a_1 + k_2a_2 + \cdots + k_na_n$$

$$y = k_1'a_1 + k_2'a_2 + \cdots + k_n'a_n$$

then

$$x + y = (k_1 + k_1')a_1 + \cdots + (k_n + k_n')a_n$$

and

$$xy = l_1a_1 + l_2a_2 + \cdots + l_na_n$$

where x and y have been multiplied together as polynomials, the product a_ia_j being determined by the law of multiplication of the group. See also **Algebraic Equations** and terms listed under **Mathetmatics.**

ALGEBRAIC EQUATIONS. An equation, or set of simultaneous equations, in which the unknowns occur as rational functions only. Hence the equations are expressible by equating polynomials to zero. Here the case of a single equation in one unknown is considered:

$$a_0P(x) \equiv a_0x^n + a_1x^{n-1} + \cdots + a_n = 0, \quad a_0 \neq 0$$

where the a_i do not depend upon x, and are called the coefficients of the equation, while n is the degree. This equation is equivalent to

$$P(x) \equiv x^n - c_1x^{n-1} + \cdots + (-1)^nC_n = 0$$

$$c_i = (-1)^ia_i/a_0$$

where the c_i are the elementary symmetric functions of the roots x_i. See also **Symmetric Function.**

The *remainder theorem* states that if $a_0P(x)$ is divided by $x - r$, the remainder is $a_0P(r)$:

$$a_0P(x) \equiv (x - r)Q(x) + a_0P(r)$$

the *factor theorem* is a corollary and states that if r is a root of $P(x) = 0$, then $x - r$ divides $P(x)$. The "fundamental theorem of algebra" states that every algebraic equation has a root, real or complex. These theorems imply that $P(x)$ can be factored completely:

$$P(x) = (x - x_1)(x - x_2) \cdots (x - x_n)$$

each x_i being a root. If $x_i = x_j$ for some $i \neq j$, then $x_i = x_j$ is a double root; if $x_i = x_j = x_k$, a triple root, ... Counting a root x_i of multiplicity m as being m coincident roots, one says that an algebraic equation of degree n has exactly n roots (neither more nor less).

A Taylor series expansion gives

$$P(x - r) = P(r) + (x - r)P'(r) + \cdots + (x - r)^nP^{(n)}(r)/n!$$

Hence r is a root of multiplicity m if and only if

$$0 = P(r) = P'(r) = \cdots = P^{(m-1)}(r) \neq P^{(m)}(r)$$

hence r satisfies the derived equations

$$P^{(i)}(x) = 0, \quad i = 0, 1, \ldots, m-1$$

In general, setting $y = x - r$, if

$$b_i = P^{(n-1)}(r)/(n-i)!$$

then any root of

$$y^n + b_1 y^{n-1} + \cdots + b_n = 0$$

is r less than a root x_i, hence the roots are said to have been reduced by r. Repeated synthetic division can be applied to the original equation to obtain the b_i, since $b_n = P(r)$ is the remainder after dividing $P(x)$ by $x - r$; b_{n-1} that after dividing the quotient by $x - r$; . . . . See also **Synthetic Division.**

Other useful transformations are the following. The roots of

$$a_0 y^n - a_1 y^{n-1} + \cdots + (-1)^n a_n = 0$$

are the negatives of the roots of the original; those of

$$a_0 y^n + a_1 \alpha y^{n-1} + \cdots + a_n \alpha^n = 0$$

are α times the roots of the original; those of

$$a_n y^n + a_{n-1} y^{n-1} + \cdots + a_0 = 0$$

are the reciprocals of those of the original. These can be derived by setting $x = -y$, $x = y/\alpha$, and $x = 1/y$.

When the coefficients a_i are integers, then any rational root when expressed as a fraction p/q in lowest terms is such that p divides a_n and q divides a_0 without remainder. In particular, any integral root must be a divisor of a_n. In principle all rational roots can be found exactly, and once any root r is known, all other roots must satisfy $Q(x) = 0$ where $Q = P/(x - r)$. For irrational roots, see **Budan Theorem; Graeffe Method; Iteration (Method of); Newton's Formula for Interpolation;** and **Sturm Theorem.**

While these methods, except for Horner's, apply as well to complex roots as to real, it may be convenient to evaluate $P(z)$ with $z = x + iy$, and write

$$P(z) = R(x, y) + iJ(x, y)$$

after collecting real and pure imaginary terms. Then

$$R(x, y) = J(x, y) = 0$$

See also **Bernoulli Method** and terms listed under **Mathematics.**

ALGEBRAIC GEOMETRY. Geometry.

ALGEBRAIC NUMBER THEORY. Number Theory.

ALGEBRAIC TOPOLOGY. Topology.

ALGICIDE. A substance, natural or synthetic, used for destroying or controlling algae. The term is also sometimes used to describe chemicals used for controlling aquatic vegetation, although these materials are more properly classified as aquatic herbicides. See **Herbicide.**

ALGIN. A hydrophilic colloidal polysaccharide obtained from several species of brown algae. The term is used both in reference to the pure substance, alginic acid, extracted from the algae and also to the salts of this acid such as sodium or ammonium alginate, in which forms it is used commercially. The alginates currently find a large number of applications in the paint, rubber, pharmaceutical, food, and other industries. See also **Gums and Mucilages.**

ALGOL (β Persei). One of the first variable stars to be recognized as such. The first scientific notice of this variability was made by Montanari in 1670, but it is quite evident that the changes in the light of this star were noticed long before then. The name Algol, which signifies "demon star" probably was suggested because of the star's peculiar behavior. Algol is an eclipsing binary and the first star of this type to be explained. Because of its great brightness, it has been extensively observed with all types of stellar photometers, and the characteristics of its light curve are known. Algol is also a spectroscopic binary and, from the determination of the orbital elements from light variability as well as from the spectroscopic data, the physical characteristics of the component parts may be determined. See also **Binary Stars; Eclipsing Binary;** and **Spectroscopic Binaries.**

ALGONKIAN. A term applied to rock formations of late pre-Cambrian (Proterozoic) date in the Great Lakes region, and to the unit of time represented by these formations. Some American geologists use the term as a synonym for late pre-Cambrian.

ALGORITHM. A term derived from the word algorism, which meant the art of computing with Arabic numerals. The term algorithm is now used (1) to denote any method of computation, whether algebraic or numerical, or (2) any method of computation consisting of a comparatively small number of steps; the steps to be taken in a preassigned order and usually involving iteration, which are specifically adapted to the solution of a problem of some particular type. The best known is Euclid's algorithm for finding the highest common factor of two given numbers.

In computer terminology, an algorithm is a detailed logical procedure which represents the solution of a particular problem. Most commonly, the term is used to indicate an analysis procedure such as that used for the evaluation of a square root or for sorting a data file. Programming algorithms are widely published in the computer field literature. See also terms listed under **Mathematics.**

ALIASING ERROR (Data Acquisition System). The Shannon-Nyquist sampling theorem states that in order to reconstruct a signal containing frequency components in the spectrum 0 to f_m Hz from sampled data, samples must be taken at a rate of at least $2f_m$ samples per second. Failure to obtain data at this rate converts high-frequency components into low-frequency components. Thus, the reconstructed signal contains low-frequency energy not present in the original signal. Figure 1 illustrates a 5-Hz signal sampled at a 4-samples-per-second rate. The reconstructed signal has a frequency of 1 Hz. An *aliasing error* is an error which can be introduced into sampled data in a digital-data-acquisition system as the result of violation of the basic sampling theorem.

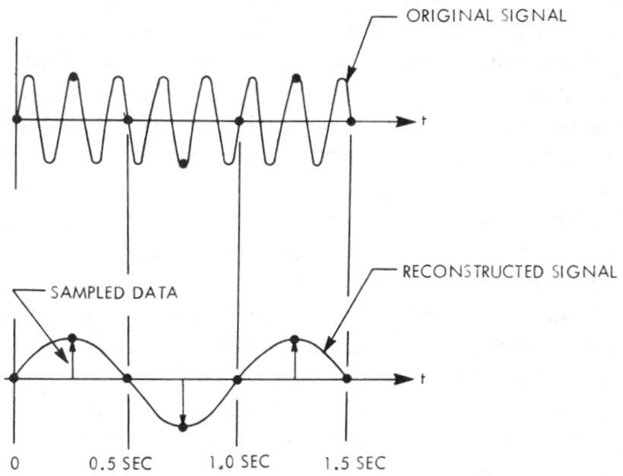

Fig. 1. Example of aliasing error in a sampled-data system.

The term *aliasing* results from the interpretation that the high-frequency components take the "alias" of a lower-frequency component. An equivalent term *folding error* arises from the interpretation that the frequency spectrum of the signal is folded such that high-frequency components appear in a lower-frequency spectrum.

Sampling a continuous signal is equivalent to modulating the signal with a series of uniformly spaced impulses as indicated in Fig. 2. If a signal containing frequency components up to a frequency f_m is modulated by a carrier of frequency f_s, the frequency spectrum of the resulting sampled signal consists of the original spectrum, plus harmonic spectra centered on the sampling frequency and its harmonics as shown in Fig. 1. This is the same effect as the modulation of an rf carrier with an audio signal in amplitude-modulated radio, except that there are an infinite number of carrier frequencies because the

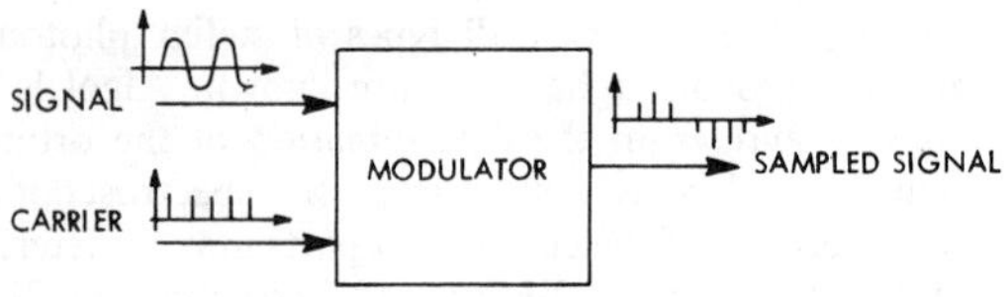

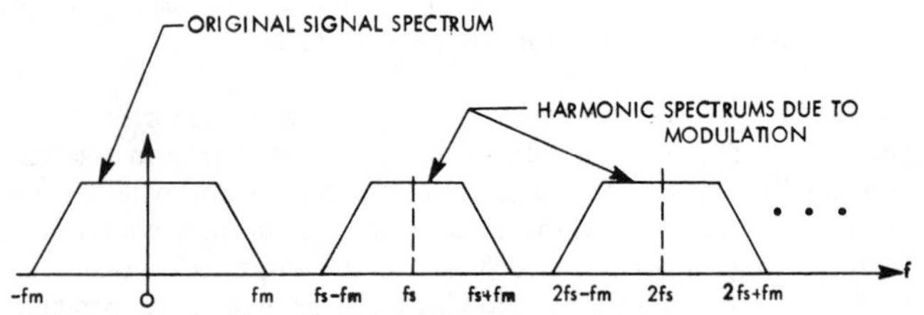

Fig. 2. Frequency-domain sampled spectra.

frequency spectrum of a train of equally spaced impulses (the modulating signal) consists of an infinite number of frequency components at f_s and its harmonics.

Figure 1 shows that if the sampling frequency f_s is less than $2f_m$, the lower sideband of the first-harmonic spectrum overlaps the original signal spectrum. Inasmuch as mathematical signal-reconstruction methods are equivalent to low-pass filtering in the signal domain, it is evident that the signal energy from the first-harmonic spectrum which overlaps the original spectrum will be passed by the low-pass filter and, therefore, will affect the nature of the reconstructed signal.

In the foregoing description, the signal was assumed to have no frequency components above the frequency f_m. In practice, signals are not definitively band-limited and high-frequency components are present in any real signal. In the majority of process-control applications, the signals are nominally band-limited by the low-pass characteristics of the process or the signal transducers. In connection with such signals, the sampling theorem frequently is stated so as to require sampling at twice the highest *significant* signal frequency. In practice, selection of the sampling frequency depends upon the interpretation of the term significant. A rule of thumb is that the sampling rate should be 5 to 10 times the highest frequency of interest. Although not technically perfect, the rule is valuable inasmuch as signals are usually band-limited to reject frequencies above those of interest. The 5 or 10 multiplier provides a safety factor to account for the finite rolloff rate of most low-pass systems. Where accurate reconstruction of the sampled signal is needed, an estimate of the aliasing error should be made.

An estimate of the aliasing error can be made by using a graphical technique, based on a frequency-spectrum diagram of the type shown in Fig. 2. The frequency spectrum of the original signal and the spectra of the modulation harmonics are drawn as indicated. At each frequency in the bandpass of the reconstruction filter or its mathematical equivalent, the amplitude of the original frequency components and all contributions from aliased harmonics are summed. This sum is an estimate of the magnitude of this frequency component in the reconstructed signal. The estimate is conservative inasmuch as the true contribution is represented by the vector sum of the components, whereas the estimate is based upon the algebraic sum. See also terms listed under **Data Processing.**

Thomas J. Harrison, International Business Machines Corporation, Boca Raton, Florida.

ALICYCLIC COMPOUND. Compound (Chemical); Organic Chemistry.

ALIDADE. Eccentricity Correction.

ALIENATION (Coefficient of). A term occurring in psychology. If the coefficient of correlation between two variables is r, the coefficient of alienation is defined as $(1 - r^2)^{1/2}$.

ALIGNMENT CHART. Nomograms or calculating charts used to represent formulas containing three or more variables. Graduated lines represent the variables; an index line or isopleth passing through points of two scales of known values will intersect the third scale to give the solution of the problem. The nomogram shown will give the solution of the equation $0.785D^2S = T$, where D is the root diameter of an American Standard screw, S is the unit tensile strength, and T is the total stress.

In addition to the accompanying three-parallel-line chart, alignment charts may take other forms, such as straight lines forming a Z shape, or the lines may be curved. The type of chart depends on the form of the mathematical equation represented.

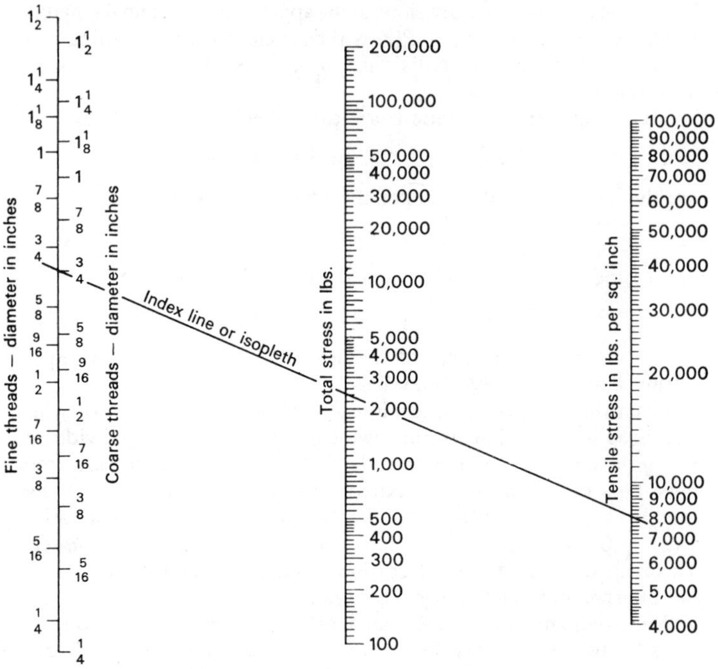

Nomogram for calculating thread diameter and tensile strength and total stress for American Standard screws.

ALIMENTARY CANAL. Digestive System (Human).

ALIMENTARY TRACT. The structures through which nourishment passes during the process of digestion and elimination. These include the mouth, pharynx, esophagus, stomach, the small intestine, which includes duodenum, jejunum, and ileum, and the large intestine, which includes the cecum, colon, rectum, and anus. See also diagram under **Digestive System (Human).**

ALIMENTATION. In glaciology, the combined processes which serve to increase the mass of a glacier or snowfield; the opposite of ablation. The deposition of snow is the major form of glacial alimentation, but other forms of precipitation along with sublimation, refreezing of melted water, etc., also contribute. The additional mass produced by alimentation is termed accumulation.

ALIPHATIC COMPOUND. An organic compound that can be regarded as a derivative of methane, CH_4. Most aliphatic compounds are open carbon chains, straight or branched; saturated, or unsaturated. Originally, the term was used to denote the higher (fatty) acids of the $C_nH_{2n}O_2$ series. The word is derived from the Greek term for oil (fatty). See also **Compound (Chemical); Organic Chemistry.**

ALIZARIN. Rubiaceae.

ALKALI. A term that was originally applied to the hydroxides and carbonates of sodium and potassium but since has been extended to include the hydroxides and carbonates of the other alkali metals and ammonium. Alkali hydroxides are characterized by ability to form soluble soaps with fatty acids, to restore color to litmus which has been reddened by acids, and to unite with carbon dioxide to form soluble compounds. See also **Acids and Bases.**

ALKALI METALS. The elements of group 1a of the periodic classification. In order of increasing atomic number, they are hydrogen, lithium, sodium, potassium, rubidium, cesium, and francium. With exception of hydrogen which is a gas and which frequently imparts a quality of acidity to its compounds, the other members of the group display rather striking similarities of chemical behavior, all reactive with H_2O to form strongly alkaline solutions. The elements in the group, including hydrogen, are characterized by a valence of one, having one electron in an outer shell available for reaction. Because of their chemical similarities, these elements, along with *ammonium* and sometimes magnesium, are considered the sixth group in classical qualitative chemical analysis separations.

ALKALINE EARTHS. The elements of group 2a of the periodic classification. In order of increasing atomic number, they are beryllium, magnesium, calcium, strontium, barium, and radium. The members of the group display rather striking similarities of chemical behavior, including stable oxides and carbonates, with hydroxides that are less alkaline than those of group 1a. The elements of the group are characterized by a valence of two, having two electrons in an outer shell available for reaction. Because of their chemical similarities, these elements are considered the fifth group in classical qualitative chemical analysis separations.

ALKALINE PRIMARY CELL. Battery.

ALKALINITY (Boiler Water). Feedwater (Boiler).

ALKALI ROCKS. Igneous rocks which contain a relatively high amount of alkalis in the form of soda amphiboles, *soda* pyroxenes, or felspathoids, are said to be alkaline, or alkalic. Igneous rocks in which the proportions of both lime and alkalis are high, as combined in the minerals, feldspar, hornblende, and augite, are said to be calcalkalic.

ALKALI-VAPOR MAGNETOMETER. Magnetometer.

ALKALOIDS. The term, alkaloid, which was first proposed by the pharmacist, W. Meissner, in 1819, and means "alkali-like," is applied to basic, nitrogen-containing compounds of plant origin. Two further qualifications usually are added to this definition: (1) the compounds have complex molecular structures; and (2) manifest significant pharmacological activity. Such compounds occur only in certain genera and families, rarely being universally distributed in larger groups of plants. Many widely distributed bases of plant origin, such as methyltrimethyl- and other open-chain simple alkylamines, the cholines, and the phenylalkylamines, are not classed as alkaloids. Alkaloids usually have a rather complex structure with the nitrogen atom involved in a heterocyclic ring. However, thiamine, a heterocyclic nitrogenous base, is not regarded as an alkaloid mainly because of its almost universal distribution in living matter. Colchicine, on the other hand, is classed as an alkaloid even though it is not basic and its nitrogen atom is not incorporated into a heterocyclic ring. It apparently qualifies as an alkaloid because of its particular pharmacological activity and limited distribution in the plant world.

Over 2000 alkaloids are known and it is estimated that they are present in only 10–15% of all vascular plants. They are rarely found in cryptogamia (exception, ergot alkaloids), gymnosperms, or monocotyledons. They occur abundantly in certain dicotyledons and particularly in the following families: *Apocynaceaae* (dogbane, quebracho, pereiro bark); *Papaveraceae* (poppies, chelidonium); *Papilionaceae* (lupins, butterfly-shaped flowers); *Ranunculaceae* (aconitum, delphinium); *Rubiaceae* (cinchona bark, ipecacuanha); *Rutaceae* (citrus, fagara); and *Solanaceae* (tobacco, deadly nightshade, tomato, potato, thorn apple). Well-characterized alkaloids have been isolated from the roots, seeds, leaves or bark of some 40 plant families. *Papaveraceae* is an unusual family, in that all of its species contain alkaloids.

Brief descriptions in alphabetical order of alkaloids of commercial or medical importance or of societal concern (alkaloid narcotics) are given later in this entry. See also **Amphetamine; Drug Addiction; Morphine;** and **Pyridine and Derivatives.**

The nomenclature of alkaloids has not been systemized, both because of the complexity of the compounds and for historical reasons. The two commonly used systems classify alkaloids either according to the plant genera in which they occur, or on the basis of similarity of molecular structure. Important classes of alkaloids containing generically related members are the aconitum, cinchona, ephedra, lupin, opium, rauwolfia, senecio, solanum, and strychnos alkaloids. Chemically derived alkaloid names are based upon the skeletal feature which members of a group posses in common. Thus, indole alkaloids (e.g., psilocybin, the active principle of Mexican hallucinogenic mushrooms) contain an indole or modified indole nucleus, and pyrrolidine alkaloids (e.g., hygrine) contain the pyrrolidine ring system. Other examples of this type of classification include the pyridine, quinoline, isoquinoline, imidazole, pyridine-pyrrolidine, and piperidine-pyrrolidine type alkaloids. Several alkaloids are summarized along these general terms in the accompanying table.

The beginning of alkaloid chemistry is usually considered to be 1805 when F. W. Sertürner first isolated morphine. He prepared several salts of morphine and demonstrated that it was the principle responsible for the physiological effect of opium. Alkaloid research has continued to date, but because most likely plant sources have been investigated and because a large number of synthetic drugs serve medical and other needs more effectively, the greatest emphasis has been placed upon the synthetics.

Sometimes, there is confusion between alkaloids and narcotics. It should be stressed that all alkaloids are not narcotics; and all narcotics are not alkaloids. A narcotic has the general definition of a drug which produces sleep or stupor, and also relieves pain. Many alkaloids do not meet these specifications.

The molecular complexity of the alkaloids is demonstrated by the accompanying figure. Alkaloids react as bases to form salts. The salts used especially for crystallization purposes are the hydrochlorides, sulfates, and oxalates which are generally soluble in water or alcohol, insoluble in ether, chloroform, carbon tetrachloride, or amyl alcohol. Alkaloid salts unite with mercury, gold, and platinum chlorides. Free alkaloids lack characteristic color reactions but react with certain reagents, as follows, with (1) iodine in potassium iodide solution, forming chocolate brown precipitate; (2) mercuric iodide in potassium iodide solution (potassium mercuriiodide), forming precipitate; (3) potassium iodobismuthate, forming orange-red precipitate; (4) bromine-saturated concentrated hydrobromic acid forming yellow precipitate; (5) tannic acid, forming precipitate; (6) molybdophosphoric acid, forming precipitate; (7) tungstophosphoric acid, forming precipitate; (8) gold(III) chloride, forming crystalline precipitate of characteristic melting point; (9) platinum(IV) chloride, forming crystalline precipitate of characteristic melting point; (10) picric acid, forming precipitate; (11) perchloric acid, forming precipitate. Many alkaloids form more or less characteristic colors with acids, solutions of acidic salts, etc.

The function of alkaloids in the source plant has not been fully explained. Some authorities simply regard them as by-products of the plant metabolism. Others conceive of alkaloids as reservoirs for protein synthesis; as protective materials discouraging animal or insect attacks; as plant stimulants or regulators in such activities as growth, metabolism, and reproduction; as detoxicating agents, which render harmless (by processes such as methylation, condensation, and ring closure) substances whose accumulation might otherwise cause damage to the plant. While these theories are of interest, it is also of interest to observe that from 85–90% of all plants manage well without the presence of alkaloids in their structures.

Adrenaline®. See Epinephrine later in this entry.

Atropine, also known as daturine, $C_{17}H_{23}NO_3$ (see structural formula in accompanying diagram), white, crystalline substance, optically inactive, but usually contains levorotatory hyoscyamine. Compound is soluble in alcohol, ether, chloroform, and glycerol; slightly soluble in water; mp 114–116°C. Atropine is prepared by extraction from *Datura stramonium,* or synthesized. The compound is toxic and allergenic. Atropine is used in medicine and is an antidote for cholinesterase-inhibiting compounds, such as organophosphorus insecticides and certain nerve gases. Atropine is commonly offered as the sulfate. Atropine is used in connection with the treatment of disturbances of cardiac rhythm and conductance, notably in the therapy of sinus bradycardia and sick sinus syndrome. Atropine is also used in some cases of heart block. In particularly high doses, atropine may induce ventricular tachycardia in an ischemic myocardium. Atropine is frequently one of several components in brand name prescription drugs.

Caffeine, also known as theine, or methyltheobromine, 1,2,7-trimethyl xanthine (see structural formula in accompanying diagram), white, fleecy or long, flexible crystals. Caffeine effloresces in air and commences losing water at 80°C. Soluble in chloroform, slightly soluble in water and alcohol, very slightly soluble in ether, mp 236.8°C, odorless, bitter taste. Solutions are neutral to litmus paper.

Caffeine is derived by extraction of coffee beans, tea leaves, and kola nuts. It is also prepared synthetically. Much of the caffeine of commerce is a by-product of decaffeinized coffee manufacture. See also **Coffee.** The compound is purified by a series of recrystallizations. Caffeine finds use in medicine and in soft drinks. Caffeine is also available as the hydrobromide and as sodium benzoate, which is a

GENERAL CLASSIFICATION OF ALKALOIDS

General Class	Examples
Derivatives of aryl-substituted amines	**Adrenaline, amphetamine, ephedrine, phenylephrine tyramine**
Derivatives of pyrrole	Carpaine, hygrine, nicotine
Derivatives of imidazole	Pilocarpine
Derivatives of pyridine and piperidine	Anabasine, coniine, ricinine
Containing fusion of two piperidine rings	Isopelletierine, pseudopelletierine
Pyrrole rings fused with other rings	Gelsemine, physostigmine, vasicine, yohimbine
Aporphone alkaloids	Apomorphine corydine, isothebaine
Berberine alkaloids	Berberine, emetine
Bis-benzylisoquinoline alkaloids	Bebeering, trilobine
Cinchona alkaloids	Cinchonine, quinidine, quinine
Cryptopine alkaloids	Cryptopine, protopine
Isoquinoline alkaloids	Anhalidine, pellotine, sarsoline
Lupine alkaloids	Lupanine, sparteine
Morphine and related alkaloids	Codeine, morphine, thebaine
Papaverine alkaloids	Codamine, homolaudanosine, papeverine
Phthalide isoquinoline alkaloids (also known as narcotine alkaloids)	Hydrastine, narceine, narcotine
Quinoline alkaloids	Dictamine, galipoline, lycorine
Tropine alkaloids	Atropine, cocaine, ecgonine, scopolamine, tropine
Other alkaloids	Brucine, solanidine, strychnine

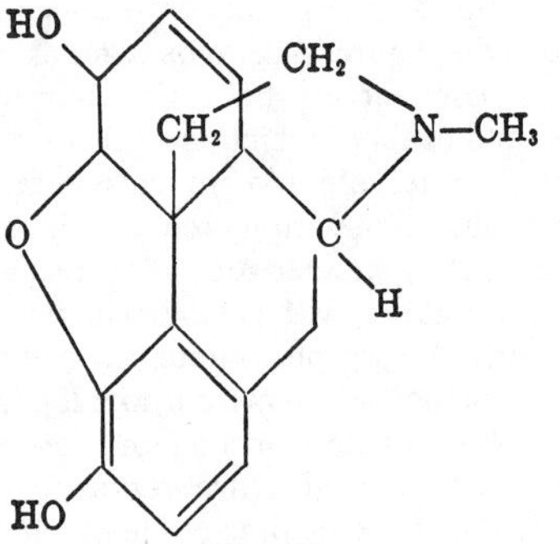
Morphine

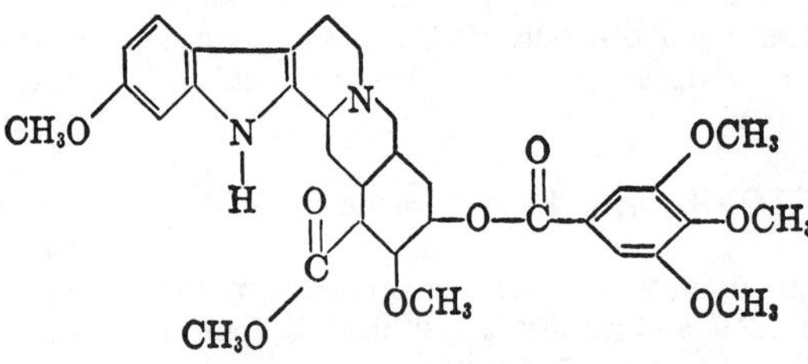
Reserpine

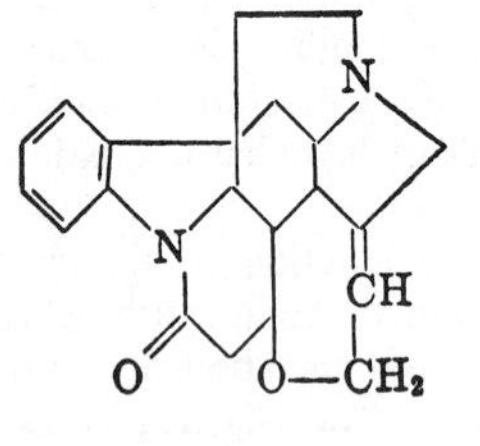
Strychnine

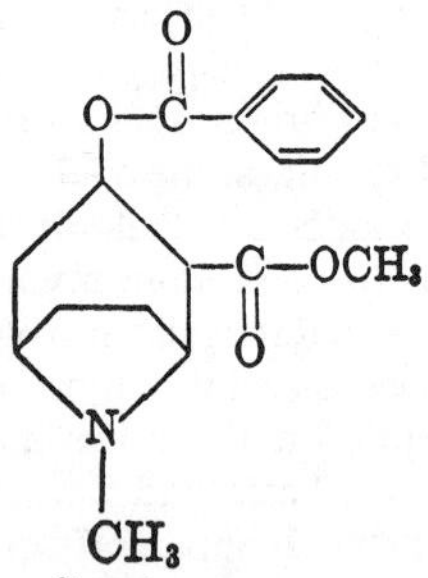
Cocaine

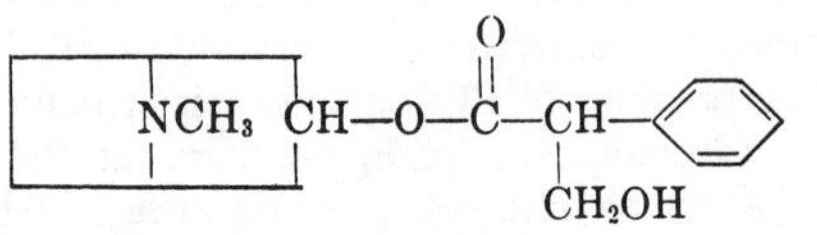
Atropine

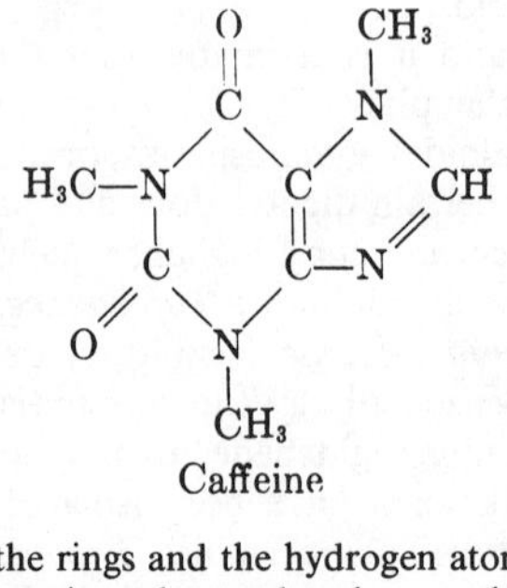
Caffeine

Structures of representative alkaloids. The carbon atoms in the rings and the hydrogen atoms attached to them are not designated by letter symbols. However, there is understood to be a carbon atom at each corner (except for the cross-over in the structure of morphine) and each carbon atom has four bonds, so that any bonds not shown or represented by attached groups are joined to hydrogen atoms.

mixture of caffeine and sodium benzoate, containing 47–50% anhydrous caffeine and 50–53% sodium benzoate. This mixture is more soluble in water than pure caffeine. A number of nonprescription (pain relief) drugs contain caffeine as one of several ingredients. Caffeine is a known cardiac stimulant and in some persons who consume significant amounts, caffeine can produce ventricular premature beats.

Cocaine (also known as methylbenzoylepgonine), $C_{17}H_{21}NO_4$, is a colorless-to-white cyrstalline substance, usually reduced to powder. Cocaine is soluble in alcohol, chloroform, and ether, slightly soluble in water, giving a solution slightly alkaline to litmus. The hydrochloride is levorotatory, mp 98°C. Cocaine is derived by extraction of the leaves of coca (*Erythroxylon*) with sodium carbonate solution, followed by treatment with dilute acid and extraction with ether. The solvent is evaporated after which the substance is redissolved and subsequently crystallized. Cocaine also is prepared synthetically from the alkaloid ecgonine. Cocaine is *highly toxic* and *habit forming.* While there are some medical uses of cocaine, usage must always be under the direction of a physician. It is classified as a narcotic in most countries. Society's major concern with cocaine is its use (increasing in recent years) as a narcotic.

Cocaine has been known as a very dangerous material since the early 1900s. When use of it as a narcotic increased during the early 1970s, serious misconceptions concerning its "safety" as compared with many other narcotics led and continue to lead to many deaths from its use. For example, health authorities in Dade County, Florida reported in 1977 that over 50% of drug-related overdose deaths were attributable to cocaine (Wetli and Wright, 1979).

Addicts use cocaine intravenously or by snorting the powder. After intravenous injections, coma and respiratory depression can occur rapidly. It has been reported that fatalities associated with snorting usually occur shortly after the abrupt onset of major motor seizures, which may develop within minutes to an hour after several nasal ingestions. Similar results occur if the substance is taken by mouth. Treatment is directed toward ventilatory support and control of seizures—although in many instances a victim may not be discovered in time to prevent death. It is interesting to note that cocaine smugglers, who have placed cocaine-filled condoms in their rectum or alimentary tract, have died (Suarez et al, 1977). The structural formula of cocaine is given in the accompanying diagram.

Codeine, also known as methylmorphine, $C_{18}H_{21}NO_3 \cdot H_2O$, is a colorless white crystalline substance, mp 154.9°C, slightly soluble in water, soluble in alcohol and chloroform, effloresces slowly in dry air. Codeine is derived from opium by extraction or by the methylation of morphine. For medical use, codeine is usually offered as the dichloride, phosphate, or sulfate. Codeine is *habit forming.* Codeine is known to exacerbate *urticaria* (familiarly known as *hives*). Since codeine is incorporated in numerous prescription medicines for headache, heartburn, fatigue, coughing, and relief of aches and pains, persons with a history of urticaria should make this fact known to their physician. Codeine is sometimes used in cases of acute *pericarditis* to relieve severe chest pains in early phases of disease. Codeine is sometimes used in drug therapy of renal (kidney) diseases.

Colchicine, an alkaloid plant hormone, $C_{22}H_{25}NO_6$, is yellow crystalline or powdered, nearly odorless, mp 135–150°C, soluble in water, alcohol, and chloroform, moderately soluble in ether. Solutions are levorotatory and deteriorate under light. The substance is highly toxic (0.02 gram may be fatal if ingested). Colchicine is extracted from the plant *Colchicum autumnale* after which it is crystallized. The compound also has been synthesized. Biologists have used colchicine to induce chromosome doubling in plants. Colchicine finds a number of uses in medicine.

Although colchicine has been known for many years, interest in the drug has been revitalized in recent years as the result of the discovery that it interferes with cell division by destroying the spindle mechanism. The two chromatids which represent one chromosome at the metaphase stage fail to separate and do not migrate to the poles (ends) of the cell. Each chromatid becomes a chromosome in situ. The entire group of new chromosomes now form a resting nucleus and the next cell division reveals twice as many chromosomes as before. The cell has changed from the diploid to the tetraploid condition. Applied to germinating seeds or growing stem tips in concentrations of about 1 gram in 10,000 cubic centimeters of water for 4 or 5 days, colchicine may thus double the chromosome number of many or all of the cells, producing a tetraploid plant or shoot. Offspring from such plants may be wholly tetraploid and breed true. Tetraploid plants are larger than diploid plants and often more valuable. The alkaloid has also been used to double the chromosome number of sterile hybrids produced by crossing widely separated species of plants. Such plants, after colchicine treatment, contain in each cell two complete diploid sets of chromosomes, one from each of the parent species, and become fertile, pure-breeding hybrid species.

In medicine, colchicine is probably best known for its use in connection with the treatment of gout. Acute attacks of gout are characteristically and specificially aborted by colchicine. The response noted after administration of the drug also can be useful in diagnosing gout cases where synovial fluid cannot be aspirated and examined for the presence of typical urate crystals. However, colchicine does not affect the course of acute synovitis in rheumatoid arthritis.

Kaplan (1960) observed that colchicine may produce objective improvement in the periarthritis associated with *sarcoidosis* (presence of noncaseating granulomas in tissue). Colchicine is sometimes used in the treatment of *scleroderma* (deposition of fibrous connective tissues in skin or other organs); it may assist in preventing attacks of Mediterranean fever; and it is sometimes used as part of drug therapy for some renal (kidney) diseases.

Colchicine can cause diarrhea as the result of mucosal damage and it has been established that colchicine interferes with the absorption of vitamin B_{12}.

Emetine, an alkaloid from ipecac, $C_{29}H_{40}O_4N_2$, is a white powder, mp 74°C, with a very bitter taste. The substance is soluble in alcohol and ether, slightly soluble in water. Emetine darkens upon exposure to light. The compound is derived by extraction from the root of *Cephalis ipecacuanha* (ipecac). It is also made synthetically. Medically, ipecac is useful as an emetic (induces vomiting) for emergency use in the treatment of drug overdosage and in certain cases of poisoning. Ipecac should not be administered to persons in an unconscious state. It should be noted that emesis is not the proper treatment in all cases of potential poisoning. It should not be induced when such substances as petroleum distillates, strong alkali, acids, or strychnine are ingested.

Ephedrine, 1-phenyl-2-methylaminopropanol, $C_6H_5CH(OH)CH(NHCH_3)CH_3$, is a white-to-colorless granular substance, unctuous (greasy) to the touch, and hygroscopic. The compound gradually decomposes upon exposure to light. Soluble in water, alcohol, ether, chloroform, and oils, mp 33–40°C, by 255°C, and decomposes above this temperature. Ephedrine is isolated from stems or leaves of *Ephedra,* especially Ma huang (found in China and India). Medically, it is usually offered as the hydrochloride. In the treatment of bronchial asthma, ephedrine is known as a *beta agonist.* Compounds of this type reduce obstruction by activating the enzyme adenylate cyclase. This increases intracellular concentrations of cAMP (cyclic 3′5′-adenosine monophosphate) in bronchial smooth muscle and mast cells. Ephedrine is most useful for the treatment of mild asthma. In severe asthma, ephedrine rarely maintains completely normal airway dynamics over long periods. Ephedrine also has been used in the treatment of cerebral transient ischemic attacks, particularly with patients with vertabrobasilar artery insufficiency who have symptoms associated with relatively low blood pressure, or with postural changes in blood pressure. Ephedrine sulfate also has been used in drug therapy in connection with urticaria (hives).

Epinephrine, a hormone having a benzenoid structure, $C_9H_{13}O_3N$, also called adrenaline. It can be obtained by extraction from the adrenal glands of cattle and also prepared synthetically. Its effect on body metabolism is pronounced, causing an increase in blood pressure and rate of heartbeat. Under normal conditions, its rate of release into the system is constant, but emotional stresses, such as fear or anger rapidly increase the output and result in temporarily heightened metabolic activity. Epinephrine is used for the symptomatic treatment of bronchial asthma and reversible bronchospasm associated with chronic bronchitis and emphysema. The drug acts on both alpha and beta receptor sites. Beta stimulation provides bronchodilator action by relaxing bronchial muscle. Alpha stimulation increases vital capacity by reducing congestion of the bronchial mucosa and by constricting pulmonary vessels.

Epinephrine is also used in the management of anesthetic procedures in connection with noncardiac surgery of patients with active ischemic heart disease. The drug is useful in the treatment of severe urticarial (hives) attacks, especially those accompanied by angioedema.

Epinephrine has numerous effects on intermediary metabolism. Among these are promotion of hepatic glycogenolysis, inhibition of hepatic gluconeognesis, and inhibition of insulin release. The drug also promotes the release of free fatty acids from triglyceride stores in adipose tissues. Epinephrine produces numerous cardiovascular effects. Epinephrine is particularly useful in treating conditions of immediate hypersensitivity—interactions between antigen and antibody. These mechanisms cause attacks of anaphylaxis, hay fever, hives and allergic asthma. Anaphylaxis can occur after bee and wasp stings, venoms, etc. Although the mechanism is not fully understood, epinephrine can play a lifesaving role in the treatment of acute systemic anaphylaxis.

In some instances, epinephrine can be a cause of a blood condition involving the leukocytes and known as neutrophilia. In very rare cases, an intramuscular injection of epinephrine can be a cause of clostridial myonecrosis (gas gangrene).

Heroin, diacetylmorphine $C_{17}H_{17}NO(C_2H_3O_2)_2$, is a white, essentially odorless, crystalline powder with bitter taste, soluble in alcohol, mp 173°C. Heroin is derived by the acetylization of morphine. The substance is highly toxic and is a habit-forming narcotic. One-sixth grain (0.0108 gram) can be fatal. Although emergency facility personnel in some areas during recent years have come to regard heroin overdosage as approaching epidemic statistics, it is nevertheless estimated that the majority of persons with heroin overdose die before reaching a hospital. The initial crisis of an overdose is a severe respiratory depression and sometimes *apnea* (cessation of breathing). In emergency situations, the victim may be ventilated with a self-inflating resuscitative bag with delivery of 100% oxygen. Then, an endotracheal tube attached to a mechnical ventilator may be inserted. Naloxone (*Narcan*®), a narcotic antagonist, then may be administered intravenously, often with repeated dosages over short intervals, until an improvement is noted in the respiratory rate or sensorial level of the victim. If a victim does not respond, this is usually indication that the situation is not opiate-related, or that other drugs also have been taken. Inasmuch as the antagonizing action of naloxone persists for only a few hours, a heroin overdose patient should be observed in the hospital for an indeterminate period. In heroin overdose cases, pulmonary edema (as the result of altered capillary permeability) may occur. This is directly associated with the overdose and not with subsequent treatment. Aside from the severe overdose situation, use of the drug causes or contributes to a number of ailments. These include chronic renal (kidney) failure and nephrotic syndrome. Septic arthritis caused by *Pseudomonas* and *Serratia* infections, is sometimes found as the result of intravenous heroin abuse. Drug-induced immune platelet destruction also may occur.

Morphine. See separate entry on **Morphine.**

Neo-Synephrine®. See Phenylephrine hydrochloride later in this entry.

Nicotine, beta-pyridyl-alpha-N-methylpyrrolidine, $C_5H_4NC_4H_7NCH_3$, is a thick, water-white levorotatory oil that turns brown upon exposure to air. The compound is hygroscopic, soluble in alcohol, chloroform, ether, kerosine, water, and oils, bp 247°C, at which point it decomposes. Specific gravity is 1.00924. Nicotine is combustible with an autoignition temperature of 243°C. Nicotine is derived by distilling tobacco with milk of lime and extracting with ether. Nicotine is used in medicine, as an insecticide, and as a tanning agent. Nicotine is commercially available as the dihydrochloride, salicylate, sulfate, and bitartrate. Nicotinic acid (pyridine-3-carboxylic acid) is a vitamin in the B complex. See also **Vitamin.**

Phenylephrine Hydrochloride, *l*-1-(meta-hydroxyphenyl-2-methylaminoethanol hydrochloride, $HOC_6H_4CH(OH)CH_2NHCN_3 \cdot HC1$, white or nearly white crystalline substance, odorless, bitter taste. Solutions are acid to litmus paper, freely soluble in water and in alcohol, mp 140–145°C. Levorotatory in solution. Phenylephrine hydrochloride is used medically as a vasoconstrictor and pressor drug. It is chemically related to epinephrine and ephedrine. Actions are usually longer lasting than the latter two drugs. The action of phenylephrine hydrochloride contrasts sharply with epinephrine and ephedrine, in that its action on the heart is to slow the rate and to increase the stroke output, inducing no disturbance in the rhythm of the pulse. In therapeutic doses, it produces little if any stimulation of either the spinal cord or cerebrum. The drug is intended for the maintenance of an adequate level of blood pressure during spinal and inhalation anesthesia and for the treatment of vascular failure in shock, shocklike states, and drug-induced hypotension, or hypersensitivity. It is also used to overcome paroxysmal supraventricular tachycardia, to prolong spinal anesthesia, and as a vasoconstrictor in regional analgesia. Caution is required in the administration of phenylephrine hydrochloride to elderly persons, or in patients with hyperthyroidism, bradycardia, partial heart block, myocardial disease, or severe arteriosclerosis. The brand name *Neo-Synephrine®* is also used to designate another product (nose drops) which does not contain phenylephrine hydrochloride. The nose drops contain xylometazoline hydrochloride.

Quinine, $C_{20}H_{24}N_2O_2 \cdot H_2O$, a bulky, white, amorphous powder or crystalline substance, with very bitter taste. It is odorless and levorotatory. Soluble in alcohol, ether, chloroform, carbon disulfide, oils, glycerol, and acids; very slightly soluble in water. Quinine is derived from finely ground cinchona bark mixed with lime. This mixture is extracted with hot, high-boiling paraffin oil. The solution is filtered, shaken with dilute sulfuric acid and then neutralized while hot with sodium carbonate. Upon cooling, quinine sulfate crystallizes out. Pure quinine is obtained by treating the sulfate with ammonia. In addition to medical uses, quinine and its salts are used in soft drinks and other beverages.

Quinine derivatives are used in therapy for mytonic dystrophy (usually weakness and wasting of facial muscles); in the treatment of certain renal (kidney) diseases. Quinine and derivatives are best known for their use in connection with malaria. Acute attacks of malaria are usually treated with oral chloroquine phosphate. The drug is given intramuscularly to patients who cannot tolerate oral medication. Combined therapy is indicated for treating *P. falciparum* infections, using quinine sulfate and pyrimethamine. A weekly oral dose of chloroquinone phosphate is frequently prescribed for persons who travel in malarious regions. The drug is taken one week prior to travel into such areas and continued for six weeks after leaving the region. Chloroquine phosphate has not proved fully satisfactory in the treatment of babesiosis, a malarialike illness caused by a parasite.

Strychnine, $C_{21}H_{24}ON_2$, hard, white crystals or powder of a bitter taste. Soluble in chloroform, slightly soluble in alcohol and benzene, slightly soluble in water and ether, mp 268–290°C, bp 270°C (5 millimeters pressure). Strychnine is obtained by extraction of the seeds of *Nux vomica* with acetic acid, followed by filtration, precipitation by an alkali, followed by final filtration. The compound is highly toxic by ingestion and inhalation. The phosphate finds limited medical use. Strychnine is also used in rodent poisons. Strychnine acts as a powerful stimulant to the central nervous system. At one time, strychnine was used in a very carefully controlled way in the treatment of some cardiac disorders. Acute strychnine poisoning resembles fully developed generalized tetanus.

References

Kaplan, H.: "Sarcoid Arthritis with a Response to Colchicine: Report of Two Cases," *N. Engl. J. Med.*, **263**, 778 (1960).

Pelletier, S. W.: "The Chemistry of Certain Imines Related to Diterpene Alkaloids," *Experimentia*, **20**, 1–10 (1964).

Raffauf, R. F.: "A Handbook of Alkaloids and Alkalid-Containing Plants," Wiley, New York, 1970.

Rao, T. K. S., et al.: "Natural History of Heroin-Associated Nephropathy," *N. Engl. J. Med.*, **290**, 19 (1974).

Suarez, C. A., et al.: "Cocaine-Condom Ingestion: Surgical Treatment," *J. Amer. Med. Assn.*, **238**, 1391 (1977).

Swan, G. A.: "An Introduction to Alkaloids," Halsted Press, London, 1967.

Wetli, C. V., and R. K. Wright: "Death Caused by Recreational Cocaine Use," *J. Amer. Med. Assn.*, **241**, 2519 (1979).

ALKALOIDS (Food Poisons). **Foodborne Diseases.**

ALKALOSIS. A condition of excess alkalinity (or depletion of acid) in the body, in which the acid-base balance of the body is upset. The hydrogen ion concentration of the blood drops below the normal level, increasing the pH value of the blood above the normal 7.4. The condition can result from the ingestion or formation in the body of an excess of alkali, or of loss of acid. Common causes of alkalosis include: (1) overbreathing (hyperventilation) where a person may breathe too deeply for too long a period, consequently washing out carbon dioxide from the blood, (2) ingestion of excessive alkali, as for example an over-dosage of sodium bicarbonate possibly taken for the relief of gastric distress, and (3) excessive vomiting, which leads to loss of chloride and retention of sodium ions. The usual, mild symptoms of alkalosis are restlessness, possible numbness or tingling of the extremities (hands and feet), and generally increased muscular irritability. Only in extreme cases, tetany (muscle spasm) and convulsions may be evidenced.

See also **Acid-Base Regulation (Blood); Blood; Kidney and Urinary Tract;** and **Potassium and Sodium (In Biological Systems).**

ALKANE. One of the group of hydrocarbons of the paraffin series, e.g., methane, ethane, and propane. See also **Organic Chemistry.**

ALKENE. One of a group of hydrocarbons having one double bond and the type formula C_nH_{2n}, e.g., ethylene and propylene. See also **Organic Chemistry.**

ALKYD PAINT. Paint.

ALKYD RESINS. The esterification of a polybasic acid with a polyhydric alcohol yields a thermosetting hydroxycarboxylic resin, commonly referred to as an alkyd resin. Some common uses of alkyds are military switchgear, electrical terminal strips, electrical relay housings and bases, and television tuner segments. The resin frequently is combined with organic and inorganic fillers. These impart desired electrical and physical properties and advantageously influence the molding characteristics. Among the advantages of alkyds are rapid curing, with no volatiles emitted during the cure cycle; low molding pressures; and high production rates on compression or transfer presses and in injection molding machines. The resins have very good dimensional stability and electrical properties.

The mineral-filled grades, sometimes modified with cellulose for reducing specific gravity and cost, are used in small switch housings, automotive ignition parts, and electronic component bases.

Alkyd resins are furnished in three major forms: (1) *fibrous*, in which the resins are compounded with long glass fibers (about 1/2-inch; 12 millimeters) and have medium strength; (2) *rope*, which is a medium-impact material and conveniently handled and processed; and (3) *granular*, in which the resins are compounded with other fibers, such as glass, asbestos, and cellulose (length about 1/16-inch; 2 millimeters). A commonly used member of the alkyd resin family is made from phthalic anhydride and glycerol. These resins are hard and possess very good stability. Where maleic acid is used as a starting ingredient, the resin has a higher melting point. Use of azelaic acid produces a softer and less brittle resin. Very tough and stable alkyds result from the use of adipic and other long-chain dibasic acids. Pentaerythritol may be substituted for glycerol as a starting ingredient.

Alkyd resins are extensively used in paints and coatings. Some advantages include good gloss retention and fast drying characteristics. However, most unmodified alkyds have low chemical and alkali resistance. Modification with esterified rosin and phenolic resins improves hardness and chemical resistance. Styrene and vinyl toluene improve hardness and toughness. For high-temperature coatings (up to about 450°F; 232°C), copolymers of silicones and alkyds are used. Such coatings include stove and heating equipment finishes. To obtain a good initial gloss, improved adhesion, and exterior durability, acrylic monomers can be copolymerized with oils to modify alkyd resins. Aromatic acids, such as benzoic or butylbenzoic, also have been used with alkyd resins in coatings.

ALKYL. A generic name for any organic group or radical formed from a hydrocarbon by elimination of one atom of hydrogen and so producing a univalent unit. The term is usually restricted to those radicals derived from the aliphatic hydrocarbons, those owing their origin to the aromatic compounds being termed "aryl." Usually, saturated radicals, such as methyl, ethyl, propyl, etc., containing, respectively, one hydrogen atom less than the corresponding saturated hydrocarbons—methane, ethane, propane, etc.—are understood.

ALKYLATING AGENTS. Cancer and Oncology.

ALKYLATION. Addition of an alkyl group. These reactions are important throughout synthetic organic chemistry; for example, in the production of gasolines with high antiknock ratings for automobiles or for use in aircraft.

The nature of the products of these reactions, as well as the yields, depend upon the catalysts and physical conditions. The reactions below have been written to show two combination reactions of two isobutene molecules, one yielding diisobutene, which reduces to isooctane, and the other yielding a trimethylpentane by a direct reduction reaction.

Specifically, the term is applied to various methods, including both thermal and catalytic processes, for bringing about the union of paraffin hydrocarbons with olefins. The process is especially effective in yielding gasolines of high octane number and low boiling range (aviation fuels).

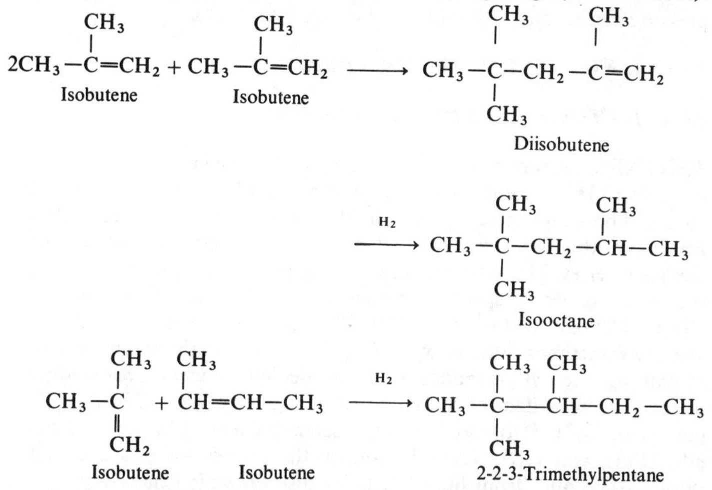

In the petroleum industry, catalytic cracking units provide the major source of olefinic fuels for alkylation. A feedstock from a catalytic cracking units it typified by a C_3/C_4 charge with an approximate composition of: propane, 12.7%; propylene, 23.6%; isobutane, 25.0%; *n*-butane, 6.9%; isobutylene, 8.8%; 1-butylene, 6.9%; and 2-butylene, 16.1%. The butylenes will produce alkylates with octane numbers approximately three units higher than those from propylene.

One possible arrangement for a hydrofluoric acid alkylation unit is shown schematically in the accompanying figure. Feedstocks are pretreated, mainly to remove sulfur compounds. The hydrocarbons and acid are intimately contacted in the reactor to form an emulsion, within which the reaction occurs. The reaction is exothermic and temperature must be controlled by cooling water. After reaction, the emulsion is allowed to separate in a settler, the hydrocarbon phase rising to the top. The acid phase is recycled. Hydrocarbons

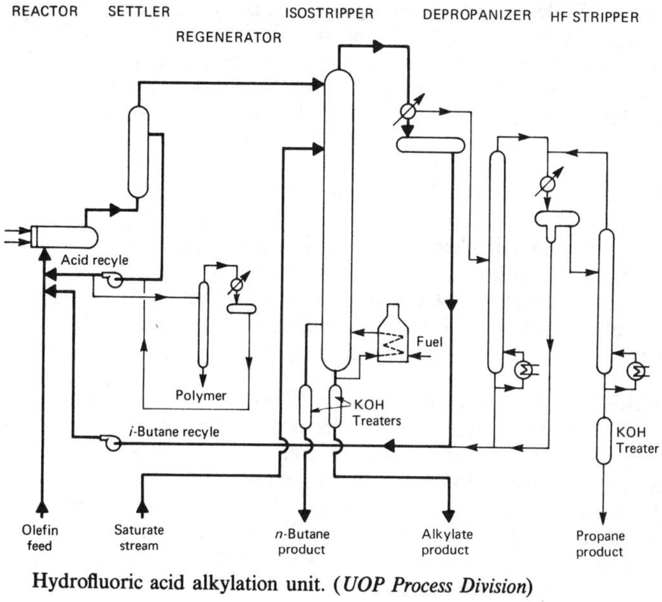

Hydrofluoric acid alkylation unit. (*UOP Process Division*)

from the settler pass to a fractionator which produces an overhead stream rich in isobutane. The isobutane is recycled to the reactor. The alkylate is the bottom product of the fractionater (isostripper). If the olefin feed contains propylene and propane, some of the isostripper overhead goes to a depropanizer where propane is separated as an overhead product. A hydrofluoric acid (HF) stripper is required to recover the acid so that it may be recycled to the reactor. HF alkylation is conducted at temperature in the range of 24–38°C (75–100°F).

Sulfuric acid alkylation also is used. In addition to the type of acid catalyst used, the processes differ in the way of producing the emulsion, increasing the interfacial surface for the reaction. There also are important differences in the manner in which the heat of reaction is removed. Often, a refrigerated cascade reactor is used. In other designs, a portion of the reactor effluent is vaporized by pressure reduction to provide cooling for the reactor.

ALKYL ETHOXYLATES. **Detergents.**

ALKYL ETHOXY SULFATE. **Detergents.**

ALKYNES. A series of unsaturated hydrocarbons having the general formula C_nH_{2n-2}, and containing a triple bond between two carbon atoms. The simplest compound of this series is acetylene HC⋮CH. Formerly, the series was named after this compound, namely the *acetylene* series. The latter term remains in popular usage. Particularly, the older names of specific compounds, such as acetylene, allylene CH_2C⋮CH, and crotonylene CH_3C⋮CCH_3, persist. These compounds also are sometimes called *acetylenic hydrocarbons.* In the alkyne system of naming, the "yl" termination of the alcohol radical corresponding to the carbon content of the alkyne is changed to "yne." Thus, C_2H_2 (acetylene by the former system) becomes *ethyne* (the "eth" from ethyl(C_2)); and C_4H_6 (crotonylene by the former system) becomes *butyne* (the "but" from butyl(C_4)). See also **Organic Chemistry.**

ALLANITE. Allanite is a rather rare monoclinic mineral of somewhat variable but quite complex chemical composition, perhaps represented satisfactorily by the formula (Ce, Ca, Y)$_2$(Al, Fe)$_3$Si$_3$O$_{12}$ (OH). The color of the fresh mineral is black but it is usually brown or yellow with a coating of some alteration product; often the altered crystals have the appearance of small rusty nails. It occurs characteristically in plutonic rocks like granite, syenite or diorite and is found in large masses in pegmatites. Localities in the United States are Essex and Orange Counties, New York, Franklin, New Jersey, Amherst County, Virginia, and Llano County, Texas. The slender prismatic crystals are sometimes called orthite. Allanite was named for its discoverer, T. Allan. Orthite was so named from the Greek word meaning straight, in reference to the straight prisms, a common habit of this mineral.

ALLANTOIS. A sac-like outgrowth of the hind gut of the embryo found only in reptiles, birds and mammals. In reptiles and birds it serves as a respiratory organ and receives waste matter, and in mammals it forms part of the placenta through which all interchange with the blood of the mother during embryonic development is carried out.

ALLEGHENY OROGENY. The term for an event which caused deformation of the rocks of the Valley and Ridge province, and those of the adjacent Allegheny Plateau in the central and southern Appalachians. It is believed that most of the orogeny occurred late in the Paleozoic, but some phases may have extended into the early Triassic. Use of this term is preferred to the more inclusive term, Appalachian Revolution.

ALLELE. Also termed *allelomorph,* one of two or more forms of a gene that occupies a particular locus on a chromosome. In humans there is a gene at a particular locus on a chromosome which produces an enzyme essential to the breakdown of the amino acid phenylalanine. An allele of this gene may sometimes be present at this same locus which does not produce the enzyme. Since there are two of each kind of chromosome in a cell, a person will normally have two genes at this locus, one on each of the two chromosomes. If at least one of these genes is the allele for enzyme production, the person will be normal. If both genes are the allele for lack of enzyme production, the phenylalanine will not be broken down and the person will suffer from phenylketonuria (PKU). Hence, we say that the gene for enzyme production is the dominant allele and the gene for no enzyme is the recessive allele.

In many cases more than two alleles are known for a particular locus on a chromosome. These are called *multiple alleles.* One locus on the X-chromosome of the fruit fly, *Drosophila,* includes a number of alleles which can cause the eye to range in color from white to a deep red. The gene for red (wild type) eye is dominant over the other alleles, but there is some intermediate inheritance when two of the other alleles are present. An individual normally will carry no more than two alleles of a series.

Ann C. Vickery, University of South Florida, College of Medicine, Tampa, Florida.

ALLELOPATHIC SUBSTANCE. A material contained within a plant that tends to suppress the growth of other plant species. The alkaloids present in several seed-bearing plants are believed to play an allelopathic role. Other suspected allelopathic substances contained in some plants include phenolic acids, flavonoids, terpenoid substances, steroids, and organic cyanides.

ALLEN'S RULE. **Adaptation (Ecology).**

ALLERGY. A reaction to a specific substance in an individual who is sensitive to that substance. If the material is borne by the wind and produces symptoms of allergy when it comes into contact with the mucous membranes of the eyes and respiratory tract, the victim is said to have *hay fever.* There are two main types: (1) *seasonal hay fever* is the most common and occurs during the spring and summer seasons as the result of pollen from various trees, grasses, and weeds (the most common cause); and (2) *nonseasonal or perennial hay fever* that may be caused by allergic reactions to house pets, foods, dust, and numerous other substances. In most cases, the causative substance is inhaled.

It is believed that, in part, the reason some people react to various allergens while others do not is due to hereditary factors. Hay fever is not inherited, but the tendency to develop an allergic disease may be inherited. Other contributing factors include psychic stress, infections, and endocrine disturbances. Any of these factors may trigger an attack. It should be stressed that the term *hay fever* is a misnomer, in that the condition is not ordinarily associated with hay or a fever. The term was used by an English physician (Bostock), himself a victim, in a report in 1812. His symptoms occurred during the haying season.

Nearly all cases of seasonal hay fever are caused by pollen. The pollen from many sweet-scented flowers is disseminated by insects, but that causing hay fever is usually spread by the wind. Since many plants produce large quantities of pollen, a sensitive person can easily become overexposed and experience an immediate reaction. Seasonal hay fever can be caused by three different groups of plants, each of which has a somewhat different season. Trees produce pollen that causes hay fever during the months of April and May in the Northern Hemisphere (October and November in the Southern Hemisphere). Various grasses are responsible for much of the hay fever that occurs during the first half of May to the first part of July (first half of November to first part of January in Southern Hemisphere). From the middle of August to October (middle of February to April in Southern Hemisphere), weeds are active pollen producers.

Usually, the least severe and least common form of seasonal hay fever is that which is induced by pollen from trees. It is estimated that 10% of all seasonal hay fever is caused by tree pollen. The oak tree is the most common tree causing hay fever and is found widely in the United States. Three other offenders, the cottonwood, the cedar, and the poplar, grow profusely in the southern and southwestern regions of the United States. Other trees with allergy-producing pollen include the birch, alder, hickory, black walnut, beech, maple, hackberry, sycamore, mulberry, and elm.

Grasses appear to account for about 35% of all cases of seasonal hay fever. The three most common grasses involved are timothy, Bermuda, and June (blue grass). There are eighty varieties of grasses found in various sections of the United States that have a common antigen. Therefore, a person sensitive to one type of grass is usually sensitive to all types of grasses.

Weeds are prolific pollen producers, some varieties producing more than 100,000 pollen grains from a single plant. The most causative agents include various species of ragweed and the thistle. Other weeds involved, but not on a wide scale, include goosefoot, buckwheat, marsh elder, rabbit bush, cocklebur, hemp, and pigweed.

Nonseasonal hay fever may be caused by the hair, feathers, or dander of a household or farm pet. Some individuals are sensitive to feathers in pillows or to kapok, the fibers used as filling for mattresses. House dust, especially in the bedroom, is a common causative agent. Workers in mills where wheat or corn is ground often inhale the flourlike powder, which produces irritation of the nasal mucous membranes. Reproductive cells of some fungi, when inhaled, may stimulate an allergic response.

Nonseasonal hay fever may be caused by certain foods, in particular eggs, chocolate, milk, coffee, or shellfish. Aspirin and quinine may be causative agents. Nonseasonal hay fever may be continuous or spasmodic, depending upon the length of contact with the exciting factor. The symptoms of nonseasonal hay fever tend to be less severe than those of the seasonal variety. At their mildest, they may consist of slight nasal congestion with sniffing, a tendency to an itchy nose, postnasal drip, and mouth breathing.

While the symptoms of hay fever are not difficult to recognize, the actual type of hay fever is not so easily determined. The prior history of a patient is of prime importance. Were there allergic reactions during childhood? Family history is helpful to determine any disposition toward the disease. In particular, details concerning the patient's living habits and possible exposure to various causative agents are important.

Skin tests (also known as "scratch" tests or intracutaneous tests) are frequently made to obtain needed information. A series of small cuts are made in the skin and a small amount of various suspected materials is dropped into each cut and allowed to remain there for a short period. If no reaction occurs after half an hour, the material is removed and the reaction is regarded as negative. If the test is positive, a large, red, itching wheal will appear. The tests are not infallible. Some individuals have more sensitive skin than others. Consequently, false positive reactions are not uncommon. A second type of test, used less frequently, is an eye test. A drop of solution of pollen extract is dropped into the eye. If the reaction is positive, a condition similar to the eye symptoms of hay fever results.

In lieu of seeking areas where the pollen count is low (often difficult for a variety of employment, economic, family and other reasons), air conditioning of the home or office is helpful. A variety of drugs provide relief of hay fever symptoms. Antihistamines are often used and help a high percentage of individuals. Ephedrine-like drugs, taken at night and either alone or in combination and in lessening early morning symptoms. Corticosteroids are useful in severe cases where other agents are not effective.

Preventive treatment also is available. It is designed to acclimate a person's body to the irritant, once identified. The treatment is commenced approximately three months prior to the usual onset of an attack, known from past history of the patient. The treatment consists of the injection of the pollen extract at intervals of about a week. The dosage is gradually increased until the largest dose is given at about the time the symptoms usually start. From then on, the dosage remains the same until the end of the season. At the beginning of treatment, there may be a mild reaction on the spot where the injection was given. This consists of itching and swelling. Subsequent dosage is regulated by the severity of this reaction. The pollen extract treatment has been beneficial in many cases.

The treatment of patients with nonseasonal hay fever consists of completely avoiding the substance or substances which cause the attack. Should contact be absolutely necessary, the physician may prescribe the allergen extract treatment, particularly if the patient's symptoms are so severe as to incapacitate him in his work.

In some allergic conditions there may be upsets of the digestive system, and occasionally severe headaches may result. In all of these cases, there may be some accompanying change in the skin, but in other commonly encountered allergies, the skin changes constitute about the only symptoms. The specific skin symptom may bear very little relation to the cause of the allergy; a particular antigen produces varying types of responses in the skin of different individuals.

One of the most common skin changes associated with allergy is simply a reddening (*erythema* or *hypermia*), caused by increased amounts of blood in the lower layers of the skin due to localized capillary dilatation. Reddened areas of this type may be restricted to a small area of the body, or may be general over its surface. They turn white when subjected to pressure from a finger, seldom are long-lasting, and either disappear within a few days, or progress into some other type of symptom.

Hives (*urticaria*) is a common skin condition in which whitish or reddish, slightly elevated areas of the skin appear. These wheals may be small, like pimples (*papules*), or much larger patches or streaks (*welts*). They generally cover the entire body, being most common on areas covered by clothing. Hives are caused by the accumulation of tissue fluids (*edema*) beneath the epidermis in areas seen as wheals. The condition generally arises rapidly, may last for an hour or so, and then disappears as quickly as it came, if its cause has been removed. See also **Hives.**

Another symptom frequently associated with allergy is known as *eczema*. There is a reddening of the skin, followed by the appearance of minute blisters or *vesicles*. These vesicles become larger and are generally accompanied by intense itching. In acute cases, these blisters break and exude a fluid which forms a crust on the skin. The crust then flakes off, frequently as the result of a secondary inflammation of the skin. Eczema may cover any area of the body, and is one of the most severe of all allergic symptoms. Eczema-type reactions of the skin may result also from some infections and as the result of various nervous conditions.

Other symptoms occasionally seen as the result of an allergy include *nodules*, which are small hard bodies beneath the skin, and large blisters (*blebs* or *bullae*). As the result of the various skin changes which occur, secondary lesions eventually may develop. These include abrasions or erosions, fissures or cracks, ulcers, and scars. These secondary lesions are seldom encountered when the patient receives prompt treatment and the cause of the allergy is determined and removed.

Food allergies in infants frequently result in a severe eczema, and are most often caused by egg white, milk, wheat, oats, barley, and corn. Since eczema may result the first time an infant eats egg white or some other of these foods, it seems possible that sensitization of a child may have occurred while it was receiving its nourishment through the placenta before birth. Infantile eczema most often appears in the second or third month of life and may disappear spontaneously by the end of the second year, with no remaining signs of the food hypersensitivity. The condition may appear again or become worse following vaccination, colds, or eruption of the teeth. Sensitivity to egg, wheat, and milk usually occurs less frequently with increasing age, and disappears almost completely between the fourth and twelfth years.

A large number of chemical substances when taken into the body or applied to the body's surface are capable of producing severe allergic symptoms. Not only are such skin conditions encountered as the result of some medicine to which the body has become sensitized, but they also occur as the result of contact with various industrial chemicals.

Skin eruptions caused by drugs differ somewhat from other allergies, in that they frequently manifest brighter colors, appear suddenly, occur symmetrically on the body, are frequently extensive, and do not generally produce any other body disturbances. Most symptoms disappear after administration of the drug is stopped. Skin eruptions caused by iodides and bromides disappear more slowly and those caused by arsenic hypersensitivity may appear long after the drug has been taken and may last indefinitely. Hypersensitivity to phenolphthalein (used in some laxatives) also may produce an inflammation which lasts long after administration of the drug has been stopped.

Among the more common drugs that may cause eruptions might be listed acetanilide, amidopyrine, antipyrine, arsenic compounds, aspirin, atabrine, barbituric acid derivatives, benzoic acid, benzocaine, opium and morphine, penicillin, phenobarbital, phenolphthalein, quinine, salicylic acid, sulfonamides, and turpentine. Except for reactions to penicillin, it is evident that allergy to any one of these drugs is a relatively rare condition, when one considers the number of persons to whom they are administered without ill effects. Included among the various medicinal preparations which are capable of producing allergies should be mentioned the various serums and other animal products. When various immunizing serums, such as tetanus antitoxin, are repeatedly injected into an individual, they occasionally produce a sensitive condition as the result of the development of antibodies against the proteins in the serum. In some acute cases, the entire body may react violently to a further administration of the same serum. The dangerous condition which occurs within a few moments in such cases is known as *shock*. Modern methods of preparing the sera for injection have caused a marked decrease in the incidence of this condition.

Workers sometimes develop a hypersensitivity to materials to which they are constantly exposed, as bakers to flour, barbers to quinine (hair tonics), dentists to Novocaine, painters to linseed oil, and so on. Various soaps and detergents are also common allergens, although these agents are more often responsible for *primary irritant dermatitis*, a condition easily confused with true allergy. Toilet preparations, cosmetics, clothing, and a host of other substances can set up an allergic reaction in some people. Insect bite hypersensitivity is common. In some individuals, a simple mosquito bite may produce a large and painful swelling out of all proportion to that seen in most other persons. Bites or stings by bees, wasps, bedbugs, lice, fleas, gnats, caterpillars, and various marine fishes and other animals may produce extreme reactions in some few individuals who have previously been sensitized to the allergenic materials of the particular species.

Heat, cold, and light may be the direct cause of burns, chapping, and sunburn, but in some sensitive persons, they may produce allergic skin changes. These usually take the form of hives. In most cases, the symptoms subside rapidly after the cause has been removed.

Mental and emotionally induced allergic symptoms also may appear. The mechanism is not well understood, but it is believed that strong emotions may release various chemical substances into the bloodstream, substances which are capable of sensitizing the body and which act as allergens.

See also **Immune System and Immunology.**

References

Breneman, J. C.: "Basics of Food Allergy," Charles C. Thomas, Springfield, Illinois, 1978.

Dickey, L. D.: "Clinical Ecology," Charles C. Thomas, Springfield, Illinois, 1976.

Gerrad, J. W.: "Understanding Allergies," Charles C. Thomas, Springfield, Illinois, 1977.

Gerrard, J. W.: "Food Allergy," Charles C. Thomas, Springfield, Illinois, 1980.

McGovern, J. P., Smolensky, M. H., and A. Reinberg: "Chronobiology in Allergy and Immunology," Charles C. Thomas, Springfield, Illinois, 1977.

ALLIGATION. A simple mathematical method for calculating the correct proportioning of the ingredients of a mixture or, in general, the value of a property of a mixture from the values of that property in its components. It is based upon the formula

$$P_{xy} = \frac{xX + yY}{x + y}$$

in which P_{xy} is the value of a property of the mixture, X and Y are its values in the components, and x and y are the proportions of the components. It assumes, of course, no change in properties on mixing.

ALLIGATOR. Crocodiles and Alligators.

ALLIUM (*Liliaceae*). A large genus whose species are found widely. Some 75 species are found in North America, especially in the western states. All are bulbous plants with flat or tubular leaves, and with spherical heads or umbels of variously colored flowers. Particularly important cultivated species are the onion, *Allium cepa*; leek, *Allium porrum*; garlic, *Allium sativum*; and chives, *Allium schoenoprasum*. One European species now extensively introduced in the United States is the common weed, field garlic, *Allium vineale*, which (if eaten by cows) noticeably flavors milk and butter.

ALL-LATITUDE COMPASS. Compass (Navigation).

ALLOBAR. A form of an element differing in atomic weight from the naturally occurring form, hence a form of element differing in isotopic composition from the naturally occurring form.

ALLOCHROMATIC. With reference to a mineral that, in its purest state, is colorless, but that many have color due to submicroscopic inclusions, or to the presence of a closely related element that has become part of the chemical structure of the mineral. With reference to a crystal that may have photoelectric properties due to microscopic particles occurring in the crystal, either present naturally, or as the result of radiation.

ALLOCHROMY. Any fluorescence, or reradiation of light, in which the wavelength (and hence color) of the emitted light differs from that of the absorbed light.

ALLOCHTHONOUS. A term proposed by Gümbel in 1888 for sedimentary rocks whose constituents have been transported and deposited at some distance from their place of origin. The bulk of the sedimentary rocks are of this type. The term is now used most commonly in reference to masses of rock transported considerable distances by tectonic movements. See **Overthrust.**

ALLOGENIC (Ecology). Successive ecologic events or conditions that result from factors which arise from outside the natural community and thus alter the more localized situation. The term may be applied, for example, to an allogenic drought of very long duration.

ALLOGENIC (Geology). Minerals and rock constituents derived from pre-existing rocks that have been transported from their original site. The term also applies to a stream (allogenic stream) which is fed by water from a distant terrain. An example would be a stream that originates in a humid or glacial region and that later flows through an arid or desert region.

ALLOGYRIC BIREFRINGENCE. A beam of plane-polarized light may be regarded as the resultant of two equal beams of circularly polarized light, one right-handed and the other left-handed. The phenomenon of optical rotation may be represented by assuming that in optically active media, circularly polarized light is transmitted unchanged, but the velocity of left-handed circularly polarized light is not the same as that of right-handed circularly polarized light. Fresnel demonstrated this difference directly, and the phenomenon is called allogyric birefringence.

ALLOMERISM. A property of substances that differ in chemical composition but have the same crystalline form.

ALLOMORPHISM. A property of substances that differ in crystalline form but have the same chemical composition.

ALLOTROPES. Chemical Elements.

ALLOTYPE. An animal or plant fossil selected, as a species or subspecies, as illustrating morphological details not shown in the holotype.

ALLOWED BANDS (Solids). Solids (Band Theory).

ALLOXAN. Diabetes Mellitus.

ALLOYS (Intermediate Compound). Intermetallic Compound.

ALLOYS. Fundamentally, an alloy is an intentional mixture of two or more chemical elements which have metallic properties. Most metals are soluble in one another in the liquid state and alloying procedures usually involve melting; however, alloying by treatment in the solid state without melting is accomplished by the methods of powder metallurgy. When molten alloys solidify they may remain soluble in one another or may separate into intimate mechanical mixtures of the pure constituent metals. More often there is partial solubility in the solid state and the structure consists of a mixture of the saturated solid solutions. Another important type of solid phase is the intermetallic compound which is characterized by hardness and brittleness and usually has only limited solid solubility with the other phases present. Many possibilities exist and these can be protrayed by diagrams showing the relationships between the solid and liquid phases at various temperatures under equilibrium conditions.

It is well-known that many important alloy combinations have properties which are not easy to predict on the basis of the properties of the constituent metals. For example, copper and nickel, both having good electrical conductivity, form solid-solution type alloys having very low conductivity, or high resistivity, making them useful as electrical resistance wires. In some cases very small amounts of an alloying element produce remarkable changes in properties, as in steel containing less than 1% carbon with the balance principally iron. Steels and the age-hardening alloys depend on heat treatment to develop special properties such as great strength and hardness. Other properties which can be developed to a much higher degree in alloys than in pure metals include corrosion-resistance, oxidation-resistance at elevated temperatures, abrasion- or wear-resistance, good bearing characteristics, creep strength at elevated temperatures, and impact toughness. However, solid state physics has been successful in explaining many of these properties of metal and alloys.

There are various types of alloys. Thus, the atoms of one metal may be able to replace the atoms of the other on its lattice sites, forming a substitutional alloy, or solid solution. If the sizes of the atoms, and their preferred structures, are similar, such a system may form a continuous series of solutions; otherwise, the miscibility may be limited. Solid solutions, at certain definite atomic proportions, are capable of undergoing an order-disorder transition into a state where the atoms of one metal are not distributed at random through the lattice sites of the other, but form a superlattice. Again, in certain alloy systems, intermetallic compounds may occur, with certain highly complicated lattice structures, forming distinct crystal phases. It is also possible for light, small atoms to fit into the interstitial positions in a lattice of a heavy metal, forming an interstitial compound.

In this encyclopedia alloys of chemical elements of alloying importance are discussed under that particular element, or in an entry immediately following. Commercially, the major categories of alloys include:

CAST FERROUS METALS

Gray, Ductile, and High-Alloy Irons

In gray iron, most of the contained carbon is in the form of graphite flakes, dispersed throughout the iron. In ductile iron, the major form of contained carbon is graphite spheres which are visible as dots on a ground surface. In white iron, practically all contained carbon is combined with iron as iron carbide (cementite), a very hard material. In malleable iron, the carbon is present as graphite nodules. High-alloy irons usually contain an alloy content in excess of 3%.

Malleable Iron

The two main varieties of malleable iron are ferritic and pearlitic, the former more machinable and more ductile; the latter stronger and harder. Carbon in malleable iron ranges between 2.30 and 2.65%. Ranges of other constituents are: manganese, 0.30 to 0.40%; silicon, 1.00 to 1.50%; sulfur, 0.07 to 0.15%; and phosphorus, 0.05 to 0.12%.

Carbon and Low-Alloy Steels

Low-carbon cast steels have a carbon content less than 0.20%; medium-carbon steels, 0.20 to 0.50%; and high-carbon steels have in excess of 0.50% carbon. Ranges of other constituents are: manganese, 0.50 to 1.00%; silicon, 0.25 to 0.80%; sulfur, 0.060% maximum; and phosphorus, 0.050% maximum.

Low-alloy steels have a carbon content generally less than 0.40% and contain small amounts of other elements, depending upon the desired end-properties. Elements added include aluminum, boron, chromium, cobalt, copper, manganese, molybdenum, nickel, silicon, titanium, tungsten, and vanadium.

High-Alloy Steels

When "high-alloy" is used to describe steel castings, it generally means that the castings contain a minimum of 8% nickel and/or chromium. Commonly thought of as stainless steels, nevertheless *cast grades* should be specified by ACI (Alloy Casting Institute) designations and not by the designations that apply to similar *wrought alloys.*

WROUGHT FERROUS METALS

Carbon Steels

These steels account for over 90% of all steel production. There are numerous varieties, depending upon carbon content and method of production. In one classification, there are *killed* steels, *semikilled* steels, *rimmed* steels, and *capped* steels. These are described in considerable detail under **Iron Metals, Alloys, and Steels.**

High-Strength Low-Alloy Steels

There are several varieties, with high-yield strength depending mainly on the precipitation of martensitic structures from an austenitic field during quenching. Small additions of alloy elements, such as manganese and copper, are dissolved in a ferritic structure to obtain high strength and corrosion resistance.

Low and Medium-Alloy Steels

The two basic types are (1) *through* hardenable, and (2) *surface* hardenable. Subcategories of surface hardenable alloys include carburizing alloys, flame and induction-hardening alloys, and nitriding alloys.

Stainless Steels

A stainless steel is defined as iron-chromium alloy that contains at least 11.5% chromium. There are three major categories: (1) austenic, (2) ferritic, and (3) martensitic, depending upon the metallurgical structure. There are scores of varieties. Type 302 is the base alloy for austenitic stainless steels. Representative stainless steels in this category provide some insight as to why so many varieties are made and of how rather small changes in composition and production can bring about significant differences in the final properties of the various stainless steels. A slightly lower carbon content improves weldability and inhibits carbide formation. An increase in nickel content lowers the work hardening. By increasing both chromium and nickel, better corrosion and scaling resistance is achieved. The addition of sulfur or selenium increases machinability. The addition of silicon increases scaling resistance at high temperature. Small amounts of molybdenum improve resistance to pitting corrosion and temperature strength.

High-Temperature, High-Strength, Iron-Base Alloys

There are two general objectives in making these alloys: (1) they can be strengthened by a martensitic type of transformation, and (2) they will remain austenitic regardless of heat treatment and derive their strength from cold working or precipitation hardening. Again, there are numerous types. Considering the main types, the carbon content may range from 0.05% to 1.10%; manganese, 0.20 to 1.75%; silicon, 0.20 to 0.90%; chromium, 1.00 to 20.75%; nickel, 0 to 44.30%; cobalt, 0 to 19.50%; molybdenum, 0 to 6.00%; vanadium, 0 to 1.9%; tungsten, 0 to 6.35%; copper, 0 to 3.30%; columbium (niobium), 0 to 1.15%; tantalum, 0 to $< 1\%$; aluminum, 0 to 1.17%; and titanium, 0 to 3%.

Ultrahigh-Strength Steels

Normally a steel is considered in this category if it has a yield strength of 160,000 psi or more. The first of these steels to be produced was a chromium-molybdenum alloy steel, shortly followed by a stronger chromium-nickel-molybdenum grade.

Free-Machining Steels

Normally, the carbon content is kept under 0.10%, but as much as 0.25% carbon has little deleterious effect on machinability. Aluminum and silicon are held to a minimum (aluminum not used as a deoxidizer where machinability is extremely important). Lead, sulfur, bismuth, selenium, and tellurium (0.04%) improve machinability when in the proper combination. Sulfur improves machinability by combining with any manganese and oxygen present to form oxysulfides.

NONFERROUS METALS

Aluminum Alloys

These alloys are available as wrought or cast alloys. The principal metals alloyed with aluminum include copper, manganese, silicon, magnesium, and zinc. These alloys are discussed in considerable detail under **Aluminum Alloys.**

Copper Alloys

These alloys are available as wrought or cast alloys. The principal wrought copper alloys are the brasses, leaded brasses, phosphor bronzes, aluminum bronzes, silicon bronzes, beryllium coppers, cupro nickels, and nickel silvers. The major cast copper alloys include the red and yellow brasses, manganese, tin, aluminum, and silicon bronzes, beryllium coppers, and nickel silvers. The chemical compositions range widely. For example, a leaded brass will contain 60% copper, 36 to 40% zinc, and lead up to 4%; a beryllium copper is nearly all copper, containing 2.1% beryllium, 0.5% cobalt, or nickel, or in another formulation, 0.65% beryllium, and 2.5% cobalt.

Nickel Alloys

Although nickel is present in varying amounts in stainless steel *commercially* a high-nickel stainless steel is not categorized as a nickel *alloy,* but rather as a stainless steel. Most nickel alloys are proprietary formulations and hence designated by trade names, such as *Duranickel, Monel* (several), *Hastelloy* (several), *Waspaloy, Rene 41, Inco, Inconel* (several), and *Illium* G. The nickel content will range from about 30% to nearly 95%.

Magnesium Alloys

It is the combination of low density and good mechanical strength which provides magnesium alloys with a high strength-to-weight ratio. Again, these generally are proprietary formulations. Aluminum, manganese, thorium, zinc, zirconium, and some of the rare-earth metals are alloyed with magnesium.

Zinc Alloys

Zinc alloys are available as die-casting alloys or wrought alloys. The principal alloys used for die casting contain low percentages of magnesium, from 3.5 to 4.3% aluminum, and carefully controlled amounts of iron, lead, cadmium, and tin.

Titanium Alloys

The titanium-base alloys are considerably stronger than aluminum alloys and superior to most alloy steels in several respects. Several types are available. Alloying metals include aluminum, vanadium, tin, copper, molybdenum, and chromium.

References

Baker, H., and D. Benjamin (editors): "Glossary of Metallurgical Terms and Engineering Tables," American Society for Metals, Metals Park, Ohio, 1979.

Bardes, B. P. (editor): "Metals Handbook," 9th Edition, Vol. 1 (Properties and Selection: Irons and Steels); and Vol. 2 (Properties and Selection: Nonferrous Alloys and Pure Metals), American Society for Metals, Metals Park, Ohio, 1979.

Chin, G. Y.: "New Magnetic Alloys," *Science,* **208**, 888–894 (1980).

Coyne, J. E.: "Technology is Key to Production of Superalloy Disc Forgings," *Metal Progress,* **118**, 6, 34–41 (1980).

Eiselstein, H. S.: "Monel Alloys Mark 75th Anniversary," *Metal Progress,* **118**, 8, 11–16 (1980).

Furrer, A. Editor: "Crystal Field Effects in Metals and Alloys," Plenum, New York, 1977.

Ketcham, S. J., and E. J. Jankowsky: "How Aluminum Alloys Fare in Shipboard Exposure Tests," *Metal Progress,* **119**, 4, 38–44 (1981).

Peckner, D., and I. M. Bernstein: "Handbook of Stainless Steels," McGraw-Hill, New York, 1977.

Pickering, F. B.: "The Metallurgical Evolution of Stainless Steels," American Society for Metals, Metals Park, Ohio, 1979.

Rashid, M. S.: "High-Strength, Low-Alloy Steels," *Science,* **208**, 862–869 (1980).

Schetky, L. M.: "Shape-Memory Alloys," *Sci. Amer.,* **241**, 5, 74–82 (1979).

Schillmoller, C. M., and H. P. Klein: "Selecting and Using Some High Technology Stainless Steels," *Metal Progress,* **119**, 2, 22–29 (1981).

Smith, D. F., and E. F. Clatworthy: "The Development of High Strength, Low Expansion Alloys," *Metal Progress,* **119**, 4, 32–35 (1981).

Smith, R. D.: "Temper Designations for Copper and Copper Alloys," *Metal Progress,* **119**, 2, 47–49 (1981).

Staff: "Unified Numbering System for Metals and Alloys," 2nd Edition, Society of Automotive Engineers, Warrendale, Pennsylvania, 1977.

Staff: "Source Book on Selection and Fabrication of Aluminum Alloys," American Society for Metals, Metals Park, Ohio, 1978.

Staff: "Source Book on Copper and Copper Alloys," American Society for Metals, Metals Park, Ohio, 1979.

Staff: "Worldwide Guide to Equivalent Irons and Steels," and "Worldwide Guide to Equivalent Nonferrous Metals and Alloys," American Society for Metals, Metals Park, Ohio, 1979.

Staff: "Trends in Superalloy Technology," *Metal Progress,* **119**, 1, 44–46 (1981).

Tien, J. K., et al. (editors): "Superalloys 1980," American Society for Metals, Metals Park, Ohio, 1980.

ALLSPICE. Sometimes called *pimento,* allspice is prepared from the dried, pea-size, unripened, dark reddish-brown berry of a West Indian evergreen tree (*Pimenta officinalis* L.). The tree, which reaches a height of about 20 feet (6 meters) is an evergreen and a member of the Myriaecea (myrtle) family, of which probably the best known members are bayberry trees and shrubs.

The word *pimento* is not to be confused with the word *pimentio* (see **Pepper**). The substances are not related. Also, to contribute to word confusion, allspice is sometimes called Jamaica pepper, a term which tends to associate it with the more familiar black or white pepper of the family *Piperiaceae.* There is no such association.

Allspice is well named because its essence and taste resemble that of a mixture of cinnamon, nutmeg, and cloves. Pimento oil is a fragrant essential oil distilled from allspice berries. It contains eugenol and cineol which have a carnationlike essence.

Allspice finds numerous uses in food products, including:

Bakery products—special breads and rolls; muffins; coffee cakes; spice cakes; fruit cakes; fruit, chocolate, and custard cream pies.
Beverages—cordials and liqueurs.
Condiments—catsup, chili sauce.
Confections—licorice.
Ethnic dishes—German.
Fruits and fruit-based products—apples; apple sauce; apple drink; apricots; cranberry drink; cranberry sauce; peaches; pears; plums; preserves; spiced fruits; stewed fruits.
Sauces—tomato sauce.
Meats and meat dishes—bologna; beef; frankfurters; hamburger; head cheese; meatloaf; mincemeat; pork; sausage.
Pickles.
Poultry.
Soups—beef; consommé; tomato; vegetable.
Vegetables—beans, beets; cole slaw; spinach; squash; sweet potatoes; tomatoes; turnips.

ALLSPICE TREE. Of the family *Myraceae* (myrtle family), the allspice tree (*Pimenta officinalis*) is a small tree native to the West Indies and Central America. The tree is known for its fruit which, when dried, is commercially marketed as a spice of the name *allspice.* The name was given by early users who concluded that the flavors of cinnamon, cloves, and nutmeg were all present in the one spice. Jamaica produces most of the allspice of commerce. The tree grows to a height of from 30 to 40 feet (9 to 12 meters), has leathery leaves, fragrant with the oil which they contain, and produces small, white flowers borne in axillary cymes. The fruit contains one or two seeds. Before ripening, the fruits are gathered, rapidly dried, sometimes further processed, and then marketed as allspice.

ALLUVIAL FAN. Also called *subaerial delta,* a cone-shaped to delta-shaped collection of coarsely graded sediments deposited by intermittent streams that debouch from steep valleys onto a relatively gentle slope or plain. See accompanying figure. Alluvial fans may extend for many miles. Confluent fans may eventually cover and fill relatively large intermontane basins. An alluvial-fan shoreline is one in which an alluvial fan is built out into a lake or sea.

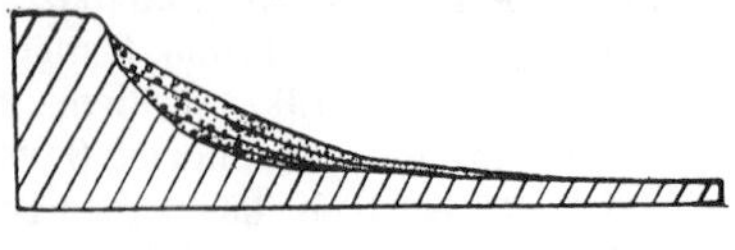

Cross section of an alluvial fan or subaerial delta.

ALLUVIUM. A general term used to designated the sand, silt, and mud deposited by a stream, along its bank or upon its floodplain, during periods of high water. The word is derived from the Latin *ad*, to; and *luo*, wash. When alluvium is relatively fine-textured and contains sufficient organic matter it forms soil. Some of the oldest and richest agricultural regions are the great delta areas, such as the Nile and Euphrates.

ALLYL CHLORIDE. Chlorinated Organics.

ALLYLIC RESINS. Among the thermosetting allylic resins are diallyl phthalate, diallyl isophthalate, diallyl maleate, and diallyl chlorendate. The allylic monomers serve as nonvolatile cross-linking agents in polyester compounds. Allylic resins find numerous applications because of their good temperature range, dimensional stability, and fine electrical properties. Allylic prepolymers are used in compounds for either compression or transfer molding. Generally, these prepolymers may be classified as medium-to-soft-flow molding materials. The addition of fibrous fillers improve strength, but at the sacrifice of electrical properties. The use of *Orlon* fibers enhances electrical properties, even under adverse humidities. *Dacron* provides impact resistance and strength in thin sections. The allylic prepolymers are especially suited for electronic gear that is subjected to adverse environmental conditions. Because of their chemical inertness, the prepolymers also are used for molding pump impellers and other chemical equipment.

Usually allylic resin parts are compression molded when long fiber glass fillers are used. Allylic molding compounds flow readily around inserts, fill complex cavities, knit well, and have low mold shrinkage. Parts can have small draft angles without causing mold-release problems. In addition to good chemical resistance, fungus resistance, thermal stability, high heat distortion temperature, and moisture resistance are among the advantages of the allylic resins. By comparison with polyesters, allylic systems have no styrene odor, low toxicity, low evaporation losses during evacuation cycles, and no subsequent bleed out.

ALLYL REARRANGEMENT. Rearrangement (Organic Chemistry).

ALMAGEST. The name assigned by the Arabs to the great treatise on science compiled by Ptolemy during the second century. The very name Almagest, which is a hybrid combination of the Greek superlative ($\mu\epsilon\gamma\iota\sigma\tau\eta$) with the Arabic article (al), indicates the importance of this work to the early astronomers.

The Almagest is a collection of treatises on a variety of scientific subjects. In it is to be found the complete exposition of the Ptolemaic system for the structure of the universe. Perhaps the best-known section of the Almagest is that dealing with the stars and the constellations. This section was taken from the works of Hipparchus and incorporated in the Almagest by Ptolemy, with some improvements and additions. It is in this catalogue that we first find the brightnesses of the stars divided into six magnitudes, a system that has persisted down to modern times. The positions of the stars given in the Almagest have proven of some little value in determining the constants of precession and, also, the proper motions of the stars.

ALMANAC (Astronomical). For the work of every person engaged in astronomy, whether as an astronomer in an observatory, as a navigator on a ship at sea or in the air, or as a surveyor in the field, tables of certain astronomical data are indispensable. Many, in fact most, of these tables change from year to year. Among such material may be listed the positions of the sun, moon, and planets for every day in the year; accurate positions of stars to be used for determination of local time; tables for computing precession, nutation, aberration, etc. Such material is computed and published in almanacs several years in advance so that ships embarking on long voyages can have the data at hand when they leave port.

In addition to the ephemerides and data listed above, astronomical almanacs also contain descriptions of such phenomena as eclipses of the sun and moon, occultations of stars by the moon, eclipses and configurations of the satellites of Jupiter, etc. An examination of the preface for the American Ephemeris and Nautical Almanac or the publications in Great Britain of H.M. Stationery Office for any year will show how the work for that particular year was distributed.

ALMANDINE. Carbuncle; Garnet.

ALMOND. Rose Family.

ALNICO MAGNET. Magnetism.

ALOE (*Liliaceae*). A large genus of plants characteristic of drier parts of Africa, especially the southern part. Because of their ornamental appearance, with stiff habit and spiny-margined leaves, many of them are grown in cultivation. The rather small yellow or red flowers are borne in large masses. Many species yield from the crushed leaves a purgative juice, which is called aloes, and which has been used extensively by eastern people.

ALOPECIA. Loss of body hair. The affliction may be the natural accompaniment of (1) inheritance- or age-induced processes. (2) inflammatory changes in the hair follicles, or (3) from a toxic process. All body hair may be affected. The causation is not fully understood. When the hair loss is confined to small patches on the scalp or bearded portion of the face, the condition is referred to as *alopecia areata.* Although this occurs in all age groups, it is much more common among young adults and is not confined to males. The condition also may occur in infants. The condition can be transient or permanent and the occurrence or reoccurrence is difficult to predict. Statistically, there appears to be no relationship with inheritance, environment, or disposition of the affected person. When large areas of the hairy portions of the body are affected, the term *alopecia generalisata* is used. When all body hair is lost, the affliction is termed *alopecia universalis.* Corticosteroids have been used with limited success toward restoration of hair in some instances with exception of ordinary baldness, which is termed *alopecia prematura.* The latter condition affects large numbers of males and hair loss may commence as early as age 17 and progress steadily for a number of years. This form of alopecia definitely appears to be hereditary, even the patterns of baldness showing a resemblance from one generation to the next. The basic cause of the inheritance mechanism is not understood. Preventive measures, such as meticulous scalp hygiene, have no effect on the unremitting progress of the condition. No successful drugs or therapies have been developed to date to naturally restore hair lost by alopecia prematura.

ALPACA. Camels and Llamas.

ALPHA-ADRENERGIC RECEPTORS. Ischemic Heart Disease.

ALPHA CELLS (Pancreas). Diabetes Mellitus.

ALPHA CENTAURI. Ranking third in apparent brightness among the stars, Alpha Centauri has a true brightness value of 1.5 as compared with unity for the sun. Alpha Centauri is a yellow, spectral type G star and is one of the terminal stars in the pattern of the constellation Centaurus located south of the ecliptic. In the mid-1800s, the South African astronomer, Thomas Henderson, determined that Alpha Centauri is the nearest star to the sun, an estimated 4.3 light years distant. Actually, Alpha Centauri is a double star, with a third star, Proxima Centauri, revolving around the two stars. Alpha Centauri and another star in the constellation Centaurus, Beta Centauri, form a line which points quite closely to the south pole of the celestial sphere. See also **Constellations;** and **Star.**

ALPHA CHAMBER. A counter tube or counting chamber for the detection of alpha particles; often operated in the nonmultiplying (ionization chamber) or proportional region with pulse height selection to discriminate against pulses due to beta or gamma rays and to pass only those due to alpha particles.

ALPHA CRUCIS. Ranking thirteenth in apparent brightness among the stars, Alpha Crucis has a true brightness value of 4,000 as compared with unity for the sun. Alpha Crucis is a blue-white, spectral type B star and is located in the constellation Crux (Southern Cross) south of the ecliptic. Estimated distance from the earth is 400 light years. See also **Constellations.**

ALPHA CUTOFF. The frequency at which the alpha (current amplification) of a transistor has fallen to 0.7 (3 decibels) of its low-frequency value.

ALPHA DECAY. The process that occurs when alpha particles are emitted by radioactive nuclei. The name *alpha particle* was applied in the earlier years of radioactivity investigations, before it was fully understood what alpha particles are. It is known now, of course, that alpha particles are the same as helium nuclei. When a radioactive nucleus emits an alpha particle, its atomic number decreases by $Z = 2$ and its mass number by $A = 4$. It is, therefore, a spontaneous nuclear reaction of the form

$$A_Z \rightarrow {}^{A-4}Z - 2 + {}^4\text{He}.$$

The entire energy released by the transition is carried away by the product nuclei. Therefore, a spectrum of alpha-particle numbers as a function of energy shows a series of distinct peaks, each corresponding to a single alpha-particle transition. To conserve both energy and momentum, the energy must be shared by the two product nuclei, with the daughter nucleus (${}^{A-4}Z - 2$) recoiling away from the direction of emission of the alpha particle. If E_x and M_x are, respectively, the kinetic energy and mass of the alpha particle and E_R and M_R the kinetic energy and mass of the recoiling product nucleus, the transition energy is $Q = E_\alpha + E_R$; and the kinetic energy of the emitted alpha particle is $E_\alpha = [M_R\ (M_\alpha + M_R)]Q$.

Almost all radioactive nuclides that emit alpha particles are in the upper end of the periodic table, with atomic numbers greater than 82 (lead), but a few alpha-particle emitting nuclides are scattered through lower atomic numbers. The reason why alpha-particle emitters are limited to nuclides with larger mass numbers is that generally only in this region is alpha-particle emission energetically possible. Most radioactive nuclides with smaller mass numbers emit beta-particle radiation.

See **Particles (Subatomic);** and **Radioactivity.**

ALPHA EMITTER. A radionuclide that undergoes a transformation by alpha-particle emission.

ALPHANUMERIC CODE. **Code (Computer System).**

ALPHA-PARTICLE BINDING ENERGY. **Binding Energy.**

ALPHA WAVES (Sleep). **Sleep.**

ALPHERATZ (α Andromedae). A star formerly allotted to the constellation Pegasus by the Arabs. It is situated at the northeast corner of the great square of Pegasus. The star is a spectroscopic binary with a period of approximately 100 days. See also **Constellations.**

ALPINE (Ecology). Characteristic of mountainous regions which occur between the timberline and the snowline. The term is used with reference to the flora, climate, relief, ecology, etc. of such regions. In a less restrictive sense, the term pertains to high elevations and cold climates.

ALPINE FIR. **Fir Trees.**

ALPINE OROGENY. The term for the relatively young orogenic events which occurred in southern Europe and Asia, during which the rocks of the Alps and the remainder of the Alpide orogenic belt were strongly deformed. Most geologists restrict the era to the Tertiary, ending during the Miocene or Pliocene.

ALTAIR (α Aquilae). A star that forms, with β and γ of the constellation Aquila, a well-known line of stars sometimes referred to as the shaft of Aquila. These stars are a conspicuous feature of the early autumn sky. Ranking twelfth in apparent brightness among the stars, Altair has a true brightness value of 11 as compared with unity for the sun. Estimated distance from the earth is $16\frac{1}{2}$ light years. Altair is classified as a white star of spectral type A. See also **Constellations;** and **Star.**

ALTAZIMUTH. An instrument so mounted that it may be rotated about a horizontal and a vertical axis (i.e., rotated in altitude and azimuth); the earliest type of mounting for astronomical telescopes. Perhaps the most familiar altazimuth instrument is the ordinary surveyor's transit or theodolite.

The great advantage of the altazimuth is the ease with which it may be set up. If the instrument has been properly constructed, the horizontal and vertical axes will be strictly perpendicular to each other, and the only necessary adjustment will be to level the horizontal axis.

The altazimuth instrument is used in the field for laying down azimuth lines and for determining latitude and longitude by measuring altitudes of celestial objects. A few large, fixed altazimuth instruments are in use in observatories for accurate determination of declinations of stars, but for this purpose, the meridian circle is most commonly used.

For ordinary astronomical observing, the altazimuth instrument is less convenient than the equatorial because of the fact that the diurnal motion of the celestial sphere is parallel to the equator rather than to the horizon, with the result that the instrument has to be moved about both axes to follow the celestial objects. See also **Theodolite.**

ALTERNATING CURRENT CIRCUITS. A circuit is an interconnection of electrical elements consisting of energy sources, resistors, capacitors, and inductors having self and mutual inductance. See also **Electric Circuits.** The steady-state response of a circuit or electrical network to sinusoidal sources obtained in complex representation. The complex form of a sinusoidal voltage

$$e(t) = E_m \cos(\omega t + \alpha)$$

is

$$E_m \epsilon^{j(\omega t + \alpha)} = (E_m \epsilon^{j\alpha})(\epsilon^{j\omega t})$$
$$= \mathbf{E}_m \epsilon^{j\omega t}, \qquad (j = \sqrt{-1})$$

The complex $\mathbf{E}_m = E_m \epsilon^{j\alpha} = E_m \angle \alpha$ is called the phasor voltage in terms of its maximum value. $\mathbf{E} = \mathbf{E}_m/\sqrt{2} = .707 E_m \angle \alpha$, the phasor voltage in terms of its rms or effective value, is called simply the phasor voltage. It can be seen that the instantaneous voltage is the real part of its complex representation. The complex representation of the sinusoidal current

$$i(t) = I_m \cos(\omega + \beta) \text{ is } I_m \epsilon^{j(\omega t + \beta)} = (I_m \epsilon^{j\beta})\epsilon^{j\omega t} = \mathbf{I}_m \epsilon^{j\omega t}$$

where $\mathbf{I}_m$, the phasor current in terms of its maximum is $I_m \epsilon^{j\beta} = I_m \angle \beta$. The phasor current $\mathbf{I} = \mathbf{I}_m/\sqrt{2} = .707 I_m \angle \beta$. Again, the instantaneous value of the current $i(t) =$ real part of $\mathbf{I}_m \epsilon^{j\omega t}$. Phasors are expressed in either polar form, magnitude and angle, or rectangular form, real and imaginary parts.

For an inductance, the voltage

$$e(t) = L \frac{di(t)}{dt}$$

Now $i(t)$ in complex representation is $\mathbf{I}_m \epsilon^{j\omega t}$, so $e(t)$ in complex representation is $\mathbf{I}_m (j\omega L) \epsilon^{j\omega t}$ and

$$\mathbf{E}_m = \mathbf{I}_m (j\omega L)$$

The ratio of the two phasors,

$$\frac{\mathbf{E}_m}{\mathbf{I}_m} = \frac{\mathbf{E}}{\mathbf{I}} = j\omega L = \omega L \angle \frac{\pi}{2} = (\omega L)\epsilon^{j\omega/2}$$

The instantaneous voltage $e(t) = E_m \cos(\omega t + \alpha) =$ real part of

$$\mathbf{I}_m (\omega t) \epsilon^{j(\omega t + \pi/2)} = \mathbf{I}_m (\omega L) \cos\left(\omega t + \beta + \frac{\pi}{2}\right)$$

So

$$E_m = I_m(\omega L)$$

and

$$\alpha = \beta + \frac{\pi}{2}$$

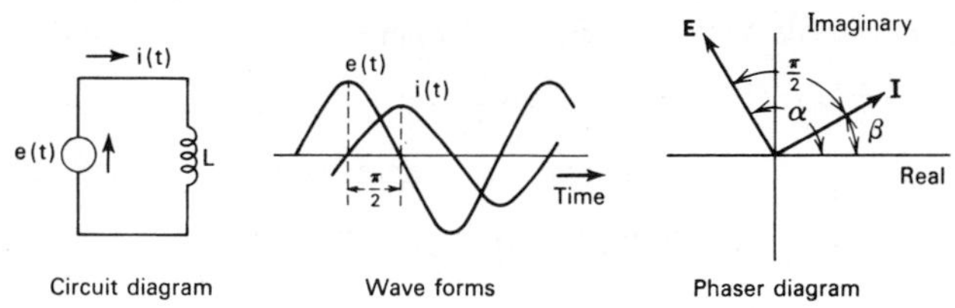

Fig. 1. Case where sinusoidal current lags the voltage by $\pi/2$ radians.

Figure 1 shows the circuit diagram, the sinusoidal waveforms and the phasor diagram. As $E_m = I_m(\omega L)$ then $E_{rms} = I_{rms}(\omega L)$ or the rms value of the voltage across an inductance is equal to $(\omega L) = X_L$, the inductive reactance (given in ohms) multiplied by the rms value of the current through the inductance. The sinusoidal current lags the voltage by $\pi/2$ radians.

For a capacitance, the current

$$i(t) = C\frac{de(t)}{dt}$$

In complex representation,

$$\mathbf{I}_m\epsilon^{j\omega t} = \mathbf{E}_m(j\omega C)\epsilon^{j\omega t}$$

or

$$\mathbf{I}_m = \mathbf{E}_m(j\omega C)$$

$$\frac{\mathbf{E}_m}{\mathbf{I}_m} = \frac{\mathbf{E}}{\mathbf{I}} = \frac{1}{j\omega C} = -\frac{j}{\omega C} = jX_C = \frac{1}{\omega C}\angle\frac{\pi}{2}$$

The instantaneous voltage $e(t) = I_m(1/\omega C)\cos(\omega t + \beta - \pi/2)$. Therefore $\alpha = \beta - \pi/2$ and $E_{rms} = I_{rms}(-X_C)$, where X_C, the capacitive reactance is $-1/\omega C$, or the rms value of the voltage across a capacitor is equal to the magnitude of the capacitive reactive multiplied by the rms current. The sinusoidal current leads the voltage by $\pi/2$ radians. Figure 2 illustrates this case.

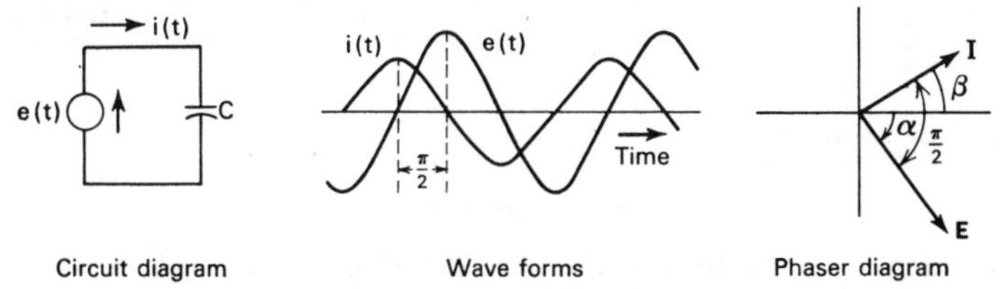

Fig. 2. Case where sinusoidal current leads the voltage by $\pi/2$ radians.

For a resistance, the voltage, $e(t) = Ri(t)$. In the complex representation,

$$\mathbf{E}_m\epsilon^{j\omega t} = R\mathbf{I}_m\epsilon^{j\omega t}$$

or

$$\mathbf{E}_m = \mathbf{I}_m R, \qquad \frac{\mathbf{E}_m}{\mathbf{I}_m} = \frac{\mathbf{E}}{\mathbf{I}} = R$$

The instantaneous voltage $e(t) = I_m R\cos(\omega t + \beta)$, where $E_m = I_m R$ and $\alpha = \beta$.

The rms value of the voltage across a resistor is equal to the product of the current through the resistance by the value of the resistance. The sinusoidal current is in phase with the voltage. Figure 3 illustrates this case.

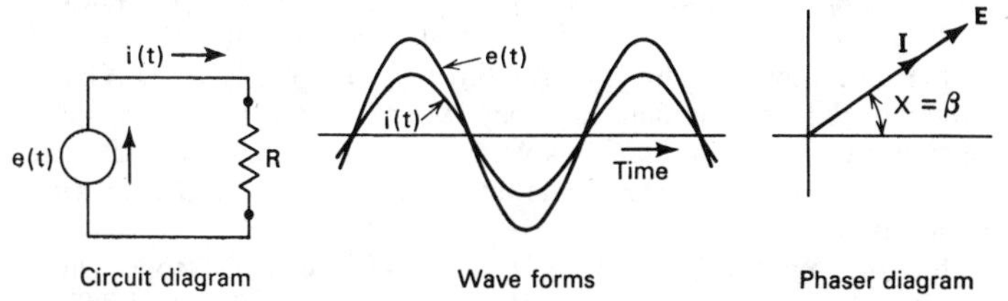

Fig. 3. Case where sinusoidal current is in phase with the voltage.

The responses of a network to sinusoidal sources are determined directly from the general but simple problem of the steady-state response of a network to sinusoidal sources.

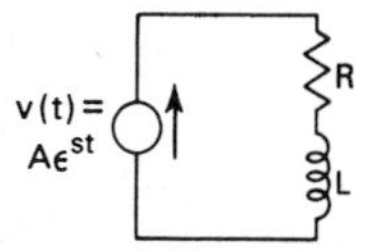

Fig. 4. Circuit with an exponential voltage source.

Thus for the circuit shown in Fig. 4 with an exponential voltage source $A\epsilon^{st}$, the application of Kirchhoff's laws gives

$$Ri + L\frac{di}{dt} = A\epsilon^{st}$$

As the particular integral is the steady state solution,

$$i(t) = \frac{A\epsilon^{st}}{R + sL}$$

For sinusoidal response, using complex representation

$$A \text{ is } \mathbf{E}_m, \qquad s = j\omega$$

and

$$i(t) \text{ is } \mathbf{I}_m\epsilon^{j\omega t}$$

The equation for the current to the network becomes

$$\mathbf{I}_m\epsilon^{j\omega t} = \frac{\mathbf{E}_m\epsilon^{j\omega t}}{R + j\omega L}$$

or

$$\mathbf{I}_m = \frac{\mathbf{E}_m}{R + j\omega L}, \qquad \mathbf{I} = \frac{\mathbf{E}}{R + j\omega L} = \frac{\mathbf{E}}{\mathbf{Z}} = \mathbf{EY}$$

The impedance of the network (sometimes referred to as the complex impedance) is

$$\mathbf{Z} = \frac{\mathbf{E}_m}{\mathbf{I}_m} = \frac{\mathbf{E}}{\mathbf{I}}$$

For this network, the impedance is $R + j\omega L$ or its real part is the resistance R and its imaginary part, the inductive reactance ωL. The admittance of the network, $\mathbf{Y}$, is equal to $\mathbf{I}_m/\mathbf{E}_m = \mathbf{I}/\mathbf{E} = 1/\mathbf{Z}$. For this network, the admittance is $1/(R = j\omega L)$. As

$$\mathbf{I}_m = \frac{\mathbf{E}_m}{R + j\omega L} = \frac{E_m\angle\alpha}{\sqrt{R^2 + (\omega L)^2}\angle\theta} = \frac{E_m\angle(\alpha - \theta)}{\sqrt{R^2 + (\omega L)^2}}$$

So the instantaneous current

$$i(t) = \frac{E_m}{\sqrt{R^2 + (\omega L)^2}}\cos(\omega t + \alpha + \theta), \qquad \theta = \tan^{-1}\frac{\omega L}{R}$$

$$= I_m\cos(\omega t + \beta)$$

$$I_m = \frac{E_m}{\sqrt{R^2 + (\omega L)^2}}$$

or

$$I_{rms} = \frac{E_{rms}}{\sqrt{R^2 + (\omega L)^2}}$$

and $\beta = \alpha - \theta$ or the current lags the voltage by the angle θ. Figure 5 shows the waveform and the phasor diagram.

Fig. 5. Waveform and phasor diagram for case of Fig. 4.

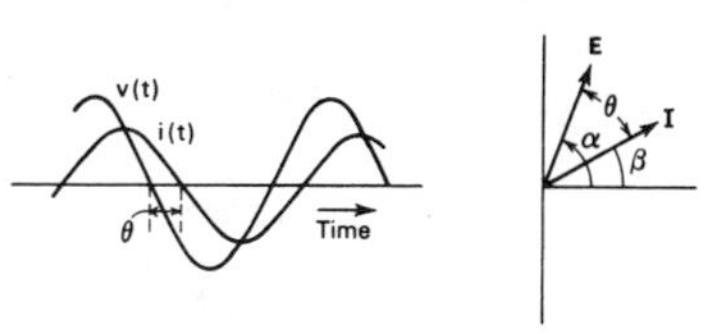

For the series R-L-C circuit,

$$Ri + L\frac{di}{dt} + \frac{1}{C}\int i\,dt = A\epsilon^{st}$$

The particular integral is

$$i(t)=\frac{A\epsilon^{st}}{R+sL+\frac{1}{sC}}$$

This relation may be written directly from the network, labelling each element as shown in Fig. 6 and calculating the current as if it were

Fig. 6. Example of labeling each element in network.

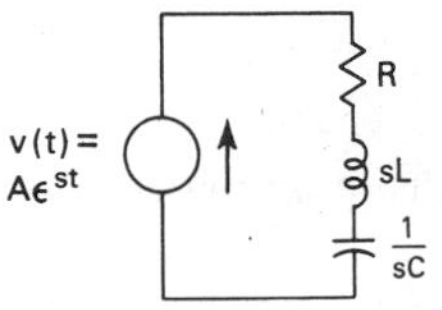

a dc circuit with each element equivalent to a resistor. For the sinusoidal response

$$\mathbf{I}=\frac{\mathbf{E}}{R+j\omega L+1/j\omega C}=\frac{\mathbf{E}}{R+j(X_L+X_C)}=\frac{\mathbf{E}}{\mathbf{Z}}$$

and

$$\mathbf{Z}=R+j(X_L+X_C)=\sqrt{R^2+(X_L+X_C)^2}\,\angle\,\theta$$

where

$$\theta=\tan^{-1}\frac{X_L+X_C}{R}$$

and

$$i(t)=\frac{E_m}{\sqrt{R^2+(X_L+X_C)^2}}\cos(\omega t+\alpha-\theta)$$

So

$$I_{rms}=\frac{E_{rms}}{\sqrt{R^2+(X_L+X_C)^2}} \quad \text{and} \quad \beta=\alpha-\theta$$

When $|X_L| > |X_C|$, θ is positive and the current in the circuit lags the impressed voltage, and when $|X_L| < |X_C|$, θ is negative and the current leads the impressed voltage.

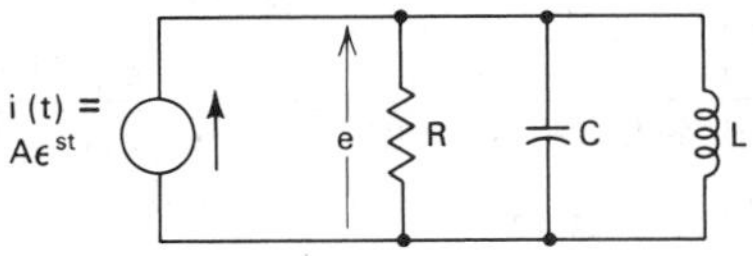

Fig. 7. Parallel *R-L-C* circuit with a current source.

For the parallel *R-L-C* circuit with a current source as shown in Fig. 7, the dual of circuit shown in Fig. 6,

$$\frac{e}{R}+C\frac{de}{dt}+\frac{1}{L}\int e\,dt=A\epsilon^{st}$$

The particular integral is

$$e(t)=\frac{A\epsilon^{st}}{\frac{1}{R}+sC=\frac{1}{sL}}$$

In complex representation,

$$A=\mathbf{I}_m, \qquad s=j\omega$$

and $e(t)$ is $\mathbf{E}_m\epsilon^{j\omega t}$. So

$$\mathbf{E}_m=\frac{\mathbf{I}_m}{\frac{1}{R}+j\omega C+\frac{1}{j\omega L}}=\frac{\mathbf{I}_m}{\mathbf{Y}}$$

or

$$\mathbf{E}=\frac{\mathbf{I}}{\frac{1}{R}+j\left(\omega C-\frac{1}{\omega L}\right)}=\frac{\mathbf{I}}{G+j(B_C+B_L)}=\frac{\mathbf{I}}{\mathbf{Y}}$$

The admittance $\mathbf{Y}=G+j(B_C+B_L)$ where

$$G=\text{conductance}=1/R$$

$$B_C=\text{capacitive susceptance}=\omega C$$

$$B_L=\text{inductive susceptance}=-\frac{I}{\omega L}$$

$$e(t)=\frac{I_m}{\sqrt{G^2+(B_C+B_L)^2}}$$

$$\theta=\tan^{-1}\frac{B_C+B_L}{G}$$

The techniques for solving ac network problems are exactly the same as those for networks with exponential sources but with $j\omega$ replacing s. For impedances in series

$$Z_t=\sum_{i=1}^{n} Z_i$$

and for admittances in parallel,

$$Y_t=\sum_{i=1}^{n} Y_i$$

Also, the same mesh and nodal analysis may be applied as in dc circuits with resistances and conductances being replaced by impedances and admittances in complex representation.

The response of a linear network to a periodic source function that is not sinusoidal may be determined by describing the function as a sum of sine waves in a Fourier analysis. The sum of responses to each sine wave is the total response.

The instantaneous power supplied to an ac circuit is $p = ei$, and the average power is

$$P_{av}=\frac{1}{T}\int_0^T ei\,dt$$

For sine waves

$$P_{av}=E_{rms}I_{rms}\cos\theta$$

The angle θ is the *phase angle* between the current and voltage. The expression $\cos\theta$ is called the *power factor* of the circuit. In terms of voltage and current phasors, the average power is

$$P_{av}=\text{Real part of } \mathbf{EI^*}$$
$$=\text{Real part of } \mathbf{E^*I}$$

where $\mathbf{I^*}$ and $\mathbf{E^*}$ are the conjugates of phasors $\mathbf{I}$ and $\mathbf{E}$ respectively. See **Direct Current Circuits; Electric Circuits;** and **Resonance.**

ALTERNATING CURRENTS. Electric currents which vary periodically with time. These currents are produced by impressing periodic voltages on an electric system. Alternating voltages are produced by rotating generators called alternators or by electronic devices called oscillators, signal generators or wave form generators. Various alternating waveforms are shown in the accompanying figure. One complete set of positive and negative values is called a *cycle*. The time T of a complete cycle is the *period* and is given in seconds. The *frequency* of a wave is the number of cycles occurring in one second. Thus f, the frequency, $= 1/T$ cycles per second. The angular frequency of the periodic wave is defined as 2π times the frequency or ω (Greek letters omega), the angular frequency, $= 2\pi f$ radians per second. For the sine wave, the instantaneous voltage $e(t) = E_m\cos(\omega t+\alpha)$. The angle α is called the phase angle and is determined by the position of the origin.

The instrument for measuring waveforms is the oscilloscope which describes a plot on the face of a cathode-ray tube. The vertical deflection of the describing electron beam is proportional to the voltage on the vertical presentation terminals and the horizontal deflection is proportional to the linear sweep-voltage on the vertical presentation terminus. When the wave being observed is repetitive at a fast enough rate, the display appears as a steady figure.

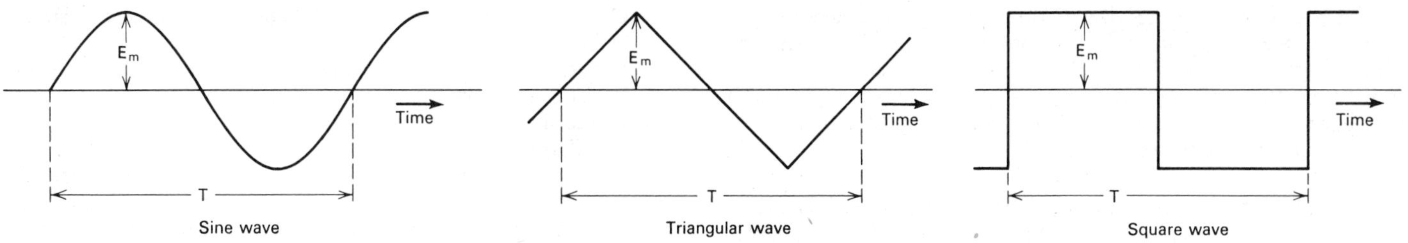

Fundamental waveforms.

The average value of the sine wave voltage for a complete cycle is 0. The average value for half a cycle is $2E_m/\pi$. The rms (root-mean-square) value of the sine wave voltage is $\sqrt{2}E_m/2 = .707E_m$. See accompanying figure.

Through common usage, the expression alternating current or ac refers to a current or a voltage that has a waveform closely approximating the sine wave. There are many advantages to the use of sinusoidal voltages such as:

1. Efficient transmission of electrical energy over long distances by stepping up the voltage with static transformers.
2. Simplification in design and reduction in cost of generators and motors.
3. Basic waveform for signal processing in the generation of modulated carriers for communication systems.
4. Signal analysis using the Fourier methods. See **Frequency Measurement.**
5. Systems analysis in terms of frequency and phase response.

See also **Oscilloscope; Signal Generator.**

ALTERNATING GRADIENT FOCUSING. This principle, discovered independently by Christofilos in 1950 and by Courant, Livingston, and Snyder in 1952, is based on the observation from geometrical optics that a combination of a convex lens and a concave lens of equal focal length have a net focusing effect. This type of focusing of moving charged particles is accomplished by appropriately shaped magnetic fields. But, since magnetic fields cannot focus simultaneously in both the vertical and horizontal planes that can be constructed through a line representing the direction of motion of a beam of charged particles, alternate lenses that focus first in the horizontal direction and then in the vertical direction are needed. This can be accomplished for an external beam of charged particles by pairs of quadrupole magnets. The same particle has been applied to the magnetic fields in particle accelerators. In a synchrotron, for example, a section of magnet focuses axially and defocuses radially, while the following section defocuses axially and focuses radially. Sector focusing in an AVF cyclotron accomplishes the same purpose.

ALTERNATING-GRADIENT SYNCHROTRON (AGS). Particles (Subatomic).

ALTERNATING GROUP. The group of even permutations of *n* objects. See **Permutation.**

ALTERNATING SERIES. A series whose terms are alternately positive and negative. See **Series.**

ALTERNATION OF GENERATIONS. A term used to describe the alternation of two distinct body forms, one reproducing sexually and one reproducing asexually, in the life cycle of a plant or animal. A typical example is to be found in the hydrozoan coelenterate, *Obelia.* See Fig. 1. One generation, the hydroid stage, reproduces asexually by budding and forms a colony consisting of a stalk bearing many polyps. As the colony grows older, reproductive polyps are formed which produce buds that grow into small jellyfish known as medusa. Some polyps produce male and others produce female medusae. These medusae break free of the polyp and swim in the water. The males produce sperms which swim to the eggs produced by the females. The zygote grows into a ciliated larva which swims around for a time and then settles down and becomes attached. It then grows into a hydroid stage.

Parasitic flatworms, such as flukes and tapeworms, and the malarial parasite (*Plasmodium*) are among the other animals that have alternation of generations.

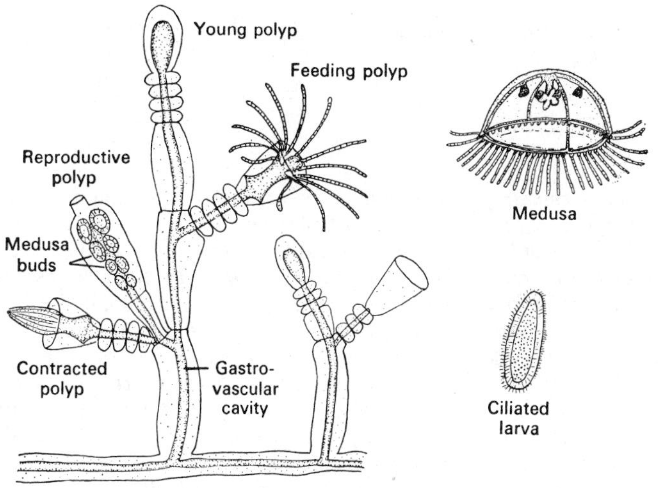

Fig. 1. The life cycle of *Obelia.*

In all except some of the simplest plants there is alternation of generations. The sporophyte generation reproduces asexually by means of spores. These spores are produced by a series of two special cell divisions known as meiosis which reduces the chromosome number to half, the haploid number. A spore grows into a gametophyte generation which will have the haploid number. Gametes are produced by the gametophyte, and when these unite a zygote is formed with the diploid number. The zygote then grows into a sporophyte generation.

As plants become larger and more complex, there is a tendency

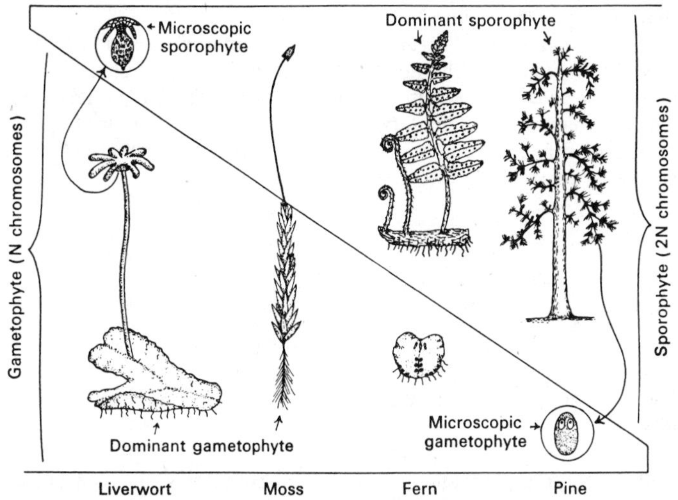

Fig. 2. Alternation of generations is demonstrated by the gradual shift of emphasis from the dominant gametophyte generation (exemplified by the liverworts) to the dominant sporophyte in the seed plants (exemplified by the pine). (*Winchester, "Biology and Its Relation to Mankind," Van Nostrand Reinhold*)

toward a reduction in the size of the gametophyte generation. In liverworts, the sporophyte is microscopic in size; in the mosses the gametophyte is still dominant, but there is a clearly visible sporophyte; in the ferns the sporophyte is dominant, but the gametophyte is still present as a distinct and separate plant; in the seed plants the gametophyte is microscopic in size. See Fig. 2.

ALTERNATOR. An electromotive force is generated in a conductor when it is moved so as to cut the lines of force between the poles of a magnet. The elementary principle of a simple two-pole, single-phase alternator is shown in Fig. 1. When the magnet revolves it will carry with it lines of force which will cut the conductor, which is a wire loop embedded in the stationary portion called the armature, and will generate ac.

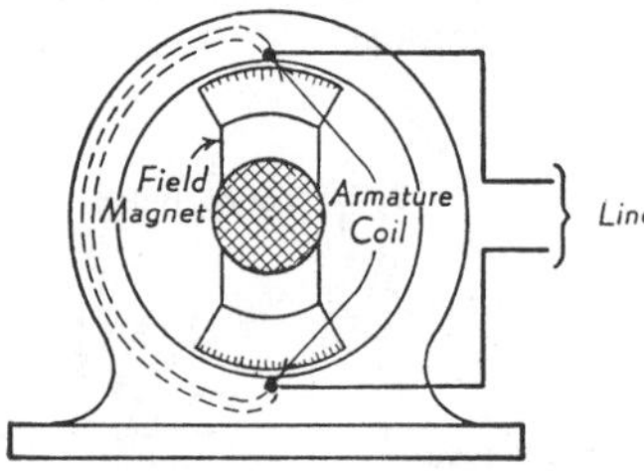

Fig. 1. Elementary alternator.

This elementary principle must be expanded in several directions if a practical generator of ac is to be had. First, the rotating part, or rotor, must have magnetic strength in excess of that which could be obtained from a simple permanent magnet. In other words, the poles must be formed by electromagnets whose excitation, in the form of dc, must be carried to the rotor through slip-ring connections. The rotor is called the field, and the current it uses is called the field current. In high-speed steam turbine-driven alternators as few as two poles are often used, while in slow-speed water-turbine units the number is frequently nearly 100. The stationary part, called the stator, or armature, usually has three sets of overlapping coils, connected in three separate circuits. These three circuits, or phases, are usually connected in one or the other ways shown in Fig. 2. The Y connection is preferred because of the usefulness of the neutral point, and the fact that the line voltage is $\sqrt{3}$ times the phase voltage, whereas it is only equal to the phase voltage in Δ connection. The neutral point is connected to the fourth wire of a four-wire, three-phase system, and left unconnected, or grounded, in the three-wire system. Several advantages are realized by making the rotating part the field, and the stationary part the armature. The ac may be generated at very high voltages because it is not necessary to connect it through movable contacts as would be the case if the armature revolved. It is not necessary to conduct high-load currents through slip rings and brushes if the armature is fixed. The armature conductors can be very rigidly braced in position, and may be much better disposed than if they were required to be in the rotor.

Engine and hydraulic turbine-driven alternators are in the slow-speed class, and are characterized by large diameter, short length, and many poles. The steam turbine-driven alternator is a high-speed machine having a length larger than its diameter. Standard speeds of turbine-driven alternators range from 1200 to 3600 rpm, with 1800 rpm very common practice.

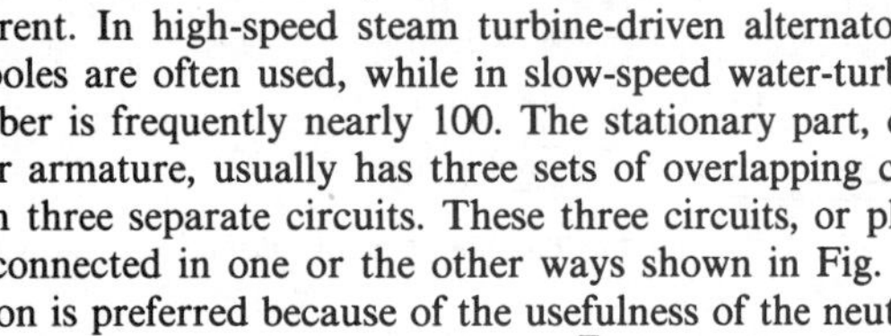

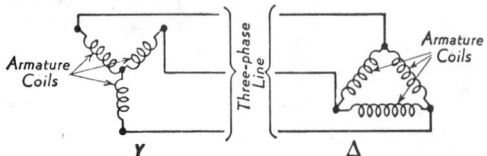

Fig. 2. Comparison of Y and Δ connections of three-phase alternator windings.

Basically, the alternator is a device for converting mechanical into electrical energy. While it is able to do this with a high degree of efficiency, it does suffer the following losses:

1. Friction and windage from bearings, brushes, and fan action of the rotor.
2. Core loss, which is the result of eddy currents and hysteresis in the iron core.
3. Resistance heating loss in the armature and field conductors.
4. Resistance loss of the field rheostat.
5. Exciter loss.
6. Ventilation loss.

Practically all generator losses appear as heat in and about the windings, and to maintain these at a safe working temperature, a cooling medium must be employed. Air has been the medium generally used. The rotor may or may not be able to produce its own fan action, depending on the size, speed, and construction of the rotor. Ventilating air frequently has to be brought through a duct and discharged through the alternator. Hydrogen is rapidly becoming a common cooling medium as it is much more effective and produces less windage loss. For its use the generator must be totally enclosed and gas-tight. To supply the dc for the field, a source of dc at 110–250 volts is necessary. This is delivered from a small dc generator called the exciter. The exciter may be driven from an extension of the alternator shaft, or it may be driven by some independent means, such as motor or engine.

When alternators are operated in parallel, the division of load between them is not accomplished by changing the generated voltage, as in dc generators, but by changing power input from the prime mover through the adjustment of the engine or turbine governor. Before being paralleled, two alternators must have the same phase sequence and frequency. Their voltages must be equal, and in phase; that is, with the peaks of wave forms coincident in time and direction.

ALTERNATOR (Fluidic). **Fluidics.**

ALTIMETRY. The technique of using air pressure to measure altitude. In altimetry there is a precisely ordered atmosphere called the *Standard Atmosphere* in which there is an exclusive value of altitude associated with a given value of air pressure. An altimeter converts static pressure into altitude. It is this precise and mathematically rigid relationship between static air pressure and altitude that the altimeter incorporates into its mechanism.

The standard atmosphere is defined by the following parameters:

(1) Zero pressure altitude corresponds to:

29.9213 inches of mercury
1013.250 millibars
760.000 millimeters of mercury

(2) Temperature at zero pressure altitude is 15°C which is equivalent to 288.16 K.
(3) Density of air at zero pressure altitude is 0.0012250 gram per cubic centimeter.
(4) The rate of temperature decrease with altitude is 6.5° per kilometer from zero pressure altitude up to 11 kilometers.
(5) The temperature at 11 kilometers is −56.5°C.
(6) The temperature from 11 kilometers up to 25 kilometers is constant at −56.5°C.
(7) The acceleration of gravity is everywhere 980.665 centimeters/sec^2.*
(8) There is no water vapor in the physical make-up of the atmosphere.*
(9) There is perfect obedience to the perfect gas law and the hydrostatic equation.*

In this concept of standard atmosphere, there is a unique value of density, pressure, and temperature at each altitude. The relationship between pressure and height is given by:

$$H = \frac{T_0}{\beta}\left[1 - \left(\frac{P}{P_0}\right)^{R\beta/g}\right]$$

where P_0 = the reference pressure which *in practice* is usually one of the following:

* It should be stressed that the foregoing statements are part of the arbitrary, but important definition of standard atmosphere and should be interpreted in this context. For a more detailed definition of standard atmosphere, see **Atmosphere (Earth).**

(1) Mean sea level pressure
(2) Field pressure
(3) 29.92 inches of mercury (1013.3 millibars)
T_0 = the standard atmosphere temperature at P_0
β = the temperature rate of decrease with altitude
R = the gas constant for standard air
g = acceleration due to gravity
P = the sensed and measured pressure by the altimeter
H = the altitude corresponding to P in the standard atmosphere

Millibars of pressure are used in much of international aviation instead of inches of mercury, but the latter is still the common usage in much of North America.

Toussaint's Formula. A rule proposed by Toussaint for the linear decrease of temperature with height in the atmosphere for which the temperature at mean sea level is 15°C is given by:

$$t = 15 - 0.0065z$$

where t is the temperature in degrees centigrade and z is the geometric height in meters above mean sea level. Toussaint's formula is used to determine temperature below 11,000 meters in the ICAO standard atmosphere.

Field Pressure. This is the atmospheric pressure measured and observed at the elevation of the airport or weather observatory in the real atmosphere by use of a mercurial barometer or a calibrated aneroid instrument. All pressure-related functions in altimetry must start with field pressure.

Field pressure *reduction to mean sea level* relates a calculated atmospheric pressure at sea level to field pressure by application of standard atmosphere relations. Field elevation above mean sea level is known and field pressure is measured. The pressure increment dictated by the standard atmosphere is added to the field pressure to obtain a value for mean sea level pressure.

Mean Sea Level Pressure. This is used extensively in altimeter-measured altitudes above sea level up to 18,000 feet (5486 meters). Mean sea level pressure is also known as *altimeter setting,* to which altimeters in airborne aircraft are set while flying below 18,000 feet (5486 meters).

Pressure Altitude. This is height in the standard atmosphere measured positively upward from the zero elevation of 29.92 inches of mercury (1013.3 millibars). An altimeter carrying a reference pressure of 29.92 inches of mercury (1013.3 millibars) will display pressure altitude.

Altimeter. A very sensitive instrument that converts static air pressure to altitude according to the precise relationship of the pressure and height in the standard atmosphere. There is a *reference pressure* adjustable setting on the altimeter that can be set to any reference pressure within the operating range of the altimeter. The altimeter measures altitude positively upward (negatively downward) from the level of the reference setting. In practice, three reference settings are predominantly used:

(1) *Reference setting at field pressure* will cause the altimeter to measure altitude positively upward from the level of the airport. When sitting on the airport, an aircraft altimeter will read zero.

(2) *Reference setting at mean sea level pressure,* also called "altimeter setting," will cause the altimeter to read altitude above mean sea level.

(3) *Reference setting at 29.92 inches of mercury* (1013.3 millibars) will cause the altimeter to measure altitude positively upward from the level where a pressure of 29.92 inches of mercury (1013.3 millibars) prevails in the real atmosphere. Altitude readings obtained with the reference set at 29.92 inches of mercury (1013.3 millibars) are known as "pressure altitude."

The altimeter shown in the accompanying diagram is simply an aneroid barometer calibrated to read in feet of altitude instead of barometric pressure. It thus reads pressure, not density, altitude. The heart of this instrument is a circular evacuated capsule, one face of which is fastened to the case, while the other is attached through a system of multiplying levers (terminating in a chain wrapping around the spindle) to a spindle carrying a pointer which moves over the scale. A U-shaped spring holds the capsule distended and atmospheric pressure tends to collapse it.

When mounted in an airplane which is climbing to higher altitudes, the capsule is subjected to decreasing atmospheric pressure, allowing the spring to distend the capsule and move the pointer. A small hairspring keeps the mechanism taut and the pointer will move upon the slightest movement of the capsule face in either direction.

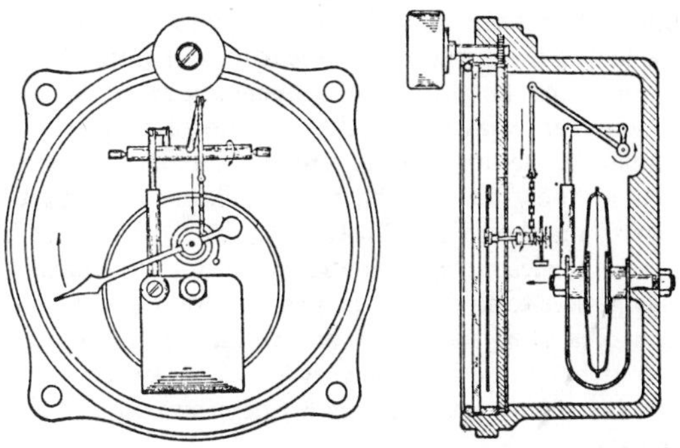

Simple altimeter.

The instrument may be made to indicate altitude above any given point, e.g., the airport runway, by rotating the dial containing the scale until the pointer registers with zero altitude, this being done when the airplane is on the runway. A knob having a pinion engaging a gear on the periphery of the dial makes this adjustment.

Vertical Separation. The vertical separation of airborne aircraft is accomplished by use of onboard altimeters and the application of altimetry.

(1) Below 18,000 feet (5486 meters), all aircraft have their altimeters set to a common altimeter setting (mean sea level) in a given traffic control area. If there are deviations from the real altitude above sea level, all aircraft carry the same deviation.

(2) Above 18,000 feet (5486 meters), all aircraft use a reference setting of 29.92 inches of mercury (1013.3 millibars) and operate on pressure altitude.

Terrain clearance when using altimeter-indicated altitude requires some precaution in air that is colder than the standard atmosphere. Mountain peaks have fixed real heights which obviously must be exceeded.

(1) If the reference setting is mean sea level pressure, the altimeter indicates altitude above sea level. When the real atmosphere is colder than the standard atmosphere, the altimeter will display a reading *too high,* i.e., the aircraft will have *less* real altitude than shown on the altimeter.

(2) If the reference setting is 29.92 inches of mercury (1013.3 millibars), the altimeter indicates altitude above the level where this value of P prevails. In addition to the cold air deviation, an additional caution is needed. When sea level pressure is less than 29.92 inches of mercury (1013.3 millibars), the altimeter starts counting altitude from *below sea level.* The altitude displayed by the altimeter will be too high by reason of the below-standard sea level pressure; and additionally may be in error because the air is colder than standard air.

(*Mean Sea Level*)—The average height of the sea surface, based upon hourly observation of the tide height on the open coast or in adjacent waters that have free access to the sea. These observations are to have been made over a "considerable" period of time. In the United States, mean sea level (MSL) is defined as the average height of the surface of the sea for all stages of the tide over a 19-year period.

(*Mean Sea Level Pressure*)—The value assigned to surface pressure of the standard atmosphere; 29.92 inches of mercury; or 1013.3 millibars.

Other Forms of Altimeters. A radio altimeter is an electronic instrument that may operate on one of the following principles: (1) pulse radar techniques which measure height in terms of the transit time of the radar pulse; (2) continuous-wave radar which measures height in terms of the phase difference between the transmitted and received signals; and (3) the variation of the electrical capacitance between aircraft and ground.

For references see entries on **Climate;** and **Meteorology.**

Peter E. Kraght, Certified Consulting Meteorologist, Mabank, Texas.

ALTITUDE. The term altitude is used synonymously for height, or distance above the surface of the earth. The altimeter is used in aviation to measure the altitude of a plane above sea level.

In astronomy and navigation the altitude of a celestial object is that coordinate in the horizontal system of spherical coordinates that is measured in the plane of the vertical circle through the object from the horizon to the object. Altitude is probably the most frequently measured of all celestial coordinates since it is universally used in celestial navigation to find lines of position for the determination of the location of a ship at sea, or an aircraft in the air. Altitude is also measured by geodetic surveyors for accurate determination of latitude and is used by astronomers for determination of the declinations of the stars.

The principal points of high and low elevation on the earth's surface are given in the entry on **Earth.**

ALTITUDE-TEMPERATURE AND PRESSURE RELATIONSHIPS (Earth's Atmosphere). Atmosphere (Earth).

ALTOCUMULUS. Clouds and Cloud Formation.

ALTOSTRATUS. Clouds and Cloud Formation.

ALUM. A series (*alums*) of usually isomorphous crystalline (most commonly octahedral) compounds in which sulfate is usually the negative ion (though it may often be replaced by selenate, and in some compounds partly by such complexes as $BeF_4^=$ or $ZnCl_4^=$), and two positive ions are present. Alums have the general formula $M^IM^{III}(AX_4)\cdot 12H_2O$. The first is potassium or a higher alkali metal (sodium alums are rare) or thallium(I) or ammonium (or a subsituted ammonium ion) and the second is a tripositive ion of relatively small ionic radius (0.5–0.7Å). The tripositive ions of larger ionic radius such as the Lanthanides form double sulfates, but not alums. The tripositive ions in alums, in roughly the order of number of known compounds, are aluminum, chromium, iron, manganese, vanadium, titanium, cobalt, gallium, rhodium, iridium, and indium.

The term *alum,* itself, refers to potassium alum, potassium aluminum sulfate, $KAl(SO_4)_2\cdot 12H_2O$. Other common alums are ferric ammonium alum, $NH_4Fe(SO_2)_2\cdot 12H_2O$, and sodium chrome alum, $NaCr(SO_4)_2\cdot 12H_2O$.

ALUMEL. Nickel.

ALUMINA. Adsorption Operations; Bauxite.

ALUMINA-SILICA CERAMIC FIBER INSULATION. Insulation (Thermal).

ALUMINUM. Chemical element symbol Al, at. no. 13, at. wt. 26.98, periodic table group 3b, mp 660 ± 1°C, bp 2452 ± 15°C, sp gr 2.699. In Canada and several other English-speaking nations, the spelling of the element is *aluminium.* The element is a silver-white metal, with bluish tinge, capable of taking a high polish. Commercial aluminum has a purity of 99% (minimum), is ductile and malleable, possesses good weldability, and has excellent corrosion resistance to many chemicals and most common substances, particularly foods. The electrical conductivity of aluminum (on a volume basis) is exceeded only by silver, copper, and gold. First ionization potential, 5.984 eV; second, 18.823 eV; third, 28.44 eV. Oxidation potentials, $Al \rightarrow Al^{3+} + 3e^-$, 1.67 V; $Al + 4OH^- \rightarrow AlO_2^- + 2H_2O + 3e^-$, 2.35 V. Other important physical properties of aluminum are given under **Chemical Emements.**

Aluminum occurs abundantly in all ordinary rocks, except limestone and sandstone; is third in abundance of the elements in the earth's crust (8.1% of the solid crust), exceeded only by oxygen and silicon, with which two elements aluminum is generally found combined in nature; present in igneous rocks and clays as aluminosilicates; in the mineral cryolite in Greenland as sodium aluminum fluoride Na_3AlF_6; in the minerals corundum and emery, the gems ruby and sapphire, as aluminum oxide Al_2O_3; in the mineral bauxite in southern France, Hungary, Yugoslavia, Greece, the Guianas of South America, Arkansas, Georgia, and Alabama of the United States, Italy, the U.S.S.R., New Zealand, Australia, Venezuela, and Indonesia as hydrated oxide $Al_2O(OH)_4$; in the mineral alunite or alum stone in Utah as aluminum potassium sulfate $Al_2(SO_4)_3\cdot K_2SO_4\cdot 4Al(OH)_3$. See also **Bauxite;** and **Cryolite.**

Uses. Because of its corrosion resistance and relatively low cost, aluminum is used widely for food-processing equipment, food containers, food-packaging foils, and numerous vessels for the processing of chemicals. No ill effects on life processes have been established for aluminum. Because of the presence of aluminum in soils and rocks, there are natural traces of the element in nearly all foods. Whereas the processing of foods in copper vessels will cause destruction of vitamins, aluminum does not accelerate the degradation of vitamins. Aluminum foil has been used to cover severe burns as a means to enhance healing. Because of its comparatively good electrical conductivity compared with other metals except silver and copper, aluminum is used as an electrical conductor, particularly for high-voltage transmission lines. For the same conductance, the weight of aluminum required is about one-half that of annealed copper. The greater diameter of aluminum conductors also reduces corona loss. But, because aluminum has about 1.4 × the linear temperature coefficient of expansion as compared with annealed copper, the changes in sag of the cable are greater with temperature changes. For long spans requiring high strength, the center strand may be a steel cable, or supporting steel cables may be used. Aluminum finds use for bus bars because of its large heat-dissipating surface available for a given conductance.

Aluminum alloys readily with copper, manganese, magnesium, silicon, and zinc. Many aluminum alloys are commercially available. See also **Aluminum Alloys.**

Production. Although aluminum was predicted by Lavoisier (France) as early as 1782 when he was investigating the properties of aluminum oxide (alumina), the metal was not isolated until 1825 by H. C. Oersted (Denmark). Oersted obtained an impure aluminum metal by heating potassium amalgam with anhydrous aluminum chloride, followed by distilling off the mercury. Using similar methods, Woehler (Germany) produced an aluminum powder in 1827. In 1854, Deville (France) and Bunsen (Germany) separately but concurrently found that aluminum could be isolated by using sodium instead of potassium in the amalgam of Oersted and Woehler. Deville exhibited his product at the Paris Exposition of 1855, after which Napoleon III commissioned Deville to improve the process and lower the cost. Because of improved processing techniques, the price declined from $115 to $17 per pound ($254 to $38 per kilogram) by 1859, with numerous plants throughout France. The price was further reduced to $8 per pound ($17.60 per kilogram) by 1885. The first breakthrough for production of aluminum on a large scale and at a much lower cost occurred as the result of experimentation by Hall (Oberlin, Ohio) who found that metallic aluminum could be produced by dissolving Al_2O_3 (alumina) in molten $3NaF\cdot AlF_3$ (cryolite) at a temperature of above 960°C and then passing an electric current through the bath. Heroult (France) independent discovered the same process. Most modern aluminum production plants utilize this electrolytic process.

The major differences found in aluminum production plants relate to the design of the electrolytic cells. Basically, the cells are large carbon-lined steel boxes. The carbon lining serves as the cathode. Separate anodes are immersed in the bath. The electric circuit is completed by passage of current through the bath. Thus, the alumina is decomposed into aluminum and oxygen by action of the electric current. The molten aluminum collects at the bottom of the cell from which it is siphoned off periodically. The released oxygen combines with carbon at the electrodes to form CO_2. The bath is replenished with alumina intermittently. The carbon anodes must be replaced periodically. Additional cryolite also is required to make up for volatilization losses. The several reactions which occur during electrolysis still are not fully understood.

Energy requirements for aluminum production in modern plants approximate 6 to 8 kilowatt-hours per pound (13.2 to 17.6 kilowatt-hours per kilogram). Per weight unit of aluminum produced, 0.4 to 0.6 weight units of carbon, 1.9 weight units of alumina, and 0.1 weight unit of cryolite are required. One form of electrolytic cell is shown in Fig. 1.

As of the early 1980s, there are 32 aluminum smelters within the United States and two smelters in the planning stage. Smelters originally were located mainly for considerations of low-cost energy and transportation—and thus are found in Washington, Oregon, western Montana, Texas, New York, and along the Mississippi River. The capacity in the United States is 5.2 million short tons (about 4.9 metric tons) per year. The United States and Canada account for 30.6% of the world production, which is 19.6 million short tons (17.6 million metric tons). This figure is expected to increase to 22.2 million short tons (about 20 million metric tons) by the mid-1980s or earlier. The developing areas for new production capacity are Venezuela, New Zealand, Australia, Brazil, and possibly the Arabian Peninsula.

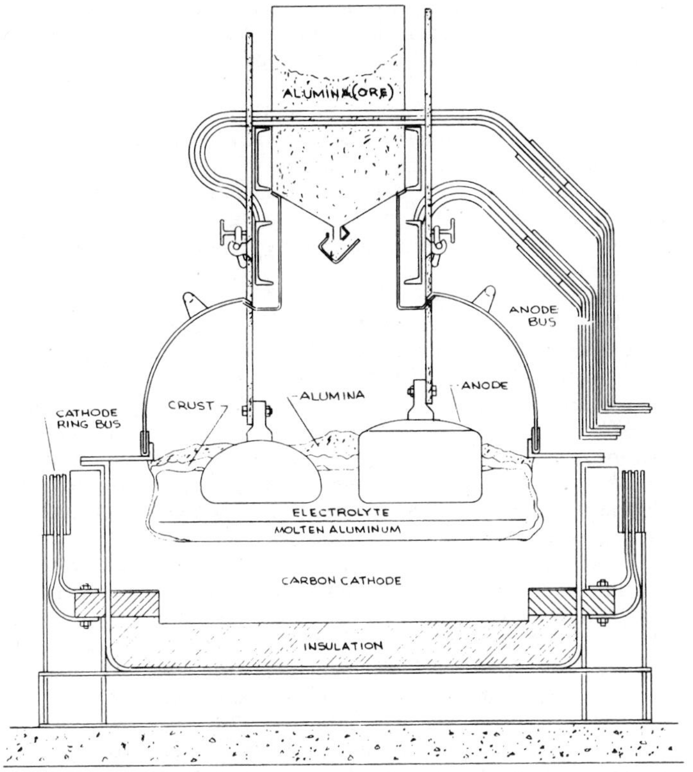

Fig. 1. Aluminum electrolytic reduction cell of the prebake type. The anodes are constructed of separate blocks of carbon which have been prebaked. There is a lead to the main bus bar from each block. (*Reynolds Metals Company*)

By 1995, Australia will probably be one of the world's three largest aluminum producers. Australia has long been known for its vast deposits of bauxite and its significant potential capacity to produce aluminum. Australia presently supplies one-third of the world's bauxite. New smelter construction is expected to boost the world production by 2 million tons (1.8 million metric tons). The rapid development in Australia can be traced to a number of circumstances, including increasing demand for the metal as compared with existing production capacity. Electrical energy costs have increased markedly, especially in regions dependent upon oil and gas. For example, in the United States, political and environmental restraints placed on new power-generating facilities have stymied growth. Australia has coal reserves estimated for 1500 years.*

It is also projected that Venezuela, because of its abundance of low-cost hydropower, will become a major aluminum producer. During the past decade, Brazil has attracted considerable interest because of its vast reserves of good-quality bauxite and its potential to develop attractive hydropower.

The People's Republic of China is planning to modernize its existing capacity of about 300,000 short tons (270,000 metric tons) to about 400,000 short tons (360,000 metric tons) during the period 1980–1985.

Thus, most of the increasing capacity for aluminum production during the 1980–1985 period will be outside the continental United States, principally in the developing countries. As of 1980, developing countries account for 14% of total free world aluminum capacity, but this is expected to increase to 22% by 1985.

Secondary Recovery. In 1979, the secondary recovery of aluminum through recycling programs in the United States was 1.75 million short tons (1.6 million metric tons), amounting to the equivalent of one-third of the primary production. Secondary recovery of scrap will continue to play a major role in the aluminum industry. Larger quantities of aluminum are being used in the beverage, automotive, and several other consumer industries, making it easy to accomplish effective recycling. See Fig. 2.

Fig. 2. Recycling of aluminum beverage cans and other scrap saves 95% of the energy required when starting from ore. Approximately 20% of the metal processed in 1980 was recycled from scrap. (*Reynolds Aluminum Recycling Company.*)

The lowest price for aluminum occurred in the early 1940s as the result of improved production methodologies and increased volume. Since then, the price has risen from the low of $0.15 per pound ($0.33 per kilogram) because of the increasing costs of raw materials and electrical energy. Because of the very large capital investment represented by an aluminum reduction plant, accompanied by very heavy electric power consumption, there has been a continuing search for lower-cost processes.

New Processes. A relatively new process may reduce the electrical energy requirements by as much as 30% and involves the elimination of fluoride in the cells. The cells are operated at a low temperature and are completely enclosed. The process combines alumina (refined from bauxite) with chlorine to form aluminum chloride, after which the chloride is electrolyzed in a closed cell, releasing molten aluminum and chlorine. The chlorine gas is recycled to the chemical reactor. A 15,000-ton (13,500-metric ton) plant scaling up this process is located in Anderson County, Texas. Reports from Alcoa indicate that the 30% less electricity required (as compared with the most efficient Hall process) previously expected has been substantiated. Aluminum production at the facility will be cut back to 7500 tons (6750 metric tons) per year, with redesign, construction, and testing of a new segment of the plant to achieve further efficiency. This program is scheduled for completion during the early 1980s.

Reynolds has announced a process based upon the electrolysis of aluminum from a cell using a fused bath of 50% sodium chloride ($NaCl$), 45% lithium chloride ($LiCl$), and 5% aluminum chloride ($AlCl_3$) at 700°C. The process requires 5 Kwh per pound (11 Kwh per kilogram) or 25% less energy than present reduction cells require. The process also requires less space for production.

Thermal Reduction Processes. Thermal processes, including the use of electric arc furnaces, make it possible to achieve a greater concentration of electric energy in the refining process. Commercial electric furnace processes are fed with bauxite and clays (kaolin or kyanite or partially refined bauxite mixed with carbon or charcoal in briquetted

* Firms participating in the development of Australian production capacity include Alcan, Alcoa, Kaiser, Comalco, Alumax, Reynolds, Pechiney, and several Japanese companies.

form). An electric dc or ac arc is used. To date, this process has been practical only for the production of impure aluminum alloys, ranging in content from 65–70% aluminum, 25–30% silicon, and 1% or more of iron. Other impurities may include titanium and the carbides and nitrides of aluminum and iron. This impure product may be refined to yield pure aluminum by means of a subsequent distillation process in which the electric-furnace product is dissolved at high temperatures in zinc, magnesium, or mercury baths and then distilled. Aluminum with a purity of 99.99+% may be produced in this manner.

A gas-reaction purification process also has attracted attention over the years. This is based on reacting aluminum trichloride gas with molten aluminum at about 1000°C to produce aluminum monochloride gas. Aluminum fluoride gas may be substituted for the aluminum monochloride gas in the process. Raw materials for the process may be scrap aluminum, aluminum from thermal-reduction processes, aluminum carbide, or aluminum nitride.

For the production of superpurity aluminum on a large scale, the Hoopes cell is used. This cell involves three layers of material. Impure (99.35 to 99.9% aluminum) metal from conventional electrolytic cells is alloyed with 33% copper (eutectic composition) which serves as the anode of the cell. A middle, fused-salt layer consists of 60% barium chloride and 40% $AlF_3 1.5NaF$ (chiolite), mp 720°C. This layer floats above the aluminum-copper alloy. The top layer consists of superpurity aluminum (99.995%). The final product usually is cast in graphite equipment because iron and other container metals readily dissolve in aluminum. For extreme-purity aluminum, zone refining is used. This process is similar to that used for the production of semiconductor chemicals and yields a product that is 99.9996% aluminum.

Aluminum Chemistry and Chemicals

Since aluminum has only three electrons in its valence shell, it tends to be an electron acceptor. Its strong tendency to form an octet is shown by the tetrahedral aluminum compounds involving sp^3 hybridization. Aluminum halides include the trifluoride, an ionic crystalline solid, the trichloride and tribromide both of which are dimeric in the vapor and in nonpolar solvents, having the halogen atoms arranged in tetrahedra about each aluminum atom, giving a bridge structure; and the triiodide. They combine readily with many other molecules, especially organic molecules having donor groups, and aluminum chloride is an active catalyst. Certain of the hydrates have a saltlike nature; thus $AlCl_3 \cdot 6H_2O$ acts as a salt, even though Al_2Cl_6 is covalent.

Common water-soluble salts include the sulfate, selenate, nitrate, and perchlorate, all of which are hydrolyzed in aqueous solution. Double salts are formed readily by the sulfate and by the chloride.

There are many fluorocomplexes of aluminum. The general formula for the fluoroaluminates is $M_x^I[Al_yF_{x+3y}]$, based upon AlF_6 octohedra, which may share corners to give other ratios of Al:F than 1:6. Chloroaluminates of the type $M^I[AlCl_4]$ are obtainable from fused melts. Aluminum ions form chloro-, bromo-, and iodo-complexes containing tetrahedral $[AlX_4]^-$ ions. However, in sodium aluminum fluoride $NaAlF_4$, the aluminum atoms are in the centers of octohedra of fluorine atoms in which the fluorine atoms are shared with neighboring aluminum atoms.

Aluminum hydroxide is about equally basic and acidic, the pK_b being about 12 and the pK_a being 12.6. Sodium aluminate seems to ionize as a uni-univalent electrolyte: $NaAl(OH)_4(H_2O)_2 \leftrightarrows Na^+ + [Al(OH)_4(H_2O)_2]^-$. The high viscosity of sodium aluminate solutions is explained by hydrogen bonding between these hydrated ions, and between them and water molecules.

By reaction of aluminum and its chloride or bromide at high temperature, there is evidence of the existence of monovalent aluminum. Here the aluminum atom is apparently in the *sp* state, with an electron pair on the side away from the chlorine atom, whereby the single pairs on the two chlorine atoms are shared to form two weak π bonds.

Aluminum forms a polymeric solid hydride, $(AlH_3)_x$, unstable above 100°C, extremely reactive, e.g., giving hydrogen explosively with H_2O and igniting explosively in air. From it are derived the tetrahydroaluminates, containing the ion AlH_4^-, which are important, powerful but selective reducing agents, e.g., reducing chlorophosphines R_2PCl to phosphines R_2PH; reactive alkyl halides and sulfonates to hydrocarbons, epoxides, and carboxylic acids; aldehydes and ketones to alcohols; nitrites, nitro compounds and N,N-dialkyl amides to amines. The salt most commonly used is lithium (tetra)hydroaluminate ("lithium aluminum hydride").

Aluminum compounds are generally made starting with bauxite, which is reactive with acids and with bases. With acids, e.g., H_2SO_4, any iron contained in the bauxite is dissolved along with the aluminum and silicon is left in the residue, whereas with bases, e.g., NaOH, any silicon is dissolved and iron left in the residue.

Acetate: Aluminum acetate $Al(C_2H_3O_2)_3$, white crystals, soluble, by reaction of aluminum hydroxide and acetic acid and then crystallizing. Used (1) as a mordant in dyeing and printing textiles, (2) in the manufacture of lakes, (3) for fireproofing fabrics, (4) for waterproofing cloth.

Aluminates: Sodium aluminate $NaAlO_2$, white solid, soluble, (1) by reaction of aluminum hydroxide and NaOH solution, (2) by fusion of aluminum oxide and sodium carbonate; the solution of sodium aluminate is reactive with CO_2 to form aluminum hydroxide. Used as a mordant in the textile industry, in the manufacture of artificial zeolites, and in the hardening of building stones. See silicates below and calcium aluminates.

Alundum: See oxide (below).

Carbide: Aluminum carbide Al_4C_3, yellowish-green solid, by reaction of aluminum oxide and carbon in the electric furnace, reacts with H_2O to yield methane gas and aluminum hydroxide.

Chlorides: Aluminum chloride $AlCl_3 \cdot 6H_2O$, white crystals, soluble, by reaction of aluminum hydroxide and HCl, and then crystallizing; anhydrous aluminum chloride $AlCl_3$, white powder, fumes in air, formed by reaction of dry aluminum oxide plus carbon heated with chlorine in a furnace, used as a reagent in petroleum refining and other organic reactions.

Fluoride: Aluminum fluoride AlF_3, white solid, soluble, by reaction of aluminum hydroxide plus hydrofluoric acid and then crystallizing $2AlF_3 \cdot 7H_2O$, used in glass and porcelain ware.

Hydroxide: Aluminum hydroxide $Al(OH)_3$, white gelatinous precipitate, by reaction of soluble aluminum salt solution and an alkali hydroxide, carbonate or sulfide (sodium aluminate is formed with excess NaOH but no reaction with excess NH_4OH), upon heating aluminum hydroxide the residue formed is aluminum oxide. Used as intermediate substance in transforming bauxite into pure aluminum oxide.

Nitrate: Aluminum nitrate $Al(NO_3)_3$, white crystals, soluble, by reaction of aluminum hydroxide and HNO_3, and then crystallizing.

Oleate: Aluminum oleate $Al(C_{18}H_{33}O_2)_3$, yellowish-white powder, by reaction of aluminum hydroxide, suspended in hot H_2O, shaken with oleic acid, and then drying, the product is used (1) as a thickener for lubricating oils, (2) as a drier for paints and varnishes, (3) in waterproofing textiles, paper, leather.

Oxide: Aluminum oxide, alumina Al_2O_3, white solid, insoluble, melting point 2020°C, formed by heating aluminum hydroxide to decomposition; when bauxite is fused in the electric furnace and then cooled there results a very hard glass ("alundum"), used as an abrasive (hardness 9 Mohs scale) and heat refractory material. Aluminum oxide is the only oxide which reacts both in H_2O medium and at fusion temperature, to form salts with both acids and alkalis.

Palmitate: Aluminum palmitate $Al(C_{16}H_{31}O_3)_3$, yellowish-white powder, by reaction of aluminum hydroxide, suspended in hot H_2O, shaken with palmitic acid, and then drying, the product is used (1) as a thickener for lubricating oils, (2) as a drier for paints and varnishes, (3) in waterproofing textiles, paper, leather, (4) as a gloss for paper.

Silicates: Many complex aluminosilicates or silicoaluminates are found in nature. Of these, clay in more or less pure form, pure clay, kaolinite, kaolin, china clay $H_4Si_2Al_2O_9$ or $Al_2O_3 \cdot 2SiO_2 \cdot 2H_2O$ is of great importance. Clay is formed by the weathering of igneous rocks, and is used in the manufacture of bricks, pottery, procelain, and Portland cement.

Stearate: Aluminum stearate $Al(C_{18}H_{35}O_2)_3$, yellowish-white powder, by reaction of aluminum hydroxide suspended in hot H_2O, shaken with stearic acid, and then drying the product, used (1) as a thickener for lubricating oils, (2) as a drier for paints and varnishes, (3) in waterproofing textiles, paper, leather, (4) as a gloss for paper.

Sulfate: Aluminum sulfate $Al_2(SO_4)_3$, white solid, soluble, by reaction of aluminum hydroxide and H_2SO_4, and then crystallizing, used (1) as a clarifying agent in water purification, (2) in baking powders, (3) as a mordant in dyeing, (4) in sizing paper, (5) as a precipitating agent in sewage disposal; aluminum potassium sulfate, see **Alum** and **Alunite.**

Sulfide: Aluminum sulfide Al_2S_3, white to grayish-black solid, reactive with water to form aluminum hydroxide and H_2S, formed by heating aluminum powder and sulfur to a high temperature.

Aluminum in solution of its salts is detected by the reaction (1) with ammonium salt of aurin tricarboxylic acid ("aluminon"), which yields a red precipitate persisting in NH_4OH solution, (2) with alizarin red S, which yields a bright red precipitate persisting in acetic acid solution.

Organoaluminum Compounds: By the action of alpha olefins on AlH_3, the trialkyls of aluminum can be prepared:

$$AlH_3 + 3CH_2{=}CHR \xrightarrow{120°C} (CH_2CH_2R)_3Al$$

In the presence of ethers, magnesium-alloy alloys react with alkyl halides to yield trialkyls, R_3AlOEt_2. The trimethyl aluminum compound is a dimer, mp 15°C. Triethyl aluminum is a liquid. The alkyls react readily with H_2O:

$$Et_3Al + 3H_2O \rightarrow Al(OH)_3 + 3EtH$$

When aluminum is reacted with diphenyl mercury $(C_6H_5)_2Hg$, triphenylaluminum $(C_6H_5)_3Al$ is yielded. The aluminum organometallic alkyls and aryls act as Lewis acids to form compounds with electron-donating substances. The resulting compounds are the organometallic basis for producing polymers of the Al-N type.

References

Aluminum Association: "Aluminum Standards and Data" and "Aluminum Statistical Review" (issued periodically).

ASM: "Metals Handbook," 9th Edition, Vol. 2 (Properties and Selection: Nonferrous Alloys and Pure Metals), American Society for Metals, Metals Park, Ohio, 1979.

Reyss, J. L., Yokoyama, Y., and S. Tanaka: "Aluminum-26 in Deep-Sea Sediment," *Science*, **193**, 1119–1121 (1976).

Stoffyn, M.: "Biological Control of Dissolved Aluminum in Seawater: Experimental Evidence," *Science*, **203**, 651–653 (1979).

Van Horn, K. R., Editor: "Aluminum," Vol. 1–3, American Society for Metals, Metals Park, Ohio, 1967.

See also list of references at end of entry on **Aluminum Alloys.**

ALUMINUM ALLOYS. The intrinsic properties of aluminum—light weight, strength, good electrical conductivity, nonmagnetic properties, high reflectivity, transparency to neutrons, good thermal conductivity, weather resistance, nontoxicity, decorative appeal, weldability, machinability, corrosion resistance, fracture toughness, strength at cryogenic temperatures, and nonsparking characteristics make it one of the most versatile engineering and construction materials. The aluminum industry offers more than 300 combinations of alloys and tempers, from which engineers and designers may select the most suitable metal for their needs.

Aluminum is unique among the metals because it responds to nearly all of the known finishing processes. It can be finished in the softest, most delicate textures as exemplified by tableware and jewelry. Aluminum can be anodized and dyed to appear like gold. It can be made as specular as a silver mirror and jet black. The metal also can be anodized to an extremely hard, wear- and abrasion-resistant surface that approaches the hardness of a diamond. Aluminum is available in many convenient forms—shapes, sheet, plate, ingot, wire, rod and bar, foil, castings, forgings, powdered metals, and extrusions.

In the production of these products, new high-speed automated equipment has been developed over the past decade. Rolling speeds as high as 8000 feet (2438 meters) per minute are possible on some products. New continuous casting processes have been developed where molten metal can be converted directly into coiled sheet, using roll casting, belt casting, and articulated block casting. See Figs. 1 and 2. Many of the common alloys respond well to these new processes. Electromagnetic casting of the conventional giant ingots used for the

Fig. 2. The continuous roll casting process is the most recent, economical, and energy efficient method for producing aluminum foil feed stock and other products. Liquid metal at 1,350°F (732°C) is fed directly between water-cooled rolls to emerge as 0.4 × 78-inch (10.2 millimeter × 198.1-centimeter) strip to be coiled into 35,000-pound (15,876-kilogram) coils. (*Reynolds Metals Company*)

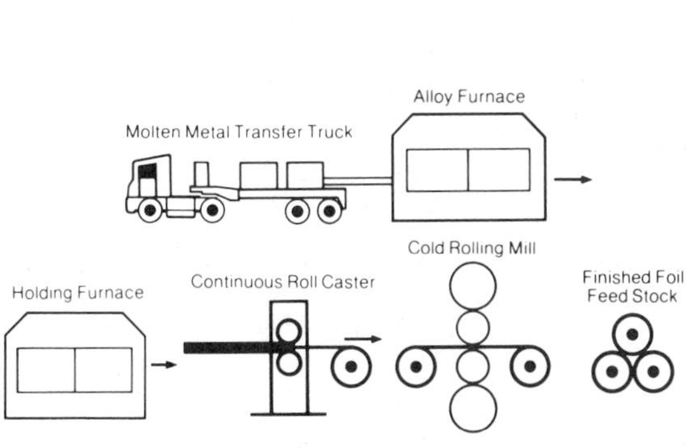

Fig. 1. Schematic portrayal of continuous aluminum roll casting process. (*Reynolds Metals Company*)

Fig. 3. Final operations on machined aluminum plates for wings of large jet aircraft. (*Reynolds Metals Company*)

conventional rolling mills has been developed in the fully automated mode. This casting process eliminates the need for scalping ingots prior to processing into final products, such as can stock, foil, auto body sheet, siding, and many other products. In this process, an electrical field rather than a solid mold wall is used to shape the ingot. The process is based upon the fact that liquids with good electrical conductivity, such as aluminum, can be constrained where an alternating magnetic field is applied. The liquid metal remains suspended by the electrical forces while it is being solidified.

The casting alloy products comprise sand castings, permanent mold castings, and die castings. Aluminum is the basic raw material for more than 20,000 businesses in the United States. Aluminum is an indispensable metal for aircraft, for example. See Fig. 3. A guide to the selection of aluminum alloys for a broad classification of uses is given in Table 1.

The free world's demand trend for aluminum over a 28-year period to 1980 shows an average growth of 7.9% per year. Projections through 1985 are estimated at a growth rate of 5% per year. Aluminum is quite well established as an energy saver and a cost-effective material in numerous applications. Particularly strong growth is occurring in the automotive and transportation fields, housing, and food packaging. For example, a number of breakthroughs in foil pouch packaging have been achieved for replacing conventional cans for shelf-stable foods. Steady growth has continued in other areas, including aircraft and space applications, electrical wire and cable, marine, architectural, and consumer durable products. See Fig. 4.

Pure aluminum is soft and ductile. In its annealed form, the tensile strength is approximately 13,000 pounds per square inch (90 megapascals). This can be work hardened by rolling, drawing into wire, or by other cold-working techniques to increase its strength to approximately 24,000 pounds per square inch (165 megapascals). Pure aluminum is particularly useful in the food and chemical industries where its resistance to corrosion and high thermal conductivity are desirable characteristics, and in the electrical industry, where its electrical conductivity of about 62% that of copper and its lightweight make it desirable for wire, cable, and bus bars. Large quantities are rolled to thin foils for the packaging of food products, for collapsible tubes, and for paste and powders used for inks and paints.

Most commercial uses require greater strength than pure alumnium affords. Higher strength is achieved by alloying other metal elements with aluminum and through processing and thermal treatments. The alloys can be classified into two categories, non-heat-treatable and heat-treatable.

Non-heat-Treatable Alloys. The initial strength of alloys in this group depend upon the hardening effect of elements such as manganese,

TABLE 1. ALUMINUM ALLOY SELECTION GUIDE

Product	Alloy	Form
Architectural and Building Products		
Awnings	3003	Sheet
Fence Wire	6061	Wire
Fittings	A514.0	Castings
Gutters	Alc. 3004	Sheet
Nails	5056, 6061	Wire
Panels	3003, 6063	Sheet and Extrusions
Roofing	Alc. 3004	Sheet
Screens	Alc. 5056	Wire
Siding	3003, Alc. 3004	Sheet
Transportation		
Aircraft	2014, 2024, 2048, 2124, 2219, 7075, 7079, 7175, 7178, 7179, 7475.	Sheet, Plate, Extrusions, and Forgings
	B295.0, 355.0, 356.0, 518.0, 520.0	Castings
Automotive		
Auto and Truck Bodies	2036, 2037, 5052, 5182, 5252, 5657, 6009, 6010, 6061, 6063, 6463, 7016, 7021, 7029, 7046, 7129	Sheet and Extrusions
Buses	2036, 5083, 5086, 5182, 5252, 5457, 5657, 6061, 6063, 6463, 7016, 7046	Sheet and Extrusions
	242.0, B295.0, 355.0, 356.0, 360.0, 380.0, 390.0, B443.0	Castings
Marine		
Barges, Small Craft, Ships, Tankers	5052, 5083, 5086, 5454, 5456, 6061, 6063	Sheet, Plate, Extrusions, and Tubing
	360.0, 413.0 B443.0	Castings
Railroad	Alc. 2024, 5052, 5083, 5086, 5454, 5456, 6061, 6063, 7005	Sheet, Plate, and Extrusions
	B295.0, 356.0, 520.0	Castings
Containers and Packaging		
Foils	1100, 1235, 3003, 5005, 8079, 8111	Foils
Cans	3003, 3004, 5182	Sheet

Product	Alloy	Form
Electrical		
Bus Bar	6063, 6201	Extrusions or Rolled Rod
Cable	E.C., 5005, 6201	Wire
Conduit	6063	Tubing
Motors	319.0, 355.0, 360.0, 380.0	Castings
Transmission Towers and Substations	6061, 6063, 7005	Extrusions
Consumer Durables		
Appliances	3003, 4343, 5052, 5252, 5357, 5457, 6063, 6463	Sheet and Extrusions
	356.0, 390.0	Castings
Cooking Utensils	1100, 3003, Alc. 3003, 3004, Alc. 3004, 5052, 5454	Sheet
	B443.0, A514.0	Castings
Furniture	3003, 3004, 5005, 5457, 6061, 6063, 6463	Sheet and Tubing
Refrigerators	3003, Alc. 3003, 3004, Alc. 3004, 4343, 5005, 5050, 5052, 5252, 5457, 6061, 6463	Sheet and Extrusions
Water Heaters	Alc. 6061, 3003	Sheet
Machinery and Equipment		
Chemical Processing	1060, 1100, 3003, Alc. 3003, 3004, Alc. 3004, 5083, 5086, 5154, 5454, 6061, 6063	Sheet, Plate, Extrusions, and Tubes
	356.0, 360.0, B514.0	Castings
Heat exchangers and Solar Panels	3003, Alc. 3003, 3004, Alc. 3004, 4343, 5005, 5052, 6061, 6951	Sheet and Tubing
Sheet for Vacuum Brazing Heat Exchangers	4003, 4004, 4005, 4044, 4104	
Irrigation	3003, Alc. 3003, 3004, Alc. 3004, 5052, 6061, 6063	Tubing
Sewage Plants	3003, Alc. 3003, 3004, Alc. 3004, 5052, 5083, 5086, 5454, 5456, 6061, 6063	Sheet, Extrusions, and Tubes
	B514.0	Castings
Screw Machine Parts	2011, 2024, 6262	Rod and Bar
Textile Machinery	2014, 2024, 6061, 6063	Extrusions, Sheet and Plate

Fig. 4. Four aluminum sections of the type shown here make up the fuel tank used for propelling the NASA space shuttle into orbit. Each section is produced from computer-machined plate and weighs 4750 pounds (2155 kilograms). (*Reynolds Metals Company*)

silicon, iron, and magnesium, singly or in various combinations (see three alloys in Table 2). The non-heat-treatable alloys are usually designated therefore in the 1xxx, 3xxx, 4xxx, and 5xxx series (Table 3). Since these alloys are work hardenable, further strengthening is made possible by various degrees of cold work denoted by the H series tempers.

Heat-Treatable Alloys. This group of alloys includes the alloying elements, copper, magnesium, zinc, and silicon (-T tempered alloys in Table 2). Since these elements singly or in combination show increasing solid solubility in aluminum with increasing temperature, it is possible to subject them to thermal treatments which will impart pronounced strengthening. The first step called heat treatment or solution heat treatment, is an elevated temperature process designed to put the soluble element or elements in solid solution. This is followed by a rapid quenching which monetarily "freezes" the structure and for a short time renders the alloy very workable. At room or elevated temperatures, the alloys are not stable after quenching, and precipitation of the constituent from the supersaturated solid solution begins. After several days at room temperature, termed aging or room temperature precipitation, the alloy is considerably stronger. Many alloys approach a stable condition at room temperature but some alloys, especailly those containing magnesium and silicon or magnesium and zinc continue to age harden for long periods of time at room temperature. By heating for a controlled time at slightly elevated temperatures, even further strengthening is possible and the properties are stabilized.

Clad Alloys. The heat-treatable alloys, in which copper or zinc are major alloying constituents, are less resistant to corrosive attack than a majority of the non-heat-treatable alloys. To increase the corrosion

TABLE 2. NOMINAL CHEMICAL COMPOSITION[1] AND TYPICAL PROPERTIES OF SOME COMMON ALUMINUM WROUGHT ALLOYS

	1100	3003	5052	2014T6[10]	2017T4[9]	2024T4[9]	6061T6[10]	7075T6[10]	6101T6[10]
Nominal chemical composition[1]	99% min. Alum.	1.2% Mn	2.5% Mg 0.25% Cr	4.4% Cu 0.8% Si 0.8% Mn 0.4% Mg	4.0% Cu 0.5% Mn 0.5% Mg	4.5% Cu 1.5% Mg 0.6% Mn	1.0% Mg 0.6% Si 0.25% Cu 0.25% Cr	5.5% Zn 2.5% Mg 1.5% Cu 0.3% Cr	0.5% Mg 0.5% Si
Tensile strength, psi	A 13,000[7]	16,000	28,000	—	—	—	—	—	—
	H 24,000[7]	29,000	42,000	70,000	62,000	68,000	45,000	83,000	32,000
Tensile strength, MPa	A 90	110	193	—	—	—	—	—	—
	H 165	200	290	483	427	469	310	572	221
Yield strength, psi[2]	A 5000	6000	13,000	—	—	—	—	—	—
	H 22,000	27,000	37,000	60,000	40,000	47,000	40,000	73,000	28,000
Yield strength, MPa	A 34	41	90	—	—	—	—	—	—
	H 152	186	255	414	276	324	276	503	193
Elongation percent in 2 in. (5.1 cm)[11]	A 45 H 15	A 40 H 10	A 30 H 8	13	22	19	17	11	15
Modulus of elasticity[3]	10	10	10.2	10.6	10.5	10.6	10	10.4	10
Brinnell hardness[8]	23–44	28–55	45–85	135	105	120	95	150	71
Melting range, °C	643–657	643–654	593–649	510–638	513–640	502–638	582–652	477–638	616–651
Melting range, °F	1190–1215	1190–1210	1100–1200	950–1180	955–1185	935–1180	1080–1250	890–1180	1140–1205
Specific gravity	2.71	2.73	2.68	2.80	2.79	2.77	2.70	2.80	2.70
Electrical resistivity[4]	2.9	3.4	4.93	4.31	5.75	5.75	4.31	5.74	3.1
Thermal conductivity[5]	0.53	A 0.46	A 0.33	0.37	0.29	0.29	0.37	0.29	0.52
SI units	221.9	A 192.6	A 138.2	154.9	121.4	121.4	154.9	121.4	217.7
Coefficient of expansion[6]	23.6	23.2	23.8	22.5	23.6	22.8	23.4	23.2	23

[1] Aluminum plus normal inpurities is the remainder.
[2] 0.2% permanent set.
[3] Multiply by 10^6.
[4] Microhms per cm (room temperature).
[5] C.g.s. units (at 100°C).
[6] Per °C (20–100°C); multiply by 10^{-6}.
[7] A = annealed; H = hard.
[8] 500 kg load, 10 mm ball.
[9] Solution heat-treated and naturally aged.
[10] Solution heat-treated and artifically aged.
[11] Round specimens, ½-m diameter.
Conversion factors used: 1 psi = 6.894757 × 10^{-3} megapascals (MPa)
C.g.s. = (cal)(cm²)/(sec)(cm)(°C)
SI unit = Watts/meter·°K
1 c.g.s. unit = 418.68 SI units

resistance of these alloys in sheet and plate form, they are often clad with a high-purity aluminum, a low-magnesium-silicon alloy, or an aluminum alloy containing 1% zinc.

The cladding, usually from 2½ to 5% of the total thickness on each side, not only protects the composite due to its own inherent excellent corrosion resistance, but also exerts a galvanic effect which further protects the core alloy.

Special composites may be obtained such as clad non-heat-treatable alloys for extra corrosion protection, for brazing purposes, and for special surface finishes. See Fig. 5.

Composites. The most commonly used metal elements alloyed with aluminum are magnesium, manganese, silicon, copper, zinc, iron, nickel, chromium, titanium, and zirconium. The strength of aluminum can be tailored to the specific end applications ranging from the very soft ductile foils to the high-strength aircraft and space alloys equal to steels in the 90,000 pounds per square inch (621 megapascals) tensile strength range. In addition to the homogeneous aluminum alloys, dispersion hardened and advanced filament-aluminum composites provide even higher strengths. Boron filaments in aluminum can provide strengths in the range of 150,000–300,000 pounds per square inch (1034–2068 megapascals) tensile strength.

Superplasticity. Eutectoid and near eutectoid alloy chemistry research has resulted in alloys exhibiting superplasticity with unusual elongation (> 1000%) and formability and with 60,000 pounds per

TABLE 3. NOMINAL CHEMICAL COMPOSITION[1] AND TYPICAL PROPERTIES OF SOME ALUMINUM CASTING ALLOYS
Properties for alloys 195, B195, 220, 355, and 356 are for the commonly used heat treatment.

	413.0[2]	B443.0[2]	208.0[3]	308.0[4]	295.0[3]	B295.0[4]	514.0[3]	518.0[2]	520.0[3]	355.0[3]	356.0[3]	380.0[2]
Nominal chemical composition	12% Si	5% Si	4% Cu 3% Si	5.5% Si 4.5% Cu	4.5% Cu 0.8% Si	4.5% Cu 2.5% Si	3.8% Mg	8% Mg	10% Mg	5% Si 1.3% Cu	7% Si 0.3% Mg 0.5% Mg	8.5% Si 3.5% Cu
Tensile strength, psi[5]	37,000	19,000	21,000	28,000	36,000	45,000	25,000	42,000	46,000	35,000	33,000	45,000
Tensile strength, MPa[5]	255	131	145	193	248	310	172	290	317	241	228	310
Yield strength, psi[5]	18,000	9000	14,000	16,000	24,000	33,000	12,000	23,000	25,000	25,000	24,000	25,000
Yield strength, MPa[5]	124	62	97	110	165	228	83	159	174	174	165	174
Elongation, percent[5]	1.8	6	2.5	2	5	5	9	7	14	2.5	4	2
Brinnell hardness[6]	—	40	55	70	75	90	50	—	75	80	70	—
Melting range, °C	574–585	577–630	521–632	—	549–646	527–627	580–640	540–621	449–621	580–627	580–610	521–588
Melting range, °F	1065–1085	1070–1165	970–1170	—	1020–1195	980–1160	1075–1185	1005–1150	840–1150	1075–1160	1075–1130	970–1090
Specific gravity	2.66	2.69	2.79	2.79	2.81	2.78	2.65	2.53	2.58	2.70	2.68	2.76
Electrical resistivity	4.40	4.66	5.56	4.66	4.66	3.45	4.93	7.10	8.22	4.79	4.42	6.50
Thermal conductivity[7]	0.37	0.35	0.29	0.34	0.35	0.45	0.33	0.24	0.21	0.34	0.36	0.26
SI units	154.9	146.5	121.4	142.4	146.5	188.4	138.2	100.5	87.9	142.4	150.7	108.9
Coefficient of expansion[8]	20.0	22.8	22.8	22.7	23.9	22.8	24.8	24.0	25.4	22.8	22.8	20.0

[1] Remainder is aluminum plus minor impurities.
[2] Die cast.
[3] Sand cast.
[4] Permanent mold cast.
[5] For separately cast test bars.
[6] 500 kg/load, 10 mm. ball.
[7] C.g.s. units.
[8] Multiply by 10^{-6}. Per °C, for temperature range 20 to 200°C.
Conversion factors used: 1 psi = 6.894757 × 10^{-3} megapascals (MPa)
SI unit = Watts/meter·°K
1 c.g.s. unit = 418.68 SI units

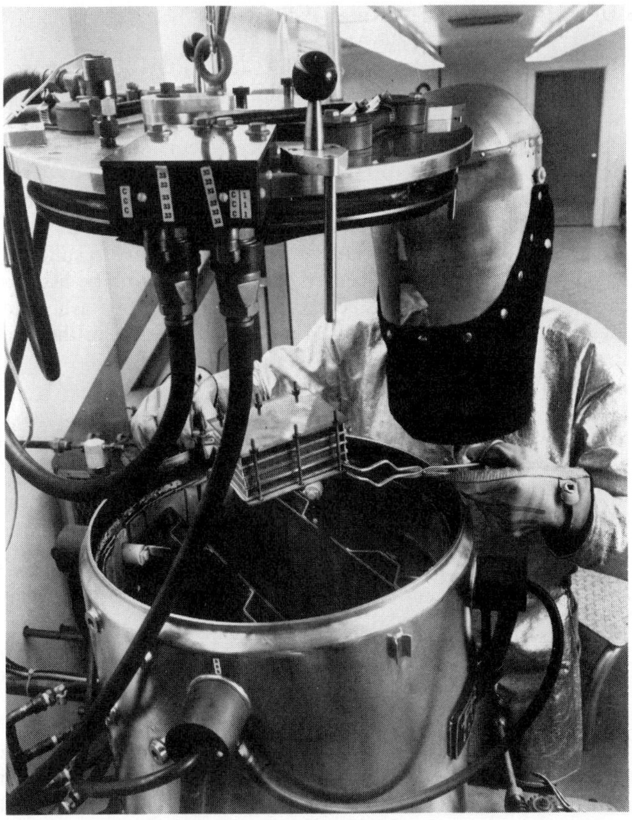

Fig. 5. Operator placing experimental assembled heat exchanger, fabricated of special aluminum alloys which respond to brazing, into vacuum tank. After pressure is reduced to 10^{-5} millimeters, radiant heaters are turned on to obtain a temperature of 1,000°F (593°C), whereupon unit is instantly brazed. (*Reynolds Metals Company, Research and Development Laboratories*)

square inch (414 megapascals) tensile strength. Three examples of these alloys are: 94.5% Al-5% Cu-0.5% Zr; 22% Al-78% Zn; and 90% Al, 5% Zn, 5% Ca.

Casting Alloys. During the last two decades, the quality of castings has been improved substantially by the development of new alloys and better liquid-metal treatment and also by improved casting techniques. Casting techniques include sand casting, permanent mold casting, pressure die casting, and others. Today sand castings can be produced in high-strength alloys and are weldable. Die casting permits large production outputs per hour on intricate pieces that can be cast to close dimensional tolerance and have excellent surface finishes; hence, require minimum machinings. Since aluminum is so simple to melt and cast, a large number of foundry shops have been established to supply the many end products made by this method of fabrication. See Table 3.

Casting Semi-solid Metal. A new casting technology is based on vigorously agitating the molten metal during solidification. A very different metal structure results when this metal is cast. The vigorously agitated liquid-solid mixture behaves as a slurry still sufficiently fluid (thixotropic) to be shaped by casting. The shaping of these metal slurries is termed "Rheocasting."

The slurry nature of "Rheocast" metal permits addition and retention of particulate nonmetal (e.g., Al_2O_3, SiC, T, C, glass beads) materials for cast composites.

The importance of this new technology has yet to be commercialized.

Alloy and Temper Designation Systems for Aluminum. The aluminum industry has standardized the designation systems for wrought aluminum alloys, casting alloys and the temper designations applicable. A system of four-digit numerical designations is used to identify wrought aluminum alloys. The first digit indicates the alloy group as shown in Table 4. The 1xxx series is for minimum aluminum purities of 99.00% and greater; the last two of the four digits indicate the minimum aluminum percentage; i.e., 1045 represents 99.45% minimum aluminum, 1100 represents 99.00% minimum aluminum. The

TABLE 4. DESIGNATIONS FOR WROUGHT ALUMINUM ALLOY GROUPS

		Alloy No.
Aluminum—99.00% minimum and greater		1xxx
	Major Alloying Element	
Aluminum Alloys grouped by Major Alloying Elements	Copper	2xxx
	Manganese	3xxx
	Silicon	4xxx
	Magnesium	5xxx
	Magnesium and Silicon	6xxx
	Zinc	7xxx
	Other Element	8xxx
Unused Series		9xxx

(1) For codification purposes an alloying element is any element which is intentionally added for any purpose other than grain refinement and for which minimum and maximum limits are specified.

(2) Standard limits for alloying elements and impurities are expressed to the following places:

Less than 1/1000%	0.000X
1/1000 up to 1/100%	0.00X
1/100 up to 1/10%	
Unalloyed aluminum made by a refining process	0.0XX
Alloys and unalloyed aluminum not made by a refining process	0.0X
1/10 through 1/2%	0.XX
Over 1/2%	0.X, X.X, etc.

2xxx through 8xxx series group aluminum alloys by major allowing elements. In these series the first digit represents the major alloying element, the second digit indicates alloy modification, while the third and fourth serve only to identify the different alloys in the group. Experimental alloys are prefixed with an X. The prefix is dropped when the alloy is no longer considered experimental.

Cast Aluminum Alloy Designation System. A four-digit number system is used for identifying aluminum alloys used for castings and foundry ingot (see Table 5). In the 1xx.x group for aluminum purities of 99.00% or greater, the second and third digit indicate the minimum aluminum percentage. The last digit to the right of the decimal point indicates the product form: 1xx.0 indicates castings and 1xx.1 indicates ingot. Special control of one or more individual elements other than aluminum is indicated by a serial letter before the numerical designation. The serial letters are assigned in alphabetical sequence starting with A but omitting I, O, Q, and X, the X being reserved for experimental alloys.

In the 2xx.x through 9xx.x alloy groups, the second two of the four digits in the designation have no special significance but serve only to identify the different aluminum alloys in the group. The last digit to the right of the decimal point indicates the product form: .0 indicates casting and .1 indicates ingot. Examples: Alloy 213.0 represents a casting of an aluminum alloy whose major alloying element is copper. Alloy C355.1 represents the third modification of the chemistry of an aluminum alloy ingot whose major alloying elements are silicon, copper, and magnesium.

Temper Designation System. A temper designation is used for all forms of wrought and cast aluminum alloys. The temper designation follows the alloy designation, the two letters being separated by a hyphen. Basic designations consist of letters followed by one or more digits. These designate specific sequences of basic treatments but only operations recognized as significantly influencing the characteristics of the product. Basic tempers are -F (as fabricated), -O annealed (wrought products only), -H strain-hardened (degree of hardness is normally quarter hard, half hard, three-quarters hard, and hard designated by the symbols H12, H14, H16, and H18, respectively). -W solution heat treated and -T thermally treated to produce stable tempers. Examples: 1100-H14 represents commercially pure aluminum cold rolled to half-hard properties. 2024-T6 represents an aluminum alloy whose principal major element is copper which has been solution heat treated and then artificially aged to develop stable full-strength properties of the alloy.

TABLE 5. DESIGNATIONS FOR CAST ALUMINUM ALLOY GROUPS

		Alloy No.
Aluminum	99.00% minimum and greater	1xx.x
	Major Alloy Element	
	Copper	2xx.x
Aluminum Alloys	Silicon, with added Copper and/or Magnesium	3xx.x
grouped	Silicon	4xx.x
by Major	Magnesium	5xx.x
Alloying	Zinc	7xx.x
Elements	Tin	8xx.x
	Other Element	9xx.x
Unused Series		6xx.x

(1) For codification purposes an alloying element is any element which is intentionally added for any purpose other than grain refinement and for which minimum and maximum limits are specified.
(2) Standard limits for alloying elements and impurities are expressed to the following places:

Less than 1/1000%	0.000X
1/1000 up to 1/100%	0.00X
1/100 up to 1/10%	
Unalloyed aluminum made by a refining process	0.0XX
Alloys and unalloyed aluminum not made by a refining process	0.0X
1/10 through 1/2%	0.XX
Over 1/2%	0.X, X.X, etc.

References

Ballance, J. B.: "Aluminum Recycling Comes of Age," *J. Metals*, **32**, 2, 55–62 (1980).
Brondyke, K. J.: "Energy Savings in Aluminum Production," *Industrial Heating*, **XLVI**, 7, 29–32 (1979).
Canby, T. Y.: "Aluminum the Magic Metal," *Nat. Geog.*, **154**, 2, 186–211 (1978).
EPA: "Environmental Considerations of Selected Energy Conserving Manufacturing Process Options," VIII (Alumina/Aluminum Industry Report). EPA-600/7-76-034h, Environmental Protection Agency, Office of Research and Development, Cincinnati, Ohio, 1976.
Froberg, H.: "Status and Outlook for the Aluminum Industry," *J. Metals*, **32**, 4, 58–59 (1980).
Mazel, J. L.: "Expansion-Minded Aluminum Producers are Eyeing Australia," *33 Metal Producing*, **18**, 1, 56–61 (1980).
Regan, R. J.: "Windfall Profits Still Elude Caribbean Bauxite," *Iron Age*, Reprint 3532, 31–33 (1980).
Staff: "Superplastic Aluminum Alloy Contains Zinc and Calcium," *Metal Progress*, **116**, 6, 73 (1979).
Staff: "Alcoa Pushes Ahead on Aluminum Process," *Chemical & Engineering News*, **57**, 23, 8 (1979).
Staff: "Successful Ocean Energy Test Stirs Interest," *Chemical & Engineering News*, **58**, 18, 24 (1980).

ALUMINUM BOROHYDRIDE. Hydride.

ALUMINUM BOROSILICATE. Dumortierite.

ALUMINUM BRONZE. Copper.

ALUMINUM FUEL. Rocket Propellants.

ALUMINUM-GALLIUM-ARSENIDE. Telephony.

ALUMINUM HYDROXIDE. Antacids.

ALUMINUM (Insulation). Insulation (Thermal).

ALUMINUM PHOSPHATE. Wavellite.

ALUMINUM SILICATE. Angelsite; Kaolinite; Pyrophyllite; Sillimanite.

ALUMINUM (Surface Treatment). Conversion Coatings.

ALUMSTONE. Alunite.

ALUM TREATMENT. Water Pollution.

ALUNITE. The mineral alunite, $KAl_3(SO_4)_2(OH)_6$, is a basic hydrous sulfate of aluminum and potassium; a variety called natroalunite is rich in soda. Alunite crystallizes in the hexagonal system and forms rhombohedrons with small angles, hence resembling cubes. It may be in fibrous or tabular forms, or massive. Hardness, 3.5–4; sp gr, 2.58–2.75; luster, vitreous to pearly; streak white; transparent to opaque; brittle; color, white to grayish or reddish.

Alunite is commonly associated with acid lavas due to sulfuric vapors often present; it may occur around fumaroles or be associated with sulfide ore bodies. It has been used as a source of potash. Alunite is found in Czechoslovakia, Italy, France, and Mexico; in the United States, in Colorado, Nevada, and Utah.

Alunite is also known as *alumstone.*

ALVEOLAR HYPER- AND HYPOVENTILATION. Carbon Dioxide; Respiratory System.

ALVEOLUS. 1. A minute sac-like chamber in a hollow organ, such as the air sacs of the lungs and the components of various glands. Its sac-like form distinguishes it from other chambers in which the wall are relatively thicker. 2. The cavity or socket in the jaw in which the root of a tooth is fixed.

ALVIN RESEARCH VESSEL. Ocean; Ocean Research Vessels.

ALZHEIMER'S DISEASE AND OTHER DEGENERATIVE MENTAL DISORDERS. The majority of cases of primary dementia are represented by Alzheimer's disease, with an estimate of one million persons in the United States afflicted by it. It is further estimated that approximately 75,000 people die of this condition each year. Commencing after 45 years of age, the incidence of the disease is divided equally among men and women. Although familial connections have been noted, these are considered rare. Often the initial symptoms are not noted by the patient, but rather are observed by friends, particularly by persons who have not seen the patient for a comparatively long time. Early symptoms include depression, moodiness, decreased sociability, and lack of drive. Hypomanic behavior may be present in some cases. As the disease progresses, there is observable intellectual decline. The latter is revealed by difficulty in making decisions and basic errors of judgment. In many instances, the patient becomes aware and hence anxious over these symptoms. There is a further development of memory loss, usually commencing with an impaired retention of new information as contrasted with older memory. There usually follows speech difficulties and minor muscular spasms (jerks). With the advent of computed tomography (CT scan), brain atrophy can be demonstrated. Pathologically, the disease features brain atrophy, senile plaques, neurofibrillary tangles, and granulovacuolar degeneration of neurons, conditions which also are found in senile dementia. Actually very little is currently known concerning the etiology and pathogenesis of the condition.

In *arteriosclerotic dementia*, many of the symptoms of which parallel those of Alzheimer's disease, the cause of the condition is well known, namely, a reduced cerebral blood flow. In Alzheimer's disease, the blood flow is normal. Arteriosclerotic dementia is sometimes called *multi-infarct dementia* because of atherosclerosis of small arteries and patchy cerebral infarction. The onset of this condition may be abrupt.

In the syndrome of *occult hydrocephalus*, where there is normal cerebrospinal fluid pressure, some researchers have found that this form of dementia may be reversible as the result of shunting. At one time, it was believed that shunting may improve Alzheimer's disease, but this has not proven so. However, it is difficult for neurosurgeons to discern situations where shunting may or may not be useful. Some neurologists prefer to use the shunt procedure only in cases where there has been a known episode of meningitis, trauma, or subarachnoid hemorrhage.

See also **Brain and Nervous System; Brain (Injury);** and **Memory.**

References

Adams, R. D., et al.: "Symptomatic Occult Hydrocephalus with "Normal" Cerebrospinal-Fluid Pressure: A Treatable Syndrome," *N. Engl. J. Med.*, **273**, 117 (1965).

Crapper, D. R., Krishnan, S. S., and S. Quittkat: "Aluminum, Neurofibrillary Degeneration and Alzheimer's Disease," *Brain*, **99**, 67 (1976).
Hachinski, V. C., et al.: "Cerebral Blood Flow in Dementia," *Arch. Neurol.*, **32**, 632 (1975).
Katzman, R.: "The Prevalence and Malignancy of Alzheimer's Disease: A Major Killer," (editorial), *Arch. Neurol.*, **33**, 217 (1976).
Messert, B., and B. B. Wannamaker: "Reappraisal of the Adult Occult Hydrocephalus Syndrome," *Neurology* (*Minneapolis*), **24**, 224 (1974).
Roth, M., Tomlinson, B. E., and G. Blessed: "Correlation Between Scores for Dementia and Counts of "Senile Plaques" in Cerebral Grey Matter of Elderly Subjects," *Nature*, **209**, 109 (1966).
Terry, R. D.: "Dementia: A Brief and Selective Review," (editorial), *Arch. Neurol.*, **33**, 1 (1976).
Tomlinson, B. E., Blessed, G., and M. Roth: "Observations on the Brains of Demented Old People," *J. Neurol. Sci.*, **11**, 205 (1970).
Wells, C. E. (editor): "Dementia. Contemporary Neurology Series," 2nd edition, F. A. Davis, Philadelphia, 1977.

AMAGAT'S LAW. **Combustion.**

AMALGAM. 1. An alloy containing mercury. Amalgams are formed by dissolving other metals in mercury, when combination takes place often with considerable evolution of heat. Amalgams are regarded as compounds of mercury with other metals, or as solutions of such compounds in mercury. It has been demonstrated that products which contain mercury and another metal in atomic proportions may be separated from amalgams. The most commonly encountered amalgams are those of gold and silver. See also **Gold; Mercury; Silver.**

2. A naturally occurring alloy of silver with mercury, also referred to as mercurian silver, silver amalgam, and argental mercury. The natural amalgam crystallizes in the isometric system; hardness, 3–3.5; sp gr, 13.75–14.1; luster, metallic; color, silver-white; streak, silver-white; opaque. Amalgam is found in Bavaria, British Columbia, Chile, Czechoslovakia, France, Norway, and Spain. In some areas, it is found in the oxidation zone of silver deposits and as scattered grains in cinnabar ores.

AMALTHEA. **Jupiter.**

AMANTADINE HYDROCHLORIDE. **Antiviral Drugs.**

AMARANTHUS. Genus of plants in the family *Amaranthaceae* (amaranth family), including many coarse and rather obnoxious pigweeds. *Amaranthus caudatus*, the love-lies-bleeding; and the *A. hypochondriacus*, the Princess feather are cultivated in gardens for their attractive green, purple, and crimson flowers. They are annuals.

AMATOL EXPLOSIVE. **Explosive.**

AMBER. Amber is a fossil resin which has been known since early times because of its property of acquiring an electric charge when rubbed. In modern times it has been used largely in the making of beads, cigarette holders, and trinkets. Its amorphous non-brittle nature permits it to be carved easily and to acquire a very smooth and attractive surface. Amber is soluble in various organic solvents, such as ethyl alcohol and ethyl ether.

It occurs in irregular masses showing a conchoidal fracture. Hardness, 2.25; sp gr, 1.09; luster, resinous; color, yellow to reddish or brownish; it may be cloudy. Some varieties will exhibit fluorescence. Amber is transparent to translucent, melts between 250 and 300°C.

Amber has been obtained for over 2,000 years from the lignite-bearing Tertiary sandstones on the coast of the Baltic Sea from Danzig to Memel; also from Denmark, Sweden and the other Baltic countries. Sicily furnishes a brownish-red amber that is fluorescent.

The association of amber with lignite or other fossil woods, as well as the beautifully preserved insects that are occasionally in it, is ample proof of its organic origin.

AMBERGRIS. A fragrant waxy substance formed in the intestine of the sperm whale and sometimes found floating in the sea. It has been used in the manufacture of perfumes to increase the persistence of the scent. See also **Whales, Dolphins, and Porpoises.**

AMBIENT CONDITIONS. These are the conditions of environment in which an instrument or device operates. As various environmental factors change, they may seriously alter the accuracy and other performance parameters of equipment. The effects of ambient changes may be combated in two basic ways: (1) *ambient compensation* in which a technique is used to offset the effects of environmental changes, such as changes in temperature; pressure; supply voltage, frequency, and waveform; shock and vibration; and position, and (2) *ambient protection* in which the equipment is protected from its environment, that is, by way of isolation, insulation, and similar techniques. When procuring instruments and other sensitive laboratory and research equipment, the expected range of ambient conditions over which the equipment will be used always should be specified, and the manufacturer's assurance of ambient compensation and/or protection should be studied carefully.

Ambient Compensation. An electrical circuit may be made relatively insensitive to temperature changes by use of components, such as *manganin* or *evanohm* wire for resistors. The resistances of these alloys are only slightly affected by temperature changes in the range of laboratory and industrial environments. Mechanical means of offsetting temperature changes include bimetals or other combinations of materials with different temperature coefficients of expansion. Where stability of dimensions must be maintained, materials with very low thermal expansion coefficients, such as *invar*, may be used in construction of the equipment.

A method employing parallel elements is used in filled-system thermometers, where a sensing bulb is connected by a capillary to a bourdon tube, bellows, or similar element with motion as its output. The system is filled with a gas or liquid having a relatively high temperature coefficient of volumetric expansion, and sealed. If the ambient temperature changes at the instrument, or along the capillary, a change in volume of the fill will cause a change in the output of the instrument, just as if the bulb temperature were changing. This leads to inaccurate readings. The effect is offset by using an auxiliary thermal system without the bulb. The output motion of the auxiliary bourdon is connected mechanically to the primary element so that it indicates an output change only when the temperature of the bulb changes. See accompanying diagram.

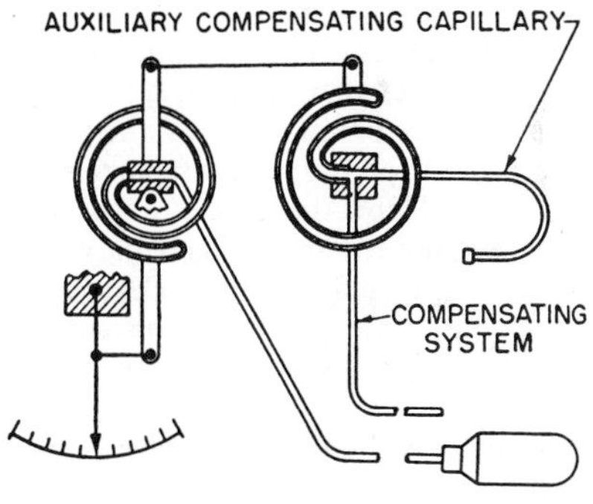

Filled-system thermometer in which auxiliary compensating capillary is used to offset ambient temperature changes.

The spring rate of a pressure-sensitive element, such as a bourdon tube, will be affected by temperature change. For most materials, the modulus of elasticity will decrease with increasing temperature, thereby causing an incorrect higher indication. Use of a bimetal to reduce the span is one common means for correcting this effect. Making the pressure element of *Ni-span C*, which has an essentially constant modulus, is another way of correcting the condition.

Manual temperature correction means are provided on several instruments, such as mercurial barometers and precision manometers. Volume changes in a fluid-filled device caused by temperature variations may be offset by a pressure-sensitive diaphragm, which is deflected by expansion of the fluid to maintain a relatively constant pressure. A similar principle is used in the hydraulic accumulator.

To prevent a pressure-sensitive element from being affected by ambient pressure changes, the element may be hermetically sealed within a rigid enclosure.

Correction curves and tables also are widely used. The output of the uncompensated instrument is read, the ambient conditions noted, and the proper correction found in a table. This technique is less convenient than automatic or even manual compensation. However, under reasonably stable environmental influences, the method may be more accurate and practical than automatic compensation, and certainly less costly.

Instruments are usually designed to tolerate some variation in supply voltage. The availability of zener diodes in a wide range of voltage ratings simplifies the problem of maintaining dc voltage levels with almost any required degree of exactness. Power supplies with ac line input and dc output are available commercially with almost any desired combination of voltage, current, range of adjustability, line and load regulation, ambient temperature compensation, ripple, transient suppression, and package configuration. Alternating current voltage regulators can be obtained to maintain constant output within the needed tolerance for input variations of any likely magnitude, with widely differing speeds of response and power capabilities. The quality of the output waveforms differs markedly, depending upon the operating principle and the magnitude and nature of the load. Care always must be exercised in determining the quality of waveform needed to assure satisfactory operation of the equipment being powered. See also **Power Supply.**

The accuracy and stability of line power frequency normally are adequate for most applications. Short-term frequency variations seldom are more than a fraction of a hertz on a 50- or 60-Hz line. Where exceptionally close regulation of frequency is needed, as in the case of very accurate timing, an electronic or tuning-fork oscillator may be built into the equipment.

Ambient Protection. Ambient condition protection may be used to isolate the instrument or device from the undesirable environment. The dividing line between this approach and that of ambient compensation is narrow and sometimes difficult to define.

Cleanliness of environment, such as air or liquid as an input to equipment, is an ambient condition difficult to measure, but often necessary for satisfactory long-term operation. Where gases and liquids are involved, it may be necessary to remove solid particles by mechanical filtering, washing, or centrifuging; moisture may have to be removed by cooling and condensing, by the use of desiccants, or by centrifuging; oil may have to be removed by chemical absorption or mechanical filtering. These problems pertain to both signal and control lines in connection with pneumatic equipment. Further, the instrument box may have to be protected from dust, oil, and chemicals in the atmosphere. Where it is not practical to seal the instrument enclosure sufficiently tight to exclude foreign materials, purging may be an attractive alternate approach. Here, the air inside the instrument case is maintained at a pressure slightly above that of the outside environment. The technique also can be used to protect electrical equipment from hazards caused by combustible gases or vapors entering the enclosure where heat or a spark might cause an explosion. See also **Intrinsic Safety.**

Changes in environmental humidity affect electrical leakage, dimensions of hygroscopic materials such as paper recorder charts, the concentration of chemical solutions, and, of course, the presence of dampness enhances the formation of rust and corrosion. Methods for overcoming humidity effects include sealing to exclude moisture, heating the equipment (maintaining a constant temperature) to lower the relative humidity, cooling by refrigeration to cause condensation on noncritical areas and permitting condensate removal, and the use of desiccants to absorb moisture. Pipes and tubing containing fluids that would freeze or become too viscous when subjected to very low ambient temperatures can be "steam traced" by paralleling the tubing with a small steam line.

Sometimes a practical solution for assuring satisfactory operation despite extremes of temperature is that of derating a component, so that its output will be adequate under the most unfavorable conditions. This applies particularly to electronic components, such as transistors, whose power dissipation decreases as the temperature rises.

The best protection against shock and vibration is careful design of the original equipment. Serious effects on the performance of transformers, capacitors, and resistors may occur unless relative motion of the materials used in constructing them is prevented, as well as motion of the entire component relative to the remainder of the circuit. Heavy parts, such as transformers, must be rigidly attached to a sufficiently rugged chassis to prevent them from breaking loose. Where shock and vibration conditions are contemplated, expensive instruments should be subjected to prescribed vibration and shock testing prior to acceptance.

Grounding may be important both for low level signals (to minimize electrical noise problems) and for high power circuits (to prevent hazards to personnel). Particular note should be taken of the distinction between line power neutral and true ground. Even a small resistance between them can produce an unacceptably high voltage difference, if there is a significant current. In many low-level instrument systems, it is vital to proper operation that a common ground or shield be used for all components of the system. This can be even more important than assuring that the ground is at true earth potential. See also **Common-Mode Voltage; Differential-Mode Voltage.**

Frequently, an expendable means may be used for protection, as exemplified by a fuse in an electrical circuit. Refinements include fast- and slow-blow fuses, and circuit breakers with various speeds of operation. Voltage limiting means are the neon lamp and the zener diode, which will pass essentially no current unless the voltage (to ground, or some other safe part of the circuit) exceeds the allowable value.

Hydraulic and pneumatic equivalents are rupture disks, which break to prevent pressure from exceeding a predetermined limit, and "hydraulic fuses," which are preloaded valves that close if the flow rate becomes too great.

Thermal devices with similar purposes are fusible links, as found in automatic sprinkler systems, and freeze-out plugs, which release if water freezes and expands, endangering a system.

AMBLYGONITE. A rather rare compound of fluorine, lithium, aluminum, and phosphorus, $(Li, Na)AlPO_4(F, OH)$. It crystallizes in the triclinic system; hardness, 5–5.6; sp gr 3.08; luster, vitreous to greasy or pearly; color, white to greenish, bluish, a yellowish or grayish; streak white; translucent to subtransparent.

Amblygonite occurs in pegmatite dikes and veins associated with other lithium minerals. It is used as a source of lithium salts. The name is derived from two Greek words meaning blunt and angle, in reference to its cleavage angle of 75° 30′.

Amblygonite is found in Saxony; France; Australia; Brazil; Varutrask, Sweden; Karibibe, S.W. Africa; and in the United States at Hebron, Paris, Greenwood, Rumford and Auburn, Maine; Branchville, Connecticut; Black Hills, South Dakota; and Pala, California.

See also **Lithium (For Thermonuclear Fusion Reactions).**

AMBLYOPIA. Vision and the Eye.

AMBLYOPSIDS. Blind-Fish.

AMBUSH BUG (*Insecta, Hemiptera*). Predacious bugs named from their habit of lying in wait for their prey in flowers, where their colors conceal them. They belong to the family *Phymatidae.* The forelegs of these predatory bugs are prehensile; the tibia can be folded back against the femur, and the tarsus is regressed. Most species are tropical, only two having penetrated as far as Europe and only one of these (*Phymata crassipes*) as far north as Germany. In the United States, the ambush bug (*Phymata fasciata*), a yellowish ambush bug which may be up to 12 millimeters (0.5 inch) long, is the most commonly encountered species, though other species are also found (for example, *Phymata erosa*).

AMEBIASIS. Also called amoebic dysentery, amebiasis is an invasion of the sigmoid colon, rectum, and/or cecum by the protozoan *Entamoeba histolytica.* A common mode of spread is by carriers—healthy individuals who pass the cysts in their stools. It is estimated that about 5% of travelers who go to regions where the disease is endemic excrete cysts, but that only about 10% of these persons have symptoms of the disease. Classical symptoms include nausea, loss of appetite and weight, and abdominal pain. Discomfort can be rather widespread, but usually is concentrated in the right lower quadrant. If the disease is allowed to go untreated, the colon may perforate and spread the disease to the peritoneum. Mortality is estimated at 3%, usually due to peritonitis.

Diagnosis involves microscopic examination of stools and sometimes sigmoidoscopy. The presence of motile forms (trophozoites) in fresh stools provides confirmation.

Therapy today is usually by administration of metronidazole (Flagyl®), a drug with less toxicity than those used a few years ago. Where parasites persist in stool samples after treatment with this drug, the use of diiodohydroxyquin (Diodoquin®) may be initiated. If still further therapy is required, a timed regimen of tetracycline, chloroquine (Aralen®), and diiodohydroxyquin will usually be used.

A serious, less frequent (about 1% of cases) complication of amebiasis is liver involvement. Migration of parasites by way of portal circulation causes liver abscesses, usually singly and in the right hepatic lobe. The symptoms parallel those of uncomplicated amebiasis, but with additional tenderness in the right upper quadrant and sometimes enlargement of the liver and accompanying local tenderness. Should abscesses rupture into the thoracic space or peritoneum, mortality rate is high (25–75%). Abdominal x-rays and liver ultrasonic scans are utilized to monitor therapy, which essentially parallels that of the uncomplicated disease.

AMEBIC MENINGOENCEPHALITIS (Primary). Primary Amebic Meningoencephalitis.

AMEIVA (*Reptilia, Sauria*). A lizard of Central and South America, small, very active, timid and mainly insectivorous. The name is that of the genus, applied as a common name to the score of included species.

AMELANCHIER. Rose Family.

AMENORRHEA. The absence of menstruation, which may be primary or secondary. Secondary amenorrhea is defined as the absence of menstruation in a woman who has had previously normal menstrual periods. The most obvious cause of this, of course, is pregnancy. The physician will probe into four possible areas of dysfunction in attempting to comprehend a particular primary or secondary amenorrhea—hypothalamic, pituitary, ovarian, and uterine defects. See **Gonads;** and **Hormones.**

AMENT. Catkin.

AMERICAN ANTELOPE. Pronghorn Antelope.

AMERICAN BISON. Bovines.

AMERICAN BLACK BEAR. Bears.

AMERICAN ELM. Elm Trees.

AMERICAN MOUNTAIN ASH. Ash Trees.

AMERICAN SHAD. Herring.

AMERICAN SWEET CHESTNUT. Chestnut Trees.

AMERICAN WINTER FLOUNDER. Flatfishes.

AMERICIUM. Chemical element symbol Am, at no. 95, at. wt. 243 (mass number of the most stable isotope), radioactive metal of the actinide series, also one of the transuranium elements. All isotopes of americium are radioactive; all must be produced synthetically. The element was discovered by G. T. Seaborg and associates at the Metallurgical Laboratory of the University of Chicago in 1945. At that time, the element was obtained by bombarding uranium-238 with helium ions to produce ^{241}Am which has a half-life of 475 years. Subsequently, ^{241}Am has been produced by bombardment of plutonium-241 with neutrons in a nuclear reactor by the series of nuclear reactions:

$$^{239}Pu + n \rightarrow {}^{240}Pu + \gamma$$
$$^{240}Pu + n \rightarrow {}^{241}Pu + \gamma$$
$$^{241}Pu + n \rightarrow {}^{241}Am + e^-$$

^{243}Am is the most stable isotope, an alpha emitter with a half-life of 7950 years. Other known isotopes are ^{237}Am, ^{238}Am, ^{240}Am, ^{241}Am, ^{242}Am, ^{244}Am, ^{245}Am, and ^{246}Am. Electronic configuration is $1s^22s^22p^63s^23p^63d^{10}4s^24p^64d^{10}4f^{14}5s^25p^65d^{10}5f^76s^26p^67s^2$. Ionic radii are: Am^{4+}, 0.85 Å; Am^{3+}, 1.00 Å.

This element exists in acidic aqueous solution in the (III), (IV), (V), and (VI) oxidation states with the ionic species probably corresponding to Am^{3+}, Am^{4+}, AmO_2^+, and AmO_2^{2+}. The oxidation potentials in acidic aqueous solution are summarized in the following diagram in which the americium(IV)-americium(V) and americium(III)-americium(V) couples are calculated from the others.

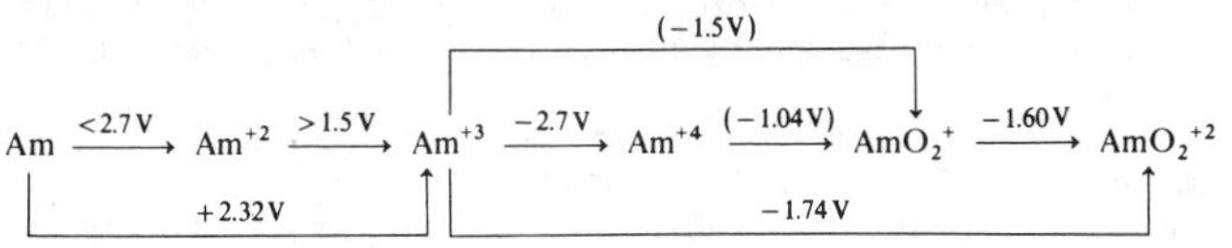

The colors of the ions are: Am^{3+}, pink; Am^{4+}, rose; AmO_2^+, yellow; and AmO_2^{2+}, rum-colored.

It can be seen that the (III) state is highly stable with respect to disproportionation in aqueous solution and is extremely difficult to oxidize or reduce. There is evidence for the existence of the (II) state since tracer amounts of americium have been reduced by sodium amalgam and precipitated with barium chloride or europium sulfate as carrier. The (IV) state is very unstable in solution: the potential for americium(III)-americium(IV) was determined by thermal measurements involving solid AmO_2. Americium can be oxidized to the (V) or (VI) state with strong oxidizing agents, and the potential for the americium(V)-americium(VI) couple was determined potentiometrically. The value of the potential for the americium(III)-americium(VI) couple has been measured as −1.9 V, but the value of −1.74 V/V is used in the above potential diagram in order to make the known disproportionation of americium(V) consistent with the other more accurately measured potentials. The value for the americium-americium(III) couple is calculated from heat of solution measurements using estimations for the entropy change.

In its precipitation reactions americium(III) is very similar to the other tripositive actinide elements and to the rare earth elements. Thus the fluoride and the oxalate are insoluble and the phosphate and iodate are only moderately soluble in acid solution, whereas the nitrates, halides, sulfates, sulfides, and perchlorates are all soluble. Americium(VI) can be precipitated with sodium acetate giving crystals isostructural with sodium uranyl acetate,

$$NaUO_2(C_2H_3O_2)_3 \cdot xH_2O$$

and the corresponding neptunium and plutonium compounds.

Of the hydrides of americium, both AmH_2 and Am_4H_{15} are black and cubic.

When americium is precipitated as the insoluble hydroxide from aqueous solution and heated in air, a black oxide is formed which corresponds almost exactly to the formula AmO_2. This may be reduced to Am_2O_3 through the action of hydrogen at elevated temperatures. The AmO_2 has the cubic fluorite type structure, isostructural with UO_2, NpO_2, and PuO_2. The sesquioxide, Am_2O_3 is allotropic, existing in a reddish brown and a tan form, both hexagonal. As in the case of the preceding actinide elements, oxides of variable composition between $AmO_{1.5}$ and AmO_2 are formed depending upon the conditions.

All four of the trihalides of americium have been prepared and identified. These are prepared by methods which are similar to those used in the preparation of the trihalides of other actinide elements. AmF_3 is pink and hexagonal, as in $AmCl_3$; $AmBr_3$ is white and orthorhombic; while a tetrafluoride, AmF_4 is tan and monoclinic.

In research at the Institute of Radiochemistry, Karlsruhe, West Germany during the early 1970s, investigators prepared alloys of americium with platinum, palladium, and iridium. These alloys were prepared by hydrogen reduction of the americium oxide in the presence of finely divided noble metals according to:

$$Am_2O_3 + 10Pt \xrightarrow[1100°C]{H_2} 2AmPt_5 + H_2O$$

The reaction is called a *coupled reaction* because the reduction of the metal oxide can be done only in the presence of noble metals. The hydrogen must be extremely pure, with an oxygen content of less than 10^{-25} torr.

References

Asprey, L. B., Stephanou, W. E., and R. A. Penneman: "Hexavalent Americium," *Amer. Chem. Soc. J.*, **73**, 5715–5717 (1951).

Asprey, L. B.: "New Compounds of Quadrivalent Americium, AmF_4, $KAmF_5$," *Amer. Chem. Soc. J.*, **76**, 2019–2020 (1954).

Cunningham, B. B.: "Chemistry of Element 96," Metallurgical Laboratory Report CS-3312, Univ. of Chicago, Chicago, 1945.

Ghiorso, A., James, R. A., Morgan, L. O., and G. T. Seaborg: "Preparation of Transplutonium Isotopes by Neutron Irradiation," *Phys. Rev.*, **78**, 4, 472 (1950).

Graf, P., et al.: "Crystal Structure and Magnetic Susceptibility of American Metal," *Amer. Chem. Soc. J.*, **78**, 2340 (1956).

Keller, C., and B. Erdmann: "Preparation and Properties of Transuranium Element—Noble Metal Alloy Phases," *Proc. of the Moscow Symposium on Chemistry of Transuranium Elements*, Pergamon, Elmsford, New York, 1976.

Roof, R. B., et al.: "High-Pressure Phase in Americium Metal," *Science*, **207**, 1353–1354 (1980).

Seaborg, G. T.: "The Chemical and Radioactive Properties of the Heavy Elements," *Chemical & Engineering News*, **23**, 2190–2193 (1945).

Seaborg, G. T., Editor: "Transuranium Elements," Dowden, Hutchinson & Ross, Stroudsburg, Pennsylvania, 1978.

Smith, J. L., and R. G. Haire: "Superconductivity of Americium," *Science*, **200**, 535–537 (1978).

AMETHYST. A purple- or violet-colored quartz having the same physical characteristics as quartz. The source of color is not definite but thought to be caused by ferric iron contamination. Oriental amethysts are purple corundum.

Amethysts are found in the Ural Mountains, India, Sri Lanka, the Malagasy Republic, Uruguay, Brazil, the Thunder Bay district of Lake Superior in Ontario, and Nova Scotia; in the United States, in Michigan, Virginia, North Carolina, Montana, and Maine.

The name amethyst is generally supposed to have been derived from the Greek word meaning not drunken. Pliny suggested that the term was applied because the amethyst approaches but is not quite the equivalent of a wine color.

See also **Quartz;** and terms listed under **Mineralogy.**

AMICI PRISM. Prism (Optics).

AMICRON. A name applied by Zsigmondy to individual disperse particles invisible under the ultramicroscope whose size is about 10^{-7}cm. They act as nuclei for the formation of submicrons which are about five times as large.

AMIDES. An amide may be defined as a compound that contains the $CO \cdot NH_2$ radical, or an acid radical(s) substituted for one or more of the hydrogen atoms of an ammonia molecule. Amides may be classified as (1) *primary amides* which contain one acyl radical, such as —$CO \cdot CH_3$ (acetyl) or —$CO \cdot C_6H_5$ (benzoyl), linked to the *amido* group (—NH_2). Thus, acetamide NH_2COCH_3 is a combination of the acetyl and amido groups; (2) *secondary amides* which contain two acyl radicals and the *imido* group (=NH_2). Diacetamide $HN(COCH_3)_2$ is an example; and (3) *tertiary amides* which contain three acyl radicals attached to the N atom. Triacetamide $N(COCH_3)_3$ is an example.

A further structural analysis will show that amides may be regarded as derivatives of corresponding acids in which the amido group substitutes for the hydroxyl radical OH of the carboxylic group COOH. Thus, in the instance of formic acid HCOOH, the amide is $HCOONH_2$ (formamide); or in the case of acetic acid CH_3COOH, the amide is CH_3CONH_2 (acetamide). Similarly, urea may be regarded as the amide of carbonic acid (theoretical) O:C, that is, NH_2CONH_2 (urea). The latter represents a dibasic acid in which two H atoms of the hydroxyl groups have been replaced by amido groups. A similar instance, malamide,

$$NH_2CO \cdot CH_2CH(OH) \cdot CONH_2$$

is derived from the diabasic acid, malic acid,

$$OHCO \cdot CH_2CH(OH) \cdot COOH.$$

Aromatic amides, sometimes referred to as *arylamides*, exhibit the same relationship. Note the relationship of benzoic acid C_6H_5COOH with benzamide $C_6H_5CONH_2$. *Thiamides* are derived from amides in which there is substitution of the O atom by a sulfur atom. Thus, acetamide $NH_2 \cdot CO \cdot CH_3$, becomes thiacetamide $NH_2 \cdot CS \cdot CH_3$; or acetanilide $C_6H_5 \cdot NH \cdot CO \cdot Ch_3$ becomes thiacetanilide $C_6H_5 \cdot NH \cdot CS \cdot CH_3$. *Sulfonamides* are derived from the sulfonic acids. Thus, benzene-sulfonic acid $C_6H_5 \cdot SO_2 \cdot OH$ becomes benzene-sulfonamide $C_6H_5 \cdot SO_2 \cdot NH_2$. See also **Sulfonamide Drugs.**

Amides may be made in a number of ways. Prominent among them is the acylation of amines. The agents commonly used are, in order of reactivity, the acid halides, acid anhydrides, and esters. Such reactions are:

$$R'COCl + HNR_2 \rightarrow R'C({=}O)NR_2 + HCl$$

$$R'C({=}O)OC({=}O)R' + HNR_2 \rightarrow R'C({=}O)NR_2 + R'COOH$$

$$R'C({=}O)OR'' + HNR_2 \rightarrow R'C({=}O)NR_2 + R''OH$$

The hydrolysis of nitriles also yields amides:

$$RCN + H_2O \xrightarrow{OH} RCONH_2$$

Amides are resonance compounds, having an ionic structure for one form:

$$R{-}C({=}O)NR_2 \qquad R{-}C({-}O^-){:}N^+R_2$$

Evidence for the ionic form is provided by the fact that the carbon-nitrogen bond (1.38 Å) is shorter than a normal C—N bond (1.47 Å) and the carbon-oxygen bond (1.28 Å) is longer than a typical carbonyl bond (1.21 Å). That is, the carbon-nitrogen bond is neither a real C—N single bond nor a C=N double bond.

The amides are sharp-melting crystalline compounds and make good derivatives for any of the acyl classes of compounds, i.e., esters, acids, acid halides, anhydrides, and lactones.

Amides undergo hydrolysis upon refluxing in H_2O. The reaction is catalyzed by acid or alkali.

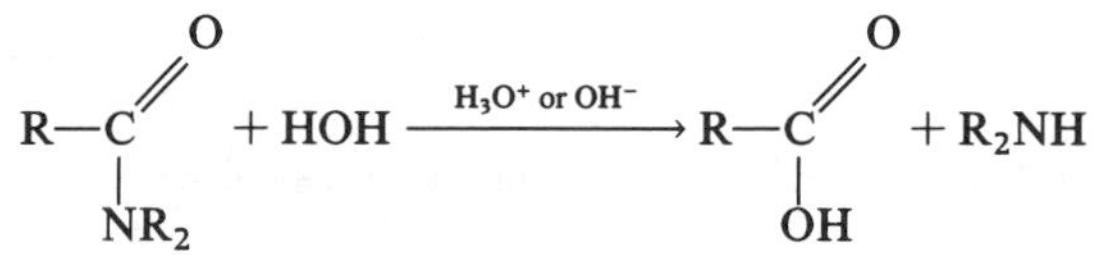

Primary amides may be dehydrated to yield nitriles.

$$R{-}CONH_2 + C_6H_5SO_2Cl \xrightarrow[70^\circ]{pyridine} R{-}CN + C_6H_5SO_3H + HCl$$

The reaction is run in pyridine solutions.

Primary and secondary amides of the type $RCONH_2$ and RCONHR react with nitrous acid in the same way as do the corresponding primary and secondary amines.

$$RCONH_2 + HONO \rightarrow RCOOH + N_2 + HOH$$

$$RCONHR + HONO \rightarrow RCON(NO)R + HOH$$

When diamides having their amide groups not far apart are heated, they lose ammonia to yield imides. See also **Imides.**

AMIKACIN. Antibiotic.

AMINATION. The process of introducing the amino group (—NH_2) into an organic compound is termed *amination.* An example is the reduction of aniline $C_6H_5 \cdot NH_2$ from nitrobenzene $C_6H_5 \cdot NO_2$. The reduction may be accomplished with iron and HCl. Only about 2% of the calculated amount of acid (to produce H_2 by reaction with iron) is required because of the fact that H_2O plus iron in the presence of ferrous chloride solution (ferrous and chloride ions) functions as the primary reducing agent. Such groups as nitroso (—NO), hydroxylamine (—NH·NH—), and azo (—N:N—) also yield amines by reduction. Amination also may be effected by the use of NH_3, a process sometimes referred to as *ammonolysis.* An example is the production of aniline from chlorobenzene:

$$C_6H_5Cl + NH_3 \rightarrow C_6H_5 \cdot NH_2 + HCl.$$

The reaction proceeds only under high pressure. In the ammonolysis of benzenoid sulfonic acid derivatives, an oxidizing agent is added to prevent the formation of soluble reduction products, such as $NaNH_4SO_4$, which commonly form. Oxygen-function compounds also may be subjected to ammonolysis: (1) methanol plus aluminum phosphate catalyst yields mono-, di-, and trimethylamines; (2) β-naphthol plus sodium ammonium sulfite catalyst (Bucherer reaction) yields β-naphthylamine; (3) ethylene oxide yields mono-, di-, and triethanolamines; (4) glucose plus nickel catalyst yields glucamine; and (5) cyclohexanone plus nickel catalyst yields cyclohexylamine.

AMINES. An amine is a derivative of NH_3 in which there is a replacement for one or more of the H atoms of NH_3 by an alkyl group, such as $—CH_3$ (methyl) or $—C_2H_5$ (ethyl); or by an aryl group, such as $—C_6H_5$ (phenyl) or $—C_{10}H_7$ (naphthyl). Mixed amines contain at least one alkyl and one aryl group as exemplified by methylphenylamine $CH_3 \cdot N(H) \cdot C_6H_5$. When one, two, and three H atoms are thus replaced, the resulting amines are known as *primary, secondary,* and *tertiary,* respectively. Thus, methylamine CH_3NH_2 is a primary amine; dimethylamine $(CH_3)_2NH$ is a secondary amine; and trimethylamine $(CH_3)_3N$ is a tertiary amine. Secondary amines sometimes are called *imines*; tertiary amines, *nitriles.*

Quaternary amines consist of four alkyl or aryl groups attached to an N atom and, therefore, may be considered substituted ammonium bases. Commonly, they are referred to in the trade as quaternary ammonium compounds. An example is tetramethyl ammonium iodide.

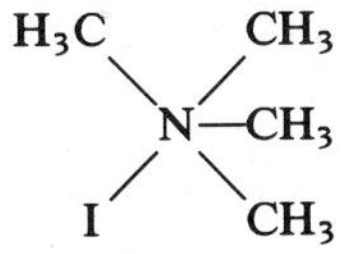

The amines and quaternary ammonium compounds, exhibiting such great versatility for forming substitution products, are very important starting and intermediate materials for industrial organic syntheses, both on a small scale for preparing rare compounds for use in research and on a tonnage basis for the preparation of resins, plastics, and other synthetics. Very important industrially are the ethanolamines which are excellent absorbents for certain materials. See also **Ethanolamines.** Hexamethylene tetramine is a high-tonnage product used in plastics production. See also **Hexamine.** Phenylamine (aniline), although not as important industrially as it was some years ago, still is produced in quantity. Melamine is produced on a large scale and is the base for a series of important resins. See also **Melamine.** There are numerous amines and quaternary ammonium compounds that are not well known because of their importance as intermediates rather than as final products. Examples along these lines may include acetonitrile and acrylonitrile. See also **Acrylonitrile.**

Primary amines react (1) with nitrous acid, yielding (a) with alkylamine, nitrogen gas plus alcohol, (b) with warm arylamine, nitrogen gas plus phenol (the amino-group of primary amines is displaced by the hydroxyl group to form alcohol or phenol) (c) with cold arylamine, diazonium compounds, (2) with acetyl chloride or benzoyl chloride, yielding substituted amides, thus, ethylamine plus acetyl chloride forms N-ethylacetamide $C_2H_5NHOCCH_3$, (3) with benzene-sulfonyl chloride $C_6H_5SO_2Cl$, yielding substituted benzene sulfonamides, thus, ethylamine forms N-ethylbenzenesulfonamide $C_6H_5SO_2—NHC_2H_5$, soluble in sodium hydroxide, (4) with chloroform $CHCl_3$ with a base, yielding isocyanides (5) with HNO_3 (concentrated), yielding nitramines, thus, ethylamine reacts to form ethylnitramine $C_2H_5—NHNO_2$.

Secondary amines react (1) with nitrous acid, yielding nitrosamines, yellow oily liquids, volatile in stream, soluble in ether. The secondary amine may be recovered by heating the nitrosamine with concentrated HCl, or hydrazines may be formed by reduction of the nitrosamines, e.g., methylaniline from methylphenylnitrosamine $CH_3(C_6H_5)NNO$, reduction yielding unsymmetrical methylphenylhydrazine, $CH_3(C_6H_5)NHNH_2$, (2) with acetyl or benzoyl chloride, yielding substituted amides, thus, diethylamine plus acetyl chloride to form N, N-diethylacetamide $(C_2H_5)_2—NOCCH_3$, (3) with benzene sulfonyl chloride, yielding substituted benzene sulfonamides, thus, diethylamine reacts to form N,N-diethylbenzenesulfonamide $C_6H_5SO_2N-(C_2H_5)_2$, insoluble in NaOH.

Tertiary amines do not react with nitrous acid, acetyl chloride, benzoyl chloride, benzenesulfonyl chloride, but react with alkyl halides to form quaternary ammonium halides, which are converted by silver hydroxide to quaternary ammonium hydroxides. Quaternary ammonium hydroxides upon heating yield (1) tertiary amine plus alcohol (or, for higher members, olefin plus water). Tertiary amines may also be formed (2) by alkylation of secondary amines, e.g., by dimethyl sulfate, (3) from amino acids by living organisms, e.g., decomposition of fish in the case of trimethylamine.

AMINES (Carcinogens). Carcinogens.

AMINO ACIDS. The scores of proteins which make up about one-half of the dry weight of the human body and that are so vital to life functions are made up of a number of amino acids in various combinations and configurations. The manner in which the complex protein structures are assembled from amino acids is described in the entry on **Protein.** For some users of this book, it may be helpful to scan that portion of the protein entry which deals with the chemical nature of proteins prior to considering the details of this immediate entry on amino acids.

Although the proteins resulting from amino acid assembly are ultimately among the most important chemicals in the animal body (as well as plants), the so-called infrastructure of the proteins is dependent upon the amino acid building blocks. Although there are many hundreds of amino acids, only about 20 of these are considered very important to living processes, of which six to ten are classified as essential. Another three or four may be classified as quasi-essential, and ten to twelve may be categorized as nonessential. As more is learned about the fundamentals of protein chemistry, the scientific importance attached to specific amino acids varies. Usually, as the learning process continues, the findings tend to increase the importance of specific amino acids. Actually, the words *essential* and *nonessential* are not very good choices for naming categories of amino acids. Generally, those amino acids which the human body cannot synthesize at all or at a rate commensurate with its needs are called *essential amino acids* (EAA). In other words, for the growth and maintenance of a normal healthy body, it is essential that these amino acids be ingested as part of the diet and in the necessary quantities. To illustrate some of the indefinite character of amino acid nomenclature, some authorities classify histidine as an essential amino acid; others do not. The fact is that histidine is essential for the normal growth of the human infant, but to date it is not regarded as essential for adults. By extension of the preceding explanation, the term nonessential is taken to mean those amino acids that are really synthesized in the body and hence need not be present in food intake. This classification of amino acids, although amenable to change as the results of new findings, has been quite convenient in planning the dietary needs of people as well as of farm animals, pets, and also in terms of those plants that are of economic importance. The classification has been particularly helpful in planning the specific nutritional content of food substances involved in various aid and related programs for the people in needy and underdeveloped areas of the world.

Food Fortification with Amino Acids. In a report of the World Health Organization, the following observation has been made: "To determine the quality of a protein, two factors have to be distinguished, namely, the proportion of essential to nonessential amino acids and, secondly, the relative amounts of the essential amino acids. . . . The best pattern of essential amino acids for meeting human requirements was that found in whole egg protein or human milk, and comparisons of protein quality should be made by reference to the essential amino acid patterns of either of these two proteins." The ratio of each essential amino acid to the total sum is given for hen's egg and human and cow's milk in Table 1.

In the human body, tyrosine and cysteine can be formed from phenylalanine and methionine, respectively. The reverse transformations do not occur Human infants have an ability to synthesize arginine

TABLE 1. REPRESENTATIVE ESSENTIAL AMINO ACID PATTERNS

A/E* RATIO (Milligrams per gram of total essential amino acids)

	Hen's Egg (Whole)	Human Milk	Cow's Milk
Total "aromatic" amino acids	195	226	197
Phenylalanine	(114)	(114)	(97)
Tyrosine	(81)	(112)	(100)
Leucine	172	184	196
Valine	141	147	137
Isoleucine	129	132	127
Lysine	125	128	155
Total "S"	107	87	65
Cystine	(46)	(43)	(17)
Methionine	(61)	(44)	(48)
Threonine	99	99	91
Tryptophan	31	34	28

SOURCE: World Health Organization; FAO Nutrition Meeting Report Series, No. 37, Geneva, 1965.

* A/E Ratio equals ten times percentage of single essential amino acid to the total essential amino acids contained.

and histidine in their bodies, but the speed of the process is slow compared with requirements.

Several essential amino acids have been shown to be the limiting factor of nutrition in plant proteins. In advanced countries, the ratio of vegetable proteins to animal proteins in foods is 1.4 : 1. In underdeveloped nations, the ratio is 3.5 : 1, which means that people in underdeveloped areas depend upon vegetable proteins. Among vegetable staple foods, wheat easily can be fortified. It is used as flour all over the world. L-Lysine hydrochloride (0.2%) is added to the flour. Wheat bread fortified with lysine is used in several areas of the world; in Japan it is supplied as a school ration.

The situation of fortification in rice is somewhat more complex. Before cooking, rice must be washed (polished) with water. In some countries, the cooking water is allowed to boil over or is discarded. This significant loss of fortified amino acids must be considered. L-Lysine hydrochloride (0.2%) and L-threonine (0.1%) are shaped like rice grain with other nutrients and enveloped in a film. The added materials must hold the initial shape and not dissolve out during boiling, but be easily freed of their coating in the digestive organs.

The amino acids are arranged in accordance with essentiality in Table 2. Each of the four amino acids at the start of the table are all limiting factors of various vegetable proteins. Chick feed usually is supplemented with fish meals, but where the latter is in limited supply, soybean meals are substituted. The demand for DL-methionine, limiting amino acid in soybean meals, is now increasing. When seed meals, such as corn and sorghum, are used as feeds for chickens or pigs, L-lysine hydrochloride must be added for fortification. Lysine production is increasing upward to the level of methionine.

Early Research and Isolation of Amino Acids. Because of such rapid studies made within the past few decades in biochemistry and nutrition, these sciences still have a challenging aura about them. But, it is interesting to note that the first two natural amino acids were isolated by Braconnot in 1820. As shown by Table 3, these two compounds were glycine and leucine. Bopp isolated tyrosine from casein in 1849. Additional amino acids were isolated during the 1880s, but the real thrust into research in this field commenced in the very late 1800s and early 1900s with the work of Emden, Fischer, Mörner, and Hopkins and Cole. It is interesting to observe that Emil Fischer (1852–1919), German chemist and pioneer in the fields of purines and polypeptides, isolated three of these important compounds, namely, proline from gelatin in 1901, valine from casein in 1901, and hydroxyproline from gelatin in 1902. As an understanding of the role of amino acids in protein formation and of the function of proteins in nutrition progressed, the pathway was prepared for further isolation of amino acids. For example, in 1907, a combined committee representing the American Society of Biological Chemists and the American Physiological Society, proposed a formal classification of proteins into three major categories: (1) simple proteins, (2) conjugated proteins, and (3) derived proteins. The last classification embraces all denatured proteins and hydrolytic products or protein breakdown and no longer is considered as a general class.

Very approximate annual worldwide production of amino acids, their current method of preparation (not exclusive), and general characteristics are given in Table 2.

The *isoelectric point* is very important in the preparation and separation of amino acids and proteins. Protein solubility varies markedly with pH and is at a minimum at the isoelectric point. By raising the salt concentration and adjusting pH to the isoelectric point, it is often possible to obtain a precipitate considerably enriched in the desired protein and to crystallize it from a heterogenous mixture.

Chemical Nature of Amino Acids

In a very general way, an amino acid is any organic acid which incorporates one or more amino groups. This definition includes a multitude of substances of most diverse structure. There are seemingly limitless related compounds of differing molecular size and constitution which incorporate varying kinds and numbers of functional groups. Most extensive study has centered around the relatively small group of alpha-amino acids which are combined in amide linkage to form proteins. With few exceptions, these compounds possess the general structure NH_2CHRCO_2H, where the amino group occupies a position on the carbon atom *alpha* to that of the carboxyl group, and where the side chain R may be of diverse composition and structure.

Few products of natural origin are as versatile in their behavior and properties as are the amino acids, and few have such a variety of biological duties to perform. Among their general characteristics would be included:

(a) Water-soluble and amphoteric electrolytes, with the ability to form acid salts and basic salts and thus act as buffers over at least two ranges of pH (hydrogen ion concentration).
(b) Dipolar ions of high electric moment with a considerable capacity to increase the dielectric constant of the medium in which they are dissolved.
(c) Compounds with reactive groups capable of a wide range of chemical alterations leading readily to a large variety of degradation, synthetic, and transformation products, such as esters, amides, amines, anhydrides, polymers, polypeptides, diketopiperazines, hydroxy acids, halogenated acids, keto acids, acylated acids, mercaptans, shorter- or longer-chained acids, and pyrrolidine and piperidine ring forms.
(d) Indispensable components of the diet of all animals including humans.
(e) Participants in crucial metabolic reactions on which life depends, and substrates for a variety of specific enzymes in vitro.
(f) Binders of metals of many kinds.
(g) Absorbers of ultraviolet and infrared radiation within specific ranges of wavelength.
(h) Possessors with one exception of optical rotatory power related to the configuration of asymmetric centers. The exception is glycine.
(i) Essential constituents of protein molecules whose biological and chemical specificities are determined in part by the number, distribution, and spatial interrelations of the amino acids of which they are composed.

They reveal at once uniformity and diversity. Uniformity, because with rare exceptions they are α-amino acids with all the physical consequences which flow from this fact and because, for those that are constituents of proteins and hence of living tissues, the same optical configuration at the α-carbon atom is common to all. Diversity, because each possesses a different side chain which confers upon it unique properties distinguishing it physically, chemically, and biologically from the others. In this duality, the array of the amino acids is a partial reflection of the larger biological world which is "always the same and yet always different."

TABLE 2. IMPORTANT NATURAL AMINO ACIDS AND PRODUCTION

Amino Acid	World Annual Production, tons	Present Mode of Manufacture	Characteristics
		ESSENTIAL AMINO ACIDS	
DL-Methionine	10^4	Synthesis from acrolein and mercaptan	First limiting amino acid for soybean
L-Lysine·HCl	10^3	Fermentation (AM)*	First limiting amino acid for all cereals
L-Threonine	10	Fermentation (AM)	Second limiting amino acid for rice
L-Tryptophan	10	Synthesis from acrylonitrile and resolution	Second limiting amino acid for corn
L-Phenylalanine	10	Synthesis from phenyl-acetaldehyde and resolution	
L-Valine	10	Fermentation (AM)	Rich in plant protein
L-Leucine	10	Extraction from protein	
L-Isoleucine	10	Fermentation (WS)**	Deficient in some cases
		QUASI-ESSENTIAL AMINO ACIDS	
L-Arginine·HCl	10^2	Synthesis from *L*-ornithine Fermentation (AM)	Essential to human infants
L-Histidine·HCl	10	Extraction from protein	
L-Tyrosine	10	Enzymation of phenol and serine	Limited substitute for phenylalanine
L-Cysteine L-Cystine	10	Extraction from human hair	Limited substitute for methionine
		NONESSENTIAL AMINO ACIDS	
L-Glutamic acid	10^5	Fermentation (WS) Synthesis from acrylonitrile and resolution	MSG, taste enhancer
Glycine	10^3	Synthesis from formaldehyde	Sweetener
DL-Alanine	10^2	Synthesis from acetaldehyde	
L-Aspartic acid	10^2	Enzymation of fumaric acid	Hygienic drug
L-Glutamine	10^2	Fermentation (WS)	Anti-gastroduodenal ulcer drug
L-Serine	< 10	Synthesis from glycolonitrile and resolution	Rich in raw silk
L-Proline	< 10	Fermentation (AM)	Rich in gelatin
L-Hydroxyproline	< 10	Extraction from gelatin	
L-Asparagine	< 10	Synthesis from L-aspartic acid	Neurotropic metabolic regulator
L-Alanine	< 10	Enzymation of L-aspartic acid	Rich in degummed white silk
L-Dihydroxy-phenylalanine	10^2	Synthesis from piperonal, vanillin, or acrylonitrile and resolution	Specific drug for Parkinson's disease
L-Citrulline	< 10	Fermentation (AM)	Ammonia detoxicant
L-Ornithine	< 10	Fermentation (AM)	

* AM, artificial mutant; ** WS, wild strain. MSG, monosodium glutamate.

Optical Properties. With the exception of glycine ($NH_2CH_2CO_2H$) and amino-malonic acid [$NH_2CH(CO_2H)_2$], all α-amino acids which are classifiable according to the general formula previously given exist in at least two different optically isometric forms. The optical isomers of a given amino acid possess identical empirical and structural formulas, and are indistinguishable from each other on the basis of their chemical and physical properties, with the singular exception of their effect on plane polarized light. This may be illustrated with the two optically active forms as shown by

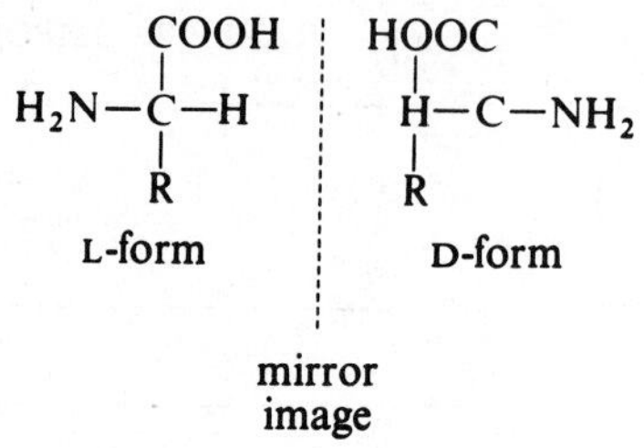

TABLE 3. FIRST ISOLATION OF AMINO ACIDS

ABBREVIATION	NAME AND FORMULA	FIRST ISOLATION AND (SOURCE)	ISOELECTRIC POINT
	Neutral Amino Acids—Aliphatic Type		
Ala	Alanine $CH_3-CH-COOH$ | NH_2	1879 by Schutzenberger 1888 by Weyl (silk fibroin)	6.0
Gly	Glycine NH_2-CH_2-COOH	1820 by Braconnot (gelatin)	6.0
Ile	Isoleucine H | $C_2H_5-C-CH-COOH$ | | H_3C NH_2	1904 by Ehrlich (fibrin)	6.0
Leu	Leucine $(CH_3)_2CH-CH_2-CH-COOH$ | NH_2	1820 by Braconnot (muscle fiber; wool)	6.0
Val	Valine $(CH_3)_2CH-CH-COOH$ | NH_2	1901 by Fischer (casein)	6.0
	Neutral Amino Acids—Hydroxy Type		
Ser	Serine $HO-CH_2-CH-COOH$ | NH_2	1865 by Cramer (sericine)	5.7
Thr	Threonine $CH_3-CH-CH-COOH$ | | OH NH_2	1925 by Gortner and Hoffman 1925 by Schryver and Buston (oat protein)	6.2
	Neutral Amino Acids—Sulfur-Containing Type		
Cys	Cysteine $HS-CH_2-CH-COOH$ | NH_2	------------------------------	5.1
Cys Cys	Cystine $(-SCH_2-CH-COOH)_2$ | NH_2	1899 by Mörner (horn) 1899 by Emden	4.6
Met	Methionine $CH_3-S-CH_2-CH_2-CH-COOH$ | NH_2	1922 by Mueller (casein)	5.7

(*continued*)

TABLE 3. FIRST ISOLATION OF AMINO ACIDS (*continued*)

Abbreviation	Name and Formula	First Isolation and (Source)	Isoelectric Point
	Neutral Amino Acids—Amide Type		
Asn	Asparagine $H_2NOC-CH_2-CH(NH_2)-COOH$	1932 by Damodaran (edestin)	5.4
Gln	Glutamine $H_2NOC-CH_2-CH_2-CH(NH_2)-COOH$	1932 by Damodaran, Jaaback, and Chibnall (gliadin)	5.7
	Neutral Amino Acids—Aromatic Type		
Phe	Phenylalanine 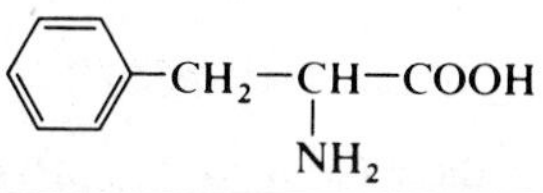	1881 by Schulze and Barbieri (lupine seedlings)	5.5
Trp	Tryptophan 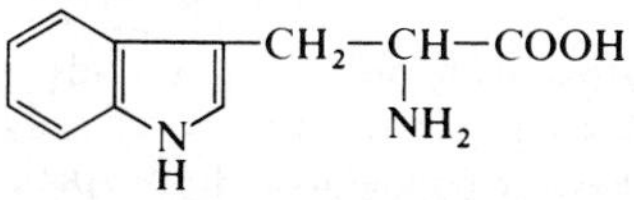	1902 by Hopkins and Cole (casein)	5.9
Tyr	Tyrosine 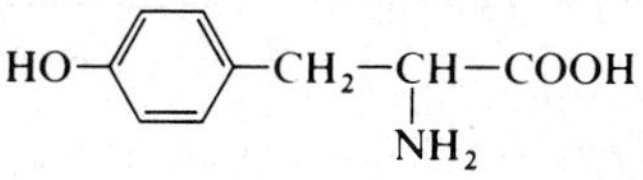	1849 by Bopp (casein)	5.7
	Acidic Amino Acids		
Asp	Aspartic Acid $HOOC-CH_2-CH(NH_2)-COOH$	1868 by Ritthausen (conglutin; legumin)	2.8
Glu	Glutamic acid $HOOC-CH_2-CH_2-CH(NH_2)-COOH$	1866 by Ritthausen (gluten-fibrin)	3.2
	Basic Amino Acids		
Arg	Arginine $H_2N-C(=NH)-NH(CH_2)_3-CH(NH_2)-COOH$	1895 by Hedin (horn)	11.2
His	Histidine 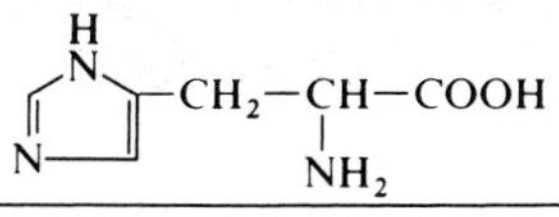	1896 by Kossel (sturine) 1896 by Hedin (various protein hydrolysates)	7.6
Lys	Lysine $H_2N-(CH_2)_4CH(NH_2)-COOH$	1889 by Dreschel (casein)	9.7
	Imino Acids		
Hyp	Hydroxyproline 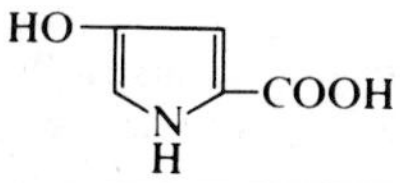	1902 by Fischer (gelatin)	5.8
Pro	Proline 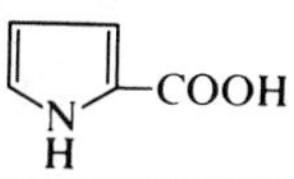	1901 by Fischer (casein)	6.3

TABLE 4. STRUCTURAL CLASSIFICATION OF AMINO ACIDS

NEUTRAL AMINO ACIDS		
Aliphatic-type	*Hydroxy-type*	*Sulfur-containing*
Glycine	Serine	Cysteine
Alanine	Threonine	Cystine
Valine		Methionine
Leucine		
Isoleucine		
Amide-type	*Aromatic-type*	
Asparagine	Phenylalanine	
Glutamine	Tryptophan	
	Tyrosine	
ACIDIC AMINO ACIDS		
Aspartic acid		
Glutamic acid		
BASIC AMINO ACIDS		
Histidine		
Lysine		
Arginine		
IMINO ACIDS		
Proline		
Hydroxyproline		

One form (L-form) exhibits the ability to rotate the plane of polarization of plane polarized light to the left (levorotatory), whereas the other form (D-form) rotates the plane to the right (dextrorotatory). Although the direction of optical rotation exhibited by these optically active forms is different, the magnitude of their respective rotations is the same. If equal amounts of *dextro* and *levo* forms are admixed, the optical effect of each isomer is neutralized by the other, and an optically inactive product known as a *racemic modification* or *racemate* is secured.

The ability of the alanine molecule, for example, to exist in two stereoisomeric forms can be attributed to the fact that the α-carbon atom of this compound is attached to four different groups which may vary in their three-dimensional spatial arrangement. Compounds of this type do not possess complete symmetry when viewed from a purely geometrical standpoint and hence are generally referred to as *asymmetric.* As a consequence of this molecular asymmetry, the four covalent bonds of an asymmetric carbon atom can be aligned in a manner such that a regular tetrahedron is formed by the straight lines connecting their ends. Hence, two different tetrahedral arrangements of the groups about the asymmetric carbon atom can be devised so that these structures relate to one another as an object relates to its mirror image, or as the right hand relates to the left hand. Molecules of this type are endowed with the property of optical activity and, together with their nonsuperimposable mirror images, are generally referred to as *enantiomorphs, enantiomers, antimers,* or *optical antipodes.*

Classification. In accordance with the structure of the R-group, the amino acids of primary importance can be classified into eight groups. Additional amino acids composing protein are not included in this classification, because they occur infrequently. See Table 4.

Normally, amino acids exist as dipolar ions. $RCH(NH_3^+)COO^-$, in a neutral state, where both amino and carboxyl groups are ionized. The dipolar form, $RCH(NH_2)COOH$ may be considered, but the dipolar form predominates for the usual monoamino monocarboxylic acid and it is estimated that these forms occur 10^5 to 10^6 times more frequently than the nonpolar forms. Amino acids decompose thermally at what might be considered a relatively high temperature (200–300°C). The compounds are practically insoluble in organic solvents, have low vapor pressure, and do not exhibit a precisely defined melting point.

The ionic states of a simple α-amino acid are given by

$$\underset{\text{(Cationic form; acidic)}}{RCH(NH_3^+)COOH} \underset{+H^+}{\overset{-H^+(K_1)}{\rightleftharpoons}}$$

$$\underset{\text{(Dipolar form; neutral)}}{RCH(NH_3^+)COO^-} \underset{+H^+}{\overset{-H^+(K_2)}{\rightleftharpoons}} \underset{\text{(Anionic form; basic)}}{RCH(NH_2)COO^-}$$

In accordance with the change of the ionic state, dissociation constants are

$$K_1(COOH) = \frac{[H^+][RCH(NH_3^+)COO^-]}{[RCH(NH_3^+)COOH]}$$

$$K_2(NH_3^+) = \frac{[H^+][RCH(NH_2)COO^-]}{[RCH(NH_3^+)COO^-]}$$

Inasmuch as $pK = -\log K$, the values for glycine are $pK_1 = 2.34$ and $pK_2 = 9.60$ (in aqueous solution at 25°C). The homologous amino acids indicate similar values. The pH at which acidic ionization balances basic ionization is termed the *isoelectric point* (pH_I), corresponding to

$$[RCH(NH_3^+)COOH] = [RCH(NH_2)COO^-]$$

Thus, from these formulas, the pH_I is

$$pH_I = 1/2(pK_1 + pK_2)$$

Formation of Salts. Amino acids have certain characteristics of both organic bases and organic acids because they are amphoteric. As amines, the amino acids form stable salts, such as hydrochlorides or aromatic sulfonic acid salts. These are used as selective precipitants of certain amino acids. As organic acids, the amino acids form complex salts with heavy metals, the less soluble salt being used for amino acid separation.

Esters. When heated with the equivalent amount of a strong acid, usually hydrochloric acid in absolute alcohol, amino acids form esters. These are obtained as hydrochlorides.

Acylation. In alkaline solution, amino acids react with acid chlorides or acid anhydrides to form acyl compounds of the type

$$\begin{array}{l} RCH\text{—}COONa \\ \;\;\;| \\ \;\;NHCOR' \end{array}$$

Van Slyke Reaction (Deamination). With excess nitrous acid, -amino acids react to form -hydroxyl acids on a quantitative basis. Nitrogen gas is generated.

$$\begin{array}{l} RCH\text{—}COOH \\ \;\;\;| \\ \;\;NH_2 \end{array} + HNO_2 \rightarrow \begin{array}{l} RCH\text{—}COOH + N_2 + H_2O \\ \;\;\;| \\ \;\;OH \end{array}$$

The reaction is completed within five minutes at room temperature. Thus, measurement of the volume of nitrogen generated can be used in amino acid determinations.

Decarboxylation. When heated with inert solvents, such as kerosene, amino acids form amines

$$\begin{array}{l} RCH\text{—}COOH \\ \;\;\;| \\ \;\;NH_2 \end{array} \rightarrow \begin{array}{l} RCH_2 + CO_2 \\ \;\;\;| \\ \;\;NH_2 \end{array}$$

Decarboxylative enzymes may react specifically with amino acids having free polar groups at the ω position. Cadaverine can be produced from lysine, histamine from histidine, and tyramine from tyrosine.

Formation of Amides. When condensed with ammonia or amines, amino acid esters form acid amides:

$$\begin{array}{l} RCH\text{—}CONH_2 \\ \;\;\;| \\ \;\;NH_2 \end{array}$$

Oxidation. Oxidizing agents easily decompose α-amino acids, forming the corresponding fatty acid with one less carbon number:

$$\begin{array}{l} RCH\text{—}COOH \\ \;\;\;| \\ \;\;NH_2 \end{array} \overset{O}{\rightarrow} RCHO \overset{O}{\rightarrow} RCOOH + NH_3 + CO_2$$

Ninhydrin Reaction. A neutral solution of an amino acid will react with ninhydrin (triketohydrindene hydrate) by heating to cause oxidative decarboxylation. The central carbonyl of the triketone is reduced to an alcohol. This alcohol further reacts with ammonia formed from the amino acid and causes a red-purplish color. Since the reaction is quantitative, measurement of the optical density of the color produced is an indication of amino acid concentration. Imino acids, such as hydroxyproline and proline, develop a yellow color in the same type of reaction.

Maillard Reaction. In amino acids, the amino group tends to form condensation products with aldehydes. This reaction is regarded as the cause of the browning reaction when an amino acid and a sugar coexist. A characteristic flavor, useful in food preparations, is evolved along with the color in this reaction.

Ion-exchange Separations. Because amino acids are amphoteric, they behave as acids or bases, depending upon the pH of the solution. This makes it possible to adsorb amino acids dissolved in water on either a strong-acid cation exchange resin; or a strong-base anion exchange resin. The affinity varies with the amino acid and the solution pH. Ion-exchange resins are widely used in amino acid separations.

Production of Amino Acids. There are three means available for making (or separating) amino acids in large quantity lots: (1) *extraction* from natural protein; (2) *fermentation*; and (3) chemical *synthesis*. During the early investigations of amino acids, the first method was widely used and still applies to four amino acids. See Table 3.

L-Leucine is easily extracted in quantity from almost any type of vegetable protein hydrolyzates. Cystine is extracted from the human-hair hydrolyzate. L-Histidine is obtainable from the blood of animals, but future yields may stem from fermentation inasmuch as some artificial mutants of bacteria have been discovered. Gelatin is the prime source of L-hydroxyproline.

Natural amino acids, normally not contained in proteins, but which are effective in medicine, include citrulline, ornithine, and dihydroxyphenylalanine. These are not listed in Tables 2 and 3. Citrulline (Cit) with an isoelectric point of 5.9 was isolated by Koga in 1914; by Odake in 1914; and by Wada in 1930. It has the formula

$$H_2N\underset{\displaystyle \overset{\|}{O}}{C}-NH-(CH_2)_3\underset{\displaystyle \overset{|}{NH_2}}{CH}-COOH$$

Dihydroxyphenylalanine (Dopa) with an isoelectric point of 5.5 was isolated by Torquati in 1913; and by Guggenheim in 1913. It has the formula

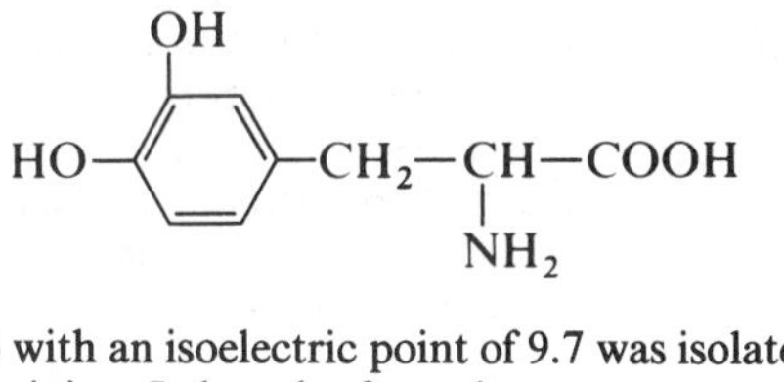

Ornithine (Orn) with an isoelectric point of 9.7 was isolated by Riesser in 1906 from arginine. It has the formula

$$\underset{\displaystyle \overset{|}{NH_2}}{CH_2}-CH_2-CH_2-\underset{\displaystyle \overset{|}{NH_2}}{CH}-COOH$$

Fermentation Methods. Numerous microorganisms can synthesize the amino acids required to support their life from a simple carbon source and an inorganic nitrogen source, such as ammonium or nitrate salts, or nitrogen gas.

Japanese microbiologists, in 1956, first succeeded in developing industrial production of L-glutamic acid by a microbiological process. As of the present, nearly all common amino acids can be produced on a low cost industrial scale by fermentation. From microbiological studies, it has been ascertained that some microbial stains isolated from natural sources serve to excrete and accumulate a large amount of a particular amino acid in the cultural broth under carefully controlled conditions. The production of glutamic acid is produced by adding a selected bacterial strain and culturing aerobically for one to two days in a chemically defined medium which contains carbon sources, such as sugar or acetate, and nitrogen sources, such as ammonium salts. About 50% (wt) of the carbon sources can be converted to glutamate.

Genetic techniques have been used to improve the ability of microorganisms to accumulate amino acids. Several amino acids are manufactured from their direct precursors by the use of microbially produced enzymes. For example, bacterial L-aspartate β-carboxylase is used for the production of L-alanine from L-aspartic acid.

In isolating the amino acids from the fermentation broth, chromatographic separations using ion-exchange resins are the most important commercial method. Precipitation with compounds which yield insoluble salts with amino acids are also used. Purification is possible by crystallization through careful adjustment of the isoelectric point, at which point the amino acid is least soluble.

There are several laboratory-size methods for synthesizing amino acids, but few of these have been scaled up for industrial production. Glycine and DL-alanine are made by the Strecker synthesis, commencing with formaldehyde and acetaldehyde, respectively. In the Strecker synthesis, aldehydes react with hydrogen cyanide and excess ammonia to give amino nitriles which, in turn, are converted into α-amino acids upon hydrolysis.

$$R\underset{\displaystyle \overset{\|}{O}}{CH} \xrightarrow{HCN} R\underset{\displaystyle \overset{|}{OH}}{CH}-CN \xrightarrow{NH_3} R\underset{\displaystyle \overset{|}{NH_2}}{CH}-CN \xrightarrow{NaOH} R\underset{\displaystyle \overset{|}{NH_2}}{CH}-COONa$$

The Hydantoin Process. Hydantoins are produced by reacting aldehydes with sodium cyanide and ammonium carbonate. Upon hydrolysis, α-amino acids will be yielded

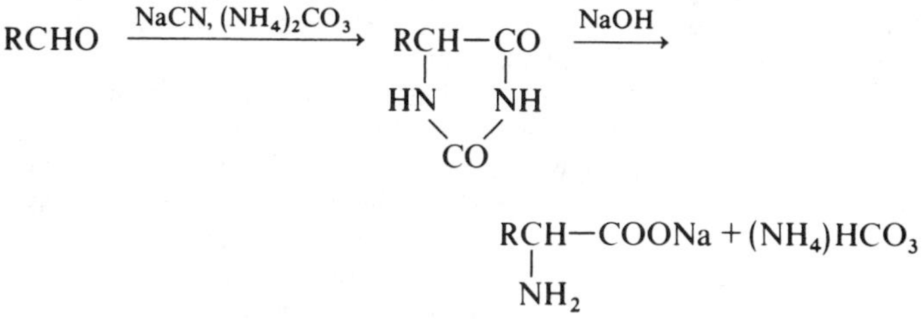

The production of α-amino acids by chemical synthesis yields a mixture of DL forms. The D-form of glutamic acid has no flavor-enhancing properties and thus requires transformation into the optically active form insofar as monosodium glutamate is concerned. The three methods for separating the optical isomers are: (1) preferential inoculation method; (2) the diasteroisomer method; and (3) the acylase method.

References

Adams, C. P.: "Nutritive Value of American Foods," Agriculture Handbook 456, U.S. Department of Agriculture, Washington, D.C., 1975.

Kolata, G. B.: "Protein Structure: Systematic Alteration of Amino Acid Sequences," *Science*, **191**, 373 (1976).

Orr, M. L., and B. K. Watt: "Amino Acid Content of Foods," Home Economics Research Report 4, U.S. Department of Agriculture, Washington, D.C. (revised periodically).

Paul, P. C., and H. H. Palomer (editors): "Food Theory and Applications," Wiley, New York, 1972.

Staff: "Amino-Acid Content of Foods and Biological Data on Proteins," Nutritional Studies No. 24, Food and Agriculture Organization (United Nations), Rome, 1976.

Staff: "Catalog of Food and Nutrition Information and Educational Materials Center," U.S. Department of Agriculture, National Agricultural Library, Beltsfille, Maryland (revised periodically).

Tabor, H., and C. W. Tabor (editors): "Metabolism of Amino Acids and Amines," Academic, New York, 1971.

Villee, C. A.: "Biology," 7th edition, Saunders, Philadelphia, 1977.

Weissbach, H., and S. Pestka, (editors): "Molecular Mechanisms of Protein Biosynthesis," Academic, New York, 1977.

Hauromi Oeda, Ajinomoto Co., Inc., Kawaski, Japan.

AMINO ACID TRANSMITTERS. Brain and Nervous System.

AMINO CARBOXYLIC ACIDS. **Carboxylic Acids.**

AMINOGLYCOSIDES. **Antibiotic.**

AMINO RESINS. A family of resins resulting from an addition reaction between formaldehyde and compounds, such as aniline, ethylene urea, dicyandiamide, melamine, sulfonamide, and urea. The resins are thermosetting and have been used for many years in such products as textile-treating agents, laminating coatings, wet-strength paper coatings, and wood adhesives. The urea and melamine compounds are the most widely used. Both of these basic resins are water white (transparent). However, the resins readily accept pigments and opacifying agents. The addition of cellulose filler can be used to reduce light transmission. Where color is unimportant, various materials are added to the melamine resin compounds, including macerated fabric, glass fiber, and wood flour. Wood flour frequently is added to the urea resins to yield a low cost industrial material.

Advantages claimed for amino resins include: (1) good electrical insulation characteristics, (2) no transfer of tastes and odors to foods, (3) self-extinguishing burning characteristics, (4) resistance to attack by oils, greases, walk alkalis and acids, and organic solvents, (5) abrasion resistance, (6) good rigidity, (7) easy fabrication by economical molding procedures, (8) excellent resistance to deformation under load, (9) good subzero characteristics with no tendency to become brittle, and (10) marked hardness.

Amino resins are fabricated principally by transfer and compression molding. Injection molding and extrusion are used on a limited scale. Urea resins are not recommended for outdoor exposure. The resins show rather high mold shrinkage and some shrinkage with age. The melamines are superior to the ureas insofar as resistance to heat and boiling water, acids, and alkalis is concerned.

Some of the hundreds of applications for amino resins include: Closures for glass, metal, and plastic containers; electrical wiring devices; appliance knobs, dials, handles, and push buttons; lamp shades and lighting diffusers; organ and piano keys; dinnerware; food service trays; food-mixer housings; switch parts; decorative buttons; meter blocks; aircraft ignition parts; heavy duty switch gear; connectors; and terminal strips. Not all of the urea or melamine amino resins are suited to all of the foregoing uses. Because of the large number of fillers and additives available, the overall range of use of this family of resins is large.

AMITOSIS. Cell division without mitosis; a splitting of the cell into two parts without the previous duplication of chromosomes and segregation of the duplicated chromosomes into two separate groups. Formerly amitosis was thought to be the method of cell division for many of the simpler, one-celled forms of life. Improved techniques of microscopic study, however, have shown that there is some form of mitosis even in these simple forms. Today amitosis is recognized as a rare and abnormal form of cell division which produces cells with a limited survival.

AMMETER. **Electrical Instruments.**

AMMINES. Dry ammonia gas reacts with dehydrated salts of some of the metals to form solid ammines. Ammines, upon warming, evolve ammonia, sometimes with final decomposition of the salt itself, in a manner analogous to the decomposition of certain hydrates. The ammines of chromium(III)(Cr^{3+}), cobalt(III), platinum(IV), and other metals have been studied in detail. Two series of ammines are shown in the accompanying diagram, the first being one in which the neutral ammonia group is replaced step by step by the negative nitro group (NO_2^-), and the second, one in which the neutral ammonia group is replaced step by step by the neutral H_2O group.

The neutral group of the complex may be replaced step by step by the following negative groups: Cl^-, Br^-, I^-, F^-, OH^-, NO_2^-, NO_3^-, CN^-, CNS^-, SO_4^{2-}, CO_3^{2-}, $C_2O_4^{2-}$; or by the following neutral groups: H_2O, NO, NO_2, SO_2, S, N_2H_4, H_2NOH, CO, C_2H_5OH, C_6H_6. All neutral groups are of substances capable of independent existence.

(a)	(b)
$[Co(NH_3)_6]Cl_3$ 410	
$[Co(NH_3)_5(NO_2)]Cl_2$ 240	
$[Co(NH_3)_4(NO_2)_2]Cl$ 95	
$[Co(NH_3)_3(NO_2)_3]$ 1.5	
$K[Co(NH_3)_2(NO_2)_4]$ 95	$[Cr(NH_3)_6]X_3$
	$[Cr(NH_3)_5(H_2O)]X_3$
$K_2[Co(NH_3)(NO_2)_5]$ 240	$[Cr(NH_3)_4(H_2O)_2]X_3$
	$[Cr(NH_3)_3(H_2O)_3]X_3$
	$[Cr(NH_3)_2(H_2O)_4]X_3$
$K_3[Co(NO_2)_6]$ 420	X = unit anion

Two series of ammines: (a) Square bracket contains the ion. The equivalent electrical conductivity is shown below each compound. The number of neutral groups, e.g., (NH_3), on metal, e.g., Co, is varied from 6 to 0. (b) The number of neutral groups is constant, but the groups are varied.

In the ammines, trivalent metals, such as cobalt(III) and chromium(III) and iron(III), possess a coordination number of 6, this number being the sum of the unit replacements on the metal in the complex ion. Since a regular octahedron has six corners equidistant from the center, it is assumed that the metal occupies the center and each of the six replacing groups occupies a corner of a regular octahedron. Support for this assumption is offered by the x-ray examination of these ammines. When there is only one of the six groups replaced by a second group, as in $[Co(NH_3)_5(NO_2)]Cl_2$, and in $[Cr(NH_3)_5(H_2O)]X_3$ the octahedral placement of groups supplies only one form, but when two of the six groups are replaced by a second group, as in $[Co(NH_3)_4(NO_2)_2]Cl$, and in $[Cr(NH_3)_4(H_2O)_2]X_3$, two different octahedral corner arrangements are possible depending upon whether the two replacing groups are adjacent (cisform) or opposite (transform). Two substances differing in physical properties and corresponding to these two forms are known. Further, when three divalent groups, e.g., $3C_2O_4^2$Â are present in the complex, two arrangements—not identical but mirror-images of each other—are possible. Two optically active substances are known in such cases corresponding to these two stereoisomeric forms.

Six is the ordinary coordination number for metallic ammines and similar complexes. Additional examples are $K_2[Pt(NH_3)_2(CN)_4]$, $[Ni(NH_3)_6]Cl_2$, $K_4[Fe(CN)_6]$, $K_3[Fe(CN)_6]$, $K_2[Fe(CN)_5(NO)]$, $K_2[SiF_6]$, $[Ca(NH_3)_6]Cl_2$. But, for the elements boron, carbon, and nitrogen four is the coordination number, e.g., $[BH_4]Cl$, $[CH_4]$, $[NH_4]Cl$, and in these substances the groups are assumed to occupy the corners of a regular tetrahedron; in $K_4[Mo(CN)_8]$ and $[Ba(NH_3)_8]Cl_2$ the coordination number is eight, and the groups are assumed to occupy the corners of a cube.

AMMONIA. Known since ancient times, ammonia, NH_3, has been commercially important for well over 100 years and has become the second largest chemical in terms of tonnage and the first chemical in value of production. The first practical plant of any magnitude was built in 1913. Worldwide production of NH_3 as of the early 1980s is estimated at 100 million metric tons per year or more, with the United States accounting for about 15% of the total production. A little over three-fourths of ammonia production in the United States is used for fertilizer, of which nearly one-third is for direct application. An estimated 5.5% of ammonia production is based in the manufacture of fibers and plastics intermediates.

Properties. At standard temperature and pressure, NH_3 is a colorless gas with a penetrating, pungent-sharp odor in small concentrations which, in heavy concentrations, produces a smothering sensation when inhaled. Formula weight is 17.03, mp −77.7°C, bp −33.35°C, and sp gr 0.817 (at −79°C) and 0.617 (at 15°C). Ammonia is very soluble in water, a saturated solution containing approximately 45% NH_3 (weight) at the freezing temperature of the solution and about 30% (weight) at standard conditions. Ammonia dissolved in water forms a strongly alkaline solution of ammonium hydroxide, NH_4OH. The univalent radical NH_4^+ behaves in many respects like K^+ and Na^+ in vigorously reacting with acids to form salts. Ammonia is an excellent nonaqueous electrolytic solvent, its ionizing power approaching that of water. Ammonia burns with a greenish-yellow flame.

Ammonia derives its name from sal ammoniac, NH_4Cl, the latter material having been produced at the Temple of Jupiter Ammon (Libya) by distilling camel dung. During the Middle Ages, NH_3 was referred to as the spirits of hartshorn because it was produced by heating the hoofs and horns of oxen. The composition of ammonia was first established by Claude Louis Berthollet (France, ca. 1777). The first significant commercial source of NH_3 (during the 1880s) was its production as a by-product in the making of manufactured gas through the destructive distillation of coal. See also **Coal Tar and Derivatives.**

Nitrogen fixation is a term assigned to the process of converting nitrogen in the air to nitrogen compounds. Although some bacteria in soil are capable of this process, N_2 as an ingredient of fertilizer is required for soils that are depleted by crop production. The production of synthetic NH_3 is the most important industrial nitrogen-fixation process. See also **Fertilizers.**

Synthesis of Ammonia

The first breakthrough in the large-scale synthesis of ammonia resulted from the work of Fritz Haber (Germany, 1913), who found that ammonia could be produced by the direct combination of two elements, nitrogen and hydrogen, ($N_2 + 3H_2 \rightleftharpoons 2NH_3$) in the presence of a catalyst (iron oxide with small quantities of cerium and chromium) at a relatively high temperature (550°C) and under a pressure of about 200 atmospheres, representing difficult processing conditions for that era. Largely because of the urgent requirements for ammonia in the manufacture of explosives during World War I, the process was adapted for industrial-quality production by Karl Bosch, who received one-half of the 1931 Nobel Prize for chemistry in recognition of these achievements. Thereafter, many improved ammonia-synthesis systems, based on the Haber-Bosch process, were commercialized, using various operating conditions and synthesis-loop designs.

The principal features of an NH_3 synthesis process system are the converter designs, operating conditions, method of product recovery, and type of recirculation equipment. Most current systems operate at or above the pressure used in the original Haber-Bosch process. Converter designs have either a single continuous catalyst bed, which may or may not have heat-exchange cooling for controlling reaction heat, or several catalyst beds with provision for temperature control between the beds.

Claude Process. The original Claude process was one of the first systems to use a high operating pressure (1000 atmospheres), achieving 40% conversion without recycling. This system used multiple converters in a series-parallel arrangement. The present Claude process[1] operates at 340–650 atmospheres, using a single converter with continuous catalyst-charged tubes externally cooled to remove the heat of reaction. Approximate hydrogen conversion is 30–34 mole percent per pass. The pressure is increased gradually to compensate for catalyst aging and loss in activity. Product recovery is by simple condensation in a water-cooled condenser. Unreacted gas is recycled by compressor.

Casale Process. This is another high-pressure conversion system, using synthesis pressures of 450–600 atmospheres, which also permits hydrogen conversions in the 30 mole percent range. As in the Claude process, the high pressure allows NH_3 to be recovered from the converter effluent by water cooling. The Casale converter uses a single catalyst bed with internal heat-exchange surfaces. Reaction rate and temperature rise across the catalyst is controlled by the internal exchanger and retaining 2–3 mole percent NH_3 in the converter feed. This eliminates the need for a mechanical recycle compressor, but requires high feed-gas pressures to supply the energy required for the ejector.

Low-Pressure Processes. Several systems use low synthesis pressures with hydrogen conversion below 30 mole percent and product recovery by water and refrigeration.

Synthesis-Gas-Production Processes. These processes were improved and developed as a result of changes in feedstock availability and economics. Before World War II, most NH_3 plants obtained H_2 by reacting coal or coke with steam in the water-gas process. A small number of plants used water electrolysis or coke-oven by-product hydrogen. The subsequent low-cost availability of natural gas brought about steam-hydrocarbon reforming as the major source of H_2 for the NH_3 synthesis gas.

Partial oxidation processes to produce H_2 from natural gas and liquid hydrocarbons were also developed after World War II and accounted for 15% of the synthetic NH_3 capacity by 1962. The steam-hydrocarbon reforming process[2] was developed in 1930. In this process, methane was mixed with an excess of steam at atmospheric pressure, and the mixture reformed inside nickel-catalyst-filled alloy furnace tubes. The heat of reaction was supplied by externally heating the catalyst-filled tubes to about 871°C. Since the late 1950s, improvements in the tubular-reforming technology and metallurgy have brought about the utilization of high-pressure (> 24 atmospheres) reforming, which cut synthesis-gas-compression costs and increased heat recovery. The first pressure reformer[3] was built in 1953. In addition, the higher pressures allowed improvements in the efficiency of synthesis-gas-purification systems. High-pressure steam-reforming technology also has been extended to cover heavier hydrocarbon gases, including propane, butane, reformer gases, and streams containing a high amount of olefins. In 1962, a process[4] for reforming straight-run liquid distillates (naphthas) was commercialized. This process is based on the use of an alkali oxide-promoted nickel catalyst[5] which permits reforming of desulfurized naphthas at low (~3.5:1) steam-to-carbon ratios, without significant carbon deposition problems.

Noncatalytic partial oxidation processes designed to produce H_2 from a wide range of hydrocarbon liquids, including heavy fuel oils, crudes, naphthas, coal tar, and pulverized bituminous coal, were commercialized in 1954[6] and 1956.[7] In both these processes, the hydrocarbon feed is oxidized and reformed in a refractory-lined pressure vessel. The required oxygen usually is supplied by an air separation plant from which nitrogen also is used as feed for the synthesis gas. The main differences between the two processes are in the reactor design, feeding method, burner design, and carbon and heat recovery. The partial oxidation processes and the steam-naphtha reforming process are favored in areas with short supplies of natural gas.

The source of nitrogen for the synthesis gas has always been air, either supplied directly from a liquid-air separation plant or by burning a small amount of the hydrogen with air in the H_2 gas. The need for air separation plants has been eliminated in modern ammonia plants by use of secondary reforming, where residual methane from the primary reformer is adiabatically reformed with sufficient air to produce a 3:1 mole ratio hydrogen-nitrogen synthesis gas.

Most ammonia plants built since the early 1960s are in the 600–1500 short tons/day (540–1350 metric tons/day) range and are based on new integrated designs that have cut the cost of ammonia manufacture in half. The plants of the early 1980s, in fact, have reached the best combination in terms of plant overall efficiency and cost by combining all the separate units (e.g., synthesis-gas preparation, purification, and ammonia synthesis) in one single train. High-pressure reforming has reduced the synthesis-gas compression load and front end plant equipment size. This compactness in design has also led to increased plant size at reduced investment and operating costs.

Use of Multistage Centrifugal Compressors. One of the major factors contributing to the improved economics of ammonia plants is the application of multistage centrifugal compressors, which have replaced the reciprocating compressors traditionally used in the synthesis feed and recycle service. A single centrifugal compressor can do the job of several banks of reciprocating compressors, thus reducing equipment cost, floor space, supporting foundations, and maintenance.

The use of multistage centrifugal compressors was made possible by redesigning the synthesis loop to operate at low pressures (150–240 atmospheres) and by increasing plant capacity to above the com-

[1] Developed by Grande Pariosse and L'Air Liquide.

[2] Originally developed by Standard Oil Company of New Jersey.

[3] Built by M. W. Kellogg, now Pullman Kellogg, for Shell Chemical Corp. (Ventura, California).

[4] M. W. Kellogg, now Pullman Kellogg, and Imperial Chemical Industries.

[5] Developed by M. W. Kellogg, now Pullman Kellogg.

[6] Texaco partial oxidation process.

[7] Shell gasification process.

pressor's minimum-flow restriction in order to obtain a reasonable compressor efficiency. (Most synthesis loops using reciprocating compressors had been operating at intermediate pressures of 300–350 atmospheres.) Centrifugal compressors capable of developing pressures up to 340 atmospheres already are being offered and used in some large-capacity (1000 short tons/day; 900 metric tons/day) plants, where the increasing compressor horsepower is partially offset by reduction of the refrigeration horsepower requirement.

An operating ammonia plant using the aforementioned improvements is shown schematically in Fig. 1. This plant[8] has a capacity of 1000 short tons/day (900 metric tons/day) and uses natural gas as feedstock. The plant can be divided into the following integrated-process sections: (a) synthesis-gas preparation; (b) synthesis-gas purification; and (c) compression and ammonia synthesis. A typical (Kellogg designed) ammonia plant is shown in Fig. 2.

Synthesis Gas Preparation. The desulfurized natural gas mixed with steam is fed to the primary reformer, where it is reacted with steam in nickel-catalyst-filled tubes to produce a major percentage of the hydrogen required. The principal reactions taking place are[9]

$$CH_4 + H_2O \rightleftharpoons CO + 3H_2 \quad (1)$$

$$\Delta H_{298} = 49.3 \text{ kcal/mole}$$

$$CO + H_2O \rightleftharpoons CO_2 + H_2 \quad (2)$$

$$\Delta H_{298} = -9.8 \text{ kcal/mole}$$

Reaction (1) is the principal reforming reaction, and reaction (2) is the water-gas shift reaction. The net reactions are highly endothermic.

[8] Designed by The M. W. Kellogg Company, now Pullman Kellogg, Houston, Texas, for which Kellogg received the 1967 Kirkpatrick Chemical Engineering Achievement Award.

[9] Heats of reaction at 198°K (25°C), 1 atmosphere pressure, gaseous substances in ideal state.

The partially reformed gas leaves the primary reformer containing approximately 10% methane, on a mole dry-gas basis, at 27–34 atmospheres and up to 816°C. The required heat of reaction is supplied by natural-gas-fired arch burners, which are designed to also burn purge and flash gases from the synthesis resection. Waste heat from the primary reformer flue gas is recovered by generating high-pressure superheated steam, which along with waste-heat process boilers and an appended auxiliary boiler assure a steam system that is always in balance, while providing high-pressure steam to compressor turbine drivers and low-pressure steam to pump drivers. Further waste heat is recovered by preheating the natural-gas-steam feed mixture, steam-air for secondary reforming, and fuel.

The primary reforming step is followed by conversion of the residual methane to hydrogen and carbon oxides over a bed of high-temperature chrome and nickel catalysts in the secondary reformer. The secondary reforming step not only achieves a great degree of overall reforming economically possible, but also reduces fuel-gas input and overall reforming costs by shifting part of the required hydrocarbon conversion from the high-cost primary reformer to the lower-cost secondary reformer. It also permits an increase in the residual methane level at the primary effluent, which results in lower operating temperatures, reduced steam requirements, and milder tube-metal conditions.

Process waste-heat boilers then cool the reformed gas to about 371°C while generating high-pressure steam. The cooled gas-stream mixture enters a two-stage shift converter. The purpose of shift conversion is to convert CO to CO_2 and produce an equivalent amount of H_2 by the reaction: $CO + H_2O \rightleftharpoons CO_2 + H_2$. Since the reaction rate in the shift converter is favored by high temperatures, but equilibrium is favored by low temperatures, two conversion stages, each with a different catalyst provide the optimum conditions for maximum CO shift. Gas from the shift converter is the raw synthesis gas, which, after purification, becomes the feed to the NH_3 synthesis section.

Purification of Synthesis Gas. This involves the removal of carbon

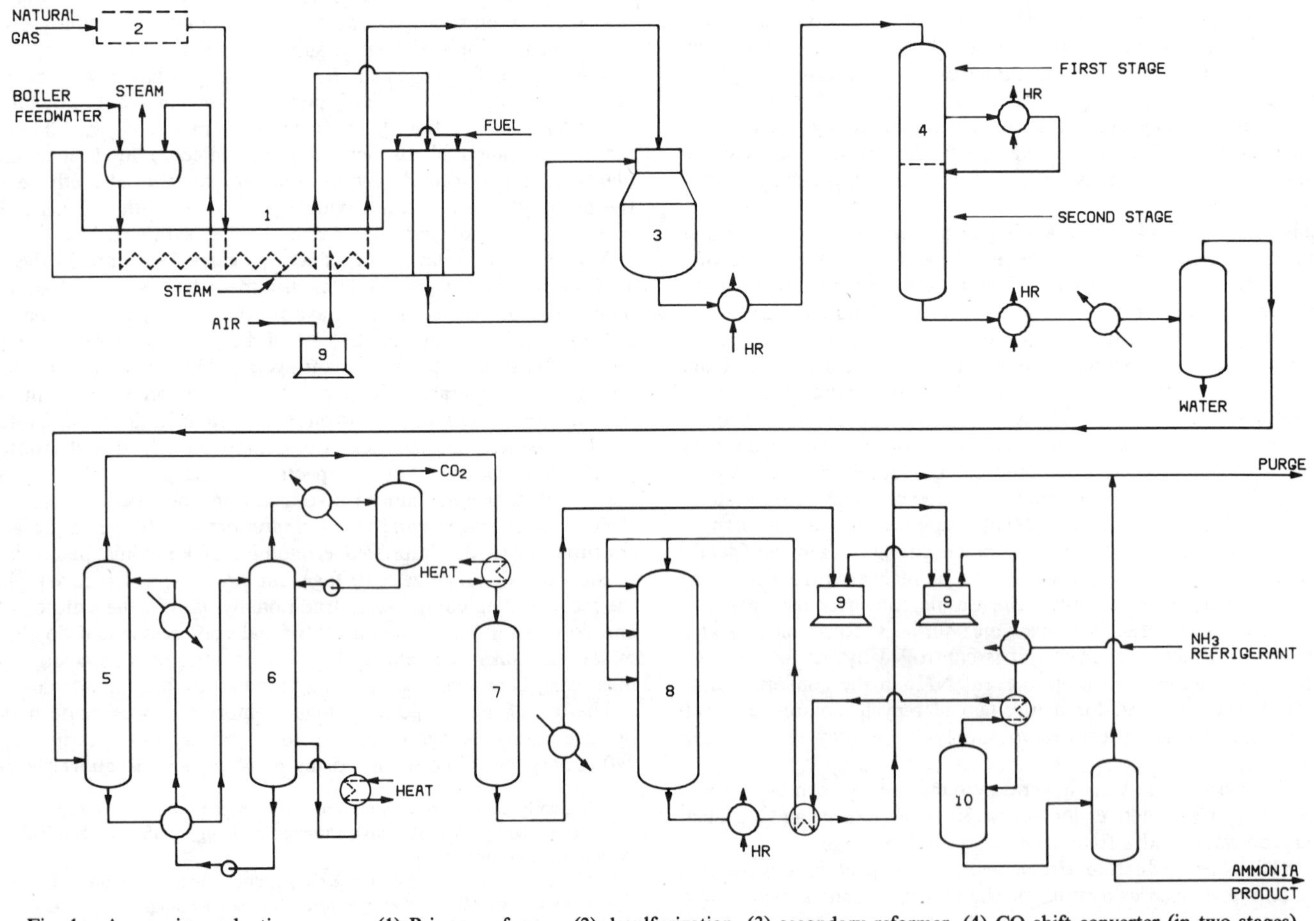

Fig. 1. Ammonia production process: (1) Primary reformer, (2) desulfurization, (3) secondary reformer, (4) CO shift converter (in two stages), (5) CO_2 absorber, (6) CO_2 stripper, (7) methanator, (8) NH_3 converter, (9) compressor, (10) separator. HR = heat recovery. (*Pullman Kellogg*)

Fig. 2. Two 1000 short tons/day (900 metric tons/day) Kellogg-designed modern ammonia plants. (*Pullman Kellogg*)

oxides to prevent poisoning of the NH_3 catalyst. An absorption process is used to remove the bulk of the CO_2, followed by methanation of the residual carbon oxides in the methanator. Modern ammonia plants use a variety of CO_2-removal processes with effective absorbent solutions. The principal absorbent solutions currently in use are hot carbonates and ethanolamines. Other solutions used include methanol, acetone, liquid nitrogen, glycols, and other organic solvents.

The partially purified synthesis gas leaves the CO_2 absorber containing approximately 0.1% CO_2 and 0.5% CO. This gas is preheated at the methanator inlet by heat exchange with the synthesis-gas compressor interstage cooler and the primary-shift converter effluent and reacted over a nickel oxide catalyst bed in the methanator. The methanation reactions are highly exothermic and are equilibrium favored by low temperatures and high pressures.

$$CO + 3H_2 \rightleftharpoons CH_4 + H_2O$$

$$CO_2 + 4H_2 \rightleftharpoons CH_4 + 2H_2O$$

The methanator effluent is cooled by heat exchange with boiler feedwater and cooling water. The synthesis gas leaves the methanator containing less than 10 parts per million (ppm) of carbon oxides.

Compression and Synthesis. The purified synthesis gas, containing H_2 and N_2 in a 3:1 mole ratio and with an inert gas (methane and argon) content of about 1.3 mole percent, is delivered to the suction of the synthesis-gas compressor. Anhydrous ammonia is catalytically synthesized in the converter. The effluent from the converter, after taking off a small purge stream, is recycled for eventual conversion to ammonia. Reaction takes place at approximately 427–482°C. Ammonia liquid, separated from the loop in the separator and from the purge, contains dissolved synthesis gas, which is released when the combined stream is flashed into the letdown drum. The flashed gas is then separated in the letdown drum and combined with the vapors from the purge separator to form a stream of purge fuel gases. Liquid ammonia in the letdown drum still contains some dissolved gases which must be disengaged. This is effected in the ammonia-refrigeration cycle.

Future Considerations in Ammonia Production. The development of a modern and more energy-efficient ammonia process has behind it a long history. Great progress has been made in the past 20 years. Presently, the demand for a more energy efficient ammonia plant is greater than before, due to the economics and short supply of light hydrocarbon feedstock. As hydrocarbon supplies diminish, costs will rise even higher. Unquestionably, the development of new ammonia processes aimed at high energy efficiency and alternate feedstocks will continue to be the primary area of ammonia technology.

In the past, coal or heavy hydrocarbon feedstock ammonia plants were not economically competitive with plants where the feedstocks were light hydrocarbons (natural gas to naphtha). Because of changing economics, however, plants that can handle heavy hydrocarbon feedstock are now attracing increasing attention. In addition, the continuous development and improvement of partial oxidation processes at higher pressure have allowed reductions in equipment size and cost. Therefore, the alternate feedback ammonia plants based on a partial oxidation process may become economically competitive in the near future.

The ultimate goal of any ammonia process will be the direct fixation of nitrogen by reaction of water with air, $1.5H_2O + 0.5N_2 \rightleftharpoons NH_3 + 0.75O_2$. The theoretical energy requirement of the reaction is about 18 MMBTU(LHV)/ST of ammonia. This feed energy requirement is the same as a natural gas feed ammonia plant. The major difference is that there is no short supply of water and air. However, the technology required for this route is not expected to be available any time soon. It is believed that, in the near future, ammonia plant designs will be based essentially on the present-day process with modifications to reduce energy consumption.

References

Allen, J. B.: "Ammonia Manufacture," *Chem. Proc. Eng.* (September 1965)

Axelrod, L. C., and T. E. O'Hare: "Production of Synthetic Ammonia," Chap. 5 in "Fertilizer Nitrogen: Its Chemistry and Technology," (V. Sauchelli, Editor), Van Nostrand, New York, 1964.

Hargette, H. L., and J. T. Berry: "Fertilizer Summary Data," Tennessee Valley Authority, Knoxville, Tennessee (published annually).

"Kirk-Othmer Encyclopedia of Chemical Technology," Chap. 470, Vol. 2, 3rd edition, Wiley, New York, 1978.

MacLean, D. L., Prince, C. E., and Y. C. Chae: "Energy-Saving Modifications in Ammonia Plants," *Chem. Eng. Progress,* **76**, 3, 98–104 (1980).

Quartulli, O. J., et al.: "Best Pressure for Ammonia Plants," *Hydrocarbon Processing,* **47**, 11 (November 1968).

Stokes, K. J.: "Compression Systems for Ammonia Plants," *Chem. Eng. Progress,* **75**, 7, 88–91 (1979).

Yost, C. C., Curtis, C. R., and C. J. Ryskamp: "Advanced Control at Sycon's Ammonia Plant," *Chem. Eng. Progress,* **76**, 4, 31–36 (1980).

J. M. Lee, Pullman Kellogg, Houston, Texas*

AMMONIA-BASED GAS TREATMENT. Pollution (Air).

AMMONIA (Fuel Cell). Fuel Cells.

AMMONIA MASER. Maser.

AMMONIA (Rocket Fuel). Rocket Propellants.

AMMONIUM CHLORIDE. NH_4Cl, formula weight 53.50, white crystalline solid, decomposes at 350°, sublimes at 520°C under controlled conditions, sp gr 1.52. Also known as *sal ammoniac,* the compound is soluble in H_2O and in aqueous solutions of NH_3; slightly soluble in methyl alcohol. Ammonium chloride is a high-tonnage chemical, finding uses as an ingredient of dry cell batteries, as a soldering flux, as a processing ingredient in textile printing and hide tanning, and as a starting material for the manufacture of other ammonium chemicals. The compound can be produced by neutralizing HCl with NH_3 gas or with liquid NH_4OH, evaporating the excess H_2O, followed by drying, crystallizing, and screening operations. The product also can be formed in the gaseous phase by reacting hydrogen chloride gas with NH_3. Ammonium chloride generally is not attractive as a source of nitrogen for fertilizers because of the build-up and damaging effects of chloride residuals in the soil. See also **Nitrogen.**

AMMONIUM COMPOUNDS. Several of the principal ammonium compounds are described in separate entries in this volume. See also **Ammonium Chloride; Ammonium Nitrate; Ammonium Phosphates;** and **Ammonium Sulfate.** The important aspects of several other ammonium compounds are summarized below.

Acetate: Ammonium acetate $NH_4C_2H_3O_2$, white solid, soluble, formed by reaction of ammonia or NH_4OH and acetic acid, reacts upon heating to yield acetamide.

Alum: Ammonium alums are those alums, such as aluminum ammonium sulfate $Al_2(NH_4)_2(SO_4)_4 \cdot 24H_2O$, ferric ammonium sulfate $Fe_2(NH_4)_2(SO_4)_4 \cdot 24H_2O$, chromium ammonium sulfate $Cr_2(NH_4)_2(SO_4)_4 \cdot 24H_2O$ where ammonium sulfate is crystallized with the heavier metal sulfate.

Benzoate: Ammonium benzoate $NH_4C_7H_5O_2$, white solid, soluble, formed by reaction of NH_4OH and benzoic acid. Used (1) as a food preservative, (2) in medicine.

Borate: Ammonium borate, ammonium tetraborate

$$(NH_4)_2B_4O_7 \cdot 4H_2O,$$

white solid, soluble, formed by reaction of NH_4OH and boric acid. Used (1) in fireproofing fabrics, (2) in medicine.

Bromide: Ammonium bromide NH_4Br, white solid, soluble, sublimes at 542°C, formed by reaction of NH_4OH and hydrobromic acid. Used in photography.

Carbonates: Ammonium carbonate, volatile $(NH_4)_2CO_3$, white solid, soluble, formed by reaction of NH_4OH and CO_2 by crystallization from dilute alcohol, loses NH_3, CO_2, and H_2O at ordinary temperatures, rapidly at 58°C; ammonium hydrogen carbonate, ammonium bicarbonate, ammonium acid carbonate NH_4HCO_3, white solid, soluble, formed by reaction of NH_4OH and excess CO_2. This salt is the important reactant in the ammonia soda process for converting sodium chloride in solution into sodium hydrogen carbonate solid.

Chloroplatinate: Ammonium chloroplatinate $(NH_4)_2PtCl_6$, yellow solid, insoluble, formed by reaction of soluble ammonium salt solutions and chloroplatinic acid. Used in the quantitative determination of ammonium.

Cobaltinitrite: Diammonium sodium cobaltinitrite

$$(NH_4)_2NaCo(NO_2)_6 \cdot H_2O$$

golden yellow precipitate, formed by reaction of sodium cobaltinitrite solution in acetic acid with soluble ammonium salt solution. Used in the detection of ammonium.

Cyanate: Ammonium cyanate NH_4CNO, white solid, soluble, formed by fractional crystallization of potassium cyanate and ammonium sulfate (ammonium cyanate is soluble in alcohol), when heated changes into urea.

Dichromate: Ammonium dichromate $(NH_4)_2Cr_2O_7$, red solid, soluble, upon heating evolves nitrogen gas and leaves a green insoluble residue of chromic oxide.

Fluoride: Ammonium fluoride NH_4F, white solid, soluble, formed by reaction of NH_4OH and hydrofluoric acid, and then evaporating. Used (1) as an antiseptic in brewing, (2) in etching glass; ammonium hydrogen fluoride, ammonium bifluoride, ammonium acid fluoride NH_4F_2, white solid, soluble.

Iodide: Ammonium iodide NH_4I, white solid, soluble, formed by reaction of NH_4OH and hydriodic acid, and then evaporating. Used (1) in photography, (2) in medicine.

Linoleate: Ammonium linoleate $NH_4C_{18}H_{31}O_2$. Used (1) as an emulsifying agent, (2) as a detergent.

Nitrite: Ammonium nitrite NH_4NO_2 when ammonium sulfate or chloride and sodium or potassium nitrite are heated, the mixture behaves like ammonium nitrite in yielding nitrogen gas.

Oxalate: Ammonium oxalate $(NH_4)_2C_2O_4$, white solid, soluble, formed by reaction of NH_4OH and oxalic acid, and then evaporating. Used as a source of oxalate; ammonium binoxalate $NH_4HC_2O_4 \cdot H_2O$, white solid, soluble.

Perchlorate: Ammonium perchlorate NH_4ClO_4, white solid, soluble, formed by reaction of NH_4OH and perchlorate acid, and then evaporating. Used in explosives and pyrotechnics.

Periodate: Ammonium periodate NH_4IO_4, white solid, moderately soluble.

Persulfate: Ammonium persulfate $(NH_4)_2S_2O_8$, white solid, soluble, formed by electrolysis of ammonium sulfate under proper conditions. Used (1) as a bleaching and oxidizing agent, (2) in electroplating, (3) in photography.

Phosphomolybdate: Ammonium phosphomolybdate

$$(NH_4)_3PO_4 \cdot 12MoO_3$$

or similar composition), yellow precipitate, soluble in alkalis, formed by excess ammonium molybdate and HNO_3 with soluble phosphate solution. Used as an important test for phosphate (similar product and reaction when arsenate replaces phosphate).

Salicylate: Ammonium salicylate $NH_4C_7H_5O_3$, white solid, soluble, formed by reaction of NH_4OH and salicylic acid, and then evaporating. Used in medicine.

Sulfide: Ammonium sulfide $(NH_4)_2S$, colorless to yellowish solution, formed by saturation with hydrogen sulfide of one-half of a solution of NH_4OH, and then mixing with the other half of the NH_4OH. Dissolves sulfur to form ammonium polysulfide, yellow solution. Used as a reagent in analytical chemistry; ammonium hydrogen sulfide, ammonium bisulfide, ammonium acid sulfide NH_4HS, colorless to yellowish solution, formed by saturation with H_2S of a solution of NH_4OH.

Tartrate: Ammonium tartrate $(NH_4)_2C_4H_4O_6$, white solid, moderately soluble, formed by reaction of NH_4OH and tartaric acid, and then evaporating. Used in the textile industry; ammonium hydro-

* This article is abstracted from a more detailed description in "The Encyclopedia of Chemistry," 4th Edition (D. M. Considine, Editor), Van Nostrand Reinhold, New York, 1983.

gen tartrate, ammonium bitartrate, ammonium acid tartrate $NH_4HC_4H_4O_6$, white solid, slightly soluble, formation sometimes used in detection of ammonium or tartrate.

Thiocyanate: Ammonium thiocyanate, ammonium sulfocyanide, ammonium rhodanate NH_4CNS, white solid, soluble, absorbs much heat on dissolving with consequent marked lowering of temperature, mp 150°C, formed by boiling ammonium cyanate solution with sulfur, and then evaporating. Used (1) as a reagent for ferric, (2) in making cooling solutions, (3) to make thiourea.

Ammonium compounds liberate NH_3 gas when warmed with NaOH solution.

AMMONIUM CYANATE. Cyanic Acid and Related Compounds; Organic Chemistry.

AMMONIUM DIHYDROGEN PHOSPHATE CRYSTAL (ADP). Piezoelectric Effect.

AMMONIUM HYDROXIDE. NH_4OH, formula weight 35.05, exists only in the form of an aqueous solution. The compound is prepared by dissolving NH_3 in H_2O and usually is referred to in industrial trade as aqua ammonia. For industrial procurements, the concentration of NH_3 in solution is normally specified in terms of the specific gravity (degrees Baumé, °Be). Common concentrations are 20° Be and 26° Be. The former is equivalent to a sp gr of 0.933, or a concentration of about 17.8% NH_3 in solution; the latter is equivalent to a sp gr of 0.897, or a concentration of about 29.4% NH_3. These figures apply at a temperature of 60°F (15.6°C). Reagent grade NH_4OH usually contains approximately 58% NH_4OH (from 28 to 30% NH_3 in solution).

Ammonium hydroxide is one of the most useful forms in which to react NH_3 (becoming the NH_4^+ radical in solution) with other materials for the creation of ammonium salts and other ammonium and nitrogen-bearing chemicals. Ammonium hydroxide is a direct ingredient of many products, including saponifiers for oils and fats, deodorants, etching compounds, and cleaning and bleaching compounds. Because aqua ammonia is reasonably inexpensive and a strongly alkaline substance, it finds wide application as a neutralizing agent. See also **Nitrogen.**

AMMONIUM MOLYBDATE. Molybdenum.

AMMONIUM NITRATE. NH_4NO_3, formula weight 80.05, colorless crystalline solid, occurs in two forms:

α-NH_4NO_3, tetragonal crystals, stable between −16°C and 32°C, sp gr 1.66.

β-NH_4NO_3, rhombic or monoclinic crystals, stable between 32°C and 84°C, sp gr 1.725.

The melting point generally ascribed to the alpha form is 169.6°C, with decomposition occurring above 210°C. Upon heating, ammonium nitrate yields nitrous oxide (N_2O) gas and can be used as an industrial source of that gas. Ammonium nitrate is soluble in H_2O, slightly soluble in ethyl alcohol, moderately soluble in methyl alcohol, and soluble in acetic acid solutions containing NH_3.

As shown in the accompanying figure, in making ammonium nitrate on a large scale, NH_3, vaporized by waste steam from neutralizer, is sparged along with HNO_3 into the neutralizer. A ratio controller automatically maintains the proper proportions of NH_3 and acid. The heat of neutralization evaporates a part of the H_2O and gives a solution of 83% NH_4NO_3. Final evaporation to above 99% for agricultural prills, or to approximately 96% for industrial prills is accomplished in a falling-film evaporator located at the top of the prilling tower. The resultant melt flows through spray nozzles and downward through the tower. Air is drawn upward by fans at the top of the tower. The melt is cooled sufficiently to solidify, forming round pellets or prills of the desired range of sizes. The prills are removed from the bottom of the tower and fed to a rotary cooler. Where industrial-type prills are produced, a pre-drier and drier precede the cooler. Fines from the rotary drums are collected in wet cyclones. This solution eventually is returned to the neutralizer. After cooling, the prills are screened to size and the over- and undersize particles are sent to a sump and returned to the neutralizer. Intermediate or product-size prills are dusted with a coating material, usually diatomaceous earth, in a rotary coating drum and sent to the bagging operation. The process can be adapted to other types of materials and mixtures of ammonium nitrate and other fertilizer materials. Mixtures include the incorporation of limestone and ammonium phosphates.

Ammonium nitrate is a very high tonnage industrial chemical, finding major applications in explosives and fertilizers, and additional uses in pyrotechnics, freezing mixtures (for obtaining low temperatures), as a slow-burning propellant for missiles (when formulated with other materials, including burning-rate catalysts), as an ingredient in rust inhibitors (especially for vapor-phase corrosion), and as a component of insecticides.

Amatol, an explosive developed by the British, is a mixture of ammonium nitrate and TNT. A special explosive for tree-trunk blasting consists of ammonium nitrate coated with TNT. In strip mining, an explosive consisting of ammonium nitrate and carbon black is used. The explosive ANFO is a mixture of ammonium nitrate and fuel oil. The typical reaction is: $3NH_4NO_3 + CH_2 \rightarrow 3N_2 + 7H_2O + CO_2 + 82$ kcal/mole. ANFO accounts for about 50% of the commercial explosives used in the United States. Slurry explosives consist of oxidizers (NH_4NO_3 and $NaNO_3$), fuels (coals, oils, aluminum, other carbonaceous materials), sensitizers (TNT, nitrostarch, and smokeless powder), and water mixed with a gelling agent to form a thick, viscous explosive with excellent water-resistant properties. Slurry explosives may be manufactured as cartridged units, or mixed on-site. Although Nobel introduced NH_4NO_3 into his dynamic formulations as early as 1875, the tremendous explosive power of the compound was not realized until the tragic Texas City, Texas disaster of 1947 when a shipload of NH_4NO_3 blew up while in harbor. See also **Explosive.**

As a fertilizer, NH_4NO_3 contains 35% nitrogen. Because of the

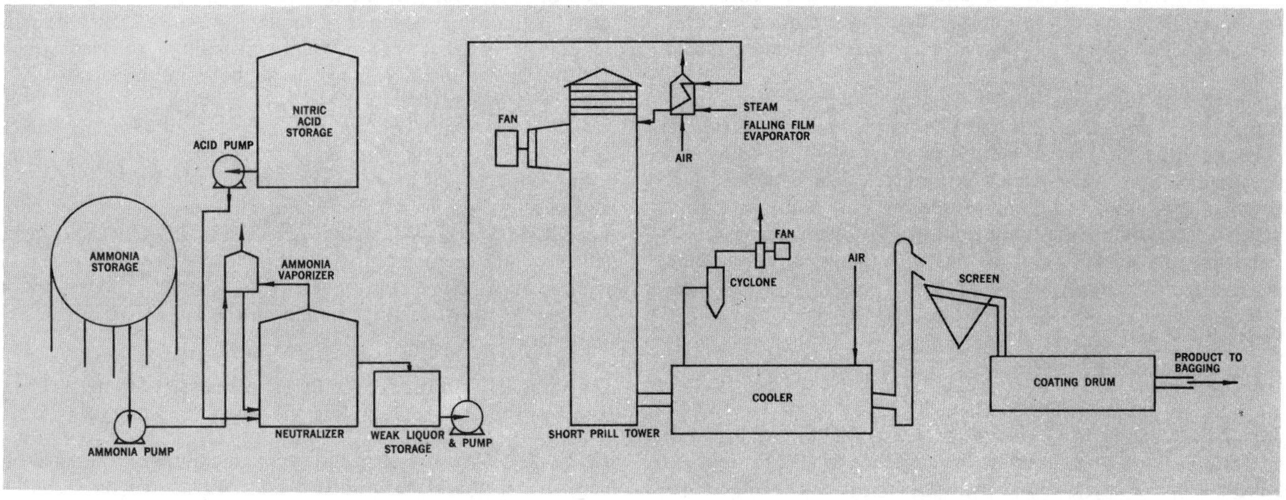

Process for making ammonium nitrate on a large scale. (*C & I/Girdler Incorporated*)

explosive nature of the compound, precautions in handling are required. This danger can be minimized by introducing calcium carbonate into the mixture, reducing the effective nitrogen content of the product to 26%. Inasmuch as NH_4NO_3 is highly hygroscopic, clay coatings and moistureproof bags are means used to preclude spoilage in storage and transportation. See also **Fertilizer;** and **Nitrogen.**

AMMONIUM PERCHLORATE ROCKET OXIDIZER. Rocket Propellants.

AMMONIUM POLYPHOSPHATE (APP) FERTILIZERS. Fertilizers.

AMMONIUM PHOSPHATES. There are two ammonium phosphates, both produced on a very high-tonnage scale.

Monoammonium phosphate, $NH_4H_2PO_4$, white crystals, sp gr 1.803
Formula weight 115.04, $N = 12.17\%$, $P_2O_5 = 61.70\%$
Diammonium phosphate, $(NH_4)_2HPO_4$, white crystals, sp gr 1.619
Formula weight 132.07, $N = 21.22\%$, $P_2O_5 = 53.74\%$

Both compounds are soluble in H_2O; insoluble in alcohol or ether. A third compound, triammonium phosphate $(NH_4)_3PO_4$ does not exist under normal conditions because, upon formation, it immediately decomposes, losing NH_3 and reverting to one of the less alkaline forms.

Large quantities of the ammonium phosphates are used as fertilizers and in fertilizer formulations. The compounds furnish both nitrogen and phosphorus essential to plant growth. The compounds also are used as fire retardants in wood building materials, paper and fabric products, and in matches to prevent afterglow. Solutions of the ammonium phosphates sometimes are air dropped to retard forest fires, serving the double purpose of fire fighting and fertilizing the soil to accelerate new plant growth. The compounds are used in baking powder formulations, as nutrients in the production of yeast, as nutritional supplements in animal feeds, for controlling the acidity of dye baths, and as a source of phosphorus in certain kinds of ceramics.

Ammonium phosphates usually are manufactured by neutralizing phosphoric acid with NH_3. Control of the pH (acidity/alkalinity) determines which of the ammonium phosphates will be produced. Pure grades can be easily made by crystallization of solutions obtained from furnace-grade phosphoric acid. Fertilizer grades, made from wet-process phosphoric acid, do not crystallize well and usually are prepared by a granulation technique. First, a highly concentrated solution or slurry is obtained by neutralization. Then the slurry is mixed with from $6\times$ to $10\times$ its weight of previously dried material, after which the mixture is dried in a rotary drier. The dry material is then screened to separate the desired product size. Oversize particles are crushed and mixed with fines from the screen operation and then returned to the granulation step where they act as nuclei for the production of further particles. Other ingredients often are added during the granulation of fertilizer grades. The ratio of nitrogen to phosphorus can be altered by the inclusion of ammonium nitrate, ammonium sulfate, or urea. Potassium salts sometimes are added to provide a 3-component fertilizer (N, P, K). A typical fertilizer grade diammonium phosphate will contain 18% N and 46% P_2O_5 (weight). See also **Fertilizer;** and **Nitrogen.**

There has been a trend toward the production of ammonium phosphates in powder form. Concentrated phosphoric acid is neutralized under pressure, and the heat of neutralization is used to remove the water in a spray tower. The powdered product then is collected at the bottom of the tower. Ammonium nitrate/ammonium phosphate combination products can be obtained either by neutralizing mixed nitric acid and phosphoric acid, or by the addition of ammonium phosphate to an ammonium nitrate melt.

AMMONIUM RADICAL. Ammonia.

AMMONIUM SULFATE. $(NH_4)_2SO_4$, formula weight 132.14, colorless crystalline solid, decomposes above 513°C, sp gr 1.769. The compound is soluble in H_2O and insoluble in alcohol. Ammonium sulfate is a high-tonnage industrial chemical, but frequently may be considered a by-product as well as intended end-product of manufacture. Large quantities of ammonium sulfate result from a variety of industrial neutralization operations required for alleviation of stream pollution by free H_2SO_4. The ammonium sulfate so produced is not always recovered and marketed. A significant commercial source of $(NH_4)_2SO_4$ is its creation as a by-product in the manufacture of caprolactam, which yields several tons of the compound per ton of caprolactam made. See also **Caprolactam.** Ammonium sulfate also is a by-product of coke oven operations where the excess NH_3 formed is neutralized with H_2SO_4 to form $(NH_4)_2SO_4$. However, as a major fertilizer and ingredient of fertilizer formulations, additional production is required, largely depending upon the proximity of consumers to by-product $(NH_4)_2SO_4$ sources. In the Meresburg reaction, natural or by-product gypsum is reacted with ammonium carbonate:

$$CaSO_4 \cdot 2H_2O + (NH_4)_2CO_3 \rightarrow CaCO_3 + (NH_4)_2SO_4 + H_2O.$$

The product is stable, free-flowing crystals. As a fertilizer, $(NH_4)_2SO_4$ has the advantage of adding sulfur to the soil as well as nitrogen. By weight, the compound contains 21% N and 24% S. Ammonium sulfate also is used in electric dry cell batteries, as a soldering liquid, as a fire retardant for fabrics and other products, and as a source of certain ammonium chemicals. See also **Fertilizer;** and **Nitrogen.**

AMMONOLYSIS. Amination; Organic Chemistry.

AMNESIA. A partial or total loss of memory of a temporary or permanent nature. The condition may result from a brain injury (See **Brain (Injury)**); or it may be symptomatic of a basic disorder of the mind. The condition is frequently associated with a dissociative reaction, but may also be evidenced in other psychoses resulting from stress, such as anxiety, phobic, obsessive-compulsive, and conversion reactions. *Retrograde amnesia* is an impaired ability to recall past events; *anterograde amnesia* is an impaired ability to learn new information; *confabulation*, sometimes associated with amnesia, is the fabrication of recent events.

In retrograde amnesia, anxiety over an event becomes so great that the individual is forced to forget it. In forgetting anxiety, the patient also forgets a multitude of necessary associations, including personal identity. Despite these shortcomings, the individual is often well oriented as to present time and place. The person simply cannot recall anything about the past. The patient's behavior appears so normal that the individual moves about freely without attracting undue notice. In some instances, the individual may wander restlessly from place to place, covering extensive regions in such travels.

Recovery of memory is sometimes spontaneous. Frequently, psychiatric help is required. This assistance may have to extend over an appreciable period. Upon regaining memory, the amnesic patient does not recall events which occurred during the period of amnesia. However, it is frequently possible to bring forth recent past events in considerable detail through hypnosis. This indicates that the loss of consciousness in dissociation is different from that of the delirious states. Patients who recover from delirium cannot recall experiences even when hypnotized.

In a dissociative reaction, there is an unconscious flight from situations of intolerable emotional stress. The reaction takes numerous forms, including some kinds of amnesia, sleep-walking, automatic writing, and the extremely rare "dual personality," in which two mental selves exist within the same body at the same time.

See also **Psychiatry.**

Amnesia may also occur as a consequence of Wernicke's encephalopathy. This condition is found in some chronic alcoholics whose diet is inadequate (frequently characterized by lack of thiamine). Neuronal degeneration results in mental confusion, disorders of eye movement (gaze), and ataxia. When the condition is left untreated, permanent neurological damage occurs, resulting in a severe impairment of memory known as Korsakoff's psychosis (or amnesic-confabulatory psychosis). See also **Memory,** including the list of references at end of that entry.

AMNIOCENTESIS. Extraction and analysis of some of the amniotic fluid from the sac surrounding the fetus. See **Embryo.**

AMNION. An accessory embryonic membrane common to reptiles, birds and mammals and a superficially similar structure found in some insects.

To the three vertebrate classes, all fundamentally terrestrial, the amnion gives the name Amniota. In this group the amnion is usually formed by the growth of folds of the somatopleure, consisting of ectoderm and somatic mesoderm, about the embryo. The union of the folds produces two membranes, the outer serosa and the inner amnion; the latter encloses the embryo and is filled with amniotic fluid which protects it against desiccation and equalizes mechanical stresses. The amnion is lined with ectoderm and covered with mesoderm, both continuous with the same tissues of the embryo itself.

In the insect egg the amion develops from folds of ectoderm which extend between the embryo and the shell of the egg and fuse to form an outer serosa and an inner amnion. These structures are not uniform in their appearance and are not persistent throughout embryonic development in all species in which they form, hence their functions are in doubt. See also **Embryo.**

AMOEBA. A genus of one-celled animals in which the body consists of a naked mass of protoplasm and the organs of locomotion are temporary blunt protuberances of cytoplasm known as pseudopodia. The large fresh water form, *Amoeba proteus*, is a typical species. When seen under the microscope, this protozoan appears as a naked bit of protoplasm surrounded only by its thin plasma membrane. When in an active state, it is constantly changing its shape as it moves about. See also **Asexual Reproduction; Cell (Biology).**

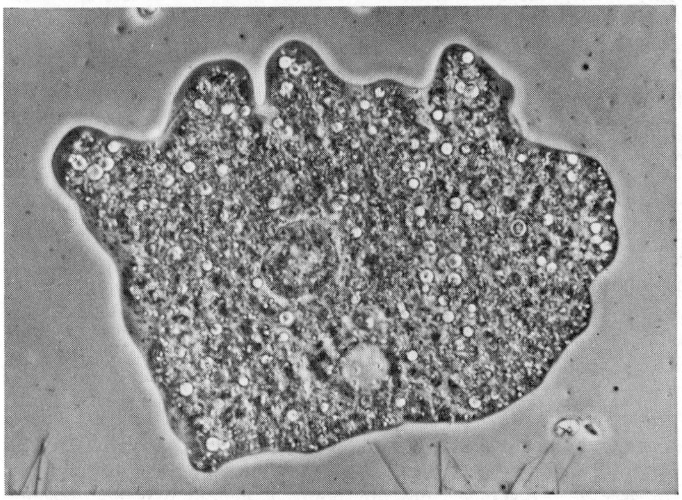

Amoeba. (*A. M. Winchester*)

AMOEBIC DYSENTERY. Amebiasis.

AMOEBOID MOVEMENT. Movement of cells by means of pseudopodia, as in *Amoeba.* The white blood cells, leucocytes, in the blood of higher animals are good examples of cells that move in this manner.

AMOEBULAE. Spores of amoeboid form which are produced by some one-celled animals.

AMOLYTIC FERMENTATION. Fermentation.

AMOR ASTEROIDS. Asteroid.

AMORPHOUS. As opposed to a crystalline substance which exhibits an orderly structure, the behavior of an amorphous substance is similar to a very viscous, inelastic liquid. Examples of amorphous substances include amber, glass, and pitch. An amorphous material may be regarded as a liquid of great viscosity and high rigidity, with physical properties the same in all directions (may be different for crystalline materials in different directions). Usually, upon heating, an amorphous solid gradually softens and acquires the characteristics of a liquid, but without a definite point of transition from solid to liquid state. In geology, an amorphous mineral lacks a crystalline structure, or has an internal arrangement so irregular that there is no characteristic external form. This does not preclude, however, the existence of any degree of order. The term amorphous is used in connection with amorphous graphite and amorphous peat, among other naturally occurring substances.

AMOSITE. Amosite is a long-fiber gray or greenish asbestiform mineral related to the cummingtonite-grunerite series, and is of economic importance. It occurs within both regional and contact metamorphic rocks in the Republic of South Africa. The name amosite is a product of the initial letters of its occurrence at the Asbestos Mines of South Africa. See also **Asbestos.**

AMPERE. Units and Standards.

AMPERE-HOUR. Units and Standards.

AMPERE PER METER. Units and Standards.

AMPERE'S LAW. This law of magnetostatics has been stated in a number of forms, one in terms of the magnetic field intensity produced by a current flowing in a thin conductor and another in terms of a line integral about a closed path. The first form, which lacks exactness, is (refer to Fig. 1) that the magnitude of the magnetic

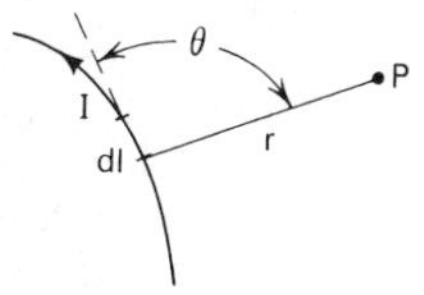

Fig. 1. Ampere's law in terms of magnetic field intensity.

field intensity dH at point P produced by a current of I amperes flowing through an element dl of a thin conducting wire at a distance r is

$$dH = \frac{Idl \cos\theta}{4\pi r^2} \text{ amperes/meter}$$

The direction of dH is given by the right hand screw rule. The total magnetic field intensity at P is the sum of the vector fields produced by all the conductor elements. The second form (refer to Fig. 2) is

Fig. 2. Ampere's law in terms of a line integral about a closed path.

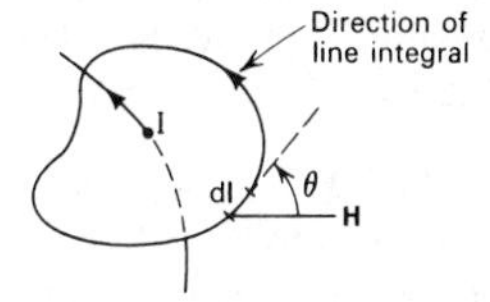

that the line integral of the magnetic field intensity around a closed path is equal to the total current which flows through any surface bounded by the closed path or $\oint H \cos\theta dl = I$. All units are in the meter-kilogram-second system.

AMPERE'S RULE. The magnetic flux generated by a current in a wire encircles the current in the counter-clockwise direction, if the current is approaching the observer.

AMPERE'S THEOREM. The magnetic field due to an electric current flowing in any circuit is equivalent at external points to that due to a simple magnetic shell, the bounding edge of which coincides with the conductor, and the strength of which is equal to the strength of the current.

AMPERE-TURN. Units and Standards.

AMPEROMETER. An instrument for the chemical analysis of electroreducible or oxidizable ions, molecules, or dissolved gases in solution. Included are the majority of metal ions and many organic substances that contain oxidizable or reducible groups. The range of concentration measurements of which the instrument is capable is from 0.01 to 1,000 ppm. This instrumental method is used for determining free available or total available chlorine, particularly in connection with water-chlorination control. In this application, the range of chlorine concentration is from 0 to 50 ppm. Free iodine also may be determined with an amperometer. The identification of unknown substances is by inference from noting the fixed potential between polarized microelectrode and reference electrode that causes oxidation or reduction of a given composition sought. Where identity is firmly established, the concentration also may be determined because the concentration is proportional to the diffusion-limited current that flows in the electrode circuit. In this latter respect, amperometry is similar to polarography. See also **Analysis (Chemical).**

AMPHETAMINE. Also called methylphenethylamine; 1-phenyl-2-aminopropane; Benzedrine; formula $C_6H_5CH_2CH(NH_2)CH_3$; *amphetamine* is a colorless, volatile liquid with a characteristic strong odor and slightly burning taste. Boils and commences decomposition at 200–203°C. Low flash point, 26.7°C. Soluble in alcohol and ether; slightly soluble in water. Amphetamine is the basis of a group of hallucinogenic, habit-forming drugs which affect the central nervous system. The drug also finds medical application, notably in appetite suppressants. It should be emphasized that administration of amphetamines for prolonged periods in connection with weight-reduction programs may lead to drug dependence. Particular attention must be paid by professionals to the possibility of persons obtaining amphetamines for nontherapeutic use or distribution to others. See also **Drug Addiction.**

Amphetamines for obesity control may be in the form of amphetamine adipate, the sulfate, the saccharate, or phosphate. Amphetamines are sympathomimetic amines with central nervous system stimulant activity. Peripheral actions include elevation of systolic and diastolic blood pressures and weak bronchodilator and respiratory stimulant action. When the drugs are used for obesity control, they are sometimes called "anorectics" or "anorexigenics." The effectiveness of amphetamines, in numerous clinical trials, as appetite depressants is not fully convincing. It has been found, for example, that the magnitude of increased weight loss of drug-treated patients over placebo-treated patients is only a fraction of a pound per week. The rate of weight loss is greatest during the first few weeks of treatment. The natural history of obesity is measured in years, whereas studies have been restricted to comparatively few weeks duration. Thus, the total impact of drug-induced weight loss over that of diet alone must be considered clinically limited. One should stress that there are varying opinions among the authorities in this area.

In connection with the treatment of narcolepsy (a sleep disorder), the use of amphetamines is not suggested because of the risks of habituation, abuse, and the development of tolerance.

AMPHIBIA. The frogs, toads, newts, salamanders and related forms. A class of the phylum *Chordata.* Since these animals live only in moist places their distribution is restricted and they are among the less familiar vertebrates.

The amphibians are distinguished by: (1) moist skin; (2) the absence of scales and claws; (3) a metamorphosis, undergone by most species during development, from an aquatic, gill-breathing larva to a semi-terrestrial, air-breathing adult stage.

While some of the salamanders are permanently aquatic and some of the tree frogs permanently terrestrial, most members of the class live near the water or in moist places and undergo the metamorphosis mentioned above.

The following orders of amphibians are recognized:

Order *Gymnophiona* (*Apoda*). Legless, worm-like animals, confined to the tropics of the Old and New Worlds.

Order *Urodela* (*Caudata*). Elongate animals with long tails and weak, short legs. The salamanders, newts, efts, hellbender, and mud puppy.

Order *Anura* (*Salientia*). Tailless species whose hind legs are the larger pair, more or less strongly developed for jumping. Most species have a larval stage known as the tadpole with a compact body and a long compressed tail but no legs until the onset of metamorphosis. The frogs and toads.

AMPHIBIA (Fossil). Fossil Amphibia.

AMPHIBIANS (Bioscience of). Herpetology.

AMPHIBIANS (Water Metabolism). Water.

AMPHIBOLE. This is the name given to a closely related group of minerals all showing in common a prismatic cleavage of 54–56° as well as similar optical characteristics and chemical composition.

The amphiboles may be said to represent chemically a series of metasilicates corresponding to the general formula $RSiO_3$ where R may be calcium, magnesium, iron, aluminum, titanium, sodium, or potassium. The crystals of the amphibole family group fall within both the monoclinic and orthorhombic systems.

There is a clear parallelism between the amphiboles and the pyroxenes. There are two basic differences between the minerals of these two family groups; amphiboles with cleavage angles of 56° and 124°, with essential OH groups in their structure; pyroxenes with cleavage angles of 87° and 93°, and being anhydrous, with no OH content. Amphibole crystals are usually long and slender and tend to be simple while pyroxene crystals tend to be complex, short, and stout prisms.

Amphibole is common in both lavas and deep-seated rocks, though less so in the basic lavas than pyroxene. Many of the amphiboles may be developed as metamorphic minerals. The following members of the amphibole group are described under their own headings: actinolite, anthophyllite, cummingtonite, glaucophane, grünerite, hornblende, riebeckite and tremolite. Amphibole was so named by Haüy from the Greek word, meaning doubtful, because of the many varieties of this mineral.

See also **Pyroxene;** and terms listed under **Mineralogy.**

AMPHIBOLITE. The amphibolites form a large group of rather important rocks of metamorphic character. As the name implies they are made up very largely of minerals of the amphibole group. There may be also a variety of other minerals present, such as quartz, feldspar, biotite, muscovite, garnet, or chlorite in greater or less amounts.

Depending upon the particular amphibole present these rocks may be light to dark green or black, the amphibole usually being in long slender prisms or laths, often quite coarse, sometimes in acicular or fibrous forms.

Because the mineral constituents are arranged parallel to the schistosity, amphibolites may have a strongly developed cleavage.

The occurrence of amphibolites accompanying gneisses, schists, and other metamorphic rocks of probable sedimentary origin strongly suggests a similar derivation. Yet some amphibolites cut other metamorphic rocks in the manner of dikes or sills. It is very likely that they have been derived from both original igneous and sedimentary rocks. Large masses of amphibolite suggest gabbroic stocks. Well-known areas in which amphibolites are found are New England, New York State, Canada, Scotland, and the Alps.

AMPHIDROMIC POINT. A geographic position in the ocean where theoretically there is no tide range and from which cotidal lines radiate in various directions. The tide amplitude presumably increases with distance from this point. A synonym is *nodal point.*

AMPHINEURA. The chitons and allied forms, a group of the phylum *Mollusca.* The more familiar members are flattened marine animals of oval outline. They have a shell composed of a series of separate plates, sometimes concealed within the body. The foot makes up most of the ventral surface and the limited mantle extends down about it to form a shallow groove. Nerve cells are in many cases distributed through the nerve cords so that the nervous system contains no ganglia. See **Ganglion.**

The group has two classes:

Class *Aplacophora* (*Solenogastres*). Worm-like animals, somewhat cylindrical and elongate. Shell lacking.

Class *Polyplacophora.* The chitons. Flattened and oval, with shell plates. They are widely distributed, chiefly in the shallow waters. Sometimes used as food.

AMPHIOXUS. Commonly used to designate any of the primitive chordates called lancelets but more accurately a genus of these animals. See **Cephalochordata.**

AMPHIPODA. An order of crustaceans including marine and freshwater species. The beach-fleas are among the few which have a common name.

AMPHIPROTIC. Capable of acting either as an acid or as a base, i.e., as a proton donor or acceptor, according to the nature of the environment. Thus, aluminum hydroxide dissolves in acids to form salts of aluminum, and it also dissolves in strong bases to form aluminates. Solvents like water which can act to give protons or accept them, are amphiprotic solvents. See **Acids and Bases; and Salt.**

AMPHISBAENA (*Reptilia, Lacertilia*). The typical genus of a snake-like family of lizards, which live in burrows, and are characterized by the absence of legs and ears, and functionless eyes.

AMPICILLIN. Antibiotic.

AMPLIFICATION. A general transmission term used to denote an increase of signal magnitude.

AMPLIFICATION FACTOR. Electron Tube; Photoemission and Photomultipliers.

AMPLIFIER. A device for increasing the strength of a signal without appreciably altering other signal characteristics, such as waveform. Amplifiers may be classified in several ways—by basic mode of operation, such as electronic (solid state or vacuum tube), magnetic, hydraulic, fluidic, and mechanical (lever systems); by application, such as data and information systems (computing, electronic data processing, communications, instrumentation) and automatic control systems. In one device or subassembly, other functions may be combined with amplification. The nomenclature of amplifiers reflects these variations in terms of numerous special designations—some by mode of operation as a transistor amplifier, magnetic amplifier, fluidic amplifier, etc.; some by use and specific function as a carrier amplifier, sample-and-hold amplifier, servoamplifier, power amplifier, etc. Amplifiers also may be classified in terms of their frequency response, or in terms of the signal frequencies they are designed to transmit. This classification is essentially one which depends on the frequency characteristics of the load impedance or coupling network used between transistor amplifier stages. Thus, there are direct current (dc) amplifiers, audio frequency amplifiers, rf amplifiers, video frequency amplifiers, etc. Also, recognizing the mechanism of amplification as the control of an output current by an input voltage or current, an amplifier may be classified in terms of the fraction of the cycle of an assumed sinusoidally varying input signal for which the output current flows. This form of description results in the designation Class A, AB, B, or C for amplifier stages. amplifier stages. Class A amplifiers are those in which the combined effect of the bias used to locate the operating point for the transistor and the input signal is such that the collector current flows during the whole input cycle. In a Class AB amplifier, the combined effect of the bias and input signal results in output current flowing for appreciably more than half, but less than the entire cycle. In a Class B amplifier, collector current is approximately zero with no input signal applied. With signal, therefore, current only flows for one-half of the input cycle. Class C amplifiers are biased so that output current does not flow in the absence of a signal, and when the input is applied, current flows in pulses which are appreciably less than half a cycle in duration.

The bias adjustments for various amplifier classes are shown in

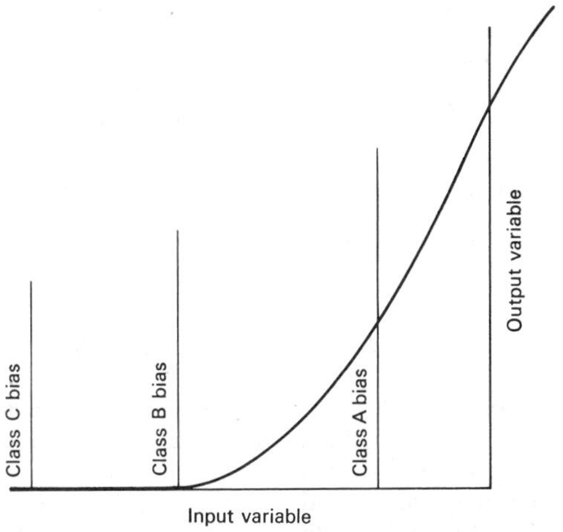

Fig. 1. Bias adjustments for various amplifier classes.

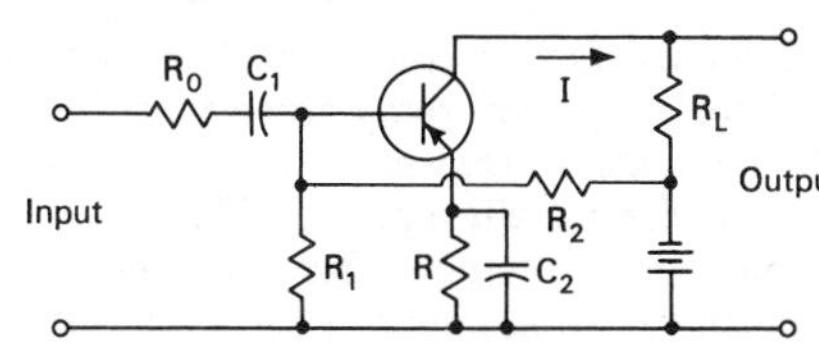

Fig. 2. Common emitter transistor amplifier connection.

Fig. 1. The diagram shows where the bias would be set for Class A, B, and C operation. Whether Class A or AB operation results in a given situation will depend on the peak-to-peak value of the input signal. The amplifier efficiency increases from Class A to B to C operation. The relation between the input signal and the output signal current becomes more nonlinear and hence the amplitude distortion is worse in the same order.

A simple transistor amplifier is shown in Fig. 2. Capacitor C_1 is a blocking capacitor to prevent modification of the bias conditions due to a dc path through the generator, R_1 and R_2 in conjunction with the battery and the dc voltage developed across R produce the desired bias current in the base lead, and C_2 is a bypass capacitor to eliminate feedback due to signal current flowing through resistor R. The application of an input signal changes the current into the base-emitter junction with a consequent change in collector current I as shown in Fig. 3. The output voltage is developed across resistor R_L as a result of the flow of collector current through it. Resistor R_0 produces a load for the source of signal voltage which is far more constant than the base to emitter resistance of the transistor alone. The input signal current is made much less dependent on the device itself by this means. When the gain obtainable from a single amplifier is not sufficient for a given application, it is necessary to cascade two or more stages. This situation requires coupling networks between amplifier stages which assume a variety of forms depending on the frequency response characteristics desired for the amplifier. A simple and common method is known as resistance coupling. This is coupling in which resistors are used as the input and output impedances of the circuits being coupled. A coupling capacitor can be used between the resistors to transfer the signal from one stage to the next.

Fig. 3. Waveform in resistance-coupled amplifier.

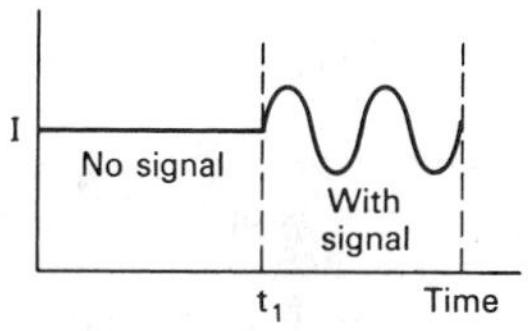

Amplifier Configurations. Several specific amplifier configurations are described briefly in the following listing, arranged alphabetically for convenience.

(*Balanced Amplifier*)—An amplifier circuit in which there are two identical signal branches connected so as to operate in phase opposition and with input and output connections each balanced to ground.

(*Booster Amplifier*)—An amplifier used in audio consoles between mixer controls and the master volume control to prevent deterioration of signal-to-noise ratio. It generally supplies sufficient gain to compensate for mixing-circuit losses.

(*Bootstrap Amplifier*)—A single-stage amplifer in which a change in input signal voltage changes the potential of the input source with respect to ground (or other reference voltage) by an amount equal to the output signal voltage. In a transistor amplifier, the input is applied between base and emitter with the output load connected between the emitter and the low potential side of the collector power supply.

(*Bridge Amplifier*)—Extensively used for instrumentation purposes, the commercial configuration generally is a direct-coupled amplifier, offering reasonably wide bandwidths up to 50 kHz at gains ranging from near unity to 1,000. See also separate entry on **Bridge Amplifier.**

(*Buffer Amplifier*)—This term is commonly applied to an amplifier stage whose main function is to isolate the oscillator of a transmitter from the main power amplifiers, but it is also applied to any amplifier which is inserted between two circuits or amplifier stages to reduce substantially the interaction of one on the other. The frequency of an oscillator depends, among other factors, upon the load which is applied to it. Since, for satisfactory communications, it is highly desirable that the oscillator frequency remain constant, buffer amplifiers are always used in broadcast transmitters and are usually used in others. Such an amplifier usually operates Class C with the signal current to the input circuit of the buffer amplifier so low that the current taken from the exciting circuit is not sufficient to change its operation appreciably from the no-load condition. The output circuit of the buffer can then supply the normal load of a conventional Class C amplifier without this load affecting the oscillator. Sometimes more than one buffer stage is used to make certain that no load is reflected back to the oscillator. See also **Buffer (Computer).**

(*Carrier Amplifier*)—A dc amplifier wherein the signal first is modulated, then demodulated during amplification. The bandwidth usually is from zero to some finite frequency. Electronic switches or electromechanical devices are used in most cases to effect the modulation. Thus, the "chopping" action accomplishes the equivalent of a square-wave modulation of the signal. See also separate entry on **Carrier Amplifier.**

(*Cascade Amplifier*)—Usually refers to a vacuum tube type amplifier consisting of a grounded-cathode input stage driving a grounded-grid output stage. Frequently used in past television and other ultra-high and very-high frequency receiver input-stages because of its low noise figure. Sometimes referred to as the "Wallman Amplifier" for its originator.

(*Cathamplifier*)—An amplifier circuit with high input impedance which permits push-pull operation from an unbalanced source.

(*Chopper Amplifier*)—In one type of chopper amplifier, the input signal is chopped (or modulated), amplified by an ac amplifier, demodulated, and filtered to provide a dc output signal. See foregoing mentioned description of carrier amplifier. In a second type, the error signal is chopped for the purpose of providing stabilization of gain and offset. The term, chopper-stabilized amplifier, may be a more apt designation. See also separate entry on **Chopper Amplifier.**

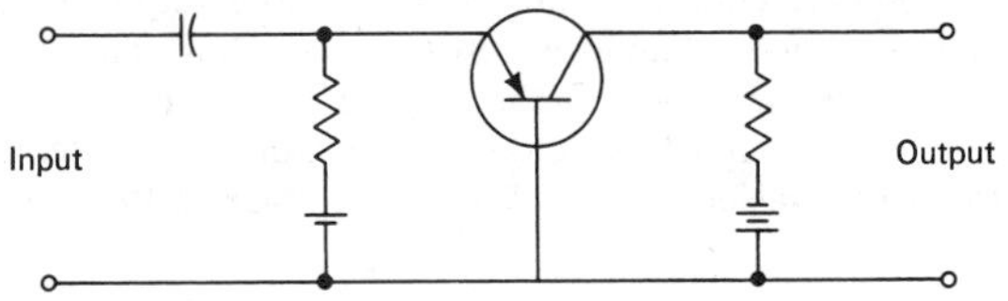

Fig. 4. Common-base transistor amplifier.

(*Common-Base Amplifier*)—A form of transistor amplifier in which the input signal is applied between emitter and base and the output signal is taken between collector and base. See Fig. 4. Characteristics associated with this connection are extremely low input impedance, a high output impedance, and a current amplification somewhat less than unity. The circuit is useful in providing impedance transformation from a low impedance source to a high impedance load.

(*Common-Emitter Amplifier*)—A form of transistor amplifier in which the input signal is applied between base and emitter and the output signal is taken between collector and emitter. This configuration is capable of providing both current gain and voltage gain exceeding unity in contrast with the characteristics of the common-base and emitter-follower amplifiers. See Fig. 5.

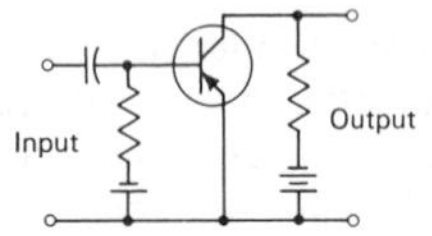

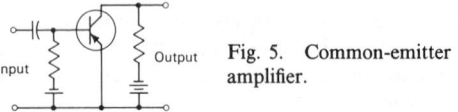

Fig. 5. Common-emitter amplifier.

(*Contact-Modulated Amplifier*)—Used for the amplification of dc and very low frequency signals. The signal source is modulated by a carrier-operated contact system (usually 60 to 400 Hz), the resulting modulated wave amplified in an ac amplifier to a suitable level, and subsequently demodulated, sometimes by the same contact system used to accomplish the original modulation.

(*Direct-Coupled Amplifier*)—An amplifier in which no coupling capacitors are employed as interstage coupling elements and which thus is capable of amplifying dc variations and ac signals of arbitrarily low frequency. A transistor direct-coupled amplifier using feedback is shown in Fig. 6. The need for batteries or Zener diodes as coupling elements has been avoided in this design by suitable choice of transistor

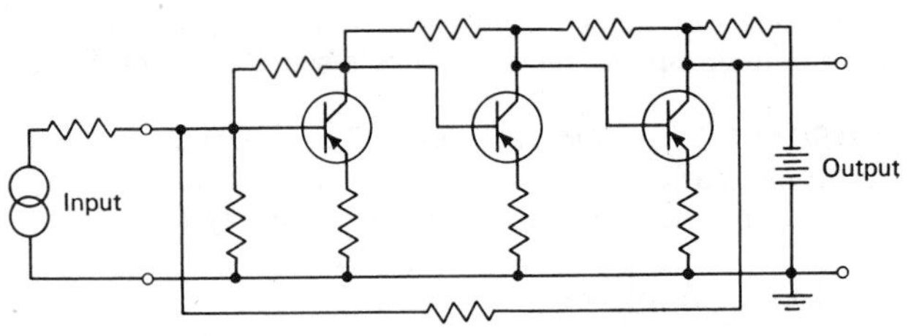

Fig. 6. Direct-coupled transistor amplifier.

operating points, values of collector and emitter resistors, and the feedback connections.

(*Distributed Amplifier*)—An amplifier consisting of components appropriately distributed along artificial transmission lines. This amplifier is capable of much greater bandwidths than a conventional amplifier, and the ordinary figure of merit or gain-bandwidth product does not apply.

(*Doherty Amplifier*)—A particular arrangement of a radio-frequency linear power amplifier wherein the amplifier is divided into two sections whose inputs and outputs are connected by quarter-wave (90°) networks, and whose operating parameters are so adjusted that, for all values of the input signal voltage up to one-half maximum amplitude, Section No. 2 is inoperative and Section No. 1 delivers all the power to the load, which presents an impedance at the output of Section No. 1 that is twice the optimum for maximum output. At one-half maximum input level, Section No. 1 is operating at peak efficiency, but is beginning to saturate. Above this level, Section No. 2 comes into operation, thereby decreasing the impedance presented to Section No. 1, which causes it to deliver additional power into the load until, at maximum signal input, both sections are operating at peak efficiency, and each section is delivering one-half the total output power to the load.

(*Double-Stream Amplifier*)—A traveling-wave amplifier in which the amplification occurs as a result of the interaction of two electron beams having different average velocities. The amplification takes place in the beam itself and is a result of what might be called electromechanical interaction.

(*Dynamo-Electric Amplifier*)—A power amplifier for dc and very low frequencies whose output circuit consists of some form of rotating armature in a controllable magnetic field. It is sometimes employed in servomechanisms and other control systems.

(*Feedback Amplifier*)—An amplifier in which feedback has been deliberately introduced to obtain certain performance characteristics. Commonly, the feedback connection is made from the output to the input of the amplifier and is employed to reduce variations in gain associated with changes in the characteristics of transistors. A representative feedback amplifier is shown in Fig. 7. In this case, the amplifier acts to maintain the output voltage constant for a constant peak amplitude input signal. The amplifier contains stabilizing networks (N_1) to prevent oscillation.

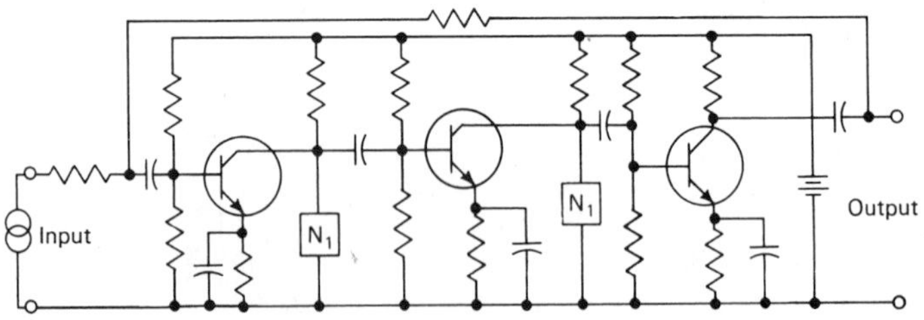

Fig. 7. Transistor feedback amplifier.

(*Final Amplifier*)—The last amplifier stage in a radio or television transmitter.

(*Floating Amplifier*)—Also known as an isolated amplifier, the design does not require that the input and output signals be referred to the same signal reference point (ground). Generally, differential-input amplifiers meet this definition. The term *floating*, however, normally excludes an amplifier where the input reference point and the output reference point are common, that is, either through a signal conductor or a power supply. Specifically, floating amplifier refers to an amplifier which includes a 4-terminal coupling device, such as a light-coupled signal-transmission element or a transformer. See also separate entry on **Floating Amplifier.**

(*Fluidic Amplifier*)—A fluid amplifier amplifies pressure, mass flow, and fluid power and, in principle, bears considerable resemblance to its electrical/electronic counterparts. Several types of fluidic amplifiers have been designed: (1) *stream interaction amplifier* which consists of a relatively powerful main stream (or jet) of fluid which has control jets of less power placed at right angles to it. Variation of the control jet power changes the direction of the main stream and can, for example, direct it to a desired outlet channel. (2) *boundary layer control amplifiers* which utilize the Coanda effect. They also have power streams directed by control streams, but in addition have boundary walls so placed that the power stream, once it is moved by the control stream, continues in its new path until it is moved again by another control stream, thus serving as a fluidic flip-flop so to speak. (3) *vortex flow amplifier* which consists of an essentially circular chamber, with the main stream entering perpendicularly to the periphery, and the control stream at the same point, but tangentially to the periphery, so it is perpendicular to the main stream. The output is discharged at the center of the chamber. The effect of the control stream is to change the angular momentum of the total flow, but the system as a whole can be used to generate any one of a number of signals. The device not only is an amplifier, but also a signal generator. (4) *turbulent flow amplifier* which operates on the turbulence produced in an otherwise laminar flow main stream by a control stream entering it at an angle. The turbulence is accompanied by a drop in pressure, which varies with the control stream velocity only up to a certain point. Thus, the device can be used not only as an amplifier, but for such logical operations as AND, OR, NOR, NAND, or flip-flop. See also separate entry on **Fluidics.**

(*Grounded-Cathode Amplifier*)—This is an electron-tube amplifier with the cathode at ground potential at the operating frequency, with input applied between the control grid and ground, and the output load connected between plate and ground.

(*Grounded-Collector Amplifier*)—An amplifier that uses a transistor in a grounded-collector connection.

(*Grounded-Gate Amplifier*)—An amplifier that uses thin-film transistors in which the gate electrode is connected to ground. The input signal is fed to the source electrode and the output is obtained from the drain electrode.

(*Grounded-Grid Amplifier*)—This is a vacuum tube amplifier circuit in which the grid is connected to ground (reference potential), the input signal is applied between cathode and ground, and the output is developed across a load impedance connected between plate and ground. The advantage is the significant reduction in interaction between input and output circuits.

(*Grounded-Plate Amplifier*)—This is an electron tube amplifier, also called cathode-follower amplifier, in which the plate is at ground potential at the operating frequency, with input applied between grid and ground, and the output load connected between cathode and ground. This amplifier is characterized by a large negative feedback and can be used as an impedance-matching device.

(*Hydraulic Amplifier*)—A power amplifier employed in some servomechanisms and control systems in which power amplification is obtained by the control of the flow of a high-pressure liquid by a valve mechanism. See also separate entry on **Hydraulic Controller.**

(*Impedance-Coupled Amplifier*)—An amplifier similar in form to a resistance-coupled amplifier in which the output voltage drop is developed across a choke or impedance coil. The choke or impedance coil can be designed to have a suitably high impedance at the frequency of the signal being applied and still have a resistance low enough so that only a relatively small voltage drop appears across it at dc. This form of coupling is usually used in amplifier design to amplify audio-frequency signals.

(*Intermediate-Frequency Amplifier*)—The amplifier used in superheterodyne receivers which amplifies the sum or difference frequency produced in the mixer or first detector by the heterodyning of the signal and oscillator frequencies.

(*Isolated Amplifier*)—See prior item in this list—Floating Amplifier.

(*Klystron Amplifier*)—A klystron tube may be used as an amplifier as well as oscillator in uhf applications. A cascade-amplifier klystron contains three resonant cavities for increased power amplification and output. The third resonator lies between the input and output resonators and has no external connection. It is excited by the bunched beam that emerges from the input-resonator gap, and it produces further bunching of the beam.

(*Light Amplifier*)—The term laser stands for light amplification by stimulated emission of radiation. A laser is an active electron device that converts input power into a very narrow, intense beam of coherent light. See also separate entry on **Laser.**

(*Linear Amplifier*)—A pulse amplifier in which the output pulse height is proportional to an input pulse height for a given pulse shape up to a point at which the amplifier overloads.

(*Linear Power Amplifier*)—A power amplifier in which the signal output voltage is directly proportional to the signal input voltage.

(*Logarithmic Amplifier*)—An amplifier whose output signal is a logarithmic function of the input signal.

(*Magnetic Amplifier*)—A device that uses saturable reactors either alone or in combination with other circuit elements to achieve amplification. A simple form of magnetic amplifier is shown in Fig. 8. An alternating voltage source is connected in series with a load resistor R_L and two coils (called gate windings) having the same number of turns (N_g). The gate windings are wound on separate cores. A third winding (N_c turns), called a control winding, is wound around the two cores containing the gate windings. The magnetic amplifier functions as an amplifier in that a signal applied to the control winding controls the flow of current in R_L produced by the source of alternating voltage. The control is effected through the action of the gate windings which act as switches which are closed for a fraction of the alternating current cycle which depends on the amplitude of the signal applied to the control winding.

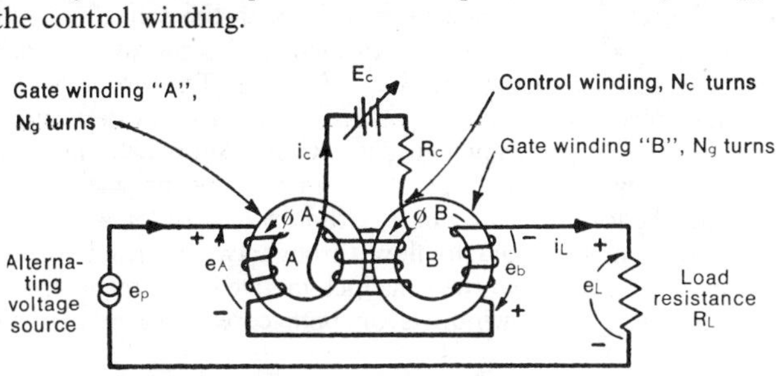

Fig. 8. Elementary magnetic amplifier.

The gate windings act as switches by virtue of the fact that their impedance is very high at one part of the cycle and very low at another part. The behavior of the impedance results from the characteristics of the magnetic material used for the cores on which the gate windings are located. The relation between the magnetomotive force produced by a coil wound on the core and the resulting magnetic flux in the core may be approximated by the graph shown in Fig. 9. Between $-M$ and $+M$ the flux curve has a very steep slope and outside of this range the graph is essentially a horizontal line. In the latter range the magnetic material is said to have reached saturation with a magnetic flux ϕ_s. The potential difference across an inductor

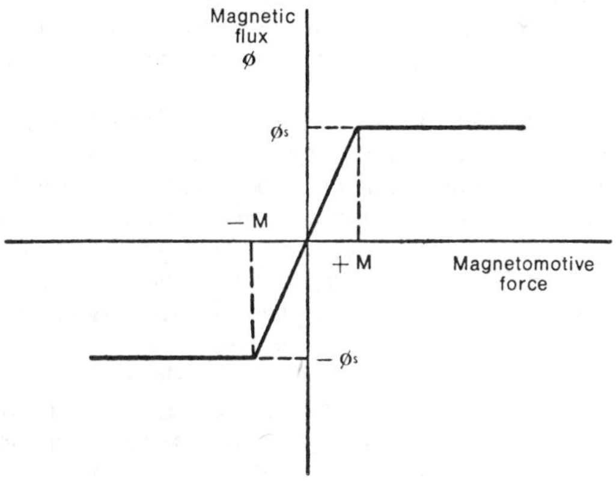

Fig. 9. Approximate relation between flux and magnetomotive force.

is directly proportional to the rate of change of the magnetic flux intercepted by the coil. Conversely, the flux intercepted may be expressed as the time integral of the potential difference. If a sinusoidal voltage is impressed across the terminals of an inductor, the value of flux that results at all instants of time may be determined from the time integral of the sinusoidal voltage.

By knowing the form of the flux-magnetomotive force (ampere-turns) relation (Fig. 9, for example) and the number of turns in the coil, the current that flows through the coil to provide the necessary ampere turns can be determined. Examination of Fig. 9 shows that, for applied voltages such that the flux produced is less than ϕ_s, a comparatively small magnetomotive force is required and hence a relatively small current will result in the coil. The inductor has a very high impedance under this condition. If the applied voltage results in a flux greater than ϕ_s, a large value of magnetomotive force, and thus a large value of current, will be required to satisfy the flux condition. The impedance of the coil for voltage amplitudes above a certain amount will thus be very low. These notions can also be expressed in terms of the equivalent inductance of the coil which is proportional to the slope of the flux-magnetomotive force curve.

From Fig. 9 it is seen that, for applied voltages resulting in operation between $-M$ and $+M$, the inductance will be very large, whereas outside of this region the inductance will be essentially zero. If a sinusoidal voltage is applied to a coil wound upon a core having the properties shown in Fig. 9, then essentially no current will flow in the coil until the amplitude of the sine wave reaches a critical value (the value corresponding to a flux ϕ_s), but for the time that the amplitude exceeds the critical value, the voltage drop across the coil will be close to zero, and the current that results will be determined principally by the amplitude of the applied voltage and the impedance of any other elements connected in series with the coil.

The ratio of power gain to time constant can be improved if positive feedback is applied to the amplifier of Fig. 8. The circuit with the feedback connection, as well as a bridge rectifier to provide a unidirectional load current, is shown in Fig. 10. For simplicity the control winding is shown as two separate windings, one on each core. In the circuit, the load current which, although it pulsates, always flows in one direction, is forced to flow through two feedback windings which are arranged so that their magnetomotive forces reenforce those created by the control windings on both cores. The ampere turns

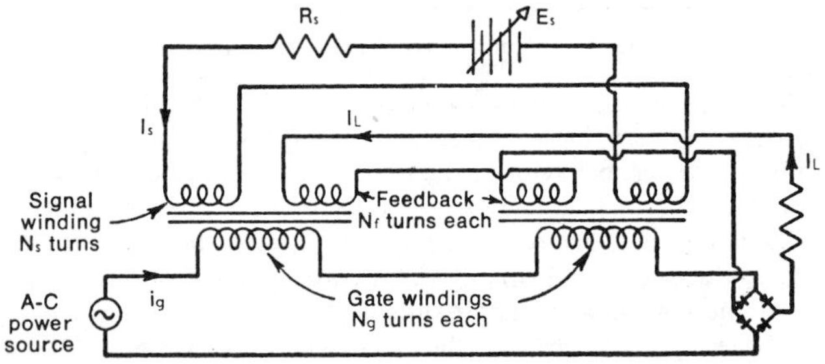

Fig. 10. Magnetic amplifier with external feedback.

contributed by control plus feedback windings are still approximately equal to those contributed by the gate windings, but because of the contribution of the feedback winding the control winding ampere turns for a given load current can be much less than without the feedback.

Two representative magnetic amplifier circuits are shown in Figs. 11 and 12.

(*Magnetron Amplifier*)—A traveling-wave magnetron used as an amplifier. The basic features of a typical structure are shown in Fig. 13. It resembles a section of a vane-type, cavity magnetron of infinite radius, excited at one end, and coupled to a load at the opposite end. A beam of electrons is projected through the space between the plane electrode and the cavity structure, which is maintained at a positive potential relative to the plane electrode. A magnetic field, normal to the plane of the paper, is adjusted so that the electrons do not strike the anode in the absence of alternating field. If the velocity of the electron beam is equal to, or nearly equal to, the phase velocity with which electromagnetic waves move down the loaded waveguide formed by the cathode and anode, energy is transferred to the electromagnetic wave from the source of direct voltage. The

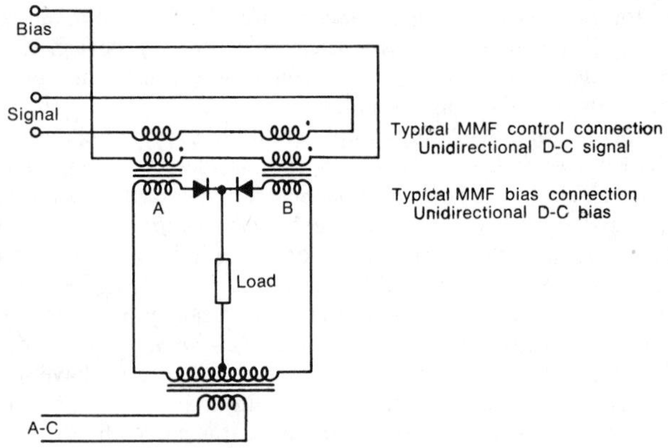

Fig. 11. Self-saturating magnetic amplifier circuit.

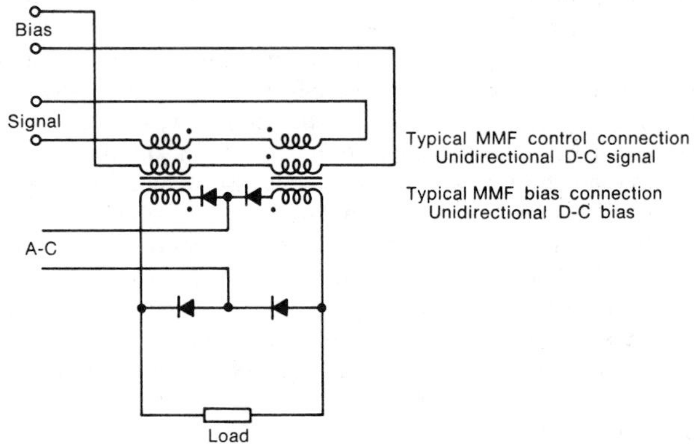

Fig. 12. Magnetic amplifier doubler circuit.

electromagnetic wave that moves from the input resonator to the output resonator, therefore, increases in amplitude, and the power output exceeds the power input.

(*Microwave Amplifier*)—The term maser stands for microwave amplification by stimulated emission of radiation. The amplification in a microwave amplifier is achieved by raising atoms or molecules of a paramagnetic material to an unstable high energy level. See also separate entry on **Maser.**

(*Modulated Amplifier*)—An amplifier stage in a transmitter in which the modulating signal is introduced and modulates the carrier.

(*Monitoring Amplifier*)—In broadcasting and recording, an amplifier with high input impedance and medium power output which is bridged across the program circuit. The available power output is used to operate a loudspeaker and/or headsets for the benefit of control personnel.

(*Neher Tetrode Amplifier*)—An integral-cavity, negative-grid tube designed to operate at a frequency of about 3,000 MHz.

(*Nonlinear Amplifier*)—An amplifier in which the output is not related to the input by a simple constant. One form is the volume-limiting amplifier, where the average gain is changed in such a manner that steady-state waveforms are accurately reproduced; another form,

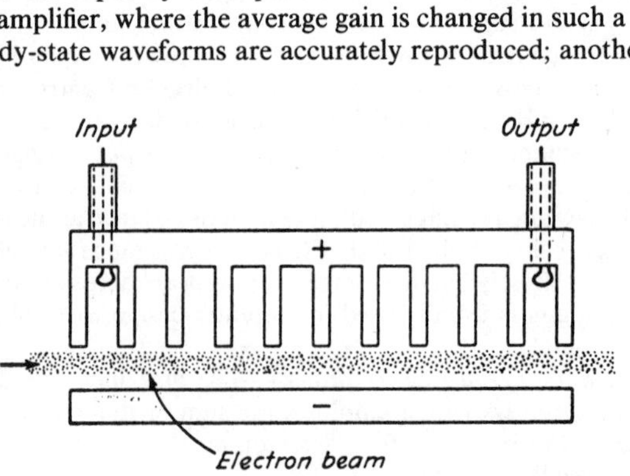

Fig. 13. Magnetron amplifier structure. Magnetic field perpendicular to plane of paper.

frequently called a clipping or over-driven amplifier, has an output which is a greatly distorted version of the input.

(*Operational Amplifier*)—Used to perform analog-computer functions, an operational amplifier is an amplifier with high dc stability and high immunity to oscillation, usually achieved by using a large amount of negative feedback. The operational-amplifier integrator shown in Fig. 14 is used in the conversion of voltage to a frequency.

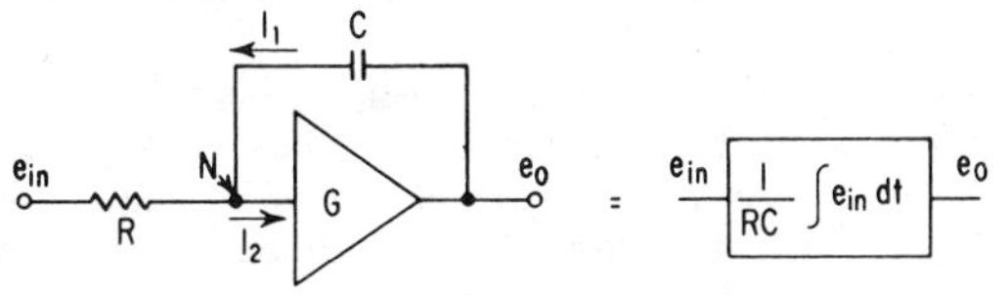

Fig. 14. Operational-amplifier integrator.

This is an amplifier with a capacitor connected between output and input and is used in digital voltmeters. As shown, there is a resistor in front of the input to the integrator. The amplifier is assumed to have a high gain. As the result of this configuration, the application of a step input results in a linear ramp output in which the slope of the ramp is proportional to the input-voltage step. The basic circuit of another integrating device making use of an operational amplifier is shown in Fig. 15. The voltage appearing at the output of the amplifier e_0 is proportional to the integral of the input voltage e_i, or $e_0 = 1/RC\int_{t_0}^{t_i} e_i\,dt$.

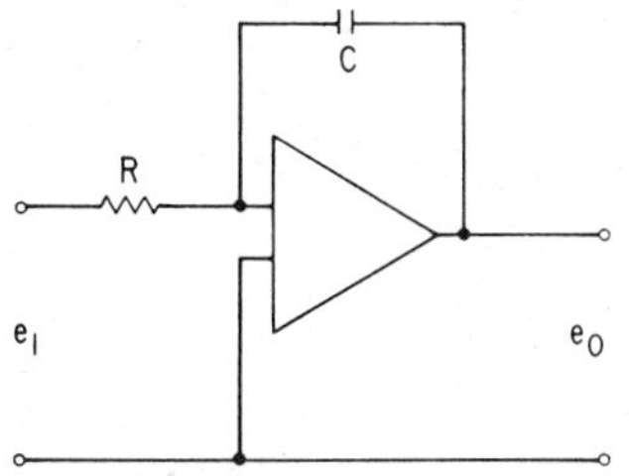

Fig. 15. Operational amplifier with capacitance feedback.

(*Paramagnetic Amplifier*)—An amplifier which increases the power level of a signal by means of the variations of an energy-storage parameter.

The classic example is the L-C resonant circuit in which the spacing of the capacitor plates is varied cyclically by a "pump" at a frequency ω_p, twice the signal frequency ω_s, which is identical to the resonant frequency of the circuit. Initial charge on the capacitor will result in a sinusoidal variation of capacitance charge as a function of time. If the capacitance is always decreased by the pump at the plus or minus peak of charge during which time charge is relatively constant, an increase in capacitor voltage must result, since $C = QV$. This increase in capacitor voltage represents an increase in the power level of the original signal, the additional energy representing the work done by the pump in increasing the separation of the capacitor plates against the forces of electrostatic attraction.

Maximum energy will be transferred from the pump frequency to the signal frequency if the pump frequency is twice the signal frequency and if the relative phase of the two is adjusted as described above. In a practical case, the separation of the two frequencies cannot be conveniently maintained constant. Therefore, the signal and pump will no longer interact favorably all the time, but will drift periodically into and out of the optimum condition. This produces a modulation of the signal frequency. resulting in frequency components in the output, among which are ω_s, ω_p, and $(\omega_p - \omega_s)$ the latter term being called the *image* or *idler* signal.

The mode of operation described above is called a *negative resistance amplifier,* since it acts to neutralize or overcome the positive resistance of the resonant circuit. Stability problems arise, since the output and input terminals are inherently the same. The closeness of the image frequency to the signal frequency makes the separation of the two difficult, one procedure being the utilization of a circulator. The pump signal may be filtered from the output by a conventional band-elimination filter.

A slightly different mode of operation, called variously the *up-conversion amplifier* or *amplifying-up converter,* differs from the negative resistance type of amplifier in these respects: (1) Since the gain can be shown to be proportional to $\omega_s + \omega_p/\omega_s$, the pump frequency is made as high as is practicable. (2) An additional frequency of interest is $(\omega_s + \omega_p)$. The signal frequency is inevitably shifted in the amplification process. (3) The up-converter is a two-port device with unconditional stability, without the requirement of a circulator.

A reverse-biased semiconductor diode provides an excellent nonlinear reactive element for use in a parametric amplifier inasmuch as it presents a very high shunt resistance and a capacitance which varies approximately inversely with the square root of the applied voltage. Thus, it is common practice to employ these elements as the capacitor C in Fig. 16. Parametric amplification can also be achieved by the utilization of the nonlinear characteristics of ferromagnetic and ferroelectric materials.

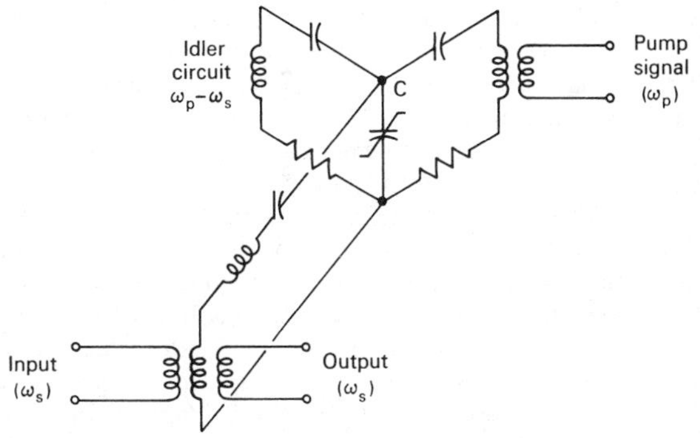

Fig. 16. Parametric amplifier.

(*Paraphase Amplifier*)—A phase-inverter amplifier used to convert a single-ended signal to a push-pull signal.

(*Pentriode Amplifier*)—A video amplifier containing a pentode which by virtue of suitable bypass and coupling devices, is made to operate as a triode (screen effectively connected to the plate) over a portion of the frequency range, and as a pentode (screen-grounded) over another part of the frequency range.

(*Pneumatic Amplifier*)—A device that increases signal strength pneumatically—as say a 3–15 psi (155–776 torr) instrument signal to several hundred psi to operate a large piece of process equipment, such as a slide valve, damper, furnace door, etc. Usually, such terms as booster, positioner, and relay, are used instead of amplifier.

(*Power Amplifier*)—An amplifier adjusted to operate with the objective of developing power in the load impedance in contrast with emphasis on production of output voltage or current. In many applications it is necessary to get power from an amplifier unit, so at least the final stage is usually adjusted to give power rather than voltage output. In radio-transmitting circuits or large power audio-amplifiers several stages may be power amplifiers. For this type of service the circuit is adjusted to give as large a current as possible through the load, which has much lower resistance than in voltage amplifiers. The transistors used are somewhat larger as a rule than those used for voltage and current amplification and are specially designed for large current outputs. Common use is made of the designations Classes A, AB, B, and C in describing the operation of power amplifiers. Since AB and B amplifiers using single tubes or transistors do not give output waves similar to the input waves, they must be operated push-pull for audio-frequency work. Class C amplifiers are used exclusively for radio-frequency work, where a tuned circuit may be used for the load. Push-pull operation may or may not be used in these applications, depending on the circumstances. For audio-frequency work the power amplifiers are normally connected to the load (loudspeaker) through a transformer, while in radio-frequency work air-cored transformers and capacitance coupling are common.

(*Preamplifier*)—This is a Class A voltage amplifier, which receives the signal from a microphone, pick-up, television camera tube, or other device supplying a low signal level and amplifies it so it can supply the input for additional amplifier circuits. Thus a preamplifier is commonly used in a radio or television studio to amplify the audio or video signal before feeding it into a mixer, line to the transmitter, or other amplifying equipment at the studio.

(*Program Amplifier*)—The amplifier following the master volume control in an audio console. Its gain brings the signal to a level suitable for transmission.

(*Push-Pull Amplifier*)—An amplifier in which the input signals applied to the transistors are opposite in phase and the output signals are combined to obtain twice the output of a single stage. Push-pull amplifiers have lower harmonic distortion than amplifiers with one transistor and with transformer-coupling they have the added advantage that the collector currents for the two devices act to magnetize the transformer core in opposite directions, tending to avoid saturation of the core. If the devices are perfectly balanced, no signal voltage exists across the bias resistor and a bypass capacitor normally used can be omitted.

The property of complementary symmetry makes it possible to dispense with a phase inverter in some push-pull transistor amplifiers. Figure 17 shows a push-pull amplifier formed from two common-collector connected transistors. Since the circuit has a low output impedance, it may be used for direct coupling to a loudspeaker without a transformer. This arrangement employs one *n-p-n* and one *p-n-p* transistor. Because of the complementary symmetry property, a positive signal current injected into the base of the *n-p-n* unit will cause its collector current to increase, whereas the same positive injected signal will cause the collector current of the *p-n-p* unit to decrease. As a result, the two base connections may be supplied from one signal source. In the figure the resistors connecting collectors and bases provide a small forward bias on the transistors. The resistance *R* is sufficiently low so that both transistors have essentially equal input currents. The transistors operate as Class B amplifiers in this circuit.

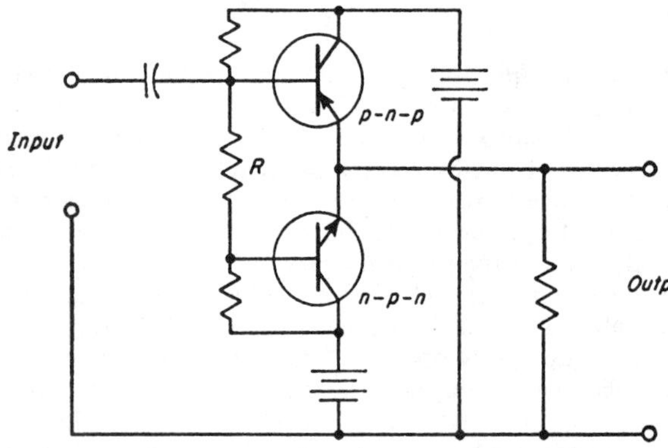

Fig. 17. Complementary symmetry push-pull amplifier.

(*Reflex Circuit Amplifier*)—A circuit in which a transistor simultaneously amplifies signals in two widely-separate frequency bands (i.e., the intermediate frequency signal of a superheterodyne and the audio-frequency output of the detector). Used very rarely because the savings of material and space are not justified by the additional complexity of operation.

(*Resistive-Wall Amplifier*)—An electron-beam amplifier in which the beam flows near a resistive wall. Gain is obtained through interaction between the stream-charge and the wall-charge, which is induced by the stream. The wall-charges act on the stream so as to cause larger and larger bunches to be formed, which result in exponential growth of the original signal with distance. Gain and gain-bandwidth are comparable to other forms of traveling wave tubes, and the stability is inherently greater.

(*Sample-and-Hold Amplifier*)—Also known as a track-and-hold amplifier, this device has an output that is proportional to the input until a "hold" signal is received. Upon receipt of that signal, the amplifier output is maintained essentially constant even though there may be changes in the input signal. See also separate entry on **Sample-and-Hold Amplifier.**

(*Servoamplifier*)—An amplifier used in a control or servosystem.

(*Single-Ended Amplifier*)—An amplifier in which each stage normally employs only one transistor, or if more than one such device is used, in which they are connected in parallel so that operation is asymmetric with respect to ground. See also separate entry on **Single-Ended Amplifier.**

(*Single-Ended Push-Pull Amplifier*)—A form of amplifier in which a pair of output terminals may have an instantaneous voltage of either polarity as may be dictated by the phase of the input signal.

(*Stagger-Tuned Amplifier*)—An amplifier incorporating staggered tuning to provide a desired bandwidth characteristic.

(*Step-Down Amplifier*)—A vacuum-tube type amplifier used to measure very high potentials which are impressed between anode and cathode, the anode being negative. The corresponding grid current is measured with the grid positive. This is sometimes called an "inverted voltmeter."

(*Transformer-Coupled Amplifier*)—An amplifier in which the interstage coupling from the output of one amplifying stage to the input of the next or to the load impedance is made using a transformer. The coupling method is used in audio-frequency amplifiers. The method is particularly advantageous in the opportunity offered for impedance matching when the load is a speaker or a low impedance transmission line. In Fig. 18 is shown the audio output stage of a transistor radio receiver. Changes in the base current resulting from the input signal cause corresponding changes in collector current which flows through the primary winding of the transformer. There results an alternating voltage across the secondary which has the same form as the input signal to the base and is capable of providing the necessary power output from the speaker. In this instance, the transformer makes possible matching the relatively low impedance of the speaker to the much higher output impedance of the transistor.

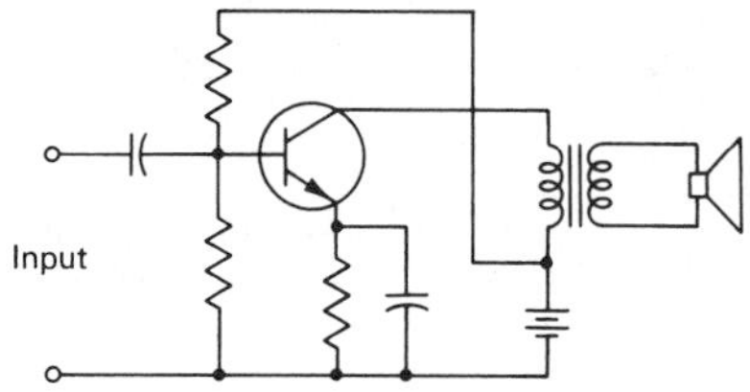

Fig. 18. Transistor audio amplifier.

(*Traveling-Wave Amplifier*)—An amplifier that uses one or more traveling-wave tubes to provide amplification of signals at frequencies of the order of thousands of megacycles.

(*Tuned Amplifier*)—An amplifier in which the load impedance consists of, generally, a parallel inductance-capacitance network, or two or more of these having electromagnetic coupling. The fact that the impedance of this network varies with frequency causes the gain of the amplifier to vary as a function of frequency in a somewhat similar manner. A tuned amplifier, employing a transistor and used for the amplification of intermediate frequency signals, is shown in Fig. 19. A tuned primary circuit is used and the transistor is connected across only a portion of the inductor. The arrangement offers advantages from the standpoint of impedance matching which permit higher gains to be achieved than would be possible otherwise in view of typical transistor characteristics.

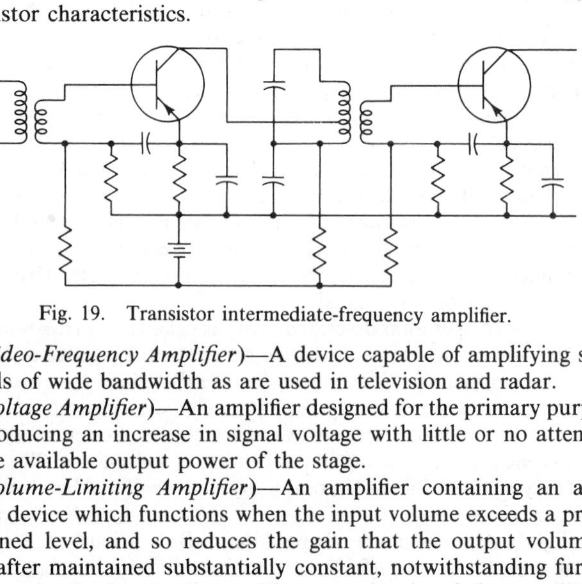

Fig. 19. Transistor intermediate-frequency amplifier.

(*Video-Frequency Amplifier*)—A device capable of amplifying such signals of wide bandwidth as are used in television and radar.

(*Voltage Amplifier*)—An amplifier designed for the primary purpose of producing an increase in signal voltage with little or no attention to the available output power of the stage.

(*Volume-Limiting Amplifier*)—An amplifier containing an automatic device which functions when the input volume exceeds a predetermined level, and so reduces the gain that the output volume is thereafter maintained substantially constant, notwithstanding further increase in the input volume. The normal gain of the amplifier is restored when the input volume returns below the predetermined limiting level.

(*Wide-Band Amplifier*)—An amplifier having uniform response over many decades of frequency. An example is the video amplifier.

References

Barna, A.: "Operational Amplifiers," Wiley, New York, 1971.
Becker, P. W.: "Design of Systems and Circuits," McGraw-Hill, New York, 1977.
Giacoletto, L. J.: "Differential Amplifiers," Wiley, New York, 1970.
Fink, D. G., Editor: "Standard Handbook for Electrical Engineers," 11th edition, McGraw-Hill, New York, 1979.
Harper, C. A.: "Handbook of Components for Electronics," McGraw-Hill, New York, 1977.
Landee, R.: "Electronics Designers' Handbook," 2nd edition, McGraw-Hill, New York, 1977.
Shrader, R. L.: "Electronics Fundamentals for Technicians," 2nd edition, McGraw-Hill, New York, 1977.
Temes, L.: "Electronic Circuits for Technicians," 2nd Edition, McGraw-Hill, New York, 1977.
Tobey, G. E., Huelsman, L. P., and G. G. Graeme: "Operational Amplifiers: Design and Application," McGraw-Hill, New York, 1971.
Zbar, P. B.: "Electricity-Electronics Fundamentals," 2nd edition, McGraw-Hill, New York, 1977.

AMPLIFIER (Analog Computer). Analog Computer.

AMPLIFIER (Darlington Pair). Darlington Pair.

AMPLIFIER (Fluidic). Fluidics.

AMPLIFIER (Maser). Maser.

AMPLIFIER (Push-Pull). Push-Pull Amplifier.

AMPLITUDE COMPARISON. The process of indicating the time at which two waveforms reach the same amplitude. It may also be considered to be the method of determining the abscissa of a waveform, given its ordinate.

AMPLITUDE DISCRIMINATOR. A circuit which performs an amplitude comparison. In addition, the sense and magnitude of the inequality of the amplitudes may be obtained.

AMPLITUDE DISTORTION. A type of distortion that occurs in an amplifier or other device when the amplitude of the output is not exactly a linear function of the input amplitude.

AMPLITUDE MODULATION. Modulation.

AMPLITUDE SEPARATION. The process of separating all values of a wave greater or less than a given amplitude, or those lying between two amplitudes.

AMPOULE. Sometimes spelled ampule, a small sealed glass container for drugs that are to be given by injection. As they are completely sealed, the contents are kept in their original sterile condition.

AMPULLA. Any flask-like dilatation, such as the small saccular outgrowth of the water vascular system of starfishes, at the inner end of the tube foot, or the dilated portion of the semicircular canals of the vertebrate ear.

AMYGDALOID. A vesicular rock, commonly a lava, whose cavities have become filled with a secondary deposit of mineral material such as quartz, calcite, and zeolites. The term is derived from the Greek word meaning almond in reference to the frequent almondlike appearance of the filled vesicles which are called amygdales or amygdules.

AMYLOPECTIN. Starch.

AMYLOSE. Starch.

ANABATIC WIND. Winds and Air Movement.

ANABOLIC AGENTS. Steroids.

ANABOLISM. Basal Metabolism.

ANACANTHINI. Codfishes.

ANACARDIACEAE. Cashew and Sumac Trees.

ANACLINAL. Used in structural geology to define a direction opposite to the dip of the strata or formations.

ANACONDA. Snakes.

ANAEROBE. An organism that can grow in the absence of free oxygen and referred to as an anaerobic organism. Subdivided into *facultative* anaerobes, which can grow and utilize oxygen when it is present; and *obligatory* anaerobes, which cannot tolerate even a trace of oxygen in their surroundings. Yeast is an example of a facultative anaerobe. Yeast can grow and utilize oxygen in its metabolism, in which case it utilizes all the energy in a carbohydrate and yields water and carbon dioxide. In the absence of oxygen, the yeast cells turn to fermentation and anaerobic metabolism wherein the carbohydrate is converted into alcohol and carbon dioxide. The bacterium which produces botulism in "preserved" foods is an obligatory anaerobe (*Clostridium botulinum*).

In addition to botulism, other species of Clostridium are the etiological agents of tetanus, gas gangrene, and a variety of animal diseases including struck, lamb dysentery, pulpy kidney, and enterotoxemia. The pathogenicity of the clostridia depends largely on the production of potent toxins. Generally, these are heat labile exotoxins and often several pharmacologically and immunologically different toxins may be produced by the same species. In recent years, certain of the toxins have been purified to the point of crystalline purity and revealed to be high-molecular-weight proteins. The availability of the pure toxins has permitted more precise studies both of the chemical nature of the toxin and also of their properties as enzymes. From such toxins, it is possible to produce efficient toxoids for use in prophylactic immunization measures for tetanus and botulism. Some clostridia, in contrast, are beneficial. In addition to the nitrogen-fixing ability of certain species, *C. sporogenes* is active in the decomposition of proteins and contributes to biodegradability of substances, particularly of plant remains. *C. acetobutyleium* is active in production of acetone and butyl alcohol by fermentation.

Glucose can be broken down through glycolysis into pyruvic acid, but in the absence of oxygen as a final acceptor for the hydrogen, the pyruvic acid cannot enter the tricarboxylic acid cycle for further breakdown. Instead the pyruvic acid itself serves as an acceptor for the hydrogen split off in glycolysis. Hence, much less energy is obtained from food in anaerobic metabolism. Lactic acid, alcohol, and some extremely poisonous substances are products of anaerobic metabolism.

In addition to the one-celled anaerobes, certain parasitic worms, such as *Ascaris*, are thought to use anaerobic metabolism, to a certain extent at least. Living in the intestine of higher animals these worms have little access to free oxygen.

The muscles of higher animals are also known to use anaerobic metabolism when the demands on the muscles for energy are greater than can be supplied through the available oxygen. Lactic acid is generated.

The diverse mechanisms that permit animal life in the absence of oxygen are described in considerable detail by P. W. Hochachka ("Living without Oxygen," Harvard Univ. Press, Cambridge, Massachusetts, 1980).

ANAL CANAL. Digestive System (Human).

ANALCIME. A common zeolite mineral, $NaAlSi_2O_6 \cdot H_2O$, a hydrous soda-aluminum silicate. It crystallizes in the isometric system, hardness, 5–5.5; specific gravity, 2.2; vitreous luster; colorless to white; but may be grayish, greenish, yellowish, or reddish. Its trapezohedral crystal resembles garnet but is softer; it is distinguished from leucite only by chemical tests.

There are many excellent European localities. Magnificent crystals occur at Mt. St. Hilaire, Quebec, Canada; in the United States at Bergen Hill and West Paterson, New Jersey, Keweenaw County, Michigan, and Jefferson County, Colorado. Nova Scotia furnishes beautiful specimens.

Analcime is a relatively common mineral and occurs with other zeolites in cavities and fissues in basic igneous rocks, occasionally in granites or gneisses. It seems to occur as a replacement and perhaps in some cases as a primary mineral crystallizing from a magma rich in soda and water vapor under pressure. The name analcime is derived from the Greek word meaning weak, in reference to the weak electric charge developed when heated or subjected to friction.

ANAL FEELERS. Posterior sensory appendages such as the anal cirri of annelid worms and the cerci of insects.

ANAL FIN. Fishes.

ANALGESIA. Anesthesia.

ANALGESICS. Drugs which diminish sensitivity to pain without impairing consciousness. These drugs act on the central nervous system and include opiates, coal tar analgesics, aminopyrine, salicylates, and phenylbutazone. Some of the analgesics also act in other ways pharmacologically. In the case of phenylbutazone, the excretion of uric acid is promoted. Salicylates reduce fever. Of the various alkaloids, only morphine and codeine are analgesics. Although still a valuable drug for some situations, particularly for relieving intense pain, morphine has several objectionable characteristics, including its toxicity and addicting nature. Codeine is a much less powerful drug and with much less objectionable aftereffects. See also **Alkaloids;** and **Morphine.**

Colchicine, an alkaloid, is used for the abatement of swelling and pain in acute attacks of gout. Normally, colchicine is not considered an analgesic, but in this circumstance, it does play this role as well as serving as an antipyretic and antiphlogistic (counteraction of fever and inflammation). See also **Alkaloids.**

For the relief of ordinary aches and pains, the coal-tar analgesics are commonly used. These include acetanilid, acetophenetidin (phenacetin), and N-acetyl-*p*-aminophenol. They frequently are mixed with caffeine and aspirin, or a barbiturate. Acetanilid is somewhat more toxic than acetophenetidin and thus is used less frequently. A side effect of N-acetyl-*p*-aminophenol may be minor gastrointestinal distress.

Aminopyrine, an analgesic and antipyretic, is used less frequently because it can cause agranulocytosis. See also **Agranulocytosis.**

Of the salicylate drugs (derivatives of salicylic acid), sodium salicylate and acetylsalicylic acid (aspirin) are the most widely used. The latter is poorly soluble in water and hydrolyzes into salicylic and acetic acids. Aspirin is combined with such compounds as phenacetin (acetophenetidin) and caffeine in a number of proprietary preparations. Aspirin is the most widely used medicine in the United States. Generally, aspirin is much more effective for pain originating in joints and muscles than in the internal organs. Aspirin also performs well in the treatment of acute rheumatic fever and is a uricosuric drug (stimulates excretion of uric acid). The latter causes symptoms of gout. Large quantities of aspirin also are consumed by arthritics. Aspirin is not totally harmless even though used widely. It contributes to accidental poisoning of children and can cause allergic reactions among some individuals.

ANALOG COMPUTER. A computer that solves problems by physical analogy. The computer translates temperature, flow, speed, altitude, voltage, and other physical variables into related electrical quantities and uses electrical equivalent circuits as an analog for the physical phenomenon being investigated. On an analog computer, for example, physical characteristics, such as weight or temperature are represented by voltage. Voltage is the electrical analog of the variable that is being analyzed. The variable itself can be electrical as well as hydraulic or pneumatic. Scale factors relate voltages in the computer to the variable in the problem being solved.

Components of the computer may be designed to operate within any fixed output voltage range. However, only two ranges are in wide use: ±100 volts and ±10 volts. All computer variables are scaled to lie within one of these ranges. For example, a temperature which varies from 0 to 1000°C is represented on a 100-volt computer by a voltage varying from 0 to 100. The scale factor would be 1/10 volt per °C.

Trends in recent years have strongly favored digital rather than analog computers for many purposes. For example, the slide rule has essentially been obsoleted by the digital pocket calculator. Because of their comparative simplicity, some analog computers have persisted. The familiar odometer in an automobile is an analog device.

ANALOG INPUT. This term is used to describe an assembly of equipment (subsystem) for performing the selection of the analog signal to be sampled, modification or conditioning of the analog signal, analog-to-digital conversion to provide a digital representation, and the necessary digital controls for subsystem control and communication with a digital computer. Generally included in this assembly are signal conditioners, such as attenuators and filters, analog signal multiplexers, amplifiers, analog-to-digital converters, and the control logic.

For data-acquisition or process-control computers, there are two fundamental analog-input subsystem types. However, within each of these basic configurations, there are numerous variations possible as dictated by the use of different kinds of multiplexers, amplifiers, and analog-to-digital converters. Cost and performance are the usual predominating factors in selection. The two major subsystem types are: (1) high-level, and (2) low-level systems.

High-Level Analog-Input Subsystems: High-level signals typically are in the range of 0 to ± 10 volts. These are connected to the subsystem input terminals of a configuration of the type shown in Fig. 1. See also **Analog-to-Digital Converter; Analog Multiplexer;** and **Signal Conditioning.** Including filtering, limiting to protect the analog-input subsystem from overvoltage, and attenuation, the signal conditioning of signals is performed on a *per-channel* basis. The multiplexer selects the signal to be converted and is under the control of the subsystem control logic. Isolation between the analog-to-digital converter and the input-signal source is provided by a buffer amplifier. A sample-and-hold amplifier may be used to reduce aperture-time errors. The buffer amplifier sometimes is included as an integral part of the analog-to-digital converter. The timing and sequencing of the subsystem operation is transmitted by the control logic, which also transmits the digital value to the digital computer.

Usually configurations of these subsystems are single-ended, although some differential systems are obtainable. The latter are more costly, but offer easier installation because noise and ground loops are less serious. Reduction of the latter problems is important particularly in the case of high-speed, high-resolution applications and where input signals must come from signal sources a considerable distance from the subsystem.

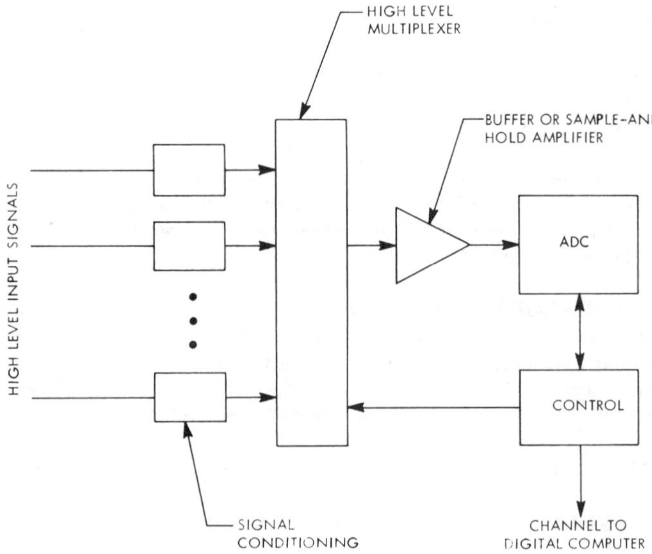

Fig. 1. High-level analog-input subsystem.

Solid-state switches, capable of sampling signals at rates in excess of 250,000 samples/second, normally are used in the analog multiplexer. Both bipolar transistors and field-effect transistors are used. Bipolar transistor switches restrict levels to a high level because of voltage-offset errors. The most common analog-to-digital converter used is of the successive-approximation type, since its speed matches the solid-state multiplexer characteristics. For lower-speed applications, ramp and intergrating-ramp analog-to-digital converters are used.

Timing and channel communications functions are provided by the control logic. The digital computer normally initiates a control word, specifying one or more input addresses, to commence operation of the subsystem. The control unit selects the appropriate multiplexer switches. After a predetermined delay time, the control unit provides a "start convert" signal to the analog-to-digital converter. The latter operates asynchronously with respect to the computer and, at the end of the conversion, signals the control logic. See also **Asynchronous.** Prior to transmitting the result back to the digital computer, the control unit may reformat the output word from the analog-to-digital converter, that is, adding parity bits or converting from parallel to a serial form. It may also collect some or all of the results in a buffer storage where they are available to the computer on demand.

The subsystem sampling speed may be as high as 100,000 samples per second at resolutions up to 14 bits. See also **Resolution (Computer System).** These characteristics are determined by the type of multiplexer, amplifier, and analog-to-digital converter used. Higher speeds are obtainable, but with reduced resolution, or using special techniques. Under controlled operating conditions, a total measurement error of less than 0.05% of full scale is obtainable.

Subsystems of the type just described are found in numerous high-speed data-acquisition applications, including hybrid computation and rocket test stand monitoring, as well as for general research. These uses typically have wide channel-bandwidth requirements, high-level signals, and/or a relatively small number of signals. Because the configuration is not adapted to the handling of low-level signals, it is not used widely for process control. For applications involving only a few data points, a high-speed low-level system can be achieved by adding a high-gain amplifier for each input signal. This is a costly approach and not practical where there are large numbers of low-level signals.

Low-Level Analog-Input Subsystems: Systems of this type are similar to the high-level system just described with the exception of the use of a time-shared low-level amplifier. See Fig. 2. In operation, the input signals are conditioned and multiplexed by a low-level multiplexer. Before conversion into a digital representation by the analog-to-digital converter, the multiplexer output is amplified by a time-shared amplifier. Measurement accuracy for the total system for a 50-millivolt signal range typically is 0.1%, with a resolution of 10 to 14 bits.

The low-level signals generally are less than 100 millivolts and often as low as 10 mV full scale. Thus, this type of subsystem usually provides a differential-signal input. See also **Differential-Mode Voltage.** The multiplexer and signal-conditioning circuits are differential. The amplifier may or may not be differential, depending on the type of multiplexer used. A single-ended amplifier may be used where the multiplexer provides common-mode isolation, such as a transformer-coupled multiplexer. The output of the amplifier and the analog-to-digital converter input are single-ended in most cases.

Differential-multiplexer configurations for low-level uses most commonly are the double-pole single-throw differential, flying-capacitor, and transformer-coupled designs. Switching devices normally are dry-reed relays, field-effect transistors, or mercury-wetted contact relays. Field-effect transistor switches can be applied for speeds in excess of 100,000 samples/second. Electromechanical devices are limited to about 300 samples/second or less. In any case, however, subsystem speed normally is limited to less than 10,000 samples/second because of amplifier performance. With the proper selection of multiplexer configuration and the capabilities of the amplifier, common-mode voltages up to several hundred volts will not affect the acceptable performance of the system. However, where solid-state switches are used in multiplexers, the tolerance is limited to something less than 30 volts of common-mode voltage. See also **Common-Mode Voltage.**

In the design of Fig. 2, a differential or single-ended amplifier may be used, depending on the multiplexer characteristics. In some cases, the amplifier will have a fixed gain. In other instances, the control logic may select the gain, either automatically or under control of the computer program. Where a differential amplifier is used, the common-mode tolerance of the amplifier is critical to determining the common-mode rejection of the subsystem. Low-level applications also require that amplifier noise be kept to a minimum.

The design configuration just described is used extensively in process control and large data-acquisition systems. Such uses typically have low channel-bandwidth requirements and a large number of low-level signals. The time-shared amplifier provides the system with a cost advantage as compared with an amplifier-per-channel type system. The low bandwidth application characteristic makes it possible to use the limited system-sampling speed in numerous applications.

For related topical coverage in this volume, see list of entries under **Data Processing.**

Thomas J. Harrison, International Business Machines Corporation, Boca Raton, Florida.

ANALOG MEMORY. **Memory Circuit (Analog).**

ANALOG MULTIPLEXER. An array of analog switches used for the selection of one of several analog signals for transmission to subsequent devices in a data acquisition subsystem. Multiplexers most typically are used for signal selection prior to amplification and conversion to digital form in time-shared subsystems. An analog multiplexer used in the analog-input subsystem of a data-acquisition system is shown schematically in Fig. 1. Multiplexers are not confined to data-acquisi-

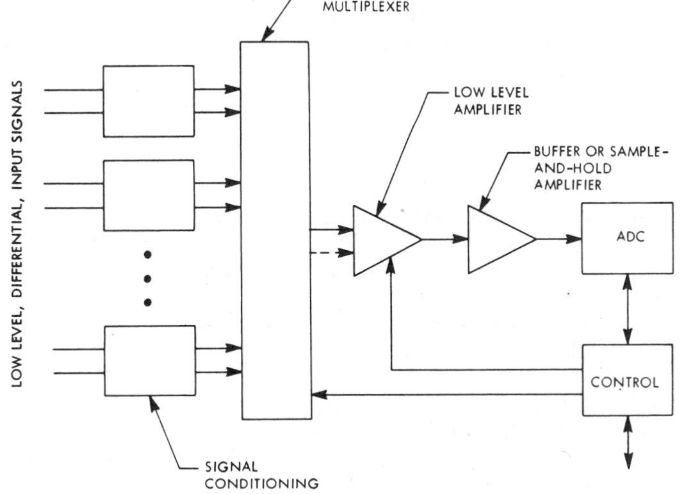

Fig. 2. Low-level analog-input subsystem.

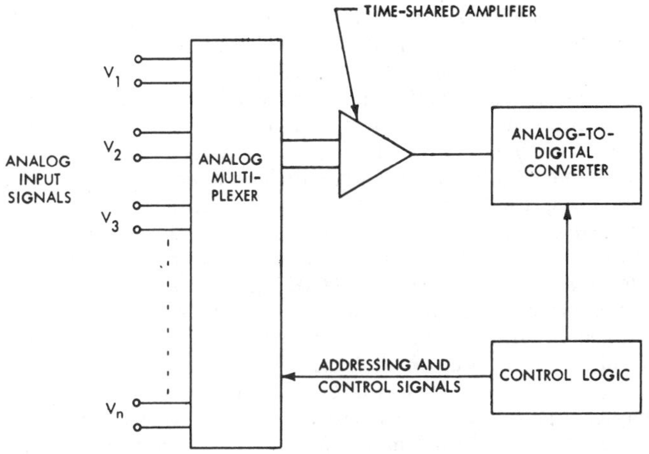

Fig. 1. Location of analog multiplexer in analog-input subsystem.

tion systems, but are used in audio-switching networks in central telephone exchanges. The particular hardware configuration varies considerably with the nature of the signals and their manipulation. See Fig. 2. See also **Analog Switch.**

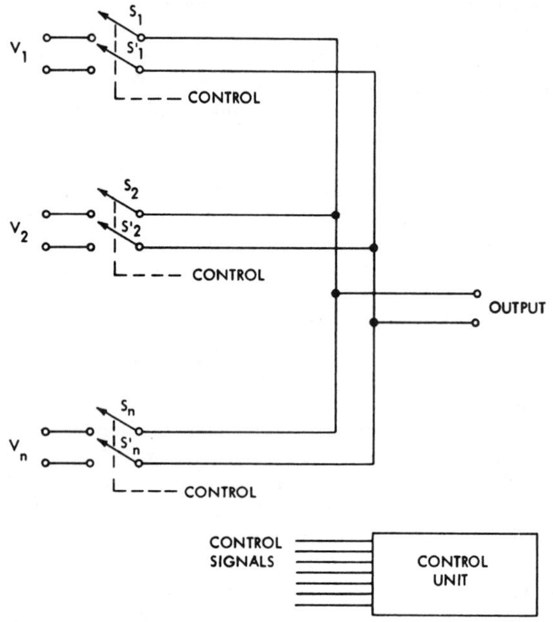

Fig. 2. Generalized configuration of analog multiplexer.

The control unit supervises switch selection. Where the multiplexer is differential, two or three switches are required for each input signal. Where the multiplexer is single-ended, only one switch per input is required, using a common ground to complete the circuit for all input signals. Although the control unit varies with particular applications, its primary function is to furnish the signals required to select or address the various switches in the multiplexer at the proper time. The logic in the control unit or a command from a computer will determine the particular address to be selected. In some cases, the switches may be selected sequentially under control of a ring counter circuit in the control unit. Or, the selection of each address may require a data word from the computer or microprogrammed control unit.

The control unit also furnishes timing for the multiplexer and associated equipment. Usually, the analog-to-digital (A/D) converter cannot commence a conversion instantly after selection of the multiplexer switch. A time delay is required to permit actuation time of the switch, for settling of the amplifier, and for the dissipation of other transients. Counters or delay circuits in the control unit usually furnish the required timing signals.

Several types of solid-state or electromechanical analog switches may be used in the multiplexer. Types of electromeromechanical switches used are mercury-wetted contact relays, dry-reed relays, and crossbar switch assemblies. Of the solid-state switches, most often field-effect transistors are used. For special applications, diodes, bipolar transistors, and silicon-controlled rectifiers may be used.

Among the major performance characteristics of an analog multiplexer are accuracy, sampling speed, noise. Common-mode rejection ratio also is important in the instance of a differential multiplexer. The features of the multiplexer switching device mainly determine sampling speed. The electromechanical devices, such as mercury-wetted contact relays and dry-reed relays, usually are limited to about 250 samples/second. Bipolar and field-effect transistors can provide sampling speeds in excess of 100,000 samples/second. Although a switch may close rapidly, a time interval must be provided to permit transients to dissipate prior to conversion of the multiplexer output signal to digital form. In the case of the mercury-wetted relay, the actual switch-closing time is in the range of 1 millisecond, but there may be a noise transient which will persist at levels in excess of 10 microvolts for perhaps 5 milliseconds or even longer. Where a noise level of this magnitude cannot be tolerated, a delay is required between switch selection and conversion initiation. Noise considerations also affect solid-state switches, but noise duration times are generally much shorter.

Careful consideration must be given to possible errors that a multiplexer can contribute to the operation of a digital-data acquisition subsystem. Leakage currents associated with the "off" switches will, in the case of solid-state switches, flow through the source impedance of the channel being sampled by the "on" switch and also through the multiplexer load impedance. Offset in the voltage being sampled thus can be caused by these currents. Even though leakage current from a single switch may be small (considerably less than a microampere), an appreciable error can result from the cumulative effect of all the switches that are connected to the common multiplexer output bus.

Inasmuch as the field-effect transistor does not show an offset voltage in the "on" condition, it may be used for multiplexing low-level signals. It should be noted, however, that some field-effect transistors show a significant "on" resistance and, because of source loading effects, can introduce errors. Further, the "on" resistance increases the effect of leakage currents as previously described. This may increase the output-signal rise time.

Where differential multiplexers are used, the common-mode rejection ratio is important. This factor is determined by leakage impedances between the switch-signal path and the drive circuit. Further, the maximum common-mode voltage which may be applied without causing damage to the multiplexer is determined by the breakdown voltage of the switch. Electromechanical devices have high drive-to-signal path and contact-breakdown voltages. Hence, these devices can be used for multiplexers that will withstand several hundreds of volts of common-mode voltage. Generally, solid-state multiplexers are limited to a common-mode voltage less than 20 to 30 volts. Improvement can be obtained through the use of isolation techniques, such as transformers.

The chances of multiplexer errors increase as the number of input channels in the multiplexer increases. Additional channels decrease the common-mode rejection ratio and also increase the offset resulting from leakage currents. *Submultiplexing* or *block switching* are techniques frequently used to minimize these conditions. Block switching entails the use of two levels of multiplexing. The first level is the same as that indicated in Fig. 2. This level furnishes selection of an input signal. The second level of multiplexing is provided at the output of the first-level multiplexer. As an example, a second-level switch may be provided for each group of 16 first-level multiplexing switches. Of course, in the selection of an input, both the first-level switches and their associated second-level switch must be actuated. The advantage of second-level multiplexing is the isolation of each block from the leakage currents and other possible disturbances which may arise from other inputs. In a system of 16 blocks of 16 channels each, for example, the leakage errors are determined mainly by the 15 "off" switches in the same block as the addressed point and the 15 "off" block or second-level switches. Hence, these errors equal those encountered in a 30-channel single-level multiplexer although a total of 256 input channels are serviced through the two-level multiplexer.

For related topical coverage in this volume, see list of entries under **Data Processing.**

Thomas J. Harrison, International Business Machines Corporation, Boca Raton, Florida.

ANALOG OUTPUT. This term is used to describe an assembly of equipment (subsystem) and operations in a process control computer or data-acquisition system with the capability to provide a continuous voltage or current output which can be controlled by a digital computer. Closed-loop control systems, for example, which utilize a digital process control computer require current signals which may have ranges of 4 to 20, 1 to 5, or 10 to 50 milliamperes. These signals, in turn, are used for controlling process actuators, such as valve positioners. For the generation of visual displays, recorders and cathode-ray tubes require similar analog signals.

The digital-to-analog (D/A) converter is the principal component of the analog-output subsystem. Simply defined, a D/A converter is an electronically controlled attenuator network and a constant reference source. Digital-input signals received by the D/A converter activate analog switches which determine the attenuation factor of a passive network. Either electromechanical or solid-state switches may be used. See also **Analog Switch.** The input to the attenuator network is a constant-voltage or -current source. The output of the network is proportional to the attenuator switch settings and, thus, to the digital-input signals.

Connection of the D/A converter to the digital computer is through control logic. The latter provides addressing for each D/A converter and also the required control and timing signals. Shown in the accompanying diagram is an analog-output system of the kind used in process

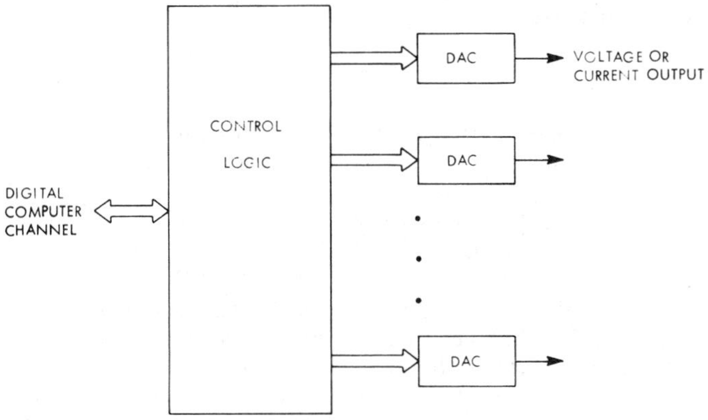

Basic analog-output subsystem.

control computer or data acquisition systems. The digital computer, via digital-control words, specifies which D/A converter is to be adjusted. Another digital data word carries the desired output value. The address information is decoded in the control logic. The digital data also are routed to the input of the addressed D/A converter. This setting of the input switches of the D/A converter provides the desired analog-output value. The kind of subsystem shown is frequently used where the analog-output value is changed often, at a rate in excess of 1000 samples/second. This subsystem configuration can adjust many D/A converters at rates exceeding 100,000 samples/second. The resolution of the D/A converters used for applications of this type usually ranges from 8 to 13 bits. The total error of the output voltage, under controlled operating conditions, may be less than 0.01%.

Thomas J. Harrison, International Business Machines Corporation, Boca Raton, Florida.

ANALOG SWITCH. This term applies to a large family of switches that are designed for switching analog signals and normally implies high accuracy and high resolution. In data-acquisition and instrumentation systems, the most important switch characteristics are speed, voltage and current errors, "on" and "off" resistance, and noise.

Electromechanical analog switches, including mercury-wetted contact and dry-reed relays, usually have a lower "on" resistance and a higher "off" resistance than most solid-state switches. Also, because of excellent isolation between drive and signal circuits, electromechanical switches have no inherent voltage offset and leakage currents. Disadvantages include their slow performance as compared with solid-state switches and the production of noise by the switching action, a condition which tends to persist for an appreciable period.

Solid-state switches on the other hand are high-speed devices and with "on" and "off" resistances, compared with electromechanical switches, as described above. Where a *pn* junction is part of the signal path, as in a bipolar transistor, the switch will show an inherent voltage offset. The leakage currents between drive and signal paths also are larger in bipolar transistors as the result of less isolation between drive and signal. Field-effect transistors display leakages comparable with those of electromechanical switches. Noise results from switching action mainly because of coupling of the drive signal into the signal path by way of interelectrode capacitance. Even though the noise magnitude may be appreciable, decay to a negligible value occurs much more rapidly than in electromechanical switches.

Use of the major classes of analog switches is summarized as follows:

Relays and Other Electromechanical Devices—used for low-level signals (less than 1 V full-scale) and multiplexing speeds generally not exceeding 250 samples/second.

Field-Effect Transistors—used for low-level applications at higher speeds. Limited in common-mode-voltage handling capability as result of lower breakdown voltages.

Bipolar Transistors—used for high-level signals, but with appropriate circuitry for compensation of offset voltages; also may be used for low-level signals. They have been used less frequently in recent years due to improvements in field-effect transistor technology.

See also **Bright Switch; Dry-Reed Relay; Field-Effect Transistor Switch (FET); Light-Coupled Switch; Mercury-Wetted Relay;** and list of entries under **Data Processing.**

Thomas J. Harrison, International Business Machines Corporation, Boca Raton, Florida.

ANALOG-TO-DIGITAL CONVERTER. Abbreviated A/D converter or ADC. A device that provides a digital representation of an analog quantity. Examples of the latter include voltage, current, or a position. There are two principal types of A/D converters used in data-acquisition systems: (1) electromechanical, and (2) electronic converters. The electromechanical types sometimes are referred to as shaft- or position-to-digital encoders. See also **Encoder.** Generally, they are comprised of a mask attached to the moving mechanical element, along with a means to read the information on the mask. Magnetic, optical, and electrical means are used for reading. Where electrical sensing is used, the mask may consist of a conducting pattern on an insulated substrate. The code represented by the mask pattern is read by means of fixed conducting brushes which are in contact with the pattern. Optical sensors use a light source and photodetectors, whereas magnetic sensors employ inductive-pickup coils.

The sensing means has a finite width. Consequently, there may be an ambiguity in the digital output. This would result, for example,where a conducting brush would be in contact with two adjacent portions of the mask pattern at the same time. Ambiguity can be avoided through the use of special codes and ingenious arrangements of the sensing detectors. Gray code, where only one bit in the digital output changes at any given time as the position of the mask is varied, may be used. A V-scan technique also can be used. In the latter, two detectors are used for each track on the mask. With appropriate decoding of the outputs from the pairs of detectors, an unambiguous digital representation will result.

Electronic A/D Converters: Two classes are used: (1) the input quantity, usually a voltage or current, is converted into another form, such as a frequency or a pulse duration. This intermediate quantity then is measured to yield a digital representation of the input signal. (2) the input is compared directly with a known reference signal which can be varied under control of the A/D converter logic. Several subclasses of A/D converters are obtainable within these two broad classes.

The principal techniques most frequently employed in process control and data-acquisition computers are (1) ramp, (2) integrating-ramp; (3) voltage-to-frequency; (4) successive-approximation; and (5) parallel-serial methods. With reference to the two broad classifications, the ramp, integrating ramp, and voltage-to-frequency methods involve the conversion of the input signal into an intermediate quantity before measurement. The successive-approximation and parallel-serial methods are direct-comparison methods.

All of the foregoing methods are described elsewhere in this book. See also **Integrating-Ramp A/D Converter; Parallel-A/D Converter; Parallel-Serial A/D Converter; Ramp A/D Converter; Successive-Approximation A/D Converter; and Voltage-to-Frequency A/D Converter.**

Numerous cost/performance trade-offs are involved in the selection of the most appropriate A/D converter for a given process control or data-acquisition system. The characteristics are summarized as follows:

Ramp and Voltage-to-Frequency Converters—relatively slow (require serial counting); used at speeds less than several thousand samples/second; usually resolutions of less than 12 bits.

Successive-Approximation Converters—useful up to about 100,000 samples/second at resolution of 16 bits. Useful up to more than 250,000 samples/second at resolution of 8 bits or less.

Parallel-Serial Converters—high-speed uses requiring conversion rates in excess of 100,000 samples/second; resolution of 8 to 14 bits.

For related topical coverage in this volume, see list of entries under **Data Processing.**

Thomas J. Harrison, International Business Machines Corporation, Boca Raton, Florida.

ANALOGY (Dynamics). Many dynamical systems, especially mechanical and acoustical ones, can be analyzed by analogy to electrical systems. This is true because these various systems possess certain common attributes such as energy, power, frequency (of vibration, rotation or alternation), and because equations for calculating these and other quantities can be written in terms of variables that are closely analogous. For example, electromotive force in an electrical system is strictly analogous to force in a mechanical rectilineal system, to torque in a mechanical rotational system and to pressure in an acoustical system. Again, current in an electrical system is closely analogous to linear velocity in a mechanical rectilineal system, to angular velocity in a mechanical rotational system and to volume current in an acoustical system.

To show the applications of these analogies the following relationships are given for the four systems cited. Note that a dot above a quantity denotes its first derivative with respect to time.

The kinetic energy T_{KE} stored in the magnetic field of the electrical circuit is

$$T_{KE} = \tfrac{1}{2}Li^2$$

where L = inductance, in abhenries, and i = current through the inductance L, in abamperes.

The kinetic energy T_{KM} stored in the mass of the mechanical rectilineal system is

$$T_{KM} = \tfrac{1}{2}m\dot{x}^2$$

where m = mass, in grams, and $\dot{x}$ = velocity of the mass m, in centimeters per second.

The kinetic energy T_{KR} stored in the moment of inertia of the mechanical rotational system is

$$T_{KR} = \tfrac{1}{2}I\dot{\phi}^2\theta$$

where I = moment of inertia, in gram (centimeter)^2, and $\dot{\phi}$ = angular velocity of I, in radians per second.

The kinetic energy T_{KA} stored in the inertance of the acoustical system is

$$T_{KA} = \tfrac{1}{2}M\dot{X}^2$$

where $M = m/S^2$, the inertance, in grams per (centimeter)^4
m = mass of air in the opening, in grams
S = cross-sectional area of the opening, in square centimeters
$\dot{X} = S\dot{x}$ = volume current, in cubic centimeters per second
$\dot{x}$ = velocity of the air particles in the opening, in centimeters per second

Similar relations hold for potential energy. Thus the potential energy V_{PE} stored in the electrical capacitance of the electrical circuit is

$$V_{PE} = \frac{1}{2}\frac{q^2}{C_E}$$

where C_E = capacitance, in abfarads, and q = charge on the capacitance, in abcoulombs.

The potential energy V_{PM} stored in the compliance or spring of the mechanical rectilineal system is

$$V_{PM} = \frac{1}{2}\frac{x^2}{C_M}$$

where $C_M = 1/s$ = compliance of the spring, in centimeters per dyne
s = stiffness of the spring, in dynes per centimeter
x = displacement, in centimeters

The potential energy V_{PR} stored in the rotational compliance or spring of the mechanical rotational system is

$$V_{PR} = \frac{1}{2}\frac{\phi^2}{C_R}$$

where C_R = rotational compliance of the spring, in radians per dyne per centimeter, and ϕ = angular displacement, in radians.

The potential energy V_{PA} stored in the acoustical capacitance of the acoustical system is

$$V_{PA} = \frac{1}{2}\frac{X^2}{C_A}$$

where X = volume displacement, in cubic centimeters
$C_A = V/\rho c^2$ = acoustical capacitance, in (centimeters)^5 per dyne
V = volume of the cavity, in cubic centimeters
ρ = density of air, in grams per cubic centimeter
c = velocity of sound, in centimeters per second

Moreover, the total energy in any of these systems can be found simply by adding the kinetic and the potential energies. These relationships are particularly useful in the facility with which relationships like Kirchhoff's law, and its mechanical and acoustical analog, D'Alembert's principle, can be used in the solution of mechanical or acoustical networks, that is, in various series of springs, gears and acoustical elements such as are found in complex trains of equipment as, for example, reduction gearing, loudspeakers, microphones, phonograph pickups, automobile suspensions, wave filters, and other electromechanical and electroacoustic transducers. Even the electrical reciprocity and superposition theorems have their analogs for mechanical and acoustical systems.

The construction of electrical analogies is frequently used in the analysis of process and process control performance. The field of fluidics has brought forth some very interesting and helpful applications of the analog technique.

ANALOGY (Physiology). The relationship between body parts having different embryonic and phylogenetic origin, but with the same function. As an example, the wings of insects and the wings of birds are analogous and illustrate the principle of analogy. These wings are used for the same function, flying, but they arise in an entirely different manner in the embryonic development. See **Homology.**

ANALYSIS (Chemical). Analytical chemistry is that branch of chemistry which is concerned with the detection and identification of the atoms, ions, or radicals (groups of atoms which react as a unit) of which a substance is composed, the compounds which they form, and the proportions of these compounds which are present in a given substance. The work of the analyst begins with sampling, since analyses are performed upon small quantities of material. The validity of the result depends upon the procurement of a sample that is representative of the bulk of material in question (which may be as large as a carload or tankload). See accompanying photo.

For clarity and convenience of description, analytical chemistry may be divided into:

1. *Qualitative chemical analysis,* in which one is concerned simply with the identification of the constituents of a compound or components of a mixture, sometimes accompanied by observations (rough estimates) of whether certain ingredients may be present in major or trace proportions.
2. *Quantitative chemical analysis,* in which one is concerned with the amounts (to varying degrees of precision) of all or frequently of only of some specific ingredients of a mixture or compound. Classically, quantitative chemical analysis is divided into (a) *gravimetric analysis* wherein weight of sample, precipitates, etc., is the underlying basis of calculation, and (b) *volumetric analysis* (titrimetric analysis) wherein solutions of known concentration are reacted in some fashion with the sample to determine the

Improved instrumentation and control techniques in the laboratory have accelerated research progress, not only by providing more complete, more accurate data faster, but by freeing scientific personnel from routine analytical tasks. The experimental set-up shown here at left is connected to a mass spectrometer for instantaneous data-taking. (*The Dow Chemical Company*)

concentration of the unknown. Obviously, the figures from either gravimetric or volumetric determinations are convertible and the two methodologies frequently are combined in a multistep analytical procedure.

See also **Titration (Potentiometric); Titration (Thermometric); Weighing.**

Analytical methods are affected not only by the type of material under examination, but also by the objectives of the analysis. *Ultimate analysis*, for example, is a term used to describe the determination of the proportions in which the elements are present in an organic substance, such as the amounts of and the ratio of carbon to hydrogen. This is normally a relatively simple procedure and often will provide answers sufficient to the need—as contrasted with the usually much more complex procedures required to identify specific compounds in a mixture of organic compounds. The term *functional group analysis*, where the analyst is concerned with the determination of specific organic groups and radicals, such as the carboxyl or carbonyl groups, also is used. Both of these terms are described under **Analysis (Organic Chemical).** In the case of solid fuels, such as coal, the term *proximate analysis* is used to describe the determination of moisture, volatile matter, fixed carbon, and ash. See also **Coal.** Although similar objectives (ultimate analysis and functional group analysis) also arise in connection with the analysis of inorganic materials, no special term applies.

Analytical chemistry also may be classified in terms of kinds of procedures in addition to those already mentioned:

3. *Classical laboratory, manual methods* conducted on a macroscale where sample quantities are in the range of grams and several milliliters. These are the techniques that developed from the earliest investigations of chemistry and which remain effective for teaching the fundamentals of analysis. However, these methods continue to be widely used in industry and research, particularly where there is a large variety of analytical work to be performed. The equipment, essentially comprised of analytical balances and laboratory glassware, tends to be of a universal nature and particularly where budgets for apparatus are limited, the relative modest cost of such equipment is attractive.
4. *Microchemical methods* which essentially extend macro-scale techniques so that they may be applied for determinations involving very small (milligram) quantities of samples. These methods have required fully new approaches or extensive modifications of macro-scale equipment. Consequently, the apparatus usually is sophisticated, relatively costly, and requires greater manipulative skills. Nevertheless, microchemical methods opened up entirely new areas of research, making possible the determination of composition where the availability of samples, as in many areas of biochemistry, was confined to very small quantities.
5. *Semi-automated apparatus* which introduces an interim step between (a) macro-scale and microchemical analysis techniques on the one hand and (b) fully instrumented and automated analytical methods on the other hand. Significant design changes in chemical balances that greatly increase the speed of weighing samples and reagents and automatic and self-refilling burettes are examples of ways in which an analytical procedure can be "tooled" to conserve manpower, reduce drudgery, and often contribute to more reliable and precise results.
6. *Analytical instrumentation* which first appeared in a major way in the 1930s and which has been expanding at a fast pace ever since, has revolutionized analytical chemistry. Instrumental procedures range from essentially the replication on a miniature scale of a highly automated laboratory analytical procedure involving chemical reagents and reactions (represented by automatic chromatographs) to numerous techniques which essentially infer the presence and quantities of substances from the manner in which the unknowns react with various forms of energy. The latter procedures are extremely *unlike* former conventional laboratory techniques. Of course, over the years, there were harbingers of later-generation analytical instruments in the form of early polariscopes and spectroscopes. Modern analytical instruments range from small, inexpensive table-top black boxes, such as simple electrometric pH and electric-conductivity devices, up to very large, costly mass spectrometers and other forms of electronic, photometric, and spectrophotometric apparatus. Analytical instrumentation also ranges widely from the standpoint of manual assistance and intervention required in its application. Usually as the equipment becomes more sophisticated, less (if any) direct attention is required in the performance of the analyses, but greater skill and time are required in the initial planning of an analytical procedure.
7. *Process analyzers* represent the ultimate extension of analytical instrumentation, moving the chemical control laboratory from a central location, dependent upon grab samples, out to the process where the chemical composition changes to be measured are occurring within process vessels and pipelines. Many of the laboratory-type analytical instruments, at least in principle, are applicable to on-line situations, but new problems arise, such as effective sampling systems, the maintenance of delicate equipment in severe environments, and arranging for back-up control when there are equipment failures. On-line analytical equipment consequently is usually costly, both in terms of initial investment and of maintenance. Nevertheless, the improved control wherein long time lags are virtually eliminated has justified the installation of tens of thousands of such systems.

Analytical Instrumentation. Chemical-composition variables are measured by observing the interactions between matter and energy. See accompanying table. That such measurements are possible stems from the fundamental that all known matter is comprised of complex, but systematic arrangements of particles which have mass and electric charge. Thus, there are neutrons which have mass but no charge; protons which have essentially the same mass as neutrons with a unit positive charge; and electrons which have a negligible mass with a unit negative charge. The neutrons and protons comprise the nuclei of atoms. Each nucleus ordinarily is provided with sufficient orbital electrons, in what is often visualized as a progressive shell-like arrangement of different energy levels, to neutralize the net positive charge on the nucleus. The total number of protons plus neutrons determines the atomic weight. The number of protons which, in turn, fixes the

ENERGY-MATTER INTERACTIONS UTILIZED IN ANALYTICAL INSTRUMENTATION

Analytical Instrumentation Techniques *See separate editorial entries with the approximate titles shown below.	Interaction with Electromagnetic Radiation			Reaction to Electric Magnetic Fields		Interaction with Other Chemicals			Interaction with Thermal and Mechanical Energy	
	Emitted Radiation		Transmission and Reflection Measurements							
	Thermally Excited	Electromagnetically Excited		Electrical Properties	Magnetic Properties	Sample or Reactant Consumed	Thermal Energy Liberated	Equilibrium Solution Potential	Thermal Energy	Mechanical Energy
Amperometer*				X						
Aulyzers, Anatomatic distillation type									X	
Analyzers, reaction-product*						X	X			
Analyzers, reagent-tape*						X				
Atomic absorption Spectrometry*	X									
Beta-ray chemical analyzers*			X							
Beta-ray spectrometry			X							
Bioluminescence and chemiluminescence meters*						X				
Boiling-point chemical analyzers*									X	
Chromatography*									X	
Colorimetry*			X							
Dielectric-constant chemical analyzers*				X						
Ebulliometer*									X	
Electric-conductivity measurements*				X						
Electrolysis-type chemical analyzers*				X						
Electrophoresis*				X						
Flame photometry and spectrometry*	X									
Fluorometers*		X								
Gamma-ray spectroscopy*			X							
Gas analyzers, combustion-type*							X			
Gas analyzers, thermal-conductivity type*									X	
Indicator, chemical*						X				
Infrared analytical techniques*			X							
Magnetic resonance spectroscopy					X					
Mass spectrometry*				X	X					
Nephelometry*			X							
Optical-emission spectrochemical analysis	X									
Orsat analyzers						X				
Oscillometers*				X						
Oxygen analyzers				X	X	X		X	X	X
pH (hydrogen ion concentration)*								X		
Polarimetry*			X							
Polarographic analyzers*								X		
Radioactivation analysis (See Radioactivity)		X								
Radiometric analysis			X							
Radioactive tracer techniques (See Radioactivity)		X								
Raman spectrometry*		X								
Redox (oxidation-reduction potential)								X		
Refractometers*			X							
Saccharimeter*			X							
Specific gravity*										X
Spectro instruments*			X							
Spectrochemical analysis, visible*			X							
Spectrography*			X							
Spectroscope*			X							
Thermoanalyzers									X	
Titrators, automatic*						X				
Titration, thermometric*						X				
Turbidimetry*			X							
Ultraviolet analytical techniques*			X							
X-ray analytical instruments*		X	X							

Note: Some kinds of analytical instruments, such as oxygen analyzers, utilize more than one of the energy-matter interactions. This table is not exhaustive. A number of techniques of lesser importance are not indicated.

number of electrons, determines the chemical properties and the physical properties, except mass, of the resulting atom.

The chemical combinations of atoms into molecules involve only the electrons and their energy states. Chemical reactions involving both structure and composition generally occur by loss, gain, or sharing of electrons among the atoms. Thus, every configuration of atoms in a molecule, crystal, solid, liquid, or gas may be represented by a specific system of electron energy states. Also, the particular physical state of the molecules, as resulting from their mutual arrangement, also is reflected upon these energy states. Fortunately, these energy states, characteristic of the composition of any particular substance, can be inferred by observing the consequences of interaction between the substance and an external source of energy.

External energy sources used in analytical instrumentation include:

1. electromagnetic radiation
2. electric or magnetic fields
3. chemical affinity or reactivity
4. thermal energy
5. mechanical energy

The interaction of electromagnetic radiation with matter yields fundamental information as the result of the fact that photons of electromagnetic radiation are emitted or absorbed whenever changes take place in the quantized energy states occupied by the electrons associated with atoms and molecules. X-rays (photons or electromagnetic wave packets with relatively high energy) penetrate deeply into electron orbits of an atom and provide, upon absorption, the large quantity of energy required to excite one of the innermost electrons. Thus, the pattern of x-ray excitation or absorption is relative to the identity of those atoms whose orbital electrons are excited, ideally suiting x-ray techniques for determining atoms and elements in dense samples. But, because of the penetrating power of x-rays, they are not suited to the excitation of low-energy states which correspond to outer-shell or valence electrons; or of the interatomic bonds which involve vibration or rotation.

In contrast, the relatively longer wavelengths of infrared radiation (photons having relatively low energy) correspond to the energy transformations involved in the vibration of atoms in a molecule as resulting from stretching or twisting of the interatomic bonds. Thus, because the penetrating power of electromagnetic radiation varies over the total spectrum, an instrumental irradiation technique can be developed for almost any analytical instrumentation requirement.

The interaction of matter with electric or magnetic fields is widely applied for determining chemical composition. The mass spectrometer, for example, which uses a combination of electric and magnetic fields to sort out constituent ions in a sample, takes full advantage of this interaction. A simple electric-conductivity apparatus determines ions in solution as the result of applying an electric potential difference across an electrolyte.

The use of chemical reactions in analytical instrumentation essentially extends the fundamental techniques of older laboratory analytical methods.

Numerous analytical instrumentation techniques involve interactions between mechanical and thermal energy with matter. All of these interactions are summarized in the accompanying table.

Qualitative Chemical Analysis. This term generally refers to the analysis of inorganic substances. Practically all of the fundamentals of inorganic chemistry are called upon in the execution of qualitative analysis. Thus, in addition to serving as an effective analytical procedure, the method is an effective teacher.

The first important step is that of putting the sample (unknown) into solution. For metals and alloys, strong acids, such as HCl, HNO_3, or aqua regia may be used. If the material is not fully dissolved by these acids, it should be fused, either with sodium carbonate (alkaline fusion) or potassium acid sulfate (acid fusion). Care should be exercised to make certain that no portion of the unknown is volatilized and thus lost during these procedures.

The next step is the detection of the cations of the metals. For this purpose, the solution should be treated with HNO_3, by evaporation and redissolving if necessary, to remove other acid radicals, so that nitrate is the only anion present in the solution. Then, a systematic procedure is followed for separation of groups of the cations. Such schemes of separation have been devised for all the metals found in nature. A shortened plan, which applies to 24 of the commonly occurring metals and ammonium, has been known and practiced for many years. This plan consists of the separation of the 24 metals into five groups:

The first is the one precipitated by HCl, consisting of silver (Ag^+), lead (Pb^{2+}) and mercury(I) or (Hg^+). Group II is that precipitated by H_2S in 3 *N* HNO_3 solution, consisting of mercury(II) or Hg^{2+}, copper (Cu^{2+}), cadmium (Cd^{2+}), lead (Pb^{2+}), which often is not completely precipitated in Group I, bismuth (Bi^{3+}), arsenic (AsO_3^{3-} or AsO_4^{3-}), antimony (Sb^{3+} or Sb^{5+}) and tin (Sn^{2+} or Sn^{4+}). The sulfides of the last three, arsenic, antimony and tin are separated from sulfide precipitates of the others by solution in alkali metal sulfides or polysulfides or in lithium hydroxide. Group III is precipitated by NH_3 and H_2S and consists of iron (Fe^{3+}), titanium (Ti^{4+}), aluminum (Al^{3+}), chromium (Cr^{3+}), manganese (Mn^{2+}), zinc (Zn^{2+}), cobalt (Co^{2+}) and nickel (Ni^{2+}). In this precipitate, the first four elements are present as their hydroxides, and the last four as the sulfides. The mixed precipitate is dissolved, the manganese separated by oxidation and precipitation as MnO_2, and the other elements separated into two subgroups by reprecipitation of the first four elements as hydroxides (by ammonia). Group IV consists of calcium (Ca^{2+}), strontium (Sr^{2+}) and barium (Ba^{2+}) which are precipitated as their carbonates. Group V consists of magnesium (Mg^{2+}), sodium (Na^+), potassium (K^+) and ammonium (NH_4^+) for which individual tests are made. Of course, there are individual tests for identifying all the elements, which must be made to confirm their presence, after further separations have been made within the groups and subgroups. Note that some details of procedure have been omitted, such as the elimination of certain interfering anions, if present, e.g., phosphate, oxalate, tartrate and citrate. Note also that in some schemes, especially those that also include the other metallic elements, magnesium is precipitated in Group IV by suitable changes in conditions.

This same scheme of separation may be adapted to all the naturally occurring metals, although with considerable modification of the details of procedure. For this purpose the metals are grouped as follows:

I. The Hydrochloric Acid Group:
   Lead, mercury(I), silver, thallium(I).

II. The Hydrogen Sulfide Group:
   Antimony, arsenic, bismuth, cadmium, copper, germanium, gold, iridium, lead, mercury, molybdenum, osmium, palladium, platinum, rhenium, rhodium, ruthenium, selenium, silver, tellurium, tin. (Partially: thallium.)
   A. Elements whose sulfides are insoluble in 3 *N* HNO_3 and in solutions of alkali sulfides:
      Bismuth, cadmium, copper, lead, mercury, silver. (Entirely or partially: osmium, palladium, rhodium, ruthenium.)
   B. Elements whose sulfides are insoluble in 3 *N* HNO_3 but are soluble in solutions of alkali sulfides:
      Antimony, arsenic, germanium, molybdenum, rhenium, selenium, tellurium, tin. (Entirely or partially: gold, iridium, platinum.)

III. The Ammonium Sulfide Group:
   Aluminum, beryllium, chromium, cobalt, gallium, indium, iron, manganese, nickel, niobium, rare earths, scandium, tantalum, thallium, thorium, titanium, uranium, vanadium, zinc, zirconium.
   A. Elements whose sulfides are soluble in 3 *N* HNO_3:
      Cobalt, gallium, indium, iron, manganese, nickel, thallium, uranium, vanadium, zinc.
   B. Elements that form hydroxides or basic compounds:
      Aluminum, beryllium, chromium, niobium, rare earths, scandium, tantalum, thorium, titanium, zirconium.

IV. The Alkaline Earths and Magnesium—The Ammonium Carbonate Group:
   Barium, calcium, magnesium, radium, strontium.

V. The Alkali Metals—The Soluble Group:
   Cesium, lithium, potassium, rubidium, sodium.

The qualitative analysis of the anions is more frequently carried out by individual tests, although there are tests which show the presence or absence of groups of them. One such system of group tests is the following:

Group I. This group consists of anions that are precipitated from a slightly basic solution by a mixture of calcium nitrate and barium nitrate. The anions are: sulfate, SO_4^{2-}; sulfite, SO_3^{2-}; thiosulfate (hyposulfite), $S_2O_3^{2-}$; carbonate, CO_3^{2-}; silicate, SiO_3^{2-}; chromate, CrO_4^{2-}; phosphate, PO_4^{3-}; arsenite, AsO_3^{3-}; arsenate, AsO_4^{3-}; borate, BO_3^{3-} (and tetraborate, $B_4O_7^{2-}$); oxalate, $C_2O_4^{2-}$; fluoride, F^-; and tartrate, $C_4H_4O_6^{2-}$.

The tartrate ion is precipitated slowly, or sometimes not at all, by the Group I reagent; hence, a negative test for Group I does not conclusively eliminate a tartrate. For this reason, tartrate is also included in Group III (soluble group). The tiosulfate ion gives a positive test for both Group I and Group II. In Group I it is precipitated as calcium thiosulfate; in Group II it is decomposed and precipitated finally as silver sulfide.

Group II. This group consists of anions that are precipitated by silver nitrate in a dilute nitric acid solution. These ions are: chloride, Cl^-; bromide, Br^-; iodide, I^-; sulfide, S^{2-}; cyanide, CN^-; ferrocyanide, $Fe(CN)_6^{4-}$; ferricyanide, $Fe(CN)_6^{3-}$; thiocyanate (sulfocyanide), SCN^-; and thiosulfate, $S_2O_3^{2-}$.

Group III (Soluble Group). This group consists of anions that are not precipitated by the previous group reagents. They are: nitrate, NO_3^-; nitrite, NO_2^-; chlorate, ClO_3^-; acetate, $C_2H_3O_2^-$; tartrate, $C_4H_4O_6^{2-}$; citrate, $C_6H_5O_7^{3-}$, and permanganate, MnO_4^-.

References

Dean, J. A.: "Chemical Separation Methods," Van Nostrand Reinhold, New York, 1969.
Dilts, R. V.: "Analytical Chemistry," Van Nostrand Reinhold, New York, 1974.
Elving, P. J., and I. M. Kolthoff (editors): "Chemical Analysis," series of monographs, Wiley, New York, various dates (1959–1972).
Gouw, T. H. (editor): "Guide to Modern Methods of Instrumental Analysis," Wiley, New York, 1972.
Kodama, K.: "Methods of Quantitative Inorganic Analysis: An Encyclopedia of Gravimetric, Titrimetric, and Colorimetric Methods," Wiley, New York, 1963.
Meites, L.: "Handbook of Analytical Chemistry," McGraw-Hill, New York, 1963.
Snell, Foster Dee: "Encyclopedia of Industrial Chemical Analysis," seventeen volumes, Wiley, New York, various dates (1966–1972).

ANALYSIS OF COVARIANCE. A generalization of *Analysis of Variance* to the case where more than one variable is observed on each member of the sample. Suppose, for example, there are two variables x and y such as the scores on a test of a group of students before undergoing a course of instruction (x) and after the course is completed (y). The final performance y will be influenced both by the course and the knowledge of the student at the outset represented by x. To disentangle these effects y is regressed on x and the residual y-bx computed, b being the regression coefficient. This residual should, in suitable circumstances, represent the effect of the course regardless of initial knowledge and the set of residuals can be subjected to variance analysis if the students are classified in any way. More generally if x is a variable unaffected by classification or treatment in an experiment its effect on y can be extracted from y and the residuals analyzed in the ordinary way; and so for several variables of type x, which can also be abstracted from y by the use of a regression equation.

Sir Maurice Kendall, International Statistical Institute, London.

ANALYSIS OF VARIANCE. In statistics, a technique for segregating the causes of variability affecting a set of observations. Consider a simple case in which a number of observations are taken on members falling into different classes (for example the yields of a number of plots of wheat, groups of which are subjected to different fertilizer treatments). The problem is whether the yields differ from group to group or, on the ohter hand, differ only as random variations from a homogeneous population. The matter is decided by comparing the sum of squares of mean yields of groups about the over-all mean of all plots with the aggregated sum of squares of observations within groups about their respective group means. On the assumption that the variation is normal (Gaussian) an exact test of significance can be applied to decide whether the difference is great enough to justify the conclusion that group differences are real.

The method can be generalized to much more elaborate situations where the classifications are more complex, especially in experimental designs which are carefully balanced so that analyses of variance are easy to apply.

The technique is frequently referred to as ANOVA.

Sir Maurice Kendall, International Statistical Institute, London.

ANALYSIS (Organic Chemical). Various techniques are used in the chemical analysis of organic substances both in microanalysis as well as in macro laboratory procedures. As contrasted with the determination of total carbon content or the amounts of other specific chemical elements, the representative analytical techniques described here are directed toward the determination of presence and amount of various functional groups (radicals). These groups also are described elsewhere in this volume and, in several instances, additional analytical procedures are related.

(1) The *carboxyl group* is determined by titration with standard sodium hydroxide solution, using phenolphthalein as the indicator, by the reaction

$$RCOOH + NaOH \rightarrow RCOONa + H_2O$$

(2) The *hydroxyl group* is determined by reaction with acetic anhydride on heating in a sealed tube by the reaction

$$ROH + (CH_3CO)_2O \rightarrow ROOCCH_3 + CH_3COOH$$

The amount of hydroxyl group present is found by titrating the resulting acetic acid (CH_3COOH) with standard sodium hydroxide, as in (1).

(3) The *acyl group* (—COOR) in esters and amides is determined by hydrolysis in alcoholic sodium hydroxide solution, followed by ion exchange with an acidic resin. The carboxylic acid formed is then titrated with standard sodium hydroxide, as in (1). The reactions are

$$\underset{\text{Ester}}{RCOOR'} + NaOH \rightarrow RCOONa + R'OH$$

or

$$\underset{\text{Amide}}{RCONHR'} + NaOH \rightarrow RCOONa + R'NH_2 + H_2O$$

$$RCOONa + \text{Resin-}SO_3H \rightarrow RCOOH + \text{Resin-}SO_3Na$$

(4) The *carbonyl group* is determined by a reaction with 2,4-dinitrophenyl hydrazine which precipitates the 2,4-dinitrophenyl hydrazone of the aldehyde or ketone, which is ten filtered off, dried, and weighed. The reaction is

$$RR'CO + H_2NNHC_6H_3(NO_2)_2 \rightarrow RR'C{:}NNHC_6H_3(NO_2)_2 + H_2O$$

(5) The *peroxy group* is determined by treatment with sodium iodide. The liberated iodine is then titrated with standard sodium thiosulfate solution. The reaction is

$$RCOO_2OCR + 2NaI \rightarrow I_2 + 2R'COONa$$

(6) The primary *amino group* is determined by treatment with nitrous acid and measurement of the nitrogen (gas) produced by the reaction

$$RNH_2 + HNO_2 \rightarrow N_2 + ROH + H_2O$$

(7) The aromatic *nitro group* is determined by its reduction with excess titanium(III) chloride. After the reaction, the unused titanium(III) ions (Ti^{3+}) are determined by titration with iron(III) sulfate or iron alum solution:

$$RNO_2 + 6TiCl_3 + 6HCl \rightarrow RNH_2 + 6TiCl_4 + 2H_2O$$

$$Ti^{3+} + Fe^{3+} \rightarrow Ti^{4+} + Fe^{2+}$$

(8) The *hydrazino group* is determined by oxidation with copper(II) sulfate solution, and measurement of the nitrogen (gas) formed. The reaction is

$$RNHNH_2 + 4CuSO_4 + H_2O \rightarrow N_2 + ROH + 2Cu_2SO_4 + H_2SO_4$$

(9) The *sulfhydryl group* is determined by reaction with iodine, which is produced in the vessel from potassium iodide, added in excess to the solution, and potassium iodate, added from a buret until the

Analytical chemist checks vacuum distillation during analytical separation. *(Olin Research Center, New Haven, Connecticut)*

completion of the reaction is shown by the permanent appearance of the blue color of starch-iodine.

$$2RSH + I_2 \rightarrow RSSR + 2HI$$

(10) *Unsaturated groups* are determined by addition of bromine, by the reaction

$$R_2C{=}CR_2' + Br_2 \rightarrow R_2CBr{-}CBrR_2'$$

The term *functional group analysis* sometimes is used to describe the foregoing kinds of analyses.

Ultimate Analysis. This term, generally limited to organic chemical analysis, denotes the determination of the proportion of each element in a given substance. The primary determination is that of carbon and hydrogen, which is conducted by mixing the sample with copper(II) oxide and heating it in a stream of oxygen to a temperature of 700 to 800°C. The carbon is converted to carbon dioxide and the hydrogen to water. These products are then absorbed by suitable reagents. For example, magnesium perchlorate dehydrate may be used to absorb water and sodium hydroxide to absorb carbon dioxide. Although the fundamental procedure is simple, a rather elaborate train of apparatus, involving both temperature and flow control, is required. See accompanying illustration.

A fundamental method for determining nitrogen in organic compounds is described under **Kjeldahl Nitrogen Determination.** In the Unterzaucher method for determining oxygen in organic substances, the sample is heated to a high temperature (approximately 1120°C) in an atmosphere of nitrogen. Under these conditions, the oxygen present combines with part of the carbon content to form carbon dioxide and with part of the hydrogen content to form water. The gases then are passed over hot carbon (1150°C), whereupon both the carbon dioxide and water are converted to carbon monoxide. The latter gas upon leaving the furnace is passed over iodine pentoxide I_2O_5 at about 110°C to form iodine by the reaction: $5CO + I_2O_5 \rightarrow I_2 + 5CO_2$. The freed iodine is titrated with a standard sodium thiosulfate solution.

See also **Analysis (Chemical); Analyzer (Reagent Tape); Chromatography; Infrared Radiation; Mass Spectroscopy; Ultraviolet Spectrometers;** and **X-Ray Analysis.**

ANALYTICAL BALANCE. Weighing.

ANALYTICAL GEOMETRY. Geometry.

ANALYTICAL INSTRUMENTATION. Analysis (Chemical).

ANALYTIC CONTINUATION. Calculation of an analytic function over some domain, from precise definition of the function over a smaller domain.

Suppose $f_1(z)$ is analytic in D_1 and $f_2(z)$ in D_2 and that D has a region in common with both D_1 and D_2. Further suppose that $f_1(z) = f_2(z)$ in D, then if $f(z)$ can be defined so that $f(z) = f_1(z)$ in D_1 and $f(z) = f_2(z)$ in D_2 the analytic continuation of either $f_1(z)$ or $f_2(z)$ in the domain $(D_1 + D_2)$ is $f(z)$. As a simple example consider the series $(1/a + z/a^2 + z^2/a^3 + \cdots)$, which represents the function $1/(a - z)$ only within C_1, a circle of radius $|a|$. Another power series of the type $[1/(a - b) + (z - b)/(a - b)^2 + (z - b)^2/(a - b)^3 + \cdots]$, however, represents the same function outside C_1 if b/a is not real and positive. This series converges at points inside another circle which has regions in common with C_1. See also **Taylor Series** and terms listed under **Mathematics.**

ANALYTIC FUNCTION. A function $f(z)$ of the complex variable $z = x + iy$ is analytic at a point on the z-plane if the function and its first derivative are finite and single-valued there. If this property applies to all points within a given region of the complex plane, $f(z)$ is an analytic function throughout the region. Any point at which the derivative fails to exist is a singularity or a singular point of the function. According to the Liouville theorem, if $f(z)$ has no singularity for z finite or infinite it is a constant.

Equivalent definitions of an analytic function are: (1) it must satisfy the Cauchy-Riemann equations and Laplace's equation; (2) it is analytic only if it may be represented by a convergent power series in some neighborhood of the given point.

Other words often used in place of analytic, and essentially equivalent, are holomorphic, meromorphic, monogenic, uniform, regular.

An analytic function of a real variable may be defined in a similar way. See also **Cauchy Theorem.**

ANALYTIC NUMBER THEORY. Number Theory.

ANALYZER (Chromatograph). Chromatography.

ANALYZER (Infrared). Infrared Radiation.

ANALYZER (Ion Microprobe). Ion Microprobe Mass Analyzer.

ANALYZER (Optics). A term applied to the Nicol prism (or other device which passes only plane polarized light) which is placed in the eyepiece of a polariscope or similar instrument.

ANALYZER (Reaction-Product). Chemical composition may be determined by the measurement of a reaction product—in an automatic fashion utilizing the basic principles of conventional qualitative and quantitative chemical analysis. Two steps usually are involved in this type of instrumental analysis: (1) the formation of a target chemical reaction, and (2) the determination of one or more of the reaction products.

Determination of a constituent in a process stream or sample by measurement of a reaction product can be represented by: $C + R \rightarrow P$, where C = constituent to be determined; R = reactant; and P = reaction product to be measured. If reactant R already is present in the sample, it is only required to expose the sample to suitable reaction conditions to form P. Under normal instrument operating conditions, the reaction of C and R may be spontaneous. In other instances, suitable conditions may have to be established either (1) to promote the desired reaction (for example, setting the proper temperature and pressure, or using a catalyst) or (2) to assure a suitable reaction rate.

Frequently, it is possible to measure the reaction product as it forms in the reaction zone. In some instances, the products and sample residue must be removed from the reaction zone before a measurement can be made. Also, the reaction product may be measured directly; or its presence may have to be inferred from a secondary reaction.

In one example, carbon monoxide in air or oxygen may be determined by combustion to carbon dioxide. The latter may be measured directly as by thermal-conductivity methods; or inferentially by absorbing the carbon dioxide in a solution and then measuring the change of that solution by electrolytic conductance.

Instruments for measuring reactions products are described under **Amperometer; Analysis (Chemical); Analyzer (Reagent-Tape); Bioluminescence; Electrolysis-Type Chemical Analyzers; Electrolytic-Conductivity Measurements; Gas Analyzers (Combustion-Type); Gas Analyzers (Thermal-Conductivity-Type); Nephelometry; Photometers; Titration (Potentiometric); Titration (Thermometric);** and **Turbidimetry.**

ANALYZER (Reagent-Tape). The key to chemical analysis by this method is a tape (paper or fabric) that has been impregnated with a chemical substance that reacts with the unknown to form a reaction product on the tape which has some special characteristic, e.g., color, increased or decreased opacity, change in electrical conductance, or increased or lessened fluorescence. Small pieces of paper treated with lead acetate, for example, have been used manually by chemists for many years to determine the presence of hydrogen sulfide in a solution or in the atmosphere. This basic concept forms the foundation for a number of sophisticated instruments that may pretreat a sample gas, pass it over a cyclically advanced tape, and, for example, photometrically sense the color of the exposed tape, to establish a relationship between color and gas concentration. Depending upon the type of reaction involved, the tape may be wet or dry and it may be advanced continuously or periodically. Obviously, there are many possible variations within the framework of this general concept.

ANALYZER (X-Ray). X-Ray Analysis.

ANAMNESTIC RESPONSE. Immune System and Immunology.

ANAMNIA. Vertebrates which do not develop an amnion during embryonic life. The group includes the cyclostomes (see **Cyclostomata**), fishes, and amphibians.

ANAMORPHISM. A term proposed by Van Hise in 1904 to designate the deep-seated constructive processes of metamorphism by which new complex (metamorphic) minerals are formed from the pre-existing simpler minerals, as contrasted with the surface alteration of rocks due to weathering and cementation, termed katamorphism.

ANAPHASE. Cell (Biology).

ANAPHYLAXIS. State of supersensitivity which may develop after a first injection of a foreign protein, such as a therapeutic or prophylactic serum. See also **Alkaloids.**

ANA-POSITION. The position of two substituent groups on atoms diagonally opposite, in α-positions on symmetrical fused rings, as the 1,5 or the 4,8 positions (which are identical) of the naphthalene ring.

ANASTIGMAT. A compound lens combination corrected so that both astigmatism and the curvature of the field are largely eliminated over a considerable area in the image plane.

ANATASE. The mineral anatase, TiO_2 crystallizing in the tetragonal system is a relatively uncommon mineral. It occurs as a trimorphous form of TiO_2 with rutile and brookite. Rutile and anatase have tetragonal crystallization; brookite, orthorhombic. It was originally named octahedrite from its pseudo-octahedral, acute pyramidal crystal habit. Hardness, 5.5–6; sp. gr. 3.82–3.97; brittle with subconchoidal fracture; color, shades of brown, into deep blue to black; also colorless, grayish, and greenish. Transparent to opaque with adamantine luster.

Anatase occurs as an accessory mineral in igneous and metamorphic rocks, gneisses, and schists. Fine crystals have been found in Arkansas in the United States, and in Switzerland.

ANATEXIS. A term proposed by Sederholm in 1907 for the supposed end-process of deep-seated metamorphism resulting in the partial or complete remelting of a specific type of rock in situ.

ANATOMY. A branch of biology dealing with structure, generally considered to be gross structure, but sometimes used to refer to microscopic structure as well. A subdivision of the more inclusive term, morphology, which includes all forms of study of structure. Human anatomy is a study of the various organs of the human body and their relationship to each other as to shape and position.

Classically, anatomy has been divided into a number of subclasses: (1) *gross anatomy* which is a study of macroscopic structure, that is, the structure which can be seen with the unaided eye; (2) *comparative anatomy* which studies the structures of animals in relation to each other, including human structure; (3) *developmental anatomy* which studies both embryonic and later development of body structures; (4) *functional anatomy* which studies the interaction of organs, particularly as they change in shape, size, pressure, temperature, and other important ways; (5) *microscopic anatomy* (histology) which investigates minute structure, particularly of cells in tissues and how cells develop into organs; and (6) *pathological anatomy* which studies diseased structures, that is, the deviation from normal structures and functions. There also are the classical subclasses of *human anatomy, animal anatomy,* and *plant anatomy.*

- Abdominal Cavity
  - Gallbladder
  - Intestines
  - Kidneys
  - Liver
  - Pancreas
  - Pelvic Cavity
    - Bladder
    - Pelvis
    - Rectum
  - Spleen
  - Stomach
- Buccal Cavity
  - Teeth
  - Tongue
- Cranial Cavity
  - Brain
- Nasal Cavity
  - Structures forming the nose
- Orbital Cavities
  - Eyes, eyeball muscles
  - Lacrimal apparatus
  - Optic nerves
- Peritoneal Cavity
- Pleural Cavities
- Spinal Canal
  - Spinal Cord
- Thoracic Cavity
  - Blood and lymph vessels
  - Esophagus
  - Heart
  - Lungs
  - Trachea
  - Thymus gland

Principal cavities of the body, indicating what they contain.

Anatomists have likened the human body (and other vertebrates) to a tube which may be referred to as the *body wall*; this tube enclosing another tube referred to as the *viscera.* The cavity between the tubes may be referred to as the *body cavity* or *celom.* The principal cavities of the human body are outlined in the foregoing list. The major systems of the body are: (1) the *circulatory* or *vascular system* (blood, blood vessesl, heart, lymphatic vessels and lymph); (2) the *digestive system* (alimentary canal, pancreas, liver, salivary gland; (3) the *endocrine system* (adrenals, parathyroids, pituitary, thyroid, portions of ovaries and testes, and other glands with ducts); (4) the *excretory system* (bladder, kidneys, ureters, urethra, respiratory systems of the skin); (5) *muscular system*; (6) *nervous system* (brain, ganglia, nerve fibers, spinal cord); (7) *reproductive system* (bulbourethral and prostate glands, penis, seminal vesicles, testes, and urethra in the male; ovaries, uterine tubes, vagina, and vulva in the female); (8) the *respiratory system* (bronchi, larynx, lungs, nose, pharynx); and (9) the *skeletal system* (bones and connective tissue).

Needless to say, the dividing line between anatomy, physiology, and other biological and medical sciences is indistinct and growing less distinct as scientists emphasize the interdisciplinary approach to their work.

ANCHOR ESCAPEMENT. Pendulum Clock.

ANCHOR RING (or Torus). A surface that has the shape of a doughnut with a hole in it. It can be generated as a surface of revolution by rotating the circle

$$(y - b)^2 + z^2 = a^2$$

around the Z-axis. Its equation, when rationalized, is of the fourth degree

$$(x^2 + y^2 + z^2 + b^2 - a^2)^2 = 4b^2(x^2 + y^2)$$

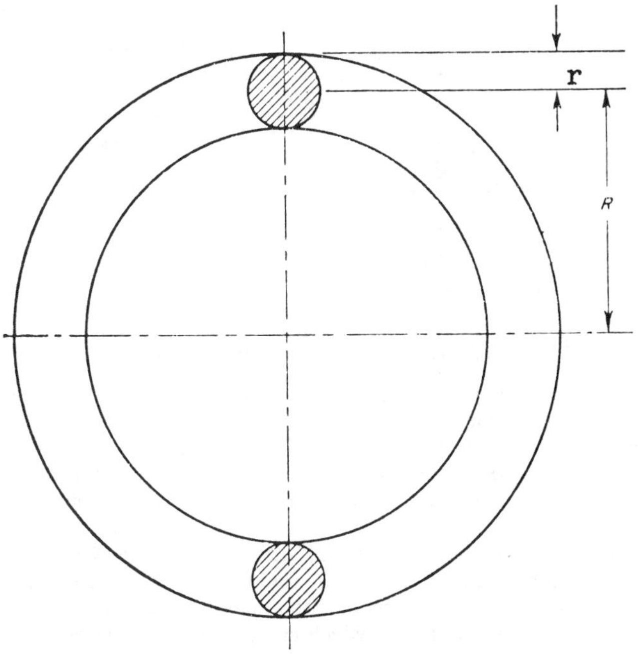

Anchor ring.

With dimensions as shown in the figure, its volume, $V = 2\pi^2Rr^2$ and its surface area, $A = 4\pi^2Rr$.

This surface is of interest in topology, where it is said to be of genus 1.

See also **Circle (Geometry); Topology;** and terms listed under **Mathematics.**

ANCHOVY AND ANCHOVETA (*Osteichthyes*). The anchovy family (*Engraulidae*) comprises 15 genera and some 100 species. These fishes are found in the tropical and temperate regions of the northern and southern hemispheres. Distribution is chiefly in the Indian and Pacific Oceans. Anchovies school along the coast and some are also found in fresh water. Because of the great masses in which they occur, the family has large commerical importance to several countries for the production of fish meal and oil. One of the main differences between the anchovy and the herring is the prominent protruding upper jaw as indicated in the accompanying illustration.

Striped anchovy (*Anchoa hepsetus*).

Anchovies. Seven species in the main genus *Engraulis* have been identified from the Pacific and Atlantic Oceans. Included is the anchovy (*Engraulis encrasicholus*) which reaches about 8 inches (20 centimeters) in length, but is usually from 4.7 to 6.3 inches (12 to 16 centimeters) in length. Coloration resembles that of the herring, with silver lateral stripes. Distribution is chiefly in the Mediterranean Sea and Black Sea as well as on the Atlantic coast of southwest Europe and north Africa. In the north, the distribution extends to the English Channel, and to the south along the African west coast from Togo to Dahomey. Anchovies are also found in the Sea of Asov, in the southern North Sea, and in small numbers as far North as Bergen, Norway.

In their chief distribution region, e.g., in the Mediterranean, anchovies migrate little. In spring and summer, they appear in great schools at the surface of the water both in the open sea and along the coast. They spawn in the Mediterranean from April to September. During this time, they are also fished. After spawning, at the commencement of winter, the adults and the subadults from the spring and summer spawn move into depths of 329 to 492 feet (100 to 150 meters). The schools probably break up at this time and the fish remain in some small region at the floor of the sea. This is evident from studies made of their stomach contents.

Anchovies in the bordering regions of their distribution are migratory. They migrate from wintering grounds in the Black Sea in early spring to the Sea of Asov, where they spawn and return to the Black Sea in the fall. In northern regions, anchovies also migrate in great schools to the north and northeast. They move through the Bristol Channel into the Irish Sea and to the west coast of Scotland, where they can be found from May to September. Spawning grounds are presumably located in this area, since specimens found there are mature for spawning.

In spring, the anchovies migrate in great numbers through the English Channel to the North Sea. They move along the French-Belgian-Dutch coast to the East Frisian coast. The spawning grounds were located here in 1929, and considerable fishing activity developed there, particularly in the Zuider Zee. Since that time, particularly after World War II, it was observed in the southeastern North Sea that the number of anchovies decreased and that they also spawned there. Eggs and larvae were found as far as the North Frisian islands. The spawning period was in warmer brackish water from June to August. Climatic changes probably acted as a factor in the spread of anchovies to the north, as was the case with sardines. In the fall, the anchovies migrate through the northern part of the English Channel along the English coast to wintering grounds off the west exit of the Channel.

The number of eggs varies between 13,000 and 20,000; they are laid in open water in groups. After a year, the fish are 3.5 to 4 inches (9 to 10 centimeters) long and spawn for the first time in the following summer at a length of about 4.75 and 5 inches (12 to 13 centimeters). The Sea of Asov anchovies, the smallest variety, grow at a slower rate. Anchovies feed on small plankton, chiefly on crustaceans of various families. Fish eggs have also been found in their stomachs. The anchovies are prey to many predatory fishes and marine birds.

The chief countries fishing anchovies include the U.S.S.R., Spain, Italy, Turkey, Yugoslavia, Greece, France, and Portugal. Anchovies are caught with drift-nets, baskets, ring-nets, and trawl nets. They are usually marketed in the salted form, in which the head and insides are removed. After an aging period of 4 to 18 months, during which time the flavor improves, the anchovies are ready for market. Part of the catch is worked into filets and conserved in oil. Anchovies are also used in the preparation of pastes and sauces.

Anchovetas. Anchovy fishing is also carried on in other parts of the world, as in the northern Pacific Ocean, on the coast of South Africa, and off Australia. The greatest fishing intensity on a single anchovy species has been carried on since the early 1950s off the coast of Peru and Chile. This species is *Engraulis ringens,* known as *anchoveta* in Spanish. The fish reaches a length of about 5.5 inches (14 centimeters). This is the most important fish in the diet of giant flocks of cormorants, pelicans, gannets, and other marine birds, which breed by the millions on the islands off western South America. The dung of these birds forms the basis of the guano industry. The anchoveta is also eaten by many other animals, including sea lions, dolphins, and predatory fishes.

As of the early 1980s, one of the world's largest fish meal and oil industries is based on these anchovetas. The anchovetas live in the Peruvian Current, ranging from central Chile (37°04′S) all along the Peruvian coast to Cabo Blanco (04°15′S) in a belt close to the shore and extending 30 miles (56 kilometers) out during the summer and 120 miles (222 kilometers) during the winter. The anchovetas spawn in both winter and summer, but with much more intensity and duration during the summer. They reach sexual maturity when approximately 1 year old when they are about 4.7 inches (12 centimeters) long. They can produce about 9000 eggs during several spawnings in the same season. It has been indicated that *E. ringens* spawns from 94°15′S to the south, and that the young anchovetas reach 3.1 and 3.5 inches (8 to 9 centimeters) in length at an age of about 6 months, being then recruited to the fishery. The diet of the species consists of 1% zooplankton and 99% phytoplankton.

The principal fishing gear is the purse-seine net. Most of the catch is reduced to fish meal and oil. Since guano is formed from anchovetas, the proper management of the anchoeva resource is of concern on 2 counts—the fish and the guano. The maximum sustainable catch of anchovetas from fishing and by birds has been estimated at about 10 million tons per year. It has been estimated that the catch by birds approximates 2.5 million tons per year. The Peruvian government has engaged in restricting the anchoveta catch to preserve a satisfactory balance between these two important resources.

Systematic studies of fish larvae have revealed that, with the decline of sardine population, its very close competitor, the anchovy has increased in abundance. It has been estimated by scientists of the California Cooperative Fishery Investigation organization that there exists off California and Baja California a standing stock of some 4 million tons of anchovies, enough to sustain a harvest of perhaps a million tons per year or more. It is believed that some reduction of the anchovy population might, at the same time, accelerate recovery of the heavily depleted sardine population.

See also **Fishes;** and **Sardine.**

ANCILLARY STATISTIC. In cases where no sufficient statistic exists, it is sometimes possible to find a set of statistics which provide no information on the parameter concerned, but which together with a suitable estimator exhaust all the information in the sample. Such statistics are called ancillary statistics; they provide information, not on the parameter, but on the accuracy with which it is estimated.

ANDALUSITE. An aluminum silicate corresponding to the formula Al_2SiO_5, and is one of a three-member polymorphous group consisting of andalusite, sillimanite, and kyanite. Andalusite occurs in contact-metamorphic shales, and in rocks of regional metamorphic origin in association with sillimanite and kyanite. Andalusite crystallizes in the orthorhombic system, developing coarse prisms of approximately square cross section, but may be massive or granular. It shows a distinct cleavage parallel to the prism; hardness 6.5–7.5; sp. gr., 3.13–3.16; vitreous luster; colorless to white, gray, brown, greenish, or reddish; streak, white; transparent to opaque.

This mineral is named for its original locality, Andalusia, Spain. A variety of andalusite, chiastolite, has carbonaceous impurities so oriented that they produce a cross or a tesselated figure at right angles to the prism. Chiastolite comes from the Greek word meaning a cross. Localities are the Urals, the Alps, the Tyrol, the Pyrenees, Australia, and Brazil; in the United States, at Standish, Maine; Sterling and Lancaster, Massachusetts; Delaware County, Pennsylvania; and Madera County, California.

When clear it is used as a gem, and it has also been used to manufacture porcelain for spark plugs.

AND (Circuit). A computer logical decision element which provides an output if and only if all the input functions are satisfied. A three-variable AND element is shown in Fig. 1. The function F is a binary 1 if, and only if, A and B and C are all 1's. When any of the input functions is 0, the output function is 0. This may be represented in Boolean algebra by $F = A \cdot B \cdot C$, or $F = ABC$. Diode and transistor-circuit schematics for the two-variable AND function are shown in Fig. 2. In modern integrated circuits, the function of the two transistors or diodes may be fabricated as a single active device. In the diode

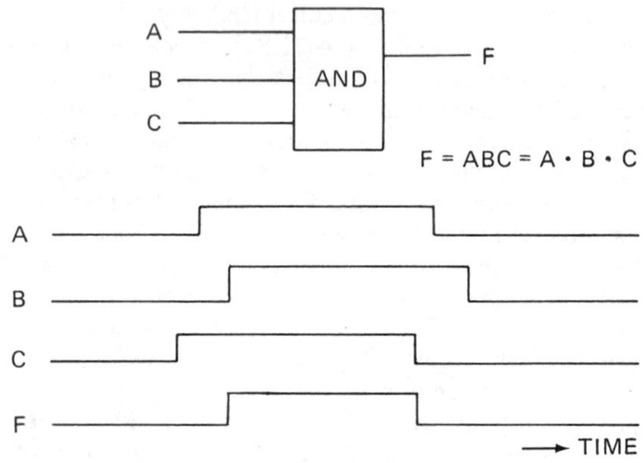

Fig. 1. AND circuit.

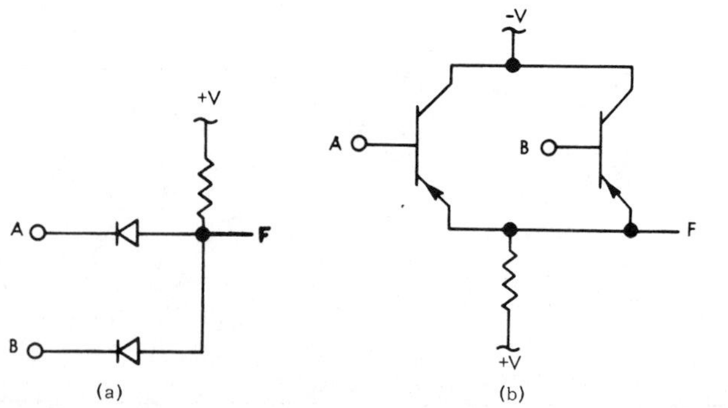

Fig. 2. (a) Diode-type AND circuit; (b) transistor-type AND circuit.

AND circuit, output F is positive only when both inputs A and B are positive. If one or both inputs are negative, one or both diodes will be forward-biased and the output will be negative. The transistor AND circuit operates in a similar manner, i.e., if an input is negative, the associated transistor will be conducting and the output will be negative.

Generally referred to as "fan in," the maximum number of input functions for which a given circuit configuration is capable is determined by the leakage current of the active element. Termed "fan out," the number of circuits which can be driven by the output is a function of current that can be supplied by the AND circuit.

For related topical coverage in this volume, see list of entries under **Data Processing.**

Thomas J. Harrison, International Business Machines Corporation, Boca Raton, Florida.

ANDERSON BRIDGE. **Bridge Circuits (Electrical).**

ANDESINE. **Feldspar.**

ANDESITE. A term originally applied to a porphyritic lava from the Andes Mountains by Leopold Van Buch. In modern terminology andesite is an extrusive igneous rock, the surface equivalent of diorite. In other words, it is composed chiefly of plagioclase, corresponding in chemical composition to oligoclase or andesine together with biotite, hornblende, or pyroxene in varying quantities.

Andesites are of rather widespread occurrence, being found in the Rocky Mountains, California, Alaska, South America, and in many other localities.

ANDRADITE. Calcium-iron garnet.

ANDROGENESIS. The development of an egg after the entry of the male germ cell without the participation of the egg nucleus.

ANDROGENS. The relation between the testis and the male secondary sex characteristics has long been known. The evidence that a chemical substance present in the testis could elicit androgenic effects was not achieved until 1908 by Walker who prepared an aqueous glycerol extract of bull testis tissue that caused growth of the capon's comb. More active extracts from bull testes were prepared in 1927 by McGee and Koch by using organic solvents. The extracts were assayed quantitatively by measuring the increase in area of the capon's comb. The discovery of androgenic activity in urine made possible the isolation of the first biologically active crystalline androgens by Butenandt et al. in 1931–1934. Androsterone and dehydroisoandrosterone were isolated from male urine. In 1935, David et al. isolated a crystalline hormone from bull testis extract. This possessed a higher biological activity than either androsterone or dehydroisoandrosterone. It was named testosterone. Testosterone was also prepared from cholesterol within months of its isolation from testular extract. See Fig. 1.

As pointed out by Liddle and Melmon (1974), adrenal androgen production also carried out in the *zona fasciculata* and in the *zona reticularis,* varies greatly at different stages of life. The fetus makes significant amounts of adrenal androgen, whereas the child makes very little. Beginning with puberty, adrenal androgen production in-

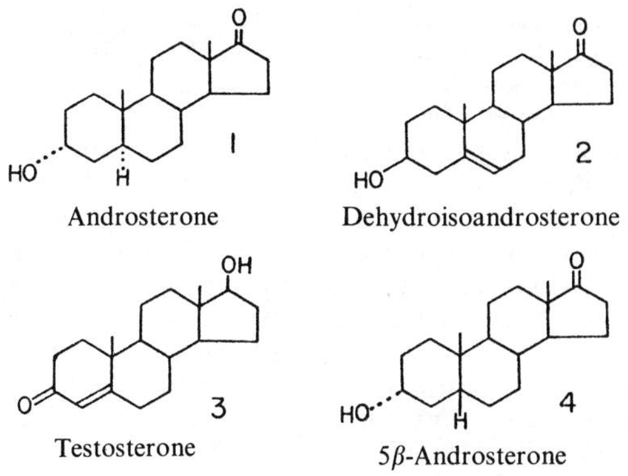

Fig. 1. Androsterone and related hormones.

creases, reaches a peak in early adulthood, and then declines to rather low levels beyond age 50. On the other hand, the secretion of ACTH, the only known control of adrenal androgen biosynthesis, shows no age-related fluctuations. The full regulation of adrenal androgen production is not understood. Adrenal androgens are relatively weak, but some serve as precursors for hepatic conversion to testosterone. Hyperfunction of this pathway in the female may lead to significant masculinization. See also **Adrenal Glands.**

The androgens stimulate the development of the male secondary structures, such as the penis, scrotum, seminal vesicles, prostate gland, vas deferens and epididymis. The deepening of the voice, the growth of pubic, axillary, body, and facial hair, as well as the development of the characteristic musculature of the human male, are also under the influence of testosterone. If the testes fail to develop or are removed prior to puberty, these changes do not occur. Thus, testosterone is essential for reproductive function of the male.

The adrenal cortex produces hydroisoandrosterone which is found in blood and urine largely conjugated as the sulfate ester. The amounts of androgen secreted by the normal adrenal cortex are insufficient to maintain reproductive function in the male. The normal human ovary and placenta also produce small amounts of androgenic steroids that serve as precursors for the estrogens in these tissues. In the human, little testosterone is excreted into the urine and virtually none into the feces. The principal metabolic transformation products are androsterone and 5β-androsterone, with small amounts of other reduced compounds. These substances are excreted in the urine in the form of esters with sulfuric acid or glycosides with glucuronic acid.

Like all other classes of steroid hormones, the androgens are synthesized from acetyl coenzyme A *via* mevalonic acid, isopentenyl pyrophosphate, farnesyl pyrophosphate, squalene, lanosterol, and cholesterol. Enzyme systems in the testis then catalyze the cleavage of the sidechain of cholesterol to pregnenolone which can give rise to testosterone by the two pathways shown in Fig. 2.

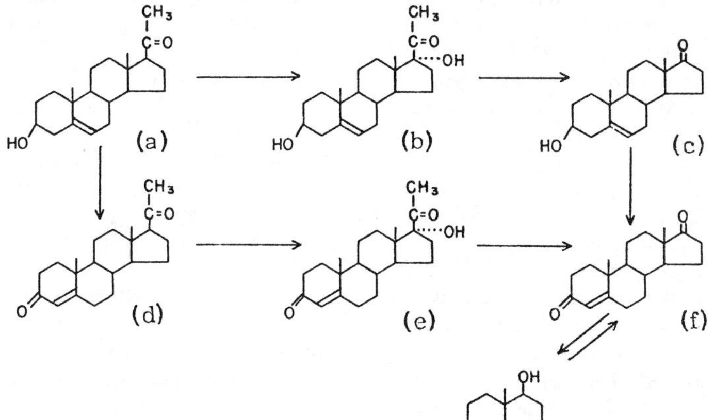

Fig. 2. Biosynthesis of testosterone: (a) Pregnenolone; (b) 17-hydroxy pregnenoline; (c) dehydroisoandrosterone; (d) progesterone; (e) 17-hydroxyprogesterone; (f) androstenedione; (g) testosterone.

Testosterone is formed by the interstitial or Leydig cells of the testes which develop under the influence of gonadotrophic hormones discharged into the bloodstream by the anterior pituitary gland. In pituitary insufficiency, this hormonal stimulus is lacking and, as a consequence, the Leydig cells do not secrete testosterone. In such instances, the male secondary sex characteristics fail to develop. However, interstitial cell tumors may occur, leading to excessive androgen production and precocious puberty. In women, tumors or excessive function of the adrenal cortex and, rarely, of the ovary, result in the production of large amounts of androgens with associated virilization.

Acne vulgaris, a chronic skin disorder, is related to androgen production. The postpubescent development of the sebaceous glands and the onset of acne are dependent upon the presence of androgens, but are not related to testosterone blood levels. Sansone et al. (1971) observed that the hypothesis of increased end-organ sensitivity, which is supported by the heightened ability of skin with acne to metabolize testosterone, may explain the lack of correlation between levels of circulating androgens and occurrence and severity of the disease.

There is androgen involvement in polycystic ovary syndrome (PCO) and hyperthecosis. In 1935, Stein and Leventhal defined a condition with hirsutism, secondary amenorrhea, and enlarged ovaries—a syndrome now referred to as PCO. In this condition, the female usually shows signs of androgen excess, including increased body hair, but true virilism, with balding and deepening of the voice, is less common. Usually, one or both ovaries are enlarged. In many patients the ovaries are cystic, with thickened capsules, yet not palpably enlarged. It has been postulated that the development of PCO commences when luteinizing hormone (LH) triggers an increase in ovarian androgen, which is converted to estrogen, causing estrogen levels (particularly estrone) to increase. This is followed by an anterior pituitary response to luteinizing hormone-releasing hormone (LRH). This completes the cycle by creating exaggerated pulsatile but surgeless LH levels. The initiating lesion remains obscure. Wedge resection of one or both ovaries has largely been replaced by the administration of the antiestrogen clomiphene. This was the first drug known to trigger ovulation in women. Patients with anovulation arising from PCO are treated with the drug primarily when fertility is desired. The hirsutism found with PCO has been difficult to treat. Wood and Boronow (1976) reported that long intervals of anovulation, such as occur in PCO, may be a prelude to endometrial carcinoma. As postulated, the link may either be continuous exposure of the endometrium to estrogen unopposed by progesterone. There is also the postulation that estrone, the estrogen that appears to be high in this disorder, may be a causative factor. A carcinogenic role has been alleged for estrone.

In a somewhat related disorder, hyperthecosis, androgen excess tends to be greater. In this condition, there is prominent luteinization of the theca, whereas the cystic development and capsular thickening of the PCO are absent. Ovarian tumors making androgen can produce the features of PCO, but they tend to have a course with sharper onset and clearer progression.

Androgens also have been used in the management and treatment of agnogenic myeloid metaplasia, aplastic anemia, breast cancer, hereditary angiodema, osteoporosis, paroxysmal nocturnal hemoglobinuria, and sideroblastic anemia.

Oral androgens as may be used in hormone therapy for the management of metastatic breast cancer are effective in about 30% of women regardless of age, but virilizing doses of the hormone are usually required. The oral androgens carry the additional risk of toxic hepatitis.

Androgens may explain some of the differences between heart diseases of males and females. See **Heart.**

References

Ginsburg, J., and C. W. H. Havard: "Polycystic Ovary Syndrome," *Br. Med. J.*, **2**, 737 (1976).

Jewelewicz, R.: "Management of Infertility Resulting from Anovulation," *Am. J. Obstet. Gynecol.*, **122**, 909 (1975).

Liddle, G. W., and K. L. Melmon: "The Adrenals: Textbook of Endocrinology," 5th edition (R. H. Williams, editor), Saunders, Philadelphia, 1974.

McGill, H. C. Jr., et al.: "The Heart is a Target Organ for Androgen," *Science*, **207**, 775–777 (1980).

Pochi, P. E., and J. S. Strauss: "Endocrinologic Control of the Development

and Activity of the Human Sebaceous Gland," *J. Invest. Dermatol.*, **62**, 191 (1974).
Sansone, G., and R. M. Reisner: *J. Invest. Dermatol.*, **56**, 366 (1971).
Stein, I. F., and M. L. Leventhal: "Amenorrhea Associated with Bilateral Polycystic Ovaries," *Am. J. Obstet. Gynecol.*, **29**, 181 (1935).
Strauss, J. S., Kligman, A. M., and P. E. Pochi: "The Effect of Androgens and Estrogens on Human Sebaceous Glands," *J. Invest. Dermatol.*, **39**, 139 (1962).

ANDROMEDA. The brighter stars of this constellation make an almost straight line between the constellations of Perseus and Pegasus. The most famous feature of the constellation is the great spiral galaxy. This is the only spiral actually visible to the naked eye, and may be distinguished as a faint blur against a moonless sky close to the faintest star in the constellation. (See map accompanying entry on **Constellations.**) The distance of this spiral from the earth is about 1.84×10^6 light years.

The bright star in Andromeda closest on the map to Perseus was called Almach by the Arabs. It is a double star, and one of the most beautiful in the sky when viewed with a small telescope. One component is a brilliant orange and the other a striking emerald color. Careful examination with a large telescope shows the green component to be also a double star.

ANDROMEDA GALAXY. Galaxy.

ANDROMEDES. Bielids.

ANDROSTERONE. Steroids.

ANECHOIC ROOM. This term means literally a room without echoes, which is actually a room in which sound reflections from the boundary surfaces have been reduced to a negligible amount.

ANEMIA (Iron Insufficiency). Iron (In Biological Systems).

ANEMIAS. The anemias comprise one of the major grops of diseases involving the blood and, in particular, the erythrocytes (red cells) and their hemoglobin, that is, the impairment of the blood's oxygen delivery system. Anemias may arise from (1) blood loss; (2) disorders of iron metabolism; (3) defects in erythrocyte production; and (4) hemolysis—destruction of erythrocytes.

A male weighing about 150 pounds (70 kilograms) will have just over 4 grams of iron in his body, with 61.7% in the form of hemoglobin; 3.5% (myoglobin); 0.2% (heme enzymes); and 34.6% in iron stores (transferrin, hemosiderin, ferritin). In a woman, there is a marked difference in the distribution of the iron. A female weighing about 132 pounds (60 kilograms) will have about 2.3 grams of iron in her body, with 81.4% in the form of hemoglobin; 5.4% (myoglobin); 0.3% (heme enzymes); and 12.9% in iron stores. The small quantity in iron stores is represented by blood loss in menstruation and in pregnancy and lactation.

Blood Loss and Iron Deficiency Anemia

Blood loss is classified as acute or chronic. In instances of blood loss from injuries, this may be immediately obvious to both patient and physician. Where there is massive acute blood loss, shock and death will occur if replacement therapy is not commenced in very short order. Where blood losses approximate one liter, there will be symptoms of incipient or overt shock. Symptoms will progress into shock, depending upon the severity of further losses. Also, in accidents, there may be deep tissue bleeding where blood losses may not be immediately apparent or measurable. For example, in the case of a fractured pelvis, where there is hemorrhage into the thigh and pelvic region, a liter or more of blood may be lost and not immediately detectable. In all severe cases of bleeding, infusions with colloid and electrolyte solutions will be commenced as soon as possible and before blood typing and cross-matching procedures can be completed. See **Blood.** Frequently, when available, a preparation known as *plasmanate* will be used. This preparation contains human plasma and a small quantity of albumin. Contrasted with dextran infusions, plasmanate does not cause platelet functional defects or red cells to aggregate. Concurrently, equal volumes of a saline-based electrolyte solution will be given. Once the steps required to restore plasma volume and electrolytes have been taken, the physician will tackle the problem of replenishing the red blood cells.

Chronic blood loss is frequently attributed to gastrointestinal tract bleeding and, in women of child-bearing age, to blood losses from the genital tract. An average woman will lose from 30 to 60 milliliters of blood per month through menstruation. During pregnancy, there is division of iron from the mother to the fetus. There are further losses during parturition and lactation. A mother may lose from 700 to 900 milligrams of iron in this way. All of these factors may be contributory to *iron deficiency anemia.* Other factors may include disturbances in the absorption of iron and deficiency of dietary iron. Less frequently, iron deficiency anemia will be a result of intravascular hemolysis resulting in iron loss through the urine (*hemoglobinuria* and *hemosiderinuria*); or even a result of loss of blood to the lungs in an uncommon condition known as idiopathic pulmonary hemosiderosis.

Iron replacement therapy involves the oral or intravenous (depending upon severity and patient reactions) administration of iron-containing compounds (commonly ferrous sulfate). The therapy is usually commenced at low levels and gradually increased so that gastrointestinal symptoms may be avoided.

In determining iron replacement requirements, numerous factors must be considered. There is no single laboratory value that defines anemia. For example, a person may lose nearly a liter of blood and show signs of impending shock from a bleeding peptic ulcer, yet samples of peripheral blood may show normal hemoglobin. This occurs because dilution of the blood (to restore loss of volume) does not occur for about 72 hours. As guidelines, the following stages of iron deficiency are used by some specialists in the field: *Normal,* a hemoglobin level of 13–15 grams per 100 milliliters; *iron deficiency without anemia,* same values; *iron deficiency with mild anemia,* a hemoglobin level of 9–10 grams per 100 milliliters; *severe iron deficiency with severe anemia,* a hemoglobin level of 6–7 grams per 100 milliliters.

Anemias from Red Blood Cell Production Disorders (Erythropoiesis)

Although anemia may result from defects in red blood cell production alone, frequently one of these disorders will be accompanied by other factors (hemolysis or blood loss) which exacerbate the anemia. Sometimes, there are also associated depression of platelet and white blood cell counts.

As mentioned in entry on **Blood,** the erythrocytes are produced in bone marrow. In what is sometimes called the *anemia of chronic disorders,* the bone marrow appears to be normal. Examination will show that there is a normal ratio of myeloid cells (precursors of the erythroid cells) in the marrow and of the erythroid cells produced. This is classified as a mild anemia and is usually presented by patients who have chronic inflammatory, infectious, or neoplastic (presence of tumors) disease. Diagnosis of the underlying condition is often difficult. Anemia is frequently found in severe renal (kidney) disease. As described in the entry on **Blood,** a chemical messenger (*erythropoietin*) is released by the kidney to signal the rate of erythrocyte generation required. Dysfunction of this system can be a causative factor. Where renal disease is accompanied by chronic uremia (blood in urine) resulting from gastrointestinal bleeding, iron deficiency will add to the complications. Drugs sometimes used in the treatment of this condition include oral androgens (oxymetholone) and intramuscular injections of nandrolone decanoate (Deca-Durabolin®), a therapy that usually requires a few months before improvements of the anemia are observed. See also **Bone.**

Also not directly marrow related are anemias resulting from starvation, such as anorexia nervosa or protein deficiency. These conditions may arise even though normal folate and vitamin B12 levels are maintained. Therapy is improvement of the diet. See **Anorexia.** Reduced red blood cell production not directly involving the marrow may also be caused by certain drugs, such as alcohol (which interferes with metabolism of folate and iron), chloramphenicol, and arsenic, among others. However, the latter drugs also can affect the marrow.

Aplastic Anemia. In this anemia, there is partial or nearly complete failure of the marrow to produce new red blood cells. This condition may arise from several causes, including ionizing irradiation, a number of chemotherapeutic drugs, as well as several diseases. There are also instances of idiopathic aplastic anemia which may be due to defective behavior of the stem cells. Benzene also has been implicated as a causative factor. Vigorous inhalation of some organic vapors (glue sniffing) can induce fatal aplastic anemia.

Among drugs that are frequently implicated in aplastic anemia are various *alkylating agents*, such as melphalan, cyclophosphamide, chlorambucil, and bisulfan; *antimetabolites*, including azathioprine, 6-mercaptopurine, 6-thioguanine, and methotrexate; and various *antitumor agents*, such as vinca alkaloids (vinblastine, vincristine), anthracyclines (daunorubicin, doxorubicin), among others. Drugs that occasionally cause aplastic anemia include arsenic, chloramphenicol, glold compounds, mesantoin, phenylbutazone, quinacrine, sulfonamides, and trimethadione. Diseases associated with aplastic anemia include viral hepatitis and paroxysmal nocturnal hemoglobinuria.

Biopsy may be required to determine aplasia of the marrow. In some cases, it may be found that the marrow has been replaced by tumors or fibrosis. Aplastic anemia has been treated with corticosteroids and splenectomy has been done, but their effectiveness has not been well documented. Aplastic anemia is a very serious disease and not always effectively treated, particularly if there are few if any surviving pluripotent stem cells. A marrow transplant may be considered. These are not always successful because of graft rejection, but the procedure may be the only remaining way of saving some patients' lives.

Pernicious Anemia. Technically termed *megaloblastic anemia*, this disorder is caused by vitamin B_{12} and folic acid deficiencies. In megaloblastic anemia, several features of the interactions between vitamin B_{12} and folic acid coenzymes are critical. See also **Folic Acid;** and **Vitamin B_{12}**. Because neither of these substances are produced by humans in adequate amounts, they must be absorbed from a good diet. Factors which cause vitamin B_{12} deficiency include: (1) Inadequate diet, particularly resulting from strict vegetarianism; (2) inadequate absorption, such as from gastric abnormalities with deficient or defective intrinsic factor, small bowel disease, and pancreatic insufficiency; (3) interference with vitamin B_{12} absorption as caused by fish tapeworm and certain drugs, such as neomycin, colchicine, para-aminosalicylic acid, and ethanol; and (4) rare congenital disorders, such as transcobalamin-II deficiency or defective intrinsic factor production. Factors which cause folic acid deficiency include: (1) Inadequate intake, as in nutritional deficiencies and alcoholism; (2) relatively inadequate intake, as may occur during pregnancy, severe hemolysis, and chronic hemodialysis; (3) inadequate absorption, as occurs in tropical sprue, Crohn's disease, lymphoma or amyloidosis of small bowel, diabetic enteropathy, and intestinal resections or diversions; and (4) interference with folic acid metabolism, as may be precipitated by drugs blocking the action of dihydrofolate reductase (methotrexate, trimethoprim, and pyrimethamine), and by other drugs, the exact mechanisms of which are not known—phenytoin, ethanol, antituberculosis drugs, and possibly oral contraceptives.

Pernicious anemia usually does not occur before middle life. It results from the disappearance of the *intrinsic factor* and with it, hydrochloric acid from gastric juices. Upon progression of the disease, certain changes can occur in the spinal cord which result in weakness and numbness of the limbs and ultimately a full loss of ability to control them. Added to weakness and pallor, the symptoms may include loss of appetite, diarrhea, nausea, sore tongue, and yellow pigmentation of the skin. Until 1926, no treatment was known. In that year Minot and Murphy, American physicians, introduced the use of dietary liver as a specific treatment for patients suffering with pernicious anemia. For this work, they received the Nobel prize in 1934.

Upon diagnosis of the disease, vitamin replacement therapy should be commenced immediately. Where the patient is symptomatic from severe anemia, packed red cells can be transferred very slowly to avoid precipitating or aggravating congestive heart failure. This will usually produce a 25% increase in oxygen-carrying capacity of the blood within a short period. Large, weekly doses of parenteral vitamin B_{12} are administered for several weeks, after which these may be scheduled on a monthly basis. Monthly doses may be required for the remainder of life. The physician will also encourage good dietary practice. Depending upon diagnosis, oral administration of folic acid may be indicated.

Sideroblastic Anemias. These comprise a heterogeneous group of disorders characterized by anemia and ineffective erythropoiesis.

Hemolytic Anemias

The anemias which result from increased red blood cell destruction are termed *hemolytic anemias*; they may be *normocytic* (red cells are of normal size) or *macrocytic* (red cells are larger than usual). Hemolysis may be caused by several differing condition. In anemias caused by hemolysis, the breakdown of red cells releases large amounts of hemoglobin end products into the plasma. These substances are converted by the liver into a number of other pigments, most of which are excreted in the bile. If production of bile pigments is excessive, some appear in body tissues, giving rise to a yellow appearance of the skin and the whites of the eyes. This condition is termed *jaundice* and is a symptom of the hemolytic anemias. The red cells are abnormally fragile and rupture easily in *hemolytic jaundice*. Thus, they are broken down by the spleen more rapidly than is usual. Such cells, without interference of the spleen, are able to function normally even though they are fragile. Thus, in some cases of hemolytic jaundice, the spleen is removed as a means of preventing too-rapid destruction of these cells. Some kinds of hemolytic anemia are inherited. Other types are acquired and may be associated with various systemic diseases. A number of drugs, physical and chemical agents, and vegetable and animal poisons have been suspect as causes. Corticosteroid therapy can be beneficial in some of these cases. Where there is no response to treatment, removal of the spleen is indicated.

Sickle Cell Anemia. This disease is caused by hemoglobin S that is inherited as a Mendelian dominant characteristic. It occurs as the *sickle trait* in 8–10% of black persons in the United States. In persons with sickle trait, the hemoglobin S concentration is less than 50% and, with rare exceptions, there are no symptoms. However, in *sickle cell anemia*, 70–98% of the hemoglobin is of the S type, leading to severe disease. The distribution of hemoglobin S in localities with a high incidence of malaria has suggested to some investigators that the sickle trait may provide some advantage to those who possess it. Studies have shown that persons with the sickle trait (but *not* with sickle cell anemia) are relatively resistant to the serious effects of falciparum malaria. It has been presumed that, in patients with the sickle trait, the parasitized cell "sickles" and thus is removed from the circulation in a sequence of events that breaks the parasite's life cycle. See **Malaria.**

Defective hemoglobin results in misshapen red cells and an inability of the blood to carry sufficient oxygen, thus producing anemia. Symptoms, as in the instances of other anemias, include general weakness and, in severe cases, headache, nausea, vomiting, fever, jaundice, and muscular and joint paints. The *sickle crisis*, which occurs in this disease, is a painful and dramatic expression of vascular occlusion. The initiating factor in the sickle crisis is not fully understood. Episodes of fever are known to predispose a patient to crisis. Studies in Ghana have shown that at the beginning of the malaria season, there are sharp increases in the incidence of sickle crisis. The traditional handling of the painful sickle crisis includes rest, the administration of drugs to relieve pain, and, if the patient is demonstrably acidotic, the administration of sodium bicarbonate in a 5% dextrose-water solution, normal saline solution, or half-normal saline solution. The bicarbonate solution is infused for a period of about 20 hours. During a crisis, the prevention of infection and other complications is very important. Past trials of oxygen therapy and urea therapy have not proved beneficial. Where a sickle crisis is of extreme severity, exchange transfusions may be indicated. As blood components are administered, part of the patient's blood will be drawn off. This procedure may be repeated three or more times, at the fastest rate permissible. Particularly in pregnant women, exchange tranfusions are commenced at the beginning of the third trimester to avoid complications in pregnancy, which may include fetal death. Transfusions may be carried out at weekly intervals. Such a program also is sometimes used prior to surgery.

Persons with sickle cell anemia should be warned about the additional dangers of high altitude and dehydration. Genetic counseling to prospective parents is universally recommended among authorities. Although much research effort has been directed toward the therapy of this disease, much fundamental information remains to be gathered.

As of the early 1980s, zinc compounds are being tested for possible efficacy in preventing or reducing the frequency of sickle crises. Various vaccines have been developed to protect sickle cell patients from infection, as, for example, the use of Pneumovax® in connection with systemic pneumococcal infections.

Research during the 1970s has led to potential new drugs in the therapy of sickle cell anemia in the early- and mid-1980s. As reported by Maugh (1981), new therapeutic agents are of three principal categories: (1) Agents that inhibit polymerization of sickle hemoglobin (HbS) by disrupting intermolecular bonding; (2) agents that inhibit polyermization by decreasing the concentration of deoxygenated HbS (deoxy HbS); and (3) agents that interact with erythrocyte membranes. New chaotropic agents as substitutes for the earlier use of urea for disrupting the hydrophobic bonding responsible for polymerization of HbS, are peptides that can bind to the surface of the HbS molecule to prevent the intermolecular contacts necessary for polymerization. Considerable research is going forth on this approach at the National Institute of Arthritis, Metabolism, and Digestive Diseases, the Massachusetts Institute of Technology, and the Weizmann Institute in Rehovot, Israel. At this time, it is in order to stress that the new agents are considered prophylactic rather than curative. As pointed out by one scientist, it is substantially more difficult to depolymerize HbS than it is to prevent it from polymerizing in the first place. Very few agents have been shown to depolymerize gelled HbS.

Hemochromatosis

In this disease, there is an increase in total-body iron content from a normal value of about 4 grams in the adult male (3 grams in adult female) to values of 15 and even 50 grams. This excessive iron is deposited in the liver, gonads, pancreas, joints, and heart. The disease is described in some detail in the entry on **Liver.**

In *Cooley's anemia*, a genetic disorder, excessive amounts of iron build up in the body. The disease, which first appears in neonatal life and continues into childhood, displays profound anemia, jaundice, and, later, enlargement of the liver and spleen. It is accompanied by numerous secondary changes in the bones, retardation of growth, increased susceptibility to infections, and cardiac hemochromatosis. Because of the seriousness of the disorder and its complications, prognosis is bleak unless exceptional and continuing therapeutic measures are taken. Iron chelating compounds, such as deferoxamine, are used to delay appearance of symptoms by increasing the iron excretion rate. Transfusions on a regular basis may be required throughout life. Splenectomy is usually indicated to relieve some of the symptoms.

References

Ammann, A. A., et al.: "Polyvalent Pneumococcal Immunization in Sickle-Cell Anemia and Patients with Splenectomy," *N. Engl. J. Med.*, **297**, 897 (1977).

Bottomley, S. S.: "Porphyrin and Iron Metabolism in Sideroblastic Anemia," *Semin. Hematol.*, **14**, 169 (1977).

Harris, J. W., and R. W. Kellermeyer: "The Red Cell: Production, Metabolism, Destruction: Normal and Abnormal," Harvard Univ. Press, Cambridge, Massachusetts, 1970.

Hillman, R. S., and C. A. Finch (editors): "The Red Cell Manual," 4th edition, F. A. Davis Co., Philadelphia, 1974.

Hoffman, A. V. (editor): "The Megaloblastic Anemias," *Clin. Haematol.*, **5**, 471 (1976).

Jacobs, A.: "Iron Overload—Clinical and Pathological Aspects," *Semin. Hematol.*, **14**, 89 (1977).

Kass, L., and B. Schnitzer: "Refractory Anemia," Charles C. Thomas, Springfield, Illinois, 1975.

Maugh, T. M., II: "Sickle Cell: Many Agents Near Trials," *Science*, **211**, 468–470 (1981).

Rawls, R. L.: "Tailor-made Drugs Treat Genetic Blood Disease (Cooley's Anemia)," *Chem. Eng. News* (May 2, 1977).

Schleicher, E. M.: "Bone Marrow Morphology and Mecanics of Biopsy," Charles C. Thomas, Springfield, Illinois, 1974.

Song, J.: "Pathology of Sickle Cell Disease," Charles C. Thomas, Springfield, Illinois, 1971.

Wormwood, M.: "The Clinical Biochemistry of Iron," *Semin. Hematol.*, **14**, 3 (1977).

ANEMOMETER. **Wind and Air Velocity Measurements.**

ANEMOPHILOUS PLANTS. **Pollination.**

ANEROID-TYPE ALTIMETER. **Altimetry.**

ANEMOTAXIS. Orientation of an insect to an air current as an inflight mechanism in seeking out a distant odor source. Some scientists now claim that insects also can follow an airborne odor trail in still air. See *Science*, **180**, 4092, 1302 (1973).

ANESTHESIA. Anesthetics are agents which, when suitably applied, cause a general or localized loss of feeling or sensation. Traditionally, anesthetics have been classified as : (1) *General anesthetics* which exert their action on the higher nerve centers and produce involuntary loss of consciousness. General anesthetics may be further classified according to their physical properties as volatile (gases or low-boiling liquids) and nonvolatile (high-boiling liquids and solids). (2) *Local, block*, or *regional anesthetics* which produce a loss of sensation within a restricted or readily predictable large segment of the body. These anesthetics may be further classified according to the manner of application. Thus, *topical anesthetics* are applied to peripheral nerve endings; *intraneural anesthetics* infiltrate the nerve fiber; *para-neural anesthetics* act around the nerve sheath; *intraspinal anesthetics* are injected into the spinal canal. Depending upon usage, some of these anesthetics are further categorized as caudal or rectal anesthetics. Anesthetics also may be categorized by their chemical structure, their immediate and residual functions or persistence, their side-effects, and their safety in usage, among other factors. It is not uncommon to use a combination of anesthetics in various procedures—for example, taking advantage of the characteristics of one compound for its rapid action coupled with excellent control, while using another compound to maintain a state of anesthesia over comparatively long periods where complex procedures may be involved. The recovery aspects of various anesthetics range rather widely. Anesthetics may also be classified in accordance with their mode of administration—inhalation, injection, surface application, etc. Some compounds possess sedative, hypnotic, and analgesic properties in addition to producing anesthesia. In some cases, minor structural alteration of some compounds may accentuate one or more of these properties while diminishing others.

General Anesthetics

Sought by surgeons for centuries, pain had been only inadequately overcome by wine, whiskey, and opium. Although ether (diethyl ether) had been discovered by Valerius Cordus in 1540 as a sleep-inducing substance, the possible use of it as an anesthetic for use in surgery did not occur until 1842 when Crawford Long, an American surgeon, first used it during an operative procedure for the removal of tumors on the neck of a patient. Long did not publish these results, however, and thus W. T. G. Morton, an American dentist, who publicly demonstrated the efficacy of ether during a tooth extraction, was generally accredited as the father of ether anesthesia. Nitrous oxide, originally prepared by Priestley in 1772 and known as "laughing gas," was first used for its anesthetic properties by Humphrey Davy in 1800. J. Y. Simpson of Edinburgh first introduced chloroform for relieving the pain in childbirth during the early 1800s. Inhalation-type general anesthetics introduced and used over various spans of time have included hydrocarbons, such as cyclopropane and ethylene; halogenated hydrocarbons, such as chloroform, ethyl chloride, and trichloroethylene; ethers, such as diethyl ether and vinyl ether; and a miscellaneous category, including tribromoethanol ("Avertin"), nitrous oxide, and barbituates.

Over the years a number of these compounds has been eliminated, mainly for reasons of unsafe usage. Some, such as ether, cyclopropane, ethylene, and trichlorethylene, are flammable and capable of producing explosive mixtures unless very carefully monitored. Others, such as chloroform, have been found carcinogenic.

As of the early 1980s and for some period in the past, *halothane* has been widely used. In this category, the proprietary compound Fluothane® is 2-bromo-2-chloro-1,1,1-trifluoroethane, with a boiling point of about 51°C (at 760 millimeters mercury pressure). The compound is nonflammable and its vapors mixed with oxygen in proportions from 0.5 to 50% (volume) are not explosive. Stability of the compound is maintained by the addition of 0.01% thymol (weight) and by storage in amber-colored bottles. Fluothane® is an inhalation anesthetic. Induction and recovery are rapid and depth of anesthesia can be rapidly altered. The compound progressively depresses respiration. There may be tachypnea with reduced tidal volume and alveolar ventilation. The compound is not an irritant to the respiratory tract, and no increase in salivary or bronchial secretions ordinarily occurs. Pharyngeal and laryngeal reflexes are rapidly obtunded. It causes bronchodilation. Hypoxia, acidosis, or apena may develop during deep anesthesia. Fluothane® is not recommended for obstetrical anesthesia except when uterine relaxation is required.

Another general, inhalation-type anesthetic in prominent usage today is *methoxyflurane,* which is 2,2-dichloro-1,1-difluoroethyl methyl ether. The proprietary compound Penthrane® has a boiling point of about 105°C. The flash point in air is 62.8°C; in oxygen (closed system), 32.8°C; and in 50% nitrous oxide with 50% oxygen, 28.2°C. The compound has a mildly pungent odor. An antioxidant, butylated hydroxytoluene (0.01% weight) is added to insure stability on standing. After surgical anesthesia with Penthrane®, analgesia and drowsiness may persist after consciousness has returned and this may obviate or reduce the need for narcotics in the immediate postoperative period. When used alone in safe concentration, the compound will not produce appreciable skeletal muscle relaxation. Thus a muscle relaxing agent, such as succinylcholine chloride, may be used as an adjunct.

Penthrane® is indicated, usually in combination with oxygen and nitrous oxide, to provide anesthesia for surgical procedures in which the total duration of Penthrane® is anticipated to be 4 hours or less, and in which the compound is not used in concentrations that will provide skeletal muscle relaxation. Penthrane® may be used alone with hand-held inhalers or in combination with oxygen and nitrous oxide for analgesia in obstetrics and in minor surgical procedures. As of 1979, the safe use of the compound other than for obstetrics has not been established with respect to adverse effects upon fetal development. Therefore, the compound is not used in women of childbearing potential and particularly during early pregnancy unless, in the judgment of the physician, the potential benefits outweigh the possible hazards.

Numerous side-effects are attributable to the aforementioned compounds and, in fact, to all kinds of anesthetics, but such effects occur in a minority of cases. The expertise of the physician and anestheologist are one's best protection against these effects. These effects are described in some detail in the *Physicians' Desk Reference* listed with other references at the end of this entry.

One of the most serious of side-effects of methoxyflurane is acute and chronic renal failure (Halpren et al., 1973).

Noted during the last decade, *malignant hyperthermia* is described as a rare and dramatic complication of anesthesia. This syndrome can develop during the induction with practically any anesthetic agent, but notably halothane or succinylcholine. Development of muscular rigidity, rapidly rising body temperature, acidosis, cardiac arrhythmias, circulatory collapse, coma, and death characterize the syndrome. A body temperature of 43.5°C (110.3°F) may develop. Present in about 80% of the cases, the muscular rigidity may disallow the use of intubation and assisted ventilation. Causation is poorly understood, but some authorities believe that this may be due to uncoupled mitochondrial respiration. There may be a genetically determined susceptibility and some authorities suggest that screening for an elevated serum CPK level may be useful in assessing risk. Mortality is high and appears to be correlated with the degree of hyperthermia. Upon indication of malignant hyperthermia, anesthesia is ceased immediately, efforts are made to cool the body, oxygen is administered, along with infusion with intravenous bicarbonate. Intravenous procaine also has been advocated.

For diagnostic and surgical procedures that do not require skeletal muscle relaxation, *ketamine hydrochloride* injection is frequently used. One of the proprietary products, Ketaject®, is a nonbarbiturate anesthetic chemically designated DL-2-(*o*-chlorophenyl)-2-(methylamino) cyclohexanone hydrochloride. It is formulated as a slightly acid (pH = 3.5–5.5) solution for intravenous or intramuscular injection. Ketaject® is a rapid-acting general anesthetic producing an anesthetic state characterized by profound analgesia, normal pharyngeal-laryngeal reflexes, normal or slightly enhanced skeletal muscle tone, cardiovascular and respiratory stimulation, and occasionally a transient and minimal respiratory depression. A patent airway is maintained partly by virtue of unimpaired pharyngeal and laryngeal reflexes. The anesthetic state produced by this compound has been termed by some authorities as "dissociative anesthesia," in that it appears to interrupt association pathways of the brain selectively before producing somesthetic sensory blockade. It may selectively depress the thalamoneocortical system before significantly obtunding the more ancient cerebral centers and pathways (reticular-activating and limbic systems). Elevation of blood pressure begins shortly after injection, reaches a maximum within a few mintues, and usually returns to preanesthetic values within 15 minutes after injection. The compound is attributed to have a wide margin of safety. Instances of gross overdoses (up to 10 times normal) have been recorded as nonfatal, but involved prolonged recovery.

Ketaject® is indicated (1) as the sole anesthetic agent for diagnostic and surgical procedures, as previously mentioned; (2) for the introduction of anesthesia prior to the administration of other general anesthetic agents; and (3) to supplement low-potency agents, such as nitrous oxide. It is contraindicated in cases where hypersensitivity to the drug has been evidenced, or where patients who have an elevated blood pressure would run a serious hazard for an additional elevation over a short time span. Because pharyngeal and laryngeal reflexes are usually active, the compound is not used alone in surgery or diagnostic procedures involving the pharynx, larynx, or bronchial tree. Another proprietary ketamine hydrochloride injection compound is known as Ketalar®.

General Anesthesia Adjuncts and Short-term Anesthetics. A number of compounds have been developed that are used to enhance and assist in the performance of general anesthetics. The expertise of the anesthesiologist is required in selecting the numerous combinations of compounds available because all such compounds are not without conflict and certain combinations may put the patient in jeopardy.

Certain *injectable barbiturates* act as an ultrashort-acting depressant of the central nervous system which induce hypnosis (within 30 to 40 seconds of intravenous injection) and anesthesia, but not analgesia. One of these compounds is thiopental sodium (Pentothal®), which is the sulfur analog of sodium pentobarbital. Chemically, the compound is sodium 5-ethyl-5-(1-methylbutyl)-2-thiobarbiturate. Recovery after a small dose is rapid, with some somnolence and retrograde amnesia. Repeated intravenous doses lead to prolonged anesthesia because fatty tissues act as a reservoir; they accumulate the drug in concentrations 6 to 12 times greater than the plasma concentrate, and then release the drug slowly to cause prolonged anesthesia. Thiopental sodium is indicated (1) as the sole anesthetic agent for brief (15-minute) procedures; (2) for induction of anesthesia prior to administration of other anesthetic agents; (3) to supplement regional anesthesia; (4) to provide hypnosis during balanced anesthesia with other agents for analgesia or muscle relaxation; (5) for the control of convulsive states during or following inhalation anesthesia, local anesthesia, or other causes; (6) in neurosurgical patients with increased intracranial pressure, if adequate ventilation is provided; and (7) for narcoanalysis and narcosynthesis in psychiatric disorders. Contraindications include (1) absence of suitable veins for intravenous administration; (2) hypersensitivity to barbiturates; (3) status asthmaticus; and (4) latent or manifest porphyria.

Another injectable barbiturate, thiamyial sodium (Surital®), has the formula, sodium-5-allyl-5-(1-methylbutyl)-2-thiobarbiturate. This compound is indicated for induction of anesthesia, for supplementing other anesthetic agents, as an intravenous anesthesia for short surgical procedures with minimal painful stimuli, and as an agent for inducing a hypnotic state. Contraindications are similar to those previously described for Pentothal®.

Injectable *succinylcholine chloride solution* is a diquaternary base consisting of the dichloride salt of the dicholine ester of succinic acid. Sterile solutions of the drug lose potency unless refrigerated. The solutions have a pH of about 4.0. Succinylcholine is indicated as an adjunct to anesthesia to induce skeletal muscle relaxation. It may be used to reduce the intensity of muscle contractions of pharmacologically or electrically induced convulsions. The safe use of succinylcholine has not been established with respect to the possible adverse effects upon fetal development. Therefore, it is not used in women of childbearing potential and particularly during early pregnancy unless in the judgment of the physician the potential benefits outweigh the potential hazards. Extra precautions are taken in considering the drug in patients with severe burns, patients who are digitalized, or those who are recovering from severe trauma because serious cardiac arrhythmias or cardiac arrest may result. Caution is also observed in patients with preexisting hyperkalemia and those who are paraplegic, who have suffered spinal cord injury, or have degenerative or dystrophic neuromuscular disease, as such patients tend to become severely hyperkalemic when succinylcholine is given.

Succinylcholine is a depolarizing skeletal muscle relaxant. Like acetylcholine, it combines with the cholinergic receptors of the motor endplate to produce depolarization followed by an initial muscle contraction often visible as fasciculations. Neuromuscular transmission is then inhibited and remains so as long as an adequate concentration of the compound remains at the receptor site. The neuromuscular block caused by succinylcholine produces a flaccid paralysis. It is hydrolyzed by plasma pseudocholinesterase at such a rate that the effect of a single paralyzing dose normally disappears within 8 to 10 minutes. Depolarization of the motor endplate gradually fades away as the concentration of succinylcholine decreases at the receptor sites. Then the endplate assumes its state of normal polarity and can again generate an electrical potential of sufficient intensity to initiate muscular contraction.

Following intravenous injection of an effective dose, complete muscular relaxation occurs within one minute, persists for about two minutes, and returns to normal within six minutes. Following intramuscular injection, onset of action may be delayed for up to three minutes.

Proprietary preparations of succinylcholine include Anectine®, Quelicin®, and Sucostrin®.

Regional Anesthetics

Several compounds are available for production of local anesthesia by caudal, epidural, or peripheral nerve block. *Bupivacaine hydrochloride,* 1-butyl-2′,6′-pipecoloxylidide hydrochloride, has an amide linkage between the aromatic nucleus and the amino or piperidine group. Anesthetics with this type of linkage are sometimes referred as amide-type local anesthetics. An injection-type anesthetic, bupivacaine hydrochloride stabilizes the neuronal membrane and prevents the initiation and transmission of nerve impulses, thereby effecting local anesthesia. The compound is not used for spinal anesthesia. Safe use of the drug in pregnant women, other than those in labor, has not been established. Onset is rapid and action is prolonged. Anesthesia may persist for several hours. The duration of action is sometimes extended by the addition of a dilute concentration of epinephrine or other appropriate vasoconstriction drugs. Analgesia often persists after sensation has returned, thereby reducing the need for additional potent analgesics. Peak blood levels of bupivacaine hydrochloride are reached within 30 to 45 minutes following injection for caudal, epidural, or peripheral nerve block, and then decline substantially during the next three to six hours.

Adverse reactions may occur from use of the drug. These are characteristic of those associated with amide-type local anesthetics and are reported in some detail by Baker (1980).

Mepivacaine hydrochloride has a similar chemical structure and is an amide-type local anesthetic, 1-methyl-2′,6′-pipercoloxylidide monohydrochloride. The compound is used much as bupivacaine hydrochloride. Safe use of the compound has not been established with respect to adverse effects on fetal development. Thus, careful consideration should be given to this before administration during pregnancy. Obstetricians are particularly careful because severe persistent hypertension, and even a rupture of a cerebral blood vessel, may occur after administration of certain oxytoxic drugs when vasopressors have already been used during labor (e.g., in the local anesthetic solution or to correct hypotension).

Prilocaine is a local anesthetic, chemically designated as 2-propylamino-*o*-propionotoludide. This drug stabilizes the neuronal membrane and prevents the initiation and transmission of nerve impulses, thereby effecting local anesthetic action. When used for infiltration anesthesia in obstetrical patients, the time of onset averages 1 to 2 minutes with an approximate duration of 60 minutes or longer. For major nerve blocks (e.g., epidural block), the onset of analgesia is approximately two minutes longer than for lidocaine (described later), whereas the duration of action is at least 30 to 60 minutes longer than for lidocaine. One proprietary preparation (Citanest®) contains no epinephrine, making it particularly useful for patients with hypertension, diabetes, thyrotoxicosis, or other cardiovascular diseases. Contraindications include use in patients with a known history of hypersensitivity to amide-type local anesthetics. The drug is not used in patients with congenital or idiopathic methemoglobinemia. Also, the drug is not used in patients with severe shock or heart block.

Etidocaine hydrochloride is another amide-type local anesthetic with the chemical structure (±)-2-(*N*-ethylpropylamino)-2′,6′butroxylidide. Its mode of operation is essentially similar to the drugs just described, stabilizing the neuronal membrane. The initial onset of sensory analgesia and motor blockade is rapid (3–5 minutes), similar to that produced by lidocaine (described later). Duration of sensory analgesia is 1.5 to 2 times longer than that of lidocaine by peridural route. Duration of analgesia in excess of nine hours is not infrequent when the compound is used for peripheral nerve blocks, such as brachial plexus blockade. The drug produces a profound degree of motor blockade and abdominal muscle relaxation when used for peridural analgesia. The compound is indicated for percutaneous infiltration anesthesia, peripheral nerve blocks, and central neural blocks, i.e., caudal or epidural blocks. Contraindications are similar to other amide-type local anesthetics.

Lidocaine hydrochloride is a local anesthetic chemically designated as diethylaminoacet-2,6-xylidide. Lidocaine is best known for its use in the control of ventricular arrhythmias in acute myocardial infarction. Lidocaine hydrochloride is indicated for production of local or regional anesthesia by infiltration techniques, including percutaneous injection and intravenous regional anesthesia, by peripheral nerve block techniques, such as brachial plexus and intercostal blocks, and by central neural techniques, including epidural and caudal blocks. The safe use of lidocaine has not been established with respect to adverse effects upon fetal development and thus it is not given to women of childbearing potential, particularly during early pregnancy. The drug is used, however, for obstetrical analgesia. Contraindications are similar to other amide-type local anesthetics. Proprietary formulations of lidocaine include LTA®, Xylocaine®, and Duranest®. Lidocaine is also used as one of several ingredients in anesthetic for surface applications.

Chloroprocaine hydrochloride, with the chemical designation, β-diethylaminoethyl-2-chloro-4-aminobenzoate hydrochloride, is used in caudal and epidural anesthesia. The parenteral administration of the drug stabilizes the neuronal membrane and prevents the initiation and transmission of nerve impulses. The decision whether or not to use chloroprocaine hydrochloride in certain instances depends upon the judgment of advantages and risks by the physician. Such situations include: (1) paracervical block when fetal distress is anticipated or when predisposing factors causative of fetal distress are present (i.e., toxemia, prematurity, diabetes, etc.); (2) injection of solutions containing epinephrine in areas where the blood supply is limited (i.e., ears, nose, digits, etc.) or when peripheral vascular disease is present; (3) serious cardiac arrhythmias, which may occur if preparations containing a vasopressor are employed in patients during or following the administration of chloroform, halothane, cyclopropane, trichloroethylene, or other related agents.

Procaine hydrochloride (Novocain®) is a local anesthetic used for major and minor surgery. This compound is indicated for the production of local anesthesia by infiltration injection, nerve block, caudal, or other epidural blocks. The anesthetic can be used safely in nearly all cases if suitable precautions are observed. However, as with any local anesthetic, uncommon idiosyncrasy manifested by the rapid appearance of symptoms, including nausea, vomiting, rapid pulse, talkativeness, syncope, respiratory difficulty, and convulsions may be encountered. Extreme caution is imperative when any local anesthetic, even in relatively small amounts, is injected into the traumatized urethra or under conditions in which trauma is likely to occur. Procaine hydrochloride does not irritate tissues and has a low relative toxicity.

Topical (Local) Anesthetics

It has long been known that lowering the temperature of a portion of the body will make the perception of pain originating in the area of lowered temperature less apparent. Likewise, it is known that pressure applied to a nerve fiber or trunk will interfere with the transmission of pain impulses originating peripherally to the site of the pressure. The first principle is still occasionally used to produce local anesthesia. Advantage is taken of the low temperatures produced by the vaporization of such compounds as ethyl chloride, ether, and solid carbon dioxide to produce local anesthesia in restricted segments of the body.

The beginning of the modern period of local anesthesia by physiologically active compounds can be traced to the isolation and characterization of the alkaloid cocaine. See also **Alkaloids.** Its local anesthetic effect on the tongue was noted by Wöhler in the mid-1800s. Its use in surgery dates from the work of Koller who, in 1884, described the use of cocaine in surgery of the eye, nose, and throat. Koller also used cocaine by infiltration techniques.

A systematic investigation of the chemical nature of cocaine led to its synthesis by Willstätter in 1902 and to the development of a large number of synthetic compounds, including the procaine group of which Novocain® is a member. As a result of much work relating to the chemical constitution of cocaine analogs and local anesthetic activity, a number of general relationships were traced. It was found that alkyl esters of aromatic acids, such as benzoic and naphthoic acids, had potential local anesthetic activity. This was found to be enhanced by the presence of substituent groups in the aromatic ring, such as alkyl, amino, hydroxy, alkoxy, and alkylthio. Isoteric compounds, such as substituted amides and amidines, as well as thiocarboxylic esters and urethanes retain the anesthesiophore.

Many of the simple alkyl esters of aromatic acids were only useful as topical anesthetics, since their water solubility was too low to permit their use in parenteral solutions. Of these, benzocaine (ethyl-4-aminobenzoate) is a prototype with the *n*-propyl and *n*-butyl esters subsequently developed. It was found that the inclusion of a tertiary or secondary aliphatic or cycloalkylamine in their structure made it possible to prepare water-soluble salts with acids which could be administered by parenteral pathways. This was also possible in the case of amides, thicarboxylic esters, and urethanes. Amidines which were of moderately high molecular weight are sufficiently basic to form salts which can be dissolved and used. A large number of individual compounds, many of which are proprietary, have been synthesized and a considerable number have been marketed. Many experimental compounds are rejected because of their locally irritant effects or systemic toxicity or both. Other compounds which, structurally, are potentially local anesthetics have found their greatest clinical use for other effects, such as antispasmodics.

A number of mildly anesthetic preparations are available for temporarily relieving pain in connection with itches, rashes, bites, sunburn, plant poisons, hives, hemorrhoids, and minor burns. These are marketed in a number of forms—ointments, creams, pastes, balms, suppositories, and sprays, among others. A survey of compounds popular as of the early 1980s contain one or more of the following anesthetic substances: alkyl and aryl phenols, benzalkonium chloride, benzocaine, benzethonium chloride, benzyl alcohol, cetyl dimethyl ethyl, cetylpyridum chloride, chloroprocaine hydrochloride, chlorothymol, cyclomethycaine sulfate, dibucaine hydrochloride, dimethisoquin hydrochloride, lidocaine hydrochloride, methylbenzethonium chloride, methyl salicylate, piperocaine hydrochloride, pramoxine hydrochloride, proparacaine hydrochloride (ophthalmic use), and sodium phenolate, among others.

References

Adriana, J.: "The Chemistry and Physics of Anesthesia," 2nd edition, Charles C. Thomas, Springfield, Illinois, 1979.

Baker, C. E., Jr.: "Physicians' Desk Reference," 34th edition, Medical Economics, Oradell, New Jersey, 1980.

Bennett, E. J.: "Fluids for Anesthesia and Surgery in the Newborn and the Infant," Charles C. Thomas, Springfield, Illinois, 1975.

Brechner, V. L.: "Pathological and Pharmacological Considerations in Anesthesiology," Charles C. Thomas, Springfield, Illinois, 1973.

Cohen, D., and J. B. Dillon: "Anesthesia for Outpatient Surgery," Charles C. Thomas, Springfield, Illinois, 1979.

deJong, R. H.: "Local Anesthetics," 2nd edition, Charles C. Thomas, Springfield, Illinois, 1977.

Eng, G. D., et al.: "Malignant Hyperthermia and Central Core Disease in a Child with Congenital Dislocating Hips," *Arch. Neurol.*, **35**, 189 (1978).

Gravenstein, J. S., et al.: "Monitoring Surgical Patients in the Operating Room," Charles C. Thomas, Springfield, Illinois, 1980.

Halpren, B. A., et al.: "Interstitial Fibrosis and Chronic Renal Failure Following Methoxyflurane Anesthesia," *J. Am. Med. Assn.*, **223**, 1239 (1973).

Harrison, G. G.: "Anaesthetic-Induced Malignant Hyperpyrexia," *Br. Med. J.*, **3**, 454 (1971).

Isaacs, H., and M. B. Barlow: "Malignant Hyperpyrexia," *J. Neurol., Neurosurg., Psychiatry*, **36**, 228 (1973).

Lorhan, P. H.: "Anesthesia for the Aged," Charles C. Thomas, Springfield, Illinois, 1971.

Moore, D. C.: "Regional Block: A Handbook for Use in the Clinical Practice of Medicine and Surgery," 4th edition, Charles C. Thomas, Springfield, Illinois, 1979.

Snow, J. C.: "Anesthesia: In Otolaryngology and Ophthalmology," Charles C. Thomas, Springfield, Illinois, 1972.

ANESTRUS. A period in which there is lack of heat (estrus) in the female animal, thus precluding breeding during that period. The periods of sexual noninterest in the female that normally occur between regular heat periods generally are not referred to as anestrus. Rather, *anestrus* applies to those factors that break up the normal cycles of heat. For example, there is lactational anestrus, which is a period following the birth of the young; or there is seasonal anestrus, which is a perfectly natural condition with some animals. For example, in some regions, seasonal anestrus occurs in sheep during late-spring and early-to-mid-summer months. Abnormal anestrus is always of concern to cattle, sheep, and swine producers because of the reduction in production of calves, piglets, and lambs during any given period.

ANEURYSM. A sac or pouch filled with blood which protrudes from the wall of an artery, a vein, or the heart. In a *true* aneurysm, the wall of the sac consists of at least one of the layers of tissue that make up the wall of the blood vessel. *False* aneurysms exist when all of the layers of the artery have ruptured, but the blood is still retained by the surrounding tissues. Occasionally, an artery and vein may be connected in such a manner that a continuing flow of blood passes from the artery to the vein. Such arteriovenous communications result from wounds, aneurysms, or congenital connections between the vessels.

Aneurysms may occur in any artery. The most common site is the large artery leading from the heart (the *aorta*). Small aneurysms sometimes develop in blood vessels as the result of injuries. The rupture of even a small aneurysm in the brain, heart, or other vital organ can be fatal.

Presently, the most common cause of aortic aneurysm is atherosclerosis, known as hardening of the arteries. Also, an injury to an arterial wall can leave it so weakened that an aneurysm eventually may occur. Infected (*mycotic*) aneurysms result from destruction of arterial walls by infectious agents. Typical of diseases that may leave a weakened blood vessel are pneumonia, streptococcal infections, and gonorrhea. Now very rare, aneurysms of the aorta formerly were common as the result of previous syphilitic infection.

Although an aneurysm may exist at any site along the aorta, the most common aneurysms are in the abdominal aorta. Symptoms depend upon size of the sac and parts of the body upon which it exerts pressure. The sacs frequently become large, sometimes larger than an orange and can cause severe crowding of the chest or abdominal cavity. Bulging of the area of the collar bone can result when the aneurysm is near the top of the aorta. Sometimes aneurysms are painful, the pain usually located in the center of the chest, or may radiate into the arms. Difficulty in breathing (*dyspnea*) is another common symptom. This results from pressure on the windpipe or smaller air passages leading to the lungs. Aneurysms may cause headaches, abdominal distress, or swelling in various parts of the body. Many aneurysms are symptomless, however, and may not be discovered except as the result of an x-ray examination for some other purpose. Angiography, in which contrast material is introduced into the blood vessels, followed by x-ray examination, is a valuable diagnostic aid.

Although many aneurysms remain small and never require treatment, generally they tend to become larger and may progress to rupture. The major treatment is complete surgical excision of the aneurysm with subsequent restoration of circulation with an implanted, pleated artificial graft. Where the artery is small, the graft may be a segment of vein from the patient. Risk of rupture and development of more hazardous problems rise with a delay in surgical procedure.

For smaller blood vessels which supply nonvital areas, aneurysms may be tied off so that blood no longer passes through the distended area. Other arteries assume the work of the vessel that is closed-off.

Most cerebral aneurysms occur in or near the circle of Willis: 30% of cases are found in the anterior communicating artery; 25% in the carotid artery; 25% in the posterior communicating artery; 2% in the basilar artery; and 2% in the vertebral artery. Several other sites are less frequently involved. See **Brain and Nervous System.** Sometimes large aneurysms will produce focal neurologic signs that lead to a correct diagnosis prior to rupture. These aneurysms may occur at any age, but middle-aged persons are most prone to them. See also **Cerebrovascular Diseases.**

ANGEL. A radar echo caused by a physical phenomenon not discernible to the eye. Angels are usually coherent echoes, whose phase and amplitude at a given range remain relatively constant, and are sometimes of great signal strength (up to 40 decibels above the noise level). They have been ascribed to insects flying through the radar beam, but have also been observed under atmospheric conditions that indicate there must be other causes. Studies indicate that a pair portion of angels are caused by strong temperature and/or moisture gradients such as might be found near the boundaries of bubbles of especially warm or moist air. They frequently occur in shallow layers at or near temperature inversions within the lowest portion of the atmosphere.

ANGELFISHES (*Osteichthyes*). Of the family Pomacanthinae, angelfishes have a powerful spine on the lower rear edge of the front gill cover; there is generally no axillary scale at the base of the pectoral fin. The spine distinguishes angelfishes from the closely related butterfly fishes (Chaetodontinae) which have no similar spine. Angelfishes are among the most beautiful of fishes and are usually found on tropical reefs. They tend to travel alone or in small groups. Although most angelfishes are much smaller, some reach a length of 2 feet (0.6 meter). These fishes are not only well known to the fishers in their native habitats, but have also become very popular in Europe and North America as aquarium fishes. They often are found on aquarium hobbyist organization emblems, since hobbyists have long been attracted to them. See accompanying figure.

The length of the angelfish *Pterophyllum scalare*, when fully developed in its natural habitat, may attain about 10 inches (25 centimeters). This fish inhabits the entire Amazon and its tributaries, resulting in the development of various geographic races differing in shape as well as coloration. Spawning usually occurs on a strong plant stem, or a large leaf. Both partners clean off the spawning surface beforehand.

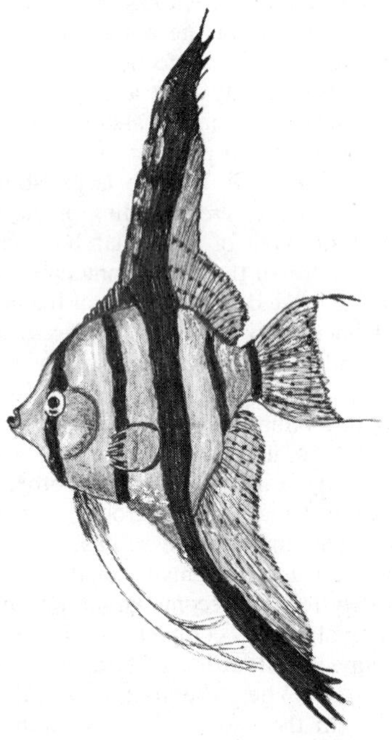

Amazon angelfish (*Pterophyllum scalare*).

Both parents also care for the young together. Scalares do not build spawning pits to which the young are brought after hatching. Instead, the young fish are carried in the parents' mouth and suspended either directly on the spawning surface, or some other cleaned area. This is done with the help of fibers projecting from adhesive glands on the top of the heads of the young. Both parents guard the developing eggs and the hatched young.

Species of angelfish include:

Blue Angelfish (*Pomacanthus semicirculatus*), tropical Indo-Pacific.
Blue-faced Angelfish (*Pomacanthus xanthometopon*), tropical Indo-Australian region.
Imperial Angelfish (*Pomacanthus imperator*), tropical Indo-Pacific.
French Angelfish (*Pomacanthus arcuatus*), tropical Atlantic, from the Florida Keys to Brazil.
Potter's Angelfish (*Centropyge potteri*), Hawaiian Islands.
Queen Angelfish (*Holacanthus ciliaris*), southwestern and northeastern Gulf of Mexico to Brazil.
Regal Angelfish (*Pygoplites diacanthus*), central Indo-Pacific (tropical).
Rock Beauty (*Holacanthus tricolor*), Florida to Brazil.

ANGEL SHARK (*Chondrichthyes*). The species *Squatina squatina* is the largest angel shark, with a maximum length of about 8 feet and weight of 160 pounds (73 kilograms). Angel sharks are intermediate between sharks and rays. Other species of the family *Squatinidae* are seldom longer than 4 to 5 feet (1.2 to 1.5 meters) at maturity. They prefer inshore waters, rarely frequenting deep waters. One exception is the Atlantic American *Squatina dumeril*, one specimen of which was found at a depth of some 4200 feet (1260 meters). Although angel sharks are not considered dangerous to swimmers, they are of a nasty disposition when out of the water and thus, if hooked, pose a danger to fishermen. They survive in captive waters for only a few weeks or months. The Pacific American *Squatina californica* inhabits waters from Alaska to lower California and like most angel sharks prefers temperate waters.

ANGINA PECTORIS. Ischemic Heart Disease.

ANGIOGRAPHY. In making an angiogram, a catheter is inserted into the individual's heart, a radiopaque medium is injected, and x-ray images and motion pictures may be made. These pictures indicate the locations where arteries are blocked and the degree to which the blockage has developed. Motion picture studies show details of heart function. Contrast angiograms are usually indicated when surgery or other therapy is being considered for persons with angina pectoris and those who are recovering from heart attack. In another version of this technique, the heart can be labeled with radioactive tracers which emit gamma rays. Images are made with a scintillation camera that is sensitive to gamma radiation. Through the use of computer techniques, three-dimensional x-ray or gamma-ray pictures of the heart can be created. Scintillation camera images have come into rather wide application as a means of diagnosing heart attacks, particularly where electrocardiograms may not be fully definitive, and where certain other clinical findings, such as serum enzymes, may not confirm heart damage. Technetium-99m, a radionuclide tracer, has a particular affinity for recently damaged heart muscle. After intravenous injection, these tracers find their way to damaged areas and emit gamma rays which are picked up as bright areas in the scintillation camera image. In persons with unstable angina pectoris, where other clinical findings are indefinite, the radioactive tracers tend to collect in the heart. Another approach is to inject radioactive tracer consisting of a monovalent cation, such as potassium-43 or thallium-201. Upon intravenous injection of these materials, the substances lodge in the heart in proportion to blood flow. Portions of the heart that are deprived of blood flow thus appear as blank regions in the scintillation camera image.

See **Heart and Circulatory System (Human); Ischemic Heart Disease;** and **Radioisotopes in Medicine.**

ANGIOMA. A tumor which is composed mainly of blood vessels (hemangioma) or of lymph vessels (lymphangioma). Both forms are ordinarily harmless: in the skin a hemangioma may appear as a disfiguring *naevus* or "port-wine stain"; lymphangiomas generally form soft bulky swellings especially about the neck.

ANGIOSPERMS. The angiosperms represent the most advanced division of the Pteropsida. They are more familiarly known as flowering plants, and the characteristic feature is the flower. The seeds are borne completely enclosed in the ovary tissue of the parent plant.

As a rule the angiosperms are land plants. A few of them have returned to the water as a habitat, but these are obviously reversions to an aqueous life. Tremendous diversity in size is found in this group; some of the so-called duckweeds are spherical masses of cells less than a millimeter in diameter; at the other end of the scale are the giant *Eucalyptus* trees, many of which are over 300 feet (90 meters) high. The variety of form shown in the angiosperms is nearly endless; each of the nearly 200,000 species has a distinct appearance. Some are tiny, herbaceous plants which live but a few weeks; others are giant trees living hundreds of years.

Included in the angiosperms are two types of plants. One, held to be the more primitive type, has a woody stem which has a much more complex structure than that found in the stems of gymnosperms. The other has an herbaceous stem, a form of stem which dies to the ground at the end of the growing season. The internal structure of angiosperm stems is much more specialized than that of gymnosperms. The xylem contains not only tracheids but also vessels and fibers. The vessels are open tubes of considerable length through which water is carried rapidly. The fibers give strength to the stem. In the phloem there are sieve tubes and companion cells, and also numerous fibers. In the woody angiosperms and in many of the herbaceous forms there is a well-developed cambium.

Reproduction in the angiosperms is described under Flower. Many of the angiosperms have highly developed mechanisms to ensure pollination and fertilization, and elaborate systems for the protection, dissemination, and germination of the seeds.

The angiosperms are separated into two large groups, the dicotyledons and the monocotyledons. The origin of the angiosperms is as yet unknown. They are known to have existed in the Jurassic period, but were not at all abundant until the Cretaceous period (see **Paleobotany**). The earliest fossil members of this group are well-differentiated plants which give little indication as to their possible ancestry. Within the group, evolution seems to be from the woody type to the herbaceous, and from plants with flowers having an indefinite number of parts arranged in spiral manner and not fused. As evolution progressed the number of flower parts became reduced and definite and finally fused. In many cases great irregularity replaced the more primitive regularity. The angiosperms are the dominant land flora of the present day.

ANGIOTENSIN. Brain and Nervous System; Hypertension (High Blood Pressure).

ANGLE (Euler). Euler Angle.

ANGLE (Mathematics). The figure obtained by drawing two straight lines, called the sides of the angle, from a point, called the vertex. In trigonometry, an angle measures the rotation of one straight line, the terminal line, about a fixed point on an initial line. It is positive if the direction of rotation is counterclockwise. A unit for measuring angles is the radian, which is that angle whose intercepted arc in a circle equals the radius of the circle. Thus π radians $= 180°$; 1 radian $= 180°/\pi = 57.29578\cdots°$ or 57° 17′ 44.6″, approximately, and 1° is approximately 0.017453 radians.

If the magnitude of an angle equals 2π radians, it is called a perigon angle and such an angle, divided into 360 equal parts, has a magnitude of 360°. A right angle equals 90° or $\pi/2$ radians; a straight angle, 180° or π radians; an acute angle is less than 90°; an obtuse angle, greater than 90° (but frequently limited to one less than 180°). A reflex angle is greater than 180° but less than 360°. Oblique angle is a general term for one not equal to 90° or 180°. Related angles are designated as follows: conjugate, if their sum equals 360°; supplementary, if the sum equals 180°; complementary, if 90°; vertical, if they have a vertex in common and the sides of one angle are prolongations of the sides of the other.

An angle inside a circle is central, if its sides are radii and its vertex is at the center of the circle; inscribed, if its sides are chords and its vertex is on the circumference of the circle.

The previous definitions refer to a plane angle, which is usually meant by the word angle, without a qualifying adjective. However, there are several other kinds of angles as the subsequent discussion will show.

The angle between two intersecting planes is called a dihedral angle. The line of intersection of the planes is the edge of the angle and the planes are the faces of the angle. Two dihedral angles are adjacent if they have a common edge and face. If two planes are parallel, the dihedral angle between them is zero; if perpendicular, the two adjacent angles are right dihedral angles. The plane angle of a dihedral angle is formed by two straight lines, one in each plane, perpendicular to the edge at the same point. The terms vertical, acute, obtuse, complementary, supplementary, etc., as used for plane angles, are also applied to dihedral angles.

Analytically, the dihedral angle can be defined as

$$\cos\theta = \lambda\lambda' + \mu\mu' + \nu\nu'$$

where the direction cosines of perpendiculars to the two intersecting planes are (λ, μ, ν) and (λ', μ', ν'). In vector notation, the relation becomes

$$\cos\theta = \csc\phi_2 \csc\phi_3(\mathbf{e}_{12} \times \mathbf{e}_{23}) \cdot (\mathbf{e}_{23} \times \mathbf{e}_{34})$$

where $\mathbf{e}_{ij}$ is a unit vector drawn from point i to point j; $\mathbf{e}_{23}$ is on the edge of the dihedral angle; $\mathbf{e}_{12}$ on one face and $\mathbf{e}_{34}$ on the other; ϕ_2 is the plane angle determined by the first vector product and ϕ_3 that angle determined by the second vector product.

If three or more planes meet at a common point, a polyhedral angle exists. The vertex, edges, faces, face angles are defined as for plane and dihedral angles. The polyhedral angle is convex if every section made by a plane cutting all of its edges is a convex polygon. It is called trihedral, tetrahedral, etc., when it has three, four faces, etc. A trihedral angle is rectangular, birectangular, trirectangular, if it has one, two, three right dihedral angles. If the vertex of a trihedral angle is at the center of a sphere, its faces intersect the sphere in great circle arcs and spherical angles are formed (see **Triangle**).

Now consider a small cone with a base of area dS and a vertex at a fixed point P. The cone will cut out an area $d\sigma$ on a sphere of radius r with center at P. The angle subtended by dS at P is defined as $d\omega = d\sigma/r^2$ and is called a solid angle. It is numerically equal to the area cut out by the same cone on a sphere of unit radius at the same point P. The unit used for measuring a solid angle is the steradian. Since the area of a sphere of unit radius equals 4π, the total solid angle about a point is 4π steradians.

For related topical coverage in this volume, see list of entries under **Mathematics.**

ANGLE MEASUREMENT. Goniometer.

ANGLE OF DEPARTURE. The angle between the line of propagation of a radiowave and the earth's surface at the point of transmission.

ANGLE OF INCIDENCE (Aircraft). Aerodynamics; Helicopters and V/STOL Craft; Supersonic Aerodynamics.

ANGLE (Of Repose). Repose, Angle of.

ANGLERFISHES (*Osteichthyes*). All members of the anglerfish order (*Lophiformes*) have a modified single movable dorsal fin ray which carries a kind of "bait" at its end; this is the basis of the name anglerfish. The "bait" is known technically as the illicium. Other unusual modifications in this group include the pectoral fins, which enable the fishes to crawl along the ground. All anglerfishes are slow-moving, almost motionless fishes, attracting their prey by means of their natural bait, at which time they suck in the prey with an extremely fast motion. The mouth acts as a giant suction trap.

Three suborders are distinguished: (1) *Goosefishes*, with 1 family; (2) *frogfishes*, with 4 families; and (3) *deepsea anglerfishes*, with 10 families. Altogether there are 225 species of anglerfishes distributed for the most part in tropical, subtropical, and temperate waters of the world's seas.

It is generally believed that anglerfishes developed from perchlike fishes which took on the froglike shape as an adaptation to their life between rocks in coastal zones. According to this viewpoint, the modified pectoral fins act to ensure balance in the breaking waves. In further developments, the goosefishes adapted to living in deep coastal waters on the floor. The frogfishes adapted to living in dense plant growths, such as Sargassum seaweed; and the deepsea anglerfishes returned to a pelagic life, losing their dependence on a supporting object.

Goosefishes. These fishes are of suborder *Lophioidei,* family *Lophiidae.* They achieve a length exceeding 4 feet (1.2 meters) and a weight of about 45 pounds (20 kilograms), and are the largest anglerfishes. It has been reported that these fishes spend most of their lives lying on the bottom, whereby they blend into their surroundings so well that they can barely be perceived. Goosefishes are flattened from the back toward the belly (i.e., not laterally). The most striking part of the body is the head, the width of which is about two-thirds its length. The mouth opening extends over almost the entire width of the head; it has many needle-sharp teeth which seem capable of holding anything they grab. To attract prey, goosefishes use their bait, the illicium, which has been formed from the first six spiny rays of the dorsal fin. This consists of a line at the end of which dangles a fleshy, sometimes worm-shaped shred of skin, which is literally dangled in the water by the goosefish. If another fish approaches with the mistaken idea that this is a worm, the goosefish waits motionless until the prey is close enough to catch. As soon as the goosefish swallows its prey, the dangling worm returns again. This method of predation must be extremely successful, because goosefishes which are caught almost always have their stomachs full, so full in fact that their contents sometimes equals one third of the body weight of the fish. The prey include many species of fishes, crustaceans, cephalopods, and other organisms. When hungry, goosefishes gather in shallow water visited by diving birds—because even small birds are eaten.

During the spawning period of the *allmouth* or *angler* (*Lophius piscatorius*), which occurs in early summer along the European coasts and between January and February in the North Atlantic Ocean, the females develop a special kind of appetite. The fish leave their grounds near the coast and move into deeper regions (3280 and 6560 feet; 1000 and 2000 meters). The eggs are released in long, glandularly secreted tubes up to 1 foot (30 centimeters) wide, $\frac{1}{4}$-inch (5 millimeters) thick and as long as 13 feet (4 meters). The eggs are situated singly or by twos in six-sided compartments in a sort of honeycomb arrangement. Eventually, the structure tears and the eggs float individually in the water until the young hatch. The larvae, which are about $\frac{1}{8}$-inch (2 millimeters) long, swimming near the surface of the water do not resemble their parents at all; they look much more like typical fish larvae. After a four-month larval period, with growth up to a length of 4 to 6 inches (10 to 15 centimeters), and a complex change in body shape, the juvenile goosefishes have the adult appearance and behavior.

In spite of their unusual appearance, goosefishes are popular as food. The head and leathery skin are removed so that little remaining in the frying pans looks like the original fish. In Europe, the fishes are sold fresh or smoked as "trout sturgeon." Insulin can be extracted from the islets of Langerhans in the pancreas.

Frogfishes. These fishes are of the suborder *Antennarioidei* and are usually small, seldom exceeding a length of 12 inches (30 centimeters). All frogfishes have a peculiar body shape, enabling them to resemble their surroundings to such a high degree that they blend into the algae "forests" like the Sargasso frogfish. This mimicry of surroundings is created by many skin folds and various skin formations, supplemented by the ability to adapt to some extent in color and pattern in the background. Frogfishes which do not swim well are prevented from falling through the Sargasso seaweed or other objects by their pectoral fins, which function much like human hands in grabbing and holding objects. The reproduction and other life habits of this suborder remain essentially unknown. Female frogfishes observed in aquariums have produced spawning tubes similar to those in goosefishes.

Deepsea Anglerfishes. These fishes are of the suborder *Ceratioidei* and are characterized by the absence of pectoral fins and the fact that only females have a fishing organ. While goosefishes and frogfishes live primarily in shallow water and may migrate into deeper water only for spawning, the approximately 120 species of deepsea anglerfishes are found at substantial oceanic depths, ranging from 985 to 13,125 feet (300 to 4000 meters). Any typical bait on the end of a fishing line would no longer be recognizable at these depths, so deepsea anglerfishes have a luminous organ at the end of the line. Production of the light is not fully understood, but some authorities attribute this to the presence of luminous bacteria. See accompanying illustration.

Deepsea anglerfish (*Linophryne arborifer*).

Parasitic Males. The ceratiid anglers (*Ceratiidae*), photocorynid anglers (*Photocorynidae*), and linophrynid anglers (*Linophrynidae*) are species with very small males which parasitize the females and stay with them. The males do have their own gill respiratory system and the necessary vessels to supply their organs with oxygen, but their food is obtained from the bloodstream of their female hosts, with certain blood vessels in the males' heads in a dependent relationship with vessels in the females. Close examination of the dwarf males shows that several systems are more or less degenerated in them, particularly the teeth and the intestinal tract. They also lack the fishing line, as do all male deepsea anglerfishes.

The biological significance of parasitic males is related to the method of propagation of the species. Despite numerous signaling systems, including the luminous organs, it remains difficult for the different sexes to find each other in the deep sea, where all sunlight is absent. This dangerous disadvantage is offset by having just one partner seek food while the other parasitizes the food gatherer. Exemplary of the size differences is the female *Ceratias hollbolli,* which is some 40 inches (103 centimeters) long, as compared with the male, which averages about 3.5 inches (9 centimeters) in length. The size differences are not this large in some species.

The female deepsea anglerfishes are very active predators which feed on large organisms. Their jaws have powerful teeth and their stomachs are so greatly distensible that they can swallow prey which is larger than they are.

Deepsea anglerfishes primarily inhabit warmer parts of the Atlantic, Pacific, and Indian Oceans. Their numbers decrease sharply in the northern and southern temperate zones. Only a few species have been found in the north Atlantic Ocean.

Researchers at the Department of Biology, California State University (Long Beach, California) studied the action of a species of Antennarius that originated in the Philippine waters. These investigators found that the pattern of movement of the illicial apparatus seems to be species-specific, ranging from simple strokes in the vertical plane to a complex triangular pattern, alternating with rapid sinusoidal thrusts. As reported by Pietsch and Grobecker (1978), "During a single luring sequence the illicium is initially brought straight forward in front of the mouth of the angler, and the bait is rapidly vibrated for 1 or 2 seconds. The bait is then held nearly motionless as the illicium is slowly laid back again onto the head and returned to its nonluring position. When the animal is sufficiently aroused, however, the bait makes a large and rapid sweeping motion that describes a nearly perfect circle. The thin membranous quality of the bait allows it to ripple while being pulled through the water, simulating the lateral undulations of a swimming fish. The lure thus provides not only a highly attractive visual cue but presumably also a low-frequency pressure stimulus for potential prey." This is an example of great energy savings in the quest for food. Further details are given in *Science*, **201**, 369–370 (1978).

In 1979, researchers at the College of Fisheries, University of Washington (Seattle, Washington) made a highspeed photographic study of the feeding actions of antennariid anglerfishes. Light cinematography at 800 and 1000 frames per second was used. It was found that single feeding events for the anglerfishes occur at speeds greater than four times those described for other fishes. Further details are given in *Science*, **205**, 1161–1162 (1979).

ANGLESITE. Naturally occurring lead sulfate ($PbSO_4$), which crystallizes in the orthorhombic system and may be found mixed with galena, from which it is usually formed by oxidation. Hardness, 3; specific gravity, 6.12–6.39; luster, adamantine to vitreous or resinous; transparent to opaque; streak, white; colorless to white or green, but may be rarely yellow or blue. This mineral is used as a source of lead.

Anglesite, whose name derives from Anglesey, England, is found in many European localities; in the United States it has been found in large crystals in the Wheatley Mine, Phoenixville, Pennsylvania, and also in Missouri, Utah, Arizona, and Idaho.

ANGLE (Slip). 1. The angle included between the direction of the applied force and the surface of shear during the plastic flow of a solid body. 2. The angle of repose. See also **Repose (Angle of).**

ANGLE (Solid). Solid Angle.

ANGLE (Trisection of). Conchoid of Nicomedes.

ANGLE-WING (*Insecta, Lepidoptera*). Butterflies of the genus *Polygonia.* Their wings are sharply angular but no more so than those of some other species.

ANGSTROM. Units and Standards.

ANGUILLID EELS. Eels.

ANGULAR DISTRIBUTION (Particle). In physics, the distribution in angle, relative to an experimentally specified direction, of the intensity of particles or photons resulting from a nuclear or an extranuclear process. Commonly, the specified direction is that of an incident beam, and the angular distribution is that of particles which are scattered or are the products of nuclear reactions. Alternatively, the specified direction might be that of an applied field, or a direction of polarization, or the direction of emission of an associated radiation.

ANGULAR FREQUENCY. Frequency.

ANGULAR GYRUS (Brain). Memory.

ANGULAR MAGNIFICATION. The ratio of the tangent of the angle with the optical axis made by a ray upon emergence from an optical instrument to the tangent of the angle for the conjugate incident ray.

ANGULAR MEASUREMENT (Eccentricity Correction). A correction for eccentricity must be applied to many types of instruments used for astronomical angular measurement. Instruments for this purpose usually consist of a circle graduated in angular units, and an arm, assumed to be concentric with the circle, that sweeps around the circle, carrying a vernier, or a measuring microscope, for the purpose of determining accurately the direction of the arm relative to the circle. It is practically a mechanical impossibility to make the centers of the circle and the measuring arm exactly coincident.

In the diagram, point *C* is the center of the circle *OMA*, graduated from *O*. *C′* is the center of the measuring arm (commonly known as the alidade). The direction *C′ M* is the actual direction of the

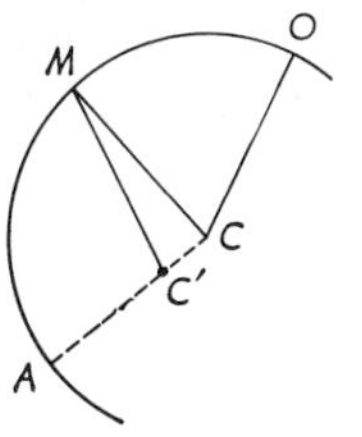

Eccentricity correction for circle with alidade.

alidade, and *OM* is the direction obtained from the circle reading (i.e., the angle *OCM*). The difference between these two directions, the angle *C′ MC*, is the eccentricity correction for the circle reading *OM*. This will be different for different circle readings, being zero for the circle reading *OA*.

An equation may be developed that will give the eccentricity correction for any circle reading as a function of the reading and three numerical constants. To determine these constants, at least three known angles must be measured with the instrument. The differences between the values obtained with the instrument and the known values of the angles are the eccentricity corrections for the circle readings. These eccentricity corrections are then used for the solution of three equations for the three constants. With the constants determined, the equation, giving the eccentricity correction for any circle reading, may be written. The results are usually tabulated, or plotted on a curve and supplied by the maker of the instrument.

ANGULAR MOMENTUM (Conservation of). Conservation Laws and Symmetry.

ANGULAR MOMENTUM (Particle). The vector product of the instantaneous values of the position vector and the linear momentum,

$$\mathbf{M} = m\mathbf{r} \times \mathbf{v}$$

For an interacting system of particles, the law of the conservation of angular momentum states that the rate of change of the total angular momentum equals the vector sum of the moments of the external forces applied to the system,

$$\frac{d}{dt}\sum_s M_s = \sum v_s \times F_s$$

where the summation is over the particles composing the system. In the absence of external forces, the angular momentum remains constant and no change of configuration can alter the total angular momentum of the system. Thus, a slowly rotating swarm of particles, like a cloud of gas in space, if it contracts under its own gravitational attraction, must rotate the more rapidly to keep its angular momentum constant. In the atmosphere, rotating air-masses tend to preserve their absolute angular momentum and the strong winds in hurricanes arise from the convergence of air into the lower levels.

If the particles are bound together in a rigid body, it is convenient to define the moment-of-inertia tensor

$$I_{ij} = \sum_s m_s(\delta_{ij} v_e v_e = v_i v_j)_s$$

where the position vectors are relative to the center of mass. Then the total angular momentum due to motion about the center of mass is

$$M_i = T_{ij}\omega_j$$

where ω_j is the instantaneous angular velocity of the rigid body (repeated tensor suffices indicate summation).

The principle of conservation applies to angular as well as to linear momentum. That is, no change of configuration within a system, uninfluenced by external forces, can alter the total angular momentum of the system. Thus, a slowly rotating swarm of particles, like a cloud of gas in space, if it contracts under its own gravitational attraction with attendant decrease in moment of inertia, must rotate the more rapidly to keep its angular momentum constant. Again, if a person, whirling about on tiptoe, with arms extended, suddenly brings the arms down to the sides, he will as suddenly begin to whirl faster, the effect being more pronounced if he holds heavy weights in his hands. Angular momentum being a vector quantity, the principle applies as well to its direction as to its magnitude. The result is that any rotating body tends to maintain the same axis of rotation, a fact well illustrated by the spinning top and by the stabilizers used on some ocean vessels.

For an elementary or other particle, angular momentum may arise from (a) rotation about an axis, (b) revolution in an orbit, or from both (a) and (b). The angular momentum of rotation is called intrinsic angular momentum, or spin. When a nucleus is considered as a single particle, its total angular momentum is also referred to as its spin. For an elementary particle, the component in a particular direction of both kinds of angular momentum is quantized; the quantum of spin angular momentum is $\frac{1}{2}\hbar$, and the quantum of orbital angular momentum is $\hbar$.

ANGULAR VELOCITY AND ANGULAR ACCELERATION. Quantities relating to rotational motion. While the use of the term "angular velocity" may be extended to any motion of a point with respect to any axis, it is commonly applied to cases of rotation. It is then the vector, whose magnitude is the time rate of change of the angle θ rotated through, i.e., $d\theta/dt$, and whose direction is arbitrarily defined as that direction of the rotation axis for which the rotation is clockwise. The usual symbol is $\boldsymbol{\omega}$ or $\boldsymbol{\Omega}$.

The concept of angular velocity is most useful in the case of rigid body motion. If a rigid body rotates about a fixed axis and the position vector of any point P with respect to any point on the axis as origin is $\mathbf{r}$, the velocity $\mathbf{v}$ of P relative to this origin is $\mathbf{v} = \boldsymbol{\omega} \times \mathbf{r}$, where $\boldsymbol{\omega}$ is the instantaneous vector angular velocity. This indeed may serve as a definition of $\boldsymbol{\omega}$.

The average angular velocity may be defined as the ratio of the angular displacement divided by the time. In general, however, this is not a vector, since a finite angular displacement is not a vector. The instantaneous angular velocity is more widely used.

Angular velocities, like linear velocities, are vectorially added; for example, if a top is spinning about an axis which is simultaneously being tipped over toward the table, the resultant angular velocity is the vector sum of the angular velocities of spin and of tipping. (This enters into the theory of precession.)

Angular acceleration is the time rate of change of the angular velocity, expressed by the vector derivative $d\boldsymbol{\omega}/dt$. Only in case the direction of the axis remains unchanged can the angular velocity and angular acceleration be treated as scalars. The effect of torque applied to a body free to rotate about an axis is to give it angular acceleration, and the opposition offered by the body to this process gives rise to the concept of moment of inertia. See also **Stroboscope; Tachometers and Speed and Velocity Measurements; Velocity (Angular).**

ANHEDRAL. Minerals in igneous rocks which are not bounded by their typical crystal faces. Such minerals are said to be anhedral or *allotriomorphic*, the latter term (now obsolete) proposed by Rosenbusch in 1887.

ANHEDRAL-CRYSTALS. Mineralogy.

ANHYDRITE. The mineral anhydrous calcium sulfate, $CaSO_4$, occurs in granular, scaly, or fibrous masses, is rarely crystallized in orthorhombic tabular or prismatic forms. Hardness, 3–3.5; sp gr, 2.9–2.98; translucent to opaque; streak white; color, white, gray, bluish, or reddish. Anhydrite has three cleavages at right angles to one another. It is similar to gypsum and occurs under the same conditions, often with the latter mineral. It is usually found in sedimentary rocks associated with limestones, salt, and gypsum, into which it changes slowly by the absorption of water. See also **Gypsum.**

Anhydrite is found in Poland, Saxony, Bavaria, Württemberg, Switzerland, and France; in the United States, in South Dakota, New Mexico, Texas, New Jersey, and Massachusetts; in Canada, in Nova Scotia, New Brunswick, and exceptional specimens from the Faraday Uranium Mine near Bancroft, Ontario.

ANIDEX FIBERS. Fibers.

ANILINE. Aniline, phenylamine, aminobenzene $C_6H_5NH_2$ is a colorless, odorous liquid, an amine, with melting point −6°C, boiling point 184°C, is slightly soluble in water, miscible in all proportions with alcohol or ether, poisonous, turns yellow to brown in the air, is a weak base forming salts with acids, e.g., anilinehydrochloride ("aniline salt," $C_6H_5NH_2 \cdot HCl$) from which aniline is reformed by addition of sodium hydroxide solution. Aniline reacts (1) with hypochlorite solution, to form a transient violet coloration, (2) with nitrous acid (a) warm, to form nitrogen gas plus phenol, (b) cold, to form diazonium salt (benzene diazonium chloride, $C_6H_5N{-}Cl$), (3) with acetyl chloride, acetic anhydride, or acetic acid glacial, to form N-phenylacetamide

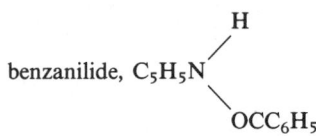

acetanilide, "antifebrin," $C_5H_6N\begin{matrix}\diagup H \\ \diagdown OCCH_3\end{matrix}$

(4) with benzoyl chloride, to form N-phenylbenzamide

benzanilide, $C_5H_5N\begin{matrix}\diagup H \\ \diagdown OCC_6H_5\end{matrix}$

(5) with benzenesulfonyl chloride, to form N-phenylbenzene sulfonamide $C_6H_5SO_2NHC_6H_5$, soluble in sodium hydroxide, (6) with chloroform $CHCl_3$ plus alcohol plus sodium hydroxide, to form phenyl isocyanide C_6H_5NC very poisonous, (7) with H_2SO_4 at 180° to 200°C, to form para-aminobenzene sulfonic acid (sulfanilic acid, $H_2N \cdot C_6H_4 \cdot SO_2H(1,4)$), (8) with HNO_3, when the amine group is protected, e.g., using acetanilide, to form mainly paranitroacetanilide $CH_3CONH \cdot C_6H_4 \cdot NO_2(1.4)$, from which paranitroaniline $H_2N \cdot C_6H_4 \cdot NO_2(1,4)$ is obtained by boiling with concentrated hydrochloric acid, (9) with chlorine in an anhydrous solvent, such as chloroform or acetic acid glacial, to form 2,4,6-trichloroaniline $(1)H_2N \cdot C_6H_2Cl_3(2,4,6)$, (10) with bromine water, to form white solid 2,4,6-tribromoaniline $(1)H_2N \cdot C_6H_2Br_3(2,4,6)$, (11) with potassium dichromate in sulfuric acid, to form aniline black dye, and, by further oxidation, benzoquinone $O{:}C_6H_4{:}O(1,4)$, (12) with potassium permanganate in sodium hydroxide, to form azobenzene $C_6H_5N{:}NC_6H_5$ along with some azoxybenzene $C_6H_5NO{:}NC_6H_5$, (13) with reducing agents, to form aminohexahydrobenzene (cyclohexylamine, $H_2N \cdot C_6H_{11}$), (14) with alkyl halides or alcohols heated, to form alkyl anilines, e.g., methylaniline $C_6H_5NHCH_3$, dimethylaniline $C_6H_5N(CH_3)_2$.

Aniline may be made (1) by the reduction, with iron or tin in HCl, of nitrobenzene, and (2) by the amination of chlorobenzene by heating with ammonia to a high temperature corresponding to a pressure of over 200 atmospheres in the presence of a catalyst (a mixture of cuprous chloride and oxide). Aniline is the end-point of reduction of most mononitrogen substituted benzene nuclei, as nitrosobenzene, beta-phenylhydroxylamine, azoxybenzene, azobenzene, hydrazobenzene. Aniline is detected by the violet coloration produced by a small amount of sodium hypochlorite.

Aniline is used (1) as a solvent, (2) in the preparation of compounds as illustrated above, (3) in the manufacture of dyes and their intermediates, (4) in the manufacture of medicinal chemicals. See also **Amines.**

ANIMAL ASSOCIATION. Gregariousness.

ANIMAL BEHAVIOR PRECURSORS. Earthquakes, Seismology, and Plate Tectonics.

ANIMALCULE. A minute animal. Applied to the protozoa and to such microscopic forms as the rotifers. See also **Rotatorias.**

Bear animalcule.

ANIMAL MANURE. **Biomass and Wastes as Energy Sources.**

ANIMAL PROTEINS. **Protein.**

ANIMAL UNIT. A term sometimes used in pasture and forage land management. One animal unit equals one mature cow, or one horse, or five sheep, or two yearling calves.

ANIMAL-UNIT-MONTH. The feed or forage needed to support one animal unit (see **Animal Unit**) for 30 days. One animal-unit-month is roughly equivalent to 0.3 ton (0.27 metric ton) of hay, or 300 pounds (135 kilograms) of total digestible nutrients.

ANION. A negatively charged atom or radical. In electrolysis, an anion is the ion which deposits on the anode; that portion of an electrolyte which carries the negative charge and travels against the conventional direction of the electric current in a cell. Within the category of anions are included the nonmetallic ions and the acid radicals, as well as the hydroxyl ion, OH^-. In electrochemical reactions, they are designated by the minus sign placed above and behind the symbol, such as Cl^- and SO_4^{2-}, the number of the minus sign indicating the magnitude, in electrons, of the electrical charge carried by the anion. In a battery, it is the deposition of negative anions that makes the anode negative. See also **Ion.**

ANION-EXCHANGE RESINS. **Ion Exchange Resins.**

ANIONIC MECHANISM. **Organic Chemistry.**

ANIONIC SURFACTANT. **Detergents.**

ANIONIC WATER. **Hydrate.**

ANISE. Of the family Umbelliferae (carrot family), the anise plant (*Pimpinella anisum*) is native to the Mediterranean region and is cultivated in Egypt, Malta, Spain, and Syria, but also in other areas of the world, such as Germany and the United States. (This plant should not be confused with *fennel* or *finocchio*, which is commonly called anise in the marketplaces of the United States.)

Frequently, the plant is grown by gardeners of small plots as part of an herb garden. The plant achieves a height of about 2 feet (0.6 meter) with several slender branches. See accompanying illustration.

Anise plant (*Pimpinella anisum*).

The aromatic, warm, and sweetish odor and taste of the seed, leaves, and stem arises from the presence of a volatile oil that contains anethole (*p*-propenyl phenylmethyl ether, $C_3H_5C_6H_4OCH_3$), the derivatives of which (anisole and anisaldehyde) are used in food flavorings, particularly bakery, liqueur, and candy products, as well as ingredients for perfumes. For commercial production of anise oil, the seeds and the dried, ripe fruit of the plant are used. Anise oil, a colorless to pale-yellow, strongly refractive liquid of characteristic odor and taste, is prepared by steam distillation of the seed and fruit. The oil contains choline which finds use in medicine as a carminative and expectorant.

The anise plant is an annual, planted directly from seed in the spring. The leaves of the plant can be used directly in salads to provide a distinctive flavor.

The fruit of a small evergreen tree (*Illicium anisatum*) of the Magnolia family is the source of *star* or *Chinese anise.* The aromatic and chemical characteristics of this plant are similar to those of *Pimpinella anisum* and thus there are similar uses for it.

ANISODESMIC STRUCTURE. A type of ionic crystal in which some of the ions tend to form tightly bound groups, e.g., nitrate and chlorate.

ANISOTROPIC MEDIUM. An anisotropic medium has different optical or other physical properties in different directions. Wood and calcite crystals are anisotropic, while fully-annealed glass and, in general, fluids at rest are isotropic.

ANISOTROPY. **Ultrasonics.**

ANKLE. In humans, a hinge joint formed by the articulation of the tibia, the malleolus of the fibula, and the convex surface of the talus. Movement of the joint provides (1) flexion or dorsiflexion (*bending the foot toward the anterior part of the leg*); (2) extension or plantar flexion (*downward* movement of the foot); (3) inversion or supination (*turning in* of the foot); and (4) eversion or pronation (*turning out* of the foot).

ANKYLOSING SPONDYLITIS. **Spondylarthropathies.**

ANNABERGITE. The mineral annabergite is a rather rare nickel arsenate with the formula $Ni_3(AsO_4)_2 \cdot 8H_2O$, crystallizing in the monoclinic system. It is of secondary origin, resulting from the alteration of pre-existing nickel minerals, commonly found as surface alteration crust on nickeline. Annabergite has been found in Saxony, France, including Annaberg, from which its name is derived, and as exceptional crystals at Laurium, Greece, and in Cobalt, Ontario, Canada.

ANNATTO FOOD COLORS. These colors are natural carotenoid colorants derived from the seed of the tropical annatto tree (*Bixa orellana*). The surface of the seeds contains a highly colored resin, consisting primarily of the carotenoid *bixin.* The bixin is extracted from the seed by a special process to produce a pure, soluble colorant. Bixin, one of the relatively few naturally occurring *cis* compounds, has a chemical structure similar to the nucleus of carotene with a free and esterified carboxyl group as end groups. Its formula is $C_{25}H_{30}O_4$.

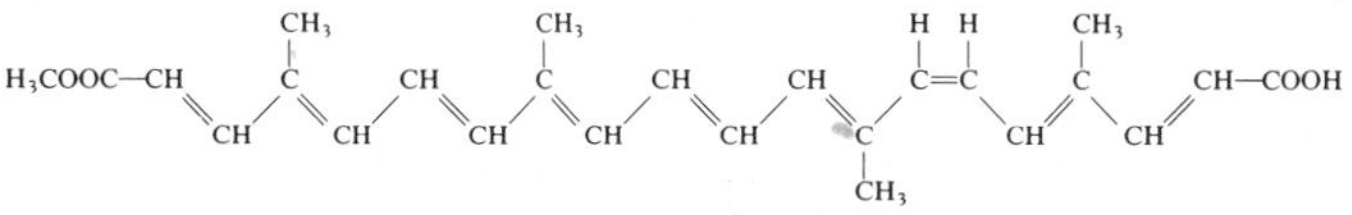

Structure of bixin.

Bixin is an oil-soluble, highly stable coloring ingredient. The saponification of the methyl ester group to form the dicarboxylic acid yields the water-soluble form of bixin, sometimes called *norbixin.* Annatto colorants date back into antiquity. The colorant has been used for centuries in connection with various textiles, medicinals, cosmetics, and foods. Annatto colors have also been used to color cheese, butter, and other dairy products for over a century. See also **Carotenoids.**

Processors make annatto colors available as a refined powder, soluble in water at pH values above 4.0 (solubility about 10 grams in

100 milliliters of distilled water at 25°C), in an acid-soluble form, in an oil-soluble form, in a water- and oil-soluble form, and in a variety of hues ranging from delicate yellows to hearty orange. Annatto extract is frequently mixed with turmeric extract to obtain various hues.

ANNEALING. The process of holding a solid material at an elevated temperature for a specified length of time in order that any metastable condition, such as frozen-in stains, dislocations, and vacancies may go into thermodynamic equilibrium. This may result in recrystallization and polygonization of cold-worked materials.

Annealing generally falls into the technology of heat treatment and varies with materials and the intended end uses of the materials, as well as the prior processing of them. In the case of nonferrous alloys, annealing is primarily a heat treatment for the purpose of removing the hardening due to cold work. Annealing also may be used with nonferrous precipitation hardening alloys to cause softening through agglomeration of the hardening constituent into fewer and larger particles.

Ferrous Metallurgy. In the case of ferrous materials, the term annealing usually implies full annealing. This heat treatment involves a change of phase inasmuch as the metal is heated into the austenitic region. Cooling slowly back to room temperature then develops a softened structure of pearlite and ferrite. The annealing of cold-worked metal is termed *process annealing,* wherein a change of phase is not involved. Annealing takes several forms in terms of the time-temperature relationships imposed upon the materials. *Box annealing, isothermal annealing, normalizing, patenting, spheroidize annealing,* and *stress relieving* are described under **Iron Metals, Alloys, and Steels.**

Annealing of Cold-Worked Metals. Ductile metals hardened by cold-working may be softened by annealing. Annealing is often an important intermediate step in producing metals by cold deformation. Thus, in the formation of fine wires through wire drawing, several intermediate anneals may be required. Annealing may also be the last production step when metal objects are desired in a final softened condition.

In general, cold working increases manyfold the dislocation density of a metal. A severely cold-worked metal may easily have a dislocation density 10^6 times greater than in the same unworked metal. Since each dislocation is surrounded by a strain field extending over long distances on an atomic scale, each dislocation contributes to the strain energy of the metal and, accordingly, to its free energy. When the metal is annealed, the free energy associated with the dislocations resulting from cold work furnishes a driving force that can effectively reduce the dislocation density back to the value that existed before deformation.

Three basic stages are generally recognized as occurring during the annealing of cold worked metals. These are recovery, recrystallization, and grain growth.

In recovery, the strain energy is lowered by the recombination of dislocations of opposite sign, or by rearrangements of dislocations into configurations of lower strain energy. A simple well-known example of this latter is the polygonization of the dislocations in a bent crystal. When a crystal is bent, the curved shape is the result of the accumulation, upon the slip planes of the crystal, of a large number of edge dislocations of the same sign. During recovery, these dislocations move from their more or less random positions along the slip planes into a set of vertical walls normal to the slip planes. This movement is accomplished by both slip and dislocation climb. The walls of dislocations that are formed in this manner constitute a form of grain boundary across which the crystal lattice is slightly rotated by the order of minutes of arc. Such boundaries are better known as subgrain boundaries. The crystalline material between these sub-boundaries is effectively free of dislocations. It is thus apparent that polygonization transforms a highly strained bent crystal into a set of small subgrains that are nearly strain-free. (See Fig. 1.)

The rate of recovery is normally highest at the start of an isothermal annealing cycle because the driving force is largest at that time. As recovery continues, the driving force diminishes as the available strain energy is used up and the rate of recovery falls continuously toward zero. A plot of the rate of recovery as a function of time yields a curve that is somewhat similar in appearance to an exponential decay curve. The rate of recovery is also temperature dependent and may

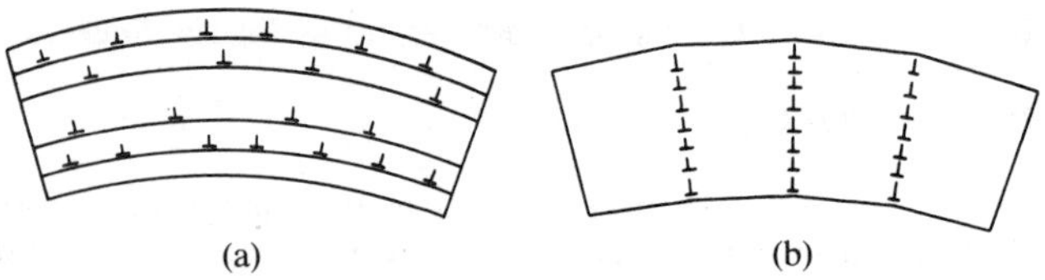

Fig. 1. Realignment of edge dislocations during polygonization: (a) excess edge dislocations that remain on active slip planes after a crystal is bent; (b) arrangement of dislocations after polygonization.

be expressed, in a number of cases, by a simple empirical equation of the form

$$1/t = Ae^{-Q/RT}$$

where t is the time to attain a certain fixed amount of recovery, A is a constant, R, the universal gas constant, T, the absolute temperature, and Q, an empirical activation energy. Because the reactions that occur during recovery are complex, it is usually not possible to attach a simple meaning to the activation energy for recovery.

Recrystallization is the process whereby the distorted grains or crystals of a cold-worked metal are reconverted into new (essentially) strain-free grains. It occurs by the nucleation of minute submicroscopic crystals that grow out into and consume the strained material surrounding them. Recrystallization is, therefore, a nucleation and growth phenomenon and, characteristically, the rate of recrystallization starts slowly, builds up to a maximum, and then diminishes back to zero. The increase in the rate at the early stages of recrystallization is due primarily to continued nucleation of new grains while the older ones continue to grow. The final falling off in the rate is the result of the progressive consumption of the material available for recrystallization. A metal is said to be completely recrystallized when all of the original deformed structure has been eliminated.

Recrystallization, like recovery, is thermally activated and occurs at a rate that grows very rapidly with increasing temperature, as may be seen in the accompanying diagram where the amount of recrystallization in copper is plotted as a function of the time for six different temperatures. (See Fig. 2.)

The driving force for recrystallization also comes from the strain energy of the excess dislocations created by cold work. It is therefore apparent that recovery and recrystallization are competitive processes. Usually a metal may undergo a considerable degree of recovery before visible evidence of recrystallization is obtained. However, since the recrystallized grains grow from very small beginnings, the recrystallization process undoubtedly is occurring long before it can be detected visually. Also, there is reason to believe that the nuclei of the recrystallized grains may be formed as a result, at least in some cases, of processes related to recovery. It should also be noted that recovery phenomena may continue to occur during recrystallization in those grains not yet consumed by the recrystallization process. The degree to which the manifestations of recrystallization and recovery appear to overlap is a function of the metal concerned and of the nature of the deformation that it has received. Under certain conditions, it is possible to have recovery occur without recrystallization. This is particularly true when the amount of deformation is insufficient to cause recrystallization, or when the type of deformation, although extensive, is very simple, as in the case of a zinc or magnesium crystal deformed only by slip on the basal plane. Examples have been observed where single crystals of these metals have been deformed in this manner

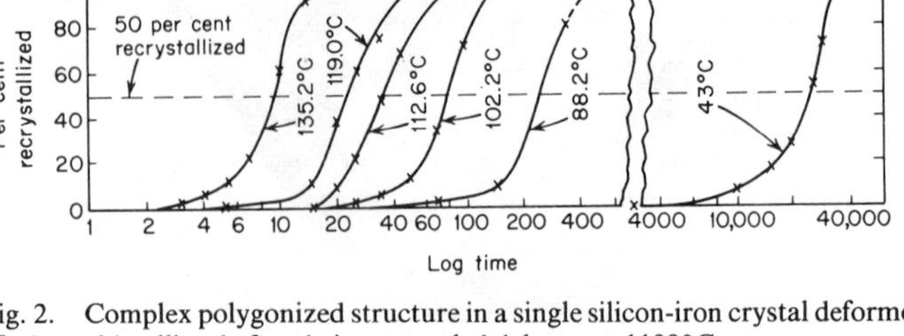

Fig. 2. Complex polygonized structure in a single silicon-iron crystal deformed 8% by cold rolling before being annealed 1 hour at 1100°C.

by as much as 700% and still failed to recrystallize on annealing.

After a metal has undergone recrystallization it can still undergo grain growth. The driving force in this case comes from the surface energy of the grain boundaries. A close analogy exists between the growth of grains in a metal during annealing and the growth of soap bubbles in a soap froth. In the soap froth, a small bubble that finds itself surrounded by larger neighbors will normally have but a few sides convex toward the bubble center. This curvature produces a small but finite excess gas pressure inside the small bubble, which causes gas to diffuse through the bubble wall into the neighboring larger bubbles. The smaller bubble, consequently, grows smaller and disappears, while the larger ones surrounding it grow in size. At the same time, the average bubble in the froth must also grow in size. The same basic phenomenon occurs during grain growth in metals. In this case, the atoms from the smaller grains move across the grain boundaries and become part of the crystals of the larger grains. At all times a geometrically similar distribution of grain sizes exists in the metal, ranging from small to large, which promotes continued grain growth. However, as the average size increases, there is a corresponding decrease in the growth rate. This is easily understood in terms of the soap froth analogy because with an increase in bubble size there is a corresponding decrease in the average bubble wall curvature and in the pressure difference across bubble walls. It may also be shown that in a soap froth the average bubble size should increase as the square root of the time. This one-half power law, however, is seldom observed during grain growth in a metal. Usually the grain growth exponent (power to which the time is raised) is much smaller than one half, signifying that the empirical growth rates are much lower than would be expected from the soap froth analogy. Several reasons may be proposed in explanation of this fact. The motion of grain boundaries is known to be easily influenced by both impurity atoms in solid solution or present as small intermetallic inclusions. In either case, the grain boundary mobility is lowered with a corresponding decrease in the value of the grain growth exponent.

Under the proper conditions, a limiting grain size may be attained in a metal at which point grain growth ceases. This is often true in very thin specimens when the average grain diameter approaches the thickness of the specimen. At this time, the grain boundary geometry becomes two- instead of three-dimensional which reduces the average curvature and hinders further growth. Alternatively, it is possible for grain boundaries to become so held up by the nonmetallic inclusions that further growth is prevented. This condition is only achieved after a critical grain size has been achieved.

Associated with the limiting grain size effect mentioned above is a phenomenon known as secondary recrystallization. Sometimes, after a limiting grain size has been attained, a few grains may begin to grow again and may obtain very large sizes. This is actually not a true recrystallization but rather an unusual manifestation of grain growth. This formation of a new set of very large grains in material where growth had apparently ceased is known as secondary recrystallization.

Annealing of Glass. As with metals, glass is fabricated at high temperatures and is annealed to relieve stresses which would develop if the glass were permitted to cool in an uncontrolled fashion. If not annealed, products made from high-expansion glasses can break spontaneously as they cool freely in air. In annealing glasses, they are raised to an annealing point temperature and then cooled gradually to a temperature that is somewhat below the strain point. Usually, the rate of cooling within this range determines the magnitude of residual stresses after the glass arrives at room temperature. Once below the strain point, the cooling rate is limited only by any transient stresses that may develop. A typical time-temperature glass-annealing curve is shown in Fig. 3. Normally, glass for optical purposes is annealed much more slowly than commercial glassware to improve the optical homogeneity of the material. In the annealing process, glass normally is heated and held at a temperature slightly higher than the annealing point temperature and controlled cooling is effected to a temperature slightly below the strain point to accommodate for differences in materials and as an extra safeguard. Controlled cooling may occur over two periods as indicated by the diagram.

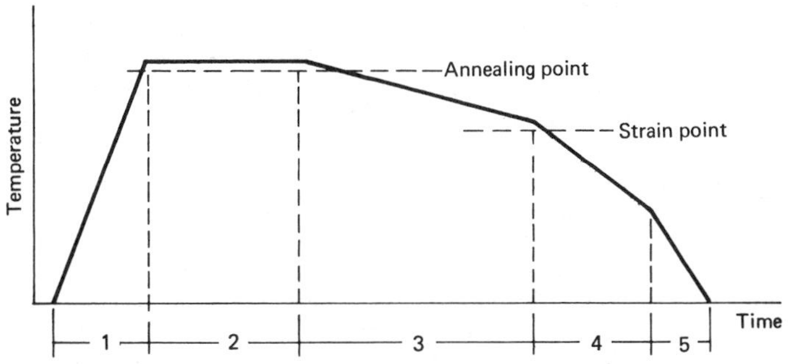

Annealing process periods: 1 – Heating to slightly above annealing point.
2 – Holding temperature for prescribed period.
3 – Controlled-cooling period (first).
4 – Controlled-cooling period (second).
5 – Natural cooling to room temperature.

Fig. 3. Typical glass annealing curve.

References

Banerjee, B. R.: "Annealing Heat Treatments," *Metal Progress*, **118**, 6, 58–64 (1980).

Bardes, B. P., Editor: "Heat Treatment of Carbon and Alloy Steels," in *Metals Handbook*, 9th edition, American Society for Metals, Metals Park, Ohio, 1979.

Chandler, H. E.: "Heat Treating Buyers Guide and Director, "American Society for Metals, Metals Park, Ohio, 1981.

McGannon, H. E.: "The Making, Shaping and Treating of Steel," 9th edition, U.S. Steel Corporation, Pittsburgh, 1971.

Staff: "Heat Treatment '79," American Society for Metals, Metals Park, Ohio, 1979.

Staff: "Heat Treating, Cleaning and Finishing," Vol. 2 of *Metals Handbook*, 8th edition, American Society for Metals, Metals Park, Ohio, 1981.

Staff: "Trends in Heat Processing Technology," *Metal Progress*, **119**, 1, 102–106 (1981).

Wilson, R.: "Metallurgy and Heat Treatment of Tool Steels," McGraw-Hill, New York, 1975.

ANNEALING (Full). **Full Annealing.**

ANNEALING (Glass). **Glass.**

ANNEALING (Semiconductor). **Semiconductor.**

ANNELIDA. The segmented worms, including earthworms and leeches. This phylum is biologically interesting because it shows in a primitive form the structural plan of the more complex animals.

The annelids are characterized by: (1) metameric segmentation; (2) a closed tubular circulatory system in most forms; (3) a coelom; (4) an excretory system with tubules opening from the coelom to the exterior in various segments; (5) a tubular alimentary tract, with regions specialized for various functions; (6) a nervous system consisting of a dorsal brain above the esophagus, connected by cords passing around the esophagus, with a ventral chain of ganglia (see **Ganglion**) below the alimentary tract; (7) setae present in many species.

The annelids are classified as follows:

Class *Archiannelida.* Small marine annelids without setae; few to many segments.

Class *Polychaeta.* Worms with setae. No suckers. External segmentation distinct and metameric. Earthworms and many aquatic species.

Class *Oligochaeta.* Earthworms and many aquatic species.

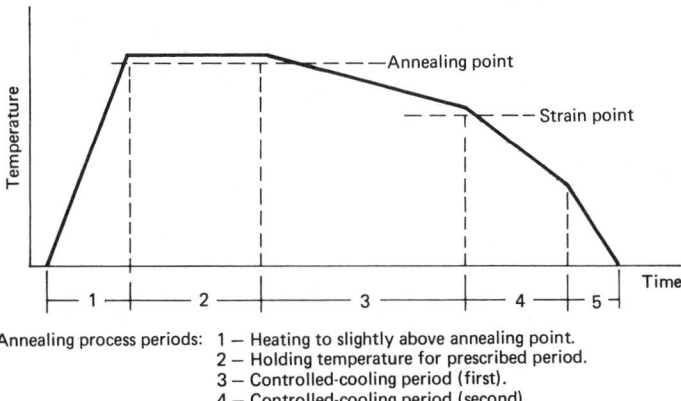

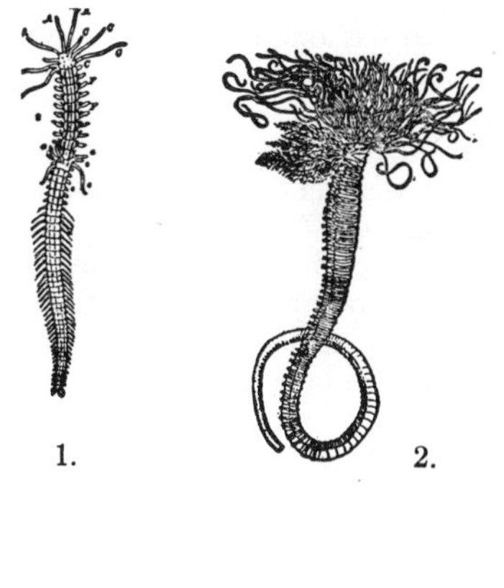

1. 2.

1. Asexual individual of *Autolytus* with male about to detach (*Verrill, "Invertebrate Animals of Vineyard Sound"*). 2. Tufted worm (*Amphitrite ornata*). (*Drawn by Verrill.*)

Class *Hirudinea*. Flattened worms without setae but with a sucker at each end of the body. External segmentation consisting of 2–14 annuli to each metamere. Mostly aquatic, a few marine and a few terrestrial. Mostly blood-sucking parasites. The leeches.

ANNIHILATION. A term used in physics to describe a process in which a particle and antiparticle combine and release the energy associated with their rest masses. The most common example is the annihilation of an electron pair. Usually the negatron and the positron of the pair first form an atom of positronium from which state they merge and are annihilated. To conserve both energy and momentum the rest mass of this particle and antiparticle is converted into two photons moving in opposite directions, each with an energy of 0.511 MeV. The energy associated with the annihilation of other particle-antiparticle pairs, such as a proton and antiproton, is much larger than for an electron pair and is carried away by pions or kaons. See also **Particle (Subatomic).**

ANNONA. Genus of the family *Annonaceae* (custard-apple family). This genus contains shrubs and small trees, many of which bear fruits that are consumed by humans. Generally, they are of relatively minor importance commercially, although some species are valued in regional markets and some are used to make purees which find use in a variety of processed foods. These fruits are composed of many individual ovaries which are more or less sunk in the fleshy receptacle and united to it and to each other. In some species, these collective fruits are quite large—up to 8 inches (20 centimeters) in length. Some of the fruits are so heavy that they drag down the branches. In tropical areas where they are grown, the small fruit trees frequently attract ants which must be destroyed, particularly during the fruiting season. The trees are susceptible to various fungus diseases, causing fruit rot. Copper-based fungicides are frequently used. *Annona* species are found in the tropics of both the northern and southern hemispheres.

Soursop. This species, *Annona muricata*, is found on the islands of the Caribbean Sea, in Florida, and in southeastern Asia, notably in Malaysia. The *A. muricata* is a small evergreen tree about the size of a peach tree. The leaves are leathery and malodorous. The fruit is large (8 inches; 20 centimeters in length) heavy, and pear-shaped, with a rough skin that has spinelike projections. The flesh is white, succulent, acidic, and is variously reported (depending upon particular variety) as having a rich flavor of wine, a taste reminiscent of the black currant, and a flavor something like that of a mango. The seeds are large and of a dark color. The tree has many branches, is decorative, and is valued as a garden ornament as well as for its fruit. In the United States, successful cultivation is confined essentially to southern Florida. The fruit can be found in the markets in the vicinity of Key West. The soursop is processed to remove seeds and fibers in preparation of a puree. This puree can be preserved by canning or freezing. It is then available for use as a base material for flavoring sherbets, ice creams, beverages, and other products. Prior to freezing, the puree is heated to a temperature of about 185°F (85°C) for a few minutes to inactivate peroxidase which, if present, may cause a pink discoloration and off-flavors. Sometimes soursop nectar is available in cans or bottles. The soursop is recommended by specialists at the Food and Agriculture Organization (United Nations) for inclusion in school and demonstration gardens in various developing tropical areas, such as found in west Africa.

Common Custard Apple. This species, *Annona reticulata*, is a deciduous tree that reaches a height of from 15 to 25 feet (4.5 to 7.5 meters). The tree is one of the more robust of the *Annona* species and is commonly found growing in the West Indies and many tropical and subtropical areas of the world. There are numerous trees of this species in southern Florida. The *A. reticulata* is an exception among the *Annona* species in that it spreads spontaneously and need not be individually planted. This widespread growth is exemplified by the profusion of the trees found in the forests of the Philippines, on the island of Guam, and in Malaysia. The fruit or custard apple is inferior in flavor and other edible characteristics as compared with the sugar apple and the cherimoya. The fruit is from 3 to 5 inches (7.5 to 12.5 centimeters) in diameter, has a smooth skin that is geometrically divided into rhomboid or hexagonal areoles. The color varies from red to reddish-brown when ripe. The pulp is semisweet with a tallowlike consistency. The seeds adhere tenaciously to the flesh.

Sweetsop or Sugar Apple. This species, *Annona squamosa*, is native to tropical America and does not do well in subtropical regions as will some of the *Annona* species. The plant is a small deciduous tree which attains a height of from 15 to 20 feet (4.5 to 6 meters). In addition to the West Indies, this species is found throughout the tropics in southern Asia and is particularly popular in India where locally it is called *custard apple*, a fact that tends to confuse it with the true custard apple previously described. The fruit is regarded more highly than *A. reticulata*. The sweetsop is much smaller than the soursop fruit, ranging from 2 to 3 inches (5 to 7.5 centimeters) in diameter and has an almost smooth, segmented skin. The flesh is fragrant and sweet to the taste. The pulp is a pale-yellow and looks very much like custard. Seeds are dark-brown. An advantage of this species is that it produces fruit throughout the year.

Cherimoya. This species, *Annona tripétala* Ait., is a larger tree, rising to a height of about 25 feet (7.5 meters) and more. It is commonly found in the Peruvian Andes and surrounding regions. The plant is also planted in Mexico, a number of Central American countries, Hawaii, India, the Canary Islands, and on the Island of Madeira. The cherimoyas of Madeira are well known and valued for their elegant flavor. In Madeira, the cherimoya plant is trained on trellises much as grapes are grown. The fruit is variously shaped—sometimes spheroidal, ovoid, conoid, and heart-shaped. A sometimes used synonym for the fruit is *bullocksheart*. The fruit has a slightly mottled, comparatively smooth surface. The pulp is white with a pleasing acidulous taste. The seeds are easily separated from the pulp, an advantage in eating the fruit fresh as well as for processing. Over the last several decades, considerable success has been enjoyed in raising this species in California.

Other varieties of *Annona* which bear edible fruit of varying quality and desirability and of very minor importance include *A. montana*, *A. purpurea*, *A. glabra* (alligator-apple), *A. diversifolia*, *A. longifolia*, among several others.

References

Bailey, L. H.: "Standard Cyclopedia of Horticulture," Macmillan, New York, 1963.

Little and Wadsworth: "Common Trees of Puerto Rico and the Virgin Islands," Agriculture Handbook 249, U.S. Forest Service, Washington, D.C., 1964.

ANNUAL. A plant which normally completes its life cycle, from seed to seed, in a single growing season. Typical annuals are corn, wheat, cucumber, and nasturtium. Annual plants are especially suited for life in regions where the growing season is short and alternates with an unfavorable cold period or dry season.

ANNUAL RING. A layer of wood added to the stem in one growing season.

In temperate climates, stem growth occurs during the warm spring and summer months. The cells formed in spring when active growth is taking place are characteristically large, while during the summer only smaller cells are added. This alternation of cells results in the formation of definite concentric rings readily seen in cross sections of woody stems. Actually the growth increment is in the form of a sheath continuous over the entire stem except at the growing tips. External conditions may have a profound effect on the appearance of the annual ring; favorable growing seasons with ample moisture result in broad rings, while seasons of drought produce narrow rings. Removal of surrounding overshading trees may result in a pronounced increase in the thickness of the annual ring. At times events such as severe defoliation by insects or cases of drought may produce two rings in one season; such rings are ordinarily not sharply distinct as are normal ones, and are called false annual rings. Counting of annual rings gives an accurate index of the age of the tree, while attention to details such as variable thickness of successive rings serves to indicate environmental changes. By careful comparison of different logs, even though they be largely reduced to charcoal, one may determine the actual year in which the ring was formed. By this means it has proved possible to establish the probable age of many ruins in the southwestern states. In tropical countries having a continuous growing season, annual rings are not formed or only slightly developed. If, however, alternating rainy and dry seasons occur, then they appear.

ANNULAR ECLIPSE. Eclipse; Sun (The).

ANNULUS. In the sporangium of many ferns, a ring of cells which have their walls characteristically thickened, and bring about the violent discharge of the spores within. In agarics, the ring of tissue which is found around the stalk in many genera, is also known as an annulus. See accompanying figure.

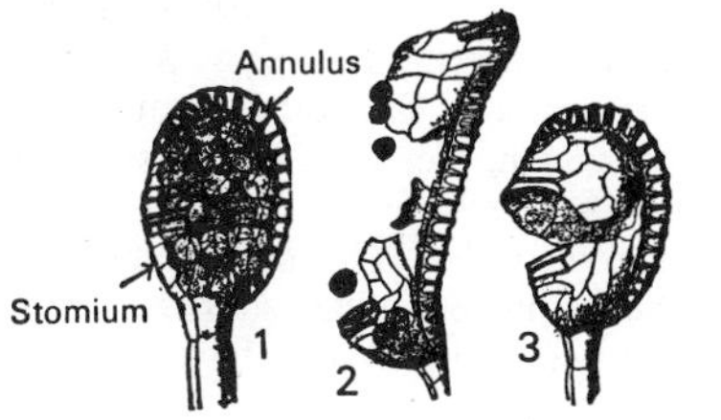

Fern sporangia: (1) unopened; (2) discharging spores; (3) empty.

ANNULUS (Geometry). Circle.

ANOA. Bovines.

ANODE. In the most general sense, an anode is the electrode via which current enters a device. The anode is the positively charged electrode of an electrolytic cell. See **Electrochemistry.** The anode (also frequently called the plate) is the principal electrode for collecting electrons in an electron tube, and is, therefore, operated at a positive potential with respect to the cathode.

ANODE (Battery). Battery.

ANODE SHEATH. In a gas discharge tube, the electron boundary which exists between the plasma and anode when the current demanded by the anode circuit is larger than the random electron current at the surface of the anode.

ANODIC OXIDATION. Since oxidation is defined not only as reaction with oxygen, but as any chemical reaction attended by removal of electrons, then when current is applied to a pair of electrodes so as to make them anode and cathode, the former can act as a continuous remover of electrons and hence bring about oxidation (while the latter will favor reduction since it supplies electrons). This anodic oxidation is utilized in industry for various purposes. One of the earliest to be discovered (H. Kolbe, 1849) was the production of hydrocarbons from aliphatic acids, or more commonly, from their alkali salts. Many other substances may be produced, on a laboratory scale or even, in some cases, on an economically sound production scale, by anodic oxidation. The process is also widely used to impart corrosion-resistant or decorative (colored) films to metal surfaces. For example, in the anodization or Eloxal process, the protection afforded by the oxide film ordinarily present on the surface of aluminum articles is considerably increased by building up this film by anodic oxidation. Also, one process for coloring the surface of aluminum, and retaining a metallic luster, is by adding substances to the metal, and subsequently oxidizing the surface anodically.

ANODIC PROTECTION. Corrosion.

ANODIZE. This term means to place a protective film on a metal surface by electrolytic or chemical action in which the metal surface is made the anode in an electrochemical process. Aluminum and magnesium parts of electronics equipment are frequently anodized.

ANODIZED COATINGS. Conversion Coatings.

ANOLIS (*Reptilia, Sauria*). The name of a genus of lizards adopted also as a common name; small, mostly brightly colored lizards of the warmer latitudes of the Americas. The little lizard sometimes sold under the name chameleon is the Carolina anolis, *Anolis carolinensis*, and is a common species of the southern United States and southward.

ANOMALODESMACEA. An order of bivalve mollusks, mostly burrowing marine species.

ANOMALOUS DISPERSION. Ordinarily the refractive index n of a medium decreases with increasing wavelength λ (see **Dispersion**). It often happens, however, that in the immediate vicinity of a certain wavelength λ_1 there is a break or discontinuity in the dispersion curve and the usual rule may be locally reversed (see figure). In some cases

Variation of refractive index with wavelength, illustrating anomalous dispersion.

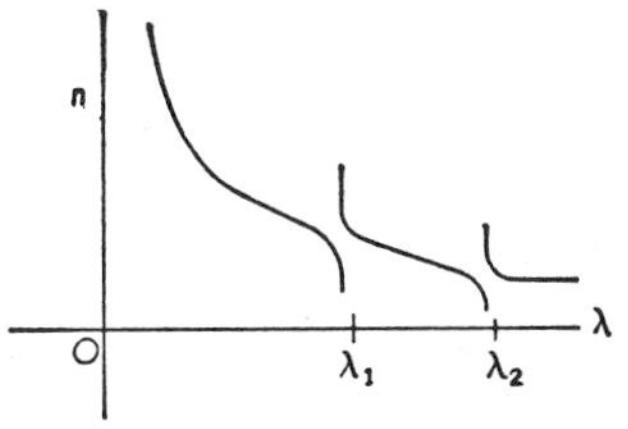

there are several such points, λ_1, λ_2, λ_3, . . . These discontinuities correspond to lines or bands in the absorption spectrum of the medium. In the Sellmeier equation

$$n = 1 + \frac{A\lambda^2}{\lambda^2 - \lambda_1^2} + \frac{B\lambda^2}{\lambda^2 - \lambda_2^2} + \ldots$$

the several fractional terms make provision for the respective discontinuities. The absorption wavelengths λ_1, λ_2, . . . and the constants A, B, . . . must be determined experimentally. It will be noticed that n becomes infinite at every wavelength λ_k and is finite at all other wavelengths. If there is pronounced absorption and anomalous dispersion in the visible range, the medium appears colored, as illustrated by transparent dyes.

ANOMALY (Oceanography). The difference between the conditions actually observed at a particular point of measurement and an ocean of standard or arbitrary temperature and salinity.

ANOPLURA. The order of insects which includes the true or sucking lice. They are wingless parasitic insects with mouths formed for piercing and sucking. See **Louse.**

ANOREXIA. A loss of appetite or distaste for food, present for short periods in connection with a variety of diseases, is known as *anorexia.* Anorexia and weight loss frequently occur among patients with cancer. The conditions may be due directly to the presence of the tumor in some cases. Anorexia may arise because of the production of anorexigenic peptides, a negative nitrogen balance as the result of unreutilized tumor amino acids, an uncoupling of oxidative phosphorylation, or a degraded glucose tolerance. Frequently, anorexia and weight loss are manifestations of the psychological and emotional stresses of a malignancy. Anorexia is almost universally present in subacute bacterial endocarditis. Usually where anorexia arises from a physical disorder that can be cured or arrested, the patient's appetite will ultimately return.

In some animals, particularly those species that hibernate, anorexia is naturally programmed. Fasting during hibernation has been recorded for centuries. However, more recent studies have shown that some animals eat very little and lose weight even when food is available. Often, this occurs when there are other more important activities which compete for the animal's time and attention. An example is that of bull seals that go without feeding for many weeks while minding their duties of defending territory and harem. Fasting associated with incubation, migration, and molting, as well as hibernation, also have been documented over a period of years. Some authorities believe that possibly through hormonal controls some species preprogram periods of anorexia that are made to coincide with other natural actions of greater importance at certain times for survival of individuals and preservation of species. In studies of this behavior, quite limited to date, no information has been uncovered that may be helpful toward understanding and treating the serious human disorder *anorexia nervosa*, which is described next. However, some early keys to improved control of obesity among humans are beginning to unfold.

Anorexia Nervosa. This disease has probably been best described as a complex psycho-endocrine disorder. This is a serious disease, with fatility rates ranging from 2 to 21%, as estimated by various authorities. Currently the disorder is poorly understood and statistics are not fully reliable or representative. For example, the rate of incidence of the disease is not accurately known.

Anorexia nervosa occurs in young women and teenage girls at ten times the rate found in males of comparable age. In 1972, Feigner and associates developed the following criteria for diagnosis of anorexia nervosa: (1) loss of 25% of body weight, (2) a desire to lose weight, (3) an onset prior to age 25, (4) presence of amenorrhea (absence of menstruation), (5) overactivity, and (6) absence of known medical or psychiatric illness. Universally present is a fear of gaining weight. The fear leads to unusual eating patterns and the avoidance of foods of high caloric value. Amenorrhea may be present at the outset or develop during the course of the disorder. In the absence of hypothalamic defects, the general view is that endocrine changes follow rather than lead the psychologic and emotional oddities of the disease. The disease has not been established as having an endocrine origin, but does cause serious endocrine consequences.

Fatness is intolerable to such patients. Sometimes a precipitating experience, difficult to comprehend by others, may be attributed by the patient as the trigger of the motivation to be thin. Most patients with anorexia nervosa are not considered psychotic. Sometimes, however, anorexia nervosa is part of the complex in a true mental illness. Patients tend to be normal in their approach to life excepting the topics of food and weight. It is not unusual for such patients to prepare nutritious meals for the consumption of others. Although the distinctive features of anorexia nervosa are easy to detect, examinations generally reveal few abnormalities except those of an endocrine nature. There are atypical cases which do make diagnosis difficult. Some of the aforementioned criteria may be absent or only partly present.

Therapy includes brief psychotherapy, isolation from family, hospitalization with tube feeding, and psychoanalysis. Numerous drugs have been used, but reliable comparative studies on their effectiveness remain to be developed. Hormonal substitution therapy has been used with limited success.

Recovery is uncertain and varies with individuals. Limited statistics indicate that about 48% of patients ultimately recover, including a return of menses, normal weight, and improved mental and psychosexual outlook. Intermediate results have been reported in 30% of cases; very poor results in 20% of cases; with 2% fatalities. However, some authorities place fatalities in the range of 15–20%.

In patients with hypothalamic defects, the most frequent mechanism of secondary amenorrhea is the absence of the LH (leutenizing hormone) required for ovulation. One of the most vulnerable components of the female reproductive endocrine system, the LH surge is inhibited in anorexia nervosa.

Delayed puberty may result from prepubertal anorexia nervosa. Persons with anorexia nervosa, as well as persons who have voluntarily (martyr syndrome) or involuntarily starved may develop leukopenia and neutropenia—low blood counts of white cells and neutrophils, respectively.

References

Andersen, A. E.: "Atypical Anorexia Nervosa," in "Anorexia Nervosa," (R. A. Vigersky, editor), Raven Press, New York, 1977.

Boyar, R. M., et al.: "Anorexia Nervosa: Immaturity of the 24-Hour Luteinizing Hormone Secretory Pattern," *N. Engl. J. Med.*, **291**, 861 (1974).

Feighner, J. P., et al.: "Diagnostic Criteria for Use in Psychiatric Research," *Arch. Gen. Psychiatry*, **26**, 57 (1972).

Halmi, K. A.: "Anorexia Nervosa: Recent Investigations," *Ann. Rev. Med.*, **29**, 137 (1978).

Mrosovsky, N., and D. F. Sherry: "Animal Anorexias," *Science*, **207**, 837–841 (1980).

Vande Wiele, R. L.: "Anorexia Nervosa and the Hypothalamus," *Hosp. Pract.*, **12**, 12, 45 (1977).

Vigersky, R. A., et al.: "Hypothalamic Dysfunction in Secondary Amenorrhea Associated with Simple Weight Loss," *N. Engl. J. Med.*, **297**, 1141 (1977).

Vigersky, R. A. (editor): "Anorexia Nervosa: III. Treatment Modalities," Raven Press, New York, 1977.

Yen, S. S. C.: "Neuroendocrine Regulation of the Menstrual Cycle," *Hosp. Pract.*, **14**, 3, 83 (1979).

ANOREXIA NERVOSA. Anorexia.

ANORTHIC CRYSTAL. Crystal.

ANORTHITE. Feldspar.

ANORTHOSITE. The name anorthosite was given by T. Sterry Hunt to rocks of gabbroid nature which were essentially free from pyroxene, hence almost wholly plagioclase *usually* labradorite. The term is derived from the French word for plagioclase, anorthose. Small quantities of pyroxene may be present as well as magnetite or ilmenite. The rock is commonly white to gray, bluish, greenish, or perhaps nearly black. A variety from the Province of Quebec is purplish-brown due to the inclusion of ilmenite dust within the feldspars. Although not a common rock in the ordinary sense of the word, occurrences of great areal extent are known in Canada, Norway, and Russia and in the United States in northern New York State and Minnesota. Opinions as to the origin of this rock differ. The development of anorthosite may have been due to the settling out of labradorite crystals from a gabbro magma as many believe, or there may have been an original anorthosite magma.

A study of anorthosite occurrences brings out two very curious circumstances, first, that there is no extrusive (lava) equivalent of anorthosite, and second, that most anorthosite masses seem to be of pre-Cambrian age.

ANOSMIA. Flavorings.

ANOVULATION. Androgens; Infertility.

ANOXEMIA. Deficiency in the oxygen content of the blood. Normally, human blood contains about 18% of oxygen, carried in combination with the hemoglobin of the red cells, and about 0.4% dissolved in the plasma. These amounts may be reduced:

1. When there is insufficient oxygen available to saturate the hemoglobin; this may occur in pulmonary diseases in which inflammatory processes interfere with the passage of oxygen into the blood, or under conditions such as are encountered in high altitude climbing, inhalation of inert gases (methane, helium), in which there is insufficient oxygen in the inhaled air, and in the late stages of cardiac and respiratory disease when insufficient air is inspired (anoxic anoxemia);
2. When the amount of hemoglobin in the blood is insufficient to carry the amount of oxygen required, as occurs in anemia from any cause, especially that following acute or chronic blood loss (anemic anoxemia).

Rapidly developing anoxemia leads, when the oxygen tension of the inspired air (normally 21%) falls to 10%, to cyanosis and to increased rate and depth of breathing; at 5% to loss of consciousness and ultimate death.

Slowly developing anoxemia may be compensated for by a process of acclimatization, not however without damage to vital tissues when the oxygen deprivation is prolonged.

ANSERIFORMES (*Aves*). A large number of goose-like or fowl-like birds which live near water and at least temporarily go into water are grouped in this order. They are ground and waterbirds; the length is 28–170 centimeters (11–67 inches), and the weight is 200–13,500 grams (7 ounces to 30 pounds). The nostrils connect with one another, and the lower mandible has a long process at the angle. The sternum has two indentations or two foramina at the rear; these are absent in the fossil giant duck (*Cnemiornis*) of the glacial period. Two pairs of muscles are located between the sternum and the trachea. The neck is extended in flight. There are 10–11 primaries, the fifth secondary is absent (diastataxic wing), and there are 12–24 tail feathers. Many down feathers are found in the fully developed plumage. The unspotted eggs are light in color. The young are nidifugous, have a dense downy plumage, and are tended for a long time by one or both parents. They are distributed over all continents except Antarctica. The *Anseriformes* are divided into two families distinguished by the absence or presence of horny lamellae in the beak: (1) screamers (*Anhimidae*), and (2) ducks and geese (*Anatidae*).

The screamers (family *Anhimidae*) are almost goose-sized birds of fowl-like appearance, with fairly thick, long legs and feet without webs. The weight is 2–3 kilograms (4.4–6.6 pounds).

There are two genera with marked differences in their internal structure, the horned screamers (*Anhima*), with 14 tail feathers, and the crested screamers (*Chauna*) with 12 tail feathers. Altogether there are three species with no subspecies: (1) the horned screamer (*Anhima cornuta*) which reaches a length of 80 centimeters (31 inches) and inhabits the flood forests of the Amazon delta, (2) the crested screamers (*Chauna torquata*) with a length of 90 centimeters (35 inches), and are found in swampy pampas areas of the La Plata States, and (3) The black-necked screamer (*Chauna chavaria*), with a length of 70 centimeters (27½ inches), is found on forest rivers of Colombia and Venezuela.

Outside the breeding season, the horned screamers live in troops of 5–10 birds. The crested screamers are, however, found in larger flocks which circle above the waters in their habitats in the evenings, calling melodiously. In contrast to the Anatidae, they can glide well. In spite of their unwebbed feet, screamers swim very well; the crested screamer will even climb onto the leaves of floating plants from the shore. These birds calmly walk about the shore or in shallow water; their food is entirely vegetarian and they obtain some of their food while swimming. They readily perch on the branches of trees and, when disturbed or pursued, generally take to trees.

All other members of this order are included in the family of ducks and geese (*Anatidae*). They have horny lamellae on the interior of the beak near the cutting edge. There are webs between the anterior toes. The upper mandible has a particularly hard process, the "nail," at its tip. There are large nasal cavities; because of this, geese breathe faster when reacting to olfactory stimuli. There are 16–25 cervical vetebrae. They cannot soar or glide to any extent; a few species are flightless, while others have a rapid flight.

There are three subfamilies: (1) the magpie goose (*Anseranatinae*) with only minute webs, (2) the geese and relatives (*Anserinae*) with larger webs and small scales on the tarsus and toes, and (3) the ducks and relatives (*Anatinae*) which also have large webs.

The Anatidae are made up of many species that are colorful and of many shapes. There is a wide range of changing forms and ways of life, from the minute African pigmy geese to the trumpeter swan, which weighs 13½ kilograms (30 pounds), and from the inconspicuous greylac goose to the colorful plumage of the king eider. Nevertheless, the different species have common characteristics which justify their grouping in one family. Thus all are water birds, even species which live mainly on land like the Cape Barren goose and the Hawaiian goose. Since, as swimmers, their plumage must always be greased, they have a particularly large preen gland.

All *Anatidae* have webbed feet. The webs extend from the second to the third and from the third to the fourth toe. Anatidae do not have to make any particular effort to remain on the surface of the water or to swim. Their buoyancy is due mainly to the air held in the plumage. They take great care not to get water under their plumage. The closed wing is covered by feathers projecting up from the side of the breast so that, of the wing feathers, only the scapulare and the primaries are generally exposed. The body plumage is continually and carefully covered with oil from the preen gland and so forms a layer impermeable to water.

A particular adaptation to the requirements of swimming is the broad cross section of the body of most *Anatidae.* This broad, bargelike body maintains its balance despite wind or waves. Another adaptation is the shortening of the thigh and tarsus. The tarsus functions like the arm of a lever in swimming, and in slow swimming is almost the only part of the leg that moves. Only in faster swimming is the thigh involved as well; it is drawn back with partially extended knee, the lower leg serving merely to transmit power.

To reduce resistance to the water when the tarsus moves forward, the webs and toes are folded together and the toes are bent. In pushing back, the toes and webs are fully extended and form an effective oarlike surface. As in walking, so in swimming the legs move alternately. Only the mute swan in its aggressive display swims with jerky movements, pushing back with both legs at the same time; however, it swims no faster by this method.

According to their manner of obtaining food, the Anatidae can be divided into several groups. Swans, shelducks, and surface feeding ducks "up end;" that is they immerse the head and neck with the rear of the body projecting almost upright above the water. In this way, they can feel over the bottom of shallow waters with the beak, and obtain food by straining it out of the water. Diving ducks also generally get their food from the bottom, but they reach greater depths and dive completely below the surface. Lastly the mergansers chase fish beneath the surface.

The true geese, swans, and whistling ducks are entirely vegetarian. Mergansers, scoters, eider ducks, and the South American torrent ducks take only animal food. The rest take both plant and animal food. The amounts of food taken are at times considerable. In the digestive tract of an eider duck, 114 mussels were found, some of which were already partially digested within their shells; the gullet and stomach of a velvet scoter contained 45 oysters.

Most of the species build their nests on the ground. Some of the genus *Tadorna,* like the common shelduck, prefer burrows in the ground as nest site, although some other species of ducks, among them the mallard, also nest on trees. The Orinoco goose, many of the *Cairini* like the maned wood duck, the mandarin, and the wood duck, the Brazilian teal, the pigmy geese, the comb duck and its relatives, the goldeneyes, and some mergansers breed preferably in tree cavities. Nests of the magpie goose, the coscoroba swan, and many diving ducks and stiff-tailed ducks are often found in dense swamp vegetation on the water.

The nest construction is simple. As far as it is possible, the birds make a hollow in the ground and pull in tems and leaves from around the nest site, as far as they can reach with their outstretched necks. All species cover incomplete clutches with plant material when leaving the nest. Shortly before the last egg is laid, the females pluck the nest-down and line the nest with it. See also **Poultry; Screamer;** and **Waterfowl.**

ANT (*Insecta, Hymenoptera*). Social insects of varied structure and habits. They may be distinguished from the related bees and wasps by the form of the slender petiole which connects thorax and abdomen; in the ants it is expanded above and looks wedgelike in profile. Ants have been known to exist on earth for some 30 to 40 million years. Evidence of their early existence is found in historic fossil Baltic amber. The average ant is about one-sixteenth inch in length, but the large Texas ant may measure up to one inch or more. Stages in the development of the ant are egg, larva, pupa, and adult. Most ants are omnivorous and wingless. See accompanying figure. There are over 5000 species of ants. Among varieties found in the United States are the army ants, carpenter ants, the leaf cutters, the dairyman, the garden (common ant), honeydew, and mound-building ants.

Ant

A sampling of ant types and habits would include: the voracious Argentine ant which steals eggs and ants from other nests, making the captive ants their slaves; the honey-pot ant which stores extra honey in its flexible body to furnish food for the young; the Legionaries ants which travel in single file to steal and store other insects for food, which live in wet areas mostly under foliage and are nearly blind; the garden ant which builds small mounds of sand in gardens or on walkways, brown in color, and with a good-natured temperament; the Texas ant known as the fungus grower in the ant world, and which builds mounds up to 12 inches (0.3 meter) in diameter, and chews leaves into a mulch for the growth of fungus which then becomes its food; the harvest ant or seed collector, one of the largest of the ants, which stores small seeds of various kinds in its mound and whose bite is painful; the carpenter ant which drills tunnels in rotting wood; the sugar ant which is extremely small, very light-brown in coloration, harmless, and which is excessively fond of sweets; the tree ant of India (*Olcophylla smaragdina*) which uses its larvae as a means of sewing leaves together to make a nest; Brazil's terrible ant which is fully 1 inch (1.5 centimeters) long, and which produces a vicious sting.

It has been determined that the behavior habits of ants are strictly acquired naturally, as in the case of the honeybee, and that no intelligence in the normal interpretation of the word is required. Ants recognize excitement and unusual conditions, however, and relay this information to other ants by stroking them with their antennae, pecking them on the head or thorax and, if a danger signal is required, the ant will open its jaws wide, or it may hold its abdomen high and run about wildly. Ants are also known to leave scent trails by pressing their stomach close to the ground. Although the scent persists for just a short time, the scent trail is refreshed by other ants repeating the same procedure.

Texas Leaf-Cutting Ant (*Atta Texana*, Buckley). Native to southern and eastern Texas (and western Louisiana), this species of ant prefers loamy soil that is adequately drained. Some nests range from 10 to 20 feet (3 to 6 meters) in depth, incorporating numerous craters, but not rising very high above ground level. Such nests are the habitat of many thousands of ants. The worker ants of this species range from $\frac{1}{16}$ to $\frac{1}{2}$ inch (1.5 to 12 millimeters) in length, they are light-brown in color, and have hardened bodies equipped with many spines on the head and thorax. Food for these ants is prepared by the workers who macerate freshly cut leaves which subsequently promotes the growth of a fungus used as food. In addition to serious pests in the garden, the ants attack field crops and, in particular, they are damaging to young pine seedlings and thus must be eradicated in areas of reforestation. Poisoned bait which the ants can carry back to their nest is effective. Argentine bait has been successful. This includes tartaric acid crystals, benzoate of soda, and sodium arsenite. These materials are boiled, along with sugar, and then mixed with honey. The resulting sirup is used as bait for ants that like sweet foods, such as the Argentine ant. The bait does not work, however, with the Pharaoh ant, the southern fire ant, or the tiny thief ant. A similar bait for protein-loving ants is prepared from thallous sulfate, groundnut (peanut), butter, and German sweet chocolate. Both of these concoctions are quite poisonous and must be prepared and used with extreme caution, marking all containers as *poison*. Placing insecticides in the normal pathways of travel of ants also is effective. Pouring liquid carbon bisulfide directly into nest openings also can be effective.

Imported Fire Ant (*Solenopsis saevissima richteri*, Forel). Native to South America, the fire ant invaded the United States at Mobile, Alabama in 1918. Since then, it has spread into more than 130 million acres (52 million hectares) in Alabama, Arkansas, Florida, Georgia, Louisiana, Mississippi, North Carolina, South Carolina, and Texas. This is a small, aggressive insect that produces a painful, burning sting. When disturbed, the ant is quick to attack both people and domestic animals. Each colony of imported fire ants builds a hard-crusted nest, or mound, sometimes 3 feet (1 meter) high and nearly 3 feet (1 meter) across. In some areas, there may be as many as 50 mounds per acre (125 per hectare), making it difficult to operate mowers and other machinery in pastures and fields, as well as lawns and park grounds.

The damage from the pest is difficult to measure in economic terms. The stings cause blisters that require as long as 10 days to heal. If the blisters break, infection may develop. Some people have been hospitalized; a few have died, primarily from allergic reaction to the stings. Some farm workers may refuse to work on land where these ants are numerous for fear of being stung while clearing clogged mower blades, handling crops, and performing other agricultural tasks.

Imported fire ants look like ordinary house and garden ants. They are from $\frac{1}{8}$- to $\frac{1}{4}$-inch (3 to 6 millimeters) long and reddish-brown or dark brown to black in color. A single mature mound contains a queen ant, several thousand winged males and females (future queens), and up to 100,000 workers. The winged forms leave the mound, most frequently in May and June, and mate in flight. Afterward, the queens land and break off their wings. They dig shallow burrows in the soil and begin to lay eggs that start new colonies. Winds and air currents may carry the new queens 12 or more miles 22+ kilometers during mating flights. They prefer to build their "homes" in open, sunny areas. Thus, the most valuable land on farms and in suburbs is often the most likely to be infested.

Cooperative programs between federal, state, county, and local governments have been established to make continuous surveys and to provide controls over the pest. Quarantines may be established to prevent further widespreading of the insect. Until the late 1970s, a control chemical, Mirex, provided effective control over the insect. Identified as a carcinogen, the compound was removed and as of late 1980 no suitable replacement has been approved.

Cornfield Ant (*Lasius alienus*, Förster). The cooperation between ant and aphid is mentioned in the entry on **Aphid.** The corn root aphid (*Anuraphis maidiradicis*, Forbes) and the cornfield ant provide an excellent example of this cooperation and thus, indirectly, the cornfield ant contributes in a major way to the damage wrought on corn by the corn root aphid. Whenever the food producer notices half-grown corn (maize) that is not doing very well, as indicated by sluggish growth and yellow- or red- tinged leaves, it will be noted that, at the base of the plant, there will be numerous blue-green aphids and a great activity of small brown cornfield ant milling about between the corn roots (which they tunnel) and a number of anthills near the plant. During the winter season, aphid eggs are collected by the ants and stored in their nests during the cold weather. The eggs are frequently moved about to assure they are obtaining the best exposure to temperature and humidity. In early spring, the aphid eggs commence to hatch. At this time, the ants carefully carry the tiny eggs to the roots of nearby smartweed plants or other grasses suitable for aphid feeding. After 2 to 3 weeks of this feeding process, the full mature aphid females commence to give birth to live female young. In turn, these females mature and give birth to others in 2 weeks or less. Reproduction is by parthenogenesis (development of the egg without fertilization). Throughout the summer, the ants care for the welfare of the aphids. Distribution of the aphids in a corn (maize) field, for example, depends almost totally upon the actions of the ants. The reward to the ant is the honeydew, a sticky sweet exudation from the anal opening of the aphid. The honeydew comprises a major part of the ant's diet. It has been observed that ants have moved over 150 feet (50 meters) from a grass meadow to a cornfield, moving their own young as well as large numbers of aphids, which the ants then turn out to pasture on young corn roots.

Thief Ant (*Solenopsis molesta*, Say). A common household pest, the thief ant prefers protein foods to sweet foods, and sometimes damages grain seed when germinating.

Recent interesting articles include: "Territorial Strategies in Ants," by B. Hölldobler and C. J. Lumsden, *Science*, **210**, 732–739 (1980), and "Canopy Orientation: A New Kind of Orientation in Ants," by B. Hölldobler, *Science*, **210**, 86–88 (1980).

ANTACIDS. These are formulations widely used in the treatment of excessive gastric secretions and peptic ulcer. Several factors determine the efficacy of antacids, including (1) the ability and capacity of the stomach to secrete acid; (2) the duration of time the antacid is retained in the stomach; and (3) the nature of the gastric response upon eating.

Five principal active ingredients are used in antacid preparations: (1) *Sodium bicarbonate* is a rapid and effective neutralizer. The compound does yield large amounts of absorbable sodium, undesirable in some persons (heart disease; hypertension). The compound also may induce milk-alkali syndrome. (2) *Calcium carbonate* is a strong, effective neutralizer, but can cause constipation, hypercalcemia, acid rebound, and milk-alkali syndrome. (3) *Aluminum hydroxide* provides slow and not potent action. The compound causes constipation, absorbs phosphates, as well as certain drugs, such as tetracyclines. (4) *Magnesium hydroxide* which provides a slow and prolonged action with no major side reactions. (5) *Magnesium trisilicate* which acts like magnesium hydroxide, but which is poorly absorbed and acts as an osmotic laxative. In cases of renal insufficiency, the serum magnesium should be monitored.

The foregoing compounds are frequently used in combination and, in some, simethicone is added to relieve flatulence. There are striking differences of commercial antacids in terms of their neutralizing capacity. Among compounds having a high neutralizing capacity are Gelusil-II®, Maalox® Therapeutic Concentrate, and Mylanta II. The relative potency of antacid tablets is less than that of liquid formulations.

The physician is concerned with at least three factors when prescribing antacids: (1) Acid rebound (associated with calcium carbonate); (2) milk-alkali syndrome (caused by ingestion of large quantities of alkali); and (3) phosphorus depletion (by aluminum salts). The mechanism of acid rebound, especially in the long-term use of calcium carbonate, is poorly understood. It has been established that there is an excessive reacidification of the antrum (pyloric gland area) a number of hours after ingestion of calcium carbonate.

The ingestion of a quart of milk or more while taking large amounts of alkali, as from antacids, sets up conditions favorable to milk-alkali syndrome. Generally, with withdrawal of the milk or the antacid, the condition is self-correcting. Symptoms of milk-alkali syndrome include nausea, vomiting, anorexia, weakness, polydipsia, and polyuria. Abnormal calcifications also may occur in the chronic stage and other symptoms include mental changes, asthenia, aching muscles, band keratopathy, and nephocalcinosis. Symptoms of milk-alkali syndrome sometimes tend to mimic hyperparathyroidism and vitamin D intoxication.

Cimetidine therapy for ulcers is described under **Ulcer.**

ANTAGONIST (Angiotensin). Hypertension (High Blood Pressure).

ANTARCTIC CONVERGENCE. A distinct, natural oceanographic boundary around the continent of Antarctica. The boundary is more or less equivalent to the 50°F (10°C) isotherm for the warmest month. It has been determined that the colder, denser Antarctic waters sink sharply below the warmer, lighter Subantarctic waters with very little mixing. The flora and fauna reflect the water and air temperature differences on either side of the boundary.

ANTARCTIC FISHES. Ice Fishes.

ANTARCTIC MAPPING (Satellite). Earth Resources Satellites and Geologic Remote Sensors.

ANTARCTIC METEORITES. Meteorite.

ANTARCTIC REGION RESEARCH. Polar Research.

ANTARCTIC WATERS. Water masses in or associated with the Antarctic Ocean, including:

Antarctic Bottom Water. An oceanic water mass arising close to the margins of the Antarctic continent. It is particularly dense due to the cold and to the fact that the surface waters freeze over in winter, leaving the salt behind to increase the density of the remaining water. Sinking to the bottom of the ocean, the water creeps north until it encounters the North Atlantic Deep Water, with which it is believed to merge.

Antarctic Circumpolar Water. Also called West Wind Drift, this is the oceanic water mass with the largest volume transport (approximately 110 × 10^6 cubic meters per second) (3,883 × 10^6 cubic feet per second) and the swiftest current. It flows from west to east through all the oceans around the Antarctic continent. The flow is locally deflected from its course, particularly by the distribution of land and sea and partly by the submarine topography. Besides the bends that are associated with the bottom topography, the effects of the distribution of land and sea and of the currents in the adjacent oceans are also evident. On its northern edge, it is continuous with the South Atlantic current, the South Pacific current and the eastward-flowing extension of the Agulhas current in the Indian Ocean. Salinity maxima occur at depths ranging from 700–1300 meters (2,310–4,290 feet), averaging 34.8%. The temperature range is 0 to 2°C (32° to 35.6°F). It is surface water in some regions; a deep mass in others.

Antarctic Intermediate Water. This is the oceanic water mass originating in the northern part of the Antarctic closest to the equator. The water moves northward as a surface current until it meets the warmer waters of the South Atlantic Ocean. Because it is colder and thus heavier, the Antarctic water sinks below the surrounding warm water to a depth of approximately 600–900 meters (1,980–2,970 feet) and continues northward until it returns to the surface between 20° and 30° north latitude.

Antarctic Surface Water. A relatively shallow oceanic water mass extending from the Antarctic convergence, where it meets the Subantarctic Water, to the shores of Antarctica. Its depth increases from about 80 meters (264 feet) in the Atlantic and 150 meters (495 feet) in the Pacific to 300–400 meters (990–1,320 feet) as it nears Antarctica. Temperature range is from −2 to 3.5°C (28.4° to 38.3°F); salinity from 34.5% to 32.8%.

Subantarctic Water. An oceanic water mass extending on the surface from the South Atlantic Central Water and the South Pacific Central Water to the Antarctic Surface Water in the south. In the Atlantic, it extends only from about 52° to 53° south latitude (from the subtropical convergence to the Antarctic convergence). In the Pacific Ocean, it covers a much larger area. Temperature range at the surface is 3 to 10°C (37.4° to 50°F); salinity from 33.8% to 34.8%.

ANTARES (*α Scorpii*). A star, whose name is derived from two Greek words signifying that it is "similar to" or the "rival of" Mars, doubtless because of its distinctly reddish hue. In fact, this reddish color has always made Antares an object of interest and importance in the ancient religions, and many of the Egyptian temples are so oriented as to indicate that this star played an important part in their ceremonials. Antares was one of the four royal stars of the Persians about 3000 B.C., and some writers claim that it is the "lance star" referred to in the 38th chapter of the Book of Job.

The diameter of Antares has been determined with the stellar interferometer and found to be about 7.5 × 10^8 kilometers, or slightly greater than the distance of Mars from the sun. It is a typical M spectral-type giant star of very low density.

Ranking sixteenth in apparent brightness among the stars, Antares has a true brightness value of 5,000 as compared with unity for the sun. Estimated distance from the earth is 400 light years. See also **Constellations;** and **Star.**

ANTEATER. Aard-Vark; Edentata; Monotremata; Pholidota.

ANTECEDENT STREAM. A stream that has maintained its consequent course in spite of localized uplifts which, if they had proceeded rapidly in relation to the cutting power of the stream, would have caused diversion of the stream. A good example of an antecedent stream valley is one which cuts across a ridge or several ridges. Excellent examples occur, in the valley and ridge province of the Appalachian Mountains. On the other hand, it has been suggested that the Appalachian antecedent stream valleys may be really superimposed. The accompanying diagram illustrates the origin of the present topography and stream pattern of the Appalachians. It is postulated that the folds were reduced to a peneplain on which were flowing a few master streams. Uplift of the peneplain caused the rejuvenation of the master streams which were able to maintain their courses across the upturned edges of the more resistant strata, while the new tributary stream pattern was largely determined by the less resistant formations.

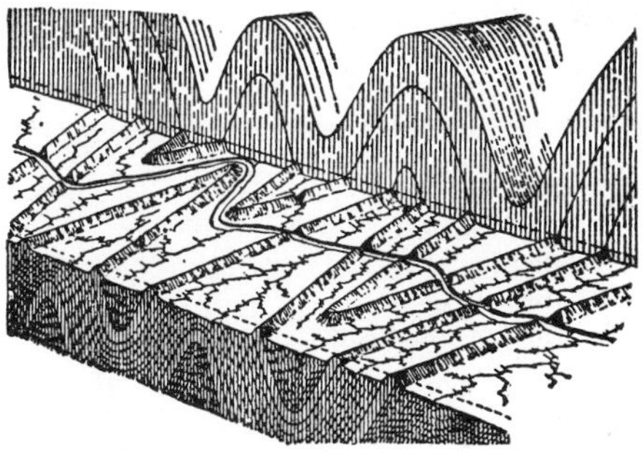

Structural and erosional history of the Appalachian Range. (*W. M. Davis*)

ANTELOPE (*Mammalia, Artiodactyla*). The antelopes (*Antelopines*) comprise one of the larger groups of the order *Artiodactyla* (even-toed hoofed mammals). In anatomy, physiology, appearance, and habits, the antelopes lie between the oxen and the sheep and goats. Many species occur in Africa and some in India and Tibet. The Pronghorn Antelope of western North America belongs to a separate group (*Antilocaprines*). The horns of the pronghorn are hollow; the horns of true antelopes are almost solid. See **Pronghorn Antelope.**

The horns, in fact, are one of the most interesting features of the antelopes. The horns consist of a structure of bone covered with keratin which is harder than bone and grows out from the animal's skull. Horns of older rams grow into nearly complete circles. It has been reported that the noise from clashing horns during fights between the males of certain species of antelopes can be heard sometimes for a distance of two miles. The shape and size of the horns frequently serve as an excellent index of the subfamily or species. For example, the Kudii has large corkscrew horns; the Giant Sable, crescent-shaped horns; the Oryx, rapierlike horns, etc. Rings on the horns form yearly, but the age is often difficult to determine with accuracy by counting the rings because there is some overlapping.

As a general observation, the various species of antelopes can run fast, but not always fast enough to avoid death on the open plains where large carnivorous animals consider the antelope good eating. Most species of antelopes are quite alert looking, with head held high, ears erect, large eyes, short hair, graceful build, and frequently attractive and distinguished horns. Size ranges from about that of a goat to as large as a horse. Some antelopes have markings on the face and head which also are indicative of species. In several species, the coats blend in well with the surrounding habitat. Some people in antelope-inhabited areas consider antelope flesh as good meat, even a delicacy.

The general organization of the Antelopines group is given in the accompanying table. Only the most important species are indicated. The following descriptive paragraphs follow the general organization of the table.

With the exception of a few species, the antelopes are almost exclusive to Africa where they prefer savanna, grassland, scrub, and semidesert as a habitat. The term *antelope* is sometimes used loosely for the related bovines and caprines.

The Horse-Antelopes are large-hoofed animals with horse-shaped bodies. The Giant Sable antelope is known only in Angola and considered quite rare. Coloration is purplish-black. The rapier-horned antelopes are small and very horselike in appearance. See accompanying figure. The Oryx is found in the desert regions of Africa and thence to Syria. All species have very long horns, straight or slightly recurved. The screw-horned antelopes are natives of the Sahara. The species is known for going long periods with absolutely no water.

Several species of the hartebeest make up a significant portion of the group of Deer-Antelopes. Generally they have large, ringed horns, irregularly spiraled with the tips pointing back. These animals are graceful and the horns of some species are described as resembling a lyre. The animals are about 4 feet high (1.2 meters) at the withers (ridge between shoulder bones). The forequarters are heavier and much higher than the hindquarters. Possibly the first hartebeest to be recognized was the Titel or Bubal, which roamed across north Africa in the days of the Roman Empire and was frequently called a horned horse. This animal, smaller than most hartebeests and with relatively

Oryx or besia antelope. (*American Museum of Natural History*)

GENERAL ORGANIZATION OF THE ANTELOPES

ANTELOPINES

HORSE-ANTELOPES (*Hippotraginae*)

Sabre-horned Antelopes (*Hippotragus*)
—Giant Sable Antelope (*H. variani*)
—Common Sable Antelope
—Roan Antelope
Rapier-horned Antelopes (*Aegoryx* and *Oryx*)
—White Oryx (*A. algazel*)
—True Oryxes
—Gemsbock
—Beisa
—Beatrix Oryx
Screw-horned Antelopes (*Addax*)

DEER-ANTELOPES (*Alcelaphinae*)

Hartebeests (*Alcelaphus*)
—Coke's Hartebeest
Damalisks (*Beatrugus* and *Damaliscus*)
—Hunter's Hartebeest (*B. hunteri*)
—Korrigum
—Topi
—Sassaby
—Blexbok
—Bontebok
Gnus (*Connochaetes* and *Gorgon*)
—White-tailed Gnus
—Brindled Gnus
—White-bearded Gnus

MARSH-ANTELOPES (*Reduncinae*)

Waterbucks (*Kobus*)
Lechwes (*Onotragus*)
Kobs (*Adenota*)
—Kob
—Puku
Reedbucks (*Redunca*)
The Rhebok (*Pelea*)

BLACKBUCK (*Antilopinae*)

PIGMY ANTELOPES (*Neotraginae*)

Klipspringers (*Oreotragus*)
Oribis (*Ourebia* and *Raphicerus*)
—Steinboks
—Grysboks
Sunis (*Neostragus*)
The Beira (*Dorcatragus*)
Dik-Diks (*Madoqua* and *Rhynchotragus*)
Royal Antelopes (*Nesotragus*)

GAZELLES (*Gazellinae*)

Impalla (*Aepyceros*)
The Gerenuk (*Litocranius*)
The Dibatag (*Ammodorcas*)
The Springbuck (*Antidorcas*)
The Addra (*Addra*)
True Gazelles (*Gazella*)
Goat-Gazelles (*Procapra*)

short thick horns, ringed and black in color, became extinct in the early 1900s. Another member of the hartebeest group is the Konzi, with small horns, pale color, and black tail and front of legs black. It is a very lively animal, living in small parties, often in company with zebra and waterbuck, usually in the flat wooded districts of Zambia and Rhodesia along the Zambezi. Another hartebeest is the Korigum of central Africa, also sometimes called the Senegal antelope. The Sassaby is sometimes called the bastard hartebeest. The animal stands about 4 feet high (1.2 meters), with horns up to 15 inches (38 centimeters) long. Coloration is deep red, blending into black on the back. Hunter's hartebeest is very rare, found only in the environs of southern Somaliland. The term hartebeests is often used in describing both hartebeests and the damalisks.

Gnus are sometimes called *wildebeests.* The animals generally have a large head, strong curved horns, an erect bristly mane and a bristly muzzle. The withers are high and the tail is hairy throughout its length. In some species, the horns resemble those of a Cape Buffalo. The animals are known for their hilarious behavior.

The Marsh-Antelopes appear and behave more like deer than horses, the qualities of both of the latter animals tending to blend in antelopes. The Waterbuck is a large antelope of southern and eastern Africa. It frequents rocky hills in the vicinity of rivers. The horns are more than 2 feet long, slightly curved and ringed almost to the tips. The Reedbuck is a comparatively small antelope. The male has small horns which turn forward. Also known by the Dutch equivalent, *reitbok.* The Rhebok also is small and is found in hilly sections of eastern and southern Africa. Some authorities have compared this animal with the chamois in its habits.

The Blackbuck, as indicated by the accompanying table, does not fit into the other large subfamilies of the antelopines. Of the antelopines, the blackbuck appears to be more closely related with the gazelles. The animal was known in Europe since the Middle Ages. Its present habitat is India and prefers the open plains from the foothills of the Himalayas to Cape Comorin; and western Pakistan to lower Assam.

As indicated by their name, the Pigmy Antelopes are very small animals, probably the smallest of the ungulates (having hoofs). The Royal antelope of west Africa is only about 8 inches (20 centimeters) high. However, the other species of pigmy antelopes are considerably larger. The Klipspringer is a great jumper, sure-footed, and known for tripping about on its toes. The Oribis is of dainty build, standing less than 2 feet high (0.6 meter), with sharply pointed horns from 4 to 5 inches (10 to 13 centimeters) in length. The color is tawny above and white below. The Sunis is black with white rings around the eyes and white underparts and ears. The horns are long and thin.

The Gazelles comprise the largest group of antelopines. To all but zoologists and keen sportsmen, most gazelles appear very much alike. They are small, delicate, with preference for the desert, grassy plains, scrub zones, and parklands. The Impala is moderately large and of several species; most of these animals have long, slender horns, slightly spiraled and ringed through most of their length. The Gerenuk inhabits eastern Africa and has a very long neck and moderate-spiraled horns, turned forward sharply at the tips. Also called Waller's gazelle. The Addra or true gazelles are found all over Africa outside the forest zones, as well as in Syria, western Arabia, the plains of India, and central Asia from Turkey eastward to the Gobi desert of Mongolia. The affinity between the antelopines and the caprines is found in the so-called Gazelle-Goats. Among these are the Chiru, the Saiga, and the Goa. The latter animal prefers the high plateaus of Tibet. The gazelle-goats travel in small herds with a diminishing population because of hunting. The gazelle-goats are not to be confused with goat-gazelles which are described under **Goats and Sheep.**

For references, see **Mammalia.**

ANTENNA (Communications). Characteristically, communication systems consist of cascaded networks, each network designed to carry out some operation on the energy conveying the information. In radio communication systems, antennas are the networks serving to transfer the signal energy from circuits to space and, conversely, from space to circuits. In circuits, the flow of energy is restricted to one or the other of two directions. The effectiveness of transfer of energy between the antenna and the adjacent circuit element is, therefore, determined solely by the terminal impedance of the antenna and that of the adjacent circuit. The knowledge of the antenna terminal impedance over the desired frequency range, therefore, fully describes the joint performance of the antenna and the circuit energy.

The relationship between the antenna and space, however, is much more complex. The distribution of the radiated energy varies with the direction in space and with the distance from the antenna. This gives rise to the directive properties of the antenna. Further, the energy is radiated in the form of an electric and a magnetic field. These are vector quantities which, at a distance from the source, are at right angles to each other and to the direction of propagation. The planes in which these vectors are located, and whether they are stationary or rotate with time determine the polarization of the radiated field. The performance of an antenna can, therefore, be fully described only by specifying several parameters, such as radiation pattern, gain, and polarization.

It is convenient, in discussing antenna properties, to consider the antenna as a radiating rather than a receiving network. The antennas are, however, linear networks and are subject to the law of reciprocity. (As used here, radiation intensity has the dimensions of power flow per unit area, normally, watts per square meter. Electric field strength, on the other hand, is in volts per meter.)

The performance of an antenna in terms of radiation pattern, gain, or polarization is the same, irrespective of whether the antenna radiates or absorbs radiation.

Except for the immediate neighborhood of the antenna, referred to as the "near-field region" of the antenna, radiated energy propagates radially from the antenna, and the radiation intensity varies inversely as the square of the distance from the antenna. This is a propagation effect. In discussing the antenna performance, it is customary to disregard this and to represent the distribution of the radiated power as a function of the two direction angles only. Such a distribution is commonly represented graphically and is then known as the radiation pattern of the antenna. See Fig. 1.

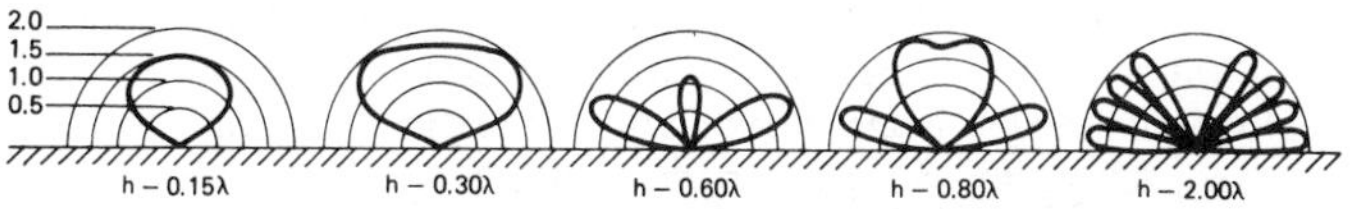

Fig. 1. Vertical polar radiation diagram in plane normal to a horizontal dipole antenna. *h* is height above ground, electrical degrees. λ is wavelength.

The radiation patterns can take a variety of forms. Sometimes they are in the form of a polar diagram, with the radial distance proportional to either field strength or intensity. The intensity may be represented linearly, as power, or logarithmically, in decibels. For representing the directive properties of an antenna in all directions, contours of equal radiation intensity may be plotted, with the two direction angles as abscissas and ordinates, respectively. See Fig. 2.

The directive properties of an antenna also lead to the concept of antenna "gain." The directive gain of an antenna in a specified direction is the radiation intensity in that direction compared to what it would be if the total radiated power were distributed equally in all directions.

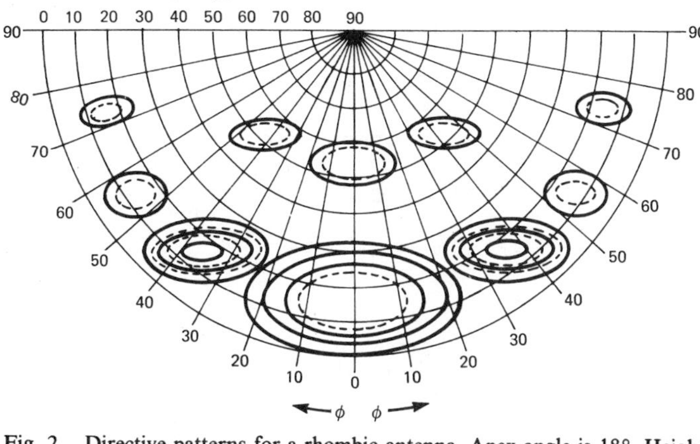

Fig. 2. Directive patterns for a rhombic antenna. Apex angle is 18°. Height above ground, *h*, is 0.8λ (wavelength).

For some applications, such as point-to-point communication, high values of antenna gain are desired because such antennas concentrate the available power, thus effectively increasing it. Conversely, in receiving applications, such antennas are more reponsive to radiation arriving from one direction. For other applications, such as broadcasting, antennas with low directivity may be desired.

The gain of an antenna is dependent principally upon the size of the antenna, expressed in wavelengths. The larger the antenna, the greater is likely to be its gain. The values of gain for different antennas range from 1.5 for an electrically small dipole to hundreds and even thousands times that. In practice, antenna gains are usually expressed logarithmically, in decibels. For the low-frequency end of the radio spectrum (15 kHz to 3 MHz), antennas, although large physically, are relatively small in terms of wavelengths. Therefore, the directive gains of these antennas seldom exceed 3 (4.8 dB). In the high-frequency band (3 to 30 MHz), which is used principally for long-distance communication, antenna gains of 10 to 100 (10 to 20 dB) are frequently encountered. At microwave frequencies, where the wavelengths are a fraction of a meter, gains of several hundred, and even thousand times (20 to over 30 dB) are common.

When an antenna has one or more of its dimensions significantly larger than a wavelength, its radiation pattern is likely to have more than one maximum. The radiation pattern, in such cases, is said to have a lobe structure. That part of the radiation pattern which encompasses the direction of the largest maximum and the radiation immediately to each side of it, is referred to as the main lobe. The radiation about the minor maxima is referred to as the secondary or side lobes. One of the frequent goals in antenna design is the reduction in the levels of secondary lobes. These may, at times, be a source of interference to other transmissions.

In common with light, radio waves consist of electric and magnetic fields at right angles to each other and to direction of propagation. In radio terminology, the orientation of the electric vector of a radio wave is taken as the direction of polarization. Thus, if the electric field vector is parallel to the ground, the radio wave is termed "horizontally polarized." Although the polarization of the energy radiated by an antenna, in general, varies with the direction, an antenna is usually designated as being horizontally (or vertically, circularly, etc.) polarized, depending on the polarization of its radiation in the direction of the main lobe maximum.

The importance of polarization in radio engineering lies principally in the different reflective properties of the ground for waves with electric field parallel to the ground and those normal to the ground. Different radio services are served best by different polarizations. Antennas for use in the low-frequency end of the radio spectrum, as previously defined, are almost invariably vertically polarized. This includes the AM broadcast band. In the high-frequency band, both horizontal and vertical polarizations are used. For FM and television

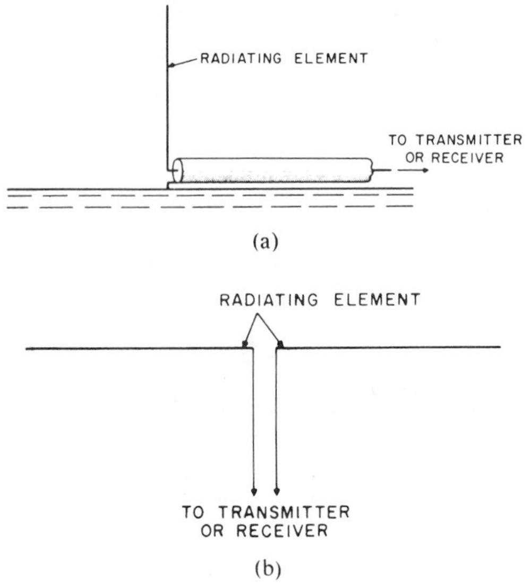

Fig. 3. Two types of elementary radiators: (a) monopole over ground; (b) dipole.

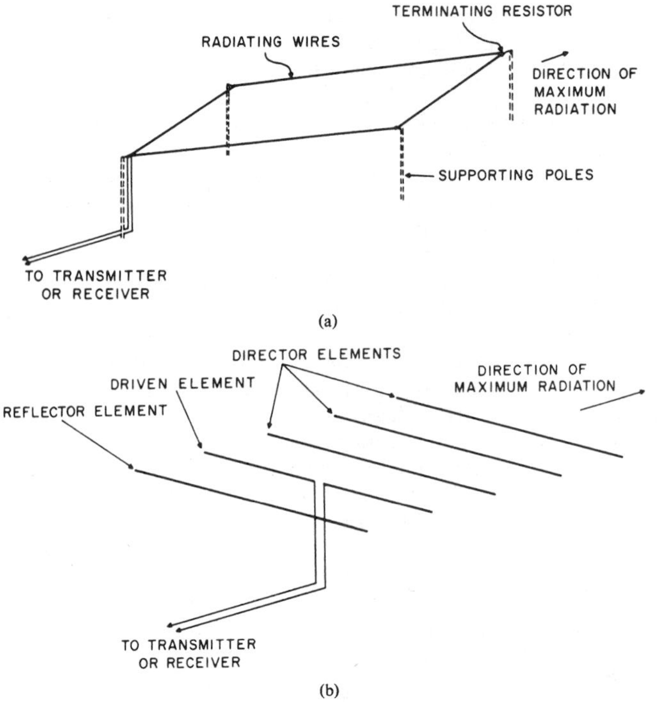

Fig. 4. Examples of directive antennas: (a) rhombic antenna; (b) Yagi antenna.

broadcast service in the United States and many other countries (but not in the United Kingdom), horizontal polarization is employed.

Antenna Configurations

The types and variations of antennas are extremely numerous. Each type has some particular advantage over the others for some specific requirement. Among some of the more important and frequently encountered requirements are those for operating bandwidth, directivity, whether high or low, and polarization. Mounting factors for receiving antennas, as in land vehicles, aircraft, and spacecraft, also pose problems of size, weight, air resistance, etc.

A few of the representative types of antennas frequently encountered in practice are shown in Figs. 3, 4, and 5. Two very elementary radiators, a monopole and a dipole, are shown in Fig. 3. A monopole in one form or another is employed almost exclusively throughout the low-frequency end of the radio spectrum. The dipole is somewhat more versatile, as it can be oriented to give either horizontal or vertical polarization. It is frequently used as an elementary radiator in large array-type antennas.

Figure 4 presents two highly directive, but otherwise radically different types of antennas. The rhombic antenna has broadband properties and is used widely in point-to-point communication service. The Yagi

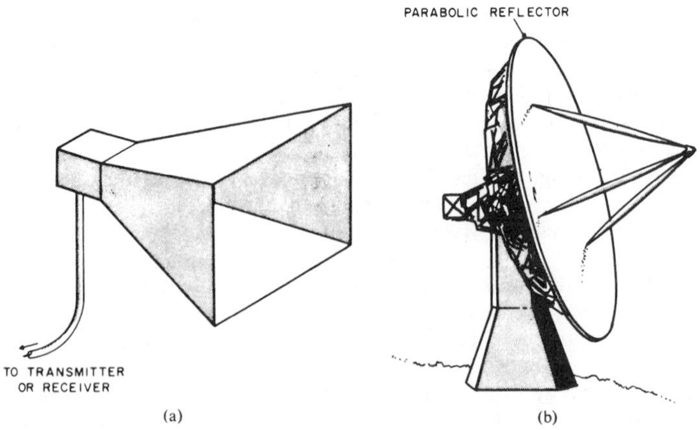

Fig. 5. Microwave-type antennas: (a) horn antenna. (b) parabolic-reflector antenna.

antenna is a relatively compact antenna with high gain for its size. Its operating frequency band is quite narrow. Two antennas used at microwave frequencies are shown in Fig. 5. The horn antenna is used generally where moderate directivity suffices. The parabolic antenna is used for high-gain applications and is a quasi-optical device.

Specific antenna configurations are described briefly in the following listing, arranged alphabetically for convenience.

(*Achromatic Antenna*)—An antenna whose characteristics are uniform over some band of frequencies.

(*Adcock Antenna*)—A form of radio antenna which in its simplest form (see Fig. 6) consists of two spaced vertical antennas. This type of antenna finds use in radio direction finding where it displays a significant advantage over the loop antenna. When the latter is used to obtain a bearing, incorrect results may be obtained because horizontally polarized downcoming radio waves will induce different voltages in the two horizontal members of the loop. In the Adcock antenna, on the other hand, horizontally polarized waves induce voltages in the horizontal elements which cancel in their effect on the output voltage.

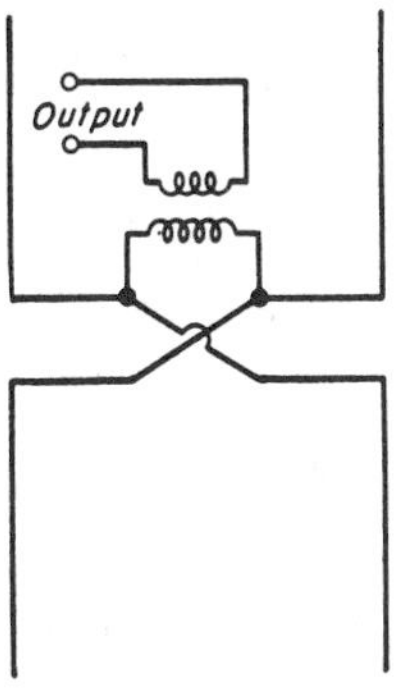

Fig.6. Elementary Adcock antenna.

(*Alford Slotted Tubular Antenna*)—A horizontally polarized antenna developed for FM broadcast work. It consists of a sheet of metal bent into the form of a cylinder which is not quite closed, hence a straight narrow slot extends the full length of the cylinder or tube. It is so dimensioned that the distribution of potential across the slot has very nearly the same phase throughout the entire length of the slot. The currents produced flow in horizontal circles around the cylinder so that the latter operates something like a stack of small, in-phase loops.

(*Aperiodic Antenna*)—A nonresonant, and thus frequently insensitive antenna.

(*Base-Loaded Antenna*)—An antenna (usually vertical) whose electrical height is increased by the addition of inductance in series with the antenna at the base.

(*Biconical Antenna*)—An antenna formed by two conical conductors, having a common axis and vertex, and excited at the vertex. When the vertex angle of one of the cones is 180°, the antenna is called a discone.

(*Broadband Antenna*)—An antenna which will function satisfactorily over a bandwidth in the order of 10% or more of its center frequency.

(*Cage Antenna*)—An antenna in which the radiating members are parallel rods arranged in a cylindrical fashion.

(*Capacitor Antenna*)—An antenna in which the capacitance between two conductors or systems of conductors is the essential characteristic. Also called a dielectric antenna.

(*Cheese Antenna*)—A cylindrical parabolic reflector enclosed by two plates perpendicular to the cylinder, so spaced as to permit the propagation of more than one mode in the desired direction of polarization. It is fed on the focal line.

(*Cloverleaf Antenna*)—An antenna for transmission or reception of horizontally polarized radiation in a nondirectional pattern in a plane normal to the axis of the antenna. Its name arises from the fact that it is comprised of a cluster of four half-wave, curved, radiating elements arranged in the pattern of a four-leaf clover.

(*Coaxial Antenna*)—An antenna comprised of a quarter wavelength extension to the inner conductor of a coaxial line, and a radiating sleeve which, in effect, is formed by folding back the outer conductor of the coaxial line for approximately one-quarter wavelength.

(*Conical Antenna*)—A wideband antenna in which the driven element or elements are conical in shape.

(*Cosecant-Squared Antenna*)—A type of radar antenna which provides constant field strength at a given altitude over an appreciable range. So-called because the power in the antenna pattern on one-way transmission decreases at a rate proportional to the square of the cosecant of the elevation angle. This radiation pattern may be approximated by using a paraboloidal reflector with several radiators arranged in a line perpendicular to the axis.

(*Dielectric Antenna*)—An antenna which employs a dielectric as the major component in producing the required radiation pattern.

(*Dipole Antenna*)—A straight radiator, usually fed in the center, and producing a maximum of radiation in the plane normal to its axis. The length specified is the overall length. Common usage in microwave antennas considers a dipole to be a metal radiating structure which supports a line current distribution similar to that of a thin straight wire, a half-wavelength long, so energized that the current has two nodes, one at each of the far ends. See Fig. 3(b).

(*Directional Antenna*)—For the usual radio-broadcast service, it is desirable to transmit the signal in all directions equally, but for special broadcast services, such as international short-wave or microwave relay transmission, it is often desirable to direct the radiation in some specific direction and avoid radiation in other directions. The need for directed radiation is even more pronounced in other types of radio service. The radio signals may be directed by the use of directional antennae, or, as often called when consisting of more than one element, directional arrays. Any antenna is directional to a certain extent, e.g., the common tower antenna for broadcast stations does not radiate directly upward, but in the sense used here a directional antenna is one having marked characteristics of this type. Basically the directional antennae all depend upon radiation from two or more components adding vectorially. If the waves radiated from various elements add in a certain direction the signal will be strong in that direction, while if they tend to cancel, or subtract, in a given direction the signal will be zero or weaker in that direction.

One of the simplest directional antennas is the loop such as used with many portable receivers. Here the two elements whose effects add vectorially are the two vertical sides of the loop. The result is a figure 8 radiation pattern, i.e., if lines are drawn to scale in various directions so their lengths represent the strength of the signal in each direction, the ends will all lie on a figure 8 curve with the antenna at the center. These antennas are used for many radio ranges, and, since the directional characteristics of any antenna are the same for transmission and reception, for radio compass use. The directional pattern may be altered by adding the radiation from a separate vertical antenna.

For broadcast use where it is necessary to decrease the signal in certain directions, usually to avoid interference with another station, systems consisting of two or more vertical antennas are quite common. By proper spacing of the elements and proper choice of the phase of the currents (which can easily be adjusted by circuit values) a wide range of radiation patterns may be obtained.

For international short-wave broadcasts and for point-to-point communication, more elaborate extensions of the same principle are used. Since the more elements in an array, the sharper the pattern, the radiation may be beamed at will, the type of service and economics being the usual limiting factors. By stacking systems one above the other in a vertical plane the radiation may be directed vertically as well as horizontally. It should be mentioned that it is not desirable to have the beam too sharp even for point-to-point service since the variations in the ionosphere may cause the signal to miss the receiver if the beam is too sharp.

Sometimes the elements of an array are not all fed directly from the transmitter, but some are fed and others pick up energy radiated by the first and reradiate it. By proper choice of the spacing and the antenna dimensions these various radiations may be made to give the desired pattern. The fed antennas are often referred to as driven antennas or elements and the others as parasitic antennas. There are many other types of directional antennas, such as rhombic, V, herringbone, binomial, broadside, continuous-spaced, linear, etc., but all depend upon vector addition of the radiation to give the pattern.

By the use of directional arrays the signal transmitted in a given direction may be increased manyfold over its value for the same transmitter with a nondirectional antenna. A measure of this is the gain of the array which is the signal with the array divided by the signal in the same direction for one element of the array serving as antenna. Where the type service permits their use, directional arrays are the most economical means of obtaining increased signal strength at the receiver. In reception the directional antenna allows the reception of a signal from the desired direction and suppresses signals and noise from other directions.

(*Discone Antenna*)—An antenna of a disk and a cone whose apex approaches and becomes common with the outer conductor of the coaxial feed at its extremity. The center conductor terminates at the center of the disk, which is perpendicular to the axis of the cone. Its most important characteristic is its ability to operate over a very wide bandwidth without a substantial change of input impedance or radiation pattern. The radiation pattern is omnidirectional in a plane perpendicular to the axis of the cone.

(*Dummy Antenna*)—A substitute for an actual antenna which is used for test purposes. In making comparative tests, calibrations, etc., on receivers, it is highly desirable to have conditions as near as possible to actual used conditions, yet have them standardized so they may be reproduced or the results on different units accurately compared. To do this a standard dummy antenna is used. The make-up of the antenna varies with different types of sets, being a series circuit with an inductance of 20 microhenries, capacitance of 200 micromicrofarads and resistance of 25 ohms for regular broadcast receivers. For auto radios, short-wave sets, etc., other circuit combinations are standard. The dummy antenna is connected between the set and the standard signal generator which supplies the radio-frequency test voltages. A dummy antenna consisting of just resistance is frequently used as a load on radio transmitters for making preliminary adjustments without radiating a signal. The output power of the transmitter is dissipated as heat in the resistance.

(*Fanned-Beam Antenna*)—A unidirectional antenna so designed that transverse cross sections of the major lobe are approximately elliptical.

(*Fishbone Antenna*)—An antenna consisting of a series of coplanar elements arranged in colinear pairs, loosely coupled to a balanced transmission line.

(*Franklin Antenna*)—An antenna consisting of a number of half-wave dipoles placed end-to-end, all operating in phase. Also called a colinear antenna.

(*Helical Antenna*)—An antenna used where circular polarization is required. The driven element consists of a helix supported above a ground plane. If the circumference of one turn is approximately one-half wavelength, the radiation is said to be in the axial mode and is directed predominately along the axis of the helix. In this mode, the antenna has good efficiency and relatively broad bandwidth.

(*Horn Antenna*)—A circular or rectangular electromagnetic horn can be used for providing unidirectional pattern coverage with either linear or circular polarization. Circular polarization can be achieved by exciting a square waveguide with a signal polarized at 45° and placing a quarter-wave plate between the feed point and the aperture. The waveguide and transition feeding it essentially determine the bandwidth of a horn. The bandwidth can be increased by using ridge waveguides and other broadbanding techniques. See Fig. 5(a).

(*Image Antenna*)—An antenna located close to the earth's surface (assumed to be a perfectly conducting plane) transmits a direct ray and a ray reflected from the earth's surface. It is convenient to represent the reflected ray as originating from an image antenna identical to the original, and located inside the earth by a distance equal to the height of the original above the earth.

(*Isotropic Antenna*)—Sometimes referred to as a unipole, this is a hypothetical antenna radiating or receiving equally in all directions. A pulsating sphere is a unipole for sound waves. In the case of electromagnetic waves, unipoles do not exist physically, but represent convenient reference antennas for expressing directive properties of actual antennas.

(*J Antenna*)—A half-wave antenna, end-fed by a parallel-wire, quarter-wave section having the configuration of a letter J.

(*Lazy H Antenna*)—An antenna array where two or more dipoles are stacked one above the other for the purpose of obtaining greater directivity.

(*Leaky-Pipe Antenna*)—External radiation is produced by providing a hole or slot in a waveguide propagating electromagnetic power. Proper choice of the size and location of a series of holes in the waveguide may lead to quite directional radiation patterns.

(*Lens Antenna*)—To satisfy high directivity requirements, a lens is often placed in front of another radiator, such as a dipole or horn. In much the same manner as an optical lens focuses light waves, these microwave lenses focus the high-frequency energy into a sharp beam.

(*Logarithmic Antenna*)—In logarithmic antennas, the radiating elements are in geometric progression in accordance with their resonant frequencies. This results in effectiveness at a wide range of frequencies. Sometimes used for special purposes in the high-frequency field. The radiating elements are folded dipoles. The aperture efficiency is comparatively low because only those elements that are fairly close to resonance will contribute to the radiation.

(*Log-Periodic Antenna*)—The geometry of this antenna repeats periodically so that the electrical properties of the antenna also repeat periodically with the logarithm of the frequency. These antennas are essentially frequency-independent and capable of bandwidths of 10:1 or greater, and with little change in patterns.

(*Loop Antenna*)—A loop antenna, when used with radio transmitters or receivers, possesses valuable directional properties.

On ships and airplanes, a loop antenna equipped with a compass card, whose 000°–180° line is parallel to the longitudinal axis of the ship, may be used to obtain relative bearings of radio stations. In using the loop, the position of minimum intensity is sought. In other words, the indicating needle on the compass card is in a direction perpendicular to the plane of the loop, and the observer rotates the antenna until the position of minimum signal intensity is found. The reading of the dial will then give a line of position through the ship and the radio station. With the simple loop there is no method for telling on which side of the instrument the station is located. To overcome this difficulty, the loop antenna is combined with a nondirectional antenna.

(*Loop-Vee® Antenna*)—A broadband antenna that produces a circularly polarized pattern. Pattern coverage of the grounded Loop-Vee is practically identical to that produced by a quarter-wave stub in conjunction with a current loop above the ground plane. The azimuth patterns are nearly perfect circles for all polarizations. In another design, the balanced Loop-Vee incorporates a second element similar to the single element of a grounded antenna and produces a butterfly-shaped pattern.

(*Marconi Antenna*)—An antenna that has one end of its radiating surface grounded (a grounded antenna). It may be considered a transmission line open-circuited at the far end and driven at the sending end, which is the junction between antenna and ground.

(*Marconi-Franklin Antenna*)—This array, one of the first used for high-speed short-wave point-to-point communication, consists of a front curtain of vertical radiators, each consisting of several cophased dipoles in series, and another curtain of reflecting wires of the same construction situated one-quarter wavelength to the rear. There are twice as many reflectors as radiators.

(*Monopole Antenna*)—Any one of several configurations of very simple design, often applied to vehicles, which protrude vertically from the vehicle, such as a whip, spike, blade, cone, or sleeve. See Fig. 3(a).

(*Multiple-Tuned Antenna*)—A low-frequency antenna having a horizontal section with a multiplicity of tuned vertical sections.

(*Musa Antenna*)—A "multiple-unit steerable antenna" consisting of a number of stationary antennas, the composite major lobe of which is electrically steerable.

(*Omnidirectional Antenna*)—An antenna producing essentially constant field strength in azimuth, and a directive radiation pattern in elevation.

(*Oscillating Doublet Antenna*)—A reference standard against which the directional characteristics of an antenna may be compared. Ideally it consists of two closely spaced charges of opposite sign, both oscillating in the same phase. Also, it may be regarded as an infinitely short, linear current-element.

(*Parabolic Antenna*)—A directional antenna using some form of a paraboloidal mirror either to convert plane waves into spherical waves or to convert spherical waves into plane waves. The reflector is fed or "illuminated" by the use of dipoles, waveguide feed system, or horns. The simple parabolic mirror is truly a broadband device. See Fig. 5(b).

(*Pencil-Beam Antenna*)—A unidirectional antenna so designed that cross sections of the major lobe by planes perpendicular to the direction of maximum radiation are approximately circular.

(*Pill-Box Antenna*)—A cylindrical, parabolic reflector enclosed by two plates perpendicular to the cylinder, so spaced as to permit the propagation of only one mode in the desired direction of polarization. It is fed on the focal line.

(*Pocket Antenna*)—A nonprotruding slot antenna developed for aircraft.

(*Quarter-Wave Antenna*)—An antenna which is electrically one-quarter of a wavelength long. It may be physically longer or shorter than one-quarter wavelength in free space.

(*Rhombic Antenna*)—An antenna composed of long-wire radiators comprising the sides of a rhombus. The antenna usually is terminated in an impedance. The sides of the rhombus, the angle between the sides, the elevation, and the termination are proportioned to give the desired directivity. See Fig. 4(a).

(*Scanning Antenna*)—A directional antenna employed in radar which mechanically or electrically causes its radiation to periodically scan a given arc or solid angle.

(*Series-Fed Vertical Antenna*)—A vertical antenna which is insulated from ground and energized at the base.

(*Shaped-Beam* (*Phase-Shaped*) *Antenna*)—A unidirectional antenna whose major lobe differs materially from that obtainable from an aperture of uniform phase. A cosec^2 θ beam is a shaped beam whose intensity in some plane varies as cosec^2 θ over a prescribed range, where θ is a polar angle in that plane. The half-power width in planes perpendicular to this plane is approximately constant for the prescribed range of θ.

(*Shunt-Fed Vertical Antenna*)—A vertical antenna connected to ground at the base and energized at a point suitably positioned above the grounding point.

(*Slot Antenna*)—A radiating element formed by a slot in a metal surface.

(*Steerable Antenna*)—A directional antenna whose major lobe can be readily shifted in direction.

(*Top-Loaded Vertical Antenna*)—A vertical antenna so constructed that, because of its greater size at the top, there results a modified current distribution giving a more desirable radiation pattern in the vertical plane. A series reactor may be connected between the enlarged portion of the antenna and the remaining structure.

(*Tridipole Antenna*)—An omnidirectional, horizontally polarized antenna consisting of three dipoles displaced from each other by 60° in the horizontal plane. The radiators are curved, causing the array to have a circular appearance.

(*Turnstile Antenna*)—An antenna composed of two dipole antennas, normal to each other, with their axes intersecting at their midpoints. Usually, the currents are equal and in phase quadrature.

(*V Antenna*)—A V-shaped arrangement of conductors, balanced-fed at the apex, and with included angle, length, and elevation proportioned to give the desired directivity.

(*Yagi Antenna*)—An array with one or more parasitic elements in addition to the driven element or elements. Currents induced in the parasitic element from the field produced by the driven antenna cause radiation in a phase (relative to the phase of the radiation from the driven unit) that is a function of both the spacing of the elements and the length of the parasitic element. If the resultant radiation pattern has its maximum in the direction of the driven element, the parasitic element is called a reflector, whereas if the maximum radiation is in the direction of the parasitic antenna, it is called a director. See Fig. 4(b).

Special Antennas

An antenna custom-designed for use with the *large radio telescope* at the Arecibo Observatory (Arecibo, Puerto Rico) is shown in Fig. 7. As illustrated further in the entry on **Radio and Radar Astronomy,**

Fig. 7. Special antenna designed for use with large radio telescope at Arecibo, Puerto Rico (operated by Cornell University). One of the three concrete towers for supporting the platform (triangular suspension system) is shown in the distance (center). Shadow of antenna platform will be noted on reflector bowl just right of center of view. Note carriage house also part of suspended transmitter system. This house weighs about 16 tons. Further illustrations are included in the entry on **Radio and Radar Astronomy.** (*The National Astronomy and Ionosphere Center, Cornell University. Photo by Russell C. Hamilton.*)

a triangular platform measuring 216 feet (66 meters) on each side, is 500 feet (152 meters) in the air and above a 1000-foot (305-meter) diameter reflector bowl. The output of the transmitter is 450,000 watts. This power is maximized by the positioning of the transmitter in a carriage house of the suspended platform, instead of on the ground. At the short wavelengths (S-Band), much power would be lost if the signal had to travel a considerable distance from the ground to the platform even through the most efficient S-band waveguide obtainable. Thus, the decision to put the transmitter on a suspended platform was made. The transmitter had to be extremely compact. The transmitter, despite its high efficiency gives off as much heat energy as it does radio energy. Heat exchangers were built on the platform to dissipate this energy, with coolant water circulating between the platform and the carriage house through a long, articulated piping system.

A special antenna used for high-frequency satellite communications studies and radio astronomy research is shown in Fig. 8.

For nearly two decades, the U.S. Navy has been considering a secure communications system for *submarines*. Particularly, interest is directed toward a system that will function when submarines are cruising at high speeds and at deep levels. Drag-type or on- or near-surface antennas mounted to submarines obviously have their disadvantages. While high-frequency (HF) and very low-frequency (VLF)

Fig. 8. Special antenna used for high-frequency satellite communications studies and radio astronomy research. This is installed at Crawford Hill, New Jersey. (*Bell Laboratories.*)

systems require the aforementioned kinds of antennas, the desire for a more secure system that could send brief but vital messages (possibly orders for a retaliatory nuclear blow) was the objective of the Navy project, first named *Sanguine.* In 1975, the project was renamed *Seafarer.* As early as 1963, Navy researchers suggested taking advantage of extremely low-frequency (ELF) radio waves, whose exceptionally long wavelengths would enable their penetration to considerable seawater depths without attenuation and resulting poor reception. ELF requires a long antenna, equal to at least a significant fraction of the transmitted wave which can be up to 2000+ miles (3218+ kilometers). In its original concept, it was proposed that five transmitters would use a grid of antennas on an area of up to several thousand square miles. It was proposed that the antenna cables be buried at a depth of 4–6 feet (1.2–1.8 meters), with positioning about 5 miles (8 kilometers) apart. It was further proposed that a length of cable, as previously mentioned, would have to be put into place. It was envisioned that an individual antenna would be from 30 to 60 miles (48 to 96 kilometers) in length and carry about 100 amperes of electric current. It was envisioned that current from a given ground terminal would pass through the earth's crust at a depth of a few miles and then back to the opposite terminal. The entire circuit would be configured as an antenna that would transmit an ELF signal to the atmosphere. This would be trapped in the ionosphere and hence travel around the earth.

Much opposition to the location of this system was shown in states such as Wisconsin, Michigan, Texas, New Mexico, and Nevada arose. Aside from other environmental considerations, the principal objections of some local citizenry were addressed mainly to the possible biological effects of this radiation—a topic generally of public concern during recent years. See **Microwave Radiation.** These fears, although lingering, were considerably assuaged by a report of the National Academy of Sciences (August 5, 1977) to the effect that *Seafarer* would be harmless if appropriate precautions were taken. Thus, subsequent environmental impact studies have been less convincing. Ultimate acceptance or rejection of the system appears to depend more upon political than technological problems.

References

Amitay, N., Galindo, V., and C. P. Wu: "Theory and Analysis of Phased Array Antennas," Wiley, New York, 1972.

Carter, L. J.: "Seafarer Project," *Science,* **197**, 964–968 (1977).

Hansen, R. C.: "Microwave Scanning Antennas," (3 volumes) Academic, New York, 1964–1966.

Harper, C. A.: "Handbook of Components for Electronics," McGraw-Hill, New York, 1977.

Rusch, W. V. T., and P. D. Potter: "Analysis of Reflector Antennas," Academic, New York, 1970.

Zbar, P. B.: "Electricity-Electronics Fundamentals," 2nd edition, McGraw-Hill, New York, 1977.

ANTENNA HEIGHT. Radio Communications.

ANTENNA (Radar). Radar.

ANTENNA (Zoology). A jointed sensory appendage of the head found in several classes of arthropoda. Crustacea have two pairs, while insects, centipedes, and millipeds have one pair. See also **Diplopoda.**

ANTERIOR LOBE. Hormones; Pituitary Gland.

ANTEROGRADE AMNESIA. Amnesia.

ANTHER. The terminal part of a stamen, containing the pollen sacs. See **Flower.**

ANTHERIDIUM. The structure which gives rise to the sperm. In the algae it is a single cell, the contents of which may become a single sperm or divide to produce many sperms. In the higher divisions of plants, the antheridium is a multicellular body which contains the sperms.

ANTHESIS. That stage in the flowering development of a plant when pollen is being produced.

ANTHOCYANINS. A group of water-soluble pigments which account for many of the red, pink, purple, and blue colors found in higher plants. Most plants contain more than one of these pigments and they occur most prevalently as glycosides. Several hundred different anthocyanins are known. Anthocyanins have been isolated and some have been found to be acylated with substituted cinnamic acids. The site of attachment of these acids to the anthocyanins has not been fully definitized. The natural role of the anthocyanins in plants to date has not been related to any factor of plant metabolism and many authorities believe that the pigments play more of an ecological role in regard to pollination and seed dispersal through their ability to act as an insect and bird attractant.

The anthocyanins are part of the larger group of aromatic oxygen-containing, heterocyclic compounds, known as flavonoids, most of which have a 2-phenylbenzopyran skeleton as their basic ring system. Although widely distributed among higher plants, including ferns and mosses, they are not found in algae, fungi, bacteria, or lichens.

There has been considerable interest and research activity in connection with anthocyanins during the past decade or so, stemming principally from the tighter restrictions, including banning, of several synthetic colorants. See also **Colorants.** Representative of the food processing industry's desire to find colorants that are beyond suspicion as health deterrents, scientists have been investigating various sources of anthocyanins. They have found that pigments from roselle plants (*Hibiscus sabdariffa*) native to the West Indies can be used for coloring apple and pectin jellies. A cranberry pomace extract has been found useful in coloring cherry pie filling. The potential of blueberry as a source of anthocyanin pigments also has been investigated. The berry is rich in nonacylated anthocyanins, but presently appears to be too costly as a coloring substitute.

Grape anthocyanins have been intensely investigated and have been found reasonably satisfactory, for example, in carbonated beverages. Although to date the grape anthocyanins are not as stable as Red No. 2, research continues, encouraged by the large amounts of grape wastes produced in the production of wine and grape juice. Red cabbage also has been seriously considered as a source of anthocyanin pigments. Because the anthocyanins are most stable at a pH range of 1.0 to 4.0, this acidity dictates the products in which they can be used.

Much more detail on this topic can be found in the references listed below. See also **Colorants (Foods); Glycosides;** and **Pigmentation (Plants).**

References

Ballinger, W. E., Maness, E. P., and L. J. Kushman: "Anthocyanins in Ripe Fruit of the Highbush Blueberry (*Vaccinium corymbosum* L.)," *J. Amer. Soc. Horticultural Sci.*, **95**, 283 (1970).

Clydesdale, F. M., et al.: "Concord Grape Pigments as Colorants for Beverages and Gelatin Desserts," *J. Food Sci.*, **43**, 6, 1687–1692 (1978).

Considine, D. M. (editor): "Foods and Food Production Encyclopedia," Van Nostrand Reinhold, New York, 1982.

Shewfelt, R. L., and E. M. Ahmed: "Anthocyanin Extracted from Red Cabbage Shows Promise as Coloring for Dry Beverage Mixes," *Food Product Development*, **11**, 4, 52–58 (1977).

Volpe, T.: "Cranberry Juice Concentrate as a Red Food Coloring," *Food Product Development*, **10**, 9, 13 (1976).

ANTHOPHYLLITE. The mineral anthophyllite is an orthorhombic amphibole essentially (Mg, Fe)$_7$Si$_8$O$_{22}$(OH)$_2$ with aluminum sometimes present. This mineral corresponds to enstatite and hypersthene in the pyroxene group. It has a prismatic cleavage; hardness, 5.5–6; sp gr, 2.8–3.57; luster, vitreous; color, gray, yellow, brown, green or brownish-green; transparent to translucent; probably always a metamorphic mineral in magnesium-rich rocks, often associated with talc; very common in schists. Found in Norway, Austria, Greenland, Pennsylvania, Georgia, and elsewhere. The name is derived from the Latin *anthophyllum*, clove, because of its usual brownish shades. See also **Amphibole.**

ANTHOXANTHINS. Pigmentation (Plants).

ANTHOZOA. The sea anemones, corals, alcyonarians and related forms. A class of the phylum *Coelenterata* in which the polyp form gains its highest development and the medusa is unknown.

Like the hydrozoan polyps, these animals have relatively thin walls, due to the thin middle layer (mesogloea), and are approximately cylindrical in form. The base is a disk by which the animal is attached to some support and the opposite end forms an oral disk bearing numerous hollow tentacles surrounding the mouth. The mouth leads into a long tube lined with ectoderm, known as the stomodaeum. In it ciliated grooves serve for the passage of currents of water into and out of the enteric cavity. In this cavity radiating partitions, the mesenteries, pass from the wall to the stomodaeum, which they hold in place. Others extend into the cavity from the wall without reaching the stomodaeum. The edges of the mesenteries bear mesenteric filaments with stinging cells. They are important in digestion and respiration. Reproductive bodies also develop in the mesenteries and slender acontia with many stinging cells arise from their edges. Muscle bands in the mesenteries and in the body wall contract the entire animal and close the margins of the oral disk in over the tentacles.

The class is divided into two subclasses:

Subclass *Alcyonaria.* Polyp with eight tentacles, pinnately branched. Colonial forms, usually supported by a hard skeleton. The sea fans, precious coral, and sea feathers.

Subclass *Zoantharia.* Colonial or solitary. Polyp with few to many tentacles, not pinnately branched. Hard deposits formed under the basal disk in some species. The stony corals and sea anemones.

ANTHRACENE. A colorless solid; melting point 218°C, blue fluorescence when pure; insoluble in water, slightly soluble in alcohol or ether, soluble in hot benzene, slightly soluble in cold benzene; transformed by sunlight into para-anthracene (C$_{14}$H$_{10}$)$_2$.

Anthracene reacts (1) with oxidizing agents, e.g., sodium dichromate plus sulfuric acid, to form anthraquinone $C_6H_4(CO)_2C_6H_4$, (2) with chlorine in water or in dilute acetic acid below 250°C to form anthraquinol and anthraquinone, at higher temperatures 9,10-dichloroanthracene. The reaction varies with the temperature and with the solvent used. The reaction has been studied using, as solvent, benzene, chloroform, alcohol, carbon disulfide, ether, glacial acetic acid, and also without solvent by heating. Bromine reacts similarly to chlorine, (3) with concentrated sulfuric acid to form various anthracene sulfonic acids. (4) with nitric acid, to form nitroanthracenes and anthraquinone, (5) with picric acid (1)HO·$C_6H_2(NO_2)_3$(2,4,6) to form red crystalline anthracene picrate, melting point 138°C.

$C_{14}H_{10}$ or 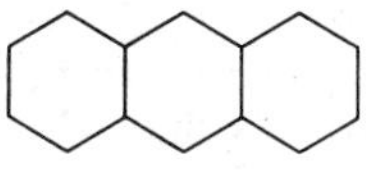

Anthracene is obtained from coal tar in the fraction distilling between 300° and 400°C. This fraction contains 5–10% anthracene from which by fractional crystallization followed by crystallization from solvents, such as oleic acid, and washing with such solvents as pyridine, relatively pure anthracene is obtained. It may be detected by the formation of a blue-violet coloration on fusion with mellitic acid. Anthracene derivatives, especially anthraquinone, are important in dye chemistry.

ANTHRACENE OIL. Coal Tar and Derivatives.

ANTHRACITE COAL. Coal.

ANTHRACITIZATION. Coal.

ANTHRACNOSE. Fungus.

ANTHRACOSIS. Pneumokonioses.

ANTHRAQUINONE. Anthraquinone (9,10) is a yellow solid;

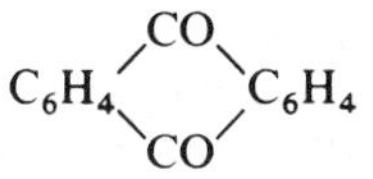

melting point 286°C; can be sublimed; forms monoxime, melting point 224°C, by heating under pressure at 180°C with hydroxylamine chloride; forms no phenylhydrazone with phenylhydrazine; with strong oxidizing agents reacts with difficulty to yield phthalic acid $C_6H_4(COOH)_2$(1,2); with reducing agents, such as sodium hyposulfite, zinc in sodium hydroxide solution, tin or stannous chloride in hydro-

anthraquinol 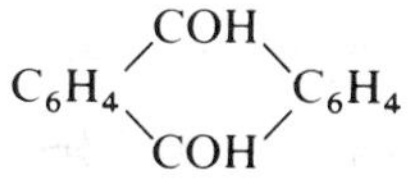

anthrone 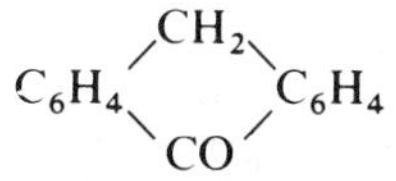

dianthrol 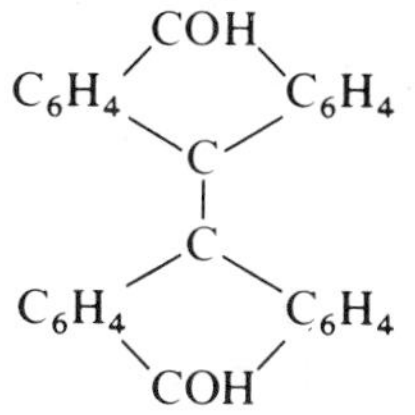

dianthrone 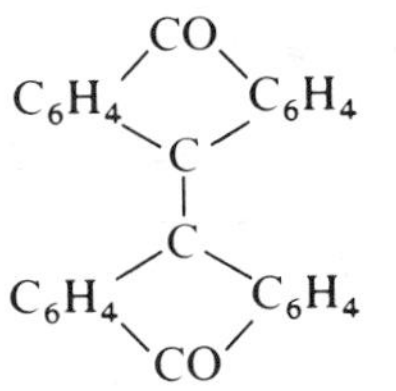

chloric acid (but not sulfurous acid), is reduced to anthraquinol, anthrone, dianthrol and dianthrone, depending on the conditions.

Anthraquinone is obtained by oxidation of anthracene using sodium dichromate plus sulfuric acid, and is purified by dissolving in concentrated sulfuric acid at 130°C and pouring into boiling water, whereupon anthraquinone separates as pure solid, and is recovered by filtration. Further purification may be accomplished by sublimation or crystallization from nitrobenzene, aniline or tetrachloroethane. Anthraquinone is used as the material from which many dyes are made, notably alizarin $C_6H_4(CO)_2C_6H_2(OH)_2$ and related substances. These are vat dyes, that is, insoluble colored substances which are readily reduced to a substance having marked affinity for the fiber to be dyed and which upon exposure to the air are readily reoxidized to the original dye. Anthraquinone may be detected by the appearance of a red color on treatment with alkali, zinc powder and water.

See also **Coal Tar and Derivatives.**

ANTHRAX. This is a highly infectious disease caused by the gram-positive *Bacillus anthracis.* The disease is of historical importance because the anthrax bacillus was the first microorganism proved definitely to be the cause of an infectious disease. Anthrax mainly is a serious disease of cattle. It is capable of transmission to humans by way of meat and animal products. In a number of countries, extensive cattle vaccination programs have been very effective. These programs have virtually eliminated reservoirs of infection in the United States, where only minor outbreaks involving very few cases may occur, usually in the Great Plains, the lower Mississippi Valley, and Texas. Infection of humans usually occurs in persons who work with animal products, such as hides. In other parts of the world, anthrax in animals persists as a major problem, notably in Asia, Africa, and South America. The Russian literature also has reported outbreaks over many years. Outbreaks were particularly severe in 1923 and 1940, but the most disastrous of all occurred in and around the city of Sverdlovsk in 1979. Over 1000 fatalities were reported in an epidemic that lasted about one month. Even though the soil of Sverdlovsk province has been known to be infected with anthrax for a century or more, the characteristics and extent of the 1979 outbreak have baffled scientists. Rather than the mild cutaneous form, or the intermediately serious intestinal form, sketchy reports from Sverdlovsk indicated that it was the inhalation form of the disease that proved so potent. Questions which remain unresolved were immediately raised concerning the possibility of anthrax spores being released to the atmosphere as the result of an accident at a biological warfare facility.

The most common form of anthrax is *cutaneous.* The cutaneous lesion is sometimes referred to as malignant pustule. The bacillus forms spores in the external environment, and in culture, but not in animal tissues. Spores can linger in the soil and in animal products for many years. A cutaneous infection is usually the result of introducing spores through an abrasion. The cutaneous form can be successfully treated with antibiotics.

The intestinal form results from ingesting anthrax-contaminated meat, notably sausage. Inhalation anthrax, while previously known, gained prominence as the result of the Sverdlovsk incident. Before antibiotics were available, mortality from cutaneous anthrax ranged up to 20%. The mortality rate for untreated inhalation anthrax and anthrax meningitis is nearly 100%.

In the early part of this century, prior to the availability of antibiotics and the initiation of cattle vaccination programs, there were about 125 cases per year of cutaneous anthrax in the United States. For several years, the number of cases per year has been 10 or fewer.

ANTHRAXOLITE. A coal-like metamorphosed bitumen, often closely associated with igneous rocks. Commonly associated with "Herkimer Diamond" type quartz crystals in dolomitic limestones in Herkimer and Montgomery counties in New York State.

ANTHROPOGENIC. Relative to the activities of humans as contrasted with the actions of natural forces and events. Thus, the addition of dust to the atmosphere from aerial crop spraying (anthropogenic) as contrasted with dust clouds produced by a volcano (natural), or the pollution arising from an oil spill (anthropogenic) versus the seepage of crude oil into the oceans from cracks in undersea rock formations (natural).

ANTHROPOGENIC CARBON DIOXIDE FLUX. Climate.

ANTHROPOIDS (*Mammalia, Primates*). This division of *Mammalia* includes a number of medium-size and large animals which have comparatively large brains and which have no tails. Zoologically, *Hominidae* are also placed in this category. The subdivisions of apes include:

- Lesser Apes (*Hylobatidae*)
  - The Siamang (*Symphalangus*)
  - Gibbons (*Hylobates*)
- Greater Apes (*Pongidae*)
  - Gorillas (*Gorilla*)
  - The Chimpanzees (*Pan*)
  - Orangutans (*Pongo*)

The lesser and greater apes are found in the forests of Africa and in the Oriental Region (eastern India to Hainan and southward to Borneo, Java, and Sumatra).

There are about six species of gibbons, of which one, the siamang, is confined to Sumatra. Of the lesser apes, the siamang is the largest, having a finger-tip-to-finger-tip spread across the chest of nearly 6 feet (1.8 meters) and a height of about 3 feet (1 meter) when sitting in an upright position. Siamangs prefer high forest country and only infrequently travel on the ground where they do so in a rather awkward manner. When required, they are good swimmers, keeping their heads well out of the water. They are known for their frightening howling which occurs at dawn and sundown. Their preferred diet is fruit. They have a coat of long, rather shaggy black hair with the exception of gray beards in the males. These animals are characterized by a pouch connected to the throat which can be inflated to appear as a rather large red balloon.

Gibbons are smaller and of more slender build than the other apes and are among the most agile of all primates in the trees. See Fig. 1. They have extremely long arms with which they swing distances up to 20 feet (6 meters) from bough to bough. Their movements are quite rapid. This arm-swinging process is known as *brachiation.* Gibbons, as do spider monkeys, have features, in addition to their long limbs, which further the effectiveness of brachiation, including: The bone structure of the hand is formed in what might be termed a "hook" or "grapple;" a reduction of the thumb, such that it does not interfere with rapid release when going from one tree limb to the next; a curling-inward position of the hand when at rest; and exceptional elongation of the fingers and palm of the hand to provide increased surface area for contacting a tree limb as a leap is being

Fig. 1. Mother gibbon and baby. (*A. M. Winchester.*)

completed. Some authorities consider the tree-swinging agility of the gibbon as unsurpassed.

A feature not found among other anthropoid apes is the presence of ischial callosities (small, naked, hardened, and thickened places) on the buttocks. For their diet, gibbons prefer leaves and fruits. The whooping noises of the siamang are also made by the gibbons, particularly for a few hours after dawn. Gibbons are well known for their intelligence, cleanliness, and gentility. Gibbons are monogamous and known for their exceptional fidelity and family cooperation. B. B. Beck (*Science*, **182**, 4112, 594–596 (1973)) reports that a bonded pair of hamadryas baboons developed use of a cooperative tool without training. The male could get food with the tool, but first had to get the tool from an adjoining cage which he could not enter. The female learned to give him the tool.

Specific variations of gibbons include: Hoolock (central Himalayas—gray to brown in color); White-handed Gibbon (Burma, Malay Peninsula—brown to black); Agile Gibbon (Sulu Archipelago and Borneo—cream-to-dark-brown color); Unkaputi (Malaysia—often taken as pets); and the Wow-Wow or Silvery Gibbon (Sumatra).

Certain generalizations can be made pertaining to all of the Greater Apes: Bodies are short and obese; small eyes, close together, pointing forward; wide and very flaring nostrils with sunken nose-bridge; protuberant muzzle; thin, mobile lips; small, fully opposed thumbs; very long arms; and short hind limbs.

Gorillas are the largest of the Greater Apes and are terrestrial. In walking, they rest partly on the backs of the bent fingers. According to Akeley's observations, they are rather poor climbers. The head of the gorilla is distinguished by the strong jaws and large teeth, the heavy ridges over the eye sockets, and the small ears. Two species are the common gorilla, *Gorilla gorilla*; and the mountain gorilla, *G. berringei.*

Gorillas live in the forests of Africa, with numbers of them found in a wide area of Cameroon and Gabon, north of the Congo, and in Katanga and Zaire. Pockets or "nations" of gorillas also are found in the northwestern portion of Nigeria, and in the volcanic mountains of Kivu. There apparently is no communication between these very isolated communities.

A large male gorilla may stand over five feet high and weigh in excess of 500 pounds (227 kilograms). The lowland dwellers usually have a rusty-gray coat, whereas the mountain dwellers have a black coat. See Fig. 2.

At one time, gorillas would not live long in captivity. This difficulty has been mastered by careful attention to diet and by protecting the animals from the respiratory diseases of humans, to which they are very susceptible. As mentioned by explorer Paul D. du Chaillu, the gorilla has a habit of beating the chest with both hands and displays a very self-centered personality. Psychologist R. M. Yerkes found the gorilla more slow in adaptability to surroundings and also lacking in initiative, when compared with orangutans and chimpanzees of the same age.

An interesting article pertaining to gorilla behavior is "Conversations with a Gorilla," by Francine Patterson (*National Geographic*, **154**, 4, 438–465 (October 1978)). During the past decade, researchers have successfully taught several chimpanzees to converse with signs. In this project, partly funded by the National Geographic Society, Koko is the first gorilla to achieve proficiency. After six years of study, Francine Patterson evaluates Koko's working vocabulary at about 375 signs.

Fig. 2. Lowland gorilla. (*New York Zoological Society Photo.*)

Even in the 1980s, there remains much to be learned concerning the gorilla, particularly pertaining to the behavior of this animal in its natural habitat. There have been numerous observations made, many scientific, some with legendary overtones. Although obviously all communities of gorillas have not been observed let alone located, it has been established that at least some of them construct platform-type homes or nests, built in trees a few feet above ground level. Some authorities have observed true knots used in the doubling over and twisting of tree and root gnarlings in the construction of such platforms. This evidences a superior intelligence which has not been observed in the comparatively few observations of animals in captivity. The strength of the gorilla always has been a topic of some controversy, although all observers agree that it is quite tremendous. There are reports of bending of heavy steel bars and gun barrels. Natives and scientific observers generally agree that gorillas prefer to spend most of their time on the ground, occasionally tree climbing for fruits, and that, when unmolested, are retiring and not ferocious as popularly reported.

Because the chimpanzees most people see on television, in the movies, or at a zoological garden are usually relatively small, sight is lost of the fact that chimpanzees can be quite large, some older male animals rivaling some of the gorillas in proportions. In their natural habitat, they can easily attain a height of five feet and a weight between 150 and 200 pounds (68 and 91 kilograms). Females are usually slightly smaller. These animals live over a large area of Africa, with particular population concentrations around Lake Victoria and in Tanzania. One authority is reported to have observed several hundred of these animals within visual range of a walking path of some 15 miles (24 kilometers) between villages in Cameroon.

In recent years, chimpanzees have become favorites for various types of animal behavior experimentation and observation. E. W. Menzel (*Science*, **182**, 4115, 943–945 (1973)) reports of a study of juvenile chimpanzees, which were carried around an outdoor field and shown up to 18 randomly placed hidden foods. The animals remembered most of the hiding places and the type of food that was in each. Their search pattern approximated an optimum routing, and they rarely rechecked a location that they had already emptied of food. As reported, the chimpanzees appeared to directly perceive the relative position of selected classes of objects and their own position in a scaled frame of reference. They proceeded on the strategy, "Do as well as you can from wherever you are," taking into account the relative preference values and spatial clusterings of the foods as well as distances. As investigator Menzel concludes, "Especially in the light of other recent research, one is struck again by the parallels between chimpanzee and human behavior, the necessity for including representational processes in any adequate formulation of learning and memory, and the apparent evolutionary independence of representational ability and verbal language."

Investigators D. M. Rumbaugh, T. V. Gill, and E. C. von Glasersfeld (*Science*, **182**, 4113, 731–733 (1973)) reported that four studies revealed that a $2\frac{1}{2}$-year-old chimpanzee, after 6 months of computer-controlled language training, proficiently read projected word-characters that constituted the beginnings of sentences and, in accordance with their meanings and serial order, either finished the sentences for reward or rejected them.

The animal is considered to be an extrovert, learning fast from certain types of information and appears to thrive on showing off. However, if it is pushed too hard in learning experiments, tantrums can be expected as well as moodiness. In nature, the male leads the family group as it travels. See Fig. 3.

In nature or captivity, there appear to be no special breeding seasons. The life span is about 35 years and adulthood is achieved at about 11 years. Chimpanzees feed on fruit, bird eggs, and plant shoots.

In semierect position, the chimpanzee walks on all fours. The arms are used for swinging. In walking, the animal leans on its hands

Fig. 3. Chimpanzee. (*A. M. Winchester.*)

Fig. 4. Adult orangutan. (*New York Zoological Society Photo.*)

propped by the knuckles, and when it stands erect, its arms reach to just below the knees.

Varied forms of chimpanzees include: The Masked Chimpanzee (*Pan satyrus verus*) found in Upper New Guinea; the Choga (*P.s. satyrus*) found in Lower Guinea; the Koola-Kamba found in Nigeria and environs; and the long-haired Eastern Chimpanzee (*P.s. schweinfurthi*) found in Tanzania and Uganda. But, as mentioned before, these animals are found in many other parts of Africa. In nature, chimpanzees build nests with half-roofs for protection against the elements. These are deftly placed in tree branches several feet above the ground.

Orangutans are found in Borneo and Sumatra. The adult animal may reach a height of 5 to $5\frac{1}{2}$ feet (1.5 to 1.7 meters) and weigh in excess of 150 pounds (68 kilograms). The sharply depressed bridge of the nose accentuates the prominence of the rounded muzzle. For some years, the zoological gardens were more successful in boarding orangutans than gorillas. However, the orangutans can be considered delicate when in captivity, particularly in northern climates. These animals, like the chimpanzee to some extent, have been found to be very intelligent, but they are less desirable as subjects because of their uncertain temper, a characteristic particularly true of older males. The term orangutan (or *Orang-Utan*) is Malayan for "man of the woods."

Orangutans are usually brownish-red in coloration, with a broad, flat face, and long, coarse hair. A sac extending from the front of the throat extends to the armpits. This feature, plus the depressed nose-bridge, the huge cheek flaps, comparatively small legs, disproportionately long and heavier forelimbs, and obese stomachs, particularly in older specimens, gives them a rather grotesque appearance. See Fig. 4. If adopted into a human household, the orangutan demonstrates amazing adaptability, often showing, as with a dog for example, definite preferences for certain people, while also demonstrating a marked antipathy for others.

In nature, the animals are vegetarians, preferring fruit (particularly of the palm nut tree), but tolerate a wider diet when in captivity. They also are "sleeping platform" builders as some of the other apes described. Even the heavier adults can travel rapidly among the tree tops and display brachiation as previously defined. See also **Mammalia;** and **Primates.**

ANTIALLATOTROPINS. Insecticide and Pesticide Technology.

ANTIANEMIA FACTOR. Folic Acid.

ANTIARRHYTHMIC AGENTS. Arrhythmias (Cardiac); Ischemic Heart Disease.

ANTIBIOTIC. A chemical substance derived from one or more types of microorganism which has the ability of inhibiting the growth of, or of killing, other microorganisms. In *bacteriostatic* drugs, very little killing, if any, of the target microorganisms occurs at the minimal inhibitory concentration (MIC). In *bactericidal* drugs, the objective is that of killing bacteria (or other target microorganisms), i.e., the minimal bactericidal concentration (MBC) is equal to or very close to the minimal inhibitory concentration (MIC). In bacteriostatic drugs, the objective is to cause bacteriostasis, a condition that is reversed if the chemotherapeutic agent is withdrawn. These rather fine dividing lines are much more clearly evident and measurable in vitro than in vivo. The principal antibiotic drugs in use today, classified by these two categories, are listed in Table 1. Sulfonamides are described in a separate entry on **Sulfonamide Drugs.**

TABLE 1. CLASSIFICATION OF DRUGS USED IN THE CHEMOTHERAPY OF MICROBIAL DISEASES

Bacteriostatic Agents	Bactericidal Agents
Sulfonamides	Penicillins (penicillin G, penicillin V, methicillin, oxacillin, cloxacillin, nafcillin, ampicillin, amoxicillin, carbenicillin)
Trimethoprim	Cephalosporins
Tetracyclines	Aminoglycosides (streptomycin, neomycin, kanamycin, gentamicin, tobramycin, amikacin)
Chloramphenicol	Vancomycin
Erythromycin	Polymyxins (polymyxin B, colistin)
Lincomycin or clindamycin	Bacitracin

Background. Although well established for about 50 years, by comparison with many other types of drugs used in medicine, the use of antibiotics as chemotherapeutic agents is relatively recent. However, the existence of antibiotic-type drugs date back over 100 years.

The first scientific demonstration of microbial antagonism was made by Pasteur and Joubert in 1877 when they observed that certain common bacteria inhibited the growth of anthrax bacilli. This basic phenomenon by which one microorganism destroys another to preserve its own life was, at that time, called *antibiosis* by Vuillemin (1889). In the decades that followed, the therapeutic efficacy of antibiotics to control infectious disease was eventually demonstrated, after which the pursuit of microbial antagonists rapidly became an organized applied science.

Pyocyanase was the first microbially derived antibiotic product to be used in treating bacterial infections in humans. Although it had only limited clinical use, it is interesting historically because it demonstrated as early as 1906 the principle of selective toxicity, i.e., specificity of action against the invading pathogen and a correlative lack of toxic action in the host. Following the decline in use of pyocyanase, it was almost a quarter of a century before interest in anti-infective agents from microbial sources was renewed.

In 1929, the British bacteriologist Alexander Fleming published his observations on the inhibition of a staphylococcus culture by growing colonies of *Penicillium notatum*. This report went largely unpursued for a decade, after which Florey and Chain reinvestigated Fleming's work and, in 1941, demonstrated the clinical usefulness of penicillin. In 1939, Dubos, by careful, well-planned studies, obtained the antibiotic tyrothricin from the soil organism *Bacillus brevis*. Although tyrothricin found only limited use, the work of Dubos on the chemical, biological, and physical properties of this antibiotic contributed immensely toward forcing a realization of the potentialities of antibiotic substances. Similarly, Waksman undertook a systematic search for antimicrobial substances in a group of soil-inhabiting microbes known as *Streptomyces* and announced the discovery of streptomycin in 1944.

The foregoing discoveries stimulated worldwide interest, and the discovery of useful new antibiotics during the period 1939–1959 was prolific. During this period, the major classes of antibacterial antibiotics were recognized. Many specific drugs which presently occupy places in therapeutic practice were discovered directly in microbial fermentations. The later work in the field had consisted principally of chemical modifications of antibiotic substances previously known.

Classification. The antibiotics comprise an unusually diverse group of substances, differing not only in chemical structure, but also in their mode of action, antibacterial spectra, origin, and other features. In spite of the variety of chemical structures involved, most antibiotics appear to arise from a limited number of biogenetic themes, and may be divided into a few main groups in accordance with their potential derivation from amino acids, sugars, and acetate or propionate units. See Table 2.

In general, antibiotics exert their toxic action on bacteria by impairing one of the following processes: (1) synthesis of the bacterial cell wall; (2) synthesis of intracellular protein; (3) synthesis of deoxyribonucleic acid; and (4) function of the cytoplasmic membrane. Grouping antibiotics by these criteria bears little resemblance to their classification according to biogenetic origin and chemical structure. Thus, while cycloserine and chloramphenicol are biogenetically derivable from a single amino acid, cycloserine inhibits cell-wall synthesis, while chloramphenicol acts on protein synthesis. In contrast, the mechanisms of action of the macrolide antibiotics and lincomycin are very similar, but their chemical structures are markedly different.

Much information concerning the use of antibiotics was known before their detailed mechanisms were explained in part. Research has shown that cell walls of bacteria are very different from the membrane surrounding mammalian cells, and the probable reason for the selectively toxic effect on bacteria rather than mammalian cells rests on this difference. The bacterial cell wall protects the microbial cell

TABLE 2. CLASSES OF ANTIBIOTICS ON BASIS OF BIOGENETIC ORIGIN AND CHEMICAL STRUCTURE

Amino Acid Units	Acetate/Propionate	Sugar Units
Amino acid cogeners	Fused-ring systems	Aminoglycosides
D-Cycloserine	Tetracyclines	Streptomycin
Chloramphenicol	Oxytetracycline	Kanamycins
Beta-Lactams	Chlortetracycline	Gentamicins
Penicillins	Steroidal antibiotics	Neomycins
Cephalosporins	Fusidic acid	Tobramycins
Polypeptides	Griseofulvin	Amikacins
Bacitracins	Antibacterial macrolides	
Polymixins	Erythromycin	
Viomycin	Oleandomycin	
Capreomycin	Leucomycins	
Vancomycin	Spiramycins	
	Polyene macrolides	
	Nystatin	
	Amphotericins	
	Ansa-macrolides	
	Rifamycins	

Note: Some of the foregoing are mainly of historical or research interest.

from the osmotic difference that exists between the inside and outside of the bacterial cell. Thus, when the cell wall is sufficiently damaged, the cell will disrupt and die unless the osmotic strength of the medium is greatly increased to minimize the osmotic difference. Under these special conditions, cells free of cell wall (protoplasts) are formed which are no longer sensitive to penicillin. Further evidence for the mechanism of action of penicillin was obtained by Park and Strominger, who observed the accumulation of a conjugate of uridine diphosphate and N-acetylmuramyl peptide in the medium of penicillin-inhibited cells. The condensation into the cell walls of this latter compound, which is a constituent of bacterial cell walls but not of mammalian cell membranes, was prevented by the drug.

Antibiotics are used to combat a large number of pathogens that range in complexity from viruses to protozoa. Bacteria constitute the largest single class of microorganisms which are susceptible to antibiotics. There are also antifungal, antiprotozoal, anthelmintic, antiviral, antineoplastic, and animal-growth promoting antibiotics. The microbial origin of antibiotics serves as yet another means of classification. For example, there are over 2000 antibiotics that have been discovered and studied since 1940, of which about 68% have been isolated from the family Streptomycetaceae, 12% from Bacillaceae, and about 20% from various fungi. These, in turn, may be further subclassified into various genera, species, and varieties of microbes. Thus, antibiotics may be classified on the basis of many different criteria, each of which represents an integral part of antibiotics technology.

Among the most important of the antibiotics used in current medical practice are: (1) the beta-lactams, which include the penicillins and cephalosporins; (2) the tetracyclines; (3) the macrolides; and (4) the aminoglycosides.

Penicillins

The penicillins are chemically characterized by a four-membered lactam ring fused to a thiazolidine ring and are differentiated by the side-chain (R) attached to the bicyclic nucleus. See Table 3. Penicillins are sometimes named by attaching the chemical name of the R-substituent as a prefix to the word penicillin. Thus, in the case where R is $C_6H_5CH_2$—, the compound may be called *benzylpenicillin*. However, this compound in commerce is more commonly referred to as *Penicillin G*. Similarly, in the case where R is $C_6H_5OCH_2$—, the compound may be called *phenoxymethylpenicillin*, although commercially it is more commonly called *Penicillin V*. Most frequently, the commercial names bear no resemblance to structure.

The naturally occurring penicillins have been discovered in the fermentation broths of *Penicillium* and *Cephalosporium* cultures.

The earliest of the penicillins, simply called penicillin, was benzylpenicillin (Penicillin G). This early product, still used, was shown to have several limitations, including acid instability, allergenicity, and susceptibility to enzymatic inactivation by penicillinases.

In 1947, it was discovered that addition of phenylacetic acid to penicillin fermentation media increased the yield of benzylpenicillin at the expense of other less desirable natural penicillins. Following this observation, a new generation of *biosynthetic* penicillins was prepared by addition of monosubstituted acetic acid derivatives to penicillin fermentations. The most important of these biosynthetic derivatives is penicillin V, obtained by adding phenoxyacetic acid to penicillin growth media. This widely used antibiotic is relatively stable in dilute acid, is not destroyed by the acidic contents of the stomach, and consequently can be effective by oral administration.

Although the biosynthetic approach created several new penicillins, it was limited in the type of side-chain (R) that could be introduced. Only derivatives with an unsubstituted methylene adjacent to the amide carbonyl (X as shown below) could be generated.

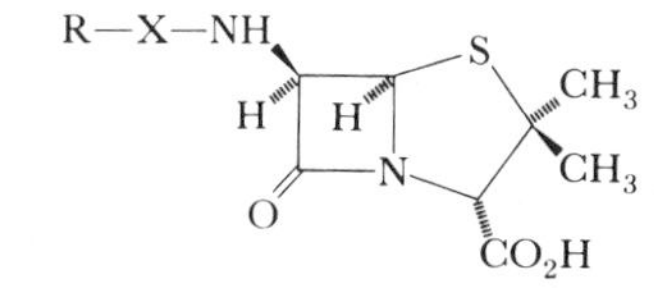

The next major breakthrough in penicillin research came in 1959 with the isolation of the penicillin nucleus 6-aminopenicillanic acid (6-APA) from fermentation mixtures to which no side-chain precursor

TABLE 3. STRUCTURES OF SOME OF THE PRINCIPAL PENICILLINS

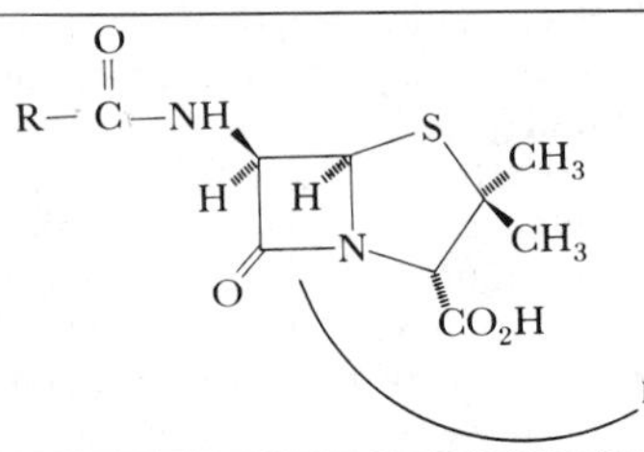

Portion of β-lactam ring cleaved by penicillinase (β-lactamase)

GENERIC OR CHEMICAL NAME	SUBSTITUENT SIDE CHAIN (R)
Penicillin G (Benzylpenicillin)	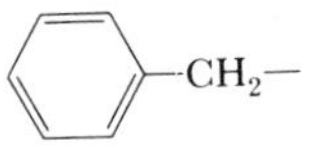
Penicillin V (Phenoxymethylpenicillin)	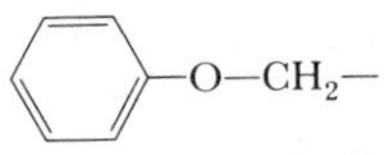
Methicillin	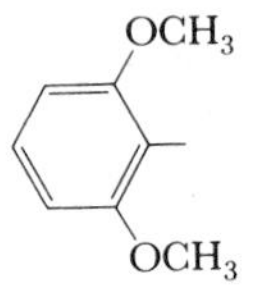
Oxacillin	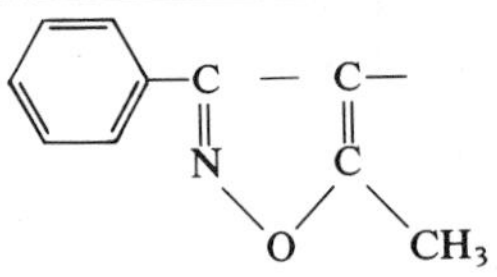
Cloxacillin	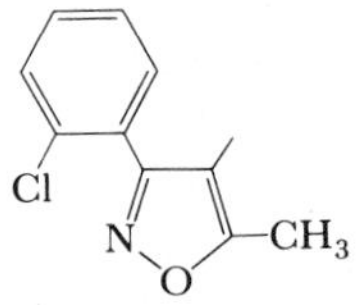
Nafcillin	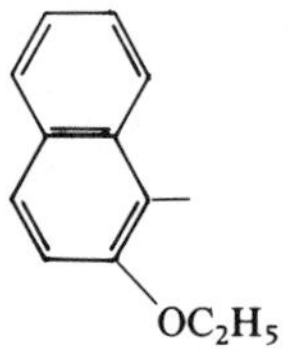
Ampicillin	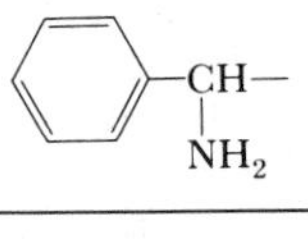
Amoxicillin	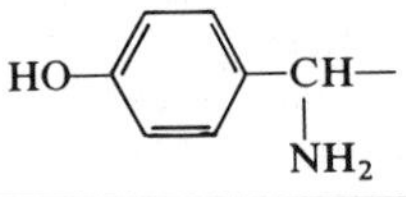
Carbenicillin	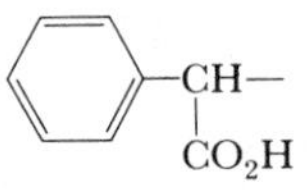

had been added. Although chemical synthesis of 6-APA and its utility for the preparation of new penicillins by acylation were announced by Sheehan in 1958, the fermentation method provided the first practical means of obtaining large quantities of 6-APA. Chemical acylation of 6-APA allowed introduction of almost unlimited varieties of side chains and gave rise to a third generation of penicillins called *semisynthetic penicillins.*

The nature of the acyl side chain has been found to have a profound effect on the properties of the penicillins, influencing such therapeutically important properties as acid stability, oral absorption, serum protein binding, penicillinase resistance, and gram-negative activity.

One of the major developments resulting from the availability of 6-APA was the creation of semisynthetic penicillins that resist destruction by the penicillinases. The empirical finding that triphenylmethylpenicillin was resistant to penicillinase led to the screening of other penicillins with sterically hindered side-chains, partly because the presence of a bulky group near the beta-lactam ring resulted in reduced affinity of these substances for the enzyme. Methicillin and cloxacillin are compounds with such side-chains which have proved clinically useful.

Administration of Penicillins. Blood levels of penicillin V are more predictable and are usually from 2 to 5 times higher than as achieved by an equal dosage of penicillin G. Although all forms of penicillin introduced since penicillin G are prescribed by weight, penicillin G is still commonly prescribed by unitage. One milligram of penicillin G = ~1600 units. Penicillin G procaine is almost painless when administered intramuscularly and is slowly absorbed, usually requiring only one injection per each 12 to 24 hours for treatment of highly susceptible infections. In the treatment of meningitis and of endocarditis due to highly susceptible organisms, high doses of penicillin G are usually administered intravenously at the shorter intervals of 2 to 4 hours because of the special problems of penetration across the blood-brain barrier and into vegetations. Where renal dysfunction is present, neurotoxicity may be produced and enhance hyperkalemia because of high potassium content. Where penicillin G may present problems, the sodium salt of penicillin G or a sodium salt of a similar penicillin, such as ampicillin or carbenicillin, may be substituted.

Ampicillin has a broader range of activity than that of penicillin G. The spectrum of ampicillin encompasses not only pneumococci, meningococci, gonococci, and a number of streptococci, but also several gram-negative bacilli, including *Haemophilus influenzae, Salmonella* species, *Shigella* species, *Escherichia coli,* and *Proteus mirabilis.* Because ampicillin is readily cleaved by β-lactamase, the drug is of no value in the treatment of infections caused by *Staphylococcus aureus* and other organisms that elaborate the enzyme. Within recent years, plasmids conferring ampicillin resistance have appeared in *Salmonella typhi, H. influenzae,* and *Neisseria gonorrhoeae.* These organisms previously were uniformly susceptible to ampicillin. The increasing use of ampicillin also has been evidenced by increasing resistance among strains of *E. coli* and nontyphoidal *Salmonella* strains.

Amoxicillin, as evident from Table 3, is structurally similar to ampicillin with exception of an OH instead of an H in one of the positions of the side-chain. Although this difference does not alter the spectrums of the two drugs, amoxicillin is better absorbed from the gastrointestinal tract, resulting in longer effective concentrations of the drug present in the circulation. The prolonged blood level and fewer problems with diarrhea sometimes permit more effective oral therapy than can be achieved with ampicillin.

Carbenicillin, in which there is a carboxyl rather than an amino substituent, has greater activity against gram-negative bacilli. These include *Pseudomonas aeruginosa,* species of *Proteus* (other than *mirabilis*), as well as some strains of *Enterobacter,* plus the strains that are susceptible to ampicillin. It has been found that over 80% of *Pseudomonas* isolates are susceptible. Carbenicillin is not absorbed from the gastrointestinal tract, although the indanyl carbenicillin ester is acid stable and thus suitable for oral administration. It is used in the treatment of urinary tract infections due to susceptible gram-negative bacilli, including *Pseudomonas.* Blood levels are too low for use of the drug in infections located elsewhere. Another semisynthetic penicillin, *ticarcillin,* has essentially the same spectrum as carbenicillin and, in the case of *Ps. aeruginosa,* is about twice as active as carbenicillin.

The penicillinase-resistant (not affected by cleavage problem previously mentioned) penicillins include *methicillin, oxacillin, nafcillin, cloxacillin,* and *dicloxacillin.* Methicillin was the first of these drugs to be introduced (early 1960s). When this drug was introduced, infections due to penicillin-resistant *Staph. aureus* had reached epidemic proportions. Methicillin is equally effective against penicillinase-producing and non-penicillinase-producing *Staph. aureus.* It is, however, less effective against pneumococci and streptococci than is penicillin G. Thus, its major use has been against penicillinase-producing *Staph. aureus.*

Methicillin is administered intravenously or intramuscularly. Because the latter route causes pain and discomfort, it is not used where large doses are required. In recent years, the semisynthetic, penicillinase-resistant *oxacillin* and *nafcillin* have markedly supplanted methicillin for many situations. These more recent antibiotics can be administered orally or parenterally. They are sometimes used in the treatment of mixed infections involving *Staph. aureus* and other gram-positive cocci or of infections of unknown etiology, but where the aforementioned organisms are suspected.

Cloxacillin and *dicloxacillin* are available as oral preparations. These compounds are sometimes preferred over oxacillin for treatment of mild staphylococcal infections. However, oxacillin and its two derivatives are usually effective in the primary treatment of soft tissue infections. They also may be used as an extension to prior intravenous therapy.

It should be mentioned that the side effects of penicillin are generally shared by the various derivatives. Allergy to one penicillin compound almost always indicates allergy to all other penicillin compounds.

Cephalosporins

These substances constitute another major class of β-lactam antibiotics and are chemically characterized by a β-lactam fused to a dihydrothiazine ring. In contrast to the penicillins, where the side-chain of the antibiotic varies, depending upon precursors present in the fermentation mixture, fermentation-derived cephalosporins contain the same side-chain. *Cephalosporin C,* the parent antibiotic of this class, is not useful clinically. For example, it is about 0.1% as active as benzylpenicillin against staphylococci. However, in early research, cephalosporin C exhibited certain interesting properties which provoked further study. Cephalosporin C was more stable toward acid than penicillin; it was unaffected by penicillinases; it exhibited appreciable activity against some gram-negative bacteria, and it appeared to have no cross-allergenicity with the penicillins. Consequently, in the late 1950s, many laboratories investigated both chemical and microbiological methods for removing the aminoadipoyl side-chain of cephalosporin C to obtain the cephalosporin nucleus, 7-aminocephalosporanic acid (7-ACA).

A practical chemical process for accomplishing this transformation was announced in 1962. Like 6-APA, previously described, 7-ACA can be readily acylated, and a large number of semisynthetic cephalosporins thus have been possible. Cephalothin was the first clinically useful broad-spectrum cephalosporin to emerge from synthetic studies. Cephaloridine soon followed cephalothin and was found to be 2 to 8 times more active than the latter against gram-positive organisms. In 1970, cephaloglycin became commercially available as the first orally effective broad-spectrum cephalosporin. Cephalexin is metabolically more stable than cephaloglycin. More recently available cephalosporins have included cephapirin, cephradine, and cefazolin.

The basic structure of the cephalosporins and structures of several of the antibiotics in this family are shown in Table 4.

Administration of Cephalosporins. The antibacterial spectrum of the current cephalosporins is fundamentally broad and includes most gram-positive bacteria (*Staph. aureus, Staph. epidermidis,* pneumococci, streptococci (except enterococci), *Actinomyces,* most gram-positive anaerobes, as well as numerous gram-negative organisms, including *E. coli, Prot. mirabilis,* and *Klebsiella.* Although the cephalosporins are resistant to the penicillinase of *Staph. aureus,* they are not effective against many gram-negative organisms because these organisms produce chromosomally determined, membrane-bound cephalosporinases.

Cephalothin, when it first became available, was used as an alternative to penicillin in treating infections arising from *Staph. aureus* and

TABLE 4. STRUCTURES OF REPRESENTATIVE CEPHALOSPORINS

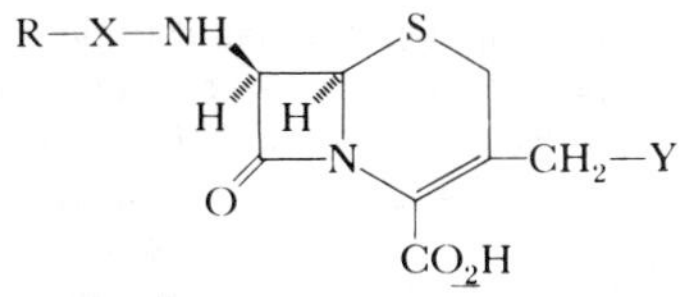

Generic or Chemical Name	R	X	Y
Cephalosporin C (parent of class—essentially inactive)	H_2N, CO_2H >—$(CH_2)_3$—	C=O	CH_3CO_2—
7-Aminocephalosporanic acid	—	H	CH_3CO_2—
Cephalothin	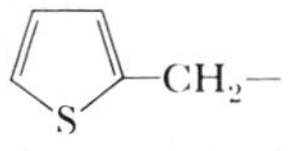	C=O	CH_3CO_2—
Cephaloridine	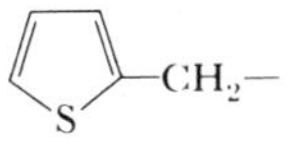	C=O	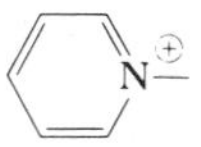
Cephaloglycin	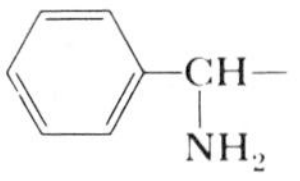	C=O	CH_3CO_2—
Cephalexin	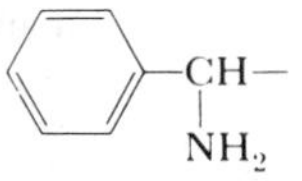	C=O	H

Note: Important cephalosporins not included in table include Cefamandole, cefazolin, Cefoxitin, cephapirin, cephradine.

some other gram-positive infections. Later, use of the drug was extended to infections arising from *K. pneumoniae*, *E. coli*, and *Prot. mirabilis*. Currently, in the treatment of major infections, cephalothin is frequently used in combination with a second antimicrobial drug. The drug usually is administered intravenously because of pain associated with intramuscular injections. Where renal dysfunction is present, the dosage must be reduced. Cephalothin is not the drug of choice in treatment of bacterial meningitis because this specific drug and the cephalosporins in general do not penetrate the cerebrospinal fluid very well. It should be noted, however, that the cephalosporins do cross the placenta and penetrate effectively into the pericardium and the joints.

Cephapirin and *cephradine* are later introductions, but cephalothin still is generally preferred because of the experience factor.

Cefazolin is also a later introduction to cephalosporin therapy. The spectrum is quite similar to cephalosporin, but two advantages are relative lack of pain on intramuscular administration and ability to use at higher blood levels.

Cefamandole and *cefoxitin* are even more recent members of this family of drugs. These drugs extend somewhat the spectrum of the cephalosporins against bacteria. They are less effective against some strains and more effective against others. For example, cefamandole is less active against *Staph. aureus* and other gram-positive cocci than cephalothin, but is more active against strains of *Proteus* species (not including *mirabilis*) and *Enterobacter*. Cefamandole is also effective against *H. influenzae*. Cefoxitin is effective against strains of *E. coli*, Klebsiella, Salmonella, and Shigella and, additionally, some strains of *Serratia* and *Proteus*, and *Bact. fragilis*. Cefoxitin is effective against enteric gram-negative bacilli and *Bacteroides*, but less active against gram-positive organisms.

The three principal oral cephalosporins are cephaloglycin, cephalexin, and cephradine. These drugs are used in the treatment of urinary tract infections that may arise from susceptible gram-negative bacilli, but frequently the drug of first choice may be sulfonamides, ampicillin, or tetracyclines.

The principal side reactions of the cephalosporins include local pain, renal impairment, and allergic reactions.

Aminoglycosides

These substances comprise a class of potent broad-spectrum antibiotics which are chemically characterized by basic carbohydrate moieties glycosidically bound to a cyclitol unit. In general, the aminoglycosides are effective against most gram-positive and gram-negative bacteria, as well as *Mycobacterium tuberculosis*. Because of their highly ionic nature, the aminoglycosides are not absorbed from the gastrointestinal tract and must be administered parenterally. In a small percentage of patients, prolonged use of this class of antibiotics can adversely affect the eighth cranial nerve, causing some impairment of hearing and balance.

Streptomycin. Discovery in 1944 of streptomycin (structure shown below) drew immediate interest because it was the least toxic of the broad-spectrum antibiotics known at that time. Indeed, streptomycin was used to treat many gram-negative microbial infections, but because of the ease with which organisms developed resistance to it during treatment, many of these applications were abandoned when the tetracyclines, discussed later, became available. Streptomycin was the first parenterally administered antibiotic active against many microorganisms, but during the last several years, its use is limited essentially to three situations: (1) the initial treatment of serious tuberculous

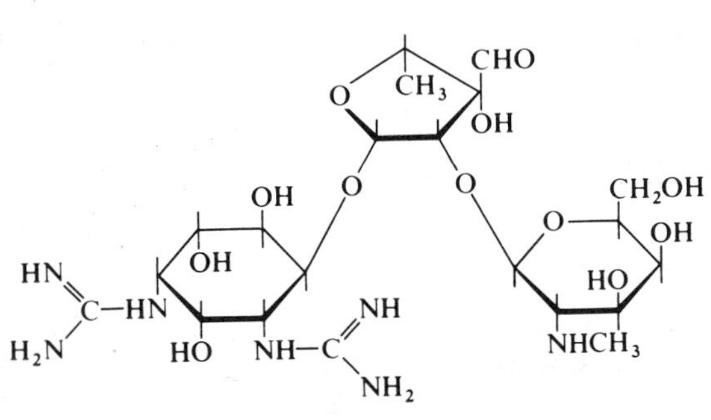

infections when the principal drugs of choice (isoniazid, rifampin) cannot be used because of their adverse effects on a particular patient; (2) treatment of enterococcal and other infections in which synergism between a penicillin and an aminoglycoside is desired; and (3) treatment of certain uncommon infections (plague and tularemia).

Kanamycin. Considerably broader in spectrum than streptomycin, kanamycin is more effective against gram-negative bacilli (other than *Pseudomonas*) and also is effective to a degree against *Staph. aureus*. However, it is ineffective against streptococci and pneumococci. The availability of penicillinase-resistant penicillins and cephalosporins essentially obsoleted kanamycin as the primary drug in the treatment of staphylococcal infections. Kanamycin has been essentially replaced by gentamicin and other aminoglycosides which are less ototoxic (adverse to hearing), and which also have a wider range of antibacterial activity.

Gentamicin. One of the successors of kanamycin, gentamicin possesses essentially the same spectrum as kanamycin, but is also active against *Pseudomonas aeruginosa*. An advantage of gentamicin is its penetration into pleural, ascitic, and synovial fluids where there is inflammation. Although not necessarily the drug of choice, gentamicin has been used in the treatment of acute cholecystitis, acute septic arthritis, anaerobic infections, *Bacillus* infections, gram-negative bacteremia, infective endocarditis, meningitis, osteomyelitis, peritonitis, staphylococcal infections, and tularemia, among others. In some situations, gentamicin acts synergistically with penicillin.

Tobramycin. Pharmacologically, tobramycin is quite similar to gentamicin. The drug is somewhat more active against *Ps. aeruginosa* than gentamicin. Tobramycin also acts synergistically with penicillin, but to a lesser degree than gentamicin.

Amikacin. This drug is a semisynthetic derivative of kanamycin. It is much less sensitive to the enzymes that inactivate aminoglycoside antibiotics. The spectrum is similar to that of gentamicin. Amikacin principally finds use in the treatment of infections arising from bacteria that are resistant to gentamicin and/or tobramycin.

Tetracyclines

These substances comprise a family of broad-spectrum antibiotics possessing a common perhydronaphthacene skeleton. They have a wider range of antimicrobial activity than other classes of clinically useful antibiotics. See Table 5. The tetracyclines are active against many species of gram-positive and gram-negative bacteria, spirochetes, rickettsiae, and some of the larger viruses.

Chlortetracycline. The first member of this class to be isolated, chlortetracycline, was discovered in 1948 among the metabolites of *Streptomyces aureofaciens*. Oxytetracycline was isolated two years later from a *S. rimosus* fermentation. Both antibiotics quickly found wide medical use, not only because they were effective orally, but because they were useful against a much wider spectrum of bacteria than penicillin G.

Chemical studies on chlortetracycline and oxytetracycline, which provided a basis for structure assignment, in general led to products with diminished or no antibacterial activity. In 1953, the first scientific reports appeared describing an active tetracycline prepared by chemical modification of a fermentation product. This was *tetracycline*, the parent member of this family of antibiotics, prepared by catalytic hydrogenolysis of chlortetracycline. It was more stable and better tolerated than its fermentation-produced progenitor, and almost completely displaced chlortetracycline from medical practice. In-terestingly, tetracycline was later found in fermentation broths of a mutant strain of *S. aureofaciens* and also may be manufactured by this method.

Following the discovery of tetracycline, useful new drugs from chemical modification of tetracycline antibiotics were slow in coming, for the complexity and chemical lability of the tetracyclines did not render them amenable to facile systematic studies of the relationships between chemical structure and biological properties. Unlike the β-lactam antibiotics, where structural modifications were being sought mainly to improve their antibacterial spectra and potency, superior semisynthetic tetracyclines were obtained, in general, with improved pharmacokinetic properties, i.e., through such factors as rate of oral absorption, degree of serum protein binding, rate of urinary excretion, and biological half-life.

An effective approach to the discovery of superior tetracycline anti-

TABLE 5. STRUCTURES OF REPRESENTATIVE TETRACYCLINES

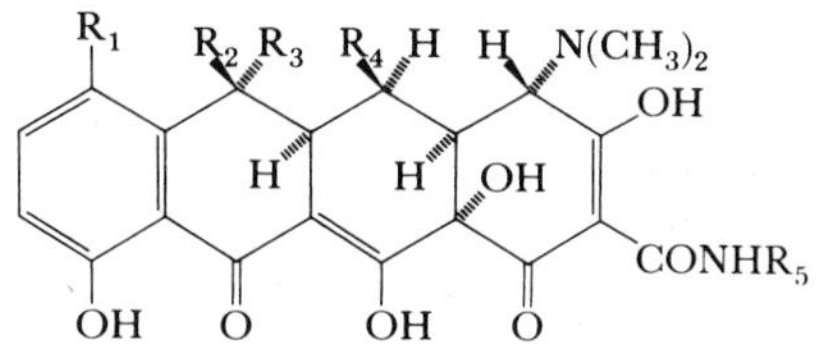

GENERIC OR CHEMICAL NAME	R_1	R_2	R_3	R_4	R_5
Tetracycline	H	OH	CH_3	H	H
Chlortetracycline	Cl	OH	CH_3	H	H
Oxytetracycline	H	OH	CH_3	OH	H
Demethylchlortetracycline	Cl	OH	H	H	H
Methacycline	H	$=CH_2$	$=CH_2$	OH	H
Doxycycline	H	H	CH_3	OH	H
Rolitetracycline	H	OH	CH_3	H	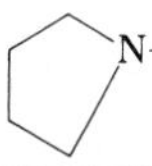 N—CH_2—

Note: Minocycline is not included in table.

biotics stemmed from studies yielding tetracyclines modified at the C-6 position. Thus, demethylchlortetracycline and methacycline were found to be somewhat superior to tetracycline in terms of a longer serum half-life, and later compounds, such as doxycycline, were shown to exhibit near-ideal pharmacokinetics. Doxycycline, among a number of other tetracyclines, remains in wide use today.

Among the many diseases treated with tetracyclines are urinary tract infections, gonorrhea, nongonococcal urethritis, Rocky Mountain spotted fever, other rickettsioses, mycoplasmal pneumonia, chlamydial diseases (psittacosis, trachoma, lymphogranuloma venereum), brucellosis, plague, cholera, granuloma inguinale, syphilis, and gonococcal pelvic inflammatory diseases, particularly in a number of instances where a patient may be allergic to penicillin. Tetracyclines also have been used in the treatment of cystic acne.

Doxycycline has been reported as effective in the prophylaxis of traveler's diarrhea in Kenya. The tetracycline minocycline is sometimes used instead of the sulfonamides against noncardial infections. It is not used in connection with meningococcal infections.

Macrolides

The macrolides comprise a family of antibiotics chemically characterized by a macrocyclic lactone to which one or more sugars are attached. The compounds are often divided into various subgroupings, but the group with antibacterial properties are known as *antibacterial macrolides.* They are distinguished chemically by having, in addition to the large lactone, various ketonic and hydroxyl functions and glycosidically bound deoxy sugars. A second grouping of commercial importance is known as the *polyene macrolides,* chemically characterized by extended conjugated double-bond systems. The polyenes are devoid of antibacterial activity, but are potent antifungal agents.

A number of the antibacterial macrolides have been found to be clinically useful chemotherapeutic substances, falling generally under the title of *medium-spectrum antibiotics.* This term is taken to mean that these substances are effective against most gram-positive bacteria and have a degree of activity against certain gram-negative organisms, such as *Haemophilus, Brucella,* and *Neisseria* species. The antibacterial macrolides also appear to inhibit certain pleuropneumonialike organisms.

Erythromycin. This is the principal drug in this category. Although available as the parent entity, semisynthetic derivatives have proved to be clinically superior to the natural cogener. Like the tetracyclines, synthetic transformations in the macrolide series have not significantly altered their antibacterial spectra, but have improved the pharmacodynamic properties. For example, the propionate ester of erythromycin lauryl sulfate (erythromycin estolate) has shown greater acid stability than the unesterified parent substance. Although the estolate appears in the blood somewhat more slowly, the peak serum levels reached are higher and persist longer than other forms of the drug. However, cholestatic hepatitis may occasionally follow administration of the estolate and, for that reason, the stearate is often preferred.

Erythromycin is effective against Group A and other nonenterococcal streptococci, *Corynebacterium diphtheriae, Legionella pneumophila, Chlamydia trachomatis, Mycoplasma pneumoniae,* and *Flavobacterium.* Because of the extensive use of erythromycin in hospitals, a number of *Staph. aureus* strains have become highly resistant to the drug. For this reason, erythromycin has been used in combination with chloramphenicol. This combination is also used in the treatment of severe sepsis when etiology is unknown and patient is allergic to penicillin.

A structural representation of erythromycin is shown below.

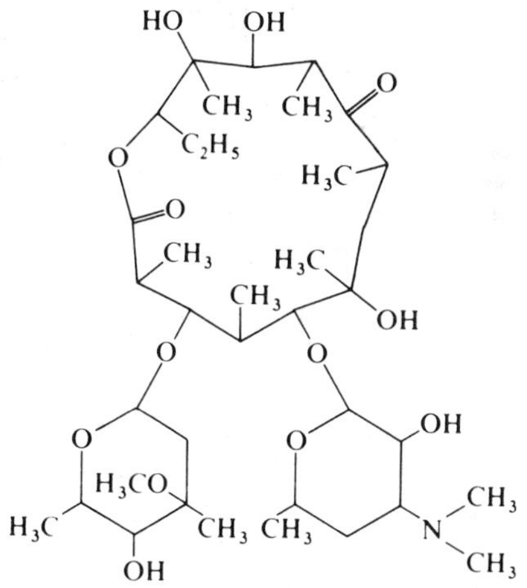

Other Antibiotics

Chloramphenicol. This compound is derived from *Streptomyces venezuelae* or by organic synthesis. It was the first substance of natural origin shown to contain an aromatic nitro group. Although the drug is a valuable broad-spectrum antibiotic, its use has been somewhat limited because of the occasional development of aplastic anemia in the patient. Thus, its use has been largely confined to its administration as the drug of choice in patients allergic to penicillin in connection with typhoid fever, nontyphoidal salmonelloses (due to ampicillin-resistant strains), *H. influenzae* meningitis, meningitis arising from *N. meningitidis,* and *Str. pneumoniae.* Because effective and safer drugs are not available, it is used for infections arising from *Bact. fragilis.* The drug is administered orally or intravenously.

Vancomycin. This is a narrow-spectrum antibiotic and produced by *Streptomyces orientalis* or synthetically. Its effectiveness is essentially confined to the treatment of streptococci (including enterococci), pneumococci, staphylococci, and a few other gram-positive bacteria. Serious side effects include possible hearing loss and renal insufficiency, particularly when the drug is administered with an aminoglycoside.

Polymyxins. This is a generic term for a series of antibiotic substances produced by strains of *Bacillus polymyxa.* Various polymyxins are differentiated by letters A, B, C, D, and E. All are active against certain gram-negative bacteria. Polymyxin B and E (colistin) have been the most important in the past, but currently are only rarely used—because they have been replaced by more effective aminoglycosides. The B and E drugs are effective against most of the common aerobic gram-negative bacilli, but not *Proteus, Providencia,* and *Serratia.* Prior to their replacement by aminoglycosides, the polymyxins were used mainly in connection with infections arising from *Ps. aeruginosa.*

Spectinomycin. This drug finds principal application in the treatment of gonorrhea. It should be noted that the antibiotic resistance among *N. gonorrhoeae* has caused a number of therapeutic problems. It has been found that only by escalating the antibiotic doses and using probenecid to retard the excretion of penicillin and ampicillin (the drugs of choice) has the continued effective use of penicillin, ampicillin, and tetracycline been possible. Even with modifications in the therapy, from 3 to 8% of cases fail to respond to the usual regimens for uncomplicated gonorrhea. Thus, the treatment of uncomplicated gonorrhea that fails to respond to the usual regimen is spectinomycin therapy.

Chemoprophylaxis with Antibiotics

In addition to their use in treating infections arising from bacteria and other microorganisms, antibiotics are also used in chemoprophylaxis, i.e., treating a patient before or shortly after the entry of pathogenic organisms. There are three common situations: (1) preventing infection following exposure to known pathogens; (2) preventing specific infections in highly susceptible individuals; and (3) preventing postoperative infectious complications.

Antibiotics in Feedstuffs

For several years, antibiotics have been used in feedstuffs, not only to lower the incidence of certain diseases in livestock, but antibiotics also play a function in the rate of growth of animals. It is estimated that, as of the late 1970s, nearly 40% of the antibacterials produced in the United States are used in animal feeds or for other nonhuman purposes. In 1977, the U.S. Food and Drug Administration commenced an investigation because it is observed that low levels of use of antibiotics in animal husbandry provide an "almost ideal" environment for breeding antibiotic-resistant strains of bacteria that could eventually infect humans. A strain resistant to both penicillin and tetracycline, for example, could overcome two of the most effective, commonly used drugs in the antibiotic therapy of humans. The problem for the authorities is to prove convincingly that such suspicions are a forecast of real truths.

Manufacture of Antibiotics

Generally, most antibiotics and the starting materials for semisynthetic antibiotics are manufactured by fermentation, with accompanying extraction, purification, crystallization, and packaging operations. Commercial fermenting vessels are stainless- or carbon-steel enclosed tanks with capacities up to several tens of thousands of gallons (many hundreds of hectoliters). Such factors as aeration, agitation, temperature, and pH must be monitored and controlled carefully.

The antibiotic-producing microorganism is grown in submerged culture in a fermentation medium which contains various carbon, nitrogen, and trace-metal sources, required by the organism for its nutrition. The organism is grown under conditions of pure culture, that is, other microorganisms excluded from the fermentation inasmuch as the latter will compete for nutrients, may contribute undesirable contamination and reduce yield of the desired product. When the fermentation has reached peak potency (varies with each product), the antibiotic may be recovered by an extraction technique, such as distribution into a water-immiscible solvent, ion-exchange chromatography, or precipitation. Following extraction, purification and crystallization are carried out by procedures compatible with the physicochemical properties of the particular antibiotic being produced. A representative flowsheet is shown in the accompanying figure.

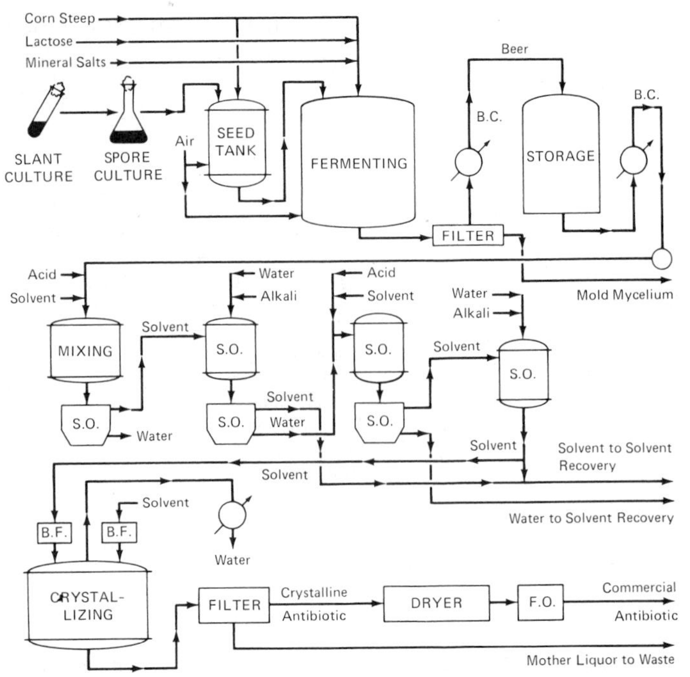

Materials flow in a representative commercial antibiotic manufacturing process. B.C. = brine cooler; S.O. = separating operation; B.F. = bacteriological filter.

References

Abramowicz, M. (editor): "Adverse Interaction of Drugs," *The Medical Letter on Drugs and Therapeutics,* **19**, 5 (1977).

Abramowicz, M. (editor): "Antimicrobial Prophylaxis: Prevention of Wound Infection and Sepsis after Surgery," *The Medical Letter on Drugs and Therapeutics,* **19**, 37 (1977).

Abramowicz, M. (editor): "The Choice of Antimicrobial Drugs," *The Medical Letter on Drugs and Therapeutics,* **20**, 1 (1978).

Braude, A. I.: "Antimicrobial Drug Therapy," Vol. 8 of "Major Problems in Internal Medicine," Saunders, Philadelphia, 1976.

Cook, F. V., and W. E. Varrar: "Vancomycin Revisited," *Ann. Intern. Med.,* **88**, 813 (1978).

Krogstad, D. J., et al.: "Plasmid-mediated Resistance to Antibiotic Synergism in Enterococci," *J. Clin. Invest.,* **61**, 1645 (1978).

Kunin, C. M., et al.: "Veterans Administration Ad Hoc Interdisciplinary Advisory Committee on Antimicrobial Drug Usage: Prophylaxis in Surgery," *J. Amer. Med. Assn.,* **237**, 1003 (1977).

Moellering, R. C., Jr., and M. N. Swartz: "Drug Therapy: The Newer Cephalosporins," *N. Engl. J. Med.,* **294**, 24 (1976).

Rahal, J. J.: "Antibiotic Combinations: The Clinical Relevance of Synergy and Antagonism," *Medicine,* **57**, 179 (1978).

Siegel, M. S., et al.: "Penicillinase-producing *Neisseria gonorrhoeae*: Results of Surveillance in the United States," *J. Infect. Dis.,* **137**, 170 (1978).

Sivonen, A., et al.: "The Effect of Chemoprophylactic Use of Rifampin and Minocycline on Rates of Carriage of *Neisseria meningitidis* in Army Recruits in Finland," *J. Infect. Dis.,* **137**, 238 (1978).

Staff: "Missed Chance" in "Science and the Citizen," *Sci. Amer.,* **239**, 5, 90–91 (1978).

ANTIBIOTICS (Foods). Antimicrobial Agents (Foods).

ANTIBODY. This article gives a generalized description of antibodies and their role in the body's immune system. More details in terms of the most recent findings in this field are given in the entry on **Immune System and Immunology.** In medicine and physiology, immunity is the ability of the body to resist invasion by pathogenic organisms and substances. Immunity may be initially in place, that is, genetically ordered for a given species. Humans are naturally immune to canine distemper; dogs are immune to measles; rats are immune to diphtheria; and domestic fowls are immune to anthrax. Many other examples could be cited. Immunity may be acquired as the result of exposure to an invasive pathogen, triggering the immune system to construct cells that will be in reserve to resist subsequent invasions by the same pathogen. Immunity also may be acquired artificially through the use of preventive immunization techniques. Immunity is effected through antibodies.

Any substance that can provoke a response by the body's immune system is called an *antigen.* This property of an antigen is referred to as *immunogenicity.* Although the first antigens to be investigated were microorganisms and proteins foreign to the body, research during recent decades has been directed toward understanding the immune response at the molecular level—for it is at this level that the actions and reactions of the immune system occur. At the molecular level, numerous previously unsuspected complexities of the immune process have been revealed and still others are only partially understood, if at all. It has been discovered that several cell types, in addition to the lymphocytes, act cooperatively in effecting what might be called the total immune response. Although the lymphocytes appear to play the dominant role in the immune system, several other cells are now known to cooperate with the lymphocytes, and the functions of these other cells are no longer considered of secondary importance. Study of the antigens at the molecular level also has contributed to a much better understanding of the immune response.

Numerous molecules can evoke an immune response, sometimes when the responses from the standpoint of protecting body functions are not immediately obvious, other than that such molecules do not meet the criteria of "self" and "nonself" described later. In a general way, it may be observed that the immune system tends to have a bias toward suspicion and may, on occasion, overreact, as in cases of autoimmunity. Currently, it is generally hypothesized that recognition of antigens at the receptor sites of the lymphocytes is based upon the shapes of molecules, reminiscent of some of the current hypotheses concerning taste and odor receptors in the tongue and nasal membranes. Until the mechanism occurring at the receptor sites is more fully explained, numerous questions as regards what molecules do and do not evoke immune response will remain unanswered.

Considerable research with synthetic polymers comprised of various amino acids has been undertaken in an effort to determine the requirements of immunogenicity. It has been established, for example, that tyrosine as well as some other aromatic amino acids will confer immunogenicity to certain polypeptides which in themselves are not or are only slightly antigenic. Further, it has been found in such cases that the antibody is directed against the polypeptide and not the amino acid. Research with synthetic polymers led investigators to the finding that immune response is under genetic control, this based upon the observations that different animal strains and species respond differently to a given polymer. However, in a descriptive fashion, this principle had been demonstrated by the different reactions to antigens by various species many decades ago.

Although not catalytic in the usual sense, certain substances, known as *adjuvants,* are capable of enhancing the immunogenicity of certain antigens. Among the adjuvants are aluminum salts, bacterial endotoxins, bacillus Calmette-Guérin (BCG), *Bordetella pertussis,* and mycobacteria. These materials and this phenomenon have been important in immunity research. Sometimes adjuvants are used clinically in connection with certain immunizations, such as against tetanus.

Where antigens are introduced into the body intravenously, they usually travel rapidly to the spleen, followed by the fast production of an antibody. Subcutaneous or intradermal injection of antigens most frequently localize in the lymph nodes and antigens that are inhaled favor local sensitization. In some cases, such as tetanus immunization, toxin produced by the bacteria may be slow and insufficient to provoke a significant immunologic reaction. Thus, the requirement for properly timed booster injections.

Clinical Use of Antigens. Without the benefit of understanding the complexities of the immune system, particularly at the molecular level, much progress was made over the years in taking advantage of certain antigens and a qualitative or descriptive understanding of the immune response. Thus, the early development of vaccines and antitoxins.

The history of the development of antitoxins in combating bacterial infection dates back to the early beginnings of organized bacteriology. Behring was the first to show that animals that were immune to diphtheria contained, in their serum, factors which were capable of neutralizing the poisonous effects of the toxins derived from the diphtheria bacillus. While this work was carried out in 1890, prior to many of the great discoveries of mass immunization, and much later the antibiotics, it is interesting to note that there remains a place for antitoxins, even though relatively limited in modern medical treatment or prophylaxis of a few diseases, such as tetanus, botulism, and diphtheria. In the case of diphtheria, equine antitoxin is the only specific treatment available. However, it is only reasonably effective if used during the first 48 hours of the onset of the disease. Trivalent (ABE) antitoxin is used in the treatment of botulism. In the treatment of tetanus, human tetanus immune globulin is preferred, but when it is not available, equine antitoxin is substituted. See **Antitoxin.**

For many years the preferred approach to immunity to infectious disease has been by development of active immunity through the injection of a vaccine. The vaccine may be either an attenuated live infectious agent, or an inactivated or killed product. In either case, protective substances called antibodies are generated in the bloodstream; these are described in the next section. Vaccines for a number of diseases have been available for many years and have assisted in the erradication of some diseases, such as smallpox. As new strains of bacteria and viruses are discovered, additional vaccines become available from time to time. See **Vaccine.**

Antibodies. Antigens are excluded from the body by skin and mucous membranes. If these barriers are penetrated, the foreign organism may be ingested by phagocytic cells (monocytes, polymorphs, macrophages) and subsequently destroyed by cytoplasmic enzymes. Some time after a foreign macromolecule has entered the body, induced mechanisms come into play. There are two basic biological manifestations of the immune reaction: (1) Immunity to infectious agents; and (2) specific hypersensitivity. Hypersensitivity, or the heightened response to an agent, can be divided into anaphylactic, allergic, and bacterial. Anaphylaxis, which can be produced by either active or passive sensitization, is a laboratory tool for studying the fundamental nature of hypersensitivity. The amounts of antigen and antibody involved, as well as the nature and source of the antibody, govern the extent of the reaction.

In immunity to infectious agents, some time after a foreign macromolecule has entered the body, induced mechanisms come into play, which result in the synthesis of specially adapted molecules (*antibodies*) capable of combining with the foreign substances which have elicited them. Most macromolecules (proteins, carbohydrates, nucleic acids, etc.) can function as antigens, provided that they are different in structure from autologous macromolecules, i.e., from the macromolecules of the responding organism.

Antibodies are proteins with a molecular weight of 150,000–1,000,000 and with electrophoretic mobility predominantly of gamma globulins. The combination between antigen and antibody results in inhibition of the biological activity of the antigen and leads to increased rate of ingestion (*opsonization*) of the antigen by phagocytic cells. In addition, combination of antigen and antibody results in the activation of a complex chain of interacting constitutive molecules—the *complement system*—leading to lysis of the cell membranes to which antibody, directed against cellular antigens, is attached.

Biochemical Individuality. This is a unique quality, genetically determined, for each individual and is exhibited with respect to: (1) The composition of blood, tissues, urine, digestive juices, cerebrospinal fluid, etc.; (2) the enzyme levels in tissues and in body fluids, particularly the blood; (3) the pharmacological responses to specific drugs and poisons; (4) the biochemical responses to bacteria, fungi, and other microorganisms; (5) the quantitative needs for specific nutrients—minerals, amino acids, vitamins, etc.—and in a number of other ways, including reactions of taste and smell and the effects of heat, cold, and electricity. Although individual *similarities* are readily apparent at the macro level, each individual must possess a highly distinctive pattern, since the differences between individuals with respect to measurable items in a potentially long list are by no means trifling. Out of these relatively small, but distinct differences has risen the concept of the so-called normal individual, against which high and low levels of response are compared when considering the "chemistry" of a given individual. The concept of individuality is covered in greater detail in the entry on **Biochemical Individuality.** Differences are particularly evident with respect to the body's immune system.

Autoimmunity. A very important characteristic of the body's immune system is a capacity to distinguish between *self* and *nonself.* In terms of the immune system, the ability to make such distinctions proceeds at the biochemical level without conscious awareness of the individual. Less than a century ago, the ability of the body to distinguish self from nonself (foreign), at the biochemical level, was considered impossible. Over the years, however, much descriptive information accumulated which, without ample explanation, proved that the body does remember at the biochemical level. Seldom, for example, have medical records shown a second infection with mumps, measles, or smallpox, once an individual survived the first attack. The question, how does the body remember? remained a mystery for many decades until the concept of the immune system was first outlined in a very general way.

When the immune system functions properly, which is the normal situation, antibodies to parts of the same body are not produced. But a condition known as *autoimmunity* can sometimes occur. In such circumstances, antibodies and sensitized (antigen-reactive) cells may be produced and directed against "self" antigens. To treat autoimmunity is one of the challenges of modern medicine. This autoimmunity mechanism may trigger some asthmatic paroxysms and is presently being considered as a suspect mechanism by cancer researchers—with considerable study being directed to tumor immunology. Some investigators hypothesize that the body system has the capacity to recognize neoplastic cells and to destroy them, but that in some individuals the process becomes inoperative, allowing the cells to multiply.

The ability of the body to tolerate self-antigens and thus to preclude autoimmunity is known as *immunologic tolerance.* As early as 1959, some scientists suggested that the antigen-specific lymphocytes which interact with self-antigens are eliminated during the prenatal state. In recent years, the concept of *suppressor mechanisms* has been well received. In this concept, the immune system, responding to an antigen, in addition to producing antibodies and sensitized cells to combat foreign substances, also initiates various suppressor mechanisms, causing some mediation of the process. Such substances are termed *mediators.* This may explain why, in some clinical situations, an early and strong immune response may be due to deficiencies in the suppressor mechanisms; or, in contrast, a deficient immune response may not be caused by the lack of particular lymphocytes, but rather by overreactive suppressor mechanisms.

A study of the body's immune system over the years has made possible the preparation of antitoxins and vaccines for preventing and treating many diseases. These studies have assisted in dealing with the problems of allergy and hypersensitivity and in finding partial or effective solutions to problems which arise from malperformance of the immune systems. Further progress during the 1980s is expected in the understanding and treatment of immune-system-related disorders, such as amyloidosis, macroglobulinemia, multiple myeloma, systemic lupus erythematosus, asthma, and various allergies, among others. A major advancement was made a few decades ago when scientists revealed the nature of factors causing hemolytic disease in infants following the first born when the mother was Rh negative and the father was Rh positive. Because of the development of RhoGam and associated procedures, this is no longer a major medical problem.

See also **Clone; and Immune System and Immunology.**

References

Capra, D., and A. B. Edmundson: "The Antibody Combining Site," *Sci. Amer.*, **223**, 2, 34–40 (1970).

Eisen, H. N.: "Immunology," Harper and Row, New York, 1974.

Fudenberg, H. H., et al. (editors): "Basic and Clinical Immunology," Lange Medical Publications, Los Altos, California, 1976.

Gell, P. G. H., Coombs, R. R. A., and P. J. Lachmann (editors): "Clinical Aspects of Immunology," Blackwell Scientific Publications, Oxford, England, 1975.

Kabat, E. A.: "Structural Concepts in Immunology and Immunochemistry," 2nd edition, Holt, Rinehart and Winston, New York, 1976.

Melchers, F., Potter, M., and N. L. Warner (editors): "Lymphocyte Hybridomas," Springer-Verlag, New York, 1978.

Nisonoff, A., Hopper, J. E., and S. B. Spring (editors): "The Antibody Molecule," Academic, New York, 1975.

Roitt, I.: "Essential Immunology," 3rd edition, Blackwell Scientific Publications, Oxford, England, 1977.

Rose, N. R.: "Autoimmune Diseases," *Sci. Amer.*, **244**, 2, 80–104 (1981).

Sela, M. (editor): "The Antigens," Academic, New York, 1973.

Staff: "Lymphocyte Function," The Cold Spring Harbor Laboratory, Cold Spring Harbor, New York, 1977.

ANTICAKING AGENTS. Some products, particularly food products that contain one or more hygroscopic substances, require the addition of an *anticaking agent* to inhibit formation of aggregates and lumps and thus retain the free-flowing characteristic of the products. Calcium phosphate, for example, is commonly used in instant breakfast drinks and lemonade and other soft-drink mixes.

The general function of an anticaking agent can be described by using silica gel as an example. Generally anticaking agents are available as very small particles (ranging from 2 to 9 micrometers in diameter). A typical application for silica gel is admixture with orange-juice crystals to assure a free-flowing product, avoiding formation of crystal cakes and hard lumps. The very high adsorption properties of the anticaking substance removes moisture that can cause fusion. The billions of extremely fine, inert particles coat and separate each grain of powder (product) to keep it free-flowing. Many anticaking agents, including silica gel, also act as dispersants for powdered products. Many food products, when stirred into water, tend to form lumps which are difficult to disperse or dissolve. The agent not only improves flow properties, but also increases speed of dispersion by keeping the food particles separated and permitting water to wet them individually instead of forming lumps. As is true with so many food additive chemicals, anticaking agents serve multiple functions. In addition to acting as an anticaking and dispersing agent, silica gel also can be used as a moisture scavenger and carrier. Some additives, when they are capable of serving several functions, may be called *conditioning agents.*

Anticaking agents commonly used include: calcium carbonate, phosphate, silicate, and stearate; cellulose (microcrystalline); kaolin; magnesium carbonate, hydroxide, oxide, silicate, and stearate; myristates; palmitates; phosphates; silica (silicon dioxide); sodium ferrocyanide; sodium silicoaluminate; and starches.

ANTICATHODE. In an x-ray tube, the target on which the electron beam is focused and from which the x-rays are radiated.

ANTICHOLINERGIC. Hallucinogens; Ulcer.

ANTICLINE. A folded structure involving bedded rocks in which the strata are arched upward so that the beds bend downward on either side. These downward-bending beds constitute the limbs of the fold.

The angle which the beds on the limbs of the fold make with the horizontal is spoken of as the dip. The term dip is also used to indicate the inclination of bedding in other structures.

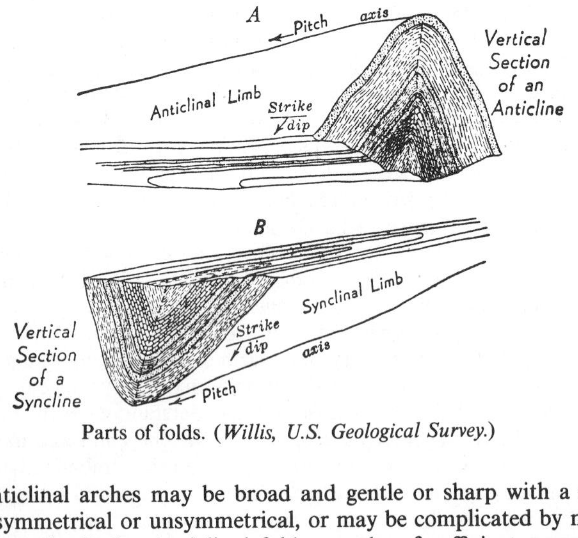

Parts of folds. (*Willis, U.S. Geological Survey.*)

Anticlinal arches may be broad and gentle or sharp with a steep dip, symmetrical or unsymmetrical, or may be complicated by minor folds on the limbs. Anticlinal folds may be of sufficient magnitude to be measured in miles, involving great thicknesses of sediments, or they may be so small as to be measured in inches.

The direction of prolongation of the fold is termed the axis of the fold, and if not exactly horizontal the angle of inclination of the top bed of the anticline is called the pitch. Plunge is used as a synonym for pitch by some geologists.

ANTICLINORIUM. A composite anticlinal structure of folded beds is called an anticlinorium; a composite synclinal structure is called a synclinorium. The latter term, however, should be applied only to the compressed sedimentary filling of a geosyncline.

Section of an anticlinorium. (*Van Hise.*)

ANTICOAGULANTS. These are substances which prevent coagulation of the blood. For blood investigations made outside the body, sodium or potassium citrates, oxalates, and fluorides are sometimes used. For blood which is to be used for transfusions, sodium citrate is used.

Organic anticoagulants are used in vivo in the treatment of numerous conditions where blood coagulation can be dangerous, as in cerebral thrombosis and coronary heart disease, among others which will be described later. The main anticoagulants used are heparin and coumarin compounds, such as warfarin.

Heparin. A complex organic acid (mucopolysaccharide) present in mammalian tissues and a strong inhibitor of blood coagulation. Although the precise formula and structure of heparin are uncertain, it has been suggested that the formula for sodium heparinate, generally the form of the drug used in anticoagulant therapy, is $(C_{12}H_{16}NS_2Na_3)_{20}$ with a molecular weight of about 12,000. The commercial drug is derived from animal livers or lungs.

Heparin is considered a hazardous drug. Heparin may be the leading cause of drug-related deaths in hospitalized patients who are relatively well (Porter and Jick, 1978). It has been reported (Bell, et al, 1976) that some patients who receive continuously infused intravenous heparin develop *thrombocytopenia* (condition where the platelet count is less than 100,000 per cubic millimeter). Some authorities believe that the risk of thrombocytopenia associated with porcine heparin may be less than the risk associated with heparin of bovine origin. (Babcock, et al, 1976; Powers, et al, 1979; Hrushesky, 1978).

Heparin, in addition to inhibiting reactions which lead to blood clotting, also inhibits the formation of fibrin clots, both in vitro and in vivo. Heparin acts at multiple sites in the normal coagulation system. Small amounts of heparin in combination with antithrombin III (heparin co-factor) can prevent the development of a hypercoagulable state by inactivating activated factor X, preventing the conversion of prothrombin to thrombin. Once a hypercoagulable state exists, larger doses of heparin, in combination with antithrombin II, can inhibit the coagulation process by inactivating thrombin and earlier clotting intermediates, thus preventing the conversion of fibrinogen to fibrin. Heparin also prevents the formation of a stable fibrin clot by inhibiting the activation of the fibrin stabilizing factor. The half-life of intravenously administered heparin is about 90 minutes.

Coumarin. Oral anticoagulants can be prepared from compounds with coumarin as a base. Coumarin has been known for well over a century and, in addition to its use pharmaceutically, it is also an excellent odor-enhancing agent. However, because of its toxicity, it is not permitted in food products in the United States (Food and Drug Administration). The commercial drug Sintrom® is 3-(alpha-acetonyl-4-nitrobenzyl)-4-hydroxycoumarin. This drug reduces the concentration of prothrombin in the blood and increases the prothrombin time by inhibiting the formation of prothrombin in the liver. The drug also interferes with the production of factors VII, IX, and X, so that their concentration in the blood is lowered during therapy. The inhibition of prothrombin involves interference with the action of vitamin K, and it has been postulated that the drug competes with vitamin K for an enzyme essential for prothrombin synthesis.

The commerical drug Dicumarol® is bis-hydroxycoumarin, $C_{19}H_{12}O_6$. The actions of this drug are similar to those just described.

Warfarin. This compound is also of the coumarin family. The formula is 3-(alpha-acetonylbenzyl)-4-hydroxycoumarin. In addition to use in anticoagulant therapy in medicine, the compound also has been used as a major ingredient in rodenticides, where the objective is to induce bleeding and, when used in heavy doses, is thus lethal. The compound can be prepared by the condensation of benzylideneacetone and 4-hydroxycoumarin.

The anticoagulant action of warfarin is through interference of the gamma-carboxylation of glutamic acid residues in the polypeptide chains of several of the vitamin K-dependent factors. The carboxylation reaction is required for the calcium-binding activity of the K-dependent factors. Because of the reserve of procoagulant proteins in the liver, usually several days are required to effect anticoagulation with warfarin.

Warfarin antagonists include vitamin K, barbiturates, gluthethimide, rifampin, and cholestyramine. Warfarin potentiators include phenylbutazone, oxyphenbutazone, anabolic steroids, clofibrate, aspirin, hepatotoxins, disulfiram, nalidixic acid, and metronidazole. In patients undergoing anticoagulation therapy with warfarin, it has been found that cimetidine (used in therapy of duodenal ulcer) may increase anticoagulant blood levels and consequently prolong the prothrombin time (Serlin et al, 1979).

Anticoagulation Therapy

Prior to administration of anticoagulant drugs, patients must be carefully evaluated. Anticoagulant drugs are to be avoided if any of the following conditions prevail: a history of abnormal bleeding, recent corticosteroid therapy, recent intraocular or intracranial bleeding, recent pericarditis, and recent peptic ulcer or esophageal bleeding. A history of the individual's use of antiplatelet agents, such as aspirin, dipyridamole, phenylbutazone, and indomethacin, should be obtained and evaluated. Anticoagulant drugs should be administered with particular care during pregnancy. Heparin does not anticoagulate the fetus because it does not cross the placenta. Warfarin, on the other hand, anticoagulates both the mother and the fetus. Problems which may arise from the administration of warfarin during pregnancy, particularly during the first trimester, are described by Hirsch and Gallus (1972) and Ravio, et al (1977).

The bile sequestrant cholestyramine is frequently used in the treatment of familial hypercholesterolemia. This drug not only binds cholesterol, but also a number of other drugs, including anticoagulants.

Deep Vein Thrombosis and Pulmonary Embolism. Prompt administration of intravenous heparin is indicated in the treatment of this condition. Heparin is fast-acting, prevents further thrombus formation, and when used in therapeutic doses, also prevents the release of serotonin and thromboxane A_2 from platelets that adhere to thrombi that embolize to the lungs. The size of the dose required varies with a number of patient conditions as described by Wessler and Gitel (1979). Several authorities are convinced that the continuous (pump-driven) infusion method is superior to intermittent injections (Salzman, et al, 1975; Glaziner and Crowell, 1976; Porter and Jick, 1977).

Heparin is usually administered for a period ranging from 7 to 10 days. Frequently, during the last half of this period of heparin therapy, oral anticoagulation will be commenced with warfarin. The time during which oral anticoagulation administration should be continued may be three months or longer after clinical evidence that the venous thrombosis has subsided; and for one year after pulmonary embolism.

Cerebral Thrombosis. Among the specific modes of treatment that have been used in anticoagulation therapy. Many authorities suggest the use of anticoagulants for an evolving stroke in an effort to arrest the propagation of thrombus. In this procedure, a lumbar puncture is usually prepared first. If there is a presence of red blood cells in the spinal fluid, this infers a hemorrhagic infarction, in which case anticoagulants are withheld for a minimum of 48 hours. If the fluid is clear, heparin can be administered by continuous intravenous drip.

Cerebral Transient Ischemic Attack (TIA). Aspirin, as a platelet-inhibiting agent, has been found effective in the medical management of TIA. As studied by the Joint Committee for Stroke Facilities in the late 1970s, the results of anticoagulant therapy for TIA were reported as vague, but possibly this is due to poorly planned tests. However, anticoagulant therapy is considered a proper mode of treatment for persons who cannot tolerate aspirin, with warfarin the drug of choice. Oral anticoagulants are not given to persons with gastrointestinal ulcerations, severe hypertension, bleeding tendencies, or renal or hepatic failure.

Coronary Heart Disease. Long-term preventive anticoagulation with warfarin and similar drugs in patients with coronary artery disease has decreased in popularity during the last few years because evidence collected over a long period of time has not shown, in a convincing way, that the therapy is of value. As of the early 1980s, the therapy of choice includes the use of platelet-inhibiting drugs, notably acetylsalicylic acid (aspirin), sulfinpyrazone, and dipyridamole. However, clinical evidence relative to this therapy thus far also has been weak.

Prevention of Thromboembolism. Anticoagulation for prevention of thromboembolism is not used to the extent that it was once employed in the 1950s. Some professionals have reexamined the data of that period, however, and have concluded that anticoagulant therapy does have value, but their observations have not been widely accepted.

In present times, because of early mobilization and shorter stays in hospital, venous thrombosis in the legs and resulting pulmonary embolism has declined to a large degree. In persons with acute myocardial infarction, prophylactic low-dose heparin has reduced the incidence of venous thrombosis in the legs. It is considered as a reasonable alternative to warfarin in selected patients. Preventive anticoagulation may be indicated in some cases to prevent strokes due to left ventricular mitral thrombi embolizing in the brain.

Prosthetic Valve Endocarditis. Anticoagulants are sometimes used in the overall treatment of PVE even though there are risks of intracerebral hemorrhage or hemorrhagic infarction. Countering this risk, however, is the risk of major thromboembolic complications involving the central nervous system that may occur in the absence of continued anticoagulant therapy.

Anticoagulant therapy is also sometimes used in cases of congestive heart failure and in the treatment of polycythemia vera (elevation of the packed cell volume or the hemoglobin level) where not contraindicated.

Massive Venous Occlusion. This may be described as a surgical emergency. Immediately after diagnosis, intravenous heparinization is started and continued during the thrombectomy.

Mini-Dose Heparin. Small subcutaneous doses of heparin have been found to be effective in high-risk postsurgical patients and in patients with acute myocardial infarction. The preventive treatment is commenced a few hours before an operative procedure and continued postoperatively for 4 to 5 days. As the result of a study in 1975, low-dose heparin prophylaxis in high-risk patients who undergo abdomino-thoracic surgery has become a widely accepted practice. However, preventive anticoagulant therapy, to date, has been unsatisfactory and controversial in the instances of hip surgery or prostatectomy.

See also **Cerebrovascular Diseases; Heart and Circulatory System (Human); Ischemic Heart Disease.**

References

Babcock, R. B., et al.: "Heparin-induced Immune Thrombocytopenia," *N. Engl. J. Med.*, **295**, 2396 (1976).

Bell, W. R., et al.: "Thrombocytopenia Occurring During the Administration of Heparin," *Ann. Intern. Med.*, **85**, 155 (1976).

Glazier, R. L., and E. B. Crowell: "Randomized Prospective Trial of Continuous vs. Intermittent Heparin Therapy," *J. Amer. Med. Assn.*, **236**, 1365 (1976).

Hirsh, J., Cade, J. F., and A. S. Gallus: "Anticoagulants in Pregnancy," *Amer. Heart J.*, **83**, 301 (1972).

Hrushesky, W. J.: "Subcutaneous Heparin-induced Thrombocytopenia," *Arch. Intern. Med.*, **138**, 1489 (1978).

Jaques, L. B.: "Heparin: An Old Drug with a New Paradigm," *Science*, **206**, 528–533 (1979).

Porter, J., and H. Jick: "Drug-related Deaths among Medical Inpatients," *J. Amer. Med. Assn.*, **237**, 879 (1977).

Powers, P. J., Cuthbert, D., and J. Hirsh: "Thrombocytopenia Found Uncommonly during Heparin Therapy," *J. Amer. Med. Assn.*, **241**, 2396 (1979).

Raivio, K. O., Ikonen, E., and S. Saarikoski: "Fetal Risks due to Warfarin Therapy during Pregnancy," *Acta Paediatr. Scand.*, **66**, 735 (1977).

Salzman, E. W., et al.: "Management of Heparin Therapy," *N. Engl. J. Med.*, **292**, 1046 (1975).

Scheinberg, P. (editor): "Cerebrovascular Diseases," Raven Press, New York, 1976.

Serlin, M. J., et al.: "Cimetidine: Interactions with Oral Anticoagulants in Man," *Lancet*, **2**, 317 (1979).

Staff: "A Randomized Trial of Aspirin and Sulfinpyrazone in Threatened Stroke," conducted by the Canadian Cooperative Study Group, *New Engl. J. Med.*, **299**, 53 (1978).

Toole, J. F.: "Cerebrovascular Disorders," McGraw-Hill, New York (1974).

Wessler, S., and S. N. Gitel: "Heparin: New Concepts Relevant to Clinical Use," *Blood*, **53**, 525 (1979).

ANTICOINCIDENCE CIRCUIT. A circuit with two input terminals which delivers an output pulse if one input terminal receives a pulse, but delivers no output pulse if pulses are received by both input terminals simultaneously or within an assignable time interval.

ANTICOINCIDENCE COUNTER. An arrangement of counters and associated circuits which will record a count if and only if an ionizing particle passes through certain of the counters but not through the others.

ANTICONVULSANT AGENTS. Seizure (Neurological).

ANTICORONA. Atmospheric Optical Phenomena.

ANTICYCLONE. Atmosphere (Earth).

ANTICYCLOGENESIS. Atmosphere (Earth).

ANTIDIURETIC HORMONE (ADH). Diabetes Insipidus; Hormones; Kidney and Urinary Tract; Water.

ANTIDOTE. An agent which inhibits or counteracts the action of a poison. There is a wide variety of poisons, such as the *corrosives* (strong acids and alkalis) which cause local destruction of tissues; irritants which produce congestion of the organ with which they come in contact; the neurotoxins which affect the nerves or some of the basic processes within the cell; hemotoxins; hepatotoxins, and nephrotoxins. Consequently, the list of effective antidotes is long, reasonably complex, and usually much less lifesaving than making immediate efforts to have the poison victim vomit and thus expell as much of the poison as may be possible. Unless the appropriate antidote is selected, administration can be harmful rather than helpful. To illustrate this, if a sleep-producing drug, such as opium or morphine, has been taken in overdosage, it is best to keep the patient awake by giving strong coffee. In contrast, in the instance of strychnine poisoning, no stimulants should be given and the patient should be kept as quiet as possible. In every type of poisoning, immediate medical aid is essential. Most local health departments have lists of antidotes for common poisons.

ANTIFERROELECTRIC. Certain crystals, such as tungstic oxide WO_3, ammonium dihydrogen phosphate $(NH_4)H_2PO_4$, sodium niobate $NaNbO_3$, and disilver trihydrogen paraperiodate $Ag_2H_3IO_6$ have been shown to exhibit spontaneous microscopic polarization similar to that in ferroelectrics, except that different types of ions are polarized in different directions, so that the megascopic spontaneous polarization is small, or even vanishing. These materials are described as being antiferroelectric. Strictly the term is applied only to substances in which the net spontaneous polarization is zero; materials in which the polarizations of the individual ions cancel only in part are often referred to as *quasi-ferroelectrics* or as *ferrielectrics.* The relation of antiferroelectrics to ferroelectrics is similar to that of antiferromagnetics to ferromagnetics. See **Antiferromagnetism; Ferromagnetism.**

ANTIFERROMAGNETISM. The observed susceptibility curves of certain substances suggest that the system has gone into a state analogous to the ferromagnetic state, but with neighboring spins antiparallel, instead of parallel. See accompanying figure. That is, such substances exhibit a paramagnetism (low positive susceptibility) that varies with temperature in a manner similar to ferromagnetism, exhibiting a Curie point. Their resulting superlattices have been observed by neutron diffraction. The interaction giving preference to the antiparallel arrangement is believed to be an exchange force, similar to that invoked in the Heisenberg theory of ferromagnetism, but opposite in sign. Evidence from face-centered crystals has suggested the importance of super-exchange between next-nearest neighbors, through the anions. The alignment of the ions can be removed by heating the crystal and the temperature at which the ordered spin arrangement breaks down is called the *Néel temperature.*

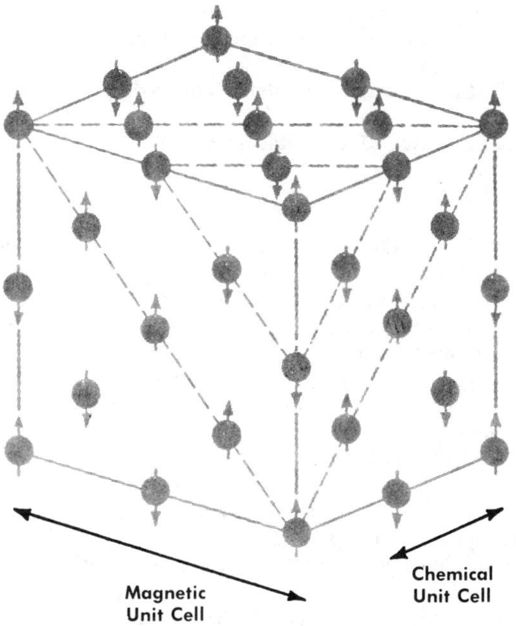

Antiferromagnetic state spin structure of manganese oxide as determined by Shull, Strauser, and Woolan.

ANTIFOAMING AGENTS. Defoaming Agents.

ANTIFOULING AGENTS. Various chemical substances added to paints and coatings to combat mildew and crustaceous formations, such as barnacles on the hull of a ship. In the past, large quantities of mercury compounds have been used in this manner. With growing environmental concern over possible mercury pollution, manufacturers have been turning to other, sometimes less efficacious compounds. Research continues to find compounds of a less toxic, but equally effective power of the mercury compounds. Bis(tributyltin) fluoride has been used on ship bottoms. See also **Mercury.**

ANTIGEN. A substance, usually a protein, a polysaccharide or a lipoid which when introduced into the body stimulates the production of antibodies. Bacteria, their toxins, red blood corpuscles, tissue extracts, pollens, dust, and many other substances may act as antigens. See **Antibody.**

ANTIGORITE. Serpentine.

ANTIGRAVITY PROPULSION. Rocket Propellants.

ANTIHEMOPHILIC FACTOR (AHF). Hemophilia.

ANTIHEMORRHAGIC VITAMIN. Vitamin K.

ANTIHISTAMINE. A synthetic substance essentially structurally analogous to histamine, the presence of which in minute amounts prevents or counteracts the action of excess histamine formed in body tissues. See also **Histamine.** Antihistamines are usually complex amines of various types. They find a number of medical uses.

In immediate hypersensitivity situations (reaction between antigen and antibody as encountered in hay fever, hives (urticaria), allergic (extrinsic) asthma, bites, drug injections, among others), antihistamines can be part of the effective therapy. Although widely used, antihistamines and steroids are not always the drugs of choice. In atopic dermatitis (chronic skin disorder), antihistamines may assist in breaking the itch-scratch cycle, particularly in persons whose sleep may be interrupted by pruritus. Antihistamine compounds for urticaria are also effective for atopic dermatitis therapy. Frequently, shifting from one antihistamine to another is effective and helps to reduce side effects of the drugs. Antihistamines are sometimes effective in the treatment of autoerythroyce purpura, a rare disease. Antihistamines are also used in connection with mild penicillin reactions, and in cases of penicillin desensitization procedures. Certain antihistamines find application to control mild parkinsonism.

Some antihistamines are particularly effective in alleviating the onset of motion sickness. Some antihistamines have been found helpful in relieving persistent, unproductive coughs that frequently accompany bronchitis or coughs associated with allergy. They are used in connection with perennial and seasonal allergic rhinitis.

Most antihistamines have anticholinergic (drying) and sedative side effects, sometimes producing marked drowsiness and reduction of mental alertness and thus should not be used by persons who operate machinery, drive vehicles, or otherwise must react quickly. Because of their similar structure, antihistamines appear to compete with histamine for cell receptor sites. Although conventional antihistaminic drugs, such as mepyramine, block the allergic and smooth muscle effects caused by histamine, the structure of these drugs is not sufficiently similar to histamine to inhibit histamine-stimulated gastric acid secretion. However, during the last few years, so-called histamine-blocking drugs have been developed which appear to be effective. It has been found that such compounds must contain the imidazole ring of histamine, with their potency enhanced by extension of the side chain. Among these new drugs are metiamide and cimetidine.

Some drugs in the antihistamine series play markedly different roles. Hydroxyzine hydrochloride and hydroxyzine pamoate have been used in the total management of anxiety, tension, and psychomotor agitation in conditions of emotional stress, usually requiring a combined approach of psychotherapy and chemotherapy. Hydroxyzine has been found to be particularly useful for making the disturbed patient more amenable to psychotherapy in long-term treatment of the psychoneurotic and the psychotic. The drug is not used as the only treatment of psychosis or of clearly demonstrated cases of depression. Hydroxyzine has also been found useful in alleviating the manifestations of anxiety and tension in acute emotional problems and in such situations as preparation for dental procedures. Hydroxyzine therapy has been used in treatment of chronic alcoholism where anxiety withdrawal symptoms or delirium tremens may be present. Hydroxyzine may potentiate narcotics and barbiturates.

Most conventional antihistamines are available for both oral and intravenous or intramuscular administration. In serious cases of urticaria (hives), for example, the injection rather than oral route is most effective. The major excretion route for most antihistamines is hepatic (liver), occurring within 4 to 15 hours.

In addition to the side effects previously mentioned, some antihistamines may cause neutropenia (neutrophil count in the blood is less than 1800 per cubic millimeter). Some antihistamines also may cause a modification of normal platelets in the blood.

Some of the more commonly used antihistamine compounds are listed below:

Ethanolamines:
- Diphenhydramine hydrochloride (Benadryl®)
- Dimenhydrinate (Dramamine®)

Ethylenediamines:
- Tripelennamine hydrochloride (Pyribenzamine®)

Alkylamines:
- Chlorpheniramine maleate (Chlor-Trimeton®)

Piperazines:
- Cyclizine hydrochloride (Marezine®)

Phenothiazines:
- Promethazine hydrochloride (Phenergan®)

Others:
- Cyproheptadine hydrochloride (Periactin®)
- Hydroxyzine hydrochloride (Atarax®)
- Hydroxyzine pamoate (Vistaril®)

References

Forno, L. S., and E. C. Alvord, Jr.: "The Pathology of Parkinsonism," in "Recent Advances in Parkinson's Disease," (F. H. McDowell and C. H. Markham, editors), F. A. Davis, Philadelphia, 1971.

Hanifin, J. M., and W. C. Lobitz: "Newer Concepts of Atopic Dermatitis," *Arch. Dermatol.*, **113**, 663 (1977).

Mathews, K. P.: "A Current View of Urticaria," in "Symposium on Allergy in Adults: Review and Outlook," (M. Samter, editor), *Med. Clin. North. Amer.* (1974).

Middleton, E. Jr., Reed, C. E., and E. F. Ellis: "Allergy Principles and Practice," C. V. Mosby, Saint Louis, Missouri, 1978.

Monroe, E. W., and H. E. Jones: "Urticaria: An Updated Review," *Arch. Dermatol.*, **113**, 80 (1977).

Rajka, G.: "Atopic Dermatitis," in "Major Problems in Dermatology," (A. Rook, editor), Vol. 3, W. A. Saunders, Philadelphia, 1975.

ANTIHUNTING CIRCUIT. A stabilizing or equalizing circuit used in a closed-loop feedback system to modify the response of the system in order that self-oscillations may be prevented.

An *antihunting transformer* is sometimes used in dc feedback systems as a stabilizing network. The primary of this transformer is in series with the load connected to the system. The secondary of the transformer has a voltage which is proportional to the derivative of the primary current, and is thus an appropriate signal to be fed back into some other part of the loop to prevent self-oscillations.

ANTIHUNT TRANSFORMER. Transformer.

ANTIHYPERTENSIVE DRUGS. Hypertension (High Blood Pressure).

ANTI-ICING FUEL ADDITIVE. Petroleum.

ANTI-INFLAMMATORY AGENTS. Rheumatoid Arthritis.

ANTIKNOCK RATING. Petroleum.

ANTILLES CURRENT. An ocean current, the northern branch of the north equatorial current flowing along the northern side of the Great Antilles carrying water that is identical with that of the Sargasso Sea. The Antilles current eventually joins the Florida current (after the latter emerges from the Straits of Florida) to form the Gulf Stream.

ANTILOCAPRINES. Pronghorn Antelope.

ANTILOGARITHM. Logarithm.

ANTIMALARIAL DRUGS. Malaria.

ANTIMATTER. Matter that consists of antiparticles. One of the great discoveries of modern physics was that for every type of elementary entity of matter and radiation (particle), there exists a corresponding conjugate type of entity (an antiparticle). In the antiparticle, certain of the particle-defining properties are identical (*conjugation invariant*), and others are reversed in sign (*conjugation reversing*). The reversed sign in a conjugation-reversing property allows one to maintain a conservation law for that property in the dramatic processes of *pair creation* and *pair annihilation* in which an antiparticle is observed to appear and disappear together with the particle to which it is conjugate. In those cases where all the conjugation-reversing properties occur with zero values, the antiparticle is identical with the particle. The progressive recognition of the existence of antiparticles was initiated by Dirac's relativistic antielectron theory in 1931, and by Anderson's independent experimental discovery of the antielectron (positron) in 1932.

Particles and (antiparticles) include: Electron (positron); Proton (antiproton); Electron neutrino (antielectron neutrino); Muon neutrino (antimuon neutrino); Neutron (antineutron); Positive pion (negative pion), etc.

The principle of charge conjugation symmetry states that if each particle in a given system is replaced by its corresponding antiparticle, then an observer will be unable to tell the difference. For example, if in a hydrogen atom, the proton is replaced by an antiproton and the electron is replaced by a positron, then this antimatter atom, if observed by persons also made of antimatter, will behave exactly like an ordinary atom. In an antimatter universe, the laws of nature could not be distinguished from the laws of an ordinary matter universe. However, it turns out that there are certain types of reactions where this rule does not hold, and these are just the types of reactions where conservation of parity breaks down. For an explanation of this, see long footnote in early portion of entry on **Particles (Subatomic).** Also see list of related entries at end of aforementioned entry.

ANTIMATTER (Celestial Objects). Gamma-Ray Astronomy.

ANTIMATTER ROCKETS. Rocket Propellants.

ANTIMERS. Amino Acids.

ANTIMETABOLITES. These substances fall into the general class of cytotoxic chemicals, i.e., agents that damage cells to which they are applied. Antimetabolites are so similar to normal enzymatic substrate molecules or metabolites as to gain entry into the cellular machinery of intermediary metabolism, but once there they differ enough to cause enzymatic inhibition. If incoporated into protein, nucleic acids, or coenzymes, for example, they will diminish the biological worth of those substances. Spectacular agents of this sort include the antifolic acids, such as aminopterin and amethopterin, various other vitamin analogs, and analogs of the naturally occurring purines, pyrimidines, nucleosides, and amino acids. Effective action against the integrity of the cell appears to be exerted at a number of points of intermediary metabolism in these multifarious antimetabolites. Of particular interest with many of them is an interference in normal nucleic acid metabolism. 5-Fluoro-2′-deoxyuridine, for example, acts to inhibit the synthesis of thymidylate, a necessary precursor of DNA, and the related 5-bromo-2′-deoxyuridine is actually incorporated into new DNA in the place of thymidine. Both of these agents increase the frequency of chromosomal disturbances. Various other base analogs, if incorporated into DNA, can lead to gene mutation by alteration of the normal sequence of nucleotides during replication through incorrect base pairing. 2-Aminopurine is an example of such a mutagen. 8-Azaguanine can be incorporated into ribonucleic acids, which are thus rendered defective. Among the actions of 6-mercaptopurine is an interference in the biochemical activity of coenzyme A, with resultant mitochondrial damage. Such amino acid analogs as β-fluorophenylalanine can effectively halt cellular activities by being incorporated into new proteins, which thereupon fail to attain their proper enzymatic or other functions.

Advantage is taken of the properties of antimetabolites in chemotherapy. In cancer chemotherapy, several antimetabolites are used. These include methotrexate, 6-mercaptopurine, 6-thioguanine, 5-fluorouracil, and cystine arabinoside. In the chemotherapy of metastatic breast cancer, 5-fluorouracil and methotrexate, in combination with cyclophosphamide, have been used. Antimetabolites, sometimes along with corticosteroids, are used in the therapy of various autoimmune diseases, such as thrombocytenic purpura, thyroiditis, Goodpasture's syndrome, among others.

Metabolites are implicated as agents that produce marrow aplasia as found in leukemia.

References

Carbone, P. P., et al.: "Chemotherapy of Disseminated Breast Cancer," *Cancer*, **39** (supplement), 2916 (1977).

Clarkson, B., Dowling, M. D., and T. S. Gee: "Treatment of Acute Leukemia in Adults," *Cancer*, **36**, 775 (1975).

Crowther, D.: "Blood and Neoplastic Diseases: A Rational Approach to the Chemotherapy of Human Malignant Disease," *Br, Med. J.*, **4**, 156 (1974).

Haskell, C. M., et al.: "Systemic Therapy for Metastatic Breast Cancer," *Ann. Intern. Med.*, **86**, 68 (1977).

Keiser, L. W., and R. L. Capizzi: "Principals of Combination Chemotherapy," in "Cancer, A Comprehensive Treatise," (J. F. Becker, editor), Vol. 5, Plenum, New York, 1977.

Lipton, A.: "Chronic Idiopathic Neutropenia: Treatment with corticosteroids and Mercaptopurine," *Arch. Intern. Med.*, **123**, 694 (1969).

Talal, N. (editor): "Autoimmunity: Genetic Immunologic, Virologic and Clinical Aspects," Academic, New York, 1978.

ANTIMICROBIAL AGENTS (Foods). These are substances added to food products, often in minute quantities, to destroy or inhibit the activity of microorganisms. These organisms are responsible for a high percentage of food spoilage and frequently severely limit the shelf life of food substances. See also **Foodborne Diseases.**

The fundamental mechanisms used by antimicrobial agents in acting against the microorganisms are (1) an adverse influence on the cellular membranes; (2) an interference with genetic mechanisms; and (3) an interference with cellular enzymes.

Among the principal antimicrobial agents used in foods are benzoic acid and sodium benzoate; the parabens; sorbic acid and sorbates; propionic acid and propionates; acetic acid and acetates; nitrates and nitrites; sulfur dioxide and sulfites; antibiotics; diethyl pyrocarbonate; epoxides; hydrogen peroxide; and phosphates. In addition to incorporation of antimicrobial agents directly into processed foods, the agents also are used in various processing stages, but are not always transferred to the food products per se.

Benzoic Acid and Sodium Benzoate. These compounds are most active against yeasts and are less effective against molds. They are best suited for foods with a natural or adjusted pH below 4.5. The average dosage in foods ranges between 0.05–0.1% (weight), depending upon product. Benzoic acid occurs naturally in cinnamon, ripe cloves, cranberries, greengage plums, and prunes. See also **Benzoic Acid.** For a number of years the sodium salt has been preferred over the acid by food processors.

Common applications for these compounds include carbonated and noncarbonated beverages, but excluding beers and wines because of their action against yeasts. They are also used in salted margarines, jams, jellies, and preserves, pie fillings, salads and salad dressings, pickles, relishes, and other condiments, as well as olives and sauerkraut. In terms of human metabolism of these substances, some authorities have suggested that benzoate is conjugated with glycine to produce hippuric acid, which is excreted, possibly accounting for 65–95% of benzoate ingested. It has been postulated that the remainder is detoxified by conjugation with glycuronic acid.

Parabens. These compounds include the methyl, ethyl, propyl, and butyl esters of para-hydroxybenzoic acid. In the United States and a number of other countries, the methyl and propyl esters are preferred, while European food processors favor the ethyl and butyl esters. The parabens were first described in 1924 as having antimicrobial activity and initially were used in cosmetic and pharmaceutical products.

The parabens are most effective against molds and yeasts, but less active against bacteria, particularly gram-negative bacteria. The antimicrobial activity of the parabens is directly related to the molecular chain length (methyl is weakest; butyl is strongest). However, the solubility of these compounds is in inverse relationship with chain length. These characteristics give rise to the use of two or more esters in combination and sometimes in combination with entirely different antimicrobial agents, such as sodium benzoate. Below a pH of 7, the parabens are only weakly effective.

The parabens are used in carbonated beverages and other soft drinks, including cider. In lieu of pasteurizing or using Millipore filtration, some brewers use the parabens for controlling secondary yeast formation. Because of their activity against yeasts, they are not used in bread and rolls, but they find use in other bakery products, such as pie crusts, certain pastries, icings, toppings, fillings, and cakes. The parabens are particularly effective in preserving fruit cakes. Usage also includes creams and pastes, fruit products, flavor extracts, pickles and olives, and artificially sweetened jams, jellies, and preserves. Average dosage ranges from 0.03–0.6% (weight).

Sorbic Acid and Sorbates. Sorbic acid and its potassium and sodium salts are effective against molds, but less effective against bacteria. These compounds may be incorporated directly into the food product, but they are frequently applied by spraying, dipping, or coating. The compounds are effective up to a pH of about 6.5. This is higher than propionates and sodium benzoate, but not so high as the parabens. Metabolism in humans parallels that of other fatty acids.

Because sorbates affect yeasts, the compounds are not directly useful in yeast-raised goods. Sorbates are particularly favored for use in chocolate syrups. They can be used in wine production in conjuction with sulfur dioxide against bacteria and are effective in inhibiting development of unwanted yeasts. They are also used in artifically sweetened jellies, jams, and preserves; in pickles and related products; in nonsalted margarines; in dried and smoked fish products; in semimoist pet foods; in dry sausage castings; in fruit-filled toaster pastries; and in cheese and cheese products. In the latter products, the agents usually are applied by dipping or spraying. Wrappers also may be impregnated with sorbates.

Propionic Acid and Propionates. The antimicrobial properties of propionic acid and its calcium and sodium salts were first noted in 1913. Today, the calcium and sodium salts are most commonly used. These compounds are more active against molds than sodium benzoate, but have little if any activity against yeasts. The propionates are well known for their effectiveness against *Bacillus messentericus*, a "rope"-forming microorganism. These compounds are effective up to a pH of 5 or slightly higher. Metabolism in the human body parallels that of other fatty acids.

An early application for the propionates was that of dipping cheddar cheese in an 8% propionic acid solution. This increased mold-free life by 4 to 5 times more than when no preservative was added. For pasteurized process cheese and cheese products, propionates can be added before or with emulsifying salts. Research has indicated that propionate-treated parchment wrappers provide protection for butter.

Use of propionates in breads can extend mold-free life by 8 days or more. Propionates are favored by bakers because of their effectiveness against ropy mold in breads up to pH levels of 6. For cakes and unleavened bakery goods, the sodium salt is usually preferred; for bread, the calcium salt is favored. This additive also contributes to the mineral enrichment of the product.

Acetic Acid and Acetates. Acetic acid (pure and as vinegar) and calcium, potassium, and sodium acetates, as well as sodium diacetate, serve as antimicrobial agents. In the United States, vinegar can contain no less than 4 grams of acetic acid per 100 milliliters of product. Acetic acid and calcium acetate are most effective against yeasts and bacteria, and to a lesser extent, molds. The diacetate is effective against both rope and mold in bread. It is interesting to note that the antimicrobial effectiveness of acetic acid and its salts is increased as the pH is lowered.

Optimal pH range varies with products and target microorganisms, but generally falls between 3.5 and 5.5. These agents are particularly effective against *Salmonella aertrycke, Staphylococcus aureus, Phytomonas phaseoli, Bacillus cereus, B. mesetericut, Saccharomyces cerevisiae,* and *Aspergillus niger.*

Unfortunately, to be effective against microorganisms in bakery products, acetic acid concentrations must be so high that an overly sour taste is imparted to the products. Sodium diacetate, however, can be used in small concentrations in bread and rolls to control rope and molds. Traditional concentrations of the acetate are 0.4 part to 100 parts of flour. During recent years, the propionates have largely displaced sodium diacetate for this use.

Vinegar or acetic acid is used in a number of products as much for its sour taste as for its antimicrobial properties. Such products include catsup, mayonnaise, pickles, salad dressing, and various condiment sauces. These agents also have been used to a lesser extent in malt syrups and concentrates, cheeses, and in the treatment of parchment wrappers for products, such as butter, to inhibit mold.

Nitrates and Nitrites. For many decades, sodium nitrate and nitrite, and potassium nitrate and nitrite have been used to cure, preserve, and provide a characteristic flavor to such meats as bacon, corned beef, frankfurters, ham, and various sausages. This tradition continues into the 1980s, but was seriously threatened in the mid- and late 1970s. Some researchers reported that N-nitrosopyrolidine (NPyr) formed in bacon upon application of heat during preparation for consumption. It was observed that there was a greater concentration of the NPyr in adipose tissue than in the lean portion. A connection was proposed that involved serious implications of the ultimate carcinogenic risk involved in meat treated with the nitrates and nitrites. Numerous tests proceeded. One of the main factors learned during this period was how little knowledge food scientists had concerning the fate of the nitrates and nitrites. Although the precursors of the nitrosamines formed under certain conditions of cooking were known, the mechanism of formation was unknown. As of the early 1980s, there is a period of quietude, with regulatory officials (United States) not fully convinced of taking serious measures that would result in serious breaks with tradition on the part of meatpackers and consumers alike. Perhaps with further investigations in a less tense atmosphere, solid scientific information will be developed to prove or disprove a number of points that remain unanswered.

Sulfur Dioxide and Sulfites. The use of sulfur dioxide gas and with it the production of sulfites differ somewhat from the other antimicrobial agents thus far described. Historical records show that burning sulfur to produce sulfur dioxide (SO_2 gas dates back to the ancient Egyptians and Romans who used it in connection with wine making). See **Sulfur.** Action by sulfur dioxide is accomplished in the gaseous phase. The effect of SO_2 is markedly determined by concentration and pH conditions of the target product. Research has demonstrated that most bacteria are inhibited by HSO_3^- at concentrations of 200 parts per million (ppm) or less. With few exceptions, yeasts are also similarly inhibited. There are, however, some strains of molds that are considerably more resistant. The sulfite salts tend to be unstable and oxidize during long periods of storage, thus decreasing the availability of SO_2. This process is aggravated by the presence of moisture.

The most effective range for optimal microbial inhibition with sulfites is a pH of 2.5 to 3. It has been found that from 2 to 4 times greater concentrations of SO_2 are needed to inhibit the growth of microorganisms at a pH of 3.5 than at a pH of 2.5. It also has been demonstrated that, at a pH of 7, SO_2 has little if any effect on yeasts and molds, even at concentrations up to 1000 ppm. The inhibitory effects against bacteria are also considerably less when pH rises above 3.5. Some researchers believe that at higher pH levels, penetration of cell walls is much more difficult.

Because residual levels of sulfites in excess of 500 ppm impart a noticeable taste to food substances, this fact, regardless of any regulations toward limiting concentration, requires SO_2 levels to be controlled.

The use of SO_2 for preserving fruit juices, sirups, concentrates, and purees is particularly attractive in regions with warm climates and where products must be stored in bulk prior to processing. In these situations, the SO_2 concentration will range between 350 and 600 ppm. High sugar concentrations require higher levels of SO_2. For optimal effectiveness, the pH of some products has to be reduced.

It is a common practice to expose many fruits to SO_2 prior to dehydration. The SO_2 also extends storage life of raw fruit prior to dehydration. The optimal temperature for exposure to the gas is from 43 to 49°C. Unlike fruit, vegetables are usually dipped in solutions of neutral sulfites and bisulfites. Suggested levels of SO_2 in some dried and dehydrated fruits and vegetables, in parts per million, are:

Apricots, peaches, and nectarines	2000
Raisins	800 to 1500
Nectarines	2000
Pears	1000
Apples	800
Cabbage	750 to 1000
Carrots and potatoes	200 to 250

It is important to note, however, that bulk-treated fruits intended for canning should not have a residual level in excess of 20 ppm SO_2 because of possible sulfide (black precipitate) forming in the can as the result of hydrogen sulfide generation.

Many countries do not allow use of SO_2 or sulfite salts for use on meats, fish, or processed meat and fish products. Where permitted, sulfite is helpful in eliminating "black spot" formation in shrimp.

A major use of sulfites is in wine making. It is used for sanitizing equipment and, prior to fermenting, the grape *musts* have to be treated with sulfites to inhibit the growth of any natural microbial flora present. This is done prior to the addition of pure cultures of the appropriate wine-making yeasts. While fermenting SO_2 also can function as an antioxidant, clarifier, and dissolving agent. Sulfur dioxide is often used after fermentation to prevent undesirable postfermentation alterations by various microorganisms. Levels of SO_2 during fermentation range from 50 to 100 ppm, depending upon condition of the grapes, temperature, pH, and sugar concentration. The wine industry uses sulfur dioxide dissolved in water, vaporized SO_2, and sulfite salts. An SO_2 level of 50–75 ppm assists the prevention of bacterial spoilage during the bulk storage of wine after fermentation.

Antibiotics. Much attention has been given to the use of antibiotics in food-associated applications since the introduction of penicillin in the 1940s. Their use has been limited. The use of antibiotics in animal feedstuffs continues to remain a topic of controversy in many countries; in other countries they have been banned. Antibiotics carry into the meat produced and further into human diets, thus possibly reducing their effectiveness in the treatment of human diseases.

Diethyl Pyrocarbonate. The preservative qualities of this compound were not recognized until the late 1930s and research on the compound continues to date. Also called pyrocarbonic acid diethyl ester, the compound is extremely effective against yeasts. It is also active against bacteria, such as *Lactobacillus pastorianus*, and various molds. The substance is generally used in still wines, fermented malt beverages, and noncarbonated soft drinks, as well as fruit-based beverages. Regulations on its use vary from one country to the next. Effective inhibition by diethyl pyrocarbonate is largely confined to acid products of low microorganism count. Some researchers point out that the pH should be less than 4 and that the microorganism count should not exceed 500 per milliliter. Some authorities observe that because of its rapid hydrolysis, no toxicity or residue problems should occur in products where the compound is permitted.

Epoxides. Two compounds are included in this category of antimicrobials. One is the gas, *ethylene oxide*; the other, *propylene oxide*, a colorless liquid with a boiling point of 35°C. Ethylene is highly reactive and must be used carefully and only with proper equipment. Somewhat less hazardous from an explosion standpoint, propylene oxide also has an explosive range of 2–22%. Consequently, these materials are usually mixed with inert substances, such as carbon dioxide or organic diluents.

Ethylene oxide is a universal antimicrobial in that it is lethal to all microorganisms. However, it is not universal from the standpoint of application. Propylene oxide is considered a broad-range microbiocide. In practically all aspects, propylene oxide is a considerably less effective agent, requiring longer exposures and greater concentrations because of its low penetrating power. However, propylene oxide is less toxic to humans.

The use of these gases (propylene oxide is volatilized) has been called "cold sterilization" and is frequently useful in sterilizing a number of low-moisture ingredients which end up on high-moisture foods. This prior sterilization lessens the total load on later thermal processing. The ability of these gases to kill microorganisms in low-moisture foods is an outstanding advantage. At the same time, macroorganisms also are killed. The gases find application in connection with spices, starches, nut meats, dried prunes, and glacé fruit. They are not used on peanuts (groundnuts).

Hydrogen Peroxide. Although usually regarded as a bleaching and oxidizing agent, this compound, H_2O_2, can be an effective antimicrobial and can be particularly useful in sterilizing processing equipment and packaging materials, notably prior to the aseptic packaging process. Regulations regarding the use of hydrogen peroxide vary from one country to the next.

Phosphates. The various phosphates are an effective multipurpose food additive chemical and functions other than their antimicrobial properties are usually given the greatest stress. The antimicrobial properties of the phosphates have been investigated over the years and are reasonably well documented.

References

Crocco, S. C.: "Sorbate Plant," *Food Engineering* (April 1977).

Gilliland, S. E., and M. L. Speck: "Inhibition of Psychotrophic Bacteria by Lactobacilli and Pediococci in Nonfermented Refrigerated Foods," *J. Food Sci.* **40**, 903 (1975).

Kraft, A. A., and C. R. Rey: "Psychotrophic Bacteria in Foods," *Food Technology*, **33**, 1, 66–71 (1979).

Miller, W. W.: "Yeasts in Food Spoilage," *Food Technology*, **33**, 1, 76–80 (1979).

Segner, W. P.: "Mesophilic Aerobic Sporeforming Bacteria in the Spoilage of Low-acid Canned Foods," *Food Technology*, **33**, 1, 55–59, 80 (1979).

Staff: "Shelf Life Increased by Sorbates," *Baking Industry* (May 1977).

Troller, J. A.: "Food Spoilage by Microorganisms Tolerating Low-a_w Environments," *Food Technology*, **33**, 1, 72–75 (9179).

ANTIMONY. Chemical element symbol Sb, at. no. 51, at. wt. 121.75, periodic table group 5a, mp 630.5°C, bp 1380°C, sp gr 6.62 (vacuum-distilled solid at 20°C) and 6.73 (single crystal). Naturally occurring isotopes are ^{121}Sb and ^{123}Sb. Antimony metal is a lustrous, silvery, blue-white solid, extremely brittle, and exhibiting a scalelike or flaky crystalline texture. The metal is easy to pulverize. The pure metal has a hardness of 3.0–3.3 on the Mohs scale and 55 on the Brinell scale. Of the more common metals, antimony is the poorest conductor

(4.5 on a scale of 100 for copper). From careful studies, it has been observed that Sb contracts upon solidification rather than expanding. The element was first described by Thölden (Valentine) in 1450.

There are two natural isotopes, ^{121}Sb and ^{123}Sb; and ten radioactive isotopes, ^{116}Sb through ^{120}Sb, ^{122}Sb, and ^{124}Sb through ^{127}Sb. ^{124}Sb is used as a radiation source in industrial instruments for the measurement of flow of slurries and interface measurements in pipelines. See also **Radioactivity.**

First ionization potential 8.64 eV; second 16.5 eV; third 25.3 eV; fourth 44.1 eV; fifth 56 eV. Oxidation potentials $Sb + H_2O \rightarrow SbO^+ + 2H^+ + 3e^-$, −0.212 V.,

$$2SbO^+ + 3H_2O \rightarrow Sb_2O_5 + 6H^+ + 4e^-,$$

−0.581 V., $Sb + 4OH^- \rightarrow SbO_2^- + 2H_2O + 3e^-$, 0.66 V. Other important physical properties of antimony are given under **Chemical Elements.**

Antimony exists in a number of allotropic forms. Gray or metallic antimony, density 6.79 g/per cm^3, is the stable form, forming rhombohedral crystals. Its vapor is that of Sb_4 up to 800°C, where dissociation to Sb_2 commences. Yellow antimony, Sb_4, density 5.3 g/cm^3 is less stable than yellow arsenic. It is produced by oxidation of stibine (see below) at very low temperatures, above which it is unstable. It changes even in the dark to black antimony at −90°C (in the light at −180°C). Black antimony, produced most readily by cooling antimony vapor or oxidizing stibine at 40°C, density 5.3, is metastable with respect to the gray form. It is also more reactive, igniting in air at room temperatures or above. Explosive antimony is produced by rapid electrodeposition of antimony from its halides. When heated or scratched, it undergoes an exothermic transformation to gray antimony. Its structure is amorphous, and differs somewhat from that of gray antimony.

Antimony is used in alloys, with lead for storage battery plates, with lead and tin in type metals and body solders, with tin and copper in bearing or antifriction metals. Antimony occurs chiefly as the sulfide (stibnite, Sb_2S_3) which is produced mainly in China, only small amounts in Mexico and Bolivia. Stibnite is (1) melted and reduced to antimony by iron metal and separated from fused ferrous sulfide (See also **Stibnite**); (2) roasted in air and sublimed antimonous oxide collected and reduced by heating to fusion with carbon and sodium carbonate.

Antimony is also leached from tetrahedryte ore and recovered by electrowinning.

Antimony is scarcely tarnished in dry air but oxidized slowly in moist air; burns at a red heat in air or oxygen with incandescence forming antimonous oxide; insoluble in HCl; converted by HNO_3 into antimonous oxide or antimonic oxide, depending upon the concentration of acid; by chlorine into trichloride or pentachloride, by NaOH solution into antimonite.

Stibine: SbH_3, is formed by hydrolysis of some metal antimonides or reduction (with hydrogen produced by addition of zinc and HCl) of antimony compounds, as in the Gutzeit test. It is decomposed by aqueous bases, in contrast with arsine. It reacts with metals at higher temperatures to give the antimonides. The antimonides of elements of group 1a, 2a, and 3a usually are stoichiometric, with antimony trivalent. With other metals, the binary compounds are essentially intermetallic, with such exceptions as the nickel series, Ni_2Sb_3, NiSb, Ni_5Sb_2 and Ni_4Sb.

Trihalides: SbF_3, $SbCl_3$, $SbBr_3$, and SbI_3, are solids, and have pyramidal structures. Except for the fluoride, which is not hydrolyzed, they undergo partial hydrolysis only (in contrast with the phosphorus trihalides) on contact with water to yield insoluble oxyhalides, either of composition SbOX or varying somewhat from this composition to give such compounds as $Sb_4O_5Cl_2$. The antimony pentahalides, SbF_5 and $SbCl_5$ can be prepared, but the pentabromide exists only in double compounds, known as bromoantimonates, those for monovalent metals being of the type $MSbBr_6$, plus water of hydration, and yielding $SbBr_6^-$ ions. $SbCl_6^-$ and SbF_6^- ions are also known. Mixture of antimony (III) chloride, $SbCl_3$, in HCl solution with antimony (V) chloride, $SbCl_5$, in equimolar proportions yields a dark colored solution. While antimony(IV) chloride cannot be isolated from it, compounds such as cesium antimony(IV) chloride, Cs_2SbCl_6 are formed by addition of cesium chloride, CsCl, and they are isomrophous with similar compounds of lead, tin, and other metals. However, tetravalent antimony should be paramagnetic because of the unpaired electron, whereas compounds of $SbCl_6^{2-}$ are diamagnetic. Therefore it may be that these compounds contain equimolar mixtures of $SbCl_6^-$ and $SbCl_6^{-3}$. The existence of these higher halide complexes with tin (and bismuth) but not with phosphorus or arsenic, may be due to steric considerations.

Antimony(III) oxide: Sb_2O_3 or Sb_4O_6, is formed by melting antimony in air, or from the hydroxide Sb $(OH)_3$. The Sb_2O_3 of commerce is produced from the oxidation of stibnite ore. Antimony is below arsenic in the periodic table, and $Sb(OH)_3$ is more definitely amphiprotic than $As(OH)_3$, forming not only antimony(III) salts and antimonites (containing the ion SbO_2^- or $Sb(OH)_4^-$), but also basic salts, especially the antimonyl salts, containing the ion SbO^+. Antimony(V) oxide, formed by oxidation of the metal with HNO_3, is less soluble in H_2O than As_2O_5. Antimonic acid cannot be obtained by hydration, and the product resulting upon hydrolysis of pentahalides has a variable H_2O content. The salts of the acid, the antimonates, are of the type $M^ISb(OH)_6$, as Pauling showed to be necessary to conform to accepted ionic radius ratios. Although the strength of antimonic acid has not been accurately determined, it appears to be comparable to acetic acid.

Antimony(IV) oxide: Obtained by heating in air the trioxide or the hydrated pentoxide.

There is a marked structural difference between the phosphates and the antimonates. Thus sodium pyroantimonate, $Na_2H_2Sb_2O_7 \cdot 5H_2O$ contains the ion $Sb(OH)_6^-$ rather than $Sb_2O_7^{4-}$, and the magnesium compound (hydated) which as a 12:1 ratio of oxygen to antimony, and would thus be a hexahydroxyantimonate, has the (X-ray determined) structure $[Mg(H_2O)_6][Sb(OH)_6]_2$.

Sulfides: Sb_2S_3 and Sb_2S_5, which may be obtained from the elements or by precipitation, respectively, of Sb(III) and Sb(V) solutions with H_2S. The Sb_2S_3 dissolves in alkaline solutions to form thioantimonites, containing the ion SbS_3^{3-}, or $Sb(SH)_6^-$, while Sb_2S_5 forms the thioantimonates, containing SbS_4^{3-}. The latter is probably present as $[SbS_2(SH)_2\text{-}(OH)_2]^{3-}$ or $[SbS_4(H_2O)_2]^{3-}$

In alloys, antimony is easily detected by its formation of a white solid upon treatment with concentrated HNO_3 and subsequent separation from tin, which is the only other metal thus forming a white solid.

Both trivalent and pentavalent antimony form several organoanti-

ANTIMONY CONTENT OF REPRESENTATIVE ANTIMONY-CONTAINING ALLOYS

Hard lead	Up to 12% Sb
Antimony reduces mp of Pb and hardens resulting alloy. Alloy has better abrasion resistance than chemical Pb at temperatures below 140°C. Alloy is age-hardenable.	
Tin-lead solders	Up to 1% Sb
Type metals	3–19% Sb
These Pb-base alloys also contain from 3–9% Sn.	
Lead-base diecasting alloys:	
[a]ASTM No. 4	14–16% Sb
ASTM No. 5	9.25–10.75% Sb
Bearing alloy	15% Sb
CT metal	12.5% Sb
Tin-free alloy	10% Sb
Babbitt (bearing) metals:	
[b] SAE 10	4–5% Sb
SAE 11	6–7.5% Sb
SAE 12	7–8.5% Sb
SAE 13	9.25–10.25% Sb
SAE 14	14–16% Sb
SAE 15	14.5–16% Sb
Britannia metal	5% Sb
This alloy also contains 93% Sn and 2% Cu. Very useful for spinning utensils.	
Pewter	Up to 7% Sb
Pewter also contains up to 20% Pb and 4% Cu with the remainder made up by Sn.	

[a] American Society for Testing and Materials
[b] Society of Automative Engineers

mony compounds. Some of these include methylstibine CH_3SbH_2 and the substitution product, methyldichlorostibine CH_3SbCl_2; phenylstibine $C_6H_5SbH_2$ and the substitution product, phenyldichlorostibine $C_6H_5SbCl_2$; methylantimony tetrachloride CH_3SbCl_4; phenylantimony tetrachloride $C_6H_5SbCl_4$; sodium methylantimonate $Na[CH_3Sb(OH)_5]$; sodium trifluoromethyl antimonate $Na[(CF_3)_3Sb(OH)_3]$; triethylstibine sulfide $(C_2H_5)_3SbS$; tetraphenylstibonium tetraphenylborate $[(C_6H_5)_4Sb][B(C_6H_5)_4]$; stibiobenzene $C_6H_5Sb{=}SbC_6H_5$; and lithium hexaphenylantimonate $LiSb(C_6H_5)_6$.

Uses: Representative alloys containing antimony are described in the accompanying table.

Metallic antimony is an effective pearlitizing agent for producing pearlitic cast iron. The principal use of antimony, however, is in the form of the oxide. Its major application is as a flame retardant for plastics and textiles. Other applications of importance are in glass, pigments, and catalysts.

Toxicity: The threshold limit value of antimony and its compounds is 0.5 milligram/cubic meter (as Sb). Antimony and its compounds used under conditions giving rise to dust, fume, and vapor should be carried out under proper ventilation. In handling antimony and its compounds, appropriate hygienic practices and good housekeeping should be observed. Stibine SbH_3, requires extreme caution in handling because it is very toxic. When using antimony and its compounds, reducing conditions, which may give rise to the undesired formation of stibine, must be avoided.

References

Carapella, S. C., Jr.: "Antimony" in "Kirk-Othmer Encyclopedia of Chemical Technology," Vol. 3, 3rd Edition, pp. 105–128, Wiley, New York, 1978.

Sneed, M. C., and R. C. Brasted: "Comprehensive Inorganic Chemistry," Vol. 5, Van Nostrand Reinhold, New York, 1956.

Wang, C. Y.: "Antimony," Chas. Griffin, London, 1952.

S. C. Carapella, Jr., ASARCO Incorporated, South Plainfield, New Jersey.

ANTIMONY ELECTRODE. An electrode of metallic antimony that has sufficient antimony oxide, Sb_2O_3, on its surface to function as an oxide electrode for the measurement of pH. With the advent of superior glass electrodes over the years, the antimony electrode is essentially of historic interest.

ANTIMONY SULFIDE. Stibnite.

ANTINODES (or Loops). The points, lines, or surfaces in a standing wave system where some characteristic of the wave field has maximum amplitude.

ANTIOXIDANT. Usually an organic compound added to various types of materials, such as rubber, natural fats and oils, food products, gasoline, and lubricating oils, for the purposes of retarding oxidation and associated deterioration, rancidity, gum formation, reduction in shelf life, etc.

Rubber antioxidants are commonly of an aromatic amine type, such as dibeta-naphthyl-para-phenylenediamine and phenyl-beta-naphthylamine. Usually, only a small fraction of a percent affords adequate protection. Some antioxidants are substitute phenolic compounds (butylated hydroxyanisole, di-tert-butyl-para-cresol, and propyl gallate).

When used in foods, antioxidants are highly regulated to extremely small percentages in most countries—down to the low fractions of one percent. Composition of the substrate, processing conditions, impurities, and desired shelf life are among the most important factors in selecting the best antioxidant system for a given food product. The desirable features of antioxidants may be summarized as (1) effectiveness at low concentrations; (2) compatibility with the substrate; (3) nontoxic to consumers; (4) stability in terms of conditions encountered in processing and storage, including temperature, radiation, pH, etc.; (5) nonvolatility and nonextractability under the conditions of use; (6) ease and safety in handling; (7) freedom from off-flavors, off-odors, and off-colors that might be imparted to the food products; and (8) cost effectiveness.

Mechanism of Oxidative Degradation. It could appear that inasmuch as oxidative degradation occurs in a variety of organic materials that are dissimilar in appearance and have entirely different applications and different properties, with degradation producing different effects, the oxidation mechanism itself might be different. Current knowledge indicates, however, that the mechanism of oxidative degradation is the same for all organic substances. They appear to degrade by the same free-radical mechanism.

Common examples of food oxidative degradation include products that contain oils and fats. For example, some antioxidants have made it possible to store groundnuts (peanuts) and other nuts, maize (corn) products, and bakery and cereal products on the shelf for periods well in excess of the four months that was considered the traditional limiting period prior to the appearance of such additives. Other examples of food products that tend to become rancid by way of oxidation include various meat-flavor stuffing mixes, cake mixes, unbaked cheesecake mix, and essentially all foods that incorporate lipids. The stability of natural fats and oils present in raw materials varies over a wide range and hence the amount of antioxidant required must be tailored to each product situation. Enzymatic "browning" is another example of oxidative degradation. The enzymes in fruits and vegetables cause apples, apricots, and potatoes, among others, to darken when they are exposed to air after being cut, bruised, or allowed to overmature. Some antioxidants can prevent or delay enzymatic browning much in the same manner as dipping freshly cut fruits in lemon, orange, or pineapple juice. Limonene and ascorbic acid naturally present in these juices serve as antioxidants. Oxidative changes may affect carbohydrate, protein, and fat substances, the primary building blocks of foodstuffs, but generally the oxidative rancidity problem results mainly from the *autoxidative* degradation of fatty (glyceridic) components.

Some authorities describe oxidation as a free-radical, chain-type reaction. At usual processing temperatures and more slowly at room temperature, organic free radicals (R•) are formed. These react with oxygen to form peroxy radicals (ROO•), which can abstract a hydrogen atom from the affected substance to form a hydroperoxide (ROOH) and another organic free radical. The cycle repeats itself with the addition of oxygen to the new free radical. The unstable hydroperoxides left along with the substance are the major source of degradation. Under the influence of heat, light, and any metals if present, the hydroperoxides decompose to form carbonyl groups. When this happens, the organic molecule breaks and splits off another organic free radical. Ultimately, this type of degradation can lead to rancidity and color deterioration in oils and fats.

An antioxidant ties up the peroxy radicals so that they are incapable of propagating the reaction chain or to decompose the hydroperoxides in such a manner that carbonyl groups and additional free radicals are not formed. The former, which are called *Chain-breaking antioxidants, free-radical scavengers,* or *inhibitors,* are usually hindered phenols or amines. The latter, called *peroxide decomposers,* are generally sulfur compounds or organophosphites. A number of antioxidants useful in rubber and plastics, for example, are not suited to food products because of their toxicity.

A mixture of two antioxidants often will display synergism. Probably the most generally effective mixtures of antioxidants are those in which one compound functions as a decomposer of peroxides (sulfides, thiodiproprionate) and the other as an inhibitor of free radicals (hindered phenols, amines). Although the latter retards the formation of reaction chains, some hydroperoxide is nevertheless formed. If this hydroperoxide then reacts with a decomposer of peroxides, instead of decomposing into free radicals, the two antioxidants act together to complement each other. Moreover, the peroxide decomposer may itself be subject to oxidation by peroxy radicals, and its efficiency will therefore be increased in the presence of an inhibitor of free radicals. In the case of phenol-sulfide mixtures, the sulfide (peroxide decomposer) also continuously regenerates the phenol (radical scavenger) to accentuate the synergistic nature of the mixture. Metal chelators or deactivators, such as citric and phosphoric acids, of prooxidant metals (iron, copper, nickel, tin), ultraviolet-light absorbers (carbon black, substitute benzophenones, benzotriazoles, and salicylates), and antiozonants (substituted phenylenediamines) also develop synergistic effects with antioxidants.

Applications of Antioxidants. The use of antioxidants in foods, pharmaceuticals, and animal feeds (direct feed additives), as well as their use in food-contact surfaces (indirect additives) is closely regulated by the governments of several countries. Antioxidants are approved only after extensive extraction, toxicological, and feeding studies. The list is relatively limited. Although antioxidants have been used for several decades and some occur naturally in food substances, intensive research in continuing, partly accelerated by the growing use of unsaturated oils in numerous food products.

Butylated hydroxyanisole (BHA) was first used in food products in 1940. This continues as one of the commonly used antioxidants, sometimes in combination with butylated hydroxytoluene (BHT), propyl gallate, citric, or phosphoric acids, to obtain a synergistic effect. In food-contact surfaces, BHT has been used by itself or in combination with thiodipropionates and/or phosphoric acids, to obtain a synergistic effect. Well over $50 million of antioxidants are produced per year commercially in the United States alone.

The value of antioxidant protection by way of natural food sources has been pointed out in the literature with considerable frequency. Among the components of soy flour known to have some antioxidant properties are isoflavones and phospholipids. Amino acids and peptides in soybean flour also possess some antioxidant activity. There also may be some antioxidant impact from aromatic amines and sulfhydryl compounds.

Rosemary and sage have been shown to have effective antioxidant properties. The extracts in the past have been of strong odor and bitter taste and thus unsuited for use in most food products. However, solvent extraction procedures have been developed to produce purified antioxidants from rosemary and sage.

For many years, in connection with certain food products, a barrier to freeze-drying has been the problems associated with the storage stability of foods that are susceptible to lipid oxidation. In order for such foods to have a reasonable shelf life and acceptable flavor characteristics, protective additives which retard oxidation, are often added before dehydration. Such antioxidants must carry through the process and not be lost because of volatilization. For these applications, BHA, BHT, and tert-butylhydroquinone (TBHQ) have been found quite effective. See also **Polymeric Food Additives.**

References

Chang, S. S., et al.: "Natural Antioxidants from Rosemary and Sage," *J. Food Sci.*, **42**, 4, 1102–1106 (1977).

Hammerschmidt, P. A., and D. E. Pratt: "Phenolic Antioxidants of Dried Soybeans," *J. Food Sci.*, **32**, 2, 556–559 (1978).

Regnarsson, J. O., et al.: "Accelerated Temperature Study of Antioxidants," *J. Food Sci.*, **42**, 6, 1536–1539 (1977).

Regnarsson, J. O., et al.: "Accelerated Shelf Life Testing for Oxidative Rancidity in Foods," *J. Food Chemistry* (1978).

Waletzko, P. T., and T. P. Labuza: "Accelerated Shelf Life Testing of an Intermediate Moisture Food in Air and in an Oxygen-free Atmosphere," *J. Food Sci.*, **41**, 1338 (1976).

ANTIOXIDANT (Fuel). Petroleum.

ANTIOXIDANT (Polymeric). Polymeric Food Additives.

ANTIOXIDANT (Rubber). Rubber (natural).

ANTIPARALLEL. Having opposite senses. Thus the vectors **A** and **B** are antiparallel, while the vectors **C** and **D** are parallel.

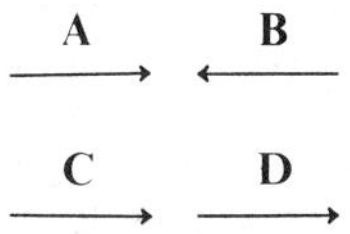

ANTIPARTICLES. One of the great discoveries of modern physics is that for every type of elementary entity of matter and radiation (particle), there exists a corresponding conjugate type of entity (antiparticle). In the antiparticle, certain of the particle-defining properties are identical (*conjugation-invariant*) and others are reversed in sign (*conjugation-reversing*). The reversed sign in a conjugation-reversing property allows one to maintain a conservation law for that property in the dramatic processes of *pair creation* and *pair annihilation* in which an antiparticle is observed to appear and disappear together with the particle to which it is conjugate. In those cases where all the conjugation-reversing properties occur with zero values, the antiparticle is identical with the particle. The progressive recognition of the existence of antiparticles was initiated by Dirac's relativistic antielectron theory in 1931, and by Anderson's independent experimental discovery of the antielectron (the positron) in 1932. See also **Particles (Subatomic).**

ANTIPHLOGISTIC. Analgesics.

ANTIPODAL CELLS. The three usually small cells which occur in the embryo sac of angiosperms at the end most distant from the micropyle. No known function has been ascribed to them.

ANTIPODE. With respect to any given point, the opposite point is known as the antipode. In terms of a sphere, Y would be the antipode of X of a line XY drawn through the center of the sphere and intersecting the two opposite surfaces. Antipodes often refer to regions rather than specific localized areas. Thus, the British Isles are approximately antipodal with Australia and New Zealand.

ANTIPODES (Optical). Amino acids.

ANTIPROTON. An elementary particle having a mass equal to that of the proton, differing from the proton only in the sign of its charge, which is negative, and a magnetic moment oppositely directed with respect to its spin. Positive identification of the antiproton was first made at the University of California. In 1959 Segre and Chamberlain received the Nobel Prize in physics for this discovery. Protons which had been accelerated to an energy of 6200 MeV in the bevatron, the proton synchrotron of the University of California Radiation Laboratory at Berkeley, were allowed to collide with a copper target. Negatively charged particles coming out in a forward direction from this collision were selected and separated in momentum by a focusing and analyzing magnet system to provide a beam of negative particles of known momentum. After a time of flight of about one-tenth of a microsecond, this beam may be expected to consist mainly of negative pions and muons, with some negative kaons (mass about 965 electron masses) and possibly negative protons. These particles were then distinguished both by measurement of their time of flight from the target (since particles of different mass have different velocities for given momentum) and by means of a device measuring the velocity of each particle passing through by the angle of its Cerenkov radiation. In this way the presence of negative particles with protonic mass (within about 10%) and distinct from the known kaons and hyperons was established. Their rate of production for the momentum and direction of this experiment was about one negative proton for every 50,000 negative pions with the same momentum and direction. See also **Particles (Subatomic);** and **Proton.**

ANTIPSYCHOTIC DRUGS. Schizophrenia.

ANTIPYRETIC. Any physical agent or drug that lowers the temperature of the body. Among antipyretics used are aspirin, antipyrine, acetanalid, and phenacetin. See also **Analgesics.**

ANTIREDEPOSITION AGENT. Detergents.

ANTIRESONANCE (or Parallel Impedance). In general, a condition of maximum impedance, as results when two or more impedors are connected in parallel, under such conditions that (for resistanceless impedors), Z approaches infinity when $\omega_1^2 = 1/LC$.

ANTISEPTIC. Antimicrobial Agents (Foods).

ANTI-SIDE-TONE (Telephone). In telephone equipment of early design, the sound going into the transmitter (mouthpiece) could be heard in the receiver. This is known as side-tone and has some objectionable features. It causes the speaker unconsciously to lower the voice and thus reduces the energy transmitted, and it also increases the effect of any local noise by causing it to appear in the receiver. It is thus desirable to eliminate or at least reduce the side-tone and special circuits and equipment have been devised. These sub-sets are known as anti-side-tone sets. Figure (a) shows a conventional connec-

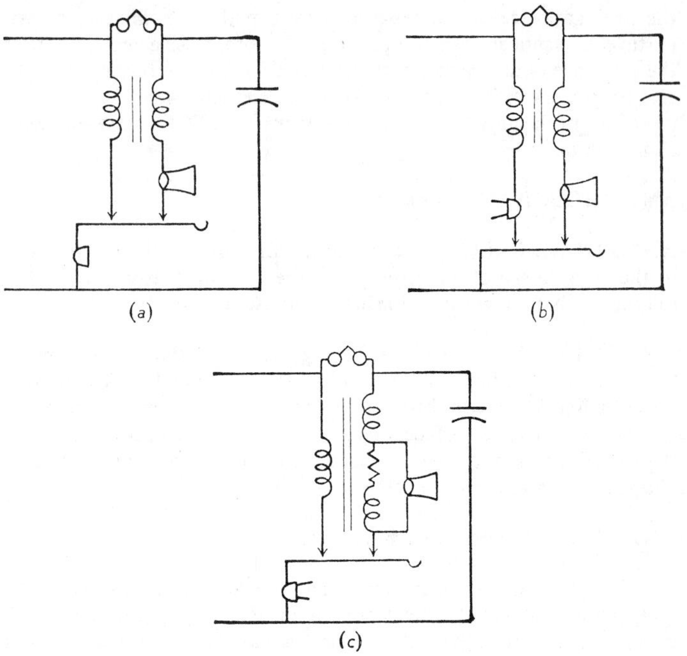

Subset connections in anti-side-tone sets.

tion, (b) a side-tone reduction circuit, and (c) an anti-side-tone circuit. In the latter, the values of the transformer windings and the resistances are so proportioned that the side-tone voltages cancel in the receiver while the incoming signal from the line is impressed on the receiver.

ANTISMOKE ADDITIVES. **Petroleum.**

ANTISOLAR POINT. The point on the celestial sphere that lies directly opposite the sun from the observer, i.e., on the line from the sun through the observer. See also **Celestial Sphere.**

ANTISTATIC ADDITIVES (Fuel). **Petroleum.**

ANTISTOKES LINES. When all the molecules are in the normal state, the transitions caused by exciting radiation are those in which the scattered or fluorescent light is of the same or lower frequency than the incident light (Stokes Lines). However if some of the atoms or molecules are in states other than normal, lines of frequencies higher than that of the incident light (Antistokes Lines) may result. See **Raman Spectrometry.**

ANTISYMMETRIC. In mathematics, a term used to denote a function which is transformed into its negative when the variables of the function are interchanged in pairs. In physical science, the term antisymmetric is applied to any physical system in which each point has properties opposite to those of a point symmetrically located with respect to it, e.g., an electric dipole is antisymmetric in its charge distribution. See **Symmetric.**

ANTITHYROID DRUGS. **Thyroid Gland.**

ANTITOXIN. (1) A substance made and elaborated in the body to neutralize a specific bacterial, plant, or animal toxin; (2) one of the class of specific antibodies. See also **Antibody.**

The history of the development of antitoxins in combating bacterial infection dates back to the early beginnings of organized bacteriology. Behring was the first to show that animals that were immune to diphtheria contained, in their serum, factors which were capable of neutralizing the poisonous effect of the toxins derived from the diphtheria bacillus. While this work was carried out in 1890, prior to many of the great discoveries of mass immunization, and much later the antibiotics, it is interesting to note that there remains a place for antitoxins in medical treatment or prophylaxis for some diseases, such as tetanus and botulism.

The more important approach to immunity to infectious disease now is the development of *active immunity* by injection of a vaccine. The vaccine may be either an attenuated live infectious agent, or an inactivated or killed product. In either case, protective substances are generated in the bloodstream called antibodies which help to neutralize the infectious agent when it is introduced. The principle of *passive immunization*, on the other hand, involves the development of the antibodies in another host and most frequently a different species as well. The antiserum or antitoxin (from the other host) is employed in preventing the onset of the disease, or in actual treatment of the active infection in subjects who have not had the advantage of becoming actively immunized due either to neglect or to the fact that an effective vaccine was not available. The use of antitoxins prepared in another species (for example, horse) is not without some element of risk.

Antitoxins are prepared by injecting the donor animals with frequent and increasing doses of toxin while maintaining a level at each injection that the animal can tolerate. The initial doses are critical since these toxins may be among the most poisonous agents known. In one technique, the toxin is diluted so that the first injection contains less than the minimum lethal dose. Other programs used toxins that are inactivated with formaldehyde so that they are no longer poisonous, but may still elicit an immune response and result in antitoxin that will neutralize the unaltered toxin. This method of inactivation also was developed prior to the twentieth century for preparing many of the important vaccines against diseases, such as influenza, tetanus, and diphtheria.

Small laboratory animals are helpful in carrying out research in this field, but in commercial production, larger animals are required. The horse proved to be very satisfactory for large-scale production of antitoxins, and is still used for the major portion of antitoxin production, although some material is also derived from cattle. The management of the animals and the schedules and dosages are perhaps as much an art as an exact science. The selection of the horses that have the best potential as antitoxin producers is also a critical factor. When a horse is receiving the maximum level of toxin in the hyperimmunization program, it may be injected in a single dose with enough toxin to be fatal for 100 million mice if the material was suitably distributed.

A means for enhancing the potency of the antitoxin in horses is to use an agent called an *adjuvant.* Several adjuvants have been used, including tapioca, mineral oil, and aluminum hydroxide. The mechanism by which these agents increase the intensity of the immune response is not fully understood, but local inflammatory reaction and the resulting slower release of the injected material from the original site appear to play a role. Adjuvants also are sometimes used with vaccines for human use. It has been recorded that one horse, during an eleven-year period, gave 657 gallons (~25 hectoliters) of blood from which tetanus antitoxin and, at different times, pneumococcus antiserum, was prepared. The volume of serum removed from the horse can be increased by the return of the red cells after removal of the plasma or liquid component. It has been shown that if the red cell level can be maintained in human donors of special serums, they can safely give as much as one liter of serum per week.

The injection of serum components of another species into human patients has not been without problems. A condition known as serum sickness develops in an alarmingly high proportion of those treated in this way. Serum sickness is apparently due to a generalized sensitivity which develops to the foreign serum protein. The onset of illness is usually delayed for several days after the injection of antitoxin and symptoms may be quite severe. The death rate due to this condition has been estimated as high as 1 in 50,000 (study made in 1936). Parfentjev and Pope developed methods for treating the antitoxin with a protein enzyme (pepsin), with the objective of breaking down the protein molecule to prevent serum sickness, yet retain the effectiveness of the antitoxin. This procedure, however, did not prove to be a guarantee against serum sickness in all subjects. This is not a problem in connection with human donors.

See also **Immune System and Immunology.**

ANTITRADE WIND. **Winds and Air Movement.**

ANTITRIPTIC WIND. **Winds and Air Movement.**

ANTIVENIN. **Snakes.**

ANTIVIRAL DRUGS. Compounds for use in the treatment of viral infections are quite limited. A problem in the development of antiviral drugs centers around the relationship between the replication function of the virus and functions of host cells. Obviously, a drug must target the virus and virus-infected cells with no destruction of healthy cell functions. Among antiviral drugs currently used are amantadine hydrochloride, idoxuridine, and adenine arabinoside. Research activity in the antiviral drug field is vigorous because of the obvious great need for them.

Amantadine Hydrochloride. Chemically, this drug is l-adamantanamine hydrochloride. A commercial preparation is known as *Symmetrel*®. This drug is indicated in the preventing (prophylaxis) and symptomatic management of respiratory tract illness caused by influenza A virus strains. The drug is particularly considered in connection with high-risk patients, close household or hospital ward contacts of index cases, and patients with severe influenza A virus illness. In the prophylaxis of influenza due to A virus strains, early immunization as periodically recommended by public health authorities is the method of choice. When early immunization is not feasible, or when the vaccine is contraindicated or not available, amantadine hydrochloride is sometimes used chemoprophylactically with inactivated influenza A virus vaccine until protective antibody responses develop. Principal contraindications include hypersensitivity to the drug and patients with a history of epilepsy, congestive heart failure, or peripheral edema.

To date, amantadine hydrochloride has not been used extensively clinically, although excellent controlled trials indicating that it has a prophylactic effect have been reported. Infected persons have been reported to suffer less cough, sore throat, and fever than other persons not taking the drug. Reports also indicate the drug accelerates recovery from peripheral airway abnormalities in normal persons who have uncomplicated influenza.

In the treatment of mild parkinsonism, amantadine has been used to advantage, particularly in conjunction with L-dopa. It is believed that amantadine stimulates the release of dopamine from nerve terminals.

Side-effects from the continuous use of amantadine include dizziness, drowsiness, difficulty in thinking, hallucinations, convulsions, and, rarely, psychosis. The effects are more commonly experienced by middle-age groups and the elderly.

Idoxuridine. Chemically, this drug is 5-iodo-2′-deoxyuridine. A commercial preparation is known as Stoxil®. This drug is indicated for the treatment of herpes simplex karatitis. The drug inhibits replication of herpes simplex virus by irreversibly inhibiting the incorporation of thymidine into the viral DNA. Although tests with rabbits of known genetic ancestry have been completed with no malformations resulting from idoxuridine, the drug is still administered with caution in pregnancy or in women of childbearing potential. The commercial compound is available in the form of a solution or as an ophthalmic ointment, with petrolatum used as an inactive ingredient. While the compound will frequently control infection, it apparently has no effect on the accumulated scarring, vascularization, or on the resultant progressive loss of vision.

Adenine Arabinoside. Also known as vidarabine, or Vira-A®, has the empirical formula, $C_{10}H_{13}N_5O_4 \cdot H_2O$. The chemical name is 9-β-D-arabinofuranosyladenine monohydrate. This compound is indicated in the treatment of herpes simplex virus encephalitis. Controlled studies have indicated that therapy with this drug may reduce the mortality caused by herpes simplex virus encephalitis from 70 to 28%. The therapy does not appear to alter morbidity and resulting serious neurological sequelae in the comatose patient. Thus, early diagnosis and treatment are essential. Herpes simplex virus encephalitis should be suspected in patients with a history of an acute febrile encephalopathy associated with disordered mentation, altered level of consciousness, and focal cerebral signs. Licensing for this use (United States) was made in December 1978. The drug was licensed in 1977 for treatment of acute herpes simplex keratoconjunctivitis and recurrent epithelial keratitis caused by herpes simplex types 1 and 2. Experimentally, under controlled conditions, the drug has been used for treatment of herpes simplex type 1 encephalitis, herpes simplex type 2 infections, herpes zoster, smallpox, progressive multifocal encephalopathy, and chronic hepatitis B virus and cytomegalovirus infections. Large doses of adenine arabinoside administered parenterally have been noted to suppress bone marrow function, particularly if administered over long periods.

Still considered experimental among antiviral drugs are methisazone, human leukocyte interferon, acyclovir (Zovirax®), and cytosine arabinoside.

References

Baker, C. E.: "Physicians' Desk Reference," 34th Edition, Medical Economics, Oradell, New Jersey, 1980.

Bauer, D. J., et al.: "Prophylaxis of Smallpox with Methisazone," *Amer. J. Epidemiol.*, **90**, 130 (1969).

Greenberg, H. B., et al.: "Effect of Human Leukocyte Interferon on Hepatitis B Virus Infection in Patients with Chronic Active Hepatitis," *New Engl. J. Med.*, **295**, 517 (1976).

Jones, B. R., et al.: "Efficacy of Acyloguanosine Against Herpes Simplex Corneal Ulcers," *Lancet*, **1**, 243 (1979).

Merigan, T. C. (editor): "Antivirals with Clinical Potential," Univ. of Chicago Press, Chicago, 1976.

Merigan, T. C., et al.: "Human Leukocyte Interferon Therapy of Herpes Zoster in Patients with Cancer," *New Engl. J. Med.*, **298**, 981 (1978).

Pavan-Langston, D., et al. (editors): "Adenine Arabinoside: An Antiviral Agent," Raven Press, New York, 1975.

Whitley, R. J., et al.: "Adenine Arabinoside Therapy of Biopsy-Proven Herpes Simplex Encephalitis," *New Engl. J. Med.*, **297**, 289 (1977).

ANTLERITE. Antlerite is a relatively uncommon mineral found within the oxidized zones of copper deposits in arid regions. It is a basic sulfate of copper $Cu_3(SO_4)(OH)_4$, crystallizing in the orthorhombic system. Hardness of 3.5, specific gravity 3.88, with vitreous luster, and emerald-green to black-green color.

Originally found in Arizona. It is the principal copper ore mineral at Chuquicamata, Chile.

ANT LION (*Insecta, Neuroptera*). Immature insects of the family *Myrmeleonidae* which lie buried at the apex of conical pits in dry sand or dust. Ants or other small insects which enter the pit slide down the loose slope and are seized by the upturned jaws below.

These insects are black or brown in color. They develop within a cocoon made of silk threads developed from larvae at the insect's tail. The immature insect remains in the cocoon stage until it becomes an adult.

ANT-LOVING CRICKET (*Insecta, Orthoptera*). *Myrmecophilia.* Small peculiarly formed crickets which live in ant nests.

ANTONOFF RULE. The tension at the interface between two saturated liquid layers which are in equilibrium is equal to the difference between the individual surface tensions against air or vapor of the two saturated solutions. This rule is approximate only and a number of exceptions are known.

ANTOZONITE. **Fluorite.**

ANTRA. **Sinus.**

ANURA. The frogs, toads, and allied species; a division of the *Amphibia* characterized by the absence of the tail and a tadpole larval stage. Also known as the Salientia from their jumping powers.

ANUS. The terminal portion of the rectum, which forms an external opening. See also **Digestive System (Human).**

ANVIL CLOUD. **Clouds and Formation.**

ANXIETY REACTION. A type of neurosis in which there is some degree of depression accompanied by worry, apprehension, and agitation; generalized dread often accompanied by somatic symptoms. Anxiety has a beneficial function. Whether it warns of external threat or inner emotional stress, the function of anxiety is to alert the individual to danger. Some degree of anxiety is normally encountered by everyone in the process of growth, with each succeeding separation from familiar surroundings. The emotionally healthy individual is able to turn anxiety into constructive use, recognizing it as a danger signal and following its warning by effecting changes in the disturbing life situations.

The emotionally disturbed individual, however, may become the victim of a chronic state of anxiety which cannot be overcome because of inner conflicts. This constitutes the neurosis known as the anxiety reaction.

In the anxiety reaction, the individual is conscious of a state of tension, but unable to locate the source of the impending threat. This has been described by psychiatrists as "free-floating" anxiety, because it cannot be pinned down to any particular source. The individual simply cannot find peace of mind because of a pervading sense of uneasiness. The symptoms of agitation are frequently apparent to others. The person may appear frightened and excited. Among the person's physical responses may be a racing pulse, pounding heart, profuse sweating, and heavy breathing. The individual may feel weak and dizzy and complain of vague disturbance of the digestive tract. The patient sometimes will not know exactly what is the matter, but there may be a feeling of imminent death. The individual may suffer nightmares which leave the person with a feeling of nameless dread. These overt signs of apprehension indicate the presence of some emotional conflict of which the individual is not aware.

Most authorities hold that the roots of chronic anxiety are laid in childhood, often unintentionally, by overly rejecting or overly permissive handling on the part of the parents. Parents who are undemonstrative usually make a child feel unfavored and insecure. An adult can often appraise threats to emotional security and take measures to overcome them, but a child cannot do so—at least as well. The discrepancy between expectation and reality is a source of conflict and resentment in the child. The child responds by a show of hostility directed against those persons whom the child also loves. Awareness of hostility in turn creates feelings of guilt and fear. The child may anticipate punishment and yet may feel that punishment is deserved. Although the child tends to repress feelings of hostility, this may not be entirely successful in eliminating the accompanying guilt. The resulting anxiety becomes a pattern of reaction to many troublesome life situations. Emotional conflict is at the root of the anxiety reaction. According to May, anxiety always involves two contradictory desires within the individual. The anxiety reaction cannot be dispelled, therefore, without eliminating the source of inner conflict.

The association of anxiety with coronary disease has been reported by C. D. Jenkins et al. (*N. Engl. J. Med.*, **294**, 987, 1976). The use of benzodiazepines in the treatment of anxiety has been reported by J. F. Tallman et al. (*Science*, **207**, 274–281, 1980).

AORTA. The main and largest blood vessel of the arterial blood system. It arises from the left ventricle of the heart, and arching over the root of the left lung, descends along the vertebral column, passing through the chest, and pierces the diaphragm into the abdominal cavity, finally dividing into the right and left iliac arteries in the pelvis. The branches of the aorta supply oxygenated blood to every part of the body. See **Arterial and Venous Disorders;** and **Heart and Circulatory System (Human).**

AORTAL ANEURYSM. Aneurysm.

AORTIC VALVE STENOSIS AND REGURGITATION. Heart and Circulatory System (Human).

AOUDAD. Goats and Sheep.

APATITE. The mineral apatite is a phosphate of calcium with either fluorine or chlorine or sometimes both, hence the distinction between fluor-apatite and chlor-apatite. Sometimes both fluorine and chlorine are present. Most apatite is, however, fluorapatite, $Ca_5(PO_4)_3F$.

Apatite crystallizes in the hexagonal system in prismatic and tabular forms. Hardness, 4.5–5; specific gravity, 3.17–3.23; luster, vitreous to resinous; transparent to opaque; streak, white; cleavage, imperfect basal and prismatic; color, white, green, yellow, red, brown and purple; sub-conchoidal fracture. The variety called asparagus stone is yellow-green and manganapatite which is a dark bluish-green may contain as much as 10% manganese dioxide replacing the calcium. Werner devised the name apatite from the Greek word meaning to deceive, as it was frequently mistaken for beryl and other species. Apatite has been found widely distributed both geographically and petrologically as it occurs in many sorts of rocks, metamorphic limestones, gneisses, schists, granites and syenites, pegmatite veins and even with iron ores. It has been prepared artifically. It has been mined for the manufacture of fertilizers and to a slight extent for jewelry.

Apatite occurs extensively in Europe and America, especially in New England, New Jersey, New York, North Carolina, California, and in the provinces of Ontario and Quebec in Canada.

See also terms listed under **Mineralogy.**

APERTOMETER. A device designed by Abbe for measuring the numerical aperture of microscope objectives.

APERTURE. Qualitatively, any opening through which radiation or particle fluxes may pass. In optical instruments, in communications and in other fields, the term aperture has acquired various specific meanings. For example, the aperture of a lens is simply the diameter of the lens. However, the numerical aperture (N.A.) of a lens is the quantity $n \sin u$, where u is the angular radius of the lens as seen from a point on the optical axis at the object, and n is the refractive index of the medium between the object and the lens. The relative aperture (or f-number) is the ratio of the focal length of an optical system to the diameter of the entrance pupil. A related term is the aperture angle which is the angle subtended by the radius of the entrance pupil of an optical instrument at an axial object point.

The aperture of a unidirectional antenna is that portion of a plane surface near the antenna, perpendicular to the direction of maximum radiation, through which the major part of the radiation passes.

APERTURE CARD. A punched card (see **Input/Output Devices)** which may contain information solely for machine storage and retrieval, this being the normal interpretation of the term. In addition, however, an aperture card may contain information that is not entered into electronic storage, such as small reproductions of diagrams, photos, etc. Thus, some of the information on the card may be fully electronically processable, whereas other information requires manual scanning and interpretation.

APES. Anthropoids.

APHANITE. An aphanite is any fine-grained igneous rock whose constituents cannot be distinguished with the naked eye. The term is derived from the Greek, meaning invisible. The adjective aphanitic is applied to these rocks as well as their fine-grained groundmasses.

APHASIA. Loss of the ability to use words as symbols of ideas. The condition usually results from an injury or disease of the brain centers which control memory, hearing, speech, and associated centers. Aphasia is commonly present in middle artery cerebral thrombosis.

APHELION. The point in the orbit of a planet or other member of the solar system, except a satellite, where the object is most remote from the sun. It is the point on the line of apsides (see **Orbit (Astronomy)**) diametrically opposite to periphelion.

APHID OR PLANT LOUSE (*Insecta, Homoptera*). A small delicate insect with sucking mouth. The aphid generally lives on the sap of plants and is of major economic importance. The aphid is characterized by an intricate life cycle that results in a high rate of reproduction. In the temperate zone, aphids hatch in the spring from winter eggs; these individuals are females known as stem mothers. They bear living young without mating (viviparity, parthenogenesis) and these, in turn, are females capable of the same type of reproduction. Late in the season, a generation known as the *sexuparae* bears both male and female offspring, which mate to produce the eggs that pass the winter.

Winged aphids (Fig. 1) appear under conditions that demand migration from plant to plant. Experiments have shown that the appearance of wings is a response to definite environmental conditions, complex in nature, and not fully understood. The much more common wingless aphid is shown in Fig. 2.

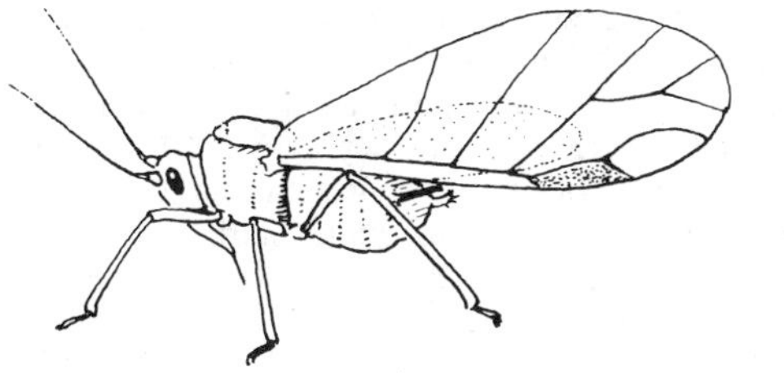

Fig. 1. Winged aphid.

The aphid injects a poisonous saliva into plants which causes discolored and unhealthy foliage and buds. A by-product of the aphid's digestion is known as honeydew, a sweet secretion arising from the ingestion of excessive sap and sugars. Ants, bees, and wasps are vigorously attracted by this honeydew. To encourage production of larger quantities of the substance, an ant has been observed to assist an aphid by carrying the aphid in its mouth from a withered plant to a healthy plant, thus encouraging the aphid to eat more and produce more honeydews. The ant also protects the aphid from intruders. The honeydew clogs the pores of plants and also nourishes damaging fungi.

Although the aphid is destructive to many crops, it is particularly damaging to certain vegetables, including bean, cabbage, cucumber, melon, pumpkin and squash; and to certain fruit and nut crops, including apple, apricot, peach, pear, pecan, and walnut. The aphid is quite damaging to a number of decorative plants.

There are many species of aphid, nearly a species for each crop attacked, indicating the high specialization created over the centuries. Generally, all aphids are tiny ($\frac{1}{16}$-inch; 1.5 millimeters) in length, ranging in color from light-green, dark-green, and black. Most varieties have a rather prominent rearward protuberance, often called a "tailpipe." Although the damage varies from one crop to the next, generally the aphid causes leaves to curl and thicken, turn yellow, and die.

Most aphids overwinter in the egg stage; eggs are laid on the twigs and bark of the trees. To kill these eggs and prevent a rapid buildup of aphids later, it is well to use a dormat oil spray. Oil sprays should be applied before any color is shown in buds of plant. Dormant oil sprays must not be used when foliage is on the tree or plant, or when the temperature is higher than 85°F (29.4°C) or lower than 35°F (2°C). When aphids become numerous during the growing season, sometimes they can be washed off with soap formulations applied under fairly high pressure, but specific treatment may involve stronger chemicals, the selection varying from one infested plant to the next.

Fig. 2. Wingless aphid feeding on rosebush. (*Bernard L. Gluck from National Audubon Society.*)

To control root-feeding aphids, such as the *woolly apple aphid*, the soil immediately under the plant should be regularly loosened and cultivated to a depth of 1 to 3 inches (2.5 to 7.5 centimeters). Attacks by the woolly aphid can be avoided by planting varieties that are resistant. Also, trees and plants that have been kept in a vigorous growing condition are less likely to suffer damage from these insects. Woolly apple aphids also feed in wounds of trunks and branches. The wounds should be treated with a wound dressing to prevent attack by aphids.

When natural enemies of the aphids are abundant, aphids usually do not require additional control measures. Before applying chemicals, check the plants carefully to see if a natural control agent, such as the larvae of lady beetles, is present. Syrphid flies and ant lions also are natural enemies of the aphid. When the spring is warm, natural enemies are usually present in sufficient numbers to control aphids completely. The greatest damage occurs after a cold spring; aphids increase more rapidly than their natural enemies in a chilly season.

Indicative of the highly specialized aphids are:

Bareberry plant louse (*Rhopalosiphium berberidis*)
Beet aphis (*Pemphigus betae*)
Birch aphis (*Callipterusbetulaecoleus*)
Black aphis (*Aphispersicae-niger*); affects peach
Bud aphis (*Siphocoryne arenae*); affects apple
Cabbage aphis (*Aphis brassicae*)
Corn-root aphis (*Aphis maidiradicis*) See entry on **Ant.**
Currant aphis (*Myzus ribis*)
Gall aphis (*Phylloxera sp.*); affects spruce
Leaf or green aphis (*Aphis pomi*); affects apple
Lettuce aphis or green fly (*Macrosiphum lactucae*)
Lupine aphis (*Macrosiphum albifrons*)
Melon aphis (*Aphis gossypii*); also affects squash, watermelon
Oleander aphis (*Aphis nerii*)
Parsnip louse (*Hyadaphis pastinacae*)
Pea aphis (*Macrosiphum pisi*); also affects bamboo
Root louse (*Aphis forbesii*); affects strawberry
Rosy aphis (*Aphis sorbi*); affects apple and quince
Spinach aphis (*Myzua persicae*); also affects green peaches
Squash aphis (*Nectarophora cucurbitae*)
Woolly aphis (*Schizoneura lanigera*); affects apple.

Oriental black citrus aphid (*Toxoptera citricada*). This insect is a serious pest of citrus in South America. The oriental black citrus aphis is black and, when fully grown, is about $\frac{1}{16}$-inch (1.5 millimeters) long. The antennae and legs are sometimes brownish, rather than black. The aphids generally are wingless. Ocassionally, winged forms appear. The winged aphids have transparent wings; the forewings are much longer than the rear wings. The aphids are most numerous in spring and early summer. Development takes about 12 days. These insects feed on leaves of citrus and cause new growth to be stunted. A heavy infestation can cause the trees to become seriously damaged. Besides damaging the trees directly, these aphids carry *tristeza*, a citrus disease that kills the roots and causes the trees to die.

APHIS-LION (*Insecta, Neuroptera*). The larva of the golden-eyes or lacewing flies (family *Chrysopidae*), so called because they feed on aphids and other small insects.

APHTHOUS ULCER. Ulcer.

APHYTIC. A paleobotanic division of geologic time, signifying the period of time that preceded the first occurrence of plant life. See also **Paleobotany.**

APICAL GROWTH. Growth at the tip of an organ, as occurs in the roots and stems of all higher plants. Examination of the stems of most plants will show that growth in length occurs only in the apical portion, and only for a relatively short period of time, usually a matter of a few weeks. This may be determined by observing the distances between successive leaves: near the growing tip the leaves are very small and close together; as one goes back along the stem the size of the leaves increases and also the distance between them; but after the leaf is mature little elongation of the stem occurs, as shown by the uniform distances between the mature leaves. Older stems increase constantly in diameter, but only exceptionally in length. One notable such exception, known as intercalary growth, is found in grasses and some other plants. Here a group of cells in the region of the node are capable for a time of active division and of causing increase in the length of older portions of the stem. Both here and in the tip of the stem increase in length is due to elongation of cells produced by an actively dividing tissue known as a meristem.

For related topical coverage in this volume, see list of entries under **Tree.**

API GRAVITY. Petroleum; Specific Gravity.

APLACOPHORA. A class of *Amphineura* (*Mollusca*) made up of animals of worm-like form with foot reduced to median ventral ridge and mantle enlarged.

APLANATIC LENS SYSTEM. A system which satisfies the equation, $ny \sin u = n'y' \sin u' =$ constant. Unprimed letters refer to object space, primed to image space. n, n' are indices of refraction; y, y' are distances of a point object and its point image from the optical axis; and u, u' are angles made by a ray with the optical axis. The equation is known as Abbe's sine condition.

A triple *aplanat* is a compound lens made of two diverging lenses of flint glass between which is cemented a converging lens of crown glass.

APLANATIC POINTS. Two points on the axis of an optical system which have the property that rays proceeding from one of them shall all converge to, or appear to diverge from, the other. The two foci of an ellipsoid of revolution are aplanatic points. An aplanatic surface is an optical surface for which two aplanatic points exist.

APLASTIC ANEMIA. Anemias.

APLITE. This term is applied to fine-grained, sometimes sugary-textured igneous rocks, composed almost wholly of quartz and feldspar. Except for size of grain, aplites resemble pegmatites both in mineral composition and in mode of occurrence in dikes and veins, save that the rare minerals often present in pegmatites are wanting here. The word aplite is derived from the Greek word meaning simple, referring to its ordinarily simple mineral composition.

APNEA (SIDS). Sudden Infant Death Syndrome (SIDS).

APOCHROMATS. Microscope.

APODEME. An internal projection of the hard outer covering (exoskeleton) of arthropods. Apodemes provide muscle attachments and in some species are extensively developed. They are collectively termed the endoskeleton.

APODES. Eels.

APODIFORMES (*Aves*). This order includes the swifts and hummingbirds which comprise those birds that have the longest primaries in proportion to their total length. Perhaps both forms represent split-offs from an ancestral "swift-sparrow" stock occurring only after the rollers and their relatives (*Coraciformes*) had already separated from the original common stock of the arboricoles, which yet more primitively had encompassed the range from the ancestral owls to the ancestral *Passeriformes.*

Swifts and hummingbirds have several characteristics in common. The wings are flat and lack, or possess only very short supporting feathers (to cover the space between wings and body). They have four complete pairs of ribs; the posterior edge of the sternum is not notched. The legs are short, the leg muscles are very weak, and the talons are pointed. The eggs are relatively elongated and are generally white.

Despite their German name "sail birds," none of the species of the swifts can actually sail or soar in the air, making use of rising convection currents. Aside from the already mentioned characteristics they share with hummingbirds, swifts have the following features. The length is 10–33 centimeters (4–13 inches). The bill is short and broad, the opening is wide, and the salivary glands are large. The tongue is small and triangular. They have 7–10 secondaries; plumage ranges from gray to black and brown (in one tree swift species, it is olive and blue), and often has white markings. The iris is brown, and the bill and feet are black. The clutch numbers from 1–6, and both sexes incubate and feed.

Swifts occur in all parts of the world except the polar regions. There are two families: 1. The Swifts (*Apodidae*); and 2. the Tree Swifts (*Hemiprocnidae*); there is a total of 77 species (from 65–81, according to other authorities).

One of the most clearly characterized, most sharply defined, and most homogeneous families is the hummingbirds (*Trochilidae*). Hummingbirds are distinguished from swifts by the following characteristics. They range in size from small to very small, with a length of 6–22 centimeters ($2\frac{1}{2}$–$8\frac{1}{2}$ inches) and a weight of 2–20 grams (0.07–0.7 ounce). The large eyes are set in wide sockets. The bill is usually long, narrow, tube-shaped, straight or curved, and very thin throughout its length. Although the neck is long and flexible, the head is usually retracted between the shoulders. The wings are long to very long. The bird has 10 well-developed primaries with the outermost almost always being the longest, but it has only 6 (rarely 7) secondaries. The origins of the powerful flight muscles are on a particularly well-developed sternum. The shins and feet are short, the leg and foot bones are thin, but the toes, of which there are 3 forward and 1 reversed, are well developed, short, with sharply pointed talons. The feet are suitable only for perching on branches, not for walking. The body plumage is moderately dense and the feathers are firm and adherent to each other, not downy.

Despite the richness of species, the family of the hummingbird is one of the most homogeneous known. The subdivision into tribes, genera, subgenera, species, and subspecies is therefore extremely difficult. There are approximately 120 genera with about 400 distinct forms, of which we consider 321 to be species. They occur only in America. See also **Swifts and Hummingbirds.**

APOENZYME. Coenzymes.

APOLLO ASTEROIDS. Asteroid.

APOLLO PROGRAM. Moon (The).

APONEUROSIS. A sheet of tough white glistening fascia or membrane which surrounds muscle and muscle fibers or connects a muscle to the part which it moves. The tensile strength and resistance of muscle tissue is dependent upon the fascial tissue around the muscular fibers. See also **Tendon.**

APOPHYLLITE. The mineral is a hydrous silicate of potassium, calcium, and fluorine, corresponding to the formula, $KCa_4Si_8O_{20}(F,OH)\cdot 8H_2O$. The true crystallographic symmetry is evident on crystals by the luster difference between the basal pinacoid facial planes and other crystal faces. Also, prism faces show vertical striations; basal planes do not. It crystallizes in the tetragonal system in square prisms resembling cubes terminated by based pinacoid or pyramids, often with both. Prism faces show vertical striations; basal pinacoid either dull or rough. Cleavage is perfect, parallel to the base; hardness, 4.5–5; sp gr, 2.3–2.4; luster, vitreous to pearly; transparent to translucent or nearly opaque; color may be white, grayish, greenish, yellowish, or reddish. This mineral was named by Haüy from the Greek words meaning from a leaf, referring to its exfoliation when heated with the blow pipe.

Apophyllite is a secondary mineral found with the zeolites and has been classed with them by some writers, but it contains no aluminum, which element is understood to be an essential in a zeolite. It occurs in cavities in basalts and less often filling openings in granites or other crystalline rocks; it also is a gangue mineral in certain ore veins.

There are many localities for apophyllite: Bohemia, Trentino, Italy, the Hartz Mountains, and Iceland. Fine specimens have been obtained from the Ghats Mountains in India. The Triassic trap rocks of New

Jersey, Connecticut, and Nova Scotia have also furnished many specimens.

APOPHYSIS. In zoology, an apophysis is a protuberance or outgrowth of an organic structure, such as a process on a bone.

In geology, an apophysis is a tongue or other direct offshoot of a larger vein or dike.

APOSTILB. Units and Standards.

APPALACHIA (Reclamation). Revegetation.

APPARENT MOLAR QUANTITY. For a solution containing n_1 moles of solvent and n_2 moles of solute, an apparent molar quantity is defined as

$$\frac{X - n_1 x_1}{n_2}$$

where X is the value of the quantity for the whole solution and x_1 the molar quantity for the pure solvent; e.g., the apparent molar volume of the solute is

$$\frac{V - n_1 v_1}{n_2}$$

V being the total volume and v_1 the volume per mole of the solvent.

See also **Molal Concentration; Molar Concentration; Mole (Stochiometry); Mole Fraction; Mole Volume.**

APPARENT SIDEREAL TIME. Time.

APPARENT SOLAR TIME. Time.

APPENDAGE. A supplementary structure attached to an organ or body. Externally many animals have appendages which serve for defense, for locomotion, for securing food, and as sensory organs.

The simplest external appendages are mere outgrowths of the body wall, either solid or hollow; the tentacles of coelenterates are an excellent example. In the annelid worms both tentacles and parapodia appear as external appendages, in the echinoderms the rays or arms may be radiating divisions of the body or appendages, and in the mollusks tentacles of complex structure occur. In all cases these structures allow greater facility of movement than is possible for the body as a whole and so compensate the lack of freedom which attends increased size and complexity or sessile habits.

The most elaborate appendages are the jointed appendages of the antropods and chordates. Although they are not fundamentally related, the appendages of both phyla are similar in principle. They consist of a series of segments connected by movable joints with each other and with the body, and are provided with muscles which operate them as a series of levers. In arthropods the skeleton is external to the muscles, in the vertebrates internal.

The jointed appendages of arthropods are specialized in various species for swimming, walking, running, jumping, feeding and grasping, and in the form of antennas and palpi as sensory organs. In the jumping appendage powerful muscles result in the extension of one segment of the leg, as in the familiar grasshoppers. Grasping is accomplished by the folding back of one segment against another in the insects or by the development of a process on the next to the last segment which works against the terminal segment in a forceps-like relation; the latter is the chelate type of appendage. Arthropod appendages were originally metameric, one pair appearing on each segment of the body. In existing species they are variably limited as described under the term *Arthropoda.*

The vertebrate appendage is also specialized for various purposes, including jumping, swimming, flight, and burrowing. The primitive form appears to be the paired fins of the fishes (*Pisces*), and the arms and legs of man are good examples of fairly specialized appendages. Two pairs, the pectoral or anterior pair and the pelvic or posterior pair are typical; only in highly specialized forms such as the snakes is this number reduced.

The vertebrate appendage differs from the arthropod appendage in the terminal series of digits. These structures have been developed in some animals so that they can be opposed to each other for grasping, as in the opposition of the human thumb to the four fingers. In other forms they have been reduced in importance.

See also **Biramous Appendage; Pentadactyl Appendage.**

APPENDECTOMY. Appendicitis.

APPENDICITIS. Inflammation of the vermiform appendix. See **Appendix.** The infection may be acute and lead to perforation of the appendix if not promptly attended, or it may subside spontaneously. Pain, tenderness, and spasm in the right lower quadrant of the abdomen are typical symptoms of appendicitis—but these are also symptoms of other conditions, such as duodenal ulcer disease, cholecystitis (inflammation of the gallbladder), gastrointestinal actinomycosis, and gastroenteritis resulting from *Salmonella* infection, among others. The symptoms of acute appendicitis develop rapidly into a severe generalized abdominal pain, and rebound tenderness is often present. In children, crying, vomiting, and refusal of food may be very early symptoms of appendicitis. Because such symptoms commonly occur from dietary indiscretions, parents may administer a laxative before consulting a physician, with the possible result that perforation of the appendix may be precipitated. Thus, a laxative or enema is not advised in the case of abdominal pain without first consulting a physician.

Immediate surgery (*appendectomy*) is indicated in acute appendicitis. Delay invites complications, such as peritonitis. See **Peritonitis.** Devoid of complications, the prognosis for successful recovery from appendicitis is excellent.

APPENDICULARIAN (*Chordata, Tunicata*). A free-swimming tailed tunicate with a permanent chordate tail. These forms make up the class *Larvacea.*

APPENDIX (or Appendix Vermiformis). A blind worm-like tubular portion of the intestine which arises from the base of the cecum. Its size and position vary greatly, the length averaging $3\frac{1}{2}$ inches (8.8 centimeters), although it has been found to vary from $\frac{3}{4}$–9 inches (1.9–22.8 centimeters). Its position often varies from the normal so that it has been reported in every possible situation in the abdomen, depending on the position of the cecum and the length of the organ and its attachments.

An appendix is found only in man, the higher apes, and the wombat, and possibly in some rodents. In herbivorous animals the cecum attains a very large size and it is thought by some that the appendix represents the degenerated remains of the herbivorous cecum. Others believe that it is a lymph organ functioning as other lymph glands in the body. By many it is considered to be in the process of gradual obliteration in the human species. It is subject to inflammatory processes because of its limited blood supply, and since it is a blind tube, it is subject to obstruction from fecal impaction. See also **Appendicitis.**

APPETITE (Loss). Anorexia.

APPLE LEAF SKELETONIZER (*Insecta, Lepidoptera*). The adult of this species (*Psorosina ammondi,* Riley), overwinters as a small dark-brown moth. Eggs are laid in early spring. The larva is a caterpillar that feeds on the underside of leaves and which later constructs shelters on the upper surface of leaves by drawing leaves together with silken webbing. The leaves become a mass of webbing and grass and are damaged by suffocation. The insect is distributed throughout the United States, but is most abundant in the central latitudes. The caterpillar is brownish-green and about $\frac{1}{2}$-inch (12 centimeters) long, with four shining black tubercles on back just to the rear of the head. This is not regarded as a serious pest because it occurs infrequently. The insect is mainly controlled by predators and parasites. When infestation is small, handpicking is effective. When infestations are large during the growing season, chemical controls, such as lime and the arsenate formulations, may be used.

APPLE REDBUG (*Insecta, Hemiptera*). A sucking insect that attacks apple, haw, and pear and which occurs east of the Mississippi River in the United States and in southeastern Canada. The adult is orange-red with dark markings, up to $\frac{1}{4}$-inch (6 millimeters) long. The nymph is bright red, somewhat smaller than the adult. This insect, *Lygidea mendax* (Reuter), punctures the fruit, causing spots and deformation. These insects pass the winter in the egg stage. The eggs are laid in the bark of branches of trees and in the bark pores. They hatch in early spring. When light to moderate infestations are found during the season, cultural precautions for the next season may be the best practice. This involves using a dormant fruit tree oil spray that will kill the overwintering eggs. Treatment is similar to that for the aphid. See **Aphid.** Where the infestation may be heavy during the growing season, control chemicals, such as parathion, may be effective. Nicotine sulfate spray is also effective if applied at the cluster-bud stage.

APPLE TREES. Rose Family.

APPROXIMATE CALCULATION. It is often necessary, especially in physical problems, to make approximate, rather than exact, calculations. Typical cases include: the solution of polynomial or transcendental equations; evaluation of integrals which cannot be given in terms of simple functions; solution of differential equations with assigned boundary conditions.

Sometimes graphical or mechanical methods are convenient and these, while limited in their precision, frequently yield results of complete satisfaction. Numerical methods are more generally applicable, for with desk calculating machines an answer can usually be obtained with any reasonable number of significant figures desired. In these cases, the procedure can be classified as one of two general classes: polynomial approximation or iteration.

The literature of this subject is very large and only a brief survey of it can be given here. Moreover, the methods involved can be modified in many different ways and there is no generally accepted name given to the various procedures.

Approximating methods also are described under **Curve Fitting; Differential Equation (Numerical Solution of); Differentiation (Numerical); Equation (Numerical Solution of); Integration (Numerical);** and **Interpolation.** Also see list of entries under **Mathematics.**

APRICOT. Rose Family.

APSIDES (Line of). Line of Apsides.

APTERYGIFORMES. Kiwi.

APTERYGOTA. The primitive wingless insects with varying number of paired appendages on the abdomen and metamorphosis slight or absent. A subclass made up of the orders *Protura, Thysanura,* and *Collembola* in which the existing species are wingless and there is no evidence to show that wings have occurred in any ancestral form.

AQUA. Latin, water, previously used widely by chemists as a term for various solutions in water, e.g., aqua calcis, lime water, calcium hydroxide solution; aqua fortis, nitric acid; aqua regia, mixture of concentrated nitric and hydrochloric acids, named "royal water" because it was the only acid then known that would attack the noble metal gold.

AQUACULTURE. This group of disciplines and practices, fundamentally centuries old in basic concept, essentially parallels agriculture (land crops and plants) and animal husbandry (livestock) in the quest to increase productivity of natural food substances, while simultaneously achieving efficiency and convenience of harvesting by way of concentrating the sources into smaller spaces. Possibly, the most useful synonym for aquaculture would be *fish farming* (where the term fish is interpreted broadly to include mollusks and crustaceans). Aquaculture can be broken down into several categories, usually in terms of environments and species involved, but possibly the most meaningful breakdown is (1) those operations directed toward the culture and harvesting of freshwater fishes; and (2) those operations directed toward marine animals, from which the terms *sea farming* or *mariculture* have developed. There are always the inevitable complications that detract from a crisp categorization—for example, the application of aquaculture to estuarine and brackish waters involving species that may be fully or partially tolerant to both fresh and saline waters. While most aquaculture operations are fully land-based or operate in the tidal zone between land and open water, there are also cultural operations that take place in naturally open and large bodies of water, such as lakes or various locations in the seas and oceans.

Bardach et al (1972) defined aquaculture as "the farming and husbandry of freshwater and marine organisms." Glude (1977) called aquaculture "the culture or husbandry of aquatic animals or plants by provate industry for commercial purposes or by public agencies for augmenting natural stocks." Brown (1977) uses the term "*sea ranching*—[a method that] involves releasing hatchery-reared animals of various sizes into marine waters for rearing and the subsequent recapture of the adult fish upon their return to the point of release." The general term, fish culture, is used by Edmunds and Lillard (1979).

Records indicate that aquaculture was practiced several thousand years ago in the Mediterranean area. It has been established that the Romans were successful in cultivating fishes in the brackish waters off the coast of Italy. Various Egyptian art forms depict the pond culture of fish. Records of the Chou and Shang Dynasties in China include reference to raising fish in ponds, projects of interest to the ruling class. Pioneered by Wen Fang in Honan Province (circa 1135–1122 B.C.), a treatise on these practices appeared in the Chinese literature as early as 475 B.C. To this day, aquaculture continues important to the food economy of China (Dill, 1967).

But, as pointed out by Sandifer and Smith (1978), despite its antiquity, aquaculture is a relative newcomer to the business of food production compared with agriculture and conventional fisheries. Consequently, aquaculture accounts for less than 1% of food production (Wheaton, 1977). Nevertheless, total food production from aquaculture exceeded 6 million metric tons in 1975 and this figure is expected to double by the mid to late 1980s (Pillay, 1976). As aptly pointed out by Moiseeve as early as 1973, it is not the current levels of production, but rather the large potential productivity (hundreds of millions of metric tons) that is of such interest to a protein-short world population. For some of the developed countries, increased protein from aquaculture is not yet in serious contention, but some authorities are quick to point out that aquaculture may be required within the relatively near future if the desire for seafood luxuries, such as shrimp and other shellfishes, continues to develop at the pace enjoyed during the last few decades. In some countries, during the period 1960 to 1976, the demand for shrimp increased by about 80%, as compared with increased shrimp landings of less than 65%. Some authorities have estimated that the demand for shrimp by the United States (domestic and imported) by 1990 may exceed the total world production of shrimp as of 1980.

Compared with its relatively small beginning and slow development for many years, during the past decade fish culture in the United States has expanded steadily. Reasons given are high fish prices, the realization that conventional sources of fish are not limitless, the increased costs of catching fish in conventional fashion, and problems of pollution etc. that have interferred with natural fish habitats. As the vital need of the developed nations for increased energy sources becomes even more critical, trade-offs favoring energy will increase—as, for example, where historical and natural fisheries may be adversely affected by thermal and chemical pollution, among other factors. Aquaculture may provide an effective means in some instances for essentially abandoning traditional fisheries where there is a serious conflict with energy production.

The Worldwide National Oceanic and Atmosphere Administration estimates that 10% of the total fish production (1977) was derived from aquacultural operations, whereas this figure for the United States was about 3%, with the activities concentrated mainly on salmon, oyster, catfish, trout, shrimp, and clam species. Predictions by authorities as to the potential growth are mixed, ranging as high as 20 million tons (United States) by the mid to late 1980s, with the accompanying creation of many new jobs.

Much of the research on aquaculture in the United States has been directed toward determination of optimum growth conditions, to the formulation of rations which provide maximum feed conversion rates, to a reduction in the time required to produce a marketable product,

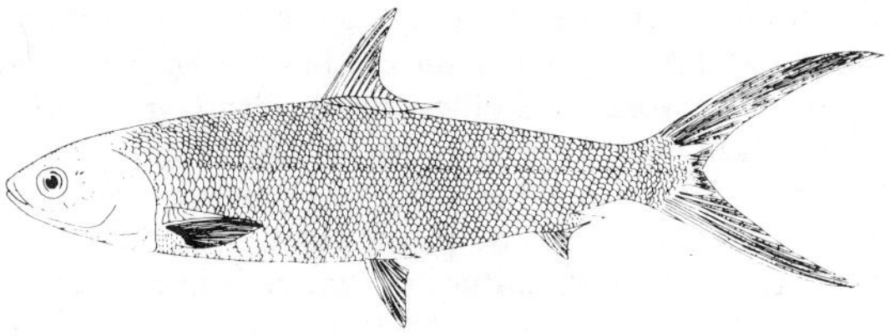

Fig. 1. Milkfish (*Chanos chanos*).

and toward closing the life cycle of the shellfish when held in captivity. Edmunds and Lillard (1979) report that recent developments in closing the life cycle for shrimp and polyculture are technological advances which dramatically increase the economic feasibility of aquaculture for the affected species. Because of little prior work in the area of assessing the sensory characteristics of cultured species, Edmunds and Lillard undertook an investigation along these lines as regards oysters, clams, and shrimps and is so reported in the listed preference. Among other observations, these researchers found that cultured shrimp were equal or superior to wild shrimp in sensory quality. Their general findings ran counter in many instances to observations that cultured fishes of the various species are frequently inferior to the wild catches. References relating to the potential of aquaculture in the United States, as listed, include: (Anon. (1) through (5); Rutherford, 1975; Shoemaker, 1975; Gallese, 1976; Twitty, 1976; and Booda, 1976).

For many years, a number of countries have developed their sea resources in coastal regions. These estuarine regions have supported the culture of fish and invertebrates in tremendous numbers as well as great fisheries. In weight yield per surface area, these estuaries are much richer in food resources than the open sea.

Indian Ocean Regions. There are extensive culture fisheries in the estuarine areas of the Indian Ocean coastal zone. These brackish water swamps are banked with dikes to take in water at high tide, and the suspended silt settles on the bottom, slowly raising its level. Within several years, the land is raised sufficiently to be used as rice paddies. When the land level is high enough, intake of tidal water is discontinued and rainwater leaches the salt from the soil. After 2 to 3 seasons, the paddy may be planted. The plot will have a canal system inside the dike; these canals are used for culturing brackish water fishes. But, during this silting period, the area is used for growing fish. The tidal water brings in the fry of commercially important fish, including prawns. Screens prevent the escape of fish or the ingress of extraneous fish, and sluice gates control the tidal flow.

The once incidental use of swampy land for raising fish promoted a system of intensive brackish water fish culture in India and Pakistan. Fish ponds of different design and shape began to be constructed for the sole purpose of commercial fish culture.

On the islands of Java and Madura (Malaysia), ponds or "tambaks" produce upwards of 40 million pounds (18 million kilograms) of fish annually. The industry is based upon the *milk fish* (*Chanos chanos*). See Fig. 1. Post larvae and juveniles are collected from inshore areas from September to December and from April to May and transported to the fish pond areas in flat, watertight bamboo baskets. The annual requirements for the tambaks of these islands is estimated at upwards of 200 million fry.

In India, grey mullet, pearl spot, prawns, or milk fish are seined from nearby areas and transported to fish ponds called "porong" ponds. Each porong-type farm has fry ponds of varying areas, ranging from about 1000 to 10,000 square feet (90+ to 900+ square meters), and rearing ponds of 10,000 to 48,500 square feet (900+ to 4500+ square meters). The irregularly shaped sections are connected by secondary sluice gates and the whole complex is controlled by a main gate located in a deep portion, having a channel in the middle. The ponds are drained and dried to eradicate predatory and weed fish and to hasten decomposition of organic matter. Then from 1.2 to 2 inches (3 to 5 centimeters) of tidal water is taken in and allowed to stand. Within 3 to 7 days, a brownish-greenish-yellowish layer of microorganisms (mainly bacteria, unicellular and filamentous blue-green algae, and diatoms) develops on the bottom. Growth and production of fish in the ponds depends upon the growth of algae. The product is marketable within 6 to 10 months. The ponds are drained at low tide to capture the fish. The tendency of the fish to swim against the current is used for partial fishing. When tidal waters are let into the ponds, the fish swim against the current and then can be led to a catching pond.

In Java and Madura, annual production approaches 25 million pounds (11.3 million kilograms) of milk fish; about 10 million pounds (4.5 million kilograms) of penaeid prawns; and 6 million pounds (2.7 million kilograms) of other fish.

The sometimes-called giant Malaysian prawn (*Macrobrachium rosenbergii* de Man) is widely distributed in the Indo-Pacific region, ranging from Australia to New Guinea to the Indus River delta (John-

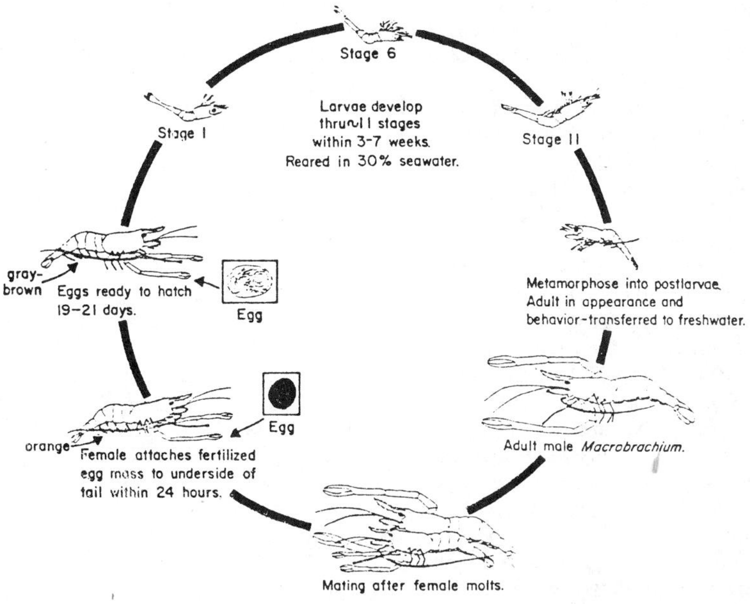

Fig. 2. Life cycle of the Malaysian prawn. (*After Sandifer and Smith reference listed.*)

son, 1960; George, 1969; Malecha, 1977). There also have been reports of populations from the Palau Islands (McVey, 1975). In their review of the aquaculture of Malaysian prawns, Sandifer and Smith (1978), observe that adults are usually found in freshwater reaches of rivers, and during the breeding season, mature females migrate downstream into estuarine areas where the eggs are hatched. Presumably, the larvae are then swept further downstream into more saline waters where they develop. But, after a few weeks after the larval phase, the postlarvae begin to migrate toward freshwater where they grow to adults. Sandifer and Smith have diagramed the life cycle of the Malaysian prawn. See Fig. 2.

Controlled-environment experiments on cultivating Malaysian prawns have been conducted in recent years by the Marine Resources Research Institute, Charleston, South Carolina and these are reported in the Sandifer and Smith reference. The outdoor climate in this area permits prawn cultivation during only the warmer half of the year. Experiments, however, have shown that one crop of marketable prawns can be reared annually in ponds in South Carolina. It has been observed that a broad area of the state's coastal plain may be suitable for prawn farming. The researchers observe that to become commercially attractive, very intensive culture systems must operate continuously at high efficiencies and produce much greater yields per unit of area than a conventional pond system. External heat is required (possibly solar) for year-round operation. Culture containers (above ground) in form of tanks may be a limiting factor. Earthen ponds are being studied. Other configurations, including Shigueno-type tanks and aquacells are under consideration. Summarizing current problems, the researchers observe prawn nutrition, behavior, physiology, and genetics; culture systems design and optimization; pilot plant demonstrations; and economic feasibility analyses. A successful controlled-environment culture of *Macrobrachium* will likely require (1) the development of more nutritionally complete, cost-effective rations; (2) the determination and maintenance of optimal conditions for prawn survival and growth under crowded conditions; (3) genetic manipulation to produce prawns better suited for intensive culture than the essentially wild animals available today; (4) reduced system costs; (5) improved management techniques; and (6) greater production efficiency.

Mosquitofish (*Nothobranchius taenipygnus*), a tiny fish a little less than 1 inch long (2.5 centimeters) at the age of 3 to 4 weeks has been suggested by some authorities as a protein source in Africa and Asia. The fish will live in a few inches of water. Areas from 6 to 12 inches (15 to 30 centimeters) allow the maximum number to be bred in the minimum volume of water. Young *Nothobranchius* feed on algae and protozoa, but older fish prefer insect larvae. Apart from some chemical fertilizers to increase the growth of algae in the feeding pools, little has been done to provide supplies of feed. It is interesting to note that in 1977, researchers of the Sutter-Yuba Mosquito Abatement District in California reported on a formulation of a specific floating feed designed for rearing of mosquitofish. It is the plan to breed these fish in sufficient numbers to release them in California rice paddies for controlling mosquitos. This is an aquaculture project with a possible double advantage.

China. An excellent summary of aquaculture in China is provided by Ryther (1979). It is estimated that freshwater fish production in China probably exceeds that of the rest of the world combined, perhaps by as much as severalfold. Various estimates have placed the total area of freshwater culture in China at about 25 million acres (10 million hectares). Yields from the smaller, more intensively managed fish ponds have been estimated to range from a low of 893 pounds per acre (1000 kilograms per hectare) to 8930 pounds per acre (10,000 kilograms per hectare), with a probable average of about 2680 pounds per acre (3200 kilograms per hectare). It has been reported that the Ching Po Fish Farm (near Shanghai) produced 5000 pounds per acre (5600 kilograms per hectare) in 1978, with an objective of reaching 6700 pounds per acre (7500 kilograms per hectare).

It has been estimated that the combined yields from aquaculture and marine fishing in China are equivalent to a per capita catch consumption of about 45 pounds (20.4 kilograms) against a backdrop of perhaps a billion people or more.

Ryther reports that the success of aquaculture relates to a number of factors: (1) multiple use of aquatic resources where several species are cultured in the same facility; (2) fish pond culture in China always has been considered an integral part of agriculture. Emphasis is given in polyculture to those species that are low in the food chain. Most of the fish species favored in Chinese pond culture are cyprinids (minnows), of which the 4 species (carp) illustrated in Fig. 3 are currently the most important. As stated earlier, it must be stressed that aquaculture in China dates back thousands of years.

By contrast, marine aquaculture or mariculture represents a much more recent endeavor, extending back only a comparatively few decades on a large scale. Culture includes species of sea plants as well as sea animals. Probably the most important cultivated marine organ-

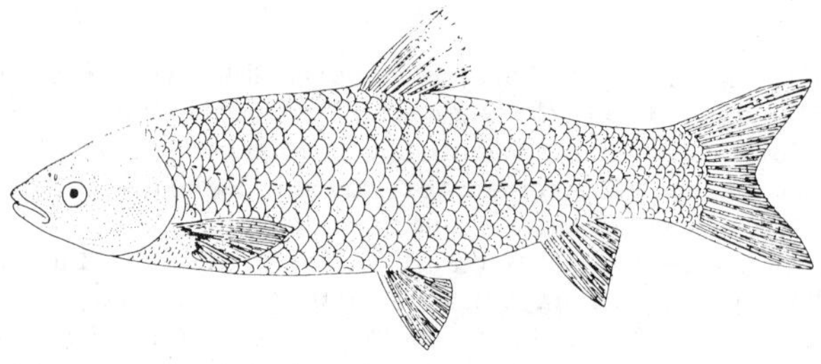

Grass carp (*Ctenopharyngodon idellus*)

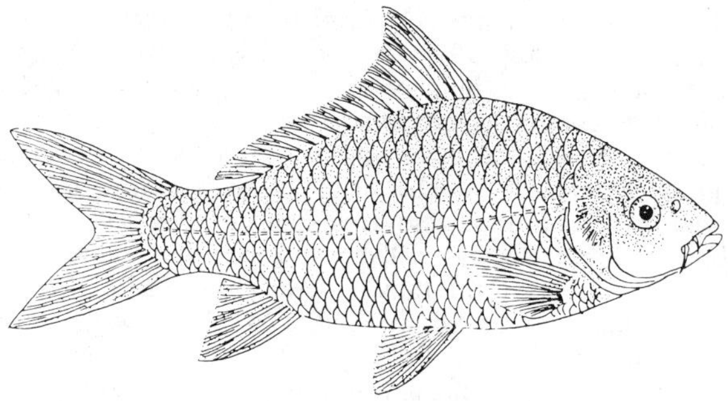

Common carp (*Cyprinus carpio L.*)

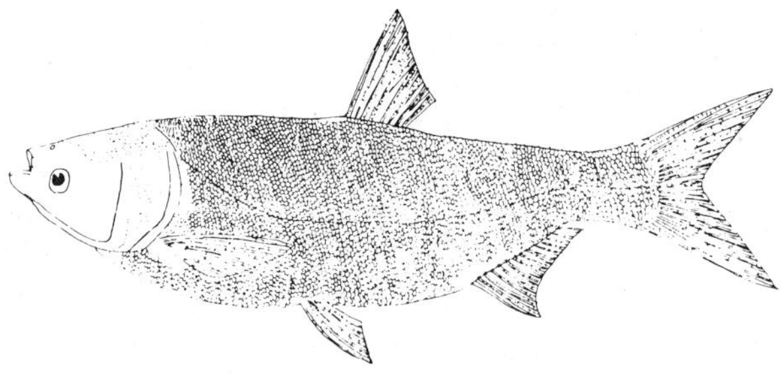

Silver carp (*Hypophthalmichthys molitrix*)

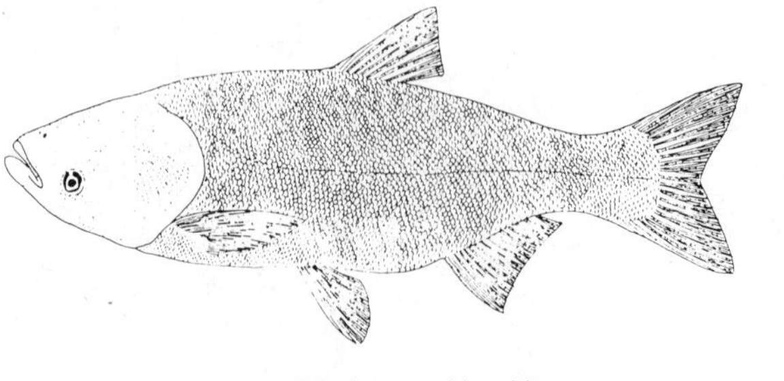

Bighead carp (*Aristichthys nobilis*)

Fig. 3. Four species of carp. (*Drawings by S. Ling, reproduced by permission of artist and publisher of reference listed.*)

ism is the brown seaweed (*Laminaria japonica*), a cold-water species of kelp. See also the entries on **Algae;** and **Seaweeds.** Other cultured species include sea cucumbers, penaeid shrimp, pearl oysters, crab, mullet, and milk fish, but all are not in a fully developed state ready for commercial production. Mariculture has lagged somewhat, possibly largely because of the heavy accent on freshwater aquaculture, which is intimately tied to agriculture and a water conservation program. Aquaculture technology in China is headquartered in the City of Tsingtao, in which are located the Institute of Oceanography (Chinese Academy of Sciences), Shantung College of Oceanography, and the Yellow Sea Fisheries Institute (National Bureau of Fisheries). A significant part of the current research is directed toward algae and seaweed production.

Europe. In Italy, the Volturno River discharges into Lake Patria. The communication of the estuary with the sea consists of a very narrow shallow watercourse. Nutrients enter the lake from volcanic subsoil. Lake Patria teems with mullet and eels. The soft, muddy subsoil makes mussel farming on the bottom impossible, but by hanging cultures above the bottom, the mussels grow and fatten rapidly.

The Oosterschelde, in the southwestern section of the Netherlands, is an estuary with a different pattern. The Oosterschelde is an embayment penetrating far into the land that receives little influx from the surrounding fertile arable land. Yet river and seawater meet under the influence of the tides. The Rhine, the Meuse, and Scheldt Rivers discharge fresh water rich in nutrients and organic materials into the North Sea. Tides and winds bring about a thorough mixing and microorganisms mineralize the remains of freshwater organisms and other organic material. It is this mixed coastal water that pours into the Oosterschelde with the tides. The high and constant salinity, together with the discharge of nutrients conducive to rich plankton development, makes good oyster water. Under natural conditions, the oyster population was limited by an unprepared bottom and by failure of the oyster larvae to settle on beds where growth and fattening would be optimal. An exploitation of the natural resource by applying techniques of cultivation has resulted in a 30-fold increase in production over what nature formerly yielded.

The Galacian bays of Spain are examples of estuaries in which shelter is a more important factor than discharge of nutrient-rich freshwater, since the seawater is sufficiently rich through upwelling. In the early 1950s, experiments were carried out to take advantage of the favorable conditions in these estuaries by growing mussels in hanging cultures following experiences gained in the Mediterranean Sea. Instead of racks, which are unsuitable where the tidal range is great, rafts were used. At first these were old boats equipped with outriggers. The success was promising and led to some 1500 installations consisting mainly of specially constructed rafts. Each raft carries 800 ropes; each rope is about 20 feet (6 meters) long and carries 100 pounds (45.4 kilograms) of marketable mussel. It takes eight months to rear a mussel seed to marketable size. See also **Mollusks.**

The Japanese use a similar technique to cultivate oysters. By growing oysters on long ropes hanging from simple rafts, the plankton of the whole water column is available to the oyster, and there can be a 50-fold increase in yield.

Another type of estuary, with extensive tidal flats and channels of varying depth through which tidal currents run, is located behind the Frisian Islands of Texel, Vlieland, and Terschelling, the Dutch Wadden Zee (western section). These flats are productive with mussels and cockles, brown shrimp, and polychaete worms of various species. Young flatfish and sole use this area as nursing grounds.

Several other examples of idealized situations for maritime culture could be described. Reference to McLeod (listed) is suggested.

Israel. Aquaculture commenced in Israel in the late 1930s and by the 1970s accounted for about 10% of the country's total annual value of agricultural production. Initial experiments utilized land and water unfit for ordinary agriculture. Productive ponds today are common throughout Israel with frequent integration of agricultural crops and fish ponds, all practicing required water conservation and reclamation. The Jordan River provides Israel with about one-third of its fresh water, the other two-thirds coming from ground water. Over 95% of the available fresh water is utilized, with emphasis being placed on use of brackish water in various agriculture/aquaculture practices. Several reasons for the development of an extensive fish culture industry in Israel include: (1) Skill in breeding and cultivation of carp was brought to Israel by immigrants from central Europe; (2) the need to provide dietary protein complicated by shortages of fresh water and good agricultural land; and (3) the eastern Mediterranean fisheries are not very productive, and projected declines in catches placed additional emphasis on expansion of intensive fish farming enterprises. In Israel, by the early 1970s, 8 pounds (3.6 kilograms) out of the 22 pounds (10 kilograms) of fish per capita consumed annually were pond-raised. The Fish Breeders Association is the governing body, controlling all aspects of freshwater fish production in Israel. All fish farmers are members of the association and pay a tax according to the area of their ponds and their production.

Fish raisers in Israel have found that only 4100 square yards (3400 square meters) of water area are required to produce one ton of fish, and portions of this can be used for irrigation. The basis of much of the technology is in controlled aeration of the pond area, thus permitting a two- to five-fold increase in fish stocking and the use of more sophisticated feeding methods. Yields of 2.8 to 12.1 tons per acre (7 to 30 tons per hectare) have been achieved by application of advanced technology as compared with yields of 1.2 to 1.6 tons per acre (3 to 4 tons per hectare) when using conventional procedures. The significant increase has been accomplished without a great change in the feed conversion ratio and total cost of production per ton. Other developments similarly have contributed to productivity of Israeli fish ponds. This has involved the use of supplemental feeding (25% protein pelleted feed), which has allowed an increase in pond stocking rate and improvement of tilapia yield in polyculture without depressing carp growth rates. Secondly, establishment of an optimum stocking density for both tilapia and mullet in the carp ponds has permitted more efficient exploitation of the pond's natural productivity.

High yields have been achieved through the use of multispecies (polyculture) culture. In polyculture, a range of fish species with a variety of feeding habits are stocked together, thus improving both the use and efficiency of conversion of the food.

The principal fish bred in the pond is carp, which feeds on grain and food wastes. Tilapis (Saint Peter's fish) and gray mullet are bred together with the carp. As noted, since each requires a different type of food (and occupies separate aquatic tropic levels), maximum use of land and water is achieved.

The species *Tilapia aurea*, introduced into carp ponds, originally presented a problem of wild spawning and resultant overcrowding and small size. Earlier investigations with tilapia were directed to finding ways to control unwanted spawning in the fish ponds. Prolific reproduction and reproduction at small sizes by tilapia results in gross overcrowding and consequent stunting of their growth. Lines of research aimed at controlling spawning included use of monosex culture (one sex of a fish present in a pond) and cultivation of tilapia in full-strength seawater where they can survive, but do not spawn. Another approach is the use of a predator fish (sea bass) that does not reproduce in fresh water and will control wild spawning.

Other cultured fishes in earlier stages of development include trout as well as (*Macrobrachium rosenbergii*), the Malaysian prawn. There is also early development of desert aquaculture in the lower Negev, including the integration of brackish pond aquaculture with intensive agricultural production in regions of high temperature and critical water availability. Cultivation of fish, such as tilapia, with variable-salinity tolerances, as well as adapted marine species, offers potential for polyculture in desert areas in other parts of the world, as well as in Israel.

Specific Aquaculture Objectives. Whether freshwater or marine species, the objectives of various countries as regards aquaculture cover a rather wide range. As of the early 1980s, these objectives in terms of species are given in the accompanying table. Only the main targets are given; the table is not meant to be all-inclusive.

Shortcutting the Conventional Food Chain. The small size of most of the food in the sea is one of the most serious constraints imposed by nature. In the ocean, unlike the land, most of the food is cycled through all but the final portions of the food chain in microscopic steps. The reasons for this are obscure, but they probably related to the absence over most of the ocean of a substrate in which large

WORLDWIDE AQUACULTURE ACTIVITY

Country	Principal Species Cultured or Under Study for Potential Culture
Australia	Rainbow trout; crayfish (marrons and yabbies); native fish (Murray cod, golden perch, silver perch, catfish)
Austria	Rainbow trout; carp; tench; grass carp; lake whitefish
Belgium and Luxembourg	Rainbow trout; brown trout; carp; tench; roach; pike; pike perch; European eel
Canada	Salmon; trout; char
China (People's Republic of)	Common carp; crucian carp; grass carp; bighead; seaweeds
Denmark	Rainbow trout; brown trout; European eel
France	Rainbow trout; brown trout; carp; European eel; coho salmon; crayfish
Germany (West)	Rainbow trout; brown trout; carp; European eel; tench; northern pike
Hungary	Carp; barbel; sturgeon; tench; bream; Danubian wels; pike-perch; pike; asp; brown trout; perch
Indonesia	Milk fish; common carp; grass carp; silver carp; kissing fourami; nilem; sepat siam; catfish; tilapia; shrimp
Ireland	Rainbow trout; brown trout; European eel
Israel	Carp; tilapia; mullet; silver carp
Italy	Rainbow trout; European eel; black bullhead; carp; mullet; sea bass; gilthead
Japan	Rainbow trout; freshwater eel; common carp; crucian carp; auy or sweetfish; loach; mullet; yellowtail; red sea bream (red porgy); horse mackerel; shrimp
Korea (South)	Carp; Japanese eel; rainbow trout; loach; grass carp; catfish; blue gill; bass
Netherlands	Roach; pike-perch; carp; pike; rainbow trout
Norway	Atlantic salmon; rainbow trout
Papua New Guinea	Rainbow trout; brown trout
Philippines	Milk fish
Portugal	Rainbow trout; brown trout; carp; steelhead trout; Atlantic salmon
Spain	Rainbow trout; Atlantic salmon; European eel
Sweden	Rainbow trout; Atlantic salmon
Switzerland	Rainbow trout; brown trout; carp; northern pike; lake whitefish; steelhead trout; char; pike-perch; perch
Taiwan	Milk fish; tilapia; Japanese eel; gray mullet; shrimp; carp
United Kingdom	Rainbow trout; carp; Atlantic salmon; turbot; Dover sole
United States	Rainbow trout; catfish; American eel; prawns; crayfish; salmon
U.S.S.R.	Sturgeon; trout; carp; Pacific salmon

plants can attach at depths shallow enough for them to receive the light required for photosynthesis.

Each of the microscopic steps up the marine food chain is perhaps only 10 to 15% efficient, so that of the great amount of potential food initially fixed by photosynthesis in the phytoplankton, only a very small portion emerges in the form of fish to be caught.

The codfish, for example, eats predatory crustaceans and gastropods and small fish, which in turn, eat small herbivorous invertebrates. Thus, each million weight unit of cod production requires 10 million weight units of smaller fish and crustaceans, which in turn, requires 100 million weight units of small herbivorous invertebrates. These, in turn, consume 10 times their own weight in plant matter and a vast amount of bottom land.

McLeod (1969) estimates that about 70% of the total annual food requirements of the adult winter flounder are met by detritus and phytoplankton-eating invertebrates. Through a series of calculations, McLeod estimates that to produce 100 weight units of winter flounder, a total of 4088 weight units of direct and indirect vegetable matter is required.

For this reason, large-scale, controlled fish farms can be achieved only in restricted lagoons and estuaries, where the migratory propensities of genetically improved fish can be curbed, and where fertilization of the water with essential nutrients is accomplished by artificially induced upwelling or by direct introduction, and where both nutrients and fish can be retained.

In shallow waters, rice paddies, or estuaries, the autotrophic (plants) and heterotrophic (animal) layers are in close contact. In the sea, the autotrophs are small and the heterotrophs are large, whereas on land, the autotrophs are large (trees) and the heterotrophs are small. Organic detritus is the chief link between these levels of primary and secondary productivity, rather than a grazing food chain as on land. Although these productive areas may be less stable than the land, the advantages in such a detritus food chain is that microbial manipulations can be used to produce protein, which is available as food. Thus, an estuary has food all the time for its populations.

Certain steps in the food chain of the sea can be circumvented to increase the ultimate yield by a factor of 6 to 10 for each step circumvented. A filter-feeding creature, like a mussel, clam, or oyster, can aggregate phytoplankton in one step into protein. Instead of moving uneconomically large masses of water to extract the plankton, efficient creatures can be used to concentrate the plankton wherever the natural movement of the water is sufficient to keep replenishing the plankton food supply. The concentration of plankton by natural advection is important in many places in the sea—at boundaries between current systems and around islands and shoals—and this can be a highly efficient concentrating mechanism. It can increase the local supply of plankton by a factor of 50 or more over the basic productivity of the contributing water.

Introduction of New Species. Closely allied to the technology of aquaculture is the introduction of new species. Reasons for introducing "new" species into a region include the market for a given species that may have to be imported from distant places to the establishment of a new industry and thus bolster the regional economy. In an excellent summary of the introduction of exotic species in aquaculture,

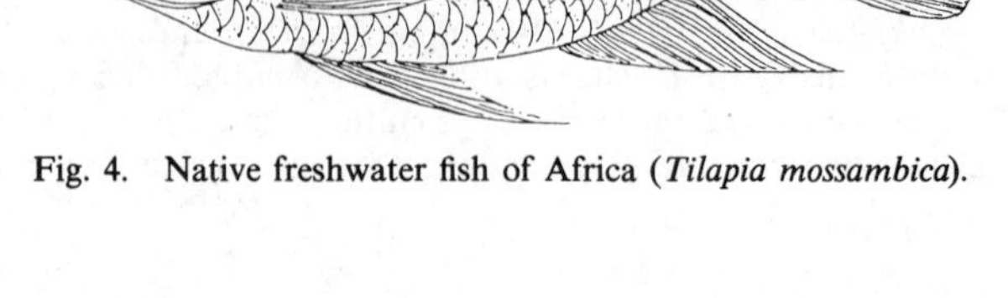

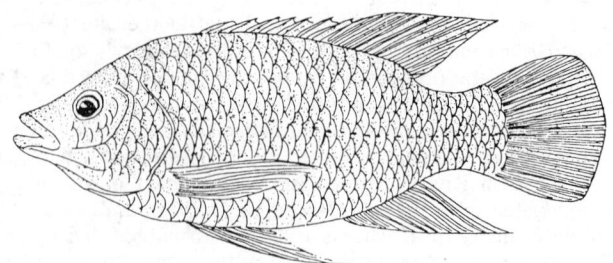

Fig. 4. Native freshwater fish of Africa (*Tilapia mossambica*).

Mann (1979) cites a number of examples, including (1) introduction of the Japanese oyster to the Pacific coast of North America and to France; and (2) introduction of the *Tilapia mossambica* (See Fig. 4), a native freshwater fish of Africa, to a number of countries, including Indonesia, Malaysia, Hong Kong, the Philippines, Vietnam, Cambodia, Laos, Taiwan, and China. The latter introduction was so successful that the fish is now one of the most important pond-cultured fish in southeast Asia. Infrequently, introductions of species can produce catastrophic instead of beneficial results and thus much trial and experimentation must go into planning such ventures. To date, as mentioned by Mann, the most disastrous introductions have occurred in terrestrial systems—rabbits in Australia, Dutch elm disease in the United States, European rats in the Pacific islands, and the giant snail, which has spread from West Africa throughout much of Asia. Guidelines for future introductions of aquatic species are being worked out by a number of organizations, including the International Council for the Exploration of the Sea (ICES).

Major Aquaculture Ventures in the United States. *Salmon.* Historically, the domestic cultivation of salmon in the Pacific Northwest has been a major aquaculture venture in the United States. Much technical attention has been directed toward maintaining salmon productivity. A new turn was taken in late 1970, when a pilot project was commenced to evaluate the rearing of Pacific salmon under captive conditions. With the cooperation of the National Marine Fisheries Service (NMFS), a joint government-industry[1] pilot program was initiated. Farms are located in the Puget Sound area of Washington. They involve two freshwater sites and two saltwater sites which sit across from each other at Manchester and Bainbridge Island, Washington.

Friedman (1978) points out that the Pacific coho salmon is an anadromous fish. It spends its early life cycle within freshwater and is not transported into saltwater until it possesses the physiological ability to osmoregulate within the saltwater environment. At this time, the coho salmon from a morphological standpoint has also lost its vertical or parr marks and has obtained a silvery coat which represents the deposition of guanine and hypoxanthine.

The coho salmon spends the following 6 to 12 months in saltwater, developing to approximately 340 grams. At this time, the salmon may be harvested as a pansized fish, or permitted to mature. If allowed to mature, these salmon then become brood stock, are spawned directly from saltwater, and are utilized as a source of future progeny. The new season commences with the return of the adult salmon to freshwater to spawn. The coho orient against the most rapidly flowing portion of a stream and will readily enter a boxlike trap wherein they are captured and retained until sufficient quantities are obtained for spawning. See also **Salmon.**

In the process, once sufficient coho salmon have been gathered within the trap, they are checked for maturity, killed, and segregated by sex, body size, and conformation. Eggs are then removed from the two ovarian sacs of the female. Milt previously artificially stripped from male coho salmon and collected within a bucket immediately contacts the micropore of the egg and enters. After ferilization, water is added to the bucket, and the excess milt is washed away and the eggs are permitted to "harden" prior to movement into the hatchery.

The eggs then are incubated. Healthy eggs under incubation are normally pink to reddish-orange and somewhat transparent. Inspections are frequently made and dead eggs are removed to prevent fungal growth. Three weeks after hatching the small coho fry will have absorbed their yolk sacs and are ready for placement within freshwater pools. They remain there until they have the physiological capacity to undergo smoltification, that is, the ability to osmoregulate in saltwater. The young salmon are vaccinated and transferred to saltwater rearing pens. Throughout the history of the pilot project, bag-type net pens supported by flotation have been a key element in the salmon-culturing program. The present net pen system can produce over 660,000 pounds (300,000 kilograms) of protein annually. To feed the salmon a total of seven feed-storage bins are servicing sites and can store approximately 300,000 pounds (136,000 kilograms). The saltwater diets are formulated by linear programming techniques.

Catfish. One of the more recent ventures in aquaculture in the United States has been that of catfish farming. The commercial production of catfish has grown from less than 400 acres (160 hectares) producing 320,000 pounds (145,000 kilograms) per year in 1960 to upwards of 60,000 acres (24,000 hectares) producing 75 million pounds (34 million kilograms) per year in the late 1970s. A primary deterrent to achieving a faster growth rate of the industry has been a persistent shortage of processing facilities.

Catfish farmers use either ponds, raceways, or cages. Three species of catfish are produced—the *channel catfish* (*Ictalurus punctatus*), the most common species raised; the *blue catfish* (*I. furcatus*), sometimes confused with the *channel catfish*; and the *white catfish* (*I. catus*). There are also bullheads (*Ictalurus* sp.), which are more difficult to manage and they often overpopulate ponds and can be quite troublesome, as well as bringing lower prices. The detailed technology for farming catfish is well described in the Grizzell reference listed.

[1] Source: Union Carbide Corporation.

References

Anon. (1): "Shrimp Farming may become Big Industry," *The Futurist*, **10**, 4, 214 (1976).

Anon. (2): "Fish Farmers get Ready to Hook into Profits," *Business Week* (June 20, 1977).

Anon. (3): "Fish Farming Continues to Gain Popularity," *Canner Packer*, **146**, 7, 54 (1977).

Anon. (4): "Global Shortage of Seafood Coming," *Sea Technology*, **18**, 9, 9 (1977).

Anon. (5): "Increased Fish Yields, Better Water Use with Polyculture Fish Farming," *Food Product Dev.*, **11**, 2, 18 (1977).

Bardach, J. E., Ryther, J. H., and W. D. McLarney: "Aquaculture. The Farming and Husbandry of Freshwater and Marine Organisms," Wiley-Interscience, New York (1972).

Booda, L. L.: "Oysters Grown in Kansas City," *Sea Technology*, **18**, 6, 16 (1977).

Brown, E. E.: "World Fish Farming: Cultivation and Economics," AVI, Westport, Connecticut (1977).

Culliton, B. J.: "Aquaculture: Appropriate Technology in China," *Science*, **206**, 539 (1979).

Dill, W. A.: "Proceedings of the World Symposium on Warm-water Pond Fish Culture," pp. 18–25, Rome (1967).

Edmunds, W. J., and D. A. Lillard: "Sensory Characteristics of Oysters, Clams, and Cultured and Wild Shrimp," *J. Food Sci.*, **44**, 2, 368–373 (1979).

FAO: "International Symposium on Finfish Nutrition and Feed Technology," European Inland Fisheries Advisory Commission (EIFAC) and Food and Agriculture Organization (United Nations), Hamburg, Germany, 1978.

Friedman, B. A.: "Domestic Culture of Salmon in the Pacific Northwest," *Food Technology*, **32**, 7, 32–34 (1978).

Gallese, L. R.: "U.S. Firms are hoping to Net Big Profits in Rapidly Expanding Fish-farm Industry," *The Wall Street Journal* (January 6, 1976).

George, M. J.: "Genus *Macrobrachium*," in "Prawn Fisheries of India," *Bull. Central Marine Fisheries Research Inst.* (Mandapam Camp, India), **14**, 178–216 (1969).

Glude, J. B. (editor): "NOAA Agriculture Plans," U.S. Department of Commerce, Washington, D.C. (1977).

Grizzell, R. A., Dillon, O. W., Jr., and E. G. Sullivan: "Catfish Farming," Farmers Bulletin 2260, U.S. Department of Agriculture, Washington, D.C., 1975.

Holden, C.: "Government Seeking Ways to Encourage Aquaculture," *Science*, **200**, 33–35 (1978).

Hunter, J. V.: "Opportunities for Crawfish in Crustacean Aquaculture," *Feedstuffs*, **49**, 14, 18–19 (1977).

Johnson, D. S.: "Some Aspects of the Distribution of Freshwater Organisms in the Indo-Pacific Area and Their Relevance to the Validity of the Concept of an Oriental Region in Zoogeography," *Proc. Cent. Bicent. Congr. Biol.*, pp. 170–181, Singapore, 1958.

Kikuchi, W. K.: "Prehistoric Hawaiian Fishponds," *Science*, **193**, 295–299 (1976).

Ling, S. W.: "Aquaculture in Southeast Asia, Univ. of Washington Press, Seattle, 1977.

Lovell, R. T.: "Fish Culture in the United States," *Science*, **207**, 1368–1372 (1979).

Malecha, S. R.: "Genetics and Selective Breeding," in "Shrimp and Prawn Farming in the Western Hemisphere," pp. 328–355, Dowden, Hutchinson & Ross, Stroudsburg, Pennsylvania, 1977.

Mann, R.: "Exotic Species in Aquaculture," *Oceanus*, **22**, 1, 29–35 (1979).

Matthes, M.: "A Bibliography of African Freshwater Fish," Food and Agriculture Organization (United Nations), Rome, 1973.

McLeod, G. C.: "Sea Farming," in "The Encyclopedia of Marine Resources," (F. E. Firth, editor), Van Nostrand Reinhold, New York, 1969.

McVey, J. P.: "New Record of *Macrobrachium rosenbergii* (de Man) in the Palau Islands," *Crustaceana*, **29**, 31–32 (1975).

Meyers, S. P.: "Use of Agricultural Wastes in Aquaculture," *Feedstuffs*, **48**, 7, 34–50 (1977).
Moiseev, P. A.: "Development of Fisheries for Traditionally Exploited Species," *J. Fish. Res. Bd. Canada*, **30**, 2109–2120 (1973).
Pillay, T. V. R.: "The State of Aquaculture, 1975," *FAO Technical Conference on Aquaculture*, Kyoto, Japan (May 26–June 2, 1976).
Ryther, J. H.: "Aquaculture in China," *Oceanus*, **22**, 1, 21–28 (1979).
Rutherford, R.: "California Shellfish Hatchery Supplies International Market," *Fish Farming Int.*, **2**, 2, 26 (1975).
Sandifer, R. A., and T. I. J. Smith: "Aquaculture of Malaysian Prawns in Controlled Environments," *Food Technology*, **32**, 7, 36–45 (1978).
Shoemaker, J.: "Happy Shrimp, Fish May Grow into PSE&G Profits," *Philadelphia Inquirer*, (June 15, 1975).
Staff: "Aquaculture in the United States," National Academy of Sciences, Washington, D.C., 1978.
Twitty, T.: "Big Scale Sea Food Farms Planned," *Orlando Sentinel Star* (February 1, 1976).
Weatherley, A. H., and B. M. G. Cogger: "Fish Culture: Problems and Prospects," *Science*, **197**, 427–430 (1977).
Wheaton, F. M.: "Aquacultural Engineering," Wiley, New York, 1977.

AQUAMARINE. A form of gem beryl. See **Beryl.**

AQUA REGIA. Also known as nitrohydrochloric acid, aqua regia is made up of three parts hydrochloric acid and one part nitric acid, each of the usual concentrated laboratory form. Aqua regia will dissolve all metals except silver. The latter is converted to silver chloride. The reaction of metals with nitrohydrochloric acid typically involves oxidation of the metal to a metallic ion and the reduction of the nitric acid to nitric oxide. Aqua regia also dissolves the common oxides and hydroxides of metals with the exception of silver, the ignited oxides of tin, aluminum, chromium, and iron, and the higher oxides of lead, cobalt, nickel, and manganese, the latter dissolving effectively in hydrochloric acid alone.

AQUARIUS (the water bearer). A relatively large constellation, important solely because of the fact that it is the eleventh sign of the zodiac. It contains no bright or particularly striking features. There is a theory that this constellation received its name because it appears over the Euphrates valley during that region's rainy season.

AQUEDUCT. An artificial conduit built to carry water is called an aqueduct. Generally speaking, aqueducts are built to convey a fresh water supply to congested districts from suitable sources more or less distant, and are therefore peculiar to cities. The first settlers in a place may depend upon local springs, streams, and wells, but with the growth of population there comes a time when these will prove inadequate, and suitable distant water supplies may have to be trapped through the medium of the aqueduct. An aqueduct may be either a pressure or grade-line type. Pressure conduits are commonly employed for small capacities or adverse topography, open or grade-line conduits for large capacity or favorable topography. Circumstances may require both types on the same project as the most economical combination. Pressure tunnels can convey water at pressures considerably above atmospheric, and are constructed with curved cross-sections. They are most frequently found in tunnel sections cut through hills and mountains, and in siphons. The principal distinguishing hydraulic characteristic of the pressure aqueduct is that it may depart from the normal open flow line both above and below the normal hydraulic gradient. However, siphon action, for sections above the hydraulic gradient, should be avoided wherever possible. Grade-line sections of aqueducts are usually built with open cut and fill construction following a hydraulic grade-line which will yield the requisite flow in the aqueduct at approximately atmospheric pressure, i.e., the fall per mile being just sufficient to overcome the friction loss in the same distance.

Some of the most important of the ancient aqueducts were those supplying the city of Rome, among which might be mentioned the Marcian, with a length of 58 miles (93 kilometers), the Julian, a length of 17 miles (27 kilometers), and the Claudian, with a length of 43 miles (69 kilometers). These were high level aqueducts of the grade-line type, principally cut and fill where possible, with grade-line tunnels for piercing hills, and resting on multiple arches when spanning valleys. These older aqueducts rarely had cross-sections greater than 30 sq. ft (2.8 sq. meters) in area. The Catskill aqueduct which conveys the water of the Ashokan Reservoir to the City of New York, approximately 100 miles (160 kilometers) away, has a capacity of 500,000,000 gallons (18,925,000 hectoliters) a day and is a splendid example of modern engineering on a large scale. It has in places cross-sections greater than 150 sq. ft. (14 sq. meters) in area, inverted siphons, one going more than 1000 feet (300 meters) below sea level, as well as a score of tunnels.

AQUEDUCT OF SYLVIUS. The portion of the central canal of the nervous system of vertebrates which lies in the mid-brain, connecting the third and fourth ventricles.

AQUEOUS HUMOR. Glaucoma; Vision and the Eye.

AQUIFERS. Hydrology.

AQUIFOLIACEAE. Box Trees and Shrubs; Holly Trees and Shrubs.

AQUILA (the eagle). A constellation lying in the Milky Way, and hence containing rich star fields when viewed with a low-powered telescope. The distinguishing feature of this constellation is the group of three stars almost in a straight line, with the bright star Altair (α Aquila) between two fainter ones. Several novae have appeared in this constellation, the most famous one being Nova Aquilae III of 1918. See map accompanying entry on **Constellations.**

ARABIAN CAMEL. Camels and Llamas.

ARABIAN PLATE. Ocean.

ARACHIDIC ACID. Also known as eicosanoic acid, formula $CH_3(2)_{18}COOH$. A widely distributed, but minor component of the fats of certain edible vegetable oils. Shining, white crystalline leaflets; soluble in ether, slightly soluble in water. Sp gr, 0.8240 (100/4°C); mp 75.4°C; bp 205°C (1 millimeter pressure). Decomposes at 328°C. Commercial product derived from groundnut (peanut) oil. Used in organic synthesis, lubricating greases, waxes, and plastics. Source of arachidyl alcohol. See also **Vegetable Oils (Edible).**

ARACHNIDA. A class of the phylum *Arthropoda* including the spiders, mites, ticks, scorpions, pseudoscorpions, whip scorpions, sun spiders and harvestmen. Next to the insects this class is probably the best known among the invertebrates.

Arachnids differ from the other members of the phylum in one or more of the following characteristics: (1) The body is usually divided into two regions, a cephalothorax and abdomen. (2) Only simple eyes are present. (3) There are no antennae. (4) The thorax bears four pairs of legs in the adult. (5) The abdomen is often unsegmented and bears no appendages. (6) The first pair of appendages are chelate grasping organs. (7) Respiration is carried on by tracheae or lung-books. Several arachnids and related forms are shown in the accompanying illustration.

Arachnids are almost exclusively terrestrial and are predominantly predacious or parasitic, although some of the mites are plant feeders and the harvestmen include vegetable materials among their food.

The development of poison glands is fairly general in the group. Spiders have such glands, opening in the jaws or chelicerae, and scorpions have a special sting at the tip of the abdomen. With the exception of the black-widow spider of the United States and a small scorpion found near Durango, the poison is not known to be harmful to man. There is some probability that these two species may sometimes inflict fatal wounds.

The secretion of silk by spiders is another salient feature of the group. Silk glands are located in the abdomen, discharging through a group of spinnerets near the posterior end of the body. The silk is used to build webs of various forms for snaring prey, for the construction of cocoons to receive the eggs, as a lining for burrows, and in some cases as a vehicle to carry the animal on currents of air.

The economic importance of arachnids is rather limited. Spider

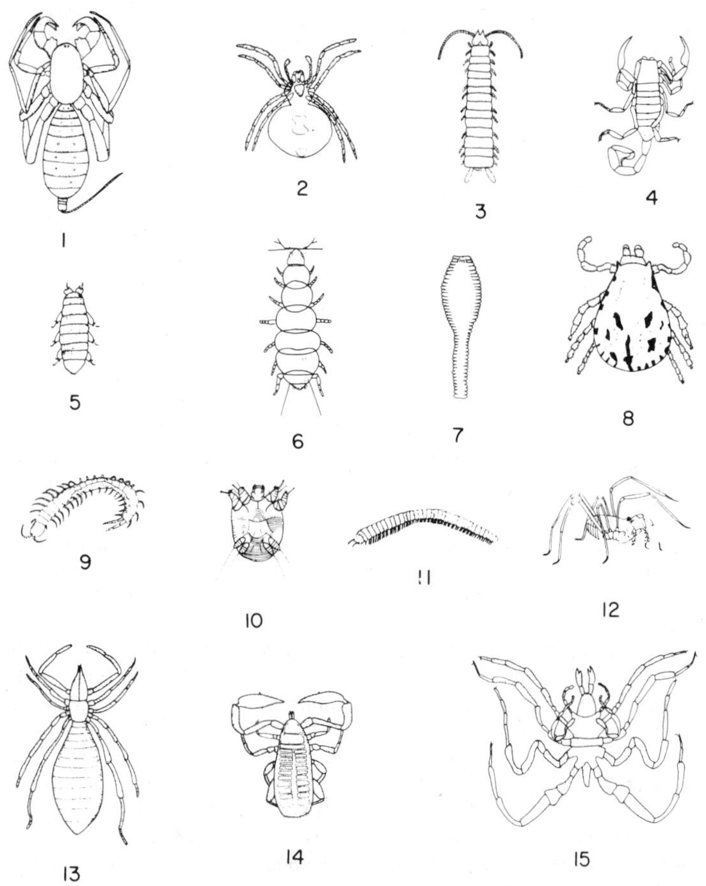

Arachnids and related forms: (1) Whip scorpion; (2) female black widow spider, *Latrodectus*, ventral view; (3) symphylid; (4) scorpion; (5) tardigrade or water bear; (6) pauropod; (7) linguatulid or tongue worm; (8) tick, *Dermacentor*; (9) centipede; (10) itch mite, *Sarcoptes*; (11) millipede; (12) opilionid; (13) solpugid; (14) pseudoscorpion; (15) pycnogonid or sea spider.

silk has been woven but it is too delicate for extensive use and is valuable only as a source of cross-hairs for optical instruments. Aside from the poisons mentioned above, the principal harm from these animals is derived from the mites and ticks. The ticks do some damage to man and domestic animals by sucking blood but their greatest damage is due to the transmission of diseases. Texas fever of cattle, Rocky Mountain spotted fever of man, and other diseases are so conveyed. Mites living in the hair follicles, the sebaceous glands, and the tissues of the skin cause such diseases as scab in sheep and itch in man. Plant-feeding species of economic importance include the bulb mite, the pear blister mite and various gall mites.

The classification of the arachnids is briefly as follows:

Order *Scorpionida.* The scorpions. Abdomen divided into a preabdomen and a slender postabdomen bearing a claw-like sting at the tip.

Order *Pedipalpi.* The whip-scorpions. Anterior pair of legs slender and antenna-like.

Order *Solpugida.* The sun spiders. Head and thorax separate.

Order *Chelonethida.* Pseudoscorpions. Very small scorpion-like animals; no postabdomen nor sting.

Order *Phalangida.* Harvestmen or daddy longlegs; commonly regarded as spiders but have a segmented abdomen broadly joined to the thorax.

Order *Araneina.* Spiders. Abdomen unsegmented and joined to the thorax by a slender waist.

Order *Acarina.* Mites and ticks. Small to moderate species with a sac-like body showing no well-marked divisions.

ARACHNOID. Brain and Nervous System.

ARAGONITE. The mineral aragonite is calcium carbonate, $CaCO_3$, chemically identical with calcite but crystallizing in the orthorhombic system, with acicular crystals. By repeated twinning, pseudo-hexagonal forms result. Aragonite may be columnar or fibrous, occasionally in branching stalactitic forms called flosferri (flowers of iron) from their association with the ores at the Carinthian iron mines. Its hardness is 3.5–4; specific gravity, 2.93–2.95; luster, vitreous to resinous; colors, white, gray, green-yellow or purple; transparent to translucent. Aragonite forms at temperatures of 80–100°C and is relatively unstable at ordinary temperatures and pressures. It alters to calcite, although very slowly. There are many localities for aragonite in Europe, Bolivia, Pennsylvania, Iowa, Missouri, South Dakota, New Mexico, Arizona and Colorado. Its name is derived from Aragon in Spain. See also **Calcite.**

ARAPAIMA. Bony Tongues.

ARAUCARIAS. Trees of the family *Araucariaceae*, sometimes referred to as the Chile pine family or Pacific pines. These are evergreen trees of usually large size and heights ranging from 80 to 200 feet (24 to 60 meters) for adult trees. Their leaves are oval, leathery, and spirally arranged. The cones are round or oval. They are most unusual trees of the southern hemisphere. The Chile pine or monkey puzzle tree (*Araucaria araucana*) rises to a height of about 80 feet (24 meters). The leaves are long, pointed, triangularly shaped and of a yellow-green color. The tree is apparently immune to known diseases and can tolerate all but the wettest or driest of soils. A group of these trees tends to purvey the appearance of a quickly assembled Hollywood set for a jungle movie. The Parana pine or candelabra tree (*A. angustifolia*) is found in southern Brazil and rises to a height of about 110 feet (33 meters). It displays a limited number of branches, has a flat crown, and long hanging branchlets. The leaves possess curved spiny tips and are scaly. The bunya-bunya tree (*A. bidwillii*) occurs in Queensland and attains a height of 150 feet (45 meters). The cones are large, up to 12 inches (30 centimeters) in length. The Norfolk Island pine (*A. excelsa*) rises to a height between 150 and 200 feet (45 and 60 meters). The leaves are small, usually overlapping on older shoots. The cones range from 3 to 4 inches (7.5 to 10 centimeters) in length and are about as broad.

It is postulated by some that a bridge between the aforementioned and most unusual trees and the Giant Sequoias is filled by the Chinese fir (*Cunninghamia lanceolata*) on the one hand and by the cryptomerias or Japanese cedars. The Chinese fir occurs in central and southern China and is a large, spreading, and weeping evergreen that rises to a height of about 150 feet (45 meters). The crytomerias are of several species, including the aforementioned Japanese cedar (*Cryptomeria japonica*), which is a narrow, conical tree (older trees are dome-topped) and rises to a height of about 150 feet (45 meters). The Chinese fir is an important timber tree in China, as is the cryptomeria in Japan.

ARAWANA. Bony Tongues.

ARBOR. A shaft or stud, usually cylindrical or conical, on which a cutting tool, a tool holder, or a part to be machined is mounted or held.

ARBORVIRUSES. Virus.

ARBORVITAE. Trees and shrubs of the genus *Thuja*, family *Cupressaceae* (cypress family). The thujas include the large eastern white cedar (*Thuja occidentalis*) and the western red cedar (*T. plicata*). Although *T. occidentalis* can attain heights in excess of 100 feet (30 meters), the tree is frequently used in landscape planning and for hedgerows. The leaves are bright green, having overlapping scales. The cones are quite small. The natural range of the species is from southern Labrador to Nova Scotia, and west into Manitoba and Minnesota, and ranging south into Pennsylvania and following the mountains into North Carolina and eastern Tennessee. Specimens in the south are considerably smaller and may be described as small trees or shrubs. The species is quite common in northern New England. Asian thujas include the Japanese arborvitae (*T. standishii*), rare outside of Japan, and the Chinese thuja (*T. orientalis*), a small, often multistemmed tree. The *T. plicata,* also known as the zebra-striped western red cedar, is also used extensively as an ornamental tree.

ARC BACK. This is the occurrence of an arc from anode to cathode in a gaseous rectifier tube. Normally such a tube has electrons flowing from the cathode to the anode but under certain conditions excessive heating of the anode, excessive voltage across the tube, or other effects may cause the anode to emit electrons and allow an arc discharge to take place in a direction opposite to the normal direction. Under many circuit conditions this may destroy the tube or it may merely open the protective devices.

ARC CUTTING. Cutting of metal by means of an arc formed between the metal and an electrode. See also arc welding under **Welding.**

ARC (Electrical). A low-voltage, high-current electrical discharge, as contrasted with a spark. The electric arc, so called because of the shape of the "flame," was discovered by Davy about 1808. It is a type of discharge between electrodes in a gas or vapor which is characterized by a relatively low voltage drop and a high current density. The two types which are of considerable practical importance are the arc in open air and the arc in gases at low pressure. The familiar carbon-arc and the electric-arc furnace are examples of the former. In this type the arc is started by impressing a voltage across the electrodes in contact and then separating them. At the instant of breaking contact the high field and current density initiate the arc. Thereafter, if the current is kept constant, the potential necessary is a linear function of the interelectrode distance. In its steady state, the arc has an intensely hot cathode which emits a plentiful supply of electrons. The energy for heating the cathode is obtained from the high current density and from the bombarding positive ions. The arc is thus apparently a thermionic phenomenon. If the carbons are impregnated with a volatile metallic salt, the result is a "flaming arc," useful in producing the arc spectrum of the metal. If a dc carbon arc is placed in parallel with a suitable condenser and an inductance, the circuit so formed may be made to oscillate, as discovered by Duddell, and to serve as a source of undamped electric waves.

The arc is one of the most serious problems in switching electrical circuits, since the separation of the switch or circuit breaker contacts establishes an arc which must be extinguished in order to break the circuit. Many schemes have been developed to accomplish this. See **Circuit Breaker.**

The mercury-arc tube is the most important example of an arc in a gas at low pressure. Here the gas is the mercury vapor, the cathode is the mercury pool, and the anode is usually carbon. The arc is initiated by breaking contact between the mercury pool and a starting electrode. In gases at low pressure an arc may be established without breaking contact between the electrodes if the impressed voltage is high enough. If the voltage impressed across the two electrodes separated by a low-pressure gas is gradually increased the current increases slowly at first due to the residual ions and electrons present in the gas. The electrons ionize the gas molecules upon colliding with them, thus giving rise to additional ions and electrons to carry the current and also to produce more ionization by collision. This process continues until suddenly the breakdown voltage is reached, when the current increases very rapidly, even if the voltage is lowered. This is the initiation of the glow discharge. If the circuit resistance does not limit the current the discharge progresses almost instantaneously into an arc discharge having the distinguishing features of the regular arc. The discharge in thyratrons and other gas-filled hot cathode tubes is often called an artificial arc discharge since it is characterized by low voltage and high current density. It is not a true arc however since the cathode heat energy is supplied by an external source and not by the discharge itself.

ARC FURNACE. A furnace heated by an electric arc. The arc may be formed either between an electrode and the metal charge to be heated, or between several arcs placed over the charge.

ARC (Geometry). Circle.

ARCH. An arch is a curved beam or truss made of wood, masonry, concrete or steel, whose supports are able to exert lateral as well as vertical forces to resist the action of any applied loads. These lateral forces are in the nature of thrusts which act inwardly toward the center of the arch span. The curvature of the beam must be in an upward direction in order to develop lateral reaction forces which will act in the required direction. The highest point of the arch is called the crown. The juncture of the arch and its foundation is called the springing.

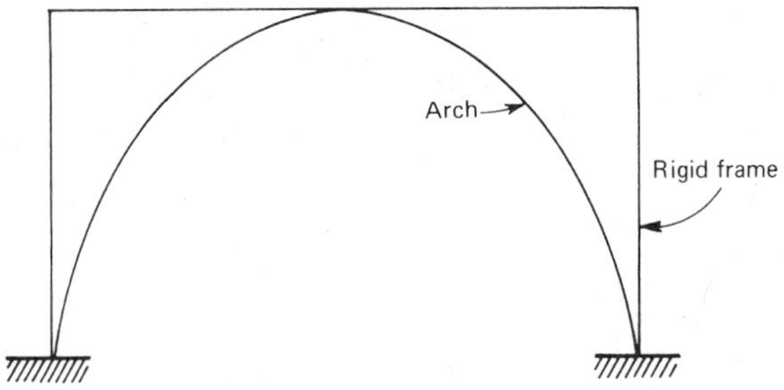

Fig. 1. Comparison of arch with rigid frame.

In an arch, loads induce both bending and direct compressive stress. Although all loads are vertical, arch reactions possess horizontal components. Also, deflections have both horizontal and vertical components.

The masonry arch was used by the Egyptians, Babylonians, Assyrians, and Greeks but received full expression in the bridges and buildings of the Romans. Such arches depended upon the compressive thrust of adjacent blocks, or *vissoirs*, for their stability. That is, the line of pressure, or pressure line, had the same shape as the arch. Under the heavy moving loads of today the pressure line can depart appreciably from the arch axis and thus large tensile forces due to bending can be developed. Masonry is unsuited for such conditions and it is necessary to use reinforced concrete or steel arches. Steel arches are of the solid-rib or trussed type.

These structures may be further classified as fixed or hinged arches. A fixed arch is a structure which is rigidly connected to its supports in such a manner that they exert vertical and lateral reactions and prevent rotation. A two-hinged arch is one which is free to rotate

Fig. 2. Gateway Arch in Saint Louis, Missouri. Structure is 630 feet (192 meters) high.

about its supports, consequently they are able to exert only vertical and lateral reactions. Three-hinged arches have an additional hinge, usually midway between the hinges at the supports. The tied arch is a structure in which the lateral forces are applied by means of a horizontal tension member connecting the ends of the arch.

From the standpoint of structure, an arch is similar to a rigid frame. See Fig. 1. Arches are not commonly precast because stacking for transportation is difficult due to the curvature. However, small spans cast at the site have been used successfully. Prestressing seldom is advantageous because arches are subjected to large compressive forces.

Arches are frequently used for very long spans and support roofs of such structures as auditoriums, exhibition halls, hangars, stadiums, sporting arenas, and transportation terminals. Typical arch bridges are shown under **Bridge (Structural).**

One of the most interesting arches is the 630-foot (192-meter) high Gateway Arch, made of stainless steel, situated in downtown St. Louis, Missouri. This catenary arch was designed by Eero Saarinen and is the tallest monument in the United States. See Fig. 2.

ARCHAEOPTERYX. Fossil Birds.

ARCHAIC. An archeological term to designate a prehistoric cultural stage following the Lithic. The Archaic stage is characterized by a foraging pattern of existence and numerous types of stone implements.

ARCH BRIDGE. Bridge (Structural).

ARCH DAM. Dams.

ARCHEGONIATES. Those plants in which the female sex organ is an archegonium are sometimes called Archegoniates.

ARCHEGONIUM. The multicellular female sex organ characteristic of all the *Embryophyta* except the Angiosperms. It consists of a swollen basal portion called the venter, and an elongated neck. The venter may be a single layer of cells, but is often many cells thick. Within the basal portion is contained the single large egg, and a second, somewhat smaller, ventral canal cell, while the elongated neck contains a single row of cells which eventually dissociate, leaving an open canal through which the sperm may pass to reach the egg.

ARCHEOCYTE. Cells of sponges which ingest and digest food, carry the products to other parts of the body, and form reproductive cells. They are amoeboid (see **Amoeba**) and are found in the mesenchyme.

ARCHEOLOGICAL DATA TECHNIQUES. Radioactivity and Other Dating Techniques.

ARCHEOPHYTIC. A paleobotanic division of geologic time. The period of initial plant evolution, more specifically algae. See also **Paleobotany.**

ARCHEOZOIC (Archean). The oldest of the five Eras of the earth's history. The rocks of this System are metamorphosed equivalents of all types of sedimentary and igneous rocks, but principally the latter. No undisputed fossils have been found in the Archean. The lower Archean (Keewatin) of North America is composed of a preponderance of metamorphosed, basaltic lava flows and tuffs with some metamorphosed sediments, such as quartzite and slate. The general character of the basal Archean proves that the oldest known rocks do not represent the original crust of the earth. The upper Archean (Laurentian) contains a preponderance of granite, gneisses, and schists in the form of batholiths intruding the Keewatin. The principal areas of Archeozoic rocks are in Canada, Finland, Scandinavia, Australia, Africa and northeastern South America. Many of the formations contain rich ore deposits, especially of gold and silver. Large amounts of graphite suggest the former existence of life. Length of time since the beginning of the Archeozoic, possibly more than 2500 million years. Owing to the lack of fossils, structural complexity, high degree of vulcanism and metamorphism, geologists have found great difficulty in deciphering the history of this earliest recognizable portion of the "crust" of the earth. The structural history of the Archean is therefore not so well known as that of the succeeding periods and intercontinental correlation is particularly difficult. On the other hand, the search for ore deposits has been an important stimulus to the study of the Archean formations, especially in Canada.

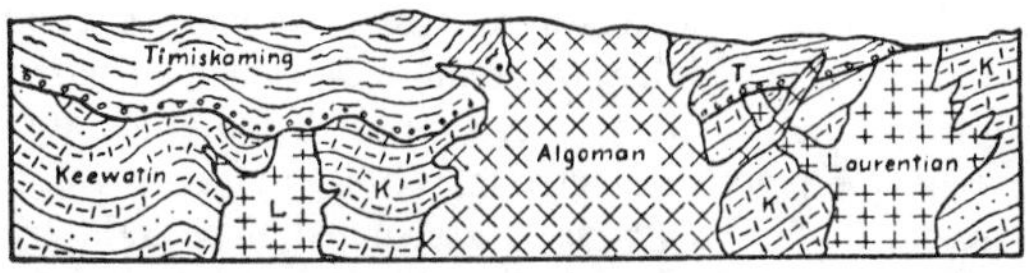

A highly generalized section, about 25 miles (40 kilometers) long, showing the relations of the Archeozoic group of rocks in the Lake Superior–Lake Huron region of Canada. The Keewatin system was moderately folded and intruded by the Laurentian granite, after which there was deep erosion. Then the Timiskaming rocks were laid down, and later, were strongly folded and intruded by the Algoman granite, after which there was another period of profound erosion, marked by the upper surface.

ARCHIANNELIDA. A class of annelid worms including small marine species of simple structure.

ARCHIMEDES SPIRAL. A special case of a spiral, represented in polar coordinates by $r = a\theta$. It may be described as the locus of a point moving with uniform velocity along the radius vector, while the radius vector also moves about the pole with constant angular velocity. The evolute of this spiral approaches asymptotically to a circle with radius a.

Another spiral, similar to this one, with polar equation, $r = a\theta^2 - b$, is known by the name of Galileo. See also **Hyperbolic Spiral.**

Spiral of Archimedes.

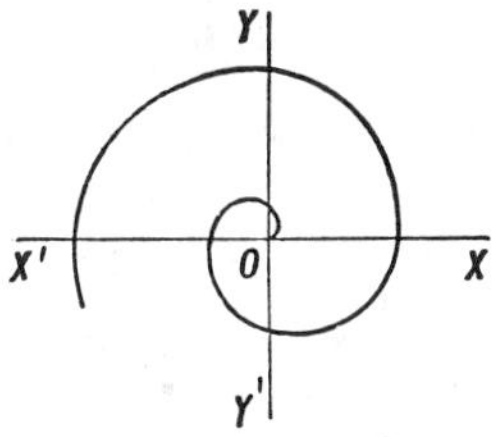

ARCHING (Bulk Solids). Feeder (Volumetric).

ARCHITECTURAL ACOUSTICS. Acoustics.

ARC LAMP. The electric-arc lamp has, as its source of illumination, an electric arc struck between two electrodes. In contrast to the incandescent lamp, in which the illumination results from a heated filament, and vapor lamps, in which the illumination is derived from a vapor made luminous by electric current, the light from an arc lamp comes from the highly incandescent crater of one of the electrodes, and from the heated, luminous, ionized gases surrounding the arc. The light from arc lamps is very much more intense than that from the incandescent type lamp. From the standpoint of current consumption, the illumination is produced efficiently. Some arc lamps may not be operated on ac, but all types are adaptable to dc. A constant-current, series type circuit is used to operate street-light arc lamps.

Open-arc type lamps can be divided into: (1) flame arcs; (2) low-intensity projector arcs; and (3) high-intensity projector arcs. The entire arc stream is made luminescent by the addition of flame materials in the flame arc light. This type of lamp is used in photography and related industrial photochemical processes because it produces an essentially continuous radiation and closely approximates natural sunshine. A special coring of rare-earth salts is used. In a special version of the flame-arc lamp, the arc is surrounded by a glass globe

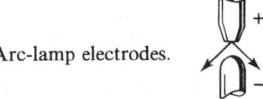

Arc-lamp electrodes.

with limited access of air, thus resulting in a nitrogen-rich atmosphere. This enhances the radiation in the violet and near-ultraviolet regions, making the lamp useful in blueprinting and related copying processes.

The incandescent tip of the positive carbon electrode, at or near its sublimation temperature, is the main light source in the low-intensity projector arc. The brightness is uniformly generated over a considerable area. This type of lamp found early use in motion picture, searchlight, and other projector systems requiring a concentrated source of light and the ability to create a well-defined narrow beam.

In the high-intensity carbon arc, rare-earth materials are included in the core of the positive electrode. These materials volatilize into the arc stream as the electrode is consumed. The light from these lamps is well suited to color motion picture photography and projection.

ARC (Mathematics). A segment or piece of a curve. See also **Circle (Geometry).**

ARC SHOOTING. (1) A method of refraction seismic prospecting in which the variation of travel time with azimuth from a shot point is used to infer geologic structure. (2) A reflection spread placed on a circle or on an arc with the center at the shot point.

ARCS OF LOWITZ. Atmospheric Optical Phenomena.

ARCTIC AIR. Air Mass; Atmosphere (Earth).

ARCTIC CIRCLE. The line of latitude 66° 32′ N (often taken as 66½° N). Along this line the sun does not set on the day of the summer solstice, about June 21 and does not rise on the day of the winter solstice, about December 22. From this line the number of annual twenty-four hour periods of continuous day or of continuous night increases northwards to about six months at the North Pole.

ARCTIC FOX. Canines.

ARCTIC FRONT. Fronts and Storms.

ARCTIC HAZE. Pollution (Air); Precipitation and Hydrometeors.

ARCTIC REGION RESEARCH. Polar Research.

ARCTIC WATERS. Water masses in or associated with the Arctic Ocean, including:

Arctic Deep Water. An oceanic water mass with a salinity of about 34.95% and temperature of about −0.85°C (30.5°F). Because of its high salinity, it is not believed that it is formed in the Arctic Ocean, but rather it is considered to be the North Atlantic Deep and Bottom Water, flowing north.

Arctic Surface Water. An oceanic water mass of low salinity, averaging from 32.0% to 33.0% in the north, but reaching very low values in summer as the rivers carrying large volumes of fresh water from melting ice decrease the salinity near the surface to values far below 30.0%. As winter approaches, the salinity increases, due to slow mixing with the Arctic Deep Water.

ARCTIC ZONE. Geographically, the area north of the Arctic Circle (66° 32′ N).

ARCTURUS (α *Bootes*). Arcturus was probably one of the first stars to be named, most likely receiving its name because of its proximity to the constellation of Ursa Major, thus being called the "watcher of the bear." It is one of the few stars named in the Bible (Job IX), although this reference is evidently to the constellation of Ursa Major rather than to the actual star itself. References to Arcturus are to be found in the writings of many of the ancient poets, including Virgil.

Arcturus is a very interesting star from the astronomical point of view, being what is known as a giant star. In appearance, it is a reddish-yellow, and the spectral type is such as to indicate that its temperature is slightly lower than that of the sun. It is one of the few stars whose diameter has actually been measured with the stellar interferometer. The angular diameter is found to be 0″.020, which, when combined with its stellar parallax of 0″.090, indicates a linear diameter of about 24 times that of our sun.

Arcturus ranks fourth in apparent brightness among the stars and has a true brightness value of 110 as compared with unity for the sun. Estimated distance from the earth is 36 light years. Arcturus is classified as an orange star of spectral type K. See also **Constellations;** and **Star.**

ARCUATE FASCICULUS (Brain). Memory.

ARC WELDING. Welding.

ARDISOLS. Soil.

AREA. If a parallelogram has sides denoted by the vectors **A** and **B**, its area is given by the vector product, **C** = **A** × **B**, which is perpendicular to the plane determined by **A** and **B**. The scalar magnitude of **C** equals that of the area and the direction of **C** is arbitrarily taken as the direction of the outward normal to the surface.

If calculus, the area of a surface may be found by integration. If $y = f(x)$ describes a curve, the area bounded by the curve, the X-axis, and the ordinates (a, b) is

$$\int_a^b f(x)\,dx$$

If a surface of revolution is generated by rotating an arc of the curve about the X-axis, its area is given by the integral

$$S = 2\pi \int_a^b y\,ds$$

where $ds^2 = dx^2 + dy^2$. A multiple integral may also be used, for an infinitesimal surface element in the XOY-plane is $dx\,dy$ and the area over a region S is

$$\iint_S dx\,dy$$

For a curved surface described by $z = f(x, y)$, the area is

$$\iint_S f(x, y)\,dx\,dy$$

Formulas for calculating the areas of major surfaces and shapes are given throughout this volume. See **Circle (Geometry)** and list of entries under Mathematics.

AREA MINING (Coal). Coal; Revegetation.

AREA SAMPLING. A method of sampling in which the domain to be examined is divided into small areas some of which are then selected to form the sample. Each area so selected may be fully inspected or may form the basis of sub-sampling. The method is particularly appropriate where no lists are available of the primary units which form the target population, e.g., individuals or dwelling units in developing countries. See also **Sampling (Statistics).**

AREA-TYPE FLOWMETERS. Flow Measurement.

ARECA PALM. Palm Trees.

ARENACEOUS. A textural term applied to sediments or sedimentary rocks which are composed of grains of sand. Psammitic has the same meaning.

ARENA VIRUSES. Virus.

ARENE COMPOUNDS. Organic Chemistry.

AREOLA. Breast.

ARÊTE. A narrow, jagged, serrate mountain crest, or a narrow, rocky, sharp-edged ridge or spur, commonly present above the snowline in rugged mountains (as the Swiss Alps) sculptured by glaciers, and resulting from the continued backward growth of the walls of adjoining cirques. (*Glossary of Geology*, American Geological Institute).

ARGALIS. **Goats and Sheep.**

ARGAND DIAGRAM. A graphical method of representing a function of a complex variable, $z = x + iy$. There are two perpendicular axes, as in the usual rectangular Cartesian coordinate system. The real part of the function is plotted on the real axis, usually the horizontal one, and the imaginary part on the imaginary or vertical axis. Points on the diagram for various values of the number pair (x, y) can then be joined to give a curve for the function of the complex plane. See also **Complex Variable;** and list of entries under **Mathematics.**

ARGENTITE. The mineral argentite, sometimes called silver glance, is naturally occurring silver sulfide, corresponding to the formula Ag_2S. It crystallizes in the isometric system in cubes, octahedrons and dodecahedrons, or may be massive. Hardness, 2–2.5; sp gr, 7.2–7.34; luster, metallic; streak, gray; color, black, blackish-gray or gray; opaque and sectile to such an extent that it cuts like wax with a knife. Heated upon charcoal it yields a malleable mass of silver. The name is derived from the Latin word for silver, *argentum.*

Localities for fine crystals are Sonora, Mexico, and Freiberg, Saxony; in the United States, at Butte, Montana; Tonopah, Nevada; and Aspen, Colorado.

Argentite is probably the most important primary silver mineral. However, it maintains its cubic (isometric) characteristic only above 179°C (354°F). Upon cooling, the inward structure inverts to a nonisometric form, usually orthorhombic, yet retaining its original outward form. It is, therefore, a paramorph after argentite, known as acanthite.

ARGILLACEOUS. This term is used to designate sedimentary rocks composed of fine particles of the nature of clay or mud. Pelitic has the same meaning.

ARGILLITE. A dense, fine-grained, hard, sedimentary rock of various colors (usually white, gray or red). Composed of minute grains of both clay and quartz. Certain types of argillites are easily confused with certain types of fine-grained acid lava flows, such as felsites, unless studied microscopically.

ARGININE. **Amino Acids.**

ARGININE-UREA CYCLE. **Urea.**

ARGON. Chemical element symbol Ar, at. no. 18, at. wt. 39.948, periodic table group 0 (inert or noble gases), mp −189.2°C, bp −185.7°C, density 1.78 g/cm^3 (solid at −233°C). Solid argon has a face-centered cubic crystal structure. At standard conditions, argon is a colorless, odorless gas and does not form stable compounds with any other element. Because of its low valence forces, argon is unable to form diatomic molecules, except in discharge tubes. It does form compounds under highly favorable conditions, as excitation in discharge tubes, or pressure in the presence of a powerful dipole. As an example of the first, argon forms amorphous compounds of the type FeA in a discharge tube having iron electrodes. An example of the second is furnished by the hydrates which argon forms with H_2O at 150 atmospheres and 0°C. Argon forms compounds, possibly clathrates, with a number of organic substances, such as a compound with hydroquinone containing 9% argon, in which the amount of argon may vary from this proportion. The compounds are made by crystallization of the aqueous solution of the hydroquinone under argon gas pressure on the order of 40 atmospheres.

Argon occurs in the atmosphere to the extent of approximately 0.935%. In terms of abundance, argon does not appear on lists of elements in the earth's crust because it does not exist in stable compounds. However, argon is 2.5× more soluble in H_2O than nitrogen and thus is found in seawater to the extent of approximately 2800 tons per cubic mile (605 metric tons per cubic kilometer). Commercial argon is derived from air by liquefaction and fractional distillation. There are three natural isotopes, ^{36}Ar, ^{38}Ar, and ^{40}Ar, and four radioactive isotopes, ^{35}Ar, ^{37}Ar, ^{39}Ar, and ^{41}Ar. The lengths of half-lives of the isotopes vary widely, the shortest ^{35}Ar with a half-life of about 2 seconds; the longest ^{39}Ar with a half-life of about 260 years. The first ionization potential of Ar is 15.755 eV; second 27.76 eV; third 40.75 eV. Other important physical characteristics of argon are given under **Chemical Elements.**

The presence of argon in air was suspected by Cavendish as early as 1785, but was not positively identified until 1894 by Lord Rayleigh and Sir William Ramsay. Argon exhibits a characteristic series of lines in the red end of the spectrum. Commercially, argon gas is used in incandescent lamps and fluorescent lamps as an inert gas to minimize vaporization of the filaments and, for this, is preferable to nitrogen. The gas also is used for shielding electrodes in arc welding. A gas of about 99.995% purity is required for lamps. Argon also has found effective use in certain lasers.

Regarding argon in meteorites, see **Krypton.** See also references listed at end of entry on **Chemical Elements.**

ARGONAUTA (*Mollusca, Cephalopoda*). The genus to which the paper nautilus belongs. This species is not a true nautilus but is more closely related to the octopus.

ARGUMENT. 1. An independent variable; e.g., in looking up quantity in a table, the number or any of the numbers which identifies the location of the desired value; or in a mathematical function the variable which when a certain value is substituted for it the value of the function is determined. 2. An operand in an operation on one or more variables.

ARGYRIA. Poisoning from the use of silver preparations over too long a period or its absorption during industrial processes, causing a ghastly bluish discoloration of the skin over the entire body.

ARID REGION (Hydrology). **Hydrology.**

ARIES (The ram). This constellation is far more famous for its classical significance than because of its appearance in the sky. It contains no bright stars and has no conspicuous features. Two thousand years ago, the vernal equinox was located in the constellation of Aries, and the symbol for the vernal equinox is the symbol for the constellation (i.e., the ram's head). Precession has caused the position of the vernal equinox to move backwards into the constellation of Pisces, so that now the "sign of the first of Aries" is to be found in that constellation. (See map accompanying entry on **Constellations.**)

ARIL. In many plants there is formed in the developing fruit an outgrowth from the funiculus, or seed stalk, one which completely or partially surrounds the seed. In the litchi nut it is the thick translucent pulp surrounding the seed; in the nutmeg it is a meshlike envelope which when removed and dried is known as mace.

ARITHMETIC. The rules for combination of two or more numbers. The operations involved are addition, subtraction, multiplication, division and the results obtained are, respectively, a sum, difference, product, quotient. Two or more of these operations can be performed successively and the final number obtained is independent of the order of the intermediate steps. Arithmetic, which is the first kind of mathematics normally studied by the beginner, is essentially the art of computation and application of this art.

See also **Progression** for meaning of arithmetic series and progression; and list of entries under **Mathematics.**

ARITHMETIC MEAN. The arithmetic mean of a number of observations x_i $(i = 1 \cdots n)$ is the simple average $\Sigma\, x_i/n$. It is commonly referred to simply as the mean.

The arithmetic mean is the most generally useful measure of location. In samples from a normal distribution, the mean is an efficient, and indeed sufficient estimate of the location parameter, though in other distributions other estimators may be preferable. If $\bar{x}$ is the mean of a sample drawn from a finite population of N members, $N\bar{x}$ is an unbiased estimate of the population total.

If κ_p is the pth cumulant of the parent distribution, that of the distribution of the mean is κ_p/n^{p-1}; in particular, the variance of the mean is σ^2/n. As this result indicates, the distribution of means from any parent distribution satisfying rather general conditions tends to normality as the sample size increases. See **Central Limit Theorem.**

A useful generalization of the arithmetic mean is the weighted mean. If each observation x_i has attached to it a weight w_i the weighted mean is given by $\Sigma\, w_i x_i/\Sigma\, w_i$, observations with large weights being given greater influence. If we have a set of estimates x_i of a quantity ξ, and if the variance of x_i is σ^2/w_i with w_i known, the estimate of ξ with least variance is the weighted means of the x_i, and the variance of this estimate is $\sigma^2/\Sigma\, w_i$.

For related topical coverage in this volume, see list of entries under **Mathematics.**

ARITHMETIC PROGRESSION. Progression.

ARITHMETIC UNIT (Computer). Central Processing Unit (Computer); Digital Computer.

ARIZONA CORAL SNAKE. Snakes.

ARIZONA SMOOTH CYPRESS. Cypress Trees.

ARKOSE. Arkose is a relatively coarse-grained feldspathic sandstone, derived from the rapid disintegration of granite or other feldspathic rock. It is characterized by its content of fresh, unaltered, euhedral feldspar. The term was proposed by Brongiart in 1823, and has been in constant use ever since. Arkose is an important type of sediment especially in relation to the study of unconformities, paleoclimatology, and "fossil" soils.

ARM. An extended lobe or appendage of a body. Its most familiar use is in application to the pectoral appendages of vertebrates when freed from the usual functions of support and locomotion, as in man and the other primates. In anatomy, the term is restricted to the region from the shoulder to the elbow, to distinguish it from the forearm. Some species are still quadrupedal on the ground, but can use their arms to some extent for handling objects.

The radiating lobes of the starfish are called arms or rays and arm is also applied to the branches of the lophophore in brachiopods and to other special structures.

ARMADILLOS. Edentata.

ARMATURE. The armature is one of the two essential parts of the dynamo electric machine. In a generator, the armature is the winding in which electromotive force (emf) is produced by magnetic induction. In the motor armature, conductors carry the input current which, in the presence of a magnetic field, produces a torque and effects the conversion of electrical into mechanical energy. In dc machines it is the rotor, but the ac armature may be rotor or stator. Larger size synchronous machines always have stationary armatures. The reluctance of the magnetic circuit to the flux which the conductors of the armature must cut in order to generate electric energy, is decreased by providing a core of soft iron or steel, on the surface of which the conductors are embedded in slots suitably provided in the core. The armature windings of a dc generator are terminated at the segments of a commutator, by means of which the alternating emf's induced in the armature are rectified and transferred by brushes from the moving rotor to stationary terminals. The conductors must be separately insulated, as must be also the commutator segments, and must be well braced and anchored in their slots to resist the electromagnetic and mechanical forces which tend to displace them.

The term armature is also applied to the moving element of a magnetically-actuated relay.

ARMATURE REACTION. This term refers to the reaction of the magnetic field produced by the current flowing in the armature conductors upon the main magnetic field of a dynamo machine. The result is a distortion of the magnetic field, the extent depending upon the reluctance of the magnetic circuit, the arrangement of the armature windings, the type field structure, and the phase angle between the armature voltage and current. In dc machines the effect is to increase the flux at some pole tips and decrease it at others, while in ac machines the effect depends upon the field structure and the phase angle of the armature current and voltage. The flux may be distorted as in the dc machine, it may be changed in magnitude but undistorted in wave form or it may be changed in magnitude and shifted in position with respect to the field windings. Armature reaction is an important factor in the speed and voltage regulation of the machines.

ARMYWORM (*Insecta, Lepidoptera*). An economically important caterpillar, *Cirphis unipuncta*, named from its habit of migrating from field to field in large numbers. When severe outbreaks occur these insects completely strip fields of grain of all kinds. When migrating they are trapped in barrier ditches dug around the fields to be protected. They are also killed by poison baits.

The true armyworm develops from a moth (*Pseudaletia unipuncta*) of the moth family *Noctuidae*. It is found in the United States and Canada, east of the Rocky Mountains. The eggs develop on the underside of leaves.

The armyworm is smooth, dark green with long white stripes extending from front to back, and often is fat and up to two inches in length. The army cutworm is sometimes referred to as an army worm. The latter is described under **Cutworm.**

Occurrence of the armyworm is subject to cycles, not fully understood or predictable. The insect is particularly damaging to maize (corn). In the young corn plants (8 inches (20 centimeters) or less in height), the armyworm devours all or nearly all of the leaves, causing the plant to expire in short order. In taller plants, some or parts of the leaves are left intact, but the insect concentrates on the center of the young stalk and damage is usually fatal to the plant—and once a field is invaded, without immediate control measures at work, all plants in the field are consumed.

The armyworm commences destructive action early in the spring. It is believed that some of these insects winter over as partially developed larvae, but that some may also winter as pupae or adults. This would explain the very early appearance of moths in the northern climes. Depending upon locale, the worms are fully grown by April or May. They pupate just below the groundlevel of the soil. A pupa is dark brown, having a length of about $\frac{3}{4}$-inch (19 millimeters), with a rather blunt head and a sharply tapering tail. The pupa stage lasts for about 2 weeks (longer if weather is unusually cold), from which the insect transforms into a rather drab gray-brown or light-brown moth having a wingspread of about 1.5 inches (about 4 centimeters). The moth can be recognized by a prominent white dot, centered in each front wing. The moth is active only at night. Light and the odor of decaying fruit attracts the moth.

The female moth lays her eggs (500 or more) in clusters or rows, preferring to place them on leaves of grasses. Pale green in color, the young worms tend to "loop" as they move about. These worms may be found by the thousands and tens of thousands in grass and fields of small grain and, frequently, are not detected until their damage has become extensive. The worms feed only at night. When fully grown, the worm is about 1.5 inches in length (4 centimeters). They maintain a green-brown color and have longitudinal stripes. After feasting for several days, the worms enter back into the soil, change to the pupal stage, and re-emerge as moths in from 2 to 4 weeks. In this fashion, there may be two to three generations per year. The larvae of the last generation of the year usually appear between mid-August and mid-September, depending upon locale and weather conditions.

Poison bran is an effective measure, particularly to protect fields which have not yet been attacked. The bran is placed in a line stretching across the probable line of march of the insect once an adjacent field has been decimated. The bait is prepared from bran that is mixed with insecticide. Blackstrap molasses or lubricating oil usually is added to provide sufficient stickiness to keep the insecticide and bran together. The worms also like molasses.

The armyworm is quite similar to the army cutworm, described in entry on **Cutworm.**

A closely related species is the *fall armyworm* (*Laphygma frugiperda*, Smith). Although widely distributed, they are notably injurious to crops in the southern United States, particularly during years when there has been a cold and wet spring. Like the armyworm and the cutworm, these insects also prefer small grains, maize (corn), sorghum, and grasses, but they also attack alfalfa, bean, cabbage, groundnut (peanut), cucumber, potato, sweet potato, spinach, and turnip. These worms are particularly fond of lawn grasses and thus not only a serious economic pest to the food producer, but to the homeowner as well.

Actually, the fall armyworm is considered a tropical insect because it cannot winter over in any area where the soil is frozen hard. Thus, a favorite winter ground is along the Gulf Coast and in southern Florida. Here, during winter, several stages of the insect may be present at the same time. After their number is increased manyfold in the spring, they swarm northward, often flying many hundreds of miles before selecting a location for their egglaying. Each female moth lays about 1000 eggs, usually on green plants. The female covers clusters of eggs with hairs from her body. Shortly thereafter, the young larva descends down through the heart of a plant and continue feeding near groundlevel until they assume a length of from 1 to 1.5 inches (2.5 to about 4 centimeters). It is at about this time that the insects are noticed as the result of the large amount of damage that becomes apparent. Unlike the armyworm and most cutworms, the worms do not take refuge in the soil during daytime, but rather they cling to parts of plants.

Quite similar in appearance to the true armyworm, the fall armyworm larvae when fully grown are a light-tan to green in color, although sometimes they are black. There are three very narrow white stripes down their back. They can be contrasted with the true armyworm by observing a white inverted "Y" design on the front of the head. Also, the tubercles are more prominent and they have more hair. The marching habit of this worm occurs during the autumn in the northern climes, but can take place in the southern states any time after the middle of summer and, if weather conditions are ideal for the insect, such marches may occur in early spring. An entire field or garden can be consumed within 36 to 48 hours. The remaining life cycle of the fall armyworm is similar to that of the true armyworm.

Another closely related species is the *beet armyworm* (*Laphygma exiqua*). This is a large caterpillar ranging up to $1\frac{1}{4}$-inch (30–32 millimeters) in length when fully mature. It is olive-green with broad light-green striping. Sugarbeet is the favorite target crop of this insect, followed by table beet and a variety of vegetables, citrus, alfalfa, and some wild grasses. This insect, native to the Orient, was first noted in California in the late 1870s and now occurs widely in the Gulf States, and from those states westward into California and northward to Nebraska and Kansas. Its habits are quite similar to those of the fall armyworm. Control is similar.

AROIDS. A large group of monocotyledonous plants, mostly tropical, having a characteristic flower habit. The numerous small inconspicuous flowers are borne on a fleshy stalk or spadix, which is surrounded, more or less completely, by a large, expanded, often brightly colored bract called a spathe. The spadix and spathe together are often but incorrectly considered to be the flower of the plant. The aroids are perennial plants, generally having tubers or rhizomes from which rise large leaves. Many tropical members are climbing plants. Well-known species are the Skunk Cabbage, whose foul-scented flowers appear so early in the spring, the Jack-in-the-Pulpit, and the wild arum, *Calla palustris*, of cold swamps, as well as the Sweet Flag, *Acorus calamus*, of the marshes. The cultivated Calla Lilies are all aroids and not lilies at all; some of them are delightfully fragrant. On the other hand, in species of *Amorphophallus*, which are sometimes seen in collections of cultivated plants, the vile odor of the flower structure prevents them from becoming popular; the spathe and spadix of some of them are of gigantic size. In the tropics several species of *Colocasia* are cultivated for the edible rhizomes which appear under the name of dasheen or taro. *Monstera* and several species of *Philodendron* are popular decorative plants in homes and in public buildings. For related topical coverage in this volume, see list of entries under **Tree.**

AROMA CHEMICALS. Flavorings.

AROMATIC AMINES (Carcinogenic). Carcinogens.

AROMATIC COMPOUND. An organic compound that incorporates a closed-chain or (ring) nucleus in its structure. This is in contrast with the aliphatic compound which is comprised of an open-chain structure. The classical example of an aromatic compound is benzene. Aromatic compounds also are sometimes referred to as benzenoids. Some ring-type compounds are not classified as aromatic. These include the cycloparaffins and cycloolefins which are considered to be derivatives of methane. See also **Compound (Chemical); Organic Chemistry.**

ARREST POINT. A temperature at which a system of more than one component that is undergoing heating or cooling absorbs or yields heat without change in temperature, thus interrupting the heating or cooling process.

ARRHENIUS EQUATION. Chemical Reaction Rate.

ARRHENIUS-GUZMAN EQUATION. A relation between the viscosity η and Kelvin temperature T, at constant pressure,

$$\eta = A \exp \frac{B}{RT}$$

where A, B are constants, and R is the gas constant; B may be identified with the *activation energy for liquid flow*.

ARRHENIUS VISCOSITY EQUATIONS. (1) Effect of temperature on viscosity, η, of a liquid

$$\frac{d}{dT}\ln(\eta v^{1/3}) = \frac{k_1}{T^2}$$

where v is the specific volume and k_1 is a constant.

(2) Viscosity of solutions, η,

$$\eta/\eta_s = A^x$$

where x is the concentration, η_s is the viscosity of the solvent and A is a constant.

(3) Viscosity of a sol, η,

$$\log \eta/\eta_\infty = kC$$

where η_∞ is the viscosity of the medium and C is the concentration of the sol-forming material.

ARRHYTHMIAS (Cardiac). An arrhythmia is a variation from what is considered normal for a rhythmic phenomenon. An analogy would be an erratic tape recorder whose speed vacillates and differs from that *one* speed which yields perfectly normal reproduction of music or voice. While the heart has a reasonably wide range in beating rate (pulse rate), depending upon the body's energy requirements (spanning from rest to heavy exertion), the rhythm of heart action is preserved even though rate may be changed. Rhythmic disturbances of the heart are called *cardiac arrhythmias.* Because, as explained in **Heart and Circulatory System (Human),** the heart is comprised of chambers (sinuses) and valves which are governed by electrical impulses, the impulses in a normal heart are programmed in just the right sequential order and at just the right instant, the timing of which is on the order of milliseconds. A crude analogy is the firing order of a multi-cylinder internal combustion engine.

Some authorities classify cardiac arrhythmias as *passive* and *active.* In some cases, both passive and active arrhythmias may be simultaneously present. The detailed etiology of these conditions is beyond the scope of this encyclopedia. The principal tool available to the physician in the analysis and diagnosis of cardiac arrhythmias is the electrocardiogram. *Bradycardia* is the term usually used to describe an abnormal slowness of the pulse rate. *Tachycardia* is the term used to describe excessive rapidity of heart action. *Heart block* is a term that signifies the interruption of muscular connection between the atrium and ventricle so that they beat independently of each other. Since there are numerous electrically conductive pathways within the heart, there are several specifically named heart blocks. These include:

Arborization heart block in which there is interference with the fine terminal subendocardial fibers of the *Purkinje system* (specialized sino-atrioventricular conduction system); *atrioventricular* heart block; *auriculoventricular* heart block; *bundle-branch* heart block, in which the two ventricles contract independently of each other; *complete heart block*, a situation in which the functional relation between the parts of the *bundle of His* (a muscular band connecting the auricles with the ventricles of the heart) is destroyed by a lesion, so that the auricles and ventricles act independently of each other; *interventricular* heart block; and *sino-auricular* heart block, in which the blocking is located between the auricles and the mouths of the great veins and coronary sinus. See Fig. 1 in entry on **Heart and Circulatory System (Human).**

Because of the several pathways of conduction, a pair of conducting leads to the electrodardiograph does not suffice. A standard arrangement of twelve leads to the instrument is commonly used. The physician will evaluate with maximum possible accuracy the risk posed by an arrhythmia, particularly a tachyarrythmia and the urgency of attempting to terminate it. Although normal hearts may tolerate tachyarrythmias for extensive periods, when there is marked underlying heart disease, such arrythmias may cause hypotension (low blood pressure), congestive heart failure, or coronary insufficiency. Tachyarrythmias are prone to degenerate into the more serious ventricular arrythmias. In addition to general cardiac health, the physician will take into consideration the heart rate, the duration of the episodes of arrythmia, and whether the arrythmia is regular or irregular.

Passive Arrythmias. Particularly in individuals who are quite physically active, *sinus bracycardia* may be present. This is defined as a resting heart (adult) which beats fewer than 60 times per minute. The rates of long-distance runners, for example, may range between 40 and 50 beats per minute. A number of conditions, however, may cause an abnormally slow discharge rate of the sinus node. This occurs during severe pain and can be induced by increased vagal tone caused by various drugs (parasympathomimetic), such as endrophonium (Tensilon®) and neostigmine (Prostignin®), and by a number of tranquilizing drugs. Increased vagal tone also can be associated with acute myocardial infarction. A number of physiologic situations will slow the sinus node, including hypothyroidism and high fever. There is also what is known as *hypersensitive carotid sinus reflex.* (The carotid artery is located in the region of the neck, face, and skull.) The carotid sinus becomes increasingly sensitive with age and a type of syncope (sudden suspension of consciousness) can be caused by twisting the neck or wearing a tight collar. See also **Syncope.**

In some patients, particularly the elderly, presenting sinus bradycardia, there may be a condition known as the *sick sinus syndrome.* The bradycardia may be punctuated by episodies of tachycardia (*bradycardia-tachycardia syndrome*). A form of sick sinus syndrome has explained some sudden deaths of young athletes, who at autopsy have exhibited an idiopathic (cause not known) obliterative disease of the artery to the sinus node.

A slow heart beat (bradycardia) may be a factor in development of congestive heart failure with patients who have myocardial disease. See also **Congestive Heart Failure.**

Active Arrythmias. Some authorities place these arrythmias into two fundamental categories—supraventricular and ventricular. Representative of *supraventricular arrythmias* are atrial premature beats and junctional premature beats; sinus tachycardia; paroxysmal supraventricular tachycardia; paroxysmal atrial tachycardia; paroxysmal (AV) atrioventricular junctional tachycardia; paroxysmal atrial tachycardia with block; multifocal atrial tachycardia; atrial flutter; and atrial fibrillation. Representative of ventricular arrythmias are ventricular premature beats, ventricular tachycardia, accelerated idioventricular rhythm, and ventricular fibrillation.

Ventricular Fibrillation. This condition may be described as an irregular twitching of the muscles in the wall of the ventricle of the heart. About two-thirds of the deaths due to heart attack are attributed to uncontrolled ventricular fibrillation (VF) which kills many patients before they reach a hospital. The probability of VF increases with the size of the dead tissue (infarct) that results when the blood flow to a portion of the heart is drastically reduced by blockage of one or more of the arteries. This is a primary motivation for finding treatments to limit infarct size. The presence of dead tissue from prior milder heart attacks will contribute to VF in a major attack. Thus, in persons with a history of heart problems, surgery to eliminate abnormal tissues that may give rise to the arrhythmia is sometimes suggested as a preventive measure. When a patient lives to reach a hospital, continuous monitoring of the heart will reveal any abnormal rhythms that presage VF and thus can be treated as soon as they occur.

Various communities have instituted emergency procedures which provide *cardiopulmonary resuscitation* (CPR), including mouth-to-mouth resuscitation and external heart massage. See **Cardiac Massage.** Therapy must be commenced within 5 minutes to prevent irreversible brain damage when the supply of blood to the brain is blocked. A major role of CPR is to identify VF and the immediate use of drug therapy. As of the 1980s, successful CPR programs require nearly an ideal environment to operate successfully, i.e., large numbers of trained personnel and sufficient units to reach a heart attack victim within 5 minutes or less; the prompt attention of bystanders (if attack occurs in public places), of family in the home, of fellow office and factory workers, etc., to take rudimentary measures and, above all, telephone for an emergency unit; adequate instrumentation, including transmission of instrumental data as well as voiced observations, in the emergency vehicle; excellent radio communication between vehicle personnel and hospital medical staff (a physician immediately available at all times) to advise most appropriate emergency procedures. The objective is to extend as much as possible the skills and means of the hospital to the patient before arrival at a hospital.

Preexcitation Syndromes. These conditions occur when all or a portion of the ventricle is activated by atrial impulses earlier than would be the case were the impulses reaching the ventricles via the normal conduction pathways. One such condition is known as the *Wolff-Parkinson-White* syndrome and is believed to result from congenital cardiac defects, frequently involving the tricuspid valve (*Ebstein's anomaly*). Recent research indicates that preexcitation syndrome can occur by a number of different pathways. Sometimes involved are the Kent bundles, the James fibers, and the septal fibers of Mahaim. Although the arrythmias (usually sporadic) associated with preexcitation syndrome can be tolerated by some patients (even at rates of 250–300 beats per minute), this tolerance is usually limited to young persons with otherwise normal hearts. In contrast, in some patients, the arrythmias can be disabling.

Pacemaker. In cases of complete or bifascicular heart block, an implanted pacemaker may be used. In the early years of pacemaker use, there were numerous problems with failures arising from electrode displacement or breakage, premature battery depletion, or faulty pulse generators. Today, the reliability of the units has markedly increased, but precaution must be exercised in selecting an instrument from the several brands available. Two techniques have been used in pacemaker implantation—insertion of an electrode through (1) a cephalic vein or (2) the external jugular vein into the apex of the right ventricle. The power source is placed subcutaneously just inferior to the clavicle; or direct epicardial electrodes are attached to the ventricle through a small midline thoracotomy. The pulse generator is placed subcutaneously in the epigastric region. Units equipped with conventional mercury batteries have an estimated life of 3–5 years, whereas the more recent lithium battery has a projected life of 8–10 years. There are also rechargeable nickel-cadmium batteries (projected life of 8–12 years) and nuclear-powered generators (10–20 years expectancy). See also **Battery.** In recent years, special clinics have been established to which a patient can transmit pacemaker impulses over the telephone, whereupon small changes in the discharge rates of the pacemaker can be detected. These small changes, as small as 25 milliseconds, would not show up on an electrocardiogram. In the *demand type* of pacemaker, the unit will not compete with the patient's own rhythm if the AV conduction should return, or with ventricular premature beats should they be present. The pacemaker can be designed and implanted in accordance with the particular problems of the patient.

Cardioversion. Certain types of arrhythmias, such as atrial flutter, atrial fibrillation, and ventricular tachycardia, that do not respond to other forms of therapy may be treated with cardioversion. In this technique, a direct-current discharge of short duration (2–3 milliseconds) is delivered to the external thorax. The very short duration of the discharge makes it possible to program the discharge at just the right instant from the electrocardiogram. The technique is not used when an arrhythmia may be due to digitalis intoxication. In most cases, some premedication may be indicated prior to the application of cardioversion.

Antiarrhythmic Agents. Some of the drugs used in the control of ventricular and atrial tachyarrhythmias include:

Quinidine—decreases automaticity, excitability, and conduction velocity. Prolongs effective refractory period and action potential duration.

Procainamide—similar in action to quinidine.

Propranolol—decreases conduction through the atrioventricular node. Effective agent for slowing ventricular response where atrial fibrillation or atrial flutter are present.

Lidocaine—effective in treatment of ventricular arrhythmias. Not considered effective in treatment of supraventricular tachyarrhythmias. See also **Anesthesia.**

Phenytoin—effective in controlling ventricular arrhythmias, particularly those due to digitalis toxicity. Not considered effective in treatment of atrial tachyarrhythmias.

Disopyramide—effectively suppresses atrial and ventricular arrhythmias. Decreases automaticity and conduction velocity. Prolongs effective refractory period.

Bretylium—for serious ventricular tachyarrhythmias. Increases ventricular fibrillation threshold. Prolongs action potential duration and effective refractory period.

Digitalis—stabilizes atrial electrical activity and assists in preventing atrial tachyarrhythmias. Particularly useful in treatment of ventricular arrhythmias arising from congestive heart failure and enlarged heart.

Some of the foregoing drugs are used in combination therapies, depending upon specific patient requirements. More recently developed drugs, some still in clinical testing, are described by Zipes (1978).

For related topics in this encyclopedia, see list of entries given at the end of entry on **Heart and Circulatory System (Human).** Particularly, see **Electrocardiography.**

References

Abbott, J. A., et al.: "Graded Exercise Testing in Patients with Sinus Node Dysfunction," *Am. J. Med.*, **62**, 330 (1977).

Akhtar, M., and A. N. Damato: "Clinical Uses of His Bundle Electrocardiography," *Am. Heart J.*, **91**, 520 (1976).

Chung, E. K.: "Wolff-Parkinson-White Syndrome," *Am. J. Med.*, **62**, 252 (1977).

Fairfax, A. J., Lambert, C. D., and A. Leatham: "Systemic Embolism in Chronic Sinoatrial Disorder," *N. Engl. J. Med.*, **195**, 190 (1976).

Ferrer, M. I.: "The Sick Sinus Syndrome," *Circulation*, **47**, 635 (1973).

Furman, S., and J. D. Fisher: "Cardiac Pacing and Pacemakers," *Am. Heart J.*, **94**, 250 (1977).

Harrison, D. C., Fitzgerald, J. W., and R. A. Winkle: "Ambulatory Electrocardiography for Diagnosis and Treatment of Cardiac Arrhythmias," *N. Engl. J. Med.*, **194**, 373 (1976).

Kastor, J. A.: "Atrioventricular Block," *N. Engl. J. Med.*, **292**, 462 (1975).

Marriott, H. J. L.: "Practical Electrocardiography," 5th edition, Williams & Wilkins, Baltimore, 1972.

Marx, J. L.: "Sudden Death: Strategies for Prevention," *Science*, **195**, 39–41 (1977).

Pollack, G. H.: "Cardiac Pacemaking: An Obligatory Role of Catecholamines?" *Science*, **196**, 731–738 (1977).

Scheinman, M. W., et al.: "Atrial Pacing in Patients with Sinus Node Dysfunction," *Am. J. Med.*, **61**, 641 (1976).

Zipes, D. P., and P. J. Troup: "New Antiarrhythmic Agents: Amiodarone, Aprindine, Disopyramide, Ethmozin, Mexiletine, Tocainide, Verapamil," *Am. J. Cardiol.*, **41**, 1005 (1978).

ARROWANA. Bony Tongues.

ARROWWOOD SHRUB. Elder Trees and Viburnums.

ARROW WORM. Small marine animals sometimes classified with the annelid worms but more often included in the separate phylum *Chaetognatha.*

ARROYO. This term is applied to dry stream channels with nearly vertical walls and flat bottoms which are characteristic of semi-arid regions. They may suddenly become filled with torrential waters after heavy rains. The word arroyo is of Spanish origin.

ARSENIC. Chemical element symbol As, at. no. 33, at. wt. 74.9216, periodic table group 5a, mp 817°C (28 atmospheres), sublimes at 618°C, density 5.72 g/cm^3. One naturally occurring stable isotope ^{75}As. Various studies indicate that arsenic exists in several allotropic forms. The metallic form has a steel-gray color in the crystalline form and is brittle. Although the red form of arsenic sulfide As_2S_2 was observed by Aristotle as early as 400 B.C., the first attempt to isolate the metal was not made until 1250 by Albert Magnus. Later documentation on the preparation of the element was given by J. Schroder and N. Lémery in the 1600s. First ionization potential 9.8 eV; second 18.63 eV; third 28.34 eV; fourth 50.1 eV; fifth 62.5 eV. Oxidation potentials $AsH_3 \rightarrow As + 3H^+ + 3e^-$, 0.54 V; $As + 2H_2O \rightarrow HAsO_2 + 3H^+ + 3e^-$, −0.2475 V; $HAsO_2 + 2H_2O \rightarrow H_3AsO_4 + 2H^+ + 2e^-$, −0.559 V; $AsO_2^- + 4OH^- \rightarrow AsO_4^{3-} + 2H_2O + 2e^-$, 0.71 V; $As + 4OH^- \rightarrow AsO_2^- + 2H_2O + 3e^-$, 0.68 V. Other important physical properties of arsenic are given under **Chemical Elements.**

Gray or metallic arsenic, density 5.73 g/cm^3, which sublimes on heating, and has the vapor composition As_4, becoming As_2 at higher temperatures, is the ordinary variety. On rapid cooling, the vapor condenses to yellow arsenic, density 1.97 g/cm^3, which reverts to the gray variety on warming. An intermediate form in the transition is black amorphous β arsenic, density 4.6–5.2, also obtained by the thermal decomposition of arsine. Brown arsenic, density 3.7–4.2, obtained by reduction of acid solutions of trivalent arsenic, is probably a finely divided form of black arsenic.

Arsenic sublimes on heating; is unchanged in dry air but a film of oxide is formed in moist air; heated in air at 180°C forms arsenic trioxide of the odor of garlic, poisonous; insoluble in HCl but soluble in concentrated HNO_3 or concentrated H_2SO_4 to form arsenic acid; soluble in hot NaOH solution; heated with chlorine forms arsenic trichloride; heated with metals forms metallic arsenides. When arsenic is heated in a tube and the vapor cooled (1) slowly (that is, in the hot part of the tube) black arsenic is formed, and this form is converted into the gray at 360°C, (2) rapidly (that is, in the cold part of the tube) yellow arsenic is formed, and this form is quickly converted into the gray by the action of light. Yellow arsenic is soluble in CS_2.

Arsenic occurs in nature as the arsenide of iron, cobalt, nickel, and as the mineral sulfides, *realgar* (arsenic monosulfide, AsS), red colored; *orpiment* (arsenic trisulfide, As_2S_3), yellow colored—these two minerals when powdered once were used as paint pigments—*arsenopyrite, mispickel* (iron arsenosulfide, FeAsS); *enargite*, Cu_3AsS_4; and *tennantite*, $Cu_8As_2S_7$.

The primary arsenic-containing material is arsenious oxide obtained by separation from roaster or smelter flue gases, and is produced in Tacoma, Washington. Metallic arsenic is obtained as sublimate by heating the oxide with carbon.

Arsine: AsH_3, is formed by hydrolysis of arsenides, or reduction (by zinc and HCl or aluminum and NaOH) of arsenic compounds, as in the Gutzeit test. It reacts with metals at higher temperatures or in solution to give the arsenides. Diarsine, As_2H_4, is produced by reduction of arsenic trichloride, $AsCl_3$, by lithium aluminum hydride, $LiAlH_4$ in ether at −190°C. It melts below −50°C, but begins to decompose into AsH_3 and brown polymeric $(AsH)_x$ about −100°C. It is more stable in the gas phase than in the solid or liquid phases.

Arsenides: These are prepared by fusion from the elements. Their properties vary across the periodic table, those of the alkalies and alkaline earths being readily hydrolyzed by H_2O or acids and are stoichiometric, while the arsenides of the other metals show an increasingly intermetallic character and resist hydrolysis.

Trihalides: The trifluoride and trichloride, AsF_3 and $AsCl_3$, are liquids at room temperature and the tribromide and triiodide, $AsBr_3$ and AsI_3 are solids, although the former melts at 31°C. Like the analogous phosphorus compounds, they have pyramidal structures. Their hydrolysis in aqueous solution is not quite complete, consistent with their greater ionic character (than the phosphorus halides), as is the fact that As^{3+} is precipitated from their solutions as the sulfide. The only stable binary pentahalogen compound of arsenic is the pentafluoride, AsF_5, a colorless gas, which like the trihalides, is less readily hydrolyzed than the corresponding phosphorus compound. A very unstable pentachloride, $AsCl_5$, has been reported. The mixed halide AsF_3Cl_2 can be made by passing chlorine into ice-cold arsenic trifluoride.

Arsenic(III) Oxide: As_4O_6, exists as tetraarsenic hexoxide in the solid state and in the vapor to above 800°C, where dissociation to As_2O_3 commences. It is somewhat soluble in H_2O (about 20 g/l at 25°C), and its solutions have some acidic properties, although the acid has not been isolated and its formula is probably not $As(OH)_3$, the form used for convenience in writing reactions. It is an amphiprotic substance, since, as stated above, As^{3+} is precipitated by H_2S from acid solutions as the sulfide, while the salts, the arsenites (containing the ion AsO_3^{3-}), are readily formed. Their solubility in H_2O varies across the periodic table, those of the alkali metals being very soluble, those of the alkaline earth metals less so, and those of the heavy metals essentially insoluble. Arsenite ion probably exists as $As(OH)_4^-$ in solution.

Arsenic(V) Oxide: As_4O_{10} is a white solid, decomposes at 315°C, isomorphous with phosphorus pentoxide, P_2O_5, but not produced by a simple oxidation of As_2O_3. It is made by dehydration of arsenic acid or $As_4O_{10} \cdot 4H_2O$. It hydrates to give arsenic acid, $H_3AsO_4 \cdot \frac{1}{2}H_2O$. This acid is only slightly weaker than phosphoric acid, which it resembles in forming a wide variety of polyacids. It also forms primary, secondary, and tertiary (ortho) arsenates. Raman spectral studies of concentrated arsenic acid solutions in H_2O have a strong band assigned to the —OH group, whence it is inferred that the acid is present in different forms in concentrated and dilute solutions. Many arsenates are converted by ignition into pyroarsenates, e.g., calcium pyroarsenate, $Ca_2As_2O_7$, and metaarsenates are also known.

Direct fusion of the elements yields a number of arsenic sulfides, including As_4S_3, As_4S_4, As_2S_3, and As_2S_5, the last two being obtained also by precipitation from arsenic(III) and arsenic(V) solutions, respectively. The trisulfide dissolves in alkali sulfide solutions to form thioarsenites:

$$As_2S_3 + 3S_2^{2-} \rightarrow 2AsS_3^{3-} + S$$

while with polysulfides it forms thioarsenates:

$$As_2S_3 + 2S_2^{2-} + S \rightarrow 2AsS_4^{2-}$$

Organoarsenic Compounds: The largest group of organic arsenic-containing compounds is the arsonic acids $RAsO(OH)_2$, where the R may be alkyl, aryl, or heterocyclic groups and their salts. In addition to specific compounds mentioned under the uses of arsenic, some organoarsenic compounds include methylarsine CH_3AsH_2; methylarsenic tetrachloride CH_3AsCl_4; diphenylarsenic peroxide $(C_6H_5)_2AsOOAs(C_6H_5)_2$; triphenylarsenic dihydroxide

$$(C_6H_5)_3As(OH)_2;$$

dimethylarsine borane $(CH_3)_2AsHBH_3$; and ethoxydichloroarsine $C_2H_5OAsCl_2$.

Uses: In the mid-1970s, the worldwide production of As_2O_3 was estimated to be about 50,000 metric tons. Future production of As_2O_3 will be influenced by the ability to handle ores in the manner required to comply with environmental restrictions. As_2O_3 is available in two grades: (1) crude, 95% As_2O_3; and (2) refined arsenic, 99% As_2O_3. Domestic supplies of the United States are supplemented by imports from Sweden, France, and Mexico. Commercial arsenic metal is produced chiefly by the United States and Sweden.

Arsenic trioxide finds major use in the preparation of other compounds, notably those used in agricultural applications. The compounds monosodium methylarsonate, disodium methylarsonate, methane arsenic acid (cacodylic acid) are used for weed control, while arsenic acid, H_3AsO_4, is used as a desiccant for the defoliation of cotton crops. Other compounds once widely used in agriculture are calcium arsenate for control of boll weevils, lead arsenate as a pesticide for fruit crops, and sodium arsenite as a herbicide and for cattle and sheep dip. In some areas, arsenilic acid has been used as a feed additive for swine and poultry. Restrictions on these compounds vary from one country and region to the next.

Refined arsenic trioxide is used both as a fining and decolorizing agent in glass. As_2O_5 and arsenic acid are used in the manufacture of chromated copper arsenate which is used extensively as a wood preservative.

Indium arsenide, gallium arsenide, and gallium arsenide phosphide find use as semiconductors. For these materials, the starting arsenic source must be extremely pure. Arsenic trichloride and arsenic hydride (very high purity) find application in the production of epitaxial gallium arsenide. Also, in various combinations with iodine, germanium, selenium, sulfur, tellurium, and thallium, arsenic will form a group of glasses with very low melting points.

The applications of arsenic as a metal are quite limited. Metallurgically, it is used mainly as an additive. The addition of from $\frac{1}{2}$ to 2% of arsenic improves the sphericity of lead shot. Arsenic in small quantities improves the properties of lead-base bearing alloys for high-temperature operation. Improvements in hardness of lead-base battery grid metal and cable-sheathing alloys can be obtained by slight additions of arsenic. Very small additions (0.02–0.05%) of arsenic to brass reduce dezincification.

Toxicity: Although metallic arsenic and arsenic trisulfide may be handled, as in the case of most arsenical compounds, skin contact should be avoided. Arsine requires extreme caution in handling because of its very high toxicity. In handling arsenic and its compounds, reducing conditions should be avoided because these may give rise to the undesired formation of arsine. Epidemiological studies indicate an association between high and lengthy exposures to inorganic arsenic compounds and cancer. Wherever arsenic and its compounds are present as dusts or vapors, proper ventilation and respirators are mandatory. Good housekeeping and appropriate hygienic practices should also be observed.

Arsenic is commonly found in small amounts in the tissues of plants and animals. A human body may contain as much as 20 mg (As_2O_3). No role in natural biological phenomena has been found for As. Although the element may be present in seawater to the extent of 0.006–0.03 ppm, it may be ten times as high in estuaries. Shellfish tend to accumulate the arsenic from the large amount of seawater with which they come in contact. Oysters may contain 3–10 ppm. However, shellfish of the same species grown in different localities show wide variations in arsenic content, suggesting that it is an accidental constituent which the organisms learn to tolerate. Its lack of function in the human body is suggested by the fact that it tends to accumulate in the hair and nails which are essentially nonliving.

References

Carapella, S. C., Jr.: "Arsenic," in "Kirk-Othmer Encyclopedia of Chemical Technology," Vol. 3, 3rd Edition, pp. 251–266, Wiley, New York, 1978.
Liddell, D. M., Editor: "Handbook of Nonferrous Metallurgy," Vol. 2, pp. 94–103, McGraw-Hill, New York, 1945.

S. C. Carapella, Jr., ASARCO Incorporated, South Plainfield, New Jersey.

ARSENIC SULFIDE. **Orpiment; Realgar.**

ARSENOPYRITE. The mineral arsenopyrite is a sulfarsenide of iron corresponding to the formula FeAsS. A variety in which some of the iron is replaced by cobalt is known as danaite. It crystallizes in the monoclinic system but twinning produces pseudo-orthorhombic crystals. Its hardness is 5.5–6; sp gr, 6.07; luster, metallic color, silvery-white to steel-gray, but usually with a yellow to gray tarnish; streak, black. Arsenopyrite is a common mineral with tin and lead ores and in pegmatites, probably having been deposited by action of both vapors and hydrothermal solutions. It is a widespread mineral, well-known deposits occurring in Austria, Saxony, Switzerland, Sweden, Norway; Cornwall and Devonshire, England; Bolivia; in the United States at Roxbury, Connecticut; Franklin, New Jersey; Paris, Maine; Emery, Montana; and Leadville, Colorado. Danaite was first found in Franconia, New Hampshire, by J. D. Dana, for whom it was later named. Arsenopyrite also is known as *mispickel*, an old German term whose exact derivation is unknown.

ARTEMISIA. Genus of the family *Carduaceae* (aster or thistle family). Many of the nearly 300 species of *Artemisia* have been cultivated. Southernwood, *Artemisia abrotanum*, is cultivated in gardens for its delicate foliage and aromatic odor. Another and a homely species, *Artemisia vulgaris* or mugwort, is also frequently cultivated, as is *Artemisia absinthium*, a native European perennial plant. All contain volatile oils. That from *Artemisia absinthium*, oil of wormwood, is a powerful drug, capable of causing violent convulsions when taken even in small doses. It is used to flavor the alcoholic beverage absinthe, a liquor capable of producing much the same effects as the drug. The sage-brushes of the western United States are all species of *Artemisia*; like other species they contain an abundance of aromatic oil. From *Artemisia dracunculus*, tarragon is obtained. This is used as a condiment, and for flavoring vinegar and mustard.

ARTERIAL AND VENOUS DISORDERS. Malfunctioning blood vessels predispose several of the major fatal and disabling diseases. These include *cardiovascular diseases*, as typified by ischemic heart disease—angina pectoris and acute myocardial infarction; and *cerebrovascular diseases*, as represented by cerebral thrombosis and cerebral hemorrhage. A general description of the blood circulatory system is given in the entry on **Heart and Circulatory System (Human).**

Blood vessels fail, fully or partially, in several ways. Particularly in connection with the arteries, a process known as *atherogenesis* may occur. This is a slow process and progresses with the age of the individual. Over a period of time, plaquelike lesions form in the arterial walls, reducing both the elasticity and useful diameter of the vessel. This, in turn, diminishes the amount of blood that can be conveyed by a given artery to the organ which it is supplying and also affects the manner in which the artery must respond in concert with the actions of the heart. Arteries so damaged are more prone to obstructing the flow of blood when thrombi or emboli are present, causing heart damage that may terminate in death (heart attack); or causing brain damage that also may terminate in death or in severe paralysis and mental degradation (stroke). Blood vessels also may leak or rupture, particularly if aneurysm is present, causing damage to surrounding tissue.

In describing vascular disorders, certain terms are frequently used. These include:

Thrombus—a blood clot formed within the heart, or in a blood vessel, remaining at its site of origin. Once a thrombus migrates by way of the bloodstream to a different site and when two or more thrombi agglomerate, the term *embolism* may be used. Thrombi and emboli also are sometimes called occlusions—as in coronary occlusion.

Spasm—coronary artery spasms are described in entry on **Ischemic Heart Disease.**

Embolism—the obstruction of a blood vessel by an embolus, often a migrated blood clot or other debris. If not absorbed, once detached from its place of origin, an embolus may circulate freely in the bloodstream until it finally reaches a vessel through which it cannot pass. An embolus on the arterial side of the heart, such as may originate in chronic valvular disease, will lodge in one of the systemic arteries of the body. On the venous side, an embolus, usually arising in some area of infection, will pass into the heart and thence to the arteries of the lungs. In either situation, the consequences are similar—the blood supply to the region is cut off and, unless adequate alternative channels are available, death or grave debility will result. In the case of the large arteries to such organs as kidneys, brain, and heart, there is only limited possibility of collateral circulation from the outset, and infarction results. An infarct is defined as an area of necrosis in a tissue caused by obstruction of circulation to the affected area. Myocardial infarction is defined as the death of an area of tissue in the heart muscle, caused by decreased blood supply to the heart. In the case of emboli in other parts of the body, unless treated, the final consequence may be gangrene.

Embolism may be caused by means other than blood clotting. Fat embolism is seen in severe bone injury, where fat globules released from the pulped bone marrow may occlude arteries in the brain and/or lungs. Air embolism may form when air is accidentally aspirated into the veins, especially those of the neck. Fragments of tumors may be carried as emboli in the circulation to distant parts. These may be too small to cause immediate effects of emboli, but instead they may form the nucleus for secondary growths at some distance from the primary tumor.

Aneurysm—a sac or pouch filled with blood which protrudes from the wall of an artery, a vein, or the heart. In a *true* aneurysm, the wall of the sac consists of at least one of the layers of tissue that make up the wall of the blood vessel. In a *false* aneurysm, all layers of the artery have ruptured, but blood is retained in connective tissue surrounding the artery. See also **Aneurysm.**

Hemorrhage—the escape of blood from a ruptured vessel. This may vary from seepage to a massive flow. See **Hemorrhage.**

Arteries

The arteries are complex, composite tubes that carry blood from the heart to the capillaries. In the adult, the largest of the arteries, the aorta and pulmonary artery, are indeed quite large—up to 1.2 inch (3 centimeters) in diameter. Likening an artery to a pipe or hose is a gross oversimplification because an artery is a dynamic structure that incorporates functions that go far beyond the simple task of a conduit. For example, the elastic properties of an artery, unlike a rigid pipe, make it possible for the artery to accommodate large and cyclic changes in blood flow that are characteristic of normal heart pumping action. With the capability of continuously changing the orifice (channel for transporting the blood), the artery assists in the outward distribution of blood from the heart through what might be called a squeezing action. Further, the action of the artery varies the resistance to flow of blood and thus the back pressure on the heart; the pliable nature of the artery resists rupturing. The muscular tissue of the artery is generously supplied with nerve fibers (*vasomotor nerves*), which in essence are a control over the relative or effective pliability of the artery.

The normal, healthy artery is composed of three coats, each of which is made up of additional layers and thus is quite unlike and much more complex than what one may envision when thinking of a pipe or tube. (1) The *inner coat* (*tunica intima*) consists of three layers—(a) endothelial cells; (b) connective tissue, which occurs only in the larger arteries; and (c) elastic fibers with microscopic perforations (*fenestrated membrane*). (2) The *middle coat* (*tunica media*) consists of muscular and elastic tissue. This middle coat provides the artery with extensible and elastic properties. These characteristics enable the artery to expand upon receiving blood at each contraction of the heart and, in turn, to squeeze the blood forward as the orifice is adjusted to receive more blood at another contraction of the heart. (3) An *external coat* (*tunica externa* or *adventita*) consists of areolar connective tissue made up of smooth muscle cells, either scattered or in the form of longitudinally arranged bundles.

When empty, the arteries do not collapse; the orifice for carrying the blood remains open. However, when an artery is cut, as from an injury, the muscular coat does contract to some extent, thus narrowing the opening of the wound and permitting a clot to form to close (plug) the orifice, an action that is mandatory to the halting of hemorrhage. In the body, a majority of the arteries are accompanied by one or more veins which are contained within a protective sheath of connective material for providing support to these vessels. Some authorities classify arteries as (1) *elastic arteries*, which are the larger vessels that carry blood from the heart to arteries of lesser diameter. The property of elasticity is a prominent feature of these vessels; and (2) *muscular arteries*, which are of smaller diameter that distribute blood to specific organs. The muscular characteristics of these vessels provide ability to reduce (by contraction) or expand (by relaxation) the blood flow in accordance with the specific demands of the organs being supplied. Arteries also have their own blood supply because of energy requirements. The arteries, capillaries, and veins so involved are identified as *vasa vasorum*. As with other blood vessels, the arteries divide and subdivide in a complex branching system. The smallest of these branches are called *arterioles*, and it is at the distal ends of these tiny arterioles that the capillaries begin. As the arteries become smaller, the proportion of smooth muscle to elastic tissue increases.

Arteriosclerosis. Commonly referred to as *hardening of the arteries*, this is a condition that exists when the walls of the blood vessels thicken and become infiltrated with excessive amounts of mineral and fatty materials. Arteriosclerosis occurs in all races and among people living in all climates. The disorder tends to progress with the age of the patient and is sometimes considered a disease of the middle-aged and elderly, but it should be quickly observed that coronary artery disease is responsible for nearly one-third of the deaths of persons between the ages of 35 and 65. The disorder has been known for centuries, although it was not even partially understood until the last half of the 1800s, when the first hypotheses as to its etiology were proposed. The complex and still only partially understood process which causes arteriosclerosis is called *atherogenesis*, particularly as the process applies to the major arteries, such as the coronary arteries. "Hardening" of the major arteries is commonly referred to as *atherosclerosis.* Because of the great importance of this disease as a predisposing condition for heart attack and stroke, intensive research into the etiology, diagnosis, treatment, and prevention of atherosclerosis is being conducted worldwide. Although considerable progress has been made in recent years, the etiology of the disease has not been fully explained and, in fact, there are several viewpoints on this topic. See also **Ischemic Heart Disease;** and **Cerebrovascular Diseases.**

Traditionally, atherogenesis has been identified as a process involving the deposition of cholesterol and other lipids in the arterial wall resulting from an abnormally high concentration of lipids in the blood. This general hypothesis was first proposed by the German physiologist, Rudolf Virchow, in the late 1800s. Virchow suggested that endothelial cell injury initiated atherogensis. During the intervening years, much clinical and experimental evidence has been collected to support the hypothesis and, in fact, during the first half of the present century, Virchow's hypothesis served as the basis for preventive cardiology. In more recent years, several other hypotheses have been proposed, but as of the early 1980s a fully satisfactory resolution of these various scientific viewpoints has not occurred.

From research on laboratory animals and human autopsies, much has been learned concerning the arterial deposits (plaques) and the structural and compositional changes which have occurred when diseased arteries are examined. Since atherogenesis is a slow process and non-invasive instrumental methods for observing living tissues over a long period are limited, a fully satisfactory step-by-step description of the total process, including the initiating event or events, remains to be delineated.

Multiple-Risk or Risk-Factor Hypothesis. Essentially by consensus among many scientists rather than by a bank of convincing scientific evidence and in lieu of a better understanding of the etiology of atherosclerosis, as of the early 1980s a multifaceted approach to the disease is considered the best guideline for practitioners in the field. The *multiple-risk* or *risk-factor hypothesis* is currently the most influential concept in preventive cardiology. The risk factors include (1) high serum concentration of blood lipids (*hyperlipidemia*) and the role of cholesterol; (2) hypertension; (3) cigarette smoking; (4) lack of exercise; (5) certain genetic disorders; and (6) excessive intake of animal proteins, among others, which appear to be predisposing factors for arteriosclerosis in susceptible individuals.

(1) *Role of Cholesterol—Hyperlipidemia.* As some authorities have observed, the correlation of high levels of serum cholesterol with arteriosclerosis is imperfect. Further, the correlation of a high level of cholesterol in the diet with high levels of serum cholesterol is equally imperfect. The linkage to cholesterol is largely based upon the observation that atherosclerotic plaques contain large deposits of cholesterol and upon data from epidemiological studies. Cholesterol is a steroid bearing an alcohol group. See also **Cholesterol.** Cholesterol is unstable in water and essentially all of it is carried in the blood as lipoproteins. It should be pointed out that some cholesterol is required by the body as a precursor for the manufacture of certain steroids, such as bile acids and several hormones. Most cholesterol synthesis occurs in the liver and, in normal persons, the amount produced is governed in some way by the dietary intake of cholesterol. Some investigators have suggested that dietary cholesterol suppresses a key regulatory enzyme (3-hydroxy 3-methylglutaryl coenzyme A reductase) which is required for cholesterol synthesis.

Lipoproteins are large, complex polar molecules. Carried on the inside are the nonpolar molecules, such as the triglycerides (esters of glycerol and three long-chain fatty acids) and esters of cholesterol. It is estimated that three-fourths of the cholesterol is carried in ester form. The remainder exists as nonesterified cholesterol contained on the surface of the lipoproteins. Within the past decade, considerable new knowledge has been gained concerning these lipoproteins and currently the compounds are put into four categories based upon size and density: (a) *chylomicrons*; (b) *very-low-density* lipoproteins (VDLs); (c) *low-density* lipoproteins (LDLs); and (d) *high-density* lipoproteins (HDLs). Investigators at the University of Texas Southwestern Medical School, among others, have found that the LDLs carry most of the cholesterol found in the blood and play a major role in cholesterol metabolism (in cells other than the liver) and participate in the formation of atherosclerotic lesions. The function of the HDLs appears to be that of transporting cholesterol from the peripheral tissues to the liver. Although the HDLs may be incorporated into LDLs or VLDLs and recycled to the peripheral tissues, some investigators believe that the HDLs may provide a means for removal of cholesterol from the tissues and thus decrease the probability of its terminal deposition as atherosclerotic plaques.

It has been found that specific receptors for the HDLs are contained in human fibroblasts and certain other cells. The interaction of the HDLs with receptors is considered by some researchers as an initial step prior to the degradation of cholesterol and thus suppress cholesterol synthesis in these cells. Individuals with a genetic disorder known as *hypercholesteremia* lack such receptors and are prone to develop atherosclerotic lesions. Studies of patients with this disorder have shown plasma cholesterol levels as much as four times the level in normal individuals. Such persons develop symptomatic atherosclerosis at an early age and frequently suffer fatal heart attacks prior to age 20 years. These findings strengthen the case that blood cholesterol is directly involved in atherogenesis.

As research continues, the explanations of the lipoproteins multiply, but currently with no common denominator within view. It is apparent that once a key is found, this will be the foundation of preventive measures. Although it is possible that such a key may be diametrically opposed to current preventive procedures, most practitioners are staying with the treatment of hyperlipidemia. Hyperlipidemia is treated primarily by dietary restriction, an approach that seems effective when elevations of both cholesterol and triglyceride levels are secondary to obesity and do not reflect a primary genetically determined form of hyperlipidemia. It has been found that weight reduction, achieved by restriction of food calories and alcohol intake, is the principal cause of the decrease in serum lipids in patients who respond to such therapy. Serum cholesterol can be reduced modestly by diets that are low in cholesterol or that substitute polyunsaturated fats (largely of vegetable origin) for saturated fats (largely of animal origin). Some authorities are quick to observe, however, that the importance of diet in prevention of coronary disease undoubtedly has been overemphasized by many physicians. Ahrens (1976) observed that "any advice to the general public to make large dietary changes now is considered premature."

Drug therapy for reduction of serum lipids has been found of limited value and is usually suggested only after dietary methods have failed. Some of the drugs, most of which have side effects, include *cholestyramine,* a high-molecular-weight resin that binds bile acids in the gastrointestinal tract; *nicotinic acid, sodium dextrothyroxine,* and *beta-sitosterol*; and *clofibrate,* used mainly for lowering triglyceride levels. The latter drug tends to increase the incidence of gallstones. *Estrogen* also lowers serum lipids. Some authorities believe that serum cholesterol is lowered because of transfer of concentration of cholesterol in the bile. It should be noted that studies have shown a low incidence of atherosclerosis in premenopausal women, which has suggested a role for estrogen in preventing atherosclerotic plaque formation.

(2) *Hypertension as a Risk Factor.* The risk of hypertension in coronary disease probably rises from the increasing physical stress placed on the arterial wall, thus enhancing the process of arterial wall injury. One hypothesis stresses how factors such as variations in intra-arterial pressure and uneven flow patterns account for the distribution of atherosclerotic plaques. A difference in arterial pressure and its effect on plaque formation is demonstrated by the fact that high-pressure systemic arteries display much more damage than that found in the low-pressure pulmonary arteries. See entry on **Hypertension.**

(3) *Smoking as a Risk Factor.* It has been established that smoking enhances platelet stickiness and causes coronary vasoconstriction produced by nicotine and hypoxemia induced by inhalation of carbon monoxide. In contrast with the risk of smoking (cumulative) in lung cancer, the risk in coronary disease appears to be reversible upon discontinuation of smoking.

(4) *Lack of Exercise as a Risk Factor.* The popularity and broad acceptance of the importance of exercise in preventing coronary artery disease greatly exceeds the scientific evidence. For example, there is no evidence that coronary collateral vessels develop or that atheromatous changes regress as a result of exercise. Exercise as a factor in

weight reduction can lead to a reduction in serum lipid levels, but as previously described, lipid levels remain a topic of controversy. **It should be noted here that the importance of exercise for the rehabilitation of patients who have had a clinical coronary disease, recently or remotely, has different goals and techniques than a program designed for a healthy person as a preventive measure.**

Although stress, personality, and behavior patterns as causative factors in coronary artery disease have received wide public recognition and recognition by some physicians as well, relatively little attention has been given to this topic by pathologists and epidemiologists, largely because these factors are difficult to study. However, some scientists have classified individuals as having a Type A, etc. behavior. Reference to Jenkins (1976) is suggested.

Response-to-Injury Hypothesis. The role of injury to endothelial cells, part of Virchow's original hypothesis on atherogenesis, has been revived in recent years and current findings are based largely upon studies of lesions of animal arteries. The present version states that plaques form in response to frequently recurring injuries to arterial walls. It has been noted that nearly any type of chronic injury to animal arteries will progress to the formation of lesions that are much like the plaques found in human arteriosclerosis. This hypothesis suggests that injury to the endothelium, causing desquamation (shedding of superficial cells), is the primary event. Smooth muscle cells then migrate from the media into the intima through fenestrae in the internal elastic lamina and there undergo active proliferation within the intima. The adherence of platelets to exposed connective tissue may cause the formation of platelet aggregates, or microthrombi. As the lesion progresses, fibrosis, lipid deposition, and calcification may ensue, thus yielding the type of complex plaque noted in diseased arteries.

Based largely upon this hypothesis, platelet-inhibiting drugs, such as *acetylsalicylic acid, sulfinpyrazone,* and *dipyridamole,* are in use and continue under investigation. Their clinical value thus far has been relatively weak. See Genton (1975) reference.

Benign Tumor (Monoclonal) Hypothesis. The suggestion that plaques are benign tumors was first made in the early 1970s. The hypothesis stems from studies made by investigators at the University of Washington School of Medicine, among others. As pointed out by Kolata (1976), this hypothesis relies on a method previously used to support the contention that benign uterine tumors made up of smooth muscle cells are derived from single cells. The method is based upon the generally accepted beliefs that only one of the two X chromosomes in a given cell of a female expresses its genes and that which X chromosome in a cell is active is decided at random during embryo development. All progeny of a particular cell express genes from the same X chromosome as their parent, but neighboring cells are often derived from different parent cells and thus often express genes from different X chromosomes. There is considerable controversy over the validity of the hypothesis but attempts to prove or disprove it, as with the other hypotheses, have opened up new perspectives in the pursuit of atherogenesis.

Buerger's Disease. This is an arterial disease (*thromboangitis obliterans*) which also may involve the veins. This is an inflammatory disease which is followed by obliteration of the lamina. This results in impairment of the circulation to the extremities, manifested by coldness, cyanosis, pain, and, if untreated and uncontrolled, eventually gangrene of the affected part. Treatment is directed toward improving the circulation of the extremities by special exercises, elevation of the affected part, drug therapy, and occasionally surgical measures which remove vessel-constricting impulses by interfering with their nervous pathways.

Arterial Transplantation. A marked advance in surgery over the past several years has been the development and refinement of arterial transplantation. Originally, sections of arteries were obtained from the bodies of healthy individuals who had died suddenly. But, because transplanted arteries rapidly underwent arteriosclerotic changes, they have been replaced by artificial vessels (Teflon or Dacron and equivalent plastic materials). Almost indestructible and readily available, these vessels have proved satisfactory. They do not cause the reaction (shock) that occurs in an individual when a part from another human or animal is placed inside his body. Pieces of vein from the patient also have been used and may further enhance long-term results.

Venous Disorders

Varicose Veins. These are veins that have lost their elasticity and, as a consequence, are irregularly enlarged and swollen. They have a dilated, lumpy, twisted and tortuous appearance. The overlying skin may be affected with ulcers. Varicose veins are most often seen in the legs of middle-aged and older persons, although certain conditions, such as pregnancy, may cause them to appear in younger adults. The dilation of the veins results from the inability of the weakened venous walls to withstand the pressure of the blood within the veins.

If the veins were simply continuous tubes running from the legs to the heart, the weight of a column of blood carried this high would press out on the leg veins when the individual stood erect. Normally, the column of blood is broken by the presence of valves which prevent the full weight of the blood from causing undue pressure on the veins in the leg. See accompanying diagram. If a vein loses its elasticity, it will become distended, and the valves will fail to close completely. The weight of the blood in that vein then presses out on the walls of the vein, causing even more distention; and a reversal of flow of blood in that vein may occur. The veins also may become swollen when a venous constriction prevents the normal emptying of the blood.

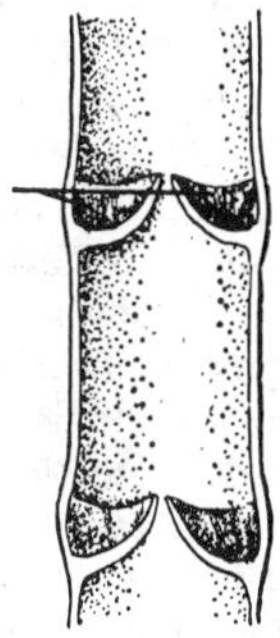

Large vein, showing how valves composed of two cusps occur along the vein and prevent the backflow of blood in these vessels.

The classical sign of varicose veins is the actual appearance in the legs of swollen, tortuous, blue veins. The enlargement may affect a short segment of a single vein, or nearly all of the veins in the entire leg. When a systemic disease is responsible for the disorder, it usually appears in both legs to an approximately equal degree. *Phlebitis* (inflammation of a vein), constriction, injury, or obstruction in the veins in one leg, however, will affect only the one leg. Aside from disfiguration, varicose veins cause appreciable physical discomfort. Both dull and stabbing pains may be noted, and the entire limb may become quite swollen. When the condition has existed for some time, the veins sometimes become toughened and thick, so that they feel firm to the touch. More often, however, they are soft and elastic, except at the hard knotty swellings which occur in the regions of the valves. Ulceration and bleeding may leave large black and blue areas beneath the skin. In men, a type of varicosity may occur in which the vein's in the scrotum are affected. In this condition, known as *varicocele,* the scrotum contains a soft, tumor-like mass of swollen venous materials. Similarly, varicosity of the veins in the rectum is known as *hemorrhoids* (piles).

In severe cases of varicose veins, it may be necessary to surgically remove portions of a vein that are particularly bothersome, or strip out the entire varicosed vein. In milder cases, merely tying off (ligate) the varicose veins to relieve pressure may suffice. Occasionally, a combination of an operative procedure and sclerosing solutions which close off small veins may be required.

Raynaud's Disease. A vascular disorder characterized by paroxysmal cyanosis of the fingers in response to cold or to emotional stimuli. Toes, ears, chin, may also be affected. The disease appears most commonly in women in the third decade of life, and familial connections have been described. The cyanosis is due to interference with the blood supply through constriction of the vessels (vasospasm) resulting from abnormalities of the nervous control of the vessel walls.

Venous Thrombosis. Two forms of venous thrombosis are presented: (1) Lesions in *superficial veins* may be caused by tightly adherent thrombi. There may be significant inflammation and in some cases cellulitis, but these thrombi rarely penetrate to deep veins to cause pulmonary emboli. Treatment includes application of local heat and possibly administration of antibiotics where infection is suspected. (2) *Deep venous thrombi* may lead to pulmonary embolism, the latter being a life-threatening situation. Some physicians describe deep venous thrombi as "silent" because they may not be detected by routine clinical examination. In fact, the primary site may be difficult to fix even after the diagnosis of pulmonary embolism. However, when present, clinical signs include distended veins, localized areas of tenderness over venous structures and, in the calf, notation of increased heat, edema, and sometimes discoloration. Some conditions mimic the symptoms of deep venous thrombi, including rupture of calf muscle fibers or of synovial cysts involving the knee. Procedures for diagnosis of deep venous thrombi include venography which will reveal any filling defects. This procedure is used with discretion because the invasion necessary for venography can in itself induce venous thrombosis. Although not applicable to the calf, noninvasive diagnostic measures include impedance plethysmography and ultrasonography. Radionuclide scanning is useful in connection with the calves, but not in the thighs. The treatment of deep vein thrombosis is the prompt administration of intravenous heparin. This drug acts rapidly in preventing further thrombus formation and prevents the release of serotonin and thromboxane A_2 from platelets adherent to those thrombi that embolize to the lungs. See **Anticoagulants.**

Pulmonary Embolism. The formation of emboli is the major consequence that is feared from deep venous thrombi. Although pulmonary emboli are sometimes overdiagnosed, statistics also show that in a high percentage of fatal cases, the disease has gone on unsuspected for quite a period of time. The usual symptoms are apprehension, weakness, faintness or syncope, dyspnea, and a substernal pain that is not distinguishable from that of myocardial infarction. Lung examination may indicate wheezes. To assist in diagnosis, the physician will order blood gases analysis and frequently an emergency perfusion lung scan. However, a number of conditions also alter pulmonary blood flow and such scans are not fully definitive. In recent years, pulmonary ateriorgraphy has become the reference standard for the diagnosis of pulmonary embolism. However, the procedure is undertaken with discretion because of risks associated with it. Other diagnostic tools include chest x-ray, electrocardiography, and serum enzyme levels. Treatment of pulmonary embolism includes drugs, such as morphine sulfate or meperidine, to relieve pain and the institution of intravenous heparin therapy. After several days, the patient may be switched to oral anticoagulants, a treatment which in some cases may extend over a year.

Massive Pulmonary Embolism. In this case, large emboli occlude the proximal pulmonary arterial circulation, producing symptoms much like those of myocardial infarction. Usually, electrocardiography will distinguish between the two conditions. Massive pulmonary embolism causes many deaths and, in any case, must be considered a life-threatening emergency. Upon diagnosis, large doses of intravenous heparin are given, along with morphine and oxygen. If there is persistence of shock for over 30 minutes, an immediate pulmonary arteriogram should be done. However, this requires the equipment and skills of an open-heart surgery facility. In severe instances and where massive pulmonary emboli are confirmed, surgery to remove the embolism may be elected to save the life of the patient.

In the management of massive pulmonary embolism and severe deep venous thrombosis, thrombolytic therapy has been relatively recently available. This involves the use streptokinase and urokinase, agents which hasten the dissolution of thrombus. Thrombolytic agents cannot be given concurrently with anticoagulants and they do have a number of side effects. Several years of additional experience will probably be required to assess the values and risks of thrombolytic agents.

See also list of related entries at end of entry on **Heart and Circulatory System (Human).**

References

Ahrens, E. H., Jr.: "The Management of Hyperlipidemia: Whether, rather than How," *Ann. Intern. Med.*, **85**, 87 (1976).

Barnes, R. W., Wu, K. K., and J. C. Hoak: "Fallibility of the Clinical Diagnosis of Venous Thrombosis," *JAMA*, **234**, 605 (1975).

Colburn, P., and V. Buonassisi: "Estrogen-Binding Sites in Endothelial Cell Cultures," *Science*, **201**, 817–819 (1978).

Coronary Drug Project Research Group: "Gallbladder Disease as a Side Effect of Drugs Influencing Lipid Metabolism," *N. Engl. J. Med.*, **296**, 1185 (1977).

Genton, E., et al.: "Platelet-inhibiting Drugs in the Prevention of Clinical Thrombotic Disease, *N. Engl. J. Med.*, **293**, 1174 (1975).

Havel, R. J.: "Classification of the Hyperlipidemias," *Ann. Rev. Med.*, **28**, 195 (1977).

Hull, R., et al.: "Combined Use of Leg Scanning and Impedance Plethysmography in Suspected Venous Thrombosis: An Alternative to Venography," *N. Engl. J. Med.*, **296**, 1497 (1977).

Jenkins, C. D.: "Recent Evidence Supporting Psychologic and Social Risk Factors for Coronary Disease," *N. Engl. J. Med.*, **294**, 987 (1976).

Kaplan, A. P., et al.: "Molecular Mechanisms of Fibrinolysis in Man," *Thromb. Haemostas.*, **39**, 263 (1978).

Kolata, G. B.: "Atherosclerotic Plaques: Competing Theories Guide Research," *Science*, **194**, 592–594 (1976).

Kolata, G. B.: "New Treatment for Coronary Artery Disease," *Science*, **206**, 917 (1979).

Marder, J. M.: "The Use of Thrombolytic Agents: Choice of Patient, Drug Administration, Laboratory Monitoring," *Ann. Intern. Med.*, **90**, 802 (1979).

Marx, J. L.: "Atherosclerosis: The Cholesterol Connection," *Science*, **194**, 711–714 (1976).

Marx, J. L., and G. B. Kolata: "Combating the #1 Killer," American Association for the Advancement of Science, Washington, D.C., 1980.

McGill, H. C., Jr.: "The Heart is a Target Organ for Androgen," *Science*, **207**, 775–776 (1980).

Richards, K. L., et al.: "Noninvasive Diagnosis of Deep Venous Thrombosis," *Arch. Intern. Med.*, **136**, 1091 (1976).

Robin, E. D.: "Overdiagnosis and Overtreatment of Pulmonary Embolism," *Ann. Intern. Med.*, **87**, 775 (1977).

Ross, R., and L. Harker: "Hyperlipidemia and Atherosclerosis," *Science*, **193**, 1094–1100 (1976).

Ross, R., and J. A. Glosmet: "The Pathogenesis of Atherosclerosis," *N. Engl. J. Med.*, **295**, 369–420 (1976).

Staff: "Vitamin B_6 versus Arteriosclerosis," *Technol. Rev.* (*MIT*), **81**, 7, 90 (1979).

Stallones, R. A.: "The Rise and Fall of Ischemic Heart Disease," *Sci. Amer.*, **243**, 5, 53–59 (1980).

Turlapaty, P. D. M. V., and B. M. Altura: "Magnesium Deficiency Produces Spasms of Coronary Arteries: Relationship to Etiology of Sudden Death Ischemic Heart Disease," *Science*, **108**, 198–200 (1980).

ARTERIOGRAPHY. Arterial and Venous Disorders.

ARTERIOLAR NEPHROSCLEROSIS. Kidney and Urinary Tract.

ARTERIOSCLEROSIS. Arterial and Venous Disorders; Ischemic Heart Disease.

ARTERIOSCLEROTIC DEMENTIA. Alzheimer's Disease and Other Degenerative Mental Disorders.

ARTERY. A vessel leading away from the heart. See **Circulatory System; and Heart and Circulatory System (Human).**

ARTHRITIS-DERMATITIS SYNDROME. Gonorrhea and Gonococcemia.

ARTHRITIS (Gouty). Gout.

ARTHRITIS (Infectious). An inflammatory joint disease caused by pyogenic (pus-forming) bacteria, mycobacteria, or fungi.

Acute Bacterial Arthritis. The *acute* form of the disease is usually caused by pyogenic bacteria, of which the most common cause is *Neisseria gonorrhoeae.* Of the remaining cases, caused by nongonococcal bacteria, *Staphylococcus aureus, Streptococcus pneumoniae,* and other strep microorganisms are involved. A wide variety of infections which cause bacteremia may predispose acute bacterial arthritis—pneumococcal pneumonia, endocarditis, skin infections, and cholangitis, among others. Uncommonly, the condition may develop as the result of joint surgery. About 10% of cases of acute bacterial arthritis are attributed to gram-negative bacili. An infection of this type is sometimes called *acute septic arthritis.* Related or predisposing factors include urinary or biliary tract infection, neoplastic (tumor-forming) diseases, diabetes mellitus, intravenous heroin abuse, and certain antibiotics and immunosuppressive drugs. In about 20% of cases, *Salmo-*

nellae are implicated; *Pseudomonas* and *Serratia* species are usually involved in cases of septic arthritis in narcotic addicts. *Haemophilius influenzae* is an uncommon agent in adults, but is a frequent cause of the disease in children. At one time, prior to tight health controls over milk and milk products, brucellosis was a major cause of spinal infectious arthritis.

Bacteria may invade the joint cavities by way of the bloodstream or they may spread from a contiguous focus of osteomyelitis. Purulent discharge may damage articular cartilage, joint capsule, and subchondral bone.

Symptoms of bacterial arthritis are hot, swollen, and painful joints. Onset may be abrupt, particularly in the presence of systemic illness. Fever of a transient or low-grade level may be present. Most frequently involved joints are the knees, followed by hips and shoulders. Symmetry is not usually present, an advantage in terms of diagnosis.

Precise diagnosis usually requires examination and culture of joint fluid. In early phases of the disease, x-rays are of little help. There are several disorders which have similar symptoms and the diagnosis may be complex. Early diagnosis, however, is extremely important if recovery without permanent joint damage is to be achieved. Choice of most effective antimicrobial and antibiotic therapy hinges on accurate diagnosis. Drainage of the affected joint is also important to reduce pressure and to remove cartilage-destroying enzymes present in the pus. Aspiration at intervals over several days may be required. Moderate movement of the joint after the initial phase is encouraged to maintain mobility of the joint. The affected joint should not be subject to weight bearing, however, until inflammation has subsided. With early diagnosis and prompt therapy, full recovery without joint damage can be expected.

Tuberculosis Arthritis. This is uncommon. The disease is caused by the presence of *Mycobacterium tuberculosis.* The joints most often involved are hips and knees in children and vertebral joints in adults. Predisposing factors include diabetes mellitus, lupus erythematosus, narcotic addiction, and intra-articular injections of corticosteroids. Often present for a long period without notice by the patient, the disease can result in serious, permanent joint damage. Diagnosis may require an open synovial biopsy. Therapy includes resting the joint and use of antimicrobial drugs, such as isoniazid and ethambutol.

Fungal Arthritis. This relatively uncommon type of infectious arthritis results from invasion by mycotic agents and varies considerably with geographical location. For example, the disease may be associated with coccidiomycosis in the southwestern United States and with blastomycosis in the southeastern United States. Amphotericin B therapy is usually indicated.

Syphilitic Arthritis. This may occur in congenital, secondary, or tertiary syphilis. It is most commonly seen in infants and very small children with congenital syphilis. In these instances it may lead to destructive osteochondritis. A more moderate form of the disease may develop in older children and adolescents with congenital syphilis. Penicillin therapy is usually indicated.

Lyme Arthritis. This relatively recently recognized disease is attributed to a virus. The disease can result in chronic joint destruction, but in most cases is self-limiting. To date, only a few hundred cases in the United States have been recorded.

References

Dryer, R. F., Goellner, P. G., and A. S. Carney: "Lyme Arthritis in Wisconsin," *JAMA*, **241**, 498 (1979).

Gifford, D. B., et al.: "Septic Arthritis due to *Pseudomonas* in Heroin Addicts," *J. Bone Joint Surg.*, **57A**, 631 (1975).

Goldenberg, D. L., and A. S. Cohen: "Acute Infectious Arthritis," *Amer. J. Med.*, **60**, 369 (1976).

McCarty, D. J. (editor): "Arthritis and Allied Conditions," 9th edition, Lea & Febiger, Philadelphia, 1979.

Steere, A. C., et al.: "Chronic Lyme Arthritis," *Ann. Intern. Med.*, **90**, 896 (1979).

Ward, J. R., and S. G. Atcheson: "Infectious Arthritis," *Med. Clin. North. Am.*, **61**, 313 (1977).

Winter, W. J., Jr., et al.: "Coccidioidal Arthritis and Its Treatment," *J. Bone Joint Surg.*, **57A**, 1152 (1975).

ARTHRITIS (Rheumatoid). Rheumatoid Arthritis.

ARTHROPODA. The largest and most diversified division of the animal kingdom, including crustaceans, horseshoe crabs (see **Xiphosura**), insects, scorpions, spiders, centipedes, millipedes (see **Diplopoda**) and other forms.

This phylum is characterized by the following structures: (1) The body is triploblastic and metameric (see **Metamere**), and is further subdivided into regions of which there may be a maximum of three: head, thorax, and abdomen. (2) Supporting structures are developed from the integument and constitute an exoskeleton made up of plates connected by flexible regions for freedom of movement. (3) The appendages are jointed, fundamentally a pair to a segment, which gives the phylum its name. (4) The circulatory system is a combination of tubes and open spaces; the coalescence of the latter to form a haemocoel is accompanied by extreme reduction of the coelom. (5) The eyes are of a form peculiar to the group. (6) The respiratory system consists of air tubes or tracheae, of gills, or of lung books formed of leaf-like expansions of the body wall located in a cavity narrowly open to the exterior. Some species have no special respiratory organs.

The phylum is divided into several classes as follows:

Class *Onychophora.* Soft-bodied worm-like animals. Commonly called Peripatus, the name of one genus.

Class *Tardigrada.* The bear animalcules.

Class *Pentastomida.* Parasites known as pentastomids or linguatulids.

Class *Pycnogonida.* The sea-spiders.

Class *Crustacea.* Crabs, lobsters, crayfishes, shrimps, barnacles, woodlice or pillbugs, and other forms.

Class *Xiphosura.* The king crab or horseshoe crab. Also given the name Palaeostraca.

Class *Arachnida.* Spiders, mites, ticks and scorpions.

Class *Diplopoda.* The millipedes.

Class *Pauropoda.* Rare forms allied to the preceding.

Class *Chilopoda.* The centipedes.

Class *Symphyla.* Rare forms allied to the preceding.

Class *Insecta.* The insects. See also **Invertebrate Paleontology.**

A very small proportion of the arthropoda, notably of the insecta, have widespread influence on the health of the human race; either by direct injury but chiefly through the destruction of growing crops and stored food and their capabilities as carriers of disease to man and domestic animals. In man, the diseases so produced are typhus and malaria, yellow fever, filariasis, bacillary dysentery and other enteric infections.

ARTHROPODA (Fossils). Invertebrate Paleontology.

ARTHROPODS (Skeletal System). Skeletal System.

ARTICHOKE. ***Compositae* or Composite Family.**

ARTICULATION (Communications). In verbal communication, the main purpose of speech is to convey thoughts. In testing speakers (human) and communication systems, a number of tests have been designed for measuring the percentage of words or individual speech sounds uttered by a speaker which are perceived correctly by listeners. For example, the Harvard PB-50 word list employs a set of phonetically balanced words and is widely used. During the testing procedure, the talkers read the word lists over the system under test to a number of listeners. As they hear them, the listeners record the words. Responses are examined and the percentage of words heard correctly is determined. This is termed *percent word articulation.*

During such tests, a number of communication system parameters are explored. Methods have been developed for computing speech intelligibility from system characteristics. The measure of intelligibility computed is termed the *articulation index* (AI). A number of observations concerning speech are taken into consideration by the articulation-index concept: (1) speech must be above threshold of audibility to be perceived; (2) noise that exceeds this threshold masks speech, effectively raising the threshold of audibility; (3) there is an upper limit to the sound pressure which the ear can utilize for perception of speech; (4) frequencies from 200 to 6100 Hz are needed for substantially perfect intelligibility; (5) speech has a 30-decibel dynamic range;

and (6) different frequency regions contribute unequally to intelligibility, but the frequency range from 200 to 6100 Hz can be divided into bands of equal contribution to speech intelligibility.

Ultimately, of course, the articulation-index concept has its foundation on an analysis of the sound pressures (signal and noise) that are produced at the listener's ear. In practice, it is expeditious to reduce all statements of sound pressure at the listener's ear to spectrum level (the sound pressure level in decibels is 0.0002 dyne per square centimeter for a 1-Hz bandwidth).

While much too complex to describe in detail here, speech intelligibility and quality are of major concern in the design of certain communications systems. The references listed contain considerable further information on this topic. See also **Telephony;** and **Voice Recognition and Synthesis.**

ARTIFICIAL ARM-HAND (Industrial). Robot.

ARTIFICIAL ARTERY. Arterial and Venous Disorders; Heart and Circulatory System (Human).

ARTIFICIAL GAS. Coal; Substitute Natural Gas.

ARTIFICIAL HEART (Total). Heart and Circulatory System (Human).

ARTIFICIAL HEART VALVE. Heart and Circulatory System (Human).

ARTIFICIAL HORIZON. A planar reflecting surface that can be adjusted to coincide with the astronomical horizon, i.e., can be made perpendicular to the zenith. This instrument is used, usually in conjunction with others, in observing celestial bodies. See also **Horizon (Astronomical); Horizon (Celestial);** and **Sextant.**

ARTIFICIAL HORIZON (Aircraft). Several ways of presenting to the pilot the information as to direction of true vertical have been used. The artificial horizon indicator shown in Fig. 1 is one acceptable

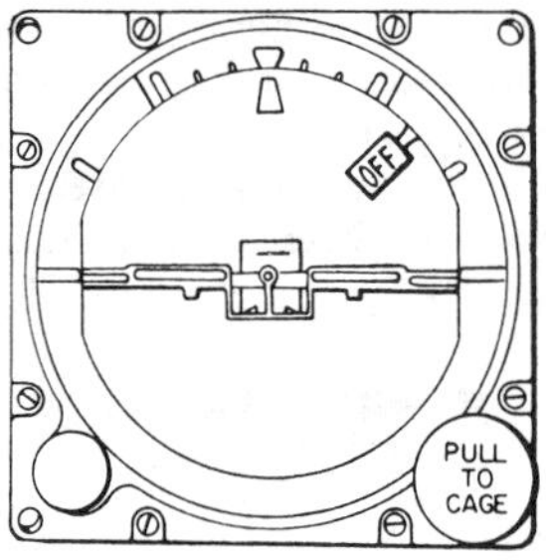

Fig. 1. Artificial horizon indicator. (*Bendix*).

means. Gyros furnish an excellent means for determining dynamic vertical. If a gyro is arranged with gimbals as shown in Fig. 2 and there is some method of detecting instantaneous or dynamic vertical by which a torque may be applied to the gyro with its downward component being directed to the rising side of the gyro wheel (Fig. 3), the spin axis will move toward the dynamical vertical. As the rate of this movement toward vertical (called the erection rate) is

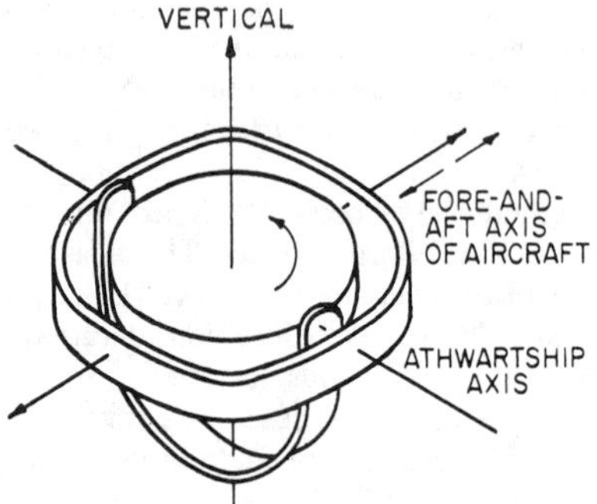

Fig. 2. Gyro in gimbals.

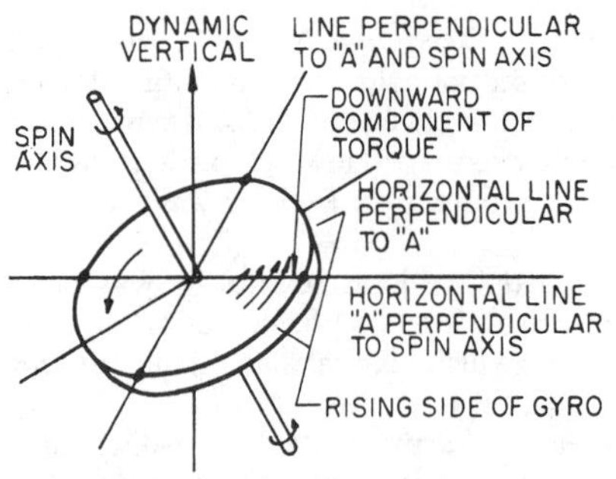

Fig. 3. Gyro vertical.

a relatively slow action, of the order of a few degrees in a minute, the spin axis indicates a long-time average of dynamic vertical which usually is very close to true vertical.

Often, the detecting means may be pendulous switches, usually electrolytic or mercury, which energize torque motors causing the gyro to erect. This system has many advantages in that the erection rate is externally adjustable if desired, and erection may be disconnected by means of switches actuated by rate gyros so that only when the dynamic vertical is near true vertical is it averaged by the gyro mechanism and a more exact true vertical indication is obtained.

If the gyro spin axis is maintained horizontal instead of vertical, it will tend to keep its spin axis pointed in one direction. By putting a degree calibration on the gyro, it forms a stable heading reference. Methods of drive, erection, and signal takeoff are similar for both gyro verticals and horizontals.

ARTIFICIAL INSEMINATION. Introduction of semen into the vagina by means of surgical instruments and procedures. Although sparingly used in *Homo sapiens,* artificial insemination is widely practiced in the breeding of horses and livestock. See also **Embryo;** and **In-Vitro Fertilization.**

ARTIFICIAL KIDNEY. Kidney and Urinary Tract.

ARTIFICIAL LARYNX. Larynx.

ARTIFICIAL LINE. An artificial line is an electrical network consisting of resistance, inductance and capacitance so connected that it has the same characteristics (electrical) as the actual transmission line. Sometimes where the artificial line is not required to duplicate exactly the actual line the inductance or capacitance may be omitted. Such a line is very valuable for making laboratory tests as it makes possible connection at points corresponding to many miles of actual line. Artificial lines are also used widely in telephone and telegraph practice to balance actual lines to give desired operating characteristics in bridge type circuits.

ARTIFICIAL RESPIRATION. A person can breathe for another person and often maintain life until medical help arrives. Many lives have been saved because someone knew how to do this. A person can stop breathing from abuse of drugs and alcohol, drowning, gas poisoning, electric shock, choking, heart failure, smothering, and numerous other causes. The fastest and best way to get air into another person is to blow air into the victim's mouth and lungs. This is called mouth-to-mouth artificial respiration. The general procedure is as follows:

1. If you can see anything in the mouth of the person who is not breathing, turn the victim's head to one side and clean out the mouth. Put the victim on his/her back. Tilt head back so that the victim's chin points up. Sometimes then the victim will start to breathe.

2. If the victim does not start to breathe without assistance when you have done this, then you will need to start blowing into the victim's mouth. Keep the victim's head tilted back and with one hand pinch the victim's nose shut to keep the air from coming out there. Make an airtight seal with your mouth over the victim's mouth and then blow into the victim's mouth until you can see the chest rise.

3. Raise your mouth after each time you blow and turn your face to the side and listen for air to come out of the victim's lungs. Blow air into the mouth about 12 times a minute for an adult. If you

ARTIODACTYLA
(Even-toed Hoofed Mammals)

- ANTELOPINES — See **Antelope.**
  - Horse-Antelopes (*Hippotraginae*)
    - Sabre-horned Antelopes (*Hippotragus*)
    - Rapier-horned Antelopes (*Aegoryx* and *Oryx*)
    - Screw-horned Antelopes (*Addax*)
  - Deer-Antelopes (*Alcelaphinae*)
    - Hartebeests (*Alcelaphus*)
    - Damalisks (*Beatragus* and *Damaliscus*)
    - Gnus (*Connochaetes* and *Gorgon*)
  - Marsh-Antelopes (*Reduncinae*)
    - Waterbucks (*Kobus*)
    - Lechwes (*Onotragus*)
    - Kobs (*Adenota*)
    - Reedbucks (*Redunca*)
    - The Rhebok (*Pelea*)
  - Blackbuck (*Antilopinae*)
  - Pigmy Antelopes (*Neotraginae*)
    - Klipspringers (*Oreotragus*)
    - Oribis (*Ourebia* and *Raphicerus*)
    - Sunis (*Nesotragus*)
    - The Beira (*Dorcatragus*)
    - Dik-Diks (*Madoqua* and *Rhynchotragus*)
    - Royal Antelopes (*Neotragus*)
  - Gazelles (*Gazellinae*)
    - Impalla (*Aepyceros*)
    - The Gerenuk (*Litocranius*)
    - The Dibatag (*Ammodorcas*)
    - The Springbuck (*Antidorcas*)
    - The Addra (*Addra*)
    - True Gazelles (*Gazella*)
    - Goat-Gazelles (*Procapra*)
- ANTILOCAPRINES — See **Pronghorn Antelope.**
  - Pronghorn Antelope
- BOVINES — See **Bovines.**
  - True Oxen (*Bovinae*)
    - Cattle (*Bos*)
    - Buffalo (*Bubalus*, *Syncerus* and *Anoa*)
    - Bison (*Bison*)
  - Deer-Oxen (*Boselaphinae*)
    - The Nilghai (*Boselaphus*)
    - The Chousingha (*Tetraceros*)
  - Twist-horned Oxen (*Strepsicerosinae*)
    - Elands (*Taurotragus*)
    - The Bongo (*Böocercus*)
- BOVINES — See **Bovines.**
    - Kudus (*Strepsiceros*)
    - Bushbucks (*Tragelaphus*)
  - Duikers (*Cephalophinae*)
    - Common Duikers (*Sylvicapra*)
    - Forest Duikers (*Cephalophus*)
    - Blue Duikers (*Philantomba*)
- CAMELINES — See **Camels and Llamas.**
    - Camels (*Camelus*)
    - Llamas (*Lama*)
    - The Vicuña (*Vicugna*)
- CAPRINES — See **Goats and Sheep.**
  - Gazelle-Goats (*Saiginae*)
    - The Chiru (*Panthalops*)
    - The Saiga (*Saiga*)
  - Rock-Goats (*Rupicaprinae*)
    - The Goral (*Naemorhedus*)
    - Serows (*Capricornis*)
    - Chamois (*Rupicapra*)
    - Rocky Mountain Goat (*Oreamnos*)
  - Ox-Goats (*Ovibovinae*)
    - Takins (*Budorcas*)
    - The Muskox (*Ovibos*)
  - True Goats (*Caprinae*)
    - Tahrs (*Hemitragus*)
    - Markhors (*Capra falconeri*)
    - The Tur (*Capra caucasica*)
    - Ibexes (*Capra ibex*, ...)
  - Sheep (*Ovinae*)
    - The Aoudad (*Ammotragus*)
    - The Bharal (*Pseudois*)
    - True Sheep (*Ovis*)
- CERVINES — See **Deer.**
  - Musk-Deer (*Moschinae*)
  - Muntjacs (*Muntiacinae*)
  - True Deer (*Cervinae*)
    - Père David's Deer (*Elaphurus*)
    - Fallow Deer (*Dama*)
    - Axis Deer (*Axis*)
    - Red Deer (*Cervus*)
  - Hollow-toothed Deer (*Odocoileinae*)
    - White-tailed Deer (*Odocoileus*)
    - Marsh Deer (*Blastocerus*)
    - The Pampas Deer (*Ozotoceros*)
    - Guemals (*Hippocamelus*)
    - Brockets (*Mazama*)
    - Pudus (*Pudua*)
  - Moose (*Alcinae*)
  - Reindeer (*Rangiferinae*)
    - Eurasian Reindeer (*Rangifer tarandus*)
    - Caribous (*Rangifer arcticus*, ...)
  - Water-Deer (*Hydropotinae*)
  - Roe Deer (*Capreolinae*)
- GIRAFFINES — See **Giraffe and Okapi.**
  - Giraffes (*Giraffinae*)
  - Okapis (*Palaeotraginae*)
- HIPPOPOTAMINES — See **Hippopotamus.**
  - Common Hippopotamus (*Hippopotamus*)
  - Pigmy Hippopotamus (*Choeropsis*)
- SUINES — See **Suines.**
  - Pigs (*Suidae*)
    - Eurasian Pigs (*Sus*)
    - African Bush-Pigs (*Potamochoerus*)
    - The Forest-Hog (*Hylochoerus*)
    - Wart-Hogs (*Phacochoerus*)
    - The Babirusa (*Babirusa*)
  - Peccaries (*Tayassuidae*)
- TRAGULINES — See **Tragulines.**
    - Oriental Chevrotains (*Tragulus*)
    - Water-Chevrotains (*Hyemoschus*)

blow every 5 seconds for adults, that will be 12 times a minute. For children, use short puffs every 3 seconds. That will be about 20 times a minute. Keep this up until medical help arrives or until the person starts to breathe unassisted.

For further information, check various publications issued by The American Red Cross.

ARTIFICIAL SWEETENERS. Flavorings.

ARTIODACTYLA (*Mammalia*). Hoofed animals which retain an even number of toes, the axis of the foot passing between the third and fourth digits. In older terminology, the term even-toed ungulates was used. Organization of the *Artiodactyla* is shown in the accompanying table, along with references to specific entries in this volume which describe the various families, subfamilies, and species found in the order of *Artiodactyla*.

For references, see **Mammalia.**

ASAFETIDA. The gum-resin exudate from the roots of a commonly occurring plant in the steppes region of Asia, upon steam distillation, yields a pale-yellow to orange-yellow liquid having a garliclike odor and slightly bitter, pungent taste. The resin from this plant (*Ferula assafoetida* L.) of the family *Umbelliferae* contains methylpropenyldisulfides, along with small amounts of vanillin. Asafetida has been used as a flavoring in a number of food products, including nonalcoholic beverages, ice creams, candies, baked goods, and condiments, among others. Most countries consider the available fluid extracts and tinctures as GRAS (generally regarded as safe).

ASBESTOS. This is the popular name for several fibrous minerals used for fireproofing and heat insulating material. Most of the commercial asbestos is a variety of serpentine called chrysotile. A fibrous kind of amphibole is called asbestus. Both terms asbestos and asbestus are derived from the Greek word meaning unquenched, which was applied to minerals that resisted fire. See also **Amphibole; Serpentine.**

Short asbestos fibers usually are compressed with binders to form insulating boards, paper, pipe covering, shingles, and various molded products where the insulative properties of asbestos are desired. The longer asbestos fibers may be woven into various fabric materials for use in brake linings, heat-resisting shields, and fireproof garments and curtains. In some cases, asbestos fibers are mixed with cotton, or with very fine metallic wires, as in the cases of clutch facings and brake linings.

When asbestos fibers are mixed with portland cement and molded under high pressure, an asbestos board is produced which is capable of withstanding temperature up to about 540°C (approximately 1000°F). Asbestos shingles for roofing and siding are made in a similar manner. When molded with cement in the form of boards, the material can be used as flooring. When asbestos is molded with asphalt, the resulting products are known as ebonized asbestos.

Asbestos paper may be made by bonding asbestos fibers with sodium silicate solution, silicone resin, or clay. In addition to excellent thermal insulation, the product has a high dielectric strength and is used for electrical insulating purposes. Being inert chemically to most corrosive materials, asbestos products are attractive for applications in the chemical and process industries, but may pose a health hazard.

Continued exposure to asbestos dust over long periods by asbestos miners and asbestos product factory workers can give rise to a serious respiratory disease sometimes referred to as asbestosis (similar to silicosis). Prolonged inhalation can cause cancer of the lung, pleura and peritoneum. Symptoms include shortness of breath and a dry cough.

See also terms listed under **Mineralogy.**

ASCARIS. Parasitic roundworms of relatively large size found in the intestines of man and other animals.

ASCENDING COLON. Digestive System (Human).

ASCENT OF SAP. All of the organs of any terrestrial plant are dependent for their existence upon water absorbed from the soil. This water, which always contains traces of solutes and hence is often referred to as sap, moves in a generally upward direction through the plant. In some of the tallest known specimens of redwood trees (*Sequoia sempervirens*) the sap must ascend to heights exceeding 350 (105 meters) feet if the topmost branch is to be kept supplied with water. The upward movement of sap in plants occurs in the xylem, which in trees and shrubs corresponds to the wood. In the trunks or larger branches of trees sap movement is confined to a few of the outermost annual rings of wood. Sap movement occurs only through the vessels and tracheids of the woody tissue.

The earlier theories of the upward movement of sap in plants mostly invoked some vaguely conceived vital activity of the cells as furnishing the motive power for sap movement. Although the vessels and tracheids through which the water moves are dead, they are always in intimate contact with living wood parenchyma and wood ray cells and it is not inconceivable that these cells might in some way motivate the upward movement of sap. However, such theories receive very little support at the present time.

It is a common observation that sap may flow from the severed stems of many kinds of plants and that this flow ("bleeding") may continue for some time. This exudation of sap results from a pressure originating in the root called *root pressure*, and the exuded sap comes from the xylem tissues. Root pressures are also present in intact plants. For several reasons, however, root pressure can be considered only a secondary mechanism of water transport in plants. In the first place, there are many species in which the phenomenon does not occur. In the second place the magnitude of measured root pressures seldom exceeds two atmospheres which could not cause a rise of sap of more than about 60 feet (18 meters). In the third place known rates of sap flow under the influence of root pressure are inadequate to compensate for many known rates of transpiration. And finally, in woody plants at least, root pressures are usually present only in the early spring; during the summer period when transpiration rates and hence rates of sap movement are greatest, root pressures are negligible or nonexistent.

The principal mechanism motivating the ascent of sap in plants is thought by most present day botanists to be dependent upon the property of *cohesion* in water. The cohesive forces between water molecules are very great. The evaporation of water from the mesophyll cells of the leaf during transpiration results in the movement of water molecules into these cells from the xylem (water-conducting tissue) of the veins. The xylem of the leaves is continuous with that of the stems which in turn is continuous with that of the root system out almost to the very tip of every rootlet. The water is apparently present in the cells and vessels of the xylem as continuous threadlike columns. As water molecules pass out of these water columns into the mesophyll cells the threads of water become taut throughout the plant. Eventually a tension of considerable magnitude may be set up in them which is transmitted from the top to the bottom of the plant. The water columns can sustain this tension only because of the high cohesive force of water. When the water in the xylem of the younger roots passes into a state of tension movement of water from the root cells into the xylem cells is induced. Loss of water from the root cells in turn causes absorption of water from the soil. Movement of water through the entire plant is thus brought about. Whenever transpiration rates are appreciable water does not, as a rule, enter the lower ends of the xylem ducts from adjacent root cells as fast as it passes from the upper ends into the mesophyll cells, hence the water is continuously under tension during periods of rapid transpiration and upward movement of sap. Calculations indicate that a cohesive force of between 30 and 50 atmospheres would be adequate to permit translocation of water to the very top of the tallest known trees by this mechanism. Experimentally determined values of the cohesive force of water are in excess of 300 atmospheres.

See also **Tree.**

ASCIDIACEA (*Chordata, Tunicata*). The tunicates, sea squirts, or ascidians (Ascidiacea), constituting a class of the subphylum *Tunicata.* They begin life as larvae which resemble tadpoles in form and later become sessile animals invested in a covering called the test or tunic. Some species are solitary and others form colonies.

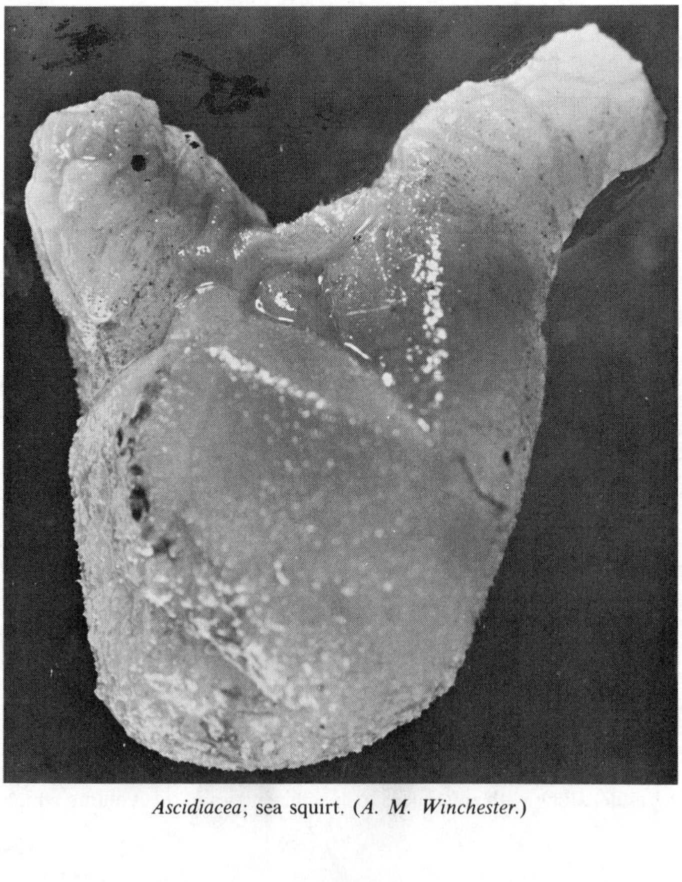

Ascidiacea; sea squirt. (*A. M. Winchester.*)

The class contains two orders:
Order *Enterogena.*
Order *Pleurogena.*

ASCITES. An abnormal accumulation of fluid in the abdominal cavity. Chronic heart failure will lead to blood being dammed up in the liver with resultant increase in its size. As a result of this congestion of the liver, the abdominal cavity may become filled with fluid (*ascites*). This fluid at times may enormously distend the abdomen. The fluid may be drained by a needle passed through the abdominal wall, or the kidneys may be forced to eliminate the fluid by the administration of diuretics.

Ascites is also common in patients with tuberculous infection of the peritoneum and in those with cancer which has spread through the peritoneal cavity. Tubercular ascites will disappear with eradication of the disease. Ascites is also a complication of portal cirrhosis; and occurs in connection with right-sided congestive heart failure.

ASCOMYCETES (Sac Fungi; Fungi). Many of the 40,000 species of fungi comprising the Ascomycetes are very common plants, but few are conspicuous. The great part of the species are small, often minute, while a few attain heights of 3 or 4 inches (7.62 or 10.2 centimeters), with a diameter of 1–2 inches (2.5 to 5.1 centimeters). Occasional individuals are even larger. All are characterized by the ascus, or spore-sac, commonly an elongate cylindrical body containing eight spores. In some species the ascus is spherical, or short cylindrical, while the number of spores may vary from two to many. Usually the asci are grouped together in a dense layer, called the hymenium. This may be composed entirely of asci, or may contain in addition numerous slender sterile filaments, called paraphyses. In some cases at least it seems the function of the paraphyses to protect the asci, since the outer tip of each paraphysis is a flattened cap which partially covers the ascus. Ascomycetes are found wherever suitable food-yielding materials exist. Many species are parasites, living on living plants; among these are species of great economic importance. Other species are saprophytes, wood-destroying species being particularly numerous.

The life-history of an Ascomycete comprises the mycelium composed of slender branching septate hyphae which penetrate throughout the substratum, and the fruiting stage in which the asci are formed. Two types of reproduction occur. One of these is the asexual type, in which asexual cells called conidia are cut off in various ways from the tips of hyphae, known as conidiophores. These conidia are single-celled spores which are disseminated by air currents. The other method of reproduction is sexual, and leads to the formation of asci. In *Pyronema confluens* this process has been carefully studied, and may be considered as typical in the main details for the process as it occurs in all the fungi of this class. The first step in this process is the formation of a multinucleate much-branched structure, which presently becomes septate. Some of the tips of this structure enlarge and become oögonia, called in this case ascogonia, while other tips become antheridia. From the oögonium a slender curved body called the trichogyne grows out. This is separated from the oögonium by a cross-wall. Since the oögonia and antheridia develop close together, the trichogyne comes in contact with the antheridium. All three bodies, oögonium, antheridium and trichogyne, are multinucleate. When the trichogyne comes in contact with the antheridium the walls between them at once break down, as does the wall between the trichogyne and the oögonium. The nuclei of the antheridium pass into the trichogyne, through it and into the oögonium. After this a new wall forms separating the trichogyne from the oögonium. In the oögonium the nuclei from the antheridium pair up with the nuclei of the oögonium, the nuclei of the trichogyne disintegrating early in the period of nuclear migration. Following the pairing of the nuclei in the ascogonium, coarse hyphae grow out from the latter. Into these the paired nuclei migrate. These coarse hyphae are the ascogenous hyphae, from which the asci eventually develop. In many Ascomycetes this process is considerably shortened, the ascogenous hyphae arising directly from the mycelium, no sex cells being formed; while other species have sex cells but no fusion, the oögonium alone developing.

The life cycles of the various ascomycetes are remarkably uniform, suggesting that they are all derived from a common ancestor. Two different views are held by botanists as to what the ancestral form may have been. According to one group, they are derived from red algae; favoring this view is the very great similarity in the development of the ascogonium and that of the carpospore formation in the algae; another favorable point is the presence of the trichogyne and the behavior of the antheridial nuclei. On the other hand, the other group holds that the ancestors of the Ascomycetes are to be found in the Phycomycetes, basing this contention on the similarity of the Phycomycete sporangium and the ascus, the latter being merely a sporangium in which the number of spores has been greatly reduced, becoming stable at eight in most species.

Many members of this class of fungi are of great importance to man because of their destructive parasitic habit. A few species are of value as food, or in the production of foodstuffs, and other products used by humans.

Among the injurious species may be mentioned the Chestnut Blight fungus, *Endothia parasitica*, a disease probably introduced from China at the beginning of the twentieth century. In China the native chestnut trees had developed immunity; this the American trees did not have, so the fungus, which attacks the cambial tissue, was particularly destructive, nearly wiping out the native chestnut trees in a few years. Another disease caused by an Ascomycete is the Brown Rot of stone fruits, caused by *Sclerotinia cinerea.* This fungus is particularly destructive in wet seasons. Often infected fruits become shriveled up and dry, in which condition they are known as "mummies." A large group of Ascomycetes are known as Powdery Mildews, because of the abundant conidiophores which are formed by the mycelium on the surface of the leaves of infected plants. Often these are so abundant as seriously to impair the functional efficiency of the leaf.

Another group of Ascomycetes contains species which are destructive and also those which are commercially of great value; these are the ubiquitous blue and green molds, species of *Aspergillus* and *Penicillium.* The destructive species attack foodstuffs everywhere, causing rotting and spoilage. Citrus fruits become covered with the bluish-green conidial masses; as does moist bread, pie crusts and many other foodstuffs. Species of the genus *Penicillium* give to Camembert and Roquefort cheese their characteristic properties. Other species of this group are the casual organisms for skin diseases of animals, including man. Another species of this genus, *Penicillium notatum*, is the source of the important drug, penicillin. Since discovery of the bactericidal effects of penicillin, commercial culture of this mold has been undertaken on a large scale. The organism is grown in sterile culture on a liquid medium consisting of water with suitable organic and inorganic substances dissolved in it. In some procedures, bottles a quart or more in capacity are used as containers for the cultures; in others, large deep tanks are used. When the tank culture method is used, forced aeration of the culture is necessary; in the bottle culture method the mold grows only on the surface of the liquid and thus obtains an adequate supply of oxygen. When the cultures have reached a suitable stage of development the mold colonies are filtered out of the liquid medium and discarded. During the growth of the organism the penicillin diffuses out of, or otherwise escapes from, the cells and hence is present in the medium. The penicillin is removed from the medium by extraction with organic solvents and suitably concentrated for medical use.

Among the largest of the Ascomycetes are species of truffles and morels, which are considered by mushroom fanciers to be particularly finely flavored. Truffles are fruit-bodies of the order Tuberales, and grow entirely underground. This makes it a matter of some difficulty to find them. Since they do not lend themselves to artificial cultivation, truffles must be sought in their wild habitat. To aid in locating them, man has trained dogs and pigs to find them by their superior sense of smell.

Another important Ascomycete is the genus *Claviceps*, which is parasitic on many grasses, including several cereal grains. This fungus forms a hard black sclerotium which is known as ergot, and which completely replaces the grain in the infected flower. The sclerotia are poisonous to livestock, causing the animals which have eaten them to become emaciated and covered with sores; another result is abortion in females.

Another very important group of Ascomycetes is the Yeasts.

ASCORBIC ACID (Degradation). **Copper (In Biological Systems).**

ASCORBIC ACID (Vitamin C). Infrequently referred to as the antiscorbutic vitamin and earlier called cevitamic acid or hexuronic acid, the present terms, *ascorbic acid* and *vitamin C*, are synonymous. Ascorbic acid was one of the first, if not the first nutrient to be associated with a major disease. Lind first described *scurvy* in 1757. However, this vitamin C deficiency disease had been recognized by Hippocrates in the 13th century and was a curse during the time of the Crusaders. In time of war, the disease killed untold numbers in armies, navies, and besieged towns. During the early days of the sailing ships, often requiring months between port calls accompanied by a lack of fresh food for long periods, the disease affected the crew as a plague. Scurvy was of some importance as recently as World War II. Currently, the disease is of prime concern in pediatrics. It is rarely seen in breast-fed children, but pasteurization of cow's milk degrades the vitamin and an addition to the diet of ascorbic acid must be provided for infants under 1 year of age.

Experimental scurvy was first produced by Holst and Frolich in 1907. About 6 months were required to produce scurvy experimentally, as individual susceptibility and the quantity of vitamin C previously stored in the body affects the onset of scurvy. The earliest sign of scurvy is usually a sallow or muddy complexion, a feeling of listlessness, general weakness, and mental depression. Soon the bones are affected and increasing pain and tenderness develop. Teeth easily decay and become loose and often fall out, while the gums bleed easily and are sore. Changes in the blood vessels occur, producing hemorrhages in different parts of the body. In infants, irritability, loss of appetite, fever, and anemia also occur. An infant between 6 and 12 months of age, who has not had sufficient intake of vitamin C (as from fruit juices, supplements, etc.) may show abnormal irritability and tenderness and pain in the legs, often accompanied by pain and swelling of joints (elbows and knees). Immediate administration of vitamin C is indicated in such cases.

In 1928, Zilva first described antiscorbutic agents in lemon juice, although the importance of fresh fruit or vegetables for preventing scurvy had been established a century or more earlier. Also in 1928, Szent-Györgyi isolated hexuronic acid (vitamin C) from lemon juice. In 1932, Waugh and King identified hexuronic acid as an antiscorbutic agent. Haworth, in 1933, established the configuration of hexuronic acid and, in that same year, Reichstein first synthesized hexuronic acid. Later in that year, Haworth and Szent-Györgyi changed the name of hexuronic acid to ascorbic acid.

In 1950, King et al, by the use of glucose labeled with radiocarbon in known positions, traced glucose through intermediate steps in the formation of ascorbic acid in plant and animal tissues, and then by using ascorbic acid with radiocarbon-labeled positions, it was possible to determine with considerable accuracy the metabolic distribution, storage, and chemical changes characteristic of the vitamin molecule. That experimentation made it clear that the carbon atoms in glucose or galactose all retain their original positions, along the carbon chain in the vitamin when it is formed biologically. No rupture or replacement in the chain during conversion was noted. It was also found that the synthesis can be considerably enhanced by feeding livestock small amounts of *chloretone* or any of a score or more of organic compounds. Reactions just described are indicated below:

Biological Role of Ascorbic Acid. Apparently all forms of life, both plant and animal, with the possible exception of simple forms, such as bacteria that have not been studied thoroughly, either synthesize the vitamin from other nutrients or require it as a nutrient. Dormant seeds contain no measurable quantity of the vitamin, but after a few hours of soaking in water, the vitamin is formed.

Ascorbic acid is easily oxidized to dehydroascorbic acid. The latter is less stable than ascorbic acid and tends to yield products, such as oxalate, threonic acid, and carbon dioxide. When administered to animals or consumed in foods, dehydroascorbic acid has nearly the same antiscorbutic activity as ascorbic acid, and it can be quantitatively reduced to ascorbic acid.

In its biochemical functions, ascorbic acid acts as a regulator in tissue respiration and tends to serve as an antioxidant in vitro by reducing oxidizing chemicals. The effectiveness of ascorbic acid as an antioxidant when added to various processed food products, such as meats, is described in entry on **Antioxidants.** In plant tissues, the related glutathione system of oxidation and reduction is fairly widely distributed and there is evidence that electron transfer reactions involving ascorbic acid are characteristic of animal systems. Peroxidase systems also may involve reactions with ascorbic acid. In plants, either of two copper-protein enzymes are commonly involved in the oxidation of ascorbic acid.

In animal tissues, it is easily demonstrated that, as the vitamin content of tissues is depleted, many enzyme systems in the body are decreased in activity. Full explanation of these decreased activities still requires further research. In the total animal and in isolated tissues from animals with scurvy, there is an accelerated rate of oxygen consumption even though the animal becomes very weak in mechanical strength and many physiologic functions are disorganized. With the onset of scurvy, the most conspicuous tissue change is the failure to maintain normal collagen. Sugar tolerance is decreased and lipid metabolism is altered. There is also marked structural disorganization in the odontoblast cells in the teeth and in bone-forming cells in skeletal structures. In parallel with the foregoing changes, there is a decrease in many hydroxylation reactions. The hydroxylation of organic compounds is one of the most characteristic features disturbed by a vitamin C deficiency. These reactions relate to the vitamin's regulation of respiration, hormone formations, and control of collagen structure.

A partial list of physiological functions that have been determined to be affected by vitamin C deficiencies include (1) absorption of iron; (2) cold tolerance, maintenance of adrenal cortex; (3) antioxidant; (4) metabolism of tryptophan, phenylalanine, and tyrosine; (5) body growth; (6) wound healing; (7) wound healing; (8) synthesis of polysaccharides and collagen; (9) formation of cartilage, dentine, bone, and teeth; and (10) maintenance of capillaries.

Requirements. Species known to require exogenous sources of ascorbic acid include the primates, guinea pig, Indian fruit bat, red vented bulbul, trypanosomes, and yeast. Species capable of endogenous sources include the remainder of vertebrates, invertebrates, plants, and some molds and bacteria. Estimates of requirements of vitamin C by humans has been approached in several ways: (1) direct observation in human studies; (2) analogy to experimentation with guinea pigs; (3) analogy to experimental studies in monkeys and other primates; and (4) analogy to animals, such as the albino rat, that normally synthesize the vitamin in accordance with physiological need. It is

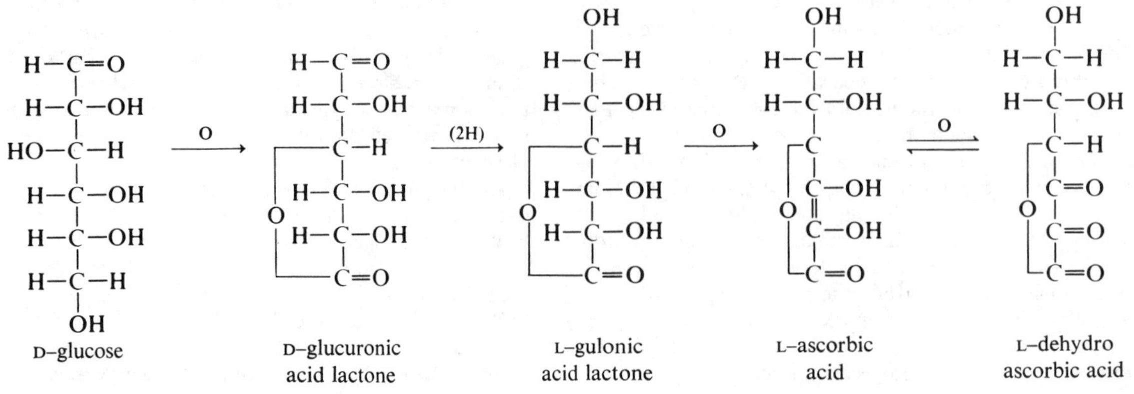

relatively easy to maintain intakes at recommended levels by use of mixed practical dietaries that include nominal quantities of fresh, canned, or frozen vegetables or fruits. Generally, ascorbic acid is considered as nontoxic to humans. Possible exceptions include kidney stones (in gouty individuals); inhibitory in excess doses on cellular level (mitosis inhibition) possible damage to beta-cells of pancreas and decreased insulin production by dehydroxyascorbic acid.

Distribution and Sources. Natural sources of vitamin C include the following:

High ascorbic acid content (100–300 milligrams/100 grams)
Broccoli, Brussels sprouts, collards, currant (black), guava, horseradish, kale, parsley, pepper (sweet), rose hips, turnip greens, walnut (green English)

Medium ascorbic acid content (50–100 milligrams/100 grams)
Beet greens, cabbage, cauliflower, chives, kohlrabi, lemon, mustard, orange, papaya, spinach, strawberry, watercress

Low ascorbic acid content (25–50 milligrams/100 grams)
Asparagus, bean (lima), cantaloupe, chard, cowpea, currant (red and white), dandelion greens, fennel, grapefruit, kumquat, lime, loganberry, mango, melon (honeydew), mint, okra, onion (spring), passion fruit, potato, radish, raspberry, rutabaga, soybean, spring greens, squash (summer), tangerine, tomato, turnip

For the species where the ascorbic acid is synthesized endogenously, the precursors include *d*-mannose, *d*-fructose, glycerol, sucrose, *d*-glucose, and *d*-galactose. Intermediates include uridine diphosphate glucose, *d*-glucuronic acid, gulon acid, *l*-gulonolactone, (Mn^{2+} cofactor). Production sites in animals are the kidney and liver in most instances. In rat, it is the intestinal bacterial supply. In plants, the production sites are found in green leaves and fruit skins. Cell sites include microsomes, mitochondria, and golgi.

Supplements. Commerically available ascorbic acid still includes isolation from natural sources, such as rose hips, but large-scale production will involve the microbiological approach, i.e., *Acetobacter suboxidans* oxidative fermentation of calcium *d*-gluconate; or the chemical approach, i.e., the oxidation of *l*-sorbose.

Bioavailability of Ascorbic Acid. The general causes of reduced availability of vitamin C include damage to adrenal cortex, presence of antagonists, and food preparation practices (oxidation, storage, leaching, cooking). Excepting the use of supplements, the almost universal requirement for fresh foods as a source of vitamin C is readily explained by the sensitivity of the vitamin to destruction by reaction with oxygen. This is accelerated by the presence of minute quantities of enzymes that occur in most living tissues, in which copper or iron is combined with a protein to form a catalyst for the oxidation reaction. Other chemicals, such as quinones or high-valence salts of manganese, chrominum, and iodine can also oxidize the vitamin readily in aqueous solutions. Most of these reactions increase rapidly in proportion to exposure to air and rising temperature. In the dry crystalline state, however, and in many dried plant tissues, particularly if acidic in reaction, the vitamin is quite stable at room temperature over a period of several months.

Freshly cut oranges or their juices may be exposed in an open glass for several hours without appreciable loss of the vitamin because of the protective effect of the acids present and the practical absence of enzymes that catalyze its destruction. In potatoes, when baked or boiled, there is a slight loss of the vitamin, but if they are whipped up with air while hot, as in the production of mashed potatoes, a large fraction of the initial vitamin content usually will be lost. In freezing foods, it is common practice to dip them in boiling water or to treat them briefly with steam to inactivate enzymes, after which they are frozen and stored at very low temperatures. In this state, the vitamin is reasonably stable. Vitamin C degradation in dehydrated food systems is described shortly.

Factors which incrase the bioavailability of ascorbic acid include the presence of antioxidants and synergists in the diet.

Vitamin C Degradation in Dehydrated Food Systems. As pointed out by J. Kirk and co-workers (Department of Food Science and Human Nutrition, Michigan State University, East Lansing, Michigan), the degradation of vitamin C in a dehydrated food system is dependent upon water activity, moisture content, and storage temperature (in containers with no headspace), and ascorbic acid destruction is dramatically increased with the presence of oxygen. Karel and Nickerson (1964), Jensen (1967), Vojnovich and Pfeifer (1970), and Lee and Labuza (1975) have studied the stability of reduced ascorbic acid in various low and intermediate moisture dehydrated foods and model systems as a function of moisture content, water activity, and storage temperature. Results reported by these investigators show that the rate of destruction of reduced ascorbic acid increased as a total moisture content and water activity increased. Lee and Labuza have interpreted the increase in destruction rates to be the result of dilution of the aqueous phase which results in a decreased viscosity and thus increased mobility of the reactants.

To further refine data in this field, Kirk et al (1977) undertook to (1) determine the rate of destruction of reduced and total ascorbic acid as a function of storage temperature, oxygen content, moisture content, and water activity in a low-moisture dehydrated food system; (2) determine the relative importance of each of these storage parameters on ascorbic acid stability; (3) determine the effect of other micronutrients (vitamins A and B_2) on the stability of ascorbic acid in the food system; and (4) predict the stability of reduced, dehydro- and total ascorbic acid using kinetic data generated from stability studies. Their general findings are described in the first sentence of this paragraph.

References

Bauernfeind, J. C., and D. M. Pinkert: "Food Processing Added Ascorbic Acid," *Advanced Food Research*, **18**, 219 (1970).

Chaberek, S., and A. E. Martell: "Organic Sequestering Agents," 2nd edition, p. 249, Wiley, New York, 1959.

Deng, J. C., et al.: "Ascorbic Acid as an Antioxidant in Fish Flesh and Its Degradation, *J. Food Sci.*, **43**, 2, 457–460 (1978).

Domah, A.A.M.B., Davidek, J., and J. Velisek: "Changes of L-Ascorbic Acid and L-Dehydroascorbic Acid during Cooking and Frying Potatoes," *Z. Lebensm. Unters. Forsch.*, **154**, 270 (1974).

Jensen, A.: "Tocopherol Content of Seaweed and Seameal. 3. Influence of Processing and Storage on the Content of Tocopherol, Carotenoids and Ascorbic Acid in Seaweed." *J. Sci. Food Agric.*, **20**, 622 (1967).

Karel, M., and J. T. R. Nickerson: "Effects of Relative Humidity, Air and Vacuum on Browning of Dehydrated Orange Juice," *Food Technology*, **18**, 104 (1964).

King, C. G.: "Ascorbic Acid (Vitamin C) and Scurvy," in "The Encyclopedia of Biochemistry" (R. J. Williams and E. M. Lansford, Jr., editors), Van Nostrand Reinhold, New York, 1967.

Kutsky, R. J.: "Handbook of Vitamins and Hormones," Van Nostrand Reinhold, New York, 1973.

Lee, S. H., and T. P. Labuza: "Destruction of Ascorbic Acid as a Function of Water Activity," *J. Food Sci.*, **40**, 370 (1975).

Linkswiler, H.: "The Effect of the Ingestion of Ascorbic Acid and Dehydroascorbic Acid Upon the Blood Levels of These Two Components in Human Subjects," *J. Nutrition*, **64**, 43 (1958).

Staff: "Food Chemicals Codex," National Academy of Sciences, Washington, D.C., 1972.

Vojnovich, C., and V. F. Pfeifer: "Stability of Ascorbic Acid in Blends with Wheat Flour, CSM, and Infant Cereals," *Cereal Science Today*, **19**, 317 (1970).

Watt, B. K., and A. L. Merrill: "Composition of Foods," *Agriculture Handbook 8*, U.S. Department of Agriculture, Washington, D.C., 1975.

ASCUS. The spore sac characteristic of the Ascomycetes. Typically each sac contains 8 spores, although there are exceptions to this. The ascus develops from an ascogenous hypha. The latter is a septate hypha which grows in the form of a hook or crosier. Each segment contains 2 nuclei; those of the apical hook-like segment divide simultaneously so that 4 nuclei are present. Septations cut off two of these nuclei, leaving a segment containing 2 nuclei which now fuse. The segment now elongates conspicuously and becomes the ascus while the fusion nucleus divides successively to form 8 nuclei which eventually become set off in 8 separate spores. Subsequent development of the segments cut off during ascus formation may lead to the development of additional asci from the same ascogenous hypha.

ASEPTIC MENINGITIS. **Meningitis.**

ASEXUAL REPRODUCTION. In asexual reproduction, a part of the parent organism becomes an organism identical with the parent.

Different types of asexual reproduction can be found in the different

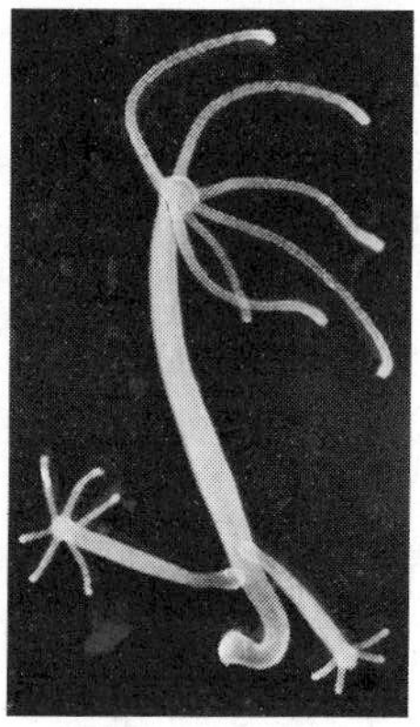

Fig. 1. Budding hydra. (*A. M. Winchester.*)

groups of plants. In bacteria, fission, a simple splitting into two equal parts, is the asexual method of reproduction. In green algae the characteristic method is by the formation of zoospores, which are unicellular motile bodies formed from the protoplast of a single cell. Each zoospore, after swimming around for a time, becomes quiet and grows to form a new individual of the parent type. In flowering plants there are many methods of asexual reproduction, some of great importance to man. A common type is found in the strawberry, where long slender branches grow out from the short stem of the plant and take root at their tip. There a new plant forms. This reproductive structure is called a stolon. Many other plants produce similar branches which run along the surface of the ground (runners) or just beneath it (rhizomes), and send up one or more new plants from the nodes. Other plants, like the tiger lily, bear small buds in the axils of their leaves: the buds or bulblets are easily detached and readily grow to form new plants. Similar bulblets are formed at the base of many bulbs. Another very common method of asexual reproduction is through the formation of suckers, branches formed at the base of the parent plants, and gradually growing to replace them. Man propagates date palms, pineapples, and bananas, for example, by means of suckers. The familiar potato tuber, the swollen tip of a rhizome, is another type of asexual reproduction which is of tremendous value to man.

All the methods so far enumerated are from the stem of the plant. However, any part of the plant may be a means of asexual renewal. Many plants are known which reproduce by means of the leaves. Several of these are now cultivated extensively as objects of beauty or of curiosity. For example, the African violet, *Saintpaulia*, and many of the ornamental begonias are readily propagated by leaves. Less widely known but equally interesting are species of *Bryophyllum* and *Kalanchoe*. In the notches of the leaves of these plants, while still attached to the parent plant or after being severed therefrom, tiny plants readily form and grow.

Many plants readily root when cut up into segments. Most people are familiar with the habit of the willow twig of striking root and growing when stuck in the ground. Equally well known is the geranium cutting, which is merely a branch removed from the parent plant and placed in a favorable environment.

Asexual reproduction is of tremendous importance to man. New and improved forms of plants are constantly being made: by asexual means they are reproduced in the great quantities necessary for commercial use. Without such reproduction their formation in quantity would be practically impossible. (See **Grafting and Budding**).

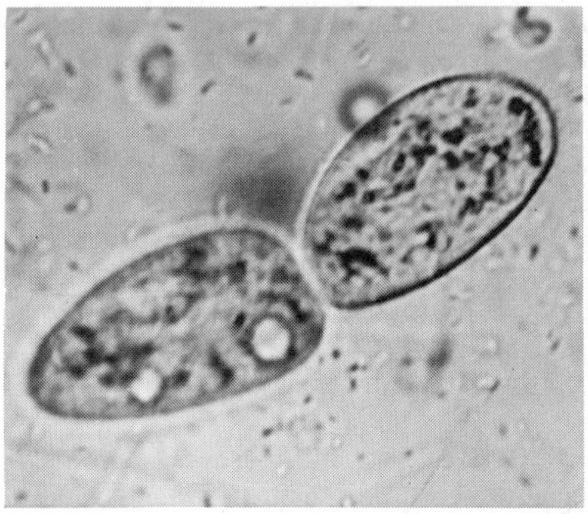

Fig. 2. Fission in paramecium. (*A. M. Winchester.*)

Asexual reproduction is also found in animals although it is restricted to some of the simpler phyla. Fission, a splitting in half, occurs in most of the *Protozoa*, such as *Amoeba* and *Paramecium*. In the class of *Protozoa* known as *Sporozoa*, one cell may break up into many reproductive bodies which may be called *merozoites* or *sporozoites*. The nucleus within a cell first divides many times and a plasma membrane forms around each nucleus and the surrounding cytoplasm, and the original cell breaks open liberating the "spores." The malarial parasite, *Plasmodium*, is a good example; merozoites are formed in red blood cells and sporozoites are formed in a mosquito.

Budding is a means of asexual reproduction found in sponges and in certain coelenterates, such as Hydra. Many of the simpler metasoans also have the power of regeneration; if cut into two or more parts, each part can grow back the missing parts, thus forming two or more complete animals. See Figs. 1 and 2.

ASH HANDLING (Boiler). Boiler; Burner.

ASH TREES. The timber and shade trees called *ashes* are members of the family *Oleaceae* (olive family); whereas the mountain ashes are members of the family *Rosaceae* (rose family). The so-called prickly ash (toothache-tree), the southern prickly ash, and the hop tree (sometimes referred to as the wafer ash) are members of the family *Rutaceae* (citrus family).

The olive ashes are of the genus *Fraxinus* and are deciduous trees (a few may be considered shrubs). Important species include:

Afghan ash	*Fraxinus xanthoxyloides*
Biltmore ash	*F. biltmoreana*
Black ash (or hoop ash)	*F. nigra*
Blue ash	*F. quadrangulata*

Record blue ash located at Danville, Kentucky. For dimensions see accompanying table. (*Kentucky Division of Forestry.*)

Carolina ash (or water ash)	*F. caroliniana*
Chinese ash	*F. chinensis*
European common ash	*F. excelsior*
Flowering ash (or Marie's ash)	*F. mariesii*
Green ash	*F. pensylvanica lanceolata*
Manna ash	*F. ornus*
Narrow-leaved ash	*F. angustifolia*
Oregon ash	*F. latifolia*
(Oxycarpa)	*F. oxycarpa*
Pumpkin ash	*F. profunda*
Red ash (or brown or river ash)	*F. pensylvanica*
Texas ash	*F. texensis*
(Tomentosa)	*F. tomentosa*
Velvet ash	*F. velutina*
Wafer ash (or hop tree)	*Ptelea trifoliata*
Weeping ash	*F. excelsior latifolia*
White ash	*F. americana*

The mountain ashes are of the genus *Sorbus* and are deciduous trees or shrubs. Important species include:

American mountain ash	*Sorbus americana*
European mountain ash (or Rowan)	*S. aucuparia*
Western mountain ash	*S. scopulina*
Notable Asian Species	
Embley mountain ash	*S. discolor*
Folgner mountain ash	*S. folgneri*
(Harrowiana)	*S. harrowiana*
(Hupehensis)	*S. hupehensis*
(Insignis)	*S. insignis*
Japanese mountain ash	*S. commixta*
Kashmir mountain ash	*S. cashmiriana*
(Pohuashaenensis)	*S. pohuashaenensis*
Sargent's mountain ash	*S. sargentiana*
Vilmorin's mountain ash	*S. vilmorinii*

As will be noted from the accompanying table of record ash trees, the white ash is a large tree and valuable for its timber. The white ash prefers moist and rich soil in a cool woods environment and particularly favors locations along rivers. The tree ranges from Newfoundland and Nova Scotia westward into Ontario and Minnesota, and southward to northern Florida and southwestward into Oklahoma and Texas. The tree frequently occurs scattered but there are large concentrations in Maine. The wood is excellent for furniture, interior finish, and implements of various types. The color ranges from pale to medium brown. The grain is close although not considered fine-grained. Commercial white ash in the green condition has a moisture content of about 43% and weighs 48 pounds per cubic foot (767 kilograms per cubic meter). After air-drying to 12% moisture content, the wood weighs 41 pounds per cubic foot (657 kilograms per cubic meter) and 1,000 board-feet (2.36 cubic meters) of nominal sizes weigh 3,420 pounds (1551 kilograms). Crushing strength of the green wood with compression applied parallel to the grain is 4,060 psi (28 MPa); of the dried wood, it is 7,280 psi (50.2 MPa). The tensile strength of the green wood with tension applied perpendicular to the grain is 580 psi (4 MPa); of the dry wood, 850 psi (5.8 MPa).

The leaves of the white ash are compound and from 7 to 12 inches (18 to 30 centimeters) long. They are of a lusterless light green, lighter and a silverish-green underneath. The flowers are green and without petals. The fruit occurs in clusters and is winged. The seed chamber is about 3/8-inch (0.9 centimeter) long and remains on bare branches well into mid-winter.

The green ash prefers the peripheries of streams and damp lowlands. It is found from Vermont southward along the mountains as far as northern Florida. Its range reaches westward to the eastern foothills of the Rocky Mountains. As shown by the accompanying table, the pumpkin ash attains great heights, normally between 60 and 100 feet (18–30 meters) for average trees. It is a slender tree with a trunk diameter usually of 3 to 4 feet (0.9 to 1.2 meters). The tree has a distinctively swollen appearing trunk at the base, giving rise to its

RECORD ASH TREES IN THE UNITED STATES[1]

Specimen	Circumference[2]		Height		Spread		Location
	(inches)	(centimeters)	(feet)	(meters)	(feet)	(meters)	
American mountain ash (1972) (*Sorbus americana*)	95	241	42	12.6	26	7.8	New York
Berlandier ash (1974) (*Fraxinus berlandierana*)	194	493	44	13.2	40	12	Texas
Black ash (1969) (*F. nigra*)	183	465	87	26.1	60	18	Ohio
Blue ash (1970) (*F. quadrangulata*)	172	437	86	25.8	71	21.3	Kentucky
Carolina ash (1973) (*F. caroliniana*)	44	118	42	12.6	18	5.4	Texas
European mountain ash (1972) (*S. aucuparia*)	65	165	42	12.6	36	10.8	Montana
Green ash (1976) (*F. pennsylvanica*)	185	470	145	43.5	70	21	Kentucky
Lowell ash (1974) (*F. anomala*)	47	119	50	15	16	4.8	Arizona
Oregon ash (1975) (*F. latifolia*)	263	668	59	17.7	11	3.3	Oregon
Pumpkin ash (1977) (*F. profunda*)	219	556	86	25.8	84	25.2	Virginia
Showy mountain ash (1972) (*S. decora*)	56	142	58	17.4	29	8.7	Michigan
Singleleaf ash (1973) (*Fraxinus anomala*)	19	48	24	7.2	23	6.9	Colorado
Texas ash (1972) (*Fraxinus texensis*)	53	135	39	11.7	34	10.2	Texas
Velvet ash (1971) (*Fraxinus velutina*)	140	356	66	19.8	65	19.5	Arizona
White ash (1976) Co-Champion (*Fraxinus americana*)	245	622	114	34.2	126	37.8	Michigan
White ash (1976) Co-Champion (*Fraxinus americana*)	288	732	80	24	82	24.6	Pennsylvania

[1] From the "Social Register of Big Trees," The American Forestry Association (by permission).
[2] At 4.5 feet (1.4 meters).

unusual name. This species likes swamps and wet areas around ponds. It ranges from western New York west to Missouri and Arkansas and southward to Florida.

The black ash, also known as hoop ash, is also a slender, very tall tree, averaging from 50 to 65 feet (15 to 19.5 meters) in height and a trunk diameter of only 1 to 2 feet (0.3 to 0.6 meter). This tree also prefers swampy country. The blue ash is valuable for timber and is found generally in the western and southwestern states. See accompanying photo. It can attain a height of close to 100 feet (30 meters), with a trunk diameter up to about 30 inches (76 centimeters), making it a slender tree. This tree is found in damp woods and prefers limestone hills. The tree occurs throughout middle America, ranging from Michigan southward into Alabama and Arkansas.

The American mountain ash is a very pretty tree, seldom exceeding 20–25 feet (6 to 7.5 meters) in height, with a trunk of from 8 to 15 inches (20.3 to 38 centimeters) in diameter. At high altitudes, it may be reduced to the status of a shrub. The tree flowers white in early spring. The fruit ranges in color from a coral red to deep scarlet. In nature, the tree is found in cool woods and along river banks, and also in swamps. The range is from Newfoundland west to middle-Canada and southward through the region of the Great Lakes and into Tennessee. The tree occurs on the mountain slopes in New England's White and Green Mountains and all along the Alleghanies as far south as North Carolina. The tree is used in landscaping.

The European mountain ash, sometimes called Rowan Tree, is used widely in parks and gardens in North America. The tree has narrow, oblong leaflets, the undersides covered with a white, hairy down. The fruit is of a bright scarlet color and is about 3/8-inch (0.9 centimeter) in diameter. There is considerably more variety in the European ashes with different colorations of bark and shape of leaves. There is a weeping ash (*Fraxinus excelsior* "Pendula") and the manna ash (*F. ornus*) which occurs in Southern Europe is known for its showy cream-colored flowers. The *F. excelsior* "Diversifolia" has a dandelion-color bark.

Several mountain ashes are native to Asia, including *Sorbus hupehensis* and *S. cashmiriana*. The *S. discolor* is well known as an excellent street tree because of its branches which point sharply upward. It has attractive large red leaves and fruit. The Chinese mountain ash, *S. vilmorinii*, is well known for its beauty of autumn coloration, with a full range from orange to purple.

The whitebeams are also of the genus *Sorbus* and thus closely related to the mountain ashes. A few highlights of the whitebeams include: *S. alinfolia* is found in Japan, crimson fruit, red-orange autumnal colors, attains a height of about 25 feet (7.5 meters); *S. aria*, a whitebeam of Europe, fragrant dull-white flowers, red fruit, brown autumnal colors, attains a height of from 30 to 45 feet (9 to 13.5 meters); *S. cuspidata*, a whitebeam of the Himalayas, white flowers, round, green fruit, large ovate leaves, attains a height up to 35 feet (10.5 meters); *S. domestica*, the service tree, occurs in southern Europe as well as western Asia and north Africa—pinnate leaves, small flowers, bears a pear-shaped fruit, may reach a height of 60 feet (18 meters); *S. hybrida*, also known as the bastard service tree, occurs in Scandinavia, white flowers, red fruit, reaches a height of 20 to 40 feet (6 to 12 meters); *S. intermedia*, the Swedish whitebeam, occurs in northwestern Europe, off-white flowers, clustered oval fruit of red coloration, attains height of about 20 feet (6 meters); *S. latifolia*, known as the service tree of Fontainebleau, ranges from Portugal to southern Germany. Rises to about 60 feet (18 meters), white flowers, brown fruit, shaggy bark; and *S. thibetica*, a whitebeam of Tibet.

For references see **Tree.**

ASPARAGINE. Amino Acids.

ASPARAGUS. Of the family *Convallariaceae* (lily-of-the-valley family), genus *Asparagus*, the familiar asparagus plant is one of about 125 species in the genus. The plant is native to temperature and tropical regions of the Old World, but widely cultivated in suitable climes throughout the world. Members of the genus are characterized by having the leaves reduced to minute scales or bristles, while small, often very leaflike branches called cladophylls function as leaves. The flowers are small, yellowish or white in color, and the fruit is a berry, often brightly colored. *Asparagus officinalis*, a native of the marshes of Europe, is the cultivated garden form with thick and fleshy young stems. Other species are widely grown for their delicate beauty, as the familiar Asparagus ferns, *Asparagus plumosus* and *Asparagus sprengeri* (which are not properly ferns at all), and the florist's smilax, *Asparagus asparagoides*. See accompanying figure.

Bunches of asparagus placed on damp moss in flat to keep them fresh for market. (*USDA photo.*)

ASPARAGUS BEETLE (*Insecta*, *Coleoptera*). An introduced European beetle, *Crioceris asparagi*, which is sometimes an important pest on asparagus. A related species, also introduced from Europe, is known as the 12-spotted asparagus beetle. They are held in check by hand-picking and by dusting plants with lead arsenate and lime.

ASPARTAME. Sweeteners.

ASPARTIC ACID. Amino Acids.

ASPECT RATIO. Aerodynamics; Supersonic Aerodynamics; Television.

ASPEN TREES. Poplar Trees.

ASPERGILLUS. Ascomycetes; Yeasts and Molds.

ASPERGILLUS NIGER. Fermentation.

ASPERGILLUS PHOENICIS. Starch.

ASPHALT (or Asphaltum). A semi-solid mixture of several hydrocarbons, probably formed because of the evaporation of the lighter and more volatile constituents. It is amorphous, of low specific gravity, 1–2, with a black or brownish-black color and pitchy luster. Notable localities for asphaltum are the Island of Trinidad and the Dead Sea region, where Lake Asphaltites were long known to the ancients. See also **Coal Tar and Derivatives; Petroleum.**

ASPHALTIC ROCK. Tar Sands.

ASPHERIC SURFACE. A surface of a lens or mirror which has been changed slightly from a spherical surface as an aid in reducing aberrations. Parabolic mirrors for telescopes and the Schmidt objective are common aspheric surfaces.

ASPHYXIA. Suffocation, the consequences of interference with the aeration of the blood, usually from interference with respiration, whether by mechanical means or by the inhalation of gases containing insufficient oxygen, although it may result from other causes which would depress the respiratory center or result in a deficiency of hemoglobin in the blood. The effects are cyanosis, increased blood pressure, violent respiratory efforts, ultimately leading to unconsciousness and death if the cause is not removed. The effects are partly due to anoxemia and partly to excess of carbon dioxide in the blood. When the blood supply to a limited portion of the body is temporarily interrupted, the term *local asphyxia* sometimes is used. See also **Anoxemia.**

ASPIRIN. Acetylsalicylic Acid; Analgesics; Anticoagulants.

ASPIRIN-INDUCED ASTHMA. Bronchial Asthma.

ASP VIPER. Snakes.

ASSASSIN BUG (*Insecta, Hemiptera*). Any bug of the large predacious species constituting the family *Reduviidae.* The assassin bug is found in the southern part of the United States and in the West Indies. Throughout the world, it is estimated that there are about 2500 species. Some species pounce upon their prey; other species stick their legs into resin from a tree and hold the sticky limbs aloft, awaiting a likely victim to come along. Some assassin bugs secrete a fluid which other insects find attractive, but the fluid has an intoxicating effect on likely victims, thus making them easy prey. An oily hair on the legs of the assassin bug helps in holding prey. The thorax of the assassin bug also produces a poisonous venom which, when injected into prey, assists in reducing the tissues of the victim to a thick juice ready for convenient consumption and assimilation.

ASSEMBLER (Computer System). A computer program which operates on symbolic input data to produce machine instructions by carrying out such functions as (1) translation of symbolic operation codes into computer instructions, (2) assigning locations in storage for successive instructions, or (3) assignment of absolute addresses for symbolic addresses. An assembler generally translates input symbolic codes into machine instructions item for item, and produces as output the same number of instructions or constants which were defined in the input symbolic codes.

Assembler language may be defined as computer language characterized by a one-to-one relationship between the statements written by the programmer and the actual machine instructions performed. The programmer thus has direct control over the efficiency and speed of the program. Usually, the language allows the use of mnemonic names instead of numerical values for the operation codes of the instructions and similarly allows the user to assign symbolic names to the locations of the instructions and data. For the first feature, the assembler contains a table of the permissible mnemonic names and their numerical equivalents. For the second feature, the assembler builds such a table on a first pass through the program statements. Then, the table is used to replace the symbolic names by their numerical values on a second pass through the program. Usually, some dummy operation codes (or pseudocodes) are needed by the assembler to pass control information to it. As an example, an origin statement is usually required as the first statement in the program. This gives the numerical value of the desired location of the first instruction or piece of data so that the assembler can, by counting the instructions and data, assign numerical values for their symbolic names.

The format of the program statements is usually rigidly specified and only one statement per input record to the assembler is permitted. A representative statement is: symbolic name, operation code (or pseudocode), modifiers and/or register addresses, symbolic name of data. The mnemonic names used for the operation codes usually are defined uniquely for a particular computer type with little standardization between computer manufacturers even for the most common operations. The programmer must learn a new language for each new machine with which he works.

An example of a program prepared in an assembler language follows. The explanatory comments following the REM (remarks) mnemonic and those to the right of the other program statements are ignored by the assembler program and thus do not affect execution of the program. See also **Language (Computer);** list of terms under **Data Processing.**

```
        ABS
        ORG   100
        REM   THIS IS AN EXAMPLE OF A PRO-
        REM   GRAM IN AN ASSEMBLER LAN-
        REM   GUAGE. "ABS", "ORG", "CALL",
        REM   "REM", "DC", "DEC" AND "END"
        REM   ARE PSEUDOCODES. THE OTHER
        REM   MNEMONICS IN THE SAME COL-
        REM   UMN ARE OPERATION CODES.
        REM   THE NAMES SUCH AS A, B,
        REM   START, AND OVER ARE SYM-
        REM   BOLIC ADDRESSES ASSIGNED BY
        REM   THE PROGRAMMER

START   SLT   32       THESE INSTRUCTIONS
        LD    A        COMPUTE A/B*C AND
        D     B        PUT THE VALUE IN D
        M     C
        SLT   1
        STO   D
        CMP   D4B1     THESE INSTRUCTIONS
        BSC   Z+       COMPARE D WITH 4 AND
        MDX   OVER     3.2 AND BRANCH OUT IF
        CMP   D32B1    IT DOES NOT LIE BE-
        BSC   Z−       TWEEN THEM
        MDX   UNDER

EXIT    CALL  EXIT     THE ASSEMBLER SETS
        REM            UP A BRANCH TO THE
        REM            SYSTEM EXIT SUBROU-
        REM            TINE (NAMED "EXIT")

OVER    LD    D1       THESE INSTRUCTIONS
        STO   EROR     SET THE VALUE OF
        MDX   EXIT     EROR TO 1 BEFORE
        REM            EXITING

UNDER   SRA   16       THESE INSTRUCTIONS
        STO   EROR     SET THE VALUE OF
        MDX   EXIT     EROR TO ZERO BEFORE
        REM            EXITING

        ORG   200      THIS SET OF PSEUDO-
A       DC    1        CODES IS USED BY THE
B       DC    2        ASSEMBLER TO SET UP
C       DC    3        THE DATA AND DATA
D       DC    0        WORKSPACE FOR THE
EROR    DEC   −1.0     ABOVE PROGRAM
D1      DEC   1.0
D4B1    DEC   4.0B1
D32B1   DEC   3.2B1
        REM            LAST STATEMENT IN
        REM            PROGRAM
        END
```

Thomas J. Harrison, International Business Machines Corporation, Boca Raton, Florida.

ASSES. Horses, Asses, and Zebras.

ASSOCIATED CORPUSCULAR EMISSION. An expression sometimes used in the definition of roentgen. Associated corpuscular emission is the full complement of secondary charged particles (usually limited to electrons) associated with an x-ray or gamma-ray beam in its passage through air. The full complement of electrons is obtained after the radiation has traversed sufficient air to bring about equilibrium between the primary photons and secondary electrons. Electronic equilibrium with the secondary photons is intentionally excluded.

ASSOCIATION (Chemical). The combination of molecules of the same substance to form larger aggregates consisting of two or more molecules. See also **Elastomers;** and **Molecule.**

Association was first thought of as a reversible reaction between like molecules that distinguished it from polymerization, which is not reversible. Association is characterized by reversibility or ease of disassociation, low energy of formation (usually about 5 and not more than 10 kcal per mole), and the coordinate covalent bond which Lewis called the acid-base bond. Association takes place between like and unlike species. The most common type of this phenomenon is hydrogen bonding. Association of like species is demonstrable by one or more of the several molecular weight methods. Association between unlike species is demonstrable by deviation of the system from Raoult's law.

The strength of the coordinate covalent bond is a function of polarity of the associating molecules. Hence, associated molecules vary in stability from very unstable to very stable. The argon-boron trifluoride complex is quite unstable, whereas calcium sulfate dihydrate (gypsum) is very stable. The bond strength associated with stability has been measured for a number of combinations. The strengths of some hydrogen bond types decrease in the order FHF, OHO, OHN, NHN, CHO, but they are dependent upon the geometry of the combination and upon the acid-base characteristics of the group. Steric effects can have a marked effect on the strength of the coordinate covalent bond. This was demonstrated by a study of the strength of the series NH_3, $C_2H_5NH_2$, $(C_2H_5)_2NH$, $(C_2H_5)_3N$ as bases toward an acid in solution and in the gaseous state and the comparison of the base strength of triethylamine and quinuclidine. The latter is, in effect, triethylamine in which the two-carbon atoms of each ethyl group are tied together by another carbon. The geometry of the ethyls around the nitrogen is drastically changed, and the cyclic is a stronger base than the triethyl compound. The factors affecting the strength of the hydrogen bond also influence the degree of association.

Association within the same species accounts for the high boiling points of water, ammonia, hydrogen fluoride, alcohols, amines, and amides. Ethyl ether and butanol contain the same number of atoms of each element, but butanol has a boiling point of 83°C above that of ethyl ether as a result of more extensive hydrogen bonding. Some substances associate completely to two or more formula weights per molecule. Carboxylic acid, by a hydrogen-to-oxygen association, form dimers with a six-membered ring. N-Unsaturated amides dimerize in the same manner, whereas N-substituted amides dimerize in a chain form in a *trans* configuration.

Hydrogen bonding is so common that coordinate bonds between other elements are sometimes overlooked. Antimony(III) halides form very few complexes with other halides, whereas aluminum halides readily form complexes. The octet of electrons is complete in all atoms of the antimony halides, but is incomplete in the aluminum atom of aluminum halides:

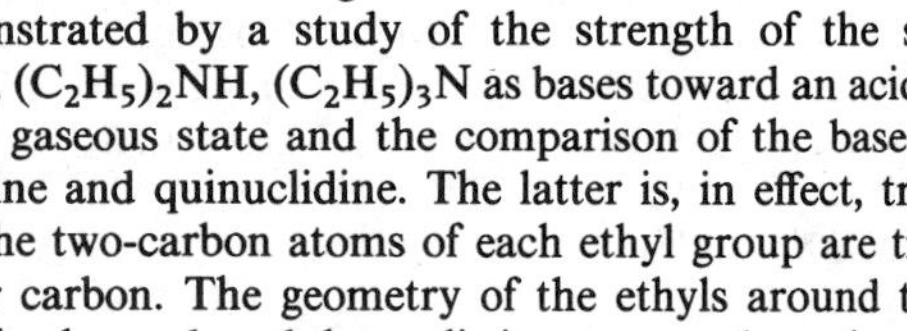

(a) (b) (c)

Aluminum can accept two electrons to complete its octet. The pair of electrons is available from the halogen. An alkali halide can supply the electrons and form a complex (c), or the electron pair may come from the halogen of another aluminum chloride. Association with other aluminum halides accounts for the higher melting point of aluminum halides over antimony(III) halides which have a formula weight of 95 or more. The association of aluminum sulfate, alkali metal sulfate, and water to form the stable alums is one of the more complex examples.

The formation of solvates is association between unlike species. Solvation is more frequent between substances of high polarity than those of low polarity. This is illustrated by the decrease in the tendency to form solvates with decrease in dipole moment and dielectric constant (shown in parentheses) for N-methylacetamide (3.59; 172), to water (1.84; 78.4), to ethanol (1.70; 24.6); to ammonia (1.48; 78.4); to ethanol (1.70; 24.6); to ammonia (1.48; 17.8); to methylcyclohexane (0; 2.02) for which few associations are known.

J. A. Riddick, Baton Rouge, Louisiana.

ASSOCIATION (Coefficient of). In statistical theory, this word is used (a) in a general sense, to denote the degree of dependence between two variables; and (b) especially to denote the relationship between two variables which are simply dichotomized. For example, if a set of n numbers is classified as A or not $-A$, and as B or not $-B$, an association table is of the following kind:

	A	NOT A	TOTALS
B	a	b	$a+b$
not $-B$	c	d	$c+d$
Totals	$a+c$	$b+d$	$a+b+c+d=n$

A coefficient of association Q is defined by

$$Q=\frac{ad-bc}{ad+bc}$$

It can vary from -1 to $+1$ according to the strength of the association. Other coefficients are sometimes used, especially

$$V=\frac{ad-bc}{\{(a+c)(b+d)(a+b)(c+d)\}^{1/2}}$$

The latter is, in fact, related to Chi-Square for the fourfold table as $V=\{\chi^2/n\}^{1/2}$.

Sir Maurice Kendall, International Statistical Institute, London.

ASSOCIATION COLLOID. **Colloid System.**

ASSOCIATION (Ecology). Central to certain concepts of ecology is the interaction between various otherwise unrelated species in a way that is beneficial to the participating parties, but not always indispensable. Ants and plant lice are sometimes associated in this way. The plant lice are guarded by the ants and sometimes carried to a good food supply whereupon the ants receive the sweet honey dew secreted by their charges. So-called cleaner fishes play useful roles in removing barnacles and other deposits from larger fishes and in recognition for these services are not eaten by the larger fishes. This type of association is also sometimes termed *commensalism*.

Plants. It is well established that plants are not distributed in nature in a haphazard fashion, but in habitats in which when certain species are present certain others usually occur also. Each such community of plants, composed of more or less the same group of species, is called a plant association. Some plant associations, such for example as the marginal rush or cattail association around a pond, may occupy only localized areas. Other associations, such as some of the grassland or desert shrub associations of western North America may occupy continuous areas totalling hundreds or thousands of square miles. In general, however, plant associations are the smaller units of vegetation occurring within a plant formation (see below) and their distribution is largely controlled by local soil and climatic conditions. Local differences in climate, in turn, are largely a function of topography. Some plant associations, such as a lichen association on a rock cliff, are relatively simple in organization. Others are relatively complex. The oak-hickory association of the eastern United States, for example, is named for the two prominent genera of trees present. Associated with the oaks and hickories, however, are occasional other large trees. In addition there are usually present smaller kinds of trees, species which constitute a shrub layer, and herbaceous species which constitute a more or less continuous ground cover.

A larger unit of vegetation than the association is the formation. A *plant formation* usually occupies thousands of square miles and its limits are controlled primarily by climatic conditions. Some of the major plant formations of North America are the tundra, the boreal forest, the hemlock-hardwood forest, the deciduous forest, the grasslands, the western coastal forest, the western mountain forest, the semi-deserts, and the tropical forests. Within each formation there are usually many different plant associations. Most plant associations are not permanent, but in the phenomenon of plant succession one association gradually replaces another. Many successions are in progress in any plant formation. The end results of the successional replacement of one plant association in turn by another is, if the process goes to completion, the establishment of a *climax association.* Such an association is a stable plant community and is not succeeded by any other association; it is the apex of the succesional process and, barring changes of climate, will continue to reproduce itself indefinitely.

Animals. While most animals are solitary, associating with others of their kind only incidentally or during the breeding season, others normally live in some relationship with members of the same or of other species.

The simplest association of members of the same species is gregariousness. Gregarious animals are not bound by the association but profit by it. Examples are the great herds of herbivorous animals such as the bison and the packs of predacious animals, such as wolves.

Colonial association may be accompanied by structural union between individuals, as in many marine polyps, or may be based on behavior, as in the social insects. The term merges with social organization. This type of association is accompanied by structural specialization of individuals for special tasks, except in human society where it depends on specialized training.

The association of individuals of different species may be the relatively loose type called commensalism in which both forms benefit but not in an essential way, or the indispensable symbiosis in which neither organism can persist without the other. An excellent example of symbiosis is the relation of termites with the protozoa found in their intestine; neither can live without the other.

An association in which one individual lives at the expense of the other is called parasitism.

Slavery is an association practiced by some of the social insects and by man; among the insects the slaves are of a different species.

Such relations as symbiosis and parasitism also occur among plants where they are exemplified by the combining of algae and fungi to form lichens and by the mistletoe, parasitic on trees. Symbiotic relations between animals and plants also occur.

See also list of entries under the entry on **Ecology.**

ASTATIC. A term meaning without orientation or directional characteristics.

ASTATIC GALVANOMETER. Galvanometer.

ASTATIC MAGNETOMETER. Magnetometer.

ASTATINE. Chemical element symbol At, at. no. 85, at. wt. 210 (mass number of the most stable isotope), periodic table group 7a, classed in the periodic system as a halogen, mp 302°C, bp 337°C. All isotopes are radioactive. This element occurs in nature only in minute amounts as a result of minor branching in the naturally occurring alpha decay series: $^{218}At(t_{1/2} = \text{ca. } 2 \text{ sec.})$ is produced to the extent of 0.03% by the beta decay of ^{218}Po(radium A), 99.97% going by alpha decay to ^{214}Pb(RaB); $^{216}At(t_{1/2} = 3 \times 10^{-4} \text{ sec.})$ 0.013% by beta decay from ^{216}Po(thorium A);

$$^{215}At(t_{1/2} = 0.018 \text{ sec.})$$

0.0005% by beta decay from ^{215}Po(actinium A). Astatine-217 ($t_{1/2} = 0.020$ second) is a principal member of the neptunium ($4n + 1$) series, all members of which occur only to that extent to which the parent ^{237}Np is produced by naturally occurring slow neutrons from uranium.

The first isotope to be discovered was ^{211}At made by Carson, Mackenzie, and Segrè by bombardment of a bismuth target with α-particles from the 60-inch cyclotron at Berkeley in 1940. The reaction is $^{209}Bi(\alpha, 2n)\ ^{211}At$. The half-life of ^{211}At is 7.2 hr. It decays in two modes, 60% by K-electron capture and 40% by α-particle emission. The longest-lived isotope is $^{210}At(t_{1/2} = 8.3 \text{ hr.})$; other isotopes having half-lives longer than 1 hr are 206, 207, 208, and 209. Various of the collateral radioactive series, involving bombardment reactions contain other astatine isotopes, such as ^{214}At, and ^{216}At. All these isotopes have half-lives that are only fractions of a second. The total number of isotopes is at least nineteen, including spallation reaction products as well as bombardment ones. They also include two short-lived isotopes, ^{215}At and ^{218}At, occurring in very small amount in the branched β-disintegration of ^{215}Po(actinium A) and ^{218}Po(radium A), respectively, as noted above.

The chemistry of astatine determined by tracer techniques, is in keeping with the regular transition of properties of the halogens. The acid properties of astatine are less marked than those of iodine, while its electropositive character is more marked than that of iodine. After reduction by SO_2 or metallic zinc, the astatine activity is carried by silver iodide or thallium iodide, so it evidently forms insoluble silver and thallium salts. This represents astatine in the univalent negative state characteristic of the halogens. However, astatine is very readily oxidized by bromine and ferric ions, giving indications of two higher oxidation states. Although there is no evidence from migration experiments of the presence of positive ions in the solution, astatine deposits on the cathode, as well as on the anode, in the electrolysis of oxidized solutions. Elemental astatine can be volatilized, although not so readily as iodine, and it has a specific affinity for metallic silver. The similarity to iodine is also shown by the observation that astatine concentrates in the thyroid glands of animals.

See list of references at end of entry on **Chemical Elements.**

ASTER. ***Compositae* or Composite Family.**

ASTER (Cell). Cell (Biology).

ASTER FAMILY. Artemisia.

ASTERISM. One of the characteristic effects sometimes observed in x-ray spectrograms. It has, roughly, the shape of a star, and commonly indicates the presence of internal stress in the material under investigation.

ASTEROID. The asteroids, sometimes called minor planets, are defined by Tver (1979)—observationally, any small interplanetary body not identified as a planet. Asteroids generally are a few kilometers in diameter or less, but can range up to about 1000 kilometers (620 miles) across. Thousands of asteroids orbit the sun between the orbits of Mars and Jupiter, and others are known elsewhere in the solar system. Most asteroids are fragments of bodies ranging up to at least 500 kilometers (310 miles) in diameter, and are sources of many meteorites. They range in composition from carbon-rich rocky bodies to metal-rich rocky bodies, according to spectroscopic and other instrumental evidence. Asteroids are numbered in order according to determinations of their orbits. They are usually named as the result of suggestions of their discoverers.

Amor Asteroids. A group of asteroids with orbits that cross the orbit of Mars (as projected on the ecliptic plane), but do not cross the orbit of the earth. Typical Amor orbits reach from the asteroid belt to a point between the earth and Mars. The group is named after the prototype asteroid, Amor.

Apollo Asteroids. A group of asteroids with orbits that cross the orbit of the earth (as projected on the ecliptic plane). Collisions with the earth are possible and apparently have occurred as described later, but these asteroids generally cross above or below the ecliptic plane, thus minimizing the possibilities of collisions. A substantial number of Apollo asteroids have been observed and their orbits have been calculated. In other instances, some have been discovered and ultimately "lost." Instrumental evidence suggests that they are rocky bodies, generally a few kilometers across. It is believed that most meteorites may be fragments broken off from them.

Trojan Asteroids. Two groups of asteroids in Jupiter's orbit about 60° ahead of and behind Jupiter. Existence of such bodies was predicted by Lagrange. Each member of the Trojan group has a period and mean distance nearly identical to those of the planet Jupiter. The group is of considerable theoretical interest because it represents examples of the solution of the three-body problem proposed by Lagrange, who proved theoretically that an object so located that it is equidistant from both Jupiter and the sun would be in a stable position, i.e., would remain there and continue to go about the sun with the same period as Jupiter. The members of the Trojan group all behave in approximately this manner, each of them being 20° of the vertex of an equilateral triangle, with the sun and Jupiter at the other vertices. They move about this vertex in a complex curve, and will remain in this vicinity unless they are perturbed by the attraction of Saturn.

Early Discoveries of Asteroids. The first asteroid discovered, Ceres, was found accidentally by Piazzi on January 1, 1801. His attention was directed to it by noticing the motion of the object through the stars. As the object approached the position of the sun, there was danger of its being lost, for the methods of orbit computation were not well developed at that time. The mathematician Gauss went to work on the problem and invented his well-known method for orbit computation, by means of which he was able to predict positions permitting the rediscovery of Ceres after it has passed the sun. Since the orbit was found to lie in the gap between the orbits of Mars and Jupiter, and the object was found to have a mean distance from the sun of 2.8 astronomical units, strong support was given by it to Bode's relation. See also **Bode's Relation.**

Up to the middle of the nineteenth century, only five more asteroids were discovered; but with the application of photography to astronomy, the discoveries became more and more frequent, until, at the present time, many hundreds are under observation. They are first detected by noticing the movement of a starlike object through the stars. Photographically, if the camera is arranged to follow the motions of the stars, the star images will appear as dots on the plate, while the asteroid image will be trailed out into a short line. The most extensive program of search for asteroids was carried on by Wolf at Heidelberg during the period following 1891. From this time on through the first two decades of the present century, Wolf and his assistants are credited with no less than 500 discoveries. Hundreds of asteroids are now picked up each year in the course of other investigations.

When an asteroid is first discovered, it is designated first by the year of discovery, then by two letters that indicate the half of the month in which the object was found, and last by the chronological order of discoveries within that half-month. After the orbit of the object has been determined, and if it proves to be a new asteroid, it is assigned a permanent number, in chronological order of discovery, and the discoverer is privileged to name the object as he may choose. In general, asteroids are given Latinized names with feminine endings.

The accompanying table gives properties of some of the better-known asteroids. The diameters of some of the brightest and, presumably, the largest asteroids have been measured with large telescopes; they vary from 700 km for Ceres down to 220 km for Juno. Estimates of the sizes of the others may be made from the amount of sunlight they reflect, after making certain assumptions regarding the reflecting power and shape of the objects. There are indications that perhaps 150 asteroids have diameters greater than 80 km, but the majority measure between 80 and 32 km in diameter, with some being even smaller than that. Masses and densities can be estimated only from statistical studies, but the indication is that the total mass of all the objects combined cannot be more than $\frac{1}{500}$ part of the mass of the earth.

There is still considerable interest in, but no definite solution for, the problem of determining the shapes of the asteroids. It is well known that the reflected sunlight from many of these objects varies in a periodic manner that can only be adequately explained on the basis of the rotation of the object. In the case of Eunomia, it has been quite definitely proven that the object must be spherical, and that the variation in light is due to different reflecting powers on different parts of the surface. On the other hand, Eros has been quite definitely proven to have a bricklike shape, with the light variations due to rotation of this irregular object.

The orbits of the asteroids have been studied carefully ever since the discovery of Ceres. See Fig. 1. In fact, this group of objects may be considered as a laboratory in which the workers in the field of celestial mechanics may test out various theories. Because their orbits lie between the orbits of Mars and Jupiter, and because their masses are very small, the asteroids have large perturbations exerted upon them by the planets; whereas the planets themselves are virtually unaffected by the asteroid attractions. Many of the methods of computing perturbations were developed as the result of research on the orbits of minor planets. One particularly interesting result is found in the case of the Trojan asteroids.

There are two theories for the origin of the asteroids. One is that

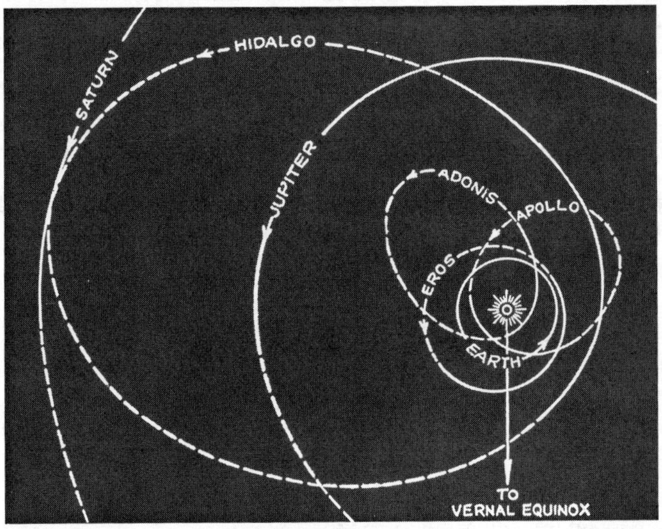

Fig. 1. Orbits of four unusual asteroids. Broken lines represent the parts of the orbits south of the plane of the ecliptic.

REPRESENTATIVE ASTEROIDS

Asteroid	Diam. (km)	Mass(10^{15}g)	Period (d)	a(A.U.)	e	i(deg.)
Ceres	700	60×10^7	1681	2.767	.08	10.6
Pallas	460	18×10^7	1684	2.767	.24	34.8
Juno	220	2×10^7	1594	2.670	.26	13.0
Vesta	380	10×10^7	1325	2.361	.09	7.1
Hebe	220	20×10^6	1380	2.426	.20	14.8
Iris	200	15×10^6	1344	2.385	.23	5.5
Hygiea	320	60×10^6	2042	3.151	.10	3.8
Eunomia	280	40×10^6	1569	2.645	.18	11.8
Psyche	280	40×10^6	1826	2.923	.14	3.1
Nemausa	80	9×10^5	1330	2.366	.06	9.9
Eros	14	5×10^3	642	1.458	.22	10.8
Davida	260	3×10^7	2072	3.182	.18	15.7
Icarus	1.4	5	408	1.077	.83	23.0
Geographos	3	50	507	1.244	.34	13.3

these objects represent a planet that was "spoiled in the making," i.e., never developed into its solid form. The other theory postulates that the asteroids represent the remains of a solidly formed planet that was later disintegrated. The latter theory is the older of the two, but has gained some considerable strength on the basis of the "families of asteroids" phenomenon. If a planet should be broken up by a series of explosions, it may be proved, theoretically, that the centers of the orbits of the asteroids (as products of any particular explosion) should lie along a line between the sun and Jupiter. In addition to the locations of the orbit centers along this line, the mean distances of the products of each explosion from Jupiter would be approximately the same; also, the inclinations of the orbit planes should be similar. From the statistical study of the orbits of asteroids, at least five asteroid families have been identified, each containing from 15 to 44 members. The theory that asteroids may be connected in some manner with comets has been gaining weight, but has yet to be proved.

In recent years, the most likely source of meteorites, primarily on the basis of rather inconclusive mineralogical evidence, has been considered to be the asteroids. Zimmerman and Wetherill (*Science*, **182**, 4107, 51–53, October 5, 1973) have made a survey of this topic. As pointed out, it is difficult to find a mechanism that will transfer an object from an orbit lying wholly between Mars and Jupiter into one crossing that of the earth. However, newly proposed mechanisms involving strong resonant perturbations by Jupiter, may explain the asteroids as a major meteorite source.

Asteroid Collisions with Earth. To some specialists in this field, it is not a question of will an asteroid collide with the earth, but of the probability of such an event. There is growing evidence of past encounters of this nature. As shown by the map of Fig. 2, craters on earth are relatively numerous. The geographical regions of central and eastern Canada embrace nearly 50% of all known ancient impact craters although the region represents only about 1% of the earth's land surface. Evidence of craters also has been found in Europe, Asia, and other areas. However, there are no known regions with the similar concentration of that in Canada. This has not been satisfactorily explained to date except in a rather qualitative way—to the effect that the Canadian scientists have conducted a much more extensive survey for craters. There is also the likely possibility that many craters are simply undiscovered in other areas because they are covered over with the debris of millions of years in the past and may be located in regions that are not particularly active in terms of other geologic interests. In commenting on the Canadian craters, as well as craters elsewhere, one scientist has observed that, on the basis of the best

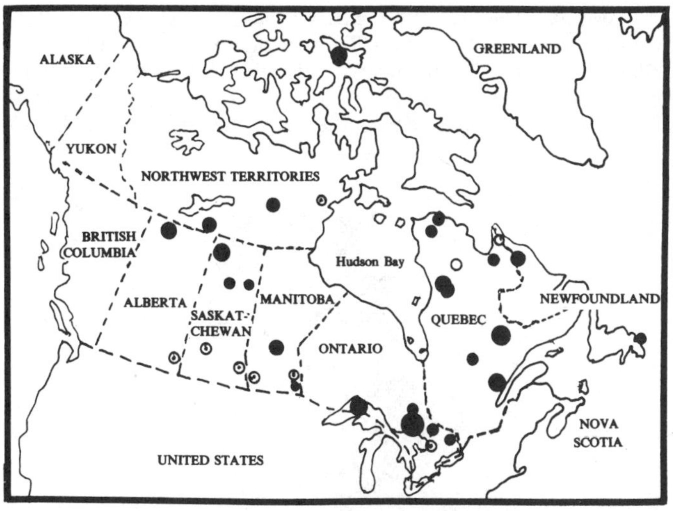

Fig. 2. Concentration of ancient (asteroid) craters in Canada, particularly in central and eastern portions. Over 20 craters (black circles) have been identified. Circles with dots identify locations of possible impact structures. These craters range in age from an estimated 1.8 billion years to less than 5 million years. The oldest crater is located near Sudbury, Ontario; the youngest crater is located in extreme northern Quebec east of Hudson Bay. Size of circles approximates size of crater (as compared with other craters). (*Drawn from data provided by Earth Physics Branch, Department of Energy, Mines and Resources, Canada.*)

available crater count, it is estimated that during the past 600 million years about 1500 Apollo objects (about 1 kilometer in diameter) or larger have struck the earth. This estimate assumes that about 70% of these bodies fell into the sea rather than on land. See also **Astrobleme.**

The relatively few craters on the earth and those many more observed on our moon and on many of the satellites of other planets are believed to have been caused by asteroid and/or comet impacts, among meteorites and other causal factors. The investigations of such impacts and their effects upon the earth not only fall within the realm of astrophysics, but of paleontology and paleogeology as well. Over the past two million years, there have been five relatively rapid environmental changes which have affected the biomass of the earth. There are hypotheses which attempt to explain periods of glaciation and intervening periods of warmer climates. That all or some of these environmental changes have occurred as the result of extraterrestrial forces has been discussed for many years. Shortly after the discovery of Ceres in 1801, there were proposals that asteroids colliding with the earth have been the principal cause of the major environmental periods, and, in particular, the most severe of these changes, which occurred at the end of the Cretaceous period and beginning of the Tertiary period, about 65 million years ago, when life epitomized by the dinosaurs became extinct. Paleontologists for several decades generally have not considered this hypothesis seriously and for understandable reasons, have opted for the more gradual causes of the Cretaceous-Tertiary extinctions, as contrasted with a single catastrophic event. The gradualists base these opinions in part upon the lack of geological and fossil evidence that would support the asteroid hypothesis. Further, no impact of an asteroid has been recorded during the time of recorded history.

In 1973, Urey speculated on the impact of a comet as the event which ended the Cretaceous period. The nuclei of comets are estimated within the same size range as an acceptable value for an impacting asteroid, i.e., from 1 to 10 kilometers (0.6–16 miles) in diameter. It is further observed, however, that comets, unlike asteroids, are composed of much ice and other substances that tend to reduce the comet size during swings close to the sun which volatilize these materials and further enrich a nebulous coma. See also **Comet.** If the estimate of the size of comets is relatively reliable, authorities suggest that observable comets have not been sufficiently abundant to produce the number of large craters on the moon, but that the abundance of Apollo objects has been sufficient to cause these craters. It has been suggested that perhaps comet impacts account for up to 35% of the larger lunar (and possibly a few earth craters), whereas the other craters have been caused by impacting asteroids and meteoroids.

The number of asteroids crossing the earth's orbit of a size sufficient to cause the five major environmental events in the earth's history (one possible example, the end of the Cretaceous period) would have required a hit about once every 100 million years. Some authorities believe that abundance of Apollo objects is sufficient to cause four collisions per every million years. Within the last few years, Alvarez and Alvarez (University of California at Berkeley) and Asaro and Michel (Lawrence Berkeley Laboratory) have located direct physical evidence (as contrasted with biological changes seen in the paleontological record) for an unusual event at exactly the time of the extinctions in the planktonic realm. A hypothesis has been developed to explain nearly all the available paleontological and physical evidence (Alvarez, et al, 1980). The Cretaceous-Tertiary boundary layer has been inspected in a number of locations, including Denmark and Italy. Deep-sea limestones exposed in New Zealand, Italy, and Denmark show iridium increases of about 20, 30, and 160 times, respectively, above the background level at precisely the time of the Cretaceous-Tertiary extinction. Field investigations indicate that this iridium is of extraterrestrial origin, but did not come from a nearby supernova. The Alvarez hypothesis accounts for the extinction and the iridium observations. Impact of a large earth-crossing asteroid would inject about 60 times the object's mass into the atmosphere as pulverized rock; a fraction of which would remain as dust in the stratosphere for several years and be distributed worldwide. The darkness resulting would suppress photosynthesis, and the expected biological consequences match quite closely the extinctions observed in the paleontological record. One prediction of this hypothesis has been verified—the chemical composition of the boundary clay (believed to have come from the stratospheric dust) is decidedly different from that of clay mixed with the Cretaceous and Tertiary limestones, which are chemically similar to each other. The research team has made four separate estimates of the diameter of the suspected asteroid, giving a value in the range of 10 ± 4 kilometers. It has been estimated that the kinetic energy of the asteroid would have been about equivalent to that of 10^8 megatons of TNT.

In an excellent summary of this hypothesis (Alvarez, 1980), the asteroid impact is compared with that of Krakatoa, an island volcano in the Sunda Strait between Java and Sumatra which erupted on August 26 and 27, 1883. See also **Volcano.** Whereas the estimated 14 cubic miles of material ejected into the atmosphere by Krakatoa required between 2 and $2\frac{1}{2}$ years to settle and to return the atmosphere to normal clarity, it is suggested that the debris from the hypothesized asteroid impact would have been greater by a factor of about 10^3 and thus would have put the earth essentially into darkness for a period of several years.

In 1963, when 10 Apollos were known, Öpik (Armagh Observatory in Ireland) concluded that there must be at least 43 Apollos and possibly many more. Since that time, an additional 28 Apollos have been discovered and the current rate of discovery is about four bodies per year. There is a general opinion as of the early 1980s that the number of Apollos, at a minimum, is well over 200. At this time, none of the known Apollo objects is on a collision course with the earth. However, both the Apollo and Amor asteroids are under continuous gravitational influence of nearby planets, particularly Jupiter, which causes the asteroidal orbits to precess. Because of precession, the major axis of an elliptical orbit gradually rotates through 360° in space. Thus, those asteroids with a perihelion inside the earth's orbit and an aphelion beyond the earth's orbit are destined at some time to be in an orbit that intersects the earth's orbit. It follows that the probability of any given Apollo to intersect the earth's orbit is once in every 5000 years. It further follows that the likelihood of the earth and asteroid being in precisly the same spot in the earth's orbit is very small—with an estimated collision probability of only about 5×10^9 per year (once in 200 million years).

Captured Asteroids. The irregular shapes and other unexpected characteristics of Deimos and Phobos, the moons of Mars, as revealed by high-resolution photos obtained by the Viking orbiters, have suggested to a number of specialists that these two bodies are indeed asteroidal satellites. If so, these are the first captured asteroids to be viewed at close range. See Figs. 3 and 4.

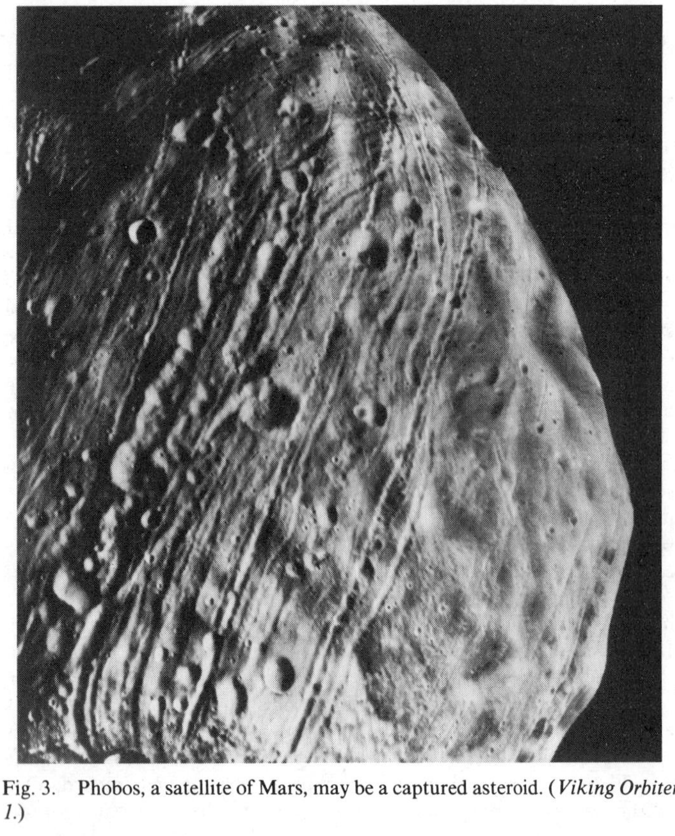

Fig. 3. Phobos, a satellite of Mars, may be a captured asteroid. (*Viking Orbiter 1.*)

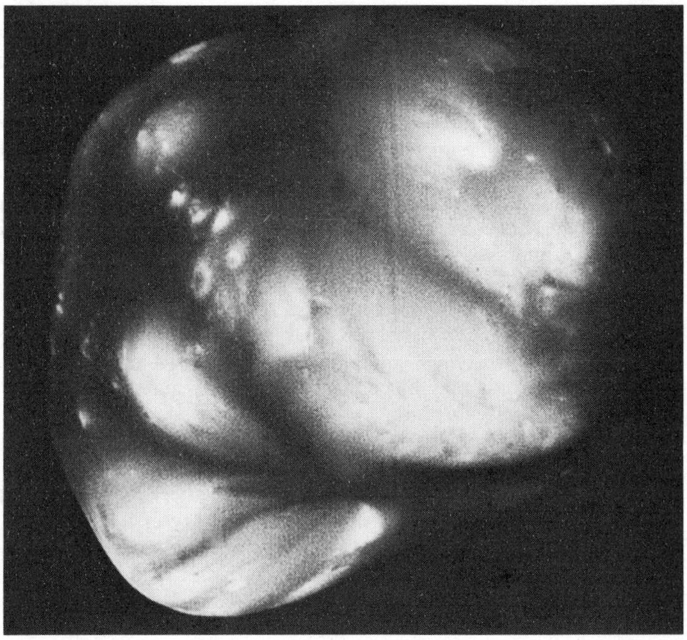

Fig. 4. Deimos, a satellite of Mars, in nearly full phase. This moon may be a captured asteroid. (*Viking Orbiter 2.*)

Binary Asteroids. Light curves obtained of 624 Hektor (Fig. 5) have suggested that the asteroid has the shape of a dumbbell or possibly a fat cigar considerably longer than it is wide. The irregular form of an asteroid is believed to result from collisions with other asteroids, but it is also regarded unlikely that a collision would produce an oblong body. Some asteroid specialists believe that Hektor may not be a dumbbell as previously proposed, but rather than Hektor is actually two asteroids in contact. Ultimately, spectroscopic observations should reveal whether Hektor is one or two objects.

In a rare event which occurred on June 7, 1978, a star was eclipsed by the asteroid 532 Herculina. While the total occultation lasted for 20.6 seconds, there were six additional dimunitions in starlight within 2 minutes of the main eclipse. These ranged from 0.5 to 4 seconds. Since the star was almost four magnitudes, atmospheric interference was ruled out. The longest secondary occultation has been confirmed by an independent observation. This occultation was caused by a secondary body about 50 kilometers in diameter and about 1000 kilometers from Herculina, whose diameter is 220 kilometers. Current thinking is that the six secondary eclipses were caused by six satellites of

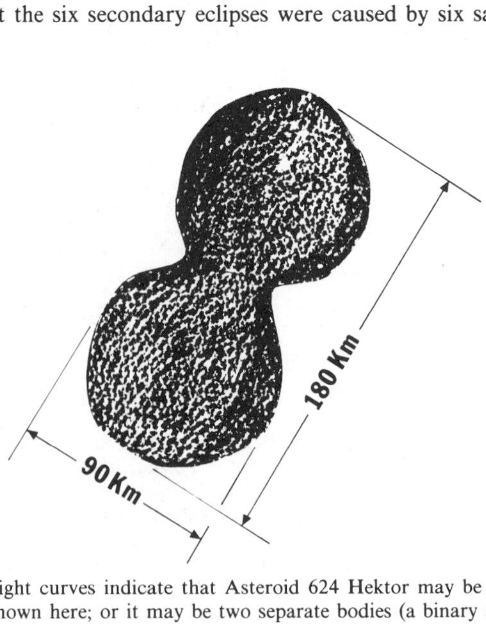

Fig. 5. Light curves indicate that Asteroid 624 Hektor may be a dumbbell shape as shown here; or it may be two separate bodies (a binary asteroid).

Herculina. The masses and distances are such that the system would be gravitationally stable.

Hirayama Families. These are groups of asteroids, usually classified as C, S, and M types, with similar orbits and suspected to be fragments from collisions between asteroid pairs. They were observed earlier in this century (Hirayama, 1918, 1929). As reported by Gradie and Zellner (1977), the comparatively recent use of reflection spectroscopy, polarimetry, and thermal radiometry methods in asteroid research, all of which provide information on the composition of these minor planetary bodies, has provided new insights as to the importance of the Hirayama families. If the foregoing definition is correct, then members of a single family should show identical compositions, assuming that the parent body was homogeneous. But, on the other hand, if the dynamical families were formed by the collisional focusing of unrelated field asteroids, then one would contemplate finding only the pattern of compositions characteristic of that region of the asteroid belt. The aforementioned investigators, who used the UBV[1] color system, have reported that the colors of minor planets indicate compositions quite distinct from those of the field population in each of the three Hirayama families. They observed that the Eos and Koronis families apparently originated from the collisional fragmentation of undifferentiated silicate bodies and the Nysa group from a geochemically differentiated parent body. Considerable research remains to further refine postulations concerning the origination of the Hirayama families.

Possible Exploitation of Asteroids. Although in a very early stage of speculation, a number of scientists have been giving consideration to the possible use of satellites as permanent sites for space stations as well as sources of minerals. Gaffey and McCord (1977) have worked out a rather elaborate, even if preliminary plan for an asteroid mining operation, including means for transporting materials to earth. Details are given in the references listed.

REFERENCES

Alvarez, L. W., et al.: "Extraterrestrial Cause for the Cretaceous-Tertiary Extinction," *Science,* **208**, 1095–1108 (1980).

Blanco, V. M., and S. W. McCuskey: "Basic Physics of the Solar System," Addison-Wesley, Reading, Massachusetts, 1961.

Chapman, C. R., Williams, J. G., and W. K. Hartmann: "The Asteroids," *Ann. Rev. Astron. and Astrophys.,* **16**, 33–75 (1978).

Davis, J.: "Asteroids," *Astronomy,* **8**, 5, 16–22 (1980).

Gaffey, M. J., and T. B. McCord: "Mining Outer Space," *Technology Review (MIT),* **79**, 7, 51–59 (1977).

Gehrels, T. (editor): "Asteroids," Univ. of Arizona Press, Tucson, Arizona (1979).

Gradie, J., and B. Zellner: "Asteroid Families: Observational Evidence for Common Origin," *Science,* **197**, 254–255 (1977).

Hartmann, W. K., and D. P. Cruikshank: "Hektor: The Largest Highly Elongated Asteroid," *Science,* **207**, 976–977 (1980).

Hirayama, K.: *Astron. J.,* **31**, 185 (1918); *Jpn. J. Astron. Geophys.,* **5**, 137 (1928).

Kerr, R. A.: "Asteroid Theory of Extinctions Strengthened," *Science,* **210**, 514–517 (1980).

King, E. A.: "Space Geology: An Introduction," Wiley, New York, 1976.

O'Leary, B.: "Mining the Apollo and Amor Asteroids," *Science,* **197**, 363–365 (1977).

Richardson, R. S.: "The Discovery of Icarus," *Sci. Amer.,* **212**, 4, 106–115 (1965).

Staff: "Close Encounters," *Sci. Amer.,* **239**, 5, 86–90 (1978).

Staff: "Hektor—A Strange Asteroid," *Astronomy,* **7**, 3, 63 (1979).

Staff: "Astounding Asteroids," *Sci. Amer.,* **240**, 5, 96–98 (1979).

Tedesco, E. F.: "Binary Asteroids: Evidence for Their Existence from Lightcurves," *Science,* **203**, 905–907 (1979).

Watson, J. T.: "Meteorites: Classification and Properties," Springer-Verlag, New York, 1974.

Wetherill, G. W.: "Apollo Objects," *Sci. Amer.,* **240**, 3, 54–65 (1979).

Wood, J. A.: "Meteorites and the Origin of Planets," McGraw-Hill, New York, 1968.

ASTEROID BELT. **Asteroid; Jupiter.**

[1] Gradie and Zellner (1977) have reported that, while UBV photometry is a coarse tool for elucidating compositional differences between asteroids, the ultraviolet-minus-blue (U-B) and the blue-minus-visual (B-V) color indices are generally sufficient to distinguish the principal C, S, and M types and recognize objects that belong to rarer or unknown types.

ASTEROIDEA. A class of the phylum *Echinodermata.* The starfishes.

The starfishes are distinguished from other echinoderms by the presence of radiating arms or rays, usually five or in multiples of five, which contain part of the internal organs and are usually not sharply separated from the central disk. There are many species but the economic importance of the group is limited. They are sometimes serious pests in oyster beds since they feed largely on shellfish.

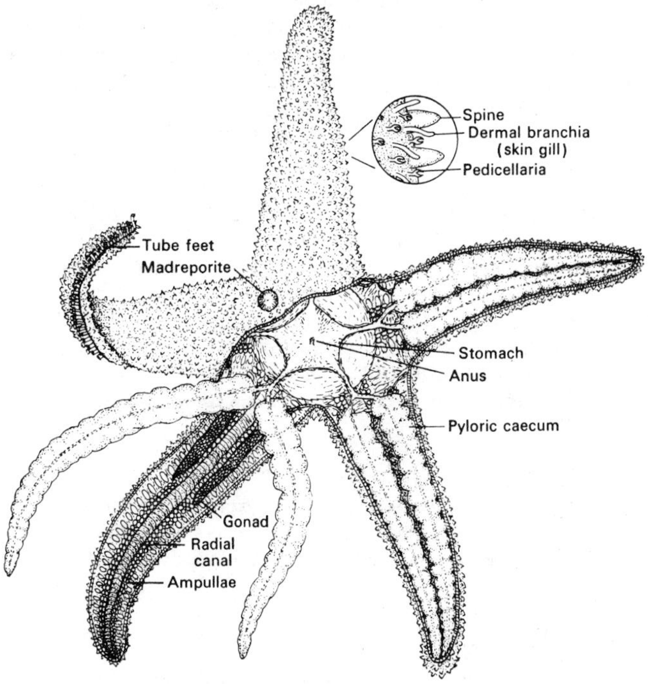

Common starfish. (*Winchester and Lovell, "Zoology," Van Nostrand Reinhold.*)

The class is divided into three orders: *Phanerozonia, Spinulosa,* and *Forcipulata;* in addition two orders—*Platyasterida* and *Hemizonida*—contain extinct asteroids.

ASTEROID (Mathematics). A higher plane curve, which is a special case of a hypocycloid. The curve is generated by a point on the circumference of a circle of radius r, which rolls around the inside of a fixed circle of radius $R = 4r$. Its parametric equations are $x = R \cos^3 \phi$, $y = R \sin^3 \phi$, and its equation in Cartesian coordinates is

$$x^{2/3} + y^{2/3} = R^{2/3}$$

The curve is symmetric to both coordinate axes. There are cusps of the first kind at the four points $(\pm R, O)$, $(O, \pm R)$, where the corresponding tangents are the X- and Y-axes. The evolute of an ellipse, which has the equation $(rx)^{2/3} + (Ry)^{2/3} = (r^2 - R^2)$ and the same general shape with four cusps, is sometimes also called an asteroid. The spelling astroid is often given.

See also **Evolute; Hypocycloid;** and terms listed under **Mathematics.**

Asteroid.

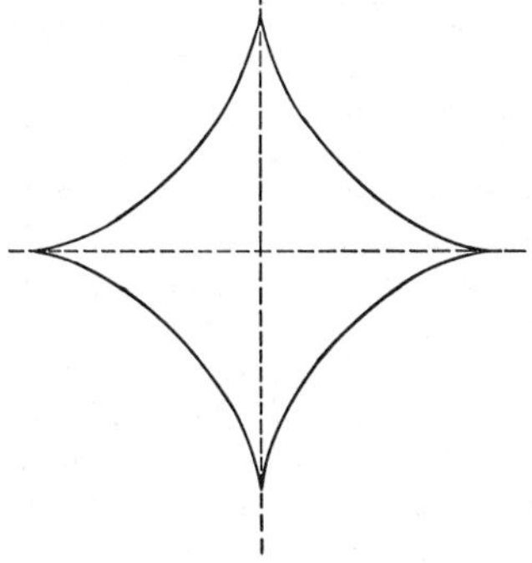

ASTHENIA. Weakness, lack or loss of strength.

ASTHENOSPHERE. A term proposed by Barrell, in 1914, for the zone beneath the relatively rigid lithosphere. The asthenosphere is considered to be the level of no strain in which there is maximum plasticity, and in which the igneous rock magmas are thought to originate. See also **Earth.**

ASTIGMATIC FOCUS. In an astigmatic system (see **Astigmatism**) some of the bundle of rays from an off-axis point meet in a line perpendicular to a plane containing the point and the optical axis. Some meet in a line (at a greater image distance) which lies in a plane containing the point and the optical axis. At all other image distances the bundle is an ellipse (or circle). The first line is called the primary or meridianal or tangential focus. The second line is called the secondary or sagittal focus.

ASTIGMATISM. This term may denote: (1) A defect in a lens including the lens of the eye, in which there is a difference in the radius of curvature of the lens as observed in one plane from that observed in another plane. (2) An aberration of a lens with spherical surfaces such that the image of a point not lying on the optical axis is a pair of short lines normal to each other and at slightly different distances from the lens. (3) In an electron-beam tube, a focus defect in which the electrons in different axial planes come to focus at different points.

ASTON WHOLE NUMBER RULE. The atomic weights of isotopes are (very nearly) whole numbers when expressed in atomic weight units, and the deviations from the whole numbers of the atomic weights of the elements are due to the presence of several isotopes with different weights.

ASTRAGALUS GUMMIFERA. Gums and Mucilages.

ASTROBLEME. A scar on the surface of the earth made by the impact of a cosmic body. The term usually connotes a so-called fossil crater of ancient origin. There are 14 large and certified meteorite craters known and undoubtedly many more that are masked by vegetation or that have been subject to subsequent alteration as the result of tectonic processes, sedimentation, and erosion. Most likely the readily visible remaining craters were created during the last million years. Over the long history of the earth, some investigators believe that many thousands of giant meteorites have impacted the surface of the planet. The moon provides strong evidence that such activity in the vicinity of the earth has been strong in the past. The moon, of course, with no atmosphere and apparent minimal tectonic activity has provided a rather ideal means for permanently recording such impacts.

Probably the most spectacular example is the Vredefort Ring in the Transvaal of South Africa. Very little of the original crater remains, but shatter cones give evidence that this probably was the greatest terrestrial explosion in relatively recent times (within last 250 million years). At one time, geologists ascribed the structure to a series of tectonic events. A shatter cone is a distinctively striated conical fragment of rock along which fracturing has occurred, ranging in length from less than a centimeter to several meters, generally found in nested or composite groups in the rocks of cryptoexplosion structures, and generally believed to have been formed by shock waves generated by meteorite impact. Shatter cones superficially resemble cone-in-cone structure in sedimentary rocks; they are most common in fine-grained homogeneous rocks, such as carbonate rocks (limestones, dolomites), but are also known from shales, sandstones, quartzites, and granites. The striated surfaces radiate outward from the apex in horsetail fashion; the apical angle varies but is close to 90 degrees. Geologists have studied the shatter cones in the area of the Vredefort Ring and confirm that, if the rocks were returned to their original positions, the shatter cones would all point inward toward the center of the ring. It is postulated that an asteroid about a mile or more in diameter struck the earth from the southwest, drilling into the earth and releasing enormous shock forces. Strata some 9 miles ($\sim$14.5 kilometers) in thickness peeled back in the fashion of a flower, opening a crater some 30 miles ($\sim$48 kilometers) in diameter and 10 miles ($\sim$16 kilometers) deep. The energy released is compared with the extent of energy required to produce the Tycho and Copernicus craters on the moon. It is further estimated that the Vredefort blast was about a million times larger than the 1883 Krakatoa volcanic explosion and probably exceeded by several thousand times the largest possible earthquake. In terms of the force of nuclear explosions, it is believed that the Vredefort blast would have been classified as a 1.5-million-megaton event.

The meteorite crater (Barringer Crater) located in Arizona is much more recent (estimated 25,000 years old) and much smaller than the Vredefort event. On a nuclear scale, as mentioned in the prior paragraph, the Barringer event would have been only a 5-megaton explosion. This crater is $\frac{3}{4}$ mile (1.2 kilometers) across and 600 feet (180 meters) deep. While not in evidence at Vredefort, coesite is found at Barringer. Coesite is a monoclinic mineral, SiO_2. It is a very dense (2.93 grams/cubic centimeter) polymorph of quartz and is stable at room temperature only at pressures above 20,000 bars. The silicon has a coordination number of 4. Coesite is found naturally only in structures that are presently best explained as impact craters, or in rocks, such as suevite, associated with such structures. Coesite is believed to be a second shock-wave product and its presence has been helpful in confirming at least five astrobleme sites. Coesite was created artificially by Loring Coes, Jr. (Norton Company, Worcester, Massachusetts) in 1953 in apparatus that produced pressures exceeding 20,000 atmospheres. See also **Meteorite Crater.**

Coesite and suevite have been found at Ries Kessel (Giant Kettle), an ancient basin formation some 17 miles (27.4 kilometers) across and located 26 miles (41.8 kilometers) from the Steinheim Basin in southern Germany. Based upon studies within the last twenty years, Ries Kessel is now considered an astrobleme. Coesite also has been found in rather large amounts of silica glass in connection with the Wabar craters in the Empty Quarter of Arabia. Similar findings have been made at the Ashanti Crater in Ghana and at the Teapot Ess Crater in Nevada (the latter created by an atomic blast at the Nevada Proving Grounds). By seeking the presence of coesite, it is believed that additional astrobleme sites will be identified.

Several fossil craters have been identified in Canada, including Carswell Lake, Keely Lake, Deep Bay, Westhawk Lake, Lac Couture, Nastapoka Arc (Hudson Bay), Clearwater Lake, Menihek Lake, Ungava Bay, Sault-Aux-Cochons, Brent, Franktown, Lake Michikamau, Manicouagan Lake, St. Lawrence Arc (New Brunswick), Mt. Canina Crater, and Holleford. The Holleford Crater is now a slight depression about $1\frac{1}{2}$ miles (2.4 kilometers) in diameter, eroded and filled with sediments. It is located in Ontario farmland and is believed to be the result of an impact some 500 million years ago. It was discovered by means of aerial photogrphy. Interesting shatter cone sites in the United States, in addition to the Barringer Crater, include Kentland, Indiana (in a large limestone quarry), Sierra Madera, Texas, Serpent Mound, Ohio, Flynn Creek and Wells Creek in Tennessee, and Crooked Creek in Missouri. Craters over one million years old are located at Boxhole, Dalgaranga, Henbury, and Wolf Creek in Australia.

Much pioneering work in recent years in connection with seeking shatter cones, coesite, and location of astroblemes has been done by Robert S. Dietz, whose writings on the subject are listed in the references.

Dietz suggests that the creation of coesite and of minute diamonds by meteorite impact opens up the new field of *impact metamorphism*, explaining that meteorite impacts are natural "experiments" in ultra-high pressures on a scale that most likely will never be equaled in the laboratory.

In the twentieth century, two great impacts have been known to occur, both in Siberia. The event at Tunguska probably was caused by the fall of a comet head. At Sikhote-Alin in 1947, a very large meteorite fell that was disintegrated in mid-air, leaving more than 100 craters on the ground. All known meteorite impacts have occurred on land, but it is highly probable that many more have fallen into the sea and thus leaving evidence very difficult for geologists to uncover with present technology.

One of the most recent impact phenomena to be reevaluated is Lonar Crater, in the Buldana District of Maharashtra, India (19°58′

N, 76°31′ E). This is an almost circular depression in the basalt flows of the Deccan Traps. The crater is 1830 meters across and nearly 150 meters deep. Most of the floor is covered by a shallow saline lake (Lonar Lake). Around most of the circumference, the rim is raised about 20 meters above the surrounding plain. A second crater appears to lie about 700 meters north of the large crater. Early investigators ascribed the formation to a volcanic explosion of subsidence. However, in 1896, Gilbert emphasized the similarity of Lonar Crater with Barringer Crater in Arizona. Recent studies of the crater are detailed by Fredriksson, Dube, Milton, and Balasundaram in *Science*, **180**, 4088, 862–864, May 25, 1973.

References

Chao, E. C. T., et al: "First Natural Occurrence of Coesite," *Science*, **132**, 220–222 (1960).

Dietz, R. S.: "Meteorite Impact Suggested by Shatter Cones in Rock," *Science*, **131**, 1781–1784 (1960).

Dietz, R. S.: "Astroblemes," *Sci. Amer.* (August 1961).

Gilbert, G. K.: "The Origin of Hypotheses, Illustrated by the Discussion of a Topographic Problem," *Science*, **3**, 1–13 (January 3, 1896).

McCall, G. J. H. (editor): "Astroblemes—Cryptoexplosion Structures," Benchmark Papers in Geology, Vol. 50, Dowden, Hutchinson & Ross, Stroudsberg, Pennsylvania, 1980.

ASTROGRAPHIC TELESCOPE. A refracting telescope designed to give a field of 10° or more. The objective is a designed compromise between the various optical aberrations at a specified wavelength. See also **Telescope.**

ASTROLABE. An ancient form of portable astronomical instrument invented during the second or third century B.C., probably either by Hipparchus or Apollonius. In its most common form, the astrolabe consists of a circular disk suspended by a ring so that it will hang in the plane of a vertical circle. A pointer, or alidade, is pivoted at the center of the disk, and angular graduations are marked about the edge. For purposes of measuring altitude, the ring is suspended by the thumb of one hand, and the other fingers of the same hand are employed to steady the disk as the alidade is moved, by the other hand, until it points directly at the object under observation. The altitude can then be read directly on the disk.

The astrolabe was used by navigators for the determination of latitude from the fifteenth century until the invention of the sextant. Since that time, it has been used as a teaching instrument in elementary classes. The astrolabe, in its modern version, is essentially the only impersonal instrument for the measurement of time and latitude that does not rely on secondary standards. The zenith telescope will do the same observational tasks as the astrolabe, but it is necessary to introduce nonfundamental stars. The modern astrolabe is free of personal errors, and gives stellar positions with an accuracy on the order of one-tenth of a second of arc.

ASTROMETRY. The branch of astronomy dealing with the positions, distances, and motions of the planets and stars; it includes determination of time and position. See also **Bonner Durchmusterung.**

ASTRONAUTICS. The science of travel in outer space.

Weight and Weightlessness. In Newtonian mechanics, weight is understood to mean the force that an object exerts upon its support. This would depend on two factors: the strength of gravity at the object's location (things weigh less on the moon) and, as Newton called it, the quantity of matter in a body (its "mass"). At any given location, where gravity is fixed, mass can be measured relative to a standard by noting the extension of a spring to which it and the standard are successively attached. Alternatively, the unknown and standard may be hung at opposite ends of a rod and the balance point noted. However, by an entirely separate experiment, mass can also be measured by noting the resistance of the object to a fixed force applied horizontally on a frictionless table. The measured acceleration provides the required basis of comparison with the standard. Needless to say, all objects measure identical accelerations when freely falling in the vertical force of gravity. This merely means that, unlike the arbitrary force we apply horizontally in the experiment above, gravity has the property of adjusting itself in just the right amount, raising or lowering its applied force, to maintain the acceleration constant.

It was well known that objects appear to increase or decrease their weight (alter the extension of the spring) if the reference frame in which the measurement takes place accelerates up or down. As gravity did not really change, however, most people were inclined to draw a distinction between *weight* defined as $m\mathbf{g}$, where m is the mass and $\mathbf{g}$ is the local gravity field, and the *appearance of weight*, the force of an object on its support as measured by the spring's extension. One way to avoid the difficulty has been to speak of an *effective* $\mathbf{g}$, which takes into consideration the frame's acceleration. For example, at the equator of the earth, we measure, say by timing the oscillation of a pendulum, the effective $\mathbf{g}$, some 0.34% less than the $\mathbf{g}$ produced by the mass of earth beneath our feet. If the earth were rotating with a period of an hour and a half instead of 24 hours, our centripetal acceleration at the equator would cause the effective $\mathbf{g}$ to vanish completely, our scales would not register, objects would be unsupported, and for all practical purposes we would be weightless.

Formally, we could state that any accelerating frame produces a local gravitational field $\mathbf{g}_{acc}$ that is equal and opposite to the acceleration. Thus, a rotating frame generates a centrifugal $\mathbf{g}_{acc}$ opposing the centripetal acceleration. We have at any point

$$\mathbf{g}_{eff} = \mathbf{g} + \mathbf{g}_{acc} \tag{1}$$

where $\mathbf{g}$ is the field produced by matter alone (e.g., the earth). By identical reasoning, an object in orbit, whether falling freely in a curved or in a straight path, will carry a reference frame in which $\mathbf{g}_{eff}$ is zero, for its acceleration will always exactly equal the local $\mathbf{g}$ by the definition of the phrase, "freely falling."

This concept was placed on a firm footing by Einstein who maintained that Eq. (1) is reasonable not only in mechanics but in all areas of physics including electromagnetic phenomena. We arrive at the inevitable conclusion that we cannot distinguish by any physical experiment between an apparent $\mathbf{g}$ accountable to an accelerating frame and a "real" $\mathbf{g}$ derived from a local accumulation of mass. This central postulate of the General Theory of Relativity also unified the two separate conceptions of mass. An object resting on a platform that is accelerating toward it will resist the acceleration in an amount depending on its inertia. It presses against the platform with a force equal to that it would have if placed at rest on the surface of a planet with local field equal and opposite to the acceleration of the frame.

General Principles of Central Force Motion. The gravitational force between point masses is inverse square, written

$$m\mathbf{g} = -\frac{\gamma m' m}{r^2}\hat{\mathbf{r}} \tag{2}$$

where the center of coordinates from which the unit vector $\hat{\mathbf{r}}$ is described lies in m', one of the masses. Thus, the force on m is directed $-\hat{\mathbf{r}}$, toward m' and is proportional to $1/r^2$ with γ the constant of proportionality. The quantity $\mathbf{g}$ is the force on m *divided by* m, (or normalized force) for which the name "gravitational field of m'" is reserved. Of course, if m were in the field of a collection of mass points, or even in a continuous distribution of mass, the summated or integrated $\mathbf{g}$ at the location of m would no longer be an inverse square function with respect to any coordinate center. However, in one special case, the inverse square functional form would be preserved: if the source mass were symmetrically distributed about the coordinate center. This would be the case if the source were a spherical shell or solid sphere, of density constant or a function only of r. The sun and earth can be regarded, at least to a first approximation, as sources of inverse square gravitational fields.

There are some important general statements we can make about the motion of an object placed with arbitrary position and velocity in a centrally directed force field, i.e., a field such as the one described, which depends only on distance from a central point (regardless of whether or not the dependence is inverse square). As the force has only a radial and no angular components, it cannot exert a torque about an axis through the center. This means that the initial angular momentum is conserved. Now angular momentum is a vector quantity and therefore is conserved both in direction and magnitude. It is defined by $\mathbf{r} \times \mathbf{p}$, where $\mathbf{r}$ is the position vector to the mass of momentum

p. The direction of the angular momentum vector is thus perpendicular to the plane containing **r** and **p**. As this direction is permanent, so also must be the plane. The planar motion of the object can be expressed in polar coordinates, so that, by writing $\mathbf{r} = r\hat{\mathbf{r}}$ and $\mathbf{p} = m(\dot{r}\hat{\mathbf{r}} + r\dot{\phi}\hat{\boldsymbol{\phi}})$, we find the specific angular momentum (angular momentum per unit mass) called h, to be

$$h = r^2\dot{\phi} \tag{3}$$

This too then must be a constant of the motion.

Consider now the rate at which area is swept out by the radius vector, dS/dt. We recall from analytic geometry that $dS = \frac{1}{2}r^2 d\phi$. Thus

$$\frac{dS}{dt} = \frac{h}{2} \tag{4}$$

so that this is a constant of the motion as well. On integration, we conclude that the size of a sector that is swept out is proportional to the time required to sweep it out. In the case of a closed orbit, the total area S would then be related to the specific angular momentum as

$$S = \frac{hT}{2} \tag{5}$$

This sector area-time relationship is Kepler's second law of planetary motion which was induced from Tycho Brahe's observation of Mars without prior knowledge of gravity and its central character.

The Laws of Kepler. Kepler stated two other laws of planetary motion: The orbits of all the planets about the sun are ellipses (a radical departure from the circles of Copernicus), and the squares of their period are proportional to the cubes of their mean distance from the sun, this mean being the semimajor axis of their ellipses. See Fig. 1. The third law pertained to the one object common to all the planets, the sun. Taken together, the three laws led Newton to the concept of gravitational force and its inverse-square form.

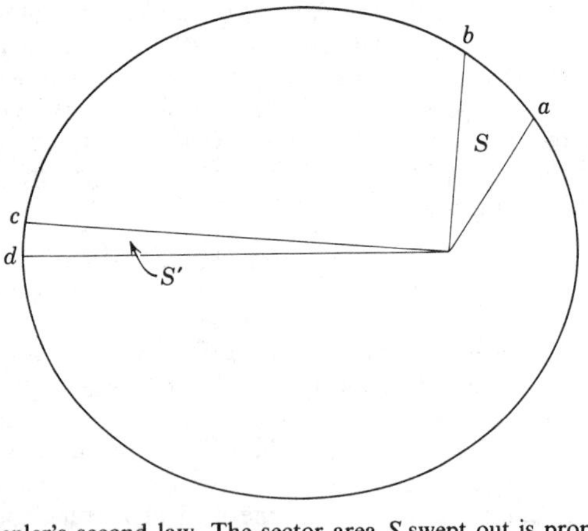

Fig. 1. Kepler's second law. The sector area S swept out is proportional to the time required for the planet to move from a to b. Thus, if $t_{cd} = t_{ab}$, then $S' = S$.

By applying Newton's law of motion $\mathbf{F} = m\mathbf{a}$, a relationship between **a**, the second derivative of the position vector, expressed in polar form, and $\mathbf{F}/m$ or **g**, as given by Eq. (2), leads to the familiar conic solution for the trajectory of an object in an inverse square field,

$$\frac{1}{r} = \frac{\gamma m'}{h^2} + A\cos(\phi - \phi_0) \tag{6}$$

where A and ϕ_0 are constants. A rotation of axis will eliminate ϕ_0, thereby aligning the coordinate axis with the conic's major axis. Also, by expressing the general conic, an ellipse or hyperbola, in terms of the usual parameters of semimajor axis a and eccentricity ϵ, we can relate the geometric parameters to the gravitational-dynamical constants, viz:

$$h = [\gamma m' a(1 - \epsilon^2)]^{1/2} \tag{7}$$

and

$$\frac{1}{r} = \frac{\gamma m'}{h^2}(1 + \epsilon\cos\phi) \tag{8}$$

Note that by substituting Eq. (7) into Eq. (5) and expressing the area of an ellipse as $S = \pi a^2(1 - \epsilon^2)^{1/2}$ we arrive at Kepler's third law,

$$T = \frac{2\pi}{(\gamma m')^{1/2}} a^{3/2} \tag{9}$$

The *energy* of the orbiting object can be calculated with ease by evaluating it at an extremal point, say the nearest point to the gravitational source, called pericenter or perifocus. As the energy is constant, it is immaterial where the calculation is made. Here the velocity has only an angular component so that the kinetic energy for a unit orbiting mass is $\frac{1}{2}v^2 = \frac{1}{2}r^2\dot{\phi}^2$. The potential energy at pericenter is $-\gamma m'/r_{pe}$ where r_{pe} is the distance of the unit mass from m', the focal point. Here $\gamma = 0$ so that by Eq. (8),

$$\frac{1}{r_{pe}} = \frac{\gamma m'}{h^2}(1 + \epsilon) \tag{10}$$

On substituting Eq. (7), we find the total kinetic and potential energy to be

$$E = \frac{\gamma m'}{2a} \tag{11}$$

Our conclusion: All objects in orbit with the same major axes have identical periods and identical energies per unit mass. Knowledge of E is invaluable in determining an object's speed when its distance from the source is known and vice versa.

In the event that the orbiting object's mass is not negligibly small compared with that of the gravitational source, one must take note that the combined center of mass, from which the acceleration is described, no longer may be assumed to lie in the center of the gravita-

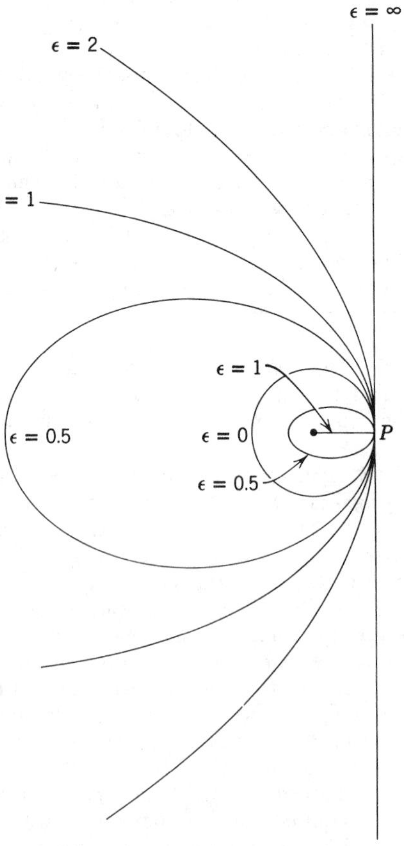

Fig. 2. Orbits of differing eccentricities and major axes which pass through a common point. Higher speeds correspond to higher energies and longer major axes.

tional source. This complicates our equations somewhat, for the accelerating force still is expressed relative to the center of the source (if spherical). The adjustment that results, when center of mass coordinates are transformed to relative coordinates in the expression for acceleration, requires our equations to take the form $\gamma(m' + m)$ wherever formerly $\gamma m'$ appeared. See Fig. 2.

Disturbances in the Central Field. The earth, of course, is spherical only to a first approximation. More accurately, it is an ellipsoid of revolution about a minor axis—an oblate spheroid. Still more accurately, it appears to be slightly pear-shaped and, in addition, its figure is distorted by continuous local variations. The spheroidal figure, nevertheless, accounts for nearly all the anomalous effects of satellite orbits. For one thing, the gravitational force on the satellite is no longer centrally directed; the excessive mass in the equatorial plane produces a force on the satellite directed out of its orbital plane. The resultant torque causes the direction of the angular momentum vector to change, i.e., the plane containing the satellite's ellipse turns. The plane turns continuously about the polar axis maintaining its angle with the axis and with the equatorial plane constant. The turning rate is greatest for low orbits and small angles of inclination with the equator. For polar satellites, the plane remains fixed. A separate effect of this equatorial bulge perturbative force is the slow turning of the ellipse's major axis *within* the orbital plane. This effect vanishes at an inclination of 63.4°; the major axis turns backward at inclinations above this angle and forward below. See Fig. 3.

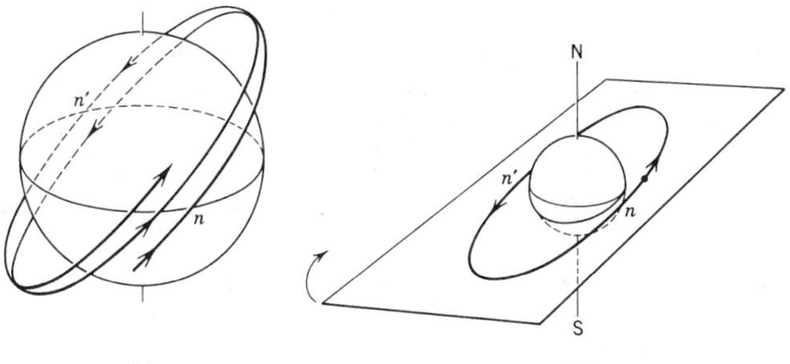

Fig. 3. The orbit of an earth satellite. The earth's equatorial bulge causes retrograde motion of the points of intersection n and n' of the orbit and equatorial plane. This can alternatively be interpreted as a retrograde motion, about the north-south axis, of the plane containing the closed orbit. The plane moves in the direction shown by the arrow in (b), maintaining a constant angle with the axis.

Rocket Propulsion. A rocket operates by the simple principle that if a small part of its total mass is ejected at high speed, the remaining mass will receive an impulse driving it in the opposite direction at a moderate speed. As δm_e, the propellant, leaves at speed v_e with respect to the rocket, the remaining rocket mass m receives a boost in speed δv such that

$$\delta m_e v_e = m\,\delta v \tag{12}$$

If additional equal propellant mass is ejected at the same speed, the boost in rocket speed is slightly greater than before as the rocket mass has been slightly depleted by the prior ejection. Indeed, if the residual rocket mass eventually were minuscule, its boost in speed could reach an enormous value. The integrated effect of these nonlinear boosts is found as

$$v_t - v_0 = v_e \log_e \frac{m_0}{m_t} \tag{13}$$

where v_0 and m_0 are the rocket speed and mass at some arbitrary initial time and v_t and m_t are the same quantities at some time t later.

From these simple considerations, it is apparent that the highest rocket velocities are attained if we could increase the propellant speed as well as the mass ratio m_0/m_t. The mass ratio can be maximized by obvious methods such as choosing a high-density propellant which cuts the tankage requirement or avoiding unnecessarily complicated apparatus for ejecting propellant at high speed. A nuclear rocket, for example, may perform well in its ability to eject propellant an order of magnitude higher in velocity than conventional chemical rockets; nevertheless, the penalty required in reactor weight and shielding severely limits its effectiveness.

Specific impulse is one performance characteristic which applies to the propellant's ability to be ejected at high speed regardless of the weight penalty required to do this. It is the impulse produced per mass of propellant ejected, or $m\,\delta v/\delta m_e$, or, by Eq. (12), simply v_e. In engineering usage, it is impulse per *weight* of propellant ejected, or v_e/g_e where g_e is the acceleration of gravity at the earth's surface. Its units are seconds, and it can be interpreted as the thrust produced by a rocket per weight of propellant ejected per second. By itself, thrust is of little importance unless it is sustained for a significant time by a large backup of propellant tankage. It is here that the mass ratio term in Eq. (13) would play an important role in any evaluation of a rocket's true performance.

Transfer Orbits. If one wishes to leave one orbit and enter another by rocket, an optimum path is generally chosen to minimize the total propellant required. Nevertheless, this should not be done at the expense of unduly long flight times, complicated guidance equipment, or high acceleration stresses. These would require unprofitable weight expenditures which would offset the frugality in propellant tankage. See Fig. 4.

Let us examine a simple but recurring example of a transfer problem, that of leaving a space platform in one circular orbit and entering another larger one concentric with the first. If the transfer path were radial or near radial (a so-called ballistic orbit) then one would have to launch at a large angle to the direction of motion of the platform, accomplished only by a velocity component opposed to the platform's motion. On reaching the outer platform, a soft landing can be made only by a substantial rocket velocity boost tangent to the orbit. Clearly, the total propellant expenditure would be far greater than one alternative of launching the rocket in the direction of motion of the first platform with just sufficient speed to reach the outer circle, timed so that the outer platform will meet the spacecraft. The transfer orbit will be an ellipse cotangent with both circles. The outer platform will be moving much faster of course at the contact point as the major axis of its orbit is much greater [see Eq. (11)], but the difference in speed is not nearly as pronounced as for the ballistic transfer case. A differential speed increment at contact completes the maneuver.

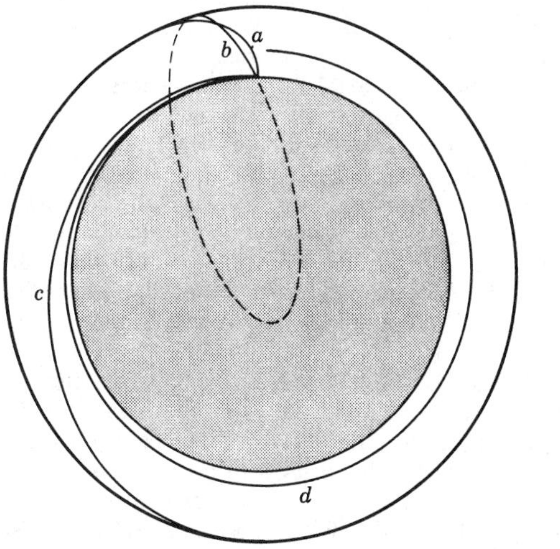

Fig. 4. Four launch trajectories into a satellite orbit about a planet. (a) If that planet has an atmosphere, the rocket may ascend in a "synergic" trajectory from the planetary surface to the final orbit, i.e., it cuts through the denser portions in an initially vertical path and gradually bends over into a horizontal path during burnout. (b) If there is no atmosphere it may ascend from the ground in a ballistic ellipse. This same ascent path may be chosen if the departure is from a parking orbit or "space platform" close to ground level. A far better choice would be (c) the Hohmann ellipse, with pericenter at the planet's surface and apocenter at the satellite orbit. Burnout time is assumed short in both this and the ballistic case. (d) A vehicle such as an ion rocket, which can sustain a microthrust for a very long time, cannot be launched from the ground but only from a parking orbit. It will spiral out to the desired altitude with few or many turns about the planet, depending on the magnitude of the thrust relative to that of the gravitational force.

The return trip, from an outer to inner circle, is made by following the second half of this cotangent ellipse, named the Hohmann transfer orbit after the German engineer who discovered its optimal property with regard to propellant expenditure. In the return case, the spacecraft is launched in opposition to the outer platform's motion. This removes kinetic energy and forces the spacecraft to fall in closer to the attractive center in order to make cotangent contact with the inner circle. The total propellant expenditure from the outer to the inner platform is the same as for the original journey.

An interesting question arises if one wishes to leave a platform for an outer orbit when it initially is in an elliptical orbit rather than a circle. Should we depart from apocenter where we are furthest from the gravitational source and closest to our destination? Or should we depart instead from some other point in the ellipse? Paradoxically, our best launch point is at pericenter, for here the largest possible amount of energy will be transferred to the spacecraft for a given expenditure of propellant. A given thrust applied for a given time interval will do more work on the spacecraft when it is moving fast, as at pericenter, for it covers a greater distance during the interval. This advantage offsets the undesirability of being at a lower potential energy point at pericenter.

Powered Trajectories. In the usual operation of a solid- or liquid-propelled rocket, the propellant is depleted in a time negligibly small compared with the total flight time. The trajectory analysis may generally be considered as that of a free orbit subject to burnout initial conditions as in the discussion above. If, however, the propellant ejection is sustained over long periods, as in an ion-propelled rocket, the trajectory analysis is necessarily complicated, for, in addition to the varying gravitational force, the vehicle, of slowly diminishing mass, is subject to a thrust which may be changing both in direction and magnitude. Even one of the simplest thrust programs, a constant thrust in the direction of motion, requires an electronic computer analysis in order to obtain the position and velocity at future times.

The continuous-thrust trajectory is a spiral with many advantages over the orbital ellipses. First, the lower sustained thrust precludes the high acceleration stresses associated with rapid-burning chemical rockets. Much of the structural weight usually needed to withstand these stresses can be replaced by propellant. Also, flights to the extremities of a gravitational region may take a shorter time in a spiral trajectory. In a long Hohmann ellipse, for example, most of the journey is made at very low speed. In a powered spiral, on the other hand, the spacecraft could be made to move fast, for the thrust, though small, is integrated over many months.

The spiral concept is ideal for rockets where very high ejection velocities are feasible by using electromagnetic or electrostatic particle accelerators, but only at the expense of a low propellant flow rate and relatively heavy power-generating equipment. However, the propellant reserve, and thrust, could then last the required long time. Such an ion rocket with its very low thrust-to-weight ratio could hardly be expected to take off from the ground, and could only take off from an orbital platform. In the vacuum of space, the ion beam meets its ideal environment.

ASTRONOMICAL CLOCK. A clock that indicates astronomical events as well as time. Historically, these clocks were developed to achieve reliable timekeeping by the mechanical simulation of the observable astronomical relationships of celestial bodies. Probably the first such clock was the great Chinese astronomical clock tower of Su Sung (1020–1101) which incorporated the first solarsideral gear and used (on the water wheel that drove the clockworks) an escapement mechanism believed to be the world's first. The first recorded astronomical clock in Europe was produced about 1330 by Richard of Wallingford of the Abbey of St. Albans in Hertfordshire, England. Better known as the "Astrarium," completed in 1364 by Giovanni de Dondi of Chioggia, Italy. An exact reproduction of this device is in the Smithsonian Institution's Museum of History and Technology, Washington, D.C. This clock indicated the movements of the Sun, Moon and the five then-known planets and displayed mean solar time and a perpetual Julian calendar for the Church's movable feasts. The outstanding modern example of an astronomical clock is Jens Olsen's "World Clock" in Copenhagen's City Hall. Detailed visual simulation of complex astronomical events within and beyond the solar system is now far more accurately achieved by planetarium projectors. See **Planetarium.**

William O. Bennett, John J. Carpenter, Frank Dostal, and E. Van Haaften, Bulova Watch Company, Inc., New York.

ASTRONOMICAL DAY SYSTEM. Time.

ASTRONOMICAL HORIZON. Horizon (Astronomical).

ASTRONOMICAL REFRACTION. Refraction (Astronomical).

ASTRONOMICAL TRIANGLE. The spherical triangle formed on the celestial sphere between the observer's meridian, the hour circle, and the vertical circle through an object. Since practically every prob-

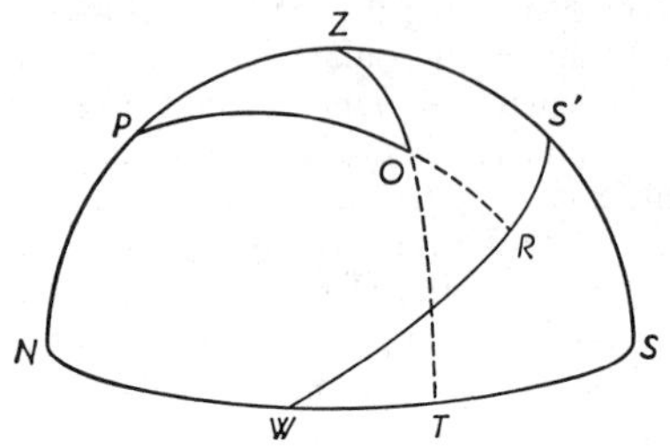

Astronomical triangle.

lem of nautical astronomy deals with the solution of this triangle, it will be well to have the definitions of the various parts well in mind. In the figure (drawn for an observer in the northern hemisphere), we have:

N, *W*, and *S*, the north, west, and south points of the observer's horizon.
S', *R*, *W*, the celestial equator.
Z, the observer's astronomical zenith.
P, the north pole of rotation.
O, the object under consideration.
PZO, the astronomical triangle, the various parts of which are defined as follows:

The vertex angles are:

ZPO = the hour angle of the object.
PZO = 180°—the astronomical azimuth of the object.
ZOP = the parallactic angle.

The sides (measured in angular units) are:

PZ = 90°—observer's astronomic latitude.
ZO = 90°—altitude of the object = zenith distance of the object.
PO = 90°—declination of the object.

See also **Celestial Sphere.**

ASTRONOMICAL UNIT. A unit of distance principally employed in expressing distances within the solar system, but also used to some extent for measuring interstellar distances. Technically defined, one astronomical unit is the mean distance of the earth from the sun. To express this in linear units, it becomes necessary to determine the distance of the earth from the sun in the units chosen or, in other words, to determine the solar parallax. The value for the length of the astronomical unit is 1.495985×10^8 kilometers (92.956×10^6 miles), and was obtained by radar astronomy. See also **Light Year;** and **Parsec.**

ASTRONOMY. Like many sciences, astronomy has become much more interdisciplinarian during the past several decades. As specialized areas of investigation grow, there is a tendency to create either subfields under the general umbrella of astronomy; or for other sciences to encroach into the classical realms of astronomy and create so-called *astro* branches. Fortunately, these divisive as well as melding actions create but few problems, most of which are confined to matters of syntax and definition.

If one takes the broad viewpoint today, astronomy could be defined as the science of matter in outer space dealing with the composition, energy, mass, relative position, size, and other chemical and physical properties of celestial bodies, including asteroids, clusters, comets, galaxies, meteoroids, natural satellites, nebulae, planets, stars, intervening space, and other forms of cosmic material and phenomena.

Several score alphabetical entries in this volume pertain to astronomical topics as indicated by the accompanying table.

ASTRONOMICAL TOPICS DESCRIBED IN THIS VOLUME

Astronomy and Related Disciplines

Astronometry
Astronomy
Astrophysics
Cosmogony
Cosmology

Catalogs and Directories

Almagest
Almanac (Astronomical)
Bonner Durchmusterung
Star Catalogue
Star Finder

Constellations

Andromeda
Aquarius (the water bearer)
Aquila (the eagle)
Aries (the ram)
Auriga (the charioteer)
Bootes (the herdsman)
Camelopardalis (the giraffe)
Cancer (the crab)
Canis Major (the great dog)
Canis Minor (the little dog)
Capricornus (the sea-goat)
Carina
Cassiopeia (the chair)
Centaurus (the centaur)
Cepheus
Cetus (the whale)
Chamaeleon
Columba (the dove)
Constellations
Corona Australis (southern crown)
Corona Borealis (northern crown)
Corvus (the crow)
Crater (Constellation)
Crux (southern cross)
Cygnus (the swan)
Draco (the dragon)
Eridanus
Gemini (the twins)
Grus (the crane)
Hercules
Horologium
Hydra (the serpent)
Indus
Leo (the lion)
Lepus (the hare)
Libra (the scales)
Lupus (the wolf)
Lyra (the harp)
Monoceros
Ophiuchus
Orion (the hunter)
Pegasus (the winged horse)
Perseus
Phoenix
Pisces (the fishes)
Piscis Austrinus
Puppis
Reticulum (Constellation)
Sagitta (the arrow)
Sagittarius (the archer)
Scorpius (the scorpion)
Sextans
Taurus (the bull)
Telescopium
Triangulum (the triangle)
Triangulum Australe
Tucana
Ursa Major (the greater bear)
Ursa Minor (the smaller bear)
Vela
Virgo (the virgin)

Coordinate Systems, Motions, Paths, Positions, Forces, Units

Antisolar Point
Aphelion
Apparent Horizon (See **Horizon (Celestial)**)
Astronomical Triangle
Astronomical Unit
Autumnal Equinox (See **Equinox**)
Azimuth (Astronomy)
Barycentric Parallax
Bode's Relation
Celestial Mechanics
Celestial Sphere
Conjunction (Astronomy)
Declination
Distance Modulus
Doppler Effects
Eccentricity Correction
Eclipse
Ecliptic
Einstein Shift
Elongation (Astronomy) (See **Conjunction (Astronomy)**)
Equatorial Coordinates (Astronomy)
Equinox
Evection
Galactic Coordinates
Geocentric Coordinates (Astronomy)
Geocentric Parallax
Horizon (Astronomical)
Horizon (Celestial)
Horizontal Coordinate System (Astronomy)
Hour Angle
Inclination
Inertial System (The Primary)
Kepler's Laws of Planetary Motions
Latitude
Librations
Light-Year
Line of Apsides
Line of Nodes
Mach Principle
Meridian
Nadir
Nutation
Occultation
Opposition (Astronomy) (See **Conjunction (Astronomy)**)
Orbit (Astronomy)
Parallax (Astronomy)
Parsec
Perihelion
Perturbation (Astronomy)
Position Angle (Stellar)
Precession (Astronomy)
Quadrature (Astronomy)
Relativity and Relativity Theory
Retrograde Motion
Secular Parallax
Secular Terms
Solar Parallax
Solstice
Sphere of Influence (Planet)
Syzygy

(*Continued*)

Coordinate Systems, Motions, Paths, Positions, Forces, Units

Three-Body Problem (Astronomy)
Transit (Astronomy)
Two-Body Problem (Astronomy)
Vertical Circle
Zenith
Zodiac

Meteorites and Comets

Ablation (Meteorite)
Achondrites
Aerolite
Astrobleme
Bennett (Comet)
Bielids
Bolide
Chondrite
Comet
Earthpoint
Encke (Comet)
Fireball
Halley's Comet
Kohoutek (Comet)
Leonids
Lyrids
Meteor
Meteorite
Meteorite Crater
Meteoroid
Meteor Shower
Meteor Trail
Micrometeorite
Perseids
Poynting-Robertson Effect
Radiant Point (Meteor)
Shooting Star
Tektite

Planets and Planetoids

Albedo (Astronomy)
Asteroid
Ceres (See **Asteroid**)
Davida (See **Asteroid**)
Earth
Earthlight
Eros
Eunomia (See **Asteroid**)
Gegenschein
Heat Balance (Planet)
Hektor (See **Asteroid**)
Jovian
Jupiter
Mars
Mercury (Planet)
Moon (The)
Neptune
Pallas (See **Asteroid**)
Planets and the Solar System
Pluto
Satellite (Planetary)
Saturn
Uranus
Venus
Vesta (See **Asteroid**)

Stars and Galaxies

Specific Stars and Galaxies
Achernar (Alpha Eridani)
Adhara (Epsilon Canis Majoris)
Aldebaran (Alpha Tauri)
Algol (Beta Persei)
Alpha Centauri
Alpha Crucis
Alpheratz (Alpha Andromedae)
Altair (Alpha Aquilae)
Antares (Alpha Scorpii)
Arcturus (Alpha Bootes)
Bellatrix (Gamma Orionis)
Beta Centauri
Beta Crucis
Betelgeuse (Alpha Orionis)
Canopus (Alpha Carinae)
Capella (Alpha Aurigae)
Castor (Alpha Geminorum)
Fomalhaut (Gamma Piscis Austrini)
Hyades
Milky Way
Mira
Mizar (Zeta Ursae Majoris)
Pleiades
Polaris (Alpha Ursae Minoris)
Pole Star (See **Vega (Alpha Lyrae)**)
Pollux (Beta Geminorum)
Praesepe (See **Cancer (The Crab)**)
Procyon (Alpha Canis Minoris)
Regulus (Alpha Leonis)
Rigel (Beta Orionis)
Shaula (Lambda Scorpi)
Sirius (Alpha Canis Majoris)
Spica (Alpha Virginis)
Sun
Vega (Alpha Lyrae)
Virgo Cluster

Types of Stars and Galaxies
Binary Stars
Black Hole
Cepheids
Cluster (Galactic)
Double Star
Eclipsing Binary
Flare Stars
Galaxies
Giant and Dwarf Stars
Magellanic Clouds
Navigators' Stars
Nebula
Neutron Stars
Nova and Supernova
Planetary Nebula (See **Nebula; Lyra (The Harp)**)
Quasars
Radio Pulsars
Ring Nebula (See **Nebula; Lyra (The Harp)**)
Seyfert Galaxy (See **Galaxy**)
Spectral Class (See **Star**)
Spectroscopic Binaries
Spiral Galaxy (See **Andromeda; Galaxy**)
Star
Universe (See **Cosmology**)
Variable Star
Visual Binaries

Stellar and Galactic Properties and Phenomena
Bailey Beads
Color Index
Color Temperature (Stellar)
Effective Temperature (Astrophysics)
Einstein Shift
Green Flash
Insolation
Interstellar Molecules
Interstellar Reddening
Limb Darkening
Mass-Luminosity Relation
Period Luminosity Law
Proper Motion (Star)
Purple Light
Radial Velocity (Star)
Red Shift
Scintillation (Astronomical)
Space Velocity (Stellar)
Spectral Equivalent Width (Stellar)
Stellar Associations

(Continued)

Stars and Galaxies

Stellar Curve of Growth
Stellar Heat Index
Stellar Luminosity
Stellar Magnitude
Stellar Parallax
Sunshine (See **Solar Energy**)
Talcott Pairs (See **Zenith Telescope**)
Twiligth
Velocity Curve (Stellar)
Zodiacal Light

Telescopes and Other Astronomical Instrumentation Systems

Alidade (See **Astrolabe**)
Altazimuth
Angular Measurement (Eccentricity Correction)
Artifical Horizon
Astrographic Telescope
Astrolabe
Augmentation
Bolometer (See **Stellar Heat Index**)
Catadioptic System (See **Maksutov-Bouwers Telescope**)
Coronagraph (See **Sun**)
Coudé
Dynamical Parallax
Eccentricity Correction
Filar Micrometer
Galilean Telescope
Gamma-Ray Astronomy
Gauss Eyepiece
Gregorian Telescope
Heliostat
Infrared Astronomy
Maksutov-Bouwers Telescope
Meridian Circle
Newtonian Telescope
Objective Prism
Photometry (See **Stellar Magnitude**)
Radio and Radar Astronomy
Ramsden Circle
Refraction (Astronomical)
Resolving Power (Telescope)
Ritchey-Chretien Telescope
Schmidt Objective
Schwarzschild Telescope
Seeing (Astronomy)
Semidiameter Correction (Sextant)
Spectroheliograph (See **Sun**)
Spectroscopic Parallax
Telecentric System
Telescope (Optical-Astronomical)
Ultraviolet Astronomy
X-Ray Astronomy
Zenith Telescope

Space-Exploration Programs
Apollo Program (See **Moon**)
Pioneer Missions to Planets
Viking Mission to Mars
Voyager Mission to Jupiter and Saturn

Time-Related Topics
Apparent Solar Time (See **Time**)
Apparent Sidereal Time (See **Time**)
Astronomical Clock
Atomic Time (See **Time**)
Calendar
Coelostat
Ephemeris Time (ET) (See **Time**)
Equation of Time
Heliostat (See **Sun**)
Mean Solar Time (See **Time**)
Saros
Sidereal Period
Synodic Period
Time
Universal Time (UT) (See **Time**)

General References

Note: See also lists of topical-related references at end of many of the entries on astronomy.

Audouze, J. (editor): "CNO Isotopes in Astrophysics," Reidel, Boston, 1977.
Brandt, J. C., and S. P. Maran: "New Horizons in Astronomy," Freeman, San Francisco, 1979.
Brecher, K., and M. Feirtag (editors): "Astronomy of the Ancients," MIT Press, Cambridge, Massachusetts, 1979.
Delsemme, A. H. (editor): "Comets, Asteroids, Meteorites," Univ. of Toledo, Toledo, Ohio, 1977.
Eddy, J.: "Probing the Mystery of the Medicine Wheels," *Nat. Geog.*, **151**, 1, 140–146 (1977).
Espenak, F.: "North American Solar Eclipses: The Next 50 Years," *Astronomy*, **8**, 11, 44–50 (1980).
Gingerich, O.: "The Basic Astronomy of Stonehenge," *Technology Review* (*MIT*), **80**, 2, 64–73 (1977).
Hartmann, W. K.: "Astronomy: The Cosmic Journey," Wadsworth, Belmont, California, 1978.
Knox, R.: "Foundations of Astronomy," Wiley, New York, 1979.
Kopal, Z.: "The Realm of the Terrestrial Planets," Wiley, New York, 1979.
Krupp, E. C.: "In Search of Ancient Astronomies," McGraw-Hill, New York, 1979.
Lang, K. R., and O. Gingerich, Editors: "Source Book in Astronomy and Astrophysics (1900–1975)," Harvard Univ. Press, Cambridge, Massachusetts, 1980.
Martin, P. G.: "Cosmic Dust," Oxford Univ. Press, New York, 1979.
Murray, B., Gulkis, S., and R. E. Edelson: "Extraterrestrial Intelligence: An Observational Approach," *Science*, **199**, 485–492 (1978).
Ney, E. P.: "Star Dust," *Science*, **195**, 541–546 (1977).
Pasachoff, J. M.: "Astronomy: from the Earth to the Universe," Saunders, Philadelphia, 1979.
Tver, D.: "Dictionary of Astronomy, Space, and Atmospheric Phenomena," Van Nostrand Reinhold, New York, 1979.
Wald, R. M.: "Space, Time, and Gravity," Univ. of Chicago Press, Chicago, 1977.
Warner, D. J.: "The Sky Explored: Celestial Cartography (1500–1800)," A. R. Liss, Inc., New York (1980).

ASTRONOMY (Neutrino). Neutrino.

ASTROPHYSICS. Commencing with the advent of photography and the study of stellar spectra in the second half of the nineteenth century, astrophysics now includes optical and radio observations of stars, clusters, interstellar material, galaxies and clusters of galaxies, and their interpretations. Radiation from these external sources provides information on the direction of the source, its velocity, composition, temperature, and other physical conditions, including magnetic fields, density, degree of ionization, and turbulence. The term *astrophysics* is generally understood to include all these aspects except the measurement of direction (positions of stars in the sky and changes due to parallax and proper motion), and the orbits of planets, asteroids and comets (celestial mechanics). Because of its proximity, the sun can be studied in more detail than other stars (solar physics); its structure and its influence on the nearby planets are closely related to geophysics and stellar astrophysics. Study of the motions of stars in pairs, groups, clusters, associations, and galaxies is the overlap of celestial mechanics with astrophysics, and the study of the distribution and patterns of motion of the distant galaxies is the overlap with cosmology.

ASYMMETRIC CARBON ATOM. Amino Acids.

ASYMMETRIC TOP. A model of a molecule which has no threefold or higher-fold axis of symmetry, so that during rotation all three principal moments of inertia are in general different. Examples are the water molecule and the ethylene molecule.

ASYMMETRY. Conservation Laws and Symmetry.

ASYMMETRY (Chemical). Asymmetry involves the presence of four different atoms or substituent groups bonded to an atom. Its existence was discovered in 1815 by the French physicist, J. B. Biot

(1774–1867). Biot found that oil of turpentine and solutions of sugar, camphor, and tartaric acid all rotate the plane of plane-polarized light when placed between two Nicol prisms. This phenomenon is called *optical rotation* and is indicated in symbols, such as: $[\alpha]_D^{20°} = +53.4$ aq., signifying that the substance gives a rotation of 53.4° to the right (clockwise, or plus) in water solution at 20°C using sodium D line as the light source. Substances in solution that rotate light to the right are designated *d* and are called *dextrorotatory*; substances rotating light to the left are designated *l* and are called *levorotatory.* See also **Isomerism.**

ASYMMETRY POTENTIAL. The potential difference between the outside and the inside surface of a hollow electrode (usually a glass electrode).

ASYMPTOTE. The limiting position of a tangent to a curve, where the point of contact is only at an infinite distance from the origin. Where there are no infinite branches, as in the cases of the circle and the ellipse, there is no real asymptote.

Suppose the equation of a given curve can be expanded in a power series

$$y = f(x) = \sum_{k=0}^{n} a_k x^k + \sum_{k=1}^{\infty} b_k / x^k = S_1 + S_2.$$

Then, if $\lim_{x\to\infty} S_2 = 0$, the equation of the asymptote is $y = S_1$. If this equation is linear, the asymptote is a straight line; otherwise, it is a more complicated curve. In the linear case, the equation of the asymptote may be written as

$$y = mx + b; \quad m = \lim_{x\to\infty} f'(x)$$

$$b = \lim_{x\to\infty} [f(x) - xf'(x)]$$

For related topical coverage in this volume, see list of entries under **Mathematics.**

ASYMPTOTIC DIRECTION. Conjugate Directions (At a Point P on a Surface).

ASYMPTOTIC RELATIVE EFFICIENCY (or ARE). The efficiency of an estimator of a statistical parameter (as compared with an optimal estimator) as the sample size on which the estimator is based tends to infinity.

ASYMPTOTIC SERIES. A divergent series of the form

$$A_0 + A_1/x + A_2/x^2 + \cdots + A_n/x^n + \cdots$$

It is an asymptotic representation of a function $f(x)$ if

$$\lim_{x\to\infty} x^n[f(x) - S_n(x)] = 0$$

for any value of n, where S_n is the sum of the first $(n + 1)$ terms of the series.

A familiar example of an asymptotis series is the Euler-Maclaurin formula, which converges for a certain number of terms and then begins to diverge. If one includes a large number of terms in this formula the successive derivatives become increasingly larger in the numerator and they increase much more rapidly than the coefficients, which occur in the denominator. However, if the summation is stopped with the term just before the smallest and not with the smallest term, the error is usually about twice the neglected term. Thus one can obtain satisfactory results in this case and with other such series when they are used with caution. Other examples are the logarithmic integral and the gamma function, both of which can be developed as asymptotic series.

An asymptotic expansion is unique; that is, a given function can be represented by only one such series. It may be integrated, two or more of them can be multiplied together, but in general it should not be differentiated.

For related topical coverage in this volume, see list of entries under **Mathematics.**

ASYNCHRONOUS. This is a term used to designate the property of a device or action whose timing is not a direct function of the clock cycles in the system. In an asynchronous situation, the time of occurrence or duration of an event or operation is unpredictable due to factors such as variable signal propagation delay or a stimulus which is not under control of the computer. See also **Synchronous.**

In terms of a computer channel, an asynchronous channel does not depend upon the computer clock pulses to control the transmission of information to and from the input or output device. Transmission of the information is under the control of interlocked control signals. Thus, when a device has data to send to the channel, the device activates a service request signal. Responding to this signal, the channel activates a "service out" signal. The latter, in turn, activates a "service in" signal in the device and also deactivates the request signal. Information then is transferred to the channel in coincidence with "service in" and the channel acknowledges receipt of the data by deactivating "service out."

Asynchronous operation also occurs in the operation of analog-to-digital subsystems. The system may issue a command to the subsystem to read an analog point and then proceed to the next sequential operation. The analog subsystem carried out the A/D conversion. When the conversion is complete, the subsystem interrupts the system to signal the completion.

For related topical coverage in this volume, see list of entries under **Data Processing.**

Asynchronous also has a broader meaning—specifically unexpected or unpredictable occurrences with respect to a program's instructions.

ATACAMITE. This mineral is a basic chloride of copper corresponding to formula $Cu_2Cl(OH)_3$. Crystallizes in thin, orthorhombic prisms, may occur massive. Hardness, 3–3.5; sp gr, 3.76–3.78; luster, adamantine to vitreous; color, green, streak, green; transparent to translucent.

It is a secondary mineral found associated with malachite and cuprite; originally found at Atacama, Chile, whence its name. Other localities are Bohemia, South Australia, and in the United States in Arizona, Utah, and Wyoming. See also **Cuprite;** and **Malachite.**

ATAVISM. The appearance through heredity of characters which have not been developed in the parents of the organism in question. The strict meaning of the word is the reappearance of grandparental characters, but it has been used also to designate the reappearance of characters from more remote generations.

ATAXIA. Lack of muscular coordination due to disease of the brain and nervous system, particularly the cerebellum or spinal cord. Occurs in cerebral palsy. Degeneration of portions of the spinal cord in later uncontrolled stages of syphilis (*neurosyphilis*) will cause loss of coordination of the limbs (*locomotor ataxia*), much less frequently seen where there is an active public health program directed to the detection, treatment, and prevention of venereal diseases.

ATAXIC. A term applied by Keyes, in 1901, to all unstratified ore deposits in contradistinction to sedimentary, stratified or eutaxic ore deposits.

ATELECTASIS. Collapse of part, or the whole, of a lung. This may be congenital, as in the stillborn infant whose lungs have never been expanded by the act of breathing; more commonly it is acquired, resulting from obstruction to a bronchus by a mucous plug, especially after surgical operations; occasionally by pressure from without as from bony deformity or tumor growth. Atelectasis is a prominent feature of adult respiratory distress syndrome (ARDS).

ATHABASCA OIL SANDS. Tar Sands.

ATHERMAL TRANSFORMATION. A reaction that occurs without thermal activation. Such a reaction also takes place without diffusion and can occur with great rapidity under the influence of a sufficiently high driving force. The martensite transformation that occurs in steel is primarily athermal, so that the amount of austenite transformed to martensite depends primarily on the temperature to which the steel is cooled and not upon the rate of cooling or the length of time the metal is held at the quenching temperature. It is necessary to note the difference between an isothermal transformation and an athermal transformation. In the former, the reaction occurs at constant temperature and depends, in general, on both diffusion and thermal activation. The transformation of austenite to pearlite can occur isothermally, with carbon atoms diffusing out of the austenite and into the cementite lamellae. See also **Iron Metals, Alloys, and Steels.**

ATHEROGENESIS. Arterial and Venous Disorders.

ATHEROSCLEROSIS. Arterial and Venous Disorders; Ischemic Heart Disease.

ATHLETE'S FOOT. Dermatitis and Dermatosis.

ATHODYD ENGINE. Airplane.

ATLANTIC CEDAR. Cedar Trees.

ATLANTIC COD. Codfishes.

ATLANTIC HERRING. Herring.

ATLANTIC MACKEREL. Mackerels.

ATLANTIC OCEAN (Hydrology). Hydrology.

ATLANTIC OCEAN (Temperatures). Atmospheric-Ocean Interface.

ATLANTIC SALMON. Salmon.

ATLANTIC SUITE. A term proposed by A. Harker, in 1896, for the chemically and structurally related igneous rocks of the Atlantic coast line. Chemically the rocks of this suite are described as alkaline and are represented by such types as granite and its magmatic relatives, as compared with the calc-alkali igneous rocks of the Pacific Suite.

ATLANTIC-TYPE MARGIN. Ocean.

ATLANTO-SCANDIAN HERRING. Herring.

ATMOLYSIS. The separation of a mixture of gases by means of their relative diffusibility through a porous partition, as burned clay. The rates of diffusion are inversely proportional to the square roots of the densities of the gases. Hydrogen, thus, is the most diffusible gas.

ATMOMETER. Precipitation and Hydrometeors.

ATMOSPHERE (Composition). Earth.

ATMOSPHERE (Earth). An envelope (actually a series of envelopes) in the form of imperfect spherical shells of various materials that are bound to the earth by gravitational force. Consisting of gases, vapors, and suspended matter, the total mass of the earth's atmosphere is estimated at approximately 5.1×10^{15} tons, or somewhat less than one-millionth part of the total mass of the earth. One-half of this total mass lies below about 5500 meters (18,000 feet). More than three-fourths of the atmosphere exists below about 10,700 meters (~ 35,000 feet). The composition of the lower layers of the atmosphere is assumed for purposes of most engineering calculations as 76.8% nitrogen and 23.2% oxygen by weight; 79.1% nitrogen and 20.9% oxygen by volume. A more precise composition of this mixture of gases, including minor constituents, is given in entry on **Air.**

The earth's atmosphere extends some 600 to 1500 kilometers into space. Two factors are involved in this great extension of the atmosphere. First, above about 100 kilometers, the atmospheric temperature increases rapidly with altitude, causing an outward expansion of the atmosphere far beyond that which would occur were the temperature within the bounds observed at the earth's surface. Second above this distance, the atmosphere is sufficiently rarefied so that the different atmospheric constituents attain diffusive equilibrium distributions in the gravitational field; the lighter constituents then predominate at the higher altitudes and extend farther into space than would an atmosphere of more massive particles. This effect is enhanced by the dissociation of some molecular species into atoms.

The composition of the atmosphere does not change much up to 100 kilometers; there is a region of maximum concentration of ozone (still a very minor constituent) near 20 to 30 kilometers; the relative concentration of water vapor falls markedly from its average sea-level value up to 10 or 15 kilometers, and the relative abundance of atomic oxygen begins to become appreciable on approaching 100 kilometers, due to photodissociation of oxygen by ultraviolet sunlight. Above 200 kilometers, atomic oxygen is the principal atmospheric constituent for several hundred kilometers. However, helium is even lighter than atomic oxygen, so its concentration falls less rapidly with altitude, and it finally replaces atomic oxygen as the principal atmospheric constituent above some altitude which varies with the sunspot cycle between 600 and 1500 kilometers. At still higher altitudes, atomic hydrogen finally displaces helium as the principal constituent. The hydrogen extends many earth radii out into space and constitutes the telluric hydrogen corona, or *geocorona.*

The temperature of the upper atmosphere, and hence its density, varies with the intensity of solar ultraviolet radiation and this, in turn, varies with the sunspot cycle and with solar activity in general. The solar radio-noise flux is a convenient index of solar activity, since it can be monitored at the earth's surface. The minimum nighttime temperature of the upper atmosphere above 300 kilometers has been expressed in terms of the 27-day average of the solar radio-noise flux at 8-centimeter wavelength. This varies from about 600 K near the minimum of the sunspot cycle to about 1400 K near the maximum of the cycle. The maximum daytime temperature is about one-third larger than the nighttime minimum.

Various properties of the earth's atmosphere are described in Tables 1 through 5 and by Figs. 1 and 2. The several layers of the atmosphere are indicated in Table 1, along with the relationship between atmospheric pressure and altitude. Atmospheric density versus altitude are given in Table 2. Geopotential altitude as related with actual altitude and the acceleration due to gravity is given in Table 3. It is interesting to note that the energy required to lift an object 2 million geometric feet is only 1.824 million times that required to lift it 1 foot above sea level—this because of the decrease in the acceleration due to gravity with altitude.

Reduction of molecular weight, indicating the change in composition of the atmosphere with increasing altitude, is shown in Table 4. The molecular weight of air is assumed essentially constant from sea level up to about 300,000 feet (91,440 meters). At altitudes higher than this, lower molecular weight is largely attributed to the dissociation of oxygen. Above an altitude of about 590,000 feet (179,832 meters), the lower molecular weight is also affected by the diffusive separation and dissociation of nitrogen.

The percent water vapor content of air at saturation versus representative temperatures and pressure altitudes is given in Table 5.

The layers of the earth's atmosphere of interest to meteorologists are the *troposphere* and the *stratosphere.* The troposphere is a thermal atmospheric region, extending from the earth's surface to the stratosphere and characterized by decreasing temperature with height, appreciable vertical wind motion, appreciable water vapor content, and containing nearly all clouds, storms, and pollutants. The thickness of the troposphere varies from as little as about 7–8 kilometers in the cold polar regions to more than 13 kilometers in the warmer, equatorial regions. Temperatures decrease to the interface between the troposphere and stratosphere. This interface is termed the *tropopause.* At the tropopause, polar temperatures average around −55°C, in equatorial regions, −80°C. Above the stratosphere are the *mesosphere* and *ionosphere,* and the outermost layer, the exosphere, gradually fades into the plasma continuum between earth and sun.

In these higher layers of the atmosphere, complex interactions between the fluxes of electromagnetic radiation of various wavelengths and corpuscular radiation from the sun on one side and the low-density concentrations of atmosphere gases on the other side take place. The particulate radiations are also governed by the earth's magnetic field. Radiations of short wavelength cause a variety of photo-

TABLE 1. ATMOSPHERIC PRESSURE VERSUS ALTITUDE ABOVE SEA LEVEL

Atmospheric Layer	Altitude Above Sea Level: feet (thousands)	Altitude Above Sea Level: meters (thousands)	Pressure: inches of mercury	Pressure: millibars
	2000	609.60	7.959×10^{-12}	269.524×10^{-12}
Mesosphere; G	1920	585.22	10^{-11}	3.4×10^{-10}
	1320	402.34	10^{-10}	3.4×10^{-9}
	1000	304.80	5.256×10^{-10}	177.989×10^{-10}
	900	274.32	10^{-9}	3.4×10^{-8}
F	640	195.07	10^{-8}	3.4×10^{-7}
Ionosphere	480	146.30	10^{-7}	3.4×10^{-6}
	400	121.92	10^{-6}	3.4×10^{-5}
	340	103.63	10^{-5}	3.4×10^{-4}
E	300	91.44	10^{-4}	3.4×10^{-3}
	260	79.25	10^{-3}	3.4×10^{-2}
D	200	60.96	10^{-2}	3.4×10^{-1}
	140	42.67	10^{-1}	3.4
Chemosphere	100	30.48	0.32	10.8
	95	28.96	0.4	13.5
	90	27.43	0.5	17.1
	85	25.91	0.64	21.6
	80	24.38	0.81	27.4
Ozonosphere	75	22.86	1.03	34.9
	70	21.34	1.31	44.4
	65	19.81	1.67	56.6
	60	18.29	2.12	71.8
	55	16.76	2.69	91.1
	50	15.24	3.42	115.8
Stratosphere	45	13.72	4.35	147.3
	40	12.19	5.54	187.6
	35	10.67	7.04	238.4
	30	9.14	8.89	301.1
	25	7.62	11.10	375.9
	20	6.10	13.75	465.6
Troposphere	15	4.57	16.89	572.0
	10	3.05	10.58	696.9
	5	1.52	24.90	843.2
	0	0	29.92	1013.3

Conversion factors used: 1 foot = 0.3048 meter

chemical reactions, the most notable of which is the creation of a layer of ozone acting as an effective absorber of solar ultraviolet and thus causing a warm layer at 30 kilometers in the atmosphere. See **Aerosol;** and **Oxygen.** The upper atmosphere, as an absorber of primary cosmic rays, shows many interesting nuclear reactions and is an important natural source of radioactive substances, including tritium and carbon 14 which are used as tracers of atmospheric motions and as criteria of age. See also **Climate;** and **Ozone.**

Most manifestations of weather take place in the troposphere. They are governed by the general atmospheric circulation which is stimulated by the differential heating between tropical and polar zones. The resulting motions in the air are subject to the laws of fluid dynamics on a rotating sphere with friction. They are characterized by turbulence of varying time and space scale. Evaporation of water (see Table 5) from the ocean and its transformation through the vapor state to droplets and ice crystals, forming clouds and precipitation, are important symptoms of the weather-producing forces.

The term *ecosphere* is sometimes used to identify that part of the lower atmosphere where unaided breathing is possible. In meteorology, the term upper atmosphere is sometimes used. That part of the atmosphere above the lower troposphere is called the *upper air,* for which no distinct lower limit is set, but the term is generally applied to the levels above 850 millibars.

The ionosphere is described in a separate entry, **Ionosphere.**

Heat Balance in the Atmosphere

Total heat received directly from the sun, at the outer limits of the atmosphere (the amount that would be received at the earth's surface if passage were unaffected by the atmosphere and clouds) is very nearly 1.94 gram-calories per square centimeter per minute. This great quantity of heat is distributed in such a way that the maximum is received directly below the sun, with a decreasing amount received as the distance from the heat equator increases. It is for this reason, of course, that tropical areas are warm; polar regions are cold.

Not all the sun's radiation is received at the earth's surface. Clouds and snow reflect about 75% of solar radiation falling upon them; land surfaces reflect an average of 10–30%; water reflects varying percentages, from 70% when the sun is only 5° high, to less than 2% when the sun is over 50° above the horizon. Some solar radiation is absorbed by the atmosphere gases and some by water vapor in the air. Another part is lost to the earth by scattering in the atmosphere. Altogether, solar radiation is distributed as follows: (1) approximately 42% is sent back into space by reflection; (2) 15% is absorbed by the atmosphere and its impurities and cloud particles; and (3) 43% is received and absorbed by the earth's surface. On cloudy days (average cloudiness is about 52%), considerably less solar radiation reaches the earth than on clear days. Loss on a clear day is approximately 17% of the total amount; but on a clouded day, the loss is about 78%. Deserts are conspicuously clear, and therefore receive a much larger percentage of the incoming solar heat than do continental west coasts, which have considerable cloud cover. Snow-covered regions lose a larger percentage of their incoming solar heat than do forest- and vegetation-covered lands. Water surfaces, averaged the world over, do not reflect a large percentage of solar heat, but water is capable of absorbing large quantities of heat with only a small temperature change. The influence of local terrain on solar radiation plays a considerable role in determining the daily and seasonal temperatures of that area.

The earth receives its heat from a number of sources: (1) about 17% is direct solar radiation; (2) 10% is sky radiation (from scattered solar radiation); (3) 70% is long-wave radiation received from the atmosphere surrounding the earth; and (4) 3% is received by contact with warm surface air currents. It should be realized, however, that all this energy, regardless of its immediate source, originates from the sun.

The fact that there is no accumulation of heat on the earth indicates a radiative heat balance. Radiation received by the earth is dispersed as follows: (1) 7% goes to space by radiation through transparent bands in the atmosphere (transparent to radiation from a black body at 300°K); (2) 78% goes to the atmosphere by radiation, where it is absorbed and redistributed; and (3) 15% is used in evaporation processes and is carried to the atmosphere, where it adds to the store of atmospheric heat. Water vapor is the principal absorber of earth radiation as it passes through the atmosphere. Carbon dioxide and ozone also have some strong absorption bands.

Those regions between approximately 35°N and 35°S receive more energy than they radiate back to space, whereas the other regions of the earth receive less energy than they radiate. The excess of energy from the subtropical and tropical zones is transferred toward the poles by both the ocean currents and the atmospheric winds. The advection

TABLE 2. ATMOSPHERIC DENSITY VERSUS ALTITUDE ABOVE SEA LEVEL

Altitude Above Sea Level		Specific Weight	
feet (thousands)	meters (thousands)	pounds per cubic foot	kilograms per cubic meter
2000	609.60	1.614×10^{-15}	25.856×10^{-13}
1000	304.80	2.374×10^{-13}	38.031×10^{-13}
100	30.48	0.00101	0.016
95	28.96	0.00129	0.021
90	27.43	0.00166	0.027
85	25.91	0.00214	0.034
80	24.38	0.00275	0.044
75	22.86	0.00350	0.056
70	21.34	0.00445	0.071
65	19.81	0.00566	0.091
60	18.29	0.00720	0.115
55	16.76	0.00915	0.146
50	15.24	0.01164	0.186
45	13.72	0.01480	0.237
40	12.19	0.01883	0.302
35	10.67	0.02370	0.380
30	9.14	0.02861	0.458
25	7.62	0.03427	0.549
20	6.10	0.04075	0.653
15	4.57	0.04812	0.771
10	3.05	0.05648	0.905
5	1.52	0.06590	1.056
0	0	0.07648	1.225

Conversion factors used: 1 foot = 0.3048 meter
1 pound/cubic foot = 16.02 kilograms/cubic meter.

TABLE 4. MOLECULAR WEIGHT OF ATMOSPHERE VERSUS ALTITUDE

Altitude Above Sea Level		
feet (thousands)	meters (thousands)	Molecular Weight
2000	609.6	15.67
1900	579.12	15.80
1800	548.64	15.96
1700	518.16	16.13
1600	487.68	16.33
1500	457.20	16.56
1400	426.72	16.82
1300	396.24	17.14
1200	365.76	17.51
1100	335.28	17.97
1000	304.8	18.54
900	274.32	19.27
800	243.8	20.24
700	213.36	21.59
600	182.88	23.60
500	152.4	24.09
400	121.92	24.76
300	91.44	28.89
0	0	28.97

Conversion factor used: 1 foot = 0.3048 meter.

TABLE 5. PERCENT WATER VAPOR CONTENT OF AIR AT SATURATION VERSUS REPRESENTATIVE TEMPERATURES AND PRESSURE ALTITUDES

Temperature °C	1,000 Millibars 370 Feet (113 Meters)	850 Millibars 4,780 Feet (1,457 Meters)	700 Millibars 9,880 Feet (3,011 Meters)	500 Millibars 18,280 Feet (5,572 Meters)
40	4.97%	5.93%	7.35%	—
30	2.76	3.28	4.03	5.79
20	1.49	1.76	2.16	3.06
10	0.77	0.91	1.12	1.57
0	0.38	0.45	0.55	0.77
−10	0.18	0.21	0.26	0.36
−20	0.08	0.09	0.11	0.16
−30	0.03	0.04	0.05	0.06
−40	0.01	0.01	0.02	0.02

TABLE 3. ACCELERATION DUE TO GRAVITY AND GEOPOTENTIAL ALTITUDE VERSUS ACTUAL ALTITUDE ABOVE SEA LEVEL

Actual Altitude Above Sea Level		Geopotential Altitude		Acceleration Due to Gravity	
feet (thousands)	meters (thousands)	feet (thousands)	meters (thousands)	feet/second/second	meters/second/second
2000	609.6	1825	556.26	26.79	8.17
1800	548.64	1657	505.05	27.26	8.31
1600	487.68	1485	452.63	27.75	8.46
1400	426.72	1310	399.29	28.25	8.61
1200	365.76	1132	345.03	28.77	8.77
1000	304.8	950	289.56	29.3	8.93
800	243.8	766	233.48	29.84	9.1
600	182.88	579	176.48	30.4	9.27
400	121.92	389	118.57	30.97	9.44
200	60.96	196	59.74	31.57	9.62
0	0	0	0	32.17	9.81

Conversion factors used: 1 foot = 0.3048 meter
1 foot/sec/sec = 0.3048 meter/sec/sec.

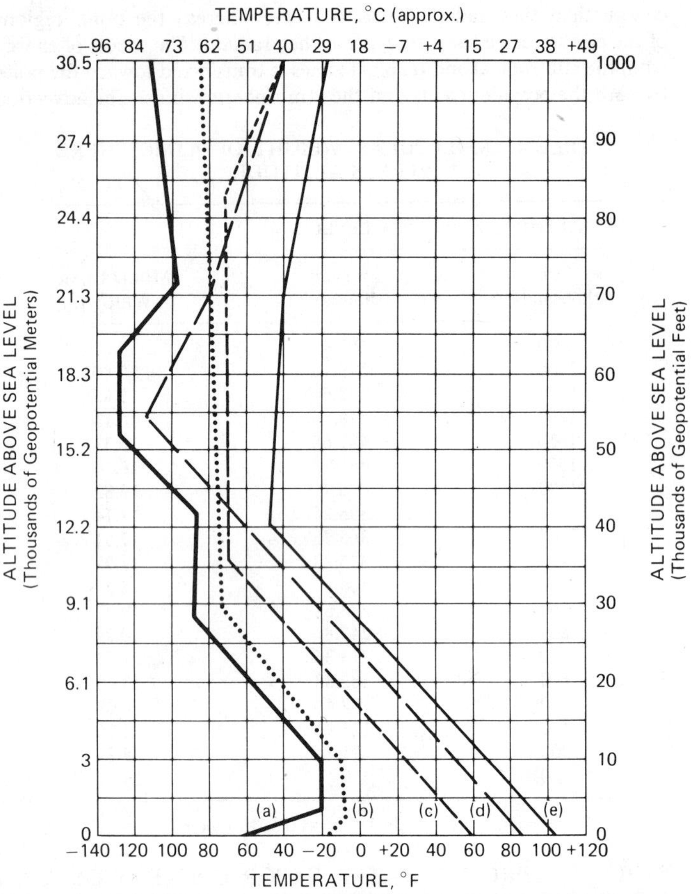

Fig. 1. Relationship between temperature and altitude: (a) cold and (e) hot are the composites of extremes of cold and hot atmospheres; (b) arctic and (d) tropical are the composites of the arctic and tropical regions; (e) is the standard atmosphere upon which altimetry is based.

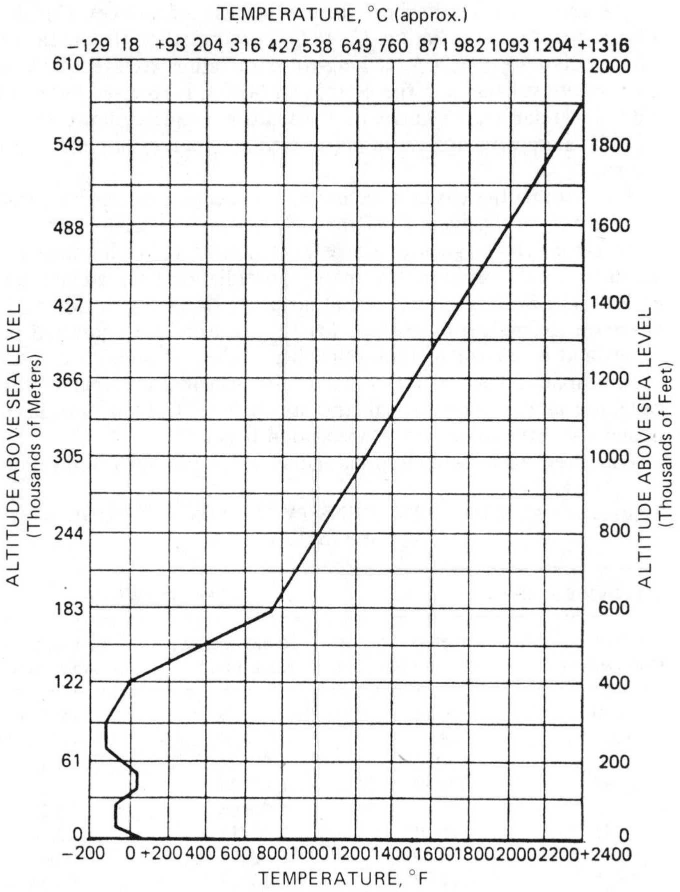

Fig. 2. Real kinetic temperature of the atmosphere, a measure of the kinetic energy of the molecules and atoms constituting the atmosphere, is plotted against altitude above sea level here. Numerical values are determined by the assumed molecular weight of the air (see Table 4), as well as assumed values of the temperature lapse rate.

of heat energy balances the differential in direct radiation. Thus, the average temperature at any point on the earth remains sensibly the same from year to year.

Elasser's radiation chart is one of the better known charts for the graphical solution of the radiative transfer problems important in meteorology. Given a radiosonde record of the vertical variation of temperature and water vapor content, such quantities can be found with this chart as the effective terrestrial radiation, net flux of infrared radiation at a cloud base or a cloud top, and radiative cooling rates. A chart of this type used widely in Europe is the Möller chart.

Atmospheric radiation is infrared radiation emitted by or being propagated through the atmosphere. Atmospheric radiation, lying almost entirely within the wavelength interval of from 3 to 80 micrometers, provides one of the most important mechanisms by which the heat balance of the earth-atmosphere system is maintained. Infrared radiation emitted by the earth's surface is partially absorbed by the water vapor of the atmosphere, which, in turn, re-emits it, partly upward, partly downward. This secondarily emitted radiation is then, in general, repeatedly absorbed and re-emitted, as the radiant energy progresses through the atmosphere. The downward flux, or counterradiation, is of basic importance in the so-called *greenhouse effect*; the upward flux is essential to the radiative balance of the planet.

Terrestrial radiation is defined as the total infrared radiation emitted from the earth's surface—to be carefully distinguished from atmospheric radiation, insolation, and effective terrestrial radiation, the latter being the difference between the outgoing infrared terrestrial radiation of the earth's surface and the downcoming infrared counterradiation from the atmosphere.

In meteorology, the cooling of the earth's surface and adjacent air, accomplished mainly at night, but whenever the earth's surface suffers a net loss of heat due to terrestrial radiation is known as *radiational cooling*. See also **Solar Energy.**

Thermodynamics of the Atmosphere

In meteorological calculations, the ideal gas law is a satisfactory approximation for the derivation of formulas for the mixture of gases that constitute the atmosphere. The derivation is:

$$PV = RT$$

where V = volume
P = pressure
R = universal gas constant
T = absolute temperature

For one gram, this becomes

$$PV = \frac{RT}{m}$$

where m is the molecular weight of the gas. For G grams, this becomes,

$$PV = \frac{GRT}{m}$$

This equation is valid for each of the constituent gases of the atmosphere.

For nitrogen,

$$P_n V = \frac{G_n RT}{m_n}$$

For oxygen,

$$P_0 V = \frac{G_0 RT}{m_0}$$

For argon,

$$P_aV=\frac{G_aRT}{m_a}$$

For water vapor,

$$P_wV=\frac{G_wRT}{m_w}$$

When there is no water vapor present in the atmosphere, these equations can be combined as follows:

$$P_tV=(P_n+P_0+P_a)V=RT\left[\frac{G_n}{m_n}+\frac{G_0}{m_0}+\frac{G_a}{m_a}\right]$$

$$=RT\left(\frac{G_t}{m_t}\right)$$

In these equations, P_t is the total pressure of the nitrogen, oxygen, and argon. Also, G_t is the total mass of the gases; and m_t is the molecular weight of the mixture, with a numerical value of 28.97.

Because water vapor is always present in varying quantities in the atmosphere, corrections in the equation of state must be made in accordance with the amount of water vapor present. Procedure is as follows:

$$PV=(P_t+P_w)V=RT\left[\frac{G_t}{m_t}+\frac{G_w}{m_w}\right]$$

$$=RT\left[\frac{G}{m_t}-\frac{G_w}{m_t}+\frac{G_w}{m_w}\right]$$

In these equations, P is the total pressure of the air gases plus the water vapor, and G is the total mass of the air gases plus the water vapor.

This equation can be rearranged and simplified:

$$PV=RT\frac{G}{m_t}\left[1-\frac{G_w}{G}\left(1-\frac{m_t}{m_w}\right)\right]$$

where m_t has a value of 28.97 and m_w has a value of 18.00. This equation is easily reduced to

$$PV=RT\frac{G}{m_t}\left[1+0.6\frac{G_w}{G}\right]$$

Virtual temperature of the air is defined as

$$T'=T\left[1+0.6\frac{G_w}{G}\right]$$

Virtual temperature is, in effect, the temperature of a mass of dry air having the same density of another mass of air containing water vapor. Virtual temperature is always greater than real temperature, except when G_w is nil.

The equation of state for real air becomes

$$PV=\frac{RT'G}{m_t}$$

If R, the universal gas constant, is made into a specific gas constant for air by letting $R/m_t=R_a$, then for one gram of air, the equation of state becomes $PV=R_aT'$.

The hydrostatic state of equilibrium of the atmosphere varies with the type of atmosphere that is under consideration.

Standard atmosphere is a term used in the following references:

1. A hypothetical vertical distribution of atmosphere temperature, pressure, and density, which, by international agreement, is taken to be representative of atmosphere for purposes of pressure altimeter calibrations, aircraft performance calculations, aircraft and missile design, ballistic tables, etc. The air is assumed to obey the ideal gas law and the hydrostatic equation, which, taken together, relate temperature, pressure, and density variations in the vertical. It is further assumed that the air contains no water vapor, and that the acceleration of gravity does not change with height. The current standard atmosphere was adopted in 1952 by the International Civil Aeronautical Organization (ICAO) and supplants the U.S. Standard Atmosphere prepared in 1925. The parametric assumptions and physical constants used in preparing the ICAO Standard Atmosphere are as follows:

(a) Zero pressure altitude corresponds to that pressure which will support a column of mercury 760 mm high. This pressure is taken to be 1.013250×10^6 dynes/cm^2, or 1013.250 mb, or 101.325 kPa (and is known as one standard atmosphere or one atmosphere).

(b) The gas constant for dry air is 2.8704×10^6 erg/gm K.

(c) The ice point at one standard atmosphere pressure is 273.16 K.

(d) The acceleration of gravity is 980.665 cm/sec^2.

(e) The temperature at zero pressure altitude is 15°C or 288.16 K.

(f) The density at zero pressure altitude is 0.0012250 gm/cm^3.

(g) The lapse rate of temperature in the troposphere is 6.5°C/km.

(h) The pressure altitude of the tropopause is 11 km.

(i) The temperature at the tropopause is −56.5°C.

2. A standard unit of atmospheric pressure; the 45° atmosphere, defined as the pressure exerted by a 760 mm column of mercury at 45° latitude at sea level at temperature 0°C (acceleration of gravity = 980.616 cm/sec^2). One 45° atmosphere equals 760 mm Hg(45°); 29.9213 in. Hg(45°); 1013.200 mb; 101.325 kPa.

Ballistics standard artillery atmosphere is composed of a set of values describing atmospheric conditions on which ballistic computations are based, namely, zero wind, pressure of 1000 millibars at the ground, temperature of 15°C, relative humidity of 78%, and a lapse rate that yields a prescribed density-altitude relationship.

Adiabatic atmosphere is characterized by a dry-adiabatic lapse rate throughout its vertical extent.

Model atmosphere is a term used for any theoretical representation of the atmosphere, with particular reference to vertical temperature and pressure distribution.

Isothermal atmosphere (or exponential atmosphere) is an atmosphere in hydrostatic equilibrium, in which the temperature is constant with height, and in which, therefore, the pressure decreases exponentially upward. In such an atmosphere, the thickness between any two levels is given by

$$Z_B-Z_A=\frac{R_dT_v}{g}\ln\frac{P_A}{P_B}$$

where R_d is the gas constant for dry air, T_v the virtual temperature (°K), g the acceleration of gravity, and P_A and P_B the pressures at the heights Z_A and Z_B, respectively. In the isothermal atmosphere, there is no finite level at which the pressure vanishes.

Polytropic atmosphere is characterized by hydrostatic equilibrium with a constant nonzero lapse rate. The vertical distribution of pressure and temperature is given by

$$\frac{p}{p_0}=\left(\frac{T}{T_0}\right)^{g/R\gamma}$$

where p is the pressure, T the Kelvin temperature, g the acceleration of gravity, R the gas constant for air, and γ the environmental lapse rate, the subscript zeros denoting values at the earth's surface.

Homogeneous atmosphere is a hypothetical atmosphere in which the density is constant with height. The lapse rate of temperature in such an atmosphere is known as the autoconvective lapse rate and is equal to g/R (or approximately 3.4°C/100 meters), where g is the acceleration of gravity and R is the gas constant for air. A homogeneous atmosphere has a finite total thickness given by R_dT_v/g, where R_d is the gas constant for dry air and T_v is the virtual temperature (°K) at the surface. For a surface temperature of 273°K, the vertical extent of the homogeneous atmosphere is approximately 8000 meters. At the top of such an atmosphere, both the pressure and absolute temperature vanish.

With respect to radio propagation, a homogeneous atmosphere is one that has a constant index of refraction, or one in which radio waves travel in straight lines at constant speed. The ideal "homogeneous atmosphere" in this sense is *free space*, which is a perfectly homogeneous medium possessing a dielectric constant of unity, and in which, as in a perfect vacuum, there is nothing to reflect, refract, or absorb energy.

Thermotropic atmosphere, a term used in numerical weather forecasting, is an atmosphere in which the parameters to be forecast are the height of one constant-pressure surface (usually 500 millibars) and one temperature (usually the mean temperature between 1000 and 500 millibars) whereby a surface prognostic chart can also be constructed.

Equivalent barotropic atmosphere is one in which the wind does not change direction with altitude and, therefore, the isotherms and isobars are everywhere parallel.

Barotropic atmosphere is one of a number in which some of the following conditions exist: (1) pressure and temperature surfaces coincide; (2) zero vertical wind shear; (3) zero vertical motion; and (4) zero horizontal velocity divergence.

Baroclinic atmosphere is one in which constant-pressure surfaces intersect constant-density surfaces, thereby creating solenoids which can cause acceleration.

Adiabatic Processes in the Atmosphere

An adiabatic process is a thermodynamic change of state of a system, in which there is no transfer of heat or mass across the boundaries of the system; where compression always results in warming, and expansion in cooling. When a parcel of air is moved from one position to another in such a manner that energy does not flow across the boundaries of the parcel, the thermal changes taking place within the parcel are said to be adiabatic changes.

Dry-adiabatic processes, during which the air involved remains unsaturated, are relatively simple. The first law of thermodynamics applied to a parcel of unsaturated air of unit mass stipulates:

$$dq = c_v\, dT + Ap\, dv$$

which, when combined with the gas equation becomes

$$dq = (c_v + AR)\, dT - \frac{ART}{p}\, dp$$

For the adiabatic process, this becomes

$$\frac{dT}{T} = \frac{AR}{c_p}\frac{dp}{p}$$

which, upon integration, becomes

$$\frac{T}{T_0} = \left(\frac{p}{p_0}\right)^{AR/c_p}$$

Dry-adiabatic horizontal transfer of a parcel from higher to lower or lower to higher pressure is of only minor consequence because of the comparatively small magnitude of pressure change. Dry-adiabatic vertical transfer of a parcel, however, is one of the important meteorological processes. Temperature decrease in a rising, and increase in a sinking, parcel amounts to very nearly 9.8°C per km, or 5.4°F per 1000 feet (304.8 meters). Dew-point changes in a vertically moving unsaturated parcel are considerably less. The dew-point decreases in rising air, and increases in sinking air, at a rate of between 1.3 and 1.8°C per kilometer (0.7 and 1.0°F per 1000 feet), depending upon air temperature.

Pseudo-adiabatic (or *saturation-* or *moist-adiabatic*) *processes* involve condensation or evaporation, and are by no means constant or simple. In a parcel rising pseudo-adiabatically, the temperature decrease is always less than the dry-adiabatic temperature change by an amount depending upon the weight of the water being condensed and the temperature at which condensation occurs. Condensation releases the latent heat of varporization within the parcel, which partially counteracts dry-adiabatic cooling. The rate of cooling in rising saturated air varies from about 2.7°C per kilometer (1.5°F per 1000 feet) in warm air at sea level to 9.7°C per kilometer (5.3°F per 1000 feet) at high altitudes in cold air, a range that is the direct result of the variance in the amount of water resident in a given mass of air at full saturation. Very cold air can retain only a slight amount of water, whereas very warm air can hold relatively large quantities. Values of resident water vapor at saturation range from 0.01% by weight in arctic air to 3% by weight in tropical air.

Sinking saturated air remains saturated only for a comparatively short distance, during which it is heated pseudo-adiabatically at a rate determined by the amount of evaporation occurring within the parcel. As soon as it becomes unsaturated, the sinking parcel descends dry-adiabatically. Foehn winds are examples of both pseudo-adiabatic and dry-adiabatic changes. Air flowing uphill is cooled pseudo-adiabatically until it reaches the hilltop. On the lee side, the air descends dry-adiabatically. Observable results of the true foehn wind are abundant clouds and rain or snow on the windward side of a mountain range, and clear, warm air on the lee side.

Dew-points in saturated air rising pseudo-adiabatically decrease at the same rate as the temperature. Dew-points in saturated sinking air increase at the same rate as the temperature until the air parcel is no longer saturated; then they rise slowly, as previously described in connection with sinking saturated air.

A large percentage of all clouds and nearly all precipitation result from adiabatic ascent of air.

Assuming increasing positive values with altitude, the following relations hold:

1. Dry-adiabatic temperature change with altitude:

$$\frac{\partial t}{\partial h} = \frac{gK}{R} = -9.8°\text{C/km}$$

where t = temperature of parcel
g = gravitational constant
$K = \dfrac{c_p - c_v}{c_p} = .288$
R = gas constant

2. Dry-adiabatic dew-point change with altitude:

$$\frac{\partial t_d}{\partial h} = -1.71\left[1 + \frac{2t_d}{237.3} - \frac{t}{273}\right] °\text{C/km}$$

where t_d = dew-point temperature in °C
t = air temperature in °C

3. Pseudo-adiabatic temperature change with altitude:

$$\frac{\partial t}{\partial h} = -g\left(\frac{A + .621\dfrac{e}{p}\dfrac{L}{Rt}}{c_p + .621\dfrac{L}{P}\dfrac{de}{dt}}\right) \text{C/km}$$

where t = temperature of the parcel in °C
g = gravitational constant
A = heat equivalent of work
e = water vapor pressure
p = air pressure
L = heat of condensation
c_p = specific heat at constant pressure for air

Virtual Temperature of Air. In meteorological calculations, it is often convenient to use, instead of the actual air temperature, the temperature which a parcel of air would have if it had the same density and pressure as the sample in question, but was entirely free from water vapor. Since dry air is denser than water vapor under the same conditions, removal of water vapor from moist air will increase its density, so that the temperature will need to be raised to obtain an equivalent density. Therefore, the virtual temperature is higher than the actual, and is given by:

$$T_v = (1 + 0.61q)T$$

where T is the absolute temperature and q the specific humidity.

Atmospheric Stability, Instability, and Equilibrium

Everywhere that air is in motion, some vertical perturbations are present. Isolated parcels and currents of air are started upwards or downwards into new environments. If the density of the environment is different from the density of the parcel after any modification caused by the change of pressure, the parcel experiences a force of buoyancy which may accelerate or retard the initial displacement. The criterion for static stability of a horizontally stratified compressible fluid is that the gradient of potential density should be negative upwards.

Stability. In meteorology, *static stability* (also called *hydrostatic stability, vertical stability,* or *convectional stability*) is the stability of an atmosphere in hydrostatic equilibrium with respect to vertical displacements, usually considered by the parcel method. The criterion for stability is that the displaced parcel be subjected to a buoyant force opposite to its displacement, e.g., that a parcel displaced upward be colder than its new environment. This is the case if $\gamma < \Gamma$, where γ is the environmental lapse rate and Γ is the process lapse rate, dry-adiabatic for unsaturated air and saturation-adiabatic for saturated air.

Neutral stability (also called *indifferent stability* or *indifferent equilibrium*) is the state of an unsaturated or saturated column of air in the atmosphere when its environmental lapse rate is equal to the dry-adiabatic lapse rate or the saturation-adiabatic lapse rate, respectively. Under such conditions, a parcel of air displaced vertically will experience no buoyant acceleration.

Hydrostatic Equilibrium. The state of a fluid whose surfaces of constant pressure and constant mass (or density) coicide and are horizontal throughout. Complete balance exists between the force of gravity and the pressure force. The relation between the pressure and the geometric height is given by the hydrostatic equation. The analysis of atmospheric stability has been developed most completely for an atmosphere in hydrostatic equilibrium. The hydrostatic equation is the form assumed by the vertical component of the vector equation of fluid motion when Coriolis, earth curvature, frictional, and vertical acceleration terms are considered negligible compared with those involving the vertical pressure force and the force of gravity. Thus,

$$\frac{\partial p}{\partial z} = -\rho g$$

where p is the pressure, ρ the density, g the acceleration of gravity, and z the geometric height.

Instability. The concept of instability is employed in many sciences. It is, in general, a property of the steady state of a system such that certain disturbances or perturbations introduced into the steady state will increase in magnitude, the maximum perturbation amplitude always remaining larger than the initial amplitude. The method of small perturbations, assuming permanent waves, is the usual method of testing for instability; unstable perturbations then usually increase exponentially with time. In meteorology, the small perturbations, may be a wave or a parcel displacement. The parcel method assumes that the environment is unaffected by the displacement of the parcel. The slice method has occasionally been used as a modification of the parcel method to gain a little information about the interaction of parcel and environment.

Absolute instability is the state of a column of air in the atmosphere when it has a superadiabatic lapse rate, i.e., greater than the dry-adiabatic lapse rate. An air parcel displaced vertically would be accelerated in the direction of the displacement. The kinetic energy of the parcel would consequently increase with the increasing distance from its level of origin.

Baroclinic instability arises from the existence of a meridional temperature gradient (and hence of a thermal wind) in an atmosphere in quasigeostrophic equilibrium and possessing static stability.

Barotropic instability arises from certain distributions of vorticity in a two-dimensional nondivergent flow. This is an *inertial instability* in that kinetic energy is the only form of energy transferred between current and perturbation. The variation of vorticity, i.e., shear, in the basic current may be concentrated in discontinuities of the horizontal wind shear (to be distinguished from *Helmholtz instability,* where the velocity itself is discontinuous) or may be continuously distributed in a curved velocity profile. A well-known necessary condition for barotropic instability is that the vorticity must change sign, i.e., vanish, at a point of maximum shear.

Colloidal instability is a property attributed to clouds (regarded in analogy to colloidal systems or aerosols) by virtue of which the particles of the cloud tend to aggregate into masses large enough to precipitate.

Conditional instability is the state of a column of air in the atmosphere when its lapse rate is less than the dry-adiabatic lapse rate but greater than the saturation-adiabatic lapse rate. With reference to the vertical displacement of an air parcel, the air will be unstable if saturated and stable if unsaturated.

Convective instability (or *potential instability*) is the state of an unsaturated layer or column of air in the atmosphere whose wet-bulb potential temperature, or equivalent potential temperature decreases with elevation.

Gravitational instability occurs in a system in which buoyancy or reduced gravity is the only restoring force on displacements.

Helmholtz instability (also called *shearing instability*) arises from a shear, or discontinuity, in current speed at the interface between two fluids in two-dimensional motion. The perturbation gains kinetic energy at the expense of that of the basic currents.

Hydrodynamic instability (or *dynamic instability*) refers to instability of parcel displacements or, more usually, of waves in a moving fluid system governed by the fundamental equations of hydrodynamics. The space scale of unstable waves is important in meteorology; thus Helmholtz, baroclinic, and barotropic instability give, in general, unstable waves of increasing length. The time scale is also important; a perturbation that grows for two days before dying out is effectively unstable for many meteorological purposes, but this is an initial value problem, and one cannot assume the existence of permanent waves. These meteorological types of hydrodynamic instability must not be confused with the phenomenon often referred to by mathematicians and physicists by the same term. A great deal of study has been devoted to the problem of the onset of turbulence in simple flows under laboratory conditions, and here viscosity is a source of instability. This is not the case in any meteorological motion yet investigated.

Inertial instability (or *dynamic instability*) is, generally, instability in which the only form of energy transferred between the steady state and the disturbance is kinetic energy. More specifically, it is the instability arising in a rotating fluid mass when the velocity distribution is such that the kinetic energy of a disturbance grows at the expense of kinetic energy of the rotation.

Rotational instability is usually synonymous with inertial instability, being, in general, any instability of a rotating fluid system.

Static instability (or *hydrostatic instability*) refers to instability of vertical displacements of a parcel in a fluid in hydrostatic equilibrium.

Thermal instability results in free convection in a fluid heated at a boundary. For the case of heating from below, the onset of convection (as opposed to conduction) is determined by a critical value of the Rayleigh number, and the linear theory admits of various convection-cell forms, including the hexagonal Benard cell. The theories of thermal instability, which are represented by sixth-order convection equations and take into account both viscosity and conductivity, must be distinguished from the theory of static instability, based on the oscillations of a parcel in an atmosphere in hydrostatic equilibrium.

Lapse Rate. This is the rate at which temperature decreases or lapses with altitude; the vertical temperature gradient. Since temperature normally decreases with altitude in the troposphere, it is convenient to assign positive values to the rate of temperature change with altitude: Lapse rate, therefore, is defined as the rate of change of temperature with altitude, and is positive when the temperature decreases. The term applies ambiguously to the environmental lapse rate and the process lapse rate (defined below), and the meaning must often be ascertained from the text.

Autoconvective lapse rate (or *autoconvection gradient*). The environmental lapse rate of temperature in an atmosphere in which the density is constant with height (homogeneous atmosphere), equal to g/R, where g is the acceleration of gravity and R the gas constant. For dry air, the autoconvective lapse rate is approximately $+3.4 \times 10^{-4}$ °C per centimeter.

Undisturbed air will remain stratified even though the lapse rate exceeds the adiabatic rate of 9.8°C per kilometer (5.5°F per 1000 feet). If, however, the lapse rate becomes sufficiently large, density of the air will increase with altitude and will overturn. This critical lapse rate is 34.17°C per kilometer (or nearly 19°F per 1000 feet). Dust devils and whirlwinds result from this steep lapse rate, which occurs at ground levels, particularly over concrete roads and the sand and rock of deserts during the heat of day.

Dry-adiabatic lapse rate. A special process lapse rate (defined below) of temperature, being the rate of decrease of temperature with height of a parcel of dry air lifted adiabatically through an atmosphere in hydrostatic equilibrium. This lapse rate is g/C_{pd}, where g is the acceleration of gravity and C_{pd} is the specific heat of dry air at constant pressure; numerically equal to 9.767°C per kilometer (about 5.4°F per 1000 feet). Potential temperature is constant with height in an atmosphere with this lapse rate.

Environmental lapse rate. The rate of decrease of temperature with elevation, $-\partial T/\partial z$, or occasionally, $\partial T/\partial p$, where p is pressure. The concept may be applied to other atmospheric variables (e.g., lapse rate of density) if these are specified. The environmental lapse rate is determined by the distribution of temperature in the vertical at a given time and place, and should be carefully distinguished from the process lapse rate (defined below), which applies to an individual air parcel.

Process lapse rate. The rate of decrease of the temperature of an air parcel as it is lifted, $-dT/dz$, or occasionally dT/dp, where p is pressure. The concept may be applied to other atmospheric variables, e.g., the process lapse rate of density. The process lapse rate is determined by the character of the fluid processes and should be carefully distinguished from the environmental lapse rate, which is determined by the distribution of temperature in space. In the atmosphere, the process lapse rate is usually assumed to be either the dry-adiabatic lapse rate or the saturation-adiabatic lapse rate (defined below).

Saturation-adiabatic lapse rate (or *moist-adiabatic lapse rate*). A special case of process lapse rate, defined as the rate of decrease of temperature with height of an air parcel lifted in a saturation-adiabatic process through an atmosphere in hydrostatic equilibrium. Owing to the release of latent heat, this lapse rate is less than the dry-adiabatic lapse rate, and the differential equation representing the process must be integrated numerically. Wet-bulb potential temperature is constant with height in an atmosphere with this lapse rate.

Superadiabatic lapse rate. An environmental lapse rate greater than the dry-adiabatic lapse rate, such that potential temperature decreases with height.

Atmospheric Inversion. This is the abnormal condition in which the temperature of the atmosphere increases with height. The term is also used for a level at which the vertical gradient of temperature changes sign. Dynamically, more importance attaches to the vertical distribution of potential temperature or potential density, and it is common to refer to the boundary between a lower region of negative gradient of potential density and an upper one of positive gradient as an inversion, whether or not the temperature gradient changes sign. An example is the tropopause.

In meteorology, *temperature inversion* refers to an atmospheric condition in which temperature increases with altitude. The temperature below the stratosphere normally decreases with altitude; thus, when it increases, normal conditions are inverted, and an inversion has occurred. Inversions in the troposphere are usually restricted to shallow layers of air, which most frequently occur in the lower 5000 feet (1524 meters) above the surface. In low latitudes, the stratosphere has a slight inversion more or less permanently.

The principal characteristics of an inversion layer is its marked static stability, so that very little turbulent exchange can occur within it. Strong wind shears often occur across inversion layers, and abrupt changes in concentrations of atmospheric particulates and atmospheric water vapor may be encountered on ascending through the inversion. When, in meteorological literature and discussion, an "inversion" is mentioned, a temperature inversion is usually meant. The following are particular types of temperature inversion:

Frontal inversion is encountered upon vertical ascent through a sloping front (or frontal zone).

Subsidence inversion is produced by the adiabatic warming of a layer of subsiding air and is enhanced by vertical mixing in the air layer below the inversion.

Surface inversion (or *ground inversion*) is a temperature inversion based at the earth's surface; that is, an increase of temperature with height beginning at the ground level. This condition is due primarily to greater radiative loss of heat at and near the surface than at levels above. Thus, surface inversions are common over land prior to sunrise and, in winter, over high-latitude continental interiors.

Trade-wind inversion (or *trade inversion*) is usually present in the trade-wind streams over the eastern portions of the tropical oceans. It is formed by broad-scale subsidence of air from high altitudes in the eastern extremities of the subtropical highs. While descending, the current meets the opposition of the low-level maritime air flowing equatorward. The inversion forms at the meeting point of these two strata, which flow horizontally in the same direction.

Circulation in the Atmosphere

The worldwide pattern of air movement is related to the global pressure belts, the rotation of the earth, the distribution of temperature over the earth, friction between the earth and the atmosphere, and the location of mountains and oceans. The major large-scale patterns of movement of the atmosphere include:

1. *General circulation,* which in the broadest sense is a complete description of atmospheric motions over the earth. These data are generated from the day-to-day patterns of flow and describe temporal as well as the mean spatial conditions, plus their variability in time and space.

2. *Planetary Circulation.* Refers specifically to (a) the system of large-scale disturbances in the troposphere when viewed on a hemispheric or world-wide scale, and (b) to the mean or time-averaged hemispheric circulation of the atmosphere; in this sense, almost synonymous with general circulation.

3. *Primary Circulation.* The prevailing fundamental atmospheric circulation on a planetary scale that must exist in response to (a) radiation differences with latitude, (b) the rotation of the earth, (c) the particular distribution of land and ocean; and which is required from the viewpoint of conservation of energy. Primary circulation and general circulation are sometimes taken synonymously. They may be distinguished, however, on the basis of approach. That is, primary circulation is the basic system of winds, of which the secondary and tertiary circulations are perturbations; while general circulation encompasses at least the secondary circulations. The latter dimension of circulation has features of cyclonic scale, while tertiary circulation is represented by such phenomena as local winds, thunderstorms, and tornadoes.

4. *Meridional Cell.* A very large-scale convection circulation in the atmosphere or ocean that takes place in a meridional plane, with northward and southward currents in opposite branches of the cell, and upward and downward motion in the equatorward and poleward ends of the cell.

5. *Zonal Flow* (or *Zonal Circulation*). The flow of air along a latitude circle; more specifically, the latitudinal (east or west) component of existing flow.

6. *Polar Vortex.* A large-scale cyclonic circulation, centered generally in the polar regions. Specifically, the vortex has two centers in the mean, one near Baffin Island and another over northwest Siberia. In the southern hemisphere, there is one center near the South Pole.

7. *Westerlies.* Specifically, the dominant west-to-east motion of the atmosphere, centered over the middle latitudes of both hemispheres. Generally, any winds with components from the west.

8. *Easterlies.* Any winds with components from the east, usually applied to broad currents or patterns of persistent easterly winds.

Forces in the Atmosphere

There are two forces that account mainly for driving the horizontal flow of air on a global scale:

1. *Coriolis Force.* This is related to the rotation of the earth and is expressed vectorially by

$$C = -2\Omega \times \mathbf{v}$$

and quantitatively by

$$C = -2\Omega v \sin\phi$$

where C = Coriolis force
Ω = angular velocity of the earth
$\mathbf{v}$ = velocity vector
v = speed of wind
ϕ = the latitude

Coriolis force acts at right angles to the direction of the wind. In the northern hemisphere, it acts toward the right; in the southern hemisphere, toward the left.

2. *Pressure Gradient Force.* This is related to the field of atmospheric pressure at a specified level at or above sea level. It is expressed vectorially by

$$P_f = -\alpha \nabla p$$

and quantitatively by

$$P_f = -\alpha \frac{\partial p}{\partial l}$$

where P_f = force resulting from the pressure gradient
α = specific volume
p = pressure gradient
l = rate of pressure change along a direction.

In meteorology, the pressure gradient is regarded as acting from high to low pressure. It is considered to be the component of the force acting at right angles to the isobars. The component parallel to isobars is zero.

Basic flow patterns in the atmosphere on the horizontal are indicated in Fig. 3. These patterns are in a steady-state (no acceleration).

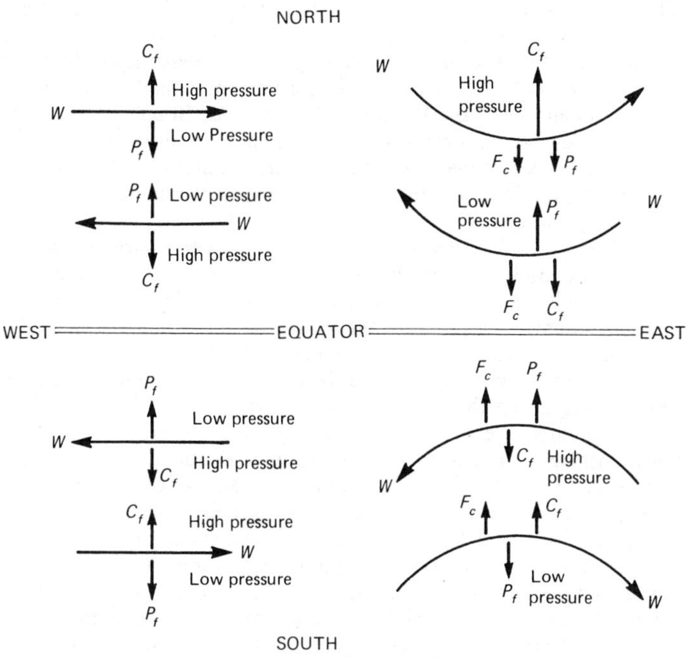

LEGEND:
W = wind direction (straight line and curved)
C_f = direction of Coriolis force
P_f = direction of pressure gradient force
F_c = direction of force associated with centripetal acceleration in the case of curved flow

Fig. 3. Basic flow patterns in the atmosphere on the horizontal. These patterns are in a steady state (no acceleration).

3. *Other Forces.* Additional forces acting on the atmosphere include:
(a) Centripetal acceleration related forces associated with curved flow.
(b) Frictional drag related forces acting near and at the surface of the earth.
(c) Isoallobaric forces related to the time rate of change of pressure field.
(d) Divergence and convergence related to vertical flow.

Pressure-Gradient Forces

Pressure Belts. Surrounding the earth at its surface and directly related to the generalized pattern of winds in the atmosphere are four alternating belts of high and low pressure. These belts are formed correspondingly in both the northern and southern hemispheres at roughly 30° intervals from equator to poles. They shift with season, on an average of 5° latitude, reaching the most northerly position in late summer; the most southerly in late winter. See Fig. 4.

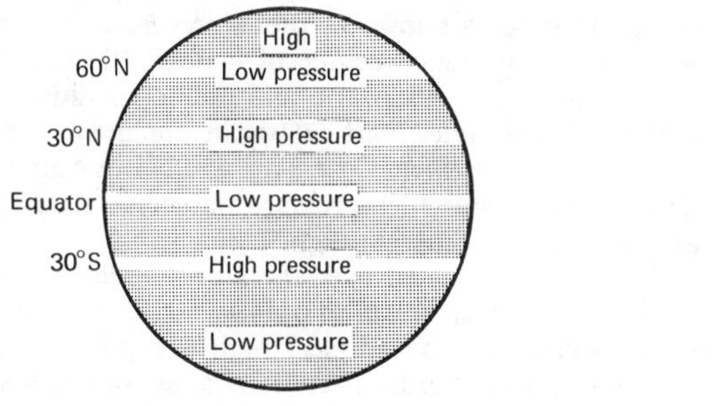

Fig. 4. Pressure belts and general circulation pattern of the air.

1. The *equatorial trough* is a quasicontinuous belt of low pressure extending north and south from the equator; it is commonly called the *doldrums,* or the *equatorial calms,* especially with reference to its light and variable winds. This entire region is one of very homogeneous air, probably the most ideally barotropic region of the atmosphere. Yet, humidity is so high that slight variations in stability cause major variations in weather. The position of the equatorial trough is fairly constant in the eastern portions of the Atlantic and Pacific; but it varies greatly with season in the western portions of those oceans and in southern Asia and the Indian Ocean. It moves into or toward the southern hemisphere.

2. The *horse latitudes* are narrow high-pressure belts over the oceans at approximately 30°–35°N and S, where winds are predominantly calm or very light, and weather is hot and dry. They are known in the northern and southern hemispheres, respectively, as the *calms of Cancer* and the *calms of Capricorn.* These latitudes mark the normal axis of the subtropical highs, and move north and south by about 5°, following the sun. In the North Atlantic Ocean, these are the latitudes of the Sargasso Sea, where surface waters converge, and which is characterized by clear, warm water, a deep blue color, and large quantities of floating Sargassum or "gulf weed." The name of the horse latitudes is believed to have originated in the days of sailing ships, when the voyage across the Atlantic in those latitudes was often prolonged by calms or baffling winds, so that water ran short, and ships carrying horses to the West Indies found it necessary to throw the horses overboard.

3. The *subpolar low-pressure belt* is located, in the mean, between 50° and 70° latitude. In the northern hemisphere, this "belt" consists of the Aleutian low and the Icelandic low. In the southern hemisphere, it is supposed to exist around the periphery of the Antarctic continent.

4. Areas of high pressure, the *polar highs,* form at the 90° poles, where the weather is violent and stormy.

Pressure Areas. These are areas within which the atmospheric pressure is either greater or smaller than other environing regions at the same altitude above sea level. In the case where the pressure is greater than other environing regions, the area of higher pressure is called a *high.* Highs are associated with anticyclonic circulation, clockwise in the northern hemisphere and counterclockwise in the southern hemisphere. In the case where the pressure is lower than other environing regions, the area of low pressure is called a *low.* Lows are associated with cyclonic circulation, counterclockwise in the northern hemisphere and clockwise in the southern hemisphere.

A further classification of pressure areas distinguishes between primary and secondary highs and lows: *Primary* (or *semipermanent*) *highs and lows* cover large areas of the earth's surface for long periods of time (sometimes for the entire year). They are the result of unequal heating of the earth's surface and the consequent movements of air. Where air rises over warmer regions, lows are likely to form; where air sinks over cooler regions, highs are likely to form. The term *center of action* refers to any one of the primary highs or lows. Fluctuations in the nature of these centers are intimately associated with relatively widespread and long-term weather changes.

Secondary highs and lows are, respectively, anticyclonic and cyclonic movements that form within the primary highs and lows. The secondary cyclonic lows are represented, generally, by the inclement weather and more-or-less violent phenomena that accompany storms. The secondary highs, unlike lows, represent a single anticyclonic air mass.

The principal semipermanent pressure areas of the northern hemisphere are:

Bermuda high: Located over the North Atlantic Ocean, and so named especially when it is located in the western part of the ocean. This same subtropical high, when displaced toward the eastern part of the Atlantic, is known as the *Azores high*. On mean charts of sea level, it is a principal center of action. When it is well-developed and extends westward, warm and humid conditions prevail over the eastern United States, particularly in summer.

North American high: Covers most of North America during winter. This relatively weak high-pressure system is not nearly so well-defined as the analogous Siberian high.

Pacific high: Located over the North Pacific Ocean and centered, in the mean, at 30–40°N and 140–150°W. On mean charts of sea-level pressure, this subtropical high is a principal center of action.

Siberian high: Forms over Siberia in winter, and is particularly apparent on mean charts of sea-level pressure. It is enhanced by surrounding mountains, which prevent the cold air from flowing away readily. In summer, the Siberian high is replaced by a low-pressure area.

Subtropical highs: Form the subtropical high-pressure belt, and include the Bermuda (and Azores) and Pacific highs.

Subpolar highs: Form over the cold continental surfaces of subpolar latitudes, principally in northern hemisphere winter. These highs typically migrate eastward and southward.

Aleutian low: Located near the Aleutian Islands on mean charts of sea-level pressure, and represents one of the main centers of action in the atmospheric circulation of the northern hemisphere. It is most intense in the winter months; in summer, it is displaced toward the North Pole and is almost nonexistent. The traveling cyclones of subpolar latitudes usually reach maximum intensity in the area of the Aleutian low.

Icelandic low: Located near Iceland, mainly between Iceland and southern Greenland, on mean charts of sea-level pressure, and is a principal center of action in the atmospheric circulation of the northern hemisphere. It is most intense during winter; in summer, it not only weakens, but also tends to split into two centers, one near Davis Strait and the other west of Iceland.

Subpolar low-pressure belt: Located, in the mean, between 50° and 70° latitude. In the northern hemisphere, this "belt" consists of the Aleutian low and the Icelandic low. In the southern hemisphere, it is supposed to exist around the periphery of the Antarctic continent.

In the southern hemisphere, there are three semipermanent high pressure centers, one each in the three oceans, the Pacific, Atlantic, and Indian Ocean. These centers are near 30°S in all cases and they do not migrate much between winter and summer. There is one semipermanent low pressure center in the southern hemisphere located over the Antarctic region. The main feature in the southern hemisphere is a zone of relatively strong westerly winds between the semipermanent high cells and the semipermanent low cells. Except for the continent of Australia, these prevailing Westerlies blow uninterrupted over ocean waters.

Commonly used general terms to describe certain types of pressure areas include:

Center of action. Any one of several large areas of high and low barometric pressure changing little in location, and persisting through a season or through the whole year. Changes in the intensity and positions of these pressure systems are associated with widespread weather changes. The term is also used to describe any region in which the variation of any meteorological element is related to weather of the following season in other regions.

Col. A relatively small area about midway between two cyclones and two anticyclones where the pressure gradient is very weak and winds are usually light and variable. It is the point of intersection between a trough and a ridge in the pressure pattern of a weather map; the point of relatively lowest pressure between the two highs and the point of relatively highest pressure between two lows.

Depression. An area of low pressure; a low or a trough. This is usually applied to a certain stage in the development of a tropical cyclone, to migratory lows and troughs, and to upper-level lows and troughs that are only weakly developed.

High. An area of high pressure, referring to a maximum of atmospheric pressure in two dimensions (closed isobars) in the synoptic surface chart, or a maximum of height (closed contours) in the constant-pressure chart. Since a high, on the synoptic chart, is always associated with anticyclonic circulation, the term can be used interchangeably with anticyclone.

Low. Also sometimes called depression. A low is an area of low pressure, referring to a minimum of atmospheric pressure in two dimensions (closed isobars) on a constant-height chart, or a minimum of height (closed contours) on a constant-pressure chart. Since a low, on a synoptic chart, is always associated with cyclonic circulation, the term can be used interchangeably with cyclone.

Ridge. Also sometimes called *wedge*, an elongated area of relatively high atmospheric pressure, almost always associated with and most clearly identified as an area of maximum anticyclonic curvature of wind flow. The locus of this maximum curvature is called the *ridge line.*

The most common use of this term is to distinguish it from the closed circulation of a high (or anticyclone); but a ridge may include a high (and an upper-air ridge may be associated with a surface high), and a high may have one or more distinct ridges radiating from its center. The opposite of a ridge is a trough.

Trough. An elongated area of relatively low atmospheric pressure. The axis of a trough is the *trough line*, along which the isobars are symmetrical and curved cyclonically. A V-shaped trough normally contains a front; a U-shaped trough generally contains no front or a very weak one. Usually there is considerable weather associated with a trough line of the V variety. A large-scale trough may include one or more lows; an upper-air trough may be associated with a lower-level low; and a low may have one or more distinct troughs radiating from it. Trough-line movements can be computed and a forecast made of future positions.

Isallobars and Isallobaric Fields. Atmospheric pressure changes at every point in the atmosphere from time to time. It is possible to measure such changes with considerable accuracy. The unit to express pressure change (or pressure tendency) is conventionally the total net change occurring in a 3-hour interval. It is customary to indicate the nature of the change, because the pressure change character may have varied during the selected time interval.

Three-hourly pressure changes are plotted on a synoptic chart (weather map) and lines drawn to join points of equal pressure change. Care is exercised, however, in noting the character of the change in judging the real value of Δp. Lines joining points of equal pressure change are isallobars. See Fig. 5. Isallobars, taken together, constitute an isallobaric field. Where there is present an isallobaric field superimposed on a pressure field, a component of the actual wind blows along the isallobaric gradient, which is directed perpendicular to the isallobars toward regions of greatest pressure fall or least pressure rise, as the case may be. Normally, the isallobaric wind component is small unless the isallobaric field is pronounced. When there is no isallobaric field, the wind is defined approximately by the orientation and spacing of isobars themselves. Isallobaric fields also are used to compute the movement of pressure areas, ridges, troughs, cols, fronts, and isobars. Centers of low pressure will tend to move toward the region of greatest pressure fall, whereas centers of high pressure tend to move toward regions of maximum pressure rise. Both cases have modifying factors, and the direction is not always exactly as would be expected from a casual glance. In all cases of computation of movement, it is necessary

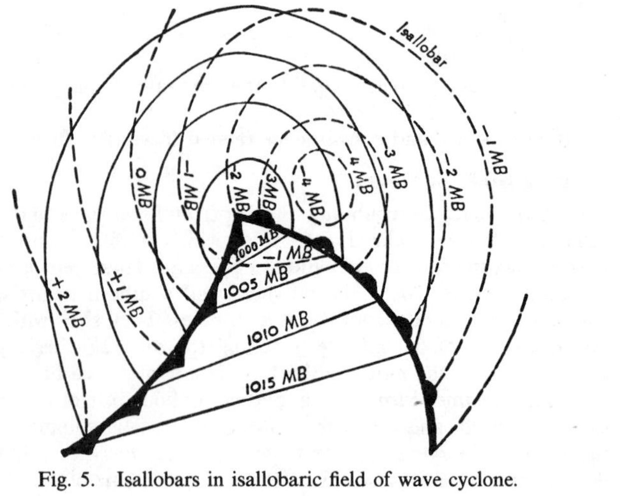

Fig. 5. Isallobars in isallobaric field of wave cyclone.

to know the isallobaric field, the pressure field, and the orientation of the system whose movement is to be computed.

In establishing an isallobaric field, it is imperative that the diurnal pressure change be accounted for and a correction applied. Diurnal pressure changes do affect movement of pressure systems, but the effect is transient, and, therefore, must be neglected in any study extending more than 6 hours into the future.

In general, isallobars and isallobaric fields are as useful in weather forecasting as the pressure field itself.

Tendency is defined as the local rate of change of a vector or scalar quantity with time at a given point in space. Thus, in symbols, $\partial p/\partial t$ is the pressure tendency, $\partial \xi/\partial t$ the vorticity tendency, etc. Because of the difficulty of measuring instantaneous variations in the atmosphere, variations are usually obtained from the difference in magnitudes over a finite period of time; and the definition of tendency is frequently broadened to include the local time variations so obtained. An example is the familiar three hourly pressure tendency given in surface weather observations.

Terms Associated with Circulation of the Atmosphere

Advection. The process of transport of an atmospheric property solely by the mass motion (velocity field) of the atmosphere; also, the rate of change of the value of the advected property at a given point. Fog drifts from one place to another and cold air moves from polar regions southward by advection. In synoptic meteorology, the term refers only to the horizontal or isobaric components of motion, that is, the wind field as shown on a synoptic chart. The distinction is made between advection and convection, the former describing the predominantly horizontal, large-scale motions of the atmosphere, while the latter describes the predominantly vertical, locally induced motions. Large-scale north-south advection is more prominent in the northern than in the southern hemisphere, but west-to-east advection is prominent on both sides of the equator.

Air Parcel. An imaginary body of air to which may be assigned any or all of the basic dynamic and thermodynamic properties of atmospheric air. A parcel is large enough to contain a very great number of molecules, but small enough so that the properties assigned to it are approximately uniform within it and so that its motions with respect to the surrounding atmosphere do not induce marked compensatory movements. It cannot be given precise numerical definition, but a cubic foot of air might fit well into most contexts where air parcels are discussed, particularly those related to static stability.

Air-Parcel Trajectory. A parcel of air located in a given pressure field will move with the gradient wind of the field (assuming steady flow). At the end of a few hours, the parcel will locate in some new region where it has been carried by the wind. If, however, the pressure field, and therefore the wind, is changing, the parcel will not move into a position indicated by the existing gradient flow. It will follow a trajectory or path dictated by successive gradient directions and velocities as indicated by synoptic charts. An approximation to its trajectory can be had by extrapolating the parcel's indicated movement for as small a time interval as practicable (usually 3 or 6 hours between synoptic charts), using successive synoptic charts. The average of the velocity vectors at the beginning and the end of a given time interval would be taken as the true velocity and direction of the parcel over that time interval. Obviously, the smaller the time interval, the more accurate the trajectory. One of the charts may be a prognostic chart for computing future trajectories. Air trajectories are valuable in estimating the influence the earth has on air as it flows over varied earth surfaces.

Anticyclone. Also termed *high,* an atmospheric circulation having a sense of rotation about the local vertical opposite to that of the earth's rotation, i.e., clockwise in the northern hemisphere; counterclockwise in the southern hemisphere, undefined at the equator; a closed circulation, whose flow is within a closed streamline. With respect to the relative direction of its rotation, it is the opposite of a cyclone.

The barometric pressure within an anticyclone is high relative to its surroundings, and a pressure gradient exists from its center toward its periphery. A well-developed anticyclone is, essentially, an air mass, whose dimensions vary from a few hundred to several thousand miles (kilometers). It is, in general, a region of slowly settling air with a descent rate of from 300–1500 feet (90–460 meters) per day. Anticyclones are migratory in the region north of 30°–40° latitude, their path usually being to the east and south. Seasonal semipermanent anticyclones develop over both North America and Eurasia during winter. A belt of permanent anticyclones, with their centers usually over the oceans, lies between 10° and 40° latitude.

Anticyclones are generally accompanied by bright, clear weather believed to be the result of descending dry air at the anticyclone center; however, rain, drizzle, and cloudy skies may develop in the southwestern and western sectors of the air mass. Anticyclones moving from the north bring cold waves in winter, and cool, clear weather at other seasons. Those moving from the south bring mild weather in winter and hot, dry spells in summer.

Anticyclogenesis is any strengthening or development of anticyclonic circulation in the atmosphere. This applies to the development of anticyclonic circulation where, previously, it was nonexistent, as well as to intensification of existing anticyclonic flow.

Convection. Atmospheric motions that are predominantly vertical, resulting in vertical transport and mixing of atmospheric properties. *Autoconvection* is the phenomenon of the spontaneous initiation of convection in an atmospheric layer in which the lapse rate is equal to or greater than the autoconvective lapse rate. The presence of viscosity, turbulence, and radiative heat transfer usually prevents the occurrence of autoconvection until the lapse rate is greater than the theoretical autoconvective lapse rate of approximately $+3.4 \times 10^{-4}$°C per centimeter.

Cyclone. An atmospheric cyclonic circulation, a closed circulation. A cyclone's direction of rotation (counterclockwise in the northern hemisphere; clockwise in the southern hemisphere) is opposite to that of an anticyclone. While modern meteorology restricts the use of the term cyclone to the so-called cyclonic-scale circulations (with wavelengths of 1000 to 2500 kilometers), it is popularly applied to the more-or-less violent small-scale circulations such as tornadoes, waterspouts, dust devils, etc. (which may, in fact, exhibit anticyclonic rotation), and even, very loosely, to any strong wind. This term was first used very generally as the generic term for all circular or highly curved wind systems. Because cyclonic circulation and relatively low atmospheric pressure usually coexist, the terms cyclone and low are used interchangeably. Also, because cyclones nearly always are accompanied by inclement, and often destructive, weather, they are frequently referred to simply as storms.

Equatorial Vortex. A closed cyclonic circulation within the equatorial trough. It develops from an equatorial wave.

Ferrel Law. When a mass of air starts to move over the earth's surface, it is deflected to the right in the northern hemisphere, and to the left in the southern hemisphere, and tends to move in a circle whose radius depends upon its velocity and its distance from the equator.

Geopotential Height. The height of a given point in the atmosphere in units proportional to the potential energy of unit mass (geopotential) at this height, relative to sea level. The relation, in the centimeter-gram-second (c.g.s.) system, between the geopotential height Z and the geometric height z is

$$Z = \frac{1}{980} \int_0^z g\, dz$$

where g is the acceleration of gravity, so that the two heights are numerically interchangeable for most meteorological purposes. Also, one geopotential meter is equal to 0.98 dynamic meter. At the present time, the geopotential height unit is used for all aerological reports, by convention of the World Meteorological Organization. See also **Geopotential.**

Geopotential Surface. Also called equigeopotential surface or level surface, this is a surface of constant geopotential, i.e., a surface along which a parcel of air can move without undergoing any changes in its potential energy. Geopotential surfaces also coincide with surfaces of constant geometric height. Because of the poleward increase of the acceleration of gravity along a constant geometric-height surface, a given geopotential surface has a smaller geometric height over the poles than over the equator.

Gradient Flow. Horizontal frictionless flow in which isobars and streamlines coincide; or equivalently, in which the tangential acceleration is everywhere zero. Important special cases of gradient flow, in which two of the normal forces predominate over the third, are: (1) *Cyclostrophic flow*, in which the centripetal acceleration exactly balances the horizontal pressure force; (2) *Geostrophic flow*, where the Coriolis force exactly balances the horizontal pressure force; (3) *Inertial flow*, which is flow in the absence of external forces; in meteorology, frictionless flow in a geopotential surface in which there is no pressure gradient, so that centripetal and Coriolis accelerations must be equal and opposite.

Inertial Force. A term used specifically in meteorology to designate a force in a given coordinate system arising from the inertia of a parcel moving with respect to another coordinate system. For example, the Coriolis acceleration on a parcel moving with respect to a coordinate system fixed in space becomes an inertial force, the Coriolis force, in a coordinate system rotating with the earth.

Rossby Number. The nondimensional ratio of the inertial force to the Coriolis force for a given flow of a rotating fluid. It may be given as

$$\mathrm{Ro} = \frac{U}{fL}$$

where U is a characteristic velocity, f the Coriolis parameter (or, if the system is cylindrical rather than spherical, twice the system's rotation rate), and L a characteristic length.

The *thermal Rossby number* is the nondimensional ratio of the inertial force due to the thermal wind and the Coriolis force in the flow of a fluid that is heated from below.

$$\mathrm{Ro}_T = \frac{U_T}{fL}$$

where f is the Coriolis parameter, L a characteristic length, and U_T a characteristic thermal wind.

Solenoid. In meteorology, a tube formed in space by the intersection of unit-interval isotimic surfaces of two scalar quantities. Solenoids formed by the intersection of surfaces of equal pressure and density are frequently referred to in meteorology. A barotropic atmosphere implies the absence of solenoids of this type, since surfaces of equal pressure and density coincide.

Tangential Acceleration. The component of the acceleration directed along the velocity vector (streamline), with magnitude equal to the rate of change of speed of the parcel dV/dt, where V is the speed. In horizontal, frictionless atmospheric flow, the tangential acceleration is balanced by the tangential pressure force

$$\frac{dV}{dt} = -\alpha \frac{\partial p}{\partial s}$$

where α is the specific volume, p the pressure, and s a coordinate along the streamline. Thus, flow without tangential acceleration is along the isobars, and the wind is the gradient wind.

Circulation Theorem

If the atmosphere is baroclinic, that is, if the surfaces of equal pressure and equal density intersect at any angle whatsoever, there is a tendency for a circulation to develop in such a manner that the atmosphere will become barotropic, that is, the surfaces of equal pressure and equal density will coincide. The atmosphere is normally baroclinic. Sea and land breezes, mountain and valley breezes are results of well-defined baroclinic states. Direction of circulation is always such that cold air flows toward warm air at the base of the circulation pattern, and warm flows toward cold at the top of the pattern. Air sinks in the cold air region and rises in the warm air region. Thus, the sea breeze blows along the surface of the earth from cold water to heated land, rises, then returns seaward, and sinks.

It is possible to compute the magnitude of the circulation from a given baroclinic state if temperature and pressure are known. Suppose a vertical plane were erected perpendicular to a shore, extending from the cold water to heated land. Lines of equal pressure and temperature drawn in this plane will produce a field of approximate parallelograms, which are known as solenoids. See Fig. 6. A tendency for circulation exists about the perimeter of each solenoid, but, in the field as a whole, this tendency is nullified in adjacent solenoids. There is no

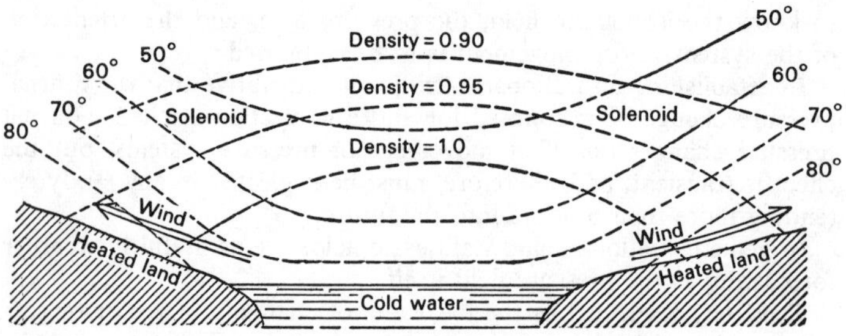

Fig. 6. Circulation theorem and solenoids.

nullification along the border solenoids of the whole field, and it is here that a circulation springs up. The number of solenoids in the field is a measure of the expected strength of the resulting circulation.

Friction Between Wind and the Earth

Even though winds may blow with considerable velocity above the earth's surface, air cannot move rapidly in shallow layers just above the earth where air is caught in the irregularities of the earth's surface. This slowing of a wind at the earth's surface is the result of friction between the moving air and the earth. The effect of friction extends to about 500 meters (> 1500 feet). The first effect of reduced velocity is a reduction in the Coriolis force and, therefore, a pressure gradient that is not balanced. Because the pressure gradient force is not balanced, air flows across isobars from high to low pressure. Thus, the unbalancing of the pressure gradient force causes air to converge toward low pressure and diverge from high pressure.

Over land, the average deflection very near the earth is up to 30° and over water it is about 20°. Reduction of wind speed depends upon the nature of the surface. Above the friction layer, the wind speed may be up to twice that at the surface.

Steady-State Wind Equation. In this equation, friction is ignored and the isallobaric and divergence-convergence factors do not come into play.

$$\frac{V^2}{r} + 2\Omega v \sin\phi = \alpha \frac{\partial p}{\partial r}$$

where v = wind speed
Ω = angular velocity of the earth
r = radius of curvature of the flow
α = specific volume, and is $1/\rho$, the density
$\frac{\partial p}{\partial r}$ = the pressure gradient along r
ϕ = the latitude

The wind speed in this relationship is given by:

$$v = -\Omega r \sin\phi \left[1 \pm \sqrt{1 + \frac{1}{r\Omega^2 \sin^2\phi}\left(\frac{1}{\rho}\right)\left(\frac{\partial p}{\partial r}\right)}\right]$$

where $\alpha = 1/\rho$ and ρ = density.

For straight-line flow,

$$v^2/r = 0,$$

and

$$v = \frac{1}{\rho}\frac{\partial p}{\partial r}\frac{1}{2\Omega \sin\phi}$$

For curved flow about an area of low pressure,

$$v = \Omega r \sin\phi \left[\sqrt{1 + \frac{1}{r\Omega^2 \sin^2\phi}\frac{1}{\rho}\frac{\partial p}{\partial r}} - 1\right]$$

which is counterclockwise in the northern hemisphere and clockwise in the southern hemisphere where ϕ is negative.

For curved flow about an area of high pressure,

$$v = -\Omega r \sin\phi \left[1 - \sqrt{1 - \frac{1}{r\Omega^2 \sin^2\phi}\frac{1}{\rho}\frac{\partial p}{\partial r}}\right]$$

which is clockwise in the northern hemisphere and counterclockwise in the southern hemisphere.

Near the equator, the Coriolis force is approximately zero, and

$$v = \sqrt{\frac{r}{\rho}\frac{\partial p}{\partial r}}$$

Atmospheric Convergence and Divergence

If an imaginary box is erected in the atmosphere near the earth's surface, in such a manner that its base, top, and sides are parallel to the air flow (i.e., the winds), it is possible to illustrate the effects of convergence and divergence. See Fig. 7. When air flows uniformly through this box, there will be no accumulation or dimunition of air inside the box. If, for any reason, however, more air flows into one end of the box than flows out the other, there is an accumulation of air, which must seek an outlet. Because the pressure is less at the top of the box than at the bottom, this accumulated air flows upward out of the box. If we followed a cube-shaped unit mass of air through this flow, it would become distorted into a rectangular prism elongated vertically. This process is called *convergence*, and results in a field

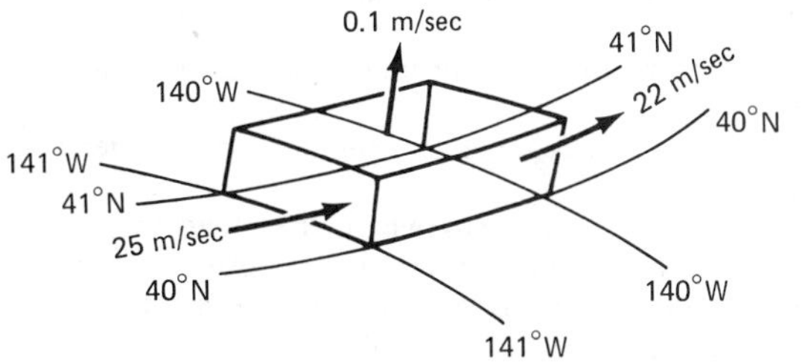

Fig. 7. Principle of convergence and divergence.

of rising air. Converging, and therefore rising air can, if the process endures over a sufficient period of time, produce clouds and precipitation. It also tends to destabilize the air. If, in the same box, less air flows into one end than flows out the other, there is an air diminution, and space is available at the top of the box for more air. One unit cube of air will be flattened into a rectangular prism elongated horizontally. This process is called *divergence*, and results in a field of sinking or subsiding air. Divergence, therefore, and subsiding air tend to stabilize the air. Clouds and turbulence diminish in regions of divergence and subsidence.

Waves in the Atmosphere

In meteorology, any pattern with some roughly identifiable periodicity in time and/or space applies to atmospheric waves in the horizontal flow pattern, i.e., in the wind field, there are wavelike disturbances, such as equatorial, easterly, frontal, Rossby, long, short, cyclone, and barotropic waves. The study of water-surface waves has bred its own special terminology, such as deep-water, shallow-water, wind, hurricane, and tidal waves. In popular terminology, a surge or influx, is often referred to as a wave, i.e., "heat wave," or "cold wave."

Barometric waves are any waves in the atmospheric pressure field, usually reserved for short-period variations not associated with cyclonic-scale motions or with atmospheric tides.

Barotropic waves occur in a two-dimensional, nondivergent flow, the driving mechanism lying in the variation of vorticity of the basic current and/or in the variation of the vorticity of the earth about the local vertical.

Cyclone waves are (1) disturbances in the lower troposphere, of wavelengths from 1000 to 2500 kilometers, recognized on synoptic charts as migratory high- and low-pressure systems, and identified with the unstable perturbations connected with baroclinic and shearing instability; and (2) frontal waves at the crests of which are centers of cyclonic circulation; therefore, the frontal waves of wave cyclones.

Easterly waves are migratory wavelike disturbances of the tropical easterlies, which move within the broad easterly current from east to west, generally more slowly than the currents in which they are imbedded. Although best described in terms of wavelike characteristics in the wind field, they also consist of weak troughs of low pressure. Easterly waves do not extend across the equatorial trough.

Equatorial waves are wavelike disturbances of the equatorial easterlies that extend across the equatorial trough.

Frontal waves are horizontal wavelike deformations of fronts in the lower levels, commonly associated with a maximum of cyclonic circulation in the adjacent flow. They may develop into wave cyclones.

Gravity wave disturbances are those in which buoyancy (or reduced gravity) acts as the restoring force on parcels displaced from hydrostatic equilibrium. There is a direct oscillator conversion between potential and kinetic energy in the wave motion.

Long Waves in the Prevailing Westerlies. There develop in the westerlies, particularly during the cold months, certain perturbations which cause the westerlies to blow alternately northward and southward in a sinusoidal wave pattern, but always with a component of velocity directed from west to east. See Fig. 8. Ridges are associated with anticyclones at ground and near-ground levels, whereas troughs are associated with cyclones. There is, therefore, a definite relation between the sinusoidal perturbations of the westerlies and large-scale surface weather phenomena. Progression eastward, or retrogression westward, of the crests and troughs is usually about the same as surface anticyclones and cyclones. Sinusoidal perturbations can, therefore, be used

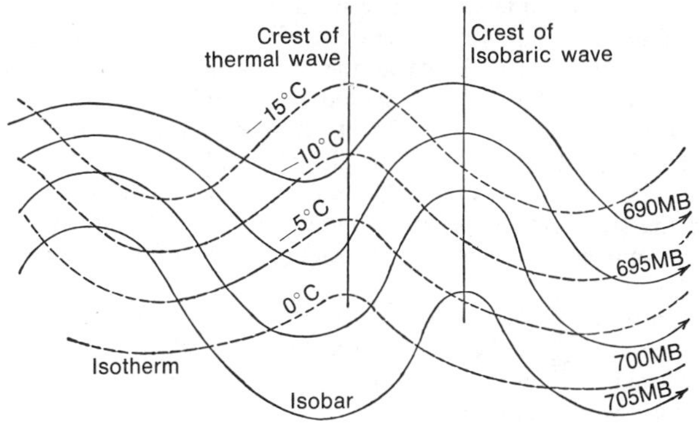

Fig. 8. Long waves in the westerlies.

for prognostic purposes computing their velocities. These velocities are given by the formula:

$$C = U - \frac{bL^2}{4\pi^2}$$

where C = the velocity of the wave
U = the west to east component of air motion
L = the wavelength of the wave
b = the rate of change in the Coriolis parameter northward

Wave velocities may be positive, zero, or negative; positive indicating easterly movement, negative, westerly movement.

Isotherms aloft also assume partial or complete sinusoidal form under some conditions. The following relations between amplitudes of the thermal and the streamline wave apply:

$$C = U\left(1 - \frac{A_s}{A_t}\right)$$

where A_s = streamline amplitude
A_t = thermal amplitude

From this, the following conclusions are possible:

1. If the amplitude of the thermal wave is greater than the streamline wave and the two are in phase, cyclones and anticyclones will have a slow eastward component of movement.
2. If the thermal wave is 180° out of phase with the streamline wave, cyclones and anticyclones move eastward rather rapidly. The smaller the amplitude of the thermal wave in relation to the streamline wave, the more rapid will be the movement of surface systems.
3. If the amplitude of the thermal wave is less than the streamline wave and the two are in phase, cyclones and anticyclones will move with a slow westward component (retrograde motion).
4. If the amplitudes of the thermal and streamline waves are the same and the waves are in phase, the surface systems will have no east-west component of movement.

The first two of these conditions are common: the latter two occur but are not usual.

Mountain Waves in the Atmosphere. These are internal waves located between the earth's surface and the base of the stratosphere. They are created when winds blow across mountainous terrain with a speed of 50 knots or more at the height of the tallest terrain and simultaneously the vertical structure of the atmosphere is favorable for wave development and growth. In North America, the most common locations for development of mountain waves are along the Rocky Mountains, over the Sierra Nevadas, and over the Pacific Coastal Range.

Within the atmosphere when mountain waves are present, the air surges upward and downward; there are vortices and the air movement is chaotic. Evidence of mountain waves is found in the presence of standing lenticular clouds, occasionally a patch of dust kicked up by a gust and, in some select sites, strong gusty winds in mountain passes and canyons.

The most serious impact of mountain waves in the atmosphere is on aircraft that happen to be in the airspace where mountain waves are present. Strong updrafts and downdrafts and moderate-to-severe turbulence are the common experiences.

Conditions under which mountain waves are generated are relatively well known and understood. The wind direction should lie between 90° and not less than 45° to the crest of the terrain. The wind speed should be 50 or more knots at the top of the terrain and should increase with altitude. There should be a stable layer of air or an inversion approximately 10,000 feet (3048 meters) above the level of high terrain. A low tropopause is also favorable.

There are many small local areas where mountain waves occur when the structure of the atmosphere and the winds are favorable. Most prominent locations in the contiguous states of the United States include:

1. Northeastern New Mexico to southeastern Wyoming, including all of central eastern Colorado. Winds are westerly.
2. Nevada-California border and over the western part of Nevada. Winds are west southwesterly.
3. Central northeast Wyoming through central Montana. Winds are southwesterly.

Periodic Changes in the Atmosphere

The atmosphere undergoes periodic recurring changes associated primarily with the relative position of the sun, but to a lesser amount on other factors. The most prominent and obvious periodic changes are the diurnal changes. Less dramatic, but nonetheless obvious are the seasonal changes. Aperiodic, but recurring changes are associated with the pattern of circulation in both hemispheres, and may also be dependent in part on the changes in ocean surface temperatures. There are also tides in the atmosphere which can be detected with very sensitive equipment.

Diurnal Changes. Changes completed within and recurring every 24 hours. The diurnal variability of nearly all meteorological elements is one of the most striking and consistent features of the study of weather. The diurnal variations of important elements at the earth's surface can be summarized as follows:

1. Atmospheric pressure varies diurnally or semidiurnally according to the effects of atmospheric tides. Surface pressure undergoes two definite periods of increase and two of decrease. Mean maximum pressure occurs approximately at ten o'clock local time, in the morning and evening; and mean minimum pressure occurs at four o'clock in the afternoon and morning. In the tropics, this surge and ebb of pressure is very pronounced and highly rhythmic.
2. Temperature tends to reach its maximum about 2–3 hours after local noon, and its minimum at sunrise. Over water, there is a minimum diurnal change as small as a fraction of a degree, and over sandy and rocky desert a maximum that sometimes amounts to 100° or more.
3. Relative humidity tends to become maximum about sunrise, and minimum in the afternoon.
4. Cloudiness and precipitation over a land surface increase by day and decrease at night; over water, the reverse is true but to a lesser extent. Over land, cumulus-type clouds tend to be maximum during afternoons and minimum at night.
5. Fogs tend to be maximum at and shortly after sunrise, and minimum in the afternoon.
6. Evaporation is markedly greater by day; and condensation is much greater at night.
7. Wind generally increases and veers by day, and decreases and backs by night; rough flying air tends to be maximum in midafternoon, and minimum at night.
8. Onshore winds tend to build up during the later morning and afternoon—then die out again in the evening and at night. Offshore winds sometimes set in during nighttime hours. Valley breezes tend to develop—blowing toward higher terrain by day and then reverse to blow downhill by night.

Seasonal Changes. These changes complete their cycle during a year and recur with some variations in intensity during the same months each year. The most prominent and obvious seasonal changes are snow in the winter and thunderstorms in the summer. Some of the more important seasonal changes include:

1. Snow occurs only in winter in most inhabited areas north of 30°N or south of 30°S.
2. The average temperature, mean maximum, and mean minimum temperatures increase to a high value during the second and third months of summer and decrease to a low value during the latter part of winter.
3. Thunderstorms increase in numbers to a maximum occurrence in summer.
4. Fog and low clouds are at a maximum during winter in temperate zone areas.
5. Tornadoes in those areas where they tend to occur are at a maximum in number and intensity during spring months.
6. Severe tropical storms tend to reach a maximum in number during autumn months.
7. Large-scale onshore monsoon circulations are predominantly a summer phenomenon whereas the offshore circulation occurs in winter.
8. Prolonged periods of rainfall (wet season) in the tropics and subtropics are associated with summer; dry periods with winter.

Aperiodic Changes. These appear to be linked to the pattern of circulation of the westerly flow in the atmosphere. There are a number of stable configurations of the west-to-east flow patterns in the atmosphere. Each one of the patterns consists of a discrete number of waves and each wave has a crest and trough. The crests and troughs tend to remain stationary for varying periods of time from weeks to months.

Air movement in the airspace between the trough and the crest of a wave (looking eastward) has a component from the equator toward the poles which tends to carry moisture. Cloud and precipitation-bearing storms predominantly are found in this zone. The airspace between the crest and the trough has an air movement component from the polar region toward the equator which has an associated suppression of cloudiness and precipitation.

A particular stable configuration may remain essentially unchanged for a substantial period of time, causing drought or cold or heat in one area and opposite conditions in another area. When the pattern breaks down and a different stable configuration is established, the newly established pattern will shift associated conditions to different areas. Shifts from one stable configuration to another are aperiodic (irregular).

Atmospheric Tide

Also known as atmospheric oscillation, the atmospheric tide is an atmospheric motion of the scale of the earth, in which vertical accelerations are neglected (but compressibility is taken into account). Both the sun and moon produce atmospheric tides, which may be thermal or gravitational. A *gravitational tide* is due to gravitational attraction of the sun or moon; the semidiurnal solar atmospheric tide is partly gravitational; the semidiurnal lunar atmospheric tide is fully gravitational. A *thermal tide,* so-called in analogy to the conventional gravitational tide, is a variation in atmospheric pressure due to the diurnal differential heating of the atmosphere by the sun.

The amplitude of the *lunar atmospheric tide* is so small (about 0.06 millibar in the tropics and 0.02 millibar in middle latitudes) that it is detected only by careful statistical analysis of a long record; the only detectable components are the 12-lunar-hour or semidiurnal, as in the oceanic tides, and two others of very nearly the same period.

The 12-hour harmonic component of the *solar atmospheric tide* is both gravitational and thermal in origin, and has, by many times, the greatest amplitude of any atmospheric tidal component (about 1.5 millibars at the equator and 0.5 millibar in middle-latitudes); 6- and 8-hour tides of small amplitude have been observed, as well as the 24-hour component, which is a thermal tide with great local variability. The fact that the 12-hour component of the solar atmospheric tide is greater than the corresponding lunar atmospheric tide is ascribed usually to a resonance in the atmosphere with a free period very close to the tidal period.

See other entries listed under **Meteorology.** For references, see **Climate;** and **Meterology.**

Peter E. Kraght, Certified Consulting Meteorologist, Mabank, Texas.

ATMOSPHERE (Lunar). Moon (The).

ATMOSPHERE-OCEAN INTERFACE. Almost 71% of the earth's atmosphere is in contact with oceanic surfaces. The ocean-air interface, therefore, plays a dominant role in determining the water content and temperature of the lower levels of the atmosphere and perhaps of the total atmosphere. The percentage of land and ocean for specified latitude belts and the temperatures in the main oceans in each zone are given in Table 1.

A broad zone lying roughly between 30°N and 30°S is approximately 75% water, with an annual average temperature of near 25°C (77°F) and nowhere below 20°C (68°F). Air in contact with this broad expanse of warm water acquires the properties of the water surface and is permeated by water vapor to considerable depths. This zone is the source region of moist, warm, unstable tropical air masses that move toward the poles as part of the general circulation of large-scale cyclones and anticyclones.

Source of Water Vapor. The air-ocean interface is the primary source of water for the atmosphere and particularly between the latitudes of 40°N and 40°S where the temperature averages 20°C (68°F) or higher. The average inches of liquid water evaporated each year into each vertical tube of air one-inch square in contact with the ocean surface is given in Table 2. This water vapor is carried aloft and transported laterally within the atmosphere. Secondary sources of water vapor entering the atmosphere include transpiration from plants, evaporation from moist soil and rock, lakes, and rivers.

Thermal Stabilization of the Atmosphere. As provided by the oceans, this is caused primarily by the large-scale uniformity of ocean temperature at the air-ocean interface. From day to day there is no appreciable change in ocean surface temperature except locally in immediate offshore waters. There is a slow seasonal change. Air in contact with the thermally stable ocean surfaces also tends to become thermally stable. Annual seasonal change in average ocean surface temperature for the Pacific, Atlantic, and Indian Oceans is indicated in Table 3. Were it not for the stabilizing influence of the oceans, the annual temperature range would be much greater than it is, i.e., warmer in summer and colder in winter.

Salt Condensation Nuclei. These nuclei required for cloud formation and precipitation originate through the air-ocean interface. Ocean spray from breaking waves creates tremendous numbers of small salt water droplets which are carried up into the atmosphere to evaporate and to leave a very small residue of sea salt. Wherever in the atmosphere the relative humidity approaches saturation values, these nuclei become the core for haze, small cloud droplets, and rain. Constituent chemical elements found in ocean water are also found in the same proportionate ratios in microscopic salt nuclei and in rain water. Over land areas, other nuclei from various sources tend to outnumber the salt nuclei.

Ocean Waves. These are predominately wind-generated. When the wind speed is near calm, the ocean surface is only rippled. As the wind speed increases, air at the ocean-air interface drags the water forward, causing waves to develop. The *fetch* is the stretch or distance over which winds can act upon the ocean surface. Fetch is limited by the distance of open water and by the distance that winds blow in one direction at a sufficient speed. Only rarely does wind direction stay in one orientation for long distances. Likewise, sustained high wind speeds only rarely extend over long distances.

TABLE 2. LIQUID WATER EVAPORATED BY OCEANS.
(Average Inches of Liquid Water Evaporated Each Year Into Each Vertical Tube of Air One-Inch Square in Contact with the Ocean Surface)

Latitude	Northern Hemisphere	Southern Hemisphere
40°	37 inches	32 inches
35°	42	39
30°	47	43
25°	51	49
20°	52	53
15°	51	53
10°	50	51
05°	43	49
00°	47	47

TABLE 3. ANNUAL SEASONAL CHANGE IN AVERAGE OCEAN SURFACE TEMPERATURES.

Latitude	Northern Hemisphere Ocean						Southern Hemisphere Ocean					
	Pacific		Atlantic		Indian		Pacific		Atlantic		Indian	
	°F	°C	°F	°C	°F	°C	°F	°C	°F	°C	°F	°C
60°			9	5			4	2.2	4	2.2	4	2.2
50°	13	7.2	11	6.1			7	3.9	5	2.7	5	2.7
40°	18	10	14	7.7			9	5	9	5	7	3.9
30°	13	7.2	13	7.2			7	3.9	11	6.1	11	6.1
20°	7	3.9	7	3.9			5	2.7	7	3.9	7	3.9
10°	4	2.2	2	1.1	5	2.7	4	2.2	6	3.4	4	2.2
00°	4	2.2	4	2.2	2	1.1	4	2.2	4	2.2	2	1.1

TABLE 1. PERCENT OF LAND AND OCEAN FOR SPECIFIED LATITUDE BELTS.

Latitude Belt	Northern Hemisphere					Southern Hemisphere				
	Percent		Average Ocean Surface Temperature, °C			Percent		Average Ocean Surface Temperature, °C		
	Water	Land	Pacific	Atlantic	Indian	Water	Land	Pacific	Atlantic	Indian
80–90	92.6	7.4					100			
70–80	71.3	28.7				24.6	75.4			
60–70	29.9	70.1		5.6		89.6	10.4	−1.3	−1.3	−1.5
50–60	42.8	57.2	5.7	8.7		99.2	0.8	5.0	1.8	1.6
40–50	47.5	52.5	10.0	13.2		96.9	3.1	11.2	8.7	8.7
30–40	57.2	42.8	18.6	20.4		88.8	10.2	17.0	16.9	17.0
20–30	62.4	37.6	23.4	24.2	26.1	76.9	23.1	21.5	21.2	22.5
10–20	73.6	26.4	26.4	25.8	27.2	78.0	22.0	25.1	23.2	25.8
00–10	77.2	22.8	27.2	26.7	27.9	76.4	23.6	26.0	25.2	27.4
00–90	60.7	39.3				80.9	19.1			

Wave height is empirically related to wind speed by:

$$H = 0.025V^2$$

where H is in feet and V is in knots. The maximum wave height is related to the fetch by:

$$H_{max} = 1.5\sqrt{F}$$

where H_{max} is in feet and F is in nautical miles. See Table 4.

Fetches greater than 1000 nautical miles are not common and those as much as 2000 miles probably do not occur. The tallest waves observed are on the order of 60 to 70 feet (18 to 21 meters). Wave heights of 50 feet (15 meters) can be expected in winds of more than 45 knots provided the fetch is sufficient. Very large waves can develop in strong winds within 12 hours when the fetch is sufficiently long. In contrast, strong winds cannot create huge waves on a short fetch.

TABLE 4. WAVE HEIGHT VERSUS WIND SPEED AND MINIMUM FETCH.

Wave Height (Feet)	Wind Speed (Knots)	Minimum Fetch Required to Attain Height (Nautical Miles)
3	10	3
10	20	48
23	30	240
42	40	760
65	50	1 860

Swells. These are more or less uniformly spaced, rounded waves that were generated as wind waves, but that have traveled well beyond the ocean areas where they were developed. The orientation of swells change only slowly, if at all, with time. Therefore, swells can be used to indicate the location of the storm that generated them. Observations of swells is usually part of a weather observation from ships at sea. The height, orientation, and speed of swells observed at a number of points in the open ocean provide meteorologists with useful information in pinpointing storm centers.

The ocean surface most often displays a chaotic mixture of swells and locally generated wind-driven waves. There may be as many as a half-dozen intermingling swells and waves in one spot, causing the sea surface to rise and fall in a most irregular manner. See list of other entries under **Meteorology.** For references, see entries on **Climate; Meteorology;** and **Ocean.**

ATMOSPHERE (Planetary). *See specific planets.*

ATMOSPHERE (Polar). Polar Research.

ATMOSPHERE (Standard). Altimetry; Atmosphere (Earth); Units and Standards.

ATMOSPHERE (Tangential Acceleration). Tangential Acceleration.

ATMOSPHERIC HEAT ENGINE. Solar Energy.

ATMOSPHERIC INTERFERENCE (or Spherics). The interference caused radio reception by natural electric disturbances in the atmosphere.

ATMOSPHERIC OPTICAL PHENOMENA. Because of varying conditions present in the atmosphere from time to time and notably the presence of ice crystals, dust particles, and other particulate matter, several interesting optical effects are the result.

Sky Color. The characteristic blue color of clear skies is due to preferential scattering of the short-wavelength components of visible sunlight by air molecules. Presence of foreign particles in the atmosphere alters the scattering processes in such a way as to reduce the blueness. Hence, spectral analysis of diffuse sky radiation provides useful information concerning the scattering particles. The study and measurement of the blueness of the sky is called *cyanometry.* Sometimes the Linke scale (or blue-sky scale) is used. The Linke scale is simply a set of eight cards of different standardized shades of blue. They are numbered 2 to 16, the odd numbers to be used by the observer when the sky color appears to lie between any of the given shades.

Halos and Coronas

A halo is any one of a large class of atmospheric phenomena appearing as colored or whitish rings and arcs about the sun or moon when seen through an ice crystal cloud, or in a sky filled with falling ice crystals. The halos exhibiting prismatic coloration are produced by refraction of light by the crystals and those exhibiting only whitish luminosity are produced by reflection from the crystal faces. The minute spicules of ice, in falling, take some definite attitude determined by their shape. Some are needlelike and assume a horizontal position; some are flat disks or stars and fall with their planes horizontal; while others, made up of both disks and rods, behave like a parachute. The sunlight is refracted by each type in a characteristic manner and dispersed into colors; it is also reflected from their external surfaces without dispersion.

Halos differ from coronas in that the former are produced by refraction and reflection due to ice crystals, whereas the latter are produced by diffraction and reflection due to water drops. A colored halo may often be distinguished from a corona in that it has the red nearest the sun or moon, whereas the corona has red in the exterior rings.

Halo Phenomena. With regards to type, orientation, motion, and solar elevation angle of ice crystals, a large variety of halos is possible, theoretically. Many varieties have been observed. Some halos theoretically predicted have not yet been reported; some that have been reported have not yet been theoretically explained, such as the Hevelian halo (described below). On rare occasions, an observer's sky will be filled with a display of four of five halo phenomena at one time, usually persisting for only a few minutes. Much supernatural lore was built up about such displays by the ancients.

By far the most common halo phenomenon is the *halo of 22°*, in the form of a prismatically colored circle of 22° angular radius around the sun or moon, exhibiting coloration from red on the inside to blue on the outside. It is produced by refraction of light that enters one prism face and leaves by the second prism face beyond, thus being refracted by an effective prism of 60° angle. In order to have a full 22° halo, the sky must be filled with hexagonal ice crystals falling with random orientations, a condition that apparently is frequently satisfied. This halo exhibits a distinct spectral pattern out to blue, due to the tipping of the crystals and consequent overlap of spectra in all but the red end. A reddish inner edge is usually all one can discern.

Closely allied to the 22° halo are the *parhelia* ("mock suns" or "sun dogs") and *paraselenae* ("mock moons"). The parhelia are two colored (reddish) luminous spots that appear at points 22° (or somewhat more) on both sides of the sun and at the same elevation as the sun. Their lunar counterpart is the weakly colored paraselenae, which are observed less frequently than the parhelia because of the moon's comparatively weak luminosity. These phenomena are produced by refraction in hexagonal crystals falling with principal axes vertical, the effective prism angle being 60°, as in the halo of 22°.

The parhelic circle ("mock sun ring") and the paraselenic circle ("mock moon ring") are halos consisting of a faint white circle passing through the sun or moon, respectively, and running parallel to the horizon for as much as 360° of azimuth. These circles are often seen in the sky along with parhelia or paraselenae, and are produced in the same manner. Parhelia and paraselenae occur at several positions along the parhelic circle other than the common 22° position, i.e., at 46°, 90°, 120°.

The *Hevelian halo* is the halo of 90°; it appears, only occasionally, as a faint, white halo on the sun or moon, and is a member of the class of halos reported but not yet fully explained.

A *sun pillar* (or a "light pillar") is a luminous streak of light, white or slightly reddened, extending above and below the sun, most frequently observed near sunrise or sunset. It may extend to about 20° above the sun, and generally ends in a point. The luminosity is thought to be produced simply as a result of reflection of sunlight from the tops and bottoms of tabular hexagonal ice crystals falling with principal axes vertical.

A *sun cross* is a rare halo phenomenon in which bands of white light intersect over the sun at right angles. It appears probable that most of such observed crosses appear merely as a result of the superposition of a parhelic circle and a sun pillar.

The *arcs of Lowitz* is a type of halo, rarely seen, in which the luminous arcs extend obliquely downward from the 22° parhelia on either side of the sun (or moon); it is concave towards the sun, with reddish inner edges; and is produced by refraction in hexagonal ice crystals that are oscillating as they descend.

The *circumhorizontal arc* is produced by refraction of light entering snow crystals, and consists of a colored arc, red on its upper margin, which extends for about 90° parallel to the horizon and lies about 46° below the sun. The *circumzenithal arc* is produced by refraction of light entering the tops of tabular ice crystals; it consists of a brightly colored arc about 90° in arc length, and is found about 46° above the sun, with its center at the zenith. It is typically very short-lived, but also very brilliant. In addition to these arcs, several types of halo arcs known generically as *tangent arcs* are occasionally formed as loci tangent to other halos, especially to the halo of 22°.

Corona. The corona consists of one or more rings located symmetrically about the sun or moon caused by diffraction of light passing through liquid water droplets. Coronas are of varied radii about the sun or moon, dependent upon the size of the water droplets. The radius of any corona is inversely related to the diameter of the water droplets causing it:

$$\sin \theta = \frac{(N + 0.22)\,L}{D}$$

where θ = angular radius of the corona ring
N = order of the corona ring (1st, 2nd, etc.)
L = wavelength of the light
D = diameter of the water droplets

The order of coloration in a corona ring is from blue on the inside to red on the outside (opposite to the coloration of halo rings).

Bishop rings are corona rings of faint reddish-brown seen in dust clouds.

Glory or Anticorona. A glory ring is observed on a cloud top of edge opposite to the position of the sun, i.e., the antisolar point. Glories are most frequently observed from aircraft flying above clouds. The shadow of the plane is in the center of the glory ring. See Fig. 1. These anticorona rings are complementary to the corona rings.

Bouguer's halo is a ring of faint white light usually about 39 feet (12 meters) in radius observed on some occasions outside the glory ring.

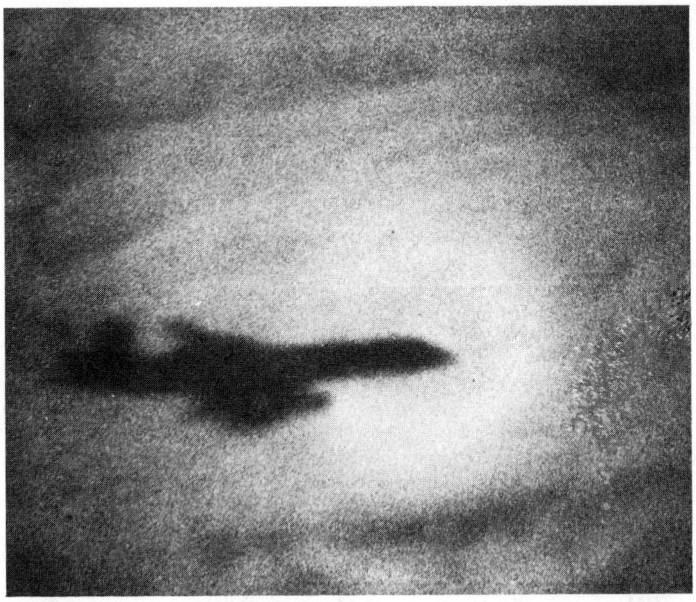

Fig. 1. Photograph of a glory ring made by a crew member of an American airliner in 1974. Note that center of glory ring is at exact position of cockpit in shadow of aircraft.

Rainbow

Looking into a "sheet" of water drops (rain, fog, spray) that is illuminated by strong white light from behind, an observer sees one, and sometimes two, concentric, spectrally colored rings, called a rainbow. If two are visible, the inner ring, called the "primary bow," is brighter and narrower than the outer ring or "secondary bow." In the primary bow, red is on the outside edge and violet on the inside edge; the order in the secondary bow is reversed. The colors are not so pure as in a spectrum because each wavelength extends over a wide radial range, the rainbow itself being made up of the fairly pronounced intensity maxima.

The colors of the rainbow are caused by the refractive dispersion of the spherical water drops. In Fig. 2 is shown the dispersion composing the primary and secondary bow. The diagram also explains why the order of colors is reversed, and shows that only the highest drops in the primary bow, and the lowest in the secondary bow, refract red light to the eye. The two internal reflections, with consequently

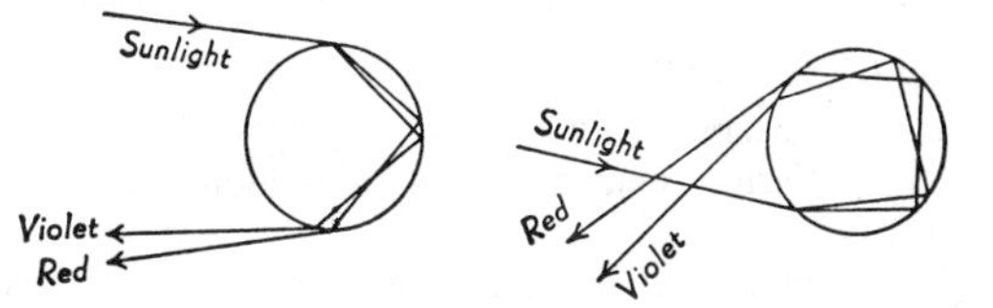

Fig. 2. Formation of primary bow (left) and secondary bow (right). Circles represent a raindrop.

greater loss of light, explain why the secondary bow is fainter. The center of the ring system is exactly opposite the source of light, so that natural rainbows are seen only when the sun is near the horizon, unless the observer is elevated high above the surrounding country and can look obliquely downward into the rain.

A *fog bow* is a type of rainbow, faintly colored, seen on fog layers whose droplets are small.

Mirage

A mirage is a curious atmospheric phenomenon caused by the total reflection of light at a layer of rarefied air. The most familiar manifestation is observed in warm weather on paved highways. The air next to the pavement becomes heated and rarefied in comparison with that above it, so that, at a sufficient angle of incidence, objects beyond the area are mirrored as if by polished silver, giving the almost irresistible impression that one is looking at a layer of water. Travelers in hot desert regions are sometimes thus deceived. Much more rarely the phenomenon appears in the air at a higher level than the observer. In either case, the images are inverted; and because of the irregular contour of the air layer, they are usually distorted. A somewhat different effect, known as *looming,* is produced by the refraction of light passing from rarefied air to a lower and denser layer. This results either in distortion, making distant objects appear grotesquely elongated vertically, or in lifting into view objects beyond the horizon. Looming effects are most frequently observed at sea.

Twilight and Afterglows

Twilight is that period between sunset and night and between night and sunrise. Civil Twilight is the period when the sun is between 0° and 6° below the horizon. Nautical Twilight is that period when the sun is between 0° and 12° below the horizon. Astronomical Twilight is that period when the sun is 0° and 18° below the horizon. Night is the period when the sun is more than 18° below the horizon.

The curtain of night is a relatively sharp slightly curved line across the sky that rises in the east in the evening. The night recedes in the east before the line of dawn, which is also a slightly curved line, appears. Both the curtain of night and the line of dawn are best observed from an aircraft flying at a high altitudes from which position both lines can be sharply delineated.

Afterglow. A broad, high arch of radiance seen occasionally in the western sky above the highest clouds in deepening twilight. It is caused by the scattering effect exerted upon the components of white light by very fine particles of material suspended in the upper atmosphere. When used in this rather broad sense, the term embraces all the complex luminosities observed in the western twilight sky, but chiefly the purple light and the bright segment.

The purple light is the faint purple glow observed on clear days over a large region of the western sky after sunset, and over the eastern sky before sunrise. The purple light first appears, in the sunset case, for example, at a solar depression of 2°; at that time, it extends from about 35° to about 50° elevation above the solar point, and has an azimuthal extent of between 40° and 80°. Maximum intensity of the glow typically occurs at the time the sun is about 4° below the horizon. Increasing depression of the sun causes the top of the purple light to descend steadily toward the western horizon. The effect disappears at solar depression angles near 7°, being replaced in the western sky by the bright segment.

Whiteout. Also termed milky weather, whiteout is an atmospheric optical phenomenon of the polar regions in which the observer appears to be engulfed in a uniformly white glow. Neither shadows, horizon, nor clouds are discernible; senses of depth and orientation are lost; only very dark, nearby objects can be seen.

Aurora and air glow are described under **Aurora and Airglow.**

Polarization of Sky Radiation

There are three commonly detectable points of zero polarization of diffuse sky radiation, neutral points, lying along the vertical circle through the sun:

Arago Point. Named for its discoverer, the Arago point is customarily located at about 20° above the antisolar point; but it lies at higher altitudes in turbid air. The latter property makes the Arago distance a useful measure of atmospheric turbidity.

Babinet Point. This point typically lies only 15° to 20° above the sun, and hence is difficult to observe because of solar glare. The existence of this neutral point was discovered by Babinet in 1840.

Brewster Point. Discovered by Brewster in 1840, this neutral point is located about 15° to 20° directly below the sun; hence, it is difficult to observe because of the glare of the sun.

Zodiacal light is not an atmospheric phenomenon. See **Zodiacal Light.**

For references see entries on **Climate;** and **Meteorology.**

Peter E. Kraght, Certified Consulting Meteorologist, Mabank, Texas.

ATMOSPHERIC PRESSURE. Also termed *barometric pressure,* atmospheric pressure is the pressure exerted by the atmosphere as a consequence of gravitational attraction on the vertical column of air lying directly above the surface of the earth upon which the pressure is effective. As with any gas, atmospheric pressure is ultimately explainable in terms of the kinetic energy of impacting constituent atmospheric gases upon the surface that experiences the pressure.

Atmospheric pressure is one of several basic meteorological parameters. It is measured fundamentally by the height of a column of mercury (or other heavy fluids) in a sealed and evacuated tube, one end of which is exposed to the air. Atmospheric pressure forces the mercury to rise in the sealed and evacuated portion of the tube to a height at which the weight of the mercury exactly balances the weight of the air column resting on the open end of the tube. Such an instrument is called a barometer.

Air pressure is expressed in several ways. The most commonly used unit in meteorology is the *millibar* in which one millibar equals 1000 dynes per square centimeter. Atmospheric pressure averages about 1013.2 millibars at sea level. The kilopascal (kPa) is also a measure of atmospheric pressure. One kPa = 10 millibars. In the kPa system, average atmospheric pressure is 101.325 kPa. The height of the mercury column in a barometer is also used, either as millimeters or inches of mercury. Average sea level pressure in this system is 760 millimeters or 29.92 inches of mercury. Pounds per square inch is used in engineering. The term atmosphere is also used, one atmosphere being average sea level air pressure.

Hydrostatic Equation. Pressure-altitude relations in the atmosphere are mathematically precise and can be determined from the hydrostatic equation:

$$\frac{\partial p}{\partial h} = -\rho g$$

and the equation of state for air,

$$p = \rho R T$$

which leads to the relation

$$p = p_0 \left[\frac{T_0 - \lambda h}{T_0}\right]^{g/R\lambda}$$

where p = pressure
p_0 = pressure at height zero
h = altitude
ρ = density
g = acceleration of gravity
T = temperature (absolute)
T_0 = temperature at height zero (also absolute)
R = universal gas constant
λ = lapse rate and is constant

In the real atmosphere, many layers of air, each having its own approximately constant lapse rate of temperature, press down on each other to create the total air pressure at the base of the bottom layer. Air pressure results from

$$p_0 = p_1 + p_2 + p_3 \cdots + p_n = \sum_1^n p_i$$

where the p's are valid at the base of their respective layers. A fictitious lapse rate can be used that will yield nearly the same results as the combined pressures of the several uniform layers.

When the atmosphere is isothermal, as it very nearly is in the stratosphere, the pressure-height relationship becomes

$$p = p_0 e^{-gh/RT}$$

These relations state that (1) pressure decreases more rapidly with altitude when the temperature is low than when temperature is high; and (2) pressure decreases more rapidly with altitude when the lapse rate is large than when it is small.

The standard atmosphere is used for calibration of altimeters. It is not very often that the real atmosphere assumes the arbitrarily assigned values of the standard atmosphere; therefore, altimeters do not often indicate exact altitude.

Total mass of air per cubic centimeter is the *density of air.* It is given by the relation,

$$\text{density} = \frac{0.0012930}{1 + 0.00367t}\left[\frac{B - 0.378e}{760}\right]$$

where t = temperature, °C
B = barometric pressure expressed in mm of mercury
e = the partial pressure of water vapor in the air

Air density at standard conditions of 0°C and 760 millimeters of mercury is 0.0012930 gram per cubic centimeter of air free from water vapor. The standard density of air at 32°F and 14.7 pounds pressure

VARIATION OF AIR PRESSURE WITH ALTITUDE IN STANDARD ATMOSPHERE AS USED IN ALTIMETRY

Altitude		Pressure		
Feet	Meters	Inches of Mercury	Millibars	Pounds per Square Inch
Sea Level	Sea Level	29.92	1013.2	14.7
1000	304.8	28.86	977.3	14.2
5000	1524	24.89	842.9	12.2
10,000	3048	20.58	696.9	10.1
15,000	4572	16.88	571.6	8.3
20,000	6096	13.75	465.6	6.8
25,000	7620	11.10	375.9	5.4
30,000	9144	8.88	300.7	4.4
40,000	12,192	5.54	187.6	2.7
50,000	15,240	3.44	116.5	1.7

is 0.081 pound per cubic foot (1.3 kilograms per cubic meter) and its composite molecular weight is 28.84.

The variation of air pressure with altitude is given in the accompanying table. See also **Atmosphere (Earth).**

Scale Height. This is a measure of the relationship between density and temperature at any point in an atmosphere; the thickness of a homogeneous atmosphere which would give the observed temperature:

$$h = kT/mg = RT/Mg$$

where k is the Boltzmann constant; T is the absolute temperature; m and M are the mean molecular mass and the molar mass of the layer; g is the acceleration of gravity; and R is the universal gas constant.

See also **Barometer.** For references see entries on **Climate;** and **Meteorology.**

Peter E. Kraght, Certified Consulting Meteorologist, Mabank, Texas.

ATMOSPHERIC RESIDUE. Petroleum.

ATMOSPHERIC TIDE. Atmosphere (Earth).

ATMOSPHERIC TURBULENCE. Air usually flows from one point to another in a turbulent manner, that is, the flow is infested with a multitude of small deviations of speed in all directions. This phenomenon has been observed for hundreds of years in the spreading and dissipation of smoke plumes. This can be demonstrated by inserting a pencil-sized smoke source into a wind stream. The thread of visible smoke does not extend downwind in a straight, thin line, but rather follows a zigzag path, spreading out and expanding downwind. This behavior demonstrates the presence of turbulence in the atmosphere.

When specific instantaneous speeds of the wind are measured at a point for a relatively large number of observations over a not-too-long period of time, and this ensemble of data is averaged, the small deviations in speed all cancel each other with the mean speed remaining.

Moving air behaves as if embedded eddies of varying sizes roll and migrate, eventually to be absorbed into another part of the main airstream, merge with other eddies, or dissipate some distance from their origin. The result of turbulent flow is the transport of atmospheric pollutants, particulates, water vapor, heat, and momentum.

The *mixing length* is a rather unprecisely defined distance over which eddies of a certain size are able to transport their own embedded properties. Thus, during its life span, an eddy rolls and migrates from one point in the airstream to another, carrying its implanted pollutants, particulates, water vapor, heat, and momentum but shedding these as it travels to become totally lost at the end of its mixing length. The concept is similar to that of the mean free path in molecular theory.

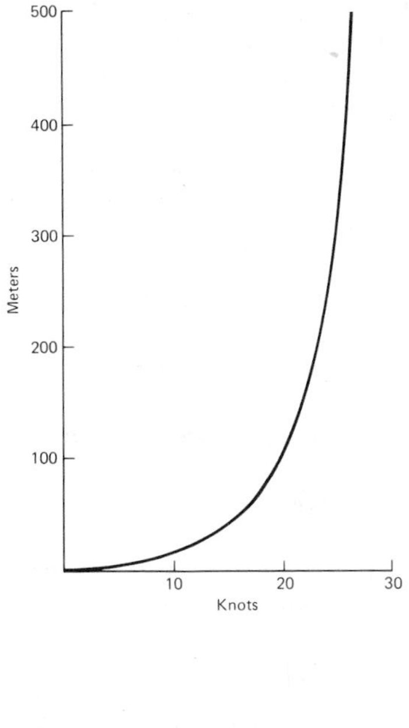

Wind speed profile based on the model: $V = 10 \log h$, where V is in knots and h is in meters.

Turbulent flow in the atmosphere has many significant meteorological consequences. Among others, the shape of the wind-speed profile just above the earth in the lower 500 meters depends in large part on the turbulent mixing in that layer. The speed profile is nearly always one in which there is a rapid increase just above the ground, but increasing less rapidly at higher levels. Exponential and logarithmic models describe the profile theoretically. See accompanying figure.

Industrial pollutants are distributed upward and laterally by atmospheric turbulence as they are carried downwind in the large-scale airflow. Acid rain and snow fall hundreds of miles from the pollutant source region. A specialized field of industrial meteorology has emerged in the past 25 years to deal with the problems of industrial pollutants in the atmosphere.

Water vapor in the atmosphere originates primarily from the oceans, rivers, and lakes, as well as from transpiration from vegetation. These sources are all at the earth's surface. Eddies in the turbulent flow of the atmosphere distribute the water vapor and thus rain and snow fall far from the source of moisture.

Condensation and sublimation nuclei are dispersed extensively throughout the tropopause by turbulent flow. Salt particles from sea spray are among the nuclei that are distributed everywhere by atmospheric turbulence. These nuclei join other factors in causing rain and snow.

Heat is transported upward by eddies in turbulent flow, thus "cooling" the earth's surface and "warming" the higher atmospheric layers.

Eddies in turbulent flow of sufficiently large dimensions can alter the longitudinal air flow past aircraft in flight, thus causing rough-flying air. Most such rough air is simply an annoyance to travelers, but in some instances, may jolt the aircraft and toss items in the cabin around.

See also **Richardson Number;** and list of entries under **Meteorology.** For references, see entries on **Climate;** and **Meteorology.**

Peter E. Kraght, Certified Consulting Meteorologist, Mabank, Texas.

ATOLL. A coral reef of ring-shape and appearing as a low, essentially circular, but sometimes elliptical or horseshoe shaped island. An atoll also may be a ring of closely spaced coral islets which encircle or nearly encircle a shallow lagoon where there is evidence of preexisting land of noncoral origin, and surrounded by deep water in the open sea. Atolls are common in the western and central Pacific Ocean.

ATOM. An atom is a basic structural unit of matter, being the smallest particle of an element that can enter into chemical combination. Each atomic species has characteristic physical and chemical properties that are determined by the number of constituent particles (protons, neutrons, and electrons) of which it is composed; especially important are the number Z of protons in the nucleus of each atom. To be electrically neutral the number of electrons in an atom must also be Z. The arrangement of these electrons in the internal structure of an atom determines its chemical properties. All atoms having the same atomic number Z have the same chemical properties, but differ in greater or lesser degree from atoms having any other value of Z. Thus, for example, all atoms of sodium ($Z = 11$) exhibit the same characteristic properties and undergo those reactions which chemists have found for the element sodium. While these reactions are similar in some degree to those reactions characteristic of certain other elements, such as potassium and lithium, they are not exactly the same and hence can be distinguished chemically (see **Chemical Elements**), so sodium has properties distinctly different from those of all other elements. Individual atoms can usually combine with other atoms of either the same or another species to form molecules.

As explained in the entry on atomic structure, atoms having the same atomic number may differ in their neutron numbers or in their nuclear excitation energies.

The term atom has a long history, which goes back as far as the Greek philosopher Democritus. The concept of the atomic nature of matter was revived near the beginning of the nineteenth century. It was used to explain and correlate advancing knowledge of chemistry and to establish many of the basic principles of chemistry, even though conclusive experimental verification for the existence of atoms was not forthcoming until late in the nineteenth century. It was on the basis of this concept that Mendeleev first prepared a periodic table. See **Periodic Table of the Elements.**

Several qualifying terms are used commonly to refer to specific types of atoms. Examples of some of the terms are given in the following paragraphs.

An *excited atom* is an atom which possesses more energy than a normal atom of that species. The additional energy commonly affects the electrons surrounding the atomic nucleus, raising them to higher energy levels.

An *ionized atom* is an ion, which is an atom that has acquired an electric charge by gain or loss of electrons surrounding its nucleus.

A *labeled atom* is a tracer which can be detected easily, and which is introduced into a system to study a process or structure. The use of those labeled atoms is discussed at length in the entry **Isotope.**

A *neutral atom* is an atom which has no overall, or resultant, electric charge.

A *normal atom* is an atom which has no overall electric charge, and in which all the electrons surrounding the nucleus are at their lowest energy levels.

A *radiating atom* is an atom which is emitting radiation during the transition of one or more of its electrons from higher to lower energy states.

A *recoil atom* is an atom which undergoes a sudden change or reversal of its direction of motion as the result of the emission by it of a particle or radiation in a nuclear reaction.

A *stripped atom* is an atomic nucleus without surrounding electrons; also called a nuclear atom. It has, of course, a positive electric charge equal to the charge on its nucleus.

Subatomic particles and organization of the atom are discussed in the entry **Particles (Subatomic).**

ATOM ENERGY LEVELS. **Energy Level.**

ATOMIC CLOCKS. These devices make use of a property that is generally found only in systems of atomic dimensions. Such systems cannot contain arbitrary amounts of energy, but are restricted to an array of allowed energy values $E_0, E_1, \ldots, E_n$. If an atomic system is to change its energy between two allowed values, it must emit (or absorb) the energy difference—as by emission (or absorption) of a quantum of electromagnetic radiation. The frequency f_{ij} of this radiation is determined by the relation

$$|E_i - E_j| = \Delta E = hf_{ij}$$

where h is Planck's constant. The rate of an atomic clock is controlled by the frequency f_{ij} association with the transition from the state of energy E_i to the state of energy E_j of a specified atomic system, such as a cesium atom or an ammonia molecule. A high-frequency electromagnetic signal is stabilized in the atomic frequency f_{ij} and a frequency converter relates the frequency f_{ij} to a set of lower frequencies which then may be used to run a conventional electric clock.

The atomic frequency f_{ij} is, according to present knowledge, free of inherent errors. It is, in particular, not subject to "aging" since any transition which the system makes puts it in a state of completely different energy, where it cannot falsify the measurement. Herein lies the principal advantage over other methods of time measurement. Two atomic clocks have exactly the same calibration so long as they are calibrated against the same atomic transition. Atomic readings made in Boulder, Colorado and Neufchâtel, Switzerland between 1960 and 1963 differed on the average by less than 3 msec, whereas the deviation of the astronomically measured time TU_2 from atomic time is of the order of 50 msec. For this reason, the atomic second was adopted as the new time unit, by the Twelfth General Conference on Weights and Measures, in October 1964. This was initially defined as the time interval spanned by 9,192,631,770 cycles of the transition frequency between two hyperfine levels of the atom of cesium 133 undisturbed by external fields. See also "Atomic Clock" in entry on **Clock.**

The accuracy of present atomic clocks is limited by the thermal noises inherent at room temperatures. Theoretically, this limitation could be removed if the clocks were maintained in an atmosphere approaching absolute zero. However, some atomic clocks, like hydrogen maser clocks, stop oscillating when they are supercooled. In 1979, scientists at the Harvard-Smithsonian Center for Astrophysics overcame this problem by coating a supercooled maser cavity with carbon tetrafluoride. With the CF_4 frozen on the interior surfaces of the cavity, the oscillating hydrogen atoms could be reflected off the walls without becoming perturbed, thus preserving the phase of the oscillations. The researchers were able to keep a hydrogen maser clock operating at temperatures somewhat above 25 K. It has been estimated that a hydrogen maser clock cooled to about 25 K could run for 300 million years before losing one second of time, a factor some six times better than present hydrogen masers. The Center for Astrophysics is interested in improved maser clocks in connection with long-baseline interferometry and satellite tracking systems. It is also envisioned that a supercooled clock put on a space probe could be helpful in research on gravity waves and possibly provide clues toward better understanding the sun's mass distribution and angular momentum.

ATOMIC DISINTEGRATION. The name sometimes given to radioactive decay of an atomic nucleus and occasionally to the breakup of a compound nucleus formed during a nuclear reaction (see **Radioactivity**).

ATOMIC ENERGY. 1. The constitutive internal energy of the atom, which would be released when the atom is formed from its constituent particles, and absorbed when it is broken up into them. This is identical in magnitude with the total binding energy and is proportional to the mass defect. 2. Sometimes this term is used to denote the energy released as the result of the disintegration of atomic nuclei, particularly in large-scale processes, but such energy is more commonly called nuclear energy. See **Nuclear Reactor.**

ATOMIC ENERGY LEVELS. 1. The values of the energy corresponding to the stationary states of an isolated atom. 2. The set of stationary states in which an atom of a particular species may be found, including the ground state, or normal state, and the excited states.

ATOMIC FREQUENCY. The vibrational frequency of an atom, used particularly with respect to the solid state.

ATOMIC HEAT. The product of the gram-atomic weight of an element and its specific heat. The result is the atomic heat capacity per gram-atom. For many solid elements, the atomic heat capacity is very nearly the same, especially at higher temperatures and is approximately equal to $3R$, where R is the gas constant (Law of Dulong and Petit).

ATOMIC HEAT OF FORMATION. Of a substance, the difference between the enthalpy of one mole of that substance and the sum of the enthalpies of its constituent atoms at the same temperature; the reference state for the atoms is chosen as the gaseous state. The atomic heat of formation at 0K is equal to the sum of all the bond energies of the molecule, or to the sum of all the dissociation energies involved in any scheme of step-by-step complete dissociation of the molecule.

ATOMIC HYDROGEN WELDING. Welding in a hydrogen atmosphere using heat from the arc between two tungsten electrodes. See also **Welding.**

ATOMIC MAGNETIC MOMENT. **Stern-Gerlach Experiment.**

ATOMIC MASS UNIT (Unified). **Units and Standards.**

ATOMIC NUCLEUS (Internal Conversion). **Internal Conversion.**

ATOMIC NUMBER. **Chemical Elements.**

ATOMIC PERCENT. The percent by atom fraction of a given element in a mixture of two or more elements.

ATOMIC PLANE. A plane passed through the atoms of a crystal space lattice, in accordance with certain rules relating its position to the crystallographic axes. See **Mineralogy.**

ATOMIC RADIUS. Chemical Elements.

ATOMIC SHELLS. Chemical Elements.

ATOMIC SPECIES. A distinctive type of atom. The basis of differentiation between atoms is (1) mass, (2) atomic number, or number of positive nuclear charges, (3) nuclear excitation energy. The reason for recognizing this third class is because certain atoms are known, chiefly among those obtained by artificial transmutation, which have the same atomic (isotopic) mass and atomic number, but differ in energetics.

ATOMIC SPECTRA. An atomic spectrum is the spectrum of radiation emitted by an excited atom, due to changes within the atom; in contrast to radiation arising from changes in the condition of a molecule. Such spectra are characterized by more or less sharply defined "lines," corresponding to pronounced maxima at certain frequencies or wavelengths, and representing radiation quanta of definite energy.

The lines are not spaced at random. In the spectrum of hydrogen, for example, there is a prominent red line (H_α) and, far from it, another (H_β) in the greenish-blue, then after a shorter wavelength interval a blue-violet line (H_γ), and after a still shorter interval another violet line (H_δ), etc. One has only to plot the frequencies of these lines as a function of their ordinal number in the sequence, to get a smooth curve which shows that they are spaced in accordance with some law. In 1885, Balmer studied these lines, now called the Balmer series, and arrived at an empirical formula which in modern notation reads

$$\nu = Rc\left(\frac{1}{n_1^2} - \frac{1}{n_2^2}\right)$$

It gives the frequency of successive lines in the Balmer series if R is the Rydberg constant, c the velocity of light, $n_1 = 2$, $n_2 = 3, 4, 5, \ldots$ As n_2 becomes large, the lines become closer together and eventually reach the series limit of $\nu = Rc/4$. Ritz, as well as Rydberg suggested that other series might occur where n_1 has other integral values. These, with their discoverers and the spectral region in which they occur are as follows:

Lyman series, far ultraviolet, $n_2 = 2, 3, 4, \ldots, n_1 = 1$
Paschen series, far infrared, $n_2 = 4, 5, 6, \ldots, n_1 = 3$
Brackett series, far infrared, $n_2 = 5, 6, 7, \ldots, n_1 = 4$
Pfund series, far infrared, $n_2 = 6, 7, 8, \ldots, n_1 = 5$

See also **Energy Level.**

ATOMIC SPECTROSCOPY. Chemical analysis by atomic absorption spectrometry involves converting the sample, at least partially, into an atomic vapor and measuring the absorbance of this atomic vapor at a selected wavelength which is characteristic for each element. The measured absorbance is proportional to concentration, and analyses are made by comparing this absorbance with that given under the same experimental conditions by reference samples of known composition. Several methods of vaporizing solids directly can be used in analytical applications. One of the first methods used was spraying a solution of the sample into a flame, giving rise to the term "absorption flame photometry." When a flame is used, the atomic absorption lines are usually so narrow (less than 0.05 Å) that a simple monochromator is not sufficient to obtain the desired resolution. Commercial atomic absorption spectrophotometers overcome this difficulty by using light sources which emit atomic spectral lines of the element to be determined under conditions which ensure that the lines in the spectrum are narrow, compared with the absorption line to be measured. With this arrangement, peak absorption can be measured, and the monochromator functions only to isolate the line to be measured from all other lines in the spectrum of the light source. See accompanying figure.

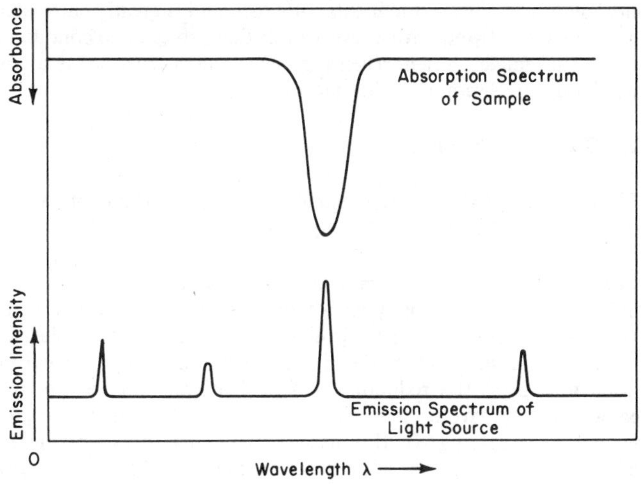

Lines emitted by the light source are much narrower than the absorption line to be measured.

Atomic spectra, which historically contributed extensively to the development of the theory of the structure of the atom and led to the discovery of the electron and nuclear spin, provide a method of measuring ionization potentials, a method for rapid and sensitive qualitative and quantitative analysis, and data for the determination of the dissociation energy of a diatomic molecule. Information about the type of coupling of electron spin and orbital momenta in the atom can be obtained with an applied magnetic field. Atomic spectra may be used to obtain information about certain regions of interstellar space from the microwave frequency emission by hydrogen and to examine discharges in thermonuclear reactions.

ATOMIC STRUCTURE. Chemical Elements; Mineraology.

ATOMIC TIME. Clock; Time.

ATOMIC WEIGHTS (Table). Chemical Elements.

ATOMIZATION. The breaking-up of a liquid into small droplets, usually in a high-speed jet or film.

ATOMIZER (Burner). Burner.

ATRIAL SEPTAL DEFECT. Heart and Circulatory System (Human).

ATRIUM. Literally, an entrance chamber, and so applied to various organs. 1. The main part of the cavity of the middle ear. 2. The vestibule of the female genital passages. 3. A chamber into which the genital organs open in the flatworms. 4. A cavity formed of folds of the body wall in Amphioxus and the tunicates, which partially surrounds the pharynx and opens to the exterior by an atriopore. 5. The chamber at the end of an air tube in the lungs, with which the ultimate air sacs or alveoli communicate. 6. The chamber of the heart in vertebrates which empties into the ventricle. In this sense the term atrium is frequently replaced by auricle, although in strict terminology the auricle refers only to a small appendage of the atrium.

ATROPHY. Physiologic or pathologic reduction in size of a mature cell or organ, usually with some degree of degeneration. Following certain diseases, types of accidents and surgery where nerves may be damaged or cut, atrophy may be temporary or permanent. Where there is traumatic injury of the nerves that cannot be repaired, atrophy is progressive. Atrophy or degeneration of the anterior lobe of the pituitary gland in adults results in Simmonds-Sheehan disease, a disorder characterized by an extreme appearance of aging. The metabolic functions of the body are affected and eventually mental functions decline as well. Pituitary atrophy is believed to result from anoxia, a lack of oxygen reaching the gland after a condition, such as postpartum hemorrhage, where excessive blood is lost. Simmonds-Sheehan disease sometimes is confused with *anorexia nervosa*, a serious nervous condition in which the patient eats little food and is greatly emaciated. A metabolic test (Metapirone test) can distinguish the two conditions. Further, in females, amenorrhea (absence of menstruation) is a constant feature of Simmonds-Sheehan disease.

ATROPINE. Alkaloids.

ATTACK ANGLE. Aerodynamics; Airplane; Helicopters and V/STOL Craft; Supersonic Aerodynamics.

ATTENUATION. 1. In its most general sense, attenuation is reduction in concentration, density or effectiveness. 2. In psychological statistics, the weakening of the correlation between two variables due to errors of measurement on them. 3. In radiation theory, attenuation is used to express the reduction in flux density, or power per unit area, with distance from the source; the reduction being due to absorption and/or scattering. In this usage, attenuation does not include the inverse-square decrease of intensity of radiation with distance from the source. 4. The same restriction applies to the use of the term in nuclear physics, where attenuation is the reduction in the intensity of radiation on passage through matter where the effect is usually due to absorption and scattering. 5. In an electric network or line, attenuation is loss, usually of current. See **Attenuation Factor; Attenuator.**

In terms of scientific instruments, the Scientific Apparatus Makers Association defines attenuation as: (1) A decrease in signal magnitude between two points, or between two frequencies; and (2) The reciprocal of gain, when the gain is less than one. Attenuation may be expressed as a dimensionless ratio, scalar ratio, or in decibels as 20 times the $\log_{10}$ of that ratio.

ATTENUATION FACTOR. 1. A measure of the opacity of a layer of material for radiation traversing it. It is equal to I_0/I, in which I_0 and I are the intensities of the incident and emergent radiation, respectively. In the usual sense of exponential absorption

$$I = I_0 e^{-\mu x}$$

where x is the thickness of the material and μ is the absorption coefficient. 2. A meaning similar to that in (1) is current in electrical circuit applications, where the attenuation factor is the ratio of the input current to the output current of a line or network.

ATTENUATION RATIO. The magnitude of the propagation constant.

ATTENUATION (Sideband). That form of attenuation in which the transmitted relative amplitude of some component(s) of a modulated signal (excluding the carrier) is smaller than that produced by the modulation process.

ATTENUATION (Underwater Sound). Sonar.

ATTENUATOR. The attenuator, often called a pad, is a network designed to introduce a definite loss in a circuit. It is designed so the impedance of the attenuator will match the impedance of the circuit to which it is connected, often being connected between two circuits of different impedance and serving as a matching network as well as an attenuator. It is distinguished from a simple resistance in that the impedance of an attenuator does not change for various values of its attenuation. It is a valuable unit in making many laboratory tests on communications equipment where it is used to adjust the outputs of two pieces of apparatus or for two different conditions so the relative merits may be determined from the attenuator setting. In much communication work it is desirable to transmit power at a higher level than will be used in order to overcome circuit noises, and then to reduce it to the proper value at the receiving end by a pad. It is usually calibrated in decibels and thus indicates the attenuation introduced by it.

Among the types of attenuators, there is the *coaxial line attenuator*, which as its name indicates, is designed for use in a coaxial line. It may be fixed or variable. One of its special types is the *chimney attenuator*, which received its name from the appearance of the stub lines. The *flap or fin attenuator* is a waveguide attenuator in which a flap or fin of conducting materials is moved into the guide in such a manner as to cause power absorption. The *transverse film attenuator* consists of a conducting film placed transverse to the axis of a waveguide.

ATTENUATOR (Analog Computer). Analog Computer.

ATTITUDE. The position or orientation of an aircraft or spacecraft, either in motion or at rest, as determined by the relationship between its axes and some reference line or plane or some fixed system of reference axes.

ATTITUDE CONTROL (Space Vehicle). Space Vehicle Guidance and Control.

ATTRITION (Geology). From the Latin *attriio* meaning a grinding or rubbing down, and used in the terminology of geological science to refer to the grinding of particles through the transporting power of wind, running water, or by the movement of glaciers.

ATTRITION MILLS. Equipment of this type is used in the process industries to reduce the size of various feeds. Attrition connotes a rubbing action, although this action usually is combined with other forces, including shear and impact. Attrition mills also are referred to as disk mills and normally comprise two vertical disks mounted on horizontal shafts, with adjustable clearance between the vertical disks. In some designs, one vertical disk may be stationary. In other designs, the two vertical disks rotate at differential speeds or in opposite directions. Material is fed to the mill so that it is subjected to a tearing or shredding action. Because of the frictional nature of the operation, temperatures build up and heat-sensitive materials cannot be size-reduced in this type of equipment. Attrition mills sometimes are used principally as mixers to provide an intimate blending of powders. Special plates are used to permit intensive blending with a minimum of grinding action. Throughput rates per horsepower required are high.

AUCTIONEERING CONTROL. A control system in which either the highest or the lowest signal from two or more input signals is automatically selected. In connection with suction and discharge pressure compressor control, normally the discharge control valve is regulated from discharge pressure. Should the suction pressure drop below its setpoint, in auctioneering control, the control will be transferred to the suction pressure controller, thus preventing excessive suction in the supply side and resultant compressor damage. In the flow and pressure control of a gas distribution system, normally the distribution valve is regulated from discharge pressure. In auctioneering control, under high demand conditions, control is transferred to the flow controller, thus limiting the maximum flow rate and the maximum pressure in the distribution system. In the temperature or vacuum and differential pressure control of a packed distillation column, the column will be controlled either from a vacuum setpoint, with the steam flow on manual for constant heat input, and the differential pressure override to prevent flooding. Or, in another configuration, regulation will be from a temperature setpoint, with vacuum on manual for constant vacuum, with differential pressure override to prevent flooding.

Terms also used to describe auctioneering control are *override control* and *limiting control.*

AUDIBILITY. The wide loudness range of the human ear is exemplified by the fact that the most intense sound that can be tolerated is a million million times greater in intensity than a sound that is just audible. This is a range of approximately 120 decibels. The decibel scale is a logarithmic ratio scale. The frequency range (*audio frequency*) of hearing is usually stated as 16 Hz to 20,000 Hz. The ear is most sensitive in the middle-frequency range of 1,000 to 6,000 Hz. Few individuals can hear above 20,000 Hz. Below 15 Hz, if detected, the sound normally is perceived not as a note, but as individual pulses.

In terms of discrimination of frequency and intensity, it is possible for about 1,400 pitches and 280 intensity levels to be distinguished.

The rather phenomenal aspects of hearing can be observed in such behavior as localization of sounds (*auditory localization*), speech perception and, in particular, the understanding of one voice in the noisy environments of many. Acoustic events that last only a few milliseconds also can usually be detected.

The instrument for measuring hearing acuity is termed an *audiometer.* See also **Auditory Organs; Hearing and the Ear.**

AUDIO FREQUENCY. **Frequency; Signal Generator.**

AUDIO FREQUENCY PEAK LIMITER. A circuit used in an audio frequency system to cut off signal peaks that exceed a predetermined value.

AUDIOGRAM. A graph showing the hearing loss, the percent of hearing loss, or the percent of hearing as a function of frequency. See also **Hearing and the Ear.**

AUDITORY ORGANS. Organs sensitive to stimulation by sound waves. True auditory organs occur in arthropods and vertebrates. In the former they vary considerably but in the latter they are the ears and can be traced through their variations to a common structural foundation.

The simplest arthropod auditory organ is known as a chordotonal organ. It consists of a nerve ending with accessory cells connected with the body wall, which is apparently the immediate source of the vibrations to which the organ responds. More elaborate auditory organs are found in grasshoppers, katydids, mosquitoes, and related species. In the grasshoppers they are located on the sides of the first abdominal segment, in the katydids in the front tibiae, and in the mosquitoes at the base of the antennae. In all forms the scolophore is the essential sensory ending; accessory structures vary to a greater degree but usually include a modification of the cuticula which serves as a resonating membrane, or tympanum.

The essential auditory portion of the vertebrate ear is the cochlea, a spiral organ of elaborate structure containing terminations of the auditory nerve. This organ is part of the inner ear. In the mammals the outer ear includes the pinna, usually called the ear, and the external auditory canal leading inward to the tympanum or ear drum which vibrates in response to sound waves. Between these two regions lies the cavity of the middle ear, derived from the pharynx and connected with it by the Eustachian tube. The middle ear is bridged by a series of small bones, the hammer, anvil, and stirrup, which convey the vibrations of the tympanum mechanically to the liquid in the inner ear. These parts are variably developed in vertebrates below the mammals, all of which have simpler ears than described. The ears of bats play a unique part in the avoidance of obstacles during flight.

See also **Hearing and the Ear; Sensory Organs.**

AUDITORY SYSTEM (Human). **Hearing and the Ear.**

AUGEN-GNEISS. A gneissoid rock that contains lenticular crystals or mineral aggregates resembling "eyes." Derived from the German *augen,* eyes.

AUGER EFFECT. A process, discovered by P. Auger, in which the energy released in the de-excitation of an excited electronic energy state of an atom is given to another one of the bound electrons rather than being released as a photon. This type of transition is usually described as radiationless. The process usually occurs only for transitions in the x-ray region of energy states. The final state corresponds to one higher degree of ionization than does the initial state. The ejected electron has kinetic energy equal to the difference between the energy of the x-ray photon of the corresponding radiative transition and the binding energy of the ejected electron.

AUGER ELECTRON (Internal Conversion). **Internal Conversion.**

AUGER MINING. **Coal.**

AUGHEY SPARK CHAMBER. An electrical spark is produced between two hollow tubular electrodes. A gas passed through the electrodes into the space where the spark occurs will be excited and caused to emit its characteristic spectrum. Traces of contamination in the gas, such as dust particles, will be efficiently excited so that their spectrum may be observed.

AUGITE. This mineral is a common monoclinic variety of pyroxene whose name is derived from the Greek word meaning luster, in reference to its shining cleavage faces. Chemcially it is a complex metasilicate of calcium, magnesium, iron and aluminum. Color, dark green to black, may be brown or even white; hardness, 5–6; specific gravity, 3.23–3.52. Augite is important as a primary mineral in the igneous rocks and also as secondary mineral. The white augite is called leucaugite from the Greek word meaning white. Chemical analysis reveals this variety as containing little or no iron. Augite is of widespread occurrence.

See also **Pyroxene.**

AUGMENTATION. A term used by astronomers to indicate the increase in apparent diameter of the moon, or of any other object close enough to the earth to be observed as a disk, as the altitude of the object increases.

In Fig. (a), we have a representation of conditions for the object, *M,* on the horizon for an observer, *O*; in Fig. (b), the object, *M,* is at the zenith on the meridian for the observer, *O′*. In both figures, *C* represents the center of the earth. The distance, *CM,* of the object from the center of the earth is assumed to be a constant. Examination of the figures will show at once that (a), gives the maximum distance of the object from the observer, whereas (b) gives the minimum value of this distance. Since apparent angular diameter of an object increases with decrease of distance, and since apparent size is defined as the apparent angular diameter, the object is seemingly larger under conditions (b) than (a), i.e., larger on the meridian than when rising.

For the sun and planets, augmentation is too small to be considered except in the most refined observations of the altitude of a limb. However, in the case of the moon, augmentation may amount to as much

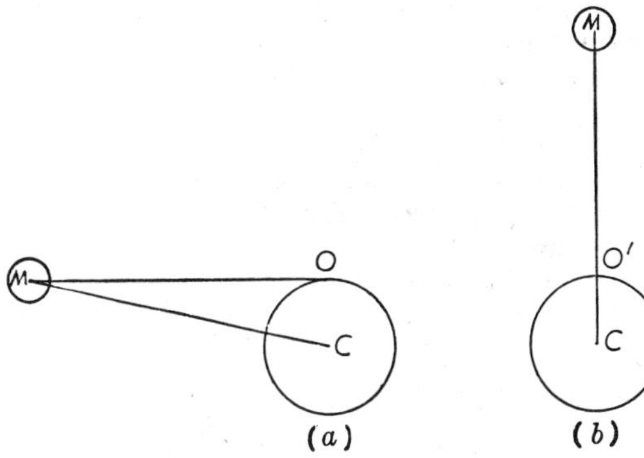

Demonstration of augmentation.

as 37″. Failure to properly correct for this effect, when a limb of the moon is observed for determination of a line of position in navigation, might introduce an error as great as 0.3 mile in the position of a ship.

AUK. **Shorebirds and Gulls.**

AUREOLE (Geology). The contact metamorphic zone of varying width that often surrounds an igneous intrusion. Such areas of contact metamorphism often contain valuable ore deposits, especially when they surround batholiths which have intruded sedimentary formations.

AURIC AND AUROUS. Prefixes often used in the naming of gold salts of valence +3(ic) and +1(ous). Thus, auric chloride; aurous nitrate, and so on.

AURICLE. **Hearing and the Ear.**

AURICLE (Heart). **Heart and Circulatory System (Human).**

AURIGA (the charioteer). This constellation is best known because it contains the bright star Capella (the she-goat) (α Aurigae) and her kids. The kids are three fainter stars, forming, to the naked eye, a small triangle, which always serves to distinguish Capella from other bright stars on a clear night. Capella is a bright star, yellowish in appearance, and of the same spectral type as our sun. The star, however, is so much larger than our sun that, in spite of its great distance (49 light-years), it appears as first magnitude, whereas the sun at the same distance would be sixth magnitude, or barely visible to the naked eye on a clear moonless night. Capella A is a spectroscopic binary with a period of 104 days. More distant are two additional M dwarf components, making Capella a four-star system, or small cluster. See map accompanying entry of **Constellations.**

AUROCHS. Bovines.

AURORA AND AIRGLOW. The visual aurora consists of luminous forms (arcs, rays, bands) in the night sky, usually confined to high latitudes and based in the ionospheric E region. See also **Ionosphere.** The airglow consists of a faint relatively uniform luminosity which is worldwide in occurrence and, except under exceptional conditions, can only be observed instrumentally. The distinction between faint aurora and bright airglow in auroral regions is not clear.

The luminosity arises from emissions of the atmospheric constituents in the atomic, molecular or ionized forms. The chief emissions in the visible region, with approximate intensities in Rayleighs for a bright aurora and temperate latitude airglow, are shown in the accompanying table. There are many other emissions in the infrared and ultraviolet. In bright aurorae, the colors can be seen visually; faint aurorae appear grayish white since they approach the color vision threshold.

An auroral arc is a narrow horizontal band of light up to hundreds of kilometers long (usually geomagnetic east-west). The term arc derives from its appearance from the earth's surface due to perspective. A band is a portion of an arc showing, distortion normal to its length. Auroral rays have been likened to searchlight beams; they lie along the geomagnetic field direction and may be several hundred kilometers long. Arcs and bands may be homogeneous or rayed. Particularly dramatic displays of the aurora are shown in Figs. 1 through 3.

Isolines of auroral occurrence are approximately centered on the geomagnetic poles. The auroral zones are defined as the regions of maximum occurrence. They are roughly circular with a radius of approximately 23° of latitude. The northern auroral zone reaches its lowest geographic latitude over eastern Canada; the southern, over the ocean south of Australia. At times of geomagnetic disturbance, the aurora appears at lower latitudes and in very great magnetic storms may be observed in the tropics. The frequency of occurrence of aurorae at lower latitudes correlates with the cycle of solar activity.

Fig. 1. A beautifully looped curtain aurora over Alaska. (*Geophysical Institute, University of Alaska.*)

Within recent years, it has been found that a relatively uniform auroral glow exists over the polar cap, extending through and beyond the classical auroral zone, on which auroral forms appear as bright patches which are visible merely because of contrast with their surroundings. Another relatively recent finding is the thesis of a local time dependence in the daily maximum of auroral occurrence which is at about 68° geomagnetic at midnight and 75 to 80° at noon. An inner auroral zone at 75 to 80° geomagnetic could possibly explain the observed results.

Many auroral forms are probably caused by the precipitation of particles (mainly electrons) into the ionosphere. Their origin is obscure, but studies suggest that they are derived from the outer regions of the magnetosphere, and are accelerated and precipitated in an irregular manner on the high latitude side of the outer radiation belt through some mechanism (e.g., turbulence) which is probably related to the solar wind. It is doubtful if precipitation of trapped particles from the outer radiation belt causes aurora directly, except in great magnetic storms.

A strong ionospheric current system is seated approximately in the classical auroral zone, but the detailed relation between aurorae and the electric currents is obscure.

The so-called radio aurora signifies the ionization in the E-layer

CHIEF AURORAL AND AIRGLOW EMISSIONS IN THE VISIBLE REGION[a]

Emission	Spectral Region or Wavelength	Approximate Height (Kilometers)	Approximate Intensity (Rayleighs)	
			Bright Aurora[b]	Nightglow
OI	5577Å	90–110	100,000	250
	6300, 6364Å	160	> 50,000	150
NII	Blue to red		25,000	
H (Balmer Series)	Red, blue	E-layer	1,000	
N_2 (1st Positive)	Red	D-layer	50,000	
(2nd Positive)	Violet	D-layer	100,000	
N_2^+ (1st Negative)	Blue-violet		165,000	
O_2^+ (1st Negative)	Red-yellow	D-layer	10,000	
NaI	5890, 5896Å	80–90		100 (winter) 20 (summer)
OH	Red-yellow	60–100		100
O_2 (Herzberg)	Blue-violet	90–100		15

[a] Adapted from Chamberlain: "Physics of the Aurora and Airglow," Academic Press, New York, 1961, to which reference should be made for data including ultraviolet and infrared emissions.

[b] International Brightness Coefficient III (brightness of moonlit cumulus clouds).

Note: Emissions are highly variable or absent with type and latitude of aurora. Heights are given only when well-defined.

Fig. 2. Auroral draperies. The drapery at the left is seen nearly edge-on. (*Hessler*: *Chamberlain, "Physics of the Aurora and Airglow," Academic Press Inc.*)

that is associated with magnetic disturbances, and gives rise to characteristic type radio reflections in the VHF (30 to 300 MHz) band and less often at higher frequencies. It has been suggested that radio aurora may be identified with the optical aurora, but little evidence exists for this.

The chief characteristic of the ionization is that it is aligned along the earth's magnetic field, the size of the irregularities ranging from meters to kilometers in length. The mechanism producing it is obscure; wind shears and particle precipitation probably contribute.

The pattern of ionization usually shows a systematic movement which in and below the auroral zone is statistically very similar to the ionospheric disturbance current system, but there are difficulties in interpreting the movement as that of the electrons in the current system. Other interpretations are that the movement is that of the ionizing sources, or even sound waves.

Frequency electromagnetic noise emission (hiss), centered around 8 kHz is observed in association with aurora. Satellite observations have sometimes shown correspondence between electron precipitation, auroral light intensity, and hiss, but at other times the correlation is poor. Theories of this noise emission all consider the interaction of a stream of particles with the surrounding plasma. Traveling wave tube amplification, Cerenkov radiation, or Doppler-shifted cyclotron generation by protons have been suggested.

The airglow is subdivided into the dayglow, twilightglow, and nightglow. The sodium intensity in the nightglow and twilightglow is highest in local winter, but seasonal and diurnal variations of the other emissions are not clear as there are marked latitude effects and distinct patchiness. The origin of the nightglow is obscure, though an important part of the oxygen red emission is excited by electron-ion recombination in the F layer. At the 85 to 100 kilometer level, there are complex chemical reactions involving oxides of nitrogen as well as the free gases and ions. The energy sources are far from understood; winds, turbulence, the quiet day ionospheric current system, thermal excitation, and even particles may contribute.

Considerable knowledge pertaining to auroral activity is being gained from photographs taken from satellites. The most illuminating aspects of recent pictures are the large field of view covering a substantial fraction of the auroral oval and coverage of formerly inaccessible areas. Feldstein and Starkov (see references) have suggested that auroral activity occurs along an oval that surrounds the north geomagnetic pole and along a similar oval around the south geomagnetic pole. The position of the oval varies with geomagnetic activity. Its geomagnetic colatitudes are about 23° on the night side and 15° on the day side during periods of moderate geomagnetic activity. Observations indicate that the aurora frequently displays an eddy-like form with a characteristic length of a few hundred kilometers. Hasegawa (see references) has suggested that this may be the result of kink instability in the field-aligned sheet current proposed by Akasofu and Meng (see references). A short summary of auroral photographic studies conducted by U.S. Air Force Weather Service satellites is given in *Science*, **183**, 4128, 951–952 (1974).

In northern latitudes, auroral displays are called *aurora borealis*, *aurora polaris*, or *northern lights*. In southern latitudes, they are called *aurora australis*.

Fig. 3. Homogeneous horseshoe band. (*Hessler*: *Chamberlain, "Physics of the Aurora and Airglow," Academic Press Inc.*)

References

Chamberlain, J. W.: "Physics of the Aurora and Airglow," Academic Press, New York, 1961.

Störmer, C.: "The Polar Aurora," Clarendon Press, Oxford, 1955.

Subcommittee of I.A.G.A. on Auroral Classification: "Atlas of Auroral Forms," Edinburgh University Press, Edinburgh, 1963.

Anger, C. D.: *Eos Trans. Am. Geophys. Union*, **53**, 108 (1972).

Feldstein, Y. I. and G. V. Starkov: Planet Space Science, **15**, 209 (1967).

Omholt, A.: "The Optical Aurora," Springer-Verlag, New York, 1971.

Hasegawa, A.: *Phys. Rev. Lett.*, **24**, 1162 (1970).

AUSCULTATION. This term is applied to the examination of the sounds within the chest, abdomen, heart, or larger blood vessels. It is carried out by listening with a stethoscope, or by applying the ear directly to the surface of the body. See also **Stethoscope.**

AUSTENITE. The solid solution based upon the face-centered cubic form of iron. The most important solute is usually carbon, but other elements may also be dissolved in the austenite. See also **Iron Metals, Alloys, and Steels.**

AUSTENITIC STEELS. Chromium; Iron Metals, Alloys, and Steels.

AUSTRALIAN GUM TREES. Eucalyptus Trees.

AUSTRALIAN PODOCARPS. Podocarps.

AUSTRALIAN REALM. Biome.

AUTHIGENOUS (or Authigenic). A geologic term proposed by Kalkovsky in 1880, meaning generated on the spot, and referring particularly to the primary and secondary minerals of igneous rocks and the cements of sedimentary rocks.

AUTOCHTHONOUS. A geologic term proposed by Gümble in 1888 for sedimentary rocks which have been formed in place. Now generally used to designate bedrock masses that have remained in place in a mountain belt, the term allochthonous denoting masses that have been moved long distances.

AUTOCLASTIC. A term proposed by Van Hise in 1894 for crush breccias or fault breccias which have been fragmented in place.

AUTOCOLLIMATING SPECTROGRAPH. Spectrography.

AUTOCOLLIMATOR. 1. A device by which a lens makes diverging light from a slit parallel, and then after the parallel light has passed through a prism to a mirror and been reflected back through the prism, the same lens brings the light to a focus at an exit slit. 2. A telescope provided with a reticle so graduated that angles subtended by distant objects may be read directly. 3. A convex mirror placed at the focus of the principal mirror of a reflecting telescope and of such curvature that the light after reflection leaves the telescope as a parallel beam.

AUTOCONVECTION. Atmosphere (Earth).

AUTOCORRELATION. The correlation of the members of a time-series among themselves.

Autocorrelation Coefficient. If ξ_t is a stationary stochastic process with mean m and variance σ^2 the autocorrelation coefficient of order k is defined by

$$\rho_k = \rho_{-k} = \frac{1}{\sigma^2} E(\xi_t - m)(\xi_{t+k} - m)$$

where the expectation relates to the joint distribution of ξ_t and ξ_{t+k}.

In a slightly more limited sense, if x_t is the realization of a stationary process with mean m and variance σ^2 the autocorrelation coefficients are given by a similar formula where the expectation is to be interpreted as

$$\lim_{n_2 - n_2 \to \infty} \frac{1}{(n_2 - n_1)} \sum_{j=n_1}^{n_2} (x_{t+j} - m)(x_{t+j+k} - m)$$

The expression is also applied to the correlations of a finite length of the realization of a series. Terminology on the subject is not standardized and some writers refer to the latter concept as serial correlation, preferring to denote the sample value by the Latin derivative "serial" and retaining the Greek derivative "auto" for the whole realization of infinite extent.

Autocorrelation Function. The graph of the autocorrelation coefficient as ordinate against the order k as abscissa is called the correlogram. When the series is continuous in time, the set of auto coefficients may be summarized in an autocorrelation function. This is the autocovariance divided by the variance, e.g., for a series with zero mean and range $a \leq t \leq b$, defined at each time point, is given by

$$\rho(\tau) = \frac{1}{b-t-a} \int_a^{b-t} u(t)u(t+\tau)\, dt \Big/ \frac{1}{a-b} \int_a^b u^2(t)\, dt$$

The limits a and b may be infinite subject to the existence of the integrals or sums involved.

The numerator of this expression is called the autocovariance function.

Sir Maurice Kendall, International Statistical Institute, London.

AUTODEPOSITION. A generic term coined to describe a fairly recent (late 1970s) development in which conversion coatings and organic coatings are applied to metal substrates in a single stage. The process is analogous to the electrodeposition of organic coatings, but in autodeposition, the coatings are deposited by means of chemical action rather than an electric current. There is no need for a separate conversion coating stage. See also entry on **Conversion Coatings.**

Autodeposition baths comprise colloidal dispersions of film-forming coating materials, such as latexes or polymer emulsions, acids, and oxidizers. Clean metal surfaces are immersed in the coating composition where the substrate metal is lightly attacked by the activating system of acid and oxidizer. This results in the formation of metal ions which overcome the stabilizing charges on the polymer particles and cause them to deposit on the metal surface. The thickness and type of conversion coating which forms simultaneously are determined by the kind and degree of activation.

Autodeposition is characterized by the growth of the coating with time of immersion and the ability to withstand water rinsing upon removal from the coating bath (without loss of coating) before fusion or curing of the coating by heat.

The elimination of the conversion coating sequence used in conventional industrial coating processes means that autodeposited coatings can be applied in fewer steps and in smaller finishing areas. There are no solvents in the commercial baths. This reduces air pollution virtually to zero and eliminates fire hazard completely. The coatings exhibit excellent adhesion, impact resistance, flexibility, and chemical resistance. There is no "throwing power" limit, such as in electrodeposition. The coatings can be applied to any partially enclosed surface that can be contacted by the bath and are not attacked by "solvent wash." As of the early 1980s, the largest application of autodeposition systems is in the automotive industry.

Wilbur S. Hall, Amchem Products, Inc., Ambler, Pennsylvania.

AUTOECOLOGY. Ecology.

AUTOERYTHROCYTE PURPURA. Purpura.

AUTOGAMY. A process of nuclear reorganization in protozoa in which the nucleus divides, each half undergoes a maturation, and the two persisting functional nuclei reunite. In the modified process known as paedogamy the individual forms a cyst within which it divides into two cells which reunite after the nuclear transformation is completed.

AUTOGENOUS. Self-generated, originating within the body. The term is usually applied to vaccines which are made from a patient's own bacteria as opposed to stock vaccines which are made from cultures grown from standard strains.

AUTOGIRO. Helicopters and V/STOL Craft.

AUTOIMMUNITY. Diabetes Mellitus; Immune System and Immunology.

AUTOINFECTION. Infection.

AUTOINTOXICATION. Seldomly used term to describe poisoning by a substance generated within the body and which the body is unable to eliminate without treatment.

AUTOIONIZATION (or Preionization). Some bound states of atoms have energies greater than the ionization energy. An atom which is in a discrete energy state above the ionization point can ionize itself automatically with no change in its angular momentum vectors if there is a continuum with exactly the same characteristics. This process is called autoionization.

AUTOLYSIS. The energy derived from biological oxidations in living cells serves to promote anabolic processes, i.e., to produce relatively complex, highly ordered molecules and structures, and thus normally keeps living cells in a steady state remote from equilibrium. In organisms that lack cellular nutrients or oxygen (or in dead organisms or cells that have been disrupted so as to destroy much subcellular organization), the opposing catabolic tendency toward equilibrium, including the tendency toward degradation of macromolecules to simpler monomeric subunits, is not counterbalanced. These degradative processes, many of them enzymatically catalyzed, are collectively termed autolysis. Autolytic processes may include, for example, hydrolysis of proteins catalyzed by proteolytic enzymes or hydrolysis of nucleic acids catalyzed by nucleases. Autolysis of tissues (e.g., liver homogenate) has sometimes been used as a method for releasing bound molecules (e.g., vitamins or coenzymes) into free soluble form.

AUTOMATIC ARM-HAND (Industrial). **Robot.**

AUTOMATIC CONTROL VALVES. **Valve (Control).**

AUTOMATIC DATA PROCESSING SYSTEM. **Data Processing.**

AUTOMATIC EXPOSURE METERING. **Photography and Imagery.**

AUTOMATIC PILOT. By combining signals from a gyro vertical as a vertical reference source, from rate gyros as needed to give stability on any or all of an aircraft's three axes, and from a directional gyro or stabilized compass for heading reference, actuators can be made to operate on the aircraft's control surfaces (ailerons, rudder, elevators) and provide automatic flight. Altitude may be controlled by signals from an altimeter. In addition, signals from ground radio navigational facilities can be used to control an aircraft automatically along the airways and to the runway. For aircraft operating over a wide range of speeds and altitudes, such as supersonic jets, some of the control parameters must be changed to reflect new flight environmental conditions. This can be done by using inputs from total and static pressure sensors on an open-loop basis, or by sensing performance response of the aircraft and adjusting the control parameters to give the desired response by self-adaptive means. Simplified block diagrams of representative automatic-pilot systems are given in the accompanying figure.

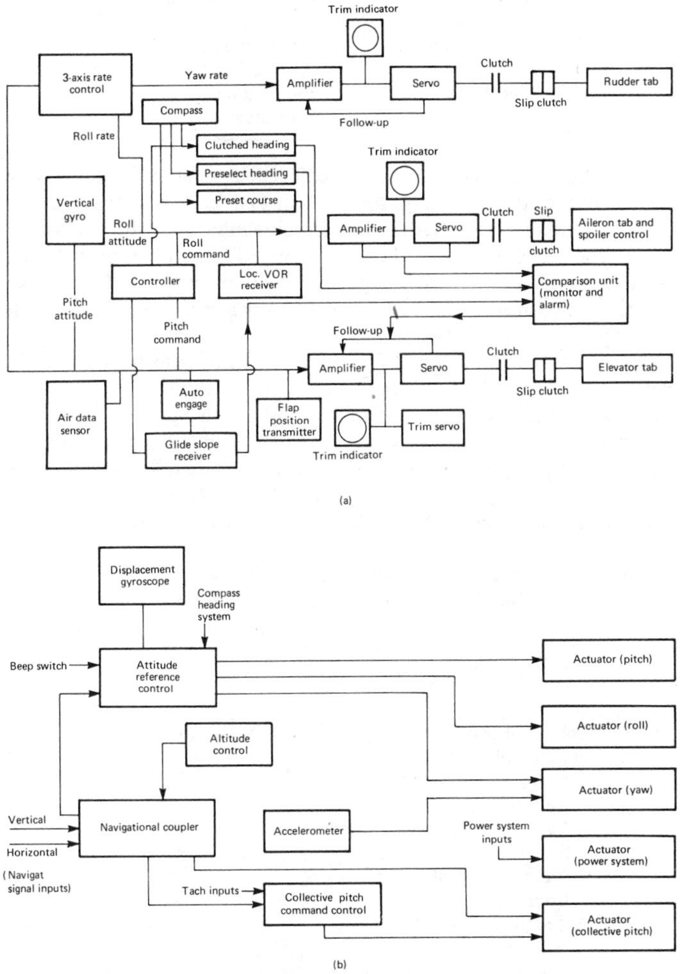

Simplified block diagrams of automatic-pilot systems.

AUTOMATIC RANGE FINDING. **Photography and Imagery.**

AUTOMATIC VOLUME CONTROL. Also called automatic gain control, a circuit incorporated in a radio receiver to maintain the amplitude of the signal applied to the detector as nearly constant as possible and thus correct for variations in the signal strength caused by fading. When a modulated radio frequency wave is detected, the output contains a dc component which has a magnitude proportional to the amplitude of the carrier impressed on the detector. Other components (af and rf) are filtered out and this dc voltage is applied as bias to the intermediate frequency and/or radio frequency amplifier stages.

The accompanying figure shows the essential components of an automatic volume control system using transistors. A dc voltage is developed across a capacitor at the output of the detector which is

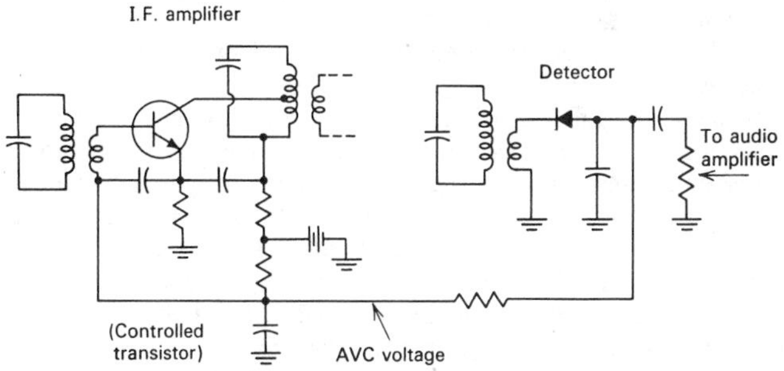

Automatic volume control system.

proportional to the carrier amplitude, and it is applied to the base of a transistor intermediate frequency (i-f) amplifier. The voltage has a polarity which reduces the emitter current as the magnitude of the rectified voltage increases. The change in emitter current reduces the gain of the i-f amplifier stage which in turn decreases the carrier signal at the detector, the net effect being a significant reduction in variations in signal level at the detector. Although control of only one state is shown, it is common practice to control several i-f amplifier stages and sometimes radio frequency amplifiers as well.

AUTOMATION. When the term *automation* appeared in the literature in the 1930s, it referred to "the automatic transfer of parts and materials from one production operation to the next." This essentially described an extension of mass production methods in use by the automotive industry, essentially an extension of mechanization of individual machines to include the mechanized interconnection of machines. Thus, for a number of years, the term was limited to those industries that specialized in the handling of discrete pieces—parts, subassemblies, assemblies, as contrasted with the handling of bulk and fluid materials as found in the process industries, typified by the chemical, petroleum, and petrochemical industries.

Actually, during the early period when automation in discrete piece manufacturing was considered newsworthy, the automaticity of the process industries far exceeded any practices found in the automotive and like industries. In retrospect, the process industries led all industries and continue to do so today in terms of automaticity. The process industries are among the least labor intensive of all industries. The term *automatic control* was used to describe automaticity in the process industries and continues, of course, in wide usage.

Over the years, automation as a term increased in scope so that today it embraces almost any system wherein the manual interface with the system is minimal. Sometimes the term is interpreted by laymen as describing a completely automatic system wherein there is no manual participation. This may be a design target, but fully automatic operation proceeds for only relatively short periods of time without manual intervention of one form or other. More practically, automation may be described as the application of measurement, information handling, and automatic control technologies to reduce, partially or extensively, the need for human intervention to make a process, machine, or almost any other kind of system proceed efficiently to produce the desired result. Even the most sophisticated systems require human programming at the start and maintenance throughout their life.

For a number of years, there was another general misinterpretation of the philosophy of automation, namely, that the primary target of turning to automation was that of reducing direct production labor costs. To be sure, this was a goal in numerous instances, but seldom the principal motivation. Automation is utilized because of human limitations in terms of human senses, reaction time, patience, reliability, and durability. Further, persons—even though qualified to accomplish many of the tasks performed by machines—prefer to turn over boring and backbreaking tasks to automation. The difficulties in finding and keeping people on tasks of this type attest to this fact. Many individuals prefer better working conditions and an opportunity to utilize their talents beyond their muscular strengths. They prefer to install, maintain, and improve machines, processes, and other systems rather than to engage in the heavy, tiring physical functions of the machine; or in the boring routine mental functions that can be handled so reliably and fast by computers.

See also **Control System (Automatic); Electronic Data Processing; Numerical Control; Robot.**

The degree of difficulty of automating an operation or process varies tremendously. In the process industries involving bulk solid materials, liquids, and gases which are transported via conveyors and pipelines, it is comparatively easy to tie various process measurements, such as temperatures, pressure, solid and liquid levels, densities, and so on, with the control of electric motor-driven conveyors and pumps or pneumatically or hydraulically operated valves, louvers, and dampers. And in the machine tool area, it is comparatively easy to preprogram the operation of a machine by reading the holes in a perforated tape. Considerably more difficulty is encountered in attempting to automate assembly operations where there is a great deal of matching of individual pieces required. An assembly operation involves a wide range of observations and judgments. Not only are these difficult to analyze in the first place, but a computer program designed to accommodate for so many variable, alternative-choice situations becomes very costly and complex. Thus, in situations which require very costly, complex approaches to their automation, there must be high production in order to economically justify automation. While there are some examples of automation wherein a good degree of flexibility has been achieved, these cases are in the minority. Usually automation does not dovetail well with complex runs of a short-term nature.

Because of these characteristics of automation, there has been a growing trend over the last several years to design for automation, that is, to make parts and assemblies less complex, easier to make and easier to put together, often of a modular nature so that the automation system can be simpler and can embrace some flexibility in handling items of a limited number of sizes, shapes, and styles. Unfortunately, an item designed with automation solely in mind is rarely of maximum appeal to the consumer and marketplace.

One of the most labor intensive industries is the manufacture of garments and apparel. The seamstress trade is progressively less attractive to the younger generation and the industry has been making efforts to automate, but the progress has been very slow. The skills of sewing have been extremely difficult to transfer to a machine control system. Unlike the metalworking field, the seamstress does not have rigid, precise pieces with which to work. Rather, the materials are soft, pliable, and from the standpoint of a machine, many more times more difficult to manipulate and control. Further, the geometry of the parts, and dependence upon the nature of the seam for shape and drape of the garment are all factors that do not enter into the assembly of hard metal parts. The extremely difficult technical problems, coupled with an industry that is not accustomed to the types of capital expenditures that are traditional in the process and automotive industries, for example, have slowed the pace of automation in the garment field.

Significant advances in the automation of laboratory (medical, microbiological, etc.) have been made in recent years.

AUTOMATISM. **Coma; Brain (Injury).**

AUTOMORPHISM. **Isomorphism; Number Theory.**

AUTOMOTIVE CARBURETOR. **Carburetor.**

AUTOMOTIVE ELECTRONICS. There have been many conferences and many papers published[1,2] on automotive electronics. Introduction of microprocessors to engine control unveiled a new era of automotive electronics. This extended application of microprocessors arrived just two years after the introduction of the first electronic spark control system (1977 Oldsmobile Toronado). Since the 1950s, twenty automotive electronic systems have been introduced, excluding communications and entertainment equipment. Sixteen of these remain and are listed in Table 1 in order of their introduction.

TABLE 1. ELECTRONIC SUBSYSTEMS IN AUTOMOBILES (Early 1980s)
(In chronological order of their introduction)

Headlight control	Wheel-lock control
Alternator rectifier	Clock
Voltage regulator	Intrusion alarm
Tachometer	Air-cushion control
Cruise control	Electronic fuel injection
Electronic ignition	Lamp timing control
Climate control	Spark timing control
Windshield wiper control	Electronic digital displays

With the rapid advances being achieved in solid-state technology, increased packaging density, lower cost, and improved reliability, it is understandable why there is considerable interest in the application of electronic technology to the automobile. The stimulation provided by new technologies and the pressures of excessive and, at times, questionable governmental regulations have accelerated the automotive application of solid-state technology.

One measure of this increase in activity is the projected automobile electronic content as a percentage of automobile cost. Sources[3,4] predict that ultimately over 10% of the cost of an automobile may be represented by electronics. That would represent a large worldwide electronic sales potential in excess of $10 billion annually, not including entertainment and communications equipment. In the United States, electronics valued at $300 (1981 $) will be installed on the average vehicle by 1985.

The size and complexity of an automobile electrical system makes it a prime candidate for the application of contemporary digital control

TABLE 2. TYPICAL ELECTRICAL SYSTEM OF LUXURY CAR (United States)*

Length of cable	1500 feet (457 meters)
Number of:	
Switches	85
Motors	15
Lamps	70
Sensors	30
Fuses and breakers	25
Connectors	110
Approximate price (1981 $)	$150–175

* Not including up to 20 electronic modules and excluding entertainment.

TABLE 3. UNITED STATES PASSENGER CAR EMISSION AND FUEL ECONOMY STANDARDS

Model Year	Test Procedure	Emission Standards				Efficiency	
		HC	CO	NOX	Particulates	mi/gal	km/lit
1981	Federal	0.41	3.4	1.5	—	22.0	9.3
	California	0.39	3.4	1.0	—		
	Altitude						
1982	Federal	0.41	3.4	1.5	0.6	24.0	10.2
	California	0.59	3.4	1.0	0.6		
	Altitude	0.58	7.8	1.0	—		
1983	Federal	0.41	3.4	1.0	0.6	26.0	11.1
	California	0.41	3.4	1.0	0.6		
	Altitude	0.51	8.1	1.0	—		
1984	Federal	0.41	3.4	1.0	0.6	27.0	11.5
	California	0.41	3.4	1.0	0.6		
	Altitude	0.41	3.4	1.0	0.6		
1985	Federal	0.41	3.4	1.0	0.2	27.5	11.7
	California	0.41	3.4	1.0	0.2		
	Altitude	0.41	3.4	1.0	0.2		

systems technology. In addition to the electronic subsystems listed in Table 1, there are many additional components as shown by Table 2. Several of these components may be replaced in the future by solid-state devices.

Emissions and Fuel Economy

Government regulations regarding emissions and fuel economy have a significant impact on the continuing application of electronics to the automobile. These regulatory requirements cannot, in general, be satisfied with current technology. The 1985 statutory emission control standards (EPA) require automobiles to be designed so that they do not exceed the exhaust emission values shown in Table 3 during the first 50,000 miles (80,450 kilometers) of vehicle operation.[5] As emission standards become more stringent, a reduction in fuel economy may be anticipated if compensating technology cannot be devised. More stringent emission standards are in direct conflict with other government regulations which require each manufacturer to improve fuel economy substantially by 1985.

Fuel economy improvements are being achieved by making vehicles smaller and lighter, by making them aerodynamically more efficient, and by improving engine and power train efficiency. It is in the latter area that electronics can be expected to be applied. See Fig. 1. It is also in this area that strict attention must be paid to meeting the stringent emission standards.

Because of pressures to reduce emissions and increase fuel economy, the highest automotive electronics priority in the United States has been, and continues to be, more precise control of the automotive power train.[6] Within three successive years, one manufacturer (Ford Motor Co.) introduced new electronic engine control (EEC) systems, each one significantly smaller, less costly, and more powerful than its predecessor. During that period, that manufacturer also developed different versions of the basic air-fuel control systems (feedback carburetor) for 4- and 6-cylinder engines.

Concurrently, another manufacturer (Chrysler Corp.) developed a lean-burn electronic control system and another manufacturer (General Motors Corp.—GM) developed a computer controlled catalytic converter (C-4) system—making marked advances in electronic power train controls.[7] The GM digital electronic fuel injection system additionally incorporates advanced self-diagnostic capability, displaying problem codes in the electronic temperature control panel, plus providing instantaneous fuel consumption data to a separate display module. Development of new and better electronic systems requires sensors (temperature, pressure, position, knock, and exhaust gas) and actuators (injectors, carburetor controllers, and EGR valve positioners) which are at least as costly in terms of effort and investment by manufacturers and their suppliers as the basic control system itself.

During the 1980s, a continuation of these trends, including faster and more powerful microprocessors, is expected. Sixteen-bit machines will replace current 8-bit designs for such future developments as sequential fuel injection. Also, simpler and lower cost single-chip microcomputers will be developed for limited control of smaller engines.

Closed-loop Feedback Systems. Expanded use of closed-loop feedback systems, including knock-detection sensors for adaptively optimized spark timing and off-stoichiometric exhaust gas sensors to enable economical lean-burn engine strategies may well provide many of the benefits of contemporary full electronic engine control systems at a fraction of present cost and complexity levels. More precise sensor technology is also anticipated to achieve direct measurement of air flow. Cost reduction and reliability improvements will remain ongoing high priorities, especially for sensors and actuators. Low-cost, reliable sensors and actuators are urgently required for the systems of the 1980s.

A closed-loop, knock-limiting system was developed by the industry which allows improvements in fuel economy while permitting an engine to operate at acceptable knock levels on current 91 octane fuels. The system is designed to control ignition timing only under knocking conditions so that optimum timing may be used under all other conditions to improve fuel economy while satisfying exhaust emission constraints.

A sensor determines engine knock and provides information to an electronic logic system which retards timing to acceptable knock levels. An acceptable system must control knock under all operating conditions of speed, load, temperature, humidity, or altitude without significant degradation in economy, driveability, performance, or emission

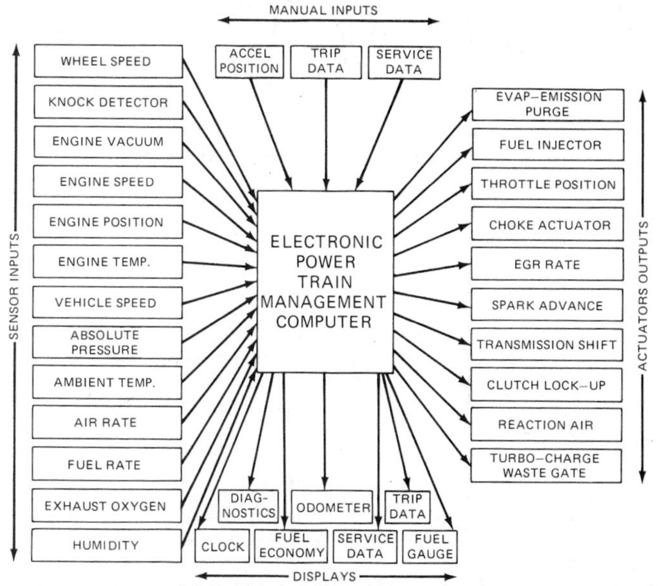

Fig. 1. Electronic power train control system for automobile.

levels. A low-cost device to sense knock at trace levels in all cylinders without mechanical or electrical interference is required.

Catalysis. The established practical method of achieving the statutory emission standards is catalysis. Systems which simultaneously control hydrocarbon (HC), carbon monoxide (CO), and nitrogen oxide (NOX) emissions with a single catalyst, currently require the air and fuel mixture to be precisely controlled to the stoichiometric ratio (14.6 : 1), at which point optimum catalyst efficiency is obtained (Fig. 2) for the three regulated byproducts of combustion. This level of control is achieved with a system in which the oxygen remaining in the engine exhaust is used as the sensing parameter. The oxygen transducer is essentially a battery cell comprised of a solid electrolyte of zirconium dioxide with platinum-plated electrodes on either side. At the stoichiometric ratio and exhaust gas temperatures, the zirconium sensor provides excellent signal characteristics (Fig. 2) which coincide with optimum catalyst efficiency for the constituents of interest. A simplified block diagram of the closed-loop system is shown in Fig. 3. Closed-loop emission control systems may be used on both electronic fuel injected and carbureted engines if a three-way catalyst is used, and considerable effort is being expended by the automobile industry to develop such systems.

Fuel Injection. This is a generic term which covers a variety of products. In diesel engines fuel injection means injection directly into each cylinder. In multipoint systems, gasoline is injected at each intake port. In throttle body injection, gasoline is injected at the intake manifold plenum. In many of the German and Japanese cars, fuel injection is continuous and mechanically controlled. In the recent fuel injected cars made in the United States, injection is intermittent and electronically controlled.

In 1981, Cadillac introduced an engine which can operate on eight, six, or four cylinders, depending on power demand.[7] Change from one operating mode to another is accomplished by selectively blocking off operation of intake and exhaust valves in given cylinders. This is done electromechanically via a microprocessor-controlled electronic system which switches solenoids mounted under the valve cover, above the rocker arms of the valves which they control.

A digital display on the dash can inform the driver of the number of cylinders on which the car is operating. By driving in such a way as to operate the engine on six or four cylinders as much of the time as possible, a careful driver can increase fuel economy.

In addition, all 1981 Cadillacs equipped with the variable-displacement V-8-6-4 engine were fitted with computerized digital fuel injection (DFI). The DFI is carefully integrated with the electronics which control the engine's valve selectors, furnishing fuel quantity precisely tailored to engine demand under all operating conditions.

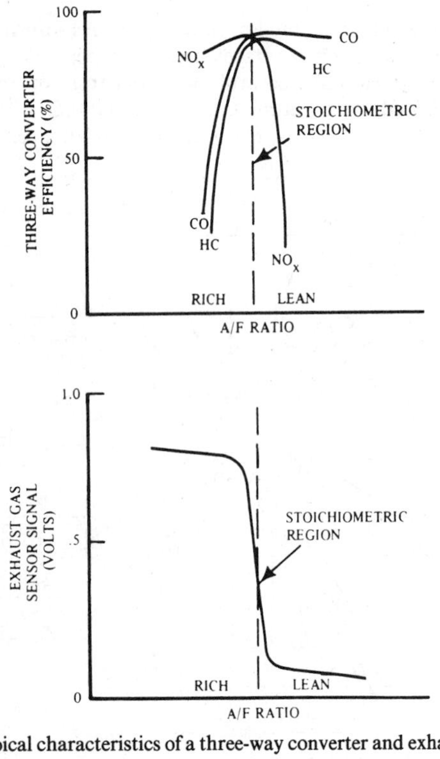

Fig. 2. Typical characteristics of a three-way converter and exhaust gas sensor. (*TRW, Inc.*)

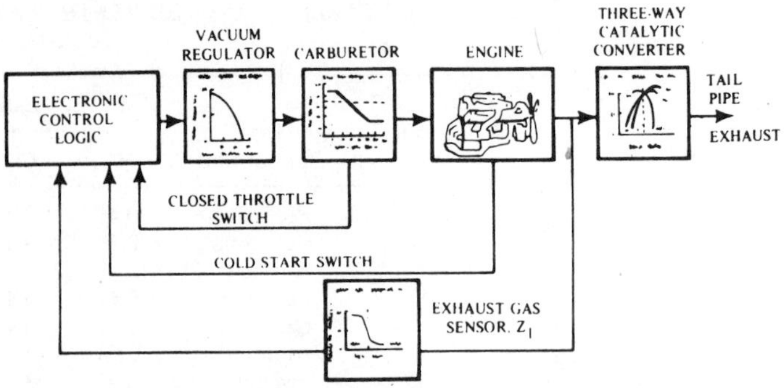

Fig. 3. Closed-loop system for controlling air-fuel mixture. (*TRW, Inc.*)

The DFI is a speed-density system which uses a throttle body injection assembly with two electronically-pulsed fuel injectors. The system's brain is an electronic control module (ECM), a digital microprocessor which provides all computational capability for the engine controls, including spark timing, fuel metering, and idle speed. The ECM also provides diagnostic capability, identifying 45 engine sensor circuits and displaying 11 pieces of sensor data.

The GM 1981 gasoline-car fleet was equipped with a computer command control (CCC) system designed to reduce emissions without penalizing fuel economy. To meet 1981 emissions standards in all 50 states, the CCC emissions system utilizing a microprocessor-based electronic control system precisely regulates air/fuel mixture and controls the air injection reactor (AIR) pump output so that optimum conversion efficiency is obtained in the catalytic converter. The system also controls engine spark advance, idle speed, transmission torque converter clutch (on rear-wheel-drive cars), and canister purge.

The CCC system contains a "Check Engine" instrument panel warning light which advises the driver of detectable malfunctions in the system. The light also assists service technicians in locating malfunctions by flashing a specific two-digit code, when requested, to isolate the potential problem areas and components on the car.

Truck Systems

Over the past few years, there have been several applications or announcements of electronic systems for trucks. In the early 1970s, government standards[8] for buses, trucks, and trailers with air brakes were announced. These were sufficiently stringent that manufacturers turned to the use of electronic wheel-lock control systems in order to meet the performance levels specified. The basic problem was that initial requirements were too stringent, and insufficient development time was allowed for by the government regulators in spite of substantial and continuing pleas on the part of the vehicle manufacturers. The resulting unacceptable reliability of the wheel-lock control systems prompted the government to suspend stopping distance requirements.

Protecting the expensive heavy-duty diesel engine has long been a requirement of the industry, and electronics has been successfully used in a shut-down system. Engine oil pressure, coolant level, and temperature are sensed; and where an unacceptable level of any of these is indicated, an alarm is first sounded and, after a period of time, the engine is shut down.

GM has developed a low tire pressure warning system for trucks, tractors, trailers, and buses. The system consists of four major components: a temperature-compensated pressure switch, an actuator-sender that replaces one wheel mounting bolt, a stationary sensor behind the mounting bolt, and a display panel. Low tire pressure lights a panel indicator and a "pop-up" button in the wheel actuator provides out-of-cab detection. The warning continues until the tire has been inflated and the "pop-up" button reset. The low tire pressure warning system maximizes tire life and fuel economy in addition to improving safety and reducing maintenance inspection time.

The recent increases in fuel prices have placed considerable interest in systems and devices which help save fuel. Some companies have introduced cruise or speed control specifically designed for heavy duty truck and bus applications. Digital microprocessor technology is utilized to control the operation of a pneumatic cylinder which is connected to the throttle and regulates vehicle speed. The action of this closed loop speed control system reduces throttle activity and is reported to increase fuel economy by up to 5%.

Speed limiting device utilizing analog electronics to sense a preset speed limit have also been introduced. When this occurs, engine fuel is reduced until vehicle speed is lowered to an established lower speed limit. If the driver attempts to exceed the top speed limit, this cycling is repeated. A fuel saving approaching 5% is reported for this device.

TRW* is pursuing a complete line of microprocessor based systems called ETEC (electronic truck engine control) which incorporates cruise control, road speed limiting, engine speed limiting, and all-speed governing into a single control package. The ETEC concept employs a serial communication bus which allows the expansion to include engine diagnostics and shutdown and vehicle operation recorder subsystems, all of which share common sensor and actuator data reducing the number of components over separately installed systems. This increases the overall system reliability and integrity.

Electronically-Controlled Diesel Injection Systems

The application of electronic controls to gasoline powered cars of the 1970s is now expanding to diesel powered cars of the 1980s.[9,10] The successful application of microprocessor technology, sensors, and actuators to gasoline cars provides a technology base for improving the performance and emissions of the diesel engine.

This is accomplished by electronically controlling the injection pump, Fig. 4(a), as opposed to mechanical control of the pump, Fig.

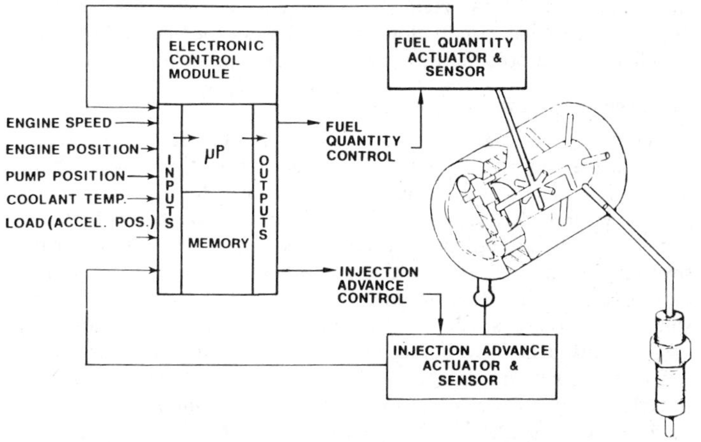

Fig. 4(a). Basic electronic diesel control system. (*TRW, Inc.*)

4(b). The open loop mechanical controls on the pump will be replaced by fuel quantity control and injection advance control actuators. Corresponding sensors for these functions will complete the closed loop system through the control module. Inputs from engine speed and position, pump position, coolant and air temperature, and accelerator pedal position complete the diesel engine control scheme.

The advantages of an electronic diesel controlled system, Fig. 4(c), in addition to the obvious increase in system precision, are numerous. The electronic control system has memory and computing capability. This makes it both simple and relatively inexpensive to add other electronically controlled subsystems such as idle speed, cruise, transmission and EGR control to the basic system. Second, because it is a closed-loop and therefore a self-compensating system, installation errors and wear decalibration can be eliminated. The additional advantages of reduced engineering development time (because software changes are much easier than pump hardware changes) and reduced manufacturing costs (because electronic PROM changes for various calibrations are simple and more cost-effective than stocking thousands of various mechanically matched parts) make electronic diesel control most enticing.

* TRW, Inc., Transportation Electrical & Electronics Operations.

Fig. 4(b). Present mechanical speed/load/advance timing map.

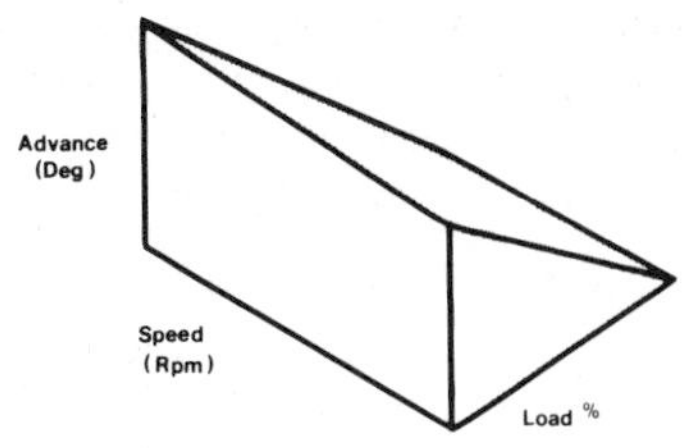

Fig. 4(c). Optimum diesel speed/load/advance timing map.

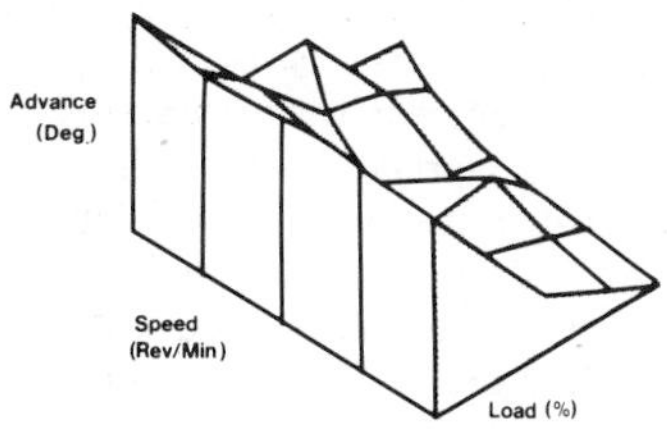

Displays

In the 1977 model year, GM introduced electronic displays integral with an automobile AM/FM radio (Fig. 5). The four-digit, seven-segment light-emitting diode (LED) displays four functions on command; namely, date, time of day, elapsed time, and radio station frequency. In addition to LEDs, electronic display technology under development around the world includes gas discharge, liquid crystal, electrochronic, and electroluminescence.

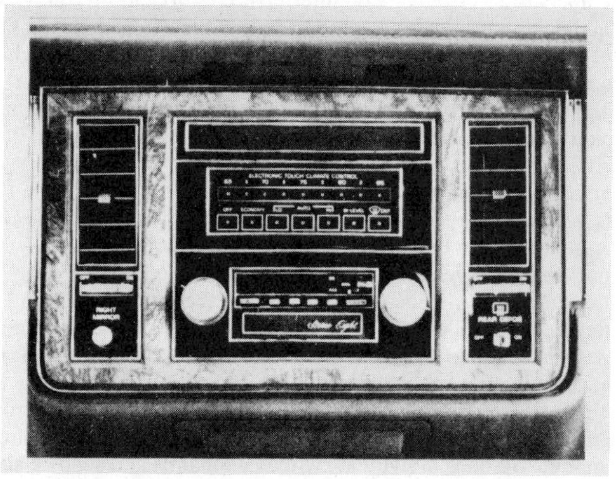

Fig. 5. GM electronic display radio and electronic automatic temperature control.

Since the late 1970s, automotive engineers have entered a new area in automotive instrumentation (Fig. 6) in which electronic technology allows redefining many historical guidelines. New display technology is permitting improvements in conventional instrumentation and sup-

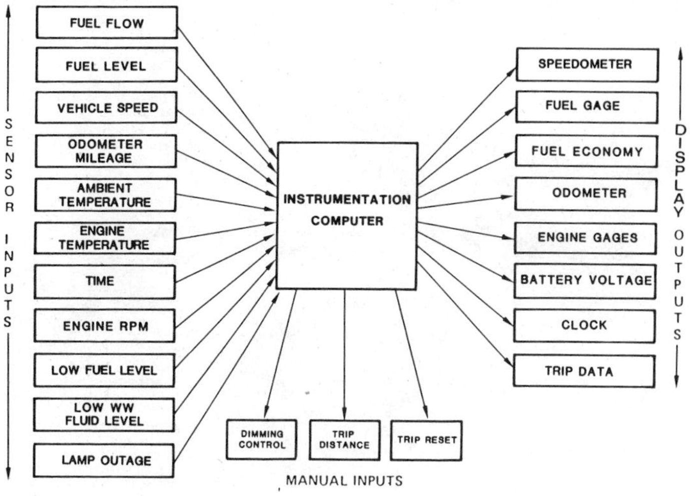

Fig. 6. Instrumentation and display system.

porting the development of completely new product features. These products can improve the overall appeal of the vehicle by adding the value of new functions, improving the performance of existing functions, and allowing the use of new and unique styling themes.

Automotive Wiring Multiplex Systems

As electrical and electronic systems are added to the automobile, it becomes progressively more difficult to meet reliability objectives because of multiple connectors.[11] Multiplex systems have the potential to improve reliability and reduce weight, and they can be designed to be immune to noise. It appears that adequate technology would be available for the use of such systems if their cost were more competitive with current wiring systems. An additional advantage of such a system is the relative ease of introducing diagnostic routines, provided low-cost transducers are available.

An interesting development is the application of fiber optics to automotive multiplexing systems. Advances in fiber optic control systems include more efficient infrared light sources, low-noise photo diodes, and rugged, efficient fiber optic cables, many of which are now available off the shelf.

Lightweight, interference-free fiber optic cables are replacing heavy, complex copper wire bundles aboard military aircraft and ships. Underground communication cables are being replaced with light pipes capable of transmitting conversations over many kilometers. Current fiber optic technology for data lines can provide a weight advantage of up to 100:1 and a cost advantage of 30:1 compared to copper lines.

The extensive advantages of this technology include the immunity of the fibers to electromagnetic interference (EMI) and to oil, grease, dirt, and acids, and the high speed of information transmission. These advantages are of considerable importance to the automotive engineer. The advent of lower-cost components, coupled with the inherent performance advantages of this technology, make it an ideal means of reducing vehicle harness weight and providing EMI-immune systems (Fig. 7).

A multiplexed control system with a single fiber optic cable (Fig. 8) has been developed to replace the multiwire harness, connectors, and printed circuitry required to interconnect steering column switching functions with vehicle power actuators and lights.

An encoder module, located at the top of the steering column (Fig. 9), interprets the following switching commands and digitally multiplexes each command signal: windshield wiper, turn signal, hazard flashers, key reminder, horn, headlamp controls, and cruise control. These signals are transmitted as light pulses through the fiber optic cable to a decoder module located at the base of the steering column. This module decodes individual switch commands and sends a signal to the appropriate power actuator, for example, the windshield wiper motor, horn relay, or lights.

An inexpensive LED on the encoder chip is electrically pulsed. The digital words are converted by the LED to light pulses which are transmitted down the cable. A photo-diode on the decoder chip receives the light pulses as they emerge from the cable and reconverts them to electrical signals, which are recognized by the decoder on individual switch commands.

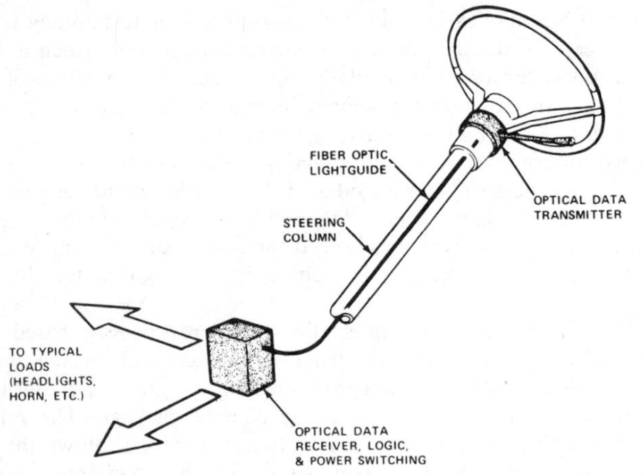

Fig. 8. Multiplexed control system, with a single fiber-optic cable. (*TRW, Inc.*)

Smaller packaging envelopes and the migration of driver controls to the steering column, for example, can complicate the electrical wiring to the extent that major manufacturing and field maintenance problems could be encountered. Proposals for more complex column control functions will further aggravate the situation. The production column shown at the bottom of Fig. 9 has controls for turn signal, head lamp beam control, cruise control, hazard warning, horn, cornering lamps, and windshield washer-wiper with analog pulse wipe. The column at the top of the figure incorporates multiplexing and features controls for rear window up and down, rear defrost, head lamp and parking lamp control, dome lamp, instrument lamp intensity, and the pulse wipers; these controls use digital positions instead of the analog system of the production column. The mechanical and electrical simplicity made possible through multiplex technology is evident.

Long-range Systems

The list of proposed electronic systems for the automobile would be extensive and exceed 100 items in addition to the systems shown in Fig. 10. Some of the advance systems which have been proposed are shown in Table 4. Many of these systems have the potential to become standard or optional automotive equipment in the next decade; several of them are under active consideration and may be seen on automobiles within this decade. In general, the near-term systems suffer from poor cost-effectiveness, generally because of the high cost of the transducers and actuators. While the utility of many of the longer-range systems can be justifiably challenged, the universal need for effective transducers and low-cost actuators cannot.

Government studies[12] continue to show that the driver is the major cause of over 80% of vehicle highway accidents.

Alcohol in particular continues to be a major cause of highway accidents. Over 50% of highway fatalities are due in part to alcohol consumption, and this percentage is expected to rise because of the projected increase in per capita alcohol consumption.[13] Drivers continue

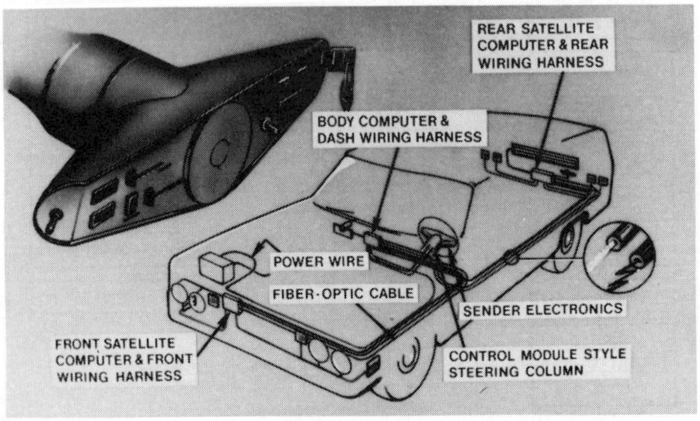

Fig. 7. Harness simplicity with optical multiplex. (*TRW, Inc.*)

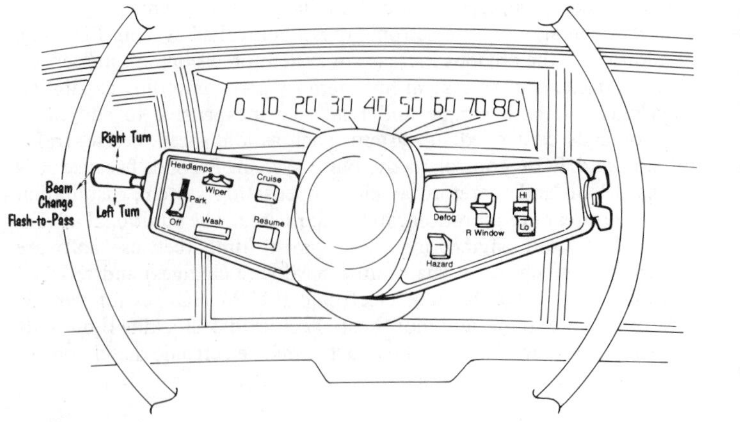

Fig. 9. Steering column control console. (*TRW, Inc.*)

TABLE 4. POTENTIAL 1990 AUTOMOTIVE ELECTRONIC SYSTEMS

COMFORT AND CONVENIENCE	SAFETY AND SECURITY	INSTRUMENTATION AND DISPLAYS	DRIVE TRAIN	ELECTRICAL
Speed Control	Flash Control	Clocks	Electronic Ignition	Alternator Diodes
Headlamp Control	Seat Belt Warning	Fuel Gages	Spark Timing	Voltage Regulator
Climate Control	Passive Restraint	Speedometer	EGR Control	Multiplex
Interval Wipers	Wheel Lock Control	Odometer	Idle Speed Control	Fiber Optics
Seat Positioner	Tire Pressure Monitor	Tachometer	Diesel Injection Control	
Illuminated Entry	Speed Limiting	Trip Computer	Gas Injection Control	
Keyless Entry	Crash Recorder	Message Center	Closed Loop Carburetor	
Garage Door Opener	Road Surface Indicator	Fuel Economy Gage	Diagnostics	
Mirror Control	Icing Control	Diagnostics	Diesel Starting Timer	
Programmed Horn	Sleep Detection	Voice Message	Evaporative Emission Purge	
Voice Actuation	Collision Avoidance	Speed Warning	Transmission Control	
Moisture-Sensitive	Blind Spot Warning	Planar Display	Clutch Lock Up	
Interval Wipers	Driver Physiology	Vehicle Data	Cooling System Control	
Advanced Speed Control	Traction Control	Computation Center	Turbo Charger Control	
			Reaction Air	
			Wide Cut Fuel	
			Programmed Combustion	
			Accessory Drive Control	
			Restart	

to run into the rear of other vehicles. In another government study[14] of rear-end collisions, 45% of the drivers in the striking vehicle claimed they never saw the vehicle struck until after the accident. Because of these conditions, it is expected that research will continue in areas aimed at reducing the driver-caused accident, with specific emphasis on research directed toward the ultimate goal of inhibiting the operation of an automobile by an intoxicated or otherwise impaired driver and toward automatic brakes.

Experiments have been conducted with electronic devices which will prevent a drunk individual from starting his car. The first device developed by General Motors was the Phystester,[15] which rejected approximately 50% of potential drivers with a blood alcohol concentration of 0.1 percent. This led to the development of a critical task tracker,[16] which gave a rejection rate of 75%. We are still seeking a device which would be 100% effective.

Accidents caused by driver inattention to vehicles stopping or slowing ahead indicate that radar-augmented braking systems may have potential for preventing highway accidents.[17] The major barrier to designing an acceptable system (Fig. 11) is the inability of the system to conduct a hazard analysis of potential obstacles. Figure 12 shows X-band radar signatures of three obstacles which are frequently involved in highway collisions. As can be seen, the small stop sign has a signal amplitude considerably higher than that of the 10-inch (25.4-centimeter) diameter tree. If the system detection threshold were set to include all three obstacles, a large number of false alarms might result, which would be unacceptable to the driver.

With the reduced cost of memories, such as charge-coupled devices and bubble memories, the potential exists to store sufficient radar signatures to support a high-speed hazard analysis. The initial application of radar to the automobile is expected to be as part of a cruise control or station-keeping system. If this is successful, it could lead to fully automatic braking systems.

Radar crash sensors for air-cushion restraint systems have been discussed[17] in the past. In this application, the severest hazard analysis

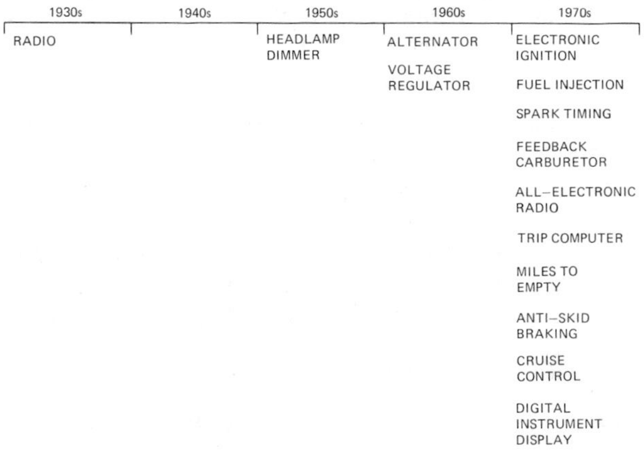

Fig. 10. Electronics applications in the automobile (1930s–1970s).

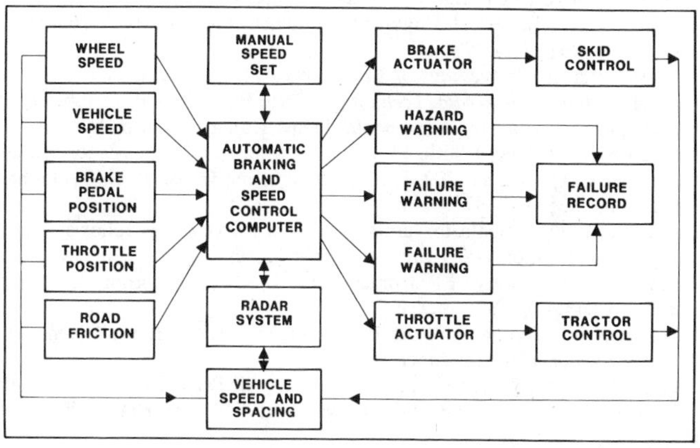

Fig. 11. Typical automatic radar braking system. (*TRW, Inc.*)

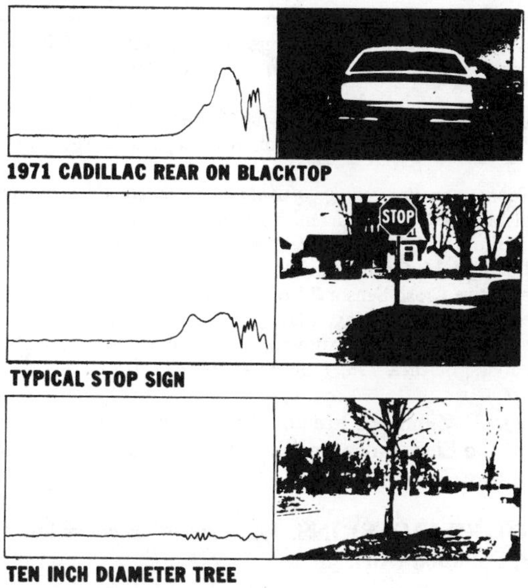

Fig. 12. X-band radar signatures of various situations. (*TRW, Inc.*)

situation exists because of the unacceptability of inadvertent or false inflations of the air cushion.

An effective collision avoidance system will be developed when pattern recognition techniques are employed. With the rapid development of solid-state radar and microprocessor technology, there is a high probability that a system will be economically feasible in the 1980s. Results obtained to date are very encouraging, though many questions and issues remain unresolved. Near-term issues of concern are:

- Evaluation of the automotive radar environment.
- Selection of the most effective pattern features which can be derived from the radar environment.
- Specification of projected system performance for various hazardous situations as well as degraded conditions.
- Development of an improved collision avoidance system concept which encompasses a greater number of driving situations.

References

1. See, for example, Society of Automotive Engineers, *Automot. Electron.*, SAE Publ No. SP-388 (January 1974); *Convergence 74*, International Colloquium on Automotive Electronic Technology, SAE Publ. No. P-57 (1974), p. 57; *Convergence 76, International Conference on Automotive Electronics and Electric Vehicles*, SAE Publ. No. P-68 (September 1976); *Automot. Electron.*, SAE Publ. No. SP-417 (February 1977); *Automot. Electron.* II, SAE Publ. No. SP-393 (February 1975); all published by the Society of Automotive Engineers, Warrendale, Pannsylvania.
2. Institution of Electrical Engineers, *Automot. Electron.* IEE Conference Publ. No. 141 (July 1976).
3. G. F. Villa, in *Proceedings of Convergence 74, International Colloquium on Automotive Electronic Technology*, SAE Publ. No. P-57 (Society of Automotive Engineers, Warrendale, Pennsylvania, 1974), p. 149.
4. Clean Air Act of 1970, Title 42, U.S. Code Emission Standards for Moving Sources, Chap. 2, Sect. 202, U.S. Govt. Printing Office, Publ. 0–445–880, Washington, D.C., 1971.
5. The metric equivalents for units used in the text are as follows: 1 mile = 1.6 km; 1 gallon = 3.7 liters; 1 inch = 2.5 cm.
6. "Status of Automotive Electronics in the U.S.A.," by Jerome G. Rivard in *Proceedings of Convergence 80.*
7. "Engineering Highlights of the 1981 Automobiles," by Larry Givens (editor), *Automot. Eng.*, (October 1980).
8. Federal Motor Vehicle Safety Standard 121, "Air Brake Systems," National Highway Safety Administration, Department of Transportation, Washington, D.C.
9. "An Overview of Electronic Controls for Passenger Car Diesel Engines," by Myron U. Trenne and Victor J. Novak, in *Proceedings of Convergence 80.*
10. "The World Automotive Electronics Market," by Trevor O. Jones and Joseph F. Ziomek, in *Proceedings of Convergence 80.*
11. "Multiplex Systems for Automobiles," by W. A. Rogers, R. A. Meade, and D. R. Kimberlin, in *Proceedings of Convergence 80.*
12. Department of Transportation, "The National Highway Safety Forecast and Assessment (A 1985 Traffic Safety Setting)," National Highway Traffic Safety Administration, Department of Transportation, Washington, D.C., 1966.
13. "Problems of Linear Closure at Urban Intersections," Contract No. CPR-11-4101, Second Quarterly Progress Report, National Highway Traffic Safety Administration, Department of Transportation, Washington, D.C., 1966.
14. "Alcohol Impairment Detection by the Phystester—Evaluation Summary," SAE Paper 730093, Society of automotive Engineers, New York, January 1973.
15. J. A. Tennant in *Proceedings of Automotive Safety Engineering Seminar*, General Motors Corp., Warren, Michigan, June 1973, p. 95.
16. D. M. Grimes and T. O. Jones, *Proc. IEEE* **62**, 804 (1974).
17. T. O. Jones, D. M. Grimes, and R. A. Dork, "A Critical Review of Radar as a Predictive Crash Sensor," SAE Paper 720424, Society of Automotive Engineers, New York, May, 1972.
18. "Some Recent and Future Automotive Electronic Developments," by Trevor O. Jones, *Science*, **195**, 1156–1160 (1977).

Joseph F. Ziomek, Director, Engineering Transportation Electrical and Electronics Operations, TRW, Inc., Farmington Hills, Michigan.

AUTOMOTIVE EMISSIONS. Automotive Electronics; Catalytic Converter; Pollution (Air).

AUTOMOTIVE FUEL (Hydrogen). Hydrogen (Fuel).

AUTOMOTIVE FUELS. Petroleum.

AUTONOMIC NERVOUS SYSTEM. The term *autonomic* signifies automatic or unconscious activity. Thus, the autonomic nervous system is sometimes referred to as the involuntary nervous system; rarely, the vegetative nervous system. A functional division of the nervous system would consist of ganglia, nerves, and plexuses, through which visceral organs, heart, blood vessels, glands, and smooth muscle receive their innervation. It is widely distributed over the body, especially in the head and neck, and in the thoracic and abdominal cavities. The autonomic system is not under voluntary control and the processes in which it is concerned are beneath consciousness for the most part. It is influenced to a great degree by the endocrine glands, particularly the adrenal and its hormone, epinephrine.

In general the autonomic nervous system may be divided into two groups both of which may send nerves to the same organs but act antagonistically, producing opposite results. One is known as the parasympathetic, which arises from the mid-brain, hind-brain, and sacral region of the cord and is stimulated by the drug pilocarpine and inhibited by atropine. The other is known as the sympathetic, which arises from the thoracic and lumbar regions of the spinal cord and is stimulated by epinephrine.

Under normal conditions there is a balance between the two systems allowing for perfect function of a bodily organ. For instance, the heart is slowed by the parasympathetic system and accelerated by the sympathetic. Movement of the stomach is increased by the parasympathetic and in inhibited by the sympathetic. The pupil of the eye is contracted by the parasympathetic and dilated by sympathetic stimuli.

Psychosomatic disturbances may take place in any of the involuntary organs of the body systems. These include the digestive, the respiratory, the heart and circulatory, the genitourinary, the endocrine system, and the skin. Gastrointestinal reactions, such as nervous diarrhea, will affect various individuals with different degrees of intensity. The emotional component of diarrhea has been recognized for centuries. It also has been recognized for centuries that certain skin disorders contain an emotional element. Eruptions arising from emotional disturbances are termed *psychogenic* skin eruptions. Most persons who show the characteristics of a nervous dermatitis, like other psychosomatic patients, appear to carry their emotional problems close to the surface of their minds, rendering them more accessible to psychiatric treatment.

As pointed out by Nauta and Feirtag (*Sci. Amer.*, **241**, 3, 109, 1979), "The autonomic nervous system is not self-governing at all. Its functions are integrated with voluntary movements no less than with motivations and affects. In short, its roots are in the brain; one's experiences from moment to moment dictate not only the contractions of one's skeletal muscles but also large functional shifts in the body's internal organs. The term autonomic has nonetheless won out in the English-speaking world. Other languages use other terms. In German one speaks of *das viszerale Nervensystem*, in French of *le système nerveux végétatif.*"

See also **Brain**, and **Nervous System.**

AUTOREGRESSION. A stochastic relation connecting the value of a variable at time t with values of the same variable at previous times. For example the linear equation

$$u_t = \alpha_1 u_{t-1} + \alpha_2 u_{t-2} + \epsilon_t \tag{1}$$

where ϵ_t is a random variable. Two common forms of autoregressive relations are the Markov scheme

$$u_t = \alpha_1 u_{t-1} + \epsilon_t \tag{2}$$

and the Yule scheme (1).

These equations bear a formal resemblance to the equations of linear regression—hence the name—but raise special problems in the estimation of the constants. They may be regarded as a class of stochastic processes.

See also **Stochastic Process.**

AUTOREGULATION (Brain). Cerebrovascular Diseases.

AUTOROTATION. Airplane; Helicopters and V/STOL Craft.

AUTOTOMY. Self-mutilation. Through the presence of a special modification near the base of the limb, some crustaceans and insects are able to drop off appendages by which they are seized. The autotomy of the arms of starfish and of the tails of lizards are other common examples. Autotomy is followed by regeneration.

AUTOTRANSFORMER. **Transformer.**

AUTOTRANSFUSION. **Blood.**

AUTOZOOID. Members of polyp colonies whose function is to feed the colony.

AUTUNITE. This mineral is a hydrous phosphate of calcium and uranium, crystallizing in the tetragonal system, usually in thin tabular crystals. Good basal cleavage; hardness, 2–2.5; specific gravity, 3.1; luster, subadamantine to pearly on the base; color, lemon yellow; streak, yellow; transparent to translucent; strongly fluorescent.

Originally from near Autun in France, whence the name, it is a secondary mineral associated commonly with uraninite. In the United States, it occurs sparsely in the pegmatites of Connecticut, New Hampshire and North Carolina. Autunite also is known as *calco-uranite.*

See also **Uraninite.**

AUXILIARY GENERATOR SET. **Gas and Expansion Turbines.**

AUXINS. **Plant Growth Modification and Regulation.**

AUXOMETER (or Auxiometer). An apparatus for measuring the magnifying power of a lens or any optical system.

AUXOSPORE. An auxospore is a special type of spore which occurs in diatoms and which seems to be a means of rejuvenating the cells. Rejuvenescence is necessary, since in the normal process of cell division one of the two daughter cells is always smaller than the parent cell. Consequently very small cells are ultimately formed. In some species of diatoms, auxospore formation is preceded by the escape of the protoplast from the walls of the cell. The free protoplast then enlarges and secretes about itself a wall. In time new valves more or less like those of the original diatom are formed. In other species of diatoms, auxospore formation is preceded by the union of the protoplasts of two similar diatom cells, the process being therefore sexual.

AVALANCHE (Electronics). The term avalanche is used in counter technology to describe the process which is essentially a cascade multiplication of ions. In this process, an ion produces another ion by collision, and the new and original ions produce still others by further collisions, resulting finally in an "avalanche" of ions (or electrons). The terms "cumulative ionization" and "cascade" are also used to describe this process.

The term avalanche or avalanche effect is sometimes applied to the Zener effect in semiconductors.

AVALANCHE (Geology). A large mass of snow, ice, soil, or rock, or mixtures of these materials, falling or sliding very rapidly under the force of gravity. Velocities may sometimes exceed 500 km/hour. Avalanches can be classified by their content, such as snow and ice avalanches, debris avalanches, soil or rock avalanches. (*Glossary of Geology,* American Geologic Institute).

AVALANCHE PHOTODIODE (APD). **Telephony.**

AVERAGE. A simple but subtle concept which attempts, in some sense, to summarize a set of numbers $x_1, \ldots, x_n$ in a single number. In statistics, the commonest forms of average are

(a) The arithmetic mean M, defined by

$$M = \frac{1}{n} \sum_{j=1}^{n} x_j$$

(b) The geometric mean G, defined by

$$\log G = \frac{1}{n} \sum_{j=1}^{n} \log x_j$$

(c) The harmonic mean H, defined by

$$\frac{1}{H} = \frac{1}{n} \sum_{j=1}^{n} \frac{1}{x_j}$$

When the individual numbers x are not regarded as of equal importance, they may be weighted by numbers $w_1, \ldots, w_n$. For example the weighted arithmetic mean is given by

$$\frac{1}{\sum_{j=1}^{n} w_j} \sum_{j=1}^{n} (w_j x_j)$$

(See **Arithmetic Mean.**)

AVERAGE DEVIATION. If $\bar{x}$ is the mean of observations $x_1, \ldots, x_n$, the mean deviation is given by

$$M.D. = \frac{1}{n} \sum_{j=1}^{n} |x_j - \bar{x}|$$

If x has a frequency distribution $f(x)$ the analogous defintion is

$$M.D. = \int_a^b f(x)|x - m|\, dx$$

where m is the mean and the distribution ranges from a to b.

Owing to its relative mathematical intractability the mean deviation is usually discarded in favor of the standard deviation.

The average deviation is a minimum when deviations are measured from the median.

AVERAGE OUTGOING QUALITY LEVEL. **Statistical Quality Control.**

AVES. **Birds.**

AVIATION WEATHER OBSERVATIONS. **Weather Observations and Forecasting.**

AVIDIN. **Egg**

AVOCADO. **Laurel Family.**

AVOCET. **Shorebirds and Gulls.**

AVOGADRO CONSTANT. The number of molecules contained in one mole or gram-molecular weight of a substance. The most recent value is $6.0220943 \times 10^{23} \pm 6.3 \times 10^{17}$. In measurements made by scientists at the National Bureau of Standards (Gaithersburg, Maryland) and announced in late-1974, the uncertainty (as compared with previous determinations) of the number has been reduced by a factor of 30.

AVOGADRO LAW. The well-recognized principle known by this name was originally a hypothesis suggested by the Italian physicist Avogadro, in 1811, to explain the puzzling rule of proportional volumes observed in chemical reactions of gases and vapors. It states simply that equal volumes of all gases and vapors at the same temperature and pressure contain the same number of molecules. Though this assumption accords with the facts and aids the kinetic theory of gases, just why it should be true is by no means self-evident, unless one starts with the much more recent Maxwell-Boltzmann law of equipartition of energy, which also requires proof. That Avogadro's law is true cannot be said to have been positively established until the experiments of J. J. Thomson, Millikan, Rutherford, and others determined the value of the electron as an electric charge and thereby made it possible to count the number of atoms of different elements in a gram. The actual number of molecules contained in one mole (gram-molecular weight) of a substance is the Avogadro constant.

At any fixed temperature and pressure, the density of carbon dioxide gas, for example, is approximately 22 times greater than the density of hydrogen gas. Thus, the mass of 1 liter of carbon dioxide is 22 times the mass of 1 liter of hydrogen gas. According to Avogadro's principle, the number of molecules in 1 liter of carbon dioxide is the same as the number of molecules in 1 liter of hydrogen. Thus, it follows that a carbon dioxide molecule must have a mass that is 22 times larger than the mass of a hydrogen molecule. Since the molecular weight of hydrogen (H_2) was set equal to 2, carbon dioxide was assigned a molecular weight of 22 × 2, or 44. Cannizzaro was the first to use gas densities to assign atomic and molecular weights. Avogadro's principle also may be used to assign molecular weights in a slightly different way. At standard temperature and pressure, the volume of a mole of any gas is 22.4 liters. The molecular weight of a gas, therefore, is the mass (in grams) of 22.4 liters of the gas under standard conditions. For most gases, the deviation from this ideal value is less than 1%. See also **Avogadro Constant.**

AVOGADRO'S LAW (Combustion). Combustion.

AVULSION. A sudden cutting off or separation of land by a flood, or by an abrupt change in the course of a stream, as by a stream breaking through a meander or by a sudden change in current whereby the stream deserts its old channel for a new one. Generally, in legal interpretation, the part thus cut off or separated belongs to the original owner.

AWANTIBO. Lemur.

AWN. Grasses.

AXES (Aircraft). Three fixed lines of reference, usually centroidal and mutually perpendicular. The horizontal axis in the plane of symmetry, usually parallel to the thrust axis, is called the longitudinal axis; the axis perpendicular to this in the plane of symmetry is called the normal axis; and the third axis perpendicular to the other two is called the lateral axis. Rotation may take place about any or all axes; translation may take place along any of the three axes. The important translational axis is the longitudinal. See also **Bank (Aircraft); Pitching Moment; Yaw (Aircraft).**

AXIAL-FLOW COMPRESSOR. Air Compression; Airplane; Gas and Expansion Turbines.

AXIAL-FLOW TURBINE (Hydro). Hydroelectric Power.

AXIAL MAGNIFICATION. The ratio of the interval between two adjacent image points on the axis of an optical instrument to the interval between the conjugate object points.

AXIAL ORGAN. An organ of peculiar structure and unknown function found near the axis of the body in all echinoderms except the sea cucumbers.

AXIAL VECTOR. Vector.

AXIL. The angle between the upper side of a leaf and the stem to which the leaf is attached. See **Stem (Plant).**

AXINITE. This mineral is an aluminum-boron-calcium silicate with iron and manganese, $(Ca, Mn, Fe)_3Al_2BSi_4O_{15}(OH)$. It crystallizes in the triclinic system, yielding broad sharp-edged forms, which has led to its name, derived from the Greek word meaning axe. It breaks with a conchoidal fracture; hardness, 6.5–7; specific gravity, 3.22–3.31; luster, vitreous; colors, brown, blue, yellow and gray; transparent to translucent.

Axinite occurs in granites or more basic rocks along contacts and in cavities in Saxony, Switzerland, France, England, Tasmania, and Japan; in the United States, in New Jersey, Pennsylvania, and California.

AXIOM. A statement of an abstract notion which is assumed without proof. Axioms constitute the unproved first principles which are used in founding a mathematical discipline. Physical science disciplines sometimes are developed using the axiomatic approach which embraces the presentation of a minimum number of statements (axioms) and derives the other relationships common in the discipline, using only these axioms and mathematical and logical processes.

AXIS. A line so situated that various parts of an object ae symmetrically located in relation to it. Also the line passing through the origin of a coordinate system which corresponds to all points of a given variable when other variables are zero. Thus, in two dimensions, the X-axis is the locus of all points whose Y-coordinate is zero. See also **Ellipse; Hyperbola; Mineralogy; Parabola.**

AXIS DEER. Deer.

AXIS (Instantaneous). In rigid body motion, a line perpendicular to the plane of the motion which passes through that point or those points of a body which are instantaneously at rest. For a cylinder rolling down an inclined plane without slipping, the instantaneous axis is the line of contact between cylinder and plane.

AXIS OF ROTATION (Fixed). The locus of points of a system along a straight line which remain stationary when the system undergoes motion of rotation.

AXIS (Optic). A direction through a doubly-refracting crystal along which no double refraction occurs. A uniaxial crystal has one such direction, a biaxial has two such directions. See **Crystal.**

AXIS (Optical). The line through the foci and the vertices of the optical surfaces. Commonly, the surfaces of lenses and mirrors are figures of revolution about the optical axis. Normally, the parts of an optical system are all coaxial.

AXOLOTL (*Amphibia, Urodela*). A salamander, *Ambystoma tigrinum*, found near Mexico City which, although related to some of the terrestrial salamanders, retains its larval form throughout life, becoming sexually mature in this stage. Under experimental conditions the animal has been caused to undergo the usual metamorphosis.

AXON. The impulse-transmitting part of a nerve cell or neuron. The other parts are the cell body, containing the nucleus, and the dendrites, branches which pick up impulses. The axon is also known as the nerve fiber. The axon of a peripheral nerve of a vertebrate animal is typically covered by an inner myelin sheath and by a thin outer cellular layer called the *neurilemma.* In certain invertebrates, such as the squid, giant axons are found which range in size from 150–700 microns in diameter. This unusually large size has made possible many fundamental measurements on the biophysical properties of the excitable cell membrane of the axon. See also **Brain and Nervous System.**

AXOPLASM. Brain and Nervous System.

AZALEA. Heather Shrubs and Trees.

AZEOTROPIC DISTILLATION. Distillation.

AZEOTROPIC SYSTEM. A system of two or more components which has a constant boiling point at a particular composition. If the constant boiling point is a minimum, the system is said to exhibit *negative azeotrophy,* if it is a maximum, *positive azeotropy.*

Consider a mixture of water and alcohol in the presence of the vapor. This system of two phases and two components is divariant (see **Phase Rule**). Now choose some fixed pressure and study the composition of the system at equilibrium as a function of temperature. The experimental results are shown schematically in the accompanying figure.

The vapor curve KLMNP gives the composition of the vapor as a function of the temperature T, and the liquid curve KRMSP gives the composition of the liquid as a function of the temperature. These two curves have a common point M. The state represented by M is that in which the two states, vapor and liquid, have the same composition x_B^a on the mole fraction scale. Because of the special properties associated with systems in this state, the Point M is called an azeotropic point and the system is said to form an azeotrope. In an azeotropic system, one phase may be transformed to the other at constant temperature, pressure and composition without affecting the equilibrium state. This property justifies the name azeotropy, which means a system which boils unchanged.

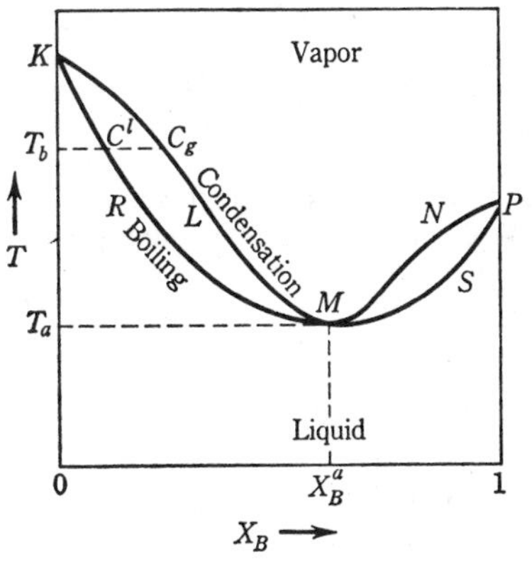

Azeotropic system.

AZIDES. The salts of hydrazoic acid are termed *azides.* Metallic azides can be prepared from barium azide and the metal sulfate, or from potassium azide and the metal perchlorate.

Soluble azides react with iron(III) salt solutions to produce a red color, similar to that of iron(III) thiocyanate. Sodium azide is not explosive, even on percussion, and nitrogen may be evolved upon heating. With iodine dissolved in cold ether silver azide forms iodine azide (IN_3), yellow explosive solid.

Sodium azide is a slow oxidizing agent. It has a selective action in inhibiting the growth of gram-negative organisms. It has been used as a component in selective media such as azide glucose broth or azide blood agar base for the isolation of mastitis and fecal *streptococci.*

A number of alkyl and aryl azides are known, such as CH_3N_3, $C_2H_5N_3$ and $C_6H_5N_3$. The nonmetallic inorganic azides include ClN_3, an explosive gas, BrN_3, an orange liquid, mp −45°C, and IN_3, a yellow solid, decomposing above −10°C. The gas FN_3 is more stable than ClN_3, decomposing only slowly at room temperature.

Lead and silver azides are widely used as initiating, or primary explosives because they can be readily detonated by heat, impact, or friction. As such, these materials, particularly lead azide, are used in blasting caps, percussion caps, and delay initiating devices. The function of the azides is similar to that of mercury fulminate or silver fulminate.

AZIMUTH (Astronomy). That coordinate of the horizontal coordinate system of a celestial object which is measured in the plane of the horizon to the point where the vertical circle of the object cuts the horizon, from the south to the right (west) through 360°. Astronomical azimuth may be computed by solving the astronomical triangle, provided three other parts are known. In most cases, the latitude of the observer, the hour angle, and declination of the object are the known parts. In case the longitude of the observer is not accurately known, the altitude of the object may be obtained and combined with latitude and declination for computing azimuth. Several of the terms used here are described elsewhere in this encyclopedia.

AZIMUTH (Navigation). The horizontal direction of a celestial point from a terrestrial point, expressed as the angular distance from a reference direction, usually measured from 0° at the reference direction clockwise through 360°. An azimuth is often designated as true, magnetic, compass, grid, or relative as the reference direction is true, magnetic, compass, grid north, or heading, respectively. Unless otherwise specified, the term is generally understood to apply to true azimuth, which may be further defined as the arc of the horizon, or the angle at the zenith, between the north part of the celestial meridian or principal vertical circle and a vertical circle, measured from 0° at the north part of the principal verticle circle clockwise through 360°. See also **Navigation.**

AZIMUTH (Surveying). The terrestrial azimuth of a mark is usually determined with an altazimuth instrument or surveyor's transit. The difference in azimuth between the vertical circle through some celestial object and the mark is measured and combined with the known azimuth of the object to obtain the azimuth of the mark. The object most commonly used for this purpose is Polaris (the North Star), since this star is within 1° of the pole of rotation, and its azimuth changes very slowly with time. Tables are published in the Nautical Almanac, and a variety of other places, which give the azimuth of Polaris in terms of local date and time. When using the North Star for ordinary surveying purposes, the local time is needed only to within about 5 minutes; but for precise geodetic work, the time must be known to within a few seconds.

Surveyors frequently run a traverse using azimuths and distances. The plotting of a traverse by this method is shown in the accompanying figure.

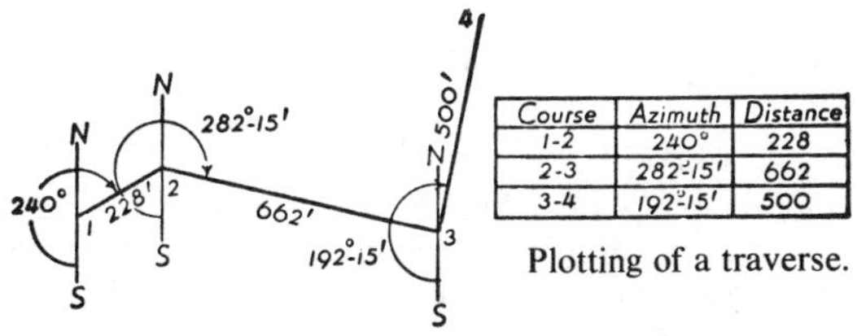

Plotting of a traverse.

Plotting of a traverse.

AZINES. The products of the reaction between an aldehyde or a ketone with hydrazine are termed *azines.* A number of dyestuffs and complex members of the pyridine family of compounds also are termed azines. See also **Pyridine.**

AZLON. Fibers.

AZO AND DIAZO COMPOUNDS. Characteristically, these are compounds containing the group —N:N— (azo) or >N⋮N (diazo). They are closely related to the substituted hydrazines. The N_2 group may be covalently attached to other groups at both ends, as in the azo compounds, or at only one end, as in the diazo compounds or diazonium salts. Although organic chemistry furnishes the most numerous examples, many inorganic azo compounds also exist.

Compounds related to aniline, either directly or by oxidation, and to nitrobenzene by reduction, are numerous and important. When nitrobenzene is reduced in the presence of hydrochloric acid by tin or iron, the product is aniline (colorless liquid); in the presence of water by zinc, the product is phenylhydroxylamine (white solid); in the presence of methyl alcohol by sodium alcoholate or by magnesium plus ammonium chloride solution, the product is azoxybenzene (pale yellow solid); by sodium stannite, or by water plus sodium amalgam, the product is azobenzene (red solid); in the presence of sodium hydroxide solution by zinc, the product is hydrazobenzene (pale yellow solid). The behavior of other nitro-compounds is similar to that of nitrobenzene.

Hydrazobenzene is converted by oxygen of the air or by ferric chloride solution into azobenzene, and by strong acids into benzidine hydrochloride (4')$H_2N \cdot C_6H_4 \cdot C_6H_4 \cdot NH_2$(4)·HCl). Benzidine is prepared by reducing nitrobenzene to hydrazobenzene as above, and then treating the product with acid. Benzidine and its toluene relative, orthotolidine, are important intermediates for dyes. The counterpart of aniline is toluidine in its three forms, *ortho, meta, para.*

Azoxybenzene is converted by distillation with iron into azobenzene (ferrous oxide also formed), and by concentrated sulfuric acid warm into *para*-hydroxyazobenzene (4)$HO \cdot C_6H_4N{:}NC_6H_5$, which is a dye.

Diazonium salts are usually colorless crystalline solids, soluble in water, moderately soluble in alcohol, and when dry are violently explosive by percussion or upon heating. These salts are generally used in cold (near 0°C) acid solution, without separation of the salt, and are prepared by reaction of the desired benzenoid primary amine with nitrous acid (from sodium nitrite plus hydrochloric acid). Alkyl amines with nitrous acid yield the corresponding alcohol.

(A) By treatment of benzene diazonium chloride

$$\left(C_6H_5\overset{\vdots}{\underset{N}{N}}{-}Cl\right)$$

solution with silver oxide, or of the diazonium sulfate solution with barium hydroxide, the hydroxide (benzene diazo hydroxide, $C_6H_5N{:}N{-}OH$) is obtained, which is intermediate in basicity between ammonium hydroxide and sodium hydroxide. Most diazo hydroxides are unstable, and are spontaneously transformed into nitrosoamines (group —NH·NO), yellow neutral compounds. With sodium hydroxide, diazonium salt solutions yield sodium benzenediazoate, more active chemically when first formed than upon standing, due to change from syn-diazoate

$$\left(\begin{array}{l} C_6H_5N \\ NaO\ddot{N} \end{array}\right)$$

which evolves nitrogen readily, to anti-diazoate

$$\left(\begin{array}{ll} C_6H_5N & \\ & \ddot{N}ONa \end{array}\right)$$

which is more stable. Sodium benzene-syn-diazoate reacts with phenols in alkaline solution to give azo-dyes, e.g., *para*-hydroxyazobenzene, wherein hydrogen *para* (or *ortho* but not *meta*) to the hydroxyl group is reactive. Other reactions of diazonium salts are:

(B) Replacement of the diazo-group with loss of nitrogen, (1) by hydroxyl-group, forming phenols by warming with water, (2) by alkoxy-group, forming ethers, by warming with an alcohol, (3) by acyl-group, forming esters, with an acid, (4) by hydrogen, forming hydrocarbons or substituted hydrocarbons, e.g., tribromobenzenediazonium chloride forms tribromobenzene, with alcohol or sodium stannite solution, (5) by chlorine, forming, for example, chlorobenzene, by warming with cuprous chloride (Sandmeyer's reaction), (6) by bromine, forming, for example, bromobenzene, by warming with cuprous bromide (Sandmeyer's reaction), (7) by iodine, forming, for example, iodobenzene, by warming a solution of the diazonium iodide, (8) by cyanide, forming, for example, cyanobenzene, by warming with cuprous cyanide (Sandmeyer's reaction), (9) by fluorine, forming, for example, fluorobenzene, by warming a solution of the diazonium borofluoride. Gatterman's modification of Sandmeyer's reactions above is to use copper powder plus the corresponding sodium or potassium salt instead of the cuprous salt. Nitro-compounds, since they are readily reduced to the corresponding amine, form an important group of compounds for the preparation of various derivatives by means of the diazo-reaction.

(C) Reduction, to form the corresponding hydrazine, for example, benzene diazonium chloride to form betaphenylhydroxylamine hydrochloride.

(D) Bromination followed by treatment with ammonia to form azides, for example, benzene diazonium bromide plus bromine forms benzene diazonium perbromide $C_6H_5Br_3$, which upon treatment with ammonia forms phenylazide

$$\left(C_6H_5N\begin{array}{l} \diagup N \\ \quad\ \cdot\cdot \\ \diagdown N \end{array}\right)$$

(E) Amines, (1) primary or (2) secondary amine to form diazo-amino-compounds, e.g., benzenediazonium chloride (a) plus aniline forms diazoaminobenzene, benzene diazoaniline $C_6H_5N{:}N—NHC_6H_5$, (b) plus methylaniline forms benzenediazomethylaniline

$$\left(C_6H_5N{:}N{-}N\begin{array}{l} \diagup C_6H_5 \\ \diagdown CH_3 \end{array}\right)$$

Diazoamino-compounds readily change into aminoazo-compounds upon standing in alcohol solution or in the presence of amine hydrochloride, thus, benzenediazoaniline changes into *para*-aminoazobenzene, benzeneazoaniline-4 $C_6H_5N{:}NC_6H_4NH_2(4)$, benzenediazomethylaniline changes into methyl-*para*-aminobenzene, benzeneazomethylaniline-4

$$\left(C_6H_5N{:}NC_6H_4N\begin{array}{l} \diagup H \\ \diagdown CH_3 \end{array}\ (4)\right)$$

(3) tertiary amine to form aminoazo-compounds directly, e.g., benzenediazonium chloride plus dimethylaniline forms dimethyl-*para*-aminoazobenzene, benzeneazodimethylaniline-4

$$\left(C_6H_5N{:}NC_6H_4N\begin{array}{l} \diagup CH_3 \\ \diagdown CH_3 \end{array}\ (4)\right)$$

In this manner are prepared the azo-dyes, which contain the chromophore azo-group —N:N— plus an auxochrome amino-group

$$\left({-NH_2},\ {-N}\begin{array}{l} \diagup H \\ \diagdown CH_3 \end{array},\ {-N}\begin{array}{l} \diagup CH_3 \\ \diagdown CH_3 \end{array}\right)$$

The simplest azo-dyes are yellow, but by increasing the number of auxochrome groups or by increasing the percentage of carbon, the color darkens to red, violet, blue, and in some cases brown. Naphthalene residues darken to red, violet, blue and finally black. These amino-azo-dyes, together with the hydroxyazo-dyes (containing auxochrome hydroxyl-group —OH), are generally only slightly soluble in water. In order that the dye may be soluble it is desirable that it contain one or more sulfonic acid groups $—SO_2OH$. This group may be introduced either by treating the dye with concentrated sulfuric acid, or by using sulfonic acid derivatives in preparing the dye, e.g., methyl orange, sodium dimethyl-*para*-aminoazobenzene-*para*-sulfonate

$$(4)(CH_3)_2NC_6H_4N{:}NC_6H_4SO_2ONa(4)$$

from dimethylaniline and diazotized sulfanilic acid (*para*-amino-benzene sulfonic acid, $(1)H_2N{\cdot}C_6H_4{\cdot}SO_2OH(4)$, and then the sodium salt made from the product. Other azo-dyes are

chrysoidine $$\left(C_6H_5N{:}NC_6H_3\begin{array}{l} \diagup NH_2(2) \\ \diagdown NH_2(4) \end{array}\right)$$

Bismarck brown $$\left((3)H_2N{\cdot}C_6H_4N{:}NC_6H_3\begin{array}{l} \diagup NH_2(2){\cdot}HCl \\ \diagdown NH_2(4) \end{array}\right)$$

Congo red $$\left(\begin{array}{l} (4)HOO_2S\diagdown \\ (1)\quad H_2N\diagup \end{array}C_{10}H_5{:}NC_6H_4\right)$$

$$C_6H_4N{\cdot}NC_{10}H_5\begin{array}{l} \diagup SO_2OH(4) \\ \diagdown NH_2(1) \end{array}$$

See also **Aniline; Dyes (Textile) Hydrazine; Nitro- and Nitroso-Compounds.**

AZOXYBENZENE. **Azo and Diazo Compounds.**

AZURITE. This mineral is a basic carbonate of copper, crystallizing in the monoclinic system, with the formula $Cu_3(CO_3)_2(OH)_2$ so called from its beautiful azure-blue color. It is a brittle mineral with a conchoidal fracture; hardness, 3.5–4; gr, 3.773; luster, vitreous; color and streak, blue; transparent to translucent. Azurite, like malachite, is a secondary mineral, but far less common than malachite. It is formed by the action of carbonated waters on compounds of copper or solutions of copper compounds, probably most abundant by rich solutions reacting with limestones.

Azurite almost always occurs associated with malachite. It is found in Siberia, Greece, Rumania, at Chessy, France (whence the name Chessylite), in Southwest Africa, Australia, and elsewhere; in the United States, at Bisbee, Arizona, and Kelly, New Mexico. Azurite also is known as *Chessylite.*

B

BABBITT METAL. Antimony; Lead; Tin.

BABESIOSIS. This is an uncommon malaria-like disease caused by an intraerythrocytic protozoan parasite, *Babesia.* The main reservoirs of human infection include rodents and, possibly, house pets. Transmission is by tick bite (family *Ixodidae*). As reported by Healy, Spielman, and Gleason (*Science*, **192**, 479–480, 1976), *Babesia microti* infection in wild mammals has been recognized in several locations in the United States and Europe. Populations of white-footed mice on Martha's Vineyard, 24 kilometers west of Nantucket, were found to be infected in 1937. Rodents of this species were infected with similar parasites around Ithaca, New York and various small rodents and rabbits carried the infection in California and in England. Thus, the parasite seems to be widespread, and this renders problematic the peculiar frequency of human infection found on Nantucket Island. However, the prevalence of infection in mice is very high on this island, greatly exceeding that reported from other locations, and this may be causally related to infection in humans. It is interesting to note that, with the exception of the aforementioned relatively isolated cases, no individuals with intact spleens located elsewhere have been infected.

The disease is characterized by fever, drenching sweats, chills, lethargy, malaise, myalgias, arthralgias, and emotional lability. Splenomegaly is occasionally present. The fact that *Babesia* organisms do not produce pigment in red blood cells is helpful in differentiating the disease from malaria. With the exception of splenectomized patients, babesiosis is generally self-limited and requires no special therapy. Chloroquine, once considered effective, is rarely used. In splenectomized patients, where the risk of the disease is much greater, a potentially toxic drug, pentamidine, has been considered.

BABINGTONITE. This mineral is a relatively rare calcium-iron-manganese silicate, occurring in small black triclinic crystals, found in Italy, Norway and in the United States at Somerville and Athol, Massachusetts, and in Passaic County, New Jersey. It was named for Dr. William Babington.

BABIRUSA. Suines.

BABOON. Monkeys and Baboons.

BACILLUS. Bacteria.

BACILLUS CEREUS GASTROENTERITIS. Foodborne Diseases.

BACILLARY DYSENTERY. Foodborne Diseases.

BACITRACIN. Antibiotic.

BACKCROSS. Plant Breeding.

BACK EMF. Electromotive Force.

BACK FOCAL LENGTH. The distance from the back surface of a lens to the second focal point. Its reciprocal is sometimes called the vertex power or the effective power of a lens. See **Mirrors and Lenses.**

BACK-GOUDSMIT EFFECT. An effect closely related to the Zeeman effect. It occurs in the spectrum of elements having a nuclear magnetic and mechanical moment. See also **Hyperfine Structure; Paschen-Back Effect.**

BACKGROUND. A general term for the totality of the effects that are always present in physical apparatus, and above which a phenomenon must show itself in order to be measured. Such unrelated effects include the unwanted counts or currents from cosmic rays in electrical apparatus for measurement of radioactivity, developable grains on photographic plates that are unrelated to the phenomena investigated, noise in acoustical apparatus, and many others.

BACKGROUND NOISE. Noise.

BACKLASH. Also termed mechanical hysteresis, backlash may be defined as that lost motion or free play that is inherent in mechanical elements, such as gears, linkages, or other mechanical transmission devices that are not rigidly connected. A physical model of backlash is shown in Fig. 1. The characteristic is shown in Fig. 2.

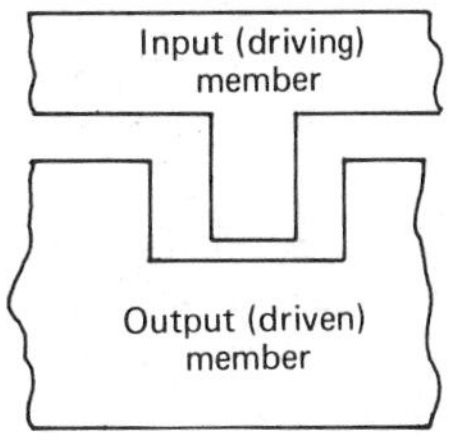

Fig. 1. Physical model of backlash.

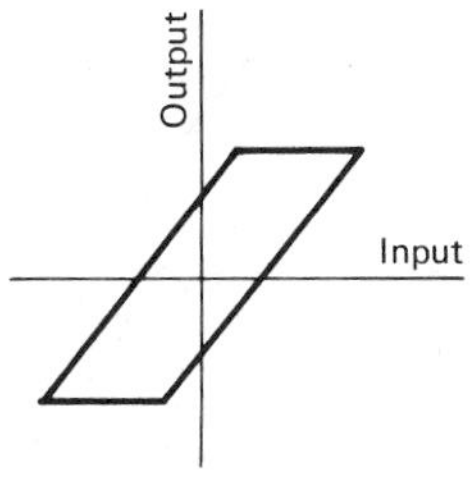

Fig. 2. Characteristic of backlash.

Backlash causes an effect similar to the hysteresis loop of Fig. 3, in that the output magnitude will assume a different value for a given value of input, depending upon whether the input is increasing or decreasing. In the classical definition of hysteresis, the output is not only dependent upon the value of input and the direction of the traverse, but also is dependent upon the history of prior excursions of the input and the span of the immediate excursion. Perhaps the most distinguishing characteristic of hysteresis is that output changes are continuous with the input so that some reversal of the output magnitude will take place for any small reversal of the input. In this sense, the output is multivalued, in that it can assume many values for a given input magnitude, depending upon the factors just described.

For pure backlash, however, the output is double-valued and is determined only by the magnitude of the input and whether it is increasing or decreasing. The *total* effect of backlash will also show

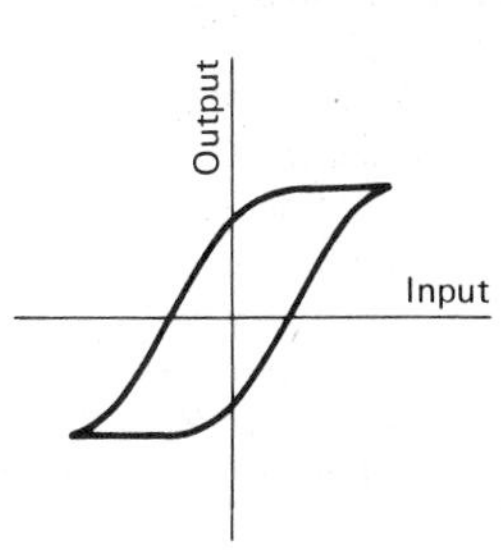

Fig. 3. Hysteresis loop.

up whenever there is any reversal of input larger (in relative values) than the magnitude of the backlash.

Most devices exhibit characteristics that are an inseparable combination of hysteresis and backlash. For this reason, it is proper to define hysteresis, for static measuring purposes, as a characteristic that includes both hysteresis error (as described above) and backlash.

The effect of backlash is of particular importance in the dynamic analysis of closed-loop systems. In this sense, the result is a discontinuous nonlinearity with a double-valued output. The describing function of such an element is a complex quantity whose magnitude and phase are both dependent upon the input signal and, if the effect is sufficiently large and not otherwise compensated, it can be a source of system instability.

In nonlinear analysis, backlash is considered to be either friction-controlled, inertia-controlled, or some combination of the two. If the output (or driven element) is considered to have friction, but no inertia, the waveform will be as shown by Fig. 4, and the peak value of the

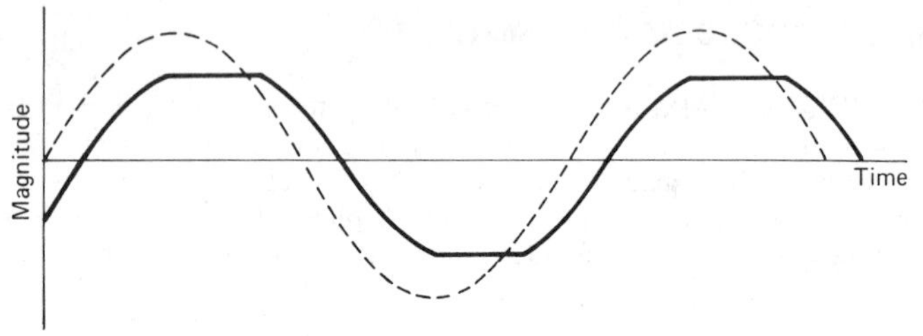

Fig. 4. Backlash waveform where output or driven element is considered to have friction, but no inertia.

output will be less than that of the input magnitude. If it is considered to have inertia and no friction, the waveform will be as shown by Fig. 5, and the peak value of the output will exceed the maximum input magnitude. In either case, there will be a phase lag which is a

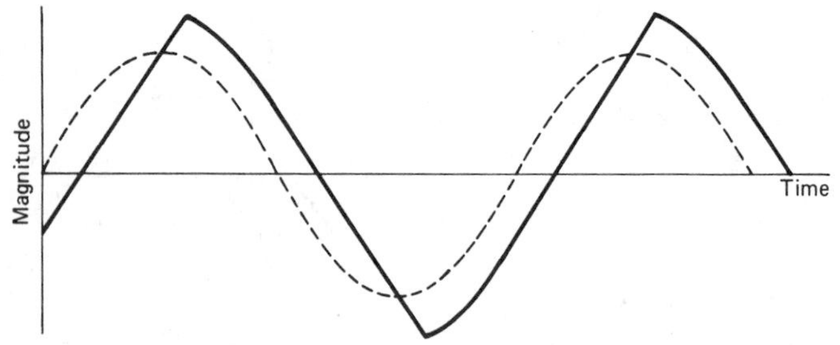

Fig. 5. Backlash waveform where output or driven element is considered to have inertia, but no friction.

function of the ratio of the backlash and the peak value of the input sine wave.

Robert L. Wilson, Honeywell Inc., Fort Washington, Pennsylvania.

BACKSCATTERING. The deflection of particles or of radiation by scattering processes through angles greater than 90° with respect to the original direction of motion.

One particular application of backscattering is the use of beta rays (electrons) to determine properties of substances. This phenomenon has been known for many years, but more recently its variations in value with differences in the atomic and molecular composition of the scattering substance have been found to give results that can be correlated with the atomic numbers of the atoms. In some cases where identical backscattering is obtained from two or more compounds (because of an accidental agreement between the total scattering of their atoms) differences in beta-ray absorption between them are usually available to provide another clue to their composition. Backscattering has also found application in measuring the thickness of coatings on materials such as paper, plastic films, and strip steel.

Backscattering also plays a part in the reception of radio waves, where it is commonly expressed in terms of a coefficient, which is said to measure the "echoing area." For an incident plane wave, the backscattering coefficient B is 4π times the ratio of the reflected power per unit solid angle (Φ) in the direction of the source divided by the power per unit area (W_i) in the incident wave:

$$B = 4\pi \frac{\Phi_r}{W_i} = 4\pi r^2 \frac{W_r}{W_i}$$

where W_r is the power per unit area at distance r. For large objects, the backscattering coefficient of an object is approximately the product of its interception area by its scattering gain in the direction of the source, where the interception area is the projected geometrical area and the scattering gain is the reradiated power gain relative to an isotropic radiator.

BACK-SWIMMER (*Insecta, Hemiptera*). An aquatic bug of boatlike form which lives in an inverted position. The hind legs are broadened by fringes and are used like oars for propulsion. Family *Notonectidae*. See illustration.

Back-swimmer

BACTEREMIA. Presence of bacteria in the blood.

BACTERIA. Microscopic, unicellular cells belonging to the kingdom Procaryotae, bounded by a membrane-wall complex and containing a variety of inclusions. Depending upon the species and cultural conditions, bacteria occur as individual cells or in clumps or chains of sister cells. Bacteria lie at the lower limits of resolution of the optical microscope. The average length lies within the range of 2 to 5 micrometers, although some are as small as 0.2 micrometer, or as large as 100 micrometers in length.

Bacteria are classified, some arbitrarily, by a descriptive array of features, one of the most common being shape. In terms of shape, the first of these are rod-shaped and are called *bacilli* (singular, *bacillus*). The bacilli often have small, whiplike structures known as flagella, with which they are able to move about. Some bacilli have oval, egg-shaped, or spherical bodies in their cells, known as spores. Under adverse conditions, dehydration, and in the presence of disinfectants, the bacteria may die, but the spores may be able to live on. The spores germinate when the conditions become favorable, and form new bacterial cells. Some are so resistant that they can withstand boiling and freezing temperatures and prolonged desiccation. See Fig. 1.

A second type of bacteria is the *cocci* (singular, *coccus*) which are spherical or ovoid in shape. The individual bacterial cells of this group may occur singly (*Micrococcus*), in chains (*Streptococcus*), in pairs

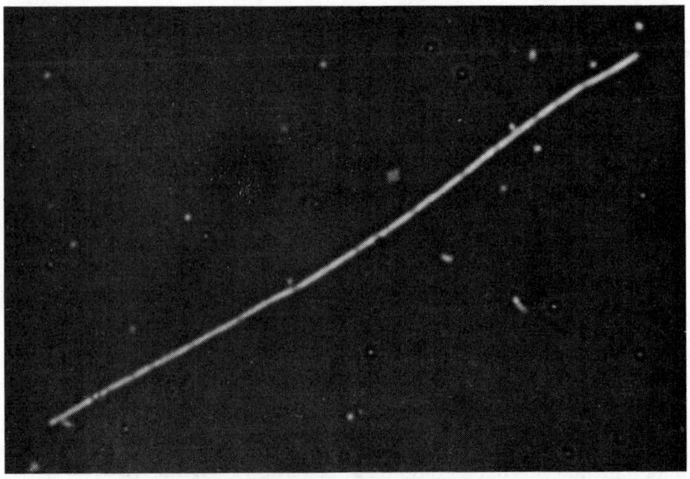

Fig. 1. Bacillus. (*A. M. Winchester.*)

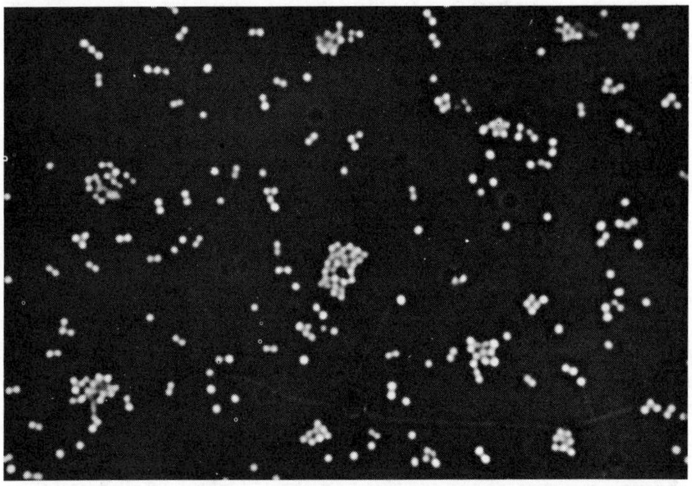

Fig. 2. Coccus. (*A. M. Winchester.*)

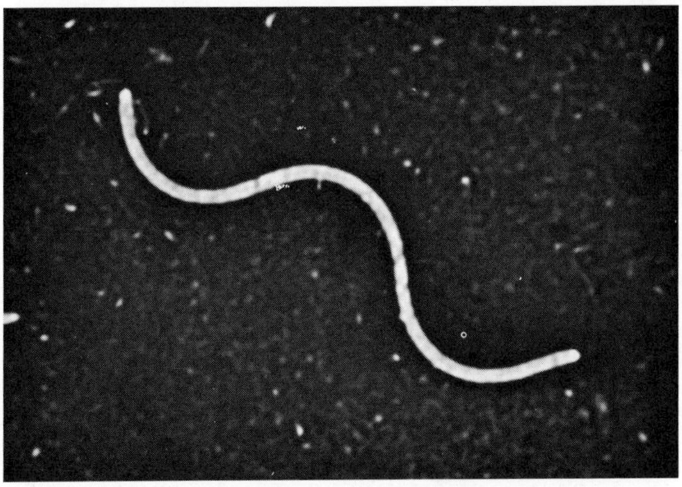

Fig. 3. Spirillum. (*A. M. Winchester.*)

(*Diplococcus*), in irregular bunches (*Staphylococcus*), and in the form of cubical packets (*Sarcina*). The coccus does not form spores and usually is nonmotile. See Fig. 2.

A third group of bacteria are the curved or bent rods. Of these, the genus *Vibrio* is composed of bacteria that are comma-shaped; and the genus *Spirillum* consists of those that are the twisted and spiral in form. All members of this group are motile, but none forms spores. However, some of these bacteria form a gelatinous capsule or covering by which they are probably protected from adverse environmental conditions. Still another group of spiral-shaped bacteria are known as the *spirochetes*, one of which is the cause of syphilis. See Fig. 3.

Bacteria also may be classified on the basis of their requirements of free atmospheric oxygen. Those which require atmospheric oxygen are called *aerobic* (air-living); those which cannot live in the presence of atmospheric oxygen are called *anaerobic*; those which do well with oxygen, but can get along without it are termed *facultative anaerobes.*

Bacteria are dependent upon the proper temperature for life and reproduction; and the various species of bacteria may differ widely in their temperature requirements. Most of the disease-producing (*pathogenic*) bacteria thrive best at body temperatures; others may live and multiply in much cooler temperatures; while still others live in hot springs. Freezing, as a rule, does not destroy bacteria, but prevents their reproduction. High temperature, conversely, quickly kills many bacteria. Most disease-producing organisms in milk, for example, may be killed by raising the temperature to 143°F (61.6°C) in the pasteurizing process. Most nonsporeforming, disease-producing microorganisms are destroyed by boiling water. In the spore stage, some bacteria must be heated to 240°F (116°C) for a considerable period of time, in order that the spores be destroyed. These high temperatures are best obtained by steam under pressure.

Bacteria may be killed by the action of chemicals called disinfectants.

PRINCIPAL BACTERIAL DISEASES

Common Name of Disease	Medical Name of Disease	Bacteria Responsible	Body Region Involved	Incubation Period
Septic sore throat	Streptococcus sore throat (Tonsillitis)	*Streptococcus* (several) species)	Throat and nasal membranes	3–5 days
Scarlet fever	Scarlatina	*Streptococcus scarlatinae*	Throat, tonsils, often other tissues	3–5 days
Pneumonia	Pneumococcal pneumonia	*Diploccus pneumoniae*	Respiratory tract, including lungs	Varies
Spinal meningitis	Epidemic meningitis	*Diploccus* (*Neisseria intracellularis*)	Respiratory tract, nervous system, sometimes blood	1–5 days
Clap	Gonorrhea	*Diploccus* (*Neisseria gonorrhoeae*)	Reproductive organs	2–8 days
Typhoid and paratyphoid fevers	Enteric fever	Short rod (*Salmonella typhosa; Salmonella paratyphi*)	Intestine	10–14 days
Bacillary dysentery	Shigellosis	Short rod (*Shigella dysenteriae*)	Intestine	1–4 days
Whooping cough	Pertussis	Small short rod (*Hemophilus pertussis*)	Respiratory tract	7–14 days
Bubonic plague	Pestis	Short rod (*Pasteurella pestis*)	Blood, spleen, liver, lymph nodes	2–10 days
Rabbit fever	Tularemia	Short rod (*Pasteurella tularensis*)	Lymph nodes, spleen, liver, kidneys, lungs	1–10 days
Undulant fever	Brucellosis	Short rod (*Brucella abortus*)	General body infection	5 days–10 weeks
Lockjaw	Tetanus	Sporeforming rod (*Clostridium tetani*)	Nervous system	2–40 days
Gas gangrene	Gas gangrene	Sporeforming rod (*Clostridium perfringens*)	Wounded areas	Varies
Botulinus	Botulism	Sporeforming rod (*Clostridium botulinum*)	Nervous system	18–66 hours
Tuberculosis	Tuberculosis	Irregular rod (*Mycobacterium tuberculosis*)	Lungs, bones, other organs	Varies
Syphilis	Lues	Spiral-shaped organism (*Treponema pallidum*)	Blood and nervous system	10–90 days
Diphtheria	Diphtheria	Irregular rod (*Corynebacterium diphtheriae*)	Respiratory tract	1–7 days

Note: Most of the diseases listed in this table are described in separate alphabetical entries in this book.

A substance which prevents infection or inhibits growth of microorganisms is called an antiseptic. Some of the most potent disinfectants are phenol and related compounds. Free chlorine gas is an excellent disinfectant, as are the hypochlorites. Tincture of iodine used on some cuts and other wounds has good disinfecting power. Bichloride of mercury and other mercury-containing compounds, mercurochrome, merithiolate, and phenylmercurinitrate also have been used as disinfectants and antiseptics. Alcohol in a 50–70% solution is a dependable disinfectant.

There has been much emphasis during the last few decades on *bacteriostatic* agents which prevent or slow down the rate of bacterial growth and reproduction, so that the natural protective mechanisms of the body can overcome the infection. These chemicals include the sulfonamide group, such as sulfathiazole, sulfadiazine, sulfanilamide, sulfasuxidine, and sulfaguanidine. Although valuable in the treatment of certain diseases, the drugs should not be taken indiscriminately, nor in conjunction with bacteriocidal agents.

Antibiotics, which are produced by other living organisms, inhibit the growth of bacteria or destroy them (bacteriocidal). There are few known bacterial diseases, the effects of which cannot be mitigated if the proper antibiotic is used early in the course of the disease. Tetanus and botulism are exceptions. These diseases are the manifestation of extremely potent toxins produced by the bacteria, rather than symptoms caused by infections of the microorganisms themselves.

Bacterial diseases may be transmitted in a number of ways. The common respiratory diseases are distributed by small droplets of sputum and nasal secretions. Sexual intercourse is the method by which venereal diseases are usually spread. Further, since many bacteria live until they are dried, diseases may be transmitted through indirect contact with persons through objects which they have handled. Some diseases are transmitted by water, milk, and foods which have become contaminated. Typhoid, cholera, and diarrheas are examples of diseases transmitted in the latter manner. Disease-producing bacteria usually cannot penetrate the unbroken skin; hence they enter by means of wounds, abrasions, or the natural openings of the body. See accompanying table.

Bacterial Genetics. The deoxyribonucleic acid (DNA) of bacteria is predominantly located in masses of variable shape, nuclear bodies (nucleoplasm or genophore), unbounded by a nuclear membrane. Bacteria thus are classified as procaryoids, in contrast to higher organisms containing nuclear membranes, the eucaryoids. In general, nongrowing, stationary-phase bacteria contain one nuclear body per cell, whereas exponentially growing, log-phase bacteria contain two or more nuclear bodies per cell. These nuclei are the sister products of a preceding nuclear division.

When bacteria are inoculated into growth medium, there is a delay (lag phase) before division and exponential growth ensue. The rate of exponential growth is a characteristic of the bacterial strain, the temperature, and the nutritional environment. The amount of DNA per nuclear body remains constant at various growth rates, although cell mass and average number of nuclei per cell are functions of the growth rate.

Most of the genetic information of bacteria is contained in a single structure of fixed DNA content, a giant circular DNA molecule that replicates semiconservatively. The enzymatic reactions involved in the biologically fundamental processes of DNA biosynthesis and genetic recombination are being elucidated in studies with bacterial systems.

Reference

Holt, J. G., Editor: "Bergey's Manual of Determinative Bacteriology," 8th edition, Williams & Wilkin, Baltimore, 1977.

This entry reviewed and updated for 6th Edition by Ann C. Vickery, Ph.D., University of South Florida, College of Medicine, Tampa, Florida.

BACTERIAL AGENTS (As Insecticides). Insecticide and Pesticide Technology.

BACTERIAL ARTHRITIS. Arthritis (Infectious).

BACTERIAL CHEMOSYNTHESIS. Ocean Resources (Energy).

BACTERIAL ENDOCARDITIS. Endocarditis.

BACTERIAL INHIBITION. Antimicrobial Agents (Foods).

BACTERIAL MENINGITIS. Meningitis.

BACTERIAL PILL. Gonorrhea and Gonococcemia.

BACTERICIDE. Pyridine and Derivatives.

BACTERICIDAL AGENT. Antibiotic.

BACTERIOPHAGE. Most bacterial viruses or bacteriophages (phages) are differentiated into head and tail. The tail parts and the head envelope consist of proteins; the envelope contains the DNA of the phage. Infection of the host bacterium begins by attachment of the tail tip to specific sites on the bacterial cell wall. The envelope then acts as a "microsyringe" and injects the phage DNA, presumably through the core of the tail, into the host. Only a small amount of protein (less than 5% of the total), of poorly understood function, is injected along with the DNA; the rest of the protein shell remains outside of the infected cell.

The essential feature of a phage infection is the injection of DNA. The DNA is the hereditary material of phage and phage infection may be considered as a genetic infection, resulting in a number of alternative interactions between host and virus.

A highly simplified definition of bacteriophage is that it is a type of virus which attacks and destroys bacteria by surrounding and absorbing them.

BACTERIOSTATIC AGENT. Antibiotic; Bacteria.

BACTRIAN CAMEL. Camels and Llamas.

BADGER. Mustelines.

BADGE READER. Input/Output Devices (Computing System).

BAD LANDS. The literal translation of the phrase *Mauvais Terre* of the French explorers who so described the highly dissected, relatively unconsolidated sandstones and shales such as occur in the western Great Plains near the Black Hills. Small areas also occur in the plateaus of the Rocky Mountain region. This type of topography develops in arid and semi-arid regions where the underlying formations are relatively soft, and, due to the climate, are not protected by a plant cover.

BAFFLE. An object, usually a partition, placed for some specific purpose in the flow path of a fluid, causing the fluid to take some prearranged and circuitous path. Thus, baffles are found in steam boilers to direct the hot gas properly back and forth over the tubes so that the gas will give up its heat to the required degree and will not short-circuit directly from the furnace to the stack. For this service the baffle is composed of refractory material similar to firebrick and will be found in longitudinal or transverse arrangement. Transverse baffling is made by building the baffle perpendicular to the tubes. Longitudinal baffles are usually precast and laid upon the tubes of the boiler, forming a baffle whose surface is parallel to the tubes.

Baffles are built in coagulation basins to impede the flow of liquids, and are also found in exhaust mufflers where their purpose is to mix the flow of gases in adjacent exhaust puffs that they may emerge from the muffler in a silent steady stream.

BAFFLE-NOZZLE SYSTEM. Pneumatic Controller; Transmission (Pneumatic).

BAGASSE. In the manufacture of sugar from sugar cane the crushed fibers from which the sap has been expressed are called bagasse. Its principal use is as a fuel to run the mills which crush the cane. For this purpose bagasse is mixed with petroleum oil. It is also used as a fertilizer and to some extent in manufacturing heavy insulation board and coarse paper.

BAGDAD BOIL. **Leishmaniasis.**

BAGRIDS. **Catfishes.**

BAG-WORM (*Insecta, Lepidoptera*). The larva of a moth which is encased in a covering of silk mixed with bits of leaves, twigs, etc. Only the head and legs protrude from the bag, hence the insect appears to be suspended from the twig on which it walks. The adult females are wingless and the eggs are deposited in the silken bag.

One species of bag-worm sometimes does great damage to evergreen trees, especially the cedars, and so has been named the evergreen bag-worm, *Thyridopteryx ephemeraeformis.* The larvae can be killed by spraying infested trees with lead arsenate and the destruction of the bags in the winter, when they contain eggs, is an important measure of control.

These insects constitute the family *Psychidae.*

BAHIAGRASS. **Grasses.**

BAHNMETAL. **Sodium.**

BAILY BEADS. During an eclipse of the sun, at the instant when the moon's edge is just tangent to the edge of the sun, i.e., at either second or third contact, the thin crescent of the disappearing sun suddenly breaks up into a number of brilliant spots known as Baily beads. These are produced because the surface of the moon is very rough, and mountains on the moon will completely cover the sun's disk while the sunlight is still coming to the earth through the valleys.

BAINITE. A product of the decomposition of austenite that usually occurs at temperatures intermediate between those that produce pearlite and those that produce martensite. Its structure consists of finely divided carbide particles in a matrix of ferrite.

BAIRSTOW METHOD. A method for finding complex roots of an algebraic equation.

Let $z^2 + az + b$ be a trial divisor of $f(z)$ and form

$$f(z) = (z^2 + az + b)^2 Q(z) + (z^2 + az + b)q(z) + r(z)$$

where

$$r(z) = r_1 z + r_0, \qquad q(z) = q_1 z + q_0$$

This means that r is the remainder after dividing f by $z^2 + az + b$, and q the remainder after dividing the quotient. Solve

$$(aq_1 - q_0)\,\delta a - q_1\,\delta b = -r_1$$
$$bq_1\,\delta a - q_0\,\delta b = -r_0$$

for δa and δb; then $z^2 + (a + \delta a)z + b + \delta b$ will be, in general, closer to a true divisor. The method is an adaptation of the Newton method to finding complex roots and was originally described by Bairstow; later by Hitchcock. See also **Algebraic Equations; Newton's Formula for Interpolation;** and classified index under **Mathematics.**

BAKING POWDER. **Leavening Agents.**

BALANCE COIL. A balance coil is a coil for supplying a three-wire circuit from a two-wire circuit. A 240-volt single-phase line, for example, can be used to supply two 120-volt circuits consisting of three wires, one of which is a common intermediate wire. The voltage between the intermediate wire and either of the outside ones is 120 volts. If the loads on the two 120-volt circuits are unequal, the voltage can be balanced by an adjustment of the central tap point at the balance coil. A balance coil is frequently an auto-transformer having only one coil, a certain portion of which is used for both a high and low tension winding. The auto-transformer has connections at the ends of the coil, and an intermediate connection.

In dc circuits the balance coil is used in conjunction with the generator to obtain a three-wire system. The generator windings are tapped at two diametrically opposite points and these taps are connected to the ends of a balance coil. The third wire for the system is obtained by connection to the center of the balance coil. Some manufacturers

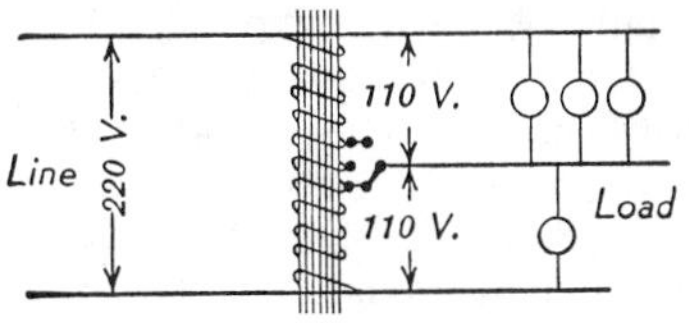

Balance coil for alternating current.

build the coil in the spider of the armature and bring out the center connection through a slip ring and brush while other manufacturers bring out the two tap connections on the armature through a pair of rings and brushes, mounting the balance coil outside the generator.

BALANCED AMPLIFIER. An amplifier circuit in which there are two identical signal branches connected so as to operate in phase opposition and with input and output connections each balanced to ground.

BALANCED DETECTOR. 1. Demodulator for frequency-modulation systems. In one form the output consists of the rectified difference of the two voltages produced across two resonant circuits, one circuit being tuned slightly above the carrier frequency and the other slightly below. 2. A detector for bridge or other null circuits, which include a balanced amplifier.

BALANCED LINE. A two-conductor balanced line is a transmission line consisting of two conductors in the presence of ground, capable of being operated in such a way that when the voltages of the two conductors at all transverse planes are equal in magnitude and opposite in polarity with respect to ground, the currents in the two conductors are equal in magnitude and opposite in direction. Currents flowing in the two conductors of a balanced line which, at every point along the line, are equal in magnitude and opposite in direction, are called balanced currents. Similarly, voltages (relative to ground) on the two conductors that are equal in magnitude and opposite in direction at every point along the line, are called balanced voltages.

BALANCED MODULATOR. A balanced modulator circuit is used to generate the sidebands of an amplitude modulated wave (see **Modulation**) and suppress the carrier. Since the power involved in a modulated wave is distributed at 100% modulation such that the carrier frequency component has $\frac{2}{3}$ and each sideband $\frac{1}{6}$ of the total, suppression of the carrier eliminates the necessity of supplying a major portion of the power usually needed. The output of such a modulator could be transmitted as only the sidebands, but in order to demodulate it a carrier wave would have to be introduced at the receiver. Such a system is known as a double sideband and suppressed carrier system. It is not used in practice because the carrier introduced at the receiver must have almost exactly the frequency of the one removed at the transmitter. Technical difficulties make this impractical. The balanced modulator output is normally fed through a filter circuit which cuts out one sideband and the other is transmitted. This gives single sideband, suppressed carrier transmission and is widely used, especially in telephony, since the frequency requirements for the carrier introduced at the receiver are not nearly so strict as the previous case. The reintroduced carrier needs to have an amplitude comparable to the received signal which is, of course, a small part of that needed at the transmitter of a radio system. The frequency band occupied by a single sideband, suppressed carrier system is somewhat less than half that needed for a full system. This is an important asset in telephone carrier circuits where the transmission line characteristics limit the total frequency band which can be transmitted. The single sideband system allows twice as many channels then on a given line.

BALANCED OSCILLATOR. An oscillator in which the impedance centers of the tank circuits are at ground potential, and the voltages between either end and their centers are equal in magnitude and opposite in phase.

BALANCE (Mechanical). The equilibrium of masses is a definition for mechanical balance. Static balance should be differentiated from dynamic balance. Static balance occurs in a system when the center of gravity of the system coincides with its reactions. For example, a rotating body in static balance has its center of gravity coincident with its axis of rotation. A system may, however, be in static balance, but become unbalanced when the system rotates. Such a system, for example, as that shown in the accompanying figure may well be in static balance, and satisfactorily pass a balance test which would consist of putting the shaft on absolutely horizontal parallel rails and trying the rotor for equilibrium in any position. But when this system rotates, the centrifugal forces of the two weights, not being in the same plane perpendicular to the axis of rotation, create a couple acting on the shaft. That couple rotates with the shaft and produces shaking forces at the journals, and vibrations in the foundation. The dynamic balancing of this system would involve the addition of a system of counter balances which, by themselves, would be in static equilibrium, but which in rotation would produce a couple equal in magnitude but opposite in direction to the one already considered.

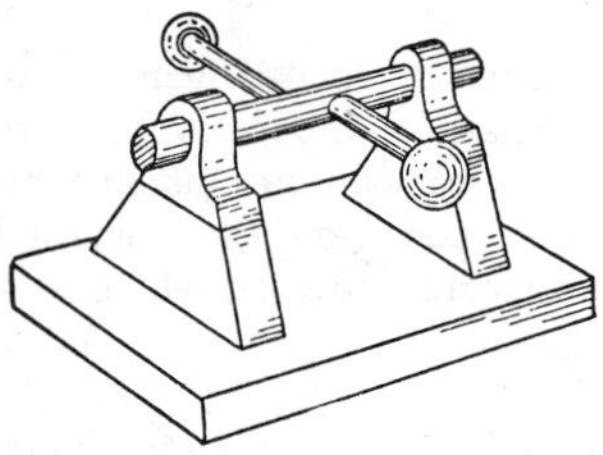

Case of static balance and dynamic unbalance.

Dynamic balance is especially important in high-speed or heavy rotating machinery, as the vibrating forces are proportional to the mass and the square of the speed of rotation. Manufacturers frequently use balancing machines for testing their product when it is especially important that it be perfectly balanced.

Reciprocating balance consists of opposing the shaking forces of a reciprocating mass by equal and opposite forces obtained from another reciprocating mass. One particularly difficult job of balancing occurs in a system consisting of both rotation and reciprocation, as exemplified by the piston, connecting rod, and crank mechanism. The difficulty lies in the fact that if perfect balance is secured in the direction of reciprocation by the employment of rotating counter balances, severe unbalance will result in a plane perpendicular to that direction. The solution of this difficulty is a compromise in which only part of the reciprocating mass is counterbalanced, thus reducing the maximum degree of unbalance, but producing a smaller unbalance in two planes.

BALANCER SET. This consists of two shunt dynamos connected in series across a two-wire dc distribution system to supply the third wire for a three-wire system. Either machine may act as motor or generator, the action being determined by the direction of unbalance of the load. For unbalanced loads one machine motors and drives the other as a generator to tend to restore the voltage balance.

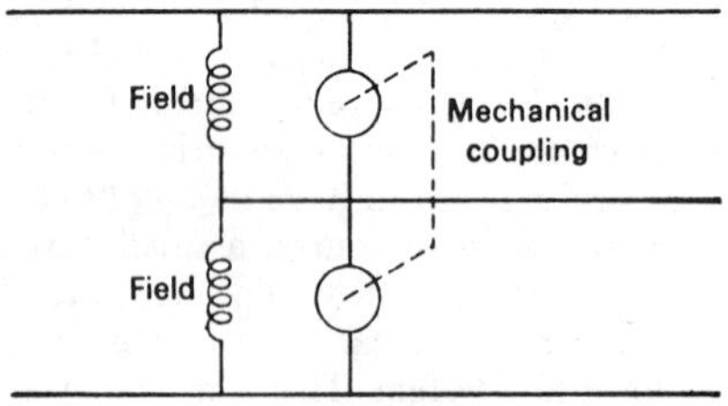

Balancer set.

BALANCE SENSING (Human). Hearing and the Ear.

BALANCE (Weighing). Weighing.

BALANCE WHEEL. Spring Clock.

BALANCING (Mechanics). A principle of statistical mechanics which shows that the steady state of affairs at equilibrium is maintained by a direct balance between the instantaneous rates of opposing processes. Thus at equilibrium the number of processes which destroy a situation A and produce a situation B is equal to the number of processes which destroy B and produce A. This principle is often valid, but not always; in the presence of a magnetic field, for instance, it does not hold for all processes A and B. It was called by Tolman the principle of *microscopic reversibility* and plays an important role in the derivation of the Onsager relations.

BALANOGLOSSUS. A genus of worm-like marine animals belonging to the lower chordates. The name is also commonly applied to any of the similar animals of several genera. *Enteropneusta.*

BALD CYPRESS. Cypress Trees.

BALD EAGLE. Eagle.

BALDNESS. Alopecia.

BALDPATE DUCK. Waterfowl.

BALLAST. Any material used for the purpose of providing stability is called ballast. Ballast is used in ships to bring the center of gravity of the vessel below a point called the metacenter, when there is a lack of cargo which would produce the same effect. Balloons and airships carry ballast which acts as a stabilizer and provides a means of controlling the rate of gain of rise as well as the altitude. Light vehicles which move at high speeds are often provided with ballast to lower the center of gravity and prevent overturning. Sand or water are very useful for ballast. Crushed stone which is placed under and between railroad ties to absorb impact and provide smooth riding conditions is also called ballast.

BALLAST EFFICIENCY. Illumination.

BALLING SCALE. Specific Gravity.

BALLISTIC MEASUREMENT. Any measurement in which an impulse is applied to the measuring device and the subsequent motion of the device is determined as a measure of the impulse. See **Ballistic Pendulum; Galvanometer.**

BALLISTIC MISSILE. A missile designed to operate primarily in accordance with the laws of ballistics. A ballistic missile is guided during a portion of its flight, usually the upward portion, and is under no thrust from its propelling system during the latter portion of its flight; it describes a trajectory similar to that of an artillery shell.

BALLISTIC PENDULUM. An instrument used for measuring the horizontal velocity component of a projectile. In its usual form it consists of a simple pendulum of mass M, and of natural frequency f. A projectile of mass m, moving with a velocity V strikes the bob and is imbedded in it. The maximum excursion X of the bob is then measured. Assuming that $M \gg m$ and that little damping is present, it may be shown by application of conservation laws that

$$V = \frac{2\pi f X M}{m}$$

BALLISTICS. Ballistics is the science which treats of the motion of masses projected into space, especially as associated with the motion of projectiles from guns, and certain aspects of the motion of rockets. The subject of ballistics is conventionally divided into three parts: interior ballistics, exterior ballistics, and terminal ballistics.

Interior ballistics is largely interwoven with the study of thermodynamics—the pressure, volume, and temperature of an expanding gas during travel of the projectile in the bore. It is concerned with the amount and combustion characteristics of gunpowder. The maximum pressures, and location of the same, stresses in the barrel, and the design of the barrel to resist these stresses may also be said to be interior ballistics.

Exterior ballistics deals with the motion of the projectile after it has left the gun. The science of exterior ballistics might be said to have been rationally developed by Newton as a by-product of his study in gravitation. If the effect of air on the motion of a projectile is omitted, then the trajectory is parabolic, since as soon as the bullet or shell leaves the muzzle of the gun, force of gravity begins to pull it toward the earth. It is therefore impossible for a bullet to travel in a straight line, and if it is to return to a target at the same elevation as the muzzle, it must have an initial upward component given by aiming the barrel somewhat above the target. If the muzzle velocity is V, and the inclination of the barrel to the horizon is i, its upward component is $V \sin i$. If the action of gravity is wholly unresisted, the time it will take to reach the top of trajectory (at which point the upward component has been reduced to zero) is the same as that required by a freely falling body to attain a velocity equal to that of the vertical component at the muzzle. Assuming a simple case where the target is at the same level above the earth's surface as the gun, it would take another equal interval of time for the projectile to move from the top of its trajectory to the target. The distance to the target would be that covered by the horizontal component $V \cos i$ in the period of time taken by the projectile in reaching the top of its trajectory, and then returning to its original level.

But the effect of air forces cannot be neglected, practically speaking, and the simple equations derived from the mechanics of a freely falling body are not applicable without considerable modification. The retarding effect of air and the effect of winds must be included, as must the spin of the projectile and its behavior under the action of these forces, as a gyroscope.

The methods of exterior ballistics are also useful in all design of rockets, where they must be considered in combination with the self-contained propulsion and control systems. This is particularly true of the (low-level) sounding rockets and ballistic missiles in which the controls shut off the propulsion system when a preset velocity and azimuth has been attained. From then on, the trajectory of the missile is determinable entirely by methods of exterior ballistics.

Terminal ballistics is concerned with fragmentation, blast, and penetration of armor and concrete.

BALLISTICS STANDARD ARTILLERY ATMOSPHERE. Atmosphere (Earth).

BALLISTIC TRAJECTORY. The trajectory followed by a body being acted upon only by gravitational forces and the resistance of the medium through which it passes. A rocket without lifting surfaces is in a ballistic trajectory after its engines cease operation.

BALLISTOCARDIOGRAPHY. Electrocardiogram.

BALLOON. A nonrigid, lighter-than-air craft, receiving its sustention from the buoyancy of the gas it contains. The shape, usually spherical, is formed by the internal pressure of the lifting gas. The lifting medium first used was air, which was expanded and made lighter than the atmosphere by heat; this hot air balloon was devised by the Montgolfiers in 1783. Coal gas and hydrogen were other early lifting media. Paper, oiled silk, goldbeater's skin, and rubber were among the materials from which the first balloons were made. Scientific balloons currently are made of plastic materials.

Balloons may be distinguished as "captive," i.e., secured by a cable to the ground, such as those used for military observation and communications purposes during World War I, or as protection against enemy bombers during World War II; or as free balloons. In addition to their value for sporting, the free balloon is widely used for making meteorological observations in the upper atmosphere. See also **Weather Observations and Forecasting.** Prior to the advent of ventures into outer space, there was considerable scientific emphasis on the use of balloons to gain scientific information at high altitudes. Captain Anderson and Major Kepner of the United States ascended to nearly 60,000 feet (18,288 meters) in a balloon in 1934. In November of 1935, Stevens and Anderson again ascended to an altitude of 73,000 feet (22,250 meters) within a period of 4 hours and during this flight gained considerable scientific information on the upper atmosphere, including air composition, spore distribution, and atmospheric effects on communications to and from the ground. These men also made cosmic ray measurements and observations on the spectral distribution of sunlight. The second flight was made in Explorer II, a balloon which had a volume of 3,700,000 cubic feet (104,784 cubic meters) when fully extended. The men were housed in a small globe-shaped gondola made of monel metal and equipped with oxygen.

An attempt to cross the Atlantic Ocean in a free balloon was made as early as 1881. This feat remained until October 1976 when Yost in his *Silver Fox* almost made it completely across the ocean. Yost flew 2740 miles (4409 kilometers) over a great-circle route from his point of launch (Milbridge, Maine) to touchdown (about 200 miles; 322 kilometers east of the Azores). The flight required 107 hours, 37 minutes and broke the previous record of 87 hours and 1896.85 miles (3052.03 kilometers) for balloons of unlimited size. The landing was short of a complete ocean crossing by about 700 miles (1126 kilometers). The *Silver Fox* was 80 feet (24.4 meters) high and contained 60,000 cubic feet (1700 cubic meters) of helium in a neoprene-coated nylon bag. The balloon and gondola combined weighed less than 2 tons, more than half of it ballast and expendable equipment. The balloon carried the American flag and the flag of the National Geographic Society, which furnished generous support for the venture.

The first successful crossing of the Atlantic Ocean occurred in August 1978. The *Double Eagle II*, with three balloonists (Abruzzo, Anderson, and Newman) aboard was launched from Presque Isle, Maine on August 11 and touched down near Miserey, France (just west of Paris) on August 17, after a flight of 137 hours, 6 minutes. The *Double Eagle II* had a capacity of 160,000 cubic feet (4531 cubic meters) of helium in a neoprene-coated nylon cloth envelope. See accompanying diagram. The Pacific Ocean was successfully crossed in November 1981.

Contemporary meteorological balloons range in capacity from 25,000 to 10,000,000 cubic feet (708 to 283,200 cubic meters) and some measure in excess of 200 feet (61 meters) in diameter, ascending to heights of 80,000 feet (24,384 meters) or more. In 1963, a balloon ascent to an altitude of 81,500 feet (24,841 meters) over New Mexico was made by a two-man crew. The unmanned 250-foot (76-meter) balloon, Stratoscope II, made observations of Mars from an altitude of 77,000 feet (23,470 meters) with a 36-inch (91-centimeter) reflecting telescope. This balloon, one of many launched from an atmospheric research center in Palestine, Texas at that time, carried a 6300-pound (2860-kilogram) instrumentation package. In another balloon, launched in 1965, telescopic pictures were obtained of Venus, at an altitude of 87,000 feet (26,518 meters). Prior to the manned space flight programs, balloons were sent aloft to obtain information on the upper atmosphere, including effects of the Van Allen radiation belt at high altitudes.

Some of the types of meteorological balloons used include:

(*Constant-Level or Constant-Pressure Balloon*)—designed to float at a constant-pressure level and used for measuring upperatmospheric conditions.

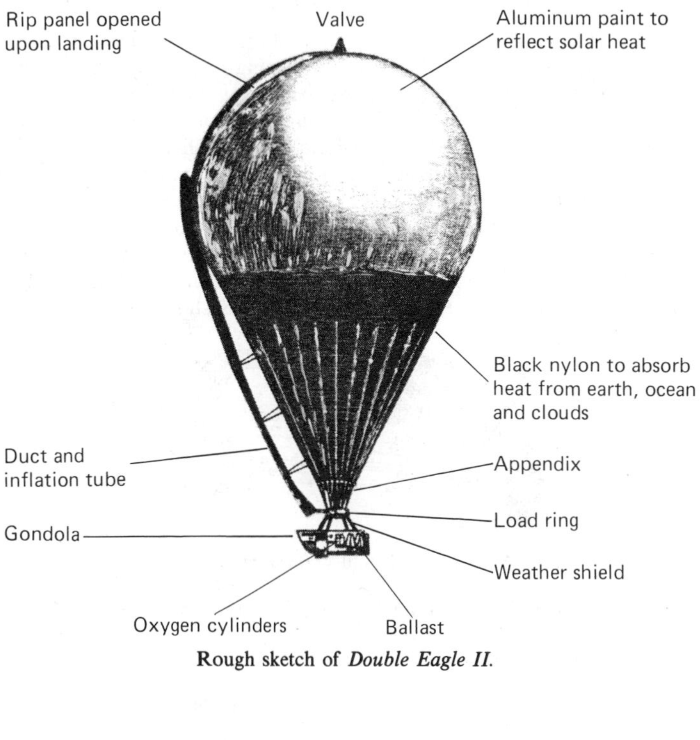

Rough sketch of *Double Eagle II.*

(*Pilot Balloon*)—used for computation of speed and direction of winds at higher levels.

(*Radiosonde Balloon*)—used to carry a radiosonde instrumentation package. Balloons have been developed which have a daytime bursting altitude of about 100,000 feet (30,480 meters) and nighttime bursting altitudes of about 80,000 feet (24,384 meters) above sea level. These balloons measure about 5 feet (1.5 meters) in diameter when first inflated, but may expand to about 20 feet (6 meters) before bursting at high altitude. The instrumentation package is returned to ground by parachute when the balloon bursts.

References

Narrative descriptions of the *Silver Fox* and *Double Eagle II* can be found in *National Geographic,* **151**, 2, 208–217 (1977); and **154**, 6, 858–882 (1978), respectively.

BALLOON (Arterial Plaque). Ischemic Heart Disease.

BALLOON-BORNE METEOROLOGICAL INSTRUMENTS. Radiosonde; Wind and Air Velocity Measurements.

BALLOON FISHES. Puffers; Porcupine Fishes.

BALLOON (Intra-Aortic). Intra-Aortic Balloon Counterpulsation; Shock Syndrome.

BALL, PEBBLE, AND ROD MILLS. Basically, all of these mills used for the size reduction of materials are comprised of a rotating drum which operates on a horizontal axis and is filled partially with a free-moving grinding medium which is harder and tougher than the material to be ground. The tumbling action of the grinding medium crushes and grinds the material by combination of attrition and impact. The grinding medium used may be a large number of round metal balls, operating in a drum with a metal lining. A conical ball mill is shown in Fig. 1. In the case of a pebble mill, a nonmetallic medium,

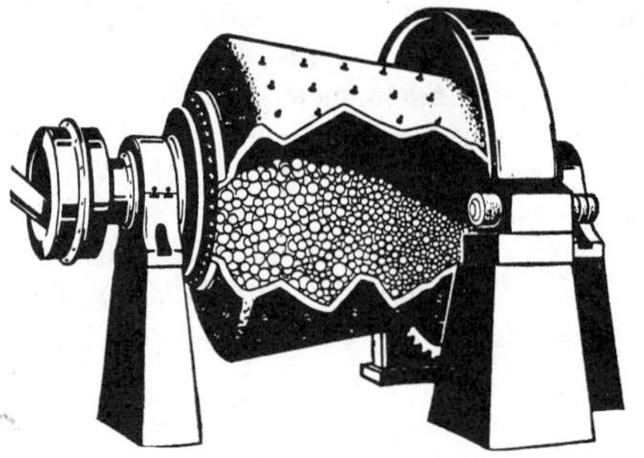

Fig. 1. Conical ball mill. (*Hardinge Co.*)

such as flint, pebbles, or even large pieces of the material being ground, is used. Instead of a metallic lining, the lining may comprise flint or porcelain blocks. In a rod mill, the grinding medium consists of a series of metallic rods essentially as long as the mill cylinder. These rods rotate freely like balls or pebbles as the mill turns. Like ball mills, rod mills have metallic linings. Feed for rotating drums varies—from a maximum of $1\frac{1}{2}$-inch ring size downward. These units can be operated either in a batch or continuous mode. Grinding generally requires several hours to assure the necessary fineness within particle-size limits. Usually, oversize material will be returned continuously or handled in a subsequent batch. A rod mill is shown in Fig. 2.

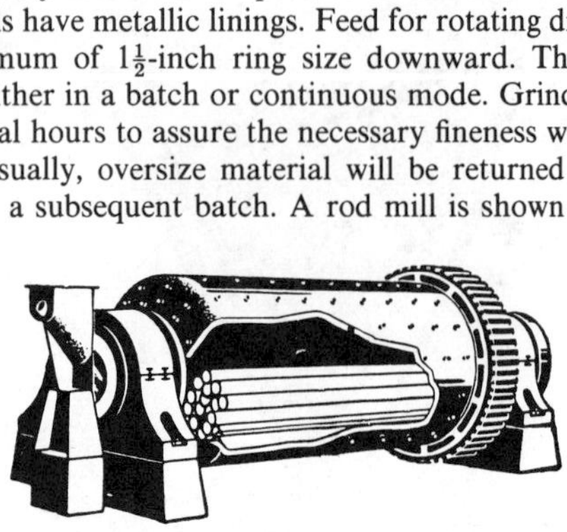

Fig. 2. Rod mill. (*Hardinge Co.*)

BALMER SERIES. Atomic Spectra; Energy Level; Stark Effect.

BALSAM FIR. Fir Trees.

BALSA TREE. Of the family *Bombaceceae* (silk cotton family), genus *Ochroma,* the balsa tree is found in the Central American tropics from southern Mexico through Ecuador and into northern Brazil. The balsa is a large, fast-growing tree, sometimes called corkwood or West Indian corkwood. Balsa wood is one of the lightest of the commercial woods, strong but pithy. At an air-dried moisture content of 12%, the specific gravity is only 0.13. The crushing strength with compression applied parallel to the grain is 1250 psi (8.6 MPa); the tensile strength with tension applied perpendicular to the grain is 115 psi (0.8 MPa). The wood from a 4-year-old tree will weigh about 8 pounds per cubic foot (128 kilograms/cubic meter). The wood becomes heavier as the tree grows older. In terms of strength/weight, balsa wood is stronger than some construction woods. It is particularly attractive when made into sandwich-type construction for its strength-weight ratio.

Balsa wood is composed largely of wood parenchyma, a tissue structure of thin-walled cells. This peculiar cellular structure makes the wood valuable for refrigeration insulation. Balsa sawdust also has been used for filling plastics because of its light weight. The cellular quality also makes the wood attractive for use in life preservers, vibration isolaters, floats, rafts, and for general use, cost permitting, where weight is the predominating factor.

Kari, a lightweight wood available in Japan from the *Paulownia tomentosa* tree competes to some extent with balsa wood. Samáuma also competes with balsa. See **Silk Cotton Trees.**

Species of balsa include: *Ochroma grandiflora* (Ecuador); *O. concolor* (Honduras and Guatemala, commercially called Barrios balsa; *O. limonensis,* called Limon balsa (Costa Rica and Panama); *O. obtusa,* called Santa Marta balsa (Colombia); and *O. velutina* or red balsa (Pacific coast of Central America).

BALUN. The name balun (balanced to unbalanced) is applied in general to devices used for the transformation from an unbalanced (coaxial) transmission line or system to a balanced (two-wire) line or system, in which the two terminals have equal impedances to ground.

BAMBOO. Of the family *Gramineae* (grass family), the tribe *Bambusae* comprises grasses which are particularly important to the economy

Medium-height bamboo (*Arundinaria simonii*). (*USDA photo.*)

of Oriental nations. The plants included in this group are of extremely variable nature, ranging from small inconspicuous species to the largest species of grass known, some having slender erect stems approaching a 100 feet (30 meters) in height. Many are clambering vines which form dense impenetrable masses. All these grasses are characterized by jointed hollow stems having solid nodes, familiar to Western people in the common bamboo fish pole. Among the Eastern peoples, bamboo is much more commonly used. The hollow stems may serve as pipes for conducting water, or as containers for storing water and other substances. Split stems may be flattened and used in constructing shelters, or boats, or furniture. Some species yield a fibrous material which is used to manufacture a kind of paper. The young stalks of certain species become an important foodstuff. See accompanying illustration.

BANACH SPACE. **Linear Topological Space.**

BANANA PLANT. Of the family *Musaceae* (banana family), genus *Musa*, the banana plant is a herbaceous perennial treelike tropical plant. Although a native of Asia, the plant is now grown in nearly all tropical areas. The fruit was first sent commercially to the United States from Central America in 1864. First found in Malaysia and the East Indies, the plant was introduced into Central and South American countries, such as Costa Rica, Honduras, Panama, Cuba, Jamaica, Haiti, Colombia, and Ecuador well over 100 years ago. As of the mid-1970s, Ecuador is the largest exporter of bananas and the fruit represents the chief export product of Honduras and Panama.

Most authorities agree that the center of origin of *Musa* genus plants was in southeastern Asia. Because of the strong association over the centuries of the banana with the tropical America, some authorities have argued that the banana originated in that region. However, wild varieties of *Musa* have been found in Asia, whereas such has not been the case in the American tropics. Efforts continue to locate new sources of propagation material for the banana. One organization devoted to this is the Plant Production and Protection Division, Food and Agricultural Organization of the United Nations in Rome. Wherever possible, "seed" sources from areas that have been free from major diseases and pests in the past are sought.

The banana is frequently referred to in ancient Hindu, Chinese, Greek, and Roman literature. Mention of the banana is found in various sacred texts of Oriental cultures. Chief of these writings are two Hindu epics, the *Mahabharata*, the work of an unknown author, and the *Ramayana* of the poet Valmiki. There are also references to the banana in certain sacred Buddhist texts. These chronicles describe a beverage derived from bananas which Buddhist monks were allowed to drink. Yang Fu, a Chinese official in the second century A.D., wrote an *Encyclopedia of Rare Things*, in which he described the banana plant. The Greek naturalist philosopher Theophrastus wrote a book on plants in the 4th century B.C. in which he described the banana. His book is considered the first scientific botanical work extant. Pliny the Elder described the banana in A.D. 77.

Although there are numerous varieties of cultivated bananas, two of the principal varieties are *Cavendish* and *Gros Michel.*

The banana and plantain are cultivated widely and are grown nearly everywhere that suitable climatic and other environmental conditions prevail. The banana is among the most important of all commercial fruits, exceeding by quite a margin the production of the apple, for example, and coming close to equaling the combined production of all citrus fruits—and, if banana and plantain production figures are combined, the banana family well exceeds combined citrus fruit production.

The banana plant grows to a height of from 10 to 30 feet (3 to 9 meters). The plant has an underground root stalk with buds which form suckers as the plants develop on the plantation. The weak suckers require frequent pruning. The plant has a thick underground rhizome from which rises what seems to be a thick, erect stem. Actually this false stem is formed by the leaf bases which grow wrapped together to a height of 10 feet (3 meters), depending upon total height of plant (variety, growing conditions, etc.). From the upper portion, the large leaf blades spread out conspicuously, sometimes to a length of 10 feet (3 meters). Up through the center of the tube formed by the leaf bases, the flower stem pushes its way and bears bunches of inconspicuous small tubular flowers in the axes of large showy bracts. After fertilization of the flowers, the bracts fall off, and the whole cluster gradually hangs over due to the weight of the developing fruit. The fruit occurs in clusters (*hands*), the individual fruit being called a *finger.* There usually are about 7 to 15 fingers per hand. When green, the banana fruit contains an abundance of starch, some of which changes to sugar as the fruit ripens. Once harvested, the plant is cut down, but by retaining a large sucker, production can be kept active over a number of years. Botanically, the banana fruit is a berry, qualifying under the definition of a fruit that has a pulpy or fleshy pericarp, that is, the wall of a mature ovary and having one or more seeds in the flesh. The latter qualification is met by those wild banana species that do have true seeds in the flesh.

The time required for the fruit to reach the harvest stage varies from 3 to 4 months after flowering, and when the fruit is harvested the false or pseudostem is cut down. The large, conspicuous leaves of a mature banana plant are the last of a long series that starts with the formation of small-scale leaves as the young shoot or sucker emerges above the ground. If the sucker is developing from a deeply buried bud, the scale leaves are followed by long narrow "sword" leaves and then, as the plant grows older, by leaves that gradually increase in size and width until the typical mature leaves are formed. In contrast, if the sucker is arising from a bud at or above the surface of the ground, it is likely to produce broad leaves very early in its development.

A healthy, mature banana plant has from 11 to 14 functioning leaves, together with the sheaths of 12 to 16 leaves that developed earlier, but those blades wither and fall as the plant develops. The last leaf to develop is short, but almost as broad as the mature leaves immediately below it. Its sheath encircles the blossom stem as the latter emerges or "shoots" above the throat of the plant and its blade droops over the blossom bud which bends downward in a few days from the weight of the rapidly developing inflorescence. The "protecting" or "follower" leaf gives the blossom bud a temporary degree of protection against the sun.

An extensive mass or matt of roots develops from the lower surface of the pseudobulb. They supply water and food materials to the aerial parts of the plant and also anchor and hold the pseudostem upright. The banana varieties grown for commercial use do not produce seeds. New plantations are established entirely from either whole pseudobulbs or portions of them containing one or more lateral buds or "eyes."

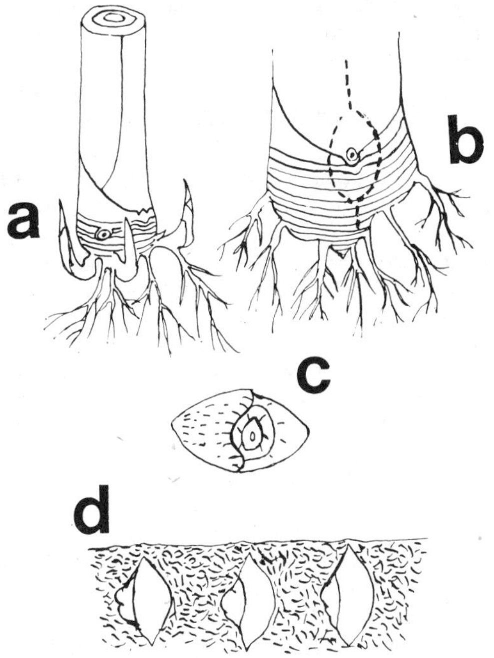

Propagation of banana plant: (a) Rhizome of banana showing suckers normally used for propagation; (b) dotted lines show how cut is made around dormant eye; (c) eye as it appears when removed from rhizome; (d) eyes planted in nursery bed, showing placement in vertical position.

Two types of suckers are formed, the broad-leafed type from buds low down on the rhizome, and the sword sucker, with narrow, upright leaves from buds nearer to the surface of the soil. Both kinds of sucker are suitable as propagating material. The sucker should be severed from the parent plant close to the base of the old pseudostem and should be lifted from the soil with the root system intact. Another method of propagation is by means of eyes. See accompanying figure.

Although there are numerous species, that commonly exported is the *Musa paradisica.*

The leaf of some species is used in making cordage from a hemplike fiber. In addition to table use, the banana is used in the preparation of flavorings. Vacuum dehydration of bananas yields banana crystals, a light-brown powder used in ice cream, bakery products, and milk-based beverages.

The banana is considered an excellent food source and contains about 460 calories per pound (1014 calories per kilogram), highest among fresh fruits. The constituents are about 75% water, 1.3% protein, 0.6% fat, and the remainder carbohydrate.

More detail on the banana plant can be found in the "Foods and Food Production Encyclopedia," (D. M. Considine, editor), Van Nostrand Reinhold, New York, 1982.

BAND EDGE ENERGY. The energy of the edge of the conduction band or valence band in a solid; that is, the minimum energy required by an electron in order that it may be free to move in a semiconductor or the maximum energy it may have as a valence electron.

BANDICOOT. Marsupialia.

BANDPASS FILTER. Filter (Communications System).

BANDSPREAD. In the finer grade of radio receivers, such as used for communications purposes, it is desirable to be able to tune rapidly over a wide frequency range, yet be able to tune carefully to rather fine limits over a small range at any selected point in the wider range. This is accomplished by the band spreading tuning provided on such receivers. The usual arrangement is to have the wider range covered by conventional tuning capacitors, but have these paralleled by small variable capacitors which are tuned by the band spread dial. Thus the complete range can be covered by one rotation of the main capacitors, while for any point at which they may be set the smaller ones can be used to give fine variation, thus giving, in effect, a spread-out tuning scale at this point. Other special circuit arrangements, such as tapping a capacitor across part of the tuning coil, are used to give the same result.

BAND THEORY OF SOLIDS. Solids (Band Theory).

BANDWIDTH. The number of cycles per second (Hertz) expressing the difference between the limiting frequencies of a frequency band.

BANDWIDTH (Effective). 1. For a specified transmission system, the bandwidth of an ideal system which (a) has uniform transmission in its pass band equal to the maximum transmission of the specified system, and (b) transmits the same power as the specified system when the two systems are receiving equal input signals having a uniform distribution of energy at all frequencies. This may be expressed mathematically as follows:

$$\text{Effective bandwidth} = \int_0^\infty G\,df$$

where f is the frequency in cycles per second and G is the ratio of the power transmission at the frequency f, to the transmission at the frequency of maximum transmission.

2. For a band-pass filter, the width of an assumed rectangular band-pass filter having the same transfer ratio at a reference frequency, and passing the same mean-square value of a hypothetical current and voltage, having even distribution of energy over all frequencies.

BANDWIDTH (Modulation). Modulation.

BANDWIDTH (Optical Fiber). Optical Fibers.

BANK (Aircraft). 1. The position of an airplane when its lateral axis is inclined to the horizontal. A right or positive bank is the position with the lateral axis inclined downward to the right. 2. To incline an airplane laterally, i.e., to rotate it about its longitudinal axis. Banking of aircraft is necessary in making a turn so that the lift force normal to the plane of the wings produces the necessary component to counteract the centrifugal force acting on the aircraft in the turn.

BANKING (Curve). Superelevation.

BANK (Production). A planned accumulation of parts or subassemblies stored along the route of a production line to serve as a buffer and thus allow for reasonable fluctuations in production rate.

BANNER CLOUD. Clouds and Cloud Formation.

BANTENG. Bovines.

BANYAN TREE. Mulberry Family.

BAOBAB TREE. Of the family *Bombacaceae* (silk cotton family), genus *Adansonia,* the baobab tree (*A. digitata*) is a familiar scene in much of the African savanna country. Because of its very large trunk, as compared with its height, and the relatively small amount of foliage, the tree sometimes has been termed pot-bellied. The trees usually occur singly and widely spaced on the savanna, thus adding to the effect of their unique shape. Travelogs of Africa almost invariably include a few views of the baobab to accentuate the African backdrop. Reasonably adult trees will attain a height of between 22 and 40 feet (6.6 and 12 meters). They live to an old age, 1000 years or more, but grow mainly in girth rather than in height. The trunk of a 30-year-old tree may be about 10 feet in diameter; a 1000-year-old tree may have a trunk diameter of 30 feet (9 meters). The roots sometimes reach out for 100 feet (30 meters) or more and the baobab is well known for its ability to store large quantities of water. In times of drought, elephants are known to chew the soft fibers of a fallen baobab to extract its watery juices. The large trunk can be easily carved to form a chamber which has been used for numerous purposes, including the storage of water by the people of the Sudan; for preserving cadavers for long periods of time without embalming by various natives. There are even unconfirmed reports that a hollowed-out baobab has been used as a temporary jail.

The tree bears long, oblong, and large fruit, sometimes called monkey bread of an acid taste. The mucilaginous pulp of the tree is sometimes eaten by natives. The bark yields a fiber which can be used for making cloth and rope and the timber can be used for various purposes. Baobab trees also are found on the island of Madagascar. These may have a girth of from 70 to 80 feet (21 to 24 meters) and hold lots of water.

The species is named for its founder, Michel Adanson, the French naturalist.

BAR. Units and Standards.

BARABOO. A baraboo is a hill, mountain or other eminence which was once buried by the deposition of sedimentary material about it and has since been exposed by the erosion of the younger beds. It is named from Baraboo, Wisconsin.

BARBARY APE. Monkeys and Baboons.

BARBARY SHEEP. Goats and Sheep.

BARBEL. Fishes.

BARBERRY SHRUB. Of the family *Berberidaceae* (barberry family), the barberries are small plants growing wild in the northern hemisphere and in parts of South America. Many of them are spiny plants which are sometimes cultivated as hedge plants. Examination shows that the leaves may vary from entire spiny-toothed structures to those reduced entirely to spines, only their position on the stem revealing the fact that they are leaves. The plants bear racemes of yellow flowers and, later, small red, yellow, or black berries, which make them attractive in appearance. The pollination in barberries is of especial interest. Each of the six stamens has a spot at the base of the filament sensitive to touch. When an insect, pushing about the base of the flower in search of nectar, touches this spot, the stamen moves violently, so that the sides of the insect's head are powdered with pollen. When the insect visits another flower, some of this pollen may be caught by the stigma, pollination thus being completed.

The American barberry (*Berberis canadensis*) ranges from Virginia southward through the Allegheny Mountains into Georgia and Missouri. The common barberry of Europe (*B. vulgaris*) was introduced into the United States many years ago as a hedge plant. Unfortunately, this species is the alternate host of *Puccinia graminis*, the organisms that causes the destructive disease, wheat rust. For many years, there has been a campaign to eradicate this species. Now much preferred as a hedge plant is the exceptionally beautiful *B. Thurnbergii*, a Japanese species, which bears bright scarlet berries. The berries of this shrub are edible and have been used in jams and jellies. Also of the genus is the trailing mahonia (*B. aquifolium*), a beautiful, low and trailing shrub found in the Rocky Mountains and Pacific northwest. The plant has leaflets something like a holly, a dark lustrous green turning to reds and yellows in the autumn. The flowers are a gold yellow, occurring in small clusters.

BARBER'S ITCH. Dermatitis and Dermatosis.

BARBET. Woodpeckers and Toucans.

BARBITURATES. Sedative drugs derived from barbituric acid. These drugs depress the central nervous system and act especially on the sleep center in the brain, thus their sedative and sometimes hypnotic effects. Barbital and phenobarbital are relatively long-lasting in effects. Other drugs, such as Amytal and Nembutal, are more powerfully hypnotic and have a shorter action. A third group, including thiopentone and hexobarbital, are very powerful and short-lived in action, and sometimes are used for the production of intravenous anesthesia. In the treatment of epilepsy, phenobarbital is sometimes used. It is an effective anticonvulsant, but produces drowsiness when given in large amounts. Dilantin suppresses grand mal convulsions and can be taken in larger doses than phenobarbital without causing lethargy. Phenobarbital and Dilantin make petit mal worse, so Tridione is sometimes used in this type of epilepsy.

Barbiturates are widely used in sleeping pills. They induce a feeling of relaxation, usually followed by sleep. Barbiturates are advantageously used to provide temporary respite in times of unusual emotional stress. They should not be taken regularly as a substitute for a cure in chronic nervous tension. This will only prolong the stress and encourage the patient to continue his reliance on drugs instead of seeking a solution to his emotional problems.

The Expert Committee on Drug Addiction of the World Health Organization advised the United Nations that barbiturates "must be considered drugs liable to produce addiction." Some people develop physical dependence on barbiturates. Others may remain able to abandon the drugs voluntarily. As in the use of other psychological supports, the need for continued barbiturates lies in the underlying personality disorder.

BARCHAN. A crescent-shaped dune or drift of wind-blown sand or snow; the arms of the crescent point downwind. Conditions under which barchans form are a moderate supply of material (sand or snow), and winds of almost constant direction and of moderate speeds.

BAR CHART. A method of representing relative magnitudes by lines or bars the lengths of which are proportional to the quantities involved. Simple barographs show individual quantities; compound barographs show complete quantities on a single bar, and are often referred to as percentage bar charts. See accompanying figure.

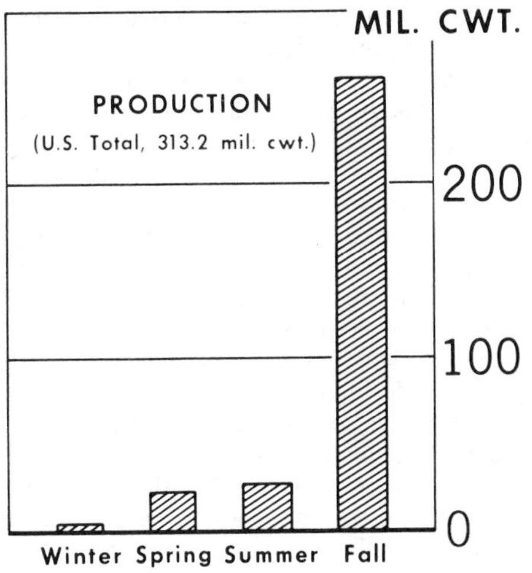

Example of bar chart. (Production of Irish potatoes in the United States—quantity versus season.)

BAR CODE. Input/Output Devices (Computing System).

BARKHAUSEN EFFECT. A series of minute "jumps" in the magnetization of iron or other ferromagnetic substance as the magnetizing force is continuously increased or decreased; discovered by H. Barkhausen in 1919. The effect may be observed by winding on the specimen, along with the magnetizing coil, a secondary coil connected to some sensitive detector of current fluctuations, such as an oscillograph or an audio amplifier. As the magnetizing current is steadily increased, the current in the secondary circuit, instead of being constant, exhibits a succession of small, sharp peaks or maxima, which the amplifier reveals by a faint clicking or snapping sound. See also **Ferromagnetism; Magnetism.**

BARITE. The mineral barite is barium sulfate, $BaSO_4$, crystallizing in the orthorhombic system. It may occur as tabular crystals, in groups, or lamellar, fibrous and massive. Barite has two perfect cleavages, basal and prismatic; hardness, 3–3.5; specific gravity, 4.5, which has led to the term heavy spar, occasionally used for this mineral. Its luster is vitreous; streak, white; color, white to gray, yellowish, blue, red and brown; transparent to opaque. It sometimes yields a fetid odor when broken or when pieces are rubbed together, due probably to the inclusion of carbonaceous matter. It is used as a source of barium compounds.

Barite is a frequently occurring gangue mineral and is found also in large masses in sedimentary rocks. It occurs in many places in Europe, including Czechoslovakia, Germany, France, Spain and England; in the United States, New York, Connecticut, Pennsylvania, Virginia, Michigan, Missouri, New Mexico, Oklahoma, Utah, Colorado, South Dakota, Georgia and Tennessee. In Canada it occurs in Ontario and in Nova Scotia.

The name of this mineral derives from the Greek word meaning heavy.

BARIUM. Chemical element symbol Ba, at. no. 56, at. wt. 137.34, periodic table group 2a (alkaline earths), mp 725°C, bp 1640°C, density 3.5 g/cm^3 (20°C). Body-centered cubic crystal form. Naturally occurring isotopes are ^{130}Ba, ^{132}Ba, ^{134}Ba, ^{135}Ba, ^{136}Ba, ^{137}Ba, and ^{138}Ba. Barium metal is comparatively soft and ductile and capable of mechanical working. Barium metal and all barium compounds are highly toxic to humans, although barium sulfate (because of its insolubility in H_2O and body fluids) can be ingested without harm and is widely used as an opaque medium in X-ray diagnostic studies of the body. First ionization potential 5.21 eV; second 9.95 eV. Oxidation potentials $Ba \rightarrow Ba^{2+} + 2e^-$, 2.90 V, $Ba + 2OH^- + 8H_2O \rightarrow Ba(OH)_2 \cdot 8H_2O + 2e^-$, 2.97 V. Other important physical properties of barium are given under **Chemical Elements.**

Barium occurs chiefly as sulfate (barite, barytes, heavy spar, $BaSO_4$), and, of less importance, carbonate (witherite, $BaCO_3$). Georgia and Tennessee are the principal producing states. The sulfate is transformed into chloride, and the electrolysis of the fused chloride yields barium metal. See also **Barite;** and **Witherite.** Barium ores are mined chiefly as a source of barium compounds because very little metallic barium is consumed commercially. The metal is obtained by thermal reduction of the oxide, using aluminum metal at a high temperature and under vacuum in a closed retort: $4BaO + 2Al \rightarrow BaOAl_2O_3 + 3Ba$. The gaseous barium produced is recovered by condensation.

As is to be expected from its high electrode potential (2.90 V) barium, like strontium and calcium, reacts readily with the halogens, oxygen and sulfur to form halides, oxide, and sulfide, as well as with nitrogen and hydrogen at higher temperatures to form the nitride and hydride. In all its stable compounds it is divalent. It reacts vigorously with water, displacing hydrogen to form the hydroxide. Barium peroxide is formed on treatment of the hydroxide with hydrogen peroxide in the cold and also by direct combination of oxygen and barium oxide or metal. The peroxide prepared in the latter way is frequently paramagnetic because of the presence of some superoxide, $Ba(O_2)_2$. Barium exhibits little tendency to form complexes, the amines formed with NH_3 being unstable and the β-diketones and alcoholates are not well characterized. Barium metal solutions in liquid NH_3 solution yield $Ba(NH_3)_6$ upon evaporation. Common compounds of barium are:

Barium acetate, $Ba(C_2H_3O_2)_2$, white crystals, solubility 76.4 g/100 ml H_2O at 26°C, formed by reaction of barium carbonate or hydroxide and acetic acid.

Barium carbide (acetylide), BaC_2, black solid, by reaction of barium oxide and carbon at electric furnace temperatures, reacts with H_2O; yielding acetylene gas and barium hydroxide.

Barium carbonate, $BaCO_3$, white solid, insoluble ($K_{sp} = 5.13 \times 10^{-9}$), formed (1) by reaction of barium salt solution and sodium carbonate or bicarbonate solution; (2) by reaction of barium hydroxide solution and CO_2. With excess CO_2 barium hydrogen carbonate, $Ba(HCO_3)_2$, solution is formed. Barium carbonate decomposes at 1450°C.

Barium chloride, $BaCl_2 \cdot 2H_2O$, white crystals, solubility 31 g/100 ml H_2O at 0°C, formed by reaction of barium carbonate or hydroxide and HCl.

Barium chromate, $BaCrO_4$, yellow precipitate, $K_{sp} = 1.17 \times 10^{-10}$, formed by reaction of barium salt solution and potassium chromate solution.

Barium cyanamide, $BaCH_2$, formed in a mixture with barium cyanide, $Ba(CN)_2$ by heating barium carbide at 800°C with nitrogen gas. Fusion of the cyanamide-cyanide mixture with sodium carbonate converts it entirely to cyanide.

Barium nitrate, $Ba(NO_3)_2$, white crystals, solubility 8.7 g/100 ml H_2O at 20°C, formed by reaction of barium carbonate or hydroxide and HNO_3.

Barium oxide, BaO, white solid, mp about 1900°C, reactive with H_2O to form barium hydroxide. Barium peroxide, $BaO_2 \cdot 8H_2O$, white precipitate, formed by reaction of barium salt solution and hydrogen or sodium peroxide, yields anhydrous barium peroxide upon heating at 100°C in a current of dry air. Anhydrous barium peroxide is also formed by heating barium oxide in air or oxygen under pressure (at somewhat over one atmosphere pressure) and temperature of 400°C.

Barium oxalate, BaC_2O_4, white precipitate, $K_{sp} = 1.1 \times 10^{-7}$, formed by reaction of barium salt solution and ammonium oxalate solution.

Barium sulfate, $BaSO_4$, white precipitate, $K_{sp} = 8.7 \times 10^{-11}$, formed by reaction of barium salt solution and H_2SO_4 or sodium sulfate solution, insoluble in acids, by heating with carbon yields barium sulfide.

Barium sulfide, BaS, grayish-white solid, formed by heating barium sulfate and carbon, reactive with H_2O to form barium hydrosulfide, $Ba(SH)_2$, solution. The latter is also made by saturation of barium hydroxide solution with H_2S. Barium polysulfides are formed by boiling barium hydrosulfide with sulfur.

Uses of Barium. The major use of barium metal for a number of years has been as a getter for oxygen in electronic vacuum tubes. A layer of the metal is deposited inside the glass envelope of the tube. Minute quantities of gases which leak into the tube react with the barium layer to form compounds. If the gases remained free, they would alter the conductance of the tube and cause deterioration of its performance. The addition of barium prolongs the useful life of vacuum tubes.

Stephen E. Hluchan, Business Manager, Calcium Metal Products, Minerals, Pigments & Metals Division, Pfizer Inc., Wallingford, Connecticut.

BARIUM CARBONATE. **Witherite.**

BARIUM SULFATE. **Barite.**

BARIUM TITANATE CRYSTAL. **Ultrasonics.**

BARK. All tissues of woody stems or roots which occur outside the cambium are collectively known as bark. In the earliest stages of its development, the stem is covered by a layer of thin epidermal cells, which may persist for some time. With increased age and growth, however, this epidermal layer is lost, and a new tissue is formed, either from the epidermal cells or from those cortical cells just beneath the epidermis. The cells which form this new protective layer are the cork cambium cells. The tissue which is formed by them is often called the cork or outer bark. If the divisions of the cork cambium cells occur fast enough the bark will remain smooth for some time. In most cases, however, an internal cork cambium forms in the deeper cortical tissues, and by producing secondary cortical tissue in isolated patches leads to the development of scales or patches of bark, with ever-deepening fissures as growth continues. In a few cases, as in the beech tree, the smooth condition persists throughout the life of the tree. In all young stems the continuity of the surface is broken by patches of loose cells, the lenticels, which permit an exchange of gases through the bark. The inner bark consists of phloem or bast, the food-conducting tissue of the plant.

BARK BEETLE (*Insecta, Coleoptera*). A small beetle of cylindrical form which burrows in the sapwood and inner bark of trees and logs, forming characteristic patterns. This habit also gives the name engraver beetle to these insects. Most of the numerous species infest forest trees but a few attack fruit trees. Together with the timber beetles and a few species which attack herbaceous plants they make up the family *Scolytidae.* Destruction of standing timber by these beetles has been estimated as high as 5 billion board feet per year.

BARKHAUSEN EFFECT. A series of minute "jumps" in the magnetization of iron or other ferromagnetic substance as the magnetizing force is continuously increased or decreased; discovered by H. Barkhausen in 1919. The effect may be observed by winding on the specimen, along with the magnetizing coil, a secondary coil connected to some sensitive detector of current fluctuations, such as an oscillograph or an audio amplifier. As the magnetizing current is steadily increased, the current in the secondary circuit, instead of being constant, exhibits a succession of small, sharp peaks or maxima, which the amplifier reveals by a faint clicking or snapping sound.

BARKHAUSEN OSCILLATOR. **Oscillator.**

BARK LOUSE (*Insecta, Homoptera*). A scale insect. These insects are minute creatures with sucking mouths. They spend most of their lives attached to the leaves, stems, or roots of plants. The name scale insect is due to the common secretion of a scale which conceals the body of the insect. There are many species, of which some, like the oyster-shell bark louse, *Lepidosaphes ulmi,* and the San Jose scale, *Comstockaspis perniciosa,* are important enemies of fruit trees.

Some scale insects produce substances of commercial value, notably cochineal and shellac.

BARKOMETER SCALE. **Specific Gravity.**

BARLEY. Of the family *Gramineae* (grass family), subfamily *Festucoideae,* tribe *Horeade,* and genus *Hordeum,* barley is an annual cereal grass of major worldwide importance, rating fourth in most lists of cereal crop statistics. Because of its outstanding tolerance to a wide range of climates and altitudes, barley is cultivated in many regions of the world, as later delineated. Barley produces stalks up to 3 feet

(0.9 meter) in height. The inflorescence is a close spike. There is much evidence of the antiquity of this plant. Remains found in the ruins of lake dwellings in Switzerland indicate an association of barley with Stone Age culture. An ancient Chinese text establishes the culture of barley as early as 2000 B.C. The plant was cultivated in ancient Egypt by the Greeks and Romans and is mentioned in *Exodus* 9:31, "And the flax and the barley was smitten: for the barley was in the ear, and the flax was boiled. But the wheat and the rie were not smitten; for they were not grown up."

Primitive peoples in Europe used barley for food, as do many modern races. However, lacking gluten, it is not well suited for making light breads, a fact that has kept barley from being of greater popularity among cultured races. Although barley is used directly in some breakfast cereals, soups, and other prepared food products, barley is consumed mainly as a source of malt (for the brewing industry) and as a major feed for animals. Barley is an excellent feed for finishing beef cattle. Its feeding value is about 90% that of maize (corn). A great advantage of barley is that it is the earliest maturing of the small grains. This makes it adaptable to a double-crop system, such as barley and soybeans, or barley and grain sorghum. Barley is an excellent winter grain crop for many regions, such as Arizona and California in the southwestern United States and in the southeastern United States, particularly in the upper Coastal Plain in the northern reaches of Georgia. Compared with most other cereal crops, barley is superior in terms of drought resistance.

A general breakdown of the uses of barley (in the United States) is: seed (10%); food products for human consumption (10%); malt production (20%); animal feeds (60%). Barley has a feeding value of about 95% that of maize (corn). Barley is used for hay in some areas and as a winter cover crop in the southeastern United States.

Linnaeus described barley in 1754 as: "Genus *Hordeum*: Spike indeterminate, dense, sometimes flattened, with brittle, less frequently tough awns. Rachis tough or brittle. Spikelets in triplets, single-flowered, but sometimes with rudiments of a second floret. Central florets fertile, sessile or nearly so, lateral florets reduced, fertile, male or sexless, sessile or on short rachillas. Glumes lanceolate or awnlike. The lemma of the fertile flowers awned, awnleted, awnless, or hooded. The back of the lemma turned from the rachis. Rachilla attached to the kernel. Kernels oblong with ventral crease, caryopsis usually adhering to lemma and palea. Annual or perennial plants."

Fig. 1. Barley plant: (a) Spike; (b) exposed upper peduncle; (c) flag leaf; (d) sheath; (e) leaf blade, third from top; (f) internode below peduncle. (*After USDA diagram.*)

To accommodate for cultivated types, Aberg and Wiebe (U.S. Department of Agriculture) added the description: "Summer or winter annuals, spikes linear or broadly linear, tough or brittle. The central florets fertile, the lateral ones fertile, male, or sexless. Glumes lanceolate, narrow or wide, projecting into a short awn. The lemma in fertile florets awned, awnleted, awnless, or hooded, the lemma in male or sexless florets without awns or hoods. Kernel weight from 20 to 80 milligrams. The chromosome number in the diploid stage 14."

Reference to Figs. 1 and 2 will assist in interpreting the foregoing technical descriptions. *Rachis* = principal stem of flower; *floret* = flower; *lemma* = sheath; *sessile* = attached to main stem; *lanceolate* = narrow and tapering (lancelike); *glume* = husk or chafflike bract; *awn* = bristly fibers or bead on head.

The culm (jointed stem) of the barley plant is made up of from five to seven hollow, cylindrical internodes that are separated by solid, somewhat larger joint nodes from which the leaves arise. The length (height) of the culm may range from about 8 inches (20 centimeters) to as much as feet (1.5 meters), depending upon variety and growing conditions. There are usually from three to six culms per plant if a normal seeding rate has been used. The leaf sheath is smooth or glabrous, although covered with hairs in a few varieties.

The botany and taxonomy of barley are complex topics because of the several species and hundreds of varieties, each exhibiting minor differences sometimes difficult to distinguish. Barley is considered by most authorities to be a plant native to the Orient, particularly of western Asia where it grows wild today. Wild species also are found in other regions of the world, including North America. When young, such wild plants can be used for forage, but as they become mature, they are regarded as rather aggressive weeds. Some of the wild species include:

- *Hordeum jubatum*, which has mature spikes with sharp-pointed joints that can be injurious to livestock.
- *H. jubatum*, *H. montansse*, and *H. nodosum*, perennial wild species.
- *H. adscendens*, *H. gussuoneanum*, *H. murinum*, and *H. pusillum*, annual species that frequently occur in pastures of the southwestern United States.
- *H. agriocrithom*, a wild species commonly found in the highlands of eastern Tibet. *H. spontaneium* is another common wild species. It is interesting to note that these two species have 14 diploid chromosomes and can be readily crossed with cultivated varieties.

In the earliest cultivated varieties of barley, the flower possessed stout barbed awns, which were disagreeable to persons who handled

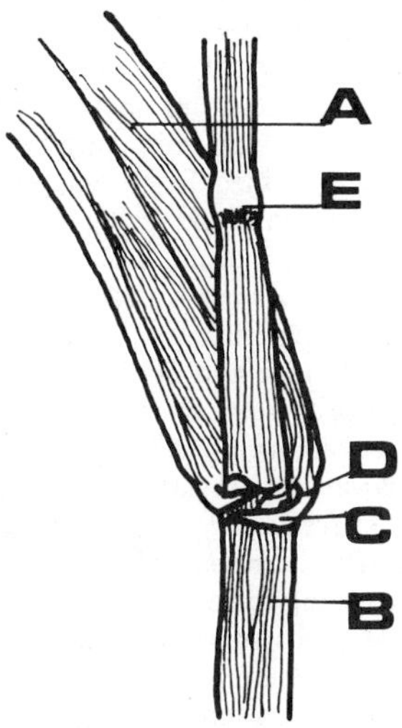

Fig. 2. Culm of barley plant partially shown: (a) Leaf blade; (b) leaf sheath; (c) auricle; (d) ligule; (e) node. (*After USDA diagram.*)

the grains and also caused serious difficulties when the grain was fed to livestock. As the result of plant breeding in the U.S.S.R. many years ago and then later carried out by scientists elsewhere, these awns were eliminated or modified, leaving a smooth-fruited variety.

In the mature barley grain, the aleurone layer is usually three cells in thickness, and the embryo is very small.

More detail on barley can be found in the "Foods and Food Production Encyclopedia," (D. M. Considine, editor), Van Nostrand Reinhold, New York, 1982.

BARLEY (Germ Plasm). Plant Breeding.

BARLOW RULE. The volumes of space occupied by the various atoms in a given molecule are approximately proportional to the valencies of the atoms; whenever an element exhibits more than one kind of valency the lowest value is generally selected.

BARN. Neutron; Units and Standards.

BARNACLE (*Crustacea, Cirripedia*). Sessile marine animals wholly unlike the more common crustaceans in appearance. Some are parasitic, some burrow in the shells of marine animals, and some attach themselves to submerged objects. The last habit is economically important since it results in the fouling of ships' bottoms and entails the expense of periodical cleaning.

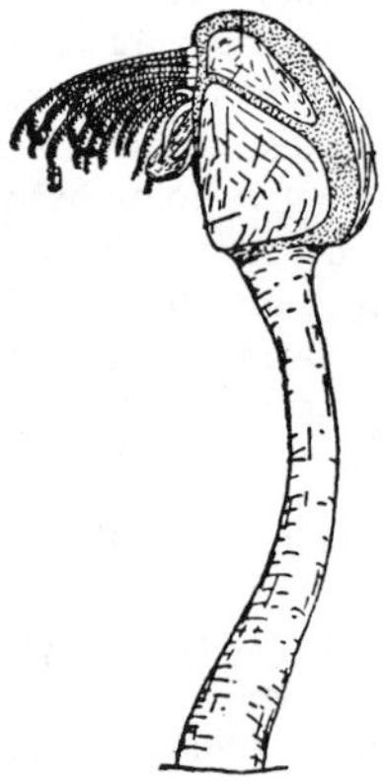

Barnacle.

BAROCLINIC ATMOSPHERE. Atmosphere (Earth).

BAROMETER. An instrument for measuring atmospheric pressure. There are two types of barometers commonly used in meteorology: (1) The mercury barometer is basically a glass tube about 3 feet (1 meter) long, closed at one end, filled with mercury, and inverted with the open end immersed in a cistern of mercury. Mercury rises in the sealed tube above the level in the cistern in direct relation to the air pressure on the cistern surface. Numerous refinements are required in a mercury barometer before it is sufficiently accurate for precise pressure measurements. Two types are shown diagrammatically in Fig. 1. (2) The aneroid barometer consists of one or more thin corrugated hollow disks (aneroid capsules), which are partially evacuated of gas, and restrained from collapsing by an external or internal spring. The deflection of the spring will be nearly proportional to the difference between the internal and external pressures. Magnification of the spring deflection is obtained both by connecting capsules in series and by mechanical linkages. When one side of a disk is fixed, the other side will move and operate a system of levers, which, in turn, cause an indicator to move over a scale. If atmospheric pressure falls, the disk expands; if atmospheric pressure increases, the disk contracts. Aneroid barometers must be calibrated against a mercury barometer and checked frequently. See also **Altimetry.**

The weight barometer is a mercury barometer that measures atmospheric pressure by weighing the mercury in the column or the cistern. A siphon barometer is so constructed that the upper and lower mercury surfaces have the same diameter.

In meteorology, barometric tendency is defined as the changes of barometric pressure within a specified time (usually 3 hours) before a weather observation.

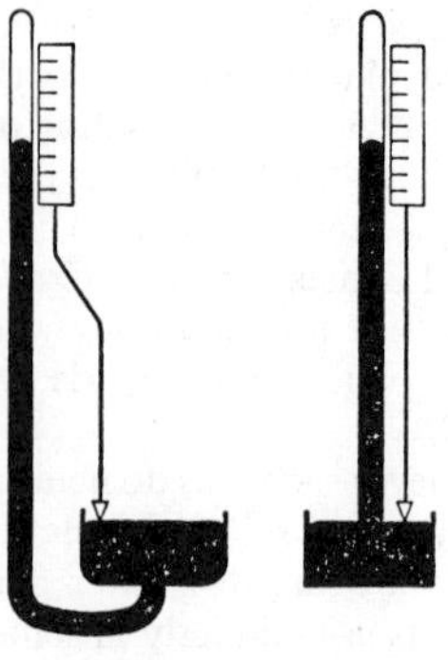

Fig. 1. Operating principle of two types of mercury barometer.

Barograph. In essence, an instrument that records barometric pressure—a recording barometer. The measurement principles are the same as the barometer, but in addition to an indicating pointer, a recording mechanism is added. An aneroid barograph is shown in Fig. 2.

Fig. 2. Aneroid barograph. Pressure can be read directly under the pen. A time record is inked onto a chart affixed to a slowly revolving drum.

BAROMETRIC HEIGHT. Pressure.

BAROMETRIC PRESSURE. Atmospheric Pressure.

BAROMETRIC WAVE. Atmosphere (Earth).

BAROTROPIC WAVE. Atmosphere (Earth).

BAROTROPY. As used in oceanography, the state of zero baroclinity, i.e., the coincidence of surfaces of constant density with those of constant pressure.

BARRACUDA (*Osteichthyes*). Of the suborder *Sphyraenoidei* and family *Sphyraenidae*, barracudas are excellent swimmers. They are chiefly found close to tropical sea coasts. From there, they migrate extensively during the summer months as far as northern or southern temperate seas. They pursue schooling fishes. The body is elongated, pikelike, and is covered with small cycloid scales. The head is very long and tapered, with a protruding lower jaw. The mouth is horizontal and can open very wide. The teeth are on the jaws and the palatine bone; they are larger in front and in the very front, barracudas have powerful grasping teeth. Like mullet, the dorsal fins are widely separated; the first dorsal fine has five strong spines. The caudal fin is forked. There is just one genus, *Sphyraena*, with 18 species.

Divers in tropical regions fear the attacks of the sometimes giant barracudas more than they fear sharks. As observed by Herald and Vogt, "Barracudas are incalculable. Although they do not sneak up on divers, they follow them everywhere. Since they perceive prey less through smell than through sight, they orient to everything which catches their attention, either through colors or unusual movements, such as those of an injured fish. Unlike shark, barracudas undertake

Great barracuda (*Sphyraena barracuda*).

just a single attack, and in attacking they leave wounds which are not torn out at the edges."

The life habits of the Pacific barracuda (*Sphyraena argentea*) are the best understood of any barracuda. This species winters off Mexico and, in spring, schools of them move up the coast to spawn. The spawning season extends from April to September. The eggs are laid at intervals and float in the water. Males attain sexual maturity in their second or third year, while females require an additional year. Barracuda diet consists chiefly of sardines. The Pacific barracuda can attain a length up to nearly 5 feet (1.5 meters). These barracudas are caught off Mexico throughout the year, and off California only in summer. The meat is considered very flavorful and is quite firm.

Other barracudas grow to be much larger. The Indo-Malaysian barracuda (*S. jello*) can reach a length up to 10 feet (3 meters). The great barracuda (*S. barracuda*) achieves equally great lengths. See accompanying illustration. The European barracuda (*S. sphyraena*) grows to about 3 feet (1 meter) in length and allegedly attacks bathers on occasion. It is considered a good market species. The southern sennet (*S. picudilla*) attains a length of about 18 inches (45 centimeters). Its habitat is the west Atlantic Ocean from the Bermudas to Brazil.

See entry on **Fishes,** which includes a list of references at the end of the entry.

BARRANCO. A deep, normally rock-walled ravine with steep sides. The term also is used to describe a gorge, small canyon, or deep cleft, gully, arroyo, or other break made by a heavy rain.

BARRED-SPIRAL GALAXY. Galaxy.

BARREL. Units and Standards.

BARRETTER. Unlike a thermistor which has a negative temperature coefficient, a barretter is a metallic resistor with a positive temperature coefficient of resistivity. Both barretters and thermistors are used in bolometers. In general, barretters are delicate and will burn out with small power overloads. Barretters are sometimes used in microwave power measuring instruments, but their characteristics are more suited to use as microwave signal detectors. A comparison of bolometers that use thermistors and barretters for microwave power measurement is given in the accompanying table.

REPRESENTATIVE BARRETTER AND THERMISTOR CHARACTERISTICS

Characteristic	Barretters	Thermistors	Remarks
Time constant	350 μs	1 s	Long time constant to measure pulsed power
Max. average power	< 10 mW	> 25 mW	Thermistor suited for pulsed power and sudden overloads
Peak overload power	25 mW	400 mW	
Power sensitivity	5 Ω/mW	35 Ω/mW	Thermistor good for low-duty-cycle pulsed power and low-level signals
Drift (200 Ω)	0.15%/°C	1.8%/°C	Barretter has accuracy and low drift
Temperature coefficient	Positive	Negative	Negative coefficient less sensitive to burnout

BARRIER BEACH. A recently emerged coast (like the present Atlantic Coastal Plain) which shallow water extending for some distance from the shore will lend itself to the formation of barrier beaches. Waves tend to stir up the bottom materials of sand and gravel to be redistributed which gradually build up a low ridge parallel to the coast. Such barrier beaches are also known as offshore bars and the quiet body of shallow water between the bar and the mainland is called a lagoon.

BARRIER LAYER. An electrical double layer formed at the junction, or surface of contact, between a metal and a semiconductor, or between two meals, for various purposes.

BARRIER LAYER CELL. The barrier layer cell is a photoelectric device which produces a small electromotive force upon the incidence of light or other radiant energy upon its barrier layer or junction.

BARRIER REEF. A narrow and long coral reef essentially parallel with the shore. The reef usually is separated from other land by a lagoon of considerable depth and width. Typically, a barrier reef will enclose much or all of a volcanic island. In the case of the Great Barrier Reef which lies off the coast of Queensland, Australia, it is a considerable distance from the continental coast.

BARRINGER CRATER. Astrobleme; Meteorite Crater.

BARTHOLIN'S GLANDS. Cowper's Glands.

BARYCENTRIC PARALLAX. A term given to a slight oscillatory motion of the earth. The center of gravity of the earth-moon system revolves about the sun in an orbit that is usually referred to as the orbit of the earth. Both the earth and moon are revolving about this center of gravity of the system as it, in turn, revolves about the sun. At the time of conjunction of the moon with the sun, the moon is inside the orbit of the system about the sun, and the center of the earth is outside. At opposition, this condition is reversed, and the center of the earth is on the inside. Hence, the center of the earth oscillates slightly back and forth across the orbit within a period of one synodic month. It is this slight oscillatory motion of the earth that is known as barycentric parallax.

Accurate measurements of barycentric parallax provide a method for the determination of the relative masses of the earth and the moon. In accordance with the principles of mechanical equilibrium, the distances of the center of gravity of the earth-moon system from the centers of mass of the two individual objects are inversely proportional to the masses of the objects. Hence, if we can determine the distance of this center of mass of the system from the centers of the earth and moon, we can at once determine the relative masses of the two objects. Accurate measurements of barycentric parallax determine the position of the center of mass of the earth-moon system as being 4,600 kilometers from the center of the earth. Since this is about 1/82.5 of the distance between the moon and the earth, we find that the mass of the moon should be 1/81.5 the mass of the earth.

BARYON NUMBER (Conservation). Conservation Laws and Symmetry.

BARYONS. A class of subatomic particles including the proton, the neutron, and several heavier particles, such as the lambda, the sigma (plus, minus, and neutral), and the omega (minus) particles. Baryons are particles that interact with the strong nuclear force. Each baryon is given a baryon number 1, each corresponding antibaryon is given a baryon number −1, while the light particles (photons, electrons, neutrinos, muons, and mesons) are given baryon number 0. The total baryon number in a given reaction is found by algebraically adding up the baryon numbers of the particles entering into the reaction. During any reaction among particles, the baryon number cannot change. This rule ensures that a proton cannot change into an electron, even though a neutron can change into a proton. Similarly, to create an antiproton in a reaction, one must simultaneously create a proton or other baryon. Baryon conservation ensures the stability of the proton against decaying into a particle of smaller mass. See also **Particles (Subatomic); Neutron;** and **Proton.**

BARYTOCALCITE. This mineral is a carbonate of barium and calcium which crystallizes in the monoclinic system but occurs massive as well. It has a perfect cleavage parallel to the prism and one, less perfect, parallel to the base; fracture, sub-conchoidal; brittle; hardness, 4; specific gravity, 3.66–3.71; luster, vitreous; color, white or gray or may be greenish or yellowish; transparent to translucent. Barytocalcite is found in Cumberland, England, associated with barite and fluorite.

BASAL CELL CARCINOMA. Dermatitis and Dermatosis.

BASAL CONGLOMERATE. A conglomerate that lies on, or occurs just above, a plane of erosion or unconformity. Such a conglomerate constitutes the first sedimentary stage in a normal cycle of sedimentation.

BASAL METABOLISM. The metabolism of a living cell or organism refers to the total turnover of chemical material and energy. It consists of *anabolism*, or assimilation, mostly of substances of high potential energy (primarily protein, fat, and carbohydrate), and *catabolism* or dissimilation. In common speech, metabolism refers to the oxidation of major foodstuffs and the concomitant release of energy. Metabolic rate refers to the metabolism in a given period of time. The "basal" metabolic rate refers to the fundamental energy requirement for maintenance and continued functioning of the organism (aside from external muscular work and work of digestion), such as respiration, contraction of the heart, function of the kidney, the liver, and of all cells in general. Basal metabolic rate (BMR) in humans refers to the determination of metabolic rate under certain standardized conditions, including complete physical rest (but not sleep), a fasting state, and an ambient temperature that does not require energy expenditure for physiological temperature regulation. Actually the BMR refers not to a "basal" rate but to a determination under these standard conditions. The BMR is below normal in sleep, starvation, anesthesia, and certain endocrine disturbances (hypothyroidism), and is elevated in fever, athletic training, under the influence of drugs (e.g., caffeine) and endocrines (adrenaline, thyroid hormones).

In studies of animals it becomes technically difficult to make observations under standard conditions which include rest. Restraining an animal increases the metabolic rate and inactivation through anesthesia lowers it; ruminants and other plant eaters cannot be brought into a fasting state unless they are deprived of food for prolonged periods of time; small animals (e.g., shrews) have such high metabolic rates that they must eat almost continuously to sustain a normal metabolic rate, etc.

Methods of Determination. In principle, the metabolic rate is determined in three different ways. (a) Determination of the energy value of all food less the energy value of excreta (mainly feces and urine) should give the energy turnover of the organism. However, the result must be corrected for any change in the composition of the body, mainly deposition or utilization of body fats. The method is cumbersome and is accurate only if the period of observation is sufficiently long. (b) Measurement of total heat production of the organism. This is fundamentally the most accurate method. The value obtained must be corrected for any external work performed, including such items as heating of the foodstuffs taken in, vaporization of water, etc. The determinations are made with the organism in a calorimeter, technically a rather difficult procedure, but it yields very accurate results. (c) The amount of oxygen used in oxidation processes can be used to determine the metabolic rate. (In theory, the carbon dioxide production could also be used, but it is less accurate, mainly because there is a large pool of carbon dioxide in the organism that undergoes changes relatively easily.) The reason that oxygen can be used is that similar amounts of heat are produced for each liter of oxygen, irrespective of whether fat, carbohydrate or protein is oxidized. The figures are: fat 4.7 kcal; carbohydrate, 5.0 kcal; and protein, 4.5 kcal, per liter oxygen. It is customary to use an average value, 4.8 kcal/liter oxygen consumed. The use of oxygen consumption for the determination of metabolic rate is so common that the two concepts have become practically synonymous. Obviously, the oxygen consumption cannot be used for determinations of metabolic rate in, for example, anaerobic organisms.

Temperature Effects. Animals whose body temperature changes with that of the environment (poikilothermic or cold-blooded animals) have a metabolic rate which depends on their temperature. In general, the metabolic rate increases, within the range tolerated by the organism, some two- or threefold for a temperature increase of 10°C. This change, designated as Q_{10}, is a term preferred by most physiologists and biologists over the use of the Arrhenius constant, which is a thermodynamically more correct way of expressing temperature dependence. Because of the temperature effect, information about metabolic rate in cold-blooded animals is meaningful only if the temperature is known.

Mammals and birds maintain a relatively constant body temperature within a wide range of ambient temperatures, and are called warm-blooded or homothermic animals. When the ambient temperature falls below a certain critical level, their metabolic rate increases so that the increased heat loss is balanced by increased heat production. Most of the increased heat production is due to involuntary muscle contractions (shivering).

Metabolic Rate in Relation to Body Size. If a uniform group of animals, such as mammals, is used for a comparison of metabolic

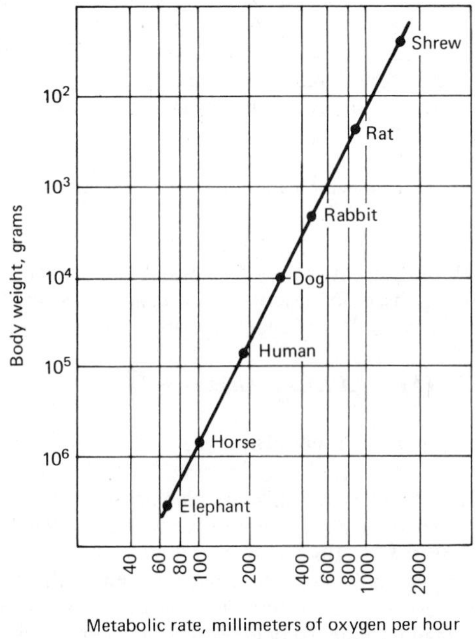

Body weight versus metabolic rate plotted on logarithmic coordinates.

rates, an interesting relationship is revealed. The smaller the animal, the higher is the metabolic rate per gram of body weight. If, on logarithmic coordinates, the metabolic rate is plotted against body size, we obtain a straight line (see accompanying figure) which corresponds to the equation: log metabolic rate $= k + 0.74$ log body weight (k being a constant whose numerical value depends on the units used). It has been suggested that this relationship expresses the need for a higher heat production in the smaller animal, which, because of its larger relative surface, must produce heat at a higher rate than a large animal in order to maintain its body temperature. However, similar relationships between metabolic rate and body size have been found in numerous groups of cold-blooded animals as well as plants, where the need for heat regulation cannot be the fundamental explanation of this interesting relationship.

BASAL PLANE. The plane normal to the principal axis, or "c" axis, in hexagonal or tetragonal crystals.

BASALT. A fine-grained to dense, sometimes porphyritic, intrusive or extrusive igneous rock, black or greenish-black in color, characterized by a preponderance of calcic plagioclase feldspars and pyroxene together with minor amounts of accessory minerals such as olivine. Glass may be present. Amygdaloidal structure is common in such cavities and beautifully crystallized species of zeolites, quartz or calcite are frequently found.

The lava flows of the Plateau of the Deccan in India, the Columbia Plateau of Washington and Oregon States, as well as the Triassic lavas of eastern North America are basalts. Perhaps the most famous basalt flow in the world is the Giants Causeway on the northern coast of Ireland, in which the vertical joints give the impression of having been artifically constructed. Pliny used the word basalt and it is said to have had an Ethiopian origin, meaning a black stone.

BASALTIC SHELL (Earth). Earth.

BASCULE BRIDGE. Bridge (Structural).

BASE (Chemistry). Acids and Bases.

BASE LEVEL. The ultimate physiographic feature of the processes of denudation is the reduction of the land surface to sea level, because it is at this level that the processes of river erosion are completely checked. As sea level is the datum plane below which stream erosion is impossible this may be taken as the theoretical lower limit of stream erosion and the level to which all the land surfaces must be brought if no other forces intervene. Actually, however, base level seems to be a limit which, however closely approached, may not in reality ever be reached. Therefore a broader use of the term base level is frequently made, base level being the lowest level to which a given stream can cut. A stream flowing into a body of water at an elevation above the ultimate datum, sea level, is said to be at a temporary base level. In any case when a stream has cut its channel to such a degree that it has just enough velocity to carry its load without further erosion being possible, it is then said to be at grade. Local or temporary base levels of relatively small area may be developed without reference to sea level. In this sense the term base level is not entirely comparable to the term peneplain.

BASE LINE. (a) A surveyed line on the earth's surface or in space, whose exact length and position have been accurately determined with more than usual care, and that serves as the origin for computing the distances and relative positions of reference to which surveys are coordinated and correlated. (b) The initial measurement in triangulation, being an accurately measured distance constituting one side of a series of connected triangles and used, together with measured angles, in computing the lengths of the other sides. (c) One of a pair of coordinated axes (along with the principal meridian) used in the United States Public Land Survey system, consisting of a line extending east and west along the true parallel of latitude passing through the initial point and along which standard township, section, and quarter-section corners are established. The base line is the line from which the survey of the meridional township boundaries and section lines is initiated and from which the townships, either north or south, are numbered. (d) An aeromagnetic profile flown at least twice in opposite directions and at the same level, in order to establish a line of reference of magnetic intensities on which to base an aeromagnetic survey. (e) The center line of location of a railway or highway; the reference line for the construction of a bridge or other engineering structure. Sometimes spelled *baseline.*

(American Geological Institute, by permission.)

BASE-LOAD PLANTS (Electric Power). Gas and Expansion Turbines; Hydroelectric Power.

BASE PERIOD. In statistics, the period of time for which data used as the base of an index-number, or other ratio, have been collected. This period is frequently one of a year, but it may be as short as one day or as long as a run of years. The length of the base period is governed by the nature of the material under review, the purpose for which the index-number (or ratio) is being compiled and the desire to use a period as free as possible from abnormal influences in order to avoid bias.

BASIC FREQUENCY. Frequency.

BASIC OXYGEN PROCESS. Iron Metals, Alloys, and Steels.

BASIC ROCK. A term applied to igneous rocks whose content of silica is less than about 52%. It is not an exact term and is gradually going out of use. It may be convenient for field use.

BASIC SALT. A compound belonging to the categories of both salts and bases, because it contains OH (hydroxyl) or O (oxide) as well as the usual positive and negative radicals of normal salts. Among the best examples are bismuth subnitrate, often written $BiONO_3$; and basic copper carbonate, $Cu_2(OH)_2CO_3$. Most basic salts are insoluble in water and many are of variable composition.

BASIC SOLVENT. A solvent that accepts protons (hydrogen ions) from the solute.

BASIDIOMYCETES. Nearly all the fungi commonly observed belong to this group, which includes toadstools, mushrooms, puff-balls, and many other forms. The characteristic feature which distinguishes them from other fungi is the basidium, typically a club-shaped structure bearing four spores at its apex. Members of this group are found wherever plant life can exist. The majority of them are saprophytes, living on dead wood and soil rich in humus; a few are parasites. The life-history of the common mushroom, one of the best known and most important, is fairly typical of the group. See Fig. 1.

The vegetative phase consists of a mycelium. This is a mass of slender much-branched threads, called hyphae, which grow throughout the substratum. The mycelium is perennial. Each hypha is a long filament composed of many segments, each containing two nuclei. The hyphae absorb from the substrate the organic materials which

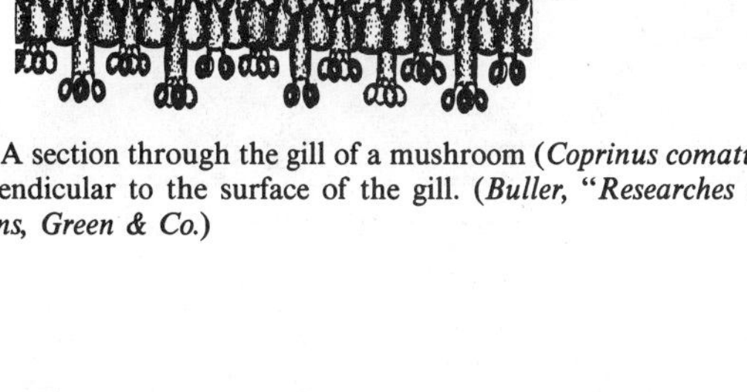

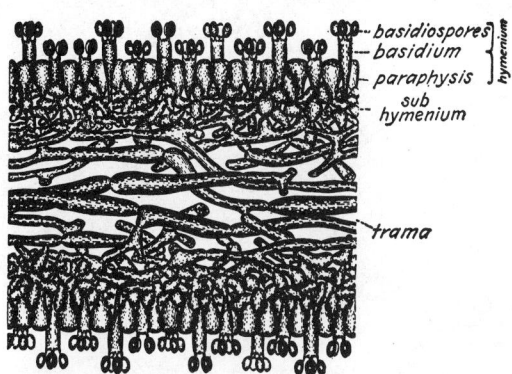

Fig. 1. A section through the gill of a mushroom (*Coprinus comatus*). Section cut perpendicular to the surface of the gill. (*Buller, "Researches on Fungi," Longmans, Green & Co.*)

Fig. 2. Development of basidium and formation of spores in mushroom: (1) Separate mycelium with binucleate cells; (2) young basidium containing two unrelated nuclei; (3, 4) fusion of nuclei in basidium; (5, 6, 7) development of four-nucleate condition in basidium; (8) formation of basidiospores that receive the four nuclei. (*Redrawn from Harper*)

the fungus needs in order to live, and convert it into other forms. Much of this food substance accumulates within the mycelium.

When sufficient material has been stored and conditions are suitable, the fungus fruits. The fruit body first appears as a small round object rising from the substratum. This elongates into a stalk bearing at its tip an umbrella-shaped cap or pileus.

On the lower surface of the pileus there are numerous thin radiating plates called gills. The lateral surfaces of the gills are formed by hyphal tips which grow perpendicular to the surface and form a compact layer called the hymenium. The hyphal tips composing the hymenium are the basidia, the reproductive structures which distinguish this group of fungi from all others. See Fig. 2. Each basidium is a cylindrical, binucleate body cut off from the tip of a hypha. The two nuclei fuse and immediately divide, usually twice, so that the basidium contains four haploid nuclei. From the outer end of the basidium four slender pegs called sterigmata develop. A small spore forms at the tip of each sterigma. Into each spore one of the nuclei of the basidium migrates. The spore is discharged from the sterigma and falls down between the gills and into the air. It is carried about by air currents and eventually falls to the ground. There the spore germinates, giving rise to a slender branching mycelium composed of uninucleate segments. This is the primary mycelium. Branches of two primary mycelia unite to form a secondary mycelium. The two nuclei present in each segment of a secondary mycelium have come from different spores. The manner in which they continue their identity during division is interesting. The two nuclei divide simultaneously. When division is about to occur a small bulge forms on the side of the hypha. See Fig. 3. One of the two nuclei enters this bulge, the other remains in the hypha. After division a cross wall forms, separating the two nuclei in the hypha. The protuberance containing the other nucleus continues to grow and forms an elbow-shaped structure, which joins the two cells of the hypha. It is called a clamp connection. One of the nuclei formed in this clamp returns to the original cell, the other passes through the clamp and into the other cell. By this means the two cells each receive a nucleus derived from one of the original nuclei. Nuclear fusion occurs only in the basidium. So there is in the mushroom an alternation of generations differing from that in most plants. The haploid or gametophyte phase consists of the primary mycelia. Following this a prolonged binucleate or dikaryon phase exists. Only in the basidium does nuclear fusion occur and reduction immediately follows. So the diploid phase or sporophyte is represented only by the basidium. There are no sex organs in this group of plants. The differences between the edible mushroom and other basidiomycetes is mainly in the structure of the fruit body, the location of the hymenium and the nature of the basidia.

The nature of the hymenium is the basis for classifying basidiomycetes. The Hymenomycetes are those in which the hymenium is exposed; in the Gasteromycetes it is formed within the fruit body. The principal order of Hymenomycetes is the *Agaricales* or agarics, or the gill fungi.

Another well-known order of basidiomycetes is the *Polyporales*; of these the polypores or Polyporaceae are best known. The distinguishing feature of these are the pits or tubes on the lower surface of the fruit body. The hymenium lines these pits. The fruit bodies of the polypores usually do not have a stalk and pileus, but form a layer spreading over the surface of rotting wood or grow out from the wood like a shelf. This shelflike habit has given to these plants the name bracket-fungi. The fruit bodies of bracket-fungi live for many years and often show distinct growth layers. In some species these are a foot or more across and several inches thick.

The other large group of basidiomycetes is the Gasteromycetes, distinguished by having the hymenium lining irregular cavities in the fruit body. Until these are fully mature, the spore-bearing parts are completely enclosed by sterile tissue. These are many different kinds of Gasteromycetes. One of the best known is the puff-balls, or Lycoperdiales. The outer wall or peridium of the puff-ball surrounds the gleba or spore-bearing tissues. When the spores are mature the peridium breaks. In some genera the wall breaks up into irregular fragments and leaves the spore-mass exposed; in others a pore is formed at the apex of the fruit body. The slightest pressure against the peridium will cause clouds of spores to puff from the pore. The number of spores formed in a single puff-ball is tremendous; 7,000,000 has been given as the number from a good-sized sporophore. Some of the puff-balls are the largest fungi known, reaching a diameter of a foot or more. If gathered before the spores are formed, most of the puff-balls are edible. None are poisonous.

Geasters or earth-stars develop very much as puff-balls do. But when they are mature the outer peridium splits into sectors which bend outward, revealing the spore-bearing part within. These fungi are frequently found growing on dry sandy soil.

Another group of Gasteromycetes includes the Bird's-nest fungi. The spore-bearing structures here are the "eggs," small oval bodies resting in the bottom of an open cup of sterile tissue. A last and curious group of Gasteromycetes is the stinkhorns, vile-smelling fungi whose spores are included in a mass of sticky stinking tissue, formed by the disintegrated glebal substance. This attracts carrion flies and other insects, which carry the spores about.

Other important families of basidiomycetes are the rusts and the smuts. See also **Rust Fungi;** and **Smuts.**

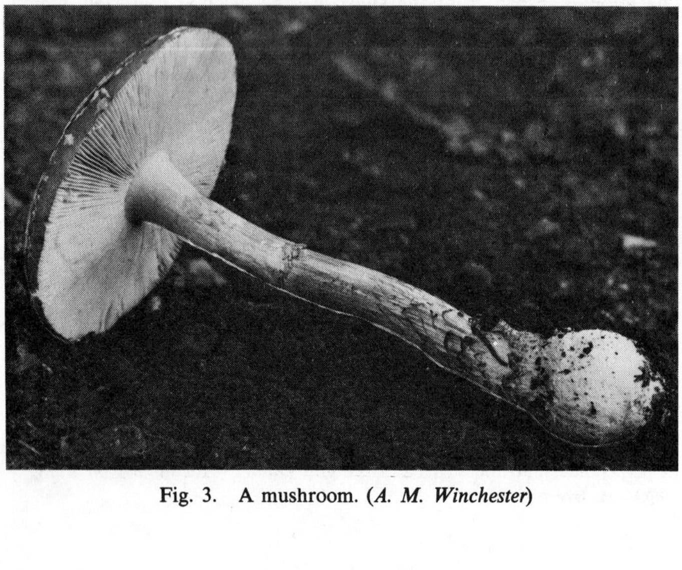

Fig. 3. A mushroom. (*A. M. Winchester*)

Fig. 4. *Agaricus*: field mushroom. (*Photograph by Brian J. Ford; copyright*)

There are several theories as to the origin of the basidiomycetes. Some maintain that they have evolved directly from certain primitive flagellates. Others derive them from the red algae. Many consider them to have descended from the ascomycetes. Adherents to this theory observe the dikaryon phase of the ascogenous hypha of the ascomycetes and note that if this phase were prolonged for some time it would be very similar to the secondary mycelium of the basidiomycete. They also note tht the asexual reproduction by conidia occurs in the basidiomycetes very much as it does in the ascomycetes.

The basidiomycetes include many important plants. Many are much sought as food. A few, and especially the field mushroom, *Agaricus campestris*, shown in Fig. 4, are extensively cultivated. See **Agarics.**

Many basidiomycetes are serious disease-producing plants, attacking trees particularly. Often their presence is unnoticed until too late; the mycelium has permeated the tissues of the host. Other basidiomycetes attack dead plants, reducing them to simpler forms. These are the basidiomycetes which cause decay. They are saprophytes.

BASIL. Flavorings; Labiatae.

BASILAR MEMBRANE. A membranous partition in the auditory chamber (cochlea) of the inner ear forming the floor of the cochlear duct. Upon it lies the organ of Corti, which contains the sensory cells for hearing. The basilar membrane increases in width as it passes from the base of the cochlea towards the apex, thus influencing the character of the vibrations with which the basilar membrane responds to sounds of different frequency.

BASILISK (*Reptilia, Sauria. Basiliscus*). A large tropical American lizard with crests on the back and tail which resemble the fins of fishes. Four species are known.

BASIN. In structural geology, a special type of folded structure in which the strata dip in toward a central point from all directions. As exposed at the surface the ground plan of a basin may be roughly circular or elliptical. Type locality, the Paris Basin, France.

BASKING SHARK. Sharks.

BASOMATOPHORA. An order of snails with eyes located at the bases of the tentacles.

BASOPHILS. Blood.

BASS (Acoustics). Those frequencies at the lower end of the audible range. A bass control is a tone control having the ability to increase or decrease the bass frequency gain of an audio amplifier. A bass compensator is an equalizer used to correct the bass response of an audio amplifier system. A bass reflex is a baffle for loudspeakers which employs a tuned port or opening which returns the sound from the rear of the speaker in an additive phase. Thus, the sound output is enhanced over a certain range of frequencies.

BASS (*Osteichthyes*). All bass are of the order *Percomorphi.* Sea bass, along with groupers, are of the family *Serranidae,* of which there are approximately 400 species. They are among the most important food fishes. Most species of bass are marine. A few inhabit brackish and fresh water. The American white and yellow bass are strictly fresh water fishes.

Sometimes called jewfishes, giant sea bass have been recorded with weights up to 800 pounds (233 kilograms). Sport fishermen's tackle records are considerably lower: 513 pounds (408 kilograms) (7 feet, 2 inches; 2.2 meters) for the *Stereolepis gigas* (California species) and 551 pounds (250 kilograms) (8 feet, 4 inches; 2.5 meters) for the Atlantic *Epinephelus itajara.* Originally, the striped bass (*Roccus saxatilus*) was found in waters along the American Atlantic coast from the Gulf of St. Lawrence to northern Florida. Two shipments of some 400 fish were moved to California where they were placed in the San Francisco area. The population exploded at such a rate that a commercial fishery for them was established only ten years later. The fish now occurs on the American west coast all the way from northern Washington to southern California. For spawning, the stripers seek

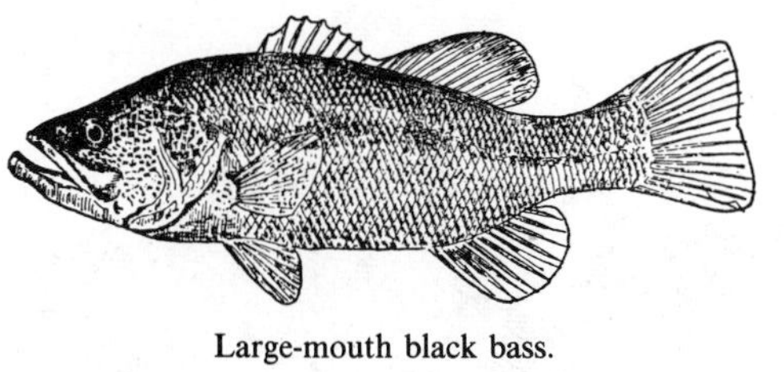

Large-mouth black bass.

fresh water. A mature female at the age of 5 years attains a length of about 2 feet (0.6 meter). Males mature earlier (2 years) and are smaller (average length of 1 foot). Records indicate the largest of the stripers to be caught was at Edenton, North Carolina (125 pounds; 57 kilograms). The largest on record on the west coast is 73 pounds (33 kilograms).

Two other species of striped bass inhabit fresh waters, mostly in the eastern United States: (1) *Roccus mississippiensis,* which has a characteristic yellow coloration of the body; and (2) *Roccus chrysops,* the white bass. Both fishes reach a length of about 18 inches (46 centimeters). Another species, *Roccus americanus,* a 12-inch (30-centimeter) perch (no stripes on side of body) originally inhabited the waters of the American Atlantic seaboard, later migrating inland.

The large-mouth bass (*Micropterus salmoides*) is a member of the family of sunfishes (*Centrarchidae*) and is one of the better known of American fresh water sporting fishes. Records indicate the largest specimen to be about $22\frac{1}{2}$ pounds (102 kilograms) ($32\frac{1}{2}$ inches (5.4 kilograms) long). The small-mouth bass (*Micropterus dolomieui*) has been recorded at 12 pounds (5.4 kilograms) (27 inches (00 centimeters) long). Both of these larger members of the centrarchids frequently feed upon smaller sunfishes. It is, therefore, usual stocking practice to include these smaller forage fishes, such as the common bluegill (*Lepomis macrochirus*) when introducing the larger bass.

The large-mouth black bass (*Micropterus* [*Huro*] *salmoides*) is shown in the accompanying figure. This fish inhabits fresh water in the eastern United States from the Great lakes region south into Mexico. The fish has been widely introduced throughout North America, Europe, and other parts of the world.

See also **Fisheries;** and **Fishes.**

BASSWOOD TREES. Known as basswoods in North America, as lindens in Germany, as tilleuls in France, and as limes in Britain, these trees are of the family *Tiliaceae* (basswood or linden or lime family). See illustration. There are several species and hybrids of interest. Generally these trees may be characterized as deciduous, favoring sun but tolerant of some shade, capable of fast growth, alternate-growing, toothed leaves, and fragrant, creamy-white but small flowers occurring in drooping clusters which are pendent from a leaflike bract. Although all species cannot be described here in detail, the more important species include:

American basswood (or American lime or linden)	*Tilia americana*
Big-leafed linden (or big-leafed lime)	*T. platyphyllos*
Chinese linden	*T. henryana*
Crimean linden	*T. X euchlora*
Downy basswood	*T. michauxii*
European linden (or common lime)	*T. X europaea*
Korean lime	*T. insularis*
Red-twigged lime	*T. platyphyllos* "Rubra"
Shrubby lime	*T. kiusiana*
Siberian lime	*T. mandshurica*
Silver linden (or silver lime)	*T. tomentosa*
Small-leaved linden (or small-leaved lime)	*T. cordata*
Weeping silver lime (or pendant silver linden)	*T. petiolaris*
White basswood	*T. heterophylla*

Linden tree.

The American basswood is native to the midwestern United States and Canada, ranging from New Brunswick in the northeast to Manitoba and the Dakotas in the west and southward into Nebraska, Oklahoma, and Kansas. It is also found in eastern Texas and in the mountains in Virginia, Georgia, and the Carolinas. The tree is particularly well known throughout New England. Normally, this species will attain a height of up to about 80 feet (24 meters), but as the accompanying table indicates there are exceptions. The tree usually is found at altitudes below about 1400 feet (420 meters). The tree has large heart-shaped leaves, 6 to 8 inches (15 to 20 centimeters) in length. The wood is of straight-grain, light-brown in color, and easily worked. The wood is used for some items of furniture, containers, and as a special millwood. When green, American basswood has a moisture content of about 105% and weighs 42 pounds per cubic foot (672.8 kilograms per cubic meter). After air-drying to 12% moisture, the weight is 26 pounds per cubic foot (416.5 kilograms per cubic meter) and 1000 board-feet (2.36 cubic meters) of nominal sizes weighs about 2170 pounds (984 kilograms). Crushing strength of the green wood when compression is applied perpendicular to the grain is 2220 pounds per square inch (15.2 MPa); of the dried wood, 4730 psi (32.6 MPa). The tensile strength with tension applied perpendicular to the grain is 280 psi (1.9 MPa) for green wood, 350 psi (2.4 MPa) for dried wood.

The white basswood is a smaller tree, ranging from 40 to 60 feet (12 to 18 meters) in height and essentially is found in the south. With preference for limestone soil, the tree ranges from southern New York through the mountains of Pennsylvania, following the Alleghenies southward into Georgia, Alabama, and central Florida. The tree also ranges westward into Kentucky and southern Indiana and Illinois. The wood is very much like that of *T. americana.* Other more regional species of American basswood include the *T. floridana* and the *T. caroliana.* A smaller tree, usually considerably less than 60 feet (18 meters) in height is the *T. michauxii* which has smaller leaves. It is found mainly in the southeastern United States, ranging westward into Texas.

Among the more well known *Tiliaceae* in Europe are the lindens of Berlin. The *T. platyphyllos* is found throughout central and southern Europe and in the British isles. These trees can attain a height up

RECORD BASSWOOD TREES IN THE UNITED STATES[1]

Specimen	Circumference[2] (inches)	(centimeters)	Height (feet)	(meters)	Spread (feet)	(meters)	Location
American basswood (1971) (*Tilia americana*)	267	678	115	34.5	76	22.8	Michigan
Carolina basswood (1973) (*Tilia caroliana*)	122	310	50	15	94	28.2	Mississippi
Florida basswood (1971) (*Tilia floridana*)	124	315	93	27.9	58	17.4	Arkansas
White basswood (1972) (*Tilia heterophylla*)	184	467	58	17.4	77	23.1	Georgia

[1] From the "Social Register of Big Trees," The American Forestry Association (by permission).
[2] At 4.5 feet (1.4 meters).

to about 135–140 feet (40.5 to 42 meters). The leaves range up to 5 or 6 inches (12.7 to 15.2 centimeters) across and are dark green. The fruit is pear-shaped. The common lime or European linden is a cross between the *T. platyphyllos* and *T. cordata*. The limes and lindens tend to be plagued with certain drawbacks, including a proliferation of suckers which occur around the base of the tree, their natural attraction to aphids which results in a production of honeydew that covers and blackens the leaves.

The wood of the lime has attractive qualities for sculpturing, the softness and workability permitting very intricate carving. Examples of carvings, as by Gibbons in the 1600s, still remain.

BAST FIBERS. In the phloem, pericycle, or cortex of many plants there occur long slender thick-walled cells with tapering ends. These cells, often occurring in groups of considerable size, are often called bast fibers. They give to the stem of the plant considerable strength. Many bast fibers are useful to man, especially in the manufacture of cloth and cordage. Linen is made from the pericyclic fibers of flax, while hemp and ramie are coarser pericyclic fibers used in making coarse sacking and cordage. The phloem fibers of jute are also used for making rope and burlap, as well as the strong webbing used in upholstering furniture. Plant anatomists call them phloem fibers, pericyclic fibers, or cortical fibers.

Characteristics of natural rope fibers are given in Tables 1 through 4. See also **Rope.**

TABLE 1. MAJOR HARD ROPEMAKING FIBERS

General Name	Other Common Names	Botanical Name	Principal Sources
Abaca	Manila, abaca	*Musa textilis*	Borneo, Costa Rica, Guatemala, Honduras, Indonesia, Panama, Philippines
Cantala	Java cantala, maguey	*Agave cantala*	Indonesia, Philippines
Henequen	Mexican, Cuban, Carrizal, and Victoria sisal	*Agave fourcroydes*	Campeche (Mexico), Cuba, Northern Mexico, Yucatan
Sisalana	Sisal	*Agave sisalana*	Brazil, East Africa, Haiti, Indonesia, Malagasy Republic, West Africa

TABLE 2. MAJOR SOFT ROPEMAKING FIBERS.

General Name	Other Common Names	Botanical Name	Principal Sources
Coir	Coconut fiber	*Cocos nucifera*	India, Indonesia, Pakistan, Philippines
Hemp (true) normally taking its name from source locality	Hemp	*Cannabis sativa*	Chile, Hungary, Italy, Poland, United States, U.S.S.R., Yugoslavia
Jute	Jute	*Corchorus*	India, Indonesia, Pakistan

TABLE 3. TENSILE STRENGTH AND SPECIFIC GRAVITY OF ROPE FIBERS.

Plant, Family, or Grade	Relative Tensile Strength	Specific Gravity
Pita floja, Aechma magdalenae	104	—
Manila Grade I, *Musa textilis*	100	1.43
Palma ixtle, *Yucca carenosana*	78	—
Tequilana, *A. tequilana*	72	—
Sisal, *Agave sisalana*	71	1.43
Zapupe maguey, *A. zapupe*	71	—
Philippine maguey, *A. cantala*	54	—
Henequen, *A. fourcroydes*	54	1.43
Jaumave ixtle, *A. heterocantha* (*funkiana*)	46	—
Mexican maguey, *A. lurida*	36	—
Tula ixtle, *A. lophanthi*	30	—
Sansevieria	23	—

TABLE 4. BREAKING STRENGTH AND WEIGHT OF HARD ROPE FIBERS.

Fiber Type	Breaking Strength Strain/Strand (grains)	Weight/Yard (grains)
Abaca (manila), *Musa textilis* (highest)	46.5	0.567
Abaca (manila), *Musa textilis* (average)	34.7	0.772
Abaca (manila), *Musa textilis* (lowest)	31.1	0.962
Sisal, *A. sisalana*	22.6	0.616
Zapupe Vincent, *A. lespinassei*	21.6	0.722
Cabuya, *Furcrea cabuya*	20.1	0.575
Phormium, *Phormium tenax*	18.9	0.660
Henequen, *Agave fourcroydes*	16.8	0.766
Cantala, *A. cantala*	9.5	0.430

Conversion factor: 1 grain = 0.0648 gram.

BASTNÂSITE. A wax-yellow reddish-brown, greasy mineral of the composition, (Ce, La)(CO_3)F, usually found in contact zones or associated with zinc lodes. Sometimes spelled *bastnaesite*. See also **Rare-Earth Elements and Metals.**

BATAGUR (*Reptilia, Chelonia*). Large fresh-water tortoises of several genera found in India and the Malayan region.

BATCH DISTILLATION. Distillation.

BATHOCHROME. A chromophore which lowers the frequency at which absorption occurs.

BATHOLITH. Batholith, by some writers spelled bathylith, is derived from the Greek meaning deep, and stone. It is a very large intrusive igneous mass with steeply inclined contacts, enlarging downward to undetermined depths. Typical batholithic rocks have a relatively coarse and even texture, such as granites and diorites. Batholiths occur as the roots or cores of folded and faulted mountain ranges and are therefore closely associated with mountain-building, although probably not the cause of the deformation but rather consequent to it. The mode of emplacement of such large cross-cutting and relatively uniform igneous rock bodies has not yet been definitely determined. The term was proposed by Zuess in 1888.

BATHYAL ZONE. Ecology.

BATHYMETRIC CHART. A map delineating the form of the bottom of a body of water, usually by means of depth contours (isobaths).

BATHYPELAGIC ZONE. A zone of the ocean into which small amounts of sunlight penetrate but in insufficient quantity to be of use to either plants or animals. See **Marine Biology.**

BATHYTHERMOGRAPH. A device for obtaining a record of temperature against depth (strictly speaking, pressure) in the upper 300 meters of the ocean, from a ship underway. For thermal element it has a xylene-filled copper coil, which actuates a stylus through a Bourdon tube. The pressure element is a copper aneroid capsule which moves a smoked glass slide at right angles to the motion of the stylus.

BATOIDEI. Skates and Rays.

BATS (*Mammalia, Chiroptera*). In terms of numbers and kinds, *Chiroptera* is second only to *Rodentia* of the living mammals. See photo. The general organization of this order is shown in the accompanying table. Bats also may be grouped in accordance with their general behavior. (1) There are the community-minded and commuting bats which live in isolated habitats a short distance from their feeding grounds, remaining in large trees or caves by day and seeking their food supply at night. Some types of bats prefer to fly out at evening en masse and return at predawn to their home base. Other bats leave their haunts on essentially an individualistic basis and return individually as their hunger has been satisfied. (2) There are the highly indi-

Bat hanging from tree bough.

vidualistic or isolationist types of bats that elect essentially to live and forage alone. It has been estimated that a group of insect-eating bats that once lived in Ney Cavern (Texas) numbered as high as 20 million. Any cave with reasonable entry can almost be counted on as a bat roost. Thousands upon thousands of bats also live together in groups of trees. Evening flights of bats have been likened to a large cloud of smoke, causing a darkening of the sky for some miles. Records also indicate that large boughs of trees have been snapped because of the weight of the bats roosting on them. Hollows in trees are also favorite haunts for bats. Since the early 1950s, there appears to be a reduction of massive bat populations occurring and several theories have been advanced. It could well be that this is a manifestation of some natural cycle not yet fully understood. A decreasing bat population, of course, is cause of some concern because they play an effective natural role in controlling insects. Some records indicate correspondence between local bat populations and the incidence of malaria in selected tropical areas. But probably much more important is the economic value of the bats in terms of killing off hordes of insects that damage agricultural crops. In contrast, however, the fruit-eating bats can be an economic menace to orchards. Certain conventional protective means, such as smoke generation or floodlighting, have proved quite ineffective against the bats.

Ages before scientists developed sonar and radar, bats had been using sound waves for communication and navigation. At one time, it was felt that this faculty was unique to bats, but later investigations showed that a number of animals also use such means, including some of the fishes (a form of underwater sonar). It can be stated with some confidence, however, that the insect-eating bats probably have developed the technique to the greatest perfection.

For decades, experimenters knew that if the eyes of a bat were blindfolded, this action seemed to have little effect on the animal's maneuverability. But many years later and within the past few decades, when experimenters taped one or both ears shut, it was obvious that the animal no longer could navigate. Experiments have been conducted in which hundreds of wires have been crisscrossed over an enclosure to demonstrate the sonic guidance of the bats. It is now known, of course, that bats emit supersonic radiation ranging up to 30,000 Hz, but that radiation as low as 200 Hz also can be emitted. This sound can be maintained for long periods throughout the flight of the animal. The sounds are emitted through the animal's nostrils and are aided by a complex flap-structure to provide precise directivity to the radiation. Echo returns from such emissions enable a bat to pick out a tiny flying insect some distance ahead. Of course, this highly refined navigation system would be ineffective if the bat could not acrobatically maneuver while in flight. This they can do in adjusting their course where signals only a few inches ahead are received by their highly complex and sensitive ears. These attributes, of course, explain why a half million or more bats can fly around in a darkened cave without fear of frequent collisions.

Although sometimes referred to as *Megachiroptera* (signifying big bats), this large group of bats is better identified as the fruit-eating bats. There are numerous types and genera. As shown in the accompanying table, there are three major classifications of the fruit-eating bats. The Fox-Bats represent the largest group as well as the largest specimens, some having a wingspan in excess of 5 feet. They are found in Australia, Indonesia, the eastern Pacific, in the Oriental Region in areas that perimeter the Indian Ocean, including much of India, and off the east coast of Africa on the Comoro Islands and Madagascar. They are not found on the African continent per se. They are of many colors, sizes, with large ears, and long tails.

There are many more varieties of insect-eating bats. The classification in detail is far from complete. In these bats, the tail can be extended or retracted. By variously manipulating the tail, the animal has an excellent steering mechanism. The Tomb-Bat (*Taphozous*) was named for the frequency with which it has been found in ancient Egyptian tombs. Vampire bats are of particular interest because they feed only on animal blood and this can range from humans on down the scale. They are quite gruesome in appearance, having naked noses, pointed ears, and spherical bodies. They move about on the ground like great spiders. They are also good fliers. There is a misconception that the vampire bat sucks blood from its victim. In actuality, the vampire bat runs on all fours to its intended victim, jumps on the

GENERAL ORGANIZATION OF BATS
(Flying Mammals)

CHIROPTERA

FRUIT-EATING BATS

- Small-tongued Fruit-Bats (*Pteropinae*)
  - Fox-Bats (*Pteropus, ...*)
  - Tailed Fruit-Bats (*Rousettus, ...*)
  - Epauletted Fruit-Bats (*Epomophorus, ...*)
  - Short-nosed Fruit-Bats (*Cynopterus, ...*)
  - Bare-backed Fruit-Bats (*Dobsonia, ...*)
  - Hammer-headed Bat (*Hypsignathus*)
- Large-tongued Fruit-Bats (*Macroglossinae*)
  - Long-tongued Fruit-Bats (*Macroglossus*)
- Tube-nosed Fruit-Bats (*Nyctimeninae*)

INSECT-EATING BATS

- Mouse-tailed Bats (*Rhinopomatidae*)
- Sheath-tailed Bats (*Emballonuridae*)
- Hare-lipped Bats (*Noctilionidae*)
- Hollow-faced Bats (*Nycteridae*)
- False-Vampire Bats (*Megadermatidae*)
- Horseshoe-nosed Bats (*Rhinolophidae* and *Hipposideridae*)
- Leaf-nosed Bats (*Phyllostomatidae*)
- Vampire Bats (*Desmodontidae*)
- Evening Bats (*Vespertillionidae*)
  - Brown Bats (Barbastelles, Serotines, ...)
  - Long-winged Bats (*Miniopterus*)
  - Tube-nosed Bats (*Harpiocephalus*)
  - Painted Bats (*Kerivoula* and *Anamygdon*)
  - Peruvian Bat (*Tomopeas*)
  - Big-eared Bats (*Antrozous* and *Nyctophilus*)
  - Mastiff-Bats (*Molossidae*)
    - Free-tailed Bat (Tadarida)
    - Naked Bat (Chiromeles)

NOTE:
(1) Fruit-eating bats are confined to tropical and subtropical areas, including Africa, India, Australia, the Oriental Region, and South Pacific islands.
(2) Insect-eating bats are found worldwide, even in some zoologically isolated areas where they might not be expected.

victim's body and slashes the skin with its two chisel-shaped upper front teeth. The bat then laps up the blood exuding from the wound. The vampire bat also is believed to spread a calming scent before the attack. Millions of these bats live in the tropical area south from the United States to Argentina. Because these bats feed occasionally on human beings, more frequently on horses, and other domestic animals, they can carry diseases, such as rabies, Murrina, Chagas disease, and other serious infectious maladies. They are seriously regarded in Central and South America. They appear to be migrating slowly northward and may one day be a threat to cattle raising in the southwest of the United States.

The Little Brown Bats are small, about $3\frac{1}{2}$ inches in length, and are found throughout the United States and temperate climates. They feed on insects during the warm season and usually hibernate in caves during the winter months. Some of the specialized bats include the Big-eared Bats, well named; the Mastiff Bats, which have peculiarly shaped ears, giving the face a mastifflike appearance; the Horseshoe-nosed Bats, which have leaflike appurtenances; and the Mexican Free-tailed Bats, which hang in colonies of hundreds of thousands in a given locale. See accompanying photo.

The Hare-lipped Bats are fish eaters and are found in tropical Central and South America. They grasp small fish with their elongated toes and claws. The Long-nosed Bats feed on nectar and pollen. Particularly in connection with night-blooming flowers, these bats play an important cross-fertilizing role.

The phrase "blind as a bat" is not entirely accurate. It is not believed that a bat can sense color, nor are its eyes large or considered important to its locomotion, as previously discussed. However, all of the elements found in a normal eye are present in the bat's eye and experiments have indicated that the ability to see does assist the bat in its general behavior.

For references, see **Mammalia.**

BATTERY. One or more galvanic cells in a finished package constitute a battery. The cells may be electrically connected in various series, parallel, and series-parallel combinations, including discrete groupings with separate terminals. In each individual cell of the battery, two electrodes of dissimilar materials are normally joined by an ion-carrying path, but are separated by an electron insulator. The electron transfer carrying path is external, via some conductor, such as a metal wire, through a using device where the useful work is done. The ion path is internal via an electrolyte which is an ionizable salt dissolved in a solvent. The electrolyte may be solid or liquid, although the former is usually limited to low discharge rates.

The active material of the solid electrodes may be solid, gaseous, or liquid and, in some special cases, such as with certain types of liquid cathode cells, the cathode serves a dual role as electrolyte solvent and depolarizer.

Control of the discharge is usually by a switch in the electron-carrying circuit, but in some special cases, it is done by making or breaking the ion-carrying path.

The cells may be rechargeable and such are often called "secondary" or "storage" batteries. Secondary batteries are reversed by passing a current through the cells in the opposite direction of current flow when the cells were working. The chemical conditions of the undischarged battery are restored and the cells are ready to be discharged again. Primary cells are meant to be discharged for exhaustion only once and then discarded. Some of the electrochemical systems used in "primary" cells are reversible, but the cells do not have the physical features needed to render them *safely* rechargeable. Various schemes or devices have been proposed over the years to recharge "primary" batteries, but generally, they are impractical and sometimes potentially *dangerous.*

Both primary and secondary types are produced in "wet-" and "dry"-cell versions. Wet cells are liquid electrolyte cells sensitive to orientation, whereas dry cells are not. Dry cells can also be made with liquid electrolytes, but in such cases, the electrolyte is immobilized by some mechanism, such as gelling or holding it in absorbent media, such as paper. Solid electrolyte cells are, by definition, also called dry cells.

As a further variation, both primary and secondary cells can be made as *reverse cells.* Activation is held off until just prior to use of the battery. This may involve holding off addition of liquid electrolyte or the solvent, such as "dry" charged car storage batteries; internal storage of liquid electrolyte or solvent apart from the active materials; or as some thermally activated batteries, heating of the batteries to change a solid inactive material into a molten active electrolyte. An example of a novel approach is a seawater-activated battery, where the salt water of the ocean serves as the electrolyte. Activation is achieved by immersing rigid electrode assemblies in seawater.

Both primary and secondary batteries are marketed in a wide variety of sizes and shapes, from the little "button" cells used in watches and hearing aids to the huge rectangular cells for missile batteries and load-leveling power storage units. Some applications also involve stringent conditions, such as very low (−50°C) or very high (above 200°C) operating temperatures, underwater usage, outer-space usage, continuous or intermittent discharge, or discharge that may be spread out over several years, or a very short period, such as a few seconds. With the larger sizes, energy per unit weight is usually the important criterion, especially in rockets or electric vehicles. With the smaller sizes, output per unit volume is usually more significant.

There are also cells which convert light and heat into electricity (solar cells and thermal cells), which are described under **Solar Energy.**

The number of electrochemical systems, the variations in cell sizes and internal configurations and the specialized applications for commercially available batteries all have increased dramatically during the past 40 years and change is continuing at a rapid pace in the early 1980s. A good example of change is the practical nonaqueous batteries now coming on the market. All of those offered thus far are with lithium anodes, but they are available in a wide range of sizes and in solid cathode, liquid cathode, and solid electrolyte versions. The rapidly changing price of raw materials is a significant factor. For example, a rise in the price of silver resulted in a switch from aqueous, silver oxide/zinc cells to nonaqueous types for many applications, such as some types of wristwatches. The development and design of galvanic cells and batteries has therefore become an increasingly complex science, far removed from the near art it was in pre-World War II years.

Cell Electrochemical Systems and Designs

The simplest cell would consist of two dissimilar metals in an ionically conductive medium, provided that the combination yielded a voltage above that needed to break down the electrolyte continuously at the positive electrode. Metal would go into solution at the negative electrode, releasing electrons to travel via the external circuit to the positive electrode, doing work during the transit. Such a cell, however, is impractical for most uses, due to shortcomings, such as gas formation, loss of electrolyte solvent, and poor storageability, among other factors. Thus, in practice, material which will go through a valency drop on electrochemical discharge is included in the positive electrode. In so doing, electrons coming from the negative electrode are accepted. This material is known as the "depolarizer." Some of the depolarizers now in use in practical cells are liquids, solids, and gases.

The depolarizer or cathode is positioned in the positive electrode in combination with some electron-carrying matrix, and usually is porous, allowing access of the ions carrying electrolyte to a large area of the depolarizer. The anode is the electrode that goes into solution in the electrolyte as ions, releasing electrons to do the work. It also can be solid, liquid, or gas, although the majority of presently marketed systems, are solid metals. The negative electrode contains the anode material and, like the cathode, it is assembled with an electron-carrying matrix and usually a design to provide a large area of contact with the electrolyte. In some cases, where the anode material is a metal, an additional conductive grid is not required. The electrolyte forms the ion bridge between the anode and the cathode materials. It is usually an ion-forming salt dissolved in a solvent. Some cell types have the electrolyte as a solid, but the majority are liquids—at least when the cell is being discharged, since the solid electrolytes accommodate only a very low rate of discharge.

Relatively few years ago, all the liquid electrolyte systems employed water as the electrolyte solvent, but in present practice, there are many nonaqueous systems in use. Thus cell systems are divided into two main categories—aqueous and nonaqueous. Electrolytes can be further subdivided into acidic, mildly acidic, and alkaline.

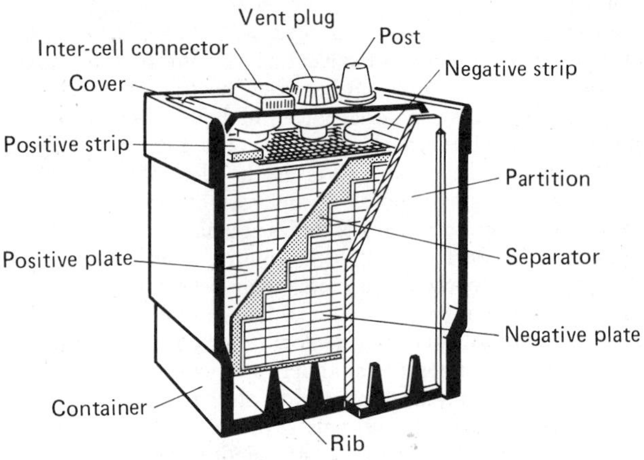

Fig. 1. Cutaway view of automotive-type storage battery cell.

Aqueous Acidic Systems. The best-known example of this type is the lead sulfuric acid, rechargeable system in wide use for starting automobiles. In the car storage battery, the configuration is usually several flat plate electrodes per cell with negative and positive electrodes interspaced, separated by porous plastic, wood, or other suitable material separators to give a high electrode interface area. In the car storage battery, the electrolyte is not immobilized—so this is a wet rather than dry cell. For present-day automobiles which operate on 12 volts nominal, six cells are connected in series per battery. A typical lead-sulfuric acid storage battery is shown in Fig. 1. This system is also marketed as a rechargeable, cylindrical dry-cell battery. The latter is achieved by winding two electrodes, one positive and one negative, separated by an absorbent material into a helix or "jelly roll" inside a cylindrical can.

Aqueous Mildly Acidic Cells. Probably the best-known cell of this type is the commonly called "carbon-zinc" or "Leclanché" cell, widely employed because of its low cost. A flashlight battery version is shown in Fig. 2. The positive electrode in this case is referred to as a "bobbin," and is a mechanical mixture of manganese dioxide powder, acetylene black particles, and an aqueous solution of zinc chloride and ammonium chloride. This electrode is molded about a carbon rod. The manganese dioxide is the depolarizer; the acetylene black is the electron-distributing conductive matrix; the salt solution is the ion-distributing electrolyte. The carbon rod is the conductive path to an outside terminal (the *cap*), since economically acceptable metals are attacked chemically by the manganese dioxide. The anode is zinc and, in this case, serves a dual role, both as the anode and as the container for the actual cell. A layer of electrolyte, gelled with a material such as starch, separates the positive and negative electrodes

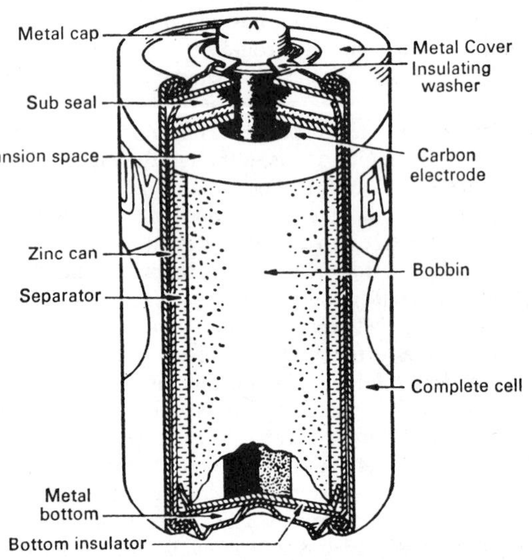

Fig. 2. Round carbon-zinc cell.

electronically, but connects them with an ion path. The actual cell is enclosed with a package imprinted with descriptive data and decorations, thus converting the single cell into a battery. The nominal voltage of a single cell is 1.5V and it discharges with a sloping voltage versus output discharge curve.

Carbon-zinc cells are marketed in many different sizes and shapes. In some cylindrical versions, the positive electrode is on the outside. These have sometimes been called "inside-out" cells. High-voltage "stacks" of cells which use flat cells have been marketed, wherein the zinc is coated on one side with an electrolyte-impervious carbon coating, serving as the positive collector for the cathode mix of the next cell. A series connection of the cells in the stack is thus achieved. Encapsulating materials are kept to a minimum in order to achieve a high-energy density. This design approach accounted for a significant proportion of the Leclanché system cells, particularly in years when there was little rural electrification, but it represents only a small portion of the market today, mainly in applications requiring small, high-voltage batteries, such as personal portable radios.

Aqueous Alkaline Systems. The majority of these systems use potassium hydroxide as the electrolyte solute, with the remainder using sodium hydroxide. Potassium hydroxide imparts higher rate capabilities, but sodium hydroxide is cheaper and easier to contain and sometimes imparts better storageability. The aqueous alkaline systems, as a rule, have higher rate capabilities, better low-temperature performance, and higher energy densities than the aqueous, mildly acidic cell designs. However, the aqueous alkaline systems are usually more complex and costly.

Aqueous alkaline systems are used in wet and dry cells and in secondary as well as primary batteries. An example of a secondary dry cell is the "jelly roll" cylindrical, "nickel-cadmium" rechargeable battery shown in Fig. 3.

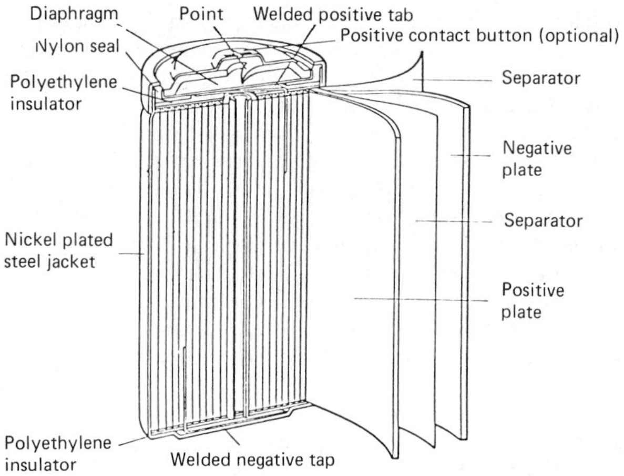

Fig. 3. Cutaway view of sealed, coiled-type, sintered nickel-cadmium cell.

As the size of batteries becomes relatively small, such as for use with electric wristwatches, the cost of the materials is a small portion of the total cost. Assembly cost for a given size remains relatively constant regardless of the aqueous system used, and since cost per hour of service usually controls these applications, systems that yield a high output per unit volume of cells, even if made with expensive electrochemical materials, become preferable. Silver oxide zinc is a good example. The miniature cell designs for the several aqueous alkaline systems in use, i.e., MnO_2/Zn, Ag_2O/Zn, AgO/Zn, HgO/Zn, CuO/Zn, and air/Zn, are quite similar. A typical miniature (sometimes called "button") silver oxide cell is shown in Fig. 4. The larger the cell, the more important the cost of the materials. Thus, in flashlight battery sizes, the MnO_2/Zn system is most common. More costly systems are usually limited to special applications, such as space vehicles. In retail stores, cells labeled simply as alkaline cells are usually MnO_2/Zn cells.

Gas Electrodes. Good examples of the wide variety of systems of this type being used or being considered for future use are the air-zinc and chlorine-zinc cells. Both of these cells use a gas as the positive

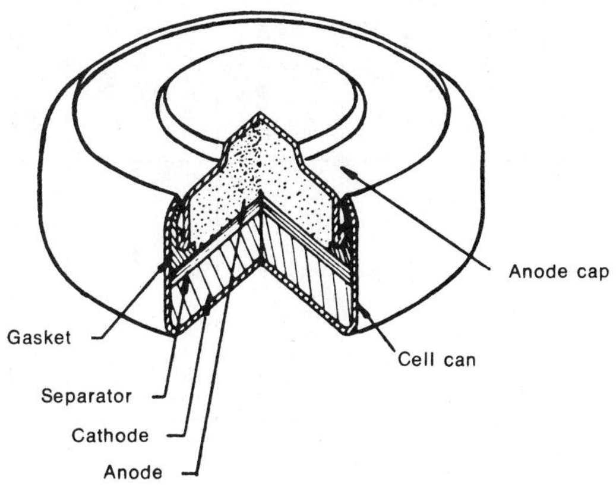

Fig. 4. Silver oxide cell.

electrode-active material. The first is an alkaline aqueous system which uses oxygen from the air, activated on a special material, to act as the oxidant or depolarizer. Button air cells for special applications, recently introduced to the market, provide significantly higher volumetric energy densities than some of the systems marketed earlier. The second is a mildly acidic aqueous system being considered for rechargeable batteries for electric cars and load-leveling in electric power generation. Chlorine is ionized at an activating electrode on discharge and formed as a gas to be transported to and stored in another location as a solid hydrate, on charge.

Nonaqueous Batteries. These cover almost as wide a range as do the aqueous systems. They employ active metals that are incompatible with water, and thus must be water-free—hence the term, *nonaqueous.* A relatively few years ago, nonaqueous batteries were unknown in the marketplace, but rapid strides have been made in recent years, due to the incentives brought about by the need for better storageability, higher energy densities, better performance at very low temperatures and, in some cases, such as silver oxide, alternatives to increasing costs of materials. The higher a metal on the electromotive (activity) series, the higher the voltage versus a given positive electrode and, therefore, the higher the watt-hours (Wh) for the same ampere-hours (Ah). With water-free systems, metals such as lithium, sodium, calcium, and magnesium, which are more negative than those compatible only in aqueous systems, can be used. To date, the nonaqueous systems reaching the market use lithium as the anode.[1]

Positive electrode materials under consideration cover a wide range. Where the battery must be substitutable for a currently available single cell aqueous type, a low-voltage match must be used. Examples are PbO, CuO, FeS, and FeS_2 versus Li. These are all solid depolarizers. For existing or new high-voltage applications, MnO_2 and fluorinated carbon are being used as solid depolarizers, and SO_2, SO_2Cl_2 and $SOCl_2$ are being used as liquid cathodes. The latter two have the advantage of being good solvents for the needed salt and of reducing

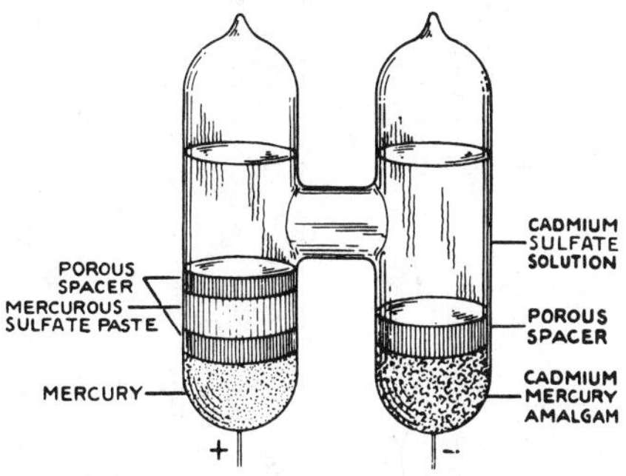

Fig. 5. Weston cadmium standard cell.

to another liquid which will also dissolve the salt. Therefore, they serve a dual role as cathode and electrolyte solvent. They are active only on a special activating electrode surface—so they do not combine directly with the lithium anode. The solid depolarizers all use organic solvents for the electrolyte. Several nonaqueous systems are being investigated for high-energy density rechargeable batteries. Those showing promise are all thermally activated types, i.e., salt is heated to the melting point to form a liquid electrolyte. This usually means that the lithium or sodium, the usual choices for the anode, are also molten. These involve high operating temperatures, such as 300°C, and thus unique design and operating problems must be solved.

The high-temperature batteries as just described use a liquid salt for the ion transporter; no solvent is used. There is a family of cells that use salt as the sole ion transporter, but the salt is solid at operating temperatures. These are known as *solid-electrolyte systems.* Because of very poor ion mobility, they are used only for extremely low current discharge, such as constant current drain over periods of years. One application, for example, is in heart pacemakers. An example is the *iodine-lithium system,* in which a layer of lithium iodide forming on the surface of the lithium serves as the separator and ion carrier. Water is a contaminant for nonaqueous systems and, therefore, special environments for cell assembly are required. These add significantly to the cost. Also, the organic solvent electrolytes are poorer in ion transportability than the aqueous electrolytes and thus are poorer on high current discharge.

Summary. Some of the systems used in commercially available batteries are described in the accompanying table. The table is abridged and included mainly to illustrate the present diversity in battery technology.

Standard Cells

Primary cells are also used as standards of electromotive force (emf) and are suitable for precise measurements, such as those made with a potentiometer which requires a constant, known emf. Over the years two principal standard cells have been used, the Clarke cell and the Weston cell. The Clarke cell has a mercury cathode and an amalgamated zinc anode. The cathode is submerged in a mercurous sulfate paste, and the electrolyte is a solution of zinc sulfate saturated at 0°C. The system is sealed in a glass tube and develops an emf of 1.440 volts at 15°C with a decrease of 0.00056 volt per °C rise in temperature. The Weston standard cell, a diagram of which is given in Fig. 5, is made in two types, the normal cell containing a saturated cadmium sulfate solution and another type which is used as a working standard in which the solution is less than saturated above 4°C. The saturated cell is the basic standard for maintaining the value of the volt, and is used in this manner in national standards laboratories. Its rather high temperature coefficient must always be taken into account. The legal voltage of the standard-cell group in the United States is 1.018636 volts absolute at 20°C and this value is accepted by international agreement.

The emf of the unsaturated Weston cell is within 1.0188 to 1.0198 volts absolute, with the exact voltage of each cell being established by comparison with the normal or saturated cell. The unsaturated cell is a useful working standard because of its negligible temperature coefficient.

[1] *Editor's Footnote*: In mid-1980, researchers at the General Motors Research Laboratories announced new findings pertaining to lithium's use in batteries. High electropositivity and low equivalent weight make lithium an ideal battery reactant, capable of supplying the specific energy needed to operate an electric vehicle. The source of the abundant energy available in lithium, however, is exactly what makes it almost impossible to manage. The challenge is to prepare alloys and find materials stable enough to contain the aggressiveness of lithium without greatly suppressing its activity. Work at GM has been related to the search for an advanced molten salt battery cell.

Specific energies greater than 180 Wh/kg, about five times that of the lead-acid battery, have been demonstrated by electrochemical cells utilizing LiCl-KCl electrolyte and electrodes of metal sulfide and lithium alloy. But operating temperatures of 723 K and the aggressive nature of the chemical reactants pose serious new challenges to cell construction materials. Of particular concern is the lithium attack upon separators and seal components. Most inorganic insulators, including the refractory oxides and nitrides, are destroyed or rendered conductive by this attack. Boron nitride, one of the more resistant materials, has been the subject of recent research. Earlier research had also indicated that silicon reduces the activity of lithium without substantially increasing its weight and produces a manageable solid at 723 K. Studies of the stability of boron nitride with Li-Si alloys of differing composition are continuing apace.

CHARACTERISTICS OF SOME COMMERICALLY AVAILABLE BATTERIES

Battery System	Basic Type	Overall Chemical Reaction	Negative Electrode	Positive Electrode	Electrolyte	Nominal Voltage per Cell	Energy Density Watt-Hours Per Pound	Energy Density Watt-Hours Per Cubic Inch	Capacity	Input for Recharging	Advantages	Disadvantages
Zinc–manganese dioxide (usually called Leclanché or carbon–zinc)	Primary	$2MnO_2 + 2NH_4Cl + Zn \rightarrow ZnCl_2 \cdot 2NH_3 + H_2O + Mn_2O_3$	Zinc	Manganese dioxide (Often natural ore)	Aqueous solution of ammonium chloride, zinc chloride	1.5	5–40	1–3	Several hundred mAh to 30 Ah*	—	Low cost; variety of shapes and sizes; excellent shelf life	Efficiency decreases at high current drains; poor low-temperature performance
Zinc–zinc chloride-manganese dioxide	Primary	$4Zn + 8MnO_2 + ZnCl_2 + 8H_2O + xH_2O \rightarrow 8MnOOH + ZnCl_2 \cdot 4Zn(OH)_2 \cdot xH_2O$	Zinc	Manganese dioxide (usually synthetic)	Aqueous solution of zinc chloride	1.5	15–30	1–3	Several hundred mAh to several Ah	—	Good service at high current drain, leak resistant, good low-temperature performance	Relatively expensive for low drains
Zinc–alkaline–manganese dioxide	Primary	$2Zn + 3MnO_2 \rightarrow 2ZnO + Mn_3O_4$	Zinc	Manganese dioxide (usually synthetic)	Aqueous solution of potassium hydroxide	1.5	20–40	2–3	Several hundred mAh to 23 Ah	—	High efficiency under moderate continuous drain conditions; good low-temperature performance; low impedance; long shelf life	Expensive for low drains
	Rechargeable	$Zn + 2MnO_2 \rightleftarrows ZnO + Mn_2O_3$					10	1.0–1.2		Approximately 100% of energy withdrawn		Rechargeable-limited cycle life; voltage-limited taper current charging
Zinc–mercuric oxide	Primary	$Zn + HgO \rightarrow ZnO + Hg$	Zinc	Mercuric oxide	Aqueous solution of potassium hydroxide	1.35	10–50	4–8	16 mAh–14 Ah	—	High service capacity/volume ratio; flat voltage discharge characteristic; good high-temperature performance; good storage life	Poor low-temperature performance on some types
Zinc–silver oxide	Primary	$Zn + Ag_2O \rightarrow ZnO + 2Ag$	Zinc	Monovalent silver oxide	Aqueous solution of potassium hydroxide or sodium hydroxide	1.5	30–60	4–8	38–190 mAh	—	Good high current pulsing; flat voltage discharge	Silver is expensive; poor storageability
	Rechargeable	$Zn + Ag \leftrightarrows ZnO + Ag$		Divalent silver oxide	Aqueous solution of potassium hydroxide	1.8/1.5 (two-step)	40–70	2–8	Vented, 5 Ah to several thousand Ah	Minimum of 110% of energy withdrawn	High-energy density; high rate capability	Expensive; two-step discharge; poor charge maintenance; limited cycle life
Lead–lead dioxide (usually called lead-acid)	Rechargeable	$2Pb + 2PbO_2 + 2H_2SO_4 + H_2O \rightleftarrows PbSO_4 + 2PbO + 3H_2O$	Lead	Lead dioxide	Aqueous solution of sulfuric acid	2	Sealed, 10–15 Vented, 7–12	0.8–1.1 0.5–2	Vented, 1–10,000 Ah	Minimum of 110% of energy withdrawn	Spill-resistant Low cost	Limited low-temperature performance; vented cells require servicing
Lithium–iron sulfide	Primary solid cathode	$FeS_2 + 4Li \rightarrow Fe + 2Li_2S$	Lithium	Iron sulfide	Nonaqueous solution of lithium salts in ethers	1.0	—	—	38 mAh to 120 mAh	—	Good storageability; lower cost substitute for silver cells.	Limited high rate capabilities

Sulfur dioxide–lithium	Primary liquid cathode	$2SO_2 + 2Li \rightarrow Li_2S_2O_4$	Lithium	Sulfur dioxide	Nonaqueous solution of lithium bromide in mixture of SO_2 and ace-tonitrile	3.0	—	—	—	—	High energy and power densities; good storageability; good at low temperatures	Complex manufacturing facilities required
Nickel–cadmium	Recharge-able	$Cd + 2NiOOH + KOH + 2H_2O \rightleftarrows Cd(OH)_2 + 2Ni(OH)_2 + KOH$	Cadmium	Nickelic hydroxide	Aqueous solution of potassium hydroxide	1.25	Sealed, 12–17 Vented, 12–20	Sealed, 1–1.5 Vented, 1–1.5	Sealed, 20 mAh–100 Ah Vented—few Ah to over 500 Ah	Sealed, minimum of 140% of energy withdrawn Vented, minimum of 125–150% of energy withdrawn	Excellent cycle life; flat voltage discharge characteristic; good high- and low-temperature performance; high resistance to shock and vibration; can be stored indefinitely in any charge state	High initial cost; only fair charge retention
Silver–cadmium	Recharge-able	$Cd + AgO + KOHÀÀH_2O \rightleftarrows Cd(OH)_2 + Ag + KOH$	Cadmium	Divalent silver oxide	Aqueous solution of potassium hydroxide	1.4	22–34	1.8–2.5	Sealed, up to 300 Ah	Minimum of 110% of energy withdrawn	Good energy/weight ratio; good charge retention; long wet-stand life	Expensive; poor low-temperature performance; two step discharge curve
Zinc–air (oxygen)	Primary	$2Zn + O_2 + 4KOH + 2H_2O \rightarrow 2K_2Zn(OH)_4$ (large cells) $2Zn + O_2 \rightarrow 2ZnO$ (small cells)	Zinc	Oxygen	Aqueous solution of potassium hydroxide	1.25	80–100	3.2	Vented, $\frac{1}{2}$–2,000 Ah	—	Flat voltage discharge characteristic; high input per unit volume in small cells	Drying out; carbonation

* Ah = ampere-hours; mAh = milliampere-hours *Source:* Union Carbide Corporation.

The two electrodes are termed negative and positive here to avoid confusion which can result from use of the terms anode and cathode, the latter often being used loosely in the battery industry. The negative electrode of the primary (or charged secondary) cell is metallic and is oxidized (increased in valence) during discharge, giving up electrons to the external circuit. The positive electrode in an aqueous system initially is an oxygen donor, usually a metal oxide, which is reduced as it receives electrons from the external circuit. Charge transfer from one electrode to the other within the cell is via the ions of the electrolyte salt. In certain cells, the electrolyte enters further into reactions and actually changes in composition during use.

References

Falk S. O., and A. J. Salkind: "Alkaline Storage Batteries," Wiley, New York, 1969.
Fleischer, A., and J. J. Lander (editors): "Zinc-Silver Oxide Batteries," Wiley, New York, 1971.
Heise, G. H., and N. C. Cahoon (editors): "The Primary Battery," Vol. 1, Wiley, New York, 1971.
Kordesch, K. V. (editor): "Batteries," Vol. 1 (Manganese Dioxide), Dekker, New York, 1974.
Murphy, D. W., and P. A. Christian: "Solid State Electrodes for High Energy Batteries," *Science*, **205**, 651–656 (1979).
Robinson, A. L.: "Advanced Storage Batteries," *Science*, **192**, 541–543 (1976).
Weinstein, J. N., and F. B. Leitz: "Electric Power from Differences in Salinity: The Dialytic Battery," *Science*, **191**, 557–559 (1976).

L. F. Urry, Battery Products Division, Union Carbide Corporation, Parma, Ohio.

BATTERY GRID METAL. Lead.

BATTERY-OPERATED CAR. Electric Car.

BAT TICK (*Insecta, Diptera*). Highly specialized flies which live as ectoparasites on bats. Like other parasitic flies, their habits are accompanied by some structural resemblance to the true ticks, hence the inaccuracy of the common name. The bat ticks belong to two families, *Streblidae* and *Nycteribiidae*. All species of the latter are wingless.

BAUD. A baud is the unit of telegraph signaling speed, derived from the duration of the shortest signaling pulse. A telegraphic speed of one baud is one pulse per second. The term "unit pulse" is often used for the same meaning as the baud. A related term, the "dot cycle," refers to an on-off or mark-space cycle in which both mark and space intervals have the same length as the unit pulse.

BAUDOT CODE. A teleprinter code that uses combinations of five and six marking and spacing intervals of equal duration. The five-unit code gives 32 possible character; and the six-unit code gives 64 possibilities. The code is used in radio and wire teleprinter operations.

BAUMÉ SCALE. Specific Gravity.

BAUXITE. There are two main types of bauxite ores used as the primary sources for aluminum metal and aluminum chemicals: $Al(OH)_3$ (gibbsite) and AlO(OH) (boehmite). Thus, bauxite is a term for a family of ores rather than a substance of one definite composition. The first bauxite ore was found near Les Baux in the south of France by P. Berthier (1821). Deposits are found worldwide except in Antarctica. Secondary sources of aluminum include a large array of clays that are rich in alumina (30–40%) found in large abundance throughout the world. Bauxite ores range in color from white to dark red or brown, largely depending upon the iron content. An average composition of the ores used by industry today would be: Al_2O_3, 35–60%; SiO_2, 1–15%; Fe_2O_3, 5–40%; TiO_2, 1–4%, H_2O; 10–35%; other substances, 0–2%.

Although developed as early as 1888 by the Austrian chemist, Karl Josef Bayer, the Bayer process still is used almost exclusively for the extraction of alumina from ores. The bauxite first is reacted under pressure with hot caustic which dissolves the $Al_2O_3 \cdot xH_2O$ to form sodium aluminate. The solution is filtered hot, then cooled and agitated with the addition of a small quantity of aluminum hydrate to enhance the precipitation of the crystalline hydrate. After filtration, the cake is kiln-dried at 1100°C to remove H_2O and yield Al_2O_3.

The purity of aluminum produced by the electrolytic process (see **Aluminum**) is determined mainly by the purity of the Al_2O_3 used. Thus, commercial grades of Al_2O_3 are 99–99.5% pure with traces of H_2O, SiO_2, Fe_2O_3, TiO_2, ZnO, and very minute quantities of other metal oxides.

As of the early 1980s, there are nine alumina plants in the United States and territories. Six plants are located on the Gulf Coast, two plants are in Arkansas, and one plant is in the Virgin Islands. Total capacity is estimated at 7.7 million short tons (approximately 6.9 million metric tons). The aluminum industry in the United States has depended on imports from the Carribean, South America, and Australia. Carribean exporting countries have high levies on bauxite exports and, politically, have moved to nationalize and expropriate the bauxite mines developed by United States industry. It is doubtful that new Bayer alumina plants will be built in the United States. The technology in the United States is focusing on extracting alumina from abundant alumina-containing clays. New Bayer plants are being built in Australia and Venezuela.

Uses and Grades of Alumina. Although aluminum production is a major consumer of alumina, the compound is used widely elsewhere. The properties of alumina and hydrated aluminas may be varied, ranging from a talclike softness to the hardness of a ruby or sapphire. Some of the uses for alumina include water purification, glassmaking, production of steel alloys, waterproofing of textiles, coatings for ceramics, abrasives and refractory materials, cosmetics, and electronics. *Hydrated aluminas* may be represented: α-$Al_2O_3 \cdot H_2O$ or α-$Al(OH)_3$. The compounds are dry, snow-white, free-flowing crystalline powders and may be obtained in a wide range of particle sizes. The compounds are widely used in the production of aluminum salts because of their reactions with strong acids and alkalies. Some of the salts prepared in this manner include aluminum chloride, aluminum phosphate, and aluminum sulfate. *Activated aluminas* are very porous aluminum oxide. λ-Al_2O_3 is made by heating the hydrate to drive off nearly all of the combined water. The final products are granules or fine powders with a large surface area to provide absorptive capacity per unit volume. Applications of the aluminas are enhanced by virtue of their inertness chemically and nontoxic qualities. They are extensively used for drying gases and for dehydrating liquids, such as alcohol, benzol, carbon tetrachloride, ethyl acetate, gasoline, toluol, and vegetable and animal oils. They also are used as filter aids in the manufacture of lubricating and other oil products. Their large surface area qualifies the aluminas as catalysts for numerous reactions. The compounds also find extensive application in ceramics, particularly in abrasive and cutting wheels, polishing compounds, additives to glass, tank linings, spark plugs, electrical substrates, and linings for high-temperature furnaces. *Corundum* is an aluminum oxide that possesses a hexagonal crystal structure. The compound is extremely hard (2000 on the Knoop scale), sp gr 3.95, and is widely used in abrasives and refractories. Corundum is manufactured by fusing alumina or bauxite in an electric arc furnace operated at about 2200°C.

See also **Aluminum; Corundum;** and terms listed under **Mineralogy.**

S. J. Sansonetti, Consultant, Reynolds Metals Company, Richmond, Virginia.

BAYBERRY SHRUBS AND TREES. Of the family *Myriaecae*, genus *Myrica*, there are three species of bayberry of interest. The *M. carolinensis* is found near the seashore from the West Indies and Florida northward to Prince Edward Island and New Brunswick. Normally, the plant may be described as a compact shrub that prefers dry, sandy, and rocky soil. However, it does occur in some of the bogs of northern New Jersey and Pennsylvania and also is found along the shore of Lake Erie. The berry occurs in crowded clusters. The berry is small, approximately $\frac{3}{16}$ inch (0.5 centimeter) in diameter, of a grayish-white color, and is waxy and resinous. The plant has long, fragrant leaves. Candles can be made from the berries. They are boiled and the waxy substance is skimmed off, after which it is melted and refined. The yield of wax is quite good, approximately 25% of the weight of the raw berries representing wax. At one time used by the settlers, the candles now are mainly associated with the Christmas holiday and are noted for a distinctive, pleasant aroma.

Also known as wax myrtle or bayberry, the *M. cerifera* is a small tree, slender of contour, that prefers sandy swamps of the southern states and coastal plains. The tree is found from Florida northward into southern Maryland and westward to the Gulf States. A large specimen, selected by The American Forestry Association for its "Social Register of Big Trees," has a circumference of 3 feet, 4 inches (0.9 meter, 10 centimeters) at $4\frac{1}{2}$ feet (1.4 meters) above ground level, a height of 29 feet (8.7 meters), and a spread of 36 feet (10.8 meters). The tree is located at McClellanville, South Carolina. This species also has small, grayish-white berries, even smaller than the *M. carolinensis.*

Also known as the California wax myrtle or bayberry, the *M. californica* is a small tree with smooth, grayish bark and branches. The tree is slender in contour, with a rounded crown. The berries are similar to those of *M. cerifera,* slightly larger, and of a dark purple coloration, but like the other bayberrys, covered with a white waxy material. The tree prefers coastal areas, such as damp slopes and sand dunes. It is found from southern California northward to Oregon. A specimen selected for the aforementioned "Social Register" is located in the Siuslaw National Forest near Reedsport, Oregon, and has a circumference of 4 feet, 4 inches (1.2 meters, 10 centimeters) at $4\frac{1}{2}$ feet (1.4 meters) above ground level, a height of 38 feet (11.4 meters), and a spread of 34 feet (10.2 meters).

Other members of the *Myricaceae* family (sometimes called Sweet Gale family) include *M. Gale* (Sweet Gale or Dutch Myrtle), a small shrub attaining a height of about 5 feet (1.5 meters) and essentially similar to the bayberrys. However, it is distributed much more widely, preferring the edges of ponds, swamps, and streams, from Labrador and Newfoundland west to the Great Lakes and northwestward to Alaska. The shrub is found in the Appalachian Mountains to the 3000-foot (900 meter) level as far south as Virginia. It is also relatively common throughout most of New York and New England. *M. asplenifolia* (sweet fern) has a nutlike fruit which, as well as the leaves, is fragrant when crushed. The shrub seldom exceeds 2 feet (0.6 meter) in height and is found in dry areas, and ranges from New Brunswick and Nova Scotia westward to Alberta and south as far as North Carolina. It is found in the White Mountains and in the Alleghany Mountains up to the 2000-foot (600 meter) level. *M. inodora* (odorless myrtle) is considerably less common than the other trees and shrubs described previously. The plant ranges in height from 5 to 18 feet (1.5 to 5.4 meters), with a grayish-white bark and leathery leaves of a deep green color. The berries are over $\frac{1}{4}$ inch (0.6 centimeter) in diameter, grayish white in color, and covered with a white wax. The plant also likes the edges of ponds and swamps and coastal climes, ranging from Florida westward to Mississippi.

The *M. cerifera* and *M. carolinensis,* previously described, as well as *M. mexicana* (grown in Mexico and Central America) and *M. ocuba* (grown in Brazil) are commercial sources of products known as vegetable tallow, bayberry tallow, myrtle wax, and ocuba wax. Ocuba wax is classified as a waxy fat, not truly a wax in the chemical sense.

BAYES' THEOREM. Let $c_1, c_2, c_3 \ldots c_s$ be some s mutually exclusive random events such that one of them must happen and let $P(c_i)$ be the probability of c_i. These events we shall describe as causes. Let E be some event which can be directly observed and let $P(E|c_i)$ be the probability that E will occur on the assumption that c_i has already happened. Let $P(c_i|E)$ be the probability of c_i after E has occurred. The probabilities $P(c_i)$ are called "a priori" probabilities, i.e., probabilities not at all dependent on the event E, where $P(c_i|E)$ are called "a posteriori" probabilities since these depend on the event E having occurred. Bayes' theorem states

$$P(c_i|E) = \frac{P(c_i)P(E|c_i)}{\sum_{j=1}^{s} P(c_j)P(E|c_j)}$$

The theorem states exactly how the probability of a certain "cause" changes as different events actually occur. The theorem was published posthumously in 1764. When all conditions of the theorem are fulfilled there is no objection to Bayes' theorem. The difficulties in applying the theorem depend upon the fact that $P(c_i)$, the "a priori" probabilities, are not known.

The theorem is nevertheless used by one school of statisticians to make inferences about different hypotheses. The procedures are called *Bayesian.* They require certain assumptions about the prior probabilities $P(c_i)$ and are justified on various grounds e.g., (1) that there is prior information enabling the probabilities to be plausibly approximated, (2) that the final results are relatively insensitive to the initial assumptions, (3) that there is no better course anyway. Another school of statistical thought prefers to avoid Bayesian methods entirely and to rely on alternative procedures which, however, also embody difficulties.

BAYOU. A stagnant or abandoned course of a meandering river. Also a general term for a stagnant inlet or outlet of a lake or bay. See **Oxbow Lake.**

BAY TREE. Laurel Family.

B COMPLEX VITAMINS. Vitamins; Vitamin B6; Vitamin B12.

BDELLOIDEA. An order of rotifers, named for their fancied resemblance to leeches in their looping method of progression.

BDELLONEMERTEA. An order of nemertine worms of broad, flat form, named from their superficial resemblance to leeches. They live in the branchial chamber of mollusks.

BEAD TEST. A test for the identification of various *metals,* made by fusing a small quantity of the sample with borax or other fluxing substance on a small loop of inert wire (usually platinum wire), to form a bead. The appearance of the bead—its color, transparency, etc.—is often characteristic of a particular metal, as well as of the flux used, and the kind of flame (oxidizing or reducing). See also **Blowpipe Analysis.**

BEAK. Birds.

BEAM BRIDGE. Bridge (Structural).

BEAM (Composite). A beam which is composed of two materials properly bonded together and having different moduli of elasticity is called a composite beam. Reinforced concrete beams, steel beams mechanically bonded to a concrete deck, and wood beams reinforced with steel plates are typical examples. It has become a rather prevalent practice in recent years to provide for composite action between the steel stringers and concrete floor of highway bridges. This permits the use of ligher beams than would otherwise be possible.

The analysis of composite beams depends on the assumption that a plane section before bending remains plane after the load is applied. Therefore the two materials must be connected in such a way that they will act as a unit. This condition is realized in the reinforced concrete beam by means of the bond (see **Stress**) between the reinforcing rods and the concrete. In the case of reinforced wood beams the parts are connected by bolts properly spaced to resist the shearing forces (see **Shear**) between the plates and the beam. Steel beams are bonded to the concrete flange by means of lugs or spirals welded into the top flange of the steel beam.

The flexure formula is applicable to composite beams if the beam is transformed into an equivalent homogeneous section by means of the transformed area method which is found in texts on strength of materials and reinforced concrete design.

BEAM (Influence Line). Influence Line.

BEAM-POWER TUBE. Electron Tube.

BEAM SPLITTER (Color Television). Television.

BEAM (Structural). Beams used for building various structures, such as bridge decks and flooring, building floors and roofs, and other applications where support over a span of distance is required, may be classified in a number of ways: (1) by cross sectional shape, such as the familiar I-beam of steel or the rectangular section of a wood beam, (2) by the material from which the beam is made, and (3) by the manner in which the beam is used, that is, the structural design configuration (method of support).

Decks, floors, and roofs often are supported on a rectangular grid made up of flexural members. Commonly, the members which provide the span between main supports are termed *girders,* essentially large heavy beams capable of carrying concentrated as well as evenly distributed loads. See also **Girder.** Beams which rest across a series of

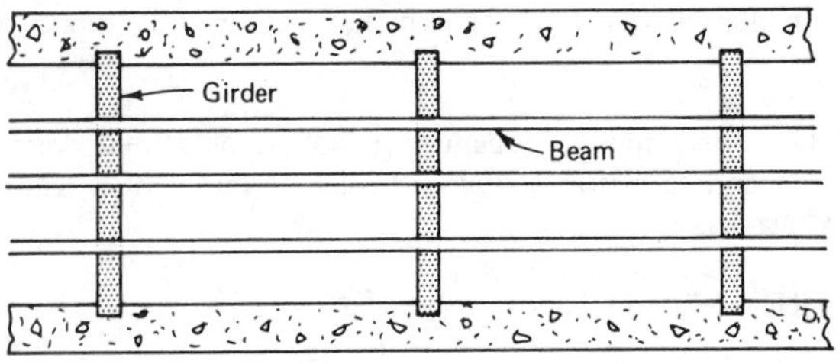

Fig. 1. Beam-and-girder framing.

girders are simply termed *beams*. See Fig. 1. The type of framing shown is commonly referred to as *beam-and-girder* framing. Special nomenclature is frequently applied for specific applications. The parallel (to flow of traffic) structural members of a bridge sometimes are called *stringers*. The transverse members supporting the floor or deck may be called *floor beams*. The grid components for a building roof often are referred to as *purlins* and *rafters*, whereas in floor supports, they may be termed *joists* and *girders*. Most frequently, beam-and-girder framing is employed for rather short spans where shallow members are desired for allowing ample headroom underneath.

The six most common methods of beam support are illustrated in Figures 2 through 7.

A *simple beam* (Fig. 2) is one which rests on two end supports in such a manner that the ends of the beam are free to rotate on the supports. The supports restrain the beam only against vertical movement. To allow for a horizontal component and also for change in length arising from a temperature change, the supports may have to prevent horizontal motion, usually accomplished by providing a horizontal restraint at one support. The term *span* is used to designate the distance between supports. The term *reaction* designates the load carried by each support.

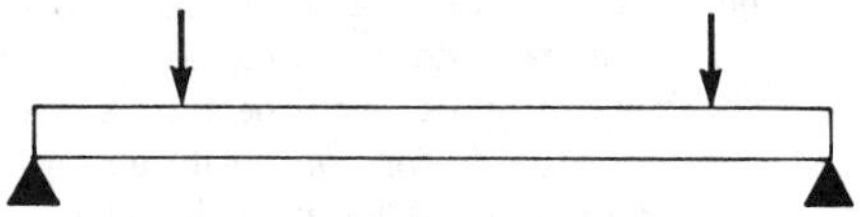

Fig. 2. Simple beam.

A *cantilever beam* (Fig. 3) is a beam which is rigidly connected at one end to a fixed support and free to move at the other end. This theoretically fixed condition rarely occurs because of deformation of the supporting material. The maximum bending moment and maximum shear occur simultaneously at the face of the support. The usefulness of this type of beam is demonstrated in structures, such as canopies, unbraced airplane wings, and cantilever retaining walls. A support of this type is called a *fixed end.*

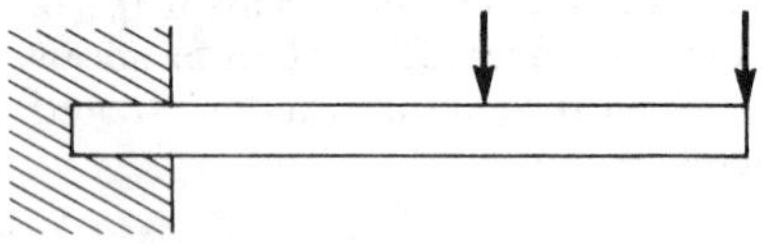

Fig. 3. Cantilever beam.

A beam with *one end fixed* (Fig. 4) results when a support is placed under the free end of a cantilever beam.

A *fixed-end beam* (Fig. 5) permits no rotation or vertical movement. In practice, a configuration of this type can seldom be obtained. The majority of support conditions fall intermediately between those applying for a simple beam and those for a fixed-end beam.

A beam with *overhangs* (Fig. 6) has supports that permit rotation, but the overhangs have a free end.

A *continuous beam* (Fig. 7) is a beam having more than two supports. A beam which is continuous over several supports offers more difficulty in analysis than would a series of freely supported beams covering the same overall span. However, it can be analyzed readily by methods, classical or numerical, available for the analysis of statically indeterminate structures. Because of the restraint at the intermediate supports, the continuous beam can carry a greater load than a simple beam of the same size and span. It is quite important to provide a firm foundation for the intermediate supports, as small deflections due to sinking of an intermediate support may introduce stresses of an entirely different nature and magnitude from those used in designing the beam.

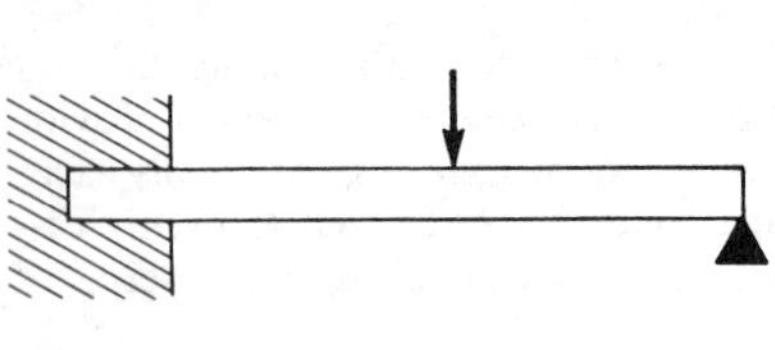

Fig. 4. Beam with one end fixed.

Fig. 5. Fixed-end beam.

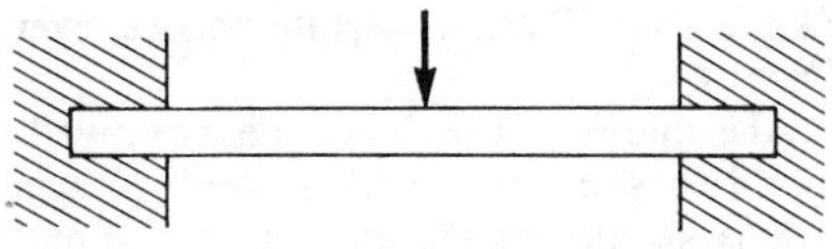

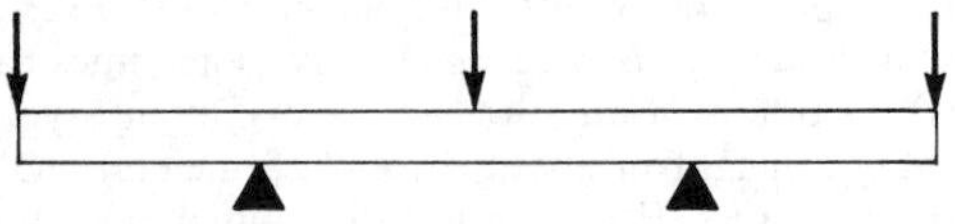

Fig. 6. Beam with overhangs.

A curved beam is a beam having a finite radius of curvature before and after the bending loads are applied. Theoretically the flexure formula is not applicable to curved beams because the unit deformation does not have a straight line variation over the depth of the beam due to the difference in the length of the various fibers between any two radial planes. Formulae for curved flexural members are given in texts on advanced strength of materials. Reliable values of extreme fiber stresses may be found by means of correction factors, applied to the flexure formual, which are also in these texts. The correction factor depends on the radius of curvature and the shape of the cross section. However, a member must have a considerable amount of curvature before there is an appreciable difference between the stresses found by the straight beam (flexure) theory and the curved beam theory. Consequently a correction factor of 1 is frequently used when the radius of curvature is large.

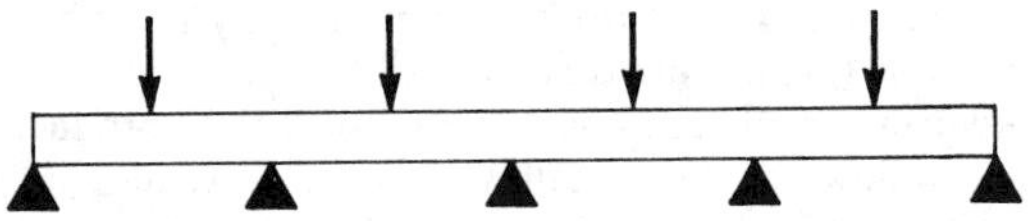

Fig. 7. Continuous beam.

The curved beam theory or a modification of this theory should be applied to the analysis of any curved flexural member of small radius of curvature even though the member cannot be classed as a beam. Hooks and chain links are typical examples.

See also **Bending Moment.**

BEAN. The fruits of many different plants and in many cases the plants themselves are called beans. Nearly all of them are members of the Leguminosae. In Europe and America, beans are principally plants of the genus *Phaseolus*, with *Phaseolus vulgaris*, the common Garden Bean, a most important species, with many varieties in cultivation. This plant is a tender annual, probably originally native in South America, and grows either as a low bush or as a twiner. Many of the varieties grown have been selected to yield a thick, rather fleshy pod with small seeds; these are string or snap beans. See accompanying figure. In some varieties chlorophyll (see **Pigmentation (Plants)**) is either entirely or largely lacking in the pods, giving the Wax or butter bean. Many other varieties are grown primarily for the dried seeds, which have a very high food value and keep exceedingly well if dried and protected from insects. Common varieties used as dry beans are Pea, Yellow Eye, and Red Kidney beans. Shell beans are those in which the nearly mature but green seed is eaten. See **Leguminoseae.**

Related to the Garden Bean is *Phaseolus multiflorus*, the Scarlet Runner Bean, which is often grown as an ornamental plant because of its showy scarlet flowers. Another species, widely grown in warmer climates, is *Phaseolus lunatus*, the Lima Bean, another plant from South America. The large flat pods of this plant contain a few large flat seeds. Varieties have been developed which can successfully mature in regions having a short growing season.

Snap bean. Variety is the *Tendercrop*, a mosaic-resistant, heavy-yielding snap bean with tender, round, green pods and a wide range of adaptability. (*USDA photo.*)

In Europe a common bean is the Broad Bean, also called Windsor or Horse Bean, *Vicia faba*. Its seeds are rich in nitrogenous compounds and rather hard to digest. These seeds are extensively used for horse food, as well as for human consumption.

In the Orient, the bean crop is almost entirely *Glycine max*, the Soy Bean. The plant is an erect bushy annual with trifoliate leaves and bears fruit (pods) in great abundance. Tremendous quantities are grown in the Orient, where large populations depend upon it as the major source of protein. The plant has been introduced into the United States, where it is extensively cultivated for several reasons. One is the fact that it will grow well on poor soils and, being a legume, will build up the fertility of the soil, because of its associated nitrogen-fixing bacteria. It is also an important forage crop. The seeds are rich in oil and have received much attention from chemists, with the result that large quantities are now used in the preparation of enamels, linoleum, inks, paints, soaps, etc. This oil is used in larger quantities than any other oil for the production of vegetable shortening. The cake remaining after the oil is pressed out is used as a stock food or as a source of industrial proteins for the production of adhesives, sizings, coatings for paper, and other products.

In Africa, species of *Dolichos* are used as food.

All the plants so far treated are legumes having "bean"-like fruit. But in the Vanilla bean one deals with an entirely different plant. Vanilla beans are the fruits of a climbing orchid growing wild in Central America and Mexico. Since prehistoric times these fruits have been used to flavor chocolate. At the present time (and with some difficulty) the plant is cultivated in several tropical countries, both in America and in the Orient. Like many other orchids, the flowers are pollinated only by certain specially adapted bees and perhaps the hummingbirds. When introduced into regions where the necessary insects are used mainly as a flavoring.

Many other plants contain vanillin, or similar substances, and so are often used in the manufacture of artificial vanilla. Among these are the seeds of a tropical South American plant, Coumarouna, which yield a product coumarin, used in perfumery. The seeds from which this coumarin is obtained are called Tonka Beans.

BEAN WEEVIL (*Insecta, Coleoptera*). The common bean weevil is a small beetle, *Mylabris obtectus*, which attacks growing beans and cowpeas in the pod. Methods of control are the same as for the pea weevil. The four-spotted bean weevil is a related species which attacks beans and peas both in the field and in storage.

BEARING MODULUS. The performance of a bearing may be measured in terms of a quasi-dimensionless function Zn/p, where Z is the absolute viscosity of the lubricant in centipoises, n is the journal speed in rpm, and p is the unit pressure on the bearing in pounds per square inch of projected area.

BEARING (Navigation). In both air and sea navigation, the term *bearing* is used as an indication of direction. Bearing is the angle, measured at the observer, or some specified point, between two lines in the plane of the horizon. See accompanying figure. Careful distinction must be made between the meaning of the term bearing when used alone and the same term when qualified, e.g., relative bearing, compass bearing, etc.

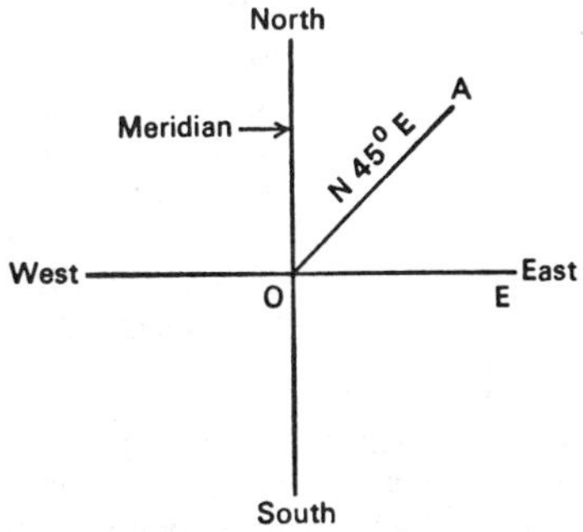

Example of navigational bearing.

The bearing of a given point from the observer is simply the true direction of that point, expressed in the conventional manner for expressing direction, i.e., to the right, through 360°, and in three digits.

If the compass is used for finding the bearing of a point from a ship, the value obtained will be the compass bearing, and the compass corrections must be applied to obtain true bearing. Methods for obtaining bearings by using the pelorus are explained in the article on that instrument.

Frequently, it is easier to use a ship's keel as a reference line, rather than north, for determining bearing. In this case, relative bearing is obtained. This is defined as the angle between the keel of the ship and a line in the plane of the horizon to the object. This angle is measured from the forward end of the keel, to the right through 360°, and expressed in three digits. To find the bearing of the object when the relative bearing is known, the heading must be added to the relative bearing. For example, a ship is heading 326° and the relative bearing of a buoy is 124°. The bearing of the buoy is $326° + 124° = 450°$, or 090°, and the buoy is due east of the ship. Radio bearings, taken with a radio direction-finder, are always given as relative bearings from the ship.

Lookouts on ships, gunners on planes, and other observers of this general type, may not have a pelorus available to determine bearings of a sighted object. Various approximate methods are in use for expressing relative bearings, depending upon the particular service. Lookouts on ships at sea frequently report relative bearings in terms of compass points. For example: an object sighted one point (11°.25) to the right of the bow would be reported as "one point off the starboard bow." An object with relative bearing 315° would be reported as "broad on the port bow," etc. In the air-services, clock numbers are sometimes used for reporting relative bearings, e.g., an object with relative bearing about 300° would be reported as "at ten o'clock." Complete descriptions of the various approximate methods used in the various services may be found in manuals of seamanship, regulations of the U.S. Air Force, etc.

In surveying, bearing is usually expressed in terms of the acute angle measured either from north or south to the east or west. For example: a bearing of 45° would be expressed as N 45° E (north 45° east); a bearing of 290° as N 707 W, etc. See figure.

The latitude of a line is its orthographic projection on the meridian; the departure is its orthographic projection on a line perpendicular to the meridian:

$$\text{latitude} = \text{length} \times \text{cosine of bearing}$$

$$\text{departure} = \text{length} \times \text{sine of bearing}$$

Latitudes are positive for north bearings and negative for south bearings. Departures are positive for east bearings and negative for west bearings. In a theoretically closed traverse the algebraic sum of both the latitudes and the departures must be zero.

The bearing of any line such as *O A* referred to a meridian through *O* is a forward bearing; if referred to a meridian through *A* it is a back bearing or reverse bearing. Forward and back bearings differ by 180°.

See also **Compass (Navigation)** and **Navigation.**

BEARING VALUE (Soil). Foundations.

BEARS (*Mammalia, Carnivora*). The general organization of Ursines may be outlined about as follows:

Brown Bears (*Ursus*)
Spectacled Bear (*Tremarctos*)
Sun-Bear (*Helarctos*)
Moon Bears (*Selenarctos*)
American Black Bear (*Euractos*)
Sloth-Bears (*Melursus*)
Polar Bears (*Thalarctos*)

Bears are found in every continent except Australia, although Africa and South America have only limited species. The term *bear* is sometimes applied to the koala (*Marsupialia*) in a superficial, erroneous fashion. The Ursines are quite closely related with the Canines. Some authorities have observed that the Ursines are huge Canines without tails. Generally, bears prefer mountainous regions and are found in the Alps, Pyrenees, Carpathians, Caucasus, the mountains and woodlands of North America, ranging from Mexico northward to the Arctic, the Andes in South America, and the hilly and mountainous areas of North Africa and southeast Asia. The Polar Bear inhabits the Arctic regions as indicated by the name. The Sloth-Bear is found in India. Bears are noted for their ability to adapt to specific regions and climes and thus, among a comparatively limited number of species (about a half-dozen), much variety is demonstrated. For example, a given species may hibernate in one region and remain active all winter in another region.

Brown Bears are often referred to by other names, reflecting the numerous variations among a given species that have arisen from years of adaptation to specific regions. The Alaskan Brown Bear, also called "Big Brownie" or the Kodiak Bear is the largest of the living bears, weighing up to 1500 or more pounds (680 kilograms). This animal also is the largest of the land carnivores. When standing erect, the height may exceed 9 feet (2.7 meters). The largest specimens are found on the islands off the coast of Alaska and notably Kodiak Island. The animal feeds extensively on salmon, captured when the fish enter the rivers for spawning. The color may range from nearly black on the dark side to nearly a blond coloration on the light side. Usually, the bear is a light brown. The Alaskan Brown Bear prefers deep underbrush. The seasonal diet of fish is augmented throughout the rest of the year with various fresh greens and even seaweed which is obtained along the shore. The animal also seeks out the burrows of mice and ground squirrels. When berries are in season, the bear relishes a number of wild berries, including cloudberries and crowberries, as well as the berries of the mountain ash. Some authorities state that just before hibernation, the bear consumes large quantities of cranberries which serve as a purgative. Hibernation commences in late October or early November when dens are dug in dry hillsides. The cubs, usually twins—occasionally triplets—are born during the cold months. Appetites are high in the early spring and at that time, the bear has been known to attack horses and cattle. Over recent years, the Alaskan Brown Bear has been overhunted, both for sport and by natives who eat the meat and utilize the hides. See accompanying figure.

Brown bear. (*A. M. Winchester.*)

Also included among the Brown Bears is the Grizzly Bear (*U. horribilis*), closely related to the Alaskan Brown Bear. The weight of an adult Grizzly Bear will range from 800 to 1400 pounds (363 to 635 kilograms). The animal is found in the wilderness of North America. So-called grizzlies are also found in eastern Asia. Other Brown Bears include the Tibetan Brown Bear (*U. pruinosus*); and the Isabelline Bear, an inhabitant of the Himalayas and known for its beautiful cream-colored coat. Other large races occur in Eastern Siberia and Kamchatka, as well as in Syria.

A general characterization of the Brown Bears would include: By nature, peaceful and not inclined to molest humans unless provoked; more dangerous than the Big Cats, however, if come upon unexpectedly or provoked; possessing a rather clumsy gait, sort of a lumbering type of walk; very swift and powerful when necessary; large eaters, of both vegetable matter and animal flesh, but with a preference for sweets, such as honey, and a relish for fish; quite immune to insects; accumulate large quantities of fat in their bodies just prior to hibernation; the young are small, seldom larger than a small rabbit.

The American Black Bear (*Euarclos*) is smaller than the Brown Bear, but is not necessarily black. There are a number of brown Black Bears, giving rise to various names including the Cinnamon Bear, the Glacier Bear, and the Kermode Bear. The American Black Bear, weighing from 200 to 500 pounds (91 to 227 kilograms), is the common bear of North America and is distributed rather widely throughout Canada, the United States, and northern Mexico. The appearance is distinguished from most other bears by the lack of a hump between the shoulders. In earlier years, it appeared that the American Black Bear might be threatened, but since the species has penetrated further into the wilderness, the trend reversed and now the population appears to be increasing slowly. This bear sleeps for long periods during winter, but authorities generally do not designate this as true hibernation. The animal is omnivorous, eating a variety of materials ranging from pine cones to wasp's nests. In captivity, the animal is dangerous to feed because it assumes that when the food runs out, the giver is holding back. Usually three cubs, about the size of a rat, are born annually.

The Spectacled Bear (*Tremarctos*) inhabits the mountainous region of the Andes in South America. Of two species, one lives in Bolivia and another in Chile. The name is derived from yellow rings around the eyes which contrast with the shaggy, black body.

The Sun-Bear (*Helarctos*) has not been fully studied. It attains a length of some $4\frac{1}{2}$ feet (1.4 meters) and has a pronounced short, wide, and flat head. They are known as excellent tree climbers.

The Moon Bear (*Selanarctos*) is found in south-central Asia, in Japan, Taiwan, as far west as Iran, and as far south as northern Pakistan and India. The animal is characterized by a V-mark on its chest and a white upper lip, both features contrasting well with the animal's black body. More than most bears, the Moon Bear is a vegetarian. However, it retains a relish for honey and sweet-tasting insects, characteristic of most bears. This bear is essentially oblivious to humans and it is reported that it can be readily approached without harm.

The Sloth-Bear (*Melursus*) is an inhabitant of India. The animal is characterized by a rather long, anteater-type snout, huge claws, and small teeth. Like the anteater, this bear is quite insectivorous. The animal has a naked face of gray color, with long, shaggy black fur over the body. The chest carriers a white marking.

The Polar Bear (*Thalarctos*), sometimes referred to as the Water Bear or Ice Bear, has a thick, white coat, including fur on the soles of the feet, and, while possessing the characteristic profile of other bears, is somewhat distinguished by a longer, more slender neck, a more pointed head, and a large, high rump. The animal may attain a weight of 1000 pounds (454 kilograms) or more. The Polar Bear is an excellent swimmer, capable of covering wide expanses of water. It is noteworthy that during the summer season, the animal usually lives on ice floes, infrequently going on land. A major dietary item is the seal. The animal is known to swim for considerable distances underwater in effecting its strategy for surprising a seal. The young are born in the spring, usually in a natural cave that has been sought out by the female, or in a cave comprised of a rock and hollowed-out snow and ice. The males, with the exception of older animals, also hibernate throughout the winter. Found throughout the Arctic region, these bears always have been important to Eskimos for food, hides, and bones, from which implements are made. For references, see **Mammalia.**

BEAT. A series of alternate maxima and minima in vibration amplitude, produced by the interference of two wave trains of different frequency. A familiar example arises in the case of musical sounds. If two musical pipes or strings of slightly different pitch are sounded together, the result is a more or less distinct throbbing, often disagreeable to the ear. The beat frequency is the difference of the two wave frequencies. Thus, if the two tones are middle-c (256) and c-sharp (271.2), there will be 15.2 beats per second. If the two tones are ultrasonic, but have a frequency difference within the audible range, the beats themselves may produce an audible "beat tone" or "beat note." A similar effect results from the simultaneous reception of two radio wave trains which are nearly, but not quite, synchronized.

When two sinusoidal signals of different frequencies are applied to a circuit whose output signal amplitude is proportional to the square or higher power of the input signal amplitude, frequency components will be produced in the output which are not all present in the input signal. Among the frequencies present will be one equal to the sum of the two original frequencies and one which has a frequency equal to the difference of the two applied frequencies. This difference frequency is known as the beat frequency. There are numerous applications of this effect, but two of the major ones in the communications fields are in the reception of continuous wave signals and in frequency shifting as in the superheterodyne receiver. Since continuous wave signals have no audio modulation superimposed on them, they cannot be made audible by ordinary detection methods. However, if the incoming signal is beat with a local signal, differing by an audible amount, the result is an audible beat frequency. Frequency shifting by use of beat frequencies is applied in the superheterodyne, in carrier telephony, frequency modulation, and numerous other circuits. In most of these applications use is made of the fact that if one of the signals is modulated, the beat frequency signal will have the same modulation.

The process wherein additional frequency components are created by passage of two signals through a nonlinear device as described above is sometimes spoken of as mixing.

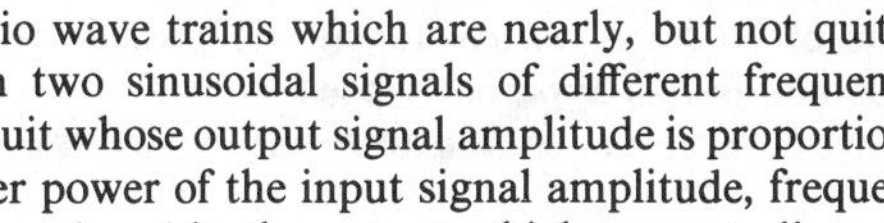

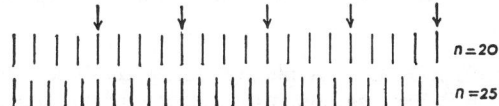

Five coincidences in unit time between wave trains of frequencies 20 and 25.

BEAT-FREQUENCY OSCILLATOR. Oscillator.

BEAUFORT WIND SCALE. Winds and Air Movement.

BEAT NOTE. The wave of difference frequency created when two sinusoidal waves of different frequencies are supplied to a nonlinear device.

BEAVER. (*Mammalia, Rodentia*). An aquatic species with a body length up to 100 centimeters (39 inches), a tail length up to 30 centimeters (12 inches), and a body weight up to 30 kilograms (66 pounds). The beaver has a heavy build and is one of the largest rodents. The body comes large as one looks at it from front to back. The hind feet have web membranes between the toes. The small forefeet are particularly good gripping tools. The feet all have five digits, all with strong nails. The second toes on the hind feet carry a weak double claw which is used for cleaning the fur. Beavers have a strong bristly hair, and their woolly hair is curled thickly. The tail is flat, 12 to 15 centimeters (5 to 6 inches) wide, and shaped like a ladle with scaly, leatherlike skin; the beaver uses it as a rudder. On land the beaver looks awkward, but in the water it is an outstanding swimmer and diver. These animals are exclusively herbivorous. In summer, they eat the juicy shrubs found on river banks, bulrushes, buds from softwood trees (especially poplars, aspens, and willows), as well as root pieces from water lilies. In winter, beavers will eat bark from shrubs and from trees they cut down. See accompanying illustration.

Beaver. (*A. M. Winchester.*)

There are several subspecies, including the Scandinavian Beaver (*Castor fiber fiber*); the Elbe Beaver (*Castor fiber albicus*); the Rhone Beaver (*Castor fiber galliae*); the Polish or Byelorussian Beaver (*Castor fiber vistulanus*); the Ural Beaver (*Castor fiber pohlei*); the Mongolian Beaver (*Castor fiber birulai*); and, from North America, the Canadian Beaver (*Castor fiber canadensis*); the Michigan Beaver (*Castor fiber michiganensis*); the Newfoundland Beaver (*Castor fiber caecator*); the Rio Grande Beaver (*Castor fiber frondator*); the Golden-Bellied Beaver (*Castor fiber subauratus*).

Beavers live much of their life in water; they settle along the banks of streams, rivers, and lakes in the thick underbrush of lowland woods. When the beaver lives near a large river full of rushing water, or when it lives near a river which is too wide to be dammed up, it builds itself a simple house in the riverbank, with at least two entrances, and sometimes four or five, all of which are underwater. The Elbe beaver often makes these simple living quarters in river banks. The entranceway rises on a diagonal from just below the water's surface to immediately below the surface of the ground. There the animals build a room which is some 4 feet (1.2 meters) wide and 15.7–19.7 inches (40–50 centimeters) high, and very carefully smoothed out on the inside. When the surface of the water rises, the floor of the room must also be raised; in order to accomplish this, the beaver gnaws or scrapes dirt from the ceiling, and this dirt then falls on the floor. Usually the ceiling is so strong that several people can stand on it together without mishap. However, if the beaver is forced to make its house somewhat higher, then it strengthens the ceiling with mounds of twigs which it lays outside over the ceiling. These can become regular mountains of twigs. As the water level rises, the height of the mound must be increased, and finally it becomes an island in the water.

If a river or lake should dry up, the entrances to the beaver mound are above the water. Beavers build dams to keep the water at a constant level and, even more, to create artificial reservoirs where they feel secure and to which they can easily transport their food. The animals build their dams by placing tree trunks and branches, which they have cut down and prepared, in the ground, perpendicular to the riverbed, and by fastening and anchoring these uprights with stones, mud, reeds, and whatever else might be at hand. The beavers will often make a support out of a tree that has fallen across or floated down the river, so they can expand their construction. The more the water behind the dam rises, filling the creek or riverbed, the higher the beavers must make their dam, extending it on the sides to the higher riverbanks. The longest dam in the Voronezh region in the U.S.S.R. is near Marinika; it is 394 feet (120 meters) long, 3.3 feet (1 meter) high, and from 23.6 to 39 inches (60 to 100 centimeters) wide. The dams across the mountain streams at the foot of the Rocky Mountains in North America are up to 10 feet (3 meters) high. The largest beaver dam of all is that on the Jefferson River, near Three Forks, Montana; one can follow it for 2297 feet (700 meters). This dam has been maintained by successive beaver generations for decades, perhaps even centuries, as long as enough food for these animals can be found in the area. There is a dam in Colorado that has been there for 70 years; trees and bushes finally grew on top of it, and the dam itself keeps getting wider and stronger. It takes a family of beavers about a week to build a dam that is 33 feet (10 meters) long. When the river has a strong current, the beavers build several auxiliary dams facing the current so that the main dam will not be destroyed. If the water breaks through the dam at a single point, the beavers will rapidly repair the leak the next night. Our expression for feverish activity, "to work like a beaver," comes about as a result of the industriousness of these animals. The dams differ according to area and subspecies. The dams in the flat swamplands of Mississippi often reach gigantic dimensions, several hundred meters in length, and the area that has been inundated provides the beaver with a large habitat.

Humans have always praised the beaver because it can fell a tree so that the crown always lies toward the water and because these animals calculate the fall of the tree so exactly that no beaver is killed in the process. In addition, beavers seem to be able to judge the direction of the tree's fall so that the crown of the tree is not caught in nearby trees and so that the tree itself does not fall in an area where it would be of no use to the beaver.

Beavers like to gnaw through tree trunks between 3.1 to 8 inches (8 to 20 centimeters) in diameter; they like aspens, willows, and poplars, and are somewhat less enthusiastic about birch and wild cherry trees. They almost completely ignore the coniferous trees, and thereby avoid harming our most important timber and hardwood trees. Two beavers often work together on a thick tree; usually one animal cuts while the other looks around. Of course, beavers prefer trees that grow close to the water. They drag thin tree trunks as well as branches they have cut from larger trees to the water by the shortest possible route. They do not cut down trees that are more than 650 feet (200 meters) from the water. After they have cut down everything within the circumference of their colony, they like to move to a new area.

Beavers place several of their harvested branches in the water around their mounds, making sure that the gnawed end is stuck firmly into the mud of the riverbed so the branch does not float away. These branches are provisions, particularly for the winter months, because in the summer these animals can also eat water plants, berries, swampwood roots, and, where possible, cultivated fruit. However, their main diet consists of fresh green bark and softwood. Hard nutrition of this type is in keeping with the strength of these animals' teeth. A single beaver exerts a chewing force of 176 pounds (80 kilograms) with its incisors; humans exert a force of only 88 pounds (40 kilograms). In addition, the beaver weighs 40–44 pounds (18–20 kilograms), 66 pounds (30 kilograms) at the very highest.

Beavers do not hibernate, although they become much less active during the cold season. One may not see them for as long as a week. When it is 59°F (15°C) outside, the air in the beaver's house remains at about the freezing point. As soon as the river freezes, the beaver swims under the ice to its storage supply of branches. Often they do not even have to make air holes in the ice, because the ice in rivers and lakes begins to freeze first at the edges, near the banks. Then, in the course of the winter, the water level sinks to some extent, especially in rivers. Thus the ice in the middle of the river sinks while the ice near the banks of the river is held up by the ground and the roots of trees, leaving safe and secure air pockets for the beaver.

Beavers are sexually mature after 3 to 4 years. Copulation takes place in the water between January and March. The male swims under the female, with his stomach facing upward. According to most observations, it appears that beavers form permanent pair bonds, although when the young are born in the beaver mound (after a gestation of 105–107 days), the male and the young from the previous year, must move out of the structure for some time. Young beavers have hair on their bodies and are able to see at birth. Their eyes are half-closed at birth and covered with a thick fluid. The incisors are already visible. Young beavers nurse for almost 2 months; they grow noticeably during this time. The beaver mother often places her tail under her stomach and lifts one leg so that the young can sit on her warm tail for support while they nurse.

Young beavers are also able to swim and dive very early. If they remain in the water, the mother often transports them back to the house forcibly. Bolau, director of the old Hamburg zoo (which is no longer in existence), once wrote that when she travels on land, the beaver mother carries her young in her outstretched forelegs and walks on her two hind legs. This description was hard to believe, but some 50 years later, this unbelievable method of carrying the young was observed and photographed in the Zurich zoo. The mother can also carry her young about by having them sit on her tail.

Usually three generations of beavers live together in one mound. The parents always drive the oldest offspring out of the den, by biting them, in order to make room for the new litter. Beavers can live to be 10 to 15 years old.

The fur of the beaver has been among the most valuable on the market and has been the chief reason for a slaughter which threatened to destroy the species. With rigid protection they have become numerous in some parts of North America during the present century so the threatened extinction has apparently been averted.

Other rodents are covered under **Rodentia;** and **Squirrels and Other Sciuromorphs.**

BECKE TEST. A microscope of moderate or high magnification is used to compare the indices of refraction of two contiguous minerals (or of a mineral and a mounting medium or immersion liquid), in a thin section or other mount. When the two substances differ substantially in refractive index, they are separated by a bright line, called the *Becke line.* The line moves toward the least refractive of the two materials when the tube of the microscope is lowered.

BECKMANN METHOD. A method of measuring elevation of the boiling-point or depression of the freezing-point of a solution. It may be used to measure concentration if the nature of the solute is known, or the molecular weight of the solute if the volume concentration is known. See also **Analysis (Chemical); Freezing-Point Depression.**

BECKMANN MIXTURE. An oxidizing reagent consisting of 60 parts potassium dichromate, 80 parts concentrated sulfuric acid, and 270 parts water, used for cleaning laboratory glassware. See also **Cleaning Solution (Glassware).**

BECKMANN REARRANGEMENT. **Rearrangement (Organic Chemistry).**

BECQUEREL EFFECT. A photographic effect discovered by E. Becquerel (1895). Experimenting with the daguerreotype process, Becquerel found that a plate will produce a direct (positive) image if exposed first to diffuse daylight. See also **Photography and Imagery.**

BEDBUG (*Insecta, Hemiptera*). A wingless, blood-sucking bug which hides during daylight in the crevices of beds and other furniture and about the woodwork of houses and seeks its victims at night. It has been found also about chicken roosts.

Corrosive sublimate (mercuric chloride) dissolved in alcohol and applied to the hiding places of the insects is one method of control.

Severe infestations are usually handled by fumigation of the entire building.

This species, *Cimex lectularius*, gives the name bedbug to the family *Cimicidae* which also contains a few species that attack bats and birds.

Still another member of this order, the large bedbug, is sometimes found in beds. It is almost an inch long and is capable of inflicting a painful wound. This species is one of the assassin bugs (family *Reduviidae*).

BEDDING. A term used by geologists to designate the natural layering or stratification usually characteristic of sediments and sedimentary rocks. Bedding is the result of the unequal rates of settling of particles of different sizes and specific gravities. In the case of very fine-grained sediments, or shales, the bedding may be shown by color bands. The thickness of a bed or stratum may vary from several feet to a fraction of an inch. Extremely thin beds are called laminae.

BEDROCK. The solid rock of the lithosphere which may be directly exposed or covered by loose unconsolidated materials such as sand, clay and soil.

BEECH TREES. Members of the family *Fagaceae* (beech family), these trees are of two principal genera: *Fagus*, large, hardy, deciduous trees; and *Nothofagus*, deciduous or evergreen shrubs and trees occurring in the southern hemisphere and sometimes collectively referred to as southern beeches. Important species of the beech family include:

American beech	*Fagus grandifolia*
Antarctic beech	*Nothofagus antarctica*
Black beech	*N. solandri*
Common or European beech	*F. sylvatica*
Copper beech	*F. s. cuprea*
Dawyck or upright beech	*F. s. "Dawyck"*
Fern-leafed beech	*F. s. heterophylla*
Mountain beech	*N. cliffortioides*
New Zealand black beech	*N. solandri*
Purple beech	*F. s. purpurea*
Red beech	*N. fusca*
Weeping beech	*F. s. purpurea pendula*

The American beech is found in the eastern part of the United States, from Maine to the Gulf Coast and throughout the midwestern states. The tree is tall, stately, and large. The height averages 60–75 feet (18–23 meters), with a trunk from 2 to 3 feet (0.6 to 0.9 meter) in diameter. The leaf is 3 to 5 inches (7.6 to 12.7 centimeters) long, dark green, sharply toothed, ovate, and deeply veined. The bud is encased in thick bronze-colored scales. The bud is slender, pointed, and about $\frac{3}{4}$ inch (2 centimeters) in length. The nut or fruit is dark brown, smooth and with a rough cap that covers nearly half of it. The kernel is bitter, but squirrels seem to like it. The flower is long, 2 to $2\frac{1}{2}$ inches (5 to 6 centimeters), light green, has a wormlike shape, and hanges in clusters from the branch of the tree. Beech wood is generally heavy, hard, tough, and of closed-grained structure. The light color is similar to that of maple. The wood has been used as veneer, for tool handles, shoe lasts, cooperage, and pulpwood. The wood is particularly suited to the production of small items, many of which now have been replaced by plastics. The moisture content of green beech wood is 54%, with a weight of about 54 pounds per cubic foot (865 kilograms per cubic meter). After air-drying to 12% moisture content, the weight is 45 pounds per cubic foot (720 kilograms per cubic meter) and 1000 board feet (2.36 cubic meter) of nominal sizes weigh 3750 pounds (1701 kilograms). The crushing strength with compression applied parallel to the grain for green wood is 2200 psi (15.2 MPa); for dried wood, 4730 psi (32.6 MPa). The tensile strength with tension applied perpendicular to the grain for green wood is 720 psi (5 MPa); for dried wood, 1010 psi (7 MPa). The record American beech, as selected by The American Forestry Association in 1976, is located in Berrien County, Michigan. Circumference at $4\frac{1}{2}$ feet (1.4 meters) above ground level is 167 inches (424 centimeters); height is 161 feet (48.3 meters); spread is 105 feet (32 meters).

The European beech (*F. sylvatica*) is not as heavy as the American species, but it is used in much the same way. The wood is a red-brown color. The tree is popular as a hedge. Although not typical, the possibilities of the tree as an effective hedge is exemplified by the huge hedge at Meikelour (Scottish Highlands), which is 95 feet (28.5 meters) high and 580 yards (530 meters) long. The European beech has done particularly well in Normandy, where specimens over a few hundred years old are found.

It is interesting to note that the beech is notable among the broad-leaved trees for its establishment in both the northern and southern hemispheres. In New Zealand, the *Nothofagus solandri* is also known as red beech, black beech, or tawhai beech. The tree is slender and has rising branches. It may attain a height of 80 feet (24 meters). The wood is strong, durable, and heavy, weighing up to 44 pounds per cubic foot (704 kilograms per cubic meter). The antarctic beech, also known as the rauli tree, is found in Chile and replaces some of the functions of oak wood in South America. It can attain a height of 100 feet (30 meters) or more.

BEE-EATER. **Kingfishers and Other Coraciiformes.**

BEEF CATTLE. **Bovines.**

BEE FLY (*Insecta, Diptera*). Flies, small to medium-sized, often beelike, whose habit of visiting flowers for pollen and nectar is like that of the bees. The larvae are parasitic, some on bees. They make up the family *Bombyliidae*.

BEE LOUSE (*Insecta, Diptera*). A minute, wingless, parasitic fly found attached to the queen and drones in honeybee colonies. The few known species constitute the family *Braulidae*, containing the genus *Braula*.

BEE-MOTH (*Insecta, Lepidoptera*). A moth whose larva eats the wax and debris in old honeycombs, spinning a silken tunnel as it goes. These insects are found chiefly in weak colonies of bees and in stored combs. They may attack beeswax products, such as comb foundation in the supplies of the apiarist, but they cannot develop on a diet of pure wax; the organic waste in old brood combs provides the necessary nitrogenous material and furnishes a favorable breeding place. The best-known form is *Galleria mellonella*.

In well-kept apiaries the bee-moth is rarely a serious pest. The maintenance of strong colonies of bees prevents its entrance and the protection of stored supplies against the entry of the adult moths safeguards them against damage. Fumigation of supplies is sometimes necessary. It may be effectively carried out with carbon disulfide but since this compound is highly explosive it must be used with due precautions.

BEER'S LAW. If the Bouguer law is applied to a solution of fixed thickness, b cm. and concentration c, the result is $\log I_0/I = abc$ which is the fundamental equation of quantitative absorptimetry. Here I/I_0 is the transmittance and the constant a, the absorptivity depends on the nature of the absorbing material, the wavelength of the incident radiation, the nature of the solvent, the temperature, and perhaps other controllable experimental conditions. When c is given in moles per liter, a is called the molar absorptivity. If other concentration units are more convenient, the numerical value of a must be changed accordingly. The product $A = abc$ is the absorbance of the sample and it equals $1/T$, the reciprocal of the transmittance. Experimental verification of Beer's law will succeed only if appropriate corrections are made or can be neglected as with the Bouguer law.

BEET (*Beta vulgaris; Chenopodiaceae*). The many varieties of beets now in cultivation are perhaps all derived from the native *Beta maritima* of southern Europe.

The most important variety is the sugar beet, which in recent years has become an important rival of the sugar cane. As a source of sugar, beets were first utilized in Germany and in France about 1800. In the United States they became important commerically only after World War I. Sugar beets are now raised commercially in several areas, California, Utah, Idaho, Oregon, and Washington; the Eastern Slope of the Rocky Mountains; Iowa and Minnesota; and in the Michigan-Ohio area.

The sugar beet is a biennial plant which during its first year of growth forms a large tapering tap root and a rosette of leaves. At the end of this first year the plant is gathered for sugar production. If allowed to grow the second year, the plant forms a branching stem and an abundance of inconspicuous flowers, utilizing the sugar stored in the root to produce them. See accompanying illustration.

The plant has an elongated tap root which tapers into a long slender root. This may penetrate 4–6 feet (1.2–1.8 meters) into the ground. In size and shape the root is extremely variable. Cut transversely, the root is seen to be composed of from six to ten or even more concentric zones. Each zone comprises a ring of conducting cells outside which is a ring of small cells in turn surrounded by a ring of large cells. The small cells are rich in sugar, while the large cells are primarily water-storage cells. The formation of these zones is a consequence of the formation of a succession of cambium rings, each of which persists for a few weeks.

In preparing the beets for sugar manufacture, the roots are first lifted from the ground, the leaves cut off, and the roots hauled to the factory. Machinery has largely replaced the hand labor formerly required for these tasks. At the factory the beets are washed thoroughly and cut into thin slices. These slices are put into hot water, which extracts the sugar. The sugar solution is next treated with lime, and then precipitated with carbon dioxide: this removes many impurities which are filtered off. The purified liquor is bleached with sulfur dioxide, and then concentrated by boiling and crystallized under a partial vacuum. From this crude product the molasses is removed by centrifuging, leaving the sugar which is dried and granulated, after which it is ready for the market. In several factories, ion exchange resins are used to remove impurities more thoroughly. This results in a higher grade of sugar.

Many of the waste products of beet sugar production are utilized. The tops are used as a stock food either in the raw condition or after preserving as ensilage. The beet pulp left after extraction of the sugar is also used as a stock food, as is the molasses from the sugar. Often the pulp and molasses are mixed before feeding. Any refuse from the factory may be used as a fertilizer.

A sugar beet of desirable shape. (*USDA photo.*)

In addition to sugar beets, several other varieties of *Beta vulgaris* are known. One of them is the common red table beet, which is eaten either boiled or pickled. When correctly grown it has a minimum of fibrous elements as well as a high sugar content. Most varieties of table beet are deep red in color, in contrast to the white-fleshed sugar beet. Another variety of beet is the Mangel-wurzel or Mangel, of which there are several varieties. They are of large size, have a sugar content varying from 4–8% and are developed principally as a stock food.

BEETLE (*Insecta, Coleoptera*). An order of insects of major economic importance in food production. Some species of beetle are quite specialized and confine their destruction to one or a narrow range of crops as, for example, the Colorado potato beetle, the cucumber beetle, the Mexican bean beetle, the rice leaf beetle, the seed corn (maize) beetle, the stored-grain beetle, and the sweet potato beetle. Other species of beetle destroy a wide variety of plant and crops. These species include the flea beetle, the ground beetle, the grub bettle, and the Japanese beetle. However, some species of ground beetle are considered economically beneficial. See accompanying illustration.

Coleoptera is the largest group in the animal kingdom, with almost 200,000 described species. It embraces almost the entire range of adaptation of the class, although very few beetles are parasitic. See also **Coleoptera.**

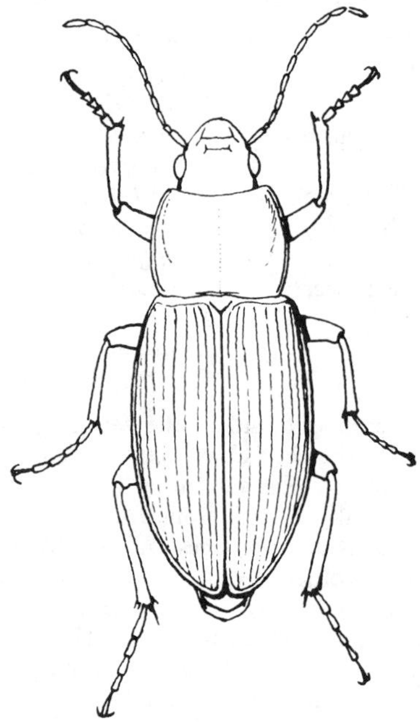

Typical ground beetle.

Beetles are usually recognized by the thickened wing covers that meet in a straight line down the middle of the back. These wing covers, or elytra, are modified forewings. In most species of beetle, they are thickened or horny, but in some they are soft. In some species, they are divergent and in some they are short, leaving much of the abdomen exposed. The typical condition of the elytra is found outside of this order only in the earwigs. Beetles have biting mouth parts and a complete metamorphosis in which the larval stage is often a grub. An excellent reference on the systematics of beetles is "Monographie der Familie Platypodidae Coleoptera," by Karl E. Schedl, published by Junk, The Hague, 1972.

Typical of the high degree of specialization found in some beetles is the furniture beetle which lives only in old wood.

Harmless and Beneficial Beetles. The familiar "lady bug" consumes only injurious insects. See also **Lady Bug.** The firefly or lightning bug is a member of *Coleoptera* and is a soft-bodied beetle with a luminous organ in the abdomen. The energy system of the lighting or flashing is of very high efficiency and involves enzyme reactions with oxygen. No economic value, beneficial or destructive, is attributed to the firefly.

BEETLE (Grain Pest). Grain-Storage Insects.

BEFORE PRESENT. B.P.

BEGGIATOA. A genus of filamentous sulfur bacteria which is capable of converting hydrogen sulfide to sulfuric acid. Sulfur granules are stored in the cells, and may be oxidized to supply the cell with the necessary energy for life. The bacteria are autotropic in form, using chemosynthesis to obtain energy.

BEL. Units and Standards.

BELLADONNA (*Atropa belladonna*; *Solanaceae*). The plant grows as a native in Europe and in parts of Asia. It is commonly known as Deadly Nightshade. It is about 3 feet (0.9 meter) tall, and has dull green leaves and purple flowers, which are followed by cherrylike red fruits. Every part of the plant contains the poisonous substance for which it is known. This drug atropine is a very poisonous alkaloid obtained principally from the roots and leaves. Belladonna is used medicinally most often as the tincture.

See also **Atropine.**

BELLATRIX (γ Orionis). Ranking twenty-fifth in apparent brightness among the stars, Bellatrix has a true brightness value of 2,300 as compared with unity for the sun. Bellatrix is a blue-white, spectral type B star and is located in the constellation Orion south of the ecliptic. Estimated distance from the earth is 360 light years. See also **Constellations.**

BELL CRANK. A means frequently used to transfer reciprocating motion at right angles is that of a rigid-angled arm pivoted to a fixed point at its vertex, and having hinge connections at its extremities. The bell crank is a type of lever, and the motions of its end are not those of reciprocation, but rather of rotation, but if comparatively long rods are hinged to it, the other ends of those rods will have similar motion. The amplitude of the reciprocation transmitted by the bell crank is directly proportional to the radii from the pivot point to the joints.

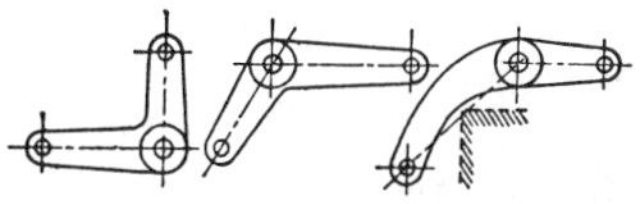

Types of bell cranks.

BELLOWS. Usually constructed of metal, this mechanical device finds wide application in mechanical, thermal, instrumentation, and control systems. A metal bellows may be defined as a flexible, expansible, and collapsible metal vessel consisting of a series of annular plates having their inner and outer circumferences joined together. The manner in which the plates are joined together depends upon the type of bellows and the method of manufacture. A seamless metal bellows is shown in the accompanying illustration. Bellows are used in thermal systems where, because of expansion and contraction resulting from temperature changes, flexibility of a leakproof nature must be imparted to piping and other connections. Bellows also are used in situations where flexibility is required because of vibration conditions. A major application for bellows is in the instrumentation field.

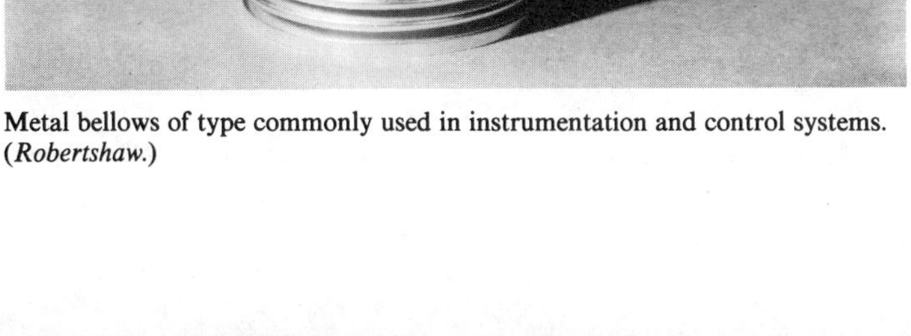

Metal bellows of type commonly used in instrumentation and control systems. (*Robertshaw.*)

Among the more important basic properties of metal bellows are (1) spring rate or flexibility, (2) pressure resistance, (3) mean effective area, and (4) life.

Spring rate. Usually, it is desirable to have the lowest possible spring rate commensurate with other properties important to the application. The spring rate is primarily regulated by the material thickness and is further affected by the material's modulus of elasticity, the depth of corrugations, and the number of free plates in the total length of the bellows.

Pressure resistance. Three generalized statements apply: The maximum pressure resistance is (1) a direct function of the square of the material thickness, (2) an inverse function of the depth of corrugation, and (3) is independent of the number of free plates except that when the total length of a bellows exceeds its diameter, internal pressure may introduce sidewise distortion or buckling.

Mean effective area. This affects the ability of the bellows to perform work from changes in pressure. Just as a pressure increase applied to the area of a piston inside a cylinder will produce motion, the pressure applied to the effective area of a bellows will produce thrust. Because of the curved sidewalls of a bellows, the effective area is not quite the exact difference between the area calculated for the extreme outside diameter and that for the extreme inside diameter. A simple expression for calculating the mean effective area of a bellows is

$$\text{Mean effective area} = 0.1963(D + d)^2$$

where D = outside diameter; d = inside diameter. Area, of course, is independent of bellows length.

Life expectancy. There are no easy rules for use in forecasting the useful life of a bellows. Factors that affect life include (1) the safe stress level, (2) the type of metal used, (3) compression or extension from the normal free length, (4) environmental conditions, including corrosive atmospheres, (5) cyclic versus fairly constant pressure changes of the application. Metals commonly used for bellows construction include brass, phosphor bronze, *Monel*, and various stainless steels. The recommendation of a multiply bellows may be in order to obtain maximum pressure resistance at the least expense of spring rate.

Bellows can be used as small pumping devices for handling gases and liquids. For example, such pumps may be used in systems for processing radioactive, hazardous, and rare gases. They also have been used as pumps in commerical aircraft for drinkwater pressurization systems. Some metal hose for nuclear applications is, in essence, a high-quality formed bellows restrained axially by steel braid and made from stainless or Inconel materials. Extremely tiny bellows have been used as sensors of intercranial fluid pressures in hydrocephalic infants. By implanting the precision miniature bellows in the skull of such infants, physicians can monitor the pressure of the fluids and possibly prevent brain damage. The fuel tank connection in the space shuttle required a three-joint flexible member providing motion compliance between the tank, structure, and related plumbing. A formed bellows is the key component. The spherical joint includes built-in stops that are internally sealed and loaded by formed bellows. Each joint must withstand pressures from 23.8 atmospheres (operating) to 95.2 atmospheres (bursting) pressures with a useful operating life of 10 years.

BELLOWS GAS METER. Flow Measurement.

BELL'S PALSY. This is a disease of the peripheral nervous system and is one of several manifestations of cranial mononeuropathy. The facial muscles innervated by the seventh cranial nerve are weakened and contribute to this condition. Onset is usually sudden and may be noticed first after a night's sleep. Sometimes patients complain of rather vague sensations or discomfort in the region of cheeks and ears. In some cases chewing and other facial motions may be difficult. Prednisone is used in the initial treatment of the disorder. With several days of treatment, the outlook for recovery are excellent. Where the facial nerve does not respond to this drug, electrical stimulation may be attempted. The disorder affects all ages. About 20 persons in 100,000 population are affected by this disorder.

BELT FEEDER (Volumetric). Feeder (Volumetric).

BELT FILTER. Filtration.

BELT (Timing). Timing Belt.

BÉNARD CONVECTION CELLS. When a layer of liquid is heated from below, the onset of convection is marked by the appearance of a regular array of hexagonal cells, the liquid rising in the center and falling near the wall of each cell. The criterion for the appearance of the cells is that the Rayleigh number should exceed 1700 (for rigid boundaries).

BENCH MARK. A bench mark is a point of known elevation and location. It is used in surveying as a vertical reference point when finding the elevation of other points of a less permanent nature. The point may be the head of a spike or bolt driven into a tree in such a manner that the top of the head is as nearly horizontal as possible, the highest point on the top of a hydrant, or the top of a flat, noncorrosive plate set securely in stone or concrete. Bench marks may be temporary or permenant depending upon their use and should be located so that they are easily accessible for instrument work.

BENDING MOMENT. The bending moment at any section of a member is equal to the algebraic sum of all the moments on one side of the section about a centroidal axis of the section. This definition assumes that all of the external forces are coplanar, that is, act in one plane. An internal resisting moment at any section is equal to the sum of the moments of the internal stresses about the gravity axis of the section. The external bending moment acting on any section is numerically equal to the internal resisting moment but acts in the opposite direction. External moments are positive or negative depending upon the direction in which they tend to rotate the section of the beam under consideration. This sign convention is entirely arbitrary although it is customary in beam analysis to assume that positive moments are those tending to shorten the top surface of the beam while negative moments are those which lengthen the top surface. The point on the longitudinal axis of a beam at which the bending moment changes sign is called the point of contraflexure or the point of inflection. Pure bending is a term used to denote the condition where the shear is zero over a finite length. It follows that the bending moment is constant over this length. Bending moments have a very important part in beam action since they cause the flexural stresses (see **Flexure**) and are the principal cause of deflections.

A graphical representation of the variation of bending moment on a beam is called a bending moment diagram. The accompanying illustration of a bending moment diagram is for an overhanging beam with a uniformly distributed load covering the entire length of the beam. The maximum bending moments are indicated on the diagram occur where the shear changes sign. See **Shear** for the shear diagram for this beam. Since the product of a force and distance is a moment, bending moments are expressed as foot-pounds, inch-pounds, etc.

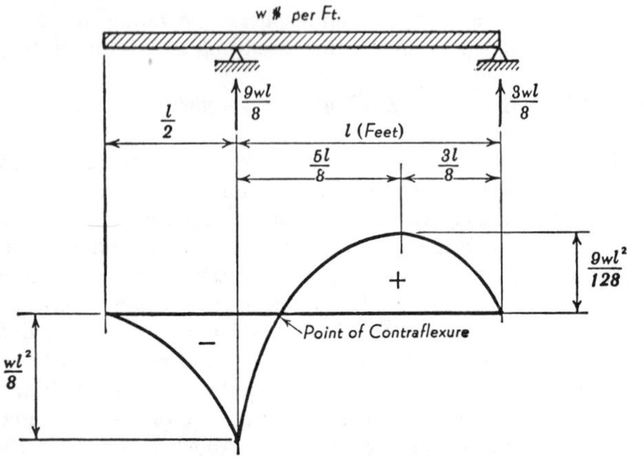

Bending moment diagram.

BENDING MOMENT (Influence Line). Influence Line.

BENDING MOMENT (Section Modulus). Section Modulus.

BENDING MOMENT (Shear). Shear.

BENDING (Unsymmetrical). Unsymmetrical Bending.

BENEDICT SOLUTION. In its original form, this was an alkaline solution of copper hydroxide and sodium citrate in sodium carbonate used either as a mild oxidizing agent or as a test for easily oxidizable groups such as aldehyde groups. The formation of cuprous oxide is a positive test, its color red but often yellow at first. Many other forms of this solution have been developed. Glucose reacts with Benedict solution to form cuprous oxide.

BENEFICIAL INSECTS. Numerous species of insects contribute in a positive way to increased quantity and quality of food production. An attempt to assign monetary values to these contributions is essentially impractical because there are too many unknown variables beyond quantification. Setting aside aesthetic, medicinal, scientific research, and scavaging roles played by insects, which are valuable but not directly related to food production, the following principal contributions of the beneficial insects include:

1. Destruction of damaging insects and other pests.
2. Pollinization of numerous food plants.
3. Improvement of soil conditions.
4. Destruction of undesired weeds.
5. Production of food products (such as honey).
6. Source of food for fishes and some farm animals and poultry.

Destruction of Damaging Insects. This is possibly the most visible role of the beneficial insects. A few examples include:

Ant-lion (Order *Neuroptera*; family *Myrmeleonidae*). See Fig. 1. Also known as the doodle bug, this is the brown-colored larva of a neuropterous insect which lives at the bottom of a conical-shaped pit in

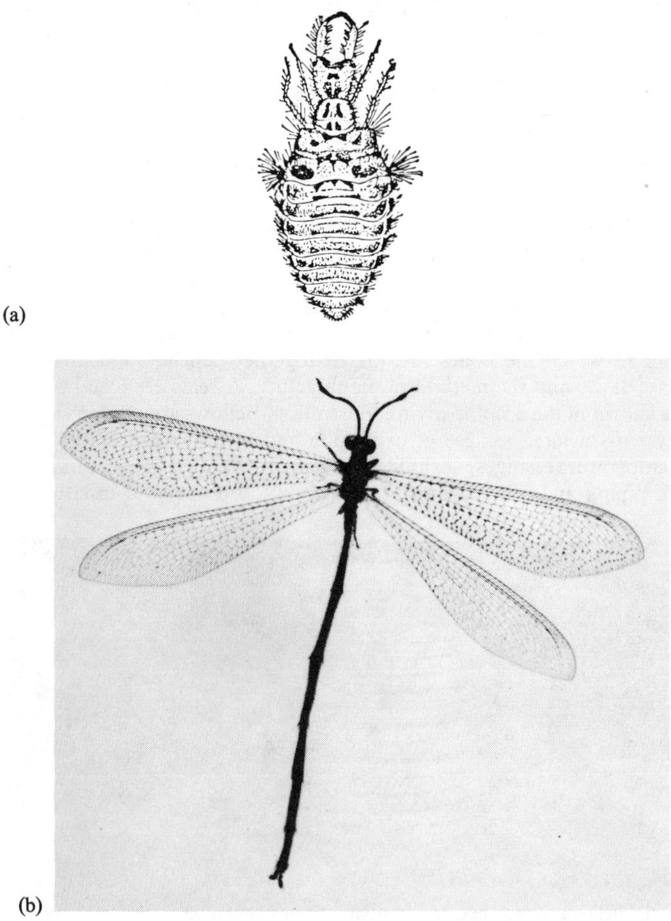

Fig. 1. (a) Ant-lion or "doodle bug;" (b) imago or adult. (*USDA photo.*)

sand or loose granular soil. The insect is equipped with powerful, spiny jaws. Once the pit is constructed, the insect, about $\frac{1}{2}$-inch (12 millimeters) in length, lies partially buried at the bottom of the pit, awaiting prey to come to investigate the pit. With extremely rapid movements, the ant-lion showers the prey with sand, causing it to fall to the bottom of the pit, at which instant the ant-lion seizes the prey with its powerful jaws. The ant-lion feeds on ants and other insects, many of which are damaging in some way to food crops. The imago adult of the ant-lion larva is a small flying insect with gauzy wings that emerges in the spring, but is seldom seen. During part of the winter, the ant-lion pupates in a silken cocoon. The ant-lion is found in many parts of the United States, but is most abundant in the southern states.

Aphid-lion (Order *Neuroptera*; family *Chrysopidae*). As with the ant-lion, it is the larva of this insect that functions in an economically beneficial way. The larva is a yellowish, or mottled red or brown creature with a long, narrow body that tapers at both ends. It is equipped with large, sickle-shaped jaws. Hairs on the body, which is about $\frac{3}{8}$-inch (9 millimeters) long, are prominent. The adult female lacewinged fly deposits her eggs singly on stems and stalks or leaves of numerous plants, including many food plants which are adversely affected by aphids. See Fig. 2. Upon hatching, the larva or aphid-lion immediately searches out aphids upon which to feed. The aphid-lion also consumes mealybugs, scales, thrips, and mites—and the eggs of numerous insects. The lacewing adult has gauzy green wings, yellow eyes, fragile antennae, and is quite fragile. Distribution of this beneficial insect is throughout the United States.

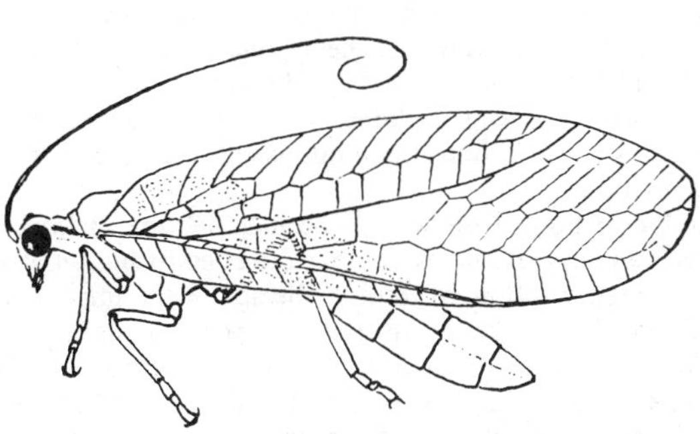

Fig. 2. Lacewing fly, the adult of the aphid-lion. (*USDA photo.*)

Assassin bug (Order *Hemiptera*; family *Reduviidae*). Some species feed on the immature forms of insects. The adult assassin bug is a light-brown insect, from $\frac{1}{2}$- to $\frac{3}{4}$-inch (12 to 19 millimeters) long. See Fig. 3. It walks over plants slowly and clumsily, holding its forelegs in prayerlike position and using them to capture and hold its prey. This insect is distributed throughout the United States. It should be added that certain species of the assassin bug should be viewed in a negative context. Some of these insects can inflict painful bites to human beings and should never be handled. One species is a carrier of Chagas' disease, which occurs in Central and South America.

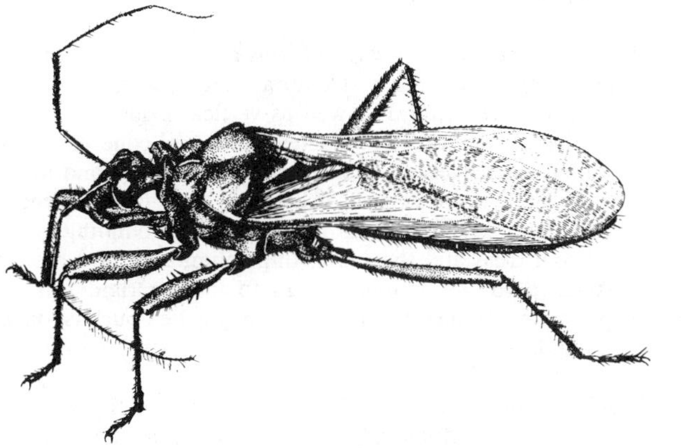

Fig. 3. Adult assassin bug. (*USDA photo.*)

Damsel bug (Order *Hemiptera*; family *Nabidae*). This insect strongly resembles the assassin bug. It is pale gray, about $\frac{3}{8}$-inch (9 millimeters) long, and uses its forelegs to capture prey. The insect feeds on aphids, fleahoppers, and small larvae of numerous insects. The front legs of the damsel bug appear to be especially modified for grasping prey. Distribution is throughout the United States.

Ground beetles (Order *Coleoptera*; family *Carabidae*). This is a very large family of long-legged beetles, most of which are predaceous and thus beneficial when in both the adult and larva stages. The adults are usually dull black or brown. The bodies are long and oval, and the heads are narrow. Ground beetles usually are seen on the ground, located under stones of loose trash. The insects hide by day and are active at night, running rapidly when disturbed. The larvae are slender, flattened bodies that taper slightly at the tail. There are two spines or bristles at the rear end. Ground beetles feed on a wide variety of caterpillars and numerous other insects. Because there are so many species, there is a wide variation in size, the length ranging from $\frac{1}{16}$-inch (1.5 millimeters) to about $\frac{2}{3}$-inch (17 millimeters). Not all beneficial species are fully colored, but may display metallic-appearing blue, green, or purple coloration. Usually the upper surface is uniform, with no markings, such as spots or stripes. Ground beetles are distributed throughout the United States.

Lady beetle (Order *Coleoptera*; family *Coccinellidae*). Also called ladybirds and ladybugs, these insects feed on aphids, spider mites, scales, mealybugs, and several other insects. The adult lady beetle is a shiny red or tan. The back is convex and the legs are short. Length ranges from $\frac{1}{16}$-inch to $\frac{1}{4}$-inch (1.5 to 6 millimeters). The body is oval-shaped and may have a black spot on the back. Lady beetles have been imported into regions to achieve specific insect control objectives. An example is the employment of the lady beetle (*Cryptolaemus montrouzieri*) to control the citrus mealybug (*Pseudococcus citri*). Distribution of the lady beetle is throughout the United States. It should be noted that not all members of the family *Coccinellidae* are beneficial. The family also included the very damaging Mexican bean beetle (*Epilachna varivestis*, Mulsant).

Praying mantis (Order *Orthoptera*; family *Mantidae*). Also called mantid, this insect is quite beneficial and is highly interesting. See Fig. 4. The praying mantis is quite large, ranging from 2.5 to 5 inches (6+ to 12+ centimeters) in length. The insect derives its name from the manner in which it holds and feeds upon its prey. The forelegs are long and powerful and the posture assumed by the insect when eating appears as though it may be praying. This insect feeds on aphids and other small insects when young. The adults can feed on

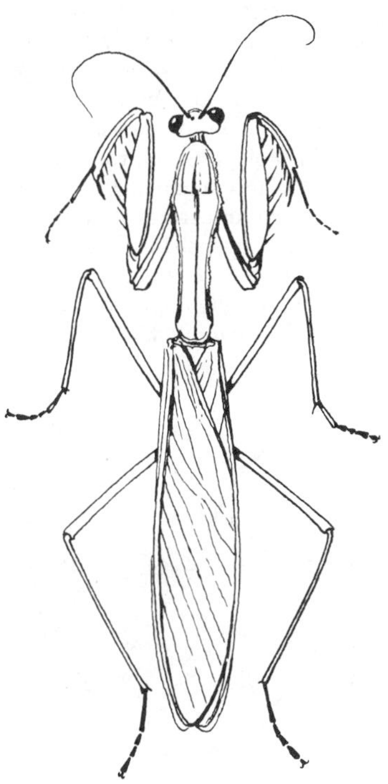

Fig. 4. Praying mantis (also called mantid). (*USDA photo.*)

rather large insects when such are available. Studies indicate that the insect is strictly carnivorous. The young mantids hatch in the spring from eggs that have been laid in masses on shrubs or tall grass and covered with a froth that hardens. The young resemble the adults, but are wingless. This beneficial insect is found throughout the United States, but is most abundant in some of the northeastern states.

Syrphid flies (Order *Diptera*; family *Syrphidae*). Also sometimes called *flower flies* or *sweet flies*, these insects are predaceous in the larval stage. Some observers report that a syrphid fly larva is satiated with juices sucked out of the aphid body. The larvae resemble slugs and are of a brown, gray, or mottled coloration. The adult fly is a bright yellow with black markings and are from $\frac{1}{4}$- to $\frac{3}{8}$-inch (6 to 9 millimeters) in length. In addition to the high rate of aphid destruction per larva, the insects are present in large numbers, at least several hundred per acre under normal circumstances. The syrphid fly larva, of course, consumes other small insects along with aphids.

Wasps and insect parasites. There are several insects, known as *entomophagous parasites*, that function in a similar manner to control insect populations. These insect parasites lay their eggs within the bodies of other insects. As the larvae from these eggs develop within the host, the host is either killed in the process, or greatly damaged, as in some cases, losing its ability to reproduce its own species.

In this group of beneficial insects are the tachinid flies (order *Diptera*; family *Tachinidae*). They appear something like a slightly overgrown housefly, but a detailed inspection reveals numerous differences, including bristles on their antennae. The adult tachinid fly lays her eggs in foliage that is frequented by other insects who may find the eggs attractive as food, or eggs may be fastened to the host insect with a sticky substance. Sometimes larvae that are hatched may be attached to the host. Once it is a part of the host, the larva consumes tissue of the host, instinctively consuming tissue that is not immediately vital to maintenance of life of the host. The host moth or butterfly will prepare a cocoon or chrysalid before it expires. The adult tachinid fly will later emerge from the cocoon, having prevented the host insect from completing its life cycle. Tachinid flies are particularly effective against armyworms. The tachinid species, *Winthermia quadripustulata*, lays eggs on the back of the armyworm. The resulting maggots penetrate through the skin of the worm and kill it within a few days. It is estimated that perhaps half of the armyworm population is destroyed in this manner. The tachinid species, *Lydella stabulans grisescens*, operates just as effectively, but in a more complex cycle, in destroying 50% or more of corn borer larvae.

Other parasitic insects include *Scelionidae* (egg parasites), very small insects that deposit their eggs inside the eggs of other insects. See Fig. 5. An example is *Eumicrosoma benefica*, which is a parasite of chinch bug eggs. The chalcid wasps, of many species, are effective against a wide range of moths and butterflies. These are described in entry on **Chalcid Wasp**. Tens of thousands of these wasps have been reared and used as biological control tools against the codling moth, oriental fruit moth, and sugarcane borer. The *Encyrtidae* are used in the biological control of certain scale insects. As an example, *Aphycus helvolus* is effective against black scale. The *Aphelinidae* are effective against many insects of *Homoptera*. As an example, *Aphelinis mali* is effective against the woolly apple aphid. And similar examples go on and on—involving the *Pteromalidae*, the *Eulophidae*, the *Ichneumonidae*, and the *Brachonidae*. The latter, for example, is effective against the tomato hornworm caterpillar.

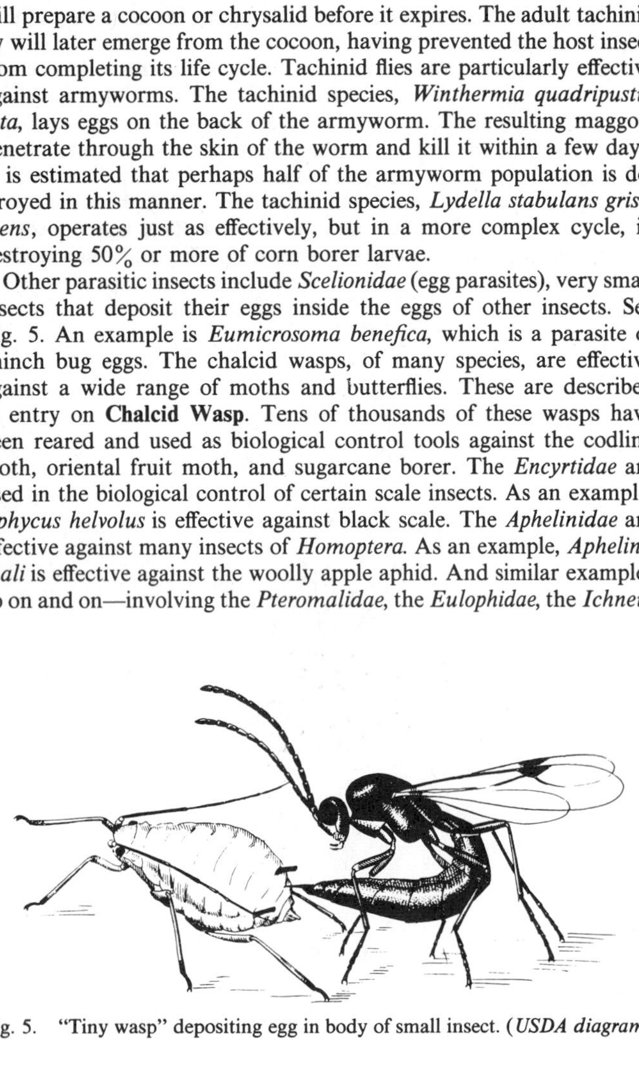

Fig. 5. "Tiny wasp" depositing egg in body of small insect. (*USDA diagram.*)

A number of wasps also function in a predaceous manner. The larger wasps, in particular, sting caterpillars, paralyzing the prey and thus making it available as feed for the young.

Pollinization of Food Plants. A second major role of insects in assisting with food production is their function as pollinators. The role of the honeybee is well understood. Plants which are fully or partially dependent upon insects for pollination are thus production of quality fruits and seeds include: alfalfa, apple, bean, blackberry, cherry, clover, cowpea, cranberry, cucumber, eggplant, fig, lemon, melon, orange, peach, pear, pepper, plum, prune, pumpkin, raspberry, soybean, squash, strawberry, sweet clover, and tomato. The dependence of fig on the fig wasp is described in entry on **Chalcid Wasp.**

Improvement of Soil Conditions. This beneficial role of insects is more difficult to assess because so many soil insects are damaging and most of the attention has been given to them. But, it is evident, of course, that in their burrowing activities, many insects reach considerable depths (up to several feet or a few meters in some cases) and thus make the soil porous and better equipped for making air, water, and nutrients available to growing plants. Some insects tend to bury organic materials in the soil. Tremendous quantities of insects perish on or in the soil, adding organic matter of their bodies to the total organic content of the soil. While living, they contribute their excreta to the soil. To comprehend these benefits, one must consider the dimensions of the phenomenon. Obviously, the contribution of a single or just a few insects is trivial to the extreme. But to this must be added the dimensions of the billions upon billions of insects that are engaged in such activities.

Production of Food Products. The manufacture of honey by the honeybees is an outstanding example of direct contribution of an insect to food for consumption by human beings. Very important is the source of food provided by insects to support populations of edible fishes. The percentage of insects in total food consumed by some fishes is often high. For example, upon examination, it has been shown that the food content of some fresh water species may run as high as 60% or more of insects consumed by the fish. The larvae maggots, and nymphs of certain insects are particularly attractive to fishes.

People of many cultures also consume insects directly as a food substance, frequently as a delicacy. An excellent, abridged summary of such foods is given in "Destructive and Useful Insects," by R. L. Metcalf (McGraw-Hill, 1962).

BENEFICIATION. **Coal; Iron.**

BENHAM TOP. A Benham top is a disc bearing characteristic black and white portions, which under certain conditions of angular velocity and illumination induces chromatic responses. See **Fechner Colors.**

BENIGN. A term signifying *nonmalignant*; favorable for recovery. Most frequently used to describe noncancerous tumors.

BENNETT (Comet). **Comet.**

BENSON-TYPE BOILER. **Boiler.**

BENT. A transverse frame which forms an integral part of a structural unit or supports another structural unit is called a bent. Bents are designed to carry lateral as well as vertical loads, and are made of structural steel, reinforced concrete, prestressed concrete, or wood. They are used extensively in connection with viaducts and industrial buildings. Viaduct bents consist of columns held firmly together by bracing in horizontal and vertical planes. Bents in buildings are composed of a roof truss and the supporting columns or of a roof girder rigidly attached to the columns so as to form a rigid frame. The bents support longitudinal beams, commonly called purlins, on which the roof is laid.

BENTHOS. The plants and animals that live on the sea bottom. These include the permanently attached or immobile forms (e.g., sponges, corals, oysters), creeping forms (e.g., crabs, snails), and the burrowing animals (e.g., worms). Barnacles, the larger seaweeds, and sea squirts are also members of this group. One of the three groups into which marine life is generally classified. See also **Ocean.**

BENTONITE. The term applied to altered fine-grained volcanic ashes which have been blown considerable distance from their origin and deposited in marine waters. The resulting material is usually a white, but sometimes a colored, clay-like sediment which may contain bits of volcanic glass but is composed mainly of colloidal silica which will absorb large quantities of water. Since bentonites are wind-blown deposits they are useful as definite datum planes in stratigraphy, especially in helping to determine the contemporaneity of the different facies of marine sediments.

BENT-TUBE BOILER. Boiler; Feedwater (Boiler).

BENZALDEHYDE. C_6H_5CHO, formula weight 106.12, colorless liquid, mp −26°C, bp 179°C, sp gr 1.046. Sometimes referred to as artificial almond oil or "oil of bitter almonds," benzaldehyde has a characteristic nutlike odor. The compound is slightly soluble in H_2O, but is miscible in all proportions with alcohol or ether. On standing in air, benzaldehyde oxidizes readily to benzoic acid. Commercially, benzaldehyde may be produced by (1) heating benzal chloride $C_6H_5CHCl_2$ with calcium hydroxide, (2) heating calcium benzoate and calcium formate:

$$(C_6H_5COO)_2Ca + (HCOO)_2Ca \rightarrow 2C_6H_5CHO + 2CaCO_3,$$

or (3) boiling glucoside amygdalin of bitter almonds with a dilute acid.

Benzaldehyde reacts with many chemicals in a marked manner, (1) with ammonio-silver nitrate ("Tollen's solution") to form metallic silver, either as a black precipitate or as an adherent mirror film on glass, but does not reduce alkaline cupric solution ("Fehling's solution"), (2) with rosaniline (fuchsine, magenta), which has been decolorized by sulfurous acid ("Schiff's solution"), the pink color of rosaniline is restored, (3) with NaOH solution, yields benzyl alcohol and sodium benzoate, (4) with NH_4OH, yields tribenzaldeamine (hydrobenzamide, $(C_6H_5CH)_3N_2$), white solid, mp 101°C, (5) with aniline, yields benzylideneaniline ("Schiff's base" $C_6H_5CH{:}NC_6H_5$), (6) with sodium cyanide in alcohol, yields benzoin $C_6H_5 \cdot CHOHCOC_6H_5$, white solid, mp 133°C, (7) with hydroxylamine hydrochloride, yields benzaldoximes $C_6H_5CH{:}NOH$, white solids, antioxime, mp 35°C, syn-oxime, mp 130°C, (8) with phenylhydrazine, yields benzaldehyde phenylhydrazone $C_6H_5CH{:}NNHC_6H_5$, pink solid, mp 156°C, (9) with concentrated HNO_3, yields metanitrobenzaldehyde ($C_6H_4CHO(NO_2)(2)$, white solid, mp 58°C, (10) with concentrated H_2SO_4 yields metabenzaldehyde sulfonic acid $C_6H_4CHO(SO_3H)(2)$, (11) with anhydrous sodium acetate and acetic anhydride at 180°C, yields sodium cinnamate C_6H_5COONa, (12) with sodium hydrogen sulfite, forms benzaldehyde sodium bisulfite $C_6H_5CHOHSO_3Na$, white solid, from which benzaldehyde is readily recoverable by treatment with sodium carbonate solution, (13) with acetaldehyde made slightly alkaline with NaOH, yields cinnamic aldehyde $C_6H_5CH{:}CHCHO$, (14) with phosphorus petachloride, yields benzylidine chloride $C_6H_5CHCl_2$.

Benzaldehyde may be detected by the appearance of a blue color on treating with acenaphthene and H_2SO_4, followed by heating. Benzaldehyde is used (1) as a flavoring material, (2) in the production of cinnamic acid, (3) in the manufacture of malachite green dye.

BENZEDRINE. Drug Addition.

BENZENE. C_6H_6, formula weight 78.11, colorless, highly flamable liquid that burns with a smoky flame, mp 5.5°C, bp 80.1°C, sp gr 0.879 (20°C referred to H_2O at 4°C). Sometimes called *benzol, phenyl hydride*, or *cyclohexatriene*, benzene is practically insoluble in water (0.07 part in 100 parts at 22°C); and fully miscible with alcohol, ether, and numerous organic liquids. Benzene is of large importance industrially, mainly as a starting ingredient of many reactions and as a solvent. The compound also is of much theoretical interest, being the simplest hydrocarbon of the aromatic group. Forming an explosive mixture with air, benzene can be used as a fuel or fuel component for internal combustion engines. Benzene is the first member of a homologous series of compounds, C_nH_{2n-6}. Methylbenzene or toluene, $C_6H_5 \cdot CH_3$, is the only homolog with the formula C_7H_8. Next in order of the homologous series, C_8H_{10}, exists in four isomeric forms, namely, ethylbenzene, $C_6H_5 \cdot C_2H_5$, and ortho-, meta-, and paradimethylbenzene, $C_6H_4(CH_3)_2$. Of the formula C_9H_{12}, eight isomerides are possible. Possible isomerides increase rapidly as the carbon count increases.

Many of the C_nH_{2n-6} series of hydrocarbons occur in coal gas and coal tar, from which they may be extracted. These were the early sources for benzene and related compounds. Much of the benzene currently is produced from petroleum sources. The separation of benzene from coal tar is difficult because of the presence of scores of isomerides with close boiling points. In one process for making high-purity benzene (99.94% or higher) from coke-oven light oil (the cut boiling between 60–150°C), the light oil and a stream of hydrogen are heated to reaction temperature and passed through fixed-bed reactors that contain a proprietary catalyst. In this reaction, the nonaromatics present are converted to light hydrocarbon gases. Sulfur compounds present are converted to H_2S. Some dealkylation of the higher aromatics present also produces benzene in addition to that contained in the feedstock. Vapors from the reactor are cooled and passed to a stabilizer tower where dissolved H_2S and light hydrocarbons (with boiling points lower than benzene) are removed. The bottoms from the stabilizer containing benzene, toluene, and xylene, are clay-treated. Then follows a series of fractionations to produce benzene, toluene, and xylene, in addition to higher-boiling hydrocarbons. If a portion of the hydrocarbons in the product fuel gas is reformed, no external hydrogen is required.

The synthetic production of benzene generally involves the dealkylation of toluene. In one noncatalytic process, a hydrogen-rich gas is mixed with liquid toluene feed and preheated prior to charging to the reactor. Toluene reacts with the hydrogen to form benzene and methane. The reaction is exothermic. Operating conditions approximate 500–1000 psi and 595–760°C. The process provides about 98% yield of benzene. The toluene is recycled.

In a catalytic dealkylation process (Fig. 1), toluene or C_8 aromatics (alkylbenzenes) are fed to a reactor, together with a hydrogen-containing gas. The hydrogen source is not critical and may be manufactured hydrogen or off-gas from a reforming or other refining unit. Effluent from the reactor, after cooling, is charged to a separator, from which hydrogen is removed and recycled to the reactor. Liquid phase from the separator is stripped of hydrocarbons (boiling lower than benzene) in a stabilizing column. One further fractionating step yields product benzene overhead. The bottoms from the tower are recycled to the reactor for dealkylation. Yields of 98% of theoretical are claimed.

In another process, mixtures of aromatics and nonaromatic hydrocarbons comprise the charge. In a first step, aromatics are continuously extracted from the feed by using an aqueous solution of *N*-methylpyrollidone. A multistage countercurrent extraction tower is used. The operation is carried out at modest temperatures and pressures. The rich aromatic extract phase then proceeds to a stripper, where pentane and a part of the benzene are removed overhead and recycled to the extractor. The bottoms from the stripper are free of nonaromatics and enter a second stripper for further separation. The distillate from the second stripper contains aromaticfree solvent, which is returned to the extractor. One or more further fractionations yield benzene, toluene, and xylenes of desired specification. A typical feedstock may

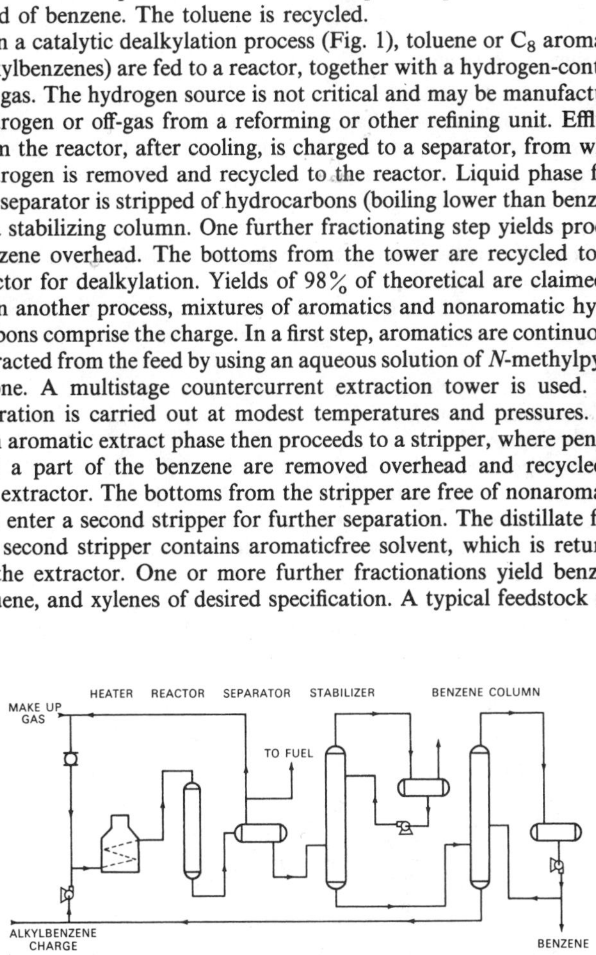

Fig. 1. Process for converting toluene or C_8 aromatics (catalytic hydrodealkylation) to high-purity benzene. (*UOP Process Division.*)

contain an aromatics mixture of about the following ranges: benzene, 26–60%; toluene, 14–22%, xylenes plus ethylbenzene, 15–50%. A similar process uses dimethyl sulfoxide as a solvent.

Styrene is a major consumer of benzene, followed by the production of cyclohexane. At one time, phenol production was the second largest consumer of benzene. Benzene and cyclohexene are closely related economically because cyclohexane can be produced by reacting benzene with hydrogen. Although the foregoing represent the major tonnage uses of benzene, the compound is critically important to the production of hundreds of other compounds. The halogen derivatives of benzene are particularly important. See also **Chlorinated Organics.** These include chlorobenzene, C_6H_5Cl; bromobenzene, C_6H_5Br; benzal chloride, $C_6H_5 \cdot CHCl_2$; benzyl chloride, $C_6H_5 \cdot CH_2Cl$; and benzotrichloride, $C_6H_5 \cdot CCl_3$. Important nitro derivatives of benzene include nitrobenzene, $C_6H_5 \cdot NO_2$; and metadinitrobenzene, $C_6H_4(NO_2)_2$. Amino compounds of large importance derived from benzene include aminobenzene, $C_6H_5 \cdot NH_2$ (aniline); diaminobenzene, $C_6H_4(NH_2)_2$; and triaminobenzene, $C_6H_3(NH_2)_3$.

Phenol, C_6H_5OH, is hydroxybenzene. Resorcinol, catechol, and quinol, $C_6H_4(OH)_2$, may be considered to be dihydroxybenzenes. Pyrogallol and phloroglucinol, $C_6H_3(OH)3$, may be considered as trihydroxybenzenes. The benzene-related alcohols, aldehydes, and ketones include benzyl alcohol, $C_6H_5 \cdot CH_2 \cdot OH$; benzaldehyde, $C_6H_5 \cdot CHO$; benzoin, $C_6H_5 \cdot CO \cdot CH(OH) \cdot C_6H_5$; salicyaldehyde, $C_6H_4(OH) \cdot CHO$; anisaldehyde, $C_6H_4(OCH_3) \cdot CHO$; acetophenone, $C_6H_5 \cdot CO \cdot CH_3$; benzophenone, $C_6H_5 \cdot CO \cdot C_6H_5$; and quinone, $C_6H_4O_2$. The benzene-related acids and salts include benzoic acid, $C_6H_5 \cdot COOH$; ethyl benzoate, $C_6H_5 \cdot COOC_2H_5$; benzoyl chloride, $C_6H_5 \cdot COCl$; benzoic anhydride, $(C_6H_4 \cdot CO)_2O$; benzamide, $C_6H_5 \cdot CO \cdot NH_2$; benzonitrile, $C_6H_5 \cdot CN$; anthranilic acid, $C_6H_4(NH_2) \cdot COOH$; phthalic acid, $C_6H_4(COOH)_2$; phthalic anhydride, $C_6H_4(CO)_2O$; phthalimide, $C_6H_4(CO)_2NH$; isophthalic acid, $C_6H_4(COOH)_2$; terephthalic acid, $C_6H_4(COOH)_2$; benzenehexacarboxylic acid, $C_6H_4(COOH)_6$; and phenylacetic acid, $C_6H_5 \cdot CH_2 \cdot COOH$.

Structure of Benzene

Benzene has long been known to consist of six C-H groups connected by valence bonds between the carbon atoms into the form of a hexagon. The precise nature of this bonding has been studied for generations. Because the studies of benzene structure were fundamental to the broader structure of organic compounds, a review of some of the earlier postulations is in order.

Among the structures suggested in the earlier days of organic chemistry are those shown in Fig. 2.

The valence bond method calculates the energy of these molecules by means of a weighted average of such configurations. The better known of these are the two *Kekulé* and the three *Dewar structures* (a, b, and c, d, e of the figure, respectively). The structure (f) proposed by *Claus* must not be considered in the valence bond method, since it can be expressed as a linear combination of the other five configurations. *Ionogenic structures* (g, h) have also been considered.

It is of interest to note that a compound of the type of structure (c) above has been prepared by replacing three of the hydrogen atoms of benzene by tertiary butyl radicals. However, this compound does not have a flat (planar) structure as shown in (c), but has the three-dimensional configuration pictured in Fig. 3. However, the presence of the Dewar-type structures in benzene itself cannot be significant,

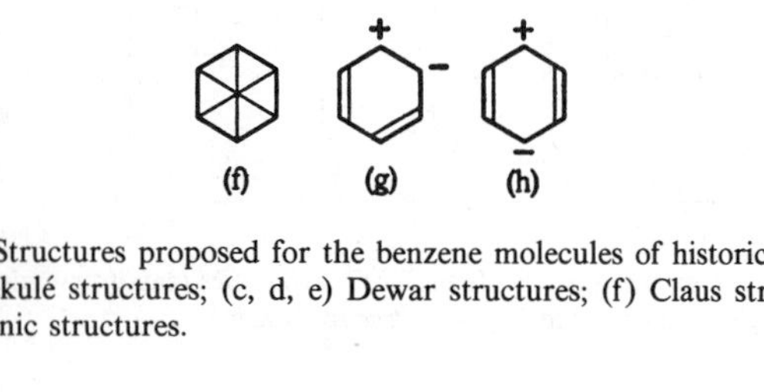

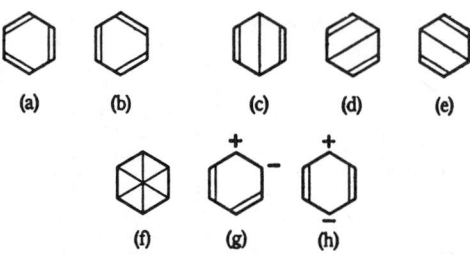

Fig. 2. Structures proposed for the benzene molecules of historical interest: (a, b) Kekulé structures; (c, d, e) Dewar structures; (f) Claus structure; (g, h) ionogenic structures.

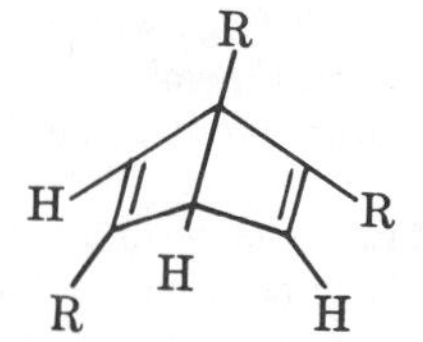

Fig. 3. Three-dimensional configuration.

since it would imply that only two disubstitution products were possible, whereas three are known.

The distance between adjacent (ortho) carbon atoms in the benzene molecule, as determined by X-ray analysis and by electron diffraction, is 1.39 Å, these distances being uniform. Since the length of a single bond is 1.54 Å and for a double bond is 1.34 Å, the benzene distance suggests a resonance hybrid between the two structures

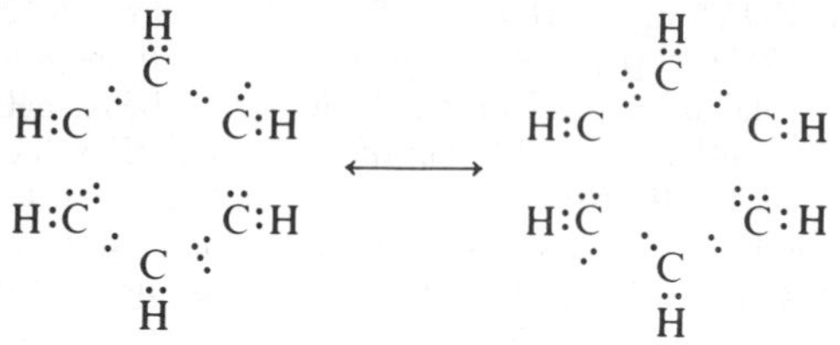

By extending the molecular orbital concept to benzene, the nature of the resonance hybrid appears as shown in Fig. 4. Three of the bonds on each carbon atom, formed through hybridized sp^2-orbitals, are sigma type molecular orbitals. One p-orbital, with one electron,

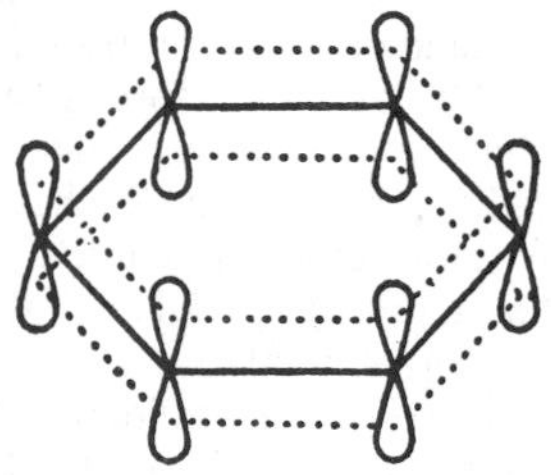

Fig. 4. Resonance hybrid model.

remains on each carbon atom. Any p-orbital can overlap the two p-orbitals on either side of it, and the dotted lines between the p-orbitals show this overlapping. (The p-orbitals are shown by the stretched "figure 8's" which are perpendicular to the plane of the sp^2-orbitals.) The fact that the p-orbitals merge to form a "π-cloud" is shown in Fig. 5.

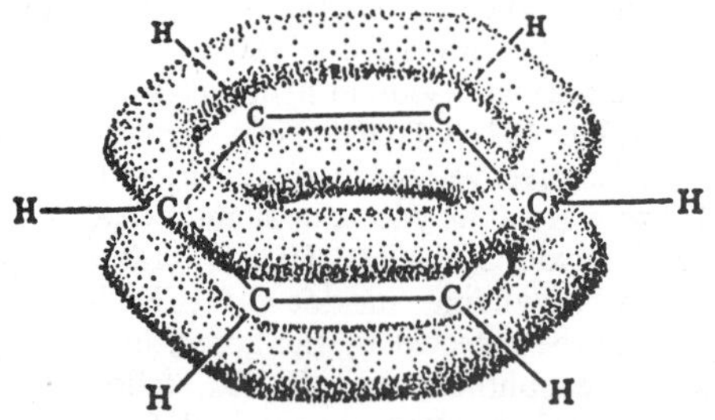

Fig. 5. "π-cloud" configuration.

Thus the pi-electrons serve to bind three carbon atoms together, resulting in increased bond energies and hence greater stability. The fact that the dotted lines are joined to form hexagons suggests that the postulated three-carbon systems may lose their identity to the extent that their bonding energy is spread out over the six carbon atoms, the point being that it is *not* restricted to a single pair.

This type of structure is represented in chemical equations by the symbol shown in Fig. 6. The relation of the positions taken by substituent or added groups are shown by the numbers in the diagram of Fig. 7. In fact, all the hydrogen atoms in benzene may be replaced

Fig. 6. Conventional symbol for benzene.

Fig. 7. Numbering system for benzene compounds.

by univalent atoms or groups forming mono-, di-, tri-, tetra-, penta-, or hexa-, derivatives according to the number of substituents. Two possibilities arise: (1) The substituents are all of one kind. Here we find but one modification of the mono-, penta-, and hexa-, derivatives and three of the di-, tri-, and tetra-, derivatives. (a) The di-derivatives are named according to the relative positions which the substituents assume around the ring as ortho-, meta-, or para-; ortho- where adjacent carbon atoms are involved, meta- when an unsubstituted carbon rests between the two substituted carbons, and para- if two unsubstituted carbon atoms occur between the substituted carbons (Fig. 8)

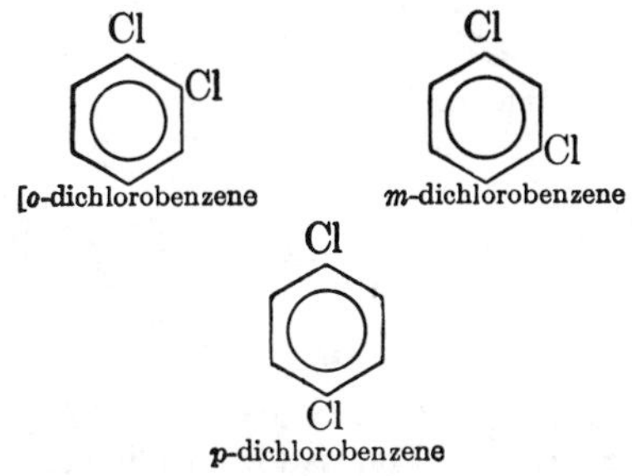

Fig. 8. Nomenclature for substituted benzene compounds with three examples of dichlorobenzene.

(b) The tri-derivatives are named adjacent (vicinal), unsymmetrical (asymmetrical), and symmetrical according as the groups take up the positions (1,2,3) (1,2,4) or (1,3,5) as shown in Fig. 9. (c) The tetra-

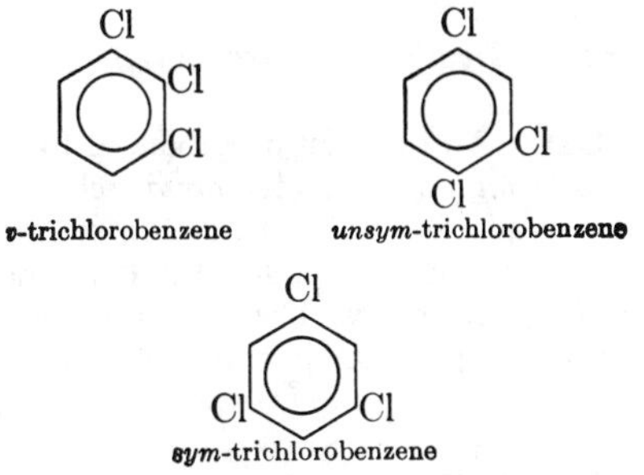

Fig. 9. Examples of tri-derivatives of benzene.

derivatives are named adjacent, symmetrical, or unsymmetrical according as the positions assumed by the substituents are (1,2,3,4), (1,2,4,5) or (1,2,3,5). (See Fig. 10.)

(2) The substituents are of two or more kinds. Here the number of isomers is greatly multiplied. (a) The di-derivatives which contain two unlike substituents may be distinguished by the prefixes ortho-, meta-, and para-, as *o*-chloronitrobenzene, or the positions of the groups may be designated by numer, as 1,2-chloronitrobenzene. (b) The system of numbering is usually followed in naming polysubstitution products of benzene, the compound being referred back to benzene or to another substance derived from benzene whose characteristic

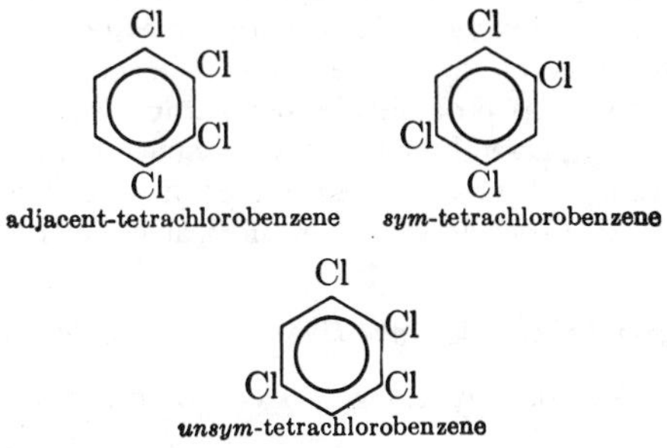

Fig. 10. Examples of tetra-derivatives of benzene.

group is present. In this latter class the group not denoted by a prefix is always regarded as assuming the 1 position; i.e., 3,6-bromonitrobenzoic acid is represented in Fig. 11.

Fig. 11. Example of polysubstitution product of benzene.

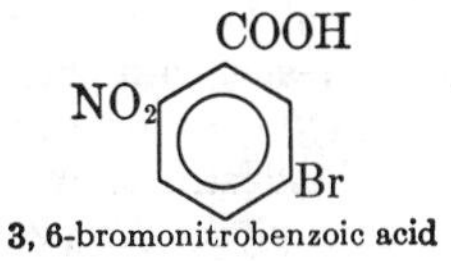

3, 6-bromonitrobenzoic acid

Benzene reacts (1) with chlorine, to form (a) substitution products (one-half of the chlorine forms hydrogen chloride) such as chlorobenzene, C_6H_5Cl; dichlorobenzene, $C_6H_4Cl_2$(1,4); and (1,2), trichlorobenzene, $C_6H_3Cl_3$(1,2,4); tetrachlorobenzene (1,2,3,5); and (b) addition products, such as benzene dichloride $C_6H_6Cl_2$; benzene tetrachloride, $C_6H_6Cl_4$; and benzene hexachloride, $C_6H_6Cl_6$. The formation of substitution products of the benzene nucleus, whether in benzene or its homologues is favored by the presence of a catalyzer, e.g., iodine, phosphorus, iron; (2) with concentrated HNO_3, to form nitrobenzene, $C_6H_5NO_2$; 1,3-dinitrobenzene, $C_6H_4(NO_2)_2$(1,3), 1,3,5-trinitrobenzene, $C_6H_3(NO_2)_3$(1,3,5); (3) with concentrated H_2SO_4, to form benzene sulfonic acid, $C_6H_5SO_3H$, benzene disulfonic acid, $C_6H_4(SO_3H)_2$(1,3), benzene trisulfonic acid, $C_6H_3(SO_3H)_3$(1,3–5); (4) with methyl chloride plus anhydrous aluminum chloride (Friedel-Crafts reaction) to form toluene, monomethyl benzene, $C_6H_5CH_3$; dimethyl benzene $C_6H_4(CH_3)_2$; trimethyl benzene, $C_6H_3(CH_3)_3$; (5) with acetyl chloride plus anhydrous aluminum chloride (Friedel-Crafts reaction) to form acetophenone (methylphenyl ketone), $C_6H_5COCH_3$; (6) with hydrogen in the presence of a catalyzer, e.g., finely divided nickel, heated, to form dihydrobenzene (cyclohexadiene), (1,3) C_6H_8; a cyclic diolefin hydrocarbon, tetrahydrobenzene, cyclohexene, C_6H_{10}; a cyclic mono-olefin, hexhydrobenzene (cyclohexane), C_6H_{12}, a cycloparaffin; (7) with ozone, to form benzene triozonide, $C_6H_6(O)_3)_3$. See also **Organic Chemistry.**

Although additional research is indicated, benzene has been implicated as a causative factor in *marrow aplasia* and some forms of *leukemia* (Forni and Vigliani, "Chemical Leukemogenesis in Man," *Series Haematologica,* **7**, 2, 214, 1974). Benzene also has been identified as a cause of *neutropenia.* See also **Carcinogens.**

BENZENE HEXACHLORIDE (BHC). A commercial mixture of isomers of 1,2,3,4,5,6-hexachlorocyclohexane. Once used extensively as an insecticide and fumigant, but now subject to regulations in a number of countries. The mixture is effective in the control of chiggers, ticks, fleas, cockroaches, lice, and *Acarus scabiei.* The mixture is sometimes called *gammexane*; more commonly called *lindane.* Quite toxic.

BENZIDINE REACTION. A test for blood (hemoglobin) made by acidifying the sample with glacial acetic acid, extracting with ether, and treating the ethereal extract with a solution of benzidine in glacial acetic acid, to which hydrogen peroxide has been added. In the absence of certain metals, a blue or green color, changing to deep blue or purple, indicates the presence of blood.

BENZIDINE REARRANGEMENT. Rearrangement (Organic Chemistry).

BENZIL REARRANGEMENT. Rearrangement (Organic Chemistry).

BENZINE. A product of petroleum boiling between 120°F and 150°F and composed of aliphatic hydrocarbons. Not to be confused with benzene, which is a single chemical compound and an aromatic hydrocarbon.

BENZOIC ACID. $C_6H_5 \cdot COOH$, formula weight 122.12, white crystalline solid, mp 121.7°C, bp 249.2°C, sublimes readily at 100°C and is volatile in steam, sp gr 1.266. Sometimes referred to as phenylformic acid, the compound is insoluble in cold H_2O, but readily soluble in hot H_2O, or in alcohol or ether. Commercially, benzoic acid finds major use as a starting or intermediate material in various industrial organic syntheses, notably in the preparation of the high-tonnage chemical, terephthalic acid. See also **Terephthalic Acid.** Benzoic acid forms benzoates: e.g., sodium benzoate, calcium benzoate which, when heated with calcium oxide, yields benzene and calcium. With phosphorus trichloride, benzoic acid forms benzoyl chloride C_6H_5COCl, an important agent for the transfer of the benzoyl group (C_6H_5CO-). Benzoic acid reacts with chlorine to form *m*-chlorobenzoic acid and reacts with HNO_3 to form *m*-nitrobenzoic acid. Benzoic acid forms a number of industrially useful esters, including: methyl benzoate, ethyl benzoate, glycol dibenzoate, and glyceryl tribenzoate.

Although benzoic acid occurs naturally in some substances, such as gum benzoin, dragon's blood resin, Peru and Tolu balsams, cranberries, and the urine of the ox and horse, the product is made on a large scale by synthesis from other materials. Benzoic acid can be prepared from toluene and air in a process that takes place in the liquid-phase in a continuous oxidation reactor operated at moderate pressure and temperature: $C_6H_5 \cdot CH_3 + 1\frac{1}{2}O_2 \rightarrow C_6H_4 \cdot COOH + 2H_2O$. The acid also can be obtained as a by-product of the manufacture of benzaldehyde from benzal chloride or benzyl chloride.

See also **Antimicrobial Agents (Foods).**

BENZOIN. The term benzoin is used in botany to denote a tree (*Styrax benzoin*) and a resin obtained from it. The former is a tall, quick-growing tree which is native in Sumatra and Java. The tree has alternate entire leaves, the lower surface of which is soft and hairy, the upper smooth. The flowers are borne in compound axillary racemes; the fruit is a drupe. The resin is obtained by making incisions in the bark. From these a thick white juice exudes and hardens. This is scraped off. It is a soft fragrant substance either white or of yellowish color. It is used in medicine as a soothing inhalant and for application to the skin. It contains about 25% benzoic acid.

Frequently confused with this is the North American shrub *Benzoin aestivale*, often called Spice bush, which blossoms very early in the spring. All parts of the shrub contain an aromatic substance which is very noticeable when the plant is bruised.

In chemistry, benzoin is a compound obtained from the resin described above, or obtained synthetically. When benzaldehyde C_6H_5CHO is warmed with sodium cyanide dissolved in alcohol, benzoin $C_6H_5COCHOHC_6H_5$ white solid, 137°C, is formed, and has the characteristics of a ketone and a secondary alcohol. Benzoin, upon reduction with sodium amalgam (sodium dissolved in mercury), forms hydrobenzoin $C_6H_5CHOHCHOHC_6H_5$ white solid, 134°C, mixed with isohydrobenzoin, 119°C.

BENZOYL CHLORIDE. Chlorinated Organics.

BENZYL BENZOATE. A water-white liquid with formula $C_6H_5CH_2OOCC_6H_5$. Sharp, burning taste with faint aromatic odor. Supercools easily. Insoluble in water and glycerin; soluble in alcohol, chloroform, and ether. 1.116–1.120 (25/25°C); bp 325°C, mp 18.8°C, flash point (~150°C). This compound is produced by the Cannizzaro reaction from benzaldehyde, by esterifying benzyl alcohol with benzoic acid, or by treating sodium benzoate with benzyl chloride. It is purified by distillation and crystallization. Benzyl benzoate is used as a fixative and solvent for musk in perfumes and flavors, as a plasticizer, miticide, and in some external medications. The compound has been found effective in the treatment of *scabies* and *pediculosis capitis* (head lice, *Pediculus humanus* var. *capitis*).

BENZYL CHLORIDE. Chlorinated Organics.

BENZYLPENICILLIN. Antibiotic.

BENZYNE. The concept of benzyne intermediates has largely stemmed from work by Dr. Georg Wittig of the University of Heidelberg, Germany, and Dr. J. D. Roberts of the California Institute of Technology. Dr. Roberts postulated that benzynes form when a substituted benzene (such as bromobenzene) reacts with a nucleophilic reagent, such as potassium amide in liquid ammonia. He and other workers have shown that strong nucleophiles add readily to arynes. If the nucleophile is attached to a side chain on the aryne, a new ring fused to the original aromatic nucleus forms by intramolecular addition. This was shown by Dr. Bunnett and Dr. B. F. Hrutfiord and also by Dr. R. Huisgen and co-workers of the University of Munich, Germany. In this work, heterocyclic compounds were usually obtained. However, in later work Dr. J. F. Bunnett and Dr. J. A. Skorez, then at Brown University, used the synthesis to obtain homocyclic ring closures, producing derivatives of such compounds as benzocyclobutene, indane, tetralin and benzocycloheptane. Their type reaction may be written as

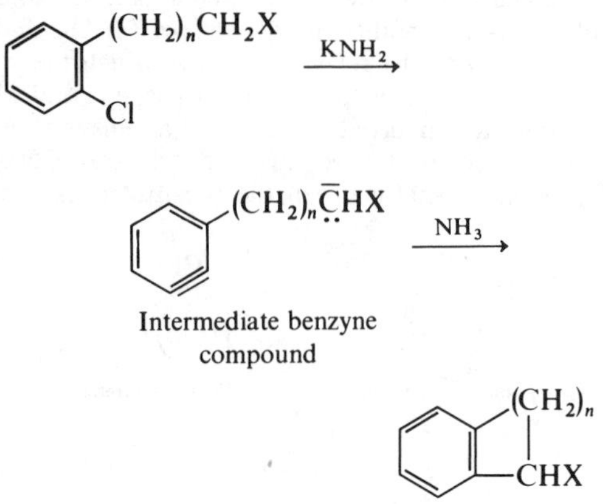

where $n = 1, 2, 3$, or 4 $X =$ cyano, acyl, carbethoxy or sulfonyl.

Dr. Lester Friedman and Francis M. Logullo prepared substituted benzynes by diazotizing substituted anthranilic acid. This is a mild, room-temperature reaction which permits simultaneous reactions of the benzynes with suitable acceptors to prepare halogen. —NO_2, —CH_3, and —OCH_3 derivatives. They have also prepared new heterocyclic arynes, such as 3-pyridyne from 3-amino-isonicotinic acid.

BERBERIDIACEAE. Barberry Shrub.

BEREAVEMENT. Depression (Psychiatric).

BERGAMOT OIL. An essential oil produced from the rind of the fruit of *Citrus aurantium*, or *C. bergamia*, relatives of the orange and lemon. The small trees are cultivated in southern Europe. The oil is expressed from the skin of the small yellow fruits and sometimes is used as a scent for cosmetics. The oil also is used sometimes as a clearing agent in the preparation of material for microscopic examination.

BERGERON CLASSIFICATION. Air Mass.

BERGERON-FINDEISEN THEORY. Precipitation and Hydrometeors.

BERGMANN'S RULE. Adaptation (Ecology).

BERIBERI. Thiamine (Vitamin B1).

BERKELEY AND HARTLEY METHOD. Osmotic Pressure.

BERKELIUM. Chemical element symbol Bk, at. no. 97, at. wt. 247 (mass number of the most stable isotope), radioactive metal of the *Actinide* series, also one of the *Transuranium* elements. All isotopes of berkelium are radioactive; all must be produced synthetically. The element was discovered by G. T. Seaborg and associates at the Metallurgical Laboratory of the University of Chicago in 1949. At that time, the element was produced by bombarding ^{241}Am with helium ions. ^{247}Bk is an alpha-emitter and may be obtained by alpha-bombardment of ^{244}Cm, ^{245}Cm, or ^{246}Cm. Other nuclides include those of mass numbers 243–246 and 248–250. Probable electronic configuration.

$$1s^22s^22p^63s^23p^63d^{10}4s^24p^64d^{10}4f^{14}5s^25p^65d^{10}5f^96s^26p^67s^2.$$

Ionic radius Bk^{+3} 0.99 Å. Longest-lived isotope, $^{247}Bk(t_{1/2} = 7000$ years).

Berkelium is known to exist in aqueous solution in two oxidation states, the (III) and the (IV) states, and the ionic species presumably correspond to Bk^{+3} and Bk^{+4}. The oxidation potential for the berkelium(III)-berkelium(IV) couple is about −1.6V on the hydrogen scale (hydrogen-hydrogen ion couple taken as zero).

The solubility properties of berkelium in its two oxidation states are entirely analogous to those of the actinide and lanthanide elements in the corresponding oxidation states. Thus in the tripositive state such compounds as the fluoride and the oxalate are insoluble in acid solution, and the tetrapositive state has such insoluble compounds as the iodate and phosphate in acid solution. The nitrate, sulfate, halides, perchlorate, and sulfide of both oxidation states are soluble.

The first compound of berkelium of proven molecular structure was isolated in 1962 by Cunningham and Wallman. A small quantity (0.004 microgram) of berkelium (as berkelium-249) dioxide was used to determine structure by X-ray diffraction.

References

Ghiorso, A., Thompson, S. G., Higgins, G. H., and G. T. Seaborg: "Transplutonium Elements in Thermonuclear Test Debris," *Phys. Rev.*, **102**, 1, 180–182 (1956).

Hulet, E. K., Thompson, S. G., Ghiorso, A., and K. Street, Jr.: "New Isotopes of Berkelium and Californium," *Phys. Rev.*, **84**, 2, 366–367 (1951).

Peterson, J. R., and B. B. Cunningham: "Crystal Structures and Lattice Parameters of the Compounds of Berkelium: I. Berkelium Dioxide and Cubic Berkelium Sesquioxide," *Inorg. Nucl. Chem. Lett.*, **3**, 9, 327–336 (1967).

Samhoun, S., and F. David: "Radiopolarography of Am, Cm, Bk, Cf, Es and Fm," *Proceedings of the 4th International Transplutonium Element Symposium*, Baden Baden, W. Germany, 1975.

Seaborg, G. T. (editor): "Transuranium Elements," Dowden, Hutchinson & Ross, Stroudsburg, Pennsylvania, 1978.

Thompson, S. G., Ghiorso, A., and G. T. Seaborg: "Element 97," *Phys. Rev.*, 77, 6, 838–839 (1950).

Thompson, S. G., and M. L. Muga: "Methods of Production and Research on Transcurium Elements," *Proc. of the Second United Nations International Conference on Peaceful Uses of Atomic Energy*, Geneva, Switzerland, 1958.

BERL SADDLE. **Absorption (Process).**

BERM (Beach). An impermanent, low, almost horizontal (or landward sloping) shelf, ledge, narrow terrace or bench on the backshore of a beach. Usually formed by material cast up and deposited by storm waves and bounded on one side by a beach scarp or beach ridge. A beach may have no berms, one, or several berms.

BERM (Geology). A relatively flat erosion surface brought to grade during a previous cycle of erosion. The term was introduced by Bascom in 1931. It was intended to replace *strath*, although as used, the term *berm* sometimes includes the shoulder of a new valley, together with the remnant of the old valley floor. The term also is used to describe a horizontal ledge of land bordering either bank of a river, as in the case of the Nile River which inundates the ledge when the river overflows.

BERM (Structural). A narrow shelf, ledge, bench or strip (horizontal) constructed along an embankment, located part way up and thus breaking the continuity of the slope. The term also is used to describe the margin, side, or shoulder of a road that is adjacent to and outside the paved portion.

BERMUDAGRASS. **Grasses.**

BERMUDA HIGH. **Atmosphere (Earth).**

BERNOULLI EQUATION. A first order nonlinear differential equation

$$y' + f(x)y = g(x)y^n$$

It may be made linear by the change of variable $y = u^{1/1-n}$ giving

$$u' + (1-n)f(x)u = (1-n)g(x)$$

The equation was offered for solution by Jacob Bernoulli in 1695. It was solved by Leibniz, as indicated here, and by John Bernoulli (1667–1748), brother of Jacob, who separated the variables by using two new ones. See **Airspeed Indicator; Bernoulli Number; Pitot Tube;** and **Polynomial.**

BERNOULLI LAW. A relationship that expresses the conservation of momentum in fluid flow. The usual form applies to the steady inviscid flow of an incompressible fluid and can be obtained by integrating the Navier-Stokes equations along a streamline,

$$gz + \frac{p}{\rho} + \tfrac{1}{2}u^2 = \text{constant on a streamline}$$

If the flow is irrotational so that the fluid velocity is the gradient of a scalar potential ϕ, the restriction to steady flow may be removed and the law is

$$\frac{\partial\phi}{\partial t} + \frac{p}{\rho} + \tfrac{1}{2}u^2 + gz = \text{constant anywhere in the fluid}$$

Lastly, if the fluid is barotropic, i.e., the density is a function of pressure alone, it takes the form,

$$\int \frac{dx}{\rho} + \tfrac{1}{2}u^2 = \text{constant}$$

valid along a streamline for steady flow. For a perfect gas, with p proportional to ρ^γ,

$$gz + \frac{\gamma}{\gamma-1}\frac{p}{\rho} + \tfrac{1}{2}u^2 = \text{constant}$$

The quantity,

$$p\left(1 + \tfrac{1}{2}\frac{\gamma-1}{\gamma}\frac{\rho u^2}{p}\right)^{\gamma/(\gamma-1)}$$

is known as the total head or stagnation pressure. For flow at small Mach numbers, i.e., $u^2 \ll \gamma p/\rho = \gamma RT$, it is $p + \tfrac{1}{2}\rho u^2$.

BERNOULLI METHOD. Given the algebraic equation

(1) $$x^n + a_1 x^{n-1} + \cdots + a_n = 0,$$

let $h_0, h_1, \ldots, h_{n-1}$ be arbitrary numbers, not all zero, and form $h_n, h_{n+1}, \ldots$, by

$$h_{n+v} + a_1 h_{n+v-1} + \cdots + a_n h_v = 0.$$

If (1) has a unique root of largest modulus, then in general the quotients $h_{\mu+1}/h_\mu$ approach that root. The method can be extended to transcendental equations. Let

(2) $$f(z) \equiv 1 + c_1 z + c_2 z^2 + \cdots$$

converge in some circle about the origin in the complex plane, and let

(3) $$g(z) \equiv g_0 + g_1 z + g_2 z^2 + \cdots$$

represent any function analytic in the same circle, and having no zero in common with $f(z)$. Let

$$h_0 = g_0,$$
$$c_1 h_0 + h_1 = g_1,$$
$$c_2 h_0 + c_1 h_1 + h_2 = g_2$$

Then if $f(z)$ has a unique zero of smallest modulus lying within that circle, then $h_\mu/h_{\mu+1}$ approaches that zero. If there are two zeros whose moduli are less than those of all others, then the roots of

$$\begin{vmatrix} z^2 & h_\mu & h_{\mu+1} \\ z & h_{\mu+1} & h_{\mu+2} \\ 1 & h_{\mu+2} & h_{\mu+3} \end{vmatrix} = 0$$

approach those zeros of $f(z)$. Likewise one can form cubics whose roots approach the three smallest roots.

BERNOULLI NUMBER AND POLYNOMIAL. A Bernoulli number is a coefficient in the power series

$$\frac{x}{e^x - 1} = \sum_{n=0}^{\infty} \frac{B_n x^n}{n!}$$

The first few numbers are $B_0 = 1$; $B_1 = -\frac{1}{2}$; $B_3 = B_5 = B_7 = \cdots = 0$; $B_2 = \frac{1}{6}$; $B_4 = -\frac{1}{30}$; $B_6 = \frac{1}{42}$; $B_8 = -\frac{1}{30}$; $B_{10} = \frac{5}{66}$. Successive values of the numbers may be found from the equation $(B + 1)^n = B^n$ by setting $B^k = B_k$. The numbers occur in the Euler-Maclaurin formula and in the Stirling formula.

The Bernoulli polynomial is a coefficient in the power series

$$\frac{t(e^{xt}-1)}{(e^t=1)} = \sum_{n=0}^{\infty} \frac{\phi_n(x)}{n!} t^n$$

The first few polynomials are $\phi_2 = x(x - 1)$; $\phi_3 = x(x^2 - 3x/2 + \frac{1}{2})$; $\phi_4 = x^2(x^2 - 2x + 1)$; $\phi_5 = x(x^4 - 5x^3/2 + 5x^2/3 - \frac{1}{6})$. If the n-th polynomial is expanded in a Maclaurin series, the coefficients are related to the Bernoulli numbers.

Both the numbers and the polynomials are often defined in other ways so that equations containing them should be carefully checked. They are used in numerical integration formulas and in the calculus of finite differences.

The Bernoulli who discovered these relations was Jacob (1654–1705), a member of a distinguished family of mathematicians and physicists. See also **Bernoulli Equation;** and **Logarithmic Spiral.**

BERNOULLI THEOREM. This theorem, due to James Bernoulli, was published posthumously in 1713. Let the probability of an event p be constant from trial to trial, and the probability of failure be denoted by q. Let the relative frequency be denoted by x/s where x is the number of successes in s trials. Let

$$P\left(\left|\frac{x}{s} - p\right|\right)$$

denote the probability of obtaining the absolute value of the deviation $x/s - p$. Bernoulli's theorem states that as the number of trials s increases, the probability of a difference in absolute value more than a stated positive amount ϵ approaches zero. In symbols,

$$\lim_{s\to\infty} P\left(\left|\frac{x}{x} - p\right| > \epsilon\right) = 0$$

The theorem does *not* state an ordinary proposition in the mathematics of limits:

$$\lim_{s\to\infty} \frac{x}{s} = p$$

Bernoulli's theorem is one of the "laws of large numbers." It gives precise form to intuitive feelings (about the limiting ratios of occurrences) which are not easy to formulate. See also **Aerodynamics.**

BERRY. A true berry consists of a fleshy fruit, derived entirely from the ovary of a flower and its contents. Usually many seeds are embedded in the flesh. Common examples are the tomato, grape, gooseberry, and currant. Frequently, as in currants and gooseberries, the calyx tube grows around the ovary wall and forms the skin of the berry. All berries of commerce do not necessarily meet this strict botanical definition. Berries are considered an excellent source of nutrients. The principal berries of commerce are blackberry and dewberry, blueberry, cranberry, currant, gooseberry, raspberry, and strawberry. Grape and tomato are not considered commercially in the same category with berries. Several berries are described in this book. Check alphabetical entries.

BERTHELOT EQUATION. A form of the equation of state, relating the pressure, volume, and temperature of a gas, and the gas constant R. The Berthelot equation is derived from the Clausius equation and is of the form

$$PV = RT\left(1 + \frac{9PT_c}{128P_cT}\left[1 - 6\frac{T_c^2}{T^2}\right]\right)$$

in which P is the pressure, V is the volume, T is the absolute temperature, R is the gas constant, T_c is the critical temperature, and P_c the critical pressure.

BERTHOLLIDE COMPOUNDS. Chemical Composition; Compound (Chemical).

BERTILLON SYSTEM. A personal identification system of principal use in criminology. It is based on physical measurements and was conceived by the French criminologist, Dr. Alphonse Bertillon, in 1886. It is used in combination with fingerprinting for identifying criminals and suspects. The system is designed to measure those characteristics which ordinarily do not change over the years, although many persons have been known to undergo various surgical procedures in order to escape their classifications in the Bertillon files. The measurements taken are those which are known to be relatively permanent after the age of 20. The principal measurement is that of head length. Each individual is classified according to small, medium, or long. Three similar divisions are made on the basis of length of middle finger, forearm, little finger, and height. The final subdivisions are made according to eye color and length of ear. Each individual's card, besides bearing these data, also presents photographs, full face and profile. Sitting height, extent of arms, length of forearm, and other measurements are also included. In addition, color of hair and location and descriptions of moles and birthmarks are noted. With such evidence, trained personnel often can identify bodies in cases where fingerprints are destroyed.

The medicolegal examination of hair has only limited value in personal identifications. Obviously, hair color and characteristics change with age; also hair from different parts of the body will differ and thus pose complications. However, evidence concerning hair is admissible in court records and is useful because it can serve as an important clue in various types of criminal investigation. Valuable clues can most often be found in crimes in which death was brought about through hand-to-hand combat, strangulation, stabbing, etc., in which cases the victim may have had an opportunity to grasp the assailant's hair. In rape cases, it is routine to examine the victim's underclothing for pubic hairs of the assailant. Sometimes hair proves valuable in hit-and-run auto cases where it may be found on part of the vehicle.

Under the microscope, hair which has been forcefully pulled out can be distinguished from hair which has fallen out naturally. Hair which has been damaged, as by a blow, will have a "bruised" appearance microscopically. Torn hair can easily be differentiated from cut hair. By examining the appearance of a single hair near a gunshot wound, it is sometimes possible to make a good estimate of the distance from which the shot was fired, since the appearance of a hair will change progressively with the amount of heat applied to it.

Fingerprints have for centuries been widely used as the basis for personal identification. Through careful research and tabulations, it has been proved that in all probability each individual's fingerprints are different from those of any one else who has ever lived or ever will. Even in the case of identical twins, fingerprints are not identical, although they may be as similar as the fingerprints from the right and left hands of the same individual. The similarity of fingerprints and palm prints is used to distinguish identical from fraternal twins in multiple births.

Fingerprints and toeprints (the fine lines seen on the inner surfaces

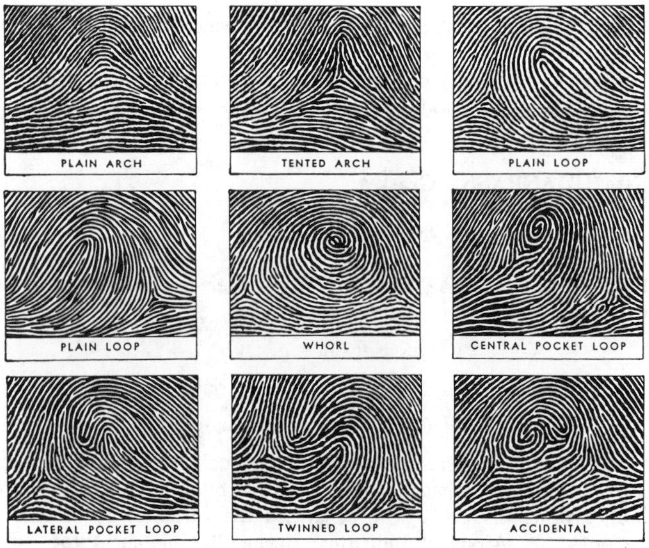

Fig. 1. Major categories of fingerprint patterns. (*Farout, Inc.*)

of the fingers and palms, toes and soles) are outward manifestations of the interweaving of two layers of the epidermis. The projections exist elsewhere in the skin, but only on the hands and feet are they so distinct.

Several systems of fingerprint classification have been developed. Four general patterns of fingerprints are widely recognized: (1) The *arched fingerprint* is composed of ridges running the width of the print, usually with more or less of an arch rising toward the tip. (2) The *loop fingerprint* is characterized by the fact that some of the ridges turn without spiraling, to return to the same side of the finger from which they came. (3) The *whorl fingerprint* is seen as a print in which some of the ridges make a spiral or complete circle. (4) The *composite fingerprint* is less definite and involves two or more of the other types, sometimes being too complex to classify, but nevertheless valuable for identification.

Pioneering work in the use of fingerprints in identification began during the mid-nineteenth century in India, through the efforts of Sir Edward Richard Henry, who developed the Henry system of fingerprint classification; and of Sir William James Herschel, son and grandson of the famous astronomers. These men, working independently in the Indian civil service, explored the method as a means of settling disputes in criminal cases. Sir Francis Galton, the famous English scientist and father of eugenics, made similar studies during the same period. If a person has been previously fingerprinted, trained personnel can match fingerprints in a matter of minutes. Fingerprints are admissible as evidence in court subject to the usual rules of evidence.

General classifications of fingerprint patterns are shown in Fig. 1. The specific characteristics of a given fingerprint are identified in Fig. 2.

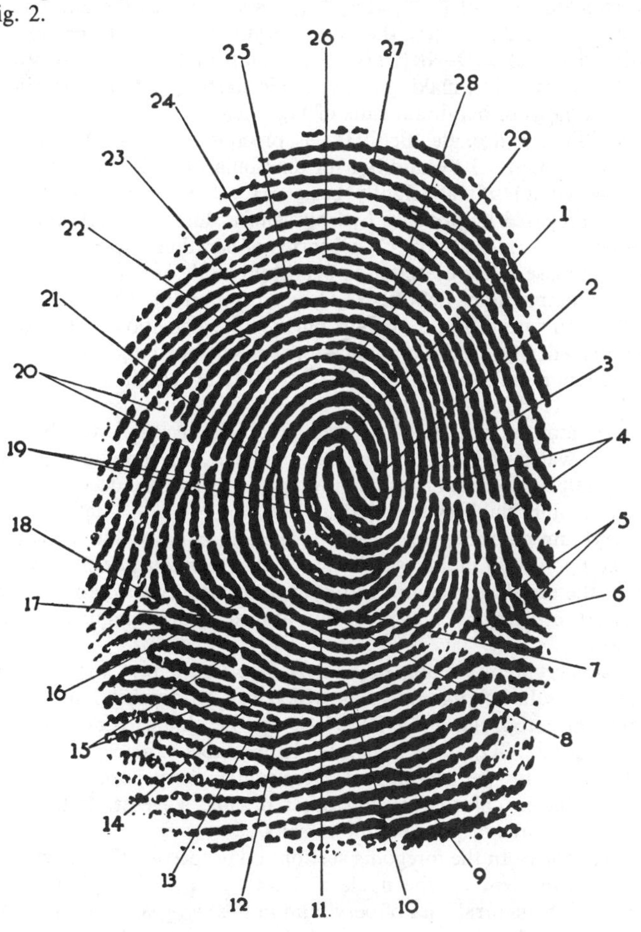

EXPLANATION OF CHARACTERISTICS ON ENLARGEMENTS

1 THE CORE
2 BIFURCATION
3 BIFURCATION
4 CICATRIX
5 ENCLOSURE
6 RIGHT DELTA
7 BIFURCATION
8 BIFURCATION
9 ABRUPT ENDING
10 ABRUPT ENDING
11 BIFURCATION
12 BIFURCATION
13 ABRUPT ENDING
14 ABRUPT ENDING
15 CICATRIX
16 BIFURCATION
17 BIFURCATION
18 LEFT DELTA
19 ENCLOSURE
20 CICATRIX
21 BIFURCATION
22 ABRUPT ENDING
23 BIFURCATION
24 BIFURCATION
25 ABRUPT ENDING
26 ABRUPT ENDING
27 ABRUPT ENDING
28 ABRUPT ENDING
29 BIFURCATION

Fig. 2. Principal characteristics of a fingerprint.

BERYL. The mineral beryl is a silicate of beryllium (glucinium) and aluminum corresponding to the formula $Be_3Al_2Si_6O_{18}$. Crystalizing in the hexagonal system the 6-sided prisms of beryl may be very small or range up to several feet in length and a yard or so in diameter. Terminated crystals are relatively rare. Its fracture is conchoidal; hardness, 7.5–8; specific gravity, 2.6–2.9; colors, emerald green, green, blue-green, blue, yellow, red, white and colorless; luster, vitreous; transparent to translucent.

Beryl has long been used as a gem, the emeralds being a rich green variety, colored probably by minute amounts of some chromium compound. A beautiful bluish sort is called aquamarine; morganite is pink, and the golden beryl is a clear bright yellow. Other shades like honey yellow and yellowish-green are common. Metallic beryllium is obtained from beryl. Its lightness and strength make it very valuable for industrial purposes.

Beryl is found in granite rocks and especially in pegmatites, but it occurs also in mica schists in the Urals. In addition to the many European localities, including Austria, Germany, and Ireland, beryls of gem quality are found in Africa, the Malagasy Republic (especially for morganite), and Brazil. The most famous place in the world for emeralds is at Muso, Colombia, South America, where they form a unique occurrence in limestones. Emeralds are also obtained in the Transvaal and near Mursinsk, in Siberia. In the United States, New England has furnished much beryl from its pegmatites, and for a long time the huge crystals from Acworth and Grafton, New Hampshire, were the largest known. Later, however, giant crystals even larger than those from New Hampshire were discovered in Albany, Maine, the largest of which was 18 feet by 4 feet, and weighed about 18 tons. Other localities are Paris and, elsewhere, in Oxford County, Maine; Royalston, Massachusetts; North Carolina; Colorado; South Dakota; and California. See also terms listed under **Mineralogy.**

Elmer B. Rowley, Union College, Schenectady, N.Y.

BERYLLIOSIS. Beryllium.

BERYLLIUM. Chemical element symbol Be, at. no. 4, at. wt. 9.0122, periodic table group 2a, mp 1287–1292 ± 3°C, bp 2970 ± 5°C. The vapor pressure at the melting point calculates to be 55 N/m^2 from the equations for the vapor pressure.

1. Solid $\log P_{(bar)} = 6.266 + 1.473 \times 10^{-4}T - 16{,}950\ T^{-1}$
2. Liquid $\log P_{(bar)} = 6.578 - 11{,}860\ T^{-1}$

Considerable variation exists in the reported specific heat data. The following appear to be representative:

Temperature °C	Specific Heat kJ/(kg·K)
−13	1630
+25	1970
100	2130
200	2340
800	1970

Similar scatter exists in the reported thermal conductivity data. An average value lies around 125 kW/(m·K).

The density of beryllium is 1.847 g/cm^3 (based upon average values of lattice parameters at 25°C ($a = 22.856$ nm and $c = 35.832$ nm). Beryllium products generally have a density around 1.850 g/cm^3 or higher because of impurities, such as aluminum and other metals, and beryllium oxide. The crystal structure is close-packed hexagonal. The alpha-form of beryllium transforms to a body-centered cubic structure at a temperature very close to the melting point.

First ionization potential 9.32 eV; second 18.4 eV. Oxidation potentials $Be \rightarrow Be^{2+} + 2e^-$, 1.70 V; $2Be + 6OH^- \rightarrow Be_2O_3^{2-} + 3H_2O + 4e^-$, 2.28 V.

All naturally occurring beryllium compounds are made up of the ^9Be isotope. Artificially produced isotopes occur during some nuclear reactor operations and include ^6Be, ^7Be, ^8Be, and ^{10}Be.

The thermal neutron absorption cross section is 0.0090 barn/atom.

The electrical conductivity of beryllium is dependent upon both temperature and metal purity. It varies at room temperature between 38–42% (International Annealed Copper Standard). Electrical resistivity of 4.266×10^{-8} ohm·m at 25°C has been reported.

Background. In 1797, Vauquelin discovered beryllium to be a constituent of the minerals beryl and emerald. Soluble compounds of the new element tasted sweet, so it was first known as glucinium from the corresponding Greek term. Quarrels over the name of the element were perpetuated by the simultaneous and independent isolations of metallic beryllium in 1828 by Wohler and Bussy. Both reduced beryllium chloride with metallic potassium in a platinum crucible. The name beryllium and symbol Be were officially recognized by the IUPAC in 1957.

Hope for the emergence of beryllium beyond the laboratory curiosity status resulted from publication of the work of the French scientist Lebeau in 1899. His paper described the electrolysis of fused sodium fluoberyllate to produce small hexagonal crystals of beryllium. Lebeau also reported the direct reduction of a beryllium oxide-copper oxide mixture with carbon to yield a beryllium-copper alloy. In 1926, Lebeau's alloy was rediscovered and found to have remarkable age-hardenable mechanical properties. A copper-beryllium alloy was first marketed in 1931, and this market remains important today.

Commercial development of beryllium in the United States was begun in 1916 by Hugh S. Cooper with the production of the first significant metallic beryllium ingot. This was followed by formation of the Brush Laboratories Company, which started its development work under the direction of Dr. C. B. Sawyer in 1921. In Germany, the Siemens-Halske Konzern began commercial development work in 1923.

Occurrence. Occurrences of beryllium in the earth's crust are widely distributed and estimates of the amount fall in the 4–6 ppm range. Forty-five beryllium-containing minerals have been identified. Only two are commercially important—beryl, $3BeO \cdot Al_2O_3 \cdot 6SiO_2$, for its high beryllium content, and bertrandite, $Be_4Si_2O_7(OH)_2$, for its large quantities located in the United States.

In 1959, beryllium was found in the rhyolitic tuffs of Spor Mountain, Utah, containing from 0.1 to 1.0% beryllium oxide. The practical processing limit requires an average beryllium oxide content of the ore of 0.6% to compete with beryl ore processing of material with more than 10% beryllium oxide. Deposits of this processable grade are adequate for the industrial requirements of the United States for several decades at present levels of consumption. Although the ore grade is much lower than that of beryl ore, the beryllium values in the rhyolitic tuffs are acid-soluble and recoverable by established processing technology.

In pure form, beryl mineral contains nearly 14% beryllium oxide, as found in its precious forms, emerald and aquamarine. Industrial grades of the mineral contain 10–12% beryllium oxide. Beryl occurs as a minor constituent of pegmatic dikes and is mined primarily as a by-product of feldspar, spodumene, and mica operations. Only the relatively large crystals are recovered by hand-picking or cobbing to supply industrial requirements of 3000 tons (2700 metric tons) per year. In 1979, the main producers were the U.S.S.R., China, Brazil, and Argentina. The opening of mines and a mill to process bertrandite-bearing ores in Utah has decreased beryl requirements in the United States to 1000–2000 tons (900–1800 metric tons) per year.

Extractive and Process Metallurgy. The production of metallic beryllium, its alloys, or its ceramic products centers around the recovery of an intermediate partially purified concentrate from ore processing. The usual intermediate is beryllium basic carbonate or hydroxide. The mill in Utah is the only one in the Western world which extracts beryllium from its ores. The processes used to extract the beryllium are based on sulfuric acid. The sulfate solutions from beryl or bertrandite sources are partially purified by solvent extraction before yielding beryllium hydroxide as the end product. The hydroxide is converted to beryllium fluoride by reaction with ammonium bifluoride. Thermal reduction with magnesium metal forms beryllium pebbles. Final purification is accomplished by vacuum melting the beryllium pebbles to remove fluorides and magnesium impurities and casting into graphite molds. Standard powder metallurgy processes are generally used to convert the cast billets to solid shapes. The prevalent final consolidation step is hot pressing.

Important Commercial Properties. Beryllium has several unique properties which have given it a position of commercial significance. Its low atomic mass, low absorption cross section, and high scattering cross section are neutronic properties of importance. These properties spurred the expansion of beryllium production beyond the pilot scale immediately after the formation of the United States Atomic Energy Commission and the initiation of nuclear reactor development programs. About 1960, structural applications using beryllium began to utilize its modulus of elasticity of 2.93×10^5 mPa, its low density, and its relatively high melting point. Beryllium has good thermal conductivity and excellent thermal capacity properties which gave rise to its use as a thermal barrier and heat sink for re-entry vehicles and other aerospace applications. The latter properties have been coupled with favorable ductility properties at elevated temperatures for the development of aircraft brakes.

Chemical Properties. Many chemical properties of beryllium resemble aluminum, and to a lesser extent, magnesium. Notable exceptions include solubility of alkali metal fluoride-beryllium fluoride complexes and the thermal stability of solutions of alkali metal beryllates.

All of the common mineral acids attack beryllium metal readily with the exception of nitric acid. It is also attacked by sodium hydroxide and potassium hydroxide, but not by ammonium hydroxide.

Beryllium interacts with most gases. Polished beryllium surfaces retain their brilliance for years on exposure to air at ambient temperatures. The oxidation rate in air increases parabolically at temperatures above 850°C with the formation of a loosely adherent, white oxide.

Compounds of Beryllium. Ammonium beryllium carbonate solutions are prepared by dissolving the hydroxide or the basic carbonate in warm (50°C) aqueous mixtures of NH_4HCO_3 and $(NH_4)_2CO_3$. After filtering to remove insoluble impurity hydroxides and adding a chelating agent, heating above 88°C evolves NH_3 and CO_2 and precipitates a high-purity, basic beryllium carbonate. If the aqueous system has the stoichiometry of $(NH_4)_4Be(CO_3)_3$, analogous to the ammonium uranyl carbonate system, the basic beryllium carbonate product of hydrolysis is $2BeCO_3 \cdot Be(OH)_2$. This compound is readily dissolved in all mineral acids, making it a valuable starting material for laboratory synthesis of beryllium salts of high purity.

Beryllium hydroxide, $Be(OH)_2$, is precipitated as an amorphous, gelatinous material by addition of ammonia or alkali to a solution of a beryllium salt at slightly basic pH values. A pure hydroxide can be prepared by pressure hydrolysis of a slurry of beryllium basic carbonate in water at 165°C. All forms of beryllium hydroxide begin to decompose in air or water to beryllium oxide at 190°C.

Beryllium sulfate, $BeSO_4 \cdot 4H_2O$, is an important salt of beryllium used as an intermediate of high purity for calcination to beryllium oxide powder for ceramic applications. A saturated aqueous solution of beryllium sulfate contains 30.5% $BeSO_4$ by weight at 30°C and 65.2% at 111°C.

Beryllium fluoride, BeF_2, is readily soluble in water, dissolving in its own water of hydration as $BeF_2 \cdot 2H_2O$. The compound cannot be crystallized from solution and is prepared by thermal decomposition of ammonium fluoberyllate, $(NH_4)_2BeF_4$.

Beryllium chloride, $BeCl_2$, with a melting point of 440°C, is used as a component of molten salt baths for electrowinning or electrorefining of the metal. The compound hydrolyzes readily with atmospheric moisture, evolving HCl, so protective atmospheres are required during processing.

Basic beryllium acetate, $Be_4O(C_2H_3O_2)_6$, is the best known of the beryllium salts of organic acids which can be divided into normal beryllium carboxylates, $Be(RCOO)_2$, and beryllium oxide carboxylates, $Be_4O(RCOO)_6$. The basic acetate is soluble in glacial acetic acid and can readily be crystallized therefrom in very pure form. It is also soluble in chloroform and other organic solvents. It has been used as a source of pure beryllium salts.

Applications. In the foregoing sections on properties, the early applications dependent on the nuclear characteristics of beryllium were described. Structural uses of beryllium in the aerospace industry have been realized because no other known material exceeds its modulus-to-weight ratio while supplying significant ductility.

Examples of applications which fully exploit the nuclear, rigidity, and high-temperature properties include heat shields, guidance system parts such as gimbals, gyroscopes, stable platforms and accelerometers, housings, mirrors, aircraft brakes, and formable grades of beryllium used in drawing and related methods of fabrication. Recent applications have been in structures which are loaded in compression. Wrought products are being developed with yield strengths approaching 690 MPa (100,000 psi) and 20% elongation at room temperature.

Additions of beryllium to commercial copper- and nickel-based alloys enable these materials to be precipitation-hardened to strengths approaching those of heat-treated steels. Yet, copper-beryllium alloys retain the corrosion resistance, electrical and thermal conductivities, and spark-resistant properties of copper-base alloys.

Beryllium oxide articles take advantage of the high thermal conductivity and heat capacity, high melting point, very low electrical conductivity, and high transparency to microwaves in microelectronic substrate applications.

Biology and Toxicology. Beryllium can be handled safely with reasonable controls, but it can cause serious illness if these controls are not observed. Skin and respiratory reactions can be experienced. There is, however, no ingestion problem.

The hazards are generally classified as (1) acute respiratory disease, (2) chronic pulmonary disease, and (3) dermatitis.

Dermatitis is produced by skin contact with soluble salts of beryllium, especially the fluoride. It is controlled by a program of good personal hygiene, frequent washing of the exposed parts of the body, as well as a program where clothing is laundered on the plant site.

Acute pulmonary disease is due exclusively to inhalation of soluble beryllium salts and is not caused by exposure to the oxide, the metal, or its alloys. The exact forms of beryllium causing the chronic pulmonary disease and the degree of exposure necessary to induce it are not precisely known. It is known that under the completely uncontrolled conditions existing in beryllium extraction plants before the establishment of air-count standards in 1949, when beryllium air-counts were in milligrams per cubic meter of air rather than micrograms, only about 1% of the exposed workers became ill. This would indicate a sensitivity of a limited number of individuals to beryllium.

Investigations by medical, toxicological, and engineering personnel led to the promulgation of safe limits of exposure by the Atomic Energy Commission in 1949. It should be noted that no disabling cases of chronic berylliosis from exposure after 1949 have been documented. The disease is believed to be avoidable when air-counts are held within average limits of 2 micrograms per cubic meter of air for an 8-hour exposure, with a maximum at any time of 25 micrograms per cubic meter of air.

The Occupational Safety and Health Administration issued a proposed new occupational standard for beryllium air-counts. This proposal is highly controversial and remains under review.

Local exhaust ventilation is the major engineering control used to limit concentrations of airborne beryllium. Modern air cleaners allow control within recommended outplant levels of 0.01 microgram beryllium per cubic meter of air, averaged over 1-month periods.

References

Ballance, J., Stonehouse, A. J., Sweeney, R., and K. Walsh: "Beryllium and Beryllium Alloys," in "Encyclopedia of Chemical Technology," 3rd edition (Kirk-Othmer, editors), Vol. 3, pp. 803–823, Wiley, New York, 1978.

Busch, L. S.: "Beryllium," in "Encyclopedia of the Chemical Elements," (C. A. Hampel, editor), pp. 49–56, Van Nostrand Reinhold, New York, 1968.

Pinto, N. P., and J. Greenspan: "Beryllium," in "Modern Materials—Advances in Development and Applications," (B. W. Gonser, editor), Vol. 6, Academic, New York, 1968.

Staff: "Beryllium in Aerospace Structures," Brush Wellman Inc., Cleveland, Ohio, 1963.

Stokinger, H. E.: "Beryllium, Its Industrial Hygiene Aspects," Academic, New York, 1966

Walsh, K., and G. Rees: "Beryllium Compounds," in "Encyclopedia of Chemical Technology," 3rd edition (Kirk-Othmer, editors), Vol. 3, pp. 824–829, Wiley, New York, 1978.

Webster, D., et al. (editors): "Beryllium Science and Technology," 2 volumes, Plenum, New York, 1979.

Kenneth A. Walsh, Ph.D., Brush Wellman Inc., Elmore, Ohio.

BERYLLIUM COPPER. **Copper.**

BERYLLIUM ROCKET FUEL. **Rocket Propellants.**

BESSEL FORMULA FOR INTERPOLATION. A central difference formula

$$y_{v+1/2} = \mu y_{1/2} + \delta y_{1/2} v + \mu\delta^2 y_{1/2} \frac{(v^2 - \frac{1}{4})}{2!} + \delta^3 y_{1/2} \frac{v(v^2 - \frac{1}{4})}{3!} + \mu\delta^4 y_{1/2} \frac{(v^2 - \frac{1}{4})(v^2 - \frac{9}{4})}{4} + \cdots$$

The independent variable x is equally spaced so that $(x_n - x_0) = nh$ and the desired value of y corresponds to $x = x_0 + hu$; $v = u - \frac{1}{2}$. The other symbols are defined as follows: $\delta^m y_{1/2} = \Delta^m_{(1-m)/2} \mu y_{1/2} = (y_0 + y_1)/2$; $\mu\delta^m y_{1/2} = (\delta^m y_0 + \delta^m y_1)/2$.

See also **Interpolation;** and list of related entries under **Mathematics.**

BESSEL FUNCTION. The differential equation

$$x^2 y'' + xy' + (x^2 - n^2)y = 0, \quad n = \text{const.}$$

is called Bessel's equation of order n. Certain of its solutions (see below) are called Bessel functions. The general solution is

$$y = AJ_n(x) + BY_n(x)$$

where

$$J_n(x) = \sum_{k=0}^{\infty} \frac{(-1)^k}{\Gamma(k+1)\Gamma(k+n+1)} \left(\frac{x}{2}\right)^{n+2k}$$

and $Y_n(x) = J_{-n}(x)$ if n is not an integer. These functions are called Bessel functions of the first kind. If n is an integer, then $J_{-n}(x) = (-1)^n J_n(x)$, so that $Y_n(x)$ is defined as

$$Y_n(x) = \lim_{k \to n} \frac{J_k(x) \cos k\pi - J_{-k}(x)}{\sin k\pi}$$

which is called a Bessel function of the second kind. The functions, much used in physics,

$$H_n^{(1)}(x) = J_n(x) + iY_n(x) \qquad i = \sqrt{-1}$$
$$H_n^{(2)}(x) = J_n(x) - iY_n(x)$$

are Bessel functions of the third kind; they are also called Hankel functions of the first and second kind, respectively. Other combinations of Bessel functions are also given names. These functions have certain standard properties of recurrence, orthogonality, etc.

BESSEL INEQUALITY. If a function $f(x)$ is expressible in terms of a set of orthogonal functions ϕ_i using expansion coefficients

$$c_i = \int f\phi_i \, dx$$

then the Bessel inequality states

$$\sum_{i=1}^{\infty} c_i^2 \leq \int f^2 \, dx$$

If the value of the integral is less than ∞, $f(x)$ is integrable square and the sum of the square of the expansion coefficients is convergent.

When the equality holds, the result is known as the Parseval theorem.

Both relations may also be written in vector form.

BESSEMER PROCESS. **Iron Metals, Alloys, and Steel.**

BETA AGONISTS. **Bronchial Asthma.**

BETA CELLS (Pancreas). **Diabetes Mellitus.**

BETA CENTAURI. Ranking eleventh in apparent brightness among the stars, Beta Centauri has a true brightness value of 5,000 as compared with unity for the sun. Beta Centauri is a blue-white, spectral type B star and is located in the constellation Centaurus south of the ecliptic. Estimated distance from the earth is 300 light years. See also **Alpha Centauri; Constellations;** and **Star.**

BETA CRUCIS. Ranking nineteenth in apparent brightness among the stars, Beta Crucis has a true brightness value of 6,000 as compared with unity for the sun. Beta Crucis is a blue-white, spectral type B star and is located in the constellation Crux (Southern Cross) south of the ecliptic. Estimated distance from the earth is 500 light years. See also **Constellations.**

BETA DECAY. The process that occurs when beta particles are emitted by radioactive nuclei. The name *beta particle* or beta radiation was applied in the early years of radioactivity investigations, before it was fully understood what beta particles are. It is known now, of course, that beta particles are electrons. When a radioactive nuclide undergoes beta decay its atomic number Z changes by $+1$ or -1, but its mass number A is unchanged. When the atomic number is increased by 1, negative beta particle (negatron) emission occurs; and when the atomic number is decreased by 1, there is positive beta particle (positron) emission or orbital electron capture.

Because atomic nuclei contain only protons and neutrons, beta particles must be created at the moment of emission, just as photons are created at the time of emission of electromagnetic radiation. Because of this creation process, the amount of energy equal to the rest energy, m_ec^2 of an electron, must be consumed when beta decay occurs. Any remaining energy can be given to the beta particle as kinetic energy. The nuclear transitions producing beta decay are between discrete energy states differing by a definite amount of energy W_0, so we expect the total energy of a beta-decay transition to be W_0. However, emitted beta particles are experimentally found to have a continuous range of total (rest plus kinetic) energies W of such magnitude that $m_ec^2 < W < W_0$, rather than all having a single energy W_0. This distribution as a function of energy (or momentum) forms what is known as a beta-ray spectrum. The shape of the spectrum depends on the sign of the charge on the beta particle (positive or negative), the energy W_0, and the degree of forbiddenness of the transition (explained below). Unless energy and momentum are not conserved in the process, the energy not carried away by the beta particle must be given to some other particle. Furthermore, since the beta particle has a spin quantum number $\frac{1}{2}$, angular momentum cannot be conserved unless another $\frac{1}{2}$ unit of angular momentum can be disposed of. Both of these possible discrepancies in the conservation laws have been taken care of in the Fermi theory of beta decay through postulation of a massless particle, a neutrino or an antineutrino, which has a spin quantum number $\frac{1}{2}$ and also carries away the remaining energy and momentum. Neutrinos were difficult to find experimentally but, even before they were experimentally detected, so much evidence had been developed to show their existence that the Fermi theory of beta decay was generally accepted.

Beta-decay processes are classified as allowed or forbidden but, as in many other physical processes, the term forbidden does not mean nonoccurrence, just a significant retardation relative to the rate for allowed transitions. The degree of forbiddenness is determined by the magnitude of the difference in angular momentum between the initial and final nuclear states as well as the parity of these states. If more than one unit of angular momentum must be carried away by the decay products ($\frac{1}{2}$ unit by the beta particle and $\frac{1}{2}$ unit by the neutrino or antineutrino), the transition must be forbidden. Allowed transitions give straight-line Kurie plots, as do some forbidden transitions but some forbidden transitions have distinct shapes other than straight lines for their Kurie plots.

A negatron emitted during beta decay has its spin aligned away from the direction of its emission (its angular momentum vector is antiparallel to its momentum vector) and hence has negative helicity, but an emitted positron has positive helicity. It is because of the absence of beta particles with both positive and negative helicity in both types of beta-emission processes that parity is not conserved in beta decay.

See also **Particles (Subatomic);** and **Radioactivity.**

BETA DECAY (Gamow-Teller Selection Rules). Gamow-Teller Selection Rules.

BETA DISTRIBUTION. Pearson Distribution.

BETA FUNCTION. Also called Euler's first integral, the beta function is defined as

$$B(r, s) = \int_0^1 t^{r-1}(1-t)^{s-1}\, dt$$

and converges for r, s positive. It is related to Euler's second integral, the gamma function, by the equation

$$B(r, s) = \frac{\Gamma(r)\Gamma(s)}{\Gamma(r+s)} = B(s, r)$$

The integral from c to x, considered as a function of x, is called the *Incomplete Beta Function.*

BETA-LACTAM RING. Antibiotic.

BETALAINES. Sometimes referred to as *beetroot pigments,* the *betalaines*[1] are made up of two main groups: (1) *betacyanins,* the principal component of which is betanin, contribute 75–95% of the total red color; and (2) *betaxanthins,* the principal component of which is vulgaxanthin-I, contribute about 95% of the yellow color. Another yellow pigment, betalamic acid, derives directly from cleavage of betanin and probably the key intermediate in the biogenesis of all betalaines.

There has been considerable interest and research activity in connection with the betalaines during the past decade or so, stemming principally from tighter restrictions, including banning, of several synthetic colorants for foods. See also **Colorants (Foods).**

The red and golden cultivars of beetroot (*Beta vulgaris* L.) appear to be excellent sources of both red and yellow, water-soluble colorants. A factor of concern in connection with the betalaines as possible coloring agents is their earthy flavor, directly reminiscent of beet taste. The principal contributor of this flavor is a substance known as *geosmin,* a complex organic alcohol. Some problems encountered to date in preparing suitable red and yellow colorants from beet raw materials, in addition to the flavor, are rather poor yields and lack of stability of the extracted substances.

References

Acree, T. E., et. al.: "Geosmin, the Earthy Component of Table Beet Odor," *J. Ag. Food Chem.*, **24**, 430 (1976).

Considine, D. M. (editor): "Foods and Food Production Encyclopedia," Van Nostrand Reinhold, New York, 1982.

Driver, M. G., and F. J. Francis: "Stability of Phytolaccanin, Betanin, and FD&C Red #2 in Dessert Gels," *J. Food Sci.*, **44**, 2, 518–520 (1979).

Pasch, J. H., and J. H. von Elbe: "Sensory Evaluation of Betanine and Concentrated Beet Juice," *J. Food Sci.*, **43**, 5, 1624–1625 (1978).

Williams, M., and G. Hrazdina: "Anthocyanins as Food Colorants," *J. Food Sci.*, **44**, 1, 66–68 (1979).

BETAMETHASONE. Steroids.

BETA-RAY CHEMICAL ANALYZERS. Instrumental beta-ray absorption techniques can be used for determining H_2 in hydrocarbons and, consequently, the hydrogen-carbon ratio. The range of concentration measurements is from 0 to 100%, although the presence of sulfur and oxygen may interfere, resulting in high H_2 readings. The measurement principle is based upon the fact that H_2 has twice as many electrons per unit weight as other atoms and, accordingly, has twice the beta-ray absorbence as carbon per unit weight. Beta-ray absorbence of the sample is measured by a null-balance ion chamber to indicate readings that are proportional to (Weight of Carbon/ml) + 2X(Weight of H_2/ml). A simultaneous density reading is proportional to (Weight of Carbon/ml) + (Weight of H_2/ml). These expressions thus permit calculation, based upon empirical calibration of the absorbence scale, of (Weight of H_2/ml) and the consequent determination of weight percent or the hydrogen-carbon ratio. Determinations require about 5 minutes, although readings can be made continuously. Less than 50 milliliters of sample are required. A convenient beta-ray source is strontium-90, with a half-life of 25 years. See also **Analysis (Chemical).**

Beta-ray backscattering techniques also are applied to determine (1) the average atomic number of a sample having a fixed thickness, and (2) the thickness of coatings having fixed composition. Elements of high atomic number scatter beta rays more intensely than those of low atomic number. The beta rays from a radioactive source strike the sample. Scattered beta rays re-emitted from the same side of the

[1] Sometimes spelled without the last *e*.

sample are detected in an ion chamber, shielded from the source, to produce a small current proportional to the backscattering effect. The sample thickness must be controlled. Sample windows must be thin and of low-atomic number material to avoid contribution to the measured effect.

BETELGEUSE (α Orionis). A star whose name is a contraction of Arabic phrase indicating that it is the "armpit of the central one," i.e., the armpit of Orion. Because of its rich reddish color, this star has frequently been referred to as the "martial one." Because it is the first star to rise of the brilliant and well-known constellation Orion, the title of "roarer" or "announcer" has been assigned to it by ancient writers.

Betelgeuse is of great interest astronomically. It is an irregular variable star. It is also one of the first stars to have its diameter measured with the stellar interferometer. The diameter is found to be variable between 12.7 and 7.8×10^8 km. At maximum diameter, the star would extend beyond the orbit of Mars, and almost to the orbit of Jupiter, if put in the sun's place.

Ranking ninth in apparent brightness among the stars, Betelgeuse has a true brightness value of 17,000 as compared with unity for the sun. Estimated distance from the earth is 500 light years. Betelgeuse is classified as a red star of special type M. See also **Constellations.**

BETEL NUT. Palm Trees.

BETULACEAE. Alder Trees.

BEUSITE. A mineral: $(Mn, Fe, Ca, Mg)_3 (PO_4)_2$.

BEVEL GEARING. Straight-tooth and spiral-tooth bevel gearing are used to transmit motion between shafts whose axes intersect. The operation of such units is analogous to that of friction cones, which may be considered to represent the pitch cones for the bevel gearing, and which correspond to pitch cylinders for spur gearing. Straight-tooth bevel gears have teeth of involute form, but the straight-line elements converge (if extended) at the intersection of the shaft axes, in contrast to the parallel-tooth elements of spur gears. There are several forms of bevel gearing; in the most important, the gear and pinion operate at a shaft axes angle of 90°. The unit is termed miter gearing if both gear and pinion have the same number of teeth, and angular gearing if the angle between the shaft axes is less than 90°. Bevel gearing in which the shaft axes angle is greater than 90° is also used to some extent. The pitch cone angles of the pinion and gear must be complementary for a shaft axes angle of 90°.

The pitch diameter of bevel gearing is measured at the large end of the pitch cones; the addendum and dedendum are not measured in the plane of the pitch circle, as in spur gearing, but are constructed perpendicular to the elements of the pitch cone on the surface of the back cone. The tooth shape is therefore dependent upon the magnitude of the back-cone radius, rather than the pitch radius, as in spur gearing.

Mortise gears have cast-iron rims with cored slots into which hard maple cogs or teeth are fitted and held in place by wedges at the back of the rim, and are designed to operate with cast iron cast tooth pinions. Cast-tooth gearing, however, is used only where the pitch-line velocity is low and where smooth action is not particularly important.

Spiral bevel gears have teeth cut in the arc of a spiral across the gear face, and bear the same relation to straight bevel gears that helical gears do to spur gears. This construction results in a larger number of teeth in contact than in straight-tooth bevel gearing, and

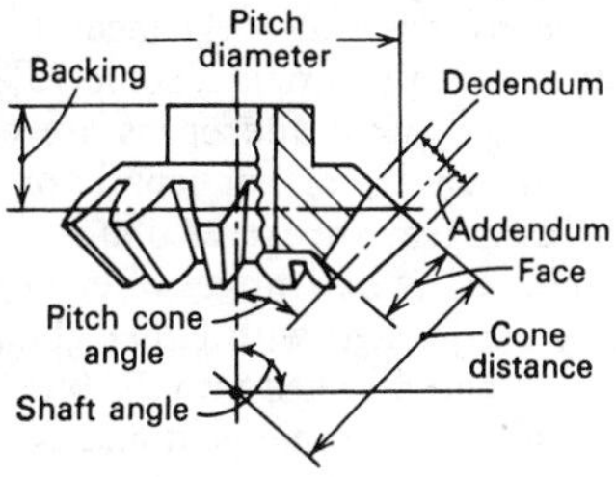

Bevel gearing.

like helical gearing, permits higher pitch-line velocities and greater load-carrying capacities for the same occupied space. They are often used to replace straight-tooth bevel gear sets.

Hypoid gearing resembles spiral bevel gearing in general appearance, but is used for transmitting power between shafts whose axes are perpendicular, but nonintersecting. They find widespread application in automobile rear-axle drives, and in numerous other instances where bevel or worm gearing is not applicable.

BHA, BHT, AND TBHQ. Antioxidant.

BHARAL. Goats and Sheep.

BIAS. A direct-current voltage applied to a transistor control electrode for establishing the desired operating point.

BIAS CELL. A small electric cell which is capable of supplying an open-circuit voltage of $1\frac{1}{2}$ or $1\frac{3}{4}$ volts indefinitely. Also used to mean the source of bias for a tube or transistor, although different numerical values than cited may be required.

BIAS (Statistics). In the theory of statistical estimation, any effect which systematically distorts an estimate from the true value, as distinct from one which may distort it on any particular occasion but averages out to zero over a long series of estimates. If t is an estimator of a parameter θ, t is said to be biassed if the *expectation* of t is not equal to θ, and unbiassed if

$$E(t) = \theta$$

BICEPS. Any of several muscles with two heads. The principal examples are (1) the biceps femoris which, in humans, lies in the back part of the thigh and flexes the lower leg, and (2) the biceps brachii of the upper arm which flexes the lower arm.

Biceps muscle contracting forearm.

BICHIRS (*Osteichthyes, Cladistia*). An air-breathing fish, *Polypterus*, found in African waters, including the Nile. One of a few living representatives of a group which was once abundant and is supposed to have been ancestral to the bony fishes. It was formerly thought that *Polypterus* belonged to the ancient and extinct paleoniscids; it probably is a link between the polypterids and the paleoniscids. Maximum size of the polypterids ranges between 2 and 3 feet. They are quite slender as indicated by the accompanying diagram. Bichirs are consumed as

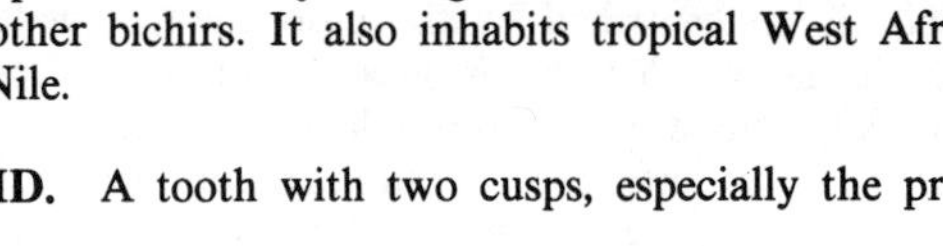

Bichir (*Polypterus senegalus*).

food by natives in some areas, but their fine bone structure makes eating quite difficult. Of the 11 species of polypterids, ten are included under *Polypterus*. The reed fish (*Erpetoichthys*) is the remaining or eleventh species. Its body is long, snakelike, but with a head much like the other bichirs. It also inhabits tropical West African rivers and the Nile.

BICUSPID. A tooth with two cusps, especially the premolars of humans.

BIELIDS. Bielids or Andromedes are the names applied to a meteor shower that is observed about November 24th of each year. The radiant point of the shower is in the constellation of Andromeda.

The history of the Bielids and their connection with Biela's comet is one of the interesting chapters in the development of meteoric astronomy. Biela's comet was first discovered in 1772, but was not found to be periodic. In 1826, Biela discovered the comet again, and it now bears his name. Its orbit is elliptical, with a period of about 6 years. In 1832, the comet passed very close to the earth. In 1845, the comet was observed to break in two, and in 1852, at the time of the predicted return, it was found that the two parts of the comet were both very faint and separated by over a million miles. They were unfavorably located relative to the sun for observation in 1859, and at the time of the return in 1866 they were not to be found.

The first mention of a swarm of meteors located on the orbit of Biela's comet is found in the display of December 5, 1741, when a brilliant shower was observed in Russia. The swarm was observed during December in 1798, 1830, and 1838. By 1838, the radiant point had been calculated and located, apparently in Cassiopeia. By 1867, the date of the shower had shifted to November, and thereafter, it has always been observed in that month. Up to 1885, many brilliant meteors were associated with the radiant point, with from two to four observers on November 27th of that year observing no less than 39,546 meteors in 4 hours and 8 minutes. Since 1899, very few members of the shower have been observed, and it is evident that perturbations have shifted the orbit of the main swarm well outside the orbit of the earth. See also **Meteorite.**

BIENNIAL. Many plants, including some of the commonest cultivated ones, require 2 years of growth to complete their life cycle. Such plants are called biennial. During the first year of growth, they commonly form a close rosette of leaves growing from a very short stem and spreading out close to the ground, or form a head (as in the cabbage), and develop a thick tap root in which is accumulated a considerable amount of food reserves. During the second year of growth, this reserve food is drawn upon to permit the development of a tall stem and flowers and fruit. Beets and carrots are examples of biennial plants in which the root rich in stored food reserves becomes an important source of food for humans.

"BIG BANG" THEORY. Cosmology; Red Shift.

BIGCONE PINE. Pine Trees.

BIGEYE TUNA. Tunas.

BIGNONIACEAE. Catalpa Trees.

"BIG TREES" (Sequoia). Giant Sequoia.

BILE. A bitter alkaline fluid secreted by the liver into the duodenum, which aids in the digestion of food. The chief components of bile are bile salts and bile pigments. Because of its strong alkalinity, bile neutralizes the acid coming into the duodenum from the stomach. The bile not only performs important functions in the process of digestion, but also serves as a vehicle for the excretion of waste products from the body.

Bile salts help in the breakdown of fat in the intestines and in fat absorption through the intestinal wall. The bile salts are injected into the digestive canal at the duodenum. They are not excreted, but are almost totally absorbed through the walls of the intestine, to be used over and over again. Bile pigments are derived from the hemoglobin of broken-down red blood cells and are excreted with the feces. When the pigments appear in excessive amounts in the blood, the mucous membranes and conjunctiva of the eye become stained a pale yellow, and the patient is said to be jaundiced.

Bile is continually secreted by the liver and stored in the gallbladder. Here the bile is concentrated by the absorption of water through the walls of the gallbladder. Bile is released from the gallbladder into the intestine when food passes through the pyloric valve from the stomach into the small intestine. Gallstones are formed of constituents of the bile which have settled out of solution. The stones vary in size, color, and structure, according to the materials composing them.

An inadequate supply of bile contributes to vitamin A deficiency because of disturbances of the intestinal tract which prevent the effective absorption of the vitamin. In an average adult, from one-half to one liter of bile is secreted every 24 hours, the quantity depending upon the amount and kind of food eaten.

Part of the cholesterol newly synthesized in the liver is excreted into bile in a free nonesterified state (in constant amount). Cholesterol in bile is normally complexed with bile salts to form soluble choleic acids. Free cholesterol is not readily soluble and with bile stasis or decreased bile salt concentration may precipitate as gallstones. Most common gallstones are built of alternating layers of cholesterol and calcium bilirubin and consist mainly (80–90%) of cholesterol. Normally, 80% of hepatic cholesterol arising from blood or lymph is metabolized to cholic acids and is eventually excreted into the bile in the form of bile salts.

The C_{24} bile acids arise from cholesterol in the liver after saturation of the steroid nucleus and reduction in length of the side chain to a 5-carbon acid; they may differ in the number of hydroxyl groups on the sterol nucleus. The four acids isolated from human bile include *cholic acid* (3,7,12-trihydroxy), as shown in Fig. 1; *deoxycholic acid*

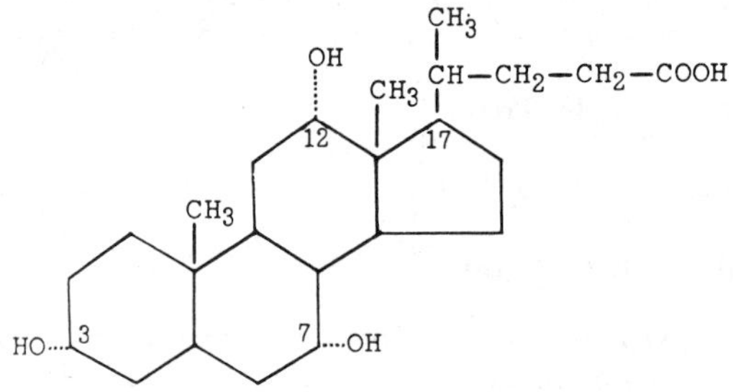

Fig. 1. Cholic acid.

(2,12-dihydroxy); *chenodeoxycholic acid* (3,7-dihydroxy); and *lithocholic acid* (3-hydroxy). The bile acids are not excreted into the bile as such, but are conjugated through the C_{24} carboxylic acid with glycine or taurine, $NH_2—CH_2—CH_2—SO_3H$. This esterification of the bile acids to soluble conjugates occurs in the microsomes and requires coenzyme A, magnesium ion, and ATP (adenosine triphosphate). Although taurocholic acid predominates at birth, the most abundant of the bile acids in the adult is glycocholic acid. In alkaline bile, the conjugated bile acids exist in their ionized form as the bile salts, glycocholate or taurocholate. Bile salts can function as effective product feedback inhibitors of hepatic cholesterol synthesis. Because of their detergent action, bile salts play an important role in the absorption of cholesterol, fats, and fat-soluble vitamins. The bile salts are believed to facilitate absorption of these compounds by the formation of micelles or aggregates of low osmotic pressure. The bile salts themselves are not absorbed during this process. Their absorption from the intestine occurs at a different site and at an entirely different rate from that of the lipids. Approximately 95% of the bile salts are reabsorbed, enter the enterohepatic circulation, and are ultimately re-excreted into the bile for further utilization in lipid absorption.

Most of the hormones that are normally conjugated in the liver to form glucuronides or sulfates, such as the steroids, thyroxine, epinephrine, and norepinephrine, are secreted into the bile, but to a varying degree, may be re-absorbed in the intestine and eventually excreted in the urine. The 17-hydroxysteroids, including cortisol, are secreted into the bile primarily as reduced glycuronide conjugates. More than 70% of these conjugates enter the enterohepatic circulation and are eventually excreted into the urine; less than 30% are found in the feces. Progesterone, after its conversion to pregnanediol, is also excreted into the bile primarily as the glucuronide, some 75% of which is eventually excreted in the feces. Most androgens are excreted as sulfates in the urine, part of which are of nonhepatic or nonbiliary origin. Significant amounts of estrogens are excreted into bile as estriol glucuronide or estrone sulfates. Many derivatives of epinephrine and norepinephrine are eventually conjugated with either glucuronide or sulfate at the 4-hydroxy position and excreted into the bile. Thyroxine is predominantly conjugated with glucuronic acid and is excreted as such into the bile. The bile, however, is not a significant route for the net disposal of thyroxine, since this hormone is rapidly re-absorbed, enters the enterohepatic circulation and is eventually excreted as urinary metabolites.

The major components of bile, the bile pigments, can account for 15–20% of the total solids. *Bilirubin* comes primarily from the degradation of heme in the reticuloendothelial system in the spleen, bone marrow, and to a lesser extent, the liver. The initial step in the metabolism of heme is the cleavage of the porphyrin ring and elimination of the alpha methylene carbon to produce an open tetrapyrrole. This may exist as a complex with iron and globin called choleglobin. After removal of the iron and globin, the resulting tetrapyrrole, biliverdin, is rapidly reduced to bilirubin, the major pigment in human bile. Not all bilirubin results from the breakdown of hemoglobin from mature red cells. The early appearance of labeled bilirubin after injection of precursor glycine-^{14}C indicates that some bilirubin (approximately 10%) may arise from: (1) the rapid breakdown of immature red cells in the bone marrow; (2) from heme that had not entered hemoglobin; or (3) from the destruction of newly formed red cells in the peripheral circulation. This "shunt" pathway for bilirubin formation may predominate in pernicious anemia and some porphyrias. A small amount of bilirubin may also arise from other heme pigments, such as myoglobin or the cytochromes. The bilirubin that enters the blood is rapidly and solely bound to albumin. Normal circulating levels of bilirubin are less than 1 mg/100 ml. Free bilirubin, which readily crosses the blood-brain barrier in the newborn, and to a lesser extent in the adult, is an effective uncoupler of oxidative phosphorylation in the brain and is highly toxic.

The hepatic transport of bilirubin from plasma to bile involves three independent, but related mechanisms, i.e., uptake, conjugation, and secretion. Plasma bilirubin is dissociated from plasma albumin in the liver and is rapidly concentrated in the cytoplasm of the hepatic cells by an unknown mechanism which precedes and is relatively independent of any subsequent hepatic conjugation. After concentration in the liver, bilirubin is conjugated with 2 moles of glucuronic acid to form bilirubin diglucuronide, the glucuronic acid moieties being attached in ester linkage to the carboxyl groups on the propionic acid side chains. See Figs. 2 and 3.

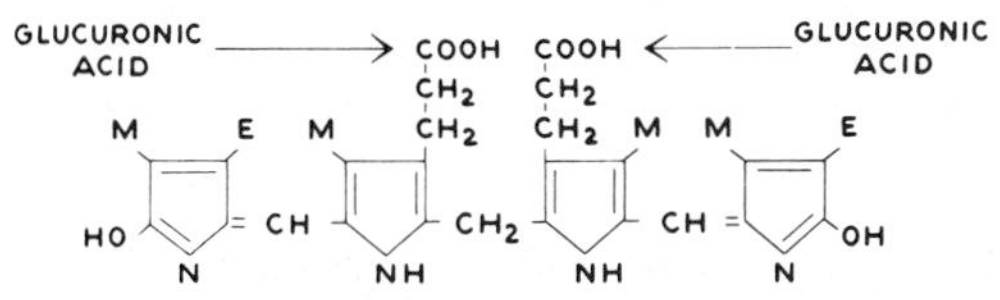

Fig. 2. Structure of "direct" bilirubin.

Glucuronyl transferase, the enzyme catalyzing the final step, is located in the smooth endoplasmic reticulum of liver and to a lesser extent in kidney and gastric mucosa, where a small amount of extrahepatic conjugation may occur. This enzyme has not been purified and it is unclear whether it nonspecifically catalyzes glucuronide conjugation of many nonbilirubinoid substrates, or is bilirubin specific and a member of a large group of closely related glucuronyl transferases. Its activity can be induced by a variety of drugs and can be inhibited with steroids or steroid glucuronides found in plasma of pregnant women.

Crigler-Najjar's disease in humans is characterized by increased levels of unconjugated bilirubin in the serum. A genetic impairment of glucuronyl transferase, the enzyme responsible for the transfer of glucuronic acid from uridine diphosphate glucuronic acid, exists not only in the liver, but in the kidney as well. Gilbert's disease is characterized by a mild increase of unconjugated bilirubin in the plasma, which may result from a partial impairment of glucuronyl transferase, from a defect in bilirubin transport in the blood, or a defect in hepatic uptake. In subjects with Dubin-Sprinz or Dubin-Johnson disease, the serum contains high levels of both unconjugated and conjugated bilirubin, and an unidentified brown pigment is present in the liver. A defect in the secretion of the bilirubin conjugates from the hepatic cell is a probable causative factor. Rotor's disease is also characterized by increased serum levels of both unconjugated and conjugated bilirubin, but it differs from Dubin-Sprinz disease in that the hepatic brown pigment is not found. The foregoing syndromes are sometimes collectively referred to as *idiopathic hyperbilirubinemia.*

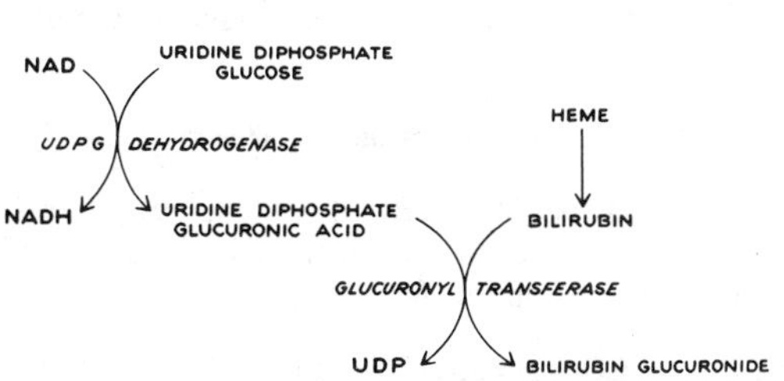

Fig. 3. Conjugation of bilirubin.

The mild nonhemolytic jaundice often present in the newborn (physiological jaundice) or the more severe jaundice and kernicterus in premature infants may result in part from an inability of the immature liver to conjugate bilirubin; low hepatic levels of both glucuronyl transferase and uridine diphosphate glucuronic acid dehydrogenase (the enzyme that catalyzes the synthesis of uridine diphosphate glucose glucuronic acid from uridine diphosphate glucose) are found in fetus and newborn. Hepatic secretion of conjugated bilirubin may also be impaired.

References

Elias, E.: "Cholangiography in the Jaundiced Patient," *Gut,* **17**, 801 (1976).

Metzger, A. L., et al.: "Diurnal Variation in Biliary Lipid Composition: Possible Role in Cholesterol Gallstone Formation," *New Engl. J. Med.,* **288**, 333 (1973).

Small, D. M.: "The Formation and Treatment of Gallstones," in "Diseases of the Liver," (L. Schiff, editor), 4th edition, Lippincott, Philadelphia, 1975.

Thistle, J. L., et al: "Chemotherapy for Gallstone Dissolution," *J. Amer. Med. Assn.,* **239**, 1041 (1978).

Warren, W. K., and E. G. C. Tan: "Diseases of the Gallbladder and Bile Ducts," in "Diseases of the Liver," (L. Schiff, editor), 4th edition, Lippincott, Philadelphia, 1975.

Wenckert, A., and B. Robertson: "The Natural Course of Gallstone Disease," *Gastroenterology,* **50**, 376 (1966).

BILIARY TRACT AND DISEASES. **Bile; Gallbladder and Biliary Tract Diseases; Jaundice; Liver; Wilson's Disease.**

BILIRUBIN. **Bile; Photochemistry and Photolysis.**

BILHARZIASIS. Also known as schistosomiasis; invasion of the body by blood flukes of the genus *Schistosomidae.* Ova from an infected person leave the body in the urine or feces and invade an intermediate host, a mollusc, in which they develop into cercariae; these escape into the water surrounding the mollusc, whence man may be infested by drinking or by washing. Chronic ulcerative lesions are produced in the lower bowel and in the ureters and bladder and accessory genitourinary organs; hematuria is a constant symptom.

Treatment of bilharziasis is difficult. Traditionally, antimony preparations have been found effective in some patients. In some instances, to treat patients with resistant strains, a mechanical filter has been surgically placed in the saphenous vein of the thigh to catch and remove eggs from the blood. If the organisms are present in the bladder for a number of years, cancer may result. One of the most dramatic examples of epidemic cancer is the high incidence of bladder cancer in Egypt, attributed to bilharziasis.

BILLET MILL. **Iron Metals, Alloys, and Steels.**

BILLET SPLIT LENS. This special lens is made by cutting a lens into two halves, parallel to the optic axis. By displacing the axes of the two halves slightly, overlapping beams for a slit source will form interference fringes. See **Interferometer.**

BILLFISHES (*Osteichthyes*). Of the family *Istiophoridae,* there are approximately ten species characterized by bills of rounded cross section and two ridges (on each side of caudal peduncle) immediately in front of the tail. It should be noted, as mentioned later, that the bill of the swordfish is flattened and there is but a single ridge on the caudal peduncle. In the overall family of billfishes are the sailfishes,

Sailfish. (*American Museum of Natural History.*)

spearfishes, and marlins. Members of this family are fish eaters. The bill is used in clublike fashion to stun their victims, such as a school of frigate mackerel, as they speed through such a group.

Because the dorsal fin is long and exceptionally high, resembling a sail as shown as the accompanying diagram, the sailfish is the most easily recognized of the istiophorids. The upper jaw of the sailfish is prolonged into a sharp sword resembling that of the related swordfish. The Pacific sailfish (*Istiophorus orientalis*) constitutes the largest species—up to about 220 pounds (100 kilograms), and 11 feet (3.4 meters) in length. Somewhat smaller, the Atlantic *I. albicans* runs about the same length, but with a maximum weight of only about 125 pounds (57 kilograms).

The black marlin of the Indo-Pacific is probably the easiest of the marlins to identify. A specimen of this fish (*Istiompax marlina*) has been recorded at 1560 pounds (708 kilograms) and a length of 14½ feet (4.4 meters), caught off Peru (Cabo Blanco). The Pacific striped marlin (*Makaira audax*) features ten or more vertical stripes on its sides. Maximum weight is about 600 pounds (272 kilograms). The Pacific blue marlin (*M. ampla*) also has vertical stripes. The blue marlin has a maximum weight of about 1400 pounds (635 kilograms). Blue marlins are also found in the Atlantic, normally no farther north than Long Island. They are quite similar to their Pacific counterparts. Also well known among sportsmen is the Atlantic white marlin (*M. albida*), with a maximum weight just in excess of 100 pounds (45.4 kilograms) and a length approaching 9 feet (2.7 meters).

Swordfish. Of the suborder *Scombroidei* (mackerels) and family *Xiphiidae*, the swordfish (*Siphias gladias* Linne) has an elongated, tuna-like body, lacking scales. Only juveniles have small, highly degenerate scales. The skin, however, has a rough texture, since it is covered by dermal teeth somewhat like those in sharks. The most distinctive feature of the species is the "sword," a long dagger-shaped beak which in adults can comprise one third of the entire length of the fish. It is formed from elongation of the upper jaw, intermaxillary, ethmoid, and vomer bones. The lower jaw is only slightly elongated. The family is represented by the sole genus and species given above. The fish attains a length of over 13 feet (4 meters) and a weight of over 660 pounds (300 kilograms). The fish is widely distributed in tropical and temperate parts of every ocean. It is also found in the Mediterranean and occasionally in the Black Sea. It occurs along Europe's Atlantic coast as far north as southern Norway, but swordfish rarely penetrate the North Sea, Scandinavian waters, and the Baltic Sea. Its back is blue-black; the sides are gray-blue, and the belly is white.

Swordfish spawning sites are located in the southern part of the Sargasso Sea and, to a lesser extent, in the Mediterranean. The swordfish is poplar wherever it occurs because of its flavorful meat, but is of major significance mainly in Japan. Sports fishermen along the Atlantic and Pacific coasts of the United States catch swordfish with hooks and harpoons. One of the kings of the sea, the swordfish puts up a mighty fight, which it often wins. Swordfish caught off the Hawaiian Islands are sometimes marketed in the United States along with swordfish from other regions. The Hawaiian swordfish normally weighs about 220 pounds (100 kilograms). See also **Fisheries;** and **Fishes.**

BIMETAL THERMOMETER. Thermostatic bimetal can be defined as a composite material, made up of strips of two or more metals fastened together, which, because of the different expansion rates of the components, tends to change its curvatuve when subjected to a change in temperature.

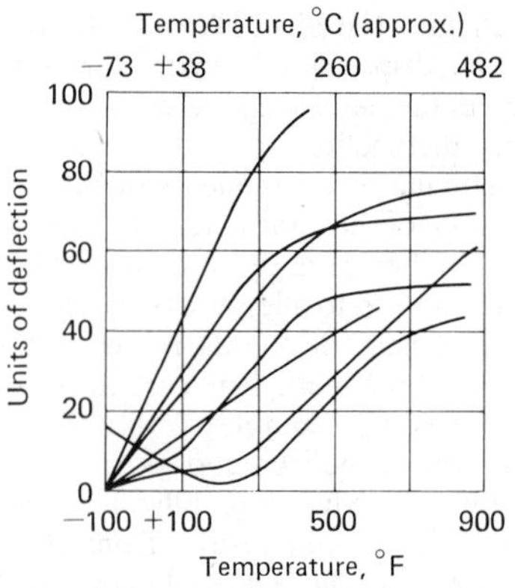

Fig. 1. Deflection characteristics of various bimetals.

With one end of a straight strip fixed, the other end deflects in direct proportion to the temperature change and the square of the length, and inversely as the thickness, throughout the linear portion of the deflection characteristic curve. If a strip of bimetal is wound into a helix or spiral and one end is fixed, the other end will rotate when heat is applied. The angular deflection varies directly with the temperature change and the length of the strip, and inversely with the thickness of the material, over the linear parts of the deflection characteristic curve. Bimetals show uniform deflection only over part of the deflection characteristic curve, as shown in Fig. 1. The three types of elements most commonly used in thermometers are shown in Fig. 2.

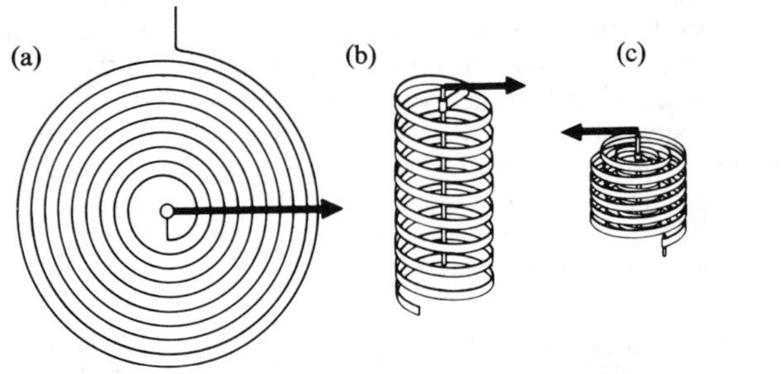

Fig. 2. Principal configurations of elements used in bimetal thermometers: (a) flat spiral, (b) single helix, (c) multiple helix.

Bimetal thermometers are made in ranges from +1000°F (538°C) down to −300°F (−184°C) and lower. However, at low temperatures, the rate of deflection drops off quite rapidly. Because of its long-term instability at high temperatures, the maximum recommended continuous use is about 800°F (427°C). Special bimetal thermometers can be obtained for continuous use up to 1200°F (649°C). Good bimetal thermometers retain their accuracy indefinitely. Usually industrial bimetal thermometers read with an accuracy of ± 1% at any point on the scale. The speed of response of bimetal thermometers is generally about the same as that for liquid-in-glass thermometers in similar ranges.

The thermostatic bimetal approach is used widely in a variety of thermostatic-type temperature-control situations, as found in heating and air-conditioning systems and in automotive cooling systems, among others.

BIMORPH CELL. Two piezoelectric plates cemented together in such a way that the application of a potential to them causes one to expand and the other to contract, thus producing a bending of the combination.

BINARY ARITHMETIC. Information Theory.

BINARY-CODED DECIMAL (BCD). A coding system for characters using six binary digits to represent each character. The system is widely used in the computer field.

BINARY COUNTER (Computer System). Counter (Computer System).

BINARY COUNTER (Fluidic). Fluidics.

BINARY ENTROPY FUNCTION. Information Theory.

BINARY ERASURE CHANNEL. Information Theory.

BINARY GRANITE. A granite containing only the essential minerals quartz and feldspar. Also used to describe granites which contain muscovite and biotite micas.

BINARY NUMBER. A number system which uses the number 2 as the base (radix) instead of 10 as used in the decimal system. The only digits are 0 and 1. The successive binary integers as 0, 1, 10, 11, 100, 101, 110, 111, 1,000 . . . etc. The binary fractions are 0.1 ($\frac{1}{2}$), 0.01 ($\frac{1}{4}$), 0.001 ($\frac{1}{8}$) . . . etc. Admittedly, the system requires many more positions to express a given number than is true of the decimal system. However, in terms of computing hardware, there are many more devices or systems which can be used to represent binary numbers than for any other base. A system with both an "on" and an "off" state, a positive and a negative state, or a high and a low state, can be used to represent binary numbers. Binary digital computers have been constructed in which mechanical, relay, electronic, and fluidic logical elements were used—along with mechanical, relay, electronic, acoustic, magnetic, photographic, and other storage elements. Most computers that are referred to as "decimal" computers are internally coded in binary.

BINARY STARS. The term *binary star* was apparently first introduced by Sir William Herschel (1802) to designate, "a real double star—the union of 2 stars that are formed together in 1 system by the laws of attraction." Binary stars are frequently classified under three types—*visual binaries, spectroscopic binaries,* and *eclipsing binaries.* Each of these types is described in separate entries in this book. A binary system also falls within the general definition of *double star.* See also **Double Star.**

A general definition of binary star (system) describes a system of two co-orbiting stars. Not many years ago, it was believed that about 25% of all stars were binary systems, with a considerable percentage, possibly as much as 10%, being multiple systems, i.e., containing more than two stars. As pointed out by Heintz in the preface of his 1979 book (see list), "Double and multiple stars are the rule in the stellar population, and single stars the minority, as the abundance of binary systems in the space surrounding the sun shows beyond doubt."

Considerable astronomical research into the presence and characteristics of binary stars has gone forward since about 1830, commencing of course with rather simple instruments. During the early part of the present century, the study of binary stars was central to much astronomical research. With the development of more advanced equipment, such as electronic detectors, very large optical telescopes, and the entry of radio astronomy in the 1950s–1960s, which greatly increased the vista of astronomical investigation, researchers generally became more intensively interested in the more remote regions of space, with relatively few scientists concentrating on binary systems. However, it is interesting to note that, during recent years, the study of binary systems has regained a high degree of interest and that such studies are now one of the important keys to the investigation of stellar evolution. See **Cosmology.**

Early investigators found a direct correlation between the period of revolution of a binary star and the eccentricity of its orbit, with systems of short periods having the smaller eccentricities. They found that there is a regular gradation from pairs with short period, in which the stars are practically in contact, to pairs so widely separated that the physical connection is only indicated by their common proper motion through space. It was also learned that, in pairs whose components are of equal brightness, both stars have the same spectral type; while in systems where the brightnesses are different, the fainter star is bluer if the brighter star is a giant, and redder if the brighter star belongs to the main sequence.

The evolution of such systems of stars is quite complicated, depending upon the exact mass ratio and stage in the nuclear history of the individual components as well as the period of the system. If two stars should become sufficiently bloated from evolution, they may form a common envelope system or contact system, in which the two stars are linked by a shared atmosphere which corotates with the pair. If only one of the stars swells far enough to feel a significant gravitational contribution from its companion, either an enhanced stellar wind or active mass streaming between the components may result (see below). Finally, a supernova explosion of one of the components may either disrupt the system or produce a bound neutron star, which will then gradually evolve from a highly eccentric orbit into a circularized path on a time scale which depends on the tidal energy dissipation within the envelope of the companion. In all cases, the processes of convection (motion of matter in the stellar interior) and nuclear burning play the major role in determining the eventual fate of the stars in a binary system.

Research was undertaken as regards the masses of these stars. Using data on gravitational attraction, the binary stars were the only stars (with exception of the sun) where masses could be determined. In the case of a visual binary star, after the orbit has been determined and the stellar parallax of the system obtained, the combined mass of the two stars may be obtained by a direct application of the Keplerian harmonic law. In the case of eclipsing binaries that are also spectroscopic binaries, it is possible to make a complete solution for the specifications (i.e., masses, densities, sizes, luminosities, and approximate shapes) of both members of the system.

For a given mass ratio and period, it is possible to compute the critical size of a gaseous body at which tidal interaction with a companion mass will literally rip matter from the stellar surface. This distance, called the Roche limit, is the most sensitive to the total mass and period of the binary. Kopal, Kuiper and Paczynski, among others, have explored the consequences of this idea quite thoroughly in a preliminary way. If the star completely fills its "Roche lobe," matter at the point along the line of centers becomes unstable and will begin to move towrd the companion, in the form of a stream. Depending upon the period and relative sizes of the components, this stream may form an accretion disk about the companion which then slowly falls onto its surface (like Algol or the binary x-ray sources) or impacts the surface forming a rapidly rotating photosphere (like U Cephei). Such a system is called "semidetached," to distinguish it from the detached systems in which the components evolve essentially as single stars. If the star should overfill its Roche lobe by a large factor, it may completely engulf the companion, forming a "contact" system like the W UMa stars. These evolve rapidly and are still not well understood. Finally, depending upon whether the companion has a magnetic field or not, material from the accretion disk, should one form, will either accrete spherically or at the magnetic poles. The latter case forms what we see as pulsed x-ray binaries when the companion is a compact object like a white dwarf or a neutron star (like Her X-1 and Cen X-3), while the former is probably responsible for the cataclysmic variables like novae (for instance, GK Per).

In searching for new pulsars (see **Radio Stars**), astronomers working with the large radio telescope in Arecibo, Puerto Rico, detected the radio pulsar PSR 1913+16. It was found that this pulsar was orbiting a massive but unseen companion. A binary system of this type had been suggested earlier as a source of intense gravitational radiation. The discovery of PSR 1913+16 was a major event in the comparatively new field of gravitational astronomy.

References

Batten, A. H.: "Binary and Multiple Systems of Stars," Pergamon, New York (1973).

Batten, A. H. (editor): "Extended Atmospheres and Circumstellar Matter in Spectroscopic Binary Systems," Reidel, Boston, 1973.

Eggleton, P., Mitton, S., and J. Whelan (editors): "Structure and Evolution of Close Binary Systems," Reidel, Boston, 1976.

Heintz, W. D.: "Binary Systems," Reidel, Boston, 1978.

Kopal, Z.: "Dynamics of Close Binary Systems," Reidel, Boston, 1978.

Manchester, R. N., and J. Taylor: "Pulsars," Freeman, San Francisco, 1977.

Popper, M., and R. Ulrich (editors): "Close Binary Stars: Observations and Interpretation," Reidel, Boston, 1980.

Slettebak, A. (editor): "Be and Shell Stars," Reidel, Boston, 1976.

Staff: "Gravity and PSR 1913+16," *Astronomy*, 7, 6, 60–61 (1979).

Steven N. Shore, Assistant Professor of Astronomy, Warner and Swasey Observatory, Case Western Reserve University, Cleveland, Ohio.

BINARY STATE. This term is applied to the property of a Boolean variable, i.e., it can assume either of two mutually exclusive alternatives called binary states. In engineering practice, the symbols 1 and 0 are assigned to the two possible values of the Boolean variable. Usually, 1 represents "Yes," "On," or "True." Similarly, 0 represents "No," "Off," or "False." In digital computer logical circuits, these two states are represented by discrete current or voltage levels.

BINARY SYMMETRIC CHANNEL. Information Theory.

BINARY SYSTEM DISTILLATION. Distillation.

BINDER (Paint). Paint.

BINDING ENERGY. This term is used in atomic physics with two closely related meanings: the binding energy of a particle (or other entity) is the energy required to remove the particle from the system to which it belongs; the binding energy of a system is the energy required to disperse the system into its constituent entities. Explicit definitions are obviously necessary.

Some explicit definitions for the binding energies of particles are the following:

1) The *electron binding energy* is the energy necessary to remove an electron from an atom. It is identical with the ionization potential.

1a) The *total electron binding energy* is the energy necessary to remove all the electrons from an atom to infinite distances, so that only the nucleus remains. It is equal to the sum of the successive ionization potentials of that atom.

2) The *proton binding energy* is the energy necessary to remove a single proton from a nucleus. Most known proton binding energies are in the range 5–12 MeV, although that for ^2H is 2.23 MeV, that for ^4He is 19.81 MeV, and those for ^5Li and ^9Be are negligible.

3) The *neutron binding energy* is the energy required to remove a single neutron form a nucleus. Most known neutron binding energies are in the ranges 5–8 MeV, though that for ^2H is 2.23 MeV, that for ^9Be is 1.67 MeV, and that for ^{12}C is 18.7 MeV.

4) The *alpha-particle binding energy* is the energy required to remove an alpha-particle from a nucleus. For most light nuclides the alpha-particle binding energy is positive and is equal to several MeV. For nuclides of mass number about 125, it is approximately zero. For nuclides of mass number about 150 to 200, it is negative by about 1 to 3 MeV, but the magnitude of the Coulomb potential barrier at the nucleus is sufficiently large that penetration by an alpha particle is so improbable that lifetimes for alpha disintegration are generally too long for detection of alpha activity. For most nuclides of mass number exceeding 200, the alpha-particle binding energy is negative by about 4 to 8 MeV, which is a negative binding energy of sufficiently large magnitude to give a measurable probability of penetration of the potential barrier by an alpha particle, hence an observable alpha activity.

Some explicit definitions for the binding energies of systems are:

(1) The *nuclear binding energy* is the energy that would be necessary to separate an atom of atomic number Z and mass number A into Z hydrogen atoms and $A - Z$ neutrons. This energy is the energy equivalent of the difference between the sum of the masses of the product hydrogen atoms and neutrons, and the mass of the atom; it includes the effect of electronic binding. (See *total electron binding energy* above.)

(2) The binding energy of a solid is the energy required to disperse a solid into its constituent atoms, against the forces of cohesion. In the case of ionic crystals, it is given by the Born-Mayer equation. See **Crystal.**

Although the concept of the atom has not been in serious question for nearly a half-century, full understanding of the forces which hold the neutrons and protons together has not yet been achieved. In 1927, Aston found that experimentally measured isotopic weights differed slightly from whole numbers. See also **Aston Whole Number Rule.** From this he was led to the concept of the *packing fraction*, which is defined as the algebraic difference between the isotopic weight and the mass number, divided by the mass number. Although the theoretical significance of the packing fraction is difficult to assess, it does lead to some interesting conclusions with respect to nuclear stability. A negative packing fraction derives from a situation where the isotopic weight is less than the mass number, inferring that in the formation of the nucleus from its constituent particles, some mass is converted into energy. Since an equivalent amount of energy would be necessary to break up the nucleus into its constituent particles again, a negative packing fraction suggests a high order of nuclear stability. By the same reasoning, a positive packing fraction indicates nuclear stability. Stable elements with mass numbers above about 175 and below about 25 have positive packing fractions. It is interesting to note that the packing fractions of both hydrogen and uranium are positive.

Actually, a comparison of the isotopic weight with the mass number (as is done in determining the packing fraction) is somewhat artificial. A rigorous determination of the mass-energy interconversion in the formation of an atom would seem to require a calculation of the difference between the sum of the masses of the constituent particles of the atom and the experimentally measured isotopic weight. The value of the mass difference thus obtained is the *mass defect.* The energy equivalent of this mass difference as derived from the Einstein equation yields a measure of the binding energy of the nucleus. Division of the binding energy of a nucleus by the number of nucleons (the total number of protons and neutrons) therein yields the binding energy per nucleon. In stable isotopes, the binding energy per nucleon decreases with increasing mass number, a fact which is important in nuclear fission. Secondly, the binding energy per nucleon derived in the manner just described is an average value, whereas each additional nucleon added to the nucleus has a binding energy less than those which preceded it. Thus, the most recently added nucleons are bound less tightly than those already present.

Additional considerations regarding nuclear stability may be gleaned from a consideration of the odd or even nature of the numbers of protons and neutrons in the nucleus. According to the Pauli exclusion principle, no two extranuclear electrons having an identical set of quantum numbers can occupy the same electron energy state. See also **Pauli Exclusion Principle.** The application of this principle to the nucleus leads to conclusions which at least are not at variance with observations of nuclear stability. Thus, it is inferred that no two nucleons possessing an identical set of quantum numbers may occupy the same nuclear energy state. It would appear, then, that both protons and neutrons which differ only in their angular momenta or spins may exist in a nuclear state. The exclusion principle requires, therefore, that only protons having opposite spins can exist in the same state. The same consideration applies to neutrons. Accordingly, two protons and two neutrons might occupy the same nuclear energy state provided the nucleons in each pair have opposite spins. Such two-proton-two-neutron groupings are termed "closed shells," and by virtue of their proton-neutron interaction, they confer exceptional stability to nuclei which are made up of them. The nuclear forces in closed shells are said to be "saturated," which means that the nucleons therein interact strongly with each other, but weakly with those in other states. Since like particles tend to complete an energy state by pairing of opposite spins, two neutrons of opposite spin, or a single neutron or proton also might exist in a particular energy state.

Any of the foregoing conditions may be achieved when the nucleus contains an even number of both protons and neutrons, or an even number of one and an odd number of the other. Since there is an excess of neutrons over protons for all but the lowest atomic number elements, in the odd-odd situation there is a deficiency of protons necessary to complete the two-proton-two-neutron quartets. It might be expected that these could be provided by the production of protons via beta decay. However, there exist only four stable nuclei of odd-odd composition, whereas there are 108 such nuclei in the even-odd form and 162 in the even-even series. It will be seen that the order of stability, and presumably the binding energy per nucleon, from greatest to smallest, seems to be even-even, even-odd, odd-odd.

Although the existence of binding energies holding the nucleus together has been demonstrated, the problem of defining the nature of these forces presents itself. Clearly, repulsive electrostatic forces must exist between protons. These are "long range" in effect. To achieve nuclear stability then, compensating attractive forces also must exist. It has been concluded that "short range" attractive forces exist between protons, between neutrons, and between protons and neutrons. The

(p-n) attractive forces are considered to be of the greatest magnitude, while the (n-n) and (p-p) forces are of less intensity, with the latter decreased by virtue of electrostatic repulsion. When the number of protons in a nucleus is greater than 20, it is found that the ratio of neutrons to protons exceeds unity. The additional short range attractive forces provided by the excess neutrons, therefore, may be considered as compensating for the long range electrostatic repulsive forces between the protons. Nevertheless, when the number of protons exceeds about 50, the short range forces are insufficient to counteract the electrostatic forces completely, with the result that the binding energy per each additional nucleon decreases.

The nature of the short range attractive forces between nucleons requires further investigation. An interpretation of them has been presented by Heisenberg in terms of wave-mechanical exchange forces. Thus, if the basic difference between the proton and neutron in a system composed of these two particles is considered to be that the former is electrically charged while the latter is not, then the transfer of the electric charge from the proton to the neutron results in an exchange of individual identity, but not a change in the system. That is to say, the system still is composed of a proton and neutron, despite the fact that the particles have exchanged their identities. Since the system itself has the same composition, it must possess the same energy after the exchange as it did before. One of the principles of wave mechanics is that, if a system may be represented by two states, each of which has the same energy, then the actual state of the system is a result of the combination, i.e., resonance, of the two separate states and is more stable than either. In the proton-neutron system, the energy difference between the "combined" state and the individual states may be considered as the "exchange energy" or "attractive force" between the particles. In an extension of Heisenberg's proposal, Yukawa postulates that the exchange energy is carried by a new particle which is termed *meson.* Particles having the properties attributed by Yukawa to mesons have been identified in cosmic rays.

With such concepts of nuclear structure and stability, however imperfect, the process of nuclear fission of uranium can be considered. Although fast neutrons (greater than 0.1 MeV) can cause fission in both uranium-235 and uranium-238, thermal neutrons (about 0.03 eV) are effective only with uranium-235. Uranium-238 is unsatisfactory as a fissionable material for most purposes, however, since it has a high probability for "resonance capture" of fast neutrons, which is a nonfission process. It is instinctive to ponder why uranium-235 fissions with thermal neutrons and uranium-238 does not. It will be recalled that the binding energy for an even-even nucleus exceeds that for an even-odd. Consequently, the addition of a neutron to uranium-235, which yields an even-even compound nucleus, will contribute a greater binding energy than in the case of uranium-238 where an even-odd compound nucleus would be produced. Calculations yield a value of 6.81 MeV for the additional neutron in the former case, and 5.31 MeV in the latter. Using Bohr and Wheeler's calculations, it is found that the activation energy for fission is 5.2 MeV for uranium-235 and 5.9 MeV for uranium-238. Thus, the binding energy for an additional neutron in uranium-235 exceeds its fission activation energy, whereas it is less in the case of uranium-238. It can be seen, then, that uranium-235 fission is energetically feasible with thermal neutrons while the fissioning of uranium-238 is not.

In considering the physical forces acting in fission, use may be made of the Bohr liquid drop model of the nucleus. Here it is assumed that in its normal energy state, a nucleus is spherical and has a homogeneously distributed electrical charge. Under the influence of the activation energy furnished by the incident neutron, however, oscillations are set up which tend to deform the nucleus. In the ellipsoid form, the distribution of the protons is such that they are concentrated in the areas of the two foci. The electrostatic forces of repulsion between the protons at the opposite ends of the ellipse may then further deform the nucleus into a dumbbell shape. From this condition, there can be no recovery, and fission results.

It will be recalled that the binding energy per nucleon decreases with increasing mass number, that is, a greater amount of energy is released in the formation of nuclei of intermediate mass number from their constituent nucleons than is the case of nuclei of high mass number. Thus, energy is released in fission because the binding energy of the high mass number uranium-236 compound nucleus is less than that of the intermediate mass number fission products which are produced. The total energy thus liberated in fission is about 200 MeV. Of this, the kinetic energy of the fission products accounts for 160 MeV. These fragments, being of significantly lower atomic number, require fewer neutrons for stability than they actually contain immediately after fission. These excess neutrons, therefore, are "boiled off" the fission fragments, the process occurring in two distinct phases. In the first phase, "prompt" neutrons of about 2-MeV energy are released within 10^{-2} seconds after fission occurs and take up about 7% of the fission energy. Subsequently, after several seconds, additional "delayed" neutrons with about 0.5-MeV energy are boiled off the fission products. See also **Energy;** and **Particles (Subatomic).**

References

Austern, N.: "Direct Nuclear Reaction Theories," (Halsted Press), John Wiley, New York, 1970.

Cool, R. L. and R. E. Marshak (editors): "Advances in Particle Physics," John Wiley, New York, 1968.

Gasiorowicz, S.: "Elementary Particle Physics," John Wiley, New York, 1966.

Omnes, R.: "Introduction to Particle Physics," John Wiley, New York, 1972.

Cohen, Bernard L.: "Concepts of Nuclear Physics," McGraw-Hill, New York, 1971.

BINDING ENERGY (Nuclide). Nuclear Reactor.

BINNACLE. The stand for supporting and protecting the compass on board a ship is known as the binnacle. This stand is usually constructed of brass, and is provided with a shaded light, which illuminates the compass during the night, but does not shine in the eyes of the helmsman.

In addition to protecting the compass from effects of weather, the binnacle also contains a number of fixed and adjustable magnets and masses of soft iron for the purpose of partially compensating for the effects of the magnetic field of the ship. The process of adjusting the various compensating devices is known as compass adjusting, and is usually carried on by experts while a ship is in port. See also **Compass (Navigation).**

BINOCULAR. An instrument composed of two similar telescopes, one for each eye, usually with focusing tubes controlled by a common screw adjustment. The ordinary opera glass is a binocular utilizing Galilean telescopes. The field glass employs erecting telescopes of the spy-glass type. A well-known modern form is the "prism binocular." The special feature of this instrument is a pair of right-angled, total reflection prisms in each telescope, which contribute three advantages. 1. The prisms, by means of two double total reflections in planes at right angles, accomplish the erection of the image without additional lenses. 2. The tube is rendered much shorter than in the ordinary field glass of equal power by the "doubling up" of the rays due to the reflections. 3. The objectives are by the same means set farther apart than the eyepieces, thus increasing the "stereo power" of the instrument as a binocular, so that objects can be seen to have depth or solidity at a greater distance than with the ordinary type.

BINOCULAR VISION. The possession of two eyes set at a distance apart, but with approximately parallel axes, enables man to obtain two views from slightly different angles, and thus to become sensible of the solidity of single objects and to get an idea of the actual distribution of different objects in space. To become vividly conscious of this faculty, one has only to look about the room for a time with a hand cupped over one eye, and then suddenly to remove the hand. If the vision is reasonably normal, it will be noticed that with one eye only the scene appears flat, like a photograph, but as soon as both eyes are used, objects spring into clear relief. In some manner the brain is able, through long experience, to blend the two different sensory pictures from the two different retinal images and to interpret the resulting sensation in terms of geometrical solidity. There is, however, a limit to the distance at which this impression is perceptible, and for very distant objects other factors must be relied upon, such as the apparent size (as of buildings or trees), or the opacity of the atmosphere (as in viewing distant mountains). In the absence of such factors, no estimate of distance can be formed; thus the stars all appear to be at the same distance. This limiting "stereoscopic radius" is for

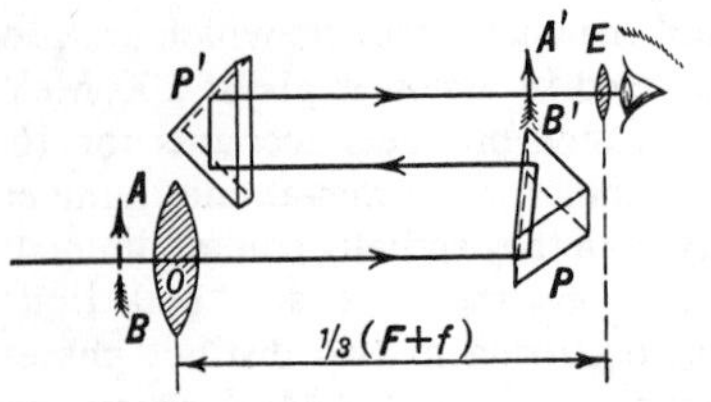

Optical system of prism binocular.

normal, unaided eyes, only a few hundred feet, but with a binocular telescope, and especially with a prism binocular, it is increased in a ratio called the "stereo power" of the instrument.

An interesting aspect of the subject is the use of binocular pictures and the stereoscope. Two photographs or drawings are prepared of the same group of objects from viewpoints approximately the same distance apart as the human eyes (~ 2.75 inches; 7 cm) and mounted side by side on a card so that each is viewed separately by the eye to which it corresponds; the observer gets the sensation of viewing a three-dimensional scene. To observation is facilitated by a pair of lenses so designed as to facilitate focusing the eyes for distance, and with a diaphragm set up between them to avoid seeing both pictures with either eye. This arrangement is the stereoscope.

BINODALS. Consider the volume-composition diagram of a binary mixture for states corresponding to the coexistence of a liquid and of a gaseous phase. If the temperature is below the critical region (see **Critical Point**) one obtains a diagram of the form represented in Fig. 1.

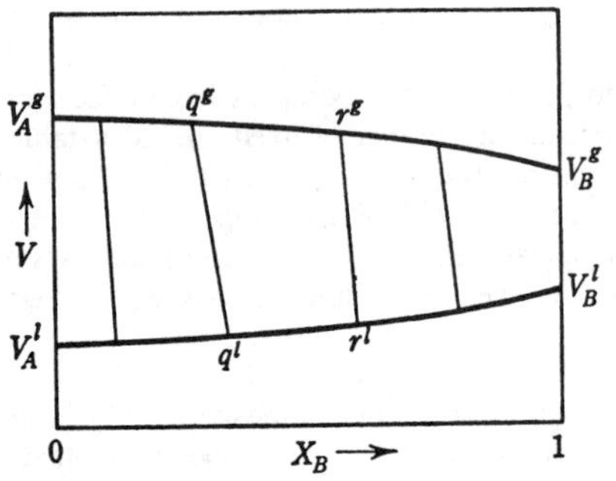

Fig. 1. Volume-composition diagram of a binary mixture below the critical region.

The line $V_A^g V_B^g$ corresponds to the molar volumes of the vapor phase, while $V_A^l V_B^l$ relates to the liquid phase. The lines such as r^l—r^g, q^l—q^g, etc., joining two phases in equilibrium are called binodals. In the critical region the diagram takes the form indicated in Fig. 2. K is the critical point. The curve V_A^g—K—V_A^l is the *saturation curve.*

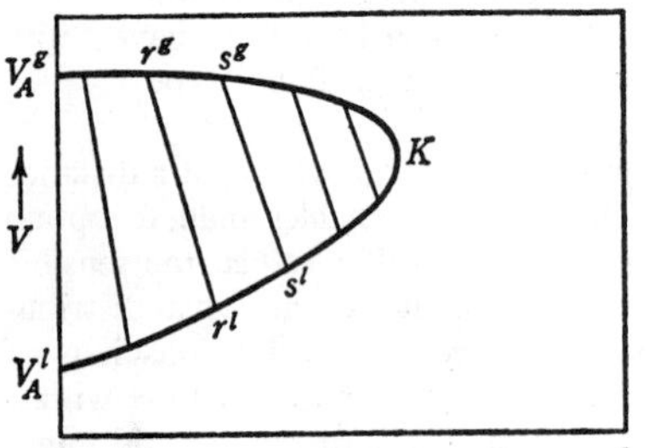

Fig. 2. Volume-composition diagram of a binary mixture in the critical region.

BINOMIAL DISTRIBUTION. If a trial can have one of two mutually exclusive results (say "success" and "failure") and if the probability p of a "success" is constant over a series of n independent trials, then the probability of obtaining r "successes" is

$$P_r = \frac{n!}{r!(n-r)!} p^r q^{n-r}$$

where $q = 1 - p$. This function is sometimes called the Bernoulli probability function, and the probability distribution of r is called the binomial distribution since the probabilities can be obtained by expanding $(q+p)^n$ by the binomial formula. The mean of the distribution is np, the variance npq. The negative binomial distribution has formally the same probability function but p and n are negative and n is not restricted to integral values.

BINOMIAL SERIES. The binomial theorem is a rule for expanding $(x + y)^n$, where n is a positive integer. The result is

$$(x+y)^n = x^n + nx^{n-1}y + \frac{n(n-1)}{2} x^{n-2}y^2 + \cdots + y^n$$

The $(k + 1)$-th term is

$$\frac{n!}{k!(n-k)!} x^{n-k}y^k$$

but the coefficient is often indicated by $\binom{n}{k}$. It is called the binomial coefficient and equals the number of combinations of n things taken k at a time. Properties of the coefficients include

$$\binom{n}{0} = \binom{n}{n} = 1; \quad \binom{n}{k} + \binom{n}{k+1} = \binom{n+1}{k+1};$$

$$\binom{n}{n-k} = \binom{n}{k}$$

If n is not a positive integer, an infinite series, called the binomial series, results. It converges absolutely for $|x| < 1$, diverges for $|x| > 1$. When $x = 1$, it converges for $n > -1$, diverges for $n \leq -1$, is absolutely convergent for $n > 0$; when $x = -1$, it is absolutely convergent for $n > 0$; divergent for $n < 0$.

If x is small enough, the quantity $(1 \pm x)^n$ may be approximated by $1 \pm nx$; $n = \pm 1, \pm 2, \pm 3, \ldots, \pm\frac{1}{2}, \pm\frac{1}{3}, \ldots, \pm\frac{3}{2}$, etc.

See also **Pascal Triangle; Taylor Series;** and terms listed under **Mathematics.**

BIN SYSTEM (Pulverized-Coal). Burner.

BIOCHEMICAL GENETICS. Genes.

BIOCHEMICAL INDIVIDUALITY. The possession of biochemical distinctiveness by individual members of a species, whether plant, animal, or human. The primary interest in such distinctiveness has centered in the human family, and in the distinctiveness within animal species as it might illuminate some of the questions on human biochemistry.

While it has been known for centuries that bloodhounds, for example, can tell individuals apart even by the attenuated odors from their bodies left on a trail, the first scientific work which hinted at the existence of substantial biochemical distinctiveness in human specimens was the discovery of blood groups by Landsteiner about 1900.

A few years later Garrod noted what he called "inborn errors of metabolism"—rare instances where individuals gave evidence of being abnormal biochemically in that they were albinos (lack of ability to produce pigment in skin, hair and eyes), or excreted some unusual substance in the urine or feces. To Garrod these observations suggested the possibility that the biochemistry of all individuals might be distinctive.

About 50 years later serious attention to the phenomenon of biochemical individuality resulted in the publication of several articles and a book on this subject. (Williams, R. J.: "Biochemical Individuality," Wiley, New York, 1956.) These reported evidence indicating that every human being, including all those designated as "normal," possesses a distinctive metabolic pattern which encompasses everything chemical that takes place in his or her body. That these patterns, like the abnormalities discussed by Garrod, have genetic roots is indicated by the pioneer explorations of Beadle and Tatum in the field of biochemical genetics in which they established the fact that the potentiality for producing enzymes resides in the genes.

Biochemical individuality, which is genetically determined, is accompanied by, and in a sense based upon, anatomical individuality, which must also have a genetic origin. Substantial differences, often of large magnitude, exist between the digestive tracts, the muscular systems, the circulatory systems, the skeletal systems, the nervous systems, and the endocrine systems of so-called normal people. Similar distinctiveness is observed at the microscopic level, for example in the size, shape and distribution of neurons in the brain and in the morphological "blood pictures," i.e., the numbers of the different types of cells in the blood.

Individuality in the biochemical realm is exhibited with respect to (1) the composition of blood, tissues, urine, digestive juices, cerebrospinal fluid, etc.; (2) the enzyme levels in tissues and in body fluids, particularly the blood; (3) the pharmacological responses to numerous specific drugs; (4) the quantitative needs for specific nutrients—minerals, amino acids, vitamins—and in miscellaneous other ways including reactions of taste and smell and the effects of heat, cold, electricity, etc. Each individual must possess a highly distinctive pattern, since the differences between individuals with respect to the measurable items in a potentially long list are by no means trifling. Often a specific value derived from one "normal" individual of a group will be several times as large as that derived from another.

The implications of this individuality are extremely broad. For medicine, they suggest that susceptibility to all disease—infective, metabolic, degenerative, mental or unclassifiable (including cancer)—probably has its roots in biochemical individuality and that the differences in responses to drugs including alcohol, caffeine, nicotine, carcinogens and morphine derivatives, which are well authenticated, have a sound and discoverable basis. It is a well-known fact that conditions which will produce disease in certain individuals will not do so in others. The basis for this observation has hitherto not been recognized; it doubtless has its roots in biochemical individuality—a development which has received little attention. People definitely possess what may be called "biochemical personalities," and it seems extremely likely that these are meaningful in connection with the numerous and increasing personality disorders and difficulties which afflict men and women in modern life.

Biochemical individuality offers a sound scientific basis for recognizing the existence in every individual of a unique makeup in the broadest sense. For centuries, "individuality" has been written about and its place in the scheme of things has been discussed, but the knowledge of what individuality consists of and how it manifests itself has indeed been scanty. Only in recent years have we had a basis for understanding it in a definitive way.

The understanding of individuality is basic to the understanding of human behavior. It is not enough to know the ways in which all human beings respond alike to certain stimuli; it is fully as important to know also why different people confronted with about the same stimulus react very differently. Since biochemistry underlies many of our moods and our reactions to different types of stimulus, and since it is in the area of biochemistry that individuality is most definitively recognized, biochemistry merits inclusion as one of the most important of the so-called behavioral sciences.

Because biochemical individuality points the way toward individuality in the broadest sense of the word, it has profound implications not only in medicine, psychiatry, and psychology but also in human relations, education, politics and even philosophy.

BIOCIDE (Fuels). Petroleum.

BIODEGRADABILITY. Detergents.

BIOFEEDBACK. A term and concept in the area of somatic medicine that grew in popularity in the late 1960s and 1970s. Biofeedback is partially an outgrowth of prior, more generalized concerns with the relationship of stress and disease and with the psychological management of serious illnesses (Jacobson, 1938). As defined by Brown (1977), "Biofeedback is an unexampled process for treating human illnesses because it evokes complex mental processes to regulate and normalize even the most complicated functions of the human body. It is called biofeedback because its effects rest upon making information about biological activities, including those of the brain, available to the mind." Experimentation and therapy have occurred in connection with muscular tension and relaxation, headaches, stuttering, hyperactivity, cardiovascular problems, among many other stress-related disorders. Cassem (1971) points out that biofeedback techniques may be helpful in a patient's recovery from serious illness, particularly when the physician (and patient) view the patient's life style as contributing to the disease process. Some physicians have suggested biofeedback, among other therapeutic measures, for the treatment of spasmodic torticollis (involuntary turning of the head). The technique also has been used in treatment of bruxism (grinding of teeth). Many words of a fictional character have been published on biofeedback. Stern and Ray (1977) concentrate on separating fact from fiction in this field.

References

Birk, L., Editor: "Biofeedback: Behavioral Medicine," Grune & Stratton, New York, 1973.

Brown, B. B.: "Stress and the Art of Biofeedback," Harper & Row, New York, 1977.

Cassem, N. H., and T. P. Hackett: "Psychiatric Consultation in a Coronary Care Unit," *Ann. Intern. Med.*, **75**, 9 (1971).

Feldman, G.: "The Effect of Biofeedback Training on Respiratory Resistance of Asthmatic Children," *Psychosomatic Medicine*, **38**, 27–34 (1976).

Finley, W. W., Smith, H., and M. Etherton: "Reduction of Seizures and Normalization of the EEG in a Severe Epileptic Following Sensorimotor Biofeedback Training; Preliminary Study," *Biological Psychology*, **2**, 189–203 (1975).

Jacobson, E.: "Progressive Relaxation: A Psychological and Clinical Investigation of Muscular States and Their Significance in Psychology and Medical Practice," Univ. of Chicago Press, Chicago, 1938.

Shapiro, D., et al.: "Effects of Feedback and Reinforcement on the Control of Human Systolic Blood Pressure," *Science*, **163**, 588–590 (1969).

Stern, R. M., and W. J. Ray: "Biofeedback," Dow Jones-Irwin, Homewood, Illinois, 1977.

Welgan, P. R.: "Learned Control of Gastric Acid Secretions in Ulcer Patients," *Psychosomatic Medicine*, **36**, 411–419 (1974).

BIOGENESIS (Sterols). Steroids.

BIOHERM. A geologic term for beds or mounds of colonial and gregarious marine fossils with calcareous shells or skeletons. Present day bioherms are usually referred to as coral reefs.

BIOLOGICAL CLOCK. Biological Timing and Rhythmicity.

BIOLOGICAL ENERGY TRANSFER. When an ionization is produced in a substance such as a protein, the net charge produced in the protein probably migrates throughout a large region of the molecule with various probabilities favoring its occurrence in one part of the molecule or another. Eventually, after approximately 10^{-14} seconds, the excess (or deficiency) of charge probably settles in an *s-s* bond or in the hydrogen atom attached to the carbon of the peptide bond which is opposite to one or other of the amino acid residues. Thus, regardless of the site of the original ionization in the molecule, there is considerable transfer of energy throughout a large portion of the molecule. However, the phrase *energy transfer* is generally meant to include those cases where it might occur in addition to this; for example, intermolecularly either between adjacent protein molecules or between protein and solvent molecules. It can also apply to excitation. See also *Active Transport* under **Cell (Biology).**

BIOLOGICAL EQUILIBRIUM. The state of coordination which maintains an animal in normal posture.

Equilibrium of aquatic animals such as the fishes is maintained by the resistance of the surrounding water in relation to specialized body form, by muscular movements of body and fins, and by the gas-filled swim bladder. The bodies of most fishes are heavier above, as is shown by their floating back downward when dead, but the combination of these factors maintains their erect position.

Terrestrial animals maintain their posture by constant muscular adjustment in response to stimuli received by sensory organs in the sole of the feet and in the muscles and tendons. A portion of the inner ear of vertebrates is also a center of equilibrium. End organs in the semicircular canals of this organ are stimulated by movement in the liquid filling the canals when the animal moves. The three canals lie in the three planes of space so that at least one is activated by any movement. The results of their reaction are transmitted to one of the lower brain centers, whence the proper impulses are relayed to the muscles.

Equilibrium in flight demands very delicate coordination of essentially the same type. In insects and bats it is supposed to be accomplished partly through delicate sense organs located in the wings.

BIOLOGICAL LEVEL. Although there is no precise scale to express the level of an organism in terms of biological organization, a few terms are used to roughly place organisms into levels, commencing at the lowest level (a bit of protoplasm surrounded by a membrane as typified by a protozoan), up through the multicellular animals, such as sponges, through the *tissue level* (evidence of differentiation of tissues in many-celled animals, through the *organ level* (animals with specific organs), and finally to the *organ-system level*, as represented by most complex animals. Thus, the term biological level is used in roughly classifying organisms in terms of complexity.

BIOLOGICALS. A catchall category of products of a biochemical nature that are sold commercially essentially for medical purposes. Generally included among the biologicals are hormones, steroids, antibiotics, vitamins, glandular extracts, and other natural or synthetic materials of a biochemical type as contrasted with inorganic chemicals and organic chemicals that normally do not enter into life processes. The production of biologicals normally is undertaken by firms also dealing with other forms of medicines and drugs. Although, because of their potency, large numbers of biologicals are not produced on a tonnage basis, their generally high costs and great variety account for a very large industry. Filling of ampoules with an antibiotic is shown in the accompanying illustration.

Filling of ampoules with an antibiotic at the Torre Annuziata plant near Naples, Italy. (*The Dow Chemical Co.*)

BIOLOGICAL TIMING AND RHYTHMICITY. Endogenous or persistent rhythmicity has been demonstrated to be widespread through the animal and plant kingdoms. Examples in plants are daily rhythms of cell division and bioluminescence in unicellular forms, and of sleep movement and growth rate in higher forms. Seeds exhibit an annual rhythm in capacity to germinate. Examples from animals are daily rhythms of skin color change and lunar tidal rhythms of spontaneous running in crabs, daily rhythms in emergence time of flies ready to leave their pupal cases, and daily rhythms of wakefulness or spontaneous activity and correlated underlying physiological phenomena in numerous kinds of animals including mammals. In organisms in which endogenous rhythms of basal metabolic fluctuations have been studied, these have displayed the natural geophysical frequencies. Some authorities assume that endogenous rhythmicity is an universal attribute of living matter.

One of the most fascinating aspects of approximately 24-hour biological rhythms (circadian rhythms) is that they persist for some time when the obvious environmental time cues—day-night illumination and temperature cycles—are precluded. The persistence of these rhythms under constant conditions is attributed to the existence of a "biological clock" resident within each organism which continues to time organismic processes.

The persistent overt physiological rhythms possess certain properties in common, quite unorthodox in terms of usual biological phenomena. The frequency or cycle duration of the rhythms is essentially independent of the level of constant temperature at which the organisms are maintained and, furthermore, drugs or other chemicals known to alter substantially general metabolic rate have little or no influence on the rhythm frequency. In addition, the patterns of activity recurring daily in constant light and temperature can be abruptly shifted to bear any desired phase relationships to the outside day-night cycles simply by subjecting the organism to a few experimental daily cycles of appropriately adjusted light or temperature changes. The new phase relationships of the rhythm may then persist in continuing constant conditions. In fact, the recurring cycles very commonly appear to shift spontaneously from day to day to yield regular cycles a little longer or shorter than the solar day. The rate and direction of such spontaneous daily phase-shifting is related to the constant temperature or light level at which the organisms are held. It has been long known that many organisms will, in response to artificial light-dark cycles, adopt "days" deviating from 24 hours by up to 4 to 6 hours, but only so long as the light cycles continue to be imposed. When, however, the aritificial light cycles deviate further from 24 hours, the organism commonly resumes its basic daily period, even in the presence of the continuing artificial cycles.

There is also a daily rhythm in the sensitivity to light and temperature as factors operating to shift phases of the persisting rhythms. There are reasons to postulate that these particular organismic rhythms are responsible for the apparent phase shifting even when the light and temperature are held constant, thereby giving rise to periods deviating from 24 hours, and responsible also for the inability of organisms to synchronize with artificial light-dark cycles when these deviate by more than a few percent from 24 hours.

There is also evidence that the same "clock system" which times the periods of persistent physiological rhythms underlies a "time sense" such as that exhibited by bees trained to feed at particular times of day. It also appears to comprise the chronometer which enables organisms as diverse as crustaceans, insects, and birds, which normally use the sun, moon, or possibly the constellations, for celestial navigation, to correct continuously for the rotation of the earth relative to each of these types of heavenly bodies, to maintain straight compass courses.

There are two basic hypotheses as to the nature of the biological clock: (1) the autonomous clock generates its own time intervals, and (2) the clock contains no independent timing mechanism, but simply indicates time from information fed to it. Palmer (1970) has suggested the analogy of these two mechanisms with a spring-driven wind-up clock and the sundial. The wind-up clock, by way of the escapement, allows energy stored in the spring to "escape" at regular intervals. In contrast, the sundial is subordinate to an external time indicator, the sun, and merely signals the time of day to an onlooker. Clearly, the escapement clock is a time-measuring device, while the sundial is simply an indicating device.

Researchers, as reflected by some of the references listed, are continuing to seek the biophysicochemical basis for the operation of either one of these mechanisms. Possibly both mechanisms play a role, depending upon species, etc. For example, the pineal gland has been implicated in some recent studies. To date, the two hypotheses have failed to establish a clearly recognizable relationship to cellular components and their functions. An attempt to overcome this difficulty has resulted in a phenomenological model developed around cycles in messenger RNA transcription. It is postulated that within every cell are hundreds of *chronons*, each of which is an unbroken polycistronic complex of DNA, 200 to 2000 cistrons in length. Messenger synthesis begins at one end of the chronon (with the initiator cistron) and proceeds sequentially along the chronon to the terminator cistron. The transcription rate of the entire chronon is prolonged because of numerous intercistronic (and interoperonic) events, which stop the transcription of one cistron and initiate transcription on the next one. It has been suggested that to accomplish this after the transcription of each cistron, the nascent mRNA passes to the cytoplasm, directs polypeptide synthesis. Certain of these products diffuse back into the nucleus, initiating messenger synthesis on the subsequent cistron of the chronon. This diffusion circuit consumes much more time than the transcription steps, thus essentially governing the rate of the transcription of the entire chronon. As a consequence, the duration of the total process is controlled primarily by diffusion effects, thereby engendering the system with the necessary approximate temperature independence. With the transcription of the terminator cistron, a period of cytoplasmic protein synthesis ensues—including the synthesis of a cistron initiator substance. The initiator substance accumulates and diffuses back into the nucleus where it reactivates the initiator cistron of the chronon, completing the feedback circuit and starting a new one. Because of the length of the chronon—which has been arrived at through natural selection—the completion of the entire loop takes approximately 24 hours. The obvious shortcoming of this model is that it does not presently provide an explanation for the phase-setting ability of light and temperature cycles. On the other hand, it has the advantage of being constructed from tangible cellular components and, therefore, many of its postulated attributes may be directly testable.

References

Bennett, M. F.: "Living Clocks in the Animal World," Charles C. Thomas, Springfield, Illinois, 1974.

Binkley, S. A., Riebman, J. B., and K. B. Reilly: "The Pineal Gland: A Biological Clock in vitro," *Science*, **202**, 1198–1200 (1978); *Sci. Amer.*, **240**, 4, 66–71 (1979).

Brown, F. A., Jr.: "Living Clock," *Science*, **130**, 1535 (1959).

Brown, F. A., Jr.: "Endogenous Rhythms," in "The Encyclopedia of the Biological Sciences," 2nd edition (P. Gray, editor), Van Nostrand Reinhold, New York, 1970.

Cloudsley-Thompson, J. L.: "Rhythmic Activity in Animal Physiology and Behavior," Academic, New York, 1961.

Delcomyn, F.: "Neural Basis of Rhythmic Behavior in Animals," *Science*, **210**, 492–498 (1980).

Edmunds, L. N. Jr., and K. J. Adams: "Clocked Cell Cycle Clocks," *Science*, **211**, 1002–1013 (1981).

Ehret, C. F., and E. Trucco: "Molecular Models for the Circadian Clock. I. The Chronon Concept," *J. Theoret. Biol.*, **15**, 240 (1967).

Enright, J. T.: "Temporal Precision in Circadian Systems: A Reliable Neuronal Clock from Unreliable Components?" *Science*, **209**, 1542–1545 (1980).

Harker, J. E.: "The Physiology of Diurnal Rhythms," Cambridge Univ. Press, Cambridge, England, 1964.

Jalife, J., and C. Antzelevitch: "Phase Resetting and Annihilation of Pacemaker Activity in Cardiac Tissue," *Science*, **206**, 695–697 (1979).

Kasal, C. A., Menaker, M., and J. Regino Perez-Polo: "Circadian Clock in Culture: N-Acetyltransferase Activity of Chick Pineal Glands Oscillates in vitro," *Science*, **203**, 656–658 (1979).

Palmer, J. D.: "Biological Clock," in "The Encyclopedia of the Biological Sciences," 2nd edition (P. Gray, editor), Van Nostrand Reinhold, New York, 1970.

Soll, D. R.: "Timers in Developing Systems," *Science*, **203**, 841–849 (1979).

Sollberger, A.: "Biological Rhythm Research," Elsevier, New York, 1965.

Underwood, H.: "Circadian Organization in Lizards: The Role of the Pineal Organ," *Science*, **195**, 587–589 (1977).

Wiesenfeld, Z., Halpern, B. P., and D. N. Tapper: "Licking Behavior: Evidence of Hypoglossal Oscillator," *Science*, **196**, 1122–1123 (1977).

BIOLOGICAL TREATMENT. Water Pollution.

BIOLOGY. The science of life. As with several of the fundamental sciences, over the last several decades, biology has been segmented into a number of fields of specialization. These include biochemistry, bioengineering, biofeedback phenomena, biomedicine, biophysics, cell biology, developmental biology, ecogenetics, evolutionary biology, marine biology, microbiology, and molecular biology, among others. Convenient umbrella terms sometimes used include the *biological sciences* and the *life sciences.*

There are hundreds of entries of varying length included throughout this encyclopedia that relate to the biological sciences. Many of these entries include lists of references for further reading.

BIOLOGY (Comparative). Comparative Biology.

BIOLUMINESCENCE. Many living organisms exhibit the unique property of producing visible light, a phenomenon referred to as bioluminescence. Known light-emitting organisms have either oxidative or peroxidative enzymes that couple the chemical energy released from the enzyme reaction to give electronic excitation of a luminescent compound. The compound that is oxidized with subsequent light emission is usually referred to as *luciferin* and the enzyme which catalyzes the reaction as *luciferase.* Most luciferins and luciferases that have been isolated from unrelated species are different in molecular structure. With one known exception, combinations of luciferin and luciferase from different species do not exhibit bioluminescence.

The light-producing reaction in a number of organisms can be represented simply by: Luciferin $+ O_2 \xrightarrow{\text{Luciferase}}$ Light. Some luminous organisms catalyzing this reaction are: (1) *Cypridina* (a crustacean); (2) *Apogon* (a fish), and (3) *Gonyaulax* (a protozoan). The latter organism is mainly responsible for the phosphorescence (so-called) of the sea.

In other instances, some luciferins must first undergo a luciferase-catalyzed activation reaction prior to their being catalytically oxidized by the enzyme to produce light. There are two well-known cases:

(1) The firefly:

$$\text{Luciferin} + \text{Aenosine Triphosphate (ATP)} \xrightarrow{\text{Luciferase; } Mg^{2+}} \text{Activated Luciferin}$$

$$\text{Activated Luciferin} + O_2 \xrightarrow{\text{Luciferase}} \text{Light}$$

(2) The sea pansy (*Renilla*):

$$\text{Luciferin} + 3',5'\text{-Diphosphoadenosine (DPA)} \xrightarrow{\text{Luciferase; } Ca^{2+}} \text{Activated Luciferin}$$

$$\text{Activated Luciferin} + O_2 \xrightarrow{\text{Luciferase}} \text{Light}$$

Both of these activation reactions are linked to adenine-containing nucleotides of great biological importance. Since the measurement of light can be made an extremely sensitive and rapid technique, the most sensitive and rapid assays known have been developed for ATP and DPA, using the foregoing luminescent systems. Nucleotide concentrations of less than 1×10^{-9} *M* are easily detectable using electronic instrumentation. Firefly luciferase-luciferin preparations for ATP assays are commercially available.

The structure of firefly luciferin has been confirmed by total synthesis. The firefly emits a yellow-green luminescence, and luciferin in this case is a benzthiazole derivative. Activation of the firefly luciferin involves the elimination of pyrophosphate from ATP with the formation of an acid anhydride linkage between the craboxyl group of luciferin and the phosphate group of adenylic acid forming luciferyl-adenylate.

All other systems that have been extensively studied emit light in the blue-green region of the spectrum. In these cases, the luciferins appear to be indole derivatives.

Some animals, such as the marine acorn worms (Balanoglossus), produce light via a peroxidation reaction and appear not to require molecular oxygen for luminescence. The luciferase in this case is a peroxidase of the classical type and catalyzes the reaction: Luciferin $+ H_2O_2 \xrightarrow{\text{Luciferase}}$ Light.

Commerically available horseradish peroxidase (crystalline) will substitute for luciferase in the foregoing reaction. In addition, a compound of known structure, 5-amino-2,3-dihydro-1,4-phthalazinedione (also known as *luminol*), will substitute for luciferin. The mechanisms appear to be the same regardless of the way in which the crosses are made. Thus, a model bioluminescent system is available and can be used as a sensitivity assay for H_2O_2 at neutral pH. The identification of luciferase as a peroxidase is of interest since this represents the only demonstration of a bioluminescent system in which the catalytic nature of a luciferase molecule has been defined.

Most of the luminescent systems mentioned appear to be under some nerve control. Normally, a luminous flash is observed after mechanical or electrical stimulation of most of the aforementioned species. A number of these also exhibit a diurnal rhythm of luminescence.

Among the lower forms of life, there are two well-known examples of luminescence which are not under nerve control, giving a continuous glow of visible light. These are the luminous bacteria, frequently found growing on dead fish, and luminous fungi which grow abundantly on rotting wood. These cells apparently depend upon the oxidation of an organic molecule and hydrogen which is transferred through diphosphopyridine nucleotide (DPN; also termed NAD, nicotinamide adenine dinucleotide) and the enzyme system to drive the luminescent reaction. Known details of these luminescent reactions are represented as follows. For bacteria:

$$\text{DPNH} + \text{H}^+ + \text{Flavin Mononucleotide (FMN)} \xrightleftharpoons{\text{Oxidase}} \text{FMNH}_2 + \text{DPN}$$

$$\text{FMNH}_2 + \text{Long-chain Aliphatic Aldehyde} + \text{O}_2 \xrightarrow{\text{Luciferase}} \text{Light}$$

and for fungi:

$$\text{DPNH} + \text{H}^+ + \text{Unknown Compound (X)} \xrightleftharpoons{\text{Oxidase}} \text{XH}_2 + \text{DPN}$$

$$\text{XH}_2 + \text{O}_2 \xrightarrow{\text{Luciferase}} \text{Light}$$

Both of these systems are apparently closely linked to respiratory processes and in this sense are analogous to one another. Luciferase from a luminous bacterium, *Photobacterium fischeri*, has been crystalled in high yield.

See also **Luminescence.**

References

Cormier, M. J. and J. R. Totter: *Ann. Rev. Biochem.* **33**, 431–458 (1964).
Dure, L. S. and M. J. Cormier: *J. Biol. Chem.*, **239**, 2351–2359 (1964).
Firth, F. E.: "The Encyclopedia of Marine Resources," Van Nostrand Reinhold, New York (1969).
Herring, P. J.: "Bioluminescence in Action," Academic, New York, 1979.
Levandowsky, M., and S. H. Hutner, Editors; "Biochemistry and Physiology of Protozoa," 3 volumes, Academic, New York (1979–1980).

BIOMASS AND WASTES AS ENERGY SOURCES. Initially, *biomass* was defined as the amount of living organisms in a particular area, stated in terms of the weight or volume of organisms per unit area or of the volume of the environment. This definition still applies very well to ecological and geophysical assessments of land areas or regions and depths of the seas and lakes. Within the past decade or two, *biomass* as a word has come to describe the exploitation of plants (terrestrial mainly, but also marine) that may be grown and harvested as crops, all or parts of which may be combusted directly for their heat energy or processed into fuels.[1] When such materials serve other primary purposes (as foods, fibers and construction members), waste is created—straw, sawdust, sewage sludge, etc.—that possesses value as an energy source. The majority of agricultural, commercial, industrial, and urban or municipal wastes are of a biological rather than mineral nature and thus fall under the umbrella of biomass. The simple burning of wood for heat illustrates one of the simplest ways to convert biomass to energy. All biomass represents an indirect form of solar energy. Biomass, as a source of energy, differs from coal, natural gas, and petroleum in one major way—biomass is renewable. Some potential biomass energy crops can be renewed as frequently as two or three times per year, depending upon location, while other materials such as trees have a renewable cycle of several years. Anthropogenic wastes are renewed on a daily basis. Interest in biomass over the last several years has stemmed from the overall concern with ultimate exhaustion of nonrenewable energy sources as well as gaining a degree of political independence by many nations that either do not have any fossil fuel resources, or that have insufficient supplies to maintain a strong economic and industrial position.

Much of the technology required to exploit biomass is available. As of the early 1980s, the principal constraints are economic and political. Aside from biomass wastes that are created anyway (and thus available for exploitation), biomass energy technology is largely a fundamental matter of agriculture and forestry (silviculture). Since modern agriculture is by itself a large consumer of fossil fuels, it is necessary to find crops that provide ample return of energy (output) for the energy invested. When analyzed critically, some previous proposals have failed to meet this criterion. Thus, the search continues for crops and regions in which to grow them that will provide a good energy return. Biomass fuels, especially when combusted directly, also create their share of environmental problems. Biomass energy sources have an advantage, on the one hand, that fuels produced from them can be reasonably close to the consuming markets and thus minimize transportation costs, but, on the other hand, highly populated urban areas are often not adjacent to good agricultural land, thus requiring long transportation of fuels—as is traditionally the case with fossil fuels. Proposals have been made to utilize crops, such as sugarcane (quite successful to date in Brazil), cassava leaves, and pineapple, that are best grown in tropical regions, but the fuels from which would have to be transported long distances to consuming centers.

When considering the potential of biomass as an energy source, there is also a confrontation with the production of food. Agriculturists, particularly in some regions of the world, have been predicting for a number of years that the available arable land is shrinking while the population is increasing. Extensive use of good land for energy-yielding plants would, of course, aggravate this situation. Particularly in some regions of the world, there is a potential crisis in terms of available irrigation water for crops of any purpose. However, even these kinds of serious problems may be amenable to solution if a total systems approach to a biomass energy plan for a given region, i.e., considering not only the contribution of such a plan to energy needs, but also all of its positive as well as negative interactions with food production, water needs, animal feedstuff needs, economic situations (role as a buffer in terms of traditional surpluses and shortages of crops), political implications (diplomatic and military advantages of energy independence), and other factors that may or may not be cost-related.

In a study of corn (maize) biomass as a source of chemicals and fuels, Lipinsky (1978) compared the traditional usage of corn grain and corn stover with an alternative routing of materials that would essentially perform the same functions as traditional routing, but furnish chemicals and fuels as well. This is evidence of how application of a *systems approach* may overcome what initially appears to be large negative concerns over using biomass as an energy source. See Fig. 1.

Some successes have been realized in the biomass energy field. (1) most automobiles in Brazil burn *gasohol* (in Brazil, there is a 20% content of alcohol made from sugarcane). By the early 1980s, Brazil plans to convert about 80,000 automobiles for using gasohol and some 250,000 new cars by the mid-1980s will be built to use gasohol. (2) It is reported that as of 1981, over 7.5 million biogas generators (methane) have been built, with obvious emphasis upon decentralized activities. Particularly in the southern area of China, the direct burning of wood is being replaced by gas generated from human and animal wastes. (3) Countries, such as Sweden, which have no coal, oil, or natural gas, are seriously planning the expansion of biomass energy systems. Sweden, a neutral nation during World War II, was cut off from fossil fuel supplies and developed a high biomass technology for that time. Sweden, which fortunately is highly forested, is taking steps toward increasing use of wood as a basic fuel. Research has been directed toward fast-rotation trees with harvesting on a 3- to 5-year schedule. Emphasis is being placed on willow and birch trees.

During the past few years in the United States, there have been several hundred research and development projects aimed toward

[1] The generation of biomass from carbon dioxide is called "primary production" because it is the first fundamental step in turning inorganic material into organic compounds and cell constituents. This reduction of carbon dioxide uses sunlight as the source of energy. See entry on **Photosynthesis.**

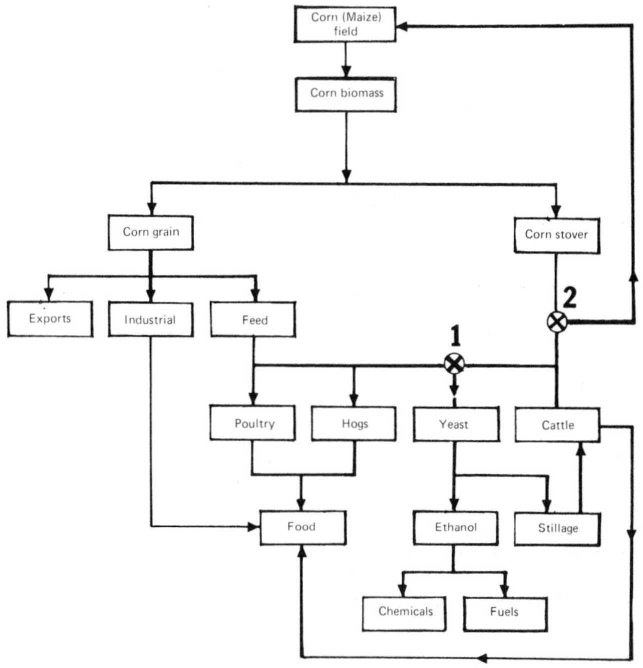

Fig. 1. Traditional procedures and alternative scheme for handling corn (maize) crops. In the conventional method, corn grain goes into (1) export, (2) industrial, and (3) animal feed channels. The feed is used for poultry, hogs, and cattle. Corn stover (stalks, leaves, etc.) is left in the field and turned under (except in no-till operations, where it is left in place) for the next crop. Thus, for the traditional method, valve 1 in the flowsheet is closed to the yeast pathway and valve 2 is closed to the return of stover to the field. In an alternative scheme (Lipinsky, 1978), some or all of the feed would be routed to the yeast pathway, with valve 1 closed to the feeding of cattle. In this scheme, stover would be used as cattle feed (preferably after treatment to improve digestibility) and part of the grain would be used in the manufacture of chemicals and fuels. Within the last few years, researchers at Purdue University have made a major breakthrough in the treatment of corn stover to obtain simple sugars for fermentation to ethanol. An organic acid treatment facilitates the attack of fungal cellulases on lignocellulose. The probability that this process would also increase the digestibility of corn stover by cattle is high because the low digestibility of raw stover arises from crystalline cellulose or cellulose protected by lignin. It is envisioned that the digestibility of corn stover could rise from 45 to 60% or more. More details are given in Lipinsky (1978).

biomass energy systems. For example, a machine has been developed that can reduce a substantial tree, including branches, to dollar-size chips in less than one minute.

Principal Biomass Materials for Energy

Biomass-to-energy systems fall into two principal categories: (1) materials for direct combustion that will generate heat for processing, for warming living and working spaces, for steam and hence also for generating electricity; and (2) materials from which both fuels and chemicals can be obtained through biochemical or thermochemical conversion processes. Resulting fuels must have ample caloric content per unit of weight and thus rich in carbon and hydrogen and poor in the content of atoms, such as oxygen and nitrogen, which do not contribute to the caloric value of the fuel. Fuels of all physical states are envisioned—solids, liquids, and gases. In searching for new biomass raw materials, scientists have found it helpful to study the various biosynthetic pathways followed by plants from seed to maturation.

In studying the potential of various biomass materials, Weisz and Marshall (1979) have compared the characteristics of several materials that have been considered for energy production. See Table 1.

Sugarcane, Pineapple, and Cassava. The optimal zone for growing sugarcane extends over a range of latitude that lies 30°N and 30°S of the equator. The growing season varies considerably—10 months or less in Louisiana and up to 2 years or longer in Hawaii, and parts of South America. Sugarcane requires considerable water—either from rain or irrigation. In several areas where rainfall is less than 1000–4000 millimeters (39–157 inches) per year, irrigation is required. Sugarcane requires a mean annual temperature of 25°C (77°F) or higher—growth essentially ceases when soil temperature drops to 16°C (61°F). Soil requirements are not rigid so long as the pH lies between 4.5–5.5. Yields of sugarcane can be increased by the application of fertilizer. The greatest experience gained from a noncereal crop as a source of biomass for fuels to date has been from sugarcane.

In recent years, cassava leaves and, even more recently, pineapple has been considered seriously as a biomass energy crop. For pineapple, the growing period extends to about 14 months, but can be as long as 3 years at the northern and southern limits of its growing area. These roughly parallel the characteristics of sugarcane. A second (ratoon) crop of pineapple can be raised in about half the time required to grow the mother plant. Cassava, on the other hand, is raised as an annual crop and is essentially limited to a region 15°N and 15°S of the equator. Cassava roots keep well if left in the soil for about 21 months to accumulate starch. Pineapple and cassava require about the same temperatures as sugarcane. Cassava is similar to sugarcane in terms of water requirements, whereas pineapple can tolerate drier conditions, although at sacrifice of yield. Cassava and pineapple require about the same soil types as sugarcane and react favorably to fertilizers.

As indicated by Table 2, a number of comparative studies have been made of these three biomass plants. As reported by Marzola and Bartholomew (1979), cassava and sugarcane are good candidates for alcohol production because of their potentially high productivity and because they accumulate large amounts of starch or sucrose. Sucrose can be fermented without pretreatment, but a preliminary degradation step is needed for starch. Sucrose levels in sugarcane stalks are low during rapid vegetative growth, but ripening of the plant with chemicals (Alexander, 1974) or by withholding nitrogen and water (Clements, 1953) results in sucrose accumulation to concentrations of 15–20%. When stored for several months, cassava roots eventually reach a starch concentration of about 33% (fresh weight basis). In contrast, sucrose and reducing sugars accumulate in pineapple fruit to a concentration of about 16%, although the plant and ratoon crop stems also contain 30–40% starch (dry weight basis) a short while after the ratoon crop fruit has been harvested. Thus, both sugar and starch substrates are available from pineapple. Pineapple is well adapted to the subhumid or semiarid tropics and thus is particularly well suited for exploiting large areas not under cultivation with other crops of commercial value.

In commenting on the potential of sugarcane as a biomass energy source, Lipinsky (1978) observes that conventional sugarcane bagasse is a source of fiber for papermaking, especially in countries that have considerable sugarcane and few pulpwood-type trees. To make paper of relatively high quality, it is necessary to depith the bagasse. Depithing is a relatively costly process with considerable loss of fibers. A relatively new process (Canadian Separator Equipment Process) generates rind fiber that already has been depithed and without injury to the fibers as the result of traditional crushing and milling operations. It is estimated that the value of rind fiber will be about double that of conventional bagasse. Thus, the new process may make it possible to generate a coproduct of significant value along with fuels from the sugarcane. Excess rind fibers could be used as a source of heat for distillation required during the manufacture of alcohol. Other byproducts, such as the leftover pith (used as an absorbant in feedstuffs made from stillage), are envisioned. As Lipinsky observes, the fact that the millable sugarcane stalk might yield so much salable coproduct that there would be no fiber left to make steam and electricity to run the process is not a serious disadvantage because very little power is required by the new process as compared with the traditional grinding and milling process.

Direct Wood Burning. Wood was the major fuel of the United States until about 1886 when the consumption of coal equaled that of wood. Oil did not appear on the chart until about 1900 and gas in 1910–1920. The use of wood tapered off while other fuels climbed at amazing rates, but wood never ceased as a factor, even if small. It is interesting to note that about a million modern woodburning stoves have been purchased since the late 1970s and that about 40% of the wood products industry is furnished by combusting bark and mill wastes. This amounts to about 1 quad (10^{15} Btu). Wood burning has been sufficiently extensive during the past few years to cause environmental concerns in some regions. Although wood has a low sulfur content and produces minimal amounts of nitrogen oxides even by the hottest

TABLE 1. COMPARISON OF BIOMASS MATERIALS

Crop	Agricultural Energy Input[1] kcal per Pound of Crop	Agricultural Energy Input[1] kcal per Kilogram of Crop	Ratio of Crop to Total Biomass (Weight)	Agricultural Energy Input kcal per Pound of Total Biomass	Agricultural Energy Input kcal per Kilogram of Total Biomass	Ratio of A to G[3]
Alfalfa	425	937	~0.9	382	842	0.21
Sorghum (dryland)	443	977	~0.9	399	880	0.22
Sorghum (irrigated)	506	1116	~0.9	455	1003	0.25
Wheat (dryland)	393	867	~0.5	200	441	0.11
Wheat (irrigated)	790	1742	~0.5	395	871	0.22
Corn (maize)	622	1372	~0.5[4]	331	730	0.18
Sugar beets[2]	320–700	706–1544	~0.6	195–420	430–926	0.15–0.23

[1] From E. S. Lipinksy, et al.: "Systems Study of Fuels from Sugarcane, Sweet Sorghum and Sugar Beets," National Technical Information Service, Springfield, Virginia, 1976.

[2] Variation over eight states of the United States.

[3] Biomass energy content assumed to be 1800 kcal/pound (3969 kcal/kilogram).

[4] Value from V. J. Kavlick, paper presented at the 42nd American Petroleum Institute Refining Department Mid-Year Meeting, Chicago, 1977.

NOTE: Format from Weisz and Marshall (1979). See references listed at end of entry. Metric equivalents added. A = Agricultural energy input; G = biomass energy content.

fires (1370°C; 2500°F), it contains air pollutants in the form of particulates, gases, and tars. Environmentalists in the New England region have estimated that the 300,000–400,000 tons of wood burned per year (New Hampshire only), if the fuel is very dry red oak, will add 1000 tons of particulates to the air; if dry white pine is burned, the total may be over 5000 tons. Since a mixture of woods usually is used, the figure lies somewhere between the two aforementioned quantities.

Fuels from Wood Wastes. Where economics may be favorable, factories that produce wood wastes may consider converting the wastes into fuels rather than using them for firing boilers. A number of processes have been suggested for producing methanol and ethanol from such wastes. Two such proprietary processes are shown schematically in Figs. 2 and 3.

Biomass from the Oceans. As pointed out by Rhyther (1980), the oceans are the largest uncultivated and underutilized pastures on earth. Seaweeds which have limited uses (hydrocolloids, such as agar, alginic acid, and carrageenan) could become a major source of biomass for energy. This topic is explored in some detail in entry on **Ocean Resources (Energy).** Although quite long-range, studies are being made of the phenomenon of bacterial chemosynthesis—a topic that has been spurred by the deep-ocean findings of the Galapagos Rift Thermal Springs. See **Ocean.** In *green plant photosynthesis,* water serves as the source of electrons and light as the source of energy. In *bacterial photosynthesis,* reduced sulfur compounds serve as the source of electrons and light as the source of energy. In *bacterial chemosynthesis,* reduced sulfur compounds serve as the source of electrons and chemical oxidation while free oxygen serves as the source of energy. Jannasch (1980) reports on very early research studies of bacterial chemosynthesis which ultimately could lead to an advantageous way of creating biomass.

Euphorbia. Considerable interest has been shown by some scientists in the genus *Euphorbia* which yields a significant quantity of a milklike emulsion of hydrocarbons in water. Natives in the forests of Brazil have been familiar with the species *Cobaifera langsdorfii* for a number of years. Trees are tapped about twice per year, with the collection of 15–20 liters of hydrocarbon. Tests have shown that the liquid can be placed directly in the fuel tank of a diesel-powered vehicle. It is estimated that a plot of 100 acres (40.4 hectares) will yield about 25 barrels of fuel per year. The Brazilian government has established experimental plantations. Very few regions in the United States, possibly including southern Florida, would support this tropical tree. In addition to work at the University of California at Berkeley and an experimental plantation operated by the Japanese in Okinawa, most attention has been concentrated in Brazil. More detail is given by Maugh (1979).

TABLE 2. COMPARISON OF THREE BIOMASS ENERGY PLANTS

Time after Planting (months)	Cassava[1] Leaves (% of Plant Dry Weight)	Sugarcane[2] Total Plant Fresh Weight (g)	Sugarcane[2] Percentage Green Top	Pineapple[3] Total Plant Dry Weight (g)	Pineapple[3] Percentage Leaves
3	16			64	83
6	22			96	88
9	8	249	32	296	90
12	2	434	21	551	88
14		433	18		
15	7			812	80
$16\frac{1}{2}$		401	16		
18	7	390	15	1176	65
20		386	13	1278	47
21	1				

[1] From C. N. Williams: *Exp. Agric.* **7**, 49 (1972).

[2] From J. L. Monteith: *Exp. Agric.* **14**, 1 (1978).

[3] From Pineapple Research Institute of Hawaii, Honolulu.

NOTE: Format from Marzola and Bartholomew (1979). See references listed at end of entry.

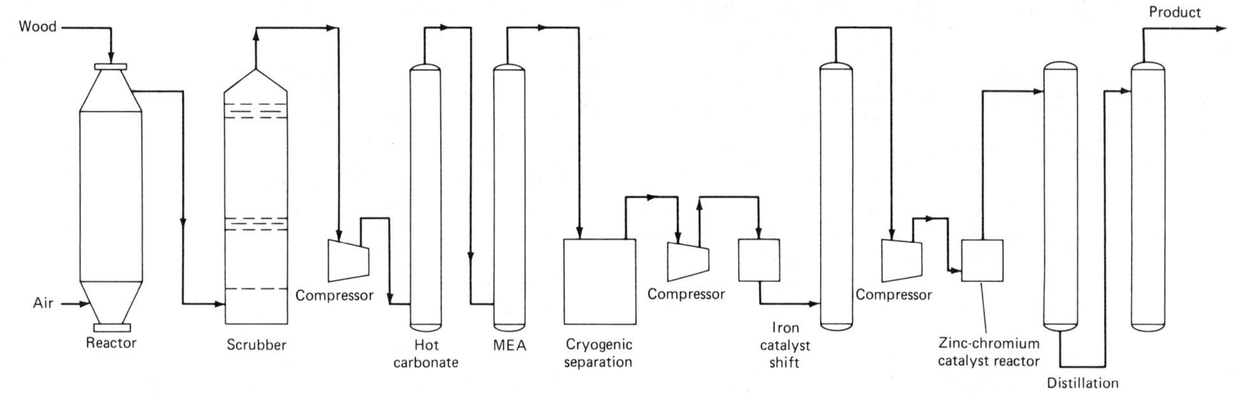

Fig. 2. Schematic flowsheet of process for producing methanol from wood wastes. MEA = monoethanolamine. (*After Hokanson and Katzen, 1978.*)

Utilization of Urban and Industrial Wastes as Energy Sources

At a minimum, the solid waste production in the United States alone is estimated at about 566 × 10^9 kilograms (624 × 10^6 tons), representing an energy content of about 2.37 × 10^{18} Calories (9.42 × 10^{15} Btu). Although impractical, if all solid wastes could be converted to useful energy, this would represent about 10% of the energy requirements of the country. The types of dry combustible solids discarded are summarized in Table 3. The quantities shown estimate only the combustible portions of the solid waste—inert materials are not included. In terms of converting waste to energy, unfortunately, solid waste materials are not available in concentrations sufficient to fuel a solid waste-energy-producing plant except in certain locations. Agricultural wastes, in particular, are not concentrated. An estimate of the quantity of concentrated waste is given in Table 4. The figures do not include wastes that are presently being recycled for use as energy sources.

Urban waste includes household, sewage, commercial, institutional, manufacturing, and demolition waste. The availability of this waste is directly related to the population living in urban areas of adequate size to support a given size system.

Manufacturing and processing wastes include all residuals generated from material inputs that leave the plant as product output. Office and packaging wastes associated with this sector are included in the urban waste sector. The majority of these wastes are from pulp and paper manufacturing, primary and secondary wood manufacturing, and the construction industry. Agricultural wastes include animal manures, crop wastes, and forest and logging residues. The only available wastes in this sector are those animal wastes generated on large feedlots and dairies, and a portion of those crop wastes (bagasse and fruit tree prunings) not readily recycled back to the soil.

The energy recovery system selected dictates the extent that solid waste must be prepared. Some systems require nothing more than the removal of massive noncombustibles, such as kitchen appliances

TABLE 3. SUMMARY OF COMBUSTIBLE SOLID WASTE DISCARDED IN THE UNITED STATES

Source of Waste	Quantity Discarded			Energy Value	
	% of Total Weight	10^9 Kilograms	10^6 Tons	10^{15} Cal	10^{15} Btu
Refuse and sewage/urban	16.3	92	101	413	1.64
Manufacturing and processing	5.8	33	36	144	0.57
Agricultural	77.9	441	487	1,817	7.21
Total	100.0	566	624	2,374	9.42

from the refuse, while other processes require extensive shredding, air classification, reshredding, and drying. In conjunction with fuel preparation, it is usually worthwhile to reclaim metals and glass for recycling.

One-stage shredding is often used to reduce waste to a nominal size as small as 1 inch (2.5 centimeters). When finer-sized fuel is required, a second shredding step is usually used after air classification has removed many of the noncombustibles. Both vertical and horizontal air classifiers depend on the heavy noncombustibles settling out by gravity in a moving air stream, while the lighter combustibles are pneumatically transferred through the air classifier. Denser combustibles, such as rubber and leather, may be removed with the heavy fraction, while some of the fine glass and metal foils are carried with the combustibles. Thus, desired separation may not always be achieved on one pass through an air classifier.

Some energy recovery systems require drying to remove excess moisture in the waste. This is required almost without exception when

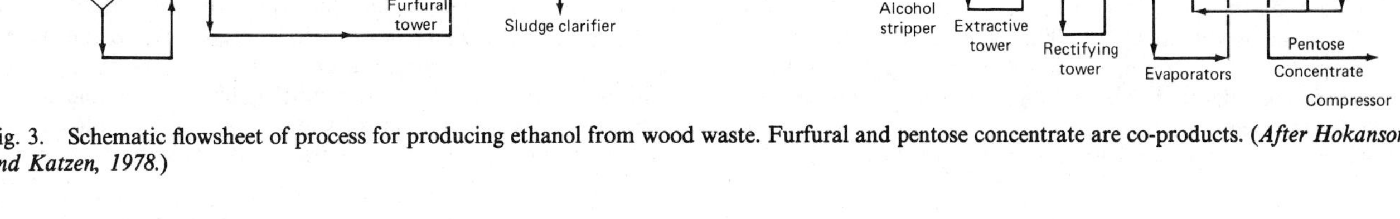

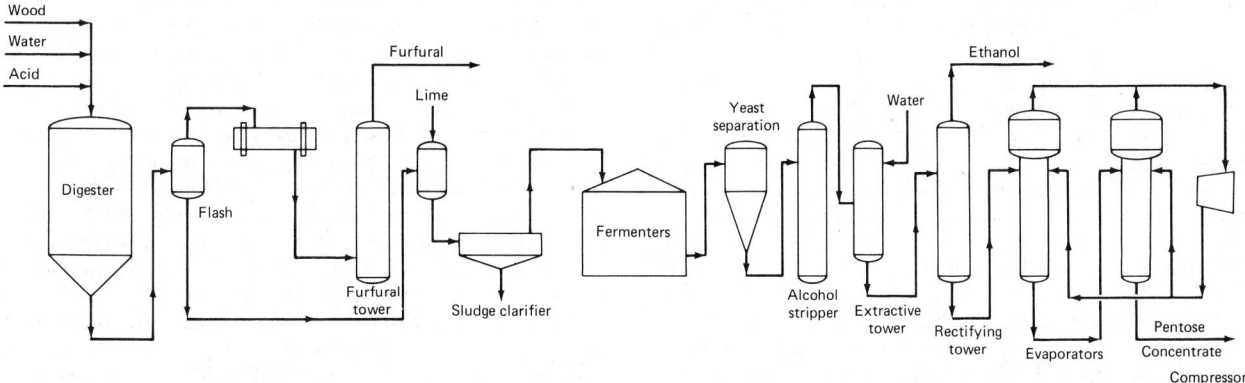

Fig. 3. Schematic flowsheet of process for producing ethanol from wood waste. Furfural and pentose concentrate are co-products. (*After Hokanson and Katzen, 1978.*)

TABLE 4. SUMMARY OF SOLID WASTE AVAILABILITY

Source of Waste	Number of Plants	Waste per Year		Energy Value per Year	
		10^9 Kilograms	10^6 Tons	10^{15} Cal	10^{12} Btu
To Solid Waste Plants capable of processing 90,700 kg (100 tons)/day					
Urban	1,598	52.9	58.3	23.8	943
Manufacturing	579	19.1	21.1	80	319
Agricultural	1,026	34.0	37.5	134	532
Sub-Total	3,203	106.0	116.9	237.8	1,794
To Solid Waste Plants capable of processing 453,000 kg (500 tons)/day					
Urban	252	41.7	46.0	187	744
Manufacturing	9	1.6	1.8	7	27
Agricultural	5	0.9	1.0	4	15
Sub-Total	266	44.2	48.8	198	786
Grand Total	3,469	150.2	165.7	435.8	2,580

NOTE: Plant size in terms of processing combustibles.

animal manures or sewage sludge is used as a fuel. Usually, waste heat from the total process can be used for the drying system.

Pyrolysis. One process uses low-temperature flash pyrolysis to produce char and a highly viscous, highly oxygenated fuel oil, having a heating value of about 10,500 Btu/pound (5838 Calories/kilogram). The system uses two stages of shredding, air classification, and drying to produce a minus-24 mesh fuel for the pyrolysis reactor. The heat required for pyrolysis is derived from the combustion of the pyrolysis off-gas and from a portion of the char produced; it is transferred by means of a heat exchanger. From the pyrolysis unit, the gases are exhausted through a cyclone to remove the char and then scrubbed to remove the oil, water, and other solids and liquids. Major properties of the pyrolytic oil include: By percent (weight): carbon, 57.5%; oxygen, 33.4%; hydrogen, 7.6%; nitrogen, 0.9%; chlorine, 0.3%; sulfur, 0.1 to 0.3%; ash, 0.2 to 0.4%. The oil has a heating value of 10,500 Btu/pound (5838 Calories/kilogram); sp gr 1.30; pour point, 32°C; flash point, 56°C; and a viscosity of 3150 (SSU at 88°C).

In another system, municipal refuse is charged at the top of a shaft furnace and is pyrolyzed as it passes downward through the furnace. Oxygen enters the furnace through tuyeres near the furnace bottom and passes upward through a 1425 to 1650°C combustion zone. The products of combustion then pass through a pyrolysis zone and exit at about 93°C. The off-gas then passes through an electrostatic precipitator to remove flyash and oil formed during pyrolysis. The latter are recycled to the furnace combustion zone. The gas then passes through an acid absorber and a condenser. The clean fuel gas has a heating value of about 300 Btu/cubic foot (2670 Calories/cubic meter) and a flame temperature equivalent to that of natural gas. The solid waste that remains is a slag at the furnace bottom.

Biological Methane Production. This process involves the anaerobic digestion of a solid waste and water or sewage sludge slurry at 60°C for five days to produce a methane-rich gas. Solid waste is prepared by shredding and air classification, followed by blending with water to produce a mixture of 10 to 20% solids concentration. The slurry is heated and placed in a mixed digester at 60°C for 5 days detention. The digester gas is drawn off and separated into carbon dioxide and methane. The spent slurry from the digester is pumped through a heat exchanger to partially heat the incoming slurry prior to filtration. The filtrate is returned to the blender and the sludge is used for landfill. Heat addition to the refuse slurry is required to maintain the required digester temperature. The process is well suited for use on sewage sludge, animal manures, and other high-moisture-content solid wastes. It is estimated that the process can reduce the volume of volatile solids by 75%, while producing about 3000 cubic feet (85 cubic meters) of methane per ton of incoming solid waste. The major residue is used for landfill or incinerated. About 10% of the methane is required to heat the digester feed.

Direct Steam Process. One process uses a rotary kiln pyrolizer followed by an afterburner and boiler to produce steam from shredded waste. The pyrolysis process in a kiln is operated countercurrently. Solid waste enters at one end and pyrolyzed residue is discharged at the other. External fuel and air are introduced at the residue discharge area and combustion products and pyrolysis gases leave the kiln at the feed opening. This arrangement causes the solid waste to be exposed to progressively higher temperatures as it passes through the kiln. The kiln off-gases pass through a refractory-lined afterburner into which air is introduced to allow complete combustion prior to passing through the waste heat boiler. A wet scrubber is used for air pollution control, while an induced draft fan is used to draw the gases through the system. One ton of solid waste, augmented by 1.25 million Btu (0.3 million Calories) from auxiliary fuel and 55 kilowatt-hours of electricity will produce about 4800 pounds (2177 kilograms) of steam at 330 psig (22.4 atmospheres) along with 200 pounds (91 kilograms) of char.

Waterwall Incinerators. These devices generate steam by burning unprepared solid waste on a grate and passing the hot products of combustion through a boiler. Numerous waterwall incinerators have been built in Europe and the United States. Unprepared refuse is taken from storage pits and charged directly into the incinerator feed hopper. From there, the refuse drops onto a feed chute and then is fed automatically onto the stoker by means of a hydraulic feed ram. Temperatures in the 870°C range effectively burn the solid waste. Before the flue gas enters the boiler, secondary air is added to produce a temperature near 1090°C. The boiler is constructed of membrane waterwalled tubes with extruded fins. After passing through the boiler, the gases travel through an economizer section and then into an electrostatic precipitator for particle removal. A typical 1000 tons/day (900 metric tons/day) waterwall incinerator produces about 300,000 pounds (136,080 kilograms) of steam per hour.

Fluid Bed/Turbine Systems. Several variations of a system involving combustion of solid wastes in a fluid bed combustor and using the gases produced for expansion through a gas turbine are available. A system of this type is shown in block diagram format in Fig. 4. Shredded and air classified waste, together with liquid waste, are fired in a fluidized bed combustor to produce hot products of combustion for expansion through the gas turbine. The turbine compressor provides pressurized air for combustion and for transporting the shredded waste from beneath the rotary air lock feeder valves into the combustor. Three stages of cyclones are used to remove fly ash from the hot gas steam prior to expansion through the gas turbine. A waste heat boiler is used to produce steam for the generation of additional electricity. The primary control loop regulates the fuel feed rate in response to turbine temperatures.

Liquid waste firing is not required, but is available as a desired option because the addition of water to the system allows more solid waste to be consumed and greater amounts of electricity to be produced

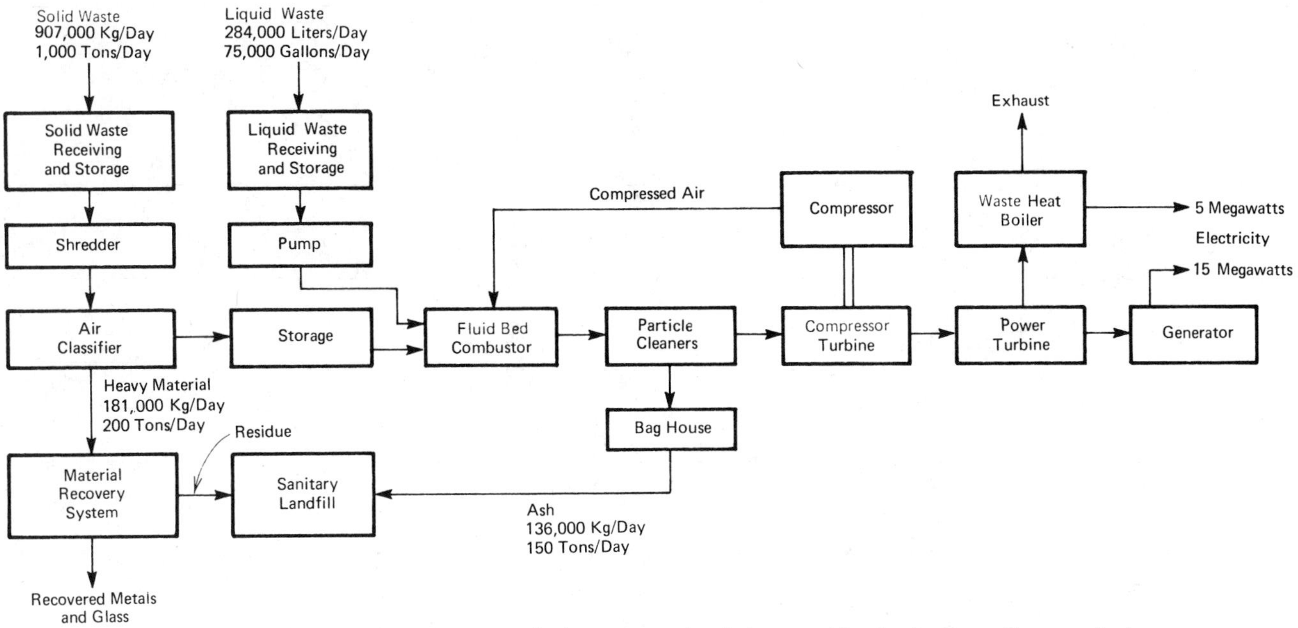

Fig. 4. Process for converting solid and liquid wastes to electrical power. (*Combustion Power Company, Inc.*)

at a lower-than-normal inlet temperature in the same size of equipment. System efficiency is reduced somewhat by the addition of water, but with a negative value fuel that is not undesirable. The energy used to evaporate the water is partially recovered by the expansion of the resulting steam through the gas turbine. The system is flexible in terms of the type of fuel that can be burned (up to 60% water) and the mix of steam and electricity produced.

References

Alexander, A. G.: "Sugarcane Physiology," Elsevier, New York, 1974.

Anderson, R. E.: "Biological Paths to Self-Reliance—Biological Solar Energy Conversion," Van Nostrand Reinhold, New York (1979).

Bolton, J. R.: "Solar Fuels," *Science*, **202**, 705–711 (1978).

Clements, H. F.: "International Society of Sugar Cane Technologists; Proceedings of the 8th Congress," Elsevier, New York, 1953.

Hammond, A. L.: "Photosynthetic Solar Energy—Biomass," *Science*, **197**, 745–746 (1977).

Hayes, D.: "Biological Sources of Commercial Energy," *Bioscience*, **27**, 540–546 (1977).

Hokanson, A. E., and R. Katzen: "Chemicals from Wood Waste," *Chem. Eng. Prog.*, **74**, 67–71 (1978).

Jannasch, H. W., and C. O. Wirsen: "Chemosynthetic Primary Production at East Pacific Seafloor Spreading Centers," *Bioscience*, **29**, 591–598 (1979).

Jannasch, H. W.: "Chemosynthetic Production of Biomass," *Oceanus*, **22**, 4, 59–62 (1980).

Kos, P., Meier, P. M., and J. M. Joyce: "Economic Analysis of the Processing and Disposal of Refuse Sludges," Curran Associates, Inc., Washington, D.C., 1974.

Lipinsky, E. S.: "Fuels from Biomass: Integration with Food and Materials Systems," *Science*, **199**, 644–651 (1978).

Lowe, R. A.: "Energy Recovery from Waste, Solid Waste as Supplementary Fuel in Power Plant Boilers," Pubn. SW-36d, U.S. Environmental Protection Agency, Washington, D.C., 1973.

Marzola, D. L., and D. P. Bartholomew: "Photosynthetic Pathway to Biomass Energy Production," *Science*, **205**, 555–559 (1979).

Mattill, J. I.: "Particulates from Wood Burning in New Hampshire," *Technology Rev.* (*MIT*), **81**, 2, 18 (1978).

Maugh, T. H., II: "Diesel Fuel Grows on Trees," *Science*, **206**, 436 (1979).

Nystrom, J. M., and S. M. Barnett: "Biochemical Engineering: Renewable Sources of Energy and Chemical Feedstocks," Amer. Inst. of Chem. Engrs., New York, 1978.

Rau, G. H., and J. I. Hedges: "Carbon-13 Depletion in a Hydrothermal Vent Mussel: Suggestion of a Chemosynthetic Food Source," *Science*, **203**, 648–649 (1979).

Rhyther, J. H., DeBoer, J. A., and B. E. Lapointe: "Cultivation of Seaweeds for Hydrocolloid, Waste Treatment and Biomass for Energy Conversion," *Proc. of 9th Intl. Seaweed Symposium*, Santa Barbara, California (A. Jensen and J. R. Stein, editors), Science Press, Princeton, New Jersey, 1979.

Sarkanen, K. V., and D. A. Tillman (editors): "Progress in Biomass Conversion," Vol. 1, Academic, New York, 1979.

Shuster, W. W., Editor: "Proceedings of 2nd Annual Fuels and Biomass Symposium, Rensselaer Polytechnic Institute, Troy, New York, June 1978.

Tillman, D. A.: "Wood as an Energy Resource," Academic, New York, 1978.

Weisz, P. B., and J. F. Marshall: "High-Grade Fuels from Biomass Farming: Potentials and Constraints," *Science*, **206**, 24–29 (1979).

BIOME. Over the earth, there are certain relatively distinct combinations of climatic conditions, life forms, and essential geologic and hydrologic features that, when taken together, form large geographic regions within which there persists a reasonably stable balance between the various natural forces and features present. Admittedly, the definition of a biome is inexact because the parameters of a biome are less than precise. Another definition of biome is that it is a climax community that characterizes a particular natural region. In ecological terms, climax refers to that final stable stage of development that a community, species, flora, or fauna attains in a given environment. Thus, the major world climaxes correspond to formations and biomes, a formation being defined as a group of associations that exist together as a result of their closely similar life pattern, habits, and climatic requirements.

Biomes do not recognize political divisons nor continental divisions, nor are their boundaries sharp, often one biome blending in with another over an extensive area. A biome, unlike a mountain range, lake, or course of a river is not measured with the precision of geodesy because it possesses fuzzy, sometimes undulating borders that generally, if illustrated on a map, will at best have wide and blended borders. In actuality, the biome is a convenient tool devised by natural scientists to map and classify what otherwise would remain blurred phenomena. And, in these respects, the concept of the biome is helpful.

The borders of biomes do correspond with geographical borders where a given biome interfaces with an ocean, the latter also considered a biome.

Land biomes include (1) the deserts, (2) tundra, (3) grassland, (4) savanna, (5) chaparral, (6) woodland, (7) coniferous forest, (8) deciduous forest, and (9) tropical forest—arranged here in order of increasing amounts of vegetation. At least two interfacing areas with the oceans are also recognized, i.e., (10) the reefs and (11) the rocky shores, which obviously are natural forms which do not satisfy the characterization of any of the aforementioned nine land biomes.

The Desert Biomes. Although the various desert biomes of the earth display considerable variation, generally a desert may be defined as "An area of low moisture due to low rainfall, i.e., less than ten inches annually, high evaporation, or extreme cold and which supports only specialized vegetation, not that typical of the latitudes in which it is located, and is generally unsuitable for human habitation under natural conditions. Deserts are not characterized by uniformity of elevation,

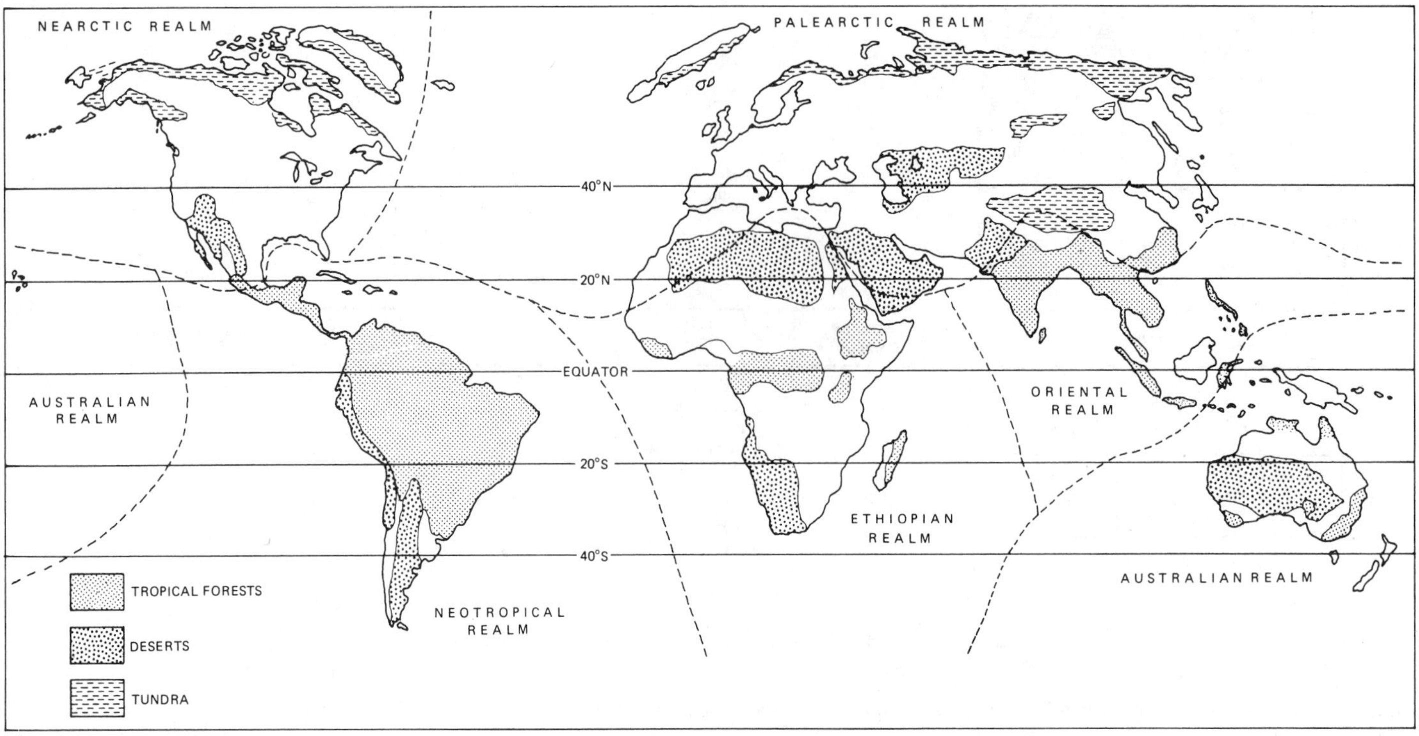

Major realms of the earth.

but wind often produces distinctive erosional features, e.g., dunes." (*American Geological Institute*).

It is estimated that deserts occupy approximately one-fifth of the earth's land surface. The principal deserts lie between 35°N and 35°S of the equator. These are regions covered by anticyclonic belts and high pressures, thus combining with other factors to result in low rainfall. See **Atmosphere (Earth);** and **Winds and Air Movement.** The major desert biomes are depicted in the accompanying figure. It is interesting to note that the major tropical forests of the earth are also located between the same aforementioned parallels of latitude, but generally considerably closer to the equator.

With reference to the figure starting with the western hemisphere, the Mojave desert of southern California is the largest in North America, with an area of approximately 13,500 square miles (34,965 square kilometers). Essentially continuous, with intervening semidesert areas, the Vizcaino desert, located in Baja California, Mexico has an area of about 6000 square miles (15,540 square kilometers). Directly south of Arizona on the Mexican mainland are semidesert areas and the Grande desert in Sonora with an area of some 2500 square miles (6475 square kilometers). Also identified as formal desert areas in the United States are the Great Salt Lake desert, located in northwestern Utah with an area of about 4000 square miles (10,360 square kilometers), the Painted desert, located in northeastern Arizona, with an area of approximately 5000 square miles (12,950 square kilometers); the Colorado-Southeastern California desert, with an area of some 4000 square miles (10,360 square kilometers); the High desert, located in central Oregon, with an area of 3000 square miles (7770 square kilometers); and the smaller Black Rock desert (600 square miles; 1554 square kilometers) and Smoke Creek desert (300 square miles; 777 square kilometers), located in northwestern Nevada.

Located in South America are the Sechura desert, situated in northwestern Peru, with an area of about 10,000 square miles (25,900 square kilometers) the Atacama desert, located in northern Chile, with an area of about 70,000 square miles (181,300 square kilometers); and the Patagonian desert of Argentina, with an estimated area of over 300,000 square miles (777,000 square kilometers).

In Africa, the Sahara, covering well over 3.5 million square miles (9 million square kilometers), represents about 32% of the total land area of that continent and spreads over parts of several African nations. Over 20% of the Sahara is located in Libya and this portion (some 650,000 miles; 1,683,500 square kilometers) is sometimes referred to as the Libyan desert. The Nubian portion of the Sahara occupies about 100,000 square miles (259,000 square kilometers). The Kalahari desert, of some 200,000 square miles (518,000 square kilometers), is located in South West Africa. It is interesting to note that the Nile River valley and environs separates the Sahara from the other desert areas of northeastern Africa and that these areas are interrupted only by the Red Sea which lies just west of the desert regions of the Arabian peninsula. Thus, with these exceptions, there is essentially continuous desert from west to east over a distance of some 5000 lineal miles (8045 kilometers).

The great desert areas of the Middle East include the Arabian desert of the Arabian peninsula which has an area of about 500,000 square miles (1,295,000 square kilometers); the Rub al Khali desert, with an area of about 250,000 square miles (647,500 square kilometers), located in southeastern Saudi Arabia; the Syrian desert, with an area of about 125,000 square miles (323,750 square kilometers), located in the northern part of the Arabian peninsula; the Nefud desert, located in the northern and central parts of Saudi Arabia, with an estimated area of 50,000 square miles (129,500 square kilometers); the Dasht-i-Lut desert of eastern Iran, with an area of approximately 20,000 square miles (51,800 square kilometers); and the Dasht-i-Kavir desert, located in north central Iran, with an area of about 18,000 square miles (46,620 square kilometers).

The Gobi desert, located in southern Mongolia, is the largest desert in Asia, with an estimated area of about 400,000 square miles (1,036,000 square kilometers). The Taklamakan desert of southern Sinkiang province (People's Republic of China) is estimated to cover an area of 125,000 square miles (323,750 square kilometers). The Thar or Indian desert, located in northwestern India has an area of about 100,000 square miles 259,000 square kilometers). The Kyzyl-Kum desert of central Turkestan (U.S.S.R.) has an area of about 90,000 square miles (233,100 square kilometers); the Kara Kum desert of southern Turkestan has an estimated area of 105,000 square miles (271,950 square kilometers); and the Peski Muyun-Kum desert of eastern Turkestan, an area of about 17,000 square miles (44,030 square kilometers).

The Australian desert is estimated to cover about 600,000 square miles; 1,554,000 square kilometers; (the total land area of Australia is 2,974,580 square miles; 7,704,162 square kilometers). Portions of the Australian desert with formal designations include: the Great Sandy desert (northwestern Australia), 160,000 square miles (414,400 square kilometers); the Great Victoria desert (southwestern), 125,000 square miles (323,750 square kilometers); the Arunta desert (central),

120,000 square miles (310,800 square kilometers); and the Gibson desert (western), 85,000 square miles (220,150 square kilometers).

Although long periods of time may elapse between rainfalls, no desert is known that is completely dry. In an exceptional rainy season, for example, parts of the Sahara may have as many as 11 days of rainfall. Most desert areas are characterized by wide temperature spans between day and night, sometimes ranging over a difference of 60°F (16°C). During the course of a year, the Gobi desert will have a variation of nearly 120 degrees. Located north of most other desert areas of the earth, the Gobi has extremely cold winter temperatures, persisting at 5°F (−15°C) (daytime) for many days and dropping at night to −30°F (−34°C).

Because of such wide temperature changes over such short periods, violent winds develop in many desert regions. Some desert winds have special names as, for example, the *simoon*, very hot and dry, which blows over the Sahara and Arabian deserts during spring and summer; the *harmattan* which carries desert air toward the Gulf of Guinea (west Africa); and the *khamsin* which blows over Egypt for many days during March, April, and May. The winds form and remove sand dunes; the terrain does not remain static. See also **Dune.**

Rivers disappear when a desert is formed. Old river beds are filled with sand. The dry beds of old river systems are known as *wadis.* Normally, the desert base immediately sucks up rainfall, but in cases of rare downpours, the wadis fill temporarily and flow violently, thus creating a hazard in the more developed areas where connecting roads have been built across desert regions. Rarely the deep waters in some desert areas are brought to the surface, such locations being known as oases. See also "Hydrology of Semiarid Areas" under entry on **Hydrology.**

The desert areas of North America support small animals, including jack rabbits, kangaroo rats, cactus mice, pocket mice, cottontails, and rock and ground squirrels. Some of these desert regions are well known for the beautiful spring flowering of desert annuals. Plants usually found in parts of these regions include Joshua trees, various cacti, the saguaro, and various low shrubs, notably creosote bush and sagebrush.

The desert regions of southern Asia in some parts may support thorny bushes, sometimes called camel sage, small clumps of wiry grass, and low-profile sagebrush. Trees occasionally found include the tamarisk, cottonwood, zag, and turai. Animals found include jerboas, sand rats, moles, hedgehogs, eagles, owls, and hawks.

The desert areas of South America are largely windswept and thus there is a very sparse covering, if any, of such plants as cacti, yucca, sagebrush, agave, cereus, creosote bush, and bunch grass. These areas support some animals, including the rhea, armadillos, guanaco, vulture, fox, and the Patagonian "hare" (tuco tuco).

The plant life of the African deserts is very scant with exception of the infrequent oases. The latter are known for their date palms. Occasional desert plants include Welwitschia, euphorbias, including plants with tuberous roots. Animals found in some areas include porcupine, gundi, rock hyrax, lizards, tenrec, springbok, and eagles.

Saltbush and bluebush are found in the Australian deserts, as well as some eucalyptus, river red gum, and acacia trees. Among animals found are spiny devil lizards, rats, mice, marsupial moles, and parakeets.

Tundra. "A treeless, level or gently undulating plain characteristic of arctic and subarctic regions. It usually has a marshy surface which supports the growth of mosses, lichens, and numerous low shrubs and is underlain by a dark, mucky soil and permafrost." (*American Geological Institute*) Permafrost may be defined as any soil, subsoil, or other surficial deposit, or even bedrock, occurring in arctic or subarctic regions at a variable depth beneath the earth's surface in which a temperature below freezing has existed continuously for a long time (from two years to tens of thousands of years). This definition is based exclusively on temperature, and disregards the texture, degree of compaction, water content, and lithologic character of the material. Its thickness ranges from over 3000 feet (914 meters) in the north to about 1 foot (0.3 meter) in the southern perimeter of a permafrost area. Permafrost underlies about one-fifth of the land area of the earth.

The regions of tundra are indicated in accompanying figure. These regions support low forms of vegetation, including mosses, sedges, scattered herbs, lichens, and stunted shrubs. Animals of the tundra include the wolf, weasel, arctic fox, arctic hare, lemming, caribou, musk ox, arctic ground squirrel, polar bear, snowy owl, and ptarmigan.

Tropical Forest. Regions of tropical forest of the earth are shown in the accompanying figure. With a plentiful supply of rainfall and warmth, the tropical forests are lushly abundant with vegetation which, in turn, assists in supporting a great variety of birds, insects, and other life forms. There are several types of tropical forests. Where a region is continuously warm and humid, as in lowlands, as found for example, in the center of Africa, certain parts of India, and in the southeastern portions of Australia and Asia, even broad-leaved trees remain green throughout the year because, in essence, the growing season does not stop for periodic breaks. Some of these trees include the ebony, mahogany, and teak. Such regions are termed *rain forests.* In other regions, because there are very sharp breaks between the dry and rainy seasons, trees will shed their leaves in systematic fashion. Because of this parallel with the patterns in temperate zones, these regions are sometimes called *winter forests.* As a tropical forest may at its outer edges begin to blend with grasslands and less-favored areas climatically, the species found in the tropical forest may follow watercourses for long distances into the grasslands. These winding, long strips of forest are sometimes called *gallery forests.* Distinctive of the tropical forests are abundant rainfall, reliable rainfall (absence of unpredictable, long dry spells), and abundant and uniform light energy (the days and nights are about equally long summer and winter). Of particular interest is the vertical layering of these forests. This represents, in essence, a struggle by the various plant species for available light. Animal communities are also divided along vertical layers. The tall trees at the very top spread a thick canopy of leaves and, with intervening vertical levels of vegetation, extremely little of the incident sunshine reaches the floor of the forest (sometimes referred to as jungle). The light at the forest floor has been likened to a kind of twilight. In a rain forest that has been left undisturbed, i.e., tall trees have not been cut down so as to dilute the effect of the covering canopy, the base of the forest is quite unlike the popular conception of a jungle. The scant light reaching the forest floor will not support twisting and tangling undergrowth—the forest floor is relatively open and clean. Fallen tree limbs and leaves are quickly consumed by insects. The rain forest displays numerous unusual and extremely interesting trees, many species of which are described in this volume. See listing under entry on **Tree.**

About half of South America is covered with tropical forests where numerous species of lichens, orchids, mosses, bromeliads, tree ferns, bamboo, lianas, cabbage palms, and numerous other forms can be found. Tree snakes, parrots, and hummingbirds are found in abundance. Other animal forms include monkeys, anteaters, coati, sloth, paca, small deer, agouti, and kinkajor. In the tropical forests of southeast Asia, some of the trees commonly found include teak, banyan, ebony, and Manila hemp. Bamboo is abundant. There are some 700 kinds of bamboo, ranging from types which grow only a few inches tall to other types that attain a height of 120 feet (36.6 meters) or more. In the forests of China, some species may grow as much as 3 feet (0.9 meter) within a 24-hour period. Animals found in the tropical forests of India and southeast Asia include the porcupine, rhinoceros, tiger, sun bear, sloth bear, antelope, and deer—with monkeys, gibbons, and orang-utans in abundance. A wide variety of lizards is found, as well as many species of pheasants and poisonous snakes.

The tropical forests of New Guinea and of relatively small areas of Australia (as compared with the vast desert regions of that continent) are of two general types—the closed-canopy rain forest; or open eucalyptus forests, where mountain ash and stunted gum trees also are found. The animal life is interesting and abundant, including numerous marsupials—kangaroos, wallabies, koala—and oppossums, the Tasmanian devil, platypus, and flying foxes.

Grasslands. In this class of biome, there is what might be termed a natural mid-form between the desert and other more richly vegetated biomes. Grasslands occur where there is inadequate rainfall and moisture to support trees, but sufficient moisture to support various kinds of grasses. The specific type of grassland mirrors the rainfall which it receives. Much of the middle portion of the United States and of Canada, in more or less of a strip north of the desert regions of the south, is or was grassland. Some authorities place North American grasslands into three categories: (1) *true prairie,* which supports blue stem and Indian grasses; (2) *short-grass plains*, where grama and buffalo grasses are found; and (3) the bunch-grass prairie. In these grasslands will be found jackrabbit, prairie dog, badger, fox, coyote, pocket gopher, pronghorn, and bison. Rattlesnakes and blue racers are prevalent.

There are large expanses of grasslands found on the other continents—the pampas of Argentina, and the great plains of southeastern Europe and Asia, sometimes termed steppes. A large portion of the former grasslands of the earth have disappeared as the result of agricultural and livestock pursuits.

Savannas. Also classified as a biome is the savanna, which may be defined as an open, grassy, essentially treeless plain, especially as found in tropical or subtropical regions. They usually are characterized by distinct wet and dry seasons. The trees and shrubs found in these areas are drought-resistant. Savannas are most prevalent in Africa and parts of Australia. A savanna has been described as being intermediate between a steppe and a forest. The savanna generally has a grassy bottom with what might be termed a sprinkling of plants and trees, but rather consistently having the aura of open country. The savannas of Africa typify what one might term the big-game landscape. Aside from the large deserts, the savanna is the most prevalent type of biome on the African continent. The typical rainfall pattern is a period of hard, soaking rain for a few months, followed by little or no rain for several months. Trees found in the African savannas include the acacia, baobab, euphorbia, and doom palm. The reasons for the formation and the continuing existence of savannas is not fully understood. Some authorities do not believe that the savanna, like the other land biomes, is essentially climate-created, but rather that the savanna can be attributed to other factors as well as climate. Factors may extend back to the primitive people who practiced shifting agriculture in these areas; or thee may be peculiarities of the soil yet not fully understood; or the grazing of native ungulate herds may be a factor. Typical of grazing animals found in the African savannas are zebras, elands, gemsbok, hartebeests, and gnus. Also found are giraffes, bush elephant, ostrich, black and white rhinoceroses, lion, wart hog, cheetah, Cape hunting dog, ground squirrels, and golden mole.

The open savanna forests which are on the fringes of the interior of Australia account for about 24% of the land area of the continent. Trees found in these areas include the eucalyptus, jarrah, wallum, ironbark, red stringybark, yellow box, coolibah, and white box. Among animal life are found the red kangaroo, emu, bandicoots, wombats, cockatoos, and parrots.

Woodland and Chaparral. These types of regions can be conveniently combined as one biome. The trees found in such regions include piñon-juniper, stands of leathery-leaved trees, such as the manzanita and chamiso. Chaparral is defined as a thicket of shrubs, and thorny bushes. In terms of fauna, these regions support few distinctive animals, but rather some animals migrate in and out from other biomes. Regions of this type are found in central and southeastern Mexico, in Europe and Africa on either side of the Mediterranean and in the southwestern United States, notably in the hills and mountains of southern California where chaparral in long dry seasons always poses a serious fire threat.

Deciduous Forest. As implied by the name, this biome is the result of climatic and other factors which favor the growth of deciduous trees (shed their leaves in the fall), of which, of course, there are scores of species, such as the beeches, oaks, basswood, elms, and maples. Roughly, the entire eastern half of the United States was originally deciduous forest, as well as southeastern Canada and pockets in Canada as far west as the Rocky Mountains. These natural forests support numerous flowering herbs, and animals commonly found include opossum, short-tailed shrew, mice, chipmunk, white-tailed deer, red fox, black bear, moles, raccoon, and gray and fox squirrels.

Other extensive regions of deciduous forests are found in the United Kingdom, the southern portions of Scandanavia, and in a broad band extending from northern Spain through France, Belgium, the Netherlands, Germany, Czechoslovakia, Poland, and the Caucasus. Large deciduous forests are also found in several parts of the People's Republic of China (east-central), in Japan and in Korea. Deciduous forests in South America essentially are limited to Chile.

Coniferous Forest. This biome is characterized by the great predominance of needle-leaf trees, such as pine and spruce. Coniferous forests are found throughout much of Canada, from coast to coast, and along a very wide strip paralleling the west coast of the United States north of San Francisco and continuing northwestward through Alaska. These forests also predominate northern Eurasia from Scandanavia eastward across the U.S.S.R. to the Bering Sea.

Realms

Closely associated with the concept of biomes is that of the realms. The concept of zoogeographic realms dates back to the early work of A. R. Wallace, who considered the earth as divided into six land realms whose boundaries were essentially fixed by impassable barriers by virtue of climate and topology. Over the years, some shifting of these boundaries has occurred, but Wallace's realms persist as the accepted biogeographical divisions of the earth. These realms are depicted in the accompanying figure.

See also **Climate; Zoogeography;** and list of entries to be found under **Meteorology.**

References

Colinvaux, P. A.: "Introduction to Ecology," Wiley, New York, 1972.
Johnson, C. E.: "The Natural World," McGraw-Hill, New York, 1972.
Ketchum, R. M.: "The Secret Life of the Forest," McGraw-Hill, New York, 1970.
Müller, P.: "The Ranges of Animal Distribution," in "Grzimek's Encyclopedia of Ecology," Van Nostrand Reinhold, New York, 1976.
Petrov, M. P.: "Deserts of the World," Wiley, New York, 1977.
Staff: "The Global Report to the President," Superintendent of Documents, Government Printing Office, Washington, D.C., 1980.
Treshow, M.: "Environment and Plant Response," McGraw-Hill, New York, 1970.
Wendorf, F., et al.: "The Prehistory of the Egyptian Sahara," *Science*, **193**, 103–114 (1976).

BIOMEMBRANES. **Brain and Nervous System; Cell (Biology).**

BIOPSY. Removal and microscopic examination of a portion of living tissue to establish a diagnosis. Biopsies are frequently performed in connection with the diagnosis of skin cancer. The technique also is widely used in the study and diagnosis of breast cancer. Other, more complex biopsies can be performed, depending upon the location of the suspect tissue and the lack of sufficient information from other diagnostic techniques.

BIOSPHERE. All the area occupied or favorable for occupation by living organisms, including parts of the lithosphere, hydrosphere, and atmosphere. See **Ecology.** In the early 1970s, the *United Nations Economic and Social Council* (UNESCO) established a "Man and the Biosphere" (MAB) program and the United States and the U.S.S.R. established an *Environmental Agreement.* This is sometimes referred to as the *Biosphere Reserve Program.* Biosphere reserves have three objectives: (1) conservation and preservation for future use of the diversity and integrity of biotic communities of plants and animals within natural ecosystems; (2) provision for ecological and environmental research, including baseline studies; and (3) facilities for education and training. Over 20 biotic provinces and subdivisions in the contiguous United States and Alaska were selected as targets. As pointed out by Franklin (1977), the program is intended to be more than simply another program of preservation layered onto existing parks and reserves. The success of the program will depend in large measure on the overall significance of the selected reserves and the degree to which they are active sites for scientific research and monitoring. See article by J. F. Franklin, *Science*, **195**, 262–267 (1977) for more detail.

BIOSPHERE (Polar). **Polar Research.**

BIOSTRATIGRAPHY. The classification, correlation, and interpretation of stratified rocks on the basis of the fossils that they contain.

BIOSTROME. A geologic term for layers, beds, or strata composed of calcareous fossil shells which form coquina, or shell-limestone. The term biostrome is primarily intended to distinguish shell-limestone from bioherms, or typical coral reefs.

BIOSYNTHESIS (Proteins). **Protein.**

BIOSYNTHETIC DRUGS. **Antibiotic.**

BIOTIC COMMUNITY. **Ecology.**

BIOTIC POTENTIAL. A quantitative expression of the dynamic significance of various inherent vital properties of living things as a factor in the establishment of external relationships. It summarizes the reproductive and survival potentialities of the organism.

BIOTIN. Infrequently referred to as Bios IIB, protective factor X, vitamin H, egg white injury factor, and CoR. Biotin, required by most vertebrates, invertebrates, higher plants, and most fungi and bacteria, falls into the general classification of vitamins. In certain species, a deficiency of this substance is a cause of desquamation of the skin; lassitude, somnolence, and muscle pain; hyperesthesia; seborrheic dermatitis; alopecia, spastic gait and kangaroolike posture (rats and mice); dermatitis and perosis (chicks and turkeys); progressive paralysis, and K^+ deficiency (dogs); alopecia, spasticity of hind legs (pigs); and thinning and depigmentation of hair (monkeys).

Biotin reacts with an oxidized carbon fragment (denoted as CO_2) and an energy-rich compound, adenosine triphosphate (ATP), to form carboxy biotin, which is "activated carbon dioxide." Biotin is firmly bound to its enzyme protein by a peptide linkage. Structurally, biotin and carboxy biotin are:

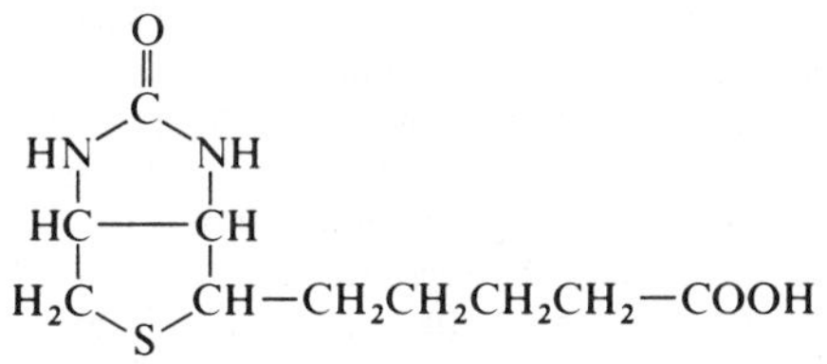

Biotin

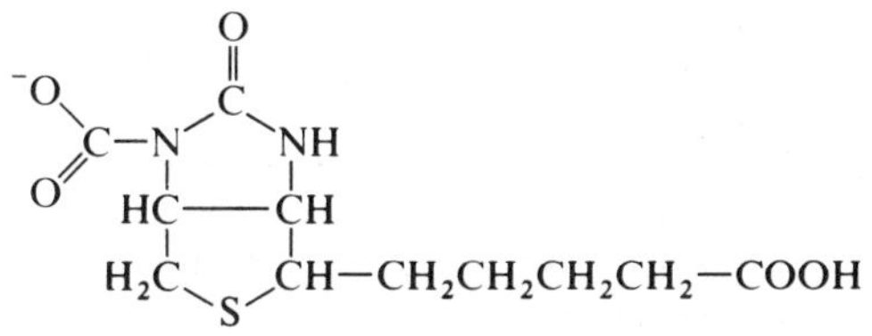

"Activated" carboxy-biotin

Biotin enzymes are believed to function primarily in reversible carboxylation-decarboxylation reactions. For example, a biotin enzyme mediates the carboxylation of propionic acid to methylmalonic acid which is subsequently converted to succinic acid, a citric acid cycle intermediate. A vitamin B_{12} coenzyme and coenzyme A are also essential to this overall reaction, again pointing out the interdependence of the B vitamin coenzymes. Another biotin enzyme-mediated reaction is the formation of malonyl-CoA by carboxylation of acetyl-CoA ("active acetate"). Malonyl-CoA is believed to be a key intermediate in fatty acid synthesis.

Bios (in Greek means life) was a word coined to describe a growth-promoting substance for yeast and discovered by Wildiers in 1901. When added in small amounts to sugar and salts medium, it permitted rapid growth of yeast even from a small seeding. Subsequent investigations proved that there was not merely a single substance involved, but that depending upon the strain of yeast and the circumstances of besting, a number of different substances could act, often synergistically, to promote the rapid growth of yeast. Pantothenic acid, biotin, inositol, thiamine, and pyridoxine all have "bios" properties when appropriately tested. Even an amino acid may be a limiting factor for yeast growth when other needs are supplied. The term "bios" has fallen from use in the literature.

Biotin required for growth and normal function by animals, yeast, and many bacteria is seldom found in deficiency in humans because the intestinal bacteria synthesize it in sufficient quantity to meet requirements. Biotin deficiency does occur, however, in animals fed raw whites of eggs. The egg white contains a protein, *avidin*, which combines with biotin and this complex is not broken down by enzymes of the gastrointestinal tract. Hence, a deficiency develops.

Biotin was first isolated in pure form in 1936 by two Dutch chemists, Koegel and Tonnis, who obtained 1.1 milligrams from 250 kilograms of dried egg yolk. They showed that the compound was necessary for the growth of yeast and gave it the name, biotin. Five years later, in America, György and co-workers found that the same compound prevented the toxicity of raw egg white in animals and, in 1942, du Vigneaud and collaborators determined the structure of the compound.

Distribution and Sources. Natural sources of biotin include the following:

High biotin content (100–400 micorgrams/100 grams)
- Lamb liver, pork liver, soyal jelly, yeast

Medium biotin content (10–100 micrograms/100 grams)
- Grains: Barley, corn (maize), oats, rice, wheat
- Meat and Fish: Beef liver, chicken, eggs, mackerel, salmon, sardines
- Nuts: Almonds, filberts, hazelnuts, groundnuts (peanuts), pecans, walnuts
- Vegetables: Cauliflower, chick-peas, cowpeas, lentils, soybeans
- Other: Chocolate, mushrooms

Low biotin content (0–10 micrograms/100 grams)
- Dairy: Cheese, milk
- Fruits: Apple, avocado, banana, cantaloupe, grape, grapefruit, orange, peach, strawberry, watermelon
- Meat and Fish: Beef, halibut, lamb, oyster, pork, tuna, veal
- Vegetables: Bean (lima), beet, beet greens, cabbage, carrot, lettuce, onion, pea, spinach, sweet corn, sweet potato, tomato

Biotin can be produced commercially by using, for starting the synthesis, meso-diamino succinic acid derivative of fumaric acid.

Determination of Biotin. Bioassay methods include the (1) rat and chick method (growth response after biotin deficiency); (2) microbiological with *L. arabinosus.* Physicochemical methods make use of polarography.

Bioavailability of Biotin. Factors which cause a decrease in bioavailability include (1) presence of avidin in food; (2) cooking losses; (3) presence of antibiotics; (4) presence of sulfa drugs; and (5) binding in foods (such as yeast). Availability can be increased by stimulating synthesis by intestinal bacteria.

Antagonists of biotin include desthiobiotin in some forms, ureylene phenyl, homobiotin, urelenecyclohexyl butyric and valeric acid, norbiotin, avidin, lysolecithin, and biotin sulfone. Synergists include vitamins B_2, B_6, B_{12}, folic acid, pantothenic acid, somatotrophin (growth hormone), and testosterone.

Precursors for the biosynthesis of biotin include pimelic acid, cysteine, and carbamyl phosphate. Desthiobiotin acts as an intermediate. In plants, the production sites are seedlings and leaves. In most animals, production is in the intestine. Storage site is the liver.

Some of the unusual features of biotin noted by investigators include (1) binding and inactivation by avidin protein found in egg white; (2) fetal tissues and cancer tissues higher in biotin than normal adult tissues; (3) biotin deficiency increases severity and duration of some diseases, notably some of the protozoan infections; and (4) oleic acid and related compounds replace biotin as unspecific stimulatory compounds in bacteria.

References

Du Vigneaud, V., et al: *J. Biol. Chem.*, **146**, 475 (1942.

Greenberg, D. M. (Editor): "Metabolic Pathways," 3rd Edition, Volumes I–IV, Academic, New York, 1967.

Haresign, W., and D. Lewis: "Recent Advances in Animal Nutrition," Butterworth Group, Woburn, Massachusetts, 1977.

Koegel, F., and B. Tonnis: *Z. Physiol. Chem.*, **242**, 43 (1963).

Kutsky, R. J.: "Handbook of Vitamins and Hormones," Van Nostrand Reinhold, New York, 1973.

Preston, R. L.: "Typical Composition of Feeds for Cattle and Sheep—1977–1978," *Feedstuffs*, A-2-A (October 3, 1977); and 3A-10A (August 18, 1978).

Staff: "Food Chemicals Codex," National Academy of Sciences, Washington, D.C., 1972.

Staff: "Nutritional Requirements of Domestic Animals," separate publications on beef cattle, dairy cattle, poultry, swine, and sheep, periodically updated, National Academy of Sciences, Washington, D.C.

White, C.: "Biotin Requirements of Broilers on Litter," World's Poultry Congress, Rio de Janeiro, Brazil, 1978.

Wood, H. G.: "Biotin," in "The Encyclopedia of Biochemistry," (R. J. Williams and E. M. Lansford, Jr., editors), Van Nostrand Reinhold, New York, 1967.

BIOTITE. A common silicate mineral containing potassium, magnesium, iron, and aluminum. Biotite is found in granitic rocks, gneisses, and schists. Although actually monoclinic, it often assumes a pseudo-hexagonal form. Like others of the mica group, it shows a highly perfect basal cleavage. Hardness, 2.5–3; sp gr, 2.7–3.4; luster, pearly to vitreous or sometimes submetallic when very black in color; cleavage sheets are elastic; color, greenish to brown or black; transparent to opaque. A general formula is $K(Mg,Fe)_3(Al,Fe)Si_3O_{10}(OH,F)_2$.

Biotite is occasionally found in large sheets, especially in pegmatite veins. It also occurs as a contact metamorphic mineral or the product of the alteration of hornblende, augite, wernerite, and similar minerals.

Biotite is found in the lavas of Vesuvius, at Monzoni, and in many other European localities; in the United States, especially in the pegmatites of New England, Virginia, and North Carolina, and the granite of Pikes Peak, Colorado. The mineral was named in honor of the French physicist, J. B. Biot. Biotite is also known as *iron mica.*

BIPHENYLS. A lay term usually used for polychlorinated biphenyls (PCBs), which have been implicated in several instances of food contamination. In 1973, food and feedstuffs were contaminated with polybrominated biphenyls in Michigan. Similarly, contamination of the James River was found in 1975. In a more recent instance in Montana, a power transformer stored in a shed at a packing company leaked PCB coolant into slaughterhouse sewage. These wastes were collected, rendered, and added to animal feed for distribution throughout nine states. A complete listing of the various PCBs can be found in "Dangerous Properties of Industrial Materials," 5th edition (I. N. Sax, editor), Van Nostrand Reinhold, New York, 1979.

BIPOLAR COORDINATE. Choose two points $\pm a$ on the X-axis of a rectangular coordinate system. Then any point in the XY-plane could be measured in either of two polar coordinate systems where the poles are at $x = \pm a$; the polar axis in both systems is the X-axis; the two polar angles are θ_1, θ_2; the two radius vectors are r_1, r_2. Define the parameters $\xi = \theta_1 - \theta_2$; $\eta = \ln r_2/r_1$ and, in terms of these parameters,

$$x = \frac{a \sinh \eta}{\cosh \eta - \cos \xi}; \quad y = \frac{a \sin \xi}{\cosh \eta - \cos \xi}$$

which are families of circles along the X- and Y-axes, respectively. Translation of these circles along the Z-axis produces the curvilinear coordinate system known as bipolar coordinates. They are families of right circular cylindrical surfaces with axes parallel to the Z-axis and centers at $y = 0$, $x = 0$, respectively (ξ, η = constant), where

$$0 \leq \xi \leq 2\pi; \quad -\infty \leq \eta \leq \infty$$

The third surface is a plane perpendicular to the Z-axis (z = constant).

See also **Coordinate System;** and terms listed under **Mathematics.**

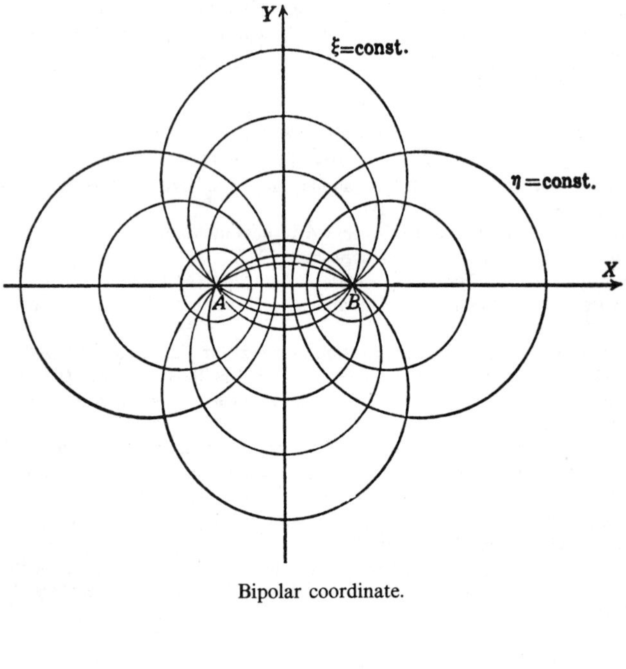

Bipolar coordinate.

BIPOLAR MICROPHONE. Microphone.

BIPOLAR TRANSISTOR SWITCH. Analog Switch.

BIPRISM. An optical prism with apex angle only slightly less than 180° used to produce a double virtual image of a point source. Interference fringes will appear on a screen placed before the biprism, as in the Young interference experiment. If the incident light is made parallel by a lens, placed between the source and the biprism, ϕ is the apex angle of the biprism, n is its refractive index, and b is the linear distance between consecutive fringes on the screen

$$\lambda = 2(n - 1)b\phi$$

where λ is the wavelength of the incident light. The device may thus be used as a method for measuring the wavelength.

BIQUADRATIC EQUATION. An algebraic equation of the fourth degree in one or more variables, also called a quartic equation. If, as is usually the case, only one variable occurs, its general form is $a_0x^4 + a_1x^3 + a_2x^2 + a_3x + a_4 = 0$. The transformation $x = y - a_1/4a_0$ removes the term in x^3, so that the equation becomes $a_0y^4 + b_2y^2 + b_3y + b_4 = 0$. The standard form is $y^4 + py^2 + qy + r = 0$, where $p = b_2/a_0$, $q = b_3/a_0$, $r = b_4/a_0$.

The roots of this equation can be determined from an associated cubic equation, as first shown by L. Ferrari (1545). Consider $z^3 + c_1z^2 + c_2z + c_3 = 0$, where $c_1 = p/2$; $c_2 = (p^2 - 4r)/16$; $c_3 = -q^2/64$, called the cubic resolvent. Suppose its roots are $z_i = Y_i^2 i = 1, 2, 3$. Then the four roots of the biquadratic equations are $y_{1,2} = Y_1 \pm Y_2 \pm Y_3$; $y_{3,4} = -Y_1 \pm Y_2 \mp Y_3$. Either of the two possible values of Y_1 and Y_2, as found from $Y_i = \pm\sqrt{z_i}$, may be taken but Y_3 must satisfy the relation $Y_3 = -q/8Y_1Y_2$.

The nature of the roots can be predicted without obtaining them. In the following cases $D = 16p^4r - 4p^3q^2 - 128p^2r^3 + 144pq^2r + 256r^3 - 27q^4$ is called the discriminant.

(a) $q \neq 0$; p, q, r real; $z_1z_2z_3 = q^2/64 > 0$.

1. z_1, z_2, z_3 all positive; four real roots; $D \geq 0$; $p < 0$, $(q^2 - 4r) > 0$; no more than two roots are equal.

2. z_1 positive; z_2, z_3 negative; two pairs of conjugate imaginary roots; $D < 0$.

3. z_1; positive; z_2 and z_3 conjugate imaginary; y_1 and y_2 real; y_3 and y_4 conjugate imaginary.

(b) $q = 0$. The cubic resolvent has one zero root, therefore the biquadratic has two pairs of equal roots, but with opposite signs.

Application of these equations to calculate the roots of a biquadratic equation would be extremely laborious and anyone would be foolish to use them for this purpose. Approximate methods, as in the case of the cubic equation, are much more satisfactory. See **Approximate Calculation;** and other related terms listed under **Mathematics.**

BI-QUARTZ. By placing two adjoining pieces of equal thickness of quartz, one dextro-, the other laevo-rotatory, over the analyzer in a polariscope, the accuracy of setting can be increased. Such a double block is called a bi-quartz. See also **Polarized Light.**

BIQUINARY CODE. Code (Computer System).

BIRAMOUS APPENDAGE. The primitive jointed appendage of the anthropods, still found in various form in the crustaceans.

The appendage consists of a single basal portion called the protopodite which is usually divided into a proximal coxopodite and a distal basipodite. It may bear on its outer margin one or several lobes called epipodites. From the protopodite two branches arise, an inner endopodite and an outer exopodite; this characteristic of the appendage is responsible for the name biramous. The endopodite is divided into five or less segments, named in order from the base the ischiopodite, meropodite, carpopodite, propodite, and dactylopodite. The exopodite is much less uniform and is often lacking.

These appendages have become modified and specialized for many functions in the existing crustaceans, as is nicely demonstrated by the appendages of the crayfish and lobster. In these animals they form sensory organs (antennae), mouth parts (jaws and accessory appendages), walking legs, swimmerets, accessory reproductive organs, and broad, flat swimming appendages. They are also regarded as the form from which the simpler jointed appendages of insects and other arthropods have been evolved.

BIRCH TREES. Members of the family *Betulaceae* (birch family), these trees are of several species. These are deciduous shrubs or trees known for their ornamental bark. They are hardy and grow fast, particularly when young. Important species of the birch family include:

Black or river birch	*Betula nigra*
Canoe birch	*B. papyrifera*
Cherry birch	*B. lenta*
Chinese paper birch	*B. albo-sinensis*
Common birch	*B. pubescens*
Forrest's birch	*B. forrestii*
Gray birch	*B. populifolia*
Himalayan birch	*B. jaquemontii*
Japanese silver birch	*B. platyphylla japonica*
Monarch birch	*B. maximowicziana*
Poplar-leaved birch	*B. populifolia*
Russian rock birch	*B. ermanii*
Silver birch	*B. pendula*
Southern white Chinese birch	*B. albo-sinensis septentrionalis*
Swedish birch	*B. pendula "Dalecarlica"*
Yellow birch	*B. lutea*

Various species of birch trees are found in Europe, North America, northeastern Asia, North America, particularly northern and eastern, but with some varieties extending westward and into Alaska. Some species were introduced into North America from Europe. Birches tend to be quite slender, the trunk diameter of some 50-year-old trees not exceeding 15 inches (38 centimeters). However, the trees tend to be stouter where they are found in warmer areas. The flower of both sexes is found on the same tree. It is spikelike with clusters close to the branch. There are no petals; the catkins mature in early spring. The fruit is a hard nut in the catkins.

The paper birch (also known as white birch or canoe birch) is possibly the best known of the birches in North America. It is a superior tree among the various species. The tree is well known for its flexible, smooth and white bark, which is probably best known for its use by American Indians in the construction of canoes. The paper birch normally ranges up to 70 feet (21 meters) in height, but as shown by the accompanying table, can reach close to 100 feet (30 meters) in height. The leaves are a dark green, pointed, ovalate, and coarsely toothed. The tree ranges westward from Newfoundland to Hudson Bay (southern portion) and to the Alaskan coast. In the west, it is found as far south as Washington and Montana. In the eastern states, it is found commonly in New England and northern New York State, and ranging down into Connecticut. The pale-brown, close-grained wood finds use as pulp wood and for making numerous small items such as spools and shoe lasts—some of these uses having been replaced by plastics and other materials in recent years. In the green state, paper birch has a moisture content of 65% and weighs 50 pounds per cubic foot (801 kilograms per cubic meter). When air dried to 12%, the weight is 44 pounds per cubic foot (705 kilograms per cubic meter) and 1000 boardfeet (2.36 cubic meter) of nominal sizes weigh 3160 pounds (1433 kilograms). The crushing strength of the green wood with compression applied parallel to the grain is 2360 per square inch (16.3 MPa); 5690 psi (39.3 MPa) for the dried wood. The tensile strength of the green wood with tension applied perpendicular to the grain is 380 psi (2.6 MPa).

The sweet birch (or black birch or cherry birch) is found in midwestern Canada and the United States eastward to the Atlantic. It is found in the Alleghany Mountains, extending southward into Kentucky, Tennessee, and even in parts of western Florida. The tree is dense with foliage, having a rounded top. The bark is cherry in color. The wood is heavy and strong. At one time, the wood was extensively used in Nova Scotia and New Brunswick by the shipbuilding industry.

The gray birch (*B. populifolia*) is common in the United States. The bark is white, spotted with dark scars. The tree tends to grow in clumps. In burned-out areas, the gray birch is often the first tree to be seen making a start out of the charred ground. The tree, also known as the white birch, Oldfield birch, or poplar birch, is relatively small (20–30 feet in height; 6 to 9 meters) with a slender trunk (from 6 to 10 inches in diameter; 15 to 25 centimeters). The tree frequently is found in what might be termed waste land—swampy areas, rocky slopes and pastures. It is found throughout New England and northern New York and up into the region of the lower St. Lawrence. The tree has a preference for coastal areas—along the Atlantic, eastern Great Lakes, and rivers. The wood finds limited commercial use.

The river birch (*B. nigra*) is native to the eastern United States. The tree is pyramidal in shape, with a reddish-brown bark. It is frequently found near water or streams and has a shallow root system. Probably the most abundant of the birches in the United States is the yellow birch (*B. alleghaniensis*), found from Newfoundland to the Gulf Coast. The wood is of a deeper brown color than most other species of birch. The wood is strong, hard, tough, close-grained, and can be polished to appear like cherry or mahogany. The wood

RECORD BIRCH TREES IN THE UNITED STATES[1]

Specimen	Circumference[2]		Height		Spread		Location
	(inches)	(centimeters)	(feet)	(meters)	(feet)	(meters)	
Gray birch (1976) (*Betula populifolia*)	47	119	55	16.5	20	6	New Hampshire
Mountain Paper birch (1973) (*Betula papyrifera*)	83	211	78	23.4	70	21	Michigan
Northwestern Paper birch (1975) (*Betula papyrifera*) (Co-Champion)	46	117	66	19.8	30	9	Oregon
Northwestern Paper birch (1970) (*Betula papyrifera*) (Co-Champion)	46	117	65	19.5	32	9.6	Idaho
Paper birch (1972) (*Betula papyrifera*)	217	551	96	28.8	83	24.9	Maine
River birch (1974) (*Betula nigra*)	161	409	95	28.5	90	27	Virginia
Sweet birch (1961) (*Betula lenta*)	182	462	70	21	87	26.1	New Hampshire
Water birch (1973) (*Betula occidentalis*)	(at 3 feet) 111	282	53	15.9	42	12.6	Oregon
Western Paper birch (1977) (*Betula papyrifera*)	135	343	50	15	45	13.5	Washington
Yellow birch (1973) (*Betula alleghaniensis*)	169	429	114	34.2	101	30.3	Michigan

[1] From the "Social Register of Big Trees," The American Forestry Association (by permission).
[2] At 4.5 feet (1.4 meters).

is used for furniture, handles, trim, and paneling. The green wood has a moisture content of 62% and weighs 57 pounds per cubic foot (913 kilograms per cubic meter). When air-dried to 12% moisture content, the weight is 44 pounds per cubic foot (705 kilograms per cubic meter) and 1000 board feet (2.36 cubic meters) of nominal sizes weigh 3670 pounds (1665 kilograms). The crushing strength of the green wood with compression applied parallel to the grain is 3510 psi (24 MPa); of the dried wood, 8310 psi (57 MPa). The tensile strength with tension applied perpendicular to the green is 430 psi (3 MPa) for the green wood; 930 psi (6.4 MPa) for the dried wood.

Some of the species just described, notably the gray and river birches, are short lived.

The common birches of Europe include the *B. pendula*, capable of attaining a height of 75 feet (22.5 meters) and having a white bark; and the *B. pubescens*, the common white birch (or downy birch) of both Europe and northern Asia. The latter tree has a white, peeling bark, but is not quite as showy as its American counterpart. The tree can attain a height of about 65 feet (19.5 meters). Various species of birches also are found in Japan, Manchuria, the region of the Himalayas, and China. Generally, the birches prefer the northern climates.

BIRD LOUSE. (*Insecta, Mallophaga*). A wingless ectoparasitic insect with biting mouth parts. Most species of bird lice live among the feathers of birds and eat bits of feather and other debris. Although a few species are found on mammals, the prevailing type of host has given its name to the entire order; the name biting lice is also distinctive. See illustration.

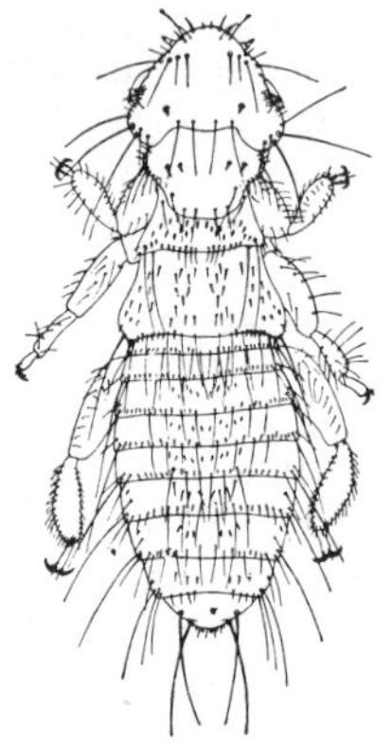

Bird louse.

The bird lice which affect poultry are economically important. Even though they do not suck blood like other parasites, the irritation resulting from their presence in large numbers is serious to the birds. Various measures of control have been devised, among them white-washing roosts, oiling perches with kerosine, and the use of various insect powders in nests and on the birds themselves. The maintenance of clean surroundings for the flock is of the utmost importance in preventing severe infestation.

BIRD OF PARADISE (*Aves, Passeriformes*). Any bird of numerous beautiful species which make up the family *Paradisaeidae*. They are characterized by the gorgeous colors, remarkable displays, and bizarre forms of the plumage and are unsurpassed in splendor by any group of birds, although they are fairly near to the crows in classification. Most of the about 40 species are found in New Guinea.

The tail of the bird of paradise may be as much as 3 feet (0.9 meter) long. The plumage is fine and silky, with black and purple or gracklelike coloration. Only the males have the highly colored plumage. The female is dull and drab. These birds make a variety of noises, ranging from a weak "peep" to throaty "caws," trumpetings, snaps, and hisses. See illustration.

Although the male carves out a single territory which he defends vigorously, he is promiscuous. There is no bond or paring off preceding mating as in the instances of many birds. The female builds her own nest and feeds the young. The nest is shallow, cuplike, and found in vines—only occasionally in trees. The egg shells are irregularly streaked.

When the bird of paradise was first discovered in New Guinea (about 1522), much trading in the plumes developed. At the peak of demand by Europeans, more than 50,000 skins were sold annually. Today, this trade and practice is forbidden.

Red bird of paradise (*Paradisaea rubra*).

BIRDS. Of the phylum *Chordata*, subphylum *Vertebrata*, and class *Aves*, the birds, which are related ancestrally to the reptiles, as apparent from reptilian scales on their legs, have numerous distinctions, most notable of which is the ability to fly, and part and parcel of which is possession of feathers, a unique anatomical feature not found in any other animal. As indicated by Fig. 1, there is a tremendous variety of form, size, and habit among the 28 orders of birds. Conventionally, the birds are put into 27 orders, although some authorities put flamingos into a separate order. A listing of the orders and parenthetical definition of the content of each order is given in the footnote to Table 2. Within these orders of birds, there are over 150 families with a total of something less than 9000 species. It should also be noted that a few authorities stretch the list of orders to 29 by separating the touracos from the cuckoos. Nearly 130 of the more important or interesting species of birds are described in this volume. Table 1 lists the specific entries providing topical coverage in this volume. It is not uncommon to have the Latin family name of a particular species as the starting information. When this is the case, Table 2 provides a convenient means for going from the Latin family name to the English family name, with examples—as well as providing information on the common name of a species. The Latin family name, very helpful in literature searching, may be found in Table 3, along with the family names and orders for over 400 species of birds.

The external structural features of a "typical" bird are shown in Fig. 2. Highlights of bird anatomy and physiology include: (1) The skin is covered with feathers; (2) the jaws are ensheathed in a horny beak and bear no teeth; (3) the pectoral appendages are usually modified for flight, forming wings, although they are rudimentary in some species and aid in swimming in others; (4) the skeleton is made rigid by the fusion of bones; (5) the heart is four-chambered; and (6) the birds are warm-blooded, the blood temperature ranging from 102°F to as high as 112°F (39°C to 44.5°C).

As is true of all extensive groups, the birds are very diverse in habits. They are both herbivorous and carnivorous, and are further distinguished as seed-eaters, fruit-eaters, insect-eaters, fish-eaters, and other types. In addition to their usual ability in the air, they are also distinguished as swimmers, waders, walkers, runners, divers, and, in a few cases, burrowers. Their nesting habits vary remarkably and the construction of the nest is, in many cases, a source of wonder.

The seasonal migrations of birds are almost unique. No other group of animals is so generally characterized by this tendency. The subject has been widely studied and has aroused much speculation without being clearly understood. It is obviously correlated with seasonal variation in the food supply and with climatic conditions, and is made possible by high specialization for flight, but exact knowledge of cause and effect in migration is lacking.

The economic importance of birds is great, and is chiefly to their credit. Insect-eating species destroy countless pests and seed-eating species aid in checking the spread of many weeds, although they may also rob the farmer of a small part of his crops. Scavengers like the turkey vultures are useful, although the degree of their usefulness is difficult to estimate. On the other hand, a few hawks and owls—

Fig. 1. Very abridged representation of various bird forms.

TABLE 1. TOPICAL COVERAGE OF BIRDS IN THIS VOLUME.

Order of Birds	Title of Entry
ANSERIFORMES	**Anseriformes**
(Duck, goose, swan)	**Waterfowl**
APODIFORMES	**Apodiformes**
	Swifts and Hummingbirds
APTERYCIFORMES	**Kiwi**
CAPRIMULGIFORMES	**Caprimulgiformes**
(Frogmouth, goatsucker, nightjar, nighthawk, oilbird, potoo)	**Nightjars and Nighthawks**
CASUARIIFORMES	**Cassowaries**
	Emu
CHARADRIIFORMES	**Charadriiformes**
(Auk, gull, plover, puffin, tern)	**Waders, Shorebirds, and Gulls**
CICONIIFORMES	**Bittern**
	Ciconiiformes
	Heron
	Ibis
	Screamer
	Stork
COLIIFORMES	**Coliiformes**
	Mousebird
COLUMBIFORMES	**Columbiformes**
	Pigeons and Doves
CORACIIFORMES	**Coraciiformes**
(Bee-eater, hoopoe, hornbill, kingfisher, motmot, roller, tody)	**Kingfishers and other Coraciiformes**
CUCULIFORMES	**Cuckoos and Coucals**
	Cuculiformes
	Turacos
FALCONIFORMES	**Caracara**
	Condor
	Eagle
	Falcon
	Falconiformes
	Hawk
	Vulture
GALLIFORMES	**Curassow**
	Galliformes
	Grouse
	Hoatzins
	Jungle Fowl
	Maleo
	Megapode
	Mound Birds
	Partridge
	Peafowl
	Pheasant
	Ptarmigan
	Quail
	Tragopans
	Turkey
GAVIIFORMES	**Gaviiformes**
	Loon
GRUIFORMES	**Gruiformes**
	Rails, Coots, and Cranes
PASSERIFORMES	**Bird of Paradise**
(Perching and song birds)	**Blackbird**
	Bluebird
	Bluethroat
	Bobolink
	Bowerbird
	Broadbills
	Bulbul
	Bulifinch
	Bunting
	Canary
	Cardinal
	Chatterer
	Chickadee
	Cowbird
	Creeper
	Crow
	Finch
	Fringillidae
	Gnatcatcher
	Grackle
	Jay
	Junco
	Kingbird
	Lark
	Lyrebird
	Magpie
	Manakin
	Martin
	Meadowlark
	Myna
	Nightingale
	Nuthatch
	Oriole
	Ouzel
	Passeriformes
	Raven
	Redstart
	Robin
	Shrike
Other species are covered under entry on **Passeriformes.**	**Sparrow**
	Starling
	Swallow
	Tanger
	Thrasher
	Thrush
	Tit
	Warbler
	Waswing
	Weaverbird
	Wren
PELECANIFORMES	**Pelecaniformes**
	Pelicans and Cormorants
PHOENICOPTERIFORMES	**Flamingo**
	Phoenicopteri
PICIFORMES	**Piciformes**
	Woodpeckers and Toucans
PODICEPEDIFORMES	**Grebe**
	Podicepediformes
PROCELLARIIFORMES	**Petrels and Albatrosses**
	Procellariiformes
PSITTACIFORMES	**Parrots and Cockatoos**
(Cockatoo, kaka, kea, lory, lovebird, macaw, parakeet, parrot)	**Psittaciformes**
RHEIFORMES	**Rhea**
SPHENISCIFORMES	**Penguin**
	Sphenisciformes
STRIGIFORMES	**Owls**
	Strigiformes
STRUTHIONIFORMES	**Ostrich**
	Ratites
TINAMIFORMES	**Tinamiformes**
	Tinamous
TROGONIFORMES	**Trogon**
	Trogoniformes
FOSSIL BIRDS	**Fossil Birds**
(Archaeopteryx, Ichthyornis)	
POULTRY	**Poultry**

and only a few—do some harm by destroying useful birds and the crow is given a very bad reputation by conservation experts as a robber of the nests of other birds. It is scarcely necessary to mention the value of birds as food and game. The domestic species, chickens, ducks, turkeys, geese, are too well known as food, and their eggs are too common a culinary material to be readily overlooked.

Probably because specialization for flight overshadows other adaptations, the classification of birds has been subject to some difficulty. The birds are divided into two subclasses by some writers, the *Ratitae* including flightless birds whose sternum is without the deep keel to which the powerful flight muscles are attached, and the *Carinarae* with a keeled sternum. These divisions are not, however, clean cut.

The development of a feather indicates that it is a modified scale, like those of reptiles. A feather consists of a central axis or rachis continuous with the hollow quill which is attached to the body. The rachis bears the flat vane of the feather which is made up of many slender barbs bearing barbules along each side. The barbules of adjacent barbs interlock to form the continuous surface of flight feathers and the similar contour feathers of the body. Down feathers are of generally soft structure and lack barbules, and filoplumes are slender feathers with few barbs. The three types of feathers are shown in Fig. 3. When birds shed their feathers, they are said to *molt.* This usually happens in late summer, at which time the bird develops a new set of feathers. These are formed within the same follicles and

TABLE 2. ALPHABETICAL LIST OF FAMILY BIRD NAMES—LATIN TO ENGLISH. (Number in ( ) after Latin name indicates Order of which Family is a part—see note at end of table.)

Latin Family Name	English Name or Examples	Latin Family Name	English Name or Examples
Accipitridae (13)	Eagle, hawk, Old World vulture, harrier	*Glareolidae* (16)	Pratincole, courser
Aegothelidae (21)	Owlet nightjar	*Grallinidae* (27)	Magpie-lark
Alaudidae (27)	Lark	*Gruidae* (15)	Crane
Alcedinidae (25)	Kingfisher	*Haematopodidae* (16)	Oystercatcher
Alcidae (16)	Auk, murre, guillemot, dovekie, puffin	*Heliornithidae* (15)	Finfoot, sun grebe
Anatidae (12)	Goose, swan, duck	*Hemiprocnidae* (22)	Crested or tree swift
Anhimidae (12)	Screamer	*Hirundinidae* (27)	Swallow, martin
Anhingidae (10)	Snakebird, anhinga	*Hydrobatidae* (9)	Storm petrel
Apodidae (22)	Swift	*Hypposittinae* (27)	Coral-billed nuthatch
Apterygidae (4)	Kiwi	*Icteridae* (27)	American blackbird, oriole, troupial
Aramidae (15)	Limpkin	*Indicatoridae* (26)	Honey-guide
Ardeidae (11)	Heron, bittern, egret	*Irenidae* (27)	Fairy bluebird, iora, leafbird
Artamidae (27)	Wood-swallow	*Jacanidae* (16)	Lily-trotter, jacana
Atrichornithidae (27)	Scrub-bird	*Laniidae* (27)	Shrike, butcher-bird
Balaeniciptidae (11)	Whale-headed stork	*Laridae* (16)	Gull, tern
Bombycillidae (27)	Waxwing, silky flycatcher, hypocolius	*Leptosomatidae* (25)	Cuckoo-roller
Brachypteraciidae (25)	Ground roller	*Megapodiidae* (14)	Megapode
Bucconidae (26)	Puffbird	*Meleagrididae* (14)	Turkey
Bucerotidae (25)	Hornbill	*Meliphagidae* (27)	Honey-eater
Burhinidae (16)	Thick-knee, stone curlew	*Menuridae* (27)	Lyrebird
Callaëidae (27)	Wattled crow, huias, saddleback	*Meropidae* (25)	Bee-eater
Campephagidae (27)	Cuckoo-shrike, minivet	*Mesoenatidae* (15)	Mesite, roatelo, monia
Capitonidae (26)	Barbet	*Mimidae* (27)	Thrasher, cat bird, mockingbird
Caprimulgidae (21)	Goatsucker, nightjar, whip-poor-will	*Momotidae* (25)	Motmot
Carduelinae (27)	Crossbill, northern grosbeak, siskin, canary	*Motacillidae* (27)	Wagtail, pipit
Cariamidae (15)	Cariama, seriema	*Muscicapidae* (27)	Old World flycatcher
Casuariidae (3)	Cassowary	*Musophagidae* (19)	Turaco, plantain-eater
Catamblyrhynchidae (27)	Plush-capped finch	*Nectariniidae* (27)	Sunbird, spiderhunter
Cathartidae (13)	New World vulture	*Numididae* (14)	Guinea fowl
Certhiidae (27)	Creeper	*Nyctibiidae* (21)	Potoo, wood-nightjar
Chamaeidae (27)	Wren-tit	*Opisthocomidae* (14)	Hoatzin
Charadriidae (16)	Plover, turnstone, surf bird	*Oriolidae* (27)	Old World oriole
Chionididae (16)	Sheath-bill	*Otididae* (15)	Bustard
Ciconiidae (11)	Stork, jabiru	*Pandionidae* (13)	Osprey, fish-hawk
Cinclidae (27)	Dipper or water ouzel	*Paradisaeidae* (27)	Bird of paradise
Cochleariinae (11)	Boat-billed heron	*Paridae* (27)	Titmouse, chickadee
Coliidae (23)	Mousebird, coly	*Paradoxornithidae* (27)	Parrotbill or suthora
Columbidae (17)	Pigeon, dove	*Parulidae* (27)	Wood warbler, honeycreeper warbler
Conopophagidae (27)	Ant-pipit, gnat-eater	*Pedionominae* (15)	Plains-wanderer, collared hemipode
Coraciidae (25)	Roller	*Pelecanidae* (10)	Pelican
Corvidae (27)	Crow, magpie, jay	*Pelecanoididae* (9)	Diving petrel
Cotingidae (27)	Cotinga	*Phaëthontidae* (10)	Tropic-bird
Cracidae (14)	Curassow, guan, chachalaca	*Phalacrocoracidae* (10)	Cormorant
Cracticidae (27)	Bell magpie, Australian butcher bird, piping crow	*Phalaropodidae* (16)	Phalarope
Cuculidae (19)	Cuckoo, anis, roadrunner, coua, coucal	*Phasianidae* (14)	Pheasant, quail, peacock
Cyclarhinae (27)	Peppershrike	*Philepittidae* (27)	Asite or philepitta
Dendrocolaptidae (27)	Woodcreeper, woodhewer	*Phoeniculidae* (25)	Wood-hoopoe
Dicaeidae (27)	Flowerpecker	*Phoenicopteridae* (28)	Flamingo
Dicruridae (27)	Drongo	*Phytotomidae* (27)	Plant-cutter
Diomedeidae (9)	Albatross	*Picathartidae* (27)	Bald crow
Drepanididae (27)	Hawaiian honeycreeper	*Picidae* (26)	Woodpecker, wryneck, piculet
Dromadidae (16)	Crab-plover	*Pipridae* (27)	Manakin
Dromiceiidae (3)	Emu	*Pittidae* (27)	Pitta or jewel thrush
Dulidae (27)	Palm chat	*Ploceidae* (27)	Weaverbird
Emberizinae (27)	Ground finch, New World sparrow, Old World finch, bunting	*Podargidae* (21)	Frogmouth
Estrildidae (27)	Waxbill, grass finch, mannikin, java sparrow	*Podicepedidae* (8)	Grebe
Eurylaimidae (27)	Broadbill	*Prionopidae* (27)	Helmet shrike or wood shrike
Eurypygidae (15)	Sun-bittern	*Procellariidae* (9)	Petrel, fulmar, shearwater
Falconidae (13)	Falcon	*Prunellidae* (27)	Accentor or hedge sparrow
Formicariidae (27)	Antbird	*Pseudochelidoninae* (27)	African river martin
Fregatidae (10)	Frigate-bird	*Psittacidae* (18)	Parrot, parakeet, cockatoo, lory
Fringillidae (27)	Cardinal, bunting, sparrow	*Psophiidae* (15)	Trumpeter
Furnariidae (27)	Ovenbird	*Pteroclidae* (17)	Sand-grouse
Galbulidae (26)	Jacamar	*Ptilonorhynchidae* (27)	Bowerbird
Gaviidae (7)	Loon	*Pycnonotidae* (27)	Bulbul
Geospizinae (27)	Galapagos and Cocos Island finch	*Rallidae* (15)	Rail, gallinule, coot
		Ramphastidae (26)	Toucan
		Recurvirostridae (16)	Avocet, stilt
		Regulinae (27)	Kinglet or goldcrest
		Rheidae (3)	Rhea
		Rhinocryptidae (27)	Tapaculo
		Rhynochetidae (15)	Kagu
		Richmondeninae (27)	Cardinal, grosbeak, saltator
		Rostratulidae (16)	Painted snipe

(*continued*)

TABLE 2 (*continued*)

Latin Family Name	English Name or Examples
Rynchopidae (16)	Skimmer
Sagittariidae (13)	Secretary bird
Scolopacidae (16)	Sandpiper, snipe, woodcock
Scopidae (11)	Hammerhead
Sittidae (27)	Nuthatch
Spheniscidae (6)	Penguin
Steatornithidae (21)	Oilbird, guacharo
Stercorariidae (16)	Skua, jaeger
Strigidae (20)	Owl
Struthionidae (1)	Ostrich
Sturnidae (27)	Starling
Sulidae (10)	Gannet, booby
Sylviidae (27)	Old World warbler
Tersinidae (27)	Swallow-tanager
Tetraonidae (14)	Grouse
Thinocoridae (16)	Seed-snipe
Thraupidae (27)	Tanager, diglossa
Threskiornithidae (11)	Ibis, spoonbill
Timaliidae (27)	Babbling thrush
Tinamidae (5)	Tinamou
Todidae (25)	Tody
Trochilidae (22)	Hummingbird
Troglodytidae (27)	Wren
Trogonidae (24)	Trogon
Turdidae (27)	Thrush, blue thrush, forktail, cochoa, robin
Turnicidae (15)	Button quail
Tyrannidae (27)	Tyrant flycatcher
Tytonidae (20)	Barn owl
Upupidae (25)	Hoopoe
Vangidae (27)	Vanga shrike
Vireolaniinae (27)	Shrike-vireo
Vireonidae (27)	Vireo
Xenicidae (27)	New Zealand wren
Zeledoniidae (27)	Wren-thrush
Zosteropidae (27)	White-eye

Identification of Order of which Family is a part:

Order No.	Name of Family
1	*Struthioniformes* (Ostriches)
2	*Rheiformes* (Rheas)
3	*Casuariiformes* (Cassowaries and Emus)
4	*Apterygiformes* (Kiwis)
5	*Tinamiformes* (Tinamous)
6	*Sphenisciformes* (Penguins)
7	*Gaviiformes* (Loons)
8	*Podicepediformes* (Grebes)
9	*Procellariiformes* (Tube-nosed Swimmers)
10	*Pelecaniformes* (Totipalmate Swimmers)
11	*Ciconiiformes* (Long-legged Waders)
12	*Anseriformes* (Waterfowl and Screamers)
13	*Falconiformes* (Diurnal Birds of Prey)
14	*Galliformes* (Gallinaceous Birds and Hoatzin)
15	*Gruiformes* (Rails, Cranes, and Bustard-like Birds)
16	*Charadriiformes* (Shorebirds, Alcids, and Gull-like Birds)
17	*Columbiformes* (Pigeon-like Birds)
18	*Psittaciformes* (Parrot-like Birds)
19	*Cuculiformes* (Cuckoo-like Birds and Turacos)
20	*Strigiformes* (Owls)
21	*Caprimulgiformes* (Nightjars)
22	*Apodiformes* (Swifts and Hummingbirds)
23	*Coliiformes* (Mousebirds or Colies)
24	*Trogoniformes* (Trogons)
25	*Coraciiformes* (Kingfishers, Todies, Rollers, Hornbills, etc.)
26	*Piciformes* (Jacamars, Barbets, Toucans, Woodpeckers, etc.)
27	*Passeriformes* (Perching Birds and Higher Song Birds)
28	*Phoenicopteriformes* (Flamingos)

are from the same papillae from which the old feathers were cast away. In some species, there is another, often partial, molt just at the start of the breeding season. Often fresh coloration is shown at this time.

In anatomy, the skeleton of a bird is frequently compared with that of a reptile, the main differences being made necessary by the two differing modes of locomotion. Bipedal locomotion of the bird requires modification of the hindlimbs and pelvic girdle. For flight, the pectoral girdle and forelimbs are modified. The trunk is rigid and the sternum is characterized by a median ridge called the *keel.* Extending rearward from selected ribs are short projections (*uncinate processes*) which contribute to firming up the thoracic framework.

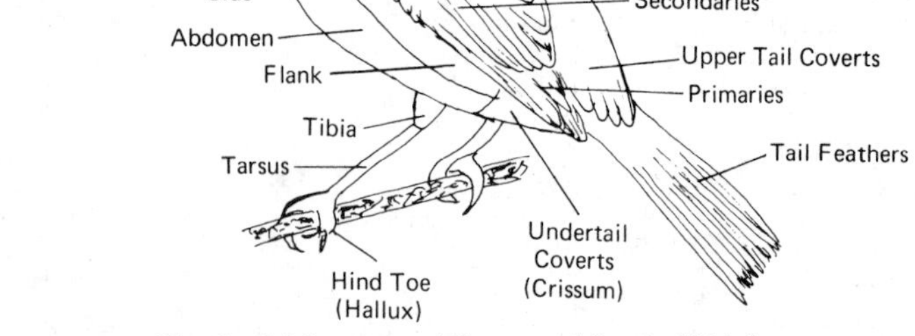

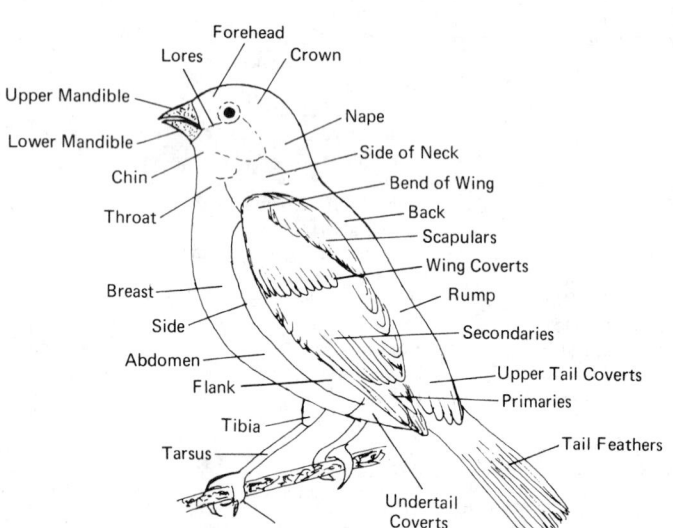

Fig. 2. Major external features of "typical" bird.

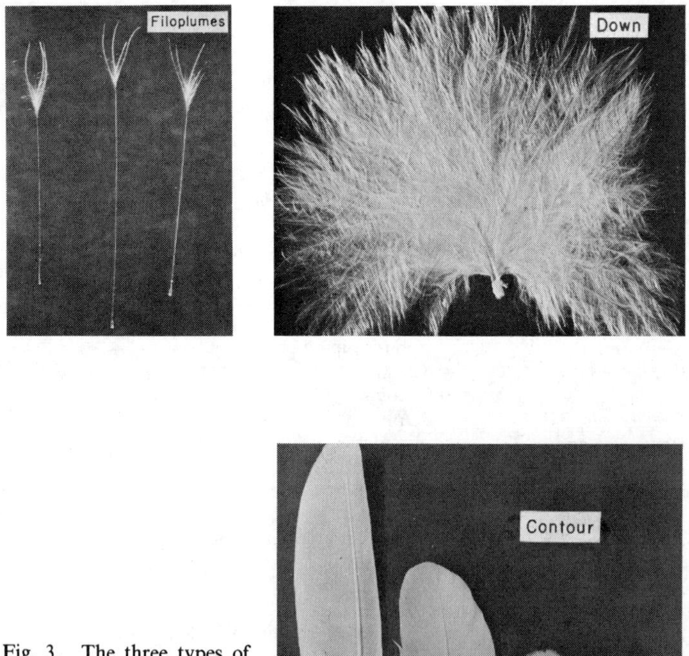

Fig. 3. The three types of bird feathers. The filoplumes and down are shown much enlarged. (*A. M. Winchester.*)

TABLE 3. ALPHABETICAL LIST OF BIRDS SHOWING LATIN FAMILY NAMES AND ORDERS.
(Orders are shown by number in ( ) after common name of bird. See note at end of Table 2 for identification of the order.)

Common Name	Family	Common Name	Family
Accentor (27)	*Prunellidae*	Crested swift (22)	*Hemiprocnidae*
Adjutant (11)	*Ciconiidae*	Crocodile Bird (16)	*Glareolidae*
African river martin (27)	*Pseudochelidoninae*	Crossbill (27)	*Carduelinae*
Albatross (9)	*Diomedeidae*	Crow (27)	*Corvidae*
American blackbird (27)	*Icteridae*	Cuckoo (19)	*Cuculidae*
Anhinga (10)	*Anhingidae*	Cuckoo-roller (25)	*Leptosomatidae*
Anis (19)	*Cuculidae*	Cuckoo-shrike (27)	*Campephagidae*
Antbird (27)	*Formicariidae*	Curassow (14)	*Cracidae*
Ant-pipit (27)	*Conopophagidae*	Curlew (11)	*Scolopacidae*
Argus (14)	*Phasianidae*	Currawong (27)	*Cracticidae*
Asite (27)	*Philepittidae*	Diamond bird (27)	*Diacaeidae*
Auk (16)	*Alcidae*	Diglossa (27)	*Thraupidae*
Australian butcher bird (27)	*Cracticidae*	Dipper (27)	*Cinclidae*
Avocet (16)	*Recurvirostridae*	Diver (7)	*Gaviidae*
Babbling thrush (27)	*Timaliidae*	Diving petrel (9)	*Pelecanoididae*
Baldcrow (27)	*Picathartidae*	Dollarbird (25)	*Coraciidae*
Baldpate (12)	*Anatidae*	Dotterel (16)	*Characriidae*
Baltimore oriole (27)	*Icteridae*	Dove (17)	*Columbidae*
Barbet (26)	*Capitonidae*	Dovekie (16)	*Alcidae*
Barn owl (20)	*Tytonidae*	Dowitcher (16)	*Scolopacidae*
Becard (27)	*Cotingidae*	Drongo (27)	*Dicruridae*
Bee-eater (25)	*Meropidae*	Duck (12)	*Anatidae*
Bellbird (27)	*Cotingidae*	Dunlin (16)	*Scolopacidae*
Bell magpie (27)	*Cracticidae*	Dunnock (27)	*Prunellidae*
Bird of paradise (27)	*Paradisaeidae*	Eagle (13)	*Accipitridae*
Bittern (11)	*Ardeidae*	Egret (11)	*Ardeidae*
Blackbird (American) (27)	*Icteridae*	Eider (12)	*Anatidae*
Bluebird (27)	*Irenidae*	Emu (3)	*Dromiceiidae*
Blue thrush (27)	*Turidae*	Fairy bluebird (27)	*Irenidae*
Boat-billed heron (11)	*Cochleariinae*	Falcon (13)	*Falconidae*
Bobolink (27)	*Icteridae*	Finch (most species) (27)	*Emberizinae*
Bob White (14)	*Phasianidae*	Finfoot (15)	*Heliornithidae*
Bonaparte's gull (16)	*Laridae*	Fire-backed pheasant (14)	*Phasianidae*
Booby (10)	*Sulidae*	Fish-hawk (13)	*Pandionidae*
Bowerbird (27)	*Ptilonorhynchidae*	Flamingo (28)	*Phoenicopteridae*
Broadbill (27)	*Eurylaimidae*	Flowerpecker (27)	*Dicaeidae*
Bulbul (27)	*Pycnonotidae*	Flycatcher, Old World (27)	*Muscicapidae*
Bunting (27)	*Emberizinae*	Forktail (27)	*Turidae*
Bustard (15)	*Otididae*	Franklin's gull (16)	*Laridae*
Butcherbird (27)	*Laniidae*	Frigate-bird (10)	*Fregatidae*
Button quail (15)	*Turnicidae*	Frogmouth (21)	*Podargidae*
California gull (16)	*Laridae*	Fulmar (9)	*Procellariidae*
Canary (27)	*Carduelinae*	Galapagos and Cocoos Island finch (27)	*Geospizinae*
Canvasback (12)	*Anatidae*	Gallinule (15)	*Rallidae*
Cardinal (27)	*Richmondeninae* also *Fringillidae*	Gambel's quail (14)	*Phasianidae*
		Gannet (10)	*Sulidae*
		Gnat-eater (27)	*Conopophagidae*
Cariama (15)	*Cariamidae*	Goatsucker (21)	*Caprimulgidae*
Cassowary (3)	*Casuariidae*	Godwit (11)	*Scolopacidae*
Cat bird (27)	*Mimidae*	Goldcrest (27)	*Regulinae*
Chachalaca (14)	*Cracidae*	Golden pheasant (14)	*Phasianidae*
Chaffinch (27)	*Fringillidae*	Goose (12)	*Anatidae*
Chat (27)	*Icteridae*	Grass finch (27)	*Estrildidae*
Chickadee (27)	*Paridae*	Gray Lag (12)	*Anatidae*
Chicken (14)	*Phasianidae*	Grebe (8)	*Podicepedidae*
Chough (27)	*Corvidae*	Greenbul (27)	*Pycnonotidae*
Cochoa (27)	*Turidae*	Greenlet (27)	*Vireonidae*
Cockatoo (18)	*Psittacidae*	Greenshank (16)	*Scolopacidae*
Cock-of-the-Rock (27)	*Cotingidae*	Grosbeak (27)	*Richmondeninae*
Collared hemipode (15)	*Pedionominae*	Ground finch (27)	*Emberizinae*
Coly (23)	*Coliidae*	Ground roller (25)	*Brachypteraciidae*
Condor (13)	*Carthartidae*	Grouse (14)	*Tetraonidae*
Coot (15)	*Rallidae*	Guacharo (21)	*Steatornithidae*
Coral-billed nuthatch (27)	*Hyposittinae*	Guan (14)	*Cracidae*
Cormorant (10)	*Phalacrocoracidae*	Guillemot (16)	*Alcidae*
Cotinga (27)	*Cotingidae*	Guinea fowl (14)	*Numididae*
Coua (19)	*Cuculidae*	Gull (16)	*Laridae*
Coucal (19)	*Cuculidae*	Hammerhead (11)	*Scopidae*
Courser (16)	*Glareolidae*	Harlequin (12)	*Anatidae*
Cowbird (27)	*Icteridae*	Harrier (Old World) (13)	*Accipitridae*
Crab-plover (16)	*Dromadidae*	Hawaiian honeycreeper (27)	*Drepanididae*
Crake (15)	*Rallidae*	Hawk (13)	*Accipitridae*
Crane (15)	*Gruidae*	Hedge sparrow (27)	*Prunellidae*
Creeper (27)	*Certhiidae*	Helmet shrike (27)	*Prionopidae*
Crested argus (14)	*Phasianidae*	Heron (11)	*Ardeidae*

(*continued*)

TABLE 3. (*continued*)

Common Name	Family	Common Name	Family
Herring gull (16)	*Laridae*	Old Squaw (12)	*Anatidae*
Hoatzin (14)	*Opisthocomidae*	Oriole (27)	*Icteridae*
Honeycreeper warbler (27)	*Parulidae*	Oriole, Old World (27)	*Oriolidae*
Honey-eater (27)	*Meliphagidae*	Osprey (13)	*Pandionidae*
Honey-guide (26)	*Indicatoridae*	Ostrich (1)	*Struthionidae*
Hoopoe (25)	*Upupidae*	Ovenbird (27)	*Furnariidae*
Hornbill (25)	*Bucerotidae*	Owl (20)	*Strigidae*
Huias (27)	*Callaeidae*	Owlet nightjar (21)	*Aegothelidae*
Hummingbird (22)	*Trochilidae*	Oyster-catcher (16)	*Haematopodidae*
Hypocolius (27)	*Bombycillidae*	Painted snipe (16)	*Rostratulidae*
Ibis (11)	*Threskiornithidae*	Palm chat (27)	*Dulidae*
Iora (27)	*Irenidae*	Parakeet (18)	*Psittacidae*
Jabiru (11)	*Ciconiidae*	Parrot (18)	*Psittacidae*
Jacamar (26)	*Galbulidae*	Parrotbill (27)	*Paradoxornithidae*
Jacana (16)	*Jacanidae*	Parson bird (27)	*Meliphagidae*
Jay (27)	*Corvidae*	Partridge (14)	*Phasianidae*
Jaeger (16)	*Stercorariidae*	Peacock (14)	*Phasianidae*
Java sparrow (27)	*Estrildidae*	Pelican (10)	*Pelecanidae*
Jewel Thrush (27)	*Pittidae*	Penguin (6)	*Spheniscidae*
Junco (27)	*Fringillidae*	Peppershrike (27)	*Cyclarhinae*
Kagu (15)	*Rhynochetidae*	Petrel (9)	*Procellariidae*
Kaka (18)	*Psittacidae*	Phalarope (16)	*Phalaropodidae*
Kea (18)	*Psittacidae*	Pheasant (14)	*Phasianidae*
Kingfisher (25)	*Alcedinidae*	Philepitta (27)	*Philepittidae*
Kinglet (27)	*Regulinae*	Piculet (26)	*Picidae*
Kite (13)	*Accipitridae*	Pigeon (17)	*Columbidae*
Kittiwake (16)	*Laridae*	Piping crow (27)	*Cracticidae*
Kiwi (4)	*Apterygidae*	Pipit (27)	*Motacillidae*
Lady Amherst (14)	*Phasianidae*	Pitta (27)	*Pittidae*
Lapwing (16)	*Charadriidae*	Plains-wanderer (15)	*Pedionominae*
Lark (27)	*Alaudidae*	Plant-cutter (27)	*Phytotomidae*
Laughing thrush (27)	*Timaliidae*	Plantain-eater (19)	*Musophagidae*
Leafbird (27)	*Irenidae*	Plover (16)	*Charadriidae*
Leatherhead (27)	*Meliphagidae*	Plush-capped finch (27)	*Catamblyrhynchidae*
Lily-trotter (16)	*Jacanidae*	Plymouth Rock (14)	*Phasianidae*
Limpkin (15)	*Aramidae*	Pochard (12)	*Anatidae*
Locust bird (16)	*Glareolidae*	Potoo (21)	*Nyctibiidae*
Log-runner (27)	*Timaliidae*	Prairie chicken (14)	*Tetraonidae*
Longclaw (27)	*Motacillidae*	Pratincole (16)	*Glareolidae*
Longspur (27)	*Fringillidae*	Ptarmigan (14)	*Tetraonidae*
Loon (7)	*Gaviidae*	Puffbird (26)	*Bucconidae*
Lorikeet (18)	*Psittacidae*	Puffin (16)	*Alcidae*
Lory (18)	*Psittacidae*	Quail (14)	*Phasianidae*
Lovebird (18)	*Psittacidae*	Rail (15)	*Rallidae*
Lyrebird (27)	*Menuridae*	Raven (27)	*Corvidae*
Macaw (18)	*Psittacidae*	Red jungle fowl (14)	*Phasianidae*
Magpie (27)	*Corvidae*	Redpoll (27)	*Fringillidae*
Magpie-lark (27)	*Grallinidae*	Redshank (16)	*Scolopacidae*
Mallard (12)	*Anatidae*	Redstart (27)	*Turdidae*
Mallee fowl (14)	*Megapodiidae*	Reeves pheasant (14)	*Phasianidae*
Manakin (27)	*Pipridae*	Rhea (3)	*Rheidae*
Mandarin (12)	*Anatidae*	Rhode Island red (14)	*Phasianidae*
Mannikin (27)	*Estrildidae*	Riflebird (27)	*Paradisaeidae*
Mannucode (27)	*Paradisaeidae*	Ring-neck (14)	*Phasianidae*
Martin (27)	*Hirundinidae*	Roadrunner (19)	*Cuculidae*
Meadowlark (27)	*Icteridae*	Roatelo (15)	*Mesoenatidae*
Megapode (14)	*Megapodiidae*	Robin (27)	*Turdidae*
Melba (27)	*Estrildidae*	Roller (25)	*Coraciidae*
Merganser (12)	*Anatidae*	Ross' gull (16)	*Laridae*
Mesite (15)	*Mesoenatidae*	Ruddy (12)	*Anatidae*
Minivet (27)	*Campephagidae*	Ruff (16)	*Scolopacidae*
Mistletoe bird (27)	*Dicaeidae*	Ruffled grouse (14)	*Tetraonidae*
Mockingbird (27)	*Mimidae*	Saddleback (27)	*Callaeidae*
Moho (27)	*Meliphagidae*	Sage grouse (14)	*Tetraonidae*
Monia (15)	*Mesoenatidae*	Saltator (27)	*Richmondeninae*
Motmot (25)	*Momotidae*	Sanderling (16)	*Scolopacidae*
Mound builder (14)	*Megapodiidae*	Sand-grouse (17)	*Pteroclidae*
Mountain quail (14)	*Phasianidae*	Sandpiper (16)	*Scolopacidae*
Mousebird (23)	*Coliidae*	Sapsucker (27)	*Picidae*
Mudlark (27)	*Grallinidae*	Scarlet tanager (27)	*Thraupidae*
Murre (16)	*Alcidae*	Scoter (12)	*Anatidae*
Muscovy (12)	*Anatidae*	Screamer (12)	*Anhimidae*
Mutton bird (9)	*Procellariidae*	Screech owl (20)	*Strigidae*
New Zealand wren (27)	*Xenicidae*	Scrub-bird (27)	*Atrichornithidae*
Nightingale (27)	*Turdidae*	Secretary bird (13)	*Sagittariidae*
Nightjar (21)	*Caprimulgidae*	Seed-snipe (16)	*Thinocoridae*
Northern grosbeak (27)	*Carduelinae*	Seriema (15)	*Cariamidae*
Nuthatch (27)	*Sittidae*	Seven Sisters (27)	*Timaliidae*
Oilbird (21)	*Steatornithidae*	Shag (10)	*Phalacrocoracidae*

(*continued*)

TABLE 3. (*continued*)

Common Name	Family	Common Name	Family
Shearwater (9)	*Procellariidae*	Towhee (27)	*Fringillidae*
Sheath-bill (16)	*Chionididae*	Tragopan (14)	*Phasianidae*
Sheldrake (12)	*Anatidae*	Tree duck (12)	*Anatidae*
Shoe-bill (11)	*Balaenicipitidae*	Tree swift (22)	*Hemiprocnidae*
Shoveller (12)	*Anatidae*	Trembler (27)	*Mimidae*
Shrike (27)	*Laniidae*	Triller (27)	*Campephagidae*
Shrike-vireo (27)	*Vireolaniinae*	Trogon (24)	*Trogonidae*
Sickle-bill (27)	*Paradisaeidae*	Tropic-bird (10)	*Phaëthontidae*
Silky Fly-catcher (27)	*Bombycillidae*	Troupial (27)	*Icteridae*
Silver-eye (27)	*Zosteropidae*	Trumpeter (15)	*Psophiidae*
Siskin (27)	*Carduelinae*	Tui (27)	*Meliphagidae*
Skimmer (16)	*Rynchopidae*	Turaco (19)	*Musophagidae*
Skua (16)	*Stercorariidae*	Turkey (14)	*Meleagrididae*
Skylark (27)	*Alaudidae*	Turnstone (16)	*Charadriidae*
Snakebird (10)	*Anhingidae*	Tyrant flycatcher (27)	*Tyrannidae*
Snipe (16)	*Scolopacidae*	Ula-ai-hawane (27)	*Drepanididae*
Snow goose (12)	*Anatidae*	Umbrella bird (27)	*Cotingidae*
Sparrow (27)	*Fringillidae*	Valley quail (14)	*Phasianidae*
Sparrow, New World (27)	*Emberizinae*	Vanga shrike (27)	*Vangidae*
Spider-hunter (27)	*Nectariniidae*	Vireo (27)	*Vireonidae*
Spine bill (27)	*Meliphagidae*	Vulture (New World) (13)	*Cathartidae*
Spoonbill (11)	*Threskiornithidae*	Vulture (Old World) (13)	*Accipitridae*
Spruce grouse (14)	*Tetraonidae*	Wagtail (27)	*Motacillidae*
Starling (27)	*Sturnidae*	Warbler (27)	*Sylviidae*
Stilt (16)	*Recurvirostridae*	Water ouzel (27)	*Cinclidae*
Stitch bird (27)	*Meliphagidae*	Wattled crow (27)	*Callaeidae*
Stone curlew (16)	*Burhinidae*	Waxbill (27)	*Estrildidae*
Stork (11)	*Ciconiidae*	Waxwing (27)	*Bombycillidae*
Storm petrel (9)	*Hydrobatidae*	Weaverbird (27)	*Ploceidae*
Sugar bird (27)	*Meliphagidae*	Whale-headed stork (11)	*Balaeniciptidae*
Sunbird (27)	*Nectariniidae*	Whip-poor-will (21)	*Caprimulgidae*
Sun-bittern (15)	*Eurypygidae*	White-eye (27)	*Zosteropidae*
Sun grebe (15)	*Heliornithidae*	White leghorn (14)	*Phasianidae*
Surf bird (16)	*Charadriidae*	Widgeon (12)	*Anatidae*
Suthora (27)	*Paradoxornithidae*	Widow bird (27)	*Ploceidae*
Swallow (27)	*Hirundinidae*	Woodcock (16)	*Scolopacidae*
Swallow-tanager (27)	*Tersinidae*	Woodcreeper (27)	*Dendrocolaptidae*
Swan (12)	*Anatidae*	Wood duck (12)	*Anatidae*
Swift (22)	*Apodidae*	Woodhewer (27)	*Dendrocolaptidae*
Tanager (27)	*Thraupidae*	Wood-hoopoe (25)	*Phoeniculidae*
Tapaculo (27)	*Rhinocryptidae*	Wood-nightjar (21)	*Nyctibiidae*
Teal (12)	*Anatidae*	Woodpecker (26)	*Picidae*
Tern (16)	*Laridae*	Wood shrike (27)	*Prionopidae*
Thermometer bird (14)	*Megapodiidae*	Wood-swallow (27)	*Artamidae*
Thick-knee (16)	*Burhinidae*	Wood warbler (27)	*Parulidae*
Thornbill (27)	*Sylviidae*	Wren (27)	*Troglodytidae*
Thrasher (27)	*Mimidae*	Wren-thrush (27)	*Zeledoniidae*
Thrush (27)	*Turidae*	Wren-tit (27)	*Chamaeidae*
Tinamou (5)	*Tinamidae*	Wryneck (26)	*Picidae*
Titmouse (27)	*Paridae*	Yellowhammer (27)	*Fringillidae*
Tody (25)	*Todidae*	Yellow-leg (11)	*Scolopacidae*
Toucan (26)	*Ramphastidae*	Yokohama chicken (14)	*Phasianidae*

The bones of birds are made very light by virtue of numerous air cavities. The skull, in particular, is very lightweight. The orbits are quite large. By extension of the facial bones, a bill is formed. There are no teeth.

In birds, the muscles, particularly of the wings, neck, tail, and legs, are extremely well developed. The *pectoralis major*, the muscle that causes the downward stroke of the wings, often will weigh as much as 20% of the total weight of the bird. The muscle which raises the wing is called the *pectoralis minor*. Together, these muscles comprise what generally is called the breast of the birds. Birds have a unique muscular mechanism which takes over when a bird assumes a squatting position for rest or sleep. In this position, certain tendons are pulled and these, in turn, flex the toes, essentially locking the bird to its perch.

Beaks of birds show a wide range of specialization, especially as related to different types of diet. Slender and elongate beaks are found in many wading species, which must reach below the surface of water for food. Hummingbirds require swordlike beaks to obtain nourishment from deep-throated flowers. The broad beak of the duck has sievelike structures at the sides and serves effectively to collect and strain out small particles from the water. The hooked beaks of birds of prey, the small and slender beaks of insectivorous birds, and the thick, strong beaks of seed-eating species, such as the parrots, indicate the wide range. See Fig. 4.

Birds are known for their high rate of metabolism. Large quantities of food are needed and digestion must be rapid. Relative to other organs of the body, the bird's heart is large. The heart may beat several hundred times per minute in a perching bird and up to a thousand times or more per minute in a canary when it is under stress.

Inspection of the wings of a bird reveals much concerning the habits and frequently the habitat of a given species. Long, narrow wings, for example, are found in the shearwaters and other ocean gliders. A shorter, more stubby wing is found in the species, such as pheasants, that require a lot of power for getting off the ground quickly. Broad, slotted wings are found among the soaring birds, such as hawks. Short, light, but strong wings are found on those birds such as swallows which require high speed and great efficiency for long migratory flights. Wing design greatly affects the bird's gliding ability—extremely accentuated in the albatross, considerably less in the pigeon, a lot less in the sparrow, and very limited in the hummingbird. On the other hand, the hummingbird has wings which can swivel nearly 180° at the shoul-

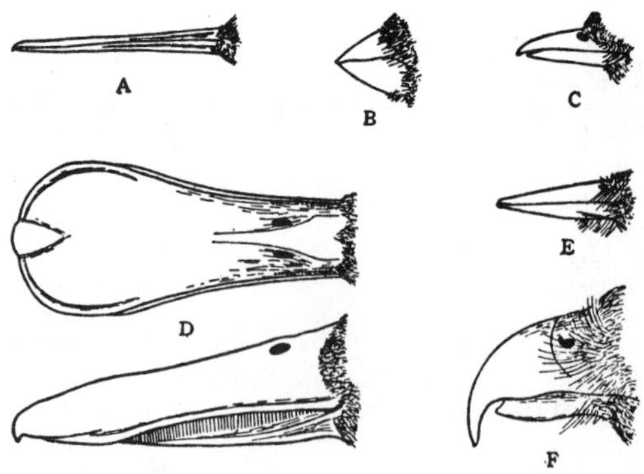

Fig. 4. The beaks of birds: (A) yellow legs, a wader; (B) cardinal, a seed-eater; (C) flycatcher, an insect eater; (D) the shoveler duck, dorsal surface above and side view showing the lateral sieve below; (E) a woodpecker's chisel-tipped beak; (F) hawk, a bird of prey.

der to provide just the right combination of lift and thrust during its wings movements to permit it to literally hang in the air in its hovering mode. Birds such as the albatross take full advantage of dynamic soaring, utilizing thermal air currents and the particular wind action next to the surface of ocean waves to maintain a gliding pattern with little flapping effort for many hours over the ocean.

Notable among the migratory birds is the Arctic tern which makes a round trip each year (about 10,000 miles in each direction; 16,090 kilometers) between the Arctic and Antarctic. The white stork is also a well-known migrator, summering in Europe and wintering in South Africa. They do not like flying over water and thus select a route to the east of the Mediterranean, or make the crossing at Gibraltar. Breeding in North America, the bobolink winters in Argentina. The route includes short hops between the islands of the Caribbean. The pectoral sandpiper prefers to breed in the tundra of the Arctic, but winters in South America, crossing mid-North America in each direction. While commonly regarded as an escape to warmer climate, the predominant motivation behind migration is that of an assured food supply during all seasons of the year. Studies indicate that migratory flights generally occur at altitudes of 3000 feet (900 meters) or less, particularly migrations involving large flocks. Single migrating birds may travel at much higher altitudes, going as high as 14,000 feet (4270 meters). There have been some radar sightings of birds at elevations up to 20,000 feet (6100 meters).

The guidance system used by birds in their rather exacting migratory patterns have been studied for a number of years, but much remains to be investigated. A large number of species travel at night and this rules out a high degree of dependence upon visual ground observations for many of them, unless, of course, there is some form of infrared "photographic" detection yet undetected. Some years ago, the concept of a pure memory system was tested by transporting two sets of birds to test their homing abilities. One set was rotated on a turntable while traveling; the other set was not. Both sets homed successfully with no measurable difference in the experiment. Golden plovers cross over 2000 miles (3218 kilometers) of open ocean between Hawaii and Alaska; the New Zealand bronze cuckoo travels about 2500 miles (4023 kilometers) over the open ocean between New Zealand, the Bismarck and Solomon Islands. Curlews travel over the ocean from Tahiti and Alaska, a distance of well over 5500 miles (8850 kilometers). In some instances, young birds make these flights for the first time without receiving directions from adults. Migratory speeds and stamina also are most impressive. In an experiment, an albatross was released over 3000 miles (4827 kilometers) from its regular habitat on Midway Island. The bird returned home in 10 days, or an average speed of 300 miles (483 kilometers) per day.

Some authorities have proposed that the earth's magnetic field may provide guidance parameters, but any connection has yet to be fully demonstrated. The use of the position of the sun by daytime migrators, in which compensation is made for the changing angle of the sun, has been demonstrated buy the late German scientist, Gustav Kramer. The use of celestial navigation by nighttime migrators also has been extensively studied, with some relatively convincing results. Numerous techniques have been developed for marking and tracking birds and perhaps one day the riddle of navigation by the migrators may yield to constant probing.

Communication between birds also has been the object of much investigation in recent years. Tape recorders have been of much assistance in learning of the qualities and range of bird sounds—calls, songs, and notes. Classification of bird sounds by bioacoustics is headquartered at a laboratory located at Cornell University, Ithaca, New York. It is estimated that over 20,000 recordings are on file in the collection. There is a similar program in Sweden.

Although some birds seem voiceless, it is usually is found that they chatter with their bills to make the necessary signals and noises required for notification of danger, in courting, and in seeking food.

As with other forms of life, generalizations can be carried only so far. Some of the specific attributes and characteristics of numerous species are described elsewhere in this volume. Consult Table 1.

References

Beebe, William: "The Bird: Its Form and Function," Dover, New York, 1973.
Crook, J. H. (editor): "Social Behavior in Birds and Mammals," Academic, New York, 1966.
Dunstan, T.: "Our Bald Eagle: Freedom's Symbol Survives," *Nat. Geog.*, **153**, 2, 187–199 (1978).
Fisher, A.: "Bird Migration," *Nat. Geog.*, **156**, 2, 154–193 (1979).
Gilliard, E. T.: "Living Birds of the World, Doubleday, Garden City, New York. 1967.
Green, P.: "Australia's Feathered Playboy," *Nat. Geog.*, **152**, 6, 865–872 (1977).
Grzimek, B.: "Grzimek's Animal Life Encyclopedia," Vol. 7–9, Van Nostrand Reinhold, New York, 1975.
Hegner, R. W., and K. A. Stiles: "College Zoology," 7th Edition, Chap. 29, Macmillan, New York, 1966.
Marshall, A. J. (editor): "Biology and Comparative Physiology of Birds," Academic, New York, 1960.
Moreau, R. E.: "The Bird Faunas of Africa and Its Island," Academic, New York, 1966.
Murton, R. K.: "The Problems of Birds as Pests," Academic, New York, 1969.
Newton, I.: "Population Ecology of Raptors," Buteo, Vermillion, South Dakota, 1979.
Peterson, R. T.: "A Field Guide to the Birds—Eastern Land and Water Birds," Houghton-Mifflin, Boston, 1947.
Zeleny, L.: "Song of Hope for the Bluebird," *Nat. Geog.*, **151**, 855–865 (1977).

BIRD SEED. **Grasses.**

BIRDS (Fossil). **Fossil Birds.**

BIRDS (Pollination). **Pollination.**

BIRDS (Water Metabolism). **Water.**

BIRTH PROCESS. A type of stochastic process describing the progress of a population for which, at each time point, there is a probability that an individual gives birth to a new individual or new individuals. Likewise, a *Death Process* is one for which an individual has a certain probability of death. More general processes can take account of birth, death, immigration and emigration.

A similar process in which individuals give rise to new ones is sometimes known as a *Branching Process.* Applications are found in many fields, from human populations to particle physics.

BICUSPID. A tooth with two cusps, especially the premolars of humans.

BISERIAL CORRELATION. Suppose we have a $2 \times q$ table of frequencies. If it is assumed that the table has arisen by grouping a sample from a bivariate normal distribution, an estimate of the correlation coefficient ρ can be obtained. Such an estimate is known as a biserial correlation coefficient. A slightly different concept which avoids the above assumptions is the point-biserial correlation coefficient. This is a measure of association between a continuous variate x and a discrete variate y which takes only two values (0 and 1 say).

BISHOP BIRD. **Weaverbird.**

BISHOP RINGS. **Atmospheric Optical Phenomena.**

BISMUTH. Chemical element symbol Bi, at. no. 83, at. wt. 208.981, periodic table group 5a, mp 271.3°C, bp 1562 ± 3°C, density 9.78 g/cm^3 (20°C). Elemental bismuth has a rhombohedral crystal structure. The metal is of a silvery-white color with limited ductility. Like gallium, bismuth is one of the few metals that increases its volume (3.32%) upon solidifying from the molten state. It is the most diamagnetic of all the metals. All isotopes of the element (^{205}Bi through ^{215}Bi) are radioactive. See also **Radioactivity.** However, the naturally occurring isotope ^{209}Bi generally is not regarded in this category because of its extremely long half-life (2 × 10^{17} years). Although described by Basil Valentine in the fifteenth century, the element was not defined as a new element until its characteristics were published in 1753 by C. Geoffroy and T. Bergman.

First ionization potential 7.287 eV; second 16.6 eV; third 25.56 eV; fourth 45.1 eV; fifth 55.7 eV. Oxidation potentials $Bi + H_2O \rightarrow BiO^+ + 2H^+ + 3e^-$, −0.32 V; $Bi + 3OH^- \rightarrow BiOOH + H_2O + 3e^-$, 0.46 V. Other important physical characteristics of bismuth are given under **Chemical Elements.**

Bismuth occurs as native bismuth in Bolivia and Saxony and frequently is associated with lead, copper, and tin ores—the sulfide (bismuthinite, bismuth glance, Bi_2S_3) is also found in nature. Separation of bismuth from lead takes place during the electrolytic refining of the latter with bismuth remaining in the anode mud, or by prometallurgical methods by which it is removed from the lead as a calcium-magnesium compound. See also **Bismuthinite.**

Alloys: Metallurgically, bismuth is used in the production of low melting point fusible alloys and as an additive to steel, cast iron, and aluminum. The fusible alloys contain about 50% bismuth in combination with lead, tin, cadmium, and indium and are used in a variety of ways, including fire-protection devices, joining and sealing hardware, and short-life dies. Because of the special volume-increase property with solidification, bismuth is used to manufacture alloys with a zero liquid-to-solid volume change. Alloy compositions are given in the accompanying table. The addition of about 0.2% bismuth, along with a similar quantity of lead, improves the machineability of aluminum. Very small quantities (0.02%) of bismuth are used in the production of melleable cast iron for stabilization of carbides upon solidification, particularly desirable for castings with heavy cross sections. Combinations of bismuth and tin and bismuth and cadmium have found use as counterelectrode alloys in the manufacture of selenium rectifiers. Bismuth telluride Bi_2Te_3 and bismuth selenide Bi_2Se_3 display thermoelectric properties. With modification, these compounds are used for certain commercial and military solid-state devices, including small units for portable power generation and refrigeration.

Chemistry and Compounds: Generally, the chemical behavior of bismuth parallels that of arsenic and antimony, but bismuth is the most metallic of the group. Bismuth is not soluble in cold H_2SO_4 or cold HCl, but is attacked by these acids when hot and also by cold aqua regia. Elemental bismuth is not attacked by cold alkalies. The metal is soluble in HNO_3 and forms nitrates. When heated with chlorine, bismuth yields a chloride.

Some of the salts of bismuth are used in medicines for the relief of digestive disorders because of the smooth, protective coating the compounds impart to irritated mucous membranes. Like barium, bismuth also is used as an aid in x-ray diagnostic procedures because of its opacity to x-rays. At one time, certain bismuth compounds were used in the treatment of syphilis. Bismuth oxychloride, which is pearlescent, finds wide use in cosmetics, imparting a frosty appearance to nail polish, eye shadow, and lipstick. Bismuth phosphomolybdate has been used as a catalyst in the production of acrylonitrile for use in synthetic fibers and paints. Bismuth oxide or subcarbonate are used as fire retardants for plastics.

SOME REPRESENTATIVE LOW-MELTING-POINT ALLOYS CONTAINING BISMUTH

Fusible alloy, melting at 96°C	53% Bi	32% Pb	15% Sn
Fusible alloy, melting at 91.5°C	52% Bi	40% Pb	8% Cd
Fusible alloy, melting at 100°C	50% Bi	30% Sn	20% Pb
Fusible alloy, melting at 70°C. (Wood's metal)	50% Bi	25% Pb	12.5% Sn 12.5% Cd
Fusible alloy, melting at 70°C. (Lipowitz' alloy)	50% Bi	27% Pb	13% Sn 10% Cd
Rose metal	50% Bi	27% Pb	23% Sn
Bismuth solder, melting at 111°C.	40% Bi	40% Pb	20% Sn

Bismuth trihalides exhibit an increased tendency toward hydrolysis, usually forming bismuthyl compounds, also called bismuth oxyhalides, which are often assumed to contain the ion BiO^+. This is not a discrete ion, however, and the crystal lattices of the "bismuthyl" compounds actually are comprised of Bi(III), O(—II) and X(—I) units. For example, BiOCl has the same crystal structure as PbFCl. The trihalides also form halobismuthates, with halogen ions, such as the chlorobismuthates, which contain the ions $BiCl_4^-$ and $BiCl_5^{2-}$. The BiI_4^- ion is precipitated analytically as the cinchonine salt.

Bismuth(III) oxide, Bi_2O_3, is the compound produced by heating the metal, or its carbonate, in air. It is definitely a basic oxide, dissolving readily in acid solutions, and unlike the arsenic or antimony compounds, not amphiprotic in solution, although it forms stoichiometric addition compounds on heating with oxides of a number of other metals. It exists in three modifications, white rhombohedral, yellow rhombohedral, and gray-black cubical. Bismuth(II) oxide, BiO, has been produced by heating the basic oxalate.

Bismuth(III) hydroxide also is not significantly amphiprotic in solution, dissolving only in acids. Its formula is given as $Bi(OH)_3$ but it is difficult to isolate, due to adsorption of acid anions and to its dehydration to BiO(OH). The actions of strong oxidants in concentrated alkalies on the hydroxide yields alkali bismuthates, such as $NaBiO_3$, sodium metabismuthate, which $NaBi(OH)_6$ is initially produced. Other metal bismuthates may be made from them or directly from the oxides and Bi_2O_3, and bismuth(V) oxide is obtained by the action of HNO_3 on the alkali bismuthates; however, some oxygen is lost and the product is a mixture of Bi_2O_5 and BiO_2.

Bismuth(III) sulfide, Bi_2S_3, is precipitated by H_2S from bismuth solutions. Complex sulfide ions form only slowly, so bismuth sulfide may be separated from the arsenic and antimony sulfides by this difference in properties. Like the oxide, bismuth sulfide forms double compounds with the sulfides of the other metals.

Bismuth forms a number of complex compounds, including the sulfatobismuthates, e.g., $NaBi(SO_4)_2$ and $Na_3Bi(SO_4)_3$; and the thiocyanatobismuthates, e.g., $Na_3Bi(SCN)_6$, by the interaction of sodium thiocyanate and $Bi(SCN)_3$. The salts of bismuth tend to lose part of their acid readily, especially on heating, to form basic salts.

True pentavalent compounds of bismuth are rare, but include bismuth pentafluoride, BiF_5 (subl. 550°C), and $KBiF_6$; pentaphenylbismuth, $(C_6H_5)_5Bi$; various compounds $(C_6H_5)_3BiX_2$, where X = F, Cl, Br, N_3, NCO, CH_3CO_2, $\frac{1}{2}CO_3$; and tetraphenylbismuthonium salts, $[(C_6H_5)_4Bi]X$, where X = Cl, $—[B(C_6H_5)_4]$, etc.

Organobismuth Compounds: Numerous bismuth organic compounds have been prepared. Some of these include methylbismuthine CH_3BiH_2; phenyldibromobismuthine $C_6H_5BiBr_2$; potassium diphenylbismuthide $K[Bi(C_6H_5)_2]$; triphenylbismuthdihydroxide $(C_6H_5)_2Bi(OH)_2$; tetraphenylbismuthonium tetraphenylborate $[(C_6H_5)_4Bi][B(C_6H_5)_4]$; and pentaphenylbismuth $(C_6H_5)_5Bi$.

References

Carapella, S. C., Jr.: "Bismuth" in "Kirk-Othmer Encyclopedia of Chemical Technology," Vol. 3, 3rd edition, pp. 921–937, Wiley, New York, 1978.

Staff: "The Bulletin of The Bismuth Institute," Bismuth Institute, Brussels, Belgium (issued periodically).

S. C. Carapella, Jr., ASARCO Incorporated, South Plainfield, New Jersey.

BISMUTH GLANCE. **Bismuthinite.**

BISMUTHINITE. A mineral containing a sulfide of bismuth, Bi_2S_3, and sometimes copper and iron; a variety from Mexico contains about 8% antimony. Bismuthinite is orthorhombic although its thin needle-like crystals are rare as it usually occurs in foliated or fibrous masses.

It has one good cleavage parallel to the prism; hardness, 2; specific gravity, 6.78; metallic luster; streak, lead gray; color, similar but often with iridescent tarnish; opaque.

Bismuthinite is a rather rare mineral although somewhat widely distributed. European localities are in Norway, Sweden, Saxony, Rumania, and England. It is found also in Bolivia, Australia, and in the United States in Utah. It is used as an ore of bismuth. Bismuthinite also is known as *bismuth glance.*

BISMUTH SOLDER. Bismuth.

BISON. Bovines.

BISPECTRUM. Spectral Analysis.

BISTABLE AMPLIFIER (Fluidics). Fluidics.

BISULFITE PROCESS. Pulp (Wood) Production and Processing.

BIT (Data System). A contraction of *bi*nary dig*it.* A single character in a binary numeral, i.e., a 1 or 0. A single pulse in a group of pulses also may be referred to as a bit. The bit is a unit of information capacity of a storage device. The capacity in bits is the logarithm to the base two of the number of possible states of the device.

Parity bit. A check bit that indicates whether the total number of binary "1" digits in a character or word (excluding the parity bit) is odd or even. If a "1" parity bit indicates an odd number of "1" digits, then a "0" bit indicates an even number of them. If the total number of "1" bits, including the parity bit, is always even, the system is called an even parity system. In an odd parity system, the total number of "1" bits, including the parity bit is always odd.

Zone bit. (1) One of the two leftmost bits in a system in which six bits are used for each character. Related to overpunch. (2) Any bit in a group of bit positions that are used to indicate a specific class of items; e.g., numbers, letters, special signs, and commands.

For related topical coverage in this volume, see list of entries under **Data processing.**

BITE (Animal). Rabies.

BITE (Snake). Snakes.

BITTERLING (*Osteichthyes*). Fishes of the general group *Cypriniformes* which also embraces other minnows, and also suckers, loaches, and hillstream fishes. The various species are thus allied to the carp. One species, *Rhodeus amarus,* lives in European waters and the others inhabit Eastern Asian waters. The Central European bitterling (*Rhodeus sericeus amarus*) is deep-bodied, attaining a length of about $3\frac{1}{2}$ inches (9 centimeters). This species displays an interesting and unusual reproductive habit, namely, that of the female developing a long ovipositor which permits deposition of the eggs into the mantle cavity of a fresh water clam or mussel. Thus, within the living clam, the eggs incubate and hatch. A related species in Japan (*Acheilognathus lanceolata*) displays similar habits.

BITTERN (*Aves, Ciconiiformes*). Wading birds allied to the herons and egrets. They have moderately long legs and a straight beak which is strong and sharp. Two species, the American, *Botaurus lentiginosus,* and least, *Ixobrychus exilis,* bitterns, occur in North America, and several others are found on other continents.

The *Botaurus lentiginosus* ranges from the Gulf of Mexico north and west to Manitoba. The bittern is found in marshy areas where there is ample vegetation for good concealment. The bird is most active at night. Freezing in position when approached, the bird holds its head high, completely still, taking advantage of the manner in which its plumage and coloring match numerous natural backgrounds. The birds are easy to lose from view.

The bittern has a booming type of cry. It is stocky and is often seen pointing its bill upward. The color is light-brown over the body with white trimming. The head is solid brown and legs are gray. Length is about 23 inches (58 centimeters); width about 35 inches (89 centimeters) with wing spread. The bittern feeds on small animals and insects found near watery areas. The neck may take the form of an S-shape when retracted for flight. The long neck enables the bill to act as a spear when spotting food. The claws have comblike serrations which are used, along with the bill, for crumbling some of the down feathers of its chest into a fine white powder which, in turn, is used for preening its feathers. This procedure is used particularly when feathered areas on the body have been soiled by fish slime. The down soaks up the slime or oil and the claws comb it out. The heron and egret also possess powder downs.

The smallest of the species is the least bittern. It possesses a rich brown plumage with white trimming, but is a weak flier. Length is about 11 inches (30 centimeters), spread is about 17 inches (43 centimeters). Bitterns are mentioned in the Scriptures. See also **Ciconiiforms.**

BITTERNUT TREE. Hickory and Wingnut Trees.

BITTER PATTERNS. A method for detecting domain boundaries at the surface of ferromagnetic crystals. If a drop of a colloidal suspension of ferromagnetic particles is placed on the surface of the crystals, the particles will collect along the domain boundaries where the field is strongest.

BITTER PECAN TREE. Hickory and Wingnut Trees.

BITUMEN. Natural flammable substances of a wide range of color, hardness, and volatility, constituted mainly of a mixture of hydrocarbons and essentially free from oxygenated bodies. Petroleums, asphalts, natural mineral waxes, and asphaltites are considered bitumens.

BITUMEN PAINTS. Paint.

BITUMINOUS FERMENTATION. Coal.

BITUMINOUS SANDS. Tar Sands.

BIVALVE. A shell composed of two distinct parts or valves. Such shells are secreted by brachiopods, in which the valves are dorsal and ventral, and by certain crustaceans (*Ostracoda*) and mollusks (*Pele cypoda*) in which the valves are lateral. The most common examples of bivalves are among the edible mollusks, including clams, oysters, and scallops. Bivalves of commercial significance are described in the entry on **Mollusks.**

BIVANE. Wind and Air Velocity Measurements.

BIXIN. Annatto Food Colors.

BLACK BASS. Sunfishes.

BLACKBERRY. Rose Family.

BLACKBIRD (*Aves, Passeriformes*). A term variously applied to different species of birds. The term sometimes is used to describe the ouzel of Europe. Several species of North American birds of the genus *Agelaius,* related to the orioles and grackles, may be called blackbirds. In the West Indies, the name is applied to the ani, a member of the order of Cuculiformes. In England, the blackbird is called the thrush (*Turdus merula*). See illustration.

In the United States, Brewer's blackbird is found in the meadows and prairies of the western states and ranges eastward to about the

Blackbird

Mississippi River and southward into Central America. The bird, from 8 to 9 inches in length (20 to 23 centimeters), appears much like a short-tailed grackle. See **Grackle.** The bird has a white eye (male). Females have dark eyes. At a distance, the bird appears all-black. Close up, purplish and greenish iridescence is noticeable. The rusty blackbird is quite similar, but prefers woodlands and swamps, whereas the Brewer's blackbird likes barnyards and fields. The song of these birds is a hoarse whistle. The redwinged blackbird (*Agelaius phoeniceus*) is quite similar with exception of red coloration on its throat and shoulders. Blackbirds tend to fly in flocks. These birds feed principally on grain. They are known for robbing the nests of other birds.

Redwing is also a term applied to a European thrush, *Turdus musicus.*

BLACK BODY. This term denotes an ideal body which would, if it existed, absorb all and reflect none of the radiation falling upon it; its reflectivity would be zero and its absorptivity would be 100%. Such a body would, when illuminated, appear perfectly black, and would be invisible except is outline might be revealed by the obscuring of objects beyond. The chief interest attached to such a body lies in the character of the radiation emitted by it when heated and the laws which govern the relations of the flux density and the spectral energy distribution of that radiation with varying temperature.

The total emission of radiant energy from a black body takes place at a rate expressed by the Stefan-Boltzmann (fourth-power) law; while its spectral energy distribution is described by Wien's laws, or more accurately by Planck's equation, as well as by a number of other empirical laws and formulas. See also **Emissivity; Thermal Radiation.**

The nearest approach to the ideal black body, experimentally, is not a sooty surface, as might be supposed, but an almost completely closed cavity in an opaque body, such as a jug. The laboratory type is usually a somewhat elongated, hollow metal cylinder, blackened inside, and completely closed except for a narrow slit in one end. When such an enclosure is heated, the radiation escaping through the opening closely resembles the ideal black-body radiation; while light or other radiation entering by the opening is almost completely trapped by multiple reflection from the walls, so that the opening usually appears intensely black. See also **Planck Radiation Formula.**

BLACKBUCK. Antelope.

BLACK DEATH. Bubonic Plague.

BLACK DIAMOND. Carbonado.

BLACKFISH (*Osteichthyes*). (1) The black sea bass, *Centropristis striatus.* See also **Bass (Osteichthyes).** (2) The Alaskan blackfish (*Dallia pectoralis*), the only representative of the family inhabiting streams and ponds of Alaska and Siberia. Chief food of natives of some parts of North Alaska. (3) A marine fish (*Centrolopus niger*) of the family *Stromateidae.*

BLACK-FLY (*Insecta, Diptera*). A minute fly whose small head and large thorax give it a hump-backed appearance. They are also called buffalo-gnats and the Indian name no-see-'em is sometimes used for the very small species. They constitute the family *Simuliidae.* Its distribution is worldwide, the larvae being attached by anal extremity to rocks in running water.

While some of these insects are harmless, others are among the most troublesome of our blood-sucking insects. Their bit is extremely irritating, considering its size, and the swarms are sometimes so numerous that their attack is serious to man and may cause the death of smaller animals, such as chicks. Certain species like *Simulum damnosum* (Theobald) of Africa transmit to man the filaria *Onchocera volvulus* (Leuckart) causing the subcutaneous disease onchoceriasis—"river-blindness." They are especially abundant in the woods, where campers and outdoor workers sometimes find it necessary to use oil of citronella, or one of the preparations containing tar, on exposed portions of the skin to prevent attack.

BLACKHAW TREES. Elder Trees and Viburnums.

BLACK HEMLOCK. Hemlock Trees.

BLACK HOLE. By definition, a black hole is a body which has become, by whatever mechanism, sufficiently compact that the escape velocity from its "surface" exceeds the speed of light. First hypothesized by Laplace in 1799, black holes were predicted by the general theory of relativity as a consequence of the distortion of the gravitational field around a massive body from the simple Newtonian inverse square law. This condition is believed to be the final state of a star, which is more massive than the upper limit for a neutron star and hence incapable of reaching hydrostatic equilibrium when its nuclear fuel has been exhausted. Under such conditions, first shown by Oppenheimer and Snyder, the collapse of the core will pass through the critical radius, which is given by:

$$R_* = 2GM/c^2$$

where G is the gravitational constant, c is the speed of light, and M is the mass. For the sun, this is of the order of 1.5 kilometers. The last stable orbit is at $3R_*$ for particles. Within this region, there is no escape—any signal will simply be directed down the hole. An observer at a large distance will, however, see progressive deceleration (a relativistic effect) of the infalling matter as the particles approach the speed of light. Once within this radius, nothing can prevent the collapse from continuing.

A famous theorem due to Israel and elaborated on by Hawing, Penrose, and Carter, the so-called "No Hair theorem," states that the only attributes which can be used to distinguish one black hole from another are mass, angular momentum, and charge. The Schwarzschild solution (found by K. Schwarzschild soon after Einstein's first paper on gravitation), was the first to demonstrate the existence of this singularity or event horizon and involved a point, nonrotating mass with no charge. The rotating case was not solved exactly until 1963 when Kerr showed that rotation introduces a second horizon, inside of which the inertial frame is dragged around with the hole. From this region, it is possible for the particle to escape from the hole and appear at infinity with some extra energy, extracted from the hole (the Penrose process).

Should the stellar collapse occur in a binary system, it may become possible for the external observer to surmise the existence of the hole by optical and x-ray observation. If mass flows from an oversized companion, and accretes onto the hole (either from a wind or forming an accretion disk) the rapid motion in the vicinity of the event horizon can raise the temperature of the matter to greater than 10^7 Kelvin. At this temperature, x-rays will be emitted with a characteristic spectrum and flickering rate. Since the mass of the emitting object can in principle also be obtained from its orbital motion (See **Binary Stars**), it is possible to choose between a neutron star and black hole as the responsible accreter; if the mass is greater than about three solar masses, it is likely a black hole.

At this writing, only one binary system clearly seems to require the presence of a black hole: Cygnus X-1, also known as HDE 226868. This star is a strong x-ray source, with a high-temperature spectrum and millisecond flickering. Bolton finds that it consists of a massive O star primary and a secondary which must be at least 16 solar masses. Attempts to explain the light curve and radial velocity variations by evoking a multiple system have so far failed to meet fairly strict observational criteria. In consequence, it is likely a firm conclusion that in this system at least we are observing a fairly low mass black hole with an age of a few million years.

The active galaxies, like *quasars,* Seyferts, and BL Lacertae galaxies, also appear to require massive (at least 10^7 solar masses) black holes in their nuclei in order to account for the X-ray, optical, and radio emissions and energies observed. There is also weak evidence that our galaxy may have a massive, but not rapidly accreting, black hole of perhaps a few million solar masses at its center. It would seem then that stellar collapse, following supernova explosions and stellar coalescence, may form massive black holes in a wide variety of galaxies, and that their presence is far more ubiquitous than might have been initially expected. See also **Quasars.**

Finally, one cosmologically interesting consequence of black-hole formation during the early stages of the big bang is connected with the process of black-hole evaporation. If holes of the order of 10^{15}

grams were formed in the early history of the universe, they may be observed now as exploding sources, emitting intense bursts of gramma radiation in a fraction of a second as they evaporate. The full and rich picture of the structure, evolution, and interaction of these objects, however, continues to be painted.

References

Blandford, R., and K. Thorne: "Black Hole Astrophysics," pp. 454–503 in "General Relativity: An Einstein Centenary Survey," (S. Hawking and W. Israel, editors), Cambridge Univ. Press, New York, 1979.
Bolton, C. T.: *Ap. J.*, **200**, 269 (1975).
Mishev, C., Thorne, K., and J. Wheeler: "Gravitation," Freeman, San Francisco, 1973.
Novikov, I., and K. Thorne: "Black Holes," page 343 (B. DeWitt and C. DeWitt, editors), Gordon and Breach, New York, 1973.
Zeldovich, Ya. B., and I. Novikov: "Relativistic Astrophysics: I. Stars and Relativity," Univ. of Chicago Press, Chicago, 1974.

Steven N. Shore, Assistant Professor of Astronomy, Warner and Swasey Observatory, Case Western Reserve University, Cleveland, Ohio.

BLACKHORSE (*Osteichthyes*). A fish (*Cycleptus elongatus*) of the Mississippi River system; also called the Missouri sucker. It attains a length of 30 inches (76 centimeters) and its flesh is excellent.

BLACK LEVEL. Cathode-Ray Tube.

BLACK LOCUST. Acacia Trees.

BLACK LUNG. Pneumokonioses.

BLACK POWDER. Explosive.

BLACK SEA TURBOT. Flatfishes.

BLACK SMOKER. Ocean; Ocean Resources (Mineral).

BLACK SPRUCE. Spruce Trees.

BLACK SQUALL. Fronts and Storms.

BLACKTONGUE. Niacin.

BLACK WIDOW SPIDER. Spider.

BLADDERNUT SHRUB. Of the family *Staphyleaceae* (bladdernut family), the American bladdernut (*Staphylea trifolia*) is a rather slender shrub, ranging from 6 to 12 feet (1.8 to 3.6 meters) in height and is rarely found as a tree (up to 25 feet; 7.5 meters). The plant has compound leaves of deep green color. The flowers are small and white. Possibly the most notable feature of the plant is the large, inflated, three-sided pods of light-brown color, 2 inches (5 centimeters) in length, and containing from one to four seeds. The seeds rattle inside the capsule when shaken. The plant is found from western Quebec westward into Ontario and Minnesota and southward to the latitude of South Carolina.

BLADDER STONES. Kidney and Urinary Tract.

BLADDER (Urinary). Kidney and Urinary Tract.

BLADDER WORM (*Platyhelminthes, Cestoda*). An immature resting stage of tapeworms consisting of a bladder-like cyst in which one or more heads are inverted. Also known as the cysticercus stage.

BLADDERWORT. Hydrophytes.

BLAGDEN LAW. The depression of the freezing point of a solution is, for small concentrations, proportional to the concentration of the dissolved substance.

BLANK CHARACTER (Computer System). Character (Computer System).

BLANKING (CRT). Cathode-Ray Tube.

BLAST FURNANCE. Iron Metals, Alloys, and Steels.

BLASTOCOELE. The first cavity formed during the embryonic development of animals. In many species the cleavage of the fertilized ovum gives rise to a hollow blastula of spheroidal form; the cavity of this structure is the blastocoele.

BLASTOMERE. Any of the cells resulting from the subdivision of the fertilized ovum during early embryonic development.

BLASTOMYCOSIS. A systemic fungus (mycotic) disease caused by the dimorphic fungus *Blastomyces dermatitidis.* The disease is found in certain parts of North, Central, and South America and in numerous areas in Africa. The disease is endemic in the southeastern and south central portions of the United States. The disease also has extended northward in several pockets along the Mississippi and Ohio Rivers and on into central Canada. In South America, the disease is caused by *Blastomyces brasiliensis.* Although the disease may affects persons of nearly any age, most of those afflicted are between 20 and 50 years of age. Incidence in males is six times that for females. Persons who work outside or who vacation a lot in areas with soils that may contain the soil sporophyte run a higher risk of becoming infected.

Blastomycosis manifests itself as a pulmonary disease with symptoms closely resembling those of tuberculosis, coccidioidomycosis, and histoplasmosis. The disease also takes a cutaneous form, but skin lesions appear to be due to metastatic infection from the primary site. The initial portal of entry is the respiratory tract. As the result of inhalation, the fungus spores are deposited in the peripheral air spaces of the lower lobes of the lung. Often minor infections will be quickly eradicated without detectable traces of the infection. But, in a certain percentage of cases, the infection may take a more serious route and this may range from mild pulmonary disease all the way to destruction and cavity formation. Metastatic spread also may include, in addition to the skin, the skeletal system, and genitalia. Less frequently, the rectum and the heart may be infected.

Diagnosis is made by growing organisms, from sputum, cutaneous lesions, etc., on the surface of Sabourand's agar slants (incubated at 30°C for 1 month). Treatment is with one of two antimicrobial agents, amphotericin B or hydroxystilbamidine isethionate. The former is usually used for patients with advanced, progressing disease, particular where several organs may be involved. The latter is more frequently used for patients with chronic dermatologic disease. This is usually considered second-line therapy. Relapses may occur within a period up to 9 years after a course of treatment.

Often acute blastomycosis will run its course without therapy, but persons who have not shown improvement within 2 weeks should be treated. In more severe, less frequent cases where host defenses are impaired, surgery may be indicated in the case of persistent pulmonary cavities and deforming orthopedic lesions that may accompany chronic infection.

BLASTULA. The stage in embryonic development which results from cleavage of the fertilized ovum and precedes the establishment of the germ layers. It is a hollow sphere in its primitive form but is modified in many animals, particularly in connection with the extensive storage of yolk in the egg, and in some of these modified forms the exact equivalent of the primitive blastula is difficult to determine.

BLANTON'S FORMULA. Trajectory (Of a Path).

BLEACHING AGENTS. The end-result of bleaching action is decolorizing although decolorizing can be accomplished by means other than bleaching. Several types of compounds are used for bleaching to satisfy a wide range of requirements. Bleaching is used in a positive way to remove color imperfections (grayness, off-whiteness, etc.) from raw materials, such as cotton, wool, and other natural fibers; to produce a white, pleasing laundered product; to bleach flour and other foodstuffs; and to bleach wood pulp prior to the preparation of paper. Bleaching also occurs in a negative fashion through the action of the rays of the sun which cause fading of large numbers of fabrics, paints, and other coatings; or the action of washing and chemicals on surfaces and fabrics. The specific use of bleaching agents in connection with detergents is described under **Detergents.**

Bleaching agents fall roughly into two categories: (1) the hypochlorite or chlorine-type bleach; and (2) the peroxy compounds.

Hypochlorites. These compounds frequently are used to provide the sanitizing and bleaching property of chlorine without requiring the handling of liquid or gaseous chlorine. The term "liquid chlorine" is used in the swimming pool trade to describe sodium hypochlorite solutions, and the term "dry chlorine" is part of the registered trademark of a proprietary calcium hypochlorite product containing 70% available chlorine. Although nonscientific terms, their usage is well established in practice.

Sodium Hypochlorite: This product generally is available in one of two strengths. The household liquid bleach contains about 5.25% (wt) NaClO. The commercial product (sometimes called 15% bleach) contains 150 grams per liter of available chlorine. This is equivalent to about 13% (wt) sodium hypochlorite.

Calcium Hypochlorites: The forerunner of bleaching agents was patented as early as 1799 and termed *bleaching powder.* It is produced by passing chlorine gas over slaked lime and the resulting powder usually contains about 30% available chlorine. Although the original compound was quite unstable, it became of immense value to the bleaching of textiles and later for sanitizing.

In the United States, bleaching powder has largely been supplanted by an improved calcium hypochlorite product containing about 70% available chlorine. The compound, available under several brand names, is essentially calcium hypochlorite dihydrate.

In another form, calcium hypochlorite, $Ca(ClO)_2$, containing from 20 to 40 grams per liter of available chlorine is commonly produced by pulp mills for pulp bleaching.

Sodium perborate is the least expensive and most commonly used of the peroxy type bleaches and is incorporated in some household and commercial detergent formulations.

Hydrogen peroxide. A significant portion of the production of hydrogen peroxide, H_2O_2, goes into the bleaching of cotton, wool, and groundwood pulp, as well as use in hair bleaching preparations.

Organic peroxides also find use as bleaching agents, notably dibenzoyl peroxide, $[C_6H_5C(O)\cdot]_2$, which is still the preferred bleaching agent for flour. Peroxyacetic acid, CH_3CO_3H, also finds use in specialized bleaching situations.

Sodium bromide, in combination with hypochlorites, is sometimes used in bleaching systems, particularly for cellulosics.

See also **Oxidation and Oxidizing Agents.**

BLEAK (*Osteichthyes*). Small fishes (*Alburnus*) of several species found in Europe and western Asia, related to the carps. Scales are used for the manufacture of artificial pearls.

BLEEDER RESISTANCE. A resistor permanently connected across the output of a power supply. The primary function of this resistor is to discharge the filter capacitors used in the supply when power is disconnected.

BLEEDING (Excessive). Hemorrhage.

BLENNIES (*Osteichthyes*). Of the suborder *Blennioidei*, blennies are generally elongated, often eellike fishes. They have been found in marine deposits as far back as the Eocene era (about 50 million years ago). Present-day blennies primarily inhabit the floors of tropical, temperate, and arctic seas, and in just a few cases, they are found in fresh water. The smallest species attain a length of an inch or less (few centimeters), while the largest blenny is the *wolffish* (*Anarrhi chas minor*), which attains a length up to 6 feet (1.8 meters). In most families, the scales are either greatly degenerate or are completely absent. In the latter case, the skin is equipped with many slime-secreting glands. The slime has the same protective function as scales.

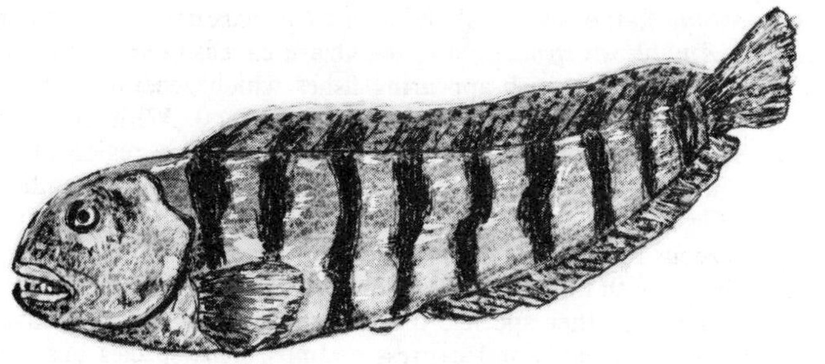

Wolffish (*Anarrhichas lupus.*)

The dorsal and anal fins of blennies are well developed; the dorsal fin extends from the rear of the body, and the anal fin extends from about the middle of the body to the caudal fin. In some species, the anal and caudal fins have fused together, forming a uniform fin seam. The pelvic fins are either degenerate or absent. The pelvic girdle, when present, is also poorly developed, and has fused with the lower part of the pectoral girdle. Of the total of 15 families, the blennies, tripterygiids, clipfishes, chaenopsids, and another family (*Congrogadi dae*) are distributed in tropical and temperate waters. Four small blenny families are found in the Australian region. The wolffishes (see accompanying illustration), the pricklebacks, and the gunnels chiefly inhabit arctic areas. Three other families, the quillfishes (*Ptili chthyidae*), the graveldivers (*Scytalinidae*), and the prowfishes (*Zapro ridae*) are only found in the north Pacific Ocean.

The blennies (*Blenniidae*) form the largest and most diverse family in the suborder. These species are found in deep water, at the water surface, and even on land. A few species have invaded fresh water habitats. Blennies are found above hard, rocky bottom, or sometimes soft mangrove mud. While most blennies are bottom dwellers, some species have taken up a free-swimming existence. One of the most prominent blenny characteristic is the presence of simple or treelike branching tentacles on the heads of many species. They may be located in front of the lowest nasal openings, above the eyes, and/or on the rear of the head. In other species, a helmet-shaped lobe of skin, the comb, is present. Tentacles and the comb can differ in size between males and females, and they may degenerate outside the spawning season. Females often lack a comb entirely.

Over the centuries, two blenny subfamilies have developed, and their differing diet forms the basis of this division. *Blenniinae* blennies feed chiefly on chaetognaths (bristle worms), crustaceans, and mollusks, eating some plant matter. Thus, their dentition and intestinal tract are well developed. These species have immovable teeth in their jaws, including larger chewing teeth at the ends. *Runula* and other genera have well-developed canine teeth in both upper and lower jaws. A few of these species have taken up a striking means of feeding; they attack larger fishes and tear pieces of their skin and fins out with their powerful teeth. The stomachless intestinal tract in *Runula* and related genera is short. The *Salariinae* blennies have fine, movable teeth with which they scrape algae from rocks. In order that their pure plant diet can be best utilized, their intestinal tract is relatively long, and leads to the anus with several convolutions.

As bottom dwellers, blennies usually lack a swim bladder. At one time, the life habits of *Blenniidae* blennies were known only through aquarium observations. In more recent times, divers have studied their actions in their natural habitats. Vision is their most important sensory modality, and it plays the most important sensory role; smell is less important; and their sense of vibration is not well developed. Blenniidae blennies can rapidly change their coloration pattern, frequently used for camouflaging. See also **Fishes.**

BLEPHARITIS. Eyelid.

BLIMP. Dirigibles and Airships.

BLIND-FISH (*Osteichthyes*). Also sometimes referred to as cave dwellers, these fishes are members of the suborder *Characoidei*, and of the family *Amblyopsidae*, of which there are three genera, including the southern cavefish (*Typhlichthys subterraneus*), the ricefish (*Cholo *gaster cornuta*), the springfish (*Chologaster agassizi*), the northern cavefish (*Amblyopsis spelaea*), and the Ozark cavefish (*Amblyopsis rosae*). These all are whitish-appearing fishes which generally reach a maximum size of about $3\frac{1}{2}$ inches (9 centimeters). With exception of one species, these fishes are found in the limestone region of the central United States, essentially between the Appalachian Mountains and the Great Plains, south of the limit of glaciation and north of the Cretaceous Mississippi embayment. However, the ricefish displays no correlation with limestone areas. It is found on the Atlantic coastal plain. Unlike the other species, the ricefish does not occur in caves, but is found in streams and cypress swamps. This species has very small functional eyes, but laboratory experiments have shown that it can obtain its food just as well without the eyes. The other amblyopsids, of course, are blind, but do display a rudimentary eye. Generally, the amblyopsids lack pigment, but it has been found experimentally that, if the southern cavefish is retained in a daylighted aquarium for a period of three months or longer, pigment coloration can be developed.

Until relatively recently, cave fishes were considered the only examples of blind characins, although there are several blind carp and catfish species. In 1965, while laborers were digging a well in Brazil, another blind characin (*Stygichthys typhlops*) was discovered. It was found at a level about 100 feet (30 meters) below ground level. This fish not only lacks eyes and normal pigmentation, but also the lateral line organ, considered so important for blind forms. The bones which normally cover the eye region have disappeared, along with most pores on the head and important sensory organs. Another special case is found in the southermost characin of all—*Gymnocharacinus*. It is completely naked, without the slightest trace of scales. For a while, this was also believed to be an exceptional case, but later a completely naked characin was discovered in Ghana.

In terms of blind fishes in general, it is now believed that the regression of the eyes and pigments is not a direct result of the darkness in which the species spends its entire life. If it is kept under daylight conditions in an aquarium and bred under these conditions for many generations, the vision still remains poor and the eyes are degenerate. One must conclude that the degeneration of the eyes and pigments is an inherited trait. In spite of this great discrepancy between the river fishes with normal vision and the blind cave fishes, they can be crossed (considered not only unusual, but quite unexpected). Thus, the courtship behavior in both forms must correspond to a high degree. The hybrid from such a cross is a mixture between the river inhabitant and the cave fish. It has small eyes, is clearly colored and completely fertile; it can be crossed with one of the parents or with another hybrid. In the latter case, a second generation is produced which varies from species with full vision to those which are completely blind. Coloration varies tremendously also in this third generation. Interestingly, there are pigmentless forms with well-developed vision and blind but fully colored fishes. Geneticists have concluded the following:

1. Development of pigmentation and the eyes proceed independently and are inherited independently.
2. The differences between river fishes and cave fishes arises from mutations. In the transition to cave life, those characteristics which have become useless degenerate by changes in the gene structure. This process in the fishes is a model for the general degeneration of organs throughout the animal kingdom, if not solely for the degeneration of pigmentation and eyes in other cave-dwelling animals.

It can be estimated how long it took for the colorless, blind-fishes to develop from the normal fishes. *Astyanax fasciatus mexicannus*, which lives aboveground, is originally from South America. It could have penetrated Mexico toward the end of the Tertiary period (about 1 million years ago) when the land bridge between South and North America was formed. It could not have invaded Mexico any earlier because it is a fresh-water fish. However, the caves into which it moved were formed during the Ice Age rainy period $\frac{1}{2}$ million years ago through an outgrowth of the calciferous stone deposits. Thus, it could have taken at most $\frac{1}{2}$ million years for the blind varieties to develop from normal fishes. The earlier cave rivers dried up during the drought of the Ice Age, so that these present-day cave fishes inhabit just a few scattered grottos.

BLINDNESS. Braille System; Vision and the Eye.

BLINDNESS (Infant). Gonorrhea.

BLIND SNAKE. Snakes.

BLIND WORM (*Amphibia, Gymnophiona*). Slender, burrowing worm-like amphibians with no trace of legs and with the tail and eyes rudimentary. They are also called caecilians.

BLISTER-BACK. Codfishes.

BLISTER BEETLE (*Insecta, Coleoptera*). Soft-bodied beetles of medium to large size. They are named from their blistering properties; when crushed on the skin even the common species are capable of raising a blister.

Blister beetles are of commercial importance as a source of a pharmaceutical preparation known as Spanish-fly, from the species of that name. This material is composed of the dried pulverized bodies of the insects and is used for producing blisters. Some of the North American species are also occasionally important enemies of plants, among them the old-fashioned potato beetle. They can be checked by the application of sprays containing arsenical poisons.

Several hundred species of blister beetles have been described. They constitute the family *Meloidae*.

BLIZZARD. Fronts and Storms.

BLOB. A radar term referring to a fairly small-scale temperature and moisture inhomogeneity produced by turbulence within the atmosphere. The resulting abnormal gradient of the refractive index can produce a radar echo of the type known as angels.

BLOCH FUNCTION. It can be shown that the wave function of an electron in a periodic lattice has the form

$$\psi = u(\mathbf{r})e^{i\mathbf{k}\cdot\mathbf{r}}$$

where $u(\mathbf{r})$ has the periodicity of the lattice (i.e., is the same in every unit cell) and k is the wave vector of the electron. Notice that this corresponds to a plane wave modulated by the periodicity of the lattice.

BLOCH WALL. This is a transition layer between adjacent ferromagnetic domains magnetized in different directions. (See accompanying figure.) The wall has a finite thickness, of the order of a few

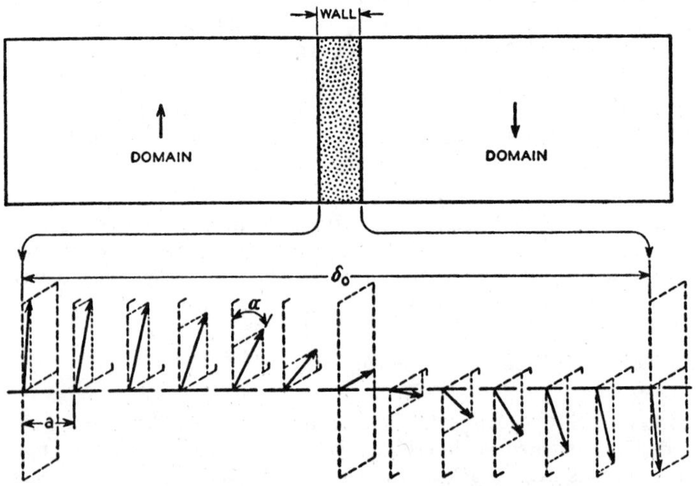

Nature of domain boundary (Bloch wall).

hundred lattice constants, as it is energetically preferable for the spin directions to change slowly from one orientation to another, going through the wall, rather than to have an abrupt discontinuity. The concept of Bloch wall is useful in solid state physics, especially in ferromagnetic theory.

BLOCK (Experimental). The name given, mainly in experimental design, to a group of items under treatment or observation. For example, a block may comprise a group of contiguous plots of land, all the animals in a litter, the results obtained by a single operator, or meteorological observations on a series of days at a given place. The general purpose of dividing all the material in an experiment into blocks (or of regarding it as so divided by the circumstances of the case) is to isolate sources of heterogeneity; the items in a block being so far as possible homogeneous and uncontrolled variation in the experimental material being measured by comparisons between blocks.

An *incomplete block* is a basic form of experimental design. If material is divided into blocks and it is desired to allocate certain treatments to the units of a block, the treatments may be too numerous for them all to appear in each block. When a block contains fewer than a complete replication of the treatments it is called incomplete. Differences between blocks are then discussed by a more elaborate analysis than is required for complete blocks. If each block contains the same number of treatments and they are arranged so that every pair of treatments occurs together in the same number of blocks, the design is said to be balanced.

The commonest type of experimental design is that known as the *randomized blocks.* Suppose that there are t treatments and that the experimental units can be divided into blocks of t units in such a way that units within a block are relatively homogeneous. Then we may allocate the treatments to the units at random, subject to the restriction that each treatment occurs just once in each block. All treatment comparisons can now be made within blocks and any differences between blocks can be eliminated from the estimate of experimental error.

BLOCK DIAGRAM (Data Processing System). A graphical presentation of hardware and paths along which data and control formation flow between various parts of a computer and/or data processing system. Through the use of numerous and varied symbols, much information can be condensed into a small space. Templates are available to facilitate the rapid construction of a block diagram. Some of the more commonly used symbols are given in Fig. 1. The use of a block diagram to represent a central processing unit of a computer is shown in Fig. 2. This overall approach to the system did not require detailed use of symbols. In a block diagram, each block generally represents a functional subdivision of the system under consideration. The function may range from a single AND block to a very complex representation of a total computer. See also **Programming Flowchart (Computer);** and for related topical coverage throughout this volume, see list of entries under **Data Processing.**

Thomas J. Harrison, International Business Machines Corporation, Boca Raton, Florida.

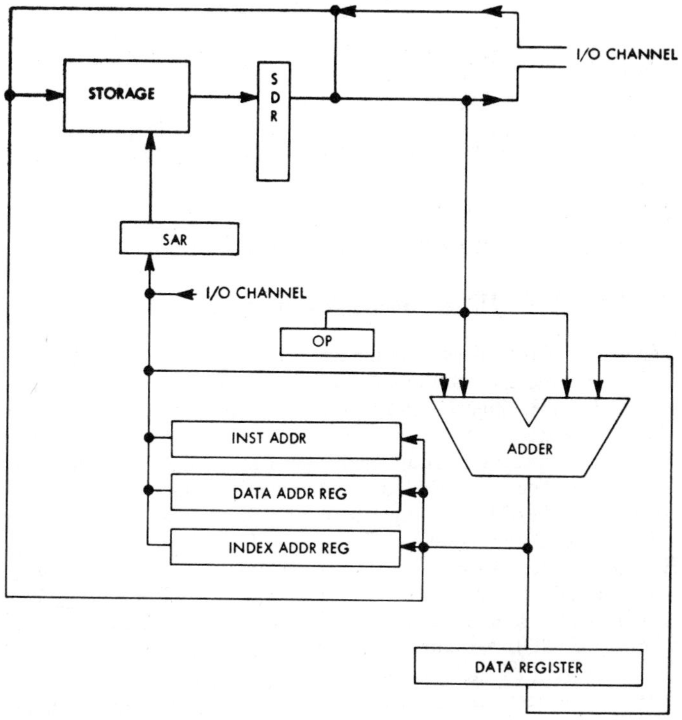

Fig. 2. Block diagram of a central processing unit of a computer.

BLOCKING CAPACITOR. This is a capacitor used at various points in an electrical circuit where it is desirable to pass alternating currents and block direct currents. It is commonly used in coupling one transistor amplifier stage to the next succeeding one. Its use prevents the dc voltage at the output of one amplifier stage from affecting the operating point of the succeeding stage. See diagram on next page.

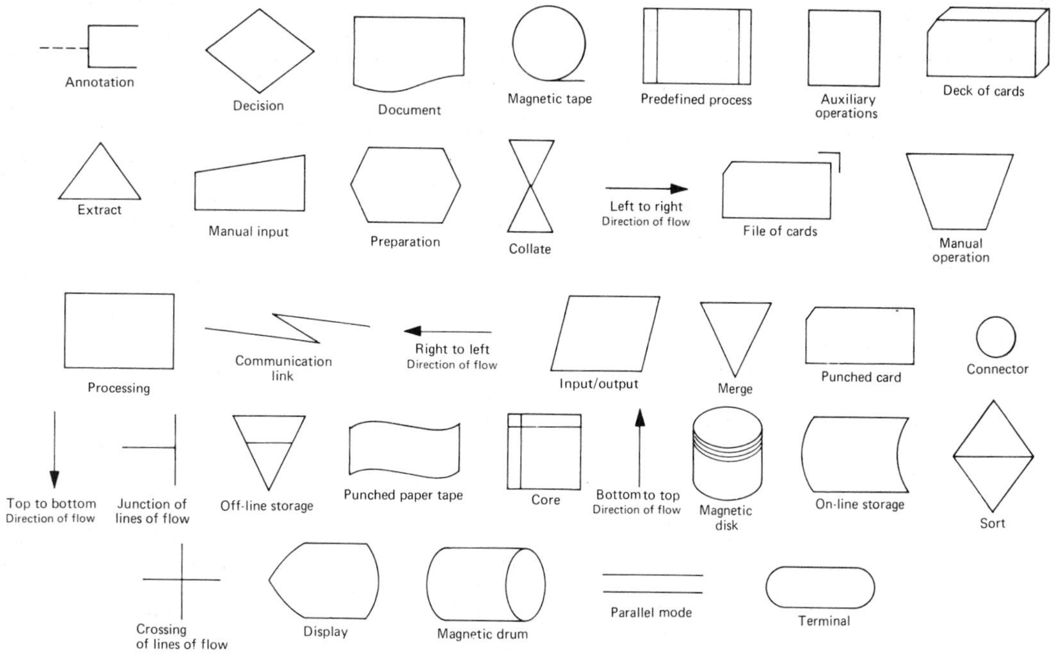

Fig. 1. Representative symbols used in preparing block diagrams.

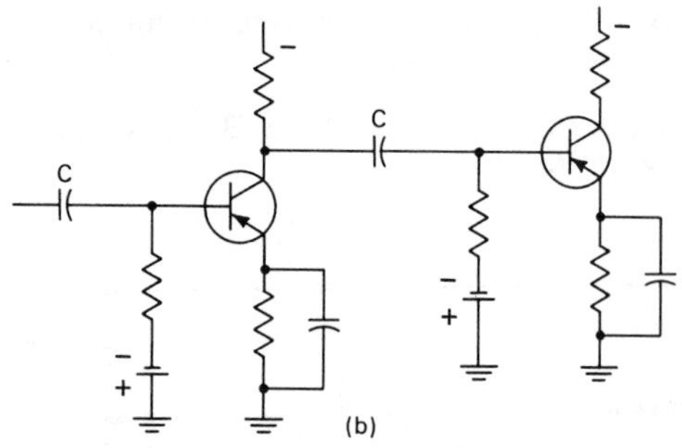

Use of blocking capacitor (C) in electronic amplifier.

BLOCKING OSCILLATOR. Oscillator.

BLOCK SWITCHING. Analog Multiplexer.

BLOOD. Classified as a major tissue of the human body, blood is a characteristically red, mobile fluid with an average specific gravity of about 1.058. Slightly sticky and somewhat viscous, blood has a viscosity between 4.5 and 5.5 times greater than that of water at the same temperature. Thus, blood flows somewhat more sluggishly than water. The odor of blood is characteristic; the taste is slightly saline. The pH of blood ranges between 7.35 and 7.45. The complex acid-base regulatory system of the blood is described in entry on **Acid-Base Regulation (Blood).** Under normal conditions, the blood circulates through the body at a temperature of 100.4°F (38.0°C). This is slightly higher than the body temperature as determined by mouth, 98.6°F (37.0°C). An adult human of average age and size has just over 6 quarts (5.7 liters) of blood.

In very general terms, blood serves as a chemical transport and communications system for the body (i.e., it carries chemical messengers as well as nutrients, wastes, etc.). Circulated by the heart through arteries, veins, and capillaries, blood carries oxygen and a variety of chemicals to all cells, acting as a delivery agent to serve the needs of the cells. Blood also takes away waste products, including carbon dioxide, from the various tissues to organs such as the kidneys and lungs which ultimately dispose these wastes to the environment. Thus, the blood serves as a collecting agent. See also **Heart and Circulatory System (Human).**

Unlike a simple liquid, such as water, or a simple solution, such as salt water, blood is a complex fluid made up of several components, each of which is, in turn, extremely complex and even today not fully understood. Many of these substances are solids in suspension. Unlike most simple liquids that are not easily changed when exposed to air or to slight alternations in their environment, the physical and biochemical properties of blood undergo marked changes (*Hemostatic responses*) when blood is taken from the body's circulatory system and, for example, placed in a test tube. Separation of blood from its usual environment immediately initiates biochemical processes which alter its properties and cause it to release its components. When so removed, blood shortly becomes viscid and forms a soft, jellylike substance, then soon separates into a firm solid mass (clot) and liquid (serum). This extremely important property of *clotting* is unique to blood among known inorganic and organic fluids and solutions. Were it not for this property, a person would bleed (*hemorrhage*) to death if a blood vessel were opened by accident or as the consequence of disease. Thus, blood may be described as a living fluid and most accurately as a living tissue, like the other tissues of the body.

Illustrative of the complex constitution of blood is the list of Table 1.

Principal Components of Human Blood

Not considering the numerous substances, other than oxygen, carried by the blood where the main function of the blood is one of transport, the main functional components of the blood are indicated by the accompanying diagram.

TABLE 1. REPRESENTATIVE CONSTITUENTS OF HUMAN BLOOD
(Values are per 100 milliliters)

Constituent	Plasma or Serum	Whole Blood	Constituent	Plasma or Serum	Whole Blood
Adenosine	1.09 mg				
Adenosine triphosphate (total)		31–57 mg	Mucopolysaccharides	175–225 mg	
Amino acids (total)		38–53 mg	Mucoproteins	86.5–96 mg	
Ammonia N	0.1–1.1 mg	0.1–0.2 mg	Nicotinic acid	0.02–0.15 mg	0.5–0.8 mg
Ascorbic acid	0.7–1.5 mg	0.1–1.3 mg	Nitrogen (total)		3.0–3.7 g
Base (total)	145–160 meq/liter		Non-protein nitrogen	18–30 mg	25–50 mg
Bicarbonate	25–30 meq/liter	19–23 meq/liter	Nucleotide (total)		31–52 mg
Bile acids		0.2–3.0 mg	Nucleotide phosphorus		2–3 mg
Biotin		0.7–1.7 μg	Oxygen (arterial)		17–22 vol %
Blood volume		2990–6980 ml	Oxygen (venous)		11–16 vol %
—adult men	33.7–43.7 ml/kg	66.2–97.7 ml/kg	pH	7.38–7.42	7.36–7.40
—adult women	32.0–42.0 ml/kg	46.3–85.5 ml/kg	Pantothenic acid	6–35 μg	15–45 μg
—infants	36.3–46.3 ml/kg	79.7–89.7 ml/kg	Polysaccharides (total)	73–131 mg	
Carbon dioxide			Protein (total)	6.0–8.0 g	19–21 g
—Arterial blood (total)		45–55 vol %	Protein (albumin)	4.0–4.8 g	
—Venous blood (total)		50–60 vol %	Protein (globulin)	1.5–3.0 g	
Cholesterol (total)	120–250 mg	115–225 mg	Purines (total)		9.5–11.5 mg
Cholesterol esters	75–150 mg	48–115 mg	Pyruvic acid	0.7–1.2 mg	0.5–1.0 mg
Cholesterol (free)	30–60 mg	82–113 mg	Riboflavin	2.6–3.7 μg	15–60 μg
Choline (total)	26–35 mg	11–31 mg	Ribonucleic acid	4–6 mg	50–80 mg
Fat (neutral)	25–260 mg	85–235 mg	Sphingomyelin	10–47 mg	150–185 mg
Fatty acids	190–450 mg	250–390 mg	Thiamine	1–9 μg	3–10.7 μg
Fibrinogen	200–400 mg	120–160 mg	Urea	28–40 mg	20–40 mg
Fructose	7–8 mg	0–5 mg	Urea N	8–28 mg	5–28 mg
Glucose (adult)	65–105 mg	80–120 mg	Uric acid (male)	2.5–7.2 mg	0.6–4.9 mg
Hemoglobin	trace	14.8–15.8 g	Vitamin A (carotenol)	15–60 μg	9–17 μg
Histamine		6.7–8.6 μg	Vitamin A (carotene)	40–540 μg	20–300 μg
Ketone bodies (total)	0.15–1.36 mg	0.23–1.00 mg	Vitamin B_{12} (cyanocobalamin)	0.01–0.07 μg	0.06–0.14 μg
Lactic acid	30–40 mg	5–40 mg	Vitamin D_2 (as calciferol)	1.7–4.1 μg	
Lecithin	100–225 mg	110–120 mg	Vitamin E	0.9–1.9 mg	
Lipids (total)	400–700 mg	445–610 mg	Water	93–95 g	81–86 g

NOTE: *Plasma* is the liquid portion of whole blood. *Serum* is the liquid portion of blood after clotting, the fibrinogen having been removed.

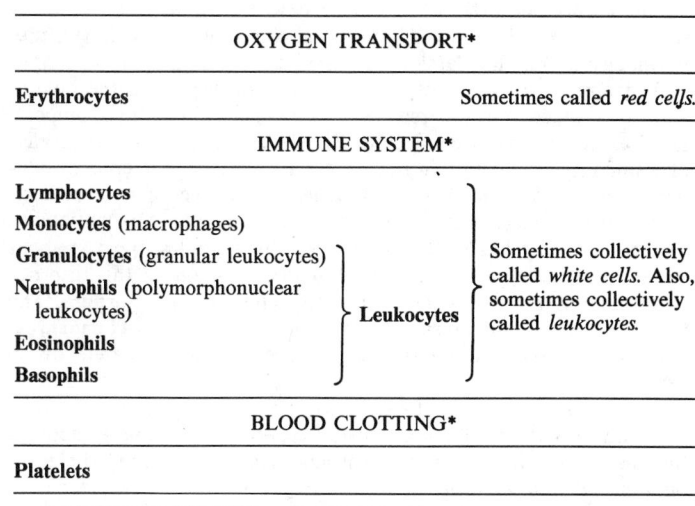

OXYGEN TRANSPORT*	
Erythrocytes	Sometimes called *red cells.*
IMMUNE SYSTEM*	
Lymphocytes	Sometimes collectively called *white cells.* Also, sometimes collectively called *leukocytes.*
Monocytes (macrophages)	
Granulocytes (granular leukocytes) — **Leukocytes**:	
Neutrophils (polymorphonuclear leukocytes)	
Eosinophils	
Basophils	
BLOOD CLOTTING*	
Platelets	
MULTIFUNCTIONAL—PROVIDES VOLUME AND FLUIDITY TO BLOOD	
Plasma	

* Predominant, but not exclusive role.

Basic components of blood.

Erythrocytes. It is estimated that an adult man will have about 5 million erythrocytes per cubic millimeter of blood. This is equivalent to about 82 billion erythrocytes per cubic inch of blood. In an adult woman, there are about 4.5 million erythrocytes per cubic millimeter. Erythrocytes are homogeneous circular disks with no nucleus. These red cells are about 0.0077 millimeter in diameter. When viewed singly by transmitted light, the erythrocyte has a yellowish red tinge, but when viewed in great numbers, the erythrocytes have the distinctly blood red coloration. Erythrocytes possess a certain degree of elasticity, so that they can pass through tiny apertures and passages on their way to reach tissue supplied by the capillaries.

The prime function of the erythrocytes is to deliver oxygen to peripheral tissues. This oxygen is furnished to these cells by an exchange-diffusion system brought about in the lungs. The color of the erythrocytes is derived from a red iron-containing pigment called *hemoglobin.* This is a conjugated protein that consists of a globin (a protein) and *hematin* (a nonprotein pigment), the latter containing iron. Hemoglobin contains 0.33% iron. When hemoglobin combines with oxygen, *oxyhemoglobin* is formed. When oxygen is given up to the tissues, it is then reduced back to hemoglobin. The erythrocytes also carry some carbon dioxide from the tissues and function to maintain a normal acid-base balance (pH) of the blood. When the hemoglobin has its full complement of oxygen, it is a bright red. This scarlet blood is found in the arteries which carry the blood to organ tissues throughout the body. As the oxyhemoglobin gives up oxygen, it takes on a darker crimson hue, and this is found in the veins which return the blood to the lungs for reoxygenation. See **Blood and Circulatory System (Human);** and **Hemoglobin.**

Megaloblasts, cells that are the precursors of erythrocytes, are noted in the blood islands of the yolk sac of the human embryo. By the end of the embryo's second month of life, manifesting the second step in the erythrocytic series, *erythroblasts* are found in the liver and spleen. These cells are somewhat smaller and possess a smaller nucleus than the megaloblasts. At about the fifth month, centers of blood formation appear in the middle regions of the bones, with an accompanying progressive expansion of the marrow cavities. At this stage the marrow assumes nearly exclusively the function of producing the erythrocytes (red cells) required by the body—a process which continues throughout the life of the individual. At the time of birth, essentially all bone marrow is engaged in blood formation (not exclusively red cells). As the individual progresses toward maturity, much of the marrow of the long bones is converted into a fatty tissue in which blood cell formation (*hematopoiesis*) is no longer apparent. In adults, bone marrow active in the formation of blood cells is found in the ribs, vertebrae, skull, and the proximal ends of the humerus (upper arm) and femur (upper thigh).

Once erythrocytes enter the blood, it is estimated that they have an average lifetime of about 120 days. In an average person, this indicates that about 1/120th or 0.83% of the red cells are destroyed each day. At least three important mechanisms are involved in the death of erythrocytes: (1) *Phagocytosis,* defined as the ingestion of solid particles by living cells—in this case, by cells of the reticuloendothelial system. (2) *Hemolysis* by specific agents in the blood plasma. The erythrocytes are protected by a membrane. If this membrane is broken, the hemoglobin goes into solution in the plasma. Numerous substances (*hemolytic agents*) may cause this action and these include hypotonic solutions, foreign blood serums, snake venom, various bacterial metabolites, chloroform, bile salts, ammonia and other alkalis, among others. In this condition, the erythrocytes no longer can serve as oxygen carriers. (3) *Mechanical damage* and destruction, brought about by simple wear and tear as the reasonably fragile red cells circulate and recirculate through the body.

The stimulus for production of new red cells is provided by *erythropoietin,* a hormone that is apparently produced by the kidneys. The actual production is accomplished almost entirely by the red portions of the bone marrow, but certain substances necessary for their manufacture must be furnished by the liver. Surplus red cells, needed to meet an emergency, are stored in the body, mainly in the spleen. The spleen also breaks down old and worn red cells, conserving the iron during the process. See **Liver;** and **Spleen.**

When a sudden loss of a large amount of blood occurs, the spleen releases large numbers of red cells to make up for the loss, and the bone marrow is stimulated to increase its rate of manufacture of blood cells. When a donor gives a pint of blood, it usually requires about seven weeks for the body reserve of red corpuscles to be replaced, although the circulating red cells may be back almost to normal within a few hours. Repeated losses of blood within a short time, however, may easily deplete the red cell reserves.

In addition to hemoglobin, it has been found that there are least two other alternative oxygen carriers—*hemerythrin* and *hemocyanin.* In overall terms, as presently understood, these carriers are minor. Unlike hemoglobin, these two blood proteins do not incorporate an iron-porphyrin ring. The three blood proteins are strikingly colored in their oxygenated states—the familiar red of hemoglobin; the unusual reddish-tinted violet of hemerythrin; and the cupric-bluish color of hemocyanin. Klotz and colleagues (Northwestern University and other locations) have made a detailed study of the alternative oxygen carriers and suggest that an understanding of the three-dimensional structure of hemerythrin and of the electronic state of the active site is approaching, in refinement, that which is currently known about hemoglobin.

White Cells. There are several types of white cells, which are sometimes collectively called *leukocytes,* although some authorities reserve that term to identify only the granulocytes. White cells are irregular in shape and size, but generally are larger than the red cells. They differ from the red cells in that each white cell contains a nucleus. Adult humans have from 5000 and 9000 leucocytes per cubic millimeter of blood. In infants, the number is essentially doubled. There is roughly a ratio of 1 white to every 700 red cells. When white cells increase in number, the condition is called *leukocytosis,* a situation that is presented in pneumonia, appendicitis, and abscesses, among other conditions. A decrease in the number of leukocytes below normal is called *leukopenia.* In *leukemia,* there is an uncontrolled increase in the number of leucocytes.

In general terms, the white cells, each type with a specific function, accomplish the following actions: (1) Protection of the body from pathogenic organisms; and (2) participation in tissue repair and regeneration. Over the years, an increasingly detailed understanding of the white cells has occurred. These matters are described in some detail in the entry on **Immune System and Immunology.** Generally, whenever bacteria or other foreign substances enter the tissues, large numbers of white cells immediately travel through the walls of the blood vessels and to the site of disturbances. They take the bacteria and any other foreign materials into their own bodies, where they are digested. White cells are able to break up and carry away even as large an object as a splinter or thorn in the skin. They also help in carrying away dead tissue and blood clots which remain after a wound. *Pus* is largely composed of white cells which have been drawn to the infected area, as well as the dead and disintegrating tissue and bacteria. During severe infections, the white cells may be increased in the blood five- or tenfold. Because of this, a white cell count is made on the blood in order to confirm diagnosis in many infections.

Lymphocytes generally comprise between 25–30% of the white cells in human blood. These immunologically active cells are comprised of several classes, each of which has specific properties and functions. Lymphocytes are derived from stem cells located in the yolk sac and fetal liver. Later, some stem cells originate from the bone marrow. These cells then differentiate into lymphocytes in the primary lymphoid organs, principally the thymus and lymph nodes.

Monocytes (macrophages) are part of the mononuclear phagocytic system. They are large, mononuclear cells and comprise 3–8% of the leukocytes found in the peripheral blood. Monocytes originate in the bone marrow. When the mature cells enter the peripheral blood, they are called monocytes; when they leave the blood and infiltrate tissues, they are called macrophages. These cells play an important role in induction of the immune response. They present antigen to the lymphocytes that bear specific receptors for the antigen and also act as effector cells, attacking certain microorganisms and neoplastic cells.

Granulocytes contain specifically identifiable granules, including the neutrophils, eosinophils, and basophils. The *neutrophils* comprise 60–70% of all leukocytes in the blood. Neutrophils arise from precursors in the bone marrow and have a half-life of 4 to 8 hours in the blood, with about a day of life in the tissues. Neutrophils hasten to inflammatory sites by a number of different and poorly understood chemotactic (response to chemical stimulation) factors. Neutrophils have a marked capacity to phagocytize and destroy microorganisms. These cells also contain a number of degradative enzymes and small proteins. The cells are endowed with receptors for IgC and for a complement component (C3b). See also **Immune System and Immunology.** The *eosinophils* are named by virtue of the fact that the granules of cytoplasm are stainable with acid dyes, such as eosin. These cells are present in small numbers (2–4% of the blood), but under certain pathological conditions they show a marked increase. The exact function of eosinophils has been a mystery for many years. Some studies commenced in the mid-1970s have indicated a number of different functions. Many eosinophils have been found in tissues at sites of immune reactions that have been triggered by IgE antibodies (as found in nasal polyps or in the bronchial wall of some patients with asthma). Eosinophils have been found to contain several enzymes that can degrade mediators of immediate hypersensitivity, such as histamine, suggesting that they may control or diminish some hypersensitivity reactions. These cells have been found associated with infections caused by helminths (worms). The *basophils* are formed in the bone marrow and have a polymorphic nucleus. They occur only to the extent of about 1% of the leukocytes. The function of these cells is poorly understood. They are known to play a role in immediate hypersensitivity reactions and in some cell-mediated delayed reactions, such as contact hypersensitivity in humans and skin graft or tumor rejections and hypersensitivity to certain microorganisms in animals.

Platelets are the smallest of the formed elements of the blood. Every cubic millimeter of blood contains about 250 million platelets, as compared with only a few thousand white cells. There are about a trillion platelets in the blood of an average human adult. Platelets are not cells, but are fragments of the giant bone-marrow cells called *megakaryocytes.* When a megakaryocyte matures, its cytoplasm (substance outside cell nucleus) breaks up, forming several thousand platelets. Platelets are roughly disk-shaped objects between one-half and one-third the diameter of a red cell, but containing only about one-thirteenth the volume of the red cell. Platelets lack DNA and have little ability to synthesize proteins. When released into the blood, they circulate and die in about ten days. However, they do possess an active metabolism to supply their energy needs.

Because platelets contain a generous amount of contractile protein (*actomyosin*), they are prone to contract much as muscles do. This phenomenon explains the shrinkage of a fresh blood clot after it stands for only a few minutes. The shrinkage plays a role in forming a hemostatic plug when a blood vessel is cut. The primary function of platelets is that of forming blood clots. When a wound occurs, numbers of platelets are attracted to the site where they activate a substance (*thrombin*) which starts the clotting process. *Prothrombin* is the precursor of thrombin. Thrombin, in addition to converting fibrinogin into fibrin, also makes the platelets sticky. Thus, when exposed to collagen and thrombin, the platelets aggregate to form a plug in the hole of an injured blood vessel. Persons with a low platelet count (*thrombocytopenia*) have a long bleeding time. Platelet counts may be low because of insufficient production in the bone marrow (from leukemia or congenital causes, or from chemotherapy used in connection with cancer), among other causes. Also, individuals may manufacture antibodies to their own platelets to the point where they are destroyed at about the same rate they are produced. A major symptom of this disorder is purpura. Aspirin may aggrevate this condition. See **Purpura.** The bleeding of hemophilia results from a different cause. See **Hemophilia.** Transfusion of blood is a major therapy used in treating platelet disorders.

Platelets not only tend to stick to one another, but to the walls of blood vessels as well. Obviously because they promote clotting, they have a key role in forming thrombi. As pointed out in the entry on **Arterial and Venous Disorders,** the dangerous consequences of thrombi are present in cardiovascular and cerebrovascular disorders. See **Cerebral Vascular Diseases;** and **Ischemic Heart Disease.** Many attempts have been made to explain the process of atherogenesis, that is, the creation of plaque which narrows arteries and, of particular concern, the coronary arteries. Recently, there has been increasing interest in the possible role of platelets in atherosclerosis. Evidence from experimentation with laboratory animals has provided some evidence of a role for platelets in this process. This is covered in some detail by Zucker (1980).

As reported by Turitto and Weiss (1980), red blood cells may have a physical and chemical effect on the interaction between platelets and blood vessel surfaces. Under flow conditions in which primarily physical effects prevail, it has been found that platelet adhesion increases fivefold as *hematocrit*[1] values increase from 10 to 40%, but undergoes no further increase from 40 to 70%, implying a sturation of the transport-enhancing capabilities of red cells. For flow conditions in which platelet surface reactivity is more dominant, platelet adhesion and thrombus formation increase monotonically as hematocrit values increase from 10 to 70%. Thus, the investigators suggest that red cells may have a significant influence on hemostasis and thrombosis; the nature of the effect is apparently related to the flow conditions.

Plasma. Normal blood plasma is a clear, slightly yellowish fluid which is approximately 55% of the total volume of the blood. The plasma is a water solution in which are transported the digested food materials from the walls of the small intestine to the body tissues, as well as the waste materials from the tissues to the kidneys. Consequently, this solution contains several hundred different substances. In addition, the plasma carries antibodies, which are responsible for immunity to disease, and hormones. The plasma transports most of the waste carbon dioxide from the tissues back to the lungs. Plasma consists of about 91% water, 7% protein material, and 0.9% various mineral salts. The remainder consists of substances already mentioned. The salts and proteins are important in keeping the proper balance between the water in the tissues and in the blood. Disturbances in this ratio may result in excessive water in the tissues (swelling or edema). The mineral salts in the plasma all serve other vital functions in the body and must be supplied through diet. See Table 2.

Some of the blood plasma, as well as some of the white cells, filters through the walls of the blood vessels and out into the tissues. This filtered plasma (lymph) is a clear and colorless fluid which returns to the blood through a series of canals referred to as the *lymphatic system.* This system contains filters (*lymph nodes*) which remove bacteria and other debris from the lymph. These nodes, especially those located in the neck, armpit, and groin, may become swollen when an infection occurs in a nearby site. Blood cots do not occur normally while the blood is in the vessels. But in an injury, one of the plasma proteins (*fibrin*) forms a mesh in which the blood cells are trapped, and this mesh is the clot. Blood serum is the yellowish fluid left after the cells and fibrin have been removed from the blood.

[1] A hematocrit is a tube calibrated to facilitate determination of the volume of erythrocytes (red cells) in centrifuged, oxalated blood, expressed as corpuscular volume percent.

TABLE 2. INORGANIC CONSTITUENTS OF HUMAN PLASMA OR SERUM.

CONSTITUENT	VALUE/100 ml
Aluminum	45 μg
Bicarbonate	24–31 meq/liter
Bromine	0.7–1.0 μg
Calcium	9.8 (8.4–11.2) mg
Chloride	369 (337–400) mg
Cobalt	10 (3.7–16.6) μg
Copper	8–16 μg
Fluorine	109 (75–145) μg
Iodine, total	7.1 (4.8–8.6) μg
Protein bound I	6.0 (3.5–8.4) μg
Thyroxine I	4–8 μg
Iron	105 (39–170) μg
Lead	2.9 μg
Magnesium	2.1 (1.6–2.6) mg
Phosphorus, total	11.4 (10.7–12.1) mg
Inorganic P	3.5 (2.7–4.3) mg
Organic P	8.2 (7–9) mg
ATP P	0.16 (0–6.4) mg
Lipid P	9.2 (6–12) mg
Nucleic acid P	0.54 (0.44–0.65) mg
Potassium	16.0 (13–19) mg
Rubidium	0.11 mg
Silicon	0.79 mg
Sodium	325 (312–338) mg
Sulfur	
Ethereal S	0.1 (0–0.19) mg
Inorganic S	0.9 (0.8–1.1) mg
Non-protein S	2.8 (2.4–3.6) mg
Organic S	1.7 (1.4–2.6) mg
Sulfate S	1.1 (0.9–1.3) mg
Tin	4 μg
Zinc	300 (0–613) μg

Blood Osmotic Pressure. The presence of solute molecules and ions in relatively high concentrations in blood establishes an osmotic pressure which tends to transport water from the exterior, through the semipermeable membranes of the blood vessel walls, into the bloodstream. This osmotic transport of water inward is opposed by the effect of hydrostatic pressure within the blood vessels, tending to force water (and soluble substances) out through the capillary walls. The loss through leakage of some of these solutes is indirectly restored through the action of the lympatic system. Among the blood constituents important in maintaining blood osmotic pressure (and thus helping to regulate the volume of fluid in the blood) are the blood proteins. Among these, the protein fraction termed *albumins*, being relatively low in molecular weight, makes the greatest contribution to the total osmotic effect.

Blood Processing and Transfusion Therapy

Blood transfusion practice has changed markedly in recent years. At one time, units of *whole blood* were administered to patients with a variety of requirements stemming from different conditions. These conditions ranged from acute blood loss (hemorrhage, bleeding from injuries, etc.) to aplastic anemia, among other blood-related problems.

The outer portion of the erythrocyte (red cell) is a very complex material composed of proteins, polysaccharides, and lipids, many of which are *antigens*, sometimes referred to as blood group substances. The presence of most, if not all, of these antigens is genetically determined, and their number is such that there may be few, if any, individuals in the world with an identical set of antigens on the red cells—monozygotic twins excepted. These differences in whole blood were learned early in the development of transfusion technology. Fundamental to the refinement of the technology was the discovery by Karl Landsteiner (Nobel Prize winner in 1930) for his observations of the four hereditary blood groups. Landsteiner developed the ABO blood-typing system which serves as a principal guideline in determining the suitability of donors and recipients. This system consists of three allelic genes, dividing all humans into four groups, A, AB, B, and O. In a few rare individuals, the presence of a suppressor gene may prevent the expression of the A, B, O group character. The products of these genes are the A, B, and O antigens or substances. These antigens not only are located on red cells, but are widely distributed in the body, occurring in the endothelium of capillaries, veins, and arteries, and in numerous cells throughout the body. In addition to a cell-associated form, these antigens occur in soluble form in many body fluids, such as the saliva, gastric juice, urine, amniotic fluid, and in very high concentrations is pseudomucinous ovarian cyst fluid. All individuals possess cell-associated A, B, and O antigens. The presence of the soluble form, however, is governed by a recessive gene called the secretor gene which exists as two alleles, *Se* and *se*. Individuals who possess at least one *Se* gene secrete the antigens, while those with two *se* genes do not. The A, B, and O antigens are not uniquely human, but are quite widely distributed in nature. They are found on primate erythrocytes and in the stomach lining of pigs and horses. Intensive investigation has produced considerable information concerning the chemical composition of these antigens. They are extremely stable substances, which is attested by the fact that they can be extracted from Egyptian mummies, thus making it possible to obtain the blood groups of this ancient people. Specific antigenic activity is associated with the carbohydrate moiety, and since the A, B, and O substances possess the same four sugars, the difference between them lies in their arrangement. Analysis of purified A, B, and O substances reveals that about three-fourths of the weight is accounted for by four sugars: L-fructose, D-galactose, *N*-acetyl-D-glucosamine, and *N*-acetyl-D-galactosamine. The reaminder consists of amino acids. See **Immune System and Immunology.**

Cross-Matching. Upon receipt of a tube of clotted blood at a blood bank for typing and cross-matching, procedures are undertaken to determine which antigens are on patient's red cells and which antibodies against red cell antigens are present in the patient's serum. Typing is routine for red blood cell antigens in the ABO system and for a single specificity in the Rh system, namely, the D phenotype.

The Rh group, so denoted because the antigen was first found in the red cells of Rhesus monkeys, is very complex, consisting of perhaps 20 antigens. The Rh_0D antigen is the most important of these antigens because of its possible involvement in the induction of *hemolytic disease of the newborn.* Today, Rh_0D immune globulin (RhoGam® and Gamulin-Rh®, among others) is available to alleviate this danger. This danger is brought about when an Rh-negative woman and an Rh-positive man have an Rh-positive child. There is the grave risk that the woman will become sensitized to the Rh factor in her infant's blood and begin to produce anti-Rh antibodies. The first child is not usually affected, but with subsequent pregnancies, the mother may send sufficient damaging antibodies into the child's blood to threaten its life. When this occurs, in the absence of using the Rh_0D immune globulin, an exchange blood transfusion with almost complete replacement of the infant's blood by Rh-negative blood of the proper ABO group is necessary.

Component Therapy. Frequently, patients do not require all of the blood components and, in fact, their presence can cause many problems. From experience with whole blood therapy over a number of years, *component therapy* emerged. In component therapy, which has many advantages, the patient is given specifically what is needed by way of blood components. Further, separate blood fractions can be stored under those special conditions best suited to assure their biological activity at the time of transfusion. Component therapy also avoids the introduction of foreign antigens and antibodies. It is seldom that fresh whole blood is the treatment of choice providing that specific components are readily available within the time needed.

Processing of Donor Blood. When donor blood is received at a processing center, it is first tested for syphilis and hepatitis B antigen. One unit of whole blood is 500 milliliters. It is then separated into: (1) A unit of *packed erythrocytes* (volume of 300 milliliters and hematocrit value of 70 to 90). This substance is storable in citrate-phosphate-dextrose at 4°C (39.2°F) for up to 3 weeks. (2) A unit of *platelets* (packet) with a volume of 50 milliliters containing about 80 billion platelets). This substance is storable (while being gently mixed) at room temperature for 2 or 3 days. (3) A unit of *cryoprecipitate* (volume of about 10 milliliters containing from 80 to 120 units of Factor VIII and from 300 to 400 milligrams of fibrinogen). This substance is stable in a frozen condition for about one year. (4) One unit of *plasma* (a volume of about 200 milliliters, from which about half of the fibrinogen has been removed). This substance contains platelets, Factor VIII, as well as all remaining procoagulants, albumin, salt, and antibodies which were a part of the original plasma. This substance may be (a) stored in the frozen state, (b) refrigerated, or (c) further processed for individual globulin classes and albumin.

Thus, somewhat analogous to obtaining specific drugs for different conditions, the physician can order up specifically those blood components required for a given need. Platelets, which survive poorly in whole blood stored in acid-citrate-dextrose solution under refrigeration, can be obtained in platelet packets as previously mentioned, or as washed platelets. Factor VIII, for the management of hemophilia, can be obtained in a lyophilized or other purified form. In whole blood, by contrast, procoagulants stored in whole blood decay so that, after a few days, the availability of Factor VIII and other allied components is extremely low. As the result of additional processing, blood centers can furnish prothrombin complex concentrate (*Proplex*), each batch bearing a specific analysis. Preparations of peripheral white blood cells are available from centers with specific blood processing equipment. Additionally, substances for expanding plasma volume, such as fresh frozen plasma, albumin solutions, and Dextran, among others, are available.

Blood Substitutes. Researchers in Japan (Fukushima Medical Center) and in other institutions in Europe and North America have been investigating substances that, in major characteristics, may serve as a substitute for blood, particularly in emergency situations where rare blood types are not immediately available to severely ill patients who require transfusions. For example, in early 1979, a Japanese patient with a rare O-negative blood was given an infusion of one liter of a new, oxygenated perfluorocarbon emulsion. This compound carried oxygen through the patient's circulatory system until the rare blood could be obtained. Later that year, eight additional patients survived infusions with artificial blood. As early as 1966, investigators at the University of Cincinnati demonstrated that life could be sustained when rodents were immersed in perflurochemicals for long periods. This class of chemicals can dissolve as much as 60% oxygen by volume, as contrasted with whole blood (20%), or salt water or blood plasma (3%). Initially, a major problem existed because pure perfluorochemicals are not miscible with blood. In the late-1960s, researchers (University of Pennsylvania; Harvard School of Public Health) demonstrated that perfluorochemicals could be emulsified. Research is continuing along these lines. Also, new chemicals of this class are being sought. Initially, the research was done with perfluorobutyltetrahydrofuran and perfluorotripopylamine, both superior carriers of oxygen, but prone to concentrate in some organs of the body, notably the liver and spleen. In 1973, perfluorodecalin was found to be completely eliminated from the body. The approach in Japan has differed somewhat, in that research has been directed to add other chemicals which will increase the half-life of the chemicals in the body. As summarized by Maugh (1979), it appears that the perfluorochemical emulsions are beginning to fulfill some of the promise they first showed and that it may only be a matter of time and research effort before they can be used to sustain life in emergency situations.

Blood Recycling. In a process known as *autotransfusion*, introduced in the early 1980s, blood lost during operative procedures, particularly in heart surgery, is recycled back to the patient. Some reports indicate the need for donor blood can be reduced by as much as 60%. Instead of discarding blood lost during surgery, as has been the traditional practice, the blood is collected in a plastic bag with a special filter to cleanse impurities before the blood is returned to the patient. The procedure has many advantages, including costs of transfusion and elimination of risks from hepatitis, errors in mismatching blood types, and other complications that may arise with donor blood. Although results appear to be positive thus far, a few additional years may be required before the procedure is fully accepted as standard practice.

Blood as an Indicator of Disorders and Diseases

Since the blood performs many services for all parts of the body, it will reflect disturbances that occur as the result of many widely divergent diseases. This had led to the development of a variety of blood tests, either to confirm a diagnosis or to follow the effectiveness of treatment in the patient. *Immunological* or *serological* tests are performed to confirm the diagnosis of selected types of infectious diseases, and are based upon the principle that in certain diseases there appear in the blood specific substances (antibodies) which are produced by the body in resisting invasion by specific disease-producing media. One of the more widely used tests is the Kolmer test for syphilis. Blood typing tests are also serological in nature. A second group of blood tests are known as *hematological.* These tests determine the number of each type of circulating blood cell (*blood count*), the total volume of red cells in a blood sample (*hematocrit*), and the hemoglobin content of the blood. A *differential* blood count is one in which selected dyes are used to distinguish better the different kinds of white blood cells. These tests are important in diagnosing and treating illnesses, such as infections, the anemias, and the leukemias.

Another group of blood tests involves *bacteriological* techniques. Blood and bone marrow samples are obtained under aseptic precautions and introduced into a variety of artificial culture media, with subsequent isolation and identification of the specific microorganism responsible for the illness. Relative susceptibility of the specific strain of bacteria to the available chemotherapeutic and antibiotic agents may then be determined and the effectiveness of such agents in sterilizing the bloodstream can be determined by further blood cultures.

Many *chemical* tests are performed on blood samples to determine the quantitiative relationships between circulating globulins, ablumin, sugar, nonprotein nitrogen, minerals, and other normal and abnormal constituents. Such chemical tests are important in diabetes, kidney diseases, the failing heart, and in pancreatic and liver diseases. In all of these disorders, pronounced changes in the relative amounts of the various chemical constituents of the blood occur. Chemical tests also may be performed on urine, spinal fluid, and saliva for some special purpose, and since most of these fluids are derived from the blood plasma, their chemical analysis frequently reflects changes in the blood itself. During prolonged therapy with certain drugs, it may be desirable to measure chemically the concentration of the drug in the blood plasma.

Sophisticated instrumentation and procedures are used in research involving blood and its functions. Phase contrast microscopy has the advantage that living cells can be studied for long periods of time; chromatin, mitochondria, centrosomes and specific granules can be seen and photographed at magnifications of 2,500×. The method is excellent for the study of granules of the matrix of cells which is unseen in traditionally fixed and stained cells. It is an excellent aid for those who wish to use the electron microscope, because areas demonstrated by light can be compared with those visualized by the electron beam. The study of blood by motion pictures (*microcinematography*) has been used for many years. With the invention of the phase microscope, this approach to the study of blood cells has been an important tool. Studies of the movements of the lymphocytes in rats showed a softening of the membrane at the forward moving end, and pseudopod formation; contractions of the cell force the inner plasma forward, while the external plasmagel remains fixed except at the posterior end, then it becomes softer and passes through the stiffer ring of plasmagel to become more gelated at the anterior end. This is an example of the type of detailed investigation that can be made with microcinematography. Using speed photography at 3,200 frames per second, the red blood cell has been observed to have an interior velocity of 30 × that of water at 38°C. In the dog's mesentery, red blood cells passing into capillaries from larger arterioles take the form of an inverted cap or parachute; when blood flow is stopped they become biconcave disks. The cup shape is suggested as bringing more surface close to the capillary endothelium.

Other blood research techniques include the use of physical and chemicals agents, ultracentrifugation, cytochemical methods, microincineration, and autoradiography.

For related topics, see the following entries: **Agranulocytosis; Anemias; Anoxemia; Arterial and Venous Disorders; Blood-Brain Barrier; Cyanosis; Gamma Globulin; Gaucher's Disease; Heart and Circulatory System (Human); Hematology; Hematoma; Hemolymph; Hemophilia; Leukemias; Leukocytosis; Leukopenia; Lipidoses; Purpura; Pus; Spleen.**

References

Albert, S. N.: "Blood Volume and Extra-cellular Fluid Volume," 2nd edition, Charles C. Thomas, Springfield, Illinois, 1971.

Bank, A., Mears, J. G., and F. Ramirez: "Disorders of Human Hemoglobin," *Science*, **207**, 486–493 (1980).
Boggs, D. R.: Neutrophils in the Blood Bank," *N. Engl. J. Med.*, **296**, 748 (1977).
Cash, J. D.: "Blood Transfusion and Blood Products," *Clin. Haematol.* (May 1976).
Grady, G. F.: "Transfusions and Hepatitis," *N. Engl. J. Med.*, **298**, 1413 (1978).
Hoffman, R., et al.: "Diamond-Blackfan Syndrome," *Science*, **193**, 899–900 (1976).
Holmsen, H., Salganicoff, L., and M. H. Fukami: "Platelet Behavior and Biochemistry," in "Haemostasis: Biochemistry, Physiology, and Pathology," (D. Ogston and B. Bennett, editors), Wiley, New York, 1977.
Klotz, I. M., et al.: "Hemerythrin: Alternative Oxygen Carrier," *Science*, **192**, 335–344 (1976).
Maines, M. D., and A. Kappas: "Metals as Regulators of Heme Metabolism," *Science*, **198**, 1215–1221 (1977).
Maugh, T. H., II: "Blood Substitute Passes Its First Test," *Science*, **206**, 205 (1979).
Moffat, K., Deatherage, J. F., and D. W. Seybert: "A Structural Model for the Kinetic Behavior of Hemoglobin," *Science*, **206**, 1035–1042 (1979).
Mustard, J. F., et al.: "Platelets, Thrombosis and Atherosclerosis," *Prog. Biochem. Pharmacol.*, **13**, 312–325 (1977).
Perutz, M. F.: "Hemoglobin Structure and Respiratory Transport," *Sci. Amer.*, **239**, 6, 92–125 (1978).
Simmons, A.: "Basic Hematology," Charles C. Thomas, Springfield, Illinois, 1973.
Staff: "Symposium on Blood Groups: Their Genetics, Function, and Relation to Disease," *Mayo Clin. Proc.*, **52**, 135 (1977).
Sussman, L. N.: "Paternity Testing by Blood Grouping," 2nd edition, Charles C. Thomas, Springfield, Illinois, 1976.
Tullis, J. L.: "Clot," Charles C. Thomas, Springfield, Illinois, 1976.
Turitto, V. T., and H. J. Weiss: "Red Blood Cells: Their Dual Role in Thrombus Formation," *Science*, **207**, 541–543 (1980).
Weiss, H. J.: "Platelet Physiology and Abnormalities of Platelet Functions," in two issues of *N. Engl. J. Med.*, **293**, 531–541; and 580–588 (1975).
Zucker, M. B.: "The Functioning of Blood Platelets," *Sci. Amer.*, **242**, 6, 86–103 (1980).

BLOOD CLOT. Arterial and Venous Disorders; Blood.

BLOOD (Fishes). Fishes.

BLOOD-BRAIN BARRIER. Many blood-borne solutes do not penetrate into central nervous tissue as rapidly as they penetrate into most other tissues. First discovered by P. Ehrlich in 1885, certain aniline dyes, when injected into the bloodstream of mice, stained most tissues of the body rapidly, but left the nervous system largely uncolored. During the ensuing half-century, the slow permeation of the brain by dyes and other histologically identifiable substances (e.g., ferricyanide and silver) was studied intensively. When these materials were placed in the cerebrospinal fluid, they entered the brain without restriction by passive diffusion through the pial surface. These observations gave rise to the erroneous concept that all metabolic exchange between blood and brain occurred via the cerebrospinal fluid. It is recognized, of course, that metabolite transfer actually occurs throughout the central nervous system vasculature, but is subject to local controlling mechanisms not found in other tissues.

Radioisotopes enabled the study of rates of exchange for many physiologically significant substances and, with few exceptions, the exchange of blood-borne solutes with the central nervous system has been found to be significantly, often orders of magnitude, slower than with other tissues. Certain metabolites and metabolic products such as glucose, oxygen, and carbon dioxide, as well as lipoid soluble substances and water itself, move rapidly between the blood and extravascular fluids of the central nervous system, but inorganic ions and most other highly dissociated compounds are very slow to equilibrate.

In attempting to evolve a general theory, the most persistent approach has been the attempt to discover physicochemical properties of molecules which determine these rates of migration. This led variously to explanations based upon electric charge, molecular size, dissociation constant, protein binding, lipoid solubility, and combinations of these. Selected series of compounds can be found which behave quite predictably according to one or more of these criteria. There is considerable similarity between blood-brain barrier permeability and cell membrane permeability, and it appears that solutes, to pass from the plasma to the extravascular fluids of the central nervous system, must for the most part pass through and not between cells.

The functional significance of the blood-brain barrier mechanism is to buffer the neuronal microenvironment against changes in plasma concentrations of various important solutes and to regulate the composition of the neuronal "atmosphere" for optimum performance. See also **Brain and Nervous System.**

BLOOD PRESSURE. The force exerted against the walls of the blood vessels by the circulating blood. Blood pressure within the arteries can be determined by using a device consisting of an elastic band around the arm, an air pump, and a column of mercury in a glass tube (manometer). The patient's age, his activity, the composition of blood, the secretion from the adrenal glands, and the thickness of the walls of the blood vessels all bear upon blood pressure.

Blood passing from the heart through the lungs has only about one-sixth of that pressure found when the blood is forced out over the body through the *aorta.* But, the pressure is sufficient to assure flow through the multitude of capillaries in the walls of the lungs. The lungs are composed of innumerable small sacs which have a supply of changing air. In the lung or pulmonary capillaries, the blood releases carbon dioxide and takes on oxygen.

The maximum pressure in the arteries is related to the contraction of the left ventricle of the heart, and is referred to as the *systolic pressure.* The minimum pressure, which exists just before the heartbeat which follows, is the *diastolic pressure.* The pressure of the blood in the smaller arterioles and in the capillaries is much less than in the arteries.

A number of factors must work together to maintain the blood pressure within normal limits. The pumping action of the heart itself is of major importance, as is the competency of the heart valves in closing so that no leakage occurs back from the arteries into the heart chambers. The elasticity of the arterial walls also influences the pressure. The resistance that the blood meets in the smaller blood vessels causes considerable variation. The amount of blood in the circulatory system and its viscosity also are factors. When any of these variables change markedly, the blood pressure may be increased or decreased. These pressure changes, in turn, may produce abnormalities in the structure and function of the heart and blood vessels. The most common variation in the blood pressure is an increase in its magnitude, which is referred to as hypertension, or high blood pressure. See also **Hypertension (High Blood Pressure); Hypotension;** and **Heart and Circulatory System (Human).**

BLOOD RECYCLING. Blood.

BLOODSTONE. A massive variety of quartz of greenish color with small spots of red jasper somewhat resembling blood drops. It is used as a semi-precious stone. When placed in water in full sunlight bloodstone will frequently give a general reddish reflection, hence the term heliotrope, derived from the Greek words meaning sun and to turn. See also **Chalcedony; Quartz.** Bloodstone also is known as *heliotrope.*

BLOOD SUGAR LEVEL. Diabetes Mellitus.

BLOOD WORM. 1. *Annelida.* certain marine worms whose bright red blood gives color to the entire body. 2. *Insecta,* Diptera. The aquatic larvae of certain midges which have hemoglobin dissolved in the plasma of the blood and so are red in color.

BLOOM. In surface-coating technology, bloom is a whitish, filmy layer which appears on films of paints, varnishes, or lacquers due to contamination from the atmosphere. The term is also applied to a filmy layer deposited on a photographic plate by tap water, which can be removed by rubbing the plate with wet cotton. The term bloom is used in metallurgy to denote a mass of malleable iron from which the slag has been removed. See also **Iron Metals, Alloys, and Steels.**

BLOOMING (CRT). Cathode-Ray Tube.

BLOOMING MILL. Iron Metals, Alloys, and Steels.

BLOWDOWN. Feedwater (Boiler).

BLOWER. A type of centrifugal air compressor, used to compress air or a gas by centrifugal force to a final pressure between 1 and 35 pounds per square inch gage. If the final pressure is below about 1 pound per square inch, the machine is known as a fan, while for pressures developed by centrifugal force above 35 pounds per square inch (2.4 atmospheres), the machine becomes a centrifugal compressor. Blowers driven at high rotative speeds (usually by steam or gas turbines) are usually called turboblowers.

BLOWER (Burner). Burner.

BLOW-FLY (*Insecta, Diptera*). Flies which deposit their eggs on meat. The name is applied to an entire family, however, containing other species which breed in dung, in wounds on living animals, and as blood-sucking parasites of nestling birds. The commoner species are also known as bluebottle flies. Family *Calliphoridae.*

The blow-fly has a black head and thorax with a steel-blue color abdomen. The eggs are long and cylindrical in shape and are deposited in stacks, many at a time. The larvae hatch in 24 hours and the insect is fully grown within 5 to 6 days. The blow-fly can spread disease by depositing infectious microorganisms on food.

BLOWHOLE. A nearly vertical hole, fissure, or natural chimney in coastal rocks, leading from the inner end of the roof of a sea cave to the ground surface above, through which incoming waves and the rising tide forcibly compress the air to rush upward or spray water to spout intermittently, often with a noise resembling a geyser outburst. It is probably formed by wave erosion concentrated along planes of weakness, as in a well-jointed rock. (*Glossary of Geology,* American Geological Institute).

BLOWING (Glass). Glass.

BLOWPIPE. A curved tube of metal or earthenware through which air may be blown and used to deflect the flame of a lamp or gas burner in such a way that an object may be heated either in the reducing or oxidizing flame.

BLOWPIPE ANALYSIS. A systematic plan of qualitative analysis based upon the use of the blowpipe, and of various bead tests, flame tests, and fusions, with or without fluxing agents, on charcoal, tile, plaster of Paris, and other surfaces. These methods have been used in the assay of ores and other metal-containing materials. One method of differentiation is by examination of the beads and other fusion products under ultraviolent light. See also **Bead Test.**

BLUE (*Insecta, Lepidoptera*). Butterflies whose prevailing color is bright blue. The famales are usually less blue than the males and some few species are not blue. With the coppers and hair-streaks they constitute the family *Lycaenidae.*

These insects are of European origin, but are now found almost worldwide. They are most abundant in the tropics. They are quite small and fragile. Their caterpillars are shaped like a sowbug, short, fat, and sometimes appear like a slug. Ants are known to defend them. The caterpillars often live in ants' nests, to which they are transported by the ants. A welcome exchange of food is the result; the ants like the caterpillar's honeydew; the caterpillars like ant larvae.

There are several hundred species of Blues. However, all species are not blue in color. Some are orange with blue spots. Among the females, many are brown or white and brown. The smallest is the pigmy Blue (*Brephidium exilis*). From wing tip to wing tip, this butterfly is only a little more than one-quarter inch in spread. The pigmy Blue exists in both North and South America.

BLUE ASBESTOS. Crocidolite.

BLUE ASH. Ash Trees.

BLUE BABY. Heart and Circulatory System (Human).

BLUEBERRY ELDER. Elder Trees and Viburnums.

BLUEBIRD (*Aves, Passeriformes*). A term variously applied to different species of birds. In North America, the term usually signifies the eastern bluebird (*Sialia sialis*) or the western mountain bluebird (*Sialia currucoides*). The term is also applied to one of the babblers of the Orient, as well as to a South African albatross (order *Procellariformes*).

The eastern bluebird is found from Newfoundland southward to Florida and the Gulf of Mexico and westward to Manitoba and the midwestern United States, notably in the Ohio Valley Region. In New England, it is usually found in the coastal areas. It is not frequently found in the immediate vicinity of the Great Lakes. The bird is a bit larger than a sparrow. It is the only blue-covered bird with a red breast. The females tend to be of paler, duller coloration. The bird feeds mostly on insects.

The male mountain bluebird is entirely blue and is about $5\frac{1}{2}$ inches (14 centimeters) in length. Although it has been observed up to 12,000 feet (3660 meters), it usually is not found at elevations exceeding 5000 feet (1525 meters). The bird has a soft, sweet, whistlelike song at dawn.

The Florida bluebird (*S. s. grata*) ranges over the southern half of that state.

BLUEBOTTLE (*Insecta, Diptera*). Large flies of shining blue, green, or purple color. They lay their eggs on meat and other foods and so are often seen in dwellings.

BLUEFIN TUNA. Tunas.

BLUEGILL (*Osteichthyes*). A fish (*Helioperca incisor*) related to the sunfishes and bass. See also **Bass (Osteichthyes).** Widely distributed east of the Rocky Mountains in lakes and the quieter parts of streams, the bluegill is esteemed as a pan fish. The fish attains a length of 10 inches (25 centimeters) or more, but in well-fished waters, rarely reaches this size. Although small, the fish readily rises to a fly and thus ranks among desirable game fishes.

BLUE GLOW. A type of luminescence emitted by certain metallic oxides, when heated. A blue glow is normally seen in electron tubes containing mercury vapor, arising from the ionization of the molecules in the mercury vapor. When a blue glow is observed near the electrodes of a conventional vacuum tube, this usually indicates that the tube contains gas and is defective. However, for some vacuum tube designs, a soft blue fluorescent glow may be normal.

BLUEGRASS. Grasses.

BLUE JAY. Jay.

BLUE SPRUCE. Spruce Trees.

BLUESTEMS. Grasses.

BLUETHROAT (*Aves, Passeriformes*). One species, *Luscinia* (*Cyanosylvia*) *svecica* is a European bird related to the warblers. It occurs sparsely in central Europe and is also found occasionally in North America, including Alaska. Its preferred habitat is the marsh, although it is found along streams and fresh-water lakes. A shy bird, it prefers the seclusion of reeds and willows along streams.

The bluethroat nests on the ground. The European species winter in Africa and southern Asia. The female and young birds have a pale blue plumage about the throat area. The feathers are black tipped. The tail is rusty-brown and is usually kept spread. The male is dark blue under its beak and over its breast, with a red spot over the beak. The feathers underneath are white with a narrow white streak running through the rusty brown plumage on its back. It is a warbler, with a gentle high-pitched tone. The bluethroat measures about 5 inches (13 centimeters) in length.

BLUFF BODY. An object immersed in fluid stream flow is said to be bluff (or blunt) if its shape promotes a rapidly increasing downstream pressure gradient in the streamline flow around it. A high adverse gradient assists the creation of a stagnation point. The streamline flow breaks loose from the surface of the body on either side, leaving a turbulent low-pressure wake. This wake causes the characteristically high drag of bluff bodies.

BLUSHING. A term applied to a surface opacity or turbidity of varnish and lacquer films. The cause of this defect is commonly rapid evaporation of solvent, or improper formulation of the product.

BOA CONSTRICTOR. Snakes.

BOAR. Suines.

BOBCAT. Cats.

BOBOLINK (*Aves, Passeriformes*). A widely distributed North American bird *Dolichonyx oryzivorus*, of which the male, marked with black, white, and yellowish, is conspicuous in the prairies and meadows where the species breeds. The female is duller and plainer. The bobolink is noted for its cheerful song. See illustration.

Bobolink

BOB-WHITE. Quail.

BODE PLOT (Frequency Response). Frequency Response.

BODE'S RELATION. In the latter part of the eighteenth century, an empirical relationship was noticed between the mean distances of the various planets from the sun. This relationship was first published by Bode, in 1772, and has since become known as Bode's relation in spite of the fact that there is certain evidence that it was known and used by Titus a number of years previous to the time of its announcement.

Bode's relation may be stated as follows: write down a series of 4's; to the first, add 0; to the second, add 3; to the third, add 6 = 3 × 2; to the fourth, 12 = 6 × 2; to the fifth, 24 = 12 × 2, etc; the resulting numbers divided by 10 will give the approximate mean distances of the planets from the sun in astronomical untis. The sequence is as follows:

Planet	Bode Distance	Mean Distance
Mercury	4 + 0 = 4	0.39
Venus	4 + 3 = 7	0.72
Earth	4 + 6 = 10	1.00
Mars	4 + 12 = 16	1.52
	4 + 24 = 28	
Jupiter	4 + 48 = 52	5.20
Saturn	4 + 96 = 100	9.54
Uranus	4 + 192 = 196	19.18
Neptune	4 + 384 = 388	30.06
Pluto	4 + 768 = 772	39.4

The value in the last column is the actual mean distance of the planet from the sun in astronomical units. Thus Bode's relation has the form

$$D = A + BC^n$$

where $A = 0.4$, $B = 0.3$, $C = 2$, and $n = -\infty, 0, 1, 2, 3, \ldots$ and yields the distance in astronomical units.

At the time that the relation was first proposed, the gap between Mars and Jupiter was not filled and no planets were known outside of Saturn. The relation predicted distances, and when Uranus was discovered with mean distance so close to the predicted value, Bode's relation was believed to be established. The discovery of the asteroid Ceres, with a mean distance of 2.77, gave further support to the validity of the relation. It is interesting to note that, in making the computations which led to the discovery of Neptune, Adams used the predicted Bode distance for the then unknown object.

During the nineteenth century, many unsuccessful attempts were made to place Bode's relation upon a theoretical foundation. The failure of the law in the cases of Neptune and Pluto has convinced most astronomers that the relation is a purely empirical relationship, more in the realm of coincidence than an actual physical law.

BODY-CENTERED STRUCTURE. A type of crystal structure in which atoms are located at the corners and center of a cubic or rectangular cell.

BODYING AND BULKING AGENTS (Foods). These terms tend to be self-defining. Additives in these classifications are frequently described together because many substances will serve one or both purposes.

Bodying Agents. The *body* of a food substance is generally associated with the textural qualities of the substance, notably with mouthfeel or chewiness. Some food products, particularly those of a fabricated nature, may possess a full complement of desirable consumer appeals (taste, odor, color, nutritive value, etc.) and yet lack the desirable textural quality of body. Thus, soups, gravies, sauces, cheese foods and spreads, dressings, snack dips, and margarines, among others, can be improved through the addition of bodying agents. For example, formulations for frozen desserts can be improved in this respect by the addition of low levels of a material such as microcrystalline cellulose (about 0.25% weight), in combination with soluble hydrocolloids, such as guar, locust bean gum, alginates, or carrageenans. See also **Gums and Mucilages.** Microcrystalline cellulose achieves about the same degree of body and substance in frozen desserts that is normally achieved only in well-emulsified products with a 2–4% higher fat content. This is the result of the ability of microcrystalline cellulose to stabilize the serum solid. Microcrystalline cellulose imparts body and smoothness to ice cream and ice milk, and tends to make them less "cold tasting." There are no off-flavors associated with the substance and frozen desserts melt to smooth, creamy consistencies. Bodying agents play an effective, if not exclusive role, in improving freeze-thaw properties of numerous products. In another example, when whey or sugar solids are used to reduce or replace portions of the milk solids nonfat (MSNF), there is a definite loss of functionality of the mix, resulting in reduced body and texture. Problems, such as stickiness, gumminess, and weak body can be corrected by the addition of a bodying agent, such as microcrystalline cellulose, in a very small amounts (0.25–0.4% weight).

Bulking Agents. These substances are added to semiliquid and solid food products to add bulk to the end product over and beyond the bulk resulting from the strict use of conventional ingredients. For example, when added as an ingredient of baked foods, microcrystalline cellulose accomplishes two functions—weight is added, thus reducing the effective caloric content of a given portion; and water is tied up so that considerably more liquid can be incorporated into the formulation. However, it should be stressed that microcrystalline cellulose is only a partial substitute for fat, which is needed for air entrapment, or flour which provides the elastic gluten structure. As an additional advantage, the cellulose also increases the fiber content of the product.

In the currently very important field of manufacturing low-calorie foods, a bulking agent essentially can be considered as a diluent even though it may play other important roles. Thus, the diet-conscious consumer can eat cookies, doughnuts, or portions of cake of traditional size and yet consume considerably fewer calories. The important factor in selecting a bulking agent for low-calorie foods is that of finding a substance that combines noncaloric qualities with other functional capabilities so that lower amounts of relatively high-calorie ingredients can be reduced or replaced without detracting drastically from the consumer appeals of the finished product.

In addition to microcrystalline cellulose, there are several other bodying and bulking compounds used. These include glycerin, methylcellulose, polyvinylpyrrolidone (PVP), sodium carboxymethylcellulose, whey solids, and xanthan gum.

References

Shama, F., and P. Sherman: "The Texture of Ice Cream: Rheological Properties of Frozen Ice Cream," *J. Food Sci.*, **31**, 699 (1966).

Shama, F., and P. Sherman: "Identification of Stimuli Controlling the Sensory Evaluation of Viscosity: Oral Methods," *J. Texture Studies*, **4**, 111 (1973).

Staff: "Food Chemicals Codex," National Academy of Sciences, Washington, D.C. (revised periodically).

Staff: "Microcrystalline Cellulose in Low Calorie Foods," FMC Corporation, Philadelphia (revised periodically).

Voisey, P. W.: "Rheology and Texture in Food Quality," AVI, Westport, Connecticut, 1976.

BOG. A waterlogged, spongy groundmass, primarily mosses, containing acidic, decaying vegetation, which may develop into peat. Also, the vegetation characteristic of this environment is, especially sphagnum, sledges, and heaths. A synonym is *peat bog*. See also **Muskeg.**

BOG LAKE. A relatively small body of open water surrounded or nearly surrounded by bogs and characterized by a false bottom of organic (peaty) material, high acidity, scarcity of aquatic fauna, and vegetation growing on a firm deposit or on a semifloating mat of peat.

BOG MANGANESE. Wad.

BOHR LIQUID DROP MODEL. Binding Energy.

BOHR MAGNETON. A unit of magnetic moment used in atomic physics, defined as

$$\mu_B = eh/4\pi m_e c = 9.27 \times 10^{-24} \text{ ampere-meter}^2$$

in which e is the electronic charge in coulombs, h is Planck's constant, and m_e is the rest mass of the electron. If the angular momentum of an orbiting electron in an atom is $L = l^* h/2\pi$, the magnetic moment established by the orbital motion of the electron is $\mu_{l^*} = l^* \mu_B$. The measured magnitude μ_s of the spin magnetic moment of an electron is, however, $2s\mu_B$, such that its gyromagnetic ratio is twice that for the orbital motion. A dimensionless multiplicative factor, called a *g*-factor, is introduced in the relationship that expresses the measured magnetic moment in terms of Bohr magnetons such that the spin magnetic moment of an electron is $\mu_s = gs\mu_B$, in which $g = 2$. A comparable insertion of a *g*-factor in the formula for the orbital magnetic moment of an electron leads to $g = 1$. See **Nuclear Magneton.**

BOHR THEORY OF ATOMIC SPECTRA. Bohr based his theory of atomic spectra upon two postulates: Postulate 1. "An atomic system can, and can only, exist permanently in a certain series of states corresponding to a discontinuous series of values for its energy, and hence any change in the energy of the system, including emission and absorption of electromagnetic radiation, must take place by a complete transition between two such states. These states will be called the stationary states of the system." Postulate 2. "That the radiation absorbed or emitted during a transition between two stationary states is monochromatic and possesses a frequency ν, given by the relation $h\nu = E_2 - E_1$," where h is the Planck constant and E_1, E_2 are energies of the two stationary states. See also **Atomic Spectrum; Quantum Mechanics; Quantum Theory.**

BOHR-WHEELER THEORY. This analytical theory of the nature of nuclear fission and the conditions under which it might occur is based upon a liquid-drop analogy. One conclusion was that if the ratio Z^2/A exceeded 45, no additional energy would be necessary to cause fission to take place, (Z = atomic number; A = mass number.) An element having a Z^2/A ratio of 45 would have an atomic number well over 100, so that elements of atomic number appreciably higher than 100 are not stable.

BOIDS. Snakes.

BOILER. In a modern steam generator, various components are arranged to absorb heat efficiently from the products of combustion. These components are generally described as boiler, superheater, reheater, economizer, and air heater.*

* Basic information for this entry from "Steam—Its Generation and Use" (39th edition), copyright The Babcock & Wilcox Co., New York, 1978.

Boiler. Boiler surface may be defined as those parts of tubes, drums and shells which are part of the boiler circulatory system and which are in contact with the hot gases on one side and water or a mixture of water and steam on the other side. Although the term boiler may refer to the overall steam generating unit, the term "boiler surface" does not include the economizer or any component other than the boiler itself. Boilers may be broadly classified as shell, fire-tube and water-tube types.

Modern boilers are of the water-tube type. The safety and dependability of operation that characterize the boilers of today had their beginning in the introduction of this boiler type. In the water-tube boiler, the water and steam are inside the tubes, and the hot gases are in contact with the outer tube surfaces. The boiler is constructed of a number of sections of tubes, headers and drums joined together in such a way that circulation of water is provided for adequate cooling of all parts, and the large indeterminate stresses of the fire-tube boilers are eliminated. With water-tube designs it is possible to protect thick drums from the hot gases and resultant high thermal stresses. With correct operation, explosive failures have been essentially eliminated with water-tube boilers. Further the water space is divided into sections so arranged that, should any section fail, no general explosion occurs and the destructive effects are limited.

The water-tube construction facilitates obtaining greater boiler capacity, and the use of higher pressure. In addition, the water-tube boiler offers greater versatility in arrangement and this permits the most efficient use of the furnace, superheater, reheater, and other heat recovery components.

Water-tube boilers may be classified as straight-tube and bent-tube. Straight-tube boilers have been supplanted by modern designs of bent-tube boilers, which are more economical and serviceable than the straight-tube designs.

The majority of fossil-fuel steam generators and all commercial nuclear steam supply systems operate at subcritical pressures. A comprehension of the boiling process is essential in the design of these units. See also entry on **Boiling.**

Factors Affecting DNB. The point of departure from nucleate boiling (DNB) is defined in entry on **Boiling.** The effect of steam quality and heat flux on the location of the DNB point are demonstrated by the figure in the entry on **Boiling.** That figure illustrates the various heat transfer regimes taking place along the length of a uniformly heated vertical tube cooled by water flowing upward. On this figure, the inner wall temperatures are plotted as functions of enthalpy and steam quality, starting with hot water, passing through the region where steam is being generated (0 to 100% quality), and finally into the superheated region.

By following the line for moderate heat flux it is seen that the metal temperature in the subcooled region is parallel to the water temperature, and only slightly above it. When boiling starts, the heat transfer coefficient increases and the metal temperature remains just above saturation temperature. Finally, at high steam quality, the DNB point is reached where the nucleate boiling process breaks down. The metal temperature increases at this point but decreases again as steam quality approaches 100%. In the superheat region the wall temperature again increases with, and approximately parallel to, the superheated steam temperature.

For the curve marked "high heat flux," the DNB point is reached at a lower steam quality, and the peak metal temperature is higher. At very high heat fluxes the DNB occurs at low steam quality and the metal temperature would be high enough to melt the tube if it were able to withstand the internal pressure without first plastically deforming and rupturing. At extremely high heat fluxes, DNB can occur in subcooled water. Avoidance of this last type of DNB is an important criterion in the design of nuclear reactors of the pressurized-water type.

Figure 1 presents the DNB phenomenon from the standpoint of a heated flow channel in which flow, pressure, and inlet temperature (inlet subcooling) remain constant. DNB is also affected by variations in mass velocity, pressure, subcooling and channel dimensions.

Many fossil-fuel boilers are designed to operate in the range between 2000 psi (136 atmospheres) and the critical pressure. In this range, pressure has an important effect, shown in Fig. 2, in that the steam quality limit for nucleate boiling falls rapidly near the critical pressure,

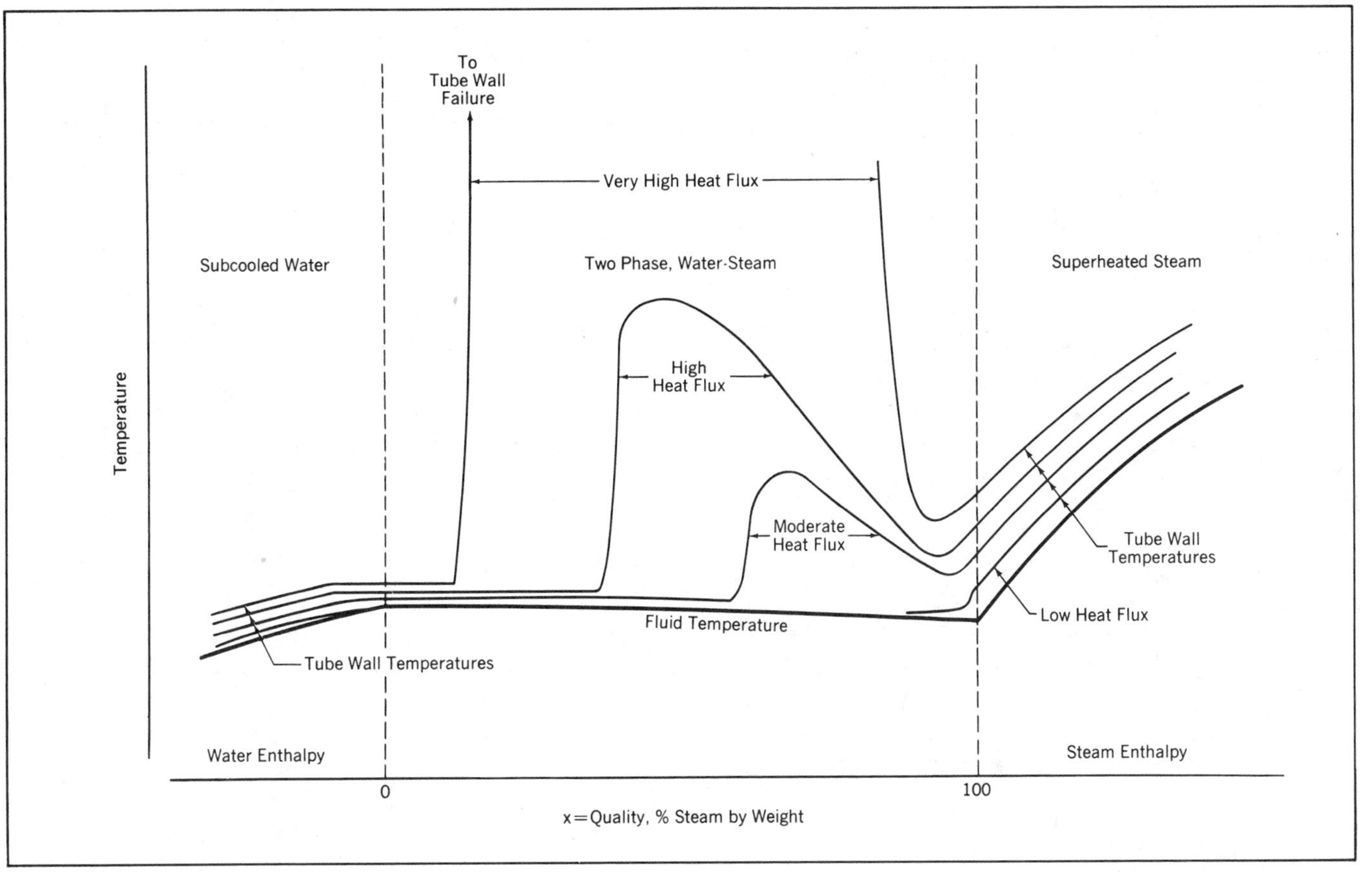

Fig. 1. Fluid and tube wall temperatures under conditions of water heating, nucleate boiling, film boiling, and superheating steam. ("*Steam—Its Generation and Use,*" *The Babcock & Wilcox Co.*)

i.e., at constant heat flux the DNB point occurs at a decreasingly low steam quality as pressure rises. Many correlations of critical heat flux or DNB have been proposed, and are satisfactory within certain limits of pressure, mass velocity and heat flux. Figure 3 is an example of a correlation which is useful in the design of fossil-fuel natural-circulation boilers. This correlation defines safe and unsafe regimes for two heat flux levels at a given pressure in terms of steam quality and mass velocity. Additional factors must be introduced when tubes are used in membrane or tangent walls or in any position other than vertical. Such factors include inside diameter of tubes and surface condition. The last of these, where the character of the inside tube surface is purposely altered, will be discussed further in the section on "Ribbed tubes."

The preceding discussion applies only to subcritical pressures. As the operating pressure is increased the various flow and boiling regimes gradually disappear. However, there are tube metal temperature excursions in low-velocity supercritical-pressure operation similar to those found in subcritical boiling. This phenomenon, known as pseudo-film boiling, is currently under intensive experimental investigation.

Ribbed Tubes. Since the 1930s, a large number of devices, including internal twisters, springs, and various grooved, ribbed, and corrugated tubes, to inhibit or delay the onset of DNB have been tried and tested.** The most satisfactory overall performance was obtained with tubes having helical ribs on the inside surface.

Steam Separation and Purity

Boilers operating below the critical point, except for once-through types, are customarily provided with a steam drum in which saturated steam is separated from the steam-water mixture discharged by the boiler tubes. Saturated steam leaves, and feedwater enters this drum through their respective nozzles (with some exceptions in multidrum boilers).

However, the primary functions of this drum are to provide a free controllable surface for separation of saturated steam from water and a housing for any mechanical separating devices. Steam drums are designed to provide the volume necessary, in combination with the controls and firing equipment, to prevent excessive rise of water into the steam separators, resulting in carry-over of water with the steam.

Solids in Boiler Water. Boiler water contains solid materials, mainly in solution. Steam contamination (solid particles in the superheated steam) comes from the boiler water, largely in the carry-over of water droplets. Therefore, in general, as boiler-water concentration increases, steam contamination may be expected to increase.

Historically the carryover of water into superheater tubes resulted in deposit of entrained solids in the superheater tubes. This caused increased tube temperatures and distortion and burnout of tubes. Therefore, it was necessary to develop devices to remove water from the steam. The need for extreme purity of steam for use in modern high-pressure turbines has provided additional incentive for reducing the carry-over of solids in steam. Troublesome deposits on turbine blades may occur with surprisingly low (0.6 ppm) total solids contamination in steam.

See **Feedwater (Boiler).**

Factors Affecting Steam Separation. Separation of steam from the mixture discharged into the drum from steam-water risers is related to both design and operating factors, some of which include:

Design factors

1. Design pressure
2. Drum size, length and diameter
3. Rate of steam generation
4. Circulation ratios—water circulated to heated tubes divided by steam generated
5. Type of arrangement of mechanical separators
6. Feedwater supply and steam discharge equipment and arrangement
7. Arrangement of downcomer and riser circuits in the steam drum

Operating factors

1. Operating pressure
2. Boiler load (steam flow)

** By the Babcock & Wilcox Company.

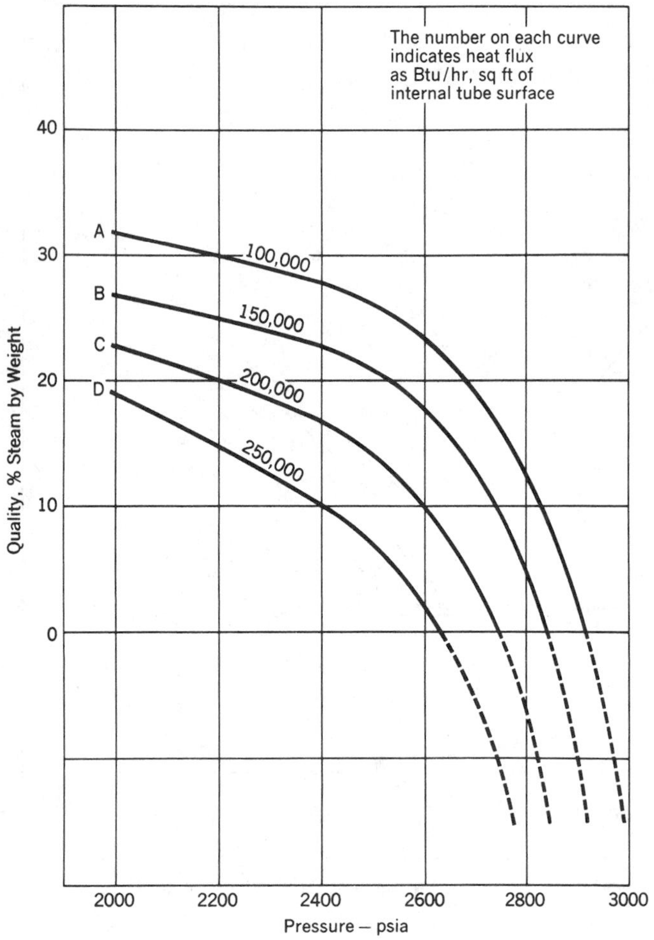

Fig. 2. Steam quality limit for nucleate boiling as a function of pressure. ("*Steam—Its Generation and Use*," *The Babcock & Wilcox Co.*)

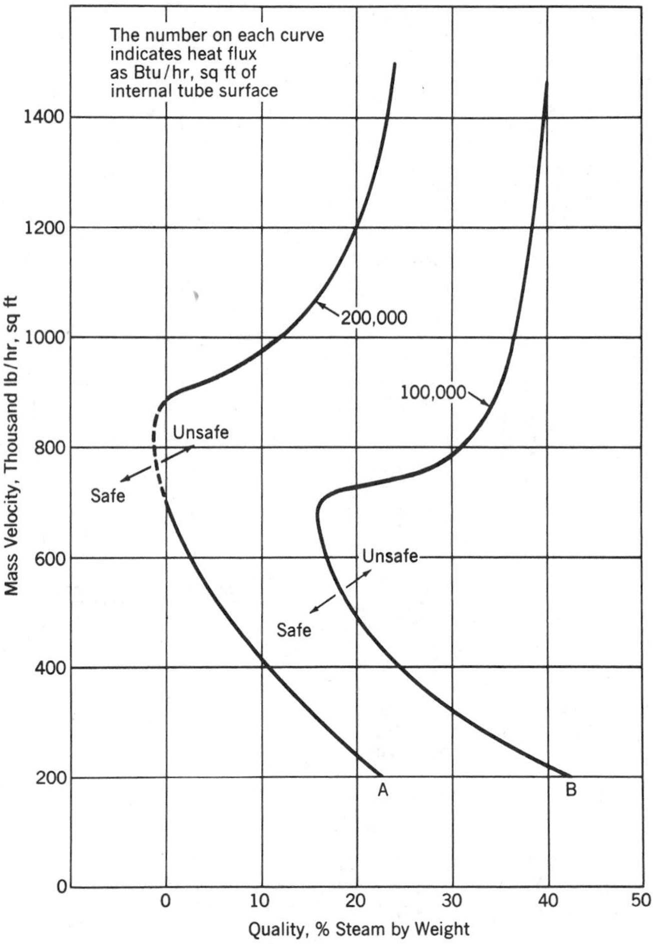

Fig. 3. Steam quality limit for nucleate boiling at 2,700 psia, as a function of mass velocity. (*The Babcock & Wilcox Co.*)

3. Type of steam load
4. Chemical analysis of boiler water
5. Water level carried

In steam drums without separation devices, where separation is by gravity only, the manner in which some of the above items affect separation is indicated in simplified form in Figs. 4 and 5.

Mechanical Steam Separators for Drums. Gravity steam separation alone is generally unsatisfactory for boilers of the usual sizes and operating requirements. Most steam drums, therefore, are fitted with some form of primary separator. Simple types of primary separators are shown in Fig. 6. These devices facilitate or supplement gravity separation. The extent and arrangement of the various baffles and deflectors should always allow for access to the drums.

In a cyclone steam separator (Babcock & Wilcox), centrifugal force many times the force of gravity is used to separate the steam from the water. Cyclones, essentially cylindrical in form, and corrugated scrubbers are the basic components of this type of separator.

The cyclones are arranged internally along the length of the drum, and the steam-water mixture is admitted tangentially. The water forms a layer against the cylinder walls, and the steam (of less density) moves to the core of the cylinder and then upward. The water flows downward in the cylinder and is discharged through an annulus at the bottom, below the drum water level. Thus, with the water returning from drum storage to the downcomers virtually free of steam bubbles, maximum net head is available for producing flow in the circuits, which is the important factor in the successful use of natural circulation. The steam moving upward from the cylinder passes through a small primary corrugated scrubber at the top of the cyclone for additional separation. Under many conditions of operation no further refinement in separation is required, although the cyclone separator is considered only as a primary separator.

When wide load fluctuations and variations in water analyses are expected, large corrugated secondary scrubbers may be installed at the top of the drum to provide nearly perfect steam separation. These scrubbers may be termed secondary separators.

The combination of cyclone separators and scrubbers described above provides the means of obtaining steam purity corresponding to less than 1.0 ppm solids content under a wire variation of operating conditions. This purity is generally adequate in commercial practice. However, further refinement in steam purification is required where it is necessary to remove boiler-water salts, such as silica, which are entrained in the steam by vaporization or solution mechanism. Washing the steam with condensate or feedwater of acceptable purity may be used for this purpose.

Steam Washing. It is often impractical to maintain boiler-water

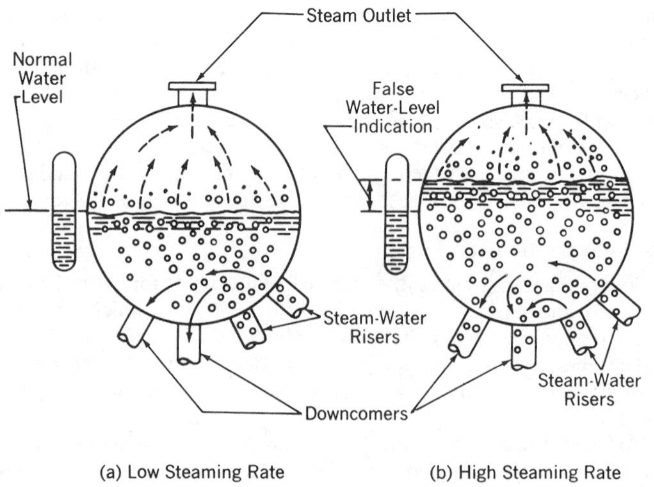

Fig. 4. Effect of rate of steam generation on steam separation in a boiler drum without separation devices. (*The Babcock & Wilcox Co.*)

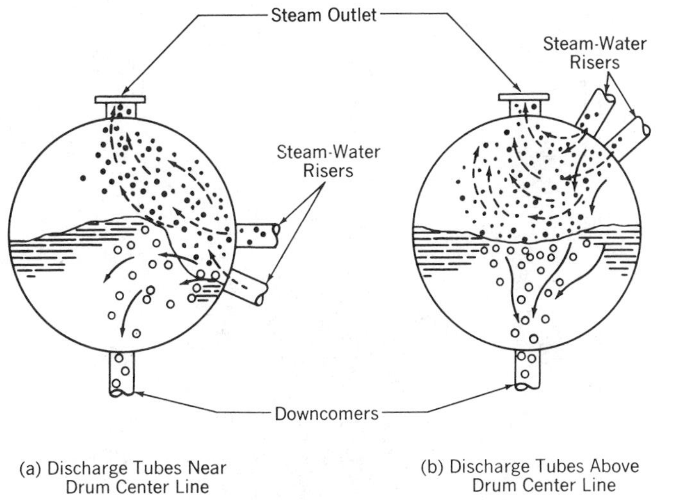

Fig. 5. Effect of location of discharge from risers on steam separation in a boiler drum without separation devices. (*The Babcock & Wilcox Co.*)

concentrations of silica sufficiently low to prevent turbine fouling, and other measures such as stem-washing are used to control this type of steam contamination.

In steam-washing, silica-laden steam is brought into intimate contact with relatively pure wash water, such as condensate or feedwater, and silica is absorbed from the steam by the fresh water.

A steam-drum arrangement employing steam-washing is shown in Fig. 7. The drum is equipped with primary mechanical separators of the centrifugal type and corrugated scrubbers. Steam leaving the primary separators flows to a steam washer arranged in the top of the steam drum. The washer consists of a rectangular column approximately the length of the steam drum. Steam passes vertically upward through a perforated plate, a pack of stainless steel wire mesh, a second perforated plate, and finally a corrugated scrubber element. Wash water enters the drum through a nozzle and flows downward through the washer, counterflow to the steam. The steam velocity through the tray perforations maintains, above each tray, a layer of wash water which is kept in violent agitation by the steam. The wire mesh provides a large surface area for achieving intimate contact between the steam and the wash water.

Bent-Tube Boilers

Many important modern designs of boilers, such as the two-drum Stirling, the Integral-Furnace, the Radiant, and the Universal Pressure

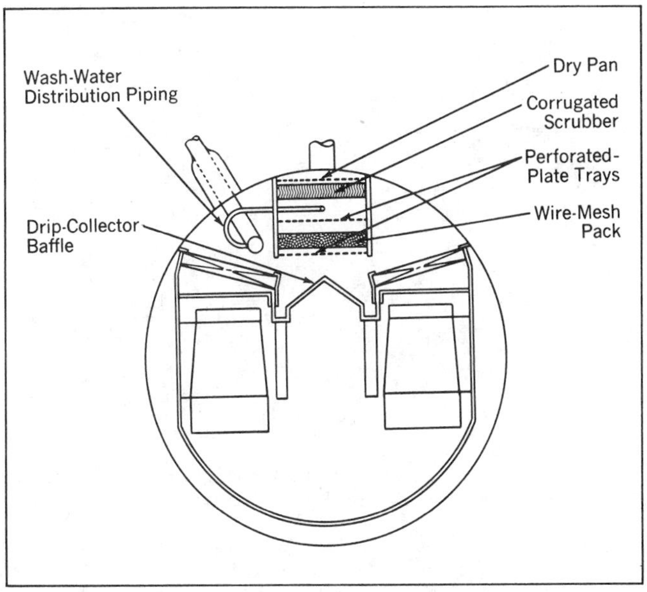

Fig. 7. Arrangement of steam-drum internals for washing silica-laden steam. (*The Babcock & Wilcox Co.*)

are included in the "bent-tube" classification. All bent-tube boilers of contemporary design, with the exception of those with stoker or flat refractory floors, have water-cooled walls and floors or hoppers. The principal types are illustrated in Figs. 8 through 14.

Integral-Furnace Boiler. This boiler is a two-drum boiler which, in the smaller capacities is adaptable to shop assembly and shipment as a package. Figure 8 shows a low-capacity Type FM Integral-Furnace boiler designed for shop assembly. This package boiler is shipped complete with support steel, casing, forced-draft fan (unmounted in larger sizes), firing equipment, and controls—ready for operation when water, fuel, and electrical connections are made. Only a stub stack is required. It is built for outputs from 8000 to 160,000 pounds (3629 to 72,576 kilograms) of steam per hour. Steam pressures range to 925 psi (63 atmospheres) and temperatures to 441°C. Units can be fired with oil, gas, or a combination of the two. Only a forced-draft fan is required, as the casing is airtight (welded) and the combustion gases are under pressure. Size of the unit is varied principally by changes in setting depth and drum length. Two combinations of width and height facilitate standardization of parts and assembly. An Integral-Furnace boiler after installation is shown in Fig. 9.

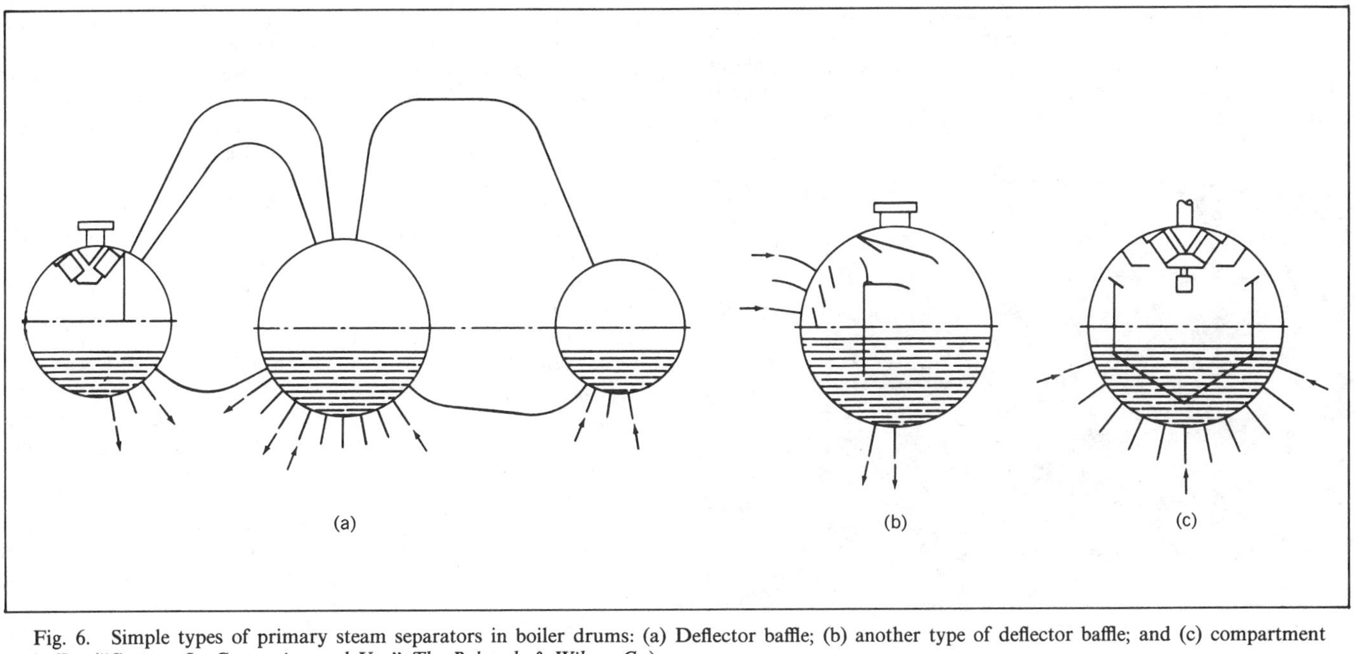

Fig. 6. Simple types of primary steam separators in boiler drums: (a) Deflector baffle; (b) another type of deflector baffle; and (c) compartment baffle. ("*Steam—Its Generation and Use,*" *The Babcock & Wilcox Co.*)

Fig. 8. Integral-furnace boiler (Type FM). Shop-assembled unit, complete and ready to operate. (*The Babcock & Wilcox Co.*)

Two-Drum Stirling Boiler. The simple arrangement possible for the connecting tubes, with one upper steam drum directly over one lower drum, led to the development of a series of designs known as the two-drum Stirling boiler. Figure 10 shows a unit of this type for cyclone-furnace firing. These designs are standardized over a wide range of capacities and pressures, with steam flows ranging from 200,000 to 1,200,000 pounds (90,720 to 544,320 kilograms) per hour, design pressures up to 1750 psi (119 atmospheres), and steam temperatures up to 538°C. The firing may be by cyclone furnace, pulverized coal, oil or gas.

The two-drum Stirling boiler is furnished for industrial and utility applications. It may be considered as a transition unit, covering an intermediate size range. Because of its versatility and economy, the boiler enjoys worldwide acceptance.

High-Pressure and High-Temperature Boilers. In the rapid development of power-plant economy, the single-boiler, single-turbine combination has been adopted for the central station and where electric power is the end product of heat transformation. There is an incentive to use very large electrical generators, since the heat rate, investment, and labor costs decrease as size increases. In the design of large boiler units for this application, the important factors are (1) high steam pressure, (2) high steam temperature, (3) bleed feedwater heating, and (4) reheat. High steam pressure means high saturation temperature and low temperature difference between steam and exit gas. High steam temperature means high initial temperature and, usually, reheating to high temperature for reuse of the steam. Bleed feedwater heating lowers the mean temperature difference in an economizer and increases the gas temperature leaving the economizer. An air heater is then required to lower the exit-gas temperature. These factors and, above all, the economic need for continuity of operation to realize an optimum return on the large investment involved have combined to produce boiler units different in many respects from earlier concepts. Thus the principle of the integrated boiler unit is firmly established for very large boilers, as well as for boilers of smaller outputs.

Fig. 9. Representative installation of an Integral-Furnace boiler. (*The Babcock & Wilcox Co.*)

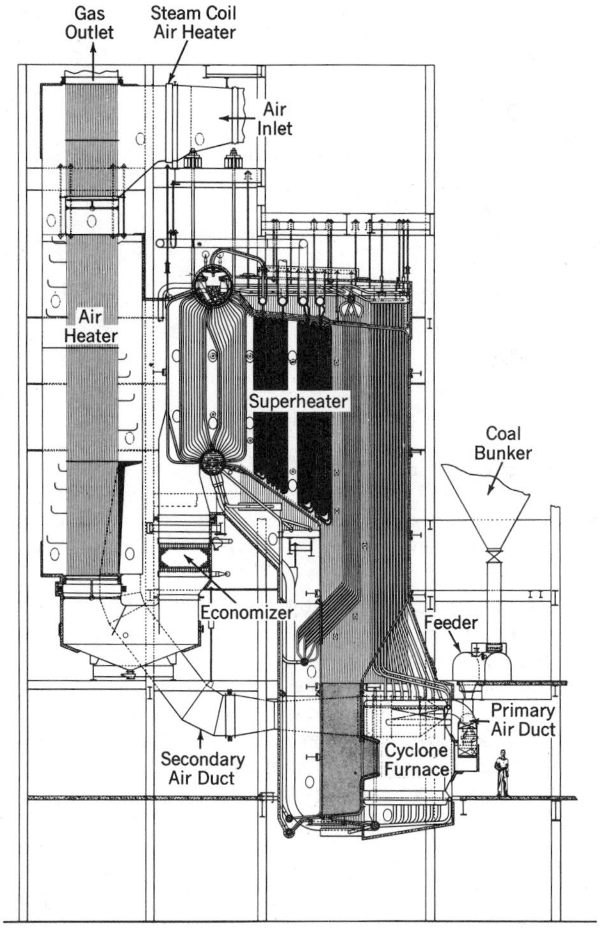

Fig. 10. Two-drum Stirling boiler for cyclone-furnace firing. (*The Babcock & Wilcox Co.*)

As steam pressures have increased, steam temperatures also have increased. This necessitates proportionally more superheating surface and less boiler surface. When pressures exceed 1500 psi (102 atmospheres) in a drum-type boiler, the heat absorbed in furnace and boiler-screen tubes is normally almost enough to generate the steam. Thus it is usually more economical to use economizer surface for any additional evaporation required as well as to raise the feed-water to saturation. All the steam is then generated in the furnace, water-cooled wall enclosures of superheater and economizer, boiler screen, division walls, and in some cases the outlet end of a steaming economizer as contrasted with only use of the boiler surfaces per se.

Radiant Boiler. This boiler, illustrated in Figs. 11 and 12, is a high-pressure, high-temperature, high-capacity boiler of the drum type. It is adaptable to pulverized-coal or cyclone-furnace firing, and also to natural gas and oil firing. Boiler convection surface is a minimum in these units.

The radiant boiler shown in Fig. 11 is a pulverized-coal-fired unit with hopper-bottom construction for dry-ash removal. It has an output of 1,750,000 pounds (793,800 kilograms) of steam per hour for continuous operation. Design pressure is 2875 psi (196 atmospheres); and primary and reheat steam temperatures are 538°C. Standard components (furnaces, superheaters, reheaters, economizers, and air heaters) are integrated to coordinate the fuel fired with the turbine throttle requirements. Standard sizes are available in reasonable increments of width and height to permit selection of economical units for the required steam conditions and capacity.

Figure 12 illustrates the El Paso-type radiant boiler, a standardized unit developed for natural gas and oil firing. This compact and economical design is suitable for these fuels because of the cleanliness of

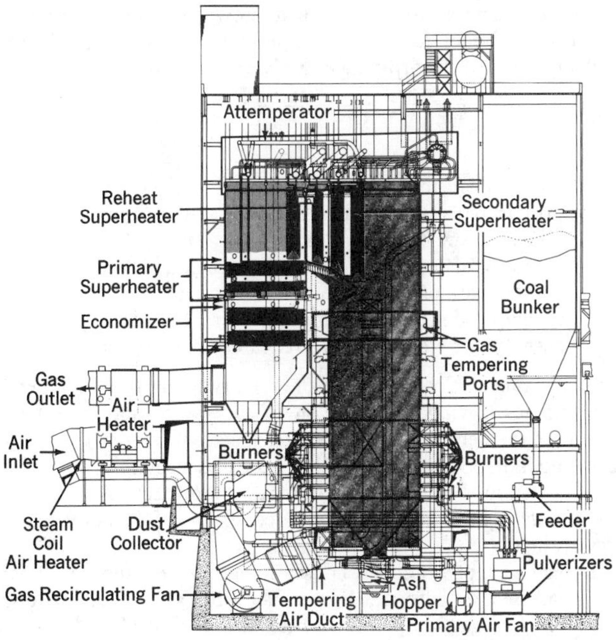

Fig. 11. Carolina-type radiant boiler for pulverized-coal firing. (*The Babcock & Wilcox Co.*)

natural gas and the relatively minor ash problems encountered with oil as compared with coal.

Universal-Pressure Boiler. This is a high-capacity, high-temperature boiler of the "once-through" or "Benson" type. See Figs. 13 and 14. Functionally applicable at any boiler pressure, it is applicable economically in the pressure range from 2000 to 4000 psi (136 to 272 atmospheres). Firing may be by coal, either pulverized or cyclone-furnace-fired, by natural gas or oil.

The working fluid is pumped into the unit as liquid, passes sequentially through all the pressure-part heating surfaces where it is converted to steam as it absorbs heat, and leaves as steam at the desired temperature. There is no recirculation of water within the unit and, for this reason, a drum is not required to separate water from steam.

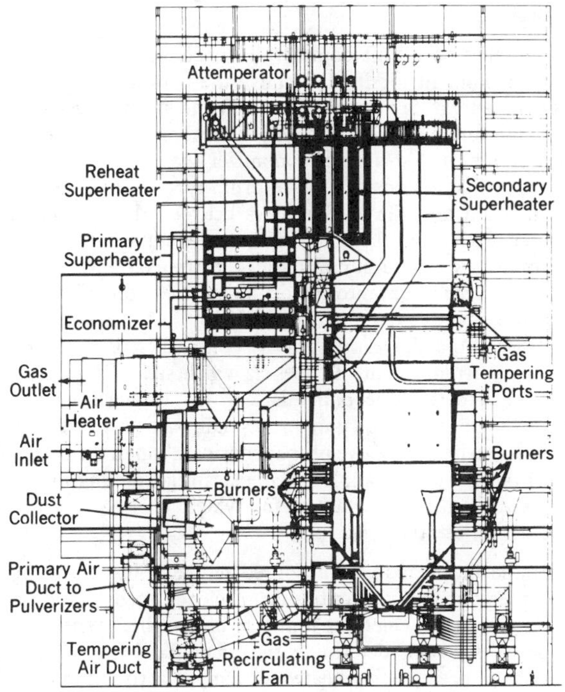

Fig. 13. Universal-Pressure boiler for pulverized-coal firing. (*The Babcock & Wilcox Co.*)

The Universal-Pressure boiler may be designed to operate at either subcritical or supercritical pressures. The size of the unit is virtually unlimited.

Figure 13 illustrates a pulverized-coal-fired Universal-Pressure boiler for a 1300 megawatt capacity electric generating unit. Figure 14 is a natural gas fired installation, supplying steam for a 750 megawatt generator.

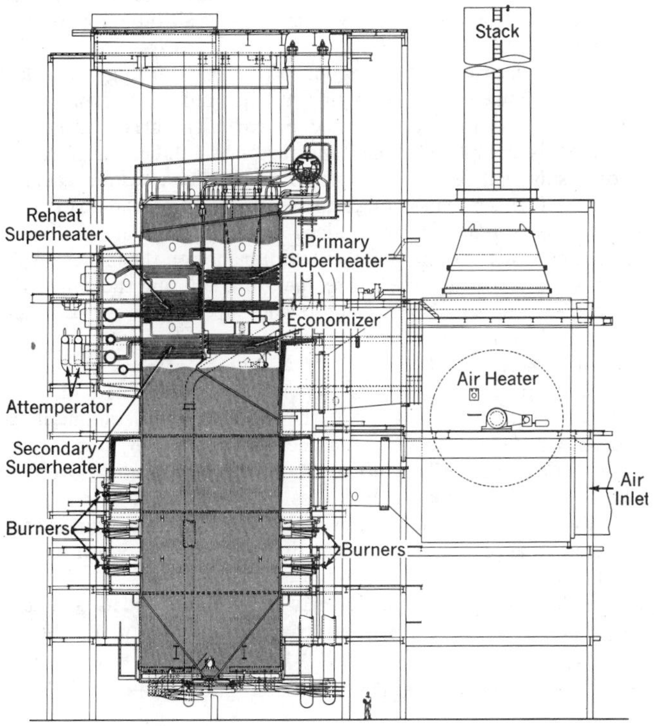

Fig. 12. El Paso type radiant boiler for natural gas and oil firing. (*The Babcock & Wilcox Co.*)

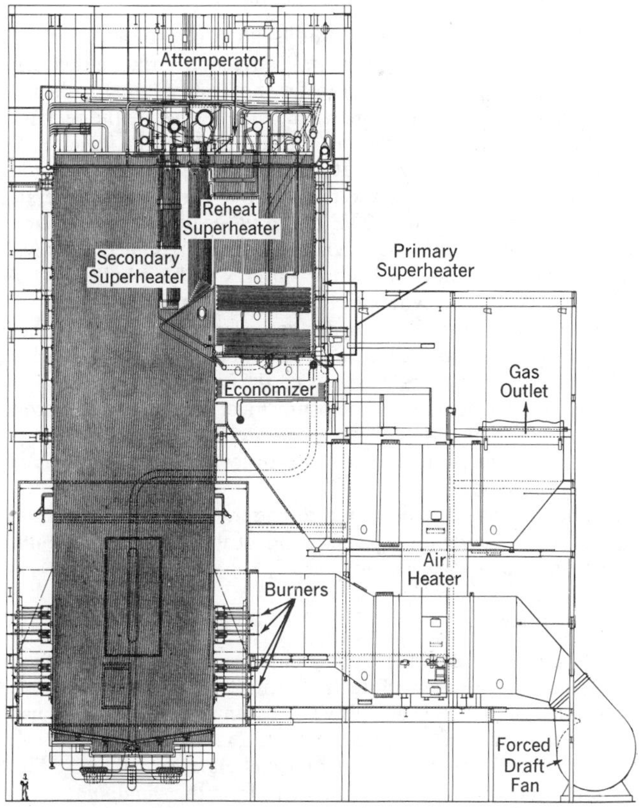

Fig. 14. Universal-Pressure boiler for natural gas firing. (*The Babcock & Wilcox Co.*)

Boiler Design

A boiler may be a unit complete in itself without auxiliary heat absorbing equipment, or it may constitute a rather small part of a large steam generating complex in which the steam is generated primarily in the furnace tubes, and the convection surface consists of a superheater, reheater, steaming economizer and air heater. In the latter case, it is possible to consider that a drum-type boiler comprises only the steam drum and the screen tubes between the furnace and the superheater. However, the furnace water-wall tubes, and usually a number of side-wall and support tubes in the convection portion of the unit, discharge steam into the drum and therefore effectively form a part of the boiler.

In the case of the Universal-Pressure boiler, there is no steam drum, but rather an arrangement of tubes in which steam is generated and superheated. Whether the boiler is a drum- or once-through type, whether it is an individual unit or a small part of a large complex, it is necessary in design to give proper consideration to the performance required from the total complex of the steam generating unit. Within this framework, the important items which must be accomplished in boiler design are the following:

1. Determine the heat to be absorbed in the boiler and other heat transfer equipment, the optimum efficiency to use, and the type of fuel or fuels for which the unit is to be designed. When a particular fuel is selected, determine the amount of fuel required, the necessary or preferred preheated air temperature and the quantities of air required and flue gas to be generated.
2. Determine the size and shape required for the furnace, giving consideration to location, the space requirements of burners or fuel bed, and incorporating sufficient furnace volume to accomplish complete combustion. Provision must also be made for proper handling of the ash contained in the fuel, and a water-cooled surface must be provided in the furnace walls to reduce the gas temperature leaving the furnace to the desired value.
3. The general disposition of convection heating surfaces must be so planned that the superheater and reheater, when provided, are located at the optimum temperature zone where the gas temperature is high enough to afford good heat transfer from the gas to the steam, yet not so high as to result in excessive tube temperatures or excessive fouling from ash in the fuel.

   While there is flexibility in the location of saturation or boiler surface, there must be enough total convection surface either before or after the superheater to transfer the heat required to heat the feedwater to saturation temperature and to generate the remainder of the steam required which is not generated in the furnace. This can be accomplished without an economizer, or an economizer can be provided to heat the feedwater to saturation temperature or even to generate up to 20% of the full-load steam requirement.

   The foregoing must be accomplished in a design that provides for proper cleanliness of heating surfaces without buildup of slag or ash deposits and without corrosion of pressure parts.
4. Pressure parts must be designed in accordance with applicable codes using approved materials with stresses not exceeding those allowable at the temperatures experienced during operation.
5. A tight boiler setting or enclosure must be constructed around the furnace, boiler, superheater, reheater and air heater, and gas-tight flues or ducts must be provided to convey the gases of combustion to the stack.
6. Supports for pressure parts and setting must be designed with adequate consideration for expansion and local requirements, including wind and earthquake loading.

Combustion Data

The basis for the designer's selection of equipment includes factors involved in the selection of fuels. In most areas, there are several fuels available and their availability and cost may be expected to change during the lifetime of the plant, with the result that the unit must be designed to burn more than one fuel. It is usually possible to determine which fuel is the most difficult from the standpoint of combustion and ash handling, and the unit, therefore, is designed for the most difficult fuel that possibly may be used.

After the steam requirements—steam flow, steam pressure, and temperature—and boiler feedwater temperature are determined, the required rate of heat absorption, q, is determined from:

$$q = w'(h_2' - h_1') + w''(h_2'' - h_1'') \qquad (1)^*$$

where:

q = rate of heat absorption, Btu/hour
w' = primary steam or feedwater flow, pounds/hour
w'' = reheat steam flow, pounds/hour
h_1' = enthalpy of feedwater entering, Btu/pound
h_2' = enthalpy of primary steam leaving superheater, Btu/pound
h_1'' = enthalpy of steam entering reheater, Btu/pound
h_2'' = enthalpy of steam leaving reheater, Btu/pound

To determine unit efficiency, it is necessary to know the temperature of the flue gas leaving the unit. This temperature may be set at the point where further addition of heating surface to reduce gas temperature would not be justified by the increased economy obtained. In the case of sulfur-bearing fuels, flue gas temperature is usually kept above the dew point to avoid sulfur corrosion of economizer or air heater surfaces.

The efficiency of combustion is 100 minus the sum of the heat losses expressed in percent. For a fuel with known characteristics and a given flue gas temperature, heat losses are evaluated.

The fuel input rate is then determined from Eqs. (1) and (2):

$$w_F = q/(Q_H \times \text{eff}) \qquad (2)$$

where

w_F = fuel input rate, pounds/hour
Q_H = high heat value of fuel, Btu/pound
eff = efficiency

From the quantity of fuel to be burned per hour, the corresponding weight of air required and the weight of combustion gases produced are determined.

Furnace Design. When pulverized-coal or cyclone-furnace firing is used, the wall(s) in which the burners or cyclones are located must be designed to accommodate them and the necessary fuel- and air-supply lines. Minimum clearances, established by experience, must be maintained between burners to avoid interference of the fuel streams from the various burners with each other. Minimum clearances must also be provided between burners and side walls and between each burner and the opposite wall to avoid flame impingement on furnace walls with consequent possible overheating of wall tubes or excessive deposits of ash or slag.

Turbulence is primarily a function of the fuel-burning equipment, and its importance lies in supplying air, not only to individual fuel particles, but also to any unburned or partially burned gases until combustion is completed. The time factor is fulfilled primarily by providing sufficient furnace volume so that the combustion gases remain in the furnace long enough to assure complete combustion.

Water-Cooled Walls. Most modern boiler furnaces have all walls water-cooled. This not only reduces maintenance on the furnace walls, but also serves to reduce the gas temperature entering the convection bank to the point where slag deposit and superheater corrosion can be controlled by sootblowers.

Handling of Ash. In the case of coal and, to a lesser extent with oil, a very important factor is the presence of ash in the fuel. If the ash is not properly considered in the design and operation, it can and does deposit not only on furnace walls and floor, but through the convection banks. This not only reduces the heat absorbed by the unit, but also increases draft loss, corrodes pressure parts, and eventually can cause shutdown of the unit for cleaning and repairs. There are two approaches to the handling of ash: (1) dry-ash furnace; and (2) slag-tap furnace.

In the dry-ash furnace, particularly applicable to coals with high ash and fusion temperatures, the furnace is provided with a hopper bottom and with sufficient cooling surface so that the ash impinging on the furnace walls or hopper bottom is solid and dry and can be

* 1 Btu = 0.2520 Calorie
1 Btu/pound = 0.556 Calorie per kilogram

removed essentially as dry particles. When pulverized coal is burned in a dry-ash furnace, about 80% of the ash is carried through the convection banks; most of the fly ash is normally removed by particulate-removal equipment located just ahead of the stack.

With many coals having low ash fusion temperatures, it is difficult to utilize a dry-bottom furnace because the slag is either molten or sticky and tends to cling and build up on the furnace walls and bottom. The slag-type furnace has been developed to handle coals of these types. The most successful form of the slag-tap furnace is that used in conjunction with cyclone-furnace firing. The furnace comprises a two-stage arrangement. In the lower part of the furnace, gas temperature is maintained high enough so that the slag drops in liquid form onto a floor where a pool of liquid slag is maintained and tapped into a slag tank containing water. In the upper part of the furnace, gases are cooled below the ash fusion point so that ash carried over into the convection banks is dry and does not adhere.

Convection Boiler Surface

The gas temperature leaving the furnace or entering the boiler depends mainly on the ratio of heat released to amount of furnace-wall cooling surface installed. Because the cost of furnace-wall cooling surface is relatively higher than that of boiler surface, the furnace size and surface are limited to the amount required to lower the gas temperature entering the convection tube banks sufficiently to avoid ash deposits.

The first few rows of tubes in the convection bank may be boiler tubes widely spaced to provide gas lanes wide enough to prevent plugging with ash and slag and to facilitate cleaning. These widely spaced boiler tubes are known as the slag screen or boiler screen. In many large units, they are used to support the furnace rear wall tubes. These screen tubes receive heat by radiation from the furnace, and by radiation and convection from the combustion gases passing through them.

In large contemporary units, the superheater generally replaces the boiler screen or, if not, is located immediately beyond it.

Design of boiler surface after the superheater will depend on the particular type of unit selected, desired gas temperature drop, and acceptable gas pressure drop (draft loss) through the boiler surface. Typical arrangements of boiler surface for various types of boilers have been illustrated. The object in the design of convection heating surfaces is to establish the combination of tube diameter, tube spacing, length of tubes, number of tubes wide and deep, and gas baffling that will give the desired gas temperature drop with the pressure drop permissible.

Heating surface and pressure drop are directly interrelated since both are primarily dependent on gas mass velocity. If either heating surface or pressure drop is increased, the other must decrease in order to maintain the desired gas temperature drop (heat transfer). Hence there is an optimum gas mass velocity which results in the optimum combination of heating surface and gas pressure drop.

For a given gas mass velocity (pounds of gas per hour per square foot of gas flow channel) or for a given gas velocity, a considerably higher gas film conductance, heat absorption, and draft loss result when the gases flow at right angles to the tubes (crossflow) than when they flow parallel to the tubes (longitudinal flow). Gas turns between tube banks generally add draft loss with little or no benefit to heat absorption and should be designed for easy flow.

From a long record of experience, given sets of conditions for each fuel to be burned have been effectively established as the conditions of economic practice. While these conditions vary as improvements occur over a period of years, at any particular time competitive economies acts to hold most of the variables involved within a fairly limited range.

Superheaters and Reheaters

Early in the eighteenth century, it was shown that substantial savings in fuel could be experienced when steam engines were run with some superheat in the steam. In the late 1800s, lubrication problems were encountered with reciprocating engines, but once these were overcome, development of superheaters continued.

Commercial development of the steam turbine hastened the general use of superheat. By 1920, steam temperatures of 343°C, representing superheats of 140°C were generally accepted. In the early 1920s, the regenerative cycle, using steam bled from turbines for feedwater heating, was developed to improve station economy without going to higher steam temperatures. At the same time, superheater development permitted raising the steam temperature to 385°C. A further gain in economy by still higher temperature was at that time limited by allowable superheater tube-metal temperature. This led to the commercial use of reheat, where the steam leaving the high-pressure stage of the turbine was reheated in a separate reheat superheater and returned at higher temperature and enthalpy to the low-pressure stage.

The first reheat unit for a central station was proposed in 1922 and went into service in 1924. It was designed for 650 psi (42 atmospheres) and operated at 550 psi and 385°C. Exhaust steam from the high-pressure turbine was reheated to 385°C at 135 psi (9.2 atmospheres). A much higher-pressure reheat unit, designed in 1924 for 1200 psi (82 atmospheres) and 371°C went into service in 1925.

Advantages of Superheat and Reheat. When saturated steam is utilized in a steam turbine, the work done results in a loss of energy by the steam and consequent condensation of a portion of the steam, even though there is a drop in pressure. The amount of work that can be done by the turbine is limited by the amount of moisture which can be handled by the turbine without excessive wear on the turbine blades. This is normally somewhere between 10 and 15% moisture. It is possible to increase the amount of work done by moisture separation between turbine stages, but this is economical only in special cases. Even with moisture separation, the total energy that can be transformed to work in the turbine is small compared with the amount of heat required to raise the water from feedwater temperature to saturation and then evaporate it. Thus, moisture constitutes the basic limitation in turbine design.

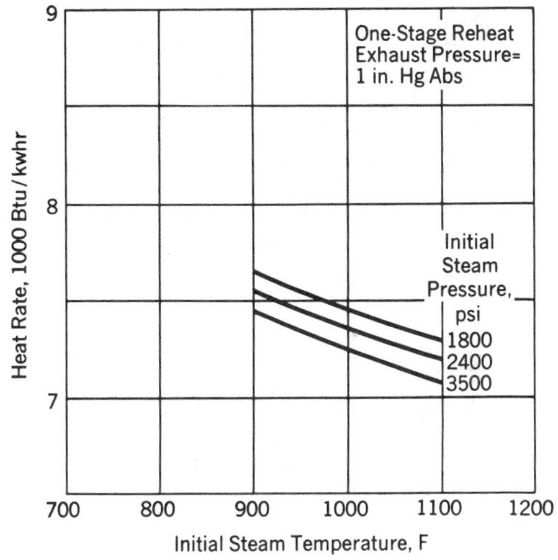

Fig. 15. Effects of changes in steam temperature and pressure on performance of ideal Rankine cycle with one-stage reheat. (*The Babcock & Wilcox Co.*)

Because a turbine generally transforms the heat of superheat into work without forming moisture, the heat of superheat is essentially all recoverable in the turbine. This is illustrated in the temperature-entropy diagram of the ideal Rankine cycle where the heat added to the right of the saturated vapor line is shown as 100% recoverable. While this is not always entirely correct, the Rankine cycle diagrams of Fig. 15 indicate that this is essentially true in practical cycles.

The foregoing factors, however, are not specifically applicable at steam pressures in the vicinity of the critical point. The term *superheat* is not quite appropriate in defining the temperature of the working fluid at or above the critical point. However, even at pressures exceeding 3208 psia (~218 atmospheres) heat added at temperatures above 374°C is essentially all recoverable in a turbine.

Types of Superheaters. The original and somewhat basic type of superheater and reheater was the convection unit for gas temperatures where heat transfer by radiation was very small. With a unit of this type, steam temperature leaving the superheater increases with boiler output because of the decreasing percentage of heat input that is absorbed in the furnace, leaving more heat available for superheater

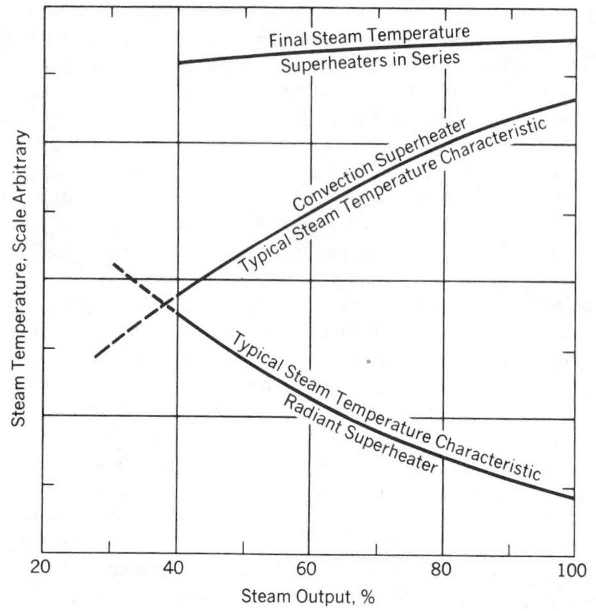

Fig. 16. A substantially uniform final steam temperature over a range of output can be attained by a series arrangement of radiant and convection superheater components. ("*Steam—Its Generation and Use,*" *The Babcock & Wilcox Co.*)

absorption. Since convection heat transfer rates are almost a direct function of output, the total absorption in the superheater per pound of steam increases with increase in boiler output. See Fig. 16. This effect is increasingly pronounced the further the superheater is removed from the furnace, that is, the lower the gas temperature entering the superheater.

Conversely, the radiant superheater receives its heat through radiation and practically none from convection. Because the heat absorption of furnace surfaces does not increase in direct proportion to boiler output, but at a considerably lesser rate, the curve of radiant superheat as a function of load slopes downward with increase in boiler output. In certain cases, the two opposite-sloping curves have been coordinated by the combination of radiant and convection superheaters to give flat superheat curves over a wide range in load as typically indicated in Fig. 16. A separately fired superheater has the characteristic that it can be fired to produce a flat superheater curve.

The early convection superheaters were placed above or behind a deep bank of boiler tubes in order to shield them from the fire or from the higher temperature gases. The greater heat absorption required in the superheater for higher steam temperatures made it necessary to move the superheater closer to the fire. This new location brought with it problems which were not apparent with the superheaters located in the original lower-gas-temperature zone. Steam- and gas-distribution difficulties and instances of general overheating of tube metal were ultimately resolved by improved superheater design, including higher mass velocity of the steam. This increased the heat conductance through the steam film, resulting in lower tube-metal temperatures, and also improved steam distribution by increasing pressure drop through the tubes.

Steam mass velocity in contemporary superheaters ranges from as low as 100,000 to 1,000,000 pounds per square foot (488,200 to 4,882,000 kilograms per square meter) per hour or higher, depending upon pressure, steam and gas temperatures, and the tolerable pressure drop in the superheater. The fundamental considerations governing superheater design apply also to reheater design. However, the pressure drop in reheaters is critical because the gain in heat rate with the reheat cycle can be fully nullified by too much pressure drop through the reheater system. Hence, steam mass flows are generally somewhat lower in the reheater.

Steam-Temperature Adjustment and Control

Improvement in the heat rate of the modern boiler unit and turbine results in large part from the high cycle efficiency possible with high steam temperatures. The importance of regulating steam temperature within narrow limits is evident from Fig. 15, which shows that a change of ~20°C corresponds to a change of about 1% in heat rate at pressures from 1800 to 3500 psi (122 to 238 atmospheres).

Other important reasons for accurate regulation of steam temperature are to prevent failures from overheating parts of the superheater, reheater, or turbine, to prevent thermal expansion from reducing turbine clearances to the danger point, and to avoid erosion from excessive moisture in the last stages of the turbine. The control of fluctuations in temperature from uncertainties of operation, such as slag or ash accumulation is important. However, superheat and reheat steam temperatures in steam generation are mainly affected by variations in steam output. See Fig. 16.

With drum-type boilers, steam output and pressure are maintained constant by firing rate, while the resulting superheat and reheat steam temperatures depend on basic design and other important operating variables, such as the ratio of convection to radiant heat-absorbing surface, excess air, feedwater temperature, changes in fuel that affect turbine characteristics and ash deposits on the heating surfaces, and the specific burner combinations in service. In the Universal-Pressure, once-through boiler which has a variable transition zone, steam output and pressure are controlled by the boiler feed pump and steam temperature by the firing rate, leaving reheat steam temperature as a dependent variable. Standard performance practice for steam generating equipment permits a tolerance of plus or minus 5.5°C in a specified steam temperature.

References

Ardell, M.: "Particulate Control for Coal-fired Boilers," *Chem. Eng. Progress*, **75**, 78–82 (1979).

Baumeister, T., (editor): "Mark's Standard Handbook for Mechanical Engineers," 8th Edition, McGraw-Hill, New York, 1978.

Criswell, R. L.: "Control Strategies for Fluidized Bed Combustion," *In-Tech.*, **27**, 1, 37–43 (1980).

Drehmel, D. C.: "Developments in Particulate Control for Coal-fired Power Plants," *Chem. Eng. Progress*, **76**, 5, 74–75 (1980).

Grey, J., Sutton, G. W., and M. Zlotnick: "Fuel Conservation and Applied Research," *Science*, **200**, 135–142 (1978).

Robnett, J. D.: "Engineering Approaches to Energy Conservation," *Chem. Eng. Progress*, **75**, 3, 59–81 (1979).

Smoot, L. D., and D. T. Pratt: "Pulverized Coal Combustion and Gasification for Continuous Flow Processes," Plenum, New York, 1979.

Staff: "Steam—Its Generation and Use," 39th edition, The Babcock & Wilcox Co., New York, 1978.

BOILING. If a liquid is heated at constant pressure in an inert atmosphere, evaporation takes place at the free surface, but bubbles may form in the interior of the liquid. This process is also known as *ebullition.* The vapor pressure inside a bubble of small diameter is considerably less than the vapor pressure over a plane surface, so that bubbles cannot persist below the temperature at which the vapor pressure equals the external pressure and do not appear until the local temperature is rather greater than this. If the whole of the liquid is above the critical temperature, usually called the "boiling-point," introduction of a source of bubbles, either deliberate or accidental, causes rapid ebullition until evaporation reduces the temperature. A strong source of heat may cause the superheating necessary for ebullition in a thin layer. The bubbles formed in the layer grow until they rise out of it and the "boiling heat-transfer" associated with the transient occurrence of bubbles either in the liquid or attached to the heated surface leads to some curious effects. Typically, normal single-phase convection is succeeded by nucleate boiling in which bubbles form on nuclei on the surface and heat transfer is much increased. Further rise of wall temperature causes partial formation of a film and a rapid decrease of heat transfer as film boiling is established.

In modern boilers,* as much as 1 to 3 cubic feet of steam per minute may be formed in a 5-foot length of $2\frac{1}{2}$-inch tubing in the furnace area. Water must be continuously supplied for this steam generation and, in many designs, excess water is provided to protect the tubes from overheating. In order to obtain information on the nature of boiling, a number of investigators have experimented

* Remaining information in this entry from "Steam—Its Generation and Use" (38th edition), copyright The Babcock & Wilcox Co., New York, 1972.

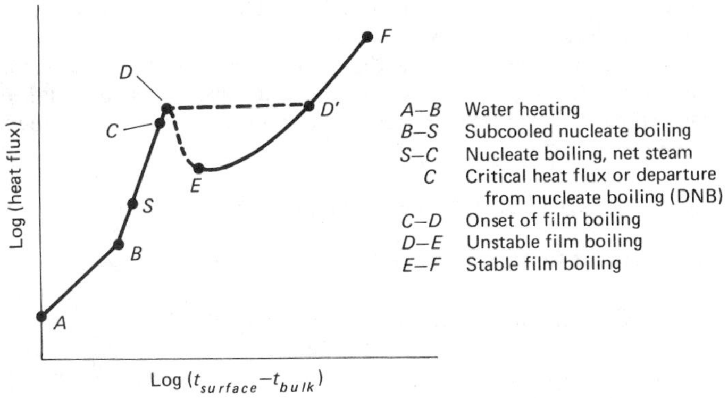

Heat transfer to water and steam in a heat flow channel. Relation of heat flux to temperature difference between channel-wall and bulk-water or steam temperature.

with electrically heated wires in a pool of water. Other investigators have performed the experiment of heating a tube or other type of flow channel cooled by a flow of water at a pressure below critical, and subjecting the tube to various levels of heat input.

The accompanying figure is a generalized curve summarizing the results of these investigations. This curve can be regarded as a general correlation of test results at a number of different heat inputs to heated wires in a pool of water or to heated tubes or flow channels. It can also be regarded as a series of different heat inputs to a single flow channel. In this case, the points on the curve represent a series of temperature differences (surface temperature minus bulk water or steam temperature) corresponding to the waer and steam conditions existing at a single location on the flow channel for different levels of heat flux or heat input. If the channel is evenly heated along its length, the location represented is the outlet end of the heated section of the channel. Absolute values on the curve are dependent on many factors, including pressure, flow-channel geometry, mass velocity, flux patterns, and degree of water subcooling.

For all heat input conditions (points on the figure), water pressure and temperature at the inlet to the channel remain constant. Hence, the amount of subcooling (saturation temperature minus water temperature) at the inlet also remains constant. Ideally, water flow through the tube is maintained at a fixed rate.

The initial heat flux at point *A* is shown increasing on a logarithmic scale for points to the right of *A*. Until point *B* is reached, the heat input is not sufficient to produce boiling.

At *B*, the local heat flux is sufficient to raise the water temperature adjacent to the heated surface to saturation temperature, or slightly above, and a change from the liquid to the vapor state occurs locally. This change is characterized by the coexistence of both phases at essentially the same temperature locally, differing only in a few degrees of liquid superheat necessary for heat transfer and by heat absorption required to overcome the molecular binding forces of the liquid phase. Here, the change of state is accompanied by ebullition of the vapor as opposed to evaporation at a free surface and the term boiling is used to describe the process. Also, the ebullition takes place at an interface other than that of the liquid and its vapor, actually at a solid-liquid interface; hence the boiling is described as "nucleate boiling."

The bulk of the water does not reach saturation temperature until the heat flux of point *S* is reached. Between *B* and *S*, the steam bubbles formed at the heated surface condense quickly in the main stream, giving up their latent heat to raise the temperature of the water. This condition is known as subcooled-nucleate or local boiling. Nucleate boiling occurs at all points up to *C*; beyond *S*, the bubbles do not collapse, since this part of the curve represents boiling with the water bulk temperature at saturation.

Both nucleate-boiling regimes, subcooled and saturated, are characterized by very high heat transfer coefficients. These are ascribed to the high secondary velocities of water caused by the liberation of surface tension energies available in the liquid-vapor-solid interfaces at the instant of bubble release from the heating surface. This is a convection-type transfer coefficient based on bubble kinetics and is also affected to some extent by bulk mass velocity, depending on the velocity range. As the result of these high heat transfer coefficients, tube- or flow-channel surface temperatures do not greatly exceed the saturation temperature.

Beyond the nucleate boiling region (*B-C* in the figure), the bubbles of steam forming on the hot tube surface begin to interfere with the flow of water to the surface and eventually coalesce to form a film of superheated steam over part or all of the heating surface. This condition is known as "film boiling." From *D* to *E* film boiling is unstable; beyond point *E* film boiling becomes stable.

In a fossil-fuel-fired boiler or in a nuclear reactor, when the local heat flux exceeds that corresponding to point *D*, the surface temperature may rise very quickly, along the horizontal dotted line in the figure, to point *D'*. If the temperature at *D'* is sufficiently high, the heating surface burns out or melts. Hence, *D* is known as the burnout point and *C*, which may be very close to it, as the point of departure from nucleate boiling (DNB), or the critical heat flux.

Stable and even unstable film boiling is acceptable in certain types of heat transfer equipment where the temperature of the heat source is within the safe operating range of the equipment, or where the boiling film heat transfer coefficient is the controlling resistance to heat flux. Steam generators for pressurized-water reactor (nuclear) systems, which are actually water-to-boiling water heat exchangers, and certain types of process heat exchange equipment are in this category.

BOILING CURVE AND CONDENSATION CURVE. Consider the phase diagram of a binary system forming a liquid and a vapor phase at constant pressure. Curve I is the boiling curve, which gives the coexistence temperature as a function of liquid composition; and curve II is the condensation curve, which gives the coexistent temperature as a function of the composition of the vapor phase. If the temperature is increased, vaporization begins when the boiling curve is crossed. Inversely, condensation begins when the temperature is decreased below the condensation curve.

Temperature composition of a liquid-vapor system at constant pressure.

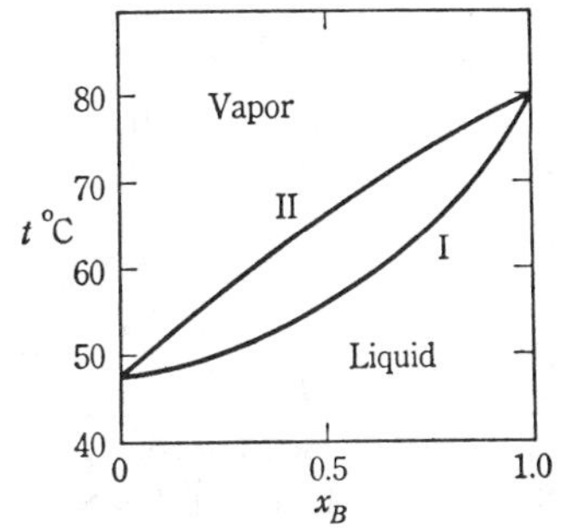

BOILING POINT. The normal boiling point of a liquid is the temperature at which its maximum or "saturated" vapor pressure is equal to the normal atmospheric pressure, 760 mm of mercury. If the pressure on the liquid varies, the actual boiling point varies in accordance with the relation between the vapor pressure and the temperature for the liquid in question. (See **Vapor.**) Water, for example, with a normal boiling point of 100°C or 212°F, boils at ordinary room temperature when the pressure is reduced to about 17 mm; and inhabitants of elevated regions often find difficulty in cooking food by boiling, because of the low boiling point. On the other hand, the boiling water and steam in a "pressure cooker" are so hot that such foods as meat and rice are cooked tender in a very short time. If a solid is dissolved in the liquid, or if another, less volatile liquid is mixed with it, the boiling point is raised to a degree expressed by the boiling point laws of Van't Hoff, Raoult, and others.

A liquid does not necessarily begin boiling when the temperature reaches the boiling point. If kept perfectly quiet, and especially if covered with a film of oil, water may be raised several degrees above its normal boiling point, before it suddenly boils with explosive violence; it then returns to its true boiling point.

To prevent this superheating, it is customary to add to laboratory distillation flasks small pieces of inert material having sharp corners; the latter favor the formation of "bubbles" of vapor when the liquid reaches the boiling point.

The *maximum boiling point* is that temperature corresponding to a definite composition of a two-component or multi-component system at which the boiling point of the system is a maximum. At this temperature the liquid and vapor have the same composition and the solution distills completely without change in temperature. Binary liquid systems which show negative deviations from Raoult's law have maximum boiling points. See **Raoult's Law; Van't Hoff Law.**

The *minimum boiling point* is that temperature corresponding to a definite composition of a two-component or multi-component liquid system at which the boiling temperature is the lowest for that particular system. At the minimum boiling point the liquid and vapor have the same composition.

BOILING-POINT CHEMICAL ANALYZERS. Essentially these instruments are for laboratory use in identifying various organic liquids, including azeotropes. They also are used for monitoring the purity of liquids and their concentrations. Some modifications of the technique make continuous, on-line process applications possible. By maintaining a known absolute pressure, the constant temperature of boiling under partial reflux of the sample is determined. Additional physical characteristics of an unknown sample may have to be measured to identify unknowns, since different substances can have the same boiling point. Once a known steady-state condition is established, however, changes of boiling point indicate the presence of impurities. The amount of impurities can be related to variation in boiling point of the sample under total reflux compared with known boiling point of the pure material. See also **Analysis (Chemical), Beckmann Method;** and **Specific Gravity.**

BOILING POINT CONSTANT. Consider a dilute solution in which all solute species may be regarded as nonvolatile. The vapor in equilibrium with the solution is then formed from the solvent only. Call T^0 the boiling point of the pure solvent at the pressure concerned, and T the boiling point of the solution. For a dilute solution, the difference

$$\theta = T - T^0$$

will be small compared with T^0.

If the solution is also ideal, one has

$$\theta = \frac{R(T^0)^2}{\Delta_e h^0} \frac{M_1}{1000} \sum_s m_s = \theta_e \sum_s m_s$$

M_1 is the molar mass of the solvent, $\Delta_e h^0$ its latent heat of vaporization in kcal per mole at temperature T^0, m_s the molality of solute s; θ_e is called the *boiling point constant*, or *ebullioscopic constant*. It depends only on the properties of the solvent. For water,

$$\theta_e = 0.51°\text{C}$$

BOILING POINT DIAGRAM. Distillation.

BOILING POINT ELEVATION. The boiling point of a solution is, in general, higher than that of pure solvent, and the elevation is proportional to the active mass of the solute for dilute (ideal) solutions.

$$\Delta T = Km$$

where ΔT is the elevation of the boiling point, K is the *boiling point constant* or the *ebullioscopic constant* and m, the molality of the solution.

There are several methods for measuring elevation: In the Beckmann method a Beckmann-type thermometer is immersed in a weighed amount of solvent and the boiling point determined by gentle heating until a steady temperature is reached. A weighed amount of solute is then added and the boiling point redetermined. The difference gives the elevation of the boiling point. The glass vessel containing the liquid is provided with a platinum wire sealed through the bottom to promote steady boiling and to prevent overheating, and reflux condensers are used to minimize loss of liquid.

In the Landsberger method, vapor from boiling solvent is passed through the solvent contained in another vessel and by giving up its latent heat will eventually raise the liquid to the boiling point. At this stage a weighed amount of solute is added to the second vessel and the boiling point is again determined.

In the Cottrell method, the thermometer is placed in the vapor phase above the surface of the liquid and the apparatus so designed that boiling liquid is pumped continuously over the bulb of the thermometer.

BOILING POINTS. Chemical Elements.

BOILING RANGE (Petroleum Fuels). Petroleum.

BOILING WATER REACTOR. Nuclear Reactor.

BOILS. A localized purulent infection of the skin and underlying tissues taking the form of nodules which initially are hard, but progress to a softer stage, ultimately rupturing and draining. Relatively small abscesses sometimes are termed furuncles. The term *carbuncle* often is used to describe larger boils, particularly those with multiple openings. Boils may occur singly or in multiples and, in some cases, may be chronic and recurrent over long periods until the underlying causes are found and corrected. The immediate cause is attributed to staphylococci, although steptococci also may be present. Factors which may contribute to a predisposition for the formation of boils include (1) inadequate hygiene, (2) prolonged, excessive sweating, particularly if unaccompanied by excellent hygiene, (3) dietary considerations, including malnutrition, (4) anemias, (5) diabetes, (6) obesity, and (7) certain drugs.

BOISE DE ROSE. An essential oil obtained from evergreen trees of the *Lauraceae* family, notably *Aniba rosaeodora* Ducke, and *Ocotea caudata* Mer.), which are found growing wild in the forests of the Amazon basis (Brazil and Peru). The chopped bark of the trees is steamed distilled to yield a colorless to pale-yellow oil. Depending upon variety of tree used, the oil has a sweet, slightly wood characteristic odor to an order in which camphoraceous notes are readily detected. The main constituent of the oil is linalool. Boise de rose oil finds extensive use in perfumery, particularly in soap and cosmetics. The oil is also used as a flavoring in chewing gum, baked goods, ice cream, candy, and beverages.

BOLE. A fine-grained, sticky, bright red laterite; the decomposition product of basic igneous rocks, such as basalt.

BOLIDE. A term occasionally applied to meteors which are observed to explode in the air and break up into two or more fragments. Such objects are frequently described as having the appearance of an exploding rocket. Not infrequently following the explosion of a bolide is heard a sharp detonation. See also **Fireball.**

BOLL WEEVIL (*Insecta, Coleoptera*). A snout-beetle or weevil, *Anthonomous grandis*, averaging about $\frac{1}{4}$ inch (6 millimeters) in length, which damages cotton. The adult punctures the cotton squares to lay its eggs and thus prevents the formation of the boll, and later in the season deposits eggs in the bolls, where its larvae damage the seeds and lint.

The species entered the United States from Mexico in the early nineties and is now an established pest in practically all cotton-growing areas, causing an annual loss in the many millions of dollars. The problem is met by various methods of keeping the insect in check. The destruction of cotton plants after the crop is harvested kills many insects, and spraying the growing plants with methyl parathion, Guthion, and Azodrin has been found effective.

Economic losses to the cotton industry caused by the boll weevil were estimated in 1974 to range from $200 to $300 million annually. Thus, the boll weevil is the most important farm pest in the United States. The U.S. Department of Agriculture (USDA) estimates that about one-third of all insecticides used in the United States are for the control of the boll weevil or for the control of other pests that would not become problems if chemicals used against the weevil did not also destroy beneficial insects. In the mid-1970s, entomologists of the USDA were pressing to have the government lead a 6- to 10-year campaign intended to rid the cottom belt of the boll weevil once and for all. The cost of the campaign is estimated at $655 million, although some entomologists believe that $1.5 billion is more realistic. The USDA states that 100% cooperation on the part of growers would be required for success.

The boll weevil has been under study since the early 1890s when first discovered in Texas, an unwelcome migrant from Mexico. The weevil's reproductive potential is enormous. A 10-acre (4-hectare) cotton patch that has a very light infestation of say 100 weevils on July 1st would have 100,000 weevils by September 1st. Economic damage usually commences when the weevils number 1000 per acre (2500 per hectare). Studies made in Texas and the Carolinas have shown that growers who have neglected to control the boll weevil have been losing up to one-half of their potential cotton crop.

For many years, the only effective boll weevil control measure was that of planting and harvesting cotton as early as possible, then destroying the stalks in order to reduce the overwintering weevil population that would threaten the crop of the following year. But such measures are only weakly effective. By the 1930s, calcium arsenate was being used effectively for control of the weevil, but that insecticide has shortcomings being dangerous to both humans and livestock. The arsenate, once applied would not disappear, but would build up steadily in the soil—to the point where the growth of some crops, such as legumes, was badly retarded. In the mid-1940s, calcium arsenate was replaced as a boll weevil control by many new families or organic compounds, notably a combination of DDT and Toxaphene. This proved effective for in-season control, but ultimately it proved even less acceptable environmentally than calcium arsenate. Because of its persistence, mobility, and tendency to build up in food chains, DDT was banned for cotton insect control in December 1972 by the Environmental Protection Agency (EPA), but such regulations, of course, are always subject to reversal as combinations of negative factors are weighed one against the other.

Organophosphate compounds, including methyl parathion, Guthion, and malathion, remained available and were already in widespread use. However, these compounds also are dangerous, particularly methyl parathion. It is a broad-spectrum pesticide which kills insects indiscriminately, the harmful and the beneficial alike.

The Boll Weevil Research Laboratory was established at Mississippi State University in 1961 and has since been undertaking intensive research on eradication and control of the boll weevil. In 1969, a special Committee on Boll Weevil Eradication developed the following strategy: (1) Normal in-season application of insecticides by the grower to control the weevil population; (2) repeated late-season insecticide treatments to kill as many weevils as possible before they achieve diapause, the state of hibernation essential to the insect's winter survival; (3) defoliation or desiccation of cotton plants before harvest and destruction of the stalks immediately afterward, the purpose being to deny the weevil a favored place to feed and produce one or more new generations; (4) the deployment in the spring, prior to the weevil's emergence from diapause, of traps containing a sex attractant called grandlure; and (5) the release of sterile boll weevils, 100 to an acre, with the aim of preventing fertile matings by the few native weevils that have survived the diapause treatment and have not been drawn to the grandlure traps. Effective use of sterile-male release techniques is possible only if the weevil population has been reduced to extremely low numbers of about two weevils per acre.

A pilot project along the foregoing lines was commenced in mid-1971. Reports of the project two years later concluded that the project had demonstrated the feasibility of eliminating the boll weevil. However, there were ambiguities in the findings. With very close control and cooperation, it is believed that this program, with improvements and refinements, can be workable and effective and, of course, very costly. One disturbing factor is that the boll weevil would have a sanctuary in Mexico and thus there would continue to be a ceaseless battle against the pest in South Texas unless those weevils that fly across the Rio Grande into Texas are quickly suppressed. To this is added the several alternative host plants, such as cienfuegosia, a wild cotton plant that grows in scattered colonies over a large area along the South Texas coast. Hibiscus plants also harbor the weevil and gardeners would have to be persuaded to stop growing it. There is also the problem of volunteer cotton which can spring up in fallow fields.

Despite these critical problems, the National Cotton Council, the USDA and other interested organizations plan to continue to push the aforementioned systematic strategy. It is planned that the campaign would advance eastward from West Texas, finally reaching Virginia and the Carolinas, involving a total of more than 10 million acres. There are outspoken critics of the program who stress that over $150 million was spent on programs to eradicate the fire ant and failed.

BOLLWORM (*Insecta, Lepidoptera*). The pink cotton bollworm is a small moth whose larva, the bollworm proper, lives in the flowers of cotton and usually prevents their maturing, and later enters the bolls and damages the seed and lint. The species lives on several other species of plants and is therefore difficult to check, although it is not yet as serious a pest as the boll weevil. The larva of another moth of larger size, more widely known as the corn earworm, also attacks cotton squares and bolls, as well as corn and tobacco, and so is known as the cotton bollworm. It is estimated to cause several millions of dollars' damage each year. Fall plowing and disking are practiced to destroy this insect in the pupal stage, which is passed in the ground, and dusting as for the weevil is effective.

BOLOMETER. A very sensitive thermometric instrument of the metallic resistance type, devised by Langley and used for measuring weak radiation. It consists of a slender strip of platinum mounted at the lower end of a long cylindrical tube having circular stops across it at intervals to screen off all radiation except that to be observed. A slight amount of radiant energy falling upon the strip causes a measurable deflection in a sensitive galvanometer in the resistance bridge coupled with it. The sensitivity of the instrument is of the same order as that of the radiomicrometer of Boys. When a bolometer tube is mounted as the receiving element of a spectroscope, instead of the usual observing telescope or camera, the instrument may be used to detect and measure the lines or bands of infrared spectra. Very sensitive bolometers have been constructed, making use of the rapid resistance change when a metal is cooled to a temperature where it exhibits superconductivity.

In microwave technology, a barretter, thermistor or other device utilizing the temperature coefficient of resistance of some substance is called a bolometer. See also **Barretter.**

BOLSON. An undrained basin in an arid region, which generally is partly filled with rock-waste washed by temporary streams from the bordering mountains.

BOLTZMANN *H*-THEOREM. **Equilibrium.**

BOLTZMANN'S DISTRIBUTION LAW. Consider a system composed of molecules, of one or more kinds, able to exchange energy at collisions but otherwise independent of one another. Evidently we cannot say anything useful or interesting about the state of a particular molecule at a particular time. We can however make useful statements about the average fraction of molecules of a given kind in a given state, or, what is the same thing, the fraction of time spent by each molecule of a given kind in a given state. If the system is maintained at a definite temperature, for example by keeping it in a thermostat, then the fraction of molecules of a given kind in a given state is determined by the energy of this state and by the temperature. In particular if we denote by i and k two completely defined states of a molecule of a given kind and by E_i and E_k the energies of these two states then the average numbers N_i and N_k of molecules in these two states are related by

$$N_i/N_k = \exp\{-\beta(E_i - E_k)\} \qquad (1)$$

where β is a parameter having a positive value determined entirely by the thermostat; i.e., β has the same value for all states of a given kind of molecule and for all kinds of molecules. In other words β has all the characteristics of temperature except that it decreases as temperature increases. If we write

$$\beta = 1/kT \qquad (2)$$

then it can be shown that T is identical with thermodynamic (or absolute) temperature and k is a universal constant whose value determines the unit of T called the degree. When k is given the value 1.38041×10^{-23} joules/degree, the temperature scale becomes the Kelvin scale defined by $T = 273.16$K at the triple point of water. Substitution of Eq. (2) into Eq. (1) leads to

$$N_i/N_k = \exp\{-(E_i - E_k)/kT\} \tag{3}$$

This fundamental relation is called Boltzmann's distribution law after the creator of statistical mechanics, Ludwig Boltzmann, (1844/1906) Professor of Physics in Leipzig, and k is called Boltzmann's constant.

We must now discuss the meaning of the words used above, "completely defined state." These words have one meaning in classical mechanics and a different, but related, meaning in quantum mechanics. Since the quantal definition is the simpler we shall discuss it first. We begin by considering a system of highly abstract "molecules" having only a single degree of freedom, for example linear oscillators. The quantum states form a simple series specified by consecutive integers called the quantum numbers. In this simple example there is no ambiguity in the meaning of "completely defined state"; each state i is completely defined by the integral value of a single quantum number. Let us now consider a "molecule" with three degrees of freedom such as a structureless particle moving in three-dimensional space. The complete specification of this particle's state requires not one but three integral quantum numbers. If the particle moves freely in a cubical box, the three quantum numbers may be associated with motion along the three directions normal to the faces of the box. The subscript labels i and k in the previous formulas are abbreviations for sets of three quantum numbers. For example i might mean (2, 5, 1) and k might mean (3, 4, 2). There can now be several states having the same energy. For a particle moving freely in a cubical box, it follows from symmetry that the states (2, 5, 1), (1, 2, 5), (5, 1, 2), (1, 5, 2), (2, 1, 5), and (5, 2, 1) all have the same energy; such an energy level is called sixfold degenerate. (One should *not* speak of a p-fold degenerate *state*, but of a p-fold degenerate energy level.) It is sometimes desirable to consider the fraction of molecules having a given energy rather than the fraction in a given state. If N_r and N_s denote the average number of molecules of a given kind having energy E_r and E_s then evidently

$$N_r/N_s = (p_r/p_s)\exp\{-(E_r - E_s)/kT\} \tag{4}$$

Alternatively, if f_r denotes the average fraction of molecules of a given kind having energy E_r, then

$$f_r = p_r \exp(-E_r/kT)/\Sigma_s\, p_s \exp(-E_s/kT) \tag{5}$$

The sum Σ_s occurring in the denominator is called the partition function.

It may happen that certain degrees of freedom are completely independent of other degrees of freedom. We call such degrees of freedom "separable." The partition function can then be separated into factors relating to the several sets of separable degrees of freedom, and Boltzmann's distribution law is applicable separately to each set of separable degrees of freedom. For example, for an electron moving freely in a rectangular box, the translational motions normal to the three pairs of faces and the fourth degree of freedom due to spin are all separable.

We shall now consider briefly the meaning of completely specified state according to classical mechanics. We know that classical mechanics is merely an approximation, sometimes good but sometimes bad, to quantum mechanics. Motion in each separable degree of freedom can be described classically by a coordinate x and its conjugate momentum p_x. If x and p_x are plotted as Cartesian coordinates, the diagram is called the phase plane. There is a simple correlation between the quantal and the classical descriptions: the density of quantum states is one per area h (Planck's constant) in the phase plane. This may be extended to several degrees of freedom. If there are f degrees of freedom, the motion is described by f coordinates $q_1, \ldots q_f$ are the conjugate momenta $p_1, \ldots p_f$. We can imagine these plotted in a $2f$ dimensional Cartesian space called phase space. There is then one quantum state per $2f$ dimensional volumes h^f of phase space. In the classical as in the quantal description there can be degenerate energy values and there can be separable degrees of freedom. The classical description is a good approximation to the quantal description when the spacing between energy levels is small compared with kT. An example of an effectively classical separable degree of freedom is the motion in a given direction of a free particle. If the linear coordinate is denoted by x and the linear momentum by p_x, then the fraction of molecules at a position between x and $x + dx$ and having a momentum between p_x and $p_x + dp_x$ is

$$\frac{\exp(-p_x^2/2mkT)\,dx\,dp_x}{\int dx \int_{-\infty}^{\infty} dp_x \exp(-p_x^2/2mkT)} \tag{6}$$

where m is the mass of the particle so that its (kinetic) energy is $p_x^2/2m$. In the classical treatment, the kinetic and potential factors are separable. Consequently the fraction of molecules, anywhere or everywhere, having momentum between p_x and $p_x + dp_x$ is

$$\exp(-p_x^2/2mkT)\,dp_x \Big/ \int_{-\infty}^{+\infty} \exp(-p_x^2/2mkT)\,dp_x \tag{7}$$

Equation (7) is called Maxwell's distribution law after Clerk Maxwell (1831/79), Professor of Physics at Cambridge (England), who obtained it in 1860 before Boltzmann in 1871 obtained his wider distribution law. Maxwell derived his distribution law from the conservation of energy together with the assumption that the motion is separable in three mutually orthogonal directions. The latter assumption was violently attacked by mathematicians, but we now recognize that the assumption is both reasonable and true.

In conclusion we must mention that a necessary condition for the validity of Eq. (3), and consequently of other formulas derived from Eq. (3) is that $N_i \ll 1$ for the state (or states) of lowest energy and *a fortiori* for all other states. When this inequality does not hold, Boltzmann's distribution law must be replaced by a more general and more precise distribution law, either that of Fermi and Dirac or that of Bose and Einstein according to the nature of the molecules. See also **Statistical Mechanics.**

BOLTZMANN TRANSPORT EQUATION. The fundamental equation describing the conservation of particles which are diffusing in a scattering, absorbing, and multiplying medium. It states that the time rate of change of particle density is equal to the rate of production, minus the rate of leakage and the rate of absorption, in the form of a partial differential equation such as

$$\frac{\partial n}{\partial t} = \text{production} - \text{leakage} - \text{absorption}$$
$$= S + D\nabla^2\phi + \phi\, \Sigma_a$$

in which S in the cgs system is in units of neutrons cm^{-3} sec^{-1}, D is the diffusion coefficient in units of cm, $\phi = nv$ is the neutron fluence in units of neutrons cm^{-2} sec^{-1}, Σ_a is the absorption cross section per unit volume in units of cm^{-1}, and ∇ is del, the vector differential operator.

BOLUS. A mass of masticated food within the mouth or alimentary canal.

BOMBACECEAE. Balsa Tree; Baobob Tree; Silk Cotton Trees.

BOMBARDIER BEETLE (*Insecta, Coleoptera*). A ground-beetle which discharges a strong-smelling volatile secretion in small jets when disturbed. Each discharge is accomplished by an audible report and a visible puff of vapor as the secretion evaporates. The material ejected contains benzoquinone and 2-methylbenzoquinone. The numerous species make up the genus *Brachinus.*

BOMBARDMENT. This term is used in atomic and nuclear physics to denote the action of directing a stream of particles or photons against a target. The term is sometimes used for irradiation in a nuclear reactor.

BOMBESIN. Brain and Nervous System.

BOMB CALORIMETER. Calorimetry.

BOMBER. Airplane.

BOND (Carbon). Organic Chemistry.

BOND (Chemical). Chemical Elements.

BONDED STRAIN GAGE. Strain Gage.

BONDI AND GOLD UNIVERSE MODEL. Cosmology; Red Shift.

BONDING (Ceramics). Ceramics.

BONDING ELECTRON. Electron.

BOND (Molecular). Molecule.

BOND (Physical). Adhesion (Physics).

BONE. The hard, calcified tissue which forms the major part of the skeletal system of the body. The bones and cartilage are referred to as *connective* or *supporting* tissue because their chief function is structural. The distinction should be made between the terms *bone* and *bones.* Bone is a *tissue,* derived from connective tissue cells which become specialized in function. Bones are *organs,* such as the skull, pubis, tibia, fibula, and so on. See also **Calcium.**

Bone tissue consists of two permanent components: (1) the specialized cells of the bone; and (2) the surrounding *matrix,* which is composed of minute fibers and a cementing substance. This cementing substance contains mineral salts, mainly calcium phosphate. Similar to bone, and comprising a portion of the skeleton is *cartilage.* Cartilage is much more elastic than bone; it is often referred to as "gristle." Some bone may begin as cartilage which is later replaced by bone tissue.

Mature bone of mammals is *lamellated,* i.e., it is made up of thin plates (*lamellae*) of bone tissue. The plates occur in bundles. This arrangement offers increased resistance to shearing forces. The shape and arrangement of the lamellae differ in the two major types of mature bone (1) *spongy,* and (2) *compact.* In spongy bone, the matrix consists of a lamellated network of interlacing walls resembling the structure of a sponge. This form can be found in the skull and ribs. In compact bone, the bundles of lamellae are arranged in vertical cylinders around a central canal. This bone is found in the long bones of the arms and legs. The blood vessels and nerves run through the central canals of compact bone and send minute extensions into the bone substance. Great numbers of these vertical cylinders are needed to make up the thickness of a typical bone.

Bone grows by the addition of new bone to old bone. In spongy bone, new bone is deposited upon the old within the meshes of the lamellated network. In compact bone, new bone is primarily laid down on the outer surface. In both types, the bone is first laid down as immature (soft) bone which gradually becomes mature bone, hard and rigid with calcification. Long, hollow bones, such as those in the arm or leg are made from compact bone. They grow in circumference by the deposition of bone on the outer surface of the shaft. At the same time, the inner cavity becomes enlarged by the resorption or eating away of bone tissue. The ends of long bones are not hollow, but consist of a spongelike section of bone covered by a layer of compact bone and capped by cartilage on the joint surface where one bone moves against another. See accompanying figure.

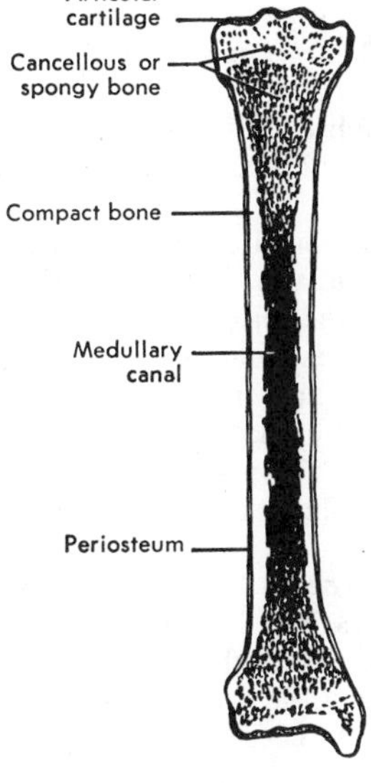

Longitudinal section through the tibia. The red marrow is in the cancellous bone and the yellow marrow in the medullary canal. The periosteum is a layer that provides a smooth protective covering around the outer surface.

The structure of the cartilage, which caps the ends of bones that rub against one another in joints is adapted to bear the strain of pressure and to facilitate the smooth gliding of the opposing surfaces during motion.

Long bones provide an example of the principle that a hollow tube is stronger than a solid one. A long bone, such as the thighbone (*femur*) is subjected to enormous stresses in the form of bending forces and in weight-bearing. It gains maximal strength with a minimal amount of material by increasing the size of the hollow center, while adding to the tissue on the outer surface at the same time.

Lengthening of long bones is accomplished by the development of bone at the ends. Between the spongelike bone ends and the shaft of the bone is an area of growth called the *epiphysis* or *epiphyseal cartilage.* Growth in length takes place only in this zone. The older cartilage becomes bone in the area next to the shaft, while new epiphyseal cartilage continues to form in the area next to the cap. When the epiphyseal area is completely replaced by bone tissue, the bone ceases to grow in length. Normally, in the human being, such growth is completed at about age 25, but physiologic disturbances may accelerate, retard, stop, or prolong growth. Hormones are important in the "sealing" of the epiphyses.

In infants and children, bones are softer than in adults, and yield readily to pressure or injury. This accounts for malformations, distortions in posture, foot defects, and other bone disorders. Young bones bend before breaking, and so-called "greenstick fractures" are common in children. In such a fracture, the shaft bends; and when the force is great enough, the bone on the convex surface breaks, much as a green twig may splinter along one side remaining intact on the other.

A deficiency of bone-making materials or disturbed processes of utilization of these materials may increase the softness and porous condition of bone. In the vitamin deficiency disease, rickets (lack of vitamin D), the shafts of long bones bend under strain, such as weight-bearing. Consequently, the patient may have curved long bones throughout his life. When vitamin C is inadequate, changes may occur at the ends of the long bones in the line between the shaft and the epiphysis and under the periosteum as a result of hemorrhages in growing children. In the adult, only the periosteal changes occur. Older persons have bones which are more porous because their bodies are not able to utilize bone-making material adequately, while absorption of bone matrix continues.

Bone is covered by a membrane called the *periosteum,* which contains the vessels for supplying some of the nourishment to the bones. The periosteum is composed of two layers of connective tissue. The outer layer is a compact arrangement of specialized fibers liberally supplied with blood vessels and nerves. Because of its nerve supply, the periosteum is sensitive and accounts for any pain or pressure felt in the bone. The structure of the inner layer of the periosteum (*cambium*) is less compact and has fewer blood vessels of its own. The periosteum adheres to the bone by means of fibers of connective tissue which anchor themselves in the bone tissue. The periosteum varies in thickness, being thinnest where the tendons of muscles are attached to bone, and thickest along the hollow portion of long bones.

Before birth and during infancy, *red marrow* is present in the long bones and in the network of spongy bones. Red marrow is one site of manufacture of the red blood cells. Blood vessels are threaded through the marrow, bringing oxygen and nutrients and taking away the waste products. The newly-made red blood cells enter the circulation by way of these vessels. When a person is about six year of age, changes begin to take place in the red marrow. *Yellow marrow,* or fatty marrow is formed, replacing the blood-making marrow; and most of the marrow cells change into fat cells. With these changes, the color of the marrow changes to yellow.

Yellow marrow is present in mature hollow bones. The formation of yellow marrow to replace red marrow takes place in a regular order beginning in the bones of the lower leg, followed by the thighbone, bones of the forearm, and finally in the bone of the upper arm. This replacement process also occurs in the epiphyses. In the adult, formation of red blood cells takes place only in the spongy flat bones of the skull, the ribs, pelvis, the bone of the spine, and the breastbone. See also **Blood.**

Biochemistry of Bone Formation. Bone formation is one of the earliest biological phenomena to be studied from a biochemical standpoint, yet its essential nature remains a subject of active research and speculation. Although the chief characteristic of bony tissue is its high content of inorganic salts, before considering the events associated with the formation of bone crystals some reference is necessary to the biogenesis of the organic portion within which the salts are deposited.

The bones originate from embryonic mesenchymal connective tissue cells which differentiate into bone-forming cells or osteoblasts. The formation of most embryonic bones occurs by calcification of a previously generated cartilaginous model, the remainder (intramembranous ossification) being the result of direct mineralization of connective tissue. In the former, the cartilage cells become hypertrophic and form centers of ossification from which cartilage is replaced centrifugally by bone cells. In the latter, calcification of the intercellular matrix occurs under the influence of osteoblasts which arise by transformation of connective tissue cells. The diaphysis, or shaft, elongates by calcification of the epiphyseal cartilage plate which is continuously regenerated by osteogenic mesenchymal cells. It increases in diameter by accretion beneath the layer of connective tissue covering the bone (the periosteum) and by concomitant removal from the endosteal surface of the marrow cavity.

Calcification results in the formation of the trabeculae of spongy bone, a form characterized by a high proportion of marrow and a profuse blood supply. Progressive deposition of new layers of bone, covered with osteoblasts, results in the generation of compact bone which is made up of units called haversion systems or osteones, each consisting of interwoven layers of bone oriented around a central vascular canal. The intercellular material is permeated by small spaces (lacunae) containing branched cells termed osteocytes. These cells are similar to osteoblasts, are rich in glycogen, and are necessary for the maintenance of bone cells. The osteones are subject to a continuing remodeling, apparently under the action of large, multinucleated osteoclasts located in tunnels which infiltrate the tissue prior to resorption.

The organic matrix of bone consists essentially of bundles of collagenous fibers imbedded in a ground substance. Although the general properties of bone collagen, which makes up over 90% of the dry fat-free organic matter, are similar to those of collagen derived from other forms of connective tissue throughout the body, the material present in osseous tissue apparently possesses some unique characteristic necessary for the nucleation of salt crystals. The ground substance is characterized chemically by the presence of mucopolysaccharides, including chondroitin sulfate, hyaluronic acid and keratosulfate, the physiological significance of which is still obscure.

The process by which bone crystals are deposited in the organic matrix, their internal structure and their chemical constitution have been under investigation for many years. The mineral consists mainly of Ca^{2+} and PO_4^{3-} ions, with smaller amounts of CO_3^{2-}, OH^-, Mg^{2+}, Na^+, F^- and citrate^{3-}. However, the concentrations of the minor ions is uncertain owing to the occurrence of surface absorption phenomena, exchange with components of the fluid medium, and the possibility that some constituents (e.g., citrate) are in a separate phase. Electron microscopy suggests that the crystals are rod-like with a diameter of about 50 Å and that they may be oriented in chains along the collagen fibrils. X-ray diffraction and chemical analysis indicate that bone mineral has the crystal lattice structure and composition of a substituted hydroxyapatite $[Ca_{10}(PO_4)_6(OH)_2]$. The architecture of the crystals provides for an enormous surface area in proportion to mass, thereby exposing the salts to intimate contact with constituents of the surrounding fluid. Exchange occurs actively, particularly in trabecular bone, not only between ions of the same species but also between dissimilar species: CO_3^{2-} for PO_4^{3-}, Sr^{2+} for Ca^{2+} and F^- for OH^-. The Ca:P molar ratio for bone is very close to the theoretical value for hydroxyapatite (1.67).

Studies on the formation of the bone salts have centered around the physiochemical concept that crystallization occurs when the concentration of Ca and P ions in the blood and circulating fluids exceeds the solubility product constant. Plasma P is present mainly as HPO_4^{2-}, in a smaller amount as $H_2PO_4^-$ and in minute concentrations as PO_4^{3-}. Observations on the calcification of cartilage *in vitro* indicate that the product $[Ca^{2+}] \times [HPO_4^{2-}]$ is the critical ion relationship in crystal formation, and it has been proposed that whereas the serum and extracellular fluid are normally undersaturated with respect to $CaHPO_4$, they are supersaturated with respect to bone salts. This proposal suggests that an ion gradient exists between the interstitial bone fluid and the extracellular fluid which is maintained by cellular activity, and it stresses the importance of the minor ions, particularly citrate^{3-}, in determining the degree of saturation. The production of citrate^{3-} by bone cells may determine whether the medium is undersaturated or supersaturated with respect to Ca^{2+} and HPO_4^{2-}, *i.e.*, whether dissolution or deposition of bone salts occurs. It is further suggested that vitamin D and parathyroid hormone may exert their influence on bone metabolism by regulating the metabolic activity of the cells and hence the production of citrate.

The mechanism by which crystal formation is initiated is still obscure. Following the discovery of phosphatase in calcifying cartilage, the view became prevalent that the local action of this enzyme on some organic phosphate ester produced a high concentration of phosphate ion which exceeded the solubility product constant of bone mineral. This theory has been largely discarded in favor of a "seeding" mechanism which assumes that some component of the organic material (presumably the collagen or the mucopolysaccharide) furnishes the seeding sites. Reconstituted collagen fibers have the ability to induce crystal formation *in vitro* from stable solutions of Ca^{2+} and HOP_4^{2-} ions; however, it is difficult to prepare collagen that is completely free of mucopolysaccharides. Once the nuclei have been formed, crystallization proceeds spontaneously. Apart from the unexplained role of phosphatase, glycolysis appears to be a necessary accompaniment of bone salt formation; inhibitors of glycolysis interrupt crystallization in a reversible manner.

The biochemical remodeling process is central to several major diseases and disorders of bone. In osteoporosis, the mass of bone is reduced per unit volume which, in connection with other less complex structural materials, would be described as a lowering of density. In osteomalacia, there is deficiency (or failure) to mineralize the newly forming organic matrix of bone. Rickets in children is a manifestation of disorder in the remodeling process that has some parallels to osteomalacia.

Bone Disorders and Diseases

Fractures. A fracture is a broken bone. In a *closed fracture*, the bone is either cracked or completely broken in two, but there is no connecting wound from the break extending through the skin. In an *open fracture*, the bone is broken and bone fragments penetrate the surface of the skin; or an external object, such as a bullet, penetrates the skin and forms a connecting wound with the broken bone. Proper handling of a fracture is essential. Rough handling may cause a simple fracture to become an open fracture; and it may cause the bone fragments to injure the blood vessels, nerves, and other tissues.

Injury to the head may result in (1) a *fracture*, in which the skull is broken or cracked, (2) a *depressed fracture*, in which the skull is broken and fragments of bone are embedded in the brain tissue, or (3) a *concussion*, in which the brain is bruised by swelling resulting from hemorrhage.

A fracture or injury to the neck or spinal column usually causes intense pain. The pain in most cases radiates outward to other parts of the body, dependent upon which of the vertebrae is affected. Fractures high on the spinal column may result in pain in the arms or chest, while fractures lower down cause pain in the abdomen or legs. When an injury has affected the spinal cord, the patient may suffer a loss of sensation and ability to move the part of the body which is supplied by nerves from the spinal column at the point of fracture and below it.

An injury is termed a *dislocation* when a bone gets out of place at a joint. The joints are flexible sacs held in place by ligaments. Ligaments are tough fibrous bands of tissue which extend from one bone to the other, entirely surrounding the joint. In a dislocation, the ligaments and sacs are partially or completely torn, the bony surfaces may be fractured, and the blood vessels and nerves may be injured or torn. Dislocations of the shoulder and fingers are most common, followed by dislocations of the jaw, elbow, kneecap, and hip.

Osteomyelitis. An infection of the bone caused by pyogenic bacteria, mycobacteria, fungi (uncommon) and viruses (rare). The disease may range from acute infection to an indolent subacute condition to a recurring chronic infection. In about one-fifth of cases, infection is by the bacteremic (blood) route and is called *hematogenous osteomyelitis,* frequently affecting children, but also quite common among older people over 50 years of age. The disease occurs to some degree among all age groups. The bones most frequently infected are the femur, tibia, and humerus, which are all long bones. Recent statistics indicate that *Staphylococcus aureus* is the causative agent in nearly half of the cases. Also important are *Mycobacterium tuberculosis* (15% of cases); enteric gram-negative bacilli (13%); *S. pneumoniae* (5%); Group A *Streptococcus* (3%). *Staphylococcus epidermis, Bacteroides* species, *Nocardia asteroides,* and *Coccidioides immitis* are less frequently implicated. Authorities observe that although tuberculosis infections generally have been declining for several years, this infection still is commonly seen in osteomyelitis. Bone infections have been increasing among heroin addicts. The marked increase in reconstructive orthopedic surgery has contributed to osteomyelitis secondary to a contiguous focus of infection.

Early symptoms of osteomyelitis parallel those of numerous infections—chills, fever, and leukocytosis (high white cell count) accompanied by tenderness, swelling, and pain in the region of infected bones. Bacteremia (bacteria in blood) is present in over half of the cases. In about one-fifth of cases, hematogenous osteomyelitis becomes chronic and may extend over months or years. *Brodie's abscess* is a subacute form of hematogenous osteomyelitis. *Vertebral body osteomyelitis* usually results from bacteremias. *S. aureus* arising from skin infections, narcotic addiction, and urinary tract infections and endocarditis is implicated in over 50% of cases. The predominant symptom of vertebral body osteomyelitis is low back pain. There also may be paraspinal muscle spasms. Subgroupings of the disease include *pyogenic osteomyelitis* (lumbar vertebrae are affected) and *spinal tuberculosis* or *Pott's disease* (thoracic vertebrae are affected). In persons with sickle cell anemia, *Salmonella* is second to *S. pneumoniae* as a cause. Syphilitic osteomyelitis is uncommon.

Treatment of acute hematogenous osteomyelitis is parenteral administration of appropriate antibiotics over a period of several weeks. This therapy is usually successful. Important to the therapy is accurate determination of the microorganism responsible for the infection.

Secondary osteomyelitis developing from a contiguous focus of infection is the other major group of osteomyelitis diseases. This type of infection may follow a puncture wound or cat bite, or may be associated with infections resulting from thermal burns and complications arising from open reduction of fractures and reconstructive orthopedic surgery. Postoperative deep wound infections following total hip replacement are not uncommon. These infections may arise from pyogenic bacteria (*S. aureus* or enteric bacilli) or from normally noninvasive bacteria, such as *S. epidermis* or diphtheroids. These infections usually appear relatively soon after surgery. At a later time, deep infections may not be apparent for several months or even years. In one statistical study, it was found that such infections arise in 2.7% of such surgical procedures. Treatment is essentially by administration of antibiotics.

Osteoporosis. A disorder, the result of which is a decrease of bone mass in the absence of a mineralization defect. The bone remodeling process previously described in this entry, rather than maintaining an equilibrium condition in this disease, is biased toward a greater rate of bone resorption than of bone formation. This suggests that the disease derives from an alteration in the relationship (both quantitatively and qualitatively) of osteoclasts with osteoblasts. In osteoporosis, radiological examination will reveal decreases in bone density, thinning of the cortexes (outer layer of bone), and a loss of trabeculae (supporting fibers). Radiological examinations, however, may not be precisely indicative of the disease, particularly because of the possible presence of osteomalacic states, multiple myeloma, metastatic neoplasms, and hyperparathyroidism. Most commonly, osteoporosis is seen in older people, particularly in women, and is the fundamental underlying cause of skeletal fractures in middle-aged and elderly women.

Involutional osteoporosis appears to be a natural accompaniment of aging, particularly of women after menopause. This is classified as a metabolic bone disease. The condition is much less commonly seen in men and consequently appears to result from the poorly understood biochemical differentiation of males and females. The condition among females bears no known racial variation. Sedentary individuals are more prone to involutional osteoporosis because exercise tends to increase bone mass in either sex. Certain factors, including dietary intake of calcium and phosphate, protein and vitamin D metabolic abnormalities, and hormones, among others, have been suggested as contributing causes, but there is no hard evidence in these areas.

Normally, throughout most of life, the remodeling processes (bone formation and resorption) occur together, but not always simultaneously. A number of authorities have suggested that resorption may be made up of short, intense periods, whereas bone formation is a slower process that continues over longer periods. In older people with osteoporosis, resorption exceeds formation, and in view of the fact that resorption may occur as the faster of the two processes, this may account for the relatively rapid onset of osteoporosis at later stages in life.

Osteoporosis occurs in a number of patients who have been on *glucocorticoid therapy* for long periods of time. Endogenous glucocorticoid excess occurs in Cushing's syndrome, causing osteoporosis among other conditions. It appears that in persons taking glucocorticoids (for example, prednisone), the severity of osteoporosis varies more closely with the duration of therapy than with the dosage. Studies indicate that glucocorticoids increase bone resorptive surfaces, while they decrease bone formation. These factors, working together, can create a rapid rate of bone loss. Halting the therapy arrests the degradation, but there is no evidence that bone restoration will occur. Sometimes vitamin D will be used to improve the intestinal absorption of calcium.

Hyperthyroidism sometimes increases bone resorption with accompanying loss of calcium and hydroxyproline through the urine.

Treatment and management of osteoporosis include mechanical devices, such as braces and corsets, analgesics to reduce pain, supervised exercises, regular periods of rest with proper postural position, and a number of drugs. Since the 1940s, estrogens have been used for treating women with the disease. Estrogens, which decrease bone resorption, have been shown to be effective in reducing fractures and the height loss frequently experienced with osteoporosis. Estrogens are used by physicians with discretion, however, because of some disadvantages (renewal of vaginal bleeding, swelling of breasts, and the possible increase of risk of endometrial carcinoma). Where osteoporosis is age-related, calcium supplements and vitamin D therapy may be beneficial. Where recurrent fractures, height loss, and spinal deformity persist, the physician may consider the administration of sodium fluoride, phosphate, or parathyroid hormone.

Paget's Disease (*Osteitis deformans*). This condition, characterized by abnormally thickened but weak bones, is believed to be caused by excessive activity of both the osteoblastic and osteoclastic cells. Some authorities postulate that an endocrine disturbance may be the root cause. The disease is found in about 4% of men over 40 years of age with an Anglo-Saxon heritage. Women are much less frequently affected. Familial connections have been observed, but no genetic details have been delineated. Normally the symptoms, including minor pain, are of such moderate proportions that the condition usually is not brought to the attention of a physician until a radiologic examination is made for other purposes—or until considerable bone deformation has taken place. In its mild form, indomethacin therapy is used. Where the disease is more serious, calcitonins and diphosphanates may be used. Where the disease seriously affects the hip, replacement arthroplasty is generally suggested.

Osteomalacia. Sometimes defined as softening of the bones due to calcium deficiency, this disease may arise from a number of causes, essentially of a metabolic nature. The disease is frequently indistinguishable clinically and radiologically from osteoporosis. In cases where there are hereditary disorders of vitamin D metabolism (hypovitaminosis D) or renal tubular acidosis, clinical manifestations are more easily identifiable. Increased renal clearance of phosphate is the principal indicator of hereditary vitamin D-resistant *rickets.* The indicated treatment for this condition is high dosages of vitamin D with a phosphate supplement. Many years ago, prior to current appreciation of the functions and importance of the various vitamins, ricketts occurred widely in infants and children, creating deformities that had to be carried throughout life. The disease is characterized by defective ossification caused by faulty deposition of calcium salts at the growing ends of the bones, generally ascribed to a deficiency of vitamin D. See also **Calcium (Biological Role of)**; and **Vitamin D.**

Traumatic Disorders. Reference to the diagram in the entry on **Skeletal System** will reveal the general structure of the spine, with the cervical vertebrae at the top of the spine and, progressing downward, the thoracic vertebrae and the lumbar vertebrae. The vertebral column, with 33 vertebrae, in a male of average height is about 28 inches (71 centimeters) long. Although the vertebrae differ in shape and size, they have a similar structure. The joints between the bodies of the vertebrae are somewhat movable and those between the arches are freely movable. *Disks* of fibrocartilage located between the vertebrae connect the bodies of the vertebrae and function essentially as shock absorbers. Because 31 nerves are associated with the spine, diseases and injuries of the spine and its vertebrae may seriously involve a few or several of these nerves. For example, in the cervical region, there are 8 pairs of nerves; in the thoracic region, 12 pairs; in the lumbar region, 5 pairs; in the sacral region, 5 pairs; and in the coccygeal region, 1 pair. The disks are subject to damage from trauma (injury) or degenerative changes. These factors may cause nerve root compression.

In *cervical disk disease,* a lateral protrusion or herniation, which most commonly occurs at the C5–6 or C6–7 vertebral spaces, may happen spontaneously, or as the result of trauma. A C5–6 disk rupture causes neck pain which is noted across the shoulders and down the arm, and may involve the thumb. Muscular weakness in the biceps may be noted. A C6–7 disk rupture causes pain and muscular weakness of the mid-arm (triceps brachii) and reduction of reflex that may extend to the index and middle fingers. Coughing, sneezing, etc., will usually accentuate these symptoms. Relief is usually provided by wearing a cervical collar adjusted to stretch the neck. The next step in therapy, if needed, is cervical traction, which normally does not require a hospital environment. In a minority of patients, surgical correction may be indicated.

In *lumbar disk disease,* the usual sites of disk protrusion are between the L4 and L5 and between the L5 and S1 interspaces. Lumbar herniation produces low back pain, which radiates and involves the thigh and calf. Such pain is accenturated by movement of the back, coughing, etc. The physician will differentiate this condition from peripheral nerve disease through a series of limb motion tests. Therapy commences with absolute bed rest in a hospital environment. Lying in a flat position is usually mandatory. Supportive measures, such as application of heat, use of analgesics to control pain, and the administration of muscle relaxants, such as diazepam, are immediately instituted. In most cases marked improvement can be expected within about three weeks. In non-responsive cases, surgical correction is indicated.

Myeloma. The stimulus for malignant conversion of the plasma cells (*myeloma*) in humans is not known. Plasma cells reside in bone marrow. Replacement of normal marrow elements may become so extensive that plasmoblasts appear in the peripheral blood, causing plasma cell leukemia. Erosion of bone may be diffusive or take the form of tumors (*plasma cytomas*). Myeloma ultimately may involve many bones of the body and migrate to other organs, such as nerves and kidneys (*multiple myeloma*), or the myeloma may be confined to one bone for long periods. There may be involvement of the axial skeleton that leads to fractures and vetebral collapse. Where bone destruction is extensive, this may be evidenced by hypercalcemia (abnormally high calcium in blood), the symptoms of which include nausea, vomiting, and somnolence. The diagnosis of multiple myeloma is complex, involving hematologic, blood chemical, urinary, and radiologic studies. Ultimate therapy and prognosis is affected by the site of the plasmocytoma. Initially, the diagnosis of multiple myeloma may be suggested by the presence of high serum proteins. Radiation and chemotherapy are important elements in current procedures for the treatment of this disease. Myeloma confined to one site may progress slowly; multiple myeloma usually takes a rather rapid course, measured in terms of months or a few years.

References

Aurbach, G. D. (editor): "Handbook of Physiology," Vol VII, American Physiological Society, Washington, D.C., 1976.

Austin, G. M. (editor): "The Spinal Cord: Basic Aspects and Surgical Considerations," 2nd edition, Charles C. Thomas, Springfield, Illinois, 1972.

Bergsagel, D. E.: "Treatment of Plasma Cell Myeloma," *Ann. Rev. Med.,* **30**, 431 (1979).

Case, D. C., Jr., Lee, D. J., III, and B. D. Clarkson: "Improved Survival Times in Multiple Myeloma," *Am. J. Med.,* **63**, 897 (1977).

DePlama, A. F., and R. H. Rothman: "The Intervertebral Disc," Saunders, Philadelphia, 1970.

Duszynski, D. O., et al.: "Early Radionuclide Diagnosis of Acute Osteomyelitis," *Radiology,* **117**, 337 (1975).

Heller, P.: "The Mechanism of the Immunologic Deficiency in Myeloma of Man and Mouse," *Blut,* **37**, 65 (1978).

Jowsey, J.: "Metabolic Diseases of Bone," Saunders, Philadelphia, 1977.

Krane, S. M., and L. V. Alvioli (editors): "Metabolic Bone Disease," Vol I, Academic, New York, 1977.

Levin, W. C.: "Symposium on Myeloma," *Arch. Intern. Med.,* **137**, 27 (1975).

McHenry, M. C., et al.: "Hematogenous Osteomyelitis: A Changing Disease," *Cleveland Clinic Q,* **42**, 125 (1975).

Mundy, G. R., et al.: "Direct Resorption of Bone by Human Monocytes," *Science,* **196**, 1109 (1977).

Owen, M.: "Cellular Dynamics of Bone. The Biochemistry and Physiology of Bone," in "Development and Growth," Vol III (G. H. Bourne, editor), Academic, New York, 1971.

Parfitt, F. B., and H. Duncan (editors): "Proceedings of the International Symposium in Clinical Aspects of Metabolic Bone Disease," American Elsevier, New York, 1973.

Rasmussen, H., and P. Bordier: "The Physiological and Cellular Basis of Metabolic Bone Disease," Williams & Wilkins, Baltimore, Maryland, 1974.

Woodruff, R. K., et al.: "Clinical Staging in Multiple Myeloma," *Br. J. Haematol.,* **42**, 199 (1979).

BONE CONDUCTION. The process by which sound is conducted to the inner ear through the cranial bones.

BONITA. Tunas.

BONNER DURCHMUSTERUNG. The name applied to the monumental catalogue of over 324,000 stars observed by that tireless observer, F. W. A. Argelander. Accompanying the catalogue is an atlas of the heavens upon which each of the catalogued stars is shown by a dot, the size of the dot being proportional to the apparent brightness of the star. The catalogue contains practically every star brighter than the tenth magnitude north of declination −2°. The catalogue is commonly referred to as the B.D., and in many astronomical writings, a particular star is referred to by its B.D. number (i.e., by the number assigned to it in the Bonner Durchmusterung).

The catalogue was continued by Schonfeld down to declination −23°, and Thome, at Cordoba, has extended it still further to −61°. It is hoped that the plan will be continued to the South Pole.

In each of the catalogues, stars are numbered in order of increasing right ascension within a particular zone of declination. Hence, a star known as CDM −48 1116 is the 1116th star in the Cordoba extension of the B.D. catalogue between declination −48° and −49°.

See also **Amalgest.**

BONNET FORM. Mean Value Theorems.

BONY FISHES. Fishes.

BONY-TAIL (*Osteichthyes*). A fish (*Gila elegans*) of the Colorado and Gila rivers, related to the minnows and chubs.

BONY TONGUES (*Osteichthyes*). Of the order *Isospondyli,* family *Osteoglossidae,* the bony tongues are fresh water fishes, apparently found only in the streams, rivers, and lakes of South America, Malaysia, Australia, and Africa. A bony plate covers the head, the eyes are large, and the body is covered with heavy scales. The *Arapaima gigas* (giant arapaima of the Amazon) is the largest of the species and, in fact, may be the largest of fresh water fishes. Records indicate attainment of lengths up to 15 feet (4.5 meters), but the average is considered to be less than 8 feet (2.4 meters). It is a favorite for aquariums. The scales are olive-green. The arapaima is well regarded

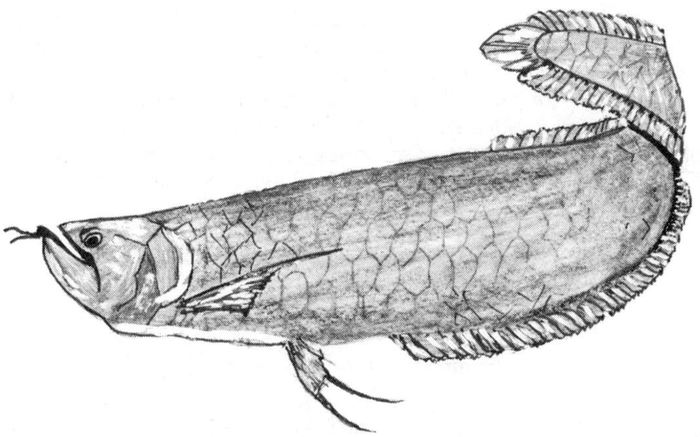

Bony tongue. Arrowana (*Osteoglossum bicirrhosum*).

as a food fish in South America with exception of Guyana where it is not esteemed. The *Clupisudis niloticus*, found in African waters, is similar to the South American arapaima, but is much smaller, rarely attaining a length in excess of 3 feet (1 meter). Both of these species are nest builders. These nests are built in the sandy bottoms of shallow areas, the fish using the fins for digging. A typical nest will have a diameter of about 20 inches (51 centimeters). The *Osteoglossum bicirrhosum* (South American arawana) grows to about 2 feet (0.6 meter) in length and appears much as the arapaimas. See accompanying illustration. It is suspected that it is a mouth breeder. Other species of bony tongues include: *Scleropages* (Asia); *Scleropages formosus* (Borneo, Malay region, Sumatra); and *S. leichardtii* (Australia and New Guinea).

BOOKKEEPING OPERATION (Computer). An operation that does not contribute directly to the result, e.g., arithmetical, logical, and transfer operations used in modifying the address section of other instructions, in counting cycles, and in rearranging data.

Similar to housekeeping operation, but generally implies an enumeration and recording of actions taken. See **Housekeeping (Computer System).**

BOOK LOUSE (*Insecta, Corrodentia*). A small insect found in old papers, books and rubbish and in collections of biological specimens. The order to which they belong is a small one containing winged species found on bark and lichens, and wingless species to which this name is applied.

Book lice must be very numerous to do appreciable damage, and since they frequent damp situations, heating and drying rooms where they are found is usually a simple method of destroying them. Severe infestations can be checked by fumigation. See also **Corredentia.**

BOOLEAN ALGEBRA. A distributive lattice which has universal bounds and has complements. (A lattice is distributive if $a(b + c) = ab + ac$ and $a \times bc = (a \times b)(a \times c)$, has universal bounds if it contains elements 0 and I with $0 \leqq a \leqq I$ for all a, and has complements if for every a there exists an a' such that $aa' = 0$ and $a + a' = I$.) Boolean algebra can be characterized in many other, equivalent ways. The subsets $a, b, \ldots$ of a set of objects S form a Boolean algebra if ab denotes the *intersection* and $a + b$ the *union* of a and b. The algebra of statements $a, b, c \ldots$ with connectives "and," "or," "not" form a Boolean algebra if $a + b$ means a or b (including the possibility of both), while ab means a and b. The simplest Boolean algebra is the one whose elements are the empty set, θ, and the set of one point I.

In practical terms of computer technology, Boolean algebra, first proposed by George Boole (1815–1864) provides a mathematical procedure for manipulating logical relations in symbolic form. Boolean variables are confined to two possible states or values. The pairs of values possible are YES and NO, ON and OFF, TRUE and FALSE. In engineering practice, it is common to employ 1 and 0 as the symbols for the Boolean variables. Inasmuch as a digital computer generally will use signals which have only two possible states or values, Boolean algebra makes it possible for designers of computers to combine these variables mathematically and to manipulate the variables in a way to obtain that minimum design which provides a desired logical function. Some of the logical operations defined in Boolean algebra and their accompanying symbols are given in the table below.

Symbol	Definition	Logical Operation
	$A \cdot A = A,\ A \cdot 0 = 0,\ A \cdot 1 = 1,\ A \cdot \bar{A} = 0$	AND
+	$A + A = A,\ A + 0 = A,\ A + 1 = 1,\ A + \bar{A} = 1$	OR
–	$A\bar{A} = 0,\ A + \bar{A} = 1$	NOT
⊕	$A \oplus B = \bar{A}B + A\bar{B} = \overline{A \odot B}$	EXCLUSIVE OR
⊙	$A \odot B = \bar{A}\bar{B} + AB = \overline{A \oplus B}$	COINCIDENCE
/	$A/B = \bar{A} + \bar{B} = \overline{AB}$	NAND (OR SHEFFER STROKE)
/	$A/B = \bar{A}\bar{B} = \overline{A + B}$	NOR (OR PEIRCE)

BOOLE'S INEQUALITY. An inequality concerning the frequencies in logical classes or equivalently of probabilities. For example if A_1, $A_2, \ldots A_k$ are compatible events (any or all can occur at any particular trial) the probability that *at least* one occurs cannot exceed the sum of the probabilities that each occurs independently of the others.

BOOM. A boom is a movable inclined arm of wood or steel used on some types of cranes or derricks to support the hoisting lines which carry the loads. The loads cause direct compression in the boom due to the manner in which the hoisting lines are connected to the member.

The word boom also describes a floating chain of logs, which is anchored in such a position in a body of water as to deflect or intercept saw logs, or to prevent floating debris from approaching water intakes to pipe lines and penstocks. Nautically, a boom is a spar holding the foot of a fore and aft sail.

BOOM (Mysterious Acoustic). Brontide.

BOOSTER (Electrical). An electrical booster is inserted in series in an electric circuit; and increases the voltage of that circuit. There are several uses to which the booster can be put. It may be employed to compensate for a line voltage drop, or it may be employed to vary voltage in such a way that constant current is maintained. The boosting of dc circuits is accomplished by rotating equipment called booster generators. If this booster is driven by an electric motor the set is called a motor-booster. The booster generator can be used to raise the line voltage at a feeder point on an electric traction system.

The booster transformer is sometimes used in alternating current circuits. On a simple single-phase circuit it boosts the line voltage by connecting the primary of the transformer across the line, and the secondary in series. There are some disadvantages to this connection, however, since blowing of a fuse, or otherwise open-circuiting the primary, leaves the transformer connected as an open-circuited series transformer, and the open-circuit voltage on the primary winding may be excessive. The induction regulator is a form of booster transformer whose effect is varied by rotating one winding with respect to the other.

The term booster is widely used as a colloquialism for an amplifier between the antenna and receiving set (especially a television set) antenna terminals.

BOOTES (the herdsman). Although not in the zodiac, Bootes is one of the earliest recorded constellations. It is readily recognized in the early summer skies from the kite-shaped configuration of stars, with the bright star Arcturus at the position of the tail of the kite. Many of the other bright stars in Bootes are double stars, several of them forming interesting objects of study with relatively small instruments. (See map accompanying entry on **Constellations.**)

BOOTSTRAP AMPLIFIER. Amplifier.

BOOTSTRAP (Computer). A technique for loading the first few instructions of a routine into storage; followed by using these instructions to bring in the remainder of the routine. The technique usually involves either entering a few instructions manually, or more commonly, using a special key on the console which invokes a routine stored in read-only storage (ROS). Also used as a verb, meaning use of a bootstrap.

BORA. Winds and Air Movement.

BORACITE. This mineral is a magnesium borate containing some chlorine, $Mg_3B_7O_{13}Cl$. It appears to be isometric but probably becomes so only at 265°C below which temperature it is believed to be orthorhombic. Its hardness is 7; specific gravity, 2.9; luster, vitreous; color, white to gray, sometimes yellow or green; translucent to subtransparent. It occurs in beds with gypsum and salt in Germany, particularly at Stassfurt in Saxony.

BORANES. Boron.

BORAX. This hydrated sodium borate mineral, $Na_2B_4O_7 \cdot 10H_2O$, is a product of evaporation from shallow lakes and plays. Borax crystallizes in the monoclinic system, usually in short prismatic crystals. Its color grades from colorless through gray, blue to greenish. Vitreous to resinous luster of translucent to opaque character. Hardness of 2–2.5, and specific gravity of 1.715.

Borax from the salt lakes of Kashmir and Tibet has been known since early history of man. India, the U.S.S.R., and Persia possess small deposits. Extensive deposits are known in the United States, notably in Lake, San Bernardino, Inyo, and Kern Counties in California, and Esmerelda and Dona Ana Counties in New Mexico.

Its use include antiseptics, medicines, a flux in smelting, soldering and welding operations, a deoxidizer in nonferrous metals, a neutron absorber for atomic energy shields, in rocket fuels, and as extremely hard abrasive boron carbide (harder than corundum). See also **Boron.**

BORDONI PEAK. A maximum in the internal friction spectrum at low temperatures found in all face-centered cubic lattices. It may be characterized by the fact that it only occurs in deformed poly- or single crystals. There are two peaks. The processes contributing to the peaks are thermally activated ones, and the peaks are removed only by annealing above the recrystallization temperature. The processes contributing to these peaks are believed to involve intrinsic dislocation loss mechanisms.

BOREAL. Tar Sands.

BORE CHECKING. Thickness Measurement and Gaging Systems.

BORER (*Insecta*). This term usually refers to the larval form of a beetle or moth, that is, a *grub.* The grubs of most species are voracious eaters of plants, with a wide range of targets, such as vegetables, fruits, nuts, grasses, grains, weeds, trees, etc. Some of the borers, such as the European corn borer, inflict many millions of dollars worth of damage to crops. Some of the borers of interest to food production and agriculture include the following:

Bronze birch borer (*Agrilus anxius* of the family *Buprestidae*, order *Coleoptera*). Attacks white or paper birches. Native to North America.

Clover root borer (*Hylastinus obscurus*, Marsham of the family *Scolytidae*, order *Coleoptera*). Prefers red and mammoth clover. Of lesser importance to white and sweet clover, alfalfa, pea, and vetch.

Currant borer (*Ramosia* or *Synanthedon tipuliformis*, Clerck of the family *Aegeriidae*, order *Lepidoptera*). Widely distributed in North America, but of European origin. Attacks currant, as well as black elder, gooseberry, and sumac. More injurious to black than to the red currant. The yellowish borers or larva (½-inch; 12 millimeters long) are found inside the canes just above groundlevel. Insect cannot be reached by contact insecticides.

European corn borer (*Pyrausta nubilalis*, Hübner of the family *Pyralididae*, order *Lepidoptera*). Although the insect feeds on a variety of herbaceous plants, it prefers corn (maize) whenever available. See Fig. 1. A native of Europe, this borer is widely distributed in Europe

Fig. 1. European corn borer. (*USDA diagram.*)

and Asia and was imported into North America in about 1917, believed to have been contained in a shipment of broomcorn from Italy or Hungary. As of the early 1950s, the insect had spread into all major corn-producing areas of the United States. Annual damage reported has been hundreds of millions of dollars on the corn crop. The insect winters as a fully developed worm or caterpillar in a borrow which it has made in the stem of a food plant. The worm ranges from ¾- to 1 inch (18 to 25 millimeters) in length and is flesh-colored, with rather small, round, brown spots on its back. The worm can be found in any part of the stem or ear, but during winter, the worm's most common location is in the cornstalk just above groundlevel. As the weather begins to warm, the caterpillar prepares a rather flimsy cocoon in the burrow and there transforms into a smooth, brown pupal stage. Adult moths emerge in late spring at which time they may migrate considerable distances. Their movement is mainly at night. When young corn plants are in abundance, the female moth (pale yellow-brown) deposits from 500 to 600 eggs on the underside of leaves of the target plant, corn (maize) when available. Until about half-grown, the young larva feeds in tight spaces between leaves, husks, or ear and stalk. After they are half-grown, they commence feeding on the stalk, the ear, and all thicker portions of the plants. To do this, they commence boring. This feeding operation continues until the larva is fully developed. Researchers have counted nearly 200 borers on a single plant and over 40 borers have been found in a single ear.

Chemical controls can be effective, but preventive measures are also very important. These include removal and burning of infested debris. Late-planting has been found of value in some areas. Planting of resistant varieties is extremely important. Rotations also are effective. Parasites have been introduced as natural enemies of the corn borer, a number of which have been established. These include *Lydella stabulans grisescens* and *Macrocentrus gifuensis.* Certain fungi, such as *Beauveria bassiana* and *Perezia pyraustae* (protozoan), also are effective.

Flatheaded apple tree borer (*Chrysobothris femorata*, Oliver, family *Buprestidae*, order *Coleoptera*). A very severe economic pest of deciduous fruit trees, as well as numerous decorative shrubs and trees. The insect is particularly damaging to young trees, during the first 2 to 3 years after planting. They are more active during dry periods. The insect mines the main trunk as well as large branches, just under the bark and penetration into the wood may be as much as 1 or 2 inches (2.5 to 5 centimeters). Such injuries are found most frequently on the sunny side of the tree. The burrows are packed with wood debris (sawdust) or excelsior fibers. All fruit-producing areas of the United States and Canada are effected. A close relative, the Pacific flatheaded borer (*Chrysobothris mali*, Horn), occurs throughout western North America and is found as far south as Arizona and Texas. The insect winters in the form of the grub (borer) stage. Length may range from ½- to 1-inch (12 to 25 millimeters). As the borer develops, it penetrates more deeply into the wood. When mature, the grub is yellowish-white, legless, with a characteristic enlargement just in back of the head. The beetle usually lies with part of its body curled to one side. The adult beetle has a very blunt head, is about ½-inch (12 millimeters) long and about ⅕-inch (5 millimeters) wide. They are a dark-green-brown color with a metallic cast and distinctly love sunlight.

Peach tree borer (*Sanninoidea* or *Conopia esitiosa*, Say, family *Aegeriidae*, order *Lepidoptera*). This is the most severe economic pest of peach. The pest also attacks apricot, cherry, nectarine, plum, and prune. The insect winters in the larval form and ranges widely in length up to about $\frac{1}{2}$-inch (12 millimeters). Some varieties may be as short as $\frac{1}{8}$-inch (3 millimeters). Size variation results from differing spans of time allowed for the worms to develop. Shortly after the soil has had opportunity to warm in the spring, the worms become active. At this time, the worms, about 1 inch (2.5 centimeters) in length, congregate under the bark of the tree and close to the ground. The worm is a dirty-white color with a dark, brownish head and has a noticeable plate behind its head. Very shortly, they convert into a cocoon to the brown pupal stage and will be found on the surface of the burrows they have made, or immediately under the soil. The range is from about 2 to 3 inches (5 to 7.5 centimeters) below groundlevel to about 1 foot (0.3 meter) above the ground. In observing an infected tree, one may see considerable bark debris at the base of the tree. The adults emerge in mid-summer, but this process may continue until late September. The moth has clear hind wings, is of a blue-black coloration with an orange crossband on the abdomen. See Fig. 2. They are somewhat wasplike in appearance and flight pattern.

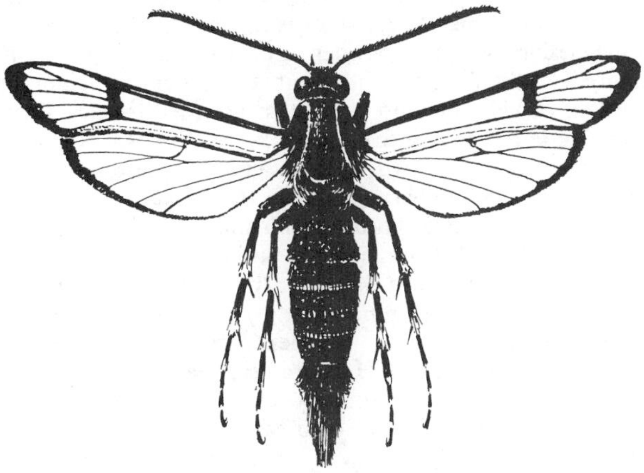

Fig. 2. Adult peach tree borer. (*USDA diagram.*)

Unless eradicated, the moths usually kill a tree within a season or two. The insect is prolific, the female laying from 200 to 800 eggs on the tree trunks or in separations in the soil just a few inches from the trunk. Certain chemical controls have proven relatively effective, including paradichlorobenzene. Treatment is similar to that for peach twig borer, described nest in this list. The peach tree borer is distributed throughout the United States and Canada. A related species is the western peach tree borer (*S. exitiosa graefi*). The lesser peach tree borer (*Synanthedon pictipes*, Grote and Robinson) is similar in its habits and destruction to deciduous trees.

Peach Twig Borer (*Anarsia lineatella* Zeller, family *Gelechiidae*, order *Lepidoptera*). The insect attacks almond, apricot, peach, and plum. The insect winters as a partially grown caterpillar. The caterpillar is brown with a black head, very small, ranging from $\frac{1}{16}$- to $\frac{1}{8}$-inch (1.5 to 3 millimeters) in length. It hides in a cocoon that is closely fixed to the tree bark, either on truck or branches. The larva leaves the cocoon at same time the tree is leafing out, and commences to feed on the tender new growth, causing wilting and destruction of twigs. The worms become about $\frac{1}{2}$-inch (12 millimeters) long and then again spin cocoons on the larger tree branches, or even the tree trunk. They become small, grey moths, with a wingspread of only about $\frac{1}{2}$-inch (12 millimeters). The female moth lays her eggs on twigs and the cycle is repeated. As many as four generations can occur per year.

Young borers can be cut from the tree with a knife. However, care must be taken not to cut away more wood than necessary. Older borers can be killed by probing with a wire with a hooked tip. The wound should be painted over to reduce damage from other insects and diseases.

Pear tree borer (*Agrilus sinuatus*, Oliver, family *Buprestidae*, order *Coleoptera*). This insect affects pear as well as some timber and shade trees. A native of Europe, the insect has been known in North America since the mid-1980s, particularly in the northeastern states. The life cycle of the insect is typical of moths. Both adult beetle and larva cause damage, the grubs burrowing into the bark and the beetles feeding on foliage.

Potato stalk borer (*Trichobaris trinotata*, Say, family *Curculionidae*, order, *Coleoptera*). An occasional pest in potato fields. The beetle larva eats out the interior of the plant stalks, usually causing the plant to die. The insect is most injurious to early potatoes. Other plants attacked include eggplant and several weeds.

Shot-hole borer (*Scolytus rugulosus*, Ratzeburg, family *Scolytidae*, order *Coleoptera*). There are numerous species of this insect. The larva is a white to yellowish-white or brown grub from 1 to $1\frac{1}{2}$ inches (2.5 to 4 centimeters) in length. See Fig. 3. The insect is very similar

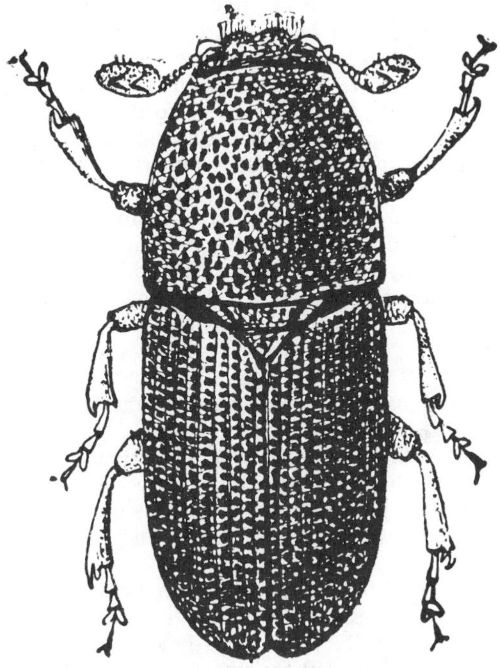

Fig. 3. Shot-hole borer. (*USDA diagram.*)

in habit to the bark beetle, attacking both trunk and branches of apple, cherry, peach, and plum trees. Serious attacks of this insect are usually limited to unhealthy and injured trees.

Squash vine borer (*Melittia cucurbitae*, Harris, family *Aegeriidae*, order *Lepidoptera*).

Strawberry crown borer (*Tyloderma fragariae*, Riley, family *Curculionidae*, order *Coleoptera*). Essentially, a pest of the midwestern United States and notably damaging in bluegrass regions of Kentucky and Tennessee. The insect winters as snout beetles about $\frac{1}{6}$-inch (4 millimeters) long among debris in strawberry patches. Damage is caused by thick-bodied, white grubs (about $\frac{1}{5}$-inch; 5 millimeters long) that tunnel through the strawberry crowns, not always killing the plants, but greatly reducing yields. Cultural practices are the best prevention.

BORE (Tidal). A large, turbulent, wall-like wave of water with a high abrupt front, caused by the meeting of two tides or by a very rapid rise or rush of the tide up a long, shallow and narrowing estuary, bay, or tidal river where the tidal range is appreciable; it can be 3–5 meters high and moves rapidly (10–15 knots) upstream with and faster than the rising tide. (*Glossary of Geology*, American Geologic Institute.) See also **Estuary.**

BORIC ACID. Boron.

BORING (Soil). Thorough site investigations of a geological and structural nature should be conducted prior to the construction of any major work that requires an excellent foundation. This practice should and often does extend to residences that are to be built on a hillside, beach, and other locations that may have a prior history of instability. Site investigations are best made prior to procurement of a site, but certainly before the commencement of construction. Test pits provide visual examination of soil in place and make it possible to manually procure an undisturbed sample of soil. Pit digging costs, however, increase with depth. Thus other methods frequently are used except for relatively shallow depths.

In *wash boring*, a hole is formed in the ground for soil sampling or rock drilling. The equipment consists of a hollow pipe called a jet pipe and a larger hollow pipe called a casing. Water under pressure is forced down the jet pipe. This water washes disintegrated material up through the space between the jet pipe and casing to the surface where it may be retained for future examination. As the material at the bottom of the casing is washed away, the casing is slowly forced downward. Information also is obtained on subsoil characteristics by noting the resistance of the casing to driving. Generally, this method is relatively unsatisfactory because the jetting disturbs the soil and the wash water leaves coarse particles behind.

Where the precise character and formation of subsurface rock formations must be known, as in the case of foundations for important dams, core boring (sometimes called core drilling) is usually a necessity. A core drill consists of a hollow cylindrical bit with its cutting edge set with hard cutting particles (such as commercial diamond particles) connected to a hollow cylindrical drill shank. The whole is rotated by mechanical power and thus cuts out a vertical, or inclined, cylinder of the rock. These cores are periodically removed and when reassembled constitute a clear and visible section of the rock structures pierced.

BORN APPROXIMATION. A method of computing approximately the wave-functions and cross-section in the quantum mechanics of collision processes, chiefly applicable when the interaction energy between the colliding particles is small compared with their kinetic energy. Thus the first Born approximation corresponds to keeping terms of first order in the interaction energy, which is treated as a perturbation to the Hamiltonian of the system.

BORNITE. Named for the German mineralogist of the eighteenth century, Ignatius von Born, this mineral is a sulfide of copper and iron corresponding to the formula Cu_5FeS_4. It is isometric with a cubic habit, although crystals are rare, usually occurring as granular or compact masses. Its fracture is conchoidal to uneven; brittle; hardness, 3; specific gravity, 5.079; color, copper-red to reddish-brown (hence the name horseflesh ore) when freshly fractured; it soon assumes an iridescent tarnish (hence the name peacock ore); luster, metallic; streak, grayish-black; opaque.

Bornite as a primary mineral has been observed in pegmatite veins and in igneous rocks and is also a common secondary mineral.

Bornite crystals have been obtained in Austria and England. As an ore it is important in Tasmania, Chile, Peru and in Montana. In the United States, bornite also has been found in Connecticut, and in Canada, in the Province of Quebec.

Bornite also is known as *peacock ore* and *horseflesh ore.*

BORN-OPPENHEIMER APPROXIMATION. An argument for calculating the force constants between atoms in a molecule or solid, based on the observation that the motion of the electrons is so rapid compared with that of the heavier nuclei that it can be assumed that the electrons follow the motion of the nuclei adiabatically. That is, one calculates the eigenvalues of energy for the electrons with the nuclei in fixed positions; the variation of this electronic energy with the configuration of the nuclei may then be treated as a contribution to the potential energy of the interatomic forces.

BORON. Chemical element symbol B, at. no. 5, at. wt. 10.81, periodic table group 3a, mp 2300°C, sublimes at approximately 2550°C, density 2.35 g/cm^3 (amorphous form). There are four principal crystal modifications of boron: (1) α-rhombohedral, (2) β-rhombohedral, (3) I-tetragonal, and (4) II-tetragonal. There are two natural isotopes, ^{10}B and ^{11}B. In 1807, Davy first produced elemental boron in amorphous form by electrolyzing boric acid. A year later, Gay-Lussac and Thénard produced elemental boron by reducing boric acid with potassium. However, it was not until 1892 that boron with a purity of over 90% was produced by Moissan who reduced the element from B_2O_3. Moissan observed that the earlier claims of producing elemental boron were in effect compounds of boron. First ionization potential 8.296 eV; second 23.98 eV; third 37.75 eV. Oxidation potential $B + 3H_2O \rightarrow H_3BO_3 + 3H^+ + 3e^-$, 0.73 V; $B + 4OH^- \rightarrow H_2BO_3^- + H_2O + 3e^-$, 2.5 V. Other important physical properties of boron are given under **Chemical Elements.**

Boron is (1) a yellowish-brown crystalline solid and (2) an amorphous greenish-brown powder. Both forms are unaffected by air at ordinary temperatures but when heated to high temperatures in air form oxide and nitride. Crystalline boron is unattacked by HCl or HNO_3, or by NaOH solution, but with fused NaOH forms sodium borate and hydrogen; reacts with magnesium but not with sodium.

Boron occurs as rasorite or kernite (sodium tetraborate tetrahydrate, $Na_2B_4O_7 \cdot 4H_2O$) and colemanite (calcium borate, $Ca_2B_6O_{11} \cdot 5H_2O$) in California, as sassolite (boric acid, H_3BO_3) in Tuscany, Italy, and also locally in Chile, Turkey, and Tibet. See also **Colemanite; Kernite;** and **Ulexite.**

Production: Commercial boron is produced in several ways. (1) Reduction with metals from the abundant B_2O_3, using lithium, sodium, potassium, magnesium, beryllium, calcium, or aluminum. The reaction is exothermic. Magnesium is the most effective reductant. With magnesium, a brown powder of approximately 90–95% purity is produced. (2) By reduction with compounds, such as calcium carbide or tungsten carbide, or with hydrogen in an electric arc furnance. The starting boron source may be B_2O_3 or BCl_3. (3) Reduction of gaseous compounds with hydrogen. In an atmosphere of a boron halide, metallic filaments or bars at a surface temperature of about 1200°C will receive depositions of boron upon admission of hydrogen to the process atmosphere. Although the deposition rate is low, boron of high purity can be obtained because careful control over the purity of the starting ingredients is possible. (4) Thermal decomposition of boron compounds, such as the boranes (very poisonous). Boranes in combination with oxygen or H_2O are very reactive. In this process, boron halides, boron sulfide, some borides, boron phosphide, sodium bornate and potassium bornate also can be decomposed thermally. (5) Electrochemical reduction of boron compounds where the smeltings of metallic fluoroborates or metallic borates are electrolytically decomposed. Boron oxide alkali metal oxide-alkali chloride compounds also can be decomposed in this manner.

Both chemical methods and float zoning are used to purify the boron product from the foregoing processes. In the latter method, a boron of 99.99% purity can be obtained.

Although the chemistry of boron is extremely interesting, there is no substantial market for elemental boron. Some boron compounds are high-tonnage products. Elemental boron has found limited use to date in semiconductor applications, although it does possess current-voltage characteristics that make it suitable for use as an electrical switching device. In a limited way, boron also is used as a dopant (*p*-type) for *p-n* junctions in silicon. The principal problem deterring the larger use of boron as a semiconductor is the high-lattice defect concentration in the crystals currently available.

Uses: Boron formed on hot tungsten wire finds use for reinforcement of metals and plastics and for weight reduction of materials and products. Deoxidizers of alloys are prepared from borides. ^{10}B and its compounds are used for neutron absorption. Various boron compounds are used as rocket fuels, diamond substitutes, and additives to aluminum alloys to improve electrical and thermal conductivity, as well as for grain refining of aluminum alloys. Boron hydrides are very sensitive to shock and can detonate easily. Boron halides are corrosive and toxic.

Biological Functions: Although boron is required by plants, there is little solid evidence to date that it is required for the nutrition of livestock or humans. Boron deficiency may alter the levels of vitamins or sugars in plants owing to the effect of boron upon the synthesis and translocation of these compounds within the plant. The addition of boron to some boron-deficient soils has increased the carotene or provitamin A concentration in carrots and alfalfa.

Like several of the other trace elements, while concentrations of very low levels are desirable, high levels of boron are toxic to plants. Different plant species vary widely in their requirement for this element and in their tolerance for high levels. Application of boron-containing fertilizer can be carefully adjusted for different crops. An application of boron-containing fertilizier to improve the yields of alfalfa or beets may be toxic to such boron-sensitive crops as tomatoes and grapes. In the southwestern United States, serious boron toxicity to plants has resulted from using irrigation waters that are high in boron.

Boric Acid: Boric oxide B_2O_3 is acidic. It exists in two forms, a glassy form obtained by high temperature dehydration of boric acid, and crystalline form obtained by slow heating of metaboric acid.

The oxyacids of boron are of two types: (A) the boric acids, based upon boric oxide, and (B) the lower oxyacids based upon boron to boron structural linkages.

The really acidic boric acids consist essentially of metaboric acid (HBO_2), a polymer, and boric or othoboric acid H_3BO_3, (pK_a = 9.24). There is no compound corresponding to the formula for tetraboric acid $H_2B_4O_7$, although there are a number of salts which may be based upon this composition. Sometimes called *boracic acid*, H_3BO_3 is a high-tonnage material, the main uses being in the medical and pharmaceutical fields. A saturated solution of H_3BO_3 contains about 2% of the compound at 0°C, increasing to about 39% at 100 °C. The compound also is soluble in alcohol. In preparations, solutions of boric acid are nonirritating and slightly astringent with antiseptic properties. Although no longer used as a preservative for meats, boric acid finds extensive use in mouthwashes, nasal sprays, and eye-hygiene formulations. Boric acid (sometimes with borax) is used as a fire-retardant. A commercial preparation of this type (*Minalith*) consists of diammonium phosphate, ammonium sulfate, sodium tetraborate, and boric acid. The tanning industry uses boric acid in the deliming of skins where calcium borates, soluble in H_2O, are formed. As sold commercially, boric acid is $B_2O_3 \cdot 3H_2O$, prepared by adding HCl or H_2SO_4 to a solution of borax.

Borates: Sodium tetraborate $Na_2B_4O_7 \cdot 10H_2O$ is a very high-tonnage material. Natural borax has a hardness of 2–2.5, mp 75°C, sp gr 1.75. An aqueous solution of borax is mildly alkaline and antiseptic. The compound finds many uses, including: (1) cleaning compounds of numerous types; (2) important ingredient of glass and ceramics, notably for heat-resistant glass where as much as 40 pounds of borax may be required per 100 pounds of finished glass; (3) source of elemental boron and other boron compounds; (4) flux for soldering and welding; (5) constituent of fertilizers; (6) filler in paper and paints; and (7) corrosion inhibitor in antifreeze formulations. Borax also is used in fire retardants.

Chemistry of Boron and Other Boron Compounds: In 1901, the German chemist, Alfred Stock, stated, "It was evident that boron, the close neighbor of carbon in the periodic system, might be expected to form a much greater variety of interesting compounds than merely boric acid and the borates, which were almost the only ones known." In 30 years of research which followed that statement, Stock synthesized almost all of the important *boranes* (hydrogen and boron). Some of these compounds now find use in glass, ceramics, synthetic lubricants, and as ingredients of high-energy rocket fuels and jet-engine and automotive fuels. Further pioneering of borohydride chemistry was carried on by Schlesinger and Burg of the University of Chicago in the late 1940s. Boron carbide B_4C is used as a neutron-absorbing material in nuclear reactors. Sodium borohydride $NaBH_4$ is applied as a reducing agent in the manufacture of certain synthetics. Although not ultimately selected, because of the greater volatility of uranium hexafluoride UF_6, both uranium borohydride $U(BH_4)_4$ and its methyl derivative $U(CH_3BH_3)_4$ were considered for use in separating the isotopes of uranium during the Manhattan Project. ^{10}B is used in brain tumor research. When injected intravenously, borax concentrates in the areas of tumors and its presence can be detected by radiation techniques. With further research, the tendency of boron to link with itself may comprise the foundation of future inorganic polymeric materials. Although they have poor mechanical strength, boron-phosphorus polymers, prepared by reacting diborane with phosphone derivatives, do exhibit excellent heat-resistance.

X-ray diffraction studies show five general types of structures in solid borates:

1. Discrete anions containing individual BO_3^{3-} groups, or a limited number of other groups combined by sharing oxygen atoms. (The simplest is $B_2O_5^{4-}$, which is called pyroborate.)
2. Extended anions in which individual BO_3 groups are linked into rings or chains, such as $B_3O_6^{3-}$ or $B_2O_4^{2-}$ (metaborate).
3. Sheet structures in which all the oxygen atoms are shared between borate groups, as in $B_5O_{10}^{5-}$ (pentaborate).
4. Structures containing the tetrahedral $B(OH)_4^-$ ion, which is the principal ion found in alkaline aqueous solutions.
5. Extended anions containing tetrahedral BO_4 units, usually linked with triangular BO_3 groups.

The lower oxyacids of boron may be considered to be derived from the various boron hydrides, whence their boron-boron linkages result. These compounds include the hypoborates, which may be produced by reactions of tetraborane with strong alkali, and which may be formulated from the structure $H_2[H_6B_2O_2]$; the subborates, derived from $H_4[B_2O_4]$, which is called subboric acid; and the borohydrates, which are derived from acids of various compositions, such as $H_2[B_4O_2]$, $H_2[B_2O_2]$ and $H_2[H_4B_2O_2]$. The last of these compounds contains a double bonded boron-boron linkage, and exhibits *cis-trans* isomerisim.

The borides are binary compounds of boron with metals or electropositive elements in general. Except in isolated cases their compositions depart from the stoichiometry of trivalent boron compounds and are determined more by the requirements of metal and boron lattices than by valencies. On the basis of composition, they may be classified into types based respectively upon zigzag chains (MB) represented by CoB; isolated boron atoms (M_2B) represented by Co_2B; double chains (M_3B_4) represented by Mo_3B_4; hexagonal layers (MB_2) represented by CoB_2; three dimensional frameworks (MB_6 or MB_{12}) represented by SiB_6 or UB_{12}. It is apparent that these borides are interstitial compounds existing primarily with the metals of main groups, II, III, IV, V, and VI.

There are at least six definitely characterized boron hydrides, as follows: diborane(6), B_2H_6; tetraborane(10) B_4H_{10}; pentaborane(9) (stable) B_5H_9; pentaborane(11) (unstable) B_5H_{11}; hexaborane(10) B_6H_{10} and decaborane(14) $B_{10}H_{14}$. In these names, note that the prefix denotes the number of boron atoms, while the figure in parentheses denotes the number of hydrogen atoms. In addition to these compounds, which are all gases or volatile liquids except decaborane(14), decomposition of the lower boron hydrides yields colorless or yellow solid boron hydrides, ranging in composition from $(BH_{1.5})_x$ to $(BH)_x$. This readiness to polymerize is evidence of the reactivity of these borane compounds, which readily form additional products with ammonia, with the amalgams of the active metals, and with many organic compounds, as well as with CO.

In addition to BH_4^- there exist a number of hydroborate anions which may be considered to be derived from real or hypothetical boron hydrides by addition of hydride ion. These include $B_2H_7^-$, formed by the reaction of B_2H_6 and BH_4^- in organic solvents, and the extremely stable ions $B_{10}H_{10}^{2-}$ and $B_{12}H_{12}^{2-}$, unaffected by either acidic or alkaline aqueous solutions or by atmospheric oxygen. Free halogens merely cause substitution of halogen for hydrogen. The structure of $B_{10}H_{10}^{2-}$ is based on the square antiprism, while that of $B_{12}H_{12}^{2-}$ is a regular icosahedron.

In 1976, the Nobel Prize for Chemistry was awarded to William Nunn Lipscomb, Jr., of Harvard University, for original research on the structure and bonding of boron hydrides and their derivatives. As pointed out by Grimes (1976), the insight into electron-deficient borane structures originally provided by Lipscomb carries over not only to the carboranes, but also to their organic cousins, the so-called "nonclassical" carbonium ions. The three-center bond descriptions given by Lipscomb to B_5H_9 and B_6H_{10} can as easily be applied to their hydrocarbon analogs, the pyramidal ions $C_5H_5^+$ and $C_6H_6^{2+}$, both presently known as alkyl derivatives. Also, molecules usually not so considered, such as metallocenes, organometallics, such as $(C_4H_4)Fe(CO)_3$ or $[(CO)_3Fe]_5C$, metal clusters and others, can be considered from the perspective of borane analogs. The boranes, once considered peculiar, over the years have provided insight to many cluster-type molecules, for which classical Lewis bond descriptions do not fit. Lipscomb's lecture given in Stockholm on December 11, 1976 provides an excellent overview of the boranes and their relatives. See Lipscomb (1977) reference.

In 1979, the Nobel Prize for Chemistry was received by Herbert C. Brown (shared with Georg Witting for research in another field) of Purdue University for the discovery of the *hydroboration reaction.* This reaction, depicted below, has made the organoboranes readily available as chemical intermediates. The boron atom adds to the less substituted carbon atom. As pointed out by Brewster and Negishi (1980), depending upon steric factors, mono-, di-, or trialkylboranes may be formed. These products comprise synthetically useful reactions whereby the boron atom is replaced, but the mono- and dialkylboranes are also useful as reducing or hydroborating agents.

$$RCH{=}CH_2 + B_2H_6 \longrightarrow (RCH_2CH_2{-})_3B \quad (1)$$

$$CH_3{-}\underset{CH_3}{\underset{|}{C}}{=}CHCH_3 + B_2H_6 \longrightarrow (CH_3{-}\underset{CH_3}{\underset{|}{CH}}{-}\underset{CH_3}{\underset{|}{CH}}{-})_2BH \quad (2)$$

$$CH_3{-}\underset{CH_3}{\underset{|}{C}}{=}\underset{CH_3}{\underset{|}{C}}{-}CH_3 + B_2H_6 \longrightarrow CH_3{-}\underset{CH_3}{\underset{|}{CH}}{-}\overset{CH_3}{\overset{|}{\underset{CH_3}{\underset{|}{C}}}}{-}BH_2 \quad (3)$$

Among the other inorganic compounds of boron are the following:

Borides: Carbon boride CB_6 and silicon borides SiB_3 and SiB_6 are hard, crystalline solids, produced in the electric furnace; magnesium boride Mg_3B_2, brown solid, by reaction of boron oxide and magnesium powder ignited, forms boron hydrides with HCl; calcium boride Ca_3B_2, forms boron hydrides and hydrogen gas with HCl.

Nitride: Boron nitride BN, white solid, insoluble, reacts with steam to form NH_3 and boric acid, formed by heating anhydrous sodium borate with ammonium chloride, or by burning boron in air.

Sulfide: Boron sulfide B_2S_3, white solid, unpleasant odor, irritating to the eyes, reactive with water to form boric acid and hydrogen sulfide, formed by reaction of boron oxide plus carbon heated in a current of CS_2 at red heat.

The great number of compounds of boron are due to the readiness with which boron atoms form, to some extent, chain structures with other boron atoms, and to a far greater extent, cyclic compounds both with other boron atoms, and with atoms of carbon, oxygen, nitrogen, phosphorus, arsenic, the halogens, and many other elements. Examples of them are shown below, beginning with the two pentaboranes B_5H_9 and B_5H_{11}:

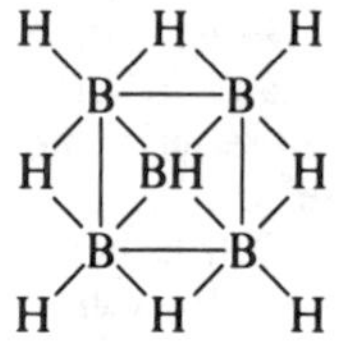

B_5H_9 Pentaborane(9)

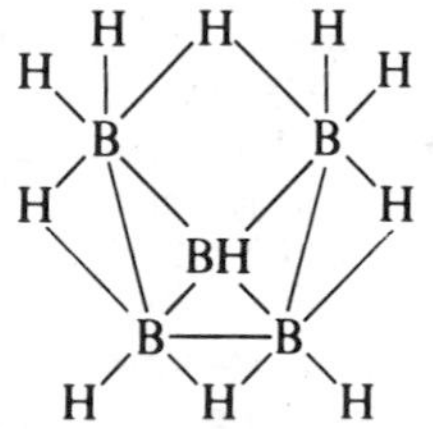

B_5H_{11} Pentaborane(11)

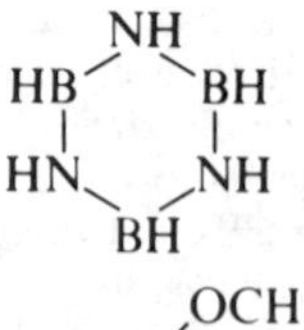

hexahydro-*s*-triazatriborine (also called *borazine*)

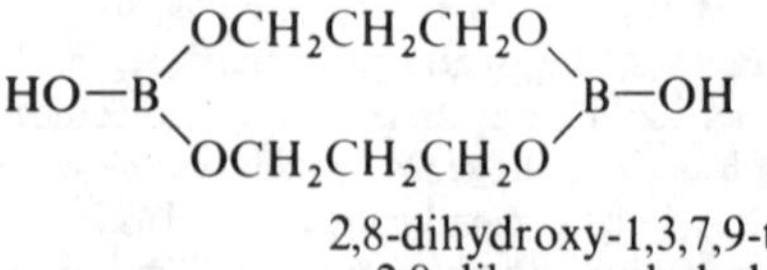

2,8-dihydroxy-1,3,7,9-tetroxa-2,8-diboracyclododecane

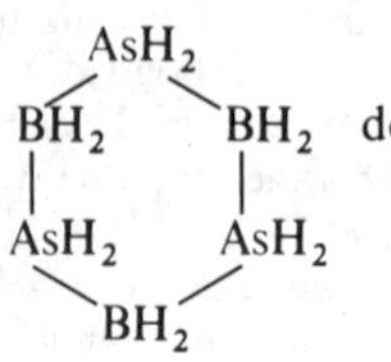

dodecahydro-*s*-triarsatriborine (also called *s*-triphosphatriborane or *borarsane*)

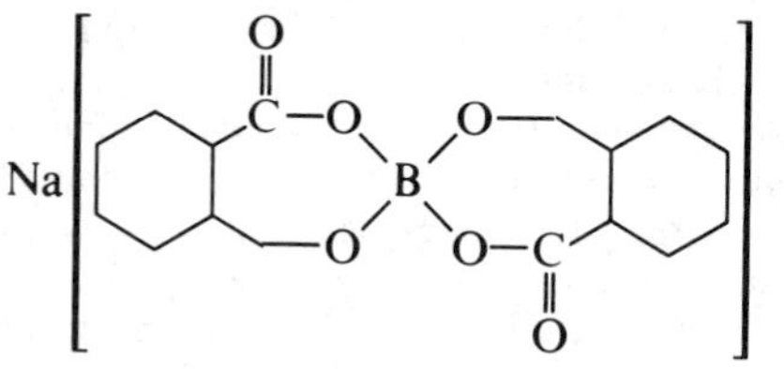

sodium bis(salicylato-*O,O′*)borate (1^-)

Halides: Since simple boron compounds have only three electron pairs in the valence shell of boron, they tend to be electron acceptors. Its simple molecules are formed by sp^2 hybrid sigma bonds lying in a plane. Its strong tendency to form an octet is shown by the tetrahedral boron compounds involving sp^3 hybridization. Boron halides include the trifluoride, BF_3, the trichloride, BCl_3, the tribromide, BBr_3 and the triiodide, BI_3, which range in mp from −127 to +43°C. Typical methods of forming the boron halides are treatment of boron oxide with hot concentrated H_2SO_4 in a reaction mixture with calcium fluoride to produce BF_3, and by heating boron, or boron oxide plus carbon with chlorine to produce the chloride.

In addition to the simple halides, boron forms fluorine complexes containing the fluoroborate ion (BF_4^-). Subhalides of boron are known (B_2X_4) of the structure:

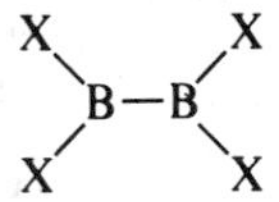

and B_4X_4 of the structure

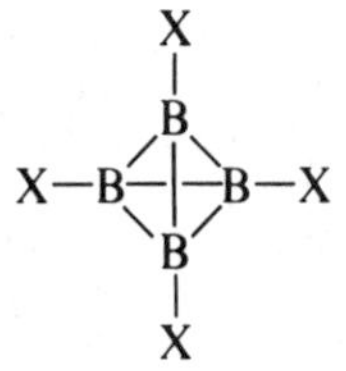

References

Armington, A. F., Bufford, J. T., and R. J. Starks: "Boron," Plenum, New York, 1965.

Brewster, J. H., and E. Negishi: "The 1979 Nobel Prize for Chemistry," *Science,* **207,** 44–46 (1980).

Brown, H. C.: "From Little Acorns to Tall Oaks: From Boranes Through Organoboranes," *Science,* **210,** 485–492 (1980).

Grimes, R. N.: "The 1976 Nobel Prize for Chemistry," *Science,* **194,** 709–710 (1976).

Lipscomb, W. N.: "The Boranes and Their Relatives," *Science,* **196,** 1047–1055 (1977).

Pulich, W. M., J.: "Photocontrol of Boron Metabolism in Sea Grasses," *Science,* **200,** 319–320 (1978).

BORON ROCKET FUEL. **Rocket Propellants.**

BOROSILICATE GLASS. **Glass.**

BORROW PIT. The section of ground from which earth is excavated for the purpose of being used as fill at some other point is called a borrow pit. Borrow pits are used extensively in connection with highway and railway construction since it is often more economical to obtain fill from some nearby source than to carry it from some distant point along the line where there is an excess of cut. When it is necessary to ascertain the amount of material taken from a borrow pit the ground surface is divided into rectangles or squares before any material has been removed. This grid system is referred to an arbitrary base line and elevations are taken on each of the corners. After the required material has been removed the system of rectangles or squares is reproduced by means of the base line and elevations are again taken on the corners. The volume of material which has been excavated then consists of the sum of the volumes of the individual prisms. See also **Earthwork.**

BOSANQUET'S LAW. **Magnetism.**

BOSE-CHAUDHURI-HOCQUENGHEM CODES. **Information Theory.**

BOSONS. Those elementary particles for which there is a symmetry under intrapair production. They obey Bose-Einstein statistics. Included are photons, pi mesons, and nuclei with an even number of particles. (Those particles for which there is antisymmetry are *fermions.*) See **Mesons; Particles (Subatomic);** and **Photon.**

Recent progress toward a complete theory of the weak interactions has led to sharper predictions for the properties of the hypothetical weak-force particles known as *intermediate bosons.* This is described by Hung and Quigg (*Science,* **210**, 1205–1211, 1980).

BOSS. The term boss or stock is used to indicate a cross-cutting mass of igneous rock which has ascended into the crust of the earth and may or may not represent the roots of volcanic conduits. Bosses are roughly circular or elliptical in ground plan and usually of greater cross-sectional area than a volcanic neck and lack pyroclastic materials. Most probably bosses are the irregular upward extensions of batholiths the main parts of which are as yet unexposed.

Boss also designates a circular projection on a casting, usually serving as the seat for a bolt head or nut.

BOSTON FERN. **Ferns.**

BOSTONITE. A rather rare rock type, dense, with an occasional feldspar phenocryst and grayish in color. it is composed almost wholly of alkaline feldspar, being analogous to aplites. The type locality is Salem Neck, Massachusetts, not many miles from Boston, for which it was named.

BOSWELLIA TREE. Of the family *Burseraceae* (torchwood family), the boswellia tree is native to the island of Socotra near Saudi Arabia. The tree is small, not attaining a height of over 12 to 20 feet (3.6 to 6 meters). The fruit is a berry about the size of an olive. The branches are short, twisted, and harsh in appearance, rising from low on the trunk. The leaves curl and are sparse. The flowers are few and appear like a red geranium blossom, but are quite fragrant. The tree was known at the time of Christ and was the source of frankincense mentioned in the Bible. The bark of the tree is filled with an amber-green resin, the source of frankincense.

BOTANY. Botany is the science which deals with plants. It is divided into many sections, each dealing with a specific part of the subject. One section, which describes plants and arranges them in classes, is called taxonomy; another section, morphology, considers the form of the various parts of a plant, while its subsections include anatomy and histology, the study of the internal structure of plants, and cytology, the study of the cell and its parts. A third, physiology, deals with the functions of the parts and the activities of the plant. In addition, one may study plant geography, or the distribution of plants on the earth; ecology, the relations of plants to each other and to their environment; phytopathology, or the diseases of plants; paleobotany, the science of fossil plants; and economic botany, which considers the uses which man has found for plants and plant products.

The science of botany is very old. Since the welfare of man is closely connected with plants, it is natural that they should receive attention early. Undoubtedly plants were known and observed by men long before the period of Greek supremacy. Various recorded observations suggest that such is true. But only with the intellectual curiosity of the Greek mind did plants receive close attention. Aristotle (384–322 B.C.) studied them attentively and cultivated many species from widely separated regions. His disciple Theophrastus (371–287 B.C.) carried on the work and wrote about them in his "Equiry into Plants," in which he describes some 500 species and gives extensive and keen observations concerning them. In Rome another naturalist, Pliny the Elder (23–79 A.D.), writes extensively on Natural History, setting forth information on some thousand species of plants. His facts are largely drawn from sources other than the plants themselves and are often grossly exaggerated. His Natural History was of immense importance, however, and largely controlled the thought of botanists for many centuries. Another ancient naturalist, Dioscorides, also studied plants. He was mainly interested in them because of the important place they held in the medical practice of that time. Indeed, the study of plants was for a long period of time considered the province of physicians and doctors, whose main interest was in plants as remedies or supposed remedies for various ills. After this, centuries followed in which little attention was given to plants; all knowledge thereof was drawn directly from the works of the ancient writers.

Beginning with the sixteenth century, however, interest in plants was revived. Men began observing the native plants around them and recording these observations, often accompanied by illustrations, in herb books or herbals. Such observations led to attempts to arrange and classify the various plants. Among the first herbals were those of Brunfels (1530) and Fuchs (1542), both of them containing excellent illustrations, but relying for their descriptions largely on the ancient writers of Greece and Rome. Hieronymus Bock (1498–1554) was another herbalist, who gave in his book extensive first-hand descriptions of the plants which he treats. William Turner and John Gerard published herbals treating of English plants. Valerius Cordus (1515–1544) gave even more complete and accurate descriptions of the plants in his books than Bock.

As a result of the work of these men and many others, came a need for a better understanding of plants and the necessity for arranging them in some sort of system other than that of size or of the alphabet. John Ray (1628–1705) advanced the problem considerably by introducing an exact concept of species, which he held to come from a single parent and to continue to produce like organisms, although he does allow some variation to occur. Ray separated flowerless plants from flowering, and divided the latter into Dicotyledons, with two seed leaves, and Monocotyledons, with only one. See also **Angiosperms;** and **Dicotyledons.**

The number of plants described was constantly increasing, rendering even more necessary a system of arranging them in order. Many systems were proposed, some having great merit. As early as 1583 Casalpino had eliminated any classification based on such variable organs as roots, stems or leaves, and had concluded that the flowers and fruit offered the only real basis. It remained for Carolus Linnaeus (1707–1778) to bring order to the situation. He invented the binomial system of nomenclature, by which each plant (and animal also) should be known by a name designating the genus and a qualifying adjective limiting the species named. His system of classification was purely artificial, being based on the number of stamens and pistils (see **Flower**), but did make it easy to refer to a description and so verify an identification. He also grouped plants and animals in larger divisions, the classes and orders. The present-day names of plants date from the time of Linnaeus. It has long been recognized that there seemed to be a natural grouping of plants; John Ray apparently understood some of the larger groups of plants. With the work of the French taxonomist A. L. de Jussieu came a definite knowledge of the natural relations of plants, which he grouped into 15 classes with about a 100 orders.

While classification and description occupied a large place in the development of botany, other branches of the science were not neglected, although of necessity many of them waited on advancement in taxonomy. The anatomy of plants was studied by Nehemiah Grew (1641–1712) in England, and Marcello Malpighi (1628–1684) in Italy, while casual observations on the internal structures of some plant substances were made by Robert Hook. The finely illustrated writings of these men established the foundations for an understanding of the internal structure of plants. Subsequent workers in this field showed the similarities existent in the internal structures of plants, and the changes which have occurred during the evolution of plants. Out of this have come the later studies of cytology and histology.

Any knowledge of the way in which the plant lives and the functions of its various parts was slow to develop. The lack of definite organs connected with such functions as digestion, circulation, respiration, etc., made the problem even more difficult. Occasional observations had been made from time to time, often leading to erroneous conclusions. With Stephen Hales (1677–1761) plant physiology became established. He first used instruments to measure various physiological activities which he studied. His observations, recorded in his "Vegetable Staticks," published in 1727, show how attentively he studied the problem of nutrition in plants and the movements of liquids within the plant. Ingen-Housz (1730–1799) gained more exact knowledge of the problem of nutrition in plants, definitely showing that the carbon in plants came from the carbon dioxide of the atmosphere. He had an accurate knowledge of the role of gases in the life of the plant. Another worker, Andrew Knight (1758–1838), studied an entirely different field, being largely interested in the problem of direction of growth of root and stem. To him is due the use of a rapidly revolving wheel to which seedlings were attached. From this experiment he determined that roots grew away from the center of the revolving wheel and stems toward the center. Out of his studies came the study of tropisms in general.

However, other branches of the science of botany have not been overlooked. The study of the distribution of plants has been pursued with great vigor, bringing to light many interesting problems, at times difficult to explain. Why should certain similar groups of plants appear in widely separated regions? At present this and many other questions are subjects for speculation and cause for further study. See **Plant Breeding.**

Another branch of botany which occupies an important position today is that of plant pathology, which treats of the diseases of plants. When a single disease such as wheat rust, attacking a single crop, causes the loss of millions of dollars in reduced harvests, and with so many crops subject to numerous diseases, this must be recognized as a study of vital importance to man. Comprehensive and exact knowledge of the disease-producing organism is necessary. Often it is obtained only after prolonged, painstaking study. Then follows the problem of treatment leading to elimination of the disease, a study in itself. Sometimes this is impracticable; it is quicker to attack the problem in another way—to attempt to develop strains of plants which are resistant or immune to the disease. In this field new problems are constantly arising, or assuming greater importance—for example, the outbreak of the Dutch Elm disease in recent years, or of the Oak Wilt disease which threatens the oak forests of America. See **Elm Trees;** and **Oak Trees.**

These and many other problems show how close is the welfare of mankind tied up with the stidy of botany and the knowledge of the many sides of that science.

Not including specific trees, plants, and plant families, the botany-related entries in this book are:

Abaca
Abscission
Achene
Adventitious Buds
Aerenchyma
Aleurone Grains
Alleopathic Substance
Angiosperm
Annual
Annual Ring
Annulus
Anther
Anthesis
Antipodal Cells
Apical Growth
Archegoniates
Archegonium
Aril
Aroids
Ascent of Sap
Axil
Bark
Bast Fibers
Berry
Biennial
Brachyblast
Bract
Bryophyllum
Bryophytes
Bud
Budding
Bulb (Botany)
Bulbil
Bundle
Calyx
Cambium (Plant)
Catkin
Chaparral
Coleoptile
Color (Plants)
Companion Cell
Conidia
Deciduous Plants
Dicotyledons
Diecious Organisms
Epiphytes
Etiolation
Euphotic Zone
Exosmosis
Ferns
Flower
Fruit
Gall (Botany)
Geotropism
Germ Plasm
Gibberellic Acid and Gibberellin Plant Growth Hormones
Grafting and Budding
Grasses
Guttation
Gymnosperms
Heterospory
Hybrid
Hydrophytes
Hydroponics
Insectivorous Plants
Leaf
Lenticels
Lichen
Monoecious Plants
Paleobotany
Parthenocarpy
Periderm
Phloem
Photoperiodism
Photosynthesis
Pigmentation (Plants)
Plant Breeding
Plant Growth Modification and Regulation
Plastids
Pollination
Respiration (Plants)
Rhizoids
Rhizome
Root (Plant)
Saprophytes
Sclerenchyma
Seed
Spore
Sporophyll
Stele
Stem (Plant)
Stolon
Stomate
Succession (Plant)
Transpiration
Vascular System (Plants)
Vernalization
Witches' Brooms
Wood
Xenia
Xerophytes
Xylem

References

Beadle, G. W.: "The Ancestry of Corn," *Sci. Amer.*, **242**, 1, 112–119 (1979).
Brown, L.: "Grasses: An Identification Guide," Houghton Mifflin, Boston, 1979.
Erickson, R. O., and W. K. Silk: "The Kinematics of Plant Growth," *Sci. Amer.*, **242**, 134–151 (1980).
Furuya, M., Katsumi, M., and A. Takimoto (editors): "Controlling Factors in Plant Development," Botanical Society of Japan, Tokyo, 1978.
Harley, J. L., and R. S. Russell, Editors: "The Soil-Root Interface," Academic, New York, 1979.
Hubbell, S. P.: "Tree Dispersion, Abundance, and Diversity in a Tropical Dry Forest," *Science*, **203**, 1299–1309 (1979).
Kirk, J. T. O., and R. A. E. Tilney-Bassett: "The Plastids," 2nd edition, Elsevier/North-Holland, New York, 1978.
Kolattukudy, P. E.: "Biopolyester Membranes of Plants: Cutin and Suberin," *Science*, **208**, 990–1000 (1980).
Levin, D. A.: "The Nature of Plant Species," *Science*, **204**, 381–384 (1979).
Lewis, W. H., and M. P. F. Elvin-Lewis: "Medical Botany: Plants Affecting Man's Health," Wiley, New York, 1978.
Maiti, I. B., and P. E. Kolattukudy: "Prevention of Fungal Infection by Plants by Specific Inhibition of Cutinase," *Science*, **205**, 507–508 (1979).
Mitchinson, G. J.: "Phyllotaxis and the Fibonacci Series," *Science*, **196**, 270–275 (1977).
Morton, J. F.: "Major Medicinal Plants: Botany, Culture and Uses," Charles C. Thomas, Springfield, Illinois, 1978.
Pinter, P. J., Jr.: "Remote Detection of Biological Stresses in Plants with Infrared Thermometry," *Science*, **205**, 585–587 (1979).
Regal, P. J.: "Ecology and Evolution of Flowering Plant Dominance," *Science*, **196**, 622–629 (1977).
Ruehle, J. L., and D. H. Marx: "Fiber, Food, Fuel, and Fungal Symbionts," *Science*, **206**, 419–422 (1979).

BOTANY (Paleo). Paleobotany.

BOT FLY (*Insecta, Diptera*). The maggots of several species of the bot fly seriously damage cattle, sheep, horses, and other farm and domestic animals. Distribution of these pests is essentially throughout the United States. The larvae live as internal parasites in mammals. The adult bot flies have, as a rule, vestigal mouth parts and attack the host only to deposit their eggs. Horses are attacked by three species of bot flies of the genus *Gasterophilus.* The *lip* or *nose bot fly* (*Gasterophilus* haemorrhoidalis, Linne) deposits its eggs on the lips, whence the larvae reach the throat or stomach. The *chin* or *throat bot fly* (*G. nasalis* or *veterinus,* Linne). The species, *G. inermis* (Brauer), attaches the larvae to the hairs of the forelegs, where they die unless the horse takes the larvae into its mouth by licking or biting the legs. The larvae develop in the alimentary tract and pass out when mature with the feces. The sheep bot fly is described in entry on **Nose Fly.**

In cattle, the larvae of the species of bot fly known as *Hypoderma lineatum* (De Villiers) is referred to generally as the common *cattle grub*, and the adult, as the *heel fly*. The larvae of the species of fly known as the bomb fly, species *Hypoderma bovis* (De Geer), is commonly referred to as the *northern cattle grub*. All species of cattle grub produce a condition sometimes called *ox warbles* because of the tumerous swellings or "warbles" produced. The adult fly may be as large as a honeybee and continues to chase and bother an animal until it finds opportunity to lay its eggs, often along the animal's back. Maggots from the eggs migrate through the animal's skin and usually ultimately find a permanent location along the back, causing a tumor, inside of which is a fat, well-nourished maggot. It is evident that hides are severely damaged from this procedure. The general health of the animal is also affected and, if a dairy cow, milk production is reduced.

The heel fly is most abundant, ranging throughout the United States. The bomb fly is most commonly found in the northeastern states and is not a pest in the southern states. However, the total range of the bomb fly is from the east to west coasts of North America, both north and south of the Canadian border.

The cattle grubs overwinter as maggots, usually in the backs of animals. After residing in the animals for about 6 months, usually in early winter, the larvae drop to the ground and pupate in the soil. The adult flies appear in the spring and commence egg-laying.

Numerous chemical formulations and methods are available for treating the animals once an infestation has occurred, but the procedures followed are complex and detailed and beyond the scope of this volume.

References

Considine, D. M. (editor): "Foods and Food Production Encyclopedia," Van Nostrand Reinhold, New York, 1982.

Metcalf, R. L.: "Destructive and Useful Insects—Their Habits and Control," McGraw-Hill, New York, 1962.

BOTHNIDS. Flatfishes.

BOTULISM. Foodborne Diseases.

BOUGAINVILLAEA. Genus of the family *Nyctaginaceae* (four-o'clock family). This is a relatively small genus of plants, natives of South America, which are frequently cultivated in the tropics and to some extent as greenhouse plants outside the tropics. However, some species do very well in subtropical areas, as in Florida, the Gulf Coast, and southern California. The flowers of the plants are small and inconspicuous, but are surrounded by showy bracts of various colors, notably pink, purple, and orange. The plants are generally cultivated in gardens and in landscaping for these brilliantly-colored bracts. *Bougainvillea spectabilis*, a heavily-thorned and climbing vine is frequently cultivated.

BOUGUER AND LAMBERT LAW. In homogeneous materials, such as glass or clear liquids, the fractional part of intensity or radiant energy absorbed is proportional to the thickness of the absorbing substance. Summing over a series of thin layers or integration over a finite thickness gives the relation

$$\log I_0/I = k_1 b$$

where I_0 is the intensity or radiant power incident on a sample b centimeters thick and I is the intensity of the transmitted beam. The constant k_1 depends on the wavelength of the incident radiation, the nature of the absorbing material and other experimental conditions. Verification of the law fails unless apporpriate corrections are made for reflection, convergence of the light beam and spectral slit width, as well as possible scattering, fluorescence, chemical reaction, inhomogeneity, and anisotropy of the sample. Formerly, the constant k_1 was called the absorption coefficient. It is now preferable to avoid this term and to call the ratio I/I_0 the transmittance. The law was first expressed by Bouguer in 1729 but it is often attributed to Lambert, who restated it in 1768.

BOUGUER'S HALO. Atmospheric Optical Phenomena.

BOUGUER'S LAW. Spectrochemical Analysis (Visible).

BOULANGERITE. A mineral compound of lead-antimony sulfide, $Pb_5Sb_4S_{11}$. Crystallizes in the monoclinic system; hardness, 2.5–3; specific gravity, 6.23; color, lead gray.

BOULDER. A large fragment of rock, usually rounded, which has been moved from its place of origin by a natural agency or has been formed in situ by weathering processes. Rather arbitrarily, 8 inches has been set as the minimum diameter for a boulder.

BOULDER CLAY. Boulder clay is a glacial deposit of clay with subangular rock fragments of different sizes.

BOUNDARY LAYER (Air). Aerodynamics; Supersonic Aerodynamics.

BOUNDARY LAYER (Fluid Flow). Fluid Flow (Boundary Layer).

BOURDON TUBE. Patented by Eugene Bourdon in 1852, the bourdon tube continues to find wide application in instruments, notably for pressure and force measurement, and for performing mechanical work in response to pressure. Filled-system thermometers also utilize bourdon tubes. Although made in various forms, the principal configurations are: (1) the "C" shape tube, (2) the helical tube, and (3) the spiral tube.

"C" Shape Bourdon Tube. This is the most common form of bourdon tube. Its use in a dial-type pressure gauge is shown in Fig. 1.

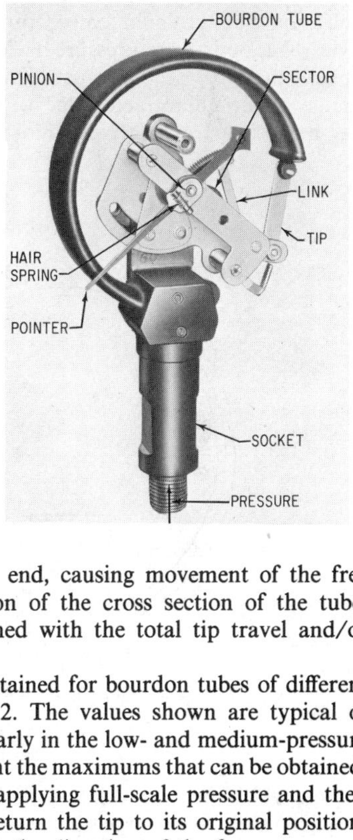

Fig. 1. Use of "C" shape bourdon tube in dial-type pressure gauge.

Pressure is applied at the fixed end, causing movement of the free end as the result of deformation of the cross section of the tube. The designer usually is concerned with the total tip travel and/or the force available at the tip.

The tip travel that may be obtained for bourdon tubes of different coiling radii is plotted in Fig. 2. The values shown are typical of what is in common use. Particularly in the low- and medium-pressure ranges, the curves do not represent the maximums that can be obtained. The tip force is determined by applying full-scale pressure and then finding the force necessary to return the tip to its original position. It can be noted from Fig. 3 that the direction of the force necessary to return the bourdon tube tip to its original position is not along the same path as the path of motion of the tip. The deflection of the tip from B to C is due to uniform loading caused by the pressure applied whereas the external force applied to the tip along B-D to return it does not load the bourdon uniformly. This fact must be considered when using bourdons in a force-balance system and accounts for the difficulty in utilizing opposing bourdon tubes as a means to measure differential pressure.

Helical Bourdon Tubes. These are used most often for high-pressure gauges—principally to permit obtaining a large tip travel without creating a high stress per unit length of tube. The direction of the motion will be an arc whose center is the center about which the helix is coiled. It must be kept in mind that the helix form introduces another

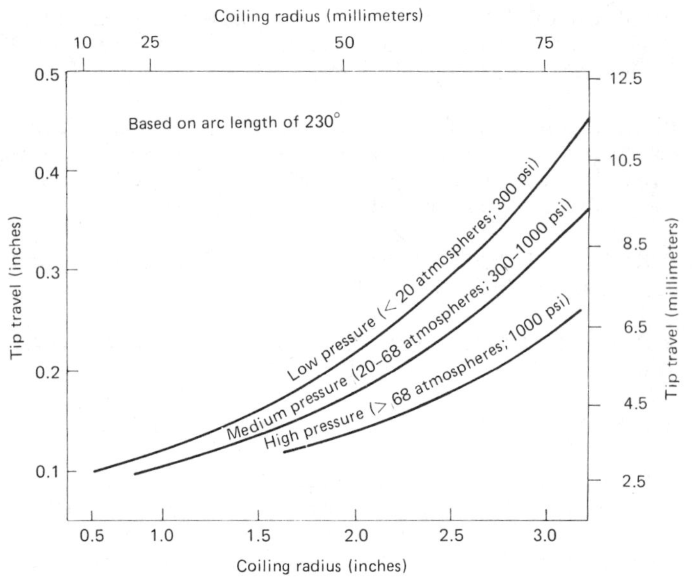

Fig. 2. Tip travel versus coiling radius (typical values).

axis of compliance so that all of the tip force will not be available unless the tip of the bourdon is constrained to move about its center.

Spiral Bourdon Tubes. Spirals often are used in liquid-filled systems, such as mercury or liquid temperature indicators. These are volumetric devices—as opposed to pressure devices—and the bourdon tube moves because the cross section must change to accommodate the volume change of the filling media due to temperature variation. The tube may be designed of relatively thin-walled tubing of a very flat cross

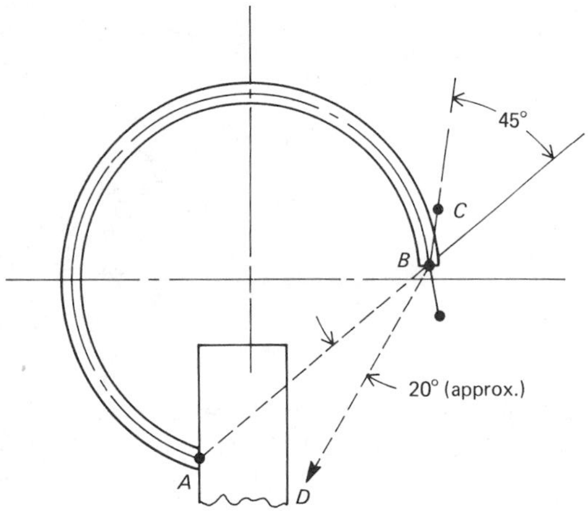

Fig. 3. Approximate direction of motion and force at tip of "C" shape bourdon tube.

section which permits winding into a compact spiral form. Spiral bourdon tubes having large tip travels also are used where it is desirable to eliminate multiplication linkages and gears. Little data are available on spiral bourdons with respect to travel and force available—hence their use generally requires cut-and-try methods.

Philip W. Harland, U.S. Gauge Division, AMETEK, Inc., Sellersville, Pennsylvania.

BOURNONITE. An antimony-copper-lead sulfide corresponding to the formula $PbCuSbS_3$. It is orthorhombic, and repeated twinning often produces crosses or wheel-shaped crystals. It is brittle; fracture, subconchoidal; hardness, 2.5–3; specific gravity, 5.83; luster, metallic; color and streak, dark gray to black; opaque.

Bournonite is found with galena, chalcopyrite, and sphalerite. There are many European localities; it was first found in Cornwall, England, by Count Bournon, for whom it was later named. Bournonite occurs in Bolivia and Peru and in the United States in Arizona, Montana, Nevada and Utah. Bournonite is also known as *wheel ore.*

BOUTONNEUSE FEVER. Rickettsial Diseases.

BOVINES (*Mammalia, Artiodactyla*). The Oxen and Duikers (*Bovines*) comprise one of the larger families of the order *Artiodactyla* (even-toed hoofed animals). The True Oxen is considered the prime origin of domestic breeds of cattle. Because of centuries of domestication and breeding, a vast series of race complexes has developed. Only the principal species of "wild" bovines are listed in Table 1. However, some of the species regarded as wild (occurring mainly in their natural habitats without supervision or procedural exploitation) also have been domesticated in certain regions to varying degrees and have been used in breeding experimentation in a continuing effort to alter and hopefully improve the strains of domestic cattle, particularly with specific regional environments or end-products in mind.

As with so many complex families of mammals, generalizations are difficult and always must be considered as such because there usually are specialized exceptions always present. Bovines graze, browse, usually moving slowly in search of vegetation for their diets. They are ruminants. They are well known for "chewing their cuds," that is, gathering food for later regurgitation and chewing at will to prepare it for final digestion and assimilation. Thus, these animals have complex stomachs, often referred to as a four-part stomach. Other than by vegetarians, people generally consider the flesh of most bovines as desirable for the human diet, both from the standpoints of taste and nutrition. The breeding, raising, and production of these animals for meat, milk, hides, and other products has been highly refined over literally centuries of time.

Like the horse, dog, goat, sheep, and other domesticates, the classification of the ancestors and currently living specimens of cattle in the natural state has posed complexities for naturalists. Until relatively recently, with the exception of swine, hippopotamuses, camels, deers, and giraffes, the remaining families of *Artiodactyla* were all grouped together in a family called *Bovidae* (or oxen). Thus, in the former classification were included cows, goats, antelopes, and gazelles, entailing some 200 species. This proved to be an untenable classification from the standpoint of scientific study and observation. *Artiodactyla* is now divided into essentially three groups: *Bovine* (oxlike); *Antelopine*; and *Caprine* (goatlike).

In Europe and the Near East until about the end of the Dark Ages, it has been quite well established that Aurochs, a wild ox (*Bos primigenius*), existed and that this animal was the major, if not the only, ancestor of domestic cattle. It is also recognized, however, that the common domesticated cattle of India (Zebu, *Bos indicus*), characterized by the highly developed dewlap and by the sharply defined hump on the withers, as well as the huge-horned Ankole, resulted from breeding of primitive domestic herds with other species of wild oxen of that time. The zebu has been introduced into the Americas for hybridization with range cattle in comparatively recent times. The primary purpose being that of developing cattle that are more tolerant of heat and more resistant of insect- and tick-borne diseases. In America, the zebu usually is referred to as Brahman cattle.

As indicated by Table 1, there are several species of wild cattle (*Bos*) identified. The Gaur is a large wild ox of India, Burma, and the Malay Peninsula. The animal has been domesticated to a relatively limited extent. These cattle are about six feet tall with white legs and gray-black/brown bodies. The horns are compressed with an elevated ridge between them. They roam in small herds in forest and hilly country. A closely related animal is the Gayal, a wild ox of northeastern India, somewhat smaller than the gaur and with less flattened horns. This animal has been domesticated to a limited extent. It has been reported that tribesmen tend them like domesticated cows, using the milk.

It is interesting to note that the Coupray was not discovered by zoologists until 1935. Found in Cambodia, it was first believed to be a hybrid, but participating ancestors could not be identified. The animal has an extensive dewlap, is grayish-black with light gray legs. The most puzzling feature of it is a cuff of shredded horn-fibers which points forward, a specialty not found in any other mammals.

TABLE 1. GENERAL ORGANIZATION OF THE OXEN AND DUIKERS

BOVINES

TRUE OXEN (*Bovinae*)	DEER-OXEN (*Boselaphinae*)
Cattle (*Bos*)	The Nilghai (*Boselaphus*)
—Gaur (*B. gaurus*)	The Chousingha (*Tetraceros*)
—Gayal or Mithan (*B. frontalis*)	
—Coupray (*B. sauveli*)	TWIST-HORNED OXEN (*Strepsicerosinae*)
—Yak (*B. grunniens*)	Elands (*Taurotragus*)
—Zebu (*B. indicus*)	The Bongo (*Böocerus*)
—Banteng (*B. banteng*)	Kudus (*Strepsiceros*)
Buffalo (*Bubalus*)	—Greater Kudu
—Water Buffalo (*B. bubalis*)	—Lesser Kudu
African Buffaloes (*Syncerus*)	Bushbucks (*Tragelaphus*)
—Forest Buffalo (*S. nanus*)	—Mountain Nyala
—Cape Buffalo (*S. caffer*)	—Nyala
Small Oxen (*Anoa*)	—Sitatunga
—Anoa (*A. depressicornis*)	—Mountain Bushbuck
—Tamareu (*A. mindorensis*)	—Common Bushbuck
Bison (*Bison*)	
—Wisent	DUIKERS (*Cephalophinae*)
—American Bison	Common Duikers (*Sylvicapra*)
—Wood Bison	Forest Duikers (*Cephalophus*)
—Plains Bison	Blue Duikers (*Philantomba*)

Regional Distribution

India	Southeast Asia	Central Asia	Africa	North America
Gayal or Mithan	Gayal or Mithan	Yak	Forest Buffalo	Wood Bison
Gaur	Gaur		Cape Buffalo	Plains Bison
Water Buffalo	Coupray		Elands	
Nilghai	Banteng		Bongo	(Extinct)
Chousingha	Anoa		Kudus	
	Tamarou		Bushbucks	Wisent (E. Europe)
			Duikers	

Well publicized and native to the high country of Tibet is the Yak. It has been domesticated by Tibetans for centuries. However, some herds still remain in the wild. The undomesticated yak is a large animal, attaining a height of well over 5 feet and a weight of more than 1000 pounds. The horns are large, and are of the characteristic curved form usually found in oxen, with smooth surfaces. The animals are marked chiefly by the long hair that clothes the flanks, legs, and tail, drooping almost to the ground. In contrast, the domesticated yak is considerably smaller, lacks horns, and has a docile disposition. Like other domestic cattle, yaks are important as beasts of burden, and as a source of meat, milk, hides, and hair. The Tibetans brew tea, using rich yak butter instead of milk.

The Banteng of southeast Asia is very buffalolike in appearance, with only a slight suggestion of a dewlap. Bantengs dwell in the forest and are considered inquisitive, although not aggressive.

Some wild Water Buffalo still exist in Borneo and other islands of Indonesia, but the animal has been domesticated for many centuries and has been widely distributed to other tropical regions. The animal is characterized by large horns which sweep back, large head, heavy body (weighs up to 1200 pounds; 543 kilograms), semiaquatic nature, and a relatively docile disposition to persons known, but not to strangers. The animal has amazing strength and is used as a general-purpose animal for farming and hauling.

The Forest Buffalo of Africa is well named for its preferred habitat and is a relatively docile animal unless attacked. There is a large variety of forest buffaloes, found in west and west central Africa. One of these is the Bush-Cow. Found along the eastern portions of Africa is the very large, strong, and aggressive ox known as the Cape Buffalo. It does not dwell in the forest, but prefers swampy areas except for occasional grazing in nearby grasslands. The Cape Buffalo may weigh up to 1500 pounds (680 kilograms) and is characterized by large, wicked horns, used for goring a victim when provoked. The animal also has a sharp rasplike tongue which can be used to tear the flesh of its victims. They are noted for their rather bad dispositions.

Two small oxen, the Anoa and the Tamareu, are native to the island Mindoro in the Philippines. The anoa is a favorite of zoological gardens.

The story of the American Bison, sometimes mistakingly called a buffalo, has been told innumerable times. They were slaughtered by the millions (estimated at 60 million) as people moved westward across America. Once near extinction, fortunately as the result of protection, the American western Plains Bison is staging a comeback. At one time, it is estimated that there were fewer than a thousand of the animals remaining. In addition to the western plains, the bison also was found in portions of the eastern United States from Lake Erie south to Georgia. The eastern form was fully extinguished by the early pioneers. The Wood Bison which is larger still exists in the wild form in northwestern Canada. See Fig. 1.

The American Bison is a large oxlike animal, weighing in excess of one ton when fully grown, has a large shaggy head, small curved horns, prefers roaming in large herds, stands about six feet high, and has humped shoulders and a small, distinguishing beard. See accompanying photo. The front legs and body of the bull are covered with rather shaggy, thick fur, but the hindquarters are fully absent of shag, giving the appearance that perhaps the rear half of his "dress" has been closely clipped.

The Deer-Oxen have been aptly described as links between the deer and the bovines. The Nilghai of India is of striking appearance, of a bluish-reddish brown coloration, with long legs and neck. Although somewhat sleek, they nevertheless retain oxlike features. They prefer open forest and park lands and are a favorite dietary item of the tigers. The Indians sometimes refer to the Nilghai as an antelope.

Relatively un-oxlike in appearance is the Chousingha, also of India, and a relative of the Nilghai. The animal appears much like a small deer. Zoologists are intrigued by the animal's four horns which are quite unusual in any mammal.

The Twist-Horned Oxen are also related to the Nilghai but are native to Africa. Sometimes this group is called the "harnessed antelopes." Of this family, the Elands prefer a habitat of plains and savan-

Fig. 1. Bison (*A. M. Winchester.*)

nas; the Kudus like mountain country; and the Bushbucks are found in forests. The eland is sometimes referred to as the largest African oxen of antelopine shape. The body is like that of a cow, but the head is smaller and the horns are moderately long, straight, and spirally twisted. There are several species of *Taurotragus.* Of the two Kudus, the Greater Kudu is the most common and prefers mountainous terrain. The animal possesses large, wide-spreading horns, which may measure up to five feet in length, twisting up to three times. The Lesser Kudu prefers the desert scrublands.

The Duikers are relatively small animals, some varieties no larger than a jack rabbit, with a few as large as a donkey. They have arched backs with rumps slightly elevated as found in the rodent. The ears are large, tail is short, but of definite oxlike design. Both sexes have simple spiked horns. Duikers are present in parts of Africa (Kenya, Angola, Rhodesia) in very great numbers in the forests and woodlands. They are such quiet, retiring animals that it is not easy to be aware of their high population.

Development of Beef Cattle Breeds

The development of modern beef breeds began in the 1600s in Europe and, in particular, the British Isles. Farmers in an area selected cattle of a kind they considered best for the locality. They continued to grow them consistently over a period of years, and these selections often resulted in the formation of a breed. Some breeds resulted from crosses of existing breeds; others from crosses of cattle that had not attained breed status. The most desirable animals tended to be gathered into a few herds that were bred by introducing little or no other stock. As they gained popularity, numbers increased and eventually a breed society was formed. In this way, highly useful and efficient kinds of animals were developed that survived as breeds. A breed may be defined as a group of animals having a common origin and possessing certain well-fixed and distinctive characteristics not common to other members of the same species; these characteristics are uniformly transmitted.

Some beef breeds are horned and some polled (hornless). Mutations have occurred in certain animals of several horned breeds, causing them and their descendants to be polled. In some cases, breeders developed these polled strains and established separate breeds. It is a moot question whether these types are truly different breeds. The question is sometimes asked: "Was the original wild ox a horned or hornless beast?" Historians are not in agreement except to state that there is considerable evidence that hornless cattle were extant in ancient times and likely arose first in the bovine world. It is possible that they were replaced by cattle with horns needed for defense against carnivorous opponents. Some scientists believe hornless cattle to be mutations, but others believe that polled animals are a reversion to the early type found in some fossil remains.

Mating Systems Used in Breeding. There are three general mating systems used to produce crossbred market cattle: (1) Mating females of one breed to males of another breed to produce F_1 market animals; (2) Terminal crossbreeding whereby F_1 females, as produced according to step (1) are mated to bulls of a third breed to produce a three-breed terminal cross; or (2) to bulls of one of the parent breeds to produce a backcross; or (3) rotational crossbreeding where a breed of sire is rotated each generation or at specific time intervals.

Heterosis, sometimes referred to as *hybrid vigor,* is a phenomenon that is most important in crossbreeding, but is essentially an extension of hybrid techniques used for centuries in connection with plants. See Table 2.

TABLE 2. HETEROSIS (HYBRID VIGOR) VERSUS CROSSBREEDING SYSTEM

	Percent of Total Possible Heterosis	
Crossbreeding System	In Brood Cow	In Calves
1. Two-breed cross	0	100
2. Three-breed terminal	100	100
Backcross	100	50
3. Two-breed rotational	67	67
Three-breed rotational	86	86

Fundamental Categories of Breeds. For convenience of classification, there are four fundamental categories of beef and dairy cattle. The categories, although commonly used, are not consistent because two of the categories reflect the origin of the cattle, and the other two reflect the purpose of the cattle.

1. British and continental European breeds (beef).
2. North American breeds (beef).
3. Dual-purpose (beef and dairy) breeds.
4. Dairy breeds.

The foregoing classification is not fully satisfactory for all of the world because it does not reflect the so-called exotic cattle, as found in Asia, and does not fully parallel cattle found elsewhere.

Detailed descriptions of the scores of breeds are well beyond the scope of this encyclopedia, but a cross section will be given briefly. Much more detail can be found in the "Foods and Food Production Encyclopedia," (D. M. Considine, editor), Van Nostrand Reinhold, New York, 1982.

Angus Cattle. Principal characteristics are: (1) black, smooth-hair coat; (2) polled; (3) generally alert and vigorous; (4) produce well-marbled beef. Angus cattle are known to have existed as early as 1523 in the county of Aberdeenshire in Scotland. In this region, the breed developed in a rigorous climate and on rolling to rough land that was not particularly fertile, except in the valleys. The first Angus bulls were imported into the United States in 1883. An Angus bull is shown in Fig. 2. In the American Southwest, the Angus cattle were crossbred with Texas Longhorn cattle. See Fig. 3.

Charolais Cattle. Principal characteristics are: (1) white, or very light straw-color coat; (2) mature purebred bull weight ranges from 2000 to 2500 pounds (907 to 1134 kilograms) or more; (3) mature cow weight ranges from 1250 to 1600 pounds (567 to 725 kilograms);

Fig. 2. Angus bull. (*USDA photo.*)

Fig. 3. Texas Longhorn steer. (*USDA photo.*)

(4) a high rate of efficiency of growth; and (5) a high percentage of lean meat with a minimum of excess fat at a young age. See Fig. 4. In France, Charolais is one of the most important beef cattle breeds. The breed did not arrive in the United States (by way of Mexico) until 1936.

Fig. 4. Charolais bull. (*USDA photo.*)

Fig. 5. Hereford bull. (*USDA photo.*)

Fig. 6. Brahman bull. (*American Brahman Breeders Association.*)

Hereford Cattle. Principal characteristics are: (1) white face, crest, dewlap, underline, and switch; white legs below the hocks and knees; red bodies; (2) medium-size horns; and (3) docile nature, easily handled. See Fig. 5. This breed originated in the County of Hereford in England. In 1817, Henry Clay (statesman from Kentucky) imported the first Herefords. The breed has been popular in the United States since the 1870s.

North American Beef Breeds. Development of beef breeds in North America has taken place mostly since the early 1900s. The Brahman

Fig. 7. Beefmaster bull. (*USDA photo.*)

was developed by combining several breeds or strains of zebu (*Bos indicus*) cattle of India. In other cases, new breeds have been developed from Brahman-European crossbred foundations. Principal characteristics: (1) distinctive appearance, a hump over shoulders, loose skin (dewlap) under throat, and large drooping ears; and (2) light gray color or red to almost black; prevailing color is light to medium-gray. Environmental adaptation, longevity, and mothering ability are the Brahman's strongest traits. A Brahman bull is shown in Fig. 6.

Fig. 8. Santa Getrudis bull. (*USDA photo.*)

Fig. 9. Japanese Wagyu cattle. (*USDA photo.*)

Beefmaster Cattle. Principal characteristics: (1) Color is variable, with more reds and duns than other colors; and (2) most animals are horned, but polled individuals do occur. See Fig. 7. Development of this breed was commenced in 1931 in Texas. The foundation herd was developed from three breeds—the Hereford, the Shorthorn, and the Brahman.

Santa Gertrudis Cattle. Principal features: (1) cherry red color; (2) the majority are horned, but polled individuals occur; (3) hides are loose, with surface area increased by neck folds and sheath or navel flap; and (4) hair is short and straight in warm climates, long in cold climates. See Fig. 8. Development of this breed dates back to the early 1900s on the King Ranch in Texas.

Fig. 10. Guernsey cow. (*American Guernsey Cattle Club.*)

Wagyu Cattle. In Japan, these cattle are the source of the well-known Kobe beef. Traditionally, about 2.5 years are required to feed a Wagyu up to time of slaughter. See Fig. 9.

Dual-Purpose (Beef and Dairy) Breeds. Among the Better-known breeds of this type the Milking Shorthorn cattle, the Red Poll cattle, the Brown Swiss cattle, and the Holstein-Friesian cattle. These breeds have reasonably good beef conformation and they are also capable of producing milk and butterfat in reasonably large quantities.

Dairy Cattle. About 70% of the dairy cattle in the United States are grades of purebreds of six breeds—Ayrshire, Brown Swiss, Guernsey, Holstein-Friesian, Jersey, and Red Danish. Two of these breeds, as previously mentioned, are dual-purpose breeds.

Guernsey Cattle. This breed originated on the island of Guernsey, off the coast of England. Over 13,000 of these animals were imported into America prior to 1914. A mature cow in milk should weigh at

Fig. 11. Jersey cow. (*American Jersey Cattle Club.*)

least 1100 pounds (499 kilograms). A mature bull in breeding condition should weigh about 1700 pounds (771 kilograms). A Guernsey cow (*Ideal's Beacon's Rosette*) produced 224,800 pounds (101,080 kilograms) or 25,912 gallons (980 hectoliters) of milk; and 10,941 pounds (4963 kilograms) of butterfat in her lifetime. See Fig. 10.

Jersey Cattle. This breed originated on the island of Jersey, off the coast of England. Jerseys were imported into the United States as early as 1800. They are a little smaller than the Guernsey cattle. A Jersey cow (*Marlu Milady's*) produced 191,226 pounds (86,760 kilograms) or 22,236 gallons (842 hectoliters) of milk; and 9444 pounds (4284 kilograms) of butterfat in her lifetime. See Fig. 11.

References

Alford, C.: "A Practical Approach to Crossbreeding," Univ. of Georgia College of Agriculture, Athens, Georgia, *Publication 35* (1978).

Clark, E.: "Mathematical Models May Help Increase Cattle Profits," *Feedstuffs* (September 25, 1978).

Considine, D. M. (editor): "Foods and Food Production Encyclopedia," Van Nostrand Reinhold, New York, 1982.

Ensminger, M. E.: "Dairy Cattle Science," The Interstate, Danville, Illinois, 1971.

Ensminger, M. E.: "Beef Cattle Science," The Interstate, Danville, Illinois, 1976.

Murphy, M. P.: "USFGC Plans New Feed Program in Japan for Wagyu Cattle," *Foreign Agr.*, U.S. Department of Agriculture, Washington, D.C. (1978).

Putnam, P. A., and E. J. Warwick: "Beef Cattle Breeds," U.S. Department of Agriculture, Washington, D.C., 1975.

Whyte, R. O.: "Milk Production in Developing Countries," Faber and Faber, London, 1977.

BOWEL DISEASES. Colitis and Other Inflammatory Bowel Diseases.

BOWEN'S DISEASE. Dermatitis and Dermatosis.

BOWERBIRD (*Aves, Passeriformes*). Birds of several species found in the Australian region. They build bowers or runs roofed with grass or sticks and decorated with bright articles of all kinds, used as playhouses and to attract females.

Bowerbirds are busy workers, neat housekeepers, and like beauty. They display most unusual and fascinating habits. They are known to build runways, sometimes 2 to 3 feet (0.6 to 0.9 meter) long and will furnish these with a colorful flooring of pebbles, bones, snails, and insect remains, all items selected for their high color and attractiveness. Sometimes bright colored feathers and orchid blossoms are strewn around. Also, the orchid stems may be used in constructing partitions of a wigwam type of design. It has been noted that fresh flowers are brought in to replace withered blossoms. It is believed that the orchid is selected because it retains its freshness and beauty over a relatively long period. These so-called bowerbird houses are associated with mating and are not the regular nests. The males dance until their death sometimes in a duel for their mate.

Species of bowerbirds include: The gardener-bird (*Anbylornis inornatus*), a species known for using moss in its house construction, sometimes banking the moss up to 18 inches (46 centimeters) in height around the trunk of a tree. Twigs are used to strengthen the walls, which are reported in some cases to have "windows." The satin-bird (*Ptilonorhynchus violaceus*) is found in southern Australia. The male has satin black plumage. The female and young are grayish-green. The hut of the satin-bird is usually dome-shaped and made of twigs a few inches long. The regent-bird (*Sericulus chrysocephalus*) is found north of Sydney, near the Brisbane River. See illustration. It mainly uses snail shells as a material of construction. The spotted bowerbird (*Chlamydera maculata*) forms runs or walkways about 3 feet long (0.9 meter) filled with attractive colored objects as previously mentioned.

Regent bowerbird (*Sericulus chrysocephalus*).

Many years ago, when the first naturalists came across the architectural skills of the bowerbirds, their houses were assumed to have been constructed by persons. But subsequent investigations over the years have demonstrated the abilities and habits of these particular birds to be unique among birds.

BOWFIN (*Osteichthyes*). The terms dogfish, grindle, spotfin, and mudfish have also been used in describing the bowfin (*Amia calva*) of the order *Protospondyli* and family *Amiidae.* In ancient times, this fish was widely distributed in the fresh waters of North America. It is now found principally in lakes and sluggish streams in the eastern and central United States. The fish is easily identifiable because of a long, spineless dorsal fin featuring about 58 rays. The bowfish has a well-developed air bladder with a cellular internal surface, thus enabling the fish to occupy waters that may contain no oxygen (or survival out of water) for as much as 24 hours. The normal bowfin weighs but a few pounds and has a length of about 2 feet (0.6 meter). However, some specimens weighing up to 8 pounds (3.6 kilograms) and 3 feet (0.9 meter) in length have been recorded. Although sometimes eaten, they are not considered a highly desirable food fish.

BOW'S NOTATION. A standard method of representing, by letters of the alphabet, forces and stresses in graphical analysis. This analysis may consist of such problems as the graphical solution of stresses in simple framed structures or the determination of the resultant of an independent system of unbalanced forces lying in the same plane and having a common point of application. The accompanying figure illustrates the method of applying Bow's notation to the latter system. Let P_1, P_2, P_3 and P_4 be a system of unbalanced forces lying in the same plane and having a common point of application. Denote the space between the line of action of each force by the letters *A, B, C* and *D.* Next construct a figure called a force polygon. This is accomplished by drawing a line parallel to P_1 and laying off its magnitude to a definite scale denoting the ends of the line by the letters *a* and *b.* From point *b* lay off *bc* equal in magnitude and parallel to P_2. Repeat the operation for the other forces. Upon completion of this graphical figure it will be found, in general, that the line representing P_4 will not pass through point *a.* The distance from point *a* to end of this line, which will be lettered *e*, represents the value of the resultant of P_1, P_2, P_3 and P_4 according to the scale used. The direction of *ae* determines the line of action of the resultant. Thus, in Bow's notation a force in space is designated by the space letters on either side of it, whereas the forces as part of the force polygon are named by the letters at their extremities.

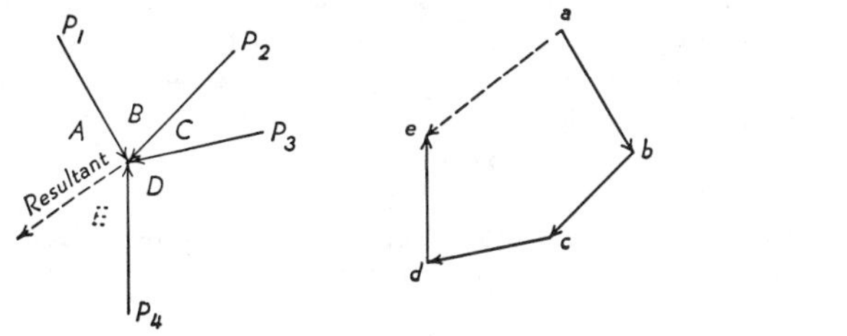

Representation of Bow's notation: (left) force system; (right) force polygon.

BOX AND PLATE GIRDER BRIDGE. Bridge (Structural).

BOX ANNEALING. Iron Metals, Alloys, and Steels.

"BOX-CAR" LENGTHENER. A pulse-lengthening circuit which lengthens a series of pulses without changing their height. Ideally, it produces flat-topped pulses prolonged throughout the interval between pulses.

BOX ELDER. Maple Tree.

BOX PRESS. Expression (Mechanical).

BOX TREES AND SHRUBS. Of the family *Aquifoliaceae* (holly family), genus *Buxus,* the common box, *Buxus sempervirens,* occurs naturally in southern Europe, north Africa, and Turkey. It is cultivated in some regions of North America. Depending upon height, the plant can be classified as an evergreen tree or small shrub. The plant can attain a height up to about 35 (10.5 meters). The box has short, oval, leathery leaves that produce a very dense foliage. The fruit is small. The flowers are of a pale green color, small, hardly noticeable. In the British Isles and Europe, the box is traditionally used for low hedges.

The wood of the box is even-grained and hard and a favorite for wood engraving blocks, rulers, instruments, and inlay work. It is called boxwood or, frequently, Turkish boxwood. The wood weighs about 65 pounds per cubic foot (1041 kilograms per cubic meter). Cape boxwood comes from *Buxus macowani,* a tree found in South Africa. The wood is somewhat softer than other boxwoods. Kamassi wood is from *Gonioma kamassi,* also a tree of South Africa. This wood is valued for making loom shuttles. Coast gray boxwood is from the *Eucalyptus bositoana* tree found in New South Wales. It is a durable wood with uniform texture, but of an interlocking grain. Maracaibo wood comes from *Casearia praecox,* a Venezuelan tree. The wood is knotless and considerable quantities are shipped in logs of about 8 feet (2.4 meters) in length and 8 inches (20 centimeters) in diameter. The wood is used for nearly all purposes served by other boxwoods, except for wood engraving blocks. Ginkgo wood comes from the large *Ginkgo biloba* tree of China and is frequently used for making chess men and chess boards. See also **Maidenhair Tree.**

BOYLE-CHARLES LAW. This law states that the product of the pressure and volume of a gas is a constant which depends only upon the temperature. This law may be stated mathematically as

$$p_2v_2 = p_1v_1[1 + a(t_2 - t_1)]$$

where p_1 and v_1 are the pressure and volume of a body of gas at temperature t_1, p_2 and v_2 are the pressure and volume of the same body of gas at another temperature t_2, and a is the volume coefficient of expansion of the gas. If the temperature is expressed in degrees absolute, this expression becomes

$$\frac{p_2v_2}{T_2} = \frac{p_1v_1}{T_1}$$

which is the ideal gas law, so-called because all real gases depart from it to a greater or lesser extent. See also **Characteristic Equation.**

BOYLE'S LAW. This law, attributed to Robert Boyle (1662) but also known as Mariotte's law, expresses the isothermal pressure-volume relation for a body of ideal gas. That is, if the gas is kept at constant temperature, the pressure and volume are in inverse proportion, or have a constant product. The law is only approximately true, even for such gases as hydrogen and helium; nevertheless it is very useful. Graphically, it is represented by an equilateral hyperbola. If the temperature is not constant, the behavior of the ideal gas must be expressed by the Boyle-Charles law.

The Boyle temperature is that temperature, for a given gas, at which Boyle's law is most closely obeyed in the lower pressure range. At this temperature, the minimum point (of inflection) in the $pV - T$ curve falls on the pV axis. See **Compression (Gas);** and **Ideal Gas Law.**

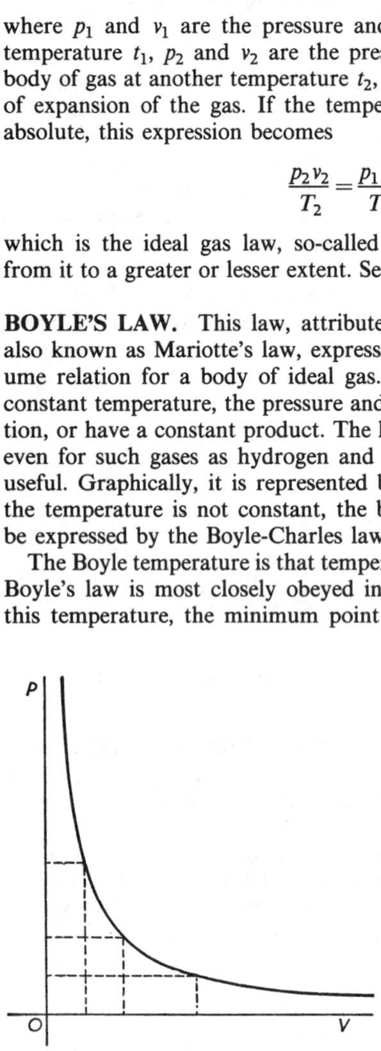

Equilateral hyperbola representing Boyle's law. The rectangular areas (PV) are all equal.

B. P. An abbreviation of "Before Present." This term is an indication of time calculation, used especially when referring to radiometric dating. When used with a carbon-14 date, for example, it indicates that the quoted data are calculated from A.D. 1950, because of the increased radioactivity introduced into the atmosphere since that time by nuclear testing.

BRACHIAL. Pertaining to the arm, from the Latin term *brachium.*

BRACHIATION. Anthropoids.

BRACHIOPODA. A phylum of marine animals which resemble the bivalve mollusks superficially. In the remote past they were much more abundant, as is shown by extensive fossil remains of many more forms than exist today.

The brachiopods are characterized by the following structures: (1) The body is enclosed by a shell consisting of dorsal and ventral valves. (2) The animal is triploblastic and coelomate but not segmented. (3) A ciliated organ called the lophophore projects about the mouth. It maintains currents of water which carry food and oxygen to the animal and wash the wastes away.

The phylum is divided into two orders:

Order *Ecardines.* Valves of shell not joined by a hinge. Anus present.

Order *Testicardines.* Valves of shell joined by a hinge. Alimentary tract without an anus.

See also **Invertebrate Paleontology.**

BRACHISTOCHRONE. The characteristic curve along which a particle will slide from one point to another under the influence of gravity in the least possible time, friction being neglected. If the particle starts from rest at the origin of a Cartesian coordinate system (it is convenient to let the Y-axis extend to the right and to measure x downward) and falls to the point (x_2, y_2) the following integral results

$$\sqrt{2g}t = \int_0^{x_2} \left(\frac{1 + y'^2}{x}\right)^{1/2} dx$$

where t is the time, g is the acceleration of gravity, and $y' = dy/dx$. It is to be minimized by methods of the calculus of variations. When the resulting differential equation is solved, the curve is found to be a cycloid. See also **Abel Equation.**

BRACHYBLAST. In numerous plants, especially in the gymnosperms, the display of leaves to light is considerably advanced by the formation of short lateral branches called brachyblasts (or short shoots). In the larch, this short shoot is well developed. In this plant, it persists year after year bearing at its tip a small group of leaves. It does not, however, increase in diameter even after several years of growth. The brachyblast develops from a bud formed in the axil of a leaf. The maidenhair tree or *Ginkgo* is another tree having well-developed short shoots. In both these plants and in many others the short shoot bears at its tip a terminal bud from which the leaves of the following year develop. In the pines, on the contrary, the short shoot is very much reduced and bears no terminal bud. In these plants the short shoot is reduced to a single bundle of leaves which persist for a year or two and then drop off completely. That this condition in pines is a reduced condition is clear from the condition found in fossil pines, which have a well-developed brachyblast, bearing many leaves and having a terminal bud.

BRACHYCEPHALIC. Short-headed. As applied to measurement of the human skull, with a width which is more than $\frac{4}{5}$ of the length.

BRACKET FUNGI (*Polyporaceae*; also called Shelf Fungi). A large group of fungi the fruit-body of which forms a characteristic shelflike outgrowth from the trunks of trees. This fruit-body arises from a mycelium of fine hyphae, which penetrate throughout the woody tissue of the host plant, from which they derive nourishment and which they slowly destroy. The fruit-bodies are often perennial, showing on sectioning the successive-growth-layers, which are added each year. See **Basidiomycetes.**

BRACKETT SERIES. **Energy Level.**

BRACKISH WATER. **Desalination; Estuary; Hydrology.**

BRACT. In many flowering plants there is found at the base of the flower stalk a small leaf, often considerably modified; this is called a bract. In many plants its minute size causes it to be overlooked; in others it is a conspicuous object. In the poinsettia, for example, the large showy red "flower" is really composed of bracts, as is also the conspicuous white petal-like structure surrounding the very small flowers of the flowering dogwood.

BRADYCARDIA. **Arrhythmias (Cardiac); Ischemic Heart Disease.**

BRAGG'S CURVE. There are two types of curves to which Bragg's name is occasionally given: 1. A graph for the average number of ions per unit distance along a beam of initially monoenergetic alpha particles, or other ionizing particles, passing through a gas. 2. A graphical relationship between the average specific ionization of an ionizing particle of a particular kind, and some other variable, such as the kinetic energy, the residual range, or the velocity of the particle.

BRAGG'S LAW. The law expressing the condition under which a crystal will reflect a beam of x-rays with maximum distinctness, at the same time giving the angle at which the reflection takes place. For x-ray reflection it is customary to use the complement of the angle of incidence and reflection, that is, the angle which the incident or the reflected beam makes the crystal planes, rather than with the normal. Let this "Bragg angle" be θ. If the planes or layers of atoms are spaced at a distance d apart, and if λ is the wavelength of the x-rays, Bragg's law is expressed by the equation

$$\sin \theta = \frac{n\lambda}{2d}$$

The condition for an intensity maximum is that n must be a whole number. For example if the planes of rock salt parallel to the natural cubical faces are spaced at $d = 2.814 \times 10^{-8}$ centimeters or 2814 x-units, and if the incident rays have a component of wavelength $\lambda =$ 714 x-units, the above equation gives $\sin \theta = 0.1269n$. Then if the crystal is rotated slowly, there will be a distinct reflection where θ reaches 7° 17′ ($n = 1$), again at 14° 42′ ($n = 2$), also at 22° 23′ ($n = 3$), etc. See also **Crystal.**

BRAGG SPECTROMETER. An instrument for the x-ray analysis of crystal structure, in which a homogeneous beam of x-rays is directed on the known face of a crystal, *C*, and the reflected beam detected in a suitably placed ionization chamber, *E*. As the crystal is rotated, the angles at which the equation expressing Bragg's law is satisfied are identified as sharp peaks in the ionization current. See accompanying figure. This is one of the early, classical instruments in the laboratory field.

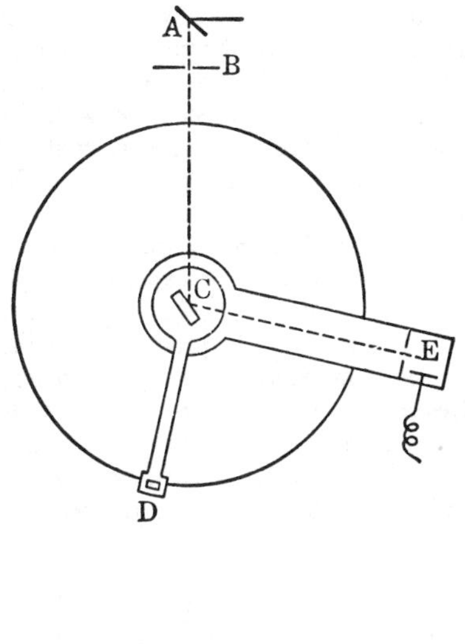

Bragg spectrometer.

BRAGG'S RULE. An empirical relationship whereby the mass stopping power of an element of alpha particles is inversely proportional to the one-half power of the atomic weight. This relationship is also stated in the form that the atomic stopping power is directly proportional to the one-half power of the atomic weight. The wide usefulness of the Bragg rule is due to the fact that it leads to relations between the stopping powers of different elements for alpha particles. It also applies to other charged particles as well as alpha particles, and to the same degree of approximation, which is about ±15%.

See **Particle (Subatomic).**

BRAHMAN CATTLE. **Bovines.**

BRAILLE SYSTEM. A printing or writing system for the blind in which letters and characters are represented by raised dots or points which are discernible to the touch. Invented by Louis Braille, a French teacher of the blind in 1829. Modified slightly since its origin, Braille writing is an almost universal system. Special books, newspapers, and periodicals are available in Braille. A second form of literature is referred to as *Moon's type.* It consists of raised lines and curves and is chiefly valuable for the small percentage of persons who do not seem able to learn Braille. Books in Moon's type are bulky and expensive and rather scarce. Braille literature, by contrast, is available in many public libraries, even of moderate size. Phonograph records and tape recordings also have augmented the Braille system of communication during the last several decades. See also **Vision and the Eye.**

BRAIN AND NERVOUS SYSTEM. Like any other tissue of the body, the brain and nervous system are made up of cells. These cells function in accordance with the same fundamentals that apply to other cells, even though the cells of the complex interwoven brain tissue and the highly specialized cells of the central and peripheral nervous systems engage in unique operations. On a comparatively large scale, investigations of the brain and of the central nervous system can be conducted much as other organs are studied. Maps have been drawn of the interconnections of the brain on a relatively massive, regional basis, but remain severely lacking in terms of the huge numbers (millions or even billions) of interconnections among extremely tiny elements that constitute the infrastructure of the brain. Both the neuroanatomy and the neurophysiology of the brain, as of the early 1980s, can be described as in a very early stage of understanding. Most scientists will agree that this statement holds even in view of the very impressive progress that has been made in brain and nervous system research over the last several years, particularly during the decade of the 1970s. Up to a point, the electrical and chemical signalling apparatus of the brain and nervous system can be measured and interpreted, but beyond a general understanding of the brain, the problem of comprehending brain function in depth is of a staggering nature—and for several reasons.

(1) *The sheer numbers* of individual elements of the brain system that exist within an organ whose mass in the average adult human is about 45 ounces (1.3 kilograms), representing an information processing system unparalleled in the universe as we know it today. It is conservatively estimated that there are on the order of 10^{11} (a hundred billion) nerve cells (neurons) in the human brain. This number approximates that of the number of stars in our galaxy. Some neuroscientists consider this figure woefully small. If the 10^{11} figure is assumed, it suggests that neurons in the embryonic and fetal stages, originating as a flat sheet of cells on the dorsal surface (*neural plate*) of the developing embryo, must develop at an average rate of more than 250,000 per minute. Once the production of neurons is completed, there is no further replication during the remainder of life, but as life progresses, large numbers of neurons are destroyed.

How can researchers "get a handle" on so many components? By comparison, the number of components in the most capable of modern computers pales. The number of synapses (*connections*) in the brain is estimated at 10^{14} (100 trillion).

(2) *The extremely small size* of the nerve cells is another problem. Research has confirmed that so many "components" are perforce extremely tiny to fit in such a relatively small volume. The typical neuron (cell) is estimated to range from 5 to 100 micrometers (thousandths of a millimeter) in diameter. There are even much tinier subelements

which make up these cells. Very important nerve actions occur in which the amount of substance involved is expressed in terms of a few hundreds of molecules.

(3) *The variations of form and function of neurons* make generalizations difficult. The neuron is not a conventionally shaped cell, but takes the form of a trunk of a tree with a well-developed root system, where the trunk may be as short as 10–12 micrometers (thousandths of a millimeter) in diameter and yet the entire cell may have a length ranging from as short as 0.1 millimeter to as long as 1 meter or more. Because neurons are tightly packed in many regions of the brain, they tend to occur in "thickets," much as the intertwining roots of closely-planted trees. But, despite the high-density packing, a very effective insulation between the intertwining fibers provides full chemical and electrical integrity of each neuron except at specific designated points (*synapses*) of connection. The fluid film which provides this protection of nerve fibers is only about 0.02 micrometer in thickness.

As will be evident from later descriptions, there are many variations in the physical and electrical and chemical characteristics among individual neurons and, to date, these variations have not been easy to classify in terms of a relatively few categories. An important aspect of progress in neuroscience during recent years has been that of identifying previously unknown neuron forms and characteristics and developing a series of generalizations that provide pathways toward a better understanding of specific information. Although from a systems standpoint, there appears to be a high degree of specificity represented by the neurons and their synapses, the tight packing of these elements into such a small volume (skull) creates a type of geometry that tends to defy the neatness desired by the human analytical mind. It is as though millions of very neatly arranged computer circuit boards or silicon chips were tightly squeezed into a plastic ball, but with the integrity of these elements still maintained. Although the analogy is not very apt, the intricacies of the brain resemble the "spaghetti" of wiring encountered in an old-fashioned radio chassis as contrasted with the extreme neatness and geometry of modern electronic equipment.

(4) *The dynamics of the brain system* can only be partially studied by examining dead brain tissue. It will be recalled that death itself is defined in terms of "brain death." It is the living, functioning brain that yields a full comprehension of brain function. Lower animals and some primates have provided leads to understanding the human brain function. Much of what has been learned concerning the nerve impulse has been gleaned from studies of the squid.

Research Breakthroughs of the Past. Because of the very high-density packing of neurons in brain tissue, a discovery by the Italian anatomist Camillo Golgi, as early as 1875, was of great note. Although the phenomenon of Golgi's brain tissue staining technique remains unexplained, the technique makes it possible to stain only small numbers of brain cells at any one time. This technique has proved invaluable to neuroanatomy and neurophysiology because the method allows specific identification and study of a few components that otherwise would be lost in a morass of thousands upon thousands of similar components. A Spanish contemporary of Golgi, Ramón y Cajal, applied Golgi's technique over a lifetime and in 1904 published *"Histologie du système nerveux de l'homme et des vertébrés."* This publication is still regarded as the greatest single work in neurobiology. In the time of Golgi and Cajal, the research tools were confined to the staining technique and the light microscope. One of Cajal's sketches of Golgi-stained nerve tissue is shown in Fig. 1. Within these limitations, Cajal proposed two important concepts: (1) Patterns do not suggest a network (particularly as suggested by a modern computer network), but rather a system comprised of specific cells which communicate with each other only at certain points (connection points or synapses). (2) Interconnections do not suggest a random arrangement, but rather a highly structured, quite specific arrangement. One might suggest that perhaps the organization of the brain may provide new concepts in computer engineering that can be reduced to practice only after a great deal more is learned of brain function and organization. Based upon present knowledge, the often suggested brain-computer analogy is of little value.

It was not until the early 1950s that a scientist in the Netherlands, Walle J. H. Nauta, developed an improved staining method. Nauta

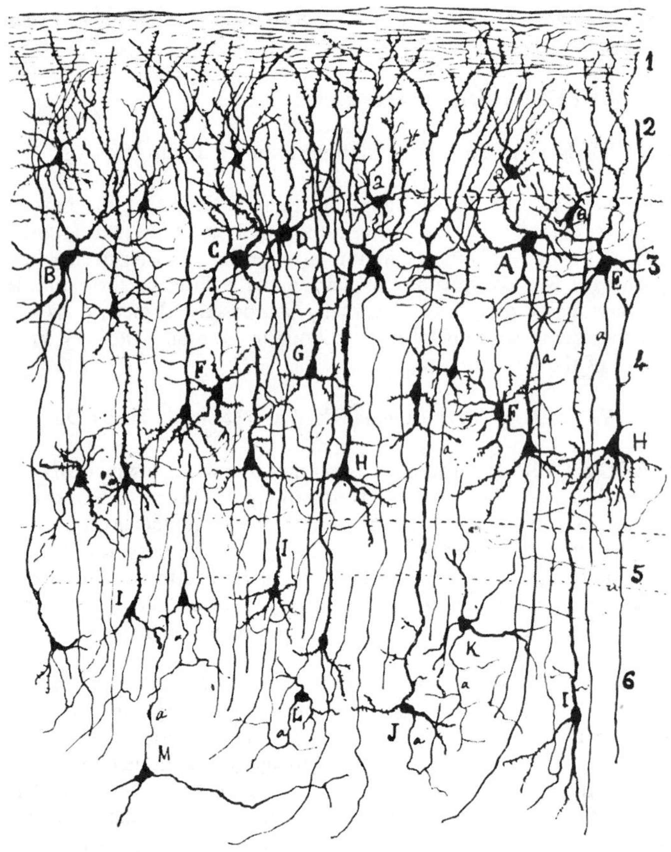

Fig. 1. Sketch from "Histologie du système nerveux de l'homme et des vertébrés," published by Ramón y Cajal in Spain in 1904. Cellular layers are identified by numerals at right-hand edge of diagram. Neurons are indicated by capital letters.

found that there is a short period of the time during which the nerve fiber degenerates when the fiber can be stained to differentiate it from living fibers. Thus, it is possible to trace that fiber to a specific neuron that may be remotely located in some other region of the brain. This technique has greatly expanded the researcher's ability to trace interconnections and thus add many details to the maps of the brain.

Since the mid-1960s, other advanced techniques have contributed in a major way to developing a better understanding of the anatomy of the brain. One of these is the injection of radioactive substances into brain tissue. Some of these substances are absorbed by cell bodies and transported to terminals; other substances are absorbed by the terminals and transported by axons to the cells. In a technique developed by Louis Sokoloff (National Institute of Mental Health), radioactively labeled deoxyglucose is used. This substance collects in nerve cells and the degree of accumulation is an indicator of a cell's activity. This technique is used in conjunction with known laboratory stimulation of various areas of the brain and thus permits brain research in real time as contrasted with examining dead tissue. For example, if the stimulation is audio, those cells which are involved in the process of hearing will be active and can be revealed by later microscopic examination. In a more recently developed technique, positron-emitting radioactive isotopes are combined with tomography which makes it possible to conduct these studies entirely outside the skull and thus makes it possible to expand the research to living animals and, in the diagnosis of mental problems, the technique may be refined to include observations of the living human brain.

The decade of the 1970s was marked by numerous discoveries pertaining to the biochemistry of the brain. For example, the known chemical transmitters has been expanded from a list of just a few to several. Much has been learned concerning the production of chemical substances within the brain, once considered a function in the exclusive province of the glands. Research commenced by neuroscientists at McGill University in the mid-1950s led to the discovery of "pleasure centers" within the brain that play a role in learning and memory. As pointed out by Routtenberg (1978), the so-called reward system can be localized in particular nerve cells and their fibers. The system is affected by drugs that interact with the substances secreted by these nerve cells. The fact that only certain nerve-fiber pathways are implicated in brain reward suggests that these pathways have a specific function. This parallels what is known of the brain's visual and movement systems, each of which has a specified set of component pathways. This research led to many new insights on the functioning of various psychoactive drugs. Neuroscientists at several institutions have found many new substances, generally referred to as neuropeptides, in brain tissue. These substances include the heavily publicized enkephalins and endorphins. Researchers have found that neuropeptides differ from previously identified transmitters in that they appear to orchestrate complex phenomena, such as thirst, memory, and sexual behavior. See also the accompanying table of research events.

VERY ABRIDGED LIST OF IMPORTANT BRAIN RESEARCH EVENTS[1]

- In 1870, G. T. Fritsch and E. Hitzig first suggested functional compartmentalization in the cerebral cortex.
- In 1875, Golgi developed a brain tissue staining technique which enables the staining of a comparatively few neurons and neuronal systems at one time, thus making it possible to study a few out of many thousands of adjacent systems.
- In 1885, P. Ehrlich discovered the blood-brain barrier, finding that many blood-borne solutes do not penetrate into central nervous system tissue as rapidly as they penetrate into most other tissues. The functional significance of the blood-brain barrier mechanism is to buffer the neuronal microenvironment against changes in plasma concentrations of various important solutes and to regulate the composition of the neuronal "atmosphere" for optimum performance. See **Blood-Brain Barrier.**
- In 1904, Cajal published the findings of a lifetime of study of brain tissue and neuronal systems as the result of combining the Golgi staining technique with light microscopy.
- In the first third of the present century, numerous investigators, among them Sir Henry Dale, Otto Loewi, A. L. Hodgkin, A. F. Huxley, B. Katz, Sir John Eccles, and S. W. Kuffler, made major contributions to an understanding of how individual neurons work. One finding was that all neurons, regardless of size and shape, appear to utilize the same two kinds of electric signals (graded potentials and action potentials). Dale initially proposed the concept that a neuron released only one transmitter chemical from all its terminals, an observation that was considered inviolable until recently. It is now suspected that some neuropeptides can coexist in the same neurons as certain transmitters, such as norepinephrine and serotonin.
- In the 1950s, E. J. Furshpan and D. D. Potter (University College London) discovered that some synapses are profoundly different from the usual chemical type, depending on the flow of current rather than diffusion of a transmitter.
- In the early 1950s, W. J. H. Nauta (then in the Netherlands) developed an improved brain tissue staining technique which differentiated degraded nerve fibers from living fibers, making it possible to identify specific fibers with remotely located neurons. This expanded the ability of researchers to provide greater detail in brain mapping.
- In the early 1950s, A. L. Hodgkin, A. F. Huxley, and B. Katz (British scientists), as the result of studying the nerve-impulse transmission in the giant axon of the squid, demonstrated that the propagation of the nerve impulse coincides with sudden changes in the permeability of the axon membrane to sodium and potassium ions.
- In the early 1950s, P. A. Weiss and colleagues (University of Chicago) discovered the phenomenon of axonal transport.
- In the mid-1950s, Sutherland and his associates (Case Western Reserve University) demonstrated that dopamine and norepinephrine, among other transmitters, increase or decrease the concentration of a second messenger substance in target cells, the latter mediating the electric or biochemical effects of the transmitter in the first messenger. (See later reference to Sutherland.)
- In the mid-1950s, neuroscientists at McGill University (Montreal) discovered pleasure centers within the brain that play a role in learning and memory.
- In the 1960s, methods for selectively staining neurons containing a particular transmitter were developed and utilized by a number of investigators. The natural transmitter substance can be converted into a fluorescent derivative that glows under ultraviolet radiation. Another method was developed that takes advantage of the high specificity of antibodies. A specific enzyme involved in the synthesis of a particular transmitter is purified from brain tissue and then injected into an experimental animal, whereupon the enzyme induces the manufacture of antibodies which, in turn, specifically combine with the enzyme. The antibodies are then purified and labeled with a fluorescent dye. Then they can be used to selectively stain neurons containing the relevant enzyme. Such techniques have greatly expanded the researcher's ability to map detailed anatomical distribution of individual transmitters. Falck (University of Lund) and Hillarp (Karolinska Institute, Sweden) first demonstrated that neurons containing monoamines fluoresce green or yellow when the transmitters are first converted into fluorescent derivatives. The best mapped transmitters are the monoamines norepinephrine, dopamine, and serotonin.
- In the early 1960s, V. P. Whittaker (University of Cambridge) and E. De Robertis (University of Buenos Aires) found that when brain tissue is gently disrupted when homogenized in a sugar solution, many nerve terminals break away from their axons. These terminals form intact, closed particles called *synaptosomes* which contain the means for synthesis, storage, release, and transmitter inactivation associated with the nerve terminal. The synaptosomes can be purified by spinning in a centrifuge. The availability of these substances made it possible for neuroscientists to study the mechanisms of synaptic transmission in vitro.
- In the 1970s, L. L. Iversen (Cambridge), T. J. Crow (London), and P. Seeman (University of Toronto), among others, revealed abnormally high concentrations of dopamine and dopamine receptors in the brains of deceased schizophrenics.
- In the 1970s, Perry (University of British Columbia) discovered that a specific deficit of gamma-aminobutyric acid (GABA) occurs in the inherited neurological syndrome known as Hungtington's chorea. See **Chorea (Huntington's).**
- In the 1970s, pioneering research efforts revealed that opiate receptors are located in those regions of the brain and spinal cord which are known to be associated with pain. S. H. Snyder and C. B. Pert (Johns Hopkins University School of Medicine), E. J. Simon (New York University School of Medicine), and L. Terenius (University of Uppsala), among others, used radioactively labeled opiate compounds to reveal these sites. This research paved the way for discovery of the enkephalins and endorphins, among other neuropeptides.
- In 1971, Bloom and co-workers (National Institute of Mental Health) demonstrated that cyclic AMP can affect signaling in neurons. A bit later, Greengard and associates (Yale University School of Medicine) showed that cyclic AMP is involved in the synaptic actions of several brain transmitters (norepinephrine, dopamine, serotonin, histamine). Greengard et al. suggested a unifying process to the effect that cyclic AMP activates specific enzymes in the target cell called *protein kinases* and, through a complex process, changes the level of excitability of the target cell.
- In 1971, Sutherland (see previous mention in this list) received the Nobel prize in physiology and medicine for identifying the second messenger substance as cyclic adenosine monophosphate (cyclic AMP).
- In 1975, J. Hughes and H. W. Kosterlitz (University of Aberdeen) first isolated the enkephalins. Shortly thereafter, the endorphins, also morphine-like compounds, were isolated from the pituitary gland.
- In the late 1970s, researchers found "substance P," associated with spinal neurons involved in pain stimuli.
- In the early 1980s, research on trophic substances, such as nerve-growth factor (NGF), accelerated into a major effort.

[1] In assembling this *very abridged* list, emphasis was given to topics which relate closely to the text of this entry.

Broad, Traditional Concepts of the Brain and Nervous System

The large, soft mass of nerve tissue making up the brain is contained within the cranium (*encephalon*). The brain consists of four major parts: (1) The *cerebrum*; (2) the *cerebellum*; (3) the *pons Varolli*; and (4) the *medulla oblongata*. The brain and spinal cord together constitute the *central nervous system*.

By way of generalization, the brain is the control station for nerve impulses. The brain is composed chiefly of nerve cells with their fibers interwoven in a complex relay system. At birth, the human brain weighs between 11 and 13 ounces (312 and 369 grams) and, as previously mentioned, attains a weight of about 45 ounces (1.3 kilogram) in the adult. The brain increases in weight until about the twentieth year, after which there is gradual loss of weight for the remainder of life. The volume of the human adult brain is 2000 cubic centimeters or greater. By comparison with other animals, the human brain is very large. See Fig. 2.

Medulla oblongata. This organ is approximately $\frac{3}{4}$ of an inch to 1 inch (19 to 25 millimeters) in length. Externally, it appears like an expanded part of the spinal cord. See Figs. 3 and 4. Internally, its structure is quite complex and consists of nerve tracts passing into the brain. From some of the nuclei come fibers that eventually emerge to form the VIIIth, IXth, Xth, XIth, and XIIth cranial nerves. Cell centers in this area also are concerned with swallowing, vomiting, breathing, speech, digestion, metabolism, and the beating of the heart. In the medulla oblongata, the large bundles of fibers, which originated in the two halves of the cerebrum and which transmit the impulses of voluntary movement, cross to the opposite sides. Thus, movement in the right arm, for example, is controlled by centers in the left half of the cerebrum.

Pons. Lying above the medulla oblongata and continuous with it is the pons. It is made up of massive bundles of fibers that start in the cerebrum and sweep backward to the cerebellum. This connection makes possible many skilled acts that require coordination of sight, hearing, muscular movement, and various other sensations. The playing of a musical instrument is an example of such an act. The pons contains a space called the *fourth ventricle*. In the floor of this ventricle

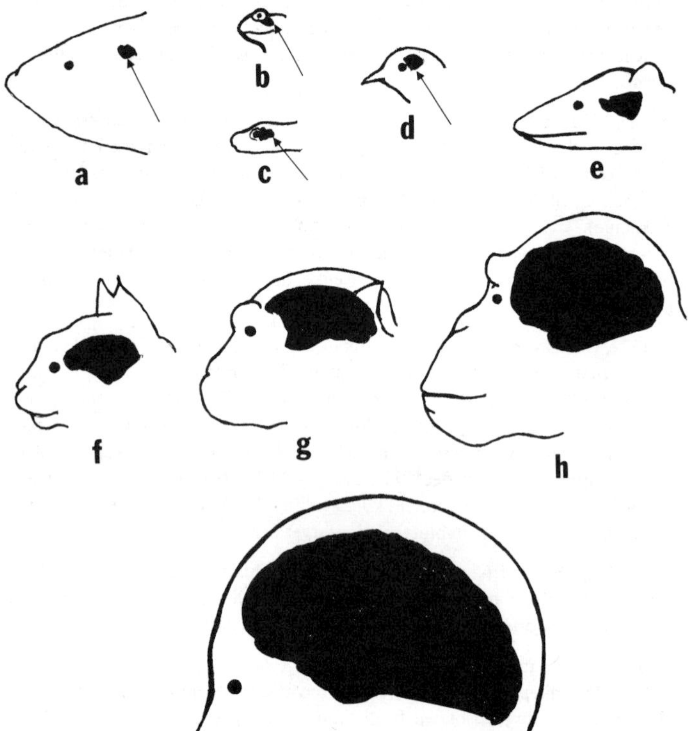

Fig. 2. Comparative sizes of the cerebrums of various animals: (a) bass, (b) leopard frog; (c) grass snake; (d) pigeon; (e) opossum; (f) cat; (g) macaque monkey; (h) chimpanzee; (i) human. (*After Hubel.*)

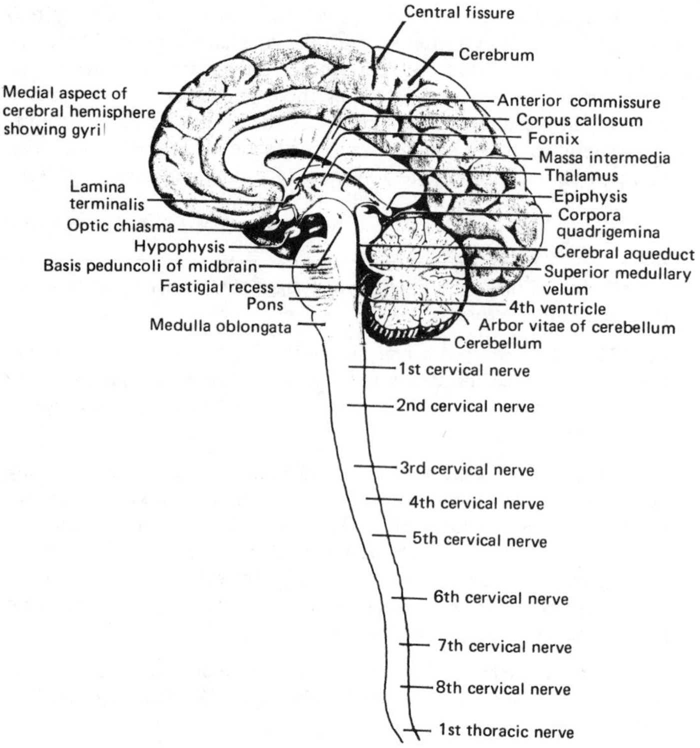

Fig. 3. Median sagittal section of human brain.

is the nucleus of the VIth cranial nerve. This nerve, which has the longest course inside the skull of all the cranial nerves, is concerned with turning the eyeball outward.

Cerebellum. The second largest part of the brain is in back of the pons and termed the cerebellum. It lies in the back of the skull. The cerebellum is made up of many narrow, leaflike folds arranged into two large masses, and of a middle portion. Rich in cells, it has many complex connections with the brain above it and with the spinal cord below. The chief function of the cerebellum is to coordinate relatively complex movements into special acts. This may be movement in different parts of the same limb; combined action of the head, limbs, and body. For example, picking up a pencil, writing with it, and laying it down again requires smooth interaction of many muscle groups. The cerebellum correlates the actions of the various groups. To do this, range, direction, rate, and force of movement must be synchronized and maintained with the movement of the eye. Disease in the cerebellum does not cause paralysis, but it does produce disturbance of muscular coordination. Tremors, staggering gait, and excessive re-

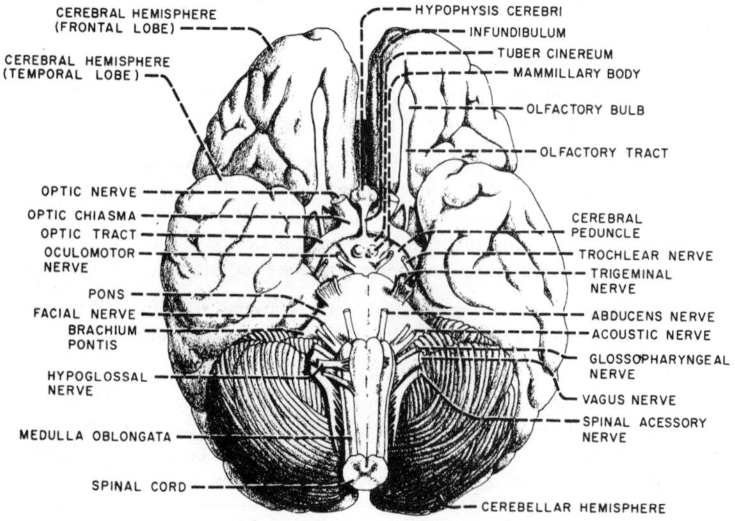

Fig. 4. Base of the human brain and cranial nerve roots.

laxation of the muscles result from disease in this part of the brain, as well as disturbances in components of muscular activity.

Midbrain. This is a small area between the pons and the cerebrum. It is an important relay station for the sensory impulses. The midbrain also governs some muscle activity of a reflex nature. Many of the involuntary acts of the eye, such as narrowing of the pupil in bright light, originate in this area. The IIIrd, IVth, and Vth cranial nerves originate from cell collections in this area of the brain.

Hypothalamus. Just above the midbrain, an important group of nuclei comprise the hypothalamus. Beneath it, the two large nerves from the eyes meet and part of their fibers cross to the opposite sides. Other cells of this region are concerned with such vital functions as regulation of body temperature, metabolism, and heart rate. Sexual development, sleep, and the body's use of fat and water are influenced by this region in the brain. There is an essential relationship between the hypothalamus and the nervous system and the endocrine system.

Thalamus. The thalamus is found next to the foregoing group of cells and contains another group of nuclei which integrate sensations of many sorts. Also, the thalamus is the site of a crude form of consciousness and plays a role in the production of emotion. When this part of the brain is diseased, spontaneous laughter or crying may occur. The crude emotional responses that arise are further elaborated and controlled by the cerebral cortex.

Cerebral Hemispheres. The two cerebral hemispheres represent 70% of the entire nervous system. This is the area of the nervous system in which all the sensory experiences are mixed and blended. Specific sensory impulses thus become associated with many others and expand the experience and consciousness. The individual's capacity for many and varied activities, and memory, emotions, and ideas is dependent upon the action of this part of the nervous system.

The surfaces of the hemispheres are marked by large, rounded folds and deep grooves. Partly on the axis of the main grooves and partly on imaginary lines, the cerebrum is divided into four lobes: *frontal, parietal, temporal,* and *occipital.* Each lobe has special functions, but these functions are only partly understood.

The occipital lobes at the back of the skull are the site where visual impressions are made. Color, size, form, movement, and distance are evaluated in this portion of the brain, leading to the identification of a particular object. Also, the differences between similar objects are discerned. For example, two objects high in the air can be recognized as a bird and an airplane on the basis of past experience. Injury to this area may cause blindness.

The temporal lobes receive the fibers concerned with hearing, speech, balance, and smell. Diseases related to these lobes cause loss of smell, or they may be responsible for imaginary smells.

The parietal lobes are concerned with taste sensations and some other sensations, such as the ability to judge weight, shape, and textures. By the action of this area, one is able to tell what various objects are by feeling, rather than seeing them.

The frontal lobes are concerned with some of the most complex abilities of the mind. Reason, emotions, and judgment have their site here. Additionally, there is a group of large cells in the posterior region of the frontal lobes which are involved with complex voluntary movements. The speech center is located here. It is found to predominate on the left side in right-handed individuals and vice versa. If the speech function is lost because of injury to only one side of the brain, it can sometimes be reacquired by reeducation. The area responsible for these complex voluntary movements is called the *motor cortex.* The muscles of the body are controlled by various areas of the motor cortex. Irritation of the cells in these zones will cause spasms of the muscles they supply. Destruction of the cells will produce complete loss of voluntary movement of the muscles. These cells also function to keep the muscles in balance between relaxation and contraction. If this region of the brain is seriously damaged or destroyed, this inhibiting power is lost. Consequently, the muscles become contracted and stiff (*spastic paralysis*).

The frontal lobes have many connections with the thalamus as well as with the other lobes of the brain. In the frontal lobes, feelings or emotions are added to the other associations. The combination of feeling and knowing determines most voluntary action of the body. Thinking, reasoning, judgment, and imagination result as the sensory and emotional associations become more complex. Disease in the frontal lobes of the cerebrum causes personality changes, errors in judgment and insight, and poor emotional control.

Twelve pairs of nerves arise from the brain proper. The Ist is associated with the sense of smell; the IInd with sight; the IIIrd and IVth with the eye muscles which move the eyeball or the muscles of the pupil; and the Vth causes the muscles of the jaw to move. The VIth is concerned with the movement of the eye to the side. The VIIth carries impulses to all the muscles of the face. The VIIIth conducts impulses having to do with hearing and with balance; the IXth transmits taste sensations from the posterior third of the tongue, and other sensations from the throat and mucous membranes; and also aids in swlallowing. The Xth is an exceedingly long nerve that extends down the neck and into the chest and abdomen; it is concerned with swallowing and talking; its action also slows the rate of the heart and regulates the movement of the stomach. It is a large part of the parasympathetic system in the upper part of the body, including the esophagus, stomach, intestines, liver, bronchi, lungs, heart, and blood vessels. The XIth supplies some of the muscles that turn the head and some of those in the neck. The XIIth is responsible for the movement of the tongue.

Spinal Cord. The nerve tracts passing to and from the brain are contained in the spinal cord, which is continuous with the lower part of the brain. It is about 18 inches (46 centimeters) long and is rounded in shape. This nerve tract is larger in the regions which give rise to the nerves to the arms and legs, since these parts have many complex functions, thus requiring a large nerve supply. From the neck to the lowest parts of the vertebral column, 31 pairs of nerves emerge from the spinal cord. Each nerve is attached to the cord by two roots. Because the spinal cord is not as long as the vertebral column, the roots of the nerves must gradually increase in length before they can emerge from between the vertebrae. These longer nerve roots collect in a mass that fills the lower end of the vertebral canal. The structure resembles a horse's tail and is called the *cauda equina.*

A cross section of the spinal cord reveals a gray figure, roughly shaped like an "H," imposed on a white background. The nerve cells make up the gray matter, while the nerve bundles form the white matter. Bundles with specific functions occupy specific areas of the spinal cord. Thus, injury to the cord will result in certain abnormal reactions which will be evident upon neurological examination. The abnormal findings will suggest where the diseased part is located.

Impulses which arise in the brain and are concerned with voluntary muscular movements are received by specific cells in the spinal cord. They relay these impulses to the nerves which control the various muscles. These cells have connections with other cells in the nervous system that act together to bring about *reflex activity.* Reflexes control the position of the head so that it automatically assumes the normal position. Withdrawal or *flexion* reflexes pull limbs away from painful or disagreeable stimuli. *Extensor* reflexes straighten out the limbs and work with the flexor reflexes. Bladder and bowel actions result from reflex action over which there is some voluntary control. In injury to the spinal cord, the tracts allowing voluntary control may be interrupted, so that action is then reflex in origin.

The cells on the front and anterior side of the spinal cord connect with cells in the cerebellum to control the direction and precision of normal muscular movement. From still another part of the brain, connecting fibers come to these cells to bring about certain automatic, associated muscular movements, as in the instance of swinging of the arms as one walks. Another function of these cells is the maintenance of the proper amount of constant contraction or *tone* of the muscles. If the muscles are too contracted, they move too slowly and rigidly. If they are too relaxed, too much stimulation is needed to make them respond.

The cells along the side of the cord send out fibers that unite with others to form the *sympathetic chain.* These cells are concerned with the action of involuntary muscles in the intestines, arteries, and other internal structures. Various glands receive fibers from these cells. Cells on the back or posterior side of the spinal cord receive the sensations of touch, pain, vibration, temperature, pressure, and postition. They then transmit these various sensations to other cells in the brain. Thus, impulses of many sorts travel down the paths in the spinal cord, while others enter it and travel upward to the brain. Still others enter and travel only part of the way up and set off the spinal reflexes.

The peripheral nervous system is composed of the cranial nerves, the spinal nerves, and the autonomic nervous system. The autonomic system supplies nerves to most of the "automatic" organs of the body, i.e., the glands, heart, blood vessels, and the involuntary muscles in the internal organs. One part of the autonomic nervous system prepares the individual for emergency situations, shifting circulating blood to skeletal and heart muscles, increasing heart and lung function, dilating the pupils of the eyes, and moistening the skin with perspiration. The other division is concerned with conserving and restoring bodily resources. Thus, it protects the eyes by causing the pupils to constrict in bright light, and prevents the heart from overexerting itself. All of the digestive processes are promoted by this division. See **Autonomic Nervous System.**

Protection of Central Nervous System. The total system is well protected by the rigid *skull* and the flexible backbone (*vertebral* or *spinal column*). The skull consists of a dome of thin, porous, but strong bones. The bones of the forehead contain the sinuses of the nose. The floor of the skull is composed of somewhat thinner bone and contains more sinuses that connect with the nasal passages. The deeper structures of the ear are embedded in the bones forming the base of the skull. The floor of the skull consists of three irregular depressions (*fossae*) that form three descending levels. The fossa in the back of the skull is the largest and deepest. The cranial nerves arise from different parts of the brain and emerge through various bony canals and openings of the skull.

Inside this hard shell, three separate tissues provide additional covering for the brain and spinal cord. The outermost, *dura mater,* consists of layers of dense, fibrous material. The outer layer of the dura adheres tightly to the bones of the skull. The inner covers the brain and spinal cord, forming a tough, saclike structure. Within the skull itself, the dura is folded into partitions which separate and support the various parts of the brain. One such fold is called the *falx cerebri,* which divides the cerebral hemispheres into right and left halves. Another fold, the *tentorium,* separates the back fossa from the vault of the skull. It provides a horizontal support for the back part of the brain and separates that part from the cerebellum. Between the layers of the dura are large blood vessels called *venous sinuses* which collect blood from the brain and return it to the heart.

The middle of the three covering tissues is a delicate membrane called the *arachnoid* (cobweb-like), a layer that encloses the brain and the spinal cord in a loose-fitting sack. The space below this layer is the *subarachnoid space.*

The third covering is a thin, delicate sheet that follows closely all the irregular surfaces and fissures of the brain and spinal cord. This is called the *pia mater* and is next to the nerve tissue proper.

Between the arachnoid and the pia mater circulates the cerebrospinal fluid, which acts as a shock absorber for the central nervous system. Also, it is postulated that this fluid helps in nourishing the nerve tissue itself. This clear, watery-appearing fluid, formed within certain cavities (*ventricles*) of the brain, flows out from small openings into the brain and spinal cord before it is absorbed back into the bloodstream. When the flow or absorption of the fluid is impaired, it accumulates in large quantities. The head enlarges and "water-head" (*hydrocephalus*) results. The spinal fluid is affected by various disorders and is easily drawn off to be examined for diagnostic purposes.

Nerves and Nervous System. A nerve is a slender cord made up of nerve fibers (neurons). In the vertebrates the nerve is surrounded by loose connective tissue called the epineurium. Each small bundle of fibers within the epineurium is surrounded by a thin, compact perineurium, and from this layer thin septa, the endoneurium, run between the irregular groups of fibers within the bundle.

Nerves form the communicating paths between the central nervous system and the various parts of the body, as well as the connections between ganglia. They have been given special names according to their anatomical distribution, and a few functional properties have been indicated by descriptive terms. Thus sensory or afferent nerves carry impulses from sense organs to the central nervous system and motor or efferent nerves lead out to muscles and other effectors. Most nerves of the body contain fibers of each kind and so are mixed nerves. Nerves arising from the brain in the vertebrates are called cranial nerves and those connected with the spinal cord are spinal nerves. Typically, all of these main nerves of the vertebrate are supposed to be based on the form of the spinal nerves, which connect with the cord by two roots. The dorsal root bears a spinal ganglion and carries all the sensory fibers and the ventral root lacks a ganglion and is motor. All sensory nerves bear a ganglion or arise from a sensory layer such as the retina of the eye and no purely motor nerves have ganglia.

The motor components of these nerves grow out from cells in the central nervous system and the sensory components grow into the central system from ganglia or from sensory cells and also out from the ganglia toward the periphery of the body.

A nerve impulse is a progressive transfer of a condition of excitation along a nerve fiber, initiated by a stimulus acting upon a sensory organ, by a cell within the nervous system, or experimentally by direct stimulation of the nerve fiber. The nerve impulse involves measurable chemical changes and energy consumption accompanied by changes in electrical potential associated with membrane depolarization. The electrical changes have been used extensively in investigation of nerve physiology and of the brain. The passage of the nerve impulse over the synapse is accompanied by the action of acetylcholine in cases which have been studied. Terminal nerve fibers may form either acetylcholine or sympathin, the latter being closely related to adrenalin. The impulse activates some organ of the body or enters the nervous system, where it is relayed to other parts. These mechanisms are further described later.

The attainment of harmonious action of all component parts of the animal and the adjustment of its behavior to environmental conditions is by way of communication and regulation through the nervous system. Coordinative processes of this type are aided by the secretion and distribution of hormones. The latter process is termed *chemical coordination.* Coordination by way of the nervous system occurs, of course, with much greater speed than actual materials can be transported from one part of the body to the next. The nervous system is responsible for the immediate adjustment of the animal to fluctuating conditions, while chemical coordination is extensively involved in the maintenance of the normal organic processes which are a more uniform part of the animal's activity. Coordination, in general, is due to close interaction of the two types: neither is wholly independent of the other.

The process of coordination through the nervous system is based on the general sensitiveness of protoplasm known as irritability and on the property of conductivity by which some result of stimulation passes rapidly through the adjacent substance. All living cells are irritable to some extent. They respond to some stimulus with characteristic activity. Sensory organs possess this property to a high degree and in addition are specialized to receive a certain kind of stimulus. Nervous tissue is specialized for ready activation and for the rapid conduction of the impulse generated within it.

The organs which receive stimuli as a special function for the benefit of the animal, whether relatively simple cells or extremely complex organs like the human eye, are known as receptors. They have special nerve endings, often associated with other structures, and are connected by afferent or sensory nerves with other parts of the nervous system or, in very simple animals, with some organ capable of acting in response to the condition from which the stimulus arose. Organs of the latter category are muscles, glands, electric organs, light organs, and some pigment cells, and are known collectively as effectors. Usually one or more cells of the nervous system intervene between the sensory cell associated with the receptor and the motor cell which communicates directly with the effector. The more complex this nervous chain, the more intricately may nerve impulses be relayed through different paths in the body, but in all cases the net result is the same. All adjustments are due to the reception of stimuli by sense organs, both internal and external, followed by appropriate reaction of other parts of the body.

The most common type of nervous system among invertebrates consists of a small brain lying above the alimentary tract near the anterior end of the body. See Fig. 5. It is connected by cords passing down around the sides of the gut with a ventral nerve cord. In the annelid worms this cord is a chain of ganglia, one lying in each segment. Conclusive evidence from embryology and minute anatomy shows that the primitive cord was a paired structure and that each segment contained a pair of ganglia connected by transverse nerves

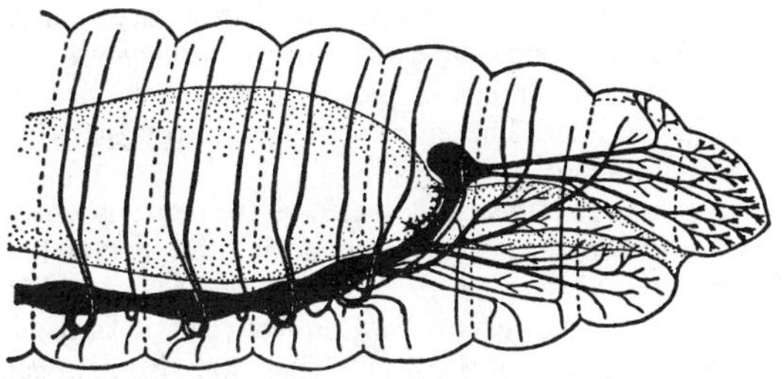

Fig. 5. Nervous system of earthworm. This is a lateral view of the arrangement of the larger nerve trunks in the left side of the anterior segments of the earthworm.

as well. In most existent species the adult shows no visible evidence of the paired condition.

In all of the more complex nervous systems, a major center, the brain, is present as a part of the central nervous system, and supplementary centers called ganglia, each containing a small number of nerve cells, are scattered through the outlying parts of the system. All of these parts are connected by nerve cords or nerves and are joined in the same way with other structures of the body.

In the arthropods a concentration of ganglia in some species has resulted in one large ganglionic mass near the anterior end of the ventral cord.

Many mollusks have a number of pairs of ganglia of about equal size as nerve centers, although the cephalopods have a concentrated brain equal to any other invertebrates.

In the vertebrates the entire central nervous system lies above the alimentary tract. The brain is large and complex in most classes. From it a spinal cord runs back along the axis of the body. Both brain and spinal cord bear nerves which extend throughout the animal. See Fig. 6. Below the spinal cord two chains of ganglia connected by slender nerves with each other and with the spinal nerves constitute the sympathetic chains of the autonomic nervous system. Nerves of this system supply the viscera and are widely distributed elsewhere in the body. In cooperation with the parasympathetic fibers of the cranial and sacral nerves, they control the functions which cannot be deliberately regulated by individual desire, such as the movements of the alimentary tract.

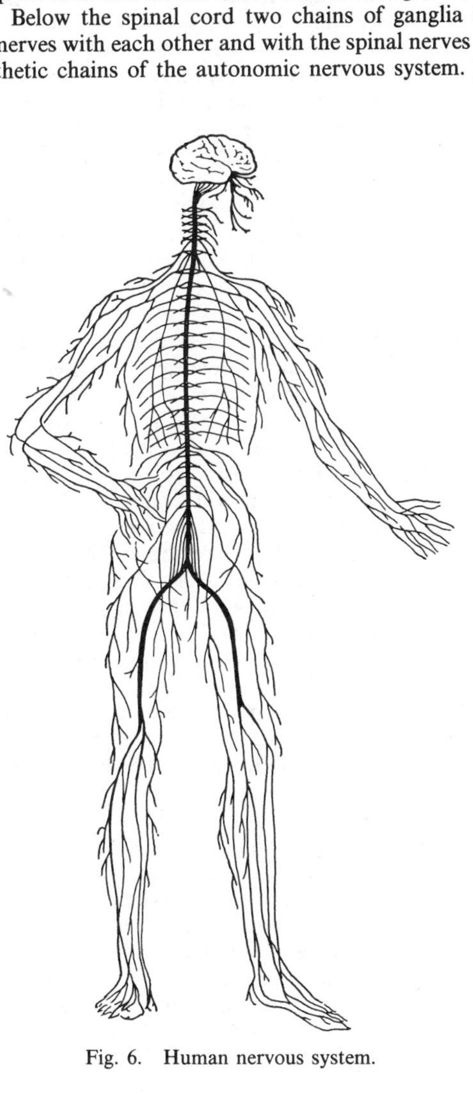

Fig. 6. Human nervous system.

The area of functional contact between two nerve cells (*synapse*) and between an axon terminal and a muscle fiber (*neuromuscular junction*) are regions of unique anatomical, physiological, and biochemical specializations which are concerned with information transfer. Since there is a wide variety of types of synapses and neuromuscular junctions, to date investigations have been relatively limited to mammalian, amphibian, and invertebrate species of select categories.

Synapses and neuromuscular junctions have been classified on a variety of morphological, functional, and biochemical bases. Anatomically, synapses have been classified as being either central (brain and spinal cord); or peripheral (autonomic ganglia, peripheral sensory ganglia and effector junctions) in position and/or axo-dentritic, axo-somatic or axo-axonic in contact relationships. Physiologically, synapses have been classified as either excitatory or inhibitory in action and/or as chemically or electrically mediated. Biochemically, synapses have been classified as either releasing acetylcholine, epinephrine, norepinephrine, or substances for which the precise chemical composition is not yet fully known. Neuromuscular junctions are classified anatomically as within the peripheral nervous system; functionally as excitatory, inhibitory or sensory; and biochemically as releasing acetylcholine.

When synapses and neuromuscular junctions are observed with the electron microscope, their fine structures are seen to be basically similar. The similarities include increased electron density of the presynaptic and postsynaptic membranes, the presence presynaptically of large numbers of mitochondria and synaptic vesicles, and an intersynaptic cleft space of 200–300Å. Synaptic vesicles are 300–600Å in diameter in synapses where acetylcholine is the mediator; and 600–1000Å in diameter in synapses where epinephrine or norepinephrine are the mediators. The accumulation of mitochondria and synaptic vesicles on the presynaptic side occurs as a result of the distal migration of materials from the cell body into the axon. Differences in the morphology of central nervous system synapses have been reported and various functional attributes have been alluded to. These include differences in presynaptic and postsynaptic interspace distances, postsynaptic membrane electron densities, and amounts and electron density of interspace substance.

Synapses transmit either electrically or chemically, and they may be either excitatory or inhibitory in action. Most synapses and neuromuscular junctions are chemically mediated and excitatory in action. Types of chemically mediated synapses include those from the mammalian and amphibian central nervous system, sympathetic and parasympathetic ganglia, and sympathetic innervation of the adrenal chromaffin cells. As previously mentioned, the three most widely known transmitter substances are acetylcholine, epinephrine, and norepinephrine.

Operation of the central nervous system depends upon two substances: the *gray matter* (nerve cells), and the *white matter* (nerve fibers given off by the cells). The function of the gray matter is the generating and dispatching of nerve impulses. The function of the white matter is the conduction of these impulses to and from the cells in the gray matter. Other cells in the nervous system have no nervous function, but instead are concerned with the support and nourishment of nerve cells.

Nerve tissue proper consists of cells giving off threadlike processes (*axons*) some of which are extremely long. The axons connect with other cells in the brain or spinal cord. These cells constantly generate, receive, or store up energy. Unlike most body tissues, nerve cells are not replaced once they are destroyed. If the cells are destroyed, their axons degenerate. Some cells concerned with generating or receiving similar impulses may be collected into definite groups called nuclei. The axons from these cells unite and form bundles of nerves which then transmit the impulses. The nerve cells are not collected into nuclei in the cerebral hemispheres, but form a uniform layer (*cortex*) of gray matter over the surfaces.

A series of highly specialized organs, called *receptors*, detect changes in and about the body. They rapidly transmit this information to definite stations within the nervous system. This is called *sensory* activity. Receptors, called *exteroceptors*, gather information from a distance, such as seeing, hearing, and smelling. *Interoceptors* detect things in contact with the body, as pain, temperature, and touch. *Proprioceptors* pick up information from within the body, giving a sense of bodily position. Fibers from the receptor organs pass into the spinal cord as a part of the nerve. Within the cord, they unite to form ascending tracts which connect with other spinal cells or enter the brain. Fibers from some special receptors, such as the eye, form nerves which enter the brain directly.

It is postulated that when a sensory impression reaches the brain, it stimulates a nerve cell which, in turn, stimulates another cell. A third cell is then stimulated and so on, until a circle has been completed, and the last cell restimulates the first one. The circuit continues to *reverberate*, thus retaining the impression which set it off. It is further postulated that these reverberating circuits hold the impressions so that they can be recalled later, or compared with other impressions. It is believed that a cell may participate in more than one circuit, thus accounting for various associations of sensory and muscular activity.

More Detailed View of Neurons and Synapses

Neurons. No two nerve cells (neurons) are exactly the same, but the majority of neurons as presently understood have three main structural features: (1) The *cell body*; (2) the *dendrites*; and (3) the *axon.* See Fig. 7.

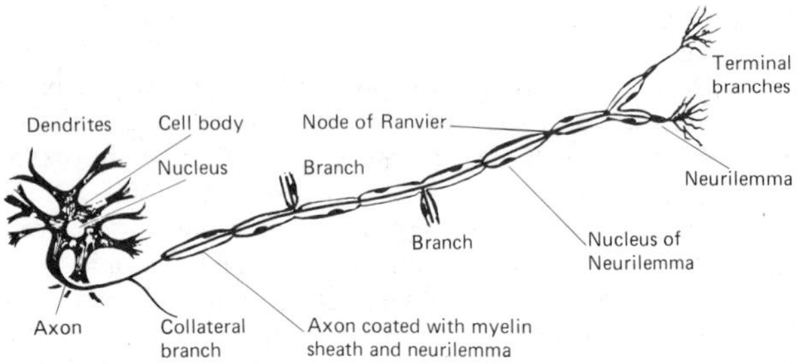

Fig. 7. Simplified presentation of principal elements of a neuron (typical of a motor neuron from ventral column of spinal cord).

Cell bodies may be spherical or pyrimidal in shape. They contain the nucleus of the neuron, that is, the biochemical substances and associated processes required to synthesize enzymes and other substances needed to maintain the life of the cell.

The dendrites (*dendrons*) most often are short and comparatively thick at their points of origin and branch out in a treelike manner. There is a variable number of dendrites with various cell bodies, ranging from a few to many in number. Generally, the dendrites are considered to be the information-receiving channels of the neuron.

The axons are threadlike channels which connect the cell body and dendrites to remote motor and other cells. In some cases, an axon may be as much as half as long as the entire human body. Generally, the axons are considered to be the information-transmitting channels of the neuron. The axons taper very little as they progress outward from the cell body, but frequently they have multiple branches. These branches are called *collaterals.* The axons are filled with a jellylike fluid called *axoplasm.* This fluid was once thought to be simply an inert mechanical support for the excitable membrane that propagates the nerve impulse. Research commenced by Weiss in the late 1940s led to the modern concept of axoplasm, which likens it to an artery for bidirectional flow of molecules moving between the cell body of the neuron and its axon terminals. A number of different systems apparently are operative within the axoplasm. For example, there is a comparatively slow system (movement of about 1 millimeter/day) which accounts for the greatest volumetric flow of substances in the axonal system. These are materials required for the growth and regeneration of the axon. There is a much faster system which transports cellular components, such as enzymes, specifically required for the manufacture of transmitters. The different rates of axonal transport are poorly understood, but in essence appear to be related to rates of demand.

The importance of axonal transport to neuron maintenance is revealed by the fact that once the central nervous system is fully developed, there is no replacement of old cells with new ones as occurs, for example, in the liver. Therefore, it is mandatory that the original cells be well maintained for long life.

While attention is usually concentrated on the neurons when describing the brain and central nervous system, it is important to mention in passing the dense network of blood vessels that furnish oxygen and other nutrients to the nerve cells. Connective tissue is also required, particularly at the surface of the brain. There are also *glial cells* which take up nearly all the volume in the nervous system that is not required by the neurons. The detailed functions of the glia are not well understood, but it can be safely generalized that the glia provide both structural and metabolic support for the intricate network of neurons.

Returning to the axons—these are jacketed by Schwann cells, a process which commences during embryonic development. A multiple layer of these cells develops and insulates the axon. This material is called *myelin* and the insulating coating is called the *myelin sheath.* It should be mentioned that some neurological disorders arise from disturbances of this sheath. In most axons, the myelin sheath is interrupted at intervals of about 1 millimeter or more. These interruptions are called the *nodes of Ranvier.* It is believed that these interruptions make it possible for extracellular fluid to directly contact the cell membrane and possibly serve as a source of certain membrane proteins, which are described later. Researchers believe that the myelin sheath functions to conserve the metabolic energy of the neuron. Research has shown that nerve fibers with the myelin sheath conduct nerve impulses at a faster rate than unmeyelinated fibers. It will be noted from Fig. 7 that some parts of the axon may be protected only by a substance known as *neurilemma* (where the myelin sheath is absent).

The importance of conserving the metabolic energy requirements of the neuron is demonstrated by the fact that, even though the human brain constitutes only about 2% of total body weight, the brain and central nervous system require at least 20% of the body's total oxygen needs when the body is at rest. This represents a use of oxygen at a rate of about 50 milliliters per minute. The brain metabolic rate is essentially constant even when the body is at rest, and some researchers suggest that during certain phases of sleep, the rate of oxygen required may rise. The total energy equivalent of brain metabolism is about 20 watts. Unlike some other cells of the body which can utilize a variety of fuels, such as sugars, fats, and amino acids, neurons require blood glucose exclusively. Since the brain depends entirely upon oxidative metabolism, consciousness is lost within 10 second and brain tissue degeneration commences upon cessation or severe dimunition of oxygen supply and blood glucose. Diabetic coma from an overdosage of insulin, which greatly lowers blood glucose levels, is exemplary of this dependency. More detail on brain metabolism is given a bit later.

Synapses. Although the detailed functioning of the neurons remains somewhat vague, it is known that a typical neuron may exchange information with a thousand or more other neurons. These connections (synapses) may arise between: (1) The axon of one neuron and the dendrite of another (apparently this occurs in the majority of cases); (2) the axon of one neuron and the axon of another; (3) the dendrite of one neuron and the dendrite of another; and (4) the axon of one neuron and the cell body of another. These four possible connections are probably not fully representative. It is this great connective potential among millions of neurons that leads to connective networks that are unknown even to the most complex of electronic circuits designed by humans and that, from the standpoint of traditional electronic connective networks, appear unworkable. The fact, of course, is that they do work.

At a point of connection (synapse), the axon normally enlarges and takes the form of a *terminal button.* This button incorporates tiny chambers (*synaptic vesicles*) which contain a *chemical transmitter.* For many years, the concept that a neuron released only one transmitter chemical from all its terminals—as proposed originally by Sir Henry Dale—was considered inviolable. Within the past few years, researchers have found that a number of neuropeptides can coexist in the same neurons as norepinephrine or serotonin. The functional significance of such a dual-transmitter system has not been established. The quantity of chemical transmitter released by the terminal button vesicles can aptly be described as vanishingly small. The quantities range from a few hundred to a few thousand molecules. As previously mentioned, transmitter chemicals may be excitatory or inhibitory.

Neuron Membranes. Like other cells of the body, neurons have membranes. In the neuron, the membrane is made up of two layers of lipid molecules. The membrane is about 5 nanometers (~0.0003 millimeter) thick. The hydrophilic ends of these molecules are oriented toward the water on the inside and outside of the cell. The hydrophobic ends are oriented away from the water and thus these form the interior of the membrane. Lipid materials tend to be relatively uniform for all cells. But membranes do differ, and these differences arise from the specific proteins that are associated with a given membrane. Membrane differences comprise just one more way in which neurons differ from each other, adding to the overall complexity of classifying and understanding them.

When the proteins are embedded in the bilayer of lipids, they are called *intrinsic proteins.* Although essentially bound, in some cells these intrinsic proteins can move about a bit by diffusion, while in other instances they are physically fixed to the membrane at specific sites. Other proteins, equally important, attach themselves to the membrane surfaces, but are not an actual part of the membrane structure.

These membrane proteins fall into at least five categories, based on their function: (1) *Pumping proteins* maintain the needed concentrations of particular molecules within the cell and thus set up the conditions necessary for a nerve impulse. The pumping proteins regulate concentrations by moving ions and molecules against concentration gradients and hence require considerable metabolic energy. (2) *Channel proteins* provide pathways (tiny orifices) through the lipid bilayer membrane and thus regulate the diffusion of specific ions through the membrane because the ions are of discrete size, some larger than others. Without these channels, the membrane would be essentially impervious. (3) *Receptor proteins* provide binding sites to enable the cell membrane to recognize and attach various specific kinds of molecules. The receptor proteins are highly specific, much as a lock is specific to one key. (4) *Enzyme proteins,* located in or on the membrane, participate in chemical reactions at the membrane surface. For example, one of these enzymes, known as adenylate cyclase, regulates the intracellular substance cyclic adenosine monophosphate (cyclic AMP). This substance serves both as a regulator and catalyst. (5) *Structural proteins* contribute to the physical support of the system.

Membrane proteins are keys to the function of the neuron because they provide the means for creating and propagating the nerve impulse. The concentrations of sodium and potassium ions on either side of the cell membrane differ markedly. In a resting state, researchers have estimated that this difference in ionic concentrations causes the interior of the axon to be about 70 millivolts negative with respect to the exterior. In terms of the sensitivity of modern instruments, this is a very significant voltage differential. As a nerve impulse propagates along the axon, the permeability of the axon membrane to both sodium and potassium ions is altered. The impulse is usually triggered in the cell body as the result of reaction to dendrite synapses, and the voltage across the membrane just ahead of the traveling impulse is lowered. This adjustment of sodium ion channeling permits sodium ions to flood into the axon. During this process, the internal potential of the membrane is changed from negative to positive. But the sodium channels remain open for but an instant, after which the potassium channels are opened to permit the outward flow of potassium ions, thus returning the original differential of −70 volts to the interior of the membrane. The sharpness of these actions appears as a spike on an oscilloscope and the phenomenon is generally called *action potential.* See also **Action Current;** and **Action Potential.** In actuality, since the axon is divided into numerous sections (Fig. 7), a wave of voltage propagates along the axon until the terminal button is reached. The action is analogous to a flame traveling along a fuse to an explosive. An explanation of this action in traditional terms is given in Fig. 8.

As previously mentioned, the channel proteins accommodate the sizes of specific ions. The sodium ion is about 30% smaller than the potassium ion. Some channel proteins are relatively undiscriminating, permitting only a slightly reduced number of sodium ions versus potassium ions to pass through the channel. Other channel proteins are highly selective, allowing, for example, fewer than 10 sodium ions to pass for every 100 potassium ions. The diameter of the former

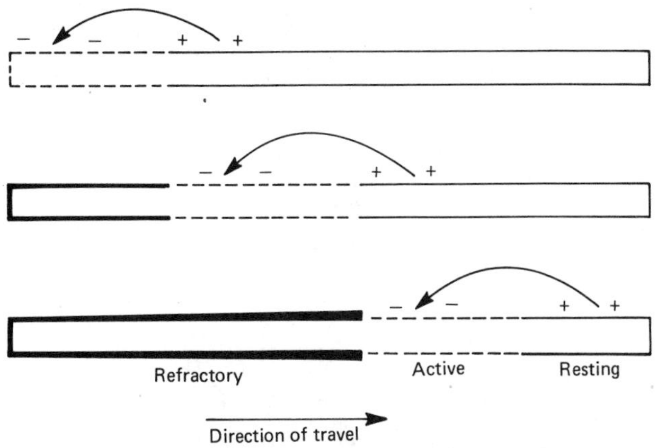

Fig. 8. Simplified representation of nerve impulse conduction.

type of channel is about 0.8 nanometer and the pore is filled with water. The opening of the second type of channel is considerably smaller and, of course, contains less water. These variations again exemplify the rather extreme degree to which neurons can be tailored to need and thus serve not simply as on-off switches, but provide shading, averaging, and tempering of input and output. The voltage-gated channels just described are traditionally named sodium or potassium channels in accordance with which ion is most readily passable. In other parts of the nervous system, calcium ions also play an important role.

The measured potential (70 millivolts) is less than the equilibrium value calculated by the Nernst equation for the potential that would be developed by a simple concentration battery formed by the unequal distribution of sodium and potassium ions. The reduced potential is probably the result of special back-pumping of the sodium (extruded) ions and the potassium (reabsored) ions against their concentration and electrochemical gradients. This pumping action proceeds simultaneously with the diffusion and thus distorts the picture of diffusion as would be determined by membrane permeability alone.

The resulting modified potential is the source of current that eddys into the point of dropped (or even reversed) potential resulting at the site of stimulation, permitting temporary free diffusion of ions and a resulting depolarization. The transient free diffusion of ions across the cell membrane carries the current into the cell. The eddy currents, in turn, act as stimuli to the surrounding membrane, which is then depolarized and draws current from more regions adjacent to the new stimulus sites. These new currents, in turn, stimulate more distant regions, and so on. Meanwhile, the original sites of stimulation and depolarization are repolarizing or recovering.

These events are shown in simplified format in Fig. 8, showing the change of resting, polarized to active, depolarized region, and the influence of the subsequent eddy current in creating a progression of the active region from left to right, tailed by a region of restored membrane. The latter is temporarily more stable than normal and, therefore, resistant to restimulation, that is, it is refractory for a short period. The eddy of current in this way spreads into a self-propagated wave, which is the conducted impulse whose minute electrical accompaniment can, after suitable amplification, be recorded as the action potential.

The channelling characteristics of the membranes of axons vary considerably, depending upon specific function. There are two principal forms—in one type, a nerve impulse opens and closes the channels in response to voltage differences across the cell membrane. Channels of this type are said to be *voltage gated.* In the other type, there is little if any response to voltage changes, but instead they respond to a particular chemical substance, as when a transmitter substance binds to a receptor region on the channel protein. These are called *chemically gated.* It has become traditional to name the chemically gated channels in accordance with the transmitter chemical involved, such as a GABA-activated channel.

Chemical Transmitters. In terms of structure, the chemical transmitters have been found to be monoamines or amino acids. The structures of the substances most extensively researched to date are given in Fig. 9. The most common of the inhibitory transmitters found in

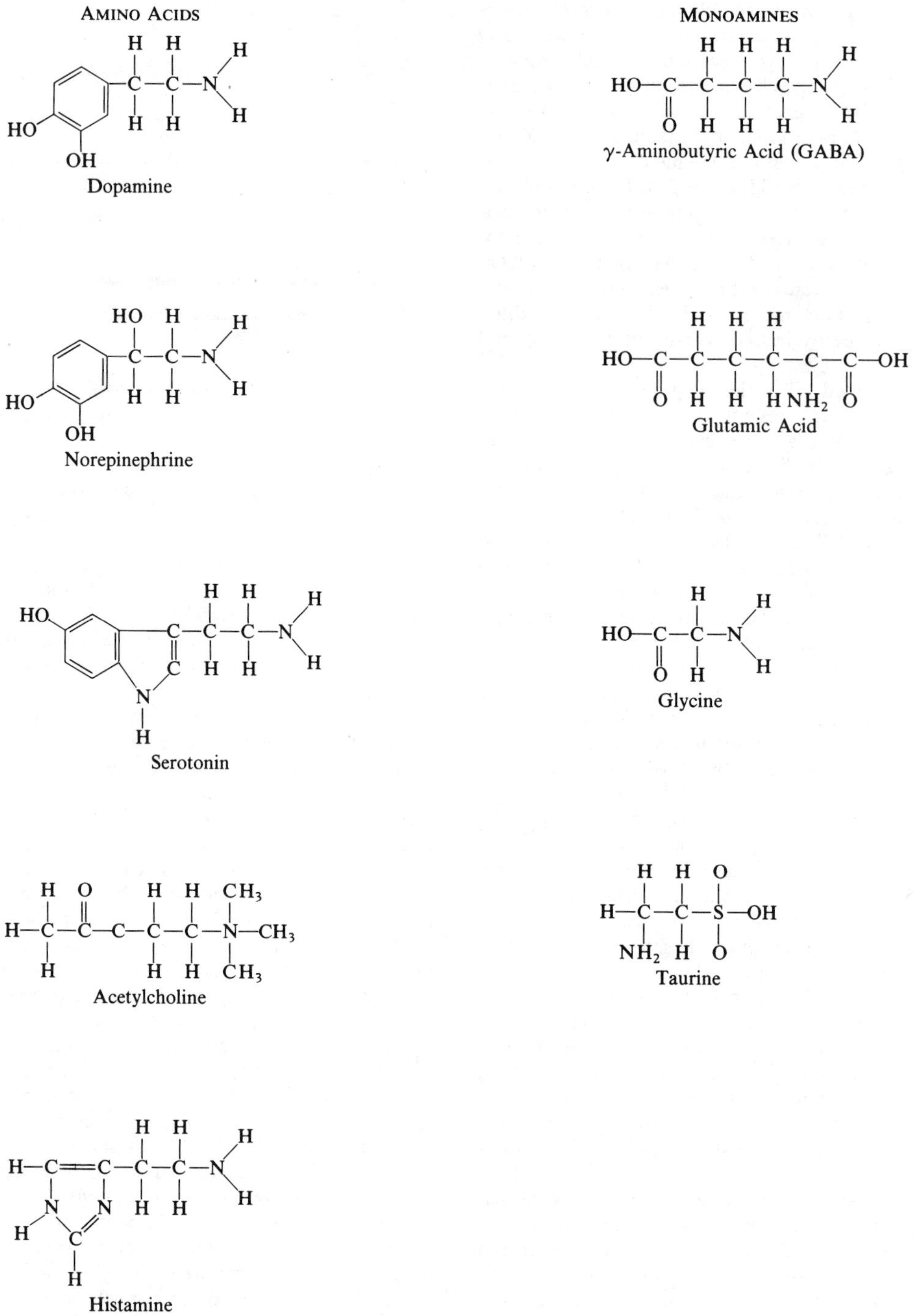

Fig. 9. Neurological system transmitter chemicals. Typically, these substances are small molecules that contain a positively charged nitrogen atom. Usually each substance characteristically inhibits or excites neurons, but in some instances the transmitter substances may play one role in one part of the brain and the other role in another part of the brain. At one time it was believed that all terminals of a given axon will transmit only one substance, but there is growing evidence that this may not always be true.

the brain is gamma-aminobutyric acid (GABA). This substance is produced almost exclusively in the brain and spinal cord. Some researchers believe that about one-third of the synapses in the brain involve GABA as a transmitter. It has been reasonably well established that a deficiency in brain GABA is the cause of Huntington's chorea. See **Chorea (Huntington's).** To date, it has not been possible to replace deficient GABA in the brain because no GABA analogues thus far developed can penetrate the blood-brain barrier. See **Blood-Brain Barrier.** Recent research has indicated that the effectiveness of certain anti-anxiety drugs, such as diazepam (Valium®) and other drugs of the benzodiazepine class, may derive their effectiveness as psychoactive drugs because of their ability to increase the effectiveness of GABA at its receptor sites.

A better understanding of the manner in which chemical transmitters are manufactured and utilized is developing, with the greatest progress taking place during the past decade. Several steps are involved in the production of these substances. One or several enzyme-catalyzed steps may be required to transform precursor molecules (which are stored in readiness for conversion) to the final transmitter chemicals. In the instance of acetylcholine, only one enzyme-catalyzed step is required. In the cases of some others, such as norepinephrine, three such steps are required, commencing the synthesis with tyrosine, which is taken up by the nerve terminal from the bloodstream. Tyrosine is converted to L-DOPA. This intermediate is then converted to dopamine and, in another step, dopamine is converted to norepinephrine. Once manufactured, vanishingly small quantities are temporarily stored for a short period (until needed) in the synaptic vesicles. The tiny quantities involve only between 10,000 and 100,000 molecules of transmitter substance. These molecules in the vesicles are protected from enzymes in the terminal that would degrade them. (The mechanism of protection is not fully understood.) A nerve impulse triggers the expulsion of the transmitter substance from the synaptic vesicles into the synaptic space; this transfer in turn activates other neurons. Few details are known concerning the transfer mechanism, but it has been established that an increase in the permeability (to calcium ions) of the nerve terminal serves to engage the release mechanism. Once released, the transmitter molecules seek out specific receptor sites and thus propagate the effects of a single nerve impulse.

Although a minor degree of interaction can take place between the electrical fields of adjacent fibers, the neural impulse conduction system is essentially an adequately well-insulated system, suited to discrete communication. The development of widespread patterns of activity reflects the functioning of interconnections at the synaptic level. It should not be surprising, since transmission is effected by specific chemicals, to find that under special circumstances of great stress, a major depot of such chemicals (adrenaline and noradrenaline) located in the medulla of the adrenal gland can pour these synaptic inhibitory chemicals into the bloodstream. Distributed in this way to all synapses, the adrenal medullary secretion can supply a cutoff influence limiting the massive discharge of impulses that stress is apt to initiate, and thus prevent it from becoming so excessive that it is detrimental. In this way, the neurohumoral transmission mechanism also affords a means of chemical homeostatic regulation.

The vulnerability of impulse transmission to chemical influences may not always serve homeostatis; for here also is where the action of poisons like mescaline and lysergic acid diethylamide takes place. They produce in humans a temporary mental derangement or psychosis. Experiments show that these substances inhibit impulse transmission at cerebral synapses and that tranquilizers prevent this effect. Such experiments in disturbed impulse transmission are developing the basis for understanding of chemically-induced psychosis and the manner of action of tranquilizers. A sufficient parallelism appears to exist between the experimental laboratory findings and data in clinical psychosis to suggest that the latter may, also, sometimes be a disturbance of impulse transmission in the brain and that tranquilizers tend to restore synaptic equilibrium.

The Neuropeptides. During the past decade, the number of chemical-messenger systems identified has increased dramatically. This advance is highlighted by the discovery of a large family of brain chemicals known as *neuropeptides*. As shown by Fig. 10, these molecules are made up of long chains of amino acids, ranging from a few to as many as 39 amino acids. Research indicates that these substances are resident within the neurons. A few of these substances, such as corticotropin (ACTH) and vasopressin, have been known for many years and identified with the hypothalamus and pituitary gland. Probably of greatest interest to researchers are the *enkephalins* and *endor-*

Tyr-Gly-Gly-Phe-Met (Met-ENKEPHALIN)

Tyr-Gly-Gly-Phe-Leu (Leu-ENKEPHALIN)

Arg-Pro-Lys-Pro-Gln-Gln-Phe-Phe-Gly-Leu-Met-NH_2 (SUBSTANCE P)

p-Glu-Leu-Tyr-Glu-Asn-Lys-Pro-Arg-Arg-Pro-Tyr-Ile-Leu (NEUROTENSIN)

Asp-Arg-Val-Tyr-Ile-His-Pro-Phe-NH_2 (ANGIOTENSIN II)

Tyr-Gly-Gly-Phe-Met-Thr-Ser-Glu-Lys-Ser-Gln-Thr-Pro-Leu-Val-Thr-Leu-Phe-Lys-Asn-Ala-Ile-Val-Lys-Asn-Ala-His-Lys-Lys-Gly-Gln (Beta-ENDORPHIN)

Ser-Tyr-Ser-Met-Glu-His-Phe-Arg-Tyr-Gly-Lys-Pro-Val-Gly-Lys-Lys-Arg-Arg-Pro-Val-Lys-Val-Tyr-Pro-Asp-Gly-Ala-Glu-Asp-Glu-Leu-Ala-Glu-Ala-Phe-Pro-Leu-Glu-Phe (ACTH, CORTICOTROPIN)

Asp-Tyr-Met-Gly-Trp-Met-Asp-Phe-NH_2 (CHOLECYSTOKININ-LIKE PEPTIDE)

His-Ser-Asp-Ala-Val-Phe-Thr-Asp-Asn-Tyr-Thr-Arg-Leu-Arg-Lys-Gln-Met-Ala-Val-Lys-Lys-Tyr-Leu-Asn-Ser-Ile-Leu-Asn-NH_2 (VASOACTIVE INTESTINAL POLYPEPTIDE, VIP)

p-Glu-His-Pro-NH_2 (THYROTROPIN RELEASING HORMONE, TRH)

p-Glu-His-Trp-Ser-Tyr-Gly-Leu-Arg-Pro-Gly-NH_2 (LUTEINIZING-HORMONE RELEASING HORMONE, LHRH)

Ala-His (CARNOSINE)

p-Glu-Gln-Arg-Leu-Gly-Asn-Gln-Trp-Ala-Val-Gly-His-Leu-Met-NH_2 (BOMBESIN)

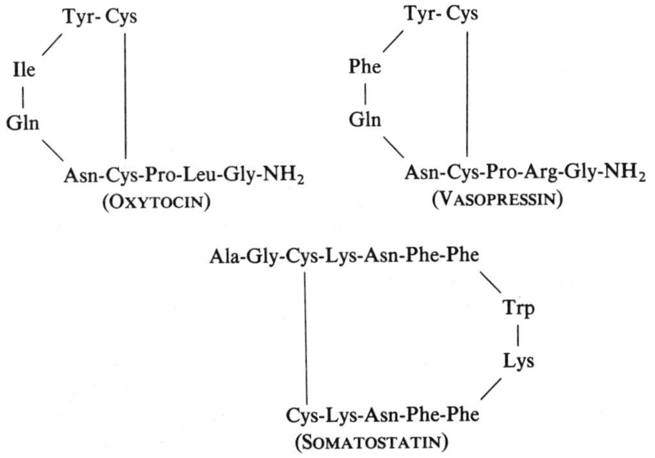

ABBREVIATIONS OF AMINO ACIDS

Ala	Alanine	Leu	Leucine
Arg	Arginine	Lys	Lysine
Asn	Asparagine	Met	Methionine
Asp	Aspartic Acid	Phe	Phenylalanine
Cys	Cysteine	Pro	Proline
Gln	Glutamine	Ser	Serine
Glu	Glutamic Acid	Thr	Threonine
Gly	Glycine	Trp	Tryptophan
His	Histidine	Tyr	Tyrosine
Ile	Isoleucine	Val	Valine

Fig. 10. Now believed to be transmitters, neuropeptides, which are short chains of amino acids found in brain tissue and notably localized in axon terminals, participate in complex mental activity, such as thirst, memory, and sexual behavior.

phins. These chemicals are strikingly similar to the opiate morphine. Identification of these substances followed the finding that specific regions of the brain possess receptor sites that bind opiates with a high affinity. These sites were revealed through the use of radioactively labeled opiate compounds. Further research showed that opiate receptors are located in those regions of the brain and spinal cord which are known to be associated with pain and emotion. Pioneering research efforts in this field included the work of S. H. Snyder and C. B. Pert (Johns Hopkins University School of Medicine), E. J. Simon (New York University School of Medicine), and L. Terenius (University of Uppsala). The enkephalins (Met- and Leu-), as shown in Fig. 10, were first isolated by J. Hughes and H. W. Kosterlitz (University of Aberdeen) in 1975. Each enkephalin contains five amino acids, one of these being methionine in one case, and leucine in the other case. Shortly after the isolation of these compounds, the endorphins, also morphinelike compounds, were isolated from the pituitary gland. Some researchers have suggested that some of the nontraditional methods used for relieving chronic pain, such as acupuncture, direct electrical stimulation of the brain, and possibly hypnosis, may be effective because these procedures may cause enkephalins or endorphins or both to be released to the brain and spinal cord. The drug naloxone (Narcan®) blocks the binding of morphine and experiments have shown that naloxone also blocks the effects of the aforementioned pain-relieving procedures; hence the tentative hypothesis.

Early research findings indicate that neuropeptides are released from axon terminals through the presence of calcium ions known to be the releasing mechanism in connection with established transmitters. Thus, some investigators believe that the neuropeptides also may be transmitters. This is particularly true of a compound simply identified as *substance P.* See Fig. 10. Some investigators have found that substance P is associated with spinal neurons involved in pain stimuli. P. T. Jessel and L. L. Iversen (Cambridge) had demonstrated that opiate drugs are able to suppress the release of substance P from sensory fibers.

Using experimental animals, researchers have shown that injection of nanogram amounts of a neuropeptide directly into brain tissue (to avoid the blood-brain barrier) causes marked behavioral changes in the subject, such as prolonged drinking behavior (at a time when the animal should not be thirsty), enhancement of sexual behavior, and improvement of memory. Some scientists have suggested, for example, that vasopressin may have beneficial effects on human patients suffering from memory loss.

One of the surprising findings of recent brain and central nervous system research is that chemical substances previously considered to be exclusive to the province of the brain have been found in organs outside the nervous system (examples: somatostatin, neurotensin, and enkephalins found in the gut); and, conversely, that other substances active in other organs, but not previously associated with the central nervous system, have been found in the latter (examples: gastrin, vasoactive intestinal polypeptide (VIP), and cholecystokinin, traditionally associated with the gastrointestinal tract, but now found in the central nervous system).

Trophic Substances. Among the most recent discoveries in brain research concerns so-called *trophic substances,* which are believed to be secreted from nerve terminals. One of these substances is nerve-growth factor (NGF). It has been established that this protein is required for the differentiation and survival of peripheral sensory and sympathetic neurons. Considerable research effort is going forward to better understand the function of trophic substances and to isolate and identify them. Substances that may be generated in the pituitary gland and transported to the brain are mentioned in the entry on **Pituitary Gland.**

Another important area of research that benefits from recent findings such as those just described is directed toward a better understanding of how pyschoactive drugs interact with the brain and central nervous system. See Fig. 11. An improved understanding will assist in the treatment of various brain disorders and of countering the effects of hallucinogenic drugs that are willfully, but tragically taken by a small number of very foolish people for "kicks." Research has progressively indicated over the years that tampering with the extremely complex brain and central nervous system is indeed attended by many risks.

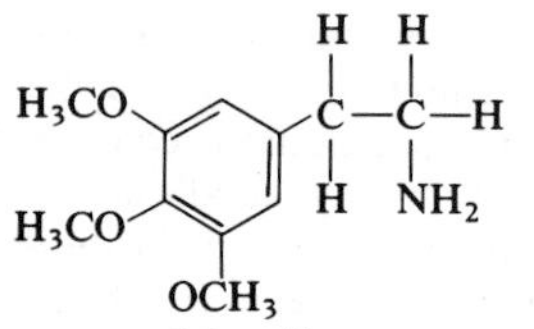

Mescaline

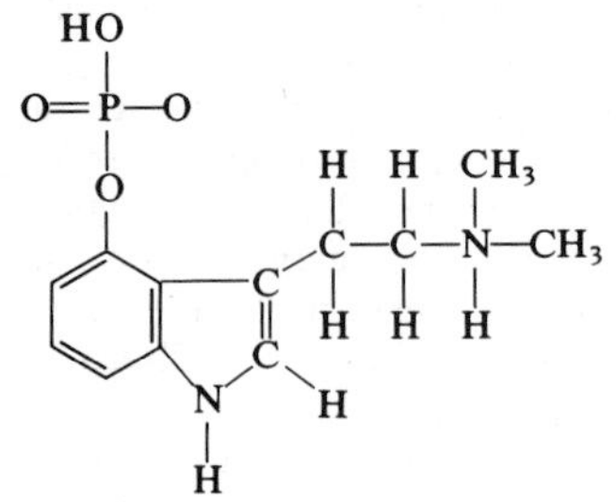

Psilocybin

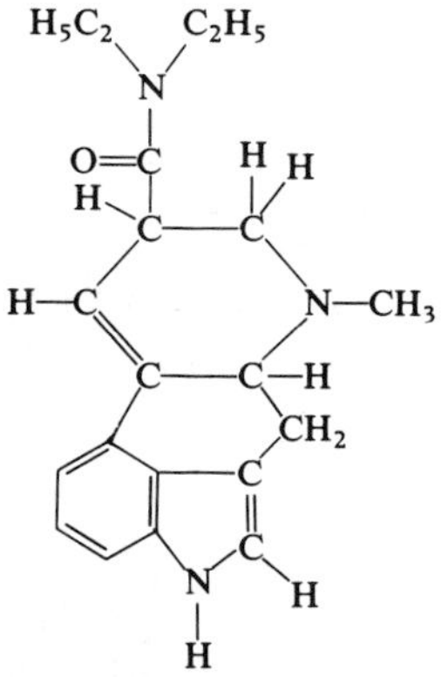

Lysergic Acid Diethylamide (LSD)

Fig. 11. Representative hallucinogenic drugs which bear structural similarities to some of the monoamine transmitters. It has been hypothesized that these similarities may cause the hallucinogens to mimic natural transmitters at synaptic receptors in the brain. Note presence of benzene-ring structure in these substances, a structure that is present in four out of the five monoamines previously shown. Also note presence of indole ring in psilocybin and lysergic acid diethylamide, a structure that is also present in the monoamines serotonin and histamine.

Brain Metabolism

The metabolism of the brain usually is measured either (1) in living organisms, or (2) in isolated systems, such as brain slices, brain homogenates, or extracts. In isolated systems, the different metabolic pathways can be closely observed, while measurements under living conditions show which of the possible capacities of the organ observed in isolated systems are operative, and at what rate, in the living brain.

The metabolism of living brain usually is measured by analyzing the arterial blood coming to, and venous blood coming from, this organ. Since cerebral circulation is very rapid (under normal conditions, about 1/6th of the total circulation goes through the brain), this method can only measure those substances that are used or produced rapidly and in significant quantities by the brain. Glucose is the main metabolite of brain and the oxidation of this compound accounts for most of the oxygen utilized by the brain. The utilization of oxygen in the brain is surprisingly high, higher than in most other organs, with about one-fifth to one-quarter of the oxygen taken up through the lungs being utilized by the brain, which comprises only 2% of body weight. About 90% of the glucose is oxidized completely to carbon dioxide, the rest mostly to lactic acid, and a small portion to pyruvic acid. The whole brain of a normal young man consumes 46 milliliters of oxygen and 76 milligrams of glucose per minute with a blood flow of 750 milliliters. These figures show that the metabolic rate of the brain is very considerable in the total body economics as far as energy and oxygen utilized, heat produced, and other factors. With physical activity, the oxygen consumption of the whole body changes significantly, but brain metabolism does not. Oxygen uptake and glucose utilization do not change significantly during mental exercise or during sleep, but are altered in extreme circumstances, such as deep anesthesia, which reduces metabolism; or high carbon dioxide content of the air, which increases cerebral blood flow.

The large amount of energy utilized by the brain is somewhat surprising since this organ does not perform work like that done by for example, muscle and kidney. The most plausible explanation is that there is continuous very high activity in the brain. Such activity can be detected by the electroencephalogram. Continuous impulse conduction involves the movement of ions, since conduction by nerve is achieved by temporary change in ion permeability.

Other components metabolized include lipids, nucleic acids, and proteins.

Cerebral metabolic rates change under a number of circumstances. There is a general decrease of oxygen utilization during development. It has been estimated that in the newborn, cerebral respiration can amount to half of that of the whole organism. Metabolism reaches a plateau in adulthood, and further decrease after 50 years of age in humans has been suggested, but not well confirmed. The metabolism of a number of compounds undergoes significant fluctuation under conditions, such as convulsions, hibernation, demyelinating diseases, and other pathological states (multiple sclerosis; myelin). A lot of interest is centered on subtle changes in metabolism which are due to changes in mental activity, such as those shown by comparison of trained versus untrained animals, or animals reared in isolation versus others reared in a complex environment. It is of interest that metabolic changes that were found and which were interpreted as changes in functional metabolism are different in the two types of cells of the brain. In functional metabolism, neurons have preference and glial cells play a supporting role.

The effect of chemical compounds on brain function is well known. Alcohol, antidepressants, hallucinatives all indicate the functional role of metabolism in the brain, particularly of compounds that inhibit cerebral protein metabolism. Compounds that inhibit the latter also inhibit the deposition of short-term memory while most do not appear to affect already deposited long-term memory.

For related information, see the following entries in this encyclopedia: **Action Current; Action Potential; Aphasia; Ataxia; Autonomic Nervous System; Axon; Blood-Brain Barrier; Brain Disorders; Brain (Injury); Cerebrospinal Fluid; Diencephalon; Electric Shock; Electroencephalogram; Ganglion; Gravireceptors; Gray Matter; Memory; Neuroses; Pain; Phrenic Nerve; Plexus; Psychiatry; Sleep; Synapse; Ventricle (Brain);** and **Seizure (Neurological).**

References

Antelman, S. M., and A. R. Caggiula: "Norepinephrine-Dopamine Interactions and Behavior," *Science*, **195**, 646–653 (1977).

Barchas, J. D., et al.: "Behavioral Neurochemistry: Neuroregulators and Behavioral States," *Science*, **200**, 964–973 (1978).

Bentley, D., and R. R. Hoy: "The Neurobiology of Cricket Song," *Sci. Amer.*, **231**, 2, 34–44 (1974).

Bunge, R., Johnson, M., and C. D. Ross: "Nature and Nurture in Development of the Autonomic Neuron," *Science*, **199**, 1409–1416 (1978).

Clarke, E. S., and C. B. O'Malley: "The Human Brain and Spinal Cord," Univ. California Press, Berkeley, California, 1968.

Cooper, J. R., Bloom, F. E., and R. H. Roth: "The Biochemical Basis of Neuropharmacology," Oxford Univ. Press, New York, 1978.

Costa, E., and M. Trabucchi, Editors: "The Endorphins," Raven, New York, 1978.

Cowan, W. M.: "The Development of the Brain," *Sci. Amer.*, **241**, 3, 112–133 (1979).

Crick, F. H. C.: "Thinking about the Brain," *Sci. Amer.*, **141**, 3, 219–232 (1979).

Evarts, E. V.: "Brain Mechanisms of Movement," *Sci. Amer.*, **241**, 3, 164–179 (1979).

Fambrough, D. M.: "Control of Acetylcholine Receptors in Skeletal Muscle," *Physiol., Rev.*, **59**, 1, 165–227 (1979).

Galaburda, A. M., et al.: "Right-Left Asymmetries in the Brain," *Science*, **199**, 852–856 (1978).

Geschwind, N.: "Specializations of the Human Brain," *Sci. Amer.*, **241**, 3, 180–199 (1979).

Granit, R.: "The Purposive Brain," MIT Press, Cambridge, Massachusetts, 1979.

Greengard, P.: "Cyclic Nucleotides, Phosphorylated Proteins, and Neuronal Function," Raven, New York, 1978.

Guillemin, R.: "Peptides in the Brain: The New Endocrinology of the Neuron," *Science*, **202**, 390–401 (1978).

Hillie, B.: "Gating in Sodium Channels of Nerve," *Ann. Rev. Physiol.*, **38**, 139–152 (1978).

Hirata, F., and J. Axelrod: "Phospholipid Methylation and Biological Signal Transmission," *Science*, **209**, 1082–1090 (1980).

Hubel, D. H.: "The Brain," *Sci. Amer.*, **241**, 3, 44–53 (1979).

Hubel, D. H., and T. N. Wiessel: "Brain Mechanisms of Vision," *Sci. Amer.*, **241**, 3, 150–162 (1979).

John, E. R., et al.: "Neurometrics," *Science*, **196**, 1393–1410 (1977).

Iversen, L. L.: "The Chemistry of the Brain," *Sci. Amer.*, **241**, 3, 134–149 (1979).

Kandel, E. R.: "Cellular Basis of Behavior: An Introduction to Behavioral Neurobiology," Freeman, San Francisco, 1976.

Kandel, E. R.: "Small Systems of Neurons," *Sci. Amer.*, **241**, 3, 66–76 (1979).

Katz, B.: "Nerve, Muscle, and Synapse," McGraw-Hill, New York, 1966.

Keynes, R. D.: "Ion Channels in the Nerve-Cell Membrane," *Sci. Amer.*, **240**, 3, 126–135 (1979).

Koshland, D. E., Jr.: "A Response Regulator Model in a Simple Sensory System," *Science*, **196**, 1055–1063 (1977).

Levi-Montalcini, R., and P. Calissano: "The Nerve-Growth Factor," *Sci. Amer.*, **240**,

Morell, P. (editor): "Myelin," Plenum Press, New York, 1978.

Morell, P., and W. T. Norton: "Myelin," *Sci. Amer.*, **242**, 5, 88–118 (1980).

Nauta, W. J. H., and M. Feirtag: "The Organization of the Brain," *Sci. Amer.*, **241**, 3, 88–111 (1979).

Nicholls, J. G., and D. Van Essen: The Nervous System of the Leech," *Sci. Amer.*, **230**, 1, 38–48 (1974).

Porter, C. G., and R. Porter: "Corticospinal Neurones: Their Role in Movement," Academic, New York, 1977.

Purves, D., and J. W. Lichtman: "Elimination of Synapses in the Developing Nervous System," *Science*, **210**, 153–157 (1980).

Rosenzweig, M. R., and E. L. Bennett: "Neural Mechanisms of Learning and Memory," MIT Press, Cambridge, Massachusetts, 1979.

Schmitt, F. O., and F. G. Worden (editors): "The Neurosciences: Fourth Study Program," MIT Press, Cambridge, Massachusetts, 1979.

Schwartz, J. H.: "The Transport of Substances in Nerve Cells,"*Sci. Amer.*, **242**, 4, 152–171 (1980).

Snyder, S. H.: "Brain Peptides as Neurotransmitters," *Science*, **209**, 976–983 (1980).

Steinbach, J. H., and C. F. Stevens: "Neuromuscular Transmission," in "Frog Neurobiology: A Handbook" (R. Llinas and W. Precht, editors), Springer-Verlag, New York, 1976.

Stevens, C. F.: "Interactions Between Intrinsic Membrane Protein and Electric Field: An Approach to Studying Nerve Excitability," *Biophys. J.*, **22**, 2, 295–306 (1978).

Stevens, C. F.: "The Neuron," *Sci. Amer.*, **241**, 3, 54–65 (1979).

Tucek, S.: "Acetylcholine Synthesis in Neurons," Chapman and Hall, London, 1978.

Warwick, R., and P. L. Williams: "Neurology in "Gray's Anatomy," 35th edition, Saunders, Philadelphia, 1973.

Wilson, R. S.: "Synchronies in Mental Development: An Epigenetic Perspective," *Science*, **202**, 939–948 (1978).

Worden, F. G., Swazey, J. P., and G. Adelman: "The Neurosciences: Paths of Discovery," MIT Press, Cambridge, Massachusetts, 1979.

BRAIN DISORDERS. Among the principal causes of brain disorders are genetic (inborn errors of metabolism, etc.), hemorrhage, pressure, displacement, inflammation, and atrophy. Several of the foregoing conditions, of course, may result from physical injury to the head.

As with other aspects of brain research, the investigation of brain disorders was hindered by the inability of researchers to explore the living brain as contrasted with examining dead brain tissue. Most of the early information was obtained at autopsy. X-rays were the first practical tool for examining the living brain. A relatively recent development, known as *pneumoencephalography*, has enhanced the value of x-ray examination. In this technique, the fluid that normally surrounds the brain is replaced with air, thus more clearly revealing structure. In still another technique, known as *cerebral angiography*, a dye opaque to x-rays is injected into the bloodstream. This enables the viewing of the pathological development of the blood vessels in the brain on the x-rays. A limitation of this method is that overlapping abnormal and normal structures are not distinguishable. The more recent computed axial tomography (CAT) technique does not have these limitations. In this technique, numerous x-ray views are taken from various angles to form a reliable, computer integrated presentation of the internal structure of the brain. Abnormal tissues, such as tumors and hemorrhage damage, are made quite visible with this method. Aside from the radiation risk associated with any x-ray procedure, the CAT scan technique can be safely used on living human patients and experimental laboratory animals. The technique serves a very useful purpose as a diagnostic as well as research tool.

Because, as discussed in the entry on **Brain and Nervous System,** there is much electrical activity occurring within the brain, this activity can be measured by picking up electric signals from the skull. By moving detector electrodes around the skull, distribution of electrical activity can be traced to specific regions and locations of the brain. See also **Electroencephalogram.**

During and since the 1960s, great progress has been made in the use of chemical indicators as criteria for brain function analysis. In a technique developed by N. A. Lassen (Bispebjerg Hospital, Copenhagen) and D. H. Ingvar (University of Copenhagen), brain blood flow and glucose consumption by the brain can be measured instrumentally and projected on a cathode-ray tube. Thus, variations in blood glucose consumption can be related to mental activities, such as reading, talking, etc. In a refinement of this technique, L. Sokoloff and associates (National Institute of Mental Health) have been able to pinpoint brain metabolic activity, greatly enhancing the mapping of brain function, both for research and diagnostic applications.

The use of various staining techniques is described in the entry on **Brain and Nervous System.** Also in that entry, mention is made of a growing understanding of the biochemistry that occurs in synapses. Disturbances of the synapses (connections) within the brain are associated with such mental disorders as schizophrenia and manic-depressive psychosis. Although they are still in a pioneering stage, it is believed that these new techniques will be invaluable to research and diagnosis in psychiatry. See accompanying illustration.

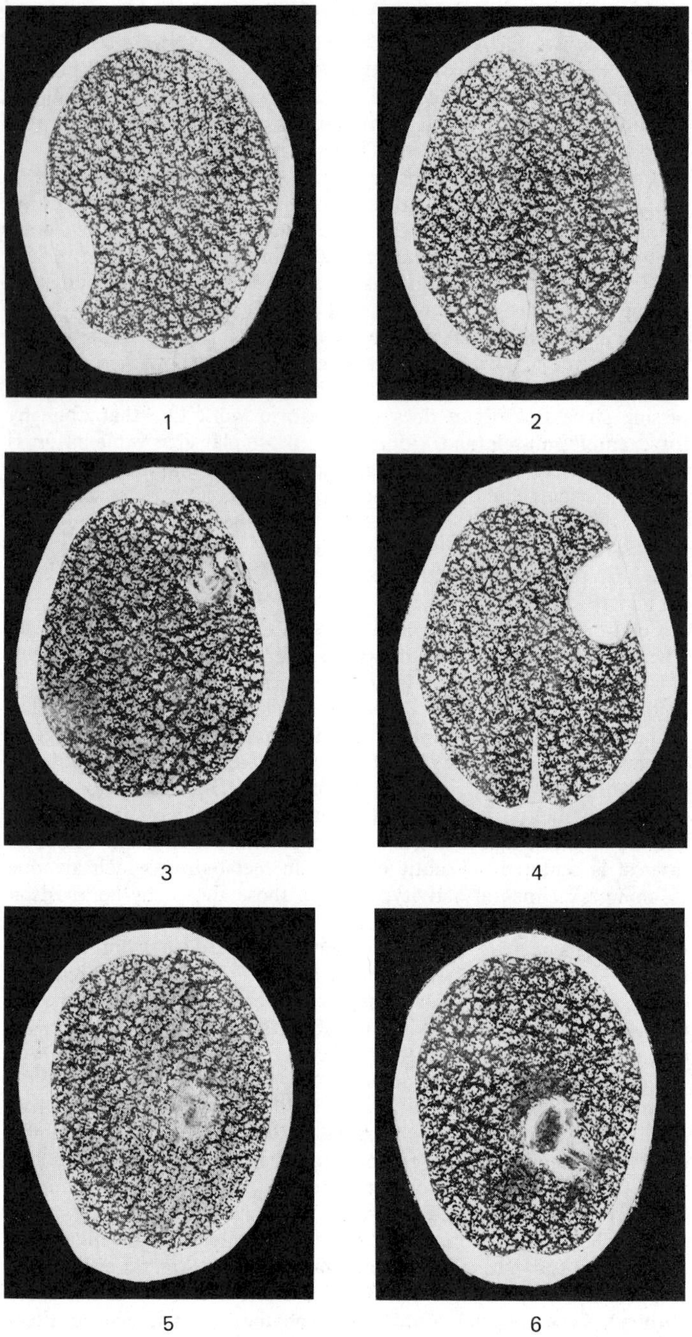

Artist's replicas of CAT scans of human brains: (1) Blood clot at left between brain and skull caused by injury. No iodine solution was required to enhance this image. (2) Tumor. Iodine solution required. (3) Meningioma (benign tumor) only shown faintly without iodine solution. (4) Same tumor enhanced by iodine solution. (5) and (6) Malignant tumor in center of view. Iodine solution used. When white ringlike zones appear in an iodine enhanced image, this usually indicates a malignant tumor.

Etiology of Brain and Nervous System Disorders

Although there are numerous causes of serious, exotic, and rare mental and nervous system disorders, by far the majority of brain disorders arise out of a deficiency of blood supply to the brain. Continuing day and night, the brain requires 20% of the body's blood supply. As pointed out in the entry on **Brain and Nervous System,** blood glucose furnished to the brain represents 20% of the body's total oxygent needs (when resting). *Atherosclerosis*, a major disorder of the blood vessels, is the main underlying cause of inadequate blood to the brain. See **Arterial and Venous Disorders.** This condition leads to a thrombus, which may progressively decrease the blood supply or cut the supply fully. Atherosclerosis also weakens blood vessels, causing them to rupture and resulting in a cerebral hemorrhage. These accidents occur over a variety of conditions and range from a small loss of blood supply or, in the case, of a weakened blood vessel, small amounts of bleeding, all the way to quickly fatal consequences. A majority of serious head injuries damage the brain blood supply, as also is the case of brain tumors. See **Brain (Injury).**

Epilepsy is a disorder of the brain that has been known since antiquity, but still is not clearly understood. Improved treatment of seizures (the symptoms of the disorder) has far outpaced an understanding of the fundamentals. Some neuroscientists believe that one major precipitating factor of epilepsy may involve neuron transmitters and, in particular, gamma-aminobutyric acid (GABA), which is an inhibitory transmitter. See **Seizure (Neurological).**

Genetic Factors. A number of enzyme catalysts are involved in brain functions. When the genetic ordering of the amino acids required to manufacture protein molecules is disturbed, a number of abnormalities, frequently resulting in mental retardation, occur. Diseases involving such enzyme deficiencies include **Galactosemia** and **Phenylketonuria,** for which there are separate entries in this encyclopedia. Genetic disorders are not limited to deficiencies of genetic material. Excesses also may cause brain disorders. Among the genetically derived mental disorders not previously mentioned are *Down's syndrome, Huntington's chorea*, and the *Lesch-Nyhan syndrome.* See **Down's Syndrome (Mongoloidism); Chorea (Huntington's);** and **Lesch-Nyhan Syndrome.** Also, there is *porphyria*, which is described in the entry on **Dermatitis and Dermatosis.**

As suggested by Linus Pauling in the late 1960s, genetic differences among people may be reflected in their minimum daily requirements for vitamins. This hypothesis has at least been partially authenticated by the successful therapeutic use of certain vitamins in the cases of a few childhood mental disorders. For major adult psychoses, however, such therapy has not provided convincing evidence to date.

For many years, neuroscientists have observed familial associations in the occurrence of schizophrenia and manic-depressive disorders. These observations have been supported by research involving identical twins. See also **Schizophrenia.**

Fetal Development. Brain and central nervous system disorders also arise from abnormalities in fetal development. For examples, see **Kernicterus;** and **Rubella.** Serious injuries to the pregnant woman may also cause fetal injury, but most frequently such situations result in a miscarriage.

Bacterial Infections. A number of brain disorders result from bacterial infections. A notable example is the brain and nervous system involvement in the later stages of syphilis. In recent years, antibiotic therapy has shortened the course of most bacterial infections and has prevented brain damage. See also "Bacterial Meningitis" in the entry on **Meningitis.**

Viral Infections. Particularly notable among the virus-caused nervous system disorders is *poliomyelitis*, which is an infection of the motor neurons. Although the incidence of this disease has been dramatically reduced, freedom from the disease is entirely dependent upon the routine use of vaccines. See **Poliomyelitis.** An example of a slow virus is the influenza virus (of the epidemic of 1918) which many years later produced *Parkinson's disease* in large numbers of individuals. See **Parkinson's Disease.** *Kuru,* a disorder thus far encountered only in tribal people of New Guinea, is a neurological disorder of viral origin. *Creutzfeldt-Jakob disease, Alzheimer's disease, progressive multifocal leukoencephalopathy* (PML) and *subacute sclerosing panencephalitis* (SSPE) also are diseases result from prior viral infections. Viruses also cause mental disorders among animals, such as chimpanzees, lower primates, sheep, and other animals. These disorders include *scrapie* and *transmissible mink encephalopathy.* See **Virus.** Also see, "Aseptic Meningitis" in the entry on **Meningitis.**

In the entry on **Brain and Nervous System,** the manner in which the brain is well protected (skull, blood-brain barrier, etc.) is discussed. Nevertheless, a number of chemical substances are highly toxic to brain tissue. There are, of course, the halucinogenic drugs, such as LSD. There are also numbers of industrial chemicals, such as mercury, manganese, and lead metals and compounds, and many identified and unidentified industrial organic chemicals, which can lead to brain disorders, some of these substances requiring several years to cause sufficient damage to be noted by the affected individual.

See also the list of other entries at end of **Brain and Nervous System.**

References

NOTE: See also the lists of references at the ends of the entries mentioned here.

Ingvar, D. H., and N. A. Lassen: "Brain Work: The Coupling of Function, Metabolism and Blood Flow in the Brain," Munksgaard, 1975.

Kety, S. S.: "The Biological Roots of Mental Illness: Their Ramifications through Cerebral Metabolism, Synaptic Activity, Genetics, and the Environment," *Harvey Lectures*, Series **71**, 1–22 (1978).

Kety, S. S.: "Disorders of the Human Brain," *Sci. Amer.*, **241**, 3, 202–214 (1979).

Kinney, D. K., and S. Matthysee: "Genetic Transmission of Schizophrenia," *Ann. Rev. Med.*, **29**, 459–473 (1978).

Penrose, L. S.: "The Biology of Mental Defect," Grune & Stratton, 1962.

Snyder, S. H.: "Madness and the Brain," McGraw-Hill, New York, 1974.

BRAIN (Fishes). Fishes.

BRAIN (Injury). Within the skull, either the brain or its covering (*dura*) may be damaged. Injury to the dura causes bleeding which may, in turn, injure the brain tissue. When bleeding is on the undersurface of the dura, it is termed a *subdural hematoma.* If the bleeding is above the dura, it is termed an *extradural hematoma.* The latter almost always occurs in the region of the temple. Individuals with hematomas have a characteristic course of symptoms. A blow on the head, for example, may or may not cause temporary loss of consciousness. This is followed by headache, which becomes increasingly severe during the next two or three hours, often followed by nausea and vomiting. There may be drowsiness, speech difficulties, and weakness in various parts of the body. If the drowsiness continues, the patient becomes stuporous and finally goes into a deep coma. Blood collects in the area of the wound. Since the skull is rigid and nonexpansile, the collecting blood can only depress the brain tissue. When this lasts for only a short time, there may be no permanent damage. When it lasts for a long period, there is usually permanent damage to the brain tissue.

Once the diagnosis of extradural clot is suspected, there should be immediate surgical exploration. This is a rather simple procedure. Two small holes are bored into the skull, and the clot is located. Then, a larger hole is made. The clot is sucked out. The bleeding artery is tied off. When treated promptly, there is an excellent chance the patient will recover.

Subdural hematomas usually develop over a period of days or weeks. Chronic subdural hematomas are more common than formerly suspected. They often follow minor head injuries and usually occur in infants and persons over 40 years of age. The bleeding is slow, and often fluid from the surrounding tissue is drawn into the clot. This results in a slowly enlarging mass which allows the brain tissue some adjustment to the increased pressure. The symptoms which appear after weeks and months are usually similar to those of brain tumor. Headache is present in most cases. Drowsiness is another conspicuous sign. Both of these symptoms may fluctuate from day to day in the same patient. Dizziness often accompanies these symptoms and vomiting may also occur. Older patients usually are confused. Personality changes are so insidious and vague that the family cannot state just what is wrong, but only that the patient is "different." There may be weakness or complete paralysis of various parts of the body. Diagnosis of this condition is sometimes difficult, and made only through an exploratory operation. The procedure may be simple, as previously described for extradural hematomas, or it may be more extensive. Recovery is good in many of these cases, particularly if the underlying brain tissue is healthy.

Injuries to the brain are extraordinarily varied in their effect. For example, the trauma of a bullet entering the brain may cause life to cease almost immediately; or the tearing and depriving brain tissues of blood and oxygen can make the difference between functional living and vegetable-like existence. In rare instances, a bullet entering the skull has been known to miss vital areas and leave the victim with little more than a severe headache. Brain injury may occur without damage to other structures of the head.

Brain injury is usually subdivided into the following classes: (1) concussion, (2) contusion, and (3) laceration. Concussion is a jarring of the brain which usually results in a transitory period of unconsciousness. It is one of the most common and mild forms of brain injury. Recovery is almost always complete. Contusion is a bruising injury to the brain. The patient's symptoms are a combination of two effects: (1) nonfunction of some nerve centers, and (2) overactivity of others which are normally inhibited by higher control centers. Disturbance of consciousness is a sign of generalized disturbance in the brain, whatever the cause. This may be a mild, transient change, or a profound and prolonged coma. A boxer's "k.o." is an example of concussion. Many fighters have no lasting effects, while others become "punch drunk" and portray unusual symptoms.

Recovery from complete loss of consciousness is attained by certain stages. The entire process may require only a few minutes. However, any of the phases may be prolonged for hours or days. In severe injury, paralysis of major brain functions, even of respiration, may occur. The latter returns quickly in nonfatal cases. Death occurs rapidly if artificial respiration is not applied in those instances when return of respiration is delayed. Deep coma is marked by flaccid paralysis and even loss of involuntary motion. As coma lightens, the patient passes into stupor, and reflex activity returns. He responds automatically to forceful commands, but is unaware of his surroundings. The next phase, excitement or delirium, is marked by extreme restlessness and confusion, and often the patient is violent. He gradually becomes quiet, but remains extremely confused mentally. In the next stage, automatism, the patient answers questions and performs simple tasks in a fairly orderly, but automatic way. The highest functions of judgment and insight are the last to return.

Laceration of the brain results in actual tearing or destruction of the brain tissue itself. Swelling of the brain occurs and probably accounts for at least part of the widespread changes that follow. Slowing of the blood flow results in poor oxygen supply, which further increases the damage.

On recovery of consciousness, there may be loss of memory (amnesia) for the accident itself. Often, this amnesia includes events that occurred before the accident (retrograde amnesia) and a variable period of time after the accident (posttraumatic amnesia). The presence of retrograde amnesia is evidence of the severity or extensiveness of the brain injury. The duration of posttraumatic amnesia varies because the patient often has isolated memories of events before the complete return of memory.

In severe injuries, if the patient arouses sufficiently to answer questions, as a rule he is fairly certain to recover from the initial generalized brain injury, but he is still liable to such complications as meningitis and hemorrhage. Personality and intellectual impairment also may occur. In general, the older the patient, the slower and less certain is the improvement. Children tolerate head injury with fewer aftereffects than adults. Some, however, show behavior disorders. In general, the duration of posttraumatic amnesia is the best single criterion for prognosis; the longer it lasts, the poorer the outlook. Improvement may continue slowly for 12 to 18 months.

Infections, such as brain abscess or meningitis, may complicate head injury. Most brain abscesses result from compound fractures or from penetrating wounds, both of which introduce bacteria into the brain. The injured brain provides an ideal place for the growth of bacteria. If the organisms are virulent, meningo-encephalitis may develop rapidly. If they are less virulent, a brain-abscess develops.

Convulsive seizures of any type may occur at any time after brain injury. The occurrence of seizures immediately following the injury does not necessarily mean that the patient will continue to have them, nor does their absence during the acute phase guarantee against them in the future. They may develop months or years after the original injury. They are more apt to occur in those injuries which produce penetration of the dura and brain damage. Retention of a foreign body of any sort leads to a higher incidence of convulsions. Laceration of the brain and small intracerebral hemorrhages also result in tissue changes which may cause convulsions.

The condition known as "punch drunk" is seen in people who have had repeated head injuries, such as professional football players and particularly boxers. The condition is thought to result from small hemorrhages throughout the brain. The changes begin gradually with the loss of dexterity, which the patient may claim is as good as ever. Lack of attention, concentration, and memory follow. Impediments of speech and glazed, staring eyes make the usually too talkative, too social person look partially drunk, hence the term for the condition. Tremor of the hands, unsteadiness of gait, and failing vision and hearing develop in severe cases. The victim is unable to engage in even simple intellectual activities and is without insight regarding his disability.

In contrast with the "punch drunk" person, the individual with post-concussion syndrome complains of greater incapacity, of which he gives little outward sign. Authorities differ as to cause. Some believe that the condition is caused by organic damage to the brain. Others believe that it is the result of psychologic factors. The condition is not related to the severity of the injury. Headache, which is quite variable in character, is the most common complaint. The patient may suffer from intolerance to cold, fatigue, and insomnia. Some memory impairment and confusion in thinking may be noted. In more severe cases, the patient may have an emotional outburst, particularly of rage. Treatment usually consists of the administration of mild sedatives and psychotherapy. See also **Headache.**

See also **Amnesia; Memory;** and other entries listed at end of entry on **Brain and Nervous System.**

BRAIN TUMOR. Cancer and Oncology.

BRAKEFIELD OSCILLATOR. Oscillator.

BRAKE HORSEPOWER. Dynamometer.

BRAN. Wheat.

BRANCH (Computer). A set of instructions that may be executed between a couple of successive decision instructions. Branching enables parts of a program to be worked on to the exclusion of other parts and provides a computer with considerable flexibility. The branch point is a junction in a computer routine where one or more of two choices is selected under control of the routine. Also refers to one instruction that controls branching.

BRANCHIAL ARTERIES. Fishes.

BRANCHING. In radioactivity, branching denotes the occurrence of more than one mode of disintegration by a radionuclide. See **Radioactivity.** The two modes operate jointly, a portion of the atoms of the radionuclide undergoing one mode, and another portion undergoing the other—both modes having characteristic rates. The branching fraction is the ratio of the number of atoms disintegrating by a particular mode to the total number of atoms disintegrating (per unit time). The branching ratio is the ratio of two specified branching fractions.

BRANCH INSTRUCTION (Computer System). Instruction (Computer System).

BRANCH POINT (Mathematics). If $f(z)$ is a multivalued function of the complex variable z and there exists a single-valued analytic function $g(z)$ such that at each z for which $f(z)$ is defined, the value of $g(z)$ coincides with one of the values of $f(z)$, then $g(z)$ is called a *branch* of $f(z)$. A curve in the domain of definition of $f(z)$ such that, if the points on this curve are removed, the remainder of the domain of definition is an open set for which there exists a branch of $f(z)$, is called a *branch cut.* A point at which a branch cut originates is called a branch point. Thus, if $f(z) = z^{1/2}$, the negative real axis, including the origin, is an example of a branch cut, while the origin itself is a branch point. See also **Node.**

BRANT. Waterfowl.

BRASS. Copper.

BRASSICA. A genus of the family *Cruciferae* (mustard family) and composed of three major groups: (1) the rapes; (2) the cabbages (or coles); and (3) the mustards. There are also numerous plants termed *cress.* See also **Cress.**

Rape. *Brassica napus* L. is characterized by foliage that is dark bluish-green and glaucous (covered with a bloom or whitish substance that rubs off) and smooth, or with a few scattered hairs near the margins. The leaves have the same general shape in all varieties. The inflorescence is an elongated raceme, the flowers large, clustered at the top but not prominently overtopping the terminal buds, often with open flowers along the axis below.

The rapes are represented by four varieties: (1) winter rape; (2) summer rape; (3) rutabaga; and (4) rape-kale. Winter rape, a biennial or winter annual, is planted for fall and winter pasture. The variety *Dwarf Essex* is planted almost exclusively so that the name is often used synonymously for winter rape. Summer rape, an annual producing comparatively little leafage, is essentially an oilseed crop. Rutabaga is grown for the young tops which are used for greens and for the tubers, which are for table use and stock feed. The rape-kales may be used as a forage crop and one variety, the *Dwarf Siberian* kale, may be planted for greens.

Cabbage or Cole. With exception of some of the kitchen kales, most of the cultivated forms of cole, or cabbage (*Brassica oleracea* L.) are characterized by foliage that is thick and somewhat leathery, glaucous, and smooth. The inflorescence is an elongated raceme with large open flowers along the axis below the terminal buds, much as in winter rape.

Mustard. The mustards may be annual or biennial, the foliage varying in shape and color from bright green and hairy to lightly glaucous and smooth. The species may be grouped roughly into three classes: (1) turnip and allies; (2) the true mustards; and (3) the oriental, or Chinese mustards. The turnip group includes three types of plants: (1) the edible turnip; (2) the so-called wild turnip; and (3) turnip-rape, annual and biennial oilseed crops. With the exception of the strap-leaved or Japanese turnips, the leaves are lyrate in form, bright green or lightly glaucous in the annual forms, sparingly to copiously hairy; the flowers are small, clustered at the top of the raceme and usually overtopping the terminal buds.

Four species are included in the true mustards: (1) Brown mustard, with several horticultural varieties; (2) black, or Trieste mustard; (3) white mustard; and (4) charlock, or wild mustard. These mustards are used chiefly in the manufacture of condiments, as table greens, and for planting as cover crops. Charlock is a widespread field weed and is sometimes screened in quantity out of grain. It finds limited use as a quick cover crop in situations where close cultivation is practiced. All four species are annuals and flower early. They are readily distinguished in early stages of growth and later by the character of the inflorescence and seed pods, or siliques.

BRAVAIS-MILLER INDICES. A modification of the Miller indices suitable for describing hexagonal crystals. In this system, three axes are taken, perpendicular to the hexagonal axis and at angles of 120° to one another. The symbols then consist of the reciprocal intercepts on these axes, followed by the reciprocal intercept on the hexagonal axis, all reduced to integers, e.g., (0001). The first three indices are not independent but must add to zero.

BRAYTON CYCLE. **Gas and Expansion Turbines; Solar Energy.**

BRAZIL-NUT TREE. Of the family *Lecythidaceae*, the giant *Bertholletia excelsa* grows in the forests of northern Brazil. The seeds of this tree are called Brazil nuts, or *Pará chestnut*; *tacarí* (in Brazil); *toura* (in French Guiana). The tree commences to bear at eight years and may yield up to a half-ton of large round fruit pods each year. Each pod contains from 18 to 24 hard-shelled kernels (the commercial nuts). The oil content of the nut is high and would be an excellent food oil except that the nuts are the more highly valued form and there is insufficient harvest to serve both purposes. The fruits develop high on the tree and fall to the ground without opening. This favors the gathering of the fruits, which are then split open and the seeds removed. Very large quantities of Brazil nuts are consumed in Brazil and the United States.

BRAZILWOOD. This term is used to describe the wood obtained from several species of tropical American trees in the family *Caesalpiniaceae* (senna family). At one time, the wood from *C. brasiliensis*, *C. drista*, and *C. echinata* was an important Brazilian export. Its principal use was as a dyewood, producing purple shades when used with a chrome mordant, and crimson shades with alum. Synthetic dyes now fill requirements for these shades. However, brazilwood extract still is used in limited quantities in connection with inks, wood stains, and silk dyeing. The wood finds continued limited demand for high-quality furniture and items such as violin cases because of the rich bright-red coloration and its capability of accepting a high polish. *Sapanwood*, also sometimes called brazilwood, is obtained from *C. sappan*, which grows in Sri Lanka, India, and Malaya.

BRAZING. Brazing, according to the American Welding Society, is a group of welding processes wherein the filler metal is a nonferrous metal or alloy whose melting point is higher than 1,000°F (538°C), but lower than that of the metals or alloys to be joined.

At one extreme, brazing is very similar to soldering and is sometimes called hard soldering. A typical example of this is silver soldering or silver brazing using alloys containing 10–80% silver, the balance principally copper and zinc. The melting points of these alloys are in the range 1,175–1,500°F (635°C–816°C), which is considerably higher than for soft solders. Silver solders are used to join both ferrous and the higher melting point nonferrous alloys. A composition containing 15% silver, 5% phosphorus, 80% copper is often used to join copper and brass.

Joints to be silver-soldered are usually designed so as to require only a thin film of filler metal which is drawn into the joint by capillary action when the solder becomes molten. This principle is also used in copper brazing, a high-temperature process in which pure copper is used to join close-fitting steel parts. Whereas most other forms of brazing depend on local heating with a torch or by other means, copper brazing is generally done in a furnace in which the part is heated to about 2,050°F (1,121°C). A reducing atmosphere is used to prevent oxidation, thus no flux is needed as in nearly all other types of brazing. The copper is applied in any convenient form such as wire or sheet and is so placed that it can flow into the joint when molten. A mechanically tight joint such as a forced fit between two cylindrical parts is the most effective in drawing the copper into the joint. The resulting thin film of bonding metal gives a high-strength connection.

Where somewhat higher temperatures can be tolerated than those used in silver soldering, brazing spelter can be used. This is an alloy containing about equal proportions of copper and zinc and melting at about 1,600°F (871°C). Like silver solders it flows readily between contacting surfaces when properly fluxed to prevent oxidation. The parts to be joined may be heated to the melting point of the spelter as in soft soldering, or may be dipped in a bath of molten brazing alloy.

Another type of brazing is similar to welding in many respects and is sometimes called bronze welding. The joints are generally of the V or fillet types in which a bead of filler metal is deposited with a torch. The technique differs from welding in that the base metal is not melted but only raised to the "tinning" temperature at which bonding takes place between the base metal and the brass or bronze filler metal by slight interdiffusion or alloying. The filler metals most widely used are brasses of the 60% copper-40% zinc type with additions of tin, iron, manganese, or silicon. For some purposes special bronzes are used such as aluminum bronze, silicon bronze, and nickel silver. The melting points of all of these alloys are higher than those of the silver solders but lower than that of pure copper. They are applied with a torch as in gas welding. Bronze welding is particularly well adapted to the repair of worn or broken parts.

BREA. **Tar Sands.**

BREAD AND BAKERY PRODUCTS. **Antimicrobial Agents (Foods); Bodying and Bulking Agents (Foods); Staling (Bakery and Food Products).**

BREADBOARD. An assembly of preliminary circuits or parts used to prove the feasibility of a device, circuit, system, or principle without regard to the final configuration or packaging of the parts.

BREADFRUIT TREE. **Mulberry Family.**

BREAKDOWN VOLTAGE. This is the voltage necessary to cause the passage of appreciable electric current without a connecting conductor. It is commonly used to express the voltage at which an insulator or insulating material fails to withstand the voltage and ceases to behave as an insulator.

BREAKPOINT INSTRUCTION (Computer System). **Instruction (Computer System).**

BREAST. The upper aspect of the chest. A mammary gland. The breasts are modified skin glands that lie in the outermost layer of connective tissue, called the fascia. In men, the breasts remain undeveloped and without specific use. In women, they are active, functioning parts of the body throughout much of life. On a well-developed, well-nourished woman who has not borne a child, the breasts may extend from the second or third rib to the sixth or seventh rib, and from the outer border of the breastbone (sternum) to the folds of the armpit. A woman who has borne children normally has somewhat larger breasts. The size and shape of the breasts in different individuals varies from round to conical. The consistency is usually firm and elastic, but varies a great deal, depending upon the presence and amount of fatty tissue. Rarely are the two breasts equal in size; the left is usually larger. There is a great divergence in breast sizes among individual women. The average breast in a woman who has not borne a child ranges from 4 to 6 inches (10 to 15 centimeters) in diameter and weighs between $2\frac{1}{2}$ ounces (71 grams) to $\frac{1}{2}$ pound (227 grams) or more. These figures depend a great deal upon age, climatic conditions, race, and general health of the individual.

The skin of the breasts is covered with tiny soft hairs associated with sebaceous glands and sweat glands like those found on the rest of the body. The skin is thin, and often superficial veins may be seen through it. The skin of the breast is elastic and flexible, despite the fact that it adheres to the fatty layer beneath it. At the tip of each breast in both men and women is a projection called the nipple, surrounded by a pigmented area (the areola) which is about $1\frac{1}{2}$ inches (3.8 centimeters) in diameter. The nipples are not in the exact middle of the breasts, but slightly to the side. The skin is wrinkled and the same color as the areola. They are usually round or cone-shaped, and the tip contains the tiny depressions which are really the openings of the milk ducts. The size of the nipple is usually directly proportionate to the size of the breast proper, but large nipples may be found on small breasts, and vice versa. In the deeper layers of the nipples, circular muscular fibers help to empty the breast of milk. When they contract, the nipple becomes harder, narrower, and more erect.

The breasts are composed primarily of a round, flattened mass of glandular tissue called the corpus mammae. This tissue is whitish or reddish-white in color and is thickest under the nipple and thinnest at the edges. The corpus mammae is a complex structure consisting of 15 to 20 separate and distinct lobes, which are separated by varying amounts of fat. They are arranged in a pattern like a wagon wheel, with the nipple as the hub. Each lobe contains a single milk duct (lactiferous duct) which opens into a tiny depression on the tip of the nipple. The lobes do not communicate with each other at any point, although two or more may have the same opening in the nipple.

The first significant changes in the female breast usually occur when the girl is 11 to 13 years of age. The activities of the gland are apparently related to changes in the reproductive system. If no function of the ovaries has been established, the breasts remain underdeveloped. During puberty, the child's breasts become more prominent, and the projection of the nipple and areola form the tip. The breasts become elastic and firm in consistency. The areola begins to attain some coloring, and the skin becomes tense; sometimes mild pain may be felt as a result of this tenseness of the skin. Between ages 14 and 16, a fat layer is deposited under the skin, softening the contour of the breast and making it more hemispherical in form. The greater part of the breast consists of this fatty layer and connective tissue. The milk glands are fully developed at this time, but only a small amount of glandular tissue has been formed; and this is found at the base and at the borders of the breast. After puberty, the amount of glandular tissue gradually increases, as well as the fat and connective tissue. Both before and after menstruation, the girl may attempt to disguise or alter the appearance of her breasts by tight, ill-fitting brassieres, or poor posture. Understanding and kindness are prerequisites for the adjustment and happiness of a young girl during this cycle.

Changes take place in the breasts during pregnancy. From five to six weeks after pregnancy begins, the breasts begin to enlarge and continue to increase rather rapidly in size until mid-pregnancy. The surface veins dilate; and if the breast has enlarged very much, bluish-white streaks may appear in the skin. The nipple becomes larger, and the size of the areola increases. The pigmentation of the areola deepens. The sebaceous glands at the base of the nipple and on the areola become more obvious. The skin covering the nipple becomes thin and may be extremely sensitive.

Even though a milklike substance (colostrum) can be squeezed from the nipples about the fifth month of pregnancy, the real production of milk does not commence until three or four days after the baby is born. See also **Colostrum.** Following birth and before the milk secretion is apparent, the breasts become more distended and tender. They may be hard and swollen, and tenderness is usually more severe in that part of the breast nearest the armpit.

Human milk is a bluish-white or slightly yellowish fluid with a characteristic odor and a rather sweetish taste. It is approximately seven parts water and one part solids and is an emulsion of fat, suspended in a solution of protein, carbohydrates, and inorganic salts. The essential food elements are present in sufficient amounts to make milk the most satisfactory food for the infant. Except for vitamins B and D, human milk also contains adequate vitamins and inorganic salts for the growing infant. Antibodies to infection are also found in breast milk. Certain drugs taken by the mother may pass into the milk, thus affecting the nursing child. Drugs which may be transmitted in this manner include iron, arsenic, lead, quinine, alcohol, and opium and its derivatives.

The breast, following the change of life, becomes quite different in appearance. Although it may retain its size (because of added fat deposits), the amount of glandular tissue diminishes, and the fibrous tissue gradually becomes more dense.

Rarely an individual will have more than the normal number of breasts. This is known as polymastia. The condition appears about twice as frequently among men as in women. These extra breasts appear more frequently on the left side of the body than on the right, and more such breasts are found below the normal breasts, than above them. They are usually in line, along the so-called "mammary line." Extremely rare is the absence of one or both breasts (amastia).

Inflammation of the breast is called mastitis. See also **Mastitis.** Persons whose health is otherwise quite normal may develop cystic nodules in their breasts. They seldom have a history of discomfort or abnormalities connected with their menstrual periods or with childbirth. The cysts associated with the disease are occasionally discovered during pregnancy, but usually appear at or near the menopause. There may be only one cyst or several. Treatment of persons with chronic cystic mastitis consists of surgical procedures or endocrine therapy. As this disease may be related to a later development of malignant conditions, careful diagnosis and continued observation are necessary.

A benign tumor is an abnormal new growth of tissue that does not spread to other body areas. Fibroadenoma is the most common benign tumor of the breast found in young females (21 to 25 years of age). They grow rapidly during pregnancy. Fibroadenomas, like other breast nodules, can be diagnosed with certainty only after sugical removal of all or some of the tissue for microscopic examination. Another tumor is intraductal papillary hyperplasia and is recognized by the discharge of blood or blood-tinged fluid from the nipple when the breast is compressed. The growths occur most often in women between the ages of 35 and 55 years. Malignant conditions of the breast are described under **Cancer and Oncology.**

Abnormal enlargement of the breasts is called hypertrophy. This condition is less common in the United States than in the tropics. It may occur in males or females, and both breasts usually are enlarged, but generally are not painful. The four most common types are: (1) infantile hypertrophy, which occurs in girls before the age of puberty; (2) gynecomastia, which occurs in males, most often at the time of adolescence; (3) virginal hypertrophy, which occurs in young females during adolescence; and (4) gravid hypertrophy, which appears during pregnancy or lactation. These conditions are treated in various ways, ranging from chemotherapy to surgical procedures. Aside from discomfort, no problems may arise in some instances.

As is so well publicized by various public organizations, frequent self-examination of the breasts is encouraged in the interest of very early detection of swelling, lumps, and abnormalities of contour and symmetry of the two breasts, any of which may indicate conditions that require immediate professional medical attention.

BREAST CANCER. **Cancer and Oncology.**

BREATHING. **Artificial Respiration; Respiratory System.**

BRECCIA. A rock formed of angular fragments in a matrix which may be of similar or different material.

Fault breccias result from the grinding action of the two fault blocks as they slide past each other. Subsequent cementation of these broken fragments may occur by means of mineral matter introduced by the ground water. Talus slopes may become buried and the talus cemented in a similar manner.

Volcanic breccias result from the cementation of fragments that have been broken by volcanic action. Sometimes the surface of a lava flow will harden while the interior will be yet liquid; the fracturing of this surface material and its subsequent cementation by the uncooled lava produces a flow breccia.

The intrusion of plutonic rocks will often shatter the invaded country rock, forming a shatter breccia. In the case of plutonic rocks partly cooled and subsequently broken by further invasions of the magma, intrusive breccias are formed.

BREECH PRESENTATION. Cesarean Section.

BREED. A type of animal produced within the species by artificial selection, distinguished by definite hereditary characteristics but usually capable of interbreeding freely with other members of the species and so maintained only through artificial control of its propagation. Thus Guernsey and Angus cattle are breeds which would soon cease to exist in a state of nature, while the Michigan beaver and the Pacific beaver are self-maintaining in nature and are called subspecies.

BREEDER REACTOR. Nuclear Reactor.

BREEDING. Heredity; Inbreeding; Plant Breeding; Poultry.

BREEZE. Winds and Air Movement.

BREMSSTRAHLUNG. A German word, meaning literally "braking radiation," which denotes the process of producing electromagnetic radiation (or the radiation itself), by the acceleration of a fast charged particle, usually an electron. A commonly occurring form of such acceleration results from deflection by another charged particle, such as a nucleus. During the bremsstrahlung-producing process the electron can give up any amount of energy ranging from near zero to its maximum kinetic energy. The resulting radiation has a continuous spectrum, as exemplified by the continuous x-ray spectra from an ordinary x-ray tube. *Outer* (or *external*) *bremsstrahlung* is a term applied in cases where the radiation is formed (the electron decelerated) in matter foreign to the source of the electron. *Inner* (or *internal*) *bremsstrahlung* is a term applied to comparatively infrequent processes occurring in beta decay, in which the bremsstrahlung is formed because of acceleration of the electron (beta particle) in the same atom in which it was formed. The abrupt change in the electric field in the region of the nucleus of the atom undergoing disintegration sometimes results in the production of a photon, in a manner similar to the emission of a photon in the ordinary (outer) bremsstrahlung process. In both negatron and positron emission the photon energy is obtained at the expense of the electron-neutron pair, and the spectral distribution decreases continuously with increasing energy of the beta particles. In electron capture, the photon energy is obtained at the expense of the neutrino, and the spectral distribution is greatest at about one-third of the normal neutrino energy, reaching zero at zero energy and at the normal neutrino energy.

In the operation of the highest energy electron synchrotrons and betatrons the acceleration required to maintain the electrons in their circular orbits is sufficiently large to produce visible bremsstrahlung.

See **Particle (Subatomic).**

BREWER'S BLACKBIRD. Blackbird.

BREWER'S SPRUCE. Spruce Trees.

BREWING (Defoaming Agents in). Defoaming Agents.

BREWSTER ANGLE. The Brewster angle, or polarizing angle, of a dielectric is that angle of incidence for which a wave polarized parallel to the plane of incidence is wholly transmitted (no reflection). An unpolarized wave incident of this angle is therefore resolved into a transmitted partly-polarized component and a reflected perpendicularly-polarized component.

BREWSTER LAW. In 1815 Sir David Brewster discovered that for any dielectric reflector there is a simple relationship between the polarizing angle for the reflected light of a particular wavelength and the refractive index of the substance for the same wavelength. The relationship is that the tangent of the polarizing angle is equal to the refractive index. For example, if the refractive index of flint glass for sodium light is 1.66 the polarizing angle for the reflection of sodium light by this glass is 50° 56′.

The law may be used to determine the refractive index of a solid which is opaque or obtainable only in a small piece, since only one small reflecting surface is required.

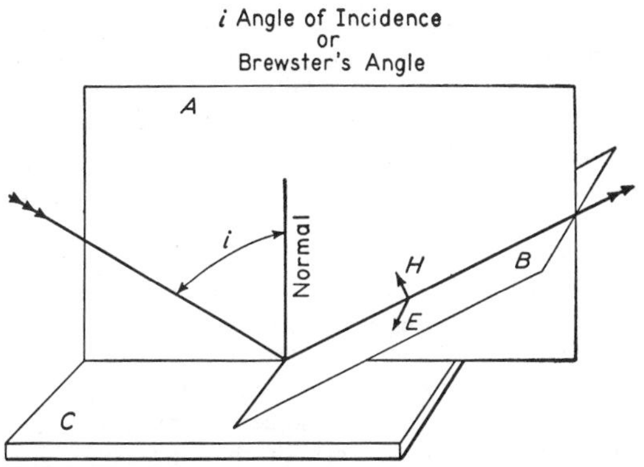

A, incident plane (plane of polarization or plane of magnetic vector, after reflection); *B*, plane of vibration (plane of electric vector, after reflection); *C*, reflecting surface (dielectric).

BRICK. Brick ordinarily refers to a rectangular prism of clay or shale which has been burned in a kiln. Clay is no longer the only material for brick manufacture, being supplemented by slag, cement and lime. However, when other than the ordinary structural clay brick is meant, a descriptive term such as fire-brick and sand-lime brick is employed. The principal classifications of brick are for structural purposes in buildings, for paving, and for lining furnaces, the latter known as refractory brick, or fire-brick. Ordinary bricks are made from a selected clay soil first by preparation of the clay by grinding and thoroughly mixing with enough water to make the mud. The bricks are then formed in the required shapes by one of several methods. In the soft mud method the prepared clay is quite plastic and the bricks are molded to shape by hand or machine. This is the principal method for making bricks by hand. The commercial manufacture of bricks is more frequently by the stiff mud process, in which the mud is less plastic, and is extruded through a die by pressure and wire-cut to the proper size. Brick made from clay that is hardly more than dampened must be formed in molds by application of a great deal of pressure. As hydraulic presses are frequently used, these dry-pressed bricks are sometimes referred to as hydraulic-pressed brick. This process gives a dense surface to the bricks which makes them suitable for facing work.

After the bricks have been molded they are air dried and piled in the kiln for burning. It is quite difficult in any other than the continuously fired kilns, in which the bricks move slowly through the kilns on conveyors, to obtain uniform characteristics in all the brick, and so the product of the ordinary brick plant consists of various grades, ranging from hard brick to softer bricks, such as salmon brick. Hard-burned brick should be used for face work exposed to the weather, and soft brick for filling, for foundations, and the like. The standard brick measures approximately $2\frac{1}{4} \times 4 \times 8$ inches ($5.7 \times 10.2 \times 20.3$ centimeters), and has a crushing strength of between 1,000 and 3,000 pounds per square inch (68 and 204 atmospheres), depending on the quality. A highly impervious and ornamental surface may be laid on brick either by salt glazing, in which salt is added during the burning process, or by the use of a "slip," which is a glaze material into which the bricks are dipped. Subsequent reheating in the kiln fuses the slip into a glazed surface integral with the brick base.

A refractory brick is built primarily to withstand temperature. Good resistance to heat flow is not to be secured simultaneously with refractoriness. Indeed, the most refractory bricks usually have the highest thermal conductivities. It is important for the refractory brick to have high resistance to erosion by ash-laden gases and to the fluxing action of molten slag. It should not spall badly under rapid temperature changes, and its structural strength should hold up well under rapid temperature changes. Fire clay bricks are made from certain clays, including a plastic clay which binds the others into brick form. The firing in the kiln is carried out at a temperature such that the brick is partly vitrified. For special purposes they may be glazed by one of the methods previously described. The fire clay brick contains 30–40% alumina and about 50% silica. Progress in the art of combustion of fuels in furnaces has advanced the service requirements of refractory brick, sometimes to the point where they are so severe that a refractory superior to fire clay is needed. High alumina bricks containing 50–

80% alumina, and correspondingly less silica, and silicon carbide, a product of the electric furnace, are typical of these super-refractories. Of course fire clay bricks are preferred wherever they give satisfactory service because they are lowest in cost of all the refractory bricks. The standard size of fire brick is $9 \times 4\frac{1}{2} \times 2\frac{1}{2}$ inches ($22.9 \times 11.4 \times 6.4$ centimeters).

BRICKWORK. When laid, bricks are bedded into a mortar which, hardening, bonds the separate bricks into a brickwork unit. A solid brick wall of more than one layer thickness has the different layers of brick bonded into each other by the use of headers, that is, brick laid perpendicular to the face of the wall. There are different systems of bonding, each of which gives a somewhat different appearance to the wall. In the common bond every fourth or fifth course is composed entirely of headers. In the English bond, every other course is a header course, while in the Flemish bond headers and stretchers alternate in each course. See **Masonry.**

The strength and durability of brickwork depend on the quality of mortar and excellence of workmanship with which the brickwork is laid. The proportions of the mortar are from one to three parts of dry sand to one part of Portland cement, depending on the strength needed. The cement mortar is much stronger than lime mortar, but the addition of a small amount of lime (see **Calcium**) to cement mortar renders it more readily worked without materially impairing its strength. In estimating brickwork, one rule is to allow 1,000 standard bricks, and $\frac{1}{2}$ cubic yard of mortar for each 2 cubic yards of brickwork in place. Some masons estimate number of bricks by assigning 7 to each superficial square foot of area of wall 1 brick thick. Brickwork varies in weight from 1.5 to 1.9 tons per cubic yard, depending on the density of the bricks used. The maximum crushing strength to which brickwork should be subjected is 170 pounds per square inch when set in cement mortar, although this may be increased to 250 pounds if the effects of eccentric loading and lateral forces are fully analyzed.

BRIDGE AMPLIFIER. A commercially available extensively used amplifier for instrumentation purposes. The commercial configuration generally is a direct-coupled amplifier, offering reasonably wide bandwidths up to 50 kHz at gains ranging from near unity to 1000. The use of four subamplifiers in a bridge-amplifier configuration is shown in the accompanying diagram. The output voltage, assuming that the open-loop gains G_1, G_2, G_3, and G_4 of the separate amplifiers are quite large, is given by

$$V_0 = \frac{R_1 + R_2}{R_1} V_1 - \frac{R_2}{R_2'} \times \frac{R_1' R_2'}{R_1} \times V_2$$

Voltage V_1 is the sum of the differential voltage $V_{\text{signal}} = V_1 - V_2$ and the common-mode voltage. Voltage V_2 is the applied common-mode voltage V_{cm}. Substituting these factors in the foregoing expression, the output voltage is given by

$$V_0 = \frac{R_1 + R_2}{R_1} V_{\text{signal}} + \left(1 - \frac{R_2 R_1'}{R_2' R_1}\right) V_{\text{cm}}$$

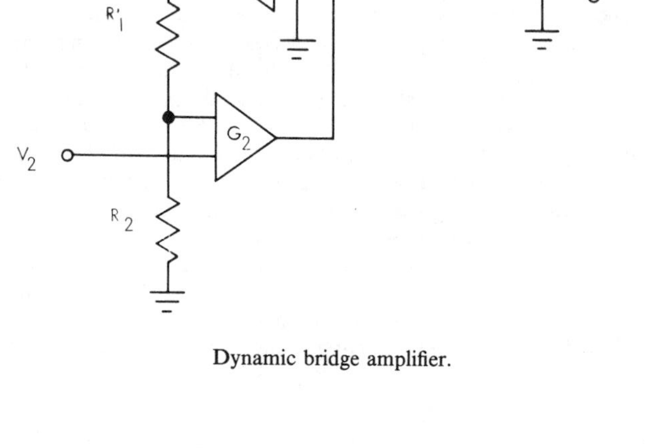

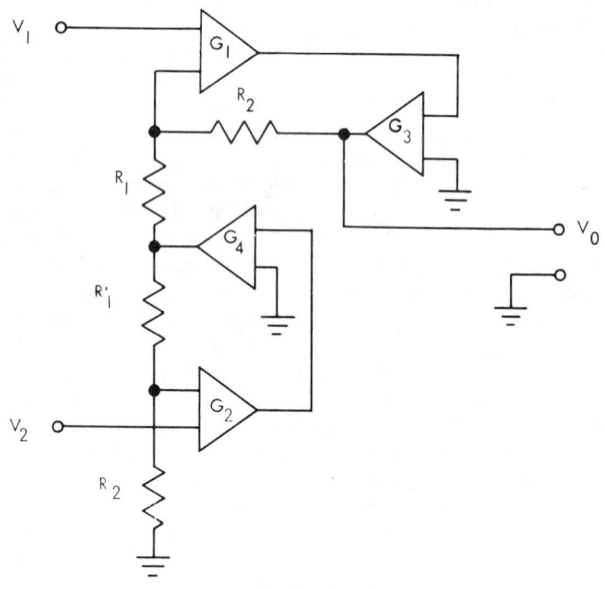

Dynamic bridge amplifier.

The closed-loop gain of the amplifier thus is $(R_1 + R_2)/R_1$. The common-mode rejection ratio is $|G/[1 - (R_2 R_1')/(R_1 R_2')]|$. If $R_1'/R_1 = R_2'/R_2$, the condition for a balanced resistive bridge, theoretically infinite common-mode rejection can be obtained. This analysis does not bring out the practical limitations of matching resistors and of other errors. Thus, the common-mode performance is finite. However, values in excess of 120 dB can be achieved. The common-mode rejection ratio of this type of amplifier is directly proportional to gain. For most differential amplifiers, the common-mode rejection ratio is largely independent of the gain.

See also **Amplifier; Analog Input;** and terms listed under **Data Processing.**

Thomas J. Harrison, International Business Machines Corporation, Boca Raton, Florida.

BRIDGE CIRCUITS (Electrical). The Wheatstone bridge circuit was discovered by S. H. Christie in 1833. Although the concept of resistance had not been formulated at that time, a short time later, in 1843, Sir Charles Wheatstone used Christie's circuit to measure resistance. Since that time, this circuit and its many modifications have been used in making precise measurements of electrical quantities. Bridge circuits are used to compare impedance elements by comparing the voltages or currents associated with them. The voltages and currents can be compared by using known impedance ratios to make potentiometers and current comparators. In essence, a bridge circuit is an arrangement of dividers for comparing the equivalent circuits of impedance elements. Therefore, bridges also can be used for the comparison of divider ratios.

Originally, *bridge* referred only to the detector connection between the divider tap points. In current practice, however, the bridge circuit includes the dividers, the generators, and the null detectors needed for furnishing power and for finding the bridge balance.

Impedance is the ratio of voltage to current at a single frequency. In a series circuit with the same current in all elements, the voltage will be proportional to the impedance. In a parallel circuit with the same voltage across all elements, the current will be inversely proportional to the impedance. These relationships make it possible to connect impedance elements in bridge circuits so that their imedances can be compared by comparing their voltage or current ratios.

Impedance elements most commonly are visualized as having only two terminals. However, terminals have stray impedances associated with them. In practice, there may be varying amounts of extension wire and contact impedances connected in series between the impedance element and its connection point in the circuit. Leakage impedances also occur from terminals to surrounding conductors. Such stray impedances frequently make it desirable to use 3- and 4-terminal impedance elements. Bridges are available for measuring the resulting 3- and 4-terminal impedance values.

A *Wheatstone bridge* suitable for both ac and dc resistance measurements is shown in Fig. 1. The bridge measures two-terminal resistance.

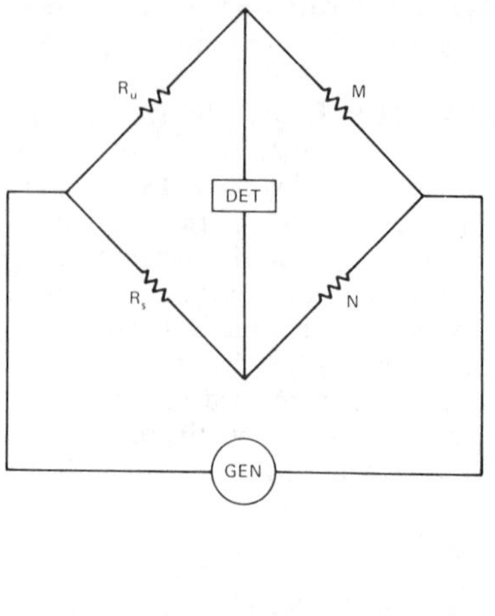

Fig. 1. Wheatstone bridge.

The balance equation is

$$R_u = R_s \left(\frac{M}{N}\right)$$

A *Kelvin bridge* for measuring four-terminal resistors is shown in Fig. 2. A bridge of this type is used for low-value resistors or precision measurements. Lead resistances of the unknown R_U, and standard R_S, resistors are included in M′, N′, Y, and the generator leads. The remaining lead resistances are L_U and L_S. When $R_U{:}R_S = M{:}N = M'{:}N' = L_U{:}L_S$, then $R_U = R_S(M/N)$. The balance equation for this bridge is

$$R_u = R_s\frac{M}{N} + \frac{N'Y}{M'+N'+Y}\left[\left(\frac{M}{N}-\frac{M'}{N'}\right) + \frac{L_u}{N}\left(\frac{M'}{N'}-\frac{L_s}{L_u}\right)\right] + L_u\frac{R_u}{N}\left(\frac{R_s}{R_u}-\frac{L_s}{L_u}\right)$$

Error term

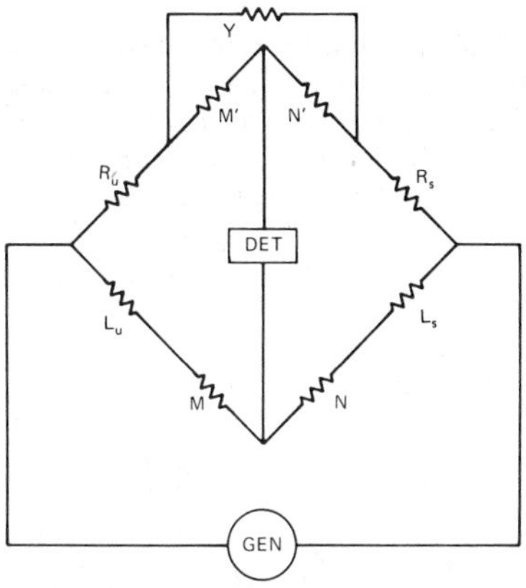

Fig. 2. Kelvin bridge.

A *Mueller bridge* for use with resistance thermometers is shown in Fig. 3. This bridge measures 4-terminal resistors by averaging two readings with lead resistance effects reversed. It should be observed that $R_U = R_S(\text{Avg})$. The balance equation is

$$R_u = \frac{R_{s(1)} + R_{s(2)}}{2}$$

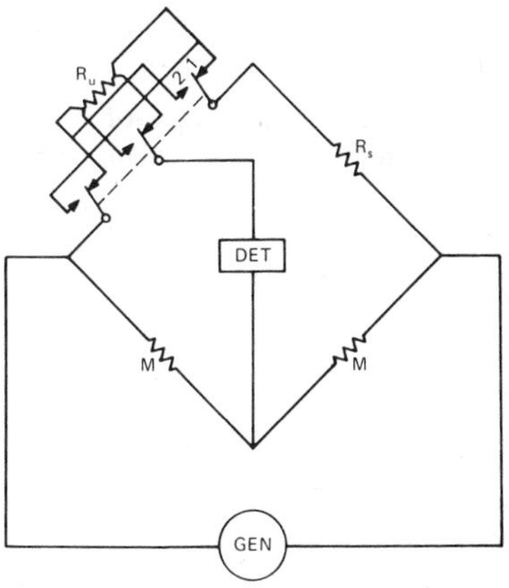

Fig. 3. Mueller bridge.

A *Smith III bridge* also for use with resistance thermometers is shown in Fig. 4. This bridge also measures 4-terminal resistors. The balance equation is

$$R_u = R_s\frac{M}{N}$$

A *Murray bridge* for use in fault location is shown in Fig. 5. This bridge measures the distance from the near end of a line to a line fault. The balance equation is

$$A = \frac{N}{M+N}$$

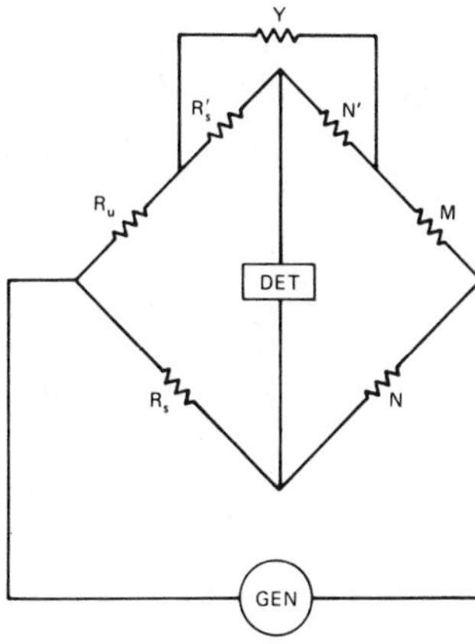

Fig. 4. Smith III bridge.

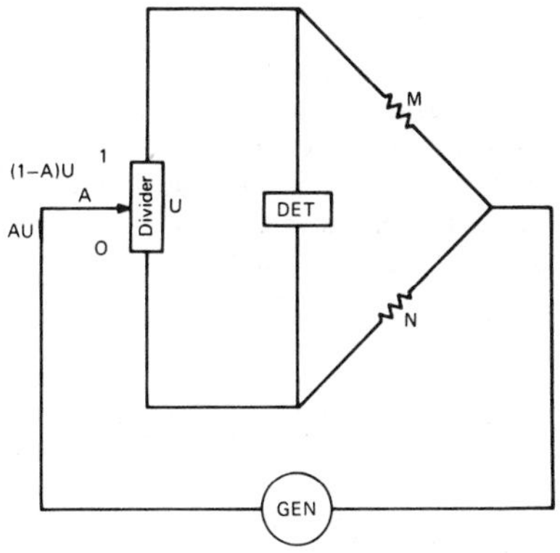

Fig. 5. Murray bridge.

The configuration of Fig. 6 measures the distance from the near end of a line to a line fault. The distance is found in terms of the line length L and the ratio M/N. The balance equation is

$$AL = \frac{2LN}{M+N}$$

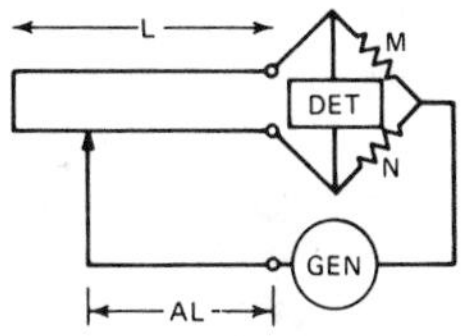

Fig. 6. Special configuration of Murray bridge.

A *Varley bridge* for use in fault location is shown in Fig. 7. The balance equation is

$$A = \frac{R}{U}\frac{M}{M+N} + \frac{MU - WN}{(M+N)U}$$

For $M/N = W/U$, $A = R/U \times M/(M+N)$; and $R = AU \times (M+N)/M$.

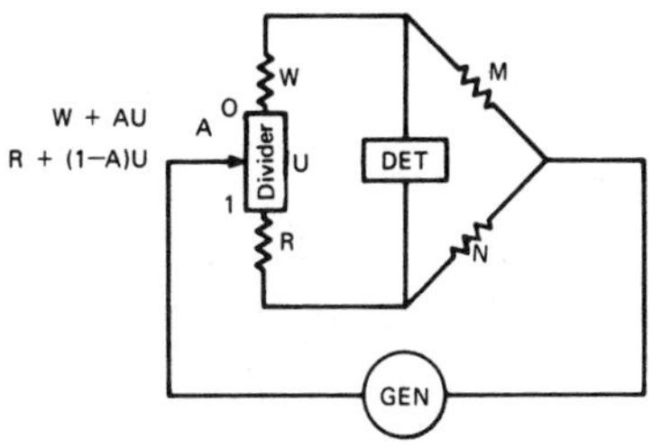

Fig. 7. Varley bridge.

In the configuration of Fig. 8, the bridge measures the distance from the far end of a line to a line fault. The distance is found in terms of the line length L, the line resistance U, the ratio M/N, and the resistance R. The balance equation is

$$AL = \frac{RLM}{(M+N)U}$$

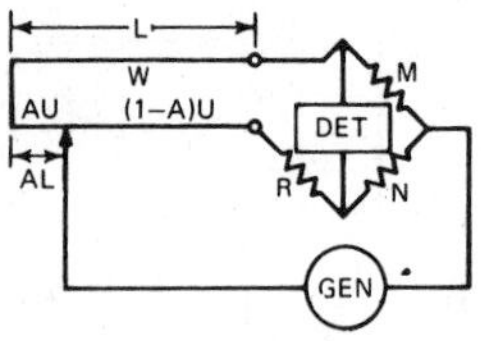

Fig. 8. Distance-measuring Varley bridge.

In the configuration of Fig. 9, the bridge measures the resistance R in terms of the divider setting A and the ratio M/N, after the equality M:N = W:U is established.

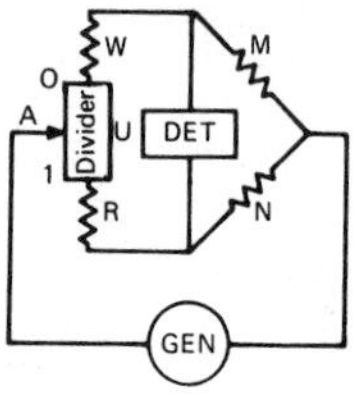

Fig. 9. Special configuration of Varley fault-locating bridge.

A general-purpose, equivalent-circuit bridge for series capacitance is shown in Fig. 10. The bridge measures unknown impedance in terms of equivalent circuit. Equivalent circuit is a capacitor and resistor in series. The bridge is used for capacitors which have a low dissipation factor. The balance equations are

$$C_u = C_s \frac{N}{M} \qquad R_u = R_d \frac{N}{M}$$

$$D_u = D_s = R_d \omega C_s$$

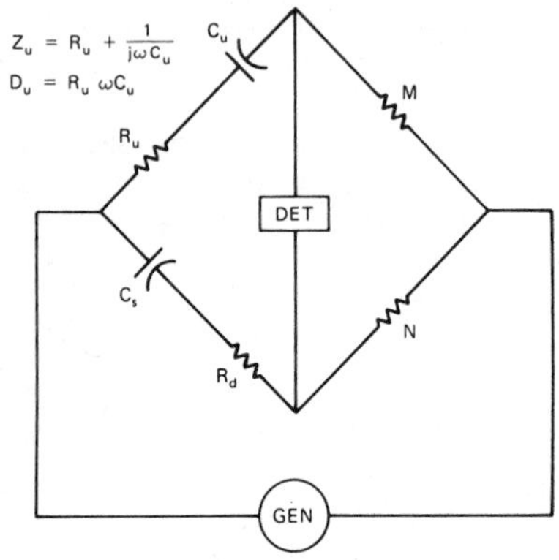

Fig. 10. General-purpose, equivalent-circuit bridge for use with capacitors which have a low dissipation factor.

The bridge shown in Fig. 11 measures unknown impedance in terms of equivalent circuit. Equivalent circuit is a capacitor and resistor in series. The balance equations are

$$C_u = C_s \frac{A}{R}(1 + D)$$

$$D_u = \omega C_s R_s$$

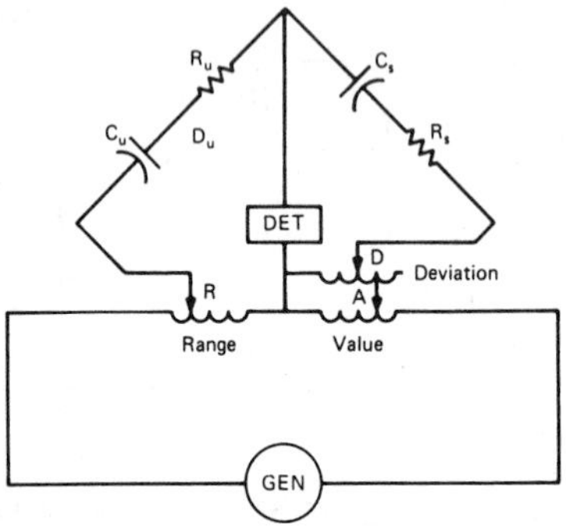

Fig. 11. Series capacitance, general-purpose, equivalent-circuit bridge.

A *Schering bridge* is shown in Fig. 12. This bridge measures unknown impedance in terms of equivalent circuit. Equivalent circuit is a capacitor and resistor in series. The bridge is used for precision capacitance measurement and high-voltage insulation leakage measurement. The balance equations are

$$C_u = R_s \frac{C_n}{M} \qquad R_u = C_s \frac{M}{C_n}$$

$$D = \omega R_s C_s$$

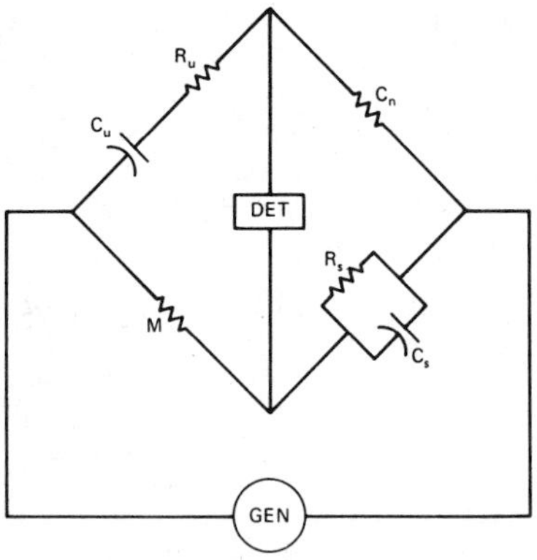

Fig. 12. Schering bridge.

A general-purpose, equivalent-circuit bridge for parallel capacitance measurements is shown in Fig. 13. This bridge measures unknown impedance in terms of equivalent circuit. Equivalent circuit is a capacitor and resistor in parallel. The bridge is used for capacitors which have a high dissipation factor. The balance equations are

$$C_u = C_s \frac{N}{M} \qquad R_u = R_Q \frac{N}{M}$$

$$D_u = \frac{1}{Q_u} = \frac{1}{R_Q \omega C_s}$$

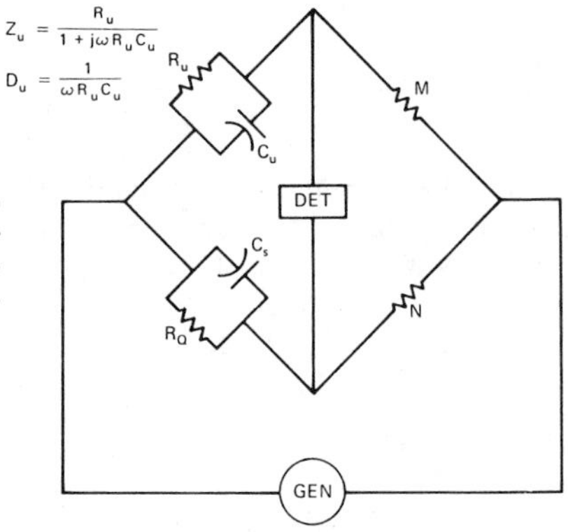

Fig. 13. Parallel capacitance, general-purpose, equivalent-circuit bridge for use with capacitors which have a high dissipation factor.

Another general-purpose, equivalent-circuit bridge for parallel capacitance measurements is shown in Fig. 14. The balance equations are

$$C_u = \frac{1}{B}(A_1 C_{s1} + A_2 C_{s2} + \cdots)$$

$$\frac{1}{R_u} = \frac{1A_R}{R_s B}$$

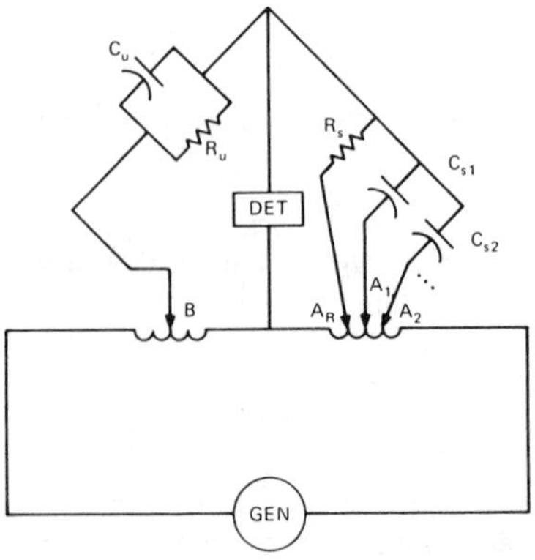

Fig. 14. Parallel capacitance, general-purpose, equivalent-circuit bridge.

A *Maxwell commutator bridge* is shown in Fig. 15. This bridge measures unknown capacitance and is used for precision measurement of capacitors with low dissipation factor. The balance equation is

$$C_u = \frac{N}{fR_s M}$$

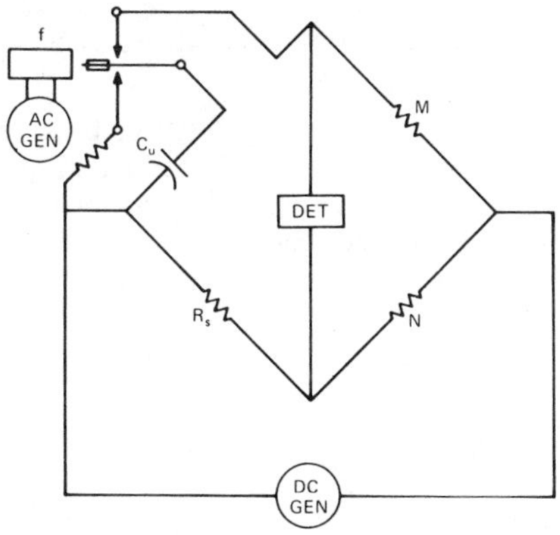

Fig. 15. Maxwell commutator bridge.

A *parallel-T bridge* is shown in Fig. 16. This bridge measures an unknown frequency or low dissipation factor capacitor. It is used at higher frequencies. The balance equations are

$$\omega^2 C_1 C_2 = \frac{2}{R_2^2} \quad \text{for } R_2 = 2R_1 \text{ and } C_2 = 2C_1$$

$$\omega^2 C_1^2 = \frac{1}{2R_1 R_2} \qquad f = \frac{1}{2\pi R_1 C_2} = \frac{1}{2\pi R_2 C_1}$$

$$C_2 R_2 = 4C_1 R_1$$

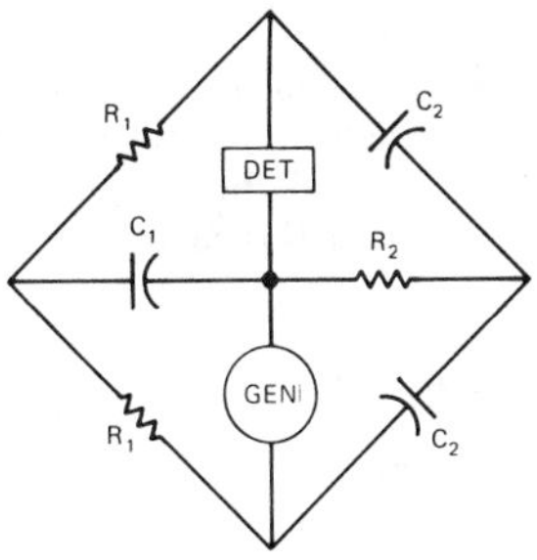

Fig. 16. Parallel-T bridge.

A *Wien bridge*, shown in Fig. 17, measures unknown frequency in terms of resistors and capacitors. The bridge is frequently used as the frequency-determining element of oscillators. Prior to development of the Schering bridge, this bridge was used for capacitance measurement. The balance equations are

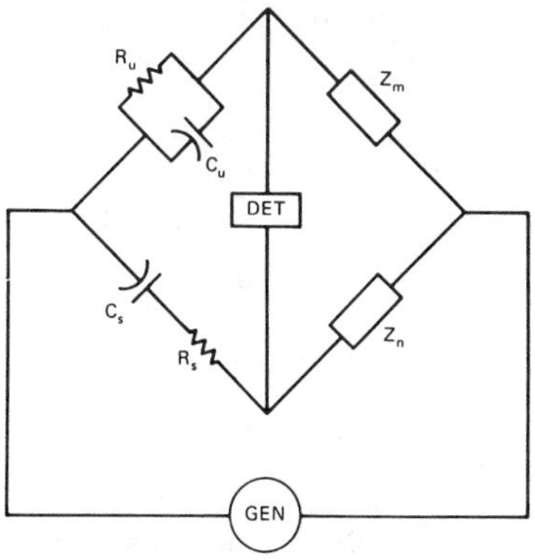

Fig. 17. Wien bridge.

$$\frac{C_u}{C_s} + \frac{R_s}{R_u} = \frac{Z_N}{Z_M} \qquad C_u = C_s \frac{Z_N}{Z_M}\left(\frac{1}{1+\omega^2 C_s^2 R_s^2}\right)$$

$$C_s = C_u \frac{Z_M}{Z_N}\left(1 + \frac{1}{\omega^2 C_u^2 R_u^2}\right)$$

$$\omega^2 = \frac{1}{R_u C_u R_s C_s} \qquad R_u = R_s \frac{Z_M}{Z_N}\left(1 + \frac{1}{\omega^2 C_s^2 R_s^2}\right)$$

$$R_s = R_u \frac{Z_N}{Z_M}\left(\frac{1}{1+\omega^2 C_u^2 R_u^2}\right) \qquad f = \frac{1}{2\pi\sqrt{R_u C_u R_s C_s}}$$

A bridge of the *Carey-Foster* and *Heydweiler* types is shown in Fig. 18. The Carey-Foster bridge measures unknown impedance in terms of equivalent circuit. Equivalent circuit is a capacitor and resistor in series. The Heydweiler bridge measures mutual inductance. The balance equations are

$$M_u = C_s R_u N \qquad C_s = \frac{M_u}{R_u N}$$

$$L_u = C_s R_u (R_s + N) \qquad R_s = \frac{N(L_u - M_u)}{M_u}$$

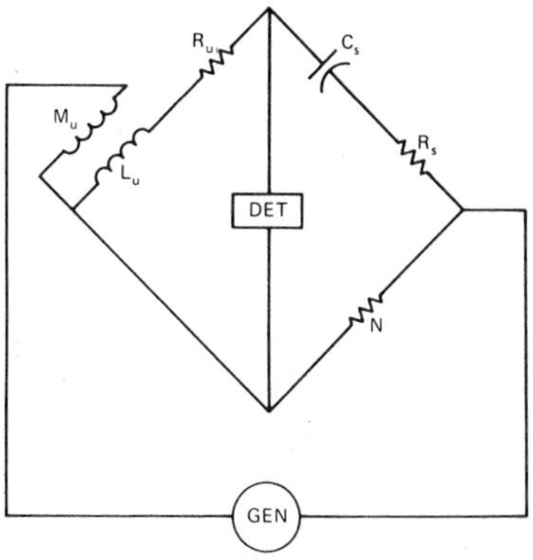

Fig. 18. General configuration of Carey-Foster; and Heydweiler bridges.

A *resonance bridge*, shown in Fig. 19, is used to measure unknown impedance in terms of equivalent circuit. Equivalent circuit is an inductor and a resistor in series. The balance equations are

$$L_u = \frac{1}{\omega^2 C_s}$$

$$R_u = R_s \frac{M}{N}$$

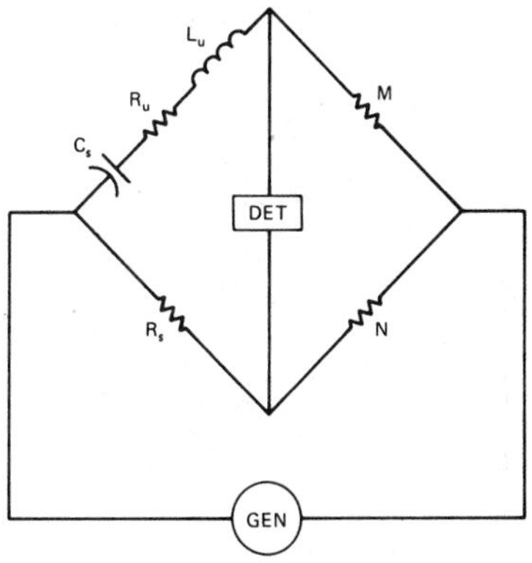

Fig. 19. Resonance bridge.

An *Owen bridge*, shown in Fig. 20, measures unknown impedance

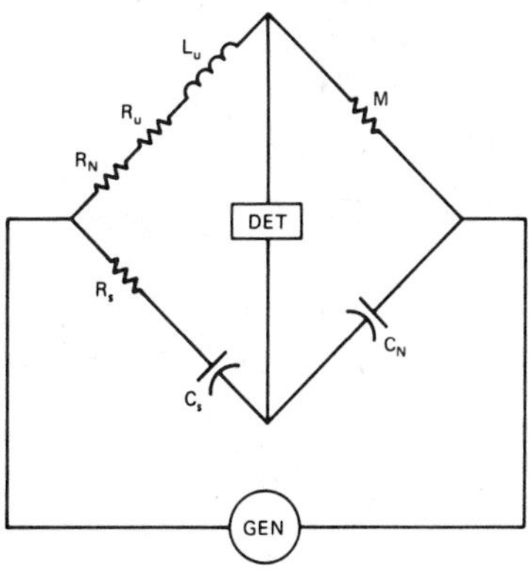

Fig. 20. Owen bridge.

in terms of equivalent circuit. Equivalent circuit is an inductor and a resistor in series. The balance equations are

$$L_u = R_s C_N M$$

$$R_u = \frac{C_N M}{C_s} - R_N$$

The *Anderson, Stroud and Oates bridges* are somewhat similar and the general configuration is shown in Fig. 21. The Anderson bridge is used for precision inductance measurement. If the generator and detector are interchanged, this becomes the Stroud and Oates bridge. The balance equations are

$$L_u = C_s M \left[R_s \left(1 + \frac{N}{R_Q} \right) + N \right] \qquad R_u = \frac{MN}{R_Q}$$

$$Q_u = \omega C_s \left[\frac{R_Q R_s}{N} + R_Q + R_s \right]$$

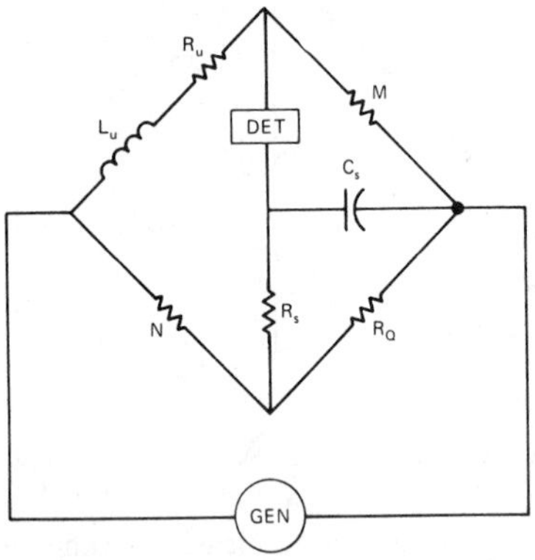

Fig. 21. General configuration of Stroud and Oates; and Anderson bridges.

A *Maxwell bridge* for measuring unknown impedance in terms of equivalent circuit is shown in Fig. 22. Equivalent circuit is an inductor and a resistor in series. This bridge is used for inductors with a low quality factor. The balance equations are

$$L_u = MNC_s \qquad R_u = \frac{MN}{R_Q}$$

$$Q_u = R_Q \omega C_s$$

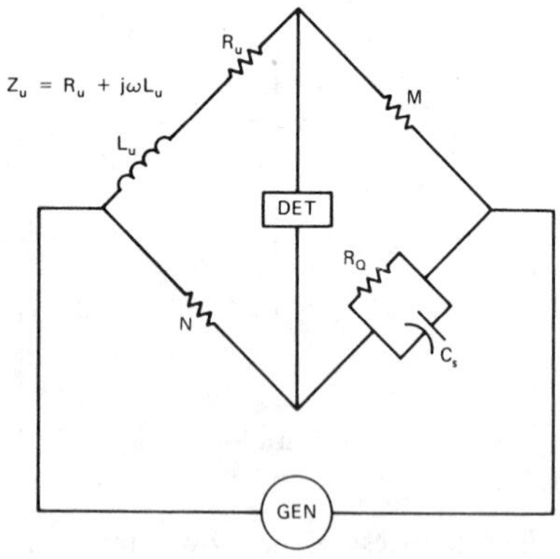

Fig. 22. Maxwell bridge.

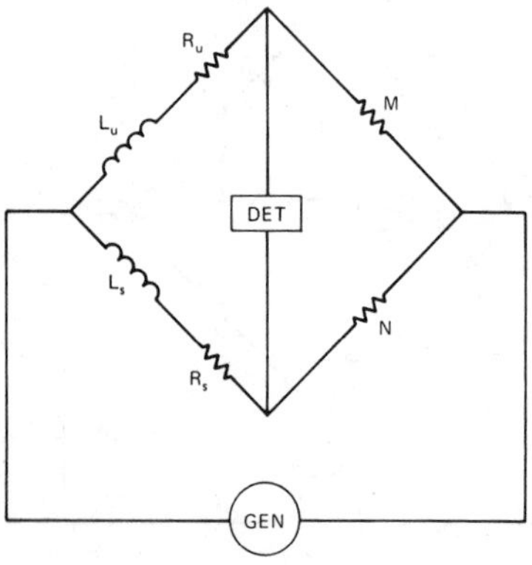

Fig. 23. Inductance comparison bridge.

An inductance-comparison bridge for use in high-frequency measurements is shown in Fig. 23. The equations of balance are

$$L_u = L_s \frac{M}{N}$$

$$R_u = R_s \frac{M}{N} \qquad Q_u = \frac{\omega L_s}{R_s}$$

A *bridged-T bridge* is shown in Fig. 24. This bridge measures unknown impedance in terms of an equivalent circuit. Equivalent circuit is an inductor and resistor in parallel. The circuit is very frequency sensitive. The bridge is used at higher frequencies. The balance equations are

$$L_u = \frac{1}{2\omega^2 C_s} \qquad D = \omega R_s C_s$$

$$R_u = \frac{1}{\omega^2 C_s^2 R_s}$$

Fig. 24. Bridged-T bridge.

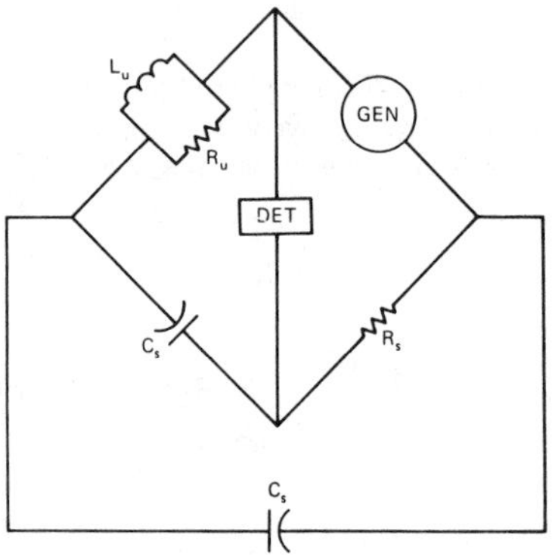

The *Hay bridge*, shown in Fig. 25, measures unknown impedance in terms of equivalent circuit as in the case of the foregoing bridge. This bridge is used for inductors with a high quality factor. The balance equations are

$$L_u = C_s MN \qquad R_u = \frac{MN}{R_D}$$

$$Q_u = \frac{1}{R_D \omega C_s}$$

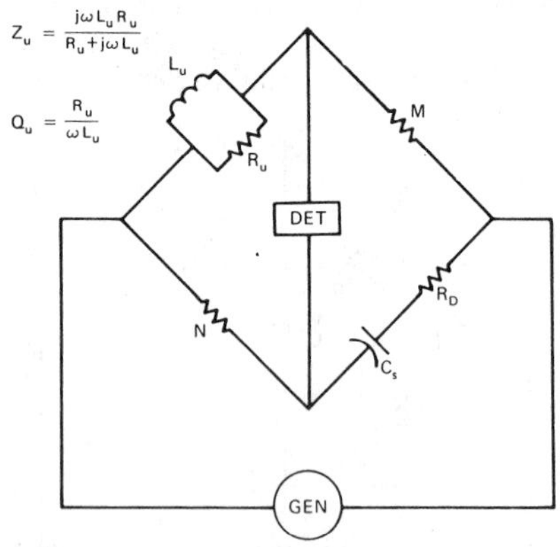

Fig. 25. Hay bridge.

The *Campbell bridge*, shown in Fig. 26, measures unknown impedance in terms of equivalent circuit. Equivalent circuit is a mutual inductor with a resistor and inductor in series on the primary side. With the switches in position 1, balance for R_U and L_U; in position 2, balance for M_U. The balance equations are

$$M_u = M_s \frac{M}{N} \qquad L_u = L_s \frac{M}{N}$$

$$R_u = R_s \frac{M}{N}$$

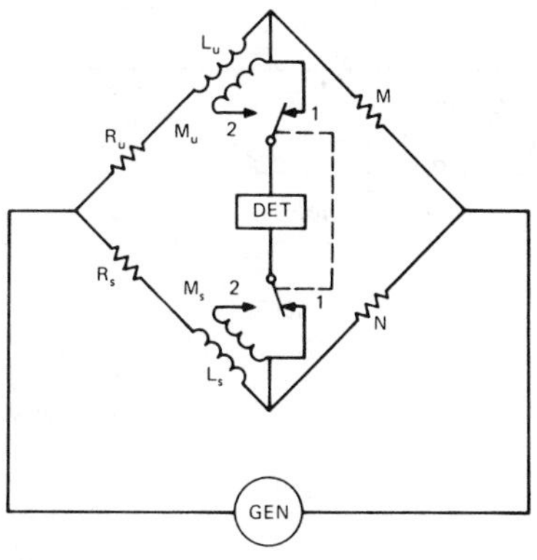

Fig. 26. Campbell bridge.

The *Heaviside bridge*, shown in Fig. 27, measures mutual inductance. The balance equations are

$$M_u = \frac{ML_s - NL_u}{M + N}$$

$$R_u = R_s \frac{M}{N}$$

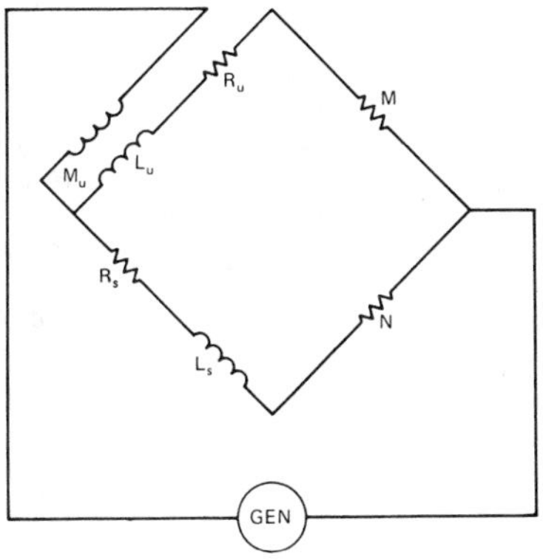

Fig. 27. Heaviside bridge.

A series-opposition bridge for measuring mutual inductance by comparing two equal mutual inductances is shown in Fig. 28. The balance equation is $M_U = M_S$.

A bridge circuit can be developed from a voltage divider. Two equal ratio voltage dividers can be bridged by a detector to compare impedances. With a constant-voltage source, the resulting circuit will maintain a fixed voltage across unknown impedances of various values.

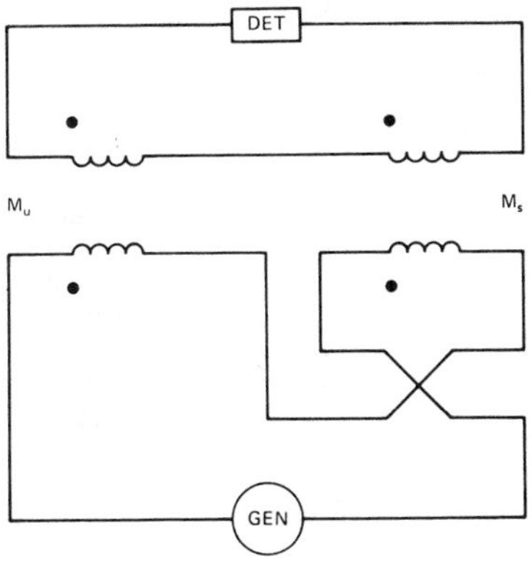

Fig. 28. Series-opposition bridge for measuring mutual inductance.

A transformer bridge can be made by using transformers to supply the bridge voltages in a desired ratio, or to compare bridge currents in a desired ratio, or both. A transformer bridge is shown in Fig. 29. Equations of balance are

$$\frac{I_{unk}}{I_{std}} = \frac{1}{A_D}$$

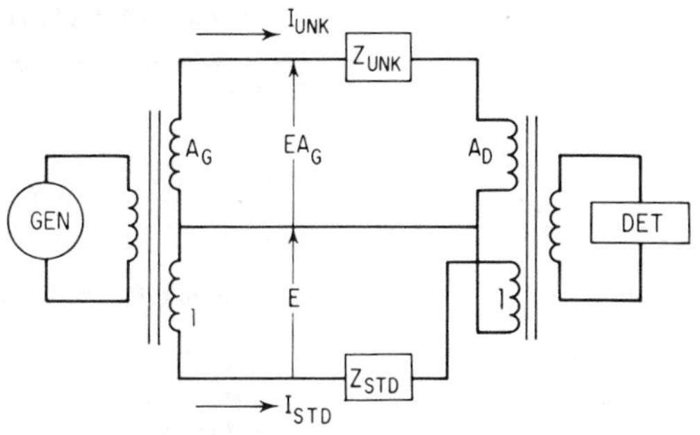

$Z_{UNK} = Z_{STD} A_G A_D$

Fig. 29. Transformer bridge.

$$I_{unk} = \frac{EA_G}{Z_{unk}} \qquad I_{std} = \frac{E}{Z_{std}}$$

$$Z_{unk} = A_G A_D Z_{std}$$

unk = unknown; std = standard

J. C. Riley, Consulting Engineer, Portland, Oregon.

BRIDGED COMPOUNDS. Organic Chemistry.

BRIDGED-T BRIDGE. Bridge Circuits (Electrical).

BRIDGE FRAME. Rigid Frame.

BRIDGE GRAFTING. Grafting and Budding.

BRIDGE (Structural). In civil engineering, a bridge is a structural unit or a series of structural units called spans designed primarily for the purpose of supporting moving loads, in addition to its own weight. The term bridge is generally associated with a structure which provides a means for foot, highway, or railroad traffic to pass over water, ground depressions or congested districts, although certain kinds of traveling cranes used for loading or unloading bulky materials such as ore or coal are sometimes referred to as bridges. All bridges are either stationary or movable. Movable spans are used in connection with low level bridges over navigable waters where these bridges interfere with shipping. See accompanying table.

There are three general types of movable spans, namely, the bascule, the vertical lift, and swing bridge. The bascule bridge pivots about a horizontal axis or rolls back on circular segments. If the entire span rotates about a horizontal axis near one end, it is called a single leaf bascule. A double leaf bascule is one which consists of two cantilevers, each of which rotates about a horizontal axis, forming a single span when closed. When the entire movable section may be lifted vertically, parallel to its original position, the bridge is called a vertical lift span. Swing bridges are those which turn in a horizontal plane about a vertical axis located at the center of the bridge. Movable bridges, when closed, are similar and perform the same service as the stationary types.

Bridges may also be classified as framed truss, beam, or suspension bridges, depending upon the way in which they support the loads. All bridges are either straight or skew. When the end supports are not on lines at right angles to the longitudinal center line of the span, the resulting structure is called a skew bridge.

Bridges are usually constructed of structural steel, reinforced concrete, or prestressed concrete, although wood is sometimes used, particularly for temporary spans. In recent times several bridges have been built of aluminum alloy. Stone or brick are occasionally used for very short spans of the arch type. Reinforced concrete particularly well adapted for use in connection with the beam or arched bridge since it can be molded into any desired form.

The oridnary framed bridge, as illustrated in Fig. 1, is composed of two vertical trusses, a floor system, upon which the roadway or railroad is directly supported, a certain amount of bracing and the end bearings.

The floor system consists of longitudinal beams called stringers

REPRESENTATIVE LARGE BRIDGES IN NORTH AMERICA

Name of Bridge	Location	Longest Span between Supports (feet)	Longest Span between Supports (meters)	Dedicated
STEEL TRUSS				
Gov. Nice Memorial	Potomac River, Maryland	800	244	1940
U.S. 60	Ohio River	800	244	1937
Owensboro, Kentucky	Ohio River	751	229	1940
I-24	Tennessee River, Kentucky	720	219	1975
U.S. 62	Green River, Kentucky	700	213	1938
U.S. 62	Cumberland River, Kentucky	700	213	1952
Jamestown	Jamestown, Rhode Island	640	195	1940
Greenville	Mississippi River, Arkansas	640	195	1940
Memphis	Mississippi River, Arkansas	621	189	1949
U.S. 22	Delaware River, New Jersey	540	165	1972
Mississippi River	Muscatine, Iowa	512	156	1972
Newport	Ohio River, Kentucky	511	156	1898
Millared E. Tydings	Susquehanna River, Maryland	490	149	1963
SIMPLE TRUSS				
Chester	Chester, West Virginia	746	227	1977
Metropolis (railroad)	Ohio River	720	219	1917
Irvin S. Cobb	Ohio River, Paducah, Kentucky	716	218	1929
Tanana River (railroad)	Nenana, Alaska	700	213	1922
MacArthur	St. Louis, Missouri	668	204	1911
Henderson (railroad)	Ohio River	665	203	1933
I-77 Ohio River	Marietta, Ohio	650	198	1967
Louisville, Kentucky	Ohio River	644	196	1919
Atchafalaya	Morgan City, Louisiana	608	185	1933
Castleton (railroad)	Hudson River, New York	598	182	1924
Elizabethtown	Great Miami River	586	179	1906
Louisville (railroad)	Ohio River	546	166	1929
Ohio River	Cincinnati, Ohio	542	165	1889
John F. Kennedy	Red River, North Dakota	279	85	1963
CONTINUOUS TRUSS				
Rocheport, Missouri	Missouri River	2500 (total length)	762	1959
Tennessee-Missouri	Mississippi River	1400	427	1971
Lyons-Fulton	Mississippi River	1340	408	1939
Astoria, Oregon	Columbia River	1232	376	1966
Marquam	Willamette River, Oregon	1044	318	1966
Caruthersville, Missouri	Mississippi River	920	280	1973
Mississippi River	Dyersburg, Tennessee	900	174	1969
Irondequoit Bay	Rochester, New York	891	272	1969
Dubuque, Iowa	Mississippi River	845	257	1943
John E. Mathews	Jacksonville, Florida	810	247	1953
Kingston-Rhinecliff	Hudson River, New York	800	244	1957
Sherman Minton	New Albany, Indiana	800	244	1961
Betsy Ross	Philadelphia, Pennsylvania	729	222	1974
Girard Point	Philadelphia, Pennsylvania	700	213	1975
I-95, Thames River	New London, Connecticut	540	165	1973
Snake River	Central Ferry, Washington	520	158	1970
CONTINUOUS BOX AND PLATE GIRDER				
Rouge River	Detroit, Michigan	1007	307	1967
Snake River	Jackson-Wilson, Wyoming	885	270	1960
Neches River	Orange County, Texas	850	259	1953
San Mateo-Hayward	San Francisco Bay, California	750	229	1967
Pennsylvania-New Jersey Turnpike	Delaware River	682	208	1956
San Diego-Coronado	San Diego Bay, California	660 (two spans)	201	1969
Ship Channel	Houston, Texas	630	192	1973
Poplar Street	Saint Louis, Missouri	600	183	1967
U.S. 64	Savannah, Tennessee	525	160	1977
McDonald-Cartier	Ottawa, Ontario	520	158	1965
Lake Konocanusa	Lincoln County, Montana	500	152	1971
Sitka Harbor	Sitka, Alaska	450	137	1972
I-430	Arkansas River	430	131	1974
I-635, Kansas City	Missouri River-Kansas-Missouri	425	130	1972
Chattanooga, Tennessee	Tennessee River	420	128	1972
Snake River	Clarkston, Washington	420	128	1978
Yukon River	North Slope Road, Alaska	410	125	1975
I-75, Tennessee River	Loudon County, Tennessee	400	122	1972
Sacramento River	Bryte, California	370	113	1971
Franklin Falls	Snoq'lmie Pass, Washington	350	107	1972
Don Pedro Reservoir	Tuolumne County, California	350	107	1971
Cumberland River	Nashville, Tennessee	330	101	1971
Copper River	Chitinia, Alaska	310	94	1971

REPRESENTATIVE LARGE BRIDGES IN NORTH AMERICA (*Continued*)

Name of Bridge	Location	Longest Span between Supports (feet)	(meters)	Dedicated
CONTINUOUS PLATE				
New Chain of Rocks	Mississippi River, Illinois	5411	1649	1965
Great Congress Gateway	Schenectady, New York	1870	570	1973
Mississippi River	La Crescent, Minnesota	450	137	1967
I-480	Missouri River, Iowa-Nebraska	425	130	1966
I-435	Missouri River, Missouri	425	130	1970
I-80	Missouri River, Iowa-Nebraska	425	130	1972
St. Croix River	Hudson, Minnesota	390	119	1971
Lafayette Street	Saint Paul, Minnesota	362	110	1968
Green River	Hendersonville, North Carolina	350	107	1970
Mississippi River	Prairie du Chien, Wisconsin	350	107	1974
Fort Smith	Arkansas River	340	104	1969
Lexington Avenue	Saint Paul, Minnesota	340	104	1964
I-BEAM GIRDER				
U.S. 31E	Rolling Fork River, Kentucky	340	104	1941
U.S. 27	Licking River, Kentucky	316	96	1948
U.S. 31E	Green River, Kentucky	316	96	1947
U.S. 62	Rolling Fork, Kentucky	240	73	1941
Licking River	Owingsville, Kentucky	240	73	1942
Fuller Warren	Jacksonville, Florida	224	68	1954
Freeway	Arkansas River	210	64	1957
STEEL ARCH				
New River Gorge	Fayetteville, West Virginia	1700	518	1977
Bayonne, New Jersey	Kill Van Kull	1652	504	1931
Fremont	Portland, Oregon	1255	383	1972
Port Mann	British Columbia	1200	366	1964
Trois-Rivières	St. Lawrence River, Quebec, Canada	1100	335	1967
Lewiston-Queenston	Niagara Falls, Ontario	1100	335	1962
Glen Canyon	Colorado River	1028	313	1959
Perrine	Twin Falls, Idaho	993	303	1976
Hell Gate	East River, New York City	977	298	1917
Rainbow	Niagara Falls, New York	950	290	1941
I-40, Mississippi River	Memphis, Tennessee (2 spans)	900	274	1972
Lake Quinisgamond	Worcester, Massachusetts	849	259	1970
Charles Braga	Somerset, Massachusetts	840	256	1966
Lincoln Trail	Ohio River, Indiana-Kentucky	825	251	1967
Sherman Minton	Louisville, Kentucky (part of total bridge)	800	244	1961
Henry Hudson	Harlem River, New York	800	244	1936
French King	Connecticut River, Massachusetts	782	238	1938
West End	Pittsburgh, Pennsylvania	778	237	1931
Piscataqua River	New Hampshire-Maine	756	230	1972
SR 156, Tennessee River	South Pittsburgh, Tennessee	750	229	1979
Cold Spring Canyon	Santa Barbara, California	700	213	1963
I-24, Paducah, Kentucky	Ohio River	700	213	1973
CONCRETE ARCH				
Selah Creek (twin)	Selah, Washington	549	167	1971
Cowitz River	Mossyrock, Washington	520	158	1968
Westinghouse	Pittsburgh, Pennsylvania	425	130	1931
Cappelen	Minneapolis, Minnesota	400	122	1923
Jack's Run	Pittsburg, Pennsylvania	400	122	1930
Elwha River	Port Angeles, Washington	380	116	1973
Bixby Creek	Monterey Coast, California	330	101	1931
Arroyo Seco	Pasadena, California	320	98	1953
Mendota	Fort Snelling, Minnesota	304	93	1927
TWIN CONCRETE TRESTLE				
Slidell, Louisiana	Lake Pontchartrain (full length of bridge)	28,547	8701	1963
SUSPENSION				
Verrazano-Narrows	New York, N.Y.	4260	1298	1964
Golden Gate	San Francisco Bay, California	4200	1280	1937
Mackinac	Straits of Mackinac, Michigan	3800	1158	1957
George Washington	Hudson River, New York	3500	1067	1931
Tacoma	Washington State	2800	853	1952
Lions Gate	Burrard Inlet, British Columbia	2778	847	1939
Transby	San Francisco Bay, California	2310 (two spans)	704	1936
Bronx-Whitestone	East River, New York	2300	701	1939
Pierre Laporte	Montreal, Quebec	2190	668	1970
Delaware Memorial (Old)	Wilmington, Delaware	2150	655	1951

REPRESENTATIVE LARGE BRIDGES IN NORTH AMERICA (*Continued*)

Name of Bridge	Location	Longest Span between Supports (feet)	Longest Span between Supports (meters)	Dedicated
Delaware Memorial (New)	Wilmington, Delaware	2150	655	1968
Walt Whitman	Philadelphia, Pennsylvania	2000	610	1957
Ambassador	Detroit-Ontario, Canada	1850	564	1929
Throgs Neck	Long Island Sound, New York	1800	549	1961
Benjamin Franklin	Philadelphia, Pennsylvania	1750	533	1926
Bear Mountain, New York	Hudson River	1632	497	1924
William Preston Lane Memorial	Sandy Point, Maryland	1600	488	1952
Williamsburg	East River, New York City	1600	488	1903
Newport	Narragansett Bay, Rhode Island	1600	488	1969
Brooklyn	East River, New York City	1595	486	1883
Mid-Hudson, New York	Poughkeepsie, New York	1500	457	1930
Vincent Thomas	Los Angeles, Harbor	1500	457	1964
Manhattan	East River, New York City	1470	448	1909
Triborough	East River, New York City	1380	421	1936
St. Johns	Portland, Oregon	1207	368	1931
Mount Hope	Rhode Island	1200	366	1929
Deer Isle	Maine	1080	329	1939
Maysville, Kentucky	Ohio River	1060	323	1931
Cincinnati	Ohio River	1057	322	1867
Dent	Clearwater County, Idaho	1050	320	1971
Miampimi	Mexico	1030	314	1900
Wheeling, West Virginia	Ohio River	1010	308	1849
CANTILEVER				
Quebec (railroad)	Quebec City, Canada	1800	549	1917
Commodore Barry	Chester, Pennsylvania	1644	501	1970
New Orleans, Louisiana	Mississippi River	1575	480	1958
Transbay	San Francisco Bay	1400	427	1936
Baton Rouge, Louisiana	Mississippi River	1235	376	1968
Tappan Zee	Hudson River	1212	369	1955
Longview, Washington	Columbia River	1200	366	1930
Queensboro	East River, New York City	1182	360	1909
Carquinez Strait	California	1100	335	1927
(parallel span)		1100	335	1958
Jacques Cartier	Montreal, Quebec	1097	334	1930
Isaiah D. Hart	Jacksonville, Florida	1088	332	1968
Richmond	San Francisco Bay	1070	326	1957
Grace Memorial	Charleston, South Carolina	1050	320	1929
Newburgh-Beacon	Hudson River, New York	1000	305	1963
Caruthersville, Missouri	Mississippi River	920	280	1975
Saint Mary's	Saint Mary's, West Virginia	900	274	1977
Vicksburg	Mississippi River	870	265	1972
DRAWBRIDGE—VERTICAL LIFT				
Houghton-Hancock	Michigan	1310	399	1959
Arthur Kill (railroad)	New York-New Jersey	558	170	1959
Cape Code Canal (railroad)	Massachusetts	544	166	1935
Delair, New Jersey (railroad)	Delaware River	542	165	1960
Marine Parkway	New York City	540	165	1937
Burlington, New Jersey	Delaware River	534	163	1931
Harry S. Truman (railroad)	Kansas City, Missouri	427	130	1945
BASCULE				
Sante Fe Railway (railroad)	Mississippi River	525	160	1926
Keokuk Municipal	Mississippi River, Iowa	377	115	1916
Elizabeth River	Chesapeake, Virginia	281	86	1969
SWING				
Douglass Memorial	Anacostia River, D.C.	386	118	1950
Lord Delaware	Mattaponi River, Virginia	252	77	1945
Eltham	Pamunkey River, Virginia	237	72	1957
SWING SPAN				
Fort Madison (railroad)	Mississippi River	525	160	1927
Willamette (railroad)	Portland, Oregon	521	159	1908
East Omaha (railroad)	Missouri River	519	158	1903
FLOATING PONTOON				
Evergreen Point	Seattle, Washington	7518	2291	1963
Lacey V. Murrow	Seattle, Washington	6561	2000	1940
Hood Canal	Port Gamble, Washington	6471	1972	1961

NOTE: Bridges that incorporate more than one type of structure will appear in two or more categories.

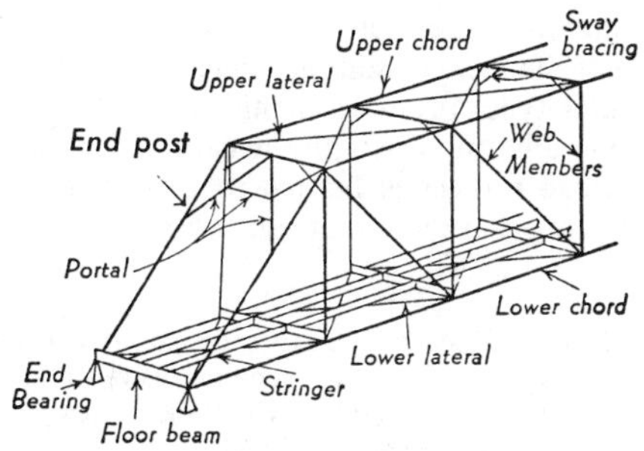

Fig. 1. Skeleton diagram of through-type truss bridge.

which transfer the effects of the moving loads to transverse beams known as floor beams. The floor beams are connected to the trusses at the lower intersection points of the truss members. Each intersection point is called a panel point. The truss is composed of an upper and lower chord and web members which are joined together in the form of triangles. Figure 2 shows some typical trusses. Since the loads are applied to the truss at the panel points, the primary stresses will be axial. When the structure deflects under load, some bending is induced in the truss members because the members are not free to rotate at the panel points. The resulting stresses are called secondary stresses. In this particular illustration the top chord will be in compression and the lower chord in tension. Some of the web members will be in tension, others in compression, but there are certain web members near the center of the span which may have either tension or compression depending upon the position of the moving loads. The bracing

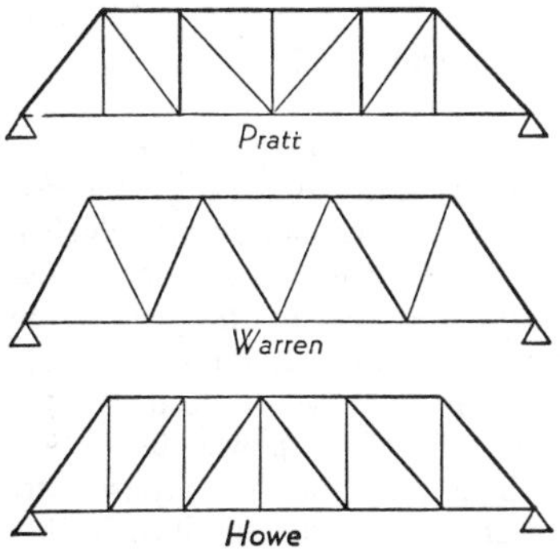

Fig. 2. Typical framed bridge trusses.

is usually made up of an upper and lower lateral system, sway frames, and portal bracing. These bracing systems resist the horizontal loads caused by wind and lateral forces, and, together with the floor system, tie the truss together forming a relatively rigid unit. Short-span framed bridges which do not require trusses sufficiently deep to allow for top chord, sway or portal bracing, because of interference with vehicular traffic are called pony truss spans. The trusses are the principal load-carrying components of the bridge, since they must support their own weight and the weight of the floor system in addition to the moving loads and wind loads. The total load on each truss is transferred through a horizontal pin to the end bearings or shoes, generally castings, which distribute the load to the supporting masonry. At one end of the bridge the bearings will be firmly fastened to the masonry, but at the other end they will be of the expansion type, which allows a limited amount of longitudinal movement to take care of the temperature changes in the structure. The separate structural members of a truss bridge are composed of rolled steel shapes or built-up sections formed by riveting two or more rolled shapes together. The truss members are connected at their intersections by riveting or bolting into gusset plates or by welding. Framed trusses are also used for industrial machines as shown in Fig. 3.

Bridge bearings are used to relieve stresses caused by temperature changes, winds, and changing soil pressures by providing slippage between the bridge and its supports. Bearings may be steel-on-steel sandwiched between lead inserts, self-lubricating bronze plates with graphite inserts, or polytetrafluoroethylene blocks that slide on stainless steel, among others. Assemblies of roller bearings and roller bearing nests may be used. Bearings must support many hundreds of

Fig. 3. Framed bridge trusses used in connection with very large settling basin. These members rotate slowly through 360°. (*Dorr-Oliver, Inc.*)

tons of weight at locations selected by the bridge designer. In recent years, more attention has been directed to the maintenance of bridge bearings and the design essentially maintenance-free bearings. Corrosion and dirt over long periods of time can cause severe damage to bridge bearings.

As pointed out by Phillips (1978), "Maintenance-free means do exist for taking up bridge movements. For example, some expansion and contraction can be absorbed within a properly designed bridge structure, and large shifts of up to three inches (7.5 centimeters) can be absorbed in flexible blocks called *elastomeric bearings.* Often made of high-quality neoprene, these pliant cushions come close to being the perfect solution to the bearing problem, because they have no moving parts to freeze, nothing to corrode, and therefore eliminate maintenance altogether."

Beam bridges are composed of two or more beams laid parallel to the direction of traffic. The roadway or track may be supported directly on the beams or by a stringer and transverse floor beam system connected to beams called girders. The construction of a simple deck plate girder bridge, frequently used for short railway spans, is shown in Fig. 4. Each girder is composed of a steel plate called the web to which are riveted four angles. Additional plates called covers are riveted to the angles. The two top or bottom angles and the attached cover plate form the flanges. The girders are stiffened laterally by the lateral system and the cross frames. A through bridge is a span of the beam or framed truss type in which the floor system is placed between the girders or trusses usually near the plane of the bottom flange or chords. In deck spans the floor system is placed between the girders or trusses near the plane of the top flanges or chords, or rests upon them.

The simplest type of suspension bridge, applicable for short spans, consists of floor system connected by hangers to two cables or chains.

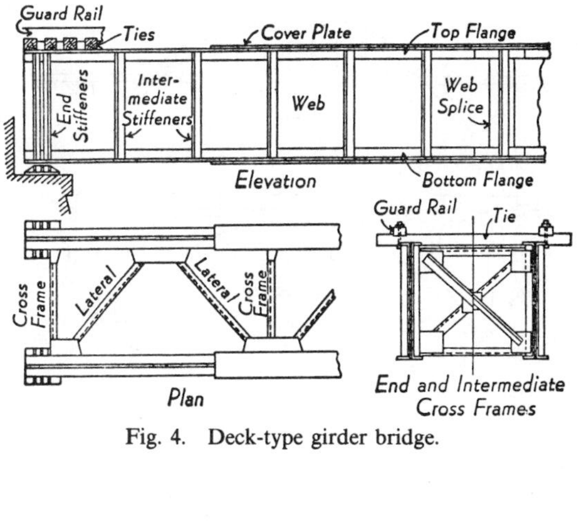

Fig. 4. Deck-type girder bridge.

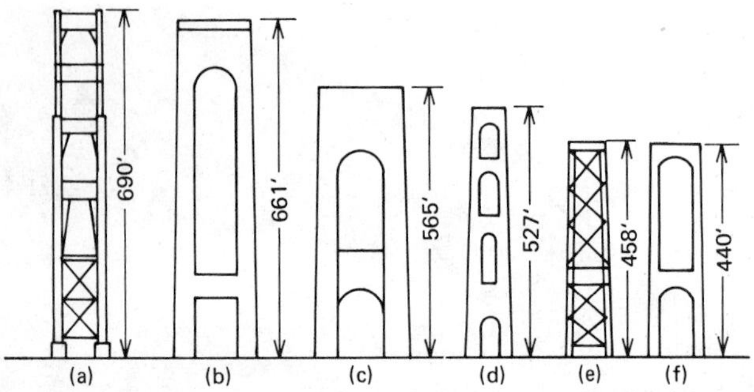

Fig. 5. Highly schematic cross sections of towers used for large suspension bridges: (a) Golden Gate; (b) Verrazano-Narrows; (c) George Washington; (d) Mackinac; (e) San Francisco-Oakland (Transbay); and (f) Delaware Memorial bridge.

The latter pass over towers and are firmly anchored at the end of the span. See Fig. 5. For longer spans it becomes necessary to connect the floor systems to stiffening trusses or girders which distribute the moving loads more uniformly to the hangers. This method of distribution reduces the distortion of the cable or chain. The hangers are formed of twisted wire ropes while the cable may consist of twisted wire ropes or a number of parallel wires securely bound together into a compact unit of circular cross-section. The chain is made up of a number of separate tension links called eye-bars. The floor system, stiffening truss and towers are constructed of rolled steel shapes.

In its simplest form the cantilever bridge consists of a suspended span and two anchor spans. Each anchor span which rests on two piers is made up of an anchor arm and a cantilever arm. The latter projects beyond the river pier to form a support for one end of the suspended span. Cantilever bridges may be either trusses or plate girders. The former are particularly well adapted to long-span construction.

A continuous bridge is one which rests on three or more supports and is capable of transmitting both shear and moment throughout its length. These bridges which are statically indeterminate structures may be plate girders or trusses. The continuous bridge is more rigid than the cantilever bridge but settlement of the supports has an effect on the stress distribution.

A bridge of two or more spans which is supported at each intermediate pier by a hinged quadrilateral is called a Wichert truss. This type of bridge is determinate and therefore the stresses are not affected by settlement of the supports.

A pontoon bridge is a floating roadway which is used to bridge narrow bodies of water. It consists of barges called pontoons which carry a roadway made up of beams which, in turn, support a plank floor. The pontoons must be firmly anchored so that they will not float out of position. The pontoon bridge is generally used for military purposes although there are instances in which this type of bridge has been constructed for ordinary vehicular and pedestrian traffic.

Rigid frame bridges are particularly popular in connection with grade separation work in highway and railways. The relatively shallow central portion of the span results in the use of less fill.

The type of bridge to be used at a particular location depends upon many considerations. Typical arched bridges are shown in Fig. 6.

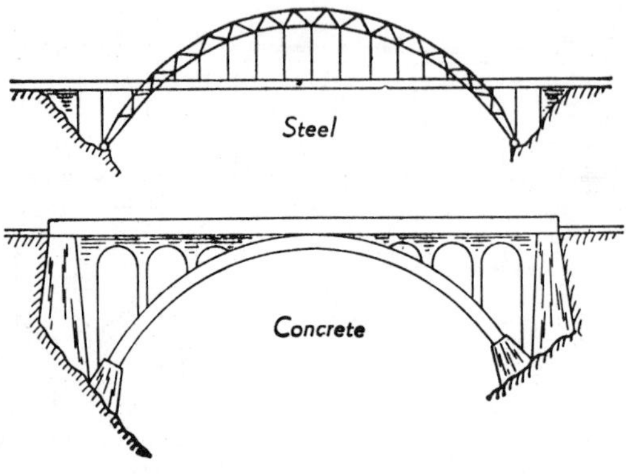

Fig. 6. Typical arched bridges.

Satisfactory foundations, possible pier locations, and access to the bridge are some of the local conditions which will influence the selection of a particular type. From an architectural standpoint the type which is used should harmonize with the natural surroundings. When the cost of a bridge project is limited to a predetermined amount certain types of bridges are automatically eliminated.

References

NOTE: Three excellent books on bridges, their design, historical development, and effect upon other industrial designs are comprehensively reviewed by Emory L. Kemp in *Science*, **208**, 727–730 (1980). The books reviewed are: Billington (1979); Rosenberg (1978); and Ruddock (1979).

Billington, D. P.: "Robert Maillart's Bridges," Princeton Univ. Press, Princeton, New Jersey, 1979.

Bessler, B., Lin, T. Y., and J. B. Scalzi: "Design of Steel Structures," 2nd edition, Wiley, New York, 1968.

Gaylord, E., and C. Gaylord: "Structural Engineering Handbook," McGraw-Hill, New York, 1968.

Merritt, F. S. (editor): "Standard Handbook for Civil Engineers," McGraw-Hill, New York, 1968.

O'Connor, C.: "Design of Bridge Superstructures," Wiley, New York, 1971.

Phillips, L. A.: "Bridge Bearings," *Technology Review (MIT)*, **80**, 3, 25 (1978).

Rosenberg, N., and W. G. Vincenti: "The Britannia Bridge," MIT Press, Cambridge, Massachusetts, 1978.

Ruddock, T.: "Arch Bridges and Their Builders," Cambridge Univ. Press, New York, 1979.

Scott, Q.: "The Eads Bridge," Univ. of Missouri Press, Columbia, Missouri, 1979.

White, R. N., Gergely, P., and R. G. Sexsmith: "Structural Engineering," Vol. 4, Wiley, New York, 1973.

BRIDGE (Truss). Truss.

BRIDGING (Bulk Solids). Feeder (Volumetric).

BRIDGING GAIN. The ratio of the power a transducer delivers to a specified load impedance under specified operating conditions, to the power dissipated in the reference impedance across which the input of the transducer is bridged. If the input and/or output power consist of more than one component, such as multifrequency signal or noise, then the particular components used and their weighting should be specified. This gain is usually expressed in decibels.

In contrast, a *bridging loss* is the ratio of the power dissipated in the reference impedance across which the input of a transducer is bridged, to the power the transducer delivers to a specified load impedance under specified operating conditions. If the input and/or output power consist of more than one component, such as multifrequency signal or noise, then the particular components used and their weighting should be specified. This loss is usually expressed in decibels. In telephone practice this term is synonymous with the insertion loss resulting from bridging an impedance across a circuit.

BRIDLED-CUP ANEMOMETER. Wind and Air Velocity Measurements.

BRIGGS LOGARITHMS. Logarithm.

BRIGHTENERS. Detergents.

BRIGHTNESS. Brightness is the attribute of visual perception in accordance with which an area appears to emit more or less light. Luminance is recommended for the photometric quantity, which has been called "brightness." Luminance is a purely photometric quantity. Use of this name permits "brightness" to be used entirely with reference to the sensory response. The photometric quantity has been often confused with the sensation merely because of the use of one name for two distinct ideas. Brightness may continue to be used properly, in nonquantitative statements, especially with reference to sensations and perceptions of light. Thus, it is correct to refer to a brightness match, even in the field of a photometer, because the sensations are matched and only by inference are the photometric quantities (luminances) equal. Likewise, a photometer in which such matches are made should be called an "equality-of-brightness" photometer. A photoelectric instrument, calibrated in foot-lamberts, should not be called a "brightness meter." If correctly calibrated, it is a "luminance meter." A troublesome paradox is eliminated by this distinction of nomenclature. The luminance of a surface may be doubled, yet it is permissible to say that the brightness is not doubled, since the sensation which is called "brightness" is generally judged to be not doubled.

BRIGHTNESS CONTROL (CRT). Cathode-Ray Tube.

BRIGHTNESS DISTINCTION. Fechner Fraction.

BRIGHTNESS METER. Photometers.

BRIGHTNESS (Stellar). Stellar Luminosity.

BRIGHTNESS (Talbot Law). Talbot Law.

BRIGHT'S KIDNEY RESEARCH. Kidney and Urinary Tract.

BRIGHT STOCK. Petroleum.

BRIGHT SWITCH. This switch invented by R. L. Bright is solid-state and consists of two bipolar transistors which are connected in the inverted connection as shown in the accompanying diagram. Inasmuch as offset voltage is reduced by the use of the inverted connection and the back-to-back transistor connection, the Bright switch more closely approximates the ideal voltage switch than a single transistor. When a bipolar transistor is driven into saturation by a signal applied between the base and emitter, an open-circuit (where the collector current is zero) voltage appears between the emitter and collector. Called the offset voltage, this voltage represents an error when the switch is used in analog-switching applications. This is because the voltage appears in series with the signal source being switched. If, on the other hand, the transistor is driven into saturation by a drive signal that is applied between the base and collector, the offset voltage magnitude is substantially reduced. The difference between the two offset voltages mainly is a function of the difference between the normal β (common-emitter current gain with the drive connected between the base and emitter) and the inverted β (common-emitter current gain with the collector functioning as the emitter).

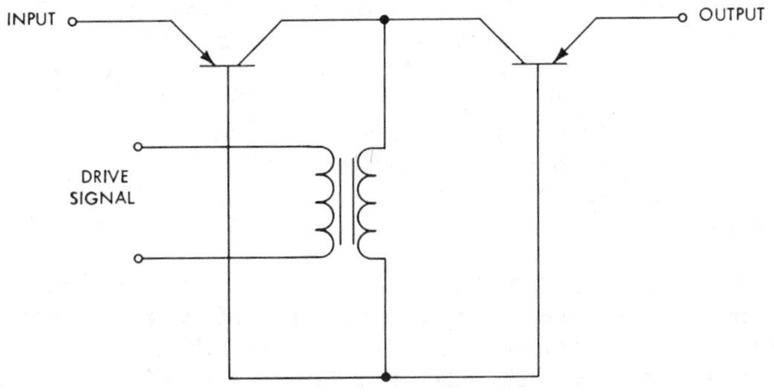

Circuit configuration of Bright switch.

The Bright switch configuration utilizes this phenomenon and the cancellation effect of connecting the transistors in a back-to-back configuration to achieve a low-offset switch. The offset of one transistor is partially cancelled by the offset voltage of the other transistor. As contrasted with an offset voltage of several millivolts in a typical silicon transistor, net offsets of much less than 50 millivolts can be achieved with the Bright switch.

See also **Analog Switch.**

BRILLIANCE. Brilliance is that attribute of any color in respect to which it may be classed as equivalent to some member of a series of grays ranging between black and white. Yellow is the most brilliant color in the spectrum of white light.

BRILLOUIN EFFECT. Upon the scattering of monochromatic radiation by certain liquids, a doublet is produced, in which the frequency of each of the two lines differs from the frequency of the original line by the same amount, one line having a higher frequency, and the other a lower frequency.

BRILLOUIN ZONE. An electron moving within an ionic crystal moves in a potential field which may be approximated to as that of a constant potential within the crystal (as in the elementary Drude-Lorentz theory), modified by a varying potential which varies as the periodicity of the lattice. The allowed solutions of the wave equation for such a system are those for which the energy lies in a series of bands, the wave vector k of the electron being imaginary at other values. The values of k at which discontinuities occur lie at the surfaces of polyhedra in k-space called Brillouin zones. The Brillouin zones may be calculated for a given lattice structure. In the study of complex metals and alloys, where there may be several overlapping bands, the geometry of the zones plays an important role. See also **Fermi Surface.**

BRILL-ZINSSER DISEASE. Rickettsial Diseases.

BRINE-FLY (*Insecta, Diptera*). Flies whose larvae live in strong briny or alkaline waters. They belong to the family *Ephydridae* which also contains species whose larvae live in fresh water and one remarkable insect which lives in pools of crude petroleum in the California oil fields.

Large quantities of the larvae of certain species are washed ashore along some of the western alkaline lakes and are gathered by the Indians as food under the native name kootsabe.

The brine-fly and the brine shrimp are among the very few life forms that inhabit the waters of the Great Salt Lake in Utah. The brine-fly can live in nearly pure salt.

BRINES (Lithium). Lithium (For Thermonuclear Fusion Reactors).

BRISTLECONE PINE. Pine Trees.

BRISTLEMOUTHS (*Osteichthyes*). Of the order *Isospondyli*, suborder *Stomiatoidea*, family *Gonostomatidae*, the bristlemouths are abundant deep sea fishes, with a herringlike appearance and incorporation of photophores (light organs) on their sides. Of the some 30-plus species, the largest does not exceed 3 inches (7.5 centimeters) in length. There remains much to be learned pertaining to the habits of the bristlemouths. Despite their abundance, the bristlemouths are seldom seen.

BRITANNIA METAL. Antimony.

BRITISH INDIAN GUM. Gums and Mucilages.

BRITISH THERMAL UNIT. Heat; Units and Standards.

BRITTLE FRACTURE. A fracture involving very little expenditure of energy. Brittle fracture usually occurs with very little accompanying plastic deformation. Brittle fractures in engineering structures have been of concern ever since it became the practice to weld large steel structures. Thus, for example, the hull of a welded ship is really one continuous piece of steel. A crack that starts in such a structure can pass completely around it, causing it to break in two. A number of failures of this type over the years have occurred. Similarly, brittle fractures have been known to travel as far as half a mile in welded gas pipelines, often with extremely high velocities. A brittle crack usually starts at a notch or stress raiser which may be due to faulty design or to accidents of construction. Most brittle fractures also occur at low ambient temperatures—the middle of winter. Finally, the metal must be subjected to a stress which furnishes the energy causing the fracture to expand.

Most brittle fractures in steel are transcrystalline, with the body-centered cubic ferrite crystals cleaving on cube planes. Cleavage is promoted by high stresses. Since plastic flow by slip tends to relieve an applied stress, the conditions that promote cleavage are normally those that restrict plastic deformation. Thus, rapid application of a load, a state of multiaxial stress, and low temperatures all limit slip deformation while encouraging cleavage. It should be noted, however, that cleavage in steels only approaches a true brittle fracture, such as that which occurs in glass, at very low temperatures like that of liquid nitrogen (−196°C). Glass at room temperature fractures without any appreciable amount of accompanying plastic deformation. In steels, some plastic deformation usually proceeds or accompanies fracture due to cleavage, the amount of deformation, however, decreasing with decreasing temperature.

In a normal steel most failures at room temperature or above are ductile fractures which involve a large expenditure of work. As the temperature is lowered, the failure becomes partly ductile and partly brittle, with a corresponding decrease in the work to cause fracture. This fact may be clearly shown by making Charpy impact tests at a number of temperatures. This test measures the energy to suddenly fracture a notched bar. The presence of the notch produces a state of multiaxial tensile stress. An important feature of the Charpy impact test is that it tends to reproduce the ductile-brittle transformation of steel in about the same temperature range as that actually observed in engineering structures. A representative curve, showing the transition from ductile to brittle behavior, as measured by the Charpy test, is shown in the first figure. (See Fig. 1.) One of its important features

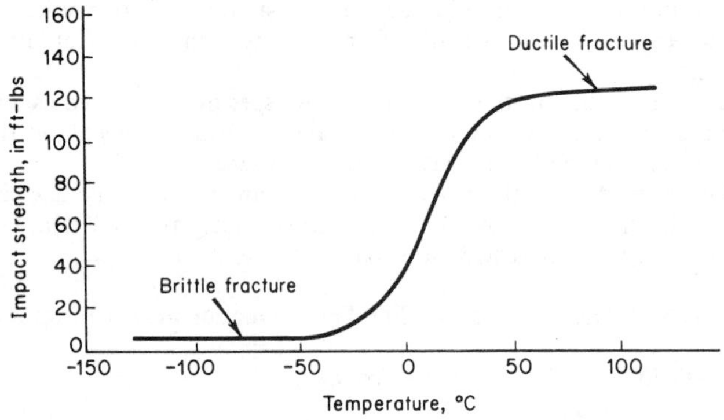

Fig. 1. Representative Charpy impact ductile to brittle fracture transition curve.

is that the transition is not sharp, but occurs over a range of temperatures. It is, therefore, necessary to arbitrarily define the transition temperature. There are several ways of doing this that are commonly employed. In one case, the transition temperature for ductile to brittle fracture is taken as the temperature at which an impact specimen fails with a half-brittle, half-ductile surface. The brittle fracture portion of the surface can always be identified by its cleavage facets which reflect light sharply. On the other hand, the ductile portion of the fracture surface is always dull and gray. A second definition uses the average energy criterion: the temperature at which the energy absorbed falls to one-half the difference between that needed to fracture a completely brittle specimen. The temperature at which a specimen fails with a fixed amount of energy, usually 15 or 20 ft-lb, is also

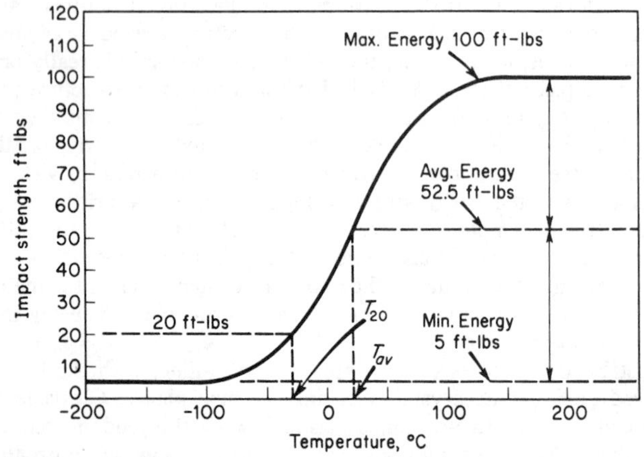

Fig. 2. The transition temperature can be defined in several ways, two of which are shown above. T_{20} is the transition temperature, using the 20-ft-lb (27.12-joule) criterion; T_{av} is the temperature for the average energy criterion.

widely employed as a basis for the transition temperature. The last two of the above criteria are illustrated in Fig. 2.

BRITTLENESS TESTING. Impact Testing.

BRITTLE STAR. Ophiuroidea.

BRIX SCALE. Specific Gravity.

BROADBAND SIGNAL SOURCE. Noise Generator.

BROAD BEAN. Bean.

BROADBILLS (*Aves, Passeriformes*). These birds belong to the suborder *Desmodactylae* and consist of only one family (*Eurylaimidae*). These birds have large heads and are generally broadbilled. There are 15 cervical vertebrae (other passeriformes have 14). There are small scales on the rear of the tarsus. The wings are short and round, and thus the birds only fly short distances. There are eight genera with 14 species, two of which are found in tropical Africa, and the rest in the Orient.

This family falls naturally into two groups. The Typical Broadbills (subfamily *Eurylaiminae*) which have very large beaks; the Green Broadbills (subfamily *Calyptomeninae*), with the single genus *Calyptomena*, with a rictal brush. The size varies from that of a sparrow to that of a jay; the body is compact, and the wings are short and round. The eyes are large, and the bill is flat, wide and hooked at the tip; the green broadbill is the only bird of the family with a smaller beak, which is covered at the base with a dense hood of feathers. The sexes are usually different in appearance.

Broadbills are mainly forest birds; some prefer mountains, and some prefer plains. Most broadbills are insectivorous; only the black-and-red broadbill eats berries and even shrimp, small fish and crabs, as well as beetles, crickets, and grasshoppers. The green broadbills, however, prefer fruit.

All species build the same type of unusual nest, which may reach a length of 6½ feet (2 meters). It usually hangs from the tip of a branch in a shady forest glade, and it is almost always over a river or some other body of water. Its shape corresponds to that of a gigantic pear, with an elongated narrow part connecting it to the end of the branch. There is usually a large projection over the entrance. The nest is built of grass, leaves, moss, rootlets, and similar materials, and it is lined with green leaves. The nest exterior is often decorated with lichens and spider webs. Broadbills usually lay 2–4 eggs; the eggs are white, light red, or cream-colored, with spots of varying density. Nothing seems to be known about the duration of incubation, the fledging period, or other details of the nest life in any of these species.

BROADENING OF SPECTRAL LINES. A spectral line emitted by an atomic or nuclear system does not consist of a single frequency, but rather of a continuous group of frequencies, which may be very narrow in its extent. The inherent width of a line is known as its natural width. A spectral line may be additionally broadened by Doppler broadening and by collision, or pressure broadening. In the latter case, the lifetime τ of an excited state may be reduced during a collision, which in turn increases the energy level width, Γ, of the excited state through the relation, $\Gamma + h/\tau$. Since the change in energy ΔE in the transition and the frequency of emitted radiation are related by $\Delta E = h\nu$, any broadening of the energy change results in a broadening of the frequency spectrum. See also **Doppler Broadening.**

BROCA'S AREA (Brain). Memory.

BROCCOLI. Brassica.

BROCHANTITE. A mineral composed of basic copper sulfate corresponding to the formula $Cu_4(SO_4)(OH)_6$, crystallizing in the monoclinic system in needle-like prisms, or forming druses or masses. Hardness, 3.5–4; specific gravity, 3.9; vitreous luster; color, green; streak, green; transparent to translucent.

Brochantite is a secondary mineral occurring in the oxidized zones with other copper minerals, and is found in the Urals, in Rumania, in Sardinia; Cornwall, England; Chile. In the United States this mineral has been found at Bisbee, Arizona, Utah, in the Tintic District and in Inyo County, California. Brochantite was named for Brochant de Villiers.

BROCKETS. **Deer.**

BRODIE'S ABSCESS. **Bone.**

BROMEGRASS. **Grasses.**

BROMIC ACID. **Bromine.**

BROMIDES. **Drug Addiction.**

BROMINE. Chemical element symbol Br, at. no. 35, at. wt. 79.904, periodic table group 7a (halogens), mp −7.2°C, bp 58.8°C, density 3.12 g/cm^3 (20°C). Bromine is one of the few elements that is liquid at standard conditions. The element volatilizes readily at room temperature to form a red vapor which is very irritating to the eyes and throat. Liquid bromine causes painful lesions upon contact with the flesh. Bromine has two stable isotopes ^{79}Br and ^{81}Br. Elemental bromine finds limited application as a chemical intermediate and as a sanitizing, disinfecting, and bleaching agent. Both the inorganic and organic compounds of the element find extensive commercial usage. Bromine was discovered in 1826 by Antoine-Jérôme Balard who identified the element as a component of seawater bitterns. Electronic configuration $1s^22s^22p^63s^23p^63d^{10}4s^24p^5$. Ionic radius Br^- 1.97 Å, Br^{7+} 0.39 Å. Covalent radius 1.193_5. First ionization potential 11.84 eV; second 19.1 eV; third 25.7 eV. Oxidation potentials $2Br^- \rightarrow Br_2(1) + 2e^-$, −1.065 V; $2Br^- \rightarrow Br_2(aq) + 2e^-$, −1.087 V; $Br^- + H_2O \rightarrow HBrO + H^+ + 2e^-$, −1.33 V; $Br^- + 3H_2O \rightarrow BrO_3^- + 6H^+ + 6e^-$, −1.44 V; $\frac{1}{2}Br_2 + 3H_2O \rightarrow BrO_3^- + 6H^+ + 5e^-$, −1.52 V; $\frac{1}{2}Br_2 + H_2O \rightarrow HBrO + H^+ + e^-$, −1.59 V; $Br^- + 6OH^- \rightarrow BrO_3^- + 3H_2O + 6e^-$, −0.61 V; $Br^- + 2OH^- \rightarrow BrO^- + H_2O + 2e^-$, −0.70 V.

Other important physical properties of bromine are described under **Chemical Elements.**

Bromine is only moderately soluble in H_2O (3.20 g/100 ml) but markedly so in nonpolar solvents, e.g., carbon tetrachloride, as is consistent with the covalent character of the Br—Br bond. It dissolves more readily in alkali bromide solutions due to the formation of the tribromide ion (Br_3^-), and in certain associated solvents, such as concentrated H_2SO_4 and ethyl alcohol. Its aqueous solution is more stable than that of chlorine, since the tendency of Br_2 to hydrolyze to unstable hypobromous acid and hydrogen bromide is less than the corresponding reaction for chlorine. Bromine exhibits in common with the other halogens a marked readiness to form singly charged negative ions, as would be expected from the fact that these atoms need only one electron to acquire an inert gas configuration. Its electron affinity (3.53 eV) is between that of chlorine and iodine. The bromides range in character from ionic to covalent compounds, many of them having bonds of intermediate nature. In addition to its negative univalence, bromine forms essentially covalent linkages with negative elements, in which it has positive valences 1, 3, and 5.

Bromine occurs as bromide in seawater (0.188% Br), in the mother liquor from salt wells of Michigan, Ohio, West Virginia, Arkansas, and in the potassium deposits of Germany and France.

Production: In the United States, nearly all bromine is derived from natural brines. The Arkansas brines which contain a minimum of 4000 ppm bromide account for over half of this production. Recovery is effected by a *steaming-out* process. After heating fresh brine, the solution is fed to the top of a tower. Chlorine and steam are injected at the bottom of the tower. The chlorine oxidizes the bromide and displaces one resultant bromine from solution. For brines of lower concentration, air instead of steam is used to sweep out the bromine vapors after chlorination.

Hydrogen Bromide and Hydrobromic Acid: HBr, is formed directly from the elements, effectively when catalyzed by sunlight, by heated charcoal or platinum, or more conveniently by hydrolysis of phosphorus tribromide. Treatment of bromides with H_2SO_4 yields mixtures of HBr and bromine. The H—Br bond is considered to be partly covalent. Hydrobromic acid is a strong acid in aqueous solution. Its salts are the bromides, all of which are water-soluble except those of copper(I), silver, gold(I), mercury(I), thallium(I) and lead(II), the divalent ions of the elements of the second and third transition series, and the salts of the heavy alkali ions with many bromo-complex anions, e.g., Cs_2PtBr_6, $RbAuBr_4$, etc. The main uses of HBr and hydrobromic acid are in the production of alkyl bromides (by replacement of alcoholic hydroxyl groups or by addition to olefins) and inorganic bromides.

Sodium Bromide: This is a high-tonnage chemical and one of the most important of the bromide salts commercially. High-purity grades are required in the formulation of silver bromide emulsions for photography. The compound, usually in combination with hypochlorites, is used as a bleach, notably for cellulosics. The production of sodium bromide simply involves the neutralization of HBr with NaOH or with sodium carbonate or bicarbonate.

Calcium Bromide: Because of its ready solubility, calcium bromide forms solutions of high density which when properly formulated are finding increasing use as functional fluids in oil well completion and packing applications.

Lithium Bromide: LiBr finds use as a desiccant in industrial air-conditioning systems.

Zinc Bromide: $ZnBr_2$ is used as a rayon-finishing agent, as a catalyst, as a gamma-radiation shield in nuclear reactor viewing windows, and as an absorbent in humidity control. It too finds use in high density formulated functional fluids in oil well applications. Zinc bromide is prepared either by the direct reduction of bromine with zinc, or by reacting HBr with zinc oxide or carbonate.

Other Bromides: Aluminum bromide $AlBr_3$ is used as a catalyst and parallels $AlCl_3$ in this role. Strontium and magnesium bromides are used to a limited extent in pharmaceutical applications. Ammonium bromide is used as a flame retardant in some paper and textile applications; potassium bromide is used in photography. Phosphorus tribromide PBr_3 and silicon tetrabromide $SiBr_4$ are used as intermediates and catalysts, notably in the production of phosphite esters.

Hypobromous Acid and Hypobromites: Hypobromous acid HOBr results from the hydrolysis of bromine with H_2O and exists only in aqueous solution. The compound finds limited use as a germicide and in water treatment; also it can be used as an oxidizing or brominating agent in the production of certain organic compound. Although hypobromous acid is low in bromine content, concentrated hypobromite solutions can be formed by adding bromine to cooled solutions of alkalis.

Bromic Acid and Bromates: Bromic acid, $HBrO_3$, can exist only in aqueous solution. Bromic acid and bromates are powerful oxidizing agents. Bromic acid decomposes into bromine, oxygen, and water. Many oxidizing agents, e.g., hydrogen peroxide, hypochlorous acid, and chlorine convert Br_2 or Br^- solutions to bromates. The decomposition reactions of bromates vary considerably. Lead(II) bromate and copper(II) bromate give the metal oxides and Br^-; silver, mercury(II) and potassium bromates give the metal ion, Br^- and oxygen, while zinc, magnesium, and aluminum bromates give the metal oxide, Br_2 and oxygen.

Halogen Compounds: Bromine forms a number of compounds with the other halogens. Its binary iodine compounds are discussed under iodine; other interhalogen compounds of bromine include bromine monochloride, bromine monofluoride, bromine trifluoride, and bromine pentafluoride. The nonexistence of higher chlorides of bromine, differing from iodine, can readily be explained in terms of the oxidation potential of Br(III) and Br(V). The monochloride, bromine chloride, BrCl, exists in pure state only at very low temperatures in the solid form. Dissociation in the gas phase is approximately 40% at 25°C and increases slowly with increasing temperature; less than 20% occurs in the liquid phase. With many substrates, bromine chloride reacts much more rapidly than does bromine itself to introduce bromine substituents. Bromine monofluoride, BrF, is also somewhat unstable, decomposing spontaneously at 50°C to Br_2, BrF_3, and BrF_5. It has never been prepared pure since it is always in equilibrium with Br_2 and BrF_5. It is a gas at room temperature, reacting readily with water, phosphorus, and the heavy metals. Bromine trifluoride, BrF_3, is much more stable than the monofluoride. It is obtained directly from the elements at 10°C, or by fluorination of univalent heavy metal bromides. It is a liquid, bp 127.6°C, mp 8.8°C. There is evidence (high Trouton constant) that it undergoes self-ionization to form BrF_2^+ and BrF_4^-. The former is found in the acidic addition products it forms with gold(III), antimony(V) and tin(IV) fluorides, BrF_2AuF_4, BrF_2SbF_6 and $(BrF_2)_2SnF_6$. The latter occurs in the tetrafluorobromates, such as $KBrF_4$ and $Ba(BrF_4)_2$. The solvent properties of BrF_3 are consistent with its indicated dissociation, i.e., reactions involving the two classes of compounds mentioned take place as if BrF_3 acts as a fluoride ion donor or acceptor as H_2O is a proton donor or acceptor in the H_2O system. For example, potassium dihydrogen phosphate, KH_2PO_4 gives KPF_6, a mixture of HNO_3 and B_2O_3 gives NO_2FB_4, etc. Bromine trifluoride fluorinates many of the metal halides and oxides. Bromine pentafluoride, BrF_5, is prepared from BrF_3 and fluorine. It is thermally stable. It is a very active fluorinating agent, converting to fluorides most metals, their oxides and other halides, and being hydrolyzed by H_2O probably to hydrofluoric and bromic acids.

The polyhalide complexes of bromine include $(PBr_4)(IBr_2)$ formed by reaction of phosphorus pentabromide and iodine monobromide, and dissociating in certain organic polar solvents to the ions PBr_4^+ and IB_2^-. Other polyhalides include NH_4IBr_2, $[(CH_3)_4N][IBr_2]$, Cs[IFBr], Rb[IClBr], and Cs[IClBr]. Most of these compounds hydrolyze readily, ionize in polar nonreacting solvents to the corresponding polyhalide ions, and decompose on heating to give the metal halide of greatest lattice energy.

Oxides: In binary combination with oxygen, bromine forms at least three compounds. Bromine(I) oxide, Br_2O, is a dark brown solid that is stable only in the dark below −40°C. It is prepared by passing dry gaseous bromine through dry mercury(II) oxide and sand. Bromine(I) oxide, in carbon tetrachloride at low temperatures, reacts with alkali hydroxide to give hypobromites. Bromine(IV) oxide, BrO_2, is obtained by reaction of the elements in a cooled electric discharge tube. It is yellow, and is stable only at low temperatures. The compound appearing in the older literature as tribromine octoxide, Br_3O_8, is actually bromine trioxide, BrO_3, or dibromine hexoxide, Br_2O_6 (cf. chlorine). It is obtained from the low temperature, low pressure reaction of ozone and bromine; it is stable only at low temperatures and is soluble in H_2O with decomposition. Bromine(VII) oxide may be present among the decomposition products of BrO_2, or BrO_3, but no other evidence for its existence has been found.

Organic Bromine Compounds: Commercially important organic bromine compounds include: (1) methyl bromide CH_3Br, formed by reacting methanol with HBr or hydrobromic acid. The compound is a highly toxic gas at standard conditions. Because of its toxicity, it is used as a soil and space fumigant. In many organic syntheses, the compound is used as a methylating agent; (2) ethylene dibromide (1,2-dibromoethane) is used in combination with lead alkyls as an antiknock agent for gasoline. The compound also is used as a fumigant; (3) methylene chlorobromide (bromochloromethane) is a low-boiling liquid of low toxicity and is useful as a fire-extinguishing agent in portable equipment and aircraft; (4) bromotrifluoromethane is increasingly employed as a fire extinguishant in permanently installed systems protecting high-cost installations, such as computer rooms, where its low toxicity and especially its freedom from corrosivity are important considerations; (5) acetylene tetrabromide (1,1,2,2-tetrabromoethane) is made by adding bromine to acetylene. The compound is comparatively dense and finds use as a gage fluid and in specific gravity separations of solids. It is also used as part of the catalyst system for the oxidation of *p*-xylene to terephthalic acid; (6) tris(2,3-dibromopropyl) phosphate may be prepared by the reaction of phosphorus oxychloride with 2,3-dibromopropanol, or by the addition of bromine to triallyl phosphate. This viscous fluid was used as a flame retardant in a number of polymer systems, but has been displaced from most, or all, of these uses because it is a mutagen and suspect carcinogen; (7) tetrabromobisphenol A is produced by the direct bromination of bisphenol A. The compound is used extensively as a flame retardant and usually is incorporated into the polymer backbone structure of epoxy resins, unsaturated polyesters, and polycarbonates; (8) tetrabromophthalic anhydride, made by the catalytic bromination of phthalic anhydride in fuming H_2SO_4, also finds use as a reactive flame retardant in the formulation of polyol systems (for polyurethane foams) and in unsaturated polyesters; (9) decabromodiphenyl ether, and others of the lower brominated diphenyl ethers, are finding increasing use as flame retardants in a variety of thermoplastic polymer systems; (10) vinyl bromide has also found major use in flame-retarding modacrylic textile fibers when introduced as a comonomer in the synthesis of the polymer itself.

Alkanes and arenes, e.g., ethane and benzene, respectively, react with bromine by *substitution* of bromine for hydrogen (hydrogen bromide also formed)—ethane to yield ethyl bromide C_2H_5Br plus further substitution products; benzene, in the presence of a catalyst, e.g., iodine, phosphorus, iron, to yield bromobenzene C_6H_5Br plus further substitution products; toluene, under like conditions to benzene, to yield orthobromotoluene and parabromotoluene $CH_3C_6H_4Br$ plus further substitution products, but at the boiling temperature, in sunlight, dry, and in the absence of a catalyst, to yield alkyl side-chain substitution products, benzyl bromide $C_6H_5CH_2Br$, benzal bromide $C_6H_5CHBr_2$, and benzotribromide $C_6H_5CBr_3$.

Alkenes, alkynes, and arenes, e.g., ethylene, acetylene and benzene, respectively, react (1) with bromine by *addition*, e.g., ethylene dibromide $C_2H_4Br_2$(1,2), acetylene tetrabromide $C_2H_2Br_4$1,1,2,2, hexabromocyclohexane $C_6H_6Br_6$; also carbon monoxide yields carbonyl bromide $COBr_2$; (2) with hypobromous acid by *addition*, e.g., olefins form, for example, ethylene bromohydrin $CH_2Br \cdot CH_2OH$; (3) with hydrogen bromide by *addition*, to form, for example, ethyl bromide $CH_3 \cdot CH_2Br$ from ethylene. When the two olefin carbons have unequal numbers of hydrogens, the carbon to which one bromide or one hydroxyl attaches can be controlled by the reaction conditions.

Oxygen-function compounds, e.g., ethyl alcohol, acetaldehyde, acetone, acetic acid, react (1) with bromine, to form *bromo-substituted* corresponding or related *compounds*, e.g., ethyl alcohol or acetaldehyde to yield bromal $CBr_3 \cdot CHO$, acetone to yield bromoacetone $CH_2Br \cdot CO \cdot CH_3$; acetic acid to yield, at the boiling temperature, dry, and in the absence of a catalyst, monobromoacetic acid $CH_2Br \cdot COOH$, dibromoacetic acid $CHBr_2 \cdot COOH$, tribromoacetic acid $CBr_3 \cdot COOH$, the substitution taking place on the alpha-carbon (the carbon next to the carboxyl-group —COOH), (2) with phosphorus bromides, to form corresponding *bromides*, e.g., ethyl bromide C_2H_5Br, ethylidene dibromide CH_3CHBr_2, acetone bromide $(CH_3)_2CBr_2$, acetyl bromide CH_3COBr, (3) with hydrobromic acid, concentrated, alcohol forms the corresponding bromide.

Bromoform is made by reaction of acetone or ethyl alcohol with sodium hypobromite; carbon tetrabromide by reaction of CS_2 plus bromine Br_2 in the presence of iron, heated; or by one reaction of bromoform with aqueous hypobromide solutions. Use is made of the diazo-reaction to introduce bromine into aryl compounds.

Many of the bromo-compounds are used as reagents or as intermediate compounds in organic chemistry. When alkyl bromocompounds are treated (1) with NaOH dissolved in alcohol, hydrogen bromide is removed, e.g., ethyl bromide $CH_3 \cdot CH_2Br$ yields ethylene $CH_2{:}CH_2$, ethyl dibromide $CH_2Br \cdot CH_2Br$ yields acetylene $CH \vdots CH$; (2) with magnesium or zinc and alcohol, bromine is removed, e.g., ethylene dibromide $CH_2Br \cdot CH_2Br$ yields ethylene $CH_2{:}CH_2$, acetylene tetrabromide $CHBr_2 \cdot CHBr_2$ yields acetylene $CH \vdots CH$.

References

Faith, W. L., Keyes, D. B., and R. L. Clark: "Industrial Chemicals," 3rd edition, Wiley, New York, 1965.

Jolles, Z. E. (editor): "Bromine and Its Compounds," Academic, London, 1966.

Robinson, A. L.: "Infrared Photochemistry—Laser-catalyzed Reactions," *Science*, **193**, 1230–1231 (1976).

Theiler, R., et al.: "Hydrocarbon Synthesis by Bromoperoxidase," *Science*, **202**, 1094–1096 (1978).

Theiler, R., et al.: "Drugs from the Sea," Univ. of Oklahoma Press, Norman, Oklahoma, 1979.

John A. Garman, Director, Research and Development, Great Lakes Chemical Corporation, West Lafayette, Indiana.

BRONCHIAL ASTHMA. A relatively common obstructive lung disease with an estimated 2.5% of the population symptomatic at any given time. Symptoms tend to range widely between patients and in any one patient. The severity of symptoms can spontaneously change rapidly in any particular patient. Although much has been learned concerning the relief of symptoms, treatment of a given patient may require a series of trials with available drugs before a reasonably effective combination of drugs can be identified. There is no known permanent cure for the disease, although some patients can be essentially symptomless for fairly long periods, only to have a spontaneous flare-up of symptoms. Generally the physician will advise the asthma patient that treatment will be over the long term. Statistics indicate that about two-thirds of asthma-prone persons will develop before the age of 5 years. The time of onset of the disease ranges widely in the other third of cases. The incidence of bronchial asthma is nearly twice as high in men as in women. Largely a result of lack of attention and treatment, approximately 5000 deaths in the United States per year are attributable to bronchial asthma. Bronchial asthma, known for centuries, remains a poorly understood disease.

Much of the knowledge pertaining to the nature of bronchial asthmas has been gained from postmortem examinations of persons who have died of status asthmaticus. The lungs are overdistended. Airways from the trachea to the respiratory bronchioles are blocked by plugs of thick, tenacious mucus. Frequently bronchiectasis (chronic necrotizing infection of the bronchi and bronchioles, accompanied by purulent exudation and very enlarged air passages) and fibrosis (fibroid tissue) are present. Emphysema is not usually indicated. Close inspection of the plugs of mucus reveals shed epithelium (Curschmann's spirals), many eosinophils, and so-called Charcot-Leyden crystals, components that are also usually found in the sputum of living patients. A heavy infiltrate of eosinophils in the airways is a prominent feature. This has led some authorities in recent years to include *eosinophilia* (excess of eosinophils) as part of the formal definition of bronchial asthma. Eosinophils make up only 2–3% of the leukocytes in the blood. See **Blood.** Total blood eosinophil counts are usually found to be in excess of $300/mm^3$ in untreated bronchial asthma patients. This involvement of the eosinophils may be a lead to a much better understanding of the etiology of bronchial asthma as research continues.

A major step forward in the understanding of bronchial asthma was made when it was recognized that this is not a single disease per se, but rather it is a group of disorders with different pathogenic mechanisms. Using this as a base, several types of bronchial asthma were identified: (1) Inherited immunologic (IgE-mediated) asthma; (2) postexercise asthma; (3) aspirin-induced asthma; (4) occupational asthma; and (5) bronchial asthma associated with system vasculitis. As more is learned of the general disease, of course, these types may be later reclassified.

IgE-mediated Asthma. Immediate hypersensitivity (reactions which may appear in the short span of seconds to minutes) occurs as the result of interaction between an antigen and an antibody. These reactions occur in anaphylaxis (susceptibility to a drug protein or toxin or toxin resulting from infection), hay fever, hives, and allergic (extrinsic) asthma. Hypersensitivity is mediated by immunolglobulin IgE, once called the reaginic antibody. Usually only traces of IgE are found in the serum of normal persons, but in some persons (atopic) there is a higher level of IgE. In these persons, there is commonly found a family history of IgE-mediated disorders. See also **Immune System and Immunology.**

The present concepts of the pathogenesis of IgE-mediated asthma are far too complex to delineate here. However, it may be said that at least three changes in the bronchi occur to create bronchial obstruction—*bronchoconstriction,* caused by increased muscle tone in the bronchial smooth muscle; *edema* of the bronchial mucosa; and secretion of thick plugs of mucus. The alveolar ducts also usually become constricted. A number of specific or classes of substances have been identified in these processes—*histamine,* which is known to constrict bronchial smooth muscle and to cause edema of bronchial mucosa, apparently as the result of increasing the permeability of small bronchial veins; a substance identified as *SRS-A,* which acts more slowly than histamine and thus prolongs the effect of histamine; a substance known as *platelet-activating factor* (PAF), which releases histamine and serotinin from platelets; and a substance referred to as *eosinophilic chemotactic factor* (ECF-A), which apparently causes the migration of eosinophils into affected regions and thus the condition of eosinophilia previously mentioned. It has been hypothesized that possibly the eosinophils, which originate in the bone marrow and reach the lungs through the blood, may assist in inactivating SRS-A. There are also secondary mediators which participate in the process. These include prostaglandins and bradykinins.

Postexercise Asthma. Rather than a specific type of bronchial asthma, as identified by some authorities, perhaps postexercise asthma is better designated as an episode which may appear in certain other types of asthma. When exercise precipitates acute attacks of asthma, this generally indicates the patient is not receiving adequate treatment.

Aspirin-Induced Asthma. In certain individuals, aspirin and various benozic acid derivatives may induce asthma that will not respond to traditional therapy. Tartrazine, used in certain food colorings, also may be implicated. Treatment is avoidance of salicylates and aspirin.

Occupational Asthma. It is believed that a number of substances may induce asthma. Examples of substances, which then repetitively inhaled may cause this type of asthma, include animal dander, exhaust gases and particulates from wood and other fuel combustion processes, castor beans, formaldehyde, grain dusts, isocyanates, metal dusts (particularly nickel, platinum, tungsten, and vanadium), plastic-generated fumes, proteolytic enzymes used in some detergents, and textile and tobacco dusts, among others. Treatment is avoidance of exposure to these substances.

Bronchial Asthma Associated with Vasculitis. A person with chronic bronchial asthma may after a period develop systemic *vasculitis* (inflammation of a vessel). Small arteries and veins, particularly involving the lungs, peripheral nerves, and skin, may be affected. This condition is sometimes referred to as the *Churg-Strauss syndrome.*

Treatment of Bronchial Asthma. The principal pharmacologic agents used in the therapy of bronchial asthma include: (1) *Beta agonists,* such as isoproterenol, salbutamol, and ephedrine, which are bronchodilators. These are available in inhaler, oral, and parenteral forms. (2) *Methylxanthine and its derivatives,* such as theophylline (aminophylline), which also are bronchodilators. These are available in oral and parenteral forms. (3) *Parasympatholytic agents,* such as atropine and S1080, which decrease acetylcholine output by the vagus nerve and thus, by a complex pathway, reduce alveolar constriction. These are available in inhaler and parenteral forms. (4) *Corticosteroids,* such as prednisone and beclomethasone. Although widely used, the mechanism of their action is poorly understood. They are available in oral, inhaler, and parenteral forms. (5) *Mast cell inhibitors,* such as cromolyn, which decreases release of mediators. These are available as inhalers.

References

Dunhill, M. S.: "The Pathology of Asthma with Special Reference to Changes in the Bronchial Mucosa," *J. Clin. Pathol.,* **13,** 27 (1960).

McFadden, E. R., Jr., R. Kiser, and W. J. de Groot: "Acute Bronchial Asthma," *N. Engl. J. Med.,* **228,** 221 (1973).

McGovern, J. P., Smolensky, M. H., and A. Reinberg: "Chronobiology in Allergy and Immunology," Charles C. Thomas, Springfield, Illinois, 1977.

Stein, M. (editor): "New Directions in Asthma," Amer. Col. Chest Physicians, Park Ridge, Illinois, 1975.

Swineford, O., Jr.: "Asthma and Hay Fever," Charles C. Thomas, Springfield, Illinois, 1971.

Thurlbeck, W. W.: "Chronic Airflow Obstruction in Lung Disease," Saunders, Philadelphia, 1976.

BRONCHIECTASIS. An inflammatory or degenerative condition of the bronchi and bronchioles in which the tubes are dilated; usually associated with abscess formation. The two main symptoms of bronchiectasis are a persistent cough and the expectoration of large amounts of sputum, sometimes foul-smelling. The condition may follow the advent of such diseases as broncho-pneumonia, tuberculosis, or lung abscess. However, the disease occurs in some patients who have no history of any prior infection. Symptoms of advanced bronchiectasis include marked weight loss, fever, loss of appetite, and in most cases, extreme weakness. A person with this disease may live for years, although he is generally uncomfortable because of the foul odor of the sputum and feeling of general malaise. If the bronchiectasis is limited to one area of the lung, surgical removal may be advised. Prolonged administration of antibiotics, use of expectorants, steam-inhalation, postural therapy to assure efficient utilization of normal pathways of bronchial discharge, relocation to dry and warm climate are among corrective measures sometimes used.

BRONCHITIS. An inflammation of the mucous membrane of the tubes leading from the windpipe to the lungs (*bronchi*). The condition usually affects the larger bronchi. When the smaller bronchi are affected, the condition is more serious. When inflammation of the smallest bronchi (or *bronchioles*) occurs, the disease is actually bronchial pneumonia. Acute bronchitis is found most often in children under three years of age, in older people, but it can occur at any age. Factors contributing to the disease include occupational conditions, diet, and general health. It appears that residence in damp, foggy climates may be a contributing factor.

Acute bronchitis is often termed "chest cold." The condition often develops after the common cold and includes chest discomfort, dry cough, fever, and loss of energy. The cough may become severe and produce mucus. Predisposing factors are exposure, chill, fatigue, malnutrition, and rickets. Physical and chemical irritants, such as tobacco smoke, strong acid fumes, ammonia, chlorine, sulfur dioxide, and bromine, may trigger the condition. The significant danger of acute bronchitis is the possible onset of pneumonia.

Generally, *chronic bronchitis* is a much more serious condition. Chronic bronchitis is now identified as one of the chronic obstructive lung diseases (COLD), which also include obstructive emphysema, bronchiolitis, cystic fibrosis, and Kartagener's syndrome, among others. Patients with this condition have a chronic cough and expectoration along with recurrent acute infections of the lower respiratory tract. The condition is especially prevalent during winter months. Chronic bronchitis usually develops over a period of years, with the tendency for an acute upper respiratory tract infection to be invariably followed by a persistent cough which hardly disappears before another episode commences. Usually, in the morning, the victim must devote considerable effort to expectoration of a thick, sticky sputum. Wheezing may be present. Later complications include shortness of breath. If left untreated, the disease may place a large strain on the heart, with resulting congestive heart failure, or an infection, such as influenza or pneumonia.

Major attention must be directed to the patient's general health. Every effort must be made to facilitate the raising of sputum and clearing of air passages. Expectorants, steam inhalation, and vasodilators are helpful. Mucolytic agents frequently are effective in loosening thick, tenacious sputum. In the instance of shortness of breath, bronchodilator aerosols may be used. In more severe cases, inhalation of oxygen is used. An important part of the therapy for bronchitis is avoidance of environmental irritants, with particular emphasis on cigarette smoking.

BRONCHODILATORS. Bronchial Asthma.

BRONCHOGENIC CARCINOMAS. Cancer and Oncology.

BRONCHOSCOPY. Internal visual examination of the bronchi by means of a tubelike instrument (bronchoscope), which contains a light. The bronchoscope is introduced into the mouth and passed through the throat into the bronchial tree.

BRONTIDE. A natural explosive noise, frequently unexplained. Brontides have been well documented and frequently are associated with seismic activity and, in some cases, as precursors to major earthquakes. Some explanations offered have included ground-to-air acoustic transmission from shallow earthquakes, as well as noises from the sudden eruption of gas from high-pressure sources in the ground. Frequently, what may appear to be a brontide will be a noise from distant thunder or artillery practice, as well as other anthropogenic causes, such as sonic booms. Brontides or natural booming noises have been reported since ancient times. Some noises have been given colloquial names. These include the "Barisal guns" heard in the Ganges delta area; "Seneca guns" in New York State; *mistpoeffers* ("fog belchers") off the coast of Belgium.

Recent attention to brontides was brought about in connection with mysterious noises noted off the eastern coast of North America during 1977 and 1978. Many thousands of persons reported these noises. Investigation by the U.S. Naval Research Laboratory accounted for about 70% of these booms as caused by supersonic aircraft. A study of Mitre Corporation concluded that the remaining booms most likely resulted from natural causes, not satisfactorily explained. Possibly some were caused by *rock bursts*, i.e., the fracture of an exposed or near-surface rock face. Where no earthquake activity can be associated with brontides, anthropogenic causes are usually suspected even though no noise-making activities of this type are identified within hearing range. It is suggested that certain favorable combinations of atmospheric conditions can enable hearing sounds as much as 100 kilometers (62 miles). Some investigators have suggested that this may explain the Belgian mistpoeffers, i.e., they may be caused by thunder or artillery fire at a considerable distance, but within audio detection range because of favorable atmospheric conditions.

In terms of earthquake precursory brontide episodes, observations were reported in the vicinity of the South Carolina earthquake of 1886 at least a year prior to the earthquake.

References

Claflin-Charlton, S., and G. J. MacDonald: "Sound and Light Phenomena: A Study of Historical and Modern Occurrences," Mitre Corporation, McLean, Virginia, 1978.

Gold, T., and S. Soter: "Brontides: Natural Explosive Noises," *Science*, **204**, 371–374 (1979).

Hill, D. P., et al.: *Bull. Seismol. Soc. Amer.*, **66**, 1159 (1976).

Staff: "Investigation of East Coast Acoustic Events," U.S. Naval Research Laboratory, Washington, D.C., 1978.

BRONZE. Copper.

BRONZE AGE. An archeological term to designate a cultural level that originally was the middle division of the so-called three-age system. The age is characterized by bronze technology. The term is principally of European interest, inasmuch as the age coincides with written history in Asian archeology and bronze was not used extensively in Africa and the Americas. It was preceded by the Stone Age and followed by the Iron Age.

BROOKHAVEN NATIONAL LABORATORY. Particles (Subatomic).

BROOKITE. Brookite, composed of titanium dioxide, TiO_2, is an orthorhombic mineral of the same chemical composition as rutile and octahedrite. It was named for the English mineralogist H. J. Brooke. See also **Rutile; Titanium Dioxide.**

BROOMCORN. Of the family *Gramineae* (grass family), genus *Sorghum, Sorghum vulgare* var. *technicum* var. A variety of tropical grasses, characterized by having an inflorescence in which the branches are very long and slender, growing in loose panicles. The plant is extremely drought resistant and so adapted for regions having an arid climate. The close-bunched stiff inflorescent-branches account for the principal use of broomcorn, i.e., the manufacture of brooms and wisk brooms, although this use has been much displaced by plastic materials. Some varieties of the sorghum grasses contain sweet juices, which can be made into a sweet syrup and which find uses like those of maple syrup. However, the syrup from any sorgo contains considerably more invert sugar. Sorghum molasses is also sold commercially.

BROWN BEAR. Bears.

BROWNIAN MOVEMENT. The random movement observed among microscopic particles suspended in a fluid medium. The phenomenon was observed in 1827 with suspensions in liquids, colloids, by Robert Brown, English botanist, who is said to have attributed it to living organisms. Not until the kinetic theory was developed was it generally understood to be due to the thermal agitation of the suspending medium. A smoke particle floating in the air, for example, is battered on all sides by the high-speed air molecules. The resultant displacement is for the most part nearly zero, but there are statistical inequalities which now and then reach such magnitude as to produce motions visible in a high-powered microscope, and which result in an irregular migration of the particle. In fact, such particles may be regarded essentially as huge molecules, with mean square speeds of thermal motion proportionately smaller as their masses are larger than that of the true molecules of the surrounding medium.

In a series of papers published from 1905 to 1908, Einstein successfully incorporated the suspended particles into the molecular-kinetic theory of heat. He treated the suspended particles as being in every way identical to the suspending molecules except for the vast difference of their size. He set forth several relationships which were capable of experimental verification and he invited experimentalists to "solve" the problem.

Several workers undertook this task. The most notable of these was Perrin. Perrin's special success was due to his technique for preparing particles to suspend which were of uniform and known size. The uniformity was achieved by fractional centrifuging, and the size was established by noting that they could be coagulated into "chains" whose length could be measured and whose "links" could be counted. The microscopic observation of these uniform particles enabled Perrin and his students to verify the Einstein results and to make four independent measurements of Avogadro's number. These results not only established an understanding of Brownian movement, but also they silenced the last critics of the atomic view of matter.

Probably the simplest example of Perrin's experiments was his test of the Law of Atmospheres. If it is assumed that air is at rest and has the same temperature from ground level upward, it can be shown that the pressure (and concentration) of the air falls off exponentially with increasing altitude. For particles of mass m and density ρ suspended in a medium of density ρ' at absolute temperature T, the ratio of the particle concentrations n_1 to n_2 at heights h_1 and h_2 is given by

$$\frac{n_1}{n_2} = \exp\left[-\frac{mg(\rho - \rho')N_0(h_1 - h_2)}{\rho RT}\right]$$

where N_0 is Avogadro's number, g is the acceleration of gravity, and R is the universal gas constant. Although the concentration of air varies slowly with height, the concentration of the relatively heavy particles varied significantly over a height change of a few millimeters. By observing the concentration variation as a function of height, all quantities in the given equation were known except Avogadro's number which could, therefore, be determined.

References

Einstein, Albert: "Investigation of the Theory of the Brownian Movement," A. D. Cowper, translator, Dover, New York, 1956.

Perrin, Jean: "Atoms," D. L. Hammick, translator, Constable, London, 1923.

Mysels, K. J.: "Introduction to Colloid Chemistry," Wiley, New York, 1959.

BROWN LUNG. **Pneumokonioses.**

BROWN ROT. **Ascomycetes.**

BROWN-TAIL MOTH (*Insecta, Lepidoptera*). A European species, *Euproctis chrysorrhea*, related to the tussock moths of this continent. It was introduced into Massachusetts during the last century and together with the gypsy moth, another introduced species, has become an important pest in the New England states and the adjacent Canadian provinces, attacking shade and fruit trees. Spraying with lead arsenate in the late summer has been found an effective method of control, as well as the collection and destruction of the winter nests in which the caterpillars hibernate.

The hairs of the larva are very irritating to the human skin.

BRUCELLOSIS. This disease, caused by infection with the gram-negative *Brucella*, is transmitted from domesticated animals to humans. The three species of *Brucella* usually implicated are *B. melitensis* (from goats); *B. suis* (from hogs); and *B. abortus* (from cattle). In recent years, *B. canis* (from dogs) also has been implicated in a few cases of brucellosis. Prior to the initiation of mandatory milk pasteurizing regulations in a number of countries several years ago, the principal cause of brucellosis was the ingestion of raw milk and butter and cheese prepared from unpasteurized milk. As recently as 1950, about 3500 cases of the disease were reported each year in the United States. The cases per year, as of the early 1980s, average about 200. The persons now at highest risk are workers in the meat-packing and livestock industries. Less than 10% of cases now result from ingestion of milk products. Only a few of these can be attributed to products made in the United States.

The microorganism generally enters through the mucous membranes of the mouth and throat, or through breaks in the skin. The organism then reduces the lymphatics, traverses the lymph node barriers, and invades the bloodstream. After this, almost any organ of the body can become involved, although lesions are most frequently seen in the spleen, liver, lymph nodes, and bone marrow. However, the heart, lungs, joints, and prostate, among other organs, can be infected. The incubation period may range from a few days to several months. Symptoms of brucellosis include chills and fever, headache, loss of weight and appetite, and myalgia. There may be pain in the region of vertebrae, or acute arthritis may be manifested. Untreated, the infection may produce complications, including osteomyelitis and localized nephritis.

Treatment is usually by tetracycline over a period of 3 weeks or longer. Trimethoprim-sulfamethoxazole also have been used. This treatment usually is quite successful. Of over 2000 cases reported since 1965, only 2 fatalities directly attributable to brucellosis have been reported. At one time, the common name for brucellosis was undulant fever or Malta fever.

BRUCINE. **Nux Vomica Tree.**

BRUCITE. The mineral brucite is magnesium hydroxide corresponding to the formula $Mg(OH)_2$; iron and manganese may occasionally be present. The crystals are usually tabular rhombohedrons of the hexagonal system; it may also occur fibrous or foliated. Brucite has one perfect cleavage parallel to the prism base; hardness, 2.5; specific gravity, 2.39; luster, pearly to vitreous; commonly white but may be gray, bluish or greenish; transparent to translucent. Brucite is a secondary mineral found with serpentine and metamorphic dolomites. It has been found in Italy, Sweden, and the Shetland Islands; and in the United States in New York, Pennsylvania, Nevada and California. Brucite was named in honor of Archibald Bruce, an American physician.

BRUCKNER CYCLE. **Climate.**

BRUNTON COMPASS. This type of compass is specially designed for geologists and fitted with a clinometer and other devices for reading both horizontal and vertical angles.

Named after its inventor David W. Brunton (1849–1927), a U.S. mining engineer, the device enables the comparatively easy determination of the strike and dip of rock formations. The device is used primarily in sketching mine workings and in preliminary topographic and geologic surveys on the surface, such as determining stratigraphic thickness and vertical elevations.

BRUSH (Electrical Machinery). A device for conducting current to or from a rotating part. The brush is stationary, and is held and guided by a fixed brush holder in which it slides freely. There may be several brushes side by side to form a single-brush set. The rotating member may be the commutator of a dc generator or motor, or it may be the slip rings of an ac motor or generator. Examples of brushes might also include those used in magnetos and static electricity machines.

Brush materials, commonly used include carbon, carbon graphite, graphite, resin-bonded graphite, metal graphite, and electrographitic substances. Important factors for designers in selecting brush materials are abrasiveness, coefficient of friction, contact drop, current capacity, hardness, ability to withstand high peripheral speeds, specific resistance, and transverse strength. Spring-leaf copper and copper gauze are rarely used except in special circumstances. Circuit connections to carbon brushes are made directly to the brush by means of short flexible cables from the external circuit because the contact surface between brush holder and brush is an unreliable conductor.

Brushes wear and must be replaced periodically. Also, their ends should fit commutators so as to make good contact over the entire brush surface. They may sometimes need periodic redressing with sandpaper in order to maintain the proper shape of contact. The most serious fault of a brush is the formation of electric arcs between the rotating member and the brush. This may be due to the condition of the brush, vibration of the brush in the holder, or improper setting of the brush. The position of the brushes on the commutator is adjusted so the coils or turns being shorted by the brushes will have a minimum voltage. This means that the coil sides will be in the position of minimum flux, which position will depend upon the load unless correction is applied. In modern machines interpole windings are used to compensate for the distortion of the field caused by the armature current so the brush position is opposite the center of the poles. In machines without some form of compensation the position will be to one side of this and is adjusted to give minimum sparking under normal load.

BRUSH TURKEY. **Megapode.**

BRUSSEL SPROUTS. **Brassica.**

BRUXISM. Grinding of the teeth while sleeping is termed *bruxism*. When this is done without awareness while awake, the habit is referred to as *bruxomania*.

BRYONIA. A perennial herb (*Bryonia alba* L.) of the *Cucurbitacea* family that grows mainly in woods, thickets, and fields in central and southern Europe, western Asia, the Far East, and North Africa. An extract from the roots is used for flavoring liqueurs, bitters, and drugs. The extract contains resin, phytosterine, bryonol, enzymes, terpenes, fatty acids, protein substances, and glucosides.

BRYOPHYLLUM. Genus of the family *Crassulaceae* (orpine family). Several tropical plants have the unusual habit of reproducing by means of their leaves. If a mature leaf is removed from one of the plants and placed on damp sand, within a short time there appear in the notches of the leaf margin, tiny roots and, later, small green plants which soon become independent of the parent leaf. See accompanying figure. Infrequently the little plants appear in the notches while

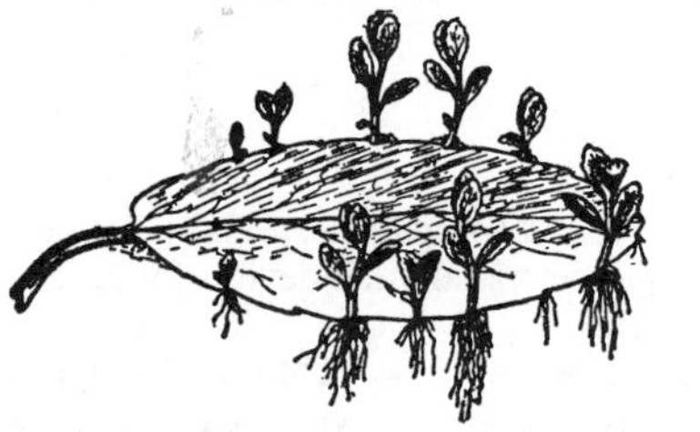

Vegetative reproduction of *Bryophyllum calycinum.* New plants develop in the notches of the leaves.

the leaf is still attached to the parent plant. The explanation of this uncommon habit is that, in the development of the leaf, certain cells in the leaf notches remain permanently embryonic. For reasons unknown, their development is inhibited so long as the leaf remains attached. Severing of the leaf removes this inhibition and the embryonic cells resume active development. See also **Asexual Reproduction.**

BRYOPHYTES. This subdivision of plants, comprising mosses and liverworts (or *Hepatics*), is a group having some 20,000 species. Most of these are terrestrial plants. The bryophytes inhabit a wide range of habitats, from dry barren rocks to submerged objects, but are most frequent where an abundance of moisture is assured. They are found on trunks and branches of trees, on the soil, and even on the leaves of some tropical plants.

The body of these plants is small and without much structural complexity. See Fig. 1. Rhizoids, slender outgrowths which serve mainly to attach the plant to its substratum and which serve only slightly as absorbing organs, are common. They may be single-celled or multicellular, but are colorless. The habit of the plant body is divese. In many hepatics it is a thin flat thallus, one to several cells thick. Often the edge of the thallus is so lobed that it appears to be differentiated into a central stem and lateral leaves. In the mosses this differentiation is much greater. There is an erect central stem which bears many thin radiating leaves. However complex the structure may be, the cells of the plant-body of a bryophyte show very little differentiation. The central cells may be longer; other cells may have thicker walls; cells nearer the surface contain more chloroplasts,

Fig. 1. (Left) The tip of a plant of *Mnium* which bears antheridia, cut lengthwise to show the antheridia. (Right) The tip of an archegonial plant of *Mnium*, cut lengthwise to show the archegonia.

but there is never any real modification of cells to form vascular elements. The latter are entirely wanting in this group, thus the group is separated from the *Tracheophyta* in the classification of plants.

The sex organs of the bryophytes are highly developed objects, which distinguish this group very sharply from the lower plants, both algae and fungi. The antheridium, in which the sperm cells are formed, is a club-shaped multicellular body. The archegonium is a flask-shaped body, also multicellular. Antheridia and archegonia may be formed on the same plant or on different plants; often they appear at different times so that self-fertilization is largely prevented. The sperm, a biciliate actively motile cell, swims to the egg, which is located in the swollen basal portion of the archegonium. There a sperm unites with the egg to fertilize the latter and incite growth of a new generation. But the plant resulting from this fertilized egg is one entirely unlike the parent plant. Commonly it is a well-developed plant, but less conspicuous than the one on which it is usually entirely dependent for its food supply. When mature it forms a large mass of small spherical cells called spores, which are freed from the body in which they are formed and carried away by currents of air. On reaching a suitable environment each spore germinates and eventually forms a plant like that which bore the sexual organs. There is then in the bryophytes a very definite alternation of generations, a haploid sexual generation called the gametophyte, bearing male and female sex organs, and a diploid asexual generation called the sporophyte, in which haploid asexual spores are formed. The gametophyte is green and carries on photosynthesis. In most cases it is terrestrial. The sporophyte, however, depends on the gametophyte for its food supply and water.

Bryophytes are separated into two classes, the *Hepaticae* or liverworts and the *Musci* or mosses. Each class is subdivided into three orders, as follows:

Class I. *Hepaticae.*
- Order 1. *Merchantiales.*
- Order 2. *Jungermanniales.*
- Order 3. *Anthocerotales.*

Class II. *Musci.*
- Order 1. *Sphagnales.*
- Order 2. *Andreaeales.*
- Order 3. *Bryales.*

The liverworts are generally considered lower in the scale of evolution than the mosses. The thallus, or plant body, of the liverworts is prostrate and flat. When it forks, the two branches are equal, a method of branching known as dichotomous. In the second order of liverworts the thallus is so divided as to appear leafy, the order often being called the leafy hepatics. All liverworts have unicellular rhizoids borne on the lower side of the thallus by which they are anchored firmly to the substratum. The sex organs are borne embedded in the body of the thallus or in special outgrowths called gametophores, which rise from the thallus. The small sporophytes are dependent on the gametophyte for their nutrients. In these plants an asexual reproduction occurs by means of small masses of cells which develop from the gametophyte to which they are attached by a very slender stalk. These small bodies, called gemmae, are easily separated from the parent plant. They develop into a new gametophyte when they are carried by wind or water to a suitable environment.

The three orders of the liverworts form an interesting series, of increasing complexity. The *Marchantiales* include forms which have

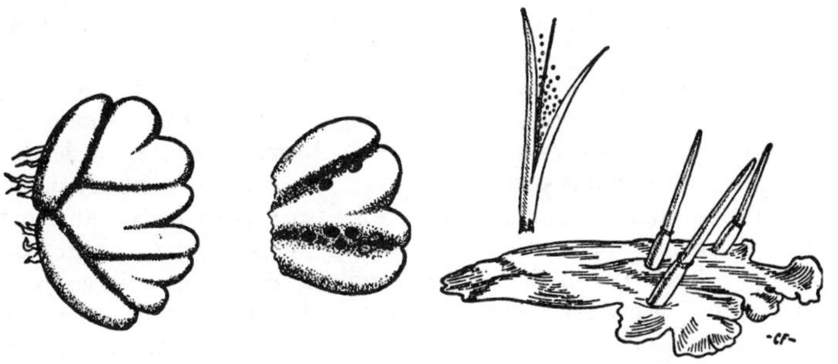

Fig. 2. (Left) Gametophytes of a liverwort, *Riccia*. The bodies embedded in the right-hand plant are sporophytes. (Right) *Anthoceros*, one of the horned liverworts. The "horns," are sporophytes. A portion of a sporophyte is shown separately to illustrate the method of liberation of spores.

a prostrate thallus, often showing a structure of considerable complexity. The sporophyte is very simple. Growth of the thallus is by repeated divisions of a single apical cell which itself sometimes forms two such cells, whose continued divisions form a dichotomous branching of the thallus. As the thallus increases in length at the apical end, death of the cells occurs at the other end, so that the plant slowly grows ahead until in time a fork is reached and the two halves separate by progressive disintegration of the older portions.

One of the simplest members of this group is *Riccia*, a small plant found either floating on still waters or growing on wet mud. Some species are thick and fleshy, others are slender much-branded bodies, having a very evident median groove. See Fig. 2.

From the lower surface single-celled rhizoids grow downward. From this surface also, thin scales or plates are developed, forming an overlapping row along the middle of the thallus. Both antheridia and archegonia are formed on the upper surface along the midrib. The antheridia have a wall a single cell in thickness, and contain many sperm mother cells, each of which divides to form two biciliate sperms. The archegonial wall is also a single cell in thickness, and encloses a row of six cells, four of which are the canal cells, the other two a ventral cell and an egg cell. When the latter is mature, the other five disintegrate, while at the same time the apical cells of the archegonium split apart, forming a canal through which the sperm swims to fuse with the egg. The sporophyte which develops from the fertilized egg remains embedded in the gametophyte thallus; when mature it is nearly all sporogenous tissue enclosed in a thin-walled capsule. A more complex member of this order is *Marchantia*. See Fig. 3. In this the gametophyte thallus is several inches long and from a half an inch to an inch (2.5 to 5 centimeters) broad. Its lower surface bears rhizoids and scales, or lamellae, quite like those of *Riccia*. The upper part of the thallus, just beneath the upper surface, contains a number of large chambers, each connected with the outside air by a

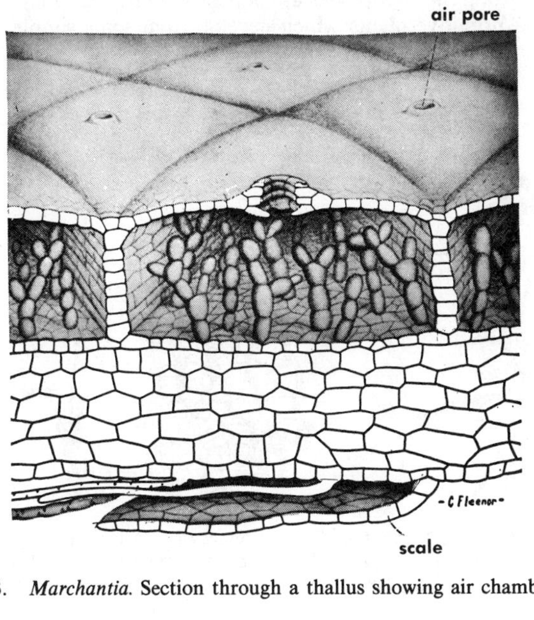

Fig. 3. *Marchantia*. Section through a thallus showing air chambers.

large pore. Gemmae are produced in cuplike organs growing out of the upper surface of the thallus. The sexual organs in this plant are not formed in the thallus, but are borne on special erect branches called gametophores. Antheridia and archegonia are borne on different plants, the plants thus being dioecious. They are quite like the antheridia and archegonia of *Riccia*. The sporophyte is considerably larger than that of *Riccia*, with a well-developed foot attaching it to the gametophyte, a short thick stalk which pushes the spore-containing capsule out from the tissues of the gametophore. Not only does this capsule contain large numbers of spores, but also, scattered among the spores, slender elongate cells called elaters, whose walls have spiral thickenings. These are affected by differences in humidity which cause the elater to twist about, apparently to stir up and loosen the spores. In *Marchantia* the gametophyte is very elaborately developed; the sporophyte, very simple.

In the second order of liverworts the thallus is so incised on its margins that it appears to bear two rows of small leaves. See Fig. 4. The *Jungermanniales* are small plants, many of them very delicate, growing in wet places, either on the ground or on rocks and tree trunks. Cellular differentiation in the thallus is very slight. In them the sporophyte is much more highly developed than in the first order. It has a long slender erect stalk which bears the capsule. When the

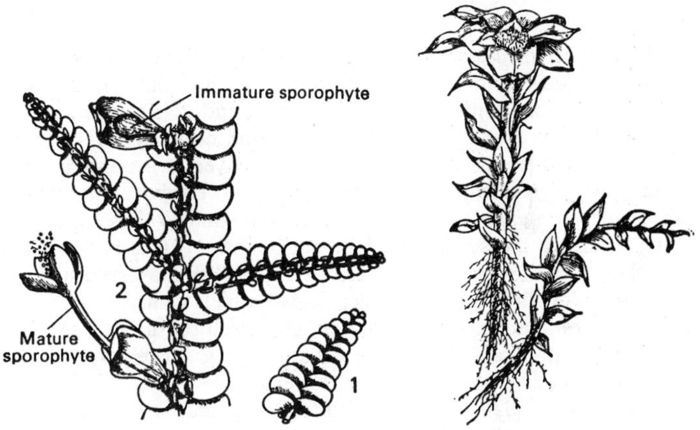

Fig. 4. (Left *Porella*, a leafy liverwort: (1) a branch seen from the upper side; (2) a portion of a plant seen from the lower side; sporophytes are also visible, attached to archegonial branches, each partially enclosed in a perianth. (Right) *Mnium*, a common genus of mosses.

latter is mature, it splits into four valves, which spread apart and free the spores within. Members of this suborder also form a series more or less as do the *Marchantiales*. The third order, the *Anthocerotales*, is a small group containing three genera. In all of them the gametophyte thallus is of a very simple type, with no great cellular differentiation, and with the sex organs always embedded. In this order the sporophyte is a most interesting object, far advanced in comparison with those of the other liverworts. It is an erect, slender, more or less cylindrical object composed of a basal foot and a long capsule. The central portion of the capsule is a rod of sterile tissue called the columella. Around it is the sporogenous tissue. The spores are formed in zones alternating with narrow bands of sterile tissue. Outside the sporogenous tissue is a wall of sterile tissue composed of chlorophyll (see **Pigmentation (Plants)**) containing cells. The epidermal portion of this wall contains many stomata (see **Stomate**). The basal portion of the sporophyte, just above the foot, is composed of meristematic cells, which by their divisions cause the capsule to elongate. When mature this capsule splits into two valves which pull apart, resembling horns. These plants, with their very simple gametophyte and very elaborate sporophyte, contrast strikingly with the *Marchantiales*. In fact, the sporophyte is so advanced that some botanists place this order in a separate class, the *Anthoceratae*.

The liverworts are of no economic importance, but are of interest because they suggest what may have been the habit of those plants which first left the water and grew on land. They have never become independent of water, since it must be present if fertilization is to occur.

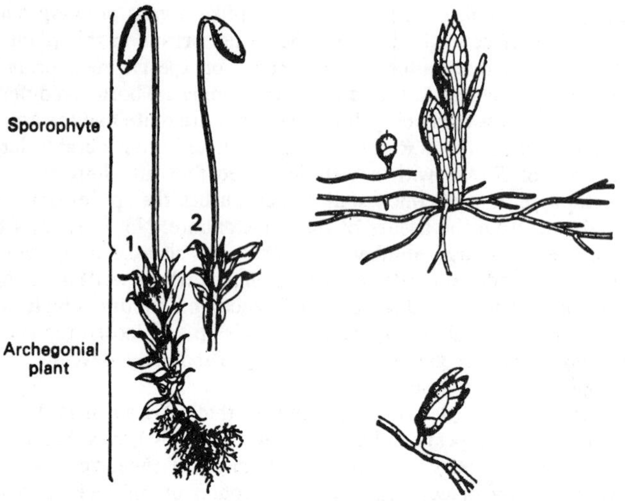

Fig. 5. (Left) (1) The sporophyte of a moss-*Mnium* attached to the gametophyte; (2) the same, the gametophyte being cut away to show the enlarged venter of the archegonium. (Right top) Moss protonema, showing the production of those buds which become the upright, leafy shoots. (Right bottom) The bud of a moss *Mnium*.

The number of mosses known is much larger than the number of liverworts. See Fig. 5. Mosses are found in many different regions, being much more abundant than liverworts, and able to grow under a wide range of environments. They are found in greatest abundance in moist shaded regions where they often cover an extensive area. Other species grow on the trunks of trees, often well above the ground, where they are exposed for long periods of time to desiccation. In tropical forests, mosses often clothe not only the trunks but also the branches of trees with a thick green covering; they may even succeed in growing on the thick evergreen leaves which characterize many tropical trees. Some species of mosses grow well on the exposed surfaces of barren rocks. A few are equatic, living entirely submerged in running water throughout their existence.

Usually moss plants are small (often tiny), and seldom exceed a few inches in length. A few genera, such as *Fontinalis,* which grows in water, and several tropical members, grow to lengths of 10–15 inches, which is very unusual in this group. Moss plants show a much higher development than hepatics. Usually the gametophyte, the part ordinarily seen and called a moss, has a very distinct, often erect stem, which bears many small radiating leaves. In each leaf there is generally a fairly evident midrib. There are many rhizoids growing from the lower part of the stem and attaching the plant to the ground. The rhizoids of mosses are longer than those of liverworts and are multicellular. There is no true vascular system in any moss, though the cells of the central portion of the stem are often much longer and more slender than those surrounding them. The sexual organs of mosses are very similar to those of hepatics and are borne at the tips of the stem or branches. Biciliate sperms are formed which must have water in which they can swim to the egg. The fertilized egg gives rise to a sporophyte which is much more highly organized than that of liverworts but still entirely dependent on the gametophyte. The basal portion of the sporophyte is the foot, a mass of cells in close contact with those of the gametophyte. Above the foot there is a stalk which in most mosses is very long and slender. It bears at its top a capsule or spore-bearing sac. This capsule has a very specialized structure. The axis of the capsule is a mass of sterile tissue called the columella. Around this the sporogenous tissue occurs, in turn surrounded by a wall many cells thick and with large cavities within it. The basal part of the capsule is also a mass of sterile tissue, often considerably swollen, known as the apophysis. The apical portion of the capsule is very complex. Over its surface is the operculum, a layer of cells, which completely covers it and which falls off like a lid when mature. Beneath this and distinct from it is the peristome, which, when mature, splits into a number of slender teeth which react to changes in humidity, rolling back when dry and closing together when wet. Surrounding the developing capsule and remaining around it for some time is a loose jacket of cells in no way connected with it. This is the calyptra, formed from cells of the gametophyte which were originally cells of the archegonal wall.

The ripe spores of mosses are shaken out through the apical opening and scattered by currents of air. On germinating, these spores do not give rise to a new moss plant directly. Instead they form a slender branching filamentous structure called a protonema which very much resembles certain kinds of algae. There are two types of cells composing the protonema: one contains many chloroplasts and so carries on photosynthesis: the other lacks chloroplasts and forms colorless rhizoids which grow downward and attach the protonema to the soil. A peculiarity of the protonema is the cross-walls between cells: they are commonly diagonal to the long axis of the filament rather than at right angles. From the cells of the protonema short erect branches ending in small buds are formed. These buds develop into erect moss plants. The life history of a moss plant shows a very distinct alternation of generations. In addition there is the juvenile phase of the gametophyte, the protonema.

The first order of mosses, the peat- or bog-mosses, contains the single genus *Sphagnum* with many species of worldwide distribution, always growing in low, wet bogs. See Fig. 6. These mosses are considered very primitive. The gametophyte has an erect stem from which arise numerous branches, all of two kinds, either spreading, or pendent against the stem. The many leaves are small and but a single cell in thickness. Some of the cells are small and elongated, forming a fine anastamosing network in the leaf. These are living cells containing chloroplasts. The openings of the network are filled by very large inflated cells with thin walls and no protoplasm. Large pores in the walls of these dead cells permit free passage of water into the cell cavity. The small sporophyte has a spherical capsule, which is black or dark brown, and a very short stalk. The sporophyte is borne at the end of a specialized structure called a pseudopodium which lifts the sporophyte above the tuft of branches at the top of the gametophyte. the protonema of *Sphagnum* is a small flat thallus resembling that of the *Anthocerotales.*

Peat-mosses are the only members of the Bryophyte group which have any commercial value. Growing slowly for long periods of time, they gradually accumulate in the wet bogs in which they are found. Gradually the lower parts amass, together with such debris as may have accumulated, forming a compact mass known as peat, and used as fuel. The ability of the large hollow cells of the leaf of *Sphagnum* to absorb and retain large quantities of water leads to the extensive use of *Sphagnum* moss as a material in which to pack live plants for shipment. For this reason also, and because *Sphagnum* is naturally a sterile substance, harboring few bacteria, certain species have been used as surgical dressings, especially in times of great need.

The *Andreaeales* is a small group of small mosses growing on the surfaces of siliceous rocks. They are unimportant.

Most mosses belong to the third order, the *Bryales,* which are the true mosses.

As a group the Bryophytes are of little importance. They are recognized as primitive plants which develop from some simple ancestral

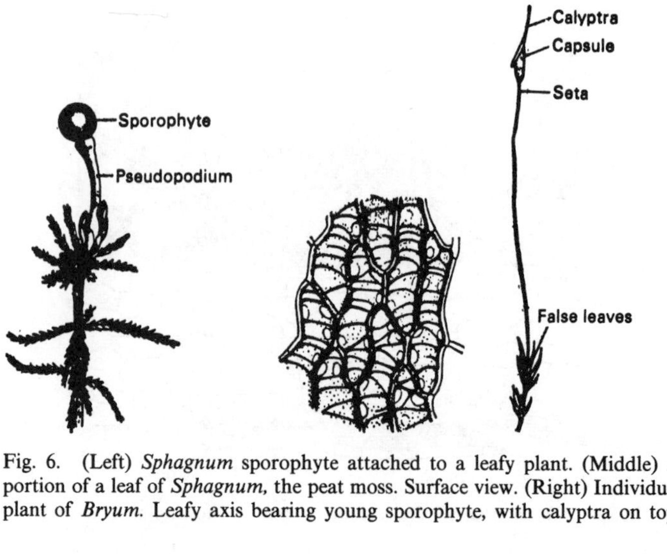

Fig. 6. (Left) *Sphagnum* sporophyte attached to a leafy plant. (Middle) A portion of a leaf of *Sphagnum,* the peat moss. Surface view. (Right) Individual plant of *Bryum.* Leafy axis bearing young sporophyte, with calyptra on top.

forms from which they have gradually diverged independently along several different lines. The existing forms do not form a single series, representing stages in the development of the most advanced forms, nor are they plants from which the higher plants have taken their origin. In this group the gametophyte appears in its most advanced form.

BRYOZOANS. **Invertebrate Paleontology.**

BUBAL. **Antelope.**

BUBBLE CAP. **Distillation.**

BUBBLE CHAMBER. A vessel filled with a transparent liquid so highly superheated that a moving ionizing particle initiates boiling in the liquid along its path. The superheating is produced by a sudden reduction of the pressure on the liquid below that at which the liquid boils. During a brief period the liquid will then boil where a disturbance is created by a charged particle but not elsewhere, so that tracks of bubbles in the liquid show the paths of charged particles. The tracks are recorded on stereophotographs in the same way as with a cloud chamber. See also **Cloud Chamber.** Bubble chambers are key instruments used in particle-physics experiments and may measure up to 12–15 feet (3.6–4.5 meters) in diameter.

Bubble chamber photographs store a large amount of data that must be extracted by first scanning the film for events that may satisfy the requirements of a particular experiment. For simple interactions, computer-guided pattern recognition methods may be applied in conjunction with human intervention and interrogation, combining the scanning and subsequent measuring processes. More often, the scanned film is transferred to measuring machines of varying complexity where positions of bubble images are measured with commensurate accuracy. As an example, on a flying-spot digitizer, a light spot, 15 micrometers in diameter, scans a photograph in a few seconds. A bubble image causes a signal in a photomultiplier, which is compared with time pulses similarly produced by a grating scanned by a synchronized light spot, locating the image with respect to fiducial marks. Hundreds of events per hour can thus be measured with excellent precision. In the spiral reader, a fine slit spirals over the photograph, again producing light signals. A flying-image digitizer will scan many different portions of a photograph simultaneously so that thousands of events can be processed per hour. The processing of bubble chamber data requires large amounts of computer time, on-line for scanning and measuring equipment, and off-line for spatial reconstruction of events from the three or more stereoscopic views, for kinematic fitting and selection of probable events, and finally for interpretation of results.

See also **Particles (Subatomic).**

BUBBLE CHARACTERISTICS. **Foam.**

BUBO. An inflamed, swollen lymph gland, particularly in the region of the armpits and groin.

BUBONIC PLAGUE. This disease, caused by infection with the gram-negative *Yersinia pestis*, is transmitted by fleas from rats to humans. Recent research has indicated that rabbits and domesticated dogs and cats may also harbor the microorganism. In the Old World in times past, the disease was commonly called *black death.* With much better control over rats, the incidence of the disease decreased. Bubonic plague is seldom seen in many of the countries in temperate climates and with advanced sanitation. Local outbreaks can occur in the United States, particularly in the region from the Pacific States eastward to Kansas.

The bacteria is present in the bloodstream of animals having the disease. Thus, a flea biting a diseased rat obtains some of the bacteria, which multiply to form a plug in the flea's gullet. This plug causes the flea to regurgitate some bacteria into the next rat it bites. When the rat dies of the plague, the fleas leave the dead animal in search of a new host. If a rat cannot be found, the flea will bite another animal or a person. If the disease is left untreated, mortality ranges from 25 to 50%.

At the site of the bite, a trivial, hardly noticeable pustule will form, but within a short period, the regional lymph node will enlarge. A very marked inflammatory reaction will involve surrounding tissue, producing a bubo. The inflamed region will be tender and usually very painful. In bubonic plague, entry of the microorganism into the bloodstream occurs early—hence the need for very early diagnosis and commencement of therapy. Incubation period of the disease extends from 2 to 10 days. Associated with the bubo and surrounding tenderness will be fever, increasing to high fever and chills, prostration, and septic shock within just a few days. A common complication is pneumonia, which progresses at exceptionally fast rates in patients and can cause respiratory failure and death within a matter of hours.

Diagnosed patients should be placed in isolation. Traditional treatment is antimicrobial therapy with large doses of intramuscular streptomycin. Tetracycline and chloramphenicol are sometimes used, as well as combined therapy with trimethoprim and sulfamethoxazole. For immunizing persons with high risk to bubonic plague exposure, a killed vaccine is available.

References

Butler, T.: "A Clinical Study of Bubonic Plague: Observations of the 1970 Vietnam Epidemic with Emphasis on Coagulation Studies, Skin Histology, and Electrocardiograms," *Am. J. Med.*, **53**, 268 (1972)

Butler, T., et al.: "*Yersinia pestis* Infection in Vietnam," *J. Infect. Dis.*, **133**, 493 (1976).

Von Reyn, C. F., et al.: "Epidemiologic and Clinical Features of an Outbreak of Bubonic Plague in New Mexico," *J. Infect. Dis.*, **136**, 489 (1977).

BUCKEYE TREES. **Horse Chestnut and Buckeye Trees.**

BUCKLEY GAGE. A very sensitive pressure gage, based on measurement of the amount of ionization produced in a gas by a specified current.

BUCKTHORN SHRUBS AND TREES. Of the family *Rhamnaceae* (buckthorn family), there are several species of shrubs, and infrequently sizeable trees. They are characterized by alternate, toothed leaves, small flowers, and, in some species, thorns. The accompanying table of record buckthorns in the United States attests to the fact, however, that some species and specimens can attain relatively significant proportions.

The Carolina buckthorn may be classified as a tall shrub or small tree, ranging in height between 10 and 30 feet (3 to 9 meters). The bark is a dark brown, reasonably smooth. This species of buckthorn does not have thorns. The leaves are large, elliptical, dark green, smooth, and only slightly pointed. The red-to-purple fruit is a little over $\frac{1}{4}$ inch (0.6 centimeter) in diameter, is of a sweetish taste, and contains three seeds. This shrub prefers wet lowlands and swampy areas, as may be found from Long Island southward to Florida, along the Ohio River valley, and as far west as Texas.

The alder buckthorn or glossy buckthorn ranges from 5 to 8 feet (1.5 to 2.4 meters) in height and was introduced into North America from Europe. The leaves are small, pale olive green, elliptical, and toothless. The fruit contains three seeds and is black, about $\frac{1}{4}$ inch (0.6 centimeter) in diameter. The shrub is reasonably local to swamps on Long Island and in northern New Jersey, although as the accompanying table indicates, is found in the midwest as well.

BUCKWHEAT (*Fagopyrum esculentum*; *Polygonaceae*). Aside from the grasses, buckwheat is the only plant used to any extent as a cereal in the United States. It is an erect branching plant from one to four feet tall, with a small root system and a smooth rather weak stem at each node of which is borne a single heart-shaped leaf. The inflorescence is a many-flowered raceme, the individual flowers being white or pink-tinged. The calyx lobes, five in number, are colored; there is no corolla. There are eight stamens and a single one-celled ovary which bears three curved styles. Cross-pollination is brought about by the numerous insect visitors attracted by the pleasant fragrance of the flowers. The mature fruit is a triangular brown or black achene. The single seed within contains an abundance of white endosperm high in starch content.

Buckwheat is an Asiatic plant which is cultivated in widely scattered regions. It grows well in cool climates, on poor soils, and where the

RECORD BUCKTHORNS IN THE UNITED STATES[1]

Specimen	Circumference[2] (inches)	Circumference[2] (centimeters)	Height (feet)	Height (meters)	Spread (feet)	Spread (meters)	Location
California buckthorn (1976) (*Rhamnus californica*)	24	61	30	9	25	7.5	California
Carolina buckthorn (1974) (*Rhamnus caroliniana*) (Co-Champion)	41	104	27	8.1	23	6.9	Virginia
Carolina buckthorn (1974) (*Rhamnus caroliniana*) (Co-Champion)	21	53	46	13.8	21	6.3	Tennessee
Cascara buckthorn (1977) (*Rhamnus purshiana*) (Co-Champion)	105 (at 29 inches)	267 (at 74 centimeters)	35	10.5	54	16.2	Oregon
Cascara buckthorn (1977) (*Rhamnus purshiana*) (Co-Champion)	99	251	37	11.1	50	15	Oregon
European buckthorn[3] (1972) (*Rhamnus cathartica*)	45	114	61	18.3	65	19.5	Michigan
Glossy buckthorn[3] (1976)	13	33	34	10.2	19	5.7	Michigan

[1] From the "Social Register of Big Trees," The American Forestry Association (by permission).
[2] At 4.5 feet (1.4 meters).
[3] Introduced

growing season is short. It matures early, producing flowers from 3–5 weeks after planting.

The plant is used as a green manure, being turned under to enrich the soil. Pancake flour is made from buckwheat seeds. The grain is also used as food for poultry and other domestic animals, either whole or divested of the hulls. Buckwheat flowers are a source of honey.

BUD. In botany, a bud is an undeveloped shoot and normally occurs in the axil of a leaf or at the tip of the stem. Once formed, a bud may remain for some time in a dormant condition, or may develop into a shoot immediately.

The buds of many woody plants, especially in temperate or cold climates, are protected by a covering of modified leaves called scales which tightly enclose the more delicate parts of the bud. Many bud scales are covered with a gummy substance, which serves as added protection. When the bud develops, the scales may enlarge somewhat but usually drop off, leaving on the surface of the growing stem a series of horizontally elongated scars. By means of these scars one can determine the age of any young branch, since each year's growth ends in the formation of a bud, the development of which causes the appearance of an additional group of bud scale scars. Continued growth of the branch causes these scars to be obliterated after a few years so that the total age of older branches cannot be determined by this means.

In many plants scales are not formed over the bud, which is then called a naked bud. The minute undeveloped leaves in such buds are often excessively hairy. Such naked buds are found in shrubs like the Sumac and Viburnums and in herbaceous plants. In many of the latter, buds are even more reduced, often consisting of undifferentiated masses of cells in the axils of leaves. A head of cabbage (see **Brassica**) is an exceptionally large terminal bud, while Brussels sprouts are large lateral buds.

Since buds are formed in the axils of leaves, their distribution on the stem is the same as that of leaves. There are alternate, opposite and whorled buds, as well as the terminal bud at the tip of the stem. In many plants buds appear in unexpected places: these are known as adventitious buds.

Often it is possible to find in a bud a remarkable series of gradations of bud scales. See Fig. 1. In the Budkeye, for example, one may observe a complete gradation from the small brown outer scale through larger scales which on unfolding become somewhat green to the inner scales of the bud, which are remarkably leaflike. Such a series suggests that the scales of the bud are in truth leaves, modified to protect the more delicate parts of the plant during unfavorable periods. See Fig. 2.

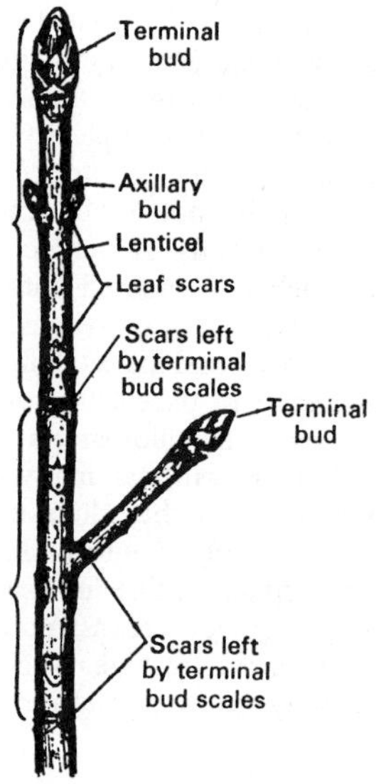

Fig. 1. Twig of buckeye (*Aesculus glabra*), showing two years of growth.

In zoology, a bud may be defined briefly as an outgrowth from the body which develops into a new individual. See **Budding:** and **Plant Growth Modification and Regulation.**

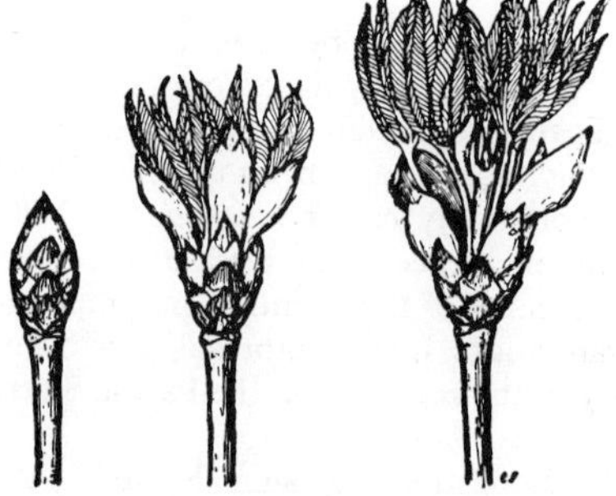

Fig. 2. A series of stages in the growth of a bud of buckeye (*Aesculus glabra*).

BUDAN THEOREM. Let $P(x) = 0$ be a polynomial of degree n with real coefficients and suppose that one wishes to locate its real roots. Choose a and b which are real numbers such that $a < b$ and neither of which is a root of the polynomial. Let V_a denote the number of variations of sign of $P(x)$, $P'(x)$, $P''(x), \ldots, P^{(n)}(x)$ for $x = a$, after vanishing terms have been deleted, and similarly V_b is the number of variations for $x = b$. Then $V_a - V_b$ is either the number of real roots $P(x) = 0$ between a and b or it exceeds the number of these roots by a positive even integer. A root of multiplicity m is here counted as m roots.

BUDDE EFFECT. The increase in volume of halogens, especially chlorine and bromine vapor, on exposure to light. It is a thermal effect, due to the heat from recombination of atoms.

BUDDING. This term is used to designate a process of asexual reproduction in which the young are formed as outgrowths of the parent body. See **Conjugation.** It is limited to animals or plants of relatively simple structure. In this process a portion of the wall of the parent cell softens and pushes out. The protuberance thus formed enlarges rapidly while at this time the nucleus of the parent cell divides. One of the resulting nuclei passes into the bud. Presently the bud becomes cut off from its parent cell and the process is repeated. Often the daughter cell starts to bud before it becomes separated from the parent, so that whole colonies of adhering cells are formed. Eventually cross walls cut off the bud from the original cell.

The term budding is also applied to a process of embryonic differentiation in which new structures are formed by outgrowth from preexisting parts. See illustration.

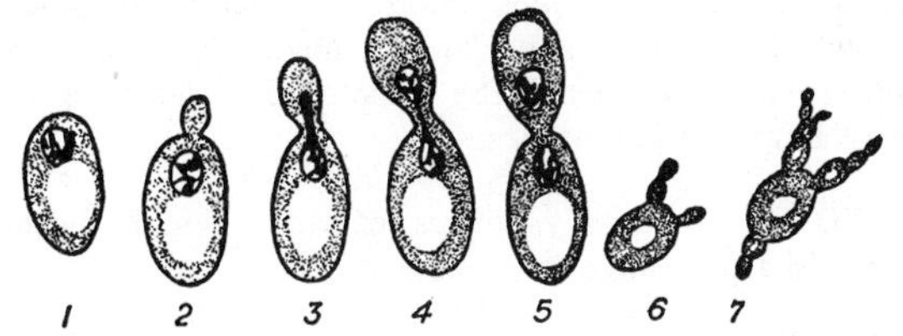

Reproduction of budding (1, 2, 3, 4, 5). Yeast cells are stained to show nucleus. (*Redrawn from Guillermond, The Yeasts, Wiley, New York*). Colony formation by rapidly growing yeast (6, 7). Nuclei are not stained.

A third use of the term budding is discussed in the entry on **Grafting and Budding.**

BUDDING (Virus). Virus.

BUDGERIGAR. Parrots and Cockatoos.

BUD MOTH (*Insecta, Lepidoptera*). This insect is an economic pest, notably on apple, pear, and pecan, and occurs throughout the United States. The apple and pear bud moth (*Tmetocera ocellana*) adult is a gray color, with a pale-beige band on forewings. The larva is brown, with black head, and ranges up to ½-inch (12 to 13 millimeters) in length. The insect eats buds, blossoms, and leaves and spins webs around them. This reduces fruit production by causing injury to terminal shoots and thus encouraging bushy growth rather than fruit production. EPN insecticide, an organic phosphorus compound, is effective against this pest. Noninsecticide control methods also can be effective. Young trees should be examined for bud moth damage in May. Dead, brown leaves are evidence of infestation by this pest. Nests can be destroyed by removal from trees, or they can be crushed on the tree to kill the enclosed caterpillars or pupae. Several parasites and predators, such as *Trichogramma* (minute wasps) and ground beetles help to control this insect. Birds also feed on the caterpillars. Mud dauber wasps sometimes store them in their shells where they are used as food for theri grubs.

The pecan bud moth (*Proteopteryx deludana*) is closely related in appearance and action and frequents pecan trees, upon which infestations can be quite damaging. Treatment is similar to that described for apple and pear. For pecan, chemical insecticides are most effective when applied just as the last of the petals are falling. A repeat application may be required.

BUERGER'S DISEASE. Arterial and Venous Disorders.

BUFFALO CARPET-MOTH (*Insecta Coleoptera*). More properly known as the carpet beetle but misnamed through the similarity of habits of its larva and those of the clothes moths.

The true carpet beetle is an introduced European species, *Anthrenus scrophulariae,* whose larva eats woolen materials of all kinds as well as furs and feathers. The adult is a compact oval insect about one-eighth inch long and marked with brick-red, black and white, and the larva is a brown hairy grub. A number of other species, native to North America, have the same habits and may be equally troublesome. The adults frequent flowers and eat pollen, hence they may migrate readily to houses. They are important pests in museums.

In the home, good housekeeping methods are usually an adequate safeguard against these pests. In special cases fumigation is necessary to destroy them but as a rule the use of sprays now supplied commercially for application to clothing and other fabrics is the only unusual measure required.

BUFFALOGRASS. Grasses.

BUFFALO FISH. Suckers.

BUFFER AMPLIFIER. Amplifier; Buffer (Computer).

BUFFER (Chemical). When acid is added to an aqueous solution, the pH (hydrogen ion concentration) falls. When alkali is added, it rises. If the original solution contains only typical salts without acidic or basic properties, this rise or fall may be very large. There are, however, many other solutions which can receive such additions without a significant change in pH. The solutes responsible for this resistance to change in pH, or the solutions themselves, are known as *buffers.* A weak acid becomes a buffer when alkali is added, and a weak base becomes a buffer upon the addition of acid. A simple buffer may be defined, in Brönsted's terminology, as a solution containing both a weak acid and its conjugate weak base.

Buffer action is explained by the mobile equilibrium of a reversible reaction:

$$A + H_2O \rightleftarrows B + H_3O^+$$

in which the base B is formed by the loss of a proton from the corresponding acid A. The acid may be a *cation,* such as NH_4^+, a *neutral molecule,* such as CH_3COOH, or an *anion,* such as $H_2PO_4^-$. When alkali is added, hydrogen ions are removed to form water, but so long as the added alkali is not in excess of the buffer acid, many of the hydrogen ions are replaced by further ionization of A to maintain the equilibrium. When acid is added, this reaction is reversed as hydrogen ions combine with B to form A.

The pH of a buffer solution may be calculated by the mass law equation

$$\mathrm{pH} = \mathrm{p}K' + \log \frac{C_B}{C_A}$$

in which pK' is the negative logarithm of the apparent ionization constant of the buffer acid and the concentrations are those of the buffer base and its conjugate acid.

A striking illustration of effective buffer action may be found in a comparison of an unbuffered solution, such as 0.1 M NaCl with a neutral phosphate buffer. In the former case, 0.01 mole of HCl will change the pH of 1 liter from 7.0 to 2.0, while 0.01 mole of NaOH will change it from 7.0 to 12.0. In the latter case, if 1 liter contains 0.06 mole of Na_2HPO_4 and 0.04 mole of NaH_2PO_4, the initial pH is given by the equation:

$$\mathrm{pH} = 6.80 + \log \frac{0.06}{0.04} = 6.80 + 0.18 = 6.98$$

After the addition of 0.01 mole of HCl, the equation becomes:

$$\mathrm{pH} = 6.80 + \log \frac{0.05}{0.05} = 6.80$$

while, after the addition of 0.01 mole of NaOH, it is

$$\mathrm{pH} = 6.80 + \log \frac{0.07}{0.03} = 6.80 + 0.37 = 7.17.$$

The buffer has reduced the change in pH from ±5.0 to less than ±0.2.

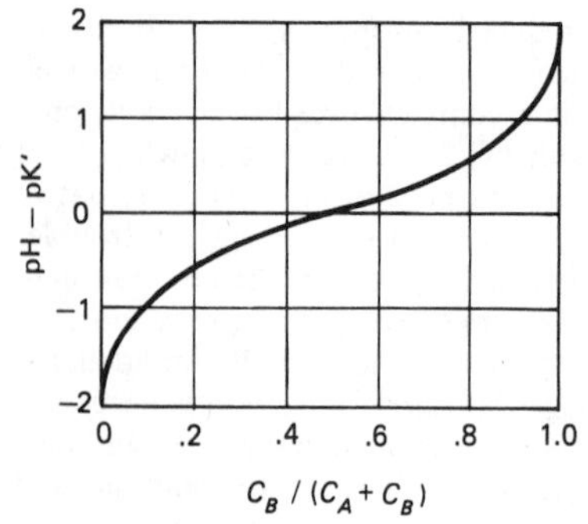

The pH of a simple buffer solution. Abscissas represent the fraction of the buffer in its more basic form. Ordinates are the difference between pH and pK'.

The accompanying diagram shows how the pH of a buffer varies with the fraction of the buffer in its more basic form. The buffer value is greatest where the slope of the curve is least. This is true at the midpoint, where $C_A = C_B$ and pH = pK'. The slope is practically the same within a range of 0.5 pH unit above and below this point, but the buffer value is slight at pH values more than 1 unit greater or less then pK'. The curve has nearly the same shape as the titration curve of a buffer acid with NaOH or the titration curve of a buffer base with HCl. Sometimes buffers are prepared by such partial titrations instead of by mixing a weak acid or base with one of its salts. Certain "universal" buffers consisting of mixed acids partly neutralized by NaOH, have titration curves which are straight over a much wider pH interval. This is also true of the titration curves of some polybasic acids, such as citric acid, with several pK' values not more than 1 or 2 units apart. Other polybasic acids, such as phosphoric acid, with pK' values farther apart, yield curves having several sections, each somewhat similar to the accompanying curve. At any pH, the buffer value is proportional to the concentration of the effective buffer substances or groups. See also **pH (Hydrogen Ion Concentration).**

The accompanying table gives approximate pK' values, obtained from the literature, for several buffer systems.

REPRESENTATIVE BUFFER SOLUTIONS

Constituents	pK'
H_3PO_4; KH_2PO_4	2.1
HCOOH; HCOONa	3.6
CH_3COOH; CH_3COONa	4.6
KH_2PO_4; Na_2HPO_4	6.8
HCl; $(CH_2OH_3)CNH_2$	8.1
$Na_2B_4O_7$; HCl or NaOH	9.2
NH_4Cl; NH_3	9.2
$NaHCO_3$; Na_2CO_3	10.0
Na_2HPO_4; NaOH	11.6

Buffer substances which occur in nature include phosphates, carbonates, and ammonium salts in the earth, proteins of plant and animal tissues, and the carbonic-acid-bicarbonate system in blood. See also **Acid-Base Regulation (Blood).**

Buffer action is especially important in biochemistry and analytical chemistry, as well as in many large-scale processes of applied chemistry. Examples of the latter include the manufacture of photographic materials, electroplating, sewage disposal, agricultural chemicals, and leather products.

BUFFER (Computer). An internal portion of a digital computer or data processing system which serves as an intermediary storage between two storage or data handling systems with different access times or formats. Buffer is also used to describe a routine for compensating for a difference in data rate, or time of occurrence of events, when transferring data from one device or task to another. For example, a buffer usually is used to connect an input or output device with the main or internal high-speed storage. The term *buffer amplifier* applies to an amplifier which provides impedance transformation between two analog circuits or devices. A buffer amplifier may be provided at the input of an analog-to-digital converter to provide a low source impedance for the analog-to-digital converter and a high load impedance for the signal source.

BUFFER STORAGE (Computer). Storage (Computer).

BUFFING. A finished process for producing a lustrous surface, usually on sheet metal. Buffing is usually effected by using a buffing wheel composed of layers of cloth sewed together to which a cake abrasive is periodically applied.

BUFFLEHEAD. Waterfowl.

BUG (*Insecta, Hemiptera*). Insects with sucking mouth parts usually arising near the front of the head, with antennae usually long but with few joints, and with wings, when present, thicker at the base and membranous at the tip, overlapping when folded to form a more or less conspicuous X on the back. The bugs are so diverse that no concise definition can be generally adequate. The term bug is not synonymous with insect.

In general terms, the head of a bug is ususlly triangular and smaller than the remainder of the body. The antennae are short, usually with 5 to 13 or more joints, and located below the eyes. When folded, the wings conceal most of the bug's body. The legs are slender and long. There are usually 11 segments to the abdomen.

There are about 50,000 living and fossil species. The oldest known fossil insect (Protocimex silurica) was obviously a bug. Parts of a wing were found in Sweden in the Upper Ordovician beds. For major families of bugs, see **Hemiptera.**

BUHRSTONE. Relatively porous, calcareous, and siliceous sandstones with sharp or angular grains.

BUILDER (Detergent). Detergents.

BULB (Botany). A thick short stem which grows many thick leaves in which food reserves are stored. In many bulbs the leaves are closely wrapped together, forming a compact body called a tunicated bulb, as is the case in the onion. In other bulbs the fleshy leaves are loosely arranged to form a scaly bulb, such as the Easter lily. Bulbs are particularly common in monocotyledonous plants, and aid the plants greatly in surviving long dry seasons. Commonly the term bulb is erroneously applied to any fleshy underground plant part, regardless of its nature, so that rhizomes, corms and tubers are all popularly classed as bulbs. The dahlia "bulb," for instance, is really a fleshy root.

BULBIL. In some plants, there occur small reproductive bodies called bulbils, or sometimes, bulblets. An example is found in the small black objects growing in the axils of the leaves of tiger lilies. These are actually buds in which the scales are very much swollen. When mature, the whole body falls to the ground and, under favorable conditions, puts out roots and, in time, grows into a new plant. Similar bodies are found in the familiar onion sets and in several sedges. Serving the same purpose are the small globose bodies which develop on the leaves of several species of ferns, as, for example, *Cystopteris bulbifera.* All of these kinds of bodies represent one method of vegetative propagation.

BULBUL (*Aves, Passeriformes*). Birds of several species found in Africa and the Oriental region, related to the babblers. They are said to be melodious singers.

The bulbul is a thrushlike bird, small, with brilliant plumage. The sexes are quite alike in plumage. The crest of the head is black. There are red spots below each eye. The throat and breast are white, with some brown on the back and wings. The tail is tipped with white. The under tail coverts are red. The bulbul attains a length of about 7 inches (18 centimeters). Diet consists of fleshy fruits, berries, and insects.

BULK (Diet). Dietary Requirements and Trends.

BULKHEAD. A bulkhead is a partition or a transverse strengthening frame. Ships' bulkheads are the important transverse partitions which subdivide the hold into separate water-tight compartments, being built from the keel to the bulkhead deck. They must be not only water-tight, but have sufficient structural strength to resist the bursting pressure to which they will be subjected when one bulkhead space is filled with water, while the adjacent one is empty.

In construction, any wall used to restrain fluid or semi-fluid pressure, such as that resulting from water or saturated earth in foundation excavations, is called a bulkhead or bulkhead wall.

BULKING AGENTS (Foods). **Bodying and Bulking Agents (Foods).**

BULK PHASE. **Gibbs Division Surface.**

BULK SAMPLING. **Sampling (Statistics).**

BULL. The male of certain animals, as domestic cattle, the bull elephant, and the bull alligator.

BULLFINCH (*Aves, Passeriformes*). Birds of northern Europe and Asia, related to the grosbeaks. The bullfinch is a favorite cage bird in Europe and can be trained to whistle. The male is plump with rich rose-colored plumage on breast and gray on back. The tail, wings, and head are covered with glossy black feathers.

BULLFROG (*Amphibia, Salientia*; Rana). Large frogs, closely related to the more familiar grass frogs and leopard frogs. The bullfrogs reach a length of 8 inches (20.3 centimeters). The flesh of the bullfrog is delicious.

BUMBLEBEE (*Insecta, Hymenoptera*). Stoutly built hairy bees of moderate to large size. Some species are colonial, building nests on the surface of the ground, while others live as parasites in the nests of other bumblebees.

Unlike the honeybee, bumblebees are not permanently colonial in temperate regions. Only the queen lives through the winter. When she emerges from hibernation in the spring she builds a nest or occupies an abandoned nest of a bird or mouse, and in it makes waxen cells in which she lays eggs and a waxen honey pot in which to store surplus food. She feeds her young until they mature, and only when they emerge as worker bees does the colony take on an organization like that of the honeybee. In the fall, males and queens appear and the colonies break up. The queens mate before hibernating. Most authorities include all of these bees in the genus *Bombus.*

The parasitic bumblebees, making up the genus *Psithyrus,* enter the nests of other bumblebees and lay their eggs to be cared for by the hosts. Their chief structural difference is the lack of pollen-gathering organs in the females.

Bumblebees are important in the cross-fertilization of red clover and other deep-throated flowers.

BUNDLE. Also often called vascular bundle or fibrovascular bundle. In most vascular plants the vascular tissues are arranged in the form of a cylinder. In many cases, notably in woody plants, this cylinder is a solid mass of cells. But in many plants, particularly in herbaceous dicotyledons and monocotyledons, the vascular tissues occur in strands which are more or less distinctly separated from one another, and are called vascular bundles. See Fig. 1. Such bundles appear as discrete objects when seen in a cross section of the stem. See Fig. 2. However,

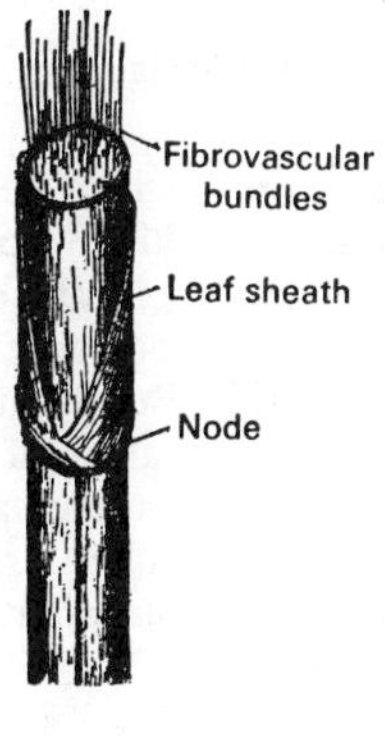

Fig. 1 Portion of corn stem (*Zea mays*), with the vascular bundles protruding, illustrating the structure of a typical monocotyledonous stem.

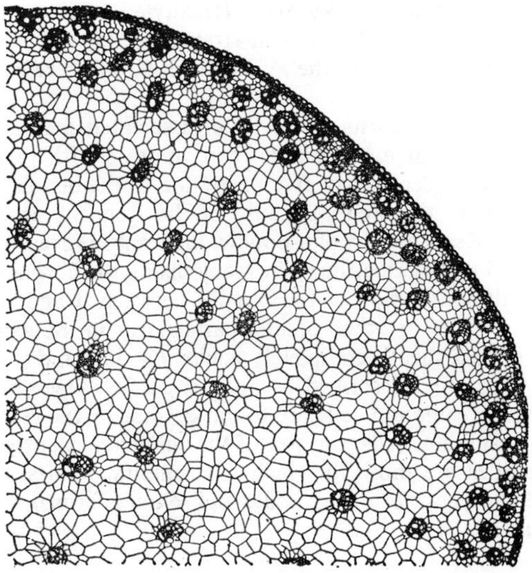

Fig. 2. Cross section of corn stem showing the distribution of fibrovascular bundles in a typical monocotyledon.

they really form a continuous conducting system which extends from a single bundle in the root through the stem and into the leaves and other parts, and becomes an elaborate system of interconnecting parts. See Fig. 3.

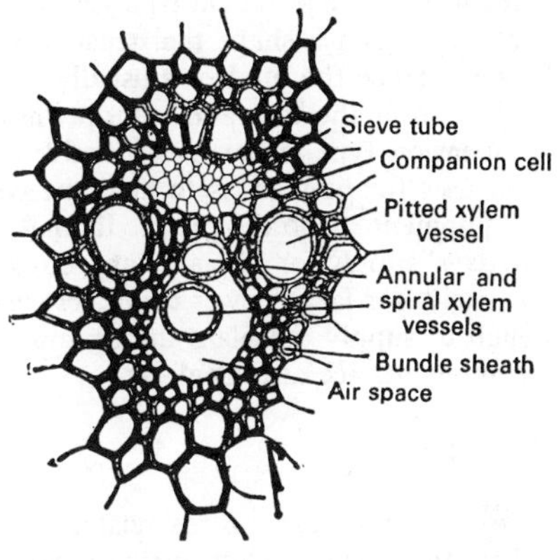

Fig. 3. Cross section of a single fibrovascular bundle of corn, *Zea mays.*

In the axis of the plant each bundle consists of masses of xylem and phloem cells which may appear in various arrangements; frequently the xylem and phloem cells appear in radially adjoining masses, forming a collateral bundle; less frequently one kind of cells is surrounded by cells of the other kind, forming a concentric bundle; in roots a third arrangement is found; the xylem occupies the center of the single bundle, nearly surrounded by strands of phloem. In cross section the xylem of the young root (see Fig. 4) is in the form

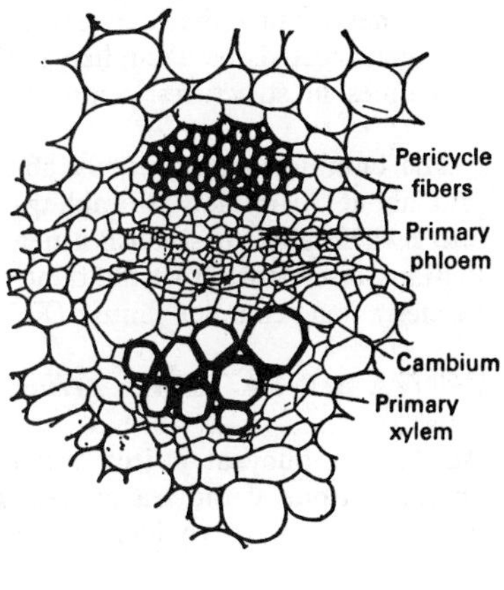

Fig. 4. Cross section of a single fibrovascular bundle of a young sunflower stem.

of a cross with four or five arms (Dicotyledons), or many arms (Monocotyledons), and the phloem strands occupy the positions between the radiating arms of the xylem. Such bundles are called radial bundles.

While a bundle consists normally of associated masses of xylem and phloem cells, in many plants there appears with them strands of fibers whose presence may give protection to the bundle, and additional rigidity to the stem. Because of the frequent occurrence of such fibrous masses as part of a bundle, the term fibrovascular bundle has been used to describe them. In many plants the fibrous cells form a mass on the outer side of the bundle, between it and the surface of the stem, while in other plants, particularly in monocotyledons, the fibers form a sheath completely encircling the bundle.

Bundles are often described as open or closed. Open bundles are those in which cambium cells are found, so that by the repeated division of the cambium cells, the size of the bundle constantly increases. Closed bundles are those in which no cambium occurs, they being composed entirely of primary tissues and so once formed remain constant in size.

BUNDLE-BRANCH HEART BLOCK. **Arrhythmias (Cardiac); Heart and Circulatory System (Human).**

BUNDLE OF HIS. **Heart and Circulatory System (Human).**

BUNKER. A bunker is any large bin, but its technical meaning is confined to receptacles for the storage of fuel or loose materials in bulk. The bunker is frequently constructed of plate steel which, for some purposes, must be lined with a substance such as concrete or wood to protect the metal. One prevalent type, the Berquist suspension bunker, has a profile of such a shape that tension is the only stress produced in the steel. Since the bunker is usually placed so that the material is deposited from the top, and withdrawn by gravity from the bottom, a suspension type is particularly advantageous. One of the most common uses for bunkers is to hold the coal required for the combustion equipment of power plants. It may be able to hold from three to five days' supply for a small plant, but space limitations will govern the capacity for large power plants. However, the bunker should hold enough to supply the plant during minor repairs to the coal conveying equipment. See also **Coal.**

BUNKER C OIL. **Petroleum.**

BUNTING (*Aves, Passeriformes*). Birds related to the finches and sparrows, including the British yellowhammer, not to be confused with the North American woodpecker which bears this name, the ortolan, and the snow bunting. In North America the indigo bunting, *Passerina cyanea*, is the most widely distributed species, ranging over the eastern half of the country and sometimes to western Texas. The lazuli, varied, painted, and lark buntings are characteristically western birds. The painted bunting, *Passerina ciris*, is also called the nonpareil (meaning no parallel—no equal). This is one of very few birds with red feathers underneath. The head is blue with red along its back. The wings and tail are brown.

Buntings are among the best of song birds. Some sing while in flight and during their courting dance. Most species are monogamous. The females incubate the eggs. The male assists in feeding the young sometimes feeding the female while she is on the rest. The nest is well constructed, of a cuplike shape, usually located in trees, but sometimes in structures. Some buntings construct roofs over their nests.

The buntings are extremely abundant. In 1951, it was estimated that about 70 million Bramblings (*Fringilla montifringilla*) ranged into Switzerland to establish a new roosting area close to Thun.

Many authorities treat the buntings, cardinals, sparrows and allied species as one complex family (*Fringillidae*).

BUNYA BUNYA TREE. **Arancarias.**

BUOY. A buoyant object, usually attached to a specific location on the bottom of the sea or to some submerged object. Buoys are used by navigators to locate channels, dangerous rocks or shoals, mooring positions, submerged wrecks, and a variety of other purposes. Various countries have devised their own systems of shapes and colors for coding buoys. In addition to the traditional uses of buoys as visual navigation markers, modern applications include: (1) Buoys equipped with electronic and acoustic transmitters to serve as location markers for ships equipped with suitable receivers; (2) buoys that support research equipment, often electronically transmitting data in connection with oceanographic and meteorological research and reporting; and (3) buoys equipped with receiving and transmitting gear as part of a communications link. In some applications, the buoy may be completely submerged, with a floating marker immediately above the submerged location.

BUOYANCY. The familiar lifting effect of a fluid upon a body wholly or partly submerged in it, known as buoyancy or buoyant force, was first closely studied by the Greek philosopher Archimedes in the third century B.C. What is now known as the "principle of Archimedes" states that the buoyant force is equal to the weight of that body of the fluid which the submerged body displaces, and may be treated as a single force acting vertically upward through the center of gravity of the displaced fluid (center of displacement). This statement applies whether the submersion is partial or complete. The principle readily follows from the consideration that if the submerged body were withdrawn and the resulting cavity allowed to fill with the fluid, the latter would be in equilibrium under the joint action of its own weight and the external forces formerly exerted by the surrounding fluid upon the submerged body.

If the buoyant force equals the weight of the submerged body and acts through its center of gravity, the body will be in equilibrium. This might be true if the body had exactly the same mean density as the fluid. It would then remain at rest when completely submerged at the proper level. A balloon, for example, may rise to a certain height and remain suspended or drift along horizontally. But a solid body completely submerged in a liquid does not come so readily to stable equilibrium, because of the very slight compressibility of liquids, the presence of highly compressible gases in pores of the body or in bubbles clinging to it, and the unequal coefficients of expansion of the body and the liquid.

If, however, the body has a lower mean density than the liquid, it will "float," partly submerged to a level, and in a position with reference to the vertical, determined by the Archimedes principle. An important phase of flotation, especially in ship design, is the degree of stability. This may be expressed in terms of the position of the "metacenter." When a boat is tipped very slightly in a given plane, the center of displacement shifts to one side. There is one point of the boat, called the metacenter, which remains vertically above the center of displacement. This point must be higher than the center of gravity, as the resulting couple then tends to restore the boat to its normal position; if it is lower, the boat is unstable and will capsize. The height of the metacenter above the center of gravity (metacentric height) is a measure of the stability in the given vertical plane. It is in general different for different planes; for example, a boat has a transverse and a longitudinal metacenter and metacentric height, corresponding, respectively, to its rolling and its pitching. Thus one may usually change seats in a rowboat without danger, while an equal shift across the boat might overturn it.

BUOYANCE (Fishes). **Fishes.**

BUPIVACAINE HYDROCHLORIDE. **Anesthesia.**

BURBLE. A separation or breakdown of the laminar flow past a body; the eddying or tubulent flow resulting from this. Burble occurs over the upper surface of an airfoil when the angle of attack has been increased to the point where the air stream no longer follows the profile of the airfoil but breaks away from it. Burble may occur also over the lower surface of an airfoil at high values of negative angles of attack. The space between the airfoil and the detached air stream is filled with eddying, burbling air, and the lift is largely lost. The airfoil is then said to have reached the burble point. This is synonymous with "stalled" in wing terminology.

BURBOT. **Codfishes.**

BURDEN (Instrument). Any circuit or device being measured will be somewhat altered by the instrument used for the measurement. Some amount of power must be transferred between the circuit and the test device in order to obtain a measurement. This characterizes the test instrument as part of the load seen by the device being tested. The power that is transferred to and from the test instrument is termed the *instrument burden.*

In circuits which have limited power supplying capabilities, this loading effect, or burdening, by the test instrument will result in erroneous readings. The test instrument chosen must, therefore, exhibit characteristics which will minimize this burdening and thus produce more dependable results. The concept of instrument burden applies to all forms of measurement and also covers cases involving transient energy transfer.

The problem of instrument burden occurs frequently in applications that require direct reading meter movements. Frequently, the meter requires a significant amount of power for its operation. This may represent a large portion of the available power in the circuit being tested and consequently will result in incorrect test results as shown in the accompanying figure.

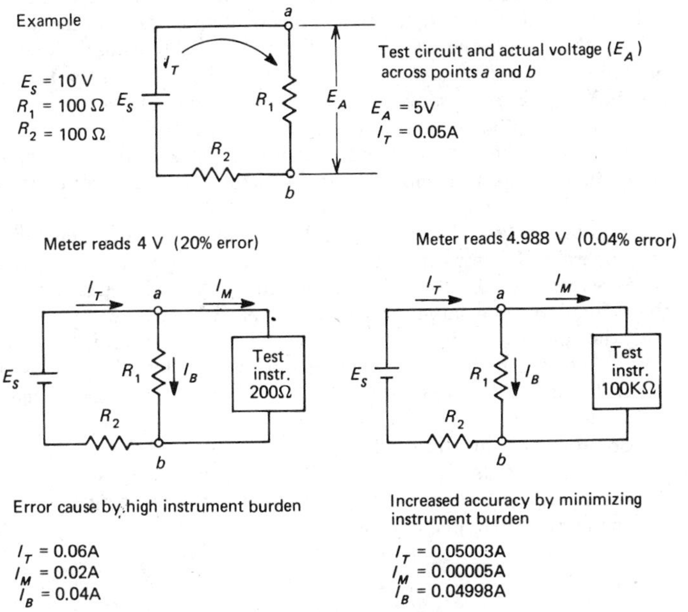

Effect of instrument burden on measurement accuracy.

BURETTE. A long, slender, gradulated glass tube used in volumetric analysis, particularly titrations. An average-size burette will contain 50 milliliters of reagent and will be graduated down to tenths of a milliliter. The amount of reagent required to effect a given reaction, such as neutralization of an acid with a base, is determined by taking the difference between the starting and ending readings of the graduations. So that the end-point can be determined with maximum precision, the bottom portion of the burette is in the form of a tapered, narrow tip, such that the exiting droplets are quite small. Flow of reagent from the burette is controlled by a ground glass stopcock which is an intimate part of the total glass assembly. In a less costly version, a short piece of rubber or plastic tubing, equipped with a pinchcock to squeeze the tubing, is attached to the lower end of the burette. Another small length of glass tubing, drawn to a fine tip for droplet control, is placed into the open (exit) end of the short tubing connection. For laboratories where numerous titrations requiring the same reagent are made routinely, a so-called automatic burette is available. Essentially, this simplifies the reagent refilling operation. See also **Titration (Potentiometric).**

BURGERS VECTOR. A vector representing the displacement of the material of the lattice required to create a dislocation. As usually defined, the Burgers vector must always be a translation vector of the lattice. The Burgers vector may be determined by comparing a path around a dislocation in the crystal with a corresponding path in a perfect part of the lattice. The difference gives the Burgers vector, which describes the magnitude and direction of slip.

BURN. The reaction of tissue to excessive exposure to heat, radiation, and corrosive materials. When a burn is caused by a hot liquid, steam, or other hot vapors, it may be termed a *scald.* Medically, burns may be classified by extent of injury: In a *first degree* burn, the skin is reddened and tender; in a *second degree* burn, the skin is blistered; in a *third degree* burn, there is extensive tissue damage, some of which may be charred and fully destroyed. When a burn is severe, the tissue may slough away, a condition known as *eschar.* In extremely severe burns, even the fat, muscles, and bone underlying the entire thickness of the skin also may be damaged. In first and second degree burns, the major factor of consideration is the extent of area damaged, determining the type and whether or not medical attention is required. A third degree burn always requires professional medical attention. In terms of relative seriousness, it was traditionally believed that a first degree burn would be fatal if it covered as much as two-thirds of the body's surface; while a second degree burn would be equally grave if it covered one-third of the suface. As of the early 1980s, for persons who are cared for in modern burn centers, statistics indicate that 44% of them will survive. Contributing factors include more rapid transportation of victims to burn centers, improved methods of restoring body fluids, and better means of controlling infection. Loss of body fluids is the most immediate danger. If not prevented, fluid loss leads to kidney failure, which in times past was usually fatal. One of the first steps taken in handling a serious burn case is administration of a basic solution of salts and water. In most cases, the patient is stabilized within one or two days.

A breakthrough in burn care occurred in the mid-1960s when water-soluble antibiotic ointments were introduced. These can be applied directly to the wound, whereupon the drug is easily absorbed. A wide range of antibiotics is used. Burn cases are isolated from other patients and placed in bacteria-free units. The main thrust of burn therapy is to close the wound as fast as possible. This reduces the risk of infection and reduces later complications. To close a burn wound, surgeons scrape away dead tissue in the burn area. This process may involve five or six operations. Then a skin graft is made, obtaining healthy tissue from the patient and using it to cover the would. A skin-typing system, similar to blood-bank typing, enables surgeons to take skin from other people, usually relatives of the victim. Artificial skin has been developed for temporarily covering the wound until the person regenerates enough natural skin. Artificial skin therapy is just getting underway and much remains to be learned about this procedure.

Once a burn patient is well on the way to recovery, the long-term prospects must be confronted. Scarring is nearly always a problem. Cosmetic surgery has accomplished much by way of restoring faces and hands. Scar tissue is also reduced by covering the burn wound early and preventing infections. As new skin grows over a wound, it contracts, sometimes locking joints or distorting the face by pulling skin down from the eyes. Early rehabilitation, including simple exercises, can contribute much toward alleviating these problems.

Statistics show that about 75,000 people in the United States alone are hospitalized each year for burns. Even with greatly improved handling of burn cases, about 10,000 persons die each year from burn injury, making it the nation's third leading cause of accidental death. Burn victims usually require months of hospital care and many suffer permanent disabilities or disfigurement.

First aid measures for burns, where immediate professional medical assistance is not available, is well covered in literature available from the American National Red Cross.

Radiation Injury. Although usually categorized as radiation burns, such injuries are quite different from thermal burns. The specific effects of radiation injury depend not only upon the exact area of the body exposed, but also upon the fact that certain kinds of tissue are more susceptible to injury than others. In humans, sensitivity of tissues to radiation decreases in the following order: (1) lymphoid tissue and bone marrow; (2) epithelial tissue, such as the testes and ovaries; (3) salivary gland; (4) skin; (5) mucous membranes; (6) endothelial cells of blood vessels and peritoneum; (7) connective tissue; (8) muscle, bone, and nerve tissue. It is in this order, therefore, that the specific effects of exposure to ionizing radiation might be expected to appear and do the most harm.

Blood changes are among the earliest to appear, and may occur as the result of doses of radiation that produce no other effect. If the white blood cells manufactured in the lymphatic tissue do not decrease in number within 72 hours following exposure, no serious dose of radiation has usually been received. Increase in the number of lymphocytes is almost the first symptom of recovery from radiation sickness. It has been recognized that leukemia, a malignant disease in which there is a considerable increase in the numbers of white blood cells, may be induced by overexposure to x-rays. The incidence of leukemia in radiologists is reported to be nine times as high as it is in other physicians. Careful studies of the incidence of leukemia among the population of cities suffering from an atomic blast indicate a significant increase in this disease among survivors. See **Blood.**

BURNER. A principal component of combustion equipment. While there are specialized burners for disposing of various kinds of waste, the usual meaning of the term applies to the burning of fuels with air to generate heat—as in the case of burning a fuel in a boiler furnace for the purpose of generating steam. See also **Boiler;** and **Combustion.**

Oil and Gas Burners

Burners are normally located in the vertical walls of the furnace.* The burners introduce the fuel and air into the furnace to sustain the exothermic chemcial reactions for the most effective release of heat. That effectiveness is judged by the following factors:

1. The rate of feed of the fuel and air shall comply with the load demand on the boiler over a predetermined operating range.
2. The efficiency of the combustion process shall be as high as possible with the minimum of unburned combustibles and minimum excess air in the products.
3. The physical size and complexity of the furnace and burners shall be as small as possible to minimize the required investment and to meet the limitations on space, weight, and flexibility imposed by the service conditions.
4. The design of the burners, including the materials used, shall provide reliable operation under specified service conditions, and shall assure meeting accepted standards on maintenance for the burners and furnaces in which they are installed.
5. Safety shall be paramount under all conditions of operation of burners, furnace, and boiler, including starting, stopping load changes, and variations in the fuel.

The normal use of a steam generator requires operation at different outputs to meet varying load demands. The specified operating range or "load range" for a burner is the ratio of full load on the burner to the minimum load at which the burner must be capable of reliable operation. For example, with a boiler of 100,000 pounds/hour capacity (steam delivered), a load range of 4 to 1 on the burners means that the unit can be operated from 100,000 pounds/hour down to 25,000 pounds/hour without changing the number of burners in operation.

Combustion air is generally delivered to the burners by fans. It is necessary to supply more than the theoretical air quantity to assure complete combustion of the fuel in the combustion chamber (furnace). The amount of excess air provided should be just enough to burn the fuel completely in order to minimize the sensible heat loss in the stack gases.

Continuity of service is enhanced by designing the furnace and arranging the burners to minimize slagging and fouling of heat-absorbing surfaces for the normal range of fuels burned.

Maintenance costs of the burner are minimized by (1) the least exposure to furnace heat, and (2) provision for replacement or repair of vulnerable parts while the unit continues in operation.

Burner Types. The most frequently used burners are the circular type. Fig. 1 shows a single circular register burner for gas and oil firing; Fig. 2 shows a circular type dual register burner for firing oil or pulverized coal. The circular type dual register burner was devel-

* Basic information for this entry from "Steam—Its Generation and Use," copyrighted by The Babcock & Wilcox Co., New York (39th Edition, 1978).

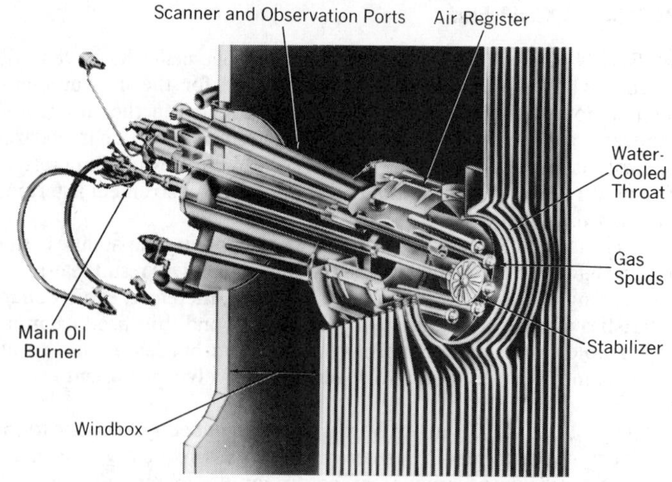

Fig. 1. Circular register burner with water-cooled throat for oil and gas firing. (*The Babcock & Wilcox Co.*)

oped for nitrogen oxides (NO_x) reduction. The maximum capacity of the individual circular burner ranges up to 300 million Btu/hour (1 Btu = 0.2520 kilogram-calories), dependent upon the atomizer used.

In both circular and cell burners the tangentially disposed "doors" built into the air register provide the turbulence necessary to mix the fuel and air and produce short, compact flames.

While the fuel is introduced to the burner in a fairly dense mixture in the center, the direction and velocity of the air, plus dispersion of the fuel, completely and thoroughly mixes it with the combustion air.

Oil Burners. In order to burn fuel oil at the high rates demanded by modern boiler units, it is necessary that the oil be atomized, that is, dispersed into the furnace as a fine mist, somewhat like a heavy fog. This exposes a large amount of oil particle surface for contact with the combustion air to assure prompt ignition and rapid combustion. There are many ways of atomizing fuel oil but the discussion here is limited to the two most popular ways, steam or air, and mechanical atomizers.

For proper atomization, oil of grades heavier than No. 2 must be heated to reduce its viscosity to 135–150 Saybolt Universal seconds. Steam or electric heaters are required to raise the oil temperature to the required degree, i.e., approximately:

57°C for No. 4 oil
85°C for No. 5 oil
93–104°C for No. 6 oil

With certain oils, better combustion is obtained at somewhat higher temperatures than required for atomization. However, oil temperature must not be raised to the point where vapor binding occurs in the

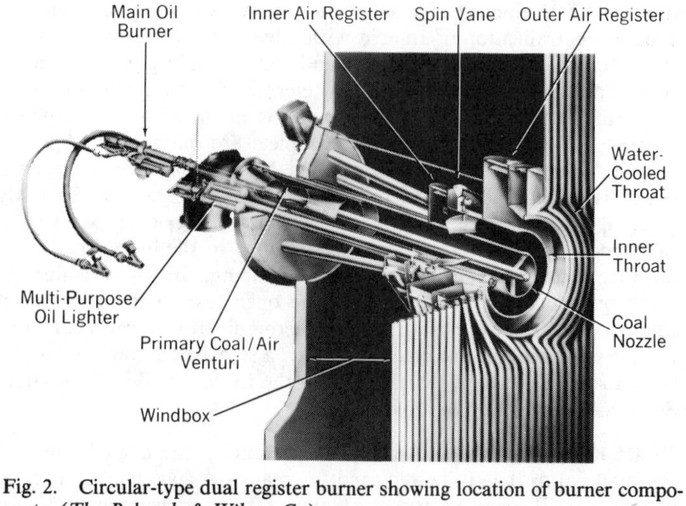

Fig. 2. Circular-type dual register burner showing location of burner components. (*The Babcock & Wilcox Co.*)

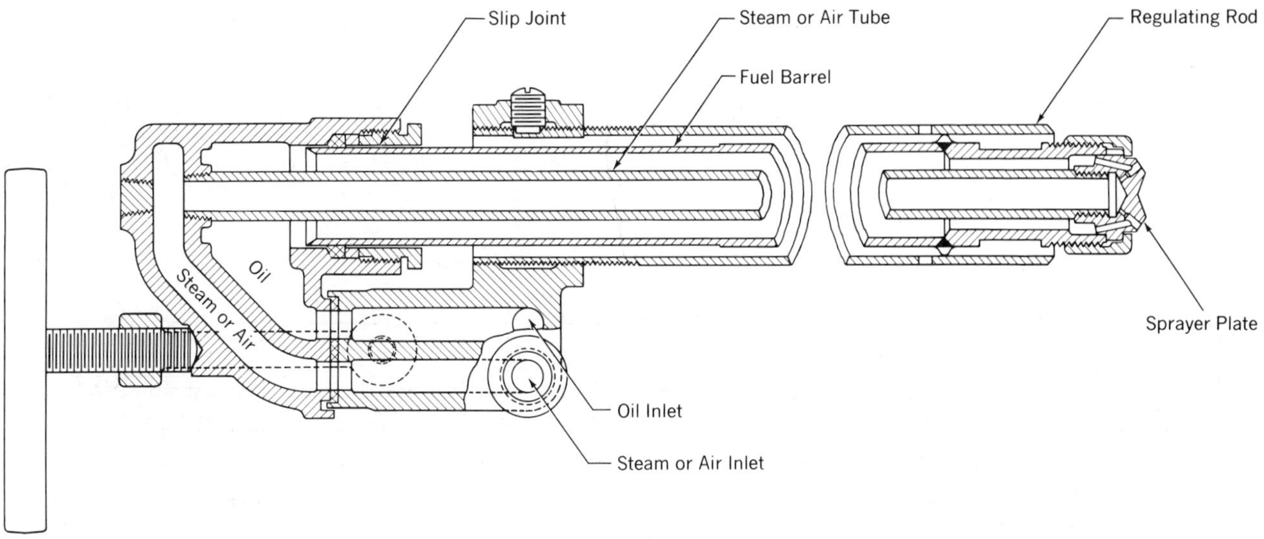

Fig. 3. Steam (or air) oil atomizer assembly. (*The Babcock & Wilcox Co.*)

pump supplying the oil, since this could cause flow interruptions followed by loss of ignition.

It is also important that oil be free from acid, grit and other foreign matter which might clog or injure burners of their control valves.

Steam or Air Atomizers. The steam (or air) atomizer (Fig. 3) is the most widely used. In general it operates on the principle of producing a steam-fuel (or air-fuel) emulsion which, when released into the furnace, atomizes the oil through the rapid expansion of the steam. The atomizing steam must be dry because moisture causes pulsations which can lead to loss of ignition. Where steam is not available, moisture-free compressed air can be substituted.

Steam atomizers are available in sizes up to 165 million Btu/hour (1 Btu = 0.2520 Calorie) input—about 9200 pounds (4173 kilograms) of oil per hour. Oil pressure is much lower than for mechanical atomizers. The steam and oil pressure required are dependent on the design of the steam atomizer. Maximum oil pressure is usually about 100 psi (6.8 atomospheres) and steam pressure is generally about 20 to 40 psi (1.4 to 2.8 atomospheres) higher than oil pressure.

The steam atomizer performs more efficiently over a wider load range than other types. It normally atomizes the fuel properly down to 20% of rated capacity, and in some instances steam atomizers have been successfully operated at 5% capacity. Frequently these extremes in range cannot be fully utilized because the temperature of the combustion space falls off to such a degree that, despite the excellent quality of atomization, there is not sufficient temperature to complete the combustion process adequately.

A disadvantage of the steam atomizer is its comsumption of steam. A good steam atomizer can operate with a steam consumption as low as 0.02 pound of steam per pound of fuel oil (2% by weight) at maximum capacity. For a large unit, this amounts to a sizable quantity of steam and consequent heat loss to the stack. When the boiler unit supplies a substantial amount of steam for a process where condensate recovery is small, the additional makeup for the steam atomizer is inconsequential. However, in a large public utility boiler where turbine losses are low and there is very little makeup, the use of atomizing steam can have a significant effect on the total makeup requirements. In spite of this, the cost of employing a compressed air supply system as a substitute for steam is not always justified.

Mechanical Atomizers. In the mechanical atomizer, the pressure of the fuel itself is used as the means of atomization. Many forms have been developed. Those with moving parts close to the furnace have lost favor because of the excessive maintenance required to keep them operational.

The return-flow atomizer (Fig. 4) is used for many marine installations and some stationary units where the use of atomizing steam is objectionable or impractical. The oil pressure required at the atomizer for maximum capacity ranges from 600 to 1000 psi (41 to 68 atmospheres), depending on capacity, load range, and fuel. The fuel flows through tangentially disposed slots in a "sprayer plate" into a "whirl" chamber, from which it issues through an orifice of the sprayer plate as a fine conical mist or spray. After the oil passes through the tangentially disposed slots, that which exceeds the boiler input requirements is returned from the base of the whirl chamber to the fuel oil system.

Mechanical atomizers are available in sizes up to 180 million Btu/hour (1 Btu = 0.2520 Calorie) input—about 10,000 pounds (4536 kilograms) oil per hour. The acceptable operating range may be as much as 10 to 1 or as little as 3 to 1 depending on the maximum oil pressure used for the system, the furnace configuration, air temperature and burner throat velocity. Return-flow atomizers are ideally suited for standard grades of fuel oil where it is desired to meet load variations without changing sprayer plates or cutting burners in and out of service.

For good performance over an operating range of 10 to 1, the combustion chamber should be of relatively small cross section at the burner zone, the oil pressure at the burners for full load should be 1000 psi (68 atmospheres), the temperature of air for combustion should be significantly above ambient throughout the load range, and the air resistance across the burner should be 8 to 12 inches (20 to 30 centimeters) of water at full load, depending upon the size of the particular burner. Departures downward from any of these values markedly affect the satisfactory load range obtainable from the burners. In a properly designed and operated system the high pressure return-flow mechanical atomizer will provide combustion efficiency comparable to that obtainable with a good steam atomizer.

Natural Gas Burners. The variable-mix multispud gas element (Fig. 1) was developed for use with circular-type burners for obtaining good ignition stability under most conditions, such as the two-stage combustion technique and vitiated air (by gas recirculation) to the burner. Distinctive features of this gas element are:

1. The individual gas spuds are removable with the boiler in service to enable cleaning or redrilling of the gas nozzles if required.
2. The individual gas spuds can be rotated to orient the gas nozzles for optimum firing conditions with the burner in service.
3. The location of the individual gas spuds with respect to the burner throat can be varied a limited distance to obtain optimum firing conditions with the burner in service.
4. The individual gas spud flame container and gas spud drilling are arranged so as to obtain optimum flame stability.

With the proper selection of control equipment, a multifuel fired burner with a variable-mix multispud type gas element is capable of changing from one fuel to another without a drop in load or boiler pressure. Simultaneous firing of natural gas and oil in the same burner is acceptable on burners equipped with variable-mix multispud type gas elements.

This type element is designed for use with natural gas or other gaseous fuels containing at least:

70% CH_4 (methane)
or 70% C_3H_8 (propane)
or 25% H_2 (hydrogen) by volume

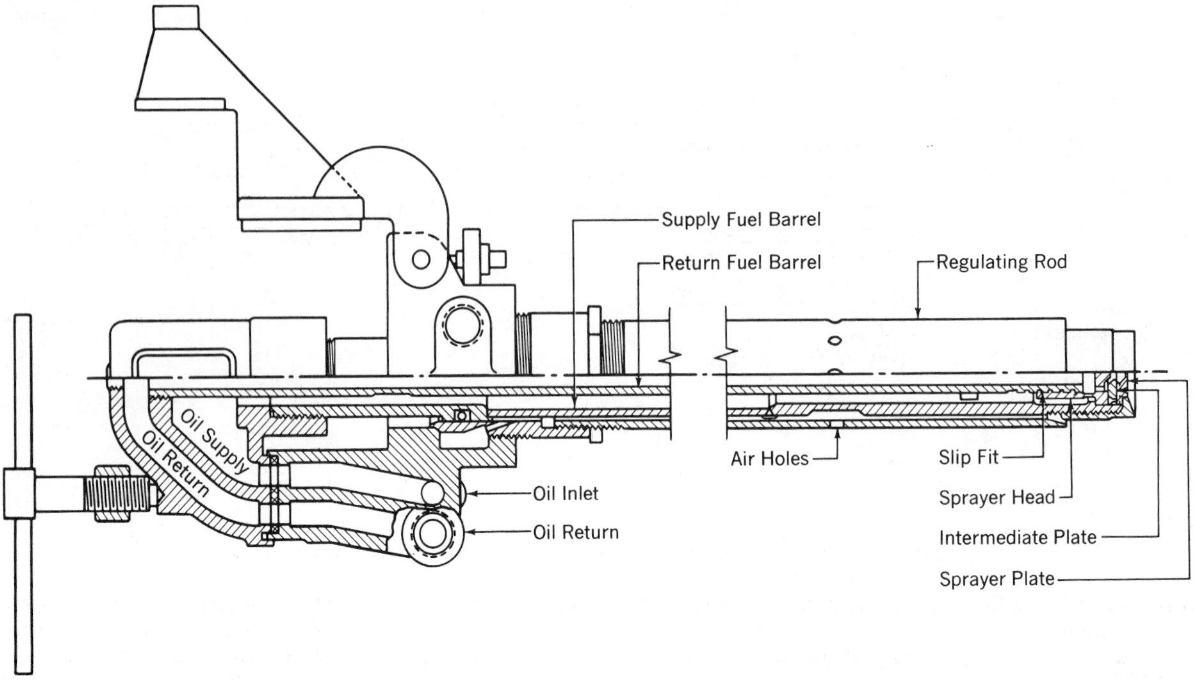

Fig. 4. Mechanical return-flow oil atomizer assembly. (*The Babcock & Wilcox Co.*)

This element is designed for a maximum input of 173 million Btu/hour per burner.

In certain respects natural gas is an ideal fuel for a burner since it requires no preparation to be suitable for rapid and intimate mixing with the combustion air flowing through the burner throat. However this characteristic of easy ignition under most operating conditions has, in some cases, led to operator carelessness with resultant damaging explosions.

To provide safe operation, ignition of a gas burner should remain close to the burner wall throughout the full range of allowable gas pressures, not only with normal air flows, but also with much more air flow through the burner than is theoretically required. Ideally it should be possible at the minimum load to pass full-load air flow through the burner, and at full load as much as 25% in excess of theoretical air without loss of ignition. With this latitude in air flow it is not likely that ignition can be lost, even momentarily, during some upset in air flow due to improper operation or error.

Burners for Other Gases. Many industrial applications utilize coke oven gas, blast furnace gas, refinery gas or other industrial by-product gases. With these gases the heat release per unit volume of fuel gas may be very different from that of natural gas. Hence, gas elements must be designed to accommodate the particular characteristics of the gas to be burned. Also burners must be designed with reference to ignition stability and load range factors which govern in each case. Other special problems may be introduced by the presence of impurities in industrial gases, such as sulfur in coke oven gas, and entrained dust in blast furnace gas.

Lighters (Ignitors) and Pilots. Equipment is available, for boiler units ranging from the smallest to the largest, that allows the boiler operator to ignite the main fuel by the simple expedient of pressing a button. This equipment ranges from spark devices that ignite fuel oil directly, to gas or light oil equipment, in itself spark-ignited, which is used for ignition of the main streams of gas and fuel oil. These devices are available with control equipment that ranges from the simplest push button requiring observation of ignition by the operator at the burner, to a fully "programmed" starting sequence, complete with interlocks and flame-sensing equipment, all remotely operated from the boiler control room.

Usually the ignition device is energized only enough to assure that the main flame is self-sustaining. With the fuel normally used in oil or natural gas burners, ignition should be self-sustaining within one or two seconds after the fuel reaches the combustion air. On a fully automated burner it is customary to allow 10 to 15 seconds "trial for ignition" so that the fuel can reach the burner after the fuel shut-off valve on the burner is opened.

In general, lighters should be used sparingly. Where special fuel is used for the lighters, it is usually more expensive that the fuel used in the regular burners. So unnecessary use adds to operating cost.

However there are applications where a continuousl burning lighter or pilot is needed. This is particularly true in the use of a by-product fuel, such as gas from a chemical process. In most installations such fuels are piped directly to the boiler house without an intervening accumulator. The quantity and quality of fuel supplied to the boiler house are subject to any malfunction in the chemcial process. Supply pressure may vary beyond the range of main burner stability or the combustible content of the gas may change. Such variations require continuous operation of a pilot. It may be necessary also to provide supplementary fuel to each burner reliably and continuously to keep the process gas burning during abnormal conditions in the process. In these instances the boiler operators have an awareness of the vagaries of the process fuel that does not exist in the operation of conventional oil- and natural-gas-fired boilers. As a consequence, they know the safety of the boiler operation is highly dependent on the reliability of the ignitor and are alert to any malfunction that may occur.

Excess Air. When firing oil or natural gas at high loads, there normally is no fuel distribution problem but distribution of combustion air in the burner windbox is not perfect, and this dictates the use of some excess air to assure complete combustion from all burners. On most units it is possible to operate with as little as 5 to 7% excess air at the furnace outlet at full load, and with extra care in design and operation, some large utility-type boilers have operated with as little as 2.5% excess air without excessive unburned combustible loss.

At partial loads on all units regardless of fuel fired, it is necessary to increase the excess air as the load is reduced. If all burners are kept in service, the furnace temperature and the air velocity, which provides burner turbulence, decrease with decrease in load. Raising the excess air tends to compensate for the deterioration of these items as the load drops. Also it may be desirable to operate with higher than normal excess air to maintain steam temperature or minimize corrosion of the "cold" end of the unit. Increased excess air is desirable even when the number of burners in use is reduced at lower loads. Burner dampers are designed not to close tightly to permit the air to protect the idle burner(s) from overheating by radiant heat from nearby operating burners.

Burner Pulsation. One of the mystifying problems associated with gas burners and, to a much less degree, with oil burners is that of burner pulsation. It appears to result from certain combinations of combustion chamber size and configuration coupled with some characteristic of the burners, perhaps too perfect mixing of fuel and air at the burner. When one or more burners on a large unit start to pulsate, it may become alarmingly violent, at times shaking the whole boiler. Making an adjustment of only one burner may start or stop pulsation. At times only minor burner adjustments eliminate the pulsation. In other instances, it is necessary to alter the burners. This may involve modifying the gas ports, impinging gas streams on one another, or using some other device that effectively alters the mixing of the gas with the air.

To avoid pulsations, burner and boiler manufacturers incorporate the latest information available into the designs of burners and furnaces. However a better understanding of this problem is needed.

Operating Range or Load Range. In the past in particular, there have been numerous claims regarding operating or load range of a particular burner. Erroneous interpretation of test results has led to performance predictions which could not be substantiated. There have been misunderstandings between vendors and users because of interpretation of results. The most persistent differences in opinion are in the acceptability of flame appearance, the length of time the unit is operated at minimum load, and the evaluation of potential hazard from deposits that may occur during the low-load tests. The following considerations are basic to understanding the difficulties arising from low-load operation.

Large boilers with air heaters have a sizeable "heat inertia" which requires more than 2 hours at full load for temperatures throughout the unit of stabilize. Conversely a significant time period is necessary for them to retrograde on dropping load.

Coal-Burning Systems

Historically in the United States, more than three-fourths of the mined tonnage of bituminous coal and lignite has been used to generate steam for electrical power. A high percentage of the coal used for the generation of steam is burned in pulverized form.

Selection of Coal-Burning Equipment. Selection of equipment for a particular installation consists of balancing the investment, operating characteristics, efficiency, and type of coal to be used—with the objective of achieving the most economical installation. Almost any coal can be burned successfully in pulverized form or on some type of stoker. The capacity limitations imposed by stokers have been overcome by the development of pulverized-coal and cyclone-furnace firing. These improved methods also provide: (1) Ability to use coal from fines up to 2 inches (5 centimeters) in maximum size; (2) improved response to load changes; (3) increase in thermal efficiency because of lower excess air for combustion and lower carbon loss than with stoker firing; (4) a reduction in labor required for operation; and (5) improved ability to burn coal in combination with oil and gas.

Experience shows that stoker firing is more economical for steam generating units of capacity less than 100,000 pounds (45,360 kilograms) of steam per hour, where the lower efficiency of a stoker can be tolerated. In larger plant, where fuel cost is a larger fraction of the operating cost, pulverized-coal or cyclone-furnace firing is more economical except in special cases. Operating characteristics may be of controlling significance in the choice of firing method. For example, where unit size is suitable for pulverized-coal, cyclone furnace, or stoker firing, the extremely wide load range of stoker firing may make it preferable. Where rotary kilns and industrial furnaces are fired by coal, the high fineness levels necessitate pulverized-coal firing.

The type of coal influences the choice of the method of firing for boiler furnaces with the primary considerations as follows:

Pulverized-coal firing: grindability, rank, moisture, volatile matter, and ash.

Stoker firing: rank of coal, volatile matter, ash, and ash-softening temperature.

Cyclone-Furnace firing: volatile matter, ash, and ash viscosity.

Convenient approximations for the selection of bituminous coals for the firing of boilers are given in Table 1.

TABLE 1. COAL CHARACTERISTICS AND METHOD OF FIRING

	Stoker	Pulverized Coal	Cyclone Furnace
Max. total moisture* (as fired), %	15–20	15	20
Min. volatile matter (dry basis), %	15	15	15
Max. total ash (dry basis), %	20	20	25
Max. sulfur (as fired), %	5	—	—

* These limits may be exceeded for lower rank, higher inherent-moisture-content coals, i.e., subbituminous and lignite.

Pulverized-Coal Systems

The function of a pulverized-coal system is to pulverize the coal, deliver it to the fuel-burning equipment, and accomplish complete combustion in the furnace with a minimum of excess air. The system must operate as a continuous process and, within specified design limitations, the coal supply or feed must be varied as rapidly and as widely as required by the combustion process.

A small portion of the air required for combustion (15 to 20% in current installations) is used to transport the coal to the burner. This is known as primary air.

In the direct-firing system, primary air is also used to dry the coal in the pulverizer. The remainder of the combustion air (80 to 85%) is introduced at the burner and is known as secondary air.

The two basic equipment components of a pulverized-coal system are:

1. The pulverizer which pulverizes the coal to the fineness required.
2. The burner which accomplishes the mixing of the pulverized-coal-primary-air mixture with secondary air in the right proportions and delivers the mixture to the furnace for combustion.

Other necessary requirements are:

3. Hot air for drying the coal for effective pulverization.
4. Fan(s) to supply air to the pulverizer and deliver the coal-air mixture to the burner(s).
5. Coal feeder to control the rate of coal feed to each pulverizer.
6. Coal and air conveying lines.
7. Pyrites reject system.
8. Measuring and control elements.

Two principal systems—the bin system and the direct-firing system—have been used for processing, distributing and burning pulverized coal. The direct-firing system is the one being installed almost exclusively today.

Bin System. Largely of historical interest, the coal in the bin system is processed at a location apart from the furnace and the end-product is pneumatically conveyed to cyclone collectors which recover the fines and clean the moisture-laden air before returning it to the atmosphere. The pulverized coal is discharged into storage bins and later conveyed pneumatically to utilization bins which may be as far as 5000 feet (1524 meters) from the point of preparation.

Direct-Firing System. The bin system has been superseded by the direct-firing system because of improvements in safety conditions, plant cleanliness, greater simplicity, lower initial investment, lower operating cost, and less space requirement.

The pulverizing equipment developed for the direct-firing system permits continuous utilization of raw coal directly from the bunkers where coal is stored in the condition in which it is received at the plant. This is accomplished by feeding the raw coal directly into the pulverizer, where it is dried as well as pulverized, and then delivering it to the burners in a single continuous operation.

Components of the direct-firing system (Fig. 5) are as follows:

1. Raw-coal feeder.
2. Source (steam or gas air heater) to supply hot primary air to the pulverizer for drying the coal.
3. Pulverizer fan, also known as the primary-air fan, arranged as a blower (or exhauster).

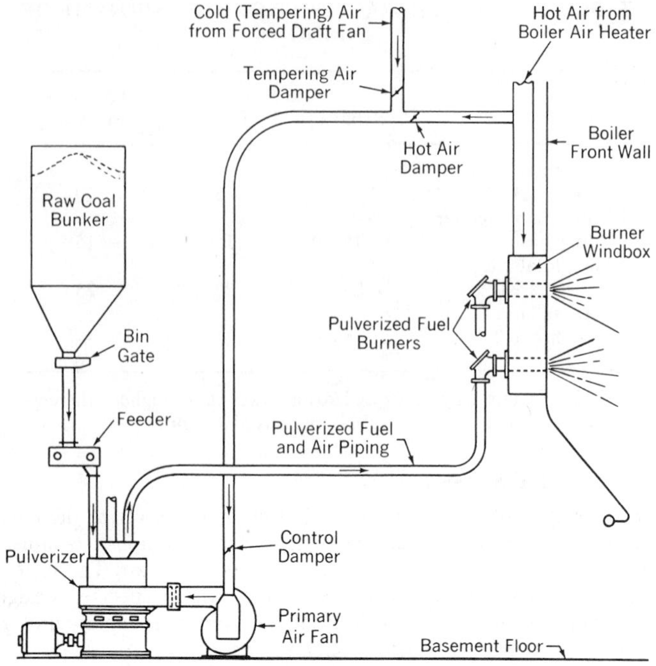

Fig. 5. Direct-firing system for pulverized coal. (*The Babcock & Wilcox Co.*)

4. Pulverizer arranged to operate under pressure (or suction).
5. Coal-and-air conveying lines.
6. Burners.

Two direct-firing methods are in use: (1) The pressure type, which is the more commonly used; and (2) the suction type. The principal differences are summarized in Table 2.

In the pressure method, the primary-air fan, located on the inlet side of the pulverizer, forces the hot primary air through the pulverizer where it picks up the pulverized coal, and delivers the proper coal-air mixture to the burners. Where a separate air heater is provided, the fan operates on cold air, forcing the air first through the air heater and then the pulverizer. In either event, the coal is delivered to the burners by a fan operating entirely on air, so that no entrained dust passes through the fan. One pulverizer generally furnishes the coal for several burners. With the pressure method, it is usual to supply each burner with a single conveying line direct from the pulverizer, thus eliminating the expense of a distributor.

In the suction method, the air and entrained coal are drawn through the pulverizer under negative pressure by an exhauster located on the outlet side of the pulverizer. With this arrangement, the fan handles a mixture of coal and air, and distribution of the mixture to more than one burner must be obtained by a distributor beyond the fan discharge.

The feeding of coal and air to the pulverizer is controlled by either of two methods: (1) The coal feed is proportioned to the load demand, and the primary-air supply is adjusted to the rate of coal feed; or (2) the primary air through the pulverizer is proportioned to the load demand, and the coal feed is adjusted to the rate of air flow. In either case, a predetermined air-coal ratio is maintained for any given load.

TABLE 2. COMPARATIVE FEATURES OF DIRECT-FIRING PRESSURE AND SUCTION SYSTEMS

Feature	Pressure System	Suction System
Type of fan	Blower	Exhauster
Location of fan	Pulverizer inlet	Pulverizer outlet
Fan construction	Standard	Explosion-proof
Fan handles	Air only	Pulverized coal and air
Relative fan efficiency	High	Low
Fan wear	Low to none	High
Pulverized coal distribution to burners	Direct distribution	Distributor required

The direct-firing system, in addition to eliminating separately fired dryers and storage facilities for pulverized coal, permits the use of inlet air temperatures to the pulverizer up to 343°C and higher for drying high-moisture coals (total moisture 20%, surface moisture 15%) or high-moisture lignites (20 to 40% total moisture) in the pulverizer.

The operating range of a pulverizer is usually not more than 3 to 1 (without change in the number of burners in service) because the air velocities in lines and other part of the system must be maintained above the minimum values to keep the coal in suspension. In practice most boiler units are provided with more than one pulverizer, each feeding multiple burners. Load variations beyond 3 to 1 are generally accommodated by shutting down (or starting up) a pulverizer and the burners it supplies.

Types of Pulverizers. The reduction of materials to a fine-particle size for countless uses is a very old art. Coal-pulverizing equipment is based, generally, on rock and mineral-ore grinding machinery. The principles involved in all pulverizing machinery are: (1) Grinding by impact; (2) by attrition; (3) by compression; or (4) by a combination of two or more of these methods. Most pulverizers involve ball-and-race or roll-and-race designs. Other types include bowl mills, tube mills, and impact mills. Pulverizer requirements may be summarized by:

1. Rapid response to load change and adaptability to automatic control.
2. Continuous service for long operating periods.
3. Maintenance of prescribed performance throughout the life of pulverizer grinding elements.
4. A wide variety of coals should be acceptable.
5. Ease of maintenance with the minimum number and variety of parts, and space adequate for access.
6. Minimum building volume required.

Exhausters and Blowers. Primary air is required for conveying the pulverized coal to the burners. In the direct-firing system the primary air is supplied through the pulverizers. With a pressure system, the primary-air fan handles clean air and is not subjected to abrasion by the pulverized coal. In this case a high-efficiency fan can be used since the conditions permit an efficient rotor design and high tip speed.

With a suction system, the fan or exhauster must handle pulverized-coal-laden air. To comply with the National Fire Protection Association requirements, the exhauster housing must be designed to withstand an explosion with the fan. Furthermore, since the exhauster is subject to excessive wear, the design is limited to a paddle-wheel type of heavy construction and hard-metal or other protective-surface coatings. All of these construction features are detrimental to the mechanical efficiency of the fan.

Power consumption of the pulverizer fan is an important item in the operating cost. For a direct-firing installation, the fan must overcome the flow resistances through the pulverizers the burner lines as well as provide the pressure required at the burner. The resistance through the pulverizer depends upon the inherent characteristics of the pulverizer and the amount of drying accomplished in it.

Pulverized-Coal Burning Equipment

As for oil and gas, the burner is the principal equipment component for the firing of pulverized coal, and much of the discussion concerning the burning of oil and gas is basically applicable to pulverized coal. However, the use of solid fuel in pulverized form presents additional problems in the design of boilers and furnaces.

As oil must be atomized to expose a large amount of oil particle surface to combustion air, so coal must be pulverized to the point where particles are small enough, i.e., surface is sufficiently large per unit of mass to assure proper combustion.

In the direct-firing system, coal is dried and delivered to the burner in suspension in the primary air, and this mixture must be adequately mixed with the secondary air at the burner.

Burner Types. As with oil and gas, the most frequently used burners are the circular type. A circular dual register type burner is shown in Fig. 6. Circular single register burners are also used. Either of these burners is designed for firing pulverized coal only. They can be used singly or in multiples. The dual register type was developed for NO_x reduction. However, either of these burner types can be equipped to fire any combination of the three principal fuels. It is

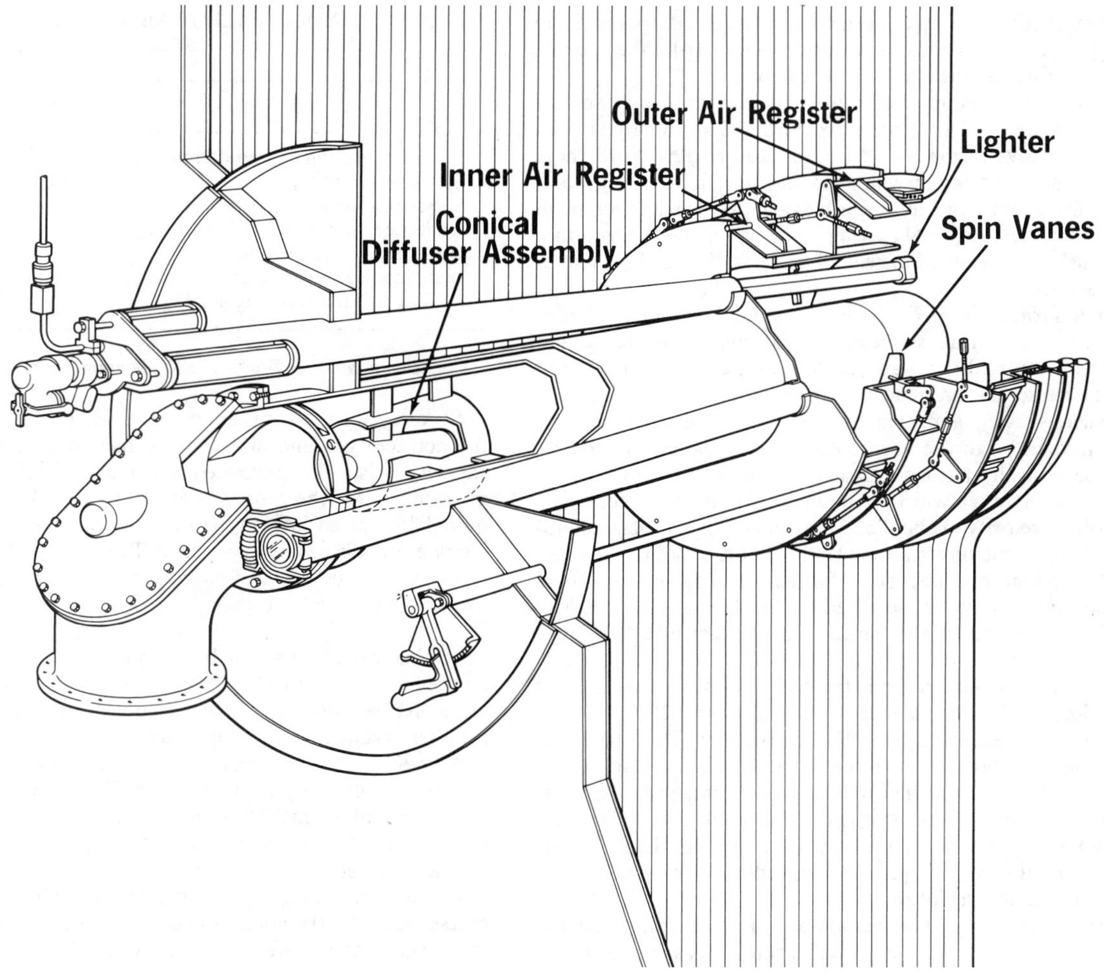

Fig. 6. Circular dual register pulverized coal burner. (*The Babcock & Wilcox Co.*)

to be stressed, however, that combination pulverized coal firing with oil in the same burner should be restricted to short emergency periods. It is not recommended for long operating periods due to possible coke formation on the pulverized-coal element.

The maximum nozzle input per individual burner is 165 million But/hour (1 Btu = 0.2520 Calorie). The secondary-air port velocity at full load on the boiler ranges from 4000 feet (~ 1200 meters) per minute for small boilers where unheated secondary air is used, to 6000 feet (~ 1800 meters) per minute for a dry-ash removal furnace with 316°C air. When circular burners are installed in slag-tap furnaces, velocities of 7500 feet (~ 2250 meters) per minute are common.

Ignition and control equipment for pulverized-coal firing is similar to that for oil and gas, although there are differences in the way it is used. In the case of pulverized coal it may be necessary to keep the ignitors operating for hours until the temperature in the combustion zone becomes high enough to assure self-sustaining ignition of the main fuel.

The self-igniting characteristics of pulverized coal vary from one fuel to another, but on most coals it should be possible to maintain ignition without auxiliary fuel down to one-third capacity of the boiler. In some instances completely reliable ignition is obtained down to a quarter load. When firing pulverized coal with volatile matter less than 25%, it may be necessary to activate the ignitors even at high loads. This can occur with any coal if it is excessivley wet, frozen, or feeding sporadically into the pulverizers. If the ignitor is not activated during the intervals when coal is not reaching the pulverizer, or is reaching it in small amounts, ignition of the burner may be lost momentarily; then upon reestablishing coal flow, the burner may ignite with explosive force from an adjacent burner.

Excess Air. Pulverized coal requires more excess air for satisfactory combustion than either oil or natural gas.

One reason for this is the inherent maldistribution of coal both to individual burner pipes and to the fuel discharge nozzles. The minimum acceptable quantity of unburned combustible is usually obtained with 15% excess air as measured at the furnace outlet at high loads. This allows for the normal maldistribution of both primary-air-coal and secondary air. Higher excess air values may be necessary to avoid slagging or fouling of the heat absorption equipment.

In the design of the burner and furnace of a pulverized-coal-fired unit, consideration must be given to the burner arrangement and furnace configuration to minimize slagging or fouling from coal ash. Increasing excess air will permit most designs to perform satisfactorily but this can be an uneconomical long-time substitute for good basic design.

In general, the pulverizer-burner combination can operate satisfactorily from full load to approximately 40% of full load with all pulverizers and burners in service. In some installations a pulverizer and set of burners, in addition to the number actually required, is provided to assure availability of the boiler unit in case of unscheduled outage of a pulverizer. Where spares are provided, it is generally most economical to operate with the greatest number of burners and pulverizers in service consistent with the capacity demand on the unit. Although the use of this excess equipment raises the minimum load which can be obtained without cutting out pulverizers and burners, other benefits offset this disadvantage.

It is easier to pick up load with an operating pulverizer than to bring an idle unit into service. Also, at high loads on a boiler unit the burner elements in idle burners deteriorate quickly because of radiant heat. Air that is admitted through idle burners to reduce over-heating does not enter into the combustion reaction but is excess air which lowers the boiler efficiency.

Cyclone Furnace. The introduction of pulverized-coal firing in the 1920s was a major advance, providing advantages over stoker firing. As of the early 1980s, pulverized-coal firing is highly developed and remains the most effective way to burn many types of coal, particularly the higher grades and ranks. However, since about 1940, the cyclone furnace has been developed and is now widely used.

The cyclone furnace is applicable to coals having a slag viscosity of 250 poises at 1427°C or lower, provided the ash analysis does not indicate excessive formation of iron or iron pyrites. With these coals, cyclone-furnace firing provides the benefits obtainable with pulverized-coal firing, plus the following advantages: (1) Reduction in fly ash content in the flue gas; (2) saving in the cost of fuel preparation, since only crushing is required instead of pulverization; and (3) reduction in furnace size.

The cyclone furnace (Fig. 7) is a water-cooled horizontal cylinder in which fuel is fired, heat is released at extremely high rates, and combustion is completed. Its water-cooled surfaces are studded, and covered with refractory over most of their area. Coal crushed in a simple crusher, so that approximately 95% will pass a four-mesh screen, is introduced into the burner end of the cyclone. About 20% of the combustion air, termed primary air, also enters the burner tangentially and imparts a whirling motion to the incoming coal. Secondary air with a velocity of about 300 feet/second (90 meters/second) is admitted in the same direction tangentially at the roof of the main barrel of the cyclone and imparts a further whirling or centrifugal action to the coal particles. A small amount of air (up to about 5%) is admitted at the center of the burner. This is known as "tertiary" air.

The combustible is burned from the fuel at heat release rates of 450,000 to 800,000 Btu/cubic foot/hour (1 Btu = 0.2520 Calorie) and gas temperatures exceeding 1649°C are developed. These temperatures are sufficiently high to melt the ash into a liquid slag, which forms a layer on the walls of the cyclone. The incoming coal particles are thrown to the walls by centrifugal force, held in the slag, and scrubbed by the high-velocity tangential secondary air. Thus, the air required to burn the coal is quickly supplied, and the products of combustion are rapidly removed.

The gaseous products of combustion are discharged through the water-cooled re-entrant throat of the cyclone into the gas-cooling boiler furnace. Molten slag in excess of the thin layer retained on the walls continually drains away from the burner end and discharges through the slag tap opening to the boiler furnace, from which it is tapped into a slag tank, solidified, and disintegrated for disposal.

The cyclone furnace is capable of burning successfully a large variety of fuels. A wide range of coals varying in rank from low volatile bituminous to lignite may be successfully burned and, in addition, other solid fuels, such as wood bark, coal chars, and petroleum coke, may be satisfactorily fired in combination with other fossil fuels. Fuel oils and gases are also suitable for firing.

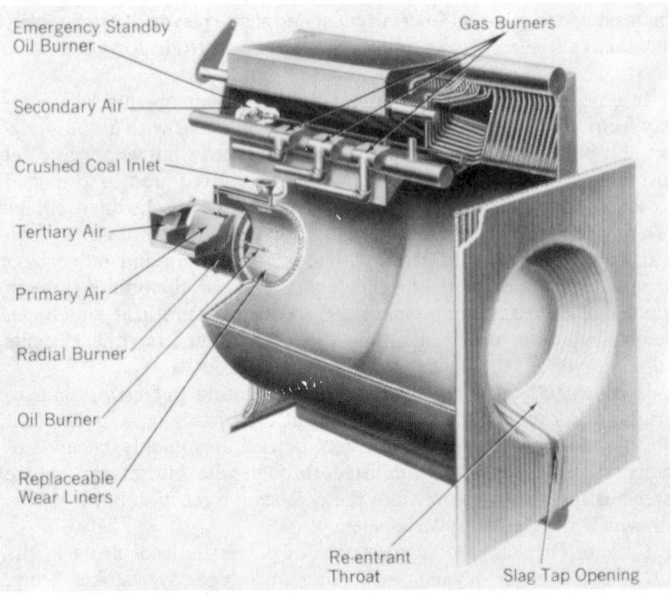

Fig. 7. Cyclone furnace. (*The Babcock & Wilcox Co.*)

TABLE 3. MAXIMUM ALLOWABLE FUEL-BURNING RATES FOR STOKERS

Type of Stoker	Btu/Square Foot/Hour
Spreader—stationary and dumping grate	450,000
Spreader—traveling grate	750,000
Spreader—vibrating grate	400,000
Underfeed—single or double retort	425,000
Underfeed—multiple retort	600,000
Water-cooled vibrating grate	400,000
Chain grate and traveling grate	500,000

Conversion factor: 1 Btu = 0.2520 kilogram-calorie.

Stokers. A successful stoker installation requires the selection of the correct type and size for the fuel to be used and the desired capacity. Also, the associated boiler unit should have the necessary instruments for the proper control of the stoker. The grate area required for a given stoker type and capacity is determined from allowable rates established by experience. Table 3 lists allowable fuel-burning rates for various types of stokers, based upon using coals suited to the stoker type in each case.

Mechanical stokers can be classified in four main groups, based on the method of introducing fuel to the furnace: (1) Spreader stokers; (2) underfeed stokers; (3) water-cooled vibrating-grate stokers; and (4) chain-grate and traveling-grate stokers.

Among these several types, the spreader stoker is the most generally used in the capacity range from 75,000 to 400,000 pounds (34,020 to 181,440 kilograms) of steam per hour, because it responds rapidly to load swings and can burn a wide range of fuels.

Underfeed stokers of the single-retort, ram-feed, side-ash-discharge type are used principally for heating and for small industrial units of less than 30,000 pounds (13,608 kilograms) of steam per hour capacity. Larger-size underfeed stokers of multiple-retort, rear-ash-discharge type have been largely displaced by spreader stokers and by the water-cooled vibrating-grate stokers in the intermediate range. Chain- and traveling-grate stokers, while still used in some areas, are gradually being displaced by the spreader and vibrating-grate types.

As the name implies, the spreader stoker projects fuel into the furnace over the fire with a uniform spreading action, permitting suspension burning of the fine fuel particles. See Fig. 8. The heavier pieces that cannot be supported in the gas flow, fall to the grate for

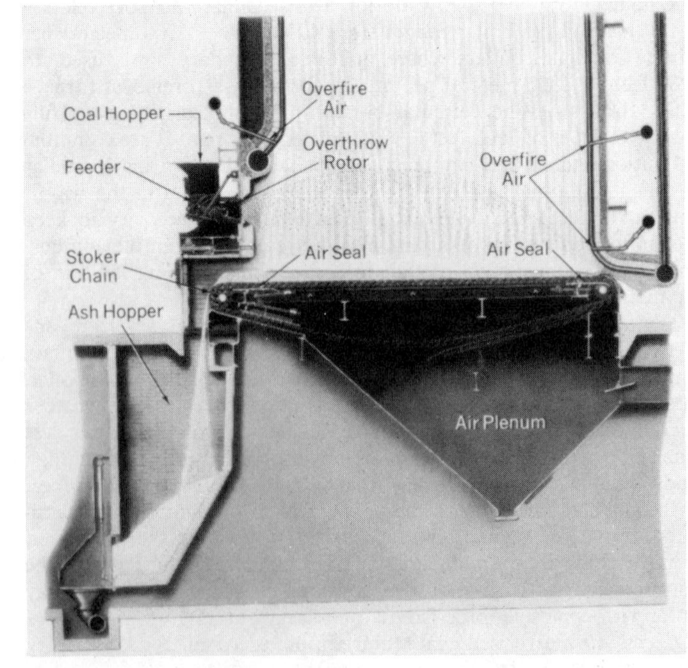

Fig. 8. Traveling-grate spreader stoker with front-ash discharge. (*The Babcock & Wilcox Co.*)

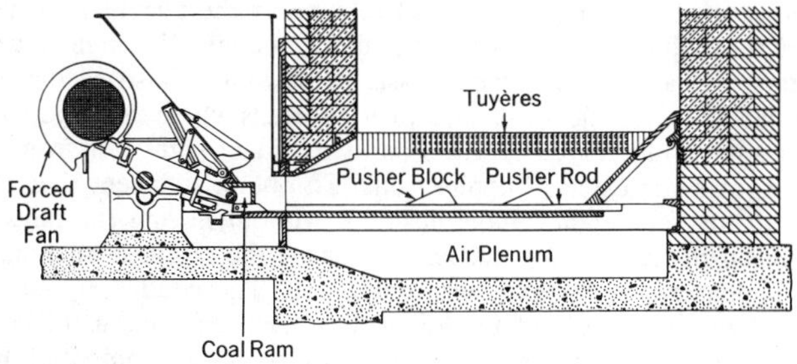

Fig. 9. Single-retort, horizontal-feed, side-ash-discharge underfeed stoker. (*The Babcock & Wilcox Co.*)

combustion in a thin fast-burning bed. This method of firing provides excellent sensitivity to load fluctuations as ignition is almost instantaneous on increase of firing rate and the thin fuel bed can be burned out rapidly when desired. All spreader stokers, and in particular, the traveling-grate spreader, have an extraordinary ability to burn fuels with a wide range of burning characteristics, including coals with cake tendencies. High-moisture, free-burning bituminous and lignite coals are commonly used, and some low-volatile fuels, such as coke breeze, have been burned in a mixture with higher volatile coal. Anthracite coal, however, is not a satisfactory fuel for spreader-stoker firing.

Underfeed stokers are generally of two types: (1) Horizontal-feed, side-ash-discharge type (Fig. 9); and (2) the gravity-feed, rear-ash discharge type (Fig. 10). In the side-ash-discharge underfeed stoker, fuel is fed from the hopper by means of a reciprocating ram to a central trough called the retort. On very small heating stokers, a screw conveys the coal from the hopper to the retort. A series of small auxiliary pushers in the bottom of the retort assist in moving the fuel rearward and, as the retort is filled, the fuel is moved upward to spread to each side over the air-admitting tuyeres and side grates.

As the fuel rises in the retort and is subjected to heat from the burning fuel above, volatile gases are distilled off and mixed with air supplied through the tuyeres above each side of the retort and through the side grates. The volatile mixture burns as it passes upward through the incandescent zone, sustaining ignition of the rising fuel. Burning continues as the incoming raw coal continually forces the fuel bed to each side. Combustion is completed by the time the bed reaches the side-dumping grates. The ash is intermittently discharged to shallow pits, quenched and removed through doors at the front of the stoker.

The single (and double) retort, horizontal-type stokers are generally limited to 25,000 to 30,000 pounds (11,340 to 13,608 kilograms) of steam per hour with burning rates of 425,000 Btu/square foot/hour (1 Btu = 0.2520 Calorie) in furnaces with water-cooled walls. For refractory-walled furnaces, the maximum rate should be reduced to about 300,000 Btu/square foot/hour. The multiple retort, rear-end-cleaning type has a retort and grate inclination of 20 to 25 degrees. This type of stoker can be designed for boiler units generating up to 500,000 pounds (226,800 kilograms) of steam per hour. Burning rates up to 600,000 Btu/square foot/hour are practicable.

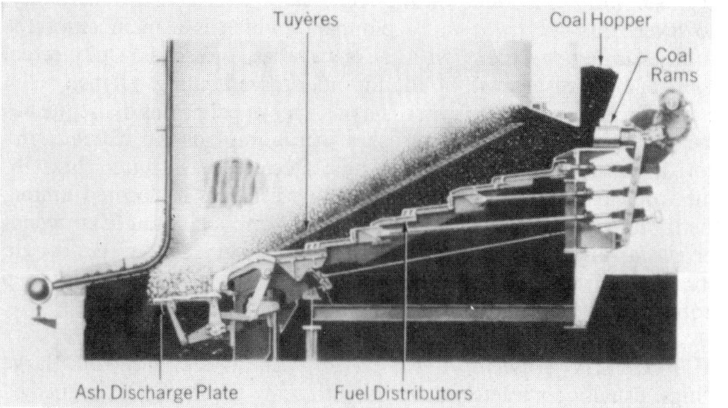

Fig. 10. Multiple-retort, gravity-feed type of rear-ash-discharge underfeed stoker. (*The Babcock & Wilcox Co.*)

Traveling-grate stokers, including the specific type known as the chain-grate stoker, have assembled links, grates, or keys joined together in endless-belt arrangements that pass over the sprockets or return bends located at the front and the rear of the furnace. Coal, fed from the hopper, onto the moving assembly, enters the furnace after passing under an adjustable grate to regulate the thickness of the fuel bed. The layer of coal on the grate, as it enters the furnace, is heated by radiation from the furnace gases and is ignited together with the hydrocarbon and other combustible gases driven off by distillation. The fuel bed continues to burn as it moves along with the bed growing progressively thinner as combustion continues. At the far end of travel, ash is discharged into the ash pit. Chain- and traveling-grate stokers can burn a wide variety of fuels, including peat, lignite, subbituminous, free-burning bituminous, anthracite, and coke breeze. These stokers are offered for maximum continuous burning rate of 425,000 Btu/square foot/hour with high-moisture (20%), high-ash (20%) bituminous coal, and 500,000 Btu/square foot/hour with lower-moisture (10%), lower-ash (8 to 12%) bituminous coal. For anthracite, the corresponding burning rate is 350,000 Btu/square foot/hour.

References

Ardell, M.: "Particulate Control for Coal-fired Boilers," *Chem. Eng. Progress*, **75**, 78–82 (1979).

Baumeister, T. (editor): "Mark's Standard Handbook for Mechanical Engineers," 8th edition, McGraw-Hill, New York, 1978.

Burkhardt, C. H.: "Domestic and Commercial Oil Burners," 3rd edition, McGraw-Hill, New York, 1969.

Criswell, R. L.: "Control Strategies for Fluidized Bed Combustion," *In-Tech.*, **27**, 1, 37–43 (1980).

Drehmel, D. C.: "Developments in Particulate Control for Coal-fired Power Plants," *Chem. Eng. Progress*, **76**, 5, 74–75 (1980).

Grey, J., Sutton, G. W., and M. Zlotnick: "Fuel Conservation and Applied Research," *Science*, **200**, 135–142 (1978).

Robnett, J. D.: "Engineering Approaches to Energy Conservation," *Chem. Eng. Progress*, **75**, 3, 59–81 (1979).

Smith, R. B.: "Heat Generation" in "Chemical Engineers' Handbook," 5th edition (R. H. Perry and C. H. Chilton, editors), McGraw-Hill, New York, 1973.

Smoot, L. D., and D. T. Pratt: "Pulverized Coal Combustion and Gasification for Continuous Flow Processes," Plenum, New York, 1979.

Staff: "Steam—Its Generation and Use," 39th edition, The Babcock & Wilcox Co., New York, 1978.

BURNER FUEL OIL. **Petroleum.**

BURNISHING. A surface-hardening or surface-finishing process for metals, effected by the application of a roller or blunt rod under pressure to the surface. It is used for gear-tooth finishing to some extent. The work is rotated between three hardened and ground burnishing gears. Burnishing will not correct errors but serves to compress the surface of the teeth and provides a slight surface hardness. As a finishing operation prior to hardening, it can also be used to remove burrs and bruises.

BURNUP FIGURE OF MERIT. **Nuclear Reactor.**

BURRFISHES. **Porcupine Fishes.**

BURROWING. The habit of living underground and also the preparation of runways and living quarters beneath the surface. Many animals, such as wolves, that are not specially developed for burrowing, prepare burrows or dens for the birth of their young and for hiding places. Others are normally at home in the ground and come to the surface only under certain conditions; here examples range from the worms, of which the earthworm is particularly well known, to highly specialized mollusks, insects, and vertebrates.

The less specialized burrowing animals cut away the earth by means of structures whose origin is not associated with this use, such as the claws of reptiles, birds, and mammals, while in many of the more specialized forms these parts are highly developed and modified and other special adaptations are evident. The earthworm merely eats its way through the earth, and some insects are capable of burrowing into the tissues of plants in the same way. The shipworm, a mollusk, has the valves of the shell adapted for cutting burrows into wood.

The moles are highly developed for burrowing by the powerful build of the fore limbs and by their large claws, and are further adjusted to life underground by their fine moisture-resisting fur. The poorly developed eyes are also correlated with conditions underground, since these animals rarely enter the light.

Mole crickets are in some ways like the moles. Their front legs are broadened and provided with clawlike processes and the body is covered with a downy vestiture which repels moisture.

Burrowing offers the animal greater safety than can be enjoyed on the surface of the earth, since no predator can compete with the highly developed burrowing animal on equal terms below the surface.

BURSA AND BURSITIS. A small sac of connective tissue, usually interposed between joints, lined by synovial membrane and filled with fluid, which reduces friction. Bursae function as lubricating buffers between moving parts, the tendons and bone, or bone and joint capsule, or between skin and bony structures. There are approximately 1,000 bursae in the human body. Bursitis is inflammation of the bursae, usually caused by overuse of a joint, trauma, or infection. Bursitis may be acute or chronic. The acute form is disabling, the involved area being swollen, tender, and very painful in motion. The bursal sac is often distended with fluid in chronic bursitis. Occasionally an acute calcium deposit also is present. The shoulder joint (acromial and subdeltoid bursae), the elbow (olecranon bursa), and the knee (prepatellar bursa) are common sites for bursitis. Treatment usually is confined to immobilization of the affected area, followed by gradually progressing exercises. In chronic cases, patients should analyze their postural habits, noting activities which may constantly aggravate a specific joint and make remedial changes. For example, shoulder bursitis may be prevented by avoiding long periods of leaning on one arm or resting an arm over a car windowsill for extensive periods.

Septic bursitis usually follows an injury and is caused by *Staphylococcus aureus* infection. There is considerable swelling, pain, and often mild fever. This condition is usually seen in the olecranon (elbow) or prepatellar (knee) bursas and occurs most frequently in men. For moderate cases, oral antibiotics and/or needle aspiration may be sufficient therapy.

BURSA COPULATRIX. A pouch found in the females of some species of insects which receives the seminal fluid of the male during copulation, before it passes to the spermatheca. A sperm pouch or reservoir for storing the spermatozoa.

BURST. In cosmic ray studies a burst is an exceptionally large electric pulse observed in an ionization chamber, signifying the simultaneous arrival or emission of several or many ionizing particles. Such an event may be caused by a cosmic-ray shower or by a spallation disintegration of the type that can produce a star. In communications a burst is a sudden increase in signal strength of waves being received by ionospheric reflection.

BUSHBUCK. **Bovines.**

BUSH COW. **Bovines.**

BUSH-CRICKET (*Insecta, Orthoptera*). Crickets of several species, most of which are found chiefly in shrubby vegetation.

BUSHING. In mechanical terminology, to bush is to reduce the size of a hole. A bushing is a hollow cylinder used as a renewable liner for a bearing or a drill jig.

A bushing is also a pipe fitting employed to reduce the size of pipe to a smaller size. When pipe is employed to contain electrical wires, the open end from which the wires emerge is often capped by a bushing which substitutes a smoothly rounded surface for the sharp edges of an unbushed conduit. The sharp edges would tend to abraid the insulation on the wires.

In electrical work, where a conductor at high voltage emerges from one insulated condition to another, an intermediate support must be provided. An electrical bushing is needed to provide the support and insulation between the conductor and the supporting surface. For example, where the conductor leaves the insulated interior of a transformer case, a bushing is provided to support the terminal where it passes through the case, and to insulate the voltage difference between the terminal and the grounded case. A bushing is also required at terminals of oil circuit breakers, and at potheads where the conductors of a multi-conductor cable are separated and brought out from the cable sheath for external connections. To obtain sufficient dielectric strength for very high voltage bushings without having the physical dimension of the bushing become excessive, the oil-filled bushing or the condenser-type bushing was developed. The condenser bushing is made of thin layers of tin foil wound between concentric layers of insulation. It is possible in this way to give uniform potential drop through the thickness of the busing.

BUS SYSTEM (Switching). **Switching (Power).**

BUSTAMITE. **Rhodonite.**

BUSTARD. **Rails, Coots, and Cranes.**

BUSTARD-QUAIL. **Rails, Coots, and Cranes.**

BUTADIENE. $CH_3CH{:}C{:}CH_2$, 1,3-butadiene (methyl-allene), formula weight 54.09, bp −4.41°C, sp gr 0.6272, insoluble in H_2O, soluble in alcohol and ether in all proportions. Butadiene is a very reactive compound, arising from its conjugated double-bond structure. Most butadiene production goes into the manufacture of polymers, notably SBR (styrene-butadiene rubber) and ABS (acrylonitrile-butadiene-styrene) plastics. Several organic synthesis, such as the Diels-Alder reaction, commence with the double-bond system provided by this compound.

Butadiene came into prominence as an important industrial chemical during World War II as the result of the natural rubber shortage. Originally, butadiene was made by the dehydrogenation of butylenes. Later, the naphtha cracking for ethylene and propylene, with a by-product C_4 stream, created another source of butadiene. The basis for one butadiene recovery process is the change in relative volatility of C_4 hydrocarbons in the presence of acetonitrile solvent. The latter makes the separation easier. The C_4 mixed charge goes to an extractive distillation column where it is separated in a solvent environment into a solvent-butadiene stream and a by-product butane-butylenes stream (overhead). The acetonitrile is recovered from the butane-butylenes. Butadiene is stripped from the fat solvent, after which it goes to a postfractionator for recovery as a 99.5% pure product.

Other solvents used in the extractive distillation include *n*-methyl pyrrolidone, dimethyl formamide, furfural, and dimethyl acetamide.

BUTANE. **Organic Chemistry.**

BUTTERFISHES (*Osteichthyes*). Members of the suborder *Stromateoidea*, these fishes are characterized by a peculiar anatomy, i.e., they incorporate an expanded and muscular esophagus, which may feature ridges, papillae, and, in some instances, teeth. Butterfishes of the family *Stromateidae* are premium food fishes. The *Poronotus triacanthus* is a 12-inch (0.3-meter) species that inhabits the American Atlantic shores. The 10-inch (25-centimeter) California pompano (*Palometa simillima*) is found on the American Pacific coast. In actuality, however, this species is *not* a pompano, which is a member of the family *Carangidae*, along with jacks, cavallas, and scads. Quite often, younger butterfishes will be found under large floating jellyfish, such as the Portuguese man-of-war. The presence of pelvic fins distinguishes the nomeids from other butterfishes. An example is the *Nomeus gronovi*, a man-of-war fish (3 inches long; 7.5 centimeters), found throughout tropical waters. Almost always, this fish will be located among the trailing tentacles of the giant jellyfishes. The squaretail (*Tetragonurus cuvieri*) is a third species of butterfish distinguished by very tough, practically irremovable scales. It is also widely distributed in deep tropical and temperate waters.

BUTTERFLY (*Insecta, Lepidoptera*). An insect with four large wings, usually completely covered with scaly vestiture. Distinguished from most other members of the order (moths and skippers) by the terminal club of the antennae.

Butterflies are found to some extent in most all regions where flowers are in abundance, but their greatest occurrence is in the tropics. There are five major families of butterflies: (1) *Hesperiidae*—small, occur worldwide, but particularly in the United States; (2) *Papilionidae*, a large family of beautifully-marked insects; (3) *Lycaenidae*, a large family of comparatively small butterflies; (4) *Lemoniidae*, a strikingly beautiful family of butterflies that occur in the tropics; and (5) *Nymphalidae*, of very ancient origin and the largest of all families of butterflies. The largest numbers of butterflies are members of the last three mentioned families.

In wing-tip spread, butterflies measure from as small as one-fourth inch to 12 inches. The bodies are slender and long. Particularly as seen in bright sunlight, butterflies are gaily colored. The wings are scaled, the scales overlapping. The name of their order (*Lepidoptera*) means "wings with scales." The wings are folded together and erect when at rest. Some butterflies can change the color of their wing spots when in danger. The two antennae are nearly hairlike and knobbed. Some butterflies have tiny brushes on their forelegs which they use for cleansing their eyes. Some species migrate to warm climates in winter. Unlike moths, butterflies do not spin cocoons. Butterflies of economic concern include:

Cabbage butterfly (*Pieris rapae*). A white butterfly with black-tipped forewings and two or three black spots on the wings of each side. Introduced from Europe in the mid-19th century, the species spread rapidly and it is found throughout the United States and much of Canada. The caterpillar feeds on all cruciferous plants, but is especially important as a pest on cabbage and cauliflower. Related caterpillars include the *potherb butterfly* (*Pieris oleracea*, Harris); the *southern cabbageworm* (*Pieris protodice*, Boisduval and LeConte); and the *Gulf white butterfly* (*Pieris monuste*, Linne). Caterpillars chew holes in leaves and attack buds to cause misshapen produce. Caterpillar droppings as well as the feeding damage can cause produce to be unmarketable. The caterpillars are usually exposed and relatively easy to control when young. Older caterpillars are difficult to kill and are usually found in protected places on the plant. Among control chemicals used are methomyl, chlordimeform, and thuricide, biotrol, and dipel (the latter contain Bacillus Thuringiensis spores as the active ingredient).

Blue butterfly (*Feniseca tarquinius*). The larva of this butterfly is an economical insect. The larva feeds on the woolly aphid.

Thistle butterfly. Feeds on Canada thistle, considered economically beneficial. But, it is also known to feed on cultivated crops when thistle is not available.

Ghiradella, Aneshansley, et al. have made a study of the ultraviolet reflection of male butterflies and this is reported in *Science*, **178**, 4066, 1214–1217 (1972). Males of the butterfly *Eurema lisa*, like many other members of the family *Pieridae*, reflect ultraviolet light. The color is structural rather than pigmentary, and originates from optical interference in a microscopic lamellar system associated with ridges on the outer scales of the wing. The dimensions and angular orientation of the lamellar system conform to predictions based on physical measurement of the spectral characteristics, including "color shifts" with varying angles of incidence, of the reflected ultraviolet light. The female lacks such scales and is consequently nonreflectant. The ultraviolet dimorphism supposedly serves as the basis for sexual recognition in courtship.

BUTTERFLY VALVE. As shown by the accompanying illustration, a butterfly valve consists of a body, a disc supported on a shaft, and a suitable packing box to allow the shaft to protrude for operation by manual and/or automatic actuators. The flow path in a butterfly valve is straight through the body with only the disc to obstruct the flow, resulting in relatively high capacity. These valves are self-cleaning, thus permitting their use for the control of heavy stocks, slurries, and sludges.

Two body styles generally are available: (1) a spool type similar to a conventional gate or globe valve; and (2) a solid ring type. The spool type requires a greater installation space and weighs more than the solid ring type. The latter type bolts between pipeline flanges, eliminates transfer of pipeline stresses to the valve body, and permits the use of lower strength and lower cost body materials.

Butterfly valves are made in many materials for a wide variety of

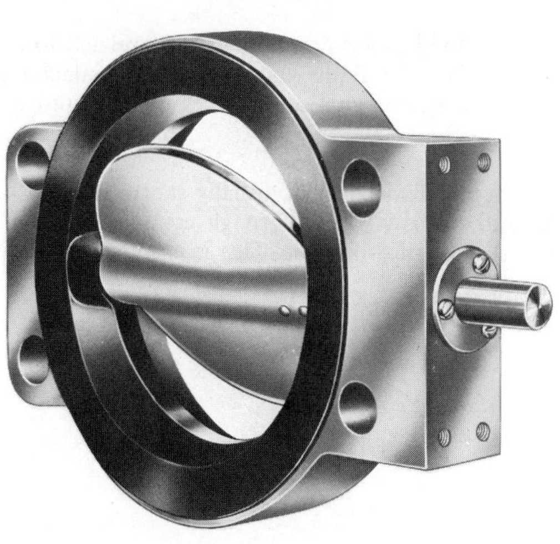

Elastomer-lined butterfly valve.

pressure, temperature, and fluid service conditions. They can be modified for tight shutoff with a soft seat or elastomer liner. Where temperatures prohibit the use of elastomers, other low-leakage designs, such as piston ring, step seated, and angle seated designs are available. Line sizes range from 1 inch (2.5 centimeters) through 108 inches (274 centimeters). Versatility of application is a feature of butterfly valves.

BUTTERNUT TREE. Walnut Trees.

BUTTONWOOD TREE. Plane Trees.

BUTYL ACETONE. Synthesis (Chemical).

BUTYLATED HYDROXYANISOLE (BHA). Antioxidant.

BUTYLATED HYDROXYTOLUENE (BHT). Antioxidant.

BUTYL RUBBER. Elastomers.

BUTYRATE PLASTICS. Cellulose Ester Plastics (Organic).

BUTYRIC FERMENTATION. Fermentation.

BUYS BALLOT'S LAW. Winds and Air Movement.

BUZZARD. Hawk.

BYPASS CAPACITOR. A capacitor placed in an electrical circuit to provide a low impedance alternative path of current flow for one of a combination of two or more signals (one of which may be dc). The most common usage is in bypassing various voltage-dropping resistors used in vacuum-tube and transistor circuits to adjust the voltages applied to the several parts of the circuits. These resistors are bypassed so there will be no, or very little, alternating signal voltage drop to produce undesirable feedback. The reactance of the bypass capacitor should be small compared to the impedance (resistance for dc) of the current path which it is desired to bypass.

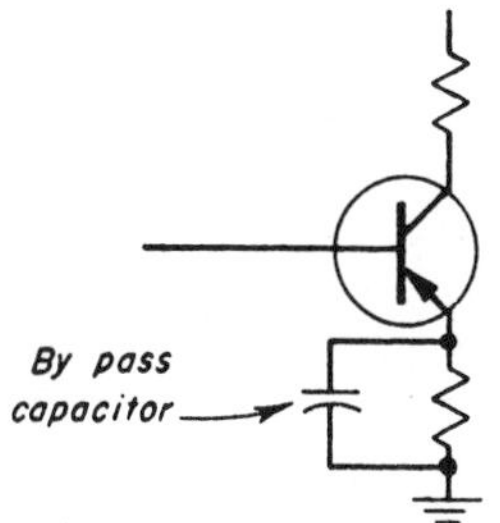

Bypass capacitor in transistor amplifier.

BYSMALITH. A plug-like igneous intrusion related to a laccolith but bounded laterally by faults due to upward "punching" rather than "pushing" of the magma as it forces its way into a series of stratified rocks.

BYTE. A group of binary digits, usually shorter than a word, and usually operated on as a unit. Through common usage, the word most often describes an 8-bit unit. This is a convenient information size to represent an alphanumeric character. Recent use, particularly in connection with communications, is toward the use of *octet* for an 8-bit byte, *sextet* for a 6-bit byte, etc. Most recent computer designs provide for instructions that operate on bytes, as well as word-oriented instructions. In addition, the capacity of storage units is often specified in terms of bytes.

BYTOWNITE. **Feldspar.**

C

CABBAGE. Brassica.

CABBAGE BUTTERFLY. Butterfly.

CABINET DRAWING. Pictorial Representation.

CABLE (Electrical). An electrical cable is one or more conductors surrounded by an insulating medium and a protective sheath. Such cables are used for the transmission of electric power and for transmission of communication signals. The power cables have relatively few conductors of heavy gauge and are insulated for high voltages. Such cables are frequently filled with oil to increase the insulation strength. The outer sheath is commonly of lead, although for submarine work this in turn is often further strengthened by a second sheath of steel strands. Communications cables usually contain many pairs of small-gauge copper conductors, paper-insulated, surrounded by a lead sheath. Sometimes the entire cable is nitrogen-filled under pressure. The various pairs of conductors are arranged by twisting and placing to minimize pick-up between them (see **Cross-Talk**). Common practice is to include two extra pairs for spares for each hundred active pairs. Cables used for submarine circuits have fewer pairs and are heavily insulated and armored to withstand the severe strains to which they may be subjected in laying and by ocean currents. The coaxial cable (see **Coaxial Line**) is a special type in which the pair of conductors is formed by a center wire and the outer sheath. In this case the sheath is copper and the insulation is often a gas with solid dielectric spacers at intervals to hold the inner conductor centered. This coaxial cable may in turn be enclosed with others in a lead sheath for protection. Coaxial cables have a wide usable frequency range and hence are used for transmission of television programs. They are also often used for radio-frequency transmission lines as the electromagnetic fields necessary for the transmission of signals penetrate the space exterior to the sheath of the cable to a negligible extent.

Detailed specifications for electrical cable are available from the American Society for Testing and Materials, Philadelphia, Pennsylvania. For information transmission by fiber optics, see also **Optical Fibers.**

CABLE (Optical Fiber). Optical Fibers; Telephony.

CABLE TELEVISION. Television.

CABLEWAY. A suspended steel cable acting as a track for aerial hoisting and conveying devices is a cableway. While occasionally used for transporting persons across deep gorges, where the amount of traffic does not warrant the building of a bridge, the cableway, in its more common application, handles construction material for building of dams, or has a permanent use in connection with the handling of material such as rock or gravel which is taken from open pits. Clear spans up to a half-mile in length are possible in a cableway. The carriage which operates on the cableway may or may not have provision for carrying passengers, depending on the purpose of the cableway. Cableways are used at winter resorts to transport skiers up steep slopes.

CACAO TREE. Of the family *Sterculiaceae* (chocolate family), there are two species of principal commercial interest: *Theobroma cacao*, native to Mexico, and *T. leiocarpum*, native to Brazil. The tree is found in its wild state, growing in the lowlands from Mexico southward to northern South America. It is of medium size with shiny, evergreen leaves about a foot long. The flowers are small. They grow from buds on the trunk or large branches of the tree. The fruits are melon-like, from 6 to 12 inches (15 to 30 centimeters) long and about 4 inches (10 centimeters) in diameter. They have a ribbed, rough surface. Each fruit contains from 20 to 50 flattened seeds or beans embedded in a gelatinous pulp. The tree is extensively cultivated in humid, tropical latitudes where rich soil is available. Important cacao bean producers include Ghana, Nigeria, and Brazil. The tree cannot tolerate sustained temperatures below 60°F (~15°C). The cultivated tree is somewhat smaller than the wild one and begins to bear within 4 to 5 years when grown from seed. Trees may continue to bear well up to an age of 50 years.

The mature pods are cut from the tree and split open. The seeds are then scooped out and fermented for 1 or 2 weeks. During fermentation, the color of the seeds darkens to a reddish tone and a rich aromatic essence develops. The pulp surrounding the seeds liquefies and runs off. After fermentation, the seeds are dried and shipped to processors throughout the world, but notably located in Europe and the United States. In the processing plants, the seeds are cleaned, after which they are roasted for 1 to 2 hours. After roasting, the seeds are cracked and the shell separated from the cotyledons. The shells may be ground up and used in the manufacture of cheap grades of cocoa, or they may be burned as fuel. From the cotyledons, an oily liquid is ground out by heated mills. This liquid hardens into familiar chocolate. If part of the oil is squeezed out and the residue ground to a powder, the product is called *cocoa.* When chocolate is mixed with sugar or flavoring is added, the product is called *bitter chocolate.* The vegetable fat removed from the pressed beans is known as cocoa butter and is used in the manufacture of various pharmaceuticals, soaps, and in the preparation of confections.

CACTUS (*Cactaceae*). The cactus is known to all as a prickly inhabitant of dry American deserts. In popular parlance, the name "cactus" applies to any fleshy spine-covered plant. But not all spiny plants are cacti, nor are all cacti characterized by spines.

With the exception of a single genus *Rhipsalis,* some of whose members are said to occur in Sri Lanka and Madagascar, all cacti are natives of America, where they are found widely scattered from latitude 59° in North America through the tropics to the southern Andean region and Argentina. They are particularly conspicuous features of the flora of dry desert regions where they are found in a wide variety of forms and sizes.

A few genera, and especially *Pereskia*, are very like ordinary mesophytic plants, having well-developed ovate leaves borne alternately on a long slender stem. But in nearly all of the Cactus Family the leaf surface is very much reduced, the leaves appearing as small fleshy bodies which last but a brief time before dropping off. In many species leaves of any recognizable kind are never formed, the green fleshy stem taking over the function of leaves completely. In these fleshy stems large amounts of water are stored, a feature which enables these plants to survive in the arid regions in which they so frequently grow. Due to the mucilaginous nature of the cell contents and to the greatly reduced surface of the plants the contained water is held most tenaciously and lost very slowly.

In their natural environment cactus plants have an extensive system of long fibrous roots which not only extend outward from the plant to considerable distances, but also penetrate the soil deeply. In cultivated plants the root system is usually greatly reduced. The stems of cacti show a variety of forms. In addition to the normal-stemmed *Pereskia*, there are the Prickly Pears, species of *Opuntia.* In most of these the stem is a series of flattened, fleshy joints often abundantly protected with bristling bunches of barbed spines. In species of *Mam-*

millaria and *Cereus* the stem is a cylindrical or globular body, often conspicuously ridged, and armed with numerous spines. In *Phyllocactus* and *Epiphyllum* the stem is flattened and largely unarmed, the small weak spines being borne in notches along the edges of the stem. The familiar night-blooming "cereus" is of this type.

One of the best known and largest of all the cacti is the sahuaro or giant cactus (*Cereus giganteus*). This species is a native of southern Arizona, northern Sonora, and extreme southeastern California. This massive cactus may grow to a height of 50 feet (15 meters) with many side branches. Some individuals probably attain an age of 200 years. In some parts of its range extensive "forests" of this species have developed.

The flowers of most species of cactus are large and brightly colored. They are regular, although in some species a definite tendency toward zygomorphic flowers is seen. The flowers are borne singly. The perianth is composed of a large number of separate members which show a gradual transition from the outer small sepals (see **Flower**) through to large brightly colored petals. The stamens are likewise numerous and have long filaments. The single compound pistil contains many ovules and in fruit becomes a many-seeded berry. In many species the fruit is edible.

Species of *Opuntia* are frequently planted in rows to form an impenetrable barrier against intruders. These plants were early introduced into the Old World and later into Australia, where in many places they have become a troublesome and almost worthless weed. Cactus plants are frequently seen in cultivation, being especially sought by those who like the bizarre effect they give. Somewhat similar in appearance are many species of *Euphorbia* from tropical Africa, and of *Stapelia*, a genus of the Milkweed Family, and likewise native to Africa. The flowers of these plants are quite unlike cactus flowers, however, so that the plants are readily distinguished when they bloom.

CADDIS FLY (or Caddice Fly). *Insecta, Trichoptera.* The adult of any species of this order. The caddis flies are slender insects with four wings, sometimes clothed with hair-like scales which give them a moth-like appearance. The mouth parts are formed for biting but are vestigial. Since the larvae are aquatic, the caddis flies are much more abundant in the vicinity of water, but they are attracted to light, often at some distance.

CADDIS WORM (*Insecta, Trichoptera*). The aquatic larva of a caddis fly. They are noteworthy for the silken webs and cases which they build, some for protection and some to catch prey. Some species spin silken nets attached to rocks in the bottom of a stream in such a position that the current washes into the wide mouth and passes out through a web at the smaller end. In this way the insect, which lives in a tube near by, snares its food. Many of the caddis worms live in cases from which only the head and legs protrude. These cases are formed of many different materials, held together by silk. Some are made of small flat pebbles, some of bits of leaves, and some of small snail shells. They are economically of some value as food for fishes.

CADELLE BEETLE. Grain Storage Insects.

CADMIUM. Chemical element symbol Cd, at. no. 48, at. wt. 112.41, periodic table group 2b, mp 320.9°C, bp 766°C, density 8.65 g/cm^3 (^{20}°C). Elemental cadmium has a hexagonal crystal structure. Cadmium is a silver-white metal, malleable and ductile, but at 80°C becomes brittle. It remains lustrous in dry air and is only slightly tarnished by air or H_2O at standard conditions. The element may be sublimed in a vacuum at a temperature of about 300°C, and when heated in air burns to form the oxide. Cadmium dissolves slowly in hot dilute HCl or H_2SO_4 and more readily in HNO_3. The element first was identified by M. Stromeyer in 1817. Naturally occurring isotopes 106, 108, 110–114, 116. ^{113}Cd is unstable with respect to beta decay (0.3 MeV) into ^{113}In ($t_{1/2} \geq 10^{13}$ years). Electronic configuration $1s^22s^22p^63s^23p^63d^{10}4s^24p^64d^{10}5s^2$. Ionic radius Cd^{2+} 0.99 Å. Metallic radius 1.489 Å. First ionization potential 8.99 eV; second, 16.84 eV; third, 38.0 eV. Oxidation potentials $Cd \rightarrow Cd^{2+} + 2e^-$, 0.402 V; $Cd + 2OH^- \rightarrow Cd(OH)_2 + 2e^-$, 0.915 V; $Cd + 4CN^- \rightarrow Cd(CN)_4 + Ze^-$, 0.90 V. Other important physical properties of cadmium are given under **Chemical Elements.**

Although ranking 57th in abundance in the earth's crust (0.15 ppm), cadmium is not encountered alone, but is always associated with zinc. The only known cadmium minerals are greenockite (sulfide) and otavite (carbonate), both minor constituents of sphalerite (zinc oxide) and smithsonite (zinc carbonate), respectively. See also **Greenockite; Smithsonite; Sphalerite Blends.**

Production: Two major processes are used for producing cadmium: (1) pyrohydrometallurgical and (2) electrolytic. Zinc blende is roasted to eliminate sulfur and to produce a zinc oxide calcine. The latter is the starting material for both processes. In the pyrohydrometallurgical process, the zinc oxide calcine is mixed with coal, pelletized, and sintered. This procedure removes volatile elements such as lead, arsenic, and the desired cadmium. From 92–94% of the cadmium is removed in this manner, the vapors being condensed and collected in an electrostatic precipitator. The fumes are leached in H_2SO_4 to which iron sulfate is added to control the arsenic content. The slurry then is oxidized, normally with sodium chlorate, after which it is neutralized with zinc oxide and filtered. The cake goes to a lead smelter, while the filtrate is charged with high-purity zinc dust to form zinc sulfate or zinc carbonate and cadmium sponge. The latter is briquetted to remove excess H_2O and melted under caustic to remove any zinc. The molten metal then is treated with zinc ammonium chloride to remove thallium, after which it is cast into various cadmium metal shapes. The process just described is known as the *melting under caustic process.* In a *distillation process,* regular rather then high-purity zinc is used to make the sponge. Then, after washing and centrifuging to remove excess H_2O, the sponge is charged to a retort. The heating and distillation process is under a reducing atmosphere. Lead and zinc present in the vapors contaminate about the last 15% of the distillate. Thus, a redistillation is required. The cadmium vapors produced are collected and handled as previously described.

Reactions which occur in the foregoing processes are: (Leaching): $CdO + H_2SO_4 \rightarrow CdSO_4 + H_2O$; (Oxidation): $3As_2O_3 + 2NaClO_3 \rightarrow 3As_2O_5 + 2NaCl$; and $6FeSO_4 + NaClO_3 + 3H_2SO_4 \rightarrow 3Fe_2(SO_4)_3 + NaCl + 3H_2O$; (Neutralization): $Fe_2(SO_4)_3 + As_2O_5 + 3ZnO + 8H_2O \rightarrow 2FeAs(OH)_8 + 3ZnSO_4$; (Cadmium Precipitation): $CdSO_4 + Zn \rightarrow Cd + ZnSO_4$; (Melting Under Caustic): $Zn + 2NaOH + \frac{1}{2}O_2 \rightarrow Na_2ZnO_2 + H_2O$.

In the electrolytic process, the calcine first is leached with H_2SO_4. Charging the resultant solution with zinc dust removes the cadmium and other metals which are more electronegative than zinc. The sponge which results is digested in H_2SO_4 and purified of all contaminants except zinc. Nearly-pure cadmium sponge is precipitated by the addition of high-purity, lead-free zinc dust. The cadmium sponge then is redigested in spent cadmium electrolyte, after which the cadmium is deposited by electrolysis onto aluminum cathodes. The metal is then stripped from the electrodes, melted, and cast into various shapes. Reactions which occur during the electrolytic process are: (Roasting): $ZnS + 1\frac{1}{2}O \rightarrow ZnO + SO_2$; (Leaching): $ZnO + H_2SO_4 \rightarrow ZnSO_4 + H_2O$; (Neutralization): $Fe_2(SO_4)_3 + 3ZnO + 3H_2O \rightarrow 2Fe(OH)_3 + 3ZnSO_4$; (Cadmium Precipitation): $CdSO_4 + Zn \rightarrow Cd + ZnSO_4$; (Electrolysis): $CdSO_4 + H_2O \rightarrow Cd + H_2SO_4 + O_2$.

Industrial specifications normally require that impurities in cadmium metal not exceed the following: zinc, 0.035%; copper, 0.015%; lead, 0.025%; tin, 0.01%; silver, 0.01%; antimony, 0.001%; arsenic, 0.003%; and tellurium, 0.003%. The metal is available in numerous forms. Electroplaters generally prefer balls 2 inches (5 centimeters) in diameter.

Uses: A major use of cadmium is for electroplating steel to improve its corrosion resistance. It is also used in low-melting-point alloys, brazing alloys, bearing alloys, nickel-cadmium batteries, and nuclear control rods, and as an alloying ingredient to copper to improve hardness. Cadmium, unfortunately, is limited in its usefulness because fumes and dusts containing cadmium are quite toxic. Melting and handling conditions that create dust or fumes must be equipped with exhaust ventilation systems.

Biological Aspects: Traditionally, cadmium has been regarded by animal feedstuff scientists as a toxic minor substance. Normally, the element is not reported in feedstuff ingredient tables because of its infrequent occurrence. When it does occur in feeds, it is believed to be derived from the fat of cereals, nuts, and vegetables. Feed researchers have found that cadmium tends to reduce iron absorption. Little information is available to indicate possible nutritional benefits of cadmium in extremely minute quantities. Cadmium-plated equipment is not used for working parts in food processing equipment.

Chemistry and Compounds: In virtually all of its compounds, cadmium exhibits the +2 oxidation state, although compounds of cadmium(I) containing the ion Cd_2^{2+}, have occasionally been reported. Cadmium hydroxide is more basic than zinc hydroxide, and only slightly amphiprotic, requiring very strong alkali to dissolve it, and forming $Cd(OH)_3^-$ or $Cd(OH)_4^{2-}$ depending upon the pH.

Cadmium oxide: CdO, formed by burning the metal in air or heating the hydroxide or carbonate, is soluble in acids, ammonia, or ammonium sulfate solution, and is more readily reduced on heating with carbon, carbon monoxide or hydrogen than zinc oxide. Cadmium suboxide, Cd_2O, formed by thermal reduction of cadmium oxalate with carbon monoxide, is believed to be a mixture of CdO and finely divided cadmium. CdO_2 and Cd_4O have been reported. Sodium hydroxide solution precipitates cadmium hydroxide, $Cd(OH)_2$, from solutions of the sulfate or nitrate, but with the chloride the $Cd(OH)_2$ precipitate is mixed with CdOHCl and other hydroxychlorides. $Cd(OH)_2$ exists in two forms, an "active" and an "inactive" one, which have different solubility products. Cadmium(I) hydroxide $Cd_2(OH)_2$, prepared by hydrolysis of Cd_2Cl_2, is, like Cd_2O, believed to be a mixture of the metal and the divalent compound. The Cd_2^{2+} ion is definitely established, however, in such compounds as $Cd_2(AlCl_4)_2$.

Cadmium halides: These compounds can be prepared by the action of the corresponding hydrohalic acids upon the carbonate; or by direct union of the elements. If bromine water is used, some hydrobromic acid must be added to prevent hydrolysis of the bromide to the oxybromide, CdOHBr.

In general, the cadmium halides show in their crystal structure the relation between polarizing effect and size of anion. The fluoride has the smallest and least polarizable anion of the four and forms a cubic structure, while the more polarizable heavy halides have hexagonal layer structures, increasingly covalent and at increasing distances apart in order down the periodic table. In solution the halides exhibit anomalous thermal and transport properties, due primarily to the presence of complex ions, such as CdI_4^{2-} and $CdBr_4^{2-}$, especially in concentrated solutions or those containing excess halide ions.

Cadmium sulfide: CdS is the most extensively used of cadmium compounds and generally is prepared by precipitation from cadmium salts. The wide range of colors, varying from lemon yellow through the oranges and deep red, coupled with the stability and intensity of these colors, qualify CdS as a most desirable pigment for paints, plastics, and other products. The range of colors of CdS precipitates results from differing conditions in their formulation, including the temperature and acidity of the salt solutions from which they are precipitated. The particular salt, such as nitrate, chloride, sulfate, etc., also affects the resulting color. The rate of addition of hydrogen sulfide to the liquor affects particle size and color of the precipitates. Cadmium sulfide is insoluble in H_2O, is dimorphous, and sublimes at 1,350°C. Several crystalline forms exist. When precipitated from normal H_2SO_4 and HNO_3 solutions, the crystals are cubic. From other media, stable alpha hexagonal and unstable beta cubic forms may be formed, these ranging in specific gravity from 3.9 to 4.5, respectively. Pigment colors are not due to crystal form, but rather derive from the particle size and dispersion of the precipitates. Of total cadmium production, pigments account for 20–25% of the total.

Other cadmium compounds used as pigments in ceramics, glass, and paints include cadmium nitrate, selenide, sulfoselenide, and tungstate. Cadmiopone ($BaSO_4$ plus CdS) ranges from yellow to crimson and is used for coloring plastics and rubber goods. Cadmium stearate, when combined with barium stearate, is widely used as a stabilizer in thermosetting plastics and accounts for well over 20% of the total cadmium produced.

Cadmium carbonate: $CdCO_3$, $pK_{sp} = 11.3$, is formed by the hydroxide upon absorption of CO_2, or upon precipitation of a cadmium salt with ammonium carbonate. With alkali carbonates, the oxycarbonates are produced.

Cadmium nitrate tetrahydrate, solubility 215 g/100 ml H_2O at 0°C, is obtained by action of HNO_3 upon the carbonate. It is ionized completely only in solutions weaker than about tenth molar. However, it does not form hydroxy compounds as readily as the zinc salt, requiring the action of NaOH, which in moderate concentration gives $Cd(NO_3)_2 \cdot 3Cd(OH)_2$ and $Cd(NO_3)_2 \cdot Cd(OH)_2$; excess sodium hydroxide precipitates the hydroxide.

Cadmium forms a wide variety of other salts, many by reaction of the metal, oxide, or carbonate with the acids, although some can be obtained only by fusion of the oxides or hydroxides. They include the antimonates (pyro and meta), the arsenates (ortho, meta, and pyro, including acid salts as well as normal), the arsenites, the borates $Cd(BO_2)_2$, $Cd_2B_6O_{11}$, $Cd_3(BO_3)_3$ and $Cd_3B_2O_6$ have been identified), the bromate, the bicarbonate, the chlorate, the chlorite, chromates and dichromates, the cyanide, the ferrate, the iodate, the molybdate, $CdMoO_4$, the nitrate, the perchlorate, various periodates, the permanganate, various phosphates (*ortho*, *meta*, and *para*, including acid salts as well as normal), the selenates and selenites, various silicates, the stannate, the sulfate (which reacts with limited amounts of NaOH or NH_3 solution to give various hydroxy sulfates), the thiosulfate, the titanate, the tungstate, and the uranate.

Cadmium arsenide, nitride, selenide, and telluride are known, the first and third obtainable from the elements, while the nitride is obtained by heating the amide (obtained by reaction of cadmium thiocyanate and potassium amide in liquid NH_3), and the telluride is obtainable by reduction of the tellurate with hydrogen. Cadmium arsenide is used as a semiconductor.

One of the features of the chemistry of cadmium is that it forms a relatively large number of complexes. A number of solid double halides of compositions $MCdX_3$, M_2CdX_4, M_3CdX_5 and M_4CdX_6 where M is an alkali metal and X a halogen are known, the last two probably existing only in the solid state. Conductance studies of solutions indicate the presence of such ions as CdX^+, CdX_3^- and CdX_4^{2-}. The donor ability of oxygen is less toward cadmium than toward zinc, fewer oxygen complexes and organic oxygen-linked complexes being known. Sulfur is a better donor than oxygen; additives of the type $(R_2S)_2 \cdot CdX_2$ are formed from dialkyl sulfides and cadmium halides. The ready reactions with NH_3, as with amines, give large numbers of complexes; those with ammonia include tetrammines and hexammines, containing $[Cd(NH_3)_4]^{2+}$ and $[Cd(NH_3)_6]^{2+}$, respectively. Ethylenediamine forms 6-coordinate compounds containing $[Cd(en)_3]^{2+}$. Prominent among the carbon donor complexes are the cyanides, principally compounds of $Cd(CN)_4^{2-}$, although $Cd(CN)_3^-$ is also known. Other carbon donor compounds are the organometallic compounds CdR_2, where R may be methyl, ethyl, propyl, butyl, isobutyl, isoamyl, amylthio, phenyl, octylthio, decylthio, and higher organic radicals.

References

Campbell, J. K., and C. F. Mills: "Effects of Dietary Cadmium and Zinc on Rats Maintained on Diets Low in Copper," *Proceedings Nutr. Society*, **33**, 1, 15A (Abstract) (1974).

Carapella, S. C., Jr.: "Cadmium," in "Metals Handbook," 9th edition, Volume 2, American Society for Metals, Metals Park, Ohio, 1979.

Chizhikov, D. M.: "Cadmium," Pergamon, London, 1966.

Chowdhury, P., and D. B. Louria: "Influence of Cadmium and Other Trace Metals on Human Alpha 1-Antitrypsin: An in vitro Study," *Science*, **191**, 480–481 (1976).

Farnsworth, M.: "Cadmium Chemicals," International Lead Zinc Research Organization, Inc., New York, 1980.

Kirchgessner, M. (editor): "Trace Element Metabolism in Man and Animals," Institut für Ernahrungsphysiologie, Technische Universität München, Freising-Weihenstephan, Germany, 1978.

Pleban, P.: "Cadmium Concentration in Blood," *Science*, **208**, 520 (1980).

Sax, N. I.: "Dangerous Properties of Industrial Materials," 5th edition, Van Nostrand Reinhold, New York, 1979.

Underwood, E. J.: "Trace Elements in Human and Animal Nutrition," 4th edition, Academic, New York, 1977.

CADMIUM COPPER. Copper.

CADMIUM/COPPER SULFIDE SOLAR CELL. Solar Energy.

CADMIUM RED LINE. A line in the spectrum of cadmium at 6438.4696 angstroms which, because it was the most narrow line known to Michelson, was used by him in measuring the standard meter and was formerly accepted as the primary standard of wavelengths.

CADMIUM SULFIDE. **Greenockite.**

CAESALPINA TREE. Of the family *Caesalpiniaceae*, this large, wide-spreading, and showy tree is found in tropical America, Sri Lanka, India, and Malaya. Some trees have been introduced into other tropical climates, but the tree is not considered plentiful. Flowers of all genera have colorful orange and red blossoms with five thin, spreading petals and long stamens. The wood is a rich red and takes a high polish. The wood also can be dyed to yield various shades of crimson and purple and thus, if available, is suited for fine items of furniture and musical instruments. Originally the tree was called the poinciana, after a governor of the French West Indies, M. dePoinci. The name was changed some years ago in honor of the Italian botanist, Andrew Cesalpino.

CAFFEINE. **Alkaloids.**

CAILLETET AND MATHIAS LAW. **Rectilinear Diameter Law.**

CAIMAN. **Crocodiles and Alligators.**

CAIRNGORM STONE. The name given to the smoky brown variety of quartz, particularly when transparent, from Cairngorm, Scotland, a well-known locality. See also **Quartz.**

CAKE MIXES. **Leavening Agents.**

CALAMINE. **Hemimorphite.**

CALANDRIA. A common device used for the heating of vacuum-evaporating apparatus, known as vacuum pans. It comprises a system of vertical steam-jacketed metal tubes open at both ends and joined by heavy metal plates so that a honeycomb structure is formed. Both the tubes and the space beneath the calandria are filled with liquid which is heated by contact with the tubes.

CALAVERITE. A gold telluride, $AuTe_2$, associated with quartz in low-temperature veins. A valuable gold ore from Kalgorrlie, Western Australia and the Cripple Creek region of Colorado. The ore occurs in bladed to lath-like monoclinic crystals with striations parallel to the long axis of the crystals. The ore has a metallic luster of brass-yellow to silver-white color, a hardness of 2.5 to 3, a specific gravity of 9.24 to 9.31, and a yellowish to greenish-gray streak.

CALCAREA. A class of the phylum porifera containing sponges whose spicules are calcareous. They are marine animals exclusively.

The sponges of this class include the simplest of the entire phylum. Some are of the ascon type, with canals passing completely through the body wall, and others are sycon sponges with two sets of canals, the incurrent leading into the body wall from the exterior and the radial leading from the body wall to the interior of the sponge. Common sponges are available to illustrate both forms, *Leucoslenia* representing the former and *Grantia* the latter. These structural differences and the examples mentioned characterize the two orders into which the class is divided:

Order Homocoela. Ascon sponges; *Leucoselenia.*
Order Heterocoela. Sycon sponges; *Grantia.*

CALCIFEROL. **Vitamins.**

CALCIFICATION (Bone). **Bone.**

CALCINATION. The subjection of a substance to a high temperature below its fusion point, often to make the substance friable. Calcination frequently is carried out in long, rotating, cylindrical vessels, known as kilns. Material so treated may (1) lose moisture, e.g., the heating of silicic acid or ferric hydroxide resulting in the formation of silicon oxide or ferric oxide, respectively, (2) lose a volatile constituent, e.g., the heating of limestone (calcium carbonate) resulting in the formation of carbon dioxide gas and calcium oxide residue—destructive distillation of many organic substances is of this type—(3) be oxidized or reduced, e.g., the heating of pyrite (iron disulfide) in air resulting in the formation of sulfur dioxide gas and ferric oxide residue. When the calcination involves oxidation, as in the preceding case, the operation is termed roasting. When heating involves reduction of metals from their ores with separation from the gangue of the liquid metal and slags, the process is termed smelting.

CALCITE. The mineral calcite, carbonate of calcium corresponding to the formula $CaCO_3$, is one of the most widely distributed minerals. Its crystals are hexagonal-rhombohedral although actual calcite rhombohedrons are rare as natural crystals. However, they show a remarkable variety of habit including acute to obtuse rhombohedrons, tabular forms, prisms, or various scalenohedrons. It may be fibrous, granular, lamellar or compact. The cleavage in three directions parallel to rhombohedron is highly perfect; fracture, conchoidal but difficult to obtain; hardness, 3; specific gravity, 2.7; luster, vitreous in crystallized varieties; color, white or colorless through shades of gray, red, yellow, green, blue, violet, brown, or even black when charged with impurities; streak, white; transparent to opaque; it may occasionally show phosphorescence or fluorescence.

Calcite is perhaps best known because of its power to produce strong double refraction of light such that objects viewed through a clear piece of calcite appear doubled in all of their parts. A beautifully transparent variety used for optical purposes comes from Iceland, for that reason is called Iceland spar.

Acute scalenohedral crystals are sometimes referred to as dogtooth spar. Calcite represents the stable form of calcium carbonate; aragonite will go over to calcite at 470°C (878°F.) Calcite is a common constituent of sedimentary rocks, as a vein mineral, and as deposits from hot springs and in caves as stalactites and stalagmites.

Localities which produce fine specimens in the United States include the Tri-State area of Missouri, Oklahoma, and Kansas, as well as Wisconsin, Tennessee, and Michigan with inclusions of native copper; several areas in Mexico, notably Charcas and San Luis Potosi; Iceland; Cumberland and Durham regions in England; and at various regions in S.W. Africa, notably Tsumeb. The exceptionally fine sand-calcite crystals from South Dakota and Fontainebleau in France are well known. See also list of terms under **Mineralogy.**

CALCITONIN. **Parathyroid Glands.**

CALCIUM. Chemical element symbol Ca, at. no. 20, at. wt. 40.08, periodic table group 2a (alkaline earths), mp 837–841°C, bp 1,484°C, density 1.54 g/cm^3 (single crystal). Elemental calcium has a face-centered cubic crystal structure when at room temperature, transforming to a body-centered cubic structure at 448°C.

Calcium is a silver-white metal, somewhat malleable and ductile; stable in dry air, but in moist air or with water reacts to form calcium hydroxide and hydrogen gas; when heated burns in air to form calcium oxide emitting a brilliant light. Discovered by Davy in 1808.

There are six stable isotopes, ^{40}Ca, ^{42}Ca, ^{43}Ca, ^{44}Ca, ^{46}Ca, and ^{48}Ca, with a predomination of ^{40}Ca. In terms of abundance, calcium ranks fifth among the elements occurring in the earth's crust, with an average of 3.64% calcium in igneous rocks. In terms of content in seawater, the element ranks seventh, with an estimated 1,900,000 tons of calcium per cubic mile (400,000 metric tons per cubic kilometer) of seawater. Electronic configuration $1s^22s^22p^63s^23p^64s^2$. Ionic radius Ca^{2+} 1.06 Å. Metallic radius 1.874 Å. First ionization potential 6.11 eV; second, 11.82 eV; third, 50.96 eV. Oxidation potentials $Ca \rightarrow Ca^{2+} + 2e^-$, 2.87 V; $Ca + 2OH^- \rightarrow Ca(OH)_2 + 2e^-$, 3.02 V.

Other important physical properties of calcium are given under **Chemical Elements.**

Calcium occurs generally in rocks, especially limestone (average 42.5% CaO) and igneous rocks; as the important minerals limestone (calcium carbonate, $CaCO_3$), gypsum (calcium sulfate dihydrate, $CaSO_4 \cdot 2H_2O$), phosphorite, phosphate rock (calcium phosphate, $Ca_3(PO_4)_2$), apatite (calcium phosphate-fluoride, $Ca_3(PO_4)_2$ plus CaF_2), fluorite, fluorspar (calcium fluoride, CaF_2); in bones and bone ash as calcium phosphate, and in egg shells and oyster shells as calcium carbonate. See also **Apatite; Calcite; Fluorite;** and **Gypsum.**

In the United States and Canada, calcium metal is produced by the thermal reduction of lime with aluminum. Before World War II, most elemental calcium was made by electrolysis of fused calcium chloride. In the thermal reduction process, lime and aluminum powder are briquetted and charged into high-temperature alloy retorts which are maintained at a vacuum of 100 μm or less. Upon heating the charge to 1,200°C, the reaction takes place slowly, releasing Ca vapor. The latter is removed continuously by condensation, thus permitting the reaction to proceed to completion. High-purity lime is required as a starting ingredient if resulting calcium metal of high purity is desired. Aluminum contamination of the resulting calcium is removed by an additional vacuum-distillation step. Other impurities, as indicated in the accompanying table, also are reduced by this distillation step.

Uses of Elemental Calcium: The very active chemical nature of calcium accounts for its major uses. Calcium is used in tonnage quantities to improve the physical properties of steel and iron. Tonnage quantities are also used in the production of automotive and industrial batteries. Other major uses include refining of lead, aluminum, thorium, uranium, samarium, and other reactive metals.

Calcium treatment of steel results in improved yields, cleanliness, and mechanical properties. Because it is a very strong deoxidizer and sulfide former, calcium will improve the deoxidation and desulfurization of steel. In addition, it alters the morphology and size of inclusions, reduces internal and surface defects, and reduces macrosegregation. Hydrogen induced cracking of line pipe steels by high-sulfur fuels is reduced with calcium treatment. Several grades of calcium treated steel are used in automotive, industrial, and aircraft applications. Oil line pipe, heavy plate, and deep drawing sheet were first treated in Japan. Additional uses have been developed in the United States and Europe.

The high vapor pressure and reactivity of calcium limited its use in steel and iron making prior to the development of injection systems and mold nodularization processes. There are two types of injections systems. One consists of the use of a holding furnace, a sealed vessel, a carrier gas, and a lance through which calcium or calcium compounds are blown into the molten metal. This system is effective for massive desulfurization of large quantities of steel. It is a ladle process. The second type of injection process is wire feeding. A steel-jacketed calcium-core wire is fed through a delivery system which drives the composite wire below the surface of the liquid metal bath. The steel jacket protects the solid metallic calcium from reacting at the surface and allows it to penetrate deep into the bath. Because the reaction occurs below the surface, high and reproducible calcium recoveries are possible. This process is used in both ladle additions and in tundish additions for continuous casting. It provides shape control, deoxidation, final desulfurization and reduction of macrosegregation.

Ladle and mold processes using calcium ferroalloys are important in the production of nodular iron castings. The principal calcium alloy used is magnesium ferrosilicon. Calcium reduces the reactivity of the alloy; with the molten iron it enhances nucleation and improves morphology. The calcium content of the alloy is proportional to the magnesium content, typically in the range of 15–50% of magnesium content. In ladle or sandwich treatment techniques, pieces of the ferroalloy are placed in a pocket cut in the refractory lining of the ladle and the molten iron is then poured into the ladle. The treated, nodularized iron is then cast from the ladle into molds.

In the mold addition process, a granular form of the alloy is placed in a small reaction chamber in the mold. The nodularization treatment occurs in the mold when the iron is cast, rather than in the ladle. The reaction is contained in the mold and high recoveries result. The production of nodular iron castings is over three million tons per year.

A calcium lead alloy is used in maintenance-free automotive and industrial batteries. The use of calcium reduces gassing and improves the life of the battery. From 0.1 to 0.5% calcium is alloyed with the lead prior to the fabrication of the battery plates either by casting or through the production of coiled sheet. With calcium present, these lead–acid batteries can be sealed and do not require the service of conventional batteries. The batteries have a higher energy-to-weight ratio. Of the battery market in the United States, over 50 million batteries per year, 40% are maintenance-free types.

Calcium is used in refining battery grade lead for removing bismuth. Calcium is also used as an electrode material in high-energy thermal batteries.

The production of samarium cobalt magnets requires the use of calcium. The reaction is

$$3SM_2O_3 + 10Co_3O_4 + 49Ca\text{ (vapor)} \xrightarrow[\Delta]{850\text{–}1150°C} 6SmCo_5 + 49CaO$$

$$0.75\text{ weight units of Ca} \longrightarrow 1\text{ weight unit of }SmCo_5$$

Samarium cobalt magnets have three to six times greater magnetic energy than alnico magnets.

Calcium serves as a reductant for such reactive metals as zirconium, thorium, vanadium, and uranium. In zirconium reduction, zirconium fluoride is reacted with calcium metal. The high heat of the reaction melts the zirconium. The zirconium ingot resulting is remelted under vacuum for purification. Thorium and uranium oxides are reduced with an excess of calcium in reactors or trays under an atmosphere of argon. The resulting metals are leached with acetic acid to remove the lime.

Calcium is also used in aluminum alloys and as an addition in a magnesium alloy used for etching. An alloy of 80% Ca–20% Mg is used to deoxidize magnesium castings. The metal also is used in the production of calcium pantothenate, a B-complex vitamin.

Chemistry and Compounds: Calcium exhibits a valence state of +2 and is slightly less active than barium and strontium in the same series. Calcium reacts readily with all halogens, oxygen, sulfur, nitrogen, phosphorus, arsenic, antimony, and hydrogen to form the halides, oxide, sulfide, nitride, phosphide, arsenide, antimonide, and hydride. It reacts vigorously with water to form the hydroxide, displacing hydrogen. Calcium oxide (quicklime) adds water readily and with the evolution of much heat (slaked lime) to form the hydroxide. Calcium hydroxide forms a peroxide on treatment with hydrogen peroxide in the cold. Calcium exhibits little tendency to form complexes, the amines formed with ammonia are unstable, although a solid of composition $Ca(NH_3)_6$ can be isolated from solutions of the metal in liquid ammonia.

Calcium acetate: $Ca(C_2H_3O_2)_2 \cdot H_2O$, white solid, solubility: at 0°C, 27.2 g; at 40°C, 24.9 g, at 80°C, 25.1 g of anhydrous salt per 100 g saturated solution, formed by reaction of calcium carbonate or hydroxide and acetic acid.

Calcium aluminates: Four in number, have been prepared by high temperature methods and identified, $3CaO \cdot Al_2O_3$, at 1,535°C, decomposes with partial fusion; $5CaO \cdot 3Al_2O_3$, mp 1,455°C, $CaO \cdot Al_2O_3$, mp 1,590°C, $3CaO \cdot 5Al_2O_3$, mp 1,720°C.

Calcium aluminosilicates: Two in number, have been prepared by high temperature methods and identified $2CaO \cdot Al_2O_3 \cdot SiO_2$, gehlinite; $CaO \cdot Al_2O_3 \cdot 2SiO_2$, anorthite.

Calcium arsenate: $Ca_3(AsO_4)_2$, white precipitate, formed by reaction of soluble calcium salt solution and sodium arsenate solution. $pK_{sp} = 18.17$.

Calcium arsenite: $Ca_3(AsO_3)_2$, white precipitate, formed by reaction of soluble calcium salt solution and sodium arsenite solution.

Calcium borates: Found in nature as the minerals colemanite, $Ca_2B_6O_{11} \cdot 5H_2O$, borocalcite, $CaB_4O_7 \cdot 4H_2O$, and pandermite $Ca_2B_6O_{11} \cdot 3H_2O$. See also **Colemanite.**

Calcium bromide: $CaBr_2 \cdot 6H_2O$, white solid, solubility 1,360 g/100 ml H_2O at 25°C, formed by reaction of calcium carbonate or hydroxide and hydrobromic acid.

Calcium carbide: CaC_2, grayish-black solid, reacts with water yielding acetylene gas and calcium hydroxide, formed at electric furnace temperature from calcium oxide and carbon.

Calcium carbonate: $CaCO_3$, found in nature as calcite, Iceland spar, marble, limestone, coral, chalk, shells of mollucks, aragonite. $pK_{sp} = 8.32$. It is (1) readily dissolved by acids forming the corresponding calcium salts, (2) converted to calcium oxide upon heating. Aragonite is an unstable form at room temperature, although no change is observable until heated, when at 470°C, it is quickly converted into calcite; calcium hydrogen carbonate, calcium bicarbonate, $Ca(HCO_3)$, known only in solution, formed by reaction of calcium carbonate and carbonic acid. See also **Aragonite; Calcite.**

Calcium chloride: $CaCl_2 \cdot 6H_2O$, white solid, solubility 536 g/100 g

H_2O at 20°C, absorbs water from moist air, formed by reaction (1) of calcium carbonate or hydroxide and HCl, (2) of calcium hydroxide and ammonium chloride.

Calcium chromate: $CaCrO_4$, yellow solid, formed by the reaction of chrome ores and calcium oxide heated to a high temperature in a current of air. $pK_{sp} = 3.15$.

Calcium citrate: $Ca_3(C_6H_5O_7)_2 \cdot 4H_2O$, white solid, solubility: at 18°C 0.085 g/100 g H_2O, formed by reaction of calcium carbonate or hydroxide and citric acid solution.

Calcium cyanamide: $CaCN_2$, white solid, formed (1) by heating cyanamide or urea with calcium oxide, sublimes at 1,050°C, (2) by heating calcium carbide at 1,100–1,200°C in a current of nitrogen. Decomposes in water with evolution of NH_3.

Calcium fluoride: CaF_2, white precipitate, formed by reaction of soluble calcium salt solution and sodium fluoride solution. $pK_{sp} = 10.40$. See also **Fluorite.**

Calcium formate: $Ca(CHO_2)_2$, white solid, solubility at 0°C 13.90 g, at 40°C 14.56 g, at 80°C 15.22 g of anhydrous salt per 100 g saturated solution, formed by reaction of calcium carbonate or hydroxide and formic acid. Calcium formate, when heated with a calcium salt of a carboxylic acid higher in the series, yields an aldehyde.

Calcium furoate: $Ca(C_4H_3O \cdot COO)_2$, formed by reaction of calcium carbonate or hydroxide and furoic acid.

Calcium hydride: CaH_2, white solid, reacts with water yielding hydrogen gas and calcium hydroxide; when electrolyzed in fused potassium lithium chloride, hydrogen is liberated at the anode.

Calcium hypochlorite: $CaOCl_2$ or $Ca(ClO)_2 \cdot 4H_2O$, white solid, contains 60%–65% "available chlorine" and sufficient calcium hydroxide to stabilize, formed by reaction of calcium hydroxide and chlorine. Very soluble in water.

Calcium hypophosphite: $Ca(H_2PO_2)_2$, white solid, solubility 15.4 g/100 g H_2O at 25°C, formed (1) by boiling calcium hydroxide suspension in water and yellow phosphorus, (2) by reaction of calcium carbonate or hydroxide and hypophosphorous acid.

Calcium iodide: CaI_2, yellowish-white solid, solubility 66 g/100 g H_2O at 10°C, formed by reaction of calcium carbonate or hydroxide and hydriodic acid. The hexahydrate, $CaI_2°6H_2O$, is soluble to the extent of 1.680 g/100 g H_2O at 30°C.

Calcium lactate: $Ca(C_3H_5O_3)_2 \cdot 5H_2O$, white solid, solubility; at 0°C 3.1 g, at 30°C 7.9 g of anhydrous salt per 100 g H_2O, formed by reaction of calcium carbonate or hydroxide and lactic acid.

Calcium malate: $CaC_4H_4O_5 \cdot 2H_2O$, white solid, solubility; at 0°C 0.670 g, at 37.5°C 1.011 g of anhydrous salt per 100 g saturated solution. Formed (1) by reaction of calcium carbonate or hydroxide and malic acid, (2) by precipitation of soluble calcium salt solution and sodium malate solution.

Calcium nitrate: $Ca(NO_3)_2 \cdot 4H_2O$, white solid, solubility 660 g/100 g H_2O at 30°C, formed by reaction of calcium carbonate or hydroxide and HNO_3.

Calcium oxalate: CaC_2O_4, white precipitate, insoluble in weak acids, but soluble in strong acids, formed by reaction of soluble, calcium salt solution and ammonium oxalate solution. Solubility at 18°C 0.0056 g anhydrous salt per liter of saturated solution.

Calcium oxide: CaO, (quicklime), white solid, mp 2,570°C, reacts with H_2O to form calcium hydroxide with the evolution of much heat; reacts with H_2O vapor and CO_2 of the atmosphere to form calcium hydroxide and carbonate mixture (slaked lime); formed by heating limestone at high temperature (800°C) and removal of CO_2. This process is conducted industrially in a lime kiln.

Tricalcium phosphate: $Ca_3(PO_4)_3$, white solid, insoluble in water; reactive with silicon oxide and carbon at electric furnace temperature yielding phosphorus vapor; reactive with H_2SO_4 to form, according to the proportions used, phosphoric acid, or dicalcium hydrogen phosphate, $CaHPO_4$, white solid, insoluble; or calcium dihydrogen phosphate, $Ca(H_2PO_4)_2 \cdot H_2O$, white solid, soluble. $pK_{sp} = 28.70$. See also **Apatite.**

Calcium silicates: Four in number, have been prepared by high temperature methods and identified, $3CaO \cdot SiO_2$, prepared by heating the constituents to a temperature below the mp (mp is 1,700°C but substance unstable); $2CaO \cdot SiO_2$, mp 2,080°C, but upon slow cooling changes to forms of different volume; $3CaO \cdot 2SiO_2$, mp 1,475°C; $CaO \cdot SiO_2$, wollastinite, mp approximately 1,400°C. See also **Clinozoisite; Datolite; Diopside; Feldspar; Lawsonite; Tremolite; Wernerite; Wollastoiite.**

Calcium sulfate: gypsum $CaSO_4 \cdot 2H_2O$ plaster of Paris $CaSO_4 \cdot \frac{1}{2}H_2O$, anhydrite $CaSO_4$, white solid, slightly soluble (about 0.2 g per 100 ml of H_2O), formed by reaction of soluble calcium salt solution with a sulfate solution. pK_{sp} of $CaSO_4 = 4.6_{25}$. See also **Anhydrite; Gypsum.**

Calcium sulfide: CaS, grayish-white solid, reactive with H_2O, formed by reaction of calcium sulfate and carbon at high temperatures. Calcium hydrogen sulfide $Ca(HS)_2$, formed in solution by saturating calcium hydroxide suspension with H_2S. pK_{sp} of CaS = 7.24.

Calcium sulfite: $CaSO_3 \cdot 2H_2O$ white precipitate, $pK_{sp} = 7.9$, formed by reaction of soluble calcium salt solution and sodium sulfite solution, or by boiling calcium hydrogen sulfite solution; calcium hydrogen sulfite, $Ca(HSO_3)_2$, formed in solution by saturating calcium hydroxide or carbonate suspension with sulfurous acid.

Calcium tartrate: $CaC_4H_4O_6 \cdot H_2O$, white solid, solubility: at 0°C 0.0875, at 80°C 0.180 g anhydrous salt in 100 ml saturated solution, formed by reaction of calcium carbonate or hydroxide and tartaric acid, or by precipitation of Ca^{2+} with a tartrate solution.

For the role of calcium in biological systems, see **Calcium (In Biological Systems).**

References

Bansal, H. N., and Z. S. Naigamwalla: "Calcium Wire Treatment of Tube Steel at Algoma," McMaster Symposium, Toronto, Canada, Paper 11, 1–17 (May 1979).

Bienvenu, Y., et al.: "Desoxydation et Desulfuration, Calcium et Baryum," *C.I.T.* (French), **6**, 1183 (1978).

Dunks, C. M., Hobman, G., and G. Mannion: *AFS Trans.* **82**, 391 (1974).

Emi, T., et al.: "Mechanism of Sulfides Precipitation during Solidification of Calcium and Rare Earth-treated Ingots" (Japan), *3rd International Iron and Steel Congress*, Chicago, Illinois, April 1978.

Faulring, G. M., Farrell, J. W., and D. C. Hilty: "Steel Flow through Nozzles—Influence of Calcium," *37th Electric Furnace Conference*, Detroit, Michigan, December 1979.

Forster, E., et al.: "Desoxidation und Entschwefelung durch Einblasen von Calciumverbindungen in Stahlschmelzen," *Stahl und Eisen*, **94**, 11, 474–485 (May 1974).

Haida, O.: "Sulphides Shape Control in Continuous Casting by Addition of Ca, R.E. (rare earths), Ca + R.E.," 95th I.S.I.J. Meeting, Paper No. 94, 1978.

Ikeshima, T.: *Transactions of the Iron and Steel Institute of Japan*, **19**, 589–594 (1979).

Kataura, Y., and D. Oeschlagel: "Die Behandlung von Stahlschmelzen mit Calcium," *Stahl und Eisen*, **100**, 1, 20–29 (1980).

Riboud, P. V., et al.: "Steel Desulphurization in the Ladle and Calcium Treatment," McMaster Symposium, Toronto, Canada, pp. 10-11-20-32, May 1979.

Scott, W., and R. A. Swift: "Advantages of Ladle Injection of Calcium and Magnesium Reagents for Steel Desulfurization," I.S.S.-A.I.M.E. Meeting, Paper 36, 128–142, 1978.

Smith, J. F.: "Calcium," in "Metals Handbook," 9th edition, Vol. 2, American Society for Metals, Metals Park, Ohio, 1979.

Staff: *Foundry Management and Technology* (January 1980).

Staff: "Proceedings of Scaninject II," 2nd International Conf. on Injection Metallurgy, Lulea, Sweden, June 12–13, 1980.

Stephen E. Hluchan, Business Manager, Calcium Metal Products, Minerals, Pigments & Metals Division, Pfizer, Inc., Wallingford, Connecticut.

CALCIUM ALGINATE. Gums and Mucilages.

CALCIUM ANTAGONIST DRUGS. Ischemic Heart Disease.

CALCIUM ARSENATE. Boll Weevil.

CALCIUM ATOM (Energy Level). Energy Level.

CALCIUM-BASED GAS TREATMENT. Pollution (Air).

CALCIUM CARBONATE. Antacids.

CALCIUM CYANAMIDE. Cyanamides.

CALCIUM DEPOSITION. **Kidney and Urinary Tract.**

CALCIUM HYPOCHLORITE. **Bleaching Agents.**

CALCIUM (In Biological Systems). The biological role and, consequently, the importance of calcium in foods for humans and feedstuffs for livestock is well established. Although about 99% of the calcium in the bodies of animals is found in bones and teeth, the element is an essential constituent of all living cells.

Various calcium salts and organic compounds fall into this category of dietary supplements and are frequently used in feeds and foods. Some of the more important additives include calcium carbonate, calcium glycerophosphate, calcium phosphate (di- and monobasic), calcium pyrophosphate, calcium sulfate, and calcium pantothenate.

Limestone is frequently used to augment animal feedstuffs. When used, it must be low in fluorine. Calcite limestone is preferred. Calcium is also supplied in the form of crushed oyster shells, marl, gypsum (calcium sulfate), bone meal, and basic slag. In compounding feedstuffs, the specific selection of calcium source is dependent upon the species to be fed. The requirements differ, for example, between cattle, swine, and poultry. The quantity required also varies with the life stage of the animal. For example, laying hens require a much higher percentage of calcium in their diet than starting poultry.

In the mammalian body, calcium is required to insure the integrity and permeability of cell membranes, to regulate nerve and muscle excitability, to help maintain normal muscular contraction, and to assure cardiac rhythmicity. Calcium plays an essential role in several of the enzymatic steps involved in blood coagulation and also activates certain other enzyme-catalyzed reactions not involved in any of the foregoing processes. Calcium is the most important element of bone salt. Together with phosphate and carbonate, calcium confers on bone most of its mechanical and structural properties.

Calcium Metabolism

The aggregate of the various processes by which calcium enters and leaves the body and its various subsystems can be summarized by the term *calcium metabolism.* The principal pathways of calcium metabolism are intake, digestion and absorption, transport within the body to various sites, deposition in and removal from bone, teeth, and other calcified structures, and excretion in urine and stool.

Pathways. The principal pathways involve three subsystems of the body: (1) the oral cavity where ingestion occurs and the gastrointestinal tract where digestion and absorption take place and from which the feces is excreted; (2) the body fluids, including blood, which transport calcium; the soft tissues and body organs to which calcium is transported and where many of its physiological functions are carried out. Some of the organs, like the kidney, the liver, and sweat glands, are also responsible for calcium excretion; (3) the skeleton, including the teeth, where calcium is deposited in the form of bone salt and from where it is removed (resorbed) after destruction of the bone salt.

Calcium Intake. This varies in different populations and is related to the food supply and to the cultural and dietary patterns of a given population. The intake of a substantial fraction of the world population falls between 400 and 1,100 mg/day, but a range encompassing 95% of all people would undoubtedly be even wider. Most populations derive half or more of their calcium intake from milk and dairy products. Calcium intakes of domestic and laboratory animals are higher than those of humans. For example, rats typically ingest 250 mg Ca/kg body weight, and cattle 100 mg/kg, whereas humans ingest only 10 mg/kg. Ingestion falls with age in all species. The average percentage concentration of minerals in the lean body mass of vertebrates ranges from 1.1 to 2.2%.

Calcium Absorption. In most animals, including the human body, this occurs mainly in the upper portion of the small intestine. The amount and, therefore, the fraction of calcium absorbed from the gut are a function of intake, age, nutritional status, and health. Generally, the fraction absorbed decreases with age and intake and as the nutritional status improves. The absolute amount absorbed increases with intake and may or may not decrease with age. The mechanisms by which calcium is absorbed are not well understood. Active transport of the ion against an electrochemical gradient seems to be involved, but not all of the calcium appears to be absorbed by ways of this process, because calcium absorption continues under conditions when active transport is severely depressed, as in vitamin D deficiency. Calcium absorption can be enhanced by the administration of large doses of vitamin D and is depressed in vitamin D deficiency. There is uncertainty regarding the effect on calcium absorption of the parathyroid hormone, the major endocrine control of the blood calcium level. Patients with hyperparathyroidism have been shown to have higher than normal absorption and patients with hypoparathyroidism to have lower than normal absorption. Similar effects have been observed in acute animal experiments, but in most of these instances a possible indirect effect has not been excluded.

Interrelationship with Phosphorus and Vitamin D. The interdependence of calcium, phosphorus, and vitamin D is exemplary of how synergistic effects can occur from combinations of feed and food components, either with a positive or negative result in the animal body. The relative concentrations (proportions) of each component in such a combination can be quite critical. Much research has gone into these particular interrelationships; much further research is required. The relationship between phosphorus and calcium nutrition has been known since the early 1840s, when Chossat in France first discovered that pigeons develop a poor bone structure when fed diets low in calcium. A few years later, the fundamental relationship of calcium and phosphorus in animal diets was developed by French and German researchers. It was not until 1922, however, with the discovery of vitamin D, that a triangular relationship was observed. See also **Bone; Phosphorus;** and **Vitamin D.**

Calcium in Blood Plasma. The concentration of calcium in the blood plasma of most mammals and many vertebrates is quite constant at about 2.5 mM (10 milligrams per 100 milliliters plasma). In the plasma, calcium exists in three forms: (1) as the free ion, (2) bound to proteins, and (3) complexed with organic (e.g., citrate) or inorganic (e.g., phosphate) acids. The free ion accounts for about 47.5% of the plasma calcium; 46% is bound to proteins; and 6.5% is in complexed form. Of the latter, phosphate and citrate account for half.

The mechanism involved in the regulation of the plasma calcium level is not fully understood. The parathyroid glands regulate both level and constancy; when these glands are removed, the plasma level drops and tends to stabilize at about 1.5 mM, but variations in calcium intake may induce fairly wide fluctuations in the plasma level. In the intact organism, wide variations in intake produce essentially no variations in the plasma calcium value which is stabilized at about 2.5 mM. The equilibrium between bone and plasma is believed to determine the level of the plasma calcium in parathyroidectomized animals, but this reasonable hypothesis requires further experimental support. See also **Blood; Endocrine System; Parathyroid Gland.**

The problem of whether parathyroid regulation is due to a single hormone with hypercalcemic properties or to two hormones, one hypocalcemic, termed calcitonin, the other hypercalcemic, termed parathyroid hormone, continues under investigation.

When the calcium ion concentration is lowered in the fluids bathing nerve axons—fluids which are in very rapid equilibrium with the blood plasma—the electrical resistance of the axon membrane is lowered, there is increased movement of sodium ions to the inside, and the ability of the nerve to return to its normal state following a discharge is slowed. Thus, on the one hand, there is hyperexcitability. But, the ability for synaptic transmission is inhibited because the rate of acetylcholine liberation is a function of the calcium ion concentration. The neuromuscular junction is affected in a similar fashion; hence, the end plate potential is lowered before the muscle membrane potential and the muscle membrane is in a hyperexcitable state. These events are reversed when the calcium ion concentration is raised above the normal in the blood plasma and in the fluids bathing muscle and nerve. It is for these reasons that hypocalcemia is associated with hyperexcitability and ultimately tetany and hypercalcemia with sluggishness and bradycardia. See also **Brain and Nervous System.**

Muscular Contraction and Relaxation. The role of calcium in this function is not fully understood. Some researchers have proposed that calcium is the link between the electrical and mechanical events in contraction. It has been shown *in vitro* that when calcium ions are applied locally, muscle fibers can be triggered to contract. It has further been postulated that relaxation of muscle fibers is brought about by an intracellular mechanism for reducing the concentration of calcium ions available to the muscle filaments. Others postulate that contraction occurs because calcium inactivates a relaxing substance which is released from the sarcoplasmic reticulum in the presence of ATP (adenosine triphosphate).

Bone. This is the most important reservoir of calcium in the animal body. Accounting for the largest portion of the body's calcium, bone calcium also constitutes about 25% (weight) of fat-free, dried bones. Calcium occurs in bone mostly in the form of a complex, apatitic salt, so named for its structural resemblance to a family of calcium phosphates of which hydroxyapatite [$Ca_{10}(PO_4)_6(OH)_2$] is the best-known mineralogical example. Since calcium occurs also as the carbonate, there is discussion as to whether bone salt contains the carbonate as a separate phase, whether some of the surface phosphate in apatite has been substituted for by carbonate, or whether bone mineral is a carbonato-apatite, such as dahlite. It is important to recognize that the crystal lattice of the bone mineral, when first laid down, does not and probably cannot have all possible calcium positions occupied. Whether stability is derived from hydrogen and/or organic bonds to which the mineral may be attached is not fully determined. It has been proposed that bone salt is a lamellar mixture of octocalcium phosphate and hydroxyapatite. This hypothesis has to account for the amount of pyrophosphate formed when bone salt is heated and also for its evolution with age, i.e., the increase with age in the calcification of bone and the corresponding drop in its induced pyrophosphate content, observations for which the apatitic structure can account. The proponents of the octocalcium phosphate hypothesis explain this by showing that octocalcium phosphate breaks down to apatite and anhydrous dicalcium phosphate which upon further heating give rise to pyrophosphate. Finally, it is postulated that octocalcium phosphate may be present in young and presumably newly formed bone, whereas in older bone an apatitic phosphate admittedly dominates the equilibrium.

Calcium enters and remains in bone as a result of calcification processes which involve two steps: (1) deposition of bone salt of a minimum calcium content and specific gravity. Deposition occurs by way of nucleation, probably an epitactic process on the collagen fibers, with the ground substance (mostly mucopolysaccharides) between the fibers exerting either a positive or an inhibitory effect on the nucleation process; and (2) subsequent further mineralization of the bone mineral, leading to an increase in its calcium content and its specific gravity.

Calcium removal, in contrast, involves destruction of the calcified structure *in toto.* There is no evidence that only particular structures are resorbed, e.g., those with a given degree of mineralization.

The amount of calcium deposited in bone at any moment may be determined from experiments with radioactive calcium. In growing individuals, it exceeds the amount removed by bone destruction. In adults, it is about the same as the amount removed. Such individuals are considered to be in "zero" calcium balance. In older persons, the amount deposited is less than the amount removed.

Because of the high incidence of osteoporosis in elderly people, a number of investigators have been looking into the effects of phosphorus consumption on calcium metabolism (Lutwak, 1969, 1974; Albanese et al., 1975; Jowsey, 1977). This particular concern comes from the fact that the American diet, for example, may have a calcium : phosphorus ratio of about 1 : 4. This ratio is much less in certain regions. The calcium : phosphorus ratio is particularly important when one of the elements is low in the diet (Wasserman, 1960). See also **Phosphorus.**

Excretion of Calcium. The principal routes of excretion are stool and urine. Calcium in the stool may be considered as made up of unabsorbed food calcium and nonreabsorbed digestive juice calcium. The latter is termed the fecal endogenous calcium. The proportion of fecal endogenous calcium to urinary calcium varies in different species. It is approximately 1 : 1 in humans and 10 : 1 in the rat and in cattle. The calcium in the urine may have a dual origin—calcium that was filtered at the glomerulus and failed to get reabsorbed along the length of the nephron, and calcium that may have originated from transtubular movement in certain regions of the nephron. The amount of calcium that may be lost in sweat can be large, but there is no convincing evidence that sweat is a habitual route of significant loss. See also **Kidney and Urinary Tract.**

Calcium Occurrence. The soils of humid regions are commonly low in calcium and ground limestone is usually applied to add the element, reduce the toxicity of aluminum and manganese, and to correct soil acidity. The soils of dry areas are frequently rich in calcium. There is little evidence to indicate a strong relationship between human nutrition and calcium excesses or deficiencies in the soil. Even with farm livestock, most calcium deficiencies are not related to levels of available calcium in the soil. The reason for this anomaly is evident when one examines some of the controls over the movement of calcium in the food chain.

At the step in the food chain when calcium moves from the soil to the plant, controls based upon the genetic nature of the plant are very important. Because of these controls, certain plant species always accumulate fairly high concentrations of calcium; while other plants accumulate rather low concentrations. Among the forage crops, red clover grown, for example, on the low-calcium soils of the northeastern United States, contains more calcium than grasses grown on the high-calcium soils of the western United States. Among the food crops, snap beans and peas normally contain about three to five times as much calcium as corn (maize) and tomatoes. Thus, the level of calcium in the diets of people or of animals depends more on what kinds of plants are included in the diet than it does on the supply of available calcium in the soil where these plants are grown.

Adding limestone to soils to correct soil acidity and to supplement available calcium will, of course, indirectly affect human and calcium nutrition, but this is a difficult quantity to measure.

References

Albanese, A. A., et al.: "Problems of Bone Health in the Elderly," *New York State J. Medicine,* **75**, 326 (1975).

Allaway, W. H.: "The Effect of Soils and Fertilizers on Human and Animal Nutrition," Agriculture Information Bulletin 378, Cornell University Agricultural Experiment Station and U.S. Department of Agriculture, Washington, D.C., 1975.

Jowsey, J.: "Osteoporosis: Dealing with a Crippling Bone Disease of the Elderly," *Geriatrics,* **32**, 41 (1977).

Kirkgessner, M. (editor): "Trace Element Metabolism in Man and Animals," Institut für Ernahrungsphysiologie, Technische Universität München, Freising-Weihenstephan, Germany, 1978.

Lutwak, L.: "Current Concepts of Bone Metabolism," *Ann. Internal Medicine,* **80**, 630 (1974).

Mahoney, A. W., and D. G. Hendricks: "Some Effects of Different Phosphate Compounds on Iron and Calcium Absorption," *J. Food Sci.,* **43**, 5, 1473–1475 (1978).

Staff: Dietary Levels of Households in the United States," U.S. Department of Agriculture, Washington, D.C. (1968).

Stewart, A. K., and A. C. Magee: "Effect of Zinc Toxicity on Calcium, Phosphorus, and Magnesium Metabolism of Young Rats," *J. Nutrition,* **82**, 287 (1964).

Underwood, E. J.: "Trace Elements in Human and Animal Nutrition," 4th edition, Academic, New York, 1977.

Wallace, G. W., et al.: "Calcium Binding and Its Effects on Properties of Food Protein Sources," *J. Food Sci.,* **42**, 2, 473–474 (1977).

Wasserman, R. H.: "Calcium and Phosphorus Interactions in Nutrition and Physiology," *Fed. Proc.,* **19**, 636 (1960).

CALCIUM METABOLISM. Parathyroid Glands.

CALCIUM OXALATE STONES. Kidney and Urinary Tract.

CALCIUM PHOSPHATE. Apatite.

CALCIUM SULFATE. Adsorption Operations.

CALCIUM TUNGSTATE. Scheelite.

CALCO-URANITE. Autunite.

CALCULATION (Approximate). Approximate Calculation.

CALCULATOR (Abacus). This scheme represents one of the first formalized approaches to counting and calculating beyond the use of fingers and toes. Essentially, the abacus is a manually manipulated digital device. Records indicate that some form of the abacus was used as early as 3,000 B.C. by the Babylonians. Formats have ranged from ruled tables to moving coins around on checkered tablecloths (from which the term British Exchequer was derived) to the currently more familiar frame-and-bead construction. Experienced operators of commercial versions of the abacus, particularly in the Orient, can add, subtract, multiply, and divide with speeds comparable to those obtainable with modern, nonelectronic adding machines. Special versions of the abacus are used in some elementary schools for teaching the fundamentals of counting and arithmetic.

The principle of the abacus is shown in the accompanying figure. Visualize a box or frame containing movable squares. In (a) the squares, all indicated by a gray tone, are in their "rest" or "zero" position. The squares along the top may be moved down into the "reckoning space" A, whereas the squares in the bottom portion of the box may be moved upward into "reckoning space" B. There is a "datum" line or bar that separates spaces A and B. The abacus is read by noting the number of squares that have been moved into the reckoning space, i.e., that make contact with the datum line. The squares in the upper portion, from right to left, represent, 5, 50, 500, 5,000 . . . etc. Note that there is only one square in each column. The squares in the lower portion, from right to left, represent 1's, 10's, 100's, 1,000's . . . etc. The extreme right-hand column permits counting from 1 to 4, depending upon how many of the squares the operator moves upward to contact the datum line.

In figures (b) through (d), the squares that have been moved into contact with the datum line, i.e., the squares to be read, are shown in black. The indication of "1" is shown in (b); of "423" in (c). In (d), the squares in the upper portion of the box are brought into play. As indicated by (e), there is no limit to the number of columns that may be used in a frame, thus permitting calculations into 8 or 10 figures, or more. Because of the limitations of squares in the columns, however, the abacus operator frequently is called upon to make minor mental calculations, i.e., to introduce a subroutine. For example, in (c), the addition of "525" to the "423" indicated is quite simple, requiring no interim calculation. There is a "5" available to be moved down; there are two remaining "20's" which can be moved up; and there is a "500" available to be moved down. Thus, the abacus will read the correct sum, i.e., "948." However, in the case of adding "107" to the "423," the operator cannot handle the "7" because only one "5" and only one "1" is available, accounting for "6" whereas "7" is required. In this case, the operator will add "10" and take away "3." There is a further problem in adding the "100" because all four of the available "100's" are in use. This can be handled by adding "500" and taking away "400." With these manipulations completed, the abacus reads the proper sum, i.e., "530."

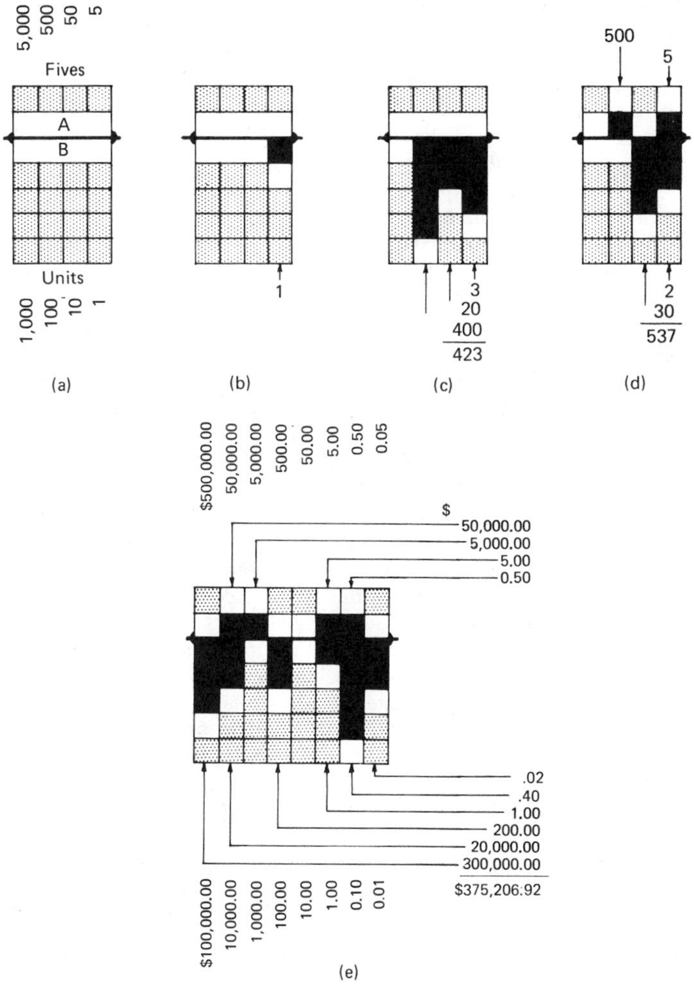

Fundamentals of the abacus.

CALCULI. A deposit from the precipitation of mineral salts in various parts of the body. Mineral salts in urine, for example, may precipitate and form calculi, commonly called stones. Calculi (singular: calculus) may be found in any part of the urinary tract—from the tubules to the orifice of the urethra. Abnormal concretions of bone or teeth are also sometimes called stones or calculi. Causes of calculus formation include decrease in intake of water over a long period, alterations of the pH of body fluids, and excessive ingestion of certain minerals.

CALCULUS. The word comes from the Latin, *calculus,* a stone or pebble used in reckoning. Sir Isaac Newton (1642–1727), the English scientist and mathematician, and Gottfried Wilhelm Leibniz (1646–1716, also called von Leibniz or spelled Leibnitz), the German philosopher and mathematician, are considered to be the founders of calculus. The unqualified word is usually taken to mean differential and integral calculus. It deals with the rate of change of a function and with the inverse process. For some of the methods and applications of calculus, consult the following topics: **Area; Curvature; Curve; Derivative (Mathematics); Differential (Mathematics); Differentiation (Mathematics); Indeterminate Form; Integral; Integration; Leibniz Rule; Length of a Curve; Limit; Mean Value Theorems; Multiple Integral; Series; Singular Point of a Function; Slope; Surface; Tangent (Geometry);** and **Volume (Geometry).**

There are several other kinds of calculus. Sometimes called the twin sister of differential calculus is the caclulus of finite differences (see **Difference**). Its principles were understood by both Newton and Leibniz. The former wrote about it in 1711 and the first book on this calculus was by Brook Taylor in 1715. It is concerned with interpolation, numerical differentiation and integration, summation of series, the solution of difference equations, and linear equations with an infinite number of unknowns.

The calculus of variations is a study of maximum and minimum properties of definite integrals. The first work on this subject was also done by Newton and, at about the same time, by the Bernoulli brothers. The founders of it as a branch of mathematics are Lagrange (1736–1813) and Euler (1707–1783). A simple case in the calculus of variations is

$$I=\int_a^b f(x, y, y')\,dx$$

where $y(x)$ is to be determined so that the integral is either a maximum or a minimum. In either case, y is said to be an extremal and the integral has a stationary value. Thus it is of a more general character than the maximum or minimum problems of differential calculus, for they require the location of a point with specified properties while in the calculus of variations, a curve or surface is sought. The subject has applications in economics, business, and other practical affairs for there one usually wishes to proceed in such a way as to secure maximum profit, minimum cost and effort, etc. See also **Abel Equation; Brachistochrome;** and **Euler Equation.**

The calculus of residues is founded on the Cauchy integral and theorem. It is applied to the evaluation of integrals in the complex variable. Suppose $f(z)$ is analytic within a region C, except for a finite number of poles, then the value of the contour integral is given by

$$\int_C f(z)\,dz = 2\pi i \sum$$

where Σ is the sum of the residues of the functions at the poles inside C. Typical integrals which may be so evaluated under certain restrictions include

$$\int_0^{2\pi} f(\cos\theta, \sin\theta)\, d\theta; \quad \int_{-\infty}^{\infty} f(x)\, dx; \quad \int_0^{\infty} x^n f(x)\, dx$$

See also list of related entries under **Mathematics.**

CALDERA. Derived from a Spanish word meaning caldron, the term caldera has been given to great crater-like depressions which are either the result of subsidence of lava within the body of a volcano or of an explosive eruption of terrific violence. Examples of these craters of explosion or subsidence are Crater Lake, Oregon, Mt. Tamboro in Indonesia, and the original *La Caldera* in the Canary Islands. Crater Lake, which occupies the caldera, is 2000 feet deep and about 25 sq. mi. in area, surrounded by cliffs whose maximum height is 2000 feet above the lake.

CALENDAR. The problem of timekeeping has always been a vexing one. There are three "natural" units, the solar day, the lunar month, and the tropical year. The normal or true solar day had to be abandoned with the improvement of mechanical timekeeping devices, and the mean solar day has been adopted as the standard short unit for keeping records. The task of the calendar builder is to combine this unit with the two longer units and, since the three are mutually incommensurable, a rigorous solution of the problem is impossible, and compromises must be made.

The fact that the economic world is largely dependent upon agriculture introduces one important restriction on the freedom of the calendar builder. The seasons should remain at approximately the same place in the completed calendar from year to year. The date upon which the sun apparently passes through the vernal equinox is of fundamental importance to the agriculturalist and, for many centuries, was considered as the time of starting a new year. One of the earliest calendars on record started the year on this date and then proceeded through ten lunar months. This calendar covered only 295.3 mean solar days, whereas the period from one passage of the sun through the vernal equinox to the next is 365.2422 days. The period between the end of one year to the beginning of the next was determined by the priesthood and by politicians, and there was conflict and confusion.

The first step toward the modern calendar was taken by Julius Caesar, with the advice of the astronomer Sosigenes. The so-called Julian calendar discards the lunar month and adopts 365.25 days as the length of the year. This year is divided into twelve periods (months) of 30 or 31 days. The normal year was 365 days in length but, to make up the extra $\frac{1}{4}$ day, an extra day was intercalated (i.e., put into the normal calendar) every four years.

Running parallel with the Julian calendar, we find the far more ancient calendar of the Jewish and Mohammedan peoples, which holds rigorously to the lunar month. Division of the number of days in the tropical year by the days in the lunar month will indicate that there are 12.36 lunar months in a tropical year. To retain the synchronism between the calendar and the seasons, this calendar is variable in the number of months it contains, and the process of intercalating months becomes very complicated. However, the Eastern calendar exerts a powerful effect upon the calendar of the Western world, because of the fact that the date of Easter is fixed by a date on the Eastern calendar.

In A.D. 325, the Christian Church took its first step in calendar building and at the Council of Nice made two decrees: a decree that the sun should pass through the vernal equinox on the 21st of March on the Julian calendar, and a second decree relative to the date for the celebration of Easter. The latter of the two decrees was within the province of the Church and can be followed; the former, however, applies to factors beyond the control of people.

It should be noted that the length of the tropical year is 0.0078 day less than the 365.25 days of the Julian calendar. This means that, after the lapse of 1000 years, the sun will pass through the vernal equinox 7.8 days earlier than the 21st of March, assuming that it was at the vernal equinox on this date in the first place. By 1582, the date of the vernal equinox was the 11th of March instead of the 21st, and Pope Gregory decided to return the sun to its proper date and to modify the calendar in such a way that the error would not reappear. The Gregorian calendar is identical with the Julian except in the fact that only such century years are leap years as are divisible by 400. This is equivalent to dropping 3 days every 400 years, leaving an average length for the year of 365.2425 days, which differs from the tropical year by only 0.0003 day. This calendar was immediately adopted by all Catholic countries, but the Greek Church and most Protestant countries refused to recognize it. The confusion following this change persisted well down into the present century (Rumania used the Julian calendar until 1919), and is still felt by historians in reading records of the early years of this country when both calendars were in use.

Within the past several decades, a movement has been underway to modify the calendar in the attempt to have dates and days of the week agree in successive years. Any such scheme involves the necessity of introducing one day each year without date or day of the week, and two such days on leap years, if the year and the seasons are to retain the present synchronism. This intercalation of a day will break the 6-day sequence between Sabbaths, an idea that is abhorrent to many religious sects. The scheme that has the most general support is one in which the year is divided into four equal quarters of 3 months each. In each quarter, the first month has 31 days and the second and third, 30 each. This gives exactly 13 weeks in each quarter, and 52 weeks in each year. The days are to be intercalated without date or day of the week between December 30 and January 1 each year and between June 30 and July 1 every leap year (e.g., the normal calendar would read Saturday, Dec. 30; New Year's Day; Sunday, Jan. 1). See also **Time.**

CALEOMETER. An electrical instrument used to measure the heat loss from a calibrated wire and useful in making a number of determinations, such as that of the variation of the concentration of one of the components of the gas surrounding the wire.

See also **Gas Analyzers (Combustion-Type);** and **Gas Analyzers (Thermal-Conductivity Type).**

CALIBRATION. With reference to industrial and scientific instruments, the Scientific Apparatus Makers Association defines *calibrate* as follows:

1. To ascertain by the use of a standard, the locations at which scale or chart graduations of a device should be placed to correspond to a series of values of the quantity which the device is to measure, receive, or transmit.
2. To adjust the output of a device, to bring it to a desired value, within a specified tolerance, for a particular value of the input.
3. To ascertain the error in the output of a device by measuring or comparing against a standard.

CALICHE (Nitrate). The gravel, rock, soil, or alluvium cemented with soluble salts of sodium in the nitrate deposits of the Atacama Desert of northern Chile and Peru. The material contains from 14 to 25% sodium nitrate, 2 to 3% potassium nitrate, and up to 1% sodium iodate, plus some sodium chloride, sulfate, and borate. At one time, this was an important natural fertilizer.

CALICHE (Soil). A commonly used term in the southwestern United States, particularly Arizona, to describe an opaque, reddish-brown to buff or white calcareous material of secondary accumulation, usually found in layers on, near, or within the surface of stony soils of arid and semiarid regions. The material also occurs as a subsoil deposit in subhumid climates. Caliche soil is composed mainly of crusts or succession of crusts of soluble calcium salts, plus gravel, sand, silt, and clay. The cementing material is essentially calcium carbonate, but magnesium carbonate, silica, or gypsum also may be present. Caliche also has been used as a term to describe the calcium carbonate cement per se. In some localities, the material is called *hardpan*, calcareous *duricrust*, *calcrete*, and *kankar* (in India).

CALIFORNIA BUCKEYE TREE. Horse Chestnut and Buckeye Trees.

CALIFORNIA NUTMEG TREE. Yew Trees.

CALIFORNIA POMPANO. Butterfishes.

CALIFORNIA VULTURE. **Condor; Vulture.**

CALIFORNIA WAX MYRTLE. **Bayberry Shrubs and Trees.**

CALIFORNIUM. Chemical element symbol Cf, at. no. 98. at. wt. 251 (mass number of the most stable isotope), radioactive metal of the *Actinide* series, also one of the *Transuranium* elements. All isotopes of californium are radioactive; all must be produced synthetically. See also **Radioactivity.** The isotope ^{245}Cf was first produced by S. G. Thompson, K. Street, Jr., A. Ghiorso, and G. T. Seaborg at the University of California at Berkeley in 1950 by bombarding microgram quantities of ^{242}Cm with helium ions. The reaction: $^{242}Cm\ (\alpha, n) \rightarrow {}^{245}Cf$. The isotope has a half-life of 44 min. A number of other isotopes of Cf have been made, one of which, ^{254}Cf, half-life 55 days, is of interest because it decays predominantly by spontaneous fission. The longest-lived isotope is $^{251}Cf(t_{1/2} =$ about 700 yrs), the next is $^{249}Cf(t_{1/2} = 470$ yrs). Except for $^{250}Cf(t_{1/2} = 10$ yrs), and $^{252}Cf(t_{1/2} = 2.2$ yrs), all other isotopes have half-lives less than one year. Several other isotopes (246, 248, 249, 250, 252) also decay by spontaneous fission, but with fission half-lives much longer than the half-lives for alpha-decay. Californium is considered to occur in its compounds only in the tripositive state.

Studied through the use of tracer quantities, the chemical properties of californium indicate that its chemical properties are analogous to the tripositive actinides and lanthanides, showing the fluoride and the oxalate to be insoluble in acid solution, and the halides, perchlorate, nitrate, sulfate and sulfide to be soluble.

Probable electronic configuration:

$$1s^2 2s^2 2p^6 3s^2 3p^6 3d^{10} 4s^2 4p^6 4d^{10} 4f^{14} 5s^2 5p^6 5d^{10} 5f^{10} 6s^2 6p^6 7s^2.$$

Ionic radius: Cf^{3+} 0.98 Å.

In 1960, Cunningham and Wallmann isolated 0.3 microgram of californium (as californium-249) oxychloride. The best isotope for the study of californium is ^{249}Cf, which can be isolated in pure form through its beta particle-emitting parent, ^{249}Bk.

Californium-252 is an intense neutron source. One gram emits 2.4 $\times\ 10^{12}$ neutrons per second. This isotope shows promise for applications in neutron activation analysis, neutron radiography, and as a portable source for field use in mineral prospecting and oil well logging. The isotope also is being investigated for medical research applications. It may find use as a neutron source for irradiation of certain tumors for which gamma-ray treatment is inadequate.

References

Choppin, G. R., G. S. Thompson, A. Ghiorso, and B. G. Harvey: "Nuclear Properties of Some Isotopes of Californium, Elements 99 and 100," *Phys. Rev.*, **94**, 4, 1080–1081 (1954).

Conway, J. G., et al.: "The Solution Absorption Spectrum of Cf^{3+}," *J. Inorg. Nucl. Chem.*, **28**, 3064–3066 (1966).

Cunningham, B. B., and T. C. Parsons: "Preparation and Determination of the Crystal Structure of Californium and Einsteinium Metals," *Lawrence Berkeley Laboratory Nuclear Chemistry Annual Report*, UCRL-20426, University of California, Berkeley, California, 1970.

Fields, P. R., et al.: "Transplutonium Elements in Thermonuclear Test Debris," *Phys. Rev.*, **102**, 1, 180–182 (1956).

Ghiorso, A., Thompson, S. G., Choppin, G. R., and B. G. Harvey: "New Isotopes of Americium, Berkelium and Californium," *Phys. Rev.*, **94**, 4, 1081 (1954).

Ghiorso, A., et al.: "New Elements Einsteinium and Fermium, Atomic Numbers 99 and 100," *Phys. Rev.*, **99**, 3, 1048–1049 (1955).

Green, J. L., and B. B. Cunningham: "Crystallography of the Compounds of Californium: I. Crystal Structure and Lattice Parameters of Californium Sesquioxide and Californium Trichloride," *Inorg. Nucl. Chem. Lett.*, **3**, 9, 343–349 (1967).

Hulet, E. K., Thompson, S. G., Ghiorso, A., and K. Street, Jr.: "New Isotopes of Berkelium and Californium," *Phys. Rev.*, **84**, 2, 366–367 (1951).

Peterson, J. R., and R. D. Baybarz: "The Stabilization of Divalent Californium in the Solid State: Californium Dibromide," *Inorg. Nucl. Chem. Lett.*, **8**, 4, 423–431 (1972).

Samhoun, K., and F. David: "Radiopolarography of Am, Cm, Bk, Cf, Es, and Fm," *Proc. 4th Int. Transplutonium Element Symp.*, Baden Baden, W. Germany, 1975.

Seaborg, G. T. (editor): "Transuranium Elements," Dowden, Hutchinson & Ross, Stroudsburg, Pennsylvania, 1978.

Thompson, S. G., Street, K., Jr., Ghiorso, A., and G. T. Seaborg: "Element 98," *Phys. Rev.*, **78**, 3, 298–299 (1950).

Thompson, S. G., and M. L. Muga: "Methods of Production and Research on Transcurium Elements," *Proc. Second United Nations Conf. Peaceful Uses of Atomic Energy*, Geneva, Switzerland, 1958.

CALIPER CHECKING. **Thickness Measurement and Gaging Systems.**

CALLA LILY. **Aroids.**

CALLA PALUSTRIS. **Aroids.**

CALLISTO. **Jupiter.**

CALLUS. In humans, an area of thickened skin, or new growth of bony tissue at the site of a fracture which has been reunited.

In plants, it is a protective tissue which occurs in many plants after injury. When the root or stem of a woody plant is wounded, exposing the tissues within, the cambium cells around the wound begin to divide rapidly, forming a protective mass of soft parenchymatous tissue. These living cells are called callus, or wound tissue, and in time will entirely close the wound if the latter is not too extensive. After the tissue is formed, cell differentiation goes on and a new phellogen layer may be formed, as well as the other tissues composing the cortex of the stem. The cambium becomes once more a continuous layer. When wounds are made in pruning, that is, when a branch is cut off, callus tissues gradually form a ring which spreads over and finally completely closes the wound.

CALM. **Winds and Air Movement.**

CALORESCENCE. A term designating the production of visible light by means of energy derived from invisible radiation of frequencies below the visible range. Tyndall found it possible to raise a piece of blackened platinum foil to a red heat by focusing upon it infrared radiation from an arc or from the sun, the visible wavelengths having been filtered out. It is to be noted that the transformation is indirect, the light being produced by heat and not by any direct stepping up of the infrared frequency. A somewhat analogous phenomenon is the production of visible sparks or the glowing of a fine platinum wire in a resonant circuit energized by long-wave Hertzian radiation.

CALORIC INTAKE. **Dietary Requirements and Trends.**

CALORIE. **Heat; Units and Standards.**

CALORIFIC VALUE. **Coal; Natural Gas; Substitute Natural Gas (SNG).**

CALORIMETRY. The study of heat as contrasted with temperature. The oxygen bomb calorimeter, which is used to determine the heat of combustion of fuels, is only one of many types of calorimeters. Steam calorimeters, for example, are used to measure heat capacities, heat of reaction, or energy changes in biological processes. Instruments for differential thermal analysis are sometimes referred to as differential scanning calorimeters. The bomb calorimeter is a batch-type instrument which requires a discrete sample and, therefore, is used only for solid and liquid materials. Gaseous fuels (nondiscrete) are analyzed in flow-type calorimeters.

One of the most important characteristics of any combustible fuel is the quantity of energy or heat that it releases as it is burned. This value is referred to as either the *heat of combustion*, or the *calorific value* of the fuel and is usually expressed in *British thermal units* (*Btu*) per pound or ton, or in *calories per gram*. The heat of combustion of solid and liquid fuels is routinely determined in order to establish the price of the fuel, as well as to serve as a basis for calculating the overall efficiency of a power generating facility or engine.

To determine the heat of combustion of a fuel, a representative sample is burned in a high-pressure oxygen atmosphere within a metal bomb or pressure vessel. The energy released by this combustion is adsorbed within the calorimeter and measured in terms of temperature change within the calorimeter. The heat of combustion of the sample is obtained by multiplying the temperature rise of the calorimeter by a previously determined energy equivalent or heat capacity for the instrument. Corrections are applied to adjust these values for any heat transfer occurring in the calorimeter as well as for any side reactions which are unique to the bomb combustion process.

The reliability of results obtained with bomb calorimetry depends upon a truly representative sample as well as a reliable calorimeter and proper operating techniques. A typical load of coal will include large lumps, fine powders, and particles varying in size between the two extremes. During loading and transit, the fines and smaller particles will work their way to the bottom of the shipment. A sample taken from the bottom would not be representative of the entire shipment inasmuch as it would be rich in fines and deficient in the larger particles. A sample from the top of the shipment obviously would be biased in favor of the larger particles. Similar problems can occur with liquid fuels.

Any oxygen bomb calorimeter consists of four essential parts: (1) A bomb or vessel in which the combustible charge is burned; (2) a bucket or container which holds the bomb as well as a measured quantity of water to absorb the heat released from the bomb and a stirring device to assure thermal equilibrium; (3) a jacket for protecting the bucket from transient thermal stresses; and (4) a calorimeter thermometer for measuring temperature changes within the bucket. The cross section of such a calorimeter is shown in Fig. 1. A photo of the actual bomb is given in Fig. 2.

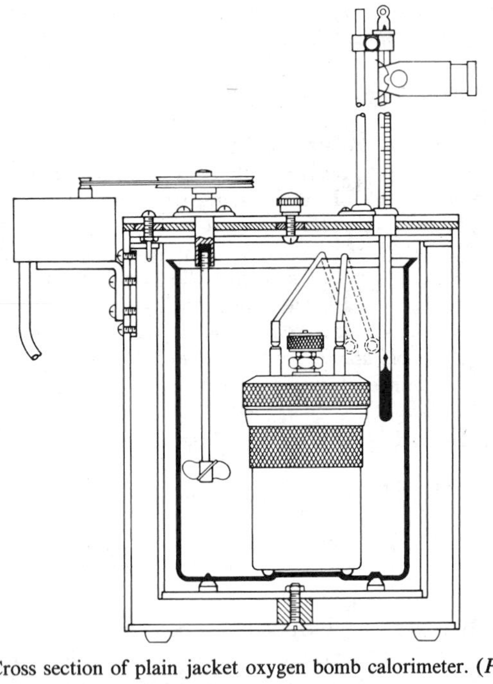

Fig. 1. Cross section of plain jacket oxygen bomb calorimeter. (*Parr Instrument Co.*)

Fig. 2. Bomb portion of oxygen bomb calorimeter. (*Parr Instrument Co.*)

The bomb consists of a strong, thick-walled, metal vessel which can be opened for inserting the sample, for cleaning, and for recovering the products of combustion. Valves must be provided for filling the bomb with oxygen under pressure and for releasing residual gases after the combustion is complete. Electrodes to carry the ignition current to the fuse wire also are required. Since an internal pressure up to 1,500 psig (102 atmospheres) can be developed during combustion, most bombs are constructed to withstand pressures of at least 3,000 psig (204 atmospheres).

In the high-pressure oxygen environment within the bomb, some of the nitrogen present will be oxidized to form nitric acid. Similarly, any sulfur contained in the sample will be converted to sulfuric acid. Because of the formation of these hot and highly corrosive acids, the bomb must be made from materials which will not be attacked by these combustion products. Until Professor S. W. Parr developed a complex nickel–chromium alloy for use in oxygen bombs in 1912, linings of platinum and gold were the only means available for protecting the inside of the bomb. While a few platinum-lined bombs are currently used for research applications, bombs for fuel testing are almost exclusively made of alloys similar to those developed by Professor Parr.

The calorimeter bucket contains the bomb plus a sufficient quantity of water to completely immerse the bomb and to absorb the heat released from the combustion within the bomb. A stirrer is used in the bucket to rapidly bring the bucket and its contents to thermal equilibrium. Bucket systems must be carefully designed. The quantity of water must be sufficient to readily absorb the heat released, but not so large as to preclude acceptable sensitivity. Likewise, the stirrer and bucket geometry must enable rapid equilibration without introducing excessive heat in the form of mechanical energy. Buckets are commonly provided with a highly polished surface to minimize absorption and emission of radiant heat.

The jacket which contains the bucket with its bomb provides a thermal shield to control heat transfer between the calorimeter bucket and its surroundings. It is not necessary to prevent this transfer if a means of precisely determining the amount of heat transferred during the determination can be established. Any effective jacket will, of course, minimize the effects of drafts, sources of radiant energy, and changes in room temperature during the test.

Two basic types of calorimeter jackets are commonly used on bomb calorimeters. The first is the *isothermal system* in which the jacket temperature remains constant while the bucket temperature rises. Isothermal jackets require that temperature readings be made to determine the net heat loss or gain from the bucket to its surroundings. Isothermal jackets are available in two styles: (1) the true isothermal jacket where the jacket is controlled to a specific temperature, and (2) the uncontrolled or plain jacket. The plain jacket assumes that the jacket temperature will remain sufficiently constant without controls. Because of its inherent simplicity, the plain jacket is popular for low-cost calorimeters.

The *adiabatic system* is the second type of jacket commonly used in bomb calorimeters. In the adiabatic system, the jacket temperature is controlled during the determination to keep it equal at all times to that of the bucket. If temperature differences between the bucket and the thermal jacket can be eliminated, there will be no heat transferred between these components and the calculations and corrections required for the isothermal systems can be eliminated. Adiabatic jackets are widely used for fuel testing since they combine high precision with speed and convenience.

The calorimetric thermometer measures temperature changes within the calorimeter bucket. It must be able to provide excellent resolution and repeatability. However, high single-point accuracy is not a requirement since it is temperature changes and not absolute temperatures that are important in calorimetry. Mercury-in-glass thermometers, platinum resistance bulbs, quartz oscillators, and thermistor systems have all been successfully used as calorimetric thermometers.

Before a material with an unknown heat of combustion can be analyzed in a bomb calorimeter, the energy equivalent or heat capacity of the calorimeter must first be determined. This value is, of course, dependent upon the heat capacities of the materials within the calorimeter; notably the metal of the bomb and bucket and the water in the bucket. Energy equivalents are determined empirically by burning a sample with a precisely known heat of combustion in the calorimeter under carefully controlled and reproducible set of operating conditions. Benzoic acid is used almost exclusively as a reference material for fuel calorimetry because it is completely combustible, nonhygroscopic, and is readily available in a very pure form.

The amount of heat introduced by the reference sample is determined by multiplying the heat of combustion of the standard material by the weight of the standard sample. If this value is divided by the net temperature rise produced in the calorimeter, the resultant is the energy equivalent under the specified operating conditions. For example, if 1.651 grams of benzoic acid with a heat of combustion of 6,318 calories per gram were burned, a total 7,361 calories would be released. If this produced a temperature rise of 3.047°C, the energy equivalent for these conditions would be 2,416 calories per degree C.

Once the energy equivalent of the calorimeter has been determined, the calorimeter can be used for actual fuel testing. A sample of known weight is burned in the calorimeter and the resulting temperature rise is measured and recorded. The total energy released by the sample is determined by multiplying the temperature rise by the energy equivalent of the calorimeter. The heat of combustion is then calculated by dividing this total energy value by the sample weight to convert to a unit weight basis.

It is important to note that the energy equivalent for any calorimeter is dependent upon a set of operating conditions and these conditions must be reproduced when the fuel sample is tested if the energy equivalent is to remain valid. A difference of 1 gram of water in the calorimeter bucket, for example, will change the energy equivalent by 1 calorie per degree C.

In a bomb combustion, the water produced by the oxidation of hydrogen condenses and liberates its latent heat of vaporization. The total heat produced is known as the gross heat of combustion at constant volume. In actual fuel-burning processes, the water escapes as a vapor and the total heat produced is known as the net heat of combustion at constant pressure. The net heat of combustion is the value of interest. It may be obtained from the gross heat of combustion and the percent hydrogen in the sample by

$$\text{Net Heat of Combustion} = \text{Gross Heat of Combustion} - 91.23 \times (\text{Weight Percent Hydrogen})$$

Because samples are completely oxidized during combustion in an oxygen bomb and because the combustion products are quantitatively retained within the bomb, procedures have been developed for determining sulfur, halogens, and other elements in conjunction with the determination of the heat of combustion.

The American Society for Testing and Materials has developed a series of standard test methods for testing both solid and liquid fuels in oxygen bomb calorimeters. ASTM Designation D271 is a standard method for laboratory sampling and analysis of coal and coke. ASTM Designation D2382 is a high-precision method for the heat of combustion of hydrocarbon fuels by the bomb calorimeter.

Calorimeters of Historical and Special Interest. Since the heat of fusion of ice is known to be very nearly 79.71 calories per gram, the heat to be measured may be applied to the melting of ice without change of temperature, and the mass of ice melted, multiplied by the heat of fusion, gives the quantity of heat. Bunsen, Lavoisier and Laplace, Black, and others devised calorimeters based upon this principle.

A steam calorimeter was perfected by J. Joly (1886) and used for the accurate determination of specific heats of solids, liquids, and gases. In principle this apparatus consists of a balance, with the specimen hung from one pan and surrounded by an enclosure which can be flooded with steam. The mass of moisture condensing on the specimen, multiplied by the heat of vaporization of water, gives the quantity of heat imparted to the specimen.

An adiabatic calorimeter is so well insulated from its surroundings that reactions or processes involving heat transfer can be studied without having any appreciable heat exchange occurring with the outside.

The Nernst calorimeter is a calorimeter in which a substance whose specific heat is to be measured is suspended in a glass or metal envelope which can be evacuated. The method is particularly suited for work at low temperatures. The envelope is immersed into a vessel containing the cooling agent, and the substance is cooled by admitting a small amount of gas into the envelope. The substance is then thermally insulated by evacuating the envelope, and the specific heat is measured by recording the increase in temperature caused by supplying a known amount of heat. In Nernst's original experiment, the same wire, attached to the substance, served as heater and resistance thermometer.

An improvement on the Nernst calorimeter is the Simon and Lange vacuum calorimeter in which the substance inside the vacuum envelope is surrounded by a shield to which a small container with the cooling agent (liquid air, liquid hydrogen or liquid helium) is attached. By pumping off the vapor from the container, a lower starting temperature can be obtained than by pumping off the vapor of the large bath of cooling agent surrounding the vacuum envelope. Moreover, by regulating the pressure in the container, the temperature difference between the substance and the shield can be kept small.

Gas Calorimeters. There are three basic classifications: (1) total calorific value types, (2) net calorific value types, and (3) inferential types. Net calorific value is less than the total calorific value by an amount equal to the latent heat of vaporization of the water formed during combustion. A net calorific value instrument uses means which give results more nearly related to the net value. Thus, these types are affected by gas composition and must be calibrated for the gas to be tested. Inferential-type instruments depend upon such characteristics as flame appearance, maximum flame temperature, specific gravity, or gas analysis as indicative of calorific value.

The most universally used instrument for gas calorimetry is the flow-type (Cutler-Hammer), in which air is used as the heat-absorbing medium. The calorific value of the gas is determined by imparting all of the heat of combustion of a metered quantity of gas to a metered quantity of air. The temperature rise of this heat-absorbing air is sensed by a pair of nickel wire resistance thermometers, forming two legs of a self-balancing wheatstone bridge-type strip-chart recorder. The scale and chart are both calibrated in Btu per cubic foot (or kilocalories per cubic meter) and the readings require no corrections or computations (to be performed on the results of the measurement) on the basis of any standard volume conditions, including the metric units mentioned. Complete information on this type of measurement is given in ASTM publications.

References

AGA: The American Gas Association, Arlington, Virginia is an excellent source of information on the properties of natural and other fuel gases and their measurement, including calorimetry. Publications are frequently updated.

ASTM: The American Society for Testing and Materials, Philadelphia, Pennsylvania, has established standards and methodologies for testing fuels of all types. The following ASTM methods and standards, periodically revised, are of particular pertinence:

Laboratory Sampling and Analysis of Coal and Coke
Method of Test for Heat of Combustion of Liquid Hydrocarbon Fuels by Bomb Calorimeter
Calorific Value of Gaseous Fuels by the Water Flow Calorimeter
Calorific Value of Gases in Natural Gas Range by Continuous Recording Calorimeter.

CALORITE. Nickel.

CALORIZING. Production of a protective coating of iron-aluminum alloy on iron or steel. The articles are ordinarily coated by heating to a high temperature in a closed container packed with powdered aluminum. Other processes include impregnation at high temperature with an aluminum chloride vapor and spraying with molten aluminum from a spray gun and then heating to a high temperature. When the aluminum coating is held at high temperatures, an iron-aluminum alloy forms which is resistant to oxidation and corrosion by hot combustion gases, especially those containing sulfur compounds which are particularly corrosive to bare iron or steel.

Steel sheets are aluminized by a hot-dip process similar to galvanizing. The principal applications for such a product are furnaces and ovens, automobile mufflers, and other equipment requiring heat and corrosion resistance. When a sheet which has been coated with aluminum by a hot-dip process is exposed to a temperature over 1,000°F (538°C), the aluminum forms an iron-aluminum alloy which is heat- and corrosion-resistant.

CALUTRON. A special type of mass spectrograph of historical interest and built by E. O. Lawrence and his co-workers at the University of California in 1945 for separating isotopes in quantity. This instrument is built around a large electromagnet, and was designed for the separation of uranium-235.

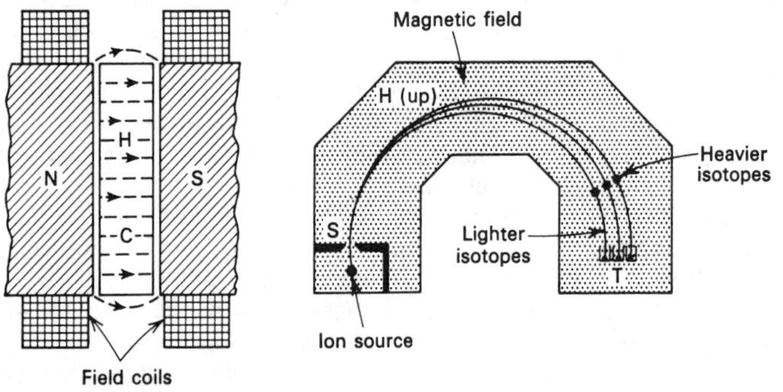

Schematic diagram of the calutron, a magnetic spectrograph for separating isotopes in quantity.

CALVING. **Ablation (Glaciology).**

CALYX. A cup-shaped or funnel-like structure, such as the body of a sea lily and the chambers branching from the principal cavity of the vertebrate kidney. Use of the term in botany is described under **Flower.**

CAM. A cam is a rotating or sliding member which imparts a desired motion or series of motions to another member. Cams are used whenever a desired motion is of such character that it cannot be obtained by using cranks or linkages. There are two important forms of cams: radial cams where the follower moves in a plane perpendicular to the axis of the shaft, and cylindrical cams where the follower moves in a plane parallel to the axis of the shaft. Each of these types may be classified further as positive motion cams in which the reciprocating motion of the follower is definitely controlled by the cam, and nonpositive-motion cams in which the follower is returned to its starting point by spring or gravity action.

Figure 1 shows a radial cam with a flat follower or cam tappet.

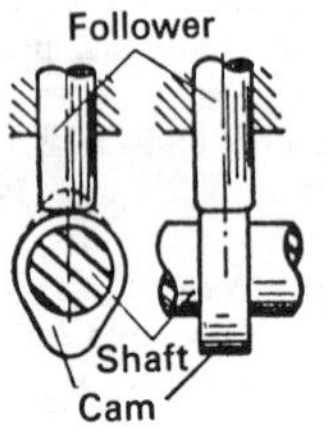

Fig. 1. Radial cam.

The cam is integral with the cam shaft. The cam profile is composed of two circular arcs connected by tangent lines. Cylindrical, helicoidal, and plane surfaces are used for cam faces whenever possible, since they are more easily and accurately manufactured than irregular curves.

The radial disk cam, at the right of Fig. 2, is similar to the cam of Fig. 1. Roller followers are preferred to flat followers because the line contact between the roller and the cam is of a rolling nature, since the sliding is transferred to the pin that carries the roller. The face cam, at the left of Fig. 2, is a positive-motion cam, but is much more difficult to manufacture than a disk cam because the cam groove must be of accurate uniform width. This face cam has a cast iron

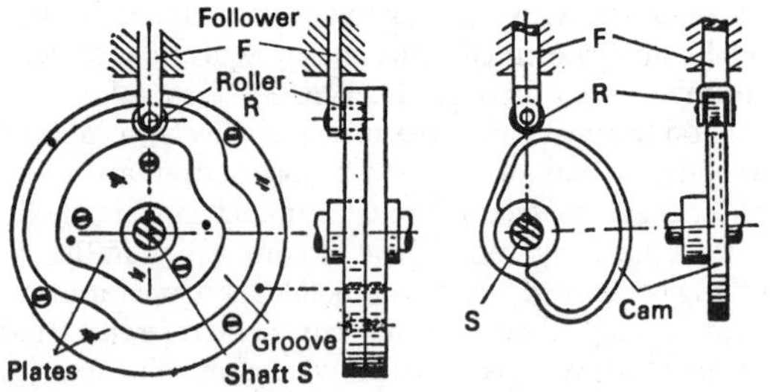

Fig. 2. (*Left*) Positive-motion cam. (*Right*) Radial disk cam.

disk on which the inner and outer hardened steel plates are screwed and dowelled.

Figure 3 shows a solid cylindrical cam with a bell-crank or lever follower for the thread-controlling function on moderate-speed sewing machines. A development or layout of a portion of a cylindrical cam

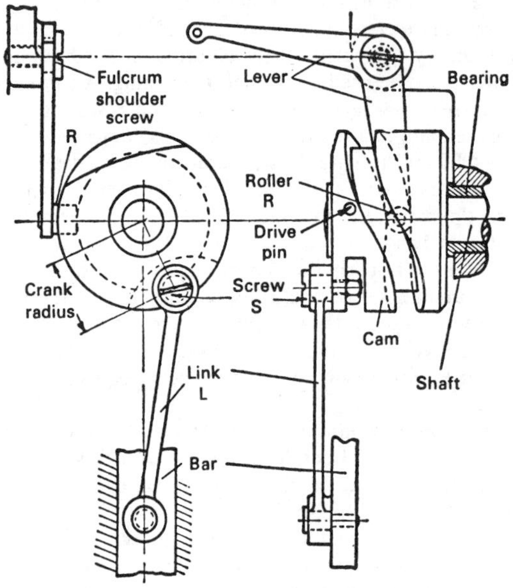

Fig. 3. Solid cylindrical cam.

is shown in Fig. 4. This development shows uniform or straight-line motion of the roller, modified by an arc equal to the roller radius at the beginning and end of each phase of motion, to permit gradual acceleration and to provide roller clearance. The drum cam may have positive motion and will therefore require a cam strap on either side of the roller, or it may be constructed with a single strap, in instances where the inertia of the slide is great enough to enable the roller to remain at rest unless acted on by the cam strap.

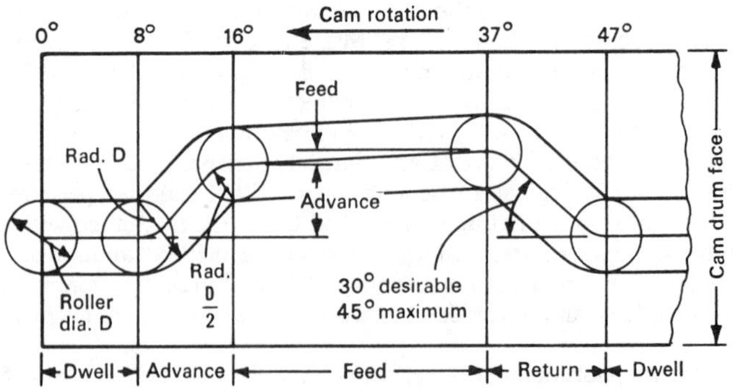

Fig. 4. Development of portion of cylindrical cam.

CAMBER. The curved line from the leading edge to the trailing edge of the airfoil is known as the camber. The curvature of the upper and lower surfaces, as well as a median line between them, is often referred to as camber or camber line.

The wheels of an aircraft landing gear are said to have camber when they make an angle with the vertical plane.

The term camber is also applied to the upward curvature which is given to bridge trusses with theoretically horizontal lower chords, bridge girders with theoretically horizontal bottom flanges, and beam bridges to compensate for the actual deflection. Although these deflections are small in a properly designed structure, they may be objectionable from the standpoint of appearance. Due to an optical illusion, these structures appear to have a pronounced downward deflection. This term is also used to denote the initial curvature which occurs in steel beams as the result of rolling.

In short-span trusses, camber is obtained by lengthening the top chords $\frac{1}{8}$ to $\frac{3}{16}$ inches for each 10 feet of length. No change is made in the lower chords and verticals, but the length of the diagonals must correspond to the new outline. Long-span trusses are cambered by increasing the geometrical length of the compression members and decreasing the geometrical length of tension members. The change in length is based on the calculated longitudinal deformation of the members under dead load and partial or full live load.

It is not customary to camber short-span girders. Long-span girders are cambered by fabricating them with an upward curvature corresponding to a predetermined amount of deflection. This is accomplished by using two or more plates for the web, spliced in such a way as to produce this curvature approximately. The straight flange angles and cover plates are then bent to the desired curvature during the fitting-up operation.

Camber may be obtained in a beam bridge by placing the beams so that the initial curvature due to rolling is upward. The initial curvature may be increased by heating the flange on the concave side with a torch.

CAMBER CHECKING. Thickness Measurement and Gaging Systems.

CAMBIUM (Bone). Bone.

CAMBIUM CELLS (Root). Root (Plant).

CAMBIUM (Plant). In Gymnosperms and dicotyledonous Angiosperms, a large part of the tissue of the stem is derived from a special layer of cells known as the cambium. The cambium originates from certain cells of the procambial strand. In the procambial strand of the stem (that part of the growing tip in which cell differentiation first takes place), cell differentiation commences at the tangential edges of the strand and progresses towards the center, forming primary xylem cells towards the center of the stem and primary phloem cells towards the surface. Some of the cells in the middle portion of the procambial strand do not differentiate into xylem or phloem, but become meristematic cells, dividing actively. These are the cambium cells. Often they begin to divide before the other cells of the procambial strand have ceased elongating.

At first the cambium is a vaguely defined layer of cells occupying the middle portion of the procambial strand. In roots the cambium appears on the inside of the primary phloem strands, which alternate with the primary xylem strands.

Gradually additional cells are formed laterally, either from those cambium cells already formed or by differentiation of parenchyma cells of the medullary ray, until a complete cylinder of cambium exists. Once formed, the cambium of woody plants persists throughout the life of the plant; in herbaceous plants its existence is rather brief, all cells of the stem becoming mature early in its development.

There are two types of cells present in the cambium of any plant. The cells of one type are isodiametric, that is, all dimensions are more or less equal; these cells give rise to the cells of the vascular rays. The other cambium cells are long cells with tapering ends; the cells which result from the division of these become either tracheids, vessels, fibers, or sieve tubes. The elongate cambium cells vary in dimensions in different plants. In various Gymnosperms they may be 3000–4000 micrometers or more in length; in dicotyledons they are much shorter, varying from 100–800 micrometers. In width cambium cells vary in different plants from 20–40 micrometers, and in thickness, or radial dimension, 5–15 micrometers. Cambium cells have a dense cytoplasm in which vacuoles are either lacking or very minute. Each cell of the cambium has a single nucleus which is usually elongated. The walls, especially the tangential ones, are very thin. Division of the cambium cells occurs in a longitudinal tangential plane, that is, the cell divides lengthwise to form two slender elongate cells, one of which lies outside the other, towards the outside of the stem or root.

It is certain that the division is always mitotic (mitosis). One of the cells resulting from this division soon begins to change its form. If this differentiating cell is on the inside of the cambium cylinder it may elongate even more, its ends sliding by and between those of other cells about it. Presently thickening of the wall occurs through deposits of cellulose which are laid down on the primary wall. The cytoplasm of the cell gradually disappears. When mature, this cell, now a tracheid, is a long slender tapering cell with thick wall and no protoplasm. In the wall are numerous simple or bordered pits, which are continuous with pits of adjoining cells.

In Gymnosperms, all elongate cells derived from the cambium become tracheids, except in those forms which have wood parenchyma cells. In these, transverse divisions occur to form a linear row of short cells. In angiosperms, other types of cells are formed. One of these, the fiber, differs little from the tracheid except that it has a thicker wall, in which there are few small pits.

The other type is quite distinct. The cambium derivative which is going to form one of these does not elongate noticeably, but does increase greatly in diameter. As it increases, a large central vacuole forms, and the nucleus moves to a position near the middle of the end wall. At that stage, the vessel appears as a series of very large vacuolate cells separated from one another by distinct end walls. When full size is reached, secondary wall thickening occurs. Then the end wall breaks down, leaving a series of cells forming a long open tube; in many plants perforations are formed in the end wall, so that direct continuity from cell to cell exists. The tremendous increase in diameter of the vessel cells causes the cells around it to be flattened and crowded into angular shapes and irregular arrangements. Once the walls have formed and the cell matured, no further change takes place. Its structure is fixed permanently.

The cells which are formed externally to the cambium become phloem cells. The manner of differentiation is not so well known in these cells as in the xylem cells. Apparently divisions of these phloem mother cells, cut off from the cambium cells, are much more frequent than are divisions of the xylem mother cells. Phloem parenchyma results from the transverse division of one of these cells to form a longitudinal series. In Angiosperms, each phloem mother cell divides unequally, cutting off a very small cell from the corner of the mother cell. The larger cell forms part of a sieve tube, the smaller becomes a companion cell. Often the companion cell divides again to form two or more companion cells associated with a single sieve tube. The cytoplasm of the companion cells remains dense, develops few vacuoles, and always has a well-developed nucleus. In the sieve tube, on the contrary, a cytoplasm becomes peripheral, and there is a large central vacuole. The nucleus has disappeared in the mature sieve tube. The end walls of the sieve tube cells are characterized by the presence of porous places called sieve plates. The pores of these sieve plates result from the enlargement or fusion of the protoplasmic strands, known as plasmodesma strands, which connect the protoplasts of adjoining cells. The enlargement of these strands causes an enlargement of the pores through which they pass, so that conspicuous connections are formed between adjacent cells. The development of sieve tubes in Gymnosperms is very similar to that in Angiosperms, but no companion cells are formed, and the pores in the sieve plates are much smaller. The development of phloem fibers is like that of xylem fibers.

It is obvious that with continued formation of xylem cells inside the cambium and consequent increase in stem diameter, the cambium is constantly being pushed outward and stretched. Gliding growth of cambium cells and those cut off from them causes increase in circumference of the cambium cylinder and so prevents any breaking of the same. For a time the phloem cells maintain their shape against the pressure of the enlarging stem within. In time, however, the older phloem cells become crushed and distorted beyond recognition.

The isodiametric cambium cells divide to form either xylem or wood ray cells inside, or phloem ray cells outside the cambium. These cells differentiate directly into ray cells.

All tissues derived from the divisions of the cambium cells are known as secondary tissues, in contrast to the primary tissues, which are formed by differentiation of the cells of the procambial strands.

Another cambium, the cork cambium (formerly called phellogen), arises in the pericycle of roots and in the outer cortex of stems. It produces the periderm. See **Bark.**

CAMBRIAN PERIOD. The earliest subdivision of the Paleozoic Era. Type locality, North Wales. The formations of this system were first studied and named by Adam Sedgwick in 1835. The Cambrian period began some 500 to 570 million years ago, and lasted for 100 million years. Cambrian formations are well exposed in North America in the Appalachians and Rocky Mountains. Important lower Cambrian beds containing the oldest known faunas occur in British Columbia. Other countries in which the Cambrian is well exposed are Sweden, Britain, Spain, Scandinavia, France, Germany, eastern China, northeastern Siberia, India (Himalayas and Salt Range), Morocco, Australia, Argentina and Antarctica. Cambrian sediments represent the earliest evidence of deposition in well-defined geosynclines, the principal types being sandstones, shales and limestones. Tillites indicate continental glaciation. The maximum thickness of 40,000 feet (12,190 meters) of Cambrian strata occurs in North America. The oldest known invertebrate fossils occur in this period, the principal types being trilobites, chitinous brachiopods, and primitive graptolites, all of which had a marine habitat. It is interesting to note that the paleontological record begins with such highly developed organisms as trilobites, whose ancestors are unknown in the pre-Cambrian formations.

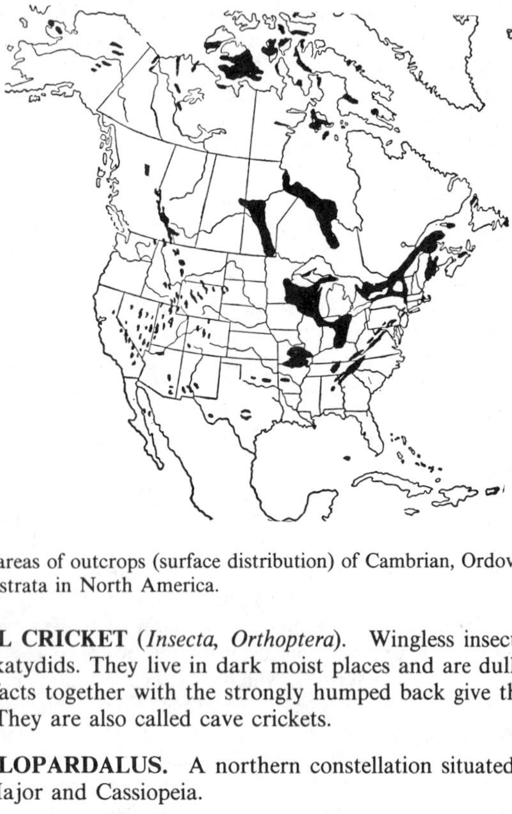

Known areas of outcrops (surface distribution) of Cambrian, Ordovician, and Silurian strata in North America.

CAMEL CRICKET (*Insecta, Orthoptera*). Wingless insects related to the katydids. They live in dark moist places and are dull colored. These facts together with the strongly humped back give them their name. They are also called cave crickets.

CAMELOPARDALUS. A northern constellation situated between Ursa Major and Cassiopeia.

CAMELS AND LLAMAS (*Mammalia, Artiodactyla*). The group of *Camelines* is one of the smaller in the order of *Artiodactyla* (even-toed hoofed animals). Included are: Camels (*Camelus*) of two species, the Bactrian (*C. bactrianus*) and the Arabian (*C. dromedarius*); the Llamas (*Lama*), including the Guanaco and the Alpaca; and the Vicuña (*Vicugna*). The extremities have only a vague resemblance to hoofs. These animals are of a most early origin and, with exception of the Chevrotains, bear little resemblance to any other living mammals. Several authorities formerly believed that the *Camelines* originated in North America. The subject now is considerably less clear. Fossilized remains indicate that there were cameline-type beasts in North America, with good indication that the so-called True Camels of the Old World and the llamas of South America stemmed from these earlier creatures.

Camels: Bactrian on left; Arabian on right. (*A. M. Winchester.*)

Camels have long legs and necks and a conspicuously humped back. See accompanying photo. They are adapted for life in arid regions, including sand deserts, by the broad feet and slit-like nostrils. Internally, the development of cells for the retention of water in one part of the stomach is especially important for life in such regions.

The Arabian camel is found both in Africa and Asia and is characterized by one large hump. There are considerable numbers of the one-humped camels roaming, unbranded and unclaimed, in the African, Arabian, and Middle East deserts, but it is not believed that these are truly wild specimens. Camels are known to detest domestication, the loss of freedom, and requirements to work; consequently, wandering away and keeping away from people is not unexpected. The Arabian camel was introduced into the southwestern states by the United States government in 1856, but after many years of apparent success, the experiment was discontinued. The animals which were freed died out after persisting for some years. The Arabian camel, often referred to as the Dromedary camel—although Dromedary applies to only a particular type—stands 7 feet (2.1 meters) high and thus is slightly taller than the Bactrian and is the faster of the two species of true camels. The Arabian camel may carry as much as 400 pounds (181 kilograms), but none go faster than 18 miles (29 kilometers) per hour and cannot endure for many hours at this speed.

When food is plentiful, fat is accumulated and later used for survival when food is scarce. The size and characteristics of the hump(s) are an index of the animal's health, stamina, and food-supply situation. The one-hump Arabians have a full, long, tall, and rigid hump when well-fed and healthy. On the other hand, in the healthy and well-fed two-humped Bactrians, the humps will be bulbous and heavy-squashy and appearing to be about to collapse because of their size and weight. As is often misunderstood, camels do *not* store water in their humps, but it has been reported that camels may be able to "manufacture" water through chemical oxidation. Special stomachs account for their water storage. Camels may live for reasonable periods without water. Reports indicate that a camel may safely lose water to an extent of 25% of its body weight. This weight can be quickly restored with a few minutes of drinking. The animals do require significant amounts of water at regular intervals. Other protective means provided for enduring arid environments include a double row of heavy eyelashes for protection against blowing sand particles. Their ear openings are protected by heavy hair and they have a very keen sense of sight and smell. During winter months, the animals grow heavy hair, while in summer most of this hair is lost.

The gestation period is 11 months. One young is produced at birth. The animals require from 10 to 12 years to reach full maturity, and their life span ranges from 30 to 40 years. During the rutting reason, camels can exhibit fits of rage and may be inordinately obstinate.

Llamas or Camelids (*Lama*) are medium sized, reaching an overall length from 125–225 centimeters (49–89 inches); the tail length is 17–25 centimeters (6.7–10 inches); the body height is 70–130 centimeters (27.5–51 inches); and the weight can reach 75 kilograms (165 pounds). The males are taller than the females. The profile of the head is straight. The large eyes have long lashes on the upper lid, the ears are long and pointed, and the lips are not too large. The long, thin neck has a slightly arched base, and is usually erect. The body has no humps, and the back is level. The round tail is rather thick, with an almost naked underside, and it is usually carried bent down and away from the body. The dense, woolly, and smooth coat has a few thin bristles which do not protect against the rain. The cutaneous foot pads are smaller than in camels, and there is a deeper cleft between the toes. The shape of the teeth is like the camels', but in the vicuña the lower incisors have smaller crowns, with open roots and continuous growth. There are two species: Guanaco (*Lama guanicoë*); and Vicuña (*Lama vicugna*).

Llamas are found in the high altitudes of western South America and on lower ground in the southern part of the continent. The animals provide wool, hides, meat, and milk and are used as beasts of burden. The adaptability of the llamas to different locales is well explained in the article by Fincher, J.: "Some Immigrant Llamas Thrive in Home of Forebears," *Smithsonian*, **10**, 9, 118–126 (December 1979).

The guanaco is the original llama and still occurs in the wild. The animal travels in large herds and prefers open country ranging from the Altiplano (15,000 feet) (4,570 meters) to the Patagonian prairies. The animal is 4 feet high (1.2 meters) at the shoulders. The legs are long and slender. The hair is long, soft, and fawn colored with some white. The animal's cry is something like the neigh of a horse. The guanaco is hunted by the Patagonian Indians and is also a favorite dietary item of the pumas. The animal is a source of food and hides for the Indians. Dried dung is used for fuel. The guanacos display a peculiar habit of going to the same place to die, a pattern that was observed years ago by Charles Darwin and W. H. Hudson.

Two domesticated animals have risen from the guanaco, known as llamas and alpacas. The animals originally were domesticated in Peru by the Spanish. They are used for riding and as beasts of burden. In early years, the male llamas were used for carrying ore and bullion for as much as 12 miles (19 kilometers) per day. They can carry a load up to about 120 pounds (54 kilograms). If overloaded, the animal automatically lies down. Female llamas provide milk and meat which resembles mutton. Llamas are of many colors and patterns. The alpaca is of more striking appearance and is found mainly at high altitudes in Bolivia and Peru. The wool is valuable. It grows about 8 inches (20 centimeters) annually and is clipped each year. The wool may be yellow-brown or gray-black and is somewhat elastic, fine, glossy, and straight.

The vicuña differs considerably from the guanacos. It was the royal animal of the Incas. Because the pelt is silky and soft, the animal was desirable to hunters. It is now under government protection. The vicuña is found in the mountains of Ecuador, Peru, and Bolivia.

For references, see **Mammalia.**

CAMERA. Photography and Imagery.

CAMERA TUBE (TV). Television.

CAMPANULARIAE (*Coelenterata, Hydrozoa*). An order made up of colonial species with two forms of polyps, one nutritive and the other reproductive, known respectively as hydranths and blastostyles, and both enveloped partially in a cupped extension of the sheath of the colony. The common and widely distributed genus *Obelia* is an example.

CAMPBELL BRIDGE. Bridge Circuits (Electrical).

CAMPHOR (*Cinnamonum camphora; Lauraceae*). A crystalline compound occurring in various parts of the wood and leaves of the camphor tree, a large evergreen tree with light green leaves growing in many warm regions of southeastern Asia, notably Taiwan. Camphor, $C_{10}H_{16}O_7$, is a white solid, mp 179°C, bp 209°C, of a characteristic pleasant odor, insoluble in H_2O, soluble in alcohol or ether. Camphor may be produced synthetically by converting pinene into bornyl chloride with HCl, thence to isobornyl acetate, thence to isoborneol, and finally oxidizing borneol to camphor. Camphor has found use in medicines, insecticides and moth preventives. Earlier uses included the manufacture of plastics and lacquers.

CAMPTONITE. A dark basaltic dike rock of the essential mineralogical composition of a diorite, requiring, however, microscopical examination for proper identification. It was named from the type locality, Campton, New Hampshire.

CANADA BALSAM. A slightly yellow, transparent, fluid resin procured from a North American species of silver fir tree. Used for mounting thin sections of rocks, and of tissues of plants and animals for microscopic examination between glass slides, and for cementing glass in optical instruments. The refractive index of Canada balsam after it has been heated varies between 1.534 and 1.540, according to A. Johannsen. See also **Resins (Natural).**

CANADA BLUEGRASS. Grasses.

CANADIAN. Geologically, a North American provincial series: Lower Ordovician (above Croixian of Cambrian; below Champlainian). The term also is an obsolete name once applied to a system of rocks between the Ozarkian below and the Ordovician above.

CANADIAN HEMLOCK. Hemlock Trees.

CANAL (Physiology). A tubular structure or passage, with specific applications in many groups of animals among which are the following: (1) the passages in the wall of a sponge, (2) slender diverticula of the enteric cavity in coelenterates and ctenophores, (3) the stone canal, ring canal, and other parts of the water vascular system in echinoderms, (4) the inguinal canal through which the testis descends from the abdomen into the scrotum in mammals.

CANANGA. The flowers of the cananga and ylang-ylang trees (*Canangium odoratum* Baill) are the source of an oil which is recovered by distillation. At one time these two trees were considered identical species, but in recent years minor differences have been noted. The recovered oil contains a multitude of organic substances, including sesquiterpenes, linalool, geranil, eugenol, methyl salicylate, among others. Canana oil or ylang-ylang oil are used extensively in perfumery, as well as for flavorings in beverages, ice creams, candies, and baked goods. The oils impart a slightly woody, floral odor with a somewhat burning taste.

CANARY (*Aves, Passeriformes*). A finch, *Serinus canarius*, native to the Canary Islands, which has been extensively used as a cage bird. The wild species is brownish with yellow markings but in captivity pure yellow strains have been developed.

The goldfinch of North America and to a lesser extent the yellow warbler are called wild canaries from their similar yellow color.

The wild canaries that are found in the Canary, Azores, and Madeira Islands have an olive coloration above and yellow below. These birds have been bred in captivity for many centuries. In nature, the canary builds a cup-like nest about 10 feet (3 meters) above the ground level in trees or shrubs. There are usually five eggs of a blue-green color with reddish-brown markings. In the wild, the canary prefers arboreal fruit and seeds.

CANCER (the crab). A small and poorly marked constellation of faint stars that is of importance principally because it is the fourth sign of the zodiac. In cancer (the crab) is found the fine cluster known as Praesepe (or the Beehive). The stars are not so numerous as in some other star clusters, but are of sufficient brightness to make this an interesting object in a small telescope. Galileo counted 36 stars with his telescope but observers using modern equipment have counted over 300. On a clear moonless night the object appears as a faint glow of light, and is frequently used by astronomers as a test of the transparency of the atmosphere. (See map accompanying entry on **Constellations.**)

CANCER AND ONCOLOGY. This entry is concerned principally with the present state of the art. Investigations of the processes of carcinogenesis are described in a separate entry. See **Cancer Research.** Hodgkin's disease and the leukemias are also described in separate entries. See **Hodgkin's Disease** and **Leukemias.** Oncology is that branch of medical science concerned with tumors. There are benign tumors and malignant (cancerous) tumors. Several of the common sites of human cancers—the gastrointestinal tract, breast, lungs, female genital tract, prostate, and skin—are described in this entry. The occurrence of cancers in a number of other organs of the body is described in entries concerned with those organs. See also **Carcinogens.**

General background. When a tumor does not spread, but remains in its original area, it is called a *benign* tumor. A benign tumor grows locally by simple expansion of the mass, whereas a cancer grows locally by extension of fingers of tissue into the normal surrounding tissues. A cancer can arise in any of the body's tissues, whenever those tissues commence to grow in an uncontrollable manner. Such a growth in the epithelial tissue (skin, as well as mucous and serous membranes) is called a *carcinoma*; a cancer in other types of tissue may be called a *sarcoma.*

A tumor is considered malignant, and therefore a cancer, if it can spread to remote areas of the body. This spread, termed *metastasis,* occurs when a minute piece of the tumor breaks off from the growth and is carried by the blood or lymph stream to other body areas or nearby lymph nodes. The metastatic cells from the original tumor attach themselves and set up a "colony," which eventually may exceed the parent growth in size and destructiveness. The most frequent sites of metastasis are the lymph nodes in the region of the tumor, the lungs, long bones, the spine and ribs, liver, skin, and brain.

Metastasis from a carcinoma usually occurs by way of the lymphatic system, whereas sarcoma spreads most often by way of the bloodstream. The primary new growth may occur in a specific organ, such as the stomach or rectum, or widely throughout the body, as in the instance of cancer of the blood (*leukemia*). In each instance, it is an abnormal growth of the tissues already present rather than of new, previously absent tissues.

Much more conclusive evidence is available pertaining to the manifestations of cancer, as the result of many years of collecting and analyzing medical statistics and of observing and treating hundreds of thousands of cases of cancer, than as to the cause(s) of cancer. Considerable evidence is available pertaining to what may be termed qualitative causes, again as the result of observation and experience, as contrasted with fundamental causes which can be explained in terms of biochemistry, molecular biology, and so on. An understanding of qualitative causations, however, in lieu of the more fundamental understanding, is extremely helpful. Some of these qualitative causations include: (1) hereditary predisposition for cancer in certain sites; (2) chronic irritation of a body area over a number of years; (3) the presence of certain substances, called *carcinogens,* which incite cancer after repeated exposure to such substances; (4) the existence of preexisting lesions, sometimes referred to as "premalignant" conditions, such as white patches on the tongue and vulva (*leukoplakia*), certain clear-colored warts on older persons (*keratosis*), large burn scars, and pendulous growths (*polyps*) in the rectum, to mention a few such lesions.

Cancer Therapy

A cure or amelioration of cancer is usually achieved by appropriate surgery, radiotherapy, or chemotherapy—these procedures used singly sometimes, but generally in combinations.

Radiation Therapy. The principles of X-rays are described in the entry on **X-ray.** Gamma rays, also used in radiation therapy, are described in the entry on **Gamma Radiation.** Gamma rays are highly penetrating and produce their biologic effects by ionization. Radiotherapy is usually carried out with megavoltage energy radiation in the form of X-rays from linear accelerators or gamma rays from cobalt-60 sources. X-rays from linear accelerators are in the range of 2 to 25 MeV and they may also produce electrons or beta particles of the same energy. Gamma rays from cobalt-60 generators have an average energy of 1.4 MeV. For effective radiation therapy, highly concentrated and purified cobalt-60 sources are required to generate sufficient gamma rays from a small focal point. Improvements in supervoltage irradiation have greatly enhanced cancer therapy. With increasing radiation energy, the skin dose is reduced and higher doses are delivered deeper into the tissues.

Chemotherapy. Chemotherapeutic drugs for cancer, as well as in other areas of medicine, are sometimes discovered by chance in the laboratory. Most frequently, drugs are developed by relating parallels in structure and performance with other proven drugs. Once a drug is considered to have potential, a fund of knowledge will be created—first by carrying out experiments with laboratory animals and, later, with selected human patients in a clinical atmosphere that has been approved by a number of authorities. Literally thousands of chemical compounds have been tested for their antitumor activity, both in vitro and in vivo. When compared with the highly successful use of chemotherapy for handling many infections (antibiotics, which were preceded by the sulfa drugs), cancer chemotherapy is somewhat less impressive. In making this comparison, however, one must take into consideration the much greater understanding of the microorganisms that produce diseases as contrasted with present knowledge of the fundamentals of carcinogenesis.

For most cancer drugs, the level of dangerous or lethal toxicity and the level of dosage required for efficient therapeutic action are disturbingly close. Thus cancer themotherapy requires the utmost care in monitoring and controlling the administration of drugs to human patients. This largely explains the sometimes extremely long periods required to develop successful regimens. To make matters more complex, a new drug with superior characteristics may appear at just about the time the regimen for an older drug has been fully refined.

In developing cancer drugs, an attempt is always made to match the performance with what is currently known about the sequence of biochemical and genetic events that occur during the development of both normal and malignant cells. Certain drugs are effective only during certain specific phases of the cell development cycle, whereas other drugs are relatively independent of the cycle. Thus, some authorities classify cancer drugs into two broad categories: (1) drugs that are *phase-specific*; and (2) drugs that are *phase-nonspecific.*

Although probably the majority of details of normal cell and malignant cell development remain to be learned, or at least may be considered to be poorly understood, for guidance at the present juncture of the technology some authorities have developed a profile of the cell cycle. It is useful to match currently used as well as experimental cancer drugs against the stages of this cycle. The cycle is divided into five phases or periods. These phases are of unequal time spans.

M phase (*mitosis*)—This is the very beginning of the cell cycle. This period lasts for only 30 to 60 minutes. No DNA synthesis is assumed during this phase. Synthetic processes are comparatively inactive. Drugs found to be effective during this phase include vincristine and vinblastine. See **Cell (Biology).**

G_1 phase—This is essentially an interval of a few to many hours and is the first time gap observed in the cell cycle. Not by confirmed observation, but by definition, it is assumed that DNA synthesis does not take place during this phase. Drugs found to be effective during this phase include actinomycin D, mitomycin, 6-mercaptopurine, and 6-thioguanine.

G_0 period—Regarded as an extension of the G_1 phase and as a "resting period." The cells are assumed not to be actively dividing during the period. It is assumed that a cell in the G_0 period may be stimulated to reenter the G_1 phase. It is believed that none of the chemotherapeutic agents are effective during this period.

S phase (*DNA synthesis*)—Regarded as the period of active DNA synthesis. A doubling of the DNA content of the cell is assumed during this period, estimated to span between 6 and 12 hours. It is assumed that once the DNA synthesis is initiated, the cell will divide. Drugs found to be effective during this phase include 6-mercaptopurine, 6-thioguanine, methotrexate, 5-fluorouracil, doxorubicin, daunorubicin, mitomycin, cyclophosphamide, and cytosine arabinoside.

G_2 phase—Estimated to span about 2 hours, this is stipulated as the second gap between DNA synthesis and cell division. Drugs found to be effective during this period include bleomycin and cyclophosphamide.

Combination chemotherapy has made major improvements in the treatment of cancer during the past several years. This is based upon the use of several drugs at one time, sometimes up to as many as six different agents. The mechanisms of action and side effects are different. By carefully determining quantities and timing of administration, the therapeutic advantages of each drug can be maximized while toxicity is minimized. This procedure, of course, requires a great deal of experience based upon empirical data gained through trials and case experimentation. Often there is only a rudimentary understanding of the mechanisms of each drug. From an overall standpoint, the multiple administration of drugs also may lessen the development of drug resistance. The prolonged administration of a single drug in ever increasing concentrations, which is retained in the environment, is believed by some authorities to be precisely that form of administration most likely to result in amplification of genes in a stable state, thereby imparting stable resistance.

There are some precautions applicable, however, to the use of combination chemotherapy. One group of researchers reported that the most frequently used antitumor agent, methotrexate, interferes with the activity of another antitumor drug, 5-fluorodeoxyuridine when given at the same time in tests on tumor cells grown in culture. Their evidence suggested that it may also interfere with the activity of the more frequently used antitumor drug, 5-fluorouracil. Methotrexate and 5-fluorouracil have been used together in therapy of many forms of cancer—breast, head, neck, liver, colon, and rectum.

Some of the most widely used of the cancer drugs are listed in the accompanying table; the structures of some of these are given in the accompanying diagram.

The first advance in cancer chemotherapy came in 1941, when researchers found that the female sex hormone, estrogen, was useful in the treatment of prostatic cancer in men. The discovery of nitrogen mustard's effectiveness as a cancer drug was the product of investigating chemical warfare agents. In 1948, the first of the antimetabolites, an antivitamin called aminopterin, was reported to be useful in the treatment of leukemia. In the following year, the effectiveness of a related drug, methotrexate, was reported. The availability of cancer drugs and their use is now widespread. Methotrexate has proven strikingly useful, particularly against a rare uterine cancer known as choriocarcinoma. Before the effectiveness of this drug was discovered, five of every six women with this or a related kind of cancer, even when treated early with surgery, died within a year of diagnosis. Similarly, the effectiveness of several other chemotherapeutic agents for different kinds of cancers has been demonstrated.

The present consensus among scientists is that these drugs interfere with cell division at the core of the cell, within the nucleus, in the chemicals that contain the cell's genetic machinery, the nucleic acids

DRUGS USED IN CHEMOTHERAPY OF CANCER

Group and Generic Name	Trade Names and Abbreviations
ALKYLATING AGENTS	
Mechlorethamine hydrochloride	Mustargen (nitrogen mustard; HN2)
Cyclophosphamide	Cytoxan, Endoxan
Chlorambucil	Leukeran
Melphalan	Alkeran (phenylalanine mustard; L-sarcolysin; L-PAM)
Busulfan	Myleran
Triethylenethiophosphoramide	Thiotepa
ANTIMETABOLITES	
Methotrexate	Methotrexate (amethopterin; MTX)
6-Mercaptopurine	Purinethol (6-MP)
6-Thioguanine	Thioguanine (6-TG)
5-Fluorouracil	Fluorouracil (5-FU)
Cytosine arabinoside	Cytosar (Ara-C)
PLANT ALKALOIDS	
Vincristine	Oncovin
Vinblastine	Velban, Velbe
ANTIBIOTICS	
Actinomycin D	Cosmegen
Doxorubicin	Adriamycin
Daunorubicin	Daunomycin
Bleomycin	Blenoxane
Mithramycin	Mithracin
Mitomycin C	Mutamycin
OTHER AGENTS	
Hydroxyurea	Hydrea
Carmustine	Bi CNU; BCNU
Lomustine	CeeNU; CCNU
Procarbazine	Matulane
Decarbazine	DITC-Dome
Cisplatin	Platinol
L-Asparaginase	Elspar
Streptozotocin	—

NOTE: See also entry on **Hormones.**

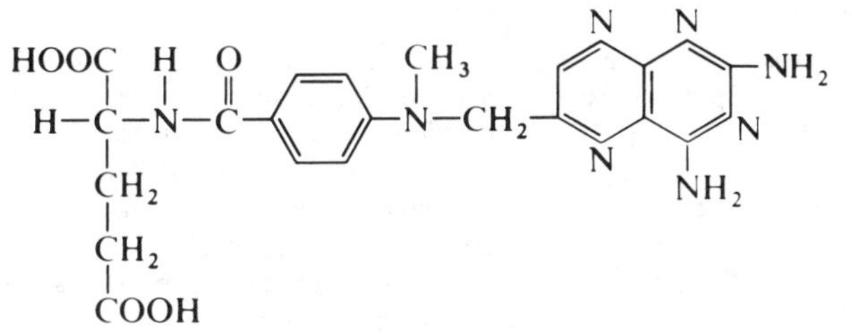

Methotrexate

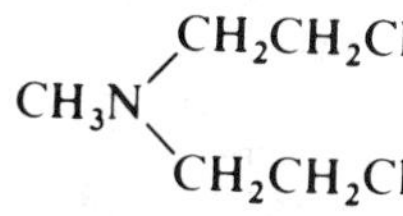

Mechlorethamine

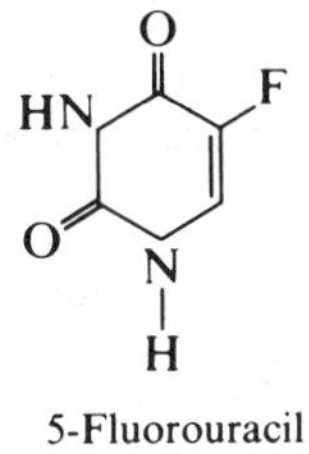

5-Fluorouracil

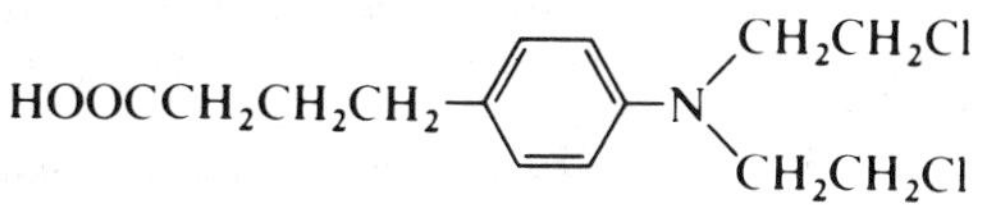

Chlorambucil

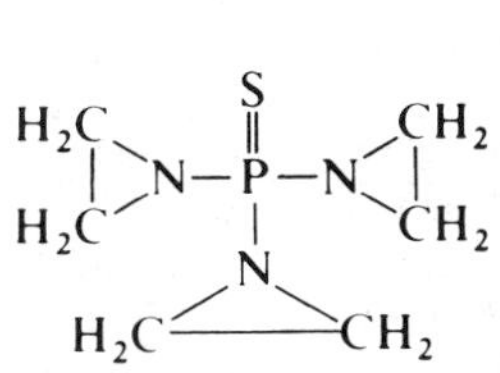

Triethylenethiophosphoramide

Chemical structures of representative drugs used in cancer chemotherapy.

DNA and RNA. Most cancer drugs are limited in their usefulness primarily by two factors: (1) *toxicity*—damage to normal cells and tissues as well as cancer cells and tissues; and (2) *decreased effectiveness*—gradual loss of drug effectiveness due to host resistance. When the drug administered is a hormone, toxicity is sometimes manifested as changes in secondary sex characteristics, such as voice and facial hair. Other drugs may temporarily produce nausea, loss of appetite, loss of hair, hypertension, or diabetes. These side effects are usually reversed when drug treatment is halted.

Chemotherapists working in the clinic can usually anticipate a drug's toxicity from data gathered in animal tests. Rats, dogs, and monkeys are ordinarily reliable indicators of the human response to potential new drugs. Initial clinical trials of a drug begin at one-tenth the dose level tolerated by the larger laboratory animals. The final dosage formula is measured in milligrams of drug per kilogram of body weight, or even more efficiently, by milligrams of drug per square meter of body surface.

As an additional measure of regulating dose, means have been devised to circulate the drugs in a closed circuit through the bloodstream of the cancer-affected region, while a tourniquet prevents the drug from reaching and damaging sensitive organs beyond the cancerous area. The drug is injected through an artery to the cancerous area and is withdrawn from a vein by special tubes, and then recirculated through the artery and vein by means of a pump oxygenator. This technique, known as *regional perfusion*, is especially adapted to treating certain cancers of the arms and legs.

Infusion methods, by which a drug is dripped slowly into a patient's bloodstream and travels throughout the entire circulatory system, have also been modified to focus drug effects on cancerous areas, for example, cancers of the head and neck. In another system, developed for treating cancer of the liver, a plastic tube carries a continuous supply of drug directly to the cancer at a uniform rate regulated by an infusion pump. The tiny pump and a 7-day supply of the prescribed drug constitute a small package that can be strapped to the chest of a nonhospitalized patient for round-the-clock treatment.

As covered further in the entry on **Cancer Research,** it is generally believed that the cancerous change is induced in the cell's nucleic acid by a chemical, by radiation, by a virus, or by some combination of these factors. Medical scientists are attempting to employ the same factors to block growth of cancer cells. As pointed out in the following paragraphs, each class of drugs functions in a somewhat different way.

Alkylating Agents. This group of quick-acting, highly reactive compounds includes nitrogen mustard and other close relatives of the wartime poison gases. Often referred to as "cell poisons," these agents are rich in electrons when they are in solution. For this reason they combine rapidly with many of the constituents of a cell. Many scientists believe that the aklylating agents exert their anticancer effects by a direct chemical interaction with the DNA of the cell.

In addition to nitrogen mustard, among the best known alkylating agents are cyclophosphamide (Endoxan, Cytoxan), chlorambucil (Leukeran), triethylene thiophosphoramide (thio-TEPA), and dimethanesulfonoxybutane (Myleran). They act primarily on tissues that are being quickly replaced, such as bone marrow and cells lining the intestine. They are used primarily in the treatment of Hodgkin's disease, lymphosarcoma, the chronic leukemias, and in some cancers of the lung, ovary, and breast.

Antimetabolites. These drugs structurally resemble metabolites, the nutrients a cell needs for growth. Antimetabolites mimic normal nutrients so closely that they are taken up by the cell through mistaken identity. Once inside the cell they interfere competitively with the production of nuclei acids and thereby prevent cell growth.

Among the antimetabolites are antagonists of purines and pyrimidines, essential components of the cell's nucleic acids. The drugs 6-mercaptopurine and 5-fluorouracil are examples, respectively, of the antipurines and antipyrimidines. 6-Mercaptopurine was developed by Elion, Burgi, and Hitchings of the Wellcome Research Laboratories; 5-fluorouracil was synthesized in the laboratory by Heidelberger and his associates at the University of Wisconsin.

One of the most widely used antimetabolites is amethopterin (methotrexate), synthesized in 1948 by Seeger and associates. By inhibiting the enzyme folic acid reductase, methotrexate acts as an antagonist to a needed B vitamin, folic acid. This, in turn, interferes with both purine and pyrimidine synthesis.

The antimetabolites are useful in leukemia and in several types of solid tumors. See also **Antimetabolites.**

Plant Akaloids. Several compounds, derived from extracts of the common periwinkle plant, seem to act through interference with a phase of cell division. Best known are vinblastine sulfate and vincristine sulfate, the latter of which is useful in treating acute lymphocytic leukemia. Both are effective in certain lymphomas.

Antibiotics. Notable in this group is actinomycin D, a drug believed to achieve its effectiveness by locking itself onto a base of the DNA molecule, thereby blocking cell growth.

Other Agents. A drug used in treating cancer of the adrenal gland, *o,p'*-DDD, is closely related chemically to the insecticide DDT. It seems to have a selective destructive effect on adrenal cells. Methylglyoxal-*bis*-(guanylhydrazone), often called Methyl-GAG, is a synthetic chemical with activity against acute myelocytic leukemia, the type of acute leukemia occurring mostly in adults.

Hormones. Also described in the entry on **Hormones,** the action of hormones in cancer chemotherapy tends to accelerate or suppress the growth of specific cells, tissues, and target organs. They are thought to derive their effectiveness by altering or reversing a hormonal imbalance in the body that encourages the cancer cells to multiply. The female hormone estrogen, for example, helps to suppress the growth of disseminated cancer of the prostate. Conversely, male hormones or androgens cause temporary regression of disease in 20% of breast cancer patients, and are especially helpful in premenopausal women.

Among the other hormonal types, corticosteroids such as cortisone and prednisone seem to suppress the growth of white blood cells known as lymphocytes. For this reason these drugs are frequently useful in acute lymphocytic leukemia, which is characterized by an abundance of abnormal lymphocytes.

Scientists are uncertain as to the exact chemical mechanism by which hormones influence the growth of cells. However, evidence is accumulating that here, too, nucleic acid may be implicated. Karlson and Clever at the Max Planck Institute of Munich have demonstrated through experiments with insects that a hormone controls certain phases of insect growth by controlling its production of the nucleic acid RNA. Studies of hydrocortisone in rat cells by Tomkins and associates at the National Institute of Arthritis and Metabolic Diseases also suggest hormonal stimulation of RNA production.

Common Sites of Cancers

Gastrointestinal Cancer. Including both sexes, gastrointestinal cancer is the most frequently occurring cancer. Of the gastrointestinal cancers, that involving the colon and rectum (colorectal) is the most prevalent. Carcinoma of the stomach is also a frequently occurring disease. This is a major problem in the Orient. In recent years in the United States, there has been a measurable decline in stomach cancer. However, growing in incidence in the United States are cancers involving the pancreas, esophagus, and liver.

Treatment of gastrointestinal cancers is essentially confined to surgery, radiation, or a combination of the two procedures. To date, chemotherapy has played a lesser role, but much research, including clinical trials of new drugs, is going forth and many authorities believe that, with time, more effective drugs and knowledge in their management of these cancers will emerge.

Diagnostic measures include X-ray examination of chest and barium opaque examinations of the gastrointestinal tract. Usually a liver scan is indicated. Laboratory analyses will include determination of the serum alkaline phosphatase level. At one time, it was believed that a carcinoembryonic antigen known as CEA could be very useful as a diagnostic indicator, but it was later found that elevated levels of this substance occur in connection with several other malignant and nonmalignant diseases. However, CEA retains some value in diagnosis when the history of the patient has been well delineated.

To aid in diagnosis, treatment, and prognosis, many physicians currently use a classification system based upon the conditions presented by the patient at the time of diagnosis.

Stage A—The tumor is confined to mucosa and no lymph nodes are involved. The 5-year survival potential ranges from 60 to 80%.

Stage B—The tumor penetrates the bowel wall to the serous coat. No lymph nodes are involved. The 5-year survival potential ranges from 25 to 65%.

Stage C—There is involvement of the lymph nodes. The 5-year survival potential ranges from 5 to 40%.

In the cases of carcinomas of the middle and lower rectum, a colostomy (fistulization of the colon, i.e., surgical creation of an opening from the body surface to the organ) may be required. Radiation therapy is useful for providing palliation for local problems—obstruction and bleeding, among others. The effectiveness of chemotherapy as an adjuvant to surgery, such as administration of the drug 5-FU (5-fluorouracil), is less certain.

Breast Cancer. This is the most common malignant tumor in women of the western world. Breast cancer has accounted for the greatest number of deaths in the 40–45 year age group. It is estimated that 5% of women in the United States will one day develop breast cancer. It is further estimated that, when treated with current technology, about one third of these women will survive for 20 years after diagnosis; about 50% of the women with cancers where there are no axillary lymph node mestastases at the time of diagnosis will have a longer period of survival. Unfortunately, the word "cured," defined as no future risk whatever of recurrence, still remains inappropriate.

Although this is a generalization, the course of an *untreated* breast cancer will commence with a small malignant tumor (or tumors) in one or both breasts. These tumors enlarge and more tumors may be formed. The cancer will spread to lymph nodes in the region of the armpit (axillary nodes) and/or clavicular nodes. Arm edema may occur. *Peau d'orange* (dimpled condition of skin resembling an orange) may be observed. Ultimately (usually in terms of a relatively few months), the cancer will further metastasize, frequently first involving the pelvic region. Metastasis will continue, ultimately leading to death. The exact course of untreated breast cancer, of course, varies considerably from one individual to the next, but the terminal nature of the disease is a common denominator.

Over a number of years, various researchers have reported that a family history of breast cancer increases an individual's risk for the disease. It has been shown that the increased risk factor varies with the number and degree of affected relatives and reaches a ninefold value for premenopausal women who have one or more first-degree relatives (mothers and sisters) with premenopausal bilateral breast cancer. Although the genetic factors involved are poorly understood, there is evidence of hormonal participation in the etiology of the disease and this further suggests that the disease may be mediated through a genetically transmitted endocrine factor. In a controlled study involving 30 young women who were genetically at risk for familial breast cancer and 30 other women fully matched for age and physical characteristics, the urine of these subjects was extracted for steroids. By radioimmunoassay, measurements were made to determine the primary glucocorticoid, androgen, and estrogen hormones and their metabolites. In the high-risk subjects, differences were found only in the case of lower values of estrone and estradiol. The investigators suggested that this endocrine abnormality may be a discriminant for identifying women at risk in the population at large.

Diagnosis and Prognosis. As an aid to diagnosis, treatment, and prognosis, many physicians use a classification scheme in which breast cancers at the *time of diagnosis* are weighted in terms of various factors. A four-stage system is widely used.

Stage I—Tumor size ranges from not palpable to less than 2 centimeters ($\frac{3}{4}$ inch). *Node* size ranges from not palpable, or palpable but clinically benign. *Metastasis* is negative.

Stage II—Tumor size ranges from not palpable to over 2 centimeters ($\frac{3}{4}$ inch), but less than 5 centimeters (2 inches). *Node* size ranges from not palpable to palpable (clinically benign) or palpable (clinically malignant). *Metastasis* is negative.

Stage III—Tumor is present (any size). *Node* (any size) and arm edema may be present. *Metastasis* is negative. This stage is broken down into four substages, but in any of these substages, both a tumor and node will be present.

Stage IV—Tumor (any size) is present. *Node* (any size) is present. *Metastasis* is positive.

These stages are used essentially as guidelines and assist in communications between diagnosticians, treating physicians, and surgeons. Identification of the stages also facilitates statistical reporting and analysis and in estimating prognosis for a given patient. As has been known for many years, the stage of the disease at the time of diagnosis and the prognosis are closely related. A study has shown that patients diagnosed and treated in Stage I have a 5-year survival potential in 85% of cases. It should be emphasized that, for study purposes, survival potential is estimated only out to 5 years. This does *not* indicate that the patient will survive only 5 years. Persons who have survived a treated cancer over a 5-year span most likely will survive (in terms of this condition) for many more years. Patients diagnosed in Stage II have a 5-year survival potential in 66% of cases; in Stage III (41%); and in Stage IV (10%). In this particular study, all patients with exception of those with Stage IV disease were treated with radical mastectomy.

Currently accepted practices for diagnosing breast cancer include mammography of both breasts, X-rays of the chest, a whole-body bone scan, and blood chemistry determinations. The latter will include serum alkaline phosphatase and serum calcium. In older women who are several years beyond the menopause, vaginal cytology for estrogen effect may be indicated.

Mammography. Introduced during the 1960s, mammography has become a widely accepted diagnostic tool in breast cancer. Almost from the outset, there have been some critics of the procedure, particularly regarding exposure to X-radiation and the risks of misdiagnosis in instances of very early breast cancers. During the 1970s, through the use of new film and a fluorescent screen with emission matched to the film sensitivity, the amount of X-radiation required to yield high contrast, low density images has been reduced to one-ninth the radiation originally required. Improvements in film and equipment have enhanced the ability of the film to show early calcifications and fibrous structures. In the early 1970s, national cancer institutions in the United States launched a breast cancer detection demonstration project which was based on the premise that if breast cancer can be discovered early enough, it can be "cured." A further premise was that, if mammography can reveal tumors too small to be felt by hand, as a screening technique, the procedure may save the lives of countless women. In a four-year program (1973–1977), mammography was used in making breast X-rays of about 270,000 women at 27 centers throughout the country. After the study, a number of questions were raised about the presumed value of early detection and whether mammography, in its ability to pick up very tiny tumors, may have revealed more than really is necessary to know. New categories of tumors, sometimes called "minimal cancers" are detected. Nearly 70 women in the program with these minimal cancers were misdiagnosed, the tumors actually being benign. Another 22 women had what pathologists called "unclear cancers," and another 374 women had minimal cancers that examiners agreed were malignant. Large numbers of women, as the result of the screening experiment, had surgery. Controversy emerged as regards the usefulness of mammography as a screening technique and, to some degree, continues into the 1980s. The objections previously mentioned are still raised by some critics. Mammography is no longer used as a mass population screening technique.

During the 1975–1977 period, the national cancer institutions reevaluated the breast cancer detection demonstration project and have since published the following series of guidelines for professionals in the field:

1. Mammography screening should be available to women over 50.
2. For women between the ages of 40 and 49, mammography should be used only for those who have had breast cancer, or who have a mother or sister who has had the disease.
3. For women between the ages of 35 and 39, mammography should be used only if a woman has previously had cancer in one breast.
4. Thermography, the examination of breast tissue by heat rather than X-ray has not been proved to be valuable, but studies of the procedure should continue.
5. Mammography should never be used to screen women under 35.

6. Mammography should be used for women of any age to aid in the diagnosis of a suspected cancer.

Surgery. In the United States, about 107,000 cases of breast cancer are diagnosed each year. Of these, about 85% are early cancer (Stages I and II). Yet approximately 30% of the mastectomies performed are radical ones, involving removal of underlying chest muscles. In a 5-year study conducted by the University of Pittsburgh, involving some 1680 patients who were randomly assigned simple or radical mastectomies, there was no significant differences detected between the survival rates of the two groups. These patients included some women whose cancers had spread to lymph nodes and who had only simple mastectomy plus radiation. Similar findings were reported by the national cancer institute in Italy. Statistics continue to be gathered.

As in the case of differing opinions in connection with mammography, some resolution regarding surgical procedures also has come out of consensus panels, made up of professionals in the field. The current essence of consensus is:

1. For Stage I and selected cases of Stage II breast cancer, total mastectomy with axillary dissection is suggested. The Halsted radical mastectomy is not advocated.
2. In advanced Stage II and in Stage III and IV diseases, radical surgery is not indicated simply because current techniques are effective in only a small percentage of patients. In individuals with a large primary lesion, external radiation therapy supplemented by iridium needle implantation is suggested.
3. A two-step procedure should be followed in nearly all cases: (a) study permanent sections of a diagnostic biopsy specimen; and (b) select the best therapeutic approach.
4. Therapeutic alternatives should be fully discussed with the patient.

It also may be observed that, until the results of trials become available, postoperative radiotherapy cannot be suggested as standard practice. Similarly, long-term studies must be completed before the lesser surgical procedures, with or without radiotherapy, can be considered a part of standard therapy in certain case categories.

Adjuvant Chemotherapy (*Chemoprophylaxis*). Over the past several years, a number of studies have been made concerning the effectiveness of postoperative and postirradiation chemotherapy to prevent recurrence of the disease. This concept dates back at least to 1957, when a drug known as thiotepa was administered after radical mastectomy. Surgeons had learned that, in the course of the operation, some cancer cells spilled into blood and lymph and thiotepa was administered on the day of surgery and for a few days after to eliminate those circulating cells. A study made at the end of a 5-year trial of this procedure showed no difference in recurrence rates between treated and control patients. In the mid-1970s, a study (the Bonadonna Study) was conducted by a cancer institute in Italy, for which three chemotherapeutic agents were administered for a year after surgery. The early results were quite positive, but, as one scientist suggested, possibly only a delay in recurrence of the disease is being observed. Long-term results will require much more time. Similar studies in the United States preceded the Bondonna Study and were reported in positive terms. One of these was a controlled trial of the use of chemotherapy as an adjuvant to surgery in women whose cancer had spread to the lymph nodes. In this study, half of the women received L-PAM (L-phenylalanine mustard) and the other half received placebos. After three years, the study report stated that treatment failures (recurrence of disease) occurred in 22% of 108 patients who received placebo and in 9.3% of 103 women given L-PAM. Controlled trials involving other drugs, singly and in combination, are continuing.

Palliation. In those unfortunate cases where metastatic breast cancer will not respond to surgery, irradiation, and adjuvant chemotherapy, palliation (supportive relief of symptoms) is the remaining objective. In these cases, three traditional approaches have been used—*chemotherapy; endocrine ablation,* which includes adrenalectomy and hypophysectomy (excision of the hypohysis cerebri, an epithelial body of dual origin at the base of the brain, attached by a stalk to the hypothalamus); and *hormone therapy.* The most effective approach depends upon many factors, such as the patient's age and menstrual status, the site of metastases, the presence of life-threatening consequences (hypercalcemia; vertebral collapse), and whether or not there is estrogen-binding receptor protein.

Male Breast Cancer. Cancer may also occur in the male breast, but this disease accounts for only about 1% or less of all breast cancers. The disease usually occurs in men between 54 and 60 years of age. The first symptom is a lump in the breast. Later symptoms include ulceration of the skin over the breast and enlarged lymph nodes in the armpit, the latter indicating an advanced stage of the disease. Upon early discovery, surgery may eradicate the growth.

Paget's Disease. Described in the entry on **Dermatitis and Dermatosis,** one form of this disease is malignant involvement of the nipple. It usually affects only one breast and is characterized by eczematoid redness, cracks, or ulcerations, and tenderness of the nipple, from which there is often an abnormal discharge. An associated breast cancer may antedate the nipple symptoms, but is usually evident a year or so after the disease of the nipple is first noticed. Persistent abnormality of the nipple, therefore, must be regarded with suspicion.

Lung Cancer. Lung cancer is a major medical problem of the western world, particularly among heavy smokers and persons who are exposed to toxic air pollutants for long periods. While the incidence of lung cancer has increased dramatically, the therapy for this disease, as contrasted with some other forms of cancer, has not kept pace. Among the major cancers, lung cancer continues to present a poor prognosis, thus giving emphasis to preventing exposure of the lungs to carcinogens, an elective option among smokers, and an increasing problem for industry and commerce in taking measures to protect workers and communities from carcinogen-containing vapors and fumes. About 100,000 deaths per year in the United States are attributed to lung cancer. It is the leading cancer in men and is rapidly increasing in frequency among women. Incidence in the early 1980s is nearly six times that of the 1950s; over 20 times that of the 1930s.

The primary types of lung cancers are bronchogenic carcinoma, bronchiolar or alveolar carcinoma, and pleural mesothelioma. See **Respiratory System.** Bronchogenic carcinomas account for 90% of all lung cancers. These are (1) *squamous* (scaly or platelike), which accounts for half of the bronchogenic carcinomas; (2) *oat cell* (undifferentiated small cell) cancer, which accounts for 20% of bronchogenic carcinomas; (3) *adenocarcinoma* (15% of cases), which is a malignant tumor composed of glandular tissue; and (4) *undifferentiated large cell tumors* (15% of cases). Each of these different forms of bronchogenic carcinomas follow different histopathologic pathways and require different treatment. Thus, correct diagnosis is extremely important. Statistics indicate that only about 5% of lung cancer patients are "cured" on the basis of 5-year survival potential. Surgery and irradiation until recently were the principal procedures used. As of the early 1980s, surgery has been notably *ineffective* in connection with the oat cell form of the disease. In contrast, this is the only major type of lung cancer that responds to single-agent chemotherapy. In about half these cases, there has been positive response to aklylating agents, the nitrosoureas, procarbazine, and vincristine. However, the response is only over the short term, with no impact on changing the ultimate course of the disease. Greater success has been obtained with multiple-drug therapy, i.e., the use of three-to-five-drug combinations. This approach has produced positive responses in a number of patients with the non-oat cell type of cancers. One combination found to have some success is cyclophosphamide, methotrexate, and vincristine. Other drugs used in various combinations include the nitrosoureas, procarbazine, blemomycin, and doxorubicin. Most physicians prefer that combination chemotherapy be under the supervision of medical oncologists and preferably as a part of clinical trials.

Many diagnosticians and physicians use a staging system for identifying the severity of lung cancer at time of diagnosis, as an aid to selecting the best therapy and for estimating prognosis. As with breast cancer, the system is based upon three factors—tumor size, involvement of nodes, and extent of metastases.

Stage I—Presence of a *tumor* less than 3 centimeters (about 1¼ inch) in diameter. *Nodes* are absent. *Metastasis* is negative.

Stage II—*Tumor* is present and larger than 3 centimeters (about 1½ inch) in diameter. *Nodes* are present. *Metastasis* is negative.

Stage III—Tumor is larger than in Stage II. Lymph *node* involvement is extensive. *Metastasis* is positive. Distant sites involved may include the vertebrae and pelvis.

In untreated lung cancers, systemic involvements may include bone lesions, airway obstruction, brain metastases, and neuropathy (inflammation of nerves due to pressure). Radiation therapy can relieve most of these symptoms for a while. In oat cell carcinoma, where life expectancy is one to two years, the incidence of brain metastases can be considerably reduced through the use of whole-brain radiation.

Where diagnosis permits, radiotherapy is usually the therapy of choice and, of course, is the logical option where patients have unresectable tumors. The decision pertaining to surgical resection (thoracotomy) is critical and often difficult to make, both for the physician and the patient. There are often borderline situations where the full extent of the disease is not known until the operative procedure is underway. Better success in lung cancer therapy is not achieved because in about half of the cases, diagnosis comes too late—at a time when the cancer is widespread—and in about 30% of cases resection is not possible at thoracotomy. Where the disease is diagnosed very early and the tumor is highly localized, the 5-year survival rate may be as high as 30%.

Gynecologic Cancers. The principal cancers of the female genital tract involve the uterus, cervix, and ovary.

In terms of improvements in cancer therapy, the management of localized carcinomas of the uterus ranks high. There has been a threefold reduction in deaths from cancer of the uterus since the 1930s. Many physicians utilize a staging classification system for cancer of the endometrium (mucous membrane lining of the uterus) as well as for other gynecologic cancers. These stages range in severity, depending upon the spread of the disease. For example, Stage 0 describes a tumor in situ, whereas Stage IV describes a situation where the carcinoma extends beyond the true pelvis or involves mucosa of bladder or rectum. Chemotherapy is the treatment of choice. Drugs used include progestational hormones. In metastatic endometrial carcinoma, a combination chemotherapy may be used.

Major improvements in the treatment of cancer of the cervix occurred during the period 1940–1960. The survival rate has remained relatively steady in recent years. Improvements are attributed to better sexual hygiene, widespread use of the Pap smear, early detection, and more effective surgical and radiation procedures. The Pap test involves the microscopic examination of cells collected from the vagina. These cells are shed from the uterus into the vagina as part of the normal life process. If examination of the smear reveals any abnormal cells, bits of tissue are taken from the cervix for further study. This widely accepted procedure, promoted for many years as part of an annual examination, was reevaluated in the late 1970s. Most authorities continue to place a respected value on the Pap test, but question the need for screening on an annual basis. For example, a Canadian task force has recommended a Pap smear for all women over 18 who have had sexual intercourse. If the first test is negative, another smear should be taken one year later to guard against any errors in the first test. If the two tests are negative, the woman is considered to be low-risk. She should continue to have Pap tests, but at 3-year rather than 1-year intervals up to age 35 years and at 5-year intervals up to age 60 years. The task force further suggested that women over 60 years old who have had repeated normal smears need have no more inasmuch as they are unlikely to develop invasive cancer during their life-span. More detail concerning this reevaluation can be found in the third Marx (1979) reference listed. The most recent approach to chemotherapy for cancer of the cervix involves the use of combinations of drugs rather than single-agent chemotherapy. The disease apparently does not respond to hormonal manipulation.

In about 50% of cases, cancer of the ovary responds to chemotherapy. There is a trend toward the use of combinations of drugs. Surgery is indicated frequently.

Prostatic Cancer. Carcinoma of the prostate ranks second as a cancer-causing death in men. Approximately 20,000 men die from this disease each year in the United States. The disease occurs at a higher frequency in black than in white men and is considered uncommon in Orientals. Sometimes the first indication of a cancerous prostate will be a routine physical examination of the rectum. Surgery is the usual procedure of choice. There are numerous variations in the profile of this disease, however, and radiation therapy can be used effectively in some patients. Radiation therapy is particularly useful for palliation of local pelvic or bone metastases. Hormonal manipulation is effective for metastatic cancer of the prostate. Chemotherapy has not been widely used in the treatment of prostatic cancer.

Skin Cancers. Whereas benign tumors of the skin may attain a certain size and then cease growth, a cancer will continue growing indefinitely, although its rate of growth may range from slow to rapid and may vary through the years. Uncontrolled skin cancer may eventually cause the death of the patient. The growth may become so large that it can destroy the nearby blood vessels supplying other parts of the body and also it may metastatize to more vital organs which, in turn, may be so damaged that the patient dies.

Most persons with skin cancer can be treated successfully if treatment is commenced early. A preliminary biopsy almost invariably is performed because of the difficulty in distinguishing skin cancer from other skin diseases. A sore or growth that persists for more than 2 to 3 weeks should be regarded with suspicion. A physician should be consulted when a sore or lump shows signs of rapid growth, or spreads, or fails to heal completely.

There are several classifications of skin cancer. Carcinoma of the skin occurs most frequently among fair-skinned persons who spend much time outdoors, particularly in a dry, sunny climate. Excessive exposure to sunlight appears to be a major factor in producing most carcinoma of the skin. There are two types of carcinoma of the skin. *Basal cell carcinoma* accounts for over half of the cases of skin cancer. The condition is made up largely of cells resembling those of the innermost cells of the stratum mucosum, the deepest layer of the epidermis. The growths occur more frequently on the face than on other areas of the body. A typical basal cell carcinoma is a hard pinkish or waxy growth, which may spread slowly, or show signs of healing with formation of a tight cluster of similar nodules around it. Its appearance is usually altered by accidental injury, bleeding, and scaling. Basal cell carcinomas may become pigmented, so that they appear more like a dark malignant mole.

Squamous cell carcinoma begins in the cells of the outermost layer of the stratum mucosum. In its early stages, a squamous cell carcinoma looks much like a basal cell carcinoma. Skin cancer of this type is found most often in persons who have been long exposed to the sun and wind. The growths usually are found on the face, the ears, and the back of the hands. Squamous cell carcinomas vary greatly in behavior; some growing rapidly, others slowly. Squamous cell lesions usually grow faster than the basal cell variety and have a tendency to metastatize. There are two varieties of squamous cell carcinoma, the *ulcerating* type and the *papillary* type (cauliflower-like in shape and structure).

Intraepidermal and superficial carcinomas of the skin are those in which the cancerous nature of the growths may remain confined to the skin for many years without metastasis or damage to nearby tissue. These diseases generally respond well to treatment.

When cancers in internal areas of the body metastatize, they may appear on the skin and are known as *metastatic carcinoma of the skin.* Secondary skin cancers usually are found on the chest, under the arms, on the abdomen, or around the genitals. The nodules may range from ivory to red in color and sometimes grow rapidly.

Surgical excision, radiation therapy, or chemotherapy are used in the treatment of skin cancer. In the case of surgical removal, skin grafting may be used if a large defect is left. Podophyllin resin, fluorouracil, colcemid, and methotrexate are among compounds which appear to selectively destroy some kinds of carcinomas while leaving the surrounding skin intact. In the treatment of malignant melanoma, extensive removal of the growth and the skin surrounding it, as well as the lymph glands that drain the involved area, usually is indicated.

Cancer arising in the skin from layers below the epidermis is termed *sarcoma.* Unlike carcinoma, the sarcomas of the skin occur in young persons. The different types of sarcoma are named in accordance with the types of cells principally involved. *Fibrosarcoma* is so named because it involves fibrous connective tissue; *fibroneurosarcoma* is made up mainly of nerve cells. Sarcomas may also arise from muscle, blood and lymph tissue, fat, and other tissues. Treatment is largely surgical, since most of these lesions are resistant to irradiation. See also **Dermatitis and Dermatosis.**

Brain Tumor. With modern brain surgery, a benign tumor can be completely removed in many cases and, when this is possible, the outlook for recovery is good. Unfortunately, the signs and symptoms of brain tumor are quite variable and depend upon the portion of the brain that is compressed or disturbed by the expanding new growth. Headache is only occasionally the first, and not the most frequent symptom. Other symptoms of brain tumor are various types of visual disturbance, incoordination, weakness, and paralysis which may affect one arm, one leg, or half the body, depending upon tumor location. Contrary to popular conceptions, personality or mental changes are the exception.

Brain tumors occur in any location of the brain within the cranial cavity and also occur at any age, including very early childhood. *Arteriography*, a technique in which the blood vessels are injected with an opaque contrast medium, is sometimes used for diagnosis. Radiograms are made of the injected area to detect the flow of the contrast medium through and around the suspected lesion. Various types of tumors and other disorders are thereby differentiated. *Radiotopography*, a technique wherein a radioisotope is injected into the area under study and a scintillation camera measures the distribution of the isotope in and around the suspected lesion, also is used.

Aside from tumors which originate in the brain, tumors arising in various parts of the body may spread to the brain by way of the blood or lymph system. Primary tumors of the lung are the most frequent to metastatize to the brain; breast tumors and gastrointestinal tumors somewhat less frequently spread to the brain, as do melanotic, thyroid, and other tumors. See also **Brain and Nervous System; Brain Disorders;** and **X-Ray Scanner (CAT).**

Other Cancers. Hodgkin's disease and leukemia are described in separate entries in this encyclopedia. Also, in many descriptions of body organs, diseases including cancer are described.

References

Baltimore, D.: "Viruses, Polymerases, and Cancer," *Science*, **192**, 632–636 (1976).

Beckler, F. F. (editor): "Mechanisms of Cancer Invasion and Metastasis," Plenum, New York, 1975.

Brouty-Boÿe, D., and B. R. Zetter: "Inhibition of Cell Motility by Interferon," *Science*, **208**, 516–518 (1980).

Cadman, E., Heimer, R., and L. Davis: "Enhanced 5-Fluorouracil Nucleotide Formation after Methotrexate Administration: Explanation for Drug Synergism," *Science*, **205**, 1135–1137 (1979).

Cairns, J.: "Cancer—Science and Society," Freeman, San Francisco, 1978.

Cameron, E., and L. Pauling: "Cancer and Vitamin C," Linus Pauling Institute, Menlo Park, California, 1980.

Carter, L. J.: "How to Assess Cancer Risks," *Science*, **204**, 811–816 (1979).

Darling, D. B. (editor): "Radiographic and Fluoroscopic Procedures," 3rd edition, Charles C. Thomas, Springfield, Illinois, 1979.

Dulbecco, R.: "From the Molecular Biology of Oncogenic DNA Viruses to Cancer," *Science*, **192**, 437–440 (1976).

Enstrom, J. E., and D. F. Austin: "Interpreting Cancer Survival Rates," *Science*, **195**, 847–851 (1977).

Ernester, V. L., Selvin, S., and W. Winkelstein, Jr.: "Cohort Mortality for Prostatic Cancer Among United States Nonwhites," *Science*, **200**, 1165–1166 (1978).

Fewer, D., Wilson, C. B., and V. A. Levin: "Brain Tumor Chemotherapy," Charles C. Thomas, Springfield, Illinois, 1976.

Fidler, I. J., and M. L. Kripke: "Metastasis Results from Preexisting Variant Cells within a Malignant Tumor," *Science*, **197**, 893–895 (1977).

Hamburger, A. W., and S. E. Salmon: "Primary Bioassay of Human Tumor Stem Cells," *Science*, **197**, 461–463 (1977).

Hiatt, H. H., Watson, J. D., and J. A. Winsten: "Origins of Human Cancer," 3 Volumes, Freeman, San Francisco, 1980.

Holden, C.: "Cancer and the Mind: How Are They Connected?" *Science*, **200**, 1363–1369 (1976).

Holden, C.: "Albert Szent-Györgyi, Electrons, and Cancer," *Science*, **203**, 522–524 (1979).

Hoppe, P. C., and C. C. Croce: "Chimeric Mice Derived from Human-Mouse Hybrid Cells," *Proc. Nat. Acad. Sci. U.S.A.*, **75**, 4, 1914–1918 (1978).

Horoszewicz, J. S., Leong, S. S., and W. A. Carter: "Noncycling Tumor Cells are Sensitive Targets for the Antiproliferative Activity of Human Interferon," *Science*, **206**, 1091–1093 (1979).

Illmensee, K., and L. C. Stevens: "Teratomas and Chimeras," *Sci. Amer.*, **250**, 4, 120–132 (1979).

Knight, E., Jr., et al.: "Human Fibroblast Interferon: Amino Acid Analysis and Amino Terminal Amino Acid Sequence," *Science*, **207**, 525–526 (1980).

Kolata, G. B.: "Genes and Cancer: The Story of Wilms Tumor," *Science*, **207**, 970–971 (1980).

Lynch, H. T.: "Cancer Genetics," Charles C. Thomas, Springfield, Illinois, 1976.

Matthews, D. A., et al.: "Dihydrofolate Reductase: X-ray Structure of the Binary Complex with Methotrexate," *Science*, **197**, 452–455 (1977).

Marx, J. L.: "Chemotherapy: Renewed Interest in Platinum Compounds," *Science*, **192**, 774–775 (1976).

Marx, J. L.: "RNA Tumor Viruses," *Science*, **199**, 161–164 (1978).

Marx, J. L.: "Crown Gall Disease: Nature as Genetic Engineer," *Science*, **203**, 254–255 (1979).

Marx, J. L.: "Interferon (Part I)," *Science*, **204**, 1183–1186 (1979).

Marx, J. L. "Interferon (Part II)," *Science*, **204**, 1293–1295 (1979).

Marx, J. L.: "The Annual Pap Smear," *Science*, **205**, 177–178 (1979).

Maugh, T. H., II: "Cancer Chemotherapy," *Science*, **194**, 310 (1976).

Maugh, T. H., II: "Biochemical Markers: Early Warning Signs of Cancer," *Science*, **197**, 543–545 (1977).

Maugh, T. H., II: "Cancer Tests Look for a Passing Grade," *Science*, **211**, 909–910 (1981).

Nicolson, G. L., and K. W. Brunson: "Specificity of Arrest, Survival, and Growth of Selected Metastatic Variant Cell Lines," *Cancer Research*, **38**, 11, 4105–4111 (1978).

Nicolson, G. L.: "Experimental Tumor Metastasis: Characteristics and Organ Specificity," *Bioscience*, **28**, 7, 441–447 (1978).

Nicolson, G. L.: "Cancer Metastasis," *Sci. Amer.*, **240**, 3, 66–76 (1979).

Nowell, P. C.: "The Clonal Evaluation of Tumor Cell Populations," *Science*, **194**, 23–28 (1976).

Rabinovitz, M., Uehara, Y., and D. T. Vistica: "Differential Competition with Cytoxic Agents: An Approach to Selectivity in Cancer Chemotherapy," *Science*, **206**, 1085–1087 (1979).

Rather, L. J.: "The Genesis of Cancer," Johns Hopkins Univ. Press, Baltimore, Maryland, 1979.

Russell, L. B. (editor): "Genetic Mosaics and Chimeras in Mammals," Plenum, New York, 1980.

Samulski, T., and P. N. Shrivastava: "Photoluminescent Thermometer Probes: Temperature Measurements in Microwave Fields (Hyperthermia)," *Science*, **208**, 193–194 (1980).

Schimke, R. T.: "Gene Amplification and Drug Resistance in Cultured Murine Cells," *Science*, **202**, 1051–1055 (1978).

Schreiber, M. H.: "Introduction to Diagnostic Radiology," Charles C. Thomas, Springfield, Illinois, 1980.

Schultz, R. M., Papamathekis, and M. A. Chirigos: "Interferon: An Inducer of Macrophage Activation by Polyanions," *Science*, **197**, 674–676 (1977).

Sherman, M. I.: "Teratocarcinoma Cells and Normal Mouse Embryogenesis," in "Concepts in Mammalian Embryogenesis," (M. I. Sherman, editor), MIT Press, Cambridge, Massachusetts, 1978.

Sklar, L. S., and H. Anisman: "Stress and Coping Factors Influence Tumor Growth," *Science*, **205**, 513–515 (1979).

Staff: "DNA Repair: New Clues to Carcinogenesis," *Science*, **200**, 518–521 (1978).

Staff: "Interfering with Cancer (Interferons)," *Sci. Amer.*, **240**, 4, 94–103 (1979).

Staff: "How Interferon Interferes," *Sci. Amer.*, **240**, 5, 90 (1979).

Staff: "Interferon Nomenclature," *Amer. Soc. Microbiol. News*, **46**, 9, 466–467 (1980).

Stewart, W. E., II: "The Interferon System," Springer-Verlag, New York, 1979.

Stewart, W. E., II: "First International Congress for Interferon Research," Washington, D.C. (November 9–12, 1980).

Stringfellow, D. A.: "Prostaglandin Restoration of the Interferon Response of Hyporeactive Animals," *Science*, **201**, 376–378 (1978).

Ting, C. C., Tsai, S. C., and M. J. Rogers: "Host Control of Tumor Growth," *Science*, **197**, 571–573 (1977).

Wallach, D. F. H.: "Membrane Molecular Biology of Neoplastic Cells," Elsevier, Amsterdam, 1975.

Wang, B. S., Onikul, S. R., and J. A. Mannick: "Prevention of Death from Metastases by Immune RNA Therapy," *Science*, **202**, 59–61 (1978).

Weaver, R.: "The Cancer Puzzle," *National Geographic*, **150**, 3, 396–399 (1976).

Weiss, L.: "A Pathobiologic Overview of Metastasis," *Seminars in Oncology*, **4**, 1, 5–19 (1977).

CANCER RESEARCH. Although cutting a wide swathe, research activities in the field of cancer fall simply into four categories: (1) What causes cancer? (2) Why do cancerous cells proliferate? (3) How do we test for cancer? and (4) How do we cure cancer?

Cancer in its diverse forms stems from mutagenic alterations of cellular deoxyribonucleic acid (DNA). DNA is not a rigid, unassailable structure, but undergoes frequent changes in its cellular environment.

Enzymes in the cell not only effect DNA replication, but also its repair, recombination, and modification. When these enzymes fail in their function, the integrity of the *genome*[1] is not maintained and, when such failures are preprogrammed by inherited genetic information, the consequences can be, and usually are, disastrous.

Three main theories of cancer genesis have been promulgated, each attempting to explain the conditions permitting or leading to cellular failure. None of these hypotheses is completely satisfactory. The somatic mutation theory assumes that the genes are not always produced in a normal, orderly fashion but, from time to time, one gene may develop a flaw which has the power to disorganize cellular functions. If this abnormal gene survives, the fault is transmitted to its descendants, leading to production of the cancerous state. An alternative theory assumes that a gene capable of causing cancerous growth is normally present, but is repressed. When restriction is removed, the cell proliferates into the cancerous state. A variation of this theory suggests that the abnormal growth characteristic of cancer cells is a reversion to an earlier evolutionary stage of growth (the α state) where cells merely proliferated rather than performed specific tasks. Regardless of the hypothesis, the ultimate trigger is *deformation of the DNA structure*. Since a single change in DNA structure may be harmful or even lethal, it is not surprising that an elaborate array of enzymatic mechanisms has evolved for repairing damaged DNA. One pathway to cancerous cell growth appears when the DNA dimerizes by cross-linking through the double helix forming bonds sufficiently strong that the enzymes responsible for repair, e.g., DNA topisomerase II which inserts negative supercoils into double-stranded DNA, are unable to become effective. Another route, as recent data suggest, requires the existence in solution of a left-handed double helix (Z-DNA) in which there is an increased accessibility of the guanine structure. There are many carcinogens which are specific for sites in the guanine structure. Any dimerization or cross-linking distorts the DNA molecule and interferes with normal synthesis of the daughter molecules.

But it is known that alterations in cell chemistry can be effected in many ways—for example, by carcinogenic chemicals, ionizing radiation, and viral activity—with some definitive evidence being seen for hereditary factors.

From 70 to 90% of human cancers are known to be determined by environmental factors, although in only a small proportion of cases do these operate as "pure" carcinogens. The known chemical carcinogens have a wide range of structures, but it has been ascertained that, when they have passed through the multistage metabolic processes, their derivatives all possess the common feature of being strongly electrophilic. Ionizing radiation is another attested carcinogen, producing skin cancers from exposure to excess sunlight or ultraviolet radiation and leukemia from atomic radiation and radioactive fission products. At the molecular level and below, ionizing radiation creates many free-radical species, and these are recognized as being of importance in an increasing number of cellular disturbances. Electron spin resonance studies have shown that resonance frequencies differ in cancerous and noncancerous tissues; this undoubtedly reflects major metabolic differences between the metalloproteins of normal and diseased tissues. From the physicochemical aspect, therefore, there is much to suggest that the strongly electrophilic carcinogenic metabolites and radiation-produced free radicals have one thing in common—sufficient chemical, or rather, electronic energy to bring about covalent bridging reactions with cellular DNA.

In contrast, virus-associated cancers have not been so readily reduced to a basic chemistry. In some animals, leukemia and some neoplasms are undoubtedly of viral origin. In humans, except for the development of warts by the papilloma virus, the association of cancer with viruses can only be considered suggestive. Much has been made of the relationship of the Epstein-Barr virus (EBV) with squamous cell carcinoma of the nasopharynx (NPC), but most recent studies indicate that the good relationship seen between EBV antibodies and undifferentiated NPC disappears as the NPC becomes well differentiated. Even the classical association of EBV with Burkitt's lymphoma increasingly appears to be adventitious, since the EBV genome has been found in the parotid and salivary glands of healthy people. And again, despite the increasing amount of data showing the presence of herpes simplex virus type 2 (HSV2)-induced antigens in cervical carcinoma in situ, definitive association of HSV2 with cervical and vulval carcinoma has yet to be demonstrated.

Investigations of the frequency with which hepatitis B antigen is to be found in the serum of patients with hepatic (liver) cancer have given conflicting results, and evidence relating to the role of viral infection in hepatic carcinogenesis must be regarded as debatable. Similarly questionable is the association of Kaposi's sarcoma to cytomegalovirus. That some products of viral chemistry, or even viruses themselves, can affect DNA polymerization sufficiently to induce the cancerous state is conceivable, but, in the absence of definitive data, such chemical reaction remains dubious. Extension of the electron spin resonance studies previously mentioned to cover virus-infected tissues might well provide some deeper insight to the basic chemistry involved.

The old question of inheritability of cancer is becoming more answerable. Definitive evidence has been presented for a heredity factor in retinoblastoma, familial polyposis of the colon, leukemia, and chronic lymphatic leukemia. Race has been increasingly determined as significant in NPC, breast, and liver cancers, and again in Kaposi's sarcoma. Chromosomal translocation has been established in many types of cancer, including leukemia and adenocarcinoma. It would seem, therefore, that the disease of cellular proliferation which manifests itself as cancer has as its base a genetic inheritance of chromosome variation, implying innate susceptibility to development of specific cancers with the cellular breakdown, e.g., by free radicals, then taking the path dictated by chromosomal variation.

The fact that some malignancies show no visible cytogenic alterations indicates that chromosomal change is not a prerequisite for the neoplastic state. But there is considerable evidence relating chromosomal changes in a nonspecific fashion to the malignant condition. There has been found, for example, an intriguing correlation between chromosomes known to possess the immunoglobulin heavy chain gene and the involvement of these in structural changes in malignant cells producing IgG and translocations involving chromosome 14.

Most cancer patients are not killed by their primary tumor. They succumb instead to metastases—colonies disseminated to anatomical sites distant from the original growth. The initial spread of growth into surrounding tissues appears to be related to various combinations of mechanical cell motility and enzymatic mechanisms. The external cell membrane is of critical importance and changes in this, mediated through cell-to-cell contact, cause deficiencies in normal growth control mechanisms. Malignant cells induce a dramatic change in the adhesion of endothelial cells—these retract, leaving cell-free spaces through which the tumor cells can encroach or escape. This encroachment is inevitably associated with further growth of the neoplasm, which then requires further fuel for more growth. The blood supply of an established organ is very stable, but that of a tumor is transient. It must recruit new blood vessels to maintain and develop its growth. For this, it excretes a tumor angiogenesis factor (TAF) which induces existing blood vessels to grow toward and infiltrate the tumor. Isolation, purification, and chemical identification of TAF would provide not only a strong weapon against cancer, but one of use in noncancerous repair processes. The TAF has many inhibitory agents which restrict tumor growth and starve out both the original tumor and its metastases, but their characterization and purification are proving extremely difficult.

Extension of the malignancy can occur through several mechanisms—lymphatic drainage, seeding, and passage through the blood stream. The last mentioned is probably the major factor and is probably permitted by enzymatic breakdown of the blood vessels attracted by TAF. But blood-borne metastasis is a highly selective process, and only a small percentage of malignant cells entering the blood stream survive to form distant metastases. An area of continuing investigation is that of seeking the reason for preferential metastasis: why does cancer of the prostate metastatize to bone, or breast cancer to brain or lung, or lung cancer to brain and adrenals? The answer

[1] Because the sexual process results in production of a fertilized egg or zygote, all somatic cells of the organisms derived through mitotic division of the zygote contain two chromosome sets or a diploid complement of chromosomes. Each set of chromosomes is called a *genome*. One genome is paternal origin; the other of maternal origin. See also **Deoxyribonucleic Acid (DNA); Nucleic Acids and Nucleoproteins; Somatoplasm;** and **Zygote.**

lies not in hematologic or lymph pathways, but apparently in some feature of the tumor cell membrane. With identification of this feature might come an ability to block metastatic growth of the neoplasm.

Since the development of neoplasms and their metastases challenges the body's defense mechanisms, we should expect the immune system to mount a response and, indeed, this is sometimes seen with effect—the tumor regressing and disappearing without metastatizing and without major influence of therapy. But in general this does not happen. Whether the tumor induces the production of a blocking mechanism which protects it against the cell-mediated immunity of the host or whether immunosuppression is effected by the DNA damage is unclear. Cancer patients not only have depressed cellular responses, but frequently a reduced capability to generate cell activity, which implies a general defect in this aspect of immune regulation. Since prostaglandin inhibits lymphocyte cytotoxicity directed against tumor cells, the production of prostaglandin E_2 by neoplasms may be the result of a defense mechanism whereby the tumor cells can subvert the immune response of the host.

The major focus of cancer immunology has now shifted away from arguments about immunosurveillance to the more specific question of tumor-specific antigens, but their existence remains unproven. Rudimentary classification of surface antigens is emerging, but these are restricted to individual tumors and more detailed knowledge is required before we can even consider an approach to immunological control. Probably the most basic response is that chemical, enzymatic, and immunological studies suggest that most neoplasms have a unicellular origin and clonal growth patterns. This aneuploidy and resulting cytogenetic individuality probably allow penetration of the immune defense bulwarks.

The final determination of tumor malignancy is cell proliferation and, although observation of this remains in the domain of microscopic examination of biopsy specimens or—as in the Pap test—scrapings of tissue, as increased knowledge of the biochemistry of cancer cells is beginning to permit the development of several chemical tests for indicating the bodily presence of cancerous growth. From this have come the alkaline phosphatase test for prostatic cancer, radioimmunoassay for carcinoembryonic antigen for intestinal cancer, and selective detection by monoclonal antibodies of tumor-associated antigens which are expressed on the cell surface. Metastases produce abnormal δ-glutamyl transpeptase, and cancerous tissue generally has a much decreased rate of lipid peroxidation and increased production of prostaglandin. But no one of these tests can be considered exclusively determinative of the presence of cancerous tissue. The various methods of mammography, radiography, and CAT scanning can indicate the existence of radio-opaque centers of growth, but it is microscopic pathology which makes the final diagnosis.

In this entry, we do not consider surgical or radiotherapeutic aspects of treatment research except to note that the radical surgery which was advocated even as recently as a decade ago has now become a regimen of conservative surgery allied with radio- and chemotherapy. See also **Cancer.**

The essential aim of chemotherapy is to introduce a cytotoxic chemical into the neoplastic cells and not into normal tissue. Unfortunately, the biochemistries of malignant and nonmalignant cells are so similar that the borderline of selective cytotoxicity is a very thin one. Nevertheless, over the past few years appreciable advances have been made. Anthracyclines complexed with DNA have been developed as an approach to selective drug delivery, based on the hypothesis that tumor cells have a higher degree of endocytosis than do normal cells leading to a greater uptake of the complex and subsequent release of the free drug within the tumor cell. Many variations and analogs of the anthracyclines are being screened for activity. Unfortunately, too many of these compounds have an unpleasant degree of cardiotoxicity and cause tissue necrosis if they are extravasated from the blood stream. Similar complexes of retinoids with vitamin A appear to possess a fairly pronounced anticarcinogenc action without the undue side-effects. Of particular interest is research on congeners of *cis*-platinum dichlorodiammine—*cis*-$[Pt(NH_3)_2Cl_2]$—which can cross-link the two strands of the DNA double helix and prevent replication of the malignant genome. Inasmuch as "*cis*-platinum" appears to be a very effective chemotherapeutic agent and yet is not the most active complex of this type, the future seems bright for treatment of this nature.

Combination or adjuvant chemotherapy is still the mainstay of cancer treatment, although this is still somewhat restricted by the cyto- and cardiotoxicity of the drugs. However, current research is beginning to separate the molecular basis for tumor cytotoxicity from that producing injury to normal tissues, especially the myocardium. Some progress has been made and new compounds with enhanced therapeutic indices have been recognized which appear to hold promise for the future treatment of leukemia, lymphosarcoma, and breast malignancies. The most recent weapons to be deployed against cancer are, however, of cellular origin—*monoclonal antibodies* and *interferon.*

Despite lack of success of the immune surveillance theory, immunological studies on the neoplastic state have developed the concept of *hybridoma,* in which mouse cells making antibodies against specific antigens are fused with tumor cells to yield hybrid cells possessing the immortality of the tumor cell and also the ability to produce antibody. Using these hybridomas, one can obtain large quantities of antibodies primed against a single enemy molecule or antigen. Although only a limited number of cells lead to hybridomas that actively secrete specific antibodies, the possibility exists of obtaining human B cell hybrids that continuously secrete human antibodies. Following successful application in animals, monoclonal antibodies have been used in humans with some success in obtaining cancer regression. Further work is needed, however, to determine how to use the hybridoma technique to produce immunoglobulins specific for specific types of cancer cell antigens.

Viral infections of cells produce *interferon*—in fact, three interferons (IFNs): IFN *a* obtained from leucocytes; IFN *b* obtained from fibroblasts; and IFN *immune* obtained from T cells. The cell proliferative inhibition characteristics of the IFNs have created great interest in cancer research and therapy, but activity has been hampered by inability to purify large amounts of these proteins from human cells. The therapeutic work which has been done has shown that IFN *a* inhibits the development of several cancers in animals. In humans, responsiveness to IFN treatment has been shown in osteogenic sarcoma, multiple myeloma, breast cancer, and certain types of leukemia and lymphoma. Evidence also exists that IFN *b* causes suppression of melanoma.

How interferon acts on neoplastic cells is uncertain, but there is evidence that it mobilizes macrophages and enhances their power to kill cancerous cells. It is also considered that where use of BCG (Bacillus Calmette-Guérin) and other antitumor agents have been successful in cancer therapy, it may be because of their ability to induce the production of interferon. The shortage of supply of interferon has thus far limited study, but the recent total synthesis of human leucocyte interferon genes by DNA recombinant techniques is not only expected to enable the production of IFN *a* in reasonably large quantities, but also to permit, through the construction of hybrid genes, a range of IFN molecules with different pharmacological profiles. The availability of a large pool of synthetic fragments will then greatly extend the range of IFN analogs which can be prepared for examination. See also **Recombinant DNA.**

R. C. Vickery, M.D., Ph.D., D.Sc., Hudson Laboratories, Hudson, Florida.

CANCRINITE. The mineral cancrinite is a complex hydrous silicate (see **Silicon**) corresponding approximately to the formula $(Na, K, Ca)_{6-8}(Al, Si)_{12}O_{24}(SO_4, CO_3, Cl)_{1-2} \cdot nH_2O$. It is hexagonal, with prismatic cleavage; hardness, 5–6; specific gravity, 2.42–2.50; color, white to gray or may be greenish, bluish, yellow, or flesh red; colorless streak; luster, subvitreous to greasy; transparent to translucent. Cancrinite is found only in the nephelite-syenites and related rock types and is commonly associated with sodalite. It is believed to be in part primary, having crystallized direct from the magma, and in part secondary as a result of alteration of nephelite by solutions of calcium carbonate. It is found in the Ilmen Mountains of the U.S.S.R., in Rumania, in Norway, in Canada in Hastings County, Ontario, and in the United States in Kennebec County, Maine. This mineral was named for Count Georg Cancrin, a Russian statesman who died in 1845.

CANCRUM ORIS. Pharyngitis.

CANDELA. Units and Standards.

CANDELABRA TREE. Araucarias.

CANDLEFISH. Smelts.

CANDLE-FLY. A southern name for moth, equivalent to the northern miller or moth-miller.

CANDLENUT. Euphorbiaceae.

CANGA. A Brazilian term for an iron-rich conglomerate or breccia in which the pebbles or anguclasts are hematite and itaberite cemented by hematite or limonite.

CANINES (*Mammalia, Carnivora*). The general organization of Canines may be outlined about as follows:

- True Canines (*Caninae*)
  - Wolves (*Canis*)
  - Jackals (*Thos*)
  - Foxes (*Vulpes*)
  - Fennecs (*Fennecus*)
  - Arctic Fox (*Alopex*)
  - Gray Fox (*Urocyon*)
  - South American Jackals (*Dusicyon*)
  - Maned Wolf (*Chrysocyon*)
  - Raccoon-Dog (*Nyctereutes*)
- False Canines (*Simocyoninae*)
  - Dholes (*Cuon*)
  - Bush Dogs (*Speothos*)
  - Cape Hunting Dog (*Lycaon*)
- Bat-eared Foxes (*Otocyoninae*)

In this large and diversified group we have, of course, the familiar dog (*Canis familiaris*), one of the most highly diversified of animals as a result of long selection and controlled breeding. While the dog has adapted extremely well to humans and vice versa, it is of interest to note that the dog, compared with many other mammals, is considered primitive, considering the absence of highly specialized structures. Authorities believe that the relationship of humans with dogs dates to the earliest days of human development on this planet. The simple observation is made that the early humans had to grub hard for their food and that they found that early dogs knew where the carrion was as well as other suitable food sources. The humans naturally took to following the dogs and a natural working companionship developed, with the dog ultimately occupying part of the human's shelter and behavior pattern. As with other domesticated animals, the true ancestors of the dog are difficult to trace with any degree of certainty, but many theories have been proposed. Dogs are closely related to jackals, coyotes, dingos, and wolves. It has been observed that once skinned, it is difficult to differentiate between a wolf, coyote, and domestic dog. However, it appears that the modern dog is a descendent of a wolf-like ancestor.

Over the years, over a hundred breeds of dogs have appeared. Generally, these are divided into six categories: (1) *working dogs* for hauling loads and herding grazing animals. Included are Collies, Mastiffs, Schnauzers, Sheepdogs, and the Siberian Husky; (2) *sporting dogs* (a) pointers, (b) setters, and (c) retrievers; (3) *hounds,* although sporting dogs, the hounds are in a separate category and include Bassets, Bloodhounds, Dachshunds, Foxhounds, and Greyhounds; (4) *terriers,* now essentially pets or house dogs, but once considered sporting dogs—including Fox, Welsh, Scottish Terriers, and Airedales; (5) *nonsporting dogs*—bred originally for sport and working, including Chow Chows, Poodles, Dalmatians, and Bulldogs; (6) *toy dogs,* such as Pekingese and Chihuahuas.

It is interesting to note that dogs, over a period of thousands of years, have formed in essence what some authorities term *nations*; that is, types and varieties that are best suited to given regions and that, in nature, the nations so-called are seldom found mixed. Of course, humans have moved domesticated dogs around the world so much over the centuries that interbreeding has added to the overall complexity of the situation. But, it is of note to observe that the dogs indigenous to the desert areas and temperate zones of the northern hemisphere are all wolves, whereas in the tropics of Asia and Africa, the jackals are the indigenous dogs, and whereas in South America, the indigenous dogs are the South American jackals (*Dusicyon*).

One of the earliest records of the existence of the dog was uncovered in exploring an Egyptian tomb (circa 3500 B.C.). Also, a terrier-like dog was found among the remains of a tomb (circa 3066 B.C.). The tomb of the monarch Antafee (3000 B.C.) revealed the presence of four dogs at the feet of the deceased. In diggings made in Denmark dating much further back (6000 B.C.), the bones of dogs were found.

Wolves are found in Europe, Asia, and North America. The Antarctic wolf of the Falkland Islands, a species somewhat smaller than the coyote, with a less bushy tail, black at the middle and tipped with white, is the sole representative of the southern hemisphere and most likely may be the result of an importation. Of the true wolves of the genus *Canis,* the coyote or prairie wolf of North America is the smallest form. Seven species ranging over various limited areas from Canada to Texas and from Iowa to California are known indiscriminately as coyotes. They are somewhat more cowardly than the larger wolves and, in settled areas, often persist as annoying predators, robbing poultry roosts and catching small game. The nature of their food supply seems to determine whether they live alone or in a pack. Such patterns of behavior may bear on the question of whether or not they are a threat to livestock. An article by M. Bekoff and M. C. Wells on "The Social Ecology of Coyotes," can be found in *Sci. Amer.*, **242**, 4, 120–148 (April 1980). The common coyote (*C. latrans*) occurs in the northern prairie area; the plains coyote (*C. nebracensis*) is found throughout the Great Plains; and the mountain coyote (*C. lestes*) occurs in the western mountain areas. Generally a coyote will weigh from 25 to 30 pounds (11.3 to 13.6 kilograms) and will attain a length of about 35 inches (89 centimeters), including tail. The body is buff color with white underneath. The legs are reddish-brown. The animal is considered sly and stealthy and is nocturnal. It may attain a speed of 40 miles (64 kilometers) per hour in pursuit of its prey. Favorite dietary items include the rabbit, chipmunk, small animals of almost any kind, mice, all kinds of birds, fowls, and sage hens. The coyote prefers to use burrows that already have been constructed by other animals, such as the prairie dog. A litter consists of from 6 to 8 pups. The life span is about 13 years. The coyote has a distinctive yap, whine, and howl.

Larger wolves are now common only in the wilder parts of North America, Europe, and Asia. Species found in North America include the gray wolf (*C. lycaon*), the southern wolf (*C. floridanus*), the lobo or timber wolf (*C. nubulis*), the Texas red wolf (*C. rufus*), and the American jackal (*C. frustron*) found in Texas and Oklahoma. Hounded by man from time immemorial, wolves still hang on by tooth and nail in North America's diminishing wilderness. More detailed coverage of this topic can be found in an article by L. Mech, "Where Can the Wolf Survive?" *National Geographic,* **152**, 4, 518–537 (October 1977).

There are over ten species of foxes and these occur widely in the northern hemisphere. Foxes (*Vulpes*) are of moderate size and slender build, with a sharp muzzle, long bushy tail, and unusually large ears. In addition to the habitats of the wolves, foxes are also found in North Africa north of the great desert regions. They also are found in Iceland. Foxes of certain types are prized for their pelts, the best furs coming from those of the far north. The Silver Fox is bred extensively in captivity for its fur.

The female fox is known as the vixen. The gestation period is 63 days. There usually are 4 to 5 young in a litter and the young are blind at birth with eyes opening at the end of 10 days. The fox has a cunning disposition. For example, a fox may break the line of scent when hunted by leaping onto the back of a sheep. Some of the cleverness of the fox remains to be fully investigated and explained.

Some genera of North American foxes include: Red Fox (*Vulpes*)—with its black fox, silver fox, and cross fox color phases—is bred on a very large scale in so-called fur farms and occurs naturally in all of the colder climes of the United States, Canada, into Alaska and Labrador, and southward into the New Mexican mountain ranges. The Kit Fox (*Vulpes*), also known as the Swift Fox, prefers open country of the Great Plains from Canada southward into Texas. The animal also has been found in the arid regions of Colorado and the Mojave Desert. Favorite dietary items include rats, squirrels, pocket mice, and small desert animals. The fur of the kit fox is not of economic value. The Gray Fox (*Urocyon*) inhabits a range somewhat more southerly of that of the red fox. The animal prefers wooded areas. There are geographic variations of this animal and some of these are found as far south as Florida, Mexico, and Central America. Fur of the gray fox is valued at about half of that of the red fox, but it is used for garment trimming, particularly after it has been dyed. The Arctic Fox (*Alopex*) is found north of the tree limit in the arctic tundra. A favorite food is the lemming, which they store much as a squirrel stores acorns. They also are known to trail polar bears, picking up the remains no longer of interest to the bear. However, polar bears, along with wolves, are natural enemies of the Arctic fox. These animals are known to be particularly friendly to humans, with reports of their trailing explorers and remaining just at the edge of campfires. The Blue Fox is a color phase of the Arctic fox. The blue fox also is farmed for fur on several of the Alaskan islands. By careful breeding, the blue fox will not turn white in winter.

Over eastern Europe, northern India, and Siberia, the gray-colored Hoary Fox (*V. canus*) is found. Further south and in Iraq, Iran, and western India is found the Desert Fox (*V. leucopus*). The Corsac Fox is found in central Asia and Siberia. Sometimes, in error, referred to as fennecs, a small, specialized group of foxes is found in Africa. They are small, with disproportionately large ears. The Kama (*V. cama*) is found in the Kalahari Desert and environs. Ruppell's Fennec (*F. familicus*) is well distributed through the Near and Middle East.

The Fennec is a small animal resembling a fox and also has enormous ears; *F. zerda* occurs in northern Africa, and *V. familicus* is found in Syria and adjacent regions. They are desert animals, living underground, and are nocturnal. See Fig. 1.

Fig. 1. Fennec fox, *Fennecus zerda.* (*Photo M. W. Fox.*)

As mentioned earlier, the jackals are the indigenous dogs of the tropics of Asia and Africa. There are almost innumerable species of jackals, often determined geographically. Generally the jackal is a scavenger and essentially omnivorous. They are hole dwellers, sleeping by day in their holes, or in hot weather, occupying a shallow watery spot to keep cool. They are known to travel in packs and to hunt fairly large game. They are also known to trail large cats and to create a noisy disturbance, whereupon they eat much of the cat's feast during the cat's period of confusion. The common jackal (*C.* (*Thos*) *aureus*) ranges from southeastern Europe to Sri Lanka and into northern Africa. The remaining species are African. They are approximately 2 feet (0.6 meter) in length, with a height of about 15 inches (38 centimeters) at the shoulders. The jackal has a dismaying cry, as truly characteristic to it as the hyena's cry is to that animal. The popular use of the term "jackal" does not seem to fit the behavior pattern of the actual jackal, particularly with reference to its association with the big cats. See Fig. 2.

Fig. 2. Asiatic jackal, *Canis aureus.* (*Photo M. W. Fox.*)

The cataloging and, in fact, actual discovery of the South American jackals is far from complete.

The false canines are so called because they are not quite so directly related to the true canines, but nevertheless have dog-like qualities. It is believed that this relatively small number of species essentially are relics of major branches that have faced extinction. These include the Dholes which occur in the Oriental region. The Cape Hunting Dog (*Lycaon pictus*) roams regions of Africa south and east of the Sahara. The animal has a large body with long legs, a broad flat head, and large, erect ears. It is of various colors, including brown, yellow, white and black. These animals usually hunt in packs of from 15 to 60 members and feed mostly on small animals. However, they also attack larger animals and are known to be especially damaging to sheep.

The Dingo (*Canis dingo*) is a wild dog found in forested areas of Australia. It is probably from dogs introduced to that continent many years ago. The dingo is a serious enemy of sheep and has been killed in large numbers for its depredations. The animal is about $2\frac{1}{2}$ feet in length, with a height of about 2 feet. It has large erect ears, a bushy tail, and has a tawny black coloration. The dog is considered crafty and courageous. When taken from the den at an early age, the dog can be trained to become a trustworthy pet.

For references, see **Mammalia.**

CANIS MAJOR (the great dog). Both this constellation and its companion Canis Minor, or the little dog, have been named from remote antiquity as the dogs of Orion. Sirius in Canis Major and Procyon in Canis Minor are both well-known stars, Sirius being the brightest observable. References to these stars are to be found in nearly all ancient classical literature. Sirius, in particular, was of great importance to the Egyptians, because it rose with the sun at the period when the waters of the Nile were due to rise, and was considered as a herald of the returning fertility of the valley. Sirius is not only the brightest, but also the closest star visible to the naked eye in the latitudes of Europe and North America. Intrinsically, Sirius has a brightness of more than 20 times that of our sun. Both Sirius and Procyon have faint companions, that of Sirius being particularly famous as the first of the white dwarfs discovered. (See map accompanying entry on **Constellations.**)

CANIS MINOR. **Canis Major.**

CANKER WORM (*Insecta, Lepidoptera*). Of two chief species, the canker worm is an economic pest against apple, apricot, plum, as well as elm trees. The spring canker worm (*Paleacrita vernata*) is a moth that emerges in early spring. The caterpillars have only two pairs of prolegs. The fall canker worm (*Alsophila pometaria*) is similar, but the timing of their habits is different. Wingless female moths normally emerge from the ground in late autumn, whereupon they ascend into trees and deposit their eggs, usually on smaller branches. These eggs hatch in mid-spring. The resulting blackish-yellow-striped looping caterpillars proceed to defoliate trees.

Adult male spring canker worms are gray moths; the females are wingless, plump, and gray. The larva is slender, light-to-dark brown and because of their locomotion may be described as "measuring worms." Distribution is the northeastern United States, as well as North Carolina, Missouri, Montana, Colorado, Utah, California, and Texas. Canker worms generally occur in cycles. Their destructive period usually lasts from 3 to 5 years before natural enemies and climatic conditions succeed in bringing about a reduction in numbers. This process may require 10 or more years.

Birds are the most effective natural enemies. Over 40 kinds of birds, chickadees, thrushes, and warblers in particular, feed on these caterpillars, their eggs, and the egg-laden female moths.

It is well to cultivate around the trees in the mid-summer in order to expose pupae (that live in earthen cells or cocoons near the surface) to birds and other enemies. This assists in reducing the population of the insect for the following year. Wingless female moths crawl up trunks of trees to lay their eggs. To prevent their climbing, a simple mechanical barrier can be erected by tightly wrapping a band of cotton batting several inches (12 to 15 centimeters) wide around the trunk. The batting should be tied tightly with string near the bottom and upper portion of the band. The upper portion of the batting should be turned down over the lower to form a funnel-shaped barrier. The cotton bands should be fluffed up after frequent rains to maintain their effectiveness.

Another simple mechanical barrier used about the trunk consists of a strip of mosquito wire netting. The netting should be at least as fine as 16 wires to the inch (2.5 centimeters) and about 14 inches (35 centimeters) wide. The netting should be tacked to the tree so that it fits tightly near the top and is held out about ½-inch (12 millimeters) at the bottom. Nails driven into the tree or spiral springs can be used to accomplish this. In badly infested trees, female canker worms that accumulate under this barrier should be crushed each night. When constructing barriers, all rough places on the bark should be smoothed and filled to prevent the moth from crawling under the edge of the bands. Sticky bands, such as flypaper, also will perform well. First, the sticky material is applied to a band of tarred or other heavy paper about 6 inches (15 centimeters) wide. Then the paper is tacked or tied around the trunk and all rough places filled with cotton. Fresh sticky substance should be added to the paper periodically.

CANNABIS INDICA. A variety of common hemp from which is procured the so-called hashish and marijuana, narcotic drugs. See also **Marijuana.**

CANNEL COAL. **Coal.**

CANNIZZARO METHOD. **Chemical Composition.**

CANONICAL. This term is used as an adjective to describe a standard form of a function or equation, especially when the form is simple. A canonical matrix, for example, has nonzero elements only on the main diagonal.

CANONICAL CORRELATION. Suppose there is a p-variate population with variates $x_1, x_2, \ldots, x_p$, and a q-variate population with variates $y_1, y_2, \ldots, y_q$, and suppose for definiteness that $p \geq q$. Then it is possible to find p linear combinations of the x's, $u_1, u_2, \ldots, u_p$ and q linear combinations of the y's, $v_1, v_2, \ldots, v_q$ all with zero mean and unit variance, such that

$$\begin{aligned} \mathrm{cov}(u_i u_j) &= 0 \qquad i \neq j \\ \mathrm{cov}(v_i v_j) &= 0 \qquad i \neq j \\ \mathrm{cov}(u_i v_j) &= 0 \qquad i \neq j \end{aligned}$$

where cov( ) denotes covariance. The correlation between u_i and v_i is called a canonical correlation; at most q of these correlations are non-zero. If u and v are the linear combinations corresponding to the largest canonical correlation, then v is the linear combination of the y's which can be predicted from the x's by linear regression with least residual variance (the most predictable criterion), and u is the appropriate predictor.

CANONICAL EQUATION OF MOTION. For the kth set of generalized coordinates and conjugate momenta in a conservative dynamical system, there can be written a pair of first-order partial differential equations:

$$\frac{\partial H}{\partial p_k} = \frac{dq_k}{dt}$$

$$\frac{\partial H}{\partial q_k} = -\frac{dp_k}{dt}$$

where H is the Hamiltonian function for the system and p_k and q_k are generalized momenta and coordinates, respectively. These equations are called the canonical equations of motion. See also **Hamiltonian.**

CANONICAL TRANSFORMATION. A transformation from one set of generalized coordinates and momenta to a new set such that the form of the canonical equations of motion is preserved. This usually involves finding a transformation function S which is a continuous and differentiable function of the old and new generalized coordinates and the time. The transformation can be defined by

$$L(q, \dot{q}) = L'(Q, \dot{Q}) + \frac{dS}{dt}$$

where L is the Lagrangian function in the original set of coordinates, and L' is the Lagrangian function in the transformed set of coordinates (the dot indicates the first derivative).

CANOPUS (α Carinae). Ranking second in apparent brightness among the stars, Canopus has a true brightness value of 1,500 as compared with unity for the sun. Canopus is a yellow-white, spectral F type star and is located in the constellation Carina south of the ecliptic. Estimated distance from the earth is 100 light years. See also **Constellations;** and **Star.**

CANTALOUPE. **Curcurbitaceae.**

CANTILEVER BRIDGE. **Bridge (Structural).**

CANVAS-BACK. **Waterfowl.**

CANYON. A long, deep, relatively narrow, steep-sided valley confined between lofty and precipitous walls in a plateau or mountainous area, often with a stream at the bottom; similar to, but larger than, a gorge. It is characteristic of an arid or semiarid area (such as western United States) where stream downcutting greatly exceeds weathering; e.g., Grand Canyon. ("Glossary of Geology," American Geological Institute).

CAPACITANCE. For an electrical system, capacitance may be defined as the ratio of its electric charge to the related change in potential, or by the time integral of the rate of flow of electric charge, divided by the related electric potential. By substitution of the quantities equivalent to electric charge and potential, the concept of capacitance is readily extended to nonelectrical physical systems by a corresponding change in quantities and units. Thus for a thermal system, the quantities equivalent to charge and potential difference are heat and temperature difference, so that English units of thermal "capacitance" would be Btu/°F, while for a pneumatic system the quantities would be mass and pressure, so the units would be lb/psi. The analogy between the capacitance of an electrical system, and analogous quantities in other physical systems is fundamental to the operation of an analog computer.

The electrical capacitance of such a capacitor as a conducting body that is completely isolated, i.e., far removed from other conductors, including the earth, and is surrounded by a homogeneous, perfect dielectric depends only upon the size and shape of its external surface, and upon the permittivity of the surrounding medium. Very long electric circuits, especially when the wire is surrounded by a conducting sheath, as an ocean cable, have considerable capacitance because of the capacitor-like action of wire and sheath with the insulation between them. The same is true of insulated wire wound in a close coil, adjacent turns of which, being at slightly different potential, act as the conductors of a capacitor—an effect which may be partially avoided by a criss-cross or "honeycomb" winding or by a "banked" winding (in flat spirals).

In MKSA units, the electric charge is given in coulombs, the electric potential in volts, and the electric capacitance in farads. For a capacitor made up of two conducting bodies in which charge is taken from one and placed on the other, its capacitance is the ratio of this charge to the difference of potential of the two bodies. The farad is a comparatively large unit and the electrical capacitance is often expressed in microfarads, $\mu f (1\ \mu f = 10^{-6}$ f) or picofarads, pf(1 pf $= 10^{-12}$ f).

For n conducting bodies, their respective potentials, $V_1, V_2, \ldots, V_n$, and charges, $Q_1, Q_2, \ldots, Q_n$, are expressed in terms of a set of algebraic equations

$$\begin{aligned} V_1 &= C_{11}Q_1 + C_{12}Q_2 + \cdots + C_{1n}Q_n \\ V_2 &= C_{21}Q_1 + C_{22}Q_2 + \cdots + C_{2n}Q_n \\ &\cdots\cdots\cdots\cdots\cdots\cdots \\ &\cdots\cdots\cdots\cdots\cdots\cdots \\ V_n &= C_{n1}Q_1 + C_{n2}Q_2 + \cdots + C_{nn}Q_n \end{aligned}$$

$C_{11}, C_{12}, \ldots, C_{nn}$ are called the capacitance coefficients of the set.

Among the special types of capacitance are:

1. The *lumped capacitance* such as the capacitance of a parallel plate capacitor. Its capacitance C in farads is $\epsilon_0 k_e A/d$ where ϵ_0 is the permittivity of free space and is equal to 8.84×10^{-9} farads per meter, k_e is the relative permittivity of the uniform isotropic dielectric between the plates. A is the area of the plate in square meters and d is the plate separation in meters.

2. *Distributed capacitance*, such as the capacitance per meters of a coaxial transmission line. Its capacitance in farads per meters is $2\pi\epsilon_0 k_e/\log_e (D/d)$. D is the inside diameter of the outside conductor and d is the outside diameter of the inner conductor.

3. *Stray capacitance*, such as the unintentional capacitance between a conducting wire or network component and chassis. See **Capacitor (Electrical).**

4. *Effective capacitance*, which is the total capacitance between two points on a circuit.

5. *Acoustic capacitance* has been defined as the negative imaginary part of acoustic impedance.

CAPACITANCE TRANSDUCER. A device whose capacitance is caused to change when exposed to a condition being measured. A simple configuration comprises two parallel plates that are separated by a small distance, with the space between the plates occupied by a dielectric medium. Relationships for such a device are

$$C = m(KA/d)$$

where C = capacitance, farads
m = proportionality constant
K = dielectric constant
A = effective area of plates
d = distance between plates

Picofarad (equals 10^{-12} farads) is the practical unit used.

In designing a capacitance transducer, one of three properties can be manipulated: (1) effective area, (2) dielectric constant, or (3) distance. All three methods are used and the capacitance principle is found in a number of transducers for various measurements, including displacement, position, velocity, level of fuels in vessels, acceleration, weight, force, and flow.

CAPACITANCE UNITS. Units and Standards.

CAPACITIVE LOAD. An alternating-current circuit in which the current drawn leads the voltage in phase is said to provide a capacitive load. Capacitive loading may be the result of actual capacitors, or of virtual capacitors in the form of long transmission lines, or over-excited synchronous rotating equipment. Most electrical apparatus, such as motors and coils, draws from the line a current which lags the voltage, and the use of some capacitive load is desirable in order to bring the total current and voltage more nearly in phase, and thus raise the power factor.

CAPACITOR (Edge Effect). Edge Effect.

CAPACITOR (Electrical). An arrangement of conductors and dielectrics used to secure a capacitance for the storage of electrical energy in the electric field. The energy stored in a capacitor is $W = \frac{1}{2}CV^2$ joules where C is the capacitance in farads and V is the voltage in volts. The essential feature of all capacitors is a system of conductors separated by dielectrics. The oldest form of a capacitor is the Leyden jar. Modern capacitors, both fixed and variable, are of many forms, such as those of metal foil and paraffin paper dielectric, metal foil and mica dielectric, metal foil and polystyrene dielectric, metal foil and ceramic dielectric, metal plates with vacuum, compressed gases or air as dielectric and aluminum and tantalum electrolytics. One finds these are in a variety of electrical systems, such as wire telephony, radio receiving, radio transmitting, television, computers, electrical measuring instruments, servomechanisms, devices for filtering, recording, or transcribing, ignition systems, power factor correcting devices, and motor starting systems.

CAPACITOR (Miniature). Microstructure Fabrication.

CAPACITOR MOTOR. Motor (Electric).

CAP CLOUD. Clouds and Cloud Formation.

CAPE BUFFALO. Bovines.

CAPE HUNTING DOG. Canines.

CAPELIN (*Osteichthyes*). A cold-water pelagic fish, the capelin (*Mallotus villosus*) occurs extensively in the north Atlantic and north Pacific and adjoining regions of the Arctic. The capelin is a soft-rayed fish and, together, with the smelts, comprises the family *Osmeridae.* In the eastern Atlantic, the capelin occurs abundantly from the Trondheim Fjord region of northwestern Norway to Jan Mayen, Spitzbergen, and Novaya Zemyla at the eastern extremity of the Barents Sea. The capelin also occurs sporadically in the White Sea and Kara Sea, but the central part of its range in the eastern Atlantic is the Barents Sea. Iceland also has an abundance of capelin around its shores, as does Greenland, where in the last several decades the center of the capelin distribution has moved north as far as Thule (76°N) on the west and Scoresby Sound (70°N) on the east. In the Canadian Arctic Archipelago, capelin have been reported from the Melville Peninsula, but not from Baffin Island. Individual occurrences of capelin have been reported from the Coronation Gulf, Bathurst Inlet, and the Great Fish River of the Canadian Arctic. Capelin are reportedly very common in the southern half of the Hudson Bay, but rare in the northern portion. They are not known in the western part of Hudson Strait. There is thus a gap in its distribution between eastern Hudson Strait and southern Hudson Bay.

Capelin are relatively small fish, the mature specimens being generally 7 to 8 inches (13 to 20 centimeters) in length, although individual fishes up to nearly 10 inches (24.5 centimeters) in length have been recorded. Growth is greatest during the first two years of life, after which it decreases until, in the fifth year, the size increment is negligible. During the first year, both male and female are the same size, but during the second year a differential growth rate sets in, favoring the male, which is from 0.4 to 1 inch (1 to 2.5 centimeters) larger than the female at sexual maturity. See accompanying figure.

From Saglek south along the Labrador coast, capelin occur in large quantities wherever suitable spawning beaches can be found. The Newfoundland coast, the Grand Bank, St. Pierre Bank, and the Banks of the Labrador Shelf also possess large populations of capelin, but they have not been reported from Flemish Cap where water tempera-

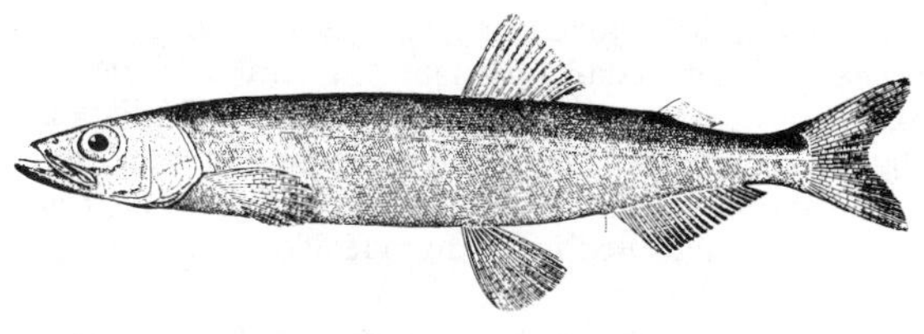

Capelin (*Mallotus villosus*).

tures are too warm. In the Gulf of Saint Lawrence, the capelin is most abundant on the northern shore, although in colder years, they also occur extensively around Gaspé. South of the Cabot Strait as far as Cape Cod, the occurrences of capelin are rare and are related and restricted to the infrequent influx of cold water into the Gulf of Maine.

In the Pacific, the distribution of the capelin extends from Cape Barrow, Alaska around the Bering Sea south along the Pacific coast to Canada to Juan De Fuca Strait. On the Asiatic coast, it extends from the Sea of Chukotsk south to Hokkaido Island, Japan, and the Tumen River in Korea.

Important Link in Food Chain. In the Barents Sea, the capelin is an abundant fish and forms an important link for many food chains. This is specially true of the cod which migrates toward the coast in pursuit of the capelin during their spawning migration in early spring, but also of the haddock and redfish. Barents Sea capelin are demersal spawners which deposit their eggs at depths between 164 and 328 feet (50 and 100 meters). The capelin of the Barents Sea are exploited commercially only during the winter–spring spawning period. Capelin represent large catches for both Norwegian and Soviet fishers. The capelin vitally affect the location and productivity of the fishers for demersal fish in the Barents Sea. Iceland, which did not initiate exploitation of its capelin resource until 1963, has expanded the catch progressively since that time. Nearly all of the Norwegian and Icelandic catch, as well as the bulk of the Soviet catch, is reduced to meal and oil. In Newfoundland, capelin have been traditionally used as a source of raw fertilizer and as bait, but these uses have declined in recent years. Although Greenland does not formally report a capelin catch, it is known that large quantities are used as food for human consumption, bait, and as a supplement in the diet of sheep and domestic cattle.

High mortality after spawning has been recorded for the capelin as a result of stranding or wounding during the act of spawning. The pelagic-living capelin enters the tidal zone in order to spawn. The fishes come to the beach in front of the crest of an advancing wave, the spawning act is completed, and they go back with the returning wave. When coming short of the reach of the returning wave, the capelins are stranded. Many others are injured during the vigorous motions that are a part of the act of spawning. In this way, very large numbers are destroyed annually in arctic regions.

See also **Fishes.**

CAPELLA (α Aurigae). The third brightest star visible in the northern latitudes, and the fifth brightest star in the celestial sphere. Capella is closer to the pole than any of the other bright stars. It has always played an important part in mythological writings, and we find it referred to on an old tablet dating back to 2000 B.C.

Astronomically, Capella is a particularly interesting star, for it is a spectroscopic binary with a period of 104 days, and the angular distance between the components has been measured with the interferometer. There are also two dwarf M stars, making Capella a four-star system. From the complete solution of the orbit, the physical characteristics of the object may be found. It is a giant star of the same spectral class as our sun.

Ranking sixth in apparent brightness among the stars, Capella has a true brightness value of 170 as compared with unity for the sun. Estimated distance from the earth is 47 light years. Capella is classified as a yellow star of spectral type G. See also **Constellations;** and **Star.**

CAPILLARITY. The name given to a class of phenomena, of which the elevating or depression of liquids in fine tubes is representative. When the interface between a liquid and a gas, or between two liquids,

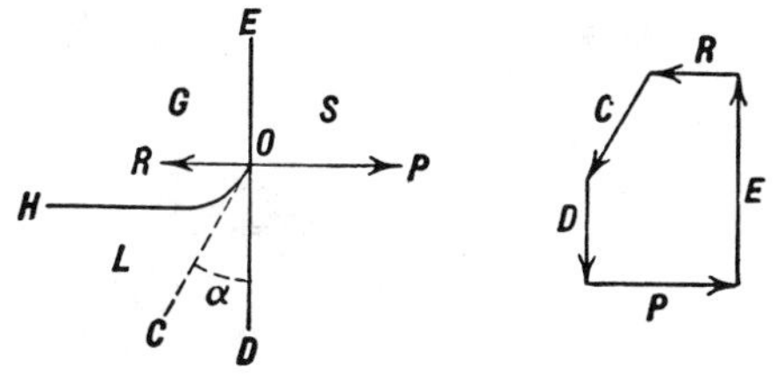

Fig. 1. Capillary force, large adhesion.

is intercepted by a solid surface, an equilibrium is established at the junction among the forces acting along the three surfaces of contact. For example, let a plate of solid *S* be dipped into a liquid *L* having gas *G* above it (see Fig. 1). A molecule at the junction *O* is acted upon by the adhesive attraction *P*, by the forces which give rise to the three surface tensions along the interfaces *OH*, *OE*, and *OD*, and by the reaction *R* of the plate *S* against which it is drawn by the adhesion. (Its weight may be considered negligible.) The flexible interface *OH* adjusts itself so that these forces come into equilibrium; unless, indeed, one of them, *E*, exceeds the sum of *D* and *C*, in which case the liquid "creeps" indefinitely along the surface as oil does over a glass or tin container. The equilibrium polygon at the right is labeled in each case to correspond with the figure representing the surfaces. The "angle of contact" α, between the liquid surface at *O* and the solid surface *OD*, is determined by the aforesaid forces acting at *O*. For most liquids against glass it is acute; for mercury against glass it is obtuse. (Fig. 2.) In special cases it may be 90°, and in others it reduces to zero.

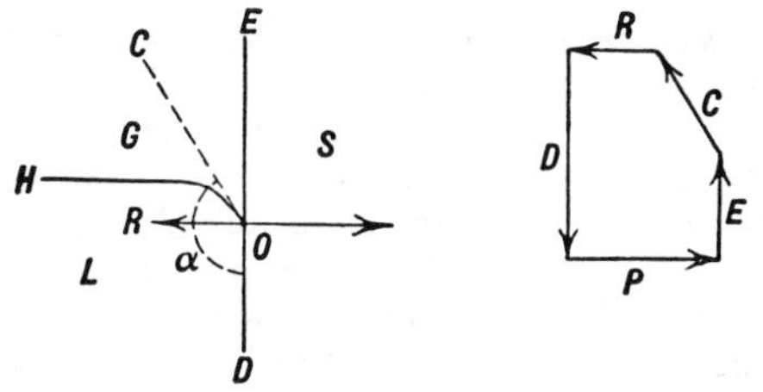

Fig. 2. Capillary force, small adhesion.

If the interface between two media *A* and *B* (Fig. 3) is curved, *A* being on the concave side, the pressure in *A* is greater than in *B* on account of the surface tension; much as the pressure inside a rubber balloon is greater than outside. We can now understand why water rises in a capillary tube. For, to secure equilibrium, the liquid must rise until the pressure inside the surface at *B*, plus the pressure due

Fig. 3. Rise of liquid in capillary tube.

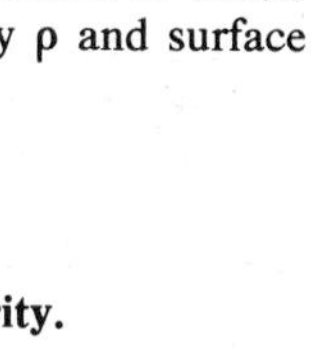

to gravity at depth *h*, makes the pressure at *L* equal to that at the surface level outside; that is, to the atmospheric pressure. Similar reasoning applies to the depression of mercury in a glass tube. For a circular tube of internal radius *r*, the distance *h* to which capillarity will elevate (or depress) a liquid of density ρ and surface tension *T* (against air) is readily shown to be

$$h = \frac{2T \cos \alpha}{r\rho g}$$

where *g* is gravity. See also **Electrocapillarity.**

CAPILLARY. 1. Hair-like, especially in application to fine tubes. 2. A minute thin-walled blood vessel intervening between the arteries and veins. See **Circulatory System.** 3. A cylindrical space of small radius, or a tube containing such a space. The numerous uses of such tubes has given rise to a number of derived terms. Thus, the capillary correction is a correction applied to mercury barometers, widebore

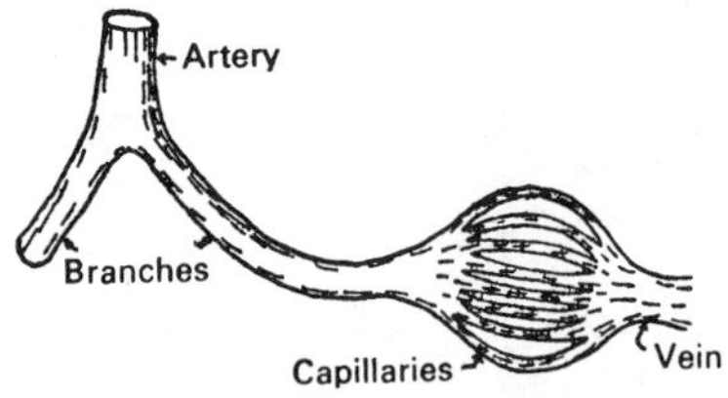

Variations in velocity of bloodflow. If a vessel divides into two branches, these will be individually of less cross section than the main trunk, but united they will exceed it. Linear velocity will be lower in the branches than in the parent stock. The sum of the cross-sectional areas of the capillaries is greater than that of the artery or vein. (*Kimber and Gray, "Textbook of Anatomy and Physiology," Macmillan.*)

thermometers, etc., for the effect of capillarity on the height of the column. Capillary pressure is a pressure due to capillary force. See **Capillarity.**

Capillary rise is the elevation of liquid in a capillary tube above the general level. Capillary separation is the separation of gases by flow through a porous medium. In a theory of this process based on the concept of momentum transfer, the actual porous medium is treated as equivalent to a bundle of parallel capillary tubes.

CAPILLARY FRINGE. Above the zone of saturation in the ground, capillary pores may exist which, if filled with water, form a zone or fringe of moisture higher than the true water table. This is the capillary fringe or zone of capillarity.

CAPILLARY PYRITES. Millerite.

CAPILLARY SYSTEM (Instrument). Capillary, or small bore tubing, has several uses in instruments. Sometimes the capillarity, or capillary attraction, plays an important functional part, as when such tubing is used in inking systems. In other cases, the capillary is utilized only to provide an optimum volume for the hydraulic transmission of a fluid in amounts proportional to the changes in some measured variable. Capillary tubing also is used to create a resistance to flow between two portions of an instrument so as to improve performance.

When used in inking systems, the fluid and the bore must be selected carefully to provide the desired level of rise. The adhesive forces between the ink and capillary wall must be sufficiently greater than the cohesive forces within the fluid to provide the transport of ink along the desired length of tubing.

Filled-system thermometers depend on the dimensions of metal capillary to provide suitable performance characteristics. The usual objective is to transmit pressure or volume changes with maximum speed and this favors large-bore capillary tubing, but this objective must be compromised inasmuch as added volume increases the effects of temperature changes along the capillary on the instrument reading, thus causing errors in the signal.

Similar to the use of capillary tubing in filled-system thermometers is its use in filled pressure and load or force measuring systems. In the filled pressure system, a flexible seal (diaphragm or bellows) separates the process fluid from the system fluid. Because the seal is flexible, the process pressure is carried by the transmitting fluid from the remote seal to an element such as a bourdon spring which converts the pressure change to a usable motion or force. This isolates all parts of the system except the diaphragm seal from the process which may contain solids, be very corrosive, or at an elevated temperature. Volumetric load or force measuring systems differ from pressure systems primarily in that the sensing element is a relatively rigid structure which changes length and volume as the force is applied. The capillary thus carries a volume change to the bourdon element, and any pressure changes resulting are incidental to the function of the system.

In any of the foregoing systems, the capillary produces an error in the signal when the tubing is subjected to ambient temperature changes.

CAPRIC ACID. Also called decanoic, decoic, and decyclic acid, formula $CH_3(CH_2)_8COOH$. The acid occurs as a glyceride in natural oils. Usual form is white crystals having an unpleasant odor. Soluble in most organic solvents and dilute nitric acid; insoluble in water. Specific gravity 0.8858 (40°C); mp 32.5°C; bp 270°C. Combustible. A component of some edible vegetable oils. See also **Vegetable Oils (Edible).** Capric acid is derived from the fractional distillation of coconut oil fatty acids. The acid is used in esters for perfumes; fruit flavors; a base for wetting agents; as an intermediate in organic synthesis; plasticizer; resins; and used in food-grade additives.

CAPRICORNUS (the sea-goat). A constellation of small stars, not at all striking in appearance, but important because it is the tenth sign of the zodiac. The star Alpha (named Giedi) is one of the more remarkable stars, actually being made up of six components. (See map accompanying entry on **Constellations.**)

CAPRIFOLIACEAE. Elder Trees and Viburnums.

CAPRIMULGIFORMES (*Aves*). This order of birds comprises crepuscular (birds active in the twilight) and nocturnal birds, loosely called the goatsuckers. The beak is broad, cleft rearward beyond the eyes; the head is flat with large eyes at the sides. The feathers are soft and have bark-colored background and markings. There are 10 primaries and 10 tailfeathers, and 14 cervical vertebrae; the thoracic vertebrae are unfused. They scratch their heads by bringing the foot over the wing. Incubation lasts only 16–17 days and is shared by both parents. The newly hatched young are covered with woolly down; they can see, and are ambulatory within a few days. Except for one species, their diet is vegetarian.

The 5 families comprise 22 genera and 96 species: 1. Oilbirds (*Steatornithidae*); 2. Potoos (*Nyctibiidae*); 3. Podargues (*Podargidae*); 4. Owlet Nighthars (*Aegothelidae*); and 5. Nightjars (*Caprimulgidae*). The length of the last two families reaches 20–55 centimeters (8–21½ inches); the length measured to the tip of the much elongated lateral tailfeathers is 80 centimeters (31 inches).

The Oilbirds (family *Steatornicidae*) comprise only one species, the Oilbird (*Steatornis caripensis*). The length is 45 centimeters (18 inches), the wingspread is 113 centimeters (44 inches), and the weight is 400 grams (14 ounces). The plumage is stiffer than that of the other goatsuckers; the tail feathers are staggered, and there are 10 rectrices. The upper mandible of the stout beak is curved like that of a predator. The legs are very short and without horny scales; the first toe points diagonally forward. The feet are weak and not suited for grasping a branch.

This family of goatsuckers is the smallest and most remarkable one. Oilbirds live in holes, fly out at night, and, unlike all the representatives of the order, feed exclusively on fruits.

The Podargues (family *Podargidae*) are rather large goatsuckers of the Australian-Papuan region and of the southeasterly parts of the Indo-Malayan islands. The length is 21.6–53.4 centimeters (8½–21 inches). Their beak is large, boat-shaped, and thickly horny. There are two genera: 1. Podargues proper (*Podargus*) with three species, among them the Australian Tawny Frogmouth (*Podargus strigoides*); and 2. Frogmouths (*Batrachostomus*) with nine species, including the Javanese Frogmouths (*Batrachostomus javensis*).

Frogmouths, like nightjars, are softly feathered. Their gray-brown and red-brown marbled color pattern is concealing. They differ considerably from nightjars in their physical structure and their behavior. The arrangement of the feathers in the podargues and in the nightjars is not markedly different. While the tongue in nightjars is more or less degenerate, in the podargues, it is firm, thick-skinned, and leaf-shaped.

The differences between the two genera (*Podargus* and *Batrachostomus*) are only minor. The moult in podargues is "staggered," as it is in potoos and oilbirds. In true goatsuckers and owlet nightjars, on the other hand, the primaries are moulted medially-laterally. Compared to nightjars, the most important differences in behavior are in regard to reproduction. Podargues proper usually lay two eggs, while frogmouths lay one pure white egg in open tree nests. Podargues build loose nests of twigs on a horizontal, forked branch, while frogmouths make a cushionlike structure from their own down and cover the outside with spiderwebs and lichen. Incubation takes about 30 days and so does the nestling period of the young, who at first are covered with long white down. Both sexes share the incubation and the care of the youngsters in the nest.

While the quiet frogmouths are to be found only in woodlands, podargues proper inhabit a variety of wooded areas, even the Australian desert with only a sparse tree cover. Nowhere are they especially common; they are also difficult to discover because of their life habits. At night they are as crowded together as are the nightjars of Africa and of tropical South America. They live in pairs and as we know they are sedentary: they neither migrate, nor roam like nomads.

The sluggish podargues and frogmouths are in no way aerial hunters, like the nightjars. Most of what they eat is taken from the ground. They fly a short distance from a tree or a pole to pick up such things as ground-dwelling scorpions, centipedes, and insects. They also consume snails, frogs, small lizards, and even birds and mice. The proper podargues also like fruit and occasionally steal grapefruit and other soft fruits from gardens.

The Potoos (family *Nyctibiidae*) are similar to the nightjars in appearance. The wings are very long and when folded reach almost to the end of the tail, which is especially long and wide. The beak extends very little beyond the outline of the head; it does, however, extend backward past the ears so that it forms a gigantic maw. The tip of the beak is free and decurved at a right angle. On the edge of the upper mandible a large tooth projects and surrounds the tender lower mandible like a clasp. The eyes are very large, have a vivid yellow color and a reddish reflection when a light is flashed on them in the night. The legs are very short, and the toes are very long and unusually broad and fleshy at the base, thus forming a large sole (which assures stable sitting on a wide surface).

There is one genus (*Nyctibius*) with five species, which resemble each other closely in their bark-colored plumage. Among them is the Common Potoo (*Nyctibius griseus*). The length is 35 centimeters (14 inches) and the weight is 160 grams ($5\frac{1}{2}$ ounces). It occurs from Mexico to Argentina. The Great Potoo (*Nyctibius grandis*) has a length of 55 centimeters ($21\frac{1}{2}$ inches) and weighs about 550 grams (19 ounces). It occurs from Guatemala to Brazil. During daylight the potoos sit immovably on a tree, not crouching like goatsuckers, but upright. They look so much like the jags on a branch that they can afford to sit on a bare branch in full view, or on a tree stump, or even on a picket fence, looking as if they were part of it. Neither the hot sun nor the rain bothers them. Now and then they droop a little, but immediately straighten up again when something near them stirs. They watch their surroundings through narrow eye slits.

The Owlet Nightjars (family *Aegothelidae*) are closely related to the podargues of the family *Podargidae*. They are smaller than podargues and have a stockier build. The length is 19–28 centimeters ($7\frac{1}{2}$–11 inches). The beak resembles that of a frogmouth, but is shorter and softer than those of podargues and frogmouths, and is extensively hidden by the forehead plumage. On the forehead and between the eye and beak are stiff and partly erectile bristles, and on the chin are some softer ones that are recurved. The coloration of the plumage resembles that of podargues and goatsuckers.

There is only one genus of Owlet Nightjars (*Aegotheles*) with seven species; one of them, the Australian Owlet Nightjar (*Aegotheles cristatus*), is widespread throughout Australia. Four species occur only in New Guinea, one in New Caledonia.

Owlet nightjars inhabit wooded areas and are less sluggish than the podargues. They sit on tree branches in an erect, owl-like posture. They never assume the stiff, "part of the tree" position during the day, but rather stay in tree hollows, from which one can easily chase them by bumping against the tree or shaking the branches. Their feeding habits place them, as indicated by the structure of the palate, between podargues and goatsuckers. Their flight course is straighter than that of goatsuckers; it lacks the latter's characteristic bends and turns. While in flight they catch flying insects. Examinations of the content of their gizzards have shown that ground-living animals, such as weevils, centipedes, and ants constitute the main part of their diet. They brood in tree hollows, occasionally in ground holes in a river bank, or in buildings. Although they do not build regular nests, the eggs are placed on a soft layer of dry leaves or mammal hairs. Like the eggs of the tawny frogmouth, the egg shell is pure white with occasional brown dots. The clutch consists of three to five eggs. The nestlings, like those of the tawny frogmouth, are covered with pure white down.

The last family of the goatsuckers or nightjars, the Goatsuckers proper (*Caprimulgidae*), derives its name from a legend reaching back to antiquity. Since these birds often flutter around grazing animals at night in order to catch insects near them, it was believed that they sucked the milk from goats. Actually, their very broad and short beak, which can be opened wide, indicates their manner of providing sustenance: they catch hawkmoths and beetles mostly in flight. Their soft plumage with its owl-like markings assures effective camouflage and is superbly adapted to the ground, which is, aside from their hunting flights, their main field of activity. The nightjar's long, slender wings lend flexibility and speed to its flight, and extensions radiating from the vanes make it noiseless.

The length is 20–41 centimeters (8–16 inches). They generally rest during the day and close their eyes down to a slit; they are active at night. They are ground-brooders and lay no nest bedding; both parents relieve each other in the care of the clutch and the young.

There are 17 genera with 69 species; the most important are: the Nightjar (*Caprimulgus europaeus*); the Standard-Wing Nightjar (*Macrodipteryx longipennis*); the Pennant-Wing Nightjar (*Semeiophorus vexillarius*); the White-Throated Poor-Will (*Phalaenoptilus nuttallii*); the Common Nighthawk (*Chordeiles minor*); and the Pauraque (*Nyctidromus albicollis*). See also **Nightjars and Nighthawks.**

CAPRINES. **Goats and Sheep.**

CAP ROCK. **Petroleum.**

CAPROIC ACID. Also called hexanoic, hexylic, or hexoic acid, formula $CH_3(CH_2)_4COOH$. Present in milk fats to extent of about 2%. Also a constituent of some edible vegetable oils. See **Vegetable Oils (Edible).** The acid is oily, colorless or slightly yellow, and liquid at room temperature. Odor is that of Limburger cheese. Soluble in alcohol and ether; slightly soluble in water. Specific gravity 0.9276 (20.4°C); mp −4.0°C; bp 205°C. Combustible. Caproic acid is derived from the crude fermentation of butyric acid; or by fractional distillation of natural fatty acids. Used in various flavorings; manufacture of rubber chemicals; varnish driers; resins; pharmaceuticals.

CAPROLACTAM. $NH(CH_2)_5CO$, formula weight 112.15, liquid ingredient used in the manufacture of type 5 nylon. See also **Fibers.** Several hundred million pounds of the compound are produced annually. There are a number of proprietary processes for caprolactam production. In one process, the chargestock is nitration-grade toluene, air, hydrogen, anhydrous NH_3, and H_2SO_4. The toluene is oxidized to yield a 30% solution of benzoic acid, plus intermediates and by-products. Pure benzoic acid, after fractionation, is hydrogenated with a palladium catalyst in stirred reactors operated at about 170°C under a pressure of 10 atmospheres. The resultant product, cyclohexanecarboxylic acid is mixed with H_2SO_4 and then reacted with nitrosylsulfuric acid to yield caprolactam. The nitrosylsulfuric acid is produced by absorbing mixed nitrogen oxides N_2O_3 in H_2SO_4: $N_2O_3 + H_2SO_4 \rightarrow SO_3 + 2NOHSO_4$. The resulting acid solution is neutralized with NH_3 to yield $(NH_4)_2SO_4$ and a layer of crude caprolactam which is further purified. The overall process reaction is:

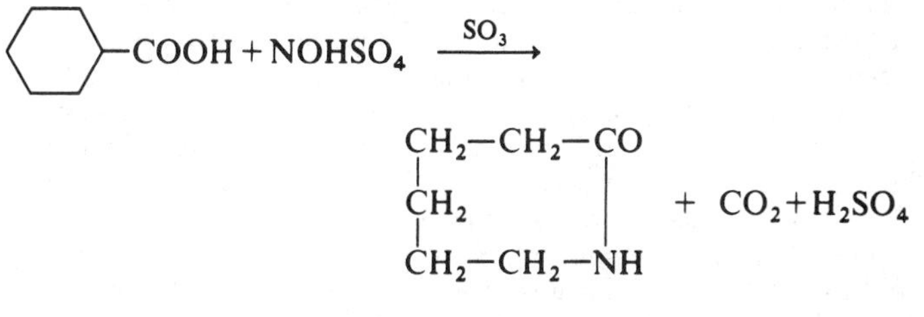

A comparatively recent process utilizes a photochemical reaction in which cyclohexane is converted into cyclohexanone oxime hydrochloride: $C_6H_{12} + NOCl \xrightarrow[\text{light}]{\text{HCl}} C_6H_{10}NOH \cdot 2HCl$. The yield of cyclohexanone is estimated at about 86% by weight. Then, in a Beckmann rearrangement, the cyclohexanone oxime hydrochloride is converted to ε-caprolactam:

$$C_6H_{10}NOH \cdot 2HCl \xrightarrow{H_2SO_4} \underbrace{CH_2{-}(CH_2)_4{-}C{=}O}_{\text{—NH—}} + 2HCl$$

To obtain the nitrosyl chloride for producing the oxime, the following reactions are required: (1) NH_3 is burned in air to produce NO_x: $2NH_3 + 3O_2 \rightarrow N_2O_3 + 3H_2O$; (2) nitrosylsulfuric acid is made by reacting nitrogen trioxide with H_2SO_4: $2H_2SO_4 + N_2O_3 \rightarrow 2HNOSO_4 + H_2O$; (3) nitrosyl chloride then is made by adding HCl to the nitrosylsulfuric acid: $HNOSO_4 + HCl \rightarrow NOCl + H_2SO_4$. High-pressure mercury lamps are used to effect the photochemical reaction. Any radiation of a wavelength shorter than 3,650 Å must be filtered out to avoid formation of tarry products. Critical factors that determine the yield of the process include temperature, the distance between the lamps and reactor, and the volume of the reactor. The crude caprolactam solution is neutralized with NH_3. The resulting mixture separates into an upper layer of crude caprolactam; the lower layer contains aqueous $(NH_4)_2SO_4$. An advantage claimed for this process is that only about one-half as much by-product $(NH_4)_2SO_4$ is formed as compared to other processes.

CAPTURE-RECAPTURE METHOD. A method of estimating the size of populations of wild animals. A number are captured, marked and released unharmed. Later, a further sample is captured and the number of marked animals in it provides the basis of the estimate.

CAPTURE RELEASE SAMPLING. **Sampling (Statistics).**

CAPUCHIN MONKEY. **Monkeys and Baboons.**

CAPYBARA. **Rodentia.**

CARACARA (*Aves, Falconiformes*). South American birds of several species related to the hawks. They eat carrion but also catch living prey and sometimes rob other birds of their prey. One species, Audubon's caracara, *Polyborus cheriway*, occurs in the extreme southern parts of the United States. The chimachima is found from Panama to southern Brazil. The chimango is found in Tierra del Fuego and the southern part of the continent. See also **Falconiformes.**

CARANGIDS (*Osteichthyes*). Of the order *Percomorphi*, suborder *Percoidea*, the family *Carangidae*, the carangids are very fast and many of the species are excellent food fishes. They are well distributed worldwide in tropical and temperate waters. There are about 200 species, most of which are shaped something like the yellow jack (*Gnathanodon speciosus*). The latter fish occurs in the Indo-Pacific and attains a length of about 3 feet (0.9 meter). It possesses several vertical greenish stripes on a pale yellow body. The tail fin is sharply forked. The jack mackerel (*Trachurus symmetricus*) frequents the waters of the American Pacific coast. These fish are characterized by a sharp ridge adjacent the caudal peduncle of the tail. This is formed by a series of bony plates, sometimes called scutes.

Another interesting carangid is the Atlantic pompano (*Trachinotus carolinus*), a valuable food fish. Possessed of spectacular blue coloration on the back, the dirigible-shaped *Elagatis bipinnulatus* (Indo-Pacific rainbow runner) attains a length of about 4 feet (1.2 meters). The yellowtail (*Seriola dorsalis*) is a highly regarded sporting fish in the waters of Mexico and southern California and attains a length in excess of 3 feet (0.9 meter). The *Naucrates ductor* is the legendary pilot fish, reputed to lead ships and swimmers to safety. The legend has no foundation. The species of jack fishes frequenting the waters of the Palmyra Islands are reputed to be poisonous. However, a related species (*Caranx melampygus*), known as the black ulua, is widely sold in the Hawaiian Islands. The jacks and cavallas found in the Philippines are considered of premium commercial value, particularly when taken from the freshwater lakes on their return to the sea. Jacks are also found in New Guinea.

CARAPACE. A shield-like covering of the upper part of the body. In the crustaceans it is the body wall of the thorax and in the turtles and tortoises it is a complex structure made up of bony plates, including flattened ribs and vertebrae, covered with thin horny plates. The armor of the armadillo, composed of many bony plates developed from the skin and covered with horny plates, is also called a carapace.

CARAPATO (*Arachnida, Acarina*). Ticks of two species, found in tropical Africa and Central America, respectively. The African species is also called the tampan.

The wounds produced by these creatures are severe in themselves but their transmission of the germs of relapsing fever is a much greater danger.

CARAWAY. **Umbelliferae.**

CARBAMATES. Derivatives of the hypothetical carbamic acid, H_2NCOOH, which does not exist. The ethyl derivative urethane, is prepared by heating urea in alcohol under pressure, by the reaction $H_2NC({=}O)NH_2 + C_2H_5OH \rightarrow H_2NCOOC_2H_5 + NH_3$. Other derivatives are shown in the accompanying table.

CARBAMIC ACID. **Herbicide; Insecticide.**

CARBANION. An ion of the general formula $\mathrm{B{-}\overset{A}{\underset{D}{C}}{:}^{-}}$, where A, B and D are substituent groups. Their importance in elucidating the mechanism of organic reactions is because a considerable proportion of all organic reactions involve carbanions, as others do carbonium ions and carbon free radicals (including carbene radicals). Many carbanion reactions involve removal of a proton from a carboxylic acid to form a carbanion. Many electrophilic substitution reactions involve carbanions. Carbanions are strong bases or nucleophiles. Many electrophilic substitution reactions that have carbanion intermediates are base-catalyzed since the basic reagent produces the basic carbanion. Because of the negative charge on carbanions, their structures are affected by cations, by attached substituents and particularly by the solvent.

CARBENE. The name quite generally used for the methylene radical, $:CH_2$. It is formed during a number of reactions. Thus the flash photochemical decomposition of ketene ($CH_2{=}C{=}O$) has been shown to proceed in two stages. The first yields carbon monoxide and $:CH_2$, the latter then reacting with more ketene to form ethylene and carbon monoxide. Carbene reacts by insertion into a C—H bond to form a C—CH_3 bond. Thus carbene generated from ketene reacts with propane to form *n*-butane and isobutane. Carbene generated by pyrolysis of diazomethane reacts with diethyl ether to form ethylpropyl ether and ethylisopropyl ether.

Substituted carbenes are also known; chloroform reacts with potassium *t*-butoxide to form dichlorocarbene $:CCl_2$, which adds to double or triple carbon-carbon bonds to form cyclopropane derivatives.

CARBENICILLIN. **Antibiotic.**

CARBIDES. **Carbon; Iron Metals, Alloys, and Steels.**

CARBOCYCLIC COMPOUNDS. **Organic Chemistry.**

CARBOHYDRATE (Nucleic Acids). **Nucleic Acids and Nucleoproteins.**

CARBOHYDRATES. These are compounds of carbon, hydrogen, and oxygen that contain the saccharose grouping (below), or its first reaction product, and in which the ratio of hydrogen to oxygen is the same as in water.

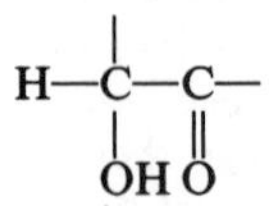

Carbohydrates are the most abundant class of organic compounds, representing about three-fourths of the dry weight of all vegetation. Carbohydrates are also widely distributed in animals and lower life forms. These compounds comprise one of the three major components (others are protein and fat) of the human diet, and indeed that of

REPRESENTATIVE CARBAMATES

Carbamate	Formula	Melting Point °C	Boiling Point °C
1. Carbamic acid (not isolated)2.0			
2. Methyl carbamate	$OC\begin{Bmatrix}OCH_3\\NH_2\end{Bmatrix}$	54	177
3. Ethyl carbamate (urethane)	$OC\begin{Bmatrix}OC_2H_5\\NH_2\end{Bmatrix}$	49	184
4. Propylcarbamate (norm.)	$OC\begin{Bmatrix}OC_3H_7\\NH_2\end{Bmatrix}$	60	200
5. Phenylcarbamate	$OC\begin{Bmatrix}OC_6H_5\\NH_2\end{Bmatrix}$	142	
6. Benzylcarbamate	$OC\begin{Bmatrix}OCH_2C_6H_5\\NH_2\end{Bmatrix}$	86	dec.
7. Ethyl-N-methyl carbamate (N-methylurethane)	$OC\begin{Bmatrix}OC_2H_5\\NHCH_3\end{Bmatrix}$	—	170
8. Ethyl-N-ethyl carbamate (N-ethylurethane)	$OC\begin{Bmatrix}OC_2H_5\\NHC_2H_5\end{Bmatrix}$	—	175
9. Ethyl-N-normal-propyl carbamate (N-normal-propylurethane)	$OC\begin{Bmatrix}OC_2H_5\\NHC_3H_7\end{Bmatrix}$	—	192
10. Ethyl-N-phenyl carbamate (N-phenylurethane)	$OC\begin{Bmatrix}OC_2H_5\\NHC_6H_5\end{Bmatrix}$	52	237
11. Ethyl-N,N-diphenyl carbamate (N,N-diphenylurethane)	$OC\begin{Bmatrix}OC_2H_5\\N(C_6H_5)_2\end{Bmatrix}$	72	> 360
12. Thiourethane	$OC\begin{Bmatrix}SC_2H_5\\NH_2\end{Bmatrix}$	108	subl.

most other animals. In a nutrition-conscious era, advocates for both more and fewer carbohydrate calories in the human diet can be found.

Classification of Carbohydrates. Because carbohydrates as components of foods and feedstuffs are not limited to just a few specific classes or types, but essentially run the gamut of the carbohydrate spectrum, it is in order here to review briefly the organization of carbohydrate chemistry, with some examples from the various classes. See also entry on **Organic Chemistry.**

Elementary Terminology. A term synonymous with carbohydrate is *saccharide* (sometimes *saccharose*). When referring to saccharides, the basic molecular formula is considered to be $C_6H_{12}O_6$. Compounds with this general formula, such as glucose, mannose, and galactose, are known as *monosaccharides* because they contain one $C_6H_{12}O_6$. A *disaccharide*, as typified by sucrose, lactose, and maltose, has the general molecular formula, $C_{12}H_{22}O_{11}$ and may be considered as containing two $C_6H_{12}O_6$ groupings that have been joined by one atom of oxygen, with the elimination of one molecule of water. Similarly, the *trisaccharides*, such as raffinose, have the molecular formula, $C_{18}H_{32}O_{16}$. Any larger molecules of the $C_x(H_2O)_y$ configuration are termed *polysaccharides*, and include the starches, celluloses, dextrin, and glycogen. See also **Starch.** An *oligosachharide* is a carbohydrate containing from two up to ten simple sugars linked together (e.g., sucrose, composed of dextrose and fructose). Beyond ten, the term *polysaccharide* is used. Gums and mucilages are complex carbohydrates. See **Gums and Mucilages.**

Both the terms carbohydrate and saccharide are significant only by way of classifying these compounds, because neither term appears in whole or in part in any of the widely used names of these compounds. About the only point of nomenclature enjoyed in common by several of the saccharides is the termination *-ose*, as found, for example, in cellulose, dextrose, sucrose, and glucose. Any saccharides having the structure of an aldehyde is termed as *aldose*; any saccharide with the structure of a ketone is termed a *ketose.* For those saccharides that contain 4–6 carbons, the number of carbons forms a nomenclature base, as a *tetrose*, $C_4H_8O_4$, a *pentose*, $C_5H_{10}O_5$, and a *hexose*, $C_6H_{12}O_6$.

To be consistent with the relationship between a mono- and a disaccharide, some authorities do not term a tetrose or a pentose a monosaccharide. By combining the *ald-* and *ket-* prefixes, certain compounds then may be called aldohexoses, such as glucose and galactose, or ketohexoses, such as fructose and sorbose.

The mono-, di-, and trisaccharides are also commonly termed *sugars.* A sugar generally is considered to possess the properties of a crystalline solid with a relatively low melting point (below 150°C), of being soluble in water, and of possessing a sweet taste. Thus, the common names of several saccharides incorporate the term sugar, preceded by the common raw source of the substance, as glucose (grape sugar), sucrose (cane or beet sugar), maltose (malt sugar), and lactose (milk sugar). The crosscurrents of the nomenclature employed for the carbohydrates will be evident from the accompanying table.

Important Carbohydrates in Foods and Biological Systems

The properties of several carbohydrates that are of particular importance in foods and biological systems are described in the following paragraphs.

Glucose. This may be considered the key carbohydrate. It is the leading member of the aldohexose group, and is formed as one of the products or the only product when the following carbohydrates are hydrolyzed, sucrose, lactose, maltose, cellulose, glycogen. In many of its properties and its structural forms, it is representative of the sugars, and it is therefore discussed in detail here. Glucose is a colorless solid ($C_6H_{12}O_6$), less sweet than sucrose, soluble in water from which it may be crystallized $C_6H_{12}O_6 \cdot H_2O$. Glucose reacts (1) with alkaline cupric salt solution (Fehling's solution or Benedict's solution) to form cuprous oxide, (2) with ammonio-silver salt solution (Tollens' solution) to form finely divided or mirror film of silver, (3) with phenylhydrazine in acetic acid, to form glucose phenylhydrazone $CH_2OH(CHOH)_4CH{:}NNHC_6H_5$, white solid, melting point alpha 159–160°C, beta 140–141°C, with excess phenylhydrazine to form glucosazone

$$CH_2OH(CHOH)_3C{:}(NNHC_6H_5)\cdot CH{:}NNHC_6H_5$$

yellow solid, melting point 205°C decom., (4) with acetic anhydride, to form glucose pentacetate $C_5H_6(OOCCH_3)_5CHO$, melting point alpha 112 to 113°C, beta 131 to 134°C, (5) with sodium amalgam, to form sorbitol $CH_2OH(CHOH)_4CH_2OH$, (6) with hydriodic acid, to form 2-iodo-normal-hexane $CH_3(CH_2)_3CHICH_3$, (7) with sodium hydroxide solution, to form yellowish-brown solutions upon warming,

CLASSES OF CARBOHYDRATES (With examples)

Monosaccharides (sugars):
crystalline solids, soluble in water, sweet taste; those that occur in nature are hydrolyzed by certain enzymes.
Tetrose, $C_4H_8O_4$
1. Erythrose
Pentoses, $C_5H_{10}O_5$
2. Arabinose
By boiling gum arabic, cherry gum, corn pith, elder pith with dilute sulfuric acid.
3. Xylose
By boiling substances mentioned under arabinose above.
4. Ribose
5. Lyxose
Hexoses, $C_6H_{12}O_6$
Aldohexoses
6. Glucose, dextrose ("grape sugar"), melting point 146°C (anhydrous). With the enzyme zymase (of yeast) yields ethyl alcohol plus carbon dioxide. Specific rotatory power—see glucose below.
7. Galactose
Specific rotatory power +83.9°.
8. Mannose
Specific rotatory power +14.1°.
9. Gulose
10. Idose
11. Talose
12. Altrose
13. Allose
Ketohexoses
14. Fructose, levulose ("fruit sugar"), melting point 95°C. Specific rotatory power −88.5°.
15. Sorbose
16. Tagatose
Disaccharides (sugars), $C_{12}H_{22}O_{11}$:
crystalline solids, soluble in water, sweet taste.
17. Sucrose ("cane sugar," "beet sugar"), melting point 170–186°C (decomposes). With the enzyme invertase, yields glucose plus fructose. Specific rotatory power +66.4°.
18. Lactose ("milk sugar"), melting point 202°C (anhydrous). With the enzyme lactase yields glucose plus galactose. Specific rotatory power +52.4°.
19. Maltose ("malt sugar"), melting point of $C_{12}H_{22}O_{11}\cdot H_2O$: 100°C. With the enzyme maltase yields glucose plus glucose. Specific rotatory power +138.5°.
20. Melibiose
With enzymes or dilute acid yields glucose plus galactose.
21. Cellobiose
With the enzymes maltase, or cellase, yields glucose plus glucose.
22. Trehalose
Trisaccharide, $C_{18}H_{32}O_{16}$:
crystalline solid, soluble in water, tasteless.
23. Raffinose, melitose, melting point 118°C (anhydrous). With the enzyme invertase, yields fructose plus melibiose. With the enzyme emulsin, yields sucrose plus galactose.
Polysaccharides (non-sugars), $(C_6H_{10}O_5)_n$:
noncrystalline solids, insoluble in water, tasteless.
24. Starches
With the enzyme diastase yield maltose.
25. Celluloses
With hydrochloric acid, heated, yield glucose.
With acetic anhydride plus concentrated sulfuric acid, yield cellobiose.
26. Dextrin
With the enzyme diastase yields maltose.
With the enzyme maltase or with acids yields glucose.
27. Inulin, melting point 178°C (decom.) $(C_6H_{10}O_5)_n$.
With the enzyme inulase (but not with diastase) yields fructose.
28. Glycogen, melting point 240°C.
With the enzyme diastase (or ptyalin), yields glucose plus maltose.
29. Pentosans

(8) with calcium hydroxide solution, to form calcium glucosate $CH_2OH(CHOH)_4COCa(OH)$, slightly soluble solid from which glucose is recoverable by action of carbon dioxide (calcium carbonate formed simultaneously). Strontium hydroxide and barium hydroxide react similarly. Any of these three reactions may be utilized to recover glucose, with the limitation that barium soluble compounds are poisonous, (9) with hydroxylamine hydrochloride, to form glucoseoxime $CH_2OH(CHOH)_4CH{:}NOH$, melting point 138°C, (10) with hydrocyanic acid, to form glucosecyanhydrin

$$CH_2OH(CHOH)_4CHOHCN,$$

(11) by oxidation, to yield with bromine gluconic acid $CH_2OH(CHOH)_4COOH$, and with nitric acid saccharic acid $COOH(CHOH)_4COOH$, (12) with alpha-naphthol dissolved in chloroform and then forming a layer of concentrated sulfuric acid beneath the mixture, to form a red coloration at the junction of the two liquid layers (Molisch's test for carbohydrates). Upon standing, the color changes to purple. (13) With methyl alcohol in the presence of hydrogen chloride, to form methyl glucoside (methyl ether of glucose). See also **Glycosides.**

If a sample of glucose is recrystallized from water, it is found that a freshly prepared aqueous solution of this sample has a specific rotation of +113°, and upon standing, the value steadily changes to +52° and remains there. On the other hand, if a sample of the same glucose is recrystallized from pyridine, a freshly prepared aqueous solution has a specific rotation of +19°, which steadily increases upon standing and levels off at a constant value of +52°. This changing of optical rotation with time is referred to as mutarotation. The fact that the two portions of glucose when recrystallized from different solvents mutarotate and stop at the same position suggests the formation of some equilibrium mixture.

To explain this situation, it must be recognized that glucose contains an aldehyde (—CHO) group and four alcohol groups (—OH). These two kinds of groups can react to form a hemiacetal just as if they were present in different molecules (Fig. 1).

Glucose and fructose are present in sweet fruits, such as grapes and figs, and in honey. These two are the only hexoses found in nature in the free state. Glucose is normally present in human urine to the extent of about 0.1%, but in the case of those suffering from diabetes glucose is excreted in large amounts. Glucose is formed, as previously mentioned, by the reaction of polysaccharides and water, the reaction with starch in the presence of very dilute hydrochloric acid serving as the industrial source (the hydrochloric acid acts as a catalyzer, and the small percentage present is later neutralized to form sodium chloride). The solution is evaporated to a syrup or to crystallization, and is used in the manufacture of sweets, and (usually) alcohol, and in foods. The reaction of glucosides with water, by enzymes or acids, produces glucose as one of the products. With sodium hydroxide, under carefully defined conditions, glucose forms lactic acid. Glucose is used as food and for the production of alcohol (wines) from fruit juices. Glucose may be detected by formation of glucosazone, and determination of its melting point.

Industrial process for converting starch into dextrose (glucose) are described under **Starch.**

Fructose. This sugar is present with glucose in sweet fruits and honey, and may be obtained free by reaction of inulin of dahlia tubers or artichokes with water, and with glucose by reaction of sucrose with water, the product being known as invert sugar. Fructose differs

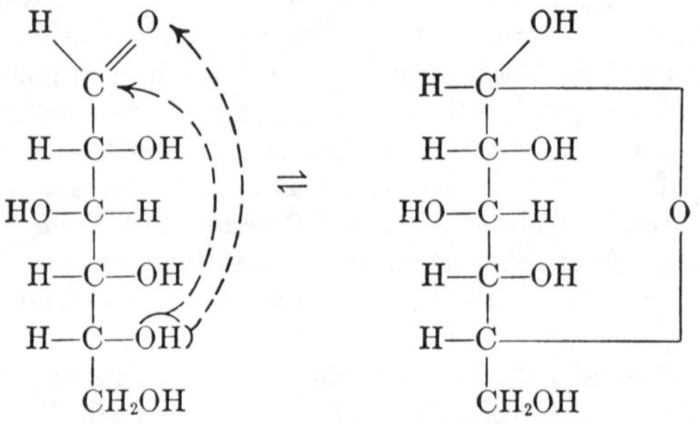

Fig. 1. Mutarotational aspects of glucose.

from glucose in structure in being a pentahydroxy-2-ketone,

$$CH_2OH(CHOH)_3COCH_2OH$$

instead of aldehyde. The specific rotary power of fructose is −88.5°. Fructose forms the same identical osazone as glucose, and sorbitol plus mannitol by reduction. Fructose may be used as sugar by diabetic patients to advantage instead of glucose or sucrose. Fructose is detected by the violet color its alkaline solution gives with meta-dinitrobenzene.

Sucrose. This is a colorless solid which when heated melts at 170–186°C, and upon cooling forms barley sugar, which gradually crystallizes. Upon heating above the melting point, it forms caramel, a brown liquid, with decomposition. Caramel is used in confectionery, and in coloring beverages and foods. At higher temperatures decomposition into gaseous and tarry substances occurs, finally leaving a residue of carbon ("sugar charcoal"). Other sugars behave similarly. Sugars are also carbonized by concentration sulfuric acid. Sucrose is very soluble in water, and is obtained from solution by crystallization, usually by vacuum evaporation. The solution has a specific rotatory power of +66.4°, does not exhibit mutarotation, but is converted by acids or invertase into invert sugar (glucose plus fructose), specific rotatory power −19.7°. Sucrose forms with calcium hydroxide calcium sucrosate, a 1% solution of sugar dissolves about 18 times as much calcium hydroxide as does pure water. This behavior is utilized to recover sugar from solutions, as in the case of glucose, and also to determine free calcium oxide in burnt lime, due to the reactivity of calcium hydroxide and non-reactivity of calcium carbonate. Sucrose is nonreactive with dilute sodium hydroxide, with phenylhydrazine, with ammonio-silver salt solution, but, when inverted to glucose plus fructose, these reactions may be obtained. Sucrose forms with acetic anhydride sucrose octaacetate. The suggested structural formula is as shown in Fig. 2. Sucrose is an important food preservative, food flavor, and a raw material for confectionery and for industrial alcohol.

Sucrose is extensively distributed in the seeds and leaves of plants, and is the most abundant of the sugars. The commercial sources of sucrose are the stems of sugar-cane (11 to 16% sucrose, average 13%), the root of the sugar-beet (average 16% sucrose, selection having raised the sucrose content from 5% to a maximum of 20%), the sap of the sugar maple, and the stems of sorghum-cane. Sucrose is pressed from the stems of sugar-cane or sorghum-cane, and extracted with the water from the sliced roots of sugar-beets. The solutions are purified, evaporated and crystallized to such a degree that commercial sucrose is practically chemically pure (about 99.8% sucrose). The purity of sugar and the concentration or strength of sugar solutions is determined by the rotatory power of the solution, the special polariscope usually used being called a saccharimeter. Sucrose is reduced with Fehling's solution only after inversion.

The sugar content of some common fruits have been reported by Kulisch:

	Sucrose	Hexoses
Apple	1.0–5.4	7.0–13.0
Apricot	6.0	2.7
Banana, ripe	5.0	10.0
Pineapple	11.3	2.0
Strawberry	6.3	5.0

See also **Beet**; and **Sugarcane.**

Lactose. This sugar is obtained from the residual water solution (whey) of milk after removal of fat and casein for making butter and cheese. Milk contains about 4.5% of lactose. Lactose forms hard gritty crystals ("sand sugar") $C_{12}H_{22}O_{11} \cdot H_2O$, loses water at 140°C, melting point 202°C (anhydrous) with decomposition; is less sweet than sucrose, reduces ammoniocupric salt solution, ammoniosilver salt solution, forms osazone, melting point 200°C, turns yellow when warmed with sodium hydroxide solution. Lactose is the source of galactose, and undergoes, with the proper enzymes, fermentation into lactic acid and butyric acid.

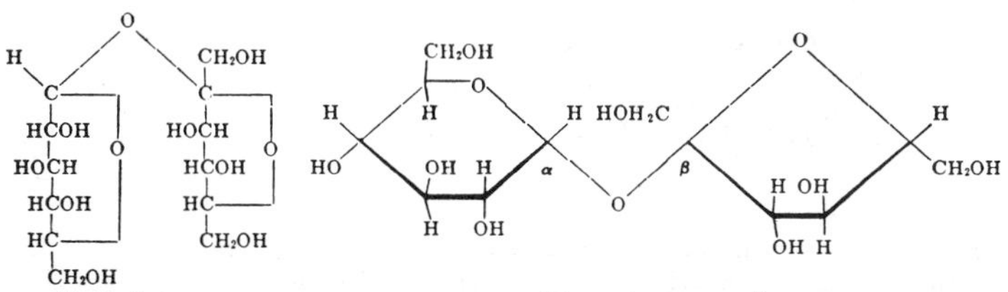

Fig. 2. Sucrose. *When the oxygen atom is drawn at the top of the furanose ring, OH groups drawn downward correspond to those on the left side of the vertical structure.

Maltose. This sugar is found in soybean, and is produced by the action of the enzyme diastase of germinated barley (malt) on starch at 50°C, and is thus an intermediate product in the transformation of starch into alcohol. Maltose $C_{12}H_{22}O_{11} \cdot H_2O$, melting point 100°C, when rapidly heated, may be crystallized from the concentrated malt syrup after removal of proteins and insoluble material. Maltose reduces ammonio-cupric salt solution, and forms osazone.

Starch. This is a white powder, odorless and tasteless, insoluble in cold water, forming an emulsion ("starch paste") or gel with hot water, the consistency of which depends upon the ratio of starch to water used. When boiled starch emulsion is cooled and treated with a solution of iodine in alcohol or potassium iodide, a blue coloration is produced, which is a sensitive and characteristic test. The blue color is associated with the adsorption of iodine on the surface of the starch, and disappears in the presence of alkalis. When boiled with dilute acid, starch is first changed into a soluble gummy mixture known as dextrin, and finally into glucose. When starch, either alone or in the presence of a slight amount of nitric acid, is heated to 120° to 200°C, dextrin is formed; at higher temperatures starch behaves similarly to sucrose. With concentrated nitric acid, starch forms esters, similar to cellulose nitrates. By the action of the enzyme diastase, starch is converted into maltose, which with the enzyme maltase yields glucose. Starch is nonreactive with ammonio-cupric salt solution, and with phenylhydrazine. See also **Starch.**

Dextrin. This is a white-to-yellow solid, forming an adhesive with water, non-reactive with ammonio-cupric salt solution, reactive with iodine in alcohol or potassium iodide, usually forming red, brown, or blue color. Formed when starch is (1) heated to 120° to 200°C either alone or in the presence of a slight amount of nitric acid. Dextrin is formed when bread is toasted and is present in well-baked bread crust, and on the surface of starched goods that have been ironed hot. Dextrin is used in adhesives.

Inulin. This is a white solid, soluble in warm water, specific rotatory power −40°, with iodine in alcohol or potassium iodide gives yellow color. Inulin is present in tubers of dahlia to the extent of about 10%. Inulin reacts with water in the presence of the enzyme inulase or of acids to form fructose. The enzyme diastase does not produce this change.

Glycogen. Also known as *animal starch,* this is a white solid, soluble in water, specific rotatory power +197°, with iodine in alcohol or potassium iodide solution, forming brown color. Glycogen is found as reserve carbohydrates in the animal body, more particularly in the liver. Horseflesh, oysters and beef are sources of glycogen.

Pentosans. These compounds are polysaccharides which may be considered as anhydrides of pentose sugars, after the manner of the hexosans, sucrose, starch, from glucose, fructose. When pentosans or pentoses are heated with hydrochloric or sulfuric acid, furfural $C_4H_3O \cdot CHO$ is formed, and addition of aniline produces a red color. Pentosans are present in gummy carbohydrates, in bran of wheat seed, and in woods.

By means of the cyanhydrin reaction, higher sugars of the heptose, octose, and nonose types have been prepared. A monosaccharide such as an aldohexose may be converted into the next lower monosaccharide, such as an aldopentose, by oxidation to the acid, which corresponds to the aldohexose, then treating the calcium salt solution of this acid with a solution of ferrous acetate plus hydrogen peroxide. Carbon dioxide is evolved and aldopentose formed.

For a description of cellulose, see **Cellulose.**

Carbohydrate Metabolism

Carbohydrates are utilized by the cells as a source of energy and as precursors for the manufacture of many of their structural and metabolic components. In the mammal, for example, D-glucose is the carbohydrate primarily used for this purpose. Certain microorganisms, in contrast, can grow on a medium containing some other hexose or a pentose as the principal source of carbon. Green plants obtain their carbohydrates by photosynthesis, while animals receive most of their carbohydrates by ingestion and digestion. See also **Photosynthesis.**

The complete oxidation of glucose to carbon dioxide and water yields 689 kcal of heat per mole of glucose. When this oxidation occurs in a cell, the energy is not all dissipated as heat. Some of the evolved energy is conserved in biochemically utilizable form of "high-energy" phosphates, such as adenosine triphosphate (ATP) and guanosine triphosphate (GTP). In addition to enzymes concerned with energy metabolism, there are enzymes in biological systems which catalyze the transformation of glucose into various carbohydrates, fatty acids, steroids, amino acids, nucleic acid components, and other necessary biochemical substances. The entire network of reactions involving compounds which interconvert carbohydrates constitutes *carbohydrate metabolism.* By convention, some reactions involving compounds which are not carbohydrates, but which are derived from them, may also be included in this area of metabolism.

Anaerobic Oxidation of Glucose. Historically, the first system of carbohydrate metabolism to be studied was the conversion by yeast of glucose to alcohol (fermentation) according to the equation: $C_6H_{12}O_6 \rightarrow 2CH_3CH_2OH + 2CO_2$. The biochemical process is complex, involving the successive catalytic actions of 12 enzymes and known as the *Embden–Meyerhof pathway.* This series of reactions is summarized in the entry on **Glycolysis.**

In order for the cell to carry out a "controlled" oxidation of D-glucose and conserve some of the energy derived from the process, it is first necessary to add phosphate to the hexose with the expenditure of energy. The necessary energy and the phosphate per se is supplied by ATP in two separate reactions of the system. Since each molecule of glucose can yield two molecules of triose phosphate for oxidation, the conversion of glucose to pyruvic acid nets two molecules of ATP per molecule of hexose utilized.

Approximately 30% of the evolved energy is conserved as ATP, but only about 8% of the total energy in glucose is made available in this anaerobic oxidation of glucose to pyruvic acid. Since nicotinamide adenine dinucleotide (NAD^+), also called diphosphoryidine nucleotide (DPN^+), which is involved in the oxidation of glyceraldehyde-3-phosphate, is present in the cell in small quantities only, this coenzyme must constantly be regenerated for the oxidative process to continue. This regeneration is accomplished by the reduction of *acetaldehyde* to *ethanol.* Since oxygen plays no role in this process, the system can obviously proceed anaerobically. In fact, the presence of oxygen decreases the net disappearance of glucose (*Pasteur effect*).

Fermentation occurs in many microorganisms, but not all organisms reoxidize the reduced nicotinamide adenine dinucleotide (NADH) through the formation of ethanol. In certain organisms, for example, *pyruvic acid* is converted to *acetoin* which is then reduced with NADH to 2,3-butylene glycol. In other organisms and in animal tissues, NADH is oxidized in the reduction of *pyruvic acid* to *lactic acid.* In insects, and possibly in some animal tissues, the reduction of *dihydroxyacetone phosphate* to *alpha-glycerol phosphate* may serve to regenerate NAD^+. The conversion of glucose to lactic acid in animal tissues is termed *glycolysis.* This term arose from the initial understanding that this process was markedly different from the microbial fermentation process. Fermentation and glycolysis are now known to differ primarily in the further anaerobic utilization of pyruvic acid.

Aerobic Oxidation of Pyruvic Acid. Pyruvic acid can be oxidized completely to carbon dioxide and water in a cyclic enzymatic system known as the *Krebs citric acid cycle,* or the *tricarboxylic acid cycle* (*TCA cycle*). In this system, a two-carbon unit in the form of acetyl coenzyme A (acetyl = CoA), derived from the NAD^+ mediated oxidative decarboxylation of pyruvic acid in the presence of coenzyme A, is condensed with oxalacetic acid to form citric acid. This tricarboxylic acid is then converted back to oxalacetic acid in a stepwise manner with the formation of $2CO_2$ and $2H_2O$. In addition to this formation of CO_2, one reduced nicotinamide adenine dinucleotide phosphate (NADPH), two NADH, one reduced flavin, and one GTP arise per two-carbon unit oxidized in the cycle. Since in the aerobic oxidation of the reduced flavin and the reduced nicotinamide adenine nucleotides, ATP is formed, the oxidation of a molecule of "acetate" results in the conservation of energy in the form of 12 molecules of triphosphate. In the complete oxidation of glucose through glycolysis and the citric acid cycle, about 40% of the energy originally present in the glucose can be retained as triphosphate. The ubiquitous distribution of this cycle in nature suggests that the citric acid cycle is a major energy-yielding pathway in biological systems.

Certain microorganisms have a modification of this cycle in which isocitric acid is cleaved to succinic acid and glyoxylic acid. The latter acid is condensed with acetyl-CoA to form malic acid. In this modification (the *glyoxylic acid cycle*), oxalsuccinic acid and alpha-ketoglutaric acid are not involved. This is sometimes referred to as the "glyoxylate shunt" pathway.

Since in the citric acid cycle there is no net production of its intermediates, mechanisms must be available for their continual production. In the absence of a supply of oxalacetic acid, "acetate" cannot enter the cycle. Intermediates for the cycle can arise from the carboxylation of pyruvic acid with CO_2 (e.g., to form malic acid), the addition of CO_2 to phosphenolpyruvic acid to yield oxalacetic acid, the formation of succinic acid from propionic acid plus CO_2, and the conversion of glutamic acid and aspartic acid to alpha-ketoglutaric acid and oxalacetic acid, respectively. See Fig. 3.

The utilization of carbohydrate intermediates for the biosynthesis of amino acids, fatty acids, steroids, etc. occurs at various stages of the cycle and its related reactions. See Fig. 4. See also **Coenzymes.**

Other Carbohydrate Interconversions. Two systems, as shown in Fig. 5, are available for the synthesis of ribose-5-phosphate, a precursor of the pentose moiety of ribonucleic acid, ATP, and other substances. The formation of ribose-5-phosphate from glucose-6-phosphate by formation and decarboxylation of 6-phosphogluconic acid and isomerization of the resulting ribulose-5-phosphate is termed the *hexose monophosphate oxidative pathway.* The scheme, together with the system involving the enzymes *transketolase* and *transaldolase* (which also can synthesize pentose) that act to form hexose phosphate from pentose

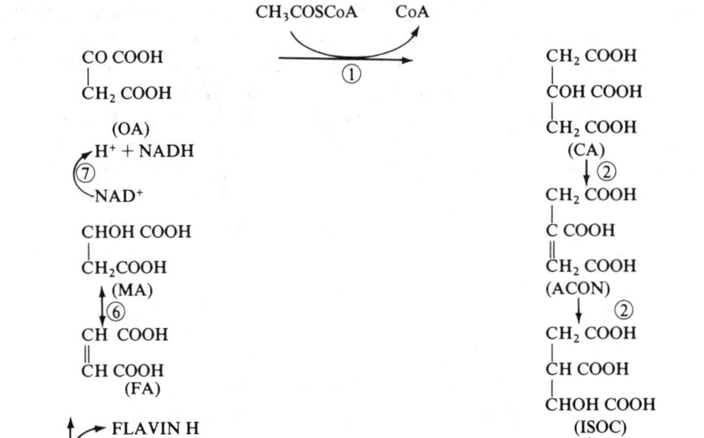

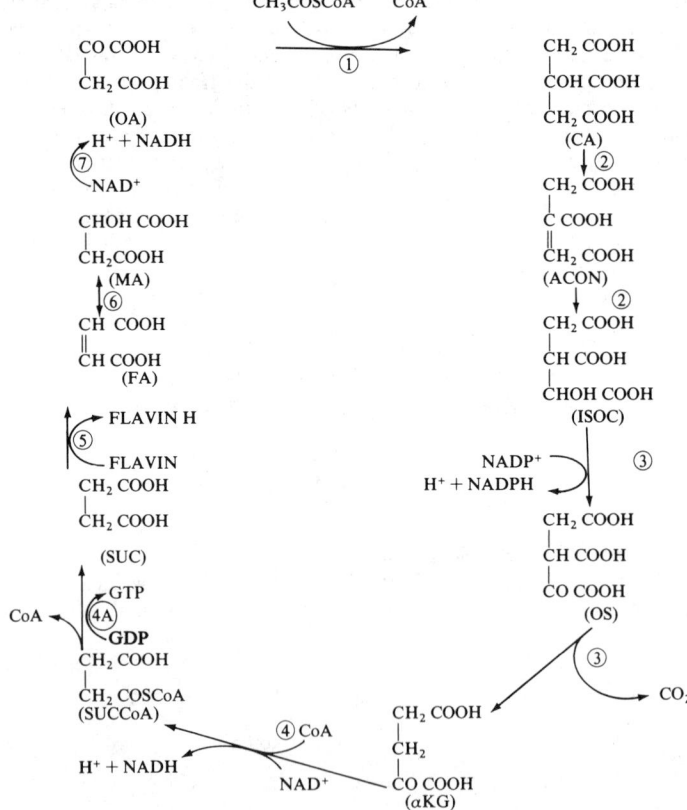

Fig. 3. Krebs citric acid cycle. *Enzymes involved*: (1) Condensing enzyme; (2) aconitase; (3) isocitric acid; (4) α-ketoglutaric acid dehydrogenase; (4A) succinic acid thiokinase; (5) succinic acid dehydrogenase; (6) fumarase; (7) malaic acid dehydrogenase. *Abbreviations*: CA = citric acid; ACON = *cis*-aconitic acid; KG = α-ketoglutaric acid; SUC = succinic acid; FA = fumaric acid; MA = malic acid; OA = oxalacetic acid.

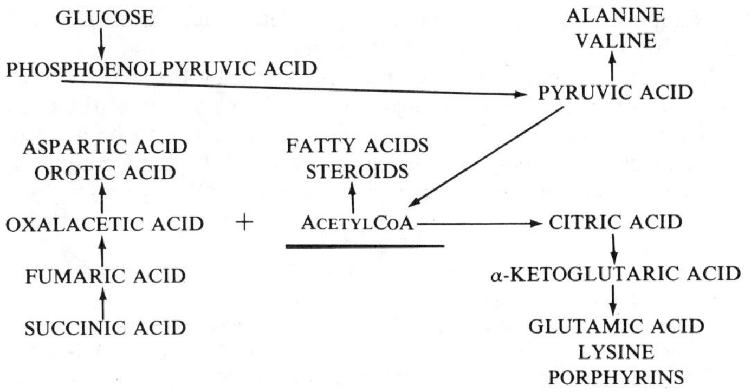

Fig. 4. Representative conversions of carbohydrates to other substances.

phosphate, is called the *pentose phosphate cycle.* This cycle represents an alternative pathway to glycolysis for the formation of triose phosphate from glucose-6-phosphate. The relative importance of the two pathways seems to be different among the various organisms and tissues.

In a certain group of bacteria, still another pathway (*Entner–Doudoroff pathway*) for the utilization of glucose has been studied. Here glucose-6-phosphate is oxidized to 6-phosphogluconic acid which is dehydrated to 2-keto-3-deoxy-6-phosphogluconic acid. This substance is then split to pyruvic acid and glyceraldehyde-3-phosphate (which also can be converted to pyruvic acid).

The formation of deoxyribose, the pentose moiety of deoxyribonucleic acid, can occur directly from ribose while the latter is in the form of a nucleotide diphosphate. Deoxyribose-5-phosphate can also be formed by condensation of acetaldehyde and glyceraldehyde-3-phosphate.

Transglycosylation. An enzymatic process, transglycosylation, plays an important role in carbohydrate metabolism. Figure 6 represents the formation of the disaccharide, sucrose, as an example of this mechanism. In the upper reaction of Fig. 6, glucose-1-phosphate is the glycosyl donor and fructose is the acceptor. In the lower reaction, the sugar nucleotide, uridine diphosphoglucose (UDP-glucose) is the glycosyl donor. With UDP-glucose as donor and glucose-6-phosphate as acceptor, trehalose-6-phosphate may be formed. Polysaccharides may also be formed by this process. The donor residues provided by sugar nucleotides are added to preexisting polysaccharide chains (known as "primers") acting as glycosyl acceptors. In the formation of glycogen, for example, UDP-glucose donates the glucose moiety which is added to the end of a previously synthesized chain by a 1,4-linkage, thereby lengthening the chain by one glucose unit.

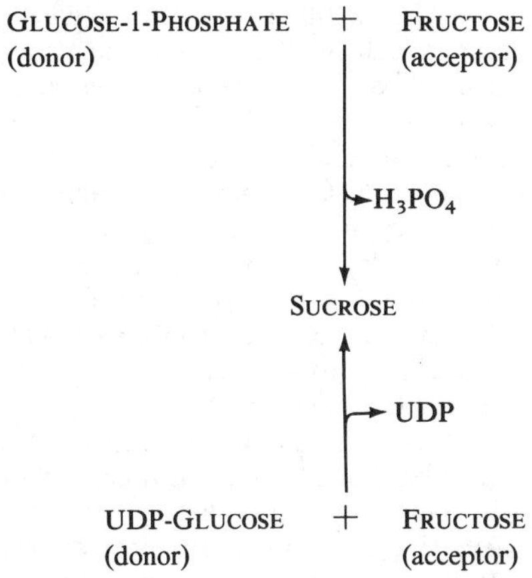

Fig. 6. Examples of transglycosylation.

Digestion, Absorption and Storage of Carbohydrates

In the mammal, complex polysaccharides which are susceptible to such treatment, are hydrolyzed by successive exposure to the amylase of the saliva, the acid of the stomach, and the disaccharidases (e.g., maltase, invertase, amylase, etc.) by exposure to juices of the small intestine. The last mechanism is very important. Absorption of the resulting monosaccharides occurs primarily in the upper part of the small intestine, from which the sugars are carried to the liver by the portal system. The absorption across the intestinal mucosa occurs by a combination of active transport and diffusion. For glucose, the active transport mechanism appears to involve phosphorylation. The details are not yet fully understood. Agents which inhibit respiration (e.g., azide, fluoracetic acid, etc.) and phosphorylation (e.g., phlorizin), and those which uncouple oxidation from phosphorylation (e.g., dinitrophenol) interfere with the absorption of glucose. See also **Phosphorylation (Oxidative).** Once the various monosaccharides pass through the mucosa, interconversion of the other sugars to glucose can begin, although the liver is probably the chief site for such conversions. Even though many organs and tissues store carbohydrates as glycogen for their own use, the liver provides the main source of glucose for all tissues through conversion of its glycogen (and other substances) to glucose-6-phosphate, hydrolysis of this ester by the specific liver glucose-6-phosphatase, and transport of the free glucose in the bloodstream throughout the body.

A common cause of *osmotic diarrhea* is the ingestion of carbohydrates that a person cannot digest. Retention of a disaccharide, such as lactose or sucrose, within the intestinal lumen occurs because of the absence of the appropriate disaccharidase at the intestinal surface

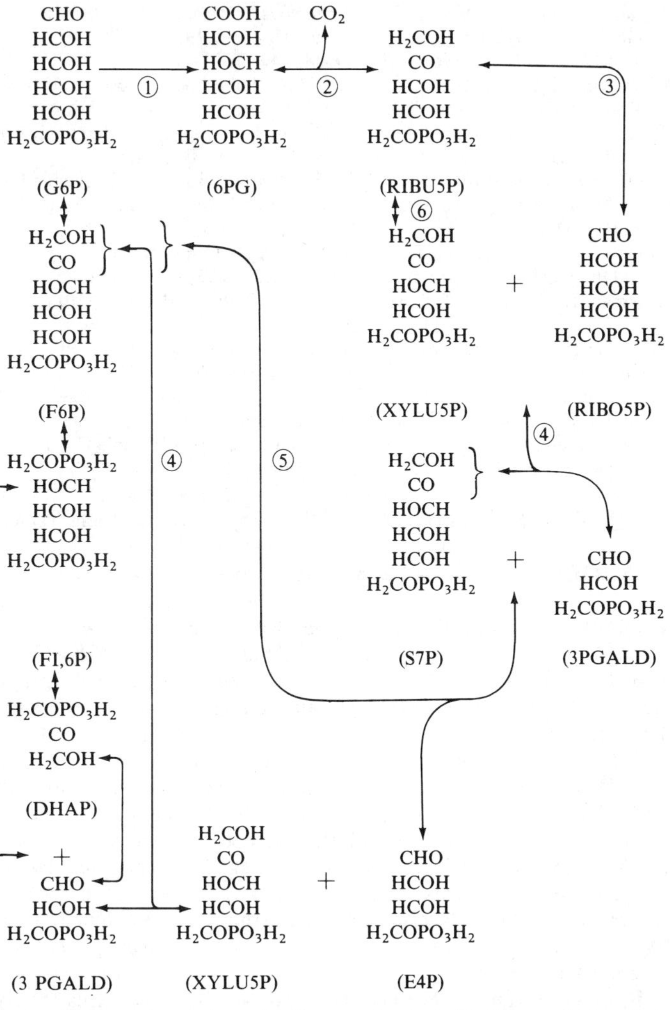

Fig. 5. Pentose phosphate cycle. *Enzymes involved*: (1) Glucose-6-phosphate dehydrogenase; (2) 6-phosphogluconic acid dehydrogenase; (3) pentose phosphate isomerase; (4) transketolase; (5) transaldolase; (6) pentose phosphate epimerase. *Abbreviations*: G6P = glucose-6-phosphate; 6PG = 6-phosphogluconic acid; RIBU5P = ribulose-5-phosphate; 3PGALD = glyceraldehyde-3-phosphate; E4P = erythrose-4-phosphate; F1,6P = fructose-1-6-diphosphate; DHAP = dihydroxyacetone phosphate; F6P = fructose-6-phosphate. Enzymes not named are those of glycolysis. $NADP^+$ is reduced in reactions (1) and (2).

membrane. Unless they are converted to monosaccharides, these sugars cannot be transported. Their retention in the lumen can cause a significant diarrheal water loss per day. As an added complication, bacteria in the lower small intestine and colon may catabolize the 12-carbon sugars to 3-carbon fragments, further aggravating the osmotic effect. Infants and very young children usually have sufficient lactase and sucrase, but there is a tendency among some people to become lactase deficient between the ages of 3 and 14 years. Once such a condition is fully recognized, the ingestion of milk and other dairy products should be eliminated. It should be pointed out, however, that the so-called *irritable bowel syndrom* is attributable to lactase deficiency in a relatively small percentage of cases. Carbohydrates that can cause diarrhea in some persons include lactose and sucrose, already mentioned; stachyose and raffinose contained in many legumes; mannitol and sorbitol, contained in artificial sweeteners (which contain sugar alcohols); glucose and galactose present in all dietary sugars; and lactulose, contained in nondietary disaccharides as parts of certain medications. See also **Diarrhea.**

Endocrine Influences. A number of hormones are known to influence carbohydrate metabolism in the mammal. Insulin seems to increase oxidation of glucose, lipogenesis, and glycogenesis. Its primary mode of action may be to facilitate the entry of glucose into the cell. The extremely important role of carbohydrate metabolism in connection with diabetes is described in entry on **Diabetes Mellitus.**

Vitamin Influences. The involvement of NAD^+ and $NADP^+$ in many carbohydrate reactions explains the importance of nicotinamide in carbohydrate metabolism. Thiamine, in the form of thiamine pyrophosphate (cocarboxylase), is the cofactor necessary in the decarboxylation of pyruvic acid, in the *trans*-ketolase-catalyzed reactions of the pentose phosphate cycle, and in the decarboxylation of alpha-ketoglutaric acid in the citric acid cycle, among other reactions. Biotin is a bound cofactor in the fixation of carbon dioxide to form oxalacetic acid from pyruvic acid. Pantothenic acid is a part of the CoA molecule. There are separate alphabetical entries in this volume on the various specific vitamins as well as a review entry on **Vitamin.**

Photosynthesis. The formation of carbohydrates in green plants by the process of photosynthesis is described in the entry on **Photosynthesis.** The synthetic mechanism involves the addition of carbon dioxide to ribulose-1,5-diphosphate and the subsequent formation of two molecules of 3-phosphoglyceric acid which are reduced to glyceraldehyde-3-phosphate. The triose phosphates are utilized to again from ribulose-5-phosphates by enzymes of the pentose phosphate cycle. Phosphorylation of ribulose-5-phosphate with ATP regenerates ribulose-1,5-diphosphate to accept another molecule of carbon dioxide. See also **Phosphorylation (Photosynthetic).**

Carbohydrates in Foods

Sugar is discussed in several entries in this volume, including **Beet; Fiber; Gums and Mucilages;** and **Sugarcane.**

Statistics on the carbohydrate content of diets of various peoples throughout the world have not been very reliable because of the scores of variables involved, the great difficulties in establishing reliable sampling procedures, lack of past records, among other factors. One summary, for example, that breaks down food energy from protein, fat, and carbohydrates shows a downward trend for carbohydrates in the American diet—from 56% in 1911 to 46% in the mid-1970s. These figures were based upon U.S. Department of Agriculture statistics of food disappearance at the retail level, but they do not take into consideration food spoilage, cooking waste, plate waste, and other factors which affect actual consumption. Since protein remained quite constant at 11–12% throughout this time span, the drop in carbohydrates was made up by an increase in fats—from 32% in 1911 to 42% in the mid-1970s. In another study, of the 46% carbohydrate energy intake as of 1977, 24% is attributed to sugar and 22% to complex carbohydrates. In a controversial U.S. government study, which attempts to set new dietary goals for the nation, it was suggested that the traditional 12% protein be retained, but that fat be reduced from 42% and carbohydrates upped to 58%, but with a major difference, namely, cutting the sugar portion of carbohydrates from 24% to 15%. Thus, the dietary goal would require 40–45% complex carbohydrates in the diet. It has been suggested that to achieve the projected carbohydrate goals, there would have to be a 66% increase in the consumption of grain products; a 25% increase of vegetables and fruit; and a 50% reduction in sugar and sweets.

Even though much visibility has been given by the various news media to the dietary role of sugar, it is obvious that, as of the early 1980s, a great deal of fundamental research remains to be done to prove or disprove many conclusions, often conflicting and confusing, in order to establish reliable dietary guidance in this area.

References

Celender, I. M., et al.: "Dietary Trends and Nutritional Status in the United States," *Food Technology*, **32**, 9, 39–41 (1978).

Cornblath, M., and R. Schwartz: "Disorders of Carbohydrate Metabolism in Infancy" in "Major Problems in Clinical Pediatrics," Vol. 3 (A. J. Schaffer and M. Markowitz, editors), Saunders, Philadelphia, 1976.

Gray, G. M.: "Intestinal Digestion and Maldigestion of Dietary Carbohydrates," *Ann. Rev. Med.*, **22**, 391 (1971).

Greenberg, D. M.: "Metabolic Pathways," 3rd edition, Vol. 1, Academic, New York, 1967.

Hollingsworth, D. F.: "Translating Nutrition into Diet," (The British Nutritional Foundation), *Food Technology*, **31**, 2, 38–44 (1977).

Hood, L. F., Wardrip, E. K., and G. N. Bollenback: "Carbohydrates and Health," AVI, Westport, Connecticut, 1977.

Mottram, R. F.: "Human Nutrition," 3rd edition, Food and Nutrition Press, Westport, Connecticut, 1979.

Ralph, C. L.: "Carbohydrate Metabolism," in "The Encyclopedia of the Biological Sciences," (P. Gray, editor), Van Nostrand Reinhold, New York, 1970.

Scala, J.: "Responsibilities of the Food Industry to Ensure an Optimum Diet," *Food Technology*, **32**, 9, 77–79 (1978).

Segal, S., and L. F. Hood, Co-Chairpersons: "Carbohydrates," *Institute of Food Technologists Symp.*, Saint Louis, Missouri, 1979.

Sharon, N.: "Carbohydrates," *Sci. Amer.*, **243**, 5, 90–117 (1980).

Siperstein, M. D., Foster, D. W., et al.: "Control of Blood Glucose and Diabetic Vascular Disease" (editorial), *N. Engl. J. Med.*, **296**, 1060 (1977).

Staff: "Food and Nutrient Intake of Individuals in the United States," Report 11, U.S. Department of Agriculture, Washington, D.C., 1972.

Staff: "A Guide for Professionals: The Effective Application of Exchange Lists for Meal Planning," Amer. Diabetes Assn. and Amer. Dietetic Assn., New York, 1977.

Sussman, K. E.: "Juvenile-Type Diabetes and Its Complications. Theoretical and Practical Considerations," Charles C. Thomas, Springfield, Illinois, 1971.

Sussman, K. E., and R. J. S. Metz (editors): "Diabetes Mellitus," 4th edition, Amer. Diabetes Assn., New York, 1975.

Sutherland, H. W., and J. M. Stowers: "Carbohydrate Metabolism in Pregnancy and the Newborn," Churchill Livingstone, New York, 1975.

Watt, B. K., and A. L. Merrill: "Composition of Foods," Agriculture Handbook 8, U.S. Department of Agriculture, Washington, D.C., 1975.

Weser, E., et al.: "Lactose Deficiency in Patients with 'Irritable-Colon' Syndrome," *N. Engl. J. Med.*, **273**, 1070 (1965).

CARBOHYDRATES (In Diet). Dietary Requirements and Trends.

CARBOLIC OIL. Coal Tar and Derivatives.

CARBON. Chemical element symbol C, at. no. 6, at. wt. 12.011, periodic table group 4a, mp 3,550°C (approximate), bp 4289°C (approximate), density 3.52 g/cm^3 (diamond at 20°C), 2.25 g/cm^3 (graphite at 20°C). The specific gravity of amorphous carbon at 20°C ranges from 1.8 to 2.1. There are two stable isotopes of the element, ^{12}C and ^{13}C, and four known radioactive isotopes ^{10}C, ^{11}C, ^{14}C, and ^{15}C. ^{14}C occurs in nature as the result of interaction of cosmic rays with ^{14}N. Inasmuch as the half-life of ^{14}C has been established (about 5,760 years), the occurrence of this isotope in ancient documents, artifacts, and materials makes it a useful diagnostic tool in archeological investigations. See also **Radioactivity.** The first ionization potential of carbon is 11.264 eV; second, 24.28 eV; third, 47.7 eV; fourth, 64.19 eV. Other important physical characteristics of carbon are given under **Chemical Elements.**

Principal Elemental Forms: There are two allotropic forms (1) diamond, and (2) graphite. Diamond, the hardest of natural substances, consists of a lattice of carbon atoms arranged in a tetrahedral structure at equal distance apart (1.544 Å) and bonded by electron pairs in localized molecular orbitals formed by overlapping of the sp^3 hybrids. In graphite, one of the softest substances, the carbon atoms are arranged in laminar sheets, 3.40 Å apart and composed of carbon atoms in hexagonal arrangement 1.42 Å apart, with each atom bonded to three others in its sheet by electron pairs in localized molecular orbitals formed by overlapping of the sp^2 hybrids. The remaining *p*-electrons form a mobile system of nonlocalized π-bonds that permits of electrical conductivity within the lamina. The various carbon blacks formed by such methods as combustion of carbon-containing materials with sufficient oxygen are found to have x-ray diffraction patterns suggestive of graphite, but with more diffuse rings, indicating a much lower degree of crystallinity. When carbon black is heated its diffraction pattern develops new rings indicative of a structure more like graphite. At the same time, its properties as an absorbent deteriorate.

The uses of carbon are dependent upon the form and variety: diamonds for jewels and as abrasives, graphite in lubricants and as an electrical conductor, cocoanut charcoal for adsorbing gases at low temperature in an enclosed space to produce a high vacuum, activated carbon to absorb color from solutions and to remove odor from water, coke and wood charcoal as fuels. See also **Carbon Black; Coal; Diamond;** and **Graphite.**

New Views of High-Temperature Behavior: The probable importance to high-temperature behavior of the —C≡C— bond, most familiarly encountered in acetylene (itself stable at high temperatures), was not proposed until the late 1960s (El Gorsey and Donnay, 1968; Sladkov and Koudrayatsev, 1969). It was proposed that high-temperature carbon forms are made up of chains of —C≡C— units, called *carbynes* by the Soviet scientists. A proposed mechanism for the transformation of a graphite basal plane sheet of atoms into (—C≡C—) units is shown in Fig. 1. It will be noted that at high temperatures, a single

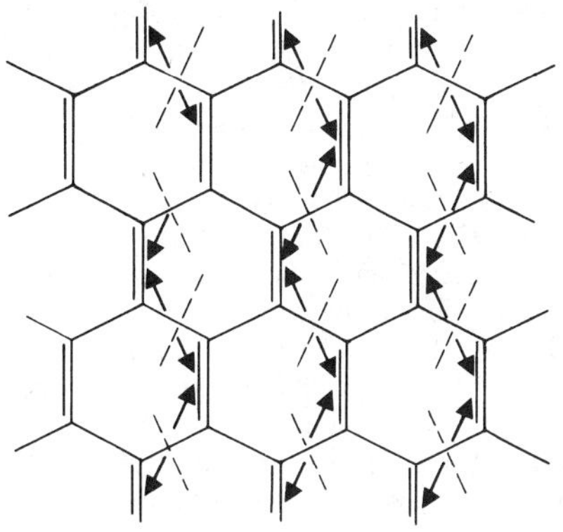

Fig. 1. Possible mechanism for transformation of a graphite basal plane sheet of atoms into carbyne chains. (*After Whittaker, 1978.*)

bond in the structure may break. This shifts an electron into each of the adjacent double bonds, forming a triple bond. Completion of the process transforms the sheet of atoms into a chain of carbynes. The chains, of course, can be variously stacked. Kasatochkin et al. (1973) reported at least five such forms. As explained at a conference on carbon held in Irvine, California in 1977, the transformation from the carbyne form to graphite involves a reaction between acetylene-like molecules (acting rapidly and exothermically), whereas the reverse reaction (breaking of single bonds) can be expected to be a much slower process. Thus, it is observed that the conventional carbon phase diagram may be lacking because it does not consider carbyne forms. Whittaker (1978) pointed out that for years it has been difficult to reconcile high-pressure results with the low-pressure data on the vapor pressure of carbon. As shown by Fig. 2, there is inclusion of a region for carbynes which accommodates experimental findings. Whittaker further notes: (1) Graphite is not stable above 2600 K at any pressure; (2) the solid–liquid–vapor triple point occurs at 3800 K and 2×10^4 Pa; and (3) carbyne forms are stable between 2600 and 3800 K, and their stability region extends to the diamond transition line.

Compounds of Carbon: With exception of hydrogen, carbon forms the largest number of known compounds (hundreds of thousands) among the chemical elements. Traditionally, carbon compounds fall into two fundamental classes (1) *inorganic* compounds and (2) *organic* compounds, although the line of demarcation is not always precise, differing from one authority to the next.

The main subclasses of inorganic carbon compounds include:

1. The *carbon oxides,* notably CO (carbon monoxide) and CO_2 (carbon dioxide). (In attempts to resolve spectral discrepancies from observations of the planet Mars, it appears that the red coloration of the planet may be due at least partially to the presence of

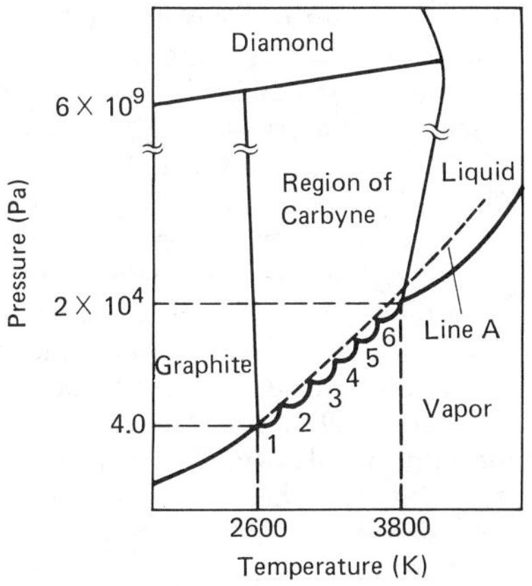

Fig. 2. Proposed carbon phase diagram to accommodate region of carbyne. Dashed line A is vapor pressure graphite would have if it were the stable from above 2600 K. (*After Whittaker, 1978.*)

carbon suboxide C_3O_2 in the Martian atmosphere. It is believed that the linear molecules of C_3O_2 polymerize, forming several heavier molecules having color gradations ranging from pale yellow to orange, reddish brown, violet, and nearly black. *Science,* **166**, 1141–1142 (November 28, 1969).
2. The *carbonates* —CO_3, which occur widely in nature—minerals, rocks, ores, and mineral waters—and include such compounds as Na_2CO_3, $CaCO_3$ (limestone), $MgCO_3$ (magnesite), $MgCa(CO_3)_2$ (dolomite), etc.
3. Some carbon–sulfur compounds, such as CS_2 (carbon disulfide), the thiocyanides, and thiocyanates —CNS, such as HCNS (thiocyanic acid), $Pb(CNS)_2$ (lead thiocyanate), etc.
4. The *carbides,* such as Na_2C_2, Cu_2C_2, WC, ZrC, etc.
5. The *carbonyls* —CO, such as $Cr(CO)_6$, $Fe(CO)_5$, $Ni(CO)_4$, etc.
6. The *halides,* such as CCl_4 (carbon tetrachloride), CBr_4, etc.

The subclasses of organic compounds comprise the realm of organic chemistry and are described under **Organic Chemistry.** There are several subclasses of organic compounds that include oxygen along with hydrogen and carbon in their structure—e.g., acid anhydrides, alcohols, aldehydes, carbohydrates, carboxylic acids, esters, ethers, fatty acids, furans, ketones, lactides, lactones, phenols, quinones, and terpenes. Some of the main subclasses of nitrogen-bearing organic compounds include the amides, amines, amino acids, anilides, azo and diazo compounds, carbamates, cyanamides, hydrazines, polypeptides, proteins, purines, pyridines, pyrroles, quaternary ammonium compounds, semicarbazones, ureas, and ureides. The addition of the halogens to the structure yields chlorine organics, brominated compounds, fluorocarbons, etc. Most of the metals combine with carbon compounds to form organometallics. Sulfur-bearing organics include the sulfonic acids, sulfonyls, sulfones, thioalcohols, thioaldehydes, sulfoxides, etc. Silicones are silicon-bearing carbon compounds. See also **Chlorinated Organics.**

Carbides: As might be expected from its position in the periodic table, carbon forms binary compounds with the metals in which it exhibits a negative valence, and binary compounds with the non-metals in which it exhibits a positive valence. A convenient classification of the binary compounds of carbon is into ionic or salt-like carbides, intermediate carbides, interstitial carbides, and covalent binary carbon compounds.

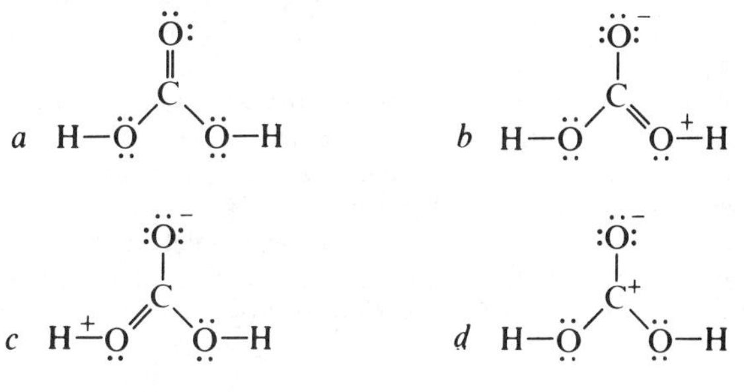

The ionic or salt-like carbides are formed directly from the elements, or from metallic oxides and carbon, carbon monoxide, or hydrocarbons. This last reaction is reversible, and this group of carbides may be further subdivided into acetylides, e.g., Li_2C_2, Na_2C_2, K_2C_2, Rb_2C_2, Cs_2C_2, Cu_2C_2, Ag_2C_2, Au_2C_2, BeC_2, Mg_2C_2, CaC_2, SrC_2, BaC_2, ZzC_2, CdC_2, Al_2C_6, Ce_2C_6, and ThC_4; methanides, e.g., Be_2C and Al_4C_3; and the allylides, primarily magnesium allylide, Mg_2C_3, according to the hydrocarbon or the principal hydrocarbon formed upon hydrolysis. By the term intermediate carbides is meant compounds intermediate in character between the ionic carbides and the interstitial carbides. The intermediate carbides, such as Cr_3C_2, Mn_3C, Fe_3C, Co_3C, and Ni_3C are similar to the ionic carbides in that they react with water or dilute acids to give hydrocarbons, and they resemble the interstitial carbides in their electrical conductivity, opacity, and metallic luster. The interstitial carbides have these properties, and are uniformly chemically inert. They include those having cubic close-packed structures, such as TiC, ZrC, HfC, VC, NbC, TaC, MoC, and WC, and those having hexagonal close-packed structures such as V_2C, Mo_2C, and W_2C. In both, the carbon atoms occupy interstitial positions in the crystal lattices of the metals, giving hardness, high melting points, and chemical inertness, as well as electrical conductivity with a positive temperature coefficient and metallic luster. The covalent binary compounds of carbon range in character from hard, chemically inert solids, such as silicon carbide, SiC, to volatile liquids, such as carbon disulfide and carbon tetrachloride, CS_2 and CCl_4, and even to gases such as carbon tetrafluoride, carbon dioxide and methane, CF_4, CO_2 and CH_4, varying in thermal stability. With several of these elements carbon forms a series of compounds, or as with hydrogen, a number of series of hydrocarbons, consisting of both compounds based upon chains and branched chains of carbon atoms, variously saturated (i.e., joined by single, double, or triple bonds), and also of ring-connected carbon atoms, with or without side chains, with varying degrees of saturation, and capable of replacement of the hydrogen atoms with other atoms or radicals.

Carbonates: Carbonic acid H_2CO_3 is present to the extent of 0.27% of the total CO_2 present in the solution that is formed by dissolving CO_2 in H_2O at room temperature. The CO_2 may be expelled fully upon boiling. The solution reacts with alkalis to form carbonates, e.g., sodium carbonate, sodium hydrogen carbonate, calcium carbonate, calcium hydrogen carbonate. The acid ionization constant usually cited for carbonic acid (4.2×10^{-7}) is actually for the equilibrium $CO_2(aq) + H_2O \rightleftharpoons H^+ + HCO_3^-$. The true ionization constant, i.e., for the equilibrium $H_2CO_3 \rightleftharpoons H^+ + HCO_3^-$ is about 1.5×10^{-4}. The carbonate ion is a resonance hybrid of the three structures shown *a*, *b*, and *c* as well as structures of the type *d* which give a partial ionic character to bonds. This resonance is somewhat inhibited in the acid and its esters, but is complete, or much more nearly complete, in many other derivatives and in the carbonate ion. Esters of both metacarbonic, $(RO)_2CO$, and orthocarbonic acid, $(RO)_4C$, are known. The esters also exhibit resonance.

Metallic carbonates are (1) soluble in H_2O, e.g., sodium carbonate, potassium carbonate, ammonium carbonate (2) insoluble in H_2O and excess alkali carbonate, e.g., calcium carbonate, strontium carbonate, barium carbonate, magnesium carbonate, ferrous carbonate (3) insoluble in H_2O but soluble in excess alkali carbonate forming carbonate complexes, e.g., compounds of uranium and ytterbium $U(CO_3)_2$, UO_2CO_3, $Yb_2(CO_3)_3$. Metallic bicarbonates are known in solution and on warming are converted into ordinary or normal carbonates, e.g., bicarbonates of sodium, potassium, calcium, barium. These are preferably named as "hydrogen carbonates," e.g., $NaHCO_3$ = sodium hydrogen carbonate. Basic carbonates are important in such cases as lead ("white lead"), zinc, magnesium, and copper. Carbonates of very weak bases, such as aluminum, iron(III), and chromium(III), are now known.

The carbonates are found in nature as the carbonates, calcite, iceland spar, limestone and various forms of impure calcium carbonate $CaCO_3$, as magnesite (magnesium carbonate, $MgCO_3$), as dolomite (various compositions of calcium and magnesium carbonates), as witherite $SrCO_3$, as strontianite $SrCO_3$, as azurite and malachite (various compositions of cupric hydroxycarbonates), in various natural waters as carbonic acid, calcium and magnesium hydrogen carbonates, in blood, as sodium hydrogen carbonate.

Many esters of carbonic acid are known, e.g., diethyl carbonate, ethyl ester of metacarbonic acid, $(C_2H_5O)_2CO$, made by reaction of ethyl alcohol and carbonyl chloride; dimethyl carbonate, $(CH_3O)_2CO$; methyl ethyl carbonate, $(CH_3O)CO(OC_2H_5)$; dipropyl carbonate, $(C_3H_7O)_2CO$; tetraethyl carbonate, ethyl ester of orthocarbonic acid, $(C_2H_5O)_4C$, bp 158°C.

Peroxycarbonic acid exists only in its compounds. Alkali peroxycarbonates are obtained by electrolysis of concentrated solutions of the carbonates, the anodic reaction being written as

$$2CO_3^{2-} \longrightarrow C_2O_6^{2-} + 2e^-$$

The peroxycarbonates are relatively stable only in concentrated alkaline solutions. On dilution they decompose to give the bicarbonate and hydrogen peroxide

$$Na_2C_2O_6 + 2H_2O \longrightarrow 2NaHCO_3 + H_2O_2$$

when acidified, the peroxycarbonate ion gives, correspondingly, CO_2 and hydrogen peroxide

$$C_2O_6^{2-} + 2H^+ \longrightarrow 2CO_2 + H_2O_2$$

Carbonyls: The metal carbonyls are strongly covalent in character, as shown by their volatility, their solubility in many nonpolar solvents, and their insolubility in polar solvents. They also behave in many reactions like mixtures of carbon monoxide, CO, and the metal. Those of group 6b elements, $Cr(CO)_6$, $Mo(CO)_6$, and $W(CO)_6$ are more stable and less reactive than the others, especially those of group 8 elements. Group 7b carbonyls are $Mn_2(CO)_{10}$, $Tc_2(CO)_{10}$, and $Re_2(CO)_{10}$, while group 8 elements form $Fe(CO)_5$, $Fe_2(CO)_9$, $Fe_3(CO)_{12}$, $Co_2(CO)_8$, $Co_4(CO)_{12}$, $Ni(CO)_4$, $Ru(CO)_5$, $Ru_2(CO)_9$, $Ru_2(CO)_{12}$, $Rh_2(CO)_8$, $Rh_3(CO)_9$, (and multiples), $Rh_4(CO)_{11}$, (and multiples), $Os(CO)_5$, $Os_2(CO)_9$, $Ir_2(CO)_8$, and $Ir_3(CO)_9$ (and multiples). The carbonyls form a wide variety of addition compounds; they are dissolve in alcoholic potassium hydroxide or other strong alkalies to form hydrides which are acids, and can be used to form a wide variety of more complex compounds. Although $H_2Fe(CO)_4$ is a moderately weak acid, $pK_1 = 4.44$, $pK_2 = 14.0$, $HCo(CO)_4$ appears to be comparable with HCl in acidity. The carbonyl compounds have zero charge number on the metal. The mononuclear carbonyls are spin-paired complexes, and are formed only by metals having even atomic numbers. However, metals having odd atomic numbers can form carbonyl compounds with other atoms or radicals, as exemplified by the nitrosyl compound of cobalt carbonyl, $Co(CO)_2NO$, where the —NO radical contributes the electron necessary to complete the 3*d* level of the cobalt atom. More than one NO group may occur in a metal carbonyl, as, for example, in $Fe(CO)_2(NO)_2$. This is isostructural with $Co(CO)_3NO$ and $Ni(CO)_4$.

Halides: The four tetrahalides of carbon are symmetrical, planar compounds, with the general property of marked stability to chemical reactions, although the tetraiodide undergoes slow hydrolysis in contact with water to form iodoform and iodine. It also decomposes under the action of light and heat. The stability of these four compounds decreases in order of descending periodic table position. Their properties are given below:

Name	Formula	MP	BP
Carbon tetrafluoride	CF_4	−184°C	−128°C
Carbon tetrachloride	CCl_4	−23.0°	76.8°
Carbon tetrabromide	CBr_4	{α48.4°, β90.1°}	189.5°
Carbon tetraiodide	CI_4	171° dec.	

The same relation of reactivity and stability to periodic position is exhibited by such other carbon halides as hexachloroethane $CCl_3 \cdot CCl_3$ and hexabromoethane, $Br_3 \cdot CBr_3$, as well as by hexachloroethylene, $CCl_2 = CCl_2$ and hexabromoethylene, $CBr_2 = CBr_2$. Carbon also forms halides containing more than one halogen. See also **Carbon Tetrachloride.**

It is well established that hydrogen forms more than one covalent binary compound with carbon. Fluorine behaves similarly. Thus, fluorine forms CF_4, C_2F_4, C_2F_6, C_3F_8 and many higher homologs, as well as the definitely interstitial compound $(CF)_n$. The other halogens form some similar compounds, although to more limited extent, and various polyhalogen compounds have been prepared. They exhibit the maximum covalency of four and are therefore inert to hydrolysis and most other low temperature chemical reactions.

Carbon Oxides: See **Carbon Dioxide; Carbon Monoxide; Carbon Suboxide.**

New Viewpoint on Structure of Carbon Compounds: For over a century, a principle stated by H. van't Hoff and Joseph A. LeBel (1874) has permeated organic chemistry, namely, that a carbon atom with substituents bonded to it prefers a tetrahedral geometry.[1] Experimental work until the mid-1970s came up with no fundamental exceptions to the principle. Organic compounds in which all four substituents lie in a plane had not been found. But, as the result of work by Schleyer (University of Erlangen-Nürnberg) and Pople (Carnegie-Mellon University), aided by advanced computer techniques in making molecular orbital calculations, a number of familiar organic compounds have been proposed as having preferred configurations that differ markedly from accepted norms. As pointed out by Maugh (1976), the technique used by Schleyer and Pople essentially comprises the selection of two or more likely geometries for a molecule, followed by calculation of their relative energies. If all appropriate structures have been included in the calculations, then the geometry with the lowest calculated energy should be the most stable. The approach has been validated for numerous molecules whose structures are well known. This is exemplified for calculations of the methane molecule. A tetrahedral configuration is approximately 150 kcal per mole lower than the energy calculated for a structure in which all five atoms lie in the same plane. On the other hand, the energy is less for a planar configuration of CF_2Li_2 than for the tetrahedral configuration. Although practical application of this new information remains to be developed, it is believed that this new structural perspective will contribute to a better understanding of many organic reactions.

References

El Gorsey, A., and G. Donnay: *Science*, **161**, 363 (1968)

Field, J. E. (editor): "The Properties of Diamond," Academic, New York, 1979.

Kasatochkin, V. I., et al.: *Carbon*, **11**, 70 (1973).

Maugh, T. H., II: "Unusual Structures Predicted for Carbon Compounds," *Science*, **194**, 413 (1976).

Sladkov, A. M., and Y. P. Koudrayatsev: *Priroda*, **5**, 37 (1969).

Whittaker, A. G.: "Carbon: A New View of Its High-temperature Behavior," *Science*, **200**, 763–764 (1978).

CARBONADO. The mineral carbonado is an opaque massive black variety of diamond, often crystalline to granular or compact and without cleavage. In thin splinters it appears greenish-black by transmitted light. It is found chiefly in Bahia, Brazil. Carbonado is used for rock-drilling apparatus.

Carbonado also is known as *black diamond.*

CARBONATED DRINKS (Additives). Acidulants and Alkalizers (Foods); Antimicrobial Agents (Foods).

CARBONATES. Carbon.

CARBON BLACK. Finely divided carbonaceous pigments of a wide variety are termed carbon blacks. Over 90% of the carbon black manufactured is consumed as reinforcing and compounding agents for rubber, mainly for motor vehicle and aircraft tires. Most users of tires do not realize that the effective use of these agents extends the life of a tire in normal usage by eight to ten times. The addition of as little as 1 to 2% carbon black to plastics greatly minimizes the effects of sunlight in degrading the materials. Most carbon blacks are derived from the pyrolysis of hydrocarbon gases and oils. The permanent and penetratingly deep black coloration obtainable with carbon blacks also makes the materials attractive for paints, inks, protective coatings, and as colorants for paper and plastics.

Two properties of carbon blacks are most significant for commercial applications: (1) particle size and (2) surface area. The particle sizes range from 100 to 5,000 micrometers. Surface areas will range from 6 to 1,100 m^2/g of material. Under electron microscopic examination, the carbon particles appear as rough spheres, usually as clusters of spheres rather than as individual spheres. The clustering characteristics stem from both chemical and physical bonding forces. Classically, the arrangement of the carbon particles may be likened to hexagonal nets of carbon atoms which are *paracrystalline* in nature. The particle size and surface area characteristics essentially are at the microscopic level—hence control over carbon black production is exacting. In terms of coloration, for example, the human eye can resolve 260 shades of blackness. The blackest of commercially produced carbon particles will have a diameter of about 100 micrometers. The grayest particle will have a diameter of about 5,000 micrometers. The blackness characteristic sometimes is referred to as *masstone* (mass-tone). The particles with the smaller diameters and hence greater surface area exhibit the highest masstone.

Lampblacks have been made for many centuries. Early methods involved the burning of petroleum-like substances or coal-tar residues with a minimum of air, thus producing large amounts of unoxidized carbon particles. The earlier settling chambers in which the particles collected have been replaced by cyclones, bag filters, or electrical precipitators. Modern installations use oil furnaces to create the particles.

Channel or *impingement carbons* are produced from burning natural gas (sometimes containing oil vapors) in many hundreds of small burners. The flames from the burners impinge upon flat surfaces called channels. The carbon deposits are periodically removed by scraping into a collector. The burning equipment is contained within a large burner house which has means for carefully regulating both bottom and top drafting of air.

Thermal blacks also are derived from natural gas, but by thermal decomposition completely in the absence of air. Large furnaces first are preheated to a temperature ranging from 1,100–1,650°C. When the checkerwork is at the proper temperature, natural gas is bled into the furnace, whereupon the gas decomposes into carbon and hydrogen. This is a batch process, requiring pairs of furnaces, one furnace preheating, while the other furnace is decomposing the gas feed. Frequently, the hydrogen by-product is recycled as fuel to heat the furnaces. Where very fine thermal blacks are produced, the by-product hydrogen is used as a diluent for the gas feed.

Furnace carbons also are derived from natural gas, but in a process in which a slight excess of air is introduced to support combustion. The hydrocarbon feedstock or liquid oil is injected into the furnace at a location where the so-called blast-flame gases are circulating at their greatest velocity. Injection of the feed at this point causes an instant high rise in temperature which results in practically instantaneous decomposition of the feed into carbon black. For coarse particles, the oil/air ratio is greater, furnace gas velocities are lower, and residence time in the furnace is longer. There is a wide range of furnace carbon particle sizes. The very fine particles go into tire treads, whereas the coarser particles are used in tire carcasses.

Acetylene black is derived from feeding acetylene into high-temperature retorts whereupon the acetylene dissociates into carbon and hydrogen. This reaction is exothermic (other carbon black processes are endothermic). Temperature control of the furnace is effected by throttling the acetylene feed.

CARBON-BUTTON MICROPHONE. Microphone.

CARBON COMPOUNDS. Organic Chemistry.

CARBON CYCLE (Nuclear). In physics and astronomy, a series of thermonuclear reactions, releasing great quantities of energy (by conversion from mass and by radiation) that are believed to furnish the energy radiated by some of the stars. This scheme was developed from theoretical considerations by H. A. Bethe in 1939 (and simultaneously by C. F. von Weizsäcker). Various possibilities were tried but the following series was the only one that gave results in agreement with the experimental facts:

[1] Double-bonded molecules, such as ethylene, and triple-bonded molecules, such as acetylene, conventionally have been considered as generally linear.

$$C^{12} + H^1 \rightarrow N^{13} + \gamma$$
$$N^{13} \rightarrow C^{13} + e^+ + \nu$$
$$C^{13} + H^1 \rightarrow N^{14} + \gamma$$
$$N^{14} + H^1 \rightarrow O^{15} + \gamma$$
$$O^{15} \rightarrow N^{15} + e^+ + \nu$$
$$N^{15} + H^1 \rightarrow C^{12} + He^4$$

where e^+ indicates a position, ν indicates a neutrino, and γ indicates a gamma ray.

The overall reaction results in the production of a helium atom, two positrons, two neutrinos and 4×10^{-5} ergs from four protons, the carbon atom that reacted initially being regenerated at the end of the process. There are, of course, other probable side reactions.

The mass rate of energy generation is given by

$$\epsilon_C = \rho X(100\ \alpha_N) f_N E_N$$

where ρ is the density, X is the fraction of hydrogen by mass, α_N is the fraction of nitrogen by mass, f_N is the shielding factor for nitrogen, and E_N is a function dependent upon temperature.

CARBON DATING. Radioactivity and Other Dating Techniques.

CARBON DIOXIDE. CO_2, formula weight 44.01, colorless, odorless, nontoxic gas at standard conditions. High concentrations of the gas do cause stupefaction and suffocation because of the displacement of ample oxygen for breathing. Density 1.9769 g/l (0°C, 760 torr), sp gr 1.53 (air = 1.00), mp −56.6°C (5.2 atmospheres), solid CO_2 sublimes at −79°C (760 torr), critical pressure 73 atmospheres, critical temperature 31°C. Carbon dioxide is soluble in H_2O (approximately 1 volume CO_2 in 1 volume H_2O at 15°C, 760 torr), soluble in alcohol, and is rapidly absorbed by most alkaline solutions. The solubility of CO_2 in H_2O for various pressures and temperatures is given in the accompanying table.

SOLUBILITY OF CARBON DIOXIDE IN WATER

Pressure (atmospheres)	Parts (weight) CO_2 Soluble in 100 Parts Water				
	18°C	35°C	50°C	75°C	100°C
25	3.7	2.6	1.9	1.4	1.1
50	6.3	4.4	4.0	2.5	2.0
75	6.7	5.5	4.5	3.4	2.8
100	6.8	5.8	5.1	4.1	3.5
200	—	6.3	5.8	5.3	5.1
300	7.4	—	6.2	5.8	5.7
400	7.8	7.1	6.6	6.3	6.4
700	—	—	7.6	7.4	7.6

Carbon dioxide plays several roles: (1) as a *raw material* for several processes, as in the Solvay process for the manufacture of sodium bicarbonate and sodium carbonate, (2) as a *by-product* from many processes, notably as a product of combustion of fossil fuels, (3) as an *ingredient* of products, for example, carbonated beverages, (4) as a *product* for direct consumption, for example, CO_2 fire extinguishers and dry ice refrigerants, and (5) as a *pollutant* of the atmosphere. Although not toxic, the presence of CO_2 in the atmosphere disturbs the environmental energy balance. The latter aspects of CO_2 are discussed under **Pollution (Air).** Normally, CO_2 is present in the air at sea level to the extent of about 0.05% by weight.

Solid carbon dioxide (dry ice) is an effective refrigerant for transportation uses. Refrigeration of moving vehicles may be derived from (1) mechanical systems which, of course, require a continuous input of energy, (2) water ice and ice–salt mixes which require water (often briny) removal, and are corrosive and subject to algae formations, and (3) dry ice, the end-product of which is simply gaseous CO_2, which is easily removed. To maintain a cool temperature in a railroad refrigerator car for a trip between California and New York, about 1,000 pounds of dry ice would be required. To maintain the same conditions with water ice and salt would require 10,000 pounds of ice. The fact that CO_2 is heavier than air makes it particularly effective for fighting fires in low places, such as pipe trenches and hard-to-reach low corners and basements, where the CO_2 tends to roll under the air required to maintain combustion. Both manually- and automatically-controlled CO_2 fire-fighting systems are available. These can be actuated by heat-sensitive systems—just as a conventional water-sprinkling system. CO_2 is effective for fires involving electrical and electronic gear because, if a fire is not fully out-of-hand, the CO_2 often can quickly quench the fire source without leaving any residual damage, as often is the disastrous consequences of using water or sand.

Although carbon dioxide must be generated on site for some processes, there is a trend toward CO_2 recovery where it is a major reaction byproduct and, in the past, vented to the atmosphere. For example, very large quantities of CO_2 are generated by various fermentation processes and in cement production. If the CO_2 must be removed from stack gases because of pollution control regulations, it is only one more step to purify the gas and sell it, usually in compressed liquid form. There are, of course, several economic tradeoffs which must be considered. Where the gas is recovered, it usually is first absorbed in sodium or potassium carbonate solutions, followed by steam-heating the solutions to free a reasonably pure CO_2. The last step is compression of the gas into steel cylinders. The ethanolamines also are excellent absorbents of CO_2.

In some dry ice manufacturing plants, CO_2 is produced by burning fuel oil. The heat generated is used to strip the resulting CO_2 from the monoethanolamine absorbing solution. The gas then is scrubbed with potassium permanganate solution to remove odorous impurities prior to compressing.

Structure. Carbon dioxide offers one of the classical examples of the role of resonance in interpreting the structure and properties of substances. The heat of formation from the atoms of the linkage of a >C=O bond is 150 kcal per mole, but the heat of formation of CO_2 from one C and two O atoms is not 2 × 150 = 300 kcal, but 336 kcal per mole. Moreover, the observed C=O bond length in carbon dioxide is not the usual 1.22 Å, but 1.15 Å. The greater than expected heat of formation, and the shorter than expected bond length are explained by assigning to CO_2 not only the structure O=C=O, but two others also, $^-$O—C≡O$^+$ and $^+$O≡C—O$^-$, the structural picture being a hybrid (not an interchange) of the three, and more stable than any one alone. The contributions of the last two are equal in amount, so the molecule has a dipole moment of zero. The molecular orbital explanation is in terms of two fractional nonlocalized π-bonds.

Direct measurement gave a value of -4×10^{-26} electrostatic unit for the molecular quadrupole moment of CO_2. This corresponds to a charge on each oxygen atom of 0.32 electronic charge.

Carbon Dioxide in Biological Systems. Carbon dioxide, which is a byproduct of the metabolic activity of all cells, is one of the most important chemical regulators in the human body. It can be said that human life without carbon dioxide would be impossible. In less specialized forms of life, carbon dioxide is essentially a waste product. In the more highly developed animals, such as humans, the gas is used to regulate the activity of the heart, the blood vessels, and the respiratory system.

As mentioned, CO_2 is normally present in air at sea level at about 0.05% (weight). A poorly ventilated room may contain as much as 1% (volume). Concentrations of the gas from about 0.1–1% (volume) induce languor and headaches; concentrations of 8–10% (volume) bring about death by asphyxiation. High concentrations of the gas are toxic. See also **Basal Metabolism.**

As a general rule, the respiration of individual cells decreases as the concentration of carbon dioxide in the medium increases. Fish show a lessened capacity to extract oxygen from their environment with increasing amounts of carbon dioxide present. On the other hand, many invertebrates show marked increases in respiratory rate (or ventilation) with increased amounts of the gas in their surroundings.

Photosynthetic and autotrophic bacteria reduce carbon dioxide which is assimilated into complex molecules for use in synthesizing various cellular constituents. The gas is apparently assimilated, at least to a small extent, by the heterotrophic bacteria. Certainly it is required for any growth in these forms. Many pathogenic bacteria required increased carbon dioxide tension for growth immediately after they are isolated from the body. The production of hemolysins and like substances is greatly enhanced by adding 10–20% of CO_2 in the air which comes in contact with the cultures.

The oxygen dissociation curve for blood is shifted to the right when the partial pressure of carbon dioxide is increased. This is referred to as the "Bohr Effect." It means that for a given partial pressure of oxygen, hemoglobin holds less oxygen at high concentration of carbon dioxide than at a lower concentration. It is evident, then, that the production of carbon dioxide by actively metabolizing tissues favors the release of oxygen from the blood to the cells where it is urgently needed. Moreover, at the alveolar surfaces in the lungs, the blood is losing carbon dioxide rapidly, which loss favors the combination of oxygen with hemoglobin. In males, the average amount of CO_2 in the alveolar air is about 5.5% (volume); during the breathing cycle, this concentration varies only slightly. In females and children, somewhat lower mean values obtain.

In every 100 milliliters of arterial blood, there is a total of 48 milliliters of free and combined CO_2. In venous blood of resting humans, there is about 5 milliliters more than this. Only about 1/20 of the carbon dioxide is uncombined, a fact which indicates that there is a specialized mechanism, aside from simple solution, for the transport of CO_2 in the blood.

About 20% of the CO_2 in the blood is carried in combination with hemoglobin as *carbaminohemoglobin.* The balance of the combined carbon dioxide is carried as bicarbonate. A CO_2 dissociation curve for blood can be prepared just as for oxygen, but the shape is not the same as for the latter. As the partial pressure of CO_2 in the air increases, the amount in the blood increases; the increase is practically linear in the higher ranges. Oxygen exerts a negative effect on the amount of CO_2 which can be taken up by the blood.

In working muscles large amounts of CO_2 are produced. This causes local vasodilation. The diffusion of some of the CO_2 into the bloodstream slightly raises the concentration there. It circulates through the body and the capillaries of the vasoconstrictor center, where it excites the cells of the center, resulting in an increase of constrictor discharges. Regarding the stimulating effect of CO_2 on cardiac output, it is evident that a most effective mechanism exists for increasing circulation through active muscles. More blood is pumped by the heart per minute and the arterial pressure is increased by the general vasoconstriction; blood is forced from the inactive regions, under increased pressure, through the widely dilated vessels of the active muscles.

The partial pressure of CO_2 is important in connection with a number of physiological problems. For example, respiratory acidosis is the result of an abnormally high p_aCO_2. The value of arterial pCO_2 varies directly with changes in the metabolic production of CO_2 and indirectly with the amount of alveolar ventilation. The problem is more commonly the result of decreased alveolar ventilation caused by abnormally low CO_2 excretion by the lungs (alveolar *hypo*ventilation).

On the other hand, primary respiratory alkalosis occurs as a result of alveolar *hyper*ventilation. This condition is associated with a number of pulmonary diseases, but also may appear during pregnancy, liver disease, and salicylate intoxication, among others. The sequence of events proceeds along these lines: (1) Ventilation removes CO_2 faster than the gas is produced by metabolism, causing a decrease in pCO_2 in the blood and body fluids, including a reduction of venous pCO_2. This reduces the gradient for excretion of CO_2 by the lungs. (2) Pulmonary excretion and metabolic production ultimately balance out at a lower pCO_2 level for all body fluids. (3) The lower pCO_2 level causes a lower carbonic acid concentration and consequently an increase in pH. The latter is relative to the reduced level of pCO_2, but the pH change also alters bicarbonate concentration. The steplike process is quite complex. See also **Blood.**

Narcosis due to CO_2 is characterized by mental disturbances which may range from confusion, mania, or drowsiness to deep coma, headache, sweating, muscle twitching, increased intracranial pressure, pounding pulse, low blood pressure, hypothermia, and sometimes papilloedema. The basic mechanisms by which carbon dioxide induces narcosis is probably through interference with the intracellular enzyme systems, which are all sensitive to pH changes.

See also **Photosynthesis.**

CARBON DIOXIDE COOLANT. Nuclear Reactor.

CARBON DIOXIDE (Critical Point). Critical Point.

CARBON DIOXIDE (Polar Atmosphere). Polar Research.

CARBON DIOXIDE POLLUTION. Climate.

CARBON GROUP (The). The elements of group 4a of the periodic classification sometimes are referred to as the Carbon Group. In order of increasing atomic number, they are carbon, silicon, germanium, tin, and lead. The elements of this group are characterized by the presence of four electrons in an outer shell. The similarities of chemical behavior among the elements of this group are less striking than that for some of the other groups, e.g., the close parallels of the alkali metals or alkaline earths. However, as more knowledge is gained of silicon, including the element's ability to form "carbon-like" chains with alternating silicon and oxygen atoms, to polymerize, and to form silicones, silanes, etc., the similarity of silicon and carbon emerges more sharply. The semiconductor properties of silicon and germanium in this group are striking, but of course such properties are not limited to elements in this group. Although some of the elements of the group have valences in addition to +4, all do have the +4 valence in common. Unlike the alkali metals or alkaline earths, for example, the elements of the carbon group are not so similar chemically that they comprise a separate group in classical qualitative chemical analysis separations.

CARBONIC ACID DERIVATIVES. Insecticide.

CARBONITRIDING. A surface hardening process for steels involving the introduction of carbon and nitrogen into steels by heating in a suitable atmosphere containing various combinations of hydrocarbons, ammonia, and carbon monoxide followed by a quenching to harden the case.

CARBONIUM ION. An ion of the general formula

$$\begin{array}{c} A \\ | \\ B\text{—}C^+ \\ | \\ D \end{array}$$

, where A, B and D are substituent groups. Its importance in elucidating the mechanism of organic reactions is because a considerable proportion of all organic reactions involve carbonium ions, as others do carbanions and carbon free radicals (including carbene radicals). Nucleophilic substitution at saturated carbon atoms includes most of carbonium ion chemistry. Carbonium ions are usually powerful acids or electrophiles, and thus many nucleophilic substitution reactions that involve carbonium ions are acid-catalyzed. For example, the tertiary-butyl carbonium ion offers a clear understanding of the probable course of the conversion of isobutylene to its dimers and trimers.

$$(CH_3)_2C{=}CH_2 + H^+ \leftrightarrows (CH_3)_3C^+$$

$$(CH_3)_3C^+ + (CH_3)_2C{=}CH_2 \leftrightarrows (CH_3)_2\overset{+}{C}\text{—}CH_2C(CH_3)_3$$

The larger carbonium ion thus formed cannot continue to exist, but may depolymerize, unite with the catalyst, or stabilize itself by the attraction of an electron pair from a carbon atom adjacent to the electronically deficient carbon (C^+) with its proton. This establishes a double bond involving the formerly deficient atom. Thus a proton is expelled to the catalyst or attracted to the catalyst. If this takes place with one of the methyl groups, the product is $CH_2{=}\underset{\displaystyle CH_3}{\underset{|}{C}}\text{—}CH_2C(CH_3)_3$. If the methylene group is involved, the product is $(CH_3)_2C{=}CHC(CH_3)_3$.

CARBONIZATION (Coal). Coal.

CARBON MONOXIDE. CO, formula weight 28.01, colorless, odorless, very toxic gas at standard conditions, density 1.2504 g/l (0°C, 760 torr), sp gr 0.968 (air = 1.000), mp −207°C, bp −192°C, critical temperature −139°C, critical pressure 35 atmospheres. Carbon monoxide is virtually insoluble in H_2O (0.0044 part CO in 100 parts H_2O at 50°C). The gas is soluble in alcohol or solutions of cupric chloride. Because carbon monoxide has an affinity for blood hemoglobin that is 300 times that of oxygen, exposure to the gas greatly reduces or fully hinders the ability of hemoglobin to carry oxygen throughout the body, causing death in excessive concentrations. Engines and stoves in poorly ventilated areas are especially hazardous.

Carbon monoxide plays several roles: (1) as a *raw material* for chemical processes (a) particularly as an effective reducing agent in various metal smelting operations, (b) in the manufacture of formates: $CO + NaOH \rightarrow HCOONa$, (c) in the production of carbonyls, such as $Ni(CO)_4$ and $Fe(CO)_5$, which are useful intermediate compounds in the separation of certain metals, (d) in combination with chlorine to form $COCl_2$ (phosgene), (e) as an ingredient of several synthesis gases, as for the production of methanol and ammonia; (2) as a *fuel* where CO is a major ingredient of such artificial fuels as coal gas, producer gas, blast-furnace gas, and water gas; (3) as a *by-product* of numerous chemical reactions, notably combustion processes where there is insufficient oxygen for complete combustion—the fumes from internal-combustion engines may contain in excess of 7% CO, and (4) as a dangerous *air pollutant*, particularly in industrial areas and where there are high concentrations of automotive vehicles and aircraft. The latter aspects of CO are discussed under **Pollution (Air).**

Summary of Chemical Reactivity: Chemically, carbon monoxide (1) reacts with oxygen to form CO_2 accompanied by a transparent blue flame and the evolution of heat, but the fuel value is low (320 Btu per ft^3), (2) reactive with chlorine, forming carbonyl chloride $COCl_2$ in the presence of light and a catalyzer, (3) reactive with sulfur vapor at a red heat, forming carbonyl sulfide COS, (4) reactive with hydrogen, forming methyl alcohol, CH_3OH or methane CH_4 in the presence of a catalyzer, (5) reactive with nickel (also iron, cobalt, molybdenum, ruthenium, rhodium, osmium, and iridium) to form nickel carbonyl, $Ni(CO)_4$ (and carbonyls of the other metals named), (6) reactive with fused NaOH, forming sodium formate, HCOONa, (7) reactive with cuprous salt dissolved in either ammonia solution or concentrated HCl, which solutions are utilized in the estimation of carbon monoxide in mixtures of gases, e.g., flue gases of combustion, coal gas, exhaust gases of internal combustion engines, (8) reactive with iodine pentoxide at 150°C. For the reaction of carbon monoxide with oxygen to form CO_2 finely divided iron or palladium wire is used as a catalyzer; for the reaction of carbon monoxide with H_2O vapor to form CO_2 plus hydrogen ("water gas reaction") important studies have been made of the conditions; and for the reaction of CO_2 plus carbon (hot) similar important studies have been made (at 675°C, 50% CO_2 plus 50% CO; at 900°C, 5% CO_2 plus 95% CO). The reaction of carbon plus oxygen at such a temperature as produces carbon monoxide (say 900°C, 95% CO plus 5% CO_2) and *evolves heat*; while the reaction of carbon plus CO_2, producing carbon monoxide at the same temperature *absorbs heat.* Accordingly it is possible to arrange the oxygen (free or as air) and CO_2 supply ratio in such a way that the desired temperature may be continuously maintained. The reduction of CO_2 by iron forms carbon monoxide plus ferrous oxide.

In valence bond terms, carbon monoxide is considered as a resonance compound with the structures

$$\overset{+}{:C}:\ddot{\underset{..}{O}}:^- \qquad :C::\ddot{O}: \qquad :C::\underset{..}{O}: \qquad \overset{-}{:C}:::\overset{+}{O}:$$

In molecular orbital terms the CO molecule is described as $CO(KK(z\sigma)^2(y\sigma)^2(x\sigma)^2(w\pi)^4)$, one $(z\sigma)$ pair being formed from the oxygen 2*s* electrons, and one $(y\sigma)$ pair held by the carbon *sp* hybrid. This $(y\sigma)^2$ pair offsets the dipole moment of the π electrons, and also accounts for the readiness with which the CO molecule coordinates with metals to form the carbonyls.

CARBON MONOXIDE (Pollution). Pollution (Air).

CARBON STAR. Infrared Astronomy.

CARBON STEELS. Iron Metals, Alloys, and Steels.

CARBON SUBOXIDE. C_3O_2, formula weight 68.03, colorless, toxic, gas at room temperature, very unpleasant odor, sp gr 2.10 (air = 1.00), 1.24 (liquid at −87°C), mp −107°C, bp 7°C (760 torr), burns with a blue smoky flame, producing CO_2. When condensed to liquid, the oxide slowly changes at ordinary temperature to a dark red solid, soluble in water to a red solution. Reacts with water to form malonic acid, with hydrogen chloride to form malonyl chloride, with ammonia to form malonamide. Made by heating malonic acid or its ester at 300°C under diminished pressure, and separation from simultaneously formed carbon dioxide and ethylene by condensation and fractional distillation.

Carbon suboxide has a linear structure, probably a resonance of four structures of which the last two below probably make a smaller contribution to the normal state of the molecule than the first two.

$$:\ddot{O}::C::C::C::\underset{..}{O}:$$
$$:\underset{..}{O}::C::C::C::\ddot{O}:$$
$$\overset{+}{:O}:::C:C:::C:\underset{..}{\ddot{O}}:^-$$
$$^-:\underset{..}{\ddot{O}}:C:::C:C:::\overset{+}{O}:$$

CARBON TETRACHLORIDE. CCl_4, formula weight 82.82, heavy, colorless, nonflammable, noncombustible liquid, mp −23°C, bp 76.75°C, sp gr 1.588 (25°C/25°C), vapor density 5.32 (air = 1.00), critical temperature 283.2°C, critical pressure 661 atmospheres, solubility 0.08 g in 100 g H_2O, odor threshold 80 ppm. Dry carbon tetrachloride is noncorrosive to common metals except aluminum. When wet, CCl_4 hydrolyzes and is corrosive to iron, copper, nickel, and alloys containing those elements. About 90% of all CCl_4 manufactured goes into the production of chlorofluorocarbons:

$$2CCl_4 + 3HF \xrightarrow{\text{catalyst}} CCl_2F_2 + CCl_3F + 3HCl.$$

Carbon tetrachloride was first made by chlorinating chloroform (1839). Later, CCl_4 was made by chlorinating carbon disulfide CS_2 in the first commercial process, developed by Müller and Dubois (1893). Large-scale production commenced in the early 1900s at which time carbon tetrachloride became a popular metal-degreasing solvent, dry-cleaning fluid, fabric-spotting fluid, grain fumigant, and fire extinguishing fluid. Many of these uses now have been displaced by other less toxic chlorinated hydrocarbons. The carbon disulfide process consists of: (1) $3C + 6S \rightarrow 3CS_2$; (2) $2CS_2 + 6Cl_2 \rightarrow 2CCl_4 + 2S_2Cl_2$; (3) $CS_2 + 2S_2Cl_2 \rightarrow CCl_4 + 6S$. The reaction must be carried out in a lead-lined reactor in a solution of CCl_4 at 30°C in the presence of iron filings as catalyst. The chlorination of methane is now the principal production route to CCl_4: $CH_4 + Cl_2 \rightarrow CH_3Cl + CH_2Cl_2 + CHCl_3 + CCl_4 + HCl +$ excess CH_4. The reaction is carried out in the liquid phase at about 35°C. Ultraviolet light is used as a catalyst. The same reaction can be carried out at 475°C without catalyst. The unreacted methane and partially-chlorinated products are recycled to control the yield of CCl_4.

Toxicity: The experimental exposure of laboratory animals to the vapors of CCl_4 has shown it to be very toxic by inhalation at concentrations which are easily obtainable at ambient temperatures. An overexposure to carbon tetrachloride has been known to cause acute but temporary loss of renal function.

CARBONYL IRON. Iron Metals, Alloys, and Steels.

CARBONYLS. Carbon.

CARBONYLS (Chlorinated). Chlorinated Organics.

CARBON-ZINC BATTERY. Battery.

CARBORUNDUM. Silicon.

CARBOXYLIC ACIDS. The general formula for a carboxylic acid is

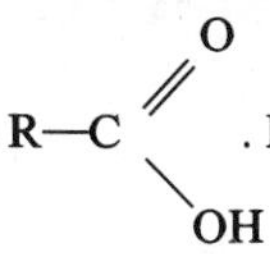

. In terms of structure, a carboxylic acid may be aliphatic, carbocyclic, or heterocyclic:

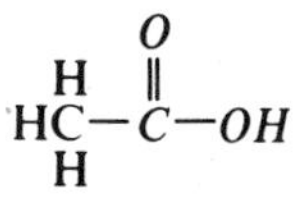

Aliphatic
acetic acid

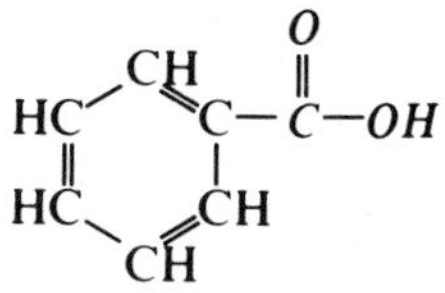

Carbocyclic or aromatic
benzoic acid

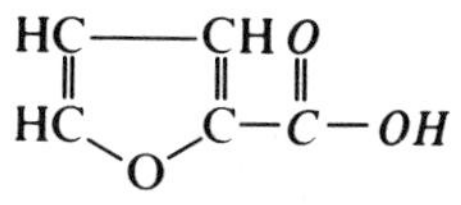

Heterocyclic
pyromucic or furoic acid

Or, a carboxylic acid may be classified in terms of the number of carboxyl (—COOH) groups which it contains. If one carboxyl group, it is designated as *mono*carboxylic; if two groups, as *di*carboxylic; if three groups, as *tri*carboxylic; and if four groups, as *tetra*carboxylic:

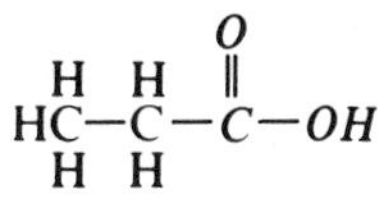

Propionic acid

(*mono*)

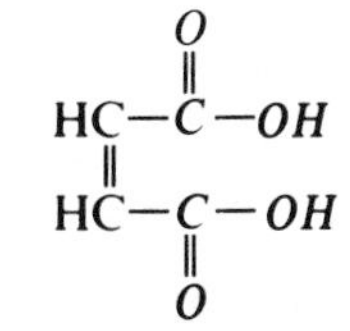

Maleic acid or
cis-ethylene dicarboxylic acid

(*di*)

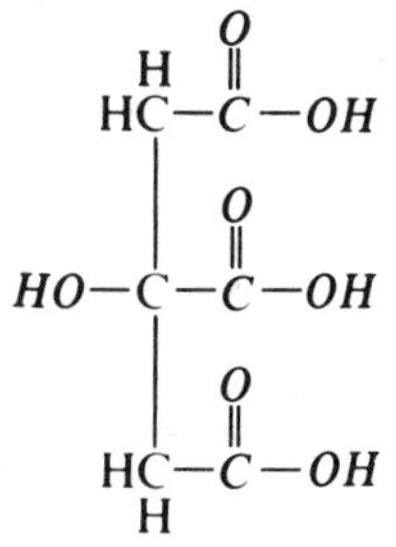

Citric acid

(*tri*)

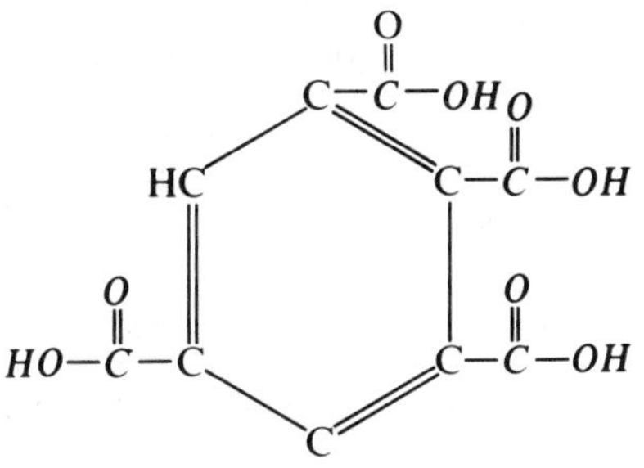

1,2,3,5-Benzenetetracarboxylic acid
or mellophanic acid

(*tetra*)

When a carboxylic acid contains a hydroxyl group in addition to that of the principal —COOH grouping, the term *hydroxy* is sometimes used. If there is only one additional hydroxyl group, the acid may be designated simply as a *hydroxycarboxylic* acid; if two groups, a *di*hydroxycarboxylic acid; if three groups, a *tri*hydroxycarboxylic acid.

A carboxylic acid may be classified in accordance with the number of available hydrogens for salt formation. If only one hydrogen is available, the acid is *monobasic*; if two hydrogens are available, the acid is *dibasic*; if three or more hydrogens are available, the acid is *polybasic*.

A carboxylic acid also may be classified from the standpoint of other groups which it contains. An *aldehydic* carboxylic acid contains the CHO group. An example is glyoxalic acid, CHO·COOH. An *amino* carboxylic acid contains the NH_2 group. An example is carbamic or amino-formic acid, NH_2COOH. A *ketonic* carboxylic acid contains the CO group. An example is benzoylacetic acid,

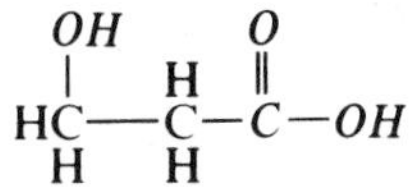

Hydracrylic acid or
β-hydroxypropionic acid
A hydroxymono-carboxylic acid

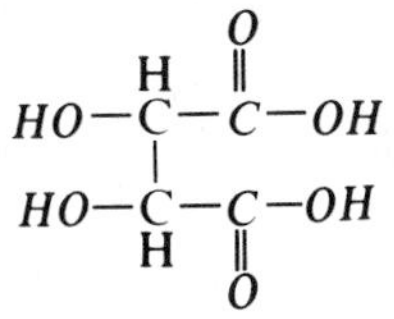

Tartaric acid
A dihydroxycarboxylic acid

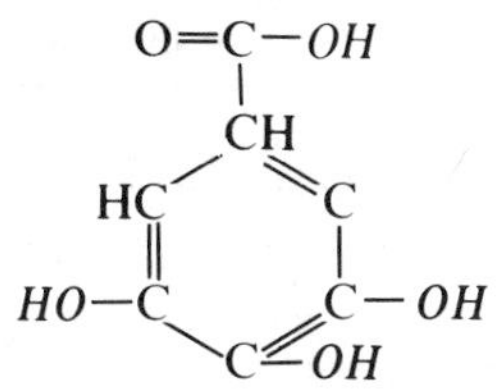

Gallic or pyrogallol carboxylic acid
or 3,4,5-trihydroxybenzoic acid
A trihydroxymonocarboxylic acid

$C_6H_5 \cdot CO \cdot CH_2 \cdot COOH$. In the case of a *phenolic* carboxylic acid, the acid is structurally derived from benzoic acid, with uniting of the OH group with a carbon of the nucleus.

There are several homologous series of carboxylic acids, including:

$C_nH_{2n}O_2$	Saturated monobasic fatty acids
$C_nH_{2n-2}O_2$	Unsaturated monobasic fatty acids
$C_nH_{2n-4}O_2$	Propioloic acid series
$C_nH_{2n}(COOH)_2$	Dicarboxylic acids, where $n = 0$ for oxalic acid
$C_nH_{2n}(OH)(COOH)$	Hydroxymonocarboxylic acids, where $n = 0$ for carbonic acid

Fatty Acids. The simplest or lowest member of the fatty acid series is formic acid, HCOOH, followed by acetic acid, CH_3COOH, propionic acid with three carbons, butyric acid with four carbons, valeric acid with five carbons, and upward to palmitic acid with sixteen carbons, stearic acid with eighteen carbons; and melissic acid with thirty carbons. Fatty acids are considered to be the oxidation product of saturated primary alcohols. These acids are stable, being very difficult (with the exception of formic acid) to convert to simpler compounds; they easily undergo double decomposition because of the carboxyl group; they combine with alcohols to form esters and water; they yield halogen-substitution products; they convert to acid chlorides when reacted with phosphorus pentachloride; and their acidic qualities decrease as their formula weight increases.

Monohydroxy Fatty Acids. Structurally, these acids may be considered as the monohydroxy derivatives of the fatty acids. Included among these acids are hydroxyacetic acid (glycollic acid) and β-hydroxypropionic acid (β-lactic acid). These acids generally are syrupy liquids that tend to give up water readily and form crystalline anhydrides, they decompose when volatilized, and they are soluble in water and usually in alcohol and ether.

Polyhydric Monobasic Acids. Structurally, these acids are considered to be the oxidation products of polyhydric alcohols. However, a number of them can be formed from the oxidation of sugars. The careful oxidation of glycerol will yield a syrupy liquid, glyceric acid, an example of a dihydroxymonobasic carboxylic acid.

Aromatic Carboxylic Acids. In many ways, these acids are similar to the fatty acids. Generally, they are crystalline solids which are only slightly soluble in water, but most often they dissolve easily in alcohol or ether. The simpler aromatic acids may be distilled (or sublimed) without decomposition. The more complex acids, such as the phenolic and polycarboxylic aromatic acids, break down when heated, yielding carbon dioxide and a simpler compound. As an example, salicylic acid degrades to carbon dioxide and phenol. In nature, the aromatic acids are found in balsams, animal organisms, and resins.

The monobasic saturated aromatic acids include benzoic, hippuric, toluic acids (three structures), phenylacetic, phenylchloracetic, and dimethylbenzoic acid. Among the monobasic unsaturated acids are cinnamic, atropic, and phenylpropionic acids. The saturated phenolic acids include gallic and salicylic acids. The alcohol acids include amygdalic, tropic, and mandelic acids. One example of an unsaturated monobasic phenolic acid is coumaric acid.

Formation of Carboxylic Acids. Commercially, these acids are produced in several ways: (1) oxidation of relevant alcohol—e.g., acetic acid from ethyl alcohol; (2) oxidation of relevant aldehyde—e.g., acetic acid from acetaldehyde; (3) bacterial fermentation of dilute alcohols; (4) reacting a methyl ketone with sodium hypochlorite (haloform reac-

SELECTED PROPERTIES OF REPRESENTATIVE CARBOXYLIC ACIDS

Carboxylic Acid	Formula	Formula Wt.	Sp Gr	MP, °C	BP, °C
Abietic	$C_{20}H_{35}O_2$	302.44	...	182	
Acetic	CH_3COOH	60.03	...	16.6	118.1
Acetoacetic	CH_3COCH_2COOH	102.06	...	...	100d
Acrylic	$CH_2:CHCOOH$	72.03	1.062	12.3	141.9
Adipic	$HOOC(CH_2)_4COOH$	146.14	1.360	151–153	265
Angelic	$CH_3CH:C(CH_3)COOH$	100.06	0.983	45	185
p-Anisic	$CH_3OC_6H_4COOH$	152.14	1.385	184.2	275–280
o-Anthranilic	$H_2NC_6H_4COOH$	137.13	...	144–145	s
Arabonic	$HOOC(CHOH)_3CH_2OH$	166.08	...	89	
Arachidic	$CH_3(CH_2)_{18}COOH$	312.52	...	77	328
Atropic	$C_6H_5CH(:CH)COOH$	148.15	...	106–107	267d
Behenic	$C_{21}H_{43}COOH$	340.34	...	84	306
1,2,3,4-Benzenetetracarboxylic	$C_6H_2(COOH)_4$	254.08	...	237d	
Benzoic	C_6H_5COOH	122.05	1.266	121.7	249.2
Benzoylacetic	$C_6H_5COCH_2COOH$	164.06	...	104	
m-Benzoylbenzoic	$C_6H_5COC_6H_4COOH$	226.08	...	162	
n-Butyric	$CH_3(CH_2)_2COOH$	88.06	0.959	−7.9	163.5
Isobutyric	$(CH_3)_2CHCOOH$	88.06	0.949	−47	154.4
Camphoric (*d*-)	$C_{10}H_{16}O$	152.23	0.999	178–179	209.1
Capric	$CH_3(CH_2)_8COOH$	172.26	0.889	31.5	268–270
Isocaproic	$(CH_3)_2CH(CH_2)_2COOH$	116.09	0.925	−35	207.7
n-Caproic	$CH_3(CH_2)_4COOH$	116.09	0.945	−1	205
n-Caprylic	$CH_3(CH_2)_6COOH$	144.21	0.910	16	237.5
Cerotic	$C_{25}H_{51}COOH$	396.41	0.836	82.5	d
Chloroacetic	$ClCH_2COOH$	94.50	1.370	61.2	189.5
o-Chlorobenzoic	ClC_6H_4COOH	156.57	1.544	141–142	
α-Chloropropionic	$CH_3 \cdot CHCl \cdot COOH$	108.53	1.306	< −20	186
Cinnamic (*cis*-)	$C_6H_5CH:CHCOOH$	148.15	1.284	68	125
Citric	$C_3H_4(OH)(COOH)_3$	192.12	1.542	153	d
α-Crotonic	$CH_3CH:CHCOOH$	86.09	0.964	72	189
p-Cumic	$(CH_3)_2CH \cdot C_6H_4COOH$	164.20	1.162	116–117	s
Cyanoacetic	$CH_2(CN)COOH$	85.06	...	65.6	108
Dichloroacetic	$Cl_2CH \cdot COOH$	128.95	1.560	9.7	194.4
Diphenylacetic	$(C_6H_5)_2CHCOOH$	212.09	...	148	
Formic	$HCOOH$	46.02	1.226	8.4	100.5
Fumaric (*trans*-)	$HOOC \cdot CH:CH \cdot COOH$	116.07	1.635	286–287	290
Furoic	$C_4H_3O \cdot COOH$	112.03	...	133–134	230–232
Gallic (3,4,5)	$(HO)_3C_6H_2COOH \cdot H_2O$	188.06	1.694	220d	
Gluconic	$C_5H_6(OH)_5COOH$	196.09	...	...	s
Glutaric	$HOOC(CH_2)_3COOH$	132.06	1.429	97.5	200
Glyceric	$CH_2OHCHOHCOOH$	106.05			
Glycollic	$CH_2OH \cdot COOH$	76.03	...	79	d
Glyoxalic	$HCOCOOH$	74.02	...		
Heptoic	$CH_3(CH_2)_5COOH$	130.19	0.918	−10	221–222
Hippuric	$C_6H_5CONHCH_2COOH$	179.08	1.371	187–188	d
Hydracrylic	CH_2OHCH_2COOH	90.08	...	...	d
α-Hydroxybutyric	$CH_3CH_2CH(OH)COOH$	104.06	...	42.5	260
Lactic	$CH_3CHOHCOOH$	90.05	1.248	18	122
Lauric	$CH_3(CH_2)_{10}COOH$	200.19	0.869	48	225
Levulinic	$CH_3CO(CH_2)_2COOH$	116.06	1.140	33.5	245–246
Lignoceric	$C_{23}H_{47}COOH$	368.37	...	81	
Linoleic	$C_{17}H_{31}COOH$	280.25	0.903	−9.5	229–230
Maleic	$HOOC \cdot CH:CH \cdot COOH$	116.03	1.609	130.5	135d
Malic (*dl*)	$HOOCCH_2CH(OH)COOH$	134.05	1.601	128–129	150d
Malonic	$HOOCCH_2COOH$	104.03	1.631	135.6	d
Mandelic (*dl*)	$C_6H_5CH(OH)COOH$	152.06	1.300	118.1	d
Margaric	$CH_3(CH_2)_{15}COOH$	270.44	0.853	60–61	227
Melissic	$C_{29}H_{59}COOH$	452.47	...	91	
Mellitic	$C_6(COOH)_6$	342.05	...	286	d
Mesoxalic	$HOOC \cdot CO \cdot COOH \cdot H_2O$	136.01	...	120	
α-Methylacrylic	$CH_2:C(CH_3)COOH$	86.09	1.015	15–16	161–163
Mucic	$(\cdot CHOHCHOHCO_2H)_2$	210.14	...	206–214d	
Myristic	$CH_3(CH_2)_{12}COOH$	228.36	0.853	57–58	250.5
Nicotinic (3-)	C_5H_4NCOOH	123.11	...	235.2	s
m-Nitrobenzoic	$NO_2 \cdot C_6H_4COOH$	167.12	1.494	140–141	
Nondecyclic	$CH_3(CH_2)_{17}COOH$	298.30	...	66.5	299
Oleic	$CH_3(CH_2)_7CH:CH(CH_2)_7COOH$	282.27	0.895	14	286
Oxalic	$HOOCCOOH$ (anhydrous)	108.03	1.653	189	
Palmitic	$CH_3(CH_2)_{14}COOH$	256.25	0.853	64	380
Pentadecyclic	$C_{14}H_{29}COOH$	242.39	...	52	257
Phenylacetic	$C_6H_5CH_2COOH$	136.14	1.081	76–77	265.5
Phthalic	$C_6H_4(COOH)_2$	166.13	1.593	208	d
Isophthalic	$C_6H_4(COOH)_2$	166.13	...	330	s
Pimelic	$HOOC(CH_2)_5COOH$	160.09	...	103	272

SELECTED PROPERTIES OF REPRESENTATIVE CARBOXYLIC ACIDS—*continued*

Carboxylic Acid	Formula	Formula Wt.	Sp Gr	MP, °C	BP, °C
Propargylic	$CH_3:CCOOH$	70.02	1.139	9	144d
Propionic	CH_3CH_2COOH	74.05	0.992	−22	141.1
Pyruvic	$CH_3COCOOH$	88.06	1.267	13.6	165
Ricinoleic	$C_{17}H_{32}(OH)COOH$	298.45	0.954	4–5	226–228
Saccharic (*d*)	$HOOC(CHOH)_4COOH$	210.08			
Isosaccharic	$HOOCCH(CHOH)_2CHCOOH$	192.06	...	185	d
Salicylic	HOC_6H_4COOH	138.05	1.443	159	s
Sebacic	$HOOC(CH_2)_8COOH$	202.14	...	−127	295
Stearic	$CH_3(CH_2)_{16}COOH$	284.28	0.847	69.3	383
Suberic	$HOOC(CH_2)_6COOH$	174.19	1.266	140–144	279
Succinic	$HOOC(CH_2)_2COOH$	118.05	1.564	185	235
Tartaric (*d* or *l*)	$(CHOHCO_2H)_2$	150.09	1.760	168–170	d
Tartronic	$CH(OH)(COOH)_2 \cdot \frac{1}{2}H_2O$	129.07	...	155.8d	s
Terephthalic	$C_6H_4(COOH)_2$	166.13	1.510	s	
o-Toluic	$CH_3 \cdot C_6H_4 \cdot COOH$	136.14	1.062	104–105	259
Triphenylacetic	$(C_6H_5)_3CCOOH$	288.12	...	265	
n-Valeric	$CH_3(CH_2)_3COOH$	102.08	0.942	−59	187
Isovaleric	$(CH_3)_2CHCH_2COOH$	102.08	0.937	−37.6	176.7
Vinylacetic	$CH_2:CH \cdot CH_2 \cdot COOH$	86.09	1.013	−39	163

d (in acid column) = dextrorotatory; d (in mp or bp column) = decomposes; *dl* = dextrolevorotatory; *l* = levorotatory; *m* = meta; *o* = ortho; *p* = para; *s* = sublimes.

tion); (5) carbonation of Grignard reagents; (6) hydrolysis of nitriles; (7) malonic ester synthesis route; (8) oxidation of relevant alkylaromatic—e.g., benzoic acid from toluene; (9) reaction of an alkali metal phenolic with carbon dioxide; and (10) hydrocarboxylation of olefins—e.g., butyric acid from propylene.

Selected properties of representative carboxylic acids are given in the accompanying table. See also **Organic Chemistry.**

Duane B. Priddy, The Dow Chemical Company, Midland, Michigan.

CARBUNCLE (Geology). A term applied to that variety of garnet, almandine, which was much used formerly for jewelry, when cut en cabochon. It is derived from the Latin, *carbunculus*, a small spark, in reference to the glowing effect of that style of cutting. In the early part of the Christian Era, the term seems to have been used for red stones of all sorts.

CARBURETOR. The fuel for an internal combustion engine must be well mixed with the air required for combustion. This is particularly true of the Otto cycle engine inasmuch as thorough distribution of particles of fuel in the air is essential to the rapid and complete explosive combustion of the fuel in that cycle. One of the most effective means of mixing the particles of a liquid fuel with air is by vaporization. The vaporizing and mixing of a liquid fuel with air in the correct proportions is called *carburetion*, and the device required to do this, a *carburetor.*

With ordinary gasoline as the fuel, this mixture should be maintained at a theoretical 15.2 weight units of air per weight unit of gasoline. The simple Venturi tube carburetor will not do that, but rather will give an increasingly richer mixture as the velocity of air through the Venturi increases. The principal reason for this characteristic will be found in the variability of the coefficients of discharge. The coefficient of discharge for air through a Venturi tube is approximately constant, but that of gasoline from a jet increases with increasing pressure differential across the jet. Thus, numerous schemes, the descriptions of which are beyond the scope of this volume, have been devised to compensate for this shortcoming of the Venturi tube.

Some of the objectives which must be met in carburetor design include:

1. Variable power over a range from idling to maximum horsepower at full rated speed.
2. Speed variations, ranging from nearly constant speed under governor control to speed controllable from idling speed to a maximum speed of several thousand revolutions per minute. Rapid acceleration at varying rates may also be an imposed requirement.
3. Thermal efficiency, which may vary from a factor of little importance for some types of engines to one of major importance in others.
4. Operation when idling, which may vary from rough and irregular to perfectly smooth.
5. Requirements imposed by the mobility of the engine. If stationary, there is no trouble with surge of fuel in supply chambers. Some engines, as on automobiles, have the carburetors subject to moderate accelerations, but remaining in approximately level position. Aircraft engine carburetors are subject to great accelerations, and those which are employed on acrobatic or combat aircraft may be required to operate in all possible physical positions.
6. Carburetion of fuels of varying volatility.

Although a mixture of 15.2 air to 1 gasoline is theoretically correct, mixtures as rich as 9:1 or as lean as 20:1 are explosive. However, the rich mixtures are uneconomical, and the lean mixtures must be employed cautiously at maximum power to prevent detonation and overheating. Thus, rich mixtures are indicated near full power operation. Also, on account of the proportionately high dilution of incoming fuel stream by unscavenged products of combustion at the previous cycle at light loads (under quantity control), a rich mixture is needed at low power output. Between these needs for rich mixture, the carburetor should deliver a fairly uniform lean mixture for the sake of fuel economy. The accompanying figure illustrates a desirable performance for the carburetor of a gasoline engine. By employing multiple jets, adjustable orifices and other intricacies, commercial carburetors attain mixture control approximating this desired performance at con-

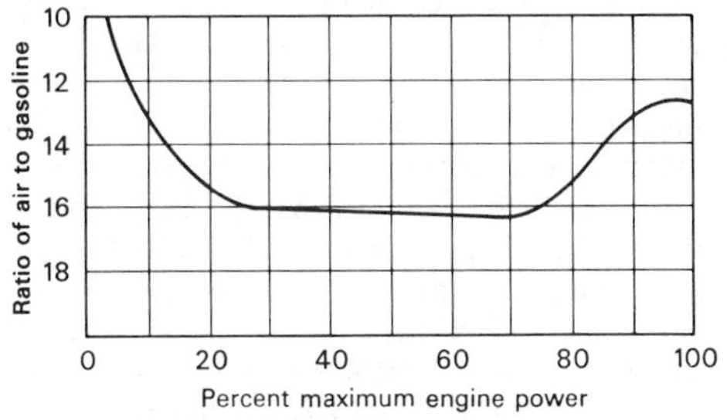

Optimal carburetor performance.

siderable sacrifice of simplicity. Increasing air pollution standards have exacted even greater demands on carburetor performance. The advent of solid-state electronic systems and their application in automotive vehicles has helped to establish and maintain optimum operating conditions. See also **Automotive Electronics.**

CARBURIZING. Machine parts requiring high strength, hardness, and toughness can often be made by either of two methods, one based on the use of a medium-carbon steel (0.30–0.50% carbon) heat treated to the required properties, and the other based on the use of a low-carbon steel (0.08–0.25% carbon) carburized to give a high-carbon surface layer and then heat treated. The carburized part will have a harder, more wear-resistant surface and a tougher core than the heat-treated medium-carbon steel. Transmission gears, camshafts, and piston pins are typical parts which can be made advantageously of carburizing grade steels.

The process consists of heating the fully machined part in an atmosphere rich in carbon monoxide or hydrocarbon gases at a temperature in the range 1650–1800°F (899–982°C). Reactions at the surface of the metal liberate atomic carbon which is readily dissolved by the steel and diffuses inward from the surface. In a typical carburized case a depth of penetration of 0.05 inch (0.13 centimeter) was obtained in 4 hours at 1700°F (927°C). The maximum carbon content at the surface was 1.10%. Shallow cases under 0.02 inch (0.05 centimeter) are useful for many purposes and very deep cases over 0.10 inch (0.25 centimeter) thick are required for gears for heavy machinery and for armor plate.

The process is most often carried out in sealed containers in which the parts are packed in carburizing compound consisting of a mixture of charcoal, coke, and other carbonaceous solids, together with barium carbonate and other compounds which act as energizers. At high temperatures these solids burn slowly, maintaining a supply of carbon monoxide. Carburizing is also carried out in batch-type and continuous-type furnaces in an atmosphere of natural gas, propane, butane, or specially mixed gases. Liquid baths consisting mainly of molten cyanide and chloride salts are also used for surface hardening. These baths supply both nitrogen and carbon to the surface of the steel, and where nitrogen is the principal hardener the process is known as cyaniding. Nitrogen hardens steel by forming hard compounds with iron and with certain alloying elements that may be present such as aluminum, chromium, and vanadium. See **Nitriding.** In general, the salt-bath methods give shallower but harder cases than regular solid-pack carburizing. The pieces are quenched for hardening directly from the bath.

Carburized steels may also be quenched in oil or water directly from the box or furnace, or they may be cooled and reheated for hardening. A low temperature tempering treatment is given for relief of quenching stresses. A surface hardness of 60 Rockwell "C" is readily obtained, and when medium alloy steels of fine grain size are used, the strength and ductility of the core is exceptionally high, for example, 165,000 psi (11,224 atmospheres) tensile strength and 18% elongation.

CARBYNE. Carbon.

CARCAJOU. Mustelines.

CARCINOGENS. Cancer-causing substances are called *carcinogens.* During the past decade, there has been a growing awareness of the presence of carcinogenic materials in the environment, both air and water. The causes of carcinogenesis are still poorly understood. Because the numbers of substances with which a person comes in contact are in the tens of thousands and because science has not achieved that position of understanding where carcinogens can be easily grouped, such as by chemical structure or other properties, the testing of so many substances becomes a formidable task. Understood even less are the long-term effects of these substances in their possible propensity to cause genetic errors that ultimately lead to carcinogenesis. Some progress, albeit a beginning, in classification has been made and reference to this is cited here. But, by and large, it would be an overemphasis to state that a great deal of progress has been made as of the early 1980s. Further complicating the social, economic, and scientific interfaces are compounds which are known to be carcinogenic above certain levels, but which are of great value, if not indispensable, to many aspects of society. Some chemical pesticides, for example, obviously fall into that category where risks and benefits must be weighed. See also **Cancer and Oncology;** and **Cancer Research.**

Until the mid-1970s, most work in chemical carcinogenesis was dominated by experiments, i.e., the testing of suspect chemicals on the skins or in the diets of animals and waiting to see if the chemicals induced tumor formation. Such work is valuable in identifying certain materials that should be removed from the environment and otherwise avoided by humans, but has not led in a major way to the understanding of the mechanism or chemistry of chemical carcinogenesis. Starting in a rather small way in the early 1970s, new emphasis now is being placed on the molecular biology of carcinogenesis. Some of the first steps in the interaction between carcinogen and cell have been demonstrated, including observations that most carcinogens must be activated by the host's metabolism. Cell culture methodologies have been developed which enable the transformation of healthy cells into malignant cells by chemicals. These methods have several advantages over work in vivo. A major step, of course, remains—the transformation of human cells by chemicals in culture.

There is some evidence that the form of the chemical carcinogen that ultimately reacts with cellular macromolecules must contain a reactive electrophilic center, that is, an electron-deficient atom which can attack the numerous electron-rich centers in polynucleotides and proteins. As examples, significant electrophilic centers include free radicals, carbonium ions, epoxides, the nitrogen in esters of hydroxylamines and hydroxamic acids, and some metal cations. It is believed that carcinogens which in themselves are not electrophiles are metabolized to electrophilic derivatives that then become the "ultimate" carcinogens.

The high incidence of skin cancer in coal tar workers was recognized as early as 1880. The carcinogenic activity of coal tar was demonstrated in 1915, when Yamagiwa and Ichikawa obtained epitheliomas (malignant tumor originating from epithelial cells) by its prolonged application to the ears of rabbits. Identification of the active material (in 1933) as the polycyclic aromatic hydrocarbon 3,4-benzopyrene (III, Fig. 1) is due to Cook, Kennaway, Hieger and their co-workers. This discovery was followed up by the synthesis and testing of a considerable variety of polycyclic aromatic hydrocarbons. All compounds of this class may be regarded as composed of condensed benzene rings. The arrangement of the hexagonal rings in various patterns results in a variety of compounds having different physical, chemical, and biological properties. However, not all polycyclic aromatic hydrocarbons possess carcinogenic activity; certain requirements of molecular geometry must be met.

For maximum activity, the molecule must have (Fig. 1): (a) an optimum size; (b) a coplanar molecular configuration, meaning that all hexagonal rings must lie flatly in one plane; in fact, hydrogenation of many of the active hydrocarbons results in buckled molecular conformation and this is concomitant with partial or total loss of activity; (c) at least one meso-phenanthrenic double bond, also called the K-region (indicated by arrows in Fig. 1) of high π-electron density (i.e., of high chemical reactivity). In addition to III, 1,2,5,6-dibenzanthracene (IV), and 20-methylcholanthrene (V) are commonly used to study the experimental induction of tumors. The activity of most hydrocarbon carcinogens was tested on the skin of mice and the subcutaneous connective tissue of mice and rats. There is a vast body of evidence indicating that 3,4-benzopyrene and other carcinogenic hydrocarbons are formed during pyrogenation or incomplete burning of almost any kind of organic material. For example, carcinogenic hydrocarbons have been identified in overheated fats, broiled and smoked meats, coffee, burnt sugar, rubber, commercial paraffin oils and solids, soot, the tar contained in the exhaust fumes of internal combustion engines, cigarette smoke, etc.

Attention to the carcinogenic aromatic amines was drawn by the high incidence of urinary bladder tumors in dye works exposed to 2-naphthylamine (VII, Fig. 2), and benzidine (IX). The carcinogenic activity of VII, IX, and 4-aminobiphenyl (X) toward the bladder of the dog and the mouse has been demonstrated. In the rat, however, there is a change in target specificity, and tumors are induced by IX and X in the liver, mammary gland, ear duct, and small intestine. Carcinogenic activity is considerably heightened in 2-acetylamino-

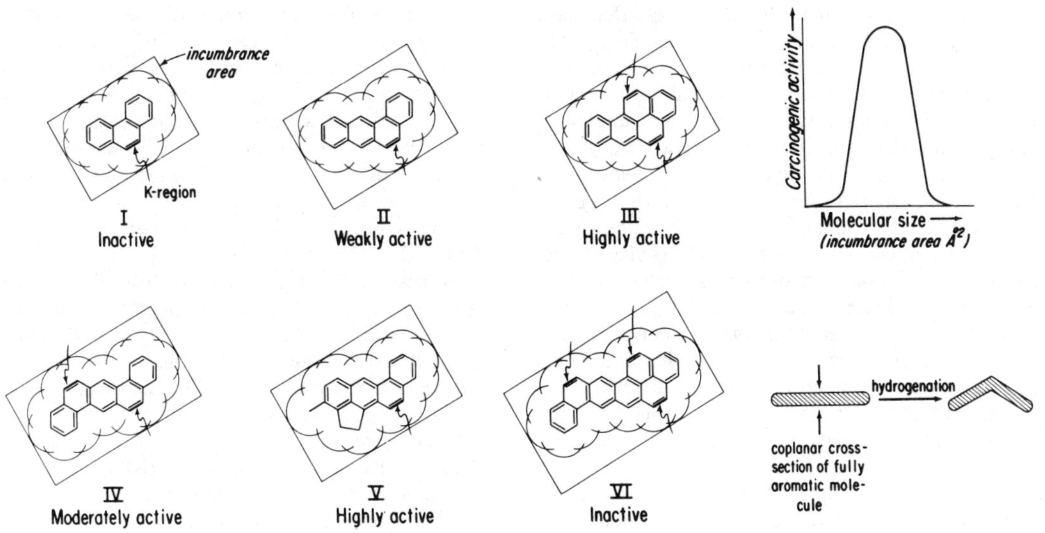

Fig. 1. Polycyclic aromatic hydrocarbons.

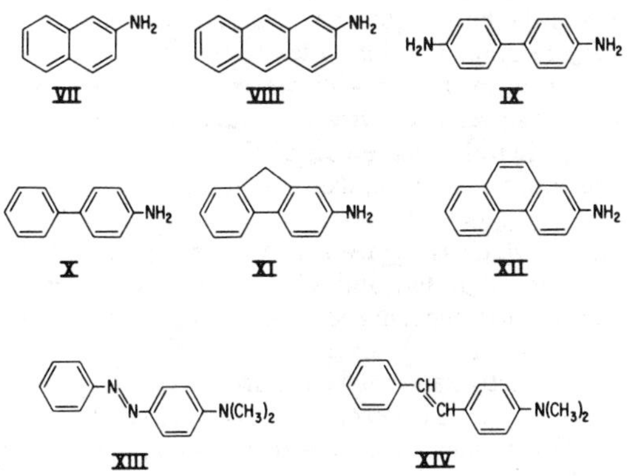

Fig. 2. Carcinogenic aromatic amines.

fluorene (XI) without change of target specificity. Increased activity is due to the fact that XI is more coplanar than X, because of the internuclear methylene (—CH_2—) bridge in the former. 2-Acetylaminofluorene was proposed as an insecticide before its carcinogenic activity was accidentally discovered; it is a ubiquitous, potent carcinogen in a variety of species. Changing the internuclear bridge of XI to —CH=CH—, as in 2-aminophenanthrene (XII), causes a shift in target specificity; thus, in the rat, XII is inactive toward the liver, but in addition to inducing tumors in the mammary gland, ear duct, and small intestine, it produces leukemia. Compound XII represents a structural link between the aromatic amine and polycyclic hydrocarbon carcinogens (compare XII with I); it is also interesting in this respect that 2-aminoanthracene (VIII), which is a higher homologue of VII, is inactive toward the bladder, but is able to induce skin tumors in rats.

4-Dimethylaminoazobenzene (XIII) is the parent compound of the aminoazo dye carcinogens; it is also known in the earlier literature as Butter Yellow, because it was used to color butter and vegetable oils before its carcinogenic activity was discovered. Many derivatives of XIII have been prepared and tested for carcinogenic activity. In the rat, the aminoazo dye carcinogens, administered in the diet, specifically induce hepatomas. Tumor induction by most of the aminoazo dyes is delayed or inhibited by high dietary levels of riboflavin (vitamin B_2) or protein. Replacement of the —N=N— azo linkage by —CH=CH—, as in 4-dimethylaminostilbene (XIV), results in widening the target tissue spectrum; XIV induces tumors in the liver, mammary gland, and ear duct. Mice are much more resistant than rats to the carcinogenic activity of both aminoazo dyes and aminostilbenes.

Figure 3 illustrates some aliphatic carcinogens. *N*-Methyl-*bis*-β-chloroethylamine (XV), a nitrogen mustard, produces local sarcomas, lung, mammary, and hepatic tumors upon injection in mice; because of its tumor-inhibitory properties, XV has also been used in the therapeutic treatment of certain types of human cancers. Bisepoxybutane (XVI), β-propiolactone (XVII), and *N*-lauroylethyleneimine (XVIII) produce local sarcomas in rats upon injection. Ethylcarbamate (XIX), the parent compounds of several hypnotic drugs used in humans, produces malignant lung adenomas in rats, mice, and chickens. Dimethylnitrosamine (XX) is a potent carcinogen toward the liver, lung, and kidney, and ethionine (XXI) toward the liver of the rat; the former is an intermediate in the manufacture of the rocket-fuel component, dimethylhydrazine $(CH_3)_2N—NH_2$, while the latter is the *S*-ethyl analogue of the natural amino acid methionine.

Chemically identified carcinogens may be grouped in many ways, including a division into inorganic ions and organic compounds. The inorganic carcinogens are the elements beryllium, cadmium, iron, cobalt, nickel, silver, lead, zinc, and possibly arsenic; these can form coordination compounds and/or react with sulfhydryl groups. Also asbestos powder is a powerful carcinogen toward the lung upon inhalation (asbestos cancer of miners). In recent years, the characteristics of asbestos and related substances have caused much controversy in connection with the pollution of certain waters, notably Lake Superior

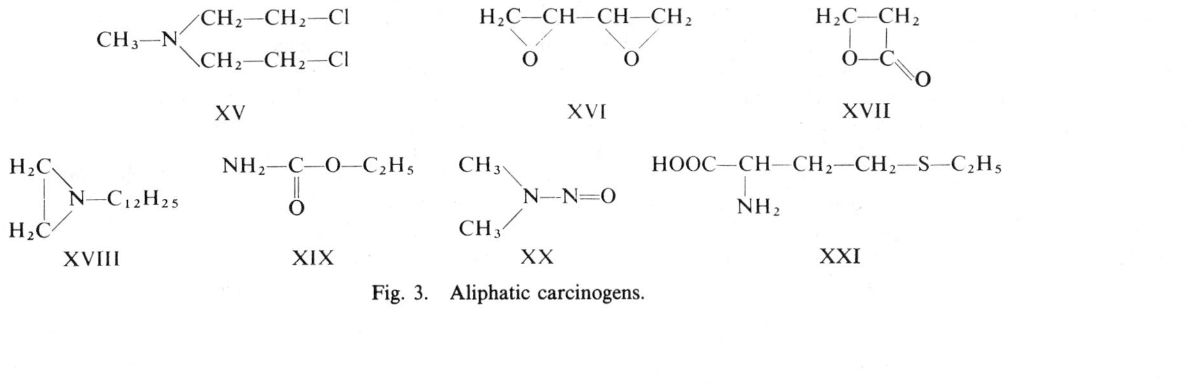

Fig. 3. Aliphatic carcinogens.

by taconite processing waste products which contain fibers that have been compared with asbestos fibers. Most likely this situation will require some years for full scientific and legal resolution.

The organic carcinogens may be subdivided in several ways, including: (a) condensed polycyclic aromatic hydrocarbons and heteroaromatic polycyclic compounds; (b) aromatic amines and *N*-aryl hydroxylamines; (c) aminoazo dyes and diarylazo compounds; (d) aminostilbenes and stilbene analogues of sex hormones. Further breaking down the aliphatic carcinogens, these include: (1) alkylating agents (such as sulfur and nitrogen mustards, derivatives of ethyleneimine, lactones, epoxides, alkane-α-ω-*bis*-methanesulfonates, certain dialkylnitrosamines, and ethionine; (b) lipophilic agents and hydrogen-bond reactors; this class comprises a wide variety of agents, such as chlorinated hydrocarbons (chloroform, carbon tetrachloride, and compounds used as pesticides under the names aldrin and dieldrin), bile acids, certain water-soluble high polymers, certain phenols, urethane and some of its derivatives, thiocarbonyls, and cycloalkynitrosamines; (c) natrually occurring carcinogens.

Until the early 1950s, the concept prevailed that the activity of carcinogenic chemicals was somehow related to the fact that they were synthetic "unnatural" substances which, since they were not present in the natural environment, were not factors of selection during developing life processes and hence contemporary living organisms were not equipped for effective metabolic "detoxification" of these compounds. However, during the last 25 years, a number of carcinogenic compounds of plant and fungal origin have been identified, including safrole in sassafras; capsaicine in chili peppers; various tannins; cycasin in the cycad groundnut; parasorbic acid in mountain ash berry; pyrollizidine alkaloids in *Senecio* shrubbery; patulin, griseofulvin, penicillin G, aflatoxin, and actinomycin produced by various molds; and many others too numerous to detail. The number and variety of identified naturally occurring carcinogens continues to increase at a rapid rate.

Carcinogen Mechanisms. The biochemical pathways in the cell are closely interconnected and are in a state of dynamic equilibrium (homeostasis). This equilibrium is maintained by feedback relationships existing between a great number of pathways. Chemical communication between subcellular organelles, such as the nucleus (within which the chromosomes contain the genetic blueprints for cell reproduction and the synthetic processes of cell life), the mitochondria (the powerhouse of the cell, which assures the synthesis of the universal cellular fuel, ATP, through the metabolism of carbohydrates and fatty acids), and the endoplasmic reticulum (synthesizing the proteins of the cell and assuring the metabolic breakdown—detoxification—of a multitude of endogenous and foreign compounds), depends on the constant interchange of a large variety of metabolic products and inorganic ions between them. There are probably a very great number of loci (receptor sites) upon which these regulatory chemical "stimuli" act. The receptor sites are of an enzymic and nucleic acid nature. Other control points of protein character regulate the morphology of the intracellular lipoprotein membranes which serve as "floor space" to the organized arrangements of multienzyme systems. The specificity of compounds of chemical control toward given receptor sites is due to a three-dimensional geometric "fit" following the lock and key analogy. Such is the general scheme of functional interrelationships in monocellular organisms which, hence, in a favorable medium multiply unchecked to the limit of the availability of nutrients.

In multicellular organisms, the subordination of the individual cells to the whole is assured by the existence of additional receptor sites which enable the cells to be response to chemical "stimuli" emitted by neighboring cells in the tissue and to hormonal regulation by the endocrine system in higher organisms. Hence, depending on the requirements of the moment, cells may remain stationary or may undergo cell division because of the need for repair of tissue injury, they may secrete different products, or they may perform some other specialized function depending on the nature of the particular tissue.

Carcinogenic substances are nonspecific cell poisons which cause the alterations and hence functional deletion of a large number of metabolic control sites. Present evidence suggests that these alterations are produced by the accumulation of the carcinogen in subcellular organelles, by covalent binding of the carcinogen to cellular macromolecules (proteins and nucleic acids) through metabolism, and by denaturation (i.e., destruction of the three-dimensional geometry) of the control sites through secondary valence interactions (hydrogen bonds, hydrophobic bonding, etc.) with the carcinogen. Early stages of tumor induction generally coincide with extensive cell death (necrosis) in the target tissues because a number of the biochemical lesions cause the irreversible blocking of metabolic pathways essential for cell life. However, because of the random distribution of the biochemical lesions in the cell population, in a small number of cells vital pathways are only slightly damaged and the lesions involve those sites and pathways which are not essential for cell life proper, but are necessary for organismic control. Thus, due to the action of the carcinogen, these cells escape physiological control and revert to a simpler, less specialized cell type (i.e., dedifferentiate). Such cells respond to continuous nutrition with continuous growth, which is an essential characteristic of malignant tumor cells.

Testing. Because of their short life span (average 3 years) small rodents (mice, rats, and hamsters) are frequently used for the testing of chemicals for carcinogenic activity; occasionally testing is done with rabbits, dogs, fowls, monkeys, etc. While a great variety of ways of administration have been used, a common method is to introduce substances to be tested in the following ways: (a) skin painting; small volumes of solution of the substance in an inactive solvent (e.g., benzene) are applied to the shaved surface of the skin (generally of mice, in the interscapular region) daily or at longer intervals; (b) subcutaneous injection of the pure substance or its solution (once or at repeated intervals); (c) feeding; the substance is mixed in the diet at given levels, or dissolved in the drinking water. Testing of new substances for possible carcinogenic activity is conducted for a minimum of 1 year to be meaningful. At the end of the testing period, all animals are autopsied, and all tumors and dubious tissues examined histopathologically.

A carcinogen which is highly active in one species may be totally inactive in another species, and vice versa. The susceptibility of a species to a given carcinogen also depends on the genetic strain, sex, and dietary conditions. Moreover, carcinogenic substances generally show a rather selective specificity toward certain target tissues; e.g., certain compounds produce exclusively hepatomas in the susceptible species. For these reasons, no chemical compound may be stated safely to the devoid of carcinogenic activity toward humans unless it has been found inactive when tested in a variety of mammalian species and by a variety of routes of administration for a length of time corresponding to half the life span of each species.

The foregoing observation emphasizes one of the main problems in cancer research, namely, *extrapolation* of research findings. To be fully safe, literally tens of thousands of commonly encountered natural materials and synthetic materials covering the complete spectrum of products with which people are in contact over their lifespan would have to be tested and regarded suspect until thoroughly tested on the species of most importance to humans, namely, the testing of reactions among people themselves. But this alone would not suffice because, as stressed throughout biochemical studies of human systems, there is individuality (see also **Biochemical Individuality**). The problem of developing improved (vastly improved) testing systems and attention to the problem of extrapolation of findings rivals in difficulty the basic problem of identifying the nature of cancer itself.

The interesting observation has been made that, because of the vast amounts of money going into cancer research, the field has become a large business for numerous suppliers. Animal cells now can be procured by the kilogram from suppliers. Because of this availability (mostly two kinds, 3T3 and W138), much information has been accumulated concerning the biology of these cells. But is much of what has been learned in this regard meaningful? Researchers have found that viruses and chemicals can transform 3T3 cells to a neoplastic state and that these cells can produce tumors when inoculated in suitable hosts. It should be noted, however, that the tumors so produced are sarcomas (derived from fibroblasts), which are very rare in human beings. Ninety percent of human tumors are carcinomas. With the exception that epithelial cells and fibroblasts are both animal cells, they have little in common. They stem from two embryonic sources with different functions, and the tumors they produce are different as well.

See also **Wastes (Toxic).**

References

Ames, B. N., McCann, J., and E. Yamasaki: "Methods for Detecting Carcinogens and Mutagens with the Salmonella/Mammalian-Microsome Mutagenecity Test," *Mutation Res.*, **31**, 6, 347–364 (1975).

Ames, B. N.: "Identifying Environmental Chemicals Causing Mutations and Cancer," *Science*, **204**, 587–593 (1979).

Devoret, R.: "Bacterial Tests for Potential Carcinogens," *Sci. Amer.*, **241**, 2, 40–49 (1979).

Fajen, J. M., et al.: "*N*-Nitrosamines in the Rubber and Tire Industry," *Science*, **205**, 1262–1264 (1979).

Gelboin, H. V. (editor): "Polycyclic Hydrocarbons and Cancer," Academic, New York, 1978.

Gori, G. B.: "The Regulation of Carcinogenic Hazards," *Science*, **208**, 256–261 (1980).

Hartline, B. K.: "Cancer and Environment," *Science*, **205**, 1363–1364 (1979).

Kolata, G. B.: "Testing for Cancer Risk," *Science*, **207**, 967–969 (1980).

Land, C. E.: "Estimating Cancer Risks from Low Doses of Ionizing Radiation," *Science*, **209**, 1197–1203 (1980).

Moreau, P., Bailone, A., and R. Devoret: "Prophage Gamma: Induction in *Escherichia coli*—A Highly Sensitive Test for Potential Carcinogens," *Proc. Nat. Acad. Sci. U.S.A.*, **73**, 10, 3700–3704 (1976).

Sax, N. I.: "Dangerous Properties of Industrial Materials," 5th edition, Van Nostrand Reinhold, New York, 1979.

Stein, M. W., and E. B. Sansone: "Degradation of Chemical Carcinogens: An Annotated Bibliography," Van Nostrand Reinhold, New York, 1979.

Witkin, E. M.: "Ultraviolet Mutagenesis and Inductible DNA Repair in *Escherichia coli*," *Bacteriol. Rev.*, **40**, 4, 869–907 (1976).

CARCINOMA. **Cancer and Oncology.**

CARDAMOM (*Elettaria Cardamomum*; *Zingiberaceae*). Cardamom is prepared from the seeds of a leafy-stemmed perennial monocotyledon growing from 5–9 feet (1.5–2.7 meters) in height. The flowers, white with purple-striped perianth parts, are borne on leafless stems which rise from the thick fleshy rhizomes apart from the leafy stems. The angular seeds are borne in 3-celled fruits. The dried seeds are used in India and elsewhere in tropical Asia as a highly flavored spice.

CARDAN'S FORMULA. **Cubic Equation.**

CARDIAC. Pertaining to the heart. A *cardiologist* is a physician who specializes in the diagnosis and treatment of heart diseases and disorders. See **Heart and Circulatory System (Human).**

CARDIAC ARRHYTHMIAS. **Arrhythmias (Cardiac); Ischemic Heart Disease.**

CARDIAC MASSAGE. When the heart collapses completely, the patient loses respiration, the pulse stops in all major blood vessels, and the pupils become dilated. In case of collapse, mouth-to-mouth artificial respiration should be given along with external cardiac massage. These are measures which can be taken prior to arrival of emergency forces, such as paramedics, a physician, or attendants at a hospital.

To give cardiac massage, the patient must be placed on a firm surface, such as the floor. If the rescuer is alone, the patient's lungs should be filled with three or four rapid mouth-to-mouth respirations before attempting to massage. Ideally, both artificial respiration and massage are performed at the same time by two rescuers. There is no need to coordinate the two activities between the rescuers. In massage, the rescuer puts one hand across the lower sternum (breastbone of the patient). The other hand is place on top to make a right angle. The full weight of the rescuer is applied rhythmically through the heel of the hands, at about one thrust per second. The sternum moves in about 4 or 5 centimeters (1.5 to 2 inches), compressing the heart. When pressure is lifted, blood reenters the heart. If the rescuer is alone, several rapid mouth-to-mouth respirations should be given to the patient every 30 seconds.

If massage is successful, gasping and some movement may occur and the pupils will constrict. If no signs of reviving occur after 3 to 4 minutes, a sharp blow to the sternum should be tried. In any case, artificial respiration and external massage should be continued until medical help arrives. The description given here is intended to provide only a general concept of what can be done. The reader is urged to discuss these procedures with expert local rescue people and to refer to publications directed specifically to first aid measures, because one really never knows just when such knowledge could save a life.

See also list of entries in entry on **Heart and Circulatory System (Human).**

CARDINAL (*Aves, Passeriformes*). Also known as the cardinal grosbeak or redbird, the cardinal is found both in North and South America. See illustration. The *Richmondena cardinalis* is found in the United States east of the plain states. It ranges northward as far as southern New York, Ontario, and Minnesota. The male is bright red with a small amount of black on its vermillion-colored head and beak. The female is rather drab in appearance. The bird is about $6\frac{1}{2}$ inches long ($16\frac{1}{2}$ centimeters), with a $3\frac{1}{2}$-inch (9-centimeter) tail, straight back, good posture, and proud, trim, and alert in appearance. It is a solitary nester. The female builds the nest, feeds the young, and incubates the eggs. The male helps with the feeding, sometimes feeding the female while she is on the nest. The birds often breed twice in a season. The egg is pale blue with brown spots and usually is one of four.

Cardinal

The cardinal is distinguished for its clear, loud, and sweet song. It is essentially nonmigratory. Other cardinals include the Florida cardinal (*R.c.floridana*), the Louisiana cardinal (*R.c.magnirostris*), found in southern Louisiana, and the gray-tailed cardinal (*R.c.canicauda*), which occurs in central and southern Texas.

CARDINAL NUMBER. It is difficult to give a satisfactory definition of a cardinal number, but this difficulty is not of great importance, since it is clear under what condition two sets contain the same cardinal number of objects, namely that they can be put into one-to-one correspondence with each other. This cardinal number is called the *power* of the set. An infinite set that can be put into one-to-one correspondence with the set of positive integers is said to be countable (or denumerable, or enumerable). A cardinal number which is not finite is called transfinite.

CARDINAL TETRA. **Characids.**

CARDIOGENIC SHOCK. **Shock Syndrome.**

CARDIOGRAPHY. **Electrocardiogram.**

CARDIOID. A higher plane curve, which is a special case of the limaçon. Its equations are

$$(x^2 + y^2 - 2ax)^2 = 4a^2(x^2 + y^2)$$

$$r = 2a(\cos\theta \pm 1)$$

$$x = a(2\cos\phi - \cos 2\phi)$$

$$y = a(2\sin\phi - \sin 2\phi)$$

where ϕ is a parameter in the last case.

The cardioid is also an epicycloid in which the radius of the fixed circle equals the radius of the rolling circle. The name comes from its heart-like shape. It has been used in the classical problem of trisecting an angle.

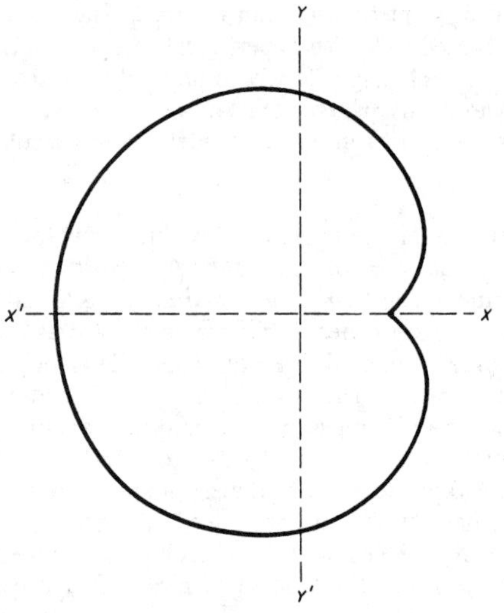

Cardioid.

See also **Epicycloid; Limacon;** and terms listed under **Mathematics.**

CARDIOID MICROPHONE. Microphone.

CARDIOMYOPATHIES. Heart and Circulatory System (Human).

CARDIOPULMONARY BYPASS. Heart and Circulatory System (Human).

CARDIOPULMONARY RESUSCITATION (CPR). Arrhythmias (Cardiac).

CARDIOVASCULAR MEDICINE. Arterial and Venous Disorders; Congestive Heart Failure; Heart and Circulatory System (Human); Ischemic Heart Disease.

CARDIOVERSION. Arrhythmias (Cardiac).

CARD READER. Input/Output Devices (Computing System).

CARDUACEAE. Artemisia.

CARDUELINAE. Fringillidae.

CAREY-FOSTER BRIDGE. Bridge Circuits (Electrical).

CARIBBEAN CURRENT. An ocean current flowing westward through the Caribbean Sea. It is formed by the commingling of part of the waters of the north equatorial current with those of the Guiana current. It flows through the Caribbean Sea as a strong current and continues with increased speed through the Yucatan Channel; there it bends sharply to the right and flows eastward with great speed out through the Straits of Florida to form the Florida current.

CARIBOUS. Deer.

CARIES AND CARIOLOGY. *Caries* is the decay of a bone or tooth; a progressive decalcification and proteolysis of the enamel and dentin. *Cariology* is the study of tooth decay. Dental caries is a process in which bacteria adhere to the tooth surface, especially in pits and other harbored areas, to form plaques. Plaque is made up of microbes that are able to attach to the teeth's surface because the bacteria secrete a sticky slime called *zooglea* (living glue). Plaque is also known as the microcosm. The microcosm keeps out substances that might harm the bacteria. Water, mouthwash, and saliva have little ability to penetrate the sticky mass, but sugar and fermentable carbohydrates penetrate easily. These foods are sources of energy for the caries bacteria.

These bacteria with their enzymes are capable of acting on fermentable foods to form acids. When sugar or carbohydrates contact the plaque, acids are produced within a few minutes. The concentration continues for several minutes and by the end of a half-hour, the concentration may be sufficient to dissolve enamel.

When acid concentration is sufficient to react with the inorganic salts of the tooth, there is partial decalcification of tooth substance. This produces a porous, opaque, white spot within the enamel substance. The process of acid formation and decalcification continues until all fermentable food is used and the acids are neutralized by saliva and minerals of the tooth substance. Decalcification stops when the acids are neutralized until more fermentable substance is brought into the plaque, upon which the cycle is repeated. Organic material of the tooth is said to be destroyed by *proteolytic* bacteria normally present in the plaque.

Any condition that leads to the formation of a bacterial film upon the tooth's surface will predispose to dental caries if acid-producing bacteria are present. Most people who appear to be resistant to dental caries can be shown to have very low intake of carbohydrates. Often these apparently resistant individuals become susceptible when they eat carbohydrates frequently. Eating in between meals (snacking) of carbohydrates is particularly favorable to microbial growth. Those microbes dependent upon carbohydrates then begin to grow and crowd out nonacid-producing bacteria (*acidogenic*). Any condition that diminishes salivary flow, thereby contributing to poor natural cleansing of the teeth and a diminished quantity of saliva in the mouth, will elevate the incidence of carious lesions. This has been observed frequently by the rapid production of decay in patients who have received radium or deep X-ray therapy for mouth cancer.

The first sign of dental decay is a white spot in the enamel. As demineralization continues, a hole (*cavity*) is produced. When a cavity forms, the area becomes more difficult to clean and the microbes flourish.

Restoration through the use of fillings is the most successful means of stopping a carious lesion. Amalgams, cast gold inlays, and gold foil have served for years as effective agents for repair. When the diseased portion of the tooth is completely removed and the remaining tooth substance cleaned and prepared to receive a filling, the caries will usually be arrested. If a cavity is not filled when it is small, decay progresses through the enamel and dentin of the tooth until the dental pulp is reached. Then, the patient experiences excruciating pain, for which there is no permanent relief until the pulp dies, or the tooth is extracted.

Although it has been known since 1924 that bacteria cause tooth decay, it it only within recent years that research findings have become more quantitative and less qualitative. Largely from experiments with laboratory rats, researchers have identified the bacterium *Streptococcus mutans* as the principal agent involved in tooth decay. *S. mutans*, as compared with other mouth bacteria, is outstanding in its ability to produce an acidic environment. Researchers at the National Institute of Dental Research have found that plaque has an electrical charge distribution that appears to contribute to damage promoted by the bacteria. The electrical field permits sucrose from food to diffuse into the plaque and thus to nourish the dense pockets of bacteria, but at the same time prevents outward diffusion of the large quantities of acid produced. Thus, in essence, a sponge of acid nestles directly against the tooth surface.

In 1977, researchers at the Forsyth Dental Center (Boston) isolated a mutant strain of *S. mutans* which generates much less acid. The mutant is deficient in the enzyme that forms lactic acid. A possible long-range concept leading from this research would be replacement of the more virulent *S. mutans* with the mutant. The concept of replacing harmful bacteria with more benign strains, of course, is not new in medicine. Research is also being conducted in Sweden wherein applications of a powerful mouthwash would destroy the virulent *S. mutans*, making colonization sites available for exposure to the mutant strains. Dental researchers also have been working for a number of years on development of a vaccine against caries-causing bacteria. But before this concept can be perfected, a much better understanding of the immune system present in saliva is needed. In past experimentation, young, bacteria-free laboratory animals injected with killed *S. mutans* exhibited significantly less disease than control animals where each group received a high-sugar diet and exposure to the virulent bacteria.

Another pathway of research is to attack the glucosyl-transferase enzymes which are responsible for converting sugar from food to the sticky material by which bacteria adhere to teeth and form plaque. It has been found that antibodies to glucosyl-transferase enzymes impede the ability of *S. mutans* to stick to a hard surface. As one authority has observed, oral immunization studies are needed to produce a form acceptable for human use.

A traditional approach to cavity control has been administration of an antiseptic to limit the number of microorganisms in the mouth. A major problem with this approach is the limited time available for the antiseptic to be effective because the saliva clears the mouth within a few minutes of the antiseptic rinse. Investigators are now thinking in terms of controlled-release antiseptics. Under test at the National Institute of Dental Research is a small device that can be attached to a back tooth for releasing fluoride, the latter acting not only to kill bacteria, but to increase decay resistance as well. Promising results have been found on tests with dogs and monkeys. Slow-release capsules are also under consideration.

Research also is being directed toward a better understanding of saliva, which is known to retard tooth decay, as evidenced by the large incidence of tooth decay in persons with xerostomia (dry mouth). A small molecule made up of just four amino acids and known as *sialin* was first isolated by Kleinberg (State University of New York at Stony Brook) in 1972. The concept of incorporating sialin in mouthwash and toothpaste, as well as in soft drinks and candy, is under active consideration.

Another traditional approach, of course, is that of substituting nonutritive sweeteners for sucrose and other natural carbohydrates. See **Sweeteners.**

Trace elements in the diet also have been studied. Investigations to date have shown that fluorine and phosphorus are strongly *cariostatic*; molybdenum, vanadium, strontium, boron, and lithium are mildly cariostatic; and selenium, magnesium, cadmium, copper, lead, silicon, and manganese are *cariogenic.* These correlations were developed at the Eastman Dental Center in Rochester, New York.

The need for improved decay prevention is obvious from the statistics. In a U.S. Navy study during the late 1970s, it was shown that out of 270,000 sailors surveyed, only 360 were found to be caries-free. Statistical studies have shown that there are wide geographical differences in the incidence of tooth decay. Decay rates are very high in the New England states; high in Illinois, Minnesota, Ohio, Oregon, Pennsylvania, Washington, and Wisconsin; moderate in Idaho, Louisiana, Montana, North and South Dakota, and Utah; and low in all other states with exception of Arkansas, Colorado, Oklahoma, and Texas, where the rates are very low. This study suggests that regions with acid soils are caries-prone, whereas those with alkaline soils are caries-low.

It has been estimated that there are about one billion unfilled cavities in the United States alone.

CARINA. A southern constellation which once, with Puppis and Vela, was part of a superconstellation known as Argo Navis. The bright star Canopus is contained in Carina.

CARLI INDEX. Index Number.

CARNALLITE. This mineral is a product of evaporation of saline deposits rich in potash content, as a hydrated chloride of potassium and magnesium, $KMgCl_3 \cdot 6H_2O$. Hardness, 2.5; specific gravity, 1.602. It crystallizes in the orthorhombic system usually as massive, granular aggregates. Luster greasy, with indistinct cleavage and conchoidal fracture. Color grades from colorless to white, into reddish from included hematite scales. Transparent to translucent with bitter taste, deliquesces readily in moist environment.

Found associated with sylvite, halite and polyhalite at Stassfurt, Germany; Abyssinia; the U.S.S.R.; and in southeastern New Mexico and adjacent areas in Texas. It is an important source of potash for use in fertilizers. See also **Potassium.**

CARNELIAN. The mineral carnelian is a red or reddish-brown chalcedony; the word is derived from the Latin word meaning flesh, in reference to the flesh color sometimes exhibited. See also **Chalcedony.**

CARNIVORA (*Mammalia*). Flesh-eating mammals, mostly predacious in habits, although some are omnivorous and some eat carrion. They have four or five toes on each foot, are armed with claws, the canine teeth are prominent, and the premolar and molar teeth are formed for cutting. The major species, families, etc. are listed in the accompanying table. In terms of numbers of living types, the *Carnivora*

CARNIVORA
(Flesh-eating Mammals)

	In this Encyclopedia
FELINES	See **Cats**
Great Cats (*Panthera*)	
Lions (*Panthera leo*)	
Tigers (*Panthera tigris*)	
Leopards (*Panthera pardus*)	
Snow Leopard (*Panthera uncia*)	
Jaguar (*Panthera onca*)	
Cats (*Profelis*)	
Pumas (*Profelis concolor*)	
Clouded Leopard (*Profelis nebulosa*)	
Golden Cats (*Profelis temmincki* and *aurata*)	
Lesser Cats (*Felis*)	
Ocelots (*Felis pardalis, ...*)	
Leopard-Cats (*Felis bengalensis, ...*)	
Tabby-Cats (*Felis lybica, ...*)	
Desert Cats (*Felis manul, ...*)	
Plain Cats (*Felis planiceps* and *badius*)	
Marbled Cats (*Felis marmorata*)	
Lynxes (*Lynx*)	
Jungle-Cats (*Lynx chaus*)	
Caracals (*Lynx caracal*)	
Northern Lynxes (*Lynx lynx, ...*)	
Bobcats (*Lynx rufa, ...*)	
Servals (*Leptailurus*)	
Jaguarondis (*Herpailurus*)	
Cheetahs (*Acinonyx*)	
VIVERRINES	See **Viverrines**
Civets (*Viverrinae*)	
True Civets (*Viverra* and *Civettictis*)	
Rasse (*Viverricula*)	
Genets (*Genetta*)	
African Linsang (*Poiana*)	
Linsangs (*Prionodon, ...*)	
Water-Civet (*Osbornictis*)	
Palm-Civets (*Paradoxurinae*)	
Musangs (*Paradoxurus*)	
Masked Palm-Civets (*Paguma*)	
Small-toothed Palm-Civets (*Arctogalidia*)	
Celebesean Palm-Civet (*Macrogalidia*)	
Binturong (*Arctictis*)	
West African False Palm-Civet (*Nandinia*)	
Hemigales (*Hemigalinae*)	
Hemigales (*Hemigale, ...*)	
Otter-Civet (*Cynogale*)	
Fanaloka (*Fossa*)	
Anteater-Civet (*Eupleres*)	
Galidines (*Galidiinae*)	
Fossas (*Cryptoproctinae*)	
Mongooses (*Herpestinae*)	
True Mongooses (*Herpestes*)	
Banded Mongoose (*Mungos*)	
Dwarf Mongooses (*Helogale*)	
Marsh Mongooses (*Atilax*)	
Cusimanses (*Crossarchus*)	
White-tailed Mongooses (*Ichneumia*)	
VIVERRINES (*Continued*)	
Mongooses (*Continued*)	
Bushy-tailed Mongooses (*Cynictis*)	
Dog-Mongooses (*Bdeogale*)	
Xenogales (*Xenogale*)	
Meerkat (*Suricata*)	
HYAENINES	See **Hyena**
Aard-Wolf (*Protelinae*)	
Hyaenas (*Hyaeniae*)	
Striped Hyaena (*Hyaena*)	
Spotted Hyaenas (*Crocuta*)	

CARNIVORA (*Continued*)

	In this Encyclopedia
PROCYONINES	See **Raccoons and Pandas**
Raccoons (*Procyoninae*)	
North American Raccoons (*Procyon*)	
Crab-eating Raccoons (*Euprocyon*)	
Coatimundis (*Nasua*)	
Mountain Coati (*Nasuella*)	
Cacomixtles (*Bassariscus*)	
Cuataquil (*Bassaricyon*)	
Kinkajous (*Cercoleptinae*)	
Pandas (*Ailurinae*)	
Lesser Panda (*Ailurus*)	
Giant Panda (*Ailuropoda*)	
CANINES	See **Canines**
True Canines (*Caninae*)	
Wolves (*Canis*)	
Jackals (*Thos*)	
Foxes (*Vulpes*)	
Fennecs (*Fennecus*)	
Arctic Fox (*Alopex*)	
Gray Fox (*Urocyon*)	
South American Jackals (*Dusicyon*)	
Maned Wolf (*Chrysocyon*)	
Raccoon-Dog (*Nyctereutes*)	
False Canines (*Simnocyoninae*)	
Dholes (*Cuon*)	
Cape Hunting Dog (*Lycaon*)	
Bush-Dogs (*Speothos*)	
Bat-eared Foxes (*Otocyoninae*)	
URSINES	See **Bears**
Common Bears (*Ursus, ...*)	
Brown Bears (*Ursus*)	
Spectacled Bear (*Tremarctos*)	
Sun-Bear (*Helarctos*)	
Moon Bears (*Selenarctos*)	
American Black Bear (*Euarctos*)	
Sloth-Bears (*Melursus*)	
Polar Bears (*Thalarctos*)	
MUSTELINES	See **Mustelines**
Weasels (*Mustelinae*)	
True Weasels (*Mustela, ...*)	
Polecats (*Putorius*)	
Minks (*Lutreola*)	
Martens (*Martes, ...*)	
Tayras (*Tayra*)	
Grisons (*Grison, ...*)	
Striped Weasels (*Poecilogale* and *Poecilictis*)	
Zorilles (*Zorilla*)	
Badgers (*Melinae, ...*)	
Wolverines (*Gulo*)	
Ratels (*Mellivora*)	
Eurasian Badgers (*Meles*)	
Sand-Badgers (*Arctonyx*)	
Teledu (*Mydaus*)	
American Badgers (*Taxidea*)	
Tree-Badgers (*Helictis*)	
Skunks (*Mephitinae*)	
Hog-nosed Skunks (*Conepatus*)	
Striped Skunks (*Mephitis*)	
Spotted Skunks (*Spilogale*)	
Otters (*Lutrinae*)	
Common Otters (*Lutra*)	
Simung (*Lutrogale*)	
Clawless Otters (*Amblonyx*)	
Small-clawed Otters (*Aonyx* and *Paraonyx*)	
Saro (*Pteroneura*)	
Sea-Otter (*Enhydra*)	

represent the fourth largest order of *Mammalia*, exceeded by *Rodentia*, *Chiroptera*, and *Artiodactyla*. Members of *Carnivora*, with exception of Australia and the oceanic islands, are widely distributed throughout the world. Specific references on the accompanying table indicate the titles of entries in this Encyclopedia where detailed information on specific varieties can be found.

CARNOSINE. **Brain and Nervous System.**

CARNOT CYCLE. An ideal cycle of four reversible changes in the physical condition of a substance; useful in thermodynamic theory. Starting with specified values of the variable temperature, specific volume, and pressure, the substance undergoes in succession (1) an isothermal (constant temperature) expansion, (2) an adiabatic expansion (see **Adiabatic Processes**), and (3) an isothermal compression to such a point that (4) a further adiabatic compression will return the substance to its original condition. These changes are represented on the volume-pressure diagram respectively by *ab, bc, cd,* and *da* in Fig. 1. Or the cycle may be reversed: *a d c b a.*

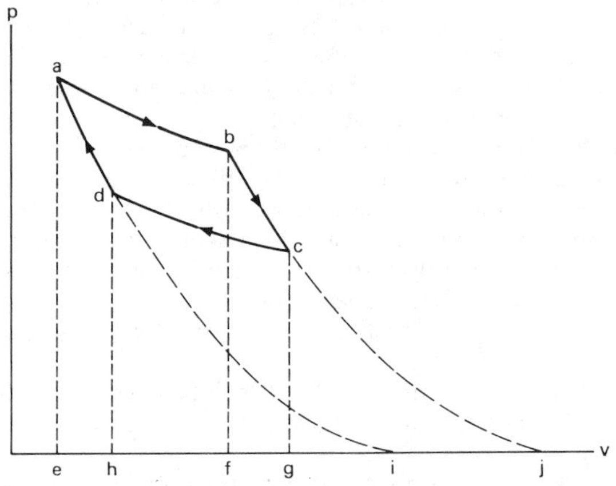

Fig. 1. Carnot cycle on *vp* diagram; *ab* and *cd*, isothermals; *bc* and *da*, adiabatics which, for some theoretical purposes, are produced to infinity.

In the forward (clockwise) case, heat is taken in from a hot source and work is done by the hot substance during the high-temperature expansion *ab*; also additional work is done at the expense of the thermal energy of the substance during the further expansion *bc*. Then a less amount of work is done on the cooled substance, and a less amount of heat discharged to the cool surroundings, during the low-temperature compression *cd*; and finally, by the further application of work during the compression *da*, the substance is raised to its original high temperature. The net result of all this is that a quantity of heat has been taken from a hot source and a portion of it imparted to something colder (a "sink"), while the balance is transformed into mechanical work represented by the area *abcd*. Thus, the forward Carnot cycle can be used for the production of power. If the cycle takes place in the counter-clockwise direction, heat is transferred from the colder to the warmer surroundings at the expense of the net amount of energy which must be supplied during the process (also represented by area *abcd*). It can thus serve as a refrigerating cycle.

The temperature-entropy diagram for the Carnot cycle, corresponding to the pressure-volume diagram is shown in Fig. 2.

It should be noted that the efficiency of the forward cycle is highest when T_1 is as high as possible. Since, in practice, T_0 will always be fixed by the temperature of the surrounding atmosphere, a high efficiency corresponds to a large difference $T_1 - T_0$. In contrast, a high

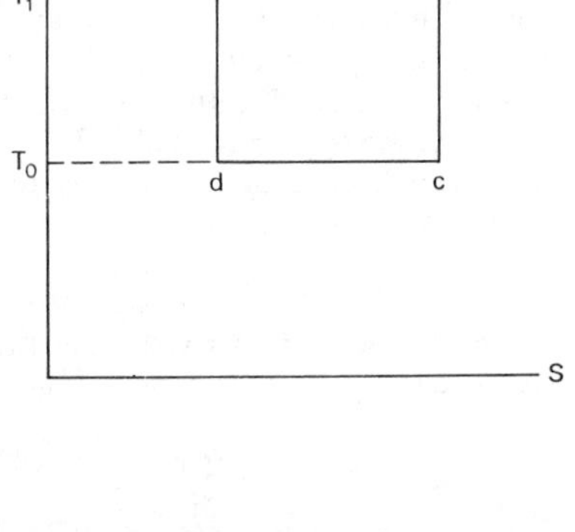

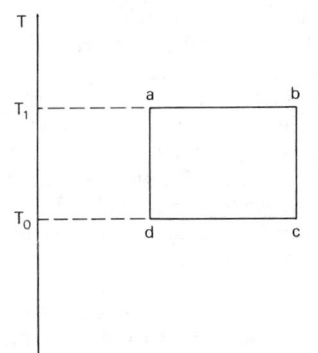

Fig. 2. Carnot cycle temperature-entropy diagram.

coefficient of performance, or a high effectiveness of a heat pump corresponds to a small difference $T_1 - T_0$.

It would appear that decreasing T_0 for a power cycle below that of the surrounding atmosphere is advantageous in that the efficiency η is increased. However, it must be realized that this can only be achieved at the expense of work in operating a refrigerator, and no advantage is gained.

CARNOT ENGINE. **Absolute Zero; Solar Energy.**

CARNOTITE. This mineral is a vanadate of potassium and uranium with small amounts of radium. Its formula may be written $K_2(UO_2)_2(VO_4)_2 \cdot 3H_2O$. The amount of water, however, seems to be variable. It occurs as a lemon-yellow earthy powder disseminated through cross-bedded sandstones with rich concentrations around petrified and carbonized trees. Soft; sp gr 4.7. It was mined in Colorado and Utah as a source of radium. Other localities are in Arizona, Pennsylvania, and Zaire.

CARNOT THEOREM. No engine operating between two given temperatures can be more efficient than a perfectly reversible engine operating between the same temperatures. See **Carnot Cycle.**

CAROLINA HEMLOCK. **Hemlock Trees.**

CAROTENES. **Carotenoids; Vitamins.**

CAROTENOID PIGMENTS. **Pigmentation (Plants).**

CAROTENOIDS. Lipid-soluble, yellow-to-orange-red pigments universally present in the photosynthetic tissues of higher plants, algae, and the photosynthetic bacteria. They are spasmodically distributed in flowers, fruit, and roots of higher plants, in fungi, and in bacteria. They are synthesized *de novo* in plants. Carotenoids are also widely, but spasmodically distributed in animals, especially marine invertebrates, where they tend to accumulate in gonads, skin, and feathers. All carotenoids found in animals are ultimately derived from plants or protistan carotenoids, although because of metabolic alteration of the ingested pigments, some carotenoids found in animals are not found in plants and protista.

Carotenoids are tetraterpenoids, consisting of eight isoprenoid 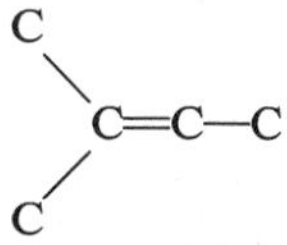 residues, and can be regarded as being synthesized by the tail-to-tail dimerization of two 20-carbon units, themselves each produced by the head-to-tail condensation of four isoprenoid units. Hydrocarbon carotenoids are termed *carotenes* and oxygenated carotenoids are known as *xanthophylls.* The structure of the best-known carotene, β-carotene, is

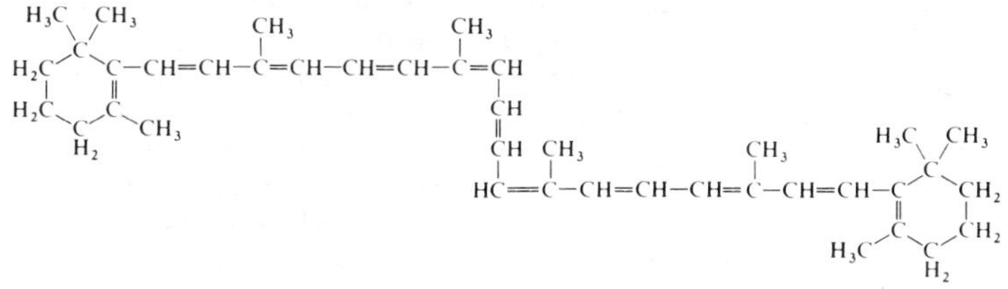

α-Carotene is widely distributed in trace amounts, together with β-carotene in leaves; γ-carotene is found in many fungi, and lycopene is the main pigment of many fruits, such as the tomato.

See also **Annatto Food Colors;** and **Pigmentation (Plants).**

CARP (*Osteichthyes*). Carp are members, along with minnows, of the family *Cyprinidae* (group *Cypriniformes*). The common carp is an introduced species, indigenous to eastern Asia, but now thoroughly acclimated in the rivers and lakes of North America and Europe. It is coarse and bony, but widely used as food. The goldfish or golden carp is a related species native to China and Japan. Many strange varieties have been developed in captivity and the species is thriving in some lakes and streams in the eastern United States.

Sports fishermen often consider the carp in extremely negative terms because in some areas the carp has literally taken over habitats, thus excluding more desirable edible and sporting species, usually present prior to the introduction of the carp. On the other hand, the carp is easily cultivated and thus in some areas of the world, the fish is a blessing as a food source. It is notable that a planted and well-fertilized pond will produce more than a half-ton of fish per acre (560 kilograms per hectare) within a reasonable time. In comparison, less than 200 pounds (91 kilograms) of black bass can be produced under the same circumstances within the equivalent period of time. Originally, *Cyprinus carpio* (the common carp) came from the Black and Caspian Seas and environs. It is now found in most of the temperate waters worldwide. The carp is characterized by its four barbels, a pair at each side of the mouth. These features are lacking in the similar goldfish. The number of scales also varies between carp and goldfishes. The Japanese golden carp is dramatic to view and is considered a show fish. Most common goldfish varieties have stemmed from the so-called wild goldfish (*Carassius auratus*), also sometimes referred to as the Missouri minnow, funa (in Japan), or johnny carp. This fish is brownish and quite plain in appearance, but with the physical features of a carp.

Numerous varieties of domestic goldfish have been developed over the years, notably by the Japanese. Some of these include the normal V-tail (the type most commonly sold in pet stores); the veiltail with its three-lobed tail; the blackmoor which features a coloration reminiscent of black velvet, bulbous "pop eyes," and a veil tail, the celestial telescope goldfish which has bulbous eyes having the appearance of looking upward as the fish swims forward; and lionheads, lacking the dorsal fin, with thick tumorous-like structures over the head. A Singapore carp is illustrated in the accompanying figure.

Singapore carp. (*A. M. Winchester.*)

The raising of carp for commercial purposes is discussed in some detail in entry on **Aquaculture.** See also **Fishes.**

CARPENTER-BEE (*Insecta, Hymenoptera*). A bee which excavates its nest in wood. The small carpenter-bee, *Ceratina dupla,* of North America merely digs out the pith or soft wood of a plant, such as sumac, while the large carpenter-bees, *Xylocopa,* of which there are several species, bore into solid wood, even attacking unpainted wood in construction. The larger bees are not unlike bumblebees in appearance.

CARPENTER-MOTH (*Insecta, Lepidoptera*). Moths whose larvae bore in the trunks of trees, entering the solid wood. A few species, including the locust borer, are of large size, and because of their narrow wings and long bodies may be mistaken for sphinx moths. These insects make up the small family *Cossidae.*

CARPETGRASS. **Grasses.**

CARPET-MOTH. Buffalo Carpet-Moth.

CARPINACEAE. Hornbeam Trees.

CARPSUCKER. Sucker.

CARRAGEENAN. Gums and Mucilages.

CARRIER AMPLIFIER. A dc amplifier wherein the signal first is modulated, the demodulated during amplification. Electronic switches or electromechanical devices are used in most cases to effect the modulation. Thus, the "chopping" action accomplishes the equivalent of a square-wave modulation of the signal.

The carrier technique is employed for two main purposes: (1) to reduce to a minimum the effects of zero-offset drift, which is a critical performance parameter in any dc amplifier, and (2) to provide isolation between the input and the output of the amplifier.

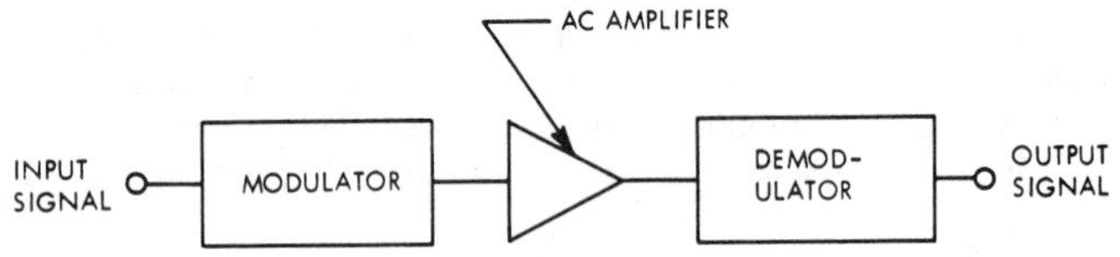

Fig. 1. Carrier amplifier.

With reference to the accompanying diagram (Fig. 1), a conceptual design is shown. The input signal first is modulated to produce an ac signal, after which the signal is amplified by an ac amplifier. Then, the output of the latter is demodulated to provide a dc output signal. Zero-offset drift in the amplification section of the amplifier does not affect the value of the output signal because only the ac component is amplified. However, offsets in the modulator can cause the equivalent of an offset in the output signal should they increase or decrease the magnitude of *both* the positive and negative peaks of the modulated signal. In most cases, if the input signal is greater than 1 V, such offsets do not create a serious problem. In the case of low-level amplifiers, however, they can cause significant errors. Because the output demodulator usually operates at a high level, demodulator offset is not considered an important limitation on overall amplifier performance. The use of carrier amplifiers designed mainly for the reduction of zero-offset drift is diminishing mainly due to the improvement of techniques and components for accomplishing low-drift direct-coupled amplifiers.

In the instance of using a carrier amplifier to provide isolation between the input and output of the amplifier, the amplifier commonly is termed a "floating amplifier." See also **Floating Amplifier.** An amplifier of this design, incorporating an overall feedback path, is shown in Fig. 2. The basic carrier concept is used—the input signal is modulated and demodulated by a chopper circuit. In this example, since the main purpose is isolation rather than reduction of drift, an ac amplifier is not used. By means of the four-terminal isolation characteristic of the transformer, the input signal can be referenced to a ground point that is independent of the output-signal reference point. Floating-carrier amplifier designs of this type are used in digital-data acquisition subsystems and instrumentation subsystems to accomplish amplification of signals under conditions where high common-mode voltages may be present. The common-mode voltage is essentially limited by the breakdown voltage of the coupling transformer. Thus, amplifiers of this design can function with up to several hundred volts of common mode, as contrasted with the usual 10 to 20 V limit inherent to most direct-coupled amplifiers as the result of the breakdown limitations of most semiconductor devices.

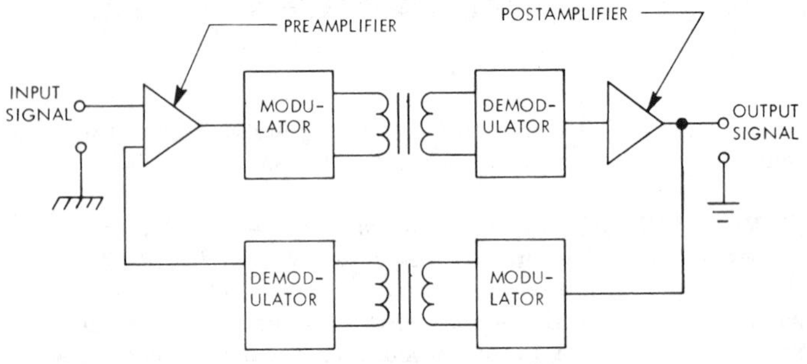

Fig. 2. Floating-carrier amplifier.

See also terms listed under **Data Processing.**

Thomas J. Harrison, International Business Machines Corporation, Boca Raton, Florida.

CARRIER-AMPLITUDE REGULATION. The change in amplitude of the carrier wave in an amplitude-modulated transmitter when modulation is applied under conditions of symmetrical modulation.

CARRIER (Communications). A wave suitable for being modulated to transmit intelligence. The modulation represents the information; the original wave is used as a "carrier" of the modulation. See also **Modulation.**

CARRIER CURRENT. Carrier current is used in connection with both power and communication circuits but, basically, the principle is the same for both systems. The term refers to the use of a relatively high-frequency ac superimposed on the ordinary circuit frequencies in order to increase the usefulness of a given transmission line. Thus in the case of power systems, carrier currents of several kHz frequency are coupled to the 60-Hz transmission lines. These carrier currents may be modulated to provide telephone communication between points on the power system or they may be used to actuate relays on the system. This latter use is known as carrier relaying. Carrier currents have greatly extended the usefulness of existing line facilities of the telephone and telegraph companies. Several carrier frequencies may be coupled to the lines already having regular voice or telegraph signals on them. Each of these carrier frequencies may be modulated with a separate voice or telegraph channel and thus a given line may carry the regular signals plus several new carrier channels, each of which is equivalent to another circuit at regular frequencies. At the receiving end, the various channels are separated by filters and the signals demodulated and then fed to conventional phone or telegraph circuits. The number of carrier channels which may be applied to a given line depends upon the characteristics of the line, varying from one or two for some lines to several hundred for the coaxial cable.

See also **Filter (Communications System).**

CARRIER (Disease). Foodborne Diseases.

CARRIER (Food Additive). A substance well named because its primary function is that of conveying and distributing other substances throughout a food substance. The role parallels that of a carrier in paint, wherein a vehicle (carrier) holds and distributes pigment throughout the entire paint product. Silica gel and magnesium carbonate serve as carriers in food substances. For example, the high porosity of silica gel enables it to adsorb internally up to three times its own weight of many liquids. This property is used to convert various liquid ingredients, such as flavors, vinegar, oils, vitamins, and other nutritional additives, into easy-to-handle powders. These powders, in turn, can be measured easily and blended effectively with other constituents to provide a uniform food substance. Advantage is taken of the properties of carriers in the convenience food field, where flavors remain entrapped inside silica particles until the food product is mixed with water, at which time the flavors are released just prior to consumption, giving the product an aura of richness and freshness.

CARRIER FREQUENCY. Also called center frequency or resting frequency, that frequency generated by an unmodulated radio, radar, or carrier communication transmitter; of the average frequency of the emitted wave when modulated by a symmetrical signal.

CARRIER (Semiconductor). Semiconductor.

CARRIER SUPPRESSION. (1) Suppression of the carrier when there is no modulation signal to be transmitted. This is practiced on ships to lessen interference between transmitter. (2) Suppression of the carrier frequency after conventional modulation at the transmitter, but with reinsertion of the carrier at the receiving end prior to demodulation.

CARRIER-TO-NOISE RATIO. The ratio of the value of the carrier to that of the noise after selection and before any nonlinear process such as amplitude limiting and detection.

CARRIER WAVE. Modulation.

CARRION BETTLE (*Insecta, Coleoptera*). Moderate to large beetles which are found above decaying flesh and to some extent about other decaying matter. Applied to members of the family *Silphidae*, although many other beetles breed in decaying matter and are found in it, both as adults and as larvae.

CARROT FAMILY. Umbelliferae.

CARTESIAN COORDINATES. Coordinate System.

CARTESIAN TENSOR. Tensor Field (Cartesian).

CARTILAGE. A supporting tissue associated with the skeleton of vertebrates. Cartilage, like the other connective and supporting tissues, contains a relatively large amount of intercellular substance in which the cells are scattered. This substance is a complex mixture of organic materials, bluish in color and translucent. It contains organic fibrils and around the cavities in which the cartilage cells lie it differs chemically as shown by its reaction to stains. The cells are rounded and may lie singly or in groups in the capsules.

Three kinds of cartilages are recognized: hyaline, elastic, and fibrocartilage. The first contains few fibrils. It is flexible, slightly elastic, and provides a support of moderate rigidity. It covers the ends of bones in movable joints as the articular cartilages, forms the rings of the trachea, and occurs in other parts of the body where such qualities are required. Elastic cartilage is similar to hyaline but has many elastic fibers in the intercellular matrix. It occurs in the pinna of the ear, where its qualities provide support and elasticity, the latter very necessary in a delicately formed projecting structure of this kind which might otherwise be easily broken. Fibrocartilage contains many inelastic white fibers which give it extreme toughness. It is associated with some joints and forms the intervertebral disks of the backbone. These disks provide very firm connections between the separate vertebrae and at the same time cushion the series.

The term cartilage is also applied to separate skeletal units formed of this material. Each cartilage is surrounded by a tough connective tissue sheath called the perichondrium.

Cartilage is a primitive skeletal material of the vertebrates. It precedes bone in embryonic development and persists in the adult skeleton in the sharks and related fishes. It is not transformed into bone but is replaced by bone in the formation of some of the parts of the skeleton. It may become rigid through the deposition of calcareous material in its matrix, particularly in old age. This calcified cartilage, while rigid like bone, does not have the minute structure of that tissue. See also **Arthritis;** and **Bone.**

Cartilage is also the term used for the internal structure of the ligament which connects the valves of the shell in some of the bivalve mollusks.

CARTILAGE FISHES. Sharks; Skates and Rays.

CARTOGRAPHIC DATA. Earth Resources Satellites and Geologic Remote Sensors.

CASALE PROCESS. Ammonia.

CASCADE. Any connected arrangement of separative elements whose result is to multiply the effect, such as isotope separation, created by the individual elements. A bubble plate-tower is a cascade whose elements are the individual plates; a plant consisting of many towers in series and parallel is similarly a cascade whose elements may be considered to be either the towers or the individual plates. Similarly, an amplifier in which each stage except the first has as its input the output of the preceding stage is spoken of as a cascade amplifier. A stage of a grounded-cathode vacuum-tube amplifier is defined as the section from a point just before the grid of one tube to that just before the grid of the next. Similarly, for grounded-emitter transistor amplifiers, it is defined as the section from a point just before the base of one transistor to that just before the base of the next.

CASCADE AMPLIFIER. Amplifier.

CASCADE AMPLIFIER (Darlington Pair). Darlington Pair.

CASCADE CONTROL. A control system in which the output signal from one controller is used as the setpoint for another controller. The application of cascade control to a stripping column is shown in the accompanying figure. In this application, the primary variable is column temperature. Measurement of column temperature is transmitted by TT to a temperature recorder controller TRC, which adjusts the reference input to the secondary steam flow control loop. Since the time response of the flow control or secondary loop is much less than that of the temperature or primary loop, steam pressure disturbances, which often occur in a situation of this kind, are corrected before they can influence column temperature. The system improves the situation in this regard, but does not accommodate other system disturbances, such as variations in feed flow, feed temperature, or feed composition. The latter variations must be corrected by the TRC in the usual fashion.

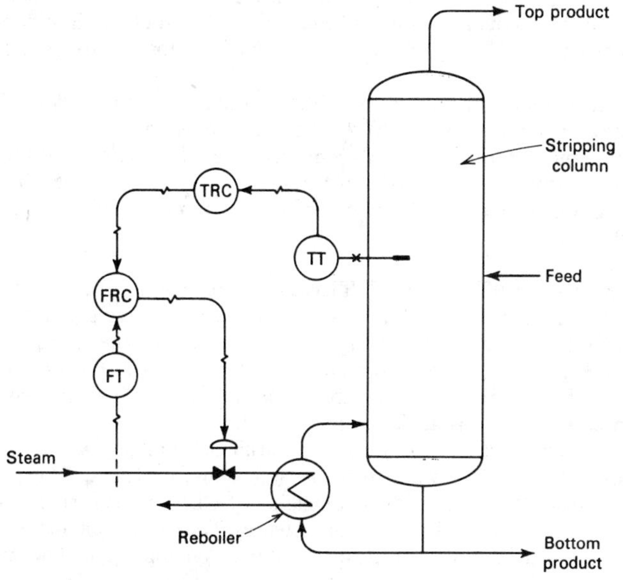

Cascade control system as applied to a stripping column.

When a cascade system is placed into operation, both controllers are initially set on manual during startup. After the process stabilizes, the secondary controller is placed on automatic and the correct control action settings are determined. When the secondary controller operation on automatic is satisfactory, the primary controller is then placed on automatic and its optimum control action settings are determined.

CASCADE COOLING. Cryogenics.

CASCADE SHOWER. A type of cosmic ray shower brought about when a high-energy electron, in passing through matter, produces one or more photons of energies of an order of magnitude of its own. These photons are converted into electron pairs by the process of pair production. Then the secondary electrons produce the same effects as the primary, so that the process continues, and the number of particles increases. This cascade shower of negatrons and positrons continues to build up until the energy level of product particles falls to a point where photon emission and pair production can no longer occur. See also **Cosmic Rays.**

CASCADE SYSTEM (LNG). **Natural Gas.**

CASCARA. A drug, used as a laxative and cathartic, obtained from the bark of a shrub, *Rhamnus purchiana*, a member of the family *Rhamnaceae* (buckthorn family). The plant is found in western North America.

CASCARILLA BARK. **Euphorbiaceae.**

CASCAVAL. **Snakes.**

CASE HARDENING. Hardening of the surface layer or case of a ferrous alloy while leaving the core or center in a softer, tougher condition. There are two basic methods of case hardening. In the first, gaseous elements such as carbon or nitrogen are introduced into the surface layer, thereby forming a hardening or hardenable alloy at the surface. Examples are carburizing, nitriding, and carbonitriding. Alternatively, the surface may be given a hardening heat treatment that does not affect the core. This may be accomplished by flame hardening or induction heating, whereby the surface is rapidly heated into the austenite range and the specimen quenched before the center has obtained a temperature high enough to allow it to be hardened.

CASEIN. Casein is the phosphoprotein of fresh milk; the rennin-coagulated product is sometimes called paracasein. British nomenclature terms the casein of fresh milk caseinogen and the coagulated product casein. As it exists in milk it is probably a salt of calcium.

Casein is not coagulated by heat. It is precipitated by acids and by rennin, a proteolytic enzyme obtained from the stomach of calves. Casein is a conjugated protein belonging to the group of phosphoproteins. The enzyme trypsin can hydrolyze off a phosphorus-containing peptone.

The commercial product which is also known as casein is used in adhesives, binders, protective coatings, and other products.

The purified material is a water-insoluble white powder. While it is also insoluble in neutral salt solutions, it is readily dispersible in dilute alkalis and in salt solutions such as those of sodium oxalate and sodium acetate.

CASHEW AND SUMAC TREES. The family *Anacardiaceae* (cashew family) is full of interesting variety, thus making generalizations difficult. Several of the species are known for their poisonous, irritating nature, such as poison ivy and poison oak. On the other hand, other species produce edible fruits and nuts, such as the mango, and cashew, and pistachio nuts.

Poison ivy (*Rhus radicans*) is a shrub or climbing woody vine, frequently found in wooded areas and along roadsides. The plant can extend itself to considerable heights by climbing tree trunks, masonry walls, and wooden screens and fences. The shrub is characterized by three pale green, ovate leaflets, smooth on top, slight fine hairs underneath on young leaves. The plant tolerates wet or dry conditions and is hardy, but generally prefers partial shade. All parts of the plant are irritating to humans. Poison oak (*Rhus toxicodendron*) also is of the cashew family and all parts of the plant are also poisonous irritants. This is an erect shrub ranging up to 20 inches (50.8 centimeters) in height, occurs in many locations, but more frequently in the southern states. The shrub is named for its oak-like leaves, which are toothless, ovate, compound, and occur in groups of three. They are of a pale green color and even a lighter green underneath. The plant frequents uncrowded woody areas and on wasteland. Its preferred regions include the coastal plains south of Maryland and New Jersey to Florida and westward into Texas. Another poisonous species is *Rhus vernix*, poison sumac, all parts of which are irritants. The shrub ranges from 5 to 10 feet (1.5 to 3 meters) in height, although it may be in the form of a small tree as high as 20 feet (6 meters). The trunk is short, with forking occurring close to the ground. The leaves are compound, smooth, toothless, sharply pointed, and light green. The plant prefers moist, swampy locations and ranges widely from southern Maine south and west to Florida and Texas.

The cashew nut is the fruit of a Brazilian tree of moderate size (*Anacardium occidentale*). The kidney-shaped nut grows at the end of a curiously enlarged fleshy peduncle which is juicy and bright yellow or red. The fleshy portion is eaten in tropical South America. The nut itself contains a biting caustic oil which is driven off by roasting. The single kernel of this fruit is the familiar cashew nut, widely distributed as a confection. The oil has been used in termite insecticides. Oil from cashew shells also has been used in compounding of rubber and plastics. The nuts range considerably in size (from 200 to 450 per pound; 441 to 992 per kilogram) and thus require grading before marketing.

The pistachio tree (*Pistacia vera*) is a small tree with deciduous pinnate leaves and is native to southwestern Asia, from which region it has spread in cultivation to the Mediterranean countries. Greece, for example, is an important producer of pistachio nuts. The apetalous flowers are unisexual and borne in panicles and the plants are dioecious. The fruit is a drupe, containing an elongated seed with a greenish kernel, having a very characteristic flavor. The kernels are used in confections, ice cream, and also eaten alone, usually after salting. Related to the true pistachio is *Pistacia lentiscus*, a shrub or small tree of the Mediterranean region with evergreen, pinnately compound leaves. From it is obtained a resin, mastic, which is often chewed by the natives of Turkey. It is used in medicine as a mild stimulant, as well as in varnishes. Another species is *Pistacia terebinthus*, a native of eastern Mediterranean countries, which yields China turpentine.

The mango tree (*Manaifera indica*), also of the cashew family, is a long-lived tree, which often develops a massive trunk and widely spreading branches. The lanceolate leaves of the mango are evergreen and about 4 inches (10 centimeters) in length. The flowers are numerous, small, pink, and borne in racemes. The ovoid fruits, 1 to 5 inches (2.5 to 12.7 centimeters) in diameter, are one-seeded berries having

RECORD SUMAC TREES IN THE UNITED STATES[1]

Specimen	Circumference[2] (inches)	Circumference[2] (centimeters)	Height (feet)	Height (meters)	Spread (feet)	Spread (meters)	Location
Dwarf sumac (1972) (*Rhus copallina var. leucantha*)	13	33	22	6.6	15	4.5	Florida
Evergreen sumac (1975) (*Rhus virens*)	22	56	17	5.1	22	6.6	Texas
Prairie sumac (1977) (*Rhus lanceolata*)	45	114	29	8.7	23	6.9	Texas
Shining sumac (1974) (*Rhus copallina*)	31	79	55	16.5	22	6.6	Mississippi
Staghorn sumac (1972) (*Rhus typhina*)	27	68.5	49	14.7	30	9	Michigan
Sugar sumac (1977) (*Rhus ovata*)	57 (at 18 inches)	144 (at 45.7 centimeters)	20	6	32	9.6	Arizona

[1] From the "Social Register of Big Trees," The American Forestry Association (by permission).
[2] At 4.5 feet (1.4 meters).

a thick, rough, greenish rind and a pleasantly aromatic, orange-colored flesh that is esteemed by many people. This fruit may be eaten fresh, or in salads. Reproduction is either by seedlings, which do not always come true, or by grafting. The tree is extensively cultivated in tropical regions, and was introduced a number of years ago to Florida and southern California. Probably it was first introduced into tropical America (Jamaica) in 1782. For successful growth, hot moist weather is necessary, followed by a short dry period for successful ripening of the fruit. Although the fruit is usually the important consideration of this tree, it also can make an excellent shade tree. The wood is soft, easily worked and, when available, can be used for constructing boats, canoes, and light buildings.

The South American paper tree (*Schinus molle*), another member of the cashew family, has been introduced into the Mediterranean region and in the warmer areas of North America. This tree, of a somewhat drooping contour, with small red berries, often gnarled trunk and branches, can make an interesting garden tree. However, generous space must be allowed because the tree can attain a height of nearly 40 feet (12 meters) within a 20-year period. This tree is not to be confused with the genera of plants (*Piper* and *Capsicum*), the sources of commercial paper.

Smoke trees are of the genus *Cotinus. C. coggyria* is frequently found in the gardens of Europe for its purple decor. An American counterpart, the *C. obovatus*, is a highly colorful plant and often used in gardens and landscaping.

Tung oil, a powerful drying oil, is obtained from the seeds of *Aleurites fordii* and closely related species, also in the family *Anacardiaceae.* This is a tree of China. The sap of *Rhus verniciflua* yields a furniture lacquer.

Some species of sumac find acceptance in gardens, notably the varieties shown on the accompanying table. Although the various sumacs, such as the staghorn and dwarf sumacs, are generally considered as shrubs, the dimensions shown in the table indicate the large proportions they can assume when situated in very favorable conditions.

CASINGHEAD GASOLINE. Natural Gas.

CASSAVA. Euphorbiaceae.

CASSAVA (As Energy Source). Biomass and Wastes as Energy Sources.

CASSINI DIVISION. Saturn.

CASSIOPEIA (the chair). One of the most widely known and striking constellations of the northern latitudes. Cassiopeia is easily recognized by the five bright stars forming an irregular W, some observers seeing not only a W, but also a chair. Since this object is circumpolar for most northern countries (i.e., remains above the horizon at all hours every night, and is easily recognized, it is frequently used as a rough indicator of sidereal time. The leading bright star of W (the star Beta Cassiopeiae) lies almost in zero hours right ascension. Hence, a line drawn through Polaris and Beta Cassiopeiae must pass close to the vernal equinox. The hour angle of this line must be equal to sidereal time. Thus, when Beta Cassiopeiae is on the meridian directly above the pole, the sidereal time is zero; when it is on the meridian directly below the pole, the sidereal time is 12 hours, etc.

One of the brightest novae on record appeared in this constellation in 1572, and was observed and recorded by Tycho Brahe. (See map accompanying entry on **Constellations.**)

CASSITERITE. The mineral cassiterite, chemically tin dioxide, SnO_2, is almost the sole ore of tin. It is a noticeably heavy mineral crystallizing in the tetragonal system, as low pyramids, prisms, often very slender, and as twinned forms. It is a brittle mineral, hardness, 6.0–7.0; specific gravity, 6.99; luster, adamantine; color; generally brown to black, but may be red, gray to white, or yellow; streak whitish, grayish, or brownish; may be almost transparent to opaque. A fibrous variety somewhat resembling wood is called wood tin. Cassiterite occurs in widely scattered areas, but deposits of a size to be commercially important are few. It is associated with granites and rhyolites.

Cassiterite is heavily concentrated in bands and layers of varying thickness, forming economically valuable deposits, such as those found in the Malay States of southeastern Asia; Bolivia, Nigeria, and the Belgian Congo are also major producers of tin ore. Cassiterite is also known as tin stone.

CASSOWARIES (*Aves, Casuariiformes, Casuriidae*). A family of birds closely related to the emus; they inhabit the primeval forests of North Australia and New Guinea as well as some of its islands. See accompanying illustration.

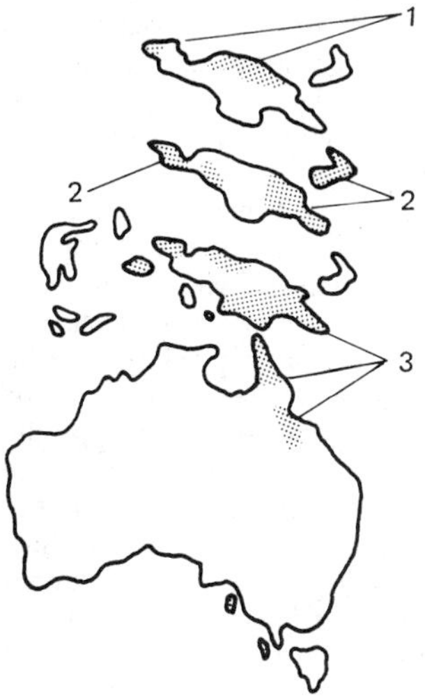

Areas inhabited by the Cassowaries: (1) one-wattled cassowary (*Casuarius unappendiculatus*); (2) Bennett's cassowary (*Casuarius bennetti*); (3) Australian cassowary (*Casuarius casuarius*).

There is only one genus (*Casuarius*). With a height at the back of up to 100 centimeters (39 inches) and a weight of 85 kilograms (187 pounds), it is the heaviest bird next to the ostrich. The legs are very strong. There are three toes, the claws of the inner toe being up to 10 centimeters (4 inches) long and straight. The feathers, like those of the emus (see also **Emu**), have an aftershaft of equal length; the flight feathers are reduced to mere rods of thick keratin. On the head they have a helmetlike, horny structure. The head and neck are bare of feathers; instead some have skin folds on the neck. The species are distinguished according to the shape of the helmet and the form of the skin folds of the neck. The bare skin differs in color in the various species and subspecies, and can be bright red, yellow, blue and/or white. Males and females are similarly colored. The chicks have a yellow-brown downy plumage with dark brown longitudinal stripes, but after a few months they become uniformly brown. The eggs average 135 × 90 milimeters (5.3 × 3.5 inches) and weigh 650 grams (23 ounces); the surface is slightly wrinkled, and the color is a shiny grass green, which later darkens somewhat.

There are 3 species: (a) Australian Cassowary (*Casuarius casuarius*); (b) One-Wattled Cassowary (*Casuarius unappendiculatus*); (c) Bennet's Cassowary (*Casuarius bennetti*).

See also **Ratites.**

CASPID. Virus.

CASTELLANUS. Clouds and Cloud Formation.

CASTING. Casting is the process of producing metal shapes by pouring molten metal into molds of the required form where it is allowed to solidify. The metal part formed as a result of this operation is called a casting. The art of casting is one of the oldest methods for making metal parts and is still extensively used in spite of more modern developments such as forging, rolling, and extrusion.

The production of a casting involves the use of a pattern, usually of wood or metal, which is similar in shape to the desired finished piece and slightly larger in all dimensions to allow for shrinkage of the metal upon solidification. The pattern is bedded down in a special damp sand by an operation called molding. When the pattern is removed it leaves an impression of the shape of the desired casting. This impression is completely surrounded by sand and provided with openings called gates through which the molten metal enters. After pouring and cooling the mold is broken open and the casting removed. All adhering sand particles together with any extraneous projections such as those left by the gate system are removed after which the casting is machined to the required finish.

The term is also applied to the casting of pig iron in blast furnace practice and the casting of ingots in steel-mill practice.

Centrifugal casting is applicable to the production of pipe and tubing, wheels, gear blanks, and other castings having rotational symmetry. While the mold is rotated on a horizontal axis for pipe and tubing, and on a vertical axis for wheels and gear blanks, a measured amount of molten metal is added. The mold may be sand or water-cooled metal for more rapid solidification. Centrifugal castings have good structure and density.

Metal molds are also used for making die castings and permanent mold castings. In the latter process a permanent metal mold is filled by gravity in the usual manner, while in die casting considerable pressure is exerted on the molten metal, insuring rapid and complete filling of the mold. Die-casting machines are highly mechanized for rapid and nearly automatic operation. The product is characterized by high dimensional accuracy and clear reproduction of mold details including screw threads, holes, and intricate sections, all of which greatly reduces the machining required. The process is limited in its application by the high cost of making alloy steel dies or molds. The lower melting zinc alloys and aluminum alloys are most successfully die cast; however, certain brasses and bronzes can also be die cast. Tin- and lead-base alloys are easily die cast but have limited application.

The zinc-base die-casting alloys are the most widely used. A typical composition is 1.0% copper, 3.9% aluminum, 0.06% magnesium, balance zinc. This alloy has a strength of about 45,000 psi (3,061 atmospheres) with 3% elongation in 2 inches (5 centimeters). Typical applications are carburetors, fuel pumps, tools, typewriter frames, instrument cases, and hardware which is often finished by chromium plating.

The investment or "lost wax" process has lately been revived as a method of making precision castings of metals such as steel and zinc having too high a melting point for die casting. A wax pattern is made in a die-casting machine, sprayed with a highly refractory slurry, dried, and embedded in sand. The mold passes through a furnace where the wax is melted or burned out, and the mold baked. The casting is then poured into the cavity left by the melting out of the wax, resulting in castings which rival die castings for dimensional accuracy.

See also **Die Casting; Iron Metals, Alloys, and Steels; Zinc.**

CASTING (Die). Die Casting.

CAST IRON. Iron Metals, Alloys, and Steels.

CASTOR (α Geminorum). The fainter star of the twins. Since these two stars are always considered together in the ancient literatures, the history and astrological significance will be found discussed under Pollux, the brighter of the two.

Astronomically, Castor is a very remarkable star. It was discovered, in 1719, to be a visual binary, with the magnitudes of the components 2.8 and 2.0. The separation is about 6 inches, and the star is certainly a true binary, but the period, probably of the order of magnitude of 350 years, has not yet been accurately determined. Each of the two components of the binary system is also a spectroscopic binary. Castor has a faint companion, separated from it by about 72 inches, but having the same parallax and proper motion. This companion is also a spectroscopic binary, with a period of slightly less than 1 day.

Sir William Herschel observed the binary nature of Castor as early as 1803. Later observations indicated that actually there is a group of six stars in the system. Ranking twenty-fourth in apparent brightness among the stars, Castor has a true brightness value of 27 as compared with unity for the sun. Castor is a white, spectral type A star and is located in the constellation Gemini, a zodiacal constellation. Estimated distance from the earth is 45 light years. See also **Constellations.**

CASTOR BEAN. Euphorbiaceae.

CASTOR OIL (*Ricinus communis*; *Euphorbiaceae*). Castor oil is obtained from a short-lived perennial tree which occurs wild in tropical Africa and perhaps in India. Cultivation of the tree is widespread not only in the tropics but also in temperate regions, where it is often grown as an ornamental plant. In the tropics it becomes a tree 36 feet tall, with large coarse leaves often of reddish color, and green flowers. An annual herbaceous variety is grown widely and produces a superior oil. The seeds, borne three in each of the smooth or prickly capsules, have a hard mottled shell. These seeds are ejected violently from the mature fruit.

The principal use of the plant is for the oil which is contained in the seeds. This oil is pressed out without heating the seeds. The particular properties make this a valuable oil for specialized uses, such as low temperature lubrication. It is an important constituent of hydraulic brake fluid and other fluids where the degree of compressibility is important. Castor oil also finds medical uses, as an ingredient of special soaps, and in the preparation of some textile dyes. Ricin, an alkaloid present in castor oil, also has been used in insecticides. Prior to the preparation of refined castor oil for medical purposes, ricin must be removed.

CAST STEELS. Iron Metals, Alloys, and Steels.

CATACLASTIC. As proposed by Teall in 1887, this term has the same meaning as crush breccias. This term is also applied to the deformation and granulation of minerals such as may take place during dynamic metamorphism.

CATACLYSM. Any of a number of geologic events, such as an exceptionally violent earthquake, that causes sudden and extensive changes in the earth's surface. An overwhelming flood of water (deluge) that spreads over a wide area of land also is sometimes referred to as a catacylsm.

CATALINA IRONWOOD TREE. Hornbeam Trees.

CATALPA TREES. Of the family *Bignoniaceae* (bignonia or trumpet creeper family), catalpa trees are of the genus *Catalpa*. These are American and Asiatic trees although they were introduced into Europe many years ago. There are two principal species in America: a northern catalpa (*C. speciosa*) and a southern catalpa (*C. bignonioides*). The southern catalpa is the most common in Europe. Several hybrid catalpas involving the crossing of American and Chinese species have been produced. These include the "J. C. Teas" (*C.* × *erubescens*); a purplish-colored cross (*C.* × *e.* 'Purpurea'); and the golden cultivar (*C.b.* 'Aurea'). Some gardeners consider the latter species the most spectacular of all yellow-leafed trees.

C. speciosa is also called the catawba tree, cigar tree, and hardy catalpa. The tree can attain heights approaching 100 feet (30 meters). Selected by The American Forestry Association in 1972 for its "Social Record of Big Trees," is the specimen located in Lansing, Michigan, with a circumference at 4½ feet (1.4 meters) above ground level of 18 feet, 10 inches (5.74 meters), a height of 94 feet (28.2 meters), and a spread of 85 feet (25.5 meters). This northern species does best in the Ohio basin, becoming a somewhat smaller tree in the eastern states. Catalpa flowers are trumpet-shaped with fluted edges, 4 to 5 inches (10 to 12.7 centimeters) long, snow white or pink-tinged, with purple veins and they are in clusters. The fruit is pod-shaped, tapered at each end (thus the name cigar), approximately 8 to 12 inches (20 to 30 centimeters) in length. The leaves are heart-shaped, quite large, 5 to 7 inches (12.7 to 17.8 centimeters), in length, toothless, with an extended sharp point. The upper side of the leaf is light green; the underside is of a slightly lighter color and covered with hair-velvet. With proper moisture, the tree is fast-growing. The wood is quite light, weighing about 26 pounds per cubic foot (416 kilograms per cubic meter).

C. bignonioides, sometimes referred to in the United States as the common catalpa, is usually smaller, less hardy, and ranges from Pennsylvania south to Florida and the southeastern states. Aside from its lesser height and stature, the tree is similar in many other respects to the northern species. The record specimen of this species is located in Water Valley, Mississippi and has a circumference (at $4\frac{1}{2}$ feet; 1.4 meters) of 18 feet, 5 inches (5.49 meters) a height of 83 feet (24.9 meters), and a spread of 58 feet (17.4 meters). The golden *C. bignonioides* 'Aurea' previously mentioned is also called the golden Indian bean tree.

Of a different genus (*Paulownia*), but related to the catalpas in the bignonia family is the so-called empress tree (*Paulownia tomentosa*). This tree is a native of the Far East and is known for its lightweight wood (15 to 16 pounds per cubic foot; 240 to 256 kilograms per cubic meter) and also as a garden tree. The tree has been introduced into Europe and North America and, under proper conditions, does quite well. The record Paulownia in the United States is located in Philadelphia County, Pennsylvania, with a circumference (at $4\frac{1}{2}$ feet; 1.4 meters) of 20 feet, 3 inches (6.17 meters), with a height of 105 feet (31.5 meters), and a spread of 70 feet (21 meters). This tree is actually a Royal *Paulownia tomentosa*. A different species, *P. fargesii*, has recently been introduced into Europe and possibly may offer more satisfying blooms. The empress tree tends to develop flower-buds in the fall, subsequently killed off by frost.

The jacaranda (*Jacaranda acutifolia*) is also a member of the bignonia family. This tree is native to tropical America and is mainly found in the northern part of South America. The tree also grows on the southwestern coast of California, on the southern tip of Florida, in the extreme south of Texas and the nearby Gulf coast of Mexico. The tree thrives in tropical areas, but can withstand months of dry weather. The tree may be described as rather exotic in appearance, ranging to a height of 50 to 100 feet (15 to 30 meters). The leaf is doubly compounded, narrow, and sharp. There are numerous leaflets which are fern-like in appearance. The flower is showy, a bell-shaped, blue, and hangs in clusters. The blossoms are about 2 inches (5 centimeters) long. The fruit is a flat capsule. There are about 50 species of the jacaranda. Several of these trees are the sources of excellent wood used for fine cabinet work, pianos, and expensive furniture. A Brazilian wood called *caroba*, for example, comes from the *Jacaranda copia*. The wood sometimes is confused with rosewood which is obtained from various species of the genus *Dalbergia*.

CATALYSIS. The process of changing the velocity of a chemical reaction by the presence of a substance that remains apparently chemically unaffected throughout the reaction. Berzelius (1836) applied this term to those reactions that do not progress unless a catalyst is present in the mixture. "Contact actions" (Mitscherlich) and "cyclic actions" (Brodie) have been suggested as names for the phenomenon.

In general, the following rules hold true for all catalytic processes:

1. The catalyst has the same composition at the beginning as at the end of the reaction.
2. A small quantity of the catalyst is capable of effecting the transformation of an indefinitely large quantity of the reacting substance.
3. No catalytic agent has power to start a chemical reaction; it may merely modify the velocity of the reaction.
4. The catalyst has no effect upon the final state of equilibrium of the forward reaction with any opposing reactions.
5. The velocity of two inverse reactions is affected in the same degree by a catalyst.

There are two general types of catalytic processes: (1) homogeneous, and (2) heterogeneous. In the *homogeneous* type, the chemical reaction is said to take place in a single phase, usually a liquid environment. In the *heterogeneous* type, the process takes place in a multiple phase, usually in a gaseous environment, and usually in the presence of a solid catalyst phase, which is either present as an undiluted material or on the surface of an inert substance, such as charcoal, the latter called a *support*.

A catalyst must be active and selective and have chemical and physical stability. Initial consideration must be given to selectivity, which is a measure of the degree the reaction is made to go in the direction that is desired. It is also desirable to convert the reactant(s) as rapidly as possible. The catalyst must be active, for an active catalyst will have a desirably high production rate. The concept of productivity is measured by the product of activity or conversion and selectivity per unit time and may be written *Yt*. The product of conversion (*C*) times selectivity (*S*) equals yield (*Y*).

In a heterogeneous catalytic reaction (by a solid surface), the following sequence of steps occurs: (1) diffusion of reactants from the bulk or gas phase to the catalytic surface, (2) adsorption of reactants, (3) diffusion of adsorbed species to active sites, (4) electron transfer processes at active sites, (5) chemical interaction of neutral and charged species, (6) desorption of reaction products from active sites, and (7) diffusion of products into the bulk phase.

Inasmuch as catalyst activity is proportional to the ability of the catalyst to chemisorb reacting species, activity will normally increase as surface area increases. For very fast reactions (high-activity catalysts), only the external surface is usually involved, and the overall rate of reaction may become controlled by the rate of transfer of reagents to the catalyst surface. In contrast, some reactions with porous catalysts may involve lesser reaction rates than rate of external diffusion. Hence overall reaction rate will be noticeably affected by diffusion within catalyst pores.

Solid catalysts which contain networks of pores and create a large surface area will allow for diffusion of reacting molecules and for high catalytic activity. Selectivity, on the other hand, may or may not be affected adversely, depending on the nature of the reaction. As a rule, oxidation catalysts usually perform best with large-diameter pores and low surface areas. Hydrogenation catalysts perform better with high surface areas.

The concept of protonic acid (hydrogen ion) and base (hydroxyl ion) catalysis has been found applicable in discussion on heterogeneous catalysis, wherein the mechanism is seen as the transfer of a proton from the catalyst to the reactant (acid catalysis), or from the reactant to the catalyst (base catalysis). Intermediates may be of two types: (Type I) in which the reversible reaction that forms the intermediate is fast compared with further change into final products, so that the intermediate is always present in its equilibrium concentration; and (Type II) in which the intermediate is never present in appreciable concentration, and the velocity of the reaction is determined by the speed of formation of the complex. A number of apparently different catalysts behave similarly because of their acidic nature, i.e., mineral acids, Friedel-Crafts catalysts, silica-alumina materials, and zeolite-cracking catalysts.

In recent years, the electronic concepts of solids have been introduced as a means toward understanding catalytic action. It is possible to divide heterogeneous catalysts into groups based upon their electronic properties. For example, metals (conductors) can be used in hydrogenation, dehydrogenation, and hydrogenolysis reactions; metal oxides or their sulfides (semiconductors) can be used in oxidation, reduction, dehydrogenation, or cyclization reactions; and salts or acid-site (insulators) catalysts can be used in cracking, dehydration, isomerization, polymerization, alkylation, dehalogenation, halogenation, and hydrogen tranfer reactions.

The effects of different catalysts on the same reagent is shown by the reaction of ethyl alcohol. When reacted over copper (a conductor), either acetaldehyde or ethyl acetate plus hydrogen is obtained. When reacted over alumina (an insulator), ethylene plus water or diethyl ether plus water is recovered.

In general, a catalyst problem is either one of development of a new catalyst for a novel process, or of a catalyst modification for an existing process. A catalyst must be considered as part of an overall process, for when studying a chemical process, not only is the precise nature of a catalyst studied, but also the chemical-reaction mechanism, practical processing details, and production techniques. Advances in physical chemistry, instrument analysis, radioactive techniques, solid-state physics, and computer applications have assisted catalytic work to a large degree. However, published discovery of new processes and continued application of old catalysts to new reactions indicates that catalytic chemistry remains an experimental art and is dependent to a large extent on the intuition, ingenuity, and perseverance of both chemist and chemical engineer.

Chemical and petroleum processing require large production of catalysts. In cracking operations, where hydrocarbons are changed from

EXAMPLES OF CATALYTIC PROCESSES[a]

Process and Product	Catalyst	Reactants	Yield
Amination			
Amines	Al_2O_3(Co)	Alcohols + Ammonia	90+
Ammoxidation			
Acrylonitrile	Bi-Mo-P CuO, SbSn	Propylene + O_2 + NH_3	60–80
Benzonitrile	V_2O_5-Sb	Toluene + O_2 + NH_3	90+
Phthalonitrile	V_2O_5	*o*-Xylene + O_2 + NH_3	90+
Chlorination			
Chlorobenzene	Fe	Benzene + Cl_2	70–75
Chloroacetic acid	red P	Acetic Acid + Cl_2	90
Benzoyl chloride	UV Light	Toluene + Cl_2	95+
Hydration			
Acetaldehyde	Hg_2SO_4	Acetylene + H2O	95
Ethanol (alcohols)	H_3PO_4, WO_3	Ethylene + H2O	95
Dehydration			
Styrene	TiO_2	Methylethyl carbinol	80+
Ethylene (olefins)	Al_2O_3, ThO_2	Ethanol (alcohols)	90+
Acrylonitrile	Al_2O_3	Ethylene cyanohydrin	90+
Hydrogenation			
Aniline	Fe-HCl, $Cu \cdot SiO_2$	Nitrobenzene	90–95
Butanol	Co, Ni-SiO_2	Butyraldehyde	98+
Cyclohexane	Ni-Al_2O_3, PtO_2	Benzene	96+
Ethylene	Fe, Ni, Cu, Pd-$BaSO_4$	Acetylene	99
Methanol	ZnO-CrO_3, Ni-Co	Carbon monoxide	60
Dehydrogenation			
Acetaldehyde	Cu, Ag, $FeMoO_4$	Ethanol + H_2	85–95
Benzene	Nu/Al_2O_3, Pt-Al_2O_3	Cyclohexane	95+
Butadiene	Fe, Cr, K, $CaNiPO_4$	Butenes	75–85+
Butene	Cr_2O_3-Al_2O_3	Butane	
Methyl ethyl ketone	ZnO-ZnCu	Sec-Butanol	85–90
Styrene	$ZnCrO_2$-FeMgO $CaNiPO_4$	Ethyl benzene	86–92
Styrene	TiO_2	Phenyl methyl carbinol	80+
Oxidation			
Acetaldehyde	$PdCl_2$-MgO-Cu	Ethylene	95+
Acetic acid	Mn^{2+}	Acetaldehyde	88–95
Acetic acid	Co, Bi	Butane	20–40
Acetic anhydride	Cu, Co	Acetaldehyde	70–75
Acetone	Cu, Ag, Zno	Isopropanol	85–90
Adipic acid	Cu-Mn, V-Cu	Cyclohexanone	70–90
Benzoic acid	Co^{2+}	Toluene	90
Benzoic acid	Cu	Phenol	90
Benzaldehyde	UO_2-MoO_3-Cu	Toluene	30–50
Ethylene oxide	Ag, AgO	Ethylene	70
Propylene oxide	Mo, W, Ti, V	Propylene	90
Phthalic anhydride	V_2O_5-K_2SO_4	Napthalene *o*-Xylene	70–80
Maleic anhydride	Mo-V-P-Na	Benzene	85
Maleic anhydride	V-P	Butene	60
Terephthalic acid	Mn-Co	*p*-Xylene	90+
Reductive Dehydration			
Butane	Ni-Al_2O_3	Butanol + H_2	90+
Reforming			
Aromatic	Mo-Al_2O_3 Pt-Al_2O_3-halides	Naphthenes + H_2	
Desulfurization			
Butane	Co-Mo-Al_2O_3	Thiophene + H_2S + H_2	

SOURCE: Catalyst Development Corporation.
[a] Discussed in detail in Ref. 1.

such substances as tar and asphalt to fuels, such as light oils and gasoline, the consumption of silica, aluminas, processed clays, and zeolite catalysts is very high. Isomerization operations, where straight-chain paraffinic molecules are rearranged to branched-chain molecules in order to improve octane number of gasoline, consume large quantities of aluminum chloride catalyst. In alkylation reactions, where an alkyl radical is introduced into a molecule by addition or substitution into an organic compound in order to make such substances as gasoline, rubber antioxidants, dyes, flavors, and other compounds, sulfuric acid is used extensively as a catalyst.

As shown by the accompanying table, scores of reaction types and catalyst materials are involved in industrial processes.

References

Baker, R. T. K.: "Catalysts in Action," *Chem. Eng. Progress*, **73**, 4, 97–99 (1977).
Kay, E., and P. S. Bagus (editors): "Topics in Surface Chemistry," Plenum, New York, 1978.
Lundberg, W. C.: "Extending Catalyst Life," *Chem. Eng. Progress*, **75**, 6, 81–86 (1979).
Parshall, G. W.: "Organometallic Chemistry in Homogeneous Catalysis," *Science*, **208**, 1221–1224 (1980).
Robinson, A. L: "Homogeneous Catalysis: Transition Metal Clusters," *Science*, **194**, 1150–1152 (1976).
Sinfelt, J. H.: "Heterogeneous Catalysis," *Science*, **195**, 641–646 (1977).
Sleight, A. W.: "Heterogeneous Catalysts," *Science*, **208**, 895–900 (1980).
Voorhoeve, R. J. H., et al.: "Perovskite Oxides: Materials Science in Catalysis," *Science*, **195**, 827–833 (1977).

CATALYSTS. **Iridium; Osmium; Platinum and Platinum Group; Rare Earth Elements and Metals; Rhodium; Ruthenium.**

CATALYTIC CONVERTER (Internal Combustion Engine). A combination of the Clean Air Act Amendments of 1970 and the Energy Policy and Conservation Act of 1975 (United States Congress) has promoted the widespread use of catalytic aftertreatment to control automotive exhaust emissions with a concomitant increase in fuel economy. The catalytic converter, comprised of a ceramic catalyst and the necessary stainless steel hardware to ensure that the exhaust gases pass through the catalyst, permits the conventional spark-ignition automobile engine to run at near optimum efficiency to afford good fuel economy. The catalyst itself has the capability of promoting (or accelerating) the rate at which reactions occur. In the case of an oxidation catalyst, the function is to cause the carbon monoxide (CO) and hydrocarbons (HC) which result from incomplete combustion to be converted to CO_2 and water. In the case of a three-way catalyst, the oxidation reactions (HC and CO) are promoted as well as the reduction reaction of oxides of nitrogen (NO_x).

Converters now in use contain noble metals on a ceramic substrate (e.g., platinum dispersed on alumina). The converter is located in the exhaust system in one of two general locations: an underfloor location is presently employed for General Motor's pelleted converter, and close-coupled location near the manifold is presently employed for Chrysler's California light-off converters. The operating temperature range for noble metal catalysts is from 600 to 1200°F (316 to 649°C), which is similar to the exhaust pipe skin temperature range normally encountered on standard autombile engines.

Catalytic materials can be physically supported on either pelleted or monolithic substrates. In the case of the pelleted catalyst, the support is an activated alumina. A typical monolithic catalyst is composed of a channeled ceramic (cordierite) support having, for example, 300 to 400 square channels per square inch on which an activated alumina layer is applied. The active agents (platinum, palladium, rhodium, etc.) are then highly dispersed on the alumina. The average noble metal usage per vehicle is about 0.05 troy ounce of platinum metal (worth about $20.00 as of 1979) and from zero to 0.03 troy ounce of palladium.

In the case of pelleted catalyst, the pellets are confined by screens (Fig. 1); the monolithic-type catalyst (Fig. 2), being a single rigid material, needs no such confinement. The arrangement within the container, regardless of which type of catalyst is used, is intended to ensure that the exhaust gases pass through the catalyst bed without by-passing it or "channeling" along the walls.

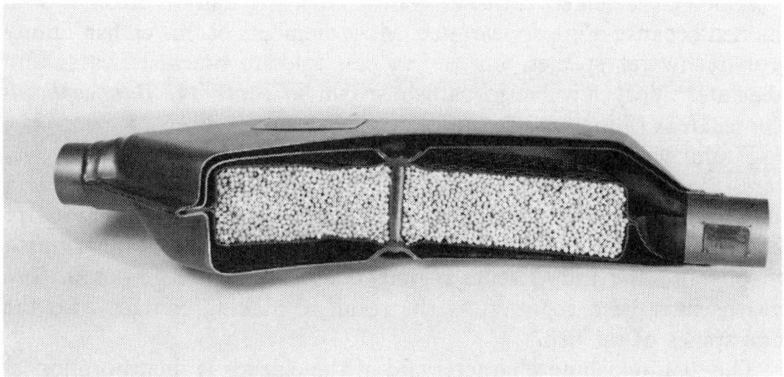

Fig. 1. UOP-designed converter to use pelleted catalyst.

Fig. 2. UOP-designed converter to use monolithic catalyst.

Exhaust emission standards for 1981 model year vehicles will likely require the use of three-way catalysts, either alone or in combination with an oxidation catalyst. Three-way catalysts are designed to operate in a very narrow range about the stoichiometric air/fuel ratio. In this range the HC and CO are subject to oxidation and the NO_x compounds undergo reduction. The downstream oxidation catalyst in a dual bed system is generally used as a "clean-up" catalyst to further control HC and CO emissions. The most common catalytic combination in three-way uses is platinum/rhodium. Current production applications use these elements in a relatively rich proportion of 10 : 1, whereas the respective mine ratio is about 19 : 1. Future applications may well employ this ratio because of cost and availability implications.

While the noble-metal-containing catalysts require the use of essentially lead-free gasoline for efficient performance and long life, an occasional tank full of leaded gasoline is not a catastrophic occurrence. The catalyst will be at least partially deactivated by the lead compounds, but the effect will be temporary. Resumed operation of the engine on unleaded gasoline will vaporize these materials and they will be vented from the tail pipe. Frequent or continued operation on leaded fuel, however, will seriously, if not permanently, damage noble metal catalysts.

The use of unleaded gasoline, in addition to preventing "poisoning" of the catalyst, has the advantage of not releasing lead oxides and lead chlorides and bromides to the atmosphere which result from burning gasoline containing lead antiknock compounds. Further, halogen compounds, such as hydrochloric and hydrobromic acids, hasten the ultimate corrosion of the exhaust train from exhaust manifold to muffler and tail pipe.

Technical Staff, UOP Inc., Des Plaines, Illinois.

CATALYTIC CRACKING. **Cracking Process.**

CATAMARAN. **Wind Power.**

CATARACT. **Vision and the Eye.**

CATAWBA TREE. **Catalpa Trees.**

CATAWBERITE. The term applied by Lieber to a metamorphic rock chiefly composed of magnetite and talc.

CATCLAW ACACIA. **Acacia Trees.**

CATENARY. The locus of the transcendental equation

$$y = \frac{a}{2}(e^{x/a} + e^{-x/a}) = a \cosh \frac{x}{a}$$

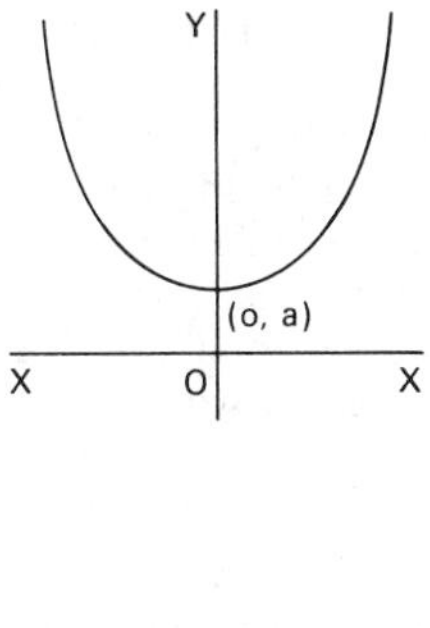

Catenary.

The curve can be generated by the focus of a parabola rolling along a straight line and its shape is that taken by a uniform, heavy flexible cable freely suspended from its ends. The involute of the catenary is called the tractrix. See also **Funicular Polygons and Catenaries; Parabola;** and **Tractrix of Huygens.**

CATENATION COMPOUND. Compound (Chemical).

CATERPILLAR. The larval form of the butterflies and moths. See accompanying illustration. These animals are of very simple construc-

Caterpillar. (*A. M. Winchester.*)

tion. Their main task is eating, and they have no organs which are not associated with this function: they have no wings, no highly developed sense organs, and only short legs. They lack compound eyes; on either side of the head they have only a row of six ocelli which are barely able to distinguish light from dark. Nevertheless, they are capable of perceiving the trunk of a tree, so that they can proceed from the ground into the foliage of the tree on which they feed. Of the mouthparts, the mandibles are always large and powerful, admirably suited for chewing up plant food, even solid wood. As weapons of defense, though, they are of little use, and even less as aggressive weapons. The remaining mouthparts are small and poorly developed; they bear sense organs for touch and taste and serve to guide the caterpillar to an appropriate food plant. The caterpillars of many lepidopteran species are restricted to certain plants and would starve before accepting any other plant as food.

In the middle of the labium of the caterpillar there is a small papilla, the spinneret, with the openings of the two silk glands. These glands are often very large, in some cases extending throughout the body. All caterpillars, without exception, are able to produce silken threads, though the amount they can produce varies. The thread extruded from a silk gland consists of two substances which harden on contact with the air. The silk marketed commercially is this substance; true silk is produced by the silkworm (*Bombyx mori*), but the silk of other caterpillars is also used. The threads are usually white but often yellowish or, in the case of many large caterpillars, even dark brown.

These silk threads play a central role in the life of a caterpillar. Many caterpillars actually live on an endless string which they produce continuously; they attach it to the substrate and clasp it with their legs. Thus they can creep about on even the smoothest surfaces. If such a caterpillar falls from its position, it spins the thread out a bit more, still hanging from it, and then uses it like a rope to climb back into place. The same technique allows it to escape from enemy attacks. Species which live in hiding, and others which build communal nests, use the thread as a guideline to and from the feeding place. The caterpillar of course uses large quantities of silk to build cocoons and other woven structures, and the threads are also used to line mines in leaves and tunnels in wood.

The caterpillar's wormlike body is often decoratively colored and sometimes oddly shaped, with various outgrowths or with wartlike verrucae, thornlike scoli, or hairs on the upper surface. Presumably all these devices serve as protection from enemies, whether by camouflaging the caterpillar of frightening the predator. Some caterpillars also have organs specially designed for defense. One of these is the eversible osmeterium of the swallowtail caterpillars, which produces a repugnant smell. Other weapons of defense are urticating hairs (hairs which cause irritation when touched) and hard, sharp bristles, whose painful stab may be accentuated by poisonous substances.

CATFISHES (*Osteichthyes*). Members of the suborder *Siluroidea*, catfishes are of many species. As a general description, they are without scales, although the skin has bony plates in some species. Barbels (feelers) occur on the head. They are represented in the fresh waters of all continents with exception of Australia and gain great diversity in the Americas. The catfishes are important food fishes in some parts of the world. In North America, the channel cats are especially desirable. They reach a large size in some of the larger rivers and lakes.

The various species of catfishes may be categorized along the following lines:

- Armored Catfishes
  - Doradid Catfishes (Family *Doradidae*)
  - Callichthyid Catfishes (Family *Callichthyidae*)
  - Loricariid Catfishes (Family *Loricariidae*)
- Naked Catfishes
  - Banjo Catfishes (Family *Aspredinidae*)
  - Ariid Marine Catfishes (Family *Arridae*)
  - Plotosid Marine Catfishes (Family *Plotosidae*)
  - Clariid Catfishes (Family *Clariidae*)
  - Silurid Catfishes (Family *Siluridae*)
  - Pimelodid Catfishes (Family *Pimelodidae*)
  - Bagrid Catfishes (Family *Bagridae*)
  - Parasite Catfishes (Family *Trichomycteridae*)
  - North American Catfishes (Family *Ictaluridae*)
  - Schilbeid Catfishes (Family *Schilbeidae*)
  - Upside-Down Catfishes (Family *Mochocidae*)
  - Electric Catfish (Family *Malapteruridae*)

There is not full agreement on methods for classifying catfishes. In some classifications, the group may range from 25 to 31 for coverage of over two thousand species.

The doradids occur in South America and are known for their heavy armor, comprised of a series of overlapping plates. One of the better known members of this family is the *Acanthodoras spinosissismus* which produces grunt-like sounds when in or out of the water. The sounds are derived from activity of its air bladder. The callichthyids are also South American, possessing a smooth armor made up of plates. A favorite of tropical-fish fanciers is the 3-inch (7.5-centimeter)—or less—*Corydoras* which is reasonably peaceful in captivity. They are not brilliant in coloration, but do display interesting patterns. Another South American variety is loricariid catfishes. They have a high dorsal fin and a V-shaped tail fin. They appear somewhat like a North American minnow and are of appropriate size for aquariums, but require special attention and diets. The heavier genus *Plecostomus* (averaging 4 to 5 inches; 10 to 12.5 centimeters in length) is popular with tropical-fish fanciers, as well as being well liked by some Indians in South America as a food.

Among the naked catfishes (no armor), the banjo catfish is well named because of its appearance. Most members of this catfish family are freshwater species, but a few can tolerate brackish water and seawater. Very few banjo catfish spawn in captivity. *Bunocephalus coracoideus* (5-inch; 12.5-centimeter fish) is an exception.

Found widely distributed throughout tropical and subtropical waters, the ariid marine catfishes are fast-moving and often travel in schools. They are frequently used as food. The *Plotosus anguillaris*, colorful and an inhabitant of tropical reefs and well distributed throughout the Indo-Pacific region, is considered dangerous in that deaths have been reported as the result of making contact with the fine spines of the fish.

The distinguishing characteristic of the clariids is incorporation of an auxiliary breathing apparatus, which permits them to live out of water for periods much longer than tolerable to other catfishes. They are found from Africa to the East Indian archipelago. More recently, they have become established in the waters around the Hawaiian Islands and Guam. Reaching an average length of about 16 inches (40 centimeters), this catfish is considered quite hardy, frequently living in captivity for several years. Also classified as clariids are the *Gymnallabes typus* and *Channallabes apus*, very strange, specialized fishes, sometimes referred to as West African eelcats. They possess very long dorsal and anal fins. When mature, they reach a length of about 1 foot (0.3 meter). The body is no thicker than an average pencil and thus appear much as eels.

The pimelodids represent the largest family of South American catfishes and inhabit waters from Mexico southward, including most of South America. There are many variations. One of particular interest is the *Typhlobagrus kronei*, a blind cave catfish that inhabits the Caverna das Areias in Sao Paulo, Brazil. Some pimelodids are of the proper size and other qualities to be of interest to tropical-fish fanciers, but they require a lot of aquarium space and are quite aggressive, often attacking other fishes.

The bagrids as a general family are much like the pimelodids just described, except they are inhabitants of the Old World. The *Leiocassis siamensis* found in Thailand and of beautiful brown coloration and yellow and white band stripping is well known for the croaking sounds which it creates. The fish is usually about 7 inches (18 centimeters) in length. The similar striped *Mystus vittatus* is found in Thailand, Burma, and India. Also among the bagrids is the unusual *Bagrichthys hypselopterus*, an inhabitant of the rivers of Borneo and Sumatra. A fully-grown fish reaches about 16 inches (40 centimeters) in length and is characterized by a dorsal fin which extends obliquely upward almost the full length of the fish. Biologically, the need for such a development remains unaccounted for even though the species have been known for over a hundred years.

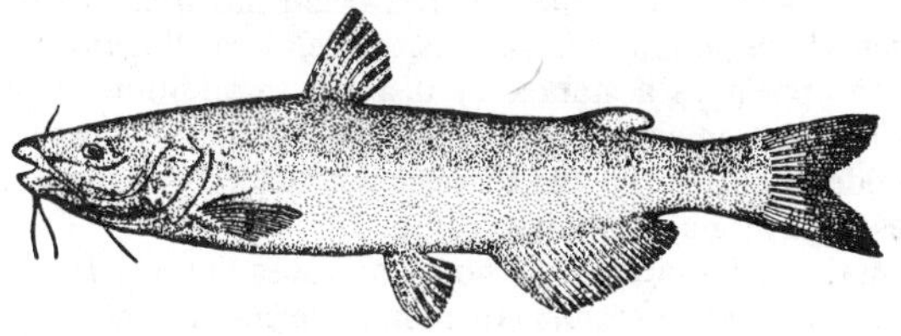

Bullhead catfish (*Ictalurus nebulosus*).

The ictalurids are what one might term the average catfish of the North American continent. Among the largest is the flathead (*Pylodictis olivaris*). This fish has been reported to weigh as much as one-hundred pounds with a length of nearly $5\frac{1}{2}$ feet (1.7 meters). This is a square-tailed species and is found widely throughout the central United States. The smaller variety of ictalurids are known as the madtoms and are dangerous because their pectoral spines and associated venom glands can cause serious and painful wounds. They are of small size, usually not exceeding 5 inches (12.5 centimeters) in length. The bullhead (*Ictalurus nebulosus*) is a favorite among fishermen. See accompanying diagram. At one time limited to the eastern United States, it is now well distributed over the western portion of North America. It is also widely found throughout the Hawaiian Islands and Europe. The brown bullheads mature at 6 inches (15 centimeters), but reach an average length of about 16 inches (40 centimeters). The brown bullheads are known for the tender care of their young. The dense school of free-swimming juveniles is amply protected by one or both patents.

The white catfish (*Ictalurus catus*), at one time was limited to the Atlantic seaboard. It is now found throughout the central United States, as well as in an increasing number of areas to which it has been introduced. It is a large fish, weighing up to close to 60 pounds (27 kilograms) and attaining a length of about 4 feet (1.2 meters). It is considered perhaps the most valuable of the catfishes in North America as a source of food.

One of the largest catfishes in the world is found among the schilbeids. This is a heavy-bodied herbivore without teeth (*Pangasianodon gigas*) which can attain a length in excess of $7\frac{1}{2}$ feet (2.3 meters) and a weight of about 250 pounds (113 kilograms). Studies have shown that spawning migrations are made up the Mekong River, possibly to Yunan province in China. Spawning may take place in Lake Tali. The Cambodian people call it the giant fish. Some other species among the schilbeids are quite small, as represented by *Etropiella debauwi* (3 inches; 7.5 centimeters long). Typical characteristics of the schilbeids include: (1) short barbels about mouth—usually two or three pairs, (2) adipose fin is quite small, (3) long anal fin distinct from tail fin, (4) short, high dorsal fin, and (5) a forked tail.

Members of the catfish family *Mochocidae* frequently reverse their swimming position; hence the "upside-down" name for them. They inhabit tropical African fresh waters, where they also are known as "squeakers" because of grunting noises sometimes made by rotation of the dorsal and pectoral spines in their sockets. The variety *Synodontis nigriventris* (speckled brown) has become attractive to tropical-fish hobbyists.

The *Malapterurus electricus* (electric catfish) is the only known species of catfish to prossess electrogenic powers. It is a very pungnacious fish. It had a special regard among ancient Egyptians who inscribed likenesses of the fish in their various art forms. The electric catfish is found in the Nile valley and in tropical central Africa. When mature, it can measure up to 4 feet in length and weigh up to 50 pounds (23 kilograms). It is believed that these fish can discharge up to 100 volts in one major jolt. It is interesting to note that the electrical polarity of the electric catfish differs from that of the South American electric eel. In the eel, the charge is positive on the head; negative on the tail. This situation is reversed in the electric catfish. It is not believed that the electric catfish uses the electric organs as a means of detection, as is true of some other electrogenic fishes. The *Malapterurus* survives well in captivity if not overfed.

The raising of catfish for commercial purposes is discussed in some detail in entry on **Aquaculture.** See also **Fishes;** and **Plecostomus.**

CATHETER. A tube for removing or injecting fluids through a natural body passage; made of plastic, rubber, glass, metal, or other appropriate materials.

CATHETERIZATION (Cardiac). Angiography.

CATHETOMETER. A form of optical comparator used for the accurate measurement of vertical distances. Some cathetometers also have been adapted for horizontal measurements. More sophisticated instruments are available which measure two coordinates in a vertical plane. Cathetometers are used whenever the object or action is not accessible by ordinary means, or when other methods of measurement introduce errors due to parallax or physical contact. Cathetometers are well suited to inspection and layout work, especially in inspection departments, model shops, and industrial research laboratories.

The vertical cathetometer essentially consists of a telescope (or for close work, a microscope) that is horizontally mounted on a guide bar whose length is parallel to the displacement to be measured. The height of the telescope and hence that of the object is read on a precision scale attached to the guide bar. For most precise measurements, a separate standard scale, supported at the same distance as the object and as close to the object as possible, is used. The height of the object then is determined by reading the scale through the telescope with the aid of a filar micrometer eyepiece.

A typical precision cathetometer, which utilizes an accurately calibrated guide bar, will have a measuring range of 100 centimeters and can be read to 0.01 millimeter. The typical precision micrometer slide cathetometer will have a measuring range of 100 millimeters and can be read to 0.001 millimeter. Typically, a coordinate cathetometer will have a measuring range of 30 inches (76 centimeters) in both the vertical and horizontal dimensions and can be read to 0.001 inch (0.025 millimeter).

CATHODE. 1. In general, the electrode at which positive current leaves a device which employs electrical conduction other than that through solids. 2. In an electron tube, the electrode through which a primary stream of electrons enters the interelectrode space. 3. The negative terminal of an electroplating cell (i.e., the electrode from which electrons enter the cell, and thus at which positively charged ions (cations) are discharged). 4. The positive terminal of a battery. See also **Battery.**

CATHODE DARK SPACE. In a gas discharge tube, the dark band between the cathode glow and the negative glow. Also known as Crookes dark space or Hittorf dark space.

CATHODE (Electron Tube). Electron Tube.

CATHODE GLOW. At sufficiently high voltage, a glow exists about the negative terminal of an arc. By operating the arc at low pressure (in partial vacuum, as in a gas discharge tube), this glow may fill much of the tube, lying between the cathode dark space and the Aston dark space. A substance placed on the cathode will produce its characteristic spectrum in the cathode glow. Also called "*Glimmschicht method.*" However, in many discharges both the Aston dark space and the cathode glow will be absent or indiscernible.

CATHODE RAY. A stream of electrons usually associated with their emission from a heated filament in a tube; or their emission by the cathode of a gas-discharge tube upon bombardment of the cathode by positive ions. After the discovery of the cathode ray in high-vacuum discharge tubes by Plücker in 1858, there developed, with the experiments of Goldstein, Crookes, Hertz, Lenard, and Schuster, a controversy over the nature of the rays. The British physicists thought they were negatively-charged particles. A predominately German school held that the rays were a peculiar form of electromagnetic rays. The controversy provides a classic "case history" of the typical scientific controversy in which two quite different models both explain most, but not all, of the observable facts.

The proponents of each model designed ingenious experiments and in some cases were so trapped in their preconceptions that they badly misinterpreted their observations. The Germans were especially impressed by the fact that the rays could go through thin foils—something no known particles could do. The British were firm in pointing out that the rays could be deflected by magnetic fields—something not possible with electromagnetic waves. Hertz, in what he thought was a crucial experiment, was unable to detect deflection of the rays by electric fields, but this very phenomenon was demonstrated by J. J. Thomson and made the basis for his conclusive experiments that the rays had velocities less than that of light. Thomson showed, further, that if one assumed that the rays were composed of particles, then the particles had the same ratio of charge to mass regardless of the cathode material or the nature of the residual gas. Perrin's classic experiment, meanwhile, proved that the rays did indeed convey negative charge. In the decade between 1896 and 1906, Thomson and others showed that negatively-charged particles from sources other than cathode rays had the same ratio of charge to mass; the negative particles emitted by hot filaments in the Edison effect, the beta rays emitted by some radioactive materials, and the negative particles emitted in the photoelectric effect that had so ironically been discovered by Heinrich Hertz in his great experiment which demonstrated the electromagnetic rays predicted by Maxwell's equations.

An emission from the cathode in a vacuum tube becomes more conspicuous as the tube is cleared of gas molecules with diminishing pressure. At pressures of 0.01 millimeter of mercury or lower, the rays leave the cathode normally to its surface and move in straight lines across the tube as demonstrated by early experiments with the Crookes tube. By using a concave cathode, they may be brought to a focus, and any obstacle placed at the focus becomes intensely hot. Thomson determined the charge-mass ratio, known now to be about 1.76×10^8 coulombs per gram. The rays move with speeds varying with the voltage, but commonly of the order of one-third the speed of light.

Lenard showed in 1898 that cathode rays will penetrate through thin aluminum or gold leaf and can thus be allowed to pass outside the tube. Electrons so escaping are termed Lenard rays.

Numerous electronic devices take advantage of cathode ray phenomena, including cathode-ray tubes, used in oscilloscopes, television receivers, in connection with computer display systems, and in telecommunications systems.

CATHODE-RAY TUBE. A special form of vacuum tube (CRT) used in a large variety of electronic applications, e.g., the television receiver picture tube, oscilloscope tube, and as a display device for numerous process control and data processing instrumentation systems. A fundamental function of the CRT is to convert information contained in an input signal to electron beam energy and finally to convert that energy into light energy to provide a visual information output. As will be noted from Fig. 1, a basic cathode ray tube is

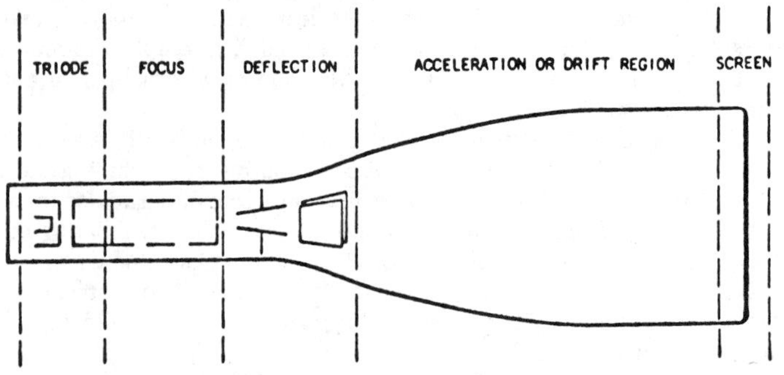

Fig. 1. Principal sections of a cathode-ray tube.

divided into five sections. Electrons are emitted from a thermionic cathode and controlled by the triode section. The electrons are then formed into a beam and accelerated in the focus section. The deflection section deflects the beam, typically on vertical and horizontal axes, by internal electrostatic deflection plates, or by external electromagnetic deflection coils. The acceleration or drift area controls (often with some further acceleration) the electron beams until the energy arrives at the CRT screen. The electrons upon striking a light-emitting phosphor coated on the inside face of the CRT screen cause the phosphor to fluoresce and emit visible light. The phosphors used in CRTs have the characteristic of phosphorescence, i.e., emitting light energy for a short interval after the electron beam has been removed. It is this effect which permits image persistence, thus allowing a repetitive pattern to appear as a stationary display. In addition to presenting x and y information on the deflection plates, a cathode ray tube utilizes the cathode or grid of the gun to present z axis information (intensity). A representative gun is shown in Fig. 2.

Two major subgroups of cathode-ray tubes are: (1) *monoaccelerators*, and (2) *postaccelerators*. In monoaccelerators, generally a high voltage of from 3 to 4 kV is applied to the second (focus) anode. In postaccelerators, from 10 to 14 kV will be applied to a high-voltage electrode near the CRT screen. The latter tubes typically have a higher light output inasmuch as the light output from a phosphor increases with voltage through which the electrons have been accelerated. Postacceleration also permits the deflection region to be maintained at a relatively low voltage, thus helping deflection sensitivity.

Postacceleration tubes are usually "aluminized" with a thin coating of aluminum. This acts as a mirror and reflects to the screen light energy that would otherwise be lost.

Phosphors. Originally, natural substances were used in cathode ray tubes for converting energy of the scanning electron beam into light. It was during this early period that the word "phosphor" was coined. Synthetic phosphors have been used for many years. They are usually

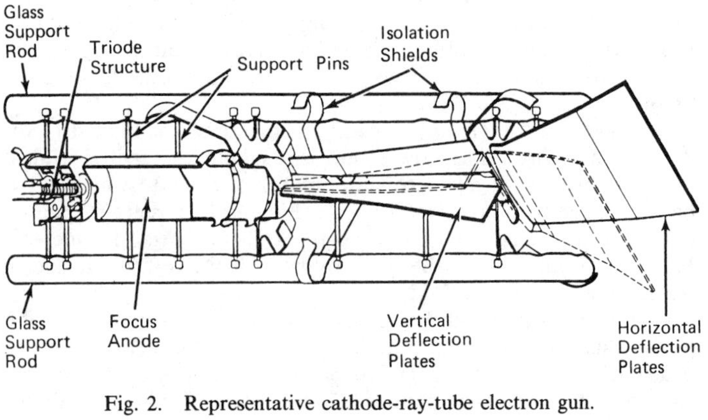

Fig. 2. Representative cathode-ray-tube electron gun.

zinc, cadmium, calcium, and magnesium compounds (as sulfides, selenides, silicates, and tungstates). The materials must withstand "bake-out" temperatures of 400°C or greater. They must have a low vapor pressure and an ability to hold up over long periods of time against the bombardment of electrons. Variation in the quality of specific phosphors is obtained through the use of accelerators, notably copper, silver, magnesium, chromium, and bismuth, among others. The activators enable a selection of efficiency, color of luminescence, and decay time.

It is estimated that there are over 50 commercial phosphors from which to select. A significant percentage of these compounds are made up of Group II and Group VI elements. Zinc sulfide activated by magnesium produces a blue emission, whereas zinc and/or cadmium sulfide activated by copper or aluminum produces a green emission. Zinc sulfides activated by silver or copper can convert up to 20% of electron beam energy to light. These compounds are important in color television tubes. Where particularly long periods of electron bombardment are involved, the compound $ZnSiO_4$ activated by manganese is well suited and thus finds wide application in oscilloscopes and aircraft instruments which require bright displays. A green luminescence is produced.

In recent years, some of the rare-earth elements, such as terbium and europium, have found use in color tubes. These compounds emit a red color that is comparably efficient with the well-established green and blue emitting compounds. Other rare-earth element compounds include La_2O_2S activated by terbium (green emission) and Y_2O_2S, also activated by terbium (white emission). An outstanding advantage of rare-earth phosphors is their ability not to become saturated at high power levels. Confinement of their emission to rather narrow bands is also advantageous in providing images with high contrast even in the presence of high ambient light levels.

The desired persistence time of phosphors varies with application. Whereas a time of 30–40 milliseconds is satisfactory for television, a longer time (up to a second or even longer) is desirable for radar displays. Zinc–cadmium sulfide activated by copper persists for a number of seconds with a yellowish-orange color. For extremely short persistance, as required in flying-spot scanners, a material such as calcium–magnesium silicate, with persistence in terms of a fraction of a microsecond, is desirable. This compound emits in the violet and ultraviolet range.

Phosphor particles range in diameter from 1 to 10 micrometers. Image resolution varies inversely with the diameter of the particle, but efficiency decreases when particles are too small.

Storage Cathode-Ray Tubes. Tubes of this type have two electron sources. There is a writing gun to provide the electrons for writing and a flood gun to provide broad coverage of low velocity of electrons that bombard the storage screen uniformly. This flood of electrons holds the writing gun information in the written mode by means of secondary emission electrons and thus maintains the stored image for an indefinite period after the writing beam has been cut off. Tubes of this type are used for displaying signals that occur only once (transients), or signals that have low repetition rates. Much of the need for formerly photographing transients on oscilloscope screens no longer is required with the availability of storage-type oscilloscopes. High-resolution storage is also useful for presenting graphic and alphanumeric displays in computer readout applications. This eliminates the bulk of local storage that may be required for continually refreshing displays and to provide a flicker-free display.

Storage-type cathode-ray tubes are classified as bistable or as halftone tubes. On a bistable tube, the stored display has one level of intensity. In a halftone tube, a stored signal may be displayed at different levels of intensity. The intensity of a halftone tube depends upon beam current and the time that the beam remains on a particular phosphor particle. A bistable tube either stores or does not store, with all stored events having the same intensity.

A direct-view bistable storage CRT is shown in Fig. 3. Action of the writing gun shown in (a); of the flood gun in (b). The writing gun bombards the screen. High-energy electrons light the phosphor and also knock loose many secondary electrons. The written area, losing electrons, charges positive. Electrons from the flood-gun hit unwritten areas too slowly to activate or light the phosphor. They simply accumulate, driving the area negative. But, the written area

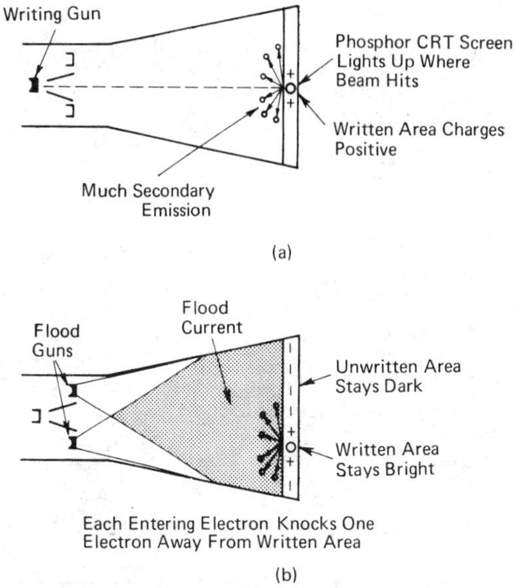

Fig. 3. Direct-view bistable storage cathode-ray tube: (a) action of writing gun; (b) action of flood gun.

(positively charged) attracts electrons at high speed, keeping the phosphor lit, as well as knocking sufficient secondaries away to hold the area positive.

Character-generation tubes pass the electron beam through an aperture of an appropriate shape. Basic methods used to generate character information include: (1) *Raster scan technique* which involves controlling the intensity of the electron beam during sweep. The process is similar to facsimile recording where the characters are generated in segments, (2) *Lissajous technique* in which the electron beam serves as a pencil, (3) *shaped-beam technique* in which the tube incorporates a number of stencil-type openings that are used to shape the electron beam.

Although not display tubes in themselves, scan conversion tubes are an important link in some display systems. For example, a scan conversion tube will convert radar blips display to a television signal for viewing on a TV screen. The scan conversion tube enables information to be put in at one rate and taken out at another rate, thus providing some storage. The technique is particularly useful for retaining aircraft locations in an air traffic control instrument so that the path of an aircraft can appear as a dotted line.

Over the past 10 to 15 years, cathode-ray tubes have displaced a number of formerly conventional display instruments, such as large panel-mounted indicators and recorders. At one time, CRTs were uncommon as parts of control panels for industrial processes. Today, CRTs associated with key measuring and controlling circuits, are central to many panels. The combination of computers, microprocessors and CRTs in desk-top computers, computer graphics, and data retrieval systems is common and growing at a rapid rate.

A computer display terminal is illustrated in Fig. 4. The screen (25-inch diagonal; 63.5 centimeters) can be used by designers of electronic circuit boards, utility networks, automotive components, schematic diagrams, and street maps, among other applications, embracing fine detail while maintaining the total picture perspective. A direct view bistable storage tube is used. With an instrument of this type, over 15,000 characters may be displayed simultaneously and may be formatted in 179 alphanumeric characters per line pages (like a printer), or as two 85-character per line pages (like an open book). The latter format is particularly useful for developing software and scanning large amounts of alphanumeric output.

A color graphics terminal with a raster scan display is shown in Fig. 5. Colors are selected from a 64-color palette with up to eight colors displayed simultaneously on the screen. The instrument uses the hue, lightness, and saturation method for specifying color. The instrument features coloring vectors, characters, symbols, and filling polygons. The firmware enables another color to border the polygon. Eight colors are displayable at once on the screen (13-inch diagonal;

Fig. 4. Computer display terminal features a direct-view bistable storage tube. The instrument is suited for displaying the highly complex graphics found in contour mapping, seismic analysis, energy field modeling, among other applications. (*Tektronix, Inc.*)

33 centimenters). Up to 120 user-defined patterns can be programmed. Additional solid color mixtures are created by alternating pixels of the eight colors within the patterns. Patterns can also be stripes, plaids, and polkadots. Vectors can be written with these patterns and polygons can be filled with them. Similar sophisticated instruments using CRTs are shown in Figures 6 and 7. An instrument designed for use with an electron pen is shown in Fig. 8.

Glossary of Terms. Some terms frequently used in connection with cathode-ray tubes and CRT instruments and displays include:

(*Angle of Deflection*)—The angle through which the beam is deflected.

(*Angle of Divergence*)—The maximum angle of deflection experienced by electrons in an electron beam due to debunching.

Fig. 5. Color graphics terminal. The computer terminal with a raster scan display has capabilities of scrolling graphics and alphanumerics, with color. (*Tektronix, Inc.*)

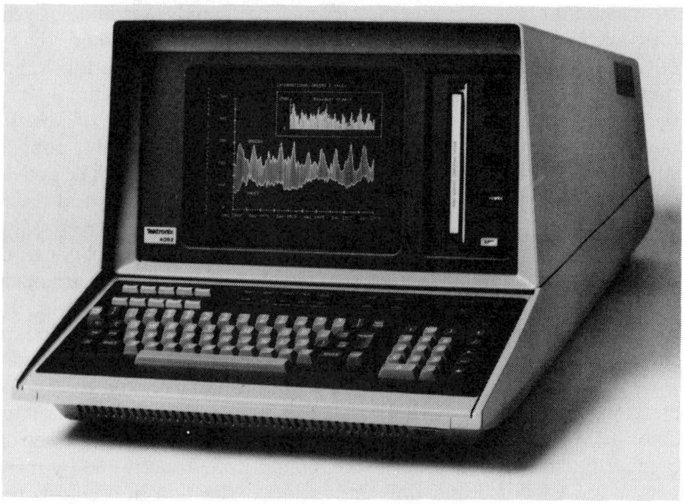

Fig. 6. Graphic computing system featuring CRT. (*Tektronix, Inc.*)

(*Black Level*)—In television, that level of the picture signal corresponding to the maximum limit of black peaks.

(*Blanking*)—In television, the substitution for the picture signal, during prescribed intervals, of a signal whose instantaneous amplitude is such as to make the return trace invisible. The term is also applied in connection with laboratory cathode-ray oscilloscopes.

(*Blooming*)—The mushrooming of an electron beam (with consequent defocusing) produced by too high a setting of the brightness control.

(*Brightness Control*)—The manual bias control of a cathode-ray tube. The brightness controls affects both the average brightness and the contrast of the picture.

(*Cathode Disintegration*)—The destruction of the active area of a cathode by positive-ion bombardment.

(*Cathodoluminescence*)—The excitation of luminescence in a solid through the action of an electron beam impinging on the luminescent material or phosphor. This is the type of luminescence present in television picture tubes, in radar cathode-ray tubes, and in oscilloscopes.

(*Cathodophosphorescence*)—Phosphorescence resulting from cathode-ray bombardment.

(*Damping Tube*)—A tube used with magnetic deflecting-coils to prevent any transient oscillations from being set up in the tube or its associated circuits.

(*Dark Trace Tube*)—A cathode-ray tube, on which the face is bright, and signals are displayed as dark traces or dark blips.

(*Deflection Sensitivity*)—1. Of an electrostatic-deflection cathode-ray

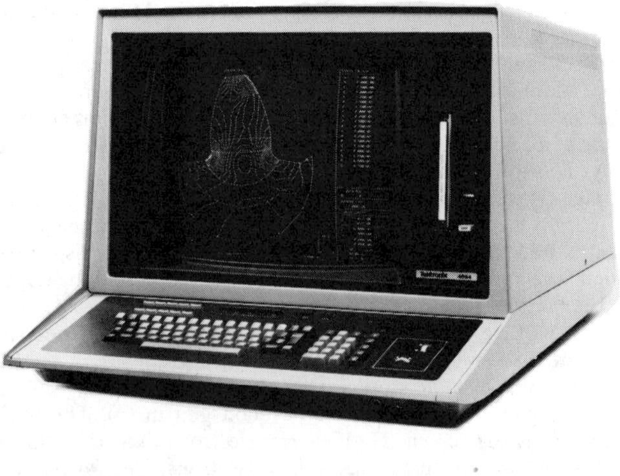

Fig. 7. Graphic computing system featuring CRT, engineered for uses where complex diagrams and high-density interactive graphics are required. (*Tektronix, Inc.*)

Fig. 8. Use of electronic pen to instruct data processing system to display more detailed breakdown report of item to which operator points pen.

tube, the quotient of the spot displacement by the change in deflecting potential. 2. Of a magnetic-deflection cathode-ray tube, the quotient of the spot displacement by the change is deflecting magnetic field. 3. Of a magnetic-deflection cathode-ray tube and yoke assembly, the quotient of the spot displacement by the change in deflecting-coil current. Deflection sensitivity is usually expressed in millimeters per volt applied between the deflecting electrodes, or in millimeters per gauss of the deflecting magnetic field.

(*Electron Image Tube*)—A cathode-ray tube used to increase the brightness or size of an image or to produce a visible image from invisible radiation, such as infrared. A large, light-sensitive cold cathode serves as the focal plane for the optical image. The resulting emission from the cathode is accelerated through an appropriate lens system before striking a fluorescent screen, where it produces an enlarged and brightened reproduction of the original image. This device has been used in electron microscopes and telescopes, infrared miscroscopes and telescopes, and fluoroscope intensifiers.

(*Grass*)—The pattern on the cathode-ray tube display of a radar or similar system, which is produced by the random noise output of the receiver.

(*Holding Beam*)—A diffuse beam of electrons for regenerating the charges retained on the dielectric surface of an electrostatic memory or storage tube.

(*Horizontal Blanking*)—The interruption of the electron beam of a cathode-ray tube during horizontal retrace.

(*Horizontal Centering Control*)—A control that enables the operator to move a cathode-ray image in a right or left direction across the screen.

(*Horizontal Deflecting Electrodes*)—The pair of electrodes located in the vertical plane in an electrostatic-deflection cathode-ray tube which is used to produce beam deflection in the horizontal plane.

(*Horizontal Line Frequency*)—In television, the number of horizontal lines per second: 15,750 for standard black-and-white television in the United States.

(*Horizontal Hold Control*)—The control which varies the free-running period of the horizontal-deflection oscillator in a television receiver.

(*Horizontal Resolution*)—In television, the number of light variations or picture elements along a line which can be distinguished from each other.

(*Horizontal Retrace*)—In cathode-ray equipment with linear, horizontal time-bass, the rapid right-to-left motion of the electron beam at the end of each sweep.

(*Horizontal Sweep*)—Sweep of an electron beam in the horizontal plane.

(*Ion Burn*)—A deactivation of a small spot of the phosphor of a cathode-ray tube, caused by bombardment by heavy negative ions in the beam. The effect is noticeable only in magnetic-deflection systems, since an electrostatic deflection system deflects the negative ion through the same deflection angle as the electrons. Magnetic-deflection tubes require an ion trap to prevent permanent damage.

(*Radarscope*)—The CRT indicator of a radar apparatus on which echoes from targets detected by the radar by visually displayed. The A scope is a type of radar indicator that presents the signal strength of a target signal and range of a target in rectangular coordinates. The R scope gives information similar to that of the A scope on an expanded horizontal scale. It takes a limited portion of the A scope presentation at any range and expands the horizontal coordinate so that a more detailed study that of portion may be made. It is distinguished from the A scope in that the zero range of the A scope is always presented.

(*Retrace Line*)—The line traced by the electron beam in a cathode-ray tube in going from the end of one line or field to the start of the next line or field.

(*Scanning—"Flying Spot"*)—The subject is illuminated by a "flying spot" light source of constant intensity, developed on the face of a cathode-ray tube with a short-persistence phosphor. The spot of light is made to follow the conventional raster pattern so that a phototube receiving transmitted or reflected light from the subject will have a signal output proportional to subject brightness and subject position as required.

References

Bailey, S. J.: "Computer Graphics: Hot Line between Process and Operator," *Control Eng.*, **26**, 7, 46–48 (1979).

Landee, R.: "Electronics Designers' Handbook," 2nd edition, McGraw-Hill, New York, 1977.

Morris, J. G.: "Using Color in Industrial Control Graphics," *Control Eng.*, **26**, 7, 41–43 (1979).

Staff: "Selecting CRT-based Process Interfaces," *Instrumentation Technol.*, **26**, 2, 28–33 (1979).

Umbers, I. G.: "Facing Up to CRT Communication," *Process Engineer* (April 1978).

CATHODIC ELECTRODEPOSITION. **Conversion Coatings.**

CATHODIC PROTECTION. **Corrosion.**

CATHODOLUMINESCENCE. **Cathods-Ray Tube; Luminescence.**

CATION. A positively charged ion. Cations are those ions that are deposited, or which tend to be deposited, on the cathode. They travel in the nominal direction of the current. In electrochemical reactions they are designed by a dot or a plus sign placed above and behind the atomic or radical symbol as $H^{\cdot}$ or H^{+}, the number of dots or plus signs indicating the valence of the ion.

In electrolysis, the cathode is negative, and attracts cations. In a battery, the transfer of charges of cations to the cathode makes it the positive terminal.

CATION-EXCHANGE RESINS. **Ion Exchange Resins.**

CATIONIC MECHANISM. **Organic Chemistry.**

CATIONIC SURFACTANT. **Detergents.**

CATIONIC WATER. **Hydrate.**

CATION TRANSPORT. **Cell (Biology).**

CATKIN. An inflorescence, also called ament, composed of many flowers, aggregated into long, often tassel-like masses. The perianth is completely lacking, or may be present in a scale-like form. The flowers of willows (pussy willows), poplars, alders, beeches, oaks, and

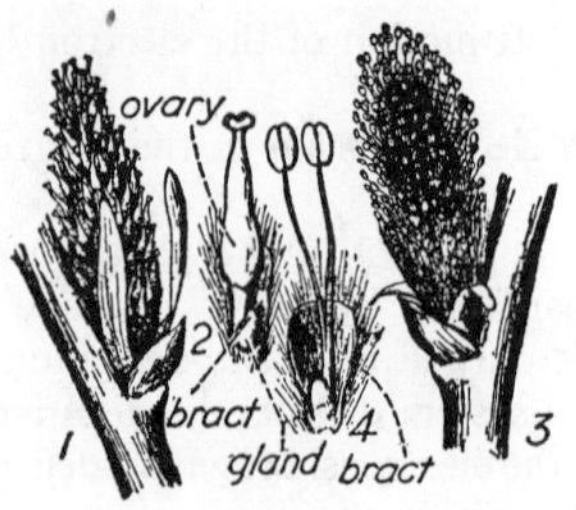

Flowers of willow, *Salix*: (1) pistillate catkin; (2) a single pistillate flower; (3) staminate catkin; (4) a single staminate flower.

birches are familiar examples. Most of them are wind-pollinated flowers. See also **Flower.** See accompanying diagram.

CATS (*Mammalia, Carnivora*). Over fifty species of cats (*Felines*) have been described. They are all assigned to the single family (*Felidae*). The general organization of the cats is shown by the accompanying table. The position of the cats with reference to other families in the order *Carnivora* is given in the entry on **Carnivora.** Cats, of course, are flesh-eating mammals; they have simple dentition; the claws are sharp, curved, and retractible. In most species, the claws can be withdrawn completely into sheaths. Considering the domestic cats as well as the lesser cats and lynxes, it can be stated that cats occur essentially worldwide with the exception of Australia and the oceanic islands.

Included among the Great Cats (*Panthera*) are the lion, tiger, leopard, snow leopard, and jaguar. The lion is certainly one of the better known species of large cats and sometimes is referred to as the "King of the Beasts" (see Fig. 1). This reference, however, is highly debatable and possibly the term may have stemmed from their most impressive appearance. They are exceeded in size by the north Manchurian tigers, particularly when sans manes. It has been said that children terrify lions and that they are easily disturbed by comparatively innocuous events and situations. One authority has mentioned the flapping of laundry on a clothes line as frightening to these animals. Although a female lion (females usually do the killing) may claim an antelope or two per month, lions have been known to lie down peacefully among antelopes during the day. While highly publicized, the man-hunting, man-killing escapades of lions would seem to occur most

GENERAL ORGANIZATION OF THE CATS

FELINES

Great Cats (*Panthera*)
Lions (*Panthera leo*)
Tigers (*Panthera tigris*)
Leopards (*Panthera pardus*)
Snow Leopard (*Panthera uncia*)
Jaguar (*Panthera onca*)
Lesser Cats (*Felis*)
Ocelots (*Felis pardalis, ...*)
Leopard-Cats (*Felis bengalensis, ...*)
Tabby-Cats (*Felis lybica, ...*)
Desert Cats (*Felis manul, ...*)
Plain Cats (*Felis planiceps* and *badius*)
Marbled Cats (*Felis marmorata*)
Other Cats (*Profelis*)
Pumas (*Profelis concolor*)
Clouded Leopard (*Profelis nebulosa*)
Golden Cats (*Profelis temmincki* and *aurata*)
Lynxes (*Lynx*)
Jungle-Cats (*Lynx chaus*)
Caracals (*Lynx caracal*)
Northern Lynxes (*Lynx lynx, ...*)
Bobcats (*Lynx rufa, ...*)
Servals (*Leptailurus*)
Jaguarondis (*Herpailurus*)
Cheetahs (*Acinonyx*)

Fig. 1. Lioness and cubs. (*A. M. Winchester.*)

infrequently, probably precipitated by a maddening disease or extreme hunger. Unmated juveniles are often identified in such rare escapades. It is recorded that the lion will not approach within 20 feet (6 meters) the very small zorille (similar to the American skunk) even though the animal may be casually inspecting fresh kill in which the lion(s) is deeply interested. Of course, this observation may be a credit to the lion's judgment rather than a criticism of lack of courage.

The lion reaches a length of some 10 feet (3 meters) from tip to tip and a weight of about 500 pounds (227 kilograms). The animal is uniformly tawny (brownish-yellow) in color, but the shades range from light yellows to dark brown. The male usually has a full mane, but this also is a variable characteristic. Some males with no mane have been found. The lion is nocturnal in habit and is generally regarded as shy, unless provoked. Possibly to a degree more noticeable than in many other mammals, a lion shows age, its physical condition showing up in contour, posture, and condition of mane (in the male). The older lion usually subsists on small rodents, scorpions, and fairly small creatures, a limited diet which may contribute to the aging process. Lions hunt in groups, sharing the animals that they kill, and sometimes eat the carcasses of animals that they have not killed, even when badly decomposed.

The lion thrives and reproduces well in captivity and there is rarely a short supply. Lion populations in Africa are high and frequently are regarded as a serious menace to domestic stock. Unlike some popular portrayals of lion habitats, the lion does not inhabit forests or jungles, but prefers grassy plains, savannas, scrubland, and even semi-desert areas. At one time, lions were found in eastern Europe, the Near East, and the northern portions of Africa, and across to India. The population in India today is quite limited, but lions are found widely in the habitats which they prefer in Africa south of the desert belt.

As do all of the Great Cats, the lion has some small bones called hyoids located at the base of the tongue, which gives the animal the ability to make roaring sounds, as contrasted with the lesser sounds of the lesser cats. It is interesting to note that lions and tigers can interbreed, producing *Ligers* (lion is male parent) or *Tigons* (tiger is male parent).

The tiger rivals the lion in size and strength. The fur normally varies from reddish to brownish yellow, with transverse black stripes and a black-ringed rail (see Fig. 2). The total length of adult males, including the tail, is from 9 to 10 feet (2.7 to 3 meters). The tiger is an Asiatic animal, found chiefly in the warm southern countries, but also northward into Turkestan and southern Siberia. It is by no means a tropical species. In fact, the tiger originated as an Arctic animal, coming from eastern Siberia. The largest tigers are found in the colder areas of Siberia in an area between the Altai and Stanovoi Mountains. Inasmuch as the tiger is from colder climes, it is not unusual for tigers to bathe as a means of keeping cool in the warmer regions. A litter of tigers usually contains five babies, but often only two are permitted to live, the parents sometimes eating the others. The young travel with the mother for at least a year. Occasionally tigers have become man-eaters, but generally they are considered timid and make great efforts to stay out of the realms of humans. The tigers in India

Fig. 2. Tigers. (*A. M. Winchester.*)

prefer the great nilghai as a favorite dietary item, but when necessary can subsist on smaller creatures, including mice, locusts, and fish. The abilities of the tiger to climb trees has been overstated. They are not considered good climbers except in emergency situations for escape. Tigers are known for rather poor vision and sense of smell, but do possess excellent hearing capabilities. Thus, the prey they are seeking are generally safe if they remain motionless and quiet. In killing, the tiger leaps on and hugs the victim, biting at the throat. Tigers are known to consume 200 pounds (91 kilograms) of flesh within a short period, followed by huge quantities of water. Among their natural enemies are packs of feral dogs, water buffaloes, elephants, and wild dholes.

The leopard is found throughout most of Africa with exception of the big desert areas. They are considered far too numerous in some of the cultivated and industrialized areas of Africa. As shown in Fig. 3, the leopard is marked with black rings and spots, although a black variety occurs in which the spots are faintly traceable. The basic coloration of most leopards is tawny. Although numerous albino tigers have been reported, there are no records of albino leopards. As compared with the lions and tigers, the leopard is faster and less fearless in most situations and tend to be much less discriminate. The animal tends to attack monkeys, baboons, even humans when opportune rather than seeking certain types of dietary favorites. Leopards will often eat only the choicest parts of a meal at first, dragging the remains to a thicket or even hide it in the branches of a tree for later consumption, a practice known as hoarding. Closely related to the common leopard is the snow leopard which lives at high altitudes in central Asia. This leopard is also spotted, but the basic color is a grayish-white. The skull is characteristically shaped. Another relative is the clouded leopard, which is found in southeastern Asia. Its legs are shorter than the other species, is basically gray or grayish-yellow in coloration, with what might be termed blotches of dark brown. This animal has not been fully studied, but it is believed to subsist mainly on birds and makes it home in trees. More detailed coverage on the status of the leopard can be found in a news item in *Science*, **208**, 18 April 1980.

Sometimes disputes have arisen concerning the terms *leopard* and *panther*. The terms can be used interchangeably.

The last of the Great Cats to be described here is the jaguar, a large South American cat, found chiefly in the jungles, but also in

Fig. 3. Leopard. (*A. M. Winchester.*)

Fig. 4. Adult jaguar. (*New York Zoological Society photo.*)

open country. As shown by Fig. 4, this cat is tan, marked with rings and dots of black, resembling the leopard. The jaguar is larger than the leopard and differs in details of structure and markings. The animal has a deep and hoarse cry, usually used at mating time. It feeds on wild horses, tapirs, capybara, dogs, and cattle. The animal reaches a length of about 4 feet (1.2 meters) and a height of from $2\frac{1}{2}$ to 3 feet (0.8 to 1 meter). It is heavily built, a rapid runner, graceful, and agile. It should be pointed out that there is some controversy pertaining to the aforementioned measurements, some authorities attributing sizes approaching those of tigers to this animal. The jaguar prefers the forest and is an excellent tree climber, the best of the Great Cats. It ranges from the southwestern United States southward to Argentina. The animal tends to vary its behavior with its habitat, becoming almost as water-loving as the tiger in the Amazonian region. Two to four cubs are produced annually. The cubs closely follow their mother for about 35 days.

The ocelot is a moderately large cat of South America, but also occurring in Central America and the southern part of the North American continent. The animal is tawny or reddish, marked with black spots and blotches. Reports of the animal as far north as southwestern Arkansas, Texas, and Arizona have been recorded. The adult is from 40 to 50 inches (102 to 127 centimeters) in length, not including the tail of some 13 to 15 additional inches (33 to 38 centimeters). There are two black bars on the cheeks, black spots on the head, and from four to five parallel stripes on the neck. The underpart of the animal is white. The ocelot prefers forest or brushy regions and feeds on small animals and reptiles. There usually are two young per season. Some varieties of ocelot can be tamed, but they are not necessarily fully reliable in captivity. The pelts are considered of economic value.

Leopard-cats are small spotted cats of many varieties. They are found well distributed throughout the tropical regions. These animals seldom exceed 2 feet in length (0.6 meter), rarely 3 feet (1 meter). Closely associated with this loosely-knit group of leopard-cats are the tabby cats, typified by the common wild cats of Africa (*Felis lybica* and *ocreata*). *Felis lybica* is essentially unchanged in the Abyssinian domestic cat breed. This animal is believed to be the ancestor of European house cats. The ancient Egyptians and early Greeks and Romans domesticated the animal and had a high regard for it. The animal was called the "Mu" (a familiar sounding word in terms of cats) and was trained to retrieve fallen water fowl, and to control the mice population. It is of interest to note that a popular concept to the effect that cats do not like water is untrue. However, it is true that cats do not like cold water. The true wild cat, pretty, shy, but savage when provoked, yellowish with tabby markings, blue eyes, and pink nose, still exists in the wild in isolated parts of Europe, but it is not found in Italy or the Scandanavian countries. Tabby-marked wild cats (tabby = gray or brown with dark stripes) are found in the Altai Mountains (Siberia) and over a large part of Africa. Some authorities believe that the present domestic cats of the Oriental region may have been bred from imported wild cats from Africa or Europe, or possibly have stemmed from the so-called waved cat (*F. torquata*), commonly found throughout northern India. Authorities are yet to agree upon the cause for a breed of cat from the Isle of Man and another from Korea to have lost their tails. The grass-cat (*F. pajeros*) is a tabby-like cat found in southern Argentina.

The marbled cats found in the forests of Tibet and into southeast Asia appear much like the previously mentioned clouded leopard, but they are much smaller. Authorities have yet to establish if these cats are miniaturized leopards or simply small cats that look like the clouded leopard.

The various lynxes bear no ready anatomical distinctions that make them differ from the previously described Lesser Cats. The lynx is essentially limited to the northern hemisphere and is characterized by conspicuous ear tufts and a fringe of long fur about the throat. Among the several North American species, most are called wildcats or bobcats. The Canada lynx (*Lynx canadensis*) alone is known as the lynx. Several species of lynx also occur in Europe, Africa, and Asia. The margay is a wild cat of moderate size, reddish marked with black spots, that occurs from Mexico to Paraguay. The manul is a wildcat found in Siberia, Mongolia, and Tibet. It is about the size of the domestic cat and varies from buff to silver-gray in color. It is also called Pallas' cat.

The serval is an equatorial African cat that can attain a length of 5 feet (1.5 meters). It is light tawny and spotted with black. It eats birds, lizards, and insects.

The cheetah is a large cat found in Africa and India and is marked by its slender build. It is tawny with black spots, thus somewhat resembling the leopards. The cheetah sometimes is referred to as the hunting leopard and has been tamed for use in the chase. The animal attains a length of from 3 to 4 feet (1 to 1.2 meters), with another $2\frac{1}{2}$ feet (0.8 meter) for the tail. The fur is coarse and crisp, with white underneath. The animal was domesticated hundreds of years ago in Asia for coursing game. The cheetah prefers to feed on young animals, but sometimes will tackle an adult animal to the size of a small antelope. The cheetah is swift, its legs are long and muscular, and its back is curved. Cheetahs have been recorded at speeds of over 60 miles (96.5 kilometers) per hour. The animal possesses a springiness in its running and the strides are long. Although very fast, the cheetah is best for short sprints, and can be overtaken by slower animals, such as a horse, over long distances.

Cheetahs today are outnumbered by their enemies, they are largely defenseless, and, where unprotected, they are likely headed for extinction. Cheetahs have been exterminated from large portions of their former range in Africa and the Middle East. Once plentiful in India, cheetahs have totally disappeared there, the victims of hunters and loss of habitat. Further detailed coverage of this topic can be found in an article by Frame, G. and L. Frame: "Cheetahs: In a Race for Survival," *National Geographic*, **157**, 5, 712–728 (May 1980).

Cheetahs have been used for killing coyotes in the southwestern United States, but not extensively because a pair of highly trained cheetahs is quite costly.

Pumas are of several types. The puma proper (*P. concolor*) originally inhabited the eastern half of the North American continent, ranging from Virginia into Canada. The animal is now considered extinct in the settled parts of the country, but possibly may still exist in wilder areas. This animal is of uniform tawny to brownish color. A darker species (*P. coryi*) is found in Florida; the *P. arundivaga* is found in Louisiana; and the western puma or mountain lion (*P. oregonensis*) ranges from Mexico into Canada. The eastern species has been variously called the cougar, mountain lion, panther, catamount, and painter. The puma is generally considered a menace to stock and sheep. However, it is shy in terms of the human realm.

Cougars are considered to have originated in Peru, the term being a native Brazilian name. The South American variety measures from 6 to 8 feet (1.8 to 2.4 meters) in length from tip of nose to tip of tail. The color is red-tawny, pale around the eyes, with a white throat and legs and white underneath. The animal is considered exceptionally intelligent. The cry of the cougar when it is hunting at night is described as terrifying. Cougars are considered excellent tree climbers.

For references, see **Mammalia.**

CAT SCANNER. Brain and Nervous System; X-Ray Scanner (CAT).

CAT'S-EYE. This name is applied to varieties of several mineral and gemstone species that enclose fine fibers or cellular structures in parallel arrangement, causing, particularly when cut and polished *en cabochon*, a band of reflected light to play on the surface of it. Because of fancied resemblance to the eyes of cats, such stones are called cat's-eyes, and the effect is referred to as chatoyancy. The stone is said to be chatoyant. True cat's-eye is a variety of chrysoberyl, but tourmaline and quartz are also found which show this same effect. Ordinary quartz cat's-eyes are a pale yellowish or greenish, but a beautiful golden-yellow sort is known from South Africa called tiger's-eye which probably represents a replacement of crocidolite by quartz.

When the term *tiger's eye* is used, this applies only to chrysoberyl. Other gemstones that exhibit this phenomenon include sillimanite, scapolite, cordierite, orthoclase, albite, and beryl. See also **Chrysoberyl; Crocidolite.**

CATSHARKS. Sharks.

CATTAIL. Of the family *Typhaceae*, genus *Typha*, there are several species of plants which grow in marshy places and along the margins of ponds and slow-flowing streams. Usually, they form extensive stands, crowding out nearly all other plants. The cattail plant has a thick horizontal rhizome which grows along the surface of the ground or just beneath it, generally in several inches of water. From this rhizome, the long linear leaves grow in erect bunches. The leaves of the *Typha latifolia* have widely overlapping bases and are from 3 to 6 feet (0.9 to 1.8 meters) long. The flower stem rises stiffly erect in the center of the bunch of leaves and is from 3 to 8 feet (0.9 to 2.4 meters) tall. Near its tip, the cattail bears two dense cylindrical spikes of flowers, one above the other, which are unisexual. The pistillate flowers are found below the staminate flowers. Each staminate flower consists of from 2 to 5 or more stamens surrounded by a number of hairs. Soon after the pollen grains are shed, the staminate flowers drop off, leaving the naked tip of the stem projecting above the pistillate spike. Each pistillate flower consists of a single pistil surrounded by a group of long hairs. Cattails are entirely wind-pollinated. After pollination, the ovaries develop to 1-seeded achenes surrounded by the fine hairs. These fruits form the familiar black or dark brown cattail of late summer and fall. The seeds are blown about by the wind, the long hairs greatly aiding in distribution.

The dried leaves of cattails were formerly used in making the seats of rush-bottomed chairs. The hair-covered seeds have been used to a slight extent for stuffing for pillows and small things. The entire fruiting stem is often used as an ornament.

The cattails are related to the rushes and sedges.

CATTLE. Bovines.

CAUCHY CONVERGENCE TEST. If

$$\lim_{n\to\infty} |s_n|^{1/n} < 1$$

then the infinite series

$$\sum_{n=1}^{\infty} s_n$$

converges absolutely. It diverges if the limit is greater than unity. This is also known as Cauchy's criterion of the first kind. His second test is more commonly called d'Alembert's test. See also **D'Alembert Test.**

CAUCHY DISTRIBUTION. A frequency distribution of the form

$$f(x) = \frac{1}{\pi(1+x^2)}, \quad -\infty \leq x \leq \infty$$

CAUCHY-RIEMANN EQUATION. If

$$\partial u/\partial x = \partial v/\partial y$$

and

$$\partial u/\partial y = -\partial v/\partial x$$

where u and v are both functions of x and y, these equations will be satisfied for an analytic function $(u + iv)$ of the complex variable $z = (x + iy)$. They are often used to show that such a function is analytic. See also **Laplace equation;** and terms listed under **Mathematics.**

CAUCHY THEOREM. A basic formula in the calculus of residues. If $f(z)$ is an analytic function of a complex variable z which has no singular point within or on a given closed curve C, then

$$\int_C f(z)\, dz = 0$$

where the integral is extended over the entire contour C.

An extension of the Cauchy theorem, known as the Cauchy integral formula or residue theorem, is also of importance. See **Residue.**

CAUDA EQUINA. Brain and Nervous System.

CAUDAL ANESTHETICS. Anesthesia.

CAUDAL FILAMENT. A median jointed appendage resembling an antenna and sensory in function, found in some of the primitive insects.

CAUDAL PEDUNCLE. Fishes.

CAULDRON-SUBSIDENCE. A term proposed by E. B. Bailey and other Scottish geologists for the sinking of the portion of the roof or cover of a deep-seated igneous intrusion, aided by circumferential faults.

CAULIFLOWER. Brassica.

CAULIFLOWER EAR. Hearing and the Ear.

CAUSALITY. Causality is the hypothesis that a precisely determined set of conditions will always produce precisely the same effects at a later time. Classical physics was based on firm belief both in philosophical causality and in the idea that the precise determination of the initial conditions was possible in principle. The impossibility of such precise determination is a basic result of quantum mechanics.

CAUSTIC (Chemical). A corrosive substance, almost always of an alkaline nature, such as sodium hydroxide, NaOH; potassium hydroxide, KOH; or calcium oxide, CaO. Such substances attack many metals, plastics, and other materials, including human tissue, and generally fall in the category of *corrosives.*

CAUSTIC EMBRITTLEMENT. Embrittlement.

CAUSTIC (Optical). An envelope curve giving the boundaries of an initially parallel beam after reflection of refraction by an optical system that has spherical aberration. See accompanying diagram.

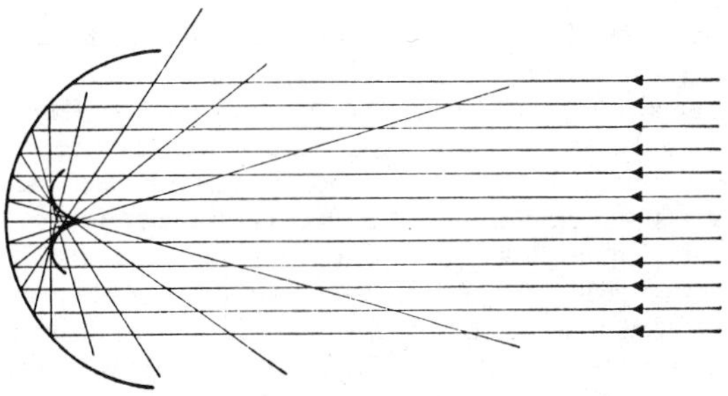

Caustic envelope curve.

CAUSTIC POTASH. Potassium.

CAUSTIC SODA. Sodium.

CAUTERIZATION. The purposeful destruction of tissue. Chemical agents, electrical and mechanical means may be used. Silver nitrate (lunar caustic) is one of the more commonly used chemical cauterizing agents. Cryogenic methods also may be used, wherein a low-temperature gas or liquid may be directed at the tissue in a carefully-controlled manner. In electrocautery, a fine platinum wire may be electrically heated by passage of a controlled current to any desired temperature.

CAVALIER DRAWING. Pictorial Representation.

CAVE. A natural opening in the earth's surface, chiefly developed in limestone regions where, because of the easy solubility of calcium carbonate, the groundwater dissolves and carries away large quantities of this otherwise resistant rock. The water enters through the joint cracks or bedding planes, passing downward by gravity until it becomes saturated with calcium carbonate or reaches the ground water level where more or less complete saturation exists. With the continued solution of the limestone, large channels and even great underground chambers are formed. The steady removal of the limestone in solution thus tends to weaken the whole formation with the result that the roofs of underground channels or chambers frequently collapse, forming depressions varying in size from a few feet (meters) in depth and of small area to those of many acres and 100 feet (30 meters) or more in depth. Such fallen-in areas are called sink holes or simply sinks. If they contain water they are then referred to as sink-hole lakes. Sometimes after the continued collapse of the roofs of caverns a small portion will remain, thus forming a natural bridge, the classical example of which is the Natural Bridge, Virginia.

Wherever limestones occur, if there is a sufficient supply of groundwater, underground drainage will develop.

Among the more famous caverns of the United States are the Luray and Shenandoah Caverns in Virginia, Mammoth Cave, Kentucky, and Carlsbad Caverns in New Mexico.

Carlsbad Caverns are on three levels and possess the largest natural cave "room" known in the world, 1500 × 300 feet (457 × 91 meters) and 300 feet (91 meters) high. Mammoth Cave has 150 miles (241 kilometers) of passageways and rooms with 200-feet (61-meters) high ceilings. The Echo River runs through the cave, 360 feet (110 meters) below ground level.

Certains parts of Florida abound in sink holes and sink-hole lakes. An important feature in caverns is the so-called rock icicles or stalactites and their associated stalagmites.

CAVE-FISH. Blind-Fish.

CAVENDISH GRAVITATIONAL METHOD. Torsion Balance.

CAVIAR. Sturgeons.

CAVITATION. Cavities may form, grow, and collapse in a liquid when variational tensile stresses are superimposed on the prevailing ambient pressure. Pure liquids have theoretical tensile strengths which are estimated on various grounds to be of the order of 300 to 1500 atmospheres (bars), but the observed tensile strengths of real liquids are much lower. It is presumed, therefore, that the observed tensile strength is a measure of the stress required to enlarge the minute cavities, or cavitation nuclei, which already exist in the liquid rather than the stress required to form new interior surfaces.

The transient cavities formed by tensile stress are unstable and would grow indefinitely if the stress were maintained. After the cavitation nuclei have been expanded to many times their original size, however, they may collapse violently if the stress is reduced or removed. The kinetic energy of the liquid that follows each inwardly collapsing interface becomes highly concentrated as the cavity collapses. If such transient cavities contain very little permanent gas, the peak pressures at collapse may reach thousands of bars, the temperature may reach thousands of degrees, and strong shock waves may be radiated to a distance of several cavity radii. Similar cavities formed in saturated liquids will usually contain more gas and their collapse will be less violent, but the peak pressures attained are sufficient to produce unique mechanical effects, such as the corrosion and pitting of metallic surfaces (as in marine propellers and sonar projectors) and the beneficial removal of embedded dirt (as in ultrasonic cleaners). In the latter case, the soil to be removed provides a prolific source of cavitation nuclei at exactly the sites where cavitation is desired.

In hydrodynamic cavitation, the tensile stress is of relatively long duration and plenty of cavitation nuclei are usually available. As a result, cavitation occurs when the total net pressure, or the stagnation pressure, becomes approximately equal to the vapor pressure of the liquid. In acoustic cavitation, the cyclic pressure required to produce cavitation is a function of the frequency, the partial pressure of any dissolved gas, and the population of cavitation nuclei. For frequencies above about 200,000 Hz, the threshold pressure for cavitation increases with the square of the frequency and is almost independent of the degree of gas saturation. For frequencies below 200,000 Hz, the threshold pressure is a function of the partial pressure of the dissolved gas. In saturated liquids at sound pressures less than a few bars, stable bubbles can grow from cavitation nuclei by the process of rectified diffusion. At higher levels of acoustic excitation, transient cavities can be formed. The threshold sound pressure at which they appear and the violence of their collapse increase as the partial pressure of the dissolved gas is lowered.

CAVITATION NUMBER. A nondimensional number whose magnitude is a measure of the likelihood of cavitating flow. If nuclei are present, flow around a bluff body is likely to cavitate when $\rho v^2/(p - p_s)$ is of order one (p and v are pressure and velocity in the stream and p_s is the saturation vapor pressure). Without nuclei in degassed liquids, the critical v value is larger and p_s is not relevant.

CAVITY FURNACE. Solar Energy.

CAVITY (Tooth). Caries and Cariology.

CAVITY-TYPE WAVEMETER. Frequency Measurement.

CAVY. Rodentia.

CAYENNE PEPPER. Flavorings.

CEBOIDS. Monkeys and Baboons.

CECA. Poultry.

CECOSTOMY. An artificial opening made through the abdominal wall into the cecum. This is done temporarily or permanently to provide an artificial anus when the large intestine is obstructed or is the site of certain chronic diseases such as colitis. It is also a frequent operation for cancerous obstruction of the large intestine, as a preliminary stage before removal of the tumor.

CECUM (or Caecum). A sac-like, blind pouch of the large intestine, situated below the level of the junction of the small intestine into the side of the large intestine. At the lower portion of the cecum, but variable in position, is the appendix. See also **Digestive System (Human).**

CEDAR ELM. Elm Trees.

CEDAR TREES. The term *cedar* is not scientifically specific, but as will be noted from the following list of cedar trees, cedars are found in several families and comparatively few are of the genus *Cedrus* (true cedars).

Cedars are characterized by having in the woody tissue an aromatic volatile oil which persists for a long time after the tree is cut down and dried. The wood of many cedars is very resistant to rotting. Therefore, cedar always has been a favored material for rail fences. The oil present in the wood is repulsive to insects. Cedars are found widely distributed in the United States, Asia, and North Africa. They grow best in warm areas, but principally in mountain climates. The trees are conifers, with cones of approximately 4 to 6 inches (10 to 15 centimeters) in length at maturity. The trees are narrow and erect. The leaves bear naked seed on cone scales. See also **Conifers.**

The weight of the wood from most cedars ranges from 22 to 33

Common Name and Family	Species
Alaska cedar (*Cupressaceae*—cypress family)	*Chamaecyparis nootkatensis*
Atlantic cedar (*Pinaceae*—pine family)	*Cedrus atlantica*
Chinese cedar (*Meliaceae*—mahogany family)	*Cedrela sinensis*
Deodar cedar (*Pinaceae*)	*Cedrus deodara*
Eastern red cedar (*Cupressaceae*)	*Juniperus virginiana*
Eastern white cedar (*Cupressaceae*)	*Thuja occidentalis*
Incense cedar (*Cupressaceae*)	*Libocedrus decurrens*
Japanese cedar (*Taxodiaceae*—swamp cypress family)	*Cryptomeria japonica*
Cedar of Goa (*Cupressaceae*)	*Cupressus lusitanica*
Cedar of Lebanon (*Pinaceae*)	*Cedrus libani*
Port Orford cedar (*Cupressaceae*)	*Chamaecyparis lawsoniana*
Spanish cedar (*Meliaceae*)	*Cedrela odorata*
Western red cedar (*Cupressaceae*)	*Thuja plicata*
White cedar (*Cupressaceae*)	*Thuja occidentalis*
Yellow cedar (*Cupressaceae*)	*Chamaecyparis nookatensis*

pounds per cubic foot (352 to 529 kilograms per cubic meter) when dried. Where plentiful, the wood is used for construction and cabinet work, interior trim, and closets and chests (with advantage of being insect repellent). The excellent condition of some temples in India 400 or more years old is exemplary of the durability of cedar wood. The Spanish cedar is related to the true mahogany tree and at one time was used extensively for making cigar boxes. This wood is frequently cut into thin sheets and applied as a veneer over cheaper woods.

Champion cedars as reported by The American Forestry Association are listed in Table 1. On the west coast of the United States, the cedar is often a companion of the giant sequoias.

The incense cedar grows along the west coast of North America from upper Mexico through California into Washington. The northern white cedar grows in the northeastern United States, Canada, and as far west as Minnesota. The western cedar is found close to the Pacific coast in northern California and into Alaska. Although not so numerous, the Atlantic white cedar is found on the eastern coast of the United States from Florida to Massachusetts. The eastern red cedar grows in abundance from Colorado to the east coast and from Florida to the Great Lakes. As will be noted by the accompanying table, there is quite a spread in the dimensions of the various types of cedars.

The cedar of Lebanon grows in the mountains of Lebanon about 100 miles (161 kilometers) along the Syrian coast and in Turkey. It is found at altitudes up to 6,000 feet (1830 meters). The tree was first grown in Berkshire (England) in about 1646. The tree still stands. In Herefordshire, a cedar of Lebanon stands 140 feet (42 meters) high and has a circumference of 38 feet (11.4 meters).

Male and female flowers are found on the same tree. The cone may take two years to ripen. It has broad scales, two seeds to each scale. The seed is about the size of a grain of wheat. The bark of the tree is thin and insect-free. The bark has shallow fissures in the pattern of squares. The heartwood is soft and of a warm-brown color. The sapwood is pale. The wood is durable and moderately strong. It is characterized by an attractive wavy pattern in the grain. The wood is in short supply, the only source being the large groves in Lebanon and on the Taurus Mountains in Turkey.

The tree grows slowly, but may attain an age of some 2,000 years.

TABLE 1. RECORD CEDAR TREES IN THE UNITED STATES[1]

Specimen	Circumference (inches)	Circumference (centimeters)	Height (feet)	Height (meters)	Spread (feet)	Spread (meters)	Location
Alaska cedar (1973) (*Chamaecyparis nootkatensis*)	373	947	134	40.2	59	17.7	Washington
Atlantic white cedar (1961) (*Chamaecyparis thyoides*)	186	472	87	26.1	—	—	Alabama
Eastern Redcedar (1970) (*Juniperus virginiana*)	203	516	42	12.6	42	12.6	South Carolina
Incense cedar (1972) (*Libocedrus decurrens*)	462	1173	152	45.6	49	14.7	California
Northern White cedar (1972) (*Thuja occidentalis*)	139	353	113	33.9	43	12.6	Michigan
Port Orford cedar (1972) (*Chamaecyparis lawsoniana*)	451	1145	219	65.7	39	11.7	Oregon
Southern Redcedar (1976) (*Juniperus siliciola*)	178	452	70	21	57	17.1	Florida
Western Redcedar (1977) (*Thuja plicata*)	732	1859	178	53.4	54	16.2	Washington

[1] From the "Social Register of Big Trees," The American Forestry Association (by permission).
[2] At 4.5 feet (1.4 meters).

At one time, grazing cattle destroyed many of the young trees. All large, old trees are now protected by the Lebanese Government. Although some of the cedars of Lebanon have survived from Biblical times, it is generally regarded as one of the least hardy of the cedars. See accompanying figure.

Cedars of Lebanon planted at Geneva, Switzerland in 1735 from seed brought from Lebanon by Bernard de Jusesieu. These trees have long borne seed rapidly, but unevenly. (*Photo by G. Pinchot.*)

The engineering characteristics of several commercial cedar woods are given in Table 2.

For references, see **Tree.**

CEDRELA. Cedar Trees.

CEDRUS. Cedar Trees.

CEFAMANDOLE. Antibiotic.

CEFAZOLIN. Antibiotic.

CEFOXITIN. Antibiotic.

CEILING (Meteorology). **Precipitation and Hydrometeors; Sky Cover.**

CEILING (Performance). Vertical operating limits of an aircraft, rocket, balloon, etc. An airplane flies by virtue of expenditure of a certain amount of power. When the power available from the engine exceeds that required to overcome the air resistance due to motion of the craft, the excess power may be used for increasing the velocity, or for climbing at a constant velocity. However, as the altitude is increased, the performance of the engine may be affected by rarefied air, with the result that the power available has reduced to the point where it is just sufficient to maintain horizontal flight. This altitude is the *absolute ceiling* of the airplane, above which it is not possible to climb. Since theoretically it requires a plane an infinite time to reach absolute ceiling, the *service ceiling* is a more practical measure of performance. This is generally considered to be the altitude at which the rate of climb has diminished to 100 feet per minute. Ceiling performance varies markedly with different types of power plants, of course.

TABLE 2. ENGINEERING DATA ON CEDAR TREES

Common Name for Species	Green Condition			Air-Dried 12% Moisture		Maximum Crushing Strength (Parallel to Grain)				Maximum Tensile Strength (Perpendicular to Grain)			
	Moisture Content (Percent)	Weight/ Cu. Foot (Pounds)	Weight/ Cu. Meter (Kilograms)	Weight/ Cu. Foot (Pounds)	Weight/ Cu. Meter (Kilograms)	Green		Dry		Green		Dry	
						(Psi)	(MPa)	(Psi)	(MPa)	(Psi)	(MPa)	(Psi)	(MPa)
Eastern red cedar	35	37	592	33	529	3570	24.6	6020	41.5	330	2.3	—	—
Incense cedar	108	45	721	26	417	3150	21.7	5200	35.9	280	1.9	270	1.9
Northern white cedar	55	28	449	22	352	1990	13.7	3960	27.3	240	1.7	240	1.7
Port Orford cedar	43	56	897	29	465	3130	21.6	6470	44.6	180	1.2	400	2.8
Western red cedar	37	27	433	23	368	2750	19.0	5020	34.6	230	1.6	220	1.4

SOURCE: U.S. Forest Products Laboratory.

CELERY. Of the family *Umbilliferae* (carrot family), wild celery (*Apium graveolens*) has been known since ancient times. This biennial or annual herb is believed to be native of the Mediterranean area and cultivated there prior to the Christian era. The plant grows wild in the marshes of western Europe and has a rank taste and strong characteristic odor. Through centuries of cultivation, these undesirable characteristics have been eliminated, and the leaf-stalk has been greatly enlarged. The improved, present-day table celery is *Apium graveolens* L. var. *dulce* Pers.

The wild celery plant has numerous leaf stalks, odd-pinnate leaves, and branching, leafy flower stalks, which may achieve a height of 2 to 3 feet (0.6 to 1 meter) or more. The leaf stalks in the cultivated plant are much more solid, less stringy, and possess a much improved flavor and odor.

Occasionally, celery plants develop as annuals and produce seedstalks the first season. The condition is called *premature seeding* or *bolting*. The affected plants are unmarketable unless seedstalk elongation is slow and occurs when plants have almost reached harvest stage. It is a potential problem in any California celery crop, for example, which matures from mid-March through late June. Exposure of plants to relatively low temperatures (about 40 to 55°F; 4.4 to 12.8°C) for as little as 10 days or so near the lower temperature results in bolting when subsequent favorable conditions predominate for the rest of the growing period. Celery varieties differ in their susceptibility to bolting.

Celeriac. (*USDA photo.*)

In the normal development of the cultivated celery plant, a well-developed root system, a short, fleshy crown stem, and a rosette of leaves are produced the first year. The thick, fleshy petioles, or leaf stalks, of the rosette leaves comprise the principal edible portion of the plant. Commercial handlers of celery usually refer to petioles as *ribs*, *shanks*, or *stems*, and to the whole marketable plants as *stalks* or *heads*. During its second year, the main stem elongates and branches to produce a seedstalk and eventually a shrubby plant about 3 feet (0.9 meter) or more in height. The plant bears compound clusters of small white flowers, which produce seeds toward the end of the flowering season.

Celery produces a well-developed root system (consisting of a tap root and laterals) when the crop is grown from seed to market maturity without transplanting. When transplanting is practiced, the tap root is destroyed and the root system is comprised of a large number of lateral fibrous roots growing from the base of the plant. A large part of this system occupies the upper 6 inches (15 centimeters) of soil with many of the roots within 2 to 3 inches (5 to 7.5 centimeters) of the surface. Some roots penetrate to a depth of 2 feet (0.6 meter) or more.

Cultivation of the plant for food was first recorded in France in 1623. By the early part of the 18th century, there had been improvement of the wild type of celery previously transported to Italy, France, and England, and as early as 1726, the plant was being used in England to flavor soups and stews. Celery was first cultivated in Europe for medicinal purposes.

Celery varieties grown in the United States belong either to the golden (or yellow class), or to the green class. Commercial celery production in California and Florida is of the green class. The yellow class is grown in other states and as a garden plant in several areas of the country.

Celeriac. Also known as turnip-rooted celery, this plant has been developed for the root instead of the top. See accompanying illustration. Its culture is the same as that of celery, and the enlarged roots can be used anytime after they become sufficiently large. The late-summer crop of celeriac may be stored for winter use. In areas having mild winters, the roots may be left in the ground and covered with a mulch of several inches (centimeters) of straw or leaves, or they may be lifted, packed in moist sand, and stored in a cool cellar.

More detail on celery can be found in the "Foods and Food Production Encyclopedia," (D. M. Considine, editor), Van Nostrand Reinhold, New York, 1982.

CELERY SEED. Flavorings.

CELESTIAL COORDINATES (Horizontal). Horizontal Coordinate System.

CELESTIAL GUIDANCE. Space Vehicle Guidance and Control.

CELESTIAL HORIZON. Horizon (Celestial).

CELESTIAL MECHANICS. The term celestial mechanics is applied to that field of astronomical study and research which deals with the motions of two or more bodies in space under the influence of their mutual gravitational attractions. The fundamental elements of the subject are found in the Newtonian law of universal gravitation, the laws of motion, and the Keplerian laws of planetary motion. In the classical theory, we find space of three dimensions treated, with time considered as an independent variable. Within recent years, some slight modifications of the classical theory, particularly when the time interval is very long or velocities and accelerations are very high, have become necessary on account of the theory of relativity. Under the general heading of celestial mechanics, we find such problems discussed as the development of the various methods for orbit computation, methods for computing perturbations, and solutions of the three-body problem.

See also **Three-Body Problem (Astronomy)**; and **Two-Body Problem (Astronomy).**

CELESTIAL NAVIGATION. Navigation.

CELESTIAL POSITION ANGLE. **Position Angle (Stellar).**

CELESTIAL SPHERE. From the earliest historical records down through the seventeenth century, in all descriptions of the structure of the universe, the concept persisted of the heavens being a sphere on which all the so-called "fixed stars" were projected. Even in modern times, such a concept is very convenient for discussing the common motions of the stars and the individual motions of the different members of the solar system. This celestial sphere may be defined as a sphere of infinite radius with the center located within the solar system. The reference frames for all systems of astronomical spherical coordinates are established on the celestial sphere.

When projecting the different members of the solar system onto the celestial sphere, it becomes necessary to restrict the location of the center to some particular point within the solar system. If the center of the celestial sphere is considered to be a point on the surface of the earth, we have systems of apparent coordinates; if the center is at the center of the earth, we have geocentric coordinates; if at the center of the sun, heliocentric coordinates; if at the center of Jupiter, Jovicentric coordinates; etc.

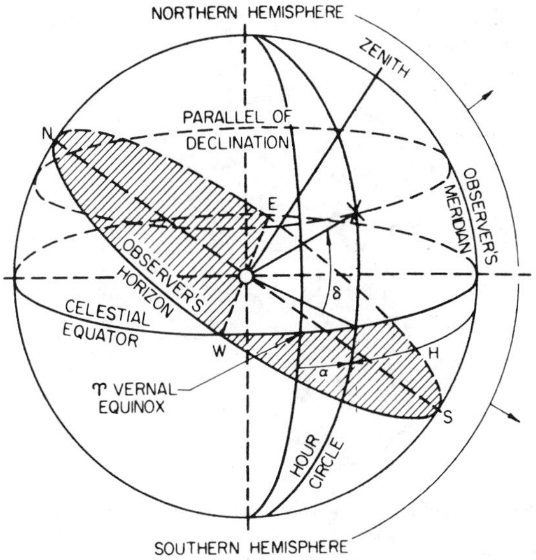

Celestial sphere. Right ascension α is measured (eastward) along celestial equator from vernal equinox to hour circle of X. Declination δ is measured north or south along hour circle from celestial equator to X. Hour angle H is measured (westward) along celestial equator from observer's meridian to hour circle of X.

Due to the fact that the earth is actually rotating about an axis, the celestial sphere is apparently rotating, as seen from the earth, about an axis parallel to the axis of the earth and with the same angular velocity as the earth, but, of course, in the opposite direction. For purposes of convenience, it is customary to refer to this apparent rotation simply as the rotation of the celestial sphere. All systems of spherical coordinates that are established on the sphere, such as the equatorial, galactic, ecliptic, etc., rotate with the sphere, while the horizontal system apparently remains fixed in space. See also **Ecliptic.**

CELESTITE. The mineral celestite (also known as celestine) is composed of strontium sulfate, $SrSO_4$, occasionally with calcium and barium. It crystallizes in the orthorhombic system in tabular or prismatic crystals. More rarely it may be pyramidal or simply fibrous or granular. Two essentially perfect cleavages may be observed, one parallel to the base, the other parallel to the prism. Its fracture is uneven; hardness, 3–3.5; specific gravity, 3.97; luster, vitreous; color, white, but may be slightly reddish or bluish; transparent to translucent.

Celestite may occur with gypsum and salt associated with beds of limestone, or by itself in large, commercially important veins. It sometimes occurs with sulfur in volcanic localities and is often a gangue mineral in veins of galena, sphalerite and similar metallic minerals. In Europe there are many localities for fine crystals, especially in England. In the United States celestite is found in New York, Pennsylvania, West Virginia, Tennessee, Kansas, Colorado, and California. The first celestite described was the delicate blue material from Blair County, Pennsylvania. Its "celestial" tints suggested the name. Celestite resembles barite.

CELL (Biology). All animals and plants are made up of cells. Some lower forms of life, such as bacteria, may be nothing more than a single cell, while larger plants and animals are built up by gluing together many of these cells into a larger mass. The human body itself begins as a single cell. This cell divides to form two cells, and these divide in turn to form four, and so on, until the complete body, consisting of billions of cells, results. As the human embryo forms, like cells organize the body according to the genetic "blueprint" each cell carries within itself. The size of cells varies greatly, but most of them are so small that a million of them would not be much larger than the head of an ordinary pin. Each cell may be thought of having a life of its own. After a human being dies, it may require hours, even days, before all the cells of the body are dead. When a large number of cells with the same special function work together, they are called a *tissue.* A body tissue is made up of billions of cells which look more or less alike and all of which contribute the same general type of special service to the body. The five different tissue types are: (1) *epithelial tissue* (surface of the body and linings of various internal tubes and cavities); (2) *connective or supporting tissue* (bones and cartilage); (3) *muscle tissue*; (4) *nerve tissue*; and (5) the *blood* and *lymph.* When several kinds of tissue are grouped together, they are called an *organ*; and when a number of organs work together as a unit in the body, they are referred to as a *system.*

The major activity of all cells is to transform energy. A typical cell is shown in Fig. 1. Cells gather energy from the breakdown of food at the molecular level. Cells use energy to grow, to eliminate waste material, to reproduce by splitting in two (*mitosis*), to move about the body, and to perform their specialized tasks. From the single-celled fertilized egg comes specialized cells that make up the complex body. As the body grows, cells specialize more and more. The cells have differentiated to a point where, in the adult, certain kinds of cells no longer reproduce. Reproduction is left to the eggs and spermatozoa. See also **Gamete.** The specialized cells perform the work of muscles, arteries, lungs, kidneys, and so on. See also **Carcinogens.**

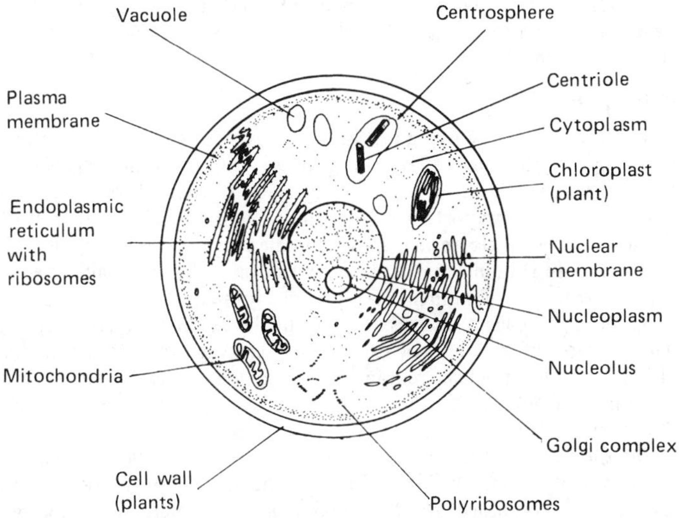

Fig. 1. Schematic of typical cell showing major parts of internal structure. In the cytoplasm, *mitochondria* are organelles in which energy is produced. Ribosomes associated with endoplasmic reticulum or as polyribosomes are active in protein manufacture. *Vacuoles* are storage regions while the *Golgi complex* functions in secretion. The *centrosphere,* containing *centrioles,* functions in normal cell division. Nuclear materials composed of *chromatin* are suspended in the nucleoplasm. Plant cells also possess a rigid cell wall and *chloroplasts* which contain the photosynthetic apparatus.

As indicated by Fig. 1, all cells contain a *nucleus* that is a center for reproduction and carries the *genetic code.* The nucleus contains deoxyribonucleic acid (DNA) which produces ribonucleic acid (RNA). The RNA organizes the essential amino acids into the proteins necessary for life. The nucleus is made up of protoplasm, termed *nucleoplasm.* Outside of the nucleus, the remainder of the cell is composed of large, complex molecules and membrane. It contains protoplasm, termed *cytoplasm.* Within this network are lysosomes, pockets of digestive enzymes that break up big molecules of fat and protein. The food is passed on to the *mitochondria* where further digestion occurs leading to production of energy in the form of ATP. Ribosomes, associated either with the endoplasmic reticulum, or in clusters referred to as polyribosomes, are sites of protein synthesis in the cell cytoplasm. Ribosomes consist of RNA and protein.

Although the cells of the body may resemble each other in some respects, their appearance may vary greatly if they come from parts of the body that perform vastly different tasks. Some nerve cells which transmit messages from one part of the body to another may have specialized projections which are as much as a yard long. They are so fine and threadlike, however, that they are invisible to the unaided eye. Some of the white blood cells behave like independent little animals, frequently leaving the blood and traveling throughout the other areas of the body. These cells are very important in bodily defense against invading microorganisms, especially at sites of inflammation.

In the course of a normal lifetime, the ability of the cells of the tissues (in most cases) to replace themselves as they become worn out continues without interruption. A red blood cell normally survives in the blood stream for about 3 or 4 months before it becomes worn out or is destroyed and must be replaced. Over a period of many years, however, the restorative ability of the body generally falls behind. This is part of the aging process. Occasionally, a small area of the body may lose control over its normally systematic and careful replacement of cells and start making new cells at an uncontrolled and rapid rate. Usually, such an occurrence subsides after a short time and no harm is done, but if this process continues, the result is called a *cancer.*

Cell Division

The division of one living cell into two is one of the most important of biological phenomena. By this process, continuity of a species is ensured and mutation of a species is made possible. Moreover, cell division plays an important role in the growth and differentiation of tissues in embryonic forms, in wound healing, in the formation of tumors, and in the normal replacement of old cells in certain tissues, such as the skin of humans. As previously mentioned, a living cell consists of cytoplasm and a body within the cytoplasm, the nucleus. The division of a cell involves not only cleavage of cytoplasm, but also a complex nuclear reorganization in which replicated genetic material, chromatin, of the mother nucleus is distributed to each daughter nucleus.

Interphase. After cell cleavage and prior to visible nuclear changes of the next division, a cell is said to be in *interphase.* In this condition, the nucleus is bounded by a thin double-layered structure, the nuclear membrane. In the interphase nucleus, chromatin is present in the form of very fine, extended threads, and present also is at least one spherical body known as the nucleolus. See also **Chromatin.** The chromatin and the nucleolus are immersed in a clear, homogeneous liquid, the nuclear sap, which has a viscosity only a few times greater than that of water. Within the cytoplasm of a cell in interphase are a number of components, such as mitochondria, lysosomes, plastids, vacuoles, endoplasmic reticulum with associated ribosomes and Golgi body, and centrioles (see Fig. 2). Centrioles are always present in animal cells, but they have not been detected in the cells of higher plants.

During interphase, a cell prepares for the ensuing division. During this period, replication of the chromatin threads occurs, forming sister threads. Chromatin consists essentially of two substances, a basic protein *histone,* and deoxyribonucleic acid (DNA). It should be pointed out that a nucleic acid is a substance of great molecular weight made up of many units of nucleotides. A nucleotide consists of phosphoric acid, a 5-carbon sugar, and an organic base, either a purine or a pyrimidine. In the case of DNA, the 5-carbon sugar is deoxyribose, the purine bases are adenine and guanine, and the pyrimidine bases are cytosine and thymine.

The synthesis of new DNA within an interphase nucleus can be marked by adding to the environment of a cell a precursor of DNA, thymidine, labeled with radioactive hydrogen (tritium). Radioactive thymidine passes into the cells and can be detected by autoradiography. In a cell not destined to divide, DNA synthesis does not occur, and in a cell in which DNA synthesis does occur, division typically takes place. If DNA synthesis is blocked by treatment of cells with deuterium oxide, division is inhibited; when the block is removed, DNA synthesis proceeds and division occurs.

Toward the end of interphase, the adjacent two pairs of centrioles begin to move in opposite directions. When they come to rest, they will form the poles of a structure known as the spindle. The centrioles are by no means simple structures. From studies with the electron microscope, it is known that a centriole is a cylindrical body made up of parallel, tubule-like structures. Often the centrioles of a pair lie at right angles to each other. Radiating from the region around a centriole pair is a system of fibers, the *aster.* The chemistry of centrioles requires much further investigation. There is some evidence that they contain ribonucleic acid (RNA). It should be pointed out that RNA differs from DNA in that the 5-carbon sugar is ribose rather than deoxyribose and the pyrimidine base, thymine, is replaced by the base, *uracil.*

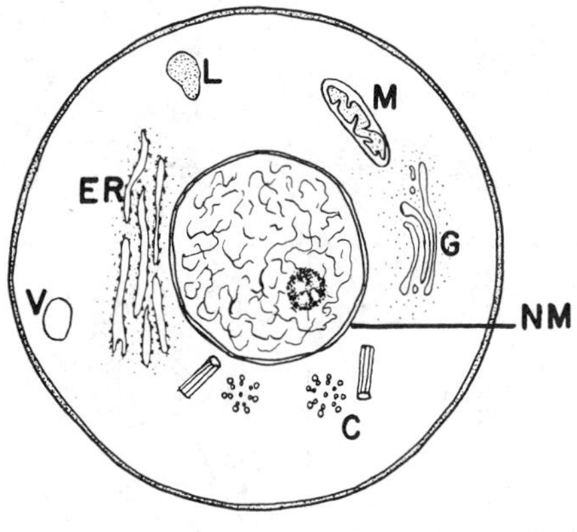

Fig. 2. Structures of interphase cell. C, centriole; ER, endoplasmic reticulum; G, Golgi complex; L, lyosome; M, mitochondrion; NM, nuclear membrane surrounding chromatin threads and nucleolus; V, vacuole.

Prophase. About the time the centrioles begin to move, other events are initiated, including the dissolution of the nuclear membrane, disappearance of the nucleolus, condensation of the chromatin threads by coiling, and spindle formation. This stage of the division process is called *prophase* (see Fig. 3). Presumably, the nuclear membrane is a lipoprotein similar to that of the cell membrane. Dissolution of the nuclear membrane apparently is brought about by a calcium-ion activated proteolytic enzyme.

The nucleolus contains protein and ribonucleic acid. RNA is not synthesized in the nucleolus, but accumulates here after being synthesized by DNA. During prophase, RNA accumulates on the coiling chromatin threads, the chromosomes, and it has been suggested that the accumulation of RNA on the chromosomes is the result of blockage of the transfer of RNA from the chromosomes to the nucleolus. Late in prophase, the nucleolus disappears and its RNA passes into the cytoplasm.

During the prophase, a dual system of fibers forms between the two poles from protein synthesized during interphase. One system, the primary spindle, consists of fibers extending from pole to pole; the other system, the chromosomal spindle, consists of fibers extending from the poles to the equatorial region between the poles.

Metaphase. Later in prophase, the tightly coiled chromosomes begin a movement which will culminate in the alignment of the chromosomes in a narrow band in the equatorial plane of the spindle. When the chromosomes reach this position on the equatorial plate, the cell is said to be in *metaphase.* Each of the sister chromosomes of a pair is connected to a chromosomal spindle fiber of its pole by means of a body on the chromosome called the *kinetochore.* By this time, chromo-

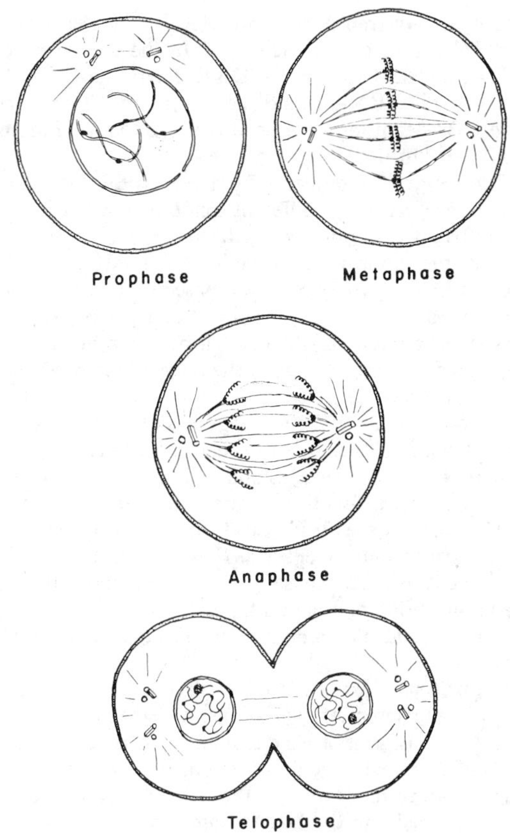

Fig. 3. Stages of cell division.

somes, spindle fibers, centrioles, and asters have become surrounded by a gelatinous matrix. This whole structure, known as the *mitotic apparatus*, differs physically from the remainder of the cytoplasm. For example, the mitotic apparatus can be dislodged from its normal position and moved about the cell by means of ultrasonic vibration applied to the cell surface.

Anaphase. The end of metaphase and the beginning of *anaphase* is marked by the separation of sister chromosomes and the beginning of movement of the sister chromosomes toward opposite poles. During anaphase, RNA is lost from the chromosomes, and toward the end of anaphase, there is an accumulation of ribonucleoprotein in the equatorial region between the two chromosome groups. The source of this equatorial RNA is under investigation. It may represent RNA lost from the chromosomes; it may orginate form other sources.

Telophase. During the final stage of the division process, *telophase*, the cell body divides into two, a process called *cytokinesis*, and each daughter cell completes processes which restore it to an interphase cell. The mitotic apparatus disappears, the chromosomes become attenuated, the centrioles duplicate and split, the nuclear membrane becomes reconstituted, and the nucleolus reappears.

During anaphase, nucleolar material can be detected among the chromosomes, and by the end of telophase, this material has been accumulated in the body known as the nucleolus. Under investigation is what fraction of RNA in the telophase nucleolus represents RNA lost to the cytoplasm from the prophase nucleolus. At least part of the RNA in the telophase nucleolus would appear to be newly synthesized.

Although the centriole has the usual capability of reproducing itself, much remains to be found concerning its chemical makeup. Some evidence indicates that centrioles contain RNA.

Various schemes have been proposed to explain the furrowing or cleavage of the cell body. According to one hypothesis, cleavage is brought about by synthesis of new material in the region of the furrow. Another hypothesis proposes that there is a contraction or constriction of cytoplasmic substance in the equatorial region of the cell. The "expanding membrane" hypothesis maintains that furrowing is caused by the expansion of folded protein molecules in the cortical region of the cell.

The events occurring at the furrowing cell surface and those preceding them in the interior of the cell are related. Furrowing always occurs in the equatorial plane between the poles of the spindle. Presumably the interior-to-surface messenger is a chemical substance(s), but the nature and source remain to be determined.

Although the division process requires energy, the active phases of the mitotic cycles are not marked by great metabolic or respiratory activity. On the contrary, the period between active phases, interphase, is the time of high metabolic and respiratory activity. Thus, many investigators believe that the energy required for division is obtained from a "reservoir" prepared during interphase. The nature of the "reservoir" remains to be studied further. Conceivably, a high-energy compound such as adenosine triphosphate may be involved, but evidence for this is not strong.

As a result of cell division, each daughter cell has a full complement of chromosomes (diploid) just as the mother cell. In reproductive organs during the production of germ cells (sperm and eggs of animals and spores of plants), the cells undergo two divisions, called meiotic divisions, which result in gametes each having only half the chromosome complement (haploid) of a somatic cell.

Chromosomes. Each plant or animal species possesses a characteristic although somewhat variable chromosome number which can easily be visualized in the form of a *karyotype*, the number and form of chromosomes present at metaphase. Originally, chromosome was a term used by microscopists to describe the type of cellular organelles which could be observed microscopically during cell division processes and which had strong affinity toward basic dyes. As genetic studies advanced, it was recognized that these cellular components represented the vehicles which carried the hereditary determinants (*genes*). In more recent years, the term chromosomes has been used to describe the "gene carrier" of any cell, whether it is microscopically visible or not. Thus, the term has lost its morphological connotation, because the chromosomes of most microorganisms, such as bacteria, and viruses, are not always detectable by standard microscopes. The chemical composition of these two types of chromosomes is also quite different. In the following description, chromosome refers to the classic definition, namely, the chromosome of higher organisms.

Investigations into the chemistry of chromosomes have been impeded by (1) a chromosome is not composed of a single type of compound, but is a composite of several species of complex macromolecules, whose interrelationships have not been well established; (2) no satisfactory method is readily available to isolate chromosomes in quantities to facilitate chemical analysis; and (3) chromosomes can be observed microscopically only when a cell enters division stages (mitosis or meiosis), which represent a small fraction of the life span of a cell. During the rest of the time, the cell is in interphase when the chromosomes are invisible. The chromosomes are "decondensed" and are enclosed in a nuclear envelope. Isolation of chemical constituents, therefore, can be done only from isolated nuclei. Further, it is not fully established that the deoxyribonucleoprotein so isolated equates to the chromosomes.

Some information on the chemical composition of chromosomes has been obtained by cytochemical analysis. The Feulgen reaction has demonstrated that DNA is the major component of chromosomes. Alkaline fast green staining shows that there is also a basic protein component (histone) is the chromosomes. In some viruses, the genetic determinants are in the form of RNA rather than DNA. However, DNA appears to be the hereditary material of the overwhelming majority of life forms, including numerous viruses and bacteria, protozoa, and all higher plants and animals.

In bacteria, and viruses, each "organism" possesses one chromosome, and each chromosome is a single, circular molecule. There is no evidence of proteins closely associated with DNA, such as nucleoproteins in the case of higher life forms. A possibility exists, however, that amino acids or small peptides may be present to interrupt the continuity of the long DNA molecule. Bendich and collaborators have postulated that amino acids may serve as punctuation points for the genetic messages. In the chromosomes of higher forms, more than one DNA molecule per chromosome is probable.

Replication and RNA Synthesis. DNA molecules have two functions: (1) replicating themselves; and (2) providing a template for RNA synthesis. Ample evidence has been accumulated to show that DNA replicates itself in a semiconservative manner. That is to say, each of the complementary strands of the double-helical structure synthesizes its complementary strand, resulting in two double helices each containing one old strand and one new strand. Apparently, the chromosomes replicate the same way.

The bacterial chromosome replicates itself from a predetermined point and proceeds around the circular chromosome with an apparent constant rate until the entire ring is duplicated. At that time, the two rings separate and each may begin another generation of replication. In higher plants and animals, the chromosomal replication may begin at multiple sites and is asynchronous among different chromosomes of the same cell. The asynchrony is not lost even after an artificial arrest of the DNA synthesis by analogues or inhibitors. Under the influence of arresting agents, the DNA synthesis process ceases. When thymidine is later introduced to the arrested cells, DNA replication resumes at the point where it was stopped. The asynchronous pattern is not altered. This asynchrony of chromosome replication has much genetic significance.

As important as self-replication, the second function of DNA is to serve as template for RNA synthesis. The biosynthesized messenger, mRNA, copies the base sequence of the DNA so that it transcribes the information carried in the original genetic code. The function of mRNA is to carry the genetic message from DNA to the cytoplasm for translation of the message into protein. In *eukaryotic cells*, each mRNA strand codes for one protein, whereas in some bacteria and viruses, mRNA forms polycystronic strands which code for more than one protein. Chromosomes synthesize DNA only when they are not condensed, i.e., when they are in the interphase stage.

Control of Genetic Expression. All cells of an organism find their lineage from a single fertilized cell, the *zygote*. The original zygote thus contains all the genetic information required for the development and maintenance of this organism. A great deal of genetic information for a specific function necessary for one type of cell will be useless in other cells. In other words, not all the DNA molecules should actively synthesize RNA at one time. In fact, a specialized cell should have many inactivated DNAs whose information belongs to other cell types. Otherwise development cannot proceed in an orderly manner. Thus, some mechanisms must be operative to control the RNA synthetic activity of the chromosomes.

In the chromosomes of higher forms, a considerable quantity of protein is present. One group of proteins is basic, known as the histones. Histones are basic because of their high content of arginine and lysine. It has been suggested (Stedman and Stedman) that histones may be the agents which regulate the activity of the genes. It also has been found (Huang and Bonner) that histones inhibit the *in vitro* synthesis of DNA-dependent RNA. Other studies indicate that various fractions of histones differ in their affinity to DNA.

Genes. These may be defined as segments of genetic material which determine the sequence of amino acids in specific polypeptides, such that there is a one-to-one relation between gene and polypeptide. This definition applies at least to those genes called *structural* genes because they determine the primary structure of proteins. More generally, genes are the physical units of heredity. Structural genes in all organisms appear to be composed of nucleic acids. In the RNA viruses, the genes are RNA only, but in all other organisms, the DNA viruses and the cellular forms which all possess both DNA and RNA, the gene material is either known to be DNA or assumed to be DNA.

The genes of viruses and bacteria appear to consist of nucleic acid unaccompanied by closely-bound protein. Ordinarily, this naked nucleic acid is in the two-stranded condition; exceptions are known among both the RNA and DNA viruses, some of which possess single-stranded genetic material. In those organisms with true nuclei, the genetic material is always double-stranded DNA associated with protein ordinarily of the histone type. The function of the protein is not considered to be genetic. It probably controls DNA in its role of determining protein structure. Also, it may serve to hold genes together and attached to the chromosomes of which they are a part.

Structural genes carry out their role of dictating protein structure by producing a messenger RNA (mRNA) which is a single strand of RNA containing nucleotide bases complementary to one of the strands of the double-stranded DNA of the gene from which it is copied or "transcribed." The evidence is that the same DNA strand of a gene is always transcribed into mRNA. In this way, only one kind of mRNA is made for each gene. In the transcription process, the C, T, A, and G bases of the DNA determine G, A, U, and C, respectively, in the mRNA strand. Transcription effectively constitutes *gene action*. By definition, if a gene is not actively forming mRNA, it is inactive or "turned off."

Each kind of gene is different from every other gene in its DNA sequence. Hence, as many different kinds of mRNA are formed as there are different genes in the organism.

After their formation to the gene level, the mRNA strands attach to ribosomes in the cytoplasm. The process of protein biosynthesis then commences. The significant point is that the sequence of nucleotide bases (the "genetic code") in a particular gene is reflected in a specific sequence of amino acids in the polypeptide produced through the protein synthetic mechanism.

The one-to-one relation between gene and polypeptide is a more accurate statement of the situation than the earlier *one gene–one enzyme* hypothesis. It has been established that a number of proteins are constituted in their functional state of subunits which are polypeptides. When subunits are all identical, the *one gene–one protein* statement holds with certain exceptions. However, proteins such as vertebrate lactic acid dehydrogenase (LDH) and hemoglobin are known to be made up of different subunits.

The term *cistron* also is sometimes used as the name for a structural gene.

Mutation of Genes. Mutation takes place in all types of organisms and is the origin of hereditary variations. A change in the base sequence of the DNA constituting a gene results in an inherited alteration in the code and is called a gene mutation. Changes in base sequence may conceivably result from: (1) the deletion or addition of one or more nucleotide pairs in the DNA chain; (2) changes in one or more bases along the chain; or (3) inversion of a segment of the chain. When the mutation involves the chromosome structure, it is considered a chromosomal mutation or aberration. Somatic mutations are not transmitted from generation to generation, but may produce severe changes in the organism depending upon the type of cell affected and the time at which mutation occurs, i.e., during embryonic development.

Active Transport in Cells

Common to certain of the organelles and to the cell itself are structures referred to as *unit membranes*. Cellular membranes are generally composed of lipid and protein molecules, spatially oriented so that the inner part of the membrane is an area of interdigitated phospholipid and cholesterol ester molecules and the inside and outside coatings of the membrane are largely protein. According to the generally accepted fluid mosaic model, the lipid layers of the membrane are fluid in which (and sometimes through which) float protein molecules. These proteins serve many functions, but perhaps most important of these are their function as cell surface receptors for such various agents as hormones, viruses, and antibodies. The thickness of biomembranes varies, but many membranes appear microscopically as two dark lines of approximately 10 micrometers thickness, separated by a lighter band of about 50 micrometers. Myelin, the covering of nerve axons, is a repeating structure of lipid-protein layers of 140–170 micrometers. The molecules making up these thin structures serve to isolate the cell contents from the environment. They do not fit tightly together as a solid wall or surface, but have areas in which there are pores. These pores or spaces through biomembranes are limited in size (0.7 nanometer) and serve to effectively prevent the passage of a variety of large molecules, thus giving to most biomembranes the property of semipermeability.

Most cells contain considerably more protein than is present in the fluids bathing the cells. The presence of a high concentration of cellular protein, together with the high concentration of salts associated with the charged protein structure, causes an osmotic gradient to exist and results in the flow of solvent into the cell in an attempt to compensate for the osmotic pressure difference. Plant cell walls have rigid structural features that prevent cell wall rupture and cell death when osmotic or hydrostatic pressures are imposed. Animal cells, on the other hand, lack supportive wall structures and must depend upon other mechanisms to restore water and solute balance (and thus osmotic balance). The cell membrane then becomes a dynamic focal point of fluid and solute flow rather than just a static barrier unassociated with the life process.

Diffusion is that motion which is imparted to solutes by the random molecular movements of materials in solution. The diffusion movement of solute is increased with increasing temperature and is directly dependent upon its concentration. In dilute solutions, the diffusion of one species or particle is independent of the diffusion of another species, provided there is not interaction between the species. For a small solute molecule (or the solvent itself), the movement through the small distance involved in the thickness of biomembranes may be of a similar magnitude to the movement in free solution. Water moves rather quickly across many cell membranes. A solute may cross a cell membrane at a rate greater than it would by simple diffusion in water if, in the process of flowing rapidly through the membrane, water "drags" soluble with it. This process is known as "solvent drag."

The movement of certain molecules through cell membranes may be restricted or aided because of the lipid layers in the membrane. If a material is not small enough to diffuse through the solvent phase of the membrane, or is hydrophobic in nature, then it may still pass through the membrane phase, later leaving this phase and entering the cell. Since diffusion is related directly to concentration, high solubility of a solute in the membrane should lead to a high probability of its crossing the membrane.

Charged solutes (cations and anions) may be subjected to additional forces in their movements. If one side of the membrane, i.e., the outside of the cell, has a positive charge on it and the inside of the cell has a negative charge, a potential difference exists. Thus, a charged solute in the vicinity of the inner or outer environment of the cell will move with greater or less speed, depending upon the nature of its own charge. An area of opposite charge then can be an attracting force and cause the solute to move at a speed greater than that expected by simple diffusion. Conversely, the electrical field may serve as a barrier to an ion of the same charge as the field.

The net movement of material across a biomembrane against a concentration gradient at a rate greater than that predicted by simple diffusion or by electrical gradients is considered to be *facilitated diffusion* or *active transport.* Facilitated diffusion does not directly involve cell metabolism and does not require energy. Active transport is metabolically coupled and requires expenditure of energy in the form of ATP. Both types of processes appear to involve membrane protein carrier molecules. The descriptive term "uphill transport" has been used synonymously with active transport.

Cation Transport. Plasma membranes have in common the ability to transport alkali metal ions in the face of osmotic, electrical, and concentration gradients. By means of mechanisms not fully explained, cells may maintain a high internal concentration of K^+ and a low concentration of Na^+, while existing in an environment that has a low K^+ and a high Na^+ concentration. Research with radioisotopes has shown that this unlikely distribution is not due to a lack of movement or passage across the cell membranes, but rather it is due to selective processes that extruded ("pumped") ions from an area of low to an area of high concentration. Data supports the concept that the driving force of the "pump" is closely linked to reactions involving the use of ATP (adenosine triphosphate). This prime energy source is supplied by the catabolism of foodstuffs and the union of food hydrogen with environmental oxygen.

Sugar Transport. A variety of mechanisms operate for the movement of sugar into cells. Sugars may enter by simple diffusion, but this is a slow process without great structural specificity. This movement is always away from the region of highest concentration. Transport from a high to a low concentration is often called "downhill" transport. In the case of many cells, the movement of certain sugars appears to be much more rapid than expected from simple diffusion. In those instances where there is rapid movement, but where the movement is still "downhill," the transport is called *facilitated transport.* Facilitation of solute movement is presumed to result from the interaction of the sugar with a carrier substance in the membrane. The complex moves across the membrane, the sugar is discharged on the far side, and the carrier returns for another cycle. In the case of glucose uptake by human red cells, the exchange is up to 100 times faster than that predicted by simple diffusion. Although this facilitated transport shows specificity for structures and demonstrates saturation phenomena, the process does not seem to directly require metabolic energy.

This entry reviewed and updated for 6th edition by Ann C. Vickery, Ph.D., University of South Florida, College of Medicine, Tampa, Florida.

References

Albertsson, P. A.: "Partition of Cell Particles and Macromolecules," 2nd edition, Wiley, New York, 1972.

Bell, T. I.: "Models for the Specific Adhesion of Cells to Cells," *Science,* **200**, 618–627 (1978).

Busch, H.: "The Cell Nucleus," Academic, New York, 1978.

Change, T. M. S.: "Artificial Cells," Charles C. Thomas, Springfield, Illinois, 1979.

Cheung, W. Y.: "Calmodulin Plays a Pivotal Role in Cellular Regulation," *Science,* **207**, 19–27 (1980).

Clark, R. L., and T. L. Steck: "Morphogenesis in Dictyostelium: An Orbital Hypothesis," *Science,* **204**, 1163–1168 (1979).

DeRobertis, E. D. P., and E. M. F. De Robertis, Jr.: "Cell and Molecular Biology," 7th edition, Saunders, Philadelphia, 1980.

Dustin, P.: "Microtubules," *Sci. Amer.,* **243**, 2, 66–76 (1980).

Ebert, J. D., and T. S. Okada (editors): "Mechanisms of Cell Change," Wiley, New York, 1979.

Fox, G. E., et al.: "The Phylogeny of Prokaryotes," *Science,* **209**, 457–463 (1980).

Friedlander, M. J., Lin, C. S., and S. M. Sherman: "Structure of Physiologically Identified X and Y Cells in the Cat's Lateral Geniculate Nucleus," *Science,* **204**, 1114–1117 (1979).

Giese, A. C.: "Cell Physiology," 5th edition, Saunders, Philadelphia, 1979.

Goldman, R., Pollard, T., and J. Rosenbaum (editors): "Cell Motility," Cold Spring Harbor Laboratory, Cold Spring Harbor, New York, 1976.

Gore, R.: "The Awesome Worlds Within a Cell," *National Geographic,* **150**, 3, 355–395 (1976).

Hall, B. K.: "Developmental and Cellular Skeletal Biology," Academic, New York, 1978.

Harold, F. M: "The 1978 Nobel Prize in Chemistry," *Science,* **202**, 1174–1176 (1978).

Horan, P. K., L. L. Wheeless, Jr.: "Quantitative Single Cell Analysis and Sorting," *Science,* **198**, 149–157 (1977).

Jaeger, E. C.: "A Source-Book of Biological Names and Terms," 3rd edition, Charles C. Thomas, Springfield, Illinois, 1978.

Lazarides, E., and J. P. Revel: "The Molecular Basis of Cell Movement," *Sci. Amer.,* **240**, 5, 100–112 (1979).

Lehninger, A. L.: "Biochemistry," 3rd edition, Worth, New York, 1979.

Ling, G. N., and C. L. Walton: "What Retains Water in Living Cells?" *Science,* **191**, 293–295 (1976).

Lodish, H. F., and J. E. Rothman: "The Assembly of Cell Membranes," *Sci. Amer.,* **240**, 1, 48–63 (1979).

Margulis, L., To, L., and D. Chase: "Microtubules in Prokaryotes," *Science,* **200**, 1118–1124 (1978).

Marx, J. L.: "Cell Biology: Cell Surfaces and the Regulation of Mitosis," *Science,* **192**, 455–457 (1976).

Marx, J. L.: "Newly Made Proteins Zip through the Cell," *Science,* **207**, 164–166 (1980).

Miller, J. A.: "Puzzling Out the Cell's Power Plant," *Science News,* **116**, 184–185 (1979).

Pepe, F. A., Sanger, J. W., and V. T. Nachmias: "Motility in Cell Function," Academic, New York, 1979.

Rothman, J. E., and J. Lenard: "Membrane Asymmetry," *Science,* **195**, 743–753 (1977).

Sloat, B. F., and J. R. Pringle: "A Mutant of Yeast Defective in Cellular Morphogenesis," *Science,* **200**, 1171–1173 (1978).

Staff: "The Nobel Prizes (1978)," in "Science and the Citizen," *Sci. Amer.,* **239**, 6, 80 (1978).

Yen, A., and A. B. Pardee: "Role of Nuclear Size in Cell Growth Initiation," *Science,* **204**, 1315–1317 (1979).

CELL BURNER. **Burner.**

CELL-MEDIATED IMMUNITY. **Immune System and Immunology.**

CELLOBIOSE. **Acetate Fibers.**

CELLS (Morphogenic Induction). **Morphogenic Induction.**

CELLULITIS. An infection of the superficial fascia. See **Fascia.**

CELLULOSE. The formula for cellulose is sometimes given as $(C_6H_{10}O_5)_n$. This is an oversimplification inasmuch as the cellulose present in natural substances, such as wood and cotton fibers, usually is combined with other constituents, such as fats and gums. Cellulose is found almost exclusively in plants and accounts for about 30% of all vegetable matter. Cellulose is the principal substance of which the walls of vegetable cells are constructed. The term *cellulose* is derived from the Latin *cellula*, meaning little cell. Relatively pure cellulose can be obtained from cotton fibers (90% cellulose) and flax fibers. Very small amounts of cellulose are found in insects and none in animal tissues. Digestive juices and enzymes present in animal systems do not appear to attack cellulose and thus ingestion by humans is relatively limited. By means of other biological processes, such as amoeboid protozoa present in the digestive tract, herbivora and insects digest and absorb some cellulose.

Cellulose is a polysaccharide of glucose. See Fig. 1. Cellulose is a white solid, odorless and tasteless, insoluble in cold or hot water, and chemically nonreactive except when treated with strongly corrosive materials. If heated with water at 260°C and under rather high pressure, however, cellulose dissolves, but with decomposition. Concentrated sulfuric acid dissolves cellulose, the solution upon dilution and boiling yielding glucose. When treated with sodium hydroxide (15 to 25% NaOH) cellulose fibers swell up and upon washing and drying possess a lustrous appearance. This is the mechanism of *mercerization.* With iodine in potassium iodide solution plus zinc chloride (Schulze's solution), cellulose produces a dark blue color. When treated with an 80% sulfuric acid solution and rapidly washed and dried, cellulose yields a parchment-like surface.

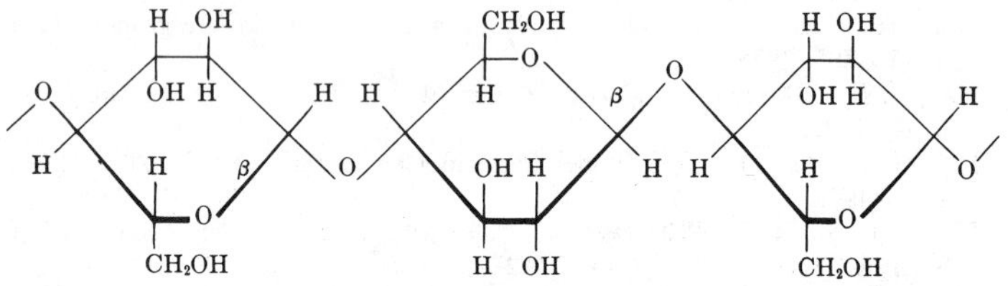

Fig. 1. A segment of the cellulose molecule.

In biochemical terms, cellulose is the name given both to a specific polysaccharide, consisting of β-D-glucose residues joined end-to-end by linkage through —O— of C_1 of one residue to C_4 of the next (see Fig. 2); and to a resistant family of polysaccharides (containing cellulose in the strict sense) isolated from plants by specific chemical treatment. Cellulose appears to be always associated in plants with other polysaccharides and polysaccharide derivatives, such as mannan, xylan, araban, galactan, polygalacturonic acid, and in woody plants, with lignin. Except in many of the fungi, and in a few seaweeds, it forms the skeletal polysaccharide walls of plant cells.

The mechanism of cellulose synthesis and microfibril orientation is not fully understood. Synthesis probably occurs from a glucose/phosphate precursor which may be guanosine diphosphate glucose. The enzyme system involved can be extracted from the plant (e.g., from *Acetobacter xylinum*) and synthesis by cell-free extracts has been achieved. The synthetic mechanism in plants higher than bacteria is thought to be located on the cell surface, and both granular aggregates and microtubules seen in the electron microscope are considered as possible sites.

The "brown rots" (e.g., *Coniphora casebella, Poria monticola*) and the "white rots" (e.g., *Polystictus versicola*) are both basidiomycetes. Both attack cellulose, but only the latter takes lignin to any large extent. "Soft rot fungi," recognized relatively recently as important in this regard are members of the *Ascomycetes* and *Fungi imperfecti* (e.g., *Chaetomium globosum*). They all attack the cellulose of wood, producing characteristic angular cavities. The evidence is that all of these fungi attack the paracrystalline component of cellulose more rapidly than the crystalline component.

Compound celluloses are widely distributed in plants, the two principal types being:

(a) Lignocelluloses, of woods, cereal straws, jute. These cellulose materials yield lignin by treatment (1) with 43% hydrochloric acid, cold for 12 hours, (2) with 8 to 12% sodium hydroxide at 140 to 160°C for 6 to 10 hours, (3) with 72% sulfuric acid at ordinary temperature for 18 hours (the common method). Wood yields about 25% of lignin by the last treatment.

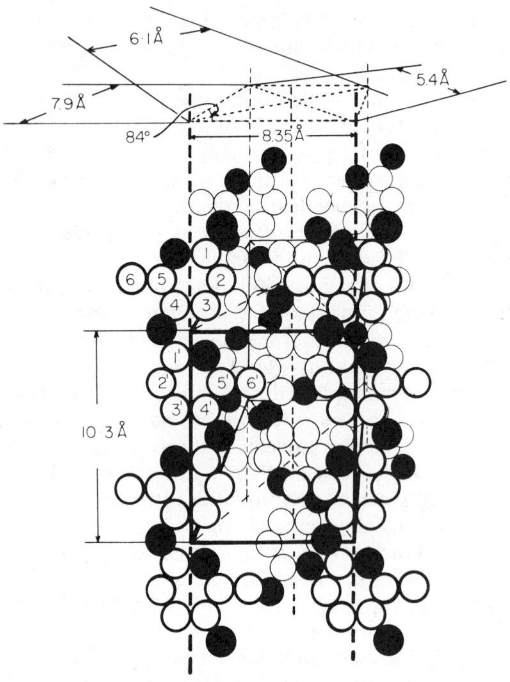

Fig. 2. Diagrammatic representation of the unit cell of cellulose. The monoclinic cell, dimensions 10.3 Å × 8.35 Å = 84°, is delineated by solid lines with one cellulose chain at each vertical edge and one (antiparallel) in the center. Open circles = carbon atoms; solid circles = oxygen atoms. For clearness of the diagram, the hydroxyl groups on carbons 2, 3, and 6 are omitted; and hydrogen atoms are omitted. Two spacing at 6.1 Å and 5.4 Å are included, inasmuch as these are strongly represented in the x-ray diffraction diagram. ("*The Encyclopedia of Biochemistry,*" *Van Nostrand Reinhold Co.*)

(b) Pectocelluloses, of flax, hemp, ramine. These cellulose materials yield pectic substances by treatment with oxalic acid or ammonium oxalate at 85°C for 24 hours, followed by carefully defined treatment with alcohol, acetic acid and calcium chloride. Pectic substances are most abundant in leaves, e.g., ivy, sycamore, and in apples or oranges, especially the white peel of the latter.

Cellulose dissolves in Schweitzer's reagent, an ammoniacal solution of cupric oxide. After treatment with an alkali, the addition of carbon disulfide causes formation of sodium xanthate, a process which is used in the production of rayon. See also **Fibers (Textile).** The action of acetic anhydride in the presence of sulfuric acid produces cellulose acetates, the basis for a line of synthetic materials. See also **Acetate Fibers; Cellulose Ester Plastics (Organic).** Nitrocelluloses are produced by the action of nitric acid and sulfuric acid on cellulose, yielding compounds which are highly flammable and explosive. See also **Explosive.**

On a heavy tonnage basis, cellulose is important as a raw material in the production of wood pulp and paper. See also **Paper;** and **Pulp (Wood) Production and Processing.**

Dietary aspects of cellulose are given in the entries on **Digestive System (Human); Digestive System (Ruminants); Feedstuffs;** and **Fibers (Dietary).**

CELLULOSE ESTER PLASTICS (Organic). The cellulosics are unique among the plastics in that the basic materials used in their manufacture are not synthetic polymers. Rather, they are derivatives of a natural polymer, *cellulose.* See also **Cellulose.** The preparation of an organic cellulose ester plastic involves the formation of a suitable cellulose derivative, followed by processing steps that convert the cellulose derivative into a plastic.

Cellulose, with its many hydroxyl groups, can react with organic reagents such as acids, anhydrides, and acid chlorides to form organic esters. The first reported organic ester of cellulose was cellulose acetate, prepared by Schützenberger in 1865 by heating cotton and acetic anhydride to about 180°C in a sealed tube until the cotton dissolved.

Franchimont, in 1879, accomplished this reaction at a lower temperature with the aid of sulfuric acid as a catalyst. The product in both cases was very nearly the triester. Miles, in 1903, first described partially hydrolyzed (generally called "secondary") cellulose acetate and distinguished it from the triacetate by its acetone solubility. The solubility of secondary cellulose acetate in such inexpensive and relatively nontoxic solvents as acetone contributed greatly to the development and commercialization of this material.

Cellulose esters of the 2-, 3-, and 4-carbon acids are readily prepared by the cellulose-anhydride reaction; the acetate ester and the mixed acetate butyrate and acetate propionate esters are manufactured and used in large amounts. Esters of higher acids require different synthesis techniques and tend to be prohibitively expensive except as specialty products. Some are in commercial production, however. Cellulose acetate phthalate, for example, is manufactured for use as an enteric coating on pills.

Most commercial preparations of cellulose esters still follow, basically, the methods described by Franchimont and Miles—esterification with sulfuric acid catalyst followed by hydrolysis. The principal steps in this process are shown in Fig. 1.

Esterification. The nature of cellulose is such that its esterification does not occur randomly. Even when the DS (degree of substitution—the average number of hydroxyl groups replaced per anhydroglucose unit in the cellulose chain) is approaching 3 (complete reaction), many anhydroglucose units have a DS of zero. If the ester is recovered before the reaction is complete, the product will not be homogeneous and it will be hazy. Regardless of the desired DS of the final product, therefore, the reaction must be allowed to proceed to virtual completion if a homogeneous material is to be produced. In most processes, the ester dissolves in the reaction mixture as the reaction approaches completion.

The cellulose used to manufacture cellulose esters is highly purified

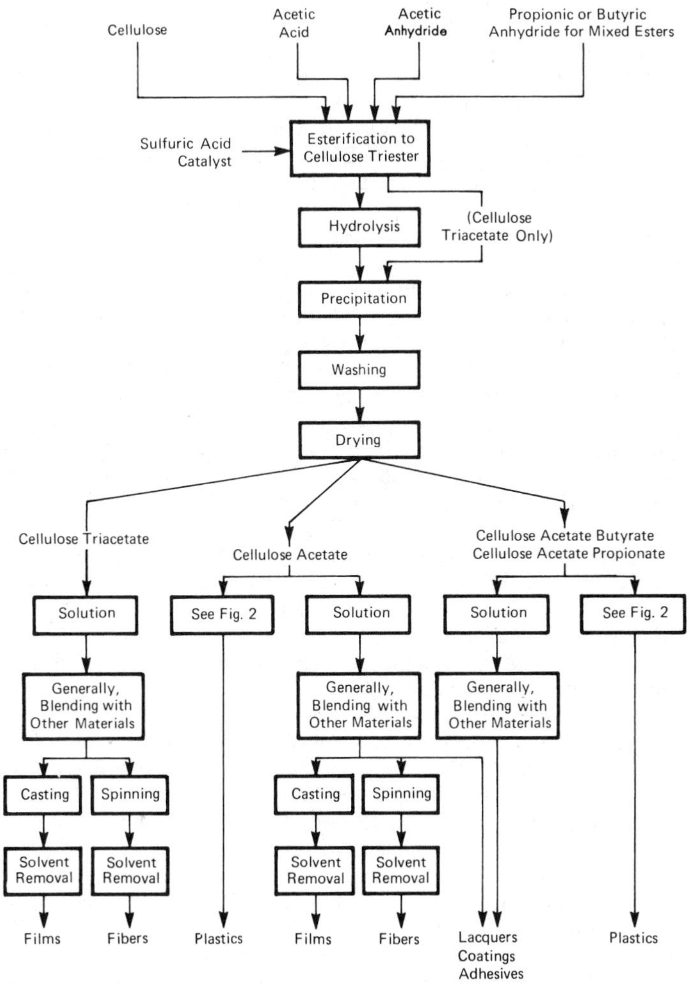

Fig. 1. Production and end-uses of organic cellulose esters.

cotton linters or wood pulp. It is generally treated to reduce its crystallinity and make it more reactive, then agitated at somewhat elevated temperatures with the appropriate acids, anhydrides, and catalyst until it dissolves. Some of the polymeric chains of the cellulose are broken during the reaction, and consequently the molecular weight decreases; thus, the catalyst concentration, reaction temperature, and reaction time must be controlled very carefully to give a product of the desired molecular weight.

Some acetate ester is recovered at the completion of reaction and marketed as commercial cellulose triacetate; it has a DS very close to 3. Other triesters have found no commercial applications.

Hydrolysis. The compatibility of a cellulose ester with other materials is influenced by the acid or acids used in its preparation, but it is controlled primarily by the hydroxyl content of the ester, which varies reciprocally with the degree of substitution. Cellulose triacetate, for example, does not dissolve in acetone; its solution requires very strong solvents such as methylene chloride. Plasticizers can be added to the triacetate in solution and will remain in the ester if the solvent is removed. Plasticizer added in this manner will affect some of the properties of the triester, but they will not appreciably reduce its softening temperature, which is higher than its decomposition temperature. The hydroxyl groups in secondary cellulose acetate, however, have for other polar materials an affinity that is lacking in the triacetate. The ester will dissolve in acetone, and plasticizers will both affect its mechanical properties and reduce its softening temperature sufficiently for the mixture to be processed as a thermoplastic. It is necessary, then, that cellulose esters to be used for plastics contain a significant number of hydroxyl groups, and these are produced by partial hydrolysis of the tiester formed in the reaction step. Since the triester is in solution, hydrolysis is random and produces a homogeneous product.

Hydrolysis is initiated by the addition of water and stopped at the desired point by neutralization of the catalyst.

Precipitation, Washing, and Drying. The cellulose ester in solution in the reaction mixture is precipitated by the addition of water. Some precipitation processes produce a flake precipitate; some produce a powder. The precipitate is removed from the slurry, washed with water until it is free of acids, and dried.

Cellulose Esters

The cellulose esters that result from the process described are chemical raw materials and are used by many branches of the chemical industry. They are generally characterized by acyl content in weight percent and viscosity in seconds, the viscosity being obtained by timing the fall of a steel ball through a solution of the cellulose ester in accordance with ASTM Method D 1343. Low-viscosity esters are generally used in solution processes; high-viscosity esters are used in the production of plastics.

Cellulose Triacetate. It has already been implied that cellulose triacetate will not produce a thermoplastic, as its softening point cannot be reduced appreciably by plasticizers. It is used in solution processes, however, to produce films and fibers. Triacetate films absorb less water than films of secondary cellulose acetate, and they are therefore more dimensionally stable in environments where the humidity is not controlled. Triacetate fibers, with a similar resistance to water, impart to fabrics wrinkle resistance, dimensional stability, and the ability to dry rapidly. Under United States federal regulations, a fiber must be made from a cellulose acetate having at least 92% of its hydroxyl groups acetylated if it is to be called "triacetate."

Cellulose Acetate. Like the triacetate, secondary cellulose acetate (CA) is used in solution processses to produce fibers and films. CA fibers were originally called "rayon," the name that was already in use for regenerated cellulose fibers. In 1951, however, the regulatory authorities formally acknowledged the chemical distinction between CA and cellulose, and the term "rayon" was reserved for fibers of regenerated cellulose. CA fibers are officially called "acetate," and they are used in a wide variety of fabrics. They also are used for cigarette filters. However, the majority of CA produced is used for manufacture of plastics.

Cellulose Acetate Butyrate and Cellulose Acetate Propionate. These two cellulose esters are somewhat similar in properties and applications. Cellulose acetate butyrate is commonly referred to in the chemical industry as CAB, while cellulose acetate propionate is simply termed "cellulose propionate" and referred to as CAP or as CP.

The major usage of CAB and CP is in plastics. Additionally, CAB and CP are mixed with a variety of synthetic polymers to produce lacquers for wood, metal, and plastics. CAB finds relatively small usage in hot-metal coatings and in optical-grade cast film used in the manufacture of sunglass lenses.

Organic Cellulose Ester Plastics

Although the first cellulose plastic (cellulose nitrate plastic—based on an *inorganic* ester of cellulose) was developed in 1865, the first *organic* cellulose ester plastic was not offered commercially until 1927. In that year, cellulose acetate plastic became available as sheets, rods, and tubes. Two years later, in 1929, it was offered in the form of granules for molding. It was the first thermoplastic sufficiently stable to be melted without excessive decomposition, and it was the first thermoplastic to be injection molded. Cellulose acetate butyrate plastic became a commercial product in 1938 and cellulose propionate plastic followed in 1945. The latter material was withdrawn after a short time because of manufacturing difficulties, but it reappeared and became firmly established in 1955.

Since the cellulose esters CA, CAB, and CP are chemical raw materials, the word "plastic" was used in the preceding paragraph to differentiate the product from the raw material. The cellulose ester plastics are commonly called simply "acetate," "butyrate," and "propionate," and these names will be used in the text that follows.

Commercial Production. In the manufacture of cellulose ester plastic, the appropriate ester is blended with plasticizer and other additives, such as stabilizers, ultraviolet inhibitors, dyes, and pigments, commonly in a large sigma-blade mixer. The mixture thus obtained is heated to its softening temperature and kneaded until it is homogeneous. This is done on hot milling rolls, in a compounding extruder, or in a Banbury mixer. The molten mass of plastic that results is formed into small rods or strips that are then cut into cylindrical or cubical pellets, which ordinarily have dimensions of about $\frac{1}{8}$ inch (3 millimeters). See Fig. 2.

Properties. The concentration of plasticizer in a cellulose ester plastic determines its "flow temperature," which in turn determines its "flow designation," as defined by ASTM Method D 569. Flow designations range from various degrees of hardness (H4, H3, H2, H) through medium-hard (MH), medium (M), and medium-soft (MS) to various degrees of softness (S, S2, S3, S4, etc). At any given flow designation, the characteristics of the plastic will vary somewhat with the identity of the plasticizer used. Some plasticizers, for example, give very hard materials, some give very low water absorption, and some permit unusual ease of processing. The flow temperature is used for quality control and the corresponding flow designation is a part of the purchase specification when material is ordered. See accompanying table.

Representative properties of general-purpose formulations of acetate, butyrate, and propionate are also shown in the table.

Depending upon plasticizer content, cellulose ester plastics range from soft, extremely tough materials to hard, strong, stiff compositions that still retain a considerable degree of toughness over a wide range

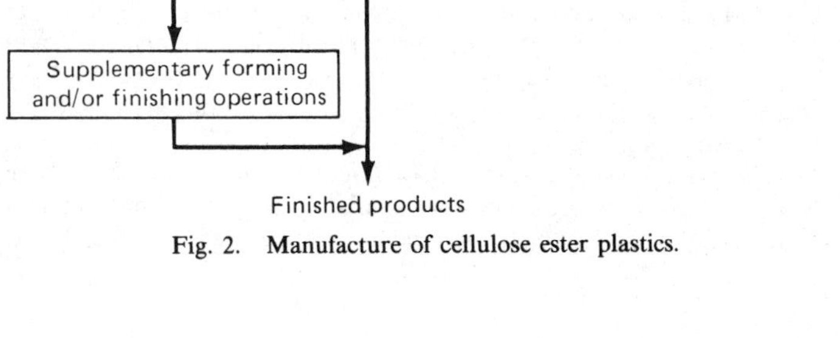

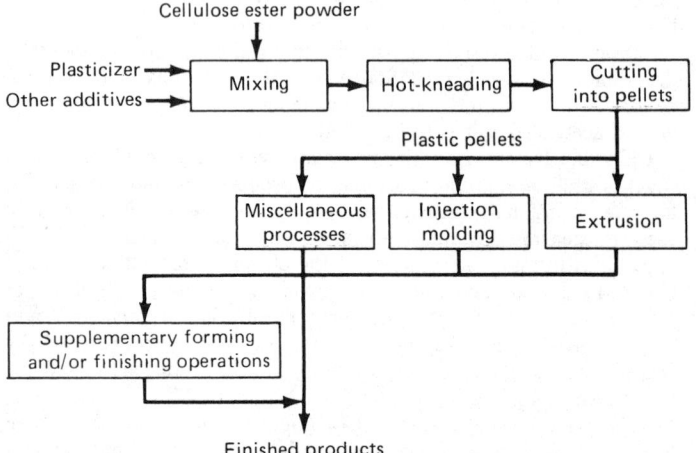

Fig. 2. Manufacture of cellulose ester plastics.

of temperatures. They are basically transparent and virtually colorless, which makes it possible for them to be manufactured in almost any desired transparent, translucent, or opaque color. They are resistant to water and aqueous salt solutions, but they are attacked by aqueous solutions that are strongly acidic or basic. They resist several types of organic solvents, such as ethers and aliphatic hydrocarbons, but they are dissolved or swollen by strongly polar liquid organic compounds such as aromatic hydrocarbons, chlorinated hydrocarbons, ketones, and esters. The susceptibility of the plastics to attack decreases as the molecular weight of the attacking compound increases.

All three cellulose ester plastics are available in formulations that meet the regulatory requirements for use in contact with food.

Butyrate and propionate are available in special formulations for continuous use outdoors, where they generally remain useful for several years. Formulations other than the special outdoor-type materials should not be used in this manner. Acetate formulations are not suggested for outdoor use, as CA does not respond well to the addition of protective compounds.

Although acetate, butyrate, and propionate resemble each other in many ways, there are a number of significant differences among them. Butyrate and propionate are generally easier to process than acetate, and this factor, also, subtracts from the price advantage of acetate. Acetate is available in very hard flows, so it can be obtained with higher stiffness, hardness, and tensile strength than can the other two cellulose ester plastics. Butyrate is available in the softest flows, so it can be obtained in the toughest and easiest-processing formulations. Butyrate and propionate are generally considered to be tougher than acetate, even though in some instances the measured impact strengths may be similar. Butyrate retains its toughness better than does propionate at low temperatures. Butyrate and propionate use higher-boiling, less-water-soluble plasticizers than does acetate, which leads to better retention of plasticizer by butyrate and propionate when exposed to elevated temperatures or to the leaching action of water. Better plasticizer retention, in turn, leads to better permanence characteristics in the plastic—i.e., smaller changes in dimensions and properties with time.

Processing. The organic cellulose ester plastics are versatile materials and can be processed by almost any hot-processing technique used for thermoplastics. The principal techniques for all three plastics are injection molding and extrusion. Blow molding is also possible. Butyrate and propionate powder are used in fluidized-bed and electrostatic coating processes, as well as in the rotational molding process.

The toughness of cellulosics nearly always enters into the selection of one of these plastics for a particular application, but if the potential toughness of cellulosics is to be realized, the materials must be processed correctly. Correct processing involves heating the plastic sufficiently for it to flow freely (*not* forcing half-melted plastic through a die or into a mold) and cooling it slowly. Fast cooling causes the outside of the finished product to harden while the inside is still molten. When the inside cools and contracts, powerful stresses form within the plastic, and these frozen-in stresses can detract very significantly from the toughness of the plastic. Sprues, runners, trim, and other scrap from molding and extrusion operations, if kept clean, can be reground, mixed with new feed stock, and reused. Butyrate and propionate are sometimes compatible with each other, but acetate is not compatible with either of the others and must not be allowed to contaminate them. Synthetic polymers must be rigorously excluded from all cellulosics.

Injection Molding. Flows of acetate and butyrate most commonly used for injection molding are MS, M, and MH; flows of propionate, H and H_2. Melt temperatures are generally in the 350 to 500°F (177 to 260°C) range; pressures, between 10,000 and 30,000 psi (680 to 2040 atmospheres) on the plastic. Molding shrinkage varies from about 0.003 to 0.008 inch per inch (millimeter per millimeter). Molding cycles can be as low as about 15 seconds for thin sections, increasing with the thickness of the molded item.

Extrusion. Extruders can process higher-melting materials than injection molding machines, and the average flow of cellulosic plastic for extrusion is 2 or 3 flows harder than the average flow used for most injection molding. Extrusion temperatures are ordinarily between about 335 and 450°F (~168 and 232°C). Extruded products are generally sized by building the die slightly oversize and drafting the extru-

REPRESENTATIVE PROPERTIES OF CELLULOSE ESTER PLASTICS

	Property, Unit [special conditions of test][a]	ASTM Test Method	Acetate H2 Flow	Acetate M Flow	Butyrate H2 Flow	Butyrate M Flow	Butyrate S2 Flow	Propionate H4 Flow	Propionate H Flow
Injection-Molded Specimens $\frac{1}{8}$-inch thick (0.32 centimeter)	Flow Temperature, °F (°C)	D 569	320(160)	293(145)	320(160)	293(145)	266(130)	338(170)	311(155)
	Specific Gravity	D 792	1.28	1.27	1.21	1.19	1.17	1.22	1.20
	Hardness, Rockwell, R Scale	D 785	85	55	96	63	—	99	51
	Tensile Strength at Yield, psi	D 638	4,800	3,400	5,400	3,800	2,250	5,200	3,600
	Tensile Strength at Fracture, psi								
	at 73°F (23°C)	D 638	6,300	4,700	6,300	5,050	3,600	6,100	4,500
	at 158°F (70°)		3,200	1,950	3,900	2,300	1,350	4,000	2,400
	Flexural Strength at Yield, psi	D 790	7,500	4,800	7,750	5,400	3,200	6,800	4,400
	Stiffness in Flexure, 10^5 psi	D 474	1.90	1.45	1.60	1.20	0.80	1.70	1.20
	Impact Strength, Izod, ft-lb/in. of notch								
	at 73°F (23°C)	D 256	3.5	5.1	4.2	6.5	9.2	2.7	9.5
	at −40°F (−40°C)	D 758	1.1	1.2	1.2	1.6	2.5	1.1	1.4
	Deformation Under Load, %								
	[24 hours at 122°F (50°)] at 1000 psi	D 621	1	2	1	3	25	1.0	2.0
	at 2000 psi		3	23	2	14	45	1.0	21
	Water Absorption, % [24-hour immersion]	D 570	2.3	2.3	1.8	1.5	1.3	2.0	1.9
	Soluble Matter Lost, %		0.2	0.6	0.1	0.1	0.1	0.1	0.1
	Weight Loss on Heating, % [72 hours at 180°F (82°C)]	See Note	1.2	3.6	0.3	0.9	2.0	0.3	1.2
Compression-Molded Specimens $\frac{1}{2}$-inch thick (1.27 centimeters)	Hardness, Rockwell, R Scale	D 785	101	75	109	89	—	107	83
	Impact Strength, Izod, ft-lb/in. of notch								
	at 73°F (23°C)	D 256	1.6	2.5	1.4	2.5	4 5	1 7	4.0
	at −40°F (−40°C)	D 758	0.5	0.6	0.6	0.9	1.3	0.6	0.8
	Compressive Strength at Yield, psi	D 695	7,200	4,900	7,800	5,500	3,400	6,600	4,400
	Deflection Temperature, °F (°C)								
	at 264-psi fiber stress	D 648	151(66)	128(53)	173(78)	143(62)	123(51)	172(78)	136(58)
	at 66-psi fiber stress		171(77)	144(62)	199(93)	165(74)	140(60)	200(93)	167(75)

[a] All tests run at 73°F (23°C) and 50 percent relative humidity unless otherwise specified.
NOTE: D 706 for Acetate; D 707 for Butyrate; D 1562 for Propionate.
Degrees Celsius are approximate.

date, but heavy sections should not be drafted more than a few percent. Excessive draft will cause stresses to form in the plastic, which can cause the product to become brittle.

Blow Molding. Processing of cellulosics by the normal extruded-parison method can be done, but it requires special techniques and is not widely practiced. Considerable success has been achieved, however, with cutting extruded butyrate pipe into short lengths, reheating, and blow molding.

Thermoforming. Thin sheeting of all three plastics is widely thermoformed for blister packaging. Heavy sheeting of special outdoor-type formulations, principally of butyrate, is formed into outdoor signs. Excellent detail can be obtained. Drawdown ratios as high as 3:1 are practical with adequate draft in the mold. Molds can be relatively inexpensive.

Cementing. Any of three organic cellulose ester plastics can be cemented to itself, and propionate and butyrate generally can be cemented to each other using solvents. A cement produced by dissolving (in an appropriate solvent) a small amount of the plastic to be cemented will ordinarily produce a better bond. Solvents most commonly used are mixtures of an active solvent, such as acetone or methylethyl ketone, a latent solvent, such as alcohol, and a diluent, such as benzene. Most solvents are flammable and toxic and must be handled accordingly.

R. P. Rich, Plastics Division, Eastman Chemical Products, Inc., Kingsport, Tennessee.

[Neither Eastman Kodak Company nor its marketing affiliates shall be responsible for the use of this information, or of any product, method or apparatus mentioned, and you must make your own determination of its suitability and completeness for your own use, for the protection of the environment, and for the health and safety of your employees and purchasers of your products. No warranty is made of the merchantability or fitness of any product; and nothing herein waives any of the Seller's conditions of sale.]

CELLULOSIC PAINTS. Paint.

CELSIUS DEGREE. Units and Standards.

CELSIUS TEMPERATURE SCALE. Temperature.

CELLULOSE (Microcrystalline). Bodying and Bulking Agents (Foods).

CELLULOSIC INSULATION. Insulation (Thermal).

CEMENT. Cement is a finely powdered substance which possesses strong adhesive powers when combined with water. Gypsum plaster (see **Calcium**), common lime, hydraulic limes, Puzzolan, natural and Portland cements are a few of the materials which are used for cementing purposes.

Portland cement, which is the most important of these materials since it is a basic ingredient of concrete, was first manufactured in England in the early part of the nineteenth century. It derived its name from the fact that this newly discovered cement resembled a building stone that was quarried near Portland, England.

There are three fundamental stages in the process of manufacture

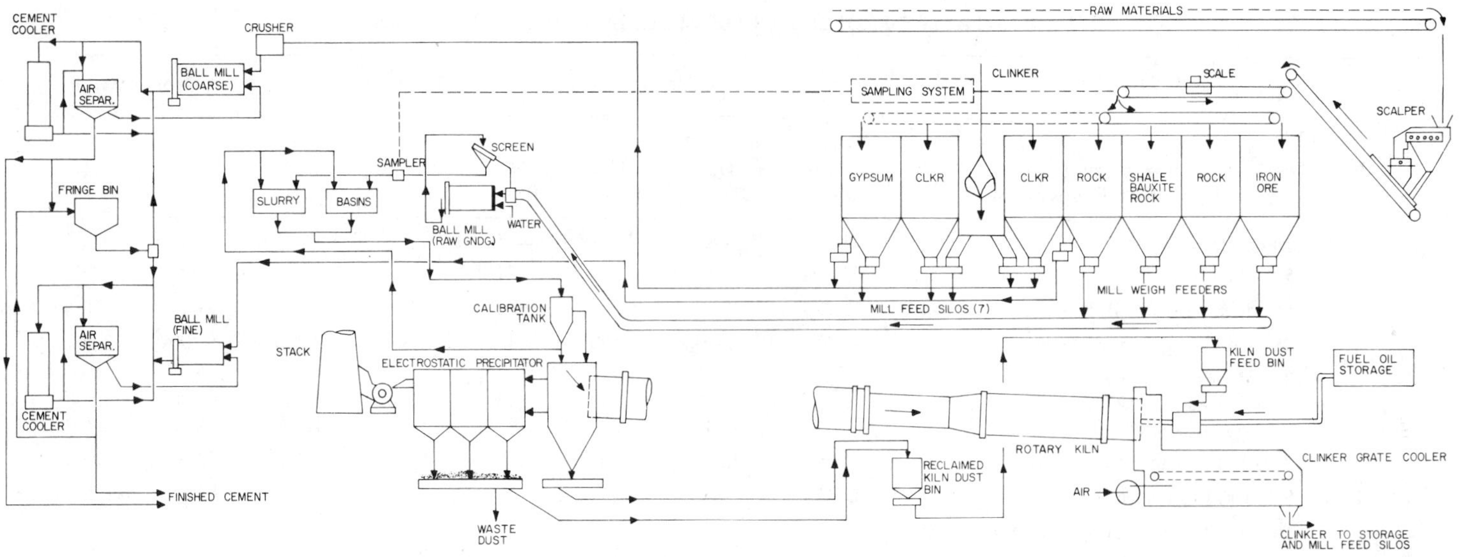

Wet process for making cement.

of Portland cement, namely, (1) preparation of the raw mixture, (2) production of the clinker, (3) preparation of the cement. Whether the process used is wet or dry, the raw materials are selected, analyzed, and mixed so that, after treatment, the product, or clinker, has a desired, narrowly specified composition. A factory analysis of slurry, where the wet process is in use, is as follows: calcium oxide 44%, aluminum oxide 3.5%, silicon oxide 14.5%, ferric oxide 3%, magnesium oxide 1.6%, loss on ignition about 33% (largely carbon dioxide), showing that the composition of the resulting burned clinker is essentially a calcium aluminosilicate. The system calcium oxide-aluminum oxide-silicon oxide has been determined by Rankin and co-workers. In some places the composition of the rock is practically of the desired composition, and in other places clay and limestone are mixed in the desired proportions.

The raw mixture is heated in a continuously operated, long, almost horizontal, slowly rotated furnace or kiln at a high temperature. The temperature is regulated so that the product consists of sintered but not fused lumps. This is clinker. Too low a temperature causes insufficient sintering, and too high a temperature results in a molten mass or glass, the product is either of these cases being valueless for cement purposes. Clinker is unaffected by water, and may be stored indefinitely without detriment. In 1824, Joseph Aspin, an English bricklayer, took out a patent for the manufacture of an improved cement, which he called Portland cement. The cement thus produced was not what is known now as Portland cement as the temperature of burning was not sufficiently high. The value of burning at a temperature sufficiently high to cause incipient fusion was soon afterwards discovered.

In order to obtain the desired setting qualities in the finished cement, there is added to the clinker about 2% of gypsum (calcium sulfate, $CaSO_4 \cdot 2H_2O$), and the mixture is pulverized very finely. For every ton of Portland cement shipped, over two and one-half tons of raw materials *and* cement clinker must be ground very finely. See Table 1.

The wet process for making cement is shown in the accompanying flowsheet. Selective quarry of raw materials for specific cement requirements is based upon chemical data. A television receiver, mounted on the quarry control panel, oversees these operations. Rocks sent to the crushing plant are reduced to pieces smaller than 2 inches. These are dropped on to a reversible shuttle conveyor that stockpiles the materials according to chemical composition in the raw materials storage located above a reclaiming tunnel. From the stockpiles, vibrating feeders then withdraw specified amounts of raw materials and discharge them over a 1,230-foot (375-meter) belt conveyor that passes through the reclaiming tunnel. A television camera at the tunnel entrance observes the materials on their way to the next crushing stage. A second television camera supervises the screening operation, and a third camera observes the unloading of crushed material into the storage silos.

Raw materials, withdrawn from the silos, are conveyed to the raw-grinding mill, and water is added. A sonic system which "listens" to the sound of steel balls impacting the inside of the mill indicates the amount and fineness of the material being ground. Feedback from this system controls the feed rate to the mill. Raw-mill discharge (a slurry of about two-thirds solids content) is screened over a 50-mesh screen cloth and pumped to two slurry basins where it is homogenized. Each basin (44 feet high × 85 feet diameter) (13.2 meters high × 25.5 meters diameter) holds about a 3-day supply for the kiln.

The rotary kiln is 510 feet (153 meters) long and is fired by oil. Maximum daily output of the kiln is 1,760 tons (1,584 metric tons)

TABLE 1. ANALYSIS OF MATERIALS USED FOR MANUFACTURE OF LIME AND CEMENT

Material	SiO_2	Fe_2O_3 Al_2O_3	CaO	MgO	CO_2	SO_3	Used for
Limestone	0.36	0.45	54.45	0.54	43.24		Portland cement
Limestone	3.30	1.30	52.15	1.58	40.98		" "
Limestone	0.74	0.13	52.94	1.87	43.68		" "
Marl	1.78	1.21	49.55	1.30	40.35		" "
Cement rock	13.44	6.60	41.84	1.94	32.94		" "
Cement rock	11.11	6.31	42.51	2.89	36.57		" "
Clay	61.09	26.97	2.51	0.65	—	1.42	" "
Clay	58.44	26.50	1.70	1.88	—		" "
Cement rock	15.37	11.38	25.50	12.35	34.20		Natural cement
Limestone	0.89	0.47	54.68	0.32	43.44		Lime
Limestone	0.78	0.48	31.15	20.78	45.76		"
Oyster shells	3.30	0.25	52.14	0.25	41.61		"

of clinker, equivalent to 9,700 barrels of cement. Operations are automatically controlled, and two television cameras, one at each end of the kiln, observe all material flow.

In its downward path through the kiln, the slurry passes first into a 91-foot long drying zone (maintained at 2,000°F) (1,093°C) and then into a hotter (above 2,800°F) (1,538°C) calcining and burning section. The drying zone is fitted with steel chains to improve fuel efficiency through better heat transfer. Feed rate of the oil is controlled by the gas temperature in the calcining zone. Thermocouples placed at intervals inside the kiln measure the temperature of solid particles and kiln gases. The critical parameters are relayed to the plant central control room.

The initial quarrying, primary, and secondary crushing and screening of raw materials are not shown in the flowsheet. See also **Gypsum.**

When Portland cement is mixed with water, the product sets in a few hours and hardens over a period of weeks. The initial setting is caused by the interaction of water and tricalcium aluminate $3CaO \cdot Al_2O_3$, present in the cement, accompanied by the separation of gelatinous hydrated product. The later hardening and the development of cohesive strength are due to the interaction of water and tricalcium silicate $3CaO \cdot SiO_2$, also present in the cement, accompanied by the separation of gelatinous hydrated product. In each case the gelatinous material surrounds and cements together the individual grains. The hydration of dicalcium silicate $2CaO \cdot SiO_2$, also present in the cement, proceeds still more slowly than that of the above compounds. The ultimate cement agent is probably gelatinous hydrated silica SiO_2. See also **Concrete.** The analyses of some typical Portland cements are given in Table 2.

Deductions regarding the mechanism of setting and hardening, and the identity of the substances concerned are the results of extensive studies, involving the use of the microscope in the examination of thin sections, on the individual compounds, the clinker, and the resulting concrete. Elaborate researches have also been conducted to determine the best way of incorporating the ingredients of concrete, the nature of the aggregate (sand, gravel, crushed rock) to be used, and the proportions of cement, water, and aggregate, in order that the resulting concrete, really an artificial rock, shall possess the greatest possible strength.

Special applications of cements and lutes are frequently demanded, as for example in floor covering, in tank lining, and in the closure of joints. The difference between a cement and a lute is that the former sets to a rigid solid mass whereas the latter retains some plasticity so that some movement of the lute is possible without cracking. A lute must have support in order that it be retained in position.

A somewhat crude though convenient classification can be made on the basis of the principal ingredients, thus, (1) Portland cement, (2) high alumina cement, (3) sodium silicate, (4) magnesium oxychloride plus copper powder, (5) litharge or red lead plus glycerol, (6) rubber latex, and (7) synthetic resins. Supplementary materials to be considered are asbestos, white lead, plaster of Paris, sulfur, graphite, sand, pitch, tar, rosin, and boiled linseed oil.

The choice to be made depends upon the kind of material to which the cement or lute must adhere; what it must withstand in the way of acid, base, sulfate, or organic liquid; also what temperature is involved; and finally the matter of resistance to vibration and shock.

Portland and high alumina cements do not withstand acids but are resistant to bases. High alumina cement attains its maximum strength more quickly than Portland, and has the extra advantage that it withstands solutions of sulfates.

Sodium silicate cement does not withstand bases, but is resistant to acids except hydrofluoric. This cement sets to a very rigid solid, so that when subjected to mechanical shock or to temperature change it is liable to crack.

A cement containing 90% magnesium oxychloride and 10% copper powder is strong, resistant to abrasion, and can be bonded to Portland cement.

Litharge or red lead plus glycerol is very commonly used, especially for pipe joints.

Rubber latex cement withstands silute acids and dilute bases, and adheres well to ceramic materials such as stoneware. This cement remains somewhat pliable, thus resisting mechanical shock and temperature change. Organic liquids in general attack this cement.

Synthetic resin cements withstand hydrochloric acid, dilute nitric acid, dilute sulfuric acid, and dilute bases, and are frequently more resistant to organic liquids than is rubber latex cement. The adherence to ceramic materials is good, and the liability to cracking less than for sodium silicate cement.

As for the miscellaneous ingredients mentioned, some are used as fillers or extenders as in the case of sand, and some are used in their own right as when pitch, tar, rosin, molten sulfur, or packed asbestos can be used. See also **Adhesives.**

CEMENT (Analysis of). X-Ray Analysis.

CEMENTATION. In geology, cementation is the process of deposition from solution of mineral matter in the interstices of rocks, and is an important factor in the consolidation of coarse-grained elastic rocks such as sandstones and conglomerates or breccias. This action is continually going on in the groundwater zone, so much so that the term zone of cementation has come into common use.

Cementation may occur in fissures or other openings of the rocks and in time all such spaces will be closed to further deposition or entrance of groundwater.

In metallurgy, cementation is the process by which one substance is caused to penetrate and change the character of another, by the action of heat, at temperatures below the melting points.

CEMENT (Insulating). Insulation (Thermal).

CEMENTITE. Iron carbide, Fe_3C, a compound which is present at room temperature in nearly all iron-carbon alloys such as steel and cast iron. It is very hard and brittle and weakly magnetic and

TABLE 2. ANALYSIS OF PORTLAND CEMENTS[a]

Where Made	Made from	SiO_2	Fe_2O_3	Al_2O_3	CaO	MgO	SO_3	Loss
New Jersey	Cement rock and	21.82	2.51	8.03	62.19	2.71	1.02	1.05
Pennsylvania	limestone	21.94	2.37	6.87	60.25	2.78	1.38	3.55
Michigan	Marl and clay	22.71	3.54	6.71	62.18	1.12	1.21	1.58
Ohio		21.86	2.45	5.91	63.09	1.16	1.59	2.98
Virginia		21.31	2.81	6.54	63.01	2.71	1.42	2.01
Missouri	Limestone and clay	23.12	2.49	6.18	63.47	0.88	1.34	1.81
Pennsylvania[b]		23.56	0.30	5.68	64.12	1.54	1.50	2.92
Illinois		22.41	2.51	8.12	62.01	1.68	1.40	1.02
Germany	Blast furnace slag	20.48	3.88	7.28	64.03	1.76	2.46	
Belgium	limestone	23.87	2.27	6.91	64.49	1.04	0.88	
France		22.30	3.50	8.50	62.80	0.45	0.70	
England		19.75	5.01	7.48	61.39	1.28	0.96	
Germany[c]	Iron ore and limestone	20.5	11.0	1.5	63.5	1.5	1.0	

[a] From Meade's "Portland cement."
[b] White Portland cement.
[c] Seawater cement.

has an orthorhombic *crystal* structure. In commercial steels, the chemical composition of cementite is usually changed by the presence of manganese and similar carbide forming elements in the steels which replace iron atoms in the compound. Cementite is a metastable phase and, under the proper conditions, decomposes to form carbon (graphite) and iron. In ordinary steels, this decomposition rarely occurs. Most cast irons, however, contain graphite. This is because the higher silicon content of the cast iron makes cementite less stable. See also **Iron Metals, Alloys, and Steels.**

CENOPHYTIC. A paleobotanic division of geologic time. This term signifies the time which has passed since the development of the angiosperms in the middle or late Cretaceous. See also **Paleobotany.**

CENOZOIC. An era of geologic time, commencing with the Tertiary period to the present. Considered to have begun about 70 million years ago, the Cenozoic is characterized paleontologically by the evolution and abundance of mammals, advanced mollusks, and birds. Paleobotanically, it is characterized by angiosperms. Informally, the Cenozoic era is sometimes referred to as the *age of mammals.*

CENTAURUS (the centaur). A large and brilliant constellation of the southern sky, which is invisible to observers in North America and Europe. Centaurus has two bright stars, frequently spoken of as the "southern pointers," since the line through them passes through the Southern Cross (the constellation Crux). The brighter of these stars (α Centauri) is not only the third brightest star in the entire sky, but also the nearest bright star to earth, at a distance of 4.3 lightyears. The only star known to be closer to the earth is a faint star known as Proxima Centauri. Alpha Centauri is a double star, and its brighter component is interesting in that it is almost a duplicate of our sun in size, temperature, and other physical characteristics. (See map accompanying entry on **Constellations.**)

CENTAURUS X-3. Neutron Stars.

CENTER-DISTANCE CHECKING. Thickness Measurement and Gaging Systems.

CENTER FREQUENCY. Frequency.

CENTER GAGE. A small gage used for checking lathe center point angles and threading tool points; also used for setting single-point threading tools on a lathe.

CENTER (Instantaneous). The point about which a body having general motion may be considered to be in pure rotation (i.e., without translation) for any instant. The instantaneous center is not necessarily on the body; in fact, it can be, in the case of rectilinear motion, infinitely distant.

CENTER OF ACTION. Atmosphere (Ocean).

CENTER OF GRAVITY. For an extended body or collection of particles subject to the earth's gravitation, the point through which the resultant force of gravity acts (i.e., the weight of the body or collection) no matter how the body is oriented. In a uniform gravitational field in which the ratio of gravitational force to mass is always the same, the center of gravity is the same as the center of mass. See also **Centroid.**

CENTER OF MASS. If we imagine a body divided into infinitesimal particles or elements of mass, and if each of these elements is acted upon in the same direction, chosen at random, by a force proportional to its mass, it is easily shown that, whatever the direction of this set of parallel forces, their resultant always passes through a certain point, which is the center of mass of the body. Since the weights of the particles of a small body constitute approximately such a system of forces, it has become customary to call this point the center of gravity, though it is in general not strictly correct to do so.

If the body is given a linear acceleration in any direction without rotation, since the inertia of each particle is proportional to its mass and acts in direct opposition to the acceleration, the resultant inertia of the whole body acts in a line through the center of mass; which is therefore properly called also the center of inertia.

If any plane is passed through the center of mass of a body, it divides the body into two parts which have the property that their mass moments with respect to the plane are equal but of opposite sign. This means that if the mass of each particle is multiplied by its distance from the plane, and the products added, the sum is numerically the same for both parts of the body. This principle may be put into mathematical form and the position of the center of mass calculated therefrom. It may be shown, for example, that the center of mass of a homogeneous right circular cone is on its axis at a distance from the apex equal to $\frac{3}{4}$ of the altitude. For a continuous body of homogeneous material, the center of mass coincides with the centroid. See also **Centroid.**

CENTER-OF-MASS SYSTEM. In general, any frame of reference moving with the center of mass of a system. Calculations in the center of mass system are often simpler than in other coordinate systems because the total momentum in the former is always zero.

CENTER OF OSCILLATION. The frequency of oscillation of a physical pendulum is given by

$$f = \frac{1}{2\pi}\sqrt{\frac{gl}{k^2 + l^2}}$$

where l is the distance from the point of suspension to the center of mass, and k is the radius of gyration. A simple pendulum having the same frequency will be of length

$$l_1 = \frac{k^2 + l^2}{l}$$

The point which lies at a distance l_1 from the point of suspension on the line through the point of suspension and center of mass is called the center of oscillation. If the physical pendulum were to be suspended from this point, its frequency would be the same as for the original point of suspension. See also **Pendulum.**

CENTER OF PERCUSSION. That point (with respect to a given point of suspension) on a rigid rod, like a tennis racket or baseball bat, such that an impulse applied perpendicular to the rod at the point produces no impulsive reaction at the original point of suspension. The center of percussion coincides with the center of oscillation.

CENTER OF PRESSURE. Aerodynamics; Supersonic Aerodynamics.

CENTER OF PRESSURE (Hydrostatic). The point of application of the resultant of all the pressure forces acting upon an exposed area is called the center of pressure. Water and air create pressures which are of importance in phases of engineering. The center of pressure on a horizontal plane immersed in water to a depth h is the center of the area, that is, the point corresponding to the center of mass of a flat, uniform plate coinciding with the area.

More generally, if a plane surface is completely immersed in a liquid at an angle θ with the horizontal, the center of pressure upon that surface lies at a depth below the top of the liquid, given by the equation $d = \sin\theta \cdot I/M$; in which I is the moment of inertia of the surface with respect to the line of intersection of its plane with the plane of the top of the liquid, and M is the moment of the surface with respect to the same intersection. For a vertical surface ($\theta = 90°$), $d = I/M$. Thus, in the case of a vertical rectangle whose upper and lower edges are horizontal and at depths h_1 and h_2, and whose width is b, we have $I = b/3(h_2^3 - h_1^3)$, while $M = b/2(h_2^2 - h_1^2)$; hence the depth of the center of pressure is

$$= \frac{2}{3}\,\frac{h_1^2 + h_1 h_2 + h_2^2}{h_1 + h_2}$$

In particular, if the upper edge is at the top of the liquid, and if the altitude of the rectangle is a, so that $h_1 = 0$ and $h_2 = a$, $d = \frac{2}{3}a$. A rectangular dam 12 feet high, and completely filled with water, would thus have a center of pressure at a depth of 8 feet.

Water which is assumed by engineers to weigh 62.5 pounds per cubic foot exerts a horizontal pressure varying uniformly from zero at the water surface to $62.5h$ at a depth h. The total horizontal water pressure acting on the vertical rectangle mentioned above at $\frac{2}{3}a$ below the water surface is $31.25h^2b$ pounds in which b is the other principal dimension or the vertical width of the rectangle. See **Hydrostatics.**

CENTER OF SYMMETRY. A point within a body or within a set of points, having such location that for every point there is a similar point on a line passing through the first point and the center of symmetry, the distance of the two points from the center being identical. If the center of symmetry is located at the origin of a Cartesian coordinate system, then for every point with coordinates x, y, z there will be an identical point with coordinates $-x$, $-y$, $-z$.

CENTER PUNCH. A hand tool with a sharp conical point, used for layout work and locating centers for drilling.

CENTIMETER-GRAM-SECOND (CGS) SYSTEM. Units and Standards.

CENTIPEDE (*Insecta, Arthropoda, Chilopoda*). Wormlike land-living tracheate animals with segmented bodies, a distinct head bearing a pair of antennae and three pairs of jaws, and a pair of jointed legs on most segments of the body. The first pair of legs are modified as poison claws for the capture of prey.

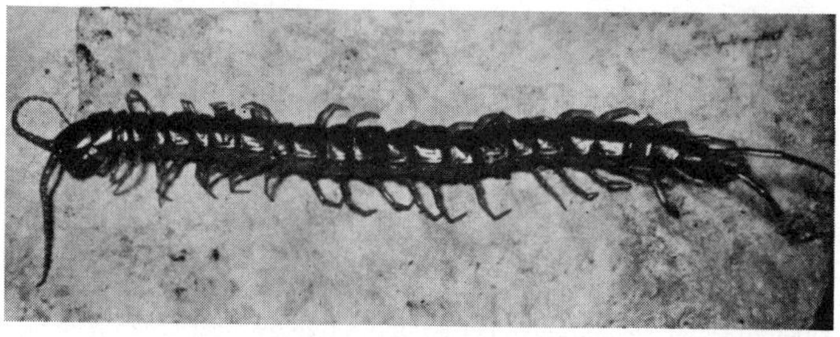

Centipede. (*A. M. Winchester.*)

The centipedes differ from the nearly related millipedes (*Diplopoda*) in the presence of poison claws, in having only one pair of legs to a segment and in the more flattened body. Each segment has a dorsal and a ventral plate connected by softer tissue and has the legs joined to the sides of the body.

The house centipede illustrated is not to be confused with the garden centipede or symphylid. The latter is damaging to vegetable plants, whereas the house centipede consumes numerous insects, including roaches, flies, and other small species, and thus is considered a beneficial insect.

CENTIPOISE. Units and Standards.

CENTI PREFIX. Units and Standards.

CENTISTOKES. Units and Standards.

CENTRAL BARRIER. Nuclear Forces.

CENTRAL FORCE. A force acting along the line joining two centers, as, for example, the electrostatic attraction between charges.

CENTRAL FORCE MOTION. Astronautics.

CENTRAL LIMIT THEOREM. The distribution of sample means usually approximates a normal distribution. The approximation becomes more accurate as the size of the sample increases. This is particularly remarkable in that the shape of the distribution of the population from which the sample was drawn has little influence over the shape of the distribution of sample means. The necessary and sufficient condition that sample means be distributed normally constitutes the central limit theorem of probability. We shall state a sufficient condition proved by Liapounoff which is extremely general. Let $x_1, x_2, \ldots, x_n$ be independent variables with means zero, possessing absolute third moments. $\mu_3^{(1)}, \mu_3^{(2)}, \ldots \mu_3^{(n)}$, respectively. (An absolute third moment is defined as the expected value $|x|^3$.)

Let σ_i^2 be the variance of x_i and

$$\sigma^2 = \sum_{i=1}^{n} \sigma_i^2$$

let

$$\alpha_3 = \sum_{i=1}^{n} \mu_3^{(i)}/\sigma^2$$

Then if $\lim_{n\to\infty} \alpha_3$ approaches zero the mean $\bar{x}$ approaches a normal distribution.

An example of a distribution which does not obey the central limit theorem is the Cauchy Distribution.

There are various generalizations of the theorem, e.g., by relaxing the conditions under which it holds, by extension to sums of correlated variables, and by extension to multivariate situations.

CENTRAL NERVOUS SYSTEM. Brain and Nervous System.

CENTRAL PROCESSING UNIT (Computer). Sometimes referred to as the *main frame*, the central processing unit (CPU) is that part of a computing system exclusive of input or output devices and, sometimes, main storage. As indicated by Fig. 1, the CPU includes the arithmetic and logical unit, channels, storage and associated registers, and controls. Information is transmitted to and from the input or output devices via the channel. Within the central processing unit, data are transmitted between storage and the channel and between each of these and the arithmetic and logical unit. In some computing systems, if the channel and storage are separate assemblages of equipment, the central processing unit includes only the arithmetic and logical unit and the instruction interpretation and execution controls.

A simplified example of a CPU data flow is given in Fig. 2. Generally, the operations performed in the CPU can be divided into (1) instructions fetch, (2) indexing (if appropriate to the instruction format), and (3) execution.

In the example shown, to initiate an operation the contents of the instruction address register is transmitted to the storage address register (SAR), and the addressed instruction is fetched from storage to the storage data register. The operation code portion of the instruction is set into the operation code register (OP), and the data address part is set into the data address register. If indexing is defined in the instruction format, the indexing address is set into the index address register.

During the indexing phase of the operation, the index address register is transmitted to the SAR and the contents of the storage location is set into the storage data register. The contents of the storage data register and contents of the data address register are summed in the adder, and the result is placed in the data address register.

In connection with an arithmetic operation, the data address register

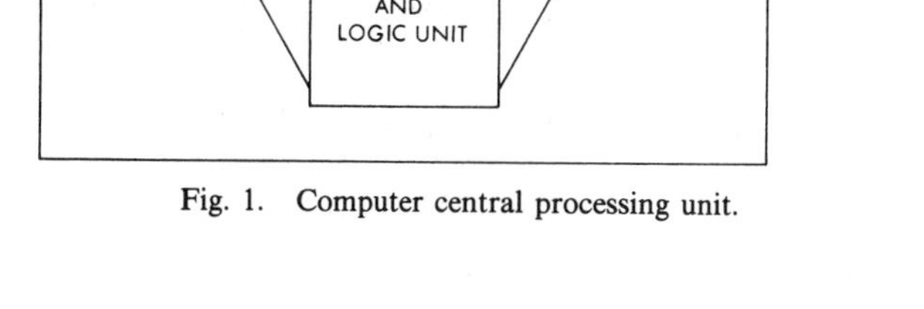

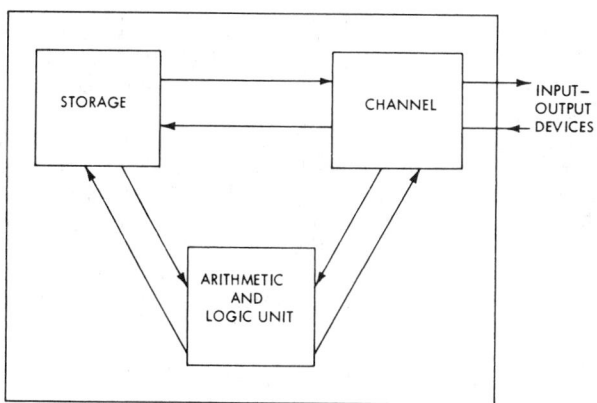

Fig. 1. Computer central processing unit.

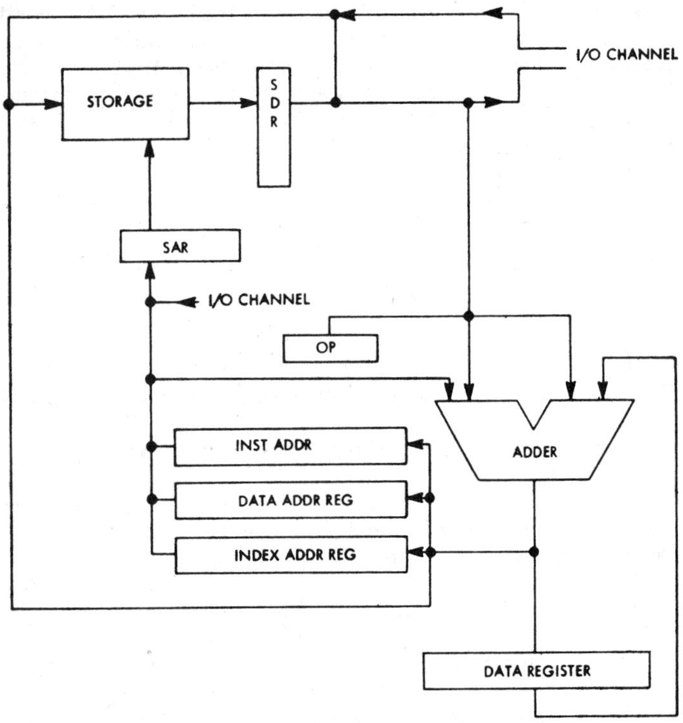

Fig. 2. Data flow in computer central processing unit.

is transmitted to the SAR, and the data in the referenced location are gated through the adder along with the contents of the data register. The result replaces the contents of the data register. If the arithmetic operation causes a carry out of the high-order position of the word being operated on, the overflow indication is saved in a condition code trigger. This condition may be tested on a subsequent operation, such as a transfer on condition. At the completion of the specified operation, the next instruction is fetched from storage and the previously described phases are repeated.

See also **Input/Output Devices; Register (Computer); Storage (Computer);** and terms listed under **Data Processing.**

CENTRECHINOIDEA (*Echinodermata, Echinoidea*). Sea urchins with a central mouth about which are gills. The commoner sea urchins of North America belong in this group.

CENTRIFUGAL COMPRESSOR. Airplane.

CENTRIFUGAL FAN. Fan.

CENTRIFUGAL FORCE. A radially outward force experienced by an observer in a reference frame which is rotating at an angular velocity ω with respect to an inertial frame (cf. Coriolis Effects, 2). The centrifugal force is the reaction to the centripetal force necessary to hold the observer at a fixed point in the rotating frame, and thus has a magnitude equal to and a direction opposite to the centripetal force.

The centrifugal force involved in the motion of a vehicle around a curve makes desirable the "superelevation" or "banking" of the roadway at an angle dependent upon the curvature and the speed, in order that the wheels of the vehicle may push perpendicularly against the pavement or track. If the speed is V and the radius of the curve is r, then the roadbed should be inclined at an angle s given by the formula: $\tan s = V^2/gr$, where g is the acceleration of gravity. For example, if $V = 40$ feet (12.19 meters) per second and $r = 1000$ feet (304.8 meters) the value of g being 32.15 feet (9.8 meters) per second per second, the formula gives $s = 2°51'$; which on a standard-gage railroad track would require the outer rail to be 2.8 inches (7.1 centimeters) above the inner.

CENTRIFUGAL GOVERNOR. Governor.

CENTRIFUGAL PUMP. Pump (Liquid).

CENTRIFUGING. A separation technique based upon the application of centrifugal force to a mixture or suspension of materials of closely similar densities. The smaller the difference in density, the greater is the force required. The equipment used (centrifuge) is a chamber revolving at high speed to impart a force up to 17,000 times that of gravity (much higher in the *ultracentrifuge*). The materials of higher density are thrown toward the outer portion of the chamber, while those of lower density are concentrated at or near the inner portion. A common cream separator is a type of centrifuge in which the flow is continuous. The centrifugal force throws the heavier milk into a different chamber from the lighter cream. Many hundreds of products, particularly in the chemical and food processing industries, require centrifugation at some stage in their manufacture. For example, important applications are found in the beet sugar and sugar cane industries, in the processing of corn (maize) products, in soybean processing, and in milk and diary products manufacture.

Basic Principle of Centrifuging. The force acting on a particle within a centrifugal field is defined by Newton's fundamental force equation, $F = ma$. Acceleration acting upon the particle, directed toward the center of rotation is $a = r\omega^2$. Therefore the centrifugal force acting on the particle is $F = mr\omega^2$, or, expressed as multiples of gravity,

$$F = 14.2 \times 10^{-6} DN^2 \ (D \text{ in inches})$$
$$= 5.59 \times 10^{-6} DN^2 \ (D \text{ in centimeters})$$

where m = mass of particle, g
a = acceleration, cm/s^2
r = radial distance of a particle in a centrifugal field from axis of rotation, cm
ω = angular velocity, rad/s
D = inner diameter of centrifugal bowl
N = bowl speed, rpm

A particle and a mixture introduced, confined, and rotated within a circular enclosure accelerates as it moves from a neutral center toward the maxium diameter (inner periphery) of the enclosure. Thus, if a mixture is introduced to the center of a 24-inch (61-centimeter) diameter solid-bowl centrifuge rotating at 1500 rpm, the particle will be caused to move at a speed of 32.2 feet (9.8 meters) per second at the center. At the maximum diameter, the particle will have a terminal velocity of 766.8 × 32.2 feet per second, or 24,690 feet (7525 meters) per second. Essentially, the separation occurs at 766.8 × g.

Industrial centrifuges are available in a variety of designs. Bowl sizes range from less than 0.5 to over 4 feet (15.2 to 122 centimeters) in diameter. Forces applied generally range from about 500 to 14,200 times the force of gravity. Operating temperatures range up to 260°C or higher; pressures up to about 8.5 atmospheres. Liquid flow rates range from 5 to 500 gallons (19 to 1893 liters) per minute; solid rates from 0.2 to 68 metric tons per hour. Depending upon design, the range of size of particles which can be handled is from 1 micrometer

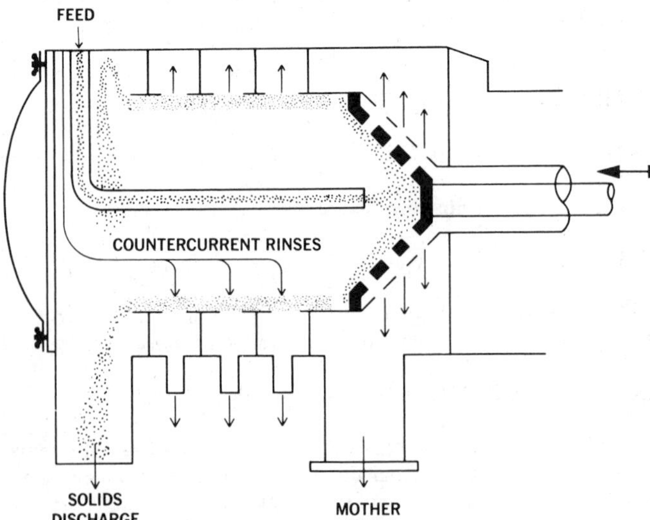

Continuous horizontal, pusher (reciprocating) filter centrifuge. Double arrows at right indicate motion of ram.

to as large as 0.6 centimeter. Centrifuges often are equipped with filters to provide a highly clarified effluent. The majority of industrial centrifuges are designed for continuous operation, although batch designs are obtainable.

One of several types of industrial designs is shown in the accompanying figure. This is a continuous horizontal, pusher (reciprocating) filter centrifuge. The feed is continuously introduced through a stationary pipe. The solids settle to the screen surface and form a filter bed through which the liquid passes. The liquid continues through the basket and is discharged from the machine, while the solids, having formed a uniform filter-bed thickness in the cylinder basket, are assisted to discharge by a pusher ram. Washing of cake solids, following the initial dewatering, is a feature of this design.

Among other types are the gas centrifuges which have been used for the separation of isotopes. In the concurrent type, one or more streams of gas enters at one end of the centrifuge and the partially-separated isotopes are removed in two or more streams at the other end. In the countercurrent type, countercurrent circulation is established in a centrifuge either thermally or mechanically. By the circulation of the gas the radial concentration gradient is converted into an axial gradient. The evaporation type operates on volatile liquids, which evaporate within the apparatus. Two streams of vapor are removed from a point near the axis of the centrifuge, having been separated by diffusion through the centrifugal field.

Ultracentrifuge. Centrifuges which operate at very high speeds and which find applications in colloid chemistry and biochemical research are sometimes called *ultracentrifuges.* In research laboratories where proteins, polymers, and other substances with high molecular weights are studied, the ultracentrifuge is effectively used. The sedimentation of large molecules in a strong centrifugal field enables the determination of both average molecular weights and the distribution of molecular weights in various systems. When a solution containing polymer or other large molecules is centrifuged at forces up to 250,000 times gravity, the molecules begin to settle, leaving pure solvent above a boundary which progressively moves toward the bottom of the cell. This boundary is a rather sharp gradient of concentrations for molecules of uniform size, such as globular proteins. For polydisperse systems, the boundary is diffuse, the lowest molecular weights lagging behind the larger molecules. An optical system can be provided for viewing this boundary, and a study as a function of time of centrifuging yields the rate of sedimentation for the single component, or for each of many components of a polydisperse system. These sedimentation rates may then be related to the corresponding molecular weights of the species present after the diffusion coefficients for each species are determined by independent experiments. Both the sedimentation and diffusion rates are affected by interactions between molecules, so that each must be studied as a function of concentration and extrapolated to infinite dilution, as is done for the colligative properties. The result of this detailed work is the distribution of molecular weights in the sample, which is available by few other methods. Extrapolation of diffusion coefficients to infinite dilution is difficult for high-molecular-weight linear polymers, and so alternate means are used to relate sedimentation constants to molecular weights in these important applications.

For research applications, ultracentrifuges may be equipped with microprocessors for programming, as well as automated handling of samples.

References

Charm, S. E.: "Fundamentals of Food Engineering," 3rd edition, AVI, Westport, Connecticut, 1978.

Considine, D. M. (editor): "The Encyclopedia of Chemistry," 4th edition, Van Nostrand Reinhold, New York, 1982.

Hall, C. W., and D. C. Davis: "Processing Equipment for Agricultural Products," AVI, Westport, Connecticut, 1979.

Staff: "Water Removal Processes: Drying and Concentration of Foods and Other Materials," Amer. Inst. of Chem. Eng., New York, 1977.

Staff: "Food, Pharmaceutical and Bioengineering," *Pubn. S–172,* Amer. Inst. of Chem. Eng., New York, 1978.

CENTRIOLE. **Cell (Biology).**

CENTRIPETAL ACCELERATION. 1. The acceleration towards the center to which any particle moving in a circular orbit is subject. This acceleration is equal to v^2/r, where v is the orbital velocity and r the radius. 2. More generally, that part of the radial component of the acceleration of a particle moving in any curved path which is equal to the magnitude of the radius vector from the instantaneous center of rotation multiplied by the square of the instantaneous angular velocity about that center. See **Coriolis Effect.**

CENTRIPETAL FORCE. The force necessary to impart centripetal acceleration to a body. For a body of mass m moving about a fixed axis at a distance r and with an angular velocity ω, the centripetal force is $-m\omega^2 r = -mv^2/r$, where v is the linear velocity of the particle and where the negative sign indicates that the force is directed radially inward. Effects such as the rupture of a rotating body or the outward skidding of an automobile in rounding a corner are often discussed as effects of centrifugal force, but are better described as being due to the insufficiency of the centripetal forces (elastic stresses or the frictional force of the wheels on the road) to maintain the masses at a constant distance from the center of rotation.

CENTRODE. The path of the instantaneous center of a plane figure having plane motion, that is, motion resulting when all points in the figure move in parallel fixed planes, is called the centrode. Any plane body having plane motion which is neither entirely rectilinear nor entirely rotative, but a combination of the two, may be considered at any instant as having rotary motion about a moving point called the instantaneous center of rotation. As shown in the illustration the

Centrode.

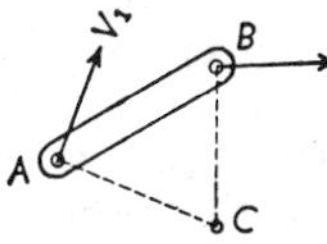

plane body AB has a motion such that the velocity of A is V_1 while the velocity of B is V_2. At the instant corresponding to the position shown for AB, the body must be rotating about a point C, which is located at the intersection of the perpendiculars to V_1 and V_2 dropped from A and B respectively. C is the instantaneous center of rotation of AB and the centrode is the path traced by the point C while AB is in motion.

CENTROID. The centroid of a given geometrical figure (curve arc, portion of a plane or curved surface, or solid) is the point whose coordinates are the mean values of the coordinates of the points of the given figure; it is independent of the choice of axes. The centroid of a geometrical figure corresponds to the center of gravity (or center of mass) of a material body of similar form.

For a plane curve arc, the centroid is given by

$$\bar{x} = \frac{\int_a^b x\,ds}{L}, \quad \bar{y} = \frac{\int_a^b y\,ds}{L}$$

$$ds = \sqrt{1 + \left(\frac{dy}{dx}\right)^2} \cdot dx = \sqrt{1 + \left(\frac{dx}{dy}\right)^2} \cdot dy$$

where L is the length of the arc, and ds is the length of an infinitesimal portion of the curve.

For a plane area, the centroid is given by

$$\bar{x} = \frac{\iint_A x\,dS}{A}, \quad \bar{y} = \frac{\iint_A y\,dS}{A}$$

where A is the area of the given region, and dS is the area of an infinitesimal portion of the surface.

For a solid, the centroid is given by

$$\bar{x} = \frac{\iiint_V x\,dV}{V}, \quad \bar{y} = \frac{\iiint_V y\,dV}{V}, \quad \bar{z} = \frac{\iiint_V z\,dV}{V}$$

where V is the volume of the given region, and dV is the volume of an infinitesimal portion of the solid.

CENTROMERE. A localized region of the chromosome to which the microtubules of the spindle apparatus attach during mitosis. It has a cuplike shape and is composed of nonchromatin material. During the process of cell division, the chromosomes become duplicated in the late resting stage (interphase). The resulting two chromatids are held together by the centromere until, at the metaphase, the centromere divides and each daughter centromere progresses, with the aid of the spindle apparatus, to opposite poles of the cell. The daughter chromosomes are pulled along behind. See **Cell (Biology).**

CENTROSOME. A clear zone surrounding one or two centrioles found in the cytoplasm near the nucleus of most animal cells. The centrosome is sometimes referred to as the microcentrum. See **Cell (Biology).**

CENTURY PLANT. Of the family *Amaryllidaceae* (amaryllis family), genus *Agave*, the century plant, *Agave americana*, the century plant is a fleshy-leaved agave native of Mexico, and well-named because of its blooming habit. The plant has a very short, thick stem which bears a rosette of thick fleshy leaves, in which are stored food reserves. Each year the plant adds three or four leaves to the rosette. After a period varying from 5 to many years (perhaps a century) vegetative growth ceases and flowering occurs. A gigantic terminal flower bud is formed and grows rapidly. It bears many flowers. After the fruit has formed the whole plant dies. It is this habit of bearing flowers only after a prolonged period of vegetative growth that has given this plant the common name, century plant. It is widely grown in cultivation, and seldom flowers under such conditions. The plant propagates vegetatively by means of suckers which grow out from the base of the stem. See also **Agave.**

CEPHALEXIN. Antibiotic.

CEPHALOCHORDATA (*Chordata*). A subphylum containing a number of species of small marine animals called lancelets. From one of the included genera the name Amphioxus has become common in laboratories to designate any lancelet, although the animals so designated more commonly belong to the genus *Branchiostoma.*

The lancelets differ from the other groups of lower chordates in their fish-like form. They are small, usually from $1\frac{1}{2}$ inches (3.8 centimeters) to rather more than 2 inches (5.1 centimeters) long, and taper toward both ends; the body is laterally compressed. Like other members of the phylum they have a dorsal nerve cord, below it a stiffening longitudinal rod, the notochord, and a pharynx perforated by many slits. A depression called the vestibule leads to the mouth. It is surrounded by a circlet of slender cirri. Cilia about the mouth carry food particles into the alimentary tract where they are caught by the mucus in a ventral groove, the endostyle, carried forward, upward, and then back to the intestine, all by ciliary action.

The lancelets are important to the evolutionist in several ways. Since they swim actively in the tides and bury themselves in the sand at other times, taking their food like sessile animals, they illustrate possible ancestral conditions for the tunicates on the one hand and the fishes on the other. The peculiar food-concentrating mechanism described above is also important; it occurs in larval lampreys and provides one of the few well-marked evidences of relationship between the vertebrates and lower chordates.

CEPHALOGLYCIN. Antibiotic.

CEPHALOPODA. A class of the phylum *Mollusca* containing the squids, cuttlefish, octopus or devilfish, and nautilus. The class is relatively small but it includes the most highly developed of unsegmented invertebrates and is of some economic importance. All of the species are marine.

These mollusks are distinguished by the following characteristics: (1) The head bears a pair of large eyes which are of the camera type like those of man although they differ fundamentally in structure. (2) The mouth is surrounded by a group of tentacles provided with many cup-like suckers and sometimes with hooks. (3) The mantle cavity opens to the exterior by a broad entrance and by a slender tube, the funnel or siphon; by forcing jets of water from the cavity through the siphon the animal propels itself through the water in the opposite direction. (4) The foot is modified to form the siphon and possibly the tentacles. (5) The mouth is provided with a sharp beak superficially like that of a bird of prey. (6) Some species have an ink-sac from which a dark fluid is discharged for concealment. (7) The shell is internal or absent in most species.

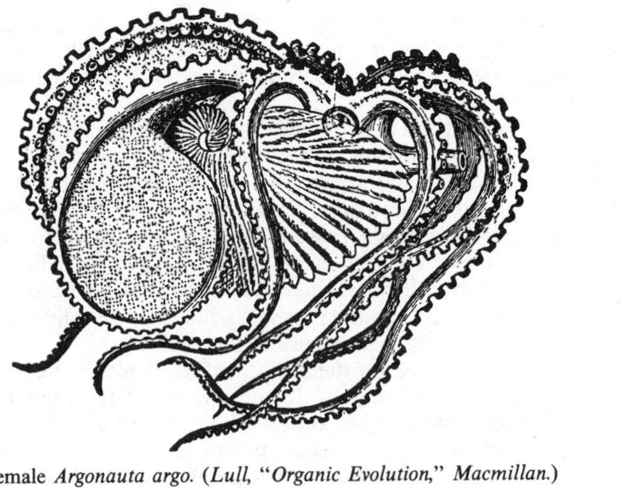

Female *Argonauta argo.* (*Lull, "Organic Evolution,"* *Macmillan.*)

Squids and cuttlefish are eaten by some peoples. The latter species furnishes the cuttlebone so familiar in bird cages and the true sepia pigment. Squids are of importance in North America chiefly as bait in the fisheries of the Atlantic Coast. The octopus has been credited with amazing feats of destruction at sea which are probably entirely imaginary. Giant squids are known to occur, with a total length of 50 feet (15 meters), and are more likely subjects of these tales, but there is no satisfactory evidence that even these creatures are greatly to be feared by man.

The class is divided into two orders. The first was abundant in the past but is now almost extinct.

Order *Tetrabranchiata.* Shell external, spiral, and divided into a series of chambers of increasing size, in the last of which the animal lives. Once abundant, now limited to four known species in the genus *Nautilus.*

Order *Dibranchiata.* Shell internal or lacking and usually straight. The squids and cuttlefishes have ten tentacles, the octopus or devilfish and the argonaut or paper nautilus only eight. See also **Invertebrate Paleontology.**

CEPHALOPODA (Fossils). Invertebrate Paleontology.

CEPHALOSPORINS. Antibiotic.

CEPHALOTHIN. Antibiotic.

CEPHALOTHORAX. The anterior body region of spiders, crustaceans and some other arthropods, consisting of the compactly associated head and thorax.

CEPHAPIRIN. Antibiotic.

CEPHEIDS. Variable, or pulsating, star types, so-named for their earliest-studied prototype, Delta Cephei. This group of variable stars is divided into two subgroups: classical cepheids, which are yellow supergiants, rare in space, and high in luminosity with periods ranging generally from 1 to 50 days (most commonly around 5 days); and type II cepheids, whose light curves have broader maxima and are more nearly symmetrical than the former, and whose periods are mostly from 12 to 20 days. A number of these stars are visible to the naked eye (e.g., Polaris, δ Cephei, η Aquilae, ζ Germinorum, β Doradus), and an interesting project is to plot their light curves as a function of time.

A star type related to the cepheids is the RR Lyrae group of variables. These stars, often called cluster variables, were first found in globular clusters, but are now recognized in greater numbers outside the clusters. They are blue giants, far outnumbering the classical cepheids, but not visible to the naked eye. Their periods range from somewhat more than an hour to about one day.

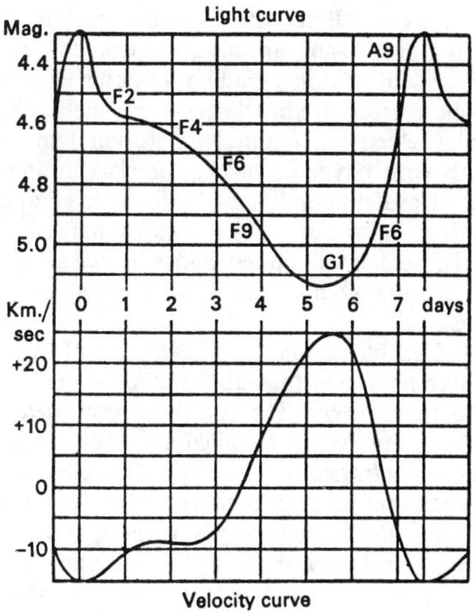

Light and velocity curves of W. Sagittarii. (*Lick Observatory.*) See also entry on **Milky Way.**

The shapes of light curves for variable stars with periods of less than 50 days are remarkably similar. Shown in the accompanying figure is the mean light curve together with the mean velocity curve as a function of period for the classical cepheid, W Sagittarii. This light curve is very typical in that the rise to maximum for almost all cepheids is rather rapid, followed by a more gradual decline to minimum. The descending portion of the curve often will show slight irregularities. The normal cepheid is usually a supergiant of spectral class F or G; it changes its spectral type slightly when going through its light variation, and is somewhat bluer at maximum than at minimum. Note that the radial velocity curve correlates with the light curve, indicating that the star is actually pulsating.

Another interesting point about these stars is that their absolute magnitudes increase as their periods grow longer, and the difference between maximum and minimum also increases.

Early development of an understanding of the cepheids is quite interesting. For many years, it was believed, because of the variations in radial velocity, that the objects were spectroscopic binaries. Shapley, Eddington, and others later showed that this is not a true explanation and proposed the alternative hypothesis that the stars are pulsating. These investigators reasoned that the star when approaching its minimum is relatively cool and is contracting under the influence of gravitational attraction. As the mass of gas contracts, the increased pressure in the interior produces a rise of temperature and eventually outward pressure due to high temperature overbalances the gravitational force and the star suddenly expands. The sudden expansion produces cooling until gravitation overbalances the outward force and contraction again takes place. It was reasoned that since at the time of expansion the outer surface will be approaching the observer and also the temperature will be higher than during the contracting stage, there is the expected correlation between brightness, spectral type, and radial velocity.

In the course of the study of typical Cepheids in the Small Magellanic Cloud, Leavitt (Harvard) found a direct correlation between the average magnitude of the Cepheids and the period of variation. Since all variables in the cloud are at approximately the same distance from earth, it appeared that the period of variation was directly correlated with intrinsic brightness or, in other words, with average absolute magnitude. Shapley, after an exhaustive study of numerous cepheids with known distances and hence known average absolute magnitudes, was able to show that the correlation between period and luminosity was apparently characteristic of all cepheids and, by plotting period against average absolute magnitude, he obtained what is commonly known as the period–luminosity curve. After this curve is established, it is possible to determine the absolute magnitude of any cepheid, no matter how distant, by simply determining the period. With the absolute magnitude and the apparent magnitude known, the calculation of the distance of the cepheid is a relatively easy task. See also **Period–Luminosity Law.**

The period–luminosity relationship was extended to the cluster-type cepheids, this serving a valuable purpose in determining distances of globular clusters. Cepheid variables have been found in distant nebulae well beyond our own galaxy and thus can be used to establish distances to objects outside the galaxy. This led to the once commonly used phrase, "measuring rods of the universe."

Within the relatively recent past, astronomers have worked on the development of a new cosmic yardstick, so to speak. Radio astronomers have found that the neutral hydrogen gas which pervades a spiral galaxy emits at a wavelength of 21 centimeters. By using the width of the 21-centimeter line, a measure of the rotation of the galaxy is gained. It is further reasoned that the rotation speed is dependent upon the mass of the galaxy. Past studies have shown that for spiral galaxies mass and optical luminosity also go together—because mass and luminosity are both largely provided by stars. Thus, it is reasoned that the width of the 21-centimeter line should be of use in determining the galaxy's intrinsic brightness. By combining the intrinsic brightness from the line width, observed optical brightness can be used for distance calculations. In 1976, the concept was initially tested on nearby galaxies, the distances of which had been previously determined by other methods. Reasonably good correlation was found when brightness in blue light was measured. But the conclusion was drawn that reliable galaxy brightness could not be derived from 21-centimeter line widths alone. Later research showed that there is much better correlation between the 21-centimeter line and the infrared intrinsic brightness. Research in this area is continuing, with the target of ultimately pinpointing the Hubble constant. This constant is discussed in entries on **Cosmology;** and **Red Shift.**

See also **Giant and Dwarf Stars.**

CEPHEUS. A constellation that is particularly interesting because it contains the remarkable variable star, δ Cephei. This star is the typical cepheid variable, one of a class that is most valuable in providing a measuring rod for probing the remote regions of space, because it obeys a relation called the period-luminosity law. (See map accompanying entry on **Constellations.**)

CEPHRADINE. Antibiotic.

CERAMICS. Derived from the Greek word *keramos* ("burnt stuff"), ceramics comprise a wide variety of materials which constitute a major industry. The principal facets of the ceramic industry, in order of increasing value of annual production, are: (1) abrasives; (2) porcelain enamel coatings; (3) refractories; (4) whitewares; (5) structural clay products; (6) electronic and technical ceramic products; and (7) glass. Glass accounts for about 45% of all ceramics produced. See also **Glass.**

Porcelain enamels are used to protect and decorate steel and aluminum metals. Actually, they are glasses especially designed to have high thermal expansions to match the base metal and to mature (to become glassy) at temperatures low enough to prevent distortion of the underlying metal sheet. Glazes perform a similar function on ceramic substrates, again, as special glasses matching the thermal expansion of the base and maturing at the desired temperature. Although not relatively large in terms of production value, the preparation of ceramic composites is important and growing. This area covers a variety of combinations, such as sapphire, Al_2O_3, whiskers in metals, metal-bonded carbides used in the machine tool industry, and directionally solidified two-phase ceramic systems. In a system of this type, an oriented fibrous second phase is grown in a primary matrix phase to maximize in a selected direction various important characteristics, such as minimum long term, high-temperature creep. An example of use is in gas turbine rotor blades.

Conventional Ceramics (*Structurals*). The essential raw material is clay. Clay is essentially a hydrated compound of aluminum and silicon $H_2Al_2Si_2O_9$, containing more or less foreign matter such as (1) ferric oxide Fe_2O_3, which contributes the reddish color frequently associated with clay, (2) silica SiO_2 as sand, (3) calcium carbonate $CaCO_3$ as limestone. Since clay is formed by the decomposition of igneous rocks, followed by transportation of the fine particles by running water and later deposition of these particles by sedimentation when the flow of water diminishes in speed, the quality of clays shows a wide range. When clay is wet it is plastic and can be shaped according to the desire and skill of the operator. The shape is retained on drying, and subsequent heating produces a coherent, hard mass, which suffers in the process more or less shrinkage and deformation depending upon the composition of the raw materials, and the method and temperature of treatment. Common bricks are made of crude materials without careful regulation of the conditions of treatment.

Bricks and plain clay products possess an earthy surface and fracture, are porous, and the strength depends upon the materials and treatment. Porcelain, on the other hand, possesses a glasslike or vitreous surface and fracture, and is not porous. Porcelain is made by mixing with the clay some powdered feldspar mineral, potassium aluminosilicate ($KAlSi_3O_8$ approximately). At the temperature of firing, feldspar undergoes a gradual change from the crystalline to the glassy state, the rate depending upon the time of heating and the temperature to which it is subjected. The fusion point of feldspar is of the order of 1300°C, whereas that of kaolin (pur clay) is of the order of 1700°C. Subjection of the porcelain raw material to the latter temperature would result in the formation of a glass. But when the temperature used is below the melting point of the clay portion and about the melting point of the feldspar, the latter produces a glass cement which binds together the particles of the former. When ground quartz SiO_2 is added to the original clay mixture the shrinkage of the material in the processes of drying and firing is reduced, the resistance to deformation during firing is increased, and the temperature coefficient of expansion of the product is affected.

The range of clay, feldspar, and quartz, as to the ratios in the mixture and as to individual composition of each (see Tables 1 and 2), as well as the available range of temperature of firing makes possible the production of products of a wide variety of physical structure. There has been proposed an arbitrary line of demarcation, namely that the unglazed product, such as has been described, which absorbs not more than 1% of its weight upon and after immersion in water, shall be termed porcelain, otherwise it shall be called earthenware. Such a nonporous material as porcelain, which includes chinaware, is also distinctly translucent in thicknesses of a few millimeters, whereas earthenware is nontranslucent and somewhat porous.

Materials that are to be glazed are dipped in a slip (the mixture of raw materials and water), dried, and refired. The glaze mixture is made up so that its fusion temperature is lower than that of the body of the ware, and the firing temperature is such that a surface of glass is formed over the body of the ware.

Designs and colors may be placed, as is commonly done, on the glaze and refired, or, as less commonly and more recently with fine effect, directly on the body under the glaze, in which case the glaze when produced covers and protects both the body of the ware and the decoration.

The properties of ceramics depend mainly on how the atoms are arranged and the interatomic bonds they form. The most important bonding force in the crystalline phases in most ceramics is ionic bonding, the metallic atoms losing an outer electron to become positive ions; and the nonmetallic atoms gaining an outer electron to form a negative ion. Ionic crystals are brittle and hard, melt at high temperatures, and have low electrical conductivity at room temperature. Compounds of metals with oxygen ions that are largely ionic are MgO, Al_2O_3, and ZrO_2. Covalent bonding is also found in ceramic crystalline materials. In this case, a pair of electrons is shared by two atoms. Covalent crystals, such as diamond and silicon carbide, have high hardness, high melting points, and low electrical conductivities at low temperatures.

The basic building block for the silicate crystal structures is the silicon–oxygen tetrahedron with a silicon atom at the center and four oxygen atoms at the corners. The silicates are classed by the types of bonding existing between the tetrahedra in their crystal structures. In orthosilicates, the tetrahedra are independent of each other. These structures make good refractories because of their high melting points. This group includes the olivine minerals, garnets, zircon, kyanite, and mullite. When the tetrahedra are joined at only one corner (oxygen atom), they form pyrosilicates, which are rare. In metasilicates, the tetrahedra share two corners to form a variety of ring or chain structures. Minerals of this type include the pyroxines, such as spodumene, and the amphiboles, such as asbestos. Sharing three corners, the tetrahedra form disilicates, which exist as sheets or planes, forming such minerals as mica. In the various forms of silica, such as quartz and crystobalite, all four tetrahedron corners are shared.

Most ceramic shapes do not consist of one single crystal, but are composed of numerous crystals joined together to form polycrystalline structures. The characteristics of the grain boundaries between crystals can influence the strength, chemical stability, and electrical properties as much as do the crystalline structures within the individual grains.

Glass is a very important part of ceramics. The glass industry is the largest single element of the entire ceramic industry, and the glassy portions of many ceramic bodies are the bond that hold many ceramics together. Probably the majority of the ceramics produced are a mixture of crystalline grains and a glassy phase. The glass frequently acts as the bond. This is the basis of the vitrified-grinding wheel industry and much of the structural and whiteware branches of the ceramics industry.

The rare-earth elements have found application in the ceramics field. In one example, a mixture of about 90% yttrium oxide powder and 10% thorium oxide powder is pressed into the desired shape and then sintered at about 2200°C. This heat treatment removes the microscopically small pores from between the powder particles. The result is a single-phase, polycrystalline material with a grain size normally between 10 and 50 micrometers in diameter. Yttrium oxide has a cubic crystal structure, thus light is not scattered at grain boundaries. This property, combined with the absence of a second phase and pores, imparts exceptional transparency (with polishing) to visible and infrared light. The transmission cutoff in the ultraviolet range occurs at 0.24 micrometers and in the infrared range at about 9 micrometers. Although the index of refraction of the ceramic is high, about 1.91 at the sodium D line wavelength, the optical dispersion of the material is very low. See also **Rare-Earth Elements and Metals.**

TABLE 1

Raw Materials	Chinaware	Earthenware
Total clay, usually blended	46.5%	50.5%
Feldspar	15	13.5
Quartz	36	36
Dolomite	2.5	—
Porosity average	0.5	8

TABLE 2. ANALYSES OF CERTAIN CERAMIC RAW MATERIALS

	Silica	Alumina	Lime	Magnesia	Iron Oxides	Soda	Potash
	%	%	%	%	%	%	%
Kaolin	58	29	0.2	0.3	1	1	1
Fire clay	61	26	0.3	0.4	1	1	1
Common brick clay	58	14	7	1.5	4	3	3
Feldspar	71	16	0.3	0.0	0.5	4	7
Quartz	100	—	—	—	—	—	—

CERCARIA. A larval form of the flukes (parasitic flatworms) with a compact body and slender tail.

CEREBELLUM. **Brain and Nervous System.**

CEREBRAL ANEURYSM. Aneurysm; Cerebral Vascular Diseases.

CEREBRAL ANGIOGRAPHY. Brain Disorders.

CEREBRAL EMBOLISM. Cerebrovascular Diseases.

CEREBRAL GANGLION. 1. The simple brain of many invertebrates. 2. Either member of the anterior pair of ganglia in the molluscan nervous system. 3. The upper posterior component of the brain of cephalopod mollusks.

CEREBRAL HEMISPHERES. Brain and Nervous System.

CEREBRAL HEMORRHAGE. Cerebrovascular Diseases.

CEREBRAL PALSY. The common name for a group of disorders resulting from brain injury, usually manifested by some type of paralysis and incoordination. The term normally denotes a condition in which the patient has lost some degree of muscular control. There is a wide range in the degree to which paralysis may be present. Conditions are caused by damage in one or more of three main areas of the brain that regulate muscular activity: (1) the *motor cortex*, damage to which results in stiffness of the muscles (*spastic paralysis*); (2) *basal ganglia*, a group of nerve cells in the brain which normally restrain certain types of muscle activity, thus injury in this area permits unplanned movements to occur. They are of two main types—slow, squirming, twisting movements that spread from the smaller joints to the larger ones without pattern (*athetosis*), most common in the arms than in the face and legs; and tremor, characterized by rhythmic motions which may range from slight shaking to violent jerking; and (3) the *cerebellar* area of the brain which controls muscle coordination as well as balance, injury of which causes the ataxic type of cerebral palsy, characterized by clumsiness and lack of balance. See also **Brain and Nervous System.**

There are numerous causes of cerebral palsy. Prior to birth, the brain may not fully develop. Injury or disease of the mother during pregnancy and antagonistic blood factors are other causes. Damage to the brain may result at birth because of delivery difficulties. In the case of a premature baby with soft bones, there may be insufficient protection to protect the brain from harmful pressures. Bleeding in the brain after birth may cause destruction of cells. Breathing difficulties at the time of birth may prevent sufficient oxygen from getting into the blood and hence supplying the brain. Nerve cells which are easily destroyed by lack of oxygen may suffer.

After birth, the baby with cerebral palsy may appear and act normally. Presence of blueness, twitching, or convulsions, however, should be regarded with suspicion. Sometimes it is difficult to make a definite diagnosis until the second 6 months of life. If watched closely, parents may detect signs as early as the first 2 or 3 months. The baby may not move much, or the legs may seem unusually stiff. Failure to follow the normal rate of babyhood accomplishments is important. Thus, a child who cannot grasp an object at 3 months, or turn over at 5 months, or sit alone at 7 months, may be showing the first signs of cerebral palsy.

The *spastic* type of cerebral palsy is the most common (approximately 65% of all cases). The child's stiff, tense muscles do not remain quiet and relaxed when not in use, as normal muscles do. They tighten up even more as the child tries to move, or if he is excited or frightened. The posture of spastic children is characteristic. The legs turn in, bending at the hips and knees, and the heels are off the ground. The arms are bent at the elbows and wrists, while the fingers are clenched. The child has trouble speaking and swallowing. The child is shy, prefers to be alone, and is generally afraid.

In the *athetoid* type of child (19% of all cases), unwanted motions begin when the child starts a planned motion. In reaching for a ball, the arm may wave about so that the hand never comes near its goal. The aimless activity affects the muscles of the throat, face, and tongue, and seriously hampers speech and swallowing. This type of child is fearless, lovable, and patient.

The *ataxic* form (about 8% of all cases) is characterized by clumsy-looking children who have a sense of balance and try to keep from falling, but they find it difficult to walk on a narrow base. Muscle coordination is lost and they have great trouble with skillful acts, such as writing and throwing a ball. Speech and swallowing are fairly normal.

Where the entire brain suffers, *rigidity* (4% of all cases) results and is often associated with severe mental deficiency. The body is rigidly arched backward, and the head is thrown back. The victims relax some in sleep. In the *tremor* type (2% of all cases), the child has control of his muscles until he starts to do something or becomes excited. Then, the vibrations become worse and interfere with the use of his hands. In severe cases, the movements are present even when the child is quiet and at rest.

Mental deficiency is seen with all types of cerebral palsy and may be mild or severe, but only one-third of affected children are below the acceptable educational levels. Ataxic or spastic children are somewhat more apt to be deficient mentally.

Treatment is slow, long, and constant. Parents should realize that the aim of treatment is not to restore to normalcy, but to make the child useful to himself and society, and therefore happier. Speech and swallowing defects are the most urgent to overcome. Muscle training is the most valuable way of treating these children, and a way in which intelligent, cooperative parents can be immensely helpful. This training is done by a physiotherapist, but many of their routines can be learned by parents and repeated at home. The child must find a sympathetic, encouraging environment in which he or she is accepted by others and given opportunity for relationships with others.

CEREBRAL THROMBOSIS. Cerebrovascular Diseases.

CEREBRAL TRANSIENT ISCHEMIC ATTACKS (TIA). Anticoagulants; Cerebral Vascular Diseases.

CEREBROSPINAL FLUID. This fluid (CSF) is a clear watery liquid that surrounds the brain and spinal cord and fills the four cavities or ventricles of the brain. See also **Brain and Nervous System.** The bulk of the fluid is thought to originate in filamentous structures, the choroid plexuses, which are situated on the walls of the ventricles. The fluid flows out of the brain through openings in the roof of the fourth ventricle to circulate over the brain and around the spinal cord in the subarachnoid space. This area is simply the space between the membrane (pia mater) directly adjacent to the nervous tissue and the membrane (arachnoid) attached to the tough outer covering of the brain (dura mater). The liquid leaves the subarachnoid spaces

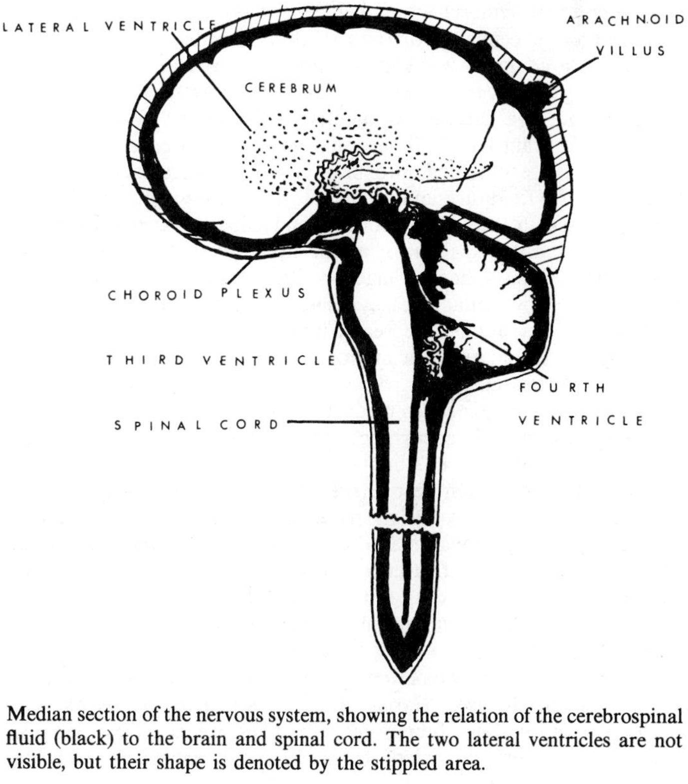

Median section of the nervous system, showing the relation of the cerebrospinal fluid (black) to the brain and spinal cord. The two lateral ventricles are not visible, but their shape is denoted by the stippled area.

to enter the blood by flowing through canals in the arachnoid villi which open into large venous channels. There are similar structures related to veins in the arachnoid membrane covering the spinal cord. These spatial relationships are shown in the accompanying diagram.

In humans, the entire volume of CSF is between 120 and 150 cubic centimeters, about 30 cc of which is in the ventricles. The rate of formation is difficult to establish, as it appears to depend upon the normal degree of absorption through the arachnoid villi. More than a liter may be formed in a day if drainage is brought about by artificial means, or by spontaneous pathological processes. Ordinarily, daily production is estimated to be from one to three times the total volume.

In addition to protecting the brain and spinal cord from mechanical shock, the CSF functions to maintain the intracranial pressure at a constant level. Thus, whenever the volume of brain tissue is altered because of differences in the volume of blood in the brain, or after surgical manipulation, the CSF volume is adjusted accordingly. No proven nutritive function and no unique biochemical action on the nervous tissue it bathes can be attributed to this fluid.

When the meninges or membranes covering the brain become infected, the protein concentration of the CSF may increase enormously and the largest protein molecules of the plasma are then found in the CSF. These disappear and the protein concentration returns to normal when the patient recovers. See also **Meningitis.**

The maintenance of such a large differential in protein concentration between the plasma and CSF indicates that a highly selective membrane exists between the blood and the CSF. In fact, the equilibration of substance between the blood and CSF does not occur with the same rapidity that it does between the blood and extracellular spaces of other body tissues. This impediment to rapid equilibration of substances between the blood and the CSF has been termed the blood–CSF barrier, although no anatomical structure can be discerned on which to explain its existence. It appears to reside in the nature of the cells of the choroid plexuses and the cells lining the surface of the brain.

CEREBROVASCULAR DISEASES. These diseases by definition involve the vascular (blood vessel) system of the brain. Like vascular diseases that occur elsewhere in the body, a major cause is atherosclerosis, a process that damages the arteries and veins and which is described in some detail in the entry on **Arterial and Venous Disorders.** The principal cerebrovascular diseases are: (1) cerebral transient ischemic attacks (TIA); (2) cerebral thrombosis; (3) cerebral embolism; and (4) cerebral hemorrhage. Other cerebrovascular diseases include hypersensitive encephalopathy and cerebral arteritis.

Stroke is a common word used to designate sudden paralysis and coma arising from cerebral effusion or extravasation of blood and a manifestation of a thrombosis or clot in the artery which supplies blood to the brain (cerebral thrombosis), or, less frequently, as the result of leakage of blood into brain tissue (cerebral hemorrhage). As used in the literature and media, stroke is not an exactly defined term and does not necessarily connote suddenness or coma. Frequent references are made to silent or minor strokes versus the fatal stroke of massive coronary hemorrhage or thrombosis. The physiological nature of cerebrovascular diseases is such that each incident is different in terms of exact site and degree of involvement.

Nearly three-quarters of a million cases of stroke are reported in the United States each year. Stroke ranks third after heart disease and cancer as a major cause of death among adults. The disease, of course, is worldwide in occurrence, but statistics indicate some differences in rate of incidence. About one-quarter of a million deaths are attributed to stroke each year. Stroke is a major debilitating disease. It is estimated that nearly half the patients hospitalized for neurological care have had cerebrovascular disease. In a major study of stroke patients, it was found that, after recovery, about 31% of persons require some assistance in providing self-care; about 20% require assistance in ambulation. Of all victims of stroke, nearly 30% are unable to handle the full load of their prior vocational responsibilities. In addition to atherosclerosis, hypertension is a major underlying cause of cerebrovascular diseases. Hypertension is estimated to increase the risk of stroke by about four times. Heart disease, diabetes mellitus, and age also increase the risk of stroke. Some authorities also implicate the use of oral contraceptives as a contributing factor to the risk of stroke.

As also pointed out in the entry on **Ischemic Heart Disease,** epidemiologists have been puzzled by the decline in incidence of these diseases, including stroke, in recent years. Historical analysis by periods of birth show no difference in the rates of decline for people born in successive 5-year periods from 1865 to 1915. However, for the period between 1948 and 1978, the decrease in the incidence of strokes brought on by obstruction of blood vessels supplying the brain was 34%; and for strokes caused by hemorrhage within the brain, the decline was 44%. The rate for subarachnoid hemorrhage (between a membrane covering the brain and brain itself) showed almost no change.

The brain has a very high metabolic rate and cannot live long when deprived of glucose and oxygen. See **Brain and Nervous System.** The brain consumes 25% of the body's total requirements for oxygen and 70% of the glucose. Research has shown that, within the brain, the rate of blood flow to a given area is related to the metabolic requirements of that area. In acute cerebrovascular disease, it has been found that a match does not always exist between blood flow and metabolic activity. To protect against the excesses of either hypertension or hypotension, a means of *autoregulation* is incorporated in the brain. This protective system helps to provide against ischemia as well as excessive pressures that may cause hemorrhage. This mechanism is not well understood, but it has been established that in focal cerebral ischemia from stroke, there is dysfunction of the autoregulation system. The degree to which this dysfunction occurs has a bearing upon the extent of recovery from ischemia.

Cerebral Transient Ischemia Attacks (TIA). In essence, these are minor or silent strokes which may develop suddenly and persist from a few minutes to a whole day. Caused by circulatory disturbances (carotid or vertebrobasilar arterial systems) in the brain, TIAs usually are not observed until the seventh decade of life. It is suggested that these attacks may arise from microemboli that rapidly fragment and thus pose no further threat. Such emboli may come from a number of sources.

Where the ophthalamic artery is involved, there may be a transient blackout of vision (up to several minutes). Some patients have described this episode as "the drawing down of a curtain." After the attack, there will be no neurological evidence of its occurrence. Mild symptoms may accompany such a TIA—numbness in the hands, short-term fuzziness of thinking, possibly some difficulty in speaking. Rarely are there disturbances of consciousness. Such attacks may occur only once or infrequently, or there may be several in a given day. The patient should consider a TIA as a warning and report such episodes to the physician. Depending upon patient, exact symptoms, and past history, the physician may decide to institute therapy to forestall the onset of a major stroke. For example, anticoagulation therapy may be indicated. See **Anticoagulants.**

It has been established that about one-third of persons who experience TIAs ultimately have a completed stroke; another one-third may continue to experience additional TIAs; for the other third, a TIA may be an isolated experience.

When the vertebrobasilar artery syndrome is involved in a TIA, the symptoms will differ. Diplopia (double vision), dizziness, nausea, vomiting, speaking difficulty, and weakness of limbs may be present.

Cerebral Thrombosis. This condition may develop suddenly or slowly. Any of the cerebral arteries may be involved, but most frequently the site of intracranial vascular thrombosis will be in the middle cerebral artery. Small areas of infarction (*lacunar sites*) also may occur where there is an occlusion in the branch of a major cerebral artery. Cerebral thrombosis frequently leads to infarction with irreversible death of brain tissue. To some extent, the damage due to dead tissue may be resolved by a full clearing of edema and perhaps functional reorganization of the brain, including new synapse formations. See **Brain and Nervous System.** The course and degree of recovery from stroke are highly variable and for the first several weeks after the episode, difficult to forecast. Treatment is generally supportive, including provision for adequate intakes by the patient of air and glucose to the brain. Thus, intravenous and nasogastric tube feeding are frequently indicated. Hypertension if severe should be treated promptly. Inasmuch as recovery is usually long and drawn out, early attention must be given to the patient's comfort and general health. Normal functions may be affected by paralysis accompanying the stroke, notably to the urinary and digestive system. Condom or catheter drainage may be provided. The patient will be turned frequently. Some physicians prefer that water mattresses be used. In an evolving stroke, many physicians use anticoagulants, but not usually in a completed stroke. Very important to the stroke patient is early commencement of physical therapy.

Cerebral Embolism. Emboli may reach the cerebrovascular system. These emboli may be blood clots, atheromatous substances (from damaged blood vessels), tumor cells, fats, foreign bodies, air, and nitrogen bubbles, among others, but in cerebral embolism, blood clots, usually originating in a diseased heart, are by far the most common. Blood clots account for 10 to 15% of all cerebrovascular accidents.

Stroke from an embolism may develop within a few seconds. Other symptoms parallel those of cerebral thrombosis. Tissue damage is highly dependent upon the final site of the embolism. If the cerebrospinal fluid is clear, some specialists use anticoagulant therapy with heparin during the acute phase of embolism. Once the patient has stabilized, the physician will attempt to determine the source of emboli and promptly initiate preventive measures against further emboli. Supportive measures and physical therapy parallel those for cerebral thrombosis.

Cerebral Hemorrhage. This condition is quite different from the similarities shown by cerebral thrombosis and cerebral embolism. In cerebral hemorrhage, bleeding into the brain tissue occurs.

In a *parenchymatous hemorrhage*, rupture may occur in small penetrating arterioles located deep within the brain. Bleeding will result in the formation of a hematoma (massive clot of extravasated blood). This will occupy space and thus displace brain structures. Often one or two hours may be required for a cerebral hemorrhage to evolve to this stage. During this time, there may be vacillating states of consciousness, almost always accompanied by severe headache and, frequently, vomiting. Deviation of the eyes may occur. Stupor and coma are not uncommon.

An intracranial hemorrhage frequently produces a bloody cerebrospinal fluid, along with increased pressure. Lumbar puncture may be indicated. Computerized axial tomography and cerebral arteriography provide more definite diagnosis when required.

Massive intracerebral hemorrhage is fatal during the acute phase in about 80% of cases. But if the hematomas are small, essentially full recovery is possible. Surgery is sometimes indicated.

In *primary subarachnoid hemorrhage* (bleeding into the cerebrospinal fluid), ruptured aneurysms are a frequent cause. Such hemorrhages also may arise from primary bleeding disorders, leukemia, overaggressive anticoagulation therapy, hemorrhagic encephalitis, and hemorrhage from tumors. Although aneurysms are prone to rupture in the middle decades of life, rupture can occur at any age. In a number of cases, there are premonitory symptoms, such as recurrent headache (during period just prior to hemorrhage) and stiff neck. However, most hemorrhages of this type occur quite suddenly, commencing with a violent headache, quickly progressing to confusion, agitation, collapse, and coma. Diagnosis is confirmed by finding increased pressure and a lot of blood in the spinal fluid. As soon as the patient's condition permits, angiography will be ordered. Neurosurgery will be weighed carefully. This early consideration of surgery is important because renewed bleeding may commence during the second week after the initial hemorrhage. A number of operative procedures have been developed and frequently prove successful.

During the early critical period of this condition, full supportive measures must be taken, with around-the-clock nursing care. Absolute bed rest is essential, with no patient self-care permitted during the critical period. Efforts may be made to induce hypotension (low blood pressure). Even with good progress, hospitalization for about 6 weeks is usually considered minimal. The patient will be advised not to resume normal activities for about 3 months.

Hypertensive Encephalopathy. This condition is believed to be caused by a spasm in a cerebral blood vessel which, in turn, precipitates a sudden increase of systemic blood pressure. The syndrome is recognized by the severity of the quickly developing symptoms—confusion, delirium, twitching muscles, and convulsions ultimately leading to stupor or coma. Sometimes, an increasingly severe headache will be a prelude. These are early morning headaches regionalized in the back of the head. Examination will show an elevation of diastolic blood pressure.

Treatment requires rapid lowering of blood pressure, with the intravenous administration of drugs, such as diazoxide or nitroprusside frequently indicated. Dexamethasone may be administered intravenously to relieve intracranial pressure as the result of cerebral edema. Convulsions may be treated with phenytoin.

Giant Cell Arteritis. In this disease, there is mononuclear and giant cell infiltration of the arterial wall, causing fibrous proliferation of elastic tissue. See **Arterial and Venous Disorders.** Blindness is a frequent accompaniment of the disease when left untreated. This disease is rarely seen in persons less than 50 years of age. Early symptoms are quite nonspecific and include headache, low-grade fever, loss of appetite and weight, fatigue, and depression, but may not be sufficiently annoying for a person to seek the counsel of a physician. Unfortunately, the disease may progress to the point where the patient has lost the sight of one eye prior to seeing a physician. Upon examination, the presence of tender nodular swelling of arteries in the scalp may be noted. Temporal artery biopsy is required to confirm giant cell arteritis. Treatment with prednisone is commenced immediately and its effect on the symptoms will be noted within a few days. Since recurrences are common in this disease, prednisone may be continued for an indefinite period. Early diagnosis can prevent blindness.

References

NOTE: See references list at ends of entries on **Arterial and Venous Disorders; Brain and Nervous System;** and **Hypertension (High Blood Pressure).**

Austin, G. M.: "Microneurosurgical Anastomoses for Cerebral Ischemia," Charles C. Thomas, Springfield, Illinois (1976).

Gresham, G. E., et al.: "Residual Disability in Survivors of Stroke—the Framingham Study," *N. Engl. J. Med.*, **293**, 954 (1975).

Marder, J. M.: "The Use of Thrombolytic Agents: Choice of Patient, Drug Administration, Laboratory Monitoring," *Ann. Intern. Med.*, **90**, 802 (1979).

Millikan, C. H., and F. H. McDowell: "Treatment of Transient Ischemic Attacks," *Stroke*, **9**, 299 (1978).

Scheinberg, P. (editor): "Princeton Conference on Cerebrovascular Diseases," Tenth Session, Raven Press, New York, 1976.

Schneider, R. C., et al.: "Correlative Neurosurgery," 3rd edition, Charles C. Thomas, Springfield, Illinois, 1980.

Smith, R. R., and J. T. Robertson: "Subarachnoid Hemorrhage and Cerebrovascular Spasm," Charles C. Thomas, Springfield, Illinois, 1975.

Toole, J. F.: "Cerebrovascular Disorders," McGraw-Hill, New York, 1974.

CEREBRUM. Brain and Nervous System.

ČERENKOV COUNTER. A detector of very fast moving mesons and electrons in which Čerenkov radiation is produced and focused onto a multiplier phototube. It measures the velocity of the particle.

ČERENKOV RADIATION. This is a feeble radiation in the visible spectrum, which occurs when a fast charged particle traverses a dielectric medium at a velocity exceeding the velocity of light in the medium. It is thus a shock-wave phenomenon, the optical analog of a sonic boom. The particle runs away from its own electromagnetic field. The radiation arises from the local and transient polarization of the medium close to the track of the particle. With reference to Fig. 1(a), consider an arbitrary element S of the medium to one side of the track AB of a fast electron, the track defining the z-axis. At a particular instant of time, when the electron is at, say, e_1, the local polarization vector P_1 will be directed along Se_1', to a point e_1' slightly behind e_1, owing to the retarded fields. As the particle goes by, the vector P_2 will turn over and, when the electron reaches e_2, will be directed to a point e_2'. The variation of P with time may be resolved into radial and axial components P_ρ and P_z, as shown in Fig. 1(b). Owing to cylindrical symmetry, this polarization, viewed at a point distant from the particle, appears as an elementary dipole lying along the z-axis, Fig. 1(c). As the particle plunges through the medium, radiation arises from the coherent growth and decay of this sequence of elementary dipoles.

Two essential features of the radiation become at once apparent. First, since it is only the P_z component which is important, the field variation (Fig. 1(b)) is that of a double δ-function. Thus, from Fourier analysis, if the circular frequency ω, we will expect a spectrum of

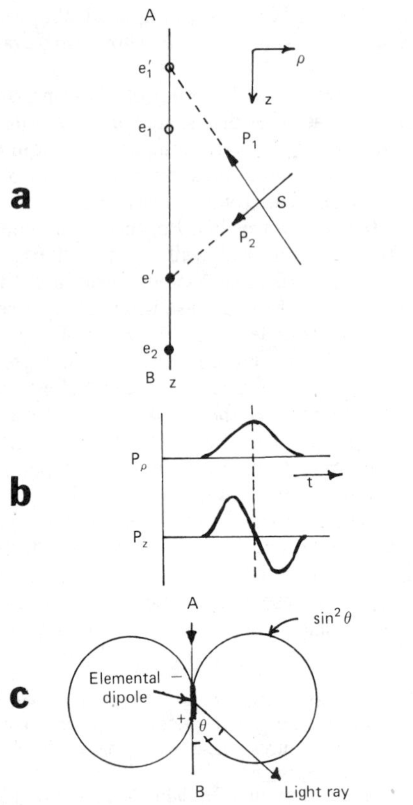

Track of particle in formation of Čerenkov radiation.

the form $\omega \cdot d\omega$, i.e., radiation which is bluer than that from an equi-energy spectrum. Secondly, since the radiating element is an axial dipole, the angular distribution, for this element alone, will be of the form $\sin^2\theta$ (Fig. 1(c)). It is important to realize that the radiation arises from the medium itself, not directly from the particle. Since the medium is stationary, the intensity and angular distributions do not contain the relativistic factor (mc^2/E); in this respect, Čerenkov radiation is essentially different from Bremsstrahlung or synchrotron radiation. (See also **Bremsstrahlung;** and **Synchrotron Radiation.**) The foregoing description applies only to one element along the track. The presence of other charged particles would add to the effect. The effect is greater when the index of refraction of the medium is large.

The phenomenon has found considerable application in the fields of high-energy nuclear physics and cosmic-ray research; in almost all practical Čerenkov counters, the light is detected by means of a photomultiplier. The unique directional and threshold properties of the radiation may be used in a number of different ways. For example, by velocity selection, it is possible to distinguish between particles of different mass having the same energy, and it is also possible to measure particle velocities directly by measuring θ. Light flashes from the night sky associated with cosmic ray showers have been attributed to the effect, and microwaves have been produced by the Čerenkov process.

CERES. Asteroid.

CERIUM. Chemical element symbol Ce, at. no. 58, at. wt. 140.12, first in the Lanthanide Series in the periodic table, mp 798°C, bp 3433°C, density 6.770 g/cm^3 (20°C). Elemental cerium has a face-centered cubic crystal structure at 25°C. Cerium is the most abundant element of the rare-earth group and is 28th in ranking of the naturally occurring elements in the earth's crust. The element is a silver-gray metal which oxidizes readily at room temperature, particularly in moist air, to form the oxide CeO_2, which is of a pale yellowish-green color. Above 300°C, the element may ignite and burn with a bright red glow. Of the nineteen isotopes of cerium, only four occur in nature, ^{136}Ce, ^{138}Ce, ^{140}Ce, and ^{142}Ce. The thermal neutron-absorption cross section of the element is low. The element has a low toxicity rating. Electronic configuration is $1s^2 2s^2 2p^6 3s^2 3p^6 3d^{10} 4s^2 4p^6 4d^{10} 4f^1 5s^2 5p^6 5d^1 6s^2$. Ionic radius Ce^{3+} 1.034 Å, Ce^{4+} 0.92 Å. Metallic radius 1.825 Å. First ionization potential 5.47 eV; second 10.85 eV. Other important physical properties of cerium are given under **Rare-Earth Elements and Metals.**

Cerium was first identified by M. H. Klaproth in 1803 and, independently, in the same year by J. J. Berzelius and W. Hisinger. The element occurs in four source minerals, allanite, bastnasite, cerite, and monazite. Bastnasite, which is a rare-earth fluorocarbonate, is found in southern California. Monazite, a phosphate that contains thorium and the light lanthanides, is distributed widely throughout the world. See also **Bastnasite;** and **Monazite.** Cerium is recovered from the minerals through an extractive process using H_2SO_4, followed by precipitation with oxalic acid which separates the light lanthanides from thorium, yttrium, and the heavy lanthanides. Cerium metal is produced from its salts, such as CeF_3 or $CeCl_3$, by thermal reduction in a tantalum or molybdenum crucible. Alternative processes include the electrolysis of $CeCl_3$ or CeO_2. The latter compounds are soluble in a complex molten halide flux. The Ce^{3+} is reduced to metal at a molybdenum electrode. The process is carried out at from 800 to 1,000°C.

A major use for CeO_2 is in decolorizing soda-lime container glass. The compound also is used for polishing gemstones and glass, notably precision optical glasses. Cerium is particularly useful in glass that is subject to α-, γ-, and x-radiation, and the impingement of light and electrons because the cerium prevents discoloration that may arise from the presence of Fe(II) by oxidizing the Fe(II) as it is formed to Fe(III). This is an important factor in color television tubes. Cerium dioxide also is used in cathodes, capacitors, phosphors, ceramic coatings, refractory oxides, semiconductors, and photochromic glasses. The compound also is used as a catalyst and as an opacifying agent in porcelain enamels. Because of its low nuclear cross section, CeO_2 may be applied as a diluent in oxide nuclear fuels.

Cerium metal finds wide application in *mischmetal,* which is a rare-earth metal comprised of 50% Ce, 25% La, 18% Nd, 5% Pr, and 2% other rare earths. This alloy is used in shell linings for military projectiles, as an alloying agent for improving the malleability of ductile iron, and in lighter "flints" where the alloy is compounded with a 30% iron alloy. The pyrophoric and incendiary nature of cerium are evident when cerium-base alloys are machined. Mischmetal also improves the creep resistance of magnesium alloys, the resistance to oxidation of nickel alloys, the hardness of copper alloys, and the strength of aluminum alloys. Both cerium metal and mischmetal are used as *getters* to remove traces of oxygen in vacuum tubes and equipment. When alloyed with cobalt, cerium is gaining importance as a magnet material. $CeCo_5$, as a permanent magnet material, has properties which exceed those of the alnicos and ferrites. Mixed rare-earth oxides and fluorides containing up to 50% cerium are used as cores for carbon arcs which, for illuminating purposes, have much greater intensity and color balance. The mixed oxides with cerium also are used as catalysts (petroleum cracking and chemical oxidation reactions) and in a variety of waterproofing agents, fungicides, and polishing materials.

See references listed at ends of entries on **Chemical Elements;** and **Rare-earth Elements and Metals.**

NOTE: This 6th edition entry was revised and updated by K. A. Gschneidner, Jr., Director, and B. Evans, Assistant Chemist, Rare-Earth Information Center, Energy and Mineral Resources Research Institute, Iowa State University, Ames, Iowa. Original 5th edition entry was prepared by J. G. Cannon, Molycorp, Inc.

CERMET. Chromium; Nuclear Reactor.

CERN. Particles (Subatomic).

CERUMEN. The waxy secretion that collects in the external ear.

CERUSSITE. The mineral cerussite, lead carbonate, $PbCO_3$, is orthorhombic with tabular, prismatic and pyramidal crystals, with twinned forms very common. If not in crystal aggregates it may occur in granular or compact masses. Cerussite is very brittle with a conchoidal fracture; hardness, 3–3.5; specific gravity, 6.55 (a heavy mineral);

luster, adamantine but may be vitreous to resinous, pearly or even submetallic. Its color is variable, white to gray, grayish-black or blue or green, transparent to translucent.

Cerussite is of secondary origin being found associated with other lead minerals, and is widely distributed. There are many European and American localities. Fine crystals have been obtained from Phoenixville, Pennsylvania; Joplin, Missouri; Leadville, Colorado; Pima County, Arizona, and Dona Ana County, New Mexico. It is an ore of lead, and frequently carries values of silver. Derived from the Latin *cerussa*, white lead.

CERVICAL DISK DISEASE. Bone.

CERVINES. Deer.

CERVIX. Any narrow or neck-like portion of an organ. The term is usually used in reference to the narrow end of the uterus that projects into the vagina.

CERVIX (Cancer). Cancer and Oncology.

CESAREAN SECTION. Surgical removal of the fetus from the uterus by means of an incision through the abdominal walls. The procedure is used when there is deformity and narrowing of the bony pelvis which does not permit delivery through the vaginal route; where a pelvic tumor blocks the birth canal; in cases where there would be difficult breech deliveries; in certain patients with conditions which make labor dangerous to the safety of the mother; and, traditionally, in the past, in patients who have previously been delivered by cesarean section. Some of these justifications are now under review.

The first record of an authentic cesarean operation on a living woman was in 1610 by Trautmann (Wittenberg, Germany). Probably other operations had occurred earlier, but were not officially recorded. Although difficult to understand, up to 1882, no sutures were used in the incision in the pregnant uterus and most of the women perished from hemorrhage. Even after sutures were used, the mortality was so great from peritonitis, shock, and hemorrhage that, in 1887, of eleven operations performed in New York City, only one mother survived. Over the intervening nearly 100 years, the success rate has improved to the point where, during the 1970s in the United States alone, the number of cesarean deliveries tripled from 5.5% to 15% of all deliveries and is continuing to increase in the early 1980s. These statistics are not sufficient for a number of authorities, however, who feel that the trend toward this procedure may be getting somewhat out of hand. Thus, in late 1980 a task force was set up by the National Institutes of Health to make an initial review of the subject.

Many of the suggestions of the first task force were essentially qualitative because, admittedly, the statistics on morbidity have not been fully or effectively compiled. Records do show, however, that in the United States in 1970, fewer than 12% of fetuses in breech presentation (buttocks or feet, rather than head first) were delivered by cesareans. By the early 1980s, this had risen to 60%. The task force indicated that term (not premature) fetuses in breech presentation can be delivered vaginally when the fetuses weigh less than 8 pounds (3.6 kilograms), when the physician is experienced in breech deliveries, when the woman has a normal pelvis, and when the fetus's head is not bent back. For premature breech babies, the task force was not definite as to whether vaginal or cesarean deliveries are safer.

The mortality rate for women who have a cesarean is twice that of the rate for vaginal delivery. Also, some authorities observe that cesarean deliveries carry a greater risk that the mother will develop a nonfatal infection after delivery.

The task force observed that traditionally 99% of women who have had a cesarean are automatically given cesareans in all future deliveries. Repeat cesareans account for 25–35% of the increase in cesarean birth rates during the past decade. Not all authorities are convinced one cesarean need follow another. The task force suggested that present horizontal incisions low on the abdomen are unlikely to rupture under the future strains of labor and a vaginal delivery. The practice many years ago involved making vertical incisions which were more vulnerable to rupture and thus established the tradition of performing cesareans on all subsequent births.

About 30% of cesareans are performed on women with prolonged or difficult labor. The task force reported that there seems to be no survival advantage for babies delivered by cesarean rather than vaginally, because of this reason. The task force suggested that physicians reconsider measures, such as sedation, stimulation of uterine contractions with oxytocin, and walking about to facilitate labor. There was some concern that prolonged or difficult labor may be overdiagnosed in many cases.

CESIUM. Chemical element symbol Cs, at. no. 55, at. wt. 132.905, periodic table group 1a, mp 28.40°C bp 678°C, density 1.88 g/cm^3 (20°C). Elemental cesium has a body-centered cubic crystal structure. Cesium is a silver-white, very soft metal, one of the softest of all metals. The element tarnishes instantly on exposure to air, soon igniting spontaneously with flame to form the oxide. Generally, the element is preserved under kerosene. Cesium reacts vigorously with H_2O, forming cesium hydroxide and hydrogen gas. The element first was identified by Bunsen and Kirchhoff in 1860 through spectroscopic observations. Cesium occurs in nature as the ^{133}Cs isotope. There are 15 radioactive isotopes ^{125}Cs through ^{132}Cs and ^{134}Cs through ^{139}Cs. The half-life of ^{137}Cs is 33 years. This isotope is used as a source of gamma radiation, particularly in radiography and therapy. See also **Radioactivity.** First ionization potential, 3.89 eV; second, 23.4 eV. Oxidation potential $Cs \rightarrow Cs^+ + e^-$, 3.02 V. Other important physical characteristics of cesium are given under **Chemical Elements.**

The main source of cesium is carnallite $KCl \cdot MgCl_2 \cdot 6H_2O$ which contains a small percentage of cesium compounds. See also **Carnallite.** Cesium also occurs in pollucite (cesium aluminosilicate, 35% Cs_2O) and lepidolite (lithium aluminosilicate). See also **Lepidolite; Pollucite.** In early processes, cesium metal was obtained by the reduction of cesium salts, such as the hydroxide or chloride. In current practice, the metal is produced by electrolyzing the cyanide. The latter compound usually is fused cesium barium cyanide mixture.

The uses for cesium and its compounds are limited. Cesium is used in photoelectric devices because of its high sensitivity to light, finding applications in television, motion picture, radar, and instrumentation equipment. Cesium also has been used in luminescent tubes and screens. Certain processes for the manufacture of synthetic resins, such as chloroprene, use cesium as a catalyst. Some interest has been indicated in cesium as a fuel for ion-propulsion engines of low thrust for spacecraft. Like sodium, cesium also has been considered as a heat-transfer medium for special applications. The function of cesium in time measurement is important. As officially defined in 1967 by the International Bureau of Weights and Measures, the atomic second is equivalent to 9,192,631,770 oscillations of the atom of ^{133}Cs. This value expresses the ephemeris time (ET) second as closely as practical in terms of an atomic standard. To derive this value, scientists at Great Britain's National Physical Laboratory and the United States Naval Observatory used a dual-rate moon-position camera and a cesium-beam clock.

Cesium forms several solid solutions with rubidium. These alloys are used as *getters* for eliminating residual gases from vacuum tubes and systems. Because of their extreme reactivity in air, the alloys are difficult to apply. For easier handling, cesium can be alloyed with calcium, barium, or strontium. The ternary alloys of cesium, aluminum, and barium or strontium are employed in photoelectric cells. Cesium alloyed with antimony, silver, bismuth, and gold also displays photoelectric properties.

Chemistry and Compounds: Cesium is more electropositive than rubidium (or the lower alkali metals) as is consistent with its position in group 1a.

Because of the ease of removal of its single 6s electron (3.89 eV) and the difficulty of removing a second electron (23.4 eV) cesium is monovalent in its compounds, which are ionic.

In its solutions in liquid NH_3, cesium is like the other alkali metals, a powerful reducing agent, so that in such solutions, titrations of cesium polysulfide with cesium are made by electrometric methods. The solubility of cesium salts in liquid NH_3 increases markedly with the radius of an anion (the chloride, CsCl, 0.0227 moles per kg, the bromide, CsBr, 0.215 moles per kg, and the iodide, CsI, 5.84 moles per kg), though the values are less than for the corresponding rubidium compounds.

As in the case of the other alkali metals, cesium forms compounds generally with the inorganic and organic anions. For a general discussion of these compounds (see also **Sodium**) because the sodium compounds differ principally in their greater extent of hydration and greater number of hydrates. However, cesium coordinates with large organic molecules, such as salicylaldehyde, even though it does not with H_2O.

One respect in which cesium (and rubidium) are outstanding among the alkali metals is the readiness with which it forms alums. Cesium alums are known for all of the trivalent cations that form alums, Al^{3+}, Cr^{3+}, Fe^{3+}, Mn^{3+}, V^{3+}, Ti^{3+}, Co^{3+}, Ga^{3+}, Rh^{3+}, Ir^{3+}, and In^{3+}.

As in the case of potassium and rubidium, cesium forms a superoxide on reaction of the metal with oxygen. The compound is orange in color and paramagnetic because it contains the O_2^- ion with an odd electron in an antibonding orbital, and has the formula CsO_2. On heating, this compound loses oxygen to form black Cs_2O_3, which contains both CsO_2 and Cs_2O_2 (peroxide), which is the product of further heating. A series of suboxides of cesium is known, Cs_7O, Cs_4O (uncertain), Cs_7O_2, Cs_3O, and Cs_2O. Moreover the normal oxide, Cs_2O, can be prepared by heating cesium nitrite with metallic cesium. It reacts explosively with oxygen to form CsO_2.

Cesium hydroxide, CsOH, is the strongest of the five alkali metal hydroxides, as would be expected from its position in the periodic table (francium hydroxide, when prepared, would be expected to be stronger). For the same reason, it has the lowest lattice energy of the five (135.6 kcal per mole).

The most numerous organic compounds of cesium are the oxygen-connected ones, such as the salts of organic acids, and the alkoxy and aryloxy compounds (alcoholates, phenates, etc.). Among the carbon-connected compounds, an ethyl cesium, CsC_2H_5, and a phenyl cesium, CsC_6H_5, have been reported.

Rogowski and Tamura (1970) studied the environmental chemistry of ^{137}Cs. Later studies by other investigators (Alberts et al., 1979) have shown that ^{137}Cs introduced into a watershed is attached to soil particles, which are removed by erosion and runoff. Some of the eroded soil particles comprise the sediments of the catchment basins in the watersheds and act as "sinks" for ^{137}Cs. Other investigators have reported an almost irreversible fixation of this element in clay interlattice sites in freshwater environments and that it is unlikely that this nuclide will be removed from these sediments under normal environmental conditions other than by exposure to solutions of high ionic strength, such as may occur in estuarine environments. Studies of ^{137}Cs have been important because the element can be introduced into a water system from a leak in a nuclear fuel element. These findings are reported in some detail by Alberts et al. in *Science*, **203**, 649–651 (1979).

See list of references at end of entry on **Chemical Elements.**

CESIUM-BEAM CLOCK. Cesium; Clock; Time.

CESTIDA. Ctenophora.

CESTODA (*Cestoidea*). The tapeworms. A class of the phylum Platyhelminthes. The tapeworms, like other members of the phylum, are flat-bodied. The body consists of two regions, a head or scolex usually bearing hooks, suckers, or both, and a strobila which, in all but the simplest species, is formed of a series of segments called proglottids. Tapeworms are parasitic in vertebrates.

In addition to the characters mentioned, tapeworms are distinguished by a complex life cycle. The adults live in the intestine of the host, absorbing food through the wall of the body since they have no alimentary tract, and produce a long succession of proglottids which break off as they mature and pass out with the feces of the host, break off and remain in the intestine, or mature while still attached to the worm. They are reproductive bodies containing the organs of both sexes, rarely those of only one. The fertilized egg becomes a simple embryo with six hooks called the onchosphere. In this stage it is taken into the alimentary tract of a new host, migrates into the blood vessels, and after drifting along the bloodstream lodges in some part of the body and develops into another form, usually vesicular, called the bladder worm. In this stage it remains inactive unless the tissue containing it is eaten by another animal, in which case it attaches itself to the intestinal wall of the new host and develops into an adult tapeworm.

Tapeworms are among the important parasites of man. Some of these species live in hogs and cattle and become established in man as the result of eating imperfectly cooked meat and one of the most dangerous species is found in the dog during its adult stage and in domestic animals and man in the bladder worm stage. The elimination of tapeworms from the human body requires the careful attention of a physician.

The tapeworms are classified as follows:

Subclass *Cestodaria.* Parasitic in fishes as adults and in annelid worms and mollusks in the early stages. No distinct scolex and no segments.
- Order *Amphilinidea.* Species of leaf-like form.
- Order *Gyrocotylidea.* Leaf-like, with a projecting organ of attachment at the posterior end.

Subclass *Cestodes.* Usually segmented. Proglottids with organs of both sexes.
- Order *Tetraphyllidea.* Scolex without retractile projections (proboscides), with four suckers or bothridia. Parasitic in cold-blooded vertebrates.
- Order *Tetrarhynchidea.* Scolex with proboscides. Parasitic in selachian fish.
- Order *Pseudophyllidea.* No proboscides; only two suckers or bothria. In vertebrates of all classes.
- Order *Cyclophyllidea.* Four suckers and usually hooks. Body elongate, with distinct segments, often numerous. Principally in warm-blooded vertebrates.

CESTODES. Hydatid Disease; Tapeworm.

CETACEA. Whales, Dolphins, and Porpoises.

CETANE NUMBER. Petroleum.

CETUS (the whale). An equatorial constellation that lies south of Pisces.

CHABAZITE. The mineral chabazite is a member of that group of hydrous silicates, the zeolites, and corresponds to the formula $CaAl_2Si_4O_{12} \cdot 6H_2O$ with sodium sometimes replacing a part of the calcium. Potassium, barium and strontium may be present in very small amounts. Chabazite is hexagonal, usually in rhombohedrons that tend to resemble cubes. It has a rhombohedral cleavage; is brittle; hardness 4–5; specific gravity 2.05–2.10; luster vitreous; color white to flesh-red; streak white; translucent to transparent. Chabazite is found in the amygdaloidal cavities of basalts often associated with other zeolites. It is occasionally found in such crystalline rocks as syenites, gneisses and schists. Chabazite is a rather common zeolite, being found in many localities in Europe. In the United States it occurs in the Triassic traps of New Jersey and Maryland. The Triassic lavas of Nova Scotia have yielded fine specimens. The name chabazite is derived from the Greek word meaning a precious stone.

CHAETOGNATHA. The arrowworms, a group of small marine animals sometimes included in the phylum *Annelida* but now more often regarded as a separate phylum.

Arrowworms are usually elongate transparent animals. They have two or three pairs of horizontal fins, one forming a caudal fin at the end of the body. The head bears a pair of eyes and a group of spine-like jaws which give the name to the phylum. The alimentary tract runs through the body as a straight tube to an anus near the caudal end. The body cavity is divided into three chambers by two transverse septa. Although the phylum includes only about 30 species, arrowworms are found from the surface to great depths and in all of the oceans.

CHAETOPODA. A division of the annelid worms including the forms which have setae set in pockets in the integument. The coelom is well developed and is at least partially divided into metameric chambers, and the external segmentation of the body is also metameric.

Most of the marine annelids such as the lobworm and clam worm and the earthworms and fresh-water annelids are included here. The leeches and a few more primitive worms make up the rest of the phylum.

By some authorities this division is called a class and is divided into two orders:

Order Polychaeta. Free-swimming and sedentary worms, mostly marine, with lobed appendages (parapodia) and many setae. Usually a head with sensory organs. This group is sometimes regarded as a class and is then divided into two orders, the *Errantia* with similar body segments, most species free-swimming or burrowing, and the *Sedentaria* with specialized body regions, living in tubes in the bottom of the ocean, or between the tides.

Order Oligochaeta. Fresh-water and terrestrial worms with few setae and with neither parapodia nor sensory appendages. The earthworms are common examples. Sometimes ranked as a class.

CHAFER (*Insecta, Coleoptera*). A name applied to certain plant-eating beetles, including the rose chafer of the United States. The adults of this species are sometimes a troublesome pest on small fruits, especially grapes. They damage the fruit itself. Spraying with lead arsenate is recommended for their control.

CHAGA'S DISEASE. Also called South American trypanosomiasis, this disease is caused by the flagellate protozoan *Trypanosoma cruzi.* The disease appears to be confined to South and Central America (35 million cases), but animal reservoirs occur widely in the southern United States. The vectors are the assassin or cone-nose bugs (*Reduviidae, Hemiptera*) of the genera *Triatoma, Rhodnius,* or *Panstrongylus.* These insects emerge from mud or thatch walls at night to suck human blood and defecate while feeding, thus contaminating the wound with the infective stage. Five to fourteen days is the usual incubation period. The trypanosome occurs in the blood stream as a motile, flagellum-bearing, spindle-like organism about 20 μm long (trypomastigote) and multiplies as a smaller, intracellular parasite (amastigote) in the cells of the reticuloendothelial system, heart, and elsewhere. The acute stages are usually seen in children, last 20–30 days, and may include fever, malaise, lymphadenopathy and hepatosplenomegaly. Unilateral palpebral (eyelid) edema (Romaña's Sign) may occur in this stage. Survivors of the acute phase may become asymptomatic or develop chronic disease, the nature of which depends on the localization of the amastigotes. Dilation of the esophagus (megaesophagus) or colon (megacolon) is due to loss of autonomic ganglia within the viscus wall. In about 50% of cases there is cardiac involvement with enlargement and eventual congestive heart failure. Neurotrophic and suprarenal types are also recognized. Diagnosis is by blood films or examination of lymph nodes or marrow biopsies. Inoculation of guinea pigs or feeding of clean reduviids (xenodiagnosis) may be necessary in chronic cases. Prognosis is poor, especially in children, but 8-aminoquinolines are of some use in controlling the blood-borne stage. Prevention is complicated by the abundance of animal reservoirs including dogs, cats, pigs, rats, oppossums, and armadillos. Control measures include residual insecticides, upgrading housing, and education of the public.

Albert L. Vincent, Assistant Professor, Department of Comprehensive Medicine, University of South Florida, Tampa, Florida.

CHAIN. A flexible connector composed of metal links, used for hoisting or for power transmission. *Coil chain* is used for hoisting and haulage, and consists of oblong links of circular sections, usually of welded wrought iron or steel. Coil chain with a stud or bridge across the center of the coil is preferred to plain coil chain in some instances, since the studs tend to prevent stretching and kinking.

Chain used for power transmission is shown in the figure. *Detachable link chain* is used for low-speed and light-load power transmission, and for conveyors and elevators of moderate capacity and length. The links can be easily detached and replaced, as illustrated. *Pintle chain* is from two to four times as strong as detachable link chain and can be used with the same sprockets. Both types of chain are usually made up of malleable iron unmachined links. They can be supplied with integral pin, plate, or scraper attachments.

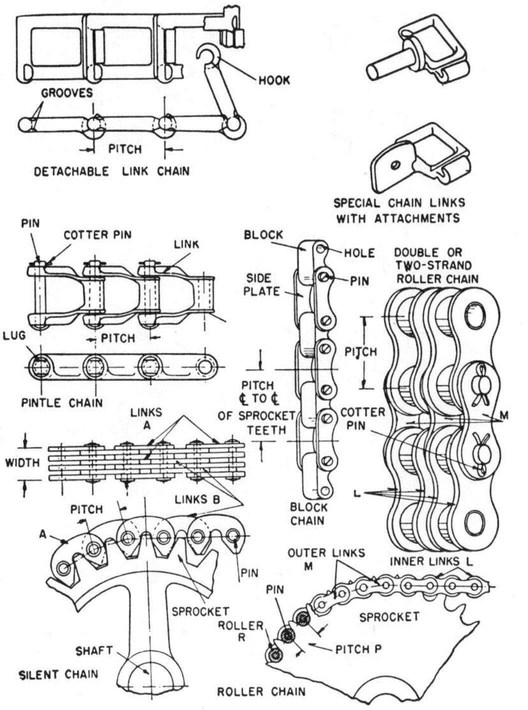

Power transmission chain configurations.

Steel block, roller, and silent chains are used where an exact speed ratio is desired and the center distance of the shafts is too large for gearing. *Block chain* is used for comparatively slow speeds and consists of blocks linked together by connecting links and pins. *Roller chain* consists of alternate links *L* and *M* held by connecting pins which are fastened by cotters. The pins also serve to carry the rollers which bear on the sprocket teeth. Roller chain can transmit more power than block chain and can operate at chain velocities up to 1200 feet (366 meters) per minute. For power requirements too great for single chains, double-, triple-, or quadruple-strand roller chains may be employed.

Silent chain is composed of alternate flat steel links *A* and *B* connected by pins. The links have straight faces in contact with the sprocket, and rotate slightly on the pins as the chain bends around the sprocket. Silent chain is used for heavy loads at speeds up to 1600 feet (488 meters) or more per minute. The silent chain is not actually quiet in operation but is much less noisy than other types of chain in use at the time of its adoption.

The speed ratio of a power chain depends upon the numbers of teeth in the driving and driven sprocket wheels; velocity ratios up to 7:1 are satisfactorily employed. Short-center drives with high ratios are usually more economical if fine-pitch chain is employed, while narrow large-pitch chain is cheaper for low-ratio long-center drives.

CHAIN-BALANCED HYDROMETER. **Specific Gravity.**

CHAIN BLOCK. Chains and sheaves may be employed in combination to produce an unusually powerful lifting mechanism. The best of these is the differential chain block, the action of which is explained in connection with the accompanying diagram. The mechanism consists of two sheaves, *A* and *B*, *A* being double sheave having diameters *R* and *r*. It will be shown that the multiplying power of this mechanism depends upon the ratio of these diameters. If they are equal, the pull *P* will not move the weight, and the efficiency of the mechanism will be 0%, but the theoretical mechanical advantage is infinity. A slight difference in radii will produce a very large lifting effort, although the efficiency may still be very low. The sheaves are made with link pockets so that the chain fits nicely into the circumference, and is restrained from slipping. Furthermore, the chain is endless, and the mechanism is self-locking by virtue of the friction intentionally allowed on the journals.

In explanation of the chain block, if the pull *P* revolves sheave *A* one revolution, the vertical chain at *a* is lowered through a distance

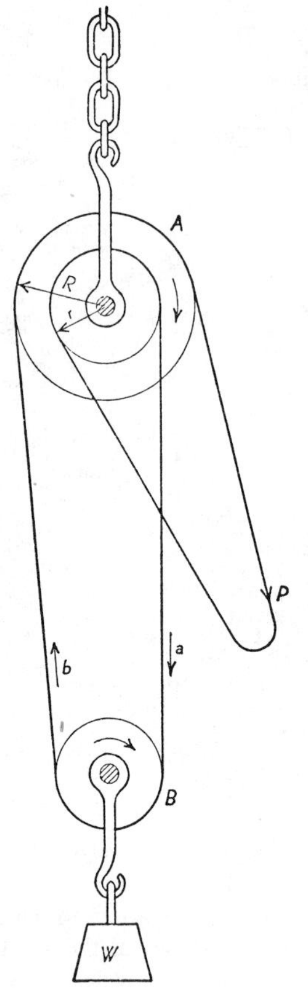

Chain block hoist.

to $2\pi r$, while the side b is raised the distance $2\pi R$. The net vertical displacement of the sheave B is $\pi(R - r)$ upward. With no friction considered, the work of lifting W through this distance must be equal to the work done by the pull P moving through $2\pi R$. Solving this equation for advantage W/P:

$$\frac{W}{P} = \frac{2R}{R - r}$$

Applying the mechanical efficiency e to this equation, the actual mechanical advantage is

$$\frac{W}{P} = \frac{2Re}{R - r}$$

These chain blocks are built in different sizes for hoisting loads from $\frac{1}{4}$ ton to 3 or 4 tons, by hand. On account of the self-locking feature depending on friction, the average mechanical efficiency of this device is only about 30%.

CHAIN-GRATE STOKER. Burner.

CHAIN MOLECULE. Identity Period.

CHAIN REACTION (Free Radical). Free Radical.

CHALAZION. Eyelid.

CHALCANTHITE. This mineral of triclinic crystallization is found only as a rare secondary mineral in the oxidized zones of sulfide copper ores within arid regions. It is a hydrous copper sulfate, $CuSO_4 \cdot 5H_2O$, and is a most unstable mineral in moist atmosphere environments, altering readily to a powder-blue dust. The mineral possesses a vitreous luster of deep azure-blue color, ranging from transparent to translucent. Chalcanthite is found in abundance only in Chuquicamata and other arid regions of Chile where it is an important copper ore.

CHALCEDONY. One of the cryptocrystalline varieties of the mineral quartz, having a waxy luster. It may be semitransparent or translucent and is usually white to gray or grayish-blue or some shade of brown, sometimes nearly black. Light colored clear red chalcedony is known as carnelian; deep reddish brown as sardonyx; a green variety colored by nickel oxide is called chrysoprase. Prase is a dull green. Plasma is a bright to emerald-green chalcedony which sometimes is found with small spots of jasper resembling blood drops; it is then referred to as blood stone or heliotrope.

Chalcedony and agate are essentially porous, which permits their being dyed various colors by artificial means. Red color if produced by iron nitrate solution; nickel nitrate produces vivid green color; ammonium bichromate produces blue-green; and ferrocyanide salts a vivid blue. The black onyx used extensively in rings is a product of soaking chalcedony in sugar solutions and later in sulfuric acid.

The term chalcedony is derived from the Greek name Chalkedon, a town in Asia Minor.

CHALCID WASP (*Insecta, Hymenoptera*). Very small insects of thousands of species, many of which are parasites on other insects and thus may be used for biologically controlling a number of damaging insect species. The tiny wasps are frequently of a shiny, metallic luster in appearance, with very simple wings showing a single vein. Only a few of the species feed on plants, notably the *fig wasp* (*Blastophaga psenes*), described below, which nevertheless is essential for the fertilization of Smyrna fig. The chalcid wasps are parasitic during their larval life, resulting from adult chalcids depositing eggs in a host. Host species are attacked by these parasites in all stages of metamorphosis, including the egg phase. A few species are important pests in wheat, but the chief economic importance of the group lies in the destruction of other pests by the parasitic species. The species *Trichogramma minutum* attacks more than 150 species of other insects and raised in the millions for use as a natural parasite against the sugarcane borer and the Angoumous grain moth. The species *Aphycus helvolus* is a parasite and thus an effective control against the black scale insects (a severe pest of citrus). The species *Aphelinis mali* is a parasite of the damaging woolly apple aphid. The *Coccophagus gurneyi* species is a parasite of the citrophilus mealybug. And the species *Pteromalus puparum* is a parasite of the imported cabbageworm. The parasitic chalcid wasps of some species, in turn, live in and destroy other beneficial parasites, but in the overall, their net economic effect on food production is considered positive.

The chalcid wasps are distributed widely throughout the United States.

Fig wasp (*Blastophaga psenes*, Linnaeus). A minute, shining, amber-brown wasp that is nearly indispensable to Calimyrna fig production because of its role in pollination or caprification.[1] No other insect successfully pollinates commercial varieties of figs. Thus, fig varieties that require pollination develop no fruit unless this wasp is present. Before caprifigs containing *Blastophaga* wasps were introduced into California from Algeria in 1899, attempts to produce Smyrna-type figs on a commercial scale had been unsuccessful.

The remarkable arrangement whereby the male fruits of various species of fig provide food and shelter for the larvae of these small wasps is one of the more complex relationships between a plant and an insect to be found in nature. The caprifig (male) tree supports a population of these wasps throughout the year. It bears three crops of caprifigs, one in winter, another in spring, and a third in summer and fall.

The spring crop of caprifigs produces an average of about 500 female wasps and about 30 wingless males per fig. This crop of caprifigs has a barrier of pollen-bearing flowers around the opening or eye. A cycle of wasp development begins with the laying of eggs in the gallflowers of a caprifig. The wasps develop to maturity inside these flowers. As the females escape through the eye, they become dusted with pollen. During the first half of June, the female wasps fly to other caprifigs and also to edible figs produced by female trees. In the caprifigs, the insects' cycle is repeated, but the attempts by females

[1] Experiments by University of California scientists have shown that a hormone spray applied to the fruit and foliage at the proper time will cause the fruit to set without pollination, but the method has not been adopted because the resulting quality of the fruit appears to be inferior. Further research is required.

to lay eggs in the galls of edible figs results only in pollination and setting of the fruit.

One disadvantage in the visits of these wasps to edible figs is that they carry into the figs the spores of a mold, *Fusarium moniliforme* disease (*endosepsis*) that cause spoilage. The insects also carry a bacterial disease, *Serratia plymuthica*, and a yeast, *Candida gulliermondii*, var. *carpophilia*.

CHALCOCITE. This mineral is cuprous sulfide, Cu_2S, crystallizing in the orthorhombic system, often in pseudo-hexagonal forms. Above a temperature of 91°C, chalcocite changes into an isometric form. It has conchoidal fracture; hardness, 2.5–3; specific gravity, 5.5–5.8; metallic luster; color dark gray to blackish-gray, frequently with bluish-green tarnish. Chalcocite is of widespread occurrence and a valuable copper ore. It seems in some cases to be definitely secondary in origin, in other cases primary. It may have been formed from bornite by the action of alkaline solutions. It sometimes carries valuable amounts of silver.

Among the many European localities might be mentioned Cornwall, England, the Ural Mountains, and Rumania. It occurs also in the Congo, South West Africa, Peru, Mexico, and Alaska. In the United States it is found at Bristol, Connecticut, in fine crystals, Montana, Tennessee, Arizona, Nevada, and California.

The word chalcocite is derived from the Greek word meaning copper. Chalcocite also is known as *copper glance*.

CHALCOGENIDE GLASSES. Semiconductor.

CHALCOPYRITE. The mineral chalcopyrite (also know as copper pyrites) is a sulfide of copper and iron corresponding to the formula $CuFeS_2$. Its tetragonal crystals are often complex with repeated twinning; massive chalcopyrite is common. It has an uneven fracture; is brittle; hardness, 3.5–4; specific gravity, 4.1–4.3; luster, metallic; color, brass-yellow, may be iridescent from tarnish; streak, greenish-black; opaque. Chalcopyrite is the most common copper-bearing mineral known and it is the most important ore of copper. It is a primary mineral in many igneous rocks and from it a host of secondary copper minerals have been derived.

Among the many localities where fine specimens of this mineral have been obtained might be mentioned: Freiburg, Saxony; Alsace; Rio Tinto, Spain; Cornwall, England; Australia; Chile, Peru, and Bolivia, South America; and in the United States, Ellenville, New York; Chester County, Pennsylvania; Joplin, Missouri; Gilpin County, Colorado; Arizona, Montana, Utah, Nevada, California, New Mexico and Tennessee. In Canada there are notable deposits of chalcopyrite in the Provinces of British Columbia, Ontario, and Quebec. The name chalcopyrite is derived from the Greek word meaning copper, and the word pyrites.

CHALK. Chalk is a soft, porous limestone of white, grayish-white or buff color made up of the minute shells of foraminifera and fragments of cocospheres. It occurs extensively in England and France and less so in the United States.

Chalk consists almost entirely of calcite which has formed principally by shallow-water accumulation of (1) calcareous tests of floating microorganisms and (2) comminuted remains of calcareous algae. The most widely distributed chalks are of Cretaceous age, as exemplified by the cliffs on both sides of the English Channel. Although an unaltered deposit, chalk masses may contain nodules of chert and pyrite.

CHALYBITE. Siderite.

CHAMAECYPARIS. Cedar Trees; Cypress Trees.

CHAMAELEON. A minor southern constellation.

CHAMELEON (*Reptilia, Sauria*). Any member of several genera of lizard-like reptiles of very peculiar form, occurring in Africa, the Oriental region, and about the Mediterranean. They are arboreal species with grasping feet, a crested head, and a long extensile tongue with a clubbed sticky tip which is used to catch insects. They are able to change color readily. The most common genus is *Chamaeleon*.

The little lizard sold at street fairs is not a true chameleon but is more closely related to the iguanas.

CHAMOIS. Goats and Sheep.

CHAMOMILE. The flowers of an annual herbaceous plant (*Matricaria chamomilla* L.) of the *Compositae* family, upon steam distillation, yield a viscous liquid of an intensely blue coloration and characteristic odor and taste. The principal constituents are chamazulene, sesquiterpene alcohols, and caprinic acid and ester. Flavorings, available as infusions, tinctures, or soft and dried fluid extract, are used in various foods, such as beverages, ice creams, candies, baked goods, and chewing gum for imparting a bitter-tonic flavor with a characteristic aroma. Traditionally the product has been most popular in Europe, where it is considered a mild sedative and digestive.

CHANCRE. Syphilis.

CHANCROID. Lymphogranuloma Venereum.

CHANNEL BASS. Croakers.

CHANNEL CAPACITY. Information Theory.

CHANNEL CARBON. Carbon Black.

CHANNEL (Computer System). That portion of the central processing unit (CPU) of a computer which connects input and output devices to the CPU. It may also execute instructions relating to the input or output devices. See also **Input/Output Devices.** The channel also provides the interfaces and associated controls for the transfer of data between storage and the input/output devices attached to the CPU. The channel generally maintains the storage address for the device in operation and includes the buffer registers needed for synchronizing with storage. A serial or selector channel may be used for high speed devices, whereas a multiplex channel normally is used for slower devices.

The multiplex channel has the capability of concurrently servicing several devices. The storage address for each operating device is controlled by the channel and is maintained in main storage or in the logic of the channel. If the address is maintained in the logic, this increases the maximum data rate of the channel and reduces the number of storage references needed to service the device. The storage-access channel is an adaptation of the multiplex channel. With this type of channel, the device transmits both the data and the storage address to the central processor when it requires servicing.

Where a series channel is used, one or more devices may be physically attached to the channel interface, but only one of the devices is logically connected at any given time. Thus, for the selected device, the full channel data-rate capacity is available. For a given device data-rate requirement, the series channel is less costly than either dedicated channels per device, or a multiplex channel. In the latter instance, the saving is realized from the fact that the maximum required data rate is determined by the data rate of a single attached device. The maximum rate of multiplex channel is the sum of the data-rate requirements of several devices, plus the data rate required for device addressing.

The term is also used for other portions of some computers. For example, the analog input channel refers to the path between the input terminals of an analog input subsystem and the analog-to-digital converter. Similarly, it may describe the logical or physical path between the source and destination of a message in a communication system. In some applications relating to data acquisition in physics experiments, it is synonymous with the quantization interval in an analog-to-digital converter.

The tracks along the length of magnetic tape used for storing digital data are also referred to as channels. See **Magnetic Tape Storage.**

For related topical coverage in this volume, see list of entries under **Data Processing.**

Thomas J. Harrison, International Business Machines Corporation, Boca Raton, Florida.

CHANNEL CONDITIONING. Telephony.

CHANNEL FREQUENCY. This term denotes the band of frequencies which is associated with a single unit of intelligence in a communications system. Thus it applies to the band of frequencies radiated by a broadcast station, or to the band of frequencies which must be handled by a carrier system to handle a single conversion. In the various systems the application of intelligence to a given frequency will generate certain other frequencies which are then associated with the original in some manner to convey the intelligence to the receiver. This band of frequencies then determines the response characteristics which the receiver (or other units of the system) must have for satisfactory results. Thus in conventional broadcasting the various stations use channels about 10 kHz wide; in frequency modulation, about 200 kHz; in television, 5 to 6 MHz; in carrier telephony, about 3 kHz.

CHANNELING. The transport of energetic ions and atoms in a lattice along directions parallel to close-packed rows of atoms. The ions or atoms move between atomic rows or planes, down what are effectively tunnels or channels through the crystal. The phenomenon is important in radiation damage studies in solids.

CHANNELING PROTEINS. Brain and Nervous System.

CHAPAPOTE. Tar Sands.

CHAPARRAL. The name applied to a plant association occurring over wide areas in western North America and composed of a mixed population of low-growing shrubs. Some stands of chaparral are dense; others are open. Chaparral is usually found between a lower zone of sagebrush or grassland and an upper zone of woodland or forest. This association is found principally in the foothills of the Coast Ranges and Sierra Nevada in California, of the southern Rocky Mountains, and other ranges in Utah and Arizona. Some types of chaparral, such as that occurring on the Coast Ranges of southern California, are composed largely of evergreen shrubs; others, such as that occurring in the southern Rocky Mountains, largely of deciduous shrubs. See also **Biome.**

CHAPMAN EQUATION. An equation expressing the viscosity of a gas in terms of certain molecular constants. This relationship has been simplified to the expression:

$$\eta = \frac{(0.499)m\bar{c}}{\sqrt{2}\pi\sigma^2(1 + c/T)}$$

in which η is the viscosity, m is the mass of a molecule, $\bar{c}$ is its average speed, σ is the collision diameter of the molecule, c is Sutherland's constant, and T is the absolute temperature.

CHAPMAN-JOUGET CONDITION. In steady-state detonation the lowest possible shock wave velocity is given by

$$v = \frac{1}{\rho_1}\left(\frac{P_2 - P_1}{\frac{1}{\rho_1} - \frac{1}{\rho_2}}\right)^{1/2}$$

where ρ_1 and ρ_2 are the densities in front of and behind the incident shock, respectively, and P_1 and P_2 are the corresponding pressures.

CHARACIDS (*Osteichthyes*). Of the order *Ostariophysi*, family *Characidae*, there are 6 families and up to 30 subfamilies. They are found only in South America and central Africa. In appearance, they resemble to some degree carps and minnows. However, jaw teeth are among the distinguishing features. There are numerous sizes and shapes among the characids, ranging from 1 inch (2.5 centimeters) or even less to a length of some 5 feet (1.5 meters). They also have a wide range of eating habits, varying from vegetarians to the omnivorous to the dangerous carnivorous. Unquestionably, the best known of the characids is the piranha. The piranha is probably feared even more than the shark as a killer.

Normally, piranhas have a diet of small fishes, usually other members of their family (characids), but will also attack large animals, including people. Piranhas are famous for their teeth and the manner in which they work in concert in consuming a victim. It has been recorded that a 100-pound (45.4-kilogram) capybara was ravaged and completely reduced to a skeleton by a group of piranhas within less than one minute. Because they are considered good food items, natives fish for them with hook and line. Obviously, strong leader line must be used because otherwise the sharp teeth of the piranha will easily cut it. The piranha must be regarded with respect when taken from the water even if it appears dead—the jaws can snap when touched to remove a finger or toe. The aggressive character of the piranha appears to be related to the strength it derives as a member of a large school because, when in captivity, singly or with just a few others, its aggressiveness is greatly moderated. In any event, however, the piranha should be regarded as a nervous and dangerous fish. See accompanying figure.

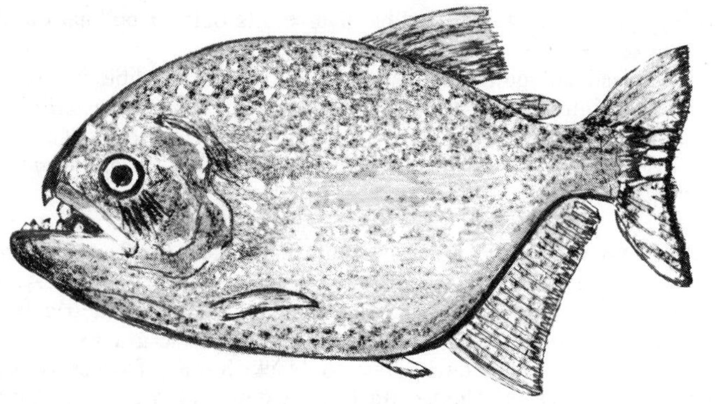

Piranha (*Serrasalmus piraya*).

The *Serrasalmus natterei* is widely distributed and is the type sometimes seen in tropical fish stores. The *S. piraya* is the largest of the dangerous piranhas and attains a length up to 2 feet (0.6 meter). It is found in eastern Brazil in the River São Francisco. Other carnivorous characids include the *Boulengerella lucius* which has the long jaw and appearance of a pike. It is noted for lurking among plants waiting for its prey. It grows to about 2 feet (0.6 meter) in length, but is not considered a danger to people. The *Hydrocyon goliath* (African Congo tiger fish) appears much as a trout and sometimes attains a weight of 125 pounds (57 kilograms). It is reputed to have attacked natives and is dangerous both in and out of the water until fully expired.

Among the very small characids are several interesting and beautiful fishes that are favorites among tropical-fish fanciers. In this group is the jewel tetra from South America (*Hyphessobrycon callistus*). Some of the common names for fishes of this type include "neon tetra," "glo-lite tetra," "head-and-tail light," and "cardinal tetra."

One of the most interesting of the South American characids is the flying hatchet fish. "Flying" is attributed to quite a few fishes, but all but the flying hatchet fish simply jump and glide, with no propulsion applied during flight. In the case of the hatchet fish, the pectoral fins are used to extend a jump into a somewhat longer flight—hence they are sometimes referred to as true flying fishes. There are nine species of hatchet fishes. They are not to be confused with the hatchet fishes of the family *Sternoptychidae* which are described in entry on **Hatchet Fishes.**

Pencil fishes (*Poecilobrycon eques*) are also small, very slender characids and favorites of tropical-fish fanciers.

CHARACTER (Computer System). One symbol of a set of elementary symbols, such as those corresponding to the keys of a typewriter, that is used for organization, representation, or control of data. Symbols may include the decimal digits 0 through 9, the letters A through Z, and any other symbol which a computer may read, store, or write. Thus, such symbols as @, #, $, /, are commonly used to expand character availability.

A *blank character* signifies an empty space on an output medium; or a lack of data on an input medium, such as an unpunched column on a punched card.

A *check character* signifies a checking operation. Such a character contains only the data needed to verify that a group of preceding characters is correct.

A *control character* controls an action rather than conveys information. A control character may initiate, modify, or stop a control operation. Actions may include the line spacing of a printer, the output hopper selection in a card punch, etc.

An *escape character* indicates that the succeeding character(s) is in a code that differs from the prior code in use.

A *special character* is not alphabetic, numeric, or blank. Thus, @, #, etc., are special characters.

See also terms listed under **Data Processing.**

CHARACTER DENSITY. Density (Data Processing).

CHARACTERISTIC EQUATION. 1. A class of equations connecting those variables, such as temperature, pressure, and volume, which define the physical condition of a given substance and are called variables of state.

The ideal gas law and the Boyle-Charles law represent approximately the behavior of all gases, but if one wishes to be accurate, some modification of these must be sought which will take into account the differences between individual gases. The best known characteristic equation for gases is that of van der Waals. Using the same notation as for the ideal gas law, this may be written

$$\left(p+\frac{n^2a}{v^2}\right)(v-nb)=nRT$$

where n is the number of moles of gas, and a and b are constants characteristic of the gas in question. They are very small; if they were zero we should have the ideal gas law. Following are their approximate values for certain gases, where a is expressed in atmosphere (liter/gram-mole)^2 and b in liter/gram-mole:

Gas	a	b
Ammonia	4.170	0.03707
Helium	0.034	0.03412
Hydrogen	0.244	0.02661
Nitrogen	1.390	0.03913
Oxygen	1.360	0.03183

Characteristic equations of this sort are also known as equations of state. See also **Berthelot Equation; Equation of State.**

2. Equations which have solutions, subject to particular boundary conditions, only for certain specific parameters occurring in them. (See **Eigenfunction;** and **Eigenvalue (Proper Value.)**) In differential equations, the complete solution includes the characteristic solution and the particular solution. The characteristic solution is obtained from the roots of the characteristic equation, and defines the transient or time response of the system. The particular solution is obtained from the forcing function or input signal and defines the steady-state response.

3. An equation in the linearized theory of hydromagnetics whose solutions show the frequencies and modes of the initial perturbations which will decay or grow exponentially in time for any given system. The solutions to this equation indicate the regions of stability for various hydromagnetic systems. See **Hydromagnetic Equations.** For related topical coverage in this volume, see list of entries under **Mathematics.**

CHARACTERISTIC FUNCTION. 1. In statistics and mathematics, if $F(x)$ is a probability distribution function, its characteristic function is given by

$$\phi(t)=\int_{-\infty}^{\infty} e^{itx}\,dF(x)$$

The characteristic function is a moment generating function,

$$\phi(t)=\sum \mu_r'(it)^r/r!$$

where μ_r' is the rth moment of $F(x)$ about the origin. The characteristic function uniquely determines the distribution function by the formula

$$F(x)-F(0)=\frac{1}{2\pi}\int_{-\infty}^{\infty}\phi(t)\frac{1-e^{-ixt}}{it}\,dt$$

Analogous functions can be defined for several variables. A most important property in the theory of sampling is that the characteristic function of a sum (convolution) of independent random variables is the product of their individual characteristic functions.

CHARACTERISTIC IMPEDANCE. This is the impedance which a transmission line would present at its input terminals if the line were infinitely long. If, instead of the line being actually infinite in length, it is finite and is terminated by an impedance equal its characteristic impedance it will behave, as far as the input is concerned, as if it were infinite. This means that there will be no reflected electrical wave, with the attendant losses, etc., at the terminal point. In electrical circuits this is an extremely important consideration as reflection means some energy which would otherwise go to the load is reflected back down the line to cause losses on the line, objectionably high voltages, echo effects and other undesired conditions. When a line is terminated in the characteristic impedance it is said to be matched, or the load matches the line. While not always attainable it is a condition highly desirable.

This term is also applied with corresponding meaning to two-port networks and to waveguides.

CHARACTERISTIC (Logarithm). Logarithm.

CHARACTERISTIC MATRIX. Consider the linear transformation $\mathbf{y} = A\mathbf{x} = \lambda\mathbf{x}$, where λ is a constant, thus A merely multiplies the vector $\mathbf{x}$ by a scalar quantity. Rewriting this equation as $[\lambda E - A]\mathbf{x} = K\mathbf{x} = 0$, we see that either $|K|$, the determinant must vanish or all x_i are zero, the latter a trivial case. The characteristic matrix of A is K, the determinant of K is the secular determinant or characteristic function of A; roots of the latter are eigenvalues, latent or characteristic roots and the corresponding values of x_i are eigenvectors.

The diagonalization of a matrix is a matter of some importance and we now show how this may be done using properties of the characteristic matrix. Let us suppose the eigenvalues of A are known and that we wish to determine a matrix X so that

$$X^{-1}AX=\Lambda=[\lambda_i\delta_{ij}]$$

Choose one eigenvalue, say λ_k, and write $A\mathbf{x} = \lambda_k\mathbf{x}$, which is a set of linear homogeneous equations in the n variables $x_1, x_2, \ldots, x_n$. Since the equations are homogeneous, one can only determine the ratio of the unknowns but, except for an arbitrary constant, they can be written as a column vector, the eigenvector $\mathbf{x}_k = \{x_{1k}, x_{2k}, \ldots, x_{nk}\}$. Continue in the same way to find the eigenvectors for the other eigenvalues. It follows that $AX = X[\lambda_i\delta_{ij}]$; hence, the collineatory transformation $X^{-1}AX = \Lambda$ does indeed diagonalize A.

The arbitrary constant in each eigenvector can be eliminated by the requirement that X be orthogonal. Such a transforming matrix is not only collineatory but also congruent. This means that $X = \tilde{X}^{-1}$ or $X\tilde{X} = \tilde{X}X = E$, where E is a unit matrix. A simple means of obtaining this property is the Schmidt process.

The preceding arguments are based on the assumption that no two or more of the eigenvalues are identical. If this is untrue, degeneracy is said to occur. Some modification of the procedure described is then necessary and the resulting matrix is not truly diagonal.

See also **Schmidt Process.** For related topical coverage in this volume, see list of entries under **Mathematics.**

CHARACTERIZATION FACTOR. A factor obtained by dividing the cube root of the boiling point (in degrees Rankine) by the specific gravity of 60°F (15.6°C). It is useful in the investigation of oils.

CHARACTER RECOGNITION. Input/Output Devices (Computing System); Magnetic Ink Character Recognition (MICR); Optical Character Recognition.

CHARADRIIFORMS (*Aves*). This order comprises a great variety of birds of distinctive sizes; it includes the extremely long-legged stilts, along with the short-legged seed snipes, the curlews and slender-beaked snipes, the puffins with their very high and laterally compressed beaks, the skuas with their hook-shaped beaks, and the skimmers, in which the lower mandible extends far beyond the upper, a unique feature among birds. Despite this external polymorphy, the taxonomic relationship of the families and genera of grallatores, gulls, and auks can be demonstrated by exact comparisons of internal details.

The waders (grallatores), shore birds, and gulls, range in length from 12 to 80 centimeters (5 to 31 inches) and weigh from 25 grams to at least 2 kilograms (0.8 ounce to $4\frac{1}{2}$ pounds). The vomer is complete and the breastbone has no inner extensions. There are 11 primaries and 12 (up to 26) rectrices. The rump feathers have aftershafts; the uropygial gland has a long feather tuft. The palates and the vocal organs are constructed alike in most species. They usually produce only one brood a year; the clutch size does not exceed four eggs, which are incubated for a period of 2 to 4 weeks or even longer. The downy chicks either leave the nest immediately or they remain there or nearby until full fledged (with the exception of some auk species). Gulls, terns, and auks breed chiefly in colonies. They feed entirely or nearly so on animals which they pick from the ground with their poker beaks or from the water by sudden plunges or wing dives. Their activity range is chiefly aquatic, in marshes, near inland waters, at the seashore, and on oceanic islands. A few species of some families inhabit dry areas, even the desert. The nasal glands, which are usually large, enable those birds that live near salt water to eliminate the salt.

There are 3 well differentiated suborders with a total of 17 families and 334 species. The ploverlike forms (*Charadrii*), include 12 families; 1. Jacanas (*Jacanidae*); 2. Phalaropes (*Phalaropodidae*); 3. Snipes, etc. (*Scolopacidae*); 4. Avocets (*Recurvirostridae*); 5. Plovers (*Charadriidae*); 6. Painted Snipes (*Rostratulidae*); 7. Oystercatchers (*Haematopodidae*); 8. Sheathbills (*Chionididae*); 9. Seed Snipes (*Thinocoridae*); 10. Coursers and European Pratincoles (*Glareolidae*); 11. Crab Plovers (*Dromadidae*); 12. Stone Curlews (*Burhinidae*). The Gull-like forms (*Lari*), include four families: 1. Skuas (*Stercorariidae*); 2. Gulls (*Laridae*); 3. Terns (*Sternidae*); 4. Skimmers (*Rynchopidae*). The Auks (*Alcae*) include one family (*Alcidae*).

Pluvialines and gulls are worldwide. Many species are pronouncedly migratory. The Arctic tern and the sandpipers share the record as long-distance travelers, for they cover 33,000–35,000 kilometers (20,506–21,749 miles) a year. The Pacific golden plover holds the record for nonstop flights (at least 3300 kilometers; 2051 miles). The auks, on the other hand, are restricted to the Northern Hemisphere. Their principal habitat is the shores and islands of northern waters that have abundant food. See also **Waders, Shorebirds, and Gulls.**

CHARGE CONJUGATION. The theoretical operation of changing the signs of all electric charges and the direction of all electromagnetic fields in a system. See also **Conservation Laws and Symmetry.**

CHARGED-COUPLED DEVICE. Invented in 1969,[1] the charge-coupled device in its configuration as of the early 1980s is a three-layered semiconductor device—one layer of metallic electrodes and another of silicon crystal, separated by an insulating layer of silicon dioxide. Charge-coupled devices (CCDs) can be fabricated with standard metal–oxide semiconductor (MOS) processing techniques. Although CCDs are of comparatively simple structure, they can perform many electronic functions that typically require more complex integrated circuits.

The CCD utilizes the familiar phenomenon found in so many microelectronic devices, namely, the ability to permit negatively charged electrons (or positively charged "holes") to move about controllably in semiconductor material. Most devices use this characteristic to change the electrical current flowing through them—as amplifiers or switches. The CCD stores and transfers information in the form of packets of electrical charge analogous to the tiny magnetic domains in bubble devices.

[1] Bell Laboratories.

When charge packets are introduced into a CCD, they can be stored in "potential wells" at the surface. These wells are actually tiny regions in which the presence or absence of charge can represent information. The packets of charge can be sequentially moved (or "coupled") from one well to the next when proper voltages are applied. In this way, the CCD can recirculate or store the charge packets of information until they are needed. Since the amount of charge in a well can be varied continuously from zero to a maximum amount, the CCD is basically an analog device and thus can be used as an efficient device for handling analog communications signals. When the packets of charge are digitized, i.e., the wells are either empty or full, the CCD can act as a digital electronic memory. If the charge packets are introduced optically instead of electronically, such as by an image focused on the light-sensitive silicon surface, the CCD can be used as an imaging device.

After over ten years of development and refining, CCDs are now used for signal processing in communications systems, for information storage in computer systems, and for image sensing in solid-state television cameras.

CCD Imaging Devices. Almost from the outset of its initial development, a very promising application has been the use of the CCD in a solid-state TV camera. When light from an image or scene is focused on the CCD, a pattern of electrical charges is created. The charges vary in proportion to the amount of light and thus serve as an accurate electrical representation of picture elements. These charges can be stored, transmitted out of the CCD chip sequentially, and later reassembled on a conventional television screen or facsimile readout. Use of the CCD eliminates the need for vacuum tubes and scanning electron beams required by the current conventional TV cameras.

CCD Memories. In commerical computers, charge packets of information can be stored, regenerated, and moved about sequentially with a CCD and thus the device can function as a shift register and can be used to construct a recirculating electronic memory. Although numerous other memory techniques are available (tape, disk, drum, ferrite core, integrated-circuit mass memories, etc.), CCD memories have made notable inroads in uses such as plug-in memories for minicomputers and refresh memories for cathode-ray tube terminals, among others. The first commercially available CCD chips (1975) had capacities of 9000 bits. Capacities have grown to 26,000 bits (1976) and 64,000 bits and greater in the last few years.

High density in electronic memories translates directly into small size and lower information-processing costs. As a result, systems designers find CCD memories attractive. One basic limitation of the CCD is that it is a serial-access device, i.e., information circulates as if in a closed pipeline and data must be retrieved from a CCD sequentially. However, in some applications, access time to a single storage site is an overriding factor and thus random access memories are required. High-speed random access memories (RAMs), such as integrated circuits with 16,000 bits per chip have much faster access time than CCDs, and the RAMs are more attractive for application in the high-speed central processors of large commercial computers and telephone switching equipment.

Analog Signal Processing. Analog applications ultimately may provide a greater demand for CCDs than the solid-state imaging device and memory applications previously described. In analog applications, the high capacity, comparatively low cost, and modest power requirements of CCD chips may be exploited in many kinds of communications devices, such as delay lines—circuits whose outputs are identical to their inputs, but delayed in time; multiplexers—circuits that combine many signals in a single transmission medium; and electrical filters—circuits that transmit some frequencies and reject others.

The voice-frequency electrical filter, a basic communications circuit, is an example of how the CCD can be used. Such filters are critical components in modern systems that carry tens of thousands of phone calls along a single transmission path. The filter confines each voice channel to a certain frequency range and helps prevent noise, crosstalk, and other kinds of interference. Telephone systems require these filters in quantities of millions of units per year. During the last decade or so, these filters have been significantly improved through developments in interrelated-circuit technologies and in piezoelectric devices. But, since CCDs utilize the same technology as conventional MOS memory circuits, it is envisioned that the CCD device can be placed on the

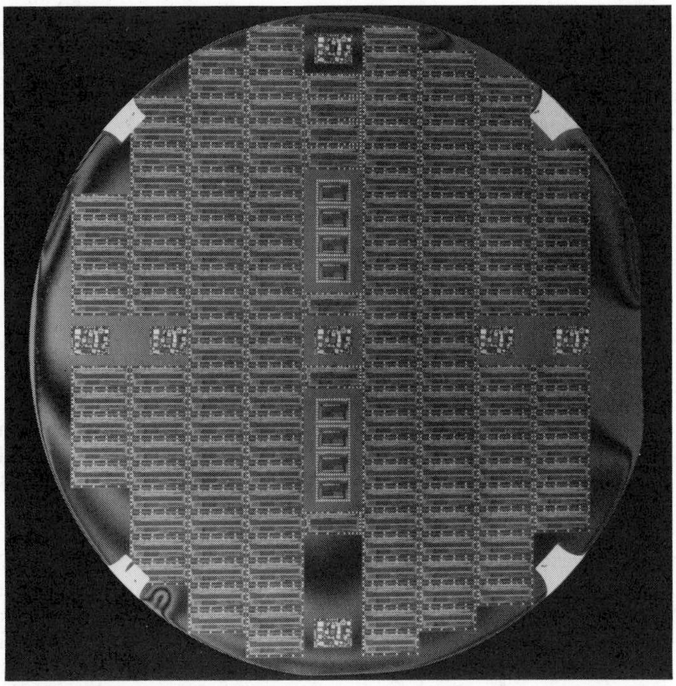

Fig. 1. A charge-coupled device (CCD) wafer, 3 inches (7.5 centimeters) in diameter, that contains 169 CCDs. Each wafer is patterned for a particular CCD filter circuit. (*Bell Laboratories*)

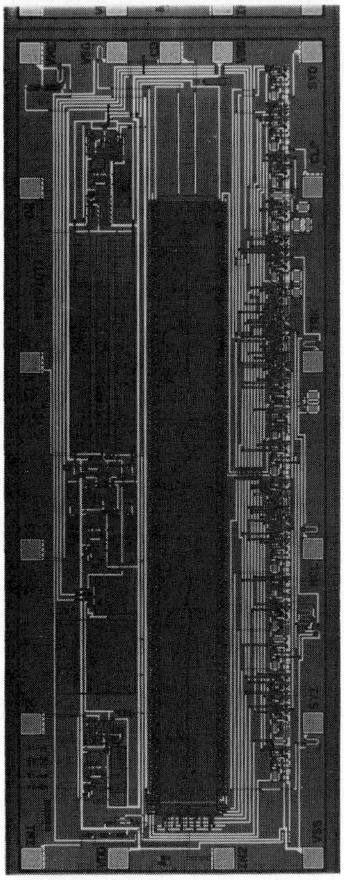

Fig. 2. Enlarged view of a CCD filter showing chip features in some detail. (*Designed by Bell Laboratories*)

same large-scale integrated circuit chip used to perform other functions in the system.[2] See Figs. 1 and 2.

See also **Signal Processing Devices.**

CHARGE (Electron). Electron Theory.

[2] In 1978, Bell Laboratories developed CCD filters for signal processing applications in Touch-Tone® receivers for use in 1979 and the 1980s.

CHARGE-MASS RATIO. This term refers to the relationship between the electric charge of a particle and its mass, so important in the physics of electrons, ions, and other electrified bodies of molecular orders.

The earliest information on the subject followed from the researches of Faraday on electrochemical equivalents. From his results it appears that in the electrolysis of chlorine, for example, 1 coulomb of negative electricity is carried by 0.00037 gram of this element, and hence that the carriers or ions have a charge-mass ratio of about 2,700 coulombs or 8.1×10^{12} electrostatic units of electricity to the gram. Similarly, 1 coulomb of positive electricity is carried by 0.0000104 gram of hydrogen, which gives about 95,700 coulombs or 2.87×10^{14} electrostatic units to the gram for hydrogen ions. This is 35 times the ratio for chlorine ions. But the atomic masses of hydrogen and chlorine are in the ratio 1:35, which means that if the carriers are atoms, the charge per carrier is the same for both elements. Bivalent elements, on the other hand, carry twice this charge per ion.

When J. J. Thomson applied a magnetic field to a stream of hydrogen canal rays, and then neutralized the resulting deflection by means of an electric field, he was able to calculate the charge-mass ratio of these particles from the curvature of the magnetically deflected stream and the values of the two field intensities. This he found to be either 95,700 coulombs per gram as in the electrolysis of hydrogen, or $\frac{1}{2}$ that value, which indicated that some of the ions were atoms and some were molecules carrying the some charge as the atoms. But when a similar test was applied to the cathode rays in a Crookes tube, the ratio was found to be about 5.303×10^{17} electrostatic units per gram, or about 1,850 times that for hydrogen atoms, whatever the nature of the cathode. We know now that this enormous difference is one of mass, not of charge; and that these experiments were the first direct revelation of the identity of the electron, the mass of which is now known to be approximately 1/1,836 of the mass of the proton (nucleus of hydrogen atom).

CHARGE TRANSFER DEVICES. Signal Processing Devices (CTD and SAW).

CHARGING CURRENT. Battery.

CHARLES LAW. Although the coefficients of expansion of different solids or of different liquids are notably different, the coefficients of expansion of all gases are nearly the same, namely, about $\frac{1}{273}$ of the volume at 0°C per centigrade degree. The law, stated by Charles in 1787 and independently by Gay-Lussac in 1802 (hence sometimes called Gay-Lussac's law) is not strictly true. Regnault obtained the following values of the volume coefficient for various gases:

Air	0.0036706
Hydrogen	0.0036613
Carbon dioxide	0.0037099
Sulfur dioxide	0.0039028
Carbon monoxide	0.0036688
Nitrous oxide	0.0037195
Cyanogen	0.0038767

None of these is far from $\frac{1}{273} = 0.003663$, which is therefore commonly taken as the expansion coefficient for gases; especially as the value for hydrogen, commonly used in the standard gas thermometer, is very near it. If the pressure as well as the volume is allowed to vary, the behavior of the ideal gas must be expressed by the Boyle-Charles law or the ideal gas law; and the behavior of a real gas by one of the other equations of state. See **Ideal Gas Law.**

CHARNOCKITE. Charnockite is a granular variety of hypersthene granite which was first described from the gravestone of Job Charnock, who founded the city of Calcutta, India, whence the derivation of the name charnockite.

CHARPY TEST. Impact Testing.

CHART (Lambert Projection). Lambert Projection.

CHAT. Warbles.

CHATTERER (*Aves, Passeriformes*). South and Central American birds making up the family *Cotingidae.* They are quite varied, some with strangely formed plumage and some beautifully colored. The family includes the umbrella bird, bell-birds, cotingas, manakins, and cocks-of-the-rock. A single species, the xantus becard, *Platypsaris aglalae,* enters the United States near the Mexican border.

These birds are near the flycatchers.

CHATTERMARK. A moon-shaped scratch or gouge on the bedrock assumed to be caused by the "chattering" action of angular boulders which are carried in the bottom of a glacier. Also, a chattermark may be defined as any mark, pit, or scratch made on a rock surface by the surface of a mass that moves over it.

CHEBYSHEV EQUATION. A special case of the Gauss hypergeometric equation

$$(1-x^2)y'' - xy' + n^2y = 0$$

where n is an integer. The name is often spelled differently, particularly Tchebycheff.

The two linearly independent solutions of the equation are known as polynomials of the first and second kind. The Chebyshev polynomials of the first kind, which are the more familiar ones, may be represented by the hypergeometric series

$$T_n(z) = F(n, -n, \tfrac{1}{2}; (1-z)/2)$$

Other definitions of them are:

$$T_n(z) = \cos(n \cos^{-1} z)$$

$$T_n(z) = z^n - \binom{n}{2} z^{n-2}(1-z^2) + \binom{n}{4} z^{n-4}(1-z^2)^2 \mp \ldots$$

$$T_n(z) = \frac{(-1)^n/(1-z^2)\, d^n}{1 \cdot 3 \cdot 5 \cdots (2n-1)\, dz^n}(1-z^2)^{n-1/2}$$

See also **Generating Function.** The polynomials are orthogonal:

$$\int_{-1}^{1} \frac{T_m(z)T_n(z)}{\sqrt{1-z^2}}\, dz = I_{mn}$$

where $I_{mn} = 0$, $m \neq n$ and $I_{mm} = \pi/2$ but $I_\infty = \pi$. Similar definitions and properties may be given for the polynomials of the second kind.

CHEBYSHEV'S INEQUALITY. Let x be a random variable, either discrete or continuous, whose probability function possesses a mean $\bar{X}$ and standard deviation σ_x. The probability of obtaining a deviation as much as or greater in absolute value than $t\sigma_x$ is less than or equal to $1/t^2$. In symbols

$$P[|x-\bar{X}| \geq t\sigma_x] \leq \frac{1}{t^2}$$

There are more general kinds of inequalities forming a family known as the Chebyshev (sometimes spelled Tchebycheff) type.

CHECK CHARACTER (Computer System). Character (Computer System).

CHECK (Computer System). A process of partial or complete testing of the correctness of computer or other data processing machine operations. Checks also may be run to verify the existence of certain prescribed conditions within the computer, or the correctness of the results produced by a program. A check of any of these conditions usually may be made automatically by the equipment, or checks may be programmed.

Automatic Check. A procedure for detecting errors that is a built-in or integral part of the normal operation of a device. Until an error is detected, automatic checking normally does not require operating system or programmer attention. For example, if the product of a multiplication is too large for the space allocated, an error condition (overflow) will be signaled.

Built-in Check. An error-detecting mechanism that requires no program or operator attention until an error is detected. The mechanism is built into the computer hardware.

Checkpoint. A point in time in a machine run at which processing is momentarily halted to perform a check or to make a magnetic tape or disk record (or equivalent) of the condition of all the variables of the machine run such as the status of input and output devices and a copy of working storage. Checkpoints are used in conjunction with a restart routine to minimize reprocessing time occasioned by functional failures.

Duplication Check. Two independent performances of the same task are completed and the results compared. This is illustrated by the operations below:

12	31	84	127
9	14	43	66
21	45	127	193

Echo Check. For checking the accuracy of transmission of data, the transmitted data are returned to the sender for comparison with the original data. Essentially a reading test, the echo check guards against malfunctions of output operations.

Modulo N Check. Same as residue check. See below.

Odd-Even Check. Same as parity check.

Parity Check. A summation check in which the binary digits, in a character or word, are added, modulo 2, and the sum checked against a single, previously computed parity digit, i.e., a check which tests whether the number of ones in a word is odd or even.

Programmed Check. A system of determining the correct program and machine functioning either by running a sample problem with similar programming and a known answer, or by using mathematical or logic checks, such as compared A × B with B × A. Also, a check system built into the program for computers that do not have automatic checking.

Reasonableness Check. Same as validity check. See below.

Residue Check. An error-detection system in which a number is divided by a quantity n and comparing the remainder with the original computer remainder. This is also termed a "modulo n" check. See above.

Sequence Check. A data processing operation designed to check the sequence of the items in a file assumed to be already in sequence.

Summation Check. A check in which groups of digits are summed, usually without regard for overflow, and that sum checked against a previously computed sum to verify that no digits have been changed since the last summation.

Validity Check. A check based upon known limits or upon given information or computer results; e.g., a calendar month will not be numbered greater than 12; a week does not have more than 168 hours, etc.

For related topical coverage in this volume, see list of entries under **Data Processing.**

Thomas J. Harrison, International Business Machines Corporation, Boca Raton, Florida.

CHECK DIGIT. Digit (Computer System).

CHEESE. Antimicrobial Agents (Foods); Ultrafiltration.

CHEETAH. Cats.

CHEILITIS. Inflammation of the lips regardless of cause. Usually due to an allergy to substances which may come in contact with the lips, including cosmetics, dentrifices, chewing gum, mouthwashes, various fruits and fruit dyes. Sometimes the inflammation may be induced by prolonged exposure to wind and sun.

CHELA. A form of grasping appendage found in lobsters, crabs and other arthropods. The large pincers of these species are the most familiar examples.

All chelate appendages have the next to the last segment prolonged into a process against which the terminal segments works to form a forceps-like organ.

CHELATES AND CHELATION. Chelation compounds are coordination compounds in which a single ligand occupies more than one coordination position. Such ligands are called chelating agents (the word being derived from the Greek, meaning crab's claw). Thus ethylenediamine $H_2N—CH_2—CH_2—NH_2$ abbreviated as en, forms a $Cr(en)_3^{3+}$ ion having three molecules of ethylenediamine, each occupying two coordination positions.

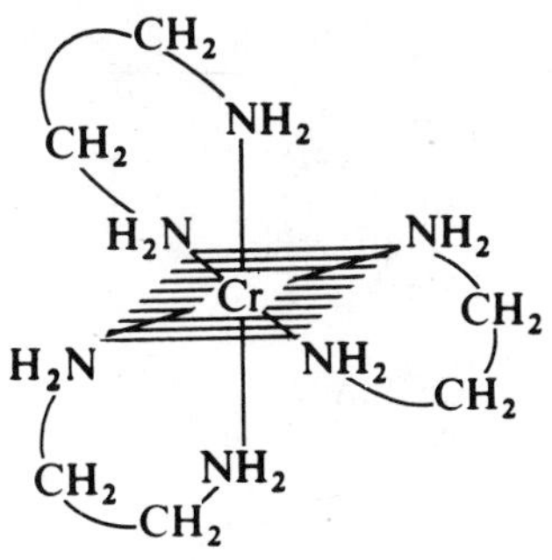

Ethylenediamine is therefore called a bidentate group, as are many other ligands, such as the β-diketones, which form chelation compounds of the type where M is a metal ion.

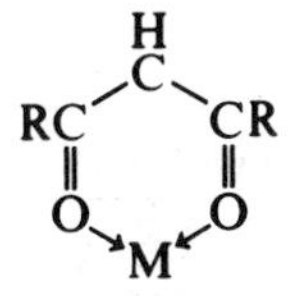

Although bidentate ligands are more common, there are polydentate ligands which occupy more than two coordination positions; ethylenediaminetetraacetic acid is such a polydentate ligand.

While ethylenediamine, and many other chelating agents, form only covalent bonds, there are others which attach by both covalent and ionic bonds. Thus glycine forms with cupric ions (Cu^{2+}) the compound copper bisaminoacetate.

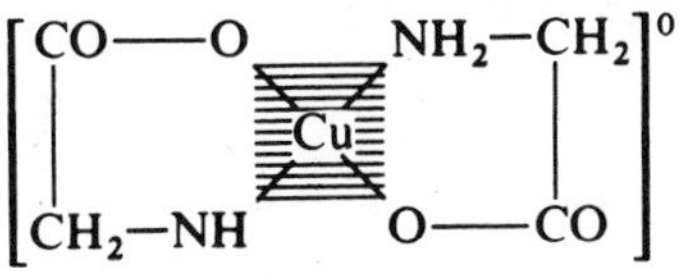

A number of synthetic chelating agents have been developed. They are substances like ethylenediaminetetraacetic acid (EDTA) and N-hydroxyethylethylenediaminetriacetic acid (HEDTA) and their salts, usually sodium salts. Many of these compounds and mixtures of these compounds are sold under trademarks.

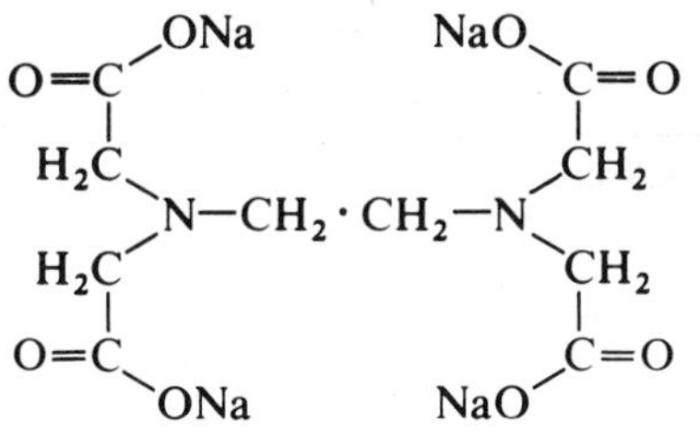

Tetrasodium Ethylenediaminetetraacetate (Tetrasodium EDTA)

Chelating agents are being used in increasing amounts for a number of important purposes. These uses may be put into two important categories: first, artificial trace metal carriers and, second, sequestering agents.

As artificial carriers for trace metals, chelating agents can be used as aids in agriculture by supplying the metals for soils which are deficient in their trace metal content. Both EDTA and HEDTA are adequate iron carriers and EDTA can be used as the carrier for bivalent copper, zinc, manganese, and cobalt. By use of such carriers certain plant deficiency diseases can be controlled. Another example of artificial carrier use is the employment of chelating agents such as the EDTA derivatives or mixtures of such derivatives with pyrophosphates or a mixture of EDTA and the sodium salt of N,N-di(2-hydroxyethyl) glycine in controlling polymerization reactions in synthetic rubber manufacture by the controlled release of trace metal catalysts.

As sequestering agents, chelating compounds have a wide variety of uses, for instance, for water softening in both soaps and synthetic detergents; in textile processing as in kier boiling operations where iron, copper, zinc, etc., ions are inactivated so that discoloration of cloth is prevented; in the stabilization of hydrogen peroxide; in boiler and heat exchanger cleaning.

Chelation in Biological Systems. Chemical reactions in biological systems are usually mediated by selective catalysts called *enzymes.* The high efficiencies and stereospecificities acheived require that enzymes have definite and characteristic geometries, whereby specific functional groups coordinated to the metal ion are held in definite spatial positions relative to each other and relative to the substances on which they exert their catalytic effects. The incorporation of metal ions into enzyme structures can assist in the maintaining of a definite geometrical relationship between ionic and polar groups, through the geometric requirements of the coordinate bonds of the metal ion. Certain metal ions may also participate in the catalytic properties of enzymes through ionic and coordinate bonding between the metal ion and electron donating groups of the enzyme and substrate, and through the ability of the metal ion to initiate oxidation-reduction reactions. Because of these chemical and steric effects, coordinated metal ions in the complex compounds that catalyze biological reactions frequently are found. See also **Metalloproteins.**

Most of the metal ions that have biological functions have a coordination number of six, with the donor groups arranged in an octahedral fashion. There are a few metals, such as Mg^{2+} and Zn^{2+}, that frequently coordinate only four donor groups tetrahedrally, and Cu^{2+}, which has four coordinations directed to the corners of a square plane with the metal ion at the center of the plane.

Many simple acid-base reactions are catalyzed by both metal ions and hydrogen ions. Because of small size, the electronic interaction of the hydrogen ion with a substrate is much greater than that of a metal ion. The latter, however, has properties not possessed by hydrogen ions, which are useful in catalysis, i.e., the ability to coordinate a large number of electron donor groups simultaneously, the specific geometric orientation of the coordinate bonds of certain metal ions, and the ability of metal ions to undergo oxidation-reduction reactions. Many of these reactions are models of the more complex catalytic effects that occur in biological systems. Since these reactions of simple coordination compounds aid in the understanding of biological reactions, a few of the more common examples are given in Table 1.

The function of the metal ions in the reactions listed is to attract electrons from the substrate. When this effect takes the form of simple polarization of the functional groups of the substrate, charge variations and electron shifts in these groups facilitate the chemical reactions listed under solvolysis and acid catalysts. When the metal ion removes completely one or more electrons from the substrate, the first step in an oxidation reaction occurs. This type of catalysis can be accomplished only by metals capable of existing in more than one valence state.

There is a saturation effect in the coordination of a metal ion by donor groups of both the enzyme and the substrate. Therefore, one would expect that the interaction of a free metal ion with the substrate would be greater than that of the metalloenzyme (in which the metal is already partially coordinated). If this were true, the metal ion would have a greater catalytic effect than the metalloenzyme. The reverse is always the case; thus far, no metal ions, or metal complex enzyme models, have been found to approach the catalytic activities of the corresponding enzyme. This high activity of the enzyme is ascribed to the special environment of the substrate around the active site of the enzyme, through which additional binding of the substrate by adjacent organic groups of the enzyme takes place.

The enzyme aconitase, which contains the Fe^{2+} ion at the reactive center, catalyzes the interconversion of citric, isocitric, and aconitic acids. The reaction has been shown to occur through the formation of a single intermediate carbonium ion structure in which the Fe^{2+} ion is always bound to the same donor atoms, while the interconversion of the substrate occurs through the migration of only protons and electrons.

Some of the more important biological reactions that are catalyzed by metal ions are summarized in Table 2.

TABLE 1. METAL ION AND METAL CHELATE CATALYSIS OF CHEMICAL REACTIONS

Solvolysis and Other Reactions Involving Acid Catalysis by the Metal Ion		
Reaction Type	**Substrate**	**Catalyst**
Solvolysis	Amino acid esters, peptides, and amides	Cu^{2+}, Co^{2+}, Mn^{2+}
	Phosphate esters	La^{3+}, Cu^{2+}, VO^{2+}
	Fluorophosphates	Cu^{2+}, UO_2^{2+} diamine-Cu(II) complexes
	Polyphosphates	Ca^{2+}, Mg^{2+}
	Schiff bases	Cu^{2+}, Ni^{2+}
Transamination	Schiff bases of pyridoxal and α-amino acids	Fe^{3+}, Cu^{2+}, Al^{3+}, Zn^{2+}, Ni^{2+}, Co^{2+}
Decarboxylation	α-Keto polycarboxylic acids (e.g., oxalacetic and oxalsuccinic acids)	Cu^{2+}, Zn^{2+}, Ni^{2+}, Co^{2+}, Mn^{2+}, Fe^{2+}
Acylation	Acetylacetone	Co(III), Rh(III) or Cr(III) chelates of acetylacetone

Catalysis of Oxidation Reactions by Electron Exchange with Metal Ions or Metal Complexes		
Reaction	**Substrate**	**Metal Ion or Complex**
Oxidation by molecular O_2	Ascorbic acid, catechols, quinoline, salacylic acid	Fe(III), Fe(III)-EDTA, Cu(II), Cu(II)-EDTA, V(IV)
Oxidation by H_2O_2	Phenol, anisole	Fe(II) (Fenton's reagent), Fe(II)-hydroquinone Fe(II)-EDTA-ascorbic acid
Formation of oxygen	Hydrogen peroxide	Fe^{3+}, Fe(III)-phthalocyanine chelate
Formation of disulfides from mercaptides	Thioglycolic acid	Fe^{3+}, Cu^{2+}

TABLE 2. BIOLOGICALLY ACTIVE METAL CHELATES

Metal	Metalloenzyme	Other Biological Functions
Mg	Polynucleotide phosphorylase, ATPase, choline acylase, deoxyribonuclease, acetate kinase, adenosine phosphokinase, fructokinase, glyceric kinase, hexokinase	Chlorophyll
Ca	α-Amylase, aldehyde dehydrogenase, lipase	
V		Green algae, blood of marine worm (ascidian)
Cr		Glucose tolerance factor
Mn	Arginase, carnosinase, prolinase, enolase, isocitricdehydrogenase, 3-phosphoglycerate kinase, glucose-1-P kinase	
Fe	Aconitase, formic hydrogenylase, phenylalanine hydroxylase, peroxidase, catalase, cytochromes	Hemoglobin, ferritin, hemosiderin, siderophilin
Co	Aspartase, acetylornithinase	Vitamin B_{12}
Cu	Lactase, phenolase, tyrosinase, uricase	Ceruloplasmin, cytochrome
Zn	Carbonic anhydrase, carboxypeptidase, alcohol dehydrogenase, glutamic dehydrogenase, acylase	
Mo	Nitrate reductase, xanthine oxidase	

CHELATES (Boiler Water). Feedwater (Boiler).

CHELATING AGENTS (Anemia). Anemias.

CHELICERAE. The first pair of appendages in spiders and related animals. They are associated with the mouth and are formed for chewing and in some cases for grasping, as in the scorpions.

CHELIPED. An appendage of the thorax formed for grasping, in the crustaceans. The chela or pincher of the lobster and crayfish.

CHEMICAL AFFINITY. The entropy production due to a chemical reaction has the form

$$\frac{d_iS}{dt}=\frac{1}{T}\mathbf{A}v\geq 0 \tag{1}$$

where **A** is the chemical affinity and v, the reaction rate. **A** is related to the characteristic functions U, H, A, G, and to the chemical potentials μ by the relations:

$$\begin{aligned}\mathbf{A}&=-\left(\frac{\partial U}{\partial \xi}\right)_{S,\,V}=-\left(\frac{\partial H}{\partial \xi}\right)_{S,\,p}\\&=-\left(\frac{\partial A}{\partial \xi}\right)_{T,\,V}=-\left(\frac{\partial G}{\partial \xi}\right)_{T,\,p}\\&=-\sum_i v_i\mu_i\end{aligned} \tag{2}$$

when ξ is the extent of reaction and v_i the stoichiometric coefficient.

The basic properties of the affinity **A** are that it is always of the same sign as the reaction rate, and that if the affinity is zero the reaction rate is also zero, i.e., the system is in equilibrium.

This definition of affinity is essentially due to De Donder and is called De Donder's fundamental inequality. In the notation used by G. N. Lewis and his school, it is supposed that ξ increases by unity, therefore the relations of (2) are written in the form:

$$\mathbf{A}=-(\Delta U)_{S.V.}=-(\Delta H)_{S.p}=-(\Delta A)_{T,\,V}=-(\Delta G)_{T,\,p}. \tag{3}$$

Note that in this entry, **A** is the affinity and A, the Helmholtz function (work function).

See also **Chemical Reaction Rate.**

CHEMICAL ANALYSIS. Analysis (Chemical).

CHEMICAL CARCINOGENS. Cancer and Oncology; Carcinogens.

CHEMICAL COMPOSITION. Matter is composed of the chemical elements, which may be in the free or elementary state, or in combination. In the former case, as exemplified by iron, tin, lead, sulfur, iodine, and the rare gases, matter commonly exhibits the properties of the atoms of the particular element, including the chemical properties whereby they combine to form molecules. Molecules may (1) be monoatomic; (2) they may consist of atoms of one element only, such as nitrogen or hydrogen molecules (N_2 or H_2), (3) they may be composed of atoms of more than one element, called compounds, which usually have distinctive properties.

The molecular formulas of gaseous compounds are obtained from a study of the composition by elements and the density, by a method introduced by the Italian chemist, Cannizzaro, in 1858. Later, in 1872, in the course of his Faraday Lecture before the Chemical Society (London) on the subject "Some Points in the Teaching of Chemistry" Cannizzaro stated that "Symbols and formulas, in my opinion, constitute the introduction, preparation, and base of the study of the transformations of matter, which is the true object of our science." The simplest way to understand the method is to arrange in tabular form (1) the individual gases, (2) the weight in grams of 1 liter (at 0°C, 760 millimeters of mercury pressure) of each gas, (3) the weight in grams of *each element* present in the above volume (1 standard liter) found by exact analysis (percentage composition by chemical elements using the methods of analytical chemistry). See Table 1.

TABLE 1. CANNIZZARO METHOD OF COMPOUND COMPUTATION

Gas	Grams per Standard Liter	Percentage Composition by Chemical Elements	Grams per Standard Liter by Chemical Elements					
			Hydrogen	Oxygen	Carbon	Nitrogen	Sulfur	Chlorine
1. Hydrogen chloride	1.639	Hydrogen 2.76% Chlorine 97.24	0.045					1.594
2. Ammonia	0.771	Hydrogen 17.75 Nitrogen 82.25	0.137			0.634		
3. Carbon dioxide	1.977	Oxygen 72.73 Carbon 27.27		1.438	0.539			
4. Carbon monoxide	1.250	Oxygen 57.14 Carbon 42.86		0.714	9.536			
5. Methane	0.717	Hydrogen 25.14 Carbon 74.86	0.180		0.537			
6. Ethylene	1.260	Hydrogen 14.38 Carbon 85.62	0.181		1.079			
7. Acetylene	1.173	Hydrogen 7.75 Carbon 92.25	0.091		1.082			
8. Oxygen	1.429	Oxygen 100.00		1.429				
9. Hydrogen	0.090	Hydrogen 100.00	0.090					
10. Nitrogen	1.251	Nitrogen 100.00				1.251		
11. Chlorine	3.214	Chlorine 100.00						3.214
12. Sulfur dioxide	2.927	Oxygen 49.95 Sulfur 50.05		1.462			1.465	
13. Hydrogen sulfide	1.539	Hydrogen 5.91 Sulfur 94.09	0.091				1.448	
14. Nitrous oxide	1.978	Oxygen 36.35 Nitrogen 63.65		0.719		1.259		
15. Nitric oxide	1.340	Oxygen 53.32 Nitrogen 46.68		0.715		0.625		
Minimum weight (approximate)			0.045	0.715	0.538	0.626	1.45	1.60

NOTE: Data are displayed in this table to illustrate the Cannizzaro method of arriving at the symbol and symbol weight of chemical elements; and the formula and formula weight of chemical compounds.

Careful examination of the figures in the last six columns reveals the experimental fact that (1) in each separate vertical column the figures represent a minimum weight or a small multiple (approximately) of this weight, (2) the smallest of the six minimum weights is that for hydrogen, namely, 0.045 gram in 1 standard liter of hydrogen chloride gas.

The next step involves changing 0.045 gram of hydrogen to exactly 1.000 gram and finding arithmetically the volume of hydrogen chloride containing this weight (1.000 gram hydrogen). The volume is found to be 22.2 standard liters.

Therefore, 1.000 gram minimum weight of hydrogen is contained in 22.2 standard liters of hydrogen chloride.

Using this standard volume of 22.2 liters, the next step is to ascertain the minimum weight of the other elements in this volume.

Chemical Element	Approximate Minimum Weight in Grams of Each of the Six Chemical Elements in the Standard Volume, 22.2 Liters
Hydrogen	1
Oxygen	16
Carbon	12
Nitrogen	14
Sulfur	32
Chlorine	35.5

Then, the abbreviation is introduced by the representation:

SYMBOL WEIGHTS OF EACH ELEMENT BY THE SYMBOLS

1 gram of hydrogen by the symbol H
16 grams of oxygen by the symbol O
12 grams of carbon by the symbol C
14 grams of nitrogen by the symbol N
32 grams of sulfur by the symbol S
35.5 grams of chlorine by the symbol Cl

By setting up again the second half of the table for the 15 gases, this time for 22.2 standard liters instead of 1 standard liter, the results obtained may be observed in Table 2.

Thus, it is seen, the chemical formulas and formula weights (last column) of 15 gaseous chemical compounds have been arrived at, using the Cannizzaro method, by purely experimental and rational means, involving no theoretical considerations. Extension of the method serves to ascertain the chemical formula of all gases and vaporizable substances. For compounds which are neither gases nor vaporizable, other methods are available. Of these the most used are those of Raoult depending upon the depression of the freezing point or the elevation of the boiling point of a compound dissolved in a given solvent.

It remains to be noted that, when there is no method available for ascertaining the formula weight of a compound, the *simplest* formula, based on chemical analysis and the use of symbol weights of the contained elements, is used, e.g., ferric oxide, Fe_2O_3, ferroferric oxide, Fe_3O_4, ferrous oxide, FeO, cupric oxide (black copper oxide), CuO, cuprous oxide (red copper oxide), Cu_2O. The customary formula of water is H_2O, which is correct at temperatures above 100°C—actually, liquid water is mainly dihydrol $(H_2O)_2$.

It should be understood from the above discussion that a chemical formula is no chance throwing together of chemical symbols, but represents the results of careful analysis, and the scrutiny and deduction of the most skillful workers in the field. On this score alone, chemical formulas demand the greatest respect in understanding and use.

Symbol weights and atomic weights are used synonymously, as are formula weights and molecular weights. Unless otherwise stated, symbol weights and formula weights are expressed in grams, and the numbers used are those taken from the accepted list of atomic weights. See **Chemical Elements.**

One formula volume of a gas is 22.242 liters. It is necessary to state that actual gases under ordinary conditions show some variation from this value, so that for accurate work the records should be consulted in each case.

Summarizing, the formula "HCl" states that "36.5 grams of hydrogen chloride gas occupies a standard volume of 22.2 liters and is composed of 1 gram of hydrogen element chemically united with 35.5 grams of chlorine element." The reason for the formulas of the simple gases, oxygen, O_2, hydrogen, H_2, nitrogen, N_2, chlorine, Cl_2, is apparent from the general method of deduction. The formula O_2 represents 22.2 liters or 32 grams of oxygen *gas*, whereas O represents 16 grams of oxygen *element* in any substance, or more precisely, 15.9994 grams.

It has become customary in chemical literature to use the formula of a substance as an accepted abbreviation for the name of the substance, especially in cases of frequent repetition.

Up to this point, the discussion in this entry has related to substances which are either elements, or single compounds of elements combined in proportions that can be represented by the ratio of small whole numbers. Such compounds are called *stoichiometric compounds* or *Daltonide compounds* (after the British chemist Dalton). There exist, however, some compounds in which the ratios of the amounts of elements present are not integral. Such compounds are called *non-stoichiometric compounds* or *Berthollide compounds* (after the French

TABLE 2. DERIVATION OF FORMULAS AND FORMULA WEIGHTS OF GASES

Gas Symbol Weight Symbol	In 22.2 Liters						Formula of Gas	Grams of Same Gas in 22.2 Liters
	1 g. H	16 g. O	12 g. C	14 g. N	32 g. S	35.5 g. Cl		
1. Hydrogen chloride	1					1	HCl	36.5
2. Ammonia	3			1			NH_3	17
3. Carbon dioxide		2	1				CO_2	44
4. Carbon monoxide		1	1				CO	28
5. Methane	4		1				CH_4	16
6. Ethylene	4		2				C_2H_4	28
7. Acetylene	2		2				C_2H_2	26
8. Oxygen		2					O_2	32
9. Hydrogen	2						H_2	2
10. Nitrogen				2			N_2	28
11. Chlorine						2	Cl_2	71
12. Sulfur dioxide		2			1		SO_2	64
13. Hydrogen sulfide	2				1		H_2S	34
14. Nitrous oxide		1		2			N_2O	44
15. Nitric oxide		1		1			NO	30

NOTE: Derivation assumes data available on the percentage composition by chemical elements of each gas and the symbols and symbol weights of the elements contained.

chemist Berthollet), and are exemplified by some oxides of the transition elements, by many intermetallic compounds, by the copper sulfide $Cu_{1.7}S$, the copper selenide $Cu_{1.6}Se$ and the cerium hydride $CeH_{2.7}$. Some such compounds vary over a range of composition, depending upon their method of preparation.

In spite of these departures of some compounds from whole number formulas, the fact remains that the great majority of compounds with which the chemist is concerned do contain their constituent elements in intergral multiples of their atomic weights. In fact, there is even a further uniformity in the behavior of many of the elements. Thus the great majority of the compounds of the alkali elements (Group 1A in the periodic table) contain equal atomic weight proportions of hydrogen or its equivalent in other elements. Thus the hydrides of this group have compositions corresponding to the formulas LiH, NaH, KH, etc.; the halogen compounds of the group have the compositions, LiF, NaF, KF, LiCl, NaCl, KCl, etc.; while their simple sulfur compounds (since in many of its compounds sulfur combines with two hydrogen equivalents as represented by the formual H_2S) have the compositions Li_2S, Na_2S, K_2S, etc. However, there also exist more complex binary sulfur compounds of these elements which contain higher proportions of sulfur, so that they combine with sulfur in more than one atomic proportion. Thus this relative combining power, which is called valence, has more than one value for many elements, but is still useful in organizing the data of chemistry. It is discussed at length in the entry on valence, and is explained in structural terms in the entry on molecule.

Radicals. In many chemical compounds there are groups of two or more elements that frequently have the properties of or enter into chemical reaction as a unit. Of those which are of outstanding importance the following are cited:

1. Ammonium NH_4— behaves as a unit in ammonium compounds and in some of these compounds is very similar to potassium K— in potassium compounds.

2. Hydroxyl —OH which behaves as a unit in bases (e.g., sodium hydroxide, NaOH), alcohols (e.g., methyl alcohol, CH_3OH), and phenols (e.g., phenol, C_6H_5OH).

3. Anion-groups of acids, their salts and their esters: Sulfate $>SO_4$, sulfite $>SO_3$, nitrate $—NO_3$, nitrite $—NO_2$, phosphate $\rightarrow PO_4$, perchlorate $—ClO_4$, chlorate $—ClO_3$, chlorite $—ClO_2$, hypochlorite —OCl, carbonate $> CO_3$, formate $—CHO_2$, acetate $—C_2H_3O_2$, palmitate $—C_{16}H_{31}O_2$, stearate $—C_{18}H_{35}O_2$, oleate $—C_{18}H_{33}O_2$, oxalate $>C_2O_4$, lactate $—C_3H_5O_3$, malate $>C_4H_4O_5$, tartrate $>C_4H_4O_6$, citrate $\rightarrow C_6H_5O_7$, benzoate $—C_7H_5O_2$, cinnamate $—C_9H_7O_2$, phthalate $>C_8H_4O_4$, salicylate $—C_7H_5O_3$.

4. Alkyl- and aryl-groups of alcohols, phenols, their esters and their alcoholates and phenolates: (a) Alkyl (non-benzenoid)-methyl CH_3—, ethyl C_2H_5—, propyl C_3H_7—, butyl C_4H_8— and similar radicals of alcohols; (b) Aryl (benzenoid)-phenyl C_6H_5—, tolyl C_7H_7—, xylyl C_8H_9—, naphthyl $C_{10}H_7$— and similar radicals of phenols.

5. Acyl-groups of organic acids: acetyl CH_3CO—, benzoyl C_6H_5CO—.

6. Miscellaneous radicals, for example, cacodyl $(CH_3)_2As$—, celebrated on account of the investigations by Bunsen (1838).

All of the above radicals are associated with a corresponding radical or element in a compound. While a radical frequently and rather generally enters into chemical reaction as a unit, it is not implied that this is always so, the stability in each case is characteristic of each radical and each reaction in which it is involved. Thus, ammonium hydroxide NH_4OH yields ammonia gas NH_3 and water H_2O at room temperature; ammonium nitrate NH_4NO_3 is decomposed, upon heating, with the accompanying disruption of both the ammonium and nitrate radicals to yield nitrous oxide N_2O gas and water H_2O.

Radicals enter widely into reactions involving electrolytic dissociation of salts, acids, bases in water solution.

Radicals exist most commonly in combination with atoms or other radicals. However, they can be produced "free," and can so exist for a finite period. Even when it is very short, the radical itself is often of great interest in elucidating reaction mechanisms. The first free radical discovered was triphenylmethyl.

Gomberg, by treating triphenylmethyl chloride in carbon dioxide, with zinc, silver, or mercury, obtained the free radical, triphenylmethyl. On dissolving the colorless solid in organic solvents a yellow solution is obtained, and the reactivity (due to unsaturation) of the yellow solution is marked towards oxygen, dissolved iodine, ether. Triphenylmethyl is present in solution in two forms, (1) monomolecular $(C_6H_5)_3C$ yellow, in equilibrium with (2) dimolecular $((C_6H_5)_3C)_2$ colorless. But tribiphenylmethyl $(C_6H_5—C_6H_4)_3C$ occurs only in the monomolecular form, purple. The action of alkali metals on ketones in some cases produces metallic ketyl (Schlenk, 1913) thus:

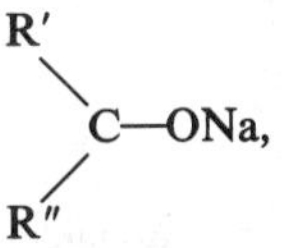

, which is a free radical, or contains trivalent carbon as does monomolecular triphenylmethyl. Many other free radicals are known. See **Free Radical.**

This entry has dealt with two types of chemical composition—elements and compounds. Many materials, including the great majority of those found in nature, are mixtures of compounds and often elements. Practically all biochemical materials and rocks are complex mixtures. Obviously the first step in the determination of the composition of such substances is their separation into the individual compounds, and elements if any, which they contain.

CHEMICAL CONCENTRATION. **Concentration (Chemical).**

CHEMICAL ELEMENTS. A chemical element may be defined as a collection of atoms of one type which cannot be decomposed into any simpler units by any chemical transformation, but which may spontaneously change into other units by radioactive processes. A chemical element is a substance that is made up of but one kind of atom. Of the over-100 chemical elements known, only 90 are found in nature. The remaining elements have been produced in nuclear reactors and particle accelerators. Theoretical physicists do not all agree, but some believe that fission-stable nuclei should exist at atomic number 114. Claims thus far have been made for the discovery, isolation, or creation of elements up to 106. The element with the highest atomic number officially named and entered into the formal table of atomic weight is lawrencium (Lr) with an atomic number of 103.

Each of the chemical elements is described in a separate alphabetical entry in this book.

Some of the principal characteristics of the elements are given in Table 1. All of the 103 elements described on this table also are explained in further detail under individual alphabetical entries throughout this volume. The lanthanide series elements also are described further under **Rare-Earth Elements and Metals.** The platinum group metals are further detailed under **Platinum Group;** the refractory metals are detailed under **Niobium.** The great versatility of carbon is described under **Organic Chemistry.** The chemical elements display a periodicity of properties when they are arranged in order of increasing atomic number. This discovery, generally attributed to Dimitri Mendeleev (1869), although some of the relationships among the elements were known earlier, led to development of the Periodic Law. The resulting matrix arrangement is called the Periodic Table. See **Periodic Table of the Elements.**

The first listing of the elements is generally attributed to Lavoisier in 1789. Of the twenty elements listed, the discovery of five was the result of research conducted by Scheele of Gothenberg. With the development of nuclear physics and the application of these principles to astronomy and cosmology, in recent years the chemical elements have been viewed from new vantage points with much concentration on physical and nuclear characteristics as well as chemical properties. An excellent summary of progress made in the study of the origin of the elements was given by Penzias in a lecture delivered in Stockholm when he received the Nobel Prize in Physics in 1978. See reference listed.

Abundance of the Chemical Elements: Considering the large number of chemical elements, it is interesting to note that insofar as the earth's crust, the oceans, and man's knowledge of the cosmos to date, there is far from a uniform distribution of the elements. In fact, the distribution is exceedingly unbalanced with, for example, only nine elements making up 99.25% of the earth's crust. In terms of abundance, many of the materials considered quite common are, in fact, scarce in terms

TABLE 1. PRINCIPAL CHARACTERISTICS OF CHEMICAL ELEMENTS

NAME	SYMBOL	ATOMIC NUMBER	ATOMIC WEIGHT	PERIODIC GROUP	VALENCY	DENSITY[b] g/cm³	MELTING POINT, °C	BOILING POINT, °C	DISCOVERY (year)
Actinium	Ac	89	227[a]	3b	3	—	1050	3500	1899
Aluminum	Al	13	26.98	3b	3	2.699	660	2467	1827
Americium	Am	95	243[a]	Actinides	3	—	990–998	2600–2608	1945
Antimony	Sb	51	121.75	5a	3,5	6.68	630.7	1587	Early
Argon	Ar	18	39.948	0	0	1.78	−189.2	−185.7	1894
Arsenic	As	33	74.9216	5a	5, ±3	5.73	613	—	Early
Astatine	At	85	210[a]	7a	—	—	302	337	1940
Barium	Ba	56	137.34	2a	2	3.5	725	1640	1808
Berkelium	Bk	97	247[a]	Actinides	3, 4	—	—	—	1950
Beryllium	Be	4	9.012	2a	2	1.848	1287–1292 ± 5	2970 ± 5	1798
Bismuth	Bi	83	208.981	5a	3, 5	9.8	271.3	1562 ± 3	1753
Boron	B	5	10.81	3a	3	2.35	2300	2550[i]	1808
Bromine	Br	35	79.904	7a	±1, 5	3.12	−7.2	58.8	1826
Cadmium	Cd	48	112.41	2b	2	8.65	321	765	1817
Calcium	Ca	20	40.08	2a	2	1.54	837–841	1484	1808
Californium	Cf	98	251[a]	Actinides	3	—	—	—	1950
Carbon	C	6	12.011	4a	± 4, 2	(c)	(c)	4827	Early
Cerium	Ce	58	140.12	Lanthanides	3, 4	6.770	798	3433	1803
Cesium	Cs	55	132.905	1a	1	1.88	28.4	678	1860
Chlorine	Cl	17	35.453	7a	±1, 5, 7	3.214[d]	−101	−34.6	1774
Chromium	Cr	24	51.996	6b	2, 3, 6	7.2	1837–1873	2671–2673	1797
Cobalt	Co	27	58.9332	8	2, 3	8.832	1495	2869–2871	1735
Copper	Cu	29	63.546	1b	1, 2	8.92	1083	2566 ± 0.5	Early
Curium	Cm	96	247[a]	Actinides	3	—	1300–1380	—	1944
Dysprosium	Dy	66	162.50	Lanthanides	3	8.551	1412	2567	1886
Einsteinium	Es	99	254[a]	Actinides	—	—	—	—	1955
Erbium	Er	68	167.26	Lanthanides	3	9.066	1529	2868	1843
Europium	Eu	63	151.96	Lanthanides	2,3	5.244	822	1529	1896
Fermium	Fm	100	257[a]	Actinides	—	—	—	—	1955
Fluorine	F	9	18.9984	7a	−1	1.696[d]	−219.62	−188.1	1771
Francium	Fr	87	223[a]	1a	1	2.4	26.28	676–678	1939
Gadolinium	Gd	64	157.25	Lanthanides	3	7.901	1313	3273	1880
Gallium	Ga	31	69.72	3a	3	5.9	29.78	2403 ± 0.5	1875
Germanium	Ge	32	72.59	4a	4	5.36	937	2830	1886
Gold	Au	79	196.967	1b	1, 3	19.32	1064.43	2805–2809	Early
Hafnium	Hf	72	178.49	4b	4	13.3	2207–2247	4601–4603	1923
Helium	He	2	4.0026	0	0	0.15–0.18	−272.2	−268.93	1895
Holmium	Ho	67	164.93	Lanthanides	3	8.795	1472	2567	1879
Hydrogen	H	1	1.0080	1a	1	0.0899	−259.14	−252.87	1766
Indium	In	49	114.82	3a	3	7.31	156.6	2078–2082	1863
Iodine	I	53	126.9045	7a	−1, 5, 7	4.94	113.5	184.35	1811
Iridium	Ir	77	192.20	8	3, 4, 6	22.42	2410	4130	1803
Iron	Fe	26	55.847	8	2, 3	7.874	1536	2745–2755	Early
Krypton	Kr	36	83.80	0	0	3.4[e]	−156.6	−152.3	1898
Lanthanum	La	57	138.91	3b	3	6.146	918	3464	1839
Lawrencium	Lr	103	257[a]	Actinides	—	—	—	—	1961
Lead	Pb	82	207.19	4a	2,4	11.35	327.5	1740	Early
Lithium	Li	3	6.939	1a	1	0.534	180.54	1315–1319	1817
Lutetium	Lu	71	174.98	Lanthanides	3	9.841	1663	3402	1907
Magnesium	Mg	12	24.312	2a	2	1.74	649	1106–1108	1755
Manganese	Mn	25	54.9380	7b	2, 3, 4, 6, 7	7.3	1241–1247	1962	1774
Mendelevium	Md	101	256[a]	Actinides	—	—	—	—	1955
Mercury	Hg	80	200.59	2b	1, 2	13.546	−38.87	356.58	Early
Molybdenum	Mo	42	95.94	6b	3, 5, 6	9.0–10.2	2610	5560	1778
Neodymium	Nd	60	144.24	Lanthanides	3	7.004	1021	3074	1885
Neon	Ne	10	20.183	0	0	1.204[f]	−248.68	−245.9	1898
Neptunium	Np	93	237.0482	Actinides	3, 4, 5, 6	18.0–20.5	629–631	3900	1940
Nickel	Ni	28	58.71	8	2, 3	8.9	1454–1456	2725–2735	1751
Niobium (Columbium)	Nb	41	92.906	5b	3, 5	8.6	2458–2478	4740–4744	1801
Nitrogen	N	7	14.0067	5a	−3, 2, 5	1.25[d]	−209.86	−195.8	1772
Nobelium	No	102	259[a]	Actinides	—	—	—	—	1957
Osmium	Os	76	190.2	8	4, 6, 8	22.5	2700 ± 5.0	>5300	1803
Oxygen	O	8	15.9994	6a	−2	1.429[d]	−218.4	−182.96	1774
Palladium	Pd	46	106.4	8	2, 4	12.16	1550–1552	3139–3141	1803
Phosphorus	P	15	30.9738	5a	±3, 5	1.82	44.1	280	1669
Platinum	Pt	78	195.09	8	2, 4	21.4	1772	3725–3925	1735
Plutonium	Pu	94	244[a]	Actinides	3, 4, 5, 6	—	—	—	1940
Polonium	Po	84	210[a]	6a	2, 4	9.4	254	962	1898
Potassium	K	19	39.098	1a	1	0.87	63.7	774	1807
Praseodymium	Pr	59	140.91	Lanthanides	3	6.773	931	3520	1879
Promethium	Pm	61	145[a]	Lanthanides	3	7.264	1042	3000	1947
Protactinium	Pa	91	231.036	Actinides	5	—	—	—	1917
Radium	Ra	88	226.026	2a	2	5	700	1140	1898
Radon	Rn	86	222[a]	0	0	9.72[d]	−71	−61.8	1900

Continued

TABLE 1—*Continued*

Name	Symbol	Atomic Number	Atomic Weight	Periodic Group	Valency	Density[b] g/cm³	Melting Point, °C	Boiling Point, °C	Discovery (year)
Rhenium	Re	75	186.2	7b	−1,4,7	20.5–21.0	3178–3182	5600–5900	1925
Rhodium	Rh	45	102.905	8	3,4	12.44	1963–1969	3625–3825	1803
Rubidium	Rb	37	85.466	1a	1	1.53	38.9	689	1861
Ruthenium	Ru	44	101.07	8	3, 4, 6, 8	12.1–12.3	2310	3900–4000	1844
Samarium	Sm	62	150.35	Lanthanides	3	7.520	1074	1794	1879
Scandium	Sc	21	44.956	3b	3	2.985	1541	2831	1879
Selenium	Se	34	78.96	6a	−2, 4, 6	4.82	217	684–686	1817
Silicon	Si	14	28.086	4a	4	2.3	1408–1412	2355	1823
Silver	Ag	47	107.868	1b	1	10.49	961.93	2212	Early
Sodium	Na	11	22.9898	1a	1	0.9721	97.82	882.9	1807
Strontium	Sr	38	87.62	2a	2	2.6	770 ± 1.0	1384	1790
Sulfur	S	16	32.064[j]	6a	−2, 4, 6	2.07[k]	112.8	444.7	Early
Tantalum	Ta	73	180.948	5b	5	16.63	2996	5325–5525	1802
Technetium	Tc	43	98.906	7b	7	11.5	2171–2172	4875–4877	1937
Tellurium	Te	52	127.60	6a	−2, 4, 6	6.25	450–452	1387–1393	1782
Terbium	Tb	65	158.92	Lanthanides	3	8.230	1365	3230	1843
Thallium	Tl	81	204.37	3a	1, 3	11.85	303.3	1460 ± 7.0	1861
Thorium	Th	90	232.038	Actinides	4	11.5–11.9	1740–1760	4780–4800	1828
Thulium	Tm	69	168.93	Lanthanides	3	9.321	1545	1950	1879
Tin	Sn	50	118.69	4a	2, 4	7.29[g]	231.97	2270	Early
Titanium	Ti	22	47.90	4b	3, 4	4.507	1650–1670	3290	1791
Tungsten	W	74	183.85	6b	6	19.3	3410	5660	1781
Uranium	U	92	238.03	Actinides	3, 4, 5, 6	19.05	1131–1133	3818	1789
Vanadium	V	23	50.941	5b	2, 4, 5	6.0–6.10	1880–2000	3390–3400	1830
Xenon	Xe	54	131.30	0	0	3.5[h]	−112	−107.1 ± 2.5	1898
Ytterbium	Yb	70	173.04	Lanthanides	2, 3	6.966	819	1196	1878
Yttrium	Y	39	88.90588.905	3b	3	4.469	1522	3338	1794
Zinc	Zn	30	65.38	2b	2	7.1	419.57	907	1746
Zirconium	Zr	40	91.22	4b	4	6.5	1853 ± 1.0	4376 ± 1.0	1789

[a] Denotes mass number of isotope of longest known half-life (or a better known one for Bk, Cf, Po, Pm and Tc).
[b] Densities for solids at 20°C unless otherwise specified.
[c] For diamond at 20°C. Graphite is 2.25 g/cm³ at 20°C.
[d] Denotes grams/liter at 0°C.
[e] Denotes solid at −273°C.
[f] Denotes density of liquid.
[g] For white tin at 15°C; 5.77 for gray tin at 13°C; 6.97 for liquid tin at melting point.
[h] For liquid at −109°C.
[i] Denotes sublimes.
[j] Atomic weight varies slightly because of naturally occurring isotopes 32, 33, 34, and 36, the total possible variation amounting to ±0.003.
[k] For rhombic sulfur; 1.96 for monoclinic sulfur; 2.046 for amorphous sulfur.

General Note: Atomic weights are believed to have the following uncertainty in specific instances: Br, ±0.002; Cl, ±0.001; Cr, ±0.001; Fe, ±0.003; Ag, ±0.003. For other elements, the last digit given for atomic weight is believed correct and reliable to ±0.5. Values are current as of early 1981. Values for all rare-earth elements furnished by Rare-Earth Information Center, Energy and Mineral Resources Research Institute, Iowa State University, Ames, Iowa.

of their percentage of the total materials in the earth's crust. In order of descending occurrence in the earth's crust, all but the very scarce elements are listed in Table 2. The occurrence of these elements in seawater also is given where data are available.

The situation alters considerably in terms of cosmic abundance of the elements, although again, the abundance is heavily slanted toward a comparatively few of the total number of elements. In 1952, Harold C. Urey made an estimate of the abundance of the chemical elements in the cosmos, wherein earlier values of V. M. Goldschmidt and Harrison S. Brown were used in the calculations. A base figure of 10,000 for silicon was used, the other elements being expressed in relationship to that base figure. The study indicated that hydrogen (3.5×10^8) is the most likely superabundant element in the cosmos, closely followed by helium (3.5×10^7). Other elements high on the list include oxygen (220,000), nitrogen (160,000), carbon (80,000), and neon (with a range estimated between 9,000 up to 240,000). Magnesium, silicon, and iron also are relatively high.

Researchers at California Institute of Technology (Pasadena, California) have been studying the occurrence of various isotopes in the earth's crust as a lead to understanding the mechanics of continental crust formation. A means for determining the time of formation of new crustal segments is paramount to an understanding of how the continental crust was evolved. In this work, they have studied samarium–neodymium and rubidium–strontium isotopic systematics. See reference listed and also entry on **Earth.**

Critical Importance of Certain Elements: The uneven distribution of the chemical elements in the earth's crust causes severe imbalances in their availability for industrial manufacturing uses. Frequently, the largest consuming nations are long distances from elemental sources, as portrayed in the case of the United States by the bar graph of Fig. 1. This clearly indicates the need for the various political entities of the world to stockpile and to project their needs many years in advance. Hence, for obvious reasons, these needs are commonly reflected in foreign policies.

There are many "hidden" uses for the elements and the compounds of which they are a part. This is evident from the list of elements used in the telephone handset as given in Table 3.

Nuclides, Isotopes, and Isobars: A nuclide may be defined as a species of atoms, with specified atomic number and mass number. The term *nuclide* should be used, *not* isotope. Different nuclides having the same atomic number are *isotopes.* Different nuclides having the same mass number are *isobars.*

A comprehensive listing of the nonradioactive nuclides of the chemical elements is given in Table 4. The isotopic abundance and mass number are given. In all but a few instances, the isotopes listed are stable. The elements also have a number of radioisotopes each, the number varying considerably from one element to the next. Extensive tables also listing the radioactive isotopes, giving lifetime, modes of decay, decay energy, particle energies, particle intensities, and thermal neutron capture cross section, among other factors, can be found in the literature. See Heath references listed at end of this entry.

An important fact pertaining to naturally occurring elements is

TABLE 2. ABUNDANCE OF THE CHEMICAL ELEMENTS
(In Grams/Metric Ton)*

Element	Terrestrial Abundance	Occurrence in Seawater	Element	Terrestrial Abundance	Occurrence in Seawater
Oxygen	466,000	850,000	Beryllium	6	—
Silicon	277,200	2.98	Praseodymium	5.53	—
Aluminum	81,300	0.01	Arsenic	5	0.003
Iron	50,000	0.01	Scandium	5	4×10^{-5}
Calcium	36,300	404	Hafnium	4.5	—
Sodium	28,300	10,550	Dysprosium	4.47	—
Potassium	25,900	380	Uranium	4	—
Magnesium	20,900	1,290	Boron	3	4.9
Titanium	4,400	0.001	Thallium	3	—
Hydrogen	1,300	108,200	Ytterbium	2.66	—
Phosphorus	1,180	0.07	Erbium	2.47	—
Manganese	1,000	0.002	Tantalum	2.1	—
Fluorine	900	1.27	Bromine	1.62	66
Sulfur	520	894	Holmium	1.15	—
Carbon	320	27.6	Europium	1.06	—
Chlorine	314	19,050	Antimony	1	0.0005
Rubidium	310	0.121	Terbium	0.91	—
Strontium	300	8.1	Lutetium	0.75	—
Barium	250	0.006	Mercury	0.50	3×10^{-5}
Zirconium	220	—	Iodine	0.30	0.05
Chromium	200	5×10^{-5}	Thulium	0.20	—
Vanadium	150	0.002	Bismuth	0.20	0.0002
Zinc	132	0.01	Cadmium	0.15	6×10^{-5}
Nickel	80	0.0005	Silver	0.10	0.0003
Copper	70	0.003	Indium	0.10	0.02
Tungsten	69	0.0001	Selenium	0.09	4×10^{-5}
Lithium	65	0.2	Argon	0.04	0.595
Nitrogen	46.3	0.51	Palladium	0.01	—
Cerium	46.1	0.00038	Tellurium	0.002	0.00001
Tin	40	0.003	Gold	0.005	4×10^{-6}
Yttrium	28.1	0.0003	Osmium	0.005	—
Niobium (Columbium)	24	5×10^{-6}	Platinum	0.005	—
Neodymium	23.9	—	Ruthenium	0.004	—
Cobalt	23	0.0005	Rhodium	0.001	—
Lanthanum	18.3	0.0003	Iridium	0.001	—
Lead	16	0.003	Neon	7×10^{-5}	0.0003
Gallium	15	3×10^{-5}	Radium	13×10^{-6}	996×10^{-13}
Molybdenum	15	0.01	Krypton	9.8×10^{-6}	0.0003
Thorium	11.5	0.0007	Xenon	1.2×10^{-6}	0.0001
Cesium	7	0.0005	Protactinium	8×10^{-7}	0.003
Germanium	7	6×10^{-5}	Actinium	3×10^{-10}	—
Samarium	6.47	—	Polonium	3×10^{-10}	—
Gadolinium	6.36	—			

* Presented in order of diminishing terrestrial abundance.

that many of them consist of several isotopes and that these isotopes are present in nearly all cases in the same proportion by weight. These constant isotopic compositions have enabled scientists over the years to analyze materials by weight, making reference to a table of atomic weights. In fact, for many years the atomic weight of an element was regarded as its most distinguishing characteristic, even though some discrepancies were noticed in the periodic table. It was not until the work of Mosely on characteristic x-ray spectra and the development of positive-ray analysis that the nuclear charge was recognized as the fundamental chemical characteristic of an element. The use of mass spectrography, with other methods of determining the masses of the atoms in a given element and the proportions in which they are present, has permitted the determination of the isotopic composition of the elements.

Sulfur is one of the few exceptions to the constancy of isotopic proportions, in that there is sufficient variation, dependent upon the source of the sulfur, to cause a variation in its atomic mass by approximately $\pm 0.01\%$. For normal stochiometric calculations, however, this small variation is unimportant.

Naturally-occurring elements which do not display any isotopic behavior include aluminum, arsenic, beryllium, bismuth, cobalt, fluorine, gold, helium, holmium, iodine, manganese, niobium (columbium), phosphorus, praseodymium, rhodium, scandium, sodium, terbium, thulium, and yttrium. Elements which have one predominating isotope (in excess of 98%) include argon, carbon, lanthanum, lutetium, nitrogen, oxygen, tantalum, and vanadium. Elements which have several isotopes and in which no one isotope is in excess of 80% of the total include antimony, barium, bromine, chlorine, copper, dysprosmium, erbium, gadolinium, gallium, germanium, hafnium, iridium, krypton, lead, magnesium, mercury, molybdenum, nickel, osmium, palladium, platinum, rhenium, rubidium, ruthenium, selenium, silver, tellurium, tin, titanium, tungsten, xenon, ytterbium, zinc, and zirconium. Tin leads with a total of ten isotopes; xenon has nine isotopes; cadmium and tellurium each have eight isotopes.

Radioactive Elements: There are (1) naturally-occurring radioactive and (2) artifically-produced radioactive elements. There are three series of naturally-occurring radioactive elements:

The actinium series: This series commences with ^{225}U and ends with the stable isotope, ^{207}Pb. The decay scheme is represented by: $^{225}U \xrightarrow{\alpha} {}^{231}Th \xrightarrow{\beta} {}^{231}Pa \xrightarrow{\alpha} {}^{227}Ac \xrightarrow{\beta \text{ and } \alpha} {}^{227}Th \xrightarrow{\alpha} {}^{223}Fr \xrightarrow{\beta} {}^{223}Ra \xrightarrow{\alpha} {}^{219}Rn \xrightarrow{\alpha} {}^{215}Po \xrightarrow{\alpha \text{ and } \beta} {}^{211}Pb \xrightarrow{\beta} {}^{215}At \xrightarrow{\alpha} {}^{211}Bi \xrightarrow{\beta \text{ and } \alpha} {}^{211}Po \xrightarrow{\alpha} {}^{207}Tl \xrightarrow{\beta} {}^{207}Pb$ (stable).

The thorium series: This series commences with ^{232}Th and ends with the stable isotope, ^{208}Pb. The decay scheme is represented by: $^{232}Th \xrightarrow{\alpha} {}^{228}Ra \xrightarrow{\beta} {}^{228}Ac \xrightarrow{\beta} {}^{228}Th \xrightarrow{\alpha} {}^{224}Ra \xrightarrow{\alpha} {}^{220}Rn \xrightarrow{\alpha} {}^{216}Po \xrightarrow{\alpha} {}^{212}Pb \xrightarrow{\beta \text{ and } \alpha} {}^{216}At \xrightarrow{\alpha} {}^{212}Bi \xrightarrow{\beta \text{ and } \alpha} {}^{212}Po \xrightarrow{\alpha} {}^{208}Tl \xrightarrow{\beta} {}^{208}Pb$ (stable).

The uranium series: This series commences with ^{238}U and ends with the stable isotope, ^{206}Pb. The decay scheme is represented by: $^{238}U \xrightarrow{\alpha} {}^{234}Th \xrightarrow{\beta} {}^{234}Pa \xrightarrow{\beta} {}^{234}U \xrightarrow{\alpha} {}^{230}Th \xrightarrow{\alpha} {}^{226}Ra \xrightarrow{\alpha} {}^{222}Ra \xrightarrow{\alpha} {}^{218}Po \xrightarrow{\alpha \text{ and } \beta} {}^{214}Pb \xrightarrow{\beta} {}^{218}At \xrightarrow{\alpha} {}^{214}Bi \xrightarrow{\beta \text{ and } \alpha} {}^{214}Po \xrightarrow{\alpha} {}^{210}Tl \xrightarrow{\beta} {}^{210}Pb \xrightarrow{\beta} {}^{210}Bi \xrightarrow{\beta \text{ and } \alpha} {}^{210}Po \xrightarrow{\alpha} {}^{206}Tl \xrightarrow{\beta} {}^{206}Pb$ (stable).

In the foregoing series, the type of radiation given off during the decay process is indicated above the arrows.

The production of artificially-produced radioactive elements dates back to the early work of Rutherford in 1919 when it was found that alpha particles reacted with nitrogen atoms to yield protons and oxygen atoms. Curie and Joliot found (1933) that when boron, magnesium, or aluminum were bombarded with alpha particles from polonium, the elements would emit neutrons, protons, and positrons. They

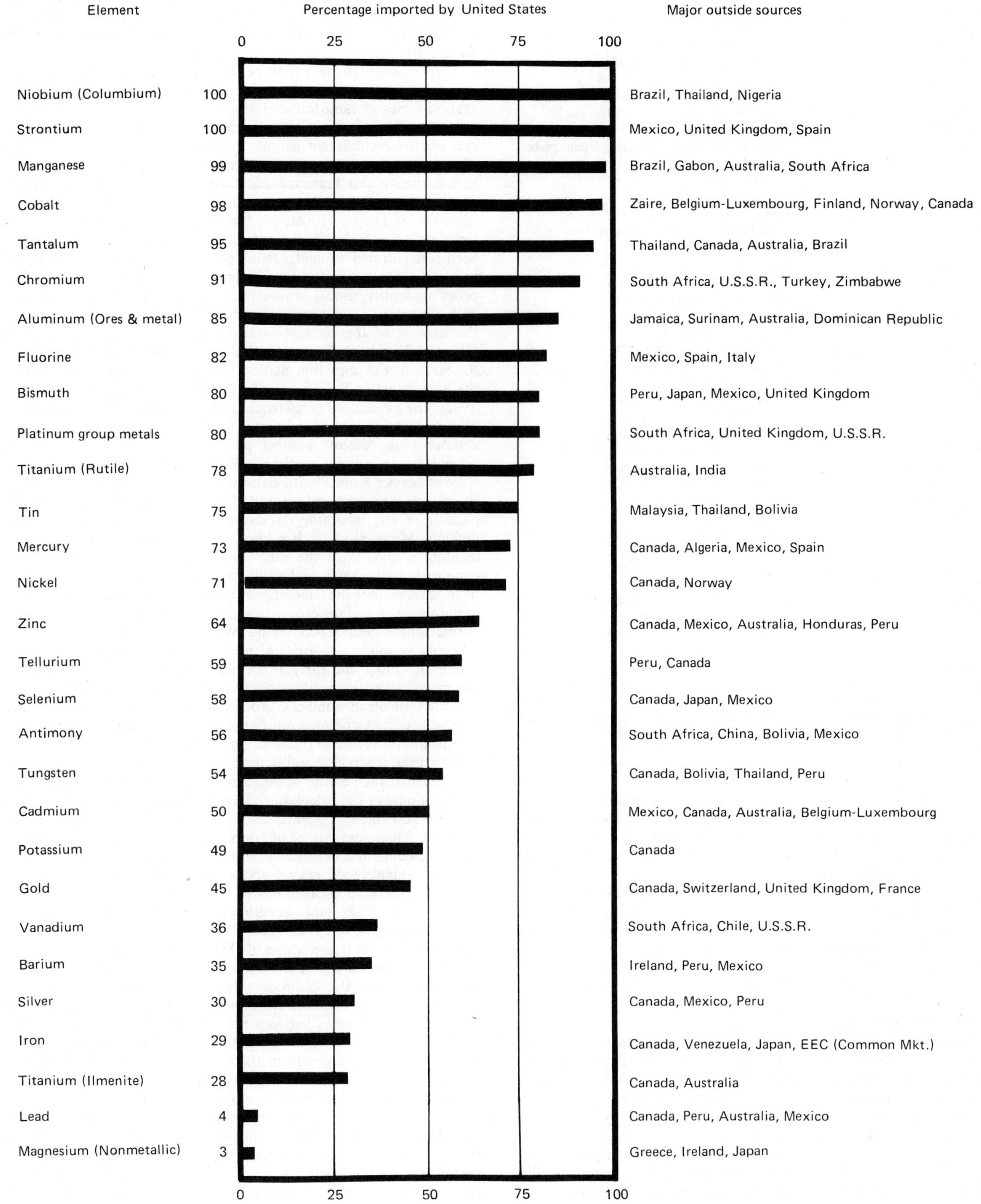

Fig. 1. Imports of strategic elements (ores) by the United States. (*U.S. Dept. Interior, Washington, D.C.*)

TABLE 3. ELEMENTS IN THE TELEPHONE HANDSET

Element	Application
Aluminum	Metal alloy in dial mechanism, transmitter, and receiver
Antimony	Alloy in dial mechanism
Arsenic	Alloy in dial mechanism
Beryllium	Alloy in dial mechanism
Bismuth	Alloy in dial mechanism
Boron	Touch-Tone® dial mechanism
Cadmium	Color in yellow plastic housing
Calcium	In lubricant for moving parts
Carbon	Plastic housing, transmitter steel parts
Chlorine	Wire insulation
Chromium	Color in green plastic housing, metal plating, stainless steel parts
Cobalt	Magnetic material in receiver
Copper	Wires, plating, brass parts
Fluorine	Plastic parts
Germanium	Transistors in some dial mechanisms
Gold	Electrical contacts
Hydrogen	Plastic housing, wire insulation
Indium	Touch-Tone® dial mechanism
Iron	Steel and magnetic materials
Krypton	Ringer in Touch-Tone® set
Lead	Solder in connections
Lithium	In lubricant for moving parts
Magnesium	Die castings in transmitter, ringer
Manganese	Steel in various parts
Mercury	Color in red plastic housing
Molybdenum	Magnet in receiver
Nickel	Magnet in receiver, stainless steel parts
Nitrogen	Hardened heat-treated steel parts
Oxygen	Plastic housing, wire insulation
Palladium	Electrical contacts
Phosphorus	Steel in various parts
Platinum	Electrical contacts
Silicon	Touch-Tone® dial mechanism
Silver	Plating
Sodium	In lubricant for moving parts
Sulfur	Steel in various parts
Tantalum	Integrated circuit in some sets (Trimline®)
Tin	Solder in connections, plating
Titanium	Color in white plastic housing
Tungsten	Lights in Princess® and key sets
Vanadium	Receiver
Zinc	Brass, die casting in transmitter, ringer

Source: Committee on Survey of Materials Science and Engineering, Appendix to the COMSAT Report (national Academy of Sciences, Washington, D.C., 1975), Volume 2.

also found that upon cessation of bombardment the emission of protons and neutrons stopped, but that the emission of positrons continued. The targets remained radioactive. They also found that the radiation emitted dropped off exponentially as would be expected from a naturally-occurring radioactive element. Further investigation indicated that nuclear reactions lead to the formation of radioactive isotopes. This and subsequent work by several investigators led to the formulation of another series of radioactive elements, namely, the Neptunium Series, which commences with ^{245}Cm (curium) and ends with the stable isotope, ^{209}Bi. The decay scheme is represented by: $^{245}Cm \xrightarrow{\alpha} {}^{241}Pu \xrightarrow{\beta} {}^{241}Am \xrightarrow{\alpha} {}^{237}Np \xrightarrow{\alpha} {}^{233}Pa \xrightarrow{\beta} {}^{233}U \xrightarrow{\alpha} {}^{229}Th \xrightarrow{\alpha} {}^{225}Ra \xrightarrow{\beta} {}^{225}Ac \xrightarrow{\alpha} {}^{221}Fr \xrightarrow{\alpha} {}^{217}At \xrightarrow{\alpha} {}^{213}Bi \xrightarrow{\beta \text{ and } \alpha} {}^{213}Po \xrightarrow{\alpha} {}^{209}Tl \xrightarrow{\beta} {}^{209}Pb \xrightarrow{\beta} {}^{209}Bi$ (stable).

A *radioactive element* is an element that disintegrates spontaneously with the emission of various rays and particles. Most commonly, the term denotes radioactive elements such as radium, radon (emanation), thorium, promethium, uranium, which occupy a definite place in the periodic table because of their atomic number. The term radioactive element is also applied to the various other nuclear species (which are produced by the disintegration of radium, uranium, etc.) including the members of uranium, actinium, thorium, and neptunium families of radioactive elements, which differ markedly in their stability, and are isotopes of elements from thallium (atomic number 81) to uranium (atomic number 92), as well as the partly artificial actinide group, which extends from actinium (atomic number 89) to lawrencium (atomic number 103), and includes the following transuranic elements: neptunium (atomic number 93), plutonium (atomic number 94), americium (atomic number 95), curium (atomic number 96), berkelium (atomic number 97), californium (atomic number 98), einsteinium (atomic number 99), fermium (atomic number 100), mendelevium (atomic number 101), nobelium (atomic number 102). The radioactive nuclides produced from nonradioactive ones are discussed under **Radioactivity.**

A radioactive element may be designated as being in a *collateral series.* In addition to the three main natural and one artificial disintegration series of radioactive elements, each has been found to have at least one parallel or collateral series. The main series and the collateral series have different parents, but become identical in the course of disintegration, when they have a member in common.

Superheavy and Transactinide Elements: Those elements with an atomic number above 103 are sometimes referred to as the *transactinide elements*; and those with atomic (proton) numbers $Z \geq 110$ are sometimes called *superheavy elements* (SHEs). Considerable research has been concentrated on synthesizing elements heavier than 103. It will be recalled that the last of the elements to be synthesized with positive proof of identity, timing of research, and place and persons associated with discovery—and thus relatively little controversy pertaining to the naming of the element—was lawrencium (103), discovered by Ghiorso, Sikkeland, Larsh, and Latimer in March 1961 at Berkeley. Research in this direction had been underway in the early 1960s at the Joint Nuclear Research Institute at Dubna (U.S.S.R.). The half-lives of the Lr isotopes range from 8 to 35 seconds, enabling researchers to use solvent extraction techniques for determining the chemical characteristics and atomic number of the element. However, with the possible exception of a predicted *island of stability,* as one goes up the scale of heavier elements, the half-lives appear to become progressively shorter, making chemical separation, needed for identifying the atomic number of any laboratory-produced superheavy nucleus, increasing difficult. Much research has been directed toward improving these techniques and, as of the early 1980s, methodology is available to cope with nuclei having a half-life of 1 second or greater. Separation techniques include the ion exchange behavior of the bromide complexes of the elements; and the ease with which the elements coprecipitate with cupric sulfide (Seaborg–Loveland–Morrissey, 1979).

Applying modern theories of nuclear structure, scientists working in the late 1960s and early 1970s made calculations that showed for superheavy elements in the vicinity of $Z = 114$ (proton number) and $N = 184$ (neutron number), ground states of nuclei were stabilized against fission. As pointed out in the aforementioned reference, "This stabilization was due to the complete filling of major proton and neutron shells in this region and is analogous to the stabilization of chemical elements, such as the noble gases by the filling of their electronic shells." Some of the calculations indicated that the half-lives of some of these superheavy nuclei might be on the order of the age of the universe. This observation, of course, was stimulating to further research. Calculations indicated that there should be an island of relative stability which would extend above the Z and N figures previously given. The calculations also showed that between the presently known elements and these stable superheavy elements, there would be an intervening region of instability.

Although beyond the scope of this volume, in the early 1930s a new mechanism for the interaction of heavy ions was discovered (See Lefort–Ngo and the Schröder–Huizenga references listed.) The method, known as *deep inelastic scattering,* involves a massive transfer of energy and nucleons between the projectile and the target. More detail is also given in the Seaborg–Loveland–Morrissey reference.

A team of researchers from the Oak Ridge National Laboratory, the University of California (Davis, California), and Florida State University reported an X-ray spectra in June 1976 that appeared to confirm the existence of superheavy elements with atomic numbers near 126. For a short period, this finding created much interest in the scientific community—mainly because the atomic numbers reported were much greater than might be expected at that stage of research, and, possibly of even greater interest, because the findings were based upon the elements being part of monazite crystals believed

TABLE 4. THE NUCLIDES (ISOTOPES AND ISOBARS)

Element	Mass No. A	Isotopic Abundance %
H	1	99.985
	2	0.015
He	3	1.3×10^{-4}
	4	~100
Li	6	7.5
	7	92.5
Be	9	100
B	10	18.7
	11	81.3
C	12	98.9
	13	1.1
N	14	99.62
	15	0.38
O	16	99.76
	17	0.04
	18	0.20
F	19	100
Ne	20	90.8
	21	<0.3
	22	8.9
Na	23	100
Mg	24	77.4
	25	11.5
	26	11.1
Al	27	100
Si	28	92.21
	29	4.70
	30	3.09
P	31	100
S	32	95.0
	33	<0.8
	34	4.2
	36	<0.02
Cl	35	75.4
	37	24.6
Ar	36	0.337
	38	0.061
	40	99.602
K	39	93.1
	40	<0.012
	41	<6.9
Ca	40	96.96
	42	0.64
	43	0.15
	44	2.06
	46	0.0033
	48	<0.019
Sc	45	100
Ti	46	7.93
	47	7.28
	48	73.94
	49	5.51
	50	5.34
V	50	0.25
	51	99.75
Cr	50	4.31
	52	83.76
	53	9.55
	54	2.38
Mn	55	100
Fe	54	5.82
	56	91.66
	57	2.19
	58	0.33
Co	59	100
Ni	58	67.88
	60	26.22
	61	1.18
	62	3.66
	64	<1.08
Cu	63	69.09
	65	30.91
Zn	64	<48.9
Zn (*cont.*)	66	27.8
	67	4.1
	68	18.6
	70	0.63
Ga	69	60.1
	71	39.9
Ge	70	20.52
	72	27.43
	73	7.76
	74	36.54
	76	<7.76
As	75	100
Se	74	0.87
	76	9.02
	77	7.58
	78	23.52
	80	49.82
	82	9.19
Br	79	50.54
	81	49.46
Kr	78	0.342
	80	2.23
	82	11.50
	83	11.48
	84	57.02
	86	<17.43
Rb	85	72.2
	87	27.8
Sr	84	0.56
	86	9.86
	87	7.02
	88	82.56
Y	89	100
Zr	90	51.5
	91	11.2
	92	17.1
	94	17.4
	96	2.8
Nb	93	100
Mo	92	15.84
	94	9.04
	95	15.72
	96	16.53
	97	9.46
	98	23.78
	100	9.63
Tc	(all radioactive)	
Ru	96	5.51
	98	1.87
	99	12.72
	100	12.62
	101	17.07
	102	31.61
	104	18.58
Rh	103	100
Pd	102	1.0
	104	11.0
	105	22.2
	106	27.2
	108	26.8
	110	11.8
Ag	107	51.4
	109	48.6
Cd	106	1.23
	108	0.88
	110	12.32
	111	12.67
	112	24.15
	113	12.21
	114	28.93
	116	7.61
In	113	4.2
	115	95.8
Sn	112	0.96
	114	0.66
	115	0.35
	116	14.30
	117	7.61
	118	24.03
	119	8.58
	120	32.85
	122	4.72
	124	5.94
Sb	121	57.25
	123	42.75
Te	120	0.088
	122	2.83
	123	0.85
	124	4.59
	125	6.93
	126	18.71
	128	<31.86
	130	<34.52
I	127	100
Xe	124	0.094
	126	0.088
	128	1.92
	129	>26.23
	130	4.05
	131	>21.14
	132	26.93
	134	10.52
	136	8.93
Cs	133	100
Ba	130	0.101
	132	0.097
	134	2.42
	135	6.59
	136	7.81
	137	>11.32
	138	>71.66
La	138	0.089
	139	99.911
Ce	136	0.19
	138	0.26
	140	88.47
	142	11.08
Pr	141	100
Nd	142	27.11
	143	12.17
	144	23.85
	145	8.30
	146	17.22
	148	5.73
	150	5.62
Pm	(all radioactive)	
Sm	147	15.0
	148	11.2
	149	13.8
	150	7.4
	152	26.8
	154	22.7
Eu	151	47.8
	153	52.2
Gd	152	0.2
	154	2.15
	155	<14.7
	156	20.5
	157	15.7
	158	<24.9
	160	21.9
Tb	159	100
Dy	156	0.052
	158	0.090
	160	2.29
	161	>18.9
Dy (*cont.*)	162	25.5
	163	<25.0
	164	28.2
Ho	165	100
Er	162	0.136
	164	1.56
	166	>33.4
	167	22.9
	168	27.1
	170	14.9
Tm	169	100
Yb	168	0.14
	170	3.03
	171	14.3
	172	>21.8
	173	16.2
	174	>31.8
	176	12.7
Lu	175	97.5
	176	2.5
Hf	174	0.18
	176	5.2
	177	>18.4
	178	27.1
	179	13.8
	180	35.3
Ta	180	0.012
	181	99.988
W	180	0.122
	182	26.20
	183	14.26
	184	<30.74
	186	<28.82
Re	185	37.1
	187	62.9
Os	184	0.018
	186	1.59
	187	1.64
	188	13.3
	189	16.1
	190	26.4
	192	<41.0
Ir	191	38.5
	193	61.5
Pt	190	0.012
	192	0.78
	194	32.8
	195	>33.7
	196	25.4
	198	7.2
Au	197	100
Hg	196	0.15
	198	10.1
	199	17.0
	200	23.3
	201	13.2
	202	<29.6
	204	6.7
Tl	203	29.5
	205	70.5
Pb	204	1.37
	206	26.26
	207	20.8
	208	>51.55
Bi	209	100

Following elements in increasing mass number are all radioactive: Po, At, Rn, Fr, Ra, Ac, Th, Pa, U, Np, Pu, Am, Cm, Bk, Cf, Es, Fm, Md, No, and Element 104 and heavier.

to be about 1 billion years old, thus giving rise to the previous mention of their life on the order of the age of the universe. Further confirmatory proof was lacking, capped by the finding of a researcher at Florida State University who showed that a gamma ray with the same energy as the X-ray peak for "element 126" is emitted when an excited praseodymium nucleus relaxes after being created from cerium during bombardment by protons. It is noteworthy that cerium is a major constituent of monazite. More details on subsequent study of the initial findings are given in the Robinson reference listed.

Element 104: As of the early 1980s, claims for discovery and thus the procedure for officially naming element 104 remain unresolved. Researchers at Dubna (U.S.S.R.), in 1964, bombarded plutonium with accelerated 113–115 MeV neon ions. During this process, an isotope that decayed by spontaneous fission was observed. It was reported that the isotope had a half-title of 0.3 ± 0.1 second and it was reasoned that the isotope was $^{260}104$, resulting from $^{242}_{94}Pu + ^{22}_{10}Ne \rightarrow ^{260}104 + 4n$. Although subsequent work toward chemically separating the new element from all others has not been conclusive, considerable evidence for evaluation has been obtained. Estimates of the half-life of the element have been reduced from the prior 0.3 second to 0.15 second. Ghiorso, Nurmia, Harris, K. Eskola, and P. Eskola (University of California, Berkeley) reported in 1969, the positive identification of two and possibly three isotopes of the element. The Berkeley discovery resulted from bombarding a target of ^{249}Cf with ^{12}C nuclei of 71 MeV, and ^{13}C nulcei of 60 MeV. The first combination resulted in the instant emission of four neutrons to produce $^{257}104$. The isotope was reported to have a half-life of 4–5 seconds. Decay was by emission of an alpha particle into ^{253}No, with a half-life of 105 seconds. In further research, several thousand atoms of $^{257}104$ and $^{259}104$ were produced. Thus far, the Dubna workers have proposed the name *kurchatovium* (Ku); and the Berkely group has suggested *rutherfordium* (Rf).

Element 105: In 1967, workers at Dubna (U.S.S.R.) reported producing a few atoms of element $^{260}105$ and $^{261}105$, as the result of bombarding ^{243}Am with ^{22}Ne. Appropriate confirmations of identification, however, were lacking, the evidence being based upon time-coincidence measurements of alpha energies. In 1970, it was reported that the Dubna researchers had investigated all the types of decay of the new element and had determined its chemical properties. As of that time, the Dubna group had not proposed a name for element 105. Ghiorso, Nurmia, Harris, K. Eskola, and P. Eskola (University of California, Berkeley) reported a positive identification of element 105. This resulted from bombarding a target of ^{249}Cf with 84 MeV nitrogen nuclei. Upon absorption of ^{15}N nuclei by a ^{249}Cf nucleus, four neutrons are emitted, forming element $^{260}105$, with a half-life of 1.6 seconds. The Berkeley group has proposed *hahnium* (Ha) as a name for the element.

Element 106: In late 1974, two groups announced the synthesis of the 14th transurium element, namely, element number 106 (eka-tungsten). The Soviet group at Dubna bombarded a target of lead atoms with ions of various weights. Argon ions were used to form a short-lived isotope of fermium which decayed by spontaneous fission. After this trial, the Soviet group used ions of titanium to produce a similarly short-lived isotope of element 104. Then, the Soviet scientists finally used ions of chromium to produce what they believe to be element 106. This presumed element is described as having 151 or 152 neutrons, which decays by spontaneous fission with a half-life of about 4 to 10 milliseconds. The Soviet scientists claim that the chromium and lead will combine to form element 106.

An American group at the University of California's Lawrence Berkeley Laboratory, using the modified super-HILAC accelerator (heavy ion linear accelerator) to bombard a target of californium-249 with ions of oxygen-18, caused the oxygen ions to combine with molecules in the target, releasing four neutrons per collision and become eka-tungsten-263. It is reported that this isotope of element 106 has a half-life of 0.9 second. Contrary to earlier predictions, it does not decay by spontaneous fission, but emits an alpha particle with an energy of 9.06 MeV to become an isotope of element 104. The previously observed daughter isotope emits an alpha particle with an energy of 8.8 MeV, becoming nobelium-255. The latter, in turn, emits an alpha particle with an energy of 8.11 MeV. The American scientists believe that observation of this complete sequence of transmutations is conclusive proof of the formation of eka-tungsten. The American team first observed element 106 in 1970, but lead impurities in the target presented difficulties in providing conclusive evidence of the existence of the new element. The HILAC accelerator was shut down shortly after the experiment and two years were required for the improvements that resulted in the super-HILAC. Then other experiments were given a higher priority. Thus there was an approximately four-year delay in the American experiments. The isotope of element 106 synthesized at Berkeley contains 157 neutrons.

As of the early 1980s, none of the researchers have suggested a name for element 106.

Element 107: Investigators at Dubna announced the synthesis of element 107 in 1976, as the result of bombarding ^{204}Bi with heavy nuclei of ^{54}Cr. Prior experiments had suggested the probable very brief (0.002 second) observation of the element. Research reports on the element remain sketchy and it is considered premature to fully presume that 107 exists.

Allotropes: Some of the elements exist in two or more modifications distinct in physical properties, and usually in some chemical properties. Allotropy in solid elements is attributed to differences in the bonding of the atoms in the solid. Various types of allotropy are known. In *enantiomorphic allotropy*, the transition from one form to another is reversible and takes place at a definite temperature, above or below which only one form is stable, e.g., the alpha and beta forms of sulfur. In *dynamic allotropy*, the transition from one form to another is reversible, but with no definite transition temperature. The proportions of the allotropes depend upon the temperature. In *monotropic allotropy*, the transition is irreversible. One allotrope is metastable at all temperatures, e.g., explosive antimony.

Examples of allotropes include:

Arsenic with four forms, metallic, yellow, gray, and brown.
Boron with two forms, crystalline and amorphous.
Carbon with three forms, amorphous, diamond, and graphite.
Phosphorus with four forms, two white forms, a violet, and a black form. Red phosphorus is a mixture of the white and violet forms.
Selenium with four forms, amorphous, two crystalline monoclinic forms (red), and the stable, crystalline gray metallic form.
Sulfur with two forms, alpha-rhombic sulfur with a density of 2.07 and a mp of 112.8°C, and beta-monoclinic sulfur with a density of 1.96 and a mp of 119°C. The beta form changes to the alpha form below 96°C.

In the case of some of the less common elements, impure forms have been mistaken in the past as allotropic forms. A number of elements that once were considered to exist in both crystalline and amorphous forms have been found to exist in only one form when perfectly pure.

Gaseous Elements: Several gaseous elements (at standard conditions of temperature and pressure) form molecules of two atoms each. These are known as *diatomic gases.* Included in this category are hydrogen, H_2, nitrogen, N_2, oxygen, O_2, and chlorine, Cl_2. The inert gases, helium, neon, argon, krypton, and xenon are *monatomic gases* and their symbols do not carry a subscript.

Atomic Structure of the Elements

The internal structure of an atom consists of electrons moving within a region having a diameter slightly greater than 10^{-8} cm and of protons and neutrons that are confined to a nucleus at the center of the electron distribution in a region having a diameter of about 10^{-12} cm. The electrons of a neutral atom are sufficient in number so that their total negative charge is equal to the positive charge on the nucleus. For example, atoms of the element hydrogen have in their neutral state a single electron moving about a nucleus which has a positive charge equal to the negative charge of an electron. Helium, which has two electrons moving about its nucleus, has a positive charge on its nucleus equal to twice one electronic charge and lithium, which has three electrons moving about its nucleus, has a positive charge on its nucleus of three electronic charges. One basis for the classification of atoms is by those numbers which correspond to the number by which the charge on the hydrogen nucleus must be multiplied to equal the nuclear charge of the atom in question. These numbers, ranging from one, for hydrogen, to 103, for lawrencium, are called atomic numbers.

The atomic number may be defined as the number of protons in an atomic nucleus, or the positive charge of the nucleus, expressed in terms of the electronic charge. Atomic number usually is denoted by the symbol Z. In the symbolic designation of individual nuclides, the atomic number sometimes is written as a subscript to the left of the chemical symbol of the atomic species, such as ${}^{16}_{8}O$ for the oxygen isotope of mass number 16. This usage is redundant, in that the chemical symbol per se specifies the atomic number of the nuclide.

Besides nuclear charge, atoms also differ in their masses. The mass is determined by the number of protons Z and the number of neutrons N in the atomic nucleus. The total number of nuclear particles, $Z + N$, in an atomic species is known as its mass number A.

The nuclear model for an atom is of relatively recent origin for, at the beginning of the 20th century, J. J. Thomson's "plum-pudding" model of an atom was the more generally accepted version. In this model Thomson supposed that the positive charge forms a plasma that is distributed throughout the atomic volume and that the electrons are mixed into this plasma with a relatively uniform distribution. E. Rutherford proposed the nuclear model on the basis of experimental work by H. Geiger and E. Marsden in which they observed that a small number of alpha particles from a naturally occurring radioactive source are scattered through angles greater than 90° by thin foils of gold and silver. Although some small angle scattering is predicted by the Thomson model, such large angle scattering is not at all expected. Large angle scattering is however completely consistent with the idea that the alpha particles interact with point positively charged objects of large mass at the center of each gold and silver atom.

Subsequently (1913), N. Bohr found that he could use the Rutherford nuclear model to explain in almost complete detail the observed spectrum of hydrogen (see **Atomic Spectra**). Bohr proposed that, in a neutral hydrogen atom, a single electron revolves in a stationary orbit around a point nucleus that has a charge $+e$. This electron is held in its orbit by the electrostatic force between the positive charge on the nucleus and its own negative charge. The stationary-orbit assumption was classically unsatisfactory but necessary to account for the behavior of the electron. If the laws of classical electrodynamics, according to which accelerated charges must be sources of electromagnetic radiation, were strictly obeyed, the electron would gradually lose energy; hence it would revolve in orbits of smaller and smaller radii and eventually fall into the nucleus. Furthermore, Bohr postulated that the magnitude of the orbital angular momentum of each stationary orbit is $L = nh/2\pi = n\hbar$, where n is an integer, h is Planck's constant, and $\hbar$ is simply an abbreviated form for $h/2\pi$.

The angular momentum of an electron moving in an orbit of the type described by Bohr is an axial vector $\mathbf{L} = \mathbf{r} \times \mathbf{p}$, formed from the radial distance $\mathbf{r}$ between electron and nucleus and the linear momentum $\mathbf{p}$ of the electron relative to a fixed nucleus. Figure 2 shows the customary method used to illustrate the axial vector $\mathbf{L}$ in terms of the orbital motion of any object, of which the electron of the Bohr atom is only one example. Although Bohr's planetary model needed only circular orbits to explain the spectral lines observed in the spectrum of a hydrogen atom, subsequent development of similar models for other atoms containing more than a single electron needed elliptically shaped orbits to explain the observed spectra. For such orbits $\mathbf{r}$ and $\mathbf{p}$ are usually not perpendicular to each other so that the magnitude $|\mathbf{L}| = |\mathbf{r} \times \mathbf{p}|$ is $rp \sin \theta$.

A significant change in the theoretical treatment of atomic structure occurred in 1924 when Louis deBroglie proposed that an electron and other atomic particles simultaneously possess both wave and particle characteristics and that an atomic particle, such as an electron, has a wavelength $\lambda = h/p = h/mv$. Shortly thereafter, C. J. Davisson and L. H. Germer showed experimentally the validity of this postulate. DeBroglie's assumption that wave characteristics are inherent in every atomic particle was quickly followed by the development of quantum mechanics. In its most simple form, quantum mechanics introduces the physical laws associated with the wave properties of electromagnetic radiation into the physical description of a system of atomic particles. By means of quantum mechanics a much more satisfactory explanation of atomic structure can be developed.

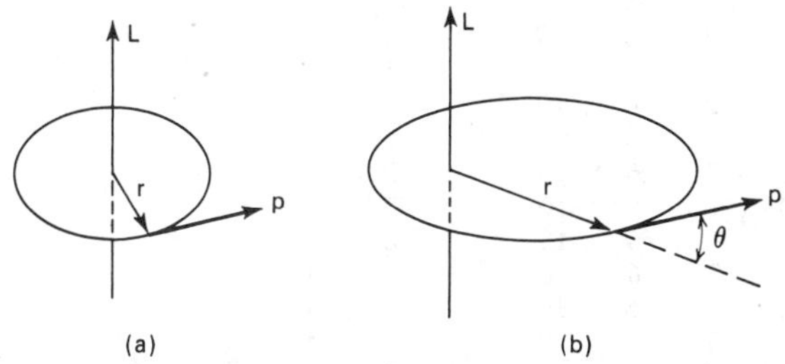

Fig. 2. Direction of angular momentum vector is perpendicular to plane formed by radial vector *r* and momentum vector *p*.

The quantity n introduced by Bohr in his description of the hydrogen atom is what is called a quantum number. These numbers enter quite naturally from quantum-mechanical descriptions of atomic energy states, including nuclear states. In quantum-mechanical descriptions of atoms, the number n characterizes a limited number of electron states that have very nearly the same energy. A group of electrons in an atom with a common value of n are usually said to be in a single shell of the atom. In the simple two-body problem of the hydrogen atom, all states having the same value of n have the same energy, but in multielectron atoms there are interactions between individual pairs of electrons as well as between electrons and the atomic nucleus. The result is a limited spread in energy between the most-tightly bound and the least-tightly bound electron with the same value of n. Except for $n = 1$, more than one orbital angular momentum state is possible for each shell. Each such state is described by a quantum number l, which can have only whole-number values between zero and $n - 1$. The number n is usually called the principal quantum number and l the orbital angular momentum, or sometimes azimuthal, quantum number. In addition to its characteristic angular momentum, each electron spins on an axis, such that it also has a spin angular momentum, described by a quantum number s. Because all electrons are identical, s has only one value, $\frac{1}{2}$.

In the development of the concepts of atomic structure much of the experimental evidence came from optical and x-ray spectroscopy. From this work certain notations have arisen that are now an accepted part of the language. For example, the $n = 1$ shell is sometimes known as the K-shell, the $n = 2$ shell as the L-shell, the $n = 3$ shell as the M-shell, etc., with consecutively following letters of the alphabet being used to designate those shells with successively higher principal quantum numbers. A Roman numeral subscript further subdivides the shells in accordance with the n, l and j quantum numbers of the electrons, as shown in Table 5.

TABLE 5. IDENTIFICATION OF ATOMIC SHELLS

X-Ray Notation	Corresponding Quantum Numbers n	l	j	X-Ray Notation	Corresponding Quantum Numbers n	l	j
K	1	0	$\frac{1}{2}$	N_I	4	0	$\frac{1}{2}$
L_I	2	0	$\frac{1}{2}$	N_{II}	4	1	$\frac{1}{2}$
L_{II}	2	1	$\frac{1}{2}$	N_{III}	4	1	$\frac{3}{2}$
L_{III}	2	1	$\frac{3}{2}$	N_{IV}	4	2	$\frac{3}{2}$
M_I	3	0	$\frac{1}{2}$	N_V	4	2	$\frac{5}{2}$
M_{II}	3	1	$\frac{1}{2}$	N_{VI}	4	3	$\frac{5}{2}$
M_{III}	3	1	$\frac{3}{2}$	N_{VII}	4	3	$\frac{7}{2}$
M_{IV}	3	2	$\frac{3}{2}$				
M_V	3	2	$\frac{5}{2}$				

A letter notation has been given to the different orbital angular momentum states by optical spectroscopists. In this notation an $l = 0$ state is called an s state (not to be confused with the s used as a quantum number for spin, and which can usually be distinguished from the way it is used) and $l = 1$ state a p state, an $l = 2$ a d state, an $l = 3$ state an f state, with consecutively higher letters above f in the alphabet designating each succeeding value of l. A number often accompanies the state designation to indicate the appropriate value of n. For example, a $3p$ level is an $n = 3$, $l = 1$ state.

Electrically neutral atoms with nuclear charge $Z > 1$ are not hydrogen-like, but have more than a single orbital electron. With more than one electron in an atom, it is necessary to determine the relationship of one electron to its neighbors. A clue to this relationship is found from the observation that the energy required to remove a second electron from an atom is greater than that required to remove the first electron and the energy so required increases with each succeeding electron that is removed. In other words, the potential energy that holds some electrons in an atom is much greater than the energy that holds other electrons in the same atom. From the results of the detailed observations, Pauli formulated a principle, the Pauli Exclusion Principle, which states that no two electrons within the same atom may have exactly the same wave function and, on this basis, that no two electrons in the same atom may have the same set of quantum numbers. Thus, if a particular configuration state in an atom is occupied by an electron, that state is forbidden to all other electrons. If each electron falls into the lowest possible energy state, the next electron to enter must remain in some higher energy state. Acceptance of this principle allows us to classify observed spectral characteristics of the radiation emitted by atoms in such a way that we can determine within reasonable limits the number of electrons in each shell or subshell of the atom producing the radiation. The electron configurations in the known atomic species are shown in Table 6, in which the atoms are ordered in accordance with their atomic numbers Z. Generally the innermost (lower quantum number) shells are filled with all the electrons allowed. For example, neon has two $1s$ electrons, two $2s$ electrons, and six $2p$ electrons, such that its $K(n=1)$ and $L(n=2)$ shells are both filled. An atom with Z having a magnitude one greater than neon, which is sodium, also has both the K and L shells filled; thus there is no space for another electron to go into either of these

TABLE 6. ELECTRON STRUCTURE OF ATOMS (NORMAL STATE)

			K	*L*		*M*			*N*				*O*				
Element	Atomic Number	Chemical Symbol	1*s*	2*s*	2*p*	3*s*	3*p*	3*d*	4*s*	4*p*	4*d*	4*f*	5*s*	5*p*	5*d*	5*f*	5*g*
Hydrogen	1	H	1														
Helium	2	He	2														
Lithium	3	Li	2	1													
Beryllium	4	Be	2	2													
Boron	5	B	2	2	1												
Carbon	6	C	2	2	2												
Nitrogen	7	N	2	2	3												
Oxygen	8	O	2	2	4												
Fluorine	9	F	2	2	5												
Neon	10	Ne	2	2	6												
Sodium	11	Na	2	2	6	1											
Magnesium	12	Mg	2	2	6	2											
Aluminum	13	Al	2	2	6	2	1										
Silicon	14	Si	2	2	6	2	2										
Phosphorus	15	P	2	2	6	2	3										
Sulfur	16	S	2	2	6	2	4										
Chlorine	17	Cl	2	2	6	2	5										
Argon	18	Ar	2	2	6	2	6										
Potassium	19	K	2	2	6	2	6		1								
Calcium	20	Ca	2	2	6	2	6		2								
Scandium	21	Sc	2	2	6	2	6	1	2								
Titanium	22	Ti	2	2	6	2	6	2	2								
Vanadium	23	V	2	2	6	2	6	3	2								
Chromium	24	Cr	2	2	6	2	6	5	1								
Manganese	25	Mn	2	2	6	2	6	5	2								
Iron	26	Fe	2	2	6	2	6	6	2								
Cobalt	27	Co	2	2	6	2	6	7	2								
Nickel	28	Ni	2	2	6	2	6	8	2								
Copper	29	Cu	2	2	6	2	6	10	1								
Zinc	30	Zn	2	2	6	2	6	10	2								
Gallium	31	Ga	2	2	6	2	6	10	2	1							
Germanium	32	Ge	2	2	6	2	6	10	2	2							
Arsenic	33	As	2	2	6	2	6	10	2	3							
Selenium	34	Se	2	2	6	2	6	10	2	4							
Bromine	35	Br	2	2	6	2	6	10	2	5							
Krypton	36	Kr	2	2	6	2	6	10	2	6							
Rubidium	37	Rb	2	2	6	2	6	10	2	6			1				
Strontium	38	Sr	2	2	6	2	6	10	2	6			2				
Yttrium	39	Y	2	2	6	2	6	10	2	6	1		2				
Zirconium	40	Zr	2	2	6	2	6	10	2	6	2		2				
Niobium	41	Nb	2	2	6	2	6	10	2	6	4		1				
Molybdenum	42	Mo	2	2	6	2	6	10	2	6	5		1				
Technetium	43	Tc	2	2	6	2	6	10	2	6	(5)		(2)				
Ruthenium	44	Ru	2	2	6	2	6	10	2	5	7		1				
Rhodium	45	Rh	2	2	6	2	6	10	2	6	8		1				
Palladium	46	Pd	2	2	6	2	6	10	2	6	10						
Silver	47	Ag	2	2	6	2	6	10	2	6	10		1				
Cadmium	48	Cd	2	2	6	2	6	10	2	6	10		2				
Indium	49	In	2	2	6	2	6	10	2	6	10		2	1			
Tin	50	Sn	2	2	6	2	6	10	2	6	10		2	2			
Antimony	51	Sb	2	2	6	2	6	10	2	6	10		2	3			
Tellurium	52	Te	2	2	6	2	6	10	2	6	10		2	4			
Iodine	53	I	2	2	6	2	6	10	2	6	10		2	5			
Xenon	54	Xe	2	2	6	2	6	10	2	6	10		2	6			

(Continued)

TABLE 6 (Continued)

			K	L	M	N				O					P						Q
ELEMENT	ATOMIC NUMBER	CHEMICAL SYMBOL				4s	4p	4d	4f	5s	5p	5d	5f	5g	6s	6p	6d	6f	6g	6h	7s
Cesium	55	Cs	2	8	18	2	6	10		2	6				1						
Barium	56	Ba	2	8	18	2	6	10		2	6				2						
Lanthanium	57	La	2	8	18	2	6	10		2	6	1			2						
Cerium	58	Ce	2	8	18	2	6	10	2	2	6				2						
Praseodymium	59	Pr	2	8	18	2	6	10	3	2	6				2						
Neodymium	60	Nd	2	8	18	2	6	10	4	2	6				2						
Promethium	61	Pm	2	8	18	2	6	10	5	2	6				2						
Samarium	62	Sm	2	8	18	2	6	10	6	2	6				2						
Europium	63	Eu	2	8	18	2	6	10	7	2	6				2						
Gadolinium	64	Gd	2	8	18	2	6	10	7	2	6	1			2						
Terbium	65	Tb	2	8	18	2	6	10	8 or 9	2	6	1 or 0			2						
Dysprosium	66	Dy	2	8	18	2	6	10	10	2	6				2						
Holmium	67	Ho	2	8	18	2	6	10	11	2	6				2						
Erbium	68	Er	2	8	18	2	6	10	12	2	6				2						
Thulium	69	Tm	2	8	18	2	6	10	13	2	6				2						
Ytterbium	70	Yb	2	8	18	2	6	10	14	2	6				2						
Lutetium	71	Lu	2	8	18	2	6	10	14	2	6	1			2						
Hafnium	72	Hf	2	8	18	2	6	10	14	2	6	2			2						
Tantalum	73	Ta	2	8	18	2	6	10	14	2	6	3			2						
Tungsten	74	W	2	8	18	2	6	10	14	2	6	4			2						
Rhenium	75	Re	2	8	18	2	6	10	14	2	6	5			2						
Osmium	76	Os	2	8	18	2	6	10	14	2	6	6			2						
Iridium	77	Ir	2	8	18	2	6	10	14	2	6	7			2						
Platinum	78	Pt	2	8	18	2	6	10	14	2	6	9			1						
Gold	79	Au	2	8	18	2	6	10	14	2	6	10			1						
Mercury	80	Hg	2	8	18	2	6	10	14	2	6	10			2						
Thallium	81	Tl	2	8	18	2	6	10	14	2	6	10			2	1					
Lead	82	Pb	2	8	18	2	6	10	14	2	6	10			2	2					
Bismuth	83	Bi	2	8	18	2	6	10	14	2	6	10			2	3					
Polonium	84	Po	2	8	18	2	6	10	14	2	6	10			2	4					
Astatine	85	At	2	8	18	2	6	10	14	2	6	10			2	5					
Radon	86	Rn	2	8	18	2	6	10	14	2	6	10			2	6					
Francium	87	Fr	2	8	18	2	6	10	14	2	6	10			2	6					(1)
Radium	88	Ra	2	8	18	2	6	10	14	2	6	10			2	6					(2)
Actinium	89	Ac	2	8	18	2	6	10	14	2	6	10			2	6	(1)				(2)
Thorium	90	Th	2	8	18	2	6	10	14	2	6	10			2	6	(2)				(2)
Protactinium	91	Pa	2	8	18	2	6	10	14	2	6	10	(2)		2	6	(1)				(2)
Uranium	92	U	2	8	18	2	6	10	14	2	6	10	(3)		2	6	(1)				(2)
Neptunium	93	Np	2	8	18	2	6	10	14	2	6	10	(5)		2	6					(2)
Plutonium	94	Pu	2	8	18	2	6	10	14	2	6	10	(6)		2	6					(2)
Americium	95	Am	2	8	18	2	6	10	14	2	6	10	(7)		2	6					(2)
Curium	96	Cm	2	8	18	2	6	10	14	2	6	10	(7)		2	6	(1)				(2)
Berkelium	97	Bk	2	8	18	2	6	10	14	2	6	10	(9)		2	6					(2)
Californium	98	Cf	2	8	18	2	6	10	14	2	6	10	(10)		2	6					(2)
Einsteinium	99	Es	2	8	18	2	6	10	14	2	6	10	(11)		2	6					(2)
Fermium	100	Fm	2	8	18	2	6	10	14	2	6	10	(12)		2	6					(2)
Mendelevium	101	Mv	2	8	18	2	6	10	14	2	6	10	(13)		2	6					(2)
Nobelium	102	No	2	8	18	2	6	10	14	2	6	10	(14)		2	6					(2)
Lawrencium	103	Lw	2	8	18	2	6	10	14	2	6	10	(14)		2	6	(1)				(2)

shells, and its additional electron must go into the $M(n = 3)$ shell. Shells with higher values of the orbital angular momentum quantum number l are sometimes partially shielded electrically from the nucleus by other electrons in such a way that their binding energies are not as great as the lower orbital angular momentum subshells associated with the next higher principal quantum number. Thus, as can be seen in Table 6, the 5s subshell is filled in atomic species of lower Z than any of those that have any 4f electrons.

The periodic table of elements can be described in terms of the similarities and differences of the angular momentum characteristics of the various atomic species. For example, similar but not identical chemical properties are observed for a number of elements that have all but one of their electrons in filled shells, with the extra electron being a single s electron. In the periodic table, lithium, sodium, potassium, rubidium, and cesium all have this characteristic and thus are placed in a single column. Other groups of atomic species that have similar electron configurations in their outermost shell are arranged in the periodic table in the same chemical grouping. See also **Periodic Table of the Elements.**

Motions of charged particles are expected to establish magnetic fields that interact with each other. As a result, coupling between spin and orbital angular momenta of a single electron or coupling between angular momenta of different electrons in a single atom is expected. The spin-orbit coupling of a single electron results in a total angular momentum state, described by a total quantum number j, such that either $j = l + \frac{1}{2}$ or $j = l - \frac{1}{2}$. Quantum-mechanical solutions show that the magnitude of the orbital angular momentum is $l^*h = [l(l+1)]^{1/2}h$, not lh, as would be expected if we had followed the simpler assumption of the Bohr model. Similarly, the magnitude of the spin and total angular momenta are $s^*h = [s(s+1)]^{1/2}h$ and $j^*h = [j(j+1)]^{1/2}h$, respectively. Coupled spin and orbital angular momenta thus do not align themselves in a linear pattern but in such

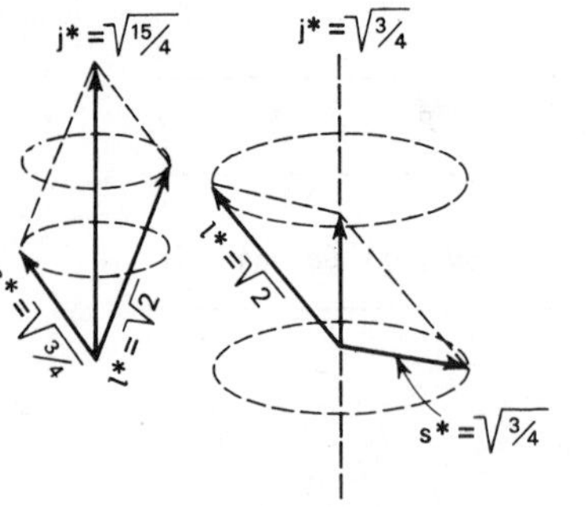

Fig. 3. Magnitudes and directions of the angular momentum vectors for an $l = 1$, $s = \frac{1}{2}$ electron in an atomic energy state.

a way that the spin and orbital angular momenta precess around the direction of the total angular momentum, as shown in Fig. 3. A similar precession around the direction of the resultant angular momentum state is found, as we shall find later, in the coupling of any two or more angular momentum states into a single common system.

Without some external influence, angular momentum states have no preferred orientation in space because the only interacting magnetic fields are entirely within the individual atom. An external magnetic field, however, may couple to the magnetic field of an atomic angular momentum state of the type described in the preceding paragraphs. In coupling to the total angular momentum, for example, only a limited number of orientations of j^*h are possible, these being such that their projections in the direction of the magnetic field have magnitudes $m_j h$ in which m_j, a magnetic quantum number, can have only those whole-number values for which $m_j \leq |j|$. Thus $2j + 1$ orientations, as illustrated in Fig. 4 for $j = \frac{3}{2}$, are possible. Note that the vector representing the total angular momentum is not parallel to the direction of the applied magnetic field. As a result, the total angular momentum vector precesses around the direction of the magnetic field such that the component of the total angular momentum that is perpendicular to the direction of the magnetic field has a time-averaged value of zero and the only component that can be observed with an external detecting device is that component parallel to the direction of the external magnetic field. A magnetic field of this type splits a set of levels characterized by a quantum number j, all of which initially have the same energy, into $2j + 1$ components. The nature of this level splitting is deduced from observations of the splitting of characteristic line spectra, in which case an originally monoenergetic radiation is split by the magnetic field into several components that are usually relatively closely spaced in wave length. The observed effect is known as the Zeeman effect. See also **Zeeman Effect.**

The use of axial vectors to describe states of an atom in terms of a coupling between the angular momentum inherent in the electrons of the atom forms what is commonly called the *vector model* of the atom. The type of coupling just described forms the basis for the *j-j* coupling model. Such coupling is found between electrons that produce the optical radiation in the higher Z atomic systems. In these systems the total angular momentum j^*h formed by the coupling of the spin and orbital angular momentum states of individual electrons are further coupled within any single shell of an atom to form a total angular momentum J^*h, characterized by a quantum number J, which then is the angular momentum for the system of several electrons. According to customary usage, a lower case letter as a designator for a quantum number indicates that the quantum number is that of a single electron, while a capital letter indicates that the quantum number represents a state formed by several electrons.

For the lower Z part of the periodic table of elements, the appropriate coupling system for angular momentum states in an atom is the *L-S*, or Russell-Saunders, coupling. In this description the orbital angular momenta of individual electrons in any single shell of an atom are coupled to form a resultant orbital angular momentum described by the quantum number L, and the spin angular momenta of the same individual electrons are coupled to form a resultant spin angular momentum, described by the quantum number S. Coupling of the individual orbital angular momentum states is not shown but possible coupling schemes for the spins of 2, 3, and 4 electrons is shown in Fig. 5. Note that as many possible couplings exist as there

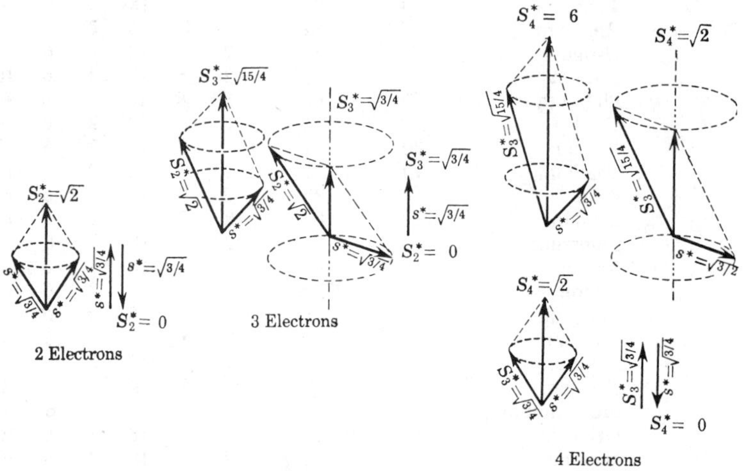

Fig. 5. Vector model coupling of the spin angular momenta of two, three, and four electrons.

are electrons. Only one of these possible schemes will be the lowest energy state, the others being at higher energies. The resultant spin and orbital angular momentum vectors for a group of electrons in a single shell of an atom can then be used to describe the coupling that gives the total angular momentum J^*h for these electrons. In Fig. 6 is shown the coupling according to the vector model between the orbital and spin angular momenta for which $L = 2$ and $S = 1$. There are $2J + 1$ possible states. All $2J + 1$ states have the same energy unless under the influence of an external magnetic field, in which case they are broken into components, each of which is designated by a magnetic quantum number M_J. In principle this splitting is the same as for the splitting of the total angular momentum states for a single electron, as shown in Fig. 4, except for the use of capital letters M_J and J.

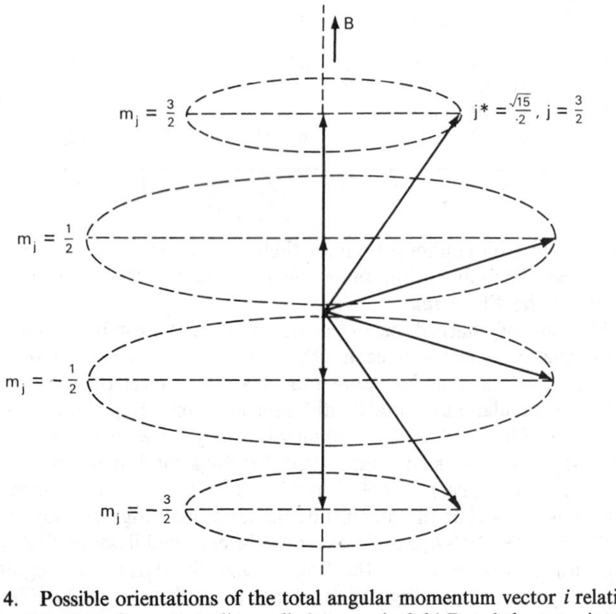

Fig. 4. Possible orientations of the total angular momentum vector i relative to the direction of an externally applied magnetic field B and the magnitudes of the associated magnetic quantum state vectors m_j.

The Pauli Exclusion Principle states that no two electrons of any single atom may simultaneously occupy a state described by only a single set of quantum numbers. Five such numbers are needed to describe fully the quantum-mechanical conditions of an electron. For *j-j* coupling this set is generally n, l, s, j, m_j, and for *L-S* coupling it is n, l, s, m_l, m_s. From the coupling of the angular momentum associated with the latter sets a full description of the multielectron state, described by n, L, S, J, M_J, is determined.

Part of the outgrowth of the determination that atomic particles have wave properties and of the subsequent development of quantum mechanics is the Heisenberg Uncertainty Principle, which states that

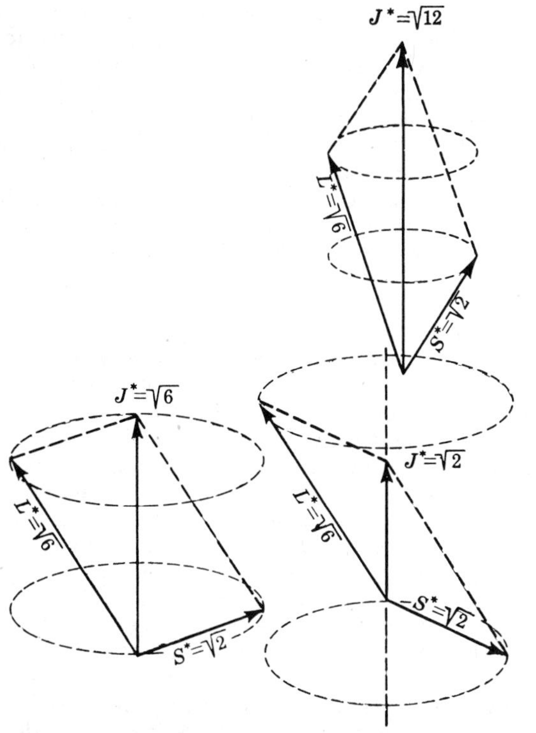

Fig. 6. Vector model coupling of spin and orbital angular momenta for which $L = 2$, $S = 1$.

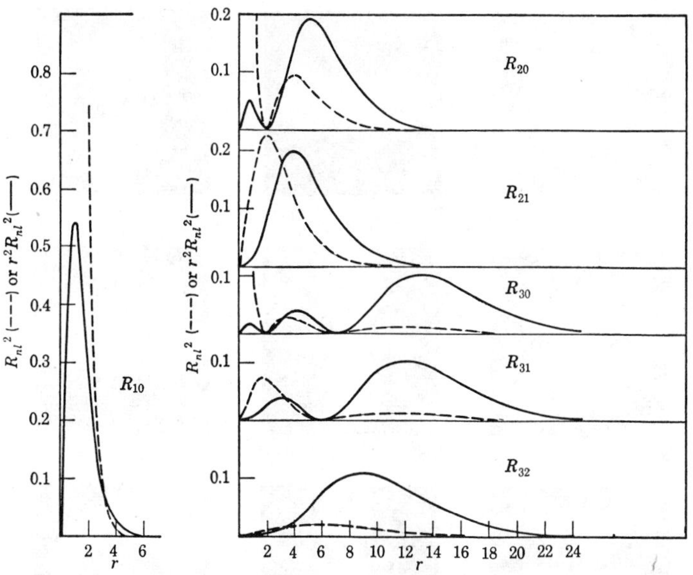

Fig. 7. Probability density distributions as a function of radial distance from the nucleus for several states of a hydrogen atom. The dashed lines are proportional to the probability for finding the electron in an incremental volume dv at the indicated radial distance. The solid lines are proportional to the probability for finding the electron in an incremental shell of volume $4\pi r^2$ dr at the indicated radius.

an electron cannot be located exactly in terms of both its space and momentum coordinates, or in terms of both its time and energy coordinates. See also **Uncertainty Principle.** If the energy of a particular atomic state is precisely defined, the time at which an electron is found in that energy state cannot be so defined. Likewise, if the momentum of an electron is precisely defined, its position in space cannot be so defined. On the other hand, the probability for finding an electron in a particular location relative to the atomic nucleus can be determined. After substitution of the appropriate potential energy terms needed to describe the atomic system, these probability density distributions may be found from solutions of the Schroedinger wave equation, which is a quantum-mechanical description of the system. This distribution, which is directly related to the wave function that describes the atomic state in a quantum-mechanical manner, must be distributed through a region of space and is hence a function of all three spatial variables, which are r, θ and ϕ in the spherical coordinate system. Because only two variables are available on the plane of a page to describe these functions, they are usually described by a series of graphs, which must then be assembled in the imagination of the reader to picture the complete three-dimensional distribution. The radial part of the distribution is dependent only on the quantum numbers n and l and is usually represented by the terminology $|R_{nl}|^2$. The radial distribution for selected states of a hydrogen atom are shown in Fig. 7. The dashed lines are proportional to the probability of finding the electron in the appropriate nl state in an incremental volume dv of constant magnitude at the indicated radial distance from the atomic nucleus. The solid lines are proportional to the probability of finding the electron in an incremental shell between the radial distances r and $r + dr$, with a volume $4\pi r^2\ dr$ at the indicated radius. Radial distances in Fig. 7 are given in units of Bohr radii, the distance from the nucleus of the $n = 1$ orbit in the Bohr planetary model of the hydrogen atom.

In the quantum-mechanical description of a hydrogen atom, the radial portion of the probability density distribution is the same in all directions from the nucleus, but only for the case $l = 0$ is the magnitude of the distribution the same in all radial directions. For all other values of l, the magnitude of the distribution is a function of the angular direction, defined by the coordinates θ and ϕ. However, as in the case of the discussion of the vector model of an atom, we cannot define an angular direction unless an axis exists to provide a reference direction. To provide this axis some force or torque external to the electron configuration must be found. The only external force strong enough to interact with the electron configuration is that provided by certain magnetic fields. The interaction with the magnetic moment of the electron configuration can then be described in terms of a magnetic quantum number m_l, which is the magnetic quantum number associated with orbital angular momentum quantum number. When $l = 1$, m_l may have any of three values, $+1$, 0, or -1. Three possible angular distributions are then possible for the electrons described by a quantum number $l = 1$, but the distributions for $m_l = +1$ and $m_l = -1$ are identical, thereby providing some simplication. If we define $\theta = 0$ in the direction of the applied magnetic field, the probability density distribution in the $r\theta$-plane for an electron of a hydrogen atom described by $l = 1$, $m_l = 0$ is given in the $2p$, $m = 0$ part of Fig. 8. This distribution is symmetric in ϕ; it may thus be rotated around an axis perpendicularly directed through the center of the figure (the $\theta = 0$ axis) to give the full three-dimensional distribution. For either $m_l = +1$ or $m_l = -1$ the distribution is given by that part of Fig. 8 labeled $2p$, $m = 1$. When rotated about the $\theta = 0$ axis the $m_l = 0$ distribution gives a dumbbell like distribution and the $|m_l| = 1$ distributions give ring-like distributions, all of which fade away to negligible magnitude at large distances from the center of the distribution. For the $2p$, $m_l = 0$ configuration, the angular distribution in the $r\theta$-plane is given by the equation $\frac{3}{2}\cos^2\theta$, and for the $2p$, $m_l = \pm 1$ configuration by the equation $\frac{3}{4}\sin^2\theta$. Since two $|m_l| = 1$ configurations exist for each $l = 1$, the distributions for all three m_l configurations, when summed, give a resultant distribution that is independent on the angle θ, as well as of the angle ϕ. This condition is reached when all possible electron states in the $2p$ subshell are filled. A similar situation, a complete symmetry of the electron distribution, exists for all filled subshells. Hence an external magnetic field interacts only with partially filled shells. In multielectron atoms only one shell is usually partially filled; thus interactions with external magnetic fields, such as those introduced artificially, by the atomic nucleus, or by neighboring atoms, occur only in one shell of the atom, sometimes called in chemical terminology the valence shell. For specific elements, the unfilled shell can be determined from Table 3. Probability density distributions for several other electron configurations besides $2p$ is given in Fig. 8.

The radial probability density distributions for individual electron configurations in atoms other than hydrogen are generally similar to those of Fig. 7 but, since the nuclear charge Z of these atoms is larger than for hydrogen, the electrostatic attractive force exerted by the nucleus on the innermost electron is stronger than for hydrogen

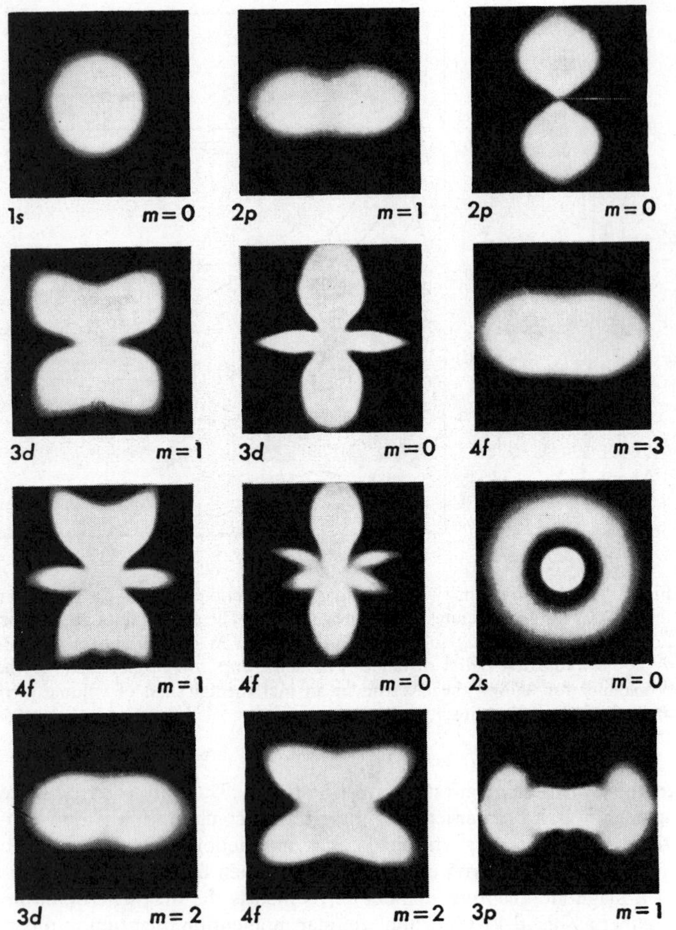

Fig. 8. Electron wave density figures representing the single electron states of the hydrogen atom.

and hence pulls its distribution closer to the nucleus. Outer electrons, however, are partially shielded electrostatically from the nucleus by the inner electrons and hence are influenced by a weaker force than that which would be provided by a bare nucleus. The least tightly bound electron is held by a force that has an effective Z of about 1 but for the more tightly bound electrons the effective Z is much higher. The probability density distribution for the least tightly bound electron then extends to a distance comparable to the distribution for hydrogen, but the main part of the distribution is much closer to the nucleus. A typical distribution for all the electrons of a multielectron atom, in this case rubidium, is shown in Fig. 9.

As stated earlier in this entry, there are three p orbitals, with magnetic quantum numbers of 0, 1 and −1. The orbital for which m = 0 corresponds to the direction of the applied magnetic field, so that its effect is zero. This direction is conventionally taken as that of

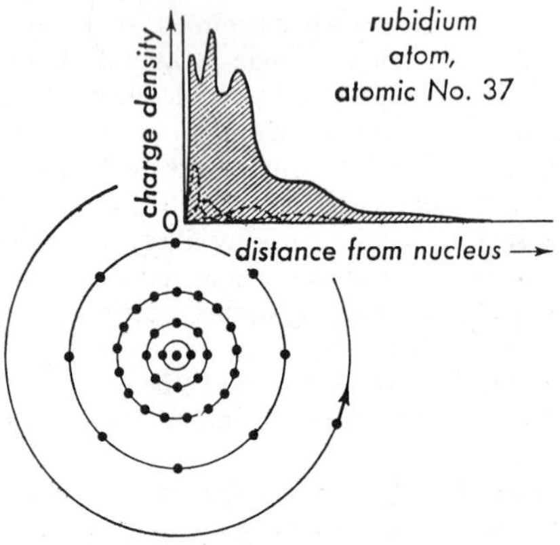

Fig. 9. Radial probability density distribution, derived from quantum-mechanical predictions, for rubidium, along with Bohr planetary model for the same atom.

the z-axis, so that p_z, which is symmetrical about that axis, coincides with p_0. The other p orbitals, p_x and p_y, are perpendicular to p_z and to each other and correspond to standing waves built up by mixing of p_{+1} and p_{-1}, which are oppositely directed waves.

An analogous situation for the d orbitals, for which l = 2, and hence m_l = 2, 1, 0, −1 and −2, gives rise to five d orbitals, designated as d_{xz}, $d_{xy}d_{yz}d_{x^2-y^2}$ and d_{z^2}.

It follows from the earlier discussion that each orbital may be occupied, in accordance with the Pauli Exclusion Principle, by two electrons of opposed spin. See also **Pauli Exclusion Principle.**

Atomic Radius

The radius of an atom may be defined as the distance of closest approach to another atom and is the distance at which the mutual repulsion of the electron clouds and the mutual attraction of the nuclear charge of each for the electrons of the other are in equilibrium under specified circumstances. If the two atoms are the same, the radius of each is one half the internuclear distance; if they are unlike, the internuclear distance is the sum of the individual radii. Atomic radii fall roughly into four categories, which, however, merge into one another. These are Van der Waals and ionic radii, which are radii of equilibrium approach for mutally nonbonded atoms (neutral and charged, respectively), and covalent and metallic radii, which are for mutually bonded atoms (where the bonding electrons are, respectively, largely localized between the bonded atoms, or highly delocalized). Radii of all types vary in essentially the same manner from one part of the periodic table to another. In general, the radii decrease from the beginning to the end of any period, the rate of decrease being less the higher the number of the period. There are discontinuities in the decrease at points at which the quantum levels become half or completely filled. For example, Mn^{2+}(r = 0.83 Å) is larger than either Cr^{2+}(0.80 Å) or Fe^{2+}(0.80 Å), and Zn^{2+}(0.75 Å) is larger than Cu^{2+}(0.72 Å). (In this description, angstroms, Å, are used for convenience of notation. 1 angstrom = 10^{-10} meter) Again, in general, the radii increase going down any column in the periodic table except in the region of hafnium. This last exception is caused by the lanthanide contraction occurring for the series of 14 elements between lanthanum and hafnium, which results in a far greater decrease between these two elements (average metallic radii: La 1.87 Å, Hf 1.58 Å) than between yttrium and zirconium (Y 1.80 Å, Zr 1.60 Å), where so such series intervenes. Because of the slower rate of decrease of the radius in the sixth period as compared to the fifth, however, the radii in the sixth period soon become larger again than those in the fifth (cf. Nb 1.429 Å, Ta 1.43 Å; Mo 1.36 Å, W 1.37 Å).

Van der Waals Radius**:** This is the radius of closest approach of one atom to another with which it does not form a chemical bond. It represents the distance at which the mutual repulsion of the nonbonding electrons of each exactly balances the attraction of the nucleus of each for the electrons of the other. See also **Van der Waals Forces.** Experimentally determined and calculated values (indicated by asterisk) of Van der Waals radii in angstrom units of a number of elements are: H 1.2, He 1.78_5, B 2.17*, C 1.85, N 1.54, 0. 140, F 1.35, Ne 1.60, Al 2.53*, Si 2.24*, P 1.9(1.96*), S 1.85, Cl 1.80, Ar 1.92, Ge 2.02, As 2.0, Se 2.00, Br 1.95, Kr 2.01, Sb 2.2, Te 2.20, I 2.15, Xe 2.20, Hg 1.50, Rn ~ 2.3*. The often cited value of 1.70 Å for the Van der Waals radius of carbon is derived from the interlayer distance in crystalline graphite (3.40 Å) and is too short because the weak interaction of the pi-electron systems draws the layers together.

Van der Waals radii of some groups are: CH_3 2.0, C_6H_5 (interplanar) 1.85, BH_4^- 2.08, SiH_3^- 2.25.

Ionic Radius**:** This is the radius of closest approach of a charged atom or group (i.e., an ion) to another atom or group with which it does not form a covalent bond. It represents the distance at which the mutal repulsion of the nonbonding electrons of each exactly balances the mutual attraction of oppositely charged ions, or the attraction of a positive ion for the electrons of a neutral atom or group or the attraction of a negative ion for the nucleus or nuclei of a neutral atom or group. Ionic radii for the cations of the least electronegative metals and of the most electronegative nonmetals are clearly defined, but for the elements of intermediate electronegativity (including the transition elements) they are much more doubtful because of the vary-

TABLE 7. IONIC CRYSTAL RADII (IN ANGSTROM UNITS)*

Element	Ion	Radius
Actinium	Ac^{3+}	1.11
Aluminum	Al^{3+}	0.57
Americium	Am^{3+}	1.00
	Am^{4+}	0.85
Antimony	Sb^{3-}	2.170
	Sb^{3+}	0.90
	(Sb^{5+}	0.62)
Arsenic	As^{3-}	1.991
	(As^{3+}	0.69)
	(As^{5+}	0.47)
Astatine	At^{-}	2.2
Barium	Ba^{2+}	1.38
Berkelium	Bk^{3+}	0.99
Beryllium	Be^{2+}	0.31
Bismuth	Bi^{3-}	2.217
	Bi^{3+}	1.20
	(Bi^{5+}	0.74)
Boron	(B^{3+}	0.20)
Bromine	Br^{-}	1.97
	(Br^{7+}	0.39)
Cadmium	Cd^{2+}	0.99
Calcium	Ca^{2+}	1.06
Californium	Cf^{3+}	0.98
Carbon	(C^{4-}	2.60)
	(C^{4+}	0.15)
Cerium	Ce^{3+}	1.034
	Ce^{4+}	0.941
Cesium	Cs^{+}	1.70
Chlorine	Cl^{-}	1.81
	(Cl^{7+}	0.26)
Chromium	Cr^{2+}	0.80
	Cr^{3+}	0.70
	(Cr^{6+}	0.52)
Cobalt	Co^{2+}	0.78
	Co^{3+}	0.65
Copper	Cu^{+}	0.96

Element	Ion	Radius
	Cu^{2+}	0.72
Curium	Cm^{3+}	1.00
Dysprosium	Dy^{3+}	0.908
Einsteinium	Es^{3+}	0.97
Erbium	Er^{3+}	0.881
Europium	Eu^{2+}	1.137
	Eu^{3+}	0.950
Fermium	Fm^{3+}	0.97
Fluorine	F^{-}	1.36
	(F^{7+}	0.07)
Francium	Fr^{+}	1.9
Gadolinium	Gd^{3+}	0.938
Gallium	Ga^{3+}	0.65
Germanium	Ge^{2+}	0.65
	(Ge^{4+}	0.55)
Gold	Au^{+}	1.37
Hafnium	Hf^{4+}	0.86
Holmium	Ho^{3+}	0.894
Hydrogen	H^{-}	2.08
Indium	In^{3+}	0.95
Iodine	I^{-}	2.16
	(I^{7+}	0.50)
Iridium	Ir^{4+}	0.66
Iron	Fe^{2+}	0.80
	Fe^{3+}	0.67
Lanthanum	La^{3+}	1.071
Lead	Pb^{2+}	1.18
	(Pb^{4+}	0.70)
Lithium	Li^{+}	0.70
Lutetium	Lu^{3+}	0.83
Magnesium	Mg^{2+}	0.75
Manganese	Mn^{2+}	0.83
	Mn^{3+}	0.52
	(Mn^{7+}	0.46)
Mendelevium	Mv^{3+}	0.96
Mercury	Hg^{2+}	1.12

Element	Ion	Radius
Molybdenum	Mo^{4+}	0.68
	(Mo^{6+}	0.65)
Neodymium	Nd^{3+}	0.995
Neptunium	Np^{3+}	1.02
	Np^{4+}	0.88
Nickel	Ni^{2+}	0.74
Niobium	Nb^{4+}	0.67
	(Nb^{5+}	0.70?)
Nitrogen	N^{3-}	1.56
	(N^{5+}	0.11)
Nobelium		
Osmium	Os^{4+}	0.65
Oxygen	O^{2-}	1.40
	(O^{6+}	0.09)
Palladium	Pd^{2+}	0.50
Phosphorus	P^{3-}	1.920
	(P^{5+}	0.34)
Platinum	Pt^{2+}	0.52
	(Pt^{4+}	0.55)
Plutonium	Pu^{3+}	1.01
	Pu^{4+}	0.86
Polonium	(Po^{4+}	0.9)
Potassium	K^{+}	1.33
Praseodymium	Pr^{3+}	1.013
	Pr^{4+}	0.87
Promethium	Pm^{3+}	0.98
Protoactinium	Pa^{3+}	1.06
Radium	Ra^{2+}	1.42
Rhenium	(Re^{6+}	0.52)
Rhodium	Rh^{3+}	0.75
	Rh^{4+}	0.65
Rubidium	Rb^{+}	1.52
Ruthenium	Ru^{4+}	0.60
Samarium	Sm^{2+}	1.143
	Sm^{3+}	0.964
Scandium	Sc^{3+}	0.83

Element	Ion	Radius
Selenium	Se^{2-}	1.960
	(Se^{6+}	0.42)
Silicon	(Si^{4+}	0.40)
Silver	Ag^{+}	0.97
Sodium	Na^{+}	1.00
Strontium	Sr^{2+}	1.18
Sulfur	S^{2-}	1.855
	(S^{6+}	0.29)
Tantalum	(Ta^{5+}	0.73)
Technetium	Tc^{4+}	0.5
Tellurium	Te^{2-}	2.21
	(Te^{4+}	0.84)
	(Te^{6+}	0.56)
Terbium	Tb^{3+}	0.923
Thallium	Tl^{+}	1.50
	Tl^{3+}	0.95
Thorium	Th^{4+}	0.95
Thulium	Tm^{3+}	0.869
Tin	Sn^{2+}	1.02
	(Sn^{4+}	0.65)
Titanium	Ti^{2+}	0.76
	(Ti^{4+}	0.60)
Tungsten	W^{4+}	0.68
	(W^{6+}	0.65)
Uranium	U^{3+}	1.04
	U^{4+}	0.89
Vanadium	V^{2+}	0.82
	V^{3+}	0.75
	(V^{5+}	0.59)
Ytterbium	Yb^{2+}	1.02
	Yb^{3+}	0.858
Yttrium	Y^{3+}	0.910
Zinc	Zn^{2+}	0.75
Zirconium	Zr^{4+}	0.80

* 1 angstrom = 10^{-10} meter.

ing degrees of covalency in the compounds of these elements. Published ionic radii for these latter elements can therefore not be considered to be accurate measures of the true sizes of the ions, but are useful for comparison of relative sizes. In particular, there are certainly no simple cations of charge greater than 4+, and the simple tetrapositive cations are limited to Th^{4+}, probably the other tetrapositive actinide cations, and possibly Ce^{4+}. "Ionic radii" cited for such "cations" as Si^{4+}, P^{5+}, S^{6+}, Cl^{7+}, etc., are fictions arrived at by subtracting from an observed interatomic distance an arbitrary anionic radius for the second atom or else are extrapolated or theoretically calculated radii. However, inasmuch as all compounds of such elements have a high degree of covalency, such radii have no real meaning in normal compounds. Simple anions of absolute charge greater than 3 undoubtedly do not exist and the existence of monatomic trinegative ions is open to question. However, in their binary compounds with the lanthanide elements, phosphorus, arsenic, antimony and possibly nitrogen and bismuth apparently have radii very close to their Van der Waals radii and may therefore be considered essentially ionic. (See Table 7.)

The monatomic ionic radii of a given element become smaller as the oxidation state increases, provided this implies actual removal of electrons. For example, the radius of Fe^{2+} is 0.80 Å, whereas that of Fe^{3+} is 0.67 Å; Cu^{+} = 0.96, Cu^{2+} = 0.72.

The ionic radii discussed above are properly "crystal radii," i.e., the radii exhibited by the ions in ionic crystals. Although these radii are probably reasonable representations of the radii of contact of the ions in solution with the nearest atoms of solvate molecules, especially for ions of low charge density, nevertheless, most ions in solution have far larger effective radii because they carry with them a sheath of solvent molecules, the tenacity and thickness of which is a function of the charge density and electronic structure of the ion. In consequence, ions of small crystal radius (e.g., Li^{+}) may act larger (e.g., have lower mobility) in solution than ions of large crystal radius (e.g., Cs^{+}).

The effective radius of an ion changes with coordination number. An ion of coordination number 4 (tetrahedral) has a radius 0.93–0.95 times as large as the same ion with coordination number 6, while an ion of coordination number 8 is about 1.03 times as large as the same ion with coordination number 6.

Effective spherical crystal radii for some polyatomic ions are given in Table 8.

TABLE 8. CRYSTAL RADII OF POLYATOMIC IONS (Å)

NH_4^+	1.48	OH^-	1.53
OH_3^+	1.35	SH^-	2.00
PH_4^+	1.61	SeH^-	2.15
BH_4^-	2.08	SiH_3^-	2.25
CN^-	1.92	TeH^-	(2.35)
NO_3^-	2.3		

* 1 angstrom = 10^{-10} meter.

***Metallic Radius*:** Metals may be considered to be composed of cations bonded together by a cement of mobile electrons which are located in the conduction bands. Since the number of available energy levels in the conduction bands is a function of the number of available orbitals of the atoms making up the metal, and since the electron population of the nonbonding orbitals of the atoms and of the conduction band is a function of the number of valence electrons of the metal, the number and distance of nearest neighbors and the strength of the metal-metal bond varies in a fairly regular way across the periodic table. In particular, bond lengths are shortest and bond strengths are greatest in the vicinity of cobalt, rhodium and iridium, where the number of valence electrons and the number of available orbitals (nine) exactly match. Below this number there are too few electrons and above, too many, for maximum sharing and use of the bonding orbitals. The interatomic distances and coordination numbers of the elements in their metallic states are given in Table 9.

***Covalent Radius*:** This is the radius of closest approach for atoms bonded together by electrons which are localized in the region between the atoms. It represents the distance at which the attraction of each nucleus for the bonding electrons is in equilibrium with the mutual repulsion of the two nuclei and the repulsion of the inner electrons of each atom for the inner electrons of the other.

TABLE 9. INTERATOMIC DISTANCES IN METALS (IN ANGSTROM UNITS)*

Element	Interatomic Distances	Coordination Number
Actinium (room temp.)	3.756	12
Aluminum (25°C)	2.8635	12 (f.c.c.)
Americium		
Antimony (25°C)	2.90, 3.36	3,3 (rhombohedral)
Arsenic	2.49_5, 3.33	3,3 (rhombohedral)
Barium (room temp.)	4.347	8 (b.c.c.)
Berkelium		
Beryllium (α-form, 20°C)	2.2260, 2.2856	6,6 (c.p. hex.)
Bismuth (25°C)	3.09_5, 3.47	3,3 (rhombohedral)
Boron	1.75–1.80	
Cadmium (21°C)	2.9788, 3.2933	6,6 (c.p. hex.)
Calcium (α-form, 18°C)	3.947	12 (f.c.c.)
(γ-form, 500°C)	3.877	8 (b.c.c.)
Californium		
Cerium (room temp.)	3.650	12 (f.c.c.)
	3.620, 3.652	6,6 (c.p. hex?)
(15,000 atm.)	3.42	(f.c.c.)
Cesium (−100°C)	5.264	8 (b.c.c.)
(−10°C)	5.309	8
Chromium (α-form, 20°C)	2.4980	8 (b.c.c.)
(β-form, > 1850°C)	2.61	12 (f.c.c.)
Cobalt (18°C)	2.5061	12 (f.c.c.)
(room temp.)	2.505–2.498, 2.505–2.507	6,6 (c.p. hex.)
Copper (20°C)	2.5560	12 (f.c.c.)
Curium		
Dysprosium (room temp.)	3.503, 3.590	6,6 (c.p. hex.)
Einsteinium		
Erbium (room temp.)	3.468, 3.559	6,6 (c.p. hex.)
Europium (room temp.)	3.989	8 (b.c.c.)
Fermium		
Francium		
Gadolinium (20°C)	3.573, 3.636	6,6 (c.p. hex.)
Gallium (20°C)	2.44_2, 2.71_2, 2.74_2, 2.80	1,2,2,2 (orthorhombic)
Germanium (20°C)	2.4498	4 (diamond)
Gold (25°C)	2.8841	12 (f.c.c.)
Hafnium (α-form, 24°C)	3.1273, 3.1947	6,6 (c.p. hex.)
Holmium (room temp.)	3.486, 3.577	6,6 (c.p. hex.)
Indium (20°C)	3.2511, 3.3730	4,8 (f.c.t.)
Iridium (room temp.)	2.714	12 (f.c.c.)
Iron (α-form, 20°C)	2.4823	8 (b.c.c.)
(γ-form, 916°C)	2.578	12 (f.c.c.)
(δ-form, 1394°C)	2.539	8 (b.c.c.)
Lanthanum (α-form, room temp.)	3.739, 3.770	6,6 (c.p. hex.)
(β-form, room temp.)	3.745	12 (f.c.c.)
Lead (25°C)	3.5003	12 (f.c.c.)
Lithium (20°C)	3.0390	8 (b.c.c.)
(78°K)	3.111, 3.116	6,6 (c.p. hex.)
Lutetium (room temp.)	3.435, 3.503	6,6 (c.p. hex.)
Magnesium (25°C)	3.1971, 3.2094	6,6 (c.p. hex.)
Manganese (γ-form, 1095°C)	2.7311	12 (f.c.c.)
(δ-form, 1134°C)	2.6679	8 (b.c.c.)
Mendelevium		
Mercury (−46°C)	3.005	6 (rhombohedral)
Molybdenum (20°C)	2.7251	8 (b.c.c.)
Neodymium (room temp.)	3.628, 3.658	6,6(?) (modified c.p. hex.)
Neptunium (α-form, 20°C)	2.60–2.64	4 (orthorhombic)
(β-form, 313°C)	2.76	4 (tetragonal)
(δ-form, 600°C)	3.05	8 (b.c.c.)
Nickel (18°C)	2.4916	12 (f.c.c.)
Niobium (20°C)	2.8584	8 (b.c.c.)
Nobelium		
Osmium (20°C)	2.6754, 2.7354	6,6 (c.p. hex.)
Palladium (25°C)	2.7511	12 (f.c.c.)
Phosphorus (black)	2.18, —	3,3 (orthorhombic)
Platinum (20°C)	2.7746	12 (f.c.c.)
Plutonium (γ-form, 235°C)	3.026, 3.159, 3.287	4,2,4 (f.c.c.)

Continued

TABLE 9—*Continued*

Element	Interatomic Distances	Coordination Number
Polonium (α-form, 10°C)	3.345	6 (cubic)
(β-form, 75°C)	3.359	6 (rhombohedral)
Potassium (78°K)	4.544	8 (b.c.c.)
Praseodymium (α-form, room temp.)	3.640, 3.673	6 (hexagonal)
(β-form, room temp.)	3.649	12 (f.c.c.)
Promethium		
Protactinium (room temp.)	3.212, 3.238	8,2 (b.c.t.)
Radium		
Rhenium (room temp.)	2.741, 2.760	6,6 (c.p. hex.)
Rhodium (20°C)	2.6901	12 (f.c.c.)
Rubidium (20°C)	4.95	8 (b.c.c.)
(−196°C)	4.860	
Ruthenium (25°C)	2.6502, 2.7058	6,6 (c.p. hex.)
Samarium		
Scandium (room temp.)	3.256, 3.309	6,6 (c.p. hex.)
(room temp.)	3.212	12 (f.c.c.?)
Selenium (20°C)	2.321, 3.464	2,4 (hexagonal)
Silicon (20°C)	2.3517	4 (diamond)
Silver (25°C)	2.8894	12 (f.c.c.)
Sodium (20°C)	3.7157	8 (b.c.c.)
Strontium (α-form, 25°C)	4.3026	12 (f.c.c.)
(β-form, 248°C)	4.32, 4.324	6,6 (c.p. hex.)
(γ-form, 614°C)	4.20	8 (b.c.c.)
Tantalum (20°C)	2.86	8 (b.c.c.)
Technetium (room temp.)	2.703, 2.735	6,6 (c.p. hex.)
Tellurium (25°C)	2.864, 3.468	2,4 (hexagonal)
Terbium (room temp.)	3.525, 3.601	6,6 (c.p. hex.)
Thallium (α-form, 18°C)	3.4076, 3.4566	6,6 (c.p. hex.)
(β-form, 262°C)	3.362	8 (b.c.c.)
Thorium (α-form, 25°C)	3.595	12 (f.c.c.)
(β-form, 1450°C)	3.56	8 (b.c.c.)
Thulium (room temp.)	3.447, 3.538	6,6 (c.p. hex.)
Tin (α-form, 20°C)	2.8099	4 (diamond)
(β-form, 25°C)	3.022, 3.181	4,2 (tetragonal)
Titanium (α-form, 25°C)	2.8956, 2.9505	6,6 (c.p. hex.)
(β-form, 900°C)	2.8636	8 (b.c.c.)
Tungsten (25°C)	2.7409	8 (b.c.c.)
Uranium (α-form, room temp.)	2.77, 2.86, 3.28, 3.37	2,2,4,4
(γ-form, 805°C)	3.058	8 (b.c.c.)
Vanadium (30°C)	2.6224	8 (b.c.c.)
Ytterbium (room temp.)	3.880	12 (f.c.c.)
Yttrium (room temp.)	3.551, 3.647	6,6 (c.p. hex.)
Zinc (25°C)	2.6649, 2.9129	6,6 (c.p. hex.)
Zirconium (α-form, 25°C)	3.1790, 3.2313	6,6 (c.p. hex.)
(β-form, 862°C)	3.1254	8 (b.c.c.)

* 1 angstrom = 10^{-10} meter.

The length of the covalent radius is a function of several factors, among which are (1) bond order (multiplicity), (2) electronegativity, (3) hybridization, (4) orbital overlap, (5) steric factors, (6) special electronic effects.

(1) The effect of bond order is exemplified by the familiar shortening of the carbon-carbon bond in ethane (1.543 Å), graphite (1.4210 Å), benzene (1.397 Å), ethylene (1.353 Å), and acetylene (1.207 Å) as the bond order goes from 1 to $1\frac{1}{3}$ to $1\frac{1}{2}$ to 2 to 3. Bond orders affect the covalent radii of other elements similarly.

The determination of the standard single bond radius is not always a simple matter. In those cases where two like atoms are joined by an unquestionable single bond (as the carbon atoms in ethane) the standard single bond radius is one-half the internuclear distance. Similarly the single bond radii for nitrogen, oxygen, and fluorine are half the interatomic distances in N_2H_4, H_2O_2 and F_2. On the other hand, acceptance of half the interatomic distances in P_4, H_2S_2 and Cl_2 for the corresponding single bond radii is open to question, since in these and similar cases the possibility exists of multiple bond formation by overlap of the electron-filled p-orbitals of each atom with the empty d-orbitals of the other. This will result in an increase in strength and a decrease in length for these bonds compared with what they would have if they were exactly single bonds. In support of this idea, it can be seen from Table 10 that although the homoatomic bond energy for silicon (where no electrons are available for multiple bonding) is less than that for carbon, the bond energies for phosphorus, sulfur, and chlorine are significantly greater than for nitrogen, oxygen, and fluorine, respectively. Representative bond strengths are given in Table 11.

TABLE 10. SINGLE BOND ENERGIES (kcal/mole)

C—C	78.9	Si—Si	53
N—N	39	P—P (in P_4)	48
O—O	35	S—S	58.1
F—F	38	Cl—Cl	57.87

In consequence of this effect, the single bond radii for such elements are better derived from the alkyl derivatives (with an electronegativity correction) in which only single bonding is possible. The single bond radii in Table 12, covalent radii of the elements, were derived insofar as possible from such compounds. The multiple bond radii were derived similarly. For example, the oxygen double bond radius may be obtained from acetone by using the double bond radius for carbon taken from ethylene and applying the electronegativity correction discussed below.

TABLE 11. REPRESENTATIVE SINGLE-BOND ENERGIES (In kcal/mole)

H—H	104.18	O—Sb	71	C—Si	72	Cl—As	70
H—B	ca 93	O—I	ca 48	C—P	63	Cl—Se	58
H—C	98.7	F—F	38	C—S	65.6	Cl—Br	52.7
H—N	93.4	F—Si	135	C—Cl	78.2	Cl—Sn	76
H—O	110.6	F—P	117	C—Zn	40	Cl—Sb	74
H—F	135	F—S	68	C—Ge	ca 44	Cl—I	51
H—Si	76	F—Cl	ca 61	C—As	48	Cl—Hg	54
H—P	ca 77	F—As	111	C—Se	58	Cl—Bi	67
H—S	83	F—Se	68	C—Br	68	K—K	12.6
H—Cl	103.1	F—Br	61	C—Cd	32	Ge—Ge	45
H—As	ca 59	F—Te	80	C—Sn	54	Ge—Br	66
H—Se	ca 66	F—I	63	C—Sb	47	Ge—I	51
H—Br	87.4	Na—Na	18.4	C—I	51	As—As	35
H—Te	ca 57	Si—Si	53	C—Hg	23	As—Br	58
H—I	71.4	Si—S	60.9	C—Pb	31	As—I	43
Li—Li	27.2	Si—Cl	91	C—Bi	31	Se—Se	41
B—C	89	Si—Br	74	N—N	39	Br—Br	46.08
B—N	106.5	Si—I	56	N—O	48	Br—Sn	65
B—O	128	P—P	48	N—F	65	Br—I	43
B—F	154	P—Cl	78	N—Cl	46	Br—Hg	44
B—Cl	109	P—Br	63	O—O	35	Rb—Rb	11.5
B—Br	90	P—I	44	O—F	45.3	Sn—Sn	39
C—C	78.9	S—S	58.1	O—Si	108	Sn—I	65
C—N	72.8	S—Cl	61	O—P	ca 80	Sb—Sb	ca 29
C—O	85.5	S—Br	ca 52	O—Cl	ca 49	I—I	36.06
C—F	116	Cl—Cl	57.87	O—As	72	I—Hg	35
C—Al	61	Cl—Ge	81	O—Br	ca 48	Cs—Cs	10.4

TABLE 12. COVALENT RADII OF THE ELEMENTS (IN ANGSTROM UNITS)*

Element	Valence	Bond Order	Hybridization	Radius
H	1	1	s	0.3754_5
Li	1	1	s	1.336
Be	2	1	s	0.86
		1	sp^3	1.07
B	1	1	p	0.79
	3	1	sp^2	0.84
		1	sp^3	0.92
C	4	1	sp	0.691
		1	sp^2	0.74
		1	sp^3	0.772
		2	sp	0.643
		2	sp^2	0.666
		3	p	0.60
		3	sp	0.602
N	3	1	p	0.73_7
	5	1	sp	0.700
		1	sp^2	0.727
		1	sp^3	0.74_9
	3	2	p	0.61
	5	2	sp	0.617
	3	3	p	0.60
	5	3	sp	0.638
O	2	1	p	0.745
		2	p	0.654
	(6)	2	sp	0.58
		3	p	0.599
F	1	1	p	0.709
Na	1	1	s	1.539
Mg	2	1	s	1.20
Al	1	1	p	1.22
	3	1	sp^2	1.24
		1	sp^3	1.26
		1	sp^3d^2	1.44
Si	4	1	sp^3	1.176
		1	sp^3d^2	1.31
		2	sp^3	1.00
P	3	1	p	1.113_5
	5	1	sp^2	1.11
		1	sp^3	1.12_5
		1	pd	1.23
P (*cont.*)	5	1	sp^3d^2	1.20
		2	sp^3	0.872
S	2,4	1	p	1.06_8
	6	1	sp^3	1.03
		1	sp^3d^2	1.10
	2	2	$p(p\pi)$	0.914
	4	2	$p(pd\pi)$	0.868
	6	2	sp^3	0.757
Cl	1	1	p	1.050
	3	1	pd	1.135
	7	2	sp^3	0.681
K	1	1	s	1.962
Ca	2	1	s	1.39
Sc				
Ti	4	1	d^3s	1.25
		1	d^5s	1.43
V	4	1	d^3s	1.10
	5	1	d^3s	1.17
		2	d^3s	1.05
Cr	3	1	d^2sp^3	1.45
	6	1	d^3s	1.16
		2	d^3s	1.03
Mn	2	1	d^2sp^3	1.6
	4	1	d^2sp^3	1.20
	7	1	d^3s	1.13
		2	d^3s	1.02
Fe	3	1	d^2sp^3	1.39
Co	2	1	sp^3	1.55
	3	1	d^2sp^3	1.35
Ni	2	1	dsp^2	1.28
		1	d^2sp^3	1.54
Cu	1	1	s	1.22
		1	sp^3	1.38
	2	1	dsp^2	1.29
		1	d^2sp^3	1.43
		1	sp	1.34
	3	1	dsp^2	1.39
Zn	2	1	sp	1.15(?)
		1	sp^3	1.34
		1	sp^3d^2	1.46

Continued

TABLE 12—*Continued*

Element	Valence	Bond Order	Hybridization	Radius
Ga	1	1	p	1.29
	3	1	sp^2	1.23
		1	sp^3	1.27
Ge	4	1	sp^3	1.225
		1	sp^3d^2	1.3
As	3	1	p	1.218
	5	1	sp^3	1.19
		1	sp^3d^2	1.33(?)
Se	2	1	p	1.21_5
	4	1	pd	1.38
	6	1	sp^3d^2	1.22(?)
	2	2	p	1.075
Br	1	1	p	1.193_5
	5	1	pd	1.28
Rb	1	1	s	2.06
Sr	2	1	s	1.49
Y	3			
Zr	4	1	d^3s	1.42
		1	d^5s	1.53
Nb	5	1	d^4s	1.37(av.)
		1	d^5s	1.5
Mo	5	1	d^4s	1.30(av.)
	6	1	d^3s	1.31
		1	d^5s	1.27
		2	d^3s	1.23
Tc				
Ru	4	1	d^2sp^3	1.38
	7	2	d^3s	1.23
	8	2	d^3s	1.14
Rh	3	1	d^2sp^3	1.48
Pd	2	1	dsp^2	1.30
Ag	1	1	s	1.32
		1	sp	1.42
		1	sp^3	1.48
Cd	2	1	sp	1.47
		1	sp^3d^2	1.62
In	1	1	p	1.47
	3	1	sp^2	1.47
		1	sp^3	1.43
		1	sp^3d^2	1.66
Sn	2	1	p	1.45
	4	1	sp^3	1.405
		1	sp^3d^2	1.47
Sb	3	1	p	1.376
Sb (*cont.*)	5	1	sp^2	1.43
		1	sp^3d^2	1.42(?)
		1	pd	1.48
Te	2	1	p	1.39
	4	1	p^3d	1.34(?)
		1	p^3d^3	1.55
	6	1	sp^3d^2	1.36(?)
	2	2	p	1.279
I	1	1	p	1.360
	5	1	pd	1.42
Cs	1	1	s	2.18
Ba	2	1	s	1.52
La				
Hf				
Ta	5	1	d^4s	1.37(av.)
		1	d^5sp	1.47(av.)
		1	d^5sp^2	1.48(av.)
W	6	1	d^5s	1.32
Re	4	1	d^2sp^3	1.44
	6	1	d^2sp^3	1.37
	7	1	d^3s	1.25
		2	d^3s	1.22
Os	4	1	d^2sp^3	1.40
	8	2	d^3s	1.171
Ir	4	1	d^2sp^3	1.50(?)
Pt	2	1	dsp^2	1.35
	4	1	d^2sp^3	1.34
Au	1	1	s	1.24
	4	1	sp^3d^2	1.51
Hg	2	1	$s(Hg_2^{2+})$	1.27
		1	sp	1.33
		1	sp^3	1.54
Tl	1	1	p	1.54
	3	1	sp^3d^2	1.58
Pb	2	1	p	1.50
	4	1	sp^3	1.44
Bi	3	1	p	1.53
		1	p^3d^3	1.58
Po	4	1	p^3d^3	1.58
Th	4	1	d^3s	1.69
U	6	1	d^2sp^3	1.50
		2	dp	1.41
Pu	4	1	d^2sp^3	1.72
Am	5	2	dp	1.42

* 1 angstrom = 10^{-10} meter.

(2) When two atoms of different electronegativity are connected by a covalent bond, the bond length is always shorter than the sum of the individual homatomic covalent radii for the atoms in question. The shortening is proportional to the difference in electronegativity of the two elements and is expressed by the relationship due to Stevenson and Schomaker.

$$R = (r_A + r_B) - 0.09|x_A - x_B|$$

where R = observed bond length
r_A, r_B = standard covalent radii of atoms A and B
x_A, x_B = electronegativities of A and B

(3) It has been pointed out by a number of workers that for bonds of the same multiplicity, the lower the average value of the l quantum number in a hybrid bonding orbital, the shorter the bond should be. For example, the carbon-carbon single bond in HC≡C—C≡CH which involves orbitals of sp hybridization is 1.37 Å long compared with the carbon-carbon bond in ethane (sp^3 hybridization) which is 1.543 Å long. Similarly the B—C bond in $B(C_6H_5)_3$ (sp^2) is shorter than in $B(C_6H_5)_4^-$ (sp^3). Thus it is found that a bond of given multiplicity has a length which is characteristic not only of the bond order, but also of the individual states of hybridization of the atoms. For example, the C—C bond in CH_3CN ($sp^3 + sp$) (1.46 Å) is almost exactly the average of the bonds in CH_3CH_3 (sp^3) and in N≡C—C≡N (sp) (1.37 Å). Though the data for other elements are less extensive than for carbon, and the interpretation frequently is much more complicated, the same general principles seem to apply to other elements as well.

(4) The effectiveness of overlap of bonding orbitals of the same symmetry appears to decrease as the principal quantum number increases and as the difference between the principal quantum numbers increases. This is reflected in the bond strengths shown in Table 11. The covalent radius of hydrogen is especially subject to effects of this kind, and has the values 0.3707, 0.362, 0.306, 0.284 and 0.293 Å respectively in H_2, HF, HCl, HBr and HI. The apparent anomaly of the P—P, S—S, and Cl—Cl bonds being stronger than the N—N, O—O, and F—F bonds has been considered in paragraph (1).

(5) In the case of very large atoms or groups bonded to small atoms, the spatial requirements of the large groups may result in a bond lengthening. For example, it is probable that the C—I bond in CI_4 is longer and weaker than in CH_3I because of the steric repulsions of the large iodine atoms. Again, the N—N bond in $[(CH_3)_3NN(CH_3)_3]^{2+}$ may be longer than in $H_3NNH_3^{2+}$.

(6) Special effects of electronic or orbital structure may result in either lengthening or shortening a bond. The first situation is exemplified by such compounds as O_2N—NO_2, O_2N—X, $(C_6H_5)_3C$—$C(C_6H_5)_3$ and the like, where the long bond is indicated in the formula. Cases of this sort involve molecules having a pi-electron sys-

tem capable of accepting additional electrons (frequently in low-lying antibonding orbitals) so that the electrons required for the bond in question are partially drained away from it, leaving the bond weak and long. Thus the N—N bond in N_2O_4 has a length of 1.75 Å and a dissociation energy of 12.9 kcal compared with 1.47 Å and 60 kcal for N_2H_4.

Bond shortening, on the other hand, may occur in compounds of the most electronegative elements, notably fluorine. Typical examples are the fluoromethanes, in which the C—F bond lengths are CH_3F 1.391 Å, CH_2F_2 1.358 Å, CHF_3 1.332 Å, and CF_4 1.323 Å. This has been explained in terms of electronegativity; i.e., that the polar C—F bond requires a high degree of *p*-character, thus releasing the *s*-orbital for the bonds to the less electronegative atoms, making them shorter and stronger. As more fluorine atoms are added, the *s*-orbital is more equally divided among the bonds, resulting in a regular shortening of the C—F bonds. This theory has been used to explain the supposed shortening of the C—C bond from 1.543 Å in C_2H_6 to 1.52 Å in C_2F_6. However, the experimental error attached to the latter value does not allow it to be considered really different from the former (a more recent value is 1.56 Å), and furthermore the effect is not observed in other halo-substituted ethanes for which more accurate data are available: e.g., the C—C bond length is CF_3CN (1.464 Å) is if anything slightly longer than that in CH_3CN (1.458 Å), and microwave determinations on C_2H_5Br, C_2H_5Cl and C_2H_5 give 1.5508, 1.5508 and 1.540, respectively, for the C—C bonds. In addition, the C—H bond lengths in the fluoromethanes appear to be essentially constant: CH_4 1.092, CH_3F 1.109, CH_2F_2 1.092 and CHF_3 1.093. It should also be noted that the C—C stretching force constants in the two cyanides (4.50×10^{-5} and 4.55×10^{-5} and 4.55×10^{-5} dyne cm^{-1}, respectively) do not differ appreciably.

A theory which more satisfactorily explains all the known facts has been suggested by J. F. A. Williams (*Trans. Faraday Soc.*, **57**, 2089 (1961)). This proposes that the highly electronegative fluorine atom drains electron density away from the carbon atom in a C—F group sufficiently to make the lobe of the σ-antibonding (σ^*) orbital which is concentrated beyond the carbon atom available for π-bonding. In CH_3F where the hydrogen atoms have no nonbonding electrons to interact with the σ^* orbital, there is little or no effect. However, in CH_2F_2, where each fluorine atom has nonbonding electrons in *p*-orbitals of favorable disposition, the *p*-electrons of each interact with the σ^* orbital associated with the other to give a p_π-σ_π^* bond, which results in a strengthening and shortening of both bonds.

This effect is observed in the shortening of such bonds as C—N in CCl_3NO_2 and $(CF_3)_3N$, C—P in $(CF_3)_3P$, C—S in $(CF_3)_2S$ and so forth. Such a mechanism, on the other hand, could not result in a shortening of the C—C bond in C_2F_6.

Table 12 gives standard covalent radii for most of the elements of the periodic table. Most of these have been calculated from the best data available in the literature using the considerations of paragraphs 1–3 above. Thus when radii of the appropriate multiplicity and hybridization are added and corrected for difference in electronegativity by the Stevenson and Schomaker relationship, the observed bond length will be obtained. For example, the O—F bond in OF_2 would have the value $0.745 + 0.709 - 0.9(0.5) = 1.409$ Å. The experimental value is 1.41 Å.

The sp^3 double bond radii for Si, P, S and Cl are taken from the paper of D. W. J. Cruickshank, *J. Chem. Soc.*, 5486 (1961). A calculation of the Cl═O bond length gives

$$0.681 + 0.654 - 0.9(0.5) = 1.290 \text{ Å}$$

while for the Cl—O bond we calculate

$$1.050 + 0.745 - 0.9(0.5) = 1.750 \text{ Å}$$

The observed Cl—O bond length in ClO_4^- (1.48 Å) indicates that it has a bond order somewhat greater than 1.5.

The hybridization designations given in the table are not intended to be exact. In particular it must be recognized that the nitrogen bonding orbitals in the ammonia molecule, for example, have considerable *s* character. Nevertheless, such orbitals, for simplicity's sake, have been designated merely as *p*. This practice has been used uniformly when a nonbonding pair of electrons is found in the valence level. In the case of "sp^3d" of hybridization, the radii are given as "sp^3d" if only average lengths were available, but are separated into sp^2 and *pd* if the data differentiated the two types of bonds. The values given for radii of the transition elements and for the less common hybridization states of the other elements, especially where the bond order is not accurately known, must be considered to be only approximate.

Valence

The capacity of an atom to combine with other atoms to form a molecule. Valence is specified as the number of hydrogen atoms or twice the number of oxygen atoms with which one atom of the element under question will combine. Thus, nitrogen has the valence 3,2,4,5 in the compounds NH_3, NO, NO_2, N_2O_5. A further distinction is made by considering positive and negative valences. If the hydrogen is assigned the valence of plus one, and oxygen that of minus two, and if the valences in a compound are made to total up to zero, we have a formal scheme of positive and negative valences. In ammonia, NH_3, the three hydrogen atoms each with a valence of plus one exactly balance the one nitrogen atom with the valence of negative three. Many atoms possess more than one valence but the principal valence is correlated with the periodic table and the atomic structure of the atom. The principal positive valence is the number of the group in which the element falls in the periodic table. Thus hydrogen is one, lithium also one, boron three, etc. The negative valence is eight minus the number of the group in the periodic table. Negative valences greater than four do not occur. For example, oxygen has the valence of eight minus six, that is two negative in H_2O (water); and nitrogen has the valence of eight minus five, that is three negative in NH_3 (ammonia).

On the basis of modern electronic theory of atomic structure we can classify the different types of valence. The guiding principle is that the atoms tend to assume an inert gas electronic structure of eight electrons is the outer shell (in the case of hydrogen it is two). To do this, the atom either loses to, gains from, or shares with other atoms, electrons. This process leads to molecule formation. The following are the principal types of valences and their electronic interpretation.

Electrovalence or polar valence is associated with a transfer of an electron from one element to the other in order to complete by such a tranfer the octet of each element. Thus in sodium chloride the sodium atom has one valence electron outside a closed octet of eight. By loss of this electron the sodium atom becomes positively charged sodium ion because the nuclear positive charge exceeds that of the electrons by one. On the other hand, the chlorine atom has a grouping of seven electrons in the outer shell. It picks up another electron to complete its outer shell to an octet, but in so doing obtains a total charge of one minus, becoming a chloride ion. The result is that in sodium chloride we are not dealing with sodium atoms and chlorine atoms but with sodium and chloride ions. This is experimentally substantiated. The forces holding the ions together are the electrostatic forces, which are equal to the product of the electronic charges on the ions divided by the product of the separation squared times the dielectric constant of the medium. Thus when the sodium chloride crystal is placed in solvent of high dielectric constant such as water, the forces between the ions are weakened and the ions float away from each other. In other words, electrolytic dissociation takes place. It must be noted that polar valences have no specific directional effects in space. The electrostatic attraction is best satisfied by a close packing of the ions. Inasmuch as there are large stray electric fields present in polar compounds, they possess a high melting point and considerable hardness.

Homopolar or covalent bonds are formed by a different mechanism. Here again we have as the basis the tendency of each atom to complete its outer shell of electrons to eight, or in the case of hydrogen to a doublet. In contrast to polar valence, in covalence we have no direct transfer of electrons, but merely a sharing. In the case of molecular hydrogen each hydrogen atom with its one electron shares this electron with the other hydrogen. The result is that each atom in the molecule has at least part of the time a complete shell of two electrons. The electrons can be visualized as traveling in orbits encompassing the two hydrogen nuclei. It is a property of the covalent bond that it is not weakened by electrolytic solvents and that it has a definite direction in space. These directional effects of covalent bonds are expressed in stereochemistry. Thus, for example, the four valence bonds of the carbon atoms are arranged to extend from the center of a tetrahedron to the four corners. Furthermore, since there is a one-to-one saturation of the electron forces, the stray electric fields are negligible, the melting points are low, and the crystals are soft.

Intermediate in properties between the electrovalent and covalent bonds discussed above is the *semi-covalent bond* (also called *dative* or *polarized ionic bond*). It is formed when both electrons that constitute the bonding pair are supplied by one of the atoms. An example is the formation of amine oxides between tertiary amines and oxygen, in which both electrons are donated by the nitrogen atom. Such bonds naturally exhibit electrical polarity. They are members of the large class of heteropolar bonds which are characterized by an unequal distribution of charge due to a displacement of the electron-pair so that the effect of the bond is to make the atoms differ in polarity. In fact, atomic bonds are best described, not qualitatively, but in terms of bond angles and distances. In water, for example, the bond angle is 109.5°, indicating that the lines joining the two hydrogen atoms to the oxygen atom meet at this angle. However, there are two special types of bonds which deserve individual mention.

The *hydrogen bond* is actually two bonds, whereby two electronegative atoms are joined through a hydrogen atom. Since a stable hydrogen atom cannot be associated with more than two electrons, the hydrogen bond may be regarded as a resonance phenomenon, whereby the hydrogen atom is periodically attached to each of the two other atoms in turn, so that its behavior is a composite of the two structures.

Another type of bond which occurs in solids is the metallic bond. It can be considered as an extreme case of sharing of electrons in that an electron gas (present in the crystal lattice) is shared not by two ions but by all the ions in the lattice. This electron gas is responsible for the metallic properties of certain solids, especially for thermal and electrical conductivity.

As a consequence of the fact that many valence bonds leave residual electrical fields, many molecules in which the "primary" valences are satisfied can combine further with other molecules or with atoms. These higher combinations enter into many important areas of chemical science. They are the basis of the formation of coordination compounds, discussed under that heading. They cause molecular association. They are responsible for the formation of hydrates. They are in many cases the binding forces in nonionic solids, and are of great importance in explaining the structure of larger material aggregates.

The foregoing discussion of valence is, of course, a simplified one. From the development of the quantum theory and its application to the structure of the atom, there has ensued a quantum theory of valence and of the structure of the molecule, discussed in this book under **Molecule.** Topics that are basically important to modern views of molecular structure include, in addition to those already indicated, the Schrodinger wave equation, the molecular orbital method (introduced in the article on **Molecule**) as well as directed valence bonds, bond energies, hybrid orbitals, the effect of Van der Waals forces, and electron-deficient molecules. Some of these subjects are clearly beyond the space available in this book and its scope of treatment. Even more so is their use in interpretation of molecular structure. (However, see **Crystal Field Theory** and **Ligand.**)

There are a number of terms used in describing the individual valence bonds. The *bond angle* is the angle between two bonds in a molecule, e.g., in water the bond angle is 109.5°, indicating that the lines joining the two hydrogen atoms to the oxygen atom meet at this angle.

The term *bond direction* arises from the fact that certain covalent bonds prefer to lie in particular directions with respect to the bonded atoms. For example, the bonds from carbon point from the center to the vertices of a regular tetrahedron.

Atoms sharing the two pairs of electrons are said to be connected by a *double bond.* The bond energy of the C—C bond is 80 kcal and that of the C═C bond is 145 kcal. The second bond is formed by *p*-electrons and, while its energy effect is considerable, it does not produce a double bond having twice the energy of the single bond. Moreover, its electrons, being less firmly held between the carbon atoms, are available for addition reactions. These bonds between *p*-electrons, or π-bonds, tend to delocalize in many cases, i.e., the electronic charges "spread" over other atoms than those furnishing them.

The diagram below, which shows the bonding of the carbon atoms in the ethylene molecule, shows the carbon atoms connected by one of the sp^2 hybrid bonds (solid line) the other two being used for the hydrogen atoms. The dotted lines show the π-bond formed between the two *p*-electrons. Since they occupy *p*-orbitals which are perpendicular to the plane of the sp^2 bonds, they cannot form a bond without considerable overlapping. From the figures of 80 and 145 kcal for single and double bonds, the π-bond accounts for 44% of the energy of the double bond, indicating extensive overlapping.

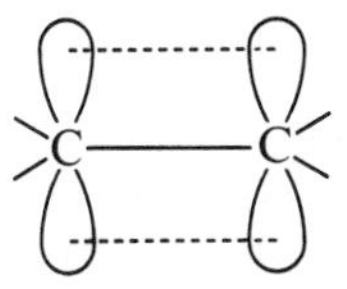

Conjugated double bonds are two double bonds in positions connecting alternate pairs of carbon atoms. For example, the compound

$$CH_2{=}CH{-}CH{=}CH_2$$

has conjugated double bonds. In addition reactions, the conjugated double bond system commonly changes to a single double bond between the second and third carbon atoms, accompanied by the addition of atoms or groups to the first and fourth carbon atoms.

Triple Bond. A single C—C bond involves *sp-sp* overlapping of orbitals, while a C—H bond is the result of *sp-s* overlapping. The other two valences on carbon atoms are represented by two remaining π-electrons, occupying mutually perpendicular *p*-orbitals. In the case of *triple bonding,* —C≡C—, overlapping of these four *p*-orbitals gives two π molecular orbitals. Thus the carbon-carbon *triple bond* is conveniently represented as —C⁚C—, in which the π-electrons are shown occupying positions on the periphery of the carbon-carbon single bond. Their mutual repulsion reduces their bonding effectiveness, so that the bond energies for single, double and triple carbon-carbon bonds are 80, 145, and 198 calories, respectively.

Bond Energies. It has been suggested that ΔH^0_{298}, the heat of formation of a molecule from its constituent atoms (see **Atomic Heat of Formation**), could be computed from a table of average bond energies, and the assumption of additivity:

$$\Delta H^0_{298} = \sum_{\substack{\text{all types}\\ \text{of bonds}}} n_{Xi-Xj} \cdot E_{Xi-Xj}.$$

n_{Xi-Xj} is the number of bonds between the two atomic species, X_i and X_j, in the molecule. E_{Xi-Xj} is the average bond energy associated with each of these bonds. Fairly accurate predictions of the heats of formations of organic molecules can be made in this way, particularly for the *larger* hydrocarbons, alcohols and other aliphatic derivatives.

The differences between the observed heats of formation of *cistrans* isomers and of branched and unbranched hydrocarbon chains show that the additivity rule is not strictly rigorous. Various improvements have been suggested.

Single bond energies are given directly by the heat of dissociation of the corresponding molecules into neutral atoms. In cases where the molecule has no independent existence, other data may often be used. For example, the complete dissociation energy of a binary compound containing more than two atoms may be divided by the number of bonds broken in the dissociation, that is, the energy of the A—B bond may be taken as $\frac{1}{2}$ the dissociation energy of A—B—A into 2A and B, or as $\frac{1}{3}$ the dissociation energy of A—B(—A)(—A) into 3A and B. This multiple bond calculation yields, of course, an average value for the energy of the bonds involved. In the simple case of the H_2O molecule, the dissociation energies of the two successive steps $H_2O \rightarrow H + OH \rightarrow H + H + O$, differ by about 10%.

Representative single bond values were given in Table 11. Summation of such energies to obtain *average* values for molecules applies only when the constituent atoms exhibit their normal covalences and is subject to the exceptions already stated.

Electrides. In late-1979, researchers at Michigan State University reported the assembly of a curious salt wherein an electron, instead of chloride for example, furnishes the negative charge. These usual compounds have been called *electrides* and apparently can be built around alkali metal ions, such as lithium, potassium, and cesium cations. In the synthesis, positively charged metal ions are trapped. Large organic molecules called *cryptate complexes* are used as the traps. Using various solvents, such as liquid ammonia, the investigators have solvated the electrons on the outside. Upon removal of the solvent, the remaining electron appears to serve as an anion to the alkali metal-cryptate complex. Most research to date has concentrated on lithium. Thus far, the electride has not been crystallized. However, several properties have been determined. For example, lithium electride, found in two forms, transmits intense blue light. The researchers observed that although electrons, as yet, have no place on the periodic table, they appear to have a demonstrated chemistry.

References

Christiansen, B., and T. H. Clack, Jr.: "A Western Perspective on Energy: A Plea for Rational Energy Planning," *Science,* **194**, 578–584 (1976).

Chynoweth, A. G.: "Electronic Materials: Functional Substitutions," *Science,* **191**, 725–732 (1976).

Harvey, B. G., et al.: "Criteria for the Discovery of Chemical Elements," *Science,* **193**, 1271–1272 (1976).

Hayes, E. T.: "Energy Implications of Materials Processing," *Science,* **191**, 661–665 (1976).

Heath, R. L.: "Table of the Isotopes," in "Handbook of Chemistry and Physics," 61st edition, CRC Press, Boca Raton, Florida, 1980.

Heath, R. L.: "Gamma Energies and Intensities of Radionuclides," in "Handbook of Chemistry and Physics," 61st edition, CRC Press, Boca Raton, Florida, 1980.

Lefort, M., Ngo, C., Peter, J., and B. Tamain: *Nucl. Phys.,* **A216**, 166 (1973).

Lefort, M., and C. Nego: *Ann. Phys.* (Paris), **3**, 5 (1978).

McCulloch, M. T., and G. J. Wasserburg: "Sm-Nd and Rb-Sr Chronology of Continental Crust Formation," *Science,* **200**, 1003–1011 (1978).

Meyerhof, W. E.: "X-rays from Coalescing Atoms," *Science,* **193**, 839–848 (1976).

Morgan, J. D., Jr.: "The Mineral Position of the United States," *Chemical Engineering Progress,* **73**, 2, 51–56 (1977).

Naldrett, A. J., and J. M. Duke: "Platinum Metals in Magmatic Sulfide Ores," *Science,* **208**, 1417–1424 (1980).

Penzias, A. A.: "The Origin of the Elements," *Les Prix Nobel en 1978,* Nobel Foundation, Stockholm; also "Nobel Lectures," (in English), Elsevier, Amsterdam and New York (1979); also reprinted in *Science,* **205**, 549–554 (1979).

Rampacek, C.: "Impact of Research and Development on Utilization of Low-Grade Resources," *Chemical Engineering Progress,* **73**, 2, 57–68 (1977).

Robinson, A. L.: "Superheavy Elements: Confirmation Fails to Materialize," *Science,* **195**, 473–474 (1977).

Schröder, W. U., and J. R. Huizenga: *Ann. Rev. Nucl. Sci.,* **27**, 465 (1977).

Seaborg, G. T. (editor): "Transuranium Elements: Products of Modern Alchemy," Academic, New York, 1979.

Seaborg, G. T., Loveland, W., and D. J. Morrissey: "Superheavy Elements: A Crossroads," *Science,* **203**, 711–717 (1979).

Staff: "Electrons as Chemical Elements," *Science News,* **116**, 427 (1979).

CHEMICAL EQUATION. By means of chemical formulas, the changes occurring during a chemical reaction can be expressed as an equation. Thus the reaction of 1 mole of sulfur with 1 mole of oxygen to produce 1 mole of sulfur dioxide is written as

$$S + O_2 \rightarrow SO_2$$

The arrow is preferred to the equality sign, which does not emphasize the direction of the reaction. In addition to the identity of the reactants and products, the equation shows the number of atoms entering into the reaction, either in the atomic state or as constituents of molecules. It also shows the number of moles of each reactant and product, so that by use of the table of atomic weights, the relative masses can be computed.

Since the principle of conservation of masses applies to chemical reactions, coefficients must often be used in writing chemical reactions so that the number of atoms of products is equal to the number of atoms of reactants. An example is the reaction of *two* moles of hydrogen with *one* mole of oxygen to form *two* moles of water

$$2H_2 + O_2 \rightarrow 2H_2O$$

In writing such equations, a convenient procedure is to write first an expression containing only the formulas, and then to add the smallest coefficients that will give the same number of atoms of products as of reactants. This operation is called balancing the equation.

Equilibrium reactions, such as that of acetic acid and ethyl alcohol to form ethyl acetate and water, which is cited in the entry on **Chemical Reaction Rate,** are indicated by use of the double arrow

$$CH_3COOH + C_2H_5OH \rightleftarrows HOH + CH_3COOC_2H_5$$

Reactions which result in the precipitation of a solid or the evolution of a gas are sometimes denoted by vertical arrows

$$AgNO_3 + NaCl \rightarrow AgCl \downarrow + NaNO_3$$

$$Na_2CO_3 + 2HCl \rightarrow CO_2 \uparrow + H_2O + 2NaCl$$

This information about the state of the products may also be denoted by writing after their formulas the expressions (s), (l), or (g).

In some cases, as in reactions in electrochemical cells or other reactions involving oxidation-reduction, the half reactions of the ions are useful. Consider the Daniell cell, which consists of a zinc electrode in a zinc sulfate solution, and a copper electrode in a copper solution, the two solutions being separated by a porous partition. The half reactions are

$$Zn \rightarrow Zn^{2+} + 2e^-$$

$$Cu^{2+} + 2e^- \rightarrow Cu$$

so that the overall reaction is

$$CuSO_4 + Zn(s) \rightarrow ZnSO_4 + Cu(s)$$

The more difficult oxidation reduction equations can often be written more easily by use of the Stock system of oxidation numbers, which are positive or negative valences or charges. Consider the reaction of potassium dichromate $K_2Cr_2O_7$ with potassium sulfite K_2SO_3 in acid solution to form chromium(III) sulfate $Cr_2[SO_4]_3$ and potassium sulfate K_2SO_4. The unbalanced expression for the ionic reaction is

$$Cr_2O_7^{2-} + SO_3^{2-} \rightarrow 2Cr^{3+} + SO_4^{2-}$$

Since the oxidation number of the combined oxygen atom is 2− throughout, that of the chromium atom in $Cr_2O_7^{2-}$ is 6+, that of the Cr^{3+} ion is obviously 3+, that of the sulfur atom is SO_3^{2-} is 4+, and that of the sulfur atom in SO_4^{2-} is 6+. The total loss in oxidation number by the two chromium atoms is therefore $(2 \times 6) - (2 \times 3) = 6+$. Since this loss must be offset by a gain made by the sulfur atoms, and since one sulfur atom gains 2+, the reaction must require 3 sulfur atoms. Therefore, the next partially balanced equation is written as

$$Cr_2O_7^{2-} + 3SO_3^{2-} \rightarrow 2Cr^{3+} + 3SO_4^{2-}$$

Counting the charges in this expression shows that there are 8 negative charges on the left-hand side and a net total of 0 charges on the right-hand side. Therefore, since the reaction occurs in acid solution, requiring that hydrogen ions are needed and must be present, $8H^+$ are added to the right-hand to balance the expression electronically, giving

$$Cr_2O_7^{2-} + 3SO_3^{2-} + 8H^+ \rightarrow 2Cr^{3+} + 3SO_4^{2-}$$

Now it is balanced in number of atoms by counting the hydrogen ions (8 on the left-hand side), and the oxygen atoms (an excess of 4 on the left-hand side). Therefore 4 H_2O is added to the right-hand side

$$Cr_2O_7^{2-} + 3SO_3^{2-} + 8H^+ \rightarrow 2Cr^{3+} + 3SO_4^{2-} + 4H_2O$$

If the molecular equation is wanted, it can be written by grouping the ions, and adding those that did not enter into the ionic equations, i.e., the potassium ions of the salts and the anions of the acid

$$K_2Cr_2O_7 + 3K_2SO_3 + 4H_2SO_4 \rightarrow Cr_2(SO_4)_3 + 4H_2O + 4K_2SO_4$$

CHEMICAL EQUILIBRIUM. The fundamental law of chemical equilibrium is that enunciated by Le Chatelier (1884), and may be stated as follows: If any stress or force is brought to bear upon a system in equilibrium, the equilibrium is displaced in a direction which tends to diminish the intensity of the stress or force. This is equivalent to the principle of least action. Its great value to the chemist is that it enables him to predict the effect upon systems in equilibrium of changes in temperature, pressure, and concentration.

The chemical system, hydrogen-nitrogen-ammonia, furnishes a notable example of the application of the principle:

$$\underbrace{\underset{1\ \text{vol.}}{\text{nitrogen}} + \underset{3\ \text{vol.}}{\text{hydrogen}}}_{4\ \text{vol.}} \leftrightarrows \underset{2\ \text{vol.}}{\text{ammonia}} + \underset{\substack{12{,}000\ \text{calories}\\ \text{per mole ammonia}}}{\text{heat}}$$

At the temperature 700°C and pressure 1 atmosphere, the equilibrium percentage of ammonia is 0.03 in the above system, and at 100 atmospheres 2.5. Increase of pressure shifts the equilibrium towards the side of the smaller total volume, at a constant temperature. Decrease of pressure shifts the equilibrium towards the side of the larger total volume, at a constant temperature. Systems of the same initial and final volumes are unaffected, as to equilibrium amounts of materials, by change of pressure.

At the pressure 100 atmospheres, and temperature 700°C, the equilibrium percentage of ammonia is 2.5 in the above system, at 600°C it is 5, at 500°C it is 10. Increase of temperature shifts the equilibrium in the direction which absorbs heat, at a constant pressure. Decrease of temperature shifts the equilibrium in the direction which evolves heat (van't Hoff's principle, 1884).

At constant pressure and temperature, the equilibrium is shifted away from the side subjected to an increase in concentration of any constituent, or towards the side subjected to a decrease in concentration of any constituent. (See **Chemical Reaction Rate.**) For the qualitative effect of temperature change, one may visualize the heat of an equilibrium reaction as material, and an increase of temperature (heat intensity) as operating to increase the concentration of "heat material" thus shifting the equilibrium away from the side of its increased concentration, and conversely. It is possible, knowing the heat of reaction, Q, on the assumption that the heat of reaction is constant between two given (absolute) temperatures, T_1 and T_2, to calculate the equilibrium constant K_2 (at T_2) when the equilibrium constant K_1 (at T_1) and the gas constant, R (equals 2 calories per mole), are known, by the application of van't Hoff's equation:

$$\log_{10} K_2 - \log_{10} K_1 = \frac{Q}{2.3 \times R}\left(\frac{1}{T_2} - \frac{1}{T_1}\right)$$

In this way the quantitative effect of temperature change on the state of equilibrium may be calculated.

In reactions of the ammonia synthesis type, to which sulfur trioxide from sulfur dioxide plus oxygen also belongs, the rate of reaction decreases with lowering of the temperature as the conversion is increased. There is, in such types of reactions, a limit to the practicable lowering of the temperature. The finding of a positive catalyzer for a given reaction of this sort permits the operation to gain the advantage of equilibrium conversion at the lower temperature as well as the increased rate of reaction at that temperature due to the presence of the catalyzer. (See **Chemical Reaction Rate.**) The time yield of product is, therefore, very important, and, with a catalyzer, the space-time yield.

Systems in equilibrium are divided into two great divisions, according to whether they are (A) homogeneous, that is, chemically and physically uniform throughout, or (B) heterogeneous, that is, not uniform throughout but consisting of two or more phases. Each phase is a homogeneous, physically distinct, and mechanically separable portion of a system. For example, ice, water, water vapor are three different phases (solid, liquid, gas) of the substance water. There can be only one gas phase of a system, and only one liquid phase where a *single* homogeneous solution is present. But the number of liquid and of solid phases in general is limited by the number of components (not constituents) of a system. The number of components is the least number of constituents, *independently* variable, and requisite to compose each and every phase. For example, the system consisting of saturated solution in water H_2O of sodium sulfate Na_2SO_4 plus solid sodium sulfate decahydrate $Na_2SO_4 \cdot 10H_2O$ plus water vapor consists of three phases, (a) gas, (b) solution, (c) solid sodium decahydrate. The *least* number of constituents, independently variable in amount *and* requisite to compose each and every phase is two, namely, Na_2SO_4 and H_2O. These, therefore, are the two components of this system. Since zero and negative as well as positive amounts of compounds are permitted in expressing the composition of each phase of any system, the three phases of this system are composed of the following components:

gas phase, zero	Na_2SO_4 plus H_2O
liquid phase,	Na_2SO_4 plus H_2O
solid phase,	Na_2SO_4 plus H_2O

The number of components in the ice-water-water vapor system is one, namely, H_2O.

To systems in which equilibrium depends solely upon the following variables, namely, (1) composition of each and every phase, (2) temperature, and (3) pressure, the phase rule (Willard Gibbs, 1874) applies: The number of variables, that is (1) the number of components, C, plus (2) temperature plus (3) pressure, above, equals the number of phases, P, plus the number of degrees of freedom, F. The number of degrees of freedom of a system is the least number of the above variables which must be arbitrarily fixed in order to define the condition of the system:

$$C + 2 = P + F$$

The phase rule applies to true equilibrium systems, where the equilibrium can be reached from either side, and, furthermore, takes no account of the time involved to attain equilibrium. The phase rule is a qualitative statement, whereas the law of mass action (concentration effect) is quantitatively applicable to those equilibrium systems where the reaction which occurs may be considered to take place in a homogeneous system, e.g., gas phase, or solution phase. (See **Chemical Reaction Rate.**).

In a one-component system, $P + F = 3$, and physical changes only occur. When only one phase is present, for example, liquid water (no vapor, no solid) the system is bivariant, that is, two variables—temperature and pressure—may be independently changed over a range. When a second phase, either vapor or solid appears through a sufficient change of temperature or pressure or both, or when two phases are originally present, the system is univariant, that is, one variable—either temperature or pressure—may be independently changed over a range. When the third phase appears or when three phases are originally present, the system is invariant, that is, a change of either temperature or pressure destroys the equilibrium, and the disappearance of one of the phases occurs. A system of one component in three phases is invariant and the conditions are represented by a point known as the triple point. The triple point for water is 0.007°C, 4.6 millimeters mercury pressure. When the total pressure is one atmosphere (760 millimeters) the equilibrium temperature of water-ice is 0.000°C, and when the water vapor pressure is one atmosphere the equilibrium temperature of water-water vapor is 100.000°C.

If, in dealing with any system, the gas phase or pressure may be neglected, on account of constancy or slightness of effect, the phase rule is simplified for practical purposes to $C + 1 = P + F$, and, if both may be neglected, to $C = P + F$.

Many two- and three-component systems have been studied and recorded in detail. The iron-carbon system is one that has attracted much attention and been of great value in iron metallurgy.

In 1977, Professor Ilya Prigogine of the Free University of Brussels, Belgium was awarded the Nobel Prize in chemistry for his central role in the advances made in irreversible thermodynamics over the last three decades. Prigogine and his associates investigated the properties of systems far from equilibrium where a variety of phenomena exist that are not possible near or at equilibrium. These include chemical systems with multiple stationary states, chemical hysteresis, nucleation processes which give rise to transitions between multiple stationary states, oscillatory systems, the formation of stable and oscillatory macroscopic spatial structures, chemical waves, and the critical behavior of fluctuations. As pointed out by I. Procaccia and J. Ross (*Science*, **198**, 716–717, 1977), the central question concerns the conditions of instability of the thermodynamic branch. The theory of stability of ordinary differential equations is well established. The problem that confronted Prigogine and his collaborators was to develop a thermodynamic theory of stability that spans the whole range of equilibrium and nonequilibrium phenomena.

CHEMICAL FORMULA. The formulas of chemistry constitute a shorthand notation used to represent the composition by weight, the molecular properties, the characteristic chemical reactions or at times even the ordering of the atoms in space of the elements which go to make up the chemical compound. Chemical formulas are classified into empirical, molecular, structural, or configurational, the order given being that of increasing content of information. The meaning of empirical and the molecular formulas is explained in the entry on **Chemical Composition,** which also describes methods for determining the formulas for some simple compounds. Their determination for compounds in general, especially if they are present in mixtures, requires considerable experimental work. The first step consists of the isolation of a pure chemical compound. Chemical purification can be obtained by methods such as crystallization, distillation, adsorption, and sublimation. Some of the criteria of purity which a substance must satisfy are constancy and sharpness of melting point and boiling point on repeated purification. As an example, let us assume that we have succeeded in purifying a solid compound which we shall call tartaric acid and whose formula we wish to determine.

The second step consists in a qualitative and quantitative analysis of the compound. In the case of tartaric acid, qualitative analysis tells us that the compound contains carbon, oxygen and hydrogen, while quantitative analysis shows that the porportions are 48 parts by weight of carbon, 96 of oxygen, and 6 of hydrogen. To obtain the empirical formula, one divides each proportion by the atomic weight of the particular element, obtaining in this way a set of numbers which can be represented by a ratio of small integers. The simplest ratio of integers is commonly used to indicate as subscripts on the right of the chemical symbol of the element to represent the empirical formula. In the case of tartaric acid, the atomic weights are approximately 12 for carbon, 16 for oxygen, and 1 for hydrogen. Dividing the percentages as determined by analysis by the atomic weights, we get:

$$\text{carbon} \quad \tfrac{48}{12} = 4.00$$
$$\text{oxygen} \quad \tfrac{96}{16} = 6.00$$
$$\text{hydrogen} \quad \tfrac{6}{1} = 6.00$$

The set of numbers is 4,6,6 and can be presented, in this case, by the ratio of integers 2:3:3. The empirical formula is therefore $C_2O_3H_3$. Empirical formula is thus only a convenient method for representing the percentage composition by weight of the different elements in the compound. The third step is the determination of the molecular weight of the compound in question. This allows us to assign to the compound a molecular formula. The molecular weight can be determined in a variety of methods, such as by the determination of the weight of 22.242 liters of the vapor of the substance at 1 atmosphere pressure and 0°C, temperature. Other methods are based on the differences in the boiling point or freezing point of solutions of known concentration and those of the pure solvent. To determine the molecular formula from the knowledge of the empirical formula and the molecular weight, the following procedure must be followed: Multiply the atomic weight of each element by its subscript, as indicated in the empirical formula, and add the result. On comparison of such a sum with the molecular weight it will be found that the molecular weight is equal to the sum times an integer. To obtain the molecular formula multiply each subscript in the empirical formula by this integer and obtain a new set of subscripts. We found the empirical formula of tartaric acid was $C_2O_3H_3$. The sum mentioned above is

$$12 \times 2 + 16 \times 3 + 1 \times 3 = 75$$

The molecular weight determined experimentally is 150. The integer multiple is 2, and the molecular formula becomes $C_4O_6H_6$.

The molecular weight of the compound can be obtained from the molecular formula by summing the products obtained by multiplication of the atomic weights of the elements times their subscripts in the molecular formula. The latter contains all the information that the empirical formula contains but in addition specifies the number of atoms in the molecule and also the molecular weight of the substance.

Important as the molecular formula is, it does not describe fully the properties, or even in some cases the identity, of chemical compounds. For example, there are two compounds we have the molecular formula C_2H_6O. They are different in all their properties, both chemical and physical. This difference is due to a difference in the manner in which the atoms are connected in the molecules of the two substances. These differences can be shown only by the use of structural formulas, such as those shown in Fig. 1, in which the valence bonds between the atom are shown. These structural formulas are determined circumstantially, that is, by the chemical reactions into which the compounds enter. (However, their arrangments have been confirmed in many cases by a direct instrumental means such as spectrometric methods, x-ray studies, etc.) These reactions differ markedly for ethyl alcohol and methyl ether. Such compounds which have the same molecular formula but differ due to the arrangements or positions of their atoms are called isomers, and the type just cited, in which the difference is in the grouping of the atoms, are called *functional isomers.* These, and many other types of isomers, are treated in the entry on **Isomerism.**

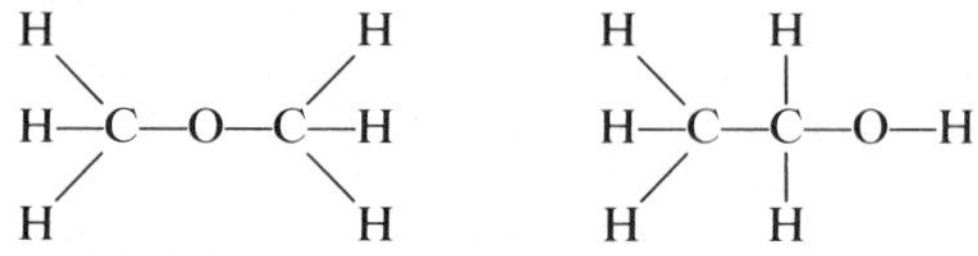

Methyl ether Ethyl alcohol

Fig. 1. Examples of structural formulas.

The structural formula is also a shorthand notation for the important chemical reactions of the compound. It can be considered as being built up of a group of organic radicals, i.e., groups of atoms which retain their individuality in the course of certain reactions. Each radical has reactions which are characteristic of its presence in the molecule. For instance, the carboxyl radical $-\overset{\overset{O}{/\!/}}{C}-OH$ will react with alkali such as sodium hydroxide to form salts $-\overset{\overset{O}{/\!/}}{C}-ONa$, with phosphorus pentachloride to form acid chlorides $-\overset{\overset{O}{/\!/}}{C}-Cl$; with alcohols to form esters; with reducing agents under certain conditions to form successively the aldehyde radical $-\overset{\overset{O}{/\!/}}{C}-H$ and the alcohol radical. Any compound which undergoes such reactions is said to contain a carboxyl group. The number of such carboxyl groups in a molecule can be determined by studying the above reactions quantitatively. On the other hand if the compound will react with sodium to give off hydrogen; with phosphorus trichloride to give a halogen substitution product which can be reduced to hydrocarbon; with an oxidating agent to give an aldehyde or ketone, with organic acids to form esters; with alcohols to form esthers; then the molecule is said to contain a hydroxyl group —OH. Analogously there are similar characteristic reactions for a variety of radicals. It often happens that the presence of one type of a radical near another type mutually influences their reactivity, but one can consider to the first approximation that the radicals act independently of each other. The structural formula is considered completely established if one can synthesize the compound by simple clear-cut reactions involving no rearrangements on the basis of the proposed formula.

Just as it was stated above that two compounds may have the same molecular formula and yet have quite different structural formulas and properties, so there are also many instances in chemistry of compounds which have the same planar structural formula and yet differ in properties. In such cases their differences in structure can be shown only by three-dimensional formulas or their projections which portray differences in the arrangement in space of the atoms or radicals that make up the molecules of the two compounds.

Thus, in cases where four different atoms or groups are attached to the same atom, it is possible to have two arrangements in space which cannot be made to coincide geometrically. This situation can be demonstrated by use of a special type of formula, shown in Fig. 2 for the two forms of the compound fluorochlorobromomethane.

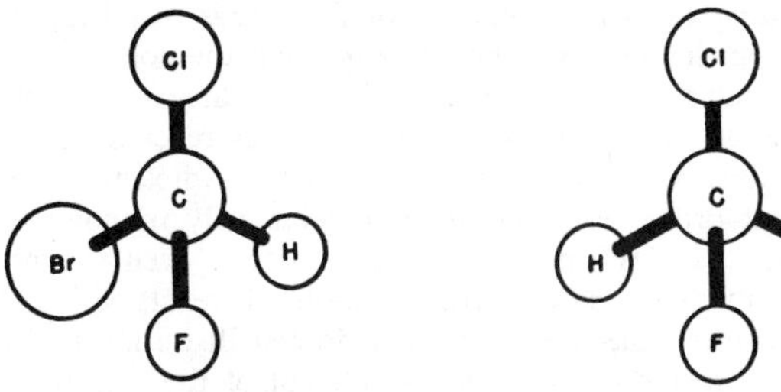

Fig. 2. Formulas to indicate spatial geometry of compounds.

This existence of two forms due to a difference in orientation in space is called *stereoisomerism*, and is discussed in the entry on isomerism. It also follows that for compounds containing more than one atom bonded to four unlike groups, the number of different forms increases rapidly, as is shown by the three possible forms of tartaric acid, HOOC—CHOH—CHOH—COOH, as portrayed by the three formulas shown in Fig. 3.

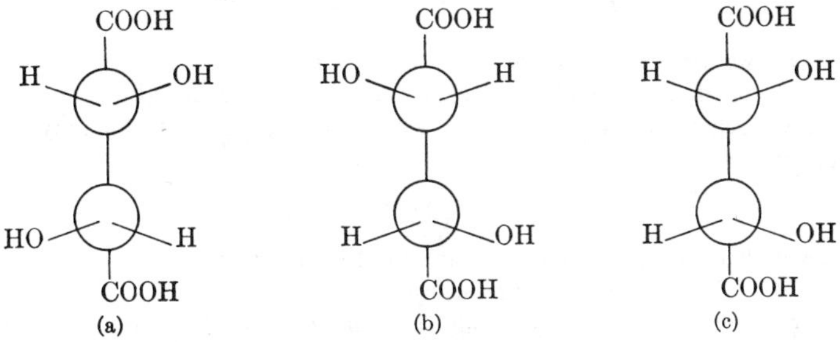

Fig. 3. Formulas to demonstrate stereoisomerism.

Still other types of formulas showing the spatial positions of atoms and groups are perspective and projection formulas, as shown in Fig. 4 for the compound 1,2-dichloroethane. These differences are due to differences in conformation, that is, to the various configurations of a molecule which differ in space by the rotation of two atoms about a single bond. (See that entry for further discussion.)

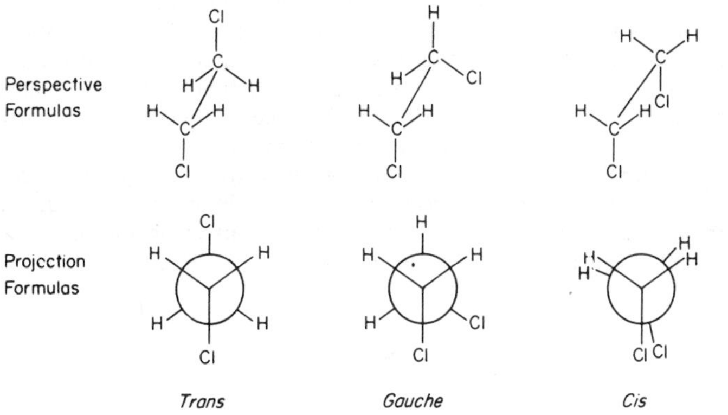

Fig. 4. Perspective and projection type formulas.

Another type of formula which is often written for compounds is the electronic formula, showing the distribution of the valence electrons among the atoms of the molecule, as shown in Fig. 5 and explained under valence and molecule.

Still other types of structures are (1) the tetrahedral models, the Brode and Boord ball-and-stick models, and the Stuart-Briegleb models. These three types are shown in Fig. 6 for the compounds methane CH_4, ethane $H_3C—CH_3$, ethylene $H_2C{=}CH_2$ and acetylene $HC{\equiv}CH$.

CHEMICAL IMAGERY. Photography and Imagery.

CHEMICAL INDICATOR. Indicator (Chemical).

CHEMICAL POTENTIAL. Ideal System.

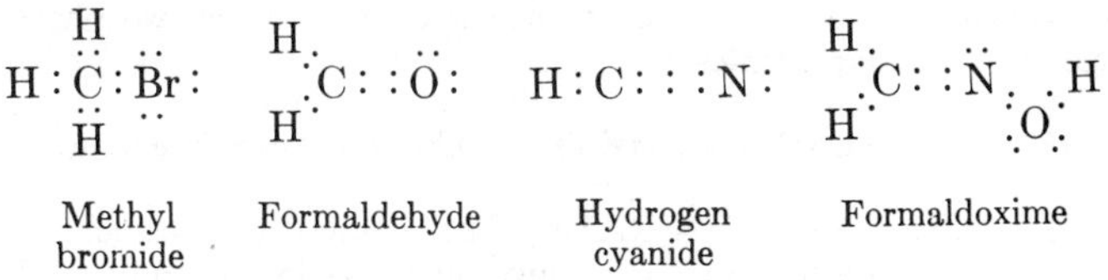

Fig. 5. Formulas to demonstrate bonding.

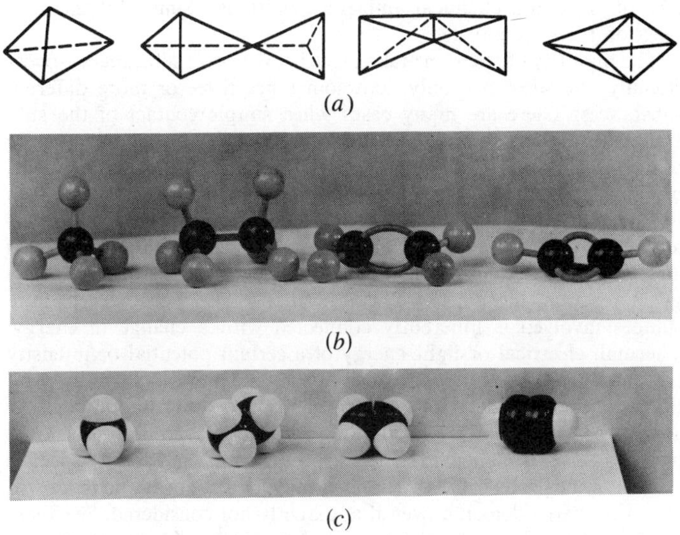

Fig. 6. (a) Tetrahedral models; (b) Brode and Boord ball-and-stick models; (c) Stuart-Briegleb molecular models.

CHEMICAL POTENTIALS. Chemical potentials are defined in terms of the entropy by the relationship

$$\mu_i = -T\left(\frac{\partial S}{\partial n_i}\right)_{U,V} \tag{1}$$

Apart from the factor T (the absolute temperature) the chemical potential is equal to the change of the entropy due to the introduction of the mole number **i** into the system, at constant total energy U and volume V. The parentheses in the above equation contain the partial derivative representing this rate of change.

Other independent variables are often much more convenient. One has also

$$\mu_i = \left(\frac{\partial U}{\partial n_i}\right)_{S,V} = \left(\frac{\partial H}{\partial n_i}\right)_{S,p} = \left(\frac{\partial A}{\partial n_i}\right)_{T,V} = \left(\frac{\partial G}{\partial n_i}\right)_{T,p} \tag{2}$$

The last member of Equation (2) shows that μ_i is the partial molar quantity associated with the Gibbs free energy, G. Euler's theorem gives then

$$G = \sum_i n_i \mu_i \tag{3}$$

The relation to chemical affinity is also very direct

$$A = -\sum_i \nu_i \mu_i \tag{4}$$

(Note that the roman capital A represents chemical affinity, while the italic capital A symbolizes the Helmholtz free energy (also called work function.) It follows that the condition for chemical equilibrium is

$$\sum_i \nu_i \mu_i = 0 \tag{5}$$

where the ν_i are stoichiometric coefficients. This formula expresses the law of mass action. Similarly the condition for two phases α and β to be in equilibrium with respect to species i is

$$\mu_i^\alpha - \mu_i^\beta \tag{6}$$

The chemical potential has then the same value in the two phases. See also **Thermodynamics.**

CHEMICAL REACTION HEAT. Gibbs-Helmholtz Equation.

CHEMICAL REACTION RATE. The chemical composition of a substance is subject to various changes under various conditions, depending upon (1) the nature of the specific substance, (2) the nature of other substances present, and (3) the environment in which it exists (the physical and chemical ambient conditions). Similarly, chemical reaction rates are affected.

The majority of reactions take place between two substances—occasionally one substance only, and sometimes three or more different substances. There are many cases when simple contact of the substances is sufficient to bring about the chemical change, e.g., the rusting of iron in oxygen. In many other cases, the change is not spontaneous, but must be induced, frequently by raising the temperature, as in the burning of fuels. The conditions that are considered important and fundamental are (1) temperature, (2) pressure, (3) medium, if any, (4) catalyzer, if any (5) electric direct current, (6) light. In a given reaction, the change in composition of the substance or substances involved is inherently connected with a change in energy. Thermal, electrical or light energy of a certain potential or intensity and in definite amounts, is requisite to initiate and carry on the reaction, and thermal, electrical or light energy of definite amount is liberated or consumed in the reaction. Every reaction, properly speaking, has both a matter and an energy aspect. While the energy aspect is frequently neglected directly, the conditions must always be in accord with the energy demand, even if apparently not considered. See **Electrochemistry; Photochemistry and Photolysis; Thermochemistry.** Chemical changes require consideration of three topics, namely, (A) natural rate of chemical reactions, (B) acceleration of the natural rate in the presence of a catalyzer, and (C) the end-point of chemical reactions.

A. *Natural Rate of Chemical Reactions.* Various factors operate to affect the rate of chemical reactions. By natural rate is understood the rate of a reaction in the absence of a catalyzer. Excluding electrochemical and photochemical reactions, and giving attention to thermochemical reactions only, there are four factors or conditions to be considered, namely, (1) concentration of constituents, (2) temperature, and (3) pressure—important where a gas is involved, (4) nature of the medium, if any.

The general mathematical definition of the rate of a chemical reaction v is

$$v = \frac{d\xi}{dt} \tag{1}$$

where ξ (Greek letter xi) is the extent of reaction, t is time, and the derivative thus represents the rate of change of the extent of reaction.

1. Relation between concentration of reactants and rate of reaction. The rate of a given reaction, at constant temperature and pressure under stated conditions of concentration of the reacting substances, is quantitatively expressible by a velocity constant, which is the fraction of the substances transformed in a unit of time. Many reactions occur instantaneously—true for most reactions in solution in inorganic chemistry—and many others are complicated in subsidiary reactions, so that the velocity constant is measurable in comparatively few cases. The principle, however, holds as stated, whether or not the desired value can be ascertained experimentally.

A simple reaction that was studied by Wilhelmy, and since then by various investigators, is the transformation (hydrolysis) of sucrose $C_{12}H_{22}O_{11}$ in water solution into glucose ($C_6H_{12}O_6$, a polyhydroxy aldehyde) plus fructose ($C_6H_{12}O_6$, a polyhydroxy ketone), which proceeds at a measurable, steady rate in the presence of acid (hydrogen ion). The rate of reaction at any instant is found to be proportional to the amount of sucrose present at that instant.

When a dilute water solution of an ester, such as methyl acetate, is similarly hydrolyzed in the presence of hydrogen ion, the reaction is of the same type. And this statement also applies to the decay of radioactive elements. One of the important radioactive constants is the half-life, that is, the time required for the decay of one-half of the element present at a given instant. See **Nuclear Reactor.**

The preceding cases are instances of *first order reactions,* that is, reactions in which the rate depends only upon the concentration of a single molecular or atomic species. They are also *monomolecular reactions,* that is, reactions in which the initial reactant is only of one species, that is, sucrose, or an ester, or a radioactive element. Note that first order reactions are not necessarily monomolecular; thus $H_2 + D_2 \leftrightarrows 2HD$, is a first order reaction, even though it is called bimolecular because a hydrogen-1 molecule reacts with a hydrogen-2 (deuterium) molecule to form hydrogen deuteride molecules.

We can now introduce a general treatment of the concept of the order of a chemical reaction. A chemical reaction is said to be of the nth order if its rate is directly proportional to the product of n concentrations. Therefore the decomposition of A, if described by the equation

$$\frac{dC_A}{dt} = -\mathrm{k}C_A \tag{2}$$

is a *first order* reaction. Similarly if it is described by the equation

$$\frac{dC_A}{dt} = -\mathrm{k}C_A C_B \quad \text{or} \quad \frac{dC_A}{dt} = -\mathrm{k}C_A^2 \tag{3}$$

it is a *second order reaction.*

The coefficient k which appears in (2) or (3) is called the *rate constant.* Generally the temperature variation of a rate constant may be expressed by

$$\mathrm{k} = Pe^{-Q/kT} \tag{4}$$

where k is the rate constant and k is the Boltzmann constant. This equation is called the *Arrhenius equation.*

We now introduce a general treatment of rate of reaction, which is often called the *absolute reaction rate theory,* because its purpose is to calculate the rate in terms of molecular quantities only.

Consider the reaction

$$A + BC \rightarrow AB + C \tag{5}$$

To simplify the discussion, assume that A, B and C always remain in a straight line. The course of the reaction may then be followed by noting the values of the two interatomic distances r_{AB} and r_{BC}. At the beginning of the reaction r_{AB} is large and r_{BC} is small while at the end of the reaction r_{AB} is small and r_{BC} is large.

Let us introduce the potential energy surface. The representative point of the system moves on this surface along the so-called *reaction coordinate.* The potential energy along the reaction coordinate is represented schematically in the figure. The maximum of the curve corresponds to a situation where three atoms are very close to one another. Moreover this point is a *maximum* along the reaction coordinate but a *minimum* for the direction normal to the reaction coordinate. Indeed the most probable path is the path involving the minimum potential energy in going from the initial to the final state.

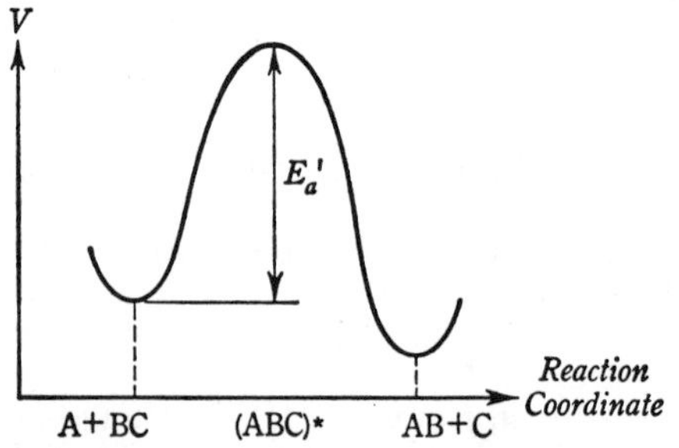

Potential energy along the reaction coordinate.

Therefore the point considered corresponds to a *saddle point* of the energy surface. It is called the *activated complex.*

One may now assume that the reaction rate is the product of the following three factors: (1) the average number of activated complexes; (2) the characteristic frequency of the activated complex (that is, the inverse of its lifetime); (3) the *transmission coefficient,* K, which is the probability that a chemical reaction takes place after the system has reached the activated state.

Moreover the number of activated complexes is calculated by the equilibrium assumption.

Using this description of the reaction process one derives the following expression for the reaction constant

$$k = K\frac{\phi_t(T)}{\phi_A(T)\phi_{BC}(T)}\frac{kT}{h}\exp\left(-\frac{E^x}{kT}\right) \quad (6)$$

Here the ϕ terms are the partition function $f(T, V)$, the volume factor being removed

$$f = V\phi \quad (7)$$

ϕ_t corresponds to the activated complex, the degree of freedom associated with the reaction coordinate being removed; k is Boltzmann's constant, h is Planck's constant, E^x is the energy associated with activated complex, or *activation energy* of the reaction.

This expression may also be written in the thermodynamic form

$$\begin{aligned} k &= K\frac{kT}{h}\exp\left(-\frac{\Delta G^\ddagger}{kT}\right) \\ &= K\frac{kT}{h}\exp[-(\Delta H^\ddagger - T\Delta S^\ddagger)/kT] \end{aligned} \quad (8)$$

where $\Delta G^\ddagger$ is a suitable free energy of activation and $\Delta H^\ddagger$, $\Delta S^\ddagger$ the corresponding enthalpy and entropy of activation.

When a reaction involves two different phases, that is, when the system is not homogeneous but heterogeneous, as in reactions between a solid phase, such as zinc or calcium carbonate, and a liquid phase, such as hydrochloric acid solution, the rate of reaction involves consideration of (1) the area of the surface of contact of the solid with the solution, and (2) the rate of diffusion from the surface of the solid, as well as (3) the concentration of hydrogen ion of the acid solution.

When the rate of a chemical process is dependent upon (1) two or more *consecutive* reactions, the observed rate is limited by the rate of the slowest reaction in the series, (2) two or more *concurrent* reactions, the products are in the same ratio at any instant only when the reactions themselves are of the same rate.

2. Relation between temperature of reactants and rate of reaction. The rate of chemical reaction is increased by an increase in temperature, as is evident from the fact that temperature occurs in the numerators in the foregoing equations.

3. Relation between pressure of reactants, if gaseous, and rate of reaction. Since pressure changes amount to concentration changes in such systems, the behavior is as described above under concentration.

4. Relation between nature of the medium and rate of reaction. Very slight changes in the nature of the medium greatly affect the rate of a chemical reaction, but attempts to relate any physical property of a solvent with the effect observed on the rate of a given reaction appear to have proved unsuccessful.

One should mention here that in reactions involving ions, the effects of electrolytes can be put into two principal categories: (a) primary salt effect and (b) secondary salt effects.

Primary salt effects refer to the effects of electrolyte concentration on the activity coefficients. Secondary salt effects are those concerned with the actual changes in concentration of the reacting species resulting from the addition of electrolytes.

Brønsted has shown that the variation in specific rate with ionic strength depends on the magnitude and sign of the ionic charges. Thus three main types of primary salt effect in reactions of two species can be distinguished. If the products of the signs of the charges are positive, the velocity of the reaction increases with increasing ionic strength:

$$Co[(NH_3)_5Br]^{2+} + Hg^{2+} \qquad (+2 \times +2 = +4)$$

or

$$S_2O_8^{2-} + I^- \qquad (-2 \times -1 = +2)$$

or

$$BrCH_2COO^- + S_2O_3^{2-} \qquad (-1 \times -2 = +2)$$

If the products of the charges are negative the velocity of the reaction decreases with increasing ionic strength:

$$Co[(NH_3)_5Br]^{2+} + OH^- \qquad (+2 \times -1 = -2)$$

or

$$H_2O_2 + H^+ + Br^- \qquad (+1 \times -1 = -1)$$

In the third category where the product of the ionic charges is zero, the ionic concentration has no or very little effect on the velocity of the reaction, particularly in dilute solution, as in the case of

$$[Cr(NH_2CONH_2)_6]^{3+} + 6H_2O \rightarrow [Cr(H_2O)_6]^{3+} + 6NH_2CONH_2 \qquad (+3 \times 0 = 0)$$

Brønsted showed the inadequacy of both the classical and the activity theories of rate of certain reactions. He proposed a new theory postulating that when ions or molecules react, they first form an unstable critical complex which then decomposes to give the reaction products. The reaction which determines the velocity of a chemical change consists in the formation of that unstable critical complex.

Summarizing, the rates of chemical reactions are subject to highly specific influences in each case, as has been abundantly demonstrated by experimental investigations, and recognized in numerous legal battles in chemical patent suits.

B. *Acceleration of the Natural Rate of Chemical Reactions* in the presence of a positive or negative catalyzer. When, in the presence of a given substance, the natural rate of a chemical reaction is changed, either increased or decreased, the given substance is called a catalyzer. Examples are numerous. (1) When a gas-lighter of the type known as platinum black, the active part of which consists of very finely divided platinum, is held in a stream of hydrogen or city gas, the gas is ignited in air. Platinum is a catalyzer for this reaction, and causes ignition to take place at a temperature much lower than by subjecting to fire. (2) The changing of sulfur dioxide into sulfur trioxide is accomplished by passing a mixture of sulfur dioxide and air (one-fifth oxygen) over asbestos coated with finely divided platinum. The temperature required is much lower by the use of platinum catalyzer than without its use. (3) Solutions of sulfites are subject to oxidation to sulfates by oxygen upon allowing to stand in air. The addition of sugar or glycerol retards the speed of this reaction. These substances act in this case as negative catalyzers. (4) The combination of nitrogen and hydrogen gases under high pressure to form ammonia gas is accomplished at a lower temperature in the presence of a catalyzer than in its absence, thus increasing the yield of ammonia (see **Equilibrium**). One of the catalyzers is composed of iron, intimately mixed with 1% aluminum oxide and 1% potassium oxide. Iron is a catalyzer for this reaction, but is more active as such in the presence of aluminum oxide and potassium oxide, which are spoken of as promoters, a sort of catalyzer of a catalyzer. (5) The hydrogenation of liquid fatty oils and of oleic acid is conducted in the presence of finely divided nickel as a catalyzer. (6) Enzymes are very specific catalyzers, "the most selective and deicate of all known catalysts (Hilditch)," at ordinary temperatures, say 25 to 30°C. Dextroglucose is converted into ethyl alcohol in the presence of the enzyme (zymase) of yeast, and ethyl alcohol into acetic acid (vinegar) in the presence of the enzyme of *Mycoderma aceti.* (7) Nitric acid reacts slowly with copper metal, but the rate of reaction is accelerated more and more as nitrogen tetroxide (catalyzer) is formed in the solution. This is an example of autocatalysis, wherein the reaction brings about the formation of its own catalyzer. (8) Arsenic-containing substances are extreme negative catalyzers, called inhibitors or poisons, of platinum catalyzer.

When the catalyzer is a solid substance, the greatest difficulty in use is to maintain a clean surface. The presence of a positive catalyzer enables a reaction to proceed more rapidly at a lower temperature than corresponds to the natural rate of the reaction. This increases the amount of substances converted in a given time, decreases the demands as to temperature resistance of materials of construction of the apparatus, and frequently makes possible a state of equilibrium more favorable to the yield of desired material.

C. *The End-point of Chemical Reactions.* If a chemically reactive system is isolated from the rest of the universe at a constant temperature and pressure, a definite end-point is often attained short of the complete transmutation of reactants into resultants. In order to be certain that this end-point (short of complete transmutation) is what is known as the equilibrium point, the equilibrium must be approached from both directions, e.g., A + B → C + D and C + D → A + B. If the equilibrium constant (see treatment below) is the same when approached from both directions, then the reaction is one of true chemical equilibrium. Such equilibrium reactions are also referred to as balanced or reversible reactions. In such reactions the extent of the chemical change is proportional to the concentrations of all the reactants—reactants and resultants being interchangeable, depending upon the direction of the reaction. (Generalization of Guldberg and Waage, 1864, called Law of Mass Action, or more correctly Law of Concentration Effect. Reaction studied by Guldberg and Waage (1867): Barium sulfate plus potassium carbonate plus barium carbonate plus potassium sulfate.)

A classical case, frequently cited, is that investigated by Berthelot in 1963. When 1 mole (60 grams) of acetic acid CH_3COOH and 1 mole (46 grams) of ethyl alcohol C_2H_5OH, both of which substances are soluble in water, are mixed, a reaction takes place which results in the formation of water and ethyl acetate ester, which is likewise in the ratio of 1 mole (18 grams) of water, and 1 mole (88 grams) of ethyl acetate $CH_3COOC_2H_5$. On the other hand, when 1 mole of water and 1 mole of ethyl acetate ester are mixed, a reaction takes place which results in the formation of acetic acid and ethyl alcohol in the ratio of 1 mole of acetic acid and 1 mole of ethyl alcohol. Three important observations have resulted from the detailed study of this reaction, namely, (1) the reaction between acetic acid and ethyl alcohol as reactants proceeds at such a rate that the fraction 0.005/5 of the amount present at any instant reacts, at 6 to 9°C, in 1 day to form equivalent amounts of water and ethyl acetate ester, (2) the reaction between water and ethyl ester as reactants proceeds at such a rate that the fraction 0.00144 of the amount present at any instant reacts, at 6 to 9°C, in 1 day to form equivalent amounts of acetic acid and ethyl alcohol, and (3) the end-point of each reaction is the same, that is, the reaction is one of true chemical equilibrium, and the resulting equilibrium mixture contains, in each case, 0.33 mole acetic acid plus 0.33 mole ethyl alcohol plus 0.67 mole water plus 0.67 mole ethyl acetate ester. This system attains practical equilibrium, at 6 to 9°C in about 1 year, at 100°C in about 8 days, and at 200°C in about 24 hours.

The equilibrium constant is calculated numerically as follows:

Equation:	CH_3COOH +	C_2H_5OH ⇆	HOH +	$CH_3COOC_2H_5$
Reaction weights:	60	46	18	88
Molar ratio at equilibrium:	0.33	0.33	0.67	0.67
Weights at equilibrium:	0.33 × 60	0.33 × 46	0.67 × 18	0.67 × 88

$$\left.\begin{array}{r}\text{Equilibrium}\\ \text{constant at 9°C}\end{array}\right\} = \frac{\text{conc. } HOH \times \text{conc. } CH_3COOC_2H_5}{\text{conc. } CH_3COOH \times \text{conc. } C_2H_5OH}$$

$$= \frac{0.67 \times 0.67}{0.33 \times 0.33} = 4$$

Knowing the equilibrium constant at any stated temperature enables one to calculate the equilibrium end-point at that temperature for any ratio of reactants. Thus, when 1 mole (60) grams of acetice acid and 10 moles (460) grams of ethyl alcohol at 9°C are taken:

$$\text{Equilibrium constant at 9°C} = 4 = \frac{X \times X}{(1 - X) \times (10 - X)}$$

where X is the number of moles of water and also the number of moles of ethyl acetate ester formed (1 mole of each is formed by reaction of 1 mole acetic acid plus 1 mole ethyl alcohol). Solution of this equation shows $X = 0.97$. Therefore, by taking the above ratio of acetic acid (1 mole) to ethyl alcohol (10 moles) 0.97 (or 97%) of the acetic acid, the excess reactant being ethyl alcohol (9 moles), is converted at equilibrium into water plus ethyl acetate ester. In practice, the reaction is conducted by the use of a catalyzer, e.g., sulfuric acid concentrated, zinc chloride.

In cases where one of two resultants can be separated from the reactants and the other resultant, by precipitation as a solid, by condensation as a liquid, or by volatilization as a gas or vapor, the yield of the desired substance from a given amount of reactants can sometimes be materially increased. In the case of heterogeneous systems, whenever a solid participant is present, the *concentration* of said solid is considered constant. The precipitation and solution of solids are in this category, as well as the reactions between a gas and a solid, e.g., the system ferroferric oxide plus hydrogen gas plus iron plus water vapor.

The effect of change of temperature on a system in chemical equilibrium is that the equilibrium point is shifted (1) towards the side *away* from that which evolves heat when the temperature is *raised*, and (2) towards the side which evolves heat when the temperature is lowered. It is *as if* the amount of heat were a *material* reactant and its concentration (temperature or intensity of heat) increased, in respect to the *direction* of the shift of the equilibrium point. The amount of the shift at constant pressure can be calculated in cases where one possesses the proper data. See **Thermochemistry.**

The effect of change of pressure on a system in chemical equilibrium is that the equilibrium point is shifted (1) towards the side possessing the smaller aggregate volume when the pressure is increased, and (2) towards the side possessing the larger aggregate volume when the pressure is decreased. The amount of the shift at constant temperature can be calculated by means of the equilibrium constant (above) recalling that increase of pressure is equivalent to increase of concentration of gases (temperature constant). When the volume of resultants equals the volume of reactants, no effect is produced on the equilibrium point by change of pressure. See **Chemical Equilibrium.**

In some chemical reactions, the use of concentrations does not give calculated results that agree with those observed, because of the departure from ideality of real gases and solutions. In such reactions, concentrations are replaced by apparent effective concentrations, or activities, as explained in that entry.

CHEMICAL ROCKETS. Rocket Propellants.

CHEMICAL SHIM. Nuclear Reactor.

CHEMICALS (Number of). With over 100 chemical elements from which chemical compounds can be built, the large numbers of atoms which may be present in various compounds, and the many ways in which the atoms can be linked (straight chains, branching chains, rings, half-rings, etc.), the number of "possible" chemical compounds is a very high number indeed and, of course, is much larger than the millions of "known" compounds, which can be described fully in terms of constituents and structure. Keeping track of so many compounds—in the interest of fundamental research and, in more recent years, the identification of chemical compounds in terms of both beneficial and adverse effects on biological systems, health, and the environment—commenced many years ago. Thousands of compounds are described in numerous handbooks, various societies have compiled lists and tabulations, as for example the Chemical Abstracts Service (CAS) of the American Chemical Society, and special encyclopedias which describe inorganic and organic compounds. Possibly the most outstanding example is the "Handbuch der Organische Chemie," first undertaken in Germany in 1918. The first 27 volumes of this book were published between 1928 and 1938. Supplements have been released periodically since that time. Part II, consisting of 29 volumes, was published between 1941 and 1957; Part III, 14 volumes, was published between 1958 and 1973. Part IV was commenced in 1972 and continues. Publisher is Springer-Verlag, Berlin.

In 1978, the CAS, in connection with the Toxic Substances Control Act, was engaged in a project for Environmental Protection Agency (United States). It was estimated that some 33,000 chemicals are in common use. The problem of identification is complicated by the fact that, because of various common and trade names for the same substance, there are approximately 183,000 designations for these 33,000 chemicals. Of all chemicals cataloged to date, by CAS and similar organizations throughout the world, it is estimated that about 3.5 to 4 million have been identified. Statistics indicate that of these about 96% are organic, i.e., they contain carbon. This translates into about 3.4 to 3.8 million organic compounds and from 100,000 to 200,000 inorganic compounds. Although essentially meaningless, a statistical summary of cataloged chemicals indicates that the "average chemical," a highly fictious entity, contains some 43 atoms, 22 of which are hydrogen. Of the organic chemicals, it is estimated that nearly 300,000 are coordination compounds and about 60,000 compounds are substances whose structures have not been completely defined. The CAS registry also lists some 72,000 alloys, 120,000 polymers, and 10,000 mixtures with definite names.

CHEMICAL SUBSTANCES INDEX. **Organic Chemistry.**

CHEMICAL TRANSMITTERS. **Brain and Nervous System.**

CHEMI-OSMOTIC THEORY. **Photosynthesis.**

CHEMISTRY. That branch of natural science which investigates the composition of all matter, and the transformations which it exhibits upon subjection to energy change. Like other basic sciences over the years, chemistry has become increasingly fractured into numerous areas of specialization. Further as all science has become more and more interdisciplinarian in nature, former sharp boundaries between chemistry and the other basic sciences have become broad zones. However, for purposes of organizing subject matter, it remains helpful to give careful and thoughtful recognition to the major fields of interest of chemistry, realizing that most fields are quite nonexclusive.

Agricultural Chemistry—Application of chemical facts and principles to the problems encountered in farming, including plant processes in their relation to the growth of crops, plant nutrition, agricultural by-product utilization, etc.

Analytical Chemistry—Determination of the qualitative and/or quantitative composition of substances and materials. This field of chemistry has been grossly altered with the development of analytical instrument techniques, such as chromatography, infrared, ultraviolet, and other types of spectroscopy, etc., that not only have improved the accuracy and sensitivity of determinations, but that also have greatly speeded up analytical procedures through the application of various degrees of automation.

Biochemistry—The chemistry of life processes.

Colloid Chemistry—The study of systems comprising a dispersion medium and a dispersed phase and possessing colloidal characteristics.

Electrochemistry—The study of electrolysis and other phenomena which relate to the passage of current through an electrolyte or are concerned with the behavior of ions in ionizing solvents. Corrosion phenomena are a concern of this field. Electrochemistry and electrometallurgy are closely allied technical areas.

Forensic Chemistry—The application of chemical knowledge to the solution of legal problems.

Industrial Chemistry—The application of the principles of chemistry to the production and utilization of chemicals in the manufacture of fuels, products, and by-products. The distinction, if any, between this field and chemical engineering is a very fine one. See also **Chemical Engineering.**

Inorganic Chemistry—The branch of chemistry dealing with salts, acids, and bases and, in general, with all compounds not containing carbon. In fact, a few carbon compounds, such as carbon monoxide, carbon dioxide, and hydrogen cyanide, are considered by many to be inorganic compounds. The chemical elements in their pure or semipure state also fall into this area. Since many of these elements are metals, the fields of interest of inorganic chemistry and metallurgy are closely allied.

Organic Chemistry—Essentially, the chemistry of carbon compounds. See also **Organic Chemistry.**

Photochemistry—The study and application of the phenomena associated with the mutual transformations of radiant energy and chemical energy. Photographic materials and processes are an important part of this field.

Stereochemistry—The study of the structure of molecules, especially concerned with those differences in chemical or physical properties of substances which are due to differences in spatial arrangement between compounds that are similar in composition.

Physical Chemistry—The study of the physical properties of chemical substances and the relations between energy and chemical change. The parameters of physical chemistry in terms of modern science are very difficult to define.

Greatly abridged, other specialized terms sometimes used to describe special fields of interest in chemistry are geochemistry (geology), microchemistry (small-scale), pathological chemistry (diseased tissue), pharmaceutical chemistry (medicinals), phytochemistry (plants), pure chemistry, stoichiometric chemistry (reactions-quantitative), structural chemistry (atomic and molecular arrangements/linkages), synthetics chemistry (building complex materials from simpler ones), theoretical chemistry, and thermochemistry (chemical thermodynamics).

CHEMORECEPTOR. Sense organ sensitive to chemical substances in the surroundings. See also **Sensory Organs.**

CHEMOPROPHYLAXIS. **Antibiotic; Cancer and Oncology.**

CHEMOSYNTHESIS (Bacterial). **Ocean Resources (Energy).**

CHEMOTAXIS. **Immune System and Immunology.**

CHEMOTAXONOMY. The application of chemical data to systematics represents a powerful approach for biologists in their attempts to answer questions about relationships among organisms. Chemotaxonomy is especially useful in classifying microorganisms, which produce primarily by asexual division, resulting in somewhat arbitrary specific designations. The chemical evidence may be equally well represented by macromolecular compounds, such as DNA, RNA and proteins, or by smaller molecules, such as the natural products of plants, e.g., alkaloids, terpenoids, and flavonoids. These latter classes of compounds have no clearly defined metabolic roles and are, therefore, often described as secondary compounds to distinguish them from the more vital substances as, for example, the protein amino acids.

The generally accepted theory for information transfer or "gene action" directly relates specific features of the chemical structures of secondary compounds, such as functional groups and number and size of rings, to segments of DNA (genes) *via* enzyme-controlled biosynthetic pathways. Since DNA is unique for each individual, we can infer that some enzymes and hence some secondary compounds may be unique for each organism. Unfortunately, for taxonomic purposes, the ability of two different organisms to synthesize the same secondary compounds can be interpreted in at least two different ways: it may reflect common ancestry or, alternatively, it may be a case of convergent evolution. Still, the careful application of chemical information to systematic problems can provide a level of insight into phylogeny not possible with morphological and cytological evidence alone.

Ann C. Vickery, Ph.D., University of South Florida, College of Medicine, Tampa, Florida.

CHEMOTHERAPY. The administration of specific substances to inhibit or destroy microorganisms, parasites, and other causes of disease and disorders, without the chemotherapeutic substances themselves causing serious harm to the tissues of the host. Such substances interact chemically in one fashion or other in target biological systems. The chemotherapeutic agents may be strictly of a chemical nature, such as the sulfonamide drugs, quinine, piperazine, tetrachlorethane, and so on, or they may be derived from and perform in a manner more closely identified with bacterial and viral processes, as in the instances of various antibiotics. In numerous instances of chemotherapy, a thorough understanding of the processes which occur is not at hand, but based upon studies with laboratory animals and further confirmation from testing programs involving human beings, many agents have come into use with varying degrees of success. Scores of examples of chemotherapy are given in this encyclopedia. Consult specific diseases and disorders.

CHEMOTHERAPY (Cancer). **Cancer and Oncology.**

CHENODEOXYCHOLIC ACID. **Bile; Gallbladder and Biliary Tract Diseases.**

CHERIMOYA. Annona.

CHERRY TREES. Rose Family.

CHERT. An impure, flinty hard rock composed chiefly of cryptocrystalline silica. Chert varies in color from gray through brown to black according to the kind and amount of coloring matter. It occurs principally as concretions, nodules or bands in limestones and dolomites, and unlike flint its fracture tends to be splintery instead of conchoidal.

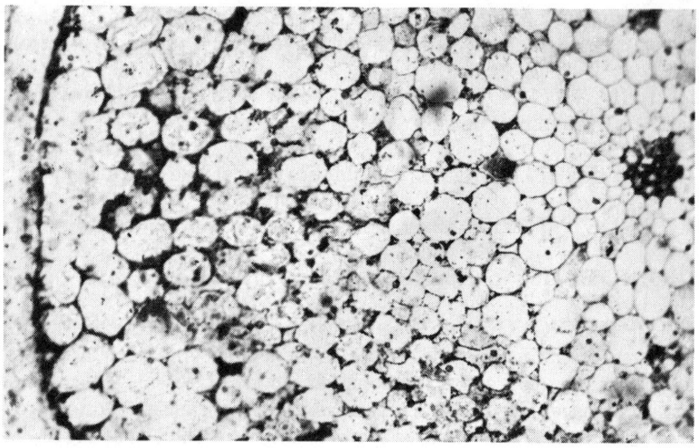

Transverse section from chert bed of Rhynie, UK. (*Photomicrograph by Brian J. Ford; copyright.*)

A great deal has been written on the occurrence and origin of chert and there is no doubt but that it may be formed in several different ways. Many of the nodular and concretionary cherts have grown around siliceous sponge spicules or radiolaria. Chert may be either sygenetic or epigenetic. The former type is supposed by some authors to be chemically precipitated from river waters on the bottom of the sea as a colloid contemporaneously with the limestones or dolomites. On the other hand certain cherts are obviously secondary although they may have been formed previous to the final lithification of the formations in which they occur. Cherts which contain relatively large amounts of iron are called Jasper.

CHERVIL. Flavorings.

CHESSYLITE. Azurite.

CHESTNUT BLIGHT. Ascomycetes; Fungus.

CHESTNUT TREES. Members of the family *Fagaceae* (beech family), these trees are of the genus *Castanea.* They are deciduous trees with toothed leaves and are quite tolerant of shade. As indicated by the names of the following species, chestnut trees are found in the eastern United States, southern Europe, northern Africa, and parts of Asia, including Korea and China.

American sweet chestnut	*Castanea dentata*
Chinese chestnut	*C. mollissima*
Japanese chestnut	*C. crenata*
Spanish or sweet chestnut	*C. sativa*

The American chestnut grows along the Alleghany Mountains and from New Hampshire southward to Georgia. It is a large tree and many years ago was one of America's most valued trees for timber. However, a fungus (*Diaporthe parasitica*) imported from Asia affected the bark of the tree. The tree was also attacked by a chestnut borer (*Ariopalus fulminans*), as a result of which a large portion of the chestnut tree population died. Some trees in Mendocino County, California were from 70 to 100 feet (21 to 30 meters) in height, had a circumference of 18 feet (5.4 meters) or more, and were over 500 years old. The present champion *C. dentata*, as listed in the "Social Register of Big Trees," issued by The American Forestry Association (1975), is located in Sherwood, Oregon and has a circumference at $4\frac{1}{2}$ feet (1.4 meters) of 221 inches (561 centimeters); a height of 66 feet (19.8 meters); and a spread of 75 feet (22.5 meters).

Chestnut wood weighs about 28 pounds per cubic foot (448.5 kilograms per cubic meter), is coarse, durable, and of a pale brown color. When available, the wood is used for posts, cross ties, veneers, and mill products.

The Japanese have successfully propagated the *C. crenata*, which is valued for its fine nuts. The tree is dwarf and is apparently free from blight attack. Orchards of these trees have been started in several parts of the world. The nut has an excellent taste.

The common horse chestnut, red horse chestnut, Baumann's horse chestnut, and Japanese horse chestnut are members of another tree family, *Hippocastanaceae* (horse chestnut family), and are closely associated with the buckeyes. See **Horse Chestnut and Buckeye Trees.**

CHEVROTAIN. Tragulines.

CHEZY FORMULA. This formula is concerned with the friction loss in water conduits. The formula is

$$V = C\sqrt{RS}$$

R is the hydraulic radius of the cross section of flow. It is the cross-sectional area of flow divided by the wetted perimeter. S is the energy loss per foot length of conduit or the drop of the water surface per foot of length of an open channel. Manning's expression for the value of C in the Chezy formula is much used. It is as follows:

$$C = \frac{1.486R^{1/6}}{n}$$

in which n is a coefficient of roughness. The values of n are the same as those used in the older Canquillet-Kutter formula for C. Values of n range from about 0.015 for vitrified sewer pipe, concrete pipe, or plank flumes, to 0.03 or more for canals and natural stream channels.

See also **Hydrokinetics.**

CHIASTOLITE. Andalusite.

CHICKADEE (*Aves, Passeriformes*). Birds of several species found in various parts of North America, all quietly colored in grays with some black markings and in some species a little white and brown. The common widely distributed species is also called the black-capped titmouse, *Penthestes atricapillus.*

Carolina chickadee.

CHICKEN. Poultry.

CHICKENPOX. Varicella. A common, communicable disease caused by a virus related to the virus that causes shingles (*Herpes zoster*). The highly contagious nature of the disease is evidenced by the fact that nearly all children have contracted the disease by the age of ten. The disease is communicated by direct contact. The incubation period is from 14 to 16 days.

In children the disease is usually quite mild. The first symptom is often the eruption which begins on the face or trunk as small, red, widely scattered itchy spots, and spreads to the rest of the body, including the scalp and mucous membranes of the mouth and pharynx. The eruption is often most abundant on the neck and shoulders. The red spots increase in size and soon are capped by small blisters of clear fluid, which progress to scab formation. The rash characteristically appears in crops over a period of 3–4 days so that early and late lesions may be seen side by side. a fever up to 101°F (38°C) may be present during the first day or two. Usually, if not infected by scratching, the lesions heal without leaving scars.

In adults, chickenpox is often a more serious disease, with severe systemic reactions and high fever.

Treatment is supportive. The patient should be isolated, and efforts made to prevent secondary infection.

CHICLE. **Gums and Mucilages; Masticatory Substances; Sapodilla.**

CHICLERO ULCER. **Leishmaniasis.**

CHICORY. ***Compositate* or Composite Family.**

CHIGGER (*Arachnida, Acarina*). This pest is of several species. It is also sometimes referred to as the jigger or the red bug.[1] It attacks humans, domestic and farm animals, poultry, some birds that nest on the ground, frogs, toads, some snakes and turtles, and rodents such as rabbits and squirrels. Because of its size, about 1/50 inch (0.5 millimeter) or less in diameter, this creature is barely visible to the naked eye. It is annoying to all of its hosts, inflicting bites that cause intense itching and small, reddish welts on the skin. The chigger is distributed throughout the United States and much of the world. In warm areas, such as Florida and southern Texas, chiggers may be present throughout the year. In most other areas of the United States, chiggers are most abundant from May to September, to the first frost. Chiggers raised experimentally complete their life cycle—from egg to egg—in about 50 days. The damaging stage is the larva, which feeds on humans and animals. The larva transforms to a nymph and the nymph to an adult, neither of which is a parasite.

The chigger larva is hairy and has three pairs of legs. Its mouth parts include two pairs of grasping palps, which are provided with forked claws. An enlarged view of the chigger larva is rather awesome, considering its real size.

A chigger attached in a pore or at the base of a hair may be so enveloped in swollen skin that it appears to be burrowing into the skin. This fact sometimes leads persons to believe, in error, that a chigger embeds itself in the skin, or that welts contain chiggers. Any welts, swelling, itching, or fever from a chigger attack will be developed within 24 hours. Chiggers attacking in large numbers can cause serious injury to poultry and sometimes can cause the death of young chickens.

Control chemicals for the chigger include diazinon, lindane, and toxaphene. An area can be checked for the presence of chiggers by placing a piece of black cardboard edgewise on the ground or floor. If chiggers are present, they will climb to the top edge and congregate there within a few minutes. If the area is large, several locations should be checked. Where chiggers are suspected, repellents can be sprayed on clothing. Deet (50%) or full-strength dimethyl phthalate or benzyl benzoate can be used. Field workers, campers, hikers, and hunters can gain considerable protection in this manner. A useful formula to obtain temporary relief from the itching caused by chigger welts is: Benzocaine (5%), methyl salicylate (2%), salicylic acid (0.5%), ethyl alcohol (73%), and water (19.5%). Proprietary ointments containing benzocaine are also useful.

The *Eutrombicula* species is most commonly encountered in North America. In the Orient, the species *Trombicula akamushi* (Brumpt) and *T. deliensis* (Walch) are responsible for spreading scrub typhus or tsutsugamushi fever. This disease is caused by *Rickettsia tsutsugamushi.*

CHIGGER BITE. **Dermatitis and Dermatosis.**

CHILBLAIN. Inflammation, accompanied by swelling, painfulness, itching, and redness of the hands or feet due to exposure to cold in persons predisposed to the condition. The causes of such predisposition are unknown.

CHILE CURRENT. **Humboldt Current.**

CHILE PINE FAMILY. **Araucarias.**

CHILE SALTPETER. **Soda Nitre.**

[1] Not to be confused with the chigoe flea. See **Flea.**

CHILL. A paroxysm of shaking or shivering, accompanied by a sense of cold and pallor of the skin. During a severe chill the temperature becomes elevated. A chill may indicate the onset of a disease, often a severe infection, notably lobar pneumonia and malaria.

Nervous chill is shaking or shivering due to excitement, fear, or anger, and is not accompanied by any rise in temperature.

See also **Fever.**

CHILLERS. **Heat Transfer.**

CHILOPODA. The centipedes, usually considered as a class of anthropods related to but distinct from the millipedes and a few rare forms, but sometimes ranked as a class in the subphylum *Myriapoda,* containing all of these forms.

These animals are elongate and slender with numerous segments, most of them bearing a single pair of appendages. They have one pair of antennae and the first pair of legs is modified to form a pair of poison claws with which they catch their prey. They are terrestrial, breathing by air tubes (tracheae) which open separately on the various segments. The body is flattened and the segments are composed of dorsal and ventral plates connected by softer lateral walls which bear the legs. Centipedes have poison glands opening through the poison claws but there is no evidence to show that they are ever dangerous to humans.

CHIMAEROIDS (*Chondrichthyes*). Otherwise termed elephant fishes, ghost sharks, or ratfishes, the chimaeroids are of the subclass *Holocephali* and are cartilage fishes. They would appear to lie midway between sharks and bony fishes, but authorities believe they had shark-like ancestors. There are three major families: (1) short-nosed chimaeras or ratfishes (*Chimaeridae*); (2) long-nosed chimaeras (*Rhinochimaeridae*); and (3) plow-nosed or elephant chimaeras (*Callorhinchidae*). The *Chimaera montrosa* is the most common ratfish of European species, occurring from Norway to the Mediterranean. The *Hydrolagus colliei* (American Pacific ratfish) frequents waters from Alaska to Lower California and seldom attains a length of 3 feet (0.9 meters). They sometimes are a nuisance to fishermen because they can fill up the trawl nets. The term ratfish comes from the long rodent-like tail.

As indicated by the name, the long-nosed rhinochimaerids have a long slender nose something like a stiletto. They prefer deep waters between 2,000 and 8,500 feet (610 and 2590 meters) and thus are difficult to obtain and specimens are quite rare. The plow-nosed chimaeras are also well-named because of their appearance. The plow or hoe shape of the nose differs drastically from the features seen in any other fishes. Only a few species have been reported. The plow-noses frequent the coastal waters of South America, South Africa, New Zealand, Tasmania, and Australia. They may attain a length of about $3\frac{1}{2}$ feet (1 meter) and a weight of 20 pounds (9 kilograms). They have a minor commercial value in the South Pacific.

CHIMERISM. **Hermaphroditism.**

CHIMNEY. A chimney is a vertical tubular structure of masonry, steel, or reinforced concrete, built for the purpose of enclosing a column of hot gas, to produce thereby a draft. Combustion requires oxygen. Air is needed to supply it. To move this air through the fuel bed, and to produce a flow of the gaseous products of combustion through the furnace and boiler, or through a stove, requires a difference of pressure, called draft. The chimney is built primarily to produce a certain available draft, although sometimes a chimney may have to be high for reasons entirely foreign to draft. In addition to the useful draft it produces, a chimney must also overcome the friction loss in the chimney itself. These losses are proportional to the cross-sectional area of the stack. Hence the problem of chimney diameter is more than the assumption of a velocity comparable to that used in actual practice; it should be such that the diameter and height it indicates result in a chimney of least cost. Ordinarily, the economic chimney gas velocities range between 20 and 40 feet per second.

A chimney produces a draft by virtue of an extremely simple principle of thermodynamics. When gas is heated it expands in volume and decreases in density, in which condition it may be displaced by a more dense gas. The light, hot flue gas is confined by the chimney. The tendency of hot gas to move up the chimney is proportional to the height of the stack, since the difference of weight of equivalent columns of air and flue gas (and this is the draft) is greater the higher the columns.

The draft of a stack is, in an elementary way, expressed by:

$$D = \text{height multiplied by difference in density of flue gas and air}$$

Many empirical formulas are advanced for computing the height of a chimney required for a given boiler. A rational scientific approach would necessarily be based on the above equation, as it truly represents the physical action actually creating the draft. When the elementary equation is written for a chimney of 100 feet incorporating certain factors needed to convert draft to inches of water, and allowing for cooling and friction in the stack, it has the form:

$$D = K(d_a - d_f) - 0.0148 d_f \sqrt{\frac{V^5}{F}}$$

where

D = effective draft per 100 feet of stack, in inches of water
K = 17.3 for masonry stacks, and 15.4 for steel stacks
d_a = density of air, in pounds per cubic foot
d_t = density of flue gas, in pounds per cubic foot
V = gas velocity in the stack, in feet per second
F = gas flow, in cubic feet per second

The second term in this equation allows for friction in the stack itself, so that D represents the effective draft per 100 feet of stack. The actual height of a chimney is obtained by dividing required draft by D and multiplying by 100. The required stack draft is the sum of all friction losses external to the stack, plus the impact loss (of gas discharged from stack), and less the effective draft furnished by fans or jets. The impact loss is $0.003\,V^2 d_f$ inches of water, the symbols having meaning as given above.

The use of perforated radial bricks has found much favor in chimney construction. Their dead air space acts somewhat as heat insulation, and they are lighter than ordinary bricks. Comparatively short stacks are frequently made of plate steel. These are lined for a portion of their height with refractory lining, and, unless very short, are braced by suitable guy wires. A tall masonry chimney must have sufficient area in any transverse section to distribute the superimposed weight sufficiently to prevent the unit pressure from exceeding the safe bearing power of the masonry material. The effect of a transverse applied load due to wind pressure cannot be neglected. The effect of wind pressure is to increase the compression in the masonry on the leeward side and decrease it on the windward side. Since tension is not permissible in masonry, not only the thickness of the wall, but the external diameter of the chimney, must be selected with due respect to strength and stability under the condition of maximum wind load. It is very important that the foundation of a tall chimney be absolutely firm and unyielding, as a very small settling on one side would throw the top of a tall chimney several inches out of line and induce an unexpected eccentric loading in the structure.

To be satisfactory, chimneys for residences must, like all other chimneys, have sufficient height for the required draft, and sufficient area to carry off the volume of gases produced. Deficiencies in either or both of these needed characteristics will be sure to cause lazy fires and smoky furnaces. Insofar as possible, a chimney should be straight and perpendicular. Necessary bends should be reduced to the minimum, and corners should be well rounded. The use of a flue lining made of jointed sections of tile is an aid to draft as well as a safeguard against the risk of fire due to defective brickwork in the chimney.

CHIMNEY ROCK. A column of rock, shaped somewhat like a chimney, which rises above its surroundings, or is isolated on the face of a steep slope. The term also is used to describe a small, weathered outlier, shaped like a sharp pinnacle. A stack formed by wave erosion also may be referred to as a chimney rock. *Pulpit rock* is synonymous.

CHIMPANZEE. **Anthropoids.**

CHINA CLAY. A commercial term, more or less identical with kaolin, as applied to the relatively pure clay concentrated by washing from a thoroughly kaolinized granite. England is the chief exporter of china clay. France has unique clays from which are made the famous Sèvres and Limoges potteries.

China-clay rock is a kaolinized granite made up chiefly of quartz and kaolin, with sometimes the presence of muscovite and tourmaline. The rock crumbles easily in the fingers. *China stone* is (1) a partially kaolinized granite, which contains quartz, kaolin, and sometimes mica and fluorite, is harder than china-clay rock and is used as a glaze in the production of china; or (2) a fine-grained, compact mudstone or limestone found in England and Wales.

CHINA TREE. **Mahogany Trees.**

CHINCH BUG (*Insecta, Hemiptera*). One of the major and most damaging of food crop pests, the chinch bug, in some years, causes many millions of dollars in damage to corn (maize) crops in the United States. The insect damages numerous crops of the grass family, including corn, small grains, and wild grasses. The insect occurs throughout the United States and southern Canada, as well as in Mexico and Central America. It is most destructive in the regions of the Mississippi, Missouri, and Ohio River valleys. Illinois, for example, has had numerous serious infestations.

The common form of chinch bug (*Blissus leucopterus*), as shown in Fig. 1, has black-and-white wings. These are strongly developed and quite useful. The hairy chinch bug (*Blissus leucopterus hirtus*, Montandon) is found more commonly in the northeastern United States. This species does not have the strong characteristic of the common form insofar as migrating from one major food source to another is concerned. The hairy chinch bug prefers continuing vegetation as provided by lawns and grass pastures to going from one plant to another, as required in a grain field.

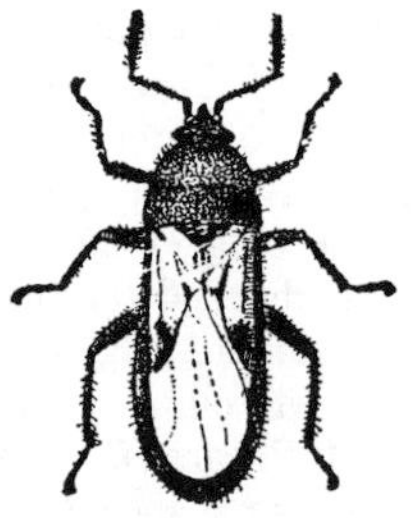

Fig. 1. Chinch bug. Actual length is about 0.25 inch (6 millimeters).

The life cycle of the common chinch bug is aptly depicted in the chart of Fig. 2.

If outbreaks of the chinch bug can be determined sufficiently early in the season, it is well to consider raising more nongrass crops during the period. The latter not only are not attacked by the chinch bug, but the areas so planted are thus subtracted from the available area for chinch bug breeding and growing. Care should be taken to avoid planting corn (maize) next to a field that is or has been planted with small grain. For example, fields of winter and spring wheat, barley, and oats are an ideal habitat for the multiplication of the chinch bug population. These matters should be taken into serious consideration in planting a year-to-year crop rotation program. Concurrent planting of soybean or cowpea with corn (maize) can produce a beneficial effect, resulting from the shade of these other plants on the base of the corn plant. The chinch bug prefers sunny locations and the absence of sunlight tends to slow down their population expansion. In the western area of the United States (west of the Mississippi River), the chinch bug usually winters in bunch grasses. Burning over such areas is sometimes done, although it is estimated that this may destroy only about one-fourth of the insects in the infested area. It is definitely a measure to be considered, provided that it can be accomplished safely.

Chinch bugs can be trapped by constructing a barrier line which the insect must cross to reach a corn (maize) field from a small grain planting. The bugs are repelled by the odor of creosote. Thus by creating a barrier line an inch or so wide on the surface of the soil,

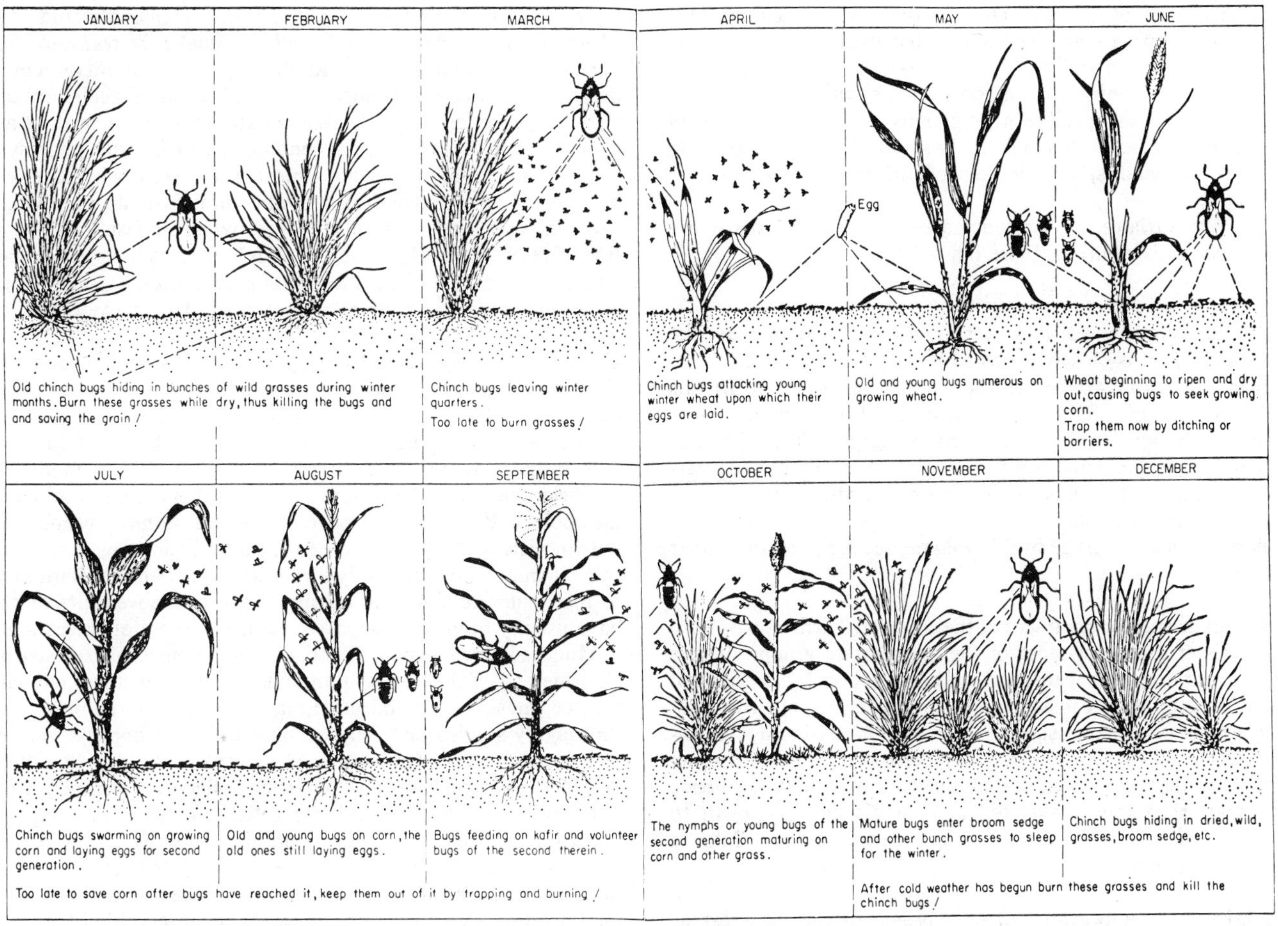
JANUARY
FEBRUARY
MARCH
APRIL
MAY
JUNE
Egg
Old chinch bugs hiding in bunches of wild grasses during winter months. Burn these grasses while dry, thus killing the bugs and and saving the grain!
Chinch bugs leaving winter quarters.
Too late to burn grasses!
Chinch bugs attacking young winter wheat upon which their eggs are laid.
Old and young bugs numerous on growing wheat.
Wheat beginning to ripen and dry out, causing bugs to seek growing corn.
Trap them now by ditching or barriers.
JULY
AUGUST
SEPTEMBER
OCTOBER
NOVEMBER
DECEMBER
Chinch bugs swarming on growing corn and laying eggs for second generation.
Old and young bugs on corn, the old ones still laying eggs.
Bugs feeding on kafir and volunteer bugs of the second therein.
Too late to save corn after bugs have reached it, keep them out of it by trapping and burning!
The nymphs or young bugs of the second generation maturing on corn and other grass.
Mature bugs enter broom sedge and other bunch grasses to sleep for the winter.
Chinch bugs hiding in dried, wild, grasses, broom sedge, etc.
After cold weather has begun burn these grasses and kill the chinch bugs!

Fig. 2. Seasonal life cycle of common chinch bug. From December to February, the bugs hibernate at edges of wooded areas, notably in clusters of grass under fallen leaves and other natural debris. In late winter and early spring (February–March), the bugs fly to young grain fields (barley, oats, wheat, etc.), a migrating process that may continue until about mid-May. In early and mid-summer (June and July), the bugs crawl *en masse* from ripening grain (such as wheat) to tender, young corn (maize) plants. It is during this migration that ditches, traps, and chemical barriers can be effective. Upon achieving adulthood (about August 1st), the bugs move throughout the corn (maize) crop, feeding and laying eggs for a second generation. The August eggs develop into adults during late September through early November and these immediately seek hibernating locations for wintering over until the following spring.

soaked with cresote, the bugs will be turned back. To establish such a line with a persistent repelling effect, several applications of the chemical may be necessary. A further method is that of digging holes on the inside of the barrier, along the border. Thousands of bugs can be trapped in these holes. Small amounts of chemical insecticide can be placed in the holes to kill the bugs as they fall in. Where the soil is gravelly, tarred felt paper strips several inches wide that have been saturated with cresote also can be used. Barriers of toxic dust also are used. These bands of chemicals, however, are easily destroyed by wind and rain.

CHINCHILLA. Rodentia.

CHINESE CABBAGE. Brassica.

CHINESE CEDAR OR TOON. Cedar Trees.

CHINESE FIR TREE. Araucarias; Fir Trees.

CHINESE PLUM YEW. Yew Trees.

CHINESE REMAINDER THEOREM. Number Theory.

CHINOOK. Winds and Air Movement.

CHINOOK SALMON. Salmon.

CHIPMUNK. Squirrels and Other Sciuromorphs.

CHIP (Silicon). Electron Beam Lithography; Microelectronics; Microstructure Fabrication; Semiconductor; Telephony.

CHIRALITY. Isomerism.

CHIROPTERA. Bats.

CHIRU. Goats and Sheep.

CHI-SQUARE. If a sample of n values is drawn from a normal distribution with variance σ^2, and if the variance is estimated as

$$s^2 = \sum (x - \bar{x})^2/(n-1)$$

then the ratio $(n-1)s^2/\sigma^2$ is known as χ^2, and has the probability function

$$P(\chi^2)\,d\chi^2 = \frac{1}{\Gamma(\frac{1}{2}\nu)}\,e^{-(1/2)\chi^2}\left(\frac{1}{2}\chi^2\right)^{\nu-2/2} d\left(\frac{1}{2}\chi^2\right) \qquad 0 \le \chi^2 \le \infty$$

when $\nu = (n-1)$ is the degrees of freedom. This is a special case of the Pearson Type III function; the probability integral is an incomplete gamma function.

If an hypothesis completely specifies the frequencies f_t to be expected in n classes, the goodness of fit of a sample providing frequencies f_o can be tested by calculating

$$\chi^2 = \sum \frac{(f_o - f_t)^2}{f_t}$$

and entering a table of χ^2 with $(n - 1)$ degrees of freedom. This quantity closely approximates the above distribution provided that none of the f_i's are too small—say, less than 5; to avoid this it is often necessary to combine two or more classes with small expectations. If the hypothesis specifies a probability distribution in which p parameters have to be estimated from the sample in order to compute the expected frequencies, the degrees of freedom are reduced to $(n - p - 1)$.

With discrete variates, χ^2 provides a sensitive test of departure from the Poisson distribution. If we put

$$\chi^2 = (n - 1)s^2/\bar{x}$$

where $\bar{x}$ is the sample mean and n the number of observations the standard distribution is closely followed for a true mean ≥ 2.

A similar test can be applied to the binomial distribution. In this form, χ^2 is sometimes referred to as the index of dispersion.

The χ^2 distribution tends to normality for large values of ν. The usual tables do not extend above $\nu = 30$; for larger values, it is ordinarily sufficient to take $\sqrt{2\chi^2}$ as normally distributed with mean $\sqrt{2\nu - 1}$ and unit standard deviation.

For related topical coverage in this volume, see list of entries under **Mathematics.**

CHITIN. An essential constituent of the cuticula of arthropods. Also found to a lesser extent in most of the invertebrate groups. It is a material of variable and intricate chemical composition. It is a principal component of the insect exoskeleton and appears in both the rigid and flexible parts, which are usually said to be chitinized or not chitinized; the degree of rigidity has recently been said to depend upon other materials deposited with the chitin. Chitin is inelastic, hence the arthropods shed the exoskeleton at intervals during growth to permit expansion during the formation of a new covering. Chitin is chemically very inactive.

CHLOASMA. The appearance of light brown patches of irregular shape and size on the skin surface, sometimes associated with endocrine imbalance. The disease usually affects only women. The patches or darkened skin are found chiefly on the face, the nipple, and the external genitalia. The condition frequently occurs in pregnant women, in which cases it disappears after the pregnancy ends. Chloasma may also appear during the menopause and accompany abnormal conditions of the ovaries and female reproductive organs. In many cases, however, the disease may occur without such internal complications.

CHLORAL. **Chlorinated Organics.**

CHLORAMPHENICOL. **Antibiotic.**

CHLORARGYRITE. Also known as horn silver, chlorargyrite is silver chloride, AgCl. The mineral crystallizes in the isometric system but is usually massive, appearing like wax or horn, hence the name. It has no cleavage, is highly sectile, yielding bright surfaces; hardness, 2.5; specific gravity, 5.55; luster, resinous to adamantine; color, gray, white to colorless. May be blue, violet-brown after exposure to light; transparent to translucent. Chlorargyrite is largely a secondary mineral, usually associated with other silver minerals as well as with compounds of lead, zinc, and copper. Saxony and the Harz Mountains are European localities. The Broken Hill district of New South Wales is a well-known occurrence, but probably the most important deposits are found in Atacama, Chile. The mineral also is found in Bolivia and Mexico. In the United States, chlorargyrite comes from Colorado, Idaho, Utah, Nevada, Arizona, and New Mexico.

CHLORELLA. A genus of unicellular green algae. Several species of this genus will grow very well in artificial media and have been used for many years in experiments on metabolism in plant cells. These green cells have been especially valuable in studying photosynthesis. Since the organism is unicellular, many of the complications caused by the complex structure of leaves are avoided.

Within the algae there is a complete nutritional spectrum from obligate heterotrophy (requirement for organic compounds) in colorless form, to obligate autotrophy (utilization of carbon dioxide). Close to the middle of the spectrum is the most commonly studied genus, *Chlorella.* Some species of *Chlorella* can make the transition between dark assimilation of glucose and photosynthetic assimilation of carbon dioxide (and vice versa) without any of the lag expected of adaptive enzyme formation. Considerable variability in metabolism of any one alga can be induced by special conditions. In *Chlorella,* photosynthetic metabolism can be switched to almost exclusive carbohydrate synthesis by a preceding period of dark starvation or light-limited growth. Nitrogen deficiency leads to shunt or overflow metabolism, and accumulation of reserve materials, such as starch, various other exotic polysaccharides, and fats. The most common forms of algae found in swimming pools are *Chlorella variegata* and *C. pyrenoidosa.*

CHLORIDE (Biological Aspects). Sodium chloride, potassium chloride, and other chloride salts, when ingested by animals from feedstuffs and humans from various food substances, reduce to a consideration of the cation involved (Na^+, K^+, etc.) and the Cl^- (chloride) ion. Generally, in terms of animal and human nutrition, more research has been conducted and more is known about the role of cations in metabolism than that of the chloride ion. Some physiologists and nutritionists in the past have described chloride as playing a "passive role" in maintaining the body's ionic and fluid balance. With exception of the "chloride shift" in venous blood, the movements of chloride have usually been considered secondary to those of the cations.

Much is known, of course, concerning the effects of excessive sodium chloride and of deficient sodium chloride in human and animal diets, but the physiological and nutritional roles of chloride have not been thoroughly studied and fully explained. C. E. Coppock, Department of Animal Science, Texas A & M University, College Station Texas; and M. J. Fettman, Cornell University College of Veterinary Medicine, undertook a study of chloride (targeted to chloride as a required nutrient for lactating dairy cows) and also carefully reviewed the prior work in this area of other researchers. (Interested readers are referred to the bibliography at end of article, "Chloride as a Required Nutrient for Lactating Dairy Cows," *Feedstuffs* (February 10, 1978).)

Despite important physiological functions and its presence in milk at about 0.11%, chloride is a neglected element in large animal nutrition. The practice of adding sodium chloride to concentrate mixtures and free-choice feeding seems to have precluded the possibility of a practical deficiency problem. When salt was omitted from the diet, researchers found that under the conditions used in their study, sodium was the first limiting element. This was true because sodium is present in most natural ingredients at much lower levels, relative to the cow's requirements, than is chloride.

Those who formulate diets usually ignore sodium levels in the natural ingredients (which are usually low in forages, but may be appreciable in certain concentrate ingredients) because of the traditional value of salt as a condiment. In addition, other sodium salts are often included in concrete mixtures: sodium sulfate as a sulfur replacement, the sodium phosphates as phosphorus supplements, and sodium bicarbonate as a buffer. Even when these supplements are used, salt is still often included because of tradition. Under many conditions, gross overfeed of both sodium and chloride occurs. High dietary levels of sodium and chloride do not increase the levels of these elements in milk. Excess will be excreted in urine and manure. Excess salt intake may result in greater water consumption, waste transport, bedding requirements, and transfer of sodium and chloride to the soil.

Of the seven macro mineral elements required by dairy cattle, five can be considered fertilizer elements (potassium, calcium, phosphorus, magnesium, and sulfur), but sodium and chloride are both toxic to plants at high concentrations and present practical problems in areas with saline soils. High salt intakes have also been shown to increase udder edema in heifers. Because of the importance of chloride in nutrition and metabolism, research is needed to define the chloride requirements of lactating cows and clarify mineral relationships, especially between chloride and potassium plus sodium.

Chloride and Plants. Chloride is one of the most recent elements to be shown essential for plant growth. In 1954, Broyer et al. presented evidence that tomato plants grown in a low-chloride solution developed wilting of leaflet blade tips, which progressed to chlorosis, bronzing, and necrosis. Growth was proportional to chloride concentration (up to a point) in the culture medium. Chloride additions to the medium prevented the deficiency symptoms and caused their disappearance in deficient plants. Later, the same team showed that often other species, including barley and alfalfa, displayed severe deficiency symptoms when grown in a low-chloride medium. Although buckwheat, corn (maize), and beans did not exhibit these obvious symptoms, yield effects were apparent.

Despite the presence of 250 parts per million (ppm) in the leaves of chloride-deficient tomato plants compared to a 0.1 ppm for molybdenum, chloride was classified as a micronutrient by plant physiologists. The essentiality of chloride escaped detection for so long because of its wide distribution in nature, the high solubility of most chloride salts, and difficulties of purification. Because chloride is a principal ion of sea water, it is picked up by winds from sea spray and carried far inland. For example, Geneva, New York has been estimated to receive 18 kilograms/hectare of chloride annually; Mount Vernon, Iowa, 73 kilograms/hectare annually. Deposits of more than 45 kilograms/hectare have been reported near coastlines. It was also suggested that leaf structures were capable of capturing airborne chloride. The popularity of potassium chloride as a potassium fertilizer and manure from cows fed excessive levels of chloride relative to their requirement are additional sources of soil chloride.

According to Stout and Jackson, chloride differs from other nutrient elements present in native rocks because it is not fixed by colloids; it is repelled by negatively charged clay surfaces, and all chloride compounds formed in soils are highly soluble. In addition to leaching, chloride is lost from soils through crop removal.

Several factors affect chloride uptake by plants: (1) age or advancing maturity shown in avocado, apricot, and grape leaves; (2) chloride concentration in the soil; (3) soil oxygen levels; (4) plant species; and (5) competition from other anions. Muraka and others have shown a specific antagonism between nitrate-nitrogen and chloride; increasing chloride levels in the growth medium reduces nitrogen uptake, but this reduction is primarily in the nitrate-nitrogen fraction, with little if any effect on the protein-nitrogen fraction. The reciprocal effect was also observed between chloride and sulfate accumulation. Chloride is essential in the plant for photosynthetic reactions in chloroplasts which produce oxygen.

Chloride is also treated as a toxic element as well as a part of saline toxicity. Over the toxic range, the reduction in growth is approximately linear. For example, at about 3500 ppm chloride in the cultural medium, alfalfa growth will be depressed to about 60% that of normal. Obviously, in areas with saline soils, there is concern about excessive levels of salt returned to the soil via manure.

Gastrointestinal Absorption of Chloride. Since 1952, when in goats, sheep, and dairy cattle, the observation was made that chloride could be absorbed from ruminal fluid into the blood against a tenfold concentration difference (normal rumen fluid chloride concentrations may range from 10 to 30 mEq/l, while those of the plasma may range from 100 to 110 mEq/l), numerous researchers have attempted to describe accurately and explain the processes by which chloride might move across the reticulorumen epithelium against its apparent chemical gradients. For a number of years, it was assumed by some workers that the observed electrical potential difference across the forestomach epithelium, making the plasma approximately 30 mV positive to the contents of the reticulorumen, could adequately account for the otherwise anomalous movements. If chloride's movement into the blood was truly attributable solely to the combined electrochemical gradient acting upon it, then its distribution across the gastric epithelium should have been describable by the Nernst equation.

In rumen-fistulated experimental animals, it has been observed that for certain distribution ratios of chloride in the ruminal fluids and blood plasma, the calculated equilibrium potential for chloride is relatively the same as that measured directly with KCl–agar bridges and calomel electrodes. However, in many circumstances, the calculated and measured values have been found to be significantly different, an observation that could only be accounted for by the presence of an active transport mechanism responsible for the movement of chloride out of its equilibrium distribution.

The active transport of chloride has also been demonstrated across the wall of the frog stomach, rat ileum, dog ileum, and the human ileum. Turnberg et al. produced a double exchange model by which bicarbonate secretion and chloride absorption are linked by an isoelectric mechanism to hydrogen ion secretion and sodium absorption across the human ileum. Chien and Stevens, in 1972, proposed a similar model of coupled transport across the reticulorumen epithelium, and further concluded that active anion and cation transport in the rumen cannot function efficiently unless both components are intact, i.e., the net transport of the body's major cations from the rumen to the blood, appeared to be dependent in part on the activity of chloride in the system.

Chloride in Cerebrospinal Fluid. A similar story has unfolded concerning the distribution and movement of chloride between the blood and cerebrospinal fluid (CSF). The first indication for the possible existence of an active mechanism responsible for the maintenance of chloride levels in the CSF came almost 40 years ago when in dogs, Hiatt demonstrated the persistence of CSF chloride concentrations at 44% of normal, despite the reduction of chloride levels to 30% of normal in all body fluids during a nitrate-induced diuresis. Over the next 20 years, researchers recorded chloride ion concentration and electrical potential differences across CSF–ECF (extracellular fluid) and CSF–blood barriers ranging from 15 to 20% and from 5 to 30 mV, respectively, in such varied subjects as dogfish, rats, cats, dogs, monkeys, and humans. In all cases, the CSF was both higher in chloride ion concentration and negative in potential with respect to the reference body fluid.

In 1970, Bourke et al. studied the distribution and kinetics of chloride in cats following the isoosmotic replacement of body chloride with isethionate via extracorporeal hemodialysis. They found that when the plasma chloride concentration was reduced by approximately 93%, the cerebral cortex, corpus callosum, and CSF chloride concentrations were reduced by approximately only 26.5, 35, and 21%, respectively. Other body tissues and fluids showed reductions in chloride concentration closer to those of the plasma (skeletal muscle and liver, 73%). The influx of chloride into the CSF at various plasma chloride concentrations was then plotted as a Lineweaver–Burk plot, and was shown to behave as a carrier-mediated process, as described by Michaelis–Menten kinetics. This information, combined with the observations made by Abbot et al. that reduction of the plasma chloride concentration by isethionate replacement did not produce a change in electrical potential "commensurate with or even in the same direction" as that expected by Nernst equation predictions, led workers in the field to ascribe the bulk of chloride movement from the blood into the CSF to an active transport process. It is possible that control of the rate of chloride transport is a factor in the regulation of the secretion of CSF, the medium that bathes, protects, and nourishes the central nervous system.

Chloride in the Humoral Regulation of Sodium and Potassium. Conventional presentations of the regulatory mechanisms involved in body fluid and electrolyte homeostasis usually have considered maintenance of the sodium/potassium ratio in the ECF both the prime means and end toward a functional electrolyte balance. Certainly, the sodium/potassium ratio provides the axis about which the body's humoral mechanism of electrolyte homeostasis revolves, represented mainly by the renin–angiotensin–aldosterone system. Aldosterone increases the activity of sodium retaining processes in the body. These include the active uptake of sodium from the gastrointestinal tract and the reabsorption of sodium from the renal tubes in exchange for potassium or hydrogen ion.

Given chloride's role in ruminal absorption and CSF secretion, perhaps its participation in the aldosterone mechanism of regulating the sodium/potassium "axis" should come as no surprise.

Upon detection of decreased blood pressure, volume, and/or sodium, the juxtaglomerular apparatus in the kidney secretes renin, an enzyme which then cleaves a decapeptide, angiotensin I, from a plasma α_2-globulin, angiotensinogen. A converting enzyme present in the plasma, and most abundant in the pulmonic circulation, cleaves a dipeptide from angiotensin I, thus forming angiotensin II. The effect of angiotensin II, potentiated by ACTH and high plasma potassium, is to induce the secretion of aldosterone by cells in the zona glomerulosa (arcuata) of the adrenal gland cortex. Aldosterone than exerts its effects on the sweat glands, salivary glands, intestinal mucosa, and distal convoluted tubules of the kidneys, promoting sodium absorption and retention. Because of its potent vascoconstrictive properties (40 times that of norepinephrine) angiotensin II can effectively reduce both renal blood flow and glomerular filtration rate, leading to an immediate decrease in excretion of water and electrolytes, before aldosterone can affect tubular sodium reabsorption.

Research has demonstrated that chloride is not only responsible in part for angiotensin II formation, but also for its deactivation or catabolism by the major angiotensinase of the body. Furthermore, chloride ion's relations to angiotensin II may not be its only route to affecting aldosterone secretion and sodium/potassium balance. Chloride's role in the metabolism of ACTH has led to support for the hypothesis that fluid and electrolyte homeostatic mechanisms may revolve not just around the sodium/potassium ratio, but also around the levels of chloride in the body.

See also **Blood; Sodium;** and **Sodium Chloride.**

CHLORINATED ORGANICS. Organic compounds containing chlorine are valued as reagents and intermediates in chemical synthesis and for their commercial and industrial importance. Several are produced in high tonnages. The large volume market for these compounds is in plastics, including vinyl chloride for polyvinyl chloride (PVC), or as a copolymer with vinyl acetate; vinylidene chloride for *Saran*; and chloroprene for neoprene. Other important uses include agricultural chemicals, solvents, plasticizers, and medicines. Uses as intermediates to produce other chemicals also are important and varied. The largest volume chlorine organic is ethylene dichloride (EDC). See Fig. 1. About 11 billion pounds (5 billion kilograms) per year of EDC are produced, but over half of this is consumed by the producers to make vinyl chloride monomer. Nearly all of the ethyl chloride produced is used in making tetraalkyl leads. Methyl chloride is the intermediate for many chemicals, as are benzyl chloride, phosgene, and chloroform.

Fig. 1. Portion of plant for oxychlorinating ethylene to ethylene dichloride (EDC). (*Ethyl Corporation.*)

The chlorine on certain compounds is used as a facile leaving group for the introduction of another functional group. The displacement of a halogen atom by a cyano group to form a nitrile is one of the most useful reactions of halogen compounds. This opens a route to carboxylic acids having one carbon atom more than the original halide, aside from the importance of the nitriles themselves. Adiponitrile, used in the manufacture of nylon, can be made by treatment of 1,4-dichlorobutane with cyanide:

$$\begin{matrix} CH_2CH_2Cl \\ | \\ CH_2CH_2Cl \end{matrix} \xrightarrow{NaCN} \begin{matrix} CH_2CH_2CN \\ | \\ CH_2CH_2CN \end{matrix}$$

Long-chain alkyl chlorides can be used for the synthesis of various amines, while benzyl chloride is used for production of quaternary ammonium compounds. Alkyl chlorides are used for the formation of organometallics, including the Grignard reagents as well as for alkylation of aromatics. One of the important reactions of phosgene is with diamines for production of diisocyanates (polyurethanes).

Synthesis of Chlorinated Organics. Chlorine derivatives or organic compounds are obtained by substitution, addition, or displacement. Substitution reactions of Cl on hydrocarbons involve radical attack to remove a hydrogen, forming the hydrocarbon radical as an intermediate: $R—H + Cl\cdot \rightarrow R\cdot + HCl$. Since a tertiary carbon radical $—\overset{|}{\underset{|}{C}}\cdot$ is most stable, it chlorinates more readily than a secondary carbon $—\overset{|}{C}H_2$ and that more readily than a methyl group. Due to inductive effects, the presence of chlorine in a molecule reduces the activity of the hydrogens on adjacent carbons more than on the chlorinated carbon. Thus, a second radical ($Cl\cdot$) will preferentially attack a hydrogen on the same carbon. For example, chlorination of ethyl chloride will produce nearly twice as much 1,1-dichloroethane as 1,2-dichloroethane. Specificity of this sort is decreased at higher temperatures, leading to more random substitution. Hydrogens further away on a longer-chained molecule are essentially unaffected by the first chlorine, therefore little selectivity occurs for subsequent substitutions.

Chlorine can be substituted onto an aromatic ring in the presence of a catalyst, such as ferric chloride, $FeCl_3$, or aluminum chloride, $AlCl_3$. The simplest case would be chlorination of benzene. Substitution of a second Cl onto the ring preferentially goes to the para position, but the ortho and meta isomers can be formed with the latter least favored. If the chlorination is carried out in the presence of a radical source, such as ultraviolet light, addition occurs instead. See *Chlorinated Aromatics* described later in this entry. When a functional group is present on the aromatic ring, Cl attack will depend upon the type of group present. Phenol and benzoic acid will chlorinate on the ring to give chlorophenol and *p*-chlorobenzoic acid. Alkyl benzenes will chlorinate on the alkyl group if a radical source is present. In the presence of an iron catalyst, the product is a mixture of the ortho- and para-chloroalkylbenzenes:

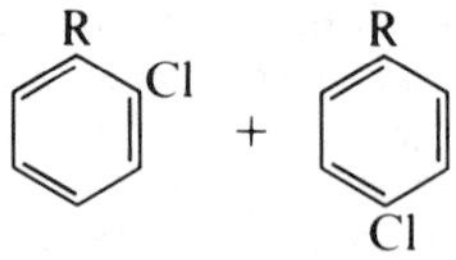

With higher aromatics, such as naphthalene, chlorine successively substitutes all of the hydrogens. The first product is α-chloronaphthalene and the final compound is perchloronaphthalene.

Chlorine addition occurs on unsaturated hydrocarbons having double or triple bonds. Addition can occur by use of Cl_2, HCl, or HOCl:

$$>C=C< + ClX \longrightarrow Cl\overset{|}{\underset{|}{C}}-\overset{|}{\underset{|}{C}}X$$

where $X = Cl$, H, or OH.

With ethylene, the products would be 1,2-dichloroethane, ethyl chloride, or ethylene chlorohydrin. When the unsaturated molecule has three or more carbons, HCl will add, preferentially, with the Cl on the carbon having the fewest hydrogens. For addition of HOCl, the opposite is favored.

Displacement occurs when a functional group is replaced. Chlorine can displace groups, such as hydroxyl, OH, in an acid-catalyzed reaction. For example, methyl chloride can be prepared from the reaction of HCl on methanol. Other alkyl chlorides can be made from their corresponding alcohols. Another type of displacement would be the exchange of one halogen for another, e.g., Cl can be substituted for bromine or iodine in a molecule. An acyl chloride can be formed by the reaction of a strong dehydrating Cl carrier, such as PCl_3, PCl_5, POCl, or $SOCl_2$, with an organic acid or its salt.

Industrially, chlorinations are carried out in five ways: (1) radical substitution of hydrogens; (2) molecular Cl addition across unsaturated (double or triple) bonds; (3) HCl addition across an unsaturated bond; (4) HCl reaction with an alcohol; and (5) oxychlorination. The latter reaction is similar to producing molecular chlorine, *in situ*, from HCl and air in the presence of a catalyst. An example of this is the production of ethylene dichloride, 1,2-dichloroethane, from ethylene, HCl, and oxygen.

Characteristics of Chlorinated Organics

The presence of Cl in an organic molecule increases the density, viscosity, and chemical reactivity, while decreasing the specific heat, solubility in water, and flammability. Chlorine is normally an excellent leaving group, particularly in base-catalyzed reactions, which makes it important for syntheses. Toxicity is the principal hazard. Threshold Limit Values, established by the American Conference of Governmental Industrial Hygienists, for tetra- and pentachloroethane are 5 ppm (vol.) in the atmosphere. Corresponding values for CCl_4, $CHCl_3$, and perchloroethylene are 10, 50, and 100 ppm, respectively. Chloroacetylenes are highly explosive, especially in contact with caustic, e.g., NaOH.

Safety and Handling. Chlorinated organics are absorbed through the skin and lungs and can seriously damage vital organs, especially the liver. Therefore, they should be handled with rubber gloves and in well-ventilated areas. When these materials are subject to burning, they have the potential of forming hydrochloric acid and phosgene, $COCl_2$, besides carbon monoxide. In highly chlorinated compounds, such as carbon tetrachloride and perchloroethylene, there is some danger of forming phosgene in a fire, or from high heat. Compounds that have sufficient hydrogen to combine with any Cl released, such as methyl chloride, vinyl chloride, and ethyl chloride, will form large amounts of hydrochloric acid. Although some small amounts of phosgene may be produced, the hydrogen chloride will naturally drive personnel away from such a fire.

Types or Families of Chlorinated Organics

Chlorinated Paraffins. Cl will displace one, two, three, or more hydrogens from the paraffins. These substitution products are referred to as *mono* ($C_nH_{2n+1}Cl$), *di* ($C_nH_{2n}Cl_2$), *tri* ($C_nH_{2n-1}Cl_3$), and so on. *Monochloro* derivatives include methyl chloride, CH_3Cl, ethyl chloride, C_2H_5Cl, and propyl chloride, C_3H_7Cl. These are also called *alkyl chlorides*. Examples of *dichloro* compounds include methylene dichloride, CH_2Cl_2, and ethylene dichloride, $C_2H_4Cl_2$. Chloroform, $CHCl_3$, and 1,1,1-trichloroethane, $C_2H_3Cl_3$, are *trichloro* derivatives, while carbon tetrachloride, CCl_4, is a *tetrachloro* molecule. When all the hydrogens are substituted by Cl, the term *perchloro* is sometimes used.

Chlorinated Carbonyls. Chlorination of an aldehyde or ketone occurs most readily on a carbon next to the carbonyl function. This is due to proton interaction with the carbonyl and is acid catalyzed. Reaction of Cl with acetone yields chloroacetone. Substitution of a second Cl on chloroacetone occurs with no preference for sites. Thus, equal amounts of 1,1-dichloro and 1,3-dichloroacetone are produced. (The opposite is true when brominating, since it is possible to form nine parts of 1,3-dibromo to one of 1,1-dibromacetone.) Chloral is produced from acetaldehyde. It is also produced by hydrolysis of trichlorodiethyl ether. Acrolein reacts with dry HCl at low temperatures to give β-chloropropionaldehyde.

The chlorinated acetones are strong lachrymators. Tear gas contains chloroacetophenone, which is also a component of the nonlethal disabling spray, Chemical Mace.

Chlorination of diketene yields α-chloroacetoacetyl chloride:

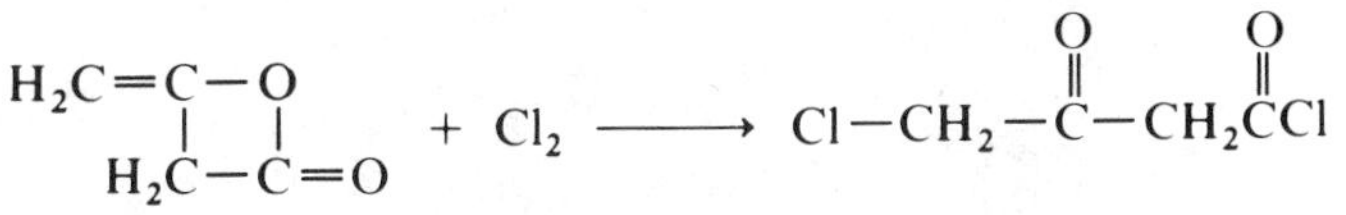

This product is both a vesicant (blistering agent) and a lachrymator.

Chlorinated Fatty Acids. Chlorination of carboxylic acids is much more difficult because the contribution of the carbonyl group toward proton removal is offset by the electron donation effect from the hydroxyl group. This hindrance is obviated by reaction with the acid chloride or anhydride. Chlorination is normally accomplished by use of a catalyst, such as phosphorus trichloride. Monochloroacetic acid is an important industrial chemical. Dichloro- and trichloroacetic acids can be produced by further chlorination, although the latter can be produced conveniently by nitric acid oxidation of chloral. Higher chlorinated fatty acids can be produced by treatment of the hydroxy carboxylic acid or ester with HCl or PCl_5:

$$CH_3CHOHCH_2COOR + PCl_5 \rightarrow CH_3CHClCH_2COOR$$

Amino fatty acids can be treated with a mixture of nitric oxide and chlorine to produce the corresponding chloroacid. Mono- and dichlorosuccinic acids are examples of chlorinated dicarboxylic acids.

Chlorinated Ethers. Ethylene chlorohydrin reacts with sulfuric acid to form β,β′-dichloroethyl ether. It is a by-product of ethylene glycol production. The chlorines on this ether are inert making it a good solvent. Further chlorination at 20–30°C gives α,β,β′-trichloro diethyl ether which hydrolyzes to chloroacetaldehyde and ethylene chlorohydrin. Ethylene and sulfur monochloride react to give β,β′-dichlorodiethyl sulfide (mustard gas), which is a thioether.

Chlorinated Aromatics. Chlorination of benzene in the presence of a catalyst ($FeCl_3$ or $AlCl_3$) yields chlorobenzene as the first product. Substitution with a second Cl yields ortho, para, or meta dichlorobenzene. Eventually all the hydrogens can be substituted to give hexachlorobenzene, C_6Cl_6. In the presence of ultraviolet light, the chlorination of benzene yields benzene hexachloride, $C_6H_6Cl_6$, a derivative of cyclohexane. Under the same conditions toluene chlorinates on the methyl group to give one, two or three substitutions (benzyl chloride, benzal chloride or benzotrichloride), while in the presence of an iron catalyst, one obtains ortho- and parachlorotoluene. See Fig. 2.

Chlorinated Heterocyclics. Substitution in pyridine is more difficult than in benzene but Cl will enter the β position slowly. Chlorine will not add to furan to give stable addition products but substitution occurs to give 2-chloro- or 3-chlorofuran, 2,5-dichlorofuran, and 2,3,5-trichlorofuran.

Important Specific Chlorinated Organic Compounds

Several thousand chlorine-containing compounds are known and have been synthesized. A select group is included for description here to provide a cross section of the most important of these compounds. The number in brackets following each heading, where appropriate, is the *Chemical Abstracts Service Registration* number.

Acetyl chloride ($CH_3\overset{\text{O}}{\overset{\|}{C}}Cl$) [75-36-5]. Acetyl chloride can be prepared by treatment of acetic acid with various reagents, such as PCl_3, $SOCl_2$ or $COCl_2$. It can be prepared by chlorination of acetic anhydride in several different ways, by reaction of methyl chloride with carbon monoxide in the presence of catalysts, by reaction of ketene ($H_2C{=}C{=}O$) with HCl, or by partial hydrolysis of 1,1,1-trichloroethane. Acetyl chloride hydrolyzes in the presence of water to give acetic acid. It reacts with ammonia and amines to give acetamides: $CH_3\overset{\text{O}}{\overset{\|}{C}}Cl + RNH_2 \rightarrow CH_3\overset{\text{O}}{\overset{\|}{C}}NHR$. Reaction with alcohols gives the corresponding acetate esters. Acetyl chloride will add across unsaturated bonds in the presence of suitable catalysts to give halogenated ketones:

Fig. 2. Industrial unit for chlorination of benzene to produce chlorobenzene. (*Dow Chemical U.S.A.*)

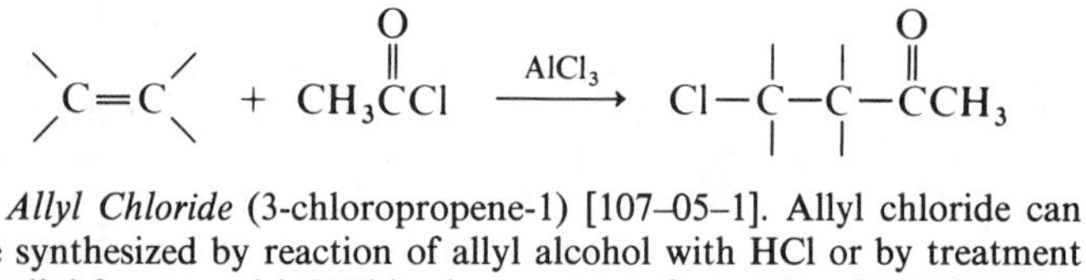

Allyl Chloride (3-chloropropene-1) [107–05–1]. Allyl chloride can be synthesized by reaction of allyl alcohol with HCl or by treatment of allyl formate with HCl in the presence of a catalyst ($ZnCl_2$). Commercial production is by chlorination of propylene at high temperatures, about 500°C, using a large excess of propylene. It is used in the synthesis of glycerol, allyl alcohol and epichlorohydrin. Since the chlorine is situated alpha to a double bond, it is particularly reactive. Thus, hydrolysis to allyl alcohol occurs rapidly in dilute caustic at about 150°C. Addition of HOCl followed by treatment with an alkali yields epichlorohydrin:

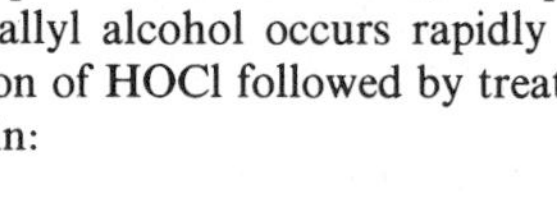

Addition of HBr in the presence of an oxidizing agent yields 1-chloro-3-bromopropane, which is used to prepare cyclopropane. Allyl chloride is one of the most toxic of the chlorinated organics.

Benzoyl Chloride [98–88–4].

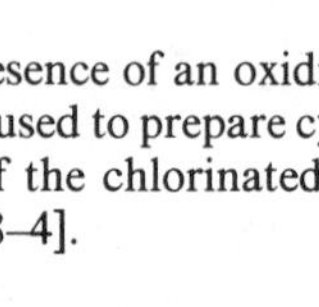

Benzoyl chloride can be prepared from benzoic acid by reaction with PCl_5 or $SOCl_2$, from benzaldehyde by treatment with $POCl_3$ or SO_2Cl_2, from benzotrichloride by partial hydrolysis in the presence of H_2SO_4 or $FeCl_3$, from benzal chloride by treatment with oxygen in a radical source, and from several other miscellaneous reactions. Benzoyl chloride can be reduced to benzaldehyde, oxidized to benzoyl peroxide, chlorinated to chlorobenzoyl chloride and sulfonated to *m*-sulfobenzoic acid. It will undergo various reactions with organic reagents. For example, it will add across an unsaturated (alkene or alkyne) bond in the presence of a catalyst to give the phenylchloroketone:

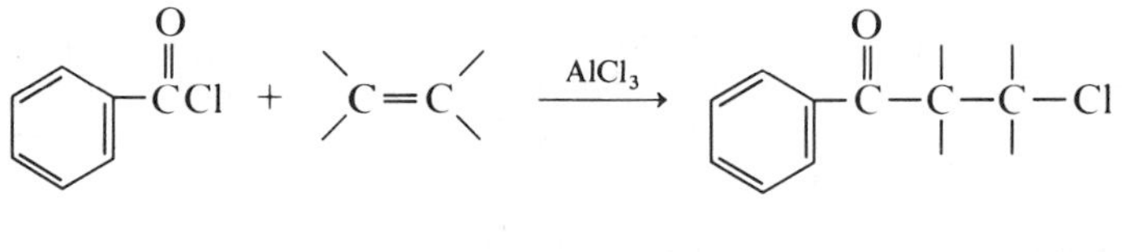

Reaction with benzene yields benzophenone while toulene gives phenyl-*p*-tolyl ketone. Reaction of benzoyl chloride with monohydric alcohols gives the corresponding alkyl ester, but with phenols the product can either be the phenylbenzoate or a phenolic ketone:

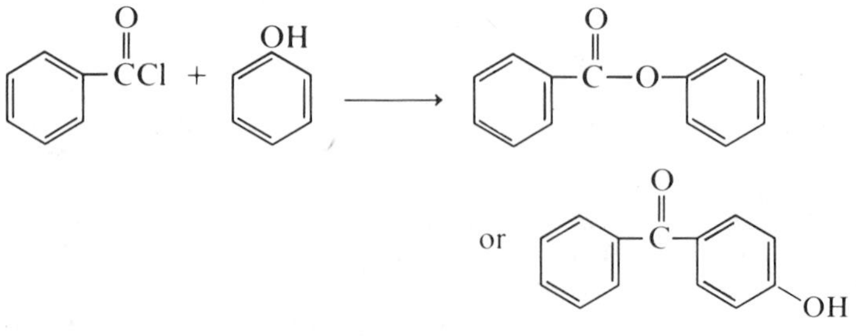

With ammonia and various primary and secondary amines the corresponding amide is formed.

Benzyl Chloride (α-chlorotoluene) [100–44–7]. Benzyl chloride can be synthesized by chloromethylation of benzene in the presence of a catalyst ($ZnCl_2$) or by treatment of benzyl alcohol with SO_2Cl_2. Commercially it is produced by chlorination of boiling toluene in the presence of light. Benzyl chloride can be oxidized to benzoic acid or benzaldehyde, or substituted to give the halogenated, sulfonated or nitrated product:

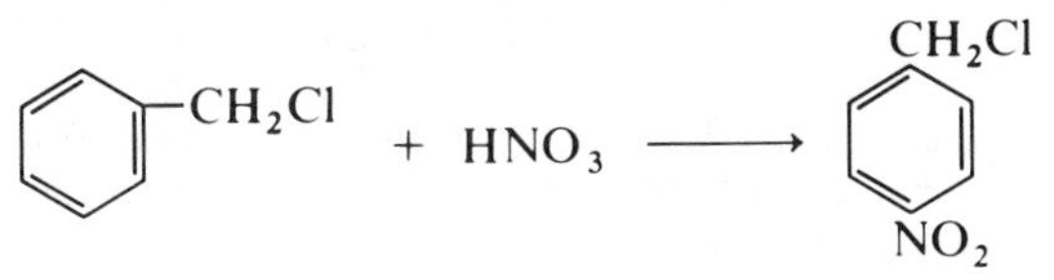

With NH_3 it yields mono-, di- or tribenzyl amine. With alcohols in base the benzylalkyl ether is formed

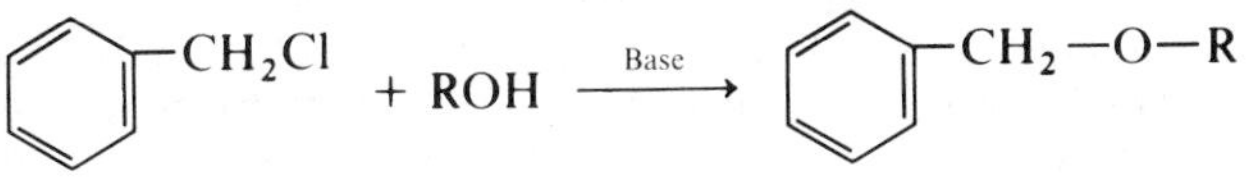

With phenols either the phenolic or nuclear hydrogens can react to give benzylaryl ether or benzylated phenols. Reaction with NaCN gives benzyl cyanide (phenylacetonitrile); with aliphatic primary amines the product is the N-alkylbenzylamine, and with aromatic primary amines N-benzylaniline is formed. Total capacity (United States) exceeds 125 million pound per year. Benzyl chloride is converted to butyl benzyl phthalate plasticizer and other chemicals.

Carbon Tetrachloride (tetrachloromethane) [56–23–5]. Carbon tetrachloride can be synthesized by the chlorination of CS_2, acetylene and other higher hydrocarbons but the primary source is the exhaustive chlorination of methane. It can be pyrolyzed to yield hexachloroethane, oxidized to phosgene and carbonylated with CO in the presence of $AlCl_3$ to give trichloroacetylchloride:

$$CCl_4 + CO \xrightarrow{AlCl_3} Cl_3C{-}\overset{\overset{\displaystyle O}{\|}}{C}Cl.$$

Some of the more important commercial uses involve fluorine displacement to yield chlorofluoromethane refrigerants, such as trichlorofluoromethane (R-11) and dichlorodifluoromethane (R-12).

Chloroacetic Acid ($ClCH_2COOH$) [79–11–8]. Chloroacetic acid can be synthesized by the radical chlorination of acetic acid, treatment of trichloroethylene with concentrated H_2SO_4, oxidation of 1,2-dichloroethane or chloroacetaldehyde, amine displacement from glycine, or chlorination of ketene. It behaves as a very strong monobasic acid and is used as a strong acid catalyst for diverse reactions. The Cl function can be displaced in base-catalyzed reactions. For example, it condenses with alkoxides to yield alkoxyacetic acids: $ClCH_2COOH + KOR \rightarrow ROCH_2COOH$. Oxidation of chloroacetic acid leads to formation of methylene chloride. Treatment with ammonia yields glycine ($ClCH_2COOH \xrightarrow{NH_3} H_2NCH_2COOH$) while use of amines leads to formation of substituted glycines. Commercially, chloroacetic acid is an intermediate in the production of herbicides (2,4-D, 2,4,5-T and others) and cellulose ethers.

Chloroacetylene ($HC{\equiv}CCl$) [593–63–51]. This compound is a gas with a very unpleasant odor. It ignites spontaneously in air and may detonate during handling. It can be synthesized by dehydrochlorination of dichloroethylenes with a strong base. It will react with silver or mercury to give explosive salts. Addition occurs across the unsaturated bond—for example, bromination yields 1-chloro-1,1,2,2-tetrabromoethane.

Chloral (trichloroacetaldehyde) [75–87–6]. Chloral can be prepared by action of Cl_2 on ethanol, chlorination of acetaldehyde, oxidation of 1,1,2-trichloroethylene in the presence of a catalyst ($FeCl_3$, $AlCl_3$, $TiCl_4$ or $SbCl_3$), and by reaction of CCl_4 with formaldehyde. Chloral can be reduced either at the —CCl_3 group or the —CHO group. In the first case the product is acetaldehyde while the second gives $Cl_3C{-}CH_2OH$. Oxidation of chloral gives trichloracetic acid. Polymerization in the presence of H_2SO_4 leads to metachloral or parachloral, depending on the temperature. It undergoes various condensation reactions with alcohols to yield hemiacetals. With organic acids and other functional groups a multitude of reactions are possible. It is used in medicine as a hypnotic.

Chlorobiphenyls. These compounds can be synthesized by direct chlorination of biphenyl in the presence of iron or other catalysts. Other means of preparation include reaction of diazotized aminobiphenyl with copper chloride. Treatment of chlorobiphenyls at elevated temperatures (300–400°C) with strong caustic yields hydroxybiphenyls. Various reactions, normal to aromatic systems, will occur—usually on the unsubstituted ring.

Chloroform (trichloromethane) [67–66–3]. Although chloroform can be prepared by various means it is almost exclusively produced by the chlorination of methane. It can be oxidized to phosgene, substituted with various halogens, nitrated to chloropicrin (Cl_3CNO_2), hydrolyzed to formic acid ($H\overset{\overset{O}{\|}}{C}OH$) and carbonylated to dichloroacetic acid. It will react with unsaturated halohydrocarbons in the presence of $AlCl_3$:

$$ClHC{=}CHCl + CHCl_3 \xrightarrow{AlCl_3} Cl{-}\underset{H}{\overset{Cl}{C}}{-}\underset{H}{\overset{Cl}{C}}{-}\underset{H}{\overset{Cl}{C}}\ Cl$$

With aromatic aldehyde or ketones base catalyzed additions occur, while it condenses with primary amines to yield isocyanides: $C_2H_5NH_2 + CHCl_3 \xrightarrow{NaOH} C_2H_5N{\equiv}C$. A special type of addition can occur when chloroform is reacted with an unsaturated molecule in the presence of potassium alkoxide or sodium hydroxide in a polymer medium. The strong base removes HCl and produces dichlorocarbene ($:CCl_2$) which adds across the double bond:

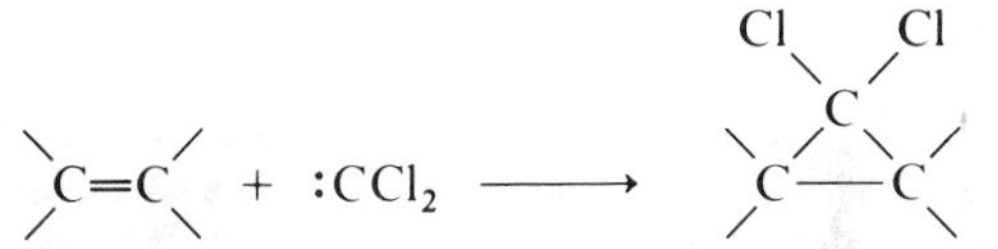

This type of synthesis is particularly useful for ring expansions with specific stereochemistry:

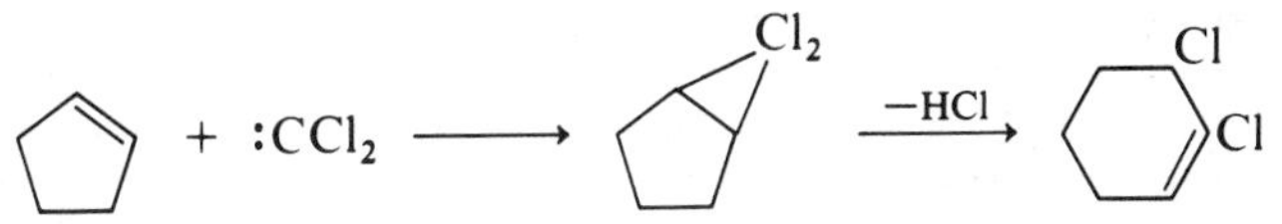

Commercially, about 60% of chloroform is used in production of fluorocarbon refrigerants and propellants. Cl is replaced by treatment with fluorinated antimony pentachloride. The product, $CHClF_2$ (R-22) is used for home air-conditioning units. It is also used as a feed for production of tetrafluoroethylene, which polymerizes to Teflon. Total demand for chloroform approximates 300 million pounds per year.

Chloronaphthalenes. These compounds can be prepared by direct chlorination of naphthalene in the liquid or vapor phase. They also can be synthesized from naphthylamines via diazotization reactions or from naphthols by treatment with PCl_5. Chloronaphthalenes can be further substituted in normal aromatic reactions e.g., halogenation, nitration and alkylation. The chloro group can be displaced to yield a naphthol, an amine, or a nitrile:

Chlorparaffins. These are produced by the random chlorination of various mixed long chain paraffins. They are used as secondary plasticizers for polyvinyl chloride, lubricating oil additives, resinous materials for coatings, and in flame-retardants.

A particularly valuable use for chlorparaffins is in preparation of linear, primarily internal olefins as feedstock for long-chain synthetic oxo-alcohols. Typically, *n*-paraffins (C_{11}–C_{14}) are chlorinated in a fluidized bed at about 300°C. Conversion is maintained low to limit multiple chlorination. After separation of the monochlorinated alkanes by distillation, dehydrochlorination over nickel acetate at 300°C yields the desired internal olefins. Unreacted paraffins are recycled.

Chloroprene (2-chlorobutadiene-1,3) [126–99–8]. Chloroprene can be synthesized by addition of HCl to vinyl acetylene ($H_2C{=}CH{-}C{\equiv}CH + HCl \rightarrow H_2C{=}CH{-}CCl{=}CH_2$) and by dehydrochlorination of dichlorobutenes or 2,2,3-trichlorobutane. It undergoes the normal addition reactions across the double bond and readily polymerizes or copolymerizes with other unsaturated compounds. These polymers resemble natural rubber but are superior in some respects, such as oil resistance (neoprene). Almost all the chloroprene produced is used for the manufacture of these polychloroprene rubbers. Chloroprene is a volatile, toxic, flammable liquid and is especially susceptible to oxidation and polymerization.

Chlorostyrene (chlorovinylbenzene) [1331–28–8]. The alpha isomer can be prepared by PCl_5 reaction on acetophenone

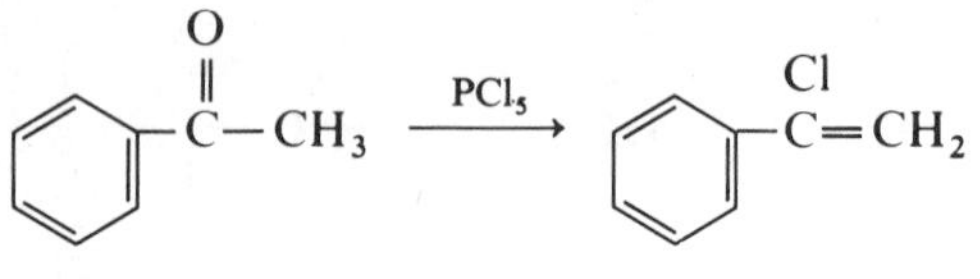

heating of acetophenone dichloride, or hydrolysis of styrene dichloride in aqueous NaOH. The beta isomer can be made by chlorination of cinnamic acid:

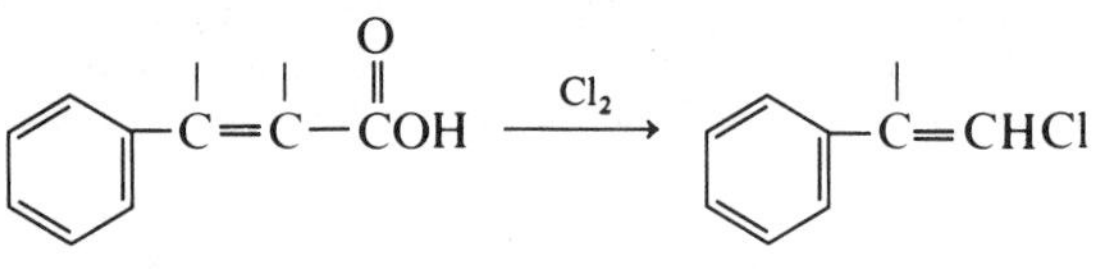

or dehydration of styrene chlorohydrin. Chlorostyrene will add Cl_2 to give α,β,β-trichloroethylbenzene or Br_2 to give the chloro-dibromo product. On treatment with alcoholic KOH the beta isomer is partly resinified.

Dichlorobenzenes [95–50–1] [106–46–7] [541–73–1]. Dichlorobenzenes are primarily produced by the chlorination of benzene in the presence of a catalyst ($FeCl_3$ or $AlCl_3$) although there are other possible synthetic routes. The two commercially important isomers are the ortho- and para-dichlorobenzenes. Further chlorination yields 1,2,4-trichlorobenzene. Dichlorobenzenes participate in normal aromatic substitution and alkylation reactions. In the presence of $CuCl_2$, ammonia will react with the dichlorobenzenes to yield chloroanilines. The ortho-dichlorobenzene is used for pesticides, moth control, as a solvent and for dyestuff manufacture. About half of the para is used as a space odorant.

Epichlorohydrin (γ-chloropropylene oxide) [106–89–8]. This compound can be prepared from 1,3-dichloropropanol-2, 2,3-dichloropropanol-1, or allyl chloride. Commercially it is prepared as an intermediate in glycerol synthesis via alkaline hydrolysis of glycerol dichlorohydrin. Both come from allyl chloride. Epichlorohydrin reacts with monohydric alcohols to give ethers by opening the oxide ring. It will react with ethers, aldehydes, ketones, organic acids and amines to give a wide variety of useful syntheses.

Commercially the most important use is production of glycerine. Large volumes are consumed in nonglycerine areas, which largely consist of the various epoxy resins. It has use as a solvent and in the production of epichlorohydrin rubber.

Ethyl Chloride (chloroethane) [75–00–3]. This compound can be synthesized by treatment of ethyl alcohol with HCl, cleavage of diethylether with HCl in the presence of a catalyst ($ZnCl_2$), chlorination of ethane or hydrochlorination of ethylene. The latter is the choice of industry. The reaction is carried out at 125°F and 125 psi in the presence of $AlCl_3$, which is dissolved in ethyl chloride. It will undergo all the reactions of a typical alkyl chloride—halogenation, hydrolysis, amination, alkylation, and will form the magnesium Grignard reagent. The compound is used in production of tetraethyllead (TEL) by reaction with sodium-lead alloy:

$$4PbNa + 4C_2H_5Cl \rightarrow Pb(C_2H_5)_4 + 3Pb + 4NaCl.$$

Ethyl cellulose is produced by treating alkali cellulose (cotton linter digested in dilute caustic) with ethyl chloride. Up to three ethyl ether stages can be made, giving various grades. These are used as synthetic gums and thickeners in the lacquer and plastics industries. Ethyl chloride is also used in the Friedel-Crafts alkylation of benzene and other aromatics. Additional uses include solvent, refrigerant, heat-transfer medium, aerosol propellant and anesthetic. Much is used captively by the producers; total demand is estimated at about 700 million pounds per year.

Ethylene Dichloride (1,2-dichloroethane) [107–06–2]. Ethylene dichloride (EDC) is produced by reacting ethylene and chlorine in the presence of ferric chloride, using the liquid product as solvent. It is also produced by oxychlorination—ethylene, hydrogen chloride, and air are reacted at about 250°C with a copper chloride catalyst. This latter is the reaction of choice only when cheap by-product HCl is available. EDC reacts with Cl_2 to give derivatives of ethylene or ethane, depending on conditions and catalysts. It will dehydrochlorinate to give vinyl chloride, which is its principal commercial use. EDC hydrolyzes to ethylene glycol and reacts with aromatic hydrocarbons in the presence of $AlCl_3$ to give polyarylethylene plastics. The largest use is for vinyl chloride; next is its use as a solvent intermediate; third is the use as a lead scavenger in lead antiknock fluids—these fluids normally contain EDC at about 30% of the weight of TEL, along with some ethylene dibromide (EDB). Other uses include the manufacture of ethylenediamine and succinic acid, by way of the nitrile. Reaction of EDC with sodium tetrasulfide is used to produce thiokol rubbers.

Methyl Chloride [74–87–3]. This compound is produced by direct chlorination of methane. Since methyl chloride adds chlorine faster than methane, the yield of methyl chloride is increased by using a large excess of methane in the feed, i.e., about ten volumes of methane to one volume of chlorine. The reaction is carried out at about 450°C with very short contact times. Methyl chloride is also commercially produced by reaction of HCl on methanol in the presence of zinc chloride. Methyl chloride is mainly used in the production of silicone resins and rubbers. Silicon metal is reacted with an excess of methyl chloride at 300°C in the presence of a copper catalyst. The product includes mono-, di-, and trichloromethyl silanes. Hydrolysis of the chloro groups converts them into the corresponding hydroxymethylsilanes. These are then polymerized to silicones. Nearly equal amounts of methyl chloride are used in making these rubbers and the other principal user, production of tetramethyllead. Production of methyl chloride approximates 440 million pounds per year.

Methylene Chloride (dichloromethane) [75–09–2]. As with the other members of the methyl series of chlorinated hydrocarbons, methylene chloride can be produced by direct chlorination of methane. The usual procedure involves a modification of the simple methane process. The product from the first chlorination passes through aqueous zinc chloride, contacting methanol at about 100°C. Thus, HCl from chlorination is used to displace the alcohol group, producing additional methyl chloride. This is further chlorinated to methylene chloride. Methylene chloride reacts violently in the presence of alkali or alkaline earth metals and will hydrolyze to formaldehyde in the presence of an aqueous base. Alkylation reactions occur at both functions, thus di-substitutions result. For example, reaction with benzene plus $AlCl_3$ yields diphenyl methane:

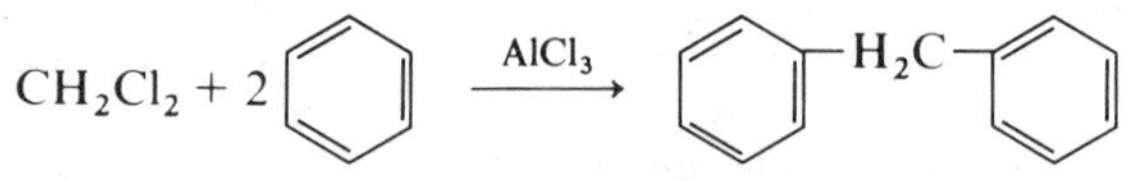

Catalyzed carbonylation (CO) reactions lead to formation of either chloroacetyl chloride or malonyl dichloride.

Methylene chloride is used in refrigeration, aerosol propellants, paint stripping, urethane foam-blowing agents, adhesive, and food extractants. It has low toxicity compared with other chlorinated hydrocarbons and has been shown to be neither mutagenic nor carcinogenic toward humans.

Monochlorobenzene (phenyl chloride) [108–90–7]. Benzene is chlorinated at 80°C in the presence of $FeCl_3$ catalyst. By using low conversions very little dichlorobenzene is produced. The chlorine on this compound is quite inactive, but hydrolysis can be effected by use of a strong caustic at high temperature and pressure, especially in the presence of a catalyst. This has been an important commercial route to phenol. With concentrated aqueous ammonia heated at high temperatures in the presence of a copper catalyst, aniline or diphenylaniline can be synthesized. Reaction with nitric acid yields chloronitrobenzenes—the para isomer predominates. Treatment with hot sulfuric acid leads to formation of *p*-chlorobenzensulfonic acid:

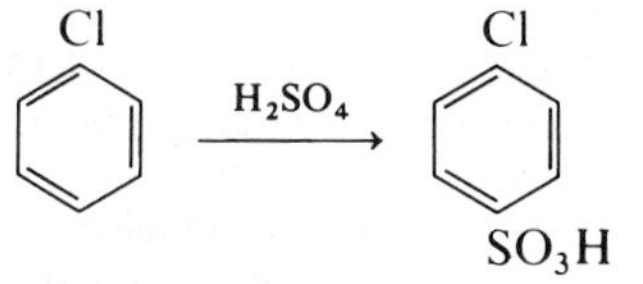

Monochlorobenzene is used commercially as a solvent and to produce phenol and nitrochlorobenzenes.

p-Nitrochlorobenzene [100–00–5].

This compound is made by the nitration of chlorobenzene and is largely used to produce *p*-nitrophenol with smaller production of *p*-nitroaniline:

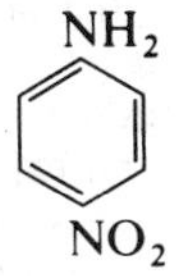

Various agricultural pesticides, rubber chemicals, phenacetin, and *p*-aminophenol consume about 30% of the total. Most of the production is used captively as an intermediate in the production of other chemicals.

Pentachlorophenol [87–86–5]. This compound can be produced by the chlorination of phenol in the presence of $AlCl_3$, or by hydrolysis of hexachlorobenzene with NaOH in methanol. Pentachlorophenol is used as a wood preservative for poles, crossarms, and pilings, and thus competes with creosote.

Vinyl Chloride [75–01–4]. This compound is produced by alkaline dehydrochlorination of ethylene dichloride, or by thermal cracking of EDC, or 1,1-dichloroethane. Vinyl chloride is polymerized in various ways to polyvinyl chloride (PVC). It is also copolymerized with various other monomers to make a variety of useful resins. The copolymers with about 3 to 20% vinyl acetate are the most important. Demand for vinyl chloride is high, approximating 7 billion pounds (3.2 billion kilograms) per year.

See also **Chlorofluorocarbons.**

References

ACS: "Chemical Abstracts Service," American Chemical Society, Washington, D.C. (Continuing).
CRC: "Handbook of Chemistry and Physics," CRC Press, Inc., Boca Raton, Florida, 1980.
Huntress, E. M.: "Organic Chlorine Compounds," Wiley, New York, 1948.
ITII: "Toxic and Hazardous Industrial Chemicals Safety Manual," The International Technical Information Institute, Tokyo, 1976.
Kirk, R. E., and D. F. Othmer: "Encyclopedia of Chemical Technology," 3rd edition, Wiley, New York, 1979.

Walter Wm. Lawrence, Jr., Ethyl Corporation, Baton Rouge, Louisiana.

CHLORINATED-RUBBER PAINTS. Paint.

CHLORINATION (Process). Chlorinated Organics.

CHLORINATION (Water). A principal means for disinfecting municipal water supplies as well as public swimming pools, and some municipal and industrial wastes is by liquid- or gas-phase chlorination. Liquid chlorine is packaged in several types of containers to accommodate a wide range of uses, which may vary from a few hundred pounds during a season (in the case of a swimming pool) to many thousands of tons per year for water supplies. Liquid chlorine is obtainable in pressurized 100- and 150-pound (~45- and 68-kilogram) cylinders, 1-ton (0.9-metric-ton) containers, and for large users, is shipped by railroad tank cars, tank barges, and tank trailers. Large users, of course, must provide local storage means.

Chlorine cylinders are equipped with a single valve. Gas is delivered when the tank is in an upright position; liquid when the cylinder is in an inverted position. However, liquid withdrawal from cylinders is not usually practiced. In the case of ton containers, two valves are provided, permitting easy withdrawal of either gaseous or liquid chlorine. Bulk shipments almost always are unloaded in the liquid phase.

In the case of gaseous withdrawal, the vaporization of the liquid chlorine lowers the temperature surrounding the valve and hence withdrawal rates are limited, ranging up to a maximum of about 1.75 pounds (0.8 kilogram) per hour for a 150-pound (~68-kilogram) cylinder; 15 pounds (6.8 kilograms) per hour for a ton container. Sometimes, cylinders are manifolded to increase the capacity of the system. In the case of liquid withdrawal, the rate ranges up to 400 pounds (181 kilograms) per hour for ton containers; up to 7,000 pounds (3175 kilograms) per hour for a tank car where discharge is from one valve. Usually the liquid is forced out of the container or tank by its own vapor pressure. However, air pressure up to 200 psi (13.6 atmospheres) may be superimposed to increase withdrawal rates.

In municipal water and wastewater treatment installations, the chlorine usually is introduced into the main water system by way of a concentrated water solution of chlorine. The schematic of such a system is shown in Fig. 1. Chlorine is metered under a vacuum created by the ejector. The chlorine is dissolved in water in the ejector, and then discharged into the water system as a high-strength solution. Where the chlorine feeding is done automatically in proportion to the flow of water being chlorinated, a system of the type shown in Fig. 2 can be used.

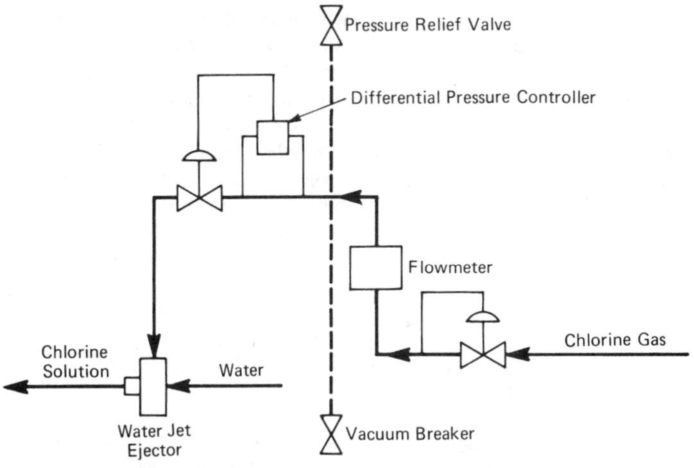

Fig. 1. Chlorinator arrangement commonly used by municipal water and waste water treatment plants.

In the control of chlorine disinfectant systems, the effective use of the chlorine for its intended purpose is assumed if the treated water considerably downstream from the chlorinator contains a residual of chlorine. Depending upon use, full-contact time may be assumed after ten minutes, or the interval may be extended to several hours. The systems also are usually carefully monitored by bacteriological testing. Normally a dose of 1 to 2 milligrams of chlorine per liter is adequate to destroy all bacteria and leave an effective residual. Residuals of 0.1 to 0.2 milligrams per liter are usually maintained in the effluent streams from water-treatment plants as a factor of safety for consumers.

Surface waters require in most instances more extensive treatment, including chlorination, than do groundwaters. By the time some river water reaches some consuming communities it will have received large inputs of organics. Because of its great oxidizing power, chlorine is highly reactive and can combine in a variety of ways with both inor-

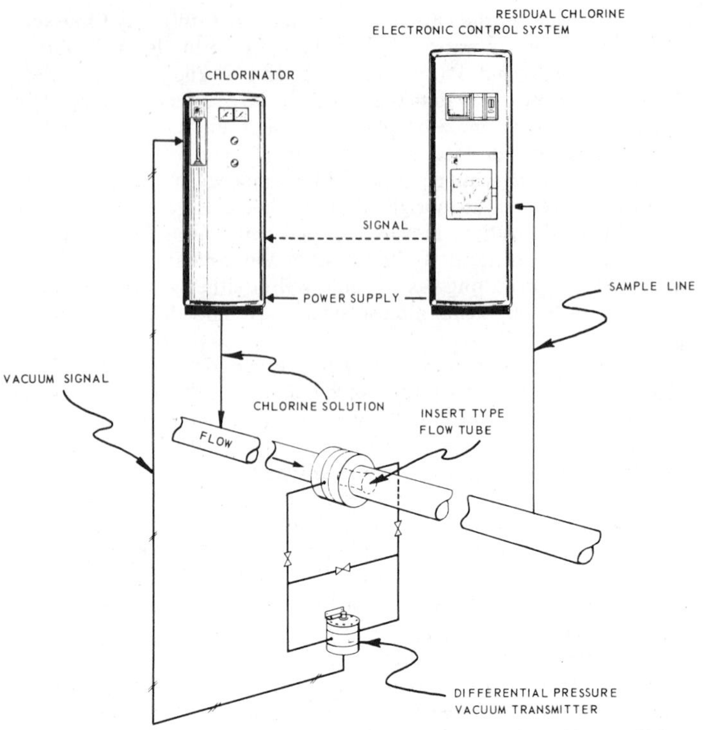

Fig. 2. Closed-loop chlorination control system. (*Fischer & Porter Co.*)

ganic and organic pollutants. Thus, there is concern over the possible formation of carcinogens or otherwise harmful compounds in waters that are heavily chlorinated, particularly waters that have been recycled a number of times along a waterway. Major halogenated compounds found in water supplies suspected of posing health hazards to humans include: (1) chloro-esters, such as *bis*-(2-chloroethyl)ether and *bis*-(2-chloroisopropyl)ether; (2) halobenzenes, such as chlorobenzenes, bromobenzenes, and chloro-bromo benzenes; and (3) haloforms, such as chloroform, bromodichloromethane, dibromochloromethane, and bromoform. More information on this topic can be found in "Chlorine in the Marine Environment," by J. C. Goldman, **Oceanus, 22,** 2, 36–43 (1979).

The fundamentals of chlorination for potable water, waste water, cooling water, industrial process water, and swimming pools are well covered in the "Handbook of Chlorination," by G. C. White (Van Nostrand Reinhold, New York, 1972).

CHLORINE. Chemical element symbol Cl, at. no. 17, at. wt. 35.435, periodic table group 7a (halogens), mp −101°C, bp −34.1°C, density (chlorine gas) 3.209 grams/liter (0°C and 1 atmosphere pressure). Chlorine gas is approximately 2.5 times heavier than air at standard conditions. Chlorine in the gaseous phase is diatomic (mol. wt. 70.906), pale greenish yellow of marked odor, irritating to the eyes and throat, poisonous. At 10°C and one atmosphere pressure 9.8 grams of Cl_2 will dissolve in one liter of water; at 30°C and one atmosphere pressure 5.6 grams will dissolve. Critical pressure is 1118.4 psia (7.7 mPa), critical temperature 144°C. CAS Registry No. 7782-50-5. Other important physical characteristics of chlorine are given under **Chemical Elements.**

Chlorine was discovered by Scheele in 1774 and confirmed as an element by Davy in 1810. It is a high-tonnage industrial chemical with many uses.

Naturally occurring isotopes* 35, 37. Electronic configuration $1s^2 2s^2 2p^6 3s^2 3p^5$. Ionic radius Cl^{7+} 0.26 Å, Cl^- 1.81 Å. Covalent radius 1.050 Å. First ionization potential 13.01 eV; second, 23.70 eV; third, 39.69 eV; fourth, 53.16 eV; fifth, 67.4 eV. Oxidation potential $ClO_3^- + H_2O \rightarrow ClO_4^- + 2H^+ + 2e^-$, −1.00 V; $HClO_2 + H_2O \rightarrow ClO_3^- + 3H^+ + 2e^-$, −1.23 V; $\frac{1}{2}Cl_2 + 4H_2O \rightarrow ClO_4^- + 8H^+ + 7e^-$, −1.34 V; $Cl^- \rightarrow \frac{1}{2}Cl_2 + e^-$, −1.3583 V; $Cl^- + 3H_2O \rightarrow ClO_3^- + 6H^+ + 6e^-$, −1.45 V; $\frac{1}{2}Cl_2 + 3H_2O \rightarrow ClO_3^- + 6H^+ + 5e^-$, −1.47 V; $Cl^- + H_2O \rightarrow HClO + H^+ + 2e^-$, −1.49 V; $Cl^- + 2H_2O \rightarrow HClO_2 + 3H^+ + 4e^-$, −1.56 V; $\frac{1}{2}Cl_2 + H_2O \rightarrow HClO + H^+ + e^-$, −1.63 V; $\frac{1}{2}Cl_2 + 2H_2O \rightarrow HClO_2 + 3H^+ + 3e^-$. −1.67 V; $ClO_3^- + 2OH^- \rightarrow ClO_4^- + H_2O + 2e^-$, −0.17 V; $ClO_2^- + 2OH^- \rightarrow ClO_3^- + H_2O + 2e^-$, −0.35 V; $ClO^- + 2OH^- \rightarrow ClO_2^- + H_2O + 2e^-$, −0.59 V; $Cl^- + 6OH^- \rightarrow ClO_3^- + 3H_2O + 6e^-$, −0.62 V; $Cl^- + 4OH \rightarrow ClO_2^- + 2H_2O + 4e^-$, −0.76 V; $Cl^- + 2OH^- \rightarrow ClO^- + H_2O + 2e^-$, −0.94 V; $ClO_2^- \rightarrow ClO_2 + e^-$, −1.15 V.

* Information in this paragraph furnished by encyclopedia staff.

Production: Most of the chlorine produced in the world is manufactured by electrolysis of sodium chloride brine. Two processes are in common use: the mercury cell process and the diaphragm cell process. Since 1969 when international concern about the effect of mercury in the environment became widespread, some mercury cell plants have been shut down. Most expansion of chlorine production has been in diaphragm cell plants. Indeed, all mercury cell plants in Japan are now required to be converted to diaphragm cell plants. However, it should be noted that existing mercury cell plants operate well within strict standards as to mercury discharge both into the air and into waterways. They produce chlorine (and caustic soda) of the most exacting quality suitable for all uses including food preparation. There is no reason to believe that mercury cell technology is obsolescent in this country or, generally speaking, worldwide.

Technology. In production of chlorine by the diaphragm cell process, salt is dissolved in water and stored as a saturated solution. Chemicals are added to adjust the pH and to precipitate impurities from both the water and the salt. Recycled salt solution is added. The precipitated impurities are removed by settling and by filtration. The purified, saturated brine is then fed to the cell which typically is a rectangular box. It uses vertical anodes (ruthenium dioxide with perhaps other rare metal oxides deposited on an expanded titanium support). The cathode is perforated metal which supports the asbestos diaphragm. This is vacuum deposited in a separate operation. The diaphragm serves to separate the anolyte (the feed brine) from the catholyte (brine containing caustic soda). Chlorine is evolved at the anode. It is collected under vacuum, washed with water to cool it, dried with concentrated sulfuric acid, and further scrubbed, if necessary. It is then compressed and sent to process as a gas or liquefied and sent to storage for transfer to shipping containers and, ultimately, shipment to consumers.

A cell of this type is called a monopolar cell. In a cell bank, several cells have their negative electrodes and their positive electrodes connected by means of external bus bars. Some companies use a bipolar cell in which the electrodes are internally connected. This results in a configuration like a plate and frame filter press.

The latest development in cell technology is the so-called membrane cell. This uses a cation exchange membrane in place of an asbestos diaphragm. It permits the passage of sodium ions into the catholyte but effectively excludes chloride ions. Thus the concept permits the production of high-purity, high-concentration sodium hydroxide directly. The chlorine side of the cell is identical to existing technology. Research in membrane cells is proceeding at a rapid rate. Some companies are known to be operating this process commercially, but as of 1981 not all problems have been solved.

In the mercury cell process chlorine is liberated from a brine solution at the anodes which are, today, typically metal anodes (Dimensionally Stable Anodes or DSA). Collection and processing of the chlorine is similar to the techniques employed when diaphragm cells are used. However, the cathode is a flowing bed of mercury. When sodium is released by electrolysis it is immediately amalgamated with the mercury. The mercury amalgam is then decomposed in a separate cell to form sodium hydroxide and the mercury is returned for reuse.

Uses: The principal use of chlorine is in the production of organic compounds. Of these the production of PVC (polyvinylchloride) is probably the single largest consumer although chlorinated solvents as a class account for larger tonnage. See **Chlorinated Organics.**

In many cases chlorine is used as a route to a final product which contains no chlorine. For instance propylene oxide has traditionally been manufactured by the chlorohydrin process. Modern technology permits abandoning this route in favor of direct oxidation, thus eliminating a need for chlorine.

Large quantities of chlorine are used in bleaching. Pulp bleaching for paper manufacture consumes about 13% of all chlorine produced in the United States. Since none of the chlorine used for this purpose winds up in the finished product, it must all be discharged as chlorides or chlorinated organics or be reprocessed. At the present time there is no proven, wholly satisfactory technique for removing chlorine compounds from pulp mill bleach plant wastes. It is doubtful that existing mills will be converted to a bleaching technique which does not require chlorine but future mills may be designed to minimize the use of chlorine.

Substantial quantities of chlorine go into household bleaches. It is used also in laundry and other commercial bleaches. It is the active element in most swimming pool sanitizers.

Large quantities of chlorine are used for treating municipal and industrial water supplies and this use will probably continue. However, some concern has been felt that traces of organic compounds in all water supplies react with the chlorine to form chlorinated organics which are suspected of being carcinogenic. Further the usefulness of chlorination of municipal wastes has been questioned in some quarters in the light of the fact that such treatment adds chlorinated organics to the waterways. See **Chlorination (Water).**

Safety and Handling: Although chlorine is a hazardous substance, it can be handled safely. All persons who handle chlorine should be thoroughly trained in its properties, in correct use of safety equipment, and in the operation of all other equipment including containers. The Chlorine Institute, 342 Madison Ave., New York, NY 10017, publishes the *Chlorine Manual* (available from the Institute at nominal cost) which provides useful information on these matters. In addition the Chlorine Institute has designed emergency kits capable of capping off certain types of leaks which can occur in chlorine containers.

There have been several recent studies of the physiologic effects of chlorine. These have considered chlorine both as an occupational exposure and as an environmental pollutant (see references). The National Institute of Occupational Safety and Health study recommended a 0.5 ppm concentration of chlorine in air for any 15-minute sampling period as the maximum permissible ceiling value. This contrasts with the generally accepted value of 1 ppm TLV (time weighted average for an eight hour exposure).

Chlorine is primarily a respiratory irritant. When the concentration in the air is sufficient, chlorine irritates the mucous membranes, the respiratory system, and the skin. It causes irritation of the eyes, coughing, and labored breathing. It may cause vomiting. In extreme cases, the difficulty of breathing may increase to the point where death can occur from suffocation. Liquid chlorine in contact with the eyes or skin will cause local irritation or severe burns.

Persons who have been overcome by chlorine should be removed to an uncontaminated area, their contaminated clothing should be removed, and they should be kept warm. Medical help should be provided. If breathing appears to have ceased artificial respiration should begin immediately. If brething is labored, the administration of oxygen may be helpful.

Chlorine Chemistry: Chlorine exhibits in common with the other halogen elements a marked readiness to form singly charged negative ions, as would be expected from the fact that these atoms need only one electron to acquire an inert gas configuration. Thus, chlorine behaves in its normal chemical reactions as an electron acceptor. While there are many compounds in which chlorine has a positive valence, there are no simple compounds of positively charged chlorine (contrast the I^+ of iodine). The positively charged chlorine forms part of a radical, as in combination with oxygen. The electron affinity of chlorine (4.02 eV) is the greatest of all the halogens, and is greater than that of oxygen.

Chlorine reacts readily with hydrogen to form hydrogen chloride, with metals and many non-metals to give chlorides, with metal oxides to give chlorides or oxychlorides, and with many salts of metals to give chlorides. These include the iodides and bromides, whose halogen is displaced by chlorine.

Four isolatable oxides of chlorine are known, Cl_2O, ClO_2, Cl_2O_6($\rightleftharpoons 2ClO_3$), and Cl_2O_7. Chlorine(I) oxide, Cl_2O, obtained by passing Cl_2 over mercury(II) oxide and sand, is a gas, bp 2°C, somewhat soluble in H_2O to form hypochlorous acid. The Cl—O—Cl bond angle is 111° and the Cl—O distance 1.71 A. Cl_2O is an active oxidizing agent. Chlorine(II) oxide, ClO, is produced by reaction of Cl_2O with atomic chlorine, or as an intermediate product in the decomposition of the Cl_2O. The ClO then decomposes into chlorine and oxygen. In view of this instability the properties of the compound are not established. Chlorine(IV) oxide, ClO_2, is produced by treatment of sodium chlorate, $NaClO_3$, with mixed HCl, oxalic acid or other mild reducing agent, and H_2SO_4 (and H_2O). Cl(IV) oxide is a greenish yellow gas, having an odd electron in its molecule and is consequently paramagnetic. Electron diffraction studies indicate its structure to be

with the odd electron in an antibonding orbital. The O—Cl—O bond angle is 116.5° and Cl—O distance 1.49 Å. It is readily hydrolyzed, but is stable when dry. The mechanism of its hydrolysis is complex, yielding all four of the oxychloric acids, and it is widely used as a heavy-duty oxidizing agent. It reacts with metal hydroxides to give the mixture of chlorate and chlorite, with metal peroxides to give chlorite and oxygen and with metals to give chlorite alone. ClO_2 is photosensitive, decomposing when illuminated at about 8°C, to give some Cl_2O_6, bp 3.5°C. Chlorine hexoxide, Cl_2O_6, has a molecular weight corresponding to the formula ClO_3—ClO_3. Its vapor pressure in the liquid state is 0.31 mm at 0°C against values of 23.7 for Cl_2O_7, 490 for ClO_2 and 699 mm for Cl_2O. This is consistent with a bi-trigonal-pyramidal structure in which the two pyramids have three oxygen atoms at their base corners, and are joined by the two chlorine atoms at the apices. In contrast, the additional oxygen atom of Cl_2O_7 would separate the two chlorine atoms, preventing close packed structure. Chlorine heptoxide, Cl_2O_7, the anhydride of perchloric acid, is obtained by heating the latter with phosphorus pentoxide, and consists of two chlorine atoms, each bonded to three oxygen atoms, and jointly bonded to the seventh. All of the oxides of chlorine are thermodynamically unstable with respect to decomposition into the elements.

Hypochlorous acid, HClO, is formed by hydrolysis of chlorine(I) oxide. It is present in aqueous solutions of chlorine because of the equilibrium

$$Cl_2 + 2H_2O \rightleftharpoons HClO + H_3O^+ + Cl^-$$

and can be freed by the addition of any substance that combines with the Cl^-, such as mercury(II) oxide, or with the H^+, such as calcium carbonate or other weak bases which do not react with HClO. HClO is a weak acid ($K = 3 \times 10^{-8}$ at 25°C). It reacts with hydrochloric acid to give chlorine and H_2O. On warming or irradiation it undergoes this reaction as well as two other decompositions, i.e., to oxygen, H^+ and Cl^-, and to ClO_3^- and H^+. The presence of oxygen favors the last reaction. HClO and its salts are strong oxidizing agents, oxidizing iodine and bromine to iodates and bromates. Covalent hypochlorites are known, such as the alkyl esters, ROCl. In common with other esters of oxidizing acids, these are unstable if R is a primary or secondary alkyl group. However, *t*-butyl hypochlorite is quite stable. Reduction of chlorine dioxide, ClO_2, with hydrogen peroxide, yields oxygen and chlorous acid, $HClO_2$, which exists only in solution. It is stronger than hypochlorous acid ($K = 1.01 \times 10^{-2}$ at 23°C). It is also a strong oxidizing agent and its sodium salt is widely used for this purpose, generally as a source of ClO_2.

Chloric acid, $HClO_3$, is readily prepared by passing chlorine into hot caustic solutions, since these conditions favor the formation of chlorate. Chloric acid is a more active oxidizing agent than HClO, reacting explosively with organic matter. Its alkali salts undergo on heating two modes of decomposition, one (catalyzed, e.g., by manganese dioxide) into the chloride and oxygen, and the other (uncatalyzed) into the chloride and perchlorate.

Perchloric acid, $HClO_4$, is obtained in anhydrous form from perchlorates by H_2SO_4 distillation, or from ammonium perchlorate by aqua regia distillation. Perchlorates are also obtained by electrolysis of chlorides or chlorates. Perchloric acid is explosive unless properly handled; it is of course a powerful oxidizing agent, but has a higher activation energy than the lower acids.

The chlorides range in character from ionic to covalent compounds, many of them having bonds of intermediate character. There are also a number of interhalogen compounds containing chlorine. Those of iodine and bromine are discussed under those entries. With fluorine, chlorine forms ClF, chlorine monofluoride, which is also obtained (along with ClF_3), when mixtures of the elements are subjected to spark discharge. It may also be obtained by heating a mixture of chlorine and chlorine trifluoride. It is a reactive gas, bp −100.8°C, and with its bond having 20–30% ionic character. Chlorine trifluoride, ClF_3, is also obtained from the elements or from chlorine monofluoride and fluorine, by varying the conditions. It is a gas, bp of liquid 11.3°C, and is a more powerful fluorinating agent than the monofluoride. Present views on its structure suggest a trigonal bi-pyramid having chlorine in the center, a fluorine atom at each apex and the third fluorine and two non-bonding pairs of electrons in the three equatorial positions, giving a T-shaped molecule. ClF_3 reacts with all elements except the noble gases, nitrogen, chromium, and certain noble metals, although some metals (e.g., copper) require elevated temperature. It does not react with oxides or salts as readily as fluorine, but nevertheless ignites such materials as asbestos.

See also **Chlorinated Organics; Halides; Hypochlorites;** and **Sodium Chloride.**

References

Sconce, J. S.: "Chlorine: Its Manufacture, Properties and Uses," ACS Monograph 154, Van Nostrand Reinhold, New York, 1962.

Somers, H. A.: "The Chlor-Alkali Industry," *Chem. Eng. Progress,* **61**, 3 (March 1965). (Covers mercury cells only.)

Staff: "Chlorine Manual," The Chlorine Institute, New York, 1969 (updated periodically).

Staff: "Exceeding All Expectations: A Short History of Chlorine," The Chlorine Institute, New York, 1968.

Staff: "Criteria for a Recommended Standard: Occupational Exposure to Chlorine," National Institute for Occupational Safety and Health, HEW Publication No. (NIOSH) 76–170, Washington, D.C., 1976.

Staff: "Medical and Biologic Effects of Environmental Pollutants: Chlorine and Hydrogen Chloride," National Academy of Sciences, Washington, D.C., 1976.

Staff: "Diaphragm Cells for Chlorine Production," Proc. Symp. held at City University, London, England, June 16–17, 1977.

Weast, R. C.: "Handbook of Chemistry and Physics," 61st edition, CRC Press, Boca Raton, Florida, 1981.

Herbert S. Hopkins, Research Center, Olin Corporation, New Haven, Connecticut.

CHLORINE (Determination of). Amperometer.

CHLORINE (Fluorides). Rocket Propellants.

CHLORINITY. A measure of the chloride content, by mass, of seawater (grams per kilogram of seawater, or per cubic mille). Originally, chlorinity was defined as the weight of chlorine in grams per kilogram of seawater after the bromides and iodides had been replaced by chlorides. To make the definition independent of atomic weights, chlorinity is now defined as 0.3285233 times the weight of silver equivalent to all the halides.

CHLORITE. Chlorite is an ubiquitous mineral usually a product of secondary origin from the alteration of silicates containing aluminum, ferrous iron, and magnesium. Pyroxenes, amphiboles, biotite garnet, and idocrase within rocks which have undergone metamorphism are common source minerals for chlorite. Distinct crystals are extremely rare; more often found as foliated masses or fine scaly aggregates. Color includes various shades of green. Hardness of 2–2.5, and specific gravity of 2.6–2.9, with vitreous to pearly luster. Individual folia characterized by flexible, not elastic property. A general formula is $(Mg,Fe^{2+},Fe^{3+}Mn)_6AlSi_3O_{10}(OH)_8$.

CHLORITE SCHIST. A schist whose color and foliation are chiefly due to the mineral chlorite. Other minerals common in this type of schist are quartz and epidote. Garnet and magnetite sometimes occur as idiomorphic crystals giving the schist a porphyroblastic texture.

CHLORITOID. A mineral which occurs [illegible]ystals, probably triclinic, foliated masses or scattered sca[illegible] a greenish-gray to greenish-black color. It is charact[illegible] intensely altered metamorphic rocks such as phyllite[illegible]mically it is a hydrous iron-aluminum silicate, $Fe_2Al_4Si_2O_{10}(OH)_4$. Ottrelite contains some manganese as well. Chloritoid was originally noted as from the Ural Mountains and named for its greenish color from the Greek word meaning green. Ottrelite was named from Ottrez in Luxemburg.

CHLOROACETIC ACID. Chlorinated Organics.

CHLOROACETYLENE. Chlorinated Organics.

CHLOROBIPHENYLS. Chlorinated Organics.

CHLOROFLUOROCARBONS. Methanes, ethanes, and ethylenes which contain at least one fluorine atom per molecule. Because of the large number of compounds in these series and to avoid the complexities of organic chemistry nomenclature insofar as the commercial user of the products is concerned, an abbreviated designation system is used. The system comprises a four-element designation, preceded by a generic term that describes the main application of the product. Thus, *Refrigerant ABCD.* Actually, this system is also complex for users—hence, a strong reliance upon strictly trade names.

In the *ABCD* system, *A* equals the number of double bonds in the molecule; *B* equals the number of carbon atoms minus 1; *C* equals 1 (one) plus the number of hydrogen atoms in the molecule; and *D* equals the number of fluorine atoms in the molecule. If *A* and *B* are zero, the digits simply are omitted. Thus, Refrigerant 12 is dichlorodifluoromethane.*

Most uses of saturated chlorofluorocarbons capitalize on the volatility, stability, and safety of this class of compounds. Refrigerant 11, Refrigerant 12, and Refrigerant 22 are applied to a variety of basic jobs. Refrigerant 11 is used in large centrifugal air-conditioning units in office buildings and industrial plants; Refrigerant 12 is usually selected for household refrigerators and freezers, as well as for automobile air conditioners; Refrigerant 22 is used extensively in residential air conditioning, where high capacity and small unit size are important. A number of binary azeotropes are used in special situations. Components of some of the common binary azeotropes, with their numerical designations, are:

Refrigerant 500	R-12/difluoroethane
Refrigerant 502	R-22/R-115
Refrigerant 503	R-13/R-23

Additionally, chlorofluorocarbons have solvent and cleaning properties that are particularly attractive in the aerospace, electronics, optical, and miniature, precision mechanism manufacturing fields. Their selective solvent properties are advantageous. Trichlorotrifluoroethane is especially suitable because of its lack of attack on paint, gaskets, and wire insulation. Binary azeotropes containing methylene chloride, ethanol, methanol, or acetone provide variations of the solvent properties obtainable with trichlorotrifluoroethane alone.

Substantial quantities of Fluorocarbon 11 and Fluorocarbon 12 are used in plastic foams. Flexible polyurethane foams are commonly expanded with Fluorocarbon 11, while rigid foams frequently are prepared from polystyrene and Fluorocarbon 12.

Dichlorodifluoromethane has been adapted to food freezing, in which food particles are frozen upon contact with boiling chlorofluorocarbon. The resulting rapid heat transfer reduces the freezing time to seconds for most foods.

Fluoroolefins, such as chlorotrifluoroethylene, tetrafluoroethylene, vinylidene fluoride, and vinyl fluoride are used extensively in the synthesis of high-performance lubricants, plastics, and elastomers.

Carbon tetrachloride or chloroform and anhydrous hydrogen fluoride are the usual starting ingredients in the manufacture of chlorofluorocarbons. A catalyst, such as AlF_3 or $SbCl_5$, is used. Examples include:

$$CCl_4 + HF \rightarrow CCl_xF_y \qquad x + y = 4$$

$$CHCl_3 + HF \rightarrow CHCl_xF_y \qquad x + y = 3$$

* These materials also are referred to by commonly known trade names, such as Freon (E. I. DuPont DeNemours & Co., Inc.); Genetron (Allied Chemical Corp.); Ucon (Union Carbide Corp.); Isotron (Pennwalt Corp.); and others.

Perchloroethylene, chlorine, and anhydrous hydrogen fluoride are used in the preparation of ethane derivatives:

$$CCl_2{=}CCl_2 + HF + Cl_2 \rightarrow \underset{\substack{\text{Fluorocarbon}\\ 113}}{CClF_2{-}CCl_2F} + \underset{\substack{\text{Fluorocarbon}\\ 114}}{[CClF_2 + CClF_2]}$$

By further fluorination, Fluorocarbon 115 and Fluorocarbon 116 can be prepared from Fluorocarbon 114:

$$\underset{\substack{\text{Fluorocarbon}\\ 114}}{[CClF_2 + CClF_2]} \xrightarrow{HF} \underset{\substack{\text{Fluorocarbon}\\ 115}}{CClF_2{-}CF_3} + \underset{\substack{\text{Fluorocarbon}\\ 116}}{CF_3{-}CF_3}$$

Such syntheses usually are accomplished in the liquid or vapor phase at moderate temperature and pressure.

Fluoroolefins result from a number of steplike reactions, typified by:

$$CClF_2{-}CCl_2F + Zn \rightarrow CClF{=}CF_2 + ZnCl_2, \text{ or}$$

$$CH{\equiv}CH + HF \rightarrow CH_2{=}CHF + CH_3{-}CHF_2, \text{ or}$$

$$CH_3{-}CClF_2 \rightarrow CH_2{=}CF_2 + HCl$$

Starting materials and byproducts are separated by fractional distillation. The products are further purified by washing, followed by drying over suitable desiccants. Because of extensive purification, chlorofluorocarbons rank among the highest-purity organic materials commercially marketed.

See also **Ozone.**

CHLOROFLUOROCARBONS (Ozone Effect). Oxygen.

CHLOROFORM. Anesthesia; Chlorinated Organics.

CHLOROMYCETIN. Antibiotic.

CHLOROPHYLL (Phosphorylation). Phosphorylation (Photosynthetic).

CHLOROPHYLL (Production). Etiolation; Magnesium (In Biological Systems).

CHLOROPHYLLS. A group of closely related green pigments occurring in leaves, bacteria, and organisms capable of photosynthesis. The major chlorophylls in land plants are designated *a* and *b*. Chlorophyll *c* occurs in certain marine organisms. Because of the overwhelming percentage of the total photosynthesis which is performed by marine organisms, it is possible that chlorophyll *c* is equivalent in importance to chlorophyll *b*. Chlorophyll *a* is several times as abundant as chlorophyll *b*. See also **Photosynthesis.**

The canonical form for chlorophyll *a* is $R{=}CH_3$ when substituted in the following formula. For chlorophyll *b*, $R{=}CHO$. These structures have been established by a long series of degradation studies mainly by R. Willstätter, Hans Fischer and their collaborators and by synthetic studies by Fischer.

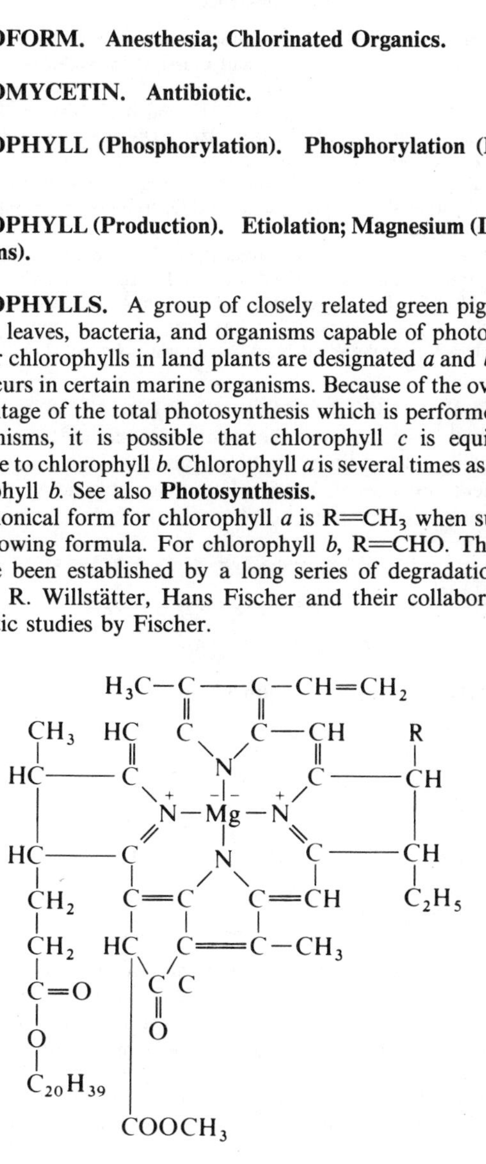

The biological significance of the chlorophylls stems from their role in photosynthesis, the process by which plants fix the sun's energy in the form of organic matter. This process corresponds to the reversal of the combustion of hydrogen. The oxygen liberated is set free in the air. Under special conditions, some organisms are also capable of liberating the hydrogen, but usually this is used for chemical reductions in the plant. Atmospheric carbon dioxide is fixed enzymatically and is thus used as the source of the carbon in the synthetic process, but is not reduced directly. The path of the carbon from carbon dioxide in photosynthesis has been elucidated largely by the studies of Calvin and his collaborators. See references.

While it is known that most of the energy fixed in photosynthesis is absorbed originally by the chlorophylls, the exact reactions which they undergo to initiate the process of reduction are not fully understood. It is known that the photosynthetic sequence requires a high degree of organization within the plant cells where it occurs and that destruction of the organization of the chloroplasts by processes like grinding are sufficient to bring photosynthesis to a stop, even when the chlorophyll and the soluble enzymes participating in the process are still presumably intact.

Chlorophyll derivatives with the phytyl group intact are oil soluble and form a series of green dyes used in the coloring of oils and waxes. The chlorophyll soaps, resulting from conbined saponification and cleavage of the isocyclic ring, form "water soluble" dyes useful in the coloring of soaps and similar products. Both the medical and cosmetic literature are replete with claims of therapeutic or physiological activity of "chlorophyll." The substances utilized in this work range from partially purified chloroplasts to mixtures of materials which have undergone deep-seated chemical alteration. Some of the types of activity claimed can be shown to be due to incidental impurities. The field for investigation of the action of pure chemical individuals produced by the action of various reagents on chlorophyll or its derivatives is largely unexplored. It is known, however, that neither chlorophyll nor hemoglobin in the diet is utilized by the body in the formation of the physiologically active pyrrole pigments. These are derived, instead, from such simple building blocks as glycine and acetate ion. Only the iron in dietary blood pigment can be utilized by the body.

The work of Granick has shown that, in the physiological processes of plants, chlorophyll is formed from protoporphyrin, which can be obtained in the laboratory by the removal of iron from hemin. The pathways to heme and to chlorophyll diverge at protoporphyrin. To form heme, an organism introduces iron into protoporphyrin. To form chlorophyll from protoporphyrin, an oxidation, a reduction, a ring closure and esterifications are performed and the magnesium is introduced. The end-product of the enzymatic synthetic chain is presumably protochlorophyll, the magnesium derivative of the porphyrin corresponding in structure to chlorophyll. The addition of the two hydrogens necessary to convert protochlorophyll to chlorophyll is accomplished under the influence of light.

For references see entry on **Photosynthesis.**

CHLOROPLAST. A plastid containing chlorophyll as found in most of the green plants. Electron microscope studies show that the chloroplasts have a complex internal structure. There are many membranous layers, known as *grana*, arranged somewhat like stacks of coins. Apparently the chlorophyll molecules are spread in single layers on the grana, thus achieving a very great surface area for trapping the energy of light in photosynthesis. See also **Photosynthesis; Pigmentation (Plants);** and **Plastids.**

CHLOROPRENE. Chlorinated Organics.

CHLOROPROCAINE HYDROCHLORIDE. Anesthesia.

CHLOROQUINE. Malaria.

CHLOROQUINE PHOSPHATE. Alkaloids.

CHLOROTHIAZIDES. Hypertension (High Blood Pressure).

CHLOROTRYPTOPHAN, D-6. Sweeteners.

CHLORTETRACYCLINE. **Antibiotic.**

CHOANOCYTE. The collar cell of sponges, bearing a high ridge surrounding a flagellum at the free end. They are located in cavities in the sponge and produce currents of water through the passages in the body wall.

CHOKE COIL. This term is applied to various types of inductances used in electrical circuits primarily to present high reactance at certain frequencies. Such coils usually have high reactance compared to their resistance and offer impedance to the flow of alternating currents by the induced counter electromotive force. Since this impedance will vary directly with frequency the choke may be designed to let certain lower frequencies through and stop or impede higher ones. An air-core choke coil is often used in electrical power circuits to block high-frequency transients produced by lightning surges. In communications circuits air-core chokes are used extensively to block radio frequencies from audio-frequency circuits or from dc parts of the circuit. In these applications they are often called radio-frequency chokes. Iron-cored chokes are frequently used in audio circuits in a similar manner. Iron-cored chokes are also important components of power supply filters as well as many wave filters.

CHOKING. **Artificial Respiration.**

CHOLECYSTITIS. **Gallbladder and Biliary Tract Diseases.**

CHOLECYSTOKININ. **Hormones.**

CHOLECYSTOKININ-LIKE PEPTIDE. **Brain and Nervous System.**

CHOLEIC ACIDS. **Bile.**

CHOLELITHIASIS. **Gallbladder and Biliary Tract Diseases.**

CHOLERA. An acute diarrheal disease caused by enterotoxin secreted by *Vibrio cholerae* organisms. These organisms invade the proximal small bowel of the human host. Without effective treatment, the mortality rate is 50% or higher in some areas. Mortality can be reduced markedly by the application of therapy, as described later. Known since antiquity, cholera frequently occurs in an epidemic or pandemic fashion, and particularly in crowded urban areas and notably in India, southeast Asia, and China. In 1961, the disease made a significant spread westward from Indonesia, reaching much of the African continent, and more recently has been reported in Italy and Spain. The first outbreak in the United States since 1911 occurred in coastal Louisiana in 1978 and was attributed to the ingestion of contaminated crab. Usually the principal path for transmitting the disease is the water supply, which can be less than sanitary in many crowded cities of the world. The disease is transmitted by the intestinal-oral pathway. Travelers to cholera-endemic areas can be protected (estimated 60–80%) by injections of cholera vaccine. Two injections about one week apart provide protection for 3–6 months. Nevertheless, travelers are advised to exert care in selecting water to drink.

During the 1879–1883 pandemic in Europe, Koch first demonstrated the causative organism. *V. cholerae* are short, comma-shaped, gram-negative bacilli and easily demonstrated in the laboratory. There are two major serotypes, Inaba and Ogawa. In addition, the El Tor strain was demonstrated in a recent cholera pandemic, causing somewhat less confidence in the degree of protection obtainable from commercial cholera vaccine.

The incubation period of the disease is short (24–72 hours). Usually the onset is sudden, commencing with diarrhea. The stools are loose and yellowish to green in color. There are no usual predromal symptoms. Inasmuch as the organisms do not invade tissue, there is no tissue inflammatory infection. Fever, chills, and the usual symptoms of infection are absent. However, there may be severe vomiting. Diarrhea becomes severe over a very short period, with the stools (sometimes called "rice water stools") becoming essentially colorless. Stool volumes become excessive and can reach a volume up to 25 liters (6.6 gallons) in one day. Without treatment, diarrhea persists for 2–4 days, with accompanying dehydration and prostration. Because of electrolyte imbalance, there may be severe cramps in the lower extremities. Death can occur within a period of hours. There are numerous secondary clinical manifestations of the disease.

Therapy consists of massive fluid replacement. Glucose is also administered orally or by nasogastric tube to reduce the volume of water output in the stools. This is effected by increasing the water absorption of the small intestine so that the net fluid loss becomes negative after a short time. Tetracycline may be administered parenterally to shorten the period of diarrhea. This therapy has been effective in saving many lives. Unfortunately, because of the nature of occurrence of the disease, many scores of cases may appear within a period of hours, far exceeding the facilities' capability for treatment and hence, in some locations, there may be many fatalities.

Inasmuch as humans are the only known hosts of the cholera-producing organism, it is logical to assume that the disease arises from fecal contamination of either water or food. Persons who are achlorhydric (as by taking generous amounts of antacids) run a higher risk of infection because the organisms naturally are sensitive to gastric acid. Persons with chronic gallbladder disease are suspected carriers of the infection. The natural immunity to cholera of persons who have recovered from the infection appears to be long-lasting. Studies made in endemic areas such as Bangladesh and India show that the incidence of the disease is eminently high in small children, being up to 10 times that of persons over 20 years old. However, in epidemics, statistics show a relatively even distribution of cases among individuals of all ages.

References

Bart, K. J., et al.: "Seroepidemiologic Studies during a Simultaneous Epidemic of Infection with El Tor, Ogawa, and Classical Inaba *Vibrio cholerae*," *J. Infect. Dis.*, **12**(suppl.), 17 (1970).
Carpenter, C. C. J., Jr., Mahmoud, A. A. F., and K. S. Warren: "Algorithms in the Diagnosis and Management of Exotic Diseases: XXVI. Cholera," *J. Infect. Dis.*, **136**, 461 (1977).
Colwell, R. R., Kaper, J., and S. W. Joseph: "*Vibrio cholerae, Vibrio parahaemolyticus*, and Other Vibrios: Occurrence and Distribution in Chesapeake Bay," *Science*, **198**, 394–396 (1977).
Hornick, R. B., et al.: "The Broad Street Pump Revisited: Response of Volunteers to Ingested Cholera Vibrios," *Bull. N.Y. Acad. Med.*, **47**, 1181 (1971).
Mosley, W. H.: "The Role of Immunity in Cholera: A Review of Epidemiological and Serological Studies," *Tex. Rep. Biol. Med.*, **27**(suppl.), 227 (1969).
Staff: "Cholera Research: What Next?" (editorial), *Lancet*, **2**, 1283 (1976).

CHOLESKY METHOD OF SOLVING EQUATIONS. The Cholesky method (also known as the square-root method) is a convenient way of solving a set of linear simultaneous equations when the matrix of coefficients on the left-hand side is symmetrical. As an example consider the three equations

$$a_{11}x_1 + a_{12}x_2 + a_{13}x_3 = y_1$$
$$a_{12}x_1 + a_{22}x_2 + a_{23}x_3 = y_2$$
$$a_{13}x_1 + a_{23}x_2 + a_{33}x_3 = y_3$$

The computational layout is shown below.

$$\begin{array}{ccccc} a_{11} & a_{12} & a_{13} & y_1 & S_1 \\ a_{12} & a_{22} & a_{23} & y_2 & S_2 \\ a_{13} & a_{23} & a_{33} & y_3 & S_3 \\ u_{11} & u_{12} & u_{13} & v_1 & S_4 \\ 0 & u_{22} & u_{23} & v_2 & S_5 \\ 0 & 0 & u_{33} & v_3 & S_6 \end{array}$$

The upper array consists of the coefficients of the equations together with the right-hand sides and a check column of S's in which each entry is the sum of all the other entries in the same row. The elements of the lower array are found row by row from the rule:

sum of products rth column with sth column of lower array
= element of upper array in rth row and sth column

Then, with

$$r=1,\ s=1 \quad \text{we have} \quad u^2_{11} = a_{11},$$
$$r=1,\ s=2 \qquad u_{11}u_{12} = a_{12}$$
$$\cdots\cdots \qquad \cdots\cdots$$
$$r=3,\ s=5 \qquad u_{13}S_4 + u_{23}S_5 + u_{33}S_6 = S_3$$

Thus

$$u_{11} = \sqrt{a_{11}}; \quad u_{12} = a_{12}/u_{11}; \ldots$$
$$u_{22} = \sqrt{(a_{22} - u^2_{12})} u_{23} = (a_{23} - u_{12}u_{13})/u_{22}; \ldots$$

and the elements of each row can be calculated successively. As a check, the first four entries in any row of the lower array should add up to the corresponding S.

The original equations have now been replaced by an equivalent set given symbolically by the lower array. These can be readily solved, working upwards from the bottom.

If there are several sets of equations with the same left-hand sides, the different columns of right-hand sides can all be included in the arrays and handled simultaneously. In particular, the inverse matrix of (a_{ij}) can be found by setting (y_1, y_2, y_3) equal successively to (1, 0, 0), (0, 1, 0) and (0, 0, 1), the columns of the unit matrix. In this case, the elements of the inverse can be evaluated directly from the columns of v's by the rule:

sum of products of pth column with qth column of v's
= element of inverse in pth row and qth column

The Cholesky method provides an alternative to the Doolittle method for solving linear equations; its principal advantage is that fewer intermediate results have to be written down.

See also **Doolittle Method of Solving Normal Equations; Matrix (Mathematics);** and terms listed under **Mathematics.**

CHOLESTERIC LIQUID CRYSTALS. Liquid Crystals.

CHOLESTEROL. Also known as cholesterin or 5-cholesten-3-beta-ol, cholesterol is the most common animal sterol, a monohydric secondary alcohol of the cyclopentenophenanthrene (4-ring fused) system containing one double bond. It occurs in part as the free sterol and in part esterified with higher fatty acids as a lipid in the human blood system. The primary precursor in biosynthesis appears to be acetic acid or sodium acetate. Cholesterol itself in the animal system is the precursor of bile acids, steroid hormones, and provitamin D3. Cholesterol is a white, or faintly yellow, almost odorless substance and may take the form of pearly granules or crystals. The substance is affected by light; mp 148.5°C; bp 360°C, but tends to decompose at lower temperatures. Specific gravity 1.067 (20/4°C); insoluble in water; slightly soluble in alcohol; soluble in fat solvents, vegetable oils, and in aqueous solutions of bile salts. Cholesterol occurs in egg yolk, liver, kidneys, saturated fats and oils. In addition to its importance in medicine and biology in general, cholesterol is used as an emulsifying agent in cosmetic and pharmaceutical products. It is the source of estradiol.

Research over the years has shown that cholesterol is carried in the bloodstream in complexes with other lipids and proteins. Based upon their density, there are four classes of lipoproteins: the chylomicrons; the VLDLs (very low density); the LDLs (low density); and the HDLs (high density). It is estimated that about 80% of the total blood cholesterol is carried by the LDLs. Most of the remainder is carried by the HDLs. Large quantities of triglycerides are carried by the chylomicrons and VLDLs, but very little cholesterol.

As early as 1951, Barr (Cornell University Medical College) observed that in males with coronary heart disease, the HDL concentrations were low. There were several confirmations of this during the 1950s and 1960s. However, for many years, the principal criteria in connection with heart attack and stroke with relation to cholesterol was considered to be total cholesterol and LDL levels, with little attention given to the HDLs. It was not until 1975 that Miller (Royal Infirmary, Edinburgh, Scotland) reported an inverse correlation between blood concentration of HDLs and total body cholesterol. Miller hypothesized that HDLs may lessen body cholesterol by facilitating its excretion. Epidemiological studies were made shortly thereafter on Japanese-Hawaiian males, Israeli males, and black sharecroppers in Georgia. These studies showed that risk of heart attack increases as blood HDL level decreases. The average value for HDL levels in human males is 45 milligrams per deciliter (55 milligrams for females). The generally higher level of HDLs in females may partially account for their lower heart attack rates. For both sexes, it has been estimated that a 5 milligram drop in the aforementioned HDL levels may increase the risk of heart attack by about 25%.

In other research efforts, Glueck and co-workers (University of Cincinnati College of Medicine) have identified two groups of people who are genetically endowed either with high HDL or low LDL concentrations and who have lifespans some 5 to 10 years greater than the average. The usual dietary and smoking restrictions for heart attack and stroke prevention do not appear to apply to these groups to any degree approaching that of the average persons. However, it also has been noted that the general dietary and nonsmoking recommendations given for many years also tend to increase HDL concentration.

In recent years, it has been found that there is little correlation between total blood cholesterol and heart attacks in people over age 50. But, the HDL level is more meaningful with this group.

Status of research in this field as of late 1979 is well summarized in *Science*, **205**, 677–679 (1979).

Cholesterol enters into several discussions in this book. In particular, see **Bile; Dietary Trends;** and **Steroids.** The structural formula for cholesterol, $C_{27}H_{15}OH$, is given in the latter entry.

CHOLESTEROL (Arteriosclerosis). Arterial and Venous Disorders.

CHOLESTEROL (Steroids). Steroids.

CHOLESTYRAMINE. Anticoagulants.

CHOLIC ACID. Bile.

CHOLINE AND CHOLINESTERASE. An enzyme (acetylcholinesterase) is specific for the hydrolysis of acetylcholine to acetic acid and choline in the animal body. It is found in the brain, nerve cells and red blood cells and is important in the mechanism of nerve action. Acetylcholine was first synthesized in 1867. It consists of a combination of choline and acetic acid in an ester linkage. The components parts of the acetylcholine molecule are both normal constituents of the animal body. Acetylcholine has the structure:

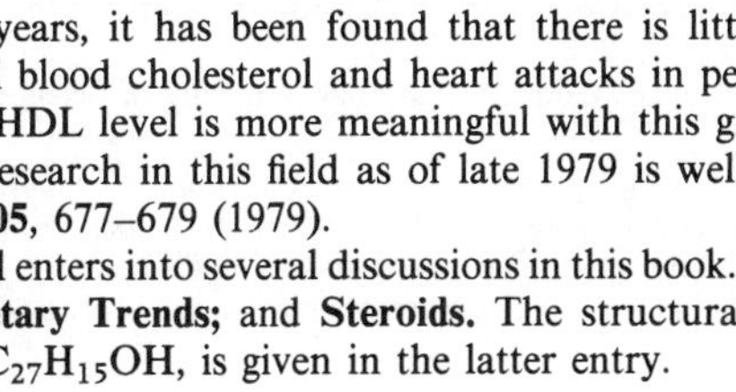

Acetylcholine assumed no importance to biologists until 1899 when Hunt identified the presence of choline in extracts of the adrenal glands and suggested that some derivative of choline was capable of causing a fall in blood pressure. This stimulated interest in studying the physiological effects of various choline derivatives and, in 1906, Hunt and Taveau found that acetylcholine was 100,000 times more effective than choline in causing a fall in blood pressure. Shortly thereafter, acetylcholine was identified in extracts of ergot, a fungus that grows on rye and other cereal grains. The first real proof of the role of acetylcholine in transmitting the effects of nerve stimulation did not come until 1921. The acetylcholine–cholinesterase system has served as the basis for the development of a number of drugs needed to alter the activity of the autonomic nervous system in certain disease states. Inhibitors of cholinesterase are used in insecticides. Parathion and malathion are examples of organic phosphate cholinesterase inhibitors that are effective for this purpose. Cholinesterase inhibitors are capable of producing poisoning and death in humans and domestic animals by the same mechanism.

Choline is an essential metabolic substance for building and maintaining cell structure. Choline is usually described along with B complex vitamins, although it is essentially a structural component of tissue rather than a metabolic catalyst. Choline is a part of the structure of phospholipids and acetylcholine.

Choline participates in normal fat metabolism and interrelates with methionine in a biochemical manipulation referred to as transmethylation. Choline, when in adequate quantity, can replace the essential amino acid methionine when the latter is in limited quantity; or the reverse may occur, that is, methionine can be dismantled to replace choline. Choline deficiencies result in numerous degradative physiologic changes in livestock. The usual dietary supplements are choline bitartrate and choline chloride.

References

DuBois, K. P.: "Acetylcholine and Cholinesterase," in "The Encyclopedia of Biochemistry," (R. J. Williams and E. M. Lansford, Jr., editors), Van Nostrand Reinhold, New York, 1967.

Haresign, W., and D. Lewis: "Recent Advances in Animal Nutrition," Butterworth Group, Woburn, Massachusetts, 1977.

Hunt, R., and R. Taveau: "On the Physiological Action of Certain Choline Derivatives and New Methods for Detecting Choline," *British Medical Journal*, **2**, 1788–1791 (1906).

Staff: "Food Chemicals Codex," National Academy of Sciences, Washington, D.C., 1972.

Staff: "Nutritional Requirements of Domestic Animals," separate publications on beef cattle, dairy cattle, poultry, swine, and sheep, National Academy of Sciences, Washington, D.C. (periodically updated).

Swan, H., and D. Lewis: "Feed Energy Sources for Livestock," Butterworth, London, 1976.

CHONDRICHTHYES. Sharks; Skates and Rays.

CHONDRITE. A term proposed by Rose in 1864, a chondrite is a stony meteorite characterized by chondrules embedded in a finely crystalline matrix consisting of orthypyroxene, olivine, and nickel-iron, with or without glass. Chondrites constitute about 80% of meteorite falls and are usually classified according to the predominant pyroxene, e.g., "enstatite chondrite," "bronzite chondrite," and "hypersthene chondrite." Chondritic meteorites have been considered as among the most primitive objects in the solar system—on the basis of their age ($\sim 4.6 \times 10^9$ years) and because of their lack of extensive chemical differentiation. However, Wetherill, Mark, and Lee-Hu (*Science*, **182**, 4109, 281–283, October 19, 1973) suggest that there is much more information concerning solar system processes contained in the strontium isotopic composition of meteorites and their constituents than has been revealed to date. Since the early 1950s, radioactive dating methods have been applied more and more to problems for establishing the chronology of the earth, moon, and solar system. Some of these methods involved a fixed percentage of error, which made it impossible to resolve events occurring within $\sim 100 \times 10^6$ years of one another early in the history of the solar system. Two methods have evolved for the resolution of the finer structure in the time scale for the formation of the solar system: (1) An approach based on the xenon daughter products of the extinct radioactive isotopes ^{129}I (half-life $= 17 \times 10^6$ years) and ^{244}Pu (half-life $= 80 \times 10^6$ years); and (2) an approach that depends on the initial isotopic abundance of radiogenic ^{87}Sr, the daughter of ^{87}Rb (half-life $= 50 \times 10^9$ years). In the latter method, the assumption is made that when formed the solar nebula had a uniform $^{87}Sr/^{86}Sr$ ratio (~ 0.698). Thus, if the solar nebula had a $^{87}Rb/^{86}Sr$ ratio, typical of chondrites (0.75), the $^{87}Sr/^{86}Sr$ ratio of the solar nebula will increase at the rate of 0.0001 per 9×10^6 years. Following this line of reasoning and as abstracted in the aforementioned paper, "A sodium-poor, calcium-rich inclusion in the carbonaceous chondrite Allende had a $^{87}Sr/^{86}Sr$ ratio at the time of its formation of 0.69880, as low a value as that found in any other meteorite. The higher $^{87}Sr/^{86}Sr$ ratios found in ordinary chondrites indicate that their formation or isotopic equilibrium occurred tens of millions of years later."

CHONDRODITE. Chondrodite, a magnesium fluosilicate mineral $Mg_5(SiO_4)_2(F, OH)_2$, crystallizing in the monoclinic system is a product of metasomatic origin in metamorphosed dolomitic limestones. Crystals are uncommon, usually occurring as discrete grains within the limestone, of light yellow to red color. Vitreous luster, translucent, with hardness of 6–6.5, and specific gravity of 3.1–3.2. This mineral is the most prominent member of minerals falling within the chondrodite group. These are norbergite, chondrodite, humite and clinohumite. Individual members of this group require optical evaluation for positive identification.

Noteworthy world occurrences include Mt. Somma, Italy; Pargas, Finland; Kafveltorp, Sweden; and in the United States at Brewster, and Warwick and Orange Counties, in New York.

CHONDROSTEL. The paddle-fishes and sturgeons. An order with the skeleton made up largely of cartilage but with some bony components.

CHONOLITH. A term proposed by R. A. Daly in 1905 for irregular igneous intrusions which according to their shapes and field relationships cannot be classified as dikes, laccoliths, batholiths, and bysmaliths.

CHOPPER. The term chopper is commonly used for two devices: 1. A device, usually mechanical, which imparts a pulsating characteristic to a current or a beam of light by a regular and frequent interruption. 2. A device which modulates a signal by opening and closing contacts periodically. The frequency of the chopper is usually greater than any frequency of interest in the signal.

CHOPPER AMPLIFIER. The term *chopper amplifier* has two connotations. In one type, the input signal is chopped (or modulated), amplified by an ac amplifier, demodulated, and filtered to provide a dc output signal. This type is described under **Carrier Amplifier.** In the second type, the error signal is chopped for the purpose of providing stabilization of gain and offset. Possibly, a more fitting term would be "chopper-stabilized amplifier."

With reference to the accompanying diagram, the unit is a chopper-stabilized amplifier with an overall gain A. The amplifier output is

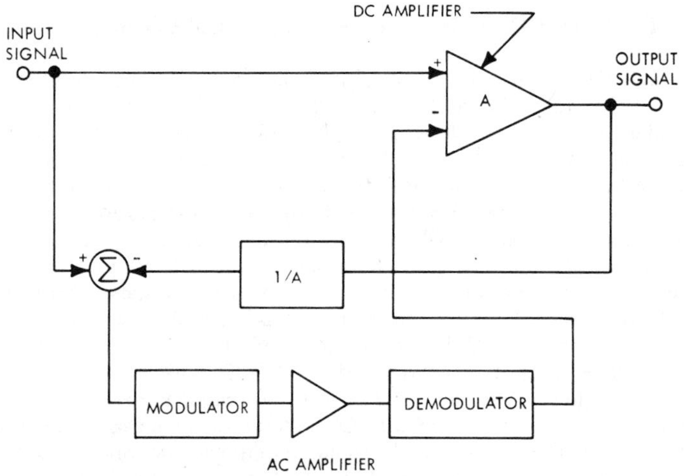

Chopper-stabilized amplifier.

attenuated by a factor of $1/A$ and then compared with the input signal. An error may exist as the result of zero-offset drift, a change in the gain of the dc amplifier, or simply noise. The error is chopped (modulated), amplified by an ac amplifier, demodulated, and then summed with the input to the dc amplifier—in a manner to compensate for the disturbance causing the error.

Excellent zero offset and gain characteristics are features of the chopper-stabilized amplifier, along with the favorable wide-bandwidth characteristics of the direct-coupled amplifier. The modulator-demodulator circuitry is relatively complex and, consequently, the design is not extensively used in digital-data acquisition and instrumentation amplifiers. Comparable performance can be obtained from conventional direct-coupled amplifiers. One negative feature of the chopper-stabilized amplifier is the frequently encountered long saturation recovery time, a deterrent for use in time-shared systems.

See also terms listed under **Data Processing.**

Thomas J. Harrison, International Business Machines Corporation, Boca Raton, Florida.

CHORD (Aircraft). The length from the leading edge to the trailing edge of an airfoil. The chord is a basic reference axis for the geometric or aerodynamic properties of an airfoil. It is normal to the span and lies in the plane of the airfoil. There are two of these reference chords. The one used for general and structural reference is the *geometric* chord. The other is an *aerodynamic chord*, being an imaginary line through the airfoil parallel to the free air stream at zero lift and passing through the trailing edge. The length of this chord is of no importance. It is useful mainly in aerodynamic studies because the lift varies directly with the angle of attack of the aerodynamic chord.

If the airfoil has a flat lower surface, an element of this surface is taken as the geometric chord. The chord length is the overall projection of the profile on this chord. In double-cambered airfoils the geometric chord is taken as the longest straight line possible between leading and trailing edges, or as a straight line joining the ends of the profile median line. The angle of attack to the geometric chord at zero lift is the angle between these chords. This may be discovered by wind tunnel tests, although empirical constructions have been devised which locate the aerodynamic chord surprisingly well. If the wing is tapered there is a tip chord and a root chord. The location of the intermediate chord on which the aerodynamic forces could be assumed to act is called the mean aerodynamic chord and is important in studies of airplane balance and stability. When the coefficient of lift may be assumed to be constant over the semispan, the mean aerodynamic chord coincides with the mean geometric chord (i.e., the centroid of the semiwing plan form). This simplification is in error if the wing has twist, or if it is rectangular, in which case the uneven downwash causes decreased lift coefficient near the tips.

See also **Airplane.**

CHORDATA. The vertebrates and a few marine animals of simpler form, including the tunicates, salpians and lancelets. Although the true vertebrates make up most of this phylum the inclusion of the other forms is scientifically accurate. With these limits the distinctive characters of the phylum are few. The animals are triploblastic, coelomate, and metameric like the higher invertebrates but differ from them in three points: (1) The skeleton is internal. In its primitive state it consists of a slender longitudinal rod lying above the alimentary tract and called the notochord. This structure is present at some stage in development in all of the included species. (2) The nervous system is entirely dorsal in position, lying in the body wall above the notochord. (3) The alimentary tract includes a chamber, the pharynx, just behind the oral cavity, whose walls are perforated by openings associated with respiration and called the gill slits or pharyngeal clefts. These openings appear or are indicated only in the embryos of terrestrial species.

It is difficult to estimate the relative importance of the phylum since man himself is one of the included species. The chordates include the most highly developed animals from the scientific point of view, and from the practical point of view they are equally important as the source of most of our animal foods, furs, feathers, wool, leather, and as beasts of burden. Man has depended on the vertebrates, indeed, for much of his progress, and has taken his domestic animals from this group.

The classification of the phylum is briefly as follows:

Subphylum *Hemichordata.* Worm-like marine animals. Balanoglossus. Also named *Enteropneusta.*

Subphylum *Urochordata.* Sessile or free-swimming forms, marine, with larvae resembling tadpoles in which the characters of the phylum are evident. Also named *Tunicata.*

- Class *Larvacea.* Small floating animals with the larval form of the subphylum.
- Class *Ascidiacea.* The tunicates. Sessile or free, named from the investing test or tunic which encloses them.
- Class *Thaliacea.* The salpians. Free-swimming.

Subphylum *Cephalochordata.* The lancelets. Small fish-like animals which swim freely and also burrow in the sand.

Subphylum *Vertebrata.* The skeleton includes cartilaginous or bony components in addition to the notochord. Also named *Craniata.*

- Class *Cyclostomata.* The round-mouth eels: lampreys and hags.
- Class *Chondrichthyes.* The cartilage fishes, such as the sharks, dogfishes, rays, and chimaeras. Sometimes called elasmobranches.
- Class *Osteichthyes.* The bony fishes. *Pisces* is an obsolete term.
- Class *Amphibia.* The salamanders, frogs, toads, etc.
- Class *Reptilia.* The lizards, snakes, turtles, crocodiles, etc.
- Class *Aves.* The birds.
- Class *Mammalia.* Popularly called animals without further qualification. They secrete milk for the nourishment of their young and the skin usually bears some hair, often a complete coat which may be in the form of fur or wool. Mice (mouse), horses, and cattle, monkeys, man, and many other forms.

CHORD (Mathematics). A segment of a straight line between two specified points of intersection of the line with a curve or surface. See also **Circle (Geometry).** In topology, the chord is an element belonging to the complement of a tree. See also **Tree (Mathematics).**

CHORDOTONAL ORGAN. An organ for the perception of vibrations—rod-like or bristle-like receptors for mechanical and sound vibrations—found in the insects where it may exist singly or in association with complex auditory organs. It consists of a nerve ending with accessory cells connected directly to the body wall or to some modified derivative of the body wall in an organ of hearing.

CHOREA (Huntington's). An inherited degenerative neurologic disorder that usually is not manifested until the adult years. In this disorder, there is progressive dementia combined with choreoathetosis, characterized by irregular movements and speech difficulties. In some patients, there is progressive rigidity rather than chorea. The latter form generally commences during childhood. The disorder progresses steadily over a number of years and creates deep depression in the patient, in some cases leading to suicide. The disorder is caused by biochemical abnormalities causing atrophy of the caudate nucleus (collection of nerve cells), among several other disturbances involving glutamic acid decarboxylase, gamma-aminobutyric acid (GABA), and reduction in activity of choline acetyl transferase. Some researchers have found a reduction of acetylcholine and serotinin receptors, but a normal number of GABA receptors.

An effective therapy remains to be developed. For controlling involuntary movements, haloperidol and chlorpromazine have been helpful in some patients. It is of great urgency for persons with a family history of this disorder to receive genetic counseling and thus avoid passing this serious disorder along to future generations. Considerable research is underway to find markers and methods of identifying people who have a potential for the disorder—both males and females who are contemplating raising a family.

CHOREA (Sydenham's). Sometimes called St. Vitus' dance, this chorea occurs in approximately 5% of cases of rheumatic fever. Onset is usually a number of months after the initial streptococcal infection and after the other usual symptoms of rheumatic fever have subsided. The patient's moves about involuntarily and purposelessly. The patient may appear clumsy and speech will be slurred. These movements disappear during sleep. Normally, the disturbance clears without residual disease. See **Rheumatic Fever.**

CHORION. 1. An accessory structure formed during embryonic development in mammals. It provides the connection with the tissues of the mother through which all interchange of materials between her blood and that of the embryo is carried on prior to birth.

The chorion is a composite structure formed of the serosa and the allantois, although the name is sometimes erroneously applied to the serosa alone. In some mammals a specialized placenta develops from part of the chorion as the persistent connection with the mother. Like all other extraembryonic membranes, this structure is discarded at birth.

2. The shell of an insect egg.

CHOROIDITIS. Uveitis.

CHOROLOGY. Zoogeography.

CHOUSINGHA. Bovines.

CHRISTMAS FERN. Ferns.

CHRISTMAS TREE (Gas Well). Natural Gas.

CHRISTOFFEL SYMBOL. One of certain quantities used in tensor analysis. They are not tensors themselves but relations involving the components and derivatives of tensors. They are of two kinds, often distinguished by bracket and brace, respectively:

$$[mn, q] = \frac{1}{2}\left(\frac{\partial g_{mq}}{\partial x^n} + \frac{\partial g_{nq}}{\partial x^m} - \frac{\partial g_{mn}}{\partial x^q}\right)$$

$$\{mn, q\} = \frac{g^{qs}}{2}\left(\frac{\partial g_{mq}}{\partial x^n} + \frac{\partial g_{nq}}{\partial x^m} - \frac{\partial g_{mn}}{\partial x^q}\right) = g^{qs}[mn, q]$$

where g_{mn} is a symmetric covariant tensor, $g^{qs} = G^{mn}/g$, with g the determinant of the components of g_{mn} and G^{mn} the cofactor of g_{mn} in g. See also **Tensor Contraction.**

CHROMA (or Munsell Chroma). The dimension of the Munsell system of color which corresponds most closely to saturation. Chroma is frequently used, particularly by English writers, as the equivalent of saturation.

CHROMATE TREATMENT. Conversion Coatings.

CHROMATIC ABERRATION. The indistinct color effects observed along the edges of images formed by a simple lens constitute what is known as the chromatic aberration of the lens. This abberation is due to the fact that the glass, or any other substance, out of which the lens is constructed produces dispersion (i.e., refracts light of different colors by different amounts). In a convergent glass lens the focal length is greater for red light than for blue, while in a divergent (concave surface) glass lens the blue focus is longer than the red. The effect for the convergent (convex surface) lens is shown, highly exaggerated, in the figure, in which the location of the image of a source, S, is shown for red light, R, to be in the plane, B, and for blue light, V, to be in the plane, A. An observer using an eyepiece focused for the plane A will see a sharp image of the source in blue light surrounded by the margin of a confused set of images of greater size in other colors.

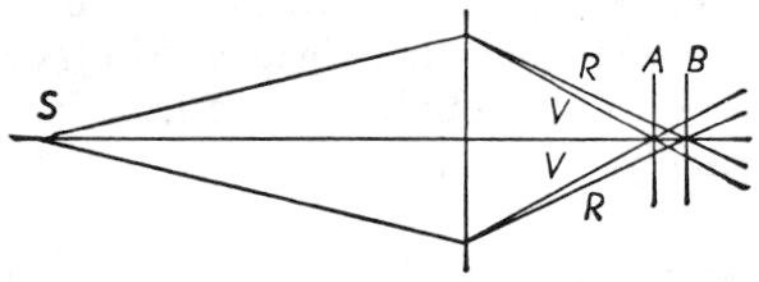

Chromatic aberration.

Chromatic aberration is very bothersome to users of lenses either for telescopic, microscopic, or photographic purposes and ever since optical instruments came into use attempts have been made to design "achromatic" lenses. For relatively short focus instruments, such as cameras or binoculars, practically complete achromatism can be obtained by using, instead of a simple convex lens, a combination of a convex and a concave lens, the two lenses being constructed of glasses of different dispersive powers. Prospective purchasers of field glasses, opera glasses or binoculars should always examine them carefully to determine whether or not the lenses are properly achromatized. A simple test is to examine the edge of a white building, which is in full sunlight, through the instrument under consideration. Move the image of the edge well over to one side of the field of view, and, if the lenses are properly figured, no color effects will appear. If, however, chromatic aberration is present, the image of the edge of the white building will be found to be bordered with a bright-colored fringe which increases in width as the image is moved closer to the edge of the field. In passing it might be said that the same test will indicate whether or not the field glasses are properly corrected for spherical aberration and other aberrations, for in a poor lens the image of the edge will become blurred and curved when moved to the side of the field of view.

In long focus instruments, such as astronomical telescopes, complete achromatism is virtually impossible. Partial achromatism may be obtained in such instruments by employing the combination of the divergent and convergent lenses of glass of different dispersive powers. [See **Dispersion (Radiation).**] Two types of partial achromatism are employed, depending upon the purposes for which the telescope is designed. In a telescope to be used for photographic purposes the colors which are most active photographically, i.e., the greens and blues, are all brought to the same focal point, whereas the reds and oranges are thrown well out of this focal plane.

In a so-called visual telescope the yellowish-green light is all brought to one sharp focus, while the blues and violets are bent well inside this visual focus. On looking at the image of a very bright object, such as the moon, a planet, or a bright star, a halo of bluish light can be observed due to the out of focus photographic light, but this halo is so diffuse that it is not objectionable when working with objects of the brilliance for which the instrument is designed. Instruments designed for visual observing (visual refractors) cannot be used satisfactorily for celestial photography without employing yellow sensitive plates and color filters to eliminate the out of focus blue and green light.

The image formed by a concave metal or silver on glass mirror is free from chromatic aberration since all colors are reflected in the same direction. This is the most important advantage of the reflecting telescope over the refracting type.

CHROMATIC DISPERSION (Optical Fiber). Optical Fibers; Telephony.

CHROMATICITY. The color quality of light definable by its chromaticity coordinates (which are based on matching a sample of light in terms of three stimuli of different, standard colors) or by its complementary or dominant wavelength and its purity taken together.

CHROMATICITY DIAGRAM. A plane diagram formed by plotting one of the three chromaticity coordinates against another. The most common chromaticity diagram at present is the CIE (x, y) diagram plotted in rectangular coordinates. It is shown in the accompanying figure.

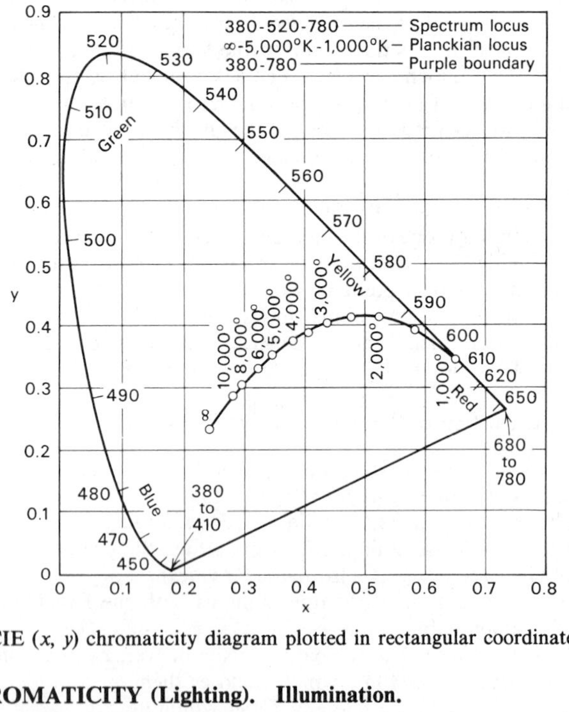

CIE (x, y) chromaticity diagram plotted in rectangular coordinates.

CHROMATICITY (Lighting). Illumination.

CHROMATICNESS. The two attributes of visual sensation, hue and saturation, taken together; as distinguished from brightness.

CHROMATIN. This term was given by Flemming to denote that substance in cell nuclei which, in the usual treatment with nuclear dyes, takes up the color. In nondividing nuclei, chromatin is distributed throughout the entire nucleus (euchromatin), but in nuclei which are undergoing cell division, chromatin is confined to the chromosomes (heterochromatin). In Flemming's time, the chemistry of the nucleus was entirely unknown. Even in view of recent knowledge, it is not fully understood as to what particular substance(s) have the special affinity for the dyes. In isolation of chromatin from pea embryos by differential centrifugation and purified by sucrose gradient centrifugation, the composition of chromatin was found to be deoxyribonucleic acid (DNA), 31%; ribonucleic acid (RNA), 17.5%; histone protein, 33%; and nonhistone protein, 18%. See also **Cell (Biology).**

CHROMATOGRAPHY. The generic name of a group of separation processes that have a common characteristic—*the separation depends upon the redistribution of the molecules of the mixture between a thin phase in contact with one or more bulk phases.* Thus, chromatography is a subclass of those separation processes which are applicable to molecular mixtures and which depend upon distribution between phases. The larger class includes distillation (vapor-liquid), solvent extraction or absorption distributions (liquid-liquid), sublimation (vapor-solid), and crystallization (liquid-solid) distributions, in all of which the phases are bulk phases. The prime difference of chromatography is that one of the phases is thin, often reaching molecular dimensions. For this reason, molecular size and shape play an exaggerated role in the separation. In essence, extremely subtle separations become possible because the limited volume in the thin film emphasizes differences in molecular size, shape, packing, and hence, orientation.

Fundamentals. A typical chromatographic separation (such as was described by Mikhael Tswett, the inventor of the method, 1906) is shown in Fig. 1, which illustrates the processes of adsorption chromatography.

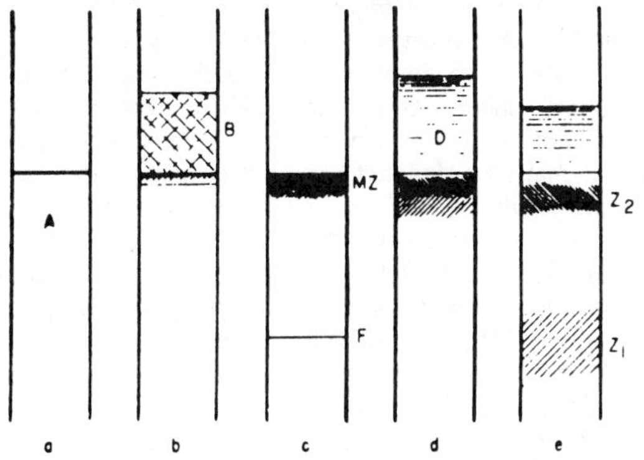

Fig. 1. Chromatography of a binary mixture. (A) Column of adsorbent in the chromatography tube. (B) Solution of mixture to be separated. MZ, mixed zone, applied to the column. F, front of empty solvent. D, developer, applied to the column. Z_1,Z_2, zones of separated components. To the packed column (a) the solution to be analyzed is applied (b). As it passes into the bed of adsorbent the components are retarded to form a mixed zone, while the solvent (also to some extent adsorbed) runs ahead (c). Developer is then applied (d), and as the development occurs the zones of separated material draw apart to produce the developed chromatogram (e).

A *column*, or *fixed bed*, of adsorbent, A, is formed in a tube by dry packing, or by pouring in a slurry from which the slurrying liquid is almost completely drained. This bulk phase will be the carrier of the thin phase, the adsorption layer. The mixture to be separated, B, dissolved in some poorly adsorbed solvent, is *applied* to the top of the column by letting the solution percolate into the bed. If the substances are adsorbed, they are removed from the solvent onto the surface of the adsorbent to form a *mixed zone*, MZ, while the solvent, emptied of solute, passes down the column. This behavior is the embodiment of the mass-action principle. The moving bulk liquid is referred to as the *mobile phase*, and the adsorbed layer as the thin *stationary phase*. The solute may move between the two in contact:

$$\text{Substances in Mobile Phase} \rightleftharpoons \text{Substances in Stationary Phase}$$

Adsorption Chromatography. When the solution is first applied to the adsorbent, the latter is *empty* of solute; there are no "substances in the stationary phase"; thus, by the principle of Le Chatelier, solvent is adsorbed and substances move out of the mobile phase, accumulating and forming the stationary phase of adsorbed solutes and solvent. But the mobile phase flows on down the tube, being replaced by fresh mixture and continually coming into contact with empty adsorbent. Thus, because of the *differential countercurrent movement* of mobile and stationary phases relative to each other, the solute is *all* removed from the solution, and forms a *mixed zone*, or *band*, at or near the top of the column. The upper part of this zone may have reached equilibrium with the solution; the lower edge may be more dilute and may already contain more of the less strongly adsorbed solute, relative to the other solutes, than was present in the original mixture: in other words, some slight separation may already have taken place. It is evident that a poorly adsorbed solvent is desirable in this process, so that it will not compete with solute for the limited surface area of the adsorbent. The *front* of the advancing solvent is shown at F. If the adsorbent is uniformly packed, the front or face of the zone will be quite regular. *The top of the bed must never be allowed to become dry*, else it will crack.

The next step is to *develop* the chromatogram. A solvent, slightly more strongly adsorbed than before, is applied to the column. Once again the principle of mass action takes effect. Empty *developer*, D, takes up solute from the stationary phase as it flows down the bed, eventually, perhaps, reaching equilibrium with the adsorbed material, but then as it reaches the front of the zone, it meets empty adsorbent, and so deposition of solute occurs as before. The developer starts empty of solute and emerges from the bed empty of solute. But in the intervening period it will have picked up solute and transported it downward. Moreover, the *less strongly adsorbed* material will be carried preferentially, since the more strongly adsorbed will occupy preferentially the available sites, or space, in the thin film of the stationary phase. One says that the more strongly adsorbed substance, Z_2—the substance preferentially distributed into the stationary phase—is *retarded* more than the other, Z_1, which *runs on ahead.* As development progresses, the zones separate until, in favorable cases, they are well and completely separated.

If development is continued, the faster zones may be *washed* successively out of the bed. Alternatively, the development may be stopped, the developer sucked from the column, and the column of adsorbent *extruded* from the tube by means of a closely fitting plunger or dowel. The zones may then be dissected out with attendant adsorbent, and the pure substance *eluted*, or *desorbed*, from each by means of a good solvent for the substance which contains, if needed, a strongly adsorbed material which *displaces* the substance from the surface. The most efficient way to carry out this step is to pack each zone as a bed in a small chromatographic tube, and apply the *eluent*, or *displacer*, to it.

For control of the process it is convenient to characterize the zone by an R_F value: the velocity of movement of the front edge of the zone divided by the velocity of movement of developer in the tube above the column of adsorbent. (Other, analogous values, are used with other chromatographic techniques).

Column Chromatography. Tswett's experiment is an example of *adsorption chromatography.* In *partition chromatography*, the stationary phase is water, held in the column by a porous solid such as silica gel or starch. The mobile phase is an immiscible liquid such as benzene or chloroform. In *reversed-phase partition chromatography*, the mobile phase is water, and the stationary phase is the immiscible organic liquid held in place by a porous hydrophobic solid such as beads of polystyrene resin, rubber, or silica gel coated with a silicone. Cations (or anions) can often be separated from each other by elution through a column of cation-exchange (or anion-exchange) resin with an aqueous salt solution as eluent. This is *ion-exchange chromatography.* Ion-exchange resins can also be used for the separation of water-soluble organic compounds with water as the eluent. This is really a type of reversed-phase partition chromatography, the resin serving as the stationary organic phase. Such separations are greatly facilitated by using an aqueous salt solution as eluent. This method is called *salting-out chromatography.* Organic compounds of insufficient solubility in water can often be separated chromatographically with an ion-exchange resin as the stationary phase and an aqueous solution of an organic solvent such as alcohol, acetone, or acetic acid as eluent. This modification is called *solubilization chromatography.*

Each of the foregoing types of chromatography can be further subdivided as follows. In *elution chromatography*, a small amount of sample solution is added to the column, previously freed from tightly sorbed material. Then a suitable solvent is passed through the column, and the sample constituents are separated into bands as in Tswett's experiment. This type is most useful in analytical chemistry because the various sample constituents can be quantitatively separated from each other. In *frontal chromatography*, no eluent is used, but the solution of the sample is passed continuously into the column. The least sorbed of the sample constituents emerges first from the column, free from the other solutes of the sample. Then a mixture of this compound and the next least readily sorbed compound emerges, then a mixture of the three least readily sorbed, and so on until finally the effluent has the same composition as the sample. *Displacement chromatography* has some characteristics of both elution and frontal chromatography. A rather large sample is first added to the column, previously freed of tenaciously sorbed compounds. Then a solution of a vigorously sorbed compound is used as eluent or displacing agent. A band of the sample constituents moves down the column in front of the highly sorbed eluent. As it moves, it is gradually separated into several bands, each containing only one constituent of the sample. At the boundaries of these bands, there is always a mixture of the solutes of the two bands. Since large quantities can be used and since a large fraction of each sample constituent can be recovered uncontaminated by the other solutes, this type of chromatography is used mostly in preparative work.

Figure 2 illustrates the behavior of a mixture of lithium, sodium, and potassium ions in the three types of ion-exchange chromatography.

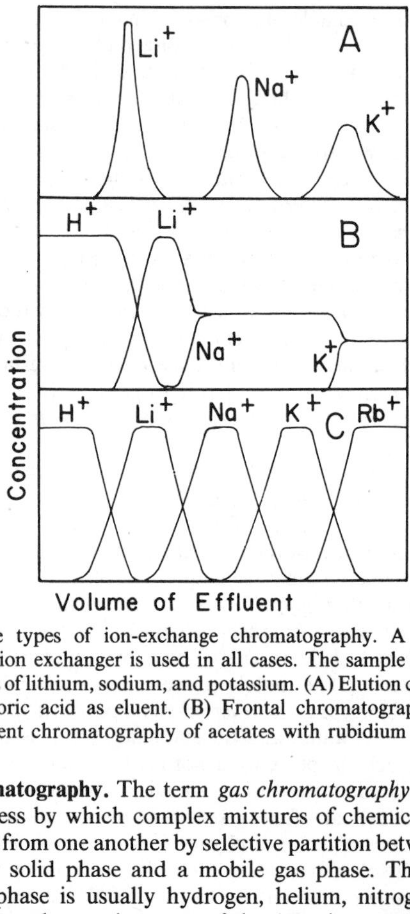

Fig. 2. Three types of ion-exchange chromatography. A hydrogen-form strong-acid cation exchanger is used in all cases. The sample is an equimolar mixture of salts of lithium, sodium, and potassium. (A) Elution chromatography with hydrochloric acid as eluent. (B) Frontal chromatography of acetates. (C) Displacement chromatography of acetates with rubidium acetate.

Gas Chromatography. The term *gas chromatography* is used to describe a process by which complex mixtures of chemical compounds are separated from one another by selective partition between a stationary liquid or solid phase and a mobile gas phase. The gas used as the moving phase is usually hydrogen, helium, nitrogen, or argon, depending upon the requirements of the detection system. Sometimes the compounds which are separated from one another are also gases, but more often they are vapors derived from solid or liquid samples. Thus this method can be used to resolve mixtures of permanent gases such as carbon dioxide, oxygen, and nitrogen at room temperature, or it can be employed to separate vapors of compounds such as steroids, fatty alcohols, and fatty acid esters at higher temperatures which are still far below their boiling points.

The stationary phase is usually contained within a glass or metal tube 0.2–1 centimeters in diameter and 0.5–10 meters long. When the stationary phase is an active solid such as molecular sieve or silica gel, the method is termed gas-solid chromatography. A very important advance was made in 1952 when James and Martin introduced gas-liquid partition chromatography. In this technique, the stationary phase is a high-boiling liquid coated on an inert solid support such as a diatomaceous earth or small glass beads. Materials which are useful as stationary liquids include silicone polymers, polyesters, polyethylene glycols and high molecular weight esters of organic acids. The separations obtained depend upon the properties of the stationary liquid phase: hydrocarbons often separate sample components in order of boiling point, and oxygenated hydrocarbons and nitriles in order of polarity, other things being equal. The stationary phase must be essentially nonvolatile at the operating temperature of the column, which will generally be between −50 and 300°C. A well-prepared column, 5 millimeters in diameter and 1 meter in length, will often provide an efficiency of 1000–2000 theoretical plates.

Detectors. The detection system which is employed will depend upon the nature of the problem. The detectors used most commonly for the analysis of inorganic gases or the isolation of relatively larger amounts of organic compounds from impurities are *thermal conductivity cells.* These can be thermistors or catharometers. Thermal detectors can measure as little as 10^{-8} gram of each component of a complex mixture. For trace analysis and most biochemical work, the *flame ionization* and *argon detectors* are much more useful. The flame ionization detector responds to as little as 10^{-11} mole of each compound in the carrier gas, has a greater dynamic range than the argon detector, and is much more sensitive to organic compounds than the thermistor or catharometer. It is insensitive to water and most other inorganic compounds, and is highly sensitive to all organic compounds containing methylene groups. Therefore, it is the method of choice for the detection of compounds of biochemical interest when semi-universal response is desired. However, in the analysis of tissue extracts, it is sometimes desirable to employ more specific detection methods to eliminate interferences and reduce background noise. For example, the *electron capture detector* can be used to detect chlorinated hydrocarbons and many conjugated unsaturated compounds with a high degree of selectivity, since it is highly sensitive to these substances while being relatively insensitive to hydrocarbons. The *microcoulometer* provides even higher selectivity since it can be used in-stream for the quantitative analysis of chlorine, sulfur or phosphorus in organic compounds. In this method, organic compounds are oxidized or reduced directly in the gas stream with oxygen or hydrogen. The HCl, SO_2, PH_3, or H_2S formed during these processes is measured electrochemically.

Capillary Columns. Chromatography can also be carried out in long capillary columns, the stationary phase consisting of a thin film of organic material coated on the internal surface of the capillary tube. The tube thus serves as the solid support. A 50-meter column can be coiled so that it is very compact. It will often provide resolutions up to 150,000 theoretical plates. For some materials such as hydrocarbons, capillary columns far exceed packed columns in their capacities for separating complex mixtures of closely related compounds. However, they are not as generally applicable as packed columns to microanalytical work since retention values are sometimes difficult to reproduce, and the total amount of material which can be applied to the column is comparatively small.

Applications. The applications of gas chromatography to biochemical problems are many and varied. Gas-solid chromatography can be used for measuring respiratory and photosynthetic gases, while gas-liquid chromatography can be used to measure the volatile compounds responsible for food aromas, terpenes and essential oils, lipids. Even nonvolatile compounds such as amino acids and oligosaccharides can be chromatographed after their conversion to volatile derivatives. Gas chromatography is sometimes used to analyze these materials directly. However, a more satisfactory approach is to first separate the sample into groups of related compounds by ancillary procedures such as distillation, liquid-liquid partition, liquid-liquid chromatography, ion-exchange chromatography, or thin-layer chromatography. Gas chromatography is then used for the final analysis. This provides for better resolution than when the method is used directly.

Permanent gases of interest to the biochemist which can be analyzed by this method include carbon dioxide, oxygen, and nitrogen. Oxygen and nitrogen are usually resolved from one another on molecular sieve. However, carbon dioxide is not eluted from this stationary phase except at high temperatures. Therefore, for a complete analysis of respiratory or photosynthetic gases, two columns are used—either in series or in parallel. Molecular sieve is used to separate oxygen from nitrogen, while silica gel is used to resolve carbon dioxide from a composite nitrogen-oxygen peak. The complete analysis can be carried out on molecular sieve alone if the temperature is programmed.

Organic compounds present in the atomosphere due to air pollutionor exudation from vegetation may also be of interest to the biochemist. Air samples are usually not analyzed for trace components directly. Instead, the air is passed through a cold trap which contains an inert (or in some cases an active) solid. Trace organic components are condensed or adsorbed on the solid. The trap is attached to the chromatograph, and its contents are vaporized by heat and flushed into the column. Stationary phases which have been used for the analysis of air pollutants include di-*n*-butylphthalate, aluminum oxide modified with propylene carbonate, squalane, and solutions of silver nitrate in ethylene glycol. This latter liquid is particularly useful for separating alkenes from alkanes and for resolving isomeric olefins.

Volatile compounds present in foods and beverages are amenable to analysis by gas-liquid chromatography. These are defined arbitrarily as materials containing one, or rarely two, functional groups, and not more than eight to ten carbon atoms. They are found in the extracts or distillates of fruits, meats, and vegetables, or in the condensates obtained on freeze-drying these products. They are usually mixtures of lower fatty acids, amines, carbonyl compounds, thiols, sulfides, alcohols, and esters. Often they are chromatographed as complex mixtures, but more meaningful results can sometimes be obtained by prefractionating them. Thus, fatty acids can be isolated as a group by distilling off neutral and basic compounds from alkaline solutions, and recovering the free acids after acidification through distillation or extraction. Carbonyl compounds can be isolated as a group by precipitation with 2,4-dinitrophenylhydrazine, while thiols and sulfides can be separated by the formation of compounds with salts of heavy metals. After isolation, the various groups of related compounds are chromatographed separately. Complex mixtures of volatiles have been chromatographed on a wide variety of liquid substrates at temperatures ranging from −80 to 200°C. No single set of conditions is satisfactory for all samples. Usually it is best to chromatograph complex mixtures on both polar and nonpolar liquids to obtain the maximum numbers of peaks. Very often, additional information can be obtained by collecting unresolved fractions and rechromatographing them on a stationary liquid having characteristics different from the ones used for the primary separation.

Gas chromatography is also an excellent tool for the separation, characterization, and quantitative analysis of essential oils. Separations of components that formerly took days by tedious chemical and physical means, or were impossible by these older methods, can be accomplished in minutes. Terpene hydrocarbons and related oxygenated terpenoids are the main constituents of essential oils. The oils may be analyzed directly, or prefractionated into hydrocarbons and oxygenated hydrocarbons. Generally, the latter is accomplished by liquid-solid chromatography on silica gel, but preparative-scale gas chromatography can also be used since the terpenes as a group are eluted well ahead of the oxygenated compounds.

Gas chromatography has proved to be more successful for the analysis of lipids than for any other group of compounds. Here again, prefractionation is desirable before injection of a sample into the chromatograph. This is usually accomplished by liquid-solid chromatography on silica gel or thin-layer chromatography. These procedures separate the sample into fractions consisting of phospholipids, sterol esters, triglycerides, and higher fatty acids. Some lipids are chromatographed directly while others are converted to volatile derivatives. The following classes of compounds are usually modified chemically before injection: phospholipids, sterol esters, higher fatty acids, O-alkyl glycerols, and higher aldehydes. Sterols and higher fatty alcohols can be chromatographed either unchanged or as derivatives. Lipids are usually chromatographed on polar and nonpolar stationary liquids to obtain optimum resolution. Chromatography of the methyl esters of the higher fatty acids on nonpolar liquids such as "Apiezon" (a high-boiling hydrocarbon) results in separation according to boiling point. By contrast, chromatography on polar phases such as polyesters [*e.g.*, poly(diethyleneglycol succinate)] leads to improved separation according to degree of unsaturation. Thus on polar liquids, saturated acids are eluted ahead of monoenes, monoenes ahead of the dienes, and dienes ahead of trienes for compounds containing the same carbon skeleton.

Compounds which are ordinarily considered to be nonvolatile can be separated by gas chromatography by first converting them to volatile derivatives. Thus, amino acids can be separated as N-acetyl alkyl esters, N-trifluoroacetyl alkyl esters, or trimethylsilyl derivatives. Di- and polybasic organic (Krebs cycle) acids can be chromatographed as their methyl esters, and oligosaccharides can be separated after converting their OH groups to OCH_3 groups. Excellent methods of separation have been developed for many of these compounds, but some problems still remain to be solved in quantitating results. It is certain that gas chromatography will continue to grow in usefulness in this field with continued advances in technology.

Process-gas Chromatography. A system for continuous, repetitive, and fully automatic on-line analysis of process streams similar in all essential elements of the basic technique to the laboratory chromatograph, but different in design and appearance. Factors affecting design include (1) the need to comply with the National Electrical Code for operation in hazardous atmospheres, (2) the need to automate the procedure, and (3) the need for ready adaptability to closed-loop process control and communication with computers. Demand for maximum reliability and minimum maintenance has emphasized simplicity of hardware and methodology. Emphasis is placed on analyzing for a few rather than a number of components and on minimizing analysis time. These design targets have resulted in extensive use of multicolumn techniques for rapid separation of selected components, with large portions of the sample being discarded. As shown in Fig. 3, the major components of a process-gas chromatograph are (1) the *analyzer*, (2) the *programmer*, and (3) one or more *recorders*. In some cases, (4) a *sample conditioning* system and (5) a *stream selector* are required. See Fig. 4.

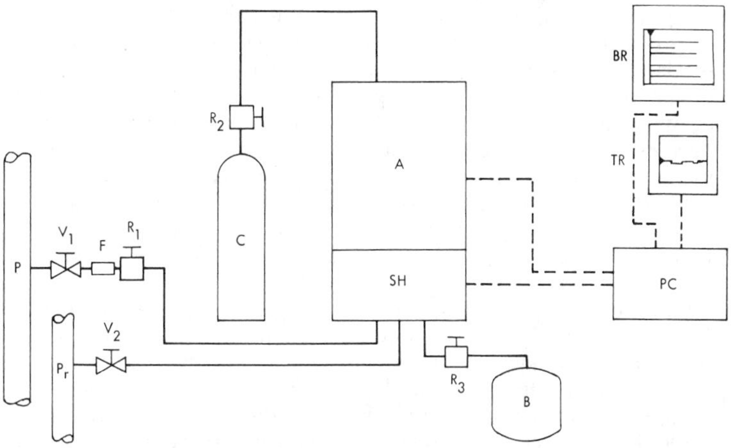

Fig. 3. Basic elements of a process-gas chromatograph. Vapor sample is continuously withdrawn at high rate from process line *P*, circulated through sample conditioner *SH*, and returned to lower pressure point P_r, through shutoff valves V_1 and V_2. Particulate matter is removed by filter *F*, and pressure reduced to constant low-level by regulator R_1. Sample conditioner contains flow control and other conditioning components and valve for switching to synthetic calibration blend *B* through pressure regulator R_3. Sample slipstream is circulated to sample valve in analyzer *A*, which also contains columns, detectors, and temperature-control system. Carrier gas *C* is controlled by regulator R_2. Programmer *PC* contains detector power supply, component attenuators, timer for controlling analyzer functions, and means for operating calibration blend valve in sample conditioner on demand. Programmer converts signal to form suitable for recording as bar graph on recorder *BR* or trend on trend recorder *TR*. (*Beckman Instruments, Inc.*)

Detectors. Thermal conductivity (for general use) and hydrogen-flame ionization (for trace organic analysis) are widely used detectors. Helium ionization (radioactive and photoionization) and thermal conductivity with amplification are used for trace inorganic and inert-gas analysis. Halogen detectors are used for applications that require sensitivity and specificity for halogen compounds.

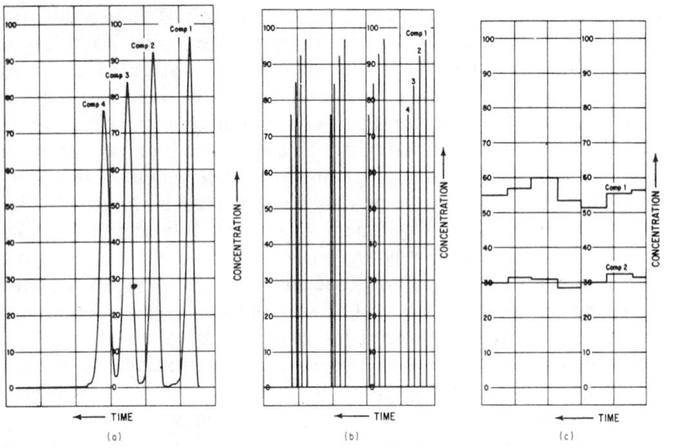

Fig. 4. Typical chromatograph readouts: (a) chromatogram; (b) bar graph; and (c) trend. (*Beckman Instruments, Inc.*)

Computer-controlled Chromatograph System. Large-process chromatograph systems may use a dedicated low-cost computer to control all analyzer functions and perform all data reduction, thereby eliminating the individual programmers. The computer monitors each individual chromatograph detector, performs integration of peak areas, corrects for zero offset and drift, identifies components from elution times, applies response factors, and computes the composition of the sample by normalization of peak areas or by comparison with calibration standards for select components. Partial or complete stream analysis may be printed out on a teletypewriter, continuously or on demand, in accordance with program format. The computer may simultaneously provide analog trend output to conventional recorders for closed-loop control. The computer may also communicate analysis results to a larger supervisory computer (e.g., for material-balance calculations or for direct digital control). High- and low-level alarms for select components may also be set by the computer. The computer program may also include subroutines for detecting and alarming in case of malfunction or failure in chromatograph analyzers.

Liquid Chromatography. This method is particularly useful for separation and analysis of high-molecular-weight compounds which are beyond the range of gas chromatography. It is generally classified according to type of stationary phase: (1) liquid-solid (adsorption), (2) liquid-liquid (partition), (3) ion-exchange, and (4) gel-filtration (gel-permeation). Each type imposes specific requirements on apparatus design. The liquid is moved through the system by gravity or constant-flow pumps. Gradient devices make stepwise or continuous changes in the composition of moving phase during analysis (gradient elution). The differential refractometer probably is the most widely used detector; other detectors are based on visible, infrared, or ultraviolet photometry. Eluting compounds may be reacted with reagents to form colored substances for ease of detection. Other detectors used include a heat-of-adsorption detector and a modified hydrogen-flame-ionization detector.

Paper Chromatography. Although paper chromatography as it is now conceived and practiced was antedated by Goppelsroeder's capillary analysis (1899) which is related to it, its modern beginnings date from a classical paper by Consden, Gordon and Martin in 1944, who successfully used paper chromatography to separate amino acids and to make possible analyses of amino acid mixtures.

The first step in making a paper chromatogram consists in applying a small drop of the solution to be analyzed (in quantitative work 5 microliters is often used) about 2–3 centimeters from the edge of a suitable piece of filter paper and allowing it to dry. Usually an organic solvent (or solvent mixture) containing water is then allowed to flow over and past the spot to a distance a few decimeters beyond. As this movement takes place, the various substances in the spot are picked up and transported by the moving solvent mixture. If the substances are extremely soluble in the solvent mixture used, they are carried along completely with the solvent front and no separation takes place. If they are extremely insoluble, the substances remain on the spot of origin; this, of course, results in no separation. If, however, the substances are soluble to an appropriate degree, each is transported to some intermediate position between the original spot and the edge of the solvent front. The choice of suitable solvent mixtures is thus crucial and depends on the nature of the substances to be separated.

Different substances have their distinctive affinities for water, for paper and for the constituents of the organic solvent mixture used. If in a given system a particular substance travels characteristically halfway to the solvent front, it is said to have a "R_F value" of 0.50. If it travels $\frac{1}{4}$ or $\frac{3}{4}$, respectively, of the distance traveled by the solvent, it is said to have an R_F value of 0.25 or 0.75.

In the original 1954 paper, the authors used 18 different solvent systems employing phenol, collidine, *n*-butanol, *t*-amyl alcohol, benzyl alcohol, *o*-, *m*-, and *p*-cresols with various additions and modifications. In a phenol-NH_3 system, for example, the various amino acids had R_F values ranging from 0.12 (aspartic acid) to 0.90 (phenylalanine). Two or more amino acids may have R_F values about the same in a given solvent system, in which case they superimpose on one another. In this case a different solvent system must be used to separate them.

Paper chromatography is particularly well adapted to amino acid analysis because after the chromatograph is complete and the solvents have been removed by evaporation, the positions of the amino acids can be made visible by spraying with ninhydrin solution which produces a typical color (in most cases blue or purple) wherever the amino acids accumulate. Various means have been used, including visual comparison with standards, to obtain quantitative results. These have been rather successful.

In the pioneering work (1944), Consden and co-workers used an apparatus in which strips of paper were hung in a vertical position and the solvent mixture was allowed to flow downward over the spot from a trough. In 1948, R. J. Williams introduced a modification which involved using a cylinder of filter paper (made by clipping together the edges of a square or rectangular sheet) on which as many as 18 spots had been placed on a line about 3 centimeters from the bottom edge. The cylinder of filter paper (about 30 centimeters in height) was then placed upright in a shallow pool of the solvent mixture, and the mixture was allowed to ascend by capillary action over the various spots, carrying the substances with various R_F values upward in vertical lines above the corresponding spots.

An interesting variation used by Martin and his co-workers was that of two-dimensional paper chromatography. In this case, a single spot is placed about 3 centimeters from the corner of a square sheet and chromatographed in one direction along the edge of the sheet. After drying, the sheet is then placed so the solvent (a different mixture from that initially used) flows at right angles to the original flow. By this means, amino acids which were not separated the first time may be well separated the second time provided the two solvent mixtures are suitably chosen. This procedure can readily be used also in the ascending technique, in which case the first cylinder used is unfastened after chromatography and drying, and clipped together into a cylinder which has an axis at right angles to first cylinder. Regardless of the particular technique used, two-dimensional chromatographs require far more paper and time, and cut down the number of analyses that can be made in a restricted laboratory.

Small temperature changes sometimes have little effect on R_F values; in other cases they may affect these values markedly. Whatever apparatus or system is used, temperature control is essential for the best results. The carrying out of paper chromatographic investigations is subject to many ramifications and modifications. The particular technique which may prove most valuable will depend to a large degree upon the nature of the particular problem at hand.

Paper chromatography is applicable to analysis on a micro scale. A fraction of a microgram of an amino acid can be detected on a chromatographic sheet, and a solution to be analyzed need not contain more than 10 micrograms/milliliter of each of the amino acids. Since paper chromatography involves little apparatus, particularly when the ascending technique is used, it has been used with more or less success in a host of laboratories, not only for amino acid separations and analyses, but for many other analyses—amines, urea, proteins, sugars, polysaccharides, phosphate esters, aliphatic acids, lipids, steroids, purines, pyrimidines, phenols, aromatic acids, porphyrins, alkaloids, pigments, antibiotics, vitamins, and inorganic ions such as chloride, sulfate, phosphate and sodium.

References

Griffin, D. E., and P. U. Webb: "Process Chromatographs and Computers in Optimizing Control Systems," *In-Tech*, **26**, 7, 47–51 (1979).

Leathard, D. A., and B. C. Shurlock: "Identification Techniques in Gas Chromatography," Wiley, New York, 1971.

Martin, F. D.: "Developments in Process Gas Chromatography," *In-Tech*, **24**, 1, 51–59 (1977).

Rich, W. E.: "Ion Chromatography," *In-Tech*, **24**, 8, 47–51 (1977).

Shellard, E. J.: "Quantitative Paper and Thin Layer Chromatography," Academic, New York, 1968.

Touchstone, J. C., and M. F. Dobbins: "Practice of Thin Layer Chromatography," Wiley, New York, 1978.

Weiss, M. D.: "Gas Chromatographs versus Mass Spectrometers on Line," *Control Eng.*, **24**, 9, 66–68 (1977).

Yancy, J. A. (editor): "Guide to Stationary Phases for Gas Chromatography," 12th edition, Analbas, Inc., North Haven, Connecticut, 1979.

Zlatkis, A., and V. Pretorius (editors): "Preparative Gas Chromatography," Wiley, New York, 1971.

CHROMATOPHORE. A general name for a definite body occurring in the cytoplasm of some cells. A characteristic of a chromatophore is that it should have a definite color, due to the pigment or pigments present in it.

In plants the most common chromatophore is the chloroplastid (see **Chloroplast**). Other chromatophores, called chromoplastids, are of various colors, including yellow, brown, orange, and red. Not all plant pigments are found in chromatophores, however. Many occur dissolved in the cell sap.

Chromatophores found in the skin of animals of several groups, including arthropoda, mollusca, and vertebrata, are large cells, often extensively branched. They are well developed in the octopus and squid, in many amphibians and fishes, and in reptiles.

Rapid changes in color such as those of squids and some lizards have been ascribed to a contraction of the chromatophores, but it now seems evident that the cell itself does not contract although the pigment within it may undergo a considerable change in distribution, revealing itself when widely distributed and otherwise concealed from view.

Starch is the first visible product of photosynthesis; in most cases it is elaborated as small grains visible within the chloroplast. In the blue green algae and certain green algae, the pigment seems to be coextensive with the general cytoplasm of the cell. In some lower forms, the pigment is found in elaborate bands or nets (chromatophores). In these, the starch may accumulate about dense proteinaceous granules called pyrenoids. See accompanying figure. This starch is known as pyrenoid starch, while the more usually occurring starch grains are called stroma starch. Both types may occur in the same chromatophore. In some red algae, starch grains regularly occur outside of the chloroplasts in the general cytoplasm of the cell. In many plants, as the leaves or fruits mature, they change from green to red to brown. This is caused by the displacement of the chlorophyll within the chloroplasts by carotenes or xanthophylls, and the change is apparently not reversible. Plastids having a preponderance of these pigments are known as chromoplasts. In potatoes or iris roots, food may be stored in colorless plastids known as leucoplasts.

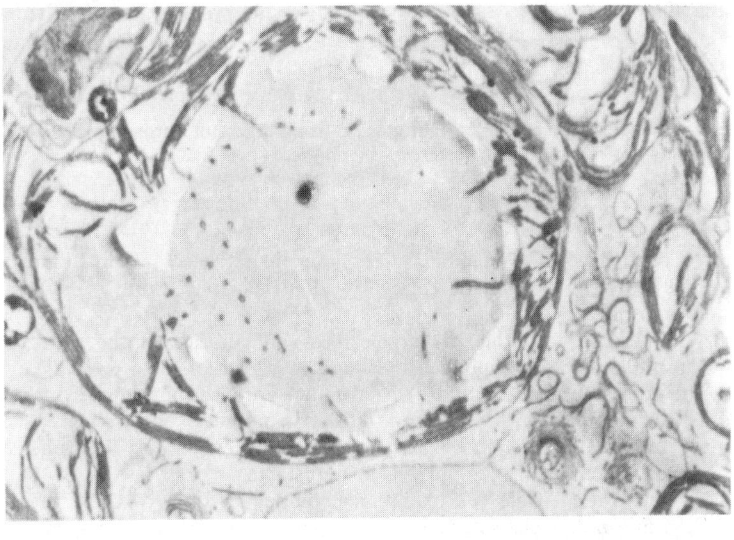

Portion of a cell from the green alga *Scenedesmus* showing the spherical gray pyrenoid surrounded by the electron transparent starch grains (~ ×8000).

CHROMINANCE. The colorimetric difference between any color and a reference color of equal luminance, the reference color having a specified chromaticity.

CHROMITE. An important mineral in the chromite series of multiple oxides. The dominant compound is $FeCr_2O_4$, but most chromite also contains magnesium Mg and aluminum Al. Crystallizes in the isometric system. Hardness, 5.5; specific gravity, 4.5–4.8; color, black. Associated with peridotite and serpentine (metamorphized peridotite). Commercial amounts occur as placer deposits in serpentine areas. In the United States chromite has been mined in Maryland, Pennsylvania, California, Montana, Oregon, Wyoming and North Carolina. Important deposits occur in Quebec, Canada. Valuable deposits occur in Asia Minor, Rhodesia, New Caledonia, Cuba, India, and the Philippines; also in New South Wales, and in the Urals associated with platinum.

CHROMIUM. Chemical element symbol Cr, at. no. 24, at. wt. 51.996, periodic table group 6b, mp 1837–1877°C, bp 2671–2673°C, density 7.2 g/cm^3. Elemental chromium has a body-centered cubic crystal structure. The metal is silver-white with a slight gray-blue tinge, very hard (9.0 on the Mhos scale), capable of taking a brilliant polish, not appreciably ductile or malleable. The element is not affected by air or H_2O at ordinary temperatures, but when heated above 200°C, chromic oxide Cr_2O_3 is formed. There are four stable isotopes ^{50}Cr, and ^{52}Cr through ^{54}Cr. Four radioactive isotopes have been identified, all with comparatively short half-lives ^{48}Cr, ^{49}Cr, ^{51}Cr, and ^{55}Cr. The element was first identified by Vauquelin in 1797.

Ionization potential 6.76 eV; second, 16.6 eV. Oxidation potentials $Cr \rightarrow Cr^{3+} + 3e^-$, 0.71 V; $Cr^{2+} \rightarrow Cr^{3+} + e^-$, 0.41 V; $2Cr^{3+} + 7H_2O \rightarrow Cr_2O_7^{2-} + 14H^+ + 6e^-$. − 1.33 V; $Cr + 3OH^- \rightarrow Cr(OH)_3 + 3e^-$, 1.3 V; $Cr + 4OH^- \rightarrow CrO_2^- + 2H_2O + 3e^-$, 1.2 V; $Cr(OH)_3 + 5OH^- \rightarrow CrO_4^{2-} + 4H_2O + 3e^-$, 0.12 V.

Other important physical properties of chromium are given under **Chemical Elements.**

Chromium occurs chiefly as chromite (ferrous chromite) $Fe(CrO_2)_2$ in Zimbabwe, the Republic of South Africa, the U.S.S.R., New Caledonia, India, Philippine Islands, Japan, Turkey, Greece, Cuba, and California. (1) Heating chromite in the electric furnace with carbon yields ferrochrome for alloys, and (2) when chromite is heated with sodium carbonate and nitrate, sodium chromate is formed, which is then extracted with H_2O. This is the substance from which chromium compounds are obtained. See also **Chromite.**

Chromium is used extensively for (1) decorative and wear-resistant electroplating, (2) many important alloys, and (3) the manufacture of numerous chemicals and refractory materials.

Alloys. In constructional steels, chromium imparts hardness by improving hardenability and promoting the formation of carbides. These steels have exceptional wear resistance and are relatively stable at elevated temperatures. In stainless and heat-resisting steels, chromium improves corrosion and heat resistance. As shown by the accompanying table, stainless steels may be grouped into three principal classes (1) austenitic, (2) martensitic, and (3) ferritic stainless steels. The fundamentals of stainless steels are further described under **Iron Metals, Alloys, and Steels.** Stainless steels, as a group of major ferrous alloys, are characterized by their high degree of resistance to chemical attack. They possess a property commonly referred to as passivity, a property eminently displayed by elemental chromium. This property is manifested in steels when the chromium content exceeds about 11%. A very marked improvement in corrosion and heat resistance is achieved when chromium is added to this or a greater extent to low-carbon steels. The addition of nickel, along with chromium, enhances these properties even more. Although a steel that contains 12% chromium will stain, it will not undergo progressive rusting in normal atmospheres. When the chromium content is increased to 18%, staining will not occur in normal atmospheres, but may stain to a limited extent in particularly bad industrial environments. However, the addition of 8% nickel, along with the 18% chromium, will make the steel stain-resistant in all but the most rigorous industrial atmospheres. Additional resistance to corrosion and heat resistance may be obtained by the presence of molybdenum and other elements in smaller quantities.

An example of the effect of chromium content on the corrosion resistance of steel is demonstrated as follows:

A low-carbon steel placed in boiling 65% HNO_3 for one month:

4.5% chromium	corrosion rate, 12.9 in.
8.0% chromium	0.14
12.0% chromium	0.01
18.0% chromium	0.003
25.0% chromium	0.0006

The corrosion resistance of iron-chromium alloys was known in England and France in the early 1800s, but passivity was not clearly recognized and reasonably understood until 1910 as the result of studies by Borchers and Monnartz in Germany. Commercial stainless steels were introduced shortly thereafter in Germany, France, England, and a bit later in the United States. Worldwide production of stainless steels now is measured in terms of millions of tons annually.

REPRESENTATIVE STAINLESS STEELS

Type	Composition and Characteristics
	Austenitic Stainless Steels
302	Basic alloy of the group: Cr, 17–19%; Ni, 6–8%
302B	Silicon (2–3%) added to increase scaling resistance
303	Sulfur (0.02–0.05%) added to improve machinability
303Se	Selenium (0.15%) added to improve machinability
304L	Extra low carbon for improved weldability
304	Lower carbon content to improve weldability and inhibit carbide formation: Cr, 18–20%; Ni, 10–12%
305	Nickel increased to lower work-hardening: Cr, 17–19%; Ni, 10–13%
308	Chromium and nickel increased to increase corrosion and scaling resistance: Cr, 19–21%; Ni, 10–12%
309S	Lower carbon content (0.08% max) for improved weldability
314	Silicon added for increased scaling resistance at high temperatures: Cr, 23–26%; Ni, 19–22%; Si, 2% max
316	Molybdenum added to improve resistance to pitting corrosion and strength at high temperatures: Cr, 16–18%; Ni, 10–14%; Mo, 2–3%
316L	Extra low carbon content for improved weldability
317	Additional molybdenum to further improve resistance to pitting corrosion: Cr, 18–20%; Ni, 11–15%; Mo, 3–4%
321	Titanium added to prevent chromium carbide precipitation: Cr, 17–19% Ni, 9–12%; C, 0.08% max; Ti, 5 × C content
347	Columbium (niobium) or tantalum added to prevent chromium carbide precipitation
	Martensitic Stainless Steels
410	Basic alloy of the group: Cr, 11.5–13.5%; Ni, 0.5% max
405	Aluminum added to prevent weld hardening
414	Nickel (2%) added to improve corrosion resistance
416	Sulfur added to improve machinability
416Se	Selenium added to improve machinability
420	Carbon increased for higher hardness: Cr, 12–14%; Ni, 0.5% max; C, over 0.15%
431	Chromium increased to further improve corrosion resistance: Cr, 15–17%; Ni, 1.25–2.50%
440A	Carbon slightly decreased to improve toughness: Cr, 10–18%; Ni, 0.5% max; C, 0.6–0.75%
440B	Carbon decreased slightly to improve toughness even more
440C	Carbon increased to increase hardness; chromium increased to make up loss in corrosion resistance: Cr, 16–18%; Ni, 0.5% max; C, 0.95–1.2%
	Ferritic Stainless Steels
430	Basic alloy of the group: Cr, 14–18%; Ni, 0.5% max
430FSe	Selenium added to improve machinability
442	Chromium increased to improve resistance to scaling and corrosion: Cr, 23–27%; Ni, 0.5% max

In addition to ferrous alloys, chromium also is added to copper, vanadium, zirconium, and other metals to form several hundred chromium-bearing alloys. Nickel-chromium-iron alloys have high electrical resistance and are used widely as electrical heating elements. *Nichrome* and *Chromel* are examples.

Chromium Plating: Although the tonnage of chromium used for electroplating is far below its comsumption for alloys, plating represents a major market for the element. In terms of special protective coatings, chromium is behind only copper, lead, and zinc in consumption. Generally, chromium plating is used for two purposes: (1) wear-resistance and (2) decorative effect, taking on polish and a much brighter surface than the other electroplated metals. The "bright work" of automotive hardware, plumbing fixtures, and electrical appliances are examples. Normally, chromium is plated over nickel where the nickel is about 100 × thicker than the chromium. In decorative plating, the thickness of the chromium plate ranges only from 0.00001 to 0.00002 inch (0.00025 to 0.0051 millimeter). Hexavalent chromium also forms protective chromate coatings over aluminum, cadmium, copper, magnesium, and zinc. These chromate conversion coatings are used on hot-dipped or electrogalvanized parts, on zinc die castings, and extensively on aluminum parts for aircraft.

In 1979, scientists at the General Motors Research Laboratories (Warren, Michigan) announced some interesting findings concerning chromium plating. The conventional process uses a bath containing Cr^{6+} ions. During plating, the Cr^{6+} ions reduce to Cr^{3+} and then to metallic Cr. Investigators have wondered for years why the process will not succeed if one commences with Cr^{3+} ions, thus avoiding the double step. The GM scientists found that starting with Cr^{3+} fails because it immediately forms a stable complex with water molecules from which Cr cannot be deposited. Commencing with Cr^{6+} succeeds because during reduction a chemical film forms around the cathode (part being plated). Since Cr^{3+} is bound in that film, it does not react with water, but, instead, plates out as chromium metal. The researchers determined all 10 steps that occur as Cr^{6+} reduces to bright Cr and pinpointed the step at which catalysis begins and also identified the active catalyst, which turned out to be the bisulfate ion, not the sulfate ion as might be expected. See accompanying diagram.

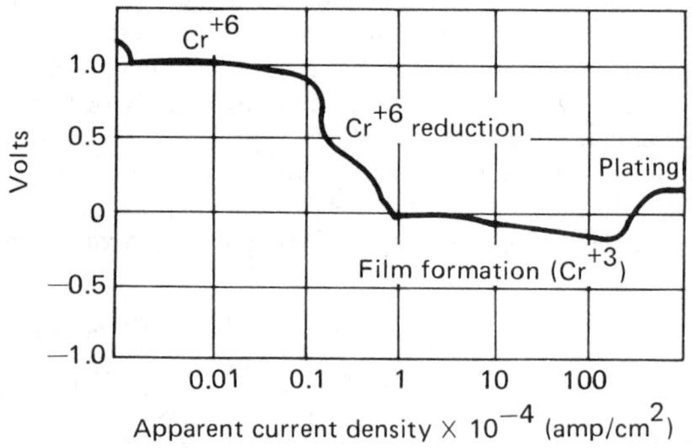

Typical polarization curve made during investigation of chromium plating process. (*General Motors Research Laboratories.*)

Chemistry and Compounds: Chromium metal is soluble in dilute HCl or H_2SO_4 and made passive by dilute or concentrated HNO_3 or concentrated H_2SO_4.

In keeping with the $3d^54s^1$ electron configuration of the atom, chromium forms compounds in which it is in the stable 6+ state. The compounds of 2+ and 3+ chromium are also quite numerous. The compounds with chromium oxidation numbers of 4+ and 5+ are unstable except under extremely alkaline conditions. The one known chromium 1+ complex is also unstable.

As is evident from the oxidation potential for Cr^{2+} to Cr^{3+}, the Cr^{2+} ion in aqueous solution is an extremely strong reducing agent, readily reacting with atmospheric oxygen. It is therefore stable in aqueous solution only in a complex or a slightly ionized salt. It may be obtained by reducing hexavalent or trivalent chromium in acid solution with one of the active metals, such as zinc. The four halogens form halides of divalent chromium, CrX_2, which dissolve in water with the evolution of heat, although the fluoride is less soluble than the other halides. The ammoniacal solution of chromium(II) sulfate, $CrSO_4$, is particularly reactive with gases, not only with oxygen like the other Cr^{2+} compounds, but also acetylene. Chromium(II) acetate is only slightly soluble in H_2O.

The Cr^{3+} ion readily forms complexes; it exists in aqueous solution as $Cr(H_2O)_6^{3+}$, and forms other complexes with anions, such as $Cr(H_2O)_5Cl^{2+}$, $Cr(H_2O)_4Cl_2^+$, etc. Thus chromium(III) halides may be crystallized in a number of forms differing in color and other properties due to variation in the bonding. Compounds of the trichloride have been reported with the following arrangements: $Cr(H_2O)_6Cl_3$, $[Cr(H_2O)_5Cl]Cl_2 \cdot H_2O$, and

$$[Cr(H_2O)_4Cl_2]Cl \cdot 2H_2O.$$

Trivalent chromium also forms double salts, notably the chromium alums, hydrated double salts of Cr(III) sulfate and the alkali metal (or thallium or ammonium) sulfates.

The three oxides of chromium, CrO, Cr_2O_3, and CrO_3, are of interest in exhibiting the transition in properties from basicity to acidity and from electrovalence to covalence, with increasing oxygen content, and the consequent increase in charge upon the Cr atom. Thus Cr_2O_3, the middle oxide, is amphiprotic, dissolving either in strong alkali hydroxide solutions or in acids.

One of the most stable of the compounds of Cr(IV) is the tetrafluoride, CrF_4, brown solid, steel blue vapor at 150°C, prepared by direct reaction of the elements; even this compound readily undergoes hydrolysis. Chromium tetrachloride, $CrCl_4$, may be prepared as a gas by reaction of Cl_2 and $CrCl_3$ at elevated temperature, but decomposes at room temperature. Chromium(V) occurs in CrF_5, fire red, volatile, and in the hypochromates, M_3CrO_4, green, which may be prepared by fusion of alkali chromate and alkali hydroxide at high temperature.

CrO_3 is the anhydride of chromic and dichromic acids, H_2CrO_4 and $H_2Cr_2O_7$, which have not been isolated, but whose anions are found in many salts. pK_{A1} and $pK_{A2} = 0.745$ and 6.49, and −1.4 and 1.64 respectively. They are strong oxidants in acid solution, and are readily obtained in basic solution by oxidation of Cr(III) compounds. The oxyhalogen compounds of chromium are chiefly the CrO_2X_2 (chromyl) compounds, which are acid halides of chromic acid. Chromyl chloride CrO_2Cl_2 deep red, liquid, is prepared by heating sodium dichromate and sodium chloride with H_2SO_4, while chromyl fluoride undergoes polymerization to a white solid. Chlorochromates, e.g., $KCrO_3Cl$, and fluorochromates, e.g., $KCrO_3F$, are also known, but are hydrolyzed in water.

Chromates and Dichromates: Sodium chromate Na_2CrO_4, potassium chromate K_2CrO_4, ammonium chromate $(NH_4)_2CrO_4$, calcium chromate $CaCrO_4$, are yellow soluble solids; barium chromate $BaCrO_4$, pale yellow, strontium chromate $SrCrO_4$, pale yellow, lead chromate $PbCrO_4$, yellow (used as a pigment, "chrome yellow"), zinc chromate $ZnCrO_4$, yellow, used as a pigment, mercurous chromate Hg_2CrO_4, yellow to red to brown, silver chromate Ag_2CrO_4, reddish-brown, are insoluble solids. Sodium dichromate $Na_2Cr_2O_7$, potassium dichromate $K_2Cr_2O_7$, readily crystallized, ammonium dichromate $(NH_4)_2Cr_2O_7$ (forming nitrogen gas and green chromic oxide solid upon heating), are red soluble solids, of important application as oxidizing agents, e.g., sulfurous acid causes reduction to chromic; silver dichromate $Ag_2Cr_2O_7$, red insoluble solid, changing to silver chromate Ag_2CrO_4 upon boiling with H_2O. Solutions of chromate in the presence of acid are changed to the corresponding dichromate.

A number of peroxychromates are known, including the deep blue, organic soluble reaction product of H_2O_2 and acid $Cr_2O_7^{2-}$ solutions, which is H_4CrO_7, i.e., $CrO_3 \cdot 2H_2O_2$.

Trivalent chromium forms many complexes. The large number is due in part to the many possible arrangements in which the same elements may be present as part of a complex anion or as the cation, as was indicated for the chloride complexes of the hydrated Cr^{3+} ion. The ligands may also be present in various proportions. As in the case with Co^{3+}, Cr^{3+} forms complexes with ammonia in the various color series, e.g., a violeo-chloride, $[Cr(NH_3)_4Cl_2]Cl$, a purpureo chloride, $[Cr(NH_3)_5Cl]Cl_2$ and a luteo chloride, $[Cr(NH_3)_6]Cl_3$. In addition to NH_3 many amines, especially ethylenediamine and its derivatives, form stable complexes with Cr^{3+}. In the oxalato complexes, the tripositive chromium ion coordinates with oxalate groups to form such anions as $[Cr(C_2O_4)_3]^{3-}$ and $[Cr(C_2O_4)_2(H_2O)_2]^-$. There are many cyano, thiocyano, nitrato, fluoro, bromo, iodo, azido, acetylido and nitro complexes.

Organometallic Compounds: These include a large number of cyclopentadienyl compounds. In addition to dicyclopentadienyl chromium itself, $(C_5H_5)_2Cr$, there are many derivatives of it with additional radicals or molecules such as $(C_5H_5)_2CrBr$, $(C_5H_5)_2CrNH_3$, $C_5H_5Cr(NO)_2CH_2Cl$, $H(C_5H_5)Cr(CO)_3$ and $C_5H_5(CO)_3CrCr(CO)_3C_5H_5$. The compounds of Cr with other hydrocarbon radicals are fewer ($Cr(C_6H_6)_2$ is known) such as $(C_6H_5)_5CrOH \cdot 4H_2O$ and

$$(C_6H_5)_3Cr \cdot 3THF \cdot (THF = \text{Tetrahydrofuran}).$$

Besides $Cr(C_6H_6)_2$, chromium hexacarbonyl, $Cr(CO)_6$, white, solid, bp (extrap.) 145.7°C, represents chromium in the zero oxidation state. This is a nonpolar compound, insoluble in H_2O, freely soluble in organic solvents. The zero-valent compound $K_6Cr(CN)_6$ has also been made by the reaction of $K_3Cr(CN)_6$ and potassium metal in liquid NH_3.

Cermets: The term *cermet* derives from the combination of ceramic and metal. Cermets are produced by powder metallurgy techniques and represent the bonding of two or more metals. They are particularly useful at high temperatures (850–1250°C). Chromium is used in several cermet combinations, including chromium-bonded aluminum oxide, metal-bonded chromium carbide, and metal-bonded chromium boride.

Biological Aspects: The first indication for a biological role of chromium in metabolism was derived from the enzymatic studies (Horecker et al., 1939). The succinate-cytochrome dehydrogenase system, an important enzyme system for the production of energy, requires certain inorganic cofactors. Of various elements tested, chromium produced the greatest increase of enzyme activity. A significant stimulation of the activity of phosphoglucomutase by chromium was described by Strickland in 1949. This system which has an important function in the early steps of carbohydrate metabolism requires magnesium and one other metal for optimal activity. Chromium was outstanding as a "second metal" because it produced the highest enzyme activation and supported a measured amount of activity even when given alone.

Chromium also stimulates fatty acid and cholesterol synthesis from acetate in liver. That chromium is an essential cofactor for the action of insulin on the rat lens was shown by Farkas in 1964. In the absence of the element, no significant insulin effect on glucose utilization of lens can be demonstrated. Chromium supplementation to the donor animals results in a significant response of lens tissue to the hormone. Numerous other findings indicate that chromium may play several vital roles in biological systems.

Chromium levels in biological matter have been studied extensively. In contrast to findings with other metals, chromium concentrations in the United States population are highest at the time of birth, with a pronounced decline during the lifetime, whereas they appear to remain high in some other countries, such as Thailand and the Philippines. These findings suggest the possibility of a relative chromium deficiency in the United States. The relationship of this to disturbances of carbohydrate metabolism in humans, while under study, remains to be established.

Not all of the various chemical forms of chromium are effective in improving sugar metabolism, and the exact nature of the compound or compounds involved in activating insulin is not established. Some of the chromium in plants may not be present in nutritionally effective forms. It has not been established that chromium is essential to plants, but high concentrations of the metal are toxic. Most agricultural crops, especially their seeds, contain only low levels of chromium.

References

Allaway, W. H.: "The Effect of Soils and Fertilizers on Human and Animal Nutrition," Agricultural Information Bulletin 378, Cornell University Agricultural Experiment Station, and U.S. Department of Agriculture, Washington, D.C., 1975.

Krichgessner, M. (editor): "Trace Element Metabolism in Man and Animals,"

Institut für Ernahrungsphysiologie, Technische Universität München, Freising-Weihenstephan, Germany, 1978.
Underwood, E. J.: "Trace Elements in Human and Animal Nutrition," 4th edition, Academic, New York, 1977.

CHROMIUM COPPER. **Copper.**

CHROMIUM STEELS. **Iron Metals, Alloys, and Steels.**

CHROMIZING. Production of a high chromium content surface layer on iron and steel by heating at high temperatures in a solid packing material containing chromium powder, or in an atmosphere containing chromium chloride. The surface layer is formed by diffusion of chromium into the iron in the same manner as carbon diffuses into iron in carburizing; however, the process is much slower and requires higher temperatures than carburizing. A similar result can be obtained by high-temperature diffusion of electrodeposited chromium. Chromized coatings have corrosion resistance and elevated temperature oxidation resistance similar to the high chromium types of stainless steels.

CHROMOGENIC COUPLERS. Couplers are used in secondary color development. See **Dyes (Textile).** The term "coupler" is applied to a large number of organic compounds which combine with a limited number of chromogenic developers to produce dye images with a wide range of color and intensity. Couplers are dye intermediates but differ from chromogenic developing agents in that they do not as a rule have the ability to develop a silver image.

Couplers may be hydroxy or amine derivatives of aromatic compounds, such as benzene, naphthalene or anthracene. Phenol, naphthol, aniline, cresol, paraaminophenol and dimethylparaphenylenediamine are examples. With these compounds coupling takes place with the hydrogen atom which is in the ortho or para position to the hydroxy or amino group on the coupler.

Compounds having active methylene $=CH_2$ groups, with strong polar groups for other valences as the cyano $-CN$, the carbonyl $=CO$, the aceto CH_3CO-, the acid ester $-COOC_2H_5$, and the phenyl, $-C_6H_5$ groups will couple. Acetoacetic ester and paranitrophenylacetonitrile are illustrations.

The methylene group may be part of a ring structure as in coumarine and indoxyl, or may be part of a heterocyclic ring attached to a phenyl group as in 1-phenyl-3-methyl-5-pyrazolone.

Compounds having a N in the ring will couple if there is a methyl $-CH_3$ group, attached to the ring in the alpha position to the nitrogen. Two illustrations for this type are picoline and 2-methyl-thiazole.

Because secondary color development has been particularly successful in the development of color materials, and because of the ease by which it is possible to control color by this method, the field of couplers has expanded rapidly.

CHROMOPHORE. Certain groups of atoms in an organic compound cause characteristic absorption of radiation irrespective of the nature of the rest of the compound. Such groups are called chromophores or color carriers.

CHROMOPHYTOSIS. **Dermatitis and Dermatosis.**

CHROMOPLASTS. **Plastids.**

CHROMOPHORIC ELECTRONS. Electrons in the double bonds of the chromophore groups. Such electrons are not bound as tightly as those of single bonds and can thus be transferred into higher energy levels with less expenditure of energy. Their electronic spectra appear at frequencies in the visible or near ultraviolet region of the spectrum.

CHROMOSOMES. **Cell (Biology); Genes; Genetics; Nucleic Acids and Nucleoproteins; Plant Breeding.**

CHROMOSPHERE. **Sun.**

CHRONIC. Of long duration, applied to a disease that is not acute.

CHRONIC RENAL FAILURE. **Kidney and Urinary Tract.**

CHRONOGRAPH. A wristwatch or pocket watch that displays time of day and incorporates the functions of the stopwatch or sports timer. The chronograph is equipped with a special internal mechanism, controlled by an external hand-operated plunger system, that permits the measurement of elapsed time—either continuously or interruptedly for periods of up to several minutes or hours. Elapsed time is displayed by separate hands on the dial; or by separate hands on subsidiary dials on the main dial. Normally, the elapsed-time reading is in fifths of a second, but if a "fast beat" (36,000 beats an hour) balance wheel movement is used, time can be shown in tenths of a second. For certain sports and industrial applications, a dial is scaled into 1/100ths of a minute instead of seconds. Certain chronographs, like some stopwatches, can display more than one elapsed time reading; others also incorporate tachometer or other special-scale dials that report computed velocities or distances.

Logically, the term chronograph is also used in connection with recording timed events. In this regard, almost all process variables, such as temperature, pressure, flow, machine operations, etc., are recorded against time. Normally, however, these instruments are not thought of as chronographs, although the term certainly is not inappropriate. More often, chronograph is applied to very special types of instruments which are precisely time-driven (where a very exact time reference is important), as in the case of seismic recorders for placing the exact time of arrival of earthquake information, for seismic mineral exploration operations, etc.

CHRONOMETER. A special type of mechanical clock used as a reference by a ship's navigator. Also known as a marine chronometer. The clock, hung on gimbals in a box to minimize the balance wheel's position error, is powered by a fusee-equipped mainspring that drives a temperature-compensated balance wheel and hairspring which work a specialized escape wheel system called a detent escapement. See also **Spring Clock.** A chronometer displays Greenwich Mean Time (GMT) and is characterized by a very constant daily rate (usually within 0.1 second per day) although the timepiece may gain or lose several seconds per day. See also **Time.**

The navigator checks the chronometer periodically (usually every week or ten days) against radio time signals to confirm its rate and error. From this information, the correct GMT can be calculated at any instant. This provides a basic time reference. Using a sextant, the navigator makes an astronomical reading, usually of the position of the sun, which is then correlated with tables given in a nautical almanac to find local mean time. Comparison of local time with the GMT readings provides the longitude.

Precision tuning fork and quartz crystal clocks now are also used by navigators. Technically, these new types of clocks are not chronometers. Consumer balance-wheel Wristwatches individually certified as "chronometers" by the Official Swiss Testing Bureaus do not fall within the historical definition of chronometer. These watches are certified to have maintained a mean accuracy of minus 3 to plus 12 seconds in five positions with a mean daily variation rate of not more than 3.2 seconds over a period of 15 days under laboratory (not use) conditions. Wristwatches with 360 Hz tuning forks maintain a mean in-use accuracy of plus or minus 2 seconds a day and have been used as navigation references by amateur navigators.

William O. Bennett, John J. Carpenter, Frank Dostal, and E. Van Haaften, Bulova Watch Company, Inc., New York.

CHRYSALIS (or Chrysalid). The third stage in the development of butterflies, also properly called the pupa. The caterpillar of a butterfly spins no cocoon but hangs itself by a silken button or by a belt and button. The skin of the pupa into which it changes is often brightly colored or protectively colored and marked, unlike the mahogany-colored pupae of most moths. Although some moths form similar naked pupae the term chrysalis is applied only to those of the butterflies.

CHRYSANTHEMUM. ***Compositae* or Composite Family.**

CHRYSOBERYL. The mineral chrysoberyl, an aluminate of beryllium corresponds to the formula $BeAl_2O_4$, crystallizes in the orthorhombic system with both contact and penetration twins common, often repeated resulting in rosetted structures. Hardness, 8.5; specific gravity, 3.75; luster vitreous; color various shades of green sometimes yellow. A variety which is red by transmitted light is known as alexandrite. Streak colorless; transparent to translucent, occasionally opalescent. Chrysoberyl also is known as *cymophane* and *golden beryl.*

Chrysoberyl occurs in granitic rocks, pegmatites and mica schists; often is found in alluvial deposits. The Ural Mountains yeild alexandrite. Other localities for chrysoberyl are Czechoslovakia; Ceylon; Rhodesia; Brazil; and the Malagasy Republic, where it occurs of gem quality in the pegmatites of that island. In the United States it is found in Maine, Connecticut, and New York. The word chrysoberyl is derived from the Greek words meaning golden and beryl. Cymophane has its derivation also from the Greek words meaning wave and appearance, in reference to the opalescence exhibited at times.

CHRYSOCOLLA. This mineral, a hydrous silicate of copper probably corresponding to the formula $Cu_2H_2Si_2O_5(OH)_4$, is perhaps a mineral gel, for it usually appears as an amorphous mass, in veins, or as incrustations. Common occurrence as massive cryptocrystalline character, possibly orthorhombic; extremely rare as small acicular crystals.

Chrysocolla is generally some shade of blue or green but if impure may be brown or black. It has a characteristic conchoidal fracture; hardness, 2–4; sp gr, 2.24; vitreous to dull luster; translucent to opaque.

Chrysocolla is a secondary mineral and associated commonly with other copper minerals of similar origin. It is one of the less important ores of copper and has a minor use as a gem stone. Among the localities for excellent specimens may be mentioned Cornwall and Cumberland, England; Congo; Chile; Lebanon and Berks Counties, Pennsylvania; the Clifton-Morenci Globe and Bisbee districts in Arizona; Dona Ana County, New Mexico, and the Tintic district, Utah.

The word chrysocolla is derived from the Greek words meaning gold and glue, formerly the name for gold solder.

CHRYSOTILE. A delicately fibrous variety of serpentine which separates easily into silky, flexible fibers of greenish or yellowish color, with formula $Mg_3Si_2O_5(OH)_4$. It crystallizes in the monoclinic system; hardness, 2.5; sp gr, 2.55. Its name is derived from the Greek words meaning gold and fibrous. Most of the common asbestos of commerce is chrysotile. It is mined in Thetford, Province of Quebec, and in the Republic of South Africa. See also **Serpentine.**

CHUB MACKEREL. Mackerels.

CHUCK. A rotating vise which may be attached to the spindle of a machine. There are two important varieties of lathe chucks, independent and universal. In general, the *independent chuck* has four jaws each of which is separately actuated and adjusted. It may be employed for almost any type of work, cylindrical, square, or irregular. In turning cylindrical work, it is necessary to adjust the jaws very carefully, and test the concentricity of the work and spindle axes with some form of indicator. When a 4-jaw independent chuck is employed for repetitive work, only two adjacent jaws are actuated as each new part is placed in the chuck after the initial adjustment and alignment of the jaws have been obtained.

Three-jaw *universal* or self-centering chucks are employed for cylindrical and hexagonal bar stock. The jaws are simultaneously advanced or retracted by turning the scroll plate in which the jaw teeth fit. On account of the curvature of the scroll, it is necessary to have separate sets of jaws for inside and outside clamping, in contrast to the independent chuck where the jaws may be reversed. Combination chucks combine an independent chuck for holding odd-shaped work; a universl chuck for self-centering and gripping round or square work; 2-jaw universal chucks have jaws to which special adapters may be fitted and are employed principally in turret lathe work.

Independent and universal chucks may be attached to a threaded adapter which screws on the spindle nose. These chucks may also be obtained with adapters to fit heavy-duty taper spindle noses; the chuck is drawn on the spindle nose by the engagement of the "pull-on" nut with the externally threaded hub of its adapter.

Air-operated chucks have a body with two jaws and work-holding adapters with an actuating wedge which closes and opens the jaws by moving parallel to the chuck axis. The actuating wedge is threaded so that a draw-rod may be attached. The other end of the draw-rod is attached to a piston operating in an air cylinder which is attached to the lathe headstock. The piston is double-acting so that the chuck jaws may be both opened and closed by the action of compressed air. The chuck is operated by a valve convenient to the machine operator. Air- and oil-operated chucks are generally employed for production work, as in turret and chucking lathes.

Draw-in chucks and collets are used for bar work, and are designed to fit on the heavy-duty spindle nose if the live center is removed. The collets, which are of various sizes, fit in the chuck and are clamped with a removable key or wrench. Drawing in the spring collet forces its outer surface against the taper on the inside of the chuck; releasing the collet causes its jaws to open by their spring action. Bars of any length may be held in the chuck, extending entirely through the hole in the spindle if necessary. Collets for all standard sizes of circular rod are available as well as collets for hexagonal bar stock and cylindrical metric sizes. Magnetic chucks of both the electrically actuated and the permanent-magnet type may also be employed on the lathe.

Three-jaw universl chucks are used on drill presses. Rotating the outer sleeve by hand or by a key fitted to the sleeve gear teeth, opens and closes the three self-centering jaws. The arbor hole in the drill chuck fits the tapered end of the drill press spindle. *Bayonet chucks* are single-purpose chucks and are used on automatic drilling machinery. They are equipped with a bayonet slot for rapid attachment and release.

CHUCK-WALLA (*Reptilia, Sauria*). A common lizard, *Sauromalus obesus*, of the southwestern deserts, ranging into Utah and Nevada. It attains a length of 11 inches (28 centimeters) and is sometimes eaten.

CHUM SALMON. Salmon.

CHURG-STRAUSS SYNDROME. Bronchial Asthma.

CICADA (*Insecta, Homoptera*). Large insects of many species. They are stoutly built and have two pairs of membranous wings which are folded roof-like over the body when at rest. They are best known for the loud songs of the males, which are produced by a pair of elaborate organs located on the under surface at the base of the abdomen. These organs have a vibrating structure controlled by special muscles and thin resonating parts which result in a peculiarly penetrating sound.

The female has a powerful ovipositor with which she punctures the twigs of trees and shrubs to deposit her eggs. The young cicada does not remain in the twig but drops to the ground and burrows, feeding on the roots of plants. In the case of one species, *Tibicina septendecim*, the duration of the larval period is unusually long and has resulted in the name seventeen-year locust. The name locust is inaccurately but very commonly applied to these insects and some are called harvest flies.

Great damage is sometimes done to young orchard trees by the breaking of twigs where the eggs have been deposited, especially after one of the great broods of the seventeen-year locust, or periodical cicada, has passed. There are approximately 20 different broods in the United States. The only effective protection is to cover young trees with inexpensive cloth when such a brood is imminent; the years of emergence are known by economic entomologists and can readily be learned for any part of the country.

In taking approximately 17 years for the nymph to fully materialize, the cicada has the longest life of any known insect. The body measures from $\frac{3}{4}$ to 1 inch (2 to 2.5 centimeters) in length; the wings are about $1\frac{1}{2}$ inches (4 centimeters) long. The eyes are large and are situated above the antennae. The insect has six legs for walking, leaping, digging, paddling, and cleaning. Most cicadas have claws on their feet with a pad between each claw. The pad is reinforced with hair moistened by glandular secretions.

CICHLIDS (*Osteichthyes*). Of the order *Percomorphi*, family *Cichlidae*, the cichlids number approximately 600 species and are found in South America, parts of the southwestern United States, and Africa. One genus, *Etroplus*, occurs in India and Ceylon. They generally are quite small and consequently many species are favorites of tropical-fish fanciers. Included are *Pterophyllum scalare* (fresh water angelfish); *Symphysodon discus* (pompadour fish; also South American discus), the habits of feeding their young being quite strange (the young obtain secretions from mucous cells of the skin of both parents as food); the genus *Tilapia*, mouthbreeders weighing from 2 to 20 pounds (0.9 to 9 kilograms); *Cichlasoma cyanoguttatum* (Rio Grande perch) occurring in Mexico and southward and attaining a length of 10 inches (25 centimeters); *Astronotus ocellatus* (oscar or peacock-eyed cichlid); and the *Steatocranus casuarius* (Congo bumphead cichlid) which is about 4 inches (10 centimeters) in length. In addition to the oscars and discus, other favorites of tropical-fish fanciers include: *Cichlasoma biocellatus* (the greenish-black Jack Dempsey); *C. festivum* (the flag cichlid); and *C. meeki* (the dramatic firemouth possessing a beautiful red chest).

The African mouthbreeders (*Tilapia mossambica*) are extremely prolific and, if competition from other species does not interfere, can become a good source of low-cost protein. They were introduced into Indonesia in the late 1930s and now are a popular pond fish.

CICONIIFORMES (*Aves*). Most birds of this order differ conspicuously by their long legs from birds living on or near the water, like the loons and grebes, and the penguins, tubenoses, and *Pelecaniformes*. Unlike the ratites, the *Ciconiiformes* cannot, however, use their legs for swift running; their gait is a much more measured walk.

Their length ranges from 30–160 centimeters (12–63 inches); in normal posture the height of the crown is 20–130 centimeters (8–51 inches). The weight is 100–6000 grams ($3\frac{1}{2}$–211 ounces). They are almost always long-legged and long-necked. They have 16–20 cervical vertebrae, and the hind toe is well developed. All species live on animal food. The crop is absent but there is a well developed proventriculus and small caeca. The eggs are usually plain-colored (exceptions are the hermit ibis and the spoonbill). The young remain in the nest for a comparatively long time.

There are five families: The Herons, Egrets, and Bitterns (*Ardeidae*); The Shoebills (*Balaenicipitidae*); The Hammerheads (*Scopidae*); The Storks (*Ciconiidae*); The Ibises (*Threskiornithidae*). Altogether there are 59 genera and 115 species.

The Heron family (*Ardeidae*) is distributed over all parts of the world. The weight varies between that of the least bittern, which weighs just over 100 grams (3.5 ounces), and the Goliath heron, which weighs 2600 grams ($5\frac{1}{2}$ pounds). There are 20–22 cervical vertebrae; the neck, which is almost immovable laterally, is S-shaped in flight and is wedged in between the breast and the wings when the bird is at rest. There are 24 genera with 63 species, found mostly in the tropics and subtropics; they do not exist in the far north or in Antarctica. See Fig. 1.

Fig. 1. Mangrove heron (*Butorides striatus*).

Herons stalk their prey carefully or stand and wait for it, their long necks in the resting position, drawn back into an S-shape. The cervical spine is so constructed that a heron can thrust its head forward in a flash to stab its prey or to seize it with the beak.

Herons are easily distinguishable from storks in flight by the way they draw their necks back. They fly well and with endurance in slow, quiet wing beats, but they do not soar like storks, even though some species know how to utilize thermals. Temperate-zone herons are generally migrants. Many species which breed in the tropics also migrate regularly according to the wet and dry seasons.

Most species of herons breed in colonies; many are also social outside of the breeding season and spend the night in communal roosts. The nests are built in trees, in shrubs or reeds, or even on rocks. Usually both parents build the nest and relieve one another during incubation. The clutch usually consists of 3–5 eggs. The eggs are white, greenish, blue, or olive-brown; a few species have spotted eggs.

The food of most species consists predominantly of fish, as well as frogs, salamander larvae, small mammals, and insects. Herons swallow their prey whole and digest fish almost completely. See also **Bittern and Heron.**

The members of the remaining four families of the *Ciconiiformes* are somewhat less adapted to life near the water than are the herons. Among them the Shoebill (*Balaeniceps rex*) differs so much from the usual type of heron and stork that it is regarded as the representative of a separate family, *Balaenicipitidae*.

The standing height of the Shoebill is about 115 centimeters (45 inches), the length is 120 centimeters (47 inches), and the wing length is 68 centimeters ($26\frac{1}{2}$ inches). The beak is extraordinarily high and wide; in connection with this, the skull is much enlarged and thus is pelicanlike. A slight crest is found at the back of the head.

The shoebill occurs only in the marshlands of tropical Africa and in general seems to be quite rare, but is more common in Uganda and the Sudan. It is easily overlooked, since it stays along river shores in dense papyrus, but is also seen on flat-flooded grassland with short grass. Its calm temperament makes it possible to approach it rather closely before it flies off. It flies well and sometimes soars in an updraft. It holds the beak pressed against the chest in flight and does this also on the ground. It lives mainly on river fish, but also on frogs and snails. The eggs are bluish white and covered with a calcareous layer; the clutch consists of 2–3 eggs.

The relationships of the Hammerheads (family *Scopidae*) are obscure, but they are generally placed with the order *Ciconiiformes*. There is only one species, the Hammerhead (*Scopus umbretta*). The length is about 50 centimeters ($19\frac{1}{2}$ inches). The head, with its medium long, laterally compressed beak and its crest pointing to the rear, looks somewhat hammerlike. The hammerhead inhabits swamp and shallow water areas, generally looks for food in the water, and occasionally swirls up the mud with its feet to stir up prey. It is found singly or in pairs; family units of up to seven birds remain together for only a short time.

Their flight is slow, as if swimming in air, and they frequently utter a shrill cry, particularly before it rains. They are active in the twilight and at night, and more rarely during the day. The nest of the hammerhead is an extraordinary, enormous structure. Its diameter is $1\frac{1}{2}$ meters (5 feet), and inside there is a small cavity only 30 centimeters (12 inches) across with an entrance that opens downward to one side. Three to six white eggs are incubated for about 30 days, and the young hammerheads leave the nest about 50 days after hatching.

The Storks (family *Ciconiidae*) are birds of medium size to very large. The beak, neck, and legs are long. The large and wide wings enable them to fly well and to soar, which saves much energy. They feed only on animals, and they build very large nests for their white eggs. There are species which have penetrated far into the temperate latitudes, and migrate far; others migrate over short distances. There are 10 genera with 18 species.

The best known species is the White Stork (*Ciconia ciconia*). Males and females look alike. The young at first have blackish beaks and legs. The length is 110 centimeters (43 inches); the wing length is 53–63 centimeters (21–25 inches); the wing span reaches over 220 centimeters (87 inches); and the weight is from 2.3–4.4 kilograms (5–10 pounds).

Storks generally feed while walking; they wander over open shallow swamps, meadows, and fields searching for food. As a rule, prey is seized with a forward thrust of the beak, but animals may also be seized under water with sideways beak movements. The diet is quite varied. Earthworms play a great role in spring, and for feeding the young. In the summer storks catch many insects, even crickets, which are difficult to catch, but mainly they feed on the more easily caught grasshoppers. Vertebrates are of lesser importance as food; however, where dead fish drift ashore, or where fish are sick and generally less mobile and hence are easily caught, storks certainly eat fish. Lizards and snakes, even vipers, are caught. Birds are occasionally taken.

Although well adapted in many respects, storks suffer greatly from cold and rainy weather. The climate, therefore, determines the northern boundary of distribution. Storks also need warm temperatures because they are dependent on thermal updrafts for soaring, as are many raptors. See also **Storks.**

The Ibises (*Threskiornithidae*) are related to the storks, with the wood ibises forming a sort of link with the true ibises. Although the ibises, with their slender curved beaks, differ strikingly from the flat-billed spoonbills, they are nevertheless closely related to them as well.

All members of the family *Threskiornithidae* are of medium size, with a length of 50–90 centimeters ($19\frac{1}{2}$–35 inches). The face and throat are more or less bare of feathers; the medium-length legs are sturdy. They are distributed over all warmer and tropical areas. Being very sociable they breed in large colonies and wander about or migrate in troops. In flight the neck is carried extended forward, as in storks.

Fig. 2. Spoonbill (*Platalea leucorodia*).

Two subfamilies are readily distinguishable by external characteristics: The Ibises (*Threskiornithinae*), with their long, narrow, and markedly down-curved beak, probe for insects, mollusks, crustacea, and worms in mud and soil; occasionally they also catch larger prey. There are 17 genera with 20 species. The Spoonbills (subfamily *Plataleinae*), with a beak which is flattened and widened at the tip, seize prey in side-to-side movements of the bill. See Fig. 2. There are two genera with six species. See also **Ibis.**

CIENEGA. A type of spring which occurs in intermontane basin deposits or bolsons, especially of semi-arid to arid regions. When the underground waterbearing stratum, or aquifer, is blocked by cemented gravels the water may be forced by hydrostatic pressure to the surface, forming the type of spring called a cienega.

CIGARETTE SMOKING. Chronic Bronchitis; Emphysema.

CIGUATERA POISONING. Foodborne Diseases.

CILIA. Fine threadlike hairs located in various parts of the body which serve as filtering mechanisms to protect body areas from foreign particles, e.g., the eyelashes.

CILIARY JUNCTION. An association of separate gill filaments in certain bivalve mollusks, which is characterized by interlocking cilia.

CILIATA. One-celled animals which have cilia during adult life but are without suctorial tentacles. They constitute a class of this name in the subphylum Ciliophora.

CILIOPHORA. A subphylum of the phylum Protozoa containing species which have cilia or cirri during some stage of life. They vary greatly in form and habits. Some are sessile, some free swimming, and some parasitic.

The following is a brief summary of a classification of ciliates now widely used:

- Class *Ciliata*. With cilia or cirri throughout life.
  - Sublcass *Protociliata*. Leaf-like species with two to many nuclei. Parasitic in the intestines of fishes and amphibians. *Opalina* and related forms.
  - Subclass *Ciliate* (*Euciliata*). With two kinds of nuclei, large and small (macronucleus and micronucleus).
    - Order *Holotrichida*. Cilia uniformly distributed. *Paramecium* and many other genera.
    - Order *Heterotrichida*. With a zone of cilia of larger size, or membranelles, associated with the mouth. *Stentor*.
    - Order *Oligotrichida*. With cilia about the mouth but few on the body.
    - Order *Hypotrichida*. Flattened, with cilia or cirri on the under surface. *Stylonychia*, etc.
    - Order *Peritrichida*. Oral end of body enlarged, ciliated, many species stalked. *Vorticella*, etc.
- Class *Suctoria*. With cilia only during early life. Adults sessile, with tentacles for ingesting food and for piercing.

CILIUM. A slender hair-like process of minute size on the surface of a cell. It is part of the living cytoplasm. In association with their minute size, cilia occur in relatively large numbers and act in unison to produce aggregate effects. They are capable of waving movement. Commonly they bend consecutively in the same direction so that a wave of movement passes along the ciliated surface, followed by the return of the cilia to the resting position and this again by their bending. This type of movement is said to be metachronal. The successive waves follow each other closely so that several may be apparent at the same time.

Cilia are found in many species of *Protozoa* and give the name *Ciliophora* to one subdivision of the phylum. Among the multicellular animals they occur on the surface of the body in a few groups (e.g., coelenterates, turbellarian worms and molluskans) during adult life and in many larvae such as those of the echinoderms, the annelid worms, coelenterates, and others. They are also found on epithelia of limited distribution in many complex animals, as in the mantle cavity of mollusks and the trachea of man.

Cilia on the surface of small animals, such as the Protozoa and larvae of greater complexity, are able by their action against a surrounding liquid to propel the animal and so serve as organs of locomotion. In animals of larger size, in those which are not surrounded by a liquid medium, on sessile forms, and where the direction of their movement is contrary to the movements of propulsion, they act to set up currents either in liquids secreted by the body or in the surrounding medium. Thus some cilia in Protozoa, coelenterates, and mollusks direct a current bearing food and oxygen into the gullet and in man cilia carry toward the throat the mucus secreted by glands in the lining of the trachea. This secretion may bear foreign particles which have been inhaled.

CIMETIDINE. Ulcer.

CINCHONA. A tall tree (*Cinchona succirubra* Pav.), or various hybrids, of the *Rubiaceae* family is the source of quinine. The tree is indigenous to Peru, but is presently cultivated throughout the Central and South American countries as well as in Madagascar and certain regions of Africa. The commonly used drug quinine is obtained from the bark of branches, generally of trees that are 15 years old or older. Cinchona and its derivatives are also widely used in the manufacture of liqueurs, some nonalcoholic beverages, ice creams, baked goods, bitters, and condiments, primarily for their characteristic bitter-tonic flavoring action. Tonic waters are prepared largely with quinine salts. In the United States, not more than 83 parts per million total chichona alkaloids can be used in finished beverages.

CINEMATOGRAPHY. **Photography and Imagery.**

CINGULUM. 1. A girdle, such as the outer ciliated ring at the anterior end of a rotifer. 2. A bundle of association fibers in the mammalian brain, partially encircling the corpus callosum not far from the median plane. 3. The basal ridge of a tooth.

CINNABAR. The mineral cinnabar, mercuric sulfide (HgS) occurs in small and often highly modified hexagonal crystals, usually of rhombohedral or tabular habit. It is found chiefly in crystalline crusts, granular or simply massive. The fracture of cinnabar is subconchoidal; hardness, 2–2.5; specific gravity, 8–8.2; luster, adamantine tending toward metallic, sometimes dull. This mineral has a characteristic cochineal-red color which, however, may be brownish at times, occasionally dull lead gray. The streak is scarlet; it is transparent to opaque.

Cinnabar occurs in veins or may be in masses in shales, slates, limestones and similar rocks due to the impregnation by mineral-bearing solutions or as replacements. The U.S.S.R., Czechoslovakia, Bohemia, Bavaria, Italy and Spain have furnished excellent specimens. The most important of the world's mercury deposits is at Almaden in Spain. Italy, Peru, Surinam, China and Mexico have commercially valuable occurrences of cinnabar. In the United States this mineral is found in California (most important deposit), Nevada, Utah, Texas and Oregon. Cinnabar is the chief ore of mercury. Its name is supposed to be of Hindu origin.

CINNAMON. **Flavorings; Laurel Family.**

CINNAMON FERN. **Ferns.**

CIPOLIN. A metamorphic rock transitional between a marble and mica-schist in which the principal mica is phlogopite.

CIPOLLETTI WEIR. **Flow Measurement.**

CIRCADIAN RHYTHMS. **Biological Timing and Rhythmicity.**

CIRCLE (Geometry). A plane curve such that all of its points are at a fixed distance, its radius, from a fixed point, its center. The line bounding the circle is its circumference. A diameter is a straight line through the center of the circle with its ends on the circumference and of length twice that of a radius of the circle. Further definitions pertaining to a circle are: secant, a straight line of any length intersecting the circumference in two points; tangent, a straight line of any length touching the circle at only one point; arc, any portion of the circumference; chord, a straight line with end points in the circumference; segment, a part of the circle bounded by an arc and its chord; semicircle, a segment equal to half of the circle; sector, a part of the area within the circle bounded by two radii and the intercepted arc; quadrant, a sector equal to one-fourth of the area within the circle. The area within a circle is commonly spoken of as "the area of the circle."

The ratio of the length of the circumference of a circle to its diameter is denoted by the Greek letter π (pi). This symbol is apt, since it recalls the efforts made by the Greeks to find an exact value for the ratio—since their mathematics was essentially geometric, this quest took the form of an effort to "square the circle," that is, to construct with straight edge and compass, a square and circle of identical areas. After the Greek period, other mathematicians sought to find an exact value, until it was proved impossible by Lindemann in 1882. In the course of the search, values were obtained to hundreds of decimal places, and in recent years the value has been extended to tens of thousands as a computer exercise, but not, however, with a view to finding an exact value.

To five decimal places, $\pi = 3.14159 \ldots$. There are a number of

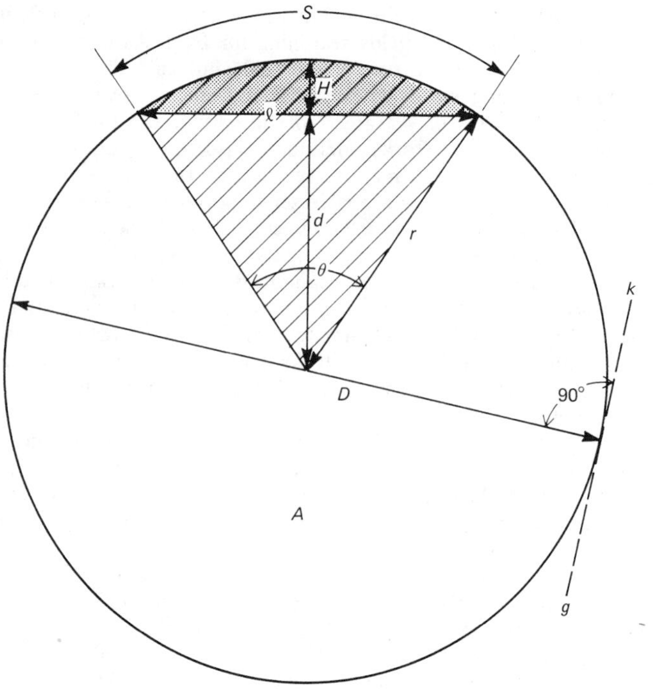

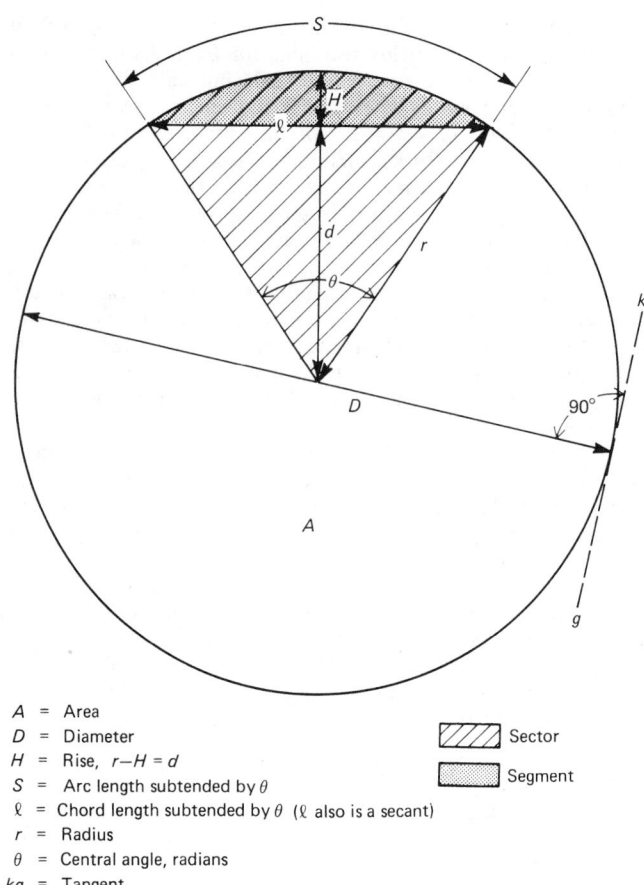

A = Area
D = Diameter
H = Rise, r–H = d
S = Arc length subtended by θ
ℓ = Chord length subtended by θ (ℓ also is a secant)
r = Radius
θ = Central angle, radians
kg = Tangent

Sector
Segment

Major characteristics of a circle.

series for π, one of the most common being $\pi = 4(1 - \frac{1}{3} + \frac{1}{5} - \frac{1}{7} \ldots)$. See also **Pi.**

A central angle has its vertex at the center of the circle and radii for its sides; an inscribed angle has its vertex on the circumference of the circle and chords for its sides.

Characteristics of a Circle (see accompanying diagram).

Arc. A portion of the circumference. $S = r\theta = \frac{1}{2}D\theta$

Area of Circle. $\pi r^2 = \frac{1}{4}\pi D^2$

Area of Sector. $\frac{1}{2}rS = \frac{1}{2}r^2\theta$

Area of Segment. $A(\text{sector}) - A(\text{triangle}) = \frac{1}{2}r^2(\theta - \sin\theta)$

$$= r^2 \cos^{-1}(r - H/r) - (r - H)\sqrt{2rH - H^2}$$

Chord. $l = 2\sqrt{r^2 - d^2} = 2r\sin(\theta/2) = 2d\tan(\theta/2)$

Circumference. The perimeter of a circle. $C = 2\pi r = \pi D$

Pi. $\pi = 3.14159\ldots$

Secant. A line cutting across a circle at two points.

Tangent. A line touching a circle at a single point.

See accompanying appended table of circumferences and areas of circles of specific diameters.

The space between the perimeters of two circles of different radii is known as *annulus*. Where the two circles are concentric, the shape commonly is referred to as a ring. The *area of the ring* (or space) between two circles of radius r_1 and r_2, where one circle fully encloses the other although the two circles need not be concentric, is $\pi(r_1 + r_2)(r_1 - r_2)$.

According to analytic geometry (see also **Conic Section**), the general equation of a circle in rectangular coordinates is given by

$$x^2 + y^2 + 2Ax + 2By + C = 0$$

provided $D = A^2 + B^2 - C$ is positive and does not vanish. Under these circumstances, the radius of the circle is $\sqrt{D}$ and its center is at $x = -A$, $y = -B$. If $D = 0$, the circle degenerates into a point; if D becomes negative the circle is imaginary.

Since the equation for the circle contains three arbitrary parameters, three conditions must be imposed in order to determine the circle uniquely. If the chosen conditions are: center $x = 0$, $y = 0$ and radius

CIRCUMFERENCES AND AREAS OF CIRCLES OF DIAMETER D

DIAMETER D	CIRCUMFERENCE	AREA
1	3.14159	0.79
2	6.28319	3.14
3	9.42478	7.07
4	12.5664	12.57
5	15.7080	19.64
6	18.8496	28.27
7	21.9911	38.49
8	25.1327	50.27
9	28.2743	63.62
10	31.4159	78.5
15	47.1239	176.7
20	62.8319	314.2
25	78.5398	490.9
30	94.2478	706.9
35	109.956	962.1
40	125.664	1256.6
45	141.372	1590.4
50	157.080	1963.5
60	188.496	2827.4
70	219.911	3848.5
80	251.327	5026.6
90	282.743	6361.7
100	314.159	7854.0
125	392.699	12271.8
150	471.239	17671.5
175	549.779	24052.8
200	628.319	31415.9
225	706.858	39760.8
250	785.398	49087.4
275	863.938	59395.7
300	942.478	70685.8
325	1021.02	82957.7
350	1099.56	96211.3
400	1256.64	125664
450	1413.72	159043
500	1570.80	196350
550	1727.88	237583
600	1884.96	282743
650	2042.04	331831
700	2199.11	384845
750	2356.19	441786
800	2513.27	502655
850	2670.35	567450
900	2827.43	636173
950	2984.51	708822
1000	3141.59	785398

r, the standard equation of the circle becomes $x^2 + y^2 = r^2$. Parametric equations are $x = r \cos \phi$, $y = r \sin \phi$. In polar coordinates, the equation is $r^2 - 2rr_1 \cos(\theta - \theta_1) + r_1^2 = a^2$, where the center of the circle is at (r_1, θ_1), its radius is a, and (r, θ) is any point on the circumference. See also **Evolute; Involute;** and terms listed under **Mathematics.**

CIRCLE OF EQUAL PROBABILITY (CEP). A measure of the accuracy with which a rocket or missile can be guided; the radius of the circle at a specific distance in which 50% of the reliable shots land. Also called *circular error probable, circle of probable error.*

CIRCUIT ANALYSIS. **Spectrum Analysis.**

CIRCUIT BREAKER. A specially designed mechanical switching device for making, carrying, and breaking electrical circuits—under normal conditions as well as performing in a special way under abnormal conditions. A circuit breaker, for example, may make and maintain circuit contact for a specified time (usually quite short) under abnormal conditions (short circuit would be an extreme case) and, if the cause of the abnormal condition has not been corrected during the short interval, then break the circuit. A fuse may be considered a form of circuit breaker with the specific purpose of providing protection against overcurrents. Ordinary switches of many designs, of course, make and break circuits routinely, but they are not designed to handle abnormal conditions and to provide special functions under such conditions. A circuit breaker need not incorporate any automatic features to qualify under the definition, but must be designed to make and break circuits under abnormal conditions quickly and without excessive arcing.

Electrical circuit breakers for both residential and industrial use are required to interrupt currents many times their rated values.* For example, the most widely used residential breakers have 15- and 20-ampere ratings. For a 15-ampere-rated breaker to pass the required tests (Underwriters Laboratory), the breaker must sustain a 15 ampere current indefinitely, must open within one hour at 18.75 amperes, must open within two minutes if the current rises to 30 amperes, and must interrupt (without change to itself) a current flowing from a source capable of 10,000 amperes. Because many loads are inductive (motors, fluorescent lights, etc.), breakers must be tested at various power factors (PF). In general, the more inductive the load, the lower the power factor and the greater the difficulty of interrupting the current. When the power factor is unity, the load is pure resistance; when a power factor is zero, the load is pure inductance. With unity PF, the power dissipated is given in watts (W), equal to the product of volts × amperes. With lower power factors, the equipment muse be rated in volt-amperes (VA, not equal to watts), in kilovolt-amperes (kVA), or in million-volt amperes (MVA). Transformers of large capacity are commonly energized by medium-voltage feeders from the power grid, often in the 13,200 volt range and usually expressed as 13.2 kilovolts (kV).

The principal categories of circuit breakers in common use are (1) *low-voltage air circuit breakers*, generally designed for use on dc circuits and low-voltage ac circuits up to about 600 volts. Air circuit breakers can be used for voltages up to about 3,000 volts for dc breaking in connection with railway circuits; (2) *oil power breakers*; (3) *oilless power breakers*.

The air cicruit breaker commonly uses two fixed terminals. A bridging member, operated by a linkage system, maintains these contacts under heavy pressure. Auxiliary arcing contacts are arranged to close before and open after the main contacts. This prevents damage of the main contacts from arcing. The auxiliary arcing contacts are usually easily replaceable. For the main contacts, most modern breakers incorporate spring-mounted, self-aligning contacts (silver) attached to a solid bridging member. The arcing contacts may be a silver-tungsten or copper-tungsten alloy although a carbon alloy was used in earlier designs. The breakers may be hand- or electrically-operated, the latter accomplished by means of solenoids or motorized mechanisms. Multiple-pole circuit breakers are widely used, with a pole for each ungrounded line of a circuit. In connection with a compound-wound generator or converter paralleled with other units, an extra pole is needed for the equalizer connection.

The so-called molded-case air circuit breaker largely has become standard in large building and industrial installations. Large units are available that will open a circuit up to 42,000 amperes at 600 volts ac, or 50,000 amperes at 250 volts dc. Smaller versions of the molded-case circuit breaker are commonly used for residence protection in place of former fuse boards.

A dead-tank type oil power breaker is comprised of a steel tank partially filled with oil. Porcelain or composition bushings pass through the cover of the device, serving as the circuit connections. Contacts located well below the oil level are bridged by a conducting crosshead, the latter being carried by a lift rod. In many designs, the rod drops by gravity after contact separation by spring action and thus the breaker is opened. To increase the rate of opening over much of the total travel, accelerating springs are commonly used. In some designs, allowance is made to provide two and up to six or more breaks per pole, thus reducing the length of stroke needed for adequate arcing distance. A special tank liner of insulating fibrous composition prevents the arc from striking the walls of the tank. In single-tank breakers, with all poles of a three-pole breaker in one tank, rating up to 69,000 volts and 3,500,000 kilovolt-amperes interrupting ratings are available.

* This description by permission from longer article by V. C. Oxley, GTE Laboratories.

For higher voltages, multitank breakers are used, in which each pole is in a separate tank. Where "isolated-phase" construction is needed, the multitank breakers are available in lower ratings.

Special arc-control devices have been developed during the last 40 years to replace plain-break breakers. Numerous schemes have been considered. In all of the commonly used breakers, advantage is taken of the oil pressure that is generated by the gas created by the arc. This pressure is used to force fresh oil through the path of the arc in sufficient quantity to provide the needed insulation at current zero, thus preventing a restrike of the arc with consequent interruption of the circuit. Special interrupting chambers contain these high pressures, thus reducing stress on the main oil tank. After cooling, the gases generated are vented to atmosphere.

Oil used in circuit breakers must have particular characteristics, including a flash point of about 133°C; a burning point of about 148°C; a freezing point of about −40°C, and a dielectric strength of at least 30,000 volts.

Since the early 1940s, the acceptance of oilless power breakers has increased greatly, to the extent that most indoor installations are of this type as well as a rapidly gaining acceptance for outdoor installations. In fact, for extra-high-voltage needs, only oilless breakers are obtainable.

In the magnetic-air circuit breaker, the main circuit is interrupted by the action of a strong magnetic field which forces the arc deep into a specially-designed arc chute. The purpose of the chute is to cool and lengthen the arc to the point where the circuit no longer can be maintained by the system voltage, thus effecting interruption. Inasmuch as the zone between the main contacts is free of ionized air by the time interruption is achieved in the arc chute, there is no problem of possible restriking of the arc. See also **Electric Shock.**

CIRCUIT (Clipping). Clipping Circuit.

CIRCUIT (Coupled). Coupled Circuit.

CIRCUIT DENSITY (Chip). Electron Beam Lithography; Microelectronics; Microstructure Fabrication; Semiconductor; Telephony.

CIRCUIT (Equivalent). Equivalent Circuit.

CIRCUIT (Integrated). Integrated Circuit.

CIRCUIT (Miniature). Microstructure Fabrication.

CIRCUIT (Resistor Transistor Logic). Resistor Transistor Logic.

CIRCUITS, FUNDAMENTAL (Mathematics). Let G be a connected graph containing v vertices and e edges and T a tree of G. The end points of each chord (with respect to T) are connected by a unique *tree path*. A chord together with its corresponding tree path forms a fundamental circuit. The number of fundamental circuits is equal to the number of chords or $e - v + 1$. If G is finite but not *connected* it consists of P maximal connected subgraphs, each of which possesses a tree and its associated fundamental circuits. See also **Graph (Mathematics); Tree (Mathematics);** and terms listed under **Mathematics.**

CIRCULAR BURNER. Burner.

CIRCULAR CURVES. From a mathematical standpoint, a circular curve is an arc having a constant radius; but it is used in civil engineering as a general heading to cover simple, compound and reversed curves.

A circular arc joining two tangents (straight lines) is called a simple curve (Fig. 1). Large radius simple curves are used in highways to provide a means of gradually changing the direction of the center line of a roadway. Simple curves connected by tangents were formerly used on railroads but they have been superseded by the combination of spiral and simple curves. This practice is also followed in modern highways built for high speeds. In Fig. 1, point A is called the point of curvature (P.C.) and point B the point of tangency (P.T.). Point C is known as the point of intersection (P.I.) of the tangents. In

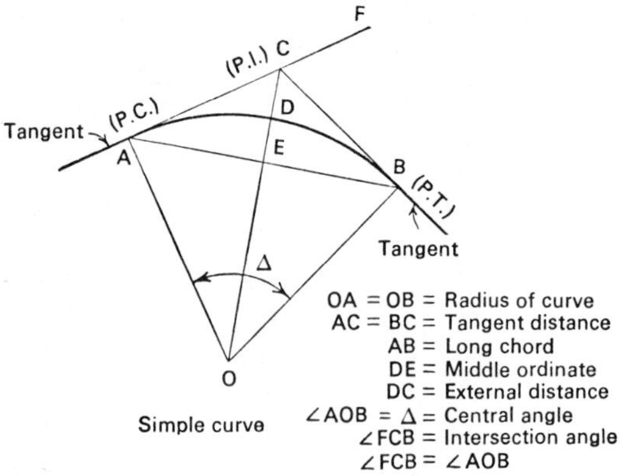

Fig. 1. Simple curve.

highway practice the length of the curve is generally represented by the length of the circular arc but in railroad practice it is given in terms of chord lengths.

A curve made up of two or more simple curves, each having a common tangent point at their junction and lying on the same side of the tangent, is called a compound curve. Compound curves have an advantage over simple curves since they may be easily adapted to the natural topography of a particular location.

A curve made up of two simple curves, having a common point of tangency at their junction and lying on opposite sides of the common tangent is called a reversed curve (Fig. 2). This type of curve is advantageous for use in connection with railroad crossovers and spur tracks but should never be employed for main lines. Reversed curves are used in highway location when the alignment requires an abrupt reversal in direction.

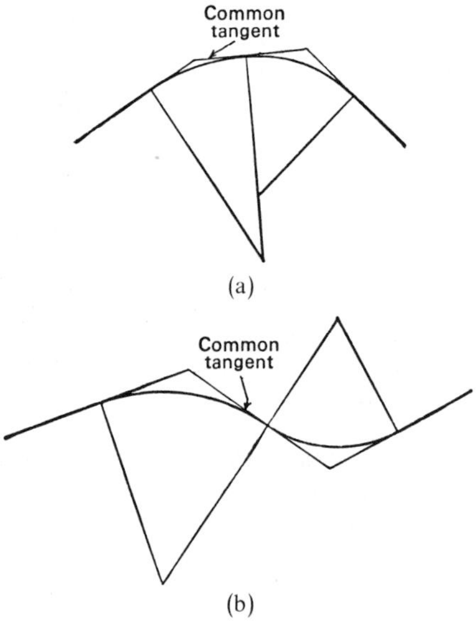

Fig. 2. (a) Compound curve; (b) reversed curve.

CIRCULAR DICHROGRAPHY. Photometers.

CIRCULAR DISTRIBUTION. A frequency distribution of a variate which ranges from 0 to 2π, so that the frequency may be regarded as distributed round the circumference of a circle. The term is used especially of phenomena which have a period of 2π (by a suitable change of scale if necessary) so that the probability density at any point a is the same as that at any point $a + 2\pi r$ for integral values of r. It is usually expressed in terms of an angle θ. The analogue of the normal distribution in this class, the so-called circular normal distribution, given by

$$f(\theta) = \exp[k \cos(\theta - \theta_0)]/2\pi I_0(k), \quad 0 \le \theta \le 2\pi$$

where I_0 is a Bessel function of the first kind of imaginary argument.

CIRCULAR MIL. Units and Standards.

CIRCULATING STORAGE (Computer). Storage (Computer).

CIRCULATION (Air). Aerodynamics; Supersonic Aerodynamics.

CIRCULATION (Atmosphere). Atmosphere (Earth).

CIRCULATION THEOREM (Meteorology). Atmosphere (Earth).

CIRCULATOR (Microwave). A nonreciprocal circuit element employing the Hall effect or Faraday rotation to produce phase shift which is a function of the direction of energy travel through the device. An application of the Faraday ferrite circulator is a radar duplexer in which the phase shift produced in the circulator can effectively switch the antenna from the transmitter to the receiver. See accompanying diagram.

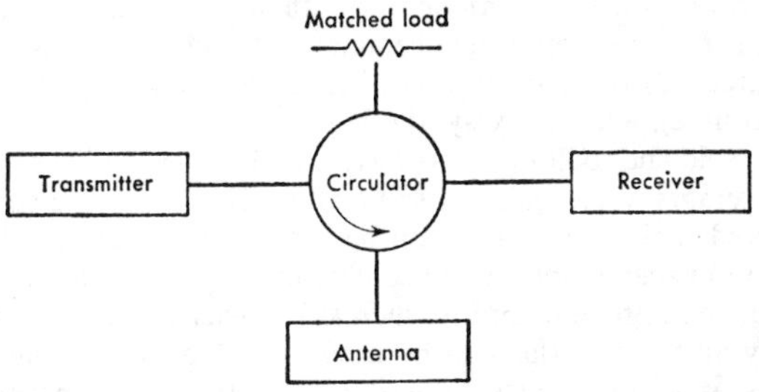

Ferrite circulator used for switching antenna from transmitter to receiver.

A circulator also may be defined as a waveguide component with several terminals so arranged that energy entering one terminal is transmitted to the next adjacent terminal in a particular direction. The terms *microwave circulator* and *ferrite circulator* also are used.

CIRCULATORY SYSTEM. The system of passages and chambers through which materials are distributed in the body in a liquid mixture (blood or hemolymph). In humans the circulatory system is highly refined and complex. See **Circulatory System (Human).**

Animals of very simple structure and small size secure adequate distribution by the diffusion of materials from cell to cell and so have no need for a special circulatory system. In more complex bodies, however, the principal centers of interchange, such as the alimentary tract, respiratory organs, and excretory system, are far removed from some of the parts that they serve; here more rapid transportation is necessary.

This need is met in some animals by the extension of centers of interchange. In the flatworms, for example, the alimentary tract branches through the body and no other structure is far from some part of it.

Other forms, as the roundworms, have extensive spaces within the body in which liquid contents are moved to some extent by the movements of the body. This type of circulation extends to animals with a true body cavity or coelom but here it is associated with the development of a closed tubular circulatory system, a condition which exists in the earthworm.

In this simple state the tubular system consists of a longitudinal vessel above the alimentary tract and others at different levels in the ventral part of the body. Other tubes running around the alimentary tract associate the longitudinal vessels, and muscular walls of certain regions, together with valves in the cavities of the tubes, propel the blood which they contain. In the earthworm the contractile vessels are principally the dorsal vessel and a series of five pairs encircling the esophagus and known as hearts. The blood flows forward in the dorsal vessel, down through the hearts, and back in the ventral vessels, reaching the dorsal vessel again after following various routes through the tissues which it serves.

In such systems as this several types of vessels are developed. Those which lead from the central pumping organ receive the blood under its highest pressure and have the strongest walls, containing elastic and muscle fibers. These vessels are called arteries. They lead into smaller branches known as arterioles and these in turn into minute vessels with very delicate walls made up chiefly of a single layer of thin cells. In the human body these delicate capillaries are from $\frac{1}{200}$ to $\frac{1}{90}$ of a millimeter in diameter. Some of the blood-fluid passes between the cells of their walls into spaces in the surrounding tissues where interchange with the various cells is possible. In some parts of the body, as in the liver of vertebrates, the ultimate tubular passages are even simpler than the capillaries and are called sinusoids. Their walls are, at least in part, merely the surrounding tissues; whether a special lining exists in some parts is disputed. From such small vessels the blood is collected into veinlets and these converge to form veins which carry it to the heart. The veins have strong walls of complex structure but in vessels of the same caliber the walls are thinner in veins than in arteries.

Some of the fluid from tissue spaces is gathered into a type of vessel resembling a vein but more delicate; it flows into trunk vessels which rejoin the veins. This system of vessels is known as the lymphatic system.

In the vertebrates the fishes present the basic plan of the circulatory system. The heart has two principal chambers, an atrium which receives the blood and a ventricle which pumps it out to the body. A single large artery, the ventral aorta, leads forward and branches into a series of afferent branchial arteries which break up into capillaries in the gills. Efferent branchial arteries lead out of the gills to the dorsal aorta which runs back to the body. The head is supplied by extensions of the dorsal aorta, the carotid arteries. From the arterial system branches conduct the blood to the capillaries of all parts of the body. The blood from the alimentary tract is collected into a hepatic portal vein which breaks up into sinusoids in the liver, and is carried thence to the heart by a hepatic vein. In a like manner some of the blood from the caudal part of the body is conveyed to the kidneys by a renal portal vein before entering the veins which carry it to the heart. From all other regions of the body the blood is collected by veins which flow directly to the heart.

The vertebrate heart is a specialized region of the tubular system whose subdivision into chambers is complicated above the fishes by some degree of longitudinal splitting. In the amphibians the atrium has become two, right and left, in the reptiles the splitting involves the ventricle, and in the birds (Aves) and mammals the division into four chambers is complete. Blood entering the right auricle comes from the body and passes on into the right ventricle to be pumped to the lungs. It returns by the pulmonary veins to the left auricle, enters the left ventricle, and is pumped to the body. In other respects the most striking difference between the circulatory system of the fish and that of man is the elimination of the renal portal system in the higher classes of vertebrates.

The circulatory system of arthropods differs from that of other complex animals in the limited extent of the closed tubes and their association with extensive blood spaces constituting a hemocoele or blood cavity and hence an open system. Insects have a single longitudinal tube lying in the dorsal region of the body. It pumps the blood into the head and is known as the heart or dorsal vessel.

CIRCULATORY SYSTEM (Fishes). Fishes.

CIRCUMCISION. The excision of the foreskin or prepuce which covers the head of the penis in the male. The operation is indicated when the foreskin is too tight, although it is commonly done for hygienic reasons.

CIRCUMFERENCE (Geometry). Circle.

CIRQUE. Topographic feature produced by a mountain glacier. A mountain glacier usually starts in some sheltered ravine, slightly below the top of the mountain. Névé (firn) is the name of the granular ice which gradually develops from the original snow. The névé or accumulation of ice is restricted to that altitude at which the average summer temperature is 32°F (0°C). This line (summer isotherm) may vary from sea level, in the polar regions, to 20,000 (6,000 meters) feet in the tropics. After the accumulation of granular ice on a slope reaches a certain thickness, the mass moves slowly downward under its own weight, by process of plastic flow. In the névé region, where the snow and ice bank rests against the sloping cliff, the ice tends to work away from the rock, forming a crevice called the bergschrund. Frost action in the region of the bergschrund causes the recession of the cliff, the broken material from which is frozen into the base of the ice and serves as tools to scour out circular basins called cirques. Small lakes and ponds which occur in cirques are called tarns. Where several cirques are developed near the summit of a mountain, the side walls of the cirques are called combs; the ridges between the cirques, cols; and the elevated terminations of the comb ridges, monuments. A triangular mountain peak, such as a Matterhorn, results from the complete cirquation of what was originally a relatively smooth-topped mountain.

CIRRHOSIS. **Liver.**

CIRRIPEDIA. Crustaceans of which the only commonly known representatives are the barnacles. The name applies to a subclass characterized by adaptations for sessile life; the included species are unlike other crustaceans in appearance.

Economically the barnacles are important because they attach themselves to the bottoms of ships and necessitate occasional removal.

Classification:

Order *Thoracica*. The common barnacles.
Order *Acrothoracica*. Small forms of which the females live in cavities in the shells of mollusks.
Order *Apoda*. One simplified form, parasitic in a common barnacle.
Order *Rhizocephala*. Parasitic, usually on crabs and related crustaceans. The adult forms root-like growths in the body of the host and is otherwise degenerate in form.
Order *Ascothoracica*. Parasitic in corals; less degenerate than the preceding.

CIRROCUMULUS. **Clouds and Cloud Formation.**

CIRROSTRATUS. **Clouds and Cloud Formation.**

CIRRUS. 1. In protozoa, a spine-like organ of locomotion formed of aggregated cilia. 2. In annelida (worms), a slender flesh process on the parapodia. 3. In meterology, see **Clouds and Cloud Formation.**

CISCOE. **Whitefishes.**

CIS-COMPOUND. **Isomerism.**

CISSOID OF DIOCLES. Draw a circle of radius a through the origin of a rectangular coordinate system with diameter $OCA = 2a$ on the X-axis. Through O take any chord OR and extend it to meet, at Q, the tangent to the circle at A. On OR choose a point P so that $PQ = OR$. As P rotates about the origin its locus is the cissoid of Diocles. Its equation is $x(x^2 + y^2) = 2ay^2$ or, in polar coordinates, $r = 2a \sin^2 \theta \sec \theta$.

The two branches of the curve are mirror images of each other with respect to the X-axis. The line $x = 2a$ is an asymptote and there is a cusp of the first kind at the origin. The name of the curve comes from a supposed resemblance to the ivy leaf (Greek, *kissoeidēs*, ivylike) for that part of the figure bounded by the semicircle BAD and the part $DOPB$. Diocles was a Greek mathematician who probably lived in the second century B.C. It is said that he used the curve to solve the classical problem of the duplication of a cube.

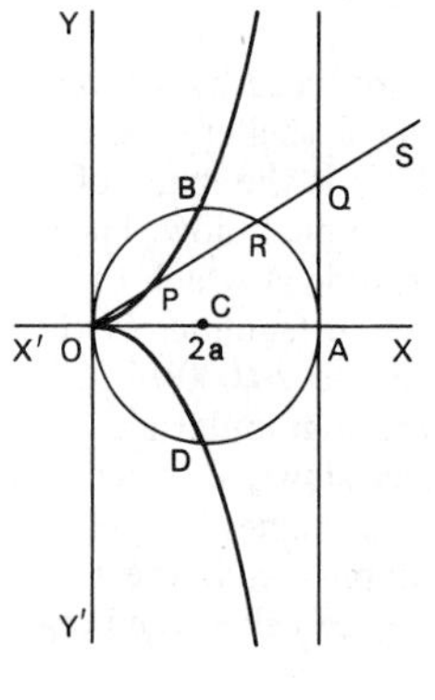

Cissoid of Diocles.

The unqualified word cissoid is sometimes taken to mean any higher plane curve constructible by a method similar to that used for this case. Other examples are the strophoid, the folium of Descartes, and the trisectrix.

See also terms listed under **Mathematics.**

CISTRON. **Cell (Biology).**

CITRATE GAS TREATMENT SYSTEM. **Pollution (Air).**

CITRIC ACID. $C_3H_4(OH)(COOH)_3$, formula weight 192.12, white crystalline solid, mp 153°C, decomposes at higher temperatures, sp gr 1.542. Citric acid is soluble in H_2O or alcohol and slightly soluble in ether. The compound is a tribasic acid, forming mono-, di-, and tri-series of salts and esters. Citric acid may be obtained (1) from some natural products, e.g., the free acid in the juice of citrus and acidic fruits, often in conjunction with malic or tartaric acid, and the juice of unripe lemons (approximately 6% citric acid) is a commercial source; (2) by fermentation of glucose (blackstrap molasses is a major source); and (3) by synthesis.

Citric acid and sodium citrate have found use as additives in effervescent beverages and medicinal salts, although excessive quantities are considered toxic. Citric acid can be used as an effective antioxidant. Because the acid is not readily soluble in fats, it is added to formulations which improve solubility, such as propylene glycol and butylated hydroxy anisol, and this can be used as a stabilizer for tallow, fats, and greases. Citric acid also is used for adjusting pH in certain electroplating baths. It finds a number of miscellaneous uses in etching, textile dyeing and printing operations.

Citrates (like tartrates) in solution change silver of ammonio-silver nitrate into metallic silver. Calcium citrate on account of its solubility characteristics, is of importance in the separation and recovery of citric acid. Calcium citrate plus dilute H_2SO_4 yields citric acid plus calcium sulfate, and the latter may be separated by filtration. Citric acid may be obtained by evaporation of the filtrate.

In most living organisms, the citric acid cycle constitutes the final common pathway in the degradation of foodstuffs and cell constituents to carbon dioxide and water. This cycle is described in the entry on **Carbohydrates.**

CITRIC ACID CYCLE. **Carbohydrates.**

CITRINE. The mineral citrine is a yellow variety of quartz sometimes used as a gem. It is often marketed under the name topaz and may mislead the unwary. Brazil and the Malagasy Republic have furnished material of excellent quality. See also **Quartz.**

CITRULLINE. **Amino Acids.**

CITRUS PSYLLA (*Insecta*, *Homopterau*). This insect is a serious pest on fruit in India. The adult citrus psylla (*Diaphorina citri*) is about $\frac{1}{8}$ inch (3 millimeters) long and mottled brown. It is covered with a waxy secretion that makes it appear dusty. The nymph is light yellow and about $\frac{1}{16}$ inch (1.5 millimeters) long. These insects suck the sap from citrus leaves, usually new growth. The leaves become covered with honeydew, on which a sooty mold grows. Prolonged feeding results in lowered yields, and if left unchecked, can cause defoliation and kill the tree. The pest is also a carrier of citrus greening disease, a virus disease of citrus.

CITRUS TREES. Citriculture is an important segment of world food production and nutrition. The principal citrus fruits with their percentage of total world production are: (1) orange, 67%, tangerine (more properly called mandarin), 14.3%; lemon and lime, 9%; grapefruit and pummelo, 7.9%; and other citrus fruits, 1.8%. Citrus fruits are of the genus *Citrus*, which is of the family *Rutaceae* (rue family) and of the subfamily *Aurantoideae. Rutaceae* is composed of a variety of trees and thorny shrubs. In all, there are 7 subfamilies in Rutaceae, comprised of about 150 genera. The subfamily *Aurantiodeae* has about 28 genera, of which six are classified as true citrus fruit trees; that is, they have a berry fruit (hesperidium) which is characterized by a juicy pulp made of vesicles filling all of the space in the segments of the fruit not occupied by seeds. The six genera are: *Citrus*, *Clymania*, *Eremocitrus*, *Fortunella*, *Microcitrus*, and *Poncirus*. Of these genera, only three are of commercial importance, namely, *Citrus*, *Fortunella*, and *Poncirus*.

Over the last 250 years, attempts to establish clearcut classifications of the various citrus fruits have met with exceptional difficulties. These difficulties, in turn, are reflected in the nomenclature used, both from a practical and a purely scientific viewpoint. The original efforts of Linnaeus in the mid-1700s were not favored with the knowledge of numerous citrus fruits, particularly those occurring in the Orient and other regions of the world whose flora had not been thoroughly studied at that time. Consequently the work of Linnaeus in the area of citrus fruits was far from complete and not fully accurate. Numerous attempts at classification during the intervening years have taken place, but as research continues right into present times, prior conclusions have been upset. Citrus fruits that have been growing in the Orient many centuries were found to be much more diverse than initially assumed. Further, citrus fruits are prone to the development of diversity through both natural and artifical cross-breeding, through bud mutations, and the development of chance seedlings.

Exemplary of the status of citrus taxonomy as of the early 1980s are persistent doubts concerning certain prior classifications. For example, there is no clear evidence that the grapefruit is a true species. The grapefruit is closely associated with the pummelo, probably originating in the West Indies a few centuries ago. It is not known whether the grapefruit arose purely as a bud sport,* or whether it is a hybrid between pummelo and some other entity. Thus, the practice of identifying grapefruit with a specific Latin name can be questioned. However, since 1930, assignment of grapefruit to *Citrus paradisi* Macf. has been accepted by a majority of botanists and horticulturists.

Similarly, the lemon is probably not a valid species. Research as recently as 1974 points to the lemon as *not* being a distinct species, nor a first-generation hybrid between citron and lime. Researchers have shown that citron is most likely one of the parents of lemon, but that an unidentified genetic source (not in the citron–lime group) may have contributed to the origin of lemon. However, as early as 1766, lemon was assigned to *Citrus limon* (L) Burm. f., an identification that generally has been accepted by citrus scientists. Questions also have been raised concerning the classification of the sweet orange as *Citrus sinensis* Osbeck, a classification dating back to 1757.

As indicated by the accompanying table, Swingle and Reece (1967) proposed 16 species of the *Citrus* genus, a classification that, at least temporarily, as been accepted by a number of citrus scientists. Eight of these species embrace the edible citrus fruits of commerce; the other species are considered inedible. However, a number of other classifications have been proposed during the last several years, including that of Tanaka (1954), who proposed as many as 159 species of *Citrus*, a scholarly classification that can be quite helpful to botanists and horticulturists, but one that is somewhat too complex for practical application by persons who are concerned mainly with the commercial aspects of citrus production.

Considering the many varieties of citrus fruits, there are remarkable similarities which the designers of citrus processing equipment can use to advantage. In terms of processing, the principal physical variables are size and shape of the whole fruit; internal dimensions, principally thickness of peel or rind, which embodies the epidermis, flavedo, oil glands, albedo and vascular bundles; number of sections; and relative size of core. The principal elements of a typical citrus fruit (shown in cross-section) are identified in Fig. 1. The presence or absence of seeds is also very important to the processor. Because the seeds contain materials that detract from the flavor of extracted juice, extraction machinery must be designed to avoid or at least minimize penetration of the outer skin of the seed that would free undesirable substances to mingle with the juice.

Discovered as a constituent of citrus fruits as early as 1841 by Bernay, the knowledge of *limonin* has progressed mainly during the past 30 to 35 years. A whole family of limonin-related compounds, called *limonoids*, have been isolated and described during this period. The structure of limonin proper was not determined until the early

* In botany, a sport is a bud variation; in biology, a form that varies markedly from the norm.

SPECIES IDENTIFIED IN GENUS CITRUS[1]

Common Name	Species Name	Year Named
Citron	*C. medica* (L.)	1753
Grapefruit	*C. paradisi* Macf.	1930
Lemon	*C. limon* (L.)	1766
Lime (common)	*C. aurantifolia* Christm.	1913
Mandarin (tangerine included)	*C. reticulata* Blanco	1837
Orange:		
Sweet orange (common orange)	*C. sinensis* Osbeck	1757
Sour orange	*C. aurantium* (L.)	1753
Indian wild orange	*C. indica* Tan.	1931
Papeda:		
Papeda	*C. micrantha* Wester	1915
Celebes papeda	*C. celebica* Koord.	1898
Ichang papeda	*C. ichangensis* Swing.	1913
Khasi papeda	*C. latipes* Tan.	1928
Maurituis papeda	*C. hystrix* D.C.	1813
Melanesian papeda	*C. jacrocarpa* Mont.	1860
Pummelo	*C. grandis* Osbeck	1765
Tachibana	*C. tachibana* Tan.	1924

[1] Adapted from Single/Reece (1967).

OTHER NAMES AND GENERA RELATED TO CITRICULTURE

Citrange	Common orange × Trifoliate orange.
Clementine	A variety of mandarin.
Fortunella	A genus of subfamily Aurantiodeae. An evergreen, unifoliate with small edible fruits. Kumquat is one of these.
Limequat	A lime-kumquat hybrid.
Ponkan	A variety of mandarin.
Poncirus	A deciduous tree with trifoliate leaves and a genus of subfamily Aurantiodeae. Fruit contains an acrid oil and is inedible.
Poncirus trifoliata	Sometimes used as a rootstock for citrus trees.
Rough lemon	A variety of lemon sometimes used as a rootstock for citrus trees.
Satsuma	A variety of mandarin.
Tangelo	Grapefruit (*C. paradisi*) × Sweet orange (*C. sinesis*). or Grapefruit (*C. paradisi*) × Pummelo (*C. grandis*). Commercially, the tangelo is classified separately, or with grapefruit. See entry on **Grapefruit.**
Tangerine	Some varieties of mandarin are sometimes called tangerine in some regions of the world.
Tangor	Mandarin (*C. reticulata* Blanco) × Sweet orange (*C. sinensis*).
Temple	Mandarin (*C. reticulata* Blanco) × Sweet orange (*C. sinensis*). Commercially, the temple is usually classified with the orange.

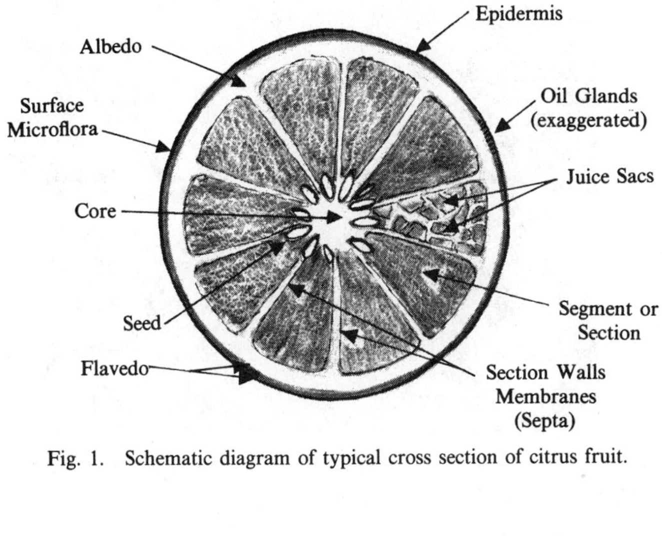

Fig. 1. Schematic diagram of typical cross section of citrus fruit.

1960s by several investigators using different analytical techniques. Research has shown that the two dozen and more known limonoids, only about 30% are bitter-tasting. Delayed bitterness in citrus juices has traditionally been attributed to the presence of limonin. Citrus scientists differentiate between the bitter aftertaste caused by limonin and the bitterness caused by *naringin* of grapefruit or the *neohesperidin* of the Seville orange. The aforementioned compounds are not universally present in citrus fruits, whereas findings to date indicate that limonin can be found in all citrus fruits.

Citron. A member of the subfamily *Aurantiodeae*, the citron (*Citrus paradisi*) is a lemonlike fruit with a thick peel and containing a relatively small amount of acid pulp. The citron is the most suitable of the various citrus fruits for preparing a candied peel used as a confection. The majority of citrons are grown in Italy, where the main variety is the *Diamante* or *Liscio di Diamante*.

Grapefruit. The grapefruit tree is a subtropical evergreen with dense foliage. The tree may reach a height of from 40 to 50 feet (12 to 15 meters) in its natural state without pruning. The tree is subtropical and sensitive to both high and low temperatures. The head of the tree varies from rounded to somewhat conical. Commercial growers hold the height to 15–25 feet (4.5–7.5 meters). Leaves of the grapefruit tree are larger than those of the sweet orange, but somewhat smaller than the closely-related pummelo (*C. grandis*). The leaves are smooth, ovate, and bluntly tipped. The flowers are large, white, and fragrant. They are borne singly or in small clusters in the axils of the leaves. The fruits are borne in grapelike clusters (possibly accounting for the name of the fruit). Generally, the fruit is oblate (globular with flattened top and bottom), although the shape varies from one variety to the next. The skin of the fruit ranges from lemon to an orangish coloration. In some varieties, the basic lemon coloration carries a pink blush coloration. The flesh is described as slightly tannish-yellow. The fruit is juicy and the juice color ranges from slightly pink to reddish of different tones, depending upon variety. Fresh juice normally has a slight bitter flavor, arising from its content of the glucoside *naringin*. The bark of the grapefruit tree is smooth and gray-brown in color.

The origin of the grapefruit is not certain and is described by some citrus scientists as remaining a puzzle. Some authorities have postulated that it may have arisen as a hybrid of the pummelo with the sweet orange. The shaddock tree, a pummelo, was observed by explorers in the very late 1600s in the West Indies, notably Jamaica.

Tangelo. This tree probably originated in China and southeastern Asia as early as 4000 years ago. It is postulated that the first tangelos resulted from insect cross-pollination of the mandarin orange and the pummelo. Possibly the first of the tangelos in Florida was the *Nocatee*, which appeared as a natural seedling in a grove located at Nocatee, Florida. Tangelo trees are vigorous growers and usually attain the size of an orange tree. They are more cold resistant then grapefruit. All are evergreen trees with unifoliolate leaves and have fragrant white flowers. Generally, tangelo fruits are about the size of the common orange, but exhibit a tendency to be slightly drawn out at the stem end (or necked). They are usually highly colored, aromatic, richly flavored, sprightly acid, with only a slightly bitter taste.

Tangelos generally are self-sterile. They produce a high percentage of nucellar embryos and thus reproduce nearly true to type by way of seed, which is not true of grapefruit.

Lemon. The lemon (*C. limon*) is probably a native of India. The trees are small and have many stout thorns. The flowers are large and purplish. The trees are considerably less hardy than orange trees. The trees require warmer winters, but less summer heat to ripen their fruit. They can survive in areas where the temperature does not drop below 44°F (6.7°C). The fruits are produced continuously throughout the year and are picked green. To allow them to ripen on the tree causes them to become bitter and unmarketable. As soon as they attain the size demanded by the market, they are picked and stored, often for several months. Coloring gradually develops during storage, or may be hastened by placing them in rooms heated above 90°F (33°C), whereupon the yellow color will develop within 4 to 5 days.

Orange. The orange (*C. sinensis*) originated in China. Some authorities believe that it was introduced into Arabia about 950 A.D. The orange tree is now widely cultivated almost wherever the climate is suitable. In the United States, there are two major growing regions. Florida produces sweet, thin-skinned juicy oranges, which ripen in early winter. They are picked before fully ripe, and mature in storage. The advent of frozen juice concentrates a number of years ago introduced a welcome element of flexibility in orange production and an effective means for matching size of crop and market demand and introduced some stability in terms of market vascillations and seasonal crop variations. See Fig. 2. California produces thick-skinned navel oranges which have a pleasing acid pulp, and ripen during the winter and spring months. The source of these navel oranges was a bud variation arising in Brazil, and later carried to California, where it

Fig. 2. Harvesting oranges for processing allows more opportunities for mechanization than picking for the fresh market. (*USDA photo.*)

was extensively propagated. Both Florida and California also produce the many-seeded Valencia oranges, which ripen from June through October, producing a crop when other varieties are not bearing. Orange trees can survive winter temperatures in the upper 40°F (4.5°C) range. The trees blossom in the spring, but the fruit requires several months to ripen, the time depending upon the climate.

Other Citrus Fruits. Lime plants (*C. aurantifolia*) are very thorny shrubs or small trees, which are not at all hardy, and so are grown mainly in tropical countries. In the United States, they are grown in California and the southern tip of Florida. The white flowers are small, as is the very acid thin-skinned fruit.

Related to the Citrus group is the genus *Fortunella*, with several species. These are the kumquats. They are small evergreen shrubs, often planted for ornament and because of the small yellow fruits, which are either eaten raw or used in preserve making.

The *Poncirus trifoliata*, or the trifoliate orange, has trifoliate leaves, and a hairy useless fruit. However, it is hardy and can be grown outdoors as far north as southern New York. Hence, it can be valuable for hybridizing with more desirable species.

CIVET. Viverrines.

CIVIL AIRCRAFT. Airplane; Helicopters and V/STOL Craft.

CLADDING (Reactor). Nuclear Reactor.

CLAIRAUT EQUATION. A differential equation of the type $f(y - xy', y') = 0$. Its solution is $f(y - Cx, C) = 0$ (see **Differential Equation of First Order, Ordinary**). It is known, however, that any equation of this general form is a family of curves and that the envelope of the family can be found in this case from the equation $\partial f/\partial C = 0$. This derivative, together with the solution to Clairaut's equation, permits the elimination of the constant of integration, C to give $y = \phi(s)$, a relation which will satisfy the differential equation. It is a singular solution because it contains no constant of integration. It is not a particular solution in the usual sense since it is not obtainable from the general solution by assigning a special value to the constant of integration. Geometrically, a singular solution is an envelope and that cannot be found from the family of curves by specifying its parameter.

Singular solutions can also occur in differential equations of higher order and in partial differential equations. Their mathematical behavior is relatively complex but they are of little interest in applied mathematics.

A simple special case is usually discussed in elementary texts and there called Clairaut's equation. It is $y = xp + p^2$, where $p = y'$, with solution $y = c_1x + c_2$.

CLAM (*Mollusca, Pelecypoda*). Any member of numerous species of bivalve mollusks; a mussel. Some species are commonly called mussels, others clams, and in some cases the names are freely interchanged. Thus the edible mussel and other members of the same family receive this name only, the little-neck clam is always a clam, and the fresh-water species are called freely both clams and mussels.

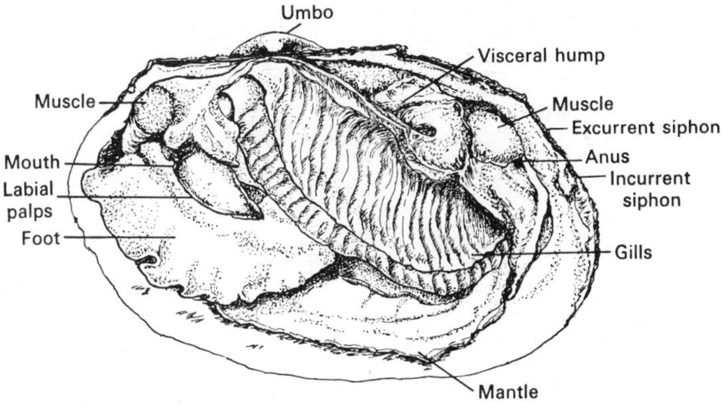

A freshwater clam with the shell and mantle removed from the left side, showing the position of the body organs within the shell.

The clams are moderately important as food. Occasionally pearls are found in them. Clams of commercial importance are described in the entry on **Mollusks.**

CLAM (Geoduck). Geoduck.

CLAM (Larva). Glochidium.

CLAPEYRON-CLAUSIUS EQUATION. This is a widely useful differential equation involving the variables associated with the transition of a pure substance from one state to another; that is, from solid to a liquid, liquid to vapor, or solid to vapor, and vice versa.

In such a system, comprising two phases of the same substance, by adding or withdrawing heat very slowly, it is possible to change one phase reversibly into the other, the system remaining at equilibrium. For such a change, thermodynamic principles indicate that the change in Gibbs free energy is zero.

Now, since by the first law of thermodynamics,

$$dq = dE + p\,dV \tag{1}$$

and since by the definition of Gibbs free energy (G),

$$G = E + pV - TS \tag{2}$$

we can differentiate this second equation and substitute dS for dq/T, obtaining

$$dG = V\,dp - S\,dt \tag{3}$$

which may be written for the phases A and B as

$$dG_A = V_A\,dp - S_A\,dT$$
$$dG_B = V_B\,dp - S_B\,dT$$

Since, as stated above, the change in Gibbs free energy is zero,

$$dG_A = dG_B \tag{4}$$

so that

$$V_A\,dp - S_A\,dt = V_B\,dp - S_B\,dt \tag{5}$$

and

$$\frac{dp}{dt} = \frac{S_B - S_A}{V_B - V_A} \tag{6}$$

Since $S_B - S_A$ is the entropy change accompanying the change of phase, we have

$$S_B - S_A = \frac{L}{T} \tag{7}$$

where L is the heat absorbed per mole in the change from A to B. Substitution of (7) in (6) gives the Clapeyron-Calusius equation:

$$\frac{dp}{dT} = \frac{L}{T(V_B - V_A)} \tag{8}$$

A similar derivation leads to another form of the equation:

$$\frac{dS}{dV} = \frac{L}{T(V_B - V_A)} \tag{9}$$

In the case of liquid ⇆ vapor transitions, we may neglect the volume of the liquid and assume the vapor to obey the ideal gas law, $pV = RT$ (both these simplifications do not usually introduce errors greater than 1–2%), and write Equation (8) as

$$\frac{d\log_e p}{dT} = \frac{L_{vap}}{RT^2}$$

where R is the gas constant.

For a transition taking place at some specified fixed temperature T, the common second member of these equations becomes a known constant, whose value represents the slope of either the T-p or the V-S transition curve, at the point of transition. In other words, it shows the rate at which one variable must change with respect to the other in order to maintain equilibrium between the two states at the given temperature.

For example, we may use Eq. (8) to calculate the change of the boiling point of a liquid with pressure in the vicinity of its normal boiling point.

Take the case of water near 100°C. The absolute temperature corresponding to 100°C is $T = 373°$. The heat of vaporization of water at this temperature is $L = 540$ cal./g. $= 2.26 \times 10^{10}$ ergs/g (see **Mechanical Equivalent of Heat**). The specific volumes of steam and of water at this temperature are respectively $V_B = 1671$ cm³/g and $V_A = 1$ cm³/g. Substituting these values in (8) we obtain $dp/dT =$ 36,260 ergs/deg cm³ = 36,260 dynes/cm² deg = 27.2 mm Hg/deg. The reciprocal of this $dT/dp = 0.368$ deg/mm. Hg, is the quantity required. That is, near normal pressure the boiling point of water rises at the rate of 0.368° per mm of pressure. This result agrees closely with experiment.

Many other applications of the Clapeyron-Clausius equation are encountered in the thermodynamic treatment of changes of state.

CLARAIN. A term proposed by Marie Stopes in 1919 for a finely banded variety of "bright" or shiny bituminous coal. In thin sections, under the microscope clarain is seen to be composed of disintegrated plant substances including bands of spore cases, which impart a yellowish to reddish color to the substance.

CLARIFIERS. Classifying (Process).

CLARIFYING AGENTS. Chemical substances used in connection with the purification of various solutions and liquors that occur during the processing of raw materials to final end-products. These agents operate in connection with mechanical equipment to bring about the removal of suspended particles that represent product impurities. Allowing such particles to settle by gravity alone would require very long periods. Clarification is part of the total process of sedimentation, which may be defined as the removal of solid particles from a liquid stream by gravitational force. The operation is effected by slowing the velocity of a feed stream in a large-volume tank so that gravitation settling can occur. Sedimentation is divided into two functions: (1) thickening, where the primary purpose is to increase the concentration of suspended solids of the feed stream, i.e., to remove liquids (this is largely a mechanical operation, assisted sometimes by clarifying agents); (2) clarification, where the purpose is to remove fine-sized particles and produce a clear effluent, i.e., to remove solids. Some equipment does both and the dividing line between thickening and clarification is not always sharp. See also **Classifying (Process).** A clarifying agent to assist in this operation must possess certain properties for acting on the suspended particles—chemical precipitation, attractive via ionic forces, absorption qualities (large surface areas plus weak forces).

In the sugar refining industry, for example, various soluble nonsugar compounds are present in sugar juices as the result of rupturing plant cells (either from pressing sugar cane stalks or by extraction of the sliced root of sugar beets). Lime is commonly used to precipitate impurities, followed by carbonation of the solution with carbon dioxide to remove residual lime as calcium carbonate. Where filtration is used, various filter aids, such as fuller's earth, may be used. Filtering-type centrifuges also may be used. Phosphates, frequently in the form of orthophosphoric acid, may be used to precipitate the calcium. Some authorities suggest the use of polyphosphates, such as superphosphate and pyrophosphate, along with lime in the clarification operation. Phosphates assist in regulating the pH for optimal precipitation of calcium.

Tannin is used as a clarifying agent in wine making. Polyvinylpyrrolidone (PVP) has been used as a clarifying agent in the food industry. Ion exchange processes are also used in various processes, along with or in lieu of conventional clarification. For example, in ion exchange, the demineralization of sugar solutions can be effected; iron can be removed from wine by substitution with hydrogen ions.

Generally, wherever practical and economical, food processors prefer to accomplish clarification without the aid of chemicals—because all or but a trace of these chemicals must be removed so that they do not reappear in the final food product.

CLARIIDS. Catfishes.

CLARKE CELL. Battery.

CLASSIC. An archeological term to designate the cultural stage following the Formative stage. The Classic stage is characterized by the rise of civilizations, such as the Mayan.

CLASSIFICATION (Genus). Genus; Taxonomy.

CLASSIFICATION (Holotype). Holotype.

CLASSIFICATION (Homeotype). Homeotype.

CLASSIFICATION (Organisms). Taxonomy.

CLASSIFICATION (Plants). Botany; Plant Kingdom; Trees.

CLASSIFYING (Process). An operation or series of operations designed to separate a mixture of substances of various sizes and specific gravities into two or more categories, the cuts or divisions being made both with reference to size and to specific gravity (density). These operations find wide application in mining, metallurgical, water and sewage treatment, as well as use in other fields, such as the chemical industry.

Flotation. This is a means of separating a relatively small particle from a liquid medium. The particle may have a specific gravity greater than, less than, or the same as the liquid from which it is floated. There are two fundamentals requirements: (1) a gas bubble and particle must come in contact with each other; and (2) the particle should have an *affinity* for attaching itself to the bubble.

To achieve the first objective, various methods of bubble production and particle agitation have been used. Since the invention of flotation by Haynes in 1860, two basic methods have emerged:

(a) *Dissolved gas—impeller agitation*, wherein gas under pressure is sparged into the bottom of a vessel in which an impeller mixes the rising bubbles with the agitated particles.

(b) *Self-induced gas—impeller agitation*, wherein the impeller is so positioned in the liquid that it inspirates ambient gas into the liquid as bubbles. These bubbles are brought into contact with the agitated particle at the impeller's most dynamic zone. This method (see Fig. 1) has become the most accepted method.

The second requirement has been served by the development of a number of chemical reagents, which fall into five basic categories: (1) collection; (2) conditioning; (3) levitation; (4) frothing; and (5) depressant.

Frothers are chemicals whose molecules contain both a polar and a nonpolar group. The purpose of a froth is to carry mineral-laden

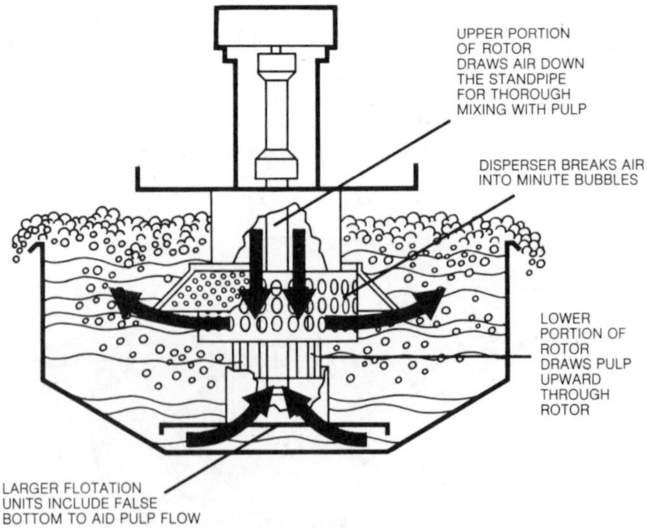
UPPER PORTION
OF ROTOR
DRAWS AIR DOWN
THE STANDPIPE
FOR THOROUGH
MIXING WITH PULP
DISPERSER BREAKS AIR
INTO MINUTE BUBBLES
LOWER
PORTION OF
ROTOR
DRAWS PULP
UPWARD
THROUGH
ROTOR
LARGER FLOTATION
UNITS INCLUDE FALSE
BOTTOM TO AID PULP FLOW

Fig. 1. Flotation cell. Upper portion of rotor draws air down the standpipe for thorough mixing with pulp. Lower portion of rotor draws pulp upward through rotor. Disperser breaks air into minute bubbles. Larger flotation units include false bottom to aid pulp flow. (*Envirotech.*)

bubbles for a period of time until the froth can be removed from the flotation machine for recovery of its mineral content. Typical frothing chemicals are alcohols, cresylic acids, eucalyptus oils, camphor oils, and pine oils, all of which are slightly soluble in water. Soluble frothers in common use include alkyl ethers and phenyl ethers of propylene and polypropylene glycols.

Collectors are chemical reagents which selectively coat the particles to be floated with a water-repellant surface which will adhere to air bubbles. Collectors generally are classified as cationic, anionic, or nonionic. Examples of collectors include the xanthates, dithiophosphates, thiocarbonilides, and thionocarbonates, all of which are anionic collectors for sulfides, a major need in ore processing. Fatty acids and soaps are anionic collectors and serve for nonsulfides. Amine salts are cationic collectors for nonsulfides.

Depressants are mainly inorganic salts, which compete with the collector for position on the sulfide surface. This permits the separation of one sulfide mineral from another. In one case, for example, in an alkaline solution, the addition of sodium cyanide prevents flotation of sphalerite and pyrite by xanthates, but not of galena, thus producing a higher grade of galena concentrates. The cyanide solution does not permanently affect the floatability of sphalerite as it can be floated by adding cupric sulfate and xanthate.

Activators are chemical reagents which alter the surface of a sulfide so that it can absorb a collector and float. Cupric sulfate is the most widely used activator. For example, xanthate as a collector will not readily float sphalerite, but the addition of cupric sulfate to the pulp changes the surface of the sphalerite particles to copper sulfide. Xanthate then will readily float the activated sphalerite as it behaves similarly to copper sulfide.

Although flotation was developed as a separation process for mineral processing and applies to the sulfides of copper, lead, zinc, iron, molybdenum, cobalt, nickel, and arsenic and to nonsulfides, such as phosphates, sodium chloride, potassium chloride, iron oxides, limestone, feldspar, fluorite, chromite, tungstates, silica, coal, and rhodochrosite, flotation also applies to nonmineral separations. Flotation is used in the water disposal field, particularly in connection with petroleum waste water cleanup.

Dense-media Separation. This operation is useful for the separation of solid particles of different densities. A liquid suspension of finely-divided high-gravity solids is prepared. Ores of different densities, when exposed to such a suspension will tend to separate by rising or settling in the liquid suspension. Numerous types of solids have been used to obtain a high-gravity medium, but the magnetic solids (ferro-silicon and magnetite) are most frequently used. These solids, alone or in combination, can provide a suitable dense medium over a gravity range of 1.25 to 3.40. Dense-media separation is applicable to any ore in which the valuable component has an appreciable gravity difference from the gangue components. In coarse-ore heavy-media separation plants, the limiting bottom size of dense-medium feed is 10 mesh and the upper size limit is 12 inches (0.3 meter). The magnetic particles of the dense medium subsequently are removed by a magnetic separator.

Jigging. In this operation, a pulsating stream of liquid flows through a bed of materials of different specific gravities, causing the heavy material to work down to the bottom of the bed and the lighter material to rise to the top. This is a very old operation used for concentrating heavy mineral from the lighter gangue. Construction costs are low, but power and water consumption are high. The process is used for the concentration of coal.

Tabling. In this concentration process, a separation between two or more minerals is effected by flowing a pulp across a riffled plane surface inclined slightly from the horizontal, differentially shaken in the direction of the long axis, and washed with an even flow of water at right angles to the direction of motion. A separation between two or more minerals depends mainly on the difference in specific gravity between the effective gravity (sp gr of mineral minus sp gr of water) of the valuable and the waste material. Tables treat metallic ores effectively in size ranges from 6 to 150 mesh, but can be used to treat lighter materials, such as coal of a considerably larger size. Dry tables also are used. The shaking motion is similar except that the direction of motion is inclined upward from the horizontal and, instead of water acting as the medium of distribution, a blast of air is driven

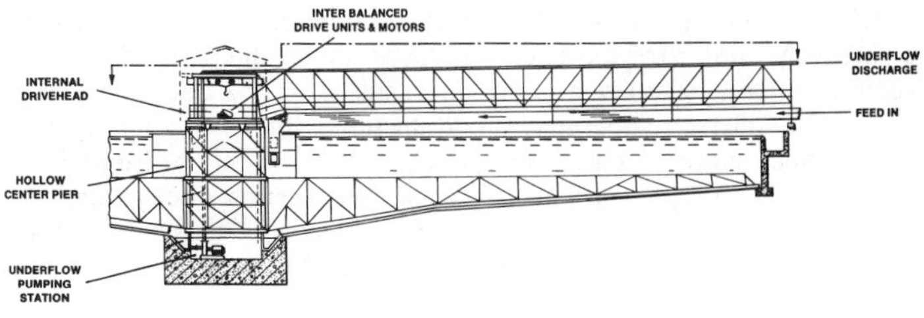

Fig. 2. Caisson-type thickener. (*Dorr-Oliver.*)

through a perforated deck. Tables also are used for selective flocculation or agglomeration of grains of one mineral in an aggregate by the addition of an agglomerating agent.

Sedimentation. This is a general term for an operation wherein suspended solids are removed from a liquid by gravitational settling. The two major forms of sedimentation equipment are (1) thickeners, and (2) clarifiers. The term *decanting* also is sometimes used to designate sedimentation.

Thickeners. The primary objective of thickening is to increase the concentration of the feedstream. The mechanical continuous thickener, equipped with sludge-raking arms, is the most common type. Usually the operation is performed in cylindrical tanks. The sludge collection system and the removal system are designed to move the settled material continuously across the tank floor to a discharge point. Feed enters through a central feed well designed to distribute around the periphery. Thickened sludge, raked toward the center by a slowing revolving mechanism, enters a central collecting trough or cone and is discharged through a spigot or removed by a sludge pump. See Figs. 2 and 3.

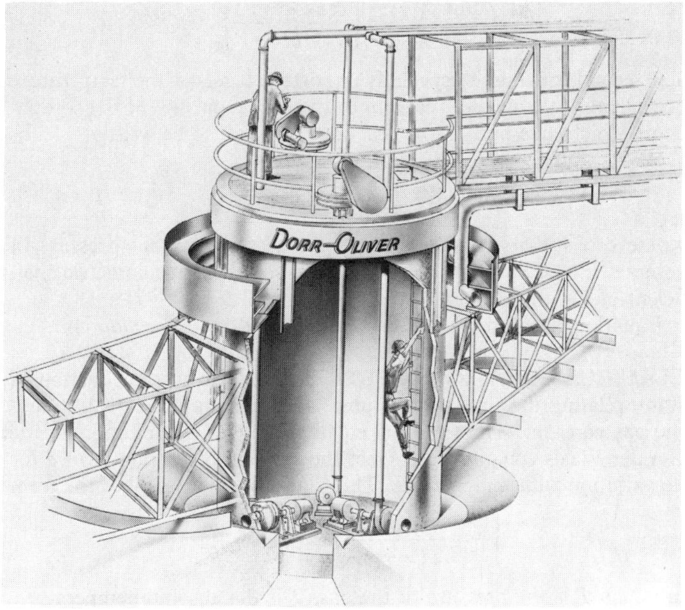

Fig. 3. Center-pier type thickener, showing pumps and access. (*Dorr-Oliver.*)

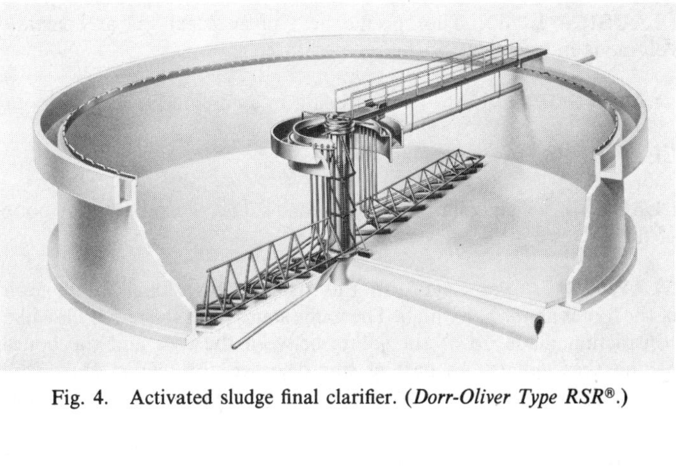

Fig. 4. Activated sludge final clarifier. (*Dorr-Oliver Type RSR®.*)

Clarifiers. The primary objective of clarifying is to free solids from a relatively dilute stream. These units operate on the basis of gravity sedimentation and utilize a raking mechanism, as in a thickener. Frequently, clarifiers are operated in conjunction with flocculation equipment, which employs chemical coagulants, such as alum, iron salts, lime, polyelectrolytes, activated silica sol, and other chemical reagents. Clarification essentially expedites the natural gravity settling process. An activated sludge final clarifier is shown in Fig. 4.

See also **Clarifying Agents.**

CLASTIC ROCK. A sedimentary rock that is entirely or chiefly composed of fragmental material. Sandstones and conglomerates are typical clastic rocks.

CLATHRATE. Compound (Chemical).

CLAUDE PROCESS. Ammonia.

CLAUSIUS-CLAPEYRON EQUATION. Clapeyron-Clausius Equation.

CLAUSIUS EQUATION. The Clausius equation(s) are partial differential equations (see **Partial Differential Equation**) which give the general relations which exist between the thermal coefficients. For the variables T, p, and ξ (extent of reaction) they are

$$\left(\frac{\partial C_{p,\,\xi}}{\partial p}\right)_{T,\,\xi} = \left[\frac{\partial(h_{T,\,\xi} + V)}{\partial T}\right]_{p,\,\xi} \tag{1}$$

$$\left[\frac{\partial h_{T,\,p}}{\partial T}\right]_{T,\,\xi} = \left(\frac{\partial C_{p,\,\xi}}{\partial \xi}\right)_{T,\,p} \tag{2}$$

$$\left(\frac{\partial h_{T,\,p}}{\partial p}\right)_{T,\,\xi} = \left[\frac{\partial(h_{T,\,\xi} + V)}{\partial \xi}\right]_{T,\,p} \tag{3}$$

The second equation is especially important; it relates the temperature coefficient of the heat of reaction to the heat capacities of the components which take part in the reaction. It may also be written

$$\frac{\partial}{\partial T} h_{T,\,p} = \sum_i \nu_i c_{p,\,i} \tag{4}$$

where $c_{p,\,i}$ is the partial molar specific heat at constant pressure of component i (see **Molar Concentration**) and ν_i the stoichiometric coefficient of i in the reaction.

Equations (2) and (4) are also called the *Kirchhoff equations.*

CLAUSIUS EQUATION OF STATE. A form of the equation of state, relating the pressure, volume, and temperature of a gas, and the gas constant. The Clausius equation applies a correction to the van der Waals equation to correct the pressure-correction term a for its variation with temperature. The Clausius equation takes the form

$$\left[P + \frac{a}{T(V+c)^2}\right](V - b) = RT$$

in which P is the pressure of the gas, T is the absolute temperature, V is the volume, R is the gas constant, b is a constant, a is a temperature-dependent constant and c is a function of a and b.

CLAUSIUS LAW. The specific heat of an ideal gas at constant volume is independent of the temperature.

CLAUSTROPHOBIA. Fear of being in a confined space.

CLAVICEPS. Ascomycetes.

CLAVICLE. The collar bone of humans. The ventral anterior bone of the pectoral girdle of vertebrates.

CLAWFOOT. Pes Cavus. In this disorder, the longitudinal arch of the foot is abnormally high. The name stems from the rather clawlike contraction made up of the joints between the toes and the bones behind these joints. As part of this disorder, the tendon above the heel (Achilles tendon) is shortened. Unfortunately, this results in additional weight being shifted to the forward part of the foot. Much of the discomfort of clawfoot can be relieved by wearing specially designed footwear.

CLAY. A very fine-grained, unconsolidated rock material which normally is plastic when wet, but becomes hard and stony when dried. Common clay essentially consists of hydrous silicates of aluminum, together with a large variety of impurities, such as hematite and limonite, which usually impart color to the clay. Geologically, clay may be defined as a rock or mineral fragment, having a diameter less than $\frac{1}{256}$ millimeter (0.00016 inch), which is about the upper limit of size of a particle that can exhibit colloidal properties. Clays are widely used in the manufacturer of tile, porcelain, earthenware, as filtering aids in oil and other processing, and as coatings for paper.

CLAY ATMOMETER. Precipitation and Hydrometeors.

CLEAN COKE PROCESS. Coal.

CLEANER FISHES. Fishes; Wrasses.

CLEANING COMPOUNDS. Detergents.

CLEANING SOLUTION (Glassware). An oxidizing reagent consisting of 60 parts potassium dichromate, 80 parts concentrated H_2SO_4, and 270 parts H_2O makes an effective cleaning agent for laboratory glassware. The solution sometimes is referred to as the Beckmann mixture.

CLEAN ROOM. Environment (Controlled).

CLEAR AIR TURBULENCE. Jet Streams.

CLEAR-WINGED MOTH (*Insecta, Lepidoptera*). A moth whose wing membranes are largely free from scales and therefore transparent. A majority belong to the families *Aegeriidae,* containing the squash-borer and other borers of economic importance, and *Sphingidae* or hawk-moths, of which the genus *Hemaris* includes such species.

CLEAVAGE (Biology). Also referred to as cytokinesis, cell cleavage is the process of segmentation and separation of two daughter cells during the process of cell division. The subdivision of the fertilized ovum which precedes the formation of germ layers as the first stage of embryonic development. In the simplest type of cleavage the egg and each cell thereafter split completely through; this process is called holoblastic cleavage. With the accumulation of yolk in the egg its division is hampered until, in the eggs of birds and reptiles, the living matter lies on one side of the yolk and cleavage merely subdivides this small mass into a layer of cells; this is meroblastic cleavage. In still other eggs with much yolk, such as those of the insects, cleavage gives rise to a layer of cells completely enclosing the yolk. The last type is called superficial cleavage. The process of cell cleavage of plant cells is different. In these, a phragmoplast is formed which is transformed into the cell plate which operates the two daughter cells. Within the cell plate is the secretory apparatus which produces the new primary cell wall. See also **Cell (Biology).**

CLEAVAGE (Cell). Cell (Biology).

CLEAVAGE (Geology). The tendency of crystalline minerals to split more easily in certain definite directions, with the development of more or less smooth surfaces called cleavage planes. Cleavage planes can only be developed parallel to some possible crystal face. Cleavage may be described either in terms of the ease with which it is developed, or its direction. Slaty cleavage, as the term implies, is the tendency for slaty rocks to split in relatively thin, flat plates. Slaty cleavage is the result of the metamorphism (foliation) of a sedimentary rock (shale or mudstone) and is not to be confused with mineral cleavage.

CLEAVAGE (Mineral). Mineraology.

CLEFT GRAFTING. Grafting and Budding.

CLEFT PALATE. Harelip.

CLETHRA TREE. Heather Shrubs and Trees.

CLICK BEETLE (*Insecta, Coleoptera*). Beetles with a peculiar junction between the first and second thoracic segments which permits them to be moved with a convulsive snap. When laid on its back the beetle uses this method of righting itself, throwing itself into the air by these snapping movements of the body. Most members of the family *Elateridae* and some of the *Eucnemidae* are click beetles. The females of these beetles cause injury to corn (maize) plants by burrowing into the soil and laying eggs around the roots.

CLIMAP. Ocean.

CLIMATE. A characteristic condition or pattern of the various elements of weather (temperature, humidity, rainfall, solar insolation, etc.) for a given geographic area or region of the earth—these conditions tending to repeat themselves in a reasonably orderly manner with time, that is, from one year to the next. Considerable variation from the mean can occur from one year to the next without suggesting a general change of the climate for a given area. For example, a warm or cool summer (that is, compared with the mean), or a rainy or dry summer or winter will fit into what may be considered reasonably normal. In other words, a lot of daily records are not broken within any given relatively normal year. Also, without considering any changes in climate, relatively abnormal situations, such as droughts, periods of intense heat, or very cold winters with heavy precipitation, may occur. Considering a century of weather, there is considerable accommodation of rather severe conditions without indicating a fundamental change in climate. Because true climatic trends are of such a slow nature, requiring essentially scores of years of measurements and interpretation of data, they are difficult to identify. Considering the current understandings of the factors and intricate systems which determine climate, perhaps the definition, "climate is what you expect; weather is what you get," is as reasonably fitting a description of climate as longer and more detailed definitions. It is not intended that this observation detract in any way from the very serious and dedicated work on the part of the numerous scientists worldwide who have been and are continuing in a very active way in attempts to build a better scientific understanding of climate, out of which may flow a reliable system of medium- and long-term climate (and weather) forecasting. Even with the assistance of computers and mathematical models, this remains a very difficult task.

Climate, of course, is very important to many human activities, but it is notably critical to agriculture and associated ventures. Unexpected variations of climate with respect to time of year are of much greater concern to growers of food than the historically established climate per se—because food production operations can be customized to the demands placed upon them by climate, but shifts in climatic patterns and those exceptional weather situations which prevail on an unscheduled, surprise basis annually cause in several regions of the world great losses of crops and subsequent food shortages. On the other hand, climate changes that tend to improve crop conditions tend to be easily overlooked.

Background. The term *climate* is derived from the Greek *klima* (meaning *inclination*), which reflects the importance attributed by the early students to the sun's influence. The relatively permanent factors that govern the general nature of the climate of a portion of the earth are known as *climatic controls.* These factors are sufficiently regular and permanent to nearly always bring weather back toward normal for the season after temporary departures, thus making possible the existence of definable types of climate. These factors include: (1) solar radiation, especially as it varies with latitude; (2) distribution of land and water masses; (3) elevation and large-scale topography; and (4) ocean currents. The general circulation of main wind systems is sometimes included, but some scientists consider this a *secondary control,* since the winds themselves are controlled by the foregoing factors. For many years, the expression of climate, other than by verbal description, has been largely by means of numerical tables and charts of average and extreme values of the climatic elements. The manipulation of these elements into various significant indices and coefficients has occurred mainly during the last few decades and has been established as a new tradition in the science of climatology, along with the introduction of statistical frequencies, deviations, correlations, etc.

Climatology is the scientific study of the climate. In addition to presentation of climatic data (*climatography*), climatology includes the analysis of the causes of differences of climate (*physical climatology*), and the application of climatic data to the solution of specific design or operation problems in which there is a direct interface between climate and human activities (*applied climatology*), of which the relationship between the climate and agricultural crops is probably the most important. Climatology may be further divided according to purpose or point of view. For example, among the many subdivisions are *agricultural climatology,* which deals in particular with the effects of climate on the health, vigor, yield, etc. of crops; *dynamic climatology* and *synoptic climatology,* which deal with the relationship between atmosphere and climate; and *paleoclimatology,* a study of the past climate throughout geologic time.

The general overall of a usually large geographic area is known as the *macroclimate.* The climate of smaller areas that may not be representative of the general climate of a large region or district are known as *mesoclimates.* Included in mesoclimatology are small valleys, forest clearings, frost hollows, and open spaces in towns, among others. Still smaller-area climates, such as a particular vineyard on a particular hillside, or of a particular field or pasture (one that may be out in the open; another next to a windbreak or partially shaded) among numerous other examples, are sometimes called *microclimates.* Crops such as grapes and citrus, notably lemon, are particularly sensitive to any alterations in their preferred microclimate.

Classification of Climates

The earliest known classification of climate, devised by the Greeks, simply divided each hemisphere into a mathematical climate of three zones; summerless; intermediate; and winterless. These zones, accounting for only the latitudinal differences in solar effect, were later labeled the torrid, temperate, and frigid zones. In the nineteenth century, Alexander Supan introduced a major improvement by basing climatic zoning on actual, rather than theoretical, temperatures; he named one hot belt, two temperate belts, and two cold caps. Supan also divided the world into thirty-four climatic provinces of essentially homogeneous climates, defining these mainly by temperature and rainfall, and partly by wind and orography. He made no attempt, however, to relate similar climates of different locations.

A basic and much used approach to climatic classification recognizes other climatic controls as well as the sun. The resulting climates, defined below, are known as polar, temperate, tropical, continental, marine, and mountain climates, and probably others, with variations. Of the major climatic classifications in use today, those of W. Köppen (1918) and C. W. Thorthwaite (1931) are referred to most often. Köppen's elaborate "geographical system of climates" is based upon annual and seasonal temperature and precipitation values; his climatic regions are given a letter code designation, and include as major categories, "tropical rainy," "temperate rainy," "snow forest," "tundra," and "perpetual frost." In 1934, Gorczynski devised a decimal number system similar to the Köppen classification.

Thornthwaite's bioclimatological system (1931) utilizes indices of precipitation effectiveness (in plant growth) to outline *Humidity provinces,* distinguished by their biological consequences of climate. This system also utilizes thermal efficiency (i.e., the effectiveness of temperature in determining the rate of plant growth), whereby *temperature provinces,* based on biological consequences of climate, are distinguished. A letter code is used to designate regions.

Thornthwaite, in 1948, introduced an approach to a "rational" classification, wherein potential evapotranspiration (i.e., the amount of moisture that, if available, would be removed from a given land area into the atomosphere by the combined process of evaporation of liquid or solid water plus transpiration from plants) is used as a measure of thermal efficiency, and is compared to precipitation to form a moisture index and to show amounts and periods of water surplus and deficiency. Definite break-points are revealed that are adaptable as climatic boundaries.

Following are definitions of some of the major categories of climatic classification:

Continental climates are characteristic of the interior of land masses of continental size. They are marked by large annual, daily, and day-to-day ranges of temperature, low relative humidity, and (generally) by a moderate or small and irregular rainfall, with the annual extremes of temperature occurring soon after the solstices. In its extreme form, a continental climate gives rise to deserts.

Marine climates (or *maritime climates*) are under the predominant influence of the sea. They are found where the prevailing winds blow onshore, as, of course, on oceanic islands, and on the western coasts of the continents in middle latitudes. They extend inland either until they meet a climatic divide or, in level country, until they become modified and gradually attain greater continentality. A marine climate is characterized by small diurnal and annual ranges of temperature, with retardation of the annual extremes until one or two months after each solstice.

Mountain climates are the climates of relatively high elevations, distinguished by the departure of their characteristics from those of surrounding lowlands. The one common basis for this distinction is that of atmospheric rarefaction; aside from this, great variety is introduced by differences in latitude, elevation, and exposure to the sun. The most common climatic results of high elevation are those of decreased pressure, reduced oxygen availability, decreased temperature, and increased insolation. On many tropical mountains, the forest zone extends into the level of average cloud height, which causes an excessive damp climate and produces the dense, rich forest growth known as the *fog forest*. The purity of high-elevation air, and the therapeutic effects of greater solar radiation, has led to the establishing of many health sanatoriums in mountain regions.

Temperature climates, very generally, are the variable climates of the middle latitudes, between the extremes of the tropical and polar climates. Included in this category are such specific divisions "temperate rainy," "snow forest," and portions of "dry" climates. Mesothermal climates, within this category, are characterized by "moderate" temperature; microthermal climates by "cool" temperatures.

Tropical climates (or *megathermal climates*), in general, are typical of equatorial and tropical regions, having continually high temperatures and considerable precipitation, at least during part of the year. These are the warmest types of climates, with a coldest-month mean temperature of 64.4°F (18°C) or higher. Included in this category are such specific divisions as "tropical rainy," "tropical rainforest," "tropical monsoon," and "tropical savannah" climates.

Polar climates (or *frost climates, arctic climates*) are, in general, found in those regions of the world where temperatures are sufficiently cold so that the annual accumulation of snow and ice is never exceeded by abalation (e.g., Antarctica, Greenland). They are characterized by a warmest-month mean temperature of less than 32°F (0°C) and include such specific divisions as "tundra," "snow forest," and "perpetual frost" climates.

The Climatic Cycle

Also known as *climatic oscillation*, the climatic cycle is a long period of climate that recurs with some regularity, but is not strictly periodic. The term "cycle" is used in climatology more loosely than in other physical sciences. It implies the occurrence of a rhythm without any real precision in the recurrence of events. It usually indicates only that the event is more probable at the peak of the cycle than at the trough. The existence of numerous cycles has been claimed, ranging from those of geological length down to a few years, the best known of which are the Brückner cycle and the sunspot cycle.

The *Brückner cycle* is an alternation of relatively cool-damp and warm-dry periods, forming an apparent cycle of about 35 years. A belief in such a cycle of 35 to 40 years in Holland was known to Sir Francis Bacon in 1625, but it was rediscovered in 1890 by E. Brückner, who regarded it was world-wide, and attached great economic importance to it. A mean cycle of 33 to 37 years (from variances of from 15 to 50 years) has been found in many meteorological and allied phenomena, including tree-ring and rainfall records. It is not known whether the Brückner cycle has any reality or is the result of statistical smoothing. It has no forecasting value.

The *sunspot cycle* refers to the rise and fall of sunspot activity, which occurs in an average cycle of 11.1 years (from variances of between about 7 and 17 years). An 11-year cycle showing correlation to sunspot activity has been found or suggested in terrestrial magnetism, frequently to aurora, and other characteristics of the ionosphere.

Climate-Oriented Terms

Continentality. The degree to which a point on the earth's surface is in all respects subject to the influence of land mass. The opposite of *oceanicity*.

Cooling Degree Days. The accumulation of degrees which the average daily temperature is above 75°F (~ 24°C). It is a measure of the air conditioning and/or refrigeration required to maintain a comfort level in homes, offices, public buildings, and factories. The annual cooling degree days are a measure of the tropical nature of a climate.

Degree Day. A measure of the departure of the daily average temperature from a preset standard. Degree-day accumulations during a season measure the requirements of fuel and energy for heating or for cooling.

Freeze. The condition that exists when, over a widespread area, the surface temperature of the air remains below freezing (32°F; 0°C) for a sufficient time to constitute the characteristic feature of the weather. This is a general term, and the time period necessary usually is considered to be two or more days; only the hardiest herbaceous crops survive, and when it cuts short a growing season, it may be termed a "killing freeze." A "hard freeze" is a freeze in which seasonal vegetation is destroyed, the ground surface is frozen solid underfoot, and heavy ice is formed on small water surfaces such as puddles and water containers. A "light freeze" is the condition that exists when the surface temperature of the air drops to below freezing for a short time period, so that only the tenderest plants and vines are adversely affected. A "dry freeze" is the freezing of the soil and terrestrial objects caused by a reduction of temperature when the adjacent air does not contain sufficient moisture for the formation of hoarfrost on exposed surfaces; with respect to vegetation alone, this is termed a "black frost."

Heating Degree Days. The accumulation of the degrees which the average daily temperature is below 65°F (~ 18°C). This is the usual, but not official accepted temperature base. For example, if on a given day, the mean temperature is 53°F, this will yield 13 degree days (65 − 53 = 13). The daily accumulations of degree days provides a measure of the fuel requirements for heating various facilities. Annual degree days thus becomes a measure of the severity of a climate.

Ice Day. In climatology, this is a day on which the maximum air temperature in a thermometer shelter does not rise above 32°F (0°C), and ice on the surface of water does not thaw.

Thermal Equator. The belt of maximum temperature surrounding the earth, which moves north and south with (but lagging) the sun's motion. It is also spoken of as the center of the area bounded by the yearly mean isotherms of 80°F (~ 27°C).

Climate Alteration

There has been much more discussion than action and much more action than results in terms of efforts to artificially alter large or even small areas of the earth's climate. A review of the entry on **Atmosphere (Earth)** and the other entries listed under **Meteorology** provides an insight to the immensity of what is involved in altering the earth's climate. It may well be, of course, that people already are well on the way toward altering the climate in what until recently were full unrecognized factors. These involve current hypotheses in connection with various types of air pollution. The earth's atmosphere and, in turn, its climate is almost wholly dependent upon the sun and it is not likely that this root cause of climate can be changed. Consequently, methods to alter climate and in essence to control climate are rooted to ways of changing the atmosphere which moderates the influence of the sun.

It is interesting to note that long-term values of temperature, cloudiness, precipitation, and sunshine have not changed appreciably within the span of climatic record keeping (a hundred years or so in a more or less formal manner). It is felt that about a half-century of record keeping is required to establish climatic trends. Thus, positive evidence of changes going on in the early 1980s probably will not be available until well after the turn of the present century.

It is also interesting to note how essentially unsuccessful scientific methods have been thus far in attempts to alter climatic factors. Cloud and hurricane seeding remains highly debatable. See **Hurricane Prediction.** Some progress has been made in providing lightning protection. Possibly the use of smudge fires to protect citrus orchards and crops from freezing, while very crude from a technological standpoint, may be one of the most successful means of highly-localized climate control on the records.

Nevertheless, scientific interest in climate control continues reasonably aggressive and some form of breakthrough may be in the offing.

"*Greenhouse*" *Effect.* Some factors that may alter the earth's climate are brought about unintentionally as the result of human (*anthropogenic*) activities. One of these is the addition of carbon dioxide to the atmosphere, first highlighted in a serious way by the British engineer, G. S. Callendar in 1938. At that time, Callendar stated, "Few of those familiar with the natural heat exchanges of the atmosphere, which go into the making of our climates and weather, would be prepared to admit that the activities of man could have any influence upon phenomena of so vast a scale." Callender estimated that about 150 billion tons of carbon dioxide, mainly as the result of burning fossil fuels over a 50-year period, had contributed this amount of atmospheric carbon dioxide by 1938. As of the early 1980s, it is estimated that carbon dioxide is being added to the atmosphere at a rate of about 18 billion tons per year. As pointed out by Brewer (1978), this annual rate of carbon dioxide addition is not large compared to the vast natural fluxes of carbon dioxide within forests and oceans, but the essential point is that the natural fluxes were in a carefully balanced steady state before the advent of industrialization. It is the new input of very old carbon that is of primary concern to scientists.

It is interesting to observe that only about half the carbon dioxide added to the atmosphere remains there. The disposition of the remaining half is not fully understood, but it is believed by many authorities that much of it must be absorbed by the oceans. The half remaining in the atmosphere has the capacity to alter climate because of the selective absorption of radiation from the sun by the CO_2 molecule. The incoming radiation from the sun is mainly of higher energy and shorter wavelength (approximately 1 micrometer). This is transmitted to the earth's surface through the atmosphere. The CO_2 molecule is transparent to these wavelengths. However, radiation emitted back from the surface of the earth to space occurs at much longer wavelengths (approximately 10 micrometers) and it so happens that both water vapor and carbon dioxide have absorption bands at these wavelengths. Thus, the earth's atmosphere tends to be heated from below. Carbon dioxide by allowing heat energy from the sun to pass through the atmosphere, but not allowing all the reradiated energy to radiate into space, causes a heating effect, popularly called the "greenhouse" effect.

Questions currently unanswered include the following: How much faster, in terms of increasing carbon dioxide content of the atmosphere at sea level, will the oceans absorb and, in effect, reduce the intensity of the "greenhouse" effect? If the oceans absorb greater concentrations of carbon dioxide, how will this affect the characteristics of the oceans themselves? Will the "greenhouse" effect be felt equally over the surface of the earth, or will there be regional differences? It is interesting to note in connection with this latter question that as of the 1980s, some differences in distribution of atmospheric carbon dioxide have been observed. Annual variations of CO_2 content vary up to 15 parts per million (ppm) at Point Barrow, Alaska, whereas the variations at the South Pole are as little as 2 ppm. An average concentration of CO_2 in the earth's atmosphere is about 330 ppm. Scientists at the Scripps Institution of Oceanography at Mauna Loa, Hawaii have shown that there is an annual cycling of CO_2 content, this cycle or ripple having an amplitude of from 4 to 6 ppm.

Many scientists hasten to observe that burning of fossil fuels is not the only factor that tends to alter the atmospheric CO_2 content. Cement production, wherein CO_2 is removed from limestone, also contributes to additions, estimated at about 2% of the total anthropogenic CO_2 flux. Large-scale changes in terrestrial vegetation also contribute to the total effect.

Some scientists estimate that the average life of a molecule of CO_2 in the atmosphere is about 10 years. After that time, it is believed to be incorporated into one of the more stable carbon reservoirs, such as the deep ocean. This interesting topic is discussed at much greater length by Brewer (1978); Broecker (1979); Machta (1977); Matthews (1976); Sagan (1979); Stuiver (1978); and Woodwell (1978).

Forecasting Climatic Trends

For medium- and long-term climate forecasting, the climatologist first must attempt to separate what may be termed *extreme differences from the norm* or *anomalies*, events which can be expected to occur infrequently, from long-term warming or cooling, moist and dry trends. As pointed out by W. O. Roberts (1978), "Our definition of anomalies, like that of climate, depends on our perceptions, our choice of time frames, and the statistical methods used. If, for example, research confirms the existence of a 22-year drought cycle for the Great Plains (United States), this should be seen as part of the 'normal' climate regime, not as an 'anomalous' or unusual event." For example, an understanding of recurrent drought cycles could discourage our holding climate responsible for great food shortfalls. Meteorologist Helmut Landsburg once commented, "I wish people would quit blaming climate or climate change for failure of agriculture in marginal lands. [Climate] is a scapegoat for faulty population and agricultural practices."

Roberts has pointed out that the most striking climate event in the United States in many years occurred in the winter of 1976–1977. (Another was the unusually hot, dry weather in many parts of North America and the very cool, wet weather of Europe during the summer of 1980. An analysis of the 1980 summer probably will not be available until the mid-1980s.) The stage was set for a severe winter in the fall of 1976, when an unusually cold body of water was detected in the northwestern Pacific Ocean. This frigid pool contrasted sharply with waters off the California coast, where temperatures were almost 3.6°F (2°C) warmer than a 30-year average. A high pressure ridge associated with the strong temperatures and pressure contrasts between the two bodies of water hovered persistently over the Rocky Mountains. The ridge steered the jet stream and associated weather systems far north of their usual paths. This brought remarkably balmy weather to Alaska. The ridge created a block, preventing moisture-laden storms from moving inland and thus causing severe drought in California and other western states. But, on the eastern coast of North America, an opposite situation prevailed. The jet stream, dynamically compensating for its northerly position in the western part of the continent, plunged southward, bringing cold Arctic air and record snows to the East. The Ohio Basin, particularly hard hit, suffered its coldest winter on record, with temperatures 9°F (5°C) or more below normal.

These anomalies caused severe dislocations. A number of states in the West suffered drought, while the East was hit with energy and other problems associated with severe winter. Some analysts estimated a loss in economic growth of some $3 billion, not including additional energy costs required in the East. Most authorities agree that even the most skillful of climatologists could not have predicted in mid-summer that such an unusual winter climate would prevail just a few months later. Some authorities believe that anomalies of this nature represent the interaction and synergistic effect of several causes. For example, the warm waters off the west coast may have deflected the faster-moving wind system in a northerly direction, which, in turn, may have reinforced the inflow of wind-driven warm water further to stabilize the anomaly (Roberts, 1978). The type of situation that occurred in the winter of 1976–1977 is shown in simplified form in Fig. 1.

Basic Approaches to Forecasting Climate. Climatologists tend to divide into two schools of philosophy as regards the proper and effective way to study and forecast climatic changes. There is the *model school*, in which researchers use large general circulation models (GCM) of the earth's atmosphere to simulate interactions in the climate/weather system. These researchers apply a fundamental and theoretical approach in which complex sets of nonlinear equations are used to describe the flow of the atmosphere. Models include in their simulations the land masses, ocean surfaces, ice bodies, and other relevant features. With a model stored in a computer, initial conditions are specified (arbitrarily or through actual observations). The state of the atmosphere at hundreds of thousands of gridpoints is simulated, out of which air flow, temperature, and moisture are calculated forward every

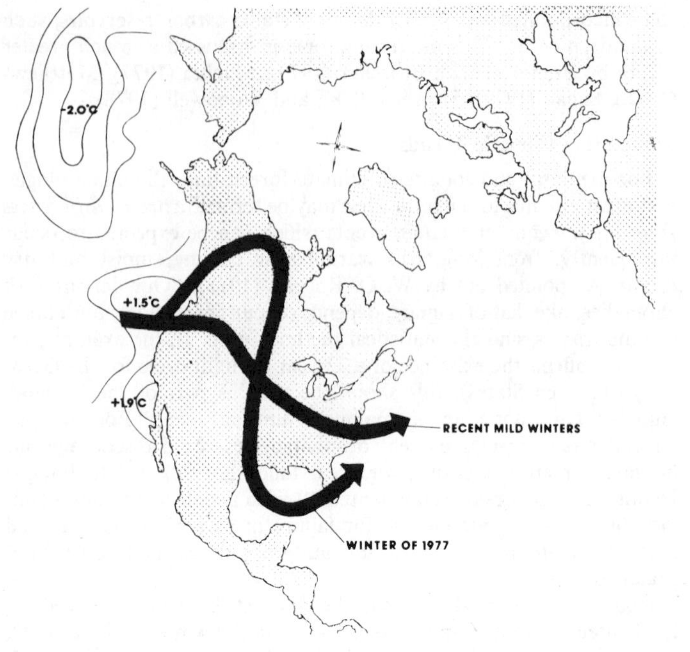

Fig. 1. Contrast of conditions that create a mild or a severe winter in parts of North America. (*Aspen Institute for Humanistic Studies.*)

few minutes. As pointed out by Roberts (1978), to be used for climate forecasts these models must be integrated forward thousands of time steps. The product is an excellent demonstration of the capabilities of the world's advanced computer systems. Most advanced nations depend on these models for day-to-day weather forecasts. Yet the models seem unlikely to achieve long-range forecasting results. Integrative schemes like these, unfortunately, begin to deteriorate beyond 3 or 4 days. And small errors in the initial conditions, considered by the model school to be irreducible "noise," gradually become as large as the deviations from the long-term averages that model proponents wish to forecast. See Fig. 2.

The second fundamental approach to forecasting climate is the *diagnostic* school, in which recent climate fluctuations are thoroughly analyzed—somewhat along the lines, although in more sophisticated form, of the analysis of the winter of 1976–1977 as previously described. Researchers examine possible connections between climate and major changes in ice masses and soil and ocean temperatures, among other parameters, that can alter major energy sources and sinks in the climate system. The effects of solar fluctuations (sunspots, storms, etc.) cannot be ignored in such analyses. Also, there are factors of dust pollution (industrial and agricultural sources, as well as natural sources such as the pollution generated by Mount Saint Helens in the State of Washington), and the effects of various chemicals, such as carbon dioxide and fluorocarbons, which probably affect climate. It is interesting to note that the last two factors (probably a fortuitous circumstance) tend to be somewhat compensating. One authority, Bryson (1977), who refers to people-created dust in the atmosphere as the "human volcano," observes that dust may lead to a global cooling. This could counter to some degree the carbon dioxide warming effect resulting from extensive combustion of fossil fuels. However, other recent research leads to the speculation that dust may enhance the carbon dioxide warming effect.

The National Climate Program Act passed by the U.S. Congress in September 1978 established a National Climate Program. With an eye toward "anticipating the effects of climate fluctuations and changes in the United States and the rest of the world," the program coordinates U.S. climate research and funding with stress on climate monitoring, the dissemination of data, the assessment of human effects on climate, and the impacts of climate on agriculture and resource management. The program is administered by the National Oceanic and Atmospheric Administration.

Historical Climatic Information

The two fundamental schools of philosophy as regards approaching climate forecasting actually are not sharply divided and liberal climatologists agree that there are attractive aspects to both schools. In any event, climatologists find that looking back for many years into climatic information can be rewarding as a way of highlighting the past magnitude of climatic changes and, in some cases, revealing characteristics of their cyclic nature. Unfortunately, reliable records in many parts of the world go back only about a century. Prior to that time, there was no established network of weather observations and the instruments used were crude. There are a few exceptions of a highly qualitative nature, such as the drought years experienced in Guatemala, records of which extend back to 1563, the first recording made by the Spanish. Considerable detail on these records and their meaning to the people of Guatemala over the centuries is given by Roberts et al. (1978). See Fig. 3.

Forecasting Activity

During the 1970s and early 1980s, the tempo of climatological research has increased. The Food and Climate Forum of the Aspen Institute for Humanistic Studies, headquartered in Boulder, Colorado, publishes a "Food and Climate Review" each year. In 1978, the Research Directorate of the National Defense University, in cooperation with the U.S. Department of Agriculture, the Defense Advanced Research Projects Agency, the National Oceanic and Atmospheric Administration, and the Institute for the Future, conducted a study of "Climate Change to the Year 2000." The study took the form of a survey of 24 climatologists from 7 countries. Individual quantitative responses to 10 major questions were weighted according to expertise and then averaged, using a method of aggregation that preserved the climatologists' collective uncertainty about future climate trends. The

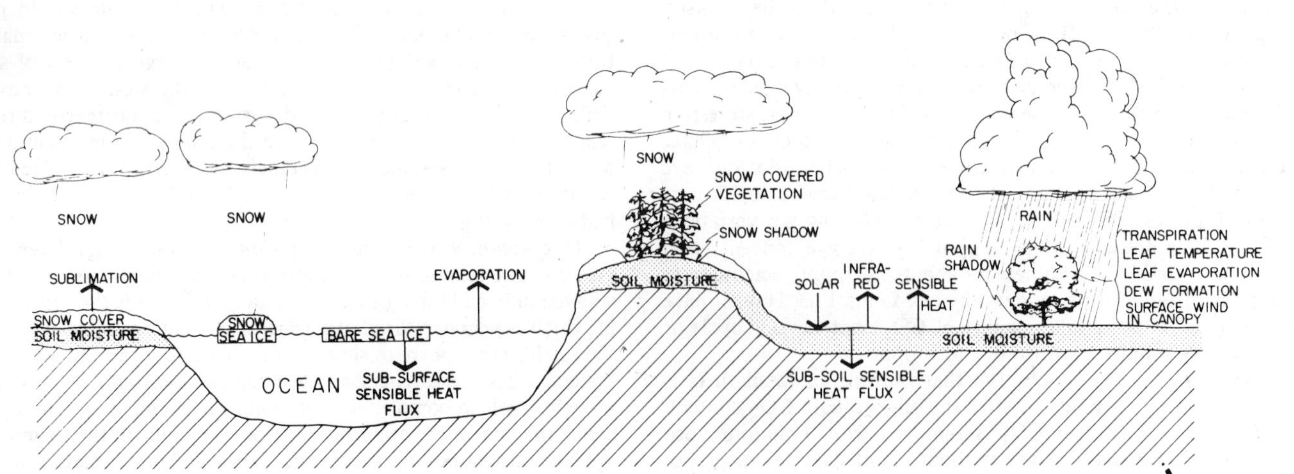

Fig. 2. Some of the many interactions that are built into advanced general circulation models (GCM) of the earth's atmosphere to simulate climate/weather system. (*Aspen Institute for Humanistic Studies.*)

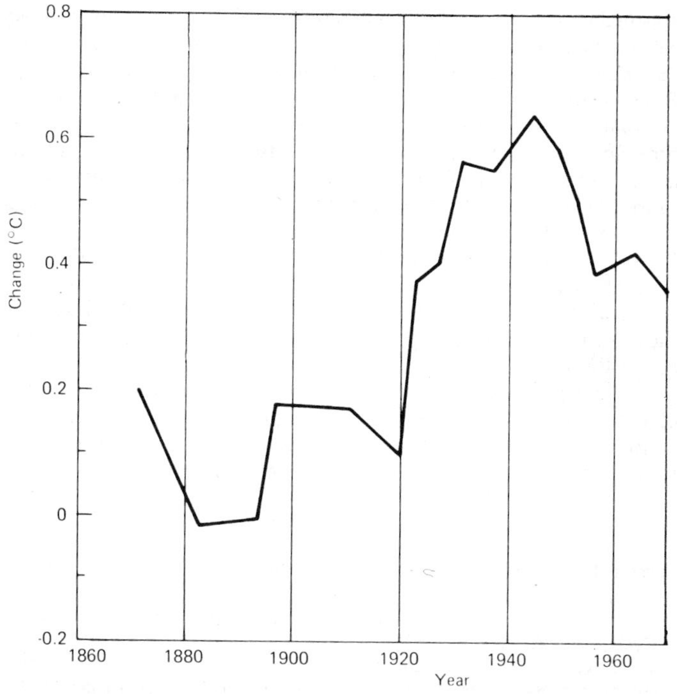

Fig. 3. Global temperature variations since 1870. "Global temperature" is used as equivalent to annual mean temperature between 0° and 80° N latitude. Period of 1880–1884 is used as zero reference base. (*U.S. National Oceanic and Atmospheric Administration.*)

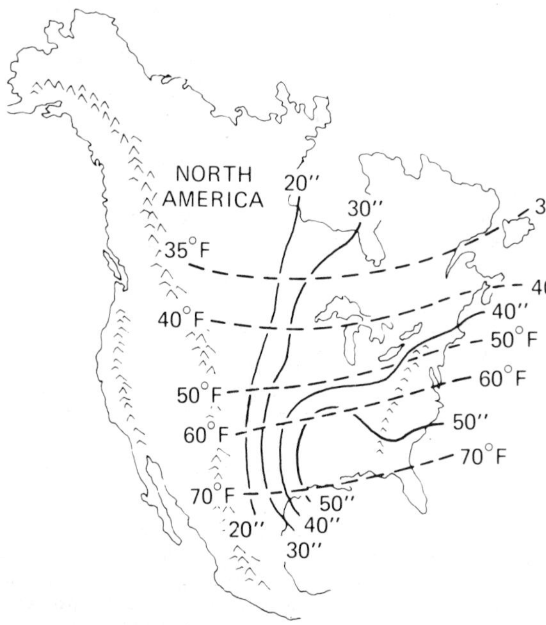

Fig. 4. In many parts of the earth, the annual isotherms and annual isohyets tend to run parallel to each other. In North America, however, the effect of the Rocky Mountains range causes the annual isotherms and isohyets to run at approximately 90° to each other, particularly in the midcontinental grasslands region. South America is similar to North America in this respect due to the Andes. Celsius equivalents of temperatures shown on map are: 35°F = 1.7°C; 40°F = 4.4°C; 50°F = 10°C; 60°F = 15.6°C; 70°F = 21.1°C. (*U.S. National Oceanic and Atmospheric Administration.*)

aggregated subjective probabilities were used to construct five possible climate scenarios for the year 2000, each having a "probability" of occurrence. Consequences of the possible climatic changes delineated in the scenarios are being considered in subsequent phases of the research.

The findings of the study, to be appreciated in the light of the research methodology used, must be gleaned directly from the 108-page report. Only a few of the observations made by particular survey participants are included here. These indicate the rather wide range of responses received from the experts.

1. The likelihood of catastrophic climatic change by the year 2000 is assessed as being small. More specifically, responses suggested only one chance in ten that the average global temperature in the next 25 years will increase by more than 1.1°F (0.6°C) relative to the global temperature of the early 1970s. Likewise, there is only one chance in ten that it will decrease by more than 0.54°F (0.3°C). The most likely event will be a climate which resembles the average of the past 30 years, arising primarily from a balancing of the warming effect of carbon dioxide with the cooling effect of a natural climate cycle. Respondents to the study tended to anticipate a slight global warming rather than a cooling. Most panelists perceived that any global temperature will be amplified at higher latitudes, particularly in the Northern Hemisphere. This magnification will be less pronounced in the Southern Hemisphere because the larger surface area of southern oceans provides more thermal inertia against change.

2. Panelist responses reflected fairly strong support for the continuation of a 20- to 22-year drought cycle in the High Plains of the United States. See Figs. 4 and 5. No causal mechanism, however, was agreed upon. For mid-latitude regions outside North America, there was more uncertainty and less support for cyclic droughts. Similarly, no periodicity was identified relative to frequency of drought in the Sahel region of Africa, or the failure of the Asian monsoons. See Fig. 6.

3. Considerable uncertainty was expressed concerning possible changes in amount and variability of precipitation—uncertainty not only with respect to magnitude of changes, but even with respect to direction of change. The uncertainty was particularly pronounced about possible changes in year-to-year variability. Some tendency was shown to associate more precipitation and decreased variability of precipitation with global warming, and less precipitation and increased variability with global cooling.

Climatic Situations in Different Countries

The following information is abstracted from the excellent report by the Aspen Institute for Humanistic Studies (Roberts et al.).

China and the U.S.S.R. Both of these countries, but to a lesser extent China, face persistent obstacles to agricultural production. Unlike the United States, which has favorable combinations of temperature, rainfall, and soil over much of the land, both the U.S.S.R. and China have relatively narrow latitudinal bands where weather and soil are suitable for cereal crop growth. Both countries have highly disparate climate regimes. In China, the semi-arid north, dominated by dry, cold, anticyclonic air, contrasts sharply with the region of abundant moisture in the south, dominated by humid equatorial and tropical air during the wet summer months. Similarly, the wet, cool climate of the U.S.S.R. west of the Urals contrasts with the drought-

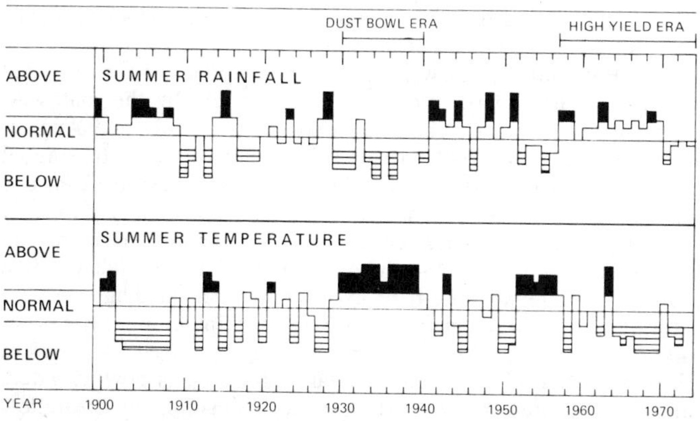

Fig. 5. The High Plains region of the United States, approximately 400 miles (644 kilometers) wide, centered on 101° W longitude from Mexico to Canada, has experienced lower than normal precipitation and attendant drought conditions approximately every 20 to 22 years, as in the mid-1930s and mid-1950s. This chart shows the 75-year record of summer average temperature and rainfall in the five major wheat-producing states of the United States. The drought period of the Dust Bowl Era (1930s) and the generally favorable conditions after the late 1950s (which continued until 1980) are shown in the diagram. The five states are Oklahoma, Kansas, Nebraska, South Dakota, and North Dakota. (*U.S. National and Atmospheric Administration, D. Gilman.*)

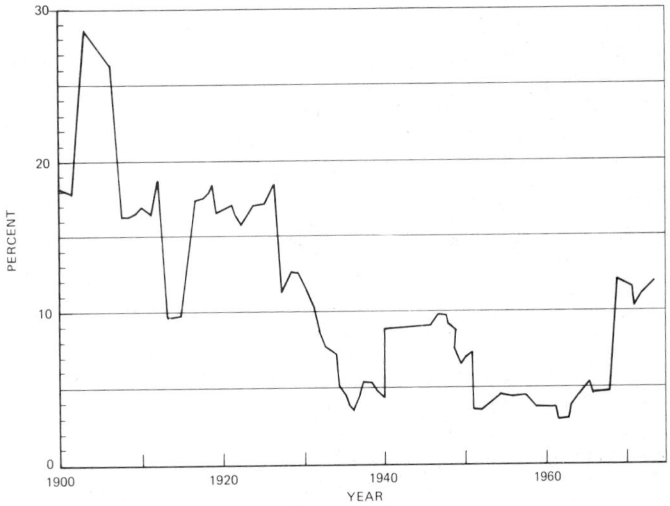

Fig. 6. Plot of data originally assembled by R. Bryson. Percent is number of weather stations in northwestern India that reported less than 50% of the normal annual rainfall in a given year (overlapping 10-year averages). This is an indication of the frequency of monsoon failures in this part of India. (*U.S. National Oceanic and Atmospheric Administration.*)

prone regions of Kazakhstan and Siberia (sometimes referred to as the "New Lands"). This frontier region first came under intense cultivation during the Khrushchev years. Under Khrushchev, a massive campaign to introduce spring wheat into marginal lands east of the Urals was inaugurated. Khrushchev banked on the contrast in climates between the New Lands and western Russia. He hoped that a climate-related shortfall in one region would be offset by a good harvest in the other, since weather conditions in the two areas are usually diametrically opposed. But introducing a single-crop monoculture into a low-precipitation region where fluctuations in rainfall are high is, according to Soviet agricultural expert Lazer Volin, "precarious insurance." The drought-related shortfall of spring wheat production in the New Lands in recent years has proved this point.

While the Soviets are trying to increase food production by expanding into marginal lands, the Chinese, particularly in recent years, are counting on increased yields per planted area through extensive multiple cropping and improved water management. China, which must feed about one-quarter of the world's population from only about 7% of the world's current farmland, has the most extensive irrigation system of any nation. Yet China is still not invulnerable to climate fluctuations, and crop shortfalls due to drought and flood are still common. But crop shortfalls due to adverse weather in 1977, for example, and other bad years would probably have been much greater if they had occurred just a decade or so ago, when irrigation systems were less extensive and sophisticated.

India. For centuries, the well-being of India's populace has depended upon the vagaries of the monsoon. Rain brought by the southwest monsoon, beginning in June and peaking during July and August, provides the Subcontinent with a large portion of its total annual water supply. The amount of rainfall which India receives from one year to the next is quite variable. Because of the complex nature of the atmospheric circulation pattern which generates the monsoon, total annual rainfall for the Subcontinent will range widely, often creating the familiar pattern of excess rains and flooding one year and little or no rain the next. This variable precipitation creates problems for agriculture which can only be partially solved by extensive irrigation. While 25% of India's total arable land is now irrigated, water supplies are not constant. Seasonal fluctuations in streamflow due to the monsoon are substantial. Three-fourths of the total arable land of India has no assured irrigation and depends entirely on the vagaries of the monsoons. Moreover, institutional factors, such as water rationing systems, often increase the instability of water supplies. In addition, not all growers have easy access to irrigation systems. Recent research indicates that while large public irrigation works, such as canals and waterways, tend to benefit all growers, smaller schemes, such as tubewell irrigation, are apt to favor the more affluent.

Further gains in India's agriculture may not be possible unless the seasonal nature of its water supplies is reduced. Some authorities advocate storing monsoon rainfall in vast underground reservoirs. Others recommend tapping the extensive groundwater resources of the Ganges Basin. Recent plans include a proposed canal linking the Brahmaputra and Ganges Rivers, which would augment the Ganges' sluggish wintertime flow and provide additional water to the crowded agricultural lands of West Bengal. While proposed plans to stabilize annual water supply face political and economic obstacles, successful water programs could have a dramatic impact upon the rural population. The Rajasthan Canal, commenced in the 1950s, irrigates 700,000 acres (280,000 hectares) of land in the dry state of Rajasthan in northern India. The canal draws its water from the rivers of the Ravi and the Beas, which also are responsible for superior agricultural conditions in the Punjab.

The foregoing examples from three major agricultural nations emphasize how extremely helpful long-range climate forecasting could be to decisions involving large costly water projects. See also **Atmosphere (Earth)** and list of entries at end of **Meteorology.**

References

Barnett, T. P.: "Ocean Temperatures: Precursors of Climate Change," *Oceanus*, **21**, 4, 27–32 (1978).

Bolin, B.: "Changes of Land Biota and Their Importance for the Carbon Cycle," *Science*, **196**, 613–615 (1977).

Borzenkova, I. I., et al.: "Change in Air Temperature in the Northern Hemisphere during the Period 1881–1975," *Meteorology and Hydrology*, **7**, 27–35 (1976).

Bretherton, F. P.: "The Problems of Climate Research," *Oceanus*, **21**, 4, 4–11 (1978).

Brewer, P. G.: "Carbon Dioxide and Climate," *Oceanus*, **21**, 4, 12–17 (1978).

Broecker, W. S., et al.: "Fate of Fossil Fuel Carbon Dioxide and the Global Carbon Content," *Science*, **206**, 409–418 (1979).

Bryan, K.: "The Ocean Heat Balance," *Oceanus*, **21**, 4, 18–26 (1978).

Bryson, R. A., and G. J. Dittberner: "A Non-equilibrium Model of Hemispheric Mean Surface Temperature," *J. Atmos. Sci.*, **33**, 2094–2106 (1976).

Fritts, H. C.: "Tree Rings and Climate," Academic, New York, 1976.

Imbrie, J.: "Geological Perspectives on our Changing Climate," *Oceanus*, **21**, 4, 65–70 (1978).

Imbrie, J., and K. P. Imbrie: "Ice Ages: Solving the Mystery," Enslow, Short Hills, New Jersey, 1979.

Imbrie, J. and J. Z. Imbrie: "Modeling the Climatic Response of Orbital Variations," *Science*, **207**, 943–953 (1980).

Jackson, I. J.: "Climate, Water and Agriculture in the Tropics," Longman Group Ltd., New York, 1977.

Lockwood, J. G.: "Causes of Climate," Wiley, New York, 1979.

Lohmann, G. P.: "Response of the Deep Sea to Ice Ages," *Oceanus*, **21**, 4, 58–64 (1978).

MacDonald, R. B., and F. G. Hall: "Global Crop Forecasting," *Science*, **208**, 670–679 (1980).

Machta, L., Hanson, K., and C. D. Keeling: "Atmospheric Carbon Dioxide and Some Interpretations," *Proc. Hawaii Meeting* (The Fate of Fossil Fuel Carbon Dioxide), 1977.

Madden, R. A., and V. Ramanathan: "Detecting Climate Changes Due to Increasing Carbon Dioxide," *Science*, **209**, 763–768 (1980).

MacLeish, W.: "Science Attempts to Fit Piece into Jigsaw of the Sea," *Smithsonian*, **10**, 1, 80–89 (1979).

Matthews, S.: "What's Happening to Our Climate?," *National Geographic*, **150**, 5, 576–620 (1976).

McLean, D. M.: "A Terminal Mesozoic 'Greenhouse,' " *Science*, **201**, 401–406 (1978).

McQuigg, J. D.: "Effective Use of Weather Information in Projects of Global Grain Production," Environmental Data and Information Service, Columbia, Missouri, March 1976.

Mercer, J. H.: "West Antarctic Ice Sheet and CO_2 Greenhouse Effect," *Nature*, **271**, 321–325 (1978).

Namais, J.: "Multiple Causes of the North American Winter 1976–77," *Monthly Weather Review*, **106**, 3 (1978).

Newman, J. A.: "Drought Impacts on American Agricultural Productivity," in "North American Droughts" (N. J. Rosenberg, editor), Westview Press, 1978.

Roberts, W. O., Slater, L. E., Jedamus, J. A., and S. K. Levin: "Food and Climate Review," The Food and Climate Forum, Aspen Institute for Humanistic Studies, Boulder, Colorado (issued annually).

Robock, A.: "The 'Little Ice Age': Northern Hemisphere Average Observations and Model Calculations," *Science*, **206**, 1402–1404 (1979).

Sagan, C., Toon, O. B., and J. B. Pollack: "Anthropogenic Albedo Changes and the Earth's Climate," *Science*, **206**, 1363–1368 (1979).

Sergin, V. Y.: "Origin and Mechanisms of Large-Scale Climatic Oscillations," *Science*, **209**, 1477–1482 (1980).
Siegenthaler, U., and H. Oeschger: "Predicting Future Atmospheric Carbon Dioxide Levels," *Science*, **199**, 388–395 (1978).
Staff: "Climate Change to the Year 2000," Research Directorate of the National Defense University, Fort Lesley J. McNair, Washington, D.C., 1978.
Staff: "Review of Exceptional Weather Events," WMO (World Meteorological Organization, United Nations), New York (Issued annually).
Staff: "Weather/Crop Assessment," U.S. Department of Agriculture, Washington, D.C. (Issued weekly).
Staff: "Major Abnormal Weather Conditions Affecting World Agriculture," Center for Climatic and Environmental Assessment (CCEA) of the Environmental Data Service, Columbia, Missouri (Issued weekly).
Stommel, H., and E. Stommel: "The Year without a Summer," *Sci. Amer.*, **240**, 6, 176–184 (1979).
Stewart, R. W.: "The Role of Sea Ice in Climate," *Oceanus*, **21**, 4, 47–57 (1978).
Stuiver, M.: "Atmospheric Carbon Dioxide and Carbon Reservoir Changes," *Science*, **199**, 253–258 (1978).
Thompson, L. M.: "World Weather Patterns and Food Supply," *J. Soil and Water Conservation*, 44–47 (January–February 1975).
Walker, J. C. G.: "Radar Measurement of the Upper Atmosphere," *Science*, **206**, 180–189 (1979).
White, R. M.: "Oceans Climate," *Oceanus*, **21**, 4, 2–3 (1978).
White, W. B., and R. L. Haney: "The Dynamics of Ocean Climate Variability," *Oceanus*, **21**, 4, 33–39 (1978).
Woodwell, G. M.: "The Carbon Dioxide Question," *Sci. Amer.*, **238**, 1, (1978).
Woodwell, G. M., et al.: "The Biota and the World Carbon Budget," *Science*, **199**, 141–146 (1978).

CLIMATE (Effect on Insects). Insecticide and Pesticide Technology.

CLIMATE (Polar). Polar Research.

CLIMATIC ALTERATION. Volcano.

CLIMAX (Ecology). The final stable stage of development that a community, species, flora, or fauna attains in a given environment. The major world climaxes correspond to formations and biomes. See also **Biome; Ecology; Formation (Ecology).**

CLIMAX VEGETATION. Ecology.

CLIMB CUTTING. There are two methods of machining surfaces with rotating multitoothed cutters; in the first method, known as climb cutting or cutting down, the work moves or feeds in the same direction as the periphery of the cutter teeth at the line of contact; in the second, known as cutting up, the work feeds in a direction opposite to that of the periphery of the cutter teeth at the line of contact. In cutting-up action, each tooth of the cutter takes a wedge-shaped chip whose initial thickness is zero and whose final thickness depends upon the rate of feed. The cutting action has a tendency to lift the work from the surface on which it is placed, and work must therefore be restrained vertically as well as longitudinally.

Climb cutting produces a chip of maximum thickness at the beginning and zero thickness at the end of the cut. As a result, the machined surface has a better finish than can be obtained by cutting up.

Climb cutting has one major disadvantage. The cutting action has a tendency to draw the work in towards the cutter. This tendency is manifested when a heavy feed is used or when the machine is not sufficiently rigid or powerful for the size of the cut. In such a case, the cutter tends to climb over the work and may thus ruin the work, the cutter, and possibly the machine. For this reason climb cutting is not employed for medium and heavy-duty service, particularly where comparatively old machine tools are used.

CLIMBING PERCH. Labyrinth Fishes.

CLINOGRAPHIC REPRESENTATION. Pictorial Representation.

CLINOHUMITE. Chondrodite.

CLINOMETER. Essentially a divided-circle instrument which simplifies the transfer of angles between planes. Bubbles or electronic levels frequently are used to establish the null setting principle of a precision clinometer, while less precise instruments utilize the comparison of an angle measured to a datum surface. Angular indications using a clinometer are established by placing the instrument successively on mutually inclined surfaces. The differences between the two observed readings is the angle between the surfaces. By this same principle, the clinometer may be moved to another surface, such as the face of a machine bed, and the readings compared. By adding an autocollimating telescope, a clinometer can be used as a goniometer.

For a single-sided divided-circle system, the composite errors of centering and scale dividing will approach 10 seconds of arc. Calibration may be accomplished by using autocollimators and precision polygons, or precision angle gage blocks. Calibrations values also may be derived by comparing readings circle right and circle left, or by evaluating the instrument on an angle generator or sine plate. Calibration factors are furnished by the manufacturer with each instrument.

CLINOZOISITE. Clinozoisite, crystallizing in the monoclinic system, is a hydrous calcium aluminium silicate $Ca_2Al_3Si_3O_{12}(OH)$. Crystals are usually of prismatic habit with striations parallel to the *b*-axis. May occur as granular or columnar masses. Hardness 6.5, sp gr 3.21–3.38, vitreous luster, transparent to translucent. Color gradations from gray through green to pink.

Clinozoisite occurs in crystalline schists which are themselves products of metamorphism from calcic feldspar-rich dark igneous rocks. Zoisite (orthorhombic) represents its dimorphous counterpart.

CLIPPER LIMITER. A transducer which gives output only when the input lies above a critical value, and a constant output for all inputs above a second higher critical value. This is sometimes called an amplitude gate, or slicer.

CLITORIS. Gonads.

CLOACA. A chamber at the end of the gut which receives the ducts of the reproductive and excretory systems. A cloaca occurs in the rotifers and in many vertebrates but in most mammals it is present only during embryonic life.

CLOCK. The measurement of time is important in so many phases of human activity that it has long engaged the inventive genius of mankind. Far in the past people discovered that the prime requisite of time-measuring devices was a process that occurred at a uniform rate. Water clocks and hour glasses were based essentially upon the flow of a known volume of homogeneous substance through an opening. While such processes are obviously subject to errors that were difficult to avoid, such as temperature effects, they measured time periods of moderate duration with sufficient accuracy for the needs of their time.

The sundial, while was also used in the earlier days, is discussed in a separate article. See **Sundial.**

The era of modern time-keeping can be said to have begun with the use of periodic processes for time measurement, and it may be traced directly to Galileo's discovery of the isochronism of the pendulum. A direct statement about the pendulum is that the period of a simple pendulum is a function of its length and not of its initial displacement. For example, if you raise the bob of a pendulum two inches, then release it and observe its period; after which you raise it four or five or six inches, and also release it and observe its period, you will find that the period remains the same. The bob merely moves more slowly over the short path than the longer one. Therefore, once men had invented a mechanism to supply the energy lost by the pendulum due to friction at its point of suspension and with the air, they were able to construct clocks as we know them today. Furthermore, when they discovered that a balance wheel turning against a spring also had a constant period they were able to produce watches as well as clocks.

The accuracy of both watches and clocks depends upon certain factors. Obviously the length of a pendulum or the dimensions of a balance wheel will change with temperature, and therefore inventors moved in the direction of greater accuracy by endeavoring to compensate for this error. One such method was the construction of these parts partly or wholly from bimetallic elements so that the overall change in dimensions with temperature would be minimized. Still another step was operation in a vacuum, to avoid the friction of the air. However, there is a practical limitation upon the extent to which the errors inherent in pendulum and balance wheel time-keepers can be minimized. See **Chronograph; Chronometer; Electric Clock; Pendulum Clock; Spring Clock;** and **Tuning Fork.**

One type of electric clock is designed to operate in synchronism with the 60-cycle current supplied by the power companies. Since their alternators must operate within a narrow frequency-range around 60 cycles per second in order to remain in synchronism, this type of electric clock is sufficiently accurate for many purposes. However, for precise scientific work a much higher degree of accuracy has been obtained from the vibrations of crystals exhibiting an electrostrictive effect. That is, any given kind of crystal, when subjected to an alternating electric field, tends to vibrate at a characteristic rate. With proper control of temperature, vibrating crystals can remain within a far smaller frequency range (i.e., inverse of period) than the devices previously mentioned in this article. Therefore, crystals are used to control oscillators, not only those which operate clocks, but those which control frequencies of broadcasting stations, and those used for other purposes.

Atomic Clock. A great advance in the accuracy of time measurement resulted from observations of various periodic processes occurring in atoms and molecules. The first atomic clock was the ammonia clock invented at the United States National Bureau of Standards in 1948. It made use of an oscillating system in the ammonia molecule (NH_3) which may be regarded as having its three hydrogen atoms occupying positions corresponding to the three lower corners of a pyramid, while the nitrogen atom is at the apex. The periodic process in question is the vibration of the nitrogen atom up and down through the base of the pyramid, so that its limiting lower position is at the apex of another pyramid below the first one and symmetrical with it. The period of this oscillator is 23,870 megacycles. Therefore, if radiowaves of this frequency are passed through a chamber filled with ammonia, they will be strongly absorbed. The design of the first atomic clock was based upon a quartz-crystal oscillator as described above, which supplied oscillatory energy through an ammonia chamber designed to absorb energy of the correct frequency, and to permit energy to pass when it varied from the correct frequency. This "passing" energy was used in a feedback circuit to correct the original oscillatory circuit. See **Crystal Oscillator.**

This ammonia clock gave far more accurate time-keeping than had ever been known before. In fact, some models have controlled frequency within three parts per billion. Yet even more accurate time-measuring devices have been developed. The cesium clock designed at the Bureau of Standards requires the use of vaporized cesium metal obtained from an electric furnace. The atomic process here involves a transition in the precession axis (hence electromagnetic field) of the outer electron of the cesium atom. The required radio frequency is 9,192,631,770 Hz (cycles per second) and the possible accuracy is to a variation of less than one part in ten billion. So unvarying, in fact, is this process that it has been made the definition of the second.

These devices have been denoted by the general term "atomic clock" which has been loosely applied to any device that depends for its constancy of rate on the frequency of a spectral line, i.e., on the energy difference of two states, for the measurement of time intervals. In some cases they are those of an atom, e.g., the two states into which the ground state of cesium is separated in a magnetic field; in other cases they are those of a molecule, e.g., in the vibration spectrum of ammonia.

Successful atomic clocks have been built employing a beam of cesium atoms. The beam is first separated into two components by an arrangement similar to that used in the Stern-Gerlach experiment. If the component made up of atoms in the lower of the two possible states is passed into a resonant chamber fed by an oscillator, the atoms will absorb energy and be raised to the higher state if the microwave frequency of the oscillator is just that for resonance absorption. A second resolution of the beam then again gives two components with opposite spin; the relative intensities of these two may be measured and may be used to keep the oscillator in exact resonance. The period of oscillation is thus matched to the frequency of the spectral line, and the number of oscillations in a time interval may be counted by electronic and/or mechanical means.

CLOCK (Biological). Biological Timing and Rhythmicity.

CLOCK CYCLE (Computer System). Cycle (Computer System).

CLONAL SELECTION. Immune System and Immunology.

CLONE. The asexual formation of descendants genetically identical to the original parent. The ability to clone plants and animals demonstrated that even highly differentiated cells contain a full set of genes necessary to form every cell type found in the complete organism. Often clones of cells are functional in character, as in the "clonal selection" theory of antibody production, contact of the protein synthesizing cells with antigen brings into existence clones of cells which produce a particular protein or proteins.

Clone may also mean a group of plants which have been produced by vegetative propagation (i.e., cutting or budding) from a single seedling. The members of a clone are identical, but cannot be grown from seed.

A recent development of great importance to many fields of science has been the production of hybridoma cells by the fusion of myeloma cells (malignant tumor cells of the immune system) with lymphocytes immunized with a particular antigen. The individual hybridoma clones can be maintained indefinitely and can be selected for the production of large amounts of identical (monoclonal) antibody against a single antigen. These highly specific monoclonal antibodies are proving to be remarkably versatile tools in many areas of biological research and clinical medicine. Monoclonal antibodies are further described by Milstein (1980). See also **Genes;** and **Recombinant DNA.**

Ann C. Vickery, Ph.D., University of South Florida, College of Medicine, Tampa, Florida.

References

Anderson, C. M., Zucker, F. H., and T. A. Steitz: "Space-Filling Models of Kinase Clefts and Conformation Changes," *Science*, **204**, 375–381 (1979).

Arehart-Treichel, J.: "Questioning the New Genetics," *Science News*, **116**, 154–156 (1979).

Chan, H. W., et al.: "Molecular Cloning of Polyoma Virus DNA in *Escherichia coli*: Lambda Phage Vector System," *Science*, **203**, 887–892 (1979).

Crick, F.: "Split Genes and RNA Splicing," *Science*, **204**, 264–271 (1979).

Davis, B. D.: "Frontiers of the Biological Sciences," *Science*, **209**, 78–89 (1980).

German, J., et al.: "Genetically Determined Sex-Reversal in 46 XY Humans," *Science*, **202**, 53–56 (1978).

Graham, J. B., and C. A. Istock: "Gene Exchange and Natural Selection Cause *Bacillus subtilis* to Evolve in Soil Culture," *Science*, **204**, 637–639 (1979).

Israel, M. A., et al.: "Molecular Cloning of Polyoma Virus DNA In *Escherichia coli*: Plasmid Vector System," *Science*, **203**, 883–886 (1979).

Linn, S.: "1978 Nobel Prize in Physiology or Medicine (Restriction Endonucleases)," *Science*, **202**, 1069–1071 (1978).

Markert, C. L., and R. M. Peters: "Manufactured Hexaparental Mice Show That Adults are Derived from Three Embryonic Cells," *Science*, **202**, 56–58 (1978).

Marx, J. L.: "Molecular Cloning: Powerful Tool for Studying Genes," *Science*, **191**, 1160–1162 (1976).

Marx, J. L.: "Gene Structure," *Science*, **199**, 517–518 (1978).

Marx, J. L.: "Function of the *src* Gene Product," *Science*, **201**, 702 (1978).

Marx, J. L.: "Successful Transplant of a Functioning Mammalian Gene," *Science*, **202**, 610 (1978).

Marx, J. L.: "Restriction Enzymes: Prenatal Diagnosis of Genetic Disease," *Science*, **202**, 1068–1069 (1978).

Marx, J. L.: "Gene Transfer," *Science*, **208**, 386–387 (1980).

Maugh, T. H., II: "The Artificial Gene: "It's Synthesized and It Works in Cells," *Science*, **194**, 44 (1976).

McKinnell, R. G.: "Cloning," University of Minnesota Press, Minneapolis, Minnesota, 1978.

Miller, J. A.: "Spliced Genes Get Down to Business," *Science News*, **117**, 202–205 (1980).

Milstein, C.: "Monoclonal Antibodies," *Sci. Amer.*, **243**, 4, 66–74 (1980).

Selden, J. R., et al.: "Genetic Basis of XX Male Syndrome and XX True Hermaphroditism: Evidence in the Dog," *Science*, **201**, 644–646 (1978).

Smithies, O.: "Cloning Human Fetal γ Globin and Mouse α-Type Globin

DNA: Characterization and Partial Sequencing," *Science*, **202**, 1284–1288 (1978).
Staff: "Gene Transfer in Mammalian Cells: Mediated by Chromosomes," *Science*, **197**, 146–148 (1977).
Staff: "Genetic Engineering," *Science*, **198**, 388 (1977).
Staff: "Missing Link (Wheat Genetics)," in "Science and the Citizen," *Sci. Amer.*, **240**, 2, 70–71 (1979).
Staff: "Charting the Human Chromosomes," *Science News*, **116**, 39 (1979).
Staff: "On the Way to a Clone," *Science News*, **116**, 68 (1979).
Staff: "Genetic Code Not Universal," *Science News*, **116**, 185 (1979).
Staff: "Cell's Gene-Copying Machinery Anchored," *Science News*, **117**, 132 (1980).
Wade, N.: "New Rulebook for Gene Splicers Faces One More Test," *Science*, **201**, 600–601 (1978).
Wade, N.: "Hybridomas: A Potent New Biotechnology," *Science*, **208**, 692–693 (1980).
Wade, N.: "Inventor of Hybridoma Technology Failed to File for Patent," *Science*, **208**, 693 (1980).

CLOSED-CIRCUIT TELEVISION. **Television.**

CLOSED-LOOP CONTROL. **Feedback Control.**

CLOSED-LOOP CONTROL (Automotive). **Automotive Electronics.**

CLOSED-LOOP GAIN. **Gain (Magnitude Ratio).**

CLOSED SYSTEM. 1. In thermodynamics, a system so chosen that no transfer of mass takes place across its boundaries; a system that can exchange energy but not matter with the outside world. 2. In mathematics, a system of differential equations and supplementary conditions such that the values of all the unknowns (dependent variables) of the system are mathematically determined for all values of the independent variables (usually space and time) to which the system applies. See also **Controlled System Automatic.**

CLOSED UNIVERSE. **Cosmology.**

CLOSTRIDIUM PERFRINGENS. **Foodborne Diseases.**

CLOT. A firm mass which results from coagulation, or change from a fluid or semi-fluid state to a semi-solid soft mass. This occurs in blood or lymph when it escapes from the blood or lymph vessels. For a discussion of the mechanism of blood clotting, see **Anticoagulants; Blood; Embolism.**

CLOT (Blood). **Blood.**

CLOTHES MOTH (*Insecta, Lepidoptera*). A moth whose larva eats dead and dry animal matter, especially fur, feathers, and wool. Three species are known. The case-bearing clothes-moth larva, *Tinea pellionella*, lives in a case made of bits of the material on which it lives, spun together with silk. The tube-building or tapestry moth, *Trichophaga tapetiella*, makes a gallery of silk and fragments as it works. The common or naked clothes-moth larva, *Tineola biselliella*, does not spin until it makes its cocoon. The adults of all three species are small, expanding usually about $\frac{1}{2}$ inch. As in the case of the buffalo carpet-moth good housekeeping is the best remedy for these pests. When present they may be killed by insecticides, the use of commercial sprays or in heavy infestations by fumigation.

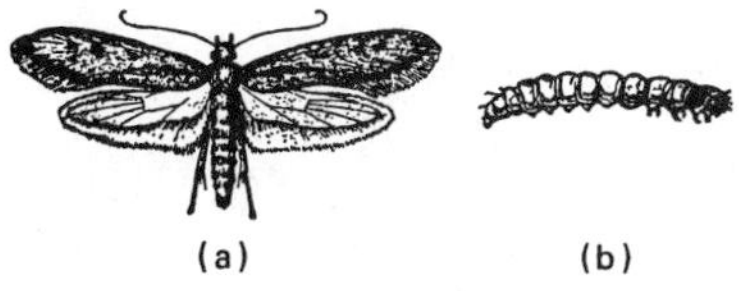

(a) Clothes moth; (b) larva of clothes moth. (*Riley, U.S. Department of Agriculture.*)

Frequently when moths become overabundant in a house it is due to some forgotten breeding place, such as feathers or old garments in storage. A search in unexpected places is often the best cure for such attacks.

CLOUDBURST. **Precipitation and Hydrometeors.**

CLOUD CHAMBER. An enclosure containing air or other gas saturated with water vapor, the cooling of which by a sudden expansion results in the formation of fog droplets upon particles of dust or other nuclei. That ions in the gas are capable of serving as condensation nuclei, even when no dust is present, was demonstrated by the experiments of C. T. R. Wilson. Thus the clouds produced are much more dense if the gas is traversed by some ionizing emission like x-rays or alpha rays. Sir J. J. Thomson utilized this effect in his early measurements of the electronic charge. One of the most striking phenomena of the Wilson cloud chamber is exhibited when single ionizing particles, such as alpha or beta particles, are allowed to traverse it just before the expansion. The path of each particle is marked by a visible white streak or "track" of mist, sometimes several centimeters in length, which soon diffuses and disappears. The study of photographs of such cloud tracks has afforded much information as to the nature and the movements of the particles producing them and the interactions between particles and radiations. See also **Bubble Chamber;** and **Particles (Subatomic).**

CLOUD-DETECTION RADAR. **Weather Observations and Forecasting.**

CLOUDED LEOPARD. **Cats.**

CLOUD-HEIGHT DETECTION. **Precipitation and Hydrometeors.**

CLOUD POINT. The temperature at which a solution becomes cloudy as it is cooled at a specified rate. The cloud point is an important property in the specification of lacquers, oils, and other industrial solutions. See also **Petroleum.**

CLOUD POINT (Fuel). **Petroleum.**

CLOUDS AND CLOUD FORMATION. A hydrometeor; large numbers of water droplets or ice crystals virtually suspended in the atmosphere. Actually, the water or ice in a cloud occupies only a small fraction of the total space appearing as a cloud; light is well-reflected from the droplets or crystals, and the cloud body appears as an opaque drifting object. Clouds differs from fog in that the latter is, by definition, in contact with the earth's surface.

Most clouds are *convective clouds*, owing their vertical development, and possibly their origin, to the process of convection, i.e., currents of moist air ascend, and adiabatic cooling and condensation occur. In rare exceptions, such as in the case of fog, which may produce stratus, the cooling may occur as a result of other processes. The ascent of air may result from vertical instability, as in most cumulus clouds; from forced lifting at a frontal surface, as in many altostratus and other stratiform clouds; from undulatory motions at inversion surfaces; or from orographic lifting.

Condensation at the point of saturation or at a low degree of supersaturation is dependent upon the presence of many condensation nuclei for water clouds, or ice nuclei for ice-crystal clouds. The size of cloud drops varies from one cloud type to another, and within any given cloud there always exists a finite range of sizes. Generally speaking, cloud drops range from between 1 and 100 micrometers in diameter, though they may be as large or larger than 200 micrometers in diameter. Larger drops within a cloud fall rapidly enough so that only very strong updrafts can sustain them. As cloud drops increase in size, they may form virga or precipitation.

Cloud Formation

Cloud formation results from the simultaneous presence in the atmosphere of several contributing factors. Because cloud particles are primarily liquid or crystalline water, the first requirement is for water substance. In fact, from a meteorological viewpoint, water vapor is the most important substance found in air. It is not only important as the raw material for clouds and rain, but also as a vehicle for the transport of energy (latent heat) and as a regulator of planetary temperatures via the greenhouse effect. Water vapor in the atmosphere has a varied distribution, ranging from zero to a volume percentage of about 4%. The water vapor content of natural air over Pittsburgh,

WATER VAPOR CONTENT OF NATURAL AIR
(Monthly Averages at Pittsburgh, Pennsylvania)

Month	Average Temperature °F	Average Temperature °C	Concentration of Water in Air: Grains per Cubic Foot	Concentration of Water in Air: Parts per Million	Volume of Water, Gallons per Hour (At air rate of 20,000 cubic feet per minute)
January	37.0	2.8	2.18	5	87
February	31.7	−0.2	1.83	4.2	73
March	47.0	8.3	3.40	7.8	136
April	51.0	10.6	3.00	6.9	120
May	61.6	16.6	4.80	11	192
June	71.6	21.4	5.94	13.6	238
July	76.2	24.4	5.60	12.8	224
August	73.6	23.1	5.16	11.8	206
September	70.4	21.3	5.68	13	227
October	56.4	13.5	4.00	9.2	160
November	40.4	4.7	2.35	5.4	94
December	36.6	2.5	2.25	5.1	90

Pennsylvania, as an example, is given on a month-to-month basis in the accompanying table.

Water vapor is concentrated in greatest quantity in the lower part of the atmosphere, with only a minute fraction occurring above the tropopause. See **Atmosphere (Earth).** The principal sources of water vapor are oceans, large lakes, and vegetation-covered land areas, especially those of the tropics and subtropics. The winds carry water vapor from the source regions elsewhere over the entire globe. Transportation of water vapor is sufficiently localized in character so that tongues and islands of moist and dry air are present everywhere in the lower hemisphere. In the northern hemisphere, moist tongues usually flow from south or west, and dry tongues from some northerly direction, except that, particularly during summer, very dry air originating in the sub-tropical anticyclones often flows from the south and west. In the temperate zone, moist tongues normally appear on the west side of anticyclones and on the east side of cyclones; and dry tongues appear on the east side of anticyclones and the west side of cyclones.

The amount of water vapor contained in the air may be expressed as the relative humidity, which is simply the fraction actually present of the amount of water required to completely saturate the air.

A second requirement is the presence of particles of solid, semisolid, or semiliquid matter that have an affinity for water. These particles act as microscopic nuclei upon which water vapor can collect and thereby permit particle growth to the necessary size for cloud droplets and crystals. Condensation nuclei present in the atmosphere include industrial combustion and other products, exhaust products from engines, and notably sea salt. Sublimation nuclei, also called freezing nuclei, in the free atmosphere consist almost entirely of ice crystals.

A third requirement for cloud formation is a refrigeration mechanism to lower the temperature of the air and thereby raise the relative humidity to saturation. Cooling mechanisms in the atmosphere are mainly thermodynamic in which ascending air is cooled very nearly 2°C for each 1000 feet (304.8 meters) of elevation. See Fig. 1. Other cooling mechanisms of lesser importance include: (1) Radiation, (2) contact cooling, and (3) mixing. Cooling beyond the initial saturation temperature forces water vapor onto growing cloud particles.

The number of condensation nuclei available to act as cores for cloud droplet growth vary from as few as one thousand to several hundred thousand per cubic centimeter. Among the available nuclei with affinity for water, some have a greater affinity than others which, therefore, causes selective growth. Clouds of liquid droplets always result when the relative humidity approaches or equals 100% when the temperature is above freezing. The relationship of cloud droplet growth with humidity is shown in Fig. 2.

Sublimation nuclei are not always present and liquid cloud droplets do form in their absence in temperatures down to at least −40°C. When sublimation nuclei are present, cloud particles consist primarily of ice crystals. See also **Precipitation and Hydrometeors.**

Cloud Seeding. During the 1940s, a discovery was made by Dr. Vincent J. Schaefer that the introduction of solid carbon dioxide into water droplet clouds having temperatures lower than 0°C caused the cloud particles to transform from liquid droplets to ice crystals. The terminology "cloud seeding" was thereby introduced. Solid carbon dioxide (dry ice) has a temperature close to −70°C and its presence in saturated air below freezing temperatures causes a multitude of micro ice crystals to be generated (sometimes called diamond dust). Water substance is transferred rapidly from the liquid droplets to the tiny ice crystals, causing the ice crystals to grow and to fall as snow. The cloud is thereby cleared away by dry ice seeding.

Later, Dr. B. Vonnegut discovered that certain other crystalline forms whose structure is very similar to water crystals cause water to collect onto these nonwater crystals. The effect was similar to that achieved by dry ice seeding. Silver iodide is the most prominent of the nonwater crystals considered in this manner. Silver iodide initiates ice crystal formation (glaciation) at about −7°C. Silver iodide is the

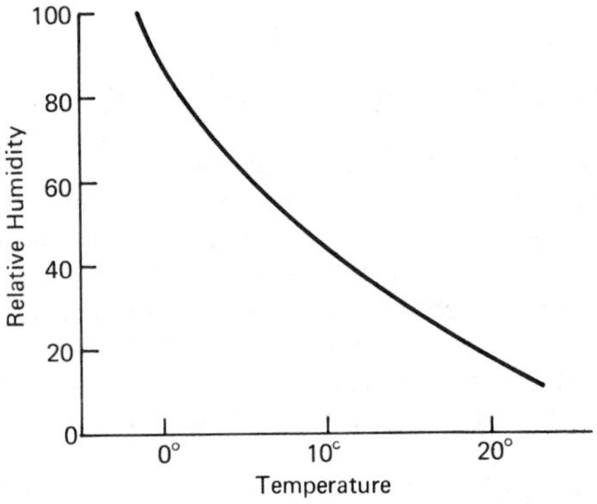

Fig. 1. Increase in relative humidity with decreasing temperature for a situation at 5000 feet (1524 meters), starting with a given water vapor content at 20°C.

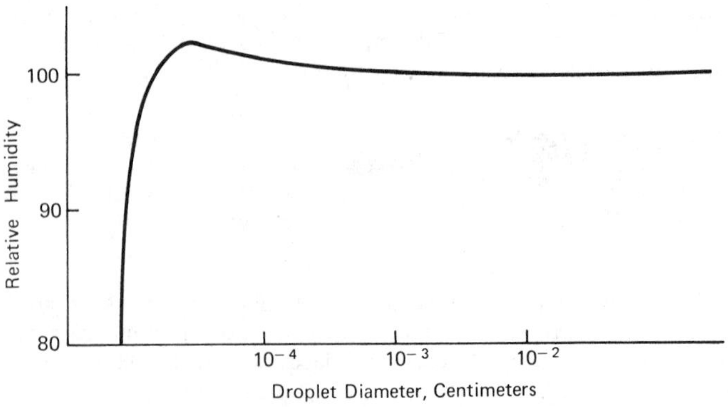

Fig. 2. Typical growth of cloud droplet in relation to relative humidity. The size of the oversaturation in particles is related to their affinity for water.

commonly used cloud seeding agent because of its low threshold temperature and because it can be introduced into clouds either by ground or airborne generators.

Glaciation of Clouds. Water droplet clouds usually have distinct, clearly delineated and sharply bordered edges. Ice crystal clouds have "soft" fuzzy edges.

When ice crystal particles are introduced into liquid droplet clouds at temperatures below freezing, the water in the liquid droplets migrates rapidly to the crystals which then grow rapidly. The transformation of a liquid droplet cloud to an ice crystal cloud is known as *glaciation.* To the observer watching this transformation, there is a rather rapid change from "hard" edged clouds to "fuzzy and milky" edged clouds. The anvil head of a thunderstorm is made of ice crystals that were glaciated in the growing cumulonimbus.

Cloud Evaporation. Any downward motion of air results in compressive heating of the air. Cloud particles in downward-moving air are in an environment of relative humidity less than 100% and lose their liquid or solid water, shrink in size quite rapidly, and assume the dimensions of nuclei. In essence, the cloud so affected disappears. This phenomenon can be observed on the lee slopes of mountains that are cloud covered on the windward side.

Cloud Classification

There are three classification schemes whereby clouds are distinguished and grouped: (1) By appearance; (2) by usual altitudes where found; and (3) by particulate composition. The scheme in general use, based on a classification system introduced by Luke Howard in 1803, is the one adopted by the World Meteorological Organization and published in the *International Cloud Atlas* (1956). This classification is based upon the determination of:

(a) Genera, the main characteristic forms of clouds. The ten cloud genera are cirrus, cirrocumulus, cirrostratus, altocumulus, altostratus, nimbostratus, stratocumulus, stratus, cumulus, and cumulonimbus. Descriptions are given later.

(b) Species, the peculiarities in shape and differences in internal structure of clouds. The fourteen cloud species are fibratus, uncinus, spissatus (false cirrus), castellanus, floccus, stratiformus, nebulosus, lenticularis, fractus, humilis, mediocris, congestus, calvus, and capillatus.

(c) Varieties, special characteristics of arrangement and transparency of clouds. The nine cloud varieties are intortus, vertebratus, undulatus, radiatus, lacunosis, duplicatus, translucidus, perlucidus, and opacus.

(d) Supplementary features and accessory clouds, appended and associated minor cloud-forms. The nine supplementary features and accessory clouds are incus, mamma, virga, praecipitatio, arcus, tuba, pileus, velum, and pannus.

(e) Mother-clouds, the clouds from which other clouds have formed.

(Note: Although these are Latin words, it is proper convention to use only the singular endings, e.g., more than one cirrus cloud is, collectively, cirrus, not cirri.)

Classification by Altitude. The three classes by altitude are distinguished as *high clouds* (cirrus, cirrocumulus, cirrostratus, occasionally altostratus, and the tips of cumulonimbus); *middle clouds* (altocumulus, altostratus, nimbostratus, and portions of cumulus and cumulonimbus); and *low clouds* (stratocumulus, stratus, most cumulus, and cumulonimbus bases, and sometimes nimbostratus).

Classification by Particulate Composition. Included are *water clouds,* composed entirely of ordinary and/or supercooled water droplets (altocumulus, nimbostratus, stratocumulus, stratus, and cumulus); *ice-crystal clouds,* composed entirely of ice crystals (cirrostratus, cirrus, occasionally altostratus, and cirrocumulus); and *mixed clouds,* a combination of water and ice crystals (cumulonimbus, usually altostratus, occasionally cirrocumulus and the water clouds).

Nimbus was a name formerly used for any rain-producing cloud, but it is not now recognized in the international cloud classification.

Popular Names. Aside from the foregoing classifications, there are various popularly recognized cloud forms or conditions, which are apt to be colorfully descriptive of the cloud's appearance or portent. *Mare's tail,* for example, describes long, detached, well-defined wisps of fibrous cirrus cloud, with feather-like tufts at one end. *Mackerel sky* is a sky with considerable cirrocumulus or, especially, small-element altocumulus clouds arranged in uniform bands similar in appearance to the scales on a mackerel. *Emissary sky* describes a sky that is often one of the first indications of the approach of a cyclonic storm, composed of isolated or small, separated groups of cirrus clouds. *Scud* is a term most often applied to low, ragged, wind-torn stratus clouds moving rapidly beneath a layer of nimbostratus. The terms "incus," "anvil cloud," or "thunderhead" are described in the definition of the genus cumulonimbus, below. "Abraham's tree" is a name given to the cloud variety radiatus. Certain clouds that form over mountain peaks or ridges are described as cap clouds, banner clouds and crest clouds.

The Ten Cloud Genera

Some of the major cloud formations are illustrated in Fig. 3. The ten genera are defined briefly in the following paragraphs.

Full discussion of all the species, varieties, supplementary features and accessory clouds would involve details beyond the scope of this article. The species *stratoformis,* for example, is the most common form of the genera altocumulus and stratocumulus, consisting of a very extensive, not necessarily continuous, horizontal layer or layers. *Translucidus,* as a variety of stratiformis, denotes the greater part of the layer, patch, or sheet as being sufficiently translucent to reveal the position of the sun or of higher clouds. The variety *perlucidus,* on the other hand, denotes distinct spaces in stratiforms, which permit the sun, moon, blue sky, or higher clouds to be seen. Similarly, distinction is made between the characteristics of cloud precipitation. *Praecipitatio* is a supplementary feature denoting precipitation that reaches the earth's surface, while *virga* refers to precipitation that evaporates before reaching the earth's surface. Through such fine distinctions, clouds can be minutely defined and classified.

Following are descriptions of the ten cloud genera:

Cirrus is a high-altitude, ice-crystal cloud in the form of white, delicate filaments, patches, or narrow bands. It is usually thin, wispy, often in streaks, and is always whitish without shadows. The term "cirrus" is frequently used for all types of cirriform clouds, i.e., cirrus, cirrocumulus, cirrostratus, and all their species and varieties.

Cirrocumulus is a small, billowed, high-altitude cloud appearing as a thick, white patch without shadows, composed of very small elements in the form of grains, ripples, etc. It may be composed of highly supercooled water droplets, as well as small ice crystals, or a mixture of both; usually, the droplets are rapidly replaced by ice crystals. Sometimes corona or irisation may be observed; mamma may appear; and small virga may fall. Cirrocumulus is most often confused with altocumulus, but differs primarily in that its constituent elements are very small and are without shadows. A cirrocumulus cloud formation indicates some instability in the layer at and above the cloud level, which permits rising currents to form the cloud parcels and descending currents to create clear spaces between them. Cirrocumulus frequently occurs in advance of a cyclonic storm.

Cirrostratus is a high-altitude, ice-crystal cloud appearing as a whitish veil, usually fibrous, but sometimes smooth, which may totally cover the sky. It is usually translucent, often produces halo phenomena, i.e., mock suns and mock moons, which are images of the real celestial bodies. Occasionally, it may be so thin and transparent that it is nearly indiscernible; at such times, the existence of a halo may be the only revealing feature. Cirrostratus often heralds the approach of a cyclonic storm, particularly in the temperate zone.

Altocumulus is a middle-altitude water or mixed cloud, whose rounded masses or rolls are usually sharply outlined. It generally has shadowed parts, varying in color from pure white to nearly black. "Mackerel sky" is an appropriate description for many altocumulus bands. Altocumulus often forms directly in clear air. With sufficiently low temperatures, ice crystals may appear in all forms of this cloud, producing showers of snow. Virga may appear, and sometimes mamma. The cloud may or may not be associated with cyclonic storms.

Altostratus is a translucent to opaque cloud composed of water droplets. It appears in the form of a dull, drab, gray or bluish sheet, through which the sun or moon might appear as seen on a ground-glass screen. It is a middle-level cloud, in contrast to the high cirrus forms, and is a precipitating cloud, often accompanied by virga and

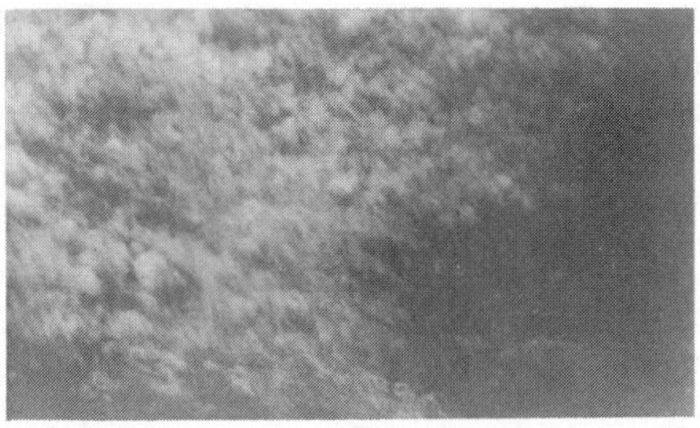

A fan shaped patch of cirrus

Trailing edge of cirrus of a snow shower

Bands of cirrus formed at the crests of gravity waves

Path of Altocumulus

Stratocumulus

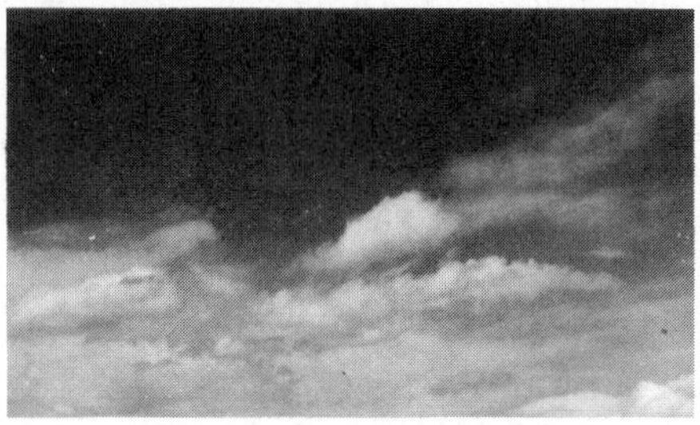

Fair weather stratocumulus with some cirrus above

Stratocumulus of an approaching rain storm

Sun illuminated fringe of a fair weather cumulus

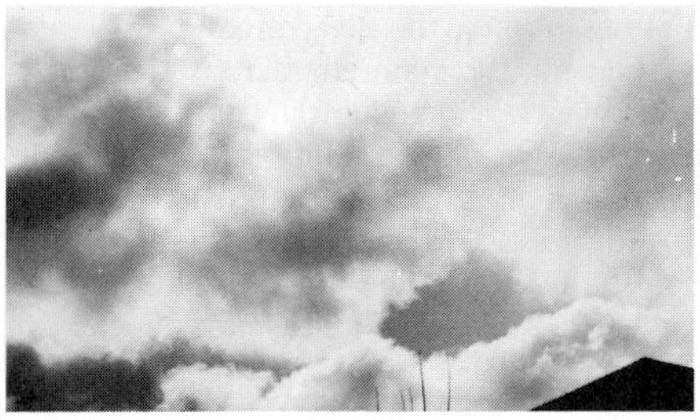

Scud and fractocumulus

Patches of cirrus associated with cumulonimbus

Cirrus anvil and shield of a cumulonimbus

Scud with a cirrus background

Low clouds on an approaching cold front

Clouds in an approaching shower

Virga of snow

Fig. 3. Various cloud formations.

mamma. Altostratus following cirrus and cirrostratus is an almost certain indication that a cyclonic disturbance is approaching.

Nimbostratus is a large middle-altitude water cloud, gray-colored, often dark and dull, and usually ragged. It is rendered diffuse by more-or-less continuously falling rain, snow, sleet, etc., of the ordinary varieties, and is not accompanied by lightning, thunder, or hail. In most cases, the precipitation reaches the ground, but not necessarily. This cloud occupies an area of large horizontal and vertical extent. Its great density and thickness (usually many thousands of feet) obscure the sun; this, plus other factors, give it the appearance of being dimly and uniformly lighted from within. Nimbostratus is most easily confused with thick masses of altostratus, stratus, or stratocumulus. Altostratus, however, is lighter in color, appears less uniform from below, and does not completely hide the sun. In case of further doubt, a cloud is called nimbostratus if precipitation from it reaches the ground.

Stratocumulus is a low-altitude water or mixed cloud, whose tesselated, rounded, or roll-shaped elements are usually arranged in orderly groups, giving the appearance of a simple wave system. It casts considerable shadow, with shades varying from very whitish in thin spots to very dark in thick spots. Occasional blue may show between the rolls. Stratocumulus frequently forms in clear air. It rarely produces precipitation, and mamma and virga may appear.

Stratus is a low-altitude water cloud in the form of a gray layer, having a rather uniform base, and often occurring as ragged patches or cloud fragments. When the sun is seen through the cloud, its outline is clearly discernible, and it may be accompanied by corona phenomena. In the immediate area of the solar disk, stratus may appear very white; away from the sun, or when the cloud is thick enough to obscure it, stratus gives off a weak, uniform luminescence. This cloud commonly develops from fog, the lower part of which evaporates, while the upper part may rise. It rarely produces precipitation, and then in the form of minute particles such as drizzle or snow grains. Fragments of stratus torn by wind or remnants of clearing stratus are usually known as fractostratus, or scud.

Cumulus is a low-altitude, water or mixed cloud in the form of billowed heaps of individual, detached elements, with flat bases and tufted tops. It has considerable shadow; its sunlit parts are mostly brilliant white; its bases are relatively dark and nearly horizontal. Size and shape vary from flat small balls of "cloud-cotton" to great towers with valleys and ravines along the sides. If precipitation occurs, it is usually of a shower nature.

The cloud is a low type, but can be found with bases from 500 to 1000 feet (152.4 to 304.8 meters) and tops as high as 20,000 feet (6096 meters). It most often forms directly in clear air as a result of convection in air of sufficiently high moisture content for a condensation level to be reached. It is composed of a great density of small water droplets, frequently supercooled; and ice-crystal formation will occur at sufficiently low temperatures, particularly in upper portions as the cloud grows vertically.

Cumulonimbus is the thunderstorm cloud; a mixed cloud of low, middle, and high altitude, with bases from 500 to 15,000 feet (152.4 to 4572 meters) and tops from 10,000 to 50,000 feet (3048 to 15,240 meters). It is exceptionally dense, tall, billowed, full of contrast from brilliant white to inky black. It occurs either as isolated clouds or as a line or wall of clouds with separated upper portions. It is vertically developed, and appears as mountains or huge towers, with at least a part of the upper portions being usually smooth, fibrous, or striated, and almost flattened. The upper portion often spreads out in the form of an anvil or a vast plume, and is known as *incus, anvil cloud,* or *thunderhead.* Cumulonimbus may be responsible for the formation of nearly all the other cloud genera.

Other Cloud Forms and Terms

Banner Cloud. Also called cloud banner, a cloud plume often observed to extend downwind from isolated mountain peaks, even on otherwise cloud-free days. The physics of the formation of such clouds is not clearly understood. Aerodynamically-induced pressure reductions in the region of separating flow to leeward from sharp peaks, combined with the cooling effect of a high peak on ambient air, may induce condensation on days when the relative humidity at peak altitude is slightly less than 100%.

Cap Cloud. An approximately stationary cloud on or hovering above an isolated mountain peak. It is formed by the cooling and condensation of humid air forced up over the peak.

Castellanus. A cloud species, found only in the genera cirrus, cirrocumulus, altocumulus, and stratocumulus, of which at least a fraction of its upper part presents some vertically-developed cumuliform protuberances (resembling rising mounds, domes, or towers, some of which are taller than they are wide), giving the cloud a crenelated or turreted appearance. The cumuliform cloud elements generally have a common base and usually seem to be arranged in lines.

Crest Cloud. A stationary cloud that forms along a mountain ridge and remains in the same position relative to the ridge. These clouds develop in a current of air rising along the ridge or mountain when air reaches its condensation level along the slope.

False Cirrus. A cloud species (*Cirrus Spissatus*) unique to the genus cirrus, of such optical thickness as to appear grayish on the side away from the sun, and to veil the sun, conceal its outline, or even hide it. These clouds often originate from the upper part of a cumulonimbus, and are often so dense that they suggest clouds of the middle level.

Floccus. A cloud species, found in the genera cirrus, cirrocumulus, altocumulus, and sometimes stratocumulus, in which each cloud element is a small tuft with a vertical development or rounded appearance, the lower part of which is more-or-less ragged and often accompanied by virga.

Fractus. Wind-torn clouds, sometimes referred to as *scud*; a cloud species, found in the genera cumulus and stratus, that presents a ragged, shredded appearance, as if torn. These clouds are irregular and generally small in size; and their characteristics change ceaselessly and often rapidly.

Lenticularis. Sometimes called lenticular cloud, a feather-shaped cloud species, found in the genera cirrocumulus, altocumulus, and rarely, stratocumulus, the elements of which have the form of more-or-less isolated, generally smooth lenses or almonds; the outlines are sharp and sometimes show irisation. These clouds usually form over mountain peaks or over terrain where the wind blows uphill; the air reaches the condensation level near the top of its flow. They also form in mid-air if there is considerable undulation in the horizontal winds and if the air rising on the crest of a wave reaches the condensation level. Clouds of this type do not travel with the wind, but remain practically stationary.

Mamma. Also called *Mammatus,* hanging protuberances, like pouches, on the undersurface of a cloud, sometimes associated with severe thunderstorms. They are indications of extreme turbulence, and occur mostly with cirrus, cirrocumulus, altocumulus, altostratus, stratocumulus, and cumulonimbus.

Noctilucent Clouds. Now rarely called luminous clouds, these are clouds of unknown composition, which occur at great heights (between 75 and 90 kilometers). They resemble thin cirrus, but usually with a bluish or silverish color, although sometimes orange to red, standing out against a dark night sky. They become more and more brilliant as the night advances, and generally more frequent and brilliant before sunrise than after sunset. These clouds have been seen rarely, and only during summer months in both hemispheres. It is thought that they may be composed of very fine cosmic dust coming from space and accumulating near a discontinuity line located at a height of about 80 kilometers.

Perlucidus. Denotes distinct spaces in stratiformis, which permit the sun, moon, blue sky, or higher clouds to be seen.

Pileus. A small accessory cloud in the form of a cap, hood, or scarf above or attached to the top of a cumuliform cloud. Several pileus clouds fairly often are observed above each other. Pileus occurs principally with cumulus and cumulonimbus clouds.

Praecipitatio. Precipitation that reaches the earth's surface.

Radiatus. A cloud variety of the genera cirrus, altocumulus, altostratus, and stratocumulus, in the form of straight parallel bands, which seem to converge toward a point on the horizon or, when crossing the entire sky, toward two opposite points. A form of raiatus, popularly known as Abraham's tree, is an assembly of long feathers and plumes of cirrus, which seem to radiate from a single point on the horizon.

Stratoformis. The most common form of the genera altocumulus and stratocumulus, consisting of a very extensive, not necessarily continuous, horizontal layer or layers.

Translucidus. A variety of stratiformis, denoting the greater part of the layer, patch, or sheet as being sufficiently translucent to reveal the position of the sun or of higher clouds.

See also other entries listed under **Meteorology.**

Peter K. Kraght, Certified Consulting Meteorologist, Mabank, Texas.

CLOUDS (Ice Content). Aircraft Icing.

CLOVER. Leguminosae.

CLOVE TREE. Of the family *Myrtaceae* (myrtle family), the clove tree (*Eugenia aromatica*) is a small evergreen tree native to the Molucca Islands. The tree is universally known for its dried flower buds, the cloves of commerce. Cultivation of the tree was introduced many years ago to numerous other tropical locations, including Zanzibar, and the Malagasy Republic. The trees grow best when near the seacoast. Reaching a height of from 24 to 40 feet (7.2 to 12 meters), the clove tree has thick shining leaves, a smooth gray bark, and flower buds and flowers of deep red color. The fragrant oil is located mainly in the leaves and flower buds, and causes the air surrounding the tree to be richly scented. Flower buds are first formed when the tree is 4 to 5 years old. The tree bears fruit in about 7 to 8 years and may produce for nearly 100 years. In gathering, the branches are pulled down and the flower buds picked off by hand, or they may be pounded from the trees with bamboo sticks. After gathering, the pedicels, or short flower stems, are picked off, and the buds dried. These buds may then be used as a spice, either whole, or ground to a powder. A single tree may yield from 8 to 12 pounds (3.6 to 5.4 kilograms) of dried fruit per year. In addition to their use as spice, an oil is distilled. The oil is rich in eugenol and is used in making artificial vanilla.

CLOWN LOACH. Loaches.

CLOXACILLIN. Antibiotic.

CLUBFOOT. Talipes. A developmental deformity of the foot, usually characterized by marked flexion, inversion, adduction of the forefoot and internal rotation of the tibia. The condition may result from an intra-uterine accident or maldevelopment, from pressure or constriction due to deficient amniotic fluid, or tumor in or around the uterus, interlocking of the feet, constriction of the umbilical cord, or pressure in the case of twins. Corrective treatment should commence shortly after birth and usually consists of progressively untwisting the foot by periodically changing plaster casts and directing the foot toward normal development. In some cases, tight tendons and stubborn contractures may require surgery. Generally, the results of treatment both cosmetically and functionally are very good. However, in older children the prognosis for a useful foot is not as promising. A similar, but less serious situation occurs much more frequently, a condition sometimes referred to as *pigeon-toes.* In this defect, the forefoot turns inward, but otherwise the foot is normal. The condition also is called *metatarsus adductus.* Treatment involves the use of plaster casts, followed by immobilization, as in a Dennis-Browne splint.

CLUPEOIDS. Herring.

CLUSIUS-DICKEL COLUMN. Diffusion.

CLUSTER ANALYSIS. In statistics, the analysis of a set of multivariate observations to see whether they cluster into groups as distinct from being more or less uniformly scattered. Various methods have been advocated for the purpose and aggregates of any size require extensive computation. Cluster analysis, as generally understood, is a form of classification but is less general in that it does not, as a rule, take account of considerations such as biological lines of descent (as in Linnean classification) or varying criteria at different levels (as in libraries).

CLUSTER (Galaxy). Galaxy.

CLUSTER (Stellar). Stellar Clusters.

CLUTTER (Radar). Radar.

CLYDESDALES HORSE. Horses, Asses, and Zebras.

CNIDOBLAST. A stinging cell of the coelenterates, found in all forms, polyps, jellyfishes, and sea anemones. It produces nematocysts which are discharged in defense and in securing food.

The *cnidocil* is a projection on the free surface of a cnidoblast which is sensitive to external stimuli.

COACERVATION. An important equilibrium state of colloidal or macromolecular systems. It may be defined as the partial miscibility of two or more optically isotropic liquids, at least one of which is in the colloidal state. For example, gum arabic shows the phenomenon of coacervation when mixed with gelatin. It also may be defined as the production, by coagulation of a hydrophilic sol, of a liquid phase, which often appears as viscous drops, instead of forming a continuous liquid phase. See also **Colloid System.**

COAGEL. A gel formed by precipitation or coagulation, as distinguished from gel formed by swelling of a solid colloid.

COAGULANTS. Water Pollution.

COAGULATION. 1. In its general scientific usage this term has two closely related meanings: (1) The process of complete or partial solidification of a colloidal solution to a gelatinous mass; or of the separation from a liquid system of a gelatinous mass. It involves the separation of the disperse from the continuous phase which fact distinguishes it from "gelation." (2) The result of an alteration of a disperse phase or of a dissolved solid which causes the separation of the system into a liquid phase and an insoluble mass, as the coagulation of egg albumin.

2. In cloud physics, coagulation is generally used synonymously with accretion. Less frequently, it refers to any process by which a cloud's numerous small cloud drops are converted into a smaller number of larger precipitation particles. When so used, the term is employed in analogy to the coagulation of any colloidal state. (See 1 above.)

3. In biological science, the term coagulation has two somewhat more specific meanings: (1) The clotting of blood or lymph. (2) The changes produced in tissue of the application of increased temperatures or by certain chemicals. See also **Anticoagulants.**

Coagulation value is the concentration of a coagulant which effects a given amount of coagulation of a colloidal, or other dispersed system.

COAGULATION (Blood). Blood.

COAGULATION (Cloud Physics). Precipitation and Hydrometeors.

COAGULATION (Hofmeister Series). A definite order of arrangement of anions and cations according to their powers of coagulation when their salts are added in quantity to lyophilic sols. Thus, the order of cations is $Mg^{2+} > Ca^{2+} > Sr^{2+} > Ba^{2+} > Li^{+} > Na^{+} > K^{+} > Rb^{+} > Cs^{+}$. The Hofmeister series is also called the *lyotropic series,* and the effect is called salting-out, a term applied strictly to the effect of electrolytes upon true solutions.

COAITA. Monkeys and Baboons.

COAL. Possibly more than any other source of energy, coal epitomizes the conflict between energy needs and environmental concerns. Coal technology has made significant strides during the past decade or so and, today, coal would be serving a greater portion of the energy needs of several countries were it not for restraints that fall outside the sphere of science and technology—and thus are not delineated in this book. Although the production and utilization of coal worldwide has been increasing over the past few years, it would appear that, as of the early 1980s, further lifting of restraints will be required before coal can fill the energy gap once envisioned for it. Although coal is not a renewable fuel, a characteristic in common with other fossil fuels, tremendous reserves of coal remain in the earth. If permitted, coal can furnish energy needs during the critical period when scientists and technologists worldwide are seeking, researching, developing, and refining what presently are considered less undesirable and, in many instances, renewable energy sources. Coal energy, it would appear, is critically needed to fill the energy gap, particularly if restraints on nuclear energy, as encountered in some countries, continue.

Nature of Coal. Containing more than 50% (weight) and 70% (volume) of carbonaceous material, including inherent moisture, coal is a readily combustible rock. Coal was formed from the compaction and induration of variously altered plant remains similar to those found in peat. Coal was formed during earlier geological periods, the process of formation acting slowly over extremely long periods of time. Coal is not a uniform substance, but reflects the conditions of its formation. These include:

1. *Differences in the kinds of plant materials* from which the coal was derived account for *different types of coal.*
2. *Differences in the degree of metamorphism* occurring during the formation of coal determine the *different ranks of coal.*
3. *Differences in the range of impurity* in coal account for the *different grades of coal.*

The fermentation of vegetable matter under conditions of no air and abundant moisture where volatiles are retained, resulting in the formation of bitumens, such as peat and coal, is known as *bituminous fermentation.* The metamorphic transformation of bituminous coal into anthracite is known as *anthracitization. Coalification* is the alteration or metamorphism of plant material into coal; the biochemical process of diagenesis and the geochemical process of metamorphism in the formation of coal. The peat-to-anthracite theory of coal formation is described as a process in which the progressive ranks of coal are indicative of the degree of coalification and, by inference, of the relative geologic age of the deposit. Peat, as the initial stage of coalification, is of recent geological age. Lignite, as an intermediate stage, is usually Tertiary or Mesozoic, and bituminous coal and anthracite, as the more advanced stages of coalification, are usually Carboniferous.

The major coals may be defined as follows:

(*Anthracite Coal*)—Coal of the highest metamorphic rank, in which the fixed carbon content is between 92 and 98%. It is hard, black, and has a semimetallic luster and semiconchoidal fracture. Anthracite ignites with difficulty and burns with a short, blue flame and without smoke. Anthracite coal is also known as hard coal, stone coal, kilkenny coal, and black coal.

(*Semianthracite Coal*)—Coal having a fixed-carbon content of between 86 and 92%. It is between bituminous coal and anthracite coal in metamorphic rank, although its physical properties more closely resemble those of anthracite.

(*Semibituminous Coal*)—Coal that ranks between bituminous coal and semianthracite. It is harder and more brittle than bituminous coal, has a high fuel ratio and burns without smoke. Semibituminous coal is also known as *metabituminous coal* which is defined as containing 89–91.2% carbon, analyzed on a dry, ash-free basis. The term *smokeless coal* also is used.

(*Bituminous Coal*)—Coal that ranks between subbituminous coal and semibituminous coal and that contains 15–20% volatile matter. It is dark brown-to-black in color and burns with a smoky flame. Bituminous coal is the most abundant rank of coal and is commonly Carboniferous in age. The most common synonym is *soft coal.*

(*Subbituminous Coal*)—A black coal intermediate in rank between lignite and bituminous coals, or in some classifications, the equivalent of *black lignite.* It is distinguished from lignite by higher carbon content and lower moisture content.

The subbituminous coals are further classified in terms of their calorific value:

Subbituminous A Coal—A type of subbituminous coal having 10,500 or more, but less than 13,000 Btu per pound (5838–7228 Calories/kg).
Subbituminous B Coal—A type of subbituminous coal having 9,500 or more, but less than 10,500 Btu per pound (5282–5838 Calories/kg).
Subbituminous C Coal—A type of subbituminous coal having 8,300 or more, but less than 9,500 Btu per pound (4615–5282 Calories/kg).

(*Lignite Coal*)—A brownish-black coal that is intermediate in coalification between peat and subbituminous coal; consolidated coal with a calorific value less than 8,300 Btu per pound (4615 Calories/kg), on a moist, mineral-matter-free basis. Synonyms include *brown lignite* and *brown coal.* Further classifications of lignite are made on the basis of calorific value:

Lignite A Coal—A lignite that contains 6,300 or more Btu per pound, but less than 8,300 Btu per pound (3503–4615 Calories/kg). Also known as *black lignite.*
Lignite B Coal—A lignite that contains less than 6,300 Btu per pound (3503 Calories/kg). Also known as brown lignite or brown coal.

(*Peat*)—This is an unconsolidated deposit of semicarbonized plant remains of a water-saturated environment, such as a bog or fen, and of persistently high moisture content (minimum of 75%). It is considered an early stage or rank in the development of coal. The carbon content is about 60%; oxygen content is about 30%. Structures of the vegetal matter can be seen. When dried, peat burns freely.

(*Peat Coal*)—This refers to two materials: (a) a coal transitional between peat and brown coal or lignite; and (b) an artificially carbonized peat that is used as a fuel.

(*Cannel Coal*)—A compact, tough *sapropelic coal* that contains spores and that is characterized by a dull-to-waxy luster, conchoidal fracture, and massiveness. It is attrital and high in volatiles. By American standards, it must contain less than 5% anthraxylon. Synonyms include *candle coal, kennel coal, cannel, cannelite, parrot coal,* and *curley cannel.* A sapropelic coal is derived from organic residues (finely divided plant material, spores, algae, etc.) in stagnant or standing bodies of water. Putrifaction is under anaerobic conditions rather than by peatification.

Ranks of Coal. Coals are classified in order to identify end-use and also to provide data useful in specifying and selecting burning and handling equipment and in the design and arrangement of heat-transfer surfaces. One classification of coal is by rank, that is, according to the degree of metamorphism, or progressive alteration, in the natural series from lignite to anthracite. Volatile matter, fixed carbon, inherent or bed moisture (equilibrated moisture at 30°C and 97% humidity), and oxygen are all indicative of rank, but no one item completely defines it. The classification of the American Society for Testing and Materials (ASTM) uses fixed carbon and calorific values, calculated on a mineral-matter-free basis, as the classifying criteria.

In establishing the rank of coals, it is necessary to use information showing an appreciable and systematic variation with age. For the older coals, a good criterion is the "dry, mineral-matter-free fixed carbon or volatile." However, this value is not suitable for designating the rank of the more recent, younger coals. A dependable means of classifying the latter is the "moist, mineral-matter-free Btu" which varies little for the older coals, but appreciably and systematically for younger coals.

Classification of major coals according to rank or age is given in Table 1. The criteria given in the prior paragraph are used in classifying the older and younger coals. Seventeen United States coals are arranged in order of the classification of Table 1 and presented in Table 2.

Classification of coals in Europe and other parts of the world differs somewhat from the American system. European classifications include: (1) the *International Classification of Hard Coals by Type*; and (2) the *International Classification of Brown Coals.* These systems were developed by a Classification Working Party established in 1949 by the Coal Committee of the Economic Commission for Europe. The term "hard coal" is defined as a coal with a clorific value of more than 10,260 Btu per pound (5705 Calories/kg) on the moist, ash-free basis. The term "brown coal" refers to a coal containing less than 10,260 Btu per pound (5705 Calories/kg). In European terminology, the term "type" is equivalent to rank in American coal classification terminology and the term "class" approximates the ASTM rank.

TABLE 1. CLASSIFICATION OF COALS BY RANK (ASTM)

Class	Group	Fixed Carbon Limits, % (Dry, Mineral-Matter-Free Basis) Equal or Greater Than	Fixed Carbon Limits, % Less Than	Volatile Matter Limits, % (Dry, Mineral-Matter-Free Basis) Greater Than	Volatile Matter Limits, % Equal or Greater Than	Calorific Value Limits, Btu/Pound (Moist*, Mineral-Matter-Free Basis) Equal or Greater Than	Calorific Value Limits, Btu/Pound Less Than	Agglomerating Character
I Anthracite	1. Meta-anthracite	98	—	—	2	—	—	Nonagglomerating
	2. Anthracite	92	98	2	8	—	—	Nonagglomerating
	3. Semianthracite[c]	86	92	8	14	—	—	Nonagglomerating
II Bituminous	1. Low volatile bituminous	78	86	14	22	—	—	Commonly Agglomerating[b]
	2. Medium volatile bituminous	69	78	22	31	—	—	Commonly Agglomerating[b]
	3. High volatile A bituminous	—	69	31	—	14,000[a]	—	Commonly Agglomerating[b]
	4. High volatile B bituminous	—	—	—	—	13,000[a]	14,000	Commonly Agglomerating[b]
	5. High volatile C bituminous	—	—	—	—	11,500	13,000	Commonly Agglomerating[b]
						10,500[b]	11,500	Agglomerating
III Subbituminous	1. Subbituminous A	—	—	—	—	10,500	11,500	Nonagglomerating
	2. Subbituminous B	—	—	—	—	9,500	10,500	Nonagglomerating
	3. Subbituminous C	—	—	—	—	8,300	9,500	Nonagglomerating
IV Lignitic	1. Lignite A	—	—	—	—	6,300	8,300	Nonagglomerating
	2. Lignite B	—	—	—	—	—	6,300	Nonagglomerating

* Moist refers to coal containing its natural inherent moisture, but not including visible water on the surface of the coal.

[a] Coals having 69% or more fixed carbon on the dry, mineral-matter-free basis are classified according to fixed carbon, regardless of calorific value.

[b] It is recognized that there may be nonagglomerating varieties in these groups of the bituminous class, and there are notable exceptions in high-volatile C bituminous group.

[c] If agglomerating, the coal is classified in the low-volatile group of the bituminous class.

The terms, *mineral-matter-free fixed carbon*; and *mineral-matter-free Btu* are defined by the following formulas:

Parr formulas

$$\text{Dry, Mm-free FC} = \frac{FC - 0.15S}{100 - (M + 1.08A + 0.55S)} \times 100, \%$$

$$\text{Dry, Mm-free VM} = 100 \times \text{Dry, Mm-free FC}, \%$$

$$\text{Moist, Mm-free Btu} = \frac{Btu - 50S}{100 - (1.08A + 0.55S)} \times 100, \text{ per pound}$$

Approximation formulas

$$\text{Dry, Mm-free FC} = \frac{FC}{100 - (M + 1.1A + 0.1S)} \times 100, \%$$

$$\text{Dry, Mm-free VM} = 100 - \text{Dry, Mm-free FC}, \%$$

$$\text{Moist, Mm-free Btu} = \frac{Btu}{100 - (1.1A + 0.1S)} \times 100, \text{ per pound}$$

Symbols Used:
Mm = mineral matter; Btu = heating value per pound; FC = fixed carbon, %; VM = volatile matter, %; M = bed moisture, %; A = ash, %; S = sulfur, %. All for coal on a moist basis.

Conversion Factor: 1 Btu/pound = 0.556 Calories/kg).

TABLE 2. REPRESENTATIVE UNITED STATES COALS ARRANGED IN ORDER OF ASTM CLASSIFICATION

Coal Rank: Class	Coal Rank: Group	State	County	Coal Analysis, Bed Moisture Basis: M	VM	FC	A	S	Btu	Rank FC	Rank Btu
I	1	Pennsylvania	Schuylkill	4.5	1.7	84.1	9.7	0.77	12,745	99.2	14,280
I	2	Pennsylvania	Lackawanna	2.5	6.2	79.4	11.9	0.60	12,925	94.1	14,880
I	3	Virginia	Montgomery	2.0	10.6	67.2	20.2	0.62	11,925	88.7	15,340
II	1	West Virginia	McDowell	1.0	16.6	77.3	5.1	0.74	14,715	82.8	15,600
II	1	Pennsylvania	Cambria	1.3	17.5	70.9	10.3	1.68	13,800	81.3	15,595
II	2	Pennsylvania	Somerset	1.5	20.8	67.5	10.2	1.68	13,720	77.5	15,485
II	2	Pennsylvania	Indiana	1.5	23.4	64.9	10.2	2.20	13,800	74.5	15,580
II	3	Pennsylvania	Westmoreland	1.5	30.7	56.6	11.2	1.82	13,325	65.8	15,230
II	3	Kentucky	Pike	2.5	36.7	57.5	3.3	0.70	14,480	61.3	15,040
II	3	Ohio	Belmont	3.6	40.0	47.3	9.1	4.00	12,850	55.4	14,380
II	4	Illinois	Williamson	5.8	36.2	46.3	11.7	2.70	11,910	57.3	13,710
II	4	Utah	Emery	5.2	38.2	50.2	6.4	0.90	12,600	57.3	13,560
II	5	Illinois	Vermilion	12.2	38.8	40.0	9.0	3.20	11,340	51.8	12,630
III	1	Montana	Musselshell	14.1	32.2	46.7	7.0	0.43	11,140	59.0	12,075
III	2	Wyoming	Sheridan	25.0	30.5	40.8	3.7	0.30	9,345	57.5	9,745
III	3	Wyoming	Campell	31.0	31.4	32.8	4.8	0.55	8,320	51.5	8,790
IV	1	North Dakota	Mercer	37.0	26.6	32.2	4.2	0.40	7,255	55.2	7,610

NOTE: Definition of coal rank is given in Table 1.

M = equilibrium moisture, %; VM = volatile matter, %; FC = fixed carbon, %; A = ash, %; S = sulfur, %; Btu = high heating value, Btu per pound; Rank FC = dry, mineral-matter-free fixed carbon, %; Rank Btu = moist, mineral-matter-free Btu per pound.

All calculations are per the Parr formulas defined in Table 1.

Conversion Factor: 1 Btu = 0.2520 Calorie.

Space does not permit a full comparison of the various systems. Reference to various ASTM publications is suggested.

The classification of coal is described in further detail later in this article under "Testing of Coal."

TABLE 3. ANTHRACITE COAL SIZES

	Diameter of Hole			
	Will Pass Through		Will Not Pass Through	
Name Used in the Trade	Inches	~ Centimeters	Inches	~ Centimeters
Broken	$4\frac{3}{8}$	11.1	$3\frac{1}{4}$ to 3	8.3 to 7.6
Egg	$3\frac{1}{4}$ to 3	8.3 to 7.6	$2\frac{7}{16}$	6.2
Stove	$2\frac{7}{16}$	6.2	$1\frac{5}{8}$	4.1
Nut	$1\frac{5}{8}$	4.1	$\frac{13}{16}$	2.1
Pea	$\frac{13}{16}$	2.1	$\frac{9}{16}$	1.4
Buckwheat	$\frac{9}{16}$	1.4	$\frac{5}{16}$	0.8
Rice	$\frac{5}{16}$	0.8	$\frac{3}{16}$	0.5

Commercial Sizes of Coal

Anthracite Coal. Standard sizes for anthracite coal are indicated in Table 3. The broken, egg, stove, nut and pea sizes are largely used for hand-fired domestic units and gas producers. Buckwheat and rice are used in mechanical types of firing equipment.

Bituminous Coal. The sizes of bituminous coal are not well standardized, but the following sizings are commonly recognized:

(*Run of Mine*)—Coal that is shipped from the mine without screening. It is used for both domestic heating and commercial steam production.

(*Run of Mine—8-inch*)—This is run-of-mine coal with oversize lumps broken up. (8 inches = 20.3 centimeters.)

(*Lump—5-inch*)—This size will not go through a 5-inch round hole. It is used for hand-firing and domestic purposes. (5 inches = 12.7 centimeters.)

(*Egg—5 by 2-inch*)—This size goes through a 5-inch hole, but is retained on 2-inch round-hole screens. It is used for hand-firing, gas producers, and domestic firing. (5 × 2 inches = 12.7 × 5.1 centimeters.)

(*Nut—2 by $1\frac{1}{4}$-inch*)—This size is used for small industrial stokers, gas producers, and hand-firing. (2 × $1\frac{1}{4}$ inches = 5.1 × 3.2 centimeters.)

(*Stoker Coal—$1\frac{1}{4}$ by $\frac{3}{4}$-inch*)—This size is largely used for small industrial stokers and domestic firing. ($1\frac{1}{4}$ × $\frac{3}{4}$ inches = 3.2 × 1.9 centimeters.)

(*Slack—$\frac{3}{4}$-inch and under*)—This is used for pulverizers, cyclone furnaces, and industrial stokers. ($\frac{3}{4}$-inch = 1.9 centimeters.)

Production and Consumption of Coal

Reliable statistics on coal production and consumption, particularly among the Communist nations, are difficult to obtain. Statistics also are usually about two years late in their compilation. Based upon several sources, including figures from the World Coal Study (Wilson, 1979; Griffith and Clark, 1979), various government statistics, and National Coal Association (U.S.) records, coal production as of the early 1980s is running at between 3500 and 4200 million metric tons per year. Of this total, approximately 72.5% is hard coal (this definition including both anthracite and bituminous coals); and 27.5% brown coals and lignites. In terms of current electrical energy production, the hard coals are of greatest importance because they furnish 25% of energy needs, whereas the brown coals and lignites furnish only about 4% of these needs. This, of course, is a reflection of the wide differences in energy content of the two kinds of coal.

There are three major hard coal producing nations: (1) the United States with an estimated 26.6% of total world production; (2) the U.S.S.R. with about 22% of the total; and (3) China with about 21.6% of the total. These three nations furnish over 70% of the world's hard coal requirements. Traditionally, Poland (8.2%), the United Kingdom (5.2%), and West Germany (3.8%) have been major suppliers. The only one of these countries that has had a reduction in coal production since the 1960s is the United Kingdom, which during an earlier period accounted for 10.7% of world hard coal production. Among the other developed nations, South Africa supplies about 3.6% of the world total; Australia, 2.8%; and France and Japan just short of 1% each. The only developing country that is a significant hard coal producer is India (4.4%). India's production since the 1960s has increased from 2.4% to the 4.4% figure of the early 1980s.

In terms of another breakdown, the Communist countries are producing 51.8% of total world hard coal production; the developed countries of the free world, 43.8%; and developing countries (India), 4.4%.

Total world production of brown coals and lignites is estimated at about 950 million metric tons per year. The major supplier is East Germany, followed by the U.S.S.R., and West Germany.

The World Coal Study, conducted under the direction of Massachusetts Institute of Technology and released in 1980, observed: (1) coal can provide the principal part of the additional energy needs of the world over the next two decades; (2) world coal production must increase by 2.5 to 3 times and world trade in coal must increase by 10 times to furnish the foregoing energy needs; (3) the United States, which possesses the largest economically and technically recoverable coal reserves in the world, will become a major exporter; and (4) as for climatic consequences of increased coal use, present knowledge of possible carbon dioxide effects on climate does not justify delaying the expansion of coal use.

In terms of coal utilization, based upon late-1970s data, the percentage use of coal varies considerably from one region or country to the next. See Table 4.

Coal Reserves

As with other fossil fuel reserves, estimates are variously made by a number of authorities and are subject to frequent revision—upward or downward, often largely dependent upon the vigor with which new exploration activities are undertaken. As pointed out in the report of the Workshop on Alternative Energy Strategies, also under the direction of the Massachusetts Institute of Technology (1977), "The true size of the world coal reserve, however, is not a critical value. Known coal resources can support any likely level of exploitation for decades to come. What is in doubt is the willingness and ability of nations to accept large increases in coal production and utilization." The Workshop involved some 70 specialists recruited from industry, government, and universities in 15 energy-consuming countries (United States, Canada, United Kingdom, France, West Germany, the Netherlands, Denmark, Sweden, Norway, Finland, Italy, Japan, Iran, Mexico, and Venezuela).

Coal reserves are usually placed in two categories: (1) *economically recoverable*, and (2) *potentially recoverable*. For comparison, coal re-

TABLE 4. ESTIMATED USAGE OF COAL

		Percent of Total Coal Consumed				
Using Region or Country	Coal as Percent of Total Energy Supply	Electricity Generation	Iron and Steel	Other Industry	Residential Commercial	Railroads
North America	19	77	10	11	2	—
Western Europe	20	57	20	9	14	—
Japan	16	18	66	5	11	—
Other non-Communist countries	19	36	16	33	8	7

TABLE 5. WORLD COAL RESERVES AND CRUDE OIL EQUIVALENTS

Continent, Country, or Region	Percent of Total Measured Reserves	Measured Reserves		Economically Recoverable Reserves	
		Billion Metric Tons	Crude Oil Equivalent Billion Barrels	Billion Metric Tons	Crude Oil Equivalent Billion Barrels
North America	30.8	410	1681	255	1046
U.S.S.R. and E. Europe	26.3	350	1435	285	1169
W. Europe	17.0	225	923	40	164
China	15.1	200	820	100	410
Australia (other Oceania)	5.7	75	308	25	103
South Africa (other Africa)	2.0	27	111	13	53
India	1.8	24	98	12	49
E. Asia	1.0	13	53	2	8
South America	0.3	4	16	2	8
Totals	100.0	1328	5445	734	3010

serves are also frequently expressed in terms of equivalent barrels of crude oil. The term *measured reserves* is also used. These are the quantities of coal in well-surveyed deposits, the extent and quality of which have been established by adequate sampling. See Table 5. The economically recoverable and the potentially recoverable reserves add up to the total measured reserves. Authorities believe, however, that the total reserves of coal on earth, including the undiscovered and the unexplored, can reasonably be estimated at 8 or 9 times the measured reserves, this giving some 11,500 billion tons of total world coal resources. Assuming that only 25% of this figure would turn out to be recoverable in a practical sense, this would provide an ultimate reserve of some 2500 billion tons of coal, equivalent to about 12,000 billion barrels of crude oil. It is interesting to note that this figure, believed to be conservative, is approximately six times greater than estimated ultimately recoverable crude oil.

It is interesting to note that exploration for coal over the years has been less wide-ranging and intensive than for gas and oil. When oil and gas were relatively inexpensive, there was no impelling incentive to undertake extensive coal exploration. Also, it should be noted that most of the known reserves today are in the Northern Hemisphere.

TABLE 6. PRINCIPAL BITUMINOUS COAL PRODUCING STATES IN THE UNITED STATES

State	Percent of Annual Total
West Virginia	20.8
Kentucky	20.4
Pennsylvania	12.8
Illinois	11.0
Ohio	8.5
Virginia	5.7
Indiana	4.4
Alabama	3.5
Tennessee	1.9
Wyoming	1.8
Montana	1.4
New Mexico	1.4
North Dakota	1.1
Colorado	0.9
Utah	0.8
Missouri	0.8
Oklahoma	0.4
Washington	0.4
Maryland	0.3
Kansas	0.2
Iowa	<0.2
Arkansas	<0.1
All others	1.3

The geology of the Southern Hemisphere suggests that this imbalance of statistical reserves exists only because of lack of more intensive exploration surveys. With the aid of the Japanese, for example, who have been interested in coal as a source of metallurgical coke for many years, substantial deposits of coal were found in New South Wales and Queensland in Australia. It is interesting to note that, although the coal reserve figures for Australia are considerably less impressive than for North America, the U.S.S.R. and Eastern Europe, Western Europe, and China, as reflected by Table 5, nevertheless Australia's coal reserves are estimated to be roughly equivalent to the oil reserves of Saudi Arabia.

Also, it is estimated that the measured reserves of coal in the United States, as of the early 1980s, are some ten times the 40 billion tons mined in the country since the earliest records of U.S. coal mining.

Coal Reserves in the United States. Although coal production now comes from 25 states, as indicated in Table 6, a coal reserve base has been established for 30 states. More than two-thirds of the reserve base is in five states—Montana, Wyoming, Illinois, West Virginia, and Pennsylvania—these same states accounting for about one-half the production as of the early 1980s. As shown by the maps in Figs. 1 and 2, coal resources in the United States are rather well dispersed. Additional deposits are located in Alaska. The Alaskan deposits are immense in quantity and are of various ranks of coal. Deposits of coal are known to occur in 37 states, in about 117 coalbeds, both thick and thin, which are found from near-surface levels to depths of several thousand feet. The measured reserves for North America of 410 billion metric tons (World Energy Conference Survey of Energy Resources, 1976) differs somewhat from the estimated coal reserves base (U.S. Bureau of Mines, 1977) for the United States of 436 billion tons. Considering the complexities and uncertainties of such estimates,

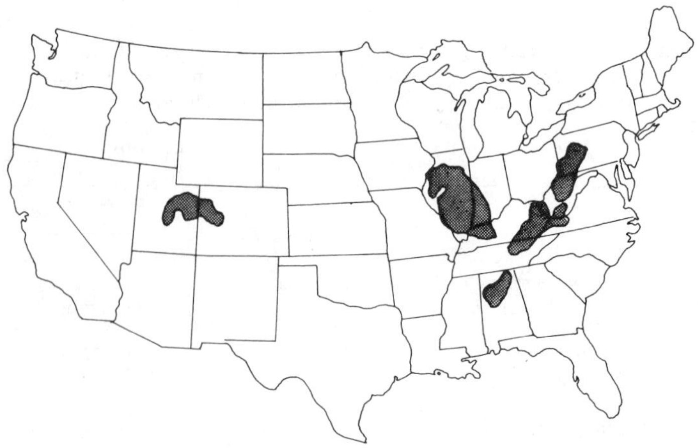

Fig. 1. Major underground coal mining regions of the United States.

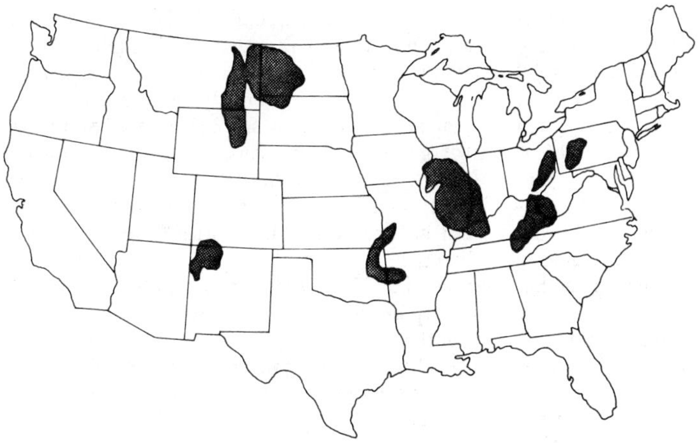

Fig. 2. Major surface coal mining regions of the United States.

this can be considered to be rather close agreement. The mix of the 436 billion tons reserves base is:

Lignite, 28 billion tons	6.5%
Subbituminous, 168 billion tons	38.5%
Bituminous, 233 billion tons	53.4%
Anthracite, 7 billion tons	1.6%

Distribution of the reserves base is:

East of Mississippi River, 202 billion tons	46.3%
Amenable to surface mining, 34 billion tons	
Minable underground, 168 billion tons	
West of Mississippi River, 234 billion tons	53.7%
Amenable to surface mining, 103 billion tons	
Minable underground, 131 billion tons	

Considering probable mining techniques to be used for the reserve base:

Surface mining, 137 billion tons	31.4%
Underground mining, 299 billion tons	68.6%

The higher ranks of coal (in terms of heating value) occur largely east of the Mississippi River. See Table 7.

The term *low-sulfur coal* is not always used in the literature in a precise fashion. At one time, the importance of sulfur content was in connection with its ranking as a coal for coking. In current usage, the term is used for evaluating coals that will meet local air quality control regulations. These regulations are based upon criteria established under the Clean Air Act of 1970 (United States) and by the various stages for Air Quality Control Regions (AQCRs) within each state. Limits on the sulfur content of coal or on the sulfur dioxide (SO_2) content in stack emissions vary widely among the AQCRs. Some authorities regard a coal as a low-sulfur fuel only if it contains less than 1% (weight) of sulfur. See Table 8.

As will be noted from the table, about 200 billion tons of the coal reserve base have been identified as low-sulfur coal. Of this amount, 167 billion tons occur west of the Mississippi River. More than one-half of the western low-sulfur reserve base consists of underground minable coal. In the eastern states, 84% of the low-sulfur coals can be recovered only by underground mining.

Premium-grade coking coals for metallurgical use occur in the eastern states, with West Virginia accounting for 54%. Eastern Kentucky and Virginia account for about 33%, with the remainder distributed in ten other states. These premium-grade coals are among the finest coals in the world. As of the early 1980s, approximately 75 million tons are used annually in coke ovens in the United States, with another 50 million tons exported for similar usage in other countries. Although the export situation probably will change within the next few years, with increasing amounts going to heating coals, nearly 80% of the United States coal exports are of these premium-grade coking coals.

Geology of Coal

Coal is interspersed as individual beds within other types of sedimentary rock beds, including sandstones, limestones, clays, shales, and mixtures of these materials. The plant material that ultimately became coal deposits was accumulated in upland bogs, coastal or near-coastal swamps, or delta plains. It is envisioned that the conditions were somewhat similar to the conditions existing today in the Okefenokee Swamp in Georgia or the Everglades of Florida. These areas may have varied from a few acres to several hundreds of square miles. Hence, the variation in the occurrence of coal as we find it today.

For the geological processes (coalification) to convert such plant materials into coal, it was necessary that the original swamps be submerged—by way of rises in sea level or land subsidence. Probably many submersive actions occurred with intermittent deposition of calcareous materials deposited from water containing muds, sands, and slimes. As the result of a series of compactions, with varying depths of burial, heat, and pressure, and length of time, the progress of coalification also varied—from peat to lignite, to subbituminous coal, to bituminous coal, possibly to anthracite.

TABLE 7. HEATING VALUES OF COALS BY SOURCE STATES

11,900–13,800 Btu/pound (6616–7673 Calories/kilogram)	9600–11,800 Btu/pound (5338–6561 Calories/kilogram)	6500–9500 Btu/pound (3614–5282 Calories/kilogram)
Alabama, Kentucky, Maryland, Ohio, Pennsylvania, Tennessee, Virginia, West Virginia	Illinois, Indiana, Michigan	Alaska, Oregon
	Arizona, Colorado, Iowa, Missouri, Montana, Texas, Washington, Wyoming	
Kansas, New Mexico, Oklahoma, Utah		

TABLE 8. UNITED STATES COAL RESERVE BASE BY AREA AND SULFUR CONTENT

	Billion Tons				
	Sulfur Range				
Location of Reserves	<1%	1.1–3.0%	3.0%	Unknown	Total
East of Mississippi River	32.9	55.5	81.4	32.5	202.3
West of Mississippi River	167.3	37.5	11.3	18.8	234.4
Total	200.2	93.0	92.7	50.8	436.7
All Sources as percent of total coal reserves	45.8	21.4	21.2	11.6	

SOURCE: U.S. Bureau of Mines.

Coals of the United States probably were formed during three major geological periods:

1. During the Pennsylvanian (Carboniferous) period which dates back approximately 300 million years. Deposits include the predominately bituminous coal beds in the Appalachian Province extending from Pennsylvania (including the anthracite beds in central Pennsylvania) into northeastern Alabama. Also the contiguous Eastern Interior Region of Illinois, southwestern Indiana, and western Kentucky; in the contiguous Western Interior Region of Iowa, Kansas, Missouri, northeastern Oklahoma and northwestern Arkansas; and in the separated central portion of Texas (excluding Texas lignite).

2. During the Cretaceous period which dates back approximately 100 million years. Deposits include the predominately bituminous and subbituminous coal beds in the Rocky Mountain Province, extending in large, separated regions from central Montana into northeastern Arizona and northwestern New Mexico.

3. During the Tertiary period which dates back approximately 65 million years. Deposits include the subbituminous coal and lignite beds in the Great Plains Province, which includes northeastern Wyoming, eastern Montana, western North Dakota, and northwestern North Dakota.

Coal beds form only a very small percentage of the total thicknesses of the overall sedimentary strata comprising the so-called "Coal Measures" in coal-bearing areas. The thicknesses of individual coal beds within the United States range from a few millimeters (horizon markers) to as much as 100 feet (30 meters) or more. The number of individual coal beds of commercial significance may range from less than 10 feet (3 meters) to over 100 feet (30 meters). The coal, however, is rarely found in full vertical sequence at any one particular spot, but usually is distributed unevenly in single beds or small groups of beds around the margin or within the interior of the generally basin-shaped areas of coal-bearing strata.

Depending on the desired or feasible rate of annual production, the amounts of coal reserves required to support a new mine designed for an economic life of 20 years or more may range from a comparatively few tons to 300 million tons or more. Such amounts are dependent upon the coal-bed thickness and the ease of mining and particularly whether surface or underground mining techniques will be required. Thus, a given mining area may range from as little as one thousand acres to as much as 50 square miles (130 square kilometers).

A variety of detrimental irregularities may accompany or interrupt an otherwise orderly accumulation of plant material either during swamp growth or shortly thereafter. While many coal beds or portions of beds are relatively low in ash content, other beds or portions of beds may contain depositional admixtures of particles of mud or silt which were washed or blown into the swamp during plant growth. Where relatively abundant, these particles serve to increase the ash content of the eventual coal bed and thus to decrease its quality correspondingly. During periods of prolonged swamp flooding, layers of mud or silt may have been deposited on the preexisting plant accumulations. Such deposition then may have been followed by additional plant accumulation. Such layers of impurities between underlying and overlying accumulations of plant materials, eventually hardening into shales or silty shales, are called *partings*. These may range from knife-edge thickness to thicknesses of up to a foot or more. Such partings within a single coal bed decrease the quality of the coal as mined and impair the mining procedures, particularly when such partings have become pyritized.

Such partings are not always evenly deposited over large portions or the entire extent of a coal-forming swamp, but may become progressively thicker toward the source of the deposited material. The partings may be wedge-shaped, with the overlying plant material occurring increasingly higher above the underlying plant material and sometimes becoming increasingly thinner to the point of disappearance. Deposition of impurities in this manner results in splitting the total thickness of the coal bed into two or more diverging *benches*. This causes mining difficulties and sometimes a bench may be so thin as not to be economically recoverable.

In some coal beds, relatively flat, lenticular masses ranging up to several feet in diameter, composed of pyrite, calcite or siderite, were formed during plant growth. Such materials (concretions) may represent the eventual immediate roof of the coal bed. These concretions impede mining operations and cause a hazard because of their tendency to drop out of the roof unexpectedly during operation of the mine.

From initial deposition and burial under overlying sedimentary materials through succeeding geological periods, coal beds are continually subject to the action of ground water. Thus, some coal beds have developed a system of essentially vertical fractures—thin cracks which are often filled with coatings of pyrite, calcite, kaolinite and other minerals deposited from ground water. Impurities from these veins lower the quality of the coal.

During the long periods since formation, many coal beds have been subject to folding and sharp deformation, resulting in specific dislocations or faults. Such shifting may range from a foot or two (less than one meter) to several hundred feet (meters) and even up to thousands of feet (meters) in linear extent. The coal and lignite fields in the Great Plains Province are relatively undisturbed. The coal-bearing strata in the Appalachian Province are relatively flat along their northwestern margin, but increase in intensity of relatively mild but significant folding toward the southeast at right angles to the regional northeast-southwest trend of the component coal fields. The coal beds in the various basins comprising the Rocky Mountain Province range from comparatively gentle slopes of but a few degrees over areas of broad extent to areas of similar extent with prevailing dips of up to 20 or 30°, along with a few areas of limited extent where the coal beds are highly deformed.

Exploration Techniques. The diamond core drill historically has been the most extensively used tool in coal exploration. Cores of coal, properly recovered, enable accurate seam descriptions and measurements; also provide material for chemical analysis. Geologging or electric logging, used for several years in the oil and gas fields, is now gaining acceptance in coal-exploration technology. The system involves hoisting a sensor up the length of a drill hole while electric pulses are transmitted through the hoist cable to a console in a truck or on the surface. Here instruments record the variations in properties of the rock strata as a function of hole depth. The electric curves usually run in coal exploration are resistivity and spontaneous potential. Radiometric or nuclear curves include gamma ray, neutron, and the density or gamma-gamma log.

The impurities in coal beds, present either as distinct partings or disseminated throughout, are composed of clay materials that have a high density and a high natural radioactivity relative to coal. Consequently, the gamma ray and density curves, invaluable for bed correlation and thickness determination, also can be used as semiquantitative indices of coal quality.

Coal Mining

Coal is produced in the United States from both underground and surface mines. Most surface-mined coal involves removal of the overburden (stripping), with auger mining contributing the balance. Surface mining has been gaining over the last several years. Surface mining is attractive due to lower investment and operating costs (less peoplepower) and because it is not subject to the health and safety problems associated with deep mining, although it is subject to increasingly stringent environmental requirements.

Underground Mining. Conventional and continuous mining provide the bulk of United States underground coal tonnages. Both systems commonly involve a room and pillar approach, but they differ in terms of machinery used and in operating sequence. Handloading, which once produced major coal tonnages, has declined rapidly since 1950 and no longer is significant. A continuous mining machine combines into one machine the work done in conventional mining by the cutter, the face drill, the loading machine, and the blasting operation. One such machine breaks or digs the coal out of the solid seam and loads it into a conveyance vehicle. Haulage and roof support, however, are accomplished in much the same manner for either conventional or continuous mining.

Although the conventional mining proportion of underground production has declined steadily since 1950, it is expected to produce an important part of total tonnage for several years into the future. Production from continuous mining has increased since 1950 and equaled conventional tonnage in 1966. A continuous mining machine is illustrated and described in Fig. 3.

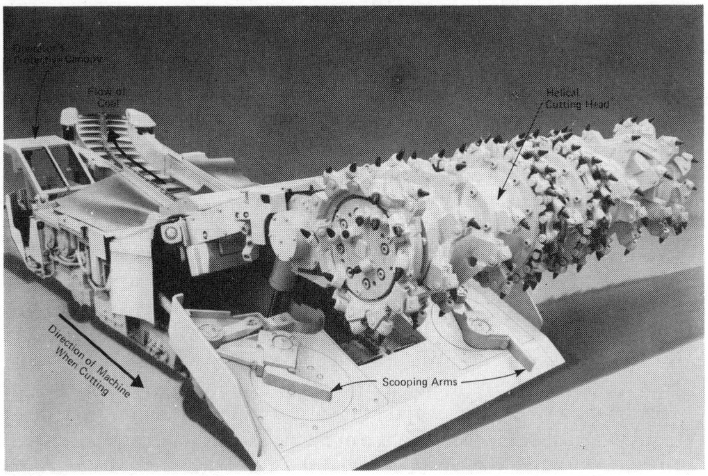

Fig. 3. Continuous mining machine. As coal accumulates on the mine floor, the helical screw effect of the cutting head constantly moves the pile toward the center of the head, contributing to fast loading and improved cleanup. Two 300-horsepower motors power all motions of the machine. A two-speed mechanical tramming system uses hydraulically-operated disk clutches. The hydraulic system of the machine operates at 1,200 psi. Safety provisions include operator's protective canopy with height control for seam variations; and a safety flapper for emergency machine shut-off and to prevent accidental motor startup. (*Jeffrey Heliminer, Jeffrey Mining Machinery Co.*)

Longwall mining, prevalent in Europe for many years, is now practiced to a limited extent in the United States. In longwall mining, a large block of coal to be mined is isolated by driving entries, or tunnels, around four sides of the block. Mining is then done across the width of the block (300 to 800 feet; 90–240 meters) a slice at a time. The mining machine is either a drum-type shearing machine or a plow which is moved along the coal face. As the coal is dug from the face, it falls to the floor where it is continuously removed by a conveyor. This method is expected to increase in acceptance in the United States because it concentrates production in a smaller area in the mine, offering better productivity and simplified ventilation.

Productivity. As a result of continued growth of continuous mining and the almost complete elimination of handloading, the average productivity of peoplepower steadily increased through the late 1960s (a 2.7% rate during the 1965–1969 period). Considering both underground and surface mining, in 1960 12.83 tons (11.55 metric tons) of coal were mined per person day; increasing to 17.52 tons (15.77 metric tons) in 1965; increasing to 19.90 tons (17.91 metric tons) in 1969; but then dropping off to some extent (18.84 tons (17.0 metric tons) in 1970) as the result of restrictions on productivity resulting from the Coal Mine Health and Safety Act of 1969. Productivity reductions at individual mines varying from 15 to 30% have been reported since enactment of the Act. Dust control provisions of the Act call for a target maximum of 2 milligrams respirable dust per cubic meter of air. Other provisions regarding roof control, ventilation, and other limitations of the mining cycle also are impacting on productivity.

Underground Roof Support. Of major importance in the safe and efficient operation of an underground coal mine is an excellent roof support system. Planning a roof support system normally involves an attempt to restore, insofar as possible, the stress equilibrium which the excavation of opening upsets. However, it is often expedient merely to delay the release of latent energy by using temporary supports. Roof failure can be one of four principal types: (1) *Falls*—consisting of the breaking away of rock from the mine roof. Their frequency and magnitude are highly variable and, to some extent, unpredictable. When unexpected falls occur in mines which are properly engineered and operated, they are usually attributable to some unforeseen structural defect in the roof itself. Falls may involve no more than a continual sloughing of small slabs of crumbly draw slate. On the other hand, it is possible for massive blocks of well indurated rock to be released. Massive falls may involve thousands of tons of material and extend into the main roof. (2) *Squeezing*—the elastic yielding of the mine roof which results from insufficient pillar strength. The magnitude of squeeze may range from a fraction of an inch to several feet. Excessive squeezing may result in general mine height reductions that are sufficient to render several sections unworkable. (3) *Bursts* (*Bumps*)—sudden explosion-like failures of coal or rocks. They are associated with the high compressive stresses which may occur in pillars or arched roof configurations. (4) *Intentional Caving*—during pillar recovery and in conjunction with longwall mining, it is imperative that the roof in the mined-out area be induced to fall. This relieves pressure on supports in the working area which otherwise would be overloaded and would fail. A pillar is a part of the coal layer that is left untouched to serve as a support.

For large mined areas having multiple openings, virtually the entire weight of overburden is supported by pillars. The so-called "room and pillar" method is the most common mining technique in the United States. The method works well, but has the drawback in that a significant portion of the coal remains in the mine. Generally where tectonic loading is negligible, pillar stress is essentially due to gravity and large horizontal stresses are not present. However, accurate prediction of stress distribution is not possible unless field measurements are made. In addition to pillars, of course, other forms of roof support are necessary.

Timber has long been an important factor in the construction of permanent and temporary roof support systems. It is relatively easy to handle and shape and is durable in the mine environment. Treated timber will provide about 10 years of service, whereas untreated materials may be expected to last about three years. Timber is further suited to roof control systems because the popping and cracking sound it emits as it begins to fail serves to warn of excessive ground pressures. Used as posts or cribs, timbers will withstand compression loads. As crossbars, timbers may provide the additional strength necessary to span an opening. Elaborate roof support structures have been built of timber, but these are not common in modern mining. However, specific timber applications include: (1) Providing additional support at important intersections and haulage ways; (2) providing temporary support in newly excavated working places; (3) providing temporary support and helping to induce the desired breakline during pillar recovery; (4) shoring up fall areas; (5) providing supplemental support where pillar dimensions have been reduced by sloughing, rolls, or by mining activity; and (6) as headers or half headers, used with roof bolts to hold loose top.

Rigid and yieldable steel arches are commonly used for roof support. Steel arch structures are generally applicable where (1) ground pressures are great; (2) where the top is extensively fractured; and (3) where ground movement (such as squeezing or heaving) is anticipated. Rigid steel arches include continuous rib configurations and rib and post configurations. The former are suitable for controlling vertical and lateral pressures, while the latter are most applicable where vertical loads only are prevalent. Common configurations of the yieldable arch include: (1) The 3-segment arch with leg segments toed in for control of vertical pressures; (2) the 3-segment arch with leg segments toed out for control of vertical and lateral pressure; and (3) the symmetrical ring (3 or more segments) for control of pressures from all directions. Some advantages of the yieldable system are: (a) Load capacities increase with the application of load; (b) joints yield at loads less than the yield load of the steel segments; (c) arches may be recovered and reshaped; and (d) they are almost maintenance free. Yieldable arches are generally used where ground pressures and movement would destroy rigid systems. The installation of yieldable structures requires more labor than that required for similar rigid support.

Masonry piers and abutments are normally used to provide permanent relatively strong support at critical locations. Grouting will seal and strengthen permanent openings. Grouting is particularly applicable to stabilizing loose or permeable rock. Lining with gunnite, concrete, or polymer resins also serves to seal and strengthen openings. For best results, lining should be accomplished as soon as possible after an opening is excavated. Lining material may be applied over reinforcing wire mesh and/or over bolted roof. Reinforced concrete lining is used to provide permanent heavy-duty support.

For the first half of the twentieth century, most mine roof was held in place by timber posts and crossbars. The theory was to wedge between the floor and roof carrying the load on the supports. In early mining, when hand tools predominated, the loading was minimized by narrow working places and arched entries. With the advent of mechanized mining in the 1930s, rooms were widened and percent recovery of coal increased. Timber crews of 4 to 6 men were required for each production section. In the 1940s, timbering machines mechanized timber installation, reducing most crews to two men, by mechanically lifting the bars and sawing the posts to length. By 1950, a new theory was promulgated and roof bolting became established. This concept was tested in southern Illinois mines—then further expanded and promoted by the U.S. Bureau of Mines.

Immediate results of the spread of roof bolting were seen in a dramatic decrease in roof-fall accidents. Production advantages were realized in faster and safer primary haulage due to the elimination of posts; greater recovery through wider working places; the mining of areas that would not have been supportable by posting; and improved ventilation.

Early roof bolts consisted of steel bars 1 inch (2.5 centimeters) in diameter, split longitudinally 6 inches (15 centimeters) to receive a steel wedge. The lower end of the bolt was threaded. Installation was effected by drilling a $1\frac{1}{4}$-inch (3.2-centimeter) hole to a predetermined depth, less than the length of the bolt. The wedge was loosely inserted in the split of the bolt and the assembly pushed into the hole with the wedge contacting the top. Through impact, the split was driven over the wedge with either side cutting into the wall of the hole, thus forming a funnel-shaped cavity and holding the bolt in place. A plate washer was then slipped over the threaded end extending from the hole, and a nut run over the threads, tensioning the bolt between the anchor point and the washer.

Both materials and tools for roof bolting rapidly improved. While the essential function of bolting is to maintain the structural integrity of the mine roof, bolts also serve to suspend loose roof material, such as draw slate, from more secure strata. In addition, holes drilled for bolt installation serve to relieve gas and hydrostatic pressure in the roof.

Underground Mining Safety and Health. Although steady technological progress has occurred for several years toward making underground coal mining a less hazardous occupation, impetus was given to safety and health programs in the United States by passage of the Federal Coal Mine Health and Safety Act of 1969. The Act embraces not only promulgation and enforcement of regulations, but technical support, worker education and training, and a broad-based research program as well. The Act has been described as the most extensive and comprehensive industrial health and safety program ever to be undertaken to promote the welfare of a single class of industrial worker. The objectives of government-funded research since 1972 essentially have been representative of the major problem areas and have included:

(*Ground Control*)—with the objective of developing technology to prevent accidental falls of roof, rib and face, and coal bumps. Study areas include: (a) Artificial support, (b) hazard detection, and (c) design of mine openings. Horizontal roof strain indicators for detection of unstable roof conditions have been tested. Other research programs have included a microseismic fracture warning system, polymeric roof bolts, and chemical impregnation techniques.

(*Fire and Explosion Prevention*)—study areas have included: (a) Ignition, (b) flame propagation, (c) fire detection and alarm, (d) suppression and extinguishment, and (e) methanometry. Devices and techniques tested have included explosion-proof bulkheads, coal dust and rock dust analyzers, ignition suppression devices for face equipment, and remote sealing techniques.

(*Industrial-type Hazards*)—with the objective of identifying hazard sources in electrical, mechanical, illumination, and non-emergency communication fields. Developments have included (a) advanced remote surveillance and communication systems, (b) portable-area illumination systems, (c) trolley-phone wireless systems, and (d) protective canopies for use on underground low-coal machines.

(*Methane Control*)—with the objective of developing safe methods for mining methane-laden coalbeds. Study areas have included: (a) Predictions of concentrations and flow; (b) control in advance of mining; and (c) control during mining. Techniques tested have included water infusion to reduce methane in the face area, degasification through vertical boreholes, the plugging of oil and gas wells which penetrate coal beds, and the complete degasification of operational mines.

(*Post-disaster Survival and Rescue*)—with the objective of developing emergency life support, communication, and rescue technology to improve the chances for a miner to survive a mine disaster. Specific examples of improved techniques include: (a) Use of survey probes at the Blacksville and Sunshine mine disasters in 1972; (b) the use of a wireless communication system for the rescue team at the Sunshine mine disaster in 1972; (c) the experimental employment of seismic methods for the location of trapped miners at the Nemacolin mine fire (1971) and the Blacksville mine fire (1972); and (d) the use of electromagnetic techniques for location of trapped miners.

(*Respirable Dust*)—with the objective or providing improvements for protecting miners from exposure to respirable coal mine dust. Study areas have included: (a) Dust formation; (b) dust control, and (c) dust measurement. Tests have included the water infusion of coal beds for control of respirable dust, the use of water-based, high-expansion foaming systems in conjunction with continuous mining machines to reduce dust at the face, the use of foam systems for dust suppression on conveyors and transfer points, and the use of prototype dust meters. See also **Pneumokonioses.**

(*Noise*)—with the assessment of permissible noise levels for communication and warning signals and the development of technology for noise abatement and control. Developments have included an audio dosimeter to replace conventional sound-level meters, discriminating earmuffs, and a noise control muffler system to reduce pneumatic drill noise.

(*Industrial Hygiene*)—with the objective of development of instrumentation for detecting and monitoring toxic gases in underground mines; and to determine the requirements for safe operation of diesel-powered equipment.

Major coal mining disasters in the United States are listed in Table 9.

Surface Mining. In general terms, surface mining (sometimes referred to as strip mining) involves removal of the overburden to expose the coal seam for subsequent loading. Surface mining can be divided into two broad classes: (1) contour mining and (2) area mining. Contour mining (in some regions called *collar mining*) is used in hilly areas where topography governs pit design. Where the terrain is steep, the recoverable reserves tend to lie in a narrow band adjacent to the coal outcrops. The coal pits are usually developed in the form of long, narrow strips, each of which follows a certain contour interval around the mountain or hill. Since the coal beds are nearly flat and the terrain is quite rough, in most cases only a few cuts can be made around the hills before the maximum economic stripping ratio is reached.

Area mining is used in flat or slightly rolling areas where the coal seams are relatively flat. The pit design is governed mainly by the equipment and the desired level of production. The pits are developed in a series of long, narrow strips. As the mining progresses, the overburden from each strip is cast back into the open pit of the previous strip. Thus, a series of parallel furrows are formed in much the same manner as a farmer plows a field. For this reason, area mining is sometimes referred to as furrow mining.

In the Appalachian region, both surface mining techniques are used. Area mining is used in Alabama, Ohio, Pennsylvania, and parts of West Virginia. Contour mining is practiced in parts of Alabama, Pennsylvania, Ohio, West Virginia, Maryland, Virginia, eastern Kentucky, and Tennessee. In the western states, surface mining is largely area mining. Auger mining of coal is frequently done in association with surface mining. In some mountain areas where surfaces lie on steep slopes, auger drills are used to remove the coal from the bank after the coal seam has been uncovered with one or two strip pits.

In 1970, about 44% of the bituminous coal and lignite produced in the United States came from strip mines. By 1972, this had increased to about 50%. The production capacity of strip mines generally ranges from 0.5 to 6.0 million tons per year, with an average output of about 2 million tons. However, western mines are scheduled to exceed 8 million tons per year. The key figure on which feasibility of surface mining depends is the stripping ratio, usually expressed in cubic yards of overburden to be removed to recover one ton of coal. The limiting ratio depends on the value of the coal; thus ratios of up to 30 to 1 have been profitably mined. Area averages are about as follows:

TABLE 9. MAJOR COAL MINE DISASTERS IN THE UNITED STATES

Year	Location	Deaths	Year	Location	Deaths
1869	Plymouth, Pa.	110	1924	Castle Gate, Utah	171
1884	Pocahontas, Va.	112	1924	Benwood, W. Va.	119
1891	Mammoth, Pa.	109	1928	Mather, Pa.	195
1892	Krebs, Okla.	100	1930	Millfield, Ohio	82
1900	Scofield, Utah	200	1932	Moweaqua, Ill.	54
1902	Coal Creek, Tenn.	184	1940	Bartley, W. Va.	91
1902	Johnstown, Pa	112	1940	Portage, Pa.	63
1903	Hanna, Wyo.	169	1943	Red Lodge, Mont.	74
1904	Cheswick, Pa.	179	1947	Centralia, Ill.	111
1905	Virginia City, Ala.	112	1951	West Frankfort, Ill.	119
1907	Monongah, W. Va.	361	1957	Bishop, Va.	37
1907	Jacobs Creek, Pa.	239	1958	Bishop, Va.	22
1908	Marianna, Pa.	154	1961	Terre Haute, Ind.	22
1909	Cherry, Ill.	259	1962	Carmichaels, Pa.	37
1911	Littleton, Ala.	128	1963	Dola, W. Va.	22
1913	Dawson, N. Mex.	263	1965	Redstone, Colo.	9
1914	Eccles, W. Va.	181	1966	Mt. Hope, W. Va.	7
1915	Layland, W. Va.	112	1968	Greenville, Ky.	9
1917	Hastings, Colo.	121	1968	Farmington, W. Va.	78
1922	Spangler, Pa.	77	1970	Hyden, Ky.	38
1923	Dawson, N. Mex.	120	**1976**	**Oven Fork, Ky.**	**26**

SOURCE: U.S. Bureau of Mines.

Kentucky	11:1
Illinois-Indiana	18:1
Ohio	15:1
Western states (overall)	6:1

Whereas 14 tons of coal (or less) per day can be produced per person in underground mining, productivity increases to about 35 tons per day for auger mining, and up to 36 tons or more per person in surface mining.

Improved Surface Mining Methods. Improvements in surface mining methods have corrected many of the problems of former traditional methods. Consolidation of extraction, backfilling, and reclamation operations into a continuous, integrated mining system has been developed during the past decade or so. With increased regulation on restoration of mining properties, mining operators have found that handling all these operations together, and thus avoiding secondary handling of the overburden, has reduced costs.

Earthmoving requirements for the newer methods include burying all acidic and toxic materials, separating and replacing topsoil and restructuring the land approximately to its original contour.

Some of the most promising of the new contour mining methods which are designed to meet all environmental regulations include (1) the *haulback*, (2) the *valley fill*, and (3) the *mountaintop-leveling methods.* Although they are significantly different in many aspects, all these methods move material laterally along the bench rather than dumping it over the outslope; and all are designed to reduce or eliminate second-handling of overburden and waste material. For comparison, the conventional method of contour mining is shown in Fig. 4.

The major operating principle of the haulback method is that all spoil except that from the initial cut is moved laterally along the bench rather than being placed on the outslope. There are two basic options with the haulback method: (1) using *scrapers* (see Fig. 5); or (2) using *trucks* (see Fig. 6). No matter which of these methods is used, nonacidic rock or clay overburden is placed over the acidic material so that trees, legumes, and other vegetation will survive when the miner completes the reclamation phase. See also **Revegetation.**

Overburden characteristics and the terrain must both be favorable to permit a scraper haulback operation. The scrapers load overburden from the advancing face of the pit and dump it at the backfill area. Ideally, scrapers should load and dump downhill for faster cycle times and less wear on the machines. When they reach acidic material overlying the coal, the scrapers load it out and spread it near the bottom of the pit so it can then be covered with nonacidic material. As the overburden dump reaches the original level of the highwall, crawlers dress the slope and restore the area to its natural contour.

The truck haulback method utilizes crawler tractors, wheel loaders, and off-highway trucks. Crawlers doze overburden off a top bench down to the loader, which dumps it into the off-highway trucks. As

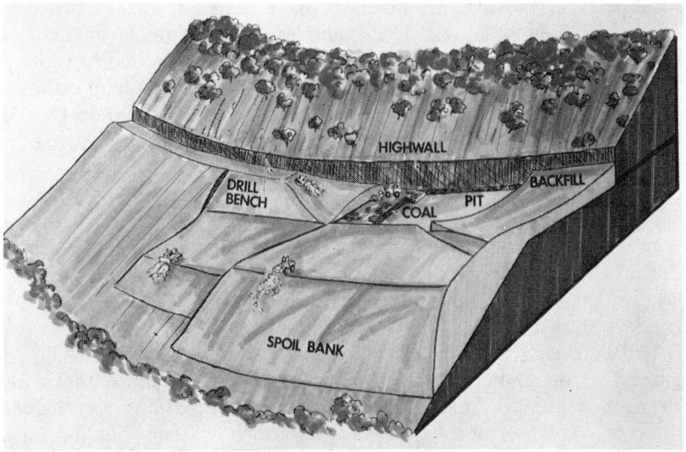

Fig. 4. Conventional method of contour surface mining. It was the predominant form of mining throughout Appalachia until the last few years when more stringent reclamation laws ushered in new integrated mining techniques. (*Caterpillar Tractor Co.*)

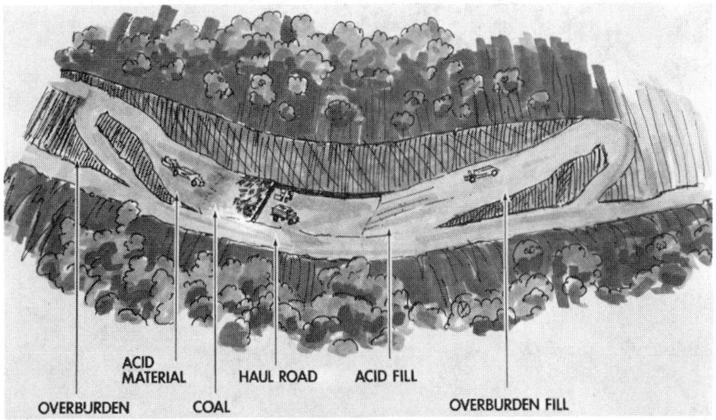

Fig. 5. One of the most promising of the new methods is the haulback. Using either scrapers or off-highway trucks, its major principle is that all spoil except that from the initial cut is moved along the bench rather than being placed on the outslope. (*Caterpillar Tractor Co.*)

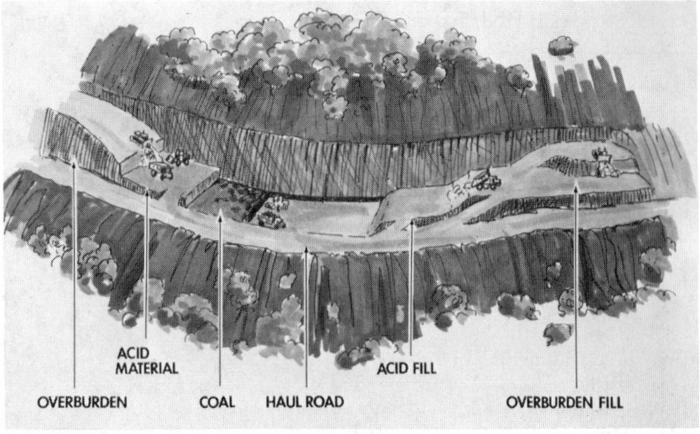

Fig. 6. The truck haulback method is used primarily in rocky overburden, which is uneconomical for loading by scrapers. With either truck or scraper haulback operations, nonacidic rock or clay overburden is placed over acidic material to facilitate revegetation. (*Caterpillar Tractor Co.*)

with scrapers, acidic material is dumped at the base of the backfill area so that it can be covered with nonacidic material. Another crawler also works the overburden area, dressing and contouring the reclaimed area.

Some miners combine scrapers and off-highway trucks in the same haulback operation. When this is done, scrapers remove the easy-to-load surface material, while trucks come in to haul away the underlying rock.

The valley fill method (also called the "head of the hollow" method) is relatively new to the Appalachian region. With it, the miner generally uses an equipment spread similar to that in a truck haulback operation. A crawler dozes overburden to the wheel loader which loads it into off-highway trucks as shown in Fig. 7. The trucks haul the overburden along the bench to a selected hollow or valley and dump it over the side, building a waste dump at bench height. Scrapers can also be used to haul the overburden in a valley fill operation. When such is the case, the scraper operator ejects overburden at bench height, and tractors doze it over the edge. Valley fill is frequently combined with the haulback method, since there is normally extra material from the initial cut as well as continuous excess from material swell which cannot be stacked directly on the bench.

The mountaintop-leveling method is quite similar to area stripping as practiced in the midwestern states. As shown in Fig. 8, a part of the mountaintop is leveled to fill in an adjacent valley. This method can be used only where topography, coal seam position, and other physical and economical factors permit. It uses similar techniques and equipment as the haulback and valley fill methods. Once the coal is loaded out, a level to gently rolling terrain results. The reclaimed

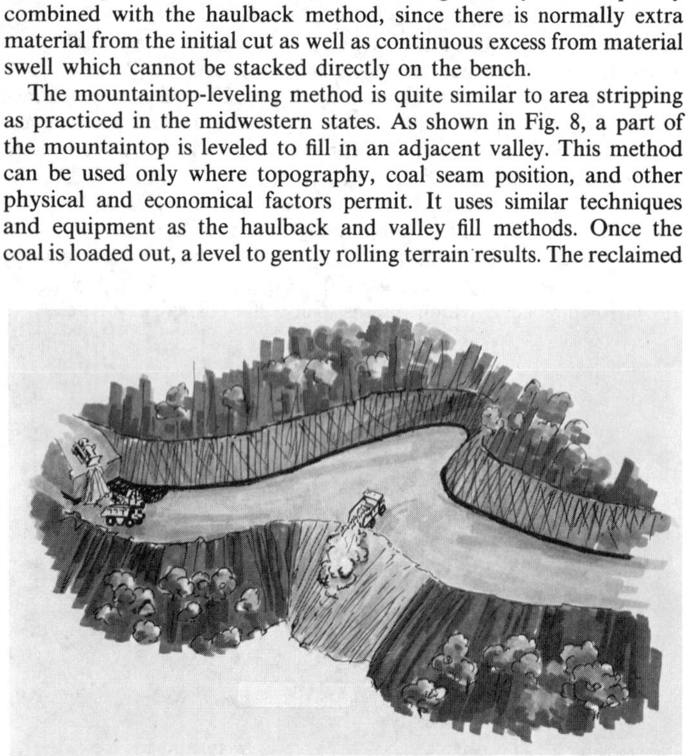

Fig. 7. In the valley fill method, the miner generally hauls overburden in trucks and constructs fills over the side at bench height as shown. When scrapers are used, the operator ejects overburden at bench height and tractors doze it over the edge. (*Caterpillar Tractor Co.*)

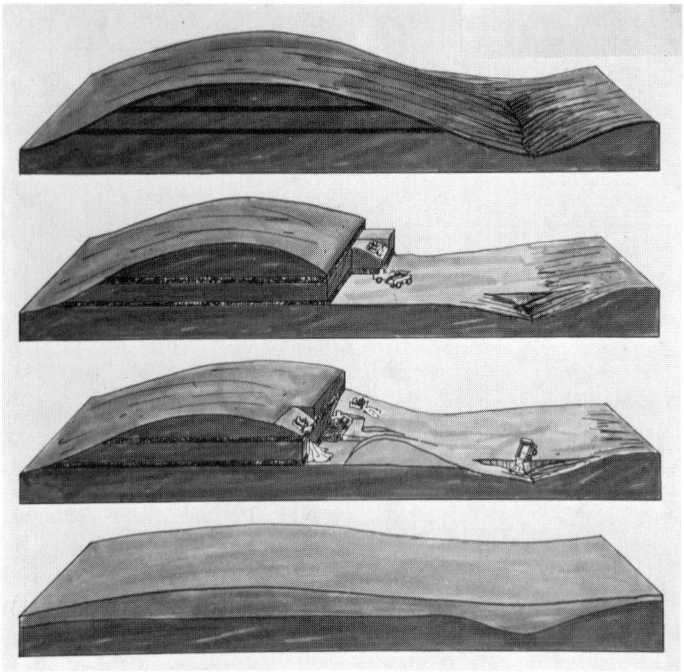

Fig. 8. In mountaintop leveling, a part of the mountaintop is moved to fill an adjacent valley and build a near-level fill in the mined area. The reclaimed flat or gently rolling land that results from this type of operation can then be used for a variety of purposes which could increase the value of the land. (*Caterpillar Tractor Co.*)

flat land, which can be more valuable than in its original contour, can then be used for grazing as well as for institutional, residential, and industrial developments.

Overburden Handling. Miners who convert their operations to some of these newer methods are making greater use of wheel loaders in load-and-carry operations and are also using two machines seldom seen previously in contour mining, i.e., wheel tractor-scrapers and off-highway trucks. Whether the miner chooses loaders, scrapers, or trucks as the prime overburden hauling unit depends upon a number of factors.

Introduction of the four-to-six-yard loaders in the early 1960s brought about the first widespread use of these units in contour mining. In the early 1980s, with the exception of crawler tractors, they are probably the most common machine in Appalachian surface mining.

Wheel loaders working a load-and-carry stripping operation with short cycle times have the obvious advantage of being the most economical of the three types of hauling units, since one machine does it all. See Fig. 9. The loader is highly mobile, capable of negotiating most grades; and it can handle a variety of materials from topsoil to shot rock. Loaders do, however, have some drawbacks in many contour mining situations. For example, they are limited to the ton-mile-per-hour rating of their tires, and they have a limited top speed. The chief limiting factor for loaders is that the one-way haul distance must be kept relatively short. This distance generally should be 700 feet or less. Over that distance, scrapers or trucks become more economical and practical for hauling overburden.

Overburden composition is the key factor in determining whether the stripping fleet consists of scrapers or of loaders with off-highway trucks. Generally scrapers are the most economical in dirt; trucks in rock. Scrapers can be divided into four basic configurations—single-engine, tandem powered, elevating, and push-pull. An off-highway truck is shown in Fig. 10; a scraper in Fig. 11.

Off-highway trucks range in capacity from about 15 tons up to giant 200-ton units. Among the key factors in selecting trucks for the newer contour mining methods, space limitations and loader matching are the most obvious. Overburden loading, hauling, and dumping take place simultaneously in many of these methods, and the small operating area rules out the use of large trucks.

Characteristics of Mining Industry. Less than 6–8% of all coal mining operators in the United States produce over 60% of the coal. About

Fig. 9. A 10-cubic-yard (7.6 cubic-meter) wheel loader working a load-and-carry operation in an Alabama surface mine. (*Caterpillar Tractor Co.*)

70% of the mining operators produce less than 10% of total coal mined. There are somewhat over 300 mines with a capacity of over 500,000 tons (450,000 metric tons) per year; about 250 mines with a capacity between 200,000 and 500,000 (180,000–450,000 metric tons) per year; slightly over 400 mines with a capacity between 100,000 and 200,000 tons (90,000–180,000 metric tons) per year; approximately 600 mines with a capacity between 50,000 and 100,000 tons (45,000–90,000 metric tons) per year; and just over 4,000 mines with a capacity under 50,000 tons (45,000 metric tons) per year. The industry trend over the past few decades has been toward increasing the number of days worked per year as well as the size of individual mines.

Fig. 10. A 50-ton (45-metric-ton) capacity off-highway truck for use in hauling overburden in a Virginia mine. (*Caterpillar Tractor Co.*)

Fig. 11. Scraper hauls overburden in Ohio surface mine. (*Caterpillar Tractor Co.*)

Coal Preparation Plants

The primary function of a coal preparation plant is to deliver a finished product which has been custom-prepared in terms of quantity and quality. Regardless of whether the coal is metallurgical quality or a power plant fuel, the coal will be scrutinized by the consumer in terms of the raw coal size, analysis, washability data, mining techniques, and the efficiency of the various types of coal washing equipment.

In 1942, 24% of the total coal production was cleaned mechanically, as compared with 65% of the production in 1965, and the trend continues into the 1980s. The amount of refuse discarded during coal preparation for the same period increased from 13 to 21%. Power plants and other consumers are using furnace and transport systems that demand very uniform feedstocks. Size, heat content, and handling characteristics are extremely important. The design and construction of the coal preparation plant require prior knowledge of the physical and chemical properties of the coal which characterize its response to crushing, screening, blending, and cleaning operations.

At the Amax Coal Company's Leahy Mine, run-of-mine coal is delivered to a 600-ton (540-metric ton) capacity truck hopper, from which it is fed to two 60-inch (1.5-meter) reciprocating feeders to a 60-inch (1.5-meter) belt that delivers the coal at a rate of 1800 tons (1620 metric tons) per hour to a rotary breaker station. See Fig. 12. A bar screen removes some of the natural 5-inch (12.5-centimeter) by 0 pieces (i.e., pieces ranging from 5 inches downward). The oversize fraction is fed to a rotary breaker, which has a diameter of 10 feet (3 meters) and a length of 24 feet (7.2 meters). The breaker is equipped with screen plates with 5.5-inch (14-centimeter) diameter holes. The rock is removed and chuted to a 100-ton (90-metric-ton) capacity bin equipped with a motorized gate. The rock is ultimately disposed of in a strip pit area.

The less than 5-inch (12.5-centimeter) fractions from the bar screen and from the rotary breaker are combined on a 48-inch (1.2-meter) belt and discharged to an 8000-ton (7200-metric ton) raw coal storage silo. The silo has a diameter of 70 feet (21 meters), walls that are 11.25 inches (28.5 centimeters) thick, and a height of 117 feet (~35 meters). Six 36-inch (1-meter) reciprocating feeders feed the 5-inch (12.5-centimeter) by 0 raw coal from the silo to a 48-inch (1.2-meter) belt that conveys the coal to the washery at a rate of 1500 tons (1350 metric tons) per hour. A belt scale provides a record of the raw coal input.

The 5-inch (12.5-centimeter) by 0 raw coal is processed in two 96-inch (2.4-meter) wide Mogul washboxes which produce clean coal, middlings, and refuse.

The refuse is dewatered on a single deck screen, which is 6 by 16

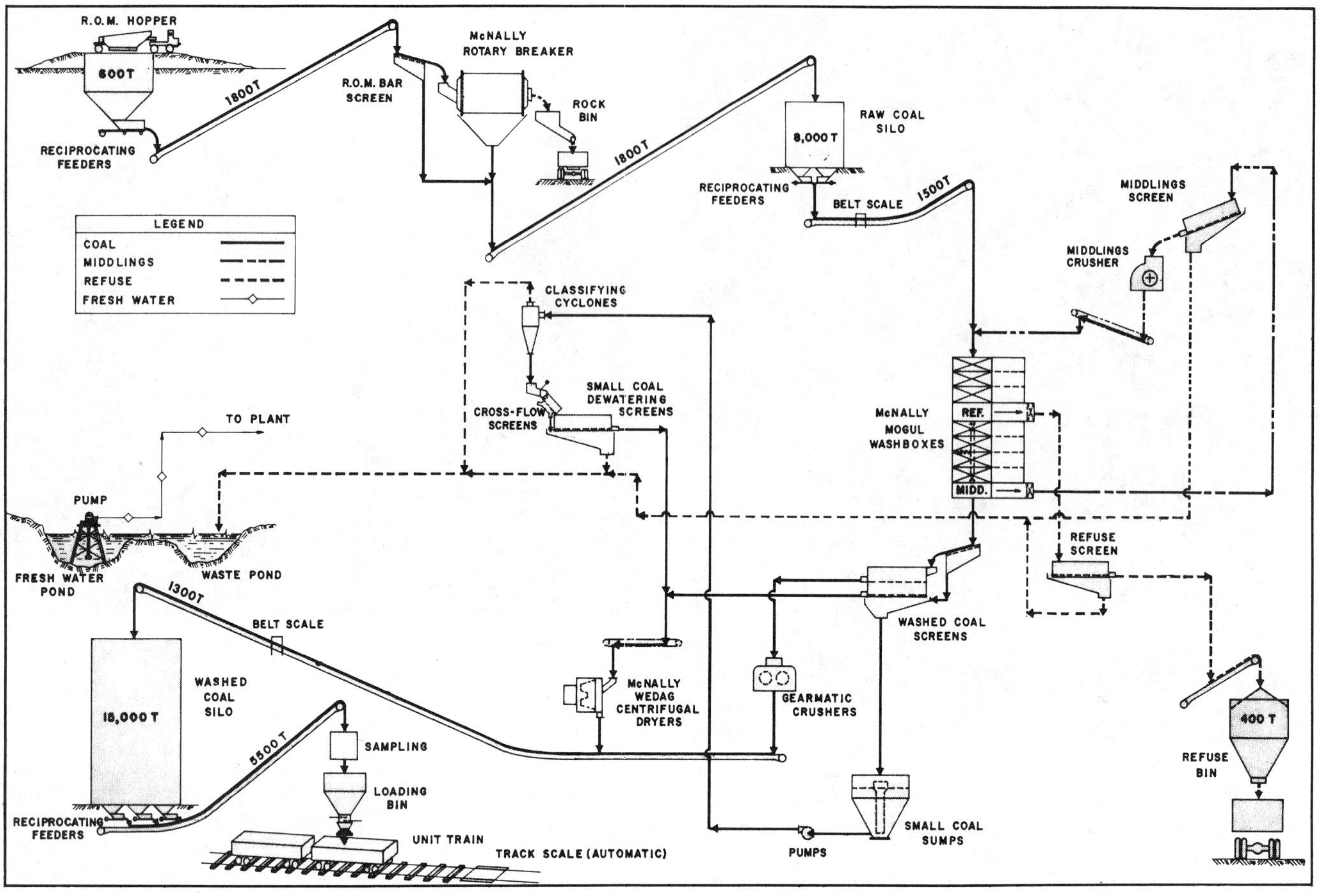

Fig. 12. Coal preparation plant at the Leahy Mine of Amax Coal Company. (*McNally Pittsburg Manufacturing Corp.*)

feet (1.8 by 4.8 meters) in dimensions and has $\frac{3}{16}$-inch (~5-millimeter) openings in the screening surface. The refuse is then discharged via a 30-inch (0.75-meter) belt to a 400-ton (360-metric ton) capacity refuse bin, from which it is removed to the strip pit area.

The middlings are dewatered on a single deck screen, which is 5 by 10 feet (1.5 by 3 meters) in dimensions and has 0.5-inch (~13-millimeter) openings in the screening surface. The 5-inch (12.5-centimeter) by 0.5-inch (~13 millimeter) screen fraction is crushed to less than 1-inch (2.5-centimeter) size, then recycled through the washboxes. The 0.5-inch (~13 millimeter) by 0 screen fraction is pumped to a waste pond where the solids settle out and the clarified water overflows to the plant fresh water supply pond.

The clean coal is discharged from the washboxes over fixed screens with $\frac{3}{16}$-inch (~5-millimeter) openings. The coal then passes to four double-deck washed coal screens. Two screens are 8 by 15 feet (2.4 by 4.5 meters) and two are 6 by 16 feet (1.8 by 4.8 meters), with 1.25-inch (3.2-centimeter) screen cloth on the top deck and $\frac{3}{8}$-inch (~1-centimeter) screen cloth on the bottom deck.

The 5-inch by 1.25-inch (12.5-centimeter by 3.2-centimeter) screen fraction is reduced by two 36 by 60-inch (0.9 by 1.5-meter) crushers to 1.5-inch (3.8-centimeter) by 0 size. The fraction then is discharged to the 48-inch (1.2-meter) belt that delivers it to a 15,000-ton (13,500-metric ton) capacity washed coal silo, which is 70 feet (21 meters) in diameter and 192 feet (~57.6 meters) in height.

The $\frac{3}{8}$-inch (~1-centimeter) by 0 fraction from the washed-coal screens is sent to two small coal sumps, from which it is pumped to eight 24-inch (0.6-meter) classifying cyclones. The cyclone underflow is discharged to four crossflow screens, after which it passes over four single-deck dewatering screens equipped with $\frac{3}{4}$-millimeter openings. Three of the screens are 6 by 16 feet (1.8 by 4.8 meters); and one is 8 by 16 feet (2.4 by 4.8 meters).

The $\frac{3}{8}$-inch (~1-centimeter) by $\frac{3}{4}$ millimeter overproduct from the dewatering screens is discharged to the distributing conveyors that feed the centrifuges. The product from these centrifuges is discharged to the washed-coal silo. A bleed of solids from the cyclone overflow, the underflow of the cross-flow screens, and the underflow of the small-coal dewatering screens are pumped to the waste pond.

In an earlier era when there was extensive use of coal for home heating and hand-fed furnaces, the coal plant operator attempted to recover as much lump as possible. In modern practice, high-quality market requirements dictate crushing to liberate the extraneous particles entrapped within the coal. Crushing allows the washing equipment to remove the extraneous material and to produce a high-quality coal. Selection of a crusher is very important because the crusher should produce a uniform top size with a minimum amount of extreme fines.

The grouping of coal particles within a range of particle sizes is one of the most important beneficiation operations performed within the coal preparation plant. Sizing is generally accomplished by passing the coal over screens which may consist of parallel bars (grizzly screens), punched steel plates with a variety of opening sizes and shapes, and woven wire cloth with square or rectangular openings. The screens may be stationary or moving. Shaking or vibrating screens assist in bringing the particles to escape-openings. Separation by differential settling in air or water currents, as well as the use of centrifugal force, also have been adopted for the separation of very fine sizes.

Since most coal impurities have specific gravities greater than coal, the density of a coal particle is a direct measure of its purity. The differences in this physical parameter is the basis for mechanical separation of coal from refuse. Both gravity and centrifugal force devices are used and these may employ air or liquids as washing media. There are relatively few instances where air washing will work because of the moisture required to be on the coal to meet mining regulations. The few cases where the raw coal is dry enough for air washing require a complete dust collection system to meet air pollution standards. Thus, few plants utilize air washing. There may be a trend to air washing in the western coal fields because of the scarcity of water.

Washing processes fall into three classes: (1) Hydraulic separation; (2) dense medium separation; and (3) centrifugal (cyclone) separation.

Hydraulic separation depends on a process called jigging, which creates a particle stratification from an alternate expansion and compaction of a bed of particles by a pulsating fluid flow. As originally developed, a basket filled with material was moved up and down in a tank filled with water. The more modern Baum jig process utilizes an air impulse concept in which the water is moved by air pressure from an adjacent sealed chamber. There are several refinements of the process, including the McNally Norton standard washer.

More accurate separations are made in dense medium vessels. Coal is slurried in a medium with a specific gravity close to that at which the separation is to be made. The lighter coal tends to float and the refuse to sink. The two fractions then can be mechanically separated. Theoretically, any size particle can be treated by the dense medium process. Practically, the sizes range from about 0.5 millimeter to about 6 inches (15 centimeters). Organic liquids, salt solutions, aerated solids and water suspensions have found use as commercial media. Water suspensions meet most practical requirements and are the least costly. The bulk of coal mechanically cleaned by the dense medium process is separated in suspensions of magnetite in water.

The use of centrifugal force as an aid in coal-refuse separation is a relatively recent addition to the coal-cleaning process. As originally developed, the device employed a dense working medium. The latest units do not employ an artificial gravity suspension and are known as hydrocyclones. Design of the unit allows the formation of a hindered settling bed as the dense particles move down the side wall under the force of gravity. Less dense particles are unable to penetrate this heavy bed and move back into the main hydraulic current and are discharged out the top of the unit.

Tables are also used to wash coal. The reciprocating action of a table stratifies the high-gravity coal particles on the bottom and the low-gravity particles rise to the upper level of the bed. As the low-gravity particles rise, they are moved across riffles which separate the high- and low-gravity material by the water flowing to the low side of the table deck. The refuse is trapped in the riffle troughs and the motion of the deck moves the refuse to discharge off the end.

For separating fine coal particles from refuse particles, flotation is often used. Finely disseminated air bubbles are passed through a coal slurry. The fine coal particles adhere to the air bubbles and rise to the top where they are removed as a concentrate while the heavy refuse particles sink and are removed by the flow of water through the flotation cell. A frothing reagent, such as methylisobutylcarbinol, is added to the feed. See also **Classifying (Process).**

Water remaining on marketable coal is a contaminant as serious as the undesirable ash. It may cause problems in handling and shipping, increase freight cost, and reduce heating value per unit weight. The difficulty of dewatering increases with increases in the surface area of the material. Several processes are used, depending upon the particle size of the coal. Vibrating-screen type centrifuges may be used. For the removal of very fine material (28-mesh and smaller), a filter process may be required. Both disk-type and drum-type filters are used. Where filters are required, filtration is usually preceded by sedimentation. Chemical flocculants sometimes are used to assist the settling process. A final reduction of moisture content frequently is accomplished by thermal drying. The use of fluidized bed coal dryers is increasing.

Transportation of Coal

Railroads handle approximately 80% of the coal mined in the United States during some part of its journey from mine to point of consumption or shipment overseas. Most American railroads offer four distinct types of service for the transportation of coal: (1) Single carload; (2) multiple carload; (3) trainload volume; and (4) unit train. Each type of freight service has its own operating characteristics which result in a distinct level of operating cost and freight rates. The first two methods are self-evident. Trainload volume is the tendering of a sufficient number of carloads of freight on one day from one origin to one destination to permit the carrier to handle the movement in a special train. The number of carloads required to form a trainload will vary from one railroad to the next. Trainload volume movements are an irregular movement on an irregular schedule. The basic operation requires simplified switching and terminal operations, resulting in economies in rail operation. Trainload volume trains use rail cars assigned to a car pool. The trains are governed by tariff provisions requiring a limited control over the loading and unloading of the railroad equipment and occasionally requiring a minimum annual volume.

A unit train movement is an integral movement of coal moving from a single origin to a single destination on a regularly scheduled train, avoiding all terminals and switching operations. Unit trains utilize specialized loading and unloading facilities and specialized railroad equipment assigned to dedicated service. The unit train movement is governed by tariff provisions requiring both controlled loading and unloading of the railroad equipment and a minimum annual tonnage. The loading of a unit train at the York Canyon Mine (New Mexico) is shown in Fig. 13. When large volumes of coal are to be loaded within a short time, some form of flood-loading is required, permitting the coal to free fall into the cars. The four basic types of flood-loading are: (1) Ground storage with loading tunnel; (2) ground storage with loading bin; (3) silo storage; and (4) silo storage with loading bin. The first type is used in the system shown in Fig. 13. A unit train being loaded from a storage silo is shown in Fig. 14.

The unloading of coal cars can be accomplished by several methods. A unit train with conventional hopper cars would be unloaded with the cars being spotted over the storage area. The gates on the cars would be opened by laborers stationed alongside the rail cars. After the first cars over the unloading area were discharged, the train would move forward until the next loaded car is moved into place. This spotting, unloading, and spotting sequence must be repeated perhaps a hundred times until the entire train is unloaded.

A unit train with quick-opening bottom-drop cars would be unloaded by having the gates on the rail cars opened by either a mechanical tripping mechanism or by an electrical device as the cars roll over the pit or tressel. After the cars are unloaded, the gates on the cars would be closed by a similar mechanical or electrical mechanism. Motion unloading systems, although costly, represent many advantages. Proceeding across the pit area at 4 to 5 miles (6.4 to 8 kilometers) per hour, a 100-car, 10,000-ton (9000-metric ton) unit train can be unloaded in 15 minutes. Considering startup time, the total unloading time may approximate one hour. The same 100-car unit train, in an efficient two-car rotary dump facility, will require from 4 to 5 hours to unload the train. If a single-car rotary dump facility were used, the unloading time will range from 8 to 12 hours. A motion unloading facility is illustrated in Fig. 15.

Coal Slurry Pipelines. The first patent covering the pumping of coal and water dates back to 1891. In 1914, the first commercial transport of coal in water was carried out in England, when a short 8-inch (20-centimeter) pipeline was used to carry coal from river barges to a power plant. Thereafter, several proposals were submitted for the long distance transport of coal from mine to market in the eastern United States, but failed to materialize for several reasons, not the least of which were technical problems. Intensive research into slurry transport was continued and, by 1957, technology and engineering had advanced to the point where the first long distance transportation of coal in water was feasible. The result was construction and operation of the Consolidation Coal Pipeline, 10 inches (~25 centimeters) in diameter and 108 miles (174 kilometers) in length, transporting 1.25 million tons (~1.1 million metric tons) of coal per year from Cadiz, Ohio, to an electrical generating station 20 miles (32 kilometers) east of Cleveland on the shores of Lake Erie. The pipeline was powered by three pump stations, spaced about 35 miles (56 kilometers) apart, where discharge pressures reached 1,000 psi (6.9 mPa). Coal with a graded size consist, 8 mesh by 0, and a concentration of 50% solids was transported. Consist means the size makeup of the solid phase of the coal slurry. The term 8 mesh by 0 indicates coal with a graded size makeup in the range of 8 mesh and zero (dust).

Although the Ohio line operated successfully, transporting 7 million tons (6.3 million metric tons) of coal, some unexpected operating problems had to be resolved. Much investigation was conducted with variables, such as size consist and slurry concentration and the resultant effect on slurry stability. After 7 years of operation, the line was shut down in 1963, when the unit train concept resulted in much lower freight rates on significantly higher tonnages of coal movement.

Economics vary in different locations, however, and slurry pipelines can be particularly attractive where no railroad facilities exist. Thus, throughout the world today there are about ten operating coal slurry pipelines. The largest in the United States is the Black Mesa Pipeline,

Fig. 13. Unit train (*Santa Fe*) being loaded at the York Canyon Mine (New Mexico). (*McNally Pittsburg Manufacturing Corp.*)

Fig. 14. Unit train is loaded with coal as it passes through base of storage silo. (*CF & I Steel Corp.*)

Fig. 15. A unit train with motion unloading of hopper cars being unloaded at Tennessee Valley Authority steam plant at Bull Run, Tennessee. (*Tennessee Valley Authority.*)

273 miles (440 kilometers) long, mostly 18-inch (~46 centimeters) diameter, but with some 12-inch (7.6-centimeters) diameter sections. The pipeline transports 5 million tons (4.5 million metric tons) per year of bituminous coal to a 1580 MW power generating station. The line originates at an altitude of 6,500 feet (1981 meters) in the northeastern part of Arizona in the heart of the Black Mesa coal fields, transports pulverized coal in a water carrier fluid across some of the most remote and rugged country of the western United States, and terminates at the Mohave Generating Station at an elevation of 700 feet (213 meters) in the southern tip of Nevada. Operating with a reliability factor of 99%, the pipeline has demonstrated the feasibility of commercial long distance transport of solids by pipeline.

A 38-mile (61-kilometer) (12-inch (30.4-centimeter) diameter) pipeline transports coal slurry from the Novovolynskaya Mine in the U.S.S.R. This line was installed in 1957. There is a 126-mile (203-kilometer) operating (10-inch (25.4-centimeter) diameter) line in Poland operated by the Polish Central Mining Industry. There is a 51-mile (82-kilometer) (15-inch; 38-centimeter diameter) line operating in France, installed in 1952, with a pumping capacity of 250 tons (225 metric tons) per hour. A 490-mile (788-kilometer) pipeline is proposed to connect East Kootenay, British Columbia, to Vancouver, British Columbia.

The feasibility of slurry transportation depends upon the resolution of a number of variables, the most important of which from a hydraulic standpoint are: (1) Size consist; (2) velocity; and (3) concentration. The selection of a proper size consist (gradation) is important in order that homogeneous flow can be achieved at prudent operating velocities. For coal slurry, such a consist is on the order of 8 mesh by 0 (approximately 0.1-inch (2.5-millimeter) particle size to dust). Homogeneous flow (solids evenly distributed across the pipe diameter) is important if excessive wear in the bottom of the pipe is to be avoided and stable operation achieved.

Of equal importance and directly related to size consist is the proper selection of velocities for transport. The velocity cannot be excessive so as to cause abrasion of pipe wall and inordinately high pressure drops. Conversely, the velocity should not be so low as to cause heterogeneous flow, with resultant excessive wear in the pipe bottom or bed formation which will cause unstable operation. Generally, practical operating velocities will be in the range of 4 to 7 feet per second. Finally, the important parameter, slurry concentration, must be considered. The relationship between concentration and viscosity for any given slurry can be determined in laboratory bench testing. Although it varies for different slurries, all systems generally demonstrate a point in inflection where a small increase in concentration causes a large viscosity increase. Hence, it is important to maintain a concentration range below the inflection point in order to provide good operation without excessive velocities. For coal, the practical concentration range appears to be from 45 to 55% solids.

Prospects for further coal movements by slurry pipelines are good. The United States has very large coal reserves, mostly in the West. The amounts are sufficient for several hundred years under reasonable usage. Western coals are low in sulfur and are less costly than most other U.S. coals to mine. Movement to major midwestern markets (800 to 1,500 miles (1287 to 2414 kilometers) distant) by slurry pipeline is economically attractive. A slurry pipeline will be attractive (1) when there is a large volume of coal (generally over 1 million tons per year) to move for a long term between stable locations; (2) when the mine or mines are in remote areas not served by existing rail lines; and (3) when the fine grind necessary for slurry transport does not conflict with end-usage. High-volume slurry pipelines may be required eventually to avoid overloading the rail lines. Once installed, a coal pipeline is less severely affected by escalation of direct costs. This is because this mode of transportation is capital intensive. About 70% of the tariff is capital related; about 15% is power related; with only 15% attributable to labor and maintenance. Coal slurry pipelines are environmentally superior to other forms of coal transportation because they are quiet and unobtrusive and because energy consumed by a coal pipeline on a ton-mile basis is significantly lower than by any other transportation mode.

The two prime disadvantages facing coal slurry pipelines are: (1) An adequate and assured water supply is required. In water-short areas of the western United States, this is a major consideration; (2) dewatering slurry for consumption at a power plant, or for transshipment by barge is required. Centrifuging is the primary method used to date. While reduction of coal particles to the very small size needed for movement by pipeline serves the requirements of a generating station for a finely-ground coal, the fine size makes dewatering difficult.

Testing of Coal

Proximate Analysis. This includes the determination of total moisture, volatile matter, and ash; and the calculation of fixed carbon for coals and cokes. The term "Proximate" should not be confused with the word "approximate," since all Proximate Analysis tests are performed according to rigid specifications and tolerances. Proximate Analysis results may be used to establish the rank of coals; to show the ratio of combustible to incombustible constituents, to provide the basis for buying and selling coal, and to evaluate for beneficiation, or other purposes.

Moisture in coal takes three forms: (1) free or adherent moisture, essentially surface water; (2) physically bound or inherent moisture (that moisture held by vapor pressure and other physical processes); and (3) chemically bound water (water of hydration or "combined" water). The ASTM defines total moisture as a loss in weight in an air atmosphere under rigidly controlled conditions of temperature, time, and air flow. *Total moisture* represents a measurement of all water not chemically combined. Total moisture is determined by a two-step procedure, involving air-drying for removal of surface moisture from the gross sample, division and reduction of the gross sample, and determination of residual moisture in the prepared sample. An algebraic calculation is used to obtain the total moisture value.

Ash is the noncombustible mineral matter left behind when coal is burned under rigidly controlled conditions of temperature, time, and atmosphere.

Total nitrogen is determined by chemical digestion (Kjeldahl-Gunning) methods.

Oxygen content is determined by calculations, subtracting total carbon, hydrogen, sulfur, nitrogen, and ash from 100%.

Chlorine is commonly included as part of the ultimate analysis.

Other important chemical and physical tests performed to characterize coal include: (1) Heating value (Btu content); (2) sulfur forms; (3) ash fusibility temperatures; (4) ash analysis; (5) trace elements; (6) free swelling index; and (7) Hardgrove grindability.

Heating value is determined by burning a coal sample in an oxygen bomb and measuring the temperature rise. See also **Calorimetry.**

Three *sulfur forms* recognized by ASTM are: (1) Sulfate sulfur, which may be in the form of calcium or iron sulfate; (2) pyritic sulfur, which is sulfur combined with iron in the form of minerals pyrite and/or marcasite; and (3) organic sulfur, which is bonded to the carbon structure. Sulfate sulfur is extracted from the coal with dilute hydrochloric acid, precipitated as barium sulfate, ignited and weighed. After sulfate removal, pyritic sulfur is extracted with nitric acid and the extracted iron is measured by redox titration. Organic sulfur is calculated by subtracting sulfate and pyritic sulfur from total sulfur.

Ash fusibility can be defined broadly as the melting temperature of the ash. Small triangular pyramides (cones) prepared from coal or coke ash pass through certain defined stages of fusing, and flow when heated at a specified rate.

Ash analysis is the term used to designate analysis of the major elements commonly found in coal and coke ash. The elements, expressed as oxides, are SiO_2, Al_2O_3, Fe_2O_3, TiO_2, CaO, MgO, Na_2O, K_2O, P_2O_5, and SO_3. Phosphorus, a trace element, has been included historically because of its importance in the subsequent steel making processes where coke is used.

Interest in *trace element analysis* has increased by environmental concerns.

Volatile matter is defined as the gaseous products, exclusive of moisture vapor, driven off during standardized test conditions. The combustible gases are carbon monoxide, hydrogen, methane, and other organic hydrocarbons. Those generally classified as noncombustible are carbon dioxide, ammonia, hydrogen sulfide, and some chlorides. Volatile matter tests are used to establish the rank of coals, to indicate coke yield upon carbonization, and to establish burning characteristics.

Fixed carbon is the solid residue, other than ash, resulting from the volatile matter test. The value is calculated by subtracting moisture, volatile matter, and ash from 100%.

Total carbon is determined by catalytic burning of the sample in oxygen to form carbon dioxide which can be readily measured.

Total hydrogen also is determined by catalytic burning of a sample in oxygen to form water. The water is absorbed in a desiccant and weighed directly. Hydrogen results as determined include the hydrogen present in both the sample moisture and water of hydration.

Importance of Testing. As previously mentioned, heating value, ash, fusibility, and other parameters are extremely important to the ultimate consumer. Advance coal testing is also very important to the layout and operation of coal preparation plants. A new dimension to coal testing has been added in recent years as various coals are considered for gasification and liquefaction processes. As will be obvious from several of the processes to be described shortly, the physical as well as the chemical properties of coal need vary but a very small amount to foul up some of the gasification and liquefaction processes. Tolerances are narrow; consistency of raw material is extremely important; very tight control over raw materials is mandatory. Variations from rigid specifications not only can adversely affect the yields and hence economic viability of coal conversion processes, but they can cause equipment damage through fouling, sometimes involving conditions that can be corrected only by shutting down operation for from a few hours to several days. The coal raw material used also can adversely affect plant effluents, upsetting delicate operating balances maintained in gas treating and recovery units, adding to the difficulty of close environmental controls.

See also **Coal Conversion Processes.**

References

Ardell, M.: "Particulate Control for Coal-Fired Boilers," *Chem. Eng. Progress*, **75**, 78–82 (1979).

Criswell, R. L.: "Control Strategies for Fluidized Bed Combustion," *In-Tech*, **27**, 1, 37–43 (1980).

Davis, J. S., and J. R. Martin: "Cryogenics for Syngas Processing," *Chem. Eng. Progress*, **76**, 2, 72–79 (1980).

Devitt, T. W., et al: "Utility Flue Gas Desulfurization in the U.S.," *Chem. Eng. Progress*, **76**, 5, 45–57 (1980).

Drehmel, D. C.: "Developments in Particulate Control for Coal-Fired Power Plants," *Chem. Eng. Progress*, **76**, 5, 74–75 (1980).

Goldberg, A. J., Hoover, L. J., and T. Surles: "Consequences of Increased Coal Utilization," *Chem. Eng. Progress*, **76**, 3, 61–64 (1980).

Gorbaty, M. L., et al.: "Coal Science: Basic Research Opportunities," *Science*, **206**, 1029–1034 (1979).

Gordon, R. L.: "The Hobbling of Coal: Policy and Regulatory Uncertainties," *Science*, **200**, 153–158 (1978).

Gray, J. A.: "Gasification of Carbonaceous Feedstocks," *Chem. Eng. Progress*, **76**, 3, 73 (1980).

Griffith, E. D., and A. W. Clarke: "World Coal Production," *Sci. Amer.*, **240**, 1, 38–47 (1979).

Grimm, C., et al.: "The Colstrip Flue Gas Cleaning System," *Chem. Eng. Progress*, **74**, 2, 51–57 (1978).

Idemura, H., Kanai, T., and H. Yanagioka: "Jet Bubbling Flue Gas Desulfurization," *Chem. Eng. Progress*, **74**, 2, 46–50 (1978).

Laseke, B. A. and T. W. Devitt: "Flue Gas Desulfurization," *Chem. Engi. Progress*, **75**, 2, 37–50 (1979).

Moll, N. G., and G. J. Quarderer: "The Dow Coal Liquefaction Process," *Chem. Eng. Progress*, **75**, 11, 46–50 (1979).

Morris, S. C., et al.: "Coal Conversion Technologies: Some Health and Environmental Effects," *Science*, **206**, 654–662 (1979).

Patterson, R. C., and S. L. Darling: "A Low-Btu Coal Gasification Scheme," *Chem. Eng. Progress*, **76**, 3, 55–60 (1980).

Princiotta, F. T.: "Advances in SO_2 Stack Gas Scrubbing," *Chem. Eng. Progress*, **74**, 2, 58–64 (1978).

Quig, R. H.: "Coal as an Energy Source: Its Impact on Engineering Planning," *Chem. Eng. Progress*, **76**, 3, 47–54 (1980).

Quillman, B.: "Present and Future Coal Utilization at Du Pont," *Chem. Eng. Progress*, **76**, 3, 43–46 (1980).

Raymond, W. J., and A. G. Sliger: "The Kellogg/Weir Scrubbing System," *Chem. Eng. Progress*, **74**, 2, 75–80 (1978).

Rochell, G. T., and C. J. King: "Alternatives for Stack Gas Desulfurization by Throwaway Scrubbing," *Chem. Eng. Progress*, **74**, 2, 65–70 (1978).

Scollon, T. R.: "An Assessment of Coal Resources," *Chem. Engi. Progress*, **73**, 6, 25–30 (1977).

Selmeczi, J. G., and D. A. Stewart: "The Thiosorbic Flue Gas Desulfurization Process," *Chem. Eng. Progress*, **74**, 2, 41–45 (1978).

Sensing, T. A., et al.: "Coal Utilization in the Chemical Industry," *Chem. Eng. Progress*, **76**, 1, 64–67 (1980).

Slack, A. V.: "Lime-Limestone Scrubbing: Design Considerations," *Chem. Eng. Progress*, **74**, 2, 71–75 (1978).

Smoot, L. D., and D. T. Pratt: "Pulverized Coal Combustion and Gasification for Continuous Flow Processes," Plenum, New York, 1979.

Squires, A. M.: "Chemicals from Coal," *Science*, **191**, 689–700 (1976).

Staff: (For background reading): Coal Research I, II, III, and IV in *Science*, **193**, 665; **193**, 750; **193**, 873; and **194**, 172 (1976).

Staff: "Energy: Global Prospects 1985–2000. Workshop on Alternative Energy Strategies," McGraw-Hill, New York, 1977.

Staff: "Energy Supply to the Year 2000: Global and National Studies Workshop on Alternative Energy Strategies," MIT Press, Cambridge, Massachusetts, 1977.

Staff: "Survey of Energy Resources," 1976 World Energy Conference, United States National Committee of the World Energy Conference, Washington, D.C., 1977.

Staff: "Conversion to Coal Means a Long Haul to Texas," *Science*, **198**, 587 (1977).

Staff: "Underground Gasification: An Alternate Way to Exploit Coal," *Science*, **198**, 1132–1134 (1977).

Staff: "Critical Issues in Coal Transportation," *ISBN 0–309–02869*, National Academy of Sciences, Washington, D.C., 1979.

Staff: "Sasol Synfuels," *Chem. Eng. Progress*, **75**, 100, 46–47 (1979).

Staff: "Sasol-II Coal Liquefaction Plant," *Chemical Engineering Progress*, **76**, 3, 85–88 (1980).

Staff: "Chemicals from Coal," *Chemical Engineering Progress*, **76**, 3, 89 (1980).

Stephens, D. R., Brandenburg, C. F., and E. L. Burwell: "Underground Coal Gasification," *Chem. Eng. Progress*, **76**, 4, 89–94 (1980).

Swabb, L. E., Jr.: "Liquid Fuels from Coal: From R & D to an Industry," *Science*, **199**, 619–622 (1978).

Thurlow, G. G.: "Producing Fuels and Chemical Feedstocks from Coal," *Chemical Engineering Progress*, **76**, 3, 81–84 (1980).

Vogt, E. V., and M. J. van der Burgt: "Status of the Shell-Koppers Process," *Chemical Engineering Progress*, **76**, 3, 65–72 (1980).

Wilson, C. L.: "Coal—Bridge to the Future: Report of the World Coal Study," Vol. 1, Ballinger, 1979.

COAL BALLS. Concretions composed of mineralized plant fragments preserved as petrifactions. Because the original structure of the plants has been so well preserved, the coal-ball flora has been of aid to paleobotanists in determining the character of the carboniferous flora.

COAL CONVERSION PROCESSES. As pointed out in the entry on **Coal,** the principal ingredient of coal is carbon and, as described in the entry on **Combustion,** it is the combination of carbon with atmospheric oxygen to produce carbon dioxide (CO_2), an exothermic reaction that releases 14,100 Btu/pound (7840 Calories/kilogram) of carbon, that provides the heat energy derived from burning coal. Depending upon the composition of the coal, other heats of reaction will occur from the combustion of hydrogen and sulfur in the coal with air, but these are secondary factors. The fixed carbon content of coal ranges from about 98% for a Class 1 anthracite or hard coal, as may be mined in Pennsylvania, to as low as about 32% for a Class 17 brown coal or lignite, as may be mined in North Dakota.

Direct use of coal, as in pulverized form for the firing of boilers in the electric utility industry, poses a number of environmental problems that can be solved only through the use of costly and elaborate antipollution measures and equipment. But, also in considering the expanded use of solid coal as a major source of energy, there are several other limitations over and beyond the environmental.

Aside from coal-powered steam locomotives and seagoing ships, which essentially were retired from most regions of the world over the past several decades, solid coal is quite unsuited for transportation energy. The energy density of raw coal means that a significant portion of the energy obtained from combusting it is required to move it (as part of a transportation vehicle). This is further amplified by the equipment required to burn coal—massive, heavy furnaces and boilers—which also have to be moved with the vehicle.

Transportation power needs dictate high energy density with fuels that are easy and convenient to handle and that can be converted to power by much smaller, lighter-weight engines. It follows, then, that for coal to be a useful energy source for transportation, it must be pre-converted in some way to overcome the aforementioned objections.

Even for stationary use, particularly where energy needs are much less concentrated than the case of a central electric power plant, such as for commercial and residential heating and large numbers of industrial plants, the conversion of coal into liquid or gaseous forms is required to provide the needed convenience, improved cleanliness, and use efficiency. Probably of equal importance is the cost of transporting large amounts of coal from sources to many tens or hundreds of thousands of locations where it can be used. With liquid or gaseous fuels, pipeline transportation, for example, becomes attractive. It is true (see entry on **Coal**) that there has been considerable attention given to coal slurry pipelines, but these are practical only for certain combinations of situations—because of the slurry preparation and handling equipment, costly and elaborate, required at both ends of the pipeline.

Since the time of Watt, the energy of coal has been converted to a gas—in the form of steam. Steam or hot water can be effectively distributed within relatively small areas, as throughout a given manufacturing plant or complex, or even a commercial and residential neighborhood. This type of district heating has been used for domestic and commercial heating in some parts of the world for many years and, in fact, is undergoing a revival in some areas. District heating (although from geothermal energy rather than from coal) has been common in parts of Iceland for many decades. But thermal losses from transporting steam or hot water limit this approach to relatively small distances.

For many years coal has also been converted into possibly the most ideal form of energy of all—electricity. Electricity has been used extensively for powering machines, lighting, communications, etc. The use of electricity for direct heating, particularly in commercial and domestic installations in many regions, has proved quite costly in recent years. However, this is not universally true. For some years, Sweden has been considering that electric heating load would constitute 50% or more of its demand and capacity for generating electricity in nuclear plants. Quite recently, Swedish authorities have been reexamining this policy. See entry on **Nuclear Reactor.**

Substitute Natural Gas. The conversion of coal into gaseous and liquid fuels is described in this immediate entry. The manufacture of substitute natural gas from hydrocarbon sources is described in the entry on **Substitute Natural Gas (SNG).**

Historical Background. With relatively few exceptions, as for example energy approaches taken in South Africa to be described later, the expanding rather than shrinking use of coal was hardly mentioned until the early 1970s when the seriousness of a world energy shortage (that is, shortage in terms of conventional sources) occurring in the near future rather than generations away finally penetrated the consciousness of most scientists and technologists and, a bit later, awakened the lay public. This process of awakening unfortunately is still, as of the early 1980s, far from complete and it may require further happenings largely of a political nature to fully consider the energy problem as fact and not fiction.

It is interesting to observe against this background, however, that the concept of synthetic fuels is far from new. In fact, in the mid-to-late 1800s and carrying into the first third of the 20th century, what are now called synfuels were used on a daily basis in many parts of the world. Thus, one might say that synthetic fuels and the role of coal in bringing them about is a situation of revival in technology rather than anything of an entirely new or pioneering nature. Petroleum fuels essentially represent a phenomenon of the automotive age (the spectacular salt dome at Spindletop, Texas was put into crude oil production in 1902). Natural gas as a major and widely distributed fuel really did not get underway on a large scale until the 1930s. Thus, for the greater part of the 20th century, with natural crude and natural gas, both excellent, convenient, generally clean and inexpensive fuels, the question could have been posed, "Who needs synthetic fuels?"

Artificial gas for use as a heating fuel and derived from coal or coke was widely used during the latter part of the nineteenth century and during the first few decades of the twentieth century. Although natural gas had been discovered and used by the Chinese some 2,000 years ago (piped from shallow wells through bamboo poles for burning under large pans for evaporation of sea water to obtain salt), the first hard evidence of commercial use of natural gas dates back to 1802 when it was used for lighting the streets of Genoa, Italy. The first natural gas utility company was formed in 1858 (The Fredonia (N.Y.) Gas Light Company). The numerous advantages of a gaseous fuel that could be piped to industrial, commercial, and residential users for heating purposes were recognized long before natural gas was found on a large scale and made available to communities hundreds and more miles from the originating wells. Thus, for many decades, artificial gas was used. The local gas utility was characterized by having one or more so-called "gas works" in which a rather poor grade of gas (on present standards) was produced essentially from coal or coke and steam. In the manufacture of producer gas, a deep hot bed of coal or coke was blasted continuously with a mixture of air and steam. The products of the process were carbon monoxide, nitrogen (from the use of air), small amounts of hydrogen, and some carbon dioxide. Because of the large percentage of nitrogen in the gas, the heating value was low [125 to 150 Btu per cubic foot (1113 to 1335 Calories per cubic meter) as compared with natural gas having a value of from 900 to 1,200 Btu per cubic foot (8,010 to 10,680 Calories per cubic meter)]. Blue water gas, carbureted water gas, and coal gas were also produced from coal or coke and, in some instances, enriched with oil and later natural gas. Because of the great availability and, at one time, apparent inexhaustible supply of natural gas in the United States (and a few other areas of the world), manufactured gases were phased out rapidly. The use of natural gas increased 730% between 1940 and 1970 in the United States, during which period the gas industry produced 313 trillion cubic feet (8.9 trillion cubic meters) of natural gas. In other areas of the world, however, where natural gas was not available locally, manufactured gas, sometimes referred to as town gas, city gas, etc., persisted. Thus, it is not surprising that the current new coal gasification technology essentially stems from Europe and the United Kingdom, where an interest in improving artificial gas manufacture continued long after such interests mainly disappeared in the United States. During the 1930s and 1940s, only a few projects for converting coal into gas were conducted in the United States—for example, the U.S. Bureau of Mines project at Louisiana, Missouri.

In the coal technology of the 1980s, rather than the term artificial gas (for gas derived from coal), SNG (meaning either substitute natural gas or synthetic natural gas) appears to be preferred.

*The South African Experience.** Governmental, economic, and scientific planners in the Republic of South Africa exhibited an uncanny early interest in coal as a major source of energy and, as a consequence, are as of the early 1980s ahead of other countries in utilizing coal conversion technology in a practical way. Of course, South Africa had the motivation because there are no signs of important oil deposits in that country. As early as 1927, a White Paper had already been published discussing the processes then available for production of oil from coal. Developments in Germany were closely followed and especially the Fischer–Tropsch process, as its operating conditions did not appear to be very extreme and the process had already been demonstrated in a number of plants. A South African mining corporation, the Anglo Transvaal Consolidated Investment Company, better known as Anglo Vaal, acquired in 1935 the South African rights to the German Fischer–Tropsch process.

During the next few years, Anglo Vaal devoted much attention to the development of a scheme for the production of oil from coal. Tenders were asked for, but because of the complications of World War II, no orders for equipment were placed. However, during the war and in the postwar years, Anglo Vaal remained in close contact with developments. In 1943, negotiations were held in America which led to the procurement of the rights to the American variation of the Fischer–Tropsch process. In 1946, a new study was undertaken, and an application was made to the government to create a suitable framework within which a long-term industry could be established. During 1947, the Liquid Fuel and Oil Act was passed and, in 1950, an agreement was reached between the South African government

* As related by Jan C. Hoogendoorn, Manager, Research and Development, South African Coal, Oil and Gas Corporation Ltd., Sasolburg, Republic of South Africa.

and Anglo Vaal in which the Anglo Vaal rights were taken over by the government. The South African Coal, Oil and Gas Corporation Ltd. was formed and incorporated under the Companies' Act as an ordinary public company. The government appoints the majority of directors, including the chairman, and the remaining directors are appointed by the Industrial Development Corporation, which is a government-owned organization with the objective, as the name implies, of stimulating industrial development in the country. Sasol operates like a normal business concern, with an autonomous board of directors, and is subject to South African company law and taxation like any other company.

A site for the plant was selected close to the banks of the Vaal River, which is South Africa's major source of water, 50 miles (80 kilometers) south of Johannesburg and on top of a vast coal field. Sasol acquired approximately 8000 acres (3200 hectares) for the plant and its township which was to have the name of Sasolburg. The site was in the middle of an area where cattle grazing and corn (maize) production were the only activities and Sasol had to create its own infrastructure from scratch.

Although it was clear that the plant would be based on the synthesis of hydrocarbons from hydrogen and carbon monoxide as invented and developed by Fischer and Tropsch, it still had to be decided which processes to choose for the individual steps in this integrated complex. For gasification, the Lurgi pressure gasification process with steam and oxygen was selected because this process had already been demonstrated in gasifiers of a smaller size. It had the advantage of being able to work on the rather low-grade, high-ash coal available to Sasol. The fact that it operated at a pressure of approximately 350 psi (2.4 mPa), which is also the desired operating pressure for the Fischer–Tropsch plant, was an additional advantage. This avoided cumbersome compression of large volumes of gas arising from low-pressure gasification. A small part of the Sasol I facility is shown in Fig. 1. Generalized flowsheet for the Sasol II facility is given in Fig. 2.

The production of synthetic motor fuels, pipeline gas, ammonia, and chemicals, where coal gasification is the key to successful production, was first commenced in 1955. Extensive experience was gained from 20 years of operation and a large expansion (Sasol II) was commenced in 1975 and completed in the spring of 1980. The newer facility processes 40,000 metric tons of coal per day and produces

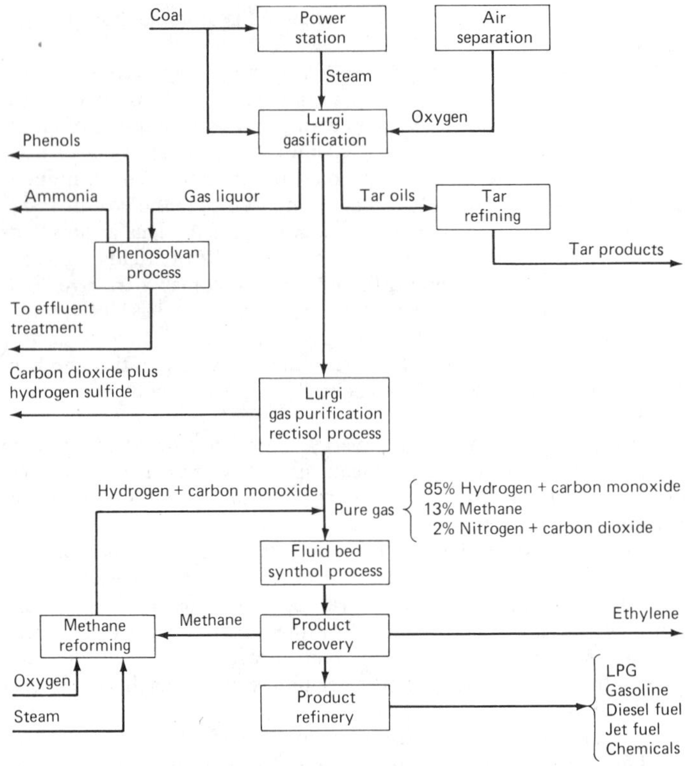

Fig. 2. Generalized flowsheet of Sasol II project. (*South Africa Coal, Oil and Gas Corporation Limited.*)

Fig. 1. *Rectisol* (gas purification) portion of coal gasification complex of Sasol I facility, located at Sasolburg, Republic of South Africa. (*South Africa Coal, Oil and Gas Corporation Limited.*)

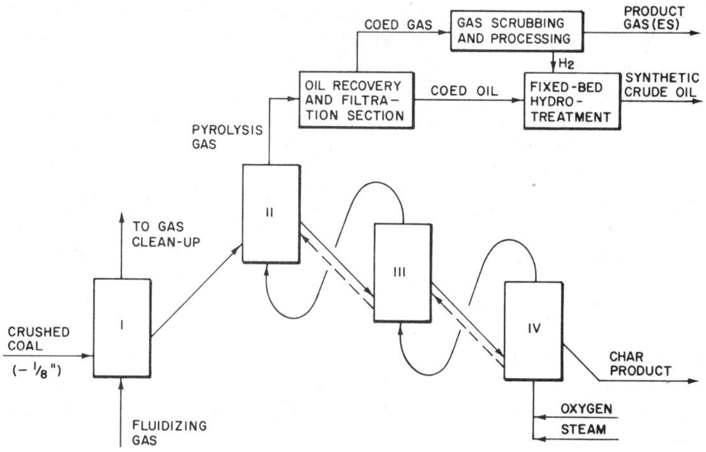

Fig. 3. COED process for coal pyrolysis. (*FMC Corp.*)

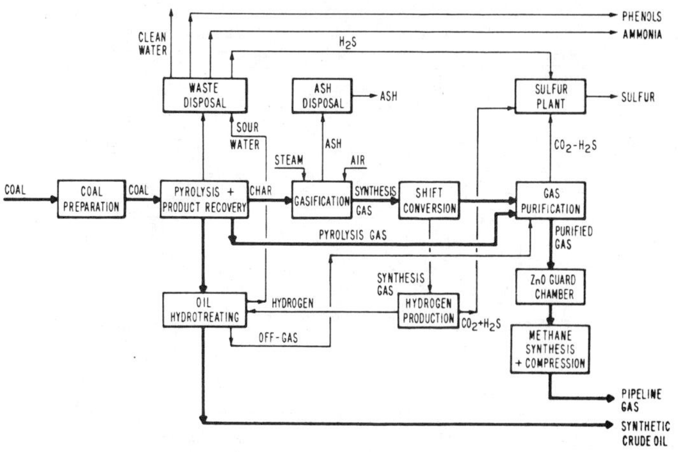

Fig. 4. COGAS pipeline gas process. (*FMC Corp.*)

the equivalent of 60,000 barrels per day of petroleumlike fuels. New mines were developed nearby to feed coal over a conveyor system into a feed preparation facility. Two huge coal piles have been built up to provide continuous operation in event of a temporary shutdown of the mines. The coal is charged to a gasification section where synthesis gas is prepared in the presence of steam and oxygen. Ash content of South African coal is relatively high (about 25%) and thus some 10,000 metric tons/day of ash are produced in the gasifiers. The gasification section contains 36 Lurgi gasifiers which produce the raw synthesis gas. These are improved versions of the gasifiers used at the original Sasol I plant. The original plant also produced town gas as well as liquids. In the new facilities, no gas will be sold. The oxygen requirements of the new facility are quite high, and six air separation plants, each with a capacity of 2500 metric tons/day, represent one of the largest oxygen facilities in the world. Liquors from the gasification section are charged to a Lurgi Phenosolvan unit where ammonia and phenols are recovered and wastewater effluents are cleaned. The raw gas has to be further treated and this is accomplished in a Lurgi Rectisol unit, where carbon dioxide, hydrogen sulfide, and other impurities are removed. The heart of the plant is the liquefaction section. This is where the synthesis gas is liquefied in the presence of an iron-based catalyst. This Fischer–Tropsch reactor technology is proprietary to Sasol but is licensed.

Methane reforming units receive methane-rich gas from a cryogenic product recovery facility and subject the gas to partial oxidation. Some of the carbon dioxide content is removed and the gas recycled to the reactors. Once liquids are recovered, the stream goes to essentially conventional refining units. The plant's production is primarily transport fuels. Most of the gasoline production is currently sold to other refineries for blending with their stocks, but a portion of the product is marketed directly to consumers.

Basic Coal Conversion Processes

There are three fundamental ways currently known for converting coal into desirable liquid and gaseous fuels: (1) Removal of carbon to alter the hydrogen-to-carbon ration[1]—known as *pyrolysis*; (2) addition of hydrogen to alter the ratio—known as *hydroliquefaction*; and (3) the *synthesis of hydrocarbons* from carbon monoxide and hydrogen—which uses Fischer–Tropsch technology. All of these methods are based upon very old chemistry and scientists continue to seek breakthroughs that may make coal conversion more attractive, both technically and economically.

Pyrolysis. Essentially straight pyrolysis is a destructive distillation process, the products of which are volatile liquid and gaseous products and solid, carbonaceous char. Much research has gone into the study of coal pyrolysis and several processes have been proposed and tested. It is not possible here to describe the various processes and their refinements except to comment on a few of them briefly.

The *COED*[2] *process* is the result of a 15-year program between the FMC Corporation and the U.S. Office of Coal Research. This process, for which a pilot plant was constructed in Princeton, New Jersey, uses a non-hydrogenative pyrolysis step, followed by catalytic hydrogenation of the liquid product. Crushed coal is pyrolyzed in three stages (at 316, 454, and 538°C). The liquid product is hydrotreated over a conventional hydrotreating catalyst at high pressure.

The COED process produces only about 1.4 barrels (0.22 cubic meter) of liquid product per ton of coal (907 kilograms); the balance is converted to char which is gasified. Hydrogen consumption in the process is 2000 standard cubic feet/ton (62 cubic meters/metric ton), and the product requires no further treatment for use as a *syncrude* (synthetic crude). To be economically viable, ways must be found to use the char in any pyrolysis process. Thus, as with the COED process, the total system encompasses much more than pyrolysis per se. The general flow of the COED process is shown in Fig. 3. In the pilot plant, a number of different coals have been tested, including high-volatile bituminous coals from Colorado, Utah, Illinois, and Kentucky; subbituminous coals from Wyoming; and lignites from North Dakota. The hydrotreating process removes the unwanted elements of nitrogen, sulfur, and oxygen from the coal oil. Also, the gravity, pour point, and viscosity are all upgraded to properties comparable to a petroleum crude oil.

In essentially an extension of the COED technology, the COGAS process produces moderate-Btu and pipeline-quality gas. A variation of the low-pressure, multistage fluidized-bed pyrolysis of coal used in the COED process is used. See Fig. 4. The hydrogen for hydrotreating is supplied by reforming a portion of the gas product. This gas is stripped of light hydrocarbons and then processed along with the synthesis gas from the char gasification. The light hydrocarbons can be marketed or blended back to increase the quality of the product gas. In-situ coal gasification also involves pyrolsis. This is described later.

Hydroliquefaction processes commence with the same initial step as in the case of pyrolysis, but differ in that hydrogen from an outside source is added to the radicals that have been generated thermally in the initial step. In straight pyrolysis, the radicals are capped by hydrogen originally in the coal, or the radicals combine with carbon, resulting in the formation of high-molecular-weight char. Although the latter two reactions also occur in hydroliquefaction, the net result is production of larger quantities of liquid and gaseous products essentially in proportion to the amount of extra hydrogen added. The process designer has a number of options, accounting at least in part for the proliferation of differing versions of hydroliquefaction proposed. The hydrogen can be furnished in molecular form, or by way of an organic donor. Catalysts may or may not be used. There are also differences in the type of solvent used. Numerous processes involving hydroliquefaction have been proposed and developed, some of them in the pilot plant or demonstration plant stage.

The *synthesis of hydrocarbons* by way of converting coal to synthesis gas (carbon monoxide and molecular hydrogen) for subsequent conversion to fuels and various chemicals is the process used at the South African facilities previously described. These involve Fischer–Tropsch technology. One problem with this system has been the need for new

[1] In raw coal, the hydrogen-to-carbon ratio is less than unity. A ratio of 1.5 to 2 or more is needed to produce desirable liquid fuels.

[2] COED = Char-Oil Energy Development (Project COED).

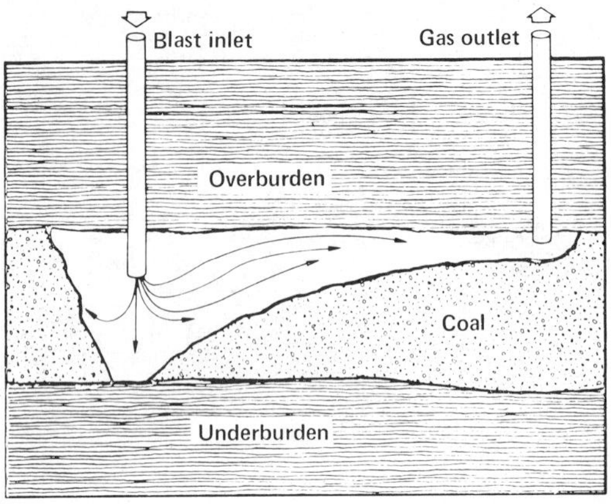

Fig. 5. Basic aspects of the percolation technique for in-situ coal gasification. (*Lawrence Livermore Laboratory.*)

(presently unknown) catalysts that would provide greater selectivity in terms of products made. There tends to be a broad distribution of products with limited selectivity among them.

Underground Coal Gasification. In-situ coal gasification was first suggested by Siemens in 1868 and the first patent on in-situ gasification was granted to Betts in 1909. The first field experimentation was conducted in England prior to World War I, and in the U.S.S.R. in 1933, where semicommercial plants were in operation. After World War II, nearly every western country with a major coal resource experimented with in-situ coal gasification. However, the leading nation in this work has been the U.S.S.R. which developed most of the technology available today.

A representative technique developed in the U.S.S.R. in the 1930s, known as the percolation method, is shown in Fig. 5. Communication is established between two boreholes either by the natural permeability of the coal, by electrolinking, or by fluid or air fracturing of the coal between the boreholes. The U.S.S.R. conducted a number of semicommercial operations between 1935 and 1965, mainly using hydraulic or air fracturing and the percolation technique, producing as much as 45 million standard cubic feet of gas. Generally, air was injected, but in some cases, oxygen or oxygen-enriched air was used. Steam was also injected in a few cases. These installations made a combustible gas of the order of 100 Btu/standard cubic foot (about 10% of the heating value of natural gas). This low-Btu gas was primarily used for electric power generation. However, the product gas flow rate fluctuated and the heating value generally decreased with time. Resource recovery was low. It was reported that for the Moscow region, over 30% of the gases was lost underground. Further, it was estimated that over half of the coal was burned to carbon dioxide and water due to short circuiting. As a consequence, the resource recovery was only 14–20%. This made for inefficient power production and the process could not compete with new oil and gas discoveries made in the U.S.S.R.

American experiments, conducted by the U.S. Bureau of Mines, during the 1950s confirmed the U.S.S.R. findings.

With sharp increases in oil and gas costs that occurred during the 1970s, underground coal gasification testing was initiated. As reported by Stephens et al. (1980), a number of major milestones have been accomplished, including an air injection producing low-heating-value gas (150 kJ/mole) at 2.4×10^5 cubic meters/day. This experiment was conducted by the Laramie Energy Technology Center and operated continuously for 55 days. Work on steam/oxygen injection producing medium-heating-value gas (234 kJ/mole) at 5×10^4 cubic meters/day was conducted by the Lawrence Livermore Laboratory, operating for 58 days of air gasification, including a 2-day oxygen test.

There also has been participation by the private sector. Basic Resources, Inc., a subsidiary of Texas Utilities Services, Inc., has a license to use Soviet technology and is developing underground gasification of lignites near Fairfield, Texas. Two tests have been completed. Atlantic Richfield has completed a successful field gasification test in a 30-meter-thick subbituminous coal near Reno Junction, Wyoming. Under the sponsorship of a consortium of private companies, Texas A&M University is carrying out underground coal gasification tests in lignite. There are also activities in Edmonton, Alberta, in West Germany, and in Belgium.

As pointed out by Stephens et al., about 1.2 trillion tons of coal are believed to be available for commercial underground gasification processes. This huge resource would more than triple proven reserves in the United States and is equivalent to more than 3×10^{21} J, after process efficiencies and inaccessibility of some of the coal is assumed. Coal seams suitable for the process are found at depths of 100 to 600 meters. However, most of the sites thus far have utilized strip-minable coal at depths of less than 60 meters. Western coal is mainly deep, thick, subbituminous or lignite and very desirable for underground gasification.

A major environmental problem essentially unique to underground gasification is surface subsidence. By wide spacing of the rows of process wells, it is believed that pillars of unburned coal will support the overburden above the gasified zones.

All coal-related references are given at the end of the main entry on **Coal.**

COALESCENCE (Cloud Physics). Precipitation and Hydrometeors.

COAL-FIRED BOILER. Boiler; Burner; Feedwater (Boiler).

COALFISH. Codfishes.

COALIFICATION. Coal.

COAL TAR AND DERIVATIVES. Coal tar constitutes the major part of the liquid condensate obtained from the "dry" distillation or carbonization of coal (mostly bituminous) to coke. The three major products of this distillation are (1) metallurgical coke, (2) gas which is suitable as a fuel after appropriate chemical treatment, and (3) vaporized materials which leave the coke oven along with the gas and which are constituted principally of ammonia liquor and coal tar. The condensables and gas impurities are separated from gas in the condensation and purification train of the coke oven plant. The purified coke oven gas is used as fuel to heat the coke ovens and steel producing furnaces. Prior to the widespread use of natural gas as a domestic fuel, coke oven gas was widely used for this purpose after additional purification.

Since metallurgical coke for use in blast furnaces is the prime product of coal carbonization, coal tar production is tied closely to the demand for metallurgical coke. Although steel production has increased progressively over many years, the demand for metallurgical coke has remained reasonably steady, for several reasons. Large improvements in blast-furnace efficiency have occurred. The amount of coke required to produce one ton of pig iron has dropped from 1760 pounds coke/ton of pig iron (878 kilograms coke/metric ton of pig iron) to 1250 pounds coke/ton of pig iron (624 kilograms coke/metric ton of pig iron) and in some modern blast furnaces, the rate is below 1000 pounds coke/ton of pig iron (500 kilograms/metric ton of pig iron). Further, coke has been partially replaced by lower-cost carbonaceous materials, such as petroleum oils, powder coal, and tar. Fundamental changes in the production of steel are expected to further reduce the need for coke.

A number of years ago, coal tar was the primary, if not the sole, source for hundreds of important organic chemicals and derivatives, notably the phenols, cresols, naphthalene, and anthracene, as well as other important coal tar end-products, such as solvent naphtha and pitch. In recent years, synthetic processes for the production of phenol, the cresols and later the xylenols, have been developed and thus, to a large extent, have pushed coal tar into the background as a source of materials for which it was at one time extremely important.

Carbonization Process. Present coking processes generally are of two types: (1) *high-temperature* (900–1200°C) carbonization for producing metallurgical coke and practically the only process practiced in the United States, and (2) *low-temperature* (500–750°C) carbonization, still practiced in some countries where there is a market for "semicoke" as a smokeless home fuel.

Currently, in the United States, slot ovens with byproduct recovery systems are used almost exclusively. These ovens are built in the shape of narrow chambers placed side by side with interspaced flues for heating. Usually, up to 90 chambers are placed together to form a battery. The chambers are charged individually with coal from the top. After carbonization is complete (14–20 hours), the chambers are discharged on one side by pushing with a ram from the opposite side. Each chamber is connected at the top to one or two collecting ducts, or *mains*, which carry the gas evolved and the distillate (tar) to suitable coolers and receivers. By high-temperature carbonization, one obtains generally:

1550 pounds coke, 11,000 standard cubic feet of coke oven gas and 10 gallons of tar from 1 ton of coal;

748 kilograms coke, 343 cubic meters of coke oven gas, and 37.9 liters of tar from 1 metric ton of coal.

Processing of Crude Tar. With reference to the accompanying diagram, the crude tar, after being separated from ammonia and other gases, is subjected to an initial distillation (called *topping*) which separates the desired chemical constitutents from the higher-boiling, more viscous tar constituents. In a typical case, the distillate from this operation (sometimes referred to as *chemical oil*) has an upper boiling point of about 250°C and contains (1) phenols (*tar acids*), (2) naphthalene, which is the most prevalent single constituent of coal tar (6–10%), (3) pyridine-type bases (*tar bases*), and (4) neutral oils. The tar acids constitute about 1.5–3% of the coal tar.

Tar Acids. These materials are recovered by extraction of the chemical oil with aqueous alkali, usually caustic solution. The aqueous layer is separated from the dephenolized (*acid-free*) oil. The phenols then are recovered in crude form by acidification (*springing*) of the aqueous solution, usually by injecting carbon dioxide, followed by gravity settling. The crude phenols then are fractionated to obtain phenol, cresols, and the higher-boiling phenols (mostly xylenols). See also **Phenol.**

Tar Bases. These materials are extracted from the dephenolized oil with aqueous solutions of mineral acids. This operation may be carried out on the entire neutral oil, or it may be done on the solvent naphtha fraction. In the latter case, only the lowest-boiling bases (picolines and lutidines) are recovered, and the higher-boiling bases (mostly quinoline and isoquinoline) can be recovered from postnapthalene fractions or left in the residue for disposal. In European practice, the topping is carried out so that several fractions are obtained: *carbolic oil*, which yields the phenols and lower-boiling bases; and *naphthalene oil*, from which naphthalene is recovered by crystallization.

The tar bases form water-soluble salts with mineral acids which are separated from the oil. They are recovered from their salts by contacting with aqueous alkali (*springing*) and separating the crude bases from the salt solution. The lutidines constitute the major part of the lower-boiling bases. See **Pyridine and Derivatives.**

Solvent Naphtha. The lower-boiling fraction of the neutral oil is a very powerful solvent, particularly for coatings containing coal tar and pitch. The material also is a source of unsaturated compounds, such as indene and, in a lesser amount, coumarone and homologues of these compounds. Resins are formed in situ from these compounds when solvent naphtha is treated with Friedel–Crafts type catalysts. See **Friedel–Crafts Reaction.** These resins are useful in the manufacture of inexpensive floor tiles and coatings. The remaining solvent is recovered by distillation and used as a solvent.

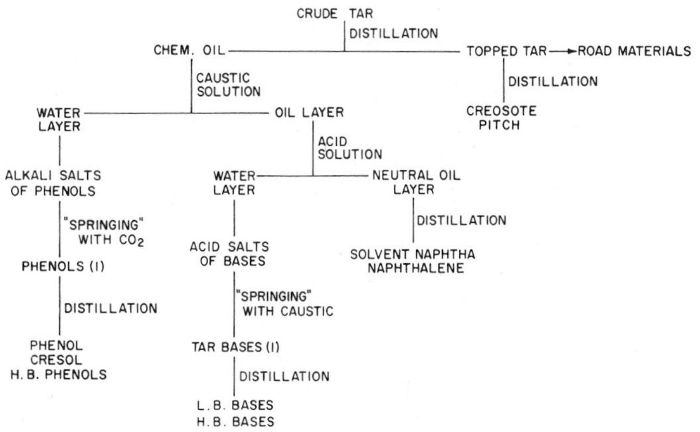

Bulk fractions from crude coal tar.

Naphthalene. This compound finds a ready market principally for the production of phthalic anhydride. There is a great variety of processes to isolate napthalene from the acid-free or neutral oils. Frequently, the naphthalene is first concentrated by distillation, and the enriched oil then is worked up by crystallation. This process is prevalent in Europe. It is also possible to isolate the naphthalene by careful fractionation. Depending on the purity desired, additional chemical treatments may be required. Naphthalene usually is traded with freezing point as a measure of purity (80.3°C) for pure naphthalene. A good quality commonly used is "78°C" naphthalene, which is about 96% pure. See also **Naphthalene.**

Topped Tar. With reference to the accompanying diagram, it will be noted that topped tar is the residue remaining from the topping operation where the chemicals are separated as the distillate. The principal use of topped tar is in road materials. A number of standard grades (RT-1 to RT-12) are available, the grade depending on the "consistency" or viscosity of the tar. Road tar has excellent weather and skid resistance, but its use is limited by availability and price as compared with asphalt. This is borne out by the respective amounts used for road building (United States) with about 90% using asphalt.

Creosote. Chemically, creosote is a mixture of a great number of compounds, almost exclusively of cyclic structure. Individual compounds present in creosote in concentrations of 2–4% are acenaphthene, fluorene, diphenylene oxide, anthracene, and carbazole. Only one compound, phenanthrene, is present in a larger concentration (12–14%). For many years, chemists in many countries have tried to isolate individual compounds and to find profitable uses for them. Most of these attempts have failed with exception of those involving anthracene. See also **Anthracene.** The principal use of creosote is for preservative of wood. Railroad ties, poles, fence posts, marine pilings, and lumber for outdoor use are impregnated with creosote in large cylindrical vessels. If properly treated, the life of the wood is greatly extended. Materials that are competitive with creosote for wood-preservation purposes include various petroleum oils, and pentachlorophenol. Pentachlorophenol is used in solutions of creosote or of petroleum oils. Blends of creosote with petroleum oils also are used for economic reasons.

Pitch. This is the residue from the processing of coal tar. Since pitch constitutes over 50% of the crude tar, its utilization has a major effect on the economics of tar processing. Coal tar contains an estimated 5,000–10,000 compounds; it is reasonable to assume that about half of this number is contained in the pitch. Of the roughly 300 compounds identified in coal tar, about one-half must be in the pitch on the basis of their boiling points. It is probable that none of these compounds is present in pitch in concentrations of more than a fraction of 1%. Coal tar pitch is a black, shiny material which is solid and brittle at low temperatures and liquid at high temperatures. Since it is composed of a great number of different compounds, many of which interact to form eutectic mixtures, it does not show a distinct melting or crystallizing point. Pitch usually is characterized by the *softening point*, which can be measured in several ways.

Because of the importance of pitch in various industries, many studies have been made to elucidate its composition. Solvent fractionation has been used to subdivide pitch into fractions by molecular weight, the higher-molecular-weight fractions requiring more powerful solvents. However, even the highest-molecular-weight compounds are only of moderate molecular weights, ranging up to 6 or 7 condensed rings with molecular weights in the range of 350–400. Constituents isolated from pitch have been identified and appear to be crystalline, well-defined substances. It is generally believed that the glasslike state of pitch is caused by association forces and by the mutual melting-point depression exerted by a large number of multiring compounds that tend to form eutectics. The uses of pitch fall in two general classes: (1) applications based upon the binder properties of carbonized pitch (*pitch coke* or *binder coke*); and (2) uses based upon the other physical qualities of pitch.

Carbon pitch is used for carbon electrodes in electrolytic reduction processes, such as aluminum reduction. Refractory pitch is used in the manufacture of refractory brick, usually burned magnesite or dolomite, the pores of which are filled with pitch by hot impregnation. Upon firing the pitch in the brick is carbonized. The remaining pitch coke retards penetration of molten metals and slags, thus prolonging the life of the brick furnace lining. Coke pitch is used in the production of foundry cores.

Roofing Pitch. A substantial amount of pitch is used as covering membrane on flat roofs on industrial plants, large office and apartment buildings, parking garages, and similar structures. Pitches of 50–60°C softening point generally are used.

George R. Romovacek, Koppers Company, Inc., Monroeville, Pennsylvania; and G. G. Lauer (Retired), Pittsburgh, Pennsylvania.

COANDA EFFECT. Fluidics.

COARCTATION (Aorta). Heart and Circulatory System (Human).

COASTAL TERRAIN HYDROLOGY. Hydrology.

COASTLINE. Ocean.

COAST REDWOOD. Redwood (Coast).

COATIMUNDIS. Raccoons.

COATING AGENTS. Substances that are used to protect the surface (and penetration through the surface) of various materials that are being processed or in final-product format. Coating agents are widely used in the pharmaceutical and food industries.

For example, in the food field, fresh produce requires very little processing. Some fruits may be dipped in 1–2% citric acid prior to freezing. This removes any residual lye from the peeling process and any further destruction of ascorbate is prevented. However, citric acid alone is not always sufficient to prevent deteriorative effects during freezing and thus a sequestrant/antioxidant may be added. A solution of 0.5% citric acid with 0.02% D-erythroascorbic acid may be used to prevent browning of some fruits during freezing and defrosting. The procedure is also useful with some vegetables. Plain wax coatings are applied to many packaged cheeses.

Dried fruits, such as raisins, prunes, and figs, require a residual moisture content because most consumers do not like a thoroughly dry or crispy product in this category. Residual moisture, of course, encourages mold and yeast growth. Protection can be gained by dipping the fruit in solutions of 2–7% potassium sorbate, this leaving a fine coating of antimicrobial agents on the fruit pieces.

Acetostearin products (di- and triglycerides) solidify into waxlike solids and are used in some protective coatings for food products. Researchers have shown these products to be effective against moisture penetration, as well as against atmospheric gases. The coating is applied by dipping; spraying in the case of nuts.

For tabletlike confections and pharmaceuticals, sucrose is one of the most common of coating materials. It is transparent, but can be made opaque, and makes an excellent film-like coating. Gums and resins are also used. Some products such as pills and candies can be multicoated by using approved colorants in the coating agents.

The film-forming characteristics of starch have been used for many years. Numerous products can be protectively and decoratively coated with starch. Starch can be added to sugar solutions to provide a less brittle and moisture-sensitive surface. Starch coatings have the advantage of being oil and grease resistant.

Methylcellulose when incorporated into sweet dough products serves to improve adherence of the glaze.

Red meats, poultry, and fish can be coated with a protective film prior to freezing by dipping into successive solutions of 10–15% sodium alginate, 3–5% calcium chloride, and 10–20% glycerol, the latter used as a plasticizer. This coating improves retention of juices upon freezing or thawing. Sodium alginate coating for sausages, alone or with ethylcellulose, prevents salt rust and increases storage stability. Researchers also have studied the effectiveness of carrageenan as a way to prevent oxidative rancidity after freezing. In a Norwegian process, fish are block-frozen in alginate jelly, forming an air-tight coating, preventing oxidative rancidity.

COATINGS. Autodeposition; Conversion Coatings; Paint.

COAXIAL CABLE. Telephony; Television.

COAXIAL LINE. A type of transmission line in which one conductor completely surrounds the other, the two being coaxial and separated by a continuous solid dielectric or by dielectric spacers with gas as the principal insulating material. Such a line is characterized by negligible external electromagnetic field and by having essentially no susceptibility to external fields from other sources. It is extensively used for radio-frequency transmission lines and is also used as a multi-channel telephone carrier and television program line. See also **Fiber Optics.**

COAXIAL LINE ATTENUATOR. Attenuator.

COBAIFERA LANGSDORFII (Fuel Source). Biomass and Wastes as Energy Sources.

COBALAMIN. Vitamins.

COBALT. Chemical element symbol Co, at. no. 27, at. wt. 58.9332, periodic table group 8, mp 1495°C, bp 2869–2871°C, density 8.832 g/cm^3. There are two allotropic modifications of cobalt, a close-packed hexagonal form (ϵ) with space group $P6_3/mmc$, stable at temperatures below 417°C; and a face-centered cubic form (α) with space group. Fm_3m stable at higher temperatures—up to the melting point. The metal is silvery gray in color. The only naturally occurring isotope ^{59}Co is stable, but the other twelve known isotopes are radioactive, their mass numbers ranging from 54 to 64. Half-lives range from 0.2 second for ^{54}Co to 5.3 years for the industrially and medically important ^{60}Co. See also **Radioactivity.**

Cobalt was identified and described by Georg Brandt in 1735, but had to wait until the last decade of the nineteenth century before the new sources of metal supply from New Caledonia and Canada stimulated its metallurgical usage.

PRINCIPAL ECONOMICALLY IMPORTANT COBALT MINERALS

Group and Name	Ideal Formula	Cobalt Content %	Source Area
Sulfides			
Linnaeite	$(Co, Cu, Ni, Fe)_3S_4$	up to ~ 48	U.S.A., Zaire, Finland
Carrolite	$CuCo_2S_4$	up to ~ 38	Zambia, Zaire, U.S.A.
Pentlandite	$(Fe, Ni, Co)_9S_8$	up to ~ 2	Canada
Cobaltiferous Pyrite	$(Fe, Co)S_2$	up to ~ 2	Canada, Finland, Zambia
Arsenides			
Skutterudite	$(Co, Ni, Fe)As_3$	up to ~ 28	Canada, Morrocco, U.S.A.
Gersdorffite	$(Ni, Co, Fe)AsS$	up to ~ 12	Canada, Zaire, U.S.A.
Oxides			
Heterogenite	$Co_2O_3H_2O$	up to ~ 57	Zaire, Zambia
Asbolite	$(Co, Ni)O2MnO_24H_2O$	up to ~ 27	New Caledonia, Canada

Note: See also **Pentlandite; Skutterudite.**

First and second ionization potentials are 7.86 eV and 17.05 eV, respectively; oxidation potential $E°$ is −0.277 V. Further general specifications are given under **Chemical Elements.**

Occurrence: The cobalt content of the earth's crust is estimated to be within the range 10 to 40 ppm. Economic concentrations of the element are the exception so that supply is governed by its by-product output from ores mined for the recovery of other elements, particularly copper and nickel. For technical reasons, the sulfides, arsenides and oxidized minerals form almost the entire economic source of the metal. Its production is restricted to a relatively few countries, the most important being the Republic of Zaire and Zambia in Africa, Canada, Finland, Morrocco, and the United States. The cobaltiferous deposits in the copper belt of central southern Africa vary in content from 1 to 30 parts of cobalt to 100 parts of copper, with an estimated hundreds of millions of tons containing 2 to 15 parts cobalt to 100 parts copper. Whereas sulfide nickel ores the world over have a cobalt content in the range 2 to 5 (very occasionally as high as 10) parts to 100 parts of nickel. Oxide nickel ores vary in cobalt content from 1 to 30 parts per 100 parts nickel with estimated thousands of millions of tons containing 5 to 15 parts cobalt per 100 nickel. The nickel lateritic deposits in the tropical and subtropical areas of the world represent a large potential future source of both nickel and cobalt.

Mining and Recovery: The economically important cobalt-containing minerals at present exploited are listed in the accompanying table. The metal extraction processes, following the usual pretreatment of the ore, are varied and complicated because the metallurgical properties of cobalt differ insufficiently from those of the associated metals and because the cobaltiferrous raw materials comprise the sulfide, arsenide, and oxide, or a mixture of these. The final refining stage invariably involves electrolysis.

Uses of Metal and Alloys: The metallurgical applications of cobalt consume approximately 70–75% of the world production; the remainder goes into chemicals. Relatively little use has been made of the pure metal, the most important being as the radioisotope ^{60}Co for teletherapy and industrial radiation processing and gamma radiography. The increasing availability of the metal in various wrought forms will stimulate its applications in those areas where the intrinsic properties of cobalt are advantageous, e.g., magnetic devices, wear resistance and bearing properties at elevated temperatures, and the manufacture of heterogeneous welding rods for hardfacing. Major uses of the metal may be classified as follows:

High-temperature Materials. To meet the demand of the gas turbine industry, alloys capable of reliable performance at high temperature and loads have been developed based on cobalt and cobalt-containing nickel and iron-based alloys. The improvement in properties at these very high temperatures (~750–1200°C) has been brought about by the addition to the oxidation and sulfidation resistant cobalt-chromium matrix of varying amounts of refractory metals, mainly tungsten, molybdenum, tantalum, and columbium (niobium) to strengthen the matrix further and promote the formation of stable carbides which make a substantial contribution to the high-temperature strength. The application of current techniques of vacuum melting and casting insure optimum material properties and improved component reliability. A new family of alloys based on Co-Fe-Cr is being successfully applied in the exacting conditions which exist in metallurgical furnaces and petrochemical plants.

Magnetic Materials. For the past 50 years, apart from iron, cobalt has been the most important constituent of permanent magnet materials. This is because cobalt additions raise the saturation magnetization and the Curie temperature to higher values than are obtained from pure iron. Present advanced Alnico type cast alloys are the results of 40 years of development and are the most widely used permanent magnet materials. More recently, permanent magnet materials based on cobalt-rare earth (e.g., samarium, praseodymium) compounds have emerged as the most powerful magnets available.

Hardfacing and Wear-resistant Alloys. These materials, essentially quaternary alloys of cobalt; chromium, tungsten (or molybdenum); and carbon, are widely used for industrial hardfacing purposes. They can be deposited by welding techniques, sprayed on as powders, or produced as separate castings. By using the weld deposition technique, a highly alloyed heat, wear, and corrosion-resistant surface can be applied to a much cheaper substrate, e.g., mild steel. Used originally as a component reclamation process, these alloys and techniques are now primary design requirements for many engineering items. The feasibility of electrodepositing composite cobalt-carbide coatings from an agitated slurry of solid carbide particles in a conventional plating bath has been demonstrated. These coatings provide excellent wear protection and are now established in the aerospace industry, while other engineering applications are being evaluated.

The demand for wear resistance allied with a low coefficient of friction has been successfully met by cobalt-molybdenum-silicon alloys. Structurally they conform to the well proven bearing concept of hard intermetallic phases dispersed through a softer cobalt matrix. They offer outstanding bearing performance in conditions of poor lubrication.

Alloy Steels. The addition of cobalt to high-speed tool steel was one of the earliest uses of cobalt. This application is represented by the super high-speed tool steel grades which contain 5–12% cobalt. Two other cutting tool materials, the nonferrous cobalt-chromium tungsten-carbon cast alloy and the cemented carbides also provide a steadily growing outlet for cobalt. Hot work die steels are another group of alloy steels which benefit from the effect cobalt has on the tempering characteristics and consequent hot strength retention. The most significant development in recent years, however, is the advent of the Ni-Co-Mo miraging family of steels which combine very high strength and toughness properties to an unusual degree. This is still an area of active development and growing engineering utilization.

Toxicity: Cobalt, like most other metals, is not entirely harmless, although it is not in any way comparable to the known toxic metals, such as mercury, cadmium, and lead. Inhalation of fine cobalt dust over long periods can cause an irritation of the respiratory organs which may result in chronic bronchitis. Complete recovery is usually achieved upon removal from the contaminated atmosphere. Cobalt salts can cause benign dermatoses, either in people new to handling them, or after prolonged exposure, usually several years.

Chemistry of Cobalt: The metal in the massive form is not attacked by air or water at temperatures below approximately 300°C; above this temperature it is oxidized in air. The metal combines readily with the halogens to form the respective halides. It combines with most of the other metalloids when heated or in the molten state. It does not combine directly with nitrogen but decomposes ammonia at elevated temperature to form a nitride. It reacts with carbon monoxide at 225–230°C to form the carbide Co_2C. Cobalt also forms intermetallic compounds with many metals, e.g., Al, Cr, Mo, Sn, Ti, V, W and Zr. Metallic cobalt is readily dissolved in dilute sulfuric, hydrochloric or nitric acids to form cobaltous salts. Like iron, cobalt is passivated by strong oxidizing agents, such as the dichromates. It is slowly attacked by ammonium hydroxide and sodium hydroxide.

Cobalt Compounds: In general the chemical properties are intermediate between those of iron and nickel. The predominant oxidation states of cobalt compounds, except for a large class of organometallic compounds, are 2+ and 3+. Common usage assigns the terms *cobaltous* and *cobaltic*, respectively, to these.

In aqueous solutions and in the absence of complexing agents, cobalt compounds are stable only in the 2+ oxidation (cobaltous) state. In the complexed state the cobaltous ion is relatively unstable being readily oxidized to the 3+ oxidation (cobaltic) state. An extremely large number of 3+ complex ions have been identified most of which are quite stable in aqueous media.

Cobalt has an electronic configuration $1s^2 2s^2 2p^6 3s^2 3p^6 3d^7 4s^2$. The two $4s$ electrons are readily removed producing the ordinary Co^{2+} ion. In principle, if the odd $3d$ electron were removed the simple cobaltic ion Co^{3+}, would be formed. This however does not occur, the ion exists only in complex ions or crystal lattices in which cases additional electron orbitals are filled.

Cobalt and oxygen form two stable oxides, cobaltous oxide CoO, stable below 200°C and above 900°C; and cobalto-cobaltic oxide Co_3O_4 which is stable below 900°C. Between 200 and 900°C the CoO oxidizes partially, or completely, to Co_3O_4.

Cobaltous oxide is usually prepared by heating the carbonate. It is insoluble in H_2O, NH_4OH, and alcohol, but dissolves in cold strong acids and in weak acids on heating. Commercial gray oxide which may contain up to 40% Co_3O_4 is used in the ceramic, glass and enamel industries and also in the production of catalysts. It is also used in the preparation of cobalt metal powder. The Co(III) oxide can also be prepared by calcining oxides, hydroxide and salts. The oxide is insoluble in H_2O and only slightly soluble in acids. The commercial black oxide consists essentially of Co_3O_4 with possibly up to 20% of CoO.

Cobalt(II) hydroxide exists in two allotropic forms, a blue α-$Co(OH)_2$ and a pink β-$Co(OH)_2$. The hydroxide is prepared by precipitation from a cobaltous salt solution by an alkali hydroxide. When the alkali is in excess the pink β-form is produced—the blue α-form is produced when the cobalt salt is in excess. The salt slowly oxidizes in air at room temperature and changes to hydrated cobaltic oxide $Co_2O_3 \cdot nH_2O$. The hydroxide is practically insoluble in H_2O and in bases, but highly soluble in mineral and organic acids. The commercial salt is used as the starting material in the preparation of drying agents.

Cobaltous halides are formed with the halogen group, but only fluorine forms a stable cobaltic compound. The other cobaltic halides are stable only in complex ions.

Commerical cobaltic fluoride is formed by the reaction of cobalt, its oxides or simple salts with ClF_3 or BrF_5 to yield a brown anhydrous salt. This is a powerful fluorinating agent readily replacing hydrogen in aliphatic and aromatic hydrocarbons. Cobaltous chloride is a pale blue compound and very hygroscopic. A series of hydrates is known, the hexahydrate is pink but becomes blue on warming. It is extremely soluble in H_2O, and in numerous organic solvents. The commercial salt is used as a starting material in the manufacture of catalysts, in electroplating, agricultural chemicals, and pharmaceuticals. Cobaltous bromide is highly hygroscopic and transforms gradually to the red hexahydrate. The salt is mainly used in the catalysts industry.

The iodide exists in two forms, the α-form, a black graphite like solid, and, iodine free β-form which is yellow-brown in color. The iodide is used as a catalyst in organic reactions.

A number of sulfides have been reported, the best characterized being Co_4S_3, Co_9S_8, CoS, Co_3S_4, and CoS_2. They are prepared either by metal or salt solution reaction with S or H_2S. The mixed sulfides of cobalt and molybdenum have catalytic properties of hydrogenation and isomerization.

The sulfate, one of the more important industrial salts of cobalt, is usually available as the red heptahydrate an efflorescent substance. On heating it converts to the dark blue monohydrate, and after prolonged heating above 250°C the red anhydrous salt. It is widely used in the electroplating and ceramic industries, and in the preparation of drying agents and agricultural pasture top-dressing.

Cobalt and nitrogen form three nitrides Co_3N_3, Co_2N, and CoN, products of metal/ammonia reaction and compound decomposition. All are gray-black or black in color.

In its commercial form cobalt nitrate appears as the red hexahydrate $Co(No_3)_2 \cdot 6H_2O$ formed by dissolving the metal oxide or carbonate in dilute HNO_3 and concentrating the solution. The compound deliquesces in moist air and effloresces in dry air. The pink anhydrous cobaltous nitrate cannot be formed by dehydrating the hexahydrate but by treating the salt with nitrogen pentoxide gas (or in solution in concentrated HNO_3). The salt is used mainly in the preparation of catalysts.

Two discrete carbides have been characterized—Co_3C and Co_2C. The former, which has the same structure as Fe_3C, has been prepared by reacting cobalt with coal gas at temperatures between 500–800°C. Co_2C is formed by the reaction of carbon monoxide at atmospheric pressure on cobalt powder at 225–230°C.

Cobaltous carbonate $CoCO_3$ is found almost pure in the mineral sphaerocobaltite in the Republic of Zaire and less extensively in Zambia. The pale-red anhydrous salt is obtained by reaction in solution of an alkaline carbonate and a cobaltous salt under a slight pressure of carbon dioxide (up to 1 atmosphere) and subsequent heating at 140°C. The commercial salt is violet-red in color, partially hydrolized with an indeterminate composition. It is insoluble in H_2O and alcohol, but dissolves easily in inorganic and organic acids, and is often used for the preparation of other salts. According to the thermal conditions it decomposes to the different types of oxides.

Cobaltous acetate $Co(CH_3CO_2)_2$ is obtained as a pink salt by the dehydration of the tetrahydrate which is prepared by dissolving the hydroxide or carbonate in acetic acid. The tetrahydrate which is the commercial form is widely used in the preparation of catalysts, e.g., OXO synthesis, and driers for inks and varnishes.

Cobaltous oxalate dihydrate $CoC_2O_42H_2O$ (pink) is obtained by adding oxalic acid or an alkaline oxalate to a cobaltous salt solution. This is the commercial form of the salt and is important as the starting material in the preparation of cobalt metal powders.

Cobalt(III) coordination compounds are the classics of coordination chemistry, many of the cobalt(III) amines having been prepared in the nineteenth century. They are invariably colored and undergo reaction slowly. Because of this many isomers have been isolated and studied. Much of the knowledge of isomerism, mechanisms of reaction, and general properties of octahedral species are based on cobalt(III) compounds. The important donor atoms (in order of decreasing tendency to complex) are nitrogen, carbon in the cyanides, oxygen, sulfur, and the halogens. The coordination number is invariably six. An extensive class of amines and compounds of the amines is known ranging from hexamine $[CoN_6]^{3+}$ through to monoamines $[CoNX_5]^{2-}$. The compounds are made in several stages by first oxidizing the cobalt(II) species and then various different substituted species are prepared by substitution reactions on the primary cobalt(III) product. When air is drawn through on aqueous solution containing cobalt(II) and ammonia the solution turns brown. From this mixture can be obtained a variety of products which depend on the initial concentrations of reactants, the pH of the solution, the anion present and the presence of heterogeneous catalysts, such as charcoal. The products are invariably colored ranging from blue-violets and green through various shades of red and brown to yellow. The absorption spectra are characteristic of the various structures. Complex cyanides of the types $[Co(CN)_6]^{3-}$ and $[Co(CN)_5X]^{3-}$ are stable and diamagnetic. The anion $[Co(CN)_6]^{3-}$ is pale yellow and is the ultimate product of the reaction when a solution of cobalt(III) cyanide in aqueous KCN is boiled in air. It is very unreactive, being untouched by chlorine, peroxide, alkali, aqueous HCl and H_2S, although it gives CO when treated with concentrated H_2SO_4.

The only cobalt(IV) compound representing this oxidation state appears to be $Cs_2[CoF_6]$ which is prepared as a yellow powder by the fluorination of Cs_2CoCl_4.

Evidence for cobalt(I) was first obtained from the electrolytic reduction of cyano-compounds and some of the reduced species have been isolated. There are also many cobalt(I) coordination compounds of the organometallic class carbonyl, isonitriles, and unsaturated hydrocarbon derivatives. The oxidation state cobalt (0) may be represented in the cyano-compound which has been formulated as $K_8[Co_2(CN)_8]$. It has been prepared as an air-sensitive brown-violet compound by reducing a liquid ammonia solution of $K_3[Co(CN)_6]$ with an excess of K metal. They only other known cobalt (0) species are organometallic compounds.

For the role of cobalt in biological systems, see **Cobalt (In Biological Systems).**

References

Bailar, J. E., et al. (editors): "Comprehensive Inorganic Chemistry," Vol. 3, Pergamon Press, Elmsford, New York, 1973.

Maykuth, D. J.: "Cobalt" in "Metals Handbook," 9th edition, Vol. 2, American Society for Metals, Metals Park, Ohio, 1979.

Smithells, C. J. (editor): "Metals Reference Book," Vol. III, Plenum, New York, 1967.

E. Williams, Cobalt Information Centre, London.

COBALT (In Biological Systems). Although cobalt is regarded as an essential element for animals, including humans, the element can perform its essential functions only after it has been incorporated into the vitamin B_{12} molecule. The microorganisms living in ruminants are the major producers of vitamin B_{12} in the food chain. Green plants do not synthesize the vitamin. The normal intake of this vitamin is by way of milk, cheese, meat, and eggs. Persons who follow a strictly vegetarian diet may become deficient in vitamin B_{12} unless supplementary sources are used. Single-stomached domestic and wild animals receive their vitamin B_{12} from animal flesh or from animal fecal material. See also **Vitamin B_{12}.**

Cobalt is present in vitamin B_{12} to the extent of about 4%. Lack of cobalt in the soil and feedstuffs prevents ruminants from synthesizing all of the vitamin B_{12} for their needs. Thus, cobalt can be added to feedstuffs as the chloride, sulfate, oxide, or carbonate. Excessive cobalt intakes are toxic, causing a reduction in feed intake and body weight, accompanied by emaciation, anemia, debility, and elevated levels of cobalt in the liver. It is of interest to note that clinical cobalt toxicity closely resembles clinical cobalt deficiency.

Cobalt is required by the microorganisms that live in nodules on the roots of legumes, such as bean and clover. They convert nitrogen from the air into chemical forms that can be used by higher plants. This is possibly the only well established and understood function of cobalt in plant growth. Legumes may grow normally and the microorganisms on their roots fix atmospheric nitrogen, even though the forage does not contain sufficient cobalt to meet the requirements of ruminants.

Areas of low cobalt content in the United States, where clovers and alfalfa are too low in cobalt content to meet requirements of cattle and sheep, include northeastern Maine, all of New Hampshire, Vermont, Massachusetts, Connecticut, and Rhode Island, much of New York with exception of the central portion, the northwestern portion of the lower peninsula of Michigan, a small area in Illinois with Peoria at its approximate center, all but eastern Iowa, and southwestern Minnesota. The low-cobalt soils of New England are primarily sandy and were formed from glacial deposits near and to the south of the White Mountains of New Hampshire. Along the south Atlantic Coastal Plain, legumes with very low concentrations of cobalt are primarily on the sandy soils formed in naturally wet areas. These soils, which are called spodosols, have light-colored subsurface layers overlying a dark-brown or dark-gray hardpan layer.

Grasses and cereal grains generally contain less than the 0.07 to 0.10 parts per million of cobalt required by ruminants. Cattle and sheep that are not fed any legumes nearly always require cobalt supplementation.

Adding cobalt to soils, either as cobalt sulfate, or as cobaltized superphosphate, can be used to increase the level of cobalt in plants and prevent cobalt deficiency in cattle and sheep. Cobalt fertilization may not be effective in preventing cobalt deficiency on alkaline soils because in these soils, the added cobalt quickly reverts to forms that are not taken up by plants. Cobalt fertilization is more common in Australia than in the United States. In the United States, cobalt is usually added to mixed feeds, mineral mixes, or salt licks.

Still another method is to place heavy ceramic "bullets" containing cobalt in the animal's rumen. These bullets remain in the rumen and slowly release cobalt to meet the animal's needs for a long period. The diets of hogs and chickens are often supplemented with concentrated forms of vitamin B_{12}.

The relationship of the levels of cobalt in soils and plants to the health of ruminants is one of the striking examples of the importance of a soil and plant relationship to animal health. When some Australian scientists discovered this relationship, new areas in several parts of the world became usable for animal production. The vitamin B_{12} formed within cattle and sheep in these new areas contributed to the vitamin B_{12} nutrition of people, even though adding cobalt to soils does not directly affect human nutrition in the absence of the production of ruminants. Cobalt-deficient grazing soils are found in Australia, New Zealand, and, in the United States, mainly in Florida, although deficient regions are found elsewhere as previously mentioned. Cobalt deficiency can result in a condition known as "pining disease," where the affected animals are quire listless. The disease is also known as "bush sickness" and "salt sickness."

References

Allaway, W. H.: "The Effect of Soils and Fertilizers on Human and Animal Nutrition," Agricultural Information Bulletin 378, Cornell University Agricultural Experiment Station and U.S. Department of Agriculture, Washington, D.C., 1975.

Kirchgessner, M. (editor): "Trace Element Metabolism in Man and Animals," Institut für Ernahrungsphysiologie, Technische Universität München, Freising-Weihenstephan, Germany (1978).

Underwood, E. J.: "Trace Elements in Human and Animal Nutrition," 4th edition, Academic, New York, 1977.

COBALTITE. The mineral cobaltite is a sulfarsenide (see **Arsenic** and **Sulfur**) of cobalt, corresponds to the formula CoAsS, crystallizing in the isometric system as cubes or pyritohedrons, also may be massive. Cobaltite has a very good cleavage parallel to the cube faces; uneven fracture; brittle; hardness, 5.5; specific gravity, 6.33; metallic luster; color, silvery-white to reddish, sometimes steel gray or violet to grayish-black; streak, grayish-black. Cobaltite is found with cobalt and nickel minerals deposited commonly by metasomatic processes. It is found in Sweden, Norway, England and the Province of Ontario. It is an ore of cobalt.

COBALT MAGNET. **Magnetism.**

COBIA (*Osteichthyes*). Of the order *Percomorphi* and family *Rachycentridae*, characterized as voracious and fast-moving. Three dark stripes on the sides of the body provide identification. Cobia is a large fish, weighing up to slightly in excess of 100 pounds (45.4 kilograms) and achieving a length of nearly 6 feet (1.8 meters). Its primary diet is fish, but it also consumes crab. Occurrence is in tropical and subtropical waters on a worldwide basis. Considered an excellent game fish, but only average for eating. The cobia is very well streamlined.

COBRAS. **Snakes.**

COCAINE. **Alkaloids; Drug Addiction.**

COCCIDIA. **Sporozoa.**

COCCIDIOMORPHA. **Sporozoa.**

COCCIDIOIDOMYCOSIS. Also known as Coccidioidal Granuloma or Valley Fever, this disease is caused by the dimorphic fungus *Coccidioides immitis.* The fungus inhabits desert soils and areas of a semiarid nature, characterized by hot, dry summers, mild winters, and moderate rainfall. These areas are found in parts of California, notably the San Joaquin Valley, in southern Arizona, Utah, New Mexico, Nevada, and southwestern Texas. There are also similar areas in Mexico and Central America where the fungus is found. The disease is contracted through the inhalation of spores (highly infectious arthrospores). Although direct transfer of the disease by human contact has not been shown, cases have been reported of persons being infected from spores contained in packages that may have been shipped hundreds of miles from the fungus' natural habitat.

In the endemic region, as many as 80% of the persons residing in the region may contract the disease during their first 5 years of residence. Small children are particularly vulnerable to infection. The primary infection is often mild and passes unnoticed. In the more serious, granulomatous form, the disease may involve bones, joints, skin, subcutaneous tissues, and internal organs. This form occurs rarely and strikes only persons of low resistance.

It is estimated that about 40% of persons who inhale *C. immitis* arthrospores will develop an influenzalike illness within 7 to 28 days after exposure. The other 60% of the persons will show a positive skin test, but no other indication of the infection. A common symptom in children is a rash, particularly on the palms and soles. The disease is treated by amphotericin B, an antimicrobial agent used in the treatment of other fungus infections, such as blastomycosis.

COCCIDIOSTAT. **Pyridine and Derivatives.**

COCCUS. One of a family of the order of *Eubacteriales* which includes bacteria whose cells are spherical in form. See **Bacteria.**

COCCYX. The lower end of the spinal column. It is composed of four rudimentary small vertebrae which are usually fused together. See **Skeletal System.**

COCHLEA. The auditory portion of the inner ear of vertebrates. See also **Hearing and the Ear.**

COCHRAN'S THEOREM. A theorem in statistics concerning quadratic forms. If $x_1, x_2, \ldots, x_n$ are independent normal variables and $Q_1, \ldots, Q_k$ are quadratic forms in those variables with ranks $n_1, \ldots, n_k$; if $\Sigma_{i=1}^{k} Q_i = \Sigma_{i=1}^{n} x_i^2$, then the necessary and sufficient condition for the Q's to be independent χ^2 variables with n_i degrees of freedom respectively is that $\Sigma_{i=1}^{k} n_i = n$. The theorem is basic to tests of significance in the analysis of variance.

COCKATIEL. **Parrots and Cockatoos.**

COCKATOOS. **Parrots and Cockatoos.**

COCKCHAFER (*Insecta, Coleoptera*). A European beetle, *Melolontha vulgaris*, related to the May beetles of North America.

COCKCROFT-WALTON GENERATOR. **Particles (Subatomic).**

COCKLE (*Mollusca, Pelecypoda; Cardium*). Marine bivalve mollusks of several species. Cockles of commercial importance are described in the entry on **Mollusks.**

COCKROACH (*Insecta, Orthoptera*). Flattened oval insects, usually brown in color. The head is almost concealed by the broad margins of the thorax and in winged species the wings overlap above the body.

Cockroaches are widely known from a few species which inhabit houses, especially where quantities of food are available. They eat almost anything used by man as food and often damage other things, such as articles made of cloth containing sizing or paste. They are especially troublesome in restaurants. The two most important pests are the Croton-bug, *Blatella germanica*, and the Oriental cockroach, *Blatta orientalis*. They can be destroyed by sprinkling borax (see **Boron**), sulfur, or pyrethrum powder liberally about their hiding places or by the use of commercial roach pastes.

The cockroaches gain their greatest development in the tropics, where species with a normal length of more than 2 inches (5 centimeters) are found.

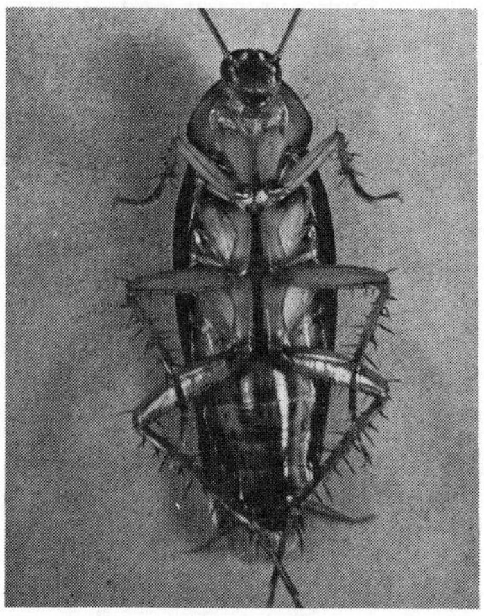

Cockroach. (*A. M. Winchester*)

COCONUT OIL. **Vegetable Oils (Edible).**

COCONUT PALM. **Palm Trees.**

COCOON STAR. **Infrared Astronomy.**

CODDINGTON EYEPIECE. An eyepiece made from a single piece of glass with a groove cut around its equator to act as a stop. The two convex surfaces are portions of the same sphere. Probably first made by Sir David Brewster.

CODE (Computer System). A code is a system of symbols for representing data or instructions in a computer or data processing machine. A machine language program sometimes is referred to as code.

Alphanumeric Code. A set of symbols consisting of the alphabet characters A through Z and the digits 0 through 9. Sometimes the definition is extended to include special characters, such as \$, %, and &. A programming system commonly will restrict the user-defined symbols to only those using alphanumeric characters and for the system to take special action on the occurrence of a nonalphanumeric character. A job currently in progress, for example, may be stopped should the \$ character be encountered.

Binary Code. (1) A coding system in which the encoding of any data is done through the use of bits; i.e., 0 or 1. (2) A code for the ten decimal digits, 0, 1, . . . , 9 in which each is represented by its binary, radix 2, equivalent; i.e., straight binary.

Biquinary Code. A two-part code in which each decimal digit is represented by the sum of the two parts, one of which has the value of decimal zero or five, and the other the values zero through four. The abacus and soroban both use biquinary codes. An example follows:

DECIMAL	BIQUINARY	INTERPRETATION
0	0 000	0 + 0
1	0 001	0 + 1
2	0 010	0 + 2
3	0 011	0 + 3
4	0 100	0 + 4
5	1 000	5 + 0
6	1 001	5 + 1
7	1 010	5 + 2
8	1 011	5 + 3
9	1 100	5 + 4

Column-binary Code. A code used with punch cards in which successive bits are represented by the presence or absence of punches on contiguous positions in successive columns as opposed to rows. Column-binary code is widely used in connection with 36-bit word computers where each group of 3 columns is used to represent a single word. Synonymous with code, chines binary.

Computer Code or *Machine Language Code.* A system of combinations of binary digits used by a given computer.

Excess-three Code. A binary coded decimal code in which each digit is represented by the binary equivalent of that number plus three; for example:

DECIMAL DIGIT	XS 3 CODE	BINARY VALUE
0	0011	3
1	0100	4
2	0101	5
3	0110	6
4	0111	7
5	1000	8
6	1001	9
7	1010	10
8	1011	11
9	1100	12

Gray Code. A binary code in which sequential numbers are represented by expressions which are the same except in one place and in that place differ by one unit; e.g.,

DECIMAL	BINARY	GRAY
0	000	000
1	001	001
2	010	011
3	011	010
4	100	110
5	101	111

thus in going from one decimal digit to the next sequential digit, only one binary digit changes its value. Synonymous with cyclic code.

Instruction Code. The list of symbols, names and definitions of the instructions which are intelligible to a given computer or computing system.

Mnemonic Operation Code. An operation code in which the names of operations are abbreviated and expressed mnemonically to facilitate remembering the operations they represent. A mnemonic code normally needs to be converted to an actual operation code by an assembler before execution by the computer. Examples of mnemonic codes are ADD for addition, CLR for clear storage and SQR for square root.

Numeric Code. A system of numerical abbreviations used in the preparation of information for input into a machine; i.e., all information is reduced to numerical quantities. Contrasted with code, alphabetic.

Symbolic Code or *Pseudo Code.* A code which expresses programs in source language; i.e., by referring to storage locations and machine operations by symbolic names and addresses which are independent of their hardware determined names and addresses.

Two-out-of-five Code. A system of encoding the decimal digits 0, 1, . . . , 9 where each digit is represented by binary digits of which 2 are zeros and 3 are ones or vice versa.

See also terms listed under **Data Processing.**

Thomas J. Harrison, International Business Machines Corporation, Boca Raton, Florida.

CODEINE. Alkaloids; Drug Addiction.

CODFISHES (*Osteichthyes*). Along with hakes and rattails, codfishes are of the order Anacanthini and specifically of the family Gadidae. All codfishes are marine with exception of the burbot, *Lota lota,* a species which ranges from the polar regions southward in North America and Eurasia. The marine codfishes prefer cold or temperate water, as contrasted with tropical climes. Their occurrence is much greater in the northern than in the southern hemisphere. Codfishes are among the greatest of world seafood sources. See also **Fisheries.**

Included in the codfish family are the pollack, the haddock, and the whiting. Because of certain anatomical differences, some investigators do not consider hakes as members of the family Gadidae, but rather as belonging to a separate family, Merlucciidae.

Atlantic Cod. This is the largest of about 150 species. The Atlantic cod may attain a length of 6 feet (1.8 meters), and weigh up to 210 pounds (95 kilograms). The commercial catch, however, usually averages from 2.5 to 25 pounds (1.1 to 11.3 kilograms). These fish spawn between January and March. A 75-pound (34-kilogram) female may lay as many as 9 million eggs. The eggs first float for a period of up to 20 days. Upon hatching, the larval forms attach themselves to floating plankton for a period of another 60 to 75 days. At the end of this period, they are about 1 inch (2.5 centimeters) in length and sink to the bottom where their subsequent growth is quite rapid. A 2-year-old cod will achieve a length of about 15 inches (38 centimeters). Normally, they are not capable of spawning until about 5 years old.

The most noticeable external characteristics are its 3 dorsal and 2 anal fins, its protruding upper jaw, its almost square tail, and a pale line running along each side of the body from head to tail. There is a fleshy barbel under the lower jaw. In most fish, the upper part of the body is thickly speckled with small, round spots somewhat darker than the body color, which may range from reddish to brown, gray, or greenish. Cod can be found from shallow water near shore, down to 250 fathoms (1500 feet; 450 meters). The cod's usual habitat is within a few fathoms of the bottom, but it also comes to the top of the water in pursuit of small fish or squid. It is most plentiful on the banks and in oceans of moderate depth. The Atlantic cods live chiefly over rocky, pebbly ground, on sand or gravel, and seldom on soft mud. They go in schools, but not in such dense bodies as mackerel and herring. The movements on and off shore and from bank to bank are due chiefly to temperature influence, the presence or absence of food, and the search for proper spawning conditions. Cod prefer temperatures of 32 to 41°F (0 to 5°C), but good catches can be made in waters up to 50°F (10°C). Cod feed on almost all types of sea life. The most important food is fish, especially herring, capelin, and sand lance, but mussels, crabs, and other bottom animals are also consumed.

The common Atlantic cod is well known on both sides of the north Atlantic Ocean. On the American coast, it is found as far north as Greenland, Davis Strait, and Hudson Strait, and south nearly to Cape Hatteras. In Europe, it is found from Novaya Zemlya, Spitzbergen, and Jan Mayen to the Bay of Biscay. It is also common near Iceland and the Faroe islands.

The cod (*Gadus callarias*), shown in Fig. 1, is caught in large quantities by British fishermen and constitutes about one-third of the total tonnage taken in British fisheries.

Whiting. This fish (*Gadus merlangus*) is essentially a near-shore fish, and occurs in large quantities around the northern coasts of Britain. It figures prominently in the landings at Scottish ports. Less esteemed than the Atlantic cod or haddock, the whiting usually finds a less favorable market.

The whiting is smaller and more short-lived than most other important codfishes. Its meat is quite popular in Great Britain and France, but less so in Germany. Its far-flung distribution extends across the entire northern and western European shelf—from the north polar cap to the Atlantic coast of Spain and the southern coast of Iceland. One subspecies, the Mediterranean cod or *molo,* extends the distribution into the Mediterranean and Black Seas. Whitings are not found in the northwestern Atlantic Ocean. They are most prevalent in the North Sea and off the western coast of England. Whiting resembles haddock somewhat in coloration, and also has a lateral black spot, which is much smaller than that of the haddock and is located at the base of the pectoral fin. As pure shelf inhabitants, whitings are found in more shallow water than haddock and are prevalent in regions with soft, muddy ground at depths of 2 and 5 feet (0.6 and 1.5 meters). Their diet consists mainly of small crustaceans and fishes. Young whitings maintain the same relationship with medusas as do haddock.

Pacific Cod. This fish (*Gadus macrocephalus*) achieves a length up to 4 feet (1.2 meters). It is believed that this species developed from the Greenland cod group, penetrating into the Pacific from the Bering Strait, and is now found on both sides of the Pacific Ocean. It is widely distributed, but as a coastal inhabitant, it does not migrate. The size and coloration are somewhat similar to that of the Atlantic cod. The maximum age is from 10 to 12 years. The diet is quite diverse, but consists primarily of crustaceans and fishes. Sexual maturity is attained after 5 to 6 years. The spawning season is in late winter. Eggs are laid in enormous numbers (similar to Atlantic cod), but instead of rising, they sink to the floor, an adaptation to the more localized nature of these codfishes.

Haddock. This fish (*Melanogrammus aeglefinus*) ranges in length up to nearly 3.5 feet (107 centimeters). See Fig. 2. Its distribution in the northern Atlantic Ocean is similar to that of the Atlantic cod. However, haddock are not usually found off Greenland, and in the northwestern Atlantic Ocean. Haddock are usually found only off the southern coast of Newfoundland, off Nova Scotia, and in the Gulf of Maine. Haddock differ from all other codfishes by a black spot above the pectoral fin. The lower jaw is very short and the barb is small. Its chief diet consists of invertebrate, bottom-dwelling organisms and herring spawn. As a pure shelf inhabitant, haddock is rarely found below 655 feet (200 meters). The oldest age of this medium-size fish is about 14 years. Like Atlantic cod, haddock undertake periodic migrations between the feeding grounds and the spawning sites. In the northeastern Atlantic Ocean, the chief spawning sites are in the northern North Sea and off the Norwegian coast. Haddock reach sexual maturity after 3 to 4 years, spawning in the spring, somewhat later than Atlantic cod. The haddock has lower fertility, but the development period of the eggs is shorter than in the Atlantic cod. The initially pelagic young are often found under the umbrellas of large medusas before changing to bottom dwelling in the autumn of their first year of life. Their length at that time is about 4 inches

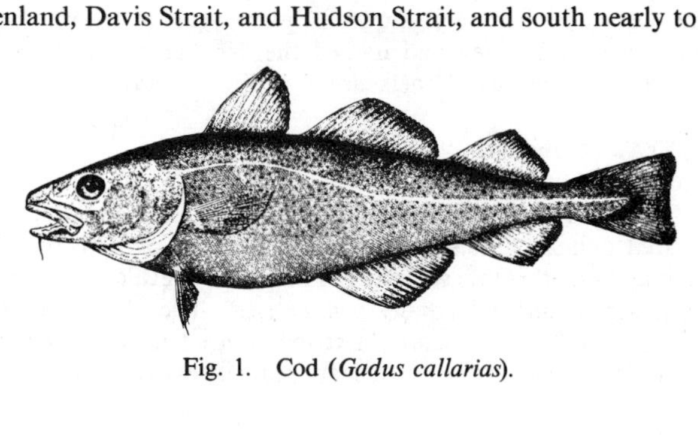

Fig. 1. Cod (*Gadus callarias*).

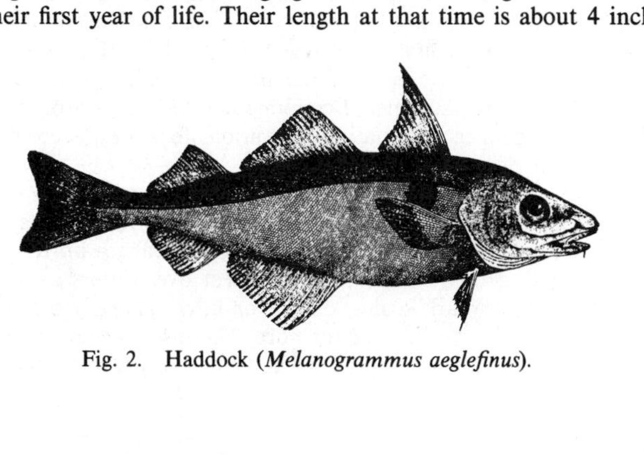

Fig. 2. Haddock (*Melanogrammus aeglefinus*).

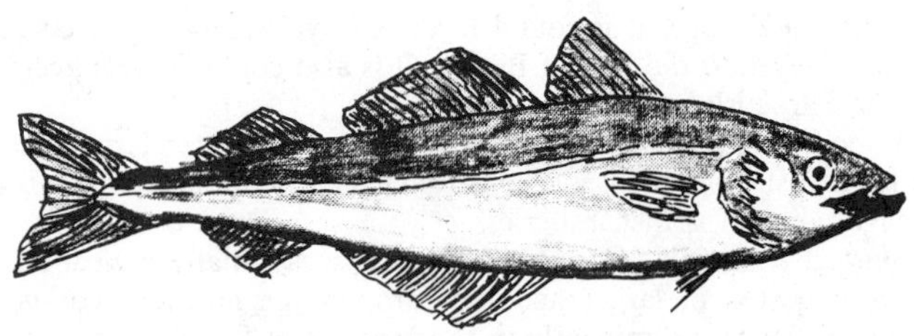

Fig. 3. Pollack (*Pollachius virens*).

(10 centimeters). Unlike other codfishes, which migrate toward the shallow coastal waters, haddock spend their first years in the open sea. Finnan haddie is smoked haddock.

Silvery Pout. This fish (*Gadiculus*) is the smallest codfish species, ranging in length up to about 6 inches (15 centimeters). Pout is found at greater depths—from 1310 to 3280 feet (400 to 1000 meters) on the continental shelf. The northern species (*Gadiculus thori*) is distributed from Norway and southern Iceland to the Bay of Biscay, while the southern species (*Gadiculus argenteus*) is found from the southernmost part of the distribution of *G. thori*, to the northern Africa coasts and the western Mediterranean.

Pollack. Also known as the *saithe* or *blister-back*, this fish (*Pollachius virens*) has about the same commercial significance as the haddock. In body shape, fin position and size, it resembles the Atlantic cod, but other characteristics clearly distinguish it. See Fig. 3. The lower jaw protrudes somewhat and has a very small barb. The dark-colored mouth is also a distinctive feature. The pollack is a pelagic predator which, as a fast and skilled swimmer, feeds chiefly on schooling fishes, especially herring. However, its extensive migrations follow less well-established routes than those of other codfishes. Its constantly changing life habits make following its growth dynamics difficult, a factor which adds to the difficulties for pollack fishers. The fish can reach an age of 18 to 20 years. The pollack attains sexual maturity in 4 to 5 years and thereafter spawns each spring in practically the same region as haddock. The eggs and larvae drift with the current and the young migrate into coastal waters after their pelagic development period. They spend their first year in coastal waters.

Alaska Pollack. This fish (*Theragra chalcogramma*) is not to be confused with the pollacks just described. The Alaska pollack has the same distribution as the Pacific cod, but is less dependent on being near the coast and may be found at depths as great as 985 feet (300 meters). This species is found in open, warmer (as compared with Pacific cod) water. Its diet consists of plankton and small schooling fishes. In the western Pacific Ocean, isolated populations have developed differences in growth and spawning seasons (October–December versus spring). In recent years, the Alaska pollack has become quite important commercially.

See also **Fishes.**

CODING THEORY. Information Theory.

CODLING MOTH (*Insecta, Lepidoptera*). The moth, *Carpocapsa pomonella*, whose larva lives in apples. An important economic species.

This insect is so serious an enemy of the apple that the successful production of the fruit depends on a program of spraying which has been carefully worked out for all parts of the country. Lead arsenate is an effective poison but its application must be regulated according to the entrance of the caterpillars into the fruit. Once they have penetrated the surface they are beyond reach of sprays. The principle of spraying is to apply a first spray when the petals of the flowers fall, and a second when the eggs of the next generation of insects are hatching later in the summer. Economic entomologists are prepared to furnish the proper information for various localities, according to the conditions of the year.

The adult codling moth is a grayish brown color, with brown wingtips, and a wingspan from $\frac{1}{2}$ to $\frac{3}{4}$ inch (12 to 18 millimeters). The larva is white to pink, with a brown head, and ranges up to $\frac{1}{2}$ inch (12 to 13 millimeters) in length. The insect overwinters as a larva in a cocoon under bark scales, debris, or litter on the ground. The larva is found in the fruit near the core. The insect occurs wherever apples are grown. Blemish marks called "stings" result when the larva chews into the fruit that has been treated with a slow-acting insecticide, such as lead arsenate. The moth emerges and lays eggs about the time the apple is in bloom. The eggs hatch in about 5 days, after which the small larvae begin to feed. There are four generations per season.

Codling moths lay eggs mostly on the upper and under surfaces of the leaves, although a few may be found on the fruit and branches. Eggs of the codling moth are parasitized by a minute wasp, *Trichogramma*, and they are attacked by mites. The larvae also have several parasites. Over a dozen species of birds are known to feed on this pest. The downy woodpecker, nuthatch, and chicadee destroy great numbers of hibernating larvae. In spite of the array of natural enemies, the codling moth remains the most destructive insect preying on apple.

Noninsecticidal control methods include the creation of a substitute location for the larvae to spin their cocoons and pupate. This can be accomplished by banding the trunks of trees by tying 6-inch (15 centimeter) strips of burlap or cardboard around them. These bands can be placed around the trunk or around large branches. Insects found in these bands should be collected and crushed immediately. To reduce this insect population further, loose bark should be scraped from the trees and any bark collected on the ground should be removed. All debris in the immediate area that may provide shelter for the insect should be removed.

COEFFICIENT. An algebraic factor. Also a quality or parameter that is characteristic of a substance or a system.

COEFFICIENT OF AGREEMENT. Agreement (Coefficient of).

COEFFICIENT OF ASSOCIATION. Association (Coefficient of).

COEFFICIENT OF CONCORDANCE. Concordance.

COEFFICIENT OF DISCHARGE. Discharge (Coefficient of).

COEFFICIENT OF VARIATION. Variation (Coefficient of).

COELACANTHS (*Osteichthyes*). Of the order *Actinistia*, the coelacanth was first discovered in the waters off South Africa in late 1938. The fish measured about 5 feet (1.5 meters) in length and weighed close to 130 pounds (59 kilograms). It previously had been assumed that fishes of this kind had been extinct for some 60 million years. Fossil records had indicated the former existence of coelacanths. Unfortunately, this first *Latimeria chalumnae* had been mounted and the internal organs destroyed before the fervid interest of scientists could intervene. After long and arduous explorations of the waters around the coast of South Africa, Professor J. L. B. Smith landed the second coelacanth. However, it differed from the first in that the first dorsal fin and middle tail fin were missing. Thus, this particular fish was considered a different species and given the name *Malania anjounae*, a name later abandoned, however, when the omissions were simply considered aberrations. The second fish was found some 1900 miles (3057 kilometers) from the site of the first catch, namely, the waters off Anjouan Island, one of a group of islands in the northern end of the Mozambique Channel, which lies between Mozambique and Madagascar.

Fortunately, only 9 months elapsed before the catch of the third coelacanth, also in the same region. Since then a number has been caught. The eighth specimen was the first one that was kept alive for nearly 18 hours. After this scientific "discovery," it was learned that the coelacanth had been known for many years by the inhabitants of Comorro Island. They had named the fish "kombessa." Perhaps many additional species of coelacanths will be discovered.

COELATA. Turbellaria.

COELENTERATA. The hydroids, jellyfishes, sea anemones, corals, and related animals. These forms make up a major division of the animal kingdom of simple structure. The body is developed from only two germ layers and is radially symmetrical. The alimentary track is sac-like, with a single opening, but sometimes has complex tubular branches. The nervous system is a scattered network of cells connected by their slender processes. Coelenterates have peculiar stinging cells (cnidoblasts) which discharge irritating nematocysts. Most species are marine and all are aquatic.

Two forms of individuals occur in this phylum, the polyp or hydroid and the medusa.

The economic importance of the phylum is limited principally to the corals. Precious coral is sold in considerable quantities and the rock corals have built up many of the oceanic islands.

The principal subdivisions of the phylum are the following:

Class Hydrozoa (*Hydromedusae*). Both polyps and medusae occur in these species, usually alternating in a reproductive cycle. Species are often colonial. In colonies individual polyps or hydranths are borne on branches of a common stalk which also gives rise to reproductive individuals or blastostyles from which medusae arise. All individuals are joined by the continuous digestive tract. Polymorphism occurs in some colonial species when polyps are modified to special functions. Most species are small but the Portuguese man-of-war and a few others are large.

Class Scyphozoa (*Scyphomedusae*). The jellyfishes. Usually free swimming species of moderate to large size. All individuals are medusae. Mostly marine.

Class Anthozoa (*Actinozoa*). Sea anemones, sea feathers, corals and allied species. Solitary or colonial marine species. The individuals are polyps. Colonies of some species build up massive hard deposits inside or outside of the body.

COELOM. The true body cavity, formed by the splitting of the middle germ layer of the body (mesoderm) and lined with a definite layer of cells. It is a perivisceral cavity.

The coelom is well developed in annelid worms, where it appears as a series of metameric chambers. It is of limited extent in other invertebrates but in the vertebrates gains great development. Here it is divided into a thoracic and an abdominal cavity and in the terrestrial species the thoracic cavity is further divided into pleural cavities containing the lungs and a pericardial cavity containing the heart. The abdominal or peritoneal cavity contains principally the greater part of the digestive tract.

Excretory organs open from the coelom in the annelids and lower vertebrates, and the liquid which it contains is apparently supplementary to the blood. In all cases the cavity furnishes a space into which developing organs may expand in complex animals.

COELOMATA. Animals which have a coelom. The term embraces the phyla Bryozoa, Brachiopoda, Phoronides, Chaetognatha, Echinodermata, Mollusca, Annelida, Arthropoda, and Chordata.

COELOMMODUCT. A tubule opening at one end into the coelom and at the other on the surface of the body. It occurs in a simple form in some annelid worms as an excretory organ and in other species is associated with the nephridial tubule to form a more complex excretory structure. Coelomoducts are regarded as the evolutionary forerunners of the excretory tubules of vertebrates. See **Excretory System.**

COELOSTAT. In many types of astronomical research, it is desirable to have the main instrument stand still. To accomplish this purpose and also allow for the apparent motion of the celestial sphere, it is necessary to reflect the light by a moving mirror from the object in question into the instrument. Such a device is known as a *coelostat.* In the coelostat, a mirror is mounted parrallel to the polar axis of an equatorial mounting. The axis is rotated by clockwork from east to west at such a rate that it would complete one rotation in 48 hours of sidereal time. Since the celestial sphere rotates from west to east once in 24 hours of sidereal time, and reflection doubles the angle, this rotation of the polar axis will compensate for the apparent rotation of the celestial sphere. With one single mirror mounted in the manner described, the direction in which the light will be reflected will depend entirely upon the declination of the object observed. To obtain any desired direction of reflection, a second mirror is employed to send the stationary beam from the first mirror in the desired direction.

In case it is desired to hold an image of the sun apparently stationary, the polar axis must be rotated once in 48 hours of solar time. Such an instrument is known as a *heliostat.* Other types of mounting, used for particular purposes, are known as *siderostats.*

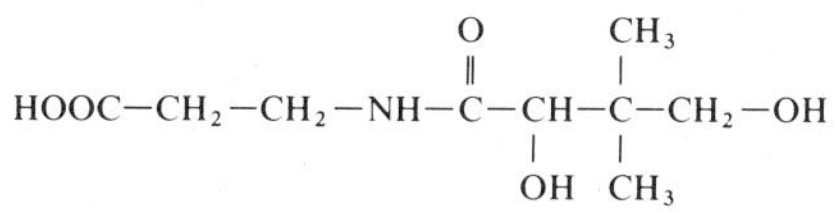

(Pantothenic acid)

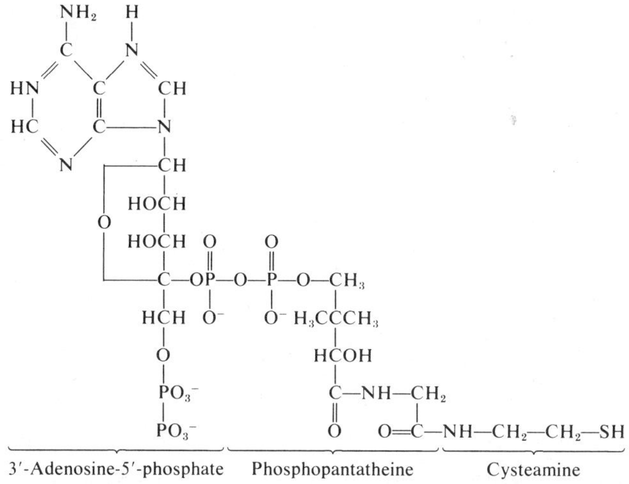

Fig. 1. Structure of CoA, composed of three parts: a nucleotide part derived from 3′-adenosine-5′-phosphate, forming a phosphodiester bond with a 4-phospho derivative of pantothenic acid, and a third part derived from the amino acid, *cysteine.* The side chain —SH group of the latter is free in this compound and is readily acylated, and thus able to act as a carrier for acyl groups in biochemical reactions in which it transfers that group between two substrates.

COENENCHYME. The middle and outer tissues of certain coelenterates (Alcyonaria); common tissue connecting polyps or zooids of a compound coral.

COENOCYTIC. A condition found in many filamentous fungi and some filamentous algae which have no cross walls to the filaments. Thus, even though the organisms are multicellular, there is no distinct separation into cells. The nuclei flow freely in the cytoplasm within the tube-like outer cell wall.

COENZYMES. A nonprotein substance that is closely associated with or bound to the protein component (*apoenzyme*) of an enzyme. Together, the coenzyme and apoenzyme form the complete enzyme known as the *holenzyme.* The presence of a coenzyme is necessary for enzyme activity. Coenzymes are organic molecules of a size intermediate between the small-molecule intermediary metabolites, which serve as the substrates of enzymatic reactions, and the macromolecular proteins. Each coenzyme acts usually as acceptor or donor of some specific type of atom or group of atoms to be removed from or added to a small-molecule substrate in a reaction catalyzed by the holoenzyme.

Coenzyme A (*CoA*). Pantothenic acid is a constituent of coenzyme A, which participates in numerous enzyme reactions. CoA (Fig. 1) was discovered as an essential cofactor for the acetylation of sulfanilamide in the liver and of choline in the brain. It has been established that CoA is involved in many biochemical reactions in the body as an "activator" of normally less reactive carbon fragments and a "transferer" of these fragments to different molecules. CoA is particularly important in the initial reaction of the citric acid cycle of carbohydrate metabolism and energy production. After oxidative decarboxylation of pyruvic acid, CoA combines with the two-carbon acetate fragment to form acetyl-CoA or "active" acetate.

$$\underset{\text{(Pyruvic acid)}}{CH_3-\overset{\overset{\displaystyle O}{\|}}{C}-COOH} + CoA \rightarrow \underset{\substack{\text{Acetyl-CoA} \\ \text{("active acetate")}}}{CH_3-\overset{\overset{\displaystyle O}{\|}}{C}-CoA} + CO_2$$

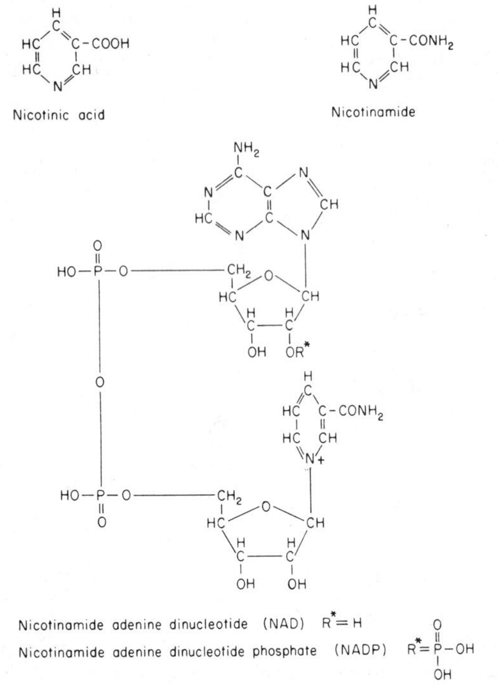

Fig. 2. Structures of nicotinic acid, nicotinamide, and nicotinamide coenzymes.

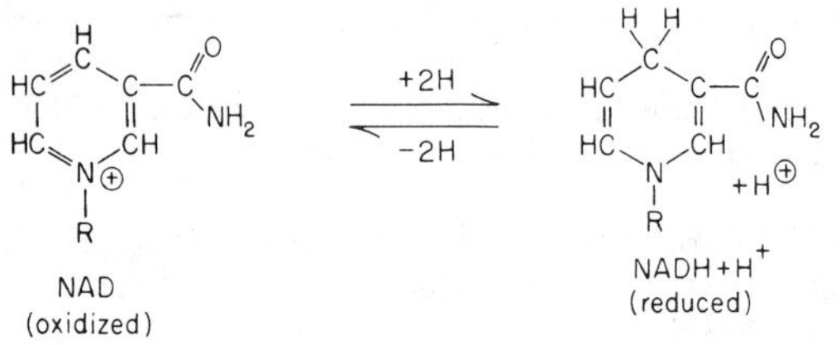

Fig. 3. Oxidized and reduced states of nicotinamide coenzymes as shown in Fig. 2.

Coenzyme A is necessary for the activation, synthesis, and degradation of fatty acids. Synthesis of cholesterol and ultimately the production of steroid hormones are also coenzyme A dependent.

Nicotinic Acid Coenzymes. Nicotinic acid can be converted to nicotinamide in the body and, in this form, is found as a component of two oxidation-reduction coenzymes (Fig. 2): *nicotinamide adenine dinucleotide* (NAD); and *nicotinamide adenine dinucleotide phosphate* (NADP). The nicotinamide portion of the coenzyme transfers hydrogens by alternating between an oxidized quaternary nitrogen and a reduced tertiary nitrogen. See Fig. 3.

Enzymes that contain NAD or NADP are usually called *dehydrogenases.* In excess of fifty NAD-dependent enzyme systems are known to exist. They participate in many biochemical reactions of lipid, carbohydrate, and protein metabolism. An example of an NAD-requiring enzyme is lactic dehydrogenase, which catalyzes the conversion of lactic acid to pyruvic acid. NADP is an essential coenzyme for glucose-6-phosphate dehydrogenase which catalyzes the oxidation of glucose-6-phosphate to 6-phosphogluconic acid. This reaction initiates metabolism of glucose by a pathway other than the citric acid cycle. The alternate route is known as the phosphogluconate oxidative pathway, or the hexose monophosphate shunt. The first step is:

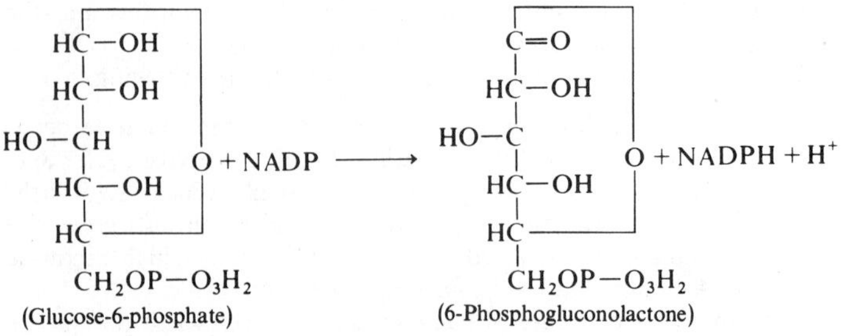

In the biological oxidation-reduction system, reduced NAD (i.e., NADH) is reoxidized to NAD by the riboflavin-containing coenzyme FAD (*flavin-adenine dinucleotide*).

Riboflavin Coenzymes. Riboflavin has been shown to be a constituent of two coenzymes: *flavin mononucleotide* (FMN) and *flavin adenine dinucleotide* (FAD). See Fig. 4. FMN was originally discovered as the coenzyme of an enzyme system that catalyzes the oxidation of the reduced nicotinamide coenzyme, NADPH, to NADP. Most of the many other riboflavin-containing enzymes contain FAD. FAD is an integral part of the biological oxidation-reduction system, where it mediates the transfer of hydrogen ions from NADH to the oxidized cytochrome system. This is illustrated in Fig. 5. FAD can also accept

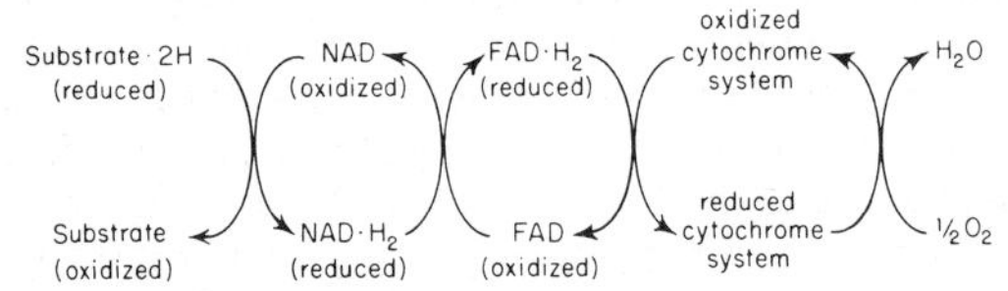

Fig. 5. Simplified representation of the biological oxidation-reduction system.

hydrogen ions directly from a metabolite and transfer them to either NAD, a metal ion, a heme derivative, or molecular oxygen. The various mechanisms of action of FAD are probably due to differences in the protein apoenzymes to which it is bound. The oxidized and reduced states of the flavin portion of FAD are shown in Fig. 6.

Decarboxylation Coenzymes. Thiamine, biotin and pyridoxine (vita-

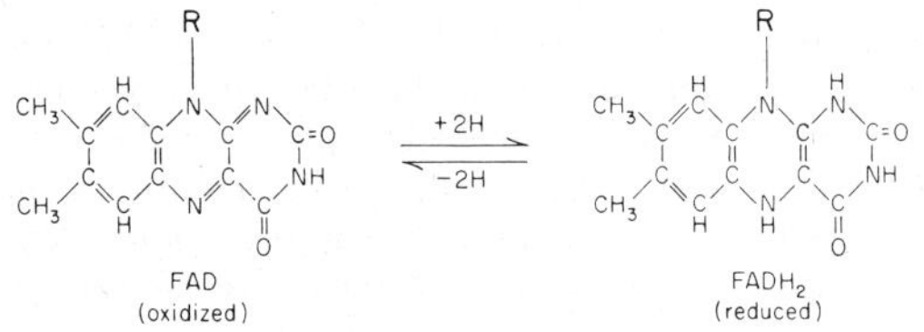

Fig. 6. Oxidized and reduced states of flavin coenzymes. R represents the remainder of the coenzyme as given in Fig. 4.

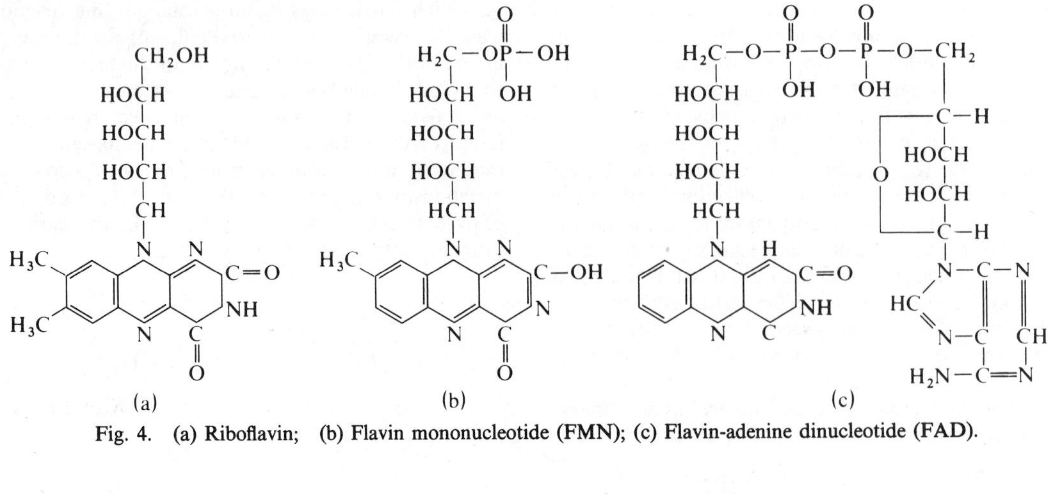

Fig. 4. (a) Riboflavin; (b) Flavin mononucleotide (FMN); (c) Flavin-adenine dinucleotide (FAD).

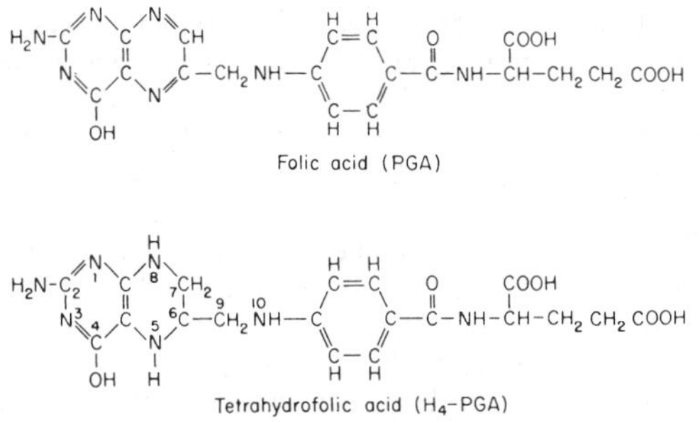

Fig. 7. Structures of folic acid and tetrahydrofolic acid.

min B) coenzymes are grouped together because they catalyze similar phenomena, i.e., the removal of a carboxyl group,—COOH, from a metabolite. However, each requires different specific circumstances. Thiamine coenzyme decarboxylates only alpha-keto acids, is frequently accompanied by dehydrogenation, and is mainly associated with carbohydrate metabolism. Biotin enzymes do not require the alpha-keto configuration, are readily reversible, and are concerned primarily with lipid metabolism. Pyridoxine coenzymes perform nonoxidative decarboxylation and are closely allied with amino acid metabolism.

Folic Acid Coenzymes. The coenzyme forms of folic acid are derivatives of tetrahydrofolic acid, FH_4. See Fig. 7. Folic acid functions as a coenzyme in enzyme reactions which involve the transfer of one-carbon fragments at various levels of oxidation. Vitamin B_2 (*cobalamin*) may be interrelated with folic acid in these reactions. Folic acid and vitamin B_{12} are also considered together since certain clinical anemias can be corrected by administration of either of the two vitamins.

Coenzyme Q. A series of quinones which are widely distributed in animals, plants, and microorganisms, these quinones have been shown to function in biological electron transport systems which are responsible for energy conversion with living cells. The nature and significance of coenzyme Q was first recognized in 1957. In structure, the coenzyme Q group closely resembles the members of the vitamin K group and the tocopherylquinones, which are derived from tocopherols (vitamin E), in that they all possess a quinone ring attached to a long hydrocarbon tail. The quinones of the coenzyme Q series which are found in various biological species differ only slightly in chemical structure and form a group of related, 2,3-dimethoxy-5-methyl-benzoquinones with a polyisoprenoid side chain in the 6-position which varies in length from 30 to 50 carbon atoms. Since each isoprenoid unit in the chain contains five carbon atoms, the number of isoprenoid units in the side chain varies from 6 to 10. The different members of the group have been designated by a subscript following the Q to denote the number of isoprenoid units in the side chain, as in coenzyme Q_{10}. The members of the group known to occur naturally are Q_6 through Q_{10}.

Coenzyme Q functions as an agent for carrying out oxidation and reduction within cells. Its primary site of function is in the terminal electron transport system where it acts as an electron or hydrogen carrier between the flavoproteins (which catalyze the oxidation of succinate and reduced pyridine nucleotides) and the cytochromes. This process is carried out in the mitochondria of cells of higher organisms. Certain bacteria and other lower organisms do not contain any coenzyme Q. It has been shown that many of these organisms contain vitamin K_2 instead and that this quinone functions in electron transport in much the same way as coenzyme Q. Similarly, plant chloroplasts do not contain coenzyme Q, but do contain *plastoquinones* which are structurally related to coenzyme Q. Plastoquinone functions in the electron transport processes involved in photosynthesis. In some organisms, coenzyme Q is present together with other quinones, such as vitamin K, tocopherylquinones, and plastoquinones; and each type of quinone can carry out different parts of the electron transport functions.

COESITE. **Astrobleme.**

COFFEE (Freeze-Dried). **Freeze-Drying.**

COFFEE TREE. Of the family *Rubiaceae* (madder family), genus *Coffea*, there are several species of this small tree or shrub, the seeds of which are obtained from a berrylike fruit. The seeds are the familiar coffee beans of commerce. The green beans are roasted, turning various shades of brown in the process. It is in this form that most people see the coffee bean. Although there are 20 or more species of the coffee tree, two species provide most of the coffee beans of commerce: *C. arabica* and *C. robusta*. The species *C. liberica* is also produced in large quantities. Because *C. arabica* is subject to attacks by insects, the latter two species were developed.

The coffee tree was originally found in Arabia, but is now grown in numerous tropical countries. In addition to a warm climate, the coffee tree requires much rainfall [70 inches (1778 millimeters) or more annually] and it cannot survive droughts. The timing of rainfall also plays an important role, with heavy rain desired early in the season when the fruit is developing and lighter rainfall when the fruit is ripening. *C. arabica* grows best at altitudes of about 2,000 feet (610 meters), whereas *C. liberica* thrives well from sea level up to about 2,000 feet (610 meters). The tree is an evergreen. The blossoms are small and white and the tree appears as covered with a light snow when in full bloom. The berries ripen between 6 and 7 months after flowering. The tree blooms and produces berries once each year, the exact time depending upon species and location. A tree will bear within 5 years from initial seeding and will yield good beans for commercial markets within 8 years. It will produce satisfactory commercial beans for a period of about 20 years thereafter and good beans, but in fewer numbers, for an additional number of years. Although some trees are greater producers, a normal tree will yield from 1 to $1\frac{1}{2}$ pounds (0.45–0.68 kilogram) of green coffee beans each year.

It is recorded that the Arabs cultivated the coffee plant as early as 600 A.D. It is first mentioned in the literature in about 900 A.D. Because of its stimulating qualities, its pleasant flavor and aroma, coffee was first recognized as a food and only later brewed as a beverage. The beverage first became popular in Arabia, gradually spreading into Turkey (circa 1554), thence to Italy (1615) and France (1644). The coffee plants were jealously guarded for many years, but over a period of years, plants were smuggled out of Arabia. The Dutch obtained a few early plants for botanical gardens in the Netherlands. After successful growth under controlled conditions, plants were sent to Java for cultivation and ultimately the coffee plant spread throughout the tropics. The first plants were brought to the Americas in about 1723. The principal countries now exporting it are Brazil, the greatest producer and supplying about 30% of the coffee consumed in the United States, Colombia, Libera, Ecuador, and other countries of Africa and the Middle East.

The tree averages from 15 to 30 feet (4.5 to 9 meters) in height. The branches start near the base of the trunk. There are wide, long

Close-up of extractors used in soluble coffee products plant. (*Stork Bowen Engineering Inc., Somerville, New Jersey.*)

leaves in clusters at the end of each branch. As the tree grows old, branching is less and leaves and berries become less plentiful. The tree is set in rows several feet apart. The native red soil in the environs of São Paulo seems particularly suited to successful coffee production. The berry is dried in large, frequently raked vats in the sun; or oven drying can be used. The ultimate producers of commercial coffees usually blend beans from various sources to obtain the desired, characteristic flavor and aroma of their brand. Mocha and Java coffees are fragrant varieties of Arabian coffee, for example. Medellin coffee from Colombia provides richness of flavor. Brazillian coffee provides an excellent base for blending. Costa Rican coffee is known for fragrance, El Salvadorian coffee for full body, and Mexican *coatepec* for a winelike flavor. Chicory, the dried, roasted, and ground root of a plant of species *Cichorium intybus*, is frequently blended in European coffees for special flavor. The amount of chicory used may range from 5 to 40%.

Over the last few decades, much research has gone into the processing and packaging of coffee, to make it more convenient to brew and to maintain its flavor or shelf life. Both freeze-drying and spray-drying have contributed to production of various instant-coffee products. See accompanying illustration. See also **Freeze-Drying.** Coffee finds limited use in the production of chemicals, plastics, and medicinals.

COFFERDAM. A cofferdam is a structure of a temporary nature, used to exclude water from an otherwise submerged area, for the purpose of preparing for foundations or for other subaqueous construction. Cofferdams are frequently required both above and below the site of a permanent dam in order to by-pass the stream through a temporary channel during the construction period. The simplest cofferdam is an earth dyke which should be used only in shallow water where there is little or no current. Facing an earth cofferdam with sand bags or constructing it entirely of sand bags will make it serviceable when there is a current which would wash away loose materials. A single line of sheet piling can be driven and braced with a form of earth or with sloping braces if the water is less than about 8 or 10 feet (2.4 or 3 meters) deep.

Sheet piling (see **Pile Foundation**) can be used to form a cofferdam around a site by erecting it in double parallel walls, the space between being filled with sand or gravel and clay. This mixture is thoroughly tamped in order to form a solid filling. This materially adds to the stability of the structure. Where a small area is to be unwatered, a single wall of sheet piling, internally braced, may be erected around the site.

When cofferdams constructed of sheet piling are to rest on hard bottom, a timber framework is required to hold the sheeting in place. These frames are generally built on shore, floated into position and sunk into place. Sheet piling is then placed around the outside and banked with earth.

After the cofferdam has been completed it must be unwatered by pumps. As sheet piling is not entirely watertight, leakage must be pumped out in order to keep the interior as dry as possible.

COFFIN STAR. Infrared Astronomy.

COHERENCE LENGTH. Superconductors.

COHERENCE (Statistics). Spectral Analysis.

COHERENCY. As applied to metallurgy, this term signifies a continuity between the lattice of a parent crystal and that of a precipitate particle embedded in the former. In general, a state of coherency can only exist as a result of strains set up in both the parent phase and the precipitate. A phase boundary in the ordinary sense does not exist in this case.

COHERENT BEAMS. Interference (Wave).

COHERENT LIGHT. Laser.

COHESION PRESSURE. Pressure.

COHO SALMON. Salmon.

COIL. This term applies to one or more turns of a conductor when wound as a definite unit of an electrical circuit. Thus there is the choke coil, or as it is sometimes called, impedance coil, as a number of turns of wire forming a coil used primarily for its reactance effect. The transformer is a unit of one or more coils used for transferring electrical energy by magnetic induction. Coils are particularly important in communications circuits where they serve in the above capacities but also form parts of the tuned circuits which make possible the complex systems. While the coil is ordinarily used for its inductive properties it inherently has both resistance and distributed capacity. The former is because of the resistance of the wire of which it is wound. The latter is due to the potential difference between turns which are separated by the turn insulation. At high frequencies this distributed capacity becomes extremely important and limits the usefulness of a given coil. Various special winding schemes have been used to minimize this effect. Electrical machines have coils as essential components; for example, field coils, and armature coils.

COIL (Distributed Capacitance). Distributed Capacitance (Coil).

COIL (Hybrid). Hybrid Transformer.

COIL (Inductance). Inductance; Induction Coil.

COINAGE METALS. Aluminum; Copper; Gold; Silver.

COINCIDENCE. A term used in counter technology to denote the occurrence of counts in two or more detectors simultaneously or within an assignable time interval. A *true coincidence* is one that is due to the detection of a single particle or of several genetically related particles. An *accidental, chance,* or *random coincidence* is one that is due to the fortuitous occurrence of unrelated counts in the separate detectors. A *delayed coincidence* is the occurrence of a count in one detector at a short, but measurable, time later than a count in another detector, the two counts being due to a genetically related occurrence such as successive events in the same nucleus.

COIR. Palm Trees.

COKE (Clean Process). Coal.

COKITE. The term applied by Lacroix in 1917 to natural coke, the result of the contact metamorphism of coal beds.

COL. Atmosphere (Earth).

COLCHICINE. Alkaloids; Gout.

COLD CATHODE. Cathode.

COLD (Common). Common Cold.

COLD FRONT. Fronts and Storms.

COLD SHORT. A metallurgical term to denote a brittle condition in a metal at temperatures below the recrystallization temperature.

COLD SORE. Dermatitis and Dermatosis.

COLD WALL. The steep water-temperature gradient between the Gulf Stream and (a) the slope water inshore of the Gulf Stream or (b) the Labrador current. See **Ocean.**

COLD WORKING. Iron Metals, Alloys, and Steels.

COLD-WORKED METAL. When metals are plastically deformed at a relatively low fraction (frequently <0.5) of their absolute melting temperatures, they are normally said to be *cold-worked.* An increase of *hardness* or strength is normally associated with such deformation. This hardening can be relatively stable as long as the metal is not heated high enough to cause extensive recovery and recrystallization to occur (see **Annealing**). In metals with high melting points, such as the alloys of iron, copper, and nickel, deformation at room temperature is normally considered to be cold-working. On the other hand, room temperature deformation of a low melting point metal such as lead is more properly designated hot-working. In this regard, it is interesting to note that lead is not normally hardened by room temperature deformation.

Cold-working is frequently used for hardening of commercial metal products. Thus, piano wire and the wire used in bridge construction obtain their hardness as a result of the final wire drawing operations. Sheet stock is often supplied in different degrees of hardness obtained by cold-rolling to the desired hardness. Mechanical working is sometimes the only feasible way to harden specific metals and alloys.

COLEMANITE. The mineral colemanite is a borate of calcium corresponding to a formula which is perhaps best represented as $Ca_2B_6O_{11} \cdot 5H_2O$. It occurs either as massive deposits or in monoclinic crystals. It has a subconchoidal fracture; hardness, 4–4.5; specific gravity, 2.42; vitreous to adamantine luster, may be colorless to milky white, grayish or yellowish; transparent to translucent. Colemanite was found originally in Death Valley, Inyo County, California, and has since been found rather widely distributed in San Bernardino, Los Angeles, Kern and Ventura Counties, California, as well as in Clark, Esmeralda and Mineral Counties in Nevada.

Colemanite was, until the discovery of kernite, the chief source of borax. Kernite, $Na_2B_4O_7 \cdot 4H_2O$, because of its easy solubility in water, has displaced very largely other boron-bearing minerals as a source of borax. Colemanite, kernite and inyoite (probably $Ca_2B_6O_{11} \cdot 5H_2O$); are lake deposits associated with other and rarer boron minerals, laid down during periods of volcanic activity or resulting from the leaching of the adjacent Tertiary sedimentary formations. Colemanite was named for Mr. William T. Coleman of San Francisco; Kernite and Inyoite were named from Kern and Inyo Counties, California.

See also terms listed under **Mineralogy.**

COLEOPTERA. The beetles. An order of insects usually recognizable by the thickened wing covers which meet in a straight line down the middle of the back. These wing covers, or elytra, are modified forewings. In most species of beetles they are thickened or horny but in some they are soft. In some species they are divergent and in some they are short, leaving much of the abdomen exposed. The typical condition of the elytra is found outside of this order only in the earwigs. Beetles have biting mouth parts and a complete metamorphosis in which the larval stage is often a grub.

This order of insects is the largest group of its rank in the animal kingdom, with almost 200,000 described species. It embraces almost the entire range of adaptation of the class, although very few beetles are parasitic. Many species are of economic importance.

An excellent reference on the systematics of beetles is "Monographie der Familie Platypodidae, Coleoptera," by Karl E. Schedl, published by Junk, The Hague, 1972.

The main families of Coleoptera include:

Burprestidae	Metallic wood borers
Cantharidae	Soldier beetles
Carabidae	Ground beetles
Cerambycidae	Long-horned beetles
Chrysomelidae	Leaf miners
Cicindelidae	Tiger beetles
Cleridae	Checkered beetles
Coccinellidae	Lady beetles
Cucujidae	Flat bark beetles
Curculionidae	Curculios, weevils, snout beetles
Dermestidae	Skin beetles
Dytiscidae	Predaceous diving beetles
Elateridae	Click beetles
Gyrinidae	Whirligig beetles or lucky beetles
Hydrophilidae	Water scavenger beetles
Lamypyridae	Fireflies or lightning beetles
Lucanidae	Stag beetles
Meloidae	Blister beetles
Mylarbridae (also called *Bruchidae*)	Bean and pea weevils
Ptinidae	Powder-post beetles, deathwatch beetles, drug store beetles
Scarabaeidea	May beetles or June bugs
Scolytidae (also called *Ipidae*)	Bark beetles
Silphidae	Carrion or burying beetles
Staphylinidae	Rove beetles
Tenebrionidae	Darkling beetles

COLEOPTILE. In the seeds of grasses the primitive bud or plumule is enclosed in a protective sheath called the *coleoptile.* During germination of the seed this coleoptile elongates, pushing its way out of the seed and up through the soil. It is very sensitive to light, growing directly toward a beam of light.

COL (Geometry). Saddle Point.

COLIC. This is a general term denoting abdominal pain which comes on quickly, is sharp and penetrating in character, intermittent, brief, and cramp-like. Biliary colic is a sharp severe pain which occurs with the passing of gallstones through the bile passages. Lead or painter's colic occurs in lead poisoning and is associated with increased intestinal peristalsis. Renal colic occurs with the passing of a stone through the ureter. Treatment of colic is symptomatic during an attack. Atropine is used to relieve muscle spasm, and morphine may be necessary for severe pain.

Commonly, the term colic is associated with periodic abdominal cramping noted in infants. Colic in a normally developing infant may arise from a variety of causes, including (1) hunger cramps confused with colic, (2) usual feeding periods may not be geared to the requirements of a particular infant, (3) a tight anal ring that makes it difficult for the infant to expel the stool, (4) sensitivity to certain foods ingested by the mother in the case of breast-fed infants, and (5) sensitivity to certain kinds of milk, the latter usually disappearing when the child is about six months old, but possibly reappearing as allergies in other forms in later years.

COLIE. Mousebird.

COLIIFORMES (*Aves*). This order of birds which is made up of the (family Coliidae) are only a little larger than finches, and are to be distinguished by long, stiff tails, prominent feather crests, and soft ragged plumage. They are extremely skillful climbers, whereby the special structure of their feet stands them in good stead: the first and the fourth toes are reversible; they can be turned forward and backward (pamprodactylous). The coloration of the males and the females is the same and they do not greatly differ in size. The eggs are relatively small and have a strangely rough, coarse-grained shell. The basic color is white. The clutch can number from two to five eggs. There is only one genus (*Colius*) with six species, all of which live exclusively in Africa south of the Sahara.

Probably this order have been given the name mousebirds not just because of their gray and brown feathers, but also because they scurry through the underbrush like mice. These birds are very sociable and can usually be seen in small family groups of 5–7, but occasionally in larger flights of 30 or more. They occur at the edges of forests, along rivers, and in areas with brushwood; they even live in cities in the Sudan and especially in South Africa. See also **Mousebird.**

COLITIS AND OTHER INFLAMMATORY BOWEL DISEASES. Based upon typical clinical findings, some authorities place inflammatory bowel disease into two major categories: (1) *ulcerative colitis,* and (2) *Crohn's disease.* There are other important diseases affecting the colon, including amebic colitis, diverticulitis, ischemic colitis, tuberculosis of the intestine, and cancer. See **Colon.**

Ulcerative Colitis. Although the incidence of ulcerative colitis is relatively high, particularly among some races and ethnic groups, the etiology of the disease is not well understood. For a number of years, some specialists considered the disease as psychosomatic, but these opinions have long been discounted. Attacks of ulcerative colitis range from mild to severe and the disease may recur and frequently worsen over a number of years. Mortality from an acute attack ranges from 0.4% (mild cases) to 15% (severe cases), depending to a large extent on the age and general health of the patient. The risk of colon cancer precipitated by ulcerative colitis increases with both the severity and duration of the disease.

Statistics show that the disease is from 2 to 4 times more prevalent among Jews than non-Jews; and about 4 times more frequent among whites than nonwhites. The disease is found in about 4 people per 100,000 population. Occurrence in females exceeds that in males. Ulcerative colitis is not considered common among persons past 60 years of age, but it is more serious and mortality is higher when it strikes elderly persons. Onset of the disease usually occurs between 25 and 40 years of age.

Early symptoms of ulcerative colitis are constipation and passage of blood or mucus with the stools. Some patients also indicate that an urgency to defecate may produce only small quantities of blood and mucus. This condition may persist for months and even years without diarrhea and any systemic symptoms. Mild ulcerative colitis appears to be limited to the distal (lower end) colon and rectum. The spread of the disease to other parts of the colon occurs only in about 15% of the cases.

In *mild ulcerative colitis* (60% of cases), the patient has intermittent diarrhea with no marked cramping or abdominal pain. There is no fever. Accurate diagnosis is largely dependent upon laboratory examinations and tests. These include sigmoidoscopy and barium-contrast x-ray examination. Tests will confirm presence of the disease, but not always its degree of severity. A general physical examination will normally yield satisfactory health. Therapy may include diphenoxylate hydrochloride with atropinesulfate (Lomotil®); or deodorized tincture of opium in water; or sulfasalazine (Azulfidine®); or corticosteroid enemas. Lomotil® acts by slowing intestinal motility. The mode of action of Azulfide® is not fully understood, but it may be related to the immunosuppressant properties.

In moderately severe ulcerative colitis (25% of cases), the patient will pass in excess of 5 stools per day. These will be watery or pasty and will contain significant quantities of blood and mucus. In addition to abdominal cramping, fatigue and intermittent low-grade fever (100.4°F; 38°C), some weight loss may be present. A physical examination will show some tenderness over the colon. There may be mild, intermittent anemia and extracolonic complications. The risk of colonic cancer is increased, particularly in a condition that has persisted for several years. Therapy consists of the use of antidiarrheal agents, sulfasalazine, and corticosteroid enemas, as well as the intermittent use of prednisone. Sometimes ACTH may be used.

In *severe ulcerative colitis* (15% of cases), hospitalization is commonly indicated. A fever between 100.4 and 104°F (38 and 40°C) may be present. The abdomen will be distended and there will be a definite tenderness over the colon. Therapy may commence as mentioned for the less severe attacks of colitis, but antidiarrheal agents are not prescribed. Prednosolone may be administered and where toxicity is present, ampicillin may be given. Where drug therapy is not fully effective, protocolectomy may be indicated.

Crohn's Disease. Similar in some respects and sometimes difficult to distinguish from other forms of colitis, Crohn's disease is also of unknown cause. There appears to be some familial involvement, although this also is unexplained. This disease, also called *granulomatous ileitis*, or simply *ileitis*, has been identified by physicians for about 50 years, but a marked distinction between it and ulcerative colitis just described has surfaced only within the last 20 years. Consequently, statistical data strictly reflective of Crohn's disease still are somewhat meager. Typical of the disease is inflammation of all layers of the bowel. Usually the disease is first evidenced by persons at about 30 years of age, but is bracketed between the ages of 20 and 40 years. The disease affects females somewhat more frequently than males. The incidence is about half that of ulcerative colitis. So-called regional ileitis (the originally named Crohn's disease) and granulomatous colitis have been identified as variations of the same syndrome.

The onset of Crohn's disease is much more subtle than ulcerative colitis and it is not uncommon for patients to delay reporting to a physician for a number of years. Some authorities have described this condition as a "smoldering" disease. Relatively mild early symptoms include lower abdominal pain, moderate diarrhea (no blood), loss of appetite, and sometimes a mild anemia. Fatigue is a common complaint. Complications may include burning of eyes and blurring of vision due to iritis, burning and urgency of urination, and arthritis, among others. In the findings of sigmoidoscopy examination, granular mucosa are usually found in Crohn's disease (but not in ulcerative colitis). The spikelike ulcerations are large (0.5–1.0 centimeter) and extend into the submucosa in Crohn's disease, whereas in ulcerative colitis, the pits are small (1 millimeter), superficial, and have been described as "collar buttons." The characteristics of inflammation also differ in the two diseases. Crohn's disease involves the nodular or stenotic distal ileum and right colon, whereas in ulcerative colitis, the sites, as previously mentioned, are the rectum and distal colon. Characteristic of Crohn's disease is the involvement of two or more separate areas (segmentation). Risk of colonic cancer from Crohn's disease (1% of cases) is less than in ulcerative colitis and does not appear to be related to the extent or duration of the disease.

Corticosteroid enemas are less effective in treatment of Crohn's disease; sulfasalazine usually effects improvement, but may require up to 4 months. Prednisone is much more effective in some patients, producing remissions within 8 weeks. It has been reported that combined sulfasalazine-prednisone therapy offers no improvement over prednisone when administered alone. It has been estimated that in the recent past about half the patients with Crohn's disease have required surgery. However, there is considerable difference of opinion as regards the long-term effectiveness of surgery. Many authorities do not suggest surgery except where there are persistent debilitating complications or uncontrollable systemic symptoms, i.e., when drug therapy is inadequate.

See also **Diarrhea;** and **Diverticulosis and Diverticulitis.** For brief descriptions of laboratory and examining procedures, see **Biopsy;** and **Sigmoidoscopy.** *Pseudomembranous enterocolitis* and *irritable bowel syndrome* (synonymous with *spastic colitis* or *mucus colitis*) are described under **Diarrhea.**

References

Brahme, F., Lindström, C., and A. Wenckert: "Crohn's Disease in a Defined Population: An Epidemiological Study of Incidence, Prevalence, Mortality, and Secular Trends in the City of Malmö, Sweden," *Gastroenterology*, **69**, 342 (1975).

Goode, A., et al.: "Use of An Elemental Diet for Long-term Nutritional Support in Crohn's Disease," *Lancet*, **1**, 122 (1976).

Kaplan, H. P., et al.: "A Controlled Evaluation of Intravenous Adrenocorticotropic Hormone and Hydrocortisone in the Treatment of Acute Colitis," *Gastroenterology*, **69**, 91 (1975).

Nugent, F. W., et al.: "Malignant Potential of Chronic Ulcerative Colitis: Preliminary Report," *Gastroenterology*, **76**, 1 (1979).

Shorter, R. G., and D. A. E. Shephard: "Frontiers in Inflammatory Bowel Disease: The Proceedings of a Conference Sponsored by the McReynolds Foundation (Parts 1 and 2), *Amer. J. Digest. Dis.*, **20**, 540, 639 (1975).

Singleton, J. W., et al.: "Trial of Sulfasalazine as Adjunctive Therapy in Crohn's Disease," *Gastroenterology*, **77**, 887 (1979).

COLLAGEN. The major protein component of connective tissue. In mammals, as much as 60% of the total body protein is collagen. It comprises most of the organic matter of skin, tendons, bones, and teeth, and occurs as fibrous inclusions in most other body structures. Collagen fibers are easily identified on the basis of the following characteristic properties: They are quite inelastic; they swell markedly when immersed in acid, alkali, or concentrated solutions of certain neutral salts and nonelectrolytes; they are quite resistant to most proteolytic enzymes, but are specifically attacked by the collagenases; they undergo thermal shrinkage to a fraction of their original length at a temperature which is characteristic of the collagen from a given animal, but this varies from one species to another; and they are converted in large part to soluble gelatin by prolonged treatment at temperatures above the thermal shrinkage level. Collagen fibers are not unique to mammals; collagen has been identified in the tissues of almost all multicellular animals, ranging from the primitive porifera and coelenterates, through the annelids and echinoderms, and up to the vertebrates.

As a protein, collagen is unusual in both chemistry and structure. Nearly one-third of its residues are glycine, and an additional 20–25% are imino acids (proline and hydroxyproline). In terms of sequence, glycine occurs regularly in essentially every third position, following as a steric requirement of the secondary-tertiary structure.

It appears that specific side-chain interactions between polar residues on adjacent collagen macromolecules are largely responsible for ordering the macromolecules into fibers. Specific cooperative interactions between functional groups on appropriately oriented macromolecules seem to be involved in the heterogeneous nucleation of hydroxyapatite crystals, and thus the initiation and control of mineralization in bones and teeth. As collagen fibers age, *in vivo*, they seem to become progressively more intermolecularly cross-linked, perhaps by the "ester-like" bonds formed. Little or no soluble collagen can be extracted from most mature connective tissue because of this extensive cross-linking, although the material can be converted into soluble gelatin by drastic thermal treatment.

Collagen and gelatin are of commercial importance. As insoluble collagen, this material may be cross-linked further by tanning and thus converted to leather. The soluble gelatins are used in the manufacture of foodstuffs, film emulsions, and glue.

See also **Bone;** and **Scleroderma.**

COLLAR. A fold or ridge of tissue more or less completely encircling the body behind its anterior end. In the snails, cuttlefishes, and related mollusks the ventral edge of the mantle is called the collar and in *Balanoglossus* (Chordata) the region of the body between the proboscis and the trunk is so named.

COLLAR CELL. A cell bearing a flagellum at one end, surrounded by a high membrane. Some of the one-celled animals and the choanocytes of sponges have this form.

COLLATERAL CIRCULATION. Auxiliary vessels for supplying blood to an area of tissue. If an artery is occluded, the collateral vessels expand to take over the task of supplying that area with blood. See also **Circulatory System (Human); Ischemic Heart Disease.**

COLLATOR. A data processing device used to combine sets or decks of cards or other information-bearing units into a desired sequence. Typically, a card collator has two input feeds so that two ordered sets may enter into the process; and four output stackers so that four ordered sets can be generated by the process. Three comparison stations are used to route the cards to one stacker or the other on the basis of comparison of criteria as specified by the collator controls. Collating is required where data from two or more physically separated files must be combined. Combining a file with names and addresses with another file containing one or more items of personal information is an example. The term *merge* usually signifies the combining of two similarly ordered sets of data into a single ordered set. The order, for example, may be alphabetical or numerical. A data set containing B, H, L, Q, and T may be combined with another set containing C, J, N, and S to produce the ordered set B, C, H, J, L, N, Q, S, and T. The combined set may be referred to as a *unified file.* The four basic operations involved in collating are merging, sequence checking, selection, and matching.

The term is also used for a program or routine which provides similar functions when applied to one or more files stored in a computer storage unit.

See also terms listed under **Data Processing.**

COLLECTING AGENTS (Chemical). Classifying (Process).

COLLECTOR (Solar Radiation). Solar energy.

COLLEMBOLA. The spring-tails. An order of primitive wingless insects characterized by a forked appendage at the tip of the body which is used in leaping. This appendage is bent forward beneath the body and when released snaps sharply down and back, projecting the animal into the air. Metamorphosis is absent. See accompanying illustration.

Spring-tails are small and delicate. They are found mostly in moist places on the ground or on bark, though a few species live in dry hot situations. The snow flea, which sometimes appears in large numbers on the surface of snow, is a spring-tail. Some species are found on the surface of water.

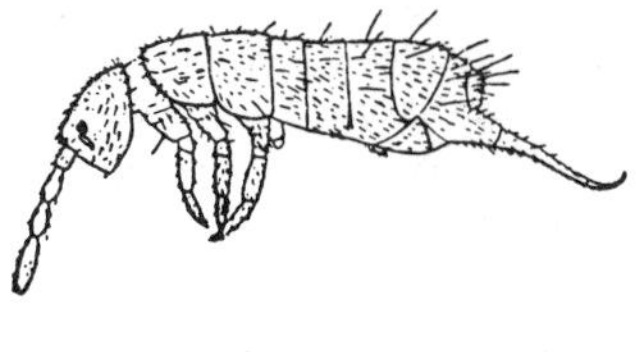

Spring-tail

COLLES' FRACTURE. A fracture of the arm near the wrist, in which the radius is broken in its lower quarter. It is one of the most common fractures and occurs usually from a fall on the outspread hand, or a direct blow against the wrist.

COLLETERIAL GLANDS. Glands associated with the female reproductive system of insects. They secrete materials which cement the eggs together or form a protective covering over them.

COLLIDINES. Pyridine and Derivatives.

COLLIDING-PARTICLE MACHINE. Particles (Subatomic).

COLLIGATIVE PROPERTY. A property, of a substance or system, which is determined by the number of particles present in the system but independent of the properties of the particles themselves.

COLLIMATOR. An optical apparatus for producing parallel rays of light. A common form consists of a converging lens, at one of whose focal points is placed a small source of light, usually a pinhole or narrow slit upon which light is focused from behind. Rays diverging from this focal point emerge from the objective lens in a parallel beam. The slit or other source is viewed through the collimator without parallax, since it appears at an infinite distance. The arrangement is very generally used on spectroscopes and spectrometers. By analogy, any arrangement of slits or apertures which limits a stream of particles to a beam in which all the particles move in the same, or nearly the same, direction is called a collimator.

Divergent rays from slit *S* rendered parallel by objective *O*.

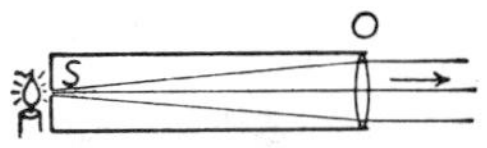

COLLIMATOR (Focal). Focal Collimator.

COLLISION. As used in physics, this term refers to any interaction between free particles, aggregates of particles, or rigid bodies in which they come near enough to exert a mutual influence, generally with exchange of energy. It does not necessarily imply actual contact. The process is always subject to conservation of momentum, and in an "elastic collision," also to conservation of energy. In the latter case, if the initial velocities are given, the velocities of the bodies after collision can be calculated by applying these two conservation principles. The subject is of special significance in atomic physics, where a collision is defined as a close approach of two or more photons, particles, atoms or nuclei during which an interchange occurs of charge, energy, momentum or other quantities. See also **Impact.**

COLLISION AVOIDANCE. Automotive Electronics.

COLLISION COEFFICIENT. In a two-body collision involving particles 1 and 2, moving in the same straight line, the coefficient of restitution is defined by

$$e = \frac{v_2 - v_1}{u_1 - u_2}$$

where $u_1 > u_2$ are the velocities with respect to a primary inertial system before collison and $v_2 > v_1$ are the corresponding velocities after collision. For a completely elastic collision $e = 1$. For an inelastic collision $d < 1$. See **Impact;** and **Restitution Coefficient.**

COLLODION EMULSION. Photography and Imagery.

COLLOID SYSTEM. Colloids are usually defined as disperse systems with at least one characteristic dimension in the range 10^{-7} to 10^{-4} centimeter. Examples include *sols* (dispersions of solid in liquid); *emulsions* (dispersion of liquids in liquids); *aerosols* (dispersions of liquids or solids in gases); *foams* (dispersion of gases in liquids or solids); and *gels* (system, such as common jelly, in which one component provides a sufficient structural framework for rigidity and other components fill the space between the structural units or spaces). All forms of colloid systems are encountered in nature. Products of a colloidal nature are commonly found in industry and are notably extensive in the food field. Foams, widely used in industrial products, but also the causes of processing problems are described in entires on **Foam;** and **Foamed Plastics.**

Early Background. Thomas Graham's investigations of diffusion (1861) led him to characterize as *crystalloids* substances, such as inorganic salts which in water solutions would diffuse through a parchment membrane; and as *colloids* (Greek word for glue) substances, such as starch and gelatin, which would not diffuse through the membrane. Sols with a given weight percent of dispersed material scatter light more strongly than a solution with the same weight percent of dissolved inorganic salt, i.e., a true solution. The Tyndall effect, in which the path of a beam of light through a turbid solution (or through dusty or smoke-filled air) is clearly defined through scattered light, is characteristic of sols. The slow diffusion and strong light scattering, together with the fact that the boiling-point elevation, freezing-point depression, and osmotic pressure caused by a given weight percent of dispersed material in sol form are much less than the corresponding magnitudes caused by the same weight percent of common inorganic salts-all of these observations indicated to early investigators that the particles dispered in the sol must be larger than those resulting from dissolving inorganic salts in water.

Development of the ultramicroscope (Siedentopf and Zsigmondy, 1903) permitted particles substantially smaller than the wavelength of light to be observed in scattered light and were thus capable of counting. Invention of the ultracentrifuge by Svedberg (1924) made it possible to cause particles in sols to sediment at observable rates, to measure these rates with reasonable precision, and to infer particle sizes from these rate measurements. The ultramicroscope and ultracentrifuge permitted validation of the early conclusions that colloidal particles are much larger than ions resulting from dissolving metal salts in water.

Svedberg found that in some sols the particles were highly uniform in size. For example, he found that the gram particle weight of insulin (a protein) was 40,900 and that apparently all insulin particles had this gram particle weight. This made it extremely likely that the insulin particles were either single molecules (albeit giant ones), or aggregates of a very definite number of smaller (but still quite large by ordinary standards) molecules. The research of Staudinger (commencing about 1920) and of Carothers (1929) opened up the field of macromolecular chemistry, leading to the recognition that giant molecules were not only abundant in nature, but could be prepared by established principles of chemistry. See also **Macromolecular Chemistry.**

It is interesting to note that Wolfgang Ostwald, in the late 1800s, stated, "There are no sharp differences between mechanical suspensions, colloidal solutions, and molecular (true) solutions. There is a gradual and continuous transition from the first through the second to the third."

Some colloidal systems are *thixotropic*, i.e., they differ in their fluid behavior from *pseudoplastic* substances in that the flow rate increases with increasing duration of agitation as well as with increased shear stress. When agitation is stopped, internal shear stress exhibits hysteresis. Upon reagitation, generally less force is required to create a given flow than is required for the first agitation. Examples of thixotropic materials include silica gel, most paints, glue, molasses, lard, fruit juice concentrates, and asphalts. By rhythmically shaking or tapping certain thixotropic suspensions, the suspensions will "set" or build up very rapidly. This type of non-Newtonian substance is said to be *rheopectic*. Bentonite sols and suspensions of gypsum in water are rheopectic. *Dilatant fluids* often are termed *inverted plastics* or inverted pseudoplastics. Initial flow under a low shear stress is at a high rate; further increases in shear stress, however, result in lower flow rate. Some liquids may change from thixotropic to dilatant or vice versa as the temperature or concentration changes. Examples of dilatant materials include quicksand, peanut (groundnut) butter, and many candy compounds. For comparison, it should be recalled that a Newtonian substance is a liquid or suspension which, when subjected to a shear stress, undergoes deformation wherein the ratio of shear rate (flow) to shear stress (force) is constant. These varying behavioral patterns of colloidal materials become important considerations in specifying pumps and other process handling equipment. See also **Gold Number.**

Sols. It is convenient to classify sols into three types: (1) *lyophilic* (solvent loving) colloids, for example, are solutions of gelatin or starch in water; (2) *association* colloids, of which a solution of soap in water at moderate concentration is an example; and (3) *lyophobic* (solvent repelling) colloids, for example, sulfur in water. Both lyophilic and association colloids can be prepared in thermodynamic equilibrium, so that when solvent is removed and then returned to the system, the original properties of the system are regained.

Lyophobic colloids are not (or at most, rarely) equilibrium systems. When solvent is removed and then returned to the system, the original dispersed material fails to redisperse, and it is usually convenient to regard such a system as one in which the dispersed particles are continuously aggregating. A lyophobic sol thus appears to be stable if the aggregation rate is slow; and unstable if it is fast. The terms lyophilic and lyophobic entered the literature before the characteristics of these systems were well understood and thus are somewhat anachronistic; they are nonetheless well-established.

Lyophilic sols are true solutions of large molecules in a solvent. Solutions of starch, proteins, or polyvinyl alcohol in water are representative of numerous examples. Properties of these solutions at equilibrium (for example, density and viscosity) are regular functions of concentration and temperature, independent of the method of preparation. The solvent–macromolecule compound system may consist of more than one phase, each phase in general containing both components. Thus, if a solid polymer is added to a solvent in an amount exceeding the solubility limit, the system will consist of a liquid phase (solvent with dissolved polymer) and a solid phase (polymer swollen with solvent, i.e., a polymer with dissolved solvent).

The foregoing characteristics also are found with solutions of small molecules. But properties of solutions, one of whose components is macromolecular, differ from those having only small molecular components in quite understandable ways. For example, where small molecules are involved, molecular distortion is minor. Quite generally, the shapes of small molecules are little affected by environment unless the small molecules react chemically. In contrast with small molecules, there is a considerable variation in polymer conformation with environment. See also **Molecule;** and **Polymer.**

A polymer dissolved in a good solvent will tend to stretch out, and the resulting entanglement of polymer chains and interference with solvent movement will lead to a high viscosity. If the solvent is a poor one, the polymer molecule will tend to form a small ball, and the viscosity for a given weight percent will be much less. A side group of a polymer may be ionizable. Ionization of this group distributes a charge along the backbone and charge repulsion causes the macromolecule to tend toward a rod shape. If there is a moderate salt concentration in the solution, the backbone charge will be partly shielded by ions from the salt of opposite charge. Thus, the tendency toward rod formation will be less pronounced. The tendency of oppositely charged macromolecules to aggregate is much greater than the tendency of oppositely charged small ions to pair simply because the charges involved are greater in the former case.

The foregoing special properties of solutions of large molecules are relatively easy to describe in qualitative terms, but a difference of a more subtle nature occurs when the system forms two liquid phases. In a macromolecular solution, both phases tend to be rich in the (small molecule) solvent, whereas in systems formed from two molecules of comparable size, one phase is rich in one component, while the other phase is rich in the second component. The formation of two liquid phases from a solvent–macromolecule system is sometimes call *coacervation*; and the phase with the higher percentage of macromolecule is sometimes called the *coacervate*. See also **Coacervation.**

Association Colloids. These are generally encountered in solutions of soaps and detergents in water. These matters become important, of course, to procedures for cleaning and sterilizing equipment (food processors, biochemical manufacturers, hospitals, etc.), but the principles also apply to other association colloids also encountered industrially, notably in the food processing field. A typical soap, such as sodium stearate, $C_{17}H_{35}COONa$, or a detergent, such as sodium dodecyl benzene sulfonate, $C_{12}H_{25}C_6H_4 \cdot SO_3Na$, consists of a long hydrocarbon tail and a polar (in the examples cited, ionizable) head group. The solubility of the soap in water is largely conferred by the head group. As the soap concentration is increased, the soap molecules tend to cluster in aggregates called *micelles*, with hydrocarbon tails in the interior of the micelles and the polar groups in contact with water. The formation of micelles is favored by the interaction between hydrocarbon tails and is opposed by charge repulsion of the polar group which are placed close together at the micelle surface. See also **Surfactants.**

Micelle formation becomes pronounced at soap concentrations exceeding the critical micelle concentration. As hydrocarbon tail length is increased, the interaction of tails is increased, and as salt concentration is increased, the repulsion of head groups is reduced because their charges are partly shielded by ions of the salt. Both of these factors favor micelle formation, causing micelles to be larger and the critical micelle concentration to be smaller. Typically, a micelle might contain about 50 soap molecules. The micelle interior is a hydrocarbon, and as such is receptive to other molecules soluble in hydrocarbons. Hence, a soap solution can 'dissolve' such molecules (taking them up in micelle interiors) even if the molecules are quite insoluble in water. This phenomenon is called *solubilization* and is a factor in detergency.

Lyophobic Sols and Aerosols. These products can be viewed most simply and, in most cases, with sufficient accuracy as two-phase systems in which the dispersed particles are steadily and irreversibly aggregating according to a second-order rate law. Thus, where C is the number of particles per cubic centimeter (an aggregate of many primary particles being counted as one particle) at time t, and where C_0 is the number of particles per cubic centimeter at zero time, and K is a constant, C depends on t according to

$$\frac{C_0}{C} - 1 = KC_0t$$

and will be one-half its value at zero time when $KC_0t = 1$. The time required for this is longer, the smaller K and the smaller C_0. If the time required is weeks, the sol will appear quite stable over a period of days. If there is no barrier to aggregation so that the particles aggregate as fast as diffusion brings them in contact, the rate constant K can be calculated approximately from diffusion theory and is $8kT/3\eta$, where k is Boltzmann's constant, T is the absolute temperature, and η the viscosity of the medium. Initial sol concentrations (particles per cubic centimeter) giving one-minute half-lives at room temperature are 1.4×10^9 in water; and 2.7×10^7 in air in the absence of aggregation barriers.

Although these numbers may appear large, they correspond to quite small volume percentages of dispersed particles. A particle of radius 5×10^{-5} centimeter is at the upper limit of the colloidal range; and 1.4×10^9 such particles occupy 0.07% of space. The behavior of smokes, fogs, and many dispersions of uncharged particles in water accords well with the rate equation and theoretical rate constant given. In contrast, the dispersed particles in many sols are electrically charged, manifesting this charge through electrophoresis (motion of colloidal particles under the influence of an electric field). In fact, Tiselius developed electrophoresis to a high degree, successfully fractionating and classifying proteins thereby. Evidently like charges on two colloidal particles will contribute to a repulsion between them, which will be greater, the greater the charge on each particle and the smaller the concentration of salts in solution (since ions from the salt will tend to mask the charges on the particle).

The theory of interaction between colloidal particles with a surface electrostatic potential (due to surface charges) surrounded by an electrical double layer (one layer of which is the layer of surface charges, the other a diffuse cloud of charges of opposite sign due to ions from salts in the solution) was developed by Derjaguin and Landau and independently by Verwey and Overbeek, and is generally known as the DLVO theory after the first letters in the names of these scientists. The DLVO theory shows how a barrier sufficient to reduce the rate constant K (and so to increase the half-life at a given initial concentration) by many powers of ten may arise from the interaction of charged particles in a solvent, and the magnitudes calculated agree rather well with experiment.

It is evident why the properties of lyophobic sols depend so critically on the chemistry of the interface between dispersed particle and solvent, for this chemistry establishes the means by which the surface charge can be established or altered. Particles of silver halide dispersed in water will acquire a positive charge if silver ion is in slight excess in the water, because the silver ion can readily add to the silver halide lattice. A negative charge is similarly acquired if the halide ion is in slight excess. The silver ions and halide ions are called *potential-determining ions* for the silver halide sol. They can lose their waters of hydration and adsorb on the particle side of the electrical double layer, conferring a charge on the particle. Hydrogen ion and hydroxyl ion are similarly potential-determining ions for many oxide sols, such as silica and alumina, including particles, such as carbon and many metals, which are ostensibly not oxides, but in fact usually have oxidized surfaces. Finally, charge can be conferred by the adsorption of charged macromolecules, such as gelatin, which are called *protective colloids.* Salts added to the sol form ions which tend to mask the particle charges and so tend to promote flocculation. The ion whose charge is opposite to the particle charge (the counter ion) is of particular importance, and the greater its charge the lower the concentration at which its flocculating effect is evident.

Emulsions. These are dispersions of one liquid in another. Most commonly, one phase is an oil which is at most slightly miscible with water. The disperse phase can either be oil (an oil-in-water emulsion) or water (a water-in-oil emulsion). For apparent stability, an emulsifying agent is almost invariably required. The emulsifying agent has an oil-soluble tail and a polar head. The emulsifying agent concentrates (adsorbs) at the interface between oil and water, lowering the interfacial tension and frequently conferring a charge on the dispersed droplets. The film of emulsificant thus formed is usually only one molecule thick, but it is essential to emulsification. Mixed emulsificants, such as a mixture of sodium stearate and octadecyl alcohol, may be more effective in emulsification than either component alone, and there is a great deal of art and experience required in the formulation of emulsions. Lecithins and some proteins are effective natural emulsificants, and a mixture of lecithin and cholesterol is an effective natural mixed emulsificant. It should be noted that the difference between solubilization (described in connection with association colloids) and emulsification is not sharp, particularly insofar as large, extensively swollen micelles and ultrafine emulsions are concerned.

Gels. These substances involve the formation of a three-dimensional structure. A gel is a colloidal disperse system in which is contained a dispersed component and a dispersion medium, both extending continuously throughout the system. Further, the system has equilibrium-elastic (time-dependent) deformation. Thus, since they have a shear modulus of rigidity, gels are like solids, but in most other physical respects, they behave like liquids. It is conceived that the three-dimentional network is kept together by bonds or junction points which essentially have an unlimited lifetime. Junction points may be described as primary valence bonds, attractive forces of long range, or secondary valence bonds which maintain an association between parts of polymer chains or form submicroscopic crystalline regions. A gel may be defined as a flocculant and gelatinous precipitate. A jelly is a transparent elastic mass. Upon standing, a gel may shrink—a process known as *syneresis.*

In 1861, Thomas Graham first used the term syneresis to describe the phenomenon of exuding small quantities of liquid by gels. By definition, syneresis is the spontaneous separation of an initially homogeneous colloid system into two phases—a coherent gel and a liquid. The liquid is actually a dilute solution whose composition depends upon the original gel. When the liquid appears, the gel contracts, but there is no net volume change. Syneresis is reversible if the colloid particles do not become too coagulated immediately after their formation.

In 1937, Heller classified three types of syneresis as to cause: (1) syneresis of desorption, caused by the particle becoming less hydrophilic with time; (2) syneresis of aggregation, whereby discrete gel particles may unite into a denser gel portion; and (3) syneresis of contraction, where a gel with fibrillar structure contracts and squeezes out the intermicellar liquid. Most commonly, syneresis is the visible manifestation of further slow coagulation which follows the initial setting of the gel, the gel-forming process itself being an enmeshing of the hydrous particles into a network. It may be further explained as the exudation of liquid held by capillary forces between the heavily hydrated particles constituting the framework of the gel. Ostwald noted that the phenomenon is one of the most characteristic of the properties of gels.

A common example of syneresis is found when a mold of gelatin remains under refrigeration for a period. A general shrinkage of the body of the gel occurs and a liquid collects around the edge of the mold. The liquid is a dilute solution of the original composition. Since the total volume of the system remains the same, syneresis should not be considered simply as the opposite of imbibition (absorption). Extending the onset of syneresis in various products, notably foods, is of obvious importance.

Dispersion Processes. 1. The simplest method of accomplishing dispersion is by grinding the solid (or liquid) material with the liquid medium until particles of the required size are ultimately obtained. The colloid mill (Plauson, 1921) is used for such purpose, as in mixing paints and pastes, regenerating milk from milk powder, dispersing cellulose in sodium hydroxide and carbon disulfide for the production of xanthates for viscose, and in emulsifying fats and waxes. See accompanying figure. 2. Zinc sulfide, cupric hexacyanoferrate(II), stannic acid, silver chloride are examples of precipitates which, when washed on the filter paper until the accompanying soluble electrolyte has been removed, form colloidal solutions and pass through the pores of the paper. Since this is usually to be avoided in practice, the washing is then done with an electrolyte which does not conflict with the treatment to follow. Frequently ammonium nitrate solution is used. 3. A peptizing agent is frequently employed. Tannin is peptized by water, and by glacial acetic acid. Soaps are peptized by water. Gelatin swells in cold water but is not peptized, but is peptized in warm water. Starch, although insoluble in cold water, behaves similarly to gelatin with warm water (63 to 74°C, depending upon the kind of starch). Cellulose nitrate swells in ethyl alcohol and not in ether, but is peptized in ethyl alcohol-ether mixture. Clay is peptized by ammonium hydroxide, and it is held by some that the action of sodium hydroxide on zinc, aluminum, and chromium hydroxides is one of peptization. 4. Water-peptizable colloidal substances such as gelatin, dextrin, gum arabic, and soap peptize many precipitates, and are often called protective colloids. Gelatin in the solution prevents the precipitation of silver dichromate upon mixing silver nitrate and potassium dichromate solutions. (See Condensation Processes, below.) 5. When dilute silver nitrate and dilute potassium bromide solutions are mixed so that there is a slight excess of either solution, silver bromide is peptized. Acheson's oil-dag and aqua-dag are suspensoids of graphite in oil or water containing a protective colloid, tannin. Oil-dag contains about 15% of a "deflocculated graphite," and is used in dilute solution in lubricating oil (about 0.1% graphite). Bearings gradually become coated with a thin layer of graphite.

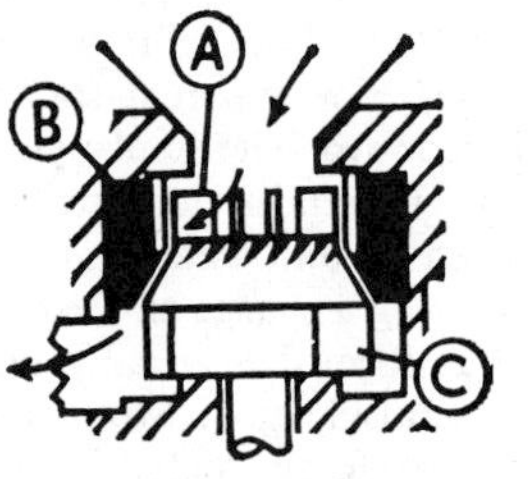

One type of colloid mill. Rotor blades *A* break up slurry. Serrations in rotor and stator provide mechanical shear and force material into adjustable gap (0.0005–0.125 inch; 0.013–3.2 millimeters) between rotor and stator *B* for intense hydraulic shear. Lower part of rotor *C* adds further whirling action.

Sonic methods also are used to create emulsions. Liquids are pumped under pressure through an orifice of special design and impinge on the edge of a blade causing it to vibrate at ultrasonic frequencies. Cavitation takes place continuously in the stream, causing violet pressure changes to be generated locally. The result is a uniform and stable emulsion and a dispersion of a very high order. See also **Cavitation.**

Condensation Processes. 1. When a solution of ferric chloride is poured into a relatively large volume of boiling water, colloidal ferric hydroxide is formed. The ferric hydroxide sol does not react with hydrogen sulfide nor with potassium hexacyanoferrate(II), and like all colloidal substances does not pass readily through animal membranes or parchment. 2. When hydrogen sulfide is passed into a solution of arsenious oxide, arsenious sulfide sol is formed which in the absence of an electrolyte may be made of the high concentration of 60 grams of arsenious sulfide per 100 grams of water. Upon addition of hydrochloric acid, arsenious sulfide coagulates and is precipitated. 3. When hydrochloric acid is added to sodium silicate solution either silicic acid sol or silicic acid gel is formed. 4. When hydrogen sulfide solution is treated with an oxidizing agent, for example, the proper concentration of nitric acid, sulfur sol is formed. 5. When gold chloride very dilute solution (0.01 to 0.001% of gold chloride) is made slightly alkaline (say by the addition of magnesium oxide) and then treated with a reducing agent, for example, formaldehyde or sodium hydrosulfite $Na_2S_2O_4$, red gold sol is formed. 6. Use of a protective colloid in solution prevents the formation of the ordinary and expected precipitate in many cases and causes the formation of the expected substance as colloidal sol. Silver nitrate (0.6 gram per liter) and potassium dichromate (0.5 gram per liter) to one of which is added 0.1 volume of hot gelatin solution (2 grams per 100 milliliters of water) are mixed with stirring silver dichromate sol is formed. 7. When an electric arc is formed under water between two metallic rods, particles of the metal of colloidal size are formed along with more or less separation of free metal. A protective colloid increases the stability. If the metal vaporizes and then condenses to the colloidal state this is strictly speaking a condensation process, if otherwise, a dispersion process.

The disappearance of the colloidal state of a substance may be accomplished in either of two directions, namely, by the colloid passing into solution or into suspension. Practically, the latter is the more important method. Coagulation, agglomeration or precipitation is readily brought about by discharge of the electric charge on the particles. Ions carrying a charge of opposite sign to that carried by the colloidal particles are active precipitants, and the higher the valency of the ion the more effective (Linder-Picton-Hardy). When the colloidal particles are made neutral the conditions are least favorable to their stability. For colloidal arsenious sulfide, which is negatively charged in water, the coagulating power of potassium iodide K^+I, calcium chloride $Ca^{2+}Cl_2$, aluminum chloride $Al^{3+}Cl_3$ is in the ratio of 1:80:1500 (Svedberg); and for colloidal ferric hydroxide, which is positively charged in water, the coagulating power of potassium chloride KCl^-, potassium sulfate $K_2SO_4^{2-}$ is in the ratio of 1:45. The active ion is carried down with the precipitated particles. Oppositely charged colloids, e.g., arsenious sulfide and ferric hydroxide, when mixed, precipitate each other. Other methods of coagulation are by migration of colloidal particles to and their discharge at electrodes, and by heating, as in the case of egg albumin. Coagulation is usually irreversible, especially when caused by electrolytes.

An interesting case, operating on a large scale in nature, of the precipitation of a colloidal system by an electrolyte is that of the action of sea water on the mud and silt of river water entering the ocean. When river water flows into the ocean the former, on account of its lower specific gravity, tends to flow over the latter and spread out in widening range. As the current diminishes some of the suspended mud and silt settles out, but the finer colloidal particles are coagulated by the electrolyte of the sea water and form deltas at the mouths of rivers.

Importance of Colloidal State. All living matter, whether animal or plant, is made up of many colloidal materials and is largely sustained by colloidal processes. Of similar importance is colloidal chemistry in everyday living, in almost all of our foods, such as proteins and starches, in our clothing, whether of natural or synthetic origin, and in our shelter materials, such as wood, bricks, concrete. When there is added to these, other common things and operations of everyday life, such as pottery and porcelain, paper, rubber and leather, and cooking and washing, where colloidal matter and processes operate, it is evident how broad is the scope and how great is the importance of the field. To these there must also be added other applications in the realm of industry, such as dyeing, printing, photography, water purification, smoke prevention, ore flotation, sewage disposal and soil preparation, paints, varnishes and lacquers, plastics, adhesives, and innumerable other operations and materials.

References

Andres, C.: "Antifoaming Agent Increases Fermentation Capactiy 20%," *Food Processing*, **38**, 5, 58–59 (1977).

Graham, H. D.: "Food Colloids," AVI, Westport, Connecticut, 1977.

Matijevic, E.: "Surface and Colloid Science," Vols. 1–5, Wiley, New York, 1969–1972.

Somorjai, G. A.: "Surface Science," *Science*, **201**, 489–497 (1978).

Staff: "Food Chemicals Codex," 2nd edition, National Academy of Sciences, Washington, D.C., 1972 (with numerous subsequent supplements).
Weiss, T. J.: "Food Oils and Their Uses," AVI, Westport, Connecticut, 1970.

NOTE: This is part of a longer article, much of which has been contributed by R. S. Hansen, that will appear in 'The Encyclopedia of Chemistry,' 4th edition (D. M. Considine, editor-in-chief), Van Nostrand Reinhold, New York, 1983.

COLLUM. 1. The dorsal plate of the first body segment in the millipedes (Diplopoda). 2. Any neck-like part or structure.

COLOBINE MONKEYS. Monkeys and Baboons.

COLOGARITHM. Logarithm.

COLON. The large intestine, which extends from the cecum to the rectum. It is divided into several parts, although the colon forms a continuous hollow muscular tube. The ascending colon extends from the lower right side of the abdomen at the termination of the small intestine, upward to the under surface of the liver, where it turns to the left and runs across the abdomen to the lower border of the spleen as the transverse colon. Beneath the spleen, it bends downward, descending along the left side of the abdomen as the descending colon. As it enters the pelvis, the colon makes a double curve, similar to the letter S. This portion is known as the sigmoid colon. The end of the sigmoid colon terminates in the rectum.

The functions of the colon are (1) final absorption of the products of digestion, (2) absorption of fluid from the feces so that this material becomes semi-solid, (3) removal of the fecal waste products into the rectum.

COLONOSCOPY. Sigmoidoscopy.

COLONY. A group of individuals of the same species living together for mutual benefit. They may be structurally united or separate and may be alike in form or of different types suited for various functions. See also **Ecology.**

COLOR. This vast subject is complicated by the distinction between the physical basis of colors and the sensations produced by them; and still further by a somewhat confused and unsettled vocabulary. Color consists of those characteristics of light other than spatial and temporal inhomogeneities, light being defined here as that aspect of radiant energy of which a human observer is aware through the visual sensations which arise from the stimulation of the retina of the eye. Color is a broad psychophysiological concept, embracing far more than the psychological sensation of hue. It includes the grays, as well as the chromatic colors; the characteristics of light constituting color may be stated in terms of the appropriate photometric quantity, dominant wavelength and purity—corresponding generally to the attributes of visual sensations, brightness, hue, and saturation.

The practical standard "white" light is direct noon sunlight. A "perfectly white" surface would reflect white light completely without any alteration. No such surface exists. Even snow does not reflect white completely, though it does reflect all visible wavelengths in the same proportion. Its color is one of the "grays" or achromatic colors, of very high "brilliance"; while that of a lead-pencil mark is a much feebler achromatic color. A "black" surface would reflect no light at all (see **Black Body**). Most colors, however, are chromatic, that is, they exhibit "hue," because their spectral energy distribution differs so much from that of white or gray that they look "reddish," "bluish," etc. Some colors of the same hue are more "brilliant" than others. Just as snow is of a more brilliant gray than graphite, so bright red is more brilliant than dark red. Further, some colors have greater purity or "saturation" than others; that is, they have more pronounced hue, or are more chromatic, and therefore differ more from a gray of the same brilliance. Thus foliage looks "greener" when freshly washed than when dusty. A chromatic color having little hue but high brilliance is a "tint," e.g., pink; while one of little hue and low brilliance, like brown, is a "shade."

We must now recognize the fact that the same color sensation can be produced by entirely different physical stimuli. Tests of a large number of observers with the spectrometer indicate that, according to the average judgment, the common names of pure spectral hues should be applied to the several wavelength ranges approximately as follows:

	Angstroms		Angstroms
Violet	3900 to 4550	Yellow	5770 to 5970
Blue	4550 to 4920	Orange	5970 to 6220
Green	4920 to 5770	Red	6220 to 7700

But the sensations produced by any of these, or by any of their tints, shades, or mixtures, can also be produced in a variety of other ways. For example, red and green light may be mixed to produce a good imitation of spectral yellow light, though no yellow wavelengths are present in the mixture.

According to the Young-Helmholtz theory, the human vision has three separate color sensations, each capable of stimulation in various degrees. It is thought that, if stimulated separately, they would prove to be the sensations produced by red, blue, and green light, respectively. But they always act together, and every color sensation is the effect of their joint stimulation in some definite proportion. A result of this is that any color can be successfully imitated by adding together red, blue, and green light with suitable relative intensities. These are therefore called "additive primaries." If added in equal intensities, they produce a sensation of white. White may, however, be produced also by adding in suitable proportions various pairs of pure spectral hues, which are "complementary" to each other; thus:

Angstroms	Angstroms
6562 and 4921	5671 and 4645
6077 and 4897	5644 and 4618
5853 and 4854	5636 and 4330
5739 and 4821	

(The third pair, for example, is a certain yellow and a certain blue.)

Color sensations are also commonly produced by removing certain components from white light, as by the use of filters. Pigments such as paint and colored inks act in this way, being selectively reflective. The complementary hues of the three additive primaries are the "subtractive primaries" blue-green (for red), yellow (for blue), and purple (for green).

Maxwell devised an ingenious graphical scheme, called the "color triangle," for representing mixtures of colored lights (not pigments). (See figure.) The three additive primaries red, blue, and green are at the vertices of an equilateral triangle, with their complementary subtractive blue-green, yellow, purple on the sides opposite. Each altitude, as PG, is taken as 100%. Any hue is represented by a point H, whose distances r, b, g, from the three sides represent the percentages of the three primaries R, B, G which must be added to imitate it ($r + b + g = 100$). Thus for the point H in the figure, $r = 60$, $b = 10$, $g = 30$, giving a reddish-yellow sensation. The point W, at the center of the triangle, corresponds to the sensation of white.

Color triangle.

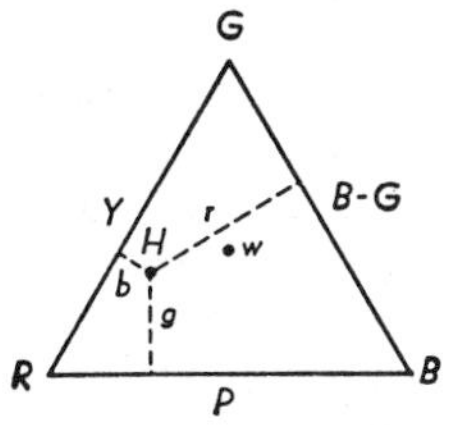

See also list of terms in entry on **Optics.**

COLORADO POTATO BEETLE (*Insecta, Coleoptera*). A leaf-eating beetle, *Leptinotarsa decemlineata*, native to the western United States. Originally it fed on a native plant but about the middle of the nineteenth century it became troublesome in potato fields and rapidly spread across the country. Both larva and adult eat the leaves of potato plants.

COLOR (Animals). The coloration of animals embraces both the colors that appear in their bodies and the patterns in which they are arranged. In many species color appears to be incidental but in others coloration has an important bearing on the life of the individual.

Colors are due in some cases to the presence of compounds which are important for other reasons. The blood of insects, for example, may be green, that of certain mollusks blue, and that of the vertebrates and annelid worms red because of their chemical composition but these colors have no important bearing on individual life. The black pigment found in eyes is a protection against random light rays but any impervious layer would be as effective; black color is unimportant in itself.

In many animals colors of this kind are supplemented by special pigments located in the superficial layers of tissue and in the vestiture, such as hair, feathers, and scales. Some animals are able to change their color by varying the expansion of the pigment. The surface of the integument in insects is often formed so that it breaks up light rays and reflects only a portion of them. The colors produced in this way are called physical colors and are usually metallic or glassy. All of these colors, whether physical or pigmental, are often arranged in intricate patterns characteristic of the species. In this arrangement details may be observed in some animals which seem to have definite value in relating the individual to its environment.

The white tail patch found in some deer and rabbits has been interpreted as a signal mark. It may serve for ready recognition in poor light and may catch the attention of other members of a group when danger is near.

Warning colors are exemplified by the brilliant red and yellow of the coral snakes. Poisonous or distasteful animals are usually avoided as food and conspicuous appearance makes it all the easier for other animals to see and avoid them. The male sex is frequently advertised by conspicuous coloration.

Concealing colors render the animal less conspicuous in its normal environment, and are essentially the same in the hunter and in the hunted. The white winter coats of the prairie hare and of the weasel are equally inconspicuous against the snow, but the one animal is helped to escape its enemies and the other to catch its prey by this means. Concealing colors may also be in patterns. The black and tawny stripes of the tiger, for example, are said to conceal it admirably in tall grasses lighted by the sun, where vertical shadows are conspicuous. This aspect of coloration, however, merges with protective mimicry, and in mimicry physical form is involved as well as coloration.

See also **Adaptation (Ecology).**

COLORANTS (Foods). Color, as a component of appearance, is important in the sensory evaluation of a food substance, including beverages. Color affects the degree of acceptability of a food product in the marketplace. Color also is a frequently useful indicator of the degree of wholesomeness of a foodstuff. Many foodstuffs are attractively colored as the result of their naturally occurring pigments. In these instances, a major objective of the food grower and processor is to protect and preserve the natural colors as long as may be required by the distribution network, considering such factors as temperature and humidity to which the food substance may be subjected before reaching the consumer.

Considering the expectations of consumers, particularly in countries that have an advanced food technology, colorants, along with flavorings and texture modifiers, are important. These factors are particularly stressed in connection with modern fabricated foods, substitute foods, and food analogues. Many years have passed since final approval was given to margarine producers to use colorants and artificial flavorings. Even though artificial flavors are used in some products, ice creams and ices, as well as soft drinks, are color matched to their fruit flavors. Yellow colorings provide a note of richness in cake mixes, eggnogs, and other products associated with their content of eggs; cheese snacks are both colored and flavored to simulate cheese; popcorn oil is usually colored; iron oxide can be added to pet foods to simulate the color of meat; the red coloration of cherries is intensified by using red colorants in the processing of maraschino cherries. Caramel colorings are widely used in both alcoholic and nonalcoholic beverages.

Particularly during the past couple of decades, regulatory agencies in various countries have scrutinized colorants (along with other food additives) against a backdrop of consumer health. Since the early 1900s, several countries have approved colorants for use in foods only after thorough physiological testing. In recent years, analytical instrumentation and research methodologies have become sophisicated and measurements of minute quantities are now practical. With improved analytical tools and a heightened awareness of the effects of foods upon health, a number of colorants that once were considered perfectly safe have come under question. In some countries, most or all synthetic colorants have been banned. Generally, the limitations on the use of colorants have become much more stringent.

In the United States, the first rather complete legislation involving such matters was the Food and Drug Act of 1906. As a result of that legislation, the list of colorants permitted was reduced to only seven days. Because the remaining seven dyes did not provide sufficient flexibility in the formulation of food products, considerable research went forth to find additional colors, not only with more desirable hues, but easier to use (solubility in oil/water, less temperature sensitivity, etc.). During the 66-year period from 1906 to 1971, several additional colors were added. In 1938, the Food, Drug, and Cosmetic Act was passed. The common names of dyes previously used were given color prefixes and numbers. For example, Amaranth became FD&C Red No. 2. Under the new act, certification became mandatory.

The color situation in the food industry was relatively without incident until the early 1950s, when, as the result of a few cases of excessive levels of usage in some candies and popcorn, two colors (FD&C Orange No. 1 and FD&C Red No. 32) were delisted. After much controversy and considerable litigation, more colors were delisted. To rectify legal complexities and unworkability of the 1938 Act, the Color Additives Amendments of 1960 were passed. Nevertheless, as pharmacological studies continued, a number of other colors were delisted during the interim. An excellent summary of the situation up to the early 1970s is contained in the Noonan reference listed. In 1973, Violet No. 1 was delisted; in 1976, Red No. 2 and Red No. 4 were delisted. As of the early 1980s, Red No. 40 is under serious scrutiny.

Current reports are available from the U.S. Food and Drug Administration; the publication "Food Colors" is released periodically by the National Academy of Sciences, Washington, D.C.; reports of the Cosmetic, Toiletries, and Fragrance Association; the Pharmaceutical Manufacturers' Association; the United Kingdom Department of Health; Food and Agriculture Organization (United Nations)—all organizations that update information as regards color additive regulations.

Natural Colorants. The use of naturally derived colorants for foods dates back many centuries. The principal concerns expressed in the foregoing paragraphs relate largely to the use of synthetic colorants. Natural colorants include the anthocyanins, annatto colors, the betalaines, the carotenoids, cochineal, saffron, turmeric, and titanium dioxide. Caramel coloring also falls into this category. Paprika is used in some foods for its coloring attributes.

Anthocyanins are water-soluble pigments which account for many of the red, pink, purple, and blue colors found in higher plants. Most plants contain more than one of these pigments and they occur most prevalently as glycosides. Several hundred different anthocyanins are known. These compounds are most stable at a pH range of 1 to 4, thus limiting the spectrum of usage. As compared with synthetic colorants, the anthocyanins produced to date generally are less stable, have less tinctorial potency, and lack some color uniformity. They are degraded by light, heat, enzymes, and interact with ascorbic acid. They also tend to form complexes with metal ions to produce off-colors. Their main advantage stems from the fact that they are naturally derived and thus not regarded with suspicion as health deterrents as are many synthetic materials. On the other hand, attempts to alter and modify them to make up for their fundamental disadvantages could also move them toward a suspicious category—a factor which researchers on anthocyanins are taking into consideration.

Numerous commercial sources for the anthocyanins have been and are continuing to be investigated. These sources include grape anthocyanins, apparently with good potential in the carbonated beverage field. Red cabbage also has been seriously considered. Roselle plants, native to the West Indies, have been studied and appear to have potential for use in apple and pectin jellies, but not for carbonated beverages, such as ginger ale. Cranberry pomace and blueberries have been investigated. Considerable interest has been shown in the red anthocyanin pigments of miracle fruit (*Synsepalum dulcificum*, Schum), a tropical plant that produces a red berry.

Betalaines are sometimes referred to as beetroot pigments. They are made up of two main groups: (1) *betacyanins*, the principal component of which is betanin; and (2) *betaxanthins*, the principal component of which is vulgaxanthin-I. The betacyanins contribute a red color, whereas the betaxanthins are yellowish. Another yellow pigment, betalamic acid, derives directly from cleavage of betanin and is probably the key intermediate in the biogenesis of all betalaines.

A factor of concern in connection with the betalaines is the earthy flavor associated with beets. To date, beets have been regarded as the primary source for these substances.

Carotenoids. These yellow-orange colorants are described in entry on **Carotenoids.**

The other natural colorants previously mentioned have been used for many years and are familiar to nearly everyone. Annatto colors are described in a separate entry, **Annatto Food Colors.**

Synthetic Colorants. Perkin, in 1856, synthesized the first synthetic dye, *mauve* or *mauveine*, by the oxidation of crude aniline. In that time and for about 80 years, coal tar was the principal source of aromatic compounds, which, in turn, were the sources of numerous synthesized dyes used primarily in textiles, but some of which were found to be adapted to other uses, including the coloring of food substances. This generally gave rise to the term "coal tar color" used commonly in the food and cosmetics industries for many years. Of course, with the development of more sophisticated organic syntheses and the petrochemical field, the association with coal tar no longer had a direct meaning. The term was finally eliminated from legislation in connection with the Color Additives Amendments of 1960. It is interesting to note, however, that prior to the first Act of 1906, it is estimated that some 80 such dyes were being used in a large number of food products, at a time when there were no regulations regarding the nature and purity of colorants used in foods.

As of the early 1980s, synthetic colorants still listed and used (although Red No. 40 is under study), are shown (structural formulas) in the accompanying diagram.

Lakes. In the United States, FD&C lakes were accepted for the approved list of certified color additives for the first time in 1959. As defined by the FDA, a lake is an "Extension on a substratum of alumina, of a salt prepared from one of the water-soluble straight colors by combining such a color with the basic radical aluminum or calcium." Because the substratum of alumina hydrate or aluminum hydroxide is insoluble, the lake provides an insoluble form of the dye, i.e., a pigment. Colors from dyes result from solution in a solvent; whereas colors from pigments result from dispersion of that pigment throughout the food substance. Prior to the acceptance of lakes, insoluble colorants were formed by absorbing them on materials (insoluble), such as cellulose, flour, and starch. Generally, these forms were inadequate because of relatively low coloring power.

When utilized in solid or semisolid vehicles, dyes must be added in solution to achieve effective coloring. Thus, color migration is a problem. Dyes can migrate with the solvent during various drying or processing operations. Because lakes are insoluble, color migration is negligible in applications where distinct interfaces are required. Striped candy pieces provide an example. Where opacity is required, titanium dioxide can be added to lakes. In high-quality lakes, nearly all particles will pass through a 325-mesh screen when wet-tested. Shades of coloration can be produced by blending the various FD&C Lake Colors.

See also **Polymeric Food Additives.**

References

Clydesdale, F. M., and F. J. Francis (co-chairpersons): "Food Colorants," Institute of Food Technologists 39th Annual Meeting, Saint Louis, Missouri, 1979.

Eagerman, B. A.: "Orange Juice Color Measurement Using General Purpose Tristimulus Colorimeters," *J. Food Sci.*, **43**, 2, 428–430 (1978).

Furia, T. E.: "Nonabsorbable, Polymeric Food Colors," *Food Technology*, **32**, 5, 26–33 (1977).

Kramer, A.: "Benefits and Risks of Color Additives," *Food Technology*, **32**, 8, 65–67 (1978).

Noonan, J.: "Color Additives in Food," in "Handbook of Food Additives" (T. E. Furia, editor), CRC Press, Boca Raton, Florida, 1972.

Riboh, M.: "Natural Color: What Works; What Doesn't?" *Food Engineering*, **49**, 5, 66 (1977).

Staff: "HT Lakes," Colorcon, Inc., West Point, Pennsylvania, 1979.

Staff: "Food Colors," National Academy of Sciences, Washington, D.C. (Issued periodically).

SULFONATED INDIGO COLORANT
FD&C Blue No. 2

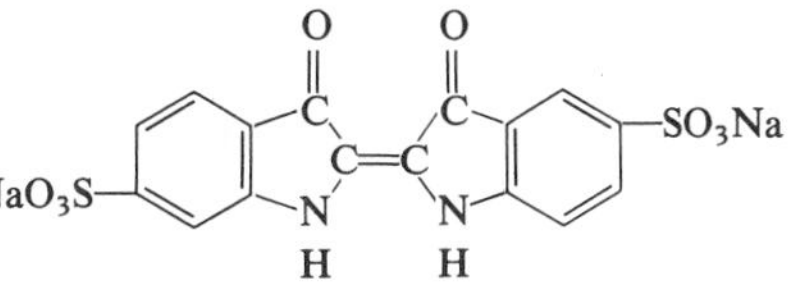

FLUORESCEIN TYPE COLORANT
FD&C Red No. 3

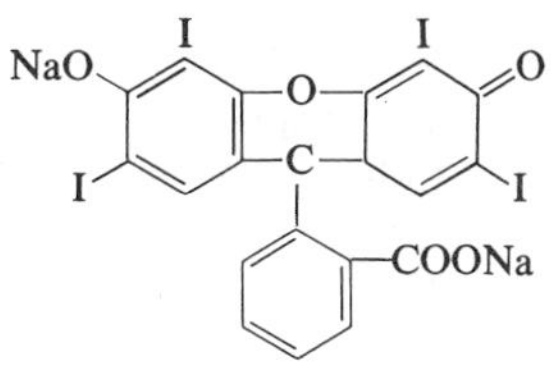

AZO COLORANTS
FD&C Red No. 40

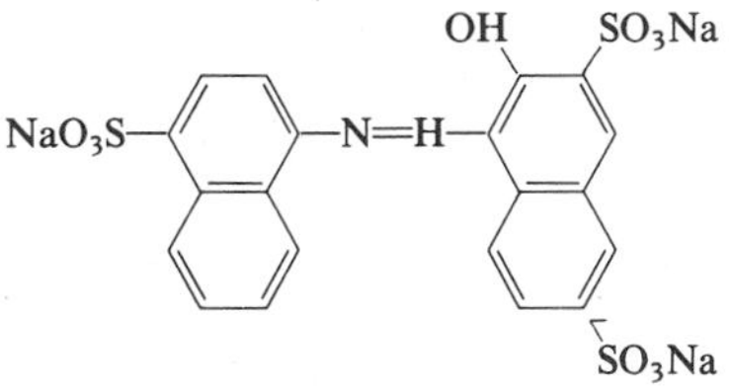

FD&C Yellow No. 5
FD&C Yellow No. 6

TRIPHENYLMETHANE COLORANTS
FD&C Blue No. 1

FD&C Green No. 3

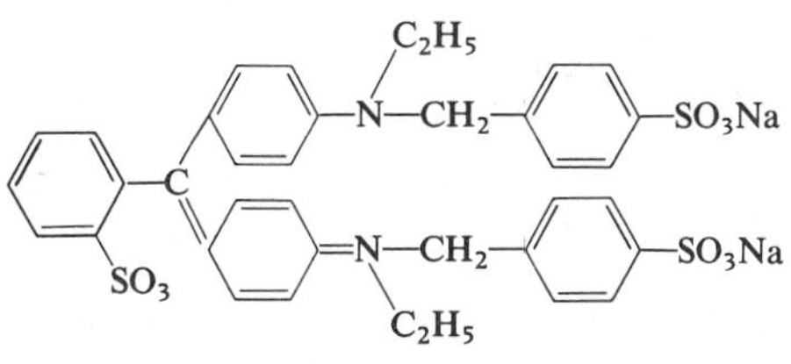

Structural formulas of several FD&C food colorants.

COLORANTS (Glass). Glass.

COLORANTS (Natural). Annato Food Colors.

COLORANTS (Polymeric). Polymeric Food Additives.

COLORATION (Animals). Pigmentation (Animals).

COLORATION (Fishes). Fishes.

COLORATION (Plants). Color (Plants); Pigmentation (Plants).

COLOR BLINDNESS. Vision and the Eye.

COLOR CENTERS. Certain crystals, such as the alkali halides, can be colored by the introduction of excess alkali metal into the lattice, or by irradiation with x-rays, energetic electrons, etc. Thus sodium chloride acquires a yellow color and potassium chloride a blue-violet color. The absorption spectra of such crystals have definite absorption bands throughout the ultraviolet, visible and near infrared regions. The term color center is applied to special electronic configurations in the solid. The simplest and best understood of these color centers is the F center. Color centers are basically lattice defects which absorb light.

COLOR ENHANCEMENT. Photography and Imagery.

COLOR (Flowers). Pollination.

COLORIMETRY. A method of chemical analysis which deals with the measurement of the light absorption by colored solutions. Since light absorption depends upon the concentration of a specific constituent in solution, colorimetry is frequently used by geologists to determine qualitatively the trace quantities of many elements.

The fundamental principle of colorimetry states that the amount of light absorbed by a given substance in solution is proportional to the intensity of incident light and to the concentration of the absorbing species. This is expressed mathematically in the *Lambert-Beer law*:

$$\log I_0/\mathrm{I} = abc$$

where

I_0 = intensity of incident light
I = intensity of transmitted light
a = absorptivity of the substance
b = light path length
c = concentration of colored substance
I/I_0 = transmittance
$\log I_0/I$ = absorbence.

The term colorimetry is generally restricted to the visual comparison and matching of the color of a standard solution with that of an unknown one, whereas *spectrophotometry* involves the use of a photoelectric cell which measures a narrow band of wavelengths for transmittance.

Visual colorimetry is a simple method and is fairly precise. Essentially it requires the matching of the color of a standard solution with that of an unknown sample so that when they become identical, they must contain the same amount of colored substance in columns of equal cross-section. At this point

$$C_x b_x = C_s b_s \quad \text{and} \quad C_x = C_s b_s/b_x$$

where

C_x = concentration of unknown solution
b_x = length of light path of unknown solution
C_s = concentration of standard solution
b_s = length of light path of standard solution

A standard series of solutions is prepared, each with a known concentration, having the same volume as the unknown, and being contained in identical flat-bottomed tubes of equal diameter (*Nessler tubes*). The solutions should be compared in daylight and examined against a white background.

A more refined method uses the Duboscq colorimeter. This instrument features a dual-matched optical system. Uniformly intense light is incident upon both colorimeter tubes and the difference in absorption of the standard and unknown solutions is compensated for by adjusting the thickness of solution through which light passes. When the two colors match $C_x = C_s b_s/b_x$.

Spectrophotometry (q.v.) is the most precise method of measuring light transmittance. It involves the use of a photoelectric cell which measures a narrow band of wavelengths for transmittance, and a prism or grating to give monochromatic light. At each wavelength, three readings are made: (1) sample cell filled with solvent, which gives the value of I_0, (2) the cell filled with standard solution, and (3) cell filled with unknown. The ratio of meter readings of the cell filled with pure solvent and sample is the I_0/I ratio. Spectrophotometry may be used analytically to measure the absorption at a single wavelength, or to determine the whole absorption spectrum of liquids and solutions in the visible and ultraviolet regions.

See also **Nephelometry; Turbidimetry.**

COLOR INDEX. The human eye is more sensitive to red light than is the photographic plate. Thus, if we have a blue and a red star of the same apparent brightness, i.e., the same visual magnitude, the blue star will appear brighter on the photographic plate, whereas the red star will appear brighter to the eye. Hence, the scale of magnitudes determined photographically for a given sequence of stars will differ from that determined visually. The difference between the photographic magnitude and the visual magnitude for any star is known as the color index of the star. Thus the relation, *color index* = $m_{pg} - m_v$.

Since the spectral type of a star is also a function of the color of the star, we should expect to find a definite relationship between the color index and the spectral type. The "zero point" for this relationship is generally fixed for an AO-type star. Techniques involving color index are quite refined, and the system of *U, B, V* photometry is a good example. If one determines the color indices $B - V$ and $U - B$, one may plot these two coordinates against each other. The main sequence then forms a somewhat skewed curve below the black-body radiation curve. It is then possible, upon obtaining these two indices for any star, to move the star along the curve for a good fit, and, in this way, to determine the spectral type and the absolute magnitude of the star. The accompanying diagram illustrates the essential features.

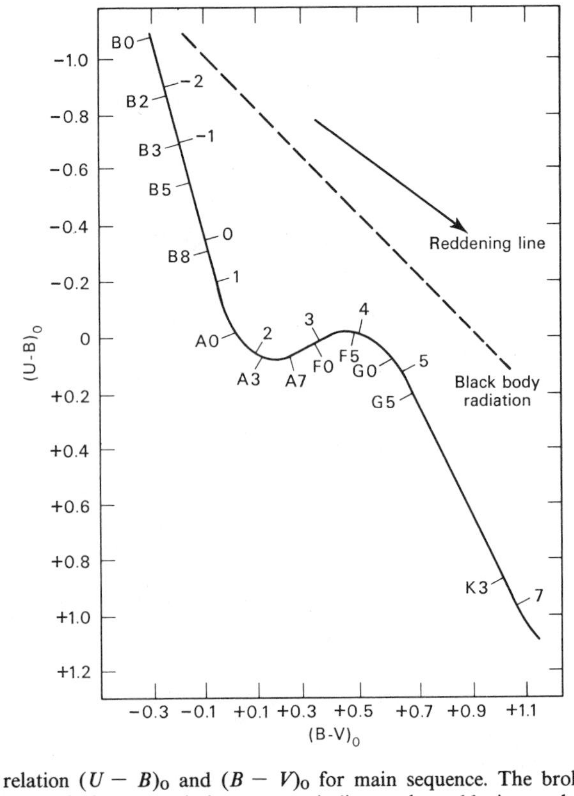

The relation $(U - B)_0$ and $(B - V)_0$ for main sequence. The broken line represents blackbody radiation. Arrow indicates the reddening path for the O stars. (*Becker, "Stars and Stellar Systems," University of Chicago Press.*)

COLOR (Light). Illumination.

COLOR MAGNITUDE DIAGRAM. Giant and Dwarf Stars.

COLOR MIXTURE CURVE. A curve representing the amount of a standard component (usually specified in wavelength) required in a three-color mixture to match a sample color, for unit flux of spectral energy. The variables plotted are usually the amount in lumens of the standard component and the indicated wavelength of the sample color matched.

COLOR (Nonspectral). 1. A color not present in the spectrum of white light. 2. A color which can be represented by a point, on the chromaticity diagram that lies on the straight line between the ends of the spectrum locus, or between that line and the achromatic point.

COLOR (Osteichthyes). Fishes.

COLOR PHOTOGRAPHY. Photography and Imagery.

COLOR (Plants). The coloration of plants may be due to pigments in plant surfaces, to the selective scattering of certain wavelengths (colors) of light by small discontinuities in the subsurface structure, to the interference of light beams reflected from interior surfaces, or to a combination of some or all of these effects.

Besides chlorophylls, two other groups of pigments contribute to the color of many plants. They are the carotenoids and the water-soluble group of anthocyanin and anthoxanthin pigments. The carotenoids range in color from yellow to red or purple. Various isomeric forms of carotene are found in carrots and other plants, and one of them is the lycopene that colors tomatoes and rose hips. Various carotenols are found in yellow corn grains and in leaves of many other plants. A carotenone is found in one of the blue-green algae.

The anthocyanins form water-soluble glycosides which are responsible for scarlet to purple or blue colorations in flowers. Similar compounds of the anthoxanthins are responsible for many pale yellow to ivory white shades. These pigments are also found often in leaves, woody tissues, and fruits.

In the leaves of higher plants, all these pigments add their colors to those of the chlorophylls, the net result depending upon the proportions and types of pigments present. The great variety of green shades observed in any wide expanse of green plants results from the various pigment mixtures modified by structural differences in leaf surfaces. In the fall after frost or following certain diseases or plant food deficiencies, leaf colors may change much due to decomposition of chlorophyll first, followed by carotenoids, permitting the remaining pigments to compose the color values. Thus a single leaf may have various color stages. These pigments all absorb much radiant energy from the sun, as do the chlorophylls, and many of them form vitamin A by changes in the animal metabolism (i.e., after they have been eaten by the animal).

See also **Annatto Food Colors; Carotenoids;** and **Pigmentation (Plants).**

COLOR PROCESS (Subtractive). Subtractive Color Process.

COLOR RENDERING INDEX. Illumination.

COLOR SATURATION. The attribute of any color perception possessing a hue, that determines the degree of its difference from the achromatic color perception most resembling it. This is a subjective term corresponding to the psychophysical term purity. The description of saturation is not commonly undertaken beyond the use of rather vague terms, such as vivid, strong, and weak. The terms brilliant, pastel, pale and deep, which are sometimes used as descriptive of saturation, have connotations descriptive also of brightness.

COLORS (Fechner). Fechner Colors.

COLOR TELEVISION. Television.

COLOR TEMPERATURE (Stellar). If a star's radiation is measured at a number of wavelengths, and a black-body curve is fitted through them, the temperature of the star can be estimated from the shape of the curve. This is called a color temperature. Using two standard colors, one can derive a series of temperatures that depend upon the color index.

COLOR THEOREM (Mapmaking). Four-Color Map Theorem.

COLOR TRANSFORMATION (Fishes). Fishes; Flatfishes.

COLOR VISION. Vision and the Eye.

COLOR (Young-Helmholtz Theory). Young-Helmholtz Theory.

COLOSTOMY. Cancer and Oncology.

COLOSTRUM. The fluid secreted by the mother's breasts a few days before and after the birth of the child, before the secretion of true milk begins. It is a thin watery fluid rich in protein, especially globulin, by means of which protein deficiencies in the serum of newborn animals may be made up. In some species there is evidence of transfer of passive immunity to the young animal, by means of antibodies carried in the globulin fraction of the protein, but such evidence for the human species is at present lacking.

COLPITTS OSCILLATOR. Oscillator.

COLUBRID SNAKES. Snakes.

COLUMBA. A southern constellation found near Canis Major.

COLUMBIA (Orbiter). Space Shuttle.

COLUMBIFORMES (*Aves*). In this order we combine several very dissimilar families. They are the pigeons (Columbidae), sandgrouse (Pteroclididae) and dodos (Raphidae), of which the last has died out, or rather, has been made extinct by man. Dodos and sandgrouse can at best be considered as very distant relatives of the pigeons.

The largest family in this order are the pigeons (family Columbidae), which are widespread in temperature and tropical regions. They are small to medium-sized birds; the length is about 15–80 centimeters (6–31 inches); they generally have small heads, plump, full-breasted bodies, and soft but very dense plumage. The aftershaft is underdeveloped or missing, but the lower part of the feathers is very downy. Head feathers sit very loosely and easily fall out. Moulting, in contrast to most other birds, is almost completely independent of brooding periods. The bills are slender with a cere at the base and a constriction at the mid-portion. There are 55–59 genera with 302 species.

All birds in this order drink by immersing the bill and sucking, a most unusual method in birds. Vocalizations are usually unmistakable cooing sounds, but also include growls, hisses and whistles. In addition there are flight sounds caused by notches on the primary wing feathers.

Pigeons are monogamous. They build weak and flimsy platform nets of twigs, straw or similar nest materials. The female sits in place and tucks the material around and under her body, while the male collects nesting material and gives it to the female. One or usually two (exceptionally three) eggs are laid, and incubation is shared by both parents. Males sit by day, and the females at night.

Most species of pigeons and doves show only a small degree of sexual dimorphism. Males are slightly larger and have slightly more conspicuous patches of display color on or near the head or anterior parts of the body.

Nearly all species are at least partly arboreal; a few live around rocky cliffs and some are nearly completely terrestrial. Pigeons are found in all parts of the earth but almost completely avoid the polar regions.

The domestic pigeon is a derivate of selective breeding in the Rock Dove (*Columba livia*). The length is about 33 centimeters (13 inches), the wing length is about 22 centimeters (7 inches), the wingspan is 63 centimeters (25 inches), the tail length is 11 centimeters (4 inches), and the body weight is 330 grams ($10\frac{1}{2}$ ounces). There are 14 subspecies of the rock doves.

Domestic pigeons which have become wild are difficult to distinguish from rock doves, insofar as they resemble the wild form. They have a larger distribution than the rock dove and thousands of them can be found in large cities in Europe, America, and Asia.

Food of the rock dove consists largely of grains containing meal and oil, available from a variety of plants, corn, and many weeds. In addition, snails, insects and occasionally worms are eaten. Small stones, sand, clay, lime, and mortar supplement the diet. The salt requirement is satisfied by feeding at dung heaps or areas near toilets or chemical refuse sites.

There are approximately 140 races of domestic pigeons with numerous coloration patterns. Most races have the four basic colors, blue, black, red, and yellow, as well as white. Earlier pigeons were divided into groups with field (hybridized) pigeons together, those of pure race together, and carrier pigeons. Another division was made on the basis of shape, color, markings and other characteristics, from which the following groups were differentiated: 1. Runts, Florentine, Maltese, and Modena pigeons; 2. Carriers, with well-developed wattles and eye ceres; 3. Show pigeons, including German beauty homers; 4. Pouters; 5. Silesian, moorheads and trumpeters; 6. Jacobins, Schmalkalden moorheads, fantails, frillbacks, and Oriental frills; 7. Tumblers and highfliers with flight peculiarities (somersault flight in the former and extensive high and elegant flight in the latter).

Carrier pigeons deserve special mention. Pigeons have been used for sending communications since earliest times. The carrier pigeon of today was created over a century ago in Belgium by breeding various races. This species can cover 800–1000 kilometers (497–621 miles) in a single day.

Pigeon breeding is considered to have begun in Egypt in the fourth century B.C. The first white pigeon was recorded in Greece in 478 B.C. Carrier pigeons came into use a short time thereafter. Innumerable pigeons were sacrificed in the Temple of Jerusalem. Some 5000 pigeons were kept in a pigeon house on the Mount of Olives. Pigeons were regarded with high esteem by the earliest peoples of the Orient. They could nest in the temples and could not be disturbed or killed.

The Mourning Dove (*Zenaidura macroura*) is the common dove of North America. The length is around 30 centimeters (12 inches) and the weight is somewhat more than 100 grams (3.5 ounces). Its predominant color is olive-gray above, shading to pinkish-gray and brownish-gray beneath.

The food of the mourning dove is extremely varied, but is almost completely vegetable matter, chiefly seeds. As is true with most other doves, this species is a feeding opportunist and will take whatever is available. See also **Pigeons and Doves;** and **Poultry.**

COLUMBITE. A mineral oxide of iron, manganese, niobium (columbium), and tantalum; $(Fe,Mn)(Nb,Ta)_2O_6$. Crystallizes in the orthorhombic system. Hardness 6; sp gr 5.20; color, red to brown.

COLUMBIUM. Niobium.

COLUMELLA. 1. The central pillar in the skeleton of some corals. 2. The central axis of the spiral shell in gasteropod shells. 3. A bone in the middle ear of amphibians, reptiles and birds.

COLUMNAR STRUCTURE. 1. In geology, prismatic columns which develop in basic lavas, due to the stress-strain relationships set up by rapid chilling. A relatively frequent phenomenon in basalt flows, such as those of the Giants' Causeway, Ireland. See also **Giant's Causeway.**

2. In metallurgy, an arrangement of coarse, parallel, elongated grains in a casting with the grain axes normal to the mold surface.

COLUMN-BINARY CODE. Code (Computer System).

COLUMN CHROMATOGRAPHY. Chromatography.

COLUMN (Data Processing). A character or digit position in a positional information format, particularly one in which characters appear in rows, and the rows are placed one above another, e.g., the right-most column in a 5-decimal place table, or in a list of data.

COLUMN (Structural). That part of a structure whose purpose is to transmit, through compression, the weight of the structure and the superimposed loads to the foundation. Other compression members are often termed columns because of the similar stress conditions. The ratio of the length of the column to the least radius of gyration of its cross section is called the slenderness ratio. This ratio affords a means of classifying columns. All the following are approximate values for convenience. A short steel column is one whose slenderness ratio does not exceed 50; an intermediate length steel column has a slenderness ratio ranging from 50 to about 200, while a long steel column may be assumed as one having a slenderness ratio greater than 200. A short concrete column is one having a ratio of unsupported length to least dimension of the cross section not greater than 10. If the ratio is greater than 10 it is a long column. Timber columns may be classed as short columns if the ratio of the length to least dimension of the cross section is equal to or less than 10. The dividing line between the intermediate and long timber columns cannot be readily evaluated. One way of defining the lower limit of long timber columns would be to set it as the smallest value of the ratio of length to least cross-sectional dimension that would just exceed a certain constant K of the material. Since K depends upon the modulus of elasticity and the allowable compressive stress parallel to the grain, it can be seen that this arbitrary limit would vary with the species of timber. The value K is given in most structural handbooks.

If the load on a column is applied through the center of gravity of its cross section, it is called an axial load. A load at any other point in the cross section is known as an eccentric load. A short column under the action of an axial load will fail by direct compression but a long column loaded in the same manner will fail by buckling (bending), the bucking effect being so large that the effect of the direct load may be neglected. The intermediate length column will fail by a combination of direct stress and bending.

In the middle of the 18th century a mathematician named Euler derived a formula which gives the maximum axial load that a long, slender ideal column can carry without buckling. An ideal column is one which is perfectly straight, homogeneous and free from initial stress. This maximum load, sometimes called the critical load, causes the column to be in a state of unstable equilibrium, that is, any increase in the loads or the introduction of the slightest lateral force will cause the column to fail by buckling. The Euler formula for columns is

$$P=\frac{K\pi^2EI}{l^2}$$

in which

$P=$ maximum or critical load
$E=$ modulus of elasticity
$I=$ moment of inertia of cross-sectional area
$l=$ unsupported length of column
$K=$ a constant whose value depends upon the conditions of end support of the column. For both ends free to turn, $K=1$; for both ends fixed, $K=4$; for one end free to turn and the other end fixed, $K=2$ approximately, and for one end fixed and the other end free to move laterally, $K=\frac{1}{4}$.

Examination of this formula reveals the following interesting facts with regard to the bearing power of columns. First, that elasticity and not compressive strength of the materials of the column determines the critical load. Secondly, the critical load is directly proportional to the moment of inertia of the cross section. The strength of a column may therefore be increased by distributing the material so as to increase the moment of inertia. This can be done without increasing the weight of the column by distributing the material as far from the principal axes of the transverse section as is possible consistent with keeping the material thick enough to prevent local buckling. This bears out the well-known fact that a tubular section is much superior to a solid section for column service. Another bit of information that may be gleaned from this equation is the effect of length upon critical load. For a given size column doubling the unsupported length quarters the allowable load. The restraint offered by the end connections of a column also affects the critical load. If the connections are perfectly rigid, the critical load will be four times that for a similar column where there is no resistance to rotation (hinged at the ends).

Since the moment of inertia of a surface is its area multiplied by the square of a length called the radius of gyration, the above formula may be rearranged as follows. Using the Euler formula for hinged ends, and substituting Ar^2 for I the following formula results:

$$\frac{P}{A}=\frac{\pi^2 E}{(l/r)^2}$$

where P/A is the allowable unit stress of the column, and the quantity l/r is the slenderness ratio.

Since the structural column is generally an intermediate length column and it is impossible to obtain an ideal column, the Euler formula has little practical value for ordinary design. Consequently, various empirical column formulae have been developed to agree with test data, all of which embody the slenderness ratio. For design, appropriate factors of safety are introduced into these formulas.

COLY. Turacos and Cuckoos.

COMA CLUSTER. Galaxy.

COMA (Diabetic). Diabetes Mellitus.

COMAGMATIC. Igneous rocks which have a common set of chemical, mineralogic, and textural features and thus considered to have derived from a common parent magma.

COMA (Optics). One of the five geometrical aberrations of a lens with spherical surfaces. Skew rays from a point object do not meet at the same point on the image plane, but rather in a pear-shaped spot (coma). The Abbe sine condition is a measure of coma. A lens system which is corrected for both spherical aberration and coma for a single object position is called aplanatic.

COMA (Physiology). A state of complete unconsciousness from which the individual cannot be aroused even by powerful stimuli. It occurs following severe head injury, in cerebro-vascular accidents (apoplexy), in acute alcoholism, in uncontrolled diabetes (diabetic acidosis), in terminal uremia, and just before death in many diseases. Coma also follows overdosage with certain drugs, particularly central nervous system depressants like morphine, barbituates, and chloral hydrate.

Recovery from complete loss of consciousness is attained by certain stages. The entire process may require only a few minutes, or any of the phases may be prolonged for hours or days. Deep coma is marked by flacid paralysis and even loss of involuntary motion. As the coma lightens, the patient passes into stupor, and reflex activity returns. He responds automatically to forceful commands, but is unaware of his surroundings. The next phase, excitement or delirium, is marked by extreme restlessness and confusion, and sometimes the patient is violent. He gradually becomes quiet, but remains extremely confused mentally. In the next stage, *automatism*, the patient answers questions and performs simple tasks in a fairly orderly but automatic way. The highest functions of the brain, judgment and insight, are the last to return.

COMBINATION. An assignment of a group of objects into two or more mutually exclusive sets. The binomial coefficient $\binom{n}{k}$ is the number of combinations or ways of selecting k objects from a set of n objects. If the k objects are permuted among themselves, no new combinations are formed, but there are $k!$ new arrangements of each combination.

COMBINATION PRINCIPLE. The principle, first recognized by Ritz, that the many frequencies exhibited by the spectrum of a substance can be regarded as differences between a comparatively few terms characteristic of the substance, taken two at a time in their various possible combinations. Ritz's statement was quite empirical, but we now understand that these terms correspond to the different possible energy levels of the atom or molecule, and that the much more numerous spectral frequencies correspond to "jumps" or transitions from one state to another with consequent release or absorption of radiation quanta. For example, if an atom had twenty possible energy states or "levels," the number of possible transitions releasing energy would be theoretically $20 \times \frac{19}{2} = 190$.

It does not follow, however, that all of the corresponding frequencies are actually found in the spectrum; some are not seen because they are forbidden transitions or for other reasons are contrary to the principles of quantum mechanics, or because they occur too seldom to produce observable spectrum lines. When the principle is applied to certain molecular spectra, slight discrepancies are found which may be explained by assuming that some of the energy levels are not single but are close doubles. Such a discrepancy is known as a "combination defect."

COMBINATORIAL TOPOLOGY. Topology.

COMBINED-CYCLE PLANTS. Gas and Expansion Turbines.

COMBINING WEIGHTS. Chemical Composition.

COMBUSTION. The rapid chemical combination of oxygen with the combustible elements of a fuel. There are three combustible chemical elements of major significance—carbon, hydrogen, and sulfur. However, as a source of heat, sulfur is of minor concern. Sulfur is of particular importance in the combustion of several fuels because of the corrosion and pollution problems which its presence creates.

Carbon and hydrogen when burned to completion with oxygen unite according to:

$$C + O_2 = CO_2 \quad + 14{,}100 \text{ Btu/pound (7840 Calories/kilogram) of carbon}$$

$$2H_2 + O_2 = 2H_2O \quad + 61{,}100 \text{ Btu/pound (33,972 Calories/kilogram) of hydrogen}$$

Air is the usual source of oxygen for boiler furnaces. These combustion reactions are exothermic as indicated by the foregoing equations.

The objective of good combustion is to release all of the indicated heat while minimizing losses from combustion imperfections and superfluous air. The combination of the combustible elements and compounds of a fuel with all the oxygen requires *temperature* high enough to ignite the constituents, mixing or *turbulence*, and sufficient *time* for complete combustion. These factors sometimes are referred to as the "three Ts" of combustion.

This description* deals with the basic chemistry necessary for understanding the phenomena of combustion in boiler furnaces. See also **Boiler.** Ability to calculate the release of heat in combustion and to determine the amount and nature of the combustion products is essential for the design, for example, of a steam generating plant and determination of its performance characteristics.

Table 1 lists the chemical elements and compounds found in fuels generally used in the commercial generation of heat with their molecular weights, heats of combustion, and other combustion constants. The term, "100% total air" used in Table 1 and figures and examples which appear elsewhere in this entry means 100% of the air theoretically required for combustion without excess. Higher percentages indicate the theoretical plus excess air, e.g., 125% total air means 100% theoretical air plus 25% excess air.

Concept of the Mole. The mass of a substance in weight units equal to its molecular weight is a mole of the substance. For example, carbon (C) has a molecular weight of 12. Therefore, a pound-mole of carbon weights 12 pounds, a gram-mole of carbon weighs 12 grams, etc. In the case of gases, the *volume* occupied by a mole is called the *molal volume* and this is constant for "ideal" gases. A pound-mole of a gas at 80°F and atmospheric pressure (1 atmosphere or 14.7 psia or 30 inches of mercury) occupies 394 cubic feet. These concepts of mass and volume are useful in combustion calculations. Pound-moles are commonly used in power plant calculations in the United States and a number of other English speaking countries. Useful metric conversions will be found in the entry on **Units and Standards.**

Fundamental Laws

Several fundamental physical laws apply to combusion calculations. These are reviewed briefly as follows:

Conservation of Matter. This is the familiar statement that "matter

* Basic information for this entry from "Steam—Its Generation and Use," copyright The Babcock & Wilcox Co., New York (39th edition, 1978).

TABLE 1. COMBUSTION CONSTANTS OF CHEMICAL ELEMENTS AND COMPOUNDS GENERALLY FOUND IN FUELS

No.	Substance	Formula	Molecular Weight	Lb per Cu Ft	Cu Ft per Lb	Sp Gr Air = 1.0000	Heat of Combustion: Btu per Cu Ft Gross (High)	Btu per Cu Ft Net (Low)	Btu per Lb Gross (High)	Btu per Lb Net (Low)	For 100% Total Air, Moles per mole of Combustible or **Volume Units of Air per Volume Units of Combustible**: Required for Combustion O_2	Required for Combustion N_2	Required for Combustion Air	Flue Products CO_2	Flue Products H_2O	Flue Products N_2	For 100% Total Air **Weight Units of Air per Weight Units of Combustible**: Required for Combustion O_2	Required for Combustion N_2	Required for Combustion Air	Flue Products CO_2	Flue Products H_2O	Flue Products N_2
1	Carbon[a]	C	12.01	—	—	—	—	—	14,093	14,093	1.0	3.76	4.76	1.0	—	3.76	2.66	8.86	11.53	3.66	—	8.86
2	Hydrogen	H_2	2.016	0.0053	187.723	0.0696	325	275	61,095	51,623	0.5	1.88	2.38	—	1.0	1.88	7.94	26.41	34.34	—	8.94	26.41
3	Oxygen	O_2	32.00	0.0846	11.819	1.1053	—	—	—	—	—	—	—	—	—	—	—	—	—	—	—	—
4	Nitrogen (atm)	N_2	28.01	0.0744	13.443	0.9718	—	—	—	—	—	—	—	—	—	—	—	—	—	—	—	—
5	Carbon monoxide	CO	28.01	0.0740	13.506	0.9672	321	321	4,347	4,347	0.5	1.88	2.38	1.0	—	1.88	0.57	1.90	2.47	1.57	—	1.90
6	Carbon dioxide	CO_2	44.01	0.1170	8.548	1.5282	—	—	—	—	—	—	—	—	—	—	—	—	—	—	—	—
Paraffin series																						
7	Methane	CH_4	16.04	0.0425	23.552	0.5543	1012	911	23,875	21,495	2.0	7.53	9.53	1.0	2.0	7.53	3.99	13.28	17.27	2.74	2.25	13.28
8	Ethane	C_2H_6	30.07	0.0803	12.455	1.0488	1773	1622	22,323	20,418	3.5	13.18	16.68	2.0	3.0	13.18	3.73	12.39	16.12	2.93	1.80	12.39
9	Propane	C_3H_8	44.09	0.1196	8.365	1.5617	2524	2322	21,669	19,937	5.0	18.82	23.82	3.0	4.0	18.82	3.63	12.07	15.70	2.99	1.63	12.07
10	*n*-Butane	C_4H_{10}	58.12	0.1582	6.321	2.0665	3271	3018	21,321	19,678	6.5	24.47	30.97	4.0	5.0	24.47	3.58	11.91	15.49	3.03	1.55	11.91
11	Isobutane	C_4H_{10}	58.12	0.1582	6.321	2.0665	3261	3009	21,271	19,628	6.5	24.47	30.97	4.0	5.0	24.47	3.58	11.91	15.49	3.03	1.55	11.91
12	*n*-Pentane	C_5H_{12}	72.15	0.1904	5.252	2.4872	4020	3717	21,095	19.507	8.0	30.11	38.11	5.0	6.0	30.11	3.55	11.81	15.35	3.05	1.50	11.81
13	Isopentane	C_5H_{12}	72.15	0.1904	5.252	2.4872	4011	3708	21,047	19.459	8.0	30.11	38.11	5.0	6.0	30.11	3.55	11.81	15.35	3.05	1.50	11.81
14	Neopentane	C_5H_{12}	72.15	0.1904	5.252	2.4872	3994	3692	20,978	19,390	8.0	30.11	38.11	5.0	6.0	30.11	3.55	11.81	15.35	3.05	1.50	11.81
15	*n*-Hexane	C_6H_{14}	86.17	0.2274	4.398	2.9704	4768	4415	20,966	19,415	9.5	35.76	45.26	6.0	7.0	35.76	3.53	11.74	15.27	3.06	1.46	11.74
Olefin series																						
16	Ethylene	C_2H_4	28.05	0.0742	13.475	0.9740	1604	1503	21,636	20,275	3.0	11.29	14.29	2.0	2.0	11.29	3.42	11.39	14.81	3.14	1.29	11.39
17	Propylene	C_3H_6	42.08	0.1110	9.007	1.4504	2340	2188	21,048	19.687	4.5	16.94	21.44	3.0	3.0	16.94	3.42	11.39	14.81	3.14	1.29	11.39
18	*n*-Butene	C_4H_8	56.10	0.1480	6.756	1.9336	3084	2885	20,854	19,493	6.0	22.59	28.59	4.0	4.0	22.59	3.42	11.39	14.81	3.14	1.29	11.39
19	Isobutene	C_4H_8	56.10	0.1480	6.756	1.9336	3069	2868	20,737	19,376	6.0	22.59	28.59	4.0	4.0	22.59	3.42	11.39	14.81	3.14	1.29	11.39
20	*n*-Pentene	C_5H_{10}	70.13	0.1852	5.400	2.4190	3837	3585	20,720	19,359	7.5	28.23	35.73	5.0	5.0	28.23	3.42	11.39	14.81	3.14	1.29	11.39
Aromatic series																						
21	Benzene	C_6H_6	78.11	0.2060	4.852	2.6920	3752	3601	18,184	17,451	7.5	28.23	35.73	6.0	3.0	28.23	3.07	10.22	13.30	3.38	0.69	10.22
22	Toluene	C_7H_8	92.13	0.2431	4.113	3.1760	4486	4285	18,501	17,672	9.0	33.88	42.88	7.0	4.0	33.88	3.13	10.40	13.53	3.34	0.78	10.40
23	Xylene	C_8H_{10}	106.16	0.2803	3.567	3.6618	5230	4980	18,650	17,760	10.5	39.52	50.02	8.0	5.0	39.52	3.17	10.53	13.70	3.32	0.85	10.53
Miscellaneous gases																						
24	Acetylene	C_2H_2	26.04	0.0697	14.344	0.9107	1477	1426	21,502	20,769	2.5	9.41	11.91	2.0	1.0	9.41	3.07	10.22	13.30	3.38	0.69	10.22
25	Naphthalene	$C_{10}H_8$	128.16	0.3384	2.955	4.4208	5854	5654	17,303	16,708	12.0	45.17	57.17	10.0	4.0	45.17	3.00	9.97	12.96	3.43	0.56	9.97
26	Methyl alcohol	CH_3OH	32.04	0.0846	11.820	1.1052	868	767	10,258	9,066	1.5	5.65	7.15	1.0	2.0	5.65	1.50	4.98	6.48	1.37	1.13	4.98
27	Ethyl alcohol	C_2H_5OH	46.07	0.1216	8.221	1.5890	1600	1449	13,161	11,917	3.0	11.29	14.29	2.0	3.0	11.29	2.08	6.93	9.02	1.92	1.17	6.93
28	Ammonia	NH_3	17.03	0.0456	21.914	0.5961	441	364	9,667	7,985	0.75	2.82	3.57	—	1.5	3.32	1.41	4.69	6.10	—	1.59	5.51
														SO_2						SO_2		
29	Sulfur[a]	S	32.06	—	—	—	—	—	3,980	3,980	1.0	3.76	4.76	1.0	—	3.76	1.00	3.29	4.29	2.00	—	3.29
30	Hydrogen sulfide	H_2S	34.08	0.0911	10.979	1.1898	646	595	7,097	6,537	1.5	5.65	7.15	1.0	1.0	5.65	1.41	4.69	6.10	1.88	0.53	4.69
31	Sulfur dioxide	SO_2	64.06	0.1733	5.770	2.2640	—	—	—	—	—	—	—	—	—	—	—	—	—	—	—	—
32	Water vapor	H_2O	18.02	0.0476	21.017	0.6215	—	—	—	—	—	—	—	—	—	—	—	—	—	—	—	—
33	Air	—	28.9	0.0766	13.063	1.0000	—	—	—	—	—	—	—	—	—	—	—	—	—	—	—	—

[a]Carbon and sulfur are considered as gases for molal calculations only.
All gas volumes corrected to 60 F and 30 in. Hg dry. (15.6°C and 101.6 kilopascals).
SOURCE: American Gas Association, Arlington, Virginia.

To convert:

lb/cu ft to kg/cu meter, multiply by	16.026
cu ft/lb to cu meters/kg, multiply by	0.0624
Btu to Calories	0.2520
Btu/cu ft to Calories/cu meter	8.898

TABLE 2. COMMON CHEMICAL REACTIONS OF COMBUSTION

COMBUSTIBLE	REACTION	MOLES	POUNDS	HEAT OF COMBUSTION (HIGH) BTU/POUND OF FUEL
Carbon (to CO)	$2C + O_2 = 2CO$	2 + 1 = 2	24 + 32 = 56	4,000
Carbon (to CO_2)	$C + O_2 = CO_2$	1 + 1 = 1	12 + 32 = 44	14,100
Carbon Monoxide	$2CO + O_2 = 2CO_2$	2 + 1 = 2	56 + 32 = 88	4,345
Hydrogen	$2H_2 + O_2 = 2H_2O$	2 + 1 = 2	4 + 32 = 36	61,100
Sulfur (to SO_2)	$S + O_2 = SO_2$	1 + 1 = 1	32 + 32 = 64	3,980
Methane	$CH_4 + 2O_2 = CO_2 + 2H_2O$	1 + 2 = 1 + 2	16 + 64 = 80	23,875
Acetylene	$2C_2H_2 + 5O_2 = 4CO_2 + 2H_2O$	2 + 5 = 4 + 2	52 + 160 = 212	21,500
Ethylene	$C_2H_4 + 3O_2 = 2CO_2 + 2H_2O$	1 + 3 = 2 + 2	28 + 96 = 124	21,635
Ethane	$2C_2H_6 + 7O_2 = 4CO_2 + 6H_2O$	2 + 7 = 4 + 6	60 + 224 = 284	22,325
Hydrogen Sulfide	$2H_2S + 3O_2 = 2SO_2 + 2H_2O$	2 + 3 = 2 + 2	68 + 96 = 164	7,100

SOURCE: "Steam—Its Generation and Use," 39th edition, The Babcock and Wilcox Company, New York, 1978.

is neither destroyed nor created." There must be a weight balance between the sum of the weights entering a process and the sum leaving. In other words, *A* pounds of fuel combined with *B* pounds of air will always result in $A + B$ pounds of products. (It should be noted that when a pound of a typical coal is burned, releasing 13,500 Btu, the quantity of mass converted to energy amounts to only 3.5×10^{-10} pound, a loss too small to be measured or considered in conventional combustion calculations. Obviously, this conversion is of significance to nuclear reactions.)

Conservation of Energy. This is the familiar statement that "energy is neither destroyed nor created." The sum of the energy (potential, kinetic, thermal, chemical, and electrical) entering a process must equal the sum of energy leaving, although the proportionate amounts of each may change. In combustion, chemical energy is exchanged for energy in the form of heat. The parenthetical observation made in the prior paragraph also applies to this relationship.

Ideal Gas Law. The volume of an ideal gas is directly proportional to its absolute temperature and inversely proportional to its absolute pressure. The proportional constant is found to be the same for one mole of any ideal gas, so this law may be expressed as:

$$v_M = \frac{RT}{p}$$

where

v_M = volume, cubic feet/mole of gas
p = absolute pressure, pounds/square foot
T = absolute temperature, degrees Rankine = °F + 460
R = universal gas constant, 1545 foot pound/mole, T

The equation states that one mole of all ideal gases occupies the same volume for the same pressure and temperature conditions—394 cubic feet at 14.7 psi and 80°F. Experiments indicate that most gases approach this ideal.

Law of Combining Weights. All substances combine in accordance with simple definite weight relationships. These relationships are exactly proportional to the molecular weights of the constituents. For example, carbon (atomic weight = 12) combines with oxygen (molecular weight = 32) to form carbon dioxide (molecular weight = 44) so that 12 pounds of carbon plus 32 pounds of oxygen unite to form 44 pounds of carbon dioxide.

Avogadro's Law. Equal volumes of different gases at the same pressure and temperature contain the same number of molecules. From the concept of the mole, a pound-mole of any substance contains a mass equal in pounds to the molecular weight of the substance. Thus the ratio of mole weight to molecular weight is a constant, and a mole of a chemically pure substance contains the same number of molecules, no matter what the substance may be. Since a mole of any ideal gas occupies the same volume at a given pressure and temperature (ideal gas law), it follows that equal volumes of different gases at the same pressure and temperature contain the same number of molecules.

Dalton's Law. The total pressure of a mixture of gases is the sum of the partial pressures which would be exerted by each of the constituents if each gas were to occupy alone the same volume as the mixture. In other words, for equal volumes, V, of three gases (A, B, and C) all at the same temperature, T, but at different pressures, P_a, P_b and P_c, when all three gases are placed in the space of the volume, V, then the resulting pressure, P, is equal to $P_a + P_b + P_c$. For gases, each gas in a mixture fills the entire volume and exerts a pressure independent of the other gases.

Amagat's Law. The total volume occupied by a mixture of gases is equal to the sum of the volumes which would be occupied by each of the constituents when at the same pressure and temperature as the mixture. This law is related to Dalton's law, but considers the additive effects of volume instead of pressure. If all three gases are at pressure, P, and temperature, T, but at volumes V_a, V_b and V_c, then, when combined so that T and P are unchanged, the volume of the mixture, $V = V_a + V_b + V_c$.

Application of Fundamental Laws

Table 2 summarizes the molecular and weight relationships between fuel and oxygen and lists the heat of combustion for the substances commonly involved in combustion. Most of the weight and volume relationships in combustion calculations can be determined by using the information in this table and the seven fundamental laws.

The data for C and H_2 can be expressed as follows:

C	+	O_2	=	CO_2*
1 molecule	+	1 molecule	→	1 molecule
1 mole	+	1 mole	=	1 mole
		1 cubic foot	→	1 cubic foot
12 pounds	+	32 pounds	=	44 pounds

$2H_2$	+	O_2	=	$2H_2O$
2 molecules	+	1 molecule	→	2 molecules
2 moles	+	1 mole	=	2 moles
2 cubic feet	+	1 cubic foot	→	2 cubic feet
4 pounds	+	32 pounds	=	36 pounds

While there is a weight balance in these equations, there is not a molecular or volume balance; for example, 2 cubic feet of H_2 unite with 1 cu ft of O_2 to form only 2 cubic feet of H_2O. This relationship is based on Avogadro's law and the law of combining weights.

The Mole in Combustion Calculations. Combustion calculations involving gaseous mixtures can be simplified by the use of the mole. Since equal volumes of gases at any given pressure and temperature contain the same number of molecules (Avogadro's law), the weights of equal volumes of gases are proportional to their molecular weights. If M is the molecular weight of the gas, 1 mole equals M pounds. Actual values are available from Table 1, e.g.:

* When 1 cubic foot of oxygen combines with carbon, it forms 1 cubic foot of carbon dioxide. If carbon were an ideal gas instead of a solid, 1 cubic foot of carbon would be required.

1 mole of O_2 = 32 pounds oxygen
1 mole of H_2 = 2 pounds hydrogen
1 mole of CH_4 = 16 pounds methane

Data from Table 1 can be used to demonstrate that the volume of 1 mole at a given pressure and temperature is approximately fixed and independent of the kind of gas.

At 60°F and atmospheric pressure (30 inches of mercury), the specific volume of oxygen is 11.819 cubic feet per pound. Therefore, 1 mole of oxygen has a volume of 32 × 11.819 = 378.21 cubic feet. Similarly, at 60°F and atmospheric pressure, the specific volume of hydrogen is 187.723 cubic feet per pound, and 1 mole has a volume of 2.016 × 187.723 = 378.45 cubic feet. This volume, usually taken as 379 cubic feet, therefore approximates the volume of 1 mole of gas at 60°F and atmospheric pressure.

The mole fraction of a component of a mixture is the number of moles of the component divided by the sum of the number of moles of all the components of the mixture. As a mole of every ideal gas occupies the same volume, it follows by Avogadro's law that in a mixture of ideal gases the mole fraction of a component will exactly equal the volume fraction:

$$\frac{\text{Moles of Component}}{\text{Total Moles}} = \frac{\text{Volume of Component}}{\text{Volume of Total Mixture}}$$

This is a valuable concept, since the volumetric analysis of a mixture of gases automatically gives the mole fractions of the different components.

In power plant practice, the practical source of oxygen is primarily air, which includes, along with the oxygen, a mixture of nitrogen, water vapor, and small amounts of inert gases, such as argon, neon, and helium. Data on the composition of air are given in Table 3.

TABLE 3. COMPOSITION OF AIR

	Composition of Dry Air	
	% by Volume	% by Weight
Oxygen, O_2	20.99	23.15
Nitrogen, N_2	78.03	76.85[a]
Inerts	0.98	—

Equivalent molecular weight of air = 29.0[a]

% Moisture = 1.3% by weight. Standard for boiler industry (ABMA)[b]

$$\begin{matrix}\text{Moles air/mole oxygen} \\ \text{Cubic feet air/cubic feet oxygen}\end{matrix} = \frac{100}{20.99} = 4.76$$

$$\text{Moles } N_2\text{/mole oxygen} = \frac{79.01}{20.99} = 3.76$$

$$\text{Pounds air (dry)/pound oxygen} = \frac{100}{23.15} = 4.32$$

$$\text{Pounds nitrogen/pound oxygen} = \frac{76.85}{23.15} = 3.32$$

[a] It is convenient in combustion calculations to account for inerts as equivalent nitrogen. The equivalent weight percentage of 76.85 and the equivalent molecular weight of 29.0 have been corrected to account for the extra weight of the inerts.

[b] Air containing 0.013 pound water/pound dry air is often referred to as standard air.

SOURCE: "Steam—Its Generation and Use," 39th edition, The Babcock and Wilcox Company, New York, 1978.

The information in Table 2 can be used for air instead of O_2 if 3.76 moles of nitrogen (N_2) are added to both left and right side of each equation for every mole of O_2 involved. For example, the burning of CO in air becomes:

$$2CO + O_2 + 3.76N_2 = 2CO_2 + 3.76N_2$$

or for methane:

$$CH_4 + 2O_2 + 2(3.76)N_2 = CO_2 + 2H_2O + 7.52N_2$$

As indicated by the following example for a fuel gas, molal calculations have a simple and direct application to gaseous fuels, where the analyses are usually reported in percent on a volume basis.

Fuel Gas Analysis (% by Volume)	
CH_4	85.3
C_2H_6	12.6
CO_2	0.1
N_2	1.7
O_2	0.3
Total	100.0

This analysis may also be expressed as 85.3 moles of CH_4 per 100 moles of fuel; 12.6 moles of C_2H_6 per moles of fuel; and so on.

The elemental breakdown of each constituent may also be designated in moles per 100 moles of fuel, as follows:

C in CH_4	= 85.3 × 1	= 85.3 moles
C in C_2H_6	= 12.6 × 2	= 25.2 moles
C in CO_2	= 0.1 × 1	= 0.1 mole
Total C per 100 moles fuel		= 110.6 moles
H_2 in CH_4	= 85.3 × 2	= 170.6 moles
H_2 in C_2H_6	= 12.6 × 3	= 37.8 moles
Total H per 100 moles fuel		= 208.4 moles
O_2 in CO_2	= 0.1 × 1	= 0.1 mole
O_2	= 0.3 × 1	= 0.3 mole
Total O_2 per 100 moles fuel		= 0.4 mole
N_2 per 100 moles fuel		= 1.7 moles

An analysis of the flue gas produced by burning a gas fuel of the composition given above could be:

Constituent	% by Volume
CO_2	10.4
O_2	2.8
N_2	86.8
Total	100.0

Analyses of flue gases are always reported on a volume basis, *dry*, when an Orsat or other type of gas analysis is used. Flue gases are cooled to room temperature and bubbled through water in most gas analyses, so that the gas becomes saturated with water vapor. This would occur even if no water vapor were formed during combustion. Proportionate parts of the water vapor content of the gas will be absorbed with the different constituents of the gas so that the resulting analysis may be safely assumed to be that of dry gas. These percentages may also be expressed as 10.4 moles CO_2, 2.8 moles O_2, and 86.8 moles N_2; each per 100 moles of dry flue gas.

For each mole of C burned, one mole of CO_2 is formed. From the fuel analysis used there are 110.6 moles C per 100 moles of fuel, and there are also 110.6 moles of CO_2 formed from the 110.6 moles C in the fuel. From the flue gas analysis, there are 100/10.4 = 9.62 moles of dry flue gas per mole of CO_2. The 100 moles of fuel will then yield 110.6 × 9.62 = 1,064 moles of dry flue gas. By the application of the mole method, an important value has been quickly determined through knowing only the flue gas analysis and the fuel analysis.

From the flue gas analysis, the molecular weight of the dry flue gas can be easily determined, as follows:

10.4 moles of CO_2 weigh		
	10.4 × 44 pounds =	457.6 pounds
2.8 moles of O_2 weigh		
	2.8 × 32 pounds =	89.6 pounds
86.8 moles of N_2 weigh		
	86.8 × 28 pounds =	2,430.4 pounds
100.0 moles of dry flue gas		2,977.6 pounds

Therefore, 1 mole equivalent of dry flue gas = 29.8 pounds, or the equivalent molecular weight of the dry flue gas = 29.8. Hence, the weight of 1,064 moles of dry flue gas is 1,064 × 29.8 = 31,700 pounds, or 100 moles of fuel yields 31,700 pounds of dry flue gas.

The weight of 100 moles of fuel is the sum of the products of each constituent in the fuel and its molecular weight.

CH_4	85.3 × 16 =	1,365
C_2H_6	12.6 × 30 =	378
CO_2	0.1 × 44 =	4.4
N_2	1.7 × 28 =	47.6
O_2	0.3 × 32 =	9.6
	100.0 moles =	1,804.6 pounds

Thus, 1,805 pounds of gas fuel yield 31,700 pounds of dry flue gas, and each pound of gas fuel yields 31,700/1,805 = 17.6 pounds dry flue gas.

Heat of Combustion

In a boiler furnance (where no mechanical work is done) the heat energy evolved from the union of combustible elements with oxygen depends on the ultimate products of combustion and not on any intermediate combinations that may occur in reaching the final result.

A simple demonstration of this law is the union of 1 pound of carbon with oxygen to produce a specific amount of heat (about 14,100 Btu, Table 2). The union may be in one step to form the gaseous product of combustion, CO_2, or under certain conditions the union may be in two steps, first to form CO, producing a much smaller amount of heat (4,345 Btu) and, second the union of the CO so obtained to form CO_2, releasing 9,755 Btu. However, the sum of the heats released in the two steps equals the 14,100 Btu evolved when carbon is burned in one step to form CO_2 as the final product.

That carbon may enter into these two combinations with oxygen is of utmost importance in the design of combustion equipment. Firing methods must assure complete mixture of fuel and oxygen to be certain that all of the carbon burns to CO_2 and not to CO. Failure to meet this requirement will result in appreciable losses in combustion efficiency and in the amount of heat released by the fuel, since only about 28% of the available heat in the carbon is released if CO is formed instead of CO_2.

Measurement of Heat of Combustion. In boiler practice, the heat of combustion of a fuel is the amount of heat, expressed in Btu, generated by the complete combustion (or oxidation) of a unit weight (1 pound in the United States) of fuel. Calorific value or "fuel Btu value" are other terms used.

The amount of heat generated by complete combustion is a constant for any given combination of combustible elements and compounds and is not affected by the manner in which the combustion takes place, provided that it is complete.

The heat of combustion of a fuel is usually determined by direct measurement in a calorimeter of the heat evolved during combustion. See entry on **Calorimetry.**

The heat of combustion of most gases encountered in boiler practice is given in Table 1. If the content of any gas mixture is known, its heat of combustion can be accurately determined by adding the products of the volume percentage of each constituent times its heat of combustion.

For accurate heat values of solid and liquid fuels calorimeter determinations are required. However, approximate heat values may be determined for most coals if the ultimate chemical analysis is known. Dulong's formula gives reasonably accurate results (within 2 to 3%) for most coals and is often used as a routine check of values determined by calorimeter:

$$\text{Btu/pound} = 14{,}544C + 62{,}028(H_2 - O_2/8) + 4050S \qquad (1)$$

In this formula, the symbols represent the proportionate parts by weight of the constituents of the fuel—carbon, hydrogen, oxygen and sulfur—as determined by an ultimate analysis; the coefficients represent the approximate heating values of the constituents in Btu per pound. The term $O_2/8$ is a correction applied to the hydrogen in the fuel to account for the hydrogen already combined with the oxygen in the form of moisture. This formula is not generally suitable for calculating the Btu values of gaseous fuels.

High and Low Heat Values. Water vapor is one of the products of combustion for all fuels which contain hydrogen. The heat content of a fuel depends on whether this water vapor is allowed to remain in the vapor state or is condensed to liquid. In the bomb calorimeter the products of combustion are cooled to the initial temperature and all of the water vapor formed during combustion is condensed to liquid. This gives the high, or gross, heat content of the fuel with the heat of vaporization included in the reported value. For the low, or net heat of combustion, it is assumed that all products of combustion remain in the gaseous state.

While the high, or gross, heat of combustion can be accurately determined by established (ASTM) procedures, direct determination of the low heat of combustion is difficult. Therefore, it is usually calculated using the following formula:

$$Q_L = Q_H - 1040\,w \qquad (2)$$

where:

Q_L = low heat of combustion of fuel, Btu/pound
Q_H = high heat of combustion of fuel, Btu/pound
w = pound of water formed per pound of fuel
1040 = factor to reduce high heat of combustion at constant volume to low heat of combustion at constant pressure

In the United States the practice is to use the high heat of combustion in boiler combustion calculations. In Europe the low heat value is used.

Ignition Temperatures

Ignition temperature may be defined as the temperature at which more heat is generated by combustion than is lost to the surroundings so that the combustion process becomes self-sustaining. The term usually applies to rapid combustion in air at atmospheric pressure.

Ignition temperatures of combustion substances vary greatly as indicated in Table 4, which lists minimum temperatures and temperature ranges in air for fuels and for the combustible constituents of fuels commonly used in the commercial generation of heat. Many factors influence ignition temperature so that any tabulation can be used only as a guide. Pressure, velocity, enclosure configuration, catalytic materi-

TABLE 4. IGNITION TEMPERATURES OF FUELS IN AIR
(Approximate Values and Ranges at Atmospheric Pressure)

Combustible	Formula	Tempfrature, °F
Sulfur	S	470
Charcoal	C	650
Fixed carbon (bituminous coal)	C	765
Fixed carbon (semibituminous coal)	C	870
Fixed carbon (anthracite)	C	840–1115
Acetylene	C_2H_2	580–825
Ethane	C_2H_6	880–1165
Ethylene	C_2H_4	900–1020
Hydrogen	H_2	1065–1095
Methane	CH_4	1170–1380
Carbon Monoxide	CO	1130–1215
Kerosine	—	490–560
Gasoline	—	500–800

als, air-fuel-mixture uniformity, and ignition source are only a few of the variables. Ignition temperature usually decreases with rising pressure and increases with increasing moisture content in the air.

The ignition temperature of the gases of a coal vary considerably and are appreciably higher than the ignition temperatures of the fixed carbon of the coal. However, the ignition temperature of coal may be considered as the ignition temperature of its fixed carbon content, since the gaseous constituents are usually distilled off but not ignited before this temperature is attained.

Adiabatic Flame Temperature

The adiabatic flame temperature is the maximum theoretical temperature which can be reached by the products of combustion of a specific fuel and air (or oxygen) combination assuming no loss of heat to the surroundings until combustion is complete. This theoretical temperature also assumes no dissociation, a phenomenon discussed later under this heading. The heat of combustion of the fuel is the major factor in the flame temperature, but increasing the temperature of the air or of the fuel will also have the effect of raising the flame temperature. As would be expected, this adiabatic temperature is a maximum with zero excess air (only enough air chemically required to combine with the fuel), since any excess is not involved in the combustion process and only dilutes the temperature of the products of combustion.

The adiabatic temperature is determined from the adiabatic enthalpy of the flue gas as follows:

$$h_g = \frac{\begin{pmatrix}\text{heat of}\\ \text{combustion}\end{pmatrix} + \begin{pmatrix}\text{sensible heat}\\ \text{in fuel}\end{pmatrix} + \begin{pmatrix}\text{sensible heat}\\ \text{in air}\end{pmatrix}}{\text{weight of products of combustion}}$$

where:

h_g = adiabatic enthalpy (adiabatic heat content of the products of combustion), Btu/pound

Knowing the moisture content of the products of combustion and its enthalpy, the theoretical flame or gas temperature can be obtained from published graphs.*

The adiabatic temperature is a fictitiously high temperature which does not exist in fact. Actual flame temperatures are lower for two main reasons:

1. Combustion is not instantaneous. Some heat is lost to the surroundings as combustion takes place. The faster the combustion occurs the less heat is lost before combustion is complete. If combustion is slow enough, the gases may be cooled sufficiently for combustion to be incomplete with some of the fuel unburned. This is related to the time factor in the "three *T*'s" of combustion mentioned previously.
2. At temperatures above 3,000°F (1,649°C), some of the CO_2 and H_2O in the flue gases dissociates, absorbing heat in the process. At 3,500°F (1,926°C), about 10% of the CO_2 in a typical flue gas dissociates to CO and O_2 with a heat absorption of 4,345 Btu/pound of CO formed, and about 3% of the H_2O dissociates to H_2 and O_2, with a heat absorption of 61,100 Btu/pound of H_2 formed. As the gas cools, the CO and H_2 dissociated recombine with the O_2 and liberate the heat absorbed in dissociation, so the heat is not lost. However, the effect is to lower the maximum actual flame temperature.

Combustion Calculations

The combustion calculations are the starting point for all design and performance determinations for boilers and their related component parts. They establish (a) the quantities of the constituents involved in the chemistry of combustion, (b) the quantity of heat released, and (c) the efficiency of the combustion process under both ideal and actual conditions.

Combustion Air. Since carbon, hydrogen, and sulfur are the only major combustible elements found in the fuels used for commercial steam generation, the air (pounds) theoretically required for the complete combustion of 1 pound of fuel is:

$$11.53C + 34.34(H_2 - O_2/8) + 4.29S \tag{3}$$

where C, H_2, O_2 and S represent the fraction by weight (percent/100) of carbon, hydrogen, oxygen and sulfur, and the constants are those given in Table 1. The factor $O_2/8$ in the term $(H_2 - O_2/8)$ is a correction for the hydrogen already combined with the O_2 in the fuel to form water vapor.

With gaseous fuels, instead of breaking down the hydrocarbons into their constituent elements, it is simpler to use the amount of air for the various compounds directly as given in Table 1. For instance, for a gaseous fuel containing the combustible gases indicated in the expression below, the theoretical air required for complete combustion (pounds air/pounds fuel) is:

$$2.47CO + 34.34H_2 + 17.27CH_4 + 13.30C_2H_2 + 14.81C_2H_4 + 16.12C_2H_6 + 6.10H_2S - 4.32O_2 \tag{4}$$

Again, the molecular symbols represent the fraction by weight of the gaseous compounds and elements.

If, as is the usual custom, the analyses of gaseous fuels are given on a volumetric basis, the cubic feet of combustion air required as given in Table 1 should be used. Thus for a gaseous fuel containing the combustible gases indicated in the following expression, the number of cubic feet of theoretical air required per cubic foot of fuel for complete combustion is:

$$2.38(CO + H_2) + 9.53CH_4 + 11.91C_2H_2 + 14.29C_2H_4 + 16.68C_2H_6 + 7.15H_2S - 4.76O_2 \tag{5}$$

where the molecular symbols represent the fraction by volume of the gaseous compounds and elements. Note that the toal air requirement is reduced if oxygen is one of the constituents of the fuel.

The products of combustion can also be determined from the data given in Table 1. Assuming complete combustion with theoretical air of the fuels ordinarily used for commerical steam generation, the products of combustion in pounds (including the nitrogen carried with the combustion air) per pound of fuel are:

$$CO_2 = 3.66C$$

$$H_2O = 8.94H_2 + H_2O^{**}$$

$$SO_2 = 2.00S$$

$$N_2 = 8.86C + 26.41(H_2 - O_2/8) + 3.29S + N_2\dagger$$

The moisture introduced with the combustion air must be added to this theoretical quantity to obtain the total weight of combustion products. The molecular symbols represent the fraction by weight of the constituents in the fuel.

Energy Losses. Not all of the Btu in the fuel are converted to heat and absorbed by the steam generation equipment. Some of the fuel may be unburned, leaving carbon in the ash or carbon may be burned incompletely to form some CO instead of all CO_2. Usually all of the H_2 in the fuel is burned. By far the greatest heat loss is the loss up the stack. Since the heat in the fuel is determined from a base of ambient temperature, all of the products of combustion must be cooled to the same temperature if all of the heat is to be utilized. Higher temperatures then represent a loss, which is the sum of four items: (1) the sensible heat in dry flue gas, (2) the sensible heat in the moisture in the air, (3) the sensible heat in the H_2O in the fuel and (4) the latent heat of the moisture in the fuel.

It is necessary practically to use more than the theoretical air requirements to assure sufficient oxygen for complete combustion. Excess air would not be required if it were possible to have an ideally perfect union of air and fuel. It is necessary, however, to keep the excess at a minimum in order to hold down the stack loss. The excess air that is not used in the combustion of the fuel leaves the unit at stack temperature. The heat required to heat this air from room temperature to stack temperature serves no purpose and is lost heat. Table 5 gives realistic values of excess air for the fuel burning equipment which

* Series of 8 graphs in "Steam—Its Generation and Use," pages 6–8 and 6–9, published by The Babcock and Wilcox Company, New York (39th edition, 1978).

** Fraction by weight of H_2O (percent/100) in the fuel as moisture.

† Fraction by weight of N_2 in the fuel as nitrogen

TABLE 5. USUAL AMOUNTS OF EXCESS AIR SUPPLIED TO FUEL-BURNING EQUIPMENT

Fuel	Type of Furnace or Burners	Excess Air % by Weight
Pulverized coal	Completely water-cooled furnace for slag-tap or dry-ash-removal	15–20
	Partially water-cooled furnace for dry-ash-removal	15–40
Crushed coal	Cyclone furnace—pressure or suction	10–15
Coal	Spreader stoker	30–60
	Water-cooled vibrating-grate stoker	30–60
	Chain-grate and traveling-grate stokers	15–50
	Underfeed stoker	20–50
Fuel oil	Oil burners, register-type	5–10
	Multifuel burners and flat-flame	10–20
Acid sludge	Cone and flat-flame type burners, steam-atomized	10–15
Natural, coke-oven, and refinery gas	Register-type burners	5–10
	Multifuel burners	7–12
Blast-furnace gas	Intertube nozzle-type burners	15–18
Wood	Dutch oven (10–23% through grates) and Hofft-type	20–25
Bagasse	All furnaces	25–35
Black liquor	Recovery furnaces for kraft and soda-pulping processes	5–7

SOURCE: "Steam—Its Generation and Use," 39th edition, The Babcock and Wilcox Company, New York, 1978.

experience has shown is required to assure complete combustion for various fuels and methods of firing.

In most furnances operating under suction, there is also some leakage of air into the setting and, consequently, the excess air leaving the furnance and the unit is greater than that at the fuel burning equipment. Another loss which must be considered is the radiation loss from the unit setting.

In summary, there are certain inherent heat losses over which there is no control, and others which are subject to some control. The inherent losses are the result of: (1) the discharge of the products of combustion at a temperature higher than ambient; and (2) the moisture content of the fuel plus the combination of some of the hydrogen with the oxygen in the fuel. The avoidable heat losses, or those which can be controlled by good design and careful operation, can be minimized by: (1) careful control of excess air; (2) tolerating virtually no unburned solid combustible matter in ash or refuse; (3) permitting no unburned gaseous combustibles in the exit gases; and (4) a well-insulated setting for the steam generating unit to reduce radiation loss.

The efficiency of combustion in a heat exchanger or boiler is 100 minus the sum of the heat losses expressed in percent. See also **Boiler;** and **Burner.**

References

Baumeister, T. (editor): "Mark's Standard Handbook for Mechanical Engineers," 8th edition, McGraw-Hill, New York, 1978.

Beer, J., and N. Chigier: "Combustion Aerodynamics," Fuel and Energy Science Monographs Series, Wiley, New York, 1972.

Grey, J., Sutton, G. W., and M. Zlotnick: "Fuel Conservation and Applied Research," *Science*, **200**, 135–142 (1978).

Johnson, S. A., and A. H. Rawdon: "NO_x Control by Furnance and Burner Design," in "Energy Technology Handbook," (D. M. Considine, editor), McGraw-Hill, New York, 1976.

Smith, R. B.: "Heat Generation," in "Chemical Engineers' Handbook," 5th edition, (R. H. Perry and C. H. Chilton, editors), McGraw-Hill, New York, 1973.

Smoot, L. D., and D. T. Pratt: "Pulverized Coal Combustion and Gasification for Continuous Flow Processes," Plenum, New York, 1979.

Staff: "Steam—Its Generation and Use," 39th edition, The Babcock & Wilcox Co., New York, 1978.

COMBUSTION (Gas Turbine). Gas and Expansion Turbines.

COMBUSTION HEAT. Calorimetry; Combustion.

COMBUSTION PROCESSES (Pollution). Pollution (Air).

COMBUSTION-TYPE GAS ANALYZERS. Gas Analyzers (Combustion-Type).

COMET. A luminous heavenly body, moving in space under the influence of the sun's gravitational field, which occasionally comes close enough to the earth to be observed. The name comes from the descriptive Latin phrase *stellae comatae* (hairy stars), and aptly describes the appearance of this class of celestial objects. The appearance of comets is so different from all other objects in the night sky that they have always attracted a great deal of attention. They were formerly regarded with superstitious awe, and were supposed to portend all sorts of calamities to the earth and its inhabitants. For these reasons, the appearance of a comet was always recorded by the ancient scribes, and we have records of comet observations going back to over more than a thousand years prior to the Christian era.

Comets take their names from their discoverers. Normally, a preliminary designation will be used. This consists of the year followed by a letter assigned in order of discovery during that year. A few years later, if all preliminary data are confirmed, a given comet will receive a permanent designation. This states the year of the comet's perihelion passage, followed by a Roman numeral giving the order of passage during that year. However, the literature will continue to identify them as well by discoverer names. See accompanying table.

Only a very small proportion of the comets that are discovered ever become visible to the naked eye, and there must be many comets that are never discovered at all. From a study of ancient records, there has been, on the average, about one comet visible to the naked eye per year. The year 1911, in which four such objects were seen, apparently holds the record for maximum number, and there are years in which no comet at all is visible to the unaided eye.

A comet consists of three general parts: the nucleus, the coma, and the tail. The relative sizes and appearances of these parts change radically as the comet approaches perihelion. It is believed that the nucleus (or head) of a comet is a mass of more or less condensed material. It looks much like an ordinary star when the comet is a long way from the sun. As the comet approaches the sun, the nucleus increases in brightness and apparently shrinks in size, although this latter characteristic may be an optical illusion, since actual measurements of the diameter of the head are practically impossible. With the approach to the sun, a hazy shell makes its appearance about the nucleus, and the so-called coma increases in size. When quite close to the sun, the coma seems to stretch out in the direction away from the sun and the tail develops in this direction. The real glory of a comet to the naked eye is this tail. (See Figs. 1, 2, 3.) It should be carefully noted that the tail points in a direction directly away from the sun and does not, in general, trail out behind the comet in its motion. Although many comets apparently never develop a tail at all, the tails that do develop have a variety of shapes and curves, ranging from short, sharply curved tails, to long, straight streamers pointing away from the sun. The tails fall into two distinct classes: those in Class I are composed of ionized molecules; those in Class II are composed of dust. Occasionally, tails are a combination of both types.

Comets are the largest members of the solar system. The nucleus may have a diameter of up to 16,000 km; the coma, a diameter ranging from 16,000 to 80,000 km; and tails have been observed with length as great as several hundred thousand km. Hence, the volume of a large comet may be greater than the volume of all other members of the solar system combined. In contrast to the enormous bulk of

COMET STATISTICS

Time of Return	Year of Discovery	Name	Period (Years)	Inclination to Ecliptic (Degrees)
1982				
January	1942	Oterma	7.88	4
April	1896	Perrine-Mrkos	6.72	18
June	1927	Gale	10.99	12
August	1862	Swift-Tuttle*	119.93	114
November	1969	Gunn	6.80	10
December	1951	Arend	7.76	22
December	1929	Neujmin-Ill	10.57	4
December	1939	Vaisala I	11.28	12
1983	1955	Harrington-Abell	7.19	17
	1786	Encke	3.30	12
	1819	Pons-Winnecke	6.34	22
	1873	Temple II	5.26	12
	1970	Kojiima	6.19	4
	1951	Dutoit-Neujmin	6.31	3
1984	1924	Wolf-Harrington	6.55	18
	1884	Wolf	8.43	27
	1955	Harrington-Abell	7.19	17
	1843	Faye	7.39	9
	1951	Arend-Rigaux	6.84	18
1985	1911	Schaumasse	8.18	12
	1948	Honda-Mrkos	5.22	13
	1933	Whipple	7.47	10
	1909	Daniel	7.09	20
1986				
January†	240 B.C.	Halley	76.1	162
	1916	Neuimin II	5.43	11
	1926	Comas-Sola	8.55	13
	1915	Taylor	6.37	15

* One appearance only.

† More detail on Halley is given in entry on **Halley's Comet.**

NOTE: The majority of comets listed will be observed only by professional astronomers.

Fig. 2. The Ikeya-Seki comet, as seen over Maryland on November 3, 1965. The length of the comet's tail was estimated at 28 million miles (~45 million kilometers).

these objects, their masses are so small that they have never been accurately measured. The only available method for determining the mass of a comet is by means of the perturbations it might produce in the orbits of the planets or asteroids. Many cases of very close approach of comets to objects of known mass have been observed, but while the comet orbit itself may be enormously perturbed, no perturbations have ever been observed in orbits of other objects. The best that can be said regarding cometary masses is that they must be less than one-hundred-thousandth of the mass of the earth.

Orbits of comets differ from orbits of the other members of the solar system in that they are, in general, much more eccentric than the planetary orbits, and while the planetary orbit planes are all nearly parallel to the plane of the ecliptic, comet orbits are found inclined through practically every angle. The great majority of the orbits that have been determined are found to be parabolic in character, which means that the objects will not return to visibility again. Two or three comets apparently have hyperbolic orbits, while others are moving about the sun in ellipses, returning to the vicinity of the sun and, hence, visibility, at periodic intervals.

Probably the most noted of all comets is Halley's comet. Halley was not its discoverer, but he was the first one to predict the period of return. This object has a period of approximately 75 years, and has been observed and recorded at practically every return for over a thousand years. The period is slightly variable due to perturbations by Jupiter. See also **Halley's Comet.**

Comet Theory

Because no cometary nucleus has been observed as more than a mere point of light, even the size of any given comet must be estimated by measuring its brightness and assuming a reflectivity (Neugenbauer and Newburn, 1978). Similarly, mass is derived from size by assuming a density. Such properties as shape, rotation, morphology, surface and internal structure remain largely unknown. Thus, for lack of information, a number of theories pertaining to the origin of comets have arisen. Some authorities believe that comets are of interstellar origin; others suggest mechanisms for continuous creation of comets within the solar system (up to the present time).

The comet Bennet was discovered at Riviera, South Africa by J. C. Bennett on December 28, 1969, and was visible to the naked

Fig. 1. Comet Arend-Roland (1956–1957).

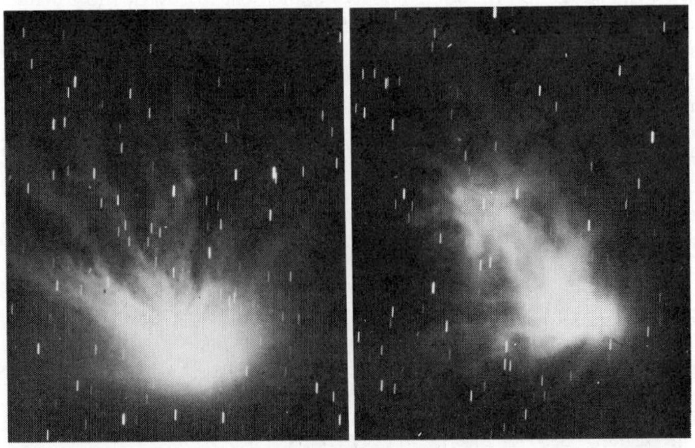

Fig. 3. Comet Humason (August 6 and 7, 1962). The comet was then very active structurally. Its tail extended almost directly away from the earth. (*U.S. Navy photo.*)

eye in the spring of 1970 when it displayed a straight narrow plasma tail and a huge moderately curved dust tail. By using photographic plates sensitive only in certain parts of the spectrum (notably in the red region) and with appropriate filters, the plasma tail can be completely suppressed. Finson and Probstein, using this technique, were able to derive significant information concerning the nature of the dust and gas production mechanisms in comets. In the Finson-Probstein model of dust comets, it is assumed that there is essentially continuous emission of solid particles of various sizes from the nucleus into the atmosphere brought about by the drag forces of the outgoing gas. Shortly after the ejection of solid particles, the solar gravity and solar radiation pressure are considered to be the only two forces that control the motion of the dust particles in the cometary tail. The trajectory of any individual particle depends upon its size and on the time, velocity, direction, and other circumstances of ejection. The observed photometric profile of the tail at any time depends upon the size distribution of the ejected particles, their emission rate as a function of time before the time of observations, and variations in the initial particle velocity with time and size. The photometric profile also depends upon the geometrical configuration of the sun, earth, and comet in space. Some authorities believe that the Finson-Probstein model is one of the most effective methods for physical research on comets. Application of this approach is described in some detail by Sekanina and Miller (1973).

As early as 1950, Whipple developed what is known as the "icy-conglomerate" model of a comet. As described by this model, a comet is sort of a "snowball" comprising frozen gases and small nonvolatile solids—the object ranging from a few hundred meters in diameter for the smallest cases to a few tens of kilometers for the largest. When the snowball approaches the sun, the volatile matter contained in its commences to sublime. This produces an extensive and very low density atmosphere (cometary atmosphere) called a **coma.** The aerodynamic drag of the released gas causes both nonvolatiles (dust) and frozen volatiles (icy grains) to lift off the surface of the comet. Even though the escape velocity is low (a few meters per second), the solids and gases escape permanently. The dust moves anti-sunward to form a tail. Gases that become ionized are subjected to much larger acceleration and, in so doing, form a separate *ion tail.* Some comets have two visible tails; others do not. The ion tail tends to be larger and thinner and more structural as compared with the essentially featureless dust tail. Usually the ion tail is bluish in color; the dust tail is yellowish—although to the eye, both tails appear colorless.

It is believed by some authorities that all or most comets lose mass as they travel through the solar system. It is assumed that each comet thus has a finite life (so many trips around the sun). What remains of a comet (remnants if any) is poorly understood.

There are about 100 known short-period comets. A short period is defined as a period of revolution around the sun of about 10 years, with periods usually lying between 3.3 and 13 years. Some authorities tend to believe that short-period comets result from the perturbing effects of Jupiter (and possibly of other planets) on the orbits of long-period comets that pass near it, especially upon those comets with perihelia (closest approach to the sun) which lie near the perturbing planet.

One argument against origination of coments in interstellar space is that, to date, no comet has been observed which approached the sun on a very hyperbolic orbit with velocity far above the velocity required to escape from the solar system. But if comets do not originate in interstellar space, from whence do they come? A general assumption is that comets originated as a part of the solar nebulae (gaseous cloud from which solar system condensed) and at a location either among the outer giant planets or even further from the sun (Pluto and beyond). Some scholars have suggested that there may be billions of comets in "cold storage" some 1000 times farther from the sun than Pluto. It is further believed that the long-period comets originate in this reservoir. This assumes that comets may have been formed from relatively small accumulations of grains of substance left over from planetary formation. Thus, it is further suggested that the interior of comets may contain some of the most primitive of solar system matter, matter that has remained unchanged by temperature, pressure, melting, resolidifying, and so on. The mechanism required to periodically release cometary material from such a reservoir and cause it to become a long-period comet is probably some kind of perturbation. But of what nature? Perhaps by an occasional passing star?

The spectra of comets have been studied for a number of years. See Fig. 4. From these observations, it has been found that comets

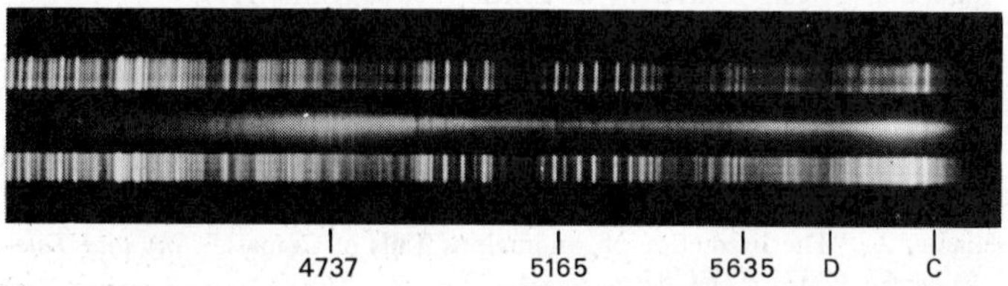

Fig. 4. Historical spectrum made of Halley's comet as early as May 23, 1910. The bright-line spectrum shows the carbon bands, indicated by their wavelengths, and the D line of sodium, which lasted only a few days. In the spectrum of sunlight reflected by the comet's head, the dark Fraunhofer lines C and F are prominent. A comparison spectrum appears above and below. (*Lowell Observatory.*)

shine partly by reflected sunlight and partly by radiation from gases. Band spectra that have been identified in the coma include those of the radicals CH, C_2, OH, NH, and CN, among others. These are transformed into stable molecules, such as CO, CO_2, and N_2, as they are driven into the tail. The first direct observation of cometary water was the detection of H_2O^+ ions in the comet Kohoutek. In 1974, radio astronomers reported molecules of water in the comet Bradfield. There is, however, no preponderance of evidence that there is consistency among the compositions or behavior of various comets. There have been a number of examples of marked departures from the "predictable" behavior of comets. Comet Tuttle-Giacobini-Kresak, in May 1973, suddenly flared to a brightness by about 9 magnitudes (factor of 4000) in less than 5 days. The comet Schwassmann-Wachmann 1 also has displayed spectacular outbursts several times per year.

Among other areas of speculation are the mechanisms which cause the atmospheric gases surrounding a comet to become ionized after they have sublimed and left the cometary nucleus. Sunlight and the solar wind are probably factors, but they do not seem sufficient to account for the phenomenon. Neugebauer and Newburn observe that it seems most likely, but by no means certain, that the ionization is produced by a cascade process in which a few ions created by sunlight and chemical reactions are accelerated and then collide with and ionize other molecules.

Whipple (1980), who has studied the comet Encke and others extensively, has observed that the puzzling changes in the orbital period of this comet, which circles the sun every 3.3 years, can be attributed to the rotation of its icy nucleus and the thrust of gases evaporated from it.

Because many authorities believe that so much can be learned from detailed observations of comets, notably because of the assumption that comets contain primitive materials and thus of great cosmological interest, a space mission to two comets was proposed in the late 1970s. The proposed plan would require a launching in July 1985. The craft would pass between the sun and Halley's comet November 1985 and would drop off a probe aimed directly at the comet's nucleus. The spacecraft would then continue and rendezvous with the comet Tempel 2 in July of 1988, where the comet could be observed from a distance of less than 2000 kilometers as it passes through perihelion. The craft would continue along with the comet and ultimately be captured by it, circling the nucleus at a distance of about 10 kilometers. After a year of circling, an attempt to "land" on the nucleus would be attempted.

Most authorities believe that there is practically no danger to the earth from comets. Some years ago, there was a suspicion that the tail of a comet might contain poisonous gases, which might cause wholesale death on earth if the earth ever came close to the tail of a comet. Such belief is without foundation, for the gas is in such a highly diffused condition that, even if it were poisonous, a lethal concentration would be highly improbable. During the return of Halley's comet in 1910, the earth passed through the tail, and not the slightest effects could be noted either in dimunition of sunlight or change in the chemical constitution of the atmosphere. A direct collision with the head of a comet would undoubtedly have a very destructive effect over the portion of the earth where the collision took place. However, from the number of comets and the average orbital characteristics, such a collision is highly improbable.

References

Delsemme, A. (editor): "Comets, Asteroids, Meteorites, Interrelations, Evolution, and Origin," University of Toledo, Toledo, Ohio, 1977.

Donn, B., et al. (editor): "The Study of Comets," NASA Rept. SP-393, National Aeronautics and Space Administration, Washington, D.C., 1976.

Marsden, B. G.: "Comets," *Ann. Rev. Astron. Astrophys.*, **12**, 1 (1974).

Neugebauer, M., and R. L. Newburn, Jr.: "The Exploration of Comets," *Technology Review (MIT)*, **80**, 5, 47–53 (1978).

Sekania, Z., et al.: *Science*, **179**, 565–567 (1973).

Sekania, Z.: "The Prediction of Anomalous Tails of Comets," *Sky and Telescope*, **47**, 6, 374–377 (1974).

Whipple, F. L.: "Background of Modern Comet Theory," *Nature*, **263**, 5572, 15–19 (1976).

Whipple, F. L., and W. F. Huebner: "Physical Processes in Comets," *Ann. Rev. Astron. Astrophys.*, **14**, 143 (1976).

Whipple, F. L.: "The Spin of Comets," *Sci. Amer.*, **243**, 3, 124–134 (1980).

COMMAND (Computer System). That which specifies the operation to be performed. In terms of a control signal, the command usually takes the form of YES (go head), or NO (do not proceed). Command should not be confused with instruction. In most computers, an instruction is one given to the central processing unit (CPU), as contrasted with a command which is an instruction to be followed by a data channel. An input command, for example, may be READ; an output command, WRITE. See also **Program (Computer);** and terms listed under **Data Processing.**

COMMAND GUIDANCE (Space Vehicle). Space Vehicle Guidance and Control.

COMMENSALISM. Association (Ecology); Symbiosis.

COMMERCIAL AIRCRAFT. Airplane.

COMMIPHORA TREE. Of the family *Burseraceae* (torchwood family), genus *Commiphora*, the *C. abyssinica* is a small, low, thorny tree that is found in the southern part of Saudi Arabia, on the island of Socotra in the Indian Ocean south of the Arabian Peninsula, and in eastern Africa. The commiphora contains a yellow-to-brown aromatic resin known as myrrh. It is postulated that the myrrh mentioned in the Bible was a mixture of this resin and labdanum, obtained from various species of the rockrose. Long before the time of Christ, the Arabians used myrrh for barter and trade with other peoples. At that time, it is believed that the commiphora tree grew in large numbers and that the trade in myrrh and of the wood of the tree was quite extensive. History records that the Romans used it, the Greeks honored Zeus by burning the resin, and that Chaldean priests used it profusely on the alters to Baal. The Egyptians also used the resin as an embalming medium.

The fruit of the tree, somewhat resembling a small cedar, is yellow, smooth, egg-shaped, and about the size of a current. The leaves are small and relatively sparse. The flower is mildly fragrant. The boswellia tree, the source of frankincense, is also of the torchwood family—so named because of the high resinous content and hence flammability of the tree. See also **Boswellia Tree.**

COMMISSURE. A transverse nerve cord or fiber tract connecting paired components of the nervous system.

COMMON-BASE AMPLIFIER. Amplifier.

COMMON BUNDLE. Heart and Circulatory System (Human).

COMMON COLD. An acute inflammation of the upper respiratory tract. Over 90% of the people in the United States suffer from colds each year. Nearly one-half of the persons in the nation have several colds during the year. The common cold is contagious and spreads easily in crowded schoolrooms and business offices. It is probably the greatest cause of absenteeism among school children and workers. It is generally believed that viruses are in the throat most of the time, but are unable to attack until body resistance is lowered. At least 30 different kinds of *rhinoviruses* and many other respiratory tract viruses are known to be causative agents. In addition, certain bacteria may be present along with the cold virus. However, these organisms are believed to be secondary invaders and not associated with the initial attack. It would appear that cold viruses weaken the tissues, making them susceptible to infection by other organisms.

Antihistamines have become popular cold remedies and sometimes relieve the symptoms of a cold in some patients, but they do not cure the cold. The common cold usually runs its course of several days. Where symptoms persist and worsen after a few days, these may indicate a more serious condition and medical attention should be sought.

The term *cold* is not to be confused with the professional use of the acronym COLD for the much more serious respiratory diseases, collectively known as *chronic obstruction lung disease* (COLD).

COMMON-EMITTER AMPLIFIER. Amplifier.

COMMON ION EFFECT. The reversal of ionization which occurs when a compound is added to a solution of a second compound with which it has a common ion, the volume being kept constant. The degree of ionization of the second compound then is lowered, i.e., it retrogresses. The common ion effect also can markedly affect solubility.

COMMON-MODE REJECTION RATIO. A parameter used to express the ability of a differential subsystem or instrument or reject the effect of a common-mode voltage applied at its input terminals. An idealized model of the common-mode rejection ratio (CMRR) is shown in the accompanying figure. Voltage V_{CM} is the common-mode voltage, V_{in} is the equivalent differential-input voltage due to V_{CM}, V_0 is the output voltage (in the case of a digital measurement subsystem or instrument, the output value), the G is the gain or conversion factor of the differential subsystem or instrument. See also **Common-Mode Voltage; Differential-Mode Voltage.**

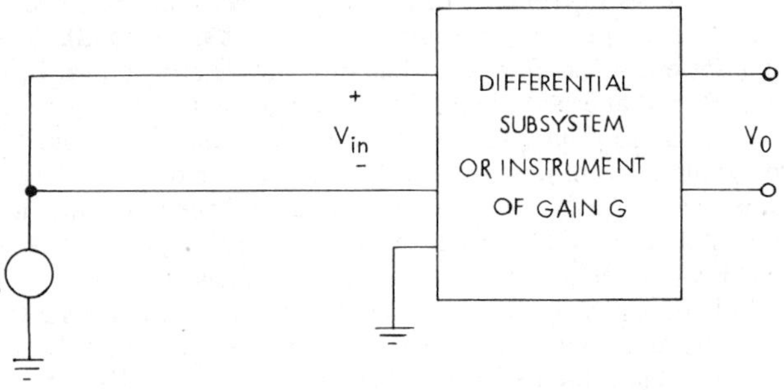

Model of common-mode rejection ratio (CMRR).

The CMRR $= V_{CM}/V_{in}$. Inasmuch as it is difficult to measure V_{in} directly, the equivalent definition, CMRR $= V_{CM}G/V_0$ may be used as the basis of measurement.

If the common-mode voltage is ac, the measurement of the CMRR is based upon the ac component in the output. Where a digital output is involved, this is the increase in the repeatability or spread in output values. Consistent units must be used, i.e., if peak-to-peak values are used for V_{CM}, then the peak-to-peak increase in the spread of output readings must be used.

Frequently, the CMRR of an instrumentation subsystem ranges from 100 to 140 dB or, when expressed as a ratio, from $10^5:1$ to $10^7:1$, with 120 dB being the most common. Definition of CMRR in decibel units follows the formula: CMRR (dB) $= 20 \log(\text{CMRR})$. A specification of $10^6:1$ signifies that every volt of common-mode voltage results in the equivalent of 1 microvolt of normal-mode voltage at the input of the subsystem. Since the CMRR of most subsystems depends upon the resistive source unbalance and, often, on the value of the common-mode voltage, a CMRR specification should include these parameters. A complete specification of the CMRR would be: CMRR $= 10^6:1$ at up to 1,000 ohms source unbalance and 200 V dc.

See also terms listed under **Data Processing.**

Thomas J. Harrison, International Business Machines Corporation, Boca Raton, Florida.

COMMON-MODE VOLTAGE. A voltage which is common to both inputs of a differential subsystem or instrument when measured with respect to a system reference point (usually ground) is termed the *common-mode voltage.* A differential subsystem used to measure a signal voltage V_S is shown in Fig. 1. The point G_1 is the ground

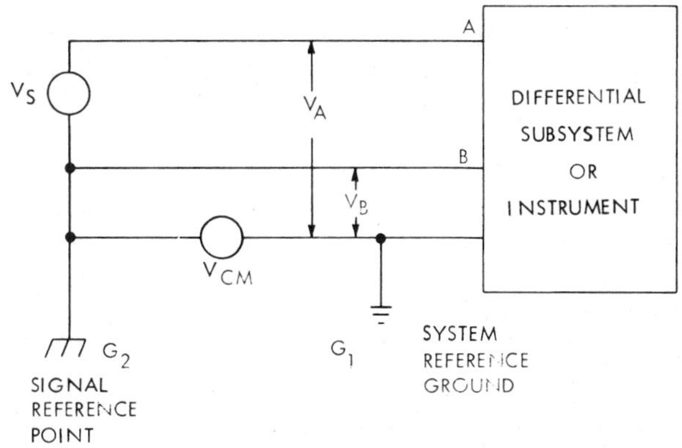

Fig. 1. Common-mode voltage.

reference point for the measurement subsystem. It will be noted that the signal voltage is referenced to G_2. The latter may or may not be at the same potential as G_1. Voltages at input terminals A and B are $V_A = V_S + V_{CM}$ and $V_B = V_{CM}$. The voltage common to both A and B is V_{CM}, the common-mode voltage. Common-mode voltage sometimes is defined as the average value of V_A and V_B, namely $(V_A + V_B)/2$. Inasmuch as common-mode voltages generally are of concern only when $V_{CM} \gg V_S$, there is little practical difference between the two definitions.

Common-mode sources cause problems in differential-measurement systems inasmuch as they can cause measurement error due to common- to normal-mode (or differential-mode) conversion. See also **Differential-Mode Voltage.** In a perfect differential-measurement system, only the value of the signal voltage V_S would be measured—with no contribution arising from the presence of a common-mode voltage V_{CM}. However, in practice some common-mode voltage usually appears as a signal voltage.

Conversion can occur as shown in Fig. 2. Resistance R_1 and R_2

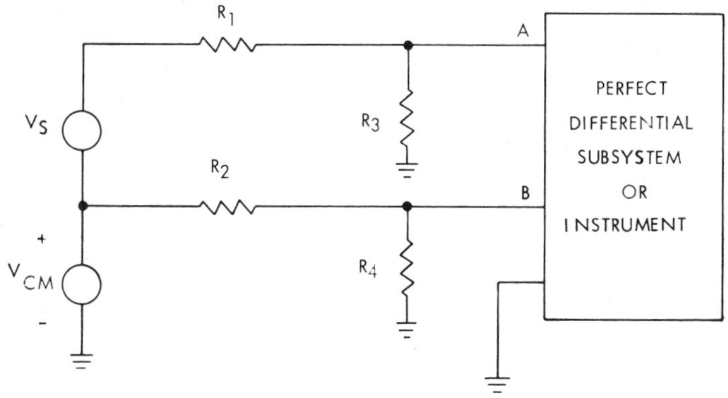

Fig. 2. Model of common- to normal-mode conversion.

are respectively representative of the input-line resistance and the source impedance of V_S. Resistances R_3 and R_4 are leakage resistances to ground in both the input lines and the differential measurement subsystem. Resistances R_1 and R_3 comprise a voltage divider, as do resistances R_2 and R_4. If the input impedance of the instrument is neglected, the differential- or normal-mode voltage, $V_A - V_B$ is

$$V_A - V_B = V_S \frac{R_3}{R_1 + R_3} + V_{CM}\left(\frac{R_3}{R_1 + R_3} - \frac{R_4}{R_2 + R_4}\right)$$

If one assumes that $R_3 \approx R_4 \gg R_2, R_2$, the expression becomes

$$V_A - V_B \approx V_S + V_{CM}\left(\frac{R_2 - R_1}{R_3}\right)$$

The desired measurement result is the value V_S. Consequently, the second term in the foregoing expression represents the error arising from common- to normal-mode conversion, commonly referred to as the *common-mode error.* The common-mode rejection ratio, which is a measure of the ability of a system to reject the effect of common-mode voltage, is the inverse of the coefficient of V_{CM}, namely, $|R_3/(R_2 - R_1)|$.

The situation shown in Fig. 2 is only one possibility whereby common-mode voltage can be converted into a normal-mode signal. Where the common-mode source is ac, resistance R_3 and R_4 usually can be neglected. However, the leakage capacitance from input terminals A and B to ground must be considered. The diagram of Fig. 2 represents an idealized version of the actual phenomenon. Lumped parameters are assumed, but in an actual situation, the reistances and capacitances will be distributed along the length of the input lines.

Common-mode voltage can result from the way a system is used, or from entirely unintentional causes. Bonding of a thermocouple to a current-carrying conductor or use of a subsystem to measure the voltage drop across an ungrounded resistor in a circuit are representative of error arising from the manner in which the system is used. Where the signal source is grounded, and where ground potential differs from that at the measurement point, the source of common-mode voltage error could be considered unintentional. Needless to say, however, the system does not differentiate between the two causes.

See also terms listed under **Data Processing.**

Thomas J. Harrison, International Business Machines Corporation, Boca Raton, Florida.

COMMUNAL ENTROPY. The contribution to the entropy of a system arising from the disorder when the molecules are sufficiently free to change positions frequently. See also **Entropy.**

COMMUNICABLE DISEASE. Contagion.

COMMUNICATION CHANNELS. Telephony.

COMMUNICATION (Fidelity). Fidelity (Communications).

COMMUNICATION (Noise Testing). Noise Generator.

COMMUNICATION (Satellite Systems). Satellites (Communication).

COMMUNICATION (Sonar). Sonar.

COMMUNICATION (Weather Data). Weather Observations and Forecasting.

COMMUNICATIONS SYSTEMS. Facsimile Transmission; Laser; Microwave Region; Microwave Tubes; Optical Fibers; Radar; Radio Communication; Satellites (Communication); Sonar; Telecommunications; Telegraphy; Telemetering (Industrial); Telephony; Television.

COMMUNITY ANTENNA TELEVISION. Television.

COMMUNITY (Biotic). Ecology.

COMMUTATION RELATIONS. Physical quantities are represented in quantum mechanics by linear operators. Two linear operators A and B do not commute in general, $AB \neq BA$. The commutation relations give an expression for the commutator $AB - BA$. Canonically conjugate dynamical variables q, p satisfy the commutation relations $qp - pq = i\hbar$, where $\hbar = h/2\pi$ and h is Planck's constant. Components M_x, M_y, M_z of an angular momentum operator satisfy the commutation relations

$$M_xM_y - M_yM_x = i\hbar M_z$$

$$M_yM_z - M_zM_y = i\hbar M_x$$

$$M_zM_x - M_xM_z = i\hbar M_y$$

See also **Commutator (Mathematics);** and list of entries under **Mathematics.**

COMMUTATIVE LAW. Arithmetic or algebraic quantities obey this law when the result of some operation is independent of the order in which it is performed. Thus, if $(a = b) = (b + a)$ or $ab = ba$, the processes are commutative. An operator or matrix is often noncommutative.

COMMUTATOR (Electric). Electric Generator.

COMMUTATOR (Mathematics). If A and B are two noncommutative operators, their commutator is

$$[A, B] = AB - BA$$

According to quantum theory, if the commutator vanishes for two operators that represent dynamical variables, then the measurement of one of these variables does not interfere with that of the other.

COMPANDOR. 1. A transmission system in which the signal-to-noise ratio is improved by signal compression before transmission, and signal expansion after reception. 2. A speech-reproducing system consisting of a compressor of the volume range at the recording end, and an expander of the volume range at the reproducing end.

COMPANION CELL. This is a very small elongate cell always found in close association with a sieve tube in higher plants. It contains a dense protoplasm with a prominent nucleus and very small vacuoles. Companion cells are assumed to function with the sieve tubes. See also **Cambium (Plant).**

COMPARATIVE BIOLOGY. In any division of biological science, the comparative treatment focuses attention upon a limited subject, such as anatomy, but introduces into its treatment data drawn from many species. This method of study has been applied widely to the vertebrates; hence, comparative anatomy is likely to mean comparative anatomy of the vertebrates unless otherwise qualified.

The comparative method is valuable in determining the evolutionary development of organs and the relationship of species.

Comparative biochemistry may be defined as the study of the nature, origin, and control of biochemical diversity. This definition suggests that (1) biochemical differences are to be found among orgnaisms, (2) such differences arise during the evolving processes of organisms, and (3) the biochemical properties of organisms are under a variety of controls in nature, and presumably may be modified when the biochemical properties and their natural controls are understood.

In the approach to comparative biochemistry, popular before the 1950s and exemplified by the books of Baldwin and Florkin (see references), biochemists studied the similarities and differences among higher organisms, mainly among animals. The compounds, such as the various phosphagens, blood-transport pigments, carotenoids, and the A vitamins, were correlated with the postulated phylogenetic position of the animals possessing these substances. The comparative aspects of nitrogenous excretion products and of salt and water balance have also been studied in great detail. Such types of studies have revealed metabolic differences among the *Metazoa*, differences related to their evolving nature and ecological niches. More recently, the distributions of the alkaloids and flavones of plants, organic acids of lichens, and the pterins of *Drosophila*, among other biochemical markers, have been analyzed in detail to help in establishing genetic and evolving interrelations.

Since about 1945, biochemistry has concentrated largely on cellular metabolism, and these studies tend to emphasize the uniformity of cellular biochemistry. Thus, almost all cells contain proteins built of the same 20 amino acids, RNA and DNA containing their characteristic constituents, common coenzymes ATP, NAD, coenzyme A, and so on. Also, many similarities could be detected in the metabolic events relating to energy and biosynthesis in many cell types, i.e., microorganisms, plants, and animals. This experience was soon interpreted by the school of Kluyver and van Niel and their students to signify a *uniformity of biochemistry*, a unity inferred to derive from a monophyletic process of organisms as they evolve from a primitive cellular type. In this view, biochemical differences among organisms are considered to reflect relatively late evolving divergences of which the metabolic differences among the *Metazoa*, e.g., patterns of nitrogen excretion, such as ammonotelism, ureotelism, and uricotelism, are clear examples. It is recognized that the biochemical differences found to exist between organisms are not only considered to be relatively late in their evolving processes, but are usually stated to represent minor alterations in the broad biochemical pattern common to many cells.

Thus comparative studies, while always remaining important as an investigative tool, have given ground to what appears to be the more important *uniformity* approach, with concentration on the fundamental aspects of biology which affect all life forms.

COMPARATOR. Generally, an instrument that facilitates the comparison of one quantity with another in a precise manner. 1. An instrument for the accurate measurement of lengths or distances (usually small). The feature common to various forms is a reading microscope or telescope arranged to travel along a scale, its axis remaining parallel to a fixed line. A typical form consists of a low-power microscope mounted on a carriage movable forward or backward by a micrometer screw. The two points or lines whose distance apart is to be measured are brought into the focal plane of the microscope, the cross-hairs of which are adjusted first upon one and then upon the other. For example, the distance to be measured may be that between two star images on an astrographic plate, or the images of two spectrum lines taken by a spectrograph; in either case the plate is simply mounted on the stage of the comparator microscope. Another familiar type is the "cathetometer," consisting of a telescope sliding on a vertical scale and provided with a vernier. This instrument is used to measure heights of liquid columns, or other differences of level. See also **Cathetometer.**

2. A circuit which compares two signals and supplies an indication of agreement or disagreement. This circuit is also known as an "add-or-subtract" circuit.

COMPARATOR AMPLIFIER. A nonlinear amplifier for sensing either the polarity or the magnitude of an input signal. The amplifier output is one of two states, usually for representing the binary states 1 and 0. Where the input signal is in excess of a predetermined level (frequently zero), the amplifier output is in one state. Where the input signal is less than the predetermined level, the output remains in the other state. In the case where the predetermined level is other than zero, a comparator amplifier may be termed a *threshold detector.* See also **Threshold Detector.**

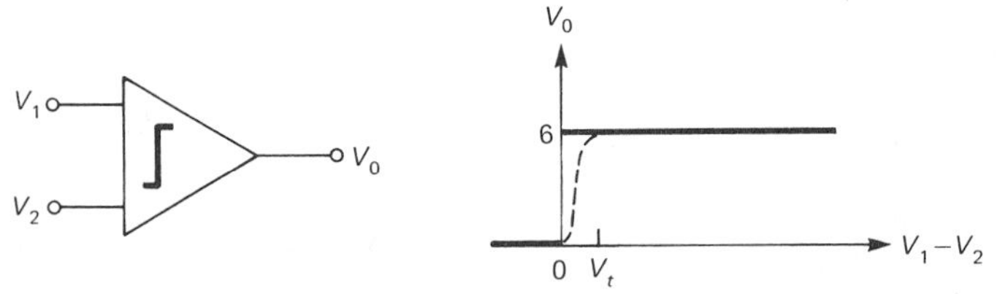

Transfer characteristics of comparator amplifier.

An ideal and a practical comparator are compared in the accompanying diagram. The ideal characteristic is given by the heavy line. In this example, the amplifier output is +6 V when the input signal is greater than zero; and zero when the input is less than zero. However, in practice, a linear region of operation, exists between these two output levels, corresponding to a small uncertainty in the performance of the comparator. Consequently, the input level must be in excess of a threshold level V_t to make certain that the output will reach the +6 V level. Within the limitations of amplifier stability factors, the threshold voltage can be decreased by increasing the gain of the amplifier. Gains in excess of 10,000 are typical of comparator circuits.

In summary, the comparator is a high-gain single-ended or differential amplifier. Usually, the comparator is configured for fast recovery from saturation, thus permitting it to follow a rapidly varying input signal. Low zero offset is required for precision applications inasmuch as zero offset is equivalent to a shift in the threshold level. Commercial monolithic integrated amplifiers configured particularly for use as comparators are available.

See also terms listed under **Data Processing.**

Thomas J. Harrison, International Business Machines Corporation, Boca Raton, Florida.

COMPASS (Navigation). An instrument used for finding direction on the surface of the earth. The oldest and most commonly used of these instruments is the magnetic compass. The directive forces are the horizontal component of terrestrial magnetism and, in a properly designed compass, the effect of the vertical component must be reduced as much as possible. The simplest form of compass is that used by surveyors, which consists of a light, thin magnet (compass needle), pivoted so that it can turn freely about an axis perpendicular to its length, and with the north-seeking end clearly marked. A compass card is mounted in the plane parallel to that in which the compass needle can turn. As shown in Fig. 1, the card incorporates "points" and other systems of marking for reading the instrument.

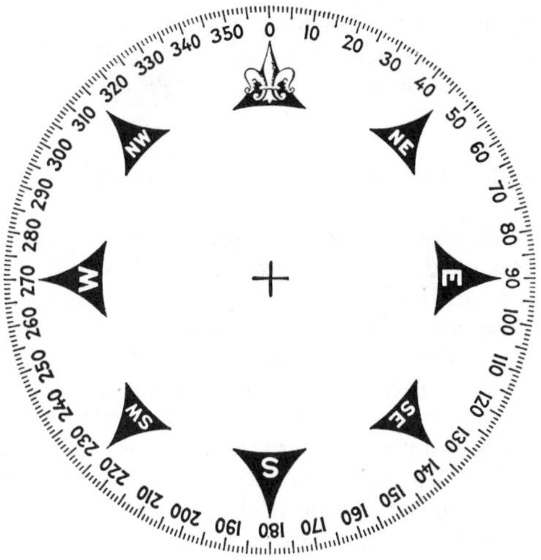

Fig. 1. Compass card.

Two principal systems are used. One system has zero both at the north and south points of the card, and reads in degrees both right and left from the zeros to the east and west points. In such a system, the northwest point would be referred to as north 45° west; and the northeast point as north 45° east. In the more widely accepted system and as shown, north is marked zero and leads to the right through the east through 360°. In this system, the northwest point is referred to as 315°; the northeast point as 045°.

For use on ships and aircraft, a much more stable system is required than that which may be satisfactory for surveyors where the environmental problems of use are minimal. A float-type magnetic compass is shown in Fig. 2. If a two-pole magnet is pivoted about a vertical axis with its poles in the horizontal plane, the magnet will align itself with the horizontal component of the earth's magnetic field and indicate magnetic north, which differs from true north by the variation angle. The latter, in turn, varies from place to place on the earth. In aircraft compasses, the magnet is attached to a float in an approximately spherical-shaped chamber that is filled with liquid, thus removing most of the weight of the magnet from its pivot to reduce pivot

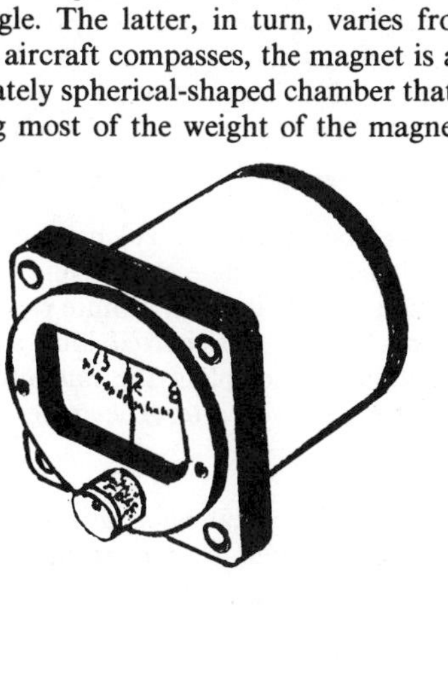

Fig. 2. Float-type magnetic compass.

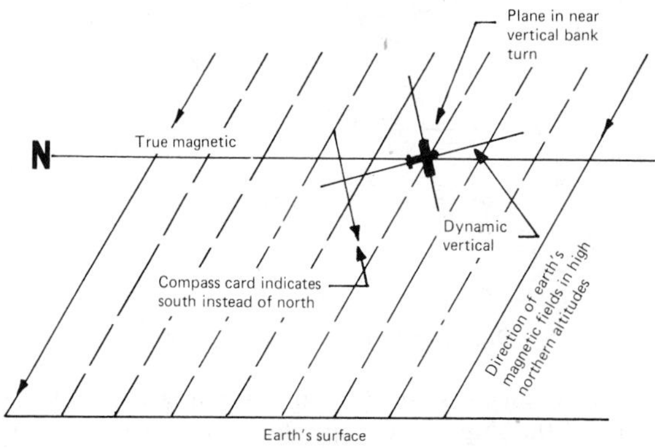

Fig. 3. Northerly-turning error.

wear and friction. The float provides a reference to dynamic vertical, and the liquid provides damping. Usually there is considerable magnetic material in an aircraft acting as a disturbing influence on the compass. Thus a compensating arrangement of small adjustable magnets usually is provided to balance out this effect.

During any acceleration, as in a turn, the float (in referencing dynamic, not true, vertical) does not align the compass magnet to read the true horizontal components of the earth's field. In fact, in a turn in one direction in high latitudes, it is possible to have the dynamic vertical at a greater angle to the true vertical than is the earth's field and thus have the compass point south instead of north. This effect, illustrated in Fig. 3, is known as the northerly-turning error. A signal can be taken from a float or pendulous-type compass and used as a slow correction on a directional gyro, and such a long-time average will be nearly correct. This is then known as a gyrosyn compass, shown in Fig. 4.

Directional gyros are used with drift rates of 0.1 to 4 degrees an hour. Correction must be applied for rotation of the earth and aircraft velocity. Such corrections, when properly applied, give a satisfactory heading reference in very high latitudes where magnetic headings are valueless. A device of this type, known as an "all-latitude" or "polar-path" compass is shown in Fig. 5.

On ships, a master gyro usually is installed in some well-protected part of the ship. First developed in 1911, the essential feature is a heavy gyroscope driven at high speed by electric power. The frame of the gyro is mounted in gimbals and a "ballistic tube" is attached, which causes the axis of rotation of the gyro to set itself parallel to the earth's axis of rotation, provided the instrument is stationary or moving in the east–west direction. Since the axis of rotation of the earth is in the plane of the meridian, the axis of the rotor will point true north on a stationary ship. A compass card may be attached to the gyro, but usually the gyro is used to operate repeaters in various locations about the ship.

The electrical circuit that operates the repeaters may be used for

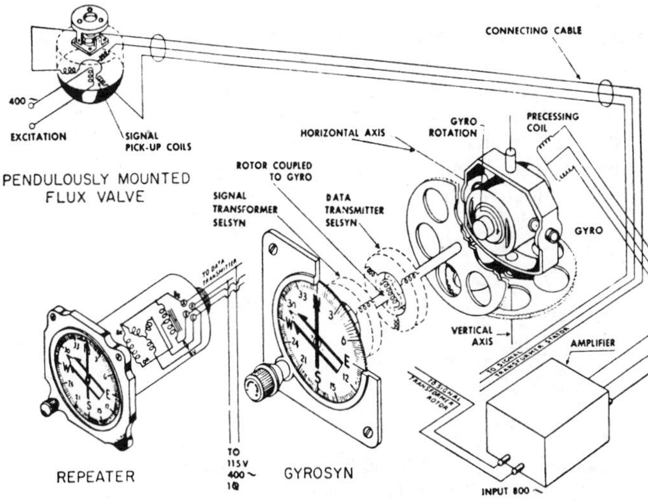

Fig. 4. Gyrosyn compass. (*Sperry Rand.*)

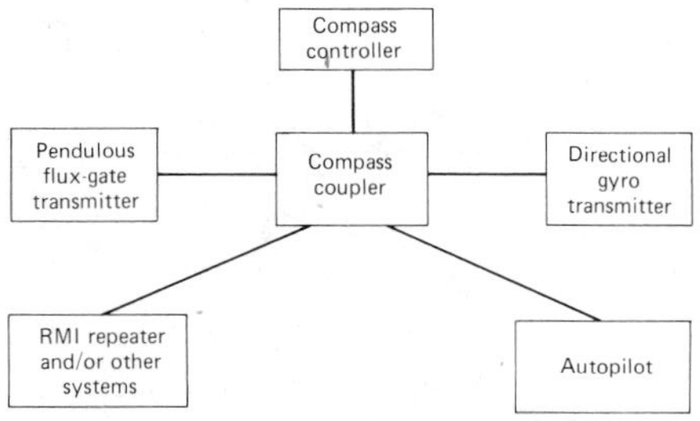

Fig. 5. Organization of polar-path compass.

a variety of other purposes, such as for keeping a constant record of the heading of the ship and for operating the steering mechanism so that the ship will be held on a predetermined heading. However, means must be made available for constant manual intervention and for minor variations, as well as major emergency intervention, by the helmsman. For security in case of possible breakdown of the gyrocompass, a ship will also carry a standard magnetic compass for which the corrections are known at all times.

A discussion of the migration of the Earth's magnetic north pole is given in entry on **Earth.**

COMPASS NORTH. North.

COMPENSATION THEOREM (Network). If a network is modified by making a change, ΔZ, in the impedance of one of its branches, the current increment thereby produced at any point in the network is equal to the current that would be produced at that point by a compensating electromotive force acting in series with the modified branch, whose value is $-i\,\Delta Z$, where i is the original current which flowed in the modified branch.

COMPENSATOR. A device, circuit, or other means provided in an instrument, subsystem, or component to counteract an effect which interferes with the major objective. There are many hundreds of forms of compensating means used in instruments and machines. Commonly encountered are ambient conditions, changes in which must be compensated. For example, an instrument or tool may be calibrated for so-called *standard conditions*, i.e., particular temperature, pressure, humidity, etc., conditions that may affect the accuracy of a measurement when the measurement is made outside the standard parameters. Barometers and altimeters, for example, require compensation for elevation above sea level. Filled-system thermometers require compensation for the height of the temperature-sensitive bulb in many instances. Compensation may be manual, in which case the operator of an instrument, tool, machine, and so on may refer to tables and correct what may be done arithmetically. More frequently, a compensator is thought of as a means for automatically and continuously applying a corrective counteraction. Thus, resistors are used in electrical-measuring circuits to counteract an excursion in ambient temperature. The following examples probe the applications of compensators in only two of many types of situations.

In optics, a form of compensator is an arrangement for measuring the phase difference between the two components of elliptically polarized light. See accompanying diagram. This is accomplished by introducing a known, opposite phase difference of equal magnitude, which reduces the existing phase difference to zero. The most familiar form, devised by Babinet, consists of two quartz wedges, with thin optic axes at right angles to each other. When passed through this apparatus

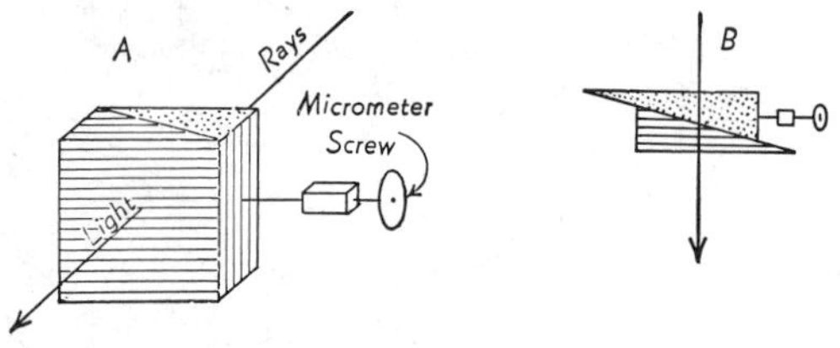

(A) Diagrammatic sketch of Babinet Compensator; angle of wedges much exaggerated. Hatching and strippling indicate direction of crystal axes. (B) Wedges displaced.

and a Nicol prism set to extinguish light plane-polarized at 45° to either axis, any given elliptically polarized light produces a system of parallel dark bands. Plane-polarized light is first used (zero phase difference), then the elliptic light of unknown phase difference, and the relative displacement of the wedges necessary to restore the bands to their original position gives the phase difference required.

In an example from the electrical field, when a motor is to be started other than by direct connection to the supply line, some means must be incorporated in the circuit to limit the current taken during the starting period. A method frequently used to start ac motors which are too large to be directly connected for starting, is to decrease the impressed voltage during the starting period. The term compensator usually refers to the automatic transformers, switches, and wiring employed to limit the starting current for the squirrel-cage type of ac motor. This equipment is enclosed in a case through which projects an operating handle which has three positions, "off," "starting," and "running." This handle controls a double-throw switch. When in starting position, the switch is thrown so as to connect the motor to the line through an auto-transformer, which reduces the voltage to an amount suitable for starting purposes. When in the running position, the switch connects the motor directly to the line through fuses, and full line voltage is impressed upon the motor windings.

See also **Ambient Conditions; Elliptical Polarization;** and **Polarized Light.**

COMPETENT (Geology). Incompetent (Geology).

COMPILER (Computer System). A program designed to translate a higher-level language into machine language. In addition to its translating function, which is similar to the process used in an assembler, the compiler program is able to replace certain items of input with a series of instructions, usually called subroutines. Thus, where an assembler translates item for item and produces as output the same number of instructions or constants which were put into it, a compiler, typically produces multiple output instructions for each input instruction or statement. The program which results from compiling is a translated and expanded version of the original.

Compiler language is characterized by the one-to-many relationship between the statements written by the programmer and the actual machine instructions executed. The programmer typically has little control over the number of machine instructions executed to perform a particular function and is dependent on the particular compiler implementation. Typically, the language is very nearly machine-independent and may be biased in its statements and features to a particular group of users with similar problems. Thus, these languages sometimes are referred to as problem-oriented languages (POL). Slightly different implementations of a given language are sometimes called "dialects." Some compiler languages in wide use are ALGOL, BASIC, COBOL, FORTRAN, and PL/I.

COMPLEMENT. 1. In decimal representation, the complement of a number x is $10^v - x$, where v is some fixed integer, positive, negative, or zero. In binary representation it is $2^\mu - x$, where μ is an integer. The subtraction of a number is thus essentially equivalent to the addition of its complement, and in computer construction it is generally easier to mechanize the formation of a complement than that of an arbitrary difference. In binary notation, the one's-complement of x is the number obtained when each digit x_i of x is replaced by $1 - x_i$. 2. The complement of a subset S of a set A is the set of all elements of A not included in S. 3. For the algebraic complement of a minor determinant, see **Minor (Of a Matrix).**

COMPLEMENTAL MALE. 1. In certain barnacles which are normally hermaphrodites a few individuals lack female organs and are called complemental males. 2. In colonies of white ants, termite reproduction is carried on principally by a highly specialized king and queen. Other sexually mature individuals resembling the immature insects are known as the second reproductive caste. The males of this caste are complemental males.

COMPLEMENTARITY PRINCIPLE. Physical phenomena may be described either in terms of particle motions characterized by a momentum p and an energy E, or in terms of waves characterized by a wavelength λ and a frequency ν. The two descriptions are connected by the equations:

$$p = h/\lambda \quad \text{and} \quad E = h\nu$$

where h is the Planck constant.

COMPLEMENTARY ANGLE. Trigonometric Function.

COMPLEMENT SYSTEM. Immune System and Immunology.

COMPLETE GRAPH. Graph (Mathematics).

COMPLEX NUMBER. Complex Variable.

COMPLEX VARIABLE. A complex number has the form $(a + ib)$, where a, b are real numbers and $i = \sqrt{-1}$. It thus consists of a real part a and a pure imaginary part ib. In the study of complex numbers they are generally regarded as an ordered pair of real numbers (a, b) subject to the following laws: (1) equality, $(a, b) = (c, d)$ if, an only if, $a = c$, $b = d$; (2) addition, $(a, b) + (c, d) = (a + c, b + d)$; (3) multiplication, $(a, b) \times (c, d) = (ac - bd, ad + bc)$.

A complex number may be represented graphically on an Argand diagram. If it is described in polar coordinates it becomes $r(\cos\theta + i\sin\theta)$, where r is the absolute value, modulus, or radius vector and the angle θ is the amplitude, argument, or phase of the number (see also **De Moivre Theorem**).

If x and y are two real variables, then $z = (x + iy)$ is a complex variable. Thus a complex number is the special case where x, y are both constants. In either case, $(u + iv)$ and $(u - iv)$ are conjugate or conjugate complex to each other. Given a complex number of a function A, the conjugate complex is obtained by changing the sign of the imaginary part. It is often indicated by the symbol $\bar{A}$ or A^*. The product of an expression and its conjugate complex is always real.

For further properties of the complex variable and its functions see: **Analytic Function; Argand Diagram; Calculus; Cauchy-Riemann Equation; Cauchy Theorem; Conformal Representation; Contour; Convolution; Laplace Equation; Laurent Series; Residue; Riemann Surface; Singular Point of a Function; Taylor Series;** and terms listed under **Mathematics.**

COMPONENT. 1. In its most general usage, one of the ingredients of a mixture, or one of the distinct molecular or atomic species composing a mixture. In physical chemistry, one among the smallest number of chemical substances which need to be specified in order to reproduce a given chemical system. 2. The projection of a vector on a particular coordinate axis or along some specified direction. 3. Component of a tensor. 4. Component of a circuit, such as a capacitor, resistor, etc.

COMPONENT ANALYSIS. This is a branch of multivariate analysis which represents a k-dimensional variation as due to a number of uncorrelated components; fewer than k if possible, but if not, in such a way that a few components account for as much of the variation as possible. The components sought in practice are linear functions of the original variates. The method of analysis into principal components is to be sharply distinguished from factor analysis, notwithstanding a formal resemblance in the mathematics of the subject. See also **Factor Analysis.**

COMPONENT THERAPY (Medical). Blood.

COMPONENTS (Electronic). Amplifier; Diode; Electron Beam Lithography; Electronics; Electron Tube; Integrated Circuit; Loudspeaker; Microelectronics; Microstructure Fabrication; Microwave Tubes; Network (Electric); Oscillator; Power Sources and Supplies; Printed Circuits and Circuit Boards; Rectifiers; Resistor; Semiconductor; Signal Conditioning; Solid-State Electronics; Transistor.

***COMPOSITAE* OR COMPOSITE FAMILY.** The largest family within the plant kingdom, embracing several thousand species within over 800 genera. The family represents the highest developmental attainment among the dicotyledonous plants (a plant having two cotyledons or seed leaves—see **Dicotyledons**). Composites are mostly herbaceous plants, distributed in nearly all parts of the earth. The few members which are shrubs or trees are mainly limited to tropical regions, especially to island floras.

The composite family is divided into two groups, distinguished by the nature of its flowers. If all the flowers of a head are ligulate, the plant is a member of the *Liguliflorae.* This group contains many common plants, including chicory, dandelion, and lettuce. The plants of this group have latex vessels usually containing a white latex. See Fig. 1. The second group, the *Tubuliflorae,* comprises all composites

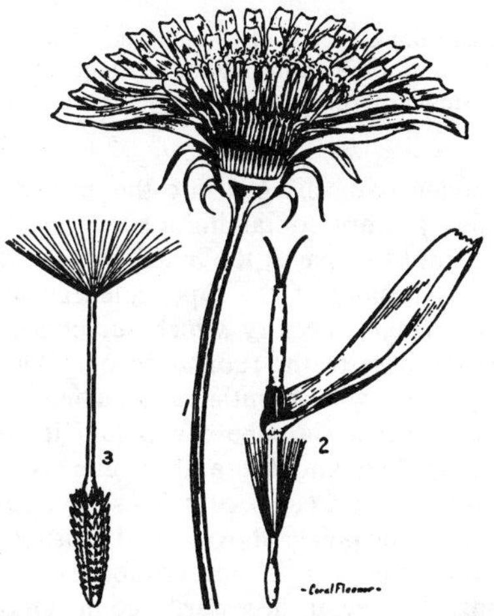

Fig. 1. Flower head and flowers of the dandelion, *Taraxacum*: (1) the inflorescence or head, composed of many flowers upon a flattened stem; (2) a single flower more enlarged; (3) a single fruit.

characterized by disk flowers. These may occupy only the central portion of the head, or the latter may be entirely of disk flowers. This group lacks latex.

The leaves of most composites are alternate, often entirely radical, although in a few cases they are opposite, as in the sunflower, *Helianthus annuus,* or whorled, as in species of *Eupatorium.* Stipules are seldom found in this family. The root is most commonly a tap-root, frequently much thickened, as in the dandelion. The inflorescence is of the type known as a head, or capitulum. Often the heads are aggregated in larger inflorescences of various types, as panicles or cymes or spikes. Commonly the single head is inaccurately regarded as a flower, rather than as an inflorescence. Surrounding the head is a group of bracts, making up the involucre. These bracts are usually green and serve to protect the flowers before they are mature and also to protect the maturing fruit. The flowers of a head are arranged on the enlarged end of the stem or axis, called the receptacle. This receptacle may be flat and disk-shaped, as in the common sunflower, conical as in the yellow daisy, *Rudbeckia hirta,* or otherwise. It may be smooth or covered with hairs or scales.

The individual flowers show considerable difference in structure. In many species the flowers of a head are all alike and all perfect. In other species they are of two kinds, one called ligulate and the other tubular. Ligulate flowers, also called ray-flowers, are irregular, but bilaterally symmetrical; tubular flowers, also called disk flowers, are regular. Both types occur in the head of the sunflower, where the tubular or disk flowers occupy the larger part of the receptacle, the ray-flowers being the conspicuous yellow flowers forming a ring around the periphery of the receptacle, just inside the involucral bracts. See Fig. 2. Each disk flower is perfect and regular. The calyx appears in different genera as bristles, bars, scales, or teeth, and sometimes is completely lacking. It is known as the pappus and occurs at the apex of the inferior ovary.

Often the pappus becomes a very important structure in the dissemination of the fruit. The corolla is tubular and 5-lobed, and inserted on the apex of the ovary. The short filaments of the five stamens are inserted on the base of the corolla tube, while the anthers are attached to each other by their edges, forming a tube which surrounds

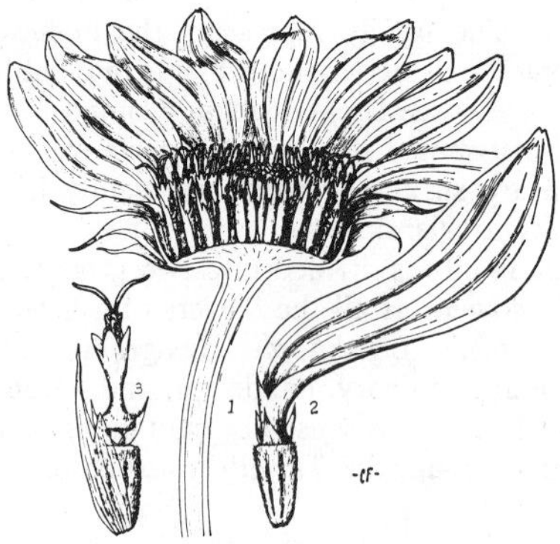

Fig. 2. Flowers and flower heads of a sunflower, *Helianthus*: (1) the flower head, cut so as to show the relation of the ray and disk flowers to the end of the stem; (2) a single ray flower; (3) a single disk flower.

the style. The pollen is discharged into the anther tube. The single pistil has an inferior ovary containing a single erect ovule, a simple style which splits into two parts, the inner surfaces of which are stigmatic. The fruit is usually of the type called an achene. Nectar is secreted in a ring-shaped nectary which surrounds the base of the style, located at the base of the tubular corolla. This nectar attracts insects, but only those with mouth parts sufficiently long to reach to the bottom of the corolla tube can obtain it. When the flower opens the pollen is shed into the anther tube. At the base of this tube the style, as yet unforked, occurs. This style elongates, pushing like a ramrod against the pollen above it, and causing an accumulation of pollen at the upper end of the anther tube.

Insects seeking the nectar necessarily come in contact with the pollen masses, which are thus likely to be transferred to another flower as the insect goes about collecting nectar. However, should insect-pollination fail, self-pollination is assured in many species, by the behavior of the styles, which emerge from the anther tube, protrude considerably, and then split and coil backwards so that the inner stigmatic surface rolls down into contact with the pollen. In some cases this self-pollination is almost the only method. Many species are self-sterile, thus requiring cross-pollination by insects.

The ligulate or ray flowers differ from the disk flowers in having the corolla of five united petals forming a tubular base which gradually emerges into a flat strap-shaped lateral structure with five teeth at its tip. Growing from the tube of the corolla are the small stigma and the five coalesced stamens. In many species where the ray flowers are marginal in the head they are entirely sterile or are pistillate.

After pollination takes place, the involucral bracts close over the head, pressing tightly against it and so protecting the developing fruit. The pappus becomes a conspicuous part of the fruit in many cases and is a very important factor in insuring scattering of the fruit. In some cases the pappus forms a parasol-like group of fine radiating hairs at the tip of a long beak of the achene. These hairs make the achene buoyant, and hence capable of being carried long distances by air currents. Often the base of the hair is hygroscopic, responding to changes in moisture, so that the parasol-like structure opens and closes with decrease or increase of humidity, which may help to loosen the achene from the receptacle, and also to shove it along over the surface of the ground, or to push it into a crack in the soil.

In *Bidens*, often called beggar-ticks or beggar-lice, the pappus is in the form of stiff usually downwardly barbed bristles which catch into the hair of passing animals or the clothing of man and so are carried about, finally breaking off and falling to the ground. In the burdock, *Arctium*, the involucral bracts form recurved hooks which serve in the same way, the entire head often breaking off and being transported. In thistles the pappus takes the form of a tuft of long silky hairs which enable the achene to float readily through the air. In many cases members of the composite family have no special pappus development, seed dispersal being entirely accidental.

The reasons which suggest that the composite family is highest in rank are found in the massing of the individual flowers into a compact head surrounded by protecting bracts, the structure of the individual flower with its inferior ovary, its united petals forming a corolla tube, the united anthers forming the anther tube, and the reduction of the number of carpels to two with but one ovule developing. See also **Flower.**

Uses of Herbs and Other Composites

Relatively few of the species of *Compositae* serve a practical purpose. Many are cultivated for ornamental reasons, the flowers often becoming very large and double and showy, as in the case of the *Chrysanthemum* and *Dahlia.* Less showy ornamentals are the *Aster, Bellis* (daisy), *Tagetes* (marigold), and *Calendula,* among others. Some composites yield oils or other substances useful to man, as *Arnica, Artemisia, Tanacetum, Calendula, Chrysanthemum* and *Helianthus.* A few are used in the diet, such as lettuce, artichokes, endive, chicory, salsify, and less frequently, dandelion.

Lettuce, *Lactuca sativa,* is an annual herb which is native to Europe, Asia and northern Africa. The plant is very leafy and contains a milk-white latex. In young plants the leaves are crowded on a short stem, forming a close rosette; in older plants the stem elongates greatly and bears a panicle of heads of yellowish flowers. The achenes are flat, ribbed, and contracted into a slender beak bearing numerous soft white or brownish pappus hairs which radiate outward like a parasol. Many varieties have been developed in cultivation. Lettuce is used almost entirely as a salad plant.

Endive, *Cichorium endivia,* is an annual or biennial herb having many basal leaves which in cultivated forms have become very much dissected and crisped. Mature plants are tall-stemmed and have purple, rarely white, flowers, all ligulate. The plant is used either as a salad plant or as a pot-herb. It is a native of India, and has long been cultivated in European countries. It is becoming more popular in the United States.

Chicory, *Cichorium intybus,* also called succory, is one of the many European plants which has become a persistent weed on introduction to North America. It is a perennial plant having a deep tap-root and a stiff tough stem two or three feet tall. Root leaves are numerous, forming a dense basal rosette: stem leaves are small, of various shapes, and clasping the stem. The flowers are usually blue, sometimes pink, or white. The plant is used as a salad plant or as a pot-herb, often being mistaken for dandelion. Its roots have been dried, ground and roasted and used as a substitute for coffee. There are several varieties in cultivation, one of which, Witloof chicory, finds considerable favor as a salad plant.

Salsify, *Tragopogon porrifolius,* or oyster plant, is a plant indigenous to southern Europe. It is a hardy biennial having a thick tap-root 8–12 inches long and an inch or two in diameter. This root is formed during the first year of growth when it bears a crowd of leaves. During the second year's growth the rather succulent branching stems grow up 2 or 3 feet tall. The stem leaves are alternate, entire and clasping, and have a smooth waxy surface. The flower heads are borne on long hollow stalks, or peduncles, and have purple flowers, all ligulate. The achenes are linear and have a long slender beak with radiating pappus hairs at its tip. The roots of the plant have a flavor suggestive of that of oysters, and are used as a cooked vegetable. A yellow-flowered species also occurs, and has been widely introduced into the United States.

Artichoke, or Jerusalem artichoke, *Helianthus tuberosus,* is sometimes called by its Italian name, girosole, meaning sunflower, which corrupted into English becomes Jerusalem artichoke. It is a perennial herbaceous plant having thick fleshy rootstocks which bear somewhat irregular tubers with very evident "eyes" (actually dormant buds). The erect stems are 6 feet or more tall, stout and branching. The leaves are simple, ovate, and long-petioled. The heads are either solitary or in corymbs, and are composed of central disk flowers and marginal ray flowers, both yellow. The achenes are thick and hairy, with two deciduous pappus scales. The tubers are used largely as stock food, especially for hogs, but are also eaten by humans.

Helianthus annuus, the common sunflower, is frequently grown, partly for ornament, partly for curiosity because of the tremendous flower heads which often contain an enormous number of flowers, and partly for the seeds. These seeds are fed to poultry and larger caged birds. From the seeds, sunflower seed oil is expressed.

Globe artichoke, *Cynara scolymus,* is another composite which is grown for food. It is a herbaceous perennial native in northern Africa. The flowers form large globular heads surrounded by several rows of fleshy bracts. The basal portions of each bract and the thick fleshy receptacle are cooked and eaten. The blanched leaves of a related species, *Cynara cardunculus*, or cardoon, are often eaten like celery.

Dandelion, *Taraxacum officinale*, is a stemless perennial herb, having a thick tap-root and a rosette of basal leaves which grow close to the ground. Contraction of the root each year keeps the leaves of that year's growth at the ground level. The heads, composed of yellow ligulate flowers, are borne singly on hollow peduncles. When the achenes are mature the peduncle elongates greatly, lifting the achenes well above the ground. Each achene is beaked and has a crown of pappus hairs which aid it in floating through the air. The dandelion is frequently used as a pot-herb, and is grown extensively. Wild plants, often pestiferous weeds in lawns, are sometimes gathered in the spring for greens. The root, containing a glucoside taraxirin, has been considered by some as a tonic. The dandelion species *Taraxacum koksaghyz* was at one time considered as a source of rubber.

Ragweed (genus *Ambrosia*) causes allergies in some people. It is widely distributed. It is a branching annual, ordinarily growing from 2 to 4 feet (0.6 to 1.2 meters) in height. The leaves are finely divided and thin. The flower heads are of two sorts, the staminate heads are borne in elongated racemes, while the pistillates are borne in clusters. Flowers are produced in late summer and autumn and, unlike most composities, are wind-pollinated.

COMPOSITE BEAM. **Beam (Composite).**

COMPOSITE COURSE. **Course (Composite).**

COMPOSITE NUMBER. **Number Theory.**

COMPOSITES (Fiber-Reinforced). **Fiber-Reinforced Composites.**

COMPOSITE SIGNAL. **Television.**

COMPOUND (Chemical). A homogeneous, pure substance, composed of two or more essentially different chemical elements, which are present in definite proportions; compounds usually possess properties differing from those of the constituent elements.

An *addition compound* is one that is formed by the junction or union of two simpler compounds. Effectively the same as a molecular compound (see definition on the following page).

An *additive compound* is formed by an additional reaction, or by the saturation of a double bond, triple bond, or more than one of them.

An *alicyclic compound* is an organic compound containing a saturated ring of carbon atoms, such as a cycloparaffin or other hydroaromatic compound. See **Hydrocarbons.**

An *Aliphatic compound* is an organic compound without ring structures, i.e., with straight chain arrangement of carbon and, possibly other, atoms. In the narrower sense, an aliphatic compound is a member of the paraffin series of hydrocarbons, or one of their derivatives.

An *aromatic compound* is an organic compound containing a ring of carbon atoms, usually unsaturated, such as a benzene, naphthalene, anthracene, and acenaphthylene ring.

An *associated compound* is a compound formed by the union of two or more molecules, usually of the same or similar chemical composition, to form a single complex molecule.

Berthollide compound. See nonstoichiometric compound in this entry.

A *binary compound* is made up of two elements in a definite molecular ratio.

A *catenation compound* has a molecular configuration resembling a linked chain, in which the atoms forming one ring pass through, but are not joined by valence forces to, the ring formed by another group of atoms. Since the two rings, while spatially interlocked, are not joined by valence forces, the application of the word compound to such aggregates may be questioned.

H. L. Frisch of Bell Laboratories has made extensive calculations of ring sizes necessary tc permit the formation of various catenation compounds. He found that 20 is the minimum number of —CH_2— groups in an alicyclic ring through which another ring can be catenated (threaded) without encountering excessively great repulsive forces from the alicyclic ring atoms. For threadings of two rings through a third ring, the probable minimum alicyclic ring size of the latter is 33 —CH_2— groups; Borromean rings, formed by the interlocking of three rings (with no two of them locked separately), require a minimum of 30 —CH_2— groups.

A laboratory preparation of the simplest of these catenation compounds, two interlocking rings, has been carried out at Bell Laboratories. They started with the dimethyl ester of a 34-carbon paraffinic dicarboxylic acid, $CH_3OOC—(CH_2)_{32}—COOCH_3$, which was reacted in a suspension of metallic sodium in xylene with acetic acid to condense the terminal ester groups to form an aceloin ring compound $O{=}C—(CH_2)_{32}—CHOH$ (ring closed between C and CHOH). Treatment of the latter with deuterated hydrochloric acid reduced it to a 34-carbon alicyclic (ring) compound,

$$HDC—CHD—CD_2—CHD—(CH_2)_{30}$$ (ring closed between HDC and $(CH_2)_{30}$)

containing five deuterium atoms. This hydrocarbon was added to the suspension of metallic sodium in xylene, and then more of the 34-carbon dimethyl ester was added. Ring formation of the latter compound to form the aceloin occurred as before, and a small percentage of the aceloin rings were found to be threaded through the deuterated rings in the solvent, yielding a catenation compound consisting of the 34-carbon aceloin ring and the 34-carbon deuterated alicyclic hydrocarbon threaded together, but without any atoms in one ring being joined by valence bonds to those in the other.

Separation and identification of this catenation compound was effected by chromatography and infrared spectroscopy, the latter being the reason why the hydrocarbon portion of the catenation compound was deuterated.

Chelation compound. See entry on **Chelates and Chelation.**

A *clathrate compound* means, literally, an enclosed compound, a term applied to a solid molecular compound in which a molecule of one component is physically enclosed in the crystal structure of a second compound, so that the properties of the aggregate are essentially those of the enclosing compound. Examples of such "cage compounds" are those of the small molecules of SO_2, CO_2, CO and the noble gases with ice and hydroquinone, which have very open crystal structures. Another example is the clathrate of benzene with nickel cyanide.

A *complex compound* is made up structurally of two or more compounds or ions. See **Ligand.**

A *condensation compound* is formed by a reaction in which the largest parts, constituting the essential structural elements, of two or more molecules combine to form a new molecule, with elimination of minor elements, such as those of water.

Coordination compound. See **Coordination Compounds.**

A *covalent compound* is formed by the sharing of electrons between atoms; as distinguished from electrovalent compounds, in which there occurs a transfer of electrons.

A *cyclic compound* has some or all of its atoms arranged in a ring structure.

An *electrovalent compound* is formed by ions, or by atoms which become ions by transfer of electrons between them. (See ionic compound below.)

An *endothermic compound* is a compound whose formation is accompanied by a positive change in heat content, i.e., by the absorption of heat.

An *epoxy compound* contains an oxygen bridge, as

$$CH_2-CH_2-CH_2-CH_2$$ (terminal CH_2 groups bridged by O)

which is 1,4 epoxy butane.

An *exothermic compound* is a compound whose formation is accompanied by a negative change in heat content, i.e., with the liberation of heat.

A *heterocyclic compound* contains one or more rings composed of atoms some of which are of dissimilar elements. A few inorganic substances fall into this classification, but by far the majority of them are carbon compounds. In organic chemistry substances of cyclic structure, as acid anhydrides, lactides, lactams, lactones, cyclic ethers, and cyclic derivatives of dicarboxylic acids which are formed by the elimination of water from aliphatic compounds, are not considered among the heterocyclic substances. Derivatives of pyridine, quinoline, thiophene, thiazole, pyrone, etc., which contain heterocyclic rings that persist in the compound through chemical reactions, are considered the true members of this class. Heterocyclic rings are known that contain nitrogen, sulfur, and oxygen members. The noncarbon members of the ring are termed "heteroatoms," and their number is indicated by the prefixes mono, di, tri, tetra, etc. The number of members in the ring may reach as high as sixteen, as in tetrasalicylide.

A *homocyclic compound* contains a homocyclic ring, i.e., a ring composed of atoms of the same element.

The term *inclusion compound* was once used for the clathrate compounds described in this entry.

In an *inner compound* an additional valence bond has been formed between two atoms of an already existing structure, usually by loss of the elements of water or other simple substance. Inner compound formation commonly results in the formation of a ring. The inner esters, inner anhydrides, and inner coordination compounds are well-known classes of inner compounds.

An *inorganic compound* means, in general, a compound that does not contain carbon atoms. Some very simple carbon compounds, such as carbon monoxide and dioxide, binary metallic carbon compounds (carbides) and carbonates, are also included in the group of inorganic compounds.

An *intermetallic compound* consists of metallic atoms only, which are joined by metallic bonds. Such compounds may be made semiconducting if the two metals between them contribute just sufficient electrons to fill the valence bond, elg., InAs. (See also **Alloys**).

An *interstitial compound* consists of a metal or metals and certain metalloid elements, in which the metalloid atoms occupy the interstices between the atoms of the metal lattice. Compounds of this type are, for example, TaC, TiC, ZrC, NbC, and similar compounds of carbon, nitrogen, boron, and hydrogen with metals.

An *ionic compound* is one of a class of compounds which are formed when atoms combine to produce molecules having stable configurations by the transfer of one or more electrons within the molecule. This type of combination is illustrated by the combination of sodium atoms and chlorine atoms to form sodium chloride. The sodium atom loses the single electron in its outer shell, and thus is left with the stable configuration of eight electrons; the chlorine atom acquires an electron to increase the number of electrons in its outer shell from seven to eight; as a result of the loss and gain of the electrons, the atoms have acquired positive and negative charges, respectively, which constitute an electrovalent bond.

A *molecular compound* is formed by the union of two or more already saturated molecules apparently in defiance of the ordinary rules of valence. The class includes double salts, salts with water of crystallization, and metal ammonium derivatives. These salts are usually formed by van der Waals attraction between the constituent molecules. They do not differ in any characteristic manner from compounds formed in strict accordance with the concept of valence. They are also called addition compounds.

A *nonpolar compound* is a compound in which the centers of positive and negative charge almost coincide, so that no permanent dipole moments are produced. The term nonpolar also applies to compounds in which the effect of oppositely directed dipole moments cancel. Nonpolar compounds may contain polar bonds, if their effect is cancelled by opposing bonds, as may occur in a perfectly symmetrical molecule. Nonpolar compounds do not ionize or conduct electricity. Most organic compounds are to be classed as nonpolar compounds.

A *nonstoichiometric compound* has a composition not in accord with the law of definite proportions, which is therefore also called a berthollide compound. Non stoichiometric compounds occur among the binary compounds of Group 6b, as examplified by $TiO_{1.8}$, $Cu_{1.7}S$ and $Cu_{1.6}Se$; among the hydrides, e.g., $CeH_{2.7}$ and especially among the intermetallic compounds.

An *organic compound* is one of the great number of compounds consisting of carbon linked in chains or rings; such compounds usually also contain hydrogen and may contain elements such as oxygen, nitrogen, sulfur, chlorine, etc. Some of the simpler carbon compounds are classified as inorganic compounds.

An *organometallic* (or *metal-organic*) *compound* is an organic compound in which one or more hydrogen atoms have been replaced by a metallic atom or atoms, usually with the establishment of a valence bond between the metal atom and a carbon atom. A metallic salt of an organic acid, in which the hydrogen atoms of a —COOH group is replaced by a metal atom, is not classified as an organometallic compound.

A *polar compound* is, in general, a compound that exhibits polarity, or local differences in electrical properties, and has a dipole moment associated with one or more of its interatomic valence bonds. Polar compounds have relatively high dielectric constants, associate readily in most cases, and include the substances that exhibit tautomerism. In the most general use of the term, polar compounds include all electrolytes, most inorganic substances, and many organic ones. Specifically, the term polar compound is frequently applied to the extreme type of polarity which arises in the presence of an electrovalent bond or, in wave-mechanical terms, to cases in which one ionic term dominates in the orbital function of the molecule. Such compounds are exemplified by the inorganic acids, bases, and salts which possess, to a greater or lesser degree the power to conduct electricity, associate, form double molecules and complex ions, etc.

In a *saturated compound* the valence of all the atoms is completely satisfied without linking any two atoms by more than one valence bond.

A *spiro-compound* contains two ring structures having one common carbon atom.

A *tracer compound* is a compound which by its ease of detection enables a reaction or process to be studied conveniently. Wide use has been made of isotopes, including radioactive isotopes of common elements, which are added in small quantities, in the form of the proper compound, to follow the course of an atom or a compound through a complicated series of reactions; or conversely to determine the properties of a tracer—that is available only in quantities too small to handle alone—by adding it to a system containing chemically related elements, and then following its course throughout a given series of reactions. Considerable use of tracer compounds is made in the study of physiological reactions.

Unsaturated compound is a term specifically applied to a carbon compound containing one or more double bonds or triple bonds. One consequence of the presence of these double bonds or triple bonds, from which a broader concept of unsaturation stems, is the relative ease with which such bonds are split, and other constituents linked to them.

See also **Organic Chemistry.**

COMPOUND DISTRIBUTION. In statistics this term occurs in three senses: (1) If two or more homogeneous populations are merged, the resulting population may be said to be compound. (2) If several distributions are convoluted they are sometimes described as compounded. (3) If a distribution of a variable x depends on a parameter θ which itself has a distribution, the resultant of integrating over the distribution of θ is sometimes called the compound distribution of x.

For example, suppose a set of observations x are each drawn from a Poisson distribution $e^{-\mu}\mu^x/x!$ where μ varies from one observation to another according to the distribution function $F(\mu)$. The resulting distribution of x is called a compound Poisson distribution. If $F(\mu)$ is a Pearson Type III distribution, x follows a negative binomial distribution.

COMPOUND (Lattice). Lattice Compounds.

COMPOUND NUCLEUS. The nucleus formed in a nuclear reaction through the amalgamation of the incident nuclear particle and the target nucleus.

COMPOUND TURBINE. Turbine (Steam).

COMPRESSED AIR (Gas Turbine). Gas and Expansion Turbines.

COMPRESSIBILITY (Air). Aerodynamics; Supersonic Aerodynamics.

COMPRESSION (Gas). The compressibility of a gas is defined as the rate of volume decrease with increasing pressure, per unit volume of the gas. The compressibility depends not only on the state of the gas, but also on the conditions under which the compression is achieved. Thus, if the temperature is kept constant during compression, the compressibility so defined is called the isothermal compressibility β_T:

$$\beta_T = -\frac{1}{V}\left(\frac{\partial V}{\partial P}\right)_T = \frac{1}{\rho}\left(\frac{\partial \rho}{\partial P}\right)_T \tag{1}$$

If the compression is carried out reversibly without heat exchange with the surroundings, the *adiabatic compressibility* at constant entropy, β_S, is obtained:

$$\beta_S = -\frac{1}{V}\left(\frac{\partial V}{\partial P}\right)_S = \frac{1}{\rho}\left(\frac{\partial \rho}{\partial P}\right)_S \tag{2}$$

Here P is the pressure, V the volume, ρ the density, T the temperature, and S the entropy.

In adiabatic compression, the temperature rises, thus the pressure increases more sharply than in isothermal compression. Therefore β_S is always smaller than β_T.

The *compressibility factor* of a gas is the ratio PV/RT. This name is not well chosen since the value of the compressibility factor by itself does not indicate the compressibility of the gas.

Experimental values for the compressibility of gases can be obtained in several ways, most of which are indirect.

Since the compressibility is proportional to the pressure derivative of the volume, any experiment that establishes the P-V-T relation of a gas with sufficient accuracy also yields data for the isothermal compressibility. For obtaining the adiabatic compressibility from the P-V-T relation, some additional information is necessary [see section (c)], for instance specific heat data in the perfect gas state of the substance considered. A more direct way of determining the adiabatic compressibility is by measuring the speed of sound v, the two quantities being related by

$$v^2 = \frac{1}{\rho \beta_S} \tag{3}$$

This relation is valid only when the compressions and expansions of the sound wave are truly reversible and adiabatic. This is the case if the frequency is fairly low and the amplitude small.

Dilute gases obey the laws of Boyle and Gay-Lussac, $PV = RT$, to a good approximation. Thus, it can readily be shown that the following relations hold for the compressibility:

$$\beta_T = 1/P = V/RT$$
$$\beta_S = 1/\gamma P = V/\gamma RT \tag{4}$$

where $\gamma = c_P/c_V$, the ratio of the specific heats at constant volume and at constant pressure, respectively, and R is the gas constant.

Compressed gases show large deviations from the behavior predicted by equation (4). This is demonstrated by the accompanying diagram, where the isothermal compressibility of argon, divided by the corresponding value for a perfect gas at the same density, is pictured as a function of density for various temperatures. It is seen, first of all, that at all temperatures the compressibility at high densities falls to a small fraction of the value for a perfect gas, and secondly, that super-critical isotherms show a maximum in the ratio $\beta/\beta_{perfect}$ as a funciton of density, which maximum is the more pronounced the closer the critical temperature. It occurs roughly at the critical density ρ_c. Since at the critical point $(\partial P/\partial V)_T$ equals zero, the isothermal compressibility becomes infinite at this point. The adiabatic compressibility, however, remains finite. Qualitatively, all gases show the same behavior as pictured for argon in the diagram.

The molecular theory can explain the general features of the compressibility in its temperature and density dependence. The pressure of the gas is caused by the impact of the molecules on the wall. If the volume is decreased at constant temperature, the average molecular speed and force of impact remain constant, but the number of collisions per unit area increases and thus the pressure rises. If the gas is compressed adiabatically, the heat of compression cannot flow off, thus

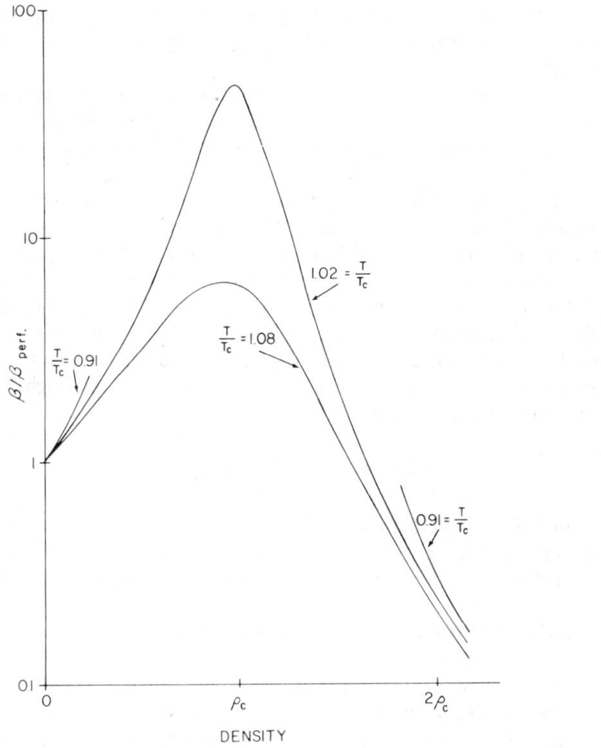

The ratio $\beta/\beta_{perf.}$ of the isothermal compressibility of argon to that of a perfect gas at the same density, as a function of the density, at 0.91, 1.02 and 1.08 times the critical temperature. The critical density is indicated by ρ_c.

the average molecular speed and force of impact increase as well, giving rise to an extra increase of pressure. Therefore $\beta_S < \beta_T$. The actual magnitude of the temperature rise depends on the internal state of the molecules: the more internal degrees of freedom available, the more energy can be taken up inside the molecule and the smaller the temperature rise on adiabatic compression. Thus for gases consisting of molecules with many internal degrees of freedom, adiabatic and isothermal compressibilities differ but little.

If the gas is assumed to consist of molecules of negligible size and without interaction, then the gas can be shown to follow the laws of Boyle and Gay-Lussac, therefore its isothermal and adiabatic compressibilities must be given by Equation (4). For a perfect gas, the percentage pressure rise is proportional to the percentage volume decrease if the change is small; thus the compressibility is inversely proportional to the pressure.

To explain the very different behavior of real gases, the model must be modified. Suppose the molecular volume is small but not negligible. In states of high compression where the total molecular volume becomes of the order of the volume available to the gas, the free space available to the molecules is only a fraction of what it would be in a perfect gas, and thus the real gas is much harder to compress than the perfect gas. This explains the low compressibility of dense gases and liquids (diagram).

Furthermore, one assumes that molecules, on approaching each other, experience a mutual attraction before they collide; this mutual attraction makes it easier to compress a real gas than a perfect gas. This explains the initial rise of the compressibility of a real gas over that of a perfect gas at temperatures not too far above the critical.

When compressed at subcritical temperatures, the gas condenses; that is, macroscopic clusters or droplets are formed under the influence of the attractive forces. During condensation, the pressure remains constant while the volume decreases, giving rise to an infinite compressibility in the two-phase region. At the critical point the system is on the verge of condensation and the compressibility is also infinite.

Theoretical predictions for the isothermal compressibility can obviously be obtained from any theory of the equation of state. If, in addition, data for the specific heat are supplied, the adiabatic compressibility can be derived in the same way. Thus the compressibility can be derived from the viral expansion of the equation of state which expresses the ratio PV/RT in a power series in the density, the coefficients being related to the interactions of groups of two, three, etc., particles.

In the dense system the convergence of the virial expansion is doubtful. In any case the higher coefficients are hard to calculate; here approximate theories have been developed, of which the cell model[5] is an example.

Many semiempirical equations of state with varying degree of theoretical foundations are in use. The van der Waals equation, a two-parameter equation which gives a qualitatively correct picture of the P-V-T relations of a gas and of the gas-liquid transition, is an example.

Modern developments are centered around the calculations of the radial distribution function $g(r)$, which is the ratio of the density of molecules at a distance r from a given molecule, to the average density in the gas. The compressibility can be expressed straightforwardly in terms of $g(r)$ as follows.

$$KT\beta_T = 1/\rho + \int_0^\infty [g(r) - 1]4\pi r^2\, dr \tag{5}$$

Approximate evaluations of the radial distribution function in dense systems are being obtained as solutions to integral equations derived from first principles under well-defined approximations.

COMPRESSION RATIO (Engine). Petroleum.

COMPRESSION (Signal). In amplitude modulation systems of radio communication, the amount of intelligence volume which can be modulated upon the carrier is limited to an amount which will give 100% modulation. Since the percentage of modulation depends directly upon the volume of the sound, it follows that, in order not to exceed the allowable modulation on very loud sounds, the percentage on most sounds will be rather low. Since the maximum use is made of the power and a higher signal-to-noise ratio is obtained for high degrees of modulation, it is very desirable to keep the level of modulation as high as possible. To do this the volume range of the original sound is compressed into a much smaller range. Thus, while a symphony orchestra may have a volume range of 100–110 dB, the range is compressed to about 40 dB for broadcast purposes. In recordings the maximum volume which may be recorded is limited by the thickness of the groove walls so the volume range is reduced here also. An expander may be used in the reproducing system of the radio circuit or the phonograph to restore the original range.

COMPRESSION (Structural). In structural engineering, compression is used to denote the type of stress which causes the fibers of a member to shorten. A compression member of a structure is subjected to a primary compressive stress. The analysis or design is the same as that for a column. See also **Column (Structural).**

When the member consists of two or more separate elements, known as ribs, the parts are connected to produce unity of action, prevent buckling of the individual elements, and also to prevent excessive distortion during fabrication, shipping, or erection. Tie plates and lacing are used for this purpose. The tie plates, also called stay plates or batten plates, which are rectangular in shape, are placed at the ends of the member and at intermediate points where the lacing is interrupted. The lacing bars, also known as lattice bars, are plates, angles, or channels placed at an angle to the longitudinal axis of the member. Lacing may be single or double as shown in the accompanying illustration of a compression member made up of two channels which are rolled steel sections. Single lacing should make an angle of not less than 60° with the longitudinal axis of the member. The angle for double lacing should be not less than 45°. The purpose of these minimum angles is to reduce the buckling tendency of the individual segments.

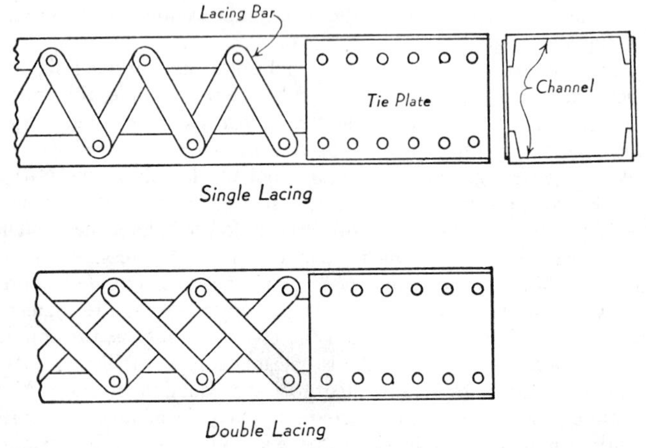

Types of compression members: (a) single lacing; (b) double lacing.

COMPRESSION SYSTEM (LNG). Natural Gas.

COMPRESSION WAVE. Aerodynamics; Supersonic Aerodynamics.

COMPRESSIVE STRESS. Stress.

COMPRESSOR (Air). Air Compression.

COMPRESSOR (Ammonia). Ammonia.

COMPTON EFFECT. This refers to the collision of a photon and a free electron in which the electron recoils and a photon of longer wavelength is emitted as indicated in Fig. 1. It is one of the most important processes by which x-rays and gamma rays interact with matter and is also one which is accurately calculable theoretically.

Barkla and other (1908) made many observations on the scattering of x-rays by different materials. The diffuse scattering was interpreted qualitatively by J. J. Thomson in terms of the interaction of electromagnetic waves with electrons which he had shown to be a constituent of all atoms. As more experiments were carried out with light elements, it was established by J. A. Gray (1920) that the diffusely scattered x-rays were less penetrating. This implied that the scattered radiation had a longer wavelength than the incident radiation. This could not be reconciled with Thomson's theory which represented x-rays as continuous electromagnetic waves with wavelengths unchanged by scattering.

The effect which now bears his name was established quantitatively by Arthur Holly Compton (1923) when he published careful spectroscopic measurements of x-rays scattered at various angles by light elements. He found that x-rays scattered at larger angles had systematically larger wavelengths. In searching for an explanation of the data, he discovered that the observations were accounted for by considering the scattering as a collision between a single photon and a single electron in which energy and momentum are conserved.

The important place which the effect occupies in the development of physics lies in his interpretation of the effect in terms of the newly emerging quantum theory. The essential duality of waves and particles was demonstrated in an especially clear way, since the collision con-

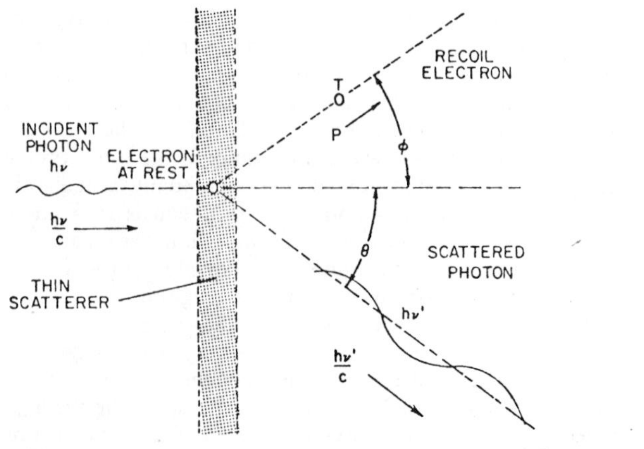

Fig. 1. Diagram showing the initial and final energies and momenta for Compton scattering.

served energy and momentum while both the incident and scattered x-rays revealed wave-like properties by their scattering from a crystal. In recognition for this contribution, Compton was awarded the Nobel Prize in 1927.

A complete theory for the effect was worked out in 1928 by Klein and Nishina using Dirac's relativistic theory of the electron. The calculation was one of the brilliant successes of the Dirac theory. It represents quantitatively, within the experimental uncertainties, all phenomena associated with the scattering of photons by electrons for energies up to several billion electron volts. Because of the confidence with which photon interaction with electrons can be interpreted, the Compton effect has been important in the analysis of the energy and the polarization of gamma rays from many sources.

Kinematics. The relations between the energies and directions of the incident and scattered photons and the recoil electron are determined by the conservation of energy and of the components of momentum parallel and at right angles to the incident beam. In the usual case, where the electron is initially at rest and the energy and momentum of the incident photon are $h\nu$ and $(h\nu/c)$, the equations are:

$$h\nu = h\nu' + T \tag{1}$$

$$\frac{h\nu}{c} = \frac{h\nu'}{c}\cos\theta + p\cos\phi \tag{2}$$

$$0 = \frac{h\nu'}{c}\sin\theta - p\sin\phi \tag{3}$$

where c is the velocity of light, h is Planck's constant, and the angles are those indicated in Fig. 1. The relativistic relation between the kinetic energy T of the recoiling electron and its momentum p is

$$pc = \sqrt{T(T+2mc^2)} \tag{4}$$

where m is the mass of the electron. These equations can be combined to obtain relations which are useful in the interpretation of data. The Compton shift is

$$\lambda' - \lambda = \frac{c}{\nu'} - \frac{c}{\nu} = \frac{h}{mc}(1-\cos\theta) \tag{5}$$

This relation was first found experimentally by Compton, who noted that the shift in wavelength $(\lambda' - \lambda)$ depended on the angle, but not on the wavelength, of the incident photon. The quantity (h/mc), which is the shift at 90°, is called the Compton wavelength of the electron and is one of the useful constants (2.4262×10^{-10} cm).

$$h\nu' = \frac{mc^2}{1-\cos\theta + \dfrac{mc^2}{h\nu}} \tag{6}$$

In this form, the energy of the scattered photon is seen to vary from that of the incident photon at 0° to less than $(mc^2/2)$ at 180°. At high energies the angle θ for which $h\nu'$ is $(h\nu/2)$ is approximately $2(mc^2/h\nu)$ radians.

The kinetic energy of the recoiling electron is

$$T = \frac{h\nu(1-\cos\theta)}{(1-\cos\theta) + \dfrac{mc^2}{h\nu}} \tag{7}$$

The relation between the scattering angles of the electron and photon is

$$\cot\phi = \left(1+\frac{h\nu}{mc^2}\right)\left(\frac{1-\cos\theta}{\sin\theta}\right) \tag{8}$$

Graphs of these kinematic relations and of the scattering cross section are given by Evans[2] and by Nelms.[3]

Scattering of Unpolarized Radiation. The differential cross section for the scattering of unpolarized radiation at an angle θ is given by the Klein and Nishina equation.

$$\frac{d\sigma}{d\Omega} = \frac{r_0^2}{2}\left(\frac{\nu'}{\nu}\right)^2\left(\frac{\nu}{\nu'} + \frac{\nu'}{\nu} - \sin^2\theta\right) \tag{9}$$

where r_0 is the electron radius $= e^2/mc^2 = 2.8177 \times 10^{-13}$ cm, and ν' is obtained from Equation (6). The cross section is shown as

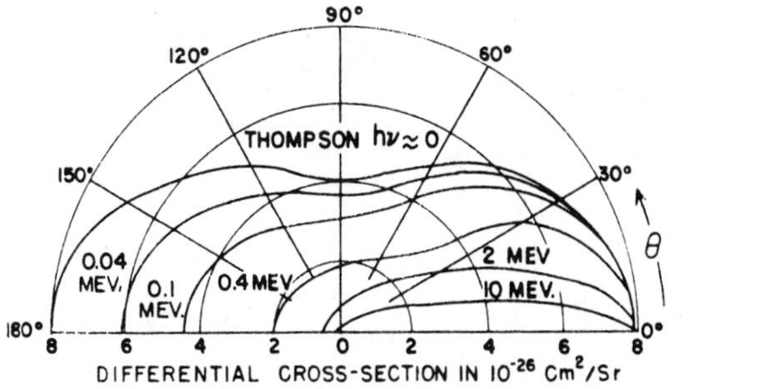

Fig. 2. Differential cross section for photons scattered at angles, θ, for a number of incident energies.

a function of θ for several energies in Fig. 2. The classical Thomson cross section $r_0^2\,(1 + \cos^2\theta)/2$ can be seen to hold for low energies where $\nu' \approx \nu$.

The total cross section obtained by integrating this cross section over angle is important in the attenuation of well-defined beams in passing through a material. The relative importance of Compton scattering as compared to the photoelectric effect and pair production is illustrated for aluminum in Fig. 3 where the attenuation coefficient

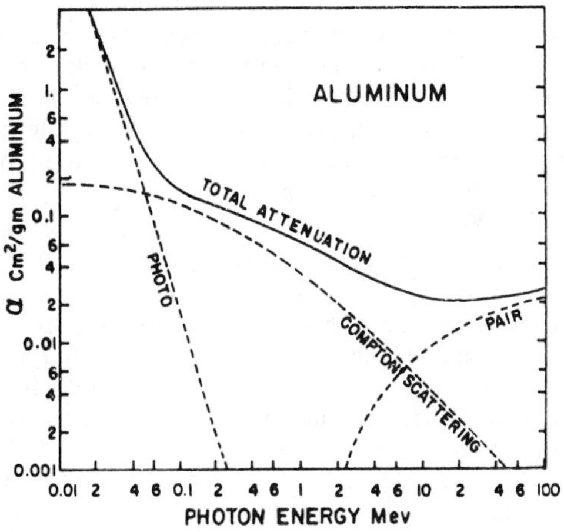

Fig. 3. The attenuation coefficients, α, for the absorption of photons in aluminum as a function of energy. The broken lines represent the separate contributions of the photoelectric effect, the Compton effect, and pair production to the absorption.

α is shown as a function of energy. The fraction of the photons surviving without an interaction upon passing through x g/cm² of aluminum is $e^{-\alpha x}$. The Compton effect is the major one between 0.5 and 2 MeV. Extensive tables and graphs for other elements are available.[2,4]

In detectors whose response is proportional to the energy deposited by the recoil electrons, the distribution of electron energies associated with a photon of known energy is of interest. The distribution is given by the relation

$$\frac{d\sigma}{dT} = \frac{\pi r_0^2 mc^2}{(h\nu)^2}\left\{2 + \left(\frac{T}{h\nu - T}\right)^2\left[\frac{(mc^2)^2}{(h\nu)^2} + \frac{h\nu - T}{h\nu} - \frac{2mc^2(h\nu - T)}{h\nu T}\right]\right\}$$

where T varies from 0 to $T_{max} = 2(h\nu)^2/(2h\nu + mc^2)$. A number of these distributions are shown in Fig. 4.

Scattering of Plane Polarized Radiation. The differential cross section for the scattering of plane polarized radiation by unoriented electrons was also derived by Klein and Nishina. It represents the probability that a photon, passing through a target containing one electron per square centimeter, will be scattered at an angle θ into a solid angle $d\Omega$ in a plane making an angle η with respect to the plane containing the electric vector of the incident wave.

$$\frac{d\sigma}{d\Omega} = \frac{r_0^2}{2}\left(\frac{\nu'}{\nu}\right)^2\left(\frac{\nu}{\nu'} + \frac{\nu'}{\nu} - 2\sin^2\theta\cos^2\eta\right) \tag{10}$$

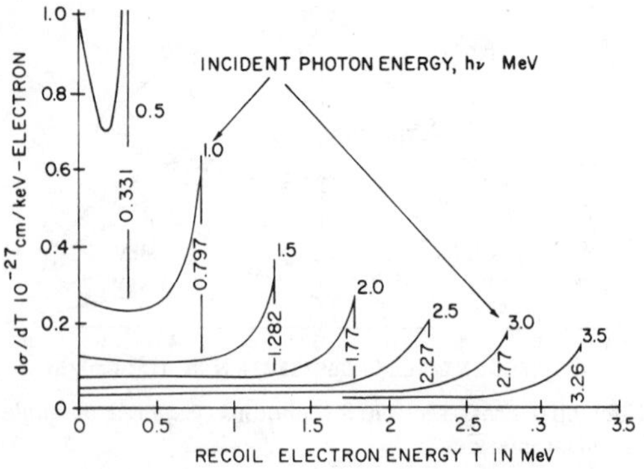

Fig. 4. The energy distribution of the Compton recoil electrons for several values of the incident photon energy *hν*. (*Based on figure in "Compton Effect" by R. D. Evans in "Handbuch der Physik," Vol. XXXIV, pp. 234–298, 1958, J. Fluge, ed., by permission of Springer-Verlag, publishers.*)

The cross section has its maximum value for $\eta = 90°$, indicating that the photon and electron tend to be scattered at right angles to the electric vector of the incident radiation.

This dependence is the basis of several instruments for determining the polarization of photons. For example, it was used by Wu and Shaknov[5] to establish the crossed polarization of the two photons emitted upon the annihilation of a positron electron pair; by Metzger and Deutsch[6] to measure the polarization of nuclear gamma rays; and by Motz[7] to study the polarization of bremsstrahlung.

Scattering of Circularly Polarized Radiation. The scattering of circularly polarized photons by electrons with spins aligned in the direction of the incident photon is represented by

$$\frac{d\sigma}{d\Omega} = r_0^2\left(\frac{\nu'}{\nu}\right)^2\left[\left(\frac{\nu}{\nu'} + \frac{\nu'}{\nu} - \sin^2\theta\right) \pm \left(\frac{\nu}{\nu'} - \frac{\nu'}{\nu}\right)\cos\theta\right] \quad (11)$$

The first term is the usual Klein-Nishina formula for unpolarized radiation. The + sign for the additional term applies to right circularly polarized photons. The ratio of the second term to the first is a measure of the sensitivity of the scattering as a detector of circularly polarized radiation and is shown in Fig. 5.

In practice, the only source of polarized electrons has been magnetized iron where 2 of the 26 electron spins can be reversed upon changing its magnetization. Although the change in the absorption or scattering is usually only a few per cent, this is often sufficient to get accurate and reliable measurements of circular polarization.

Cross sections for some practical arrangements and discussions of earlier work are presented by Tolhoek.[8] Applications to the determination of the helicities of photons, electrons, and neutrinos in confirming the two-component theory of the neutrino are reviewed in considerable detail by L. Grodzins.[9]

Proton and Deuteron Compton Effect. Particle-like scattering of high-energy photons by protons and deuterons has been observed and has been referred to as the proton and deuteron Compton effect. The kinematic equations are identical to those for electrons except that the mass is that of the proton or deuteron.

Although the cross sections are smaller than that for electrons, by the square of the ratio of the masses, the scattering is easily distinguished by the characteristically higher energy of the radiation at large angles. At energies above the pion threshold, the cross section is dominated by pion nucleon resonances. The experimental cross sections for the scattering by protons, as presented by Steining, Loh, and Deutsch,[10] are shown in Fig. 6. Some experimental results and calculations on the coherent scattering from deuterium are described by Jones, Gerber, Hanson, and Wattenberg.[11]

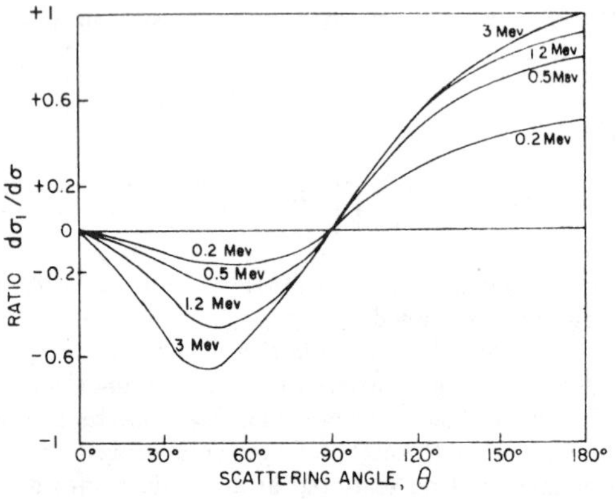

Fig. 5. The ratio of the partial cross section dependent on the spin orientation of the electrons to the average cross section as represented by the second and first terms of Eq. (10). (*Based on figure in "Compton Effect" by R. D. Evans in "Handbuch der Physik," Vol. XXXIV, pp. 234–298, 1958, J. Fluge, ed., by permission of Springer-Verlag, publishers.*)

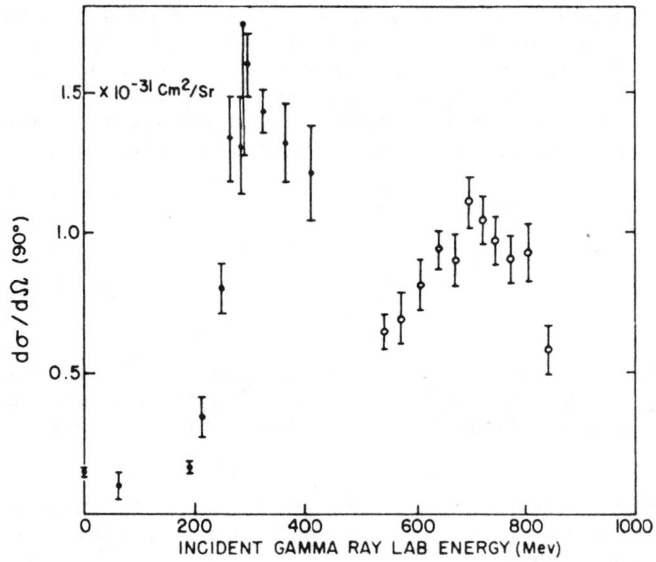

Fig. 6. The differential cross section for the scattering of high-energy photons by protons at 90° in the center of mass system. (*Based on figure in Steining, Loh and Deutsch, Phys. Rev. Letters,* **10**, *536 (1963).*)

References

1. Evans, R. D.: "The Atomic Nucleus," Chapter 23, McGraw-Hill, New York, 1955.
2. Evans, R. D.: "Compton Effect," In Flugge, S. (editor), "The Encyclopedia of Physics," vol. 34, pp. 234–298, Springer-Verlag, Berlin, 1958.
3. Nelms, A. T.: "Graphs of the Compton Energy-Angle Relationship and the Klein-Nishina Formula from 10 keV to 500 MeV," *Natl. Bur. Std. Circ.*, **542** (1953).
4. White, G. R.: "X-Ray Attenuation Coefficients from 10 keV to 100 MeV," *Natl. Bur. Std. Rept.*, **1003** (1952).
5. Wu, C. S., and I. Shaknov: "Angular Correlation of Scattered Annihilation Radiation," *Phys. Rev.*, **77**, 136 (1950).
6. Metzger, F., and M. Deutsch: *Phys. Rev.*, **78**, 551 (1950).
7. Motz, J. W.: "Bremsstrahlung Polarization Measurements for 1 MeV Electrons," *Phys. Rev.*, **78**, 551 (1950).
8. Tolhoek, H. A.: "Electron Polarization, Theory and Experiment," *Rev. Mod. Phys.*, **28**, 277 (1956).
9. Grodzins, L.: "Measurement of Helicity," in Frisch, O. R. (editor), "Progress in Nuclear Physics," Pergamon, New York, 1959.
10. Steining, R. F., Loh, E., and M. Deutsch: "The Elastic Scattering of Gamma Rays by Protons," *Phys. Rev. Letters*, **10**, 536 (1963).
11. Jones, R. S., Gerber, H. J., Hanson, A. O., and A. Wattenberg: "Deuteron Compton Effect," *Phys. Rev.*, **128**, 1357 (1962).

COMPTON RULE. An empirical relationship between thermal properties of elements, of the form

$$\frac{(\text{At. Wt.})(L_f)}{T_f} = 2$$

in which At. Wt. is the atomic weight, L_f is the heat of fusion, and T_f is the fusing point (in degrees absolute).

COMPUTED AXIAL TOMOGRAPHY (CAT). **Brain Disorders; X-Ray Scanner (CAT).**

COMPUTER. A device or machine capable of accepting information, applying prescribed processes to the information, and supplying the results of these processes. The computer usually consists of some combination of input/output devices, storage, arithmetic, and logical units, and a central control unit. In current terminology, a computer, once programmed, is essentially capable of performing all desired functions with a very minimum of human checking and intervention. Calculators also compute, but traditionally have required manual step-by-step guidance. With the advent of solid-state desk calculators, some of the functions which previously required manual attention are permanently wired into the device and thus calculators are getting closer to meeting the definition of a computer.

In terms of fundamental technique used, computers are classified as analog or digital. Although analog computers are widely used in the form of devices and subsystems, such as electronic, mechanical, pneumatic, and fluidic analog computing devices and subsystems in the instrumentation and automatic control field, they may not be identified as computers per se. Certain large analog computers, identified as such, find application, but generally when the term computer is mentioned, a digital computer is inferred. Combinations of analog and digital techniques in the form of hybrid computers also are applied in specialized areas.

Digital computers range widely in size, capability, and cost. A digital computer may be identified in terms of size and capability, as for example a minicomputer, or microcomputer; or in terms of intended use, as for example data-acquisition computers, process-control computers, general-purpose computers, commercial computers, and special-purpose computers.

The several hundred computer terms described in this volume are listed under **Data Processing.** Particular reference to **Analog Computer; Digital Computer;** and **Electronic Data Processing (EDP)** also is suggested.

Thomas J. Harrison, International Business Machines Corporation, Boca Raton, Florida.

References

EDITOR'S NOTE. Several hundred books have been published covering various aspects of the computer field. It is beyond the scope of this encyclopedia to include a long bibliography. Following is a list of stimulating articles which have appeared in the late 1970s and early 1980s which may be of interest to some readers.

Abelson, P. H., et al: "Computers and Electronics," (a feature issue), *Science*, **215**, 749872 (1982).

Baker, W. O., et al.: "Computers and Research," *Science*, **195**, 1134–1139 (1977).

Balderston, F. E., Carman, J. M., and A. C. Hoggatt: "Computers in Banking and Marketing," *Science*, **195**, 1115–1119 (1977).

Davis, R. M.: "Evolution of Computers and Computing," *Science*, **195**, 1096–1102 (1977).

Dertouzos, M. L., and J. Moses (editors): "The Computer Age: A Twenty-Year View," MIT Press, Cambridge, Massachusetts, 1980.

Dinneen, G. P., and F. C. Frick: "Electronics and National Defense: A Case Study," *Science*, **195**, 1151–1155 (1977).

Farber, D., and P. Baran: "The Convergence of Computing and Telecommunications Systems," *Science*, **195**, 1166–1170 (1977).

Gelernter, H. L., et al.: "Empirical Explorations of SYNCHEM," *Science*, **197**, 1041–1049 (1977).

Heller, S. R., Milne, G. W. A., and R. J. Feldmann: "A Computer-Based Chemical Information System," *Science*, **195**, 253–259 (1977).

Hellman, M. E.: "The Mathematics of Public-Key Cryptography," *Sci. Amer.*, **241**, 2, 146–157 (1979).

Sheridan, T. B.: "Computer Control and Human Alienation," *Technology Review (MIT)*, **83**, 1, 61–72 (1980).

White, R. L., and J. D. Meindl: "The Impact of Integrated Electronics in Medicine," *Science*, **195**, 1119–1124 (1977).

COMPUTER (Airspeed). Airspeed Indicator.

COMPUTER-BASED NUMERICAL CONTROL. Numerical Control.

COMPUTER (Digital). Digital Computer.

COMPUTER ERROR PROCEDURES. Computer errors fall into three principal categories: (1) central-processor errors, (2) peripheral-device errors, and (3) program errors. Program errors by far are the most common, but sometimes they may appear to be processor or peripheral errors. If an error cannot be duplicated through the use of other programs, it is almost certainly a program error. Detection of programming errors is described under **Debugging (Computer Program).**

The procedure for isolating peripheral-device errors normally consists of retrying the operation several times and logging the failures. When the failure persists, the operator is informed and the computer may wait on operator intervention, or abort the current job and logically disconnect the device. Or, the error may be documented to the operator with a return to the current program with a special indicator to permit the application programmer to determine the error procedure to be followed. This is a good approach where two different programs, each having different error-procedure requirements, are running in the same machine.

Procedures for central-processor errors are usually more difficult. Often a "warm start" is attempted. All current jobs are aborted; all of the storage is initialized where possible, and a restart from a previously defined state is attempted. If the first warm start fails, another cycle usually will be attempted. It is unlikely that a machine which cannot be warm-started can be safely used without operator intervention.

See also **Diagnostics (Computer System);** terms listed under **Data Processing.**

COMPUTER GRAPHICS. For some tasks, a graphic input to the computer is the best or possibly the only way to communicate, particularly in connection with computer-aided design or simulation. In chemical research, new compounds may be studied visually; molecular structures can be computer-generated with certain parameters controlled by the operator. Studies can be conducted in three dimensions; a new automobile design can be drawn on a cathode-ray tube and then rotated to be looked at from all sides. Generating visual data requires considerable software as well as hardware. One way to ease the computer load and also to provide a more realistic picture is to project slide or film images onto the face of a cathode-ray tube and superimpose computer-generated information. This technique was pioneered for military command and control systems for displaying terrain maps and to superimpose troop concentations and equipment locations. The technique had many other uses, such as aircraft-flight-plan analysis. See also **Cathode-Ray Tube.** Much simplification has been accomplished in recent years through the use of microprocessors.

A generalized block diagram for a cathode-ray tube display system is shown in Fig. 1. Entry into the system is at the operator's console, which contains an alphanumeric keyboard for writing instructions or entering data, a function selector to make inquiries or select operating modes, and a communications device. Light pens are sometimes used. With the proper device, the operator can point to information on the screen, designate a location at which information should appear, enter information, or possibly draw a line.

The operator can communicate with a digital computer that is part of the console or one that is located remotely. Simple data-retrieval systems require only a memory bank from which data are extracted, edited, and reentered. Computer information enters the display console through an interface unit which conditions the input to ensure compatible logic levels and data word lengths. Additional signals are provided by the computer for sequencing computer signals with display-generation circuits. Timing and mode control ensures the proper positioning of characters and symbols on the screen.

Among the types of displays that can be generated are point plotting, alphanumeric characters, and graphic or pictorial views. The simplest function is point plotting. This requires binary X and Y addressing to position the CRT beam. Two output functions are needed to represent the plot adequately: position generation and intensity generation (blanking).

A modern desktop computing system (4051 Graphic Computing System), shown in Fig. 2, is a microprocessor-based data system offering compact, local computing and calculating power with high-resolution graphics. The instrument shown comes with 8K bytes memory (expandable to 32K) and offers flexible data communications interfacing.

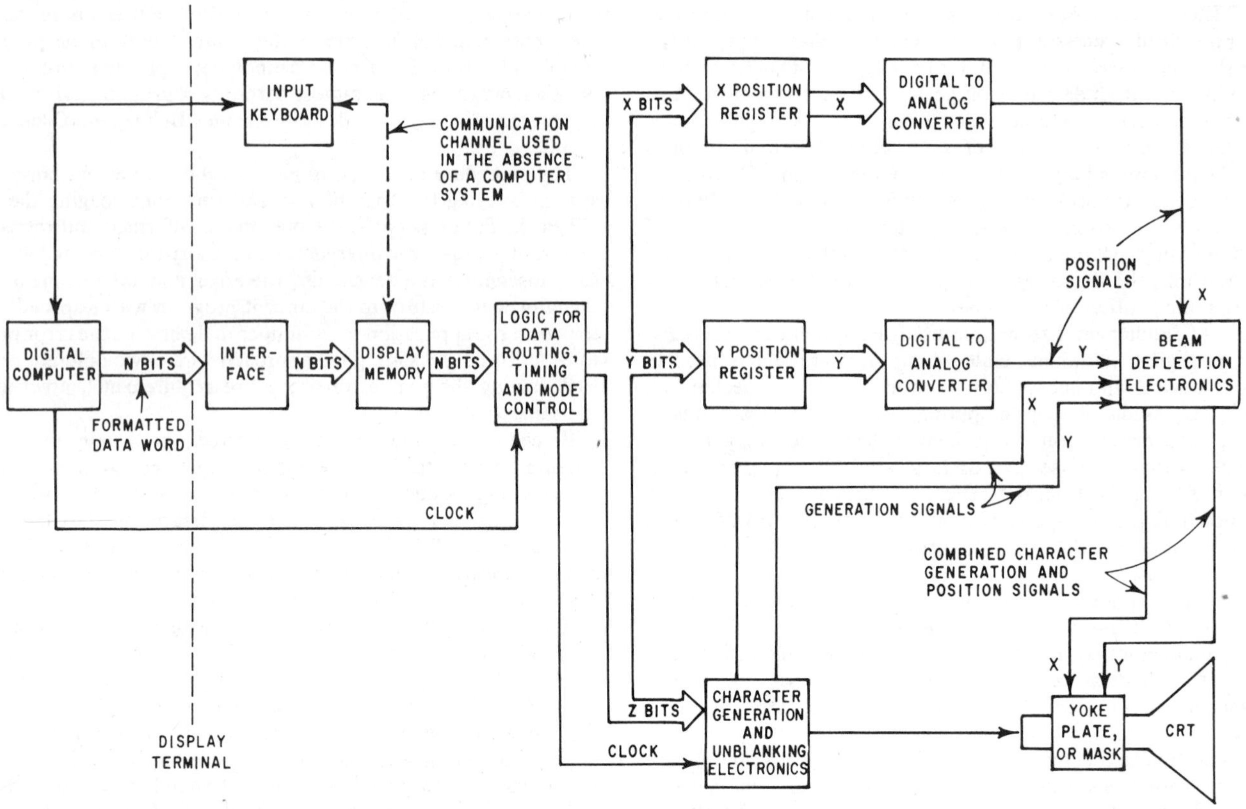

Fig. 1. Block diagram of cathode-ray-tube display system. The system can be simplified to provide local computing with use of a microprocessor-based data system.

Fig. 2. Computer graphics system, with a microprocessor based data system, allowing local computing and calculating power with high-resolution graphics. (*Tektronix, Inc., 4051 System.*)

COMPUTER (Human-Machine Communication). Telephony; Voice Recognition.

COMPUTER (Hurricane Forecasting). Hurricane Prediction.

COMPUTERIZED DIGITAL FUEL INJECTION. Automotive Electronics.

COMPUTER OPERATING SYSTEM. Generally defined as a group of interrelated programs to be used on a computer system in order to increase the utility of the hardware and software. There is a wide range in size and complexity and in application of an operating system. The need for operating systems arose from the desire to obtain the maximum amount of service from a computer. A first step was the simple monitor systems providing a smooth job-to-job transition. Modern operating systems contain coordinated programs to control input/output scheduling, task scheduling, error detection and recovery, data management, debugging, multiprogramming, and on-line diagnostics.

Operating system programs fall into two main categories: (1) control programs, and (2) processing programs:

- Control Programs
  - Data management, including all input/output
  - Job management
  - Task management
- Processing Programs
  - Language translators
  - Service programs
    - Linking programs
    - Sort/merge
    - Utilities
    - System library routines
  - User-written programs

Control programs provide the structure and environment in which work may be accomplished more efficiently. These are the components of the supervisor portion of the system. Processing programs are programs that have some specific objective that is unrelated to controlling the system work. These programs use the services of the supervisory programs rather than operating as a part of them.

Data management is involved with the movement of data to and from all input/output devices and all storage facilities. This area of the system embraces the functions referred to as the input/output control system (IOCS), which frequently is segmented into two parts, the physical IOCS and the logical IOCS.

Physical IOCS is concerned with device and channel operations, error procedures, queue processing, and, generally, all operations concerned with transmitting physical data segments from storage to external devices. *Logical IOCS* is concerned with data organization, buffer handling, data-referencing mechanisms, logical-device reference, and device independence.

Job management involves the movement of control cards of commands through the system input device, their initial interpretation, and the scheduling of jobs so indicated. Other concerns are job queues, priority scheduling of the jobs, and job accounting functions.

Task management is concerned with the order in which work is performed in the system. These include management of the control facilities, i.e., central processing unit, storage, input/output channels, and devices, in accordance with some task priority scheme.

Language translators allow users to concentrate on solving logical problems with a minimum of thought given to detailed hardware requirements. In this category are higher-level languages, report generators, and special translation programs.

Service programs are needed to facilitate system operation, or to provide auxiliary functions. Sort programs, program-linking functions, tape-print and file duplication, and system libraries are within this group.

User-written programs are programs prepared specifically to assist the user in the accomplishment of his objectives.

See also **Program (Computer);** terms listed under **Data Processing.**

Thomas J. Harrison, International Business Machines Corporation, Boca Raton, Florida.

COMPUTER PROGRAMMING FLOW CHART. Programming Flow Chart (Computer).

COMPUTER (Weather Forecasting). Weather Observations and Forecasting.

COMSAT AND COMSTAR SATELLITES. Satellites (Communication).

CONCENTRATION CELL. Galvanic Cell.

CONCENTRATION CELL CORROSION. Corrosion.

CONCENTRATION (Chemical). The quantity of matter or of a particular type of matter that exists in a unit volume, as the strength of a solution in mass of solute per unit mass of solution; or in the number of moles, hydrogen ions, etc., contained per unit volume or per unit mass.

The most commonly used method of expressing concentration is by stating the percentage, that is, parts by weight of the given substance in 100 parts by weight of the stated material. There are several exceptions to this method of expressing concentration, and the units used should be carefully recorded or observed according as the reader is operator or reader respectively. Ethyl alcohol, in water mixtures, is commonly reported as a stated percent by volume, where 50.0% by volume is equivalent to 42.47% by weight. Beverage and industrial alcohol also is rated in terms of proof, 200 proof indicating pure ethyl alcohol, that is, absolute alcohol containing no water. See also **Ethyl Alcohol.** Gases in a mixture are commonly reported by volume percent, thus nitrogen in air 78.0% by volume (equivalent to 75.5% by weight).

The concentrations of substances in solution are expressed variously, thus, percent by weight of the actual material stated, percent by weight of a material calculated chemically from the actual material, grams of the actual material (or a material calculated chemically from this) per 100 milliliters of solution, gram moles (the formula weight taken in grams) of the actual material per liter of solution (this is molar or formal concentration and the abbreviations, M or F, respectively, are used to express it), gram equivalents (the equivalent weight taken in grams of the actual material per liter of solution (this is normal concentration and the abbreviation, N, is used to express it).

Since the concentration is proportional in many individual cases to an easily determined physical constant, such as specific gravity (e.g., of solutions), the index of refraction, specific rotatory power (e.g., sugar solutions, terpenes), such constants are frequently used to ascertain and express concentration data.

See also **Demal Solution; Gram-Equivalent; Gram-Molecular Weight; Mole Fraction; Mole Volume; Molal Concentration; Molar Concentration;** and **Mole (Stochiometry); Normal Concentration.**

CONCENTRATION (Hydrogen Ion). pH (Hydrogen Ion Concentration).

CONCENTRATION (Process). In the processing of various materials, it frequently is required to increase the proportion of one material in a mixture or solution by removing all or part of one or several other components. A simple example is the retrieval of impure sodium chloride from seawater by solar evaporation of the water. Concentration is a prime reason for several of the chemical engineering unit operations described elsewhere in this volume, such as *solid–solid* separations, as by screening, jigging, tabling, flotation, sublimation, and freeze-drying; or *liquid–solid* separations, as by filtering, centrifuging, drying, evaporating, crystallizing, leaching, expressing, and prilling; or *gas–solid* separations, as by gravity settling, electrical precipitation, cyclone and impingement settling; or *gas–gas* separations, as by absorption, adsorption, gaseous diffusion, chromatography, and electromagnetic methods; or *liquid–liquid* separations, as by distillation, dialysis, and extraction; or *liquid–gas* separations, as by drying, boiling, and condensing.

The process of concentration normally connotes an increase in the proportion of one material rather than a full separation of all other materials from the target material. Raw ores, for example, frequently will contain only a small percentage of the desired mineral. Thus, an ore may be concentrated from a fraction of 1% (wt) to 50% (wt) or more. The gaseous diffusion separation of $^{235}UF_6$ (required for nuclear fission reactors) from the natural occurring $^{238}UF_6$ (comprising 99.3%, wt) of the starting materials is an extreme example of concentration. Over 4,000 diffusion stages are required in a large atomic fuels plant to effect this concentration. Prior concentration of ores is known as *beneficiation.* This is described under **Iron.** See also **Classifying (Process).**

CONCENTRATION (Statistics). As used in statistics, consider: if a variable $x \geq 0$ has frequency function $f(x)$, and distribution function

$$F(x) = \int_0^x f(t)\,dt$$

the incomplete first moment is defined as

$$\phi(x) = \frac{1}{\mu_1'} \int_0^x tf(t)\,dt$$

where μ_1' is the first complete moment $\phi(\infty)$. The graph of ϕ as ordinate against F as abscissa is called the *curve of concentration.* The coefficient G defined by

$$G = \frac{\text{M.D.}}{2\mu_1'}$$

where M.D. is the mean difference, is called the *coefficient of concentration.*

The concentration curve is convex to the F axis and the more it deviates from the straight line joining (0, 0) to (1, 1), the greater is the amount of concentration of total frequency in the lower part of the range of x.

CONCENTRATOR (Solar Radiation). Solar Energy.

CONCENTRICITY CHECKING. Thickness Measurement and Gaging Systems.

CONCH (*Mollusca, Gasteropoda*). Any of numerous species of large marine mollusks (*Mollusca*). The spiral shells have a long aperture, in many species beautifully colored. The shells are used for ornaments and as horns, and the animals are sometimes eaten. See also **Mollusks.**

CONCHOIDAL FRACTURE. Mineralogy.

CONCHOID OF NICOMEDES. A rational algebraic curve of fourth order. Its equation in rectangular coordinates is $x^2y^2 = (a + y)^2(b^2 - y^2)$ and in polar coordinates, $r = a \csc \theta \pm b$. The curve can be produced geometrically as follows. Select a point on the negative Y-axis at a distance a from the origin. This is called the pole of the conchoid. Draw a secant line from this point with equal lengths b above and below the OX-axis. As this secant rotates about the pole, the points at its ends, P and P', generate the upper and lower branches of the conchoid. It is symmetric about the Y-axis and the X-axis is its asymptote. The appearance of the curve differs as a and b vary.

The point $(0, b - a)$ on the lower branch is a node, cusp, or conjugate point as $a \lesseqgtr b$ (see **Singular Point of a Curve**).

The name of the curve comes from its shell-like shape (*L. concha*, shell). The one described here was discovered by Nicomedes, a Greek mathematician who lived about 240 B.C., and who used it to solve the Delian problems of angle trisection and cube duplication.

Actually, conchoid is often used to describe a general class of algebraic curves which can be constructed in a similar way. They can be generalized by using a secant through a conic section. The limaçon, for example, can be produced in this way. See also terms listed under **Mathematics.**

CONCHOLOGY. The study of molluscan shells. The word is sometimes applied to the study of mollusks generally but this field is more accurately called malacology.

CONCH SHELL. Invertebrate Paleontology.

CONCORDANCE. In the theory of ranking, a coefficient measuring the agreement among a set of ranks. If m rankings of n objects are arranged one above the other and the rankings summed for each object; and if S is the sum of squares of these n sums measured from their mean value $\frac{1}{2}m(n + 1)$, the *coefficient of concordance* is given by

$$W = \frac{12S}{m^2(n^3 + n)}$$

CONCRETE. Concrete is a mixture of fine and coarse aggregates firmly bound into a monolithic mass by a cementing agent. The cement ordinarily employed for concrete is the standard Portland cement. The aggregates are usually sand and crushed stone or gravel. Crushed slag or cinders are used in special kinds of concrete. The formation of concrete can be thought of as a process in which the voids between the particles of coarse aggregate are filled by the fine aggregate, and the whole is cemented together by the binding action of the cement. The nature of Portland cement in this respect is described under **Cement.**

Due to its strength, permanency, and relatively low cost, concrete is one of the most important building materials employed in modern construction. It is widely used for foundations of all types, buildings, bridges, dams, retaining walls, highways, and other purposes too numerous to mention. However, the success of concrete in meeting any particular set of conditions, depends upon the proper correlation of many factors bearing on the selection and mixing of the materials, the placing of the concrete, and the original design. Concrete is strong in compression, but relatively weak in tension. Therefore structures in which the concrete is likely to be in tension must be reinforced with steel rods, which carry the tension. For strong permanent concrete, the aggregates should be clean, coarse, and well graded. River or coarse sand is better than pit sand, and should always be used where possible. The table gives typical concrete mixes, with the characteristics of each.

This table is suitable for preliminary estimates only or for small amounts of concrete, but it should be remembered that research and development in the science of concrete proportioning have advanced to the point where little short of a laboratory analysis can establish the best and most economical mix for a given condition. The strength of the cementing agent is a function of the water-cement ratio, and therefore strength will vary widely depending upon the amount of water used. To obtain maximum strength, the water-cement ratio should be kept as low as possible. Type and gradation of aggregate, moisture content of the sand, and water-cement ratio are typical factors taken into account in a complete analysis for the specification of large amounts of concrete work.

Concrete should be transported rapidly from the mixer to the forms, so that no initial set will have occurred before the concrete is placed in its final position. It is necessary to place concrete in the forms with care to prevent segregation of the lighter and heavier parts. This precludes dropping the concrete into place from any height. After the "green" concrete has been poured, it should be cured, or hardened, slowly over a period of about a week, during which time it should be protected from vibration, freezing, and a too-rapid rate of drying out. Strengths of concrete are usually classified on the basis of 28-day strength. Most of its strength will be acquired in this time but there is a slow increase in strength for a much longer period thereafter.

For small projects, particularly around the home, a variety of premixed dry cements and concretes are available. These simply require the addition of the requisite amount of water and adequate mixing.

Prestressed concrete elements for use in construction has increased severalfold since World War I. See **Prestressed Concrete.** The concrete block industry also has expanded greatly during the same period, replacing brickwork and stonework for many applications because of economy and speed. In addition to the usual cement, sand, and gravel, concrete blocks are also available with several other ingredients to lower the weight and density of the blocks, increase thermal insulative qualities, and to add color. See also **Brickwork; Ceramics.**

CONCRETE BRIDGE. Bridge (Structural).

CONCRETE (Prestressed). Prestressed Concrete.

CONCRETION. A term used in petrology and geology for spheroidal or discoidal aggregates formed by the secretion of silica, calcium carbonate, gypsum, or other chemical compounds around an original nucleus. See also **Oölite;** and **Accretion (Geology).**

DATA ON CONCRETE MIXES TO YIELD 1 CUBIC YARD OF CONCRETE

Mixture	Cement, Sacks	Sand, Cu. Yd	Stone, Cu. Yd.	Application	Weight, Tons per Cu. Yd.
1 : 2 : 3	7	0.51	0.77	Roofs, sills, tanks, tunnels	2
1 : 2 : 4	6	0.44	0.88	R. C. floors, beams, and columns	2
1 : 2½ : 4	5.6	0.52	0.83	Building walls	2
1 : 3 : 5	4.7	0.52	0.86	Foundations and footings	2
			Cinders, Cu. Yd.		
1 : 2 : 4	6.6	0.49	0.98	R. C. floors	1.5
			Slag, Cu. Yd.		
1 : 2 : 4	6.6	0.49	0.98	R. C. floors	1.6

CONCURRENT OPERATION (Computer System). The performance of several actions during the same interval of time, although not necessarily simultaneously. Compare with **Parallel Operation (Computer System).** Multiprogramming is a technique that provides concurrent execution of several tasks in a computer. See **Multiprogramming (Computer System).**

CONCUSSION. Brain Injury.

CONDENSATE. A vapor may be reduced to liquid by removal of such portion of the latent heat of evaporation as it may contain. The act is called condensation, and the liquid is condensate. It is the property of vapors that the condensate is dense as compared to the vapor, which, in condensing, formed it; that is, there is considerable shrinkage of volume upon a reversion to the liquid state. The thermal condition of the condensate immediately upon formation is that of saturated liquid at the temperature of vapor. See also **Distillation.**

CONDENSATE (Boiler). Boiler; Feedwater (Boiler).

CONDENSATION (Atmospheric). Precipitation and Hydrometeors.

CONDENSATION (Colloidal). Colloid System.

CONDENSATION NUCLEI. Hygroscopic.

CONDENSATION TRAIL. Precipitation and Hydrometeors.

CONDENSER MICROPHONE. Microphone.

CONDENSER (Microscope). Microscope.

CONDENSER MODULATOR. Essentially a capacitor microphone which is driven by some form of electromechanical or electroacoustic transducer. The resulting change in capacitance is used as a variable impedance element in the modulated stage.

CONDOLUMINESCENCE. Hydrogen (Fuel).

CONDOR (*Aves, Falconiformes*). A large vulture of the Andes of Peru and Chile. This bird reaches a length of more than 4 feet (1.2 meters) and a wing expanse of 10 feet (3 meters) and is among the largest birds now existing. A smaller condor has been reported from Ecuador and the California vulture, *Gymnogyps californianus*, is also called a condor. The last species occurs in lower California, southern California, and east to Arizona, living in the mountains. It is sometimes larger than the condor of South America. All of the condors eat carrion.

While some species of condor are of particular concern to naturalists because they have, until recent years, been a disappearing form, the condors are not considered as friends by the Indians of the Andes. The birds attack deer and vicuna and these animals are of great economic importance to the natives. The birds also may light on the backs of larger cattle and, aided by the sharp, huge, curved claws and strong bill and jaws, will tear flesh from the cattle. The animals may die from these injuries and the birds return to consume the carcass. These birds are tenacious, ravenous, and tough.

The bill is long and slender; the tongue has serrated edges. The body is rather slender; the neck is bare with a white ruff. The naked head is elongated and rugose, with nostrils that can scent a carcass miles away. The condor is gluttonous and may eat until it cannot fly. Under such conditions, in order to escape danger, the bird may kick at and scratch its throat area until a portion of the food is regurgitated, thus lightening the body weight. The bird has been known to eat from 10 to 16 pounds (4.5 to 7.3 kilograms) of flesh without disturbing in any way its desire for food on a subsequent day. When the bird has overeaten, the Indians take revenge on its state of immobility at which time they can be more easily caught and killed.

The condor roosts and lives at high altitudes, far above the habitat of most mammals. However, from these high altitudes, the bird senses the presence of life below and is exceedingly keen in its senses for determining prospective dietary items. Vultures, hawks, and condors, in particular, are known to possess the keenest senses of sight and smell of any vertebrate. These birds are endowed with two focus points in each eye. For objects nearby, monocular vision is used; for distant objects, binocular vision is used. The eyes are situated laterally and are capable of numerous types of movements.

The egg is large and white, about $3\frac{1}{2}$ inches (9 centimeters) in length and hatches in about 7 weeks. The young bird is able to fly at 1 year, but remains with the parents for at least another year.

Coloration of the condor varies with species. Usually gray-black or brown with the feathers in a somewhat tile-like pattern. Some birds are brilliantly black all over. The male's wings are about half-white in most cases. The feather tips are white with spots, making the male the more ornamental of the two sexes. The plumage is hard and can provide protection from bullets or arrows that do not come in directly from short distances. The tail has 12 feathers and is short, black, tapering to a thin edge. Nests are constructed of sticks and located on the ground. Some of the largest California condors measure about 9 feet (2.7 meters) across wing tips. They weigh from 20 to 25 pounds (9 to 11 kilograms). This particular species is a dark gray-brown with white under the wings. There is a white ruff around the naked neck. These birds breed only every other year. See also **Falconiformes.**

CONDYLE. A rounded prominence on a bone, associated with a joint.

CONDUCTANCE UNITS. Units and Standards.

CONDUCTION ELECTRON. Electron.

CONDUCTION (Heat). Heat Transfer.

CONDUCTION (Nerve). Brain and Nervous System.

CONDUCTIVE DEAFNESS. Hearing and the Ear.

CONDUCTIVITY MEASUREMENT (Electrolytic). Electrolytic Conductivity Measurements; Oscillometry.

CONE. A solid formed by a closed conical surface, the lateral surface of the cone, and a plane, called the base of the cone, cutting the surface in a closed curve. The names cone and conical surface are frequently used interchangeably but it is proper to distinguish between them for the former is a solid and the latter a surface.

The perpendicular distance from the vertex of the cone to its base is the altitude and the straight line from vertex to base is the axis. If the axis is perpendicular to the base, the solid is a right cone; otherwise, it is oblique. A circular cone has a circle for a base. If, in addition, it is a right cone, it is called a right circular cone and also a cone of revolution for it may be produced by revolving a right triangle about one of its legs as an axis. The hypotenuse of the triangle in any position is the slant height of the cone.

A plane which cuts all elements of a cone divides the solid into two parts: another cone and a truncated cone. If the cutting plane is also parallel to the base, the latter is called the frustrum of a cone. It has both a lower and an upper base; its altitude, lateral surface, and slant height are defined as for a cone.

The following relations hold for a cone of revolution: $S = sC/2 = \pi Rs$; $A = \pi R(R + s)$; $V = \pi R^2h/3$ and, for the frustrum of such a cone: $S = s(C + c)/2$; $V = \pi h(R^2 + Rr + r^2)/3$. In these equations S is lateral areas; V, volume; A, total area; s, slant height; h, altitude; r, R and c, C radius and circumference of upper and lower base, respectively.

The volume of any cone is given by $V = Bh/3$ and the volume of any frustrum by $V = h(B + \sqrt{Bb} + b)/3$, where b, B refer to the areas of upper and lower bases.

See also **Conic Section; Conical Surface.**

CONE-IN-CONE STRUCTURE. A concretionary structure frequently observed in limestone, dolomites and other sedimentary rocks which originated as fine-grained sediments. As the term suggests the structure is formed by a series of concentric cones, the result of radial crystallization around a common axis, probably due to differential pressures.

CONES AND RODS (Eye). Vision and the Eye.

CONFABULATION. Amnesia.

CONFIDENCE INTERVAL. If a parameter θ of a frequency or probability distribution is estimated by a statistic t, the value of t calculated from any given sample will deviate from the value of θ. It is, however, often possible to set limits round t (dependent on the sample values) between which the value of θ may be asserted to lie with assigned probability. The range between these limits is a confidence interval. Customary probability levels for the purpose are 95% and 99%; for example, with 10 observations from a normal (Gaussian) distribution, if the sample mean is m and the sample standard deviation, calculated as $\Sigma^n (x - \bar{x})^2/(n - 1)$, is s, the true mean may be asserted to lie in the interval $m \pm 2.23s$ with confidence 95%, meaning that if we always make such an assertion in such cases we shall be right 95 times in 100.

CONFIDENCE LEVEL. Confidence Interval.

CONFLUENCE. Given a linear differential equation of Fuchsian type (see **Fuchs Theorem**), certain limiting processes may be carried out so that two or more of its singular points coalesce to form a singular point of a more complicated character. This procedure is called confluence, and refers to the rate at which adjacent flow is converging along an axis oriented normal to the flow at the point in question. Its importance arises from the fact that most of the linear differential equations of mathematical physics can be obtained by confluence from a single differential equation with five regular singular points. This means that the general solution to the former can be specialized to give solutions for all the rest. See the **Gauss Hypergeometric Equation.**

In natural coordinates, the confluence may be measured by

$$-\frac{\partial V_n}{\partial n} \quad \text{or} \quad -V\frac{\partial \psi}{\partial n}$$

where V is the speed of the wind, the n axis is oriented 90 degrees clockwise from the direction of the wind vector, V_n is the wind component in the n direction, and ψ is the wind direction, measured in degrees clockwise from a reference direction.

CONFLUENT. A term used by geologists and physiographers to designate a stream that unites with another; especially applied to streams nearly equal in size.

CONFORMAL MAP (or Isogonal Map; Orthomorphic Map). A map that preserves angles, i.e., if two curves intersect at a given angle, the images of the two curves on the map also intersect at the same angle. On such a map, at each point, the scale is the same in every direction. Shapes of small regions are preserved, but areas are only approximately preserved (the property of area conservation is peculiar to the equal-area map).

The most commonly used conformal map is probably the Lambert projection, with standard latitudes at 30° and 60°N. On the standard latitudes, the scale is exact; between them, it is decreased by not more than about 1%; outside them, distortion increases rapidly. The Mercator and stereographic projections are also conformal maps.

CONFORMAL MAPPING. A mapping of one region upon another that is one-to-one and continuous, and such that angles are preserved. Thus let arcs C_1 and C_2 in one region intersect in a point P; and let these be mapped into arcs C_1' and C_2' intersecting in P'. Let P_1 be any point on C_1, mapped into P_1' on C_1'. Let ϕ be the angle between tangents to C_1 and C_2 at P, ϕ' that between tangents to C_1' and C_2' at P'. Preservation of angles means that $\phi' = \phi$. In case the regions are two-dimensional and simply connected, they can be regarded as regions of two complex planes, and the mapping is then defined by $z' = f(z)$, where $f(z)$ is analytic in z throughout the region.

See also **Mapping;** and terms listed under **Mathematics.**

CONFORMAL REPRESENTATION. Consider the real variables x, y, u, v and suppose $w = u + iv$ is a single-valued function of the complex variable $z = x + iy$. Now plot z and w in two coordinate planes: the z-plane, with x, y as axes and the w-plane, with u, v as axes. Wherever w is analytic, each point in the z-plane is related to a point in the w-plane and one says that the z-plane is mapped onto the w-plane by the transformation $w(z)$. Any infinitesimal figure in the w-plane becomes a similar figure in the z-plane with angles and proportions preserved. This is the reason for the terms conformal representation or mapping. Changes of size, called magnification, and displacement of figures frequently occur in such mappings.

As a simple example, consider $w = z^2$ and apply it to the area shown in the figure. Polar coordinates are useful for we see that

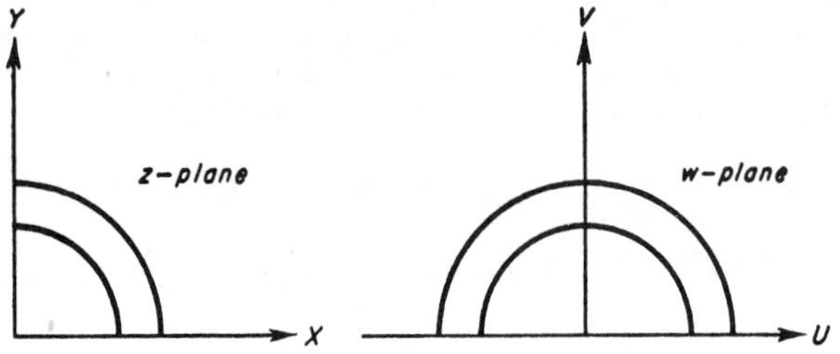

Conformal representation.

$z = re^{i\theta}$, $w = r^2e^{2i\theta}$, the modulus of w equals r^2 and its amplitude is 2θ. The two arcs in the z-plane become semicircles of radii a^2, b^2 in the w-plane; the straight lines on the X, Y-axes remain straight lines but they are now on the U-axis. The transformation is conformal everywhere within and on the boundary of the curves but the magnification is $2|z|$.

Now suppose the two arcs in the z-plane coincide ($b = 0$) so that the figure becomes one quadrant and one semicircle, respectively. The mapping is no longer conformal at the origin, which is a singular point, for the magnification vanishes there and the angles become $\pi/2$ and π, respectively.

In a conformal representation, circles always transform into circles with straight lines as limiting cases. However, a hyperbola becomes a lemniscate, showing that the degree of the equation for the curve is not an invariant of the process.

See also **Mapping; Representation (Single-Valued);** and terms listed under **Mathematics.**

CONFORMITY. With reference to industrial and scientific instruments, the Scientific Apparatus Makers Association defines conformity (of a curve) as the closeness to which it approximates a specified curve (e.g., logarithmic, parabolic, cubic, and so on). Conformity is usually measured in terms of *nonconformity* and expressed as *conformity;* e.g., the maximum deviation between an average curve and a specified curve. The average curve is determined after making two or more full range traverses in each direction. The value of conformity is referred to the output unless otherwise stated. As a performance specification, conformity should be expressed as *independent conformity, terminal-based conformity,* or *zero-based conformity.* See accompanying diagrams. When expressed simply as conformity, it is assumed to be independent conformity. See also **Linearity.**

CONGENITAL CHLORIDORRHEA. Diarrhea.

CONGENITAL DISORDER OR DISEASE. This is a condition that exists at birth, even if not immediately observable or diagnosable. The condition may be the result of genetic errors (hereditary or drug-induced), infection, etc. Several congenital conditions are described in this encyclopedia. Among the entries dealing with these disorders are: **Albinism; Anemias** (sickle cell disease); **Brain and Nervous System** (Huntington's chorea); **Colitis** (Crohn's disease); **Diabetes Mellitus; Diarrhea** (congenital chloridorrhea); **Embryo** (drugs effects, e.g., thalidomide; effects of bacteria, e.g., spirochetes of syphilis, etc., that cross the placenta); **Endocrinology** (Phenochromocytoma); **Hemophilia, Heredity; Kidney and Urinary Tract** (polycystic disease); **Lipidoses** (disturbance of lipid metabolism—Gaucher's disease, Niemann-Pick disease, Tay Sachs disease, generalized gangliosidosis); **Muscular**

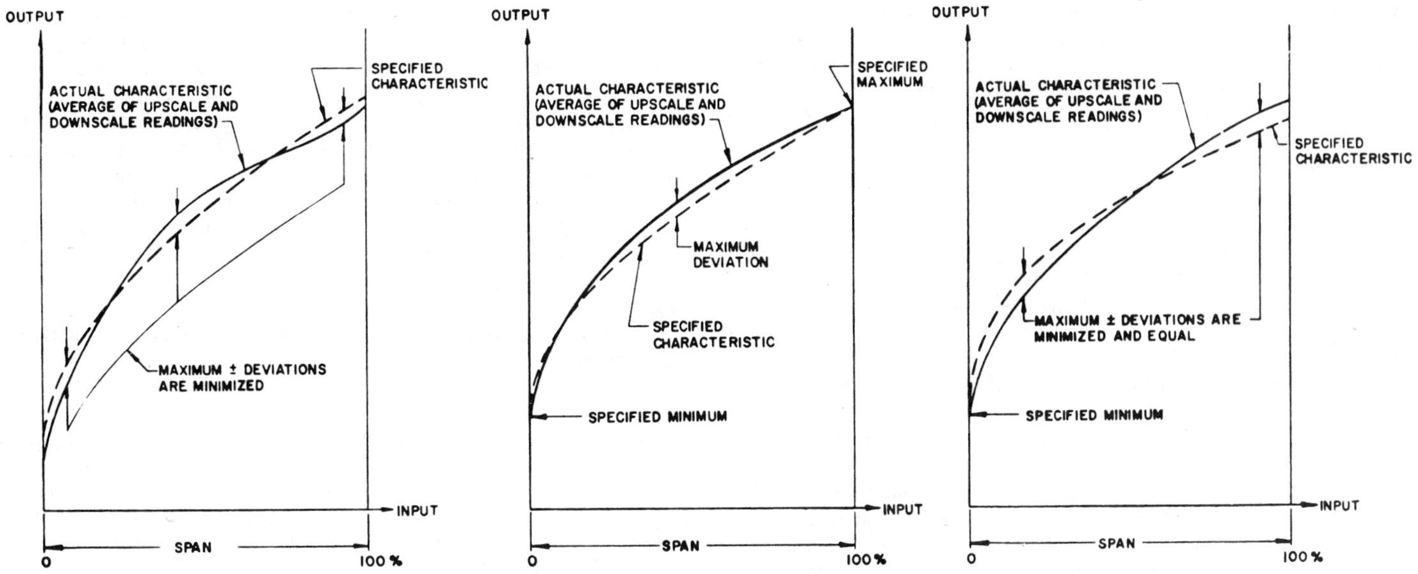

(a) Independent conformity (b) Terminal-based conformity (c) Zero-based conformity

Fundamental relationships pertaining to conformity.

Dystrophy; Myopathy; Pancreas (pancreatitis); **Porphyria; Rubella** (Down's syndrome; mongoloidism); **Respiratory System** (cystic fibrosis); and **Schizophrenia.** See also **Immune System and Immunology.**

References

Brady, R. O.: "Inherited Metabolic Diseases of the Nervous System," *Science*, **193**, 733–739 (1976).

Fuchs, F.: "Genetic Amniocentesis," *Sci. Amer.*, **242**, 6, 47–53 (1980).

Kolata, G. B.: "Thalassemias: Models of Genetic Diseases," *Science*, **210**, 300–302 (1980).

Marx, J. L.: "Restriction Enzymes: Prenatal Diagnosis of Genetic Disease," *Science*, **202**, 1068–1069 (1978).

Newman, A. H., Freeman, F. N., and K. J. Holzinger: "Twins: A Study of Heredity and Environment," Univ. Minnesota, Minneapolis, Minnesota, 1973.

Omenn, G. S.: "Prenatal Diagnosis of Genetic Disorders," *Science*, **200**, 952–958 (1978).

Scriver, C. R., et al.: "Genetics and Medicine: An Evolving Relationship," *Science*, **200**, 946–952 (1978).

Staff: "Identical Twins Reared Apart," *Science*, **207**, 1323–1328 (1980).

CONGENITAL DISORDERS (Heart). Heart and Circulatory System (Human).

CONGENITAL ICHTHYOSIS. Inbreeding.

CONGERS. Eels.

CONGESTIVE CARDIOMYOPATHY. Heart and Circulatory System (Human).

CONGESTIVE HEART FAILURE. The general term heart failure is used to describe a limited ability of the heart to furnish a sufficient supply of oxygenated blood to meet the requirements of peripheral tissues, particularly when the heart is under heavy load as in exercise conditions. The word *failure* in this context does *not* connote *complete* failure, but rather a *partial* failure of varying degrees, depending upon patient, to function in a fully normal fashion under all reasonable demand situations. The term *cardiac insufficiency* is sometimes used to identify the same situation.

Congestive heart failure (a series of disorders) can arise from a variety of root causes and is one of the most common of heart ailments. A satisfactory explanation of all of the factors which are operative in congestive heart failure would require many pages of discussion pertaining to the physiology and mechanics of the heart. Generally, it can be observed that congestive heart failure results from factors that interfere with the heart muscle's adaptive (to load) mechanisms. There is essentially a loss of strength—a decreased ability to accomplish work—of the muscle. These adaptive mechanisms involve the contractility characteristics of the heart, for it is in contracting and expanding that the heart can function as a pump. See **Heart and Circulatory System (Human).** One of the principal drugs used in the treatment of congestive heart failure is digitalis, an agent that increases the contractility of heart muscle. In a failing heart, when contractility is reduced, a lower stroke output is achieved relative to any given left ventricular filling pressure. The heart does not have full ability to match output with input. A normal heart will do more work, that is, it will increase cardiac output with an increase in left ventricular filling pressure. Further, a normal heart will change stroke volume to compensate for alterations in heart rate, that is, changes in heart rate will not significantly alter cardiac output. But, in the failing heart, in which there is a low basal cardiac output, the stroke volume will remain relatively constant and thus changes in heart rate are markedly affected by changes in cardiac output. In the total heart-circulatory system, these abnormalities cause profound problems.

Physiologists for many years have used the terms "left heart failure" and "right heart failure" to reflect two specifically different consequences. In left-side congestive heart failure, where the left ventricle is primarily involved, the left atrial pressure is elevated, leading to symptoms of pulmonary congestion and, in more advanced cases, acute pulmonary edema. In right-side congestive heart failure, the right atrial pressure is elevated, leading to symptoms of systemic venous hypertension and congestion. As congestive heart failure progresses, both sides ultimately are involved. The most common cause of right heart failure is left heart failure.

A person with congestive heart failure may have few symptoms if a normal life style is followed. But, unless diagnosed and treated, the condition tends to worsen. The condition also may be suddenly revealed as the result of some insult to the heart, such as acute myocardial infarction and, when present, congestive heart failure can seriously add to the complexities of recovery and survival.

In *left-sided congestive heart failure*, an early symptom is *dyspnea*, a feeling of breathlessness and increased effort required for breathing. While the mechanism is poorly understood, it is known that this dyspnea is related to increased interstitial pulmonary edema, which causes a stiffness of the lungs and hence a greater breathing effort. Exercise accentuates this condition. The physician will carefully differentiate this complaint from the condition of angina pectoris where a patient fears the pain of taking deep breaths and pants for air. Accurate diagnosis is of the utmost importance because the treatment of congestive heart failure differs markedly from that of angina pectoris. See also **Ischemic Heart Disease.** There are also numerous other factors, unrelated to congestive heart failure, that can contribute to dyspnea, such as lack of physical fitness, obesity, and chronic lung disease. Electrocardiography is frequently indicated.

A somewhat more definitive clinical feature of left-sided congestive heart failure is *orthopnea* (dyspnea that appears when the patient lies down). In congestive heart failure, this results from increased venous return to the heart, a condition that places extra commands on a failing left ventricle. However, persons with chronic obstructive pulmonary disease and obese people also frequently find that they can breathe easier when head and shoulders are elevated rather than in a prone position. A dry, nonproductive, hacking cough (*orthopneic cough*) that is relieved by sitting up is indicative of increased pulmonary congestion and is easily differentiated from the morning cough of bronchitis. For further confirmation, radiologic studies may be indicated. There also may be paroxysmal nocturnal dyspnea. In this instance, the patient is sharply awakened after a few hours of sleep in an episode of gasping for air.

Acute pulmonary edema is the hallmark of more mature left-sided congestion. In this state, there is movement of fluid into the interstitial and alveolar spaces. See **Respiratory System.** This condition can occur suddenly in persons with borderline pulmonary congestion with very frightening consequences—coughing, wheezing, breathlessness, intense anxiety, occasional slightly bloody sputum, and sometimes a feeling of impending death.

In *right-sided congestive heart failure*, an elevated right atrial pressure will lead to signs of systemic venous hypertension and congestion. The physician will carefully examine the neck veins. Distension of the neck veins (*hepatojugular reflux*), caused by increased venous return to the chest, is an indication of right ventricular dysfunction. In right-sided congestive heart failure, the liver usually enlarges. Edema is another common manifestation. This results from a passive venous congestion and retention of salt and water.

When congestive heart failure is first recognized, the physician will attempt to correct underlying causes. Drugs may be used to alleviate the problems of arrhythmias. See **Arrhythmias (Cardiac).** A pacemaker may be implanted in some patients; in others, consideration may be given to the surgical correction of certain congenital lesions that contribute to a failing heart. Attempts will be made to alleviate kidney problems. But if a patient still has problems after all conditions that may be reversible are treated, the long-term therapy will involve three basic approaches—*salt restriction*, *digitalis*, and *diuretics*. Because a failing circulation tends to retain salt, dietary input of salt will be controlled. This is a very difficult accomplishment with some patients. With the availability of stronger diuretics, there has been a trend toward relaxing the rules over minimizing salt intake daily. When a patient is in the hospital, an effort will be made to find a tolerable salt level with relation to the diuretics used. Of course, very tight control over salt intake is still required in the most severe cases of congestive heart failure. The role of diuretics is discussed in entry on **Diuretics.**

The effects of *digitalis* for increasing contractility and cardiac output in congestive heart failure have been known since the late 1700s when Withering first described the qualities of foxglove. The exact manner in which digitalis functions is not fully understood. However, it is suggested that the drug inhibits sodium-potassium activated ATPase (adenosine triphosphatase), thus reducing sodium and potassium transport across the plasma membrane and leading to an increase in intracellular sodium and an efflux of potassium from the cell (Smith and Haber, 1973). Accompanying the influx of sodium, there is an influx of calcium which is made available to the contractile element of the myofibril. It should be noted that, in the normal heart, digitalis not only increases contractility of the cardiac muscle, but also causes peripheral vasoconstriction, with no resulting marked change in cardiac output. On the other hand, in the failing heart of congestive heart failure, the heart is enlarged, with accompanying elevated peripheral vascular resistance. Digitalis in this instance reduces heart size, increasing cardiac output. This improves the heart's pumping efficiency, lowering oxygen consumption for a given amount of work done.

In recent years, digitalis prepared by leaf (foxglove) extraction has been largely replaced by the pure glycosides, *digitoxin* and *digoxin*. In North America, digoxin is probably the most commonly used. Since nearly 40% of the drug is excreted by the kidney each day, maintenance therapy requires replacement of that amount daily. When first administering digoxin, blood levels of the drug will be checked frequently and in about one week, an effective regimen for achieving and maintaining the proper level will be achieved. There are several side effects of the drug, known as digitalis toxicity. Symptoms include gastrointestinal disturbances, fatigue, headache, and dizziness. There may be disturbances in color vision. Sometimes the first sign of digitalis toxicity are cardiac arrhythmias, which require immediate attention and adjustment of drug blood levels, usually commencing with complete withdrawal of the drug until arrhythmias have been successfully resolved. See also **Arrhythmias (Cardiac).**

In addition to the foregoing drug therapies, in some patients the use of vasodilators may be indicated.

For associated topics, see the list at the end of the entry on **Heart and Circulatory System (Human).**

References

Brown, D. D. and R. P. Juhl: "Decreased Bioavailability of Digoxin due to Antacids and Kaolin-Pectin," *N. Engl. J. Med.*, **295**, 1034 (1976).

Burg, M. B.: "Mechanisms of Action of Diuretic Drugs," in "The Kidney," (B. M. Brenner and F. C. Rector, Jr., editors), Saunders, Philadelphia, 1976.

Frazier, H. S., and H. Yager: "The Clinical Use of Diuretics," *N. Engl. J. Med.*, **288**, 246 and 455 (1973).

Leahey, E. B., et al.: "Interaction between Quinidine and Digoxin," *JAMA*, **240**, 533 (1978).

Lindenbaum, J., et al.: "Variation in Biologic Availability of Digoxin from Four Preparations," *N. Engl. J. Med.*, **285**, 1344 (1971).

Nora, J., and A. H. Nora: "Genetics and Counseling in Cardiovascular Diseases," Charles C. Thomas, Springfield, Illinois, 1978.

Smith, T. W., and E. Haber: "Digitalis," *N. Engl. J. Med.*, **289**, 945, 1010, 1063, 1125 (four parts) (1973).

CONGLOMERATE. Conglomerate, called in older writings "pudding-stone," consists of aggregates of gravel or pebbles with a matrix of sand and cement. The proportion of pebbles and matrix may vary considerably both as to amount and actual or relative size of the component material. The common cementing materials are silica, calcite, and iron oxide.

Consolidated glacial debris called tillite, may consist of boulders of considerable size in a heterogeneous mixture of pebbles, clay and sand.

CONGRUENT. 1. Two geometric figures are congruent if they differ only in their position in space; that is, they are congruent with respect to a given geometry if they can be transformed into each other by transformations belonging to the group of the geometry; e.g., in Euclidean geometry by translations, rotations and reflections. 2. Two elements a, b of a ring are congruent *modulo m*, written $a \equiv b$ (mod m), if there exist elements p, q, r in the ring such that $a = mp + r$, $b = mq + r$; roughly speaking, if they leave the same remainder when divided by m. 3. Two square matrices A, B are congruent if A can be transformed into B by a congruent transformation; that is, if there exists a nonsingular matrix C such that $B = \hat{C}AC$, where $\hat{C}$ is the transpose of C.

See also **Matrix (Mathematics);** and terms listed under **Mathematics.**

CONICAL COORDINATE. A degenerate curvilinear coordinate system obtained from an ellipsoidal coordinate system. The surfaces are: spheres with center at the origin of a rectangular system and radii u (u = const.); two sets of conical surfaces with apexes at the origin, one along the Z-axis and the other along the X-axis (v, w = const.). Conical coordinates are related to rectangular coordinates by the equations

$$x^2 = \frac{u^2v^2w^2}{b^2c^2};$$

$$y^2 = \frac{u^2(v^2 - b^2)(w^2 - b^2)}{b^2(b^2 - c^2)}$$

$$z^2 = \frac{u^2(v^2 - c^2)(w^2 - c^2)}{c^2(c^2 - b^2)}$$

where $c^2 > v^2 > b^2 > w^2$.

CONICAL PROJECTIONS. The term applied to any one of that class of map projections in which the surface of the earth is projected from a point within the earth to the surface of one or more cones, and then developed on a plane. In the simple conical projection, a cone is placed tangent to the surface of the earth along the central parallel of latitude for the region to be mapped. The surface features are then projected onto the cone, the cone is cut along with element which represents the meridian of longitude 180° from the central longitude of the mapped region, and the cone is rolled out on a plane. The resulting graticule will show parallels of latitude as circles concentric at the pole of rotation (usually outside the limits of the map), and meridians of longitude as straight lines converging on the pole.

Along parallels of latitude and along the central meridian, the scale of distance is uniform. The map is not conformal except close to the central parallel, and the distortion of shape increases rapidly with distance from that parallel. The simple conical projection is sometimes used for maps of small countries. For relatively large areas, the polyconic and Lambert projections are preferable conical projections.

CONICAL SURFACE. A surface generated by a moving line intersecting a fixed curve and passing through a fixed point. The line is the generatrix; the curve is the directrix; the point is the vertex; the generatrix in any position is an element of the surface. Its general equation, if the vertex is taken at the origin of a Cartesian coordinate system, is $Ax^2 + By^2 + Cz^2 + 2Fyz + 2Gxz + 2Hxy = 0$.

Depending on the curve chosen as the directrix, the surface is square, rectangular, elliptical, etc. If the directrix is an ellipse with semi-axes a and b and the generatrix is taken as the line $z = c$, the resulting equation is $x^2/a^2 + y^2/b^2 - z^2/c^2 = 0$. The axis of the conical surface is then the Z-axis and the coordinate origin is a center of symmetry. Two surfaces are thus produced, one for positive values of z and its mirror image for negative z. These are the two nappes of the conical surface.

If the directrix is a circle, $a = b$ and the surface is called right circular.

A cone is not identical with a conical surface for the former is a solid bounded by a closed conical surface and a plane cutting all of its elements.

See also **Cone; Conic Section;** and terms listed under **Mathematics.**

CONIC SECTION. The curve produced by a plane cutting a right circular conical surface, provided the plane does not pass through the vertex of the cone. The possible results are: an ellipse (or a circle as a special case) if the plane cuts all elements of one nappe of the cone; a parabola if the plane cuts only one element of the cone; an hyperbola if the plane cuts both nappes of the cone. See Figures 1 through 4.

A conic section may also be defined as the locus of a point which moves so that its distance from a fixed point, the focus, is in a constant ratio, the eccentricity, to its distance from a fixed straight line, the directrix. A chord passing through a focus and perpendicular to an axis of the curve is the latus rectum.

In Cartesian coordinates, a conic section is an algebraic equation of the second degree in two variables $Ax^2 + Bxy + Cy^2 + Dx + Ey + F = 0$. The shape of the curve is determined by the parameters $A, B, \ldots, F$ and certain limiting forms, called degenerate conics,

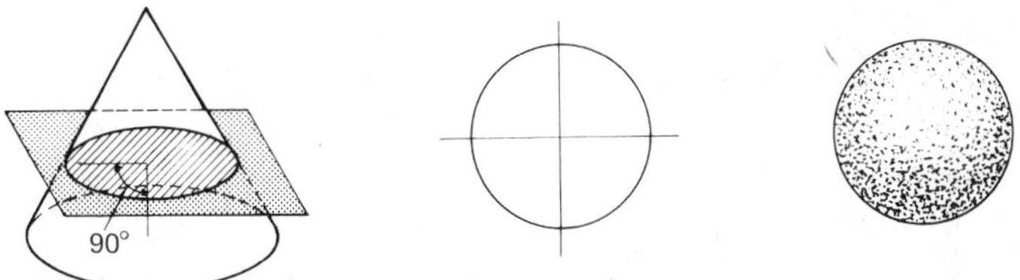

(a) Plane cuts a right circular conical surface to form circle, special case of the elipse where plane is 90° to axis of cone.

(b) Circle

(c) Sphere

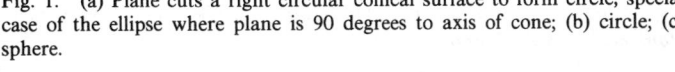

Fig. 1. (a) Plane cuts a right circular conical surface to form circle, special case of the ellipse where plane is 90 degrees to axis of cone; (b) circle; (c) sphere.

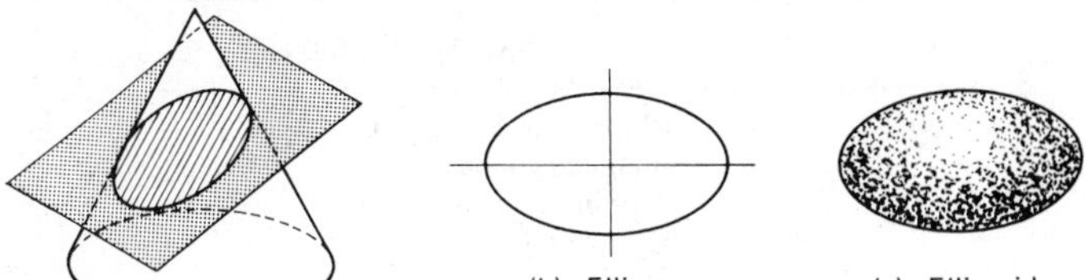

(a) Plane cuts a right circular conical surface to form an elipse.
The plane cuts all elements of one nappe of the cone.

(b) Ellipse

(c) Ellipsoid

Fig. 2. (a) Plane cuts a right circular conical surface to form an ellipse. The plane cuts all elements of one nappe of the cone; (b) ellipse; (c) ellipsoid.

may occur. These are a point, one or two straight lines, a circle, or an imaginary locus. The quantities which determine the possible cases are the invariant of the equation, I and the discriminant, Δ. These are given by the equations $I = (B^2 - 4AC)$:

$$\Delta = \frac{1}{2} \begin{vmatrix} 2A & B & D \\ B & 2C & E \\ D & E & 2F \end{vmatrix}$$

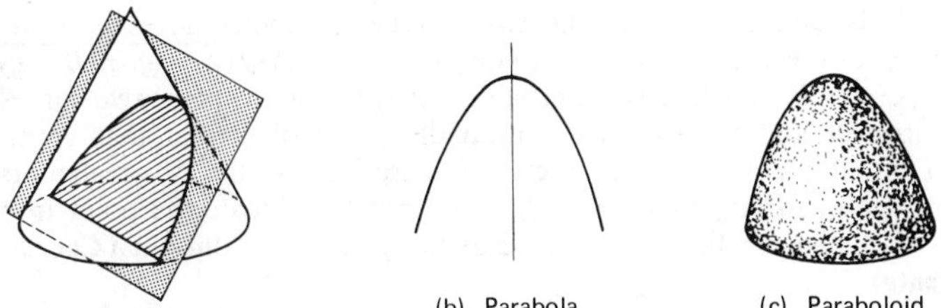

(a) Plane cuts a right circular conical surface to form a parabola. Only one element of the cone is cut.

(b) Parabola

(c) Paraboloid

Fig. 3. (a) Plane cuts a right circular conical surface to form a parabola. Only one element of the cone is cut; (b) parabola; (c) paraboloid.

The results are given in accompanying table.

The equation of a conic section in polar coordinates is $r = ep/(1 - e \cos \theta)$, where e is the eccentricity, p is the distance from the focus to the directrix, the pole is at the focus, the polar axis is perpendicular to the directrix, and (r, θ) is any point on the curve.

If the conic section is symmetric about a finite point on the plane of the curve it is a central conic. Moreover, this point called the center can be chosen at the origin of a Cartesian coordinate system and, in that case, the equation contains no terms of the first degree

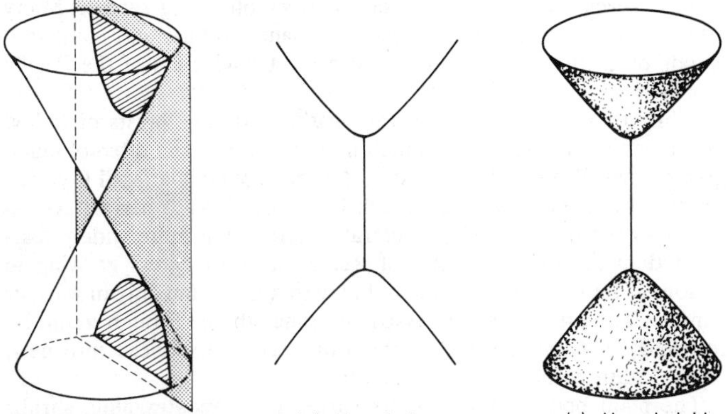

(a) Plane cut a right circular conical surface to form a hyperbola. Both nappes of the cone are cut.

(b) Hyperbola

(c) Hyperboloid

Fig. 4. (a) Plane cuts a right circular conical surface to form a hyperbola. Both nappes of the cone are cut; (b) hyperbola; (c) hyperboloid.

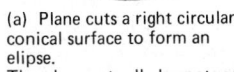

I	Δ	NATURE OF CURVE
0	$A\Delta < 0$	Ellipse; circle if $B = 0$, $A = C$
	$A\Delta > 0$	Imaginary locus
	0	A point
<0	$\neq 0$	Hyperbola
	0	Two intersecting straight lines
>0	$\neq 0$	Parabola
	0	$A \neq 0$; $H = D^2 - 4AF = 0$, one line; $H > 0$, two parallel lines; $H < 0$, imaginary locus; $A = 0$; $J = E^2 - 4CF = 0$, one line; $J > 0$, two parallel lines; $J < 0$, imaginary locus.

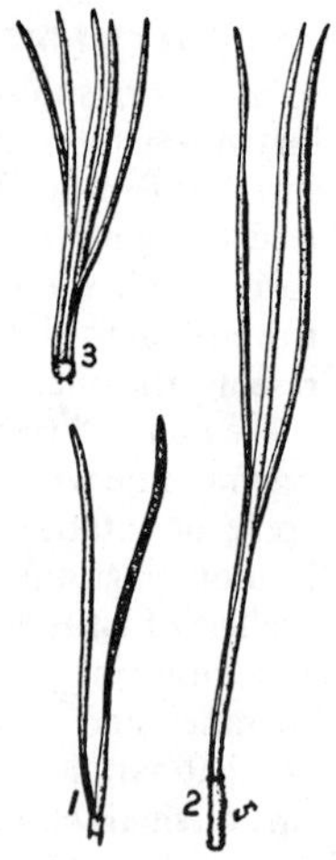

Fig. 1. Leaf clusters from different species of pine: (1) *Pinus murrayana*; (2) *Pinus ponderosa*; (3) *Pinus flexilis.*

in x and y. The central conics are the hyperbola and the ellipse but a parabola is a noncentral conic since its center is at infinity.

A system of two conics having the same foci is a system of confocal conics. Its equation in standard form is $x^2/(A - q) + y^2/(B - q) = 1$, where $A > B$ and q is a variable parameter. If q is less than B or negative, both terms on the left are positive and the resulting curves are ellipses; if $A > q > B$, the curves are hyperbolas; if $q < A$, the curves are imaginary.

It is easy to see that the two families of conics have the same foci, for the distance from the center to the focus is $\sqrt{(A - q) - (B - q)} = \sqrt{A - B}$, in both cases. Through every point in the plane two curves intersect and the curves are mutually perpendicular at that point. Thus, if the point of intersection is described by the two real roots of the quadratic equation in q, these numbers locate the point in a two-dimensional curvilinear coordinate system (see **Ellipsoidal Coordinate**).

A degenerate case of these conic families is a system of two sets of parabolas, opening in opposite directions. Since the curves are mutually perpendicular, they also may be used as a curvilinear coordinate system in two dimensions. The name parabolic coordinates is used for this system as well as for the three-dimensional case obtained by revolving the parabolas about their common axis.

See also **Cone; Conical Surface;** and terms listed under **Mathematics.**

CONIDIA. A conidium is an asexual spore, characteristic of many fungi. Commonly it is formed at the tip of a hyphal branch, which is called a conidiophore. Conidia may be one- to many-celled.

CONIFEROUS FOREST BIOME. Biome.

CONIFERS. One of the two basic groups of trees, the other being the Broad-leaves. There are about 650 species of conifers, contained within about 50 genera, and 8 families. The conifers include the dominant evergreen trees of the northern hemisphere, as well as some tropical species. Generally, conifers are trees, often of great size. Many attain a great age, as for example the Giant Sequoias and coast redwoods of California and Oregon, and the Douglas firs of the Pacific Northwest.

While they are dominant plants in the northern forests of today, in Mesozoic times they were much more numerous and often of much greater size. The petrified forest of Arizona contains a fossil Gymnosperm, *Araucarioxylon arizonicum*, from the Triassic period. As recently as the Miocene age, coniferales were much more widely scattered than at present, species of Redwood and Cypress growing in regions as far north as Greenland; today they are found only in warmer climates and often in a very restricted range there. Though naturally restricted in habitat, many of them are easily introduced into new, often distant regions, where they thrive.

The plant body of the conifers varies from low straggling shrubs to large trees. The several parts of the plant also show great diversity. Nearly all of them have tall straight stems extending to the very top of the tree, and numerous lateral branches which are progressively shorter from bottom to top of the trees, which therefore give an attractive conical shape. In many species there is a central tap root extending deep into the ground, from which smaller lateral roots arise. In other species extensive lateral roots spread out near the surface of the ground. The leaves of conifers of the northern hemisphere are either slender and needle-shaped or short and scale-like. See Fig. 1. Those of the pines are formed in fascicles or bunches of 2–5 which grow from a very short lateral branch. In spruces and firs the leaves are borne singly around the stem. In white cedar and certain other conifers the leaves are reduced to short pointed scales which are formed in pairs on opposite sides of the stem, and are pressed tightly to it. In the southern hemisphere there are several conifers with broad leaves very similar in outward appearance to those of many angiosperms. The leaves of conifers have an epidermis of thick-walled cells which give stiffness to the leaves. The stomata are sunk deep in grooves. The chlorenchyma of the leaves, the cells containing the chloroplasts, is formed of cells the walls of which are curiously infolded. These cells surround the central vein, in which are found the vascular bundles. Some species have one, others two bundles in a leaf. In the chlorenchyma, numerous resin canals are found. These are also present in the bark and in the wood of many gymnosperms. The leaves of most gymnosperms remain on the tree from 2 to 5 or more years, and even, in some species, persist for as long as 20 years. Those of the larches are deciduous, falling in the autumn, and leaving the branches bare through the winter.

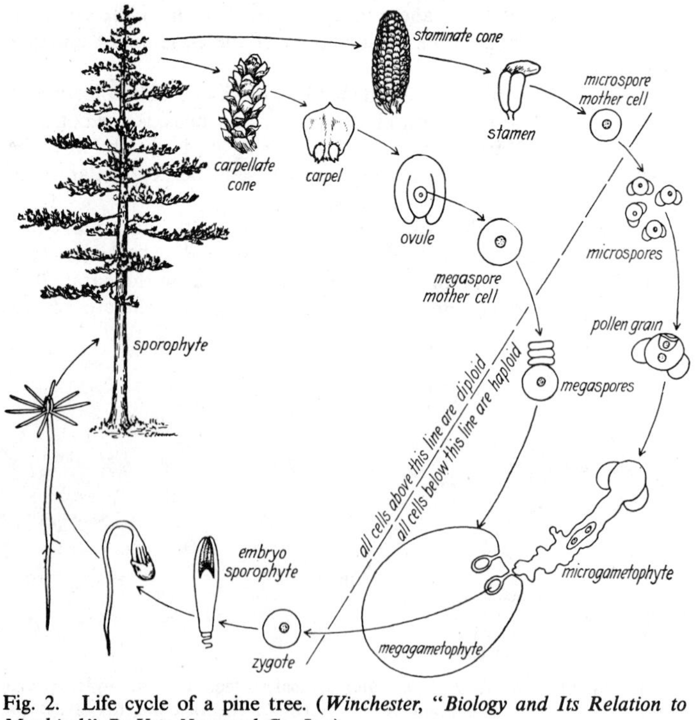

Fig. 2. Life cycle of a pine tree. (*Winchester, "Biology and Its Relation to Mankind," D. Van Nostrand Co. Inc.*)

The reproductive structures of the conifers are the cones or strobili, which are of two kinds, staminate and ovulate cones. See Fig. 2. Usually these are found on the same tree, which is therefore monoecious. The staminate cones, often called the male cones, are small and short-lived. They are generally found in clusters near the tips of the branches. A staminate cone consists of a central axis and a series of spirally arranged microsporophylls. Each microsporophyll bears two pollen-sacs or microsporangia on its lower surface. Each microsporangium contains numerous cells called microspore mother cells which divide by meiosis to form haploid microspores, or pollen grains, four from each microspore mother cell. The number of pollen grains produced is tremendous; often they are liberated from the sporangia in such quantities as to produce what are known as "sulfur showers," which cover the ground with a layer of yellow pollen, or cause the water of ponds to become turbid with the pollen grains. The pollen is carried about by the wind, often to distances of many miles. The wall of a pollen grain is composed of two layers, an inner, called intine, and an outer, exine layer. In some conifers the outer layer is separate from the inner in two places, forming conspicuous balloon-like structures which presumably are an aid in keeping the pollen grain floating in the air. Even before shedding, the nucleus of the pollen grain has divided, so that the grain at the time of shedding contains three or more cells. All but one of these are very small and last but a short time before they disintegrate, leaving thin disk-like cells pressed against one side of the pollen grain. These small cells are usually called prothallial cells, and are thought to be vestiges of extensive vegetative tissue or earlier forms. The large cell of the pollen grain divides to form two, known as the generative and tube cells.

The ovulate cones require a much longer time than the staminate to reach maturity. See Fig. 3. Often they remain on the tree for many years. The structure of the ovulate cone varies in the different genera, and is the cause of much discussion in many cases. In the pines, the genus in which the development of the ovulate structure is best known, the cone is made up of numerous scales. On the upper surface of each scale there are two ovules. The greater part of each ovule is a mass of cells called the nucellus or megasporangium. This is surrounded by an integument, a tissue which does not completely enclose the nucellus, a small opening known as the micropyle being left. It is through this opening that the pollen grains reach the surface of the megasporangium. As the ovule develops, one or sometimes more of the cells in the nucellus becomes distinct from the other cells because of its larger size and denser protoplasm. This is the megaspore mother cell. It divides by meiosis to form a row of four cells, one of which becomes the megaspore, the others degenerating. The single megaspore divides into many cells which form the female gametophyte or megagametophyte. At the end of the megagametophyte nearest the micropyle several archegonia develop. Each archegonium contains a single very large egg cell.

Fig. 3. The female (carpellate) cone of the loblolly pine, Pinus taeda; (*left*) the small spring cone; (*center*) the one-year-old cone, containing the developing seed; (*right*) the two-year-old cone, which has opened and allowed the winged seeds to escape. (*Winchester, "Biology and Its Relation to Mankind," D. Van Nostrand Co., Inc.*)

The pollen grain, carried by the wind, comes in contact with a small drop of fluid which has been secreted by cells in the region of the micropyle. As this evaporates the pollen grain is drawn down into the micropyle to the surface of the nucellus. There the pollen grain puts out a pollen tube which grows through the nucellus tissue until it reaches the tip of the megagametophyte. During this development a final division of the generative nucleus has occurred, and two male gametes or sperm nuclei are formed. When the tip of the tube reaches an archegonium these nuclei are discharged into the egg. One of the nuclei passes to the egg nucleus and fuses with it; the other disintegrates. The time required for the pollen tube to reach the egg varies greatly in different genera; in some, like the spruce and the hemlock, it is a matter of a few weeks; in others, like the pine, it is nearly a year.

The nucleus of the fertilized egg passes to the basal end, dividing twice, and forms a rosette of four nuclei. Each of these divides again, forming two tiers of four nuclei. Walls then begin to form, separating the apical tier from the other. Subsequent nuclear divisions increase the number of tiers to four, each composed of four cells. The apical tier presently divides and forms the embryo; the second tier, called the suspensor, elongates greatly, shoving the embryo down into the gametophyte tissue. There the embryo absorbs food substances from the tissues around it and matures.

The mature seed of a gymnosperm consists of a hard seed coat, developed from the integument, the nucellar tissue, and the embryo. See Fig. 4. The latter consists of a straight slender hypocotyl and two or more cotyledons, and a very small epicotyl or plumule. During its development this embryo has been surrounded by a mass of tissue called the endosperm, which is megagametophyte tissue.

Fig. 4. Pine seeds. The flattened wing attached to the main portion of the seed is a great aid to seed dispersion. (*Winchester, "Biology and Its Relation to Mankind," D. Van Nostrand Co., Inc.*)

The seeds of different conifers vary greatly in size. In some species of pine, they are large enough to be used as food for human beings. In the southwestern United States there are several species of pine which bear edible seeds. These are called piñon nuts, and are used in the same way as peanuts are. In southern Europe other species of pine yield edible seeds. One of these, called pignolea nuts, is frequently used in confectionery.

The conifers yield many other valuable products. The wood of many of them is extremely valuable, forming the principal timber used in construction work. Its use in this work is largely due to its composition. Each annual ring is composed of two distinct layers, an inner soft layer and an outer hard layer often much darker colored than the other; these give to the wood great strength and also flexibility, and allow easy driving of nails into the wood without splitting it. The wood is also used for paper pulp. Turpentine and resin are obtained from many of the conifers. Amber is a fossil resin coming from an extinct conifer. Many conifers are grown as ornamental trees, often in regions far from their natural habitat.

Numerous conifers are described in specific entries in this volume. See the following:

Arborvitae	**Hemlock Trees**	**Podocarps**
Cedar Trees	**Juniper Trees**	**Redwood (Coast)**
Cypress Trees	**Maidenhair Tree**	**Spruce Trees**
Fir Trees	**Palm Trees**	**Yew Trees**
Giant Sequoia	**Pine Trees**	

For references, see **Tree.**

CONJUGATE. 1. A special kind of singular point of a plane curve. (See **Conjugate Point.**) 2. If $u + iv$ is a complex number or a function of a complex variable, its conjugate is $u - iv$. 3. For use of the word in group theory, see **Transform.**

CONJUGATE ANGLE. Angle.

CONJUGATE DIRECTIONS (at a Point P on a Surface). The directions of the straight line joining P to a neighboring point Q on the surface and the line of intersection of the tangent planes at P and Q, in the limiting case as Q tends to coincidence with P. If these two directions are the same, the direction is said to be a *self-conjugate direction*, or *asymptotic direction*. There are two real asymptotic directions at each point of a surface for which the two principal curvatures are of opposite signs. If the two principal curvatures have the same sign then the two asymptotic directions are imaginary. If one of the principal curvatures is zero, the two asymptotic directions at the point coincide.

See also **Surface;** and terms listed under **Mathematics.**

CONJUGATE ELEMENTS OF A GROUP. In a group G an element a is said to be conjugate to an element b if there exists an element p in G such that $b = pap^{-1}$. The relation of conjugacy is easily proved to be an equivalence, so that the whole group G is thereby divided into conjugate classes. A subgroup N is called a *self-conjugate*, or *normal*, subgroup if all the conjugates of every element of n (that is, all elements of the form pap^{-1} with a in N and p in G) are included in N.

See also **Permutation Group;** and terms listed under **Mathematics.**

CONJUGATE GRADIENTS (Method of). A method of successive approximations for solving a system of linear equations which is exact after, at most, n steps. The successive approximations are so chosen that each residual is orthogonal to all preceding ones.

CONJUGATE NUMBERS. Two algebraic numbers are conjugate over a given field if they are roots of the same ireducible equation with coefficients in the field. Thus the complex numbers $a + bi$, $a - bi$ *are conjugate over the real field, since they are roots of the equation* $x^2 - 2ax + (a^2 + b^2) = 0$.

CONJUGATE POINT (Mathematics). A singular point on a curve which is the real intersection of imaginary branches of the curve. The condition which must be met is

$$\left(\frac{\partial^2 f}{\partial x\,\partial y}\right)^2 - \frac{\partial^2 f}{\partial x^2}\frac{\partial^2 f}{\partial y^2} < 0$$

All called an isolated point.

CONJUGATE SOLUTIONS. A system in equilibrium consisting of two liquid phases and two components; if the components of such a two-component system are designated as A and B, then one phase is a solution of component A in component B, and the other phase is a solution of component B in component A. Such liquids are spoken of as "partially miscible."

CONJUGATE SPACE. Linear Space (Vector Space); Linear Topological Space.

CONJUGATION. A form of sexual reproduction marked by the union of cells and the mingling of nuclear material. In the algae, *Spirogyra* (Fig. 1), one filament unites with another filament by means of conjugation tubes that connect the cells. The protoplasm of each cell becomes an isogamete. The isogametes from the cells of one fila-

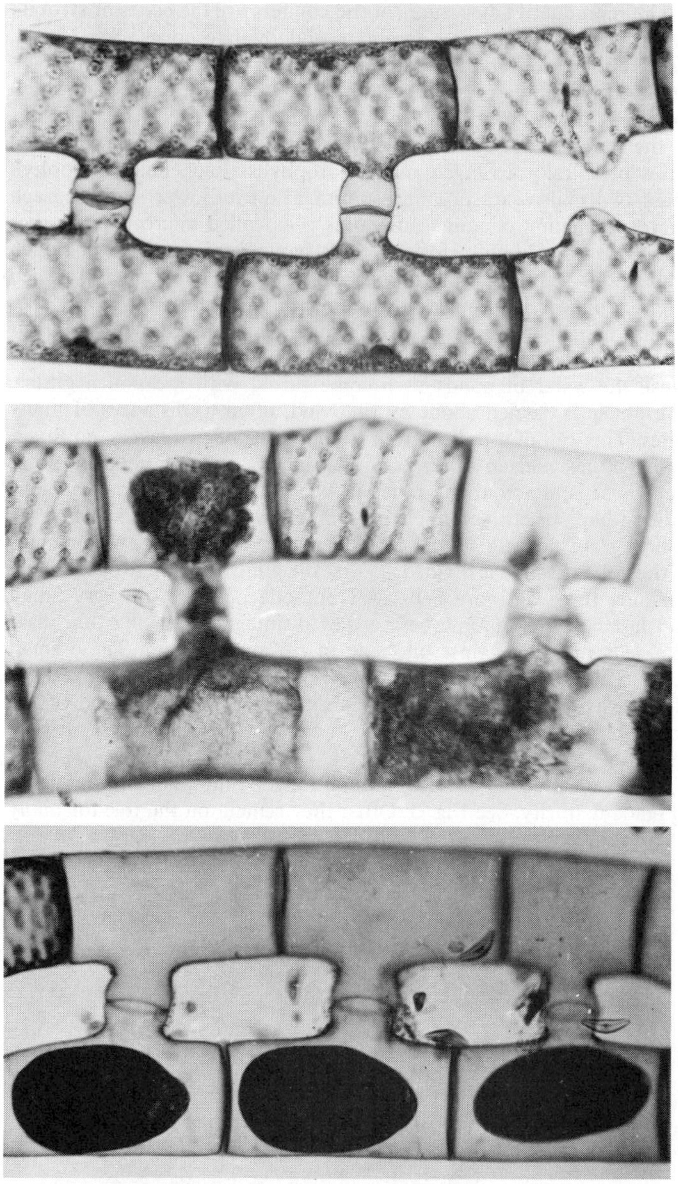

Fig. 1. Conjugation in *Spirogyra*: (top) the gametes can be seen forming in the cells and the conjugation tubes are forming; (middle) there is a union of gametes as those from one filament pass through the tube into the cells of the other filament; (bottom) the zygotes have been formed. (*A. M. Winchester*)

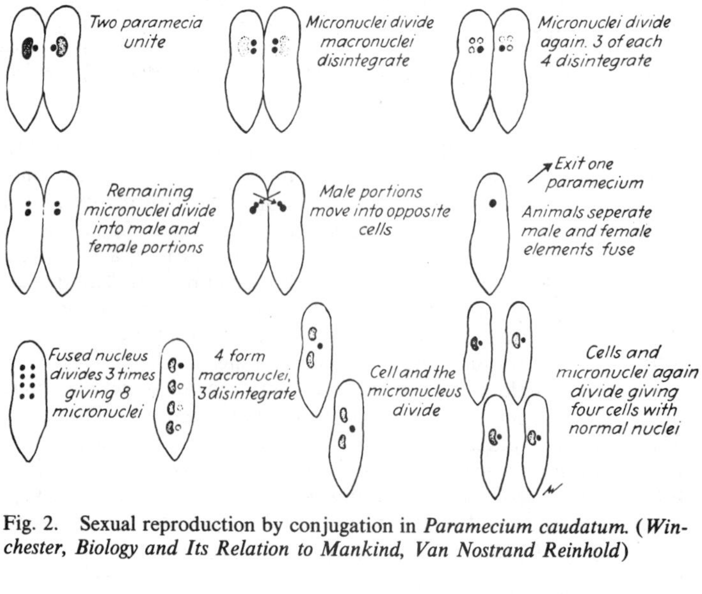

Fig. 2. Sexual reproduction by conjugation in *Paramecium caudatum.* (*Winchester, Biology and Its Relation to Mankind, Van Nostrand Reinhold*)

ment migrate down the conjugation tube into the cells of the other filament. The union of gametes forms zygotes. Conjugation is also found in yeasts, bread mold, bacteria, and other simple plants. In animals, conjugation in *Paramecium* (see Fig. 2), has been studied extensively. After a series of nuclear divisions and disintegrations, a half nucleus from one cell migrates into another cell and in turn receives a half nucleus from its partner. This achieves a variation in heredity which would not be possible if reproduction was by asexual means only.

CONJUNCTION (Astronomy). When two heavenly bodies occupy the same longitude (same degree of the Zodiac), the bodies are said to be in *conjunction*. This is a condition when the same perpendicular to the ecliptic passes through both bodies. Should both bodies have, at the same time, an identical latitude (equally far north or south of the ecliptic), then one body will fully or partially block the view of the other body from the earth. If the observer were at some other location, that is, on the moon, on a planet, or in an orbiting astronomical observatory, as examples, the phenomenon of conjunction also would occur whenever the celestial geometry from that point was just right. A conjunction as observed from the earth is termed *geocentric*; as it may be observed from the sun, *heliocentric*.

Conjunctions are most frequently mentioned in connection with the planets and asteroids. With reference to the accompanying diagram, note that the inner circle represents the orbit of an inner planet (Mercury or Venus). The middle circle is the earth's orbit. The outer circle represents the orbit of an outer planet (Mars, Jupiter, etc.). The angle at the earth between the lines drawn to a planet and the sun is known as the *elongation* of the planet. If the earth is at *E* when the inner planet is at *a*, the planet is said to be in *superior conjunction*. When the inner planet is at *b*, the planet is said to be in *inferior conjunction*. When the planet is at *g* or *h*, it is at its *greatest elongation*. The outer planet is in conjunction when at *c*; and in *opposition* when at *d*; and in *quadrature* when its elongation is 90 degrees, as noted at *e* and *f*. Elongations are termed *east* or *west*, depending upon the direction of the planet from the sun.

A *grand conjunction* occurs when 3 or more planets or stars are observed together. In explaining the Star of Bethlehem, the conjunction theory is sometimes proposed and, provided some logical adjustments in the calendar then used are made, the proposition appears feasible. A grand conjunction is mentioned in Chinese historical records as having occurred about 2500 B.C. The term conjunction is sometimes used when two bodies, such as Venus and the moon, appear in very close proximity as viewed from the earth. Astronomers normally forecast conjunctions as they will be viewed from the center of the earth.

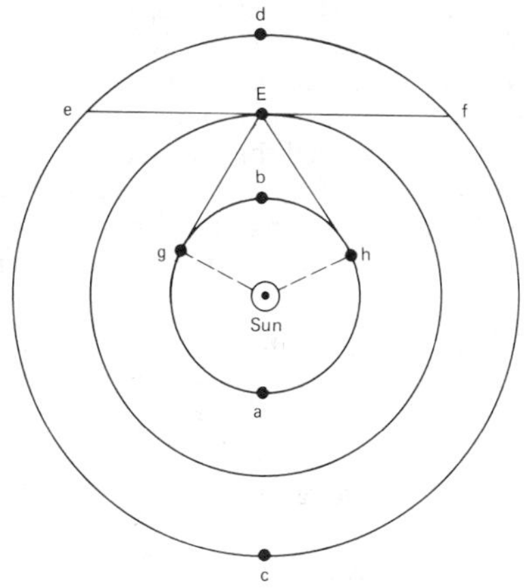

CONJUNCTIVA. The mucous membrane which covers the anterior portion of the globe of the eye, and the inner surface of the eyelids. Structurally, it completely invests the eye anteriorly.

CONJUNCTIVITIS. The general term given to any inflammation or infection of the conjunctiva. This condition has innumerable causes and constitutes the most common eye disease of the Western Hemisphere. Most cases are caused by bacterial or viral infection. However, allergy, chemical irritation, and infection by fungus or parasites are sometimes responsible. Conjunctivitis occurs in connection with a number of diseases, including ankylosing spondylitis, epidemic typhus, erythema multiforme, gonorrhea, leptospirosis, meningococcal disease, and Reiter's syndrome, among others. Pharyngoconjunctival fever (swimming-pool conjunctivitis) is usually associated with adenovirus type 3 or 7, but has been reported with other types as well. Incubation period of the disease is 2 to 8 days. The disease is quite contagious and spreads rapidly among family members. Epidemic keratoconjunctivitis and hemorrhagic cystitis are also caused by adenoviruses. Fever, photophobia and a tendency for the eyes to tear are present. Corneal erosions can occur within 2 days and may be sufficiently deep to interfere with vision. Outbreaks of this condition sometimes occur in professional ophthamological and optometrist centers. This disease was first discovered on the west coast of the United States, among shipyard workers, during World War II. It is believed that the epidemic resulted from improperly sanitized equipment used in a medical facility for treating eye injuries. Acute hemorrhagic cystitis was first observed in Japan.

Acute catarrhal conjunctivitis ("pink-eye") is a term rather loosely applied to inflammations of the conjunctiva in children, although adults are also susceptible. Pink-eye is sometimes associated with irritation from smoke, dust, wind, or intense light, as from electric arcs.

In the acute, highly contagious form of pink-eye, the eyes are red and watery at first. Then pus begins to accumulate. The eyelids may smart, burn, or itch, and become stuck together overnight by the discharge. General swelling and puffiness often surrounds the eyes.

Pink-eye usually represents an infection of the conjunctiva by pneumococci or staphylococci, but occasionally the Kochs-Weeks bacillus is responsible. Because the cornea can become involved in certain epidemic forms of the disease, medical attention should be sought whenever an eye is persistently or acutely inflamed. This condition also may mask a more serious eye disease. If pink-eye is diagnosed, scrupulous care must be taken to avoid transmitting the disease to others, or to the opposite eye if only one eye is involved. Promptly treated, conjunctivitis usually responds readily to therapy and causes no permanent eye damage.

Ophthalmia neonatorum is any inflammation of the conjuctiva in newborn infants. The condition is acquired by contact with an infected birth canal during delivery of the infant. Gonococcus is usually the infecting organism. Most areas now require the routine use of preventive measures in the delivery room. In most hospitals, two drops of a 1% silver nitrate solution are instilled into each of the infant's eyes at birth. Penicillin also can be substituted. The use of such preventive measures at birth has enormously reduced the incidence of eye damage and blindness resulting from ophthalmia neonatorum.

Trachoma is an eye disease which approaches the incidence of the common cold in some areas of the world. The condition occurs mainly under conditions of overcrowding and poor hygiene. The condition is characterized by large, clear "granulations" underneath the eyelids. Without sulfonamide or antibiotic treatment, trachoma eventually produces corneal damage and moderate to complete visual loss.

Pinguecula is a yellowish nodule of tissue which appears gradually on the conjunctivas of both eyes in some persons. These nodules are usually located on the nasal side of the iris and are relatively common among persons over 35 years of age. The nodules consist of hyaline and elastic tissue and generally no treatment is required.

See also **Vision and the Eye.**

CONNATE WATER. Water that is trapped in marine sediments at the time they are laid down in the sea is commonly called *connate water.* As the term implies, connate water is produced at the same time as the rock and constitutes a sort of fossil seawater.

When marine sediments are raised above sea level and subjected to the action of circulating meteoric water, their connate water and solutes are removed by leaching and flushing and carried back to the ocean. The flushing process however is slow in fine-grained and deeply buried sediments, and water which almost certainly owes its high content of dissolved ions to remnants of marine solutions is common in such environments. The water associated with petroleum commonly is saline and occurs in formations whose porosity and structure are generally unfavorable for extensive water circulation and coincident removal of solutes. Usually the salt dissolved in brines that occur in deeply buried rocks is considered to be of connate origin. See **Petroleum.**

Many geological processes may have modified the composition of connate brines to produce their wide range. Obviously these alterations have been extensive. Some connate waters are essentially saturated or nearly saturated solutions of sodium chloride. Others contain large proportions of calcium as well as sodium and chloride. Although the composition of connate water gives few useful clues as to the composition of the ocean in past geologic periods, the ocean has evidently been rich in chloride for a very long time and brines in which anions other than chloride are predominant cannot logically be ascribed entirely to a connate origin.

Among the processes which might be expected to alter the concentrations of ions in connate water are the precipitation of solids such as calcite on mineral surfaces, the solution of rock minerals and evaporites, sorption and desorption of ions on solids, and the differential movement of water molecules and ions through clay and shale strata. The latter effect is equivalent to the ultrafiltration effect of certain types of membranes used in the reverse osmosis method of removing solutes from water. Under the high pressures encountered at great depths, water and ion movements may be very different from the ones expected at low pressure. These effects may possibly explain the high concentrations of solutes in some connate brines and the difference in ion content between such brines and ordinary seawater. Another factor of considerable importance in many places appears to be the biochemical reduction of sulfur from S^{6+} as found in sulfate ions to sulfur in the more reduced forms of free sulfur, polysulfides, or sulfide ions. The brines encountered in oil and gas fields, and the gases themselves, are often high in hydrogen sulfide content as a result of sulfate reduction.

Analyses of a variety of connate brines have been published by White, Hem, and Waring (1963). White (1957) has suggested criteria for distinguishing connate water by means of ratios of concentrations of certain of the dissolved ions to one another. In most connate water the ratios of bromide and iodide to chlorine are relatively high and ratios of potassium and lithium to sodium are low.

CONNECTED GRAPH. Graph (Mathematics).

CONNECTING ROD. The common connecting rod is one of the four elements of the mechanism known as the slider crank chain. This mechanism consists of a base which carries two members, one of which rotates while the other reciprocates. The connecting rod connects the reciprocating and rotating elements by means of pinned or hinged joints at its ends, the same constituting the connecting rod bearings. The importance of this mechanism, and of its elements, including the connecting rod, is that it is the basis for a large number of machines of great importance to modern civilization. This is probably due to the fact that in so many cases a reciprocating motion is produced where rotary is desired, and vice versa. Among the more common illustrations of the machines of which the connecting rod is a vital and important part, are engines, pumps, compressors, punches, etc. A familiar example is the connecting rod of the gasoline engine. This engine derives its power from the push of the exploding gas against the reciprocating piston. One end of the connecting rod is joined to the piston by the wrist pin on which it has bearing. The other end of the connecting rod has a bearing on the rotating crank pin. Thus, the connecting rod has a composite motion: one end of it reciprocates, while the other rotates. It is subject not only to tension and compressive stress, but also, by virtue of its inertia, to transverse bending. Due to this latter factor, high-speed connecting rods are carefully designed so that their mass will not only be as small as possible, but so placed as to cause the least shaking forces. Commonly used sections are the solid rectangular, the I-beam, and the tubular. Almost all materials, including cast iron, brass, wood, steel, alumnium alloy, have at one time or another been used for connecting rods.

CONNECTION (Open Delta). Open Delta.

CONNECTIVE TISSUE. The connective tissues, as the name suggests, are primarily those which bind together, connect and support other structures. They are derived from the loosely arranged mesenchyme of the embryo and are characterized in the adult by the presence of various kinds of cells and of much intercellular substance, also in various forms.

In addition to bone and cartilage the connective tissues of the adult vertebrate include the loose irregularly arranged tissue which underlies the skin and occupies spaces between other organs. In this tissue lie cells which produce its own structures, blood cells, fat cells, and cells which become active in the repair of wounds. Between the cells is a soft matrix containing two kinds of fibers: white and elastic. The white fibers give tensile strength to the tissue and the elastic fibers give elasticity.

The other connective tissues are made up of certain of these structures. Thus tendons and ligaments are composed principally of parallel white fibers. Fibrous membranes may be either elastic or tough according to the fibers composing them.

Some of these connective tissues have other functions which are not associative. Adipose tissue, for example, is composed largely of cells in which fat is stored.

Some minute parts of many organs are held together by a network of reticular tissue whose fibers run in all directions among the cells of the organ.

The development of these tissues is so extensive that if all other components of the body could be removed, its gross form would still be evident.

See also **Bone;** and **Collagen.**

CONNECTIVITY. A domain bounded by a smooth curve is said to be simply connected. Any closed curve in the domain can be shrunk to a point by continuous deformation without crossing the boundary. When more than one continuous arc is required to form the boundary, the domain is multiply connected and the minimum number of arcs required to form the boundary is the connectivity.

CONRADSON CARBON TEST. This test is an indication of the percentage of carbon residue in an oil, either fuel or lubricating. Where lubricating oil is exposed to high temperatures during use, a certain amount of carbon remains after the volatile parts of it have been volatilized. This is true of the lubrication of the cylinders of the internal combustion engine, one of the objectionable features of which is the deposit of carbon in the cylinder head after continued use. Likewise with fuel oils, especially the Diesel fuel oil, carbon residue is an important characteristic. Although not very well understood at present, even by many of those associated with Diesel power development, recent research indicates that the carbon residue as indicated by tests such as the Conradson, may be expected to be an important factor in the selection of a fuel oil.

The Conradson test consists of heating, under special conditions, a weighed sample of the oil at a sufficiently high temperature to volatilize all of the volatile matter of the oil. Once the distillation is completed, there remains the carbon residue, which is determined by weighing. Although it is objected that this test fails in many important respects to duplicate the conditions under which a carbon residue is formed in internal combustion engines, it is a reasonably successful indicator of the value of an oil with respect to its carbonizing qualities.

CONSANGUINITY. The genetic relationship of igneous rocks. Related groups of rocks which have been derived from a common magma and thus form a distinct petrographic province.

CONSCIOUSNESS. Brain Injury.

CONSEQUENT STREAMS. The type of drainage pattern which develops on the initial slopes of a land surface newly exposed to erosion by running water. Some such surfaces, as in coastal plains, are underlaid by sedimentary strata gently tilted in the direction of surface slope. If one of the parallel series of tilted formations is more resistant to erosion than the others, a cliff or cuesta will be developed. Stream valleys which develop parallel to the cuesta are called subsequent, and their tributaries which cut back into the cuesta are called obsequent.

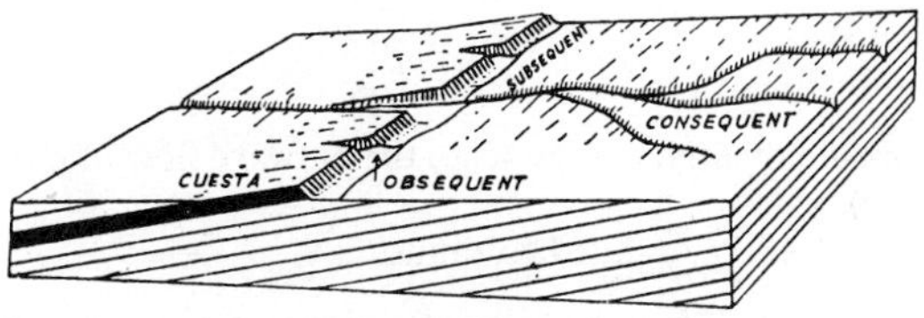

Block diagram to illustrate the meaning and relations of consequent streams, in a youthful stage of the normal cycle of erosion in a region of tilted strata. The more resistant rock layer (in black) stands out in the form of a cuesta against the erosion.

CONSERVATION LAWS AND INVARIANCE THEOREM IN QUANTUM MECHANICS. Invariance Theorem and Conservation Laws in Quantum Mechanics.

CONSERVATION LAWS AND SYMMETRY. Basic among the natural laws are the so-called conservation laws which, in essence, state that in a given physical system under specified conditions there is a certain measurable quantity that remains changeless regardless of what actions may occur within the system. Three classical laws of this type are: (1) the law of the conservation of energy; (2) the law of the conservation of momentum; and (3) the law of the conservation of angular momentum. The concept of the conservation laws dates back to the early days of science and modifications have occurred and will continue to occur as new knowledge is gained. One of the tasks of physics is to explain the detailed rationale for laws that, on the surface, have a quality of being "self-evident."

An outstanding example of how these laws are subject to modification was Einstein's elucidation of the mass-energy equivalence ($E = mc^2$). Before that, the conservation of mass and the conservation of energy were considered to be independently valid.

General Functions of Laws of Conservation. An important function of the conservation laws is that they allow predictions about the behavior of a system without going into mechanical details of what happens during the course of a reaction. The laws provide a direct connection between the state of the system before the reaction and the state after the reaction. Also, one may conclude that any action which violates one of the conservation laws must be forbidden.

Laws of this nature are postulated as a result of many measurements of the energies and momenta involved in reactions of all kinds. It is always found that within the limit of accuracy of the experiment, for example, that the amount of energy in the system after the reaction is the same as the amount of energy before the reaction. Prior to the development of elementary particle physics, there was much emphasis on transformation of energy between its various "forms," such as mechanical, electrical, and thermal energy. The present viewpoint is that the macroscopic behavior of matter is the result of interactions between elementary particles and that these elementary interactions individually obey the various conservation laws. In its elementary form, the law of conservation of energy states that when two or more particles interact, the total energy (kinetic plus potential) is always a constant. When it is said, for example, that the kinetic energy of a moving object has been transformed into the potential energy of a compressed spring, what is really meant is that on a microscopic level the atoms of the spring have been pressed closer together so that there is a greater amount of potential energy in the electric fields between the atoms of the spring. This increase of potential energy is associated with an equal decrease in the kinetic energy of the object which caused the spring to compress.

Concept of Symmetry. With the development of the Lagrangian and Hamiltonian methods of solving physical problems, and particularly with the growth of importance of quantum mechanics, it became clear that the conservation laws are closely connected with the concept of symmetry in space and time. This is based upon the ability to describe the interaction between two or more objects in terms of a potential energy function. If the potential energy of the system is known for any position of these objects in space and time, then the future motions of the objects can be predicted. Certain predictions can be made without going through a complete solution of the equations of motion. For example, it is found that if the potential energy does not depend explicitly on one of the space coordinates, then the momentum associated with that coordinate does not change, but is a constant of the motion.

Some examples of "geometrical symmetries" encountered in classical physics might include:

1. An object moves in a 3-dimensional space where its potential energy is the same at every point. The expression describing the potential does not explicitly contain the coordinates x, y, or z. That is, the system is invariant with respect to translation of the origin of the coordinate system in any direction. This symmetry is associated with conservation of linear momentum; the momentum in all three dimensions is a constant.

2. An object moves in a world which is flat so that the force of gravity is in the vertical (z) direction. The potential depends only on the height of the object above the ground, but does not depend on its location in the horizontal plane; i.e., the potential is invariant to a translation of the coordinate system in the x-y plane. There is now symmetry in two dimensions, and momentum in the x-y plane is conserved.

3. In the interaction between two spherical bodies, if the potential depends only on the distance between the bodies, then there is spherical symmetry. The system is invariant to a rotation of the coordinate system about any axis. In this case, two components of angular momentum are conserved; as the two bodies orbit around their common center of mass, the magnitude of their angular momentum is constant, while the plane of the orbit in space does not change.

4. If the interaction between two objects does not depend explicitly on the time coordinate, then the actions which take place do not depend on when one starts to measure time; i.e., the properties of the system are invariant with respect to a translation of the origin of coordinates along the time axis. This symmetry is associated with conservation of energy. Use of a 4-dimensional coordinate system allows one to associate conservation of momentum and energy in a unified manner with the geometrical symmetry of space-time.

The conservation laws and the concept of symmetry acquired importance in the area of elementary particle physics. The conservation laws act as "selection rules" to determine which reactions may take place between the many existing particles out of the very large number of otherwise conceivable reactions.

An important property of elementary particles is *parity*. Each particle has a parity number associated with it; either $+1$ or -1, depending on the type of particle. In an assembly of particles, there is a total parity, which is the sum of the individual parities. If parity is conserved, then this total parity does not change during the course of the reaction. This property of matter is associated with the so-called mirror symmetry. All the laws of nature which possess this type of symmetry are such that if the words "right" and "left" are interchanged in the statement of the law, then the behavior of the system obeying these laws is unchanged. At one time, it was thought that every natural law was out of this type. As a result of conservation of parity, it was believed that it would be impossible to describe the difference between "right" and "left" by the use of words alone. Yang and Lee (1956) pointed out that in a special class of reactions involving the "weak nuclear" interaction, parity need not be conserved. As a result of this finding, it was seen that the universe does possess an asymmetry between right and left and that it is possible to describe an experiment which will definitely distinguish between the directions "right" and "left" in the universe.

Another type of symmetry of importance in elementary particle physics is that entitled *charge conjugation*. This principle states that if each particle in a given isolated system is replaced by its corresponding antiparticle, then no difference can be observed. For example, if in a hydrogen atom, the proton is replaced by an anti-proton and the electron is replaced by a positron, then this antimatter atom will behave exactly like an ordinary atom, so long as it does not come into contact with ordinary atoms.

However, it is found that there are certain types of reactions where this rule does not hold. These are the types of reactions where conservation of parity breaks down. If one considers a piece of radioactive material emitting electrons by beta decay, the radioactive nuclei are lined up in a magnetic field which is produced by electrons traveling clockwise in a coil of wire (as seen by an observer looking down on the coil). Because of the asymmetry of the radioactive nuclei, most of the emitted electrons travel in the downward direction. If the same experiment were done with similar nuclei composed of antiparticles and the current in the magnet coil consisted of positrons instead of electrons, then the emitted positrons would be found to travel in the upward rather than in the downward direction. Thus, interchanging each particle with an antiparticle has produced a change in the experiment.

In the foregoing situation, however, the symmetry can be restored if the words "right" and "left" are interchanged in the description of the experiment at the same time each particle is exchanged with its antiparticle. In this example, this is equivalent to replacing the word "clockwise" by "counterclockwise." When this is done, the positrons are emitted in the downward direction, just as the electrons in the original experiment. The laws of nature thus have been found to be invariant to the simultaneous application of charge conjugation and mirror inversion.

Other more technical conservation laws play a role in elementary particle physics. *Conservation of baryon number* and *conservation of strangeness* are rules required to account for the fact that certain reactions involving heavy particles are forbidden. *Time reversal invariance* describes the situation that in reactions between elementary particles, it does not make any difference if the direction of the time coordinate is reversed.

Although a symmetry idea may first suggest a conservation law, the conservation law must be tested by experiment to see if the symmetry is valid.

Conservation of Energy in Thermodynamics. The principle of conservation of energy plays a fundamental role in thermodynamics and is, therefore, also called the first law of thermodynamics. In its most general form it postulates the existence of function of state, called the internal energy of the system U, such that its change per unit time is equal to some flow, called the energy flow from the surroundings.

This statement can be expressed symbolically by the formula

$$dU = d_eU \qquad \text{or} \qquad d_iU = 0 \tag{1}$$

in which d_eU is the energy received during the time dt from the outside, and d_iU is the energy "creation" inside the system.

The explicit form of the energy flow depends on the nature of the system considered.

In closed systems and in the absence of an external field, the energy U supplied from the outside during the time interval dt is equal to the sum of the heat flow dQ expressed in units of energy and the mechanical work dW performed at the boundaries of the system. If the pressure is normal to the surface, the mechanical work is simply $-p\,dV$ and the expression of the energy conservation becomes

$$dU = dQ - p\,dV \tag{2}$$

From a purely phenomenological point of view, this expression of the conservation of energy may be considered as the definition of the heat received by the system. The extension of the mechanical principle of conservation of energy to include the flow of heat is due mainly to Carnot, Joule, Helmholtz, and Clausius.

In order to express the energy conservation in continuous systems, it is useful to introduce the energy density per unit volume

$$u_v = \frac{\Delta U}{\Delta V} \tag{3}$$

In agreement with the general formulation of the principle of energy conservation

$$\frac{\partial u_c}{\partial t} = \text{div}\,\Phi[U] \tag{4}$$

where $\Phi[U]$ is the energy flow. This flow contains in general different contributions, among which are:

1. The convection flow corresponding to a center of mass motion ω, amounting to $u_v\omega$.
2. A heat flow Q.
3. A flow of energy corresponding to the pressure tensor p_{ij} in the fluid, its i component being

$$\sum_j p_{ij}\omega_j \tag{5}$$

4. A flow of potential energy (e.g., the outward flow of electromagnetic energy).
5. A flow of energy related to diffusion.

The total energy U may be split into an internal energy, a potential energy, and a macroscopic kinetic energy. Each contribution taken separately does not satisfy an equation of the simple form of Equation (4) because of possible transformation of one form of energy to another.

Conservation of Mass. This law has been put in the form that matter can neither be created nor destroyed. More accurately, the total mass of any system remains constant under all transformations. The statement, however, is subordinate to mass-energy equivalence.

Consider a homogeneous closed system contaning c components ($\gamma = 1 \ldots c$) among which a single chemical reaction is possible. In such a system any variation in the masses will result only from the chemical reaction. Thus, the change of the masses m_γ of component γ during the time interval dt can be written as

$$dm_\gamma = \nu_\gamma M_\gamma\, d\xi \tag{6}$$

where M_γ is the molar mass of component γ and ν_γ its stoichiometric coefficient in the chemical reactions.

This coefficient is generally counted positive when ν appears in the right-hand member of the reaction equation, negative when it appears in the left-hand member; ξ is the degree of advancement or extent of reaction.

The total mass of the system is given by $m = \Sigma_\gamma\, m_\gamma$. Summing over γ, the conservation of mass for a closed system is expressed by

$$dm = \left(\sum_\gamma \nu_\gamma M_\gamma\right) d\xi = 0 \tag{7}$$

The equation

$$\sum_\gamma \nu_\gamma M_\gamma = 0 \tag{8}$$

is called the equation of the chemical reaction or, more briefly, the stoichiometric equation.

Instead of the mass of the components it is often useful to consider the mole numbers $n_1 \ldots n_c$. Instead of Equation (6), this becomes

$$dn_\gamma = \nu_\gamma\, d\xi \tag{9}$$

Equations (6) and (9) are extended easily to r simultaneous reactions. The different reactions are designated by indices ($\rho = 1 \ldots r$). Instead of Equations (6) and (9), there are

$$dm_\gamma = M_\gamma \sum_{\rho=1}^{r} \nu_{\gamma\rho}\, d\xi_\rho \tag{10}$$

$$dn_\gamma = \sum_{\rho=1}^{r} \nu_{\gamma\rho}\, d\xi_\rho \tag{11}$$

where $\nu_{\gamma\rho}$ denotes the stoichiometric coefficient of γ in the reaction.

The conservation of mass in a continuous system is expressed by the equation of continuity for the density, ρ,

$$\frac{\partial\rho}{\partial t} = -\text{div}\,\rho\omega \tag{12}$$

where ω is the macroscopic velocity. This equation expresses the fact that the local change of the density is equal to the negative divergence of the flow of matter.

Equation (12) holds also for a mixture; ω is then related to the macroscopic velocities of the different components by

$$\omega = \left(\sum_\gamma \rho_\gamma\omega_\gamma\right)/\rho \tag{13}$$

Thus ω is simply the velocity of the center of gravity in an element of volume.

In general, the local change of a physical quantity is due not only to the divergence of the current which is associated with it, but a "source" term has also to be taken into account. For instance, the equation of continuity for the density ρ_γ of a component γ participating in a chemical reaction is

$$\frac{\partial \rho_\gamma}{\partial t} = -\text{div } \rho_\gamma \omega_\gamma + \nu_\gamma M_\gamma V_v \qquad (14)$$

where V_v is the rate of the chemical reaction per unit volume.

In an open system, it is possible to split the change of mass of component γ into an external part, $d_e m_\gamma$ supplied from the exterior, and an internal part, $d_i m_\gamma$ due to changes inside the system

$$dm_\gamma = d_e m_\gamma + d_i m_\gamma \qquad (15)$$

Taking into account the equations on conservation of mass in closed systems, this becomes

$$dm_\gamma = d_e m_\gamma + M_\gamma \sum_{\rho=1}^{r} \nu_{\gamma\rho}\, d\xi_\rho \qquad (16)$$

$$dn_\gamma = d_e n_\gamma + \sum_{\gamma=1}^{r} \nu_{\gamma\rho}\, d\xi_\rho \qquad (17)$$

Summing Equation (16) over γ and taking into account the stoichiometric equations $\Sigma_\gamma\ \nu_{\gamma\rho} M_\gamma = $), the total change of mass is

$$dm = d_e m \qquad (18)$$

This relation expresses the conservation of mass in open systems and indicates that the change of the total mass is equal to the mass exchanged with the outside world.

The process of splitting the total change of mass of a component into an external part and an internal part may be used on arbitrary extensive property.

Conservation of Momentum. The principle of conservation of momentum states that for a dynamical system consisting of n material particles of masses $m_1, M_2, \ldots, m_n$, respectively, and position vectors $r_1, r_2, \ldots, r_n$, respectively, if the only forces acting are the mutual interaction forces of the particles the total momentum of the system remains constant; for example

$$\sum m_i \frac{d\mathbf{r}_i}{dt} = \text{constant} \qquad (19)$$

The law of conservation of momentum is as fundamental to physics as the law of conservation of mass energy. Like that law, it holds in quantum mechanics and relativistic mechanics as well as in classical mechanics.

Conservation of Electric Charge. Since electric charge comes in discrete quantities (there is no known way of breaking an electric charge down into bundles smaller than that contained in an electron), this law deals with the counting of objects, rather than with the measurement of continuous variables, such as momentum of energy. Conservation of electric charge means that the total number of electric charges (taking positive and negative signs into account) in a closed system does not change. This, in earlier times, meant that electric charge could not be created or destroyed. The creation of charged particles is now spoken of, but the creation of a positive charge must always be accompanied by formation of an equal negative charge (e.g., an electron-positron pair is created by a photon). Conservation of electric charge is associated with a symmetry property of Maxwell's equations known as gage invariance which states that the absolute value of the electric potential (as opposed to the relative value) plays no part in physical processes. Further developments in quantum mechanics indicate that conservation of electric charge is connected with the observation that the properties of a system of particles do not depend on the phase of the wave function describing the system.

The following conservation principles apply particularly to interactions of elementary particles:

1. *The total baryon number remains constant*. A baryon is a nucleon (proton or neutron) or any particle heavier than those that can be considered to have an atomic mass number $A = 1$. Some mesons have a mass greater than the proton, but they have a mass number $A = 0$, so they are not baryons. In computing the number of baryons present in a system, each baryon counts 1; each antibaryon counts -1; and leptons and mesons count 0.

2. *The total lepton number remains constant*. Leptons consist of the neutrinos, electrons, muons, and their antiparticles. Here again, the basic principles applied are analogous to those for baryon number; particles count 1; antiparticles count -1; and baryons and mesons 0.

3. *Total strangeness quantum number.* This remains constant except as previously mentioned in connection with weak interactions.

References

Cline, D. B., and F. E. Mills (editors): "Unification of Elementary Forces and Gauge Theories," Harwood Academic, London, 1978.

d'Espagnat, B.: "The Quantum Theory and Reality," *Sci. Amer.*, **241**, 5, 158–181 (1979).

Feynman, R. P., Leighton, R. B., and M. Sands: "The Feynman Lectures on Physics," Addison-Wesley, Reading, Massachusetts, 1963.

Rothman, M. A.: "Conservation Laws and Symmetry," in "The Encyclopedia of Physics," (R. M. Besancon, editor), 2nd edition, pages 166–170, Van Nostrand Reinhold, New York, 1974.

Salam, A.: "Gauge Uninification of Fundamental Forces," (Nobel Lecture), *Science*, **210**, 723–732 (1980).

Weisskopf, V. G.: "Contemporary Frontiers in Physics," *Science*, **203**, 240–244 (1979).

Woodward, R. B., and R. Hoffman: "The Conservation of Orbital Symmetry," Academic, New York, 1969.

CONSERVATION LAWS (Fluid Flow). **Fluid Flow (Conservation Laws).**

CONSERVATION OF ENERGY. **Energy.**

CONSERVATION OF ENERGY (Combustion). **Combustion.**

CONSERVATION OF MATTER (Combustion). **Combustion.**

CONSERVATIVE SYSTEM. 1. A closed system in which mechanical energy is constant is called a conservative system, as is an open system in which all external work is available as potential and kinetic energy of the system. 2. A system of particles in which the forces acting on any particle of the system are forces which can be derived from a potential energy function. There must exist a potential energy function $V(x, y, z)$ such that the components of the resultant of these forces are given by

$$F_x = -\frac{\partial V}{\partial x}$$

$$F_y = -\frac{\partial V}{\partial y}$$

$$F_z = -\frac{\partial V}{\partial z}$$

or the resultant force vector $\mathbf{F} = -\nabla V$. The forces which satisfy this condition are called conservative forces. In a conservative system, the work required to move a particle from one point to another depends only on the positions of the two points, not on the path followed between them.

CONSISTENT STATISTIC. A statistic is said to be consistent if when calculated from the whole population it gives the correct value of the population parameter. More precisely as the size of the sample n increases then

$$\lim_{n\to\infty} P\{|t-\theta| > \epsilon\} \to 0$$

that is, the probability that the statistic t differs from the population parameter θ by more than ϵ in absolute value approaches zero. The least one should demand of a statistic is that it be consistent.

CONSOLUTE LIQUIDS. This term is applied to liquids when they are miscible in all proportions, i.e., mutually completely soluble, under some given conditions. It is not usually applied to gases because they are all miscible.

CONSOLUTE TEMPERATURE. The upper consolute temperature for two partially-miscible liquids is the critical temperature above which the two liquids are miscible in all proportions. In some systems where the mutual solubility decreases with increasing temperature over a certain temperature range, the lower consolute temperature corresponds to the critical temperature below which the two liquids are miscible in all proportions. Some systems such as methylethyl ketone and water have both upper and lower consolute temperatures.

CONSONANCE. When two or more musical tones played simultaneously produce a pleasing effect on the listener, the tones are said to be in consonance. The general condition for consonance is that the frequencies of all the sounds have the ratios of small whole numbers.

CONSTANT. An absolute constant is a single number which always has the same value. An arbitrary constant or parameter is one which has only one particular value in a given case, but may have another value in another case. For example, the equation $x^2 + y^2 = r^2$ represents a circle. If r is an absolute constant, there will be one and only one circle described by this equation. However, if r is a parameter the equation represents the entire family of circles, infinite in number, which have center at the origin of a Cartesian coordinate system. A given circle of the family is found by the choice of r, which fixes the radius of the circle.

Arbitrary constants are frequently designated by letters from the first part of the alphabet, such as *a*, *b*, *c*, etc.

CONSTANTAN. Nickel.

CONSTANT-CURRENT TRANSFORMER. This is a specially constructed transformer, sometimes called a tub transformer, built so the primary and secondary can move relative to one another under the influence of the forces set up by the load current. The currents in the primary and secondary react on one another to produce a repelling force between the two windings, the greater the current, the greater the force. The secondary is movable along the core and is partially balanced by a counterweight which is adjusted so the current forces will produce just the right reaction to hold the current essentially constant. An increase of current increases the force and hence the separation, which in turn increases the leakage flux between the windings and lowers secondary voltage. The lowered voltage causes a reduction of the current to a value which, when equilibrium is reached, is practically constant. Reduction of the load current causes an opposite sequence of reactions.

CONSTANT-DEVIATION PRISM. Prism (Optics).

CONSTANT-STRESS LAYER (Fluid). Fluid (Constant-Stress Layer).

CONSTELLATIONS. In astronomy this term is used to designate certain groupings of the stars. From earliest recorded history we find that the larger star groups (constellations), the smaller groups (asterisms such as the Pleiades), and the individual stars have received names symbolizing meteorological, religious, or mythological beliefs. The idea that the constellation names and myths are of Greek origin has been quite completely disproved. It seems highly probably that they are of Semitic or Pre-Semitic origin and that they found their way into Greece through contact with the Phoenicians (sailors who used the stars constantly in their profession).

The oldest record of an actual constellation listing is found in the Creation Legend in about 650 B.C. This Legend was recorded on Cuneiform from even earlier records. From this time onward, frequent references to the constellation legends are to be found both in poetical and historical writings. The basis for the modern constellation division is to be found in the list of 48 constellations published by Ptolemy in about 150 A.D. This list is based upon the writings of Ptolemy's predecessors, notably Hipparchus.

The boundaries of Ptolemy's constellations were very indefinite, and many visible stars were left out entirely. Furthermore, his list only covered that portion of the heavens visible from the southern Mediterranean regions. In the 1,800 years since Ptolemy's time, the list has been added to and the boundaries defined, until, at present, all stars are included in some one of the constellations. The International Astronomical Union placed the matter of defining the constellation boundaries in the hands of a special committee, and in 1930, the final list was published.

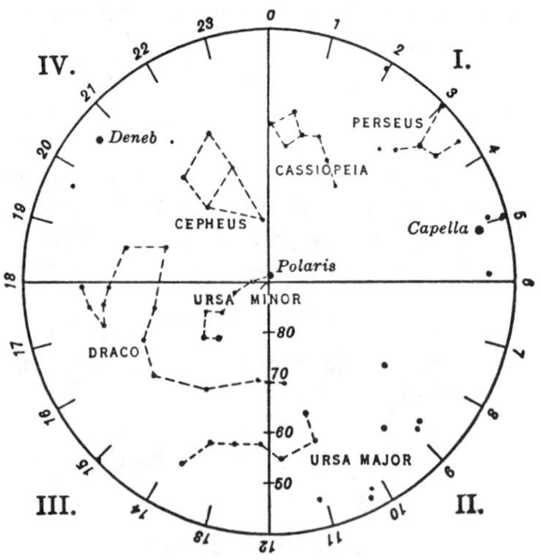

Fig. 1. Circumpolar stars.

Several attempts have been made to supplant the ancient mythological names with more modern ones (e.g., the Coelum Stellatum Christianum of Julius Schiller, in 1627, in which we find the ancient names replaced by names of various Church dignitaries), but none of the attempts has been successful.

The most important and easily recognized constellations will be found on the accompanying star maps (Figs. 1–3); the numbers, e.g., Aries—Fig. 3 (2, N), refer to the particular map and the location on that map where the constellation will be found. The list accompanying Figs. 2 and 3 contains the names of all the constellations now used; those in bold-face type are treated elsewhere in this volume in special articles, and those marked with an asterisk are the original constellations of Ptolemy.

CONSTIPATION. A sluggish action of the bowels, usually with difficulty in evacuation. Many persons have the idea that they are constipated, when actually they are not. As a result, they form the habit of taking laxatives and enemas which cause irritable bowel symptoms. The concept that defecation should occur after a meal, or that copious evacuations are necessary is unfounded, but many individuals are obsessed with notions that their bowels do not move frequently enough. A weakening of normal bowel function and the development of increasing instability in bowel habits almost always follow the habitual administration of laxative agents. The incidence of hemorrhoids is much greater in cathartic users than in other persons.

Constipation is a common complaint of peptic ulcer patients. Often, the constipation is regarded as the cause of the pain, with the result that the patient may take laxatives habitually. The resultant bowel upset may be severe.

In true constipation, the ease and sense of completeness of evacuation of the rectum and lower part of the colon is lacking. Constipation is more common among women than men. Habitual constipation occurs in young adults and becomes established in their twenties. Although practically half the population will give a history of constipation at some time, x-ray studies have shown that true constipation is not as frequent as imagined.

Many factors may be responsible for, or contribute to the production of constipation. Apart from obvious gross mechanical obstruction caused by tumor, adhesive bands, or strictures, constipation may be caused by habit, diet, or the state of the muscles of the bowel.

The act of defecating can be readily inhibited by an effort of the will, and a habit of refusing to respond to the urge to defecate is a common cause of constipation. The sensation of a full rectum usually occurs regularly at some definite time each day, and refusal to respond causes the desire to defecate to pass. The rectum becomes accustomed to the increased fecal bulk and there is retention of feces in the distal colon and rectum. This leads to excessive absorption of fluid; hence, the feces become dry, hard, and less easily expelled. As a result of continued overloading, the bowel masculature becomes sluggish in

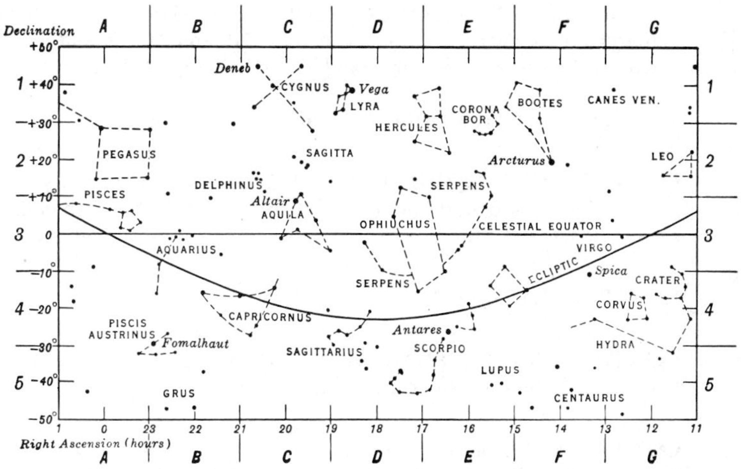

Fig. 2. Equatorial stars.

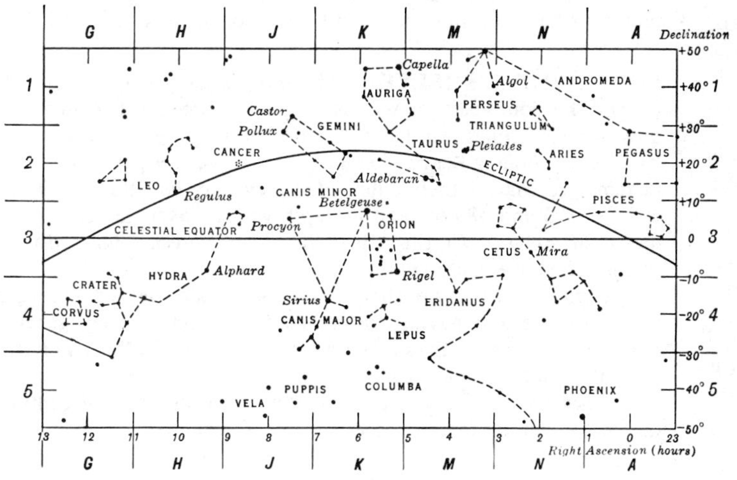

Fig. 3. Equatorial stars (continued).

LIST OF CONSTELLATIONS

NORTH OF THE ECLIPTIC

* **Andromeda**—Fig. 3 (1, N)
* **Aquila**—Fig. 2 (3, C)
* **Auriga**
* **Bootes**—Fig. 2 (1, F)
Camelopardalis
Canes Venatici
* **Cassiopeia**—Fig. 1 (I)
* **Cepheus**—Fig. 1 (IV)
Coma Berenices
* Corona Borealis
* **Cygnus**—Fig. 2 (1, C)
* Delphinus
* Draco
* Equuleus
* **Hercules**—Fig. 2 (2, E)
Lacerta
Leo Minor
Lynx
* **Lyra**—Fig. 2 (1, D)
* Ophiuchus
* **Pegasus**
* **Perseus**—Fig. 3 (1, M)
* Sagitta
* Serpens
* Triangulum
* **Ursa Major**—Fig. 1 (II)
* **Ursa Minor**—Fig. 1 (III)
Vulpecula

ZODIACAL

* **Aquarius**—Fig. 2 (3, B)
* **Aries**—Fig. 3 (2, N)
* **Cancer**—Fig. 3 (2, J)
* **Capricornus**—Fig. 2 (4, B)
* **Gemini**—Fig. 3 (2, K)
* **Leo**—Fig. 3 (2, H)
* **Libra**—Fig. 2 (4, E)
* **Pisces**—Fig. 3 (3, A)
* **Sagittarius**—Fig. 2 (4, D)
* **Scorpius**—Fig. 2 (5, E)
* **Taurus**—Fig. 3 (2, M)
* **Virgo**—Fig. 2 (3, F)

SOUTH OF THE ECLIPTIC

Antilia
Apus
* Ara
Caelum
* **Canis Major**—Fig. 3 (4, K)
* Canis Minor
Carina
* **Centaurus**—Fig. 2 (5, F)
* Cetus
Chamaeleon
Circinus
Columba
* Corona Australis
* **Corvus**—Fig. 2 (4, G)
* Crater
* **Crux** (Southern Cross)
Dorado
* Eridanus
Fornax
Grus
Horolgium
* Hydra
Hydrus
Indus
* Lepus
* Lupus
Mensa
Microscopium
Monoceros
Musca
Norma
Octans
* **Orion**—Fig. 3 (3, K)
Pavo
Phoenix
Pictor
* Piscis Austrinus
Puppis
Pyxis
Reticulum
Sculptor
Scutum
Sextans
Telescopium
Triangulum Australe
Tucana
Vela
Volans

its action, and thinning and lack of tone may result. Properly regulated habits, rather than purgatives, are more likely to correct this type of constipation.

The diet may be responsible for constipation when it contains too little of undigestible residue, or roughage, which normally stimulates intestinal activity, or when the diet is not fluid enough. The times at which meals are taken are also important, because regular meals stimulate regular emptying of the large bowel.

Senility, obesity, lesions of the central nervous system, and generalized disease may rob intestinal muscles of their normal reactivity so that the propulsive mechanism of the bowel may not function efficiently. A diet low in calcium, potassium, or vitamin B, also may cause this condition.

A segment of the large bowel may be the site of muscular spasm and cause only spasmodic contractions which have little value in moving the feces onward. Such a contracted bowel may be felt through the abdominal wall as a thick cord. This type of spasm may be caused by disease of the gallbladder, duodenum, or appendix, or it may be the result of overwork, worry, or shock.

Certain individuals are born with abnormally long colons and this condition may be accompanied by constipation. Passage of material is naturally slower through a greater length of bowel and there is more opportunity for fluid absorption. Individuals with this congenital abnormality may have the desire to defecate only once or twice a week. Distention of the elongated bowel occurs, and pressure on other parts of the bowel, bladder, or blood vessels results.

Constipation is usually relieved by adequate amounts of laxative foods in the diet. Foods which stimulate bowel action are fats, fruits, vegetables, and coarse cereals. Fruits contain a high proportion of undigestible cellulose, as well as sugars, acids, and salts which have a chemically stimulating effect on the bowel. Undigested fat supplies a mild lubrication to the feces, while partially digested fats are mildly irritating and activate the bowels.

See also **Colitis and Other Inflammatory Bowel Diseases;** and **Diarrhea.** See also the section "Role of Fiber in Diet" in the entry on **Dietary Requirements and Trends.**

CONSTITUENT. 1. In general, one of the elements or parts of a compound. 2. An identifiable component in the microstructure of an alloy. It may be a phase or a characteristic configuration of several phases.

CONSTRAINT. Any particle or collection of particles is said to be subject to constraint if the number of degrees of freedom is less than $3N$, where N is the number of particles. Specifically that property which distinguishes a mechanism from other mechanical linkages. A mechanism has constrained motion in that a motion of one part is followed by a predetermined motion of the remainder of the mechanism. To determine whether a mechanical linkage is a mechanism or not, Klein advocated applying the criterion of constraint, defined as follows:

$$J = \frac{3N - 4 + \gamma - P}{2}$$

in which J is the number of joints in the mechanism, N is the number of links in the mechanism, γ is the number of independent prismatic chains, that is, those whose joints are of the sliding type, P is the number of point or line type of contact joints in the mechanism. When this equation yields an identity, the mechanism is said to be constrained for all dimensions.

CONSTRAINT (Theorem). **Le Châtelier's Principle.**

CONSTRICTORS (Snakes). **Snakes.**

CONTACT ANGLE. A term applied to the angle formed by a liquid on the surface of a solid at the gas-solid-liquid interface, measured as the dihedral angle in the liquid. Its value depends on the relative surface energies of the three interfaces, vapor-solid, vapor-liquid and solid-liquid.

CONTACT FILTRATION. Adsorption.

CONTACT GONIOMETER. Goniometer.

CONTACT LENS. Vision and the Eye.

CONTACT POTENTIAL DIFFERENCE. In his experiments with electroscopes, Volta found that when pieces of two different metals, otherwise insulated, are brought into contact, they acquire opposite charges and maintain a difference of electrical potential even while still touching. This potential difference he found to be characteristic of the given pair of metals. Thus when the metals are iron and copper, the iron has a potential about 0.15 volt higher than the copper, while for tin and iron the difference is 0.31 volt, tin being the higher. Volta listed a series of several metals, viz., zinc, lead, tin, iron, copper, silver, gold, such that when any two are put in contact, the one first named is at the higher potential. "Volta's law," which he was not in position to demonstrate but which was established much later, states that the potential difference between any two metals in direct contact is the sum of the potential differences between intervening metals of the series. Thus for tin and copper (above) it is 0.31 volt + 0.15 volt = 0.46 volt; and it makes no difference whether the tin and copper are in direct contact or have other metals intervening between them.

The contact potential difference depends on the relative Fermi levels of the two solids. On contact, electron flow will take place until the Fermi levels of the two solids are equal. This will result in a potential difference between the two solids, called the contact potential.

A distinction must be made between the contact potentials in air and the so-called "intrinsic" contact potentials in a vacuum with all adsorbed gases removed. According to Millikan, the intrinsic potential difference between two metals A and B is expressed by $V_{AB} = h(\nu_A - \nu_B)/e$, in which h is Planck's constant, ν_A and ν_B are the critical frequencies of photoelectric emission for the two metals (see **Photoelectric Effect**), and e is the electronic charge. In any case, if the electronic work functions of the metals are p_A and p_B, the contact potential difference is $V_{AB} = (p_A - p_B)/e$. The work functions, and hence V_{AB}, are in general dependent upon the medium surrounding the metals. Accurate measurements of these potentials are, unfortunately, very difficult.

CONTAGION. The communication of disease by direct or indirect contact between a patient or carrier and a susceptible subject. Diseases so transferred are frequently termed *communicable diseases.*

It is mandatory in the United States that the following diseases be reported to local health authorities and to the Center for Disease Control (Atlanta, Georgia) on a weekly basis:

> Amebiasis, anthrax, aseptic meningitis, botulism (infantile botulism reported separately), brucellosis, chickenpox, cholera, diphtheria, hepatitis A, hepatitis B, hepatitis (unspecified), legionellosis, leprosy, leptospirosis, malaria, measles (rubeola), meningococcal infections (civilian and military separately), mumps, pertussis (whooping cough), plague, poliomyelitis, psittacosis, rabies (human), rheumatic fever (acute), rubella (German measles), rubella congenital syndrome, salmonellosis (except typhoid fever), shigellosis, syphilis (primary and secondary), tetanus, trichinosis, tularemia, typhoid fever, and typhus fever (flea-borne and tick-borne reported separately).

This responsibility rests with physicians and hospitals.

CONTAGIOUS DISTRIBUTION. A frequency or probability distribution of a compound kind (see **Compound Distribution**), so-called because it arises in the study of contagious diseases and similar events.

CONTINENTAL CLIMATE. Climate.

CONTINENTAL DISPLACEMENT (Drift). **Earthquakes, Seismology, and Plate Tectonics; Ocean.**

CONTINENTAL MARGINS. Earthquakes, Seismology, and Plate Tectonics; Ocean; Ocean Resources (Mineral); Volcano.

CONTINENTAL RISE. At the outer edges of the continental slopes, the slope lessens and decreases in degree of slope as the ocean floor is approached. The rise ranges in width from one to several hundred miles and may reach depths of 17,000 feet (5,180 meters) (approximately) where it is in contact with the floor of the ocean basin.

CONTINENTAL SHELVES. That portion of the ocean basin that fringes the continents in widths varying between a few miles and more than 200 miles (320 kilometers). They are generally very smooth, have gently sloping floors, averaging only a few feet per mile increase in depth so that at their outer edge they may be between 300 and 600 feet (90 and 180 meters) in depth with an average depth of 400 feet (120 meters). At this edge, the slope increases markedly and this is called the continental slope.

Epicontinental seas are shallow bodies of water deeper than continental shelves and having somewhat greater relief. They generally are somewhat greater than 600 feet (180 meters) in depth. Also termed epicontinental marginal seas.

Further information on the geology and composition of the continental shelves will be found in the entry on **Ocean.**

CONTINENTAL SLOPES. That part of the ocean basin found at the rims of the continental shelves. The slope is greater than the latter, averaging 3 to 6 degrees. The slope drops rapidly from depths of 300 to 600 feet (90 to 180 meters) to between 10,000–12,000 feet (3,000–3,600 meters) where the slope decreases and is replaced by the continental rise.

CONTINENTS (Statistics). **Earth.**

CONTINGENCY COEFFICIENT. Contingency Table.

CONTINUOUS CHANNEL. Information Theory.

CONTINGENCY TABLE. The members of an aggregate may be classified according to qualitative or quantitative characteristics. Where the characteristics are qualitative a classification according to two of them may be set out in a two-way table known as a contingency table. For example, if the characteristic A is p-fold and a characteristic B is q-fold then the contingency table will be one of p rows and q columns. A given cell contains the number of individuals having both characteristics corresponding to A_j and B_k.

A *contingency coefficient* is a measure of the relationship between the two qualities as exhibited by the contingency table.

CONTINUITY (Equation). In one form of the equation of continuity, the principle of the conservation of matter is stated in the following form: The rate of increase of the particles in an element of volume is equal to the net inward flow across the surfaces of the element, i.e., the negative of the divergence. In mathematical terms,

$$\frac{d\rho}{dt} + div\,\mathbf{j} = 0$$

where ρ is the density of the medium and **j** is the mass current density vector. With proper changes in the meaning of ρ and **j**, the equation expresses the conservation of other quantities, such as charge and energy.

For a fluid of constant density, conservation of mass requires that the flow field satisfies

$$\text{div}\,\mathbf{v} = \nabla \cdot \mathbf{v} = 0$$

in the Eulerian specification of the flow. The flow is *solenoidal.*

CONTINUITY OF STATE. This term is applied to a transition between two states, as between the gaseous and liquid states, in either direction, without discontinuity, or abrupt change in physical properties. Although this transition is not realizable in practice by mere pressure-volume change, it can be accomplished by some processes, as by a sequence of temperature changes in one direction at constant volume, followed by temperature changes in the other direction at constant pressure, or vice versa.

CONTINUOUS FUNCTION. A function of one variable $f(x)$ is continuous at a value $x = c$ when $f(c)$ has a definite value which is equal to the limit of $f(x)$ when $x \rightarrow c$. A function is continuous within an interval (a, b) when it is continuous at every point of the interval; at the end points it is sufficient that

$$\lim_{x \to a^+} f(x) = f(a); \qquad \lim_{x \to b^-} f(x) = f(b)$$

For the meaning of these symbols, see **Limit.**

A function has a discontinuity or is discontinuous where it is not continuous. The usual type of discontinuity is a point at which the function becomes infinite, or where it has a finite jump (also called a saltus).

Important properties of a continuous function are: (1) if it is continuous in a closed interval (a, b), then among the different values of $f(x)$ in this interval, there is a greatest value M and a least value m; (2) if its interval of continuity is again (a, b), then between every two values x_1 and x_2 of x in this interval, $f(x)$ takes at least once every value between $f(x_1)$ and $f(x_2)$.

A function of two variables, $f(x, y)$, is continuous at a point (a, b) if

$$\lim_{\substack{x \to a \\ x \to b}} f(x, y) = f(a, b)$$

See also terms listed under **Mathematics.**

CONTINUOUS INFORMATION SOURCE. Information Theory.

CONTINUOUS MINING MACHINE. Coal.

CONTINUOUS-PATH CONTROL. Numerical Control.

CONTINUOUS-PATH ROBOT. Robot.

CONTINUOUS STEEL CASTING. Iron Metals, Alloys, and Steels.

CONTINUOUS VARIABLE. A continuous variable is one which may have any value within the range of variation. Examples are the weights of objects, the ages of people, and so on. A variable which is not continuous is said to be discrete or discontinuous.

CONTINUOUS WAVE TRANSMISSION. Used for radio telegraph communication, continuous wave transmission utilizes a wave which is constant in amplitude, as opposed to the varying amplitude of a modulated wave, and which changes only during keying. For transmission of dots and dashes of telegraph code, the wave is transmitted at constant amplitude for a time corresponding to the length of the dot or dash. This type of transmission, when combined with well-designed receiving systems, produces better reception than the damped waves or interrupted continuous waves which were employed at one time. A much narrower frequency channel is required for continuous wave transmission.

CONTOUR. A smooth curve composed of arcs bound together continuously, each arc having a common tangent. A closed contour may be decomposed into a series of closed smooth curves having no singular points. If the closed contour is traversed in a counterclockwise direction with respect to some point in the domain, the sense of the contour is positive.

Integration along a closed contour C is indicated by the symbols $\int_C$ or $\oint$ and the result is a contour integral.

In the calculus of residues, the Bromwich contour is often useful. It extends from $c - i\infty$ to $c + i\infty$, where c is real and positive and the path so chosen that all singular points of the complex function are to the left of it.

In surveying terms, a contour is a means by which the three dimensions necessary to represent a point in space can be shown on a map. These three variables are: length measured north and south, length measured east and west, and height, above some arbitrary reference plane. On a topographical map height is shown by contours. See also **Topographical Mapping.** All points on a given contour line will be at the same elevation. The accompanying figure shows the relation between height of a small irregular hill, and the corresponding contour lines as drawn upon a map. Occasionally special maps, instead of showing elevation as the third variable, give line of constant magnetic variation. See **Compass (Navigation).** These lines are called isogonic lines and the map is known as an isogonic chart.

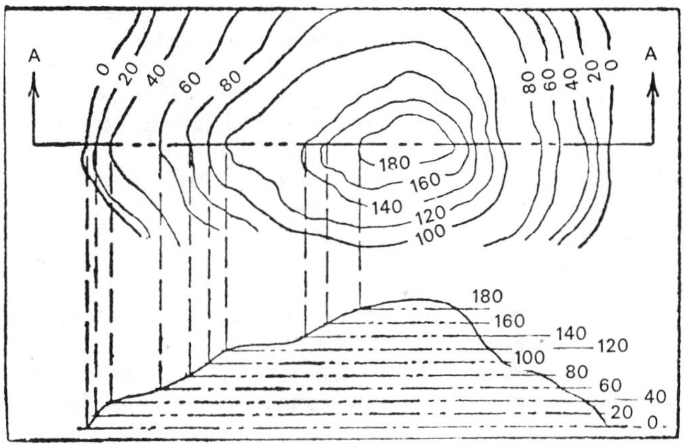

Contours.

While the contour line has its greatest use in land surveying and mapping, there are other technical uses. For example, the efficiency of a centrifugal pump varies both with the pumping head and with the discharge. Accordingly, centrifugal pump characteristics are often given with the efficiency displayed as contour lines upon a head-discharge plane.

CONTOURING CONTROLS. Numerical Control.

CONTOUR MINING (Coal). Coal; Revegetation.

CONTRACTILE FORCE (Interfacial). Interfacial Tension.

CONTRACTILE VACUOLE. A vacuole within the cytoplasm of many freshwater protozoans and a few metazoans which collects and expels excess water from the cell. In most species of *Paramecium* there are two contractile vacuoles. These can be seen in living forms under the microscope. They gradually expand and then suddenly contract as the water is forced from the cell. Water is constantly being taken into these cells by osmosis and along with food. Without the functioning of contractile vacuoles excess water would accumulate in the cell.

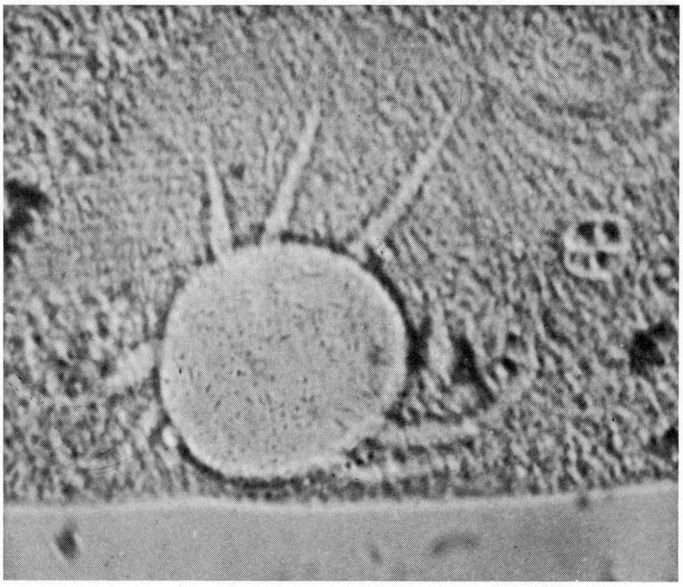

Contractile vacuole in paramecium. (*A. M. Winchester*)

CONTRACTILITY AND CONTRACTILE PROTEINS. The fundamental property of living matter on which its power of movement depends is termed contractility. In the simplest forms of living things, it is evident in the flowing movement of the material of the cell. In more complex forms, the property is centralized in muscle tissues. Muscle cells are elongate and are so arranged that necessary movements result from their shortening when stimulated.

The study of the fibrillar proteins of muscle invites interest for two reasons. On the one hand, they form the prime ingredients of the mechanism that performs a typical vital activity, movement, in its most specialized form; thus, the elucidation of their functions, and of the physical and chemical properties basic to it, represents one of the cardinal parts of molecular biology. On the other hand, the isolated proteins display such striking physical behavior that to the macromolecular physicist they are among the most fascinating materials. Unfortunately, they are difficult to obtain and exceedingly changeable and labile; thus, working with them is somewhat difficult.

The major fibrous proteins are myosin and actin. Other proteins that may be involved in myofibrillar structure are tropomyosin and paramyosin. All four proteins share certain chemical properties; among others, they are all exceptionally rich in ionizing amino acids, thus they are highly charged molecules. Myosin is an adenosinetriphosphatase. The molecular weight is not accurately known; a number of determinations cluster around 500,000. Light scattering dissymmetry suggests a mean molecular length of about 1600 micrometers, and electron microscopy suggests similar lengths.

See also **Muscle.**

CONTRAST. This term is used in physical science and statistics with at least four meanings: 1. In photography, the slope of the curve plotted between density and the log of exposure. 2. In television, the ratio between the maximum and minimum brightness values in a picture. 3. In psychophysics, the change in the response to a stimulus as a result of the proximity in space or time of other stimuli. In general, the response to a stimulus of given physical intensity is reduced if neighboring stimuli have greater intensities, and vice versa. A measure of contrast is the contrast sensitivity which may be expressed, for example, for brightness by the expression $\Delta B/B$, where B is the brightness of a background, and ΔB is the difference in brightness of a small spot that is just discernible against that background. 4. In statistics, a contrast between a set of independent observations x_i of equal accuracy is a linear function $\Lambda = \Sigma\ \lambda_i\ x_i$ in which the sum of the coefficients $\Sigma\ \lambda_i = 0$. If σ^2 is the variance of a single observation and Λ, Λ' are two contrasts with coefficients λ_i, λ'_i, then the variance of Λ is $\Sigma\ \lambda_i^2$ and the covariance of Λ and Λ' is $\Sigma\ \lambda_i \lambda'_i$. If $\Sigma\ \lambda_i \cdot \lambda'_i = 0$, the contrasts are said to be orthogonal.

CONTRAST (Statistics). A contrast between a set of independent observations x_i of equal accuracy is a linear function $\Lambda = \Sigma\ \lambda_i\ x_i$ in which the sum of the coefficients $\Sigma\ \lambda_i = 0$. If σ^2 is the variance of a single observation and Λ, Λ' are two contrasts with coefficients λ_i, λ'_i, then the variance of Λ is $\Sigma\ \lambda_i^2$ and the covariance of Λ and Λ' is $\Sigma\ \lambda_i\lambda'_i$. If $\Sigma\ \lambda_i \cdot \lambda'_i = 0$, the contrasts are said to be orthogonal.

CONTROL ACTION. Of a controller or a controlling system, the nature of the change of the output effected by the input. The output may be a signal or the value of a manipulated variable. The input may be the control loop feedback signal when the setpoint is constant, an actuating error signal, or the output of another controller.

Adaptive. Control action whereby automatic means are used to change the type of influence (or both) of control parameters in such a way as to improve the performance of the control system.

Cascade. Control action where the output of one controller is the setpoint for another controller.

Derivative (*Rate*). Control action in which the output is proportional to the rate of change of the input.

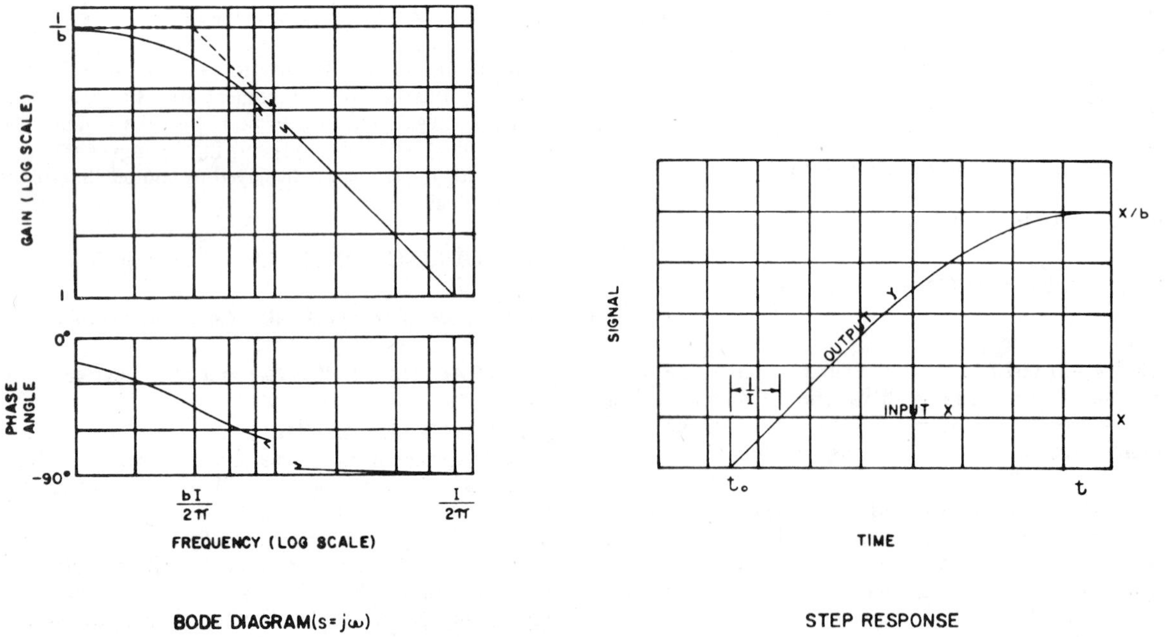

Fig. 1. Integral control action.

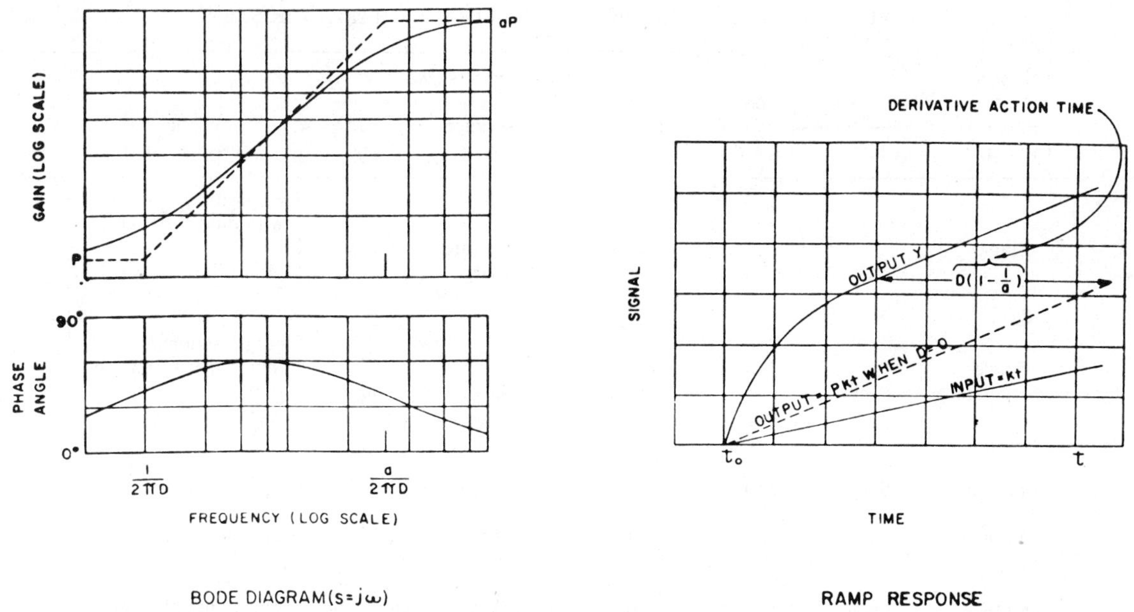

Fig. 2. Proportional-plus-derivative control action.

Direct Digital. Control action in which control is performed by a digital device which establishes the signal to the final controlling element.

Examples of possible digital (D) and analog (A) combinations for this definition are:

	Feedback Elements	Controller	Final Controlling Element
1.	D	D	D
2.	A	D	D
3.	A	D	A
4.	D	D	A

Feedback. Control action in which a measured variable is compared to its desired value to produce an actuating error signal which is acted upon in such a way as to reduce the magnitude of the error.

Feedforward. Control action in which information concerning one or more conditions that can disturb the controlled variable is converted into corrective action to minimize deviations of the controlled variable.

Feedforward control action can be combined with other types of control to anticipate and minimize deviations of the controlled variable.

High Limiting. Control action in which the output never exceeds a predetermined high limit value.

Integral (Reset). Control action in which the output is proportional to the time integral of the input, i.e., the rate of change of output is proportional to the input. See Fig. 1.

In the practical embodiment of integral control action, the relation between output and input, neglecting high frequency terms, is given by

$$\frac{Y}{X} = \pm \frac{I/s}{bI/s + 1}, \text{ where } 0 \leq b \ll 1$$

and

b = reciprocal of static gain
$I/2\pi$ = gain crossover frequency in cycles per unit time
s = complex variable
X = input transform
Y = output transform
I = integral action rate

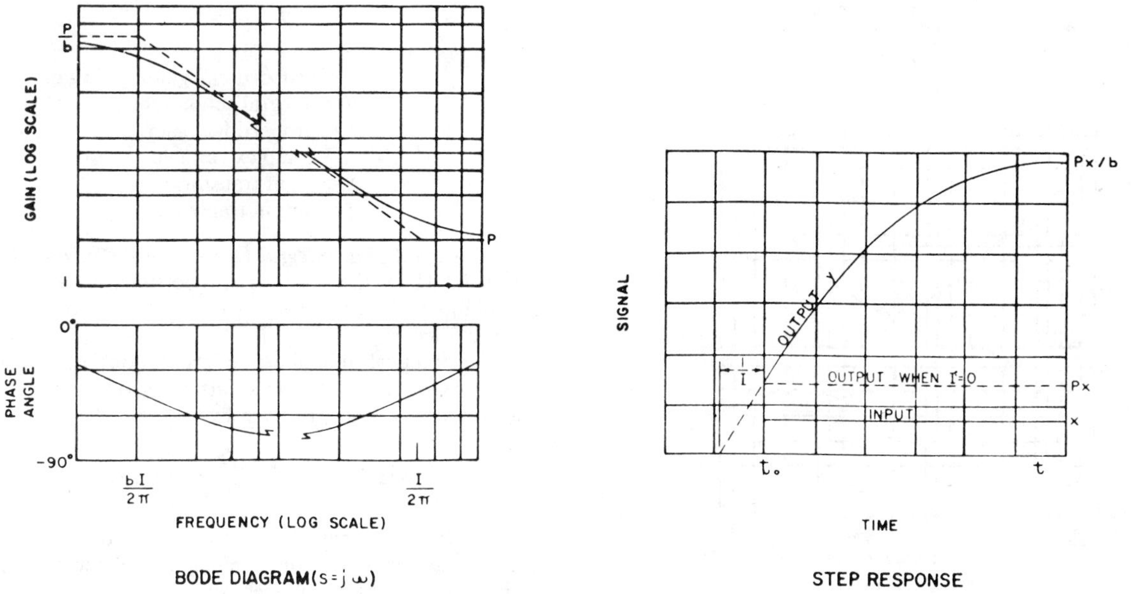

Fig. 3. Proportional-plus-integral control action.

TABLE 1. CONTROL ACTION VERSUS PROCESS CHARACTERISTICS

Number of Process Capacities	Load Changes		Process Reaction Rate	Process Lag Times		Appropriate Control Action
	Size	Speed		Resistance-capacity	Dead Time	
Single	Any Moderate	Any Slow	Slow	Moderate to large	Small	Two position. Two position with differential gap Multiposition. Time proportioning
Single (self-regulating)	Any	Slow Moderate	Fast	Small	Small	Floating actions: Single speed; multispeed Integral
Multiple	Small	Moderate	Slow to moderate	Moderate	Small	Proportional
Multiple	Small	Any	Moderate	Any	Small	Proportional plus derivative
Multiple	Large	Slow to moderate	Any	Any	Small to moderate	Proportional plus integral
Multiple	Large	Fast	Any	Any	Small	Proportional plus integral plus derivative
Any	Any	Any	Faster than that of the control system	Small or nearly zero	Small to moderate	Wideband proportional plus fast integral

Low Limiting. Control action in which the output is never less than a predetermined low limit value.

Optimizing. Control action that automatically seeks and maintains the most advantageous value of a specified variable, rather than maintaining it at one set value.

Proportional. Control action in which there is a continuous linear relation between the output and the input.

This condition applies when both the output and input are within their normal operating ranges and when operation is at a frequency below a limiting value.

Proportional plus Derivative (*Rate*). Control action in which the output is proportional to a linear combination of the input and the time rate-of-change of input. See Fig. 2.

In the practical embodiment of proportional plus derivative control action, the relationship between output and input, neglecting high frequency terms, is given by

$$\frac{Y}{X} = \pm P \frac{1 + sD}{1 + sD/a}, \text{ where } a > 1$$

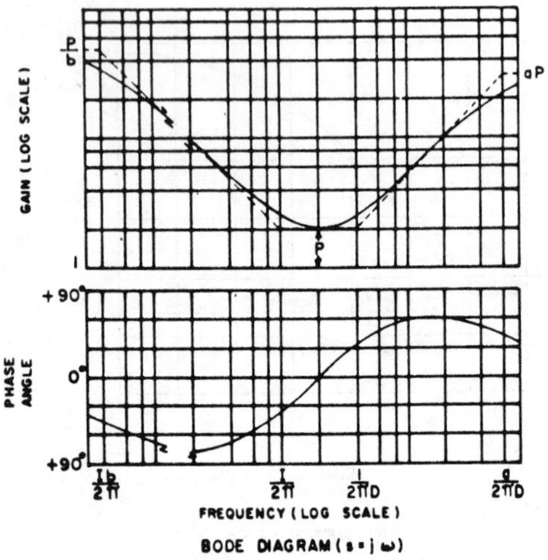

Fig. 4. Proportional-plus-integral-plus derivative control action.

and

a = derivative action gain
D = derivative action time constant
P = proportional gain
s = complex variable
X = input transform
Y = output transform

Proportional plus Integral (*Reset*). Control action in which the output is proportional to a linear combination of the input and the time integral of the input. See Fig. 3.

In the practical embodiment of proportional plus integral control action, the relation between output and input, neglecting high frequency terms, is given by

$$\frac{Y}{X} = \pm P \frac{I/s + 1}{bI/s + 1}, \text{ where } 0 \leq b \ll 1$$

and

b = proportional gain/static gain
I = integral action rate
P = proportional gain
s = complex variable
X = input transform
Y = output transform

Proportional plus Integral (*Reset*) *plus Derivative* (*Rate*). Control action in which the output is proportional to a linear combination of the input, the time integral of input and the time rate-of-change of input. See Fig. 4.

In the practical embodiment of proportional plus integral plus derivative control action, the relationship of output to input, neglecting high frequency terms, is given by

$$\frac{Y}{X} = \pm P \frac{I/s + 1 + Ds}{bI/s + 1 + Ds/a}, \text{ where } a > 1, 0 \leq b \ll 1$$

Shared Time. Control action in which one controller divides its computation or control time among several control loops rather than acting on all loops simultaneously.

Supervisory. Control action in which the control loops operate inde-

TABLE 2. CONTROL ACTION VERSUS PROCESS TYPES

Process Type	Gain (Proportional)	Integral Control Action	Derivative Control Action
Flow and liquid pressure	0.2–50 (500–2% PB)	Necessary	Not necessary
Gas pressure	20–∞ (5–0% PB)	Not necessary	Not necessary
Liquid level	2–20 (50–5% PB)	Occasionally needed	Not necessary
Temperature	1–10 (100–10% PB)	Necessary	Necessary

NOTE: PB = Proportional Band.

pendently subject to intermittent corrective action, e.g., setpoint changes from an external source.

Velocity Limiting. Control action in which the rate of change of a specified variable will not exceed a predetermined limit.

For a fundamental description of automatic control, see **Control System (Automatic).**

Each process or machine to be controlled requires individual analysis prior to firmly specifying the desired control action. However, certain generalities can be helpful and these are summarized in Tables 1 and 2.

CONTROL ALGORITHM. A general computational procedure which may include instructions, limits, and equations representing functional relationships in the controlling elements. In control applications the algorithm usually defines the functional relationship existing between the manipulated variable and the actuating or error signal.

CONTROL (Boiler). Boiler; Burner; Feedwater (Boiler).

CONTROL CHARACTER (Computer System). Character (Computer System).

CONTROL CHART (Statistical Quality Control). A graphical device used to show the results of small scale repeated sampling of a manufacturing process. It usually consists of a central horizontal line corresponding to the average value of the quantitative characteristic under investigation together with upper and lower limits between which a stated proportion of the sample statistics should fall. Any marked divergence above or below these control limits will tend to indicate that new causes are at work beyond those responsible for the random variations inherent in large scale production. A set of points outside the control limits will signal the need for special inquiries for the purpose of identifying the new factor(s) at work. The two lines are known as *upper and lower control limits*.

CONTROL COUNTER (Computer System). Counter (Computer System).

CONTROLLABLE-PITCH PROPELLER. Airplane.

CONTROLLED-ATMOSPHERE SYSTEMS (Storage). Hypobaric (Controlled-Atmosphere) Systems.

CONTROLLED RECTIFIERS. Rectifiers.

CONTROLLED THERMONUCLEAR REACTOR. Nuclear Reactor.

CONTROLLER (Automatic). A device which operates automatically to regulate a controlled variable. During the mid- and late 1970s and carrying into the 1980s, digital techniques have had a heavy impact on automatic controllers, particularly in those systems where there are numerous variables that must be controlled and related with each other to achieve an optimized control result. The fundamentals remain essentially the same, but the pathways to control, involving centralized computers, minicomputers, and microprocessors, have altered the so-called architecture of control systems, particularly those of a multiple and complex nature. A large percentage of the installed control instrumentation throughout the world, as of the early 1980s, is along conventional lines, but in terms of new installations, again of a multiple and complex nature, digital techniques are gaining on the older analog approaches at a rapid pace. There are, of course, many processing and manufacturing operations that do not require a high degree of sophistication—so that what was conventional instrumentation as of just a few years ago, will remain available for quite some time.

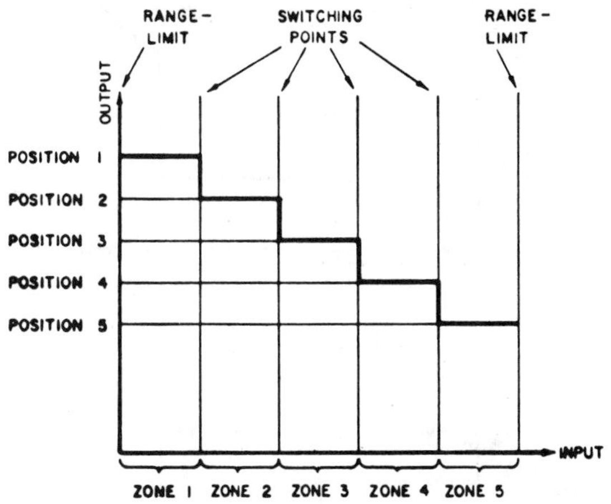

Fig. 1. Multiposition controller.

The major types of controllers may be described as follows:

Direct Acting. A controller in which the absolute value of the output signal increases as the absolute value of the input (measured variable) increases.

Floating. A controller in which the rate of change of the output is a continuous (or at least a piecewise continuous) function of the actuating error signal.

The output of the controller can remain at any value in its operating range when the actuating error signal is zero and constant. Hence the output is said to float. When the controller has integral control action only, the mode of control has been called "proportional speed floating." The use of the term integral control action is recommended as a replacement for "proportional speed floating control."

Integral (*Reset*). A controller which produces integral control action only. This type of controller also may be referred to as a proportional speed floating controller.

Multiple-speed Floating. A floating controller in which the output may change at two or more rates, each corresponding to a definite range of values of the actuating error signal.

Multiposition. A controller having two or more discrete values of output. See Fig. 1.

On-off. A multiposition controller having two discrete values of output, fully on, or fully off. See Figs. 2 and 4.

Proportional. A controller which produces proportional control action only.

Proportional plus Derivative (*Rate*). A controller which produces proportional plus derivative (rate) control action.

Proportional plus Integral (*Rate*). A controller which produces proportional plus integral (reset) control action.

Proportional plus Integral (*Reset*) *plus Derivative* (*Rate*). A controller which produces proportional plus integral (reset) plus derivative (rate) action.

Ratio. A controller that maintains a predetermined ratio between two or more variables.

Reverse Acting. A controller in which the absolute value of the output signal decreases as the absolute value of the input (measured variable) increases.

Sampling. A controller using intermittently observed values of a signal such as the setpoint signal, the actuating error signal, or the signal representing the controlled variable to effect control again.

Self-operated (*Regulator*). A controller in which all the energy to

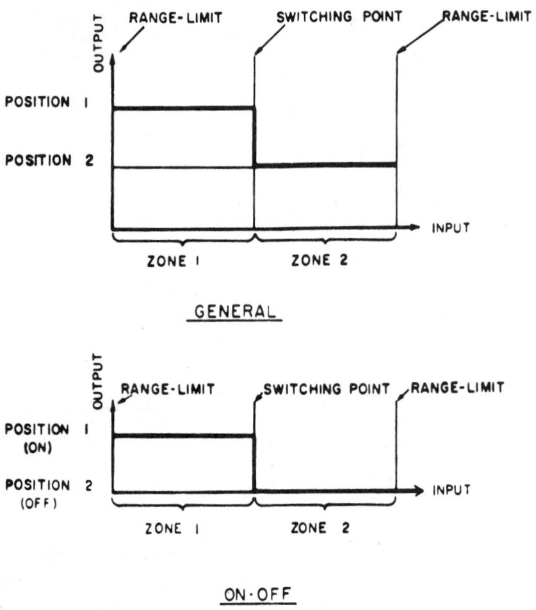

Fig. 2. Two-position controller.

operate the final controlling element is derived from the controlled system, through the sensing element.

Single-speed Floating. A controller in which the output changes at a fixed rate increasing or decreasing depending on the sign of the actuating error signal.

A neutral zone of values of the actuating error signal in which no action occurs may be used.

Three Position. A multiposition controller having three discrete values of output. See Fig. 3.

This is commonly achieved by selectively energizing a multiplicity of circuits (outputs) to establish three discrete positions of the final control element.

Time Proportioning. A controller whose output consists of periodic pulses whose duration is varied to relate, in some prescribed manner, the time average of the output to the actuating error signal.

Time Schedule. A controller in which the setpoint (or reference input signal) automatically adheres to a predetermined time schedule.

Two Position. A multiposition controller having two discrete values of output. See Figs. 2 and 4.

The control actions mentioned in the foregoing definitions are described under **Control Action.** For a fundamental description of automatic control, see **Control System (Automatic).**

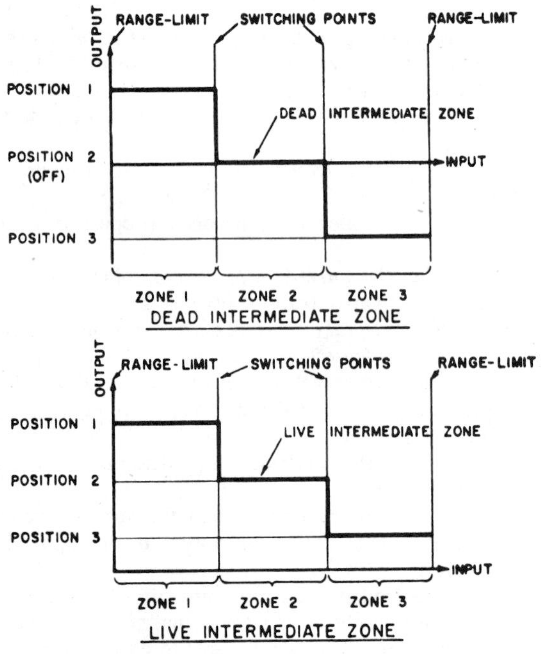

Fig. 3. Three-position controller.

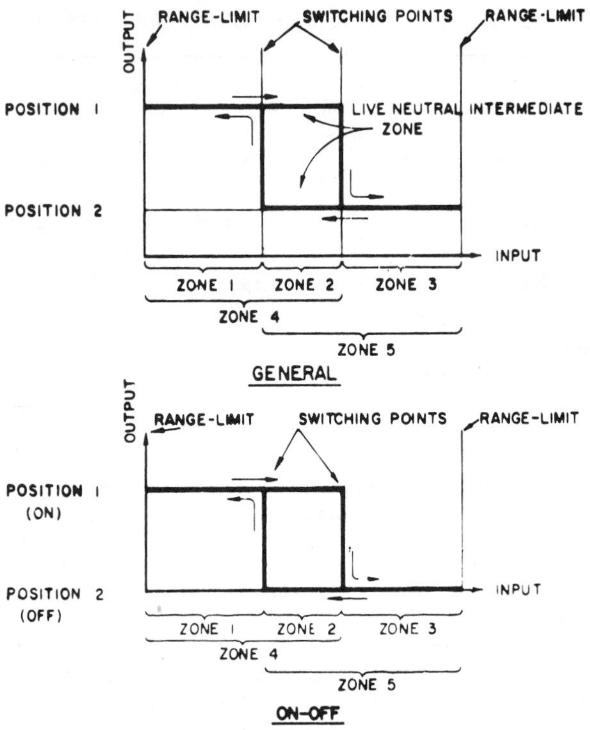

Fig. 4. Two-position controller with neutral intermediate zone.

CONTROLLER (Electric). Electrical Instruments.

CONTROLLER (Fluidic). Fluidics.

CONTROLLER (Hydraulic). Hydraulic Controller.

CONTROLLER (Pneumatic). Pneumatic Controller.

CONTROL ROD. Nuclear Reactor.

CONTROL SYSTEM (Automatic). A system in which deliberate guidance, with a minimum assistance of human intervention, is used to achieve a prescribed value of a variable (temperature, pressure, flow, etc.) or group of variables. An automatic control system may be very simple, such as the thermostatic regulation of a refrigerator, oven, or living space, to the full coordination of a complex manufacturing process. Automatic control systems are widely used in the process industries (so-called wet processes) in which variables ae controlled—as in an alkylation or catalytic cracking process found in a petroleum refinery; or in the manufacture of pulp for paper from wood feedstocks; or in the automatic batching, mixing, and melting of ingredients for making glass and other ceramics. Automatic control systems are also widely used in the discrete-piece manufacturing industries, as encountered in the automated machine shop, in engine assembly, and in the packaging of foods and other products. Automatic control systems find common application in all types of power-generating facilities—hydroelectric, geothermal, conventional fossil-fuel steam plants, solar and nuclear power facilities. Automatic control systems are also found in numerous aspects of transportation, ranging from autopilots for aircraft and automated navigation systems for ships and missile guidance to simpler applications as found in automotive electronic systems.

To explain the fundamentals of automatic control systems, the example used here is that which generally applies to the control of a process and for ease of comprehension, the hardware depicted is essentially of the analog (rather than digital) type. Control fundamentals apply to either analog or digital control systems, but to most persons it is easier to visualize the operation of an analog system—and then translate this understanding to digital modes. During the 1970s and continuing into the 1980s, the trend to digital equipment is strong. However, the amount of equipment currently installed is still preponderantly analog. As this equipment depreciates, much of it will be replaced, partially or nearly completely, by digital hardware. But, for many of the relatively simple applications of automatic control, it is expected that at least some analog hardware will be used for a long time.

A control system, for purposes of definition, can be subdivided into two systems: (1) a *controlling system*; and (2) a *controlled system*. With reference to Fig. 1, these and other definitions are described in the following paragraphs. Most definitions were initiated in the early 1970s by the Scientific Apparatus Makers Association (U.S.), with more recent attention given to this topic by the Instrument Society of America (ISA) and the American National Standards Institute (ANSI). The effects of these activities have been felt worldwide, but outstanding contributions to clarity of definitions and uniformity of standards and practices have also been made by several other countries.

Controlling System. 1. Of a *feedback control system*, that portion which compares functions of a directly controlled variable and a setpoint and adjusts a manipulated variable as a function of the difference. It includes the reference input elements, summing point, forward and final controlling elements, and feedback elements (including sensing element). 2. Of an automatic control system *without feedback*, that portion of the control system which manipulates the controlled system.

Controlled System (*or Process*). The collective functions performed in and by the equipment in which the variable(s) is(are) to be controlled. Equipment as embodied in this definition should be understood not to include any automatic control equipment. The terms *process* and *controlled system* are interchangeable.

Variable (*Directly Controlled*). In a control loop, that variable whose value is sensed to originate a feedback signal.

Variable (*Indirectly Controlled*). A variable which does not originate a feedback signal, but which is related to and influenced by the directly controlled variable.

As shown by Fig. 1, the directly controlled variable is the temperature of the fluid in the tank. An indirectly controlled variable is the vapor pressure above the level of the fluid in the tank.

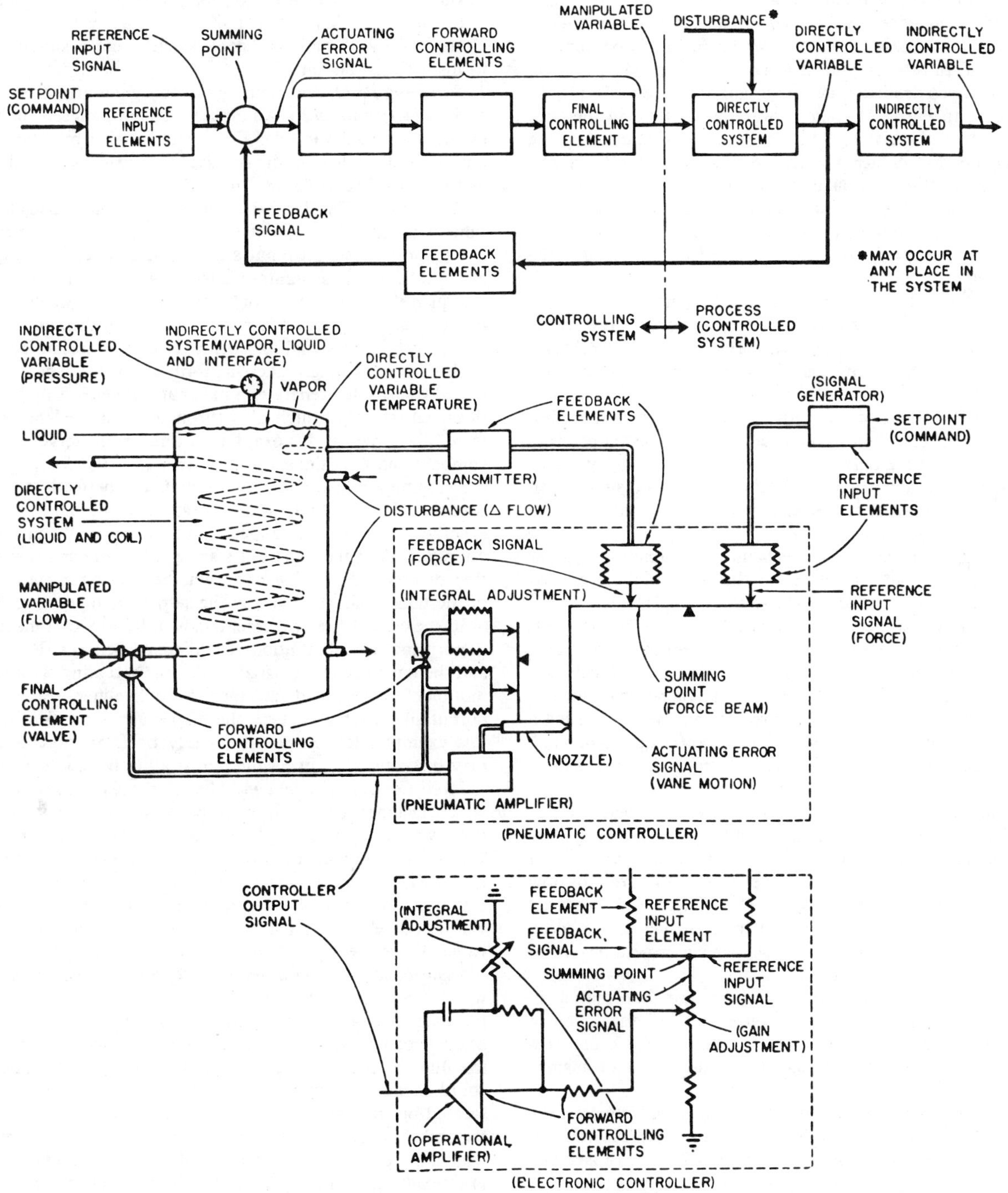

Fig. 1. Representative automatic control system, indicating major elements and terminology. For ease of comprehension, analog hardware has been featured in this example. Much of this hardware is also available in digital format. (*Scientific Apparatus Makers Association.*)

There are numerous variables which may be controlled, depending upon the process involved. A process need not be a process in the normal sense, i.e., a heating, cooling, or reacting process, but for the purposes of definition here also includes machines whose operation is dependent upon the manipulation and control of variables. Variables include temperature (as in the example of Fig. 1), pressure, the level of liquids (as in tanks, reactors, and other process vessels), the level of bulk solids (as in bins and silos), the flow of a liquid or gas in a pipeline, the flow of a bulk solid (as on a conveyor), the density or specific gravity of liquids and gases, the viscosity of fluids, the moisture content of materials, the chemical composition of materials, and several other variables, the control of which affects such factors as final product quality, yield, and safety. Some typical process operations in which control is important include drying, distilling, absorbing, adsorbing, centrifuging, filtering, heating/cooling, compressing, electroplating, reacting, power generation, and dozens of others.

Variable (Manipulated). A quantity or condition which is varied as a function of the actuating error signal so as to change the value of the directly controlled variable. In any practical control system, there may be more than one manipulated variable. Accordingly, when using the term, it is necessary to state which manipulated variable is being discussed. In process control work, the one immediately preceding the directly controlled system is usually intended.

With reference to Fig. 1, the temperature of the fluid in the tank (the directly controlled variable) is held at a steady value by adjusting (manipulating) the flow of steam to the heating coil of the tank. The flow of steam is thus the manipulated variable.

Feedback Control Action. This is a control action in which a measured variable is compared to its desired value to produce an actuating error signal which is acted upon in such a way as to reduce the magnitude of the error.

Measured Variable. The physical quantity, property, or condition which is measured. Sometimes this is referred to as the measurand. In the example of Fig. 1, the measured variable is the temperature of the fluid in the tank. In this case, it is the same as the directly controlled variable.

Desired Value. The value of the controlled variable wanted or chosen. In the example of Fig. 1, the desired value refers to the temperature of the water desired in the tank.

Signal. Information about a variable that can be transmitted. In the example, this is information pertaining to the temperature of the water in the tank.

Measured Signal. The electrical, mechanical, pneumatic, or other variable applied to the input of a device. It is the analog of the measured variable produced by a transducer (when such is used).

In the case of a filled-system thermometer for measuring temperature, as shown in Fig. 1 measuring the tank temperature, the change of temperature surrounding the sensitive thermometer bulb causes a change in the pressure within the filled system, an increasing temperature causing a greater internal system pressure and vice versa for a decreasing temperature. This pressure is constantly detected, measured, and compared by the thermometer instrument. Thus, the measured signal in this instance is pressure.

In the case of a thermocouple-type thermometer, the measured signal is an emf which is the electrical analog of the temperature applied to the thermocouple. In the case of a flowmeter, the measured signal may be a differential pressure which is the analog of the rate of flow through an orifice. In an electric tachometer system, the measured signal may be a voltage which is the electrical analog of the speed of rotation of the part coupled to the tachometer generator. In all of these examples, transducers are involved.

Transducer. An element or device which receives information in the form of one physical quantity and converts it to information in the form of the same or some other physical quantity. Types of transducers include a primary element, a signal transducer, and a transmitter.

Primary Element. The system element that quantitatively converts measured variable energy into a form suitable for measurement. In the example of Fig. 1, the thermometer bulb is the primary element.

Signal Transducer. Also called signal converter, a transducer which converts one standardized transmission signal to another, as in the case of converting from pneumatic to equivalent electric signals and vice versa.

Transmitter. A transducer which responds to a measured variable by means of a sensing element, and converts it to a standardized transmission signal which is a function only of the measurement. It will be noted from Fig. 1 that the measured signal (pressure in the thermometer system) goes directly to a transmitter. Here that signal is converted into a pneumatic signal, lying within a range of 3 to 15 psi (0.44 to 2.18 kPa).

Input Signal. A signal applied to a device, element or system.

Return Signal. In a closed loop, the signal resulting from a particular input signal, and transmitted by the loop and to be subtracted from the input signal.

Error Signal. In a closed loop, the signal resulting from subtracting a particular return signal from its corresponding input signal.

Actuating Error Signal. The reference-input signal minus the feedback signal.

Reference-Input Signal. One external to a control loop which serves as the standard of comparison for the directly controlled variable.

Closed Loop. Also termed *feedback loop*, a signal path which includes a forward path, a feedback path, and a summing point, and forms a closed circuit.

Summing Point. Any point at which signals are added algebraically.

Setpoint (Command). An input variable which sets the desired value of the controlled variable. The input variable may be manually set, automatically set, or programmed. It is expressed in the same units as the controlled variable.

Returning to Fig. 1 and relating the foregoing terms to the illustration, note that the setpoint (either manually set or remotely set by a program) causes an input signal to the system to be generated. In essence, the signal generator translates a dial setting into a corresponding quantitative electrical or pneumatic signal. In this case, a pneumatic signal is shown. After this translation, the command is termed a reference input signal and through an appropriate bellows element in this case, the signal takes the form of a force. The previously mentioned transmitter generates a like, but not necessarily an equal, signal (feedback signal), which also through a matched bellows takes the form of a force. In this example, a force beam serves to algebraically compare the two signals and in this case the force beam becomes the summing point. If the two signals are perfectly equal, no action occurs, indicating that the temperature of the fluid in the tank is at exactly the value commanded by the setpoint. When the signals are not equal, the balance assumes an unbalanced condition in one direction or other, depending upon whether the temperature of the fluid in the tank is above or below the setpoint temperature.

In the case of the instrument shown, by way of a movable vanestationary nozzle configuration, the difference between the feedback signal and the reference input signal (called the actuating error signal) causes motion of the vane and via a pneumatic amplifier creates the controller output signal which causes the final controlling element (a valve in this example) to become more fully open or more fully closed, as may be required to bring the temperature back into line.

Final Controlling Element. That element in the controlling system which changes a variable in response to the actuating error signal, synonymous in the example of Fig. 1 to the controller output signal. The valve shown in this case is a pneumatically-actuated diaphragm motor valve. In the case of an electric controller, the steam valve may be a motor-operated valve, or an electropneumatic operator which will take an electric signal and convert it for operation of a diaphragm valve. There are many possible hardware configurations.

Disturbance. An undesired change in a variable applied to a system which tends to affect adversely the value of a controlled variable. In the example, one form of disturbance could be a greater or lesser drawoff of heated fluid from the tank because of varying needs for the fluid at point of use. Or, the pressure in the steam supply line could change, altering the Btu content of the steam. If the tank were located outside, ambient conditions could change. In the long-term, the coil could scale up, thus affecting heat-transfer efficiency. In other words, there are numerous factors which motivate against maintenance of a steady state of the system. If this were not the case, an automatic control system hardly could be justified.

Because the system has capacity, an adjustment of the steam supply

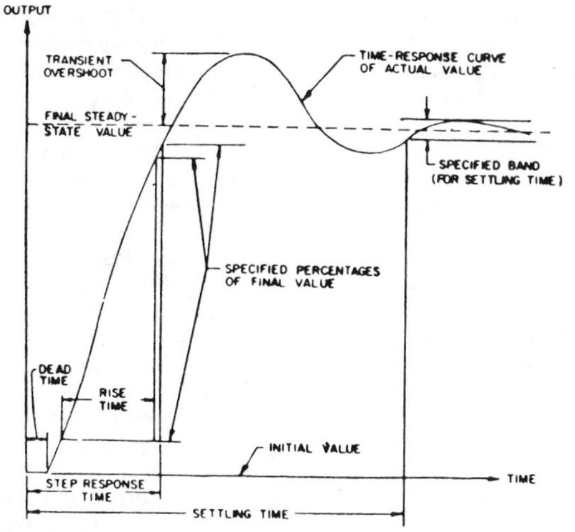

Fig. 2. Typical time response of a system to a step increase of input.

valve will not be immediately sensed by the thermometer bulb and hence a warming or cooling trend will not be corrected instantaneously.

Dead Time. The interval of time between initiation of an input change or stimulus and the start of the resulting response. This is illustrated in Fig. 2. Dead time also is referred to as *distance-velocity lag* or *transportation lag*. Terms included in Fig. 2 are defined as follows:

Rise Time. The time required for the output of a system (other than first order) to make the change from a small specified percentage (often 5 or 10) of the steady-state increment to a large specified percentage (often 90 to 95), either before or in the absence of overshoot. If the term is unqualified, response to a unit step stimulus is understood, otherwise the pattern and magnitude of the stimulus should be specified.

Step Response Time. Of a system or an element, the time required for an output to make the change from an initial value to a large specified percentage of the final steady-state value either before or in the absence of overshoot, as a result of a step change to the input. Usually stated for 90, 95, or 99% change.

Settling Time. The time required, following the initiation of a specified stimulus to a system, for the output to enter and remain within a specified narrow band centered on its steady-state value. The stimulus may be a step impulse, ramp parabola, or sinusoid. For a step or impulse, the band is often specified as $\pm 2\%$. For nonlinear behavior, both magnitude and pattern of the stimulus should be specified.

Depending upon the nature of the process and, in particular, the manner and speed with which a process responds to change and correction, the type of automatic control system best suited may range from a simple on–off controller to a proportional plus integral (reset) plus derivative (rate) controller. The cost of the control system versus the desired control results often is a tradeoff. However, it usually is best to err on the side of the more sophisticated control because even just a fraction of a percentage increase in product quality or yield multiplied by the many thousands of days the control system probably will be in use can justify the very best control system for a given situation.

Fig. 3. Plants in the process industries deal largely with liquids, gases and bulk solid materials that are handled within enclosures—vessels, pipes, tanks, conveyors, silos, and the like. The operator seldom has opportunity to see what is taking place, but must depend upon instruments to measure flow rates, liquid levels, pressures, temperatures, viscosities, chemical compositions, and numerous other variables. Further, the operator's capacity to react to a myriad of changing conditions must be grossly amplified by way of automatic control devices, involving analog and digital computers within the control loop. This installation is representative of a control room in a modern petroleum refinery, and is located at the Beaumont, Texas refinery of Mobil Oil Corporation. (*Foster Wheeler Corporation.*)

The numerous types of control actions available are defined under **Control Action.** In addition to matching the control action with the needs of the process, the type of control medium must be selected. Generally, these are of the following categories: (1) *pneumatic,* in which air pressure values comprise the control system signals, (2) *electric,* in which either millivolts or milliamps are the control system signals, and (3) *hydraulic,* usually used where powerful final controlling elements are required. Conversion of signals from one medium to the next—pneumatic to electric, electric to pneumatic, electric to hydraulic, and even double conversions, such as pneumatic to electric back to pneumatic, or electric to pneumatic and back to electric, are rather commonly encountered in control system technology. The task of selecting control system equipment is further complicated by the rather wide variety of primary elements obtainable. For example, just for the measurement of temperature, there are thermocouples, various forms of radiation and optical pyrometers, resistance thermometers, filled-system thermometers (used in the example of Fig. 1), bimetallic thermometers, and so on. However, because of the vast variety of requirements posed by thousands of different processes (and machines), it is fortunate that such a wide variety of hardware is available. See Fig. 3.

Human Factors in Control System Design. For a long period extending from the 1930s through the mid-1950s, the control room of a large manufacturing facility featured large instruments (indicators and recorders, many of which were equipped with control-adjusting means) mounted on long panels. Such panels were usually sectionalized to make it effective and convenient for two or more operators to monitor and adjust a series of related instruments. In the much earlier days of process instrumentation (prior to World War II), the majority of instruments were locally mounted out on the process so to speak, with few, if any, of the measurements transmitted to a central control room. Developments in both the fields of pneumatic and electric transmission of measurement data made it possible to put nearly all of the instruments in a central control facility. This greatly improved coordination of the process, but did introduce much additional expenditure in transmission tubing or wiring. As processes became more complex, requiring many more variables to be measured and controlled, the control rooms and instrument panels in them became exceedingly large.

As a partial answer to this problem, indicating, recording, and controlling instruments were miniaturized. This trend, of course, greatly increased the information display density on a control panel, which on the one hand, made it more physically convenient for the operators, but on the other hand, tended to contribute to operator confusion; thus there was introduced a much greater attention to the human engineering factors involved in the operator/process or machine interface. So-called graphic instrument panels were introduced in the early 1950s, in which the smaller (miniaturized) instruments were grouped on the panel within a diagram of the process or machine being controlled. This approach improved the immediate identification of measurement data with a given part of the process. In the 1960s and continuing to the present era, the use of cathode ray tube (CRT) displays has been stressed. The CRT, for example, ties in very well with measurement data in digital input formats. Further, the CRT screen can be time-shared among numerous variables. The use of the CRT increased at a rapid pace as the control systems proper became more and more digitalized, with the introduction not only of central computers, but of minicomputers, miniprocessors, and the like.

Instrumentation and automatic control professionals have become increasingly concerned with the operator/process interface. These activities have been spurred by the thorough analysis of this problem in connection with the handling of emergency situations, such as that which arose at the nuclear reactor located on Three Mile Island near Harrisburg, Pennsylvania. Further attention to this topic is given in this book. See the entry on **Human Factors Engineering.** See also the entry on **Data Processing,** which includes a long list of specific entries included in this book. See also the entries listed below:

Actuator (Control System)
Adaptive Control
Auctioneering Control
Cascade Control
Computer Process Control
Control Action
Controller (Automatic)
Dead Band
Dead Time
Degeneration (Control System)
Derivative (Rate) Control
Diaphragm Actuator and Valve
Direct Acting Controller
Electric and Electronic Controllers
Feedback
Feedback Control
Feedforward Control
Final Controlling Element
Floating Controller
Hunting
Hydraulic Controller
Hysteresis (Instrument)
Integral (Reset) Control
Intrinsic Safety
Loop Gain
Multiposition Controller
Numerical Control
Optimizing Control
Pneumatic Controller
Primary Element
Process Control
Process (Control System)
Proportional Control
Proportional Plus Derivative (Rate) Control
Proportional Plus Integral (Reset) Control
Proportional Plus Integral (Reset) Plus Derivative (Rate) Control
Range and Span (Instrument)
Ratio Controller
Remote Control
Reverse Acting Controller
Sampling Controller
Self-Operated Controller
Selsyn
Sensitivity (Instrument)
Sensor (Measurement)
Servomechanism
Settling Time
Shared Time Control
Stability (System)
Step Response Time
Supervisory Control
Time Constant
Time Proportioning Controller
Time Schedule Controllers
Timing Controllers
Transducer
Valve (Control)
Variable (Process)
Velocity Limiting Control

References

AIChE: "Industrial Process Control," Publication W-10, American Institute of Chemical Engineers, New York, 1979.
AIChE: "Chemical Process Control," Publication S-159, American Institute of Chemical Engineers, New York, 1976.
Considine, D. M. (editor): "Process Instruments and Controls Handbook," 2nd edition, McGraw-Hill, New York, 1974.
Farmer, E.: "Planning and Implementing New Control Systems in Existing Process Plants," Instrument Society of America, Research Triangle Park, North Carolina, 1979.
Hougen, J. O.: "Measurements and Control Applications," Instrument Society of America, Research Triangle Park, North Carolina, 1979.
ISA: "Standards and Practices for Instrumentation," 6th edition, Instrument Society of America, Research Triangle Park, North Carolina, 1980.
Morris, H. M.: "Operator Convenience is Key as Process Controllers Design Evolves," *Control Engineering,* **28**(3), 65–66 (1981).
Schetzen, M.: "The Volterra and Wiener Theories of Nonlinear Systems," Wiley, New York, 1980.
Shinskey, F. G.: "Distillation Control: for Productivity and Energy Conservation," McGraw-Hill, New York, 1977.
Staff: "Trends in Control," *Control Engineering,* **28**(6), 170–172 (1981).
Willard, H. H., et al.: "Instrumental Methods of Analysis," Van Nostrand Reinhold, New York, 1981.
Wise, K. D., Chen K., and R. E. Vokley: "Microcomputers: A Technology Forecast and Assessment to the Year 2000," Wiley, New York, 1980.

CONTROL TRACK. A supplementary sound track, usually placed on the same film with the sound track carrying the program material. Its purpose is to control, in some respect, the reproduction of the sound track. Ordinarily, it contains one or more tones, each of which may be modulated either as to amplitude or frequency.

CONTROL VALVE (Fluidic). Fluidics.

CONTROL VALVE (Pneumatic and Electric). Valve (Control).

CONTUSION. Brain Injury.

CONURE. Parrots and Cockatoos.

CONVALLARIACEAE. Asparagus.

CONVECTION (Heat). Heat Transfer.

CONVECTION (Meteorlogy). Atmosphere (Earth).

CONVECTION SURFACES (Boiler). Boiler; Feedwater (Boiler).

CONVERGENCE. A sequence $\{s_n\}$ converges if it has a limit. The same statement holds for an infinite series s_n if that symbol is regarded as the sum to n terms of the series $s_1 + (s_2 - s_1) + (s_3 - s_2) + \cdots + (s_n - s_{n-1})$.

There are several ways in which a sequence or series can be examined for convergence (see **Cauchy Convergence Test; D'Alembert Test; Raabe Test**).

When a series is shown to be convergent, it can be further classified as absolutely convergent, if the sum of its terms without regard to sign approaches a limit; nonabsolutely convergent (or divergent), if it is not absolutely convergent. The series $S_1 = 1 - \frac{1}{4} + \frac{1}{9} - \frac{1}{16} \pm \cdots$ and $S_2 = 1 - \frac{1}{2} + \frac{1}{3} - \frac{1}{4} \pm \cdots$ are typical examples of these two cases. If a series continues to give the same sum no matter how its terms are added, it is absolutely convergent; otherwise, conditionally convergent. Examples of these cases are $\Sigma\,(-1)^{n+1}/2^n$ and $\Sigma\,(-1)^{n-1}/n$.

The foregoing rules will not cover all cases, but are illustrative of the procedures used to study the convergence properties of series.

The term convergence is widely used in science and engineering in the sense of motion toward a point or limited area or volume, i.e., motion attended by increasing concentration.

See also terms listed under **Mathematics.**

CONVERGENCE (Meteorology). Atmosphere (Earth).

CONVERGENT ADAPTATION. Adaptation (Ecology).

CONVERSION. 1. In its most general usage, this term denotes a change, often with the force of a directed or induced change. One specific use is a change in numerical value of a quantity resulting from the use of a different unit in the same or a different system of measurement. 2. An intramolecular rearrangement of organic substances in which the relative positions of the radicals are modified. The Beckmann rearrangement is a case in point. Radicals may be transferred from carbon, oxygen, or nitrogen to carbon; from carbon or oxygen to nitrogen; from side chains to nucleus, etc. 3. The process of changing information from one form of representation to another; such as, from the language of one type of machine to that of another or from magnetic tape to the printed page. 4. The process of changing from one data processing method to another, or from one type of equipment to another; e.g., conversion from punch card equipment to magnetic tape equipment.

CONVERSION COATINGS. The industrial application of organic finishes to metals almost always requires the use of an intermediate conversion coating, particularly when the performance demands are high. Conversion coatings are formed chemically by causing the surface of the metal to be "converted" into a tightly adherent amorphous or crystalline coating, part or all of which consists of an oxidized form of the substrate metal. Conversion coatings can provide high corrosion resistance as well as strong affinity for organic coatings. They are also useful as lubricants for the drawing and forming of metals and sometimes are used for decorative purposes. The most important and widespread use of conversion coatings is on steel, zinc or galvanized steel, and aluminum alloys. The most widely used classes of conversion coatings for use on these metals are the phosphates and chromates. Depending on size, shape, volume of production, and other factors, the coating may be applied by spray, immersion, roll coat, or brush. A typical metal pretreatment sequence is: (1) cleaning; (2) rinsing; (3) conversion coating; (4) rinsing; and (5) final rinsing.

Phosphating of Steel. When a ferrous alloy is immersed in phosphoric acid, it initially forms a soluble phosphate. As the pH rises at the metal/solution interface, the phosphate becomes insoluble and crystallizes epitaxially on the substrate metal. The phosphate coating thus produced consists of a nonconductive layer of crystals that insulates the metal from any subsequently applied film and provides a topography with enhanced "tooth" for increased adhesion. The crystals insulate microanode and microcathode centers caused by stress or imperfections in the metal surface. This greatly reduces the severity of electrochemical corrosion.

Practically all phosphating processes involve patented proprietary solutions which produce superior coatings in shorter times and at lower temperatures than are obtainable with phosphoric acid alone. Treating times have been reduced from the earlier, 1–2-hour to 1–5-minute periods, while temperatures have been reduced from 98°C to 20–30°C.

Increasing energy costs have led to more widespread use of low-temperature phosphatizing baths. These comprise a lower free acid content which results in higher saturation of the primary zinc phosphate and thus provides a greater tendency to deposit at temperatures at or close to ambient temperature. In many cases, depending on the type of soil, it is difficult to reduce the cleaner temperature and, as a result, the cleaned parts tend to heat the phosphatizing stage to approximately 50°C even when no external heat is used.

Iron Phosphate Coatings. These coatings, weighing about 40–60 milligrams per square foot (430–646 milligrams per square meter), provide good paint adhesion, but inferior heat and corrosion resistance. They are used when coating performance is not very demanding. Iron phosphate coatings appear as a very thin blue or brown film to the naked eye.

Crystalline Zinc Phosphate Coatings. These coatings, weighing about 200 milligrams per square foot (2152 milligrams per square meter) when applied by spray, or as much as 3000 milligrams per square foot (32,280 milligrams per square meter) when applied by immersion, are of a medium-gray color and are used when higher quality is mandatory. For even greater corrosion resistance and paint adhesion, *microcrystalline zinc phosphate* coatings are used. They are dark in color and give coating weights of about 150–200 milligrams per square foot (1614–2152 milligrams per square meter) when sprayed and up to 1000 milligrams per square foot (10,760 milligrams per square meter) when applied by immersion. They are used as a paint base for products which are expected to last for years under varying environmental conditions.

Phosphating Chemistry. Iron and iron oxide react with phosphoric acid to form soluble primary iron phosphate, $Fe(H_2PO_4)_2$, liberating H_2 and H_2O, respectively. Because of the consumption of acid at the metal interface, there is a local rise in pH, which causes insoluble secondary iron phosphate, $FeHPO_4$, to coat the metal. The iron in the coating is supplied by the substrate.

In zinc phosphating, a small amount of iron phosphate is formed initially, but the bath contains primary zinc phosphate, $Zn(H_2PO_4)_2$, which crystallizes on the metal surface as secondary and tertiary zinc phosphates, $ZnHPO_4$ and $Zn_3\,(PO_4)_2$, respectively, when the pH rises at the metal/solution interface. The most frequently used baths contain accelerators, preferably nitrates and nitrites, which oxidize the hydrogen formed by the pickling reactions. The fundamental zinc phosphate reactions occur in three steps, all in the same bath:

(Pickling)

$$Fe^0 + 2H^+ \rightarrow Fe^{++} + H_2$$
$$3Fe^0 + 2NO_3^- + 8H^+ \rightarrow 3Fe^{++} + 2NO + 4H_2O$$

(Coating)

$$3Fe^{++} + 2H_2PO_4^- \rightarrow 4H^+ + Fe_3(PO_4)_2$$
$$3Zn^{++} + 2H_2PO_4^- \rightarrow 4H^+ + Zn_3(PO_4)_2$$

(Iron Removal)

$$4Fe(H_2PO_4)_2 + O_2 \rightarrow \underset{\text{(sludge)}}{4FePO_4} + 4H_3PO_4 + 2H_2O$$
$$Fe(H_2PO_4)_2 + NaNO_2 \rightarrow \underset{\text{(sludge)}}{FePO_4} + NO + H_2O + NaH_2PO_4$$

In the coating reaction, 3 moles of iron or zinc liberate 4 moles of hydrogen ion. However, in the pickling reaction, 8 moles of hydrogen ion are consumed for 3 moles of iron. Thus, the pH at the metal interface rises, and insoluble tertiary ferrous phosphate and zinc phosphate crystallize on the iron surface. The coating closest to the metal interface is largely iron phosphate, while that farther away is rich in zinc phosphate.

Iron buildup in the bath is objectionable; in the iron-removal equations above, it is seen that dissolved Fe^{++} can be removed by oxidation, slowly in air or more rapidly by peroxides or nitrate, as shown in the final equation. The iron removed becomes ferric phosphate, while iron in the coating is ferrous phosphate.

Accelerators speed phosphating reactions by reacting with hydrogen liberated at the metal. Were hydrogen not removed, it would form gas bubbles which would interfere with metal/solution contact. Strong oxidizer accelerators also serve to precipitate dissolved iron and, to a degree, act as metal cleaners by oxidizing residual organic soils. In spray application, mild accelerators are usually adequate for maintaining dissolved iron at safe levels because atomization of the solution permits it to absorb from air the oxygen needed for precipitation of iron.

Zinc phosphate coatings consist of varying ratios of hopeite, $Zn_3(PO_4)_2 \cdot 2H_2O$, and phosphophyllite, $Zn_2Fe(PO_4)_2 \cdot 4H_2O$, with hopeite usually predominant. Hopeite and phosphophyllite grow epitaxially on alpha-iron crystallites. Only a slight adaptation deformation is necessary for the lattice planes of both foreign phases compared with the alpha-iron lattice of the substrate. Good adhesive strength can be expected from such a bond.

The smaller the crystals, the better the adhesion: crystal-to-metal, crystal-to-crystal, and crystal-to-final-finish. Also, the smaller the crystals, the tighter the packing, the denser the coating, the less total porosity area for corrosive reactions to take place.

Crystal size depends upon such factors as growth rate, agitation, and the effects of nucleating agents and foreign atoms in the crystal lattice. Spray application provides agitation which reduces crystal size. Accelerators increase the number of nucleation sites on the substrate and result in smaller crystals. The most refined technique for producing very small crystals is the introduction of foreign elements of different atomic radii into the crystal lattice. A calcium additive in a zinc phosphating solution produces scholzite, $CaZn_2(PO_4)_2 \cdot 2H_2O$, in which the crystals may average one-twentieth the size of the finest crystals produced by other methods. It is believed that the foreign elements cause uneven growth along one crystal face, creating stresses that either stunt growth at an early stage, or rupture the crystal.

The relatively recent surge in the use of cathodic electrodeposition, especially in the automotive industry, has necessitated change in metal pretreatment. The generation of hydroxyl ions at the cathode where the organic coating is deposited tends to dissolve zinc from the conventional zinc phosphate conversion coating with consequent deleterious coating performance. By increasing the concentration of phosphoric acid in the zinc phosphatizing stage, the ratio of iron to zinc in the conversion coating is increased. Such conversion coatings show superior performance under cathodically electrodeposited coatings.

The following sequence is typical of a production spray phosphate system: (1) Cleaning for 60 seconds at 71–77°C; (2) rinsing for 15–30 seconds (hot or cold); (3) phosphating for 60 seconds at 54–60°C; (4) rinsing for 15–30 seconds (cold); and (5) chromate-rinsing for 30–45 seconds at 27–60°C. The final stage, an acid chromate rinse, is extremely important to the overall adhesion and corrosion resistance of the finish. One of its functions is to seal the pores in the phosphate coating. The effect can be demonstrated by placing a drop of chromic acid in the center of a phosphatized panel and subjecting it to corrosive exposure. Corrosion in the chromated area will be strongly inhibited.

Phosphating of Zinc. The chemistry in the phosphating of zinc alloys is similar, with the exception that iron does not play an important role in the coating or in the bath. The only cation involved is Zn. Zinc phosphate coatings are used widely in the treatment of galvanized steel for refrigerators, air conditioners, kitchen cabinets, and house and building siding.

Aluminum Pretreatment. Although Al protects itself against corrosion by forming a natural oxide, the protection is not complete. In the presence of moisture and electrolytes, Al alloys, particularly the high-copper alloys, corrode much more rapidly than pure Al.

Chemically produced oxides can be formed by treatment in 2–3% sodium carbonate containing 0.1% sodium dichromate for 10–20 minutes at 66°C, followed by "sealing" in 5% sodium dichromate at 82–88°C for 10 minutes. Such coatings are softer, more porous, and not as effective as those produced by chromic acid anodizing.

Electrically produced anodic coatings from chromic, sulfuric, or oxalic acid electrolytes are more dense and less porous. Their corrosion resistance is improved by hot-water sealing, which is even more effective with the inclusion of dichromate.

Generally, the higher the alloy content of the base aluminum, the heavier the oxide present and the more difficult it is to remove prior to effective chemical pretreatment. A properly formulated deoxidizer will remove the oxide only to the desired degree, with minimum attack on the base metal. The baths usually consist of fairly high concentrations (5–10%) of either nitric or sulfuric acid, along with chromates and either free or complex fluorides. A deoxidizer will also remove the smut that forms on etching in strong alkali. Smut consists of the alkali-insoluble alloying elements and their oxides.

Amorphous Phosphate Coatings for Aluminum. These coatings were introduced in 1945. Their simplicity of application, speed, and economy have resulted in wide commercial acceptance. They provide a continuous, uniform green coating with excellent paint-bonding properties and underfilm corrosion protection. The coatings consist of varying ratios of chromic phosphate and hydrated aluminum oxide. The bath contains hydrofluoric acid, which removes the natural oxide to permit contact of the coating-forming chemicals with the metal. The complexity of the reactions involves makes it difficult to present a simplified chemistry, but the results of many tests and analyses give the following coating composition: $x\text{CrPO}_4 \cdot y\text{Al}_2\text{O}_3 \cdot z\text{H}_2\text{O}$. The phosphate coatings vary from 10 to 300 milligrams per square foot (108 to 3228 milligrams per square meter), depending on the end use. The lower coating weights are used for paint bonding; the higher range is used for decorative purposes.

Gold-colored conversion coatings are formed in baths containing hydrofluoric acid, to remove the natural oxide, and chromic acid. The coating composition is chromic chromate plus varying amounts of hydrated aluminum oxide. Some baths also contain ferricyanide iron, which greatly accelerates the coating action and forms some chromic ferricyanide in the coating. This constitutes one of the most widely used conversion coatings on aluminum because of its high speed, excellent corrosion resistance, and high affinity for organic finishes. The hexavalent chromium content permits these coatings to withstand somewhat more severe corrosive environments than do the amorphous phosphate coatings. The baths have a pH of about 1.2–1.9 and can be applied by dip, brush, spray, or reverse roll coater.

Proprietary chromatic rinses are frequently used over conversion coatings on aluminum for increased corrosion resistance. See also **Autodeposition.**

Wilbur S. Hall, Amchem Products, Inc., Ambler, Pennsylvania.

CONVERSION ELECTRON. Internal Conversion.

CONVERSION FRACTION. Internal Conversion.

CONVERSION GAIN RATIO. In communications, the ratio of the available signal power at the output to the available signal power at the input of a frequency converter or mixer.

CONVERSION PROCESS (Coal). Coal.

CONVERSION PROCESS (SNG). Substitute Natural Gas (SNG).

CONVERSION RATIO. 1. The ratio of the number of internal conversion electrons to the number of gamma rays emitted in a given time interval by a single nuclidic species during the de-excitation of one of its excited energy states. Sometimes known as the *internalconversion coefficient.* 2. In a nuclear reactor, the number of fissionable atoms produced per fissionable atom destroyed. See **Nuclear Reactor.**

CONVERSION TIME (Computer System). The interval of time between the initiation and completion of a single analog-to-digital or digital-to-analog conversion operation. Also, the reciprocal of the conversion rate. In practice, the term usually is with reference to analog-to-digital converters or digital voltmeters. The conversion time required by an analog-to-digital converter is comprised of: (1) the time needed to reset and condition the logic, (2) a delay to allow for settling time of the input buffer amplifier, (3) a polarity-determination time, (4) the actual analog-to-digital conversion operation, and (5) any time required to transfer the digital result into an output register. Not all factors always are present. A unipolar analog-to-digital converter, for example, will involve no polarity-determination time.

Conversion time also is used in connection with data-acquisition or analog-input subsystems for process control computers. The more precise term in this case would be *measurement time.* In this case, the measurement rate may not be the reciprocal of the measurement time inasmuch as some of the operations performed may be overlapped with other operations. With some types of analog-to-digital converters, it is possible to select the next multiplexer point during the time that the previous value is being converted from analog to digital form.

Conversion or measurement time also may include the time required for such operations as multiplexer address decoding, multiplexer switch-selection and settling time, amplifier settling time, range-selection and settling time, and the time required to deactive the multiplexer switches and permit the subsystem to return to an initial state.

See also **Analog-to-Digital Converter; Digital-to-Analog Converter;** and terms listed under **Data Processing.**

Thomas J. Harrison, International Business Machines Corporation, Boca Raton, Florida.

CONVERTER (Analog-to-Digital). Analog-to-Digital Converter.

CONVERTER (Bessemer). Iron Metals, Alloys, and Steels.

CONVERTER (Catalytic). Catalytic Converter.

CONVERTER (Digital-to-Analog). Digital-to-Analog Converter.

CONVERTER (Electric). Dynamotor.

CONVERTER (Electropneumatic). Electropneumatic Converter.

CONVERTIBLE AIRCRAFT. Helicopters and V/STOL Craft.

CONVOLUTION. If f and g are both functions of the complex variable z, the convolution or *faltung* of f and g, often indicated by the symbol $(f * g)$ is the integral

$$(f * g) = \frac{1}{\sqrt{2\pi}} \int_{-\infty}^{\infty} f(t) g(z - t)\, dt$$

Let $\mathscr{F}$ be the Fourier transform of the integral, then $\mathscr{F}(f * g) = \mathscr{F}(g * f) = \mathscr{F}(f)\mathscr{F}(g)$. This theorem also applies directly to the Laplace transform and, with minor modifications, to the Mellin transform.

The German word *faltung* means folding. It is used in this case for the following reason. Let a line of length z be folded back in the middle as shown. The points opposite each other on its two parts then lie at distances of t and $(z - t)$ from the origin.

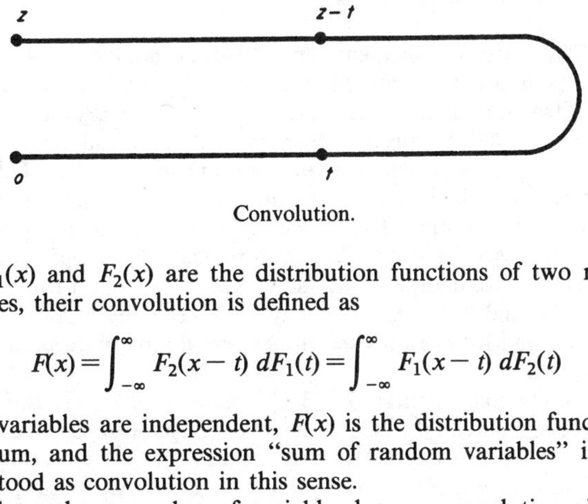

Convolution.

If $F_1(x)$ and $F_2(x)$ are the distribution functions of two random variables, their convolution is defined as

$$F(x) = \int_{-\infty}^{\infty} F_2(x - t)\, dF_1(t) = \int_{-\infty}^{\infty} F_1(x - t)\, dF_2(t)$$

If the variables are independent, $F(x)$ is the distribution function of their sum, and the expression "sum of random variables" is to be understood as convolution in this sense.

Analogously, a number of variables have a convolution given by

$$F(x) = \int_{-\infty}^{\infty} dF_1(t_1) \cdots \int_{-\infty}^{\infty} F_n(x - t_1 - \cdots - t_{n-1})\, dF_{n-1}(t_{n-1})$$

CONVOLUTIONAL CODING. Information Theory.

CONVOLUTION THEOREM. Laplace Transform.

CONVOLVULACEAE. Sweet Potato.

CONVULSION. A violent, uncontrollable contraction or series of contractions of voluntary muscles. It may or may not be accompanied by loss of consciousness. See **Seizure (Neurological).**

CONY. Hyraxes.

COOLER (Thermoelectric). Thermoelectric Cooling.

COOLING (Atmospheric). Clouds and Cloud Formation.

COOLING CURVE. A graph of the temperature of a substance plotted against time, such as is obtained for a molten alloy cooling through its solidification temperature or range of temperatures.

COOLING DEGREE DAY. Climate.

COOLING (Solar). Solar Energy.

COORDINATE GEOMETRY. Geometry.

COORDINATES (Eulerian). Eulerian Coordinates.

COORDINATES (Galactic). Galactic Coordinates.

COORDINATES (Generalized). (Also called Lagrangian Coordinates). Any set of coordinates specifying the state of the system under consideration. Usually employed in problems involving a finite number of degrees of freedom, the generalized coordinates are chosen so as to take advantage of the constraints of the system in reducing the total number of coordinates. See **Lagrangian Coordinates.**

COORDINATE SYSTEM. A coordinate is one of a set of numbers used to locate a point relative to a system of axes or surfaces, the coordinate system. The simplest and most frequently used system is called rectangular Cartesian. Choose three mutually perpendicular straight lines, the coordinate axes, intersecting at a point O, the origin. Label the axes OX, OY, OZ and imagine that OX and OY are drawn on a piece of paper, the XY-plane, so that the positive direction of OX points toward the reader's right and the positive direction of OY points toward the top of the page. The positive direction of OZ is then taken upward from the page, toward the reader. This arrangement determines a right-handed coordinate system, the most common one. If a pair of axes is exchanged, the system is then left-handed. A unit distance of convenient size is selected and this distance is marked off on each axis, repeatedly and in both positive and negative directions. Now suppose three numbers are given, one each in multiples (or fractions) of the unit distance along each axis. The numbers then locate the position of a point in space, relative to the coordinate system and they are the coordinates of the point. A suitable notation is (x, y, z). If one coordinate is zero, the point is on one of the three possible planes of the system, a coordinate surface; if two coordinates vanish, the point lies at the intersection of two coordinate surfaces, which is a coordinate axis; if all three coordinates are zero, the point is at the coordinate origin.

When a given point lies on a coordinate surface, it is customary to use the XY-plane. The horizontal coordinate is then called the *abscissa* and measured along the X-axis while the vertical, or Y-coordinate, is called the *ordinate.*

These conceptions can be generalized extensively. When the surfaces are mutually perpendicular, the system is curvilinear or orthogonal; when the surfaces are orthogonal planes, the system is rectangular; if the surfaces do not intersect at right angles, the system is said to be affine or nonorthogonal. A Cartesian coordinate system could mean a rectangular system or a system of planes not at right angles to each other. The latter is called an oblique coordinate system. However, the term Cartesian generally means rectangular Cartesian.

After Cartesian coordinates, the next most often used type is the polar coordinate in two or three dimensions, the latter being called spherical polar coordinates. The two-dimensional case is appropriate for the study of many plane curves and for the complex variable (see **Argand Diagram**). Systems of three axes in a plane or four axes on the sides of a tetrahedron, called trilinear and quadriplanar coordinates, respectively, are sometimes applicable to special problems.

Many problems in chemistry and physics involve partial differential equations and these often require for their solution a suitable coordinate system so that the method of separation of variables can be used. While the geometry or symmetry of a given problem frequently suggests the correct coordinate system, considerable attention has been paid to a more general problem about coordinate systems. One attempts to derive all systems which will yield separation of variables under certain general conditions as permitted by a partial differential equation. The solution to this problem is known for the Schrödinger wave equation of quantum mechanics and the eleven possible systems in this case are: rectangular, conical, cylindrical, ellipsoidal, elliptic cylindrical, spheroidal (either oblate or prolate), parabolic, parabolic cylindrical, paraboloidal, spherical polar, Laplace's equation also separates in these eleven systems and in bipolar and toroidal coordinates.

In addition to these coordinate systems which differ in the geometry upon which they are based, there are also the generalized coordinates of Lagrange, Lagrangian coordinates, natural coordinates, relative coordinates, inertial coordinates, and Eulerian coordinates. These systems are especially useful in mechanics, meteorology, and other physical problems.

See also terms listed under **Mathematics.**

COORDINATION. Brain and Nervous System.

COORDINATION COMPOUND (Hydrate). Hydrate.

COORDINATION COMPOUNDS. One of a number of types of complex compounds, usually derived by addition from simpler inorganic substances. Coordination compounds are essentially compounds to which atoms or groups have been added beyond the number possible on the basis of electrovalent linkages, or the usual covalent linkages, to which each of the two atoms linked donates one electron to form the duplet. The coordinate groups are linked to the atoms of the compound usually by coordinate valences, in which both the electrons in the bond are furnished by the linked atom of the coordinated group. The ammines and complex cyanides are representative of coordination compounds.

In attempting to classify coordination compounds, Sidgwick noted that the number of molecules or atoms coordinated with a metallic atom (which he called the *coordination number*) is 2, 3, 4, 5, 6, or 8, and that 2, 4, and 6 are the most common. In forming such compounds, each molecule or atom donates a pair of electrons to the metallic atom, forming a semi-covalent type of bond. Thus, in the nitrocobaltates, six nitro groups each donate a pair of electrons to a cobalt(III) ion, forming the complex ion.

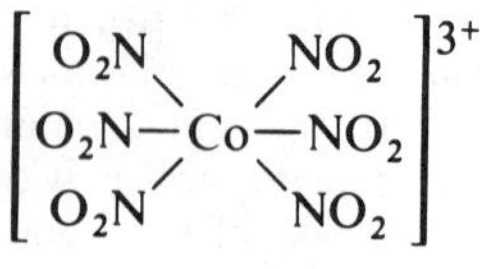

In the positive triammine cobalt(III) ion, three ammonia molecules donate one pair of electrons each to the cobalt(III) ion, to form the complex ion.

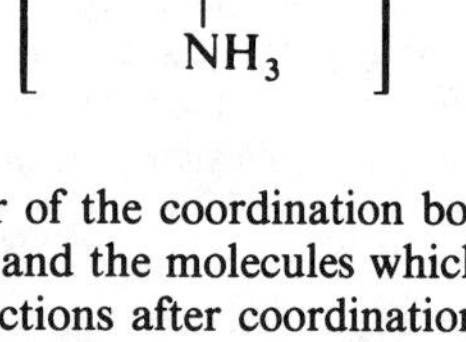

The covalent character of the coordination bond is evident from the fact that both the ions and the molecules which form it fail to exhibit their characteristic reactions after coordination.

Included in the coordination compounds are the double salts, the complex salts, the oxysalts, and the hydrates.

See also **Chelates and Chelation; Cobalt; Copper; Gold; Hydrate; Iron; Manganese; Molybdenum.**

COORDINATION NUMBER. The number of nearest neighbors of a given atom in a crystal structure. In covalent crystals, only those neighbors to which the atom is directly bonded are counted, and this number is usually 4 or less. In metals, the coordination number may be as high as 12, as in the close-packed structures.

COORDINATION POLYHEDRA. The arrangement of oxygen ions about the cation to which they are closely bonded in an ionic crystal, as, for example, the group SiO_4, which forms a tetrahedron. Such polyhedra pack as units in the crystal structure.

COOT. Rails, Coots, and Cranes.

COOTAMUNDRA WATTLE TREE. Acacia Trees.

COOTER (*Reptilia, Testudinata*). A turtle of the genus *Pseudemys.* The several species are found chiefly in the eastern and southern United States. Also called sliders.

COPAL. Copal is the hardened resin derived from several tropical trees. One of these, *Trachylobium hornemannianum,* is a large white-flowered tree of tropical east Africa. Another *Hymenaea courbaril,* a tree with large white or purplish flowers, is a native of tropical South America.

The resin may be obtained from living trees, in which case it is a soft substance naturally slow to harden. In Zanzibar, copal resin occurs in fossil form, masses of resin closely resembling amber being dug from the ground. This is the best grade of copal, known as Zanzibar copal. Similarly, in South America the resin may be dug from the ground at the base of the tree, where it slowly accumulates. Copal is used in the making of high-grade varnishes. Of late it is being supplanted with synthetic cellulose lacquers. Several other trees also yield copal resin.

COPALITE (or Copaline). The mineral copalite or "Highgate resin" is a fossil resin found in irregular fragments in the blue clay of London, England. It resembles copal, the resin of certain modern tropical trees. Copalite is pale yellow to greenish or brownish, and emits an aromatic odor when broken. It has a hardness of 1.5; a specific gravity of 1.046; burns with a very smoky yellow flame.

COPEPODA. A subclass of small crustaceans. Some species are free-swimming and others live as parasites on fishes. The latter are called fish-lice.

COPLANAR FORCE. Force.

COPPER. Chemical element symbol Cu, at. no. 29, at. wt. 63.546, periodic table group lb, mp 1083°C, bp 2566 ± 0.5°C, density 8.92 g/cm^3. Elemental copper has a face-centered cubic crystal structure. The metal is yellowish-red, soft, very malleable and ductile. Very thin sheet copper is translucent and transmits greenish-blue light. The element is unattacked by dry air, but in moist air containing CO_2, a protective greenish film of basic carbonate is formed. There are two natural isotopes ^{63}Cu and ^{65}Cu. Seven radioactive isotopes have been identified, all with comparatively short half-lives: ^{58}Cu through ^{62}Cu, ^{64}Cu, ^{66}Cu, and ^{67}Cu. Copper may have been the first metal used by humans and today ranks second, exceeded only by iron, in annual consumption.

First ionization potential 7.723 eV; second, 20.29 eV; third, 29.5 eV. Oxidation potentials: $2Cu + 2OH^- \rightarrow Cu_2O + H_2O + 2e^-$, 0.361 V; $Cu + 2OH^- \rightarrow Cu(OH)_2 + 2e^-$, 0.224 V; $Cu^+ \rightarrow Cu^{2+} + e^-$, −0.153 V; $Cu \rightarrow Cu^{2+} + 2e^-$, −0.344 V; $Cu \rightarrow Cu^+ + e^-$, −0.522 V. Other important physical properties of copper are given under **Chemical Elements.**

Copper occurs as native copper particularly in the region south of Lake Superior (often 99.9% Cu), as sulfides (chalcocite, copper glance, cuprous sulfide, Cu_2S; chalcopyrite, $CuFeS$), as oxide (cuprite, cuprous oxide, Cu_2O, red); as basic carbonates (malachite,

$CuCO_3 \cdot Cu(OH)_2$, green; azurite, $2CuCO_3 \cdot Cu(OH)_2$, blue). The copper content of its ores varies from 0.3 to 8% Cu and the average is of the order of 2.5%. The value depends largely upon the content of silver and gold. The area of production is widely distributed: in the United States, Montana, Utah, New Mexico, Arizona, Michigan, and Tennessee.

Copper is frequently detected in atmospheric particles, even in those collected at locations far removed from anthropogenic sources. Cattell and Scott (1978) reported similar enrichments found over the north Atlantic and the South Pole. Duce et al. (1975) proposed that measured atmospheric concentrations are larger than those predicted for unenriched crustal weathering or oceanic production. It was proposed that the enrichment may result from natural processes of anomalously enriched elements in aerosol particles. These may derive from low-temperature volatilization processes, such as biological methylation, or volcanism, or direct sublimation from the earth's crust, or emissions from plants, or fractionating at the air–sea interface which enriches elements in particles produced from the oceans. Atmospheric studies conducted by Cattell and Scott near the island of Tasmania in 1977 led to the conclusion that a biogenic agent may be responsible for the approximately 20,000-fold enrichment of copper during aerosol production from the ocean.

Native copper ore is crushed, concentrated by washing with water, smelted, and cast into bars. Oxide and carbonate ores are treated with carbon in a smelter. Sulfide ore treatment is complex, but in brief, consists of smelting to a matte of cuprous sulfide, ferrous sulfide, and silica, which molten matte is treated in a converter by the addition of lime and air is forced under pressure through the mass. The products are blister copper, ferrous calcium silicate slag, and SO_2. Refining is conducted by electrolysis, and the anode mud is treated to obtain the gold and silver.

See also **Azurite; Chalcocite; Chalcopyrite; Cuprite; Malachite; Mineralogy.**

Leading world producers of copper include: United States (23.9%); Chile (11.8%); Canada (11.8%); Zambia (10.9%); U.S.S.R. (10.6%); Zaire (7.0%); Peru (3.7%); Philippines (3.0%); South Africa (2.9%); Australia (2.6%); Japan (2.1%); and China (1.7%).

Copper is distinguished by several properties which contribute to its extensive use: (1) a combination of mechanical workability with corrosion resistance to many substances, (2) excellent electrical conductivity, (3) superior thermal conductivity, (4) effect as an ingredient of alloys to improve their physical and chemical properties, (5) efficiency of copper and some of its compounds as catalysts for several kinds of chemical reaction, (6) nonmagnetic characteristics, advantageous in electrical and magnetic apparatus, and (7) nonsparking characteristics, mandatory for tools for use in explosive atmospheres. There are additional attractions of copper for many other applications. The metal would be used even more widely, but for some uses, even though superior, copper cannot compete with substitute materials because of cost.

Unalloyed Copper: In the United States, the term *copper* signifies copper that contains less than 0.5% impurities or alloying elements. Copper-base alloys are those that contain no less than 40% copper. Additionally, copper appears as a minor, but important ingredient of several alloys. There are six major types of commercial, unalloyed copper. These are described briefly in Table 1.

Very-High Copper Alloys: Although not meeting the foregoing definition of copper precisely, there is a group of copper alloys which contain only a few percent of other ingredients and commonly these are also referred to as coppers, usually with the name of the other element preceding copper in the name—as chromium copper or beryllium copper. These very-high copper alloys are described briefly in Table 2.

The Brasses: There are eight principal categories of brasses, not including the leaded and alloy brasses. Brass essentially is an alloy of copper and zinc. Several of the brasses contain other ingredients, such as lead and iron. When zinc is added to copper, there is a progressive alteration of color and lowering of melting point. When the zinc content is about 10%, the metal is a bronze color; with 15% zinc, the color may be described as golden; from 20–40% zinc, there is a range of yellow colors; over 45% zinc, the color is silver-white. The melting point of a 95% copper–5% zinc brass is about 1,065°C, whereas the melting point of 50% copper–50% zinc brazing metal drops to about 880°C. The ratio of copper to zinc also progressively affects mechanical and corrosion-resistance properties. Maximum tensile strength, for example, is attained with a 55% copper content, whereas maximum ductility is attained with a 70% copper content. This exceptional range of properties accounts for the availability and demand for a wide variety of brasses. Metallurgically, brasses may be classified as (1) *alpha brass*, in which the content of zinc is less than 36% and in which the zinc is dissolved in the copper, imparting to the alloy the basic structure of copper; (2) *beta brass*, in which the content of zinc ranges between 36% and 45%. This alloy contains the CuZn as a compound and enhances the hot workability of the alloy; and (3) *gamma* brass, in which the zinc content exceeds 45% and where there are Cu_2Zn_3 crystals in the alloy. This combination does not lend itself to either hot or cold workability. Some of the important commercial brasses are described briefly in Table 3.

The Bronzes: Classically, a bronze is defined as an alloy of copper and tin, but over the years the term has taken on a much broader meaning. The term may apply to numerous copper alloys that possess a crystalline, bronze-like structure, are of a bronze color, or simply because they may contain some tin. Further, bronze generally is considered a casting metal. In contrast, brass is generally wrought. Some alloys are commercially named bronze even though they contain no tin whatsoever.

Copper Wire and Cable: *The International Annealed Copper Standard* (IACS), which sets annealed copper as having 100% electrical conductivity as a basis against which to compare other metals, alloys, and materials, is accepted internationally. Using this standard for comparison, the conductivity of copper is exceeded only by silver for which the IACS figures is 108.4%. This comparison is on the

TABLE 1. COMMERCIAL UNALLOYED COPPERS

Electrolytic Tough-Pitch Copper
Cu, 99.90%; O, 0.04% nominal; density 8.89–8.94 g/cm³, EC, 101%
Architecture: downspouts, flashing, building fronts, gutters, screening, roofing
Automotive: radiators and gaskets
Electrical: conductive-wire contacts, terminals, switch parts, bus bars
Hardware: cotter pins, nails, rivets, soldering copper, ball floats
Other: anodes, chemical process equipment, kettles, pans, printing rolls, expansion plates, rotation bands, die-pressed forgings
Deoxidized Copper
Cu, 99.90% minimum; P, 0.025% nominal; density 8.94 g/cm³, EC, 80–90%
Industrial: condensers, evaporators, heat exchangers, dairy tubes, fractionating columns, kettles, pulp and paper piping, steam and water piping, tanks
Transportation: gasoline, oil, air, and hydraulic fluid lines, oil coolers
Other: shell rotation bands, die-pressed forgings, gauge lines
Oxygen-free Copper
Cu, 99.92% minimum; no residual oxidants; density 8.89–8.94 g/cm³, EC, 101%
Electrical: conductors, electron tubes, bus bars, waveguides (for operation at high temperatures in presence of reducing gases)
Industrial: heaters, oil coolers, gasoline supply lines, radiators, refrigeration lines, water piping
Silver-bearing Copper
Cu, 99.90% minimum; 8–25 ounce (226–708 grams) Ag/ton; density 8.91 g/cm³, EC, 100–101%.
Electrical: commutator bars, heavy-duty motor windings (particularly for retention of strength at elevated temperatures)
Other: brazing solders, die-pressed forgings
Arsenical Copper
Cu, 99.68% nominal; P, 0.025% nominal; As, 0.30% nominal; density 8.94 g/cm³, EC, 90%
Industrial: heat-exchangers, boilers, radiators, condenser tubes
Free-cutting Copper
Cu, 99.4–99.5%; Te, 0.5–0.6%; density 8.94 g/cm³, EC, 90%
Industrial: electrical connectors, motor and switch parts, soldering coppers, screw-machine parts, forgings, welding-torch tips

EC, electrical conductivity (International Annealed Copper Standard).

TABLE 2. VERY HIGH COPPER ALLOYS

Cadmium Copper
Cu, 99.00–plus %; Cd, 0.6–1.0%
Cadmium toughens copper and increases resistance to fatigue; also increases softening temperature.
Electrical conductivity (fully annealed) is about 95% (IACS).
Essentially free of oxygen; not susceptible to gassing.
Uses: contact wires used in electrical transportation, notably long-span overhead electric transmission lines.

Chromium Copper
Cu, 99.50%; Cr, 0.5%
Chromium improves mechanical properties while retaining high thermal and electrical conductivities.
Strength and hardness depend on heat treatment and not cold working—hence alloy can be used up to temperature of about 450°C without danger of softening.

Tellurium Copper
Cu, 99.50%; Te, 0.5%
Tellurium increases softening temperature of work-hardened copper.
Alloy is excellent where combination of good machinability and electrical conductivity is required.
Uses: motor and switch parts, electrical connectors, screw-machine parts, electrical instrument parts.

Beryllium Copper
Type 1: Cu, 98%; Be, 2%
Type 2: Cu, 97%; Be, 0.4%; Co, 2.6%
Cobalt is added as a lower-cost substitute for beryllium.
Uses: instrument springs, bellows, diaphragms, bourdon tubes, non-sparking tools for hazardous locations.
Alloy permits springs to be shaped while soft, followed by hardening.

Selenium copper also available for combining high electrical conductivity with free-machining and hot-working properties. Alloy makes excellent copper-to-glass seals.

basis of conductivity per gram, pound, or other mass unit. Aluminum, although widely used as an electrical conductor for selected applications, has a rating on this schedule of approximately 61%, steel a rating of 11%, and nickel-chromium alloy (valued because of its high electrical resistance rather than conductivity) has a rating of 1.5%. Thus, the value of copper for electrical conductors, considering its availability and economics, is self-evident.

Processing Equipment: In terms of thermal conductivity, assigning a value og 100 to copper, the metal is exceeded only by silver which has a value of 108. Copper is followed by gold (76), aluminum (56), magnesium (41), zinc (29), nickel (15), iron (15), steel (13–17), lead (9), and antimony (5). Thermal conductivity means good heat transfer and this is extremely important in most industrial processing equipment where heating and cooling cycles are involved. This property, when combined with corrosion resistance, makes copper attractive for the construction and lining of process vessels. Deoxidized copper, admiralty brass, and arsenical copper are effective in condenser tubes operating with fresh water. Copper is less suitable for seawater because of its inability to form a protective film. Aluminum brass and 70–30 cupronickel alloy are favored for severe seawater service. Copper and copper alloys are not suited for use in oxidizing acidic solutions, in mercury, or in the presence of free NH_3. Copper vessels are used extensively in food processing and for numerous organic materials, particularly distillation columns and hardware. The relatively high cost of copper as compared with other metals, however, is always an important factor.

Piping: Copper and copper alloys are used in a wide variety of pipes and tubes, both for industrial and domestic systems. In many areas, galvanized water pipe has been almost completely replaced by copper piping in new construction. Advantages include corrosion resistance and ease of installation which offset higher costs. Because of excellent thermal conductivity, copper and brass fittings are used widely in hot water and steam-heating systems. However, the greater conductivity of bare copper pipe in long runs requires more attention to insulation covering.

Chemistry and Compounds: Copper is dissolved best by HNO_3; not attacked by cold dilute HCl or H_2SO_4, but in hot HCl dissolves to yield cuprous chloride, in hot concentrated H_2SO_4 to yield copper sulfate; attached by chlorine, especially when heated, to form cuprous and cupric chlorides; only slight action by H_2S or SO_2 at ordinary temperatures in the absence of air.

In view of its $3d^{10}4s^1$ electron configuration and the relatively small energy difference between the two levels, copper forms dipositive ions as well as monopositive ones. In fact, the former are the more stable in aqueous solution, due primarily to the larger heat of hydration of Cu^{2+} than Cu^+. Moreover, the *d*-electrons may participate in bonding, and tripositive copper, Cu(III), appears in complexes. In addition, copper forms a number of compounds essentially covalent in character, such as copper(I) oxide, Cu_2O.

This compound is less stable at room temperature than copper(II) oxide, CuO, although Cu_2O occurs in nature (as cuprite). It is the stable oxide above 1,026°C. It is prepared by fusion of copper(I) chloride, CuCl with sodium carbonate, Na_2CO_3. In its crystal, each copper atom has two colinear bonds, and each oxygen atom four tetrahedral ones; two such interpenetrating lattices constitute the structure. Copper(I) hydroxide, CuOH, is relatively stable, and is produced by electrolysis of a sodium chloride solution between copper electrodes (by action of NaOH on the cathode). Copper(II) oxide, produced by heating copper in air, is also essentially covalent (Cu—O, 1.95 Å); it has tetrahedral bonding of the oxygen atoms, and coplanar bonding of the copper atoms. Copper(II) hydroxide, $Cu(OH)_2$, is precipitated by alkali hydroxides from Cu^{2+} solutions. It is gelatinous, and its composition and solubility vary somewhat with the alkali concentration. It is thermodynamically unstable even in contact with liquid water with respect to dehydration to CuO, but this occurs only very slowly, except upon heating or when catalyzed by hypochlorite, hydrogen peroxide, etc.

Copper(I) halides are formed with chlorine, bromine and iodine, the chloride and bromide by reduction of the copper(II) halides with copper powder, and the iodide by reduction of copper(II) sulfate, $CuSO_4$, solution with potassium iodide. The fluoride appears never to have been made, despite reports to the contrary. All are insoluble in H_2O. Copper(II) fluoride, CuF_2 may be made from CuO and hydrofluoric acid at 400°C, copper(II) chloride, $CuCl_2$ by dissolving the oxide or carbonate in HCl, and copper(II) bromide, $CuBr_2$, from copper and bromine water; copper(II) iodide, CuI_2, is unstable at room temperature with respect to decomposition into CuI and iodine. The chloride and bromide are water-soluble, and ionic. The fluoride is only slightly water-soluble. Anhydrous copper(II) chloride, $CuCl_2$, is monoclinic and its structure contains infinite chain molecules formed by $CuCl_4$ groups that share opposite edges. $CuBr_2$ has a similar structure.

Complex halides of both monovalent and divalent copper ar known. The monovalent complexes are primarily of the composition $MCuX_2$, where X is a halogen atom and M usually an alkali metal, although $CuCl_3^{2-}$ ions are also known, being found in infinite chain $(CuCl_3^{2-})_n$ structures, as in crystals of Cs_2CuCl_3. The ion $CuCl_4^{3-}$ is also known. The composition of the copper(II) complex halides is primarily in terms of $CuCl_3^-$, $CuCl_4^{2-}$, or $CuBr_4^{2-}$ ions, although the corresponding complex fluoride has trivalent copper, as in K_3CuF_6. Its paramagnetic moment indicates two unpaired electrons.

Copper oxyhalides of a number of different compositions have been reported, but the most definitely established compositions are $Cu(OH)Cl$, $Cu_2(OH)_2Cl_2$, and $CuBr_2 \cdot 2Cu(OH)_2$. The property of forming basic salts is not limited to the halides of copper. Basic sulfates, such as $CuSO_4 \cdot 2Cu(OH)_2$, $CuSO_4 \cdot 3Cu(OH)_2$, $CuSO_4 \cdot 4Cu(OH)_2$, and $CuSO_4 \cdot 5Cu(OH)_2$ have been prepared, more or less hydrated. In copper(II) carbonate, the stable forms are oxycarbonates, $xCuCO_3 \cdot yCu(OH_2)$, where the $x{:}y$ ratios may be 2:1, 1:1, 2:3, 1:9, and still other values. Many of these compositions occur in minerals, such as malachite and azurite. Copper forms complexes with larger number of ions and molecules. The halogen complexes were discussed above. In general, copper tends to be 6-coordinate, as in the complex ion $[Cu(H_2O)_2\ (ethylenediamine)_2]^{2+}$. With NH_3 and many amines, stable complexes are formed, both of Cu^+, such as $[Cu(NH_3)_2]^+$ and Cu^{2+}, such as $[Cu(NH_3)_4]^{2+}$ which add halogen, pseudohalogen, and many other anions to form compounds of the composition $Cu(NH_3)_2X_2$ and $Cu(NH_3)_4X_2$.

TABLE 3. REPRESENTATIVE BRASSES AND BRONZES

Gilding Brass
Cu, 95%; Zn, 5%; Pb, 0.03% maximum; Fe, 0.05% maximum; density 8.86 g/cm³; mp 1066°C; AT, 427–788°C; HWT, 760–871°C
Coinage: coins, metals, tokens
Munitions: firing-pin support shells, bullet jackets, fuse caps, primers
Novelties: emblems, plaques, jewelry
Other: base for gold plate and for vitreous enamel

Commercial Bronze
Cu, 90%; Zn, 10%; Pb, 0.05% maximum; Fe, 0.05% maximum; density 8.80 g/cm³; mp 1043°C; AT, 427–788°C; HWT, 760–871°C
Architectural: grillwork, etching bronze, screen cloth, weather stripping
Cosmetics: lipstick cases, compacts
Hardware: kickplates, line clamps, marine hardware, escutcheons, rivets, screws
Munitions: rotating bands, primer caps
Other: costume jewelry, screen wire, ornamental trim, vitreous enamel base

Red Brass
Cu, 85%; Zn, 15%; Pb, 0.06% maximum, Fe, 0.05% maximum; density 8.75 g/cm³; mp 1027°C; AT, 427–732°C; HWT, 788–900°C
Architectural: trim, etching parts, weather stripping
Electrical: screw shells, sockets, conduit
Hardware: fasteners, fire extinguishers, eyelets
Industrial: heat-exchanger tubes, condensers, flexible hose, piping, pumps, radiator cores, pickling crates
Other: compacts, costume jewelry, dials, badges, etched articles, lipstick cases

Jewelry Bronze
Cu, 87.5%; Zn, 12.5%; Pb, 0.05% maximum; Fe, 0.10% maximum; density 8.78 g/cm³; mp 1035°C; AT, 427–760°C; HWT, 760–900°C
Architectural: angles, channels
Hardware: chains, fasteners, slide fasteners, eyelets
Novelties: costume jewelry, emblems, compacts, etched articles, lipstick cases, plaques
Other: base for gold plate

Low Brass
Cu, 80%; Zn, 20%; Pb, 0.05% maximum; Fe, 0.05% maximum; density 8.67 g/cm³; mp 999°C; AT, 427–704°C; HWT, 816–900°C
Architectural: medallions, spandrels, ornamental metalwork
Electrical: battery caps
Instruments: bellows and muscal instruments
Hardware: flexible hose, pump lines, tokens, clock dials

Cartridge Brass
Cu, 70%; Zn, 29-plus %; P, 0.07% maximum; Fe, 0.05% maximum; density 8.53 g/cm³; mp 954°C; AT, 427–760°C; HWT, 732–843°C
Automotive: radiator cores and tanks, reflectors
Electrical: flashlight shells, lamp fixtures, socket shells, screw shells, bead chain
Hardware: fasteners, pins, rivets, eyelets, springs, tubes, stampings
Munitions: various components. Note: Admiralty brass is similar: Cu, 71%; Zn, 28%; Sn, 1%

Yellow Brass
Cu, 65%; Zn, 34-plus %; Pb, 0.15% maximum; Fe, 0.05% maximum; density 8.47 g/cm³; mp 932°C; AT, 427–704°C
Architectural: grillwork
Automotive: reflectors, radiator cores and tanks
Electrical: lamp fixtures, flashlight shells, screw shells, socket shells, bead chain
Hardware: kick plates, push plates, locks, hinges, grummets, fasteners, eyelets, stencils, plumbing accessories, pins, rivets, screws, springs

Muntz Metal
Cu, 60%; Zn, 39-plus %; Pb, 0.30% maximum; Fe, 0.07% maximum; density 8.39 g/cm³; mp 904°C; AT, 427–593°C; HWT, 621–788°C
Hardware: large nuts and bolts, brazing rod, condenser plates, valve stems, hot forgings

Leaded Brasses
When lead is added to brass up to about 4%, improved machinability results. The lead has practically no effect on tensile strength or hardness. However, for cold-worked materials, lead does lower ductility and shear strength.

Phosphor Bronzes
Although tin is the primary alloying element in these alloys, their name derives from the addition of small quantities of phosphorus used as a deoxidizing agent in casting the alloys. Tensile strength ranges from moderate to very high, decreasing with amount of tin added. Tin percentage will range from 1.25 to 10%. Of the copper alloys, the phosphor bronzes are best suited for sea duty and where acid reagents may be present.

Silicon Bronzes
Most of these alloys are of proprietary compositions and are known by a variety of trade names. Silicon content ranges from 1.5 to 3.5%; usually less than 1.5% zinc content. Tin, manganese, and iron also may be added in small quantities. Because of their excellent strength, ease of welding, and corrosion resistance, the alloys have become important construction materials. As the silicon content increases, the alloys become more subject to fire cracking.

Aluminum Bronzes
The aluminum content of these alloys ranges from 4 to 10%. They are moderately hard, very ductile, and tough. The alloys resist scaling and oxidation at high temperatures because of the aluminum content. They perform well in both acids and alkalis. The alloys are good for sea duty, particularly in contact with turbulent seawater.

Nickel Silvers
Nickel essentially is added to copper-zinc alloys to enhance color. With a nickel content of about 18%, the alloy is silver-white. Also, most of the mechanical properties and corrosion resistance are improved. The alloys find wide application for operations that require ductility in the cold condition, as in stamping, spinning, deep drawing, and for articles to be plated. An alloy widely used as a spring material because of its high tensile and fatigue properties has the composition: Cu, 55%; Zn, 27%; Ni, 18%. German silver contains: Cu, 50%; Ni, 30%; Zn, 20%. It is interesting to note that the nickel silvers do not contain silver.

Cupronickels
The nickel silvers generally are classified as brasses. Cupronickels fall more into basic copper–nickel alloys. Possible minor ingredients are manganese, iron, and zinc. These alloys can be used for severe drawing, spinning, and stamping operations because they do not work harden readily. They also are extensively used for condenser tubes and plates, heat exchangers, and other process equipment.

AT, annealing temperature range.
HWT, hot-working temperature range.

Among the other copper compounds, copper(II) acetate is used as a pigment and fungicide; in its basic form it is the familiar verdigris that forms on copper surfaces in the presence of moisture and organic matter. The arsenic compounds of copper are used as insecticides and wood preservatives: copper(II) arsenite is called "Scheele's green" and copper(II) acetoarsenite $Cu(AsO_2)_2 \cdot Cu(C_2H_3O_2)_2$ is "Paris green." Copper(II) hexacyanoferrate(II), brown, is precipitated from copper(II) solutions by soluble hexacyanoferrates(II), even from very dilute solutions, and copper(II) sulfide, black, by H_2S or other soluble sulfides. Copper(II) sulfate, when hydrated, forms its characteristic blue crystals (the hydrated Cu^{2+} ion is blue).

Copper(I) cyanide dissolves in alkali cyanide solution to form cyanocuprates(I) of the general formula $M_n[Cu(CN)_{n+1}]$ where M is an alkali metal, and n ranges in value from 1 to 5. Not all of the values, of course, are found for a particular alkali or in the presence of particular anions. With sodium cyanide, NaCN, most of the complex present has an n value of 2, but if the original solute was CuCl, more of the $n = 3$ cyanocuprate(I) is present. At low temperatures, values of n of 4 and 5 are found. Copper(II) appears to coordinate four cyanide ions to form the unstable tetracyanocuprate(II) ion which decomposes at once to the copper(I) complex and cyanogen. In fact, for Cu(II) the chelated complexes are more common than the simple ones, examples being those formed with ethylenediamine and its derivatives, oxalates, catechol, and the β-diketones.

A solution of CuCl in HCl absorbs carbon monoxide, forming copper(I) carbonyl chloride, $Cu(CO)Cl \cdot H_2O$. This reaction, which is used in gas analysis, is indicative of the ability of copper to combine with carbon monoxide. Evidence for a true carbonyl is limited to the observation that if hot carbon monoxide is passed over hot copper, a metallic mirror is produced in the hotter parts of the tube. Other organometallic compounds include the very unstable methyl copper, CH_3Cu, phenyl copper, C_6H_5Cu, and bischlorocopper acetylene $C_2H_2(CuCl)_2$.

Copper industrial chemicals: Copper oxides, salts, and organocopper compounds find extensive use in industry and commerce. Some of the more important compounds are summarized:

Cupric Acetate, $Cu(C_2H_3O_2)_2 \cdot H_2O$, sp gr 1.88, mp 115°C, decomposes at 240°C, dark brown powder, slightly soluble in cold H_2O and alcohol; moderately soluble in hot H_2O and ether. Used as a fungicide, insecticide, as a catalyst, and in pigments.

Cupric Acetoarsenite (Paris Green), $(CuOAs_2O_3)_3 \cdot Cu(C_2H_3O_2)_2$, emerald green powder, very slightly soluble in cold H_2O, soluble in alcohol and potassium cyanide. Used as an insecticide, wood preservative, and paint pigment.

Cupric Acid Orthoarsenite (Scheele's Green), $CuHAsO_2$, green powder, insoluble in H_2O, soluble in alcohol, acids, and NH_4OH. Used as an insecticide and wood preservative.

Copper Carbonate (Basic), $CuCO_3 \cdot Cu(OH)_2$, dark green monoclinic crystals, insoluble in cold H_2O, decomposes in hot H_2O, soluble in potassium cyanide. Malachite, a copper ore, is of this composition. Refined compound is used as a pigment.

Cupric Hydroxide, $Cu(OH)_2$, blue, gelatinous compound, insoluble in cold H_2O, decomposes in hot H_2O, soluble in alcohol, NH_4OH, and potassium cyanide. Used as a pigment.

Cuprous Cyanide, $Cu_2(CN)_2$, white monoclinic crystals, insoluble in H_2O, soluble in HCl, NH_4OH, and potassium cyanide. Used in Sandmeyer's reaction to synthesize aryl cyanides.

Cuprous Iodide, Cu_2I_2, cubic white crystals, practically insoluble in H_2O or alcohol, soluble in NH_4OH, potassium iodide, or potassium cyanide. Used in Sandmeyer's reaction to synthesize aryl chlorides.

Cupric Oxide, CuO, black cubic crystals, insoluble in H_2O, soluble in HCl, NH_4OH, or ammonium chloride. Used as a green and blue colorant in ceramics.

Cuprous Oxide, Cu_2O, red cubic crystals, insoluble in H_2O, soluble in HCl, NH_4OH, or ammonium chloride. Cuprite, a copper ore, is of this composition. Refined compound is used in electrical rectifiers.

Cupric Sulfate, $CuSO_4 \cdot 5H_2O$, blue triclinic crystals, moderately soluble in cold H_2O, quite soluble in hot H_2O, very slightly soluble in alcohol. Used in copper plating, dyestuff manufacture, water treatment, germicides, and coppering of steels.

Cupric Chloride, $CuCl_2$, brown-yellow powder, quite soluble in cold H_2O or alcohol, very soluble in hot H_2O. Catalyst for several organic syntheses, including production of vinyl chloride monomer.

References

Blackwood, A. W., and J. E. Casteras: "Copper" in "Metals Handbook," 9th edition, Vol. 2., American Society for Metals, Metals Park, Ohio, 1979.

Cattell, F. C. R., and W. D. Scott: "Copper in Aerosol Particles Produced by the Ocean," *Science*, **202**, 429–430 (1978).

Duce, R. A., Hoffman, G. L., and W. H. Zoller: *Science*, **183**, 198 (1974).

Jovanović, B.: "The Origins of Copper Mining in Europe," *Sci. Amer.*, **242**, 5, 152–167 (1980).

Marchant, G. R., et al.: "Digital Controls for Continuous Copper Smelting," *Instrumentation Technology*, **25**, 6, 51–57 (1978).

Rosenbaum, J. B.: "Minerals Extraction and Processing," *Science*, **191**, 720–723 (1976).

Staff: "Introduction to Copper and Copper Alloys," in "Metals Handbook," 9th edition, Vol. 2, American Society for Metals, Metals Park, Ohio, 1979.

COPPER AGE. An archeological term to designate a cultural level that has been discerned between the Bronze Age and Iron Age. It is characterized by use of copper for weapons and tools.

COPPER CARBONATE. **Azurite (or Chessylite); Malachite.**

COPPER DEACTIVATOR (Fuel). **Petroleum.**

COPPER GLANCE. **Chalcocite.**

COPPERHEAD. **Snakes.**

COPPER (In Biological Systems). The activity of copper in plant metabolism manifests itself in two forms: (1) synthesis of chlorophyll, and (2) activity of enzymes. In leaves, most of the copper occurs in close association with chlorophyll, but little is known of its role in chlorophyll synthesis, other than the presence of copper is required. Copper is a definite constituent of several enzymes catalyzing oxidation-reduction reactions (oxidases), in which the activity is believed to be due to the shuttling of copper between the +1 and +2 oxidation states.

Traces of copper are required for the growth and reproduction of lower plant forms, such as algae and fungi, although larger amounts are toxic.

The effects of copper deficiency in plants are varied and include: die-back, inability to produce seed, chlorosis, and reduced photosynthetic activity. In contrast, excesses of copper in the soil are toxic, as in the application of soluble copper salts to foliage. For this reason, copper fungicides are formulated with a relatively insoluble copper compound. Their toxicity to fungi arises from the fact that the latter produce compounds, primarily hydroxy and amino acids, which can dissolve the copper compounds from the fungicide.

Copper is a necessary trace element in animal metabolism. The human adult requirement is 2 milligrams per day, and the adult human body contains 100–150 milligrams of copper, the greatest concentrations existing in the liver and bones. Blood contains a number of copper proteins, and copper is known to be necessary for the synthesis of hemoglobin, although there is no copper in the hemoglobin molecule.

Anemia can be induced in animals on a low copper diet, such as milk, and appears to be due to an impaired ability of the body to absorb iron. This anemia, however, is rare, because of the widespread occurrence of copper in foods. In locations, such as Australia and the Netherlands, diseases of cattle and sheep, involving diarrhea, anemia and nervous disorders, can be traced either to a lack of copper in the diet, or to excessive amounts of molybdenum, which inhibits the storage of copper in the liver.

Ingestion of copper sulfate by humans causes vomiting, cramps, convulsions, and as little as 27 grams of the compound may cause death. An important part of the toxicity of copper to both plants and animals is probably due to its combination with thiol groups of certain enzymes, thereby inactivating them. The effects of chronic exposure to copper in animals are cirrhosis of the liver, failure of growth, and jaundice.

Copper deficiency in plants is most frequent on organic soils, such as newly drained bogs, and on very sandy soils. The severe copper deficiency often found when bogs and marshes are first used for crop production is called *reclamation disease* in some parts of the world.

Ruminants are sensitive to copper deficiency. The symptoms of copper deficiency in animals vary with the species and age, but often the fading of brown or black hair is evident. On some acidic soils, the use of copper in fertilizers increases crop and pasture production, and the increases in level of copper in the plants help to prevent copper deficiency in the cattle and sheep. In parts of Australia, livestock production was impossible until copper fertilizers were used on the pastures. Application of copper fertilizers to alkaline soils generally does not increase the copper level in the crop. Farm animals are often supplied with copper in the form of dietary mineral supplements. Compounds used include copper gluconate, copper oxide, and copper sulfate.

Although copper fertilizers will sometimes increase crop yields and improve the nutritional quality of the crops, this practice must be used with caution and only on copper-deficient soils. Both plants and animals are subject to toxicity from excessive levels of copper. Ruminants, especially sheep, are sensitive to copper toxicity as well as to copper deficiency. Adding a copper fertilizer to a soil that naturally contains rather high levels of available copper may increase levels of the metal in the forage to the point of causing copper toxicity in grazing sheep. Copper toxicity from soils naturally high in copper occurs in Australia, but is uncommon in the United States. There are soils in the United States, however, that produce forage levels of copper close to toxicity limits, and if copper-bearing mineral supplements are inadvertently used with these forages, copper toxicity to sheep may result.

It is not easy to set a definite limit, in terms of the copper concentration in the diet, that will permit accurate predictions of the danger of copper deficiency or of copper toxicity in cattle and sheep. In particular, if the molybdenum concentration in the forage is high, extra amounts of copper are needed to prevent deficiency. Also, higher copper levels can be tolerated without danger of toxicity with molybdenum present.

Monogastric animals, including humans, are less sensitive than ruminants to either copper deficiency or toxicity. Copper deficiency in people has been found only when other complications, such as excessive bleeding, general starvation, and iron deficiency, are also present. Wilson's disease, an inherited disease of humans, prevents the loss of excess copper from the body and brings on copper toxicity. No direct relationships have been found between levels of available copper in the soil and the copper status of humans.

A number of copper-containing protein compounds are enzymes with an oxidase function (ascorbic acid oxidase, urease, etc.) and these play an important role in the biological oxidation–reduction system. There is a definite relationship of copper with iron in connection with utilization of iron in hemoglobin function.

Copper absorption is depressed by ascorbic acid, dietary phytates, cadmium, mercury, silver, and zinc. It appears that metals impede copper absorption through competition for metal-binding sites. Dietary copper, molybdenum, and sulfur are closely interrelated in optimum copper and molybdenum nutrition of ruminants. Increase pasture molybdenum content and low-pasture copper result in a condition known as "peat scours."

Copper toxicity tends to accumulate in the liver. The capacity to tolerate copper varies considerably with the species. Sheep are most susceptible. Swine have a much greater tolerance and copper may be added to the swine diet for pharmacological reasons (for example, use as an anthelminthic to control internal parasites).

Continuing research is providing a better understanding of the biological role of copper. The prooxidant and antioxidant effects of ascorbic acid and metal salts, including copper, in a beta-carotene–linoleate model system were studied by Israeli scientists Kanner, Mendel, and Budowski (1977). The interacting effects of ascorbic acid and metal ions on carotene oxidation were studied in an aqueous carotene–linoleate solution at pH of 7. Ascorbic acid at concentrations up to 10^{-3} M was a prooxidant. Fe^{3+} and, to a lesser extent, Co^{2+} acted synergistically with ascorbic acid, the prooxidant effect increasing with metal concentration. Cu^{2+} formed a prooxidant system with ascorbic acid only at low metal concentration, but as the copper concentration was raised, inversion of activity occurred and the copper–ascorbic acid system exerted a stabilizing action on carotene. Prooxidant effects were enhanced and antioxidant effects weakened in the presence of added linoleate hydroperoxides. The latter were unstable in the presence of ascorbic acid and especially ascorbic acid plus Cu^{2+}. Ascorbic acid itself became unstable in the presence of Cu^{2+}. Oxygen depletion, brought about by the rapid oxidation of ascorbic acid, may be partly responsible for the carotene-stabilizing effect of the Cu^{2+} couple. The investigators postulated that additional stabilization results from the radical-scavenging properties of copper or of a copper chelate formed by ascorbic and/or dehydroascorbic acid.

Y. C. Lee and a team of investigators at the Department of Food Science and Human Nutrition, Michigan State University, studied the kinetics of ascorbic acid stability of tomato juice as a function of temperature, pH, and metal catalyst, including copper. The rate of copper-catalyzed destruction of ascorbic acid increased as copper concentration in tomato juice increased, and was affected by pH.

Relatively recent hypotheses concerning the effect of zinc-to-copper ratios in the diet as a determining factor of plasma cholesterol levels have been made (Kleva, 1973). In 1978, L. R. Helwing, Jr. and a team at the Department of Poultry Science, Cornell University, undertook a study to establish if this hypothesis could be demonstrated in an animal system (other than rat). They selected the White Leghorn laying hen. The chicken was selected as the model system for a number of reasons, including (1) any resulting alteration in plasma cholesterol levels in a species other than rat would further support the proposal that the zinc-to-copper ratio could be involved in human cholesterol metabolism; and (2) the alteration in the cholesterol metabolism of the chicken may result in lowered egg cholesterol levels which could be beneficial to persons wishing to restrict dietary cholesterol without eliminating egg intake. As reported (Helwing et al., 1978), the researchers were unable to demonstrate any effect of the zinc-to-copper ratio upon cholesterol metabolism in the White Leghorn laying hen. The researchers suggested that this inability to support the hypothesis in a species other than rat indicates that further studies, possibly with human subjects, should be undertaken prior to accepting the concept for humans.

A study by Zenoble and Bowers of the Department of Foods and Nutrition, Kansas State University, undertaken in 1977, is exemplary of the much needed further research in determining the properties of certain elements, including copper, when contained in various food substances. Part of the study was directed at determining the effects of cooking on copper content of turkey muscle. The researchers found that copper was significantly lower in cooked than in raw breast turkey muscle, but similar in raw and cooked thigh muscle.

References

Campbell, J. K., and C. F. Mills: "Effects of Dietary Cadmium and Zinc on Rats Maintained on Diets Low in Copper," *Proc. Nutr. Society*, **33**, 1, 15A (Abstract) (1974).

Cort, W. M., Mergens, W., and A. Greene: "Stability of Alpha- and Gamma-Tocopherol: Fe^{3+} and Cu^{2+} Interactions," *J. Food Sci.*, **43**, 3, 797–800 (1978).

Hamilton, R. P., et al.: "Zinc Interference with Copper, Iron, and Manganese in Young Japanese Quail," *J. Food Sci.*, **44**, 3, 738–741 (1979).

Helwing, L. R., Jr., et al.: "Effects of Varied Zinc/Copper Ratios on Egg and Plasma Cholesterol Level in White Leghorn Hens," *J. Food Sci.*, **43**, 666–669 (1978).

Hill, C. H., and G. Matrone: "A Study of Copper and Zinc Interrelationships," *Proc. 12th World's Poultry Congress*, 219 (1962).

Kenner, J., Mendel, H., and P. Budowski: "Prooxidant and Antioxidant Effects of Ascorbic Acid and Metal Salts in a Beta-Carotene-Linoleate Model System," *J. Food Sci.*, **42**, 1, 60–64 (1977).

Kirchgessner, M. (editor): "Trace Element Metabolism in Man and Animals," Institut für Ernahrungsphysiologie, Technische Universität München, Freising-Weihenstephan, Germany, 1978.

Lee, Y. C., et al.: "Kinetics and Computer Simulation of Ascorbic Acid Stability of Tomato Juice as Functions of Temperature, pH and Metal Catalyst," *J. Food Sci.*, **42**, 3, 640–644 (1977).

Sax, N. I.: "Dangerous Properties of Industrial Materials," Van Nostrand Reinhold, New York, 1979.

Zenoble, O. C., and J. A. Bowers: "Copper, Zinc and Iron Content of Turkey Muscles," *J. Food Sci.*, **42**, 5, 1408–1412 (1977).

COPPER LOSS. This term is frequently used to denote the resistance loss in the conductors of electrical circuits or machines. In most machines there are two types of electrical losses, those caused by winding resistance, i.e., the copper loss, and those caused by the magnetic core, i.e., core loss. Copper loss is given by

$$P = I^2R$$

where I is the current (effective value for ac) and R the resistance.

COPPER OXIDE. **Cuprite; Tenorite.**

COPPER OXYCHLORIDE. **Atacamite.**

COPPER PYRITES. **Chalcopyrite.**

COPPERSMITH. **Woodpeckers and Toucans.**

COPPER SULFATE. **Brochantite; Chalcanthite.**

COPPER SULFIDE. **Bornite; Chalcocite; Covellite.**

COPRA. **Palm Trees.**

COPULATION. The act of sexual union by which the seminal fluid, containing the reproductive cells of the male, is transferred to the genital passages of the female.

The germ cells are adapted for locomotion through liquids, hence many aquatic species need only discharge them into the surrounding water simultaneously to enable them to come together for fertilization. If the egg is to develop in the body of the mother, however, or if the animal is entirely terrestrial, the liquid medium in which fertilization occurs is secreted by the body and a direct transfer from male to female is usually necessary. Artificial insemination of animals has been accomplished.

In some animals copulation is accomplished merely by the apposition of the orifices of the genital ducts, but in most cases the terminal portion of the female organs becomes a vagina for the reception of a male intromittent organ. This organ varies greatly. In some of the rotifers the pointed end of the body serves for the introduction of the germinal material, although some have a special projecting organ called the penis. Some of the roundworms have a pair of copulatory setae, which project from the alimentary tract. In the crustaceans certain paired appendages are modified for introduction into the female and in the spiders the male discharges the seminal fluid onto a web and takes it up into his palpi, which are modified for the transmission of the material to the female ducts.

Among the vertebrates most intromittent organs are in the form of a penis developed either as a projecting fold in the wall of the cloaca or as a protrusible organ associated with the urogenital passages at their caudal extremity. In the shark the pelvic fins sometimes bear lobes which are thrust into the cloaca of the female during copulation.

In humans, the terms *sexual act* and *sexual intercourse* are also used. For conception to occur, intercourse must occur during the woman's fertile period, at a time close to ovulation. Generally, this time occurs at the midpoint between successive menstrual periods.

COQUETTE. Swifts and Hummingbirds.

COQUINA. This is a Spanish word meaning little shells. It is a coarse and highly porous limestone made up of shells and shell fragments loosely cemented. It is being formed at present along the coasts of Florida, where it is frequently referred to as "beach rock." Only a few of the limestone formations of former geological periods are true coquina. In Bermuda coquina, largely of Aeolian origin, is sawed into blocks and used as a building material.

CORACIIFORMES (*Aves*). This order consists of the superfamilies kingfishers, todies, motmots, with families of bee-eaters, cuckoo-rollers, rollers, hoopoes, wood hoopoes, and hornbills. Only a few common structural characteristics and habits unite the families of this order; as a result, this order was previously subject to many different decisions concerning its relationship and position in the zoological system. Today, these birds have been combined into one order on the basis of their three partially fused anterior toes (syndactylism), because of their desmognathous palatal structure, their leg muscles, and because of particular plumage developments and arrangement. The feet are generally noticeably small. There are 2 notches in the rear edge on each side of the sternum (only 1 in hoopoes and hornbills), 10 primaries (often with a vestigial eleventh one), and 12 tail feathers (only 10 in motmots, hoopoes, and hornbills). The length is 9–105 centimeters ($3\frac{1}{2}$–41 inches). These are largely colorful tropical or subtropical land birds. The beak is often large and of peculiar shape.

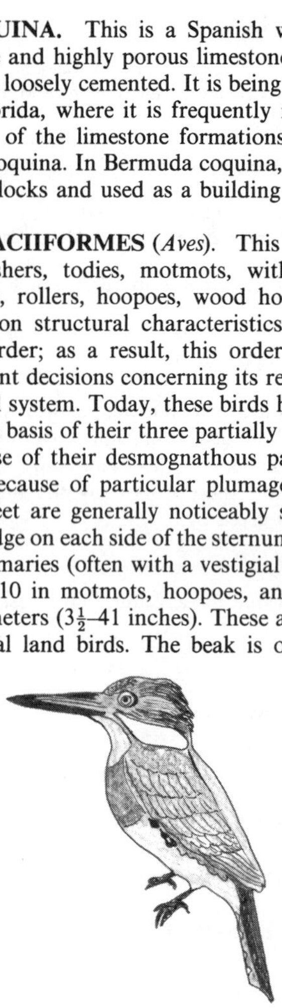

Fig. 1. Amazon kingfisher (*Chloroceryle amazona*).

Fig. 2. Turquoise-browed motmot (*Eumomota superciliosa*).

These birds are primarily meat, fish, and insect-eaters, although some also take fruit and berries. The young are naked and blind at hatching (with the exception of the hoopoes). The sexes are similar except in many of the hornbills and some kingfishers.

Most families are confined to the east of the Old World, but kingfishers are also found in the New World. Todies and motmots occur only in the New World. Bee-eaters, hoopoes, and hornbills probably originated in Africa, while the kingfishers originated in southeastern Asia. See Figs. 1 and 2.

We recognize 7 families: 1. kingfishers (*Alcedinidae*); 2. todies (*Todidae*); 3. motmots (*Momotidae*); 4. bee-eaters (*Meropidae*); 5. rollers (*Coraciidae*); 6. hoopoes (*Upupidae*); and 7. hornbills (*Bucerotidae*). There are 53 genera with 190 species. See also **Kingfishers and other Coraciiformes.**

CORAL (*Coelenterata*). The hard deposit built up by minute colonial animals called coral polyps which occur in the warmer oceans. The deposit consists principally of calcium carbonate.

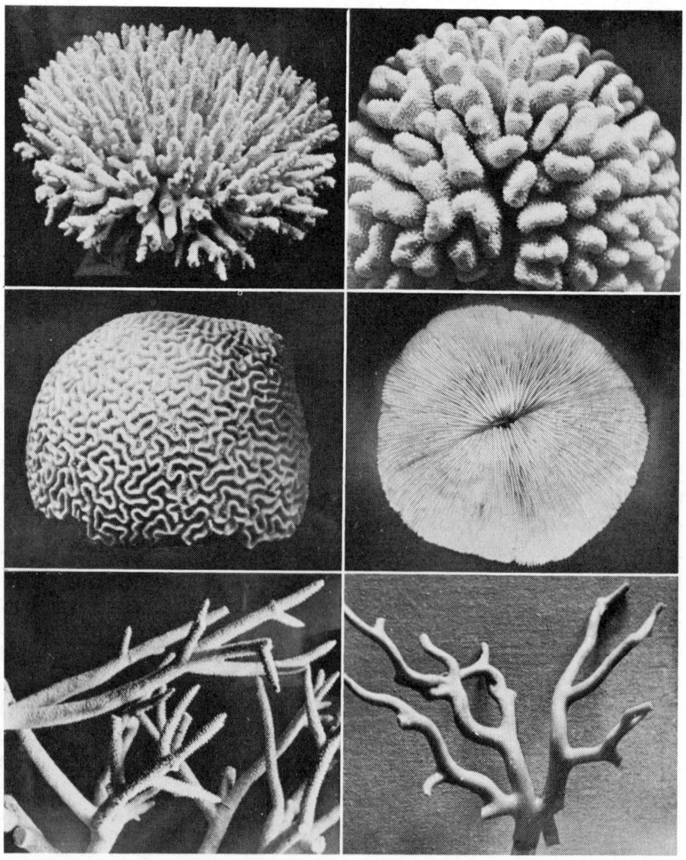

Varieties of coral. From left to right and top to bottom: Star coral, *Pacillopora grandis*; fingerlike madrepore, *Madrepore digitata*; brain coral, *Meandrina sinuosa*; mushroom coral, *Fungia dentata*; staghorn coral, *Madrepore cervicornis*; precious red coral, *Corallium rubrum*. (*A. M. Winchester.*)

The corals of the order *Hydrocorallinae* of the Class *Hydrozoa* exist as sessile colonies with a massive encrusting or branching exoskeleton with pits in the surface from which the polyps arise.

The corals of the orders *Alcyonaria* and *Zoantharia* of the Class *Anthozoa* (*Actinozoa*) are of different form and habits. Those of the alcyonarians are made up of minute spicules formed within the tissues, occasionally compacted in a hard central rod running through the entire colony and sometimes supplemented by an external covering. Red or precious coral is the hard axis of such a form and organ-pipe coral is made up of the connected tubes which once surrounded the living animals. Zoantharian corals build up hard deposits externally beneath the basal disk which attaches them to the ocean floor. As new individuals arise from the edge of the living tissue their deposits become continuous with those already laid down and so large colonies produce extensive masses of coral rock. The form of these deposits varies. Some are slender and branching and others rounded and massive. They have received common names such as staghorn coral and brain coral.

Precious coral is secured principally in the Mediterranean and is the foundation of a considerable industry in Italy. Several thousands of persons in that country work coral into beads and other ornaments and make it into jewelry.

The formation of coral islands in the warmer oceans has resulted in many habitable land masses, and in the same waters submerged reefs of this material are serious obstacles to navigation.

CORALLITE. Invertebrate Paleontology.

CORAL REEF. A complex, ecological association of benthonic (bottom-living) and attached, calcareous, shelly marine invertebrates, forming either fringing reefs, barrier reefs, or atolls. The lagoons of barrier reefs and atolls are important loci for the deposition of fine-grained calcium carbonate mud called drewite. Fossil reefs include all types of organic reefs which show a distinct ecological and structural evolution from the earliest known fossiliferous limestones to the typical atolls of the South Pacific Oceanic Islands.

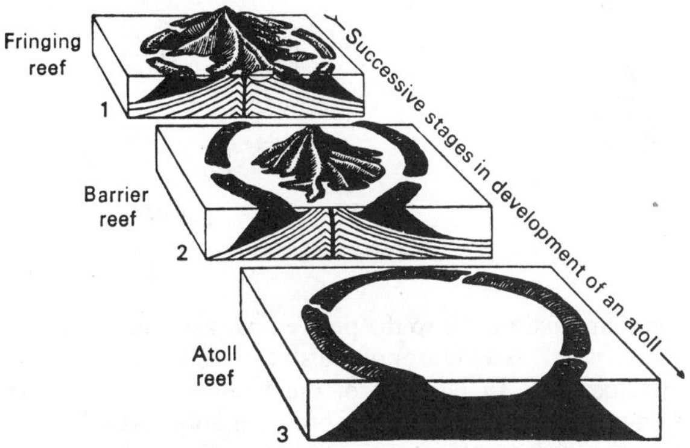

Successive stages in the development of an atoll during the subsidence of a volcanic cone. ("*Field Laboratory Manual*," *Princeton Univ. Press.*)

CORAL SNAKES. Snakes.

CORALS (Fossil). Invertebrate Paleontology.

CORAL TREE. Of the family *Leguminosae* (pea family), genus *Erythrina*, there are about 50 species of coral trees. A native tree of Brazil, the coral also is found in Africa (*E. caffra*), in India (*E. indica*), and quite commonly in the southeastern United States (*E. herbacea*), and in southern California. The height of the tree varies with climate, but some attain a height of 40 feet (12 meters) and a trunk diameter between 18 and 25 inches (46 and 64 centimeters). The leaf is trifoliolate and located at the end of the branch. Three or four large branches often grow out from the trunk up about 3 feet (0.9 meter) from the base of the tree. Smaller branches are numerous, smooth, and rather straight, gradually and slightly curving upward. These, in turn, produce a profusion of branches. The blossom is a bright red cluster of flowers located at the tip of the branch. However, some coral trees have yellow blossoms. There are from six to eight flowers in a cluster, with short stems. The flower varies from the size of a pea to 3 or 5 inches (7.5 or 12.5 centimeters) across. The tree may be described as quite spectacular when in bloom. The bark is smooth, with vertical patterning.

CORDIERITE. The mineral cordierite, composition $(Mg, Fe)_2Al_4Si_5O_{18}$ is an orthorhombic mineral frequently seen, however, in pseudo-hexagonal forms, as well as massive. It is brittle, with a subconchoidal fracture; hardness, 7–7.5; specific gravity, 2.53–2.78; luster, vitreous; color, blue of varying shades; translucent to transparent. Cordierite exhibits pleochroism (or dichroism) being dark blue, light blue and light yellow when examined by transmitted light in different directions. Hence, it is frequently called *dichroite*, and less frequently, *iolite*. It is occasionally used as a gem.

Cordierite is found as a primary mineral in the igneous rocks. It is, however, found ordinarily in gneisses, schists and in areas of contact metamorphism. Localities for good specimens are numerous in Europe, including Bavaria, Finland, Norway. It is found in Greenland, the Malagasy Republic and Ceylon, from which later place come the rolled pebbles of a rich blue color known as saphir d'eau, prized as a gem. In the United States, it is found principally in Connecticut.

Named for the French geologist, Pierre Louis Antoine Cordier, this mineral has also been called iolite from the Greek word meaning violet, and stone, as well as dichroite from the Greek meaning *two-colored*.

See also terms listed under **Mineralogy.**

CORE (Earth). Earth.

CORE LOSS. In electrical machinery, magnetic cores provide easy paths for the flux which is necessary for the proper operation of the machines. The insertion of this core is not without its drawbacks, however, as it introduces additional losses known as core losses. In spite of the additional loss present in the magnetic material the overall effect of the core is a tremendous increase in the efficiency of the machine because of the smaller currents needed to produce the desired magnetic flux. The core losses are composed of *eddy current* loss and *hysteresis* loss. The former is caused by the currents which are induced in the core material by the changing flux through it. This is, of course, much more pronounced in ac than in dc machines because of the much greater rate of change of the flux in the former. In order to reduce this loss the core of all ac machinery and much dc machinery is laminated or composed of thin sheets. The hysteresis loss comes as a result of something like viscous friction opposing the reversal of magnetic field intensity. While the two types of losses do not follow the same laws, the total loss varies approximately as the square of the frequency if other factors are held constant, or about as the 1.5 power of the flux density if it is the only variable. See also **Hysteresis (Magnetic).**

CORE (Magnetic). Electromagnetic equipment, as exemplified by the transformer, the motor, and the generator, have electrical circuits, usually of copper conductors, and magnetic circuits. The magnetic circuit follows a path largely contained in a core composed of iron or iron alloys. The purpose of the core metal is to offer the best path for the magnetic lines of flux, and its success in this respect is measured by its permeability. Cores are usually composed of a large number of thin metal laminations which are fabricated by punching from thin sheets of metal, and after being enameled are assembled to form a core. The enamel forms an insulation between laminations which reduces the eddy currents induced in the metal of the core by transformer action. Normal oxidation scale is frequently sufficient insulation for this. See also **Core Memory (Computer).**

CORE (Reactor). Nuclear Reactor.

CORE SAMPLER (Piston corer; Kullenberg corer). A long slender tube with an internal piston. The tube is lowered slowly to the ocean floor where it is suddenly released while the piston remains in place. The tube bores into the ocean floor and a core of the sea floor is drawn into the tube as a consequence of the partial vacuum created by the piston. A major tool used in oceanography.

CORE STORAGE (Computer). Also termed core memory or simply core, a core storage memory is a storage medium in which one bit of binary data is represented by the direction of magnetization in each unit of an array of ferrite toroidal rings. The ferrite material has a rectangular hysteresis loop as shown in Fig. 1 and the magnetic flux of the core remains in one of two stable states, either A or A′, when no current is flowing through the X and Y select lines, as shown in Fig. 2.

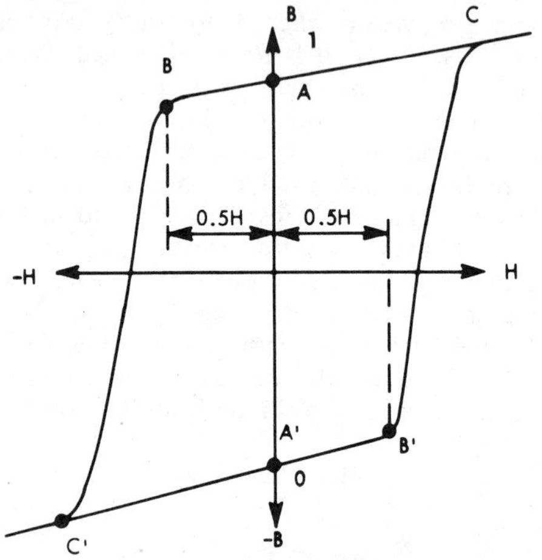

Fig. 1. Hysteresis loop of magnetic core.

When a particular core is to be read out, current is passed through the cores via the X and Y select lines in such a direction as to set the particular core to zero. Thus, if X0 and Y0 are selected, the currents produce a combined magnetizing force −H which sets the core at coordinates 00 to the flux state C′. If the core contained a "1," i.e., if the flux was at a point A on the hysteresis loop, the change in flux generates a voltage in the sense winding which is interpreted as a "1" by discriminating logic in the sense amplifier circuit. If the core is in the "0" state or A′, the change in flux is smaller, corresponding to transversal of the hysteresis loop from A′ to C′, and the voltage generated in the sense winding is ignored by the logic in the sense amplifier circuit. In this example, the cores at the other coordinates in the memory are, at most, only half selected, i.e., the magnetizing force is −0.5H which only sets the core to point B if the core is the "1" state. When the current is removed, the magnetic flux in the core returns to point A.

In order to write "1" in a core, current is passed through the X and Y select wires to produce a combined magnetizing force +H on the core located at the intersection of the two select lines. This causes the core to be set to point C and it subsequently stabilizes at point A when the current is removed. If a zero is to be set in the core, a current is driven through the inhibit winding which produces a magnetizing force of −0.5H. This, in conjunction with the magnetizing force due to the current in the select lines, limits the net magnetizing force

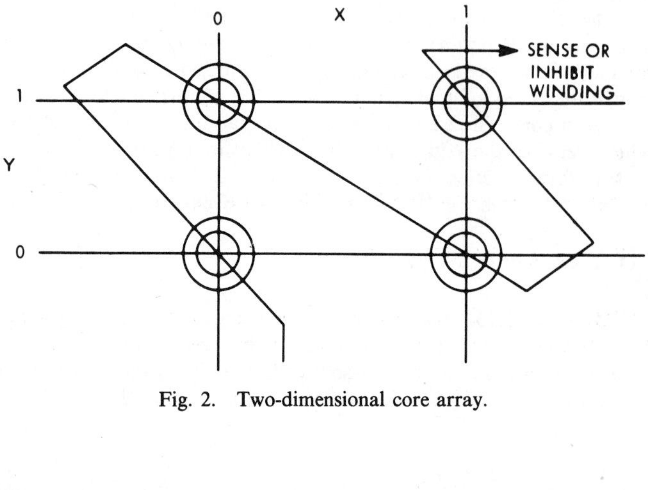

Fig. 2. Two-dimensional core array.

to 0.5H and the flux only moves on the hysteresis loop to point B′. When the currents are removed, the magnetic flux of the core reverts to the "0" state or point A′.

Although once the most widely used device for main storage, core storage has been largely displaced in practice by semiconductor storage, which is less costly, smaller, utilizes less power, and is faster. Core storage continues to be used where retention of data in main storage is required when power is removed.

Magnetic cores also find application in power and signal transformers and as transformerlike logical elements.

See also **Semiconductor Storage (Computer); Storage (Computer);** and terms listed under **Data Processing.**

Thomas J. Harrison, International Business Machines Corporation, Boca Raton, Florida.

CORIANDER. Flavorings; Umbelliferae.

CORING. A fine scale or microscopic variation in composition within the grains of a metal casting. Coring is usually associated with a solid solution phase and is caused by failure to achieve equilibrium during freezing. The alloy initially solidifies in the form of dendrites or skeleton crystals that contain a greater proportion of the higher melting components. The spaces between the dendrite arms freeze later with a composition lower in the higher melting components.

CORIOLIS EFFECT. Any object moving above the earth with constant space velocity is deflected relative to the surface of the rotating earth. This deflection was first discussed by the French scientist Coriolis about the middle of the last century, and is now usually described in terms of the Coriolis acceleration or the Coriolis force. The deflection is found to be to the right in the northern hemisphere and to the left in the southern.

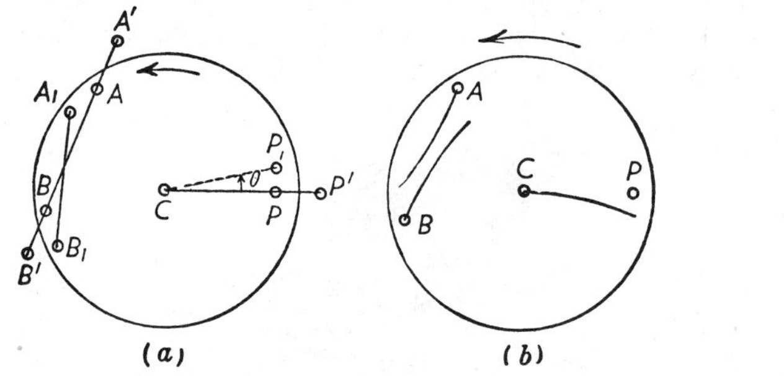

Coriolis effects.

As a first approximation to the problem we assume that an observer is at the center (C) (see figure) of a disk that is rotating with constant angular velocity ω. In part (a) of the figure, the observer can see objects off the disk and is conscious of the rotation. At a given instant, when the object P on the disk is directly in line with the point P' off the disk, the observer fires a shot at P. During the time (t) that is required for the shot to move over the distance CP at speed v ($CP = vt$), the disk will turn through the angle θ ($\theta = \omega t$). The observer notices that the bullet misses the point P, but that it hits P'. In Fig. 1(b) the observer's "world" is limited to the rotating disk and he has no way of knowing that his world is in rotation. Under these conditions when he fires a shot at P he will notice, as before, that the bullet misses P, and he will also determine that the shot follows a curve, similar to that shown, in his world. The same effects will be observed no matter where the observer is located in his world or what direction he aims his shot (e.g., from A in direction BB', or from B toward AA'). The deflection is always to the right with the direction of rotation indicated in the figure, and would be to the left if the direction of rotation were reversed.

In accordance with fundamental definitions a curved motion represents an acceleration (a) and an acceleration is the result of the action of a force (f) which is proportional to the mass (m) of the object and the speed (v) with which it is moving. Analysis of the motions shown in the figure show that the acceleration relative to the disk is given by $a = 2v\omega$, or, since $\omega = (2\pi/T)$, in which T is the period of rotation, $a = (4\pi/T)v$. The force producing this acceleration will be given by $f = k(4\pi/T)mv$, in which k is a factor of proportionality whose value is determined by the units employed.

The conditions on the rotating earth in the vicinity of either pole of rotation are comparable to the "world conditions" on the rotating disk. The sense of rotation will be opposite at the two poles and the deflection will be to the right at the north pole and to the left at the south. The angular velocity of rotation of an area at any place on the earth in latitude L is given by $\omega = (2\pi/T)\sin L$, in which T is the sidereal period of rotation of the earth or 86,163.4 seconds. Hence the Coriolis acceleration is $a_c = (4\pi/T)v \sin L$, and the Coriolis force $F_c = k(4\pi/T)mv \sin L$.

The Coriolis effects must be considered in a great variety of phenomena in which motion over the surface of the earth is involved. Among these may be listed the following: (1) Rivers in the northern hemisphere should scour their right banks more severely than the left, and the effect should be more evident for rivers in high latitudes. Studies of the banks of the Mississippi and Yukon rivers indicate the predicted results. (2) The motions of air over the earth are governed, to an appreciable extent, by the Coriolis force. (See **Winds and Air Movement.**) (3) A term, due to the Coriolis force, must be included in the equations for exterior ballistics. While the effect is negligible in small-arms fire, nevertheless, it is very important in the fire-control material for long-range guns. (4) Any level bubble, which is being carried on a ship or plane, will be deflected from its normal position. The deflection will be perpendicular to the direction of motion of the ship or plane. The correction for this effect may amount to several miles in the determination of a position of the ship or plane by methods of celestial navigation if the bubble octant is used in the necessary observations. (See **Sextant**).

CORIOLIS FORCE. Atmosphere (Earth).

CORK. Cork cells are found in the outer bark of most woody-stemmed plants, but in amounts too small and with too brittle walls to be of any use to man. But in *Quercus suber*, the Cork Oak, the cork cells become a very large part of the tissue of the bark, and have been used for centuries by man. The cork oak tree is a medium-sized tree, seldom much over 50 feet (15 meters) in height, growing in nearly all countries bordering the Mediterranean Sea. The evergreen leaves are small, $1\frac{1}{2}$–3 inches long, and about an inch wide, with slightly toothed margins. The bark of the tree soon becomes rough and deeply furrowed, but is of little value except as ground cork or as a source of tannin. When the tree is about 20 years old this first formed bark is removed, care being taken not to injure the phloem and cambium layers. Within 10 days a new cork layer has formed. This layer is the first of many layers which are removed once every 10 years or so throughout the life of the tree. Removal is generally done in the early summer at a time when hot dry winds will not cause injury to the unprotected phloem and cambium.

After removal, the cork is air-dried for a time, then boiled to soften it and to remove some of the tannin. The outer part of the bark is scraped off, and the rest pressed out flat and dried. It is then ready to ship.

The physical properties of cork account for its many uses. It is very light and buoyant, more than 50% of its volume being air, and hence is used in the manufacture of floats, life-preservers, and so forth. The living protoplasm of the cork cells dries up early in their development, leaving hollow cells, each containing a small mass of air which expands after compression. Therefore, cork is very resilient, and is frequently used as a core on which to wind yard or string in the manufacture of baseballs. In the early stages of their formation, the walls of cork cells are cellulose, but this is soon impregnated by a waterproof and non-absorbent lipoid substance, suberin. Therefore cork is used in making handles for fishing rods, shoe-soles, and cork stoppers. Since the hollow cork cells are poor conductors both of heat and sound, cork is much used as insulating material. For this use cork is ground up and then pressed into sheets with various binding materials, giving much larger sheets than can be obtained from the tree. Ground cork is also a constituent of linoleum, gaskets, and other products.

Cork is traversed by lenticels, loose masses of porous tissue, which appear as dark spots or holes in stoppers. Usually in making stoppers the bark is cut so that these will be transverse in the stopper. In making stoppers, the forms are first punched out as cylinders, and then trimmed down by machine to the required tapering shape.

CORK CELLS. Periderm.

CORKWOOD TREE. Of the family *Leitneriaceae*, the corkwood (*Leitneria floridana*) is a small deciduous tree, seldom exceeding a height of 20 to 25 feet (6 to 7.5 meters) and with a narrow trunk of about 6 inches in diameter. The tree is found in the southeastern United States and also in the regions of Arkansas and Texas. The bright green leaf is about 5 inches (12.5 centimeters) long, 2 to 3 inches (5 to 7.5 centimeters) across, and quite hairy on the underside. The fruit is of oval shape, leathery, unveined, flat, and brown in color. The wood of this tree is among the lightest, weighing about 13 pounds per cubic foot (208 kilograms per cubic meter). It finds limited application in fishing tackle. This tree is not to be confused with the balsa tree, the wood of which is even lighter. The balsa is sometimes also referred to as corkwood or Indian corkwood. See also **Balsa Tree.**

CORM. 1. In botany, a corm is a very short, thick, subterranean stem distinguished from the rhizome by erect instead of horizontal growth. Its surface shows more or less distinct nodes. From the nodes of the upper portion, buds develop. Generally, roots are formed in the lower part of the corm. In many plants the corm is surrounded by scaly leaves or leaf bases. The crocus and the gladiolus produce corms.

2. In zoology, the corm is the median branch or endopodite of the appendage of certain crustacea together with the common basal portion. When the expodite is reduced, these parts sometimes appear as the principal axis of the appendage.

CORMORANT. Pelicans and Cormorants.

CORMIDIUM. A group of individuals of various forms budded from the parent stalk of certain floating marine coelenterates.

CORN (As Energy Source). Biomass and Wastes as Energy Sources.

CORN BORER (European). European Corn Borer.

CORNCRAKE. Rails, Coots, and Cranes.

CORNEA. The transparent outer layer of the front of the eyeball. See **Eye.** It is a complicated structure composed of five layers. The cornea may be the site of inflammation or ulceration.

CORNEAGEN CELL. A kind of cell found in the eyes of some insects. It produces the transparent lenticular cornea at the outer surface of the eye and is renewed on ecdysis.

CORNEAL TRANSPLANTATION. Vision and the Eye.

CORNEAL ULCER. Vision and the Eye.

CORN EARWORM (*Insecta, Lepidoptera*). Many millions of acres of corn (maize) have been attacked and destroyed by this insect over the years in the United States and other regions of the world. In the United States, the pest is most damaging in the southern states. Although its damage is usually associated with corn, the insect is a general feeder and will attack numerous crops, such as bean, cotton, lettuce, tobacco, okra, tomato, and vetch. When found on tobacco, it may be called the tobacco budworm; or if found on tomato, the tomato fruitworm; or if on cotton, the cotton bollworm. The adult corn earworm is a moth of the family *Notodontidae.* The official species name is *Heliothis zea* or *H. armigera*, Boddie. In some regions, records indicate that nearly 100% crop loss has resulted from heavy infestations of this pest.

The insect winters over as a brownish pupa, usually found several inches in the soil below groundlevel. Moths emerge in spring and early summer. The moth is relatively large, with a wingspread of about 1.5 inches (nearly 4 centimeters). The moths are of a rather nondescript coloration, ranging from gray to brown to olive green.

Markings are irregular. After emerging, the moth consumes nectar from nearby flowers. The female deposits yellow eggs in the evening, singly, even though the total number may be as high as 3000. Depending upon latitude and prevailing seasonal temperatures, there may be as many as 3 generations of the insect per year. The eggs early in the season are placed among the curl of early-leafing plants. For subsequent generations, corn silk is one of the preferred locations. Hatching requires up to 10 days. The emerging worms feed on the corn silk. When the silk has lost much of its moisture, the worms move on to the kernels of the ear and feed on the ear for 2 to 4 weeks, during which time they molt 5 times. When fully grown the worms are striped, green-brown in color, and about 2 inches (5 centimeters) long. Their damage is extended because a single worm may move over more than one ear. When fully grown, the worms (larvae) drop to groundlevel where they excavate a cell in the soil, up to 5 inches (about 12 centimeters) deep. The pupate in this location for 1.5 to 2.5 weeks, after which they emerge as moths.

Application of a carbaryl spray (from carbaryl wettable powder) to the silk immediately after the silks appear can prevent damage to ears. Usually this treatment must be repeated at least 4 times at 2-day intervals.

Corn that is to be used for fodder adds to the difficulties of chemical controls. Fall mowing reduces the population of overwintering pupae. Within recent years, corn earworm-resistant varieties have become available and greatly reduce the damage from this insect.

CORNELL ELECTRON STORAGE RING. Particles (Subatomic).

CORNER. A corner is the point of intersection of adjacent property lines. Landed property is ordinarily bounded by broken lines meeting at the "corners" of the property. A land survey is generally a traverse with the transit stations at the corners and with the traverse lines coinciding with the property lines. The corners are indicated on a survey plat. They are frequently marked by monuments, either artificial or natural. When subdividing land in accordance with the scheme of the United States land subdivision, the corners are designated by standard monuments whose character and markings are governed by certain specific regulations.

If it is impossible to set a monument at a corner, permanent markers called "witness corners" are placed on all lines intersecting at the corner. When a property line intersects a body of water the intersection is marked by a permanent monument known as a "meander corner."

CORNER CUBE PRISM. Prism (Optics).

CORNER REFLECTOR (Optics). A reflector which consists of two plane-conducting surfaces set at an angle of 45° to 90° with the driven element on a line bisecting the angle. The reflecting surfaces are not necessarily solid, but can be made from wires spaced about 0.1 wavelength apart. In a given amount of space, the corner reflector gives better directivity than the parabolic reflector.

CORN (Hybridization). Plant Breeding.

CORN (Maize). In the United States, the word *corn* signifies Indian corn or maize, *Zea mays.* In Europe, the word is used for several cereal grains; in England for wheat; in Scotland and Ireland for oats. In its original use, corn designated a hard seed or grain.

There has been considerable speculation as to the origin of this plant. Everything indicates that it is native to America, probably originating in Mexico or Central America. The plant is not known to occur in the wild state, but a native Mexican grass, teosinte, *Euchlaena mexicana*, is a closely related grass with which corn hybridizes freely. Some botanists hold that teosinte is the ancestral grass from which corn originated. See Fig. 1.

There are many kinds of corn in cultivation, ranging from dwarf forms less than 3 feet (0.9 meter) high to giant plants 15 feet (4.6 meters) or more in height. All kinds have an extensive fibrous root system, the individual roots not only occupying the surface portion of the soil but also extending downward to depths of 6 feet (1.8 meters) or more. In addition to these normal roots, which all arise from the basal portions of the very young stem, prop roots develop from the lower nodes of the older stem. These prop roots are coarse outgrowths

Fig. 1. Probable precursor of the corn (maize) plant known as teosinte (*Euchlaena mexicana*): (1) Plant; (2) branch with staminate and pistillate inflorescences; (3) ear; (4) grains with attached stigmas; (5,6) mature grains. (*USDA diagram.*)

which radiate outward and downward until they reach the surface of the ground. During their growth in the air, their tips are protected from drying by an abundant slime coating; once they have entered the ground they branch abundantly and become like normal roots. They serve to support the plant. The stem of the corn plant is coarse and, unlike other grasses, solid throughout its length. The leaves, borne alternately on the stem, have large broad blades at the base of which is a conspicuous ligule, a membranous outgrowth which tightly invests the stem and so may serve to prevent water entering between the stem and the leaf-sheath.

The corn plant (Fig. 2) is monoecious, that is, both staminate (stamen) and pistillate (pistil) flowers are borne on the same plane. However, they are usually not borne in the same inflorescence. The staminate inflorescence or tassel appears at the top of the plant, and matures some time before the pistillate flowers do. The male flowers produce immense quantities of pollen which when ripe is shed in the air, to be carried by wind currents or gravity to the pistillate flowers. Corn pollen is a cause of hay fever (see **Allergy**). The pistillate inflorescence, or ear, is a modified branch developing in the axis of a leaf. This branch has a fleshy axis or cob on which are borne rows of pistillate flowers. These occur in two flowered spikelets, the lower flower usually being abortive, but its bracts, the lemma and palea persisting, as do the two short glumes which subtend the entire spikelet. The ovary bears a long style, commonly known as the silk of the corn. When first developed the silk is green and has a sticky surface; after pollination has occurred, the silk turns brown and dries up. The entire pistillate inflorescence is enclosed in many overlapping modified leaves known as husks, from the tip of which the silk protrudes.

The mature corn grain is variously shaped according to the kind of corn. In most species it is a flattened object with a shallow groove on one side, indicating the location of the embryo. This embryo is on one side of the grain, the rest of which is filled with a starchy substance, the endosperm. This endosperm is usually separable into

Fig. 2. Corn (maize) plant (*Zea mays*). (*USDA photo.*)

two parts, one hard and horny, called the horny endosperm; the other less firm and of lighter corn, known as the starchy endosperm. The horny portion contains more protein than does the starchy and has its starch grains more densely packed together.

The corn grain will not germinate unless the temperature is about 40°F (4.5°C) and sprouts best when the temperature is about 90°F (32°C). It is grown most successfully in regions having a deep, warm, well-drained loam soil, an abundance of rainfall and a growing season of 90 days or more, depending on the kind of corn. Improved varieties have been developed which grow satisfactorily in regions having a shorter growing season. Corn is principally grown in the Americas, especially in the United States and Argentina. It has been introduced in European countries and into Africa, but is not grown there to any great extent. In the United States the so-called corn belt grows more than half the entire world crop. This corn belt comprises the states from Ohio west to South Dakota and south to Kansas, a region having climatic conditions most favorable for this crop. Almost all of the corn produced today is hybrid corn. See also **Green Revolution.**

The uses of the corn crop are many and varied. The green plants are fed to stock directly or are stored in silos. For storage in silos the entire plant is cut down and chopped into small pieces. These are compactly stored in large tight structures of wood, concrete or other material, in which partial fermentation occurs, forming a product called ensilage. This is an important part of the ration of dairy cows.

Corn grains form a most important food product for humans and domesticated animals. The ripe ears are picked from the plants and allowed to dry. The grain is then removed from the cob and used directly as stock and poultry food or ground into coarse particles known as cracked corn, much used in feeding poultry. Ground somewhat finer, but still consisting of coarse particles, white corn becomes grits, much used in southern states. More finely ground, corn becomes corn meal, used in making corn bread and various kinds of puddings. Rarely corn is finely ground to flour. Another corn preparation is hulled corn. In making this the grain is soaked in lye which loosens the pericarp or outer portion of the grain. This is then removed and the remaining grain cooked soft. As a breakfast food corn appears principally in the form of corn flakes. In making these, clean corn grains are steamed and the hulls and embryo removed, leaving the endosperm. This is sweetened and flavored, and then cooked by steam under pressure. Following this cooking, the grains are partially dried and then passed between heavy rollers, which make them flakes. These flakes are carried to huge ovens where they are quickly toasted. Cooling follows, after which the product is packed in waterproof cases and is ready for the consumer.

In addition to use as food for man and beast, corn yields many important secondary products, such as corn starch, glucose (see **Carbohydrates**) and corn oil. Starch is prepared from corn by soaking the grains in slightly acidulated water for several days, after which they are broken up, the embryos removed and the grain ground. Following the grinding, the whole is passed through sieves which remove the pericarp. The resulting paste is allowed to flow slowly over tilted

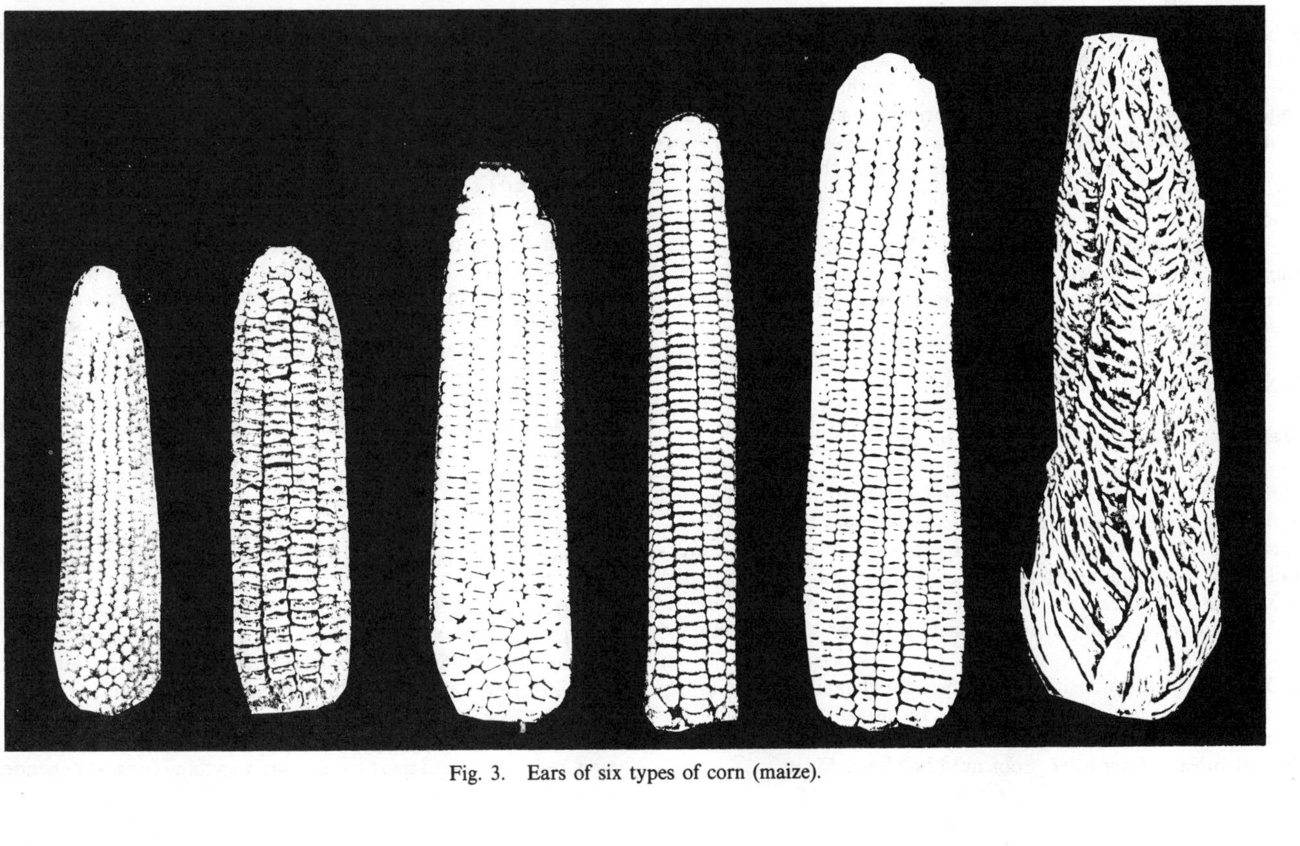

Fig. 3. Ears of six types of corn (maize).

tables, which allows the starch to settle. This starch is washed and dried, then pulverized for use. See also **Starch.**

From corn starch is prepared glucose, or corn sirup, a thick substance which results from the partial hydrolysis of starch with acid. The acid is neutralized with sodium carbonate, and the liquor resulting from neutralization is filtered. This liquor is evaporated, and again filtered, emerging as a clear thick sirup, which is boiled down even further. It now becomes commercial glucose, a thick sirupy substance about half as sweet as cane sugar, and with little flavor. It is used in making jellies and preserves, and in blending with other sweets such as cane sirup and maple sirup.

From the embryos of corn, corn oil is prepared. Corn oil is used as a cooking oil and also in making soaps and paints. The cake remaining after the oil is pressed from the embryos becomes a stock food.

Corn stalks have been used experimentally in manufacturing paper. However, the abundance of objectionable nonfibrous material will probably prevent any extensive use of the stalks in this industry. The cobs left after shelling the grain are used as a fuel to a slight extent and also in making cob-pipes. The dried husks form a stuffing for a particularly disturbing kind of mattress, and are also used to some extent as a stuffing in upholstery.

The principal varieties of corn include pod corn, a rarely grown form in which the glumes, lemma and palea of each floret are particularly well-developed, surrounding the kernel; pop corn, in which the sudden explosion of the moisture within the grain turns the latter more or less inside out; sweet corn, characterized by the high sugar content of the grains, and largely grown in home gardens; flint corn, with very hard grains containing a large amount of horny endosperm; and dent corn, so-called because the floury endosperm extends to the end of the kernel and is surrounded laterally by horny endosperm. On maturing the floury endosperm shrinks, causing an obvious dent to appear at the apex of the kernel. Dent corn is raised most extensively. See Fig. 3.

More detail on corn (maize) can be found in the "Foods and Food Production Encyclopedia" (D. M. Considine, editor), Van Nostrand Reinhold, New York, 1982.

CORN (Physiology). Dermatitis and Dermatosis.

CORN PROCESSING. Starch.

CORN (Root). Root (Plant).

CORN ROOTWORM (*Insecta, Coleoptera*). Larvae of two species of beetles. The adult of one is yellow-green with six black spots on each wing cover. The larva of this species, the southern corn rootworm, damages the roots and lower stems of various grains and grasses and sometimes kills the plants. The other species is the western corn rootworm. The adult is entirely yellowish-green and the larva burrows inside the roots of corn and stunts or kills the plant. Both beetles are about $\frac{3}{8}$ inch (1 centimeter) long.

The most effective protection against these pests is proper crop rotation. Since the southern rootworm lives on plants other than corn this treatment is supplemented by planting early or late to avoid the most serious attack of the larvae.

CORNU DOUBLE PRISM. Prism (Optics).

CORNU-JELLET PRISM. Prism (Optics).

CORNU SPIRAL. A special kind of plane curve, a spiral obtained by plotting the Fresnel integrals $C(t)$ as abscissa and $S(t)$ as ordinate. Its curvature increases proportionally to the length of arc. Also known as a clothoid, it is of interest in classical optics problems concerned with diffraction. Distances on this curve are used in computing intensities in the pattern resulting from Fresnel diffraction. See diagram.

COROLLARY. A proposition which was incidentally proved in the process of proving another proposition. In current use, a corollary is a consequence of an existing theorem (or proof) which is obvious—so that, at most, a very brief proof need be stated for it.

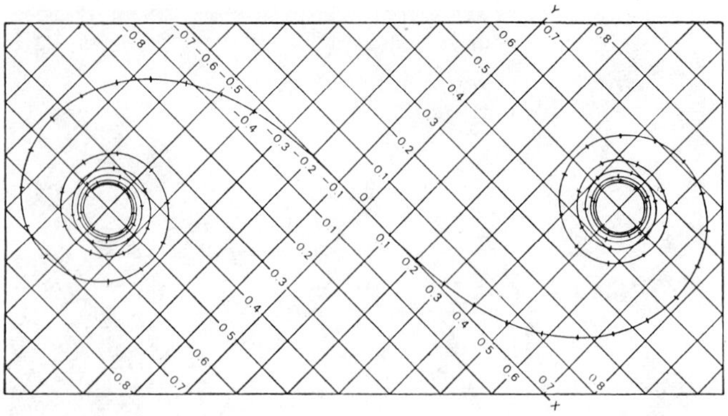

Cornu spiral.

CORONA AUSTRALIS (southern crown). A southern constellation located near Sagittarius.

CORONA BOREALIS (northern crown). A northern constellation located between Hercules and Bootes, the grouping of stars essentially forming a semicircle.

CORONA. Sun.

CORONA DISCHARGE. Saint Elmo's Fire.

CORONAGRAPH. Sun.

CORONA (Meteorology). Atmospheric Optical Phenomena.

CORONARY ANGIOGRAPHY. Angiography.

CORONARY ARTERY BYPASS. Ischemic Heart Disease.

CORONARY ARTERY SPASMS. Ischemic Heart Disease.

CORONARY CARE UNIT (CCU). Ischemic Heart Disease.

CORONARY HEART DISEASE. Anticoagulants; Ischemic Heart Disease.

CORONARY THROMBOSIS. Ischemic Heart Disease.

CORONATAE. An order of jellyfishes (*Scyphozoa*) with a lobed margin. Found in the open ocean.

CORPUS CALLOSUM. A broad band of nerve fibers within the brain, connecting the two cerebral hemispheres.

CORPUSCLE. A term applied to many minute bodies: 1. The cells of the blood or lymph. 2. The excretory unit of vertebrates, consisting of a small knot of blood vessels enveloped by a capsule (Renal corpuscle). 3. Bone cells, which are called corpuscles of Purkinje. 4. Many nerve endings including Pacinian or lamellar corpuscles, Grandry's corpuscles, tactile corpuscles. These and all other sensory corpuscles consist of nerve endings enveloped by accessory cells of various forms. They are sensory organs.

CORRASION. This term is applied to the mechanical wearing away of rocks through the agency of running water and the rock fragments which it carries in suspension.

CORRECTIONS (Navigation). Compass Corrections; Current Correction Angle (Navigation); Wind Correction Angle (Navigation).

CORRECTION TO VACUUM. The correction of wavelengths or of the speed of light or other property as measured in air to the appropriate values *in vacuo.* The index of refraction for air is about 1.000225 but is dependent on the density of the air.

CORRELATION. In general statistical usage correlation or co-relation refers to the departure of two variables from independence. In this broad sense there are several coefficients, measuring the degree of correlation, adapted to the nature of the data, e.g., *association coefficient* for dichotomous material, *contingency coefficient* for more extended classification, *rank correlation* for ranked material and so on. See also **Association (Coefficient of); Contingency Table;** and **Rank Correlation.**

In a narrower sense correlation refers to the degree of dependence of two continuous variables. Given a set of bivariate values $(x_1, y_1) \ldots (x_n, y_n)$ the correlation coefficient is given by

$$r = \frac{\sum_{i=1}^{n} (x_i - \bar{x})(y_i - \bar{y})}{\{\sum_{i=1}^{n} (x_i - x)^2 \sum_{i=1}^{n} (y_i - \bar{y})^2\}^{1/2}}$$

namely is the covariance of x and y divided by the square root of the product of their variances. It may vary between the limits ± 1. A value of zero results when the variables are independent, but strictly implies independence only when the variables are jointly distributed in the normal (Gaussian) form.

When three or more variables are jointly distributed each pair will have a correlation coefficient expressing in a summary form their degree of co-relation. If A is highly correlated with B and C is also highly correlated with B, there will in general result a correlation between A and C. Attempts to remove such part of this correlation as arise from their common dependence on B leads to the formation of a *partial* correlation coefficient between A and C, the object of which is to measure that part of their co-relation which is direct and not merely generated by their relation with a third variable. More complicated forms of partial correlation can arise, e.g., the correlation between X_1 and X_2 when the effect of their correlation with $X_3, X_4, \ldots X_n$ is removed.

The concept is easily generalized to the case where a continuous infinity of observations arises.

An obsolescent expression "correlation ratio" also occurs in older literature. The correlation ratio squared of x on y is given by

$$\eta_{xy}^2 = \sum (\bar{x}_i - \bar{x})^2 / \sum (x - \bar{x})^2$$

where $\bar{x}_i$ is the mean of x_i for a given y_i, $\bar{x}$ is the mean of all values and the summation in the numerator extends over the values of y_i and that in the denominator over all values. A closely allied ratio rises in the analysis of variance (*q.v.*).

Sir Maurice Kendall, International Statistical Institute, London.

CORRELATION (Rank). Rank Correlation.

CORRELOGRAM. The graph of the autocorrelation coefficient (or the serial correlation) as ordinate against the lag of the coefficient as abscissa. See also **Autocorrelation.**

CORRESPONDENCE PRINCIPLE. The principle that in the limit of high quantum numbers the predictions of quantum theory agree with those of classical physics.

CORRODENTIA. An order of minute insects containing the book lice and psocids. They have biting mouths and some species bear four wings. Some are found among plants and on bark and others frequent books and papers, especially in damp buildings.

CORROSION. 1. The electrochemical degradation of metals or alloys due to reaction with their environment; it is accelerated by the presence of acids or bases. In general, the corrodability of a metal or alloy depends upon its position in the activity series (electromotive force series). Corrosion products often take the form of metallic oxides; in the case of aluminum and stainless steel, this is actually beneficial, for the oxide forms a strongly adherent coating which effectively prevents further degradation. Hence, these metals are widely used for structural purposes. Probably the most familiar kind of corrosion is that of *rusting*. This is but a special case of a general classification known as atmospheric corrosion, wherein the oxygen of the atmosphere reacts with the material in question. Most metals, with exception of the noble metals, such as gold, can be oxidized by atmospheric oxygen. In the usual case, however, water vapor must be present before any appreciable oxidation can take place. With iron, for example, about 40% relative humidity is needed at ordinary temperatures before rusting will occur.

Acidic soils are highly corrosive. Sulfur is a corrosive agent in automotive fuels and in the atmosphere (SO_2) as well, and is frequently mentioned in connection with so-called acid rains. Sodium chloride in the air at locations near the sea is strongly corrosive, especially at temperatures above 70°F (21.1°C). Copper, nickel, chromium, and zinc are among the more corrosion-resistant metals and are widely used as protective coatings for other metals.

2. The term *corrosion* is also sometimes used in connection with the destruction of body tissues by strong acids and bases.

In a restricted sense, corrosion is considered to consist of the slow chemical and electrochemical reactions between metals and their environments. From a broader point of view corrosion is the slow destruction of any material by chemical agents and electrochemical reactions. This contrasts with *erosion*, which is the slow destruction of materials by mechanical agents. The character of the atmospheres to which materials are exposed may be classified as: rural, urban, industrial, urban-marine, industrial-marine, marine, tropical, and tropical-marine. In addition to these general kinds of environments, corrosion is of particular concern in the environments of chemical, petrochemical, and other processing and manufacturing environments where extremely corrosive substances may be encountered.

Metals Corrosion. The relationships between metals and hydrogen in the activity series are important because in the electrochemical processes of corrosion the discharge of hydrogen ions and the evolution of hydrogen as a gas is one of the principal cathodic reactions. The facility with which this can occur is determined by such factors as the hydrogen ion concentration (pH) of the electrolyte, the electrical potential of the corrosion cell, and the overvoltage characteristics of the cathodic surface. In a situation sometimes called *concentration cell corrosion*, two solutions of different concentrations will set up an electrical potential between them similar to that produced by a battery. If oxygen is present in the liquid and is continually replenished by contact with the air, then the oxygen concentration in the liquid will remain substantially constant. Any liquid that is contained in small holes or cracks on a metal surface will not be able to obtain oxygen from the main bulk of the solution, so when the supply in the holes and cracks is exhausted, no more oxygen can get in to replace it. Therefore, the oxygen concentration in the cracks is different from that of the main bulk of the solution and a concentration cell is set up. This minute electrical effect is sufficient to make corrosion proceed quite rapidly. A similar cell type of corrosion is that called *galvanic* or *two-metal corrosion*. Two different metals in contact will set up an electrical potential between them. If the two metals are surrounded by an electrolyte so that a closed circuit can be obtained, corrosion takes place. The magnitude of the electrical potential and, therefore, the speed and extent of the corrosion will depend upon the types of metals in each pair. In general, pairs farther apart in the activity series will corrode faster than those close together. See also **Electrochemistry.**

The electrochemical reactions in corrosion of a divalent metal may be written:

Anodic reaction: $M^0 \rightarrow M^{++} + 2$ electrons
At the cathode: (1) $2H^- + 2$ electrons $\rightarrow H_2$ gas
(2) $\frac{1}{2}O_2 + 2H^- + 2$ electrons $\rightarrow H_2O$
(3) $O_2 + 2H_2O + 2$ electrons $\rightarrow H_2O_2 + 2OH^-$
(4) $\frac{1}{2}O_2 + H_2O + 2$ electrons $\rightarrow 2(OH)^-$

It is evident that oxygen as well as hydrogen plays an important part in metal corrosion. It can accelerate corrosion by participating in cathodic reactions, or it can retard corrosion by forming protective oxides or passive films. The dual effect of oxygen is one of the factors that complicates corrosion processes, including the interpretation of observations of the process and the steps to be taken to avoid corrosion damage.

Forms of Corrosion. (1) *Pitting* resulting from local action currents, as at discontinuities in protective or passive films or under or around deposits that set up concentration cells. (2) *Stress corrosion cracking* resulting from the combined effects of corrosion by a specific environment and either applied or internal static tensile stresses. Depending upon the metal and the environment, the cracks may be either intercrystalline or transcrystalline. (3) *Corrosion fatigue*, resulting from the combined effects of corrosion and cyclic stresses. Racks of this type are characteristically transcrystalline. (4) *Intergranular corrosion* resulting from preferential attack on, or around, a phase or compound that occupies grain boundaries. (5) *Corrasion-corrosion* resulting from the combined effects of corrosion and either abrasion or attrition. The mechanism usually involves local or general removal of otherwise protective corrosion product films. Particular forms are impingement attack due to the effects of high velocity or turbulence in flowing liquids, e.g., salt water in steam condensers, or other heat exchangers, in piping systems, valves, and pumps, among others. A particularly aggressive form is associated with the severe mechanical forces that are characteristic of *cavitation phenomena.* See also **Cavitation.** (6) Uniform attack or general wastage, such as may be caused by the action of strong acids as used for pickling (scale removal) or etching. This is also characteristic of the slow corrosion of durable materials in appropriate environments, such as copper roofing in suburban atmospheres, cupronickel tubes in ships, condensers, Monel–nickel copper alloy racks for pickling steel in sulfuric acid, or stainless steel columns for handling nitric acid.

Metal Corrosion Minimization. In the most recent official assessment of the economic costs of equipment damage arising from corrosion, prepared by the National Commission on Materials Policy (U.S., 1973), it was stated that annual losses in the United States alone are on the order of $15 billion annually. Since that time and going into the 1980s, these losses have increased.

Some of the means used to combat corrosion losses include:

(1) Use of the right metal in the proper way and in the correct place. Planners tend to look too closely at first costs and not closely enough at maintenance costs; consequently, there are many applications of materials that are of lower first cost, but of severely limited life. For example, in some applications, inexpensive fasteners that will obviously corrode in a few years are used in place of stainless steel or hardened aluminum fasteners that would have cost only a few dollars more at the time of installation. When discussing applications of possible corrosion-resistant materials, it is important to define clearly the parameters of the environment of usage, such as temperature, pressure, humidity, presence of specific chemical agents, presence of living or dead organic materials, and the characteristics of associated electrical magnetic, light, and other radiation fields. Far more needs to be done on the microclimatology of environments and the specifics needed to ameliorate corrosion problems (Morgan, 1978).

(2) Protective coatings—paints, enamels, other metals, oils, greases, among others. One of the more common methods used is zinc-coating, i.e., galvanizing. Galvanized iron wire with a thin layer of zinc applied to it by dipping in molten zinc or by electrical means usually will resist corrosion for an extended period. Cadmium, nickel, tin, and chromium are metals often used as protective coatings, generally applied by electroplating. See also **Autodeposition; Conversion Coatings; Electrochemistry; Electroplating; Galvanizing;** and **Paint.**

(3) Inhibitors and neutralizers, i.e., compounds added to the environment in small concentration to form protective films which increase anodic or cathodic polarization or both, or to neutralize some corrosive constituents. For example, it is possible for corrosion to occur at many places in the piping leading to boilers or heaters, but usually it occurs in the boiler itself. The trouble is ordinarily found to be due to an acid condition of the boiler feedwater, or to dissolved oxygen contained in it. The raw water used may be acid from surface pollution or from subsurface drains. Usually this can be detected and readily remedied. A more serious factor is the oxygen dissolved in water. Under the high-temperature conditions existing in the boiler per se, this oxygen becomes extremely active in attacking metal surfaces. The operators of large, high-pressure boilers well know the necessity of removing oxygen from feedwater through the use of deactivators or deaerators. Corrosion protection of boilers in power plants is effected by installation of ion exchange resin and other purifiers in the feedwater cycle, and by monitoring them by automatic analysis. In still other installations where acidity and oxygen are a problem, neutralizing chemicals with the dosage governed by automated pH and oxygen control systems are effective. Systems of this type are also effectively used to neutralize plant effluents to streams so that water users downstream have a reasonably neutral and clean supply of water. Clean water programs not only are desirable from an environmental standpoint, but also can help in reducing the costs of corrosion.

(4) Drying of air other gases to keep humidity below the level where corrosion becomes serious.

(5) Design of hydraulic systems to avoid excessive velocities or localized turbulence or to maintain a velocity high enough to prevent the accumulation of corrosion products or other deposits that will promote localized corrosion.

(6) Various features of design and operation of structures or equipment to favor rapid drainage and drying, prevent accumulation or concentration of corrosive chemicals in crevices or low spots, hold operating stresses and temperatures within desired limits, eliminate fabricating stresses by appropriate metallurgical treatment, avoid galvanically unfavorable combinations of different metals, provide protection against stray electrical currents by appropriate insulation and electrical bonding.

(7) Heat-treating metals to leave them in optimum condition to resist corrosion.

(8) Applying protective electrical currents (*cathodic protection*) from sacrificial metals (galvanic anodes) such as zinc, magnesium, or aluminum or from some external source through a graphite, platinum, or other appropriate anode receiving current from a rectifier, generator, or battery. The location of the anodes, the magnitude of the current, and the applied voltage must be engineered so that without wasting current all surfaces that require protection will receive sufficient current to achieve this effect. Too much current may cause damage by the alkali generated by a cathodic reaction or by hydrogen evolved at the cathode, which can destroy protective films or embrittle metals. A current of 1 to 15 milliamperes per square foot (929 square centimeters) is usually required to protect bare steel area; for design purposes, the range is generally narrowed to 3 to 5 milliamperes. The potential required to produce this current flow depends, of course, on the resistivity of the electrolytic path between the electrodes.

Corrosion can also be suppressed by the controlled application of current to the metal as an anode. This is called *anodic protection.* Passivity is induced and preserved by maintaining the potential of the alloy at, or above, a critical potential in what is called the range of passivity in a potentiostatic diagram. Such diagrams are based on the relationship between applied anodic current density and the corresponding potential in the environment of interest.

When pipes and cables carrying an electric current are underground, they are commonly corroded by electrolytic action from unidirectional electric currents in the ground. Stray current from electric traction equipment is retarded by increasing the resistance of the ground circuit and by reducing the electric resistance of the track. Also, cathodic protection is widely used. An external source of dc voltage is applied so that the protected equipment (pipeline or cable, for example) becomes lower in potential than the soil that surrounds it. Thus, the buried material is the cathode rather than the anode. Usual forms of corrosion can be prevented when the preventive system causes the pipe or metal structure to be 0.25 to 0.30 volt negative with reference to the soil or liquid that may be surrounding it. The negative lead from a small generator, battery, or rectifier is connected to the metal structure; the positive lead to the ground at some distance. Or sacrificial magnesium or zinc rods sunk in the ground may be externally connected to the structure.

Atmospheric Corrosion of Metals and Nonmetals. Taking into consideration the relative order of corrodibility, it is preferable to describe corrosive damage as attributable to certain agents rather than to the indefinite characterization "smoke." Corrosive agents can be placed into four major groups, namely, oxygen and oxidants, acidic materials, salts, and alkalis.

Corrosion attributable to oxygen is deemed to result from the solution of oxygen by a thin film of liquid adjacent to the metallic surface, the transportation of the oxygen through the film, and the subsequent reaction at the surface of the metal. This explains why there is corrosive action even in relatively arid land. In a very dry atmosphere corrosion is, however, markedly reduced.

There are three principal categories of oxidizing agents which occur as air pollutants. These are ozone, nitrogen oxides and nitric acid, and organic peroxide. Many materials which are relatively resistant to attack by the free oxygen of the air are far less resistant to attack by such oxidants and peroxides. These dissolve in the surface film and thus convert metals to their oxides which react readily even with such relatively weak acids as carbonic acid and sulfurous acid. For instance, copper tarnishes rapidly forming the oxide which dissolves readily in dilute acids.

The acid components given off to the air by the various processes of combustion are sulfur dioxide and sulfurous acid, sulfuric acid, hydrogen sulfide, hydrochloric acid, carbon dioxide and carbonic acid, and tar acids. There is little doubt that the material of greatest importance in respect to atmospheric corrosion in this group is sulfur dioxide. Generally, the total acidity of the atmosphere is closely related to the sulfur dioxide content.

Other soluble acidic components such as sulfuric acid, hydrogen sulfide, hydrochloric acid, nitric acid and the like are all of minor importance. Carbon dioxide and carbonic acid play a significant role in acid decomposition.

One aspect of tar and tar acids should be noted, namely, that these are sticky and cling to the surfaces with which they come in contact. This enables the acids which such tar contains to have a prolonged corrosive action. It also increases the difficulty of removal by rain or wind or other action.

It is common to consider that certain salts have a very corrosive action. This is true in the respect that the corrodibility of marine atmospheres has been shown to be greater than rural, tropical, and urban atmospheres. For example, ammonium sulfate and ammonium chloride being salts of strong acids and a weak base, that is ammonium hydroxide, hydrolyze in water to yield the respective acids. These salts then have a corrosive action which is due actually to the acid produced in hydrolysis.

Alkalis seldom occur as air pollutants except under industrial conditions. Nevertheless, the corrosive action of alkalis should not be completely ignored. While a number of metals are relatively resistant to acid attack, they have an amphoteric action and can react with alkalies. For instance, aluminum and zinc are in this category and they are subject to corrosive attack by relatively weak alkalis.

Metals exposed to air pollution can be placed into three groups:

1. Metals that corrode rapidly because they do not form completely protective corrosion products. The major metal in this group is iron. It should be noted, however, that iron oxide Fe_2O_3 does have some protective action.
2. Metals which are initially attacked somewhat readily but subsequently become resistant to attack because of the formation of a corrosion-resisting film which hinders further attack. Among the metals in this group are aluminum, lead, zinc, brass, copper, nickel and magnesium.
3. Metals that are almost completely corrosion resistant, as for instance stainless steel of the 18/8 type, chromium plate products, monel, and gold.

Mention has already been made of the action of oxygen and oxidants on metal. It should be noted that metals react with sulfides, such as hydrogen sulfide, and are subsequently subject to additional slow attack by oxygen and oxidants. Thus, copper reacts to form sulfide and then the basic copper sulfate.

Generally, metals are resistant to attack in dry air; even in pure humid air, corrosion is slight; when, however, air pollutants are present the rate of corrosion will increase measurably, the increase being dependent upon the humidity and the character of the pollutant. Such action may be grouped as follows:

Relative Humidity		Degree of Corrosion
Less than	60	None
More than	60	Slow but definite
	80	Decided increase
Greater than	80	Very high

A factor of note is the settling and adherence of particles on metals. Particles of carbon, ammonium sulfate, and silica cause a marked increase in corrosion, and this is accentuated in atmospheres containing sulfur dioxide. The presence of such hygroscopic particles enhances the adherence of liquid and thus provides for electrochemical attack.

Stone building materials may be placed into relatively resistant and non-resistant categories. The acids of the air, such as sulfuric acid, attack carbonate-bearing stone, such as limestone, converting the calcium carbonate to calcium sulfate. The gypsum formed is dissolved by rainwater, causing pitting. Incrustations may be formed because of the crystallization of soluble salts. These break away in time and leave pitted surfaces. Other types of damage, such as porosity of the stone, are caused by analogous reactions.

Corrosion-Resistant Metals. Some concept of the economic importance of corrosion to the production and consumption of various metals can be gleaned from the accompanying table. Many mineral commodities are used in more or less direct proportion to steel production. In the case of the United States and many other countries, a large number of the mineral materials important to combatting corrosion come from distant sources. That is why many countries maintain a stockpile of such materials, particularly those that are regarded as critical or strategic. In the United States, 93 materials are officially classified for defense purposes as basic stockpile commodities. Seventy-nine of these are metals and minerals, including nearly every one of the metals with important corrosion-resistant properties.

The possible use of low-grade, currently noncommercial, mineral deposits requires constant consideration. For example, chromium has been recognized as an important strategic material ever since World War I. Over the years, the U.S. government, through the U.S. Geological Survey and the Bureau of Mines, has discovered and carefully defined numerous domestic chromium deposits. The Bureau of Mines in its metallurgical laboratories has produced acceptable chrome concentrates from these deposits as well as acceptable ferrochromium chemicals. Current Bureau of Mines research includes recovering chromium, nickel, and cobalt from laterite deposits, both domestic and from other countries, and also from flue dusts, plating wastes, and other residues (Morgan, 1978).

Corrosion Monitoring. Combatting corrosion in continuous processing plants which may be scheduled for quite infrequent but thorough equipment checking and maintenance is particularly difficult. Process downtime costs for a large unit may be several hundred thousand dollars per day in terms of lost production. To avoid excessive downtime for checking and still control the effects of corrosion (personnel and equipment safety, product quality and throughput rates, etc.) requires means to measure the status and rate of corrosion that may be taking place within the equipment. The design of corrosion monitors is among the most recent developments in overall process instrumentation. For obvious reasons, such on-line corrosion testing must be of a nondestructive nature.

In one type of monitor, changes in electrical resistance of a measuring element or probe relate to corrosion rate. The measuring element may be a wire, tube, or strip that can be inserted in a tee or an elbow in the process piping. As the measuring element corrodes, the cross-sectional area reduces and the electrical resistance increases. The thickness of the measuring element is directly proportional to a corrosion dial reading. The difference in dial readings is plotted over a period of time and from these data, corrosion rate can be determined. Probes are available in a number of different metals for different temperature ranges and corrosion conditions. In another variation, three electrode probes are used. The corrosion rate is determined by measuring electrical current flow between the test and auxiliary electrodes. That current either cathodically protects or anodically accelerates the corrosion rate of the test electrode, depending upon the flow. The current is measured on a microammeter that has been converted to read directly the corrosion rate in mils (1 mil = 0.001 inch = 25.4 micrometers) per year of the test electrode.

In another instrumental approach, a hydrogen test probe operates on the principle that hydrogen will diffuse through the thin wall of a test probe and set up a pressure within the tube. The rate at which the pressure increases is measured by a pressure gage. The rate at which hydrogen is penetrating per unit area can then be determined, using the exposed surface area and the internal volume of the probe. Beyond a certain rate, severe hydrogen damage can be anticipated.

Measurement of wall thickness can be determined using ultrasonic nondestructive testing methodology. See also **Thickness Measurement and Gaging Systems.** This method can be used to determine wall

APPLICATIONS OF CORROSION-RESISTANT METALS

Metal	Corrosion-Resistance Use	Consumption: For Corrosion Applications	Consumption: Percent of All Uses of Metal
Nickel	Alloying, 34%; high-temperature oxidation resistance, 26%; plating, 13%	144,000 MT	73
Chromium	Alloying, 57%; coatings and plating, 7%	306,000 MT	64
Titanium	Coatings, 51%; alloying, 1%	243,000 MT	52
Cadmium	Plating	2,520 MT	45
Gold	Plating and alloying	50,000 kg	35
Zinc	Galvanizing, 32%; coatings and sacrificial anodes, 3%	432,000 MT	35
Tin	Plating, tinning	19,800 MT	34
Tantalum	Alloying, 16%; cladding, 12%; high-temperature oxidation resistance, 5%	1,935 MT	33
Rare earths	Alloying	3,600 MT	30
Platinum	Resistance to chemical attack	13,375 kg	27
Silver	Alloying	1,370 kg	26
Columbium (Niobium)	Alloying; high-temperature oxidation resistance	630 MT	25
Iron oxide pigments	Coatings	30,600 MT	25
Copper	Alloying and plumbing	370,000 MT	18
Molybdenum	Alloying; coating	4,860 MT	18
Cobalt	Alloying; high-temperature oxidation resistance	1,530 MT	17
Magnesium	Alloying; sacrificial anodes	14,400 MT	15
Zirconium	Alloying; chemical resistance	340 MT	15
Thorium (ThO_2)	Alloying	31 MT	12
Hafnium	Alloying	3 MT	11
Beryllium	Alloying	4 MT	10
Lead	Pigments and plating, 8%; cable covering, 1%	108,000 MT	9
Indium	Coatings	1,555 kg	8
Aluminum	Alloying; coatings; cladding	225,000 MT	5
Manganese	Alloying; cladding	45,000 MT	4

Source: U.S. Bureau of Mines (1978).
MT = metric ton; kg = kilogram

thinning, pitting, erosion, and flaws in metals, plastics, and various rubbers and polymers. Some disadvantages of the method include the need to take many readings over a period of time to determine corrosion rate and the fact that high-temperature measurements tend to be inaccurate.

Infrared thermographic techniques can be used to identify hot spots on process equipment. The camera works on the theory that the hotter the object, the higher the frequency of radiation. For off-line corrosion monitoring, borescopes for inspecting tubes, pumps, compressors, and other equipment may be used. Spot chemical testing can indicate the presence of alloy constituents of unknown materials. Television camera and holographic techniques also have been used. The monitoring of pH is an invaluable indicator of possible corrosion problems, particularly in cooling water systems. Monitoring is usually done continuously because pH shifts can take place rapidly in many systems, particularly as a result of a process leak.

Probably the weight loss coupon approach is one of the most reliable methods and is widely used. The accuracy of the data is highly dependent on good techniques and on the statistical significance of the tests. The engineering quality data produced require the efforts of many people over a period of several weeks or months, which makes this information quite costly. However, it is the technique of first choice of many processors.

References

Arnold, C. G.: "Using Real Time Corrosion Monitors in Chemical Plants," *Chem. Eng. Progress*, **74**, 3, 26–31 (1978).

Cangi, J. W.: "Characteristics and Corrosion Properties of Cast Alloys," *Chem. Eng. Progress*, **74**, 3, 61–66 (1978).

Covington, L. C., Schutz, R. W., and I. A. Franson: "Titanium Alloys for Corrosion Resistance," *Chem. Eng. Progress*, **74**, 3, 67–69 (1978).

Draley, J. E., and J. R. Weeks: "Corrosion by Liquid Metals," Plenum, New York, 1969.

Fontana, M. G., and R. W. Staehle: "Advances in Corrosion Science and Technology," Plenum, New York, 1980.

Fontana, M. G., and N. D. Greene: "Corrosion Engineering," McGraw-Hill, New York, 1978.

Gasper, K. E.: "Non-chromate Methods of Cooling Water Treatment," *Chem. Eng. Progress*, **74**, 3, 52–56 (1978).

Harrell, J. B.: "Corrosion Monitoring in the Chemical Process Industries," *Chem. Eng. Progress*, **74**, 3, 57–60 (1978).

Leidheiser, H.: "Corrosion of Copper, Tin and Their Alloys," Wiley, New York, 1971.

Morgan, J. D., Jr.: "Supply and Demand of Corrosion-resistant Materials," *Chem. Eng. Progress*, **74**, 3, 26–31 (1978).

Rak, G. R.: "Corrosion Monitoring Equipment," *Chem. Eng. Progress*, **74**, 3, 46–51 (1978).

Schmeal, W. R., MacNab, A. J., and P. R. Rhodes: "Corrosion in Amine/Sour Gas Treating Contactors," *Chem. Eng. Progress*, **74**, 3, 37–42 (1978).

Sedricks, A. J.: "Corrosion of Stainless Steels," Wiley, New York, 1979.

CORROSION (Boiler). **Feedwater (Boiler).**

CORROSION EMBRITTLEMENT. The embrittlement or loss of ductility of metals due to corrosion, usually as a result of intergranular attack which may not readily be visible.

CORROSION FATIGUE. The condition caused by the combined action of corrosion and repeated stresses. Both factors may result in damage to and failure of metals, but when acting in concert their combined action usually accelerates the deterioration.

CORROSION INHIBITION. **Autodeposition; Conversion Coatings; Paint; Petroleum.**

CORSITE. An orbicular diorite resulting from the segregation, in rounded concentric forms, of ferro-magnesian minerals. It derives its name from its occurrence on the Island of Corsica, and is also sometimes called Napoleonite.

CORTEX. This is the outer portion of a stem or root, bounded externally by the epidermis, and internally by the cells of the pericycle. It is composed mostly of cells which are very little differentiated. Usually these are rather large, thin-walled parenchyma cells. The outer cortical cells often acquire irregularly thickened cell walls. These are the collenchyma cells. Some of the outer cortex cells may contain chloroplasts and carry on photosynthesis.

In zoology, a superficial layer of an organ. Included are such organs as the kidney, the adrenal gland, the ovary, the thymus and portions of the brain. Among these examples the cerebral cortex of the brain is the most familiar. The term designates no common characteristic of origin or structure, but only the existence of a distinctive layer at the surface of the organ involved.

CORTICOSTEROIDS. Adrenal Glands; Hormones; Steroids.

CORTICOSTERONE. Adrenal Glands; Hormones.

CORTICOTROPIN. Brain and Nervous System.

CORTI (Organ of). A center of nerve terminals located in the inner ear, adjacent to the basilar membrane. See **Hearing and the Ear.**

CORTISOL. Adrenal Glands; Corticosteroids.

CORTISONE. Steroids.

CORUNDUM. The mineral corundum, Al_2O_3, aluminum oxide, occurs as well-developed hexagonal crystals which may display prismatic, rhombohedral, pyramidal or tabular habits. The larger crystals are often rounded or barrel shaped. Corundum shows both basal and rhombohedral partings; the fracture is conchoidal; hardness, 9; specific gravity, 4.0–4.1; luster, vitreous to adamantine, may be pearly on base; transparent to translucent. Common corundum is gray, grayish-blue or brown, but may be red, yellow or whitish; it is sometimes called adamantine spar. Transparent corundum may be colorless or of various tints. The highly prized ruby is deep red; the sapphire, blue. Transparent yellow corundum is known as oriental topaz; if violet, oriental amethyst; if green, oriental emerald.

Emery is a mixture of granular corundum of dark color, magnetite and hematite, sometimes with spinel. Quartz may be present. For a long time emery was supposed to be an ore of iron. Until the introduction of artificial abrasives, emery was much used for such purposes.

Corundum is found as an accessory mineral in the crystalline rocks such as crystalline limestones and dolomites, gneisses, schists as well as in the igneous rock types granite and syenite. Corundum syenites are found in Canada, especially in the Province of Ontario. Rubies have long been mined in Upper Burma; both rubies and sapphires are found near Bangkok, Thailand. Numerous localities in India furnish gem stones of high quality.

In the United States, common corundum is found in New York, New Jersey, Pennsylvania, Virginia, North Carolina, South Carolina and Georgia; sapphires of gem quality near Helena, Montana, associated with alluvial gold in the Missouri River. From the crystalline limestones and schists of the islands of Naxos and Samos in the Grecian archipelago most of the emery of commerce comes. Other deposits are near Ephesus in Asis Minor, and in the town of Chester in Massachusetts. The word corundum comes from the Hindu, *kurand*; emery is derived from the Greek name for this substance.

See also **Bauxite;** and terms listed under **Mineralogy.**

CORVUS (the raven or crow). A small constellation containing no particular bright or interesting stars. This group of stars has long been a friend to lovers of the sea because of its resemblance to the "fore and aft" sail of a cutter. For this reason, the constellation is frequently referred to by sailors as "the cutter's mainsail." On a clear, moonless night, the resemblance to the sail is very remarkable, with even the "step" of the mast and a small "pennant" flying from the gaff being discernible. (See map accompanying entry on **Constellations.**)

CORYDALIS (*Insecta, Neuroptera*). A large gray insect with four membranous wings and in the male sex with very long slender jaws. The adult of the hellgrammite. Also called dobson fly.

In botany, the term *Corydalis* is applied to a genus of plants occurring in the north temperate regions and in South Africa, which are sometimes found in cultivation. All are herbs with small, somewhat irregular flowers.

CORYLACEAE. Hazelnut Shrubs.

CORYZA. Common Cold.

COSECANT. Trigonometric Function.

COSET. If H is a subgroup of the group G, a set of elements of the form ah, where a is in G and h runs through H is called a (left) coset of H in G and a set of the form ha is a right coset. The number of such cosets is called the *index* of H in G.

See also **Group;** and terms listed under **Mathematics.**

COSINE. Trigonometric Function.

COSINE EMISSION LAW. A law relating to the emission of radiation in different directions from a radiating surface. If a small, white-hot metal plate is viewed from a great distance, its apparent candle power, measured by a photometer is greatest when it is perpendicular to the line of sight, and reduces to practically zero when it is turned edgewise. If the observer now moves nearer, he finds that this change is due to the smaller angle subtended by the surface, that is, the smaller cross section of the beam proceeding from it in his direction; and that the apparent brightness of the surface is the same however it is turned. To apply this, let the radiating surface, of area a, be emitting a luminous flux L (lumens) in the normal direction (see figure), and, in any other direction making an angle θ with the normal, the smaller quantity L'. Then since the apparent brightness is unchanged, $L'/a' = L/a$. This gives $L'/L = a'/a$. But $a'/a = \cos\theta$, hence $L' = \cos\theta$; which means that the energy emitted in any direction is proportional to the cosine of the angle which that direction makes with the normal. This is the "cosine emission law" of Lambert. It applies to thermal radiation as well as to light, and to diffusely reflected as well as directly emitted radiation. The law is true only for a perfectly diffusing surface, strictly, for a black body, but it is a good approximation to the behavior of many surfaces.

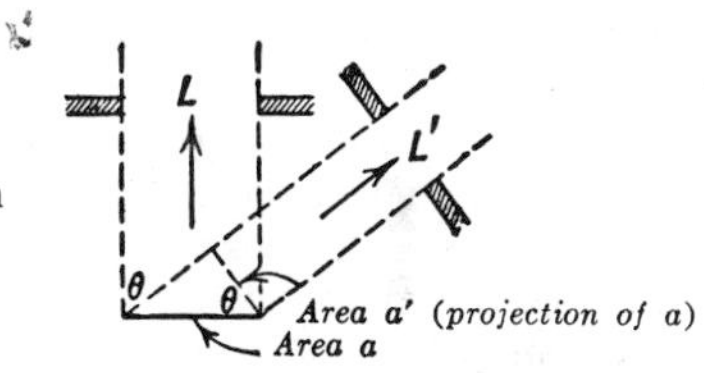

Illustration of cosine emission law.

COSINE INTEGRAL. Integral (Logarithmic).

COSMIC BACKGROUND RADIATION. Cosmology; Radio and Radar Astronomy.

COSMIC BACKGROUND EXPLORER SPACECRAFT. Cosmology.

COSMIC RADIATION (Entropy). Energy.

COSMIC RAYS. Generally, cosmic rays are divided into two classes: (1) primary, and (2) secondary. The first rays, for the most part, are energetic charged particles of extraterrestrial origin, while the latter are the products resulting from collisions of the primary cosmic rays with atoms of the earth's atmosphere.

Primary Radiation. This is characterized by at least four qualities: (1) the primary intensity is essentially constant in time; (2) isotropic in space; (3) anomalous in composition; and (4) it contains very energetic particles. Measurements indicate that with the exceptions of the local perturbations, described later under solar effects, the primary cosmic-ray intensity exhibits less than 1% variation in time. Measurements of radioactivity produced in meteors by cosmic rays show that the intensity has not appreciably changed in the last several million years.

Isotropy simply means that there is no direction or particular directions in the cosmos from which the bulk of cosmic radiation emanates, including the direction of the sun. Since there are magnetic fields of varying magnitude and direction throughout interstellar space which deflect charged particles, the isotropic nature of the primary intensity is not surprising. It could be concluded that because of collisions of the cosmic rays with these fields, the primary particles lose their "memory" so to speak of the directions to their source. Magnetic fields also act in other ways. For example, the earth's field effectively prevents particles of less than 10^8 eV energy from reaching the earth at all ($1 \text{ eV} = 1.6 \times 10^{-19}$ joule). Moreover, there is indication that cosmic rays between 10^{14} and 10^{18} eV may be very slightly guided along the field lines of the local spiral arm of our galaxy. It should be observed that the diameter of the orbit of a proton of 10^{18} eV moving in the galactic magnetic field would be comparable with the thickness of the galactic disk, so that above 10^{18} eV such guidance would not be impossible.

The total number of primary cosmic rays striking the earth's atmosphere is roughly 1 cm^{-2} sec^{-1}. These particles are mostly protons, but decreasing proportions of heavier atomic nuclei are present, ranging from helium (15% of the proton intensity for the same momentum-to-charge ratio) all the way to iron. Although there are many interesting features of the primary composition, one of the most notable is that the abundance of the elements in cosmic rays is very different from the chemical composition of the sun. Not only are cosmic rays relatively rich in heavy nuclei, but there is roughly one million times as much lithium, beryllium, and boron in cosmic rays as in the sun. These light elements presumably arise from collisions of heavy nuclei with interstellar matter. Such considerations determine a value of a few grams per square centimeter for the average amount of matter traversed by cosmic rays before reaching the earth. This, in turn, gives a mean lifetime of cosmic rays in the galaxy (density $\sim 10^{-26}$ grams per cubic centimeter) on the order of 10^8 years or $\sim 10^6$ years in the spiral arms.

The spectrum of cosmic ray energies between 10^{14} and 10^{19} eV is a power law relation given by $N(> E) = 3 \times 10^{-10} E^{-(1.7+0.1 \log_{10} E)}$ (cm^2 sec steradian)$^{-1}$, where E is the energy in 10^{15} eV. This spectrum should not be extrapolated beyond the indicated upper limit since some experiments indicate a possible change in the slope beyond this point. The maximum particle energies which have been detected are well in excess of 1 joule. These are truly phenomenal energies, equivalent to taking all of the kinetic energy of an apple dropped a distance of several meters and giving it to just one proton of the apple's atoms. Not only are some cosmic rays individually energetic, but because of their high spatial density, cosmic rays represent a large fraction of the total energy associated with astrophysical phenomena. The energy density of cosmic rays, optical photons, interstellar magnetic fields, and the turbulent motions of interstellar matter are each about equal to 1 eV/cm^3.

Origin. Since their discovery by Hess in 1911, the problem of the origin of cosmic rays has not been fully solved. The major difficulty has been that of finding a satisfactory mechanism whereby charged particles can be accelerated to the very high energies described. The earth's sun is wholly inadequate in this respect; the sun it appears cannot be considered as the sole source of cosmic rays.

Speculation considers the sources to be violently active celestial objects—exploding galaxies and exploding stars or supernovae. Supernovae are known to be rich in heavy elements. These exploding phenomena involve huge amounts of energy. This is inferred from the presence of synchrotron radiation (the emission of electromagnetic waves by electrons moving in magnetic fields) which implies in some instances electron energies as high as 10^{13} eV. These electrons are most likely the decay products of mu-mesons which are again decay products of charged pi-mesons. The pi-mesons are created in nuclear interactions, thereby suggesting the presence of very energetic nuclei, some of which could escape from the magnetic fields of the supernovae to become cosmic rays. There are possibilities of verifying such a model, for neutral pi-mesons should also be produced in the nuclear interactions. These mesons decay into gamma rays which travel in straight lines unaffected by magnetic fields. Thus, experiments to detect high-energy gamma rays coming from supernovae and radio-galaxies will assist in the cosmic-ray origin problem.

High-energy photons have been observed in the primary cosmic rays. Gamma and x-rays produced by interaction of primary particles with interstellar matter and optical photons (inverse Compton scattering) give information on the distribution and composition of matter in our galaxy. Several point sources of x-rays have been discovered, one of which is in the Crab Nebula. Processes whereby these x-rays are created are being investigated. All high-energy interactions lead ultimately to the production of neutrinos. So-called neutrino astronomy would be a useful tool for the study of astrophysical phenomena.

In mid-1967, an intense burst of gamma rays was recorded by several Vela satellites (monitors for Nuclear Test Ban Treaty). Although the information was not publicly released for some years, careful examination of this and later similar events showed that the radiation did not come from any nuclear devices, or, in fact, from within our solar system. Since that time, about five similar events have been detected each year. The events are characterized by an initial short pulse of gamma rays, ranging from 0.1 to 4 seconds, followed by one or several pulses, with the entire burst over within about 1 minute.

Unlike the radiation with energy of 50 MeV (million electron volts) studied by gamma ray astronomers, the more recently noted bursts occur between 0.1 and 1.2 MeV, with no clear distinction at the lower end of the energy range between gamma and x-rays. The intensity of the bursts leaves little doubt that the signal may simply be a fluctuation in the intensity of background noise, sometimes encountered when discrete sources of high-energy gamma rays are analyzed.

As of the end of 1973, nearly 25 such bursts had been cataloged by scientists at the Los Alamos Scientific Laboratory. There is ambiguity concerning the location of most of the bursts, but with only limited exceptions, it is evident that the bursts do not come from the earth, the sun, the planets, or the closest stars. Supernovas in our galaxy do not seem to occur with sufficient frequency to account for the bursts. Prior thinking led to the conclusion that well-known astronomical objects were not expected to produce the type of gamma ray bursts that have been observed—the recently observed bursts are less energetic, longer in duration, and multiple rather than singular.

Since the discovery of the unexplained gamma bursts was announced in June 1973, numerous theories have been proposed, including a modified theory for sources in extragalactic supernovas and an unusual concept suggesting that the gamma rays may come from the breakup of relativistic "beebees" within the solar system. These concepts are described in further detail in Reference 6.

Interesting work on the difference between the energy spectra of iron and other cosmic rays has been conducted at the Goddard Space Flight Center and is reported, in part, in Reference 7. Two source mechanisms have been proposed. One mechanism, possibly acceleration at neutron star surfaces, produces the iron; another is responsible for the rest of the primary nuclei. Within this model, observations of high-energy cosmic rays could determine whether secondary nuclei are produced in the sources or in the interstellar medium.

Solar Effects. Although the sun appears to have little effect on the high-energy cosmic-ray flux, it strongly influences the low-energy flux during the occurrence of solar flares. At such periods of solar activity, protons may be emitted by the sun with kinetic energies of nearly 10^{11} eV. The accompanying increase in the sea-level cosmic-ray intensity can be as much as 50 times larger than the normal value of about 1.0 cm^{-2} min^{-1}.

The earth is surrounded by electron ring currents circulating in the earth's magnetic field. These ring currents (Van Allen belts) contain electrons with energies ranging from 10^3 to 10^5 eV. Occasionally, the sun emits an ionized gas which moves outward with a velocity of 10^3 kilometers/second. These particles do not have enough energy to penetrate the earth's magnetic field, but they do modify the field, thereby releasing electrons trapped in the Van Allen belts. When the electrons strike the atmosphere, the resultant ionization produces auroras. The magnetic fields associated with the ionized gas also prevent low-energy galactic cosmic rays from reaching the earth. The accompanying decreases in the sea-level cosmic-ray intensity are known as Forbush decreases. Since these fluctuations in intensity are related to solar activity, their occurrence is periodic and associated with the 11-year solar sunspot cycle.

Secondary Radiation. If it were not for the secondary cosmic rays, high-energy primaries would not have been discovered. Consider a particle detector of an area of 1 cm^2 and an aperture of 1 steradian.

With such a detector, one would have to wait nearly 10^{10} years (the lifetime of the universe) before registering the passage of a 1-joule particle. However, the detection frequency is enormously enhanced by the earth's atmosphere.

The layer of air above the earth represents about 13 collision mean free paths for an incident proton. After the inevitable interaction of a primary particle with some atom high in the atmosphere, the nuclear debris so produced undergoes successive interactions with air atoms further down in the atmosphere. In each collision, pi-mesons are created with decay into mu-mesons and gamma rays. Owing to their large Lorentz factors commensurate with their high energies, the mu-mesons continue on down to sea level before decaying. The gamma rays, on the other hand, produce electron-positron pairs which, in turn, radiate more gamma rays. The huge number of electrons created in this way is called an extensive air shower (EAS). After reaching a maximum at an atmospheric depth dependent on the energy of the primary particle, the EAS slowly decays by ionization losses. Even so, energetic primaries can give rise to EAS which contain billions of electrons at sea level. The total number of EAS particles reaching sea level is nearly proportional to the primary energy, the relation being roughly 10^9 to 10^{10} primary eV per secondary electron.

As they cascade down through the atmosphere, the electrons are scattered by air atoms so that, upon reaching sea level, the EAS electrons and positrons are distributed over a large area. The density distribution of these secondaries is peaked around the shower axis (the direction of motion of the primary particle) and decreases monotonically with distance in a plane perpendicular to the shower axis. For example, a 1-joule primary can create detectable numbers of electrons per square meter even at distances of 1 kilometer from the shower axis. Thus, a small number of detectors spread out in a plane (1) can detect EAS produced by energetic primaries; (2) can give information concerning the primary energy from knowledge of the secondary electron density distribution; and (3) can be used to determine the incidence angle of the primary particle. The last measurement involves timing the arrival of the shower front (the nearly plane surface containing the majority of the secondaries and propagating along the shower axis with the velocity of light) at several distances from the shower axis. By spreading a dozen or so detectors over an area of 10^6 square meters, several showers representing primaries of 1 joule or more can be detected per year.

In order to extend the detectable upper limit of primary cosmis-ray energies, other techniques are in use which effectively increase the sensitive area of the individual detectors. These methods involve the detection of the atmospheric fluorescence produced isotropically (hence detectable over a wider area than the electrons themselves) by the secondary electrons as they pass through the atmosphere.

Among the many uses of the secondary radiation has been the discovery of new particles, notably the positron and the various mesons, and the study of their interactions with matter. In addition, some of the interactions provide remarkable clocks for finding the age of many terrestrial features. For example, the collisions of secondary neutrons with atmospheric nitrogen produce ^{14}C which combines with oxygen to form radioactive carbon dioxide, thus facilitating the familiar technique of radiocarbon dating developed by Libby.

See also **Cosmology; Neutrino;** and **Particles (Subatomic).**

References

1. "Progress in Cosmic Ray Physics," 1–3 and subsequent volumes entitled "Progress in Elementary Particle and Cosmic Ray Physics," North Holland Publishing Co., Amsterdam.
2. Hayakawa, S.: "The Origin of Cosmic Rays," *Lectures on Astrophysics and Weak Interactions II*, Brandeis University, Waltham, Massachusetts, 1963.
3. Chiu, H. Y.: "Neutrino Astrophysics," *Lectures on Astrophysics and Weak Interactions II*, Brandeis University, Waltham, Massachusetts, 1963.
4. Greisen, K.: "Cosmic Ray Showers," *Ann. Rev. Nucl. Sci.*, **10**, 63 (1960).
5. Ginzburg, V. L. and S. I. Syrovatskii: "The Origin of Cosmic Rays," Pergamon, New York, 1964.
6. Metz, W. D.: "Gamma Rays: From Neutron Stars, Supernovas, or Beebees?" *Science*, **182**, 4118, pp. 1234–1237 (1973).
7. Ramaty, R., Balasubrahmanyan, V. K., and J. F. Ormes: "Cosmic Ray Sources: Evidence for Two Acceleration Mechanisms," *Science*, **180** (4087), 731–733 (1973).
8. Parker, E. N.: "Cosmological Magnetic Fields," Oxford University Press, New York, 1979.
9. Wilson, R. M.: "The Cosmic Microwave Background Radiation," *Science*, **205**, 866–874 (1979).

COSMIC "YARDSTICK." **Cepheids; Period Luminosity Law.**

COSMOLOGY. The study of the origin and evolution of the universe as a whole. Modern cosmology is based on mathematical solutions of equations from the general theory of gravitation advanced by Albert Einstein, and on observational work in all parts of the spectrum carried out from the ground and from satellites in space.

The investigation of the structure of the universe and the Earth's role in it has been continuously carried out since the religious cosmologies of the Babylonians and Egyptians. Scientific cosmology began with the ancient Greeks, who knew of the familiar parts of earth, the sun, 5 planets other than earth, and stars. The Greek "kosmos," the root of our word "cosmology," refers to the well-ordered harmonious system composed of these elements.

The Greek scientist Aristotle and others advanced the earth-centered cosmological system that was generally accepted for over a thousand years. It was modified and elaborated on, in particular, by the Greek Alexandrian scientist Claudius Ptolemy in the second century A.D. The suggestion that had been made by Aristarchus of Samos in the third century B.C. that the sun was the center of the universe was not widely accepted.

The origin of our modern understanding of the universe comes from Nicolaus Copernicus' suggestion, in his book *De Revolutionibus* (*On the Revolutions*), published in 1543, that the sun rather than the earth is at the center of the universe. Copernicus advanced that the earth rotates on its axis and that the planets, including the earth, revolve around the sun. For a variety of religious and political reasons, these suggestions were resisted.

Fifty years after Copernicus' book the data amassed by the Danish astronomer Tycho Brahe was interpreted in the first decades of the 17th century by the German astronomer Johannes Kepler. Kepler's three laws of planetary motion—the first law stating that the orbits of the planets are ellipses with the sun at one focus and the other two laws giving rules governing the speeds of the planets in their orbits—put Copernican theory on a firm basis. Galileo Galilei, an Italian contemporary of Kepler, used the newly invented telescope to make several observations that strongly endorsed the Copernican idea that the sun rather than the earth is at the center of the Universe. He observed, for example, a set of phases of Venus that would occur only if Venus went around the sun rather than the earth. His discovery of the satellites of Jupiter indicated that not all celestial objects orbited the earth.

Later in the seventeenth century, Isaac Newton put laws governing the universe on a mathematical basis. Based on earlier ideas of Galileo and others, Newton provided laws of motion. From them, he derived Kepler's laws. Newton also realized that the force holding the moon to the earth is the same force that pulls apples to earth, and used this idea to formulate a universal law of gravity.

A cosmological question that was phrased in the 18th century and put clearly by Wilhelm Olbers in Germany in 1823, "Why is the sky dark at night?," is a precursor of modern cosmology. Olbers pointed out that if the universe extended forever and stars were distributed evenly throughout it, we would eventually see a star no matter what direction we looked. Thus the sky should be bright at night, in contradiction to what we observe. Olbers' paradox was not resolved until it was realized that the universe is expanding, as we shall see below. More recently, it has been realized that the finite ages of the stars provide another solution to Olbers' paradox.

In 1906, Albert Einstein elaborated on his special theory of relativity to show that mass and energy are equivalent, with $E = mc^2$. In the years that followed, Einstein tried to generalize his work to include the effects of gravitation, and succeeded with the general theory of relativity in 1916. This general theory is a theory of gravitation, and explains gravity as a warping of space caused by the presence of a mass. Picture a billiard table with depressions in it resulting from heavy masses causing the table to sag in certain locations. A cue ball rolled along would appear to curve as it passed through the depressions. Similarly, a light ray or a body orbiting in space would appear to curve if space itself is curved, though it is harder to visualize the four-dimensional nature of space and time that Einstein's theory involves than it is to visualize a two-dimensional surface of a billiard table warped into a third dimension.

Einstein's general theory of relativity was soon verified by an eclipse expedition, which found that the sun bent starlight that passed near it in accordance with Einstein's theory. The theory had also cleared up a problem with the orbit of Mercury that, in Einsteinian terms, resulted from Mercury travelling different distances in the space warped by the sun at different times in Mercury's elliptical orbit.

Many tests of Einstein's general theory of relativity have been carried out since, including bending of radio waves measured with interferometers and a delay of the receipt of radio signals from the Viking spacecraft in orbit around Mars. Einstein's theory continues to be more and more accurately confirmed with these experiments. The discovery in 1979 and 1980 of multiple quasars, in which the several images are apparently multiple images of a single quasar each caused by an Einsteinian bending of light, provides further confirmation. The first binary pulsar—a pulsar in orbit around another star—verifies the type of prediction general relativity made for Mercury's orbit but to a much higher degree of accuracy. The change in period of this pulsar also apparently confirms the existence of gravity waves, another consequence of Einstein's theory. Other binary pulsars are also under study.

In 1917, Einstein's own solution to the "field equations" of his general theory indicated that the universe was unstable, and would either expand or contract. Since this conclusion seemed at the time untenable because no observational evidence was known that the universe was other than static, Einstein modified his equations with an arbitrary parameter, known as the "cosmological constant," whose sole purpose was to provide a static universe. Once it became known that the universe is in fact expanding, Einstein dropped his cosmological constant, and called it his biggest mistake.

The Dutch astronomer Willem de Sitter worked out solutions to Einstein's equations in the same year as Einstein. The solutions were applicable, though, only in the case that the universe did not contain any matter. This is an interesting limiting case to consider, especially because the average density of the universe turns out to be very low when all the empty space is included along with the stars, galaxies, etc.

The solutions on which modern cosmology is developed were the work of Alexander Friedmann in the Soviet Union in 1922. Five years later, the Belgian abbé Georges Lemaître found similar solutions. Lemaître's solutions indicated that the universe expanded from a hot and dense state.

Observational work on cosmology in the early 1900s had gone on in isolation from the theoretical work. Starting in 1912, Vesto M. Slipher at the Lowell Observatory had observed the spectra of several of the spiral nebulae that were known in the sky, and discovered that they almost all had large redshifts. But the nature of the spiral nebulae was controversial, a discussion symbolized by the debate in 1920 between Harlow Shapley and Heber Curtis as to whether these spiral nebulae were within our own galaxy or were "island universes" on their own. The discussion concerned the scale of our own galaxy.

Shapley's position eventually won, though not for all the reasons advanced in the debate. Edwin Hubble, working with the largest telescope in the world, the 2.5-m (100-inch) reflector on Mount Wilson in California, was able to measure the distances to a few nearby galaxies using a method involving variable stars based on the work of Henrietta Leavitt, Shapley, and others. The work proved that the spiral "nebulae" were actually galaxies on the same scale as that of our own Milky Way Galaxy. Shapley had also found, from studies of groups of stars known as globular clusters, where the center of the Milky Way Galaxy was. His displacement of the sun from the center of our galaxy was akin to Copernicus' displacement of the earth from the center of the solar system.

Hubble went on to enlarge the work of Slipher on the spectra of galaxies, and in 1929 showed that the distance to a galaxy, measured from its redshift, is directly correlated with its velocity of recession, measured from studies of the variable stars in it. The relation is known as Hubble's law. Hubble and Milton Humason soon showed that the conclusion was valid far into space, based on studies of more distant galaxies.

Hubble's law says that the velocity at which a galaxy is receding is equal to a constant, known as Hubble's constant, times the distance to that galaxy. Hubble's constant is expressed in terms of the velocity in km/s that corresponds to each unit of distance, usually expressed in terms of megaparsecs (Mpc) away from us. One parsec, the distance from which we would see the earth and sun separated by one second of arc, is equivalent to 3.26 light years, so one megaparsec is 3.26 million light years.

The value of Hubble's constant has been of continuing interest to astronomers. Hubble's original value was 500 km/s/Mpc, but work since then based on new ways of finding or calibrating the distance scale has consistently reduced the value. The most generally used value now is that derived by Allan R. Sandage and Gustav Tammann with the 5-m Hale telescope on Palomar Mountain in California, the successor to the 2.5-m Mt. Wilson telescope as largest in the world (and still exceeded only by the Soviet 6-m telescope). Their value of 55 km/s/Mpc appears in most current articles, though many workers in the field believe that the actual value may be somewhat higher, perhaps 75 to 100. Other observational cosmologists, such as G. de Vaucouleurs of the University of Texas, have derived such higher values from their own research.

Hubble's law can be explained by a universe that is uniformly expanding everywhere. The analogy of a raisin cake rising in the oven is often given. Each raisin becomes further away from all other raisins as the cake expands, and the rate at which the farther raisins recede is proportional to their distance from the initial raisin. This is true no matter which raisin we take as our point of origin. Similarly, the universe need have no center. Hubble's law explains how even without a center, all galaxies (or, actually, clusters of galaxies) can seem to be receding with their velocities proportional to their distances.

If we follow Hubble's law backwards in time, we deduce that all the matter in the universe was together at some instant, a time that we call the "big bang." The universe has been expanding ever since this big bang. The inverse of Hubble's constant, called the "Hubble age," corresponds to the time since the big bang on the assumption that the expansion has proceeded at a constant rate. This Hubble age is 13 to 20 billion years. The Hubble age fits in well with other current work on the ages of the oldest objects we know in the universe, the stars in globular clusters.

Our universe appears to be made of matter (such as protons, neutrons, electrons) rather than antimatter (such as antiprotons, antineutrons, antielectrons). We know that matter and antimatter annihilate each other when they come in contact, and from the fact that interplanetary matter extends from the earth through the solar system and from our landings on other planets, we deduce that our solar system is made of matter rather than antimatter. Limits on the amount of radiation that might exist from annihilations in interstellar matter in our galaxy and in intergalactic matter indicate that our whole galaxy and cluster of galaxies must also be made of matter.

A possible solution to the question of why our universe is made of matter, based on the experiment, reported in 1964, of Val L. Fitch, James W. Cronin, James H. Christensen, and René Turlay that a certain kind of symmetry known as CP is violated for certain kinds of elementary particles within atoms, has been incorporated in speculative theories for how the asymmetry of matter and antimatter may have arisen. Fitch and Cronin received the 1980 Nobel Prize in physics for their work.

Basically, theoreticians believe that everything is unchanged if three quantities—charge (C), parity (P), and time (T)—are reversed simultaneously. Fitch and Cronin showed that CP together can sometimes change, which indicates that T must simultaneously change to make CPT constant. Thus the universe is not symmetric in time; if time reversal were to take place, the universe would not retrace exactly the same path it had taken with time running forward.

The asymmetry in the universe that follows from the work of Fitch and Cronin, when applied to the formation of particles of matter in the first fraction of a second after the big bang, can explain how a slight excess of matter over antimatter might have arisen. Later on, the rest of the matter and antimatter may have undergone mutual annihilation, releasing great quantities of energy. Only the result of the imbalance remains in our universe as the matter from which the celestial objects and living objects are formed.

As the universe continued to expand, it cooled. Eventually, after about a million years, it became cool enough (3000 K) so that hydrogen atoms formed and the universe became essentially transparent. From that time on, the radiation that was then present in the universe could travel long distances without being reabsorbed. As the universe continued to expand for billions of years, the temperature of the radiation dropped until today it is only about 3 K, 3 degrees above absolute zero.

In 1965, Arno Penzias and Robert W. Wilson at the Bell Laboratories detected this 3 K radiation. They found a slight amount of radiation at a certain wavelength, 7 cm, that they could not account for in any normal way. A research group at Princeton, including Robert H. Dicke, P. J. E. Peebles, P. G. Roll, and David Wilkinson had recently derived the result that such radiation would be left over from the big bang. The Bell Labs and Princeton papers were published jointly. Penzias and Wilson later received the Nobel Prize in Physics for their discovery. George Gamow, Ralph Alpher, and George Hermann had in the 1940s made a prediction that such radiation would exist, but their result had been overlooked until after the Princeton predictions and Bell Labs discovery.

The discovery of this 3 K radiation, known as the cosmic background radiation or the primordial fireball radiation, seems to have conclusively verified that a big bang did take place. It provided the effective end of an alternative theory, known as the steady-state theory, that had been advanced in the 1940s by Hermann Bondi, Fred Hoyle, and Thomas Gold. The big bang theory had used the idea known as the "cosmological principle": that the universe is isotropic and homogeneous. The steady state theory generalized this to the "perfect cosmological principle": that the universe is not only isotropic and homogeneous but also unchanging in time.

The original objection to the steady-state theory, that matter would have to be continuously created out of nothing in order to leave the density of the universe constant over time, did not prove decisive because the amount of matter that would have to be created is too small to be observationally detectable. But discoveries since, such as the discovery of quasars, tended to act against the steady-state theory. Quasars, for example, were apparently more numerous at earlier stages of the universe, which would indicate that the universe was not unchanging in time. At the time of the Penzias and Wilson discovery, few believed in the steady-state theory, and the discovery of remnant radiation from the big bang provided an effective end to the theory.

Observations of the background radiation have become more and more accurate. The wavelength range has gradually been extended so that by now the coverage shows the shape of the radiation sufficiently that it is unambiguously clear that the radiation has the shape of a black body, in agreement with the predictions. The problem is a hard one, because the shape of the spectrum approximates that of a straight line in the region that can be observed from the earth's surface. Only observations from balloons provided the shape of the spectrum in the infrared bordering on the short-wavelength radio region.

Careful observations have also been made of the isotropy of the background radiation. A high-flying airplane was used. The isotropy is very small, in agreement with a big-bang origin for the radiation. The anisotropy of one part in three thousand can be accounted for by a Doppler effect caused by the motion of the sun at a velocity of 350 km/s with respect to the background radiation. Taking out the effects of the sun's motion in our galaxy, this corresponds to our galaxy moving in the direction of the constellation Hydra with a velocity of 520 km/s.

NASA plans to launch a Cosmic Background Explorer spacecraft in 1983. It will have the ability to study the spectrum and anisotropy of the background radiation with precision higher by a factor of 10 than our current abilities allow.

Lines of evidence other than studies of the cosmic background radiation also indicate that our galaxy may have some velocity of motion through space. Work by John P. Huchra of Harvard, Marc Aaronson of the University of Arizona, Jeremy Mould, then of Kitt Peak National Observatory and the Hale Observatories, and Woodruff T. Sullivan, Robert A. Schommer, and Gregory D. Bothun of the University of Washington at Seattle, found such a velocity based on radio and infrared studies of the motions of and distances to galaxies. Since the research also indicates that the universe is expanding with a Hubble constant of about 95, it means that the universe would be only 10 billion years old. Since this is younger than some of the ages measured for globular clusters, the discrepancy must be resolved before this value can be generally accepted.

One interesting major cosmological question is what will happen to the universe in the future? Will it continue to expand forever or will the expansion ever stop and the universe contract? The first case, in which the universe will expand forever, is known as the *open universe*. It corresponds to a universe that is infinite. The second case, in which the universe will eventually contract, is known as the *closed universe*. A closed universe is finite. The case of the closed universe can be subdivided into a subcase in which this is the only cycle of expansion and contraction and into an oscillating subcase in which this cycle is only one of many.

The most obvious way to detect the difference between these cases is to observe the *deceleration parameter*, a measure of how rapidly the universe is slowing down. The latest results, using the Palomar telescope, indicate that there is in fact a slight deviation from the straight-line correspondence of Hubble's law, but we must also take into account that the farthest objects we see are so old that they may then have been of very different brightnesses from similar objects closer to us. Since we tell the distance to these farthest objects by assessing their brightnesses, this unknown effect of evolution is sufficient to indicate that the method will not in fact give us reliable results on the future course of the universe.

Other modern methods are based on assessing the amount of material in the universe, akin to seeing if there is enough gravity to slow down the expansion sufficiently to cause an eventual collapse. There is not enough visible matter by a considerable factor, so the question becomes how much matter is present in invisible form. Year by year, astronomers are able to detect more kinds of matter. For example, results from the first High Energy Astronomy Observatory showed the presence of an x-ray background that could indicate the presence of a lot of hot gas hitherto unknown. Results from the second High Energy Astronomy Observatory, known as the Einstein Observatory, on the contrary showed that most or all of the x-rays came from distant previously unknown quasars. This indicates that not enough mass is present in this form to "close" the universe. The recent indications that neutrinos may have a small mass, if verified, could provide enough mass to close the universe in that form because of the large number of neutrinos. At this writing, the results were subject to verification.

The best way to assess the future of the universe at present is to look at the abundances of the light elements in general and deuterium (heavy hydrogen) in particular. Theory indicates that the deuterium was all formed in the first few minutes of the universe, and that the amount of deuterium present is a sensitive indicator of the density of the universe at that early stage. If the universe were too dense, then all the deuterium would have quickly gone into forming helium. Only in the case where the universe was not very dense, would the amount of deuterium that we nowadays detect from radio telescopes on earth and in the ultraviolet from space satellites have survived.

Other methods, such as investigations of the velocities of galaxies due to density perturbations, are also being carried out. The preponderance of the evidence at present indicates that the universe is open.

Jay M. Pasachoff

Bibliography

Pasachoff, Jay M.: "Contemporary Astronomy," 2nd edition, Saunders, Philadelphia, 1981.

Silk, J.: "The Big Bang," Freeman, San Francisco, 1979.

Weinberg, S.: "The First Three Minutes," Basic Books, New York, 1977.

CO-SPECTRUM. **Spectral Analysis.**

COSTAL. Pertaining to a rib or the region of the ribs. Thus, coastal cartilage is the cartilaginous portion of the ribs which joins them to the sternum. The term intercostal refers to the space between the ribs.

COTANGENT. Trigonometric Function.

COT DEATH. Sudden Infant Death Syndrome (SIDS).

COTTER. A wedge-shaped metal piece for fastening two parts subjected to reciprocatory motion. The connecting rod bearing of a steam engine is often constructed with a removable cap held in place by a cotter. A cotter pin is a split pin made of semi-cylindrical bar stock, bent so that the flat surfaces are in contact, to permit insertion in a drilled hole. The cotter pin is used as a locking medium for nuts and bolts; since the pin is made of a soft steel, the split ends can be spread after it has been inserted in the hole.

COTTON (*Gossypium* species; *Malvaceae*). A natural fiber, cellulosic in composition, with the general formula $(C_6H_{10}O_5)_x$, specific gravity, 1.54; moisture regain 7–8.5% (at 70°F (21.1°C) and 65% relative humidity); and a tensile strength of 60–120 × 10^3 psi (same condition of temperature and humidity). Cotton is quite resistant to degradation by heat. After about 5 hours at 50°F (121°C), the material yellows, and above 300°F (149°C), the material decomposes. Cold concentrated acids and hot dilute acids disintegrate cotton. In alkalis, cotton mercerizes without damage. The fiber is quite resistant to most solvents.

Even with the large inroads of synthetic fibers into the textile industry, cotton remains as a major textile fiber, consumption still being expressed in terms of billions of pounds per day. As one of the most versatile of fibers, cotton blends well with other fibers. In addition to the cellulosic content (88–96% depending upon source), cotton contains proteins, pectin, sugar, and wax (about 0.4–0.8%). Much research is going forth to eliminate much of the loss and damage that occur in harvesting and ginning. These factors contribute to price instability.

From the standpoint of use in textiles, cotton is rated excellent for hand (general feel, softness, drape), pilling resistance, and stability to repeated launderings. Cotton is rated good for abrasion resistance, strength, wash and wear performance, and wrinkle resistance. Resistance to sunlight is rated only fair. Pressed-crease retention is rated poor. Safe-ironing temperature for most cotton fabrics is 425°F (219°C).

Important worldwide producers include Brazil, Egypt, India, and the United States.

Many species of cotton plants are known, some native to warm regions of America; others growing wild in tropical Asia. Cotton seeds are surrounded by an abundance of soft white fibers, which when mature are very conspicuous. These fibers are unicellular and have been known and used for hundreds of years. Cultivation of cotton in India has continued through at least 26 centuries. Even before the discovery of America, the natives of tropical America were growing Sea Island cotton and using its fibers.

The cotton plant of cultivation is a woody annual growing 3 or 4 feet (0.9 or 1.2 meters) tall. It bears large palmate leaves and showy white flowers which turn pink as they grow older. Outside the five-parted corolla and calyx is an involucre composed of three large green bracts having very irregular margins. These bracts persist throughout the growth of the fruit. The stamens of the cotton flower have their filaments united to form a hollow tube through which the stigmas must grow; this stamen structure is characteristic of the mallow family, or *Malvaceae*. The fruit is a large dehiscent capsule of ovoid shape. On dehiscence, or splitting open, at maturity, the soft white fibers which surround the five or six large dark brown or black seeds become visible. The fibers are of two kinds, one relatively long, called lint, the other short and called fuzz. The capsule of the cotton plant is generally called the cotton boll. The length of the lint fibers varies in different species: Sea Island cotton, the native American species *Gossypium barbadense*, cultivated on coastal areas of the southern states, has the longest fibers of any cotton ($1\frac{1}{2}$–$2\frac{1}{2}$ inches; 3.8–6.4 centimeters long). Egyptian cotton fibers are slightly shorter than these. Next comes upland cotton, which is mainly derived from *Gossypium hirsutum*. This is the principal cotton plant of the southern states. In these the length of the fibers varies considerably, but averages about 1 inch (2.5 centimeters). Asiatic cottons have still shorter fibers.

Cotton cultivation is limited to regions having a growing season of 6 months or more of continuous high temperature. Any soil having a proper moisture content is suitable for cotton, but a deep, well-drained loam soil is best. One requirement is that excessive rainfall shall not occur during the period when the bolls are opening, since the cotton fibers are injured by moisture at that time. During the growing season the cotton fields must be kept clear of weeds. From 5 to 6 months are required before the production of fibers begins, after which fiber production continues for another 3 months or more.

The soft, light fibers, after picking, are carted to the gins. Previous to the invention of the cotton gin by Eli Whitney in 1793, the lint was removed from the seeds by hand, a slow and laborious process. The invention of the gin greatly advanced the cotton-manufacturing industry. The cotton gin is a steel grate with narrow slits through which reach thin-notched saws. The rapid rotation of the latter causes lint to catch on their teeth. The lints are pulled through the grating, after which brushes remove them from the saws. This ginned cotton is then pressed into bales weighing about 500 pounds (227 kilograms) each.

The uses of cotton are many. First among them is the manufacture of cloth. Cloth may be woven entirely of cotton, as much is, or may be of cotton mixed with other fibers, as synthetics, linen and wool.

Cotton fibers, especially the fuzz, are frequently used to stuff mattresses, pads and upholstered furniture. Treated with chemicals which remove the thin coating of waxy substances which cover the fibers, the latter becomes absorbent cotton, which is capable of absorbing many times its weight of water.

Cotton treated in this way is almost pure cellulose, and so is in great demand by those industries using cellulose. The pure cellulose of the fiber may be dissolved and then precipitated in sheets, giving the familiar thin transparent cellophane. Or the dissolved cellulose may be pressed through fine holes and solidified, giving rayon. If treated with concentrated caustic soda, cotton fibers take on a high degree of luster. The product of this process is called mercerized cotton, after John Mercer, its discoverer.

Treated with nitric acid under various conditions, cotton yields a long series of by-products. Some of them are plastic substances. If highly nitrated, cellulose becomes gun-cotton, used in the manufacture of explosives. Collodion is one of these nitrated products. Many varnishes and lacquers are made from cotton cellulose.

Not all the derivative products of cotton come from the fibers. Some, for example, are obtained from the seeds. In preparing these, the hulls are first removed from the kernel within. These hulls are used as fuel in the ginning mill, as food for cattle, and as fertilizer. The kernels are heated and pressed to remove the oil in cotton. During this pressing the kernels are wrapped in cloth to prevent anything but oil from being expressed. The oil is purified to a soft white substance very similar to lard in appearance. Cottonseed oil is used in making salad oils, oleomargarine and soap. After the oil is expressed, the seed cake may be used as food for stock or as a fertilizer.

COTTON (Boll Weevil). Boll Weevil.

COTTONMOUTH. Snakes.

COTTON-MOUTON EFFECT. Kerr Effects.

COTTONSEED OIL. Expression (Mechanical); Vegetable Oils (Edible).

COTTONSEED PROTEIN. Protein.

COTTON STAINER (*Insecta, Hemiptera; Dysdercus*). A bug of Florida and adjacent states which punctures the bolls of cotton and causes staining of the fiber. There is another species in Egypt. It develops in groups which can be jarred from the plants into vessels containing a little kerosene.

COTTONTAIL. Rabbits and Hares.

COTTONWOOD TREES. Poplar Trees.

COTTRELL PRECIPITATOR. Electrostatic Precipitation.

COUCAL. **Turacos and Cuckoos.**

COUDÉ. A modification of the equatorial form of mounting for an astronomical telescope, originally designed for the purpose of providing a maximum amount of comfort for the observer. Now, however, the Coudé focus has found great favor where there is a need to mount very large equipment, such as special spectrographs, without flexure. The telescope itself is mounted in bearings parallel to the axis of rotation of the earth, and forms the polar axis of the instrument. The eyepiece is at the upper end, and usually projects into a closed room. Below the object glass, a mirror is mounted in such a manner that it will rotate about an axis perpendicular to the optical axis of the telescope. Hence, the mirror may be rotated about its own axis parallel to an hour circle (in the coordinate of declination), and is carried along with the rotating tube parallel to the coordinate of hour angle. The observer, seated in a comfortably heated room, looks down into the eyepiece and sees the field of view slowly rotating about the optical center of the field.

The fundamental difficulty with the Coudé mounting is that the mirror will introduce distortion. Furthermore, unless the aluminum coating on the mirror is very nearly perfect, there will be a large amount of light loss.

COUETTE FLOW. A two-dimensional steady flow without pressure gradient in the direction of flow and caused by the tangential movement of the bounding surfaces. The only practicable type is the flow between concentric rotating cylinders, although the flow between parallel planes with uniform relative velocity is used in the elementary discussion of viscosity.

COUGAR. **Cats.**

COULOMB. **Units and Standards.**

COULOMB BARRIER. **Nuclear Forces.**

COULOMB DAMPING. The dissipation of energy that occurs when a particle in a vibrating system is resisted by a force whose magnitude is a constant independent of displacement and velocity, and whose direction is opposite to the direction of the velocity of the particle. Also called *dry friction damping*.

COULOMB DEGENERACY. Identity of the energy levels of a charged particle bound in a Coulomb (electrostatic) field for different values of the orbital angular momentum, provided that the principal quantum number and spin state are the same, e.g., the $2s_{1/2}$ and $2p_{1/2}$ states of the hydrogen atom. See **Lamb Shift.**

COULOMB ENERGY. That part of the binding energy of a solid associated with the electrostatic interaction of the ions and electrons.

COULOMB EXCITATION. This terminology is commonly used to refer to the process of raising a target nucleus to an excited state through an electromagnetic interaction with a passing charged nuclear particle. If the process is purely coulomb excitation, the passing particle remains outside the range of the nuclear forces of the target nucleus, so is not strictly a nuclear reaction.

COULOMB LAW (Electrostatics). The force between two point (electric) charges in free space is a pure attraction or repulsion, and is given by

$$F = \frac{q_1 q_2}{4\pi\epsilon_0 r^2} \text{ newtons}$$

where q_1 and q_2 are the magnitudes of the charges in coulombs, r is their separation in meters and ϵ_0 is the permittivity of free space, 8.84×10^{-9} farads/meter.

COULOMETER. Also known as coulombmeter, a device for the measurement of electric current. Originally developed (1916) by the U.S. National Bureau of Standards, the *silver coulometer* consists of a small platinum vessel, acting as the cathode, into which a pure silver anode is immersed. An aqueous solution of silver nitrate (15% $AgNO_3$, wt) of very high purity is used as the electrolyte. In use, both the quantity of silver deposited and the time are carefully noted. These measurements permit a calculation of the average current strength.

The more practical form for laboratory use employs copper electrodes in a bath of copper sulfate. The thin copper cathode, between two heavy copper anodes, is removable for weighing; and since one coulomb deposits 0.000329 gram of copper, the weight of the copper deposit enables one to determine the quantity of electricity in coulombs. A solution recommended for this cell consists of 15 grams of crystalline copper sulfate, 5 grams of pure sulfuric acid, and 5 grams of pure alcohol, dissolved in 100 grams of distilled water.

The effect of current flow on electrolyte concentration also can be determined by titrating the electrolyte after electrolysis. A device of this type is known as a *titration coulometer*. Rarely, a coulometer may be referred to as a *voltameter*.

COUMARIN. **Anticoagulants; Furane and Related Compounds.**

COUMARONE-INDENE PAINTS. **Paint.**

COUNTER. A device or instrument for summing the number of pieces, events, pulses, and a variety of other phenomena and occurrences which may occur within a fixed period of time. A counter may identify and hence remember each item of concern in its effort to attain a sum, or it may determine a rate and from this infer a total. Be it ever so simple, a counter requires a memory (if direct counting is done)—the reason children use their fingers sometimes in counting. Counters as used in data processing systems are described under **Counter (Computer System).**

The counting scale is an excellent example of an inferential means for counting discrete pieces. Provided that the pieces to be counted are of reasonable uniformity and that variations of weight between individual pieces are small when compared with the weight of each piece, weighing provides a very satisfactory counting method and often one that is convenient and efficient. Scales with special readouts are furnished for parts counting.

Because of the large variety of needs for counting, numerous types of counters, ranging from purely mechanical through electric and electronic and from low cost to a high degree of sophistication, are available.

Electronic Counters. These units were initially developed for use in conjunction with Geiger or Geiger-Mueller counting tubes. Originally, electronic counters were constructed with binary counting elements. Contemporary counters utilize decimal-type counting tubes or equivalent solid-state circuitry. Electronic counters generally consist of a time-base generator, a signal gate, and decade-counting units. Frequency is measured by counting the number of input cycles over a precisely controlled period of time. The time-base generator develops control signals which are applied to the signal gate. When the first or "start" signal is received, the signal gate opens to pass input pulses from the unknown frequency source to the decade-counting units. When the second or "stop" signal is received, the signal gate closes to prevent further input pulses. Totalization of input pulses by the decade-counting units during the interval is a measure of the input-signal frequency. Frequency measurement accuracy of an electronic counter is plus or minus one count plus or minus the accuracy of the time-base generator. In some standard models, frequencies up to 500 MHz or more are measured with a 10-MHz counter and a frequency converter. With a transfer oscillator or other harmonic generator-mixer arrangements, the range is further extended to at least 40 kMHz.

Mechanical Counters. For numerous applications, simple mechanical counters are relatively low cost and most appropriate. Typically, mechanical counters have operating speeds of approximately 1,000 counts per minute or 100 revolutions per minute of the prime wheel. The life of a commercial mechanical counter can be considered to range between 20 and 100 million counts, with an average of about 50 million counts. For a counter of precise and high-quality production, the figure could go to 500 million counts. Counters of this type are widely used on a large variety of machines, ranging from printing presses, packaging equipment, conveyors, machine tools, to turnstiles, etc.

A mechanical counter is comprised of three major elements: (1) drive, (2) transfer mechanism, and (3) reset. A direct drive operates through a direct connection between the drive shaft and the first (or "unit") wheel of a counter. This type of drive provides 10 counts per revolution of the drive shaft and is frequently used when indication applications and interpolation between full-figure readings may be required. Revolution drives are often provided through 10:1 gearing between the drive shaft and the first counter wheel. One count per revolution results. Industrial applications where machine shaft revolutions are counted often require this type of input. Geared drives giving ratios other than 10 counts per revolution or 1 count per revolution can be devised and are common in certain applications. Ratchet counter drives provide counts of cycles of oscillating or reciprocating motion. The counting stroke is generally composed of "pretravel," the "counting stroke" proper, and of "overtravel." The counting stroke proper is frequently 36°, with the pretravel and overtravel something less than this amount. Return motions are frequently actuated by internal return springs. Some units count on the return stroke, using the counting (or power) stroke to cock the return spring which drives the wheels when released. Rotary ratchet drives are similar. The main difference is the elimination of mechanical stops for the counting motion.

Electromechanical Counters. The basic difference between an electromechanical counter and the previously described mechanical counters is in the input means. The electromechanical form is usually operated by an electrical impulse instead of a mechanical motion. In cases where events or units are being counted, it is usually possible to substitute an electromechanical counter for a mechanical unit by having the mechanical drive member operate a switch which is used to control on-off voltage pulses which are then applied to the counter. An advantage of this type of operation is that the counter need not be mounted in conjunction with the sensing element, but can be remotely located. Electromechanical counters are often mounted in multiple groups and are used to provide summarizing information on production statistics, or for general multiple-channel data storage and indication.

Although some electromechanical counters use iron-core solenoid driving units, it is more common to find clapper-type solenoids, such as those used in relay construction. Either single- or double-stroke drives may be used. In the single-stroke mechanism, the pull-in of the clapper rotates the lowest-order wheel by 36° through a ratchet and pawl linkage. The return stroke of the clapper is accomplished by spring action and the wheels are at rest during the return stroke. Two-stroke drives move the wheels 18° on the power stroke and the additional 18° on the return stroke.

Stepping motor drives also are used and these are capable of high speed because they have no ratchets and pawls. Electromechanical counters are basically dc operated, but models are obtainable for operating on alternating current. Direct current counters can be equipped with compact full-wave rectifiers for ac operation.

Bidirectional magnetic counters (add-and-subtract counters) also are available. They are usually equipped with two power sources, one of which adds counts and one of which subtracts counts from the total reading. Units of this type are particularly useful for stock and inventory control applications.

The "single-wheel" unit also has many applications. In its basic form, this is a counter having a single digital counting wheel and a switch capable of being actuated between the nine and zero position. This transfer switch can be used to operate a higher-order single-wheel counter, and thus units with any desired number of digits can thus be built up.

COUNTERBALANCING. Counterbalancing means simply the application of extra mass to a system in order to produce balance for the system as a whole, and to offset the unbalance arising from some particular part. Rotating machinery, especially high-speed machinery, needs to be counterbalanced if the center of gravity of the rotating mass does not lie on the axis of rotation. Hoists are frequently counterbalanced so that a descending weight will supply some of the energy required for hoisting the non-useful load. Numerous examples of counterbalancing will be found in everyday practice, but those cases associated with the counterbalancing of high-speed rotating machinery are the most imperative of solution. Counterbalancing mechanisms are very important in certain balances and scales.

COUNTERBORE. Counterboring, spot-facing, countersinking, and center drilling are hole-enlarging operations generally performed on a drill press, although they are actually reaming operations. Counterbored holes are those with cylindrical enlargements, for fillister head screws; spot-faced holes are counterbored to a depth just sufficient to clean up the surface. Countersunk holes have conical enlargements; center drilling is essentially countersinking, using an integral drill and countersink for machining center holes for turning operations on lathes.

Both counterboring and countersinking are performed with piloted cutters, but usually (except for center drilling) require separate operations for drilling the body hole before countersinking or counterboring. A subland drill is a twist drill of special design, constructed so as to provide separate cutting edges and flutes for each hole diameter, so that counterbored holes may be produced in one operation.

COUNTER (Computer System). A physical or logical device that is capable of maintaining numeric values which can be incremented or decremented by the value of another number. A counter, located in storage, may be incremented or decremented under control of the program. An example of such use would be recording the number of times a program loop has been executed. The counter location is set to the value of the number of times the sequence of instructions is to be performed. Upon completion of the sequence, the counter is decremented by one and then tested for zero. If the answer is nonzero, the sequence of instructions is repeated.

A counter also may be in the form of a circuit that records the number of times an event occurs. A counter which counts according to the binary number system is illustrated by the accompanying table. Counters also may be used to accumulate the number of times an external event takes place. The counter may be a storage location or a counter circuit which is incremented as the result of an external stimulus.

	A_3	A_2	A_1
P1	0	0	1
P2	0	1	0
P3	0	1	1
P4	1	0	0
P5	1	0	1
P6	1	1	0
P7	1	1	1
P8	0	0	0

Truth table for a binary counter. The situation illustrates a binary counter of three-stages capable of counting up to eight pulses. Each trigger changes state when a pulse is gated to its input. In the instance of trigger A_2, it receives an input pulse only when trigger A_1 and the input are both 1. Subsequently, trigger A_3 changes its state only when both triggers A_1 and A_2 and the input pulse are all 1's. This table shows the value of each trigger after each input pulse. At the eighth pulse, the counter resets to zero.

Specific counter definitions include:

The *binary counter* is (a) a counter which counts according to the binary number system, or (b) a counter capable of assuming one of two stable states.

The *control counter* records the storage location of the instruction word, which is to be operated upon following the instruction word in current use. The control counter may select storage locations in sequence, thus obtaining the next instruction word from the subsequent storage location, unless a transfer or special instruction is encountered.

The *location counter* or *instruction counter* is (a) the control section register which contains the address of the instruction currently being executed, or (b) a register in which the address of the current instruction is recorded. Synonymous with instruction counter and program address counter.

See also terms listed under **Data Processing**.

Thomas J. Harrison, International Business Machines Corporation, Boca Raton, Florida.

COUNTER EMF. Electromotive Force.

COUNTER (Fluidic). Fluidics.

COUNTER-TYPE FREQUENCY METER. Frequency Measurement.

COUNT RUMFORD CANNON-BORING EXPERIMENTS. Mechanic Equivalent of Heat.

COUNTRY ROCK. The general term used for the main mass of rock in which occur the veins, dikes or ore bodies which are of particular interest, or which are described in detail.

COUPLE. Two parallel forces of equal magnitude, but opposite direction cannot be reduced to a single force, i.e., as the resultant of two forces acting at their center of pressure. Such forces form a couple. The effect of a couple upon a body is independent of the location of that couple with respect to the body. The net action of several couples all in the same plane on a body is equal to the algebraic sum of the moments of the couples, the sign being determined from the direction of rotation which the couple tends to give. The moment of a couple is the production of the perpendicular distance between the forces and one of the forces. The action of any force, acting at any particular point on a body, upon another point lying in the plane of the force can be reproduced by another force of the same magnitude acting at the desired point plus a couple. For example (see figure), let F be

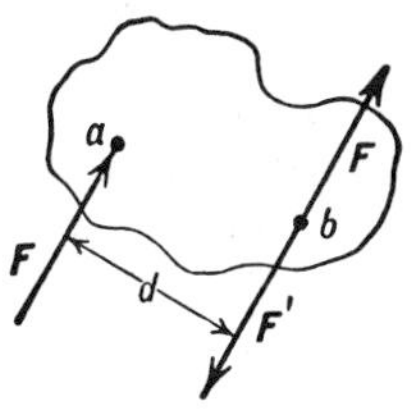

Forces acting in a couple.

any force acting at point a, and b any other point at distance d from the line of action of F. At point b place two equal and opposite forces F and F' which are parallel to the direction of the original force F. Then F' at b and F at a form a couple Fd, leaving F which acts through point b. The effect on point b of the latter force and the couple is the same as the effect of the original force F acting at a. Thus it is seen that it is possible to replace a force acting at a with an equal force acting at b, and the couple Fd, where d is the perpendicular distance from b to the force F in its original position.

COUPLED CIRCUIT. While any group of circuits which are so connected or related that effects in one produce effects in the other constitute a coupled circuit, the term is usually used to designate circuits related so ac effects are transferred but steady state dc effects are not. The two most common classifications of coupled circuits are the inductive and the capacitive coupled circuits, so named because of the primary method of transferring the effects. Capacitance coupling is used quite extensively in various vacuum tube amplifier circuits, in thyratron circuits, and similar applications where it is desired to block dc effects and transfer ac. Since the capacitor does this it may be used as the common element between the two circuits. The so-called resistance coupled amplifier is really capacitance coupled. Figure 1 shows examples of this type. Inductive coupling is the most widely used type since it is used extensively in the power field as well as in the communications and electronics fields. The ordinary power transformer is the means of inductively coupling two power circuits. The various transformers, tuning coils, etc., of radio circuits are other examples. Figure 2 shows some typical circuits.

For two circuits to be inductively coupled an inductance element

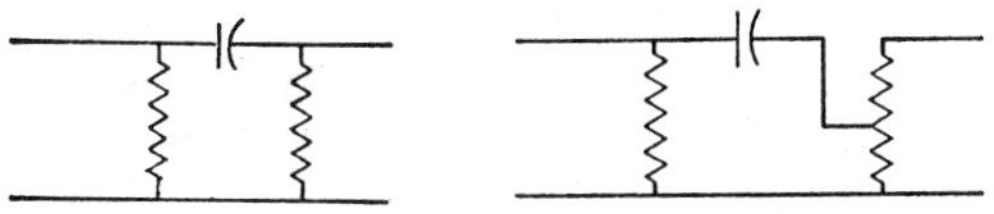

Fig. 1. Capacitance-coupled circuits.

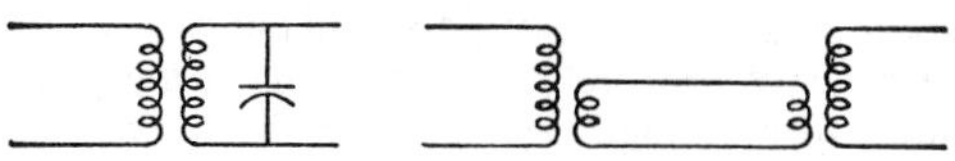

Fig. 2. Inductance-coupled circuits.

in one circuit must be so related to an inductance in the other that flux set up by one links the other. Thus a current flowing in circuit number 1 produces a flux which, in part at least, links the other. When this flux changes it produces a voltage in the second coil, and if the second circuit is closed a current flows. This flux which links both circuits is the mutual flux and the effect gives rise to mutual inductance. Mutual inductance may be defined as the flux linkages in one circuit per ampere in the other,

$$M = N_2\phi_m/I_1$$

where M is the mutual inductance, $N_2\phi_m$ the flux linkages by the mutual flux in circuit 2 and I_1 is the current in circuit 1 which caused the flux ϕ_m. M is also expressible in terms of the self-inductance of the coupled coils,

$$M = k\sqrt{L_1L_2}$$

where k is the coefficient of coupling and the L's are the respective self-inductances. The manner in which the current in the secondary circuit varies is a rather complicated function of the various circuit parameters, frequency, and primary current. However, there is a certain value of coupling which will produce a maximum secondary current for fixed values of the other parameters. This value of coupling is called the critical coupling. In the usual inductive coupled circuit used at radio frequencies where the circuits are tuned, the secondary current plotted as a function of frequency presents a single peak, increasing in value, up to the critical coupling, then presents a double peak with no increase in value as the coupling is made still closer.

Not only are coupled circuits used to transfer ac energy from one circuit to another, but by proper design may be made to match the impedances of the connected circuits also. See **Impedance Matching.**

COUPLED OSCILLATOR (Mechanical). A system with two or more components coupled by forces which can be considered either exactly or approximately as harmonic. The resultant motion of each component when the system is displaced from its equilibrium position can be considered as a linear superposition of simple harmonic oscillations with characteristic frequencies known as normal frequencies. For a nondegenerate nondissipative coupled system of n particles each having one degree of freedom, there will exist n normal frequencies. It is theoretically possible by the correct choice of initial conditions to set a coupled system into oscillation so that all the particles vibrate with only one normal frequency. Such a vibration is called a normal mode of vibration. For the n particle system there will exist n normal modes and n normal coordinates. A normal coordinate vibrates harmonically with a single normal frequency and can be found by a transformation of the actual displacement coordinates which describe the individual motion of each particle.

COUPLING (Chemical). Reactions for the formation of chemical compounds usually by establishing a valence bond between a carbon atom and a nitrogen atom, as in the following example.

$$C_6H_5N{=}NCl + H_2NC_6H_5 \rightarrow C_6H_5N{=}N{-}NHC_6H_5$$
$$\downarrow \text{ Rearrangement}$$
$$C_6H_5N{=}NC_6H_5NH_2$$

Phenols and several other organic substances are also said "to couple." Polyphenylene oxides, thermoplastic materials, are produced by means of oxidative-coupling technology.

COUPLING (Mechanical). In mechanical engineering there are a number of uses of the term coupling. A pipe coupling is a hollow cylinder with internal pipe threads, used to join two sections of externally threaded pipe, either of the same or of different sizes; in the latter case, the unit is referred to as a reducing coupling. Shaft couplings are used to connect rotating shafts, and are of two types—rigid and flexible. The sleeve coupling is an example of the former type, and consists of a hollow cylinder, usually provided with a keyway and set screws to prevent relative rotary and axial motion of the shafts. Another type of rigid coupling, the flange coupling, consists of a disk and hub on each shaft, connected by through bolts. Since perfect alignment of two theoretically collinear shafts is difficult to attain, some form of flexible coupling is usually employed for moderate or heavy duty transmissions, such as motor-generator sets, motor-driven pumps, and the like, to prevent the transmission of shock and eliminate stress reversals. There are numerous commercial forms of flexible couplings; one type is similar to a flanged coupling, but employs laminated steel pins instead of through bolts for transmitting power from one flange to the other. In another form, two sprockets of equal size are mounted—one on each shaft—and connected by means of a roller or silent chain. In other forms, the connection between the shaft flanges is effected by springs or by bolts or pins mounted in rubber.

For connecting shafts whose axes are slightly out of alignment, but approximately parallel, the Oldhams or cross-keyed coupling is used. This device consists of two coupling halves fastened to the shafts; each half has a groove or slot cut in it, and the two halves are arranged so that the grooves are perpendicular. The halves are connected by a central member with perpendicular tongues that engage the coupling half slots. In some instances, cross-keyed couplings are used as flexible couplings; in such cases, the central member is made of fiber or has leather-faced contact surfaces.

A universal joint is a rigid coupling for connecting shafts whose axes will intersect if prolonged, and for applications where the angle between the shaft axes may vary during operation. The device is usually composed of two forked coupling halves, with a central block free to oscillate about two mutually perpendicular axes lying in a plane perpendicular to the shaft axes. Universal joints operate satisfactorily when the shaft coincidence error does not exceed 10 to 15 degrees; beyond this range, the joint is likely to be quite inefficient.

COUPLING (Physics). An interaction between parts or systems. A simple illustration arising in induction heaters and other electrical apparatus is where coupling is the percentage of the total magnetic flux produced by an inductor which is useful in heating a load or charge, or which is otherwise effective. In atomic physics, the various parts of a system (e.g., electrons in an atom) each have orbital angular momentum and spin, or rotation, angular momentum. Obviously these values can be combined in various ways to obtain a resultant momentum, which is useful in computing various properties; optical, magnetic, etc., of the substance composed of these systems.

In *Russell-Saunders coupling* the orbital angular momentum and the spin angular momentum vectors of the individual particles are added separately to obtain a resultant orbital and a resultant spin angular momentum vector. The two resultants are then combined vectorially to obtain a series of allowed total angular momentum vectors.

In *spin-orbit coupling* the resultant angular momenta of the various individual particles are first found by adding the individual orbital angular momentum and spin angular momentum vectors, and then the resultants for each particle are combined to find total angular momentum vectors for the system.

Since Russell-Saunders coupling and spin-orbit coupling represent extremes, many transition cases occur, and must be reckoned with in studies of atoms. Moreover, often the spin angular momentum of the atomic nucleus must be considered. Obviously these processes are complex even for atoms; for molecules they are so involved that many different modes of coupling must be distinguished. This work was first done by Hund, who distinguished various coupling cases, which are therefore known by his name.

In *electroacoustic transducers*, performance is closely related with the tightness of coupling between mechanical and electrical aspects. Consider a piezoelectric disk which is compressed by putting in mechanical energy W_m. The appearance of surface charges shows that electrical energy W_e is stored in the self-capacitance and is available when an external circuit is connected to suitable electrodes. The ratio W_e/W_m (electromechanical coupling coefficient) sets a limit to the efficiency for a given bandwidth (frequency range). The coefficient may reach 70% for lead zirconate titanate. See also **Resonance.**

COUPON. An extra piece of metal formed on a casting, forging, or similar metal product for the purpose of furnishing material to make a metallurgical test specimen.

COUPRAY. Bovines.

COURLAN. Rails, Coots, and Cranes.

COURSE. The term course is used in a number of contradictory meanings by different writers on the general subject of navigation. The U.S. Navy has adopted two standard meanings for this term: (1) Course is the direction that a navigator desires his ship to follow for a given period of time; and (2) course is also the direction that a navigator hopes his ship has followed for a given period. Specifically, by the first definition, the navigator knows or assumes that his ship is in a given location at a given time, and he wishes to proceed to another particular location. By graphical methods, on a chart or small-area plotting sheet, or by any one of a number of standard computational methods (e.g., plane sailing, middle-latitude sailing, mercator sailing, great-circle sailing, and composite sailing), the navigator obtains the direction and length of the line joining the two points. These are known as the predicted course and distance between the two locations. By the second definition, the navigator knows, or assumes, that his ship is at a given location at a certain time. With the ship proceeding on a given heading with a known speed, the navigator may determine, by graphical or computational dead-reckoning methods, a position of the ship at the end of a definite period of time. The direction and length of the line joining the two locations are the assumed course and distance between the two points, or the dead-reckoning course and distance made good.

See also **Current Correction Angle (Navigation); Dead Reckoning; Departure (Navigation); Fix (Navigation); Great-Circle Course; Heading; Mercator Sailing; Navigation; Plane Sailing; Plotting Sheet;** and **Wind Correction Angle (Navigation).**

COURSE (Composite). The shortest track between two points on the surface of the earth is a great circle, if we neglect the slight oblateness. However, the following of such a track has two fundamental disadvantages: (1) Such a course is a rhumb line only in the particular cases of two points both on the equator, or two points on the same meridian of longitude. (2) Such a course will frequently lead the vessel into impossible positions (e.g., if the two points are in the same latitude but differ by 180° in longitude, the great-circle track between them would lead over the nearest pole).

A composite course is a combination of great-circle and rhumb line courses designed to carry a ship from one point to another by the shortest practicable path. In case the great-circle track does not lead the ship into impossible positions, the composite course is usually a series of rhumb line courses to successive positions along the great-circle track, the rhumb-line distances so figured that the course of the ship will be altered at convenient times (e.g., the changing of the watch). In case the great-circle course leads the ship into danger, the problem of computing the composite course is one of a number of compromises, which are different for every problem. A good example of such a composite course may be obtained by examining the steamer lanes across the Atlantic Ocean, which will be found on many terrestrial globes.

See also **Course; Navigation;** and **Rhumb Line.**

COURSER. Shorebirds and Gulls.

COVALENT BONDS. Chemical Elements; Mineralogy.

COVALENT RADIUS. Chemical Elements.

COVARIANCE. As an extension of the variance, the covariance of two variables x and y is the expectation of $(x - \mu_x)(y - \mu_y)$ where μ_x and μ_y are the means of x and y. For a sample of n values it is defined as $\Sigma_{i=1}^{n} (x_2 - \bar{x})(y_2 - \bar{y})/n$. Some authorities define it with a denominator of $n - 1$ instead of n. See also **Variance.**

COVARIANCE (Analysis of). Analysis of Covariance.

COVELLITE. The mineral covellite, cupric sulfide, CuS, is hexagonal, usually in thin platey crystals, but may be massive. It has a hardness of 1.5–2; specific gravity, 4.6; luster, submetallic to resinous; color, dark indigo blue, sometimes showing a purplish tarnish, or if moistened may appear purple in color. Its streak is dark gray to black; it is opaque. Covellite is found associated with chalcopyrite, bornite, and chalcocite, and is believed to be chiefly of secondary origin. Covellite occurs in Yugoslavia, Saxony, Sardinia, Argentina, Chile, Bolivia and Peru, and in the United States at Butte, Montana, and in Colorado, Wyoming, and Utah. This mineral was named for Covelli, who discovered it in the lavas of Mt. Vesuvius.

COVERING (Animal). Ingegument; Integumentary System.

COVERSINE. Trigonometric Function.

COVOLUME. The correction term applied in certain equations of state, as in that of van der Waals, to correct the volume of the gas for *the effect of* the volume of the molecules. This term is not the molecular volume itself.

COW. Bovines; Digestive System (Ruminants).

COWBIRD (*Aves, Passeriformes*). 1. The yellow wagtail of England. 2. A dark-colored bird, *Molothrus ater,* of southern Canada and the United States, related to the blackbirds, and several species of the same genus extending from Texas into South America. Most of these species deposit their eggs in the nests of other birds like the European cuckoo. They are named from their frequent association with cattle and before the settlement of North America the common species was called the buffalo bird.

COW EXPERIMENT. Gravitation.

COWPER'S GLANDS. Two small glands located beneath the male urethra which produce a mucous secretion into the urethra; analogous to Bartholin's glands in the female.

COWRY (*Mollusca, Gasteropoda*). Compactly oval shells with a long narrow aperture, smooth surface, and often bright colors.

There are many species of cowries, especially in the Pacific and Indian Oceans. The shells of some have been used as money and for decorations.

COWSHARKS. Sharks.

COXSACKIE VIRUS. This virus was so named because the first strain was isolated in Coxsackie, New York in 1969. Since that time, newly discovered enteroviruses have been identified by number rather than by specific names. Coxsackie viruses have many biological characteristics in common with echoviruses. See **Enteroviruses;** and **Virus.** Coxsackie virus is usually spread by the hand-to-mouth route and infections occur most commonly in the summer and fall. Family groupings of cases or small epidemics are frequently associated with this virus. In addition to enteric system involvement and disease, extraenteric symptoms and disease may occur in some cases of coxsackie virus infection. Coxsackie virus A16 has been implicated in Kaposi's varicelliform eruption (skin). Coxsackie virus B types have been associated with infections of the nervous system. These may be manifested in aseptic meningitis as well as involvement in encephalitis, myelitis, and ganglionitis. See **Dermatitis and Dermatosis.**

COYOTE. Canines.

COYPU. Rodentia

CPU. Central Processing Units (Computer).

CRAB (*Crustacea, Decapoda*). Crustaceans with a short broad cephalothorax and a small abdomen bent below it. The large pinchers and four pairs of legs are the only conspicuous appendages. Most species of crabs are found in or near the ocean but some are terrestrial and others live in fresh water. The land crabs deposit their eggs in the water. The many species of crabs constitute a division of the order named the *Brachyura* from the short abdomen.

A number of species of crabs are used for food. Of these the edible crab of the Atlantic, which ranges from Cape Cod to Louisiana, and the edible crab of the Pacific Coast, are the most important species.

Many crabs have received common names which apply to one species or to a group of similar species. Among these names are spider crab, hermit crab, fiddler crab and land crab. Commercially important crabs are described in the entry on **Crustaceans (Edible).**

CRAB APPLE TREES. Rose Family.

CRAB (Horseshoe). Xiphosura.

CRAB PULSAR. Radio Pulsars.

CRACKING PROCESS. A reaction in which a hydrocarbon molecule is broken or fractured into two or more smaller fragments. Sometimes the term *pyrolysis* is used for this reaction. Possibilities for cleavage of a molecule include (1) a carbon-hydrogen bond; (2) a bond between an inorganic atom and a carbon or hydrogen atom; (3) a carbon-carbon bond. Usually the objective of cracking is that of reducing the size of hydrocarbon molecules; hence the target is to fracture the carbon-carbon bonds. The main cracking processes are: (1) thermal cracking; (2) fluid catalytic cracking; and (3) hydrocracking.

Thermal Cracking. Of the thermal cracking processes, two are of major importance: (1) coking, and (2) visbreaking (viscosity breaking). Both of these processes convert nondistillable residues into more valuable products. Thermal cracking was the first of the principal cracking processes used in the petroleum industry. For increasing gasoline production and improving quality, fluid catalytic cracking has essentially replaced thermal cracking.

In *thermal coking,* heavy residual stocks are converted into gas, gasoline, distillates, and coke. Generally, the objective is that of maximizing the yield of distillates; and minimizing the production of gas, gasoline, and coke. Light distillates are used for both domestic and industrial heating oils. There are two types of thermal coking processes: (1) cyclic, semicontinuous process, sometimes referred to as *delayed coking,* decarbonizing, or low-pressure coking; and (2) a continuous fluid coking process. About 70% of the installed capacity in the United States is the delayed coking type.

As shown by Fig. 1, a delayed coking unit is comprised of three sections—a furnace, coke drums, a fractionating unit, plus coke removal and handling equipment. Usually the feedstock is charged to the lower part of the fractionator. Here the feedstock is contacted by hot vapors from the coke drum, causing any light components to be flashed from the feed before the feed joins with the recycle and charged (from the bottom of the fractionator) to the furnace. The charge in the furnace is heated to about 480°C (896°F). The heated effluent from the furnace is introduced into the bottom of one of two or more insulated vessels (coke drums) where, as the result

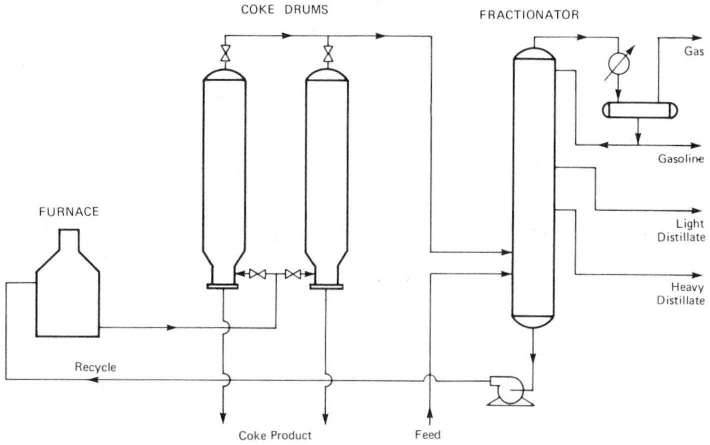

Fig. 1. Delayed coking unit shown schematically.

of its contained heat, the material cracks to form a solid coke residue. At the same time, lighter cracked products are evolved and proceed from the top of the coke drums to the fractionator. The reaction in the coke drum is endothermic and thus the temperature of the material drops to about 425°C (797°F). The cracked products leave as vapors; the coke remains in the drum. The fractionator separates the cracked vapors into several side streams as shown on the diagram. When accumulated coke reaches a certain level in the drum, that drum is temporarily taken off-stream and the flow is switched to the second drum. Prior to removal of coke, the drum is steamed to remove vapors. Water is also added to cool the coke.

The fluid coking process accomplishes the coking operation in a continuous manner. Feed is sprayed into a fluid bed of hot coke in a coking reactor. Steam introduced into the bottom of the reactor provides the fluidization energy. The cracked products are quenched in an overhead scrubber and then go to the fractionator. The coke is deposited on the particles in the reactor which commute with a heater vessel in which a portion of the coke is burned to heat up the returning coke particles to supply the energy for the coking reaction.

Fluid Catalytic Cracking. This process is used principally to create gasoline, C_3/C_4 olefins, and light distillates by the selective decomposition of heavy distillates. The process was introduced during World War II to replace earlier thermal cracking processes. Specially prepared catalysts are used. The resulting gasoline contains substantial proportions of high-octane-number hydrocarbon components, including aromatics, branched paraffins, and olefins. The cracking proceeds in accordance with the carbonium-ion mechanisms. Consequently, there are minor amounts of fragments lighter than C_3 in the products. This is to be contrasted with thermal cracking by the free-radical mechanism, wherein large proportions of fragments lighter than C_3 result. Another product of fluid catalytic cracking is *cycle oil,* a distillate that boils above gasoline. Cycle oils are withdrawn as net products and are used as components in heating oils, feedstocks to hydrocracking units, and for blending with heavy residuals as a means of reducing viscosity. The highly aromatic clarified slurry oils have been found to be useful feeds for the manufacture of carbon black.

Indicated in Fig. 2 is a representative fluid catalytic cracking unit, comprising (1) a reactor, (2) a regenerator, (3) the main fractionator, (4) an air blower or compressor, (5) a spent-catalyst stripper, (6) catalyst recovery equipment, including cyclones internal in the reactor and regenerator; and slurry settler, and possibly an electrostatic precipitator, and (7) a gas-recovery unit. The catalyst used is essentially a specially prepared composite of silica and alumina.

In operation, preheated feedstock meets a controlled stream of hot, regenerated catalyst. Vaporized oil and catalyst ascend in the riser, such that the catalyst particles are suspended in a dilute phase. Essentially all of the cracking occurs in the riser. The catalyst particles are separated from the cracked vapors at the end of the riser and the catalyst containing a coke deposit is returned to the regenerator.

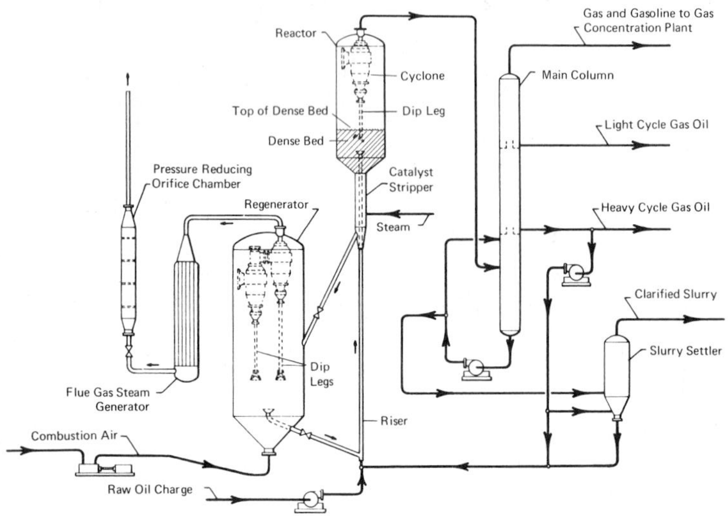

Fig. 2. Fluid catalytic cracking process. (*UOP Inc.*)

The cracked vapors pass through one or more cyclones located in the upper portion of the reactor and proceed to the fractionator (main column) that produces the side streams indicated.

Hydrocracking. Processes in this category produce gasoline and light distillates from feed distillates that are higher-boiling than the products. Hydrocracked products are not olefinic. The light gaseous hydrocarbons produced by hydrocracking are entirely paraffinic. The processes operate at elevated pressures in the presence of hydrogen and catalysts. Temperatures are usually lower than 482°C (900°F). Pressures run from 800 to 2,500 psig. Both fixed-bed and ebullating-bed configurations are used. Because carbonaceous deposits accumulate very slowly on the catalyst, the on-line periods for these units is quite long, ranging from several months to over a year. Somewhat more costly to build than fluid catalytic cracking, the hydrocracking process has the advantage that it can handle heavier and dirtier feedstocks and also may be adapted to varying product ratios of gasoline to middle distillate.

Technical Staff, UOP Inc., Des Plaines, Illinois.

CRAMÉR-RAO INEQUALITY. In statistics, an inequality giving a lower bound to the sampling variance of an estimator. If t is an estimator of a parameter θ in a frequency function f and the bias is defined as

$$b(\theta) = E(t) - \theta$$

where E denotes expectation, then the inequality states that

$$\operatorname{var} t \geq E\left\{\frac{\left(1 + \dfrac{\partial b}{\partial e}\right)^2}{\left(\dfrac{\partial \log f}{\partial \theta}\right)^2}\right\}$$

CRAMER RULE. Let

$$a_{11}x_1 + a_{12}x_2 + \cdots + a_{1n}x_n = y_1$$
$$a_{21}x_1 + a_{22}x_2 + \cdots + a_{2n}x_n = y_2$$
$$a_{n1}x_1 + a_{n2}x_2 + \cdots + a_{nn}x_n = y_n$$

be n linear equations in n unknowns with nonvanishing determinant $D = \det\{a_{ij}\}$. Let D_i denote the determinant of the matrix obtained by replacing the ith column of $\{a_{ij}\}$ by the column $y_1, y_2, \ldots, y_n$. Then Cramer's rule gives the solution of the above equations in the form $x_i = D_i/D$, $i = 1, 2, \ldots, n$. The rule is of theoretical importance, but of little value in computing practice.

CRANE (Aves). Rails, Coots, and Cranes.

CRANE FLY (*Insecta, Diptera*). Insects, *Tipulidae* (Daddy long-legs), which resemble mosquitoes in form but are usually much larger. They have a V-shaped groove across the thorax and have no scales on the wings.

The larvae of some species live in the ground and are sometimes injurious to the roots of grasses and grains. These larvae are called meadow maggots or leather jackets. They come to the surface at times and can be destroyed by the use of poison baits.

CRANE (Gantry). Gantry.

CRANIUM. Brain and Nervous System.

CRANKSHAFT. A crank is a bent arm which moves with rotary motion about its unbent end. In order to provide a support for this rotation the crank is mounted on a crankshaft. The crank and crankshaft form one of the important basic units of mechanism. This is an efficient way of transforming rotary to reciprocating motion, or vice versa. The crank alone does not accomplish this, but in conjunction with the connecting rod and slider, it forms the basis of many such important machines as engines, pumps, compressors, and a host of other mechanisms. A simple crankshaft of the overhung type is readily made from an arm (or disk) and a crank pin which is set into it at some radial distance from a crankshaft, which is the center of rotation. This type is used where a crankshaft is to accommodate but one crank, and where the crankshaft bearing surface is entirely on one side of the crank. Multiple cranks on the same crankshaft are obtained by fitting the crank pin between two crank arms, so that the connecting rod may swing freely without interference with the crankshaft. Multiple throw crankshafts are made by building up the crank pin between two cranks, and by machining from a single forged piece. It is the function of a crank to transmit the force at the crank pin into a torque at the crankshaft, the torque being equal to the component of crank pin pressure perpendicular to the crank radius multiplied by that radius. Since crank pin pressure is not always exactly perpendicular to crank radius, it follows that a crank may carry some compression or tension as well. It should not be loaded with bending by forces parallel to the crankshaft.

CRAPPIE. Sunfishes.

CRASPEDOTE. Provided with a craspedon or velum. Applied to the medusae of many species of Hydrozoa in contrast with the jellyfishes.

CRASSULACEAE. Bryophyllum.

CRATER (Constellation). A small southern constellation located near Hydra.

CRATER (Meteorite). Meteorite Crater.

CRATON. A part of the earth's crust which has attained stability and which has been little deformed for a prolonged period. The extensive central cratons of the continents (central stable regions) include both shields and platforms. Parts of the more mature Phanerozoic foldbelts have not achieved, or are approaching a cratonic condition. In terms of continental margins, a *cratonic margin* lies entirely on continental crust. Strictly speaking, they do not qualify as transition between continental and oceanic crust. They do, however, occupy large areas covered by seas. Cratonic margins also may contain thick sediment accumulations. See also **Ocean.**

CRATONIC MARGIN. Ocean.

CRAWFISH. Crustaceans (Edible).

CRAWLER TRACTOR. Coal.

CRAYFISH (*Crustacea, Decapoda*). **Crustaceans (Edible).**

CREATINE PHOSPHOKINASE (CPK). Heredity.

CREEPER (*Aves, Passeriformes*). Small insectivorous birds (Aves) which cling to the trunks of trees or cliffs in seeking food. They are found in the Northern Hemisphere and are represented in North America by the brown creeper, *Certhia familiaris*, sometimes called tree creeper. There are about five species of creepers in all, ranging from Nicaragua northward to Alaska and westward to Japan.

The brown creeper is dark brown with lighter brown on the lower half of the body. The basic brown coloration is spotted and striped with gray. The throat is white. The bill is slightly curved and slender and is especially shaped to pierce behind bark on trees. The tail is stiff and used as a brace. Often the nest will be jammed into a crevice behind scales of bark. The nest will be constructed of bits of bark, moss, or twigs. Often it will be from 20 to 40 feet (6 to 12 meters) above ground level.

The male feeds the female while she incubates the eggs. In some species, the male also helps with the incubation. The eggs hatch in about 21 days. They are white-spotted brown.

The short-tailed creeper (*C. brachydactyla*) is found in Asia, Africa, and parts of Europe. The high Himalayan mountain regions claim three species of creepers. These birds are larger, with a somewhat darker overall coloring. The bills are longer and the tails are striped. The Nepalese creeper (*C. familiaris nipalensis*) and the Sikkim creeper (*C. discolor*) are found at altitudes up to 12,000 feet (3,600 meters).

Usually the creepers are classified in the family *Certhiidae*, although some authorities believe that they should be grouped with the nuthatch family.

CREEP (Geology). The slow movement of soil and rock waste down a slope. This movement may be due to the combined influence of gravity, frost, and groundwater. Creep is a factor in soil erosion even on relatively flat slopes.

CREEPING ERUPTION. Dermatitis and Dermatosis.

CREEP (Metals). This term usually is associated with the slow, plastic deformation of metals under constant load. Continued plastic deformation, where the applied force does not change, can result from two basic causes. (1) If a metal is deformed, as in tension, its cross section is correspondingly reduced. This raises the stress level in the material and, if the rate of this increase in stress exceeds the rate of strain hardening, creep occurs. (2) Plastic flow may be promoted by thermally activated softening processes occurring in the metal that counteract the mechanisms leading to strain hardening. Thus, thermal energy may aid dislocations in cutting through one another or in passing around inclusions. Thermal energy may also provide the means for dislocations of opposite sign to move toward each other so that they can recombine. This may eliminate dislocations entrapped in each other's force field, thereby allowing additional dislocations to move out from the sources and plastic deformation to continue.

Creep of metals at high temperatures is primarily controlled by thermally activated processes. Since there are many conceivable mechanisms, creep cannot be associated with a single activation energy. However, in a given temperature range, the creep rate may be controlled by a particular mechanism and the temperature dependence of the creep rate in this range will be related to a corrresponding activation energy. Thus, several activation energies have been observed for creep in aluminum, each of which is controlling in a different temperature range. In general, the activation energy is larger the higher the temperature range in which it controls. The activation energy for creep of metals at very high temperatures (just below the melting point) often tends to be the same as that for self-diffusion. This implies that vacancy motion or dislocation climb is very important in creep at extremely high temperatures.

From a practical point of view, creep becomes an important natural phenomenon when the temperature at which a metal is loaded lies above about 0.4 to 0.5 of its melting point on an absolute scale. In some metals such as zirconium, which undergo a solid state phase change, creep becomes an important effect above about one-half of the temperature of the phase transformation. In many metals such as steel, creep is almost non-existent at room temperatures if the metal is not loaded above its annealed yield strength. However, at 900°F (482°C) steel can creep readily at very small stresses, and equipment such as boilers and tubes for petroleum cracking stills, intended to operate at high temperatures for long periods of time, must be designed on the assumption that creep will occur. Alloy steels and other materials have been developed having much higher creep strengths than carbon steel.

Creep strength is the unit stress which will produce deformation at a specified rate at a specified temperature; for example, the creep strength of a certain 0.15% carbon open-hearth steel at 1000°F (538°C) is 6,100 pounds per square inch (415 atmospheres) for a rate of 0.01% elongation in 1,000 hours. Other values for this material are 6,900 pounds per square inch (469 atmospheres) for a rate of 0.1% and 7,800 pounds per square inch (531 atmospheres) for a rate of 1.0% elongation in 1,000 hours. Creep strength values can only be determined by long-time laboratory tests under carefully controlled conditions of temperature and loading.

CREODONT. Fossil Reptilian Mammals.

CREOSOTE. Coal Tar and Derivatives.

CREST CLOUD. Clouds and Cloud Formation.

CRETACEOUS PERIOD. The last major division in the Mesozoic Era of the geologic time-scale. Type locality, chalk (creta) cliffs of

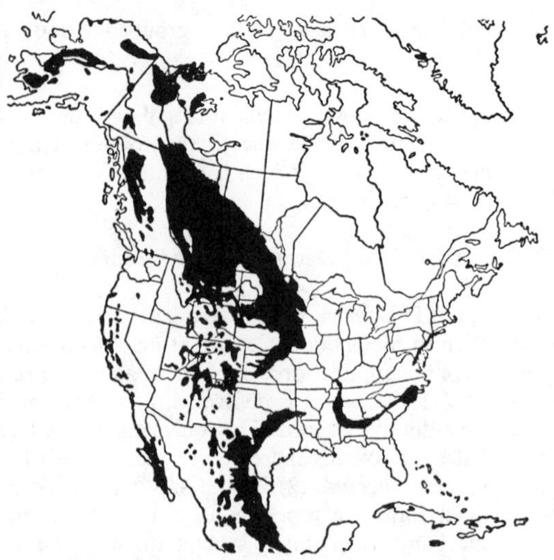

North America showing the surface distribution (areas of outcrops) of Cretaceous strata. The large and small areas in the western interior of the continent very largely represent Upper Cretaceous deposits.

the English Channel. The period was named by A. d'Halloy in 1822. In 1877, Hill proposed that the Lower Cretaceous be erected as a separate system which he called the Comanchean. The Comanchean period began about 135 million years ago and lasted about 25 million years. The Cretaceous period began about 110 million years ago, and lasted about 50 million years. The greatest thicknesses of Cretaceous strata in the United States occur in the Rocky Mountain region and in California. During the Cretaceous Period there was also an extensive overflow of the waters of the Gulf of Mexico toward the southern interior of the United States, depositing there a great series of overlapping sediments. This is the type region of the Lower Cretaceous or Comanchean. While the Atlantic and Gulf continental margins were invaded by the Cretaceous Sea, the region of the Appalachian geosyncline had been reduced to a peneplain which may have been overlapped by marine sediments. In the Great Plains region were deposited freshwater shales and sandstones with local swampy areas in which were formed numerous lignites, subsequently altered to bituminous coal. The Middle Cretaceous saw a great marine invasion of western North America from the Gulf of Mexico to the Arctic Ocean. Deposits of Cretaceous Age are well represented in central and western Europe, with fairly continuous marine deposition throughout the entire period. Cretaceous deposits occur in South America, especially in Brazil. As further evidence that there was widespread invasion of the continents by oceanic waters during this period, marine sediments occur in southwestern Asia, China, Himalayas, Japan, Siberia, and Africa. Important mineral resources are of Cretaceous Age. In the United States the Cretaceous formations contain numerous important aquifers which underlie great semi-arid areas. Bituminous coal beds of average value occur in Alaska, Australia, British Columbia and Germany. Important oil pools occur in the Gulf region. The Cretaceous clays of the eastern United States are extensively used in the manufacture of china and building materials. The sulfide copper ores of Butte, Montana, occur in igneous rocks of Cretaceous and Early Tertiary age. Lower Cretaceous plants and animals were only slightly different from those of Jurassic time. Ferns, cycads, ginkgoales and conifers still predominated, and the principal marine invertebrates were ammonites and belemnites. During the Cretaceous there was a great expansion of the pelecypods, gastropods, and the modern types of fishes. The Mesozoic reptiles had reached their climax in the early Cretaceous and, except for a few large and bizarre forms, were on their way to extinction. The most remarkable types were Triceratops (horned dinosaur), Tyrannosaurus (tyrant dinosaur), Tylosaurus (marine lizard), and Pteranodon (crested pterodactyl, or flying reptile). Since uppermost marine Cretaceous does not occur in North America, there appears to have been a pronounced period of uplift and erosion, accompanied by mountain building, with the combined growth of the Cordilleran ranges. This period of diastrophism, which closed the Mesozoic Era in the western hemisphere, is called the *Laramide Revolution.*

CRETACEOUS PERIOD (Extinction). Asteroid.

CRETINISM. Thyroid Gland.

CREUTZFELD-JAKOB DISEASE. Virus.

CREVASSE. 1. A crack or open fissure in a glacier. 2. A break in the levee of an old-age, meandering stream, often leading to disastrous floods, as in the case of the lower stretches of the Mississippi River.

CRIB DEATH. Sudden Infant Death Syndrome (SIDS).

CRICKET (*Insecta, Orthoptera*). Insects related to the katydids and grasshoppers and, like some of these forms, well known for their resonant singing. The true crickets constitute a family (*Gryllidae*) which contains the common or field crickets and in addition several other forms more or less different in appearance. The field crickets are black or brown species, some of which enter houses. Tree crickets are usually green with broad transparent wings. They frequent trees and shrubs. Mole crickets are thick-bodied brown insects whose forelegs are strongly developed for burrowing. In addition to these and a few other forms of crickets several insects belonging with the katydids among the long-horned grasshoppers bear this name. They are the cave or camel crickets, the sand cricket, and the Mormon cricket.

Crickets, like grasshoppers, beetles, cicadas, and cockroaches, are among the most ancient insect inhabitants of the earth. There are about 900 known species. The average cricket is about $\frac{1}{2}$ to 1 inch (12 to 15 millimeters) long. Its thighs are large in proportion to the rest of the body. The tarsi is in three segments. Some crickets possess tympanic membranes (on the order of the human eardrum) which are located in cavities on the front legs. Movement of a leg enables the insect to determine the direction of sound.

The male cricket produces a chirruping sound by rubbing together especially modified parts of its forewings. In many areas, the cricket appears on or about June 15 in the northern hemisphere and announces its presence with the loud "song" of the males, usually at dusk. Hoy and Paul have made a study of genetic control of song specificity in crickets which is reported in *Science,* **180** (4081), 82–83 (1973). The calling song of male field crickets is composed of stereotyped rhythmic pulse intervals, which are predictable expressions of genotype. Females identify conspecific males by their song. Two species of crickets were found to exhibit species-specific song preference, and hybrids between them preferred hybrid calls over either parental call. These results imply genetic control of song reception as well as transmission.

The *field cricket* (*Gryllus assimilis,* Fabricius) is distributed throughout the United States, southern Canada, and a large part of South America. The insect is particularly damaging to cotton in California and the Gulf states. This is caused by the insect cutting off seedling plants just above groundlevel. They devour alfalfa and grain seeds in the Great Plains states and even attack grain after it has been harvested. There is one generation per year, usually wintering over in the egg stage. However, generations of different broods overlap and thus adults may be present from spring to fall.

The *Mormon cricket* (*Anabrus simplex,* Haldeman) and the *coulee cricket* (*Peranabrus scabricollis,* Thomas) are not true crickets, but are closely related to katydids and longhorned grasshoppers. One of the first recorded outbreaks of an infestation by the Mormon cricket occurred in 1848 in the Great Salt Lake basin. This invasion was

Cricket.

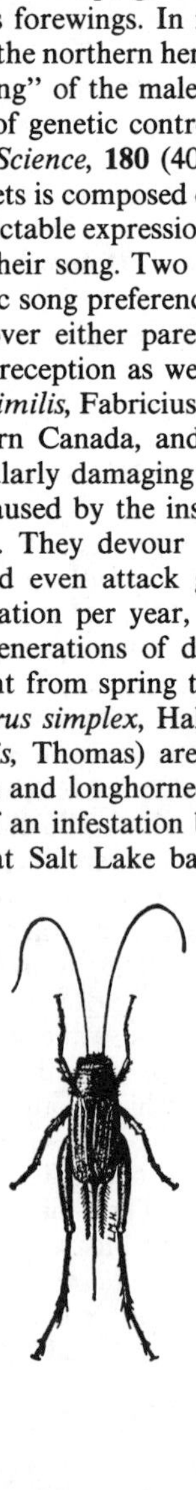

spectacularly terminated by many flocks of gulls to which the settlers later erected a monument. This cricket is found west of the Missouri River, but is most destructive in the Pacific Northwest, Nevada, Utah, Idaho, Montana, Wyoming, and Colorado. The coulee cricket has caused major damage in Montana and Washington.

These crickets attack a host of vegetable and fruit crops. Both species are cannibalistic, consuming dung and dead animals. Advantage of the fact that these insects do not fly is taken in constructing barriers, ditches, etc. to prevent their migration from one area to the next. A film of petroleum distillate on an irrigation ditch can be effective.

The *snowy tree cricket* (*Oecanthus niveus*, De Geer) and closely related tree crickets damage fruit trees and bushes. The insects drill small holes in small twigs and brambles. A rather large, pale-yellow egg (about $\frac{1}{8}$-inch long; 3 millimeters) is deposited in each hole. There may be as many as several dozen such punctures in just a few inches. These puncture wounds serve as sites for fungus and other infection. The adults puncture and eat ripe fruit, rendering it unfit for market. The tree crickets are active on apple, blackberry, cherry, loganberry, peach, plum, prune, and raspberry.

CRIGLER-NAJJAR'S DISEASE. Bile.

CRIMINOLOGY. Bertillon System.

CRINOID. Invertebrate Paleontology.

CRINOIDEA. The sea lilies, feather stars, and basket stars, a class of the phylum Echinodermata. There are now only a few hundred species of these animals although several thousand fossil species are known.

The crinoids are distinguished from the starfishes and other echinoderms by the following characteristics: (1) The body consists of a disk, arms, and a stalk. (2) The mouth and anus are both directed upward. (3) The arms bear small lateral branches and in many species fork repeatedly.

Most of the living species lose the stalk when mature and become free-swimming. These forms, known as comatulids, are found in shallower waters of the ocean, where they swim or creep by means of the arms. The stalked species are found in deep water.

The classification of crinoids is of little interest save to specialists. Many families are recognized and they are generally grouped into orders which include a primitive attached form, the stalked crinoids, and the free species.

CRITICAL COMPOSITION. Systems consisting of two liquid layers that are formed by the equilibrium between two partly-miscible liquids, frequently have a consolute temperature or a critical solution temperature, beyond which the two liquids are miscible in all proportions. At this temperature, the phase boundary disappears, and the two liquid layers merge into one. The composition of the mixture at that point is called the critical composition. There is, in some cases, a lower consolute temperature as well as an upper consolute temperature.

CRITICAL CONCENTRATION. When two immiscible liquids are heated in contact with each other their mutual solubility is usually increased until, at the critical solution temperature, they become consolute. The composition of the two solutions immediately before they become consolute is termed the critical concentration. See **Critical Solution Temperature.**

CRITICAL DENSITY. The density of a substance which is at its critical temperature and critical pressure.

CRITICAL FREQUENCY. 1. A wave radiated from the antenna of a radio transmitter spreads in various directions, the exact nature of its spread being determined by the directional characteristics of the antenna. The part which travels towards the outer atmosphere goes into the ionosphere where it acts upon the ionized particles, principally electrons, which absorb energy and re-radiate it. The net result of this action is an effective change of the index of refraction of the medium through which the wave is traveling. This causes the wave to be bent back towards the earth, the extent of the bending varying with the index. It is this which causes radio waves to be returned to the earth to give reception at points very remote from the transmitter. However, this change of index of refraction varies with frequency in such a manner that the waves are bent less and less as the frequency is raised. A critical frequency is finally reached where the wave is not bent enough to return to the earth, even for the most glancing angle of incidence which it is possible to obtain. As the frequency of the signal is progressively raised the critical frequencies for the various layers of the ionosphere are reached in turn, the wave penetrating each at its critical frequency and being refracted back to the earth by the next layer until finally all layers are penetrated and there is not returning signal. This occurs in the vicinity of 40 megacycles.

2. In a usage closely related to that in the first definition the critical frequency of a magnetohydronamic wave component is that frequency at which it is reflected by, and above which it penetrates through, an ionized medium (plasma) at vertical incidence.

CRITICAL HUMIDITY. Humidity.

CRITICALLY DAMPED. Damping.

CRITICAL MASS. Nuclear Reactor.

CRITICAL OPALESCENCE. The phenomenon sometimes produced when a homogeneous solution of two liquids at its critical composition is cooled from a temperature above its consolute temperature to one below that temperature. This phenomenon consists of a bluish haze which is believed to be due to the scattering of light brought about by local variations of density within the liquid.

CRITICAL POINT. 1. A point where two phases, which are continually approximating each other, become identical and form but one phase. With a liquid in equilibrium with its vapor, the critical point is such a combination of temperature and pressure that the specific volumes of the liquid and its vapor are identical, and there is no distinction between the two states. 2. The critical solution point is such a combination of temperature and pressure that two otherwise partially miscible liquids become consolute.

To consider in detail the critical point as defined in (1), examine the accompanying figure, which shows the family of isotherms of a pure substance in the fluid range (liquid or gas) such for example as shown in the figure for carbon dioxide.

At sufficiently high temperatures each isotherm is a continuous curve, but at low temperatures the isotherm consists of three portions.

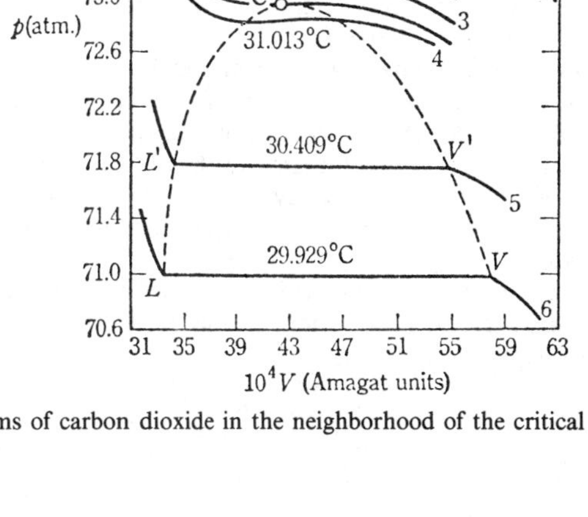

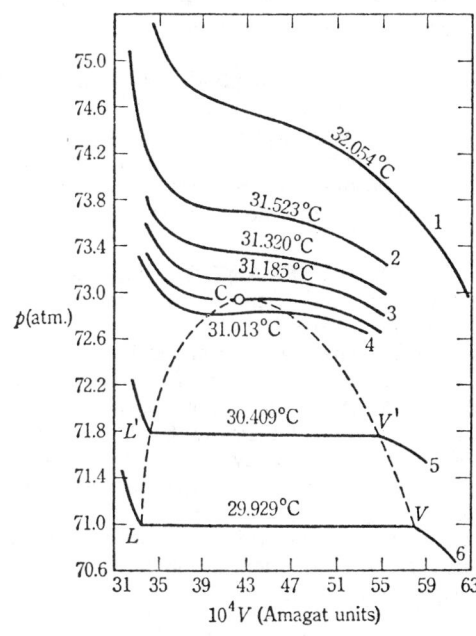

Isotherms of carbon dioxide in the neighborhood of the critical point.

The first section of the curve at high pressures corresponds to the liquid state, while that at low pressures refers to the gaseous state. These two curves are joined by a horizontal line corresponding to the simultaneous presence of two phases, liquid and gas.

The isotherm between those numbered 3 and 4 in the figure represent the transition between isotherms corresponding to the gas phase only, and those including a horizontal portion corresponding to a liquid-gas equilibrium. In this isotherm the horizontal line has contracted to a single point of inflection *C*. This is the critical point characterized by the relations

$$\left(\frac{\partial p}{\partial V}\right)_{T_c} = 0; \quad \left(\frac{\partial^2 p}{\partial V^2}\right)_{T_c} = 0; \quad \left(\frac{\partial^3 p}{\partial V^3}\right)_{T_c} < 0.$$

The curve *LL'C* gives the molar volume of the liquid. Similarly *VV'C* gives the molar volume of the gas.

At the critical point the molar volumes of the liquid and of the gas become equal. In general a critical state is characterized by the fact that the two coexistent phases (here the liquid and the vapor) are identical.

The curve *VV'CL'L* is called the *saturation curve.*

The experimental data do not indicate the existence of a critical point for the liquid-solid transition.

Above the critical point the substance can no longer exist in the liquid state. The critical temperature is thus the highest temperature at which the liquid and vapor can coexist.

Ternary Critical Point. The point where, upon adding a mutual solvent to two partially miscible liquids (as adding alcohol to ether and water), the two solutions become consolute and one phase results.

CRITICAL POTENTIAL. In atomic physics, the critical potential is used in general as a measure of the amount of energy necessary to raise an electron from a lower to a higher energy level. The term "potential" is used because the quantity of energy is measured by means of electrons accelerated by application of a known potential, the energy of the electrons being given by the product of the accelerating potential and the electronic charge.

Two kinds of critical potentials are the ionization potentials and the resonance potentials. The ionization potential represents the work necessary to remove an electron from a normal atom wherein the electron may be supposed to be in its lowest level, to an infinite distance, so that a positively charged ion results. The resonance potential is a measure of the work required to raise an electron from the lowest level to any other level, and therefore, there are first, second, etc., resonance potentials, corresponding to the transfer of an electron from the lowest level to the next level, to the next-but-one level, etc.

CRITICAL REGION. The region in the diagram of state of a substance in the neighborhood of the critical point.

CRITICAL REGION (Statistics). In the statistical theory of hypothesis testing, a region in the sample space (i.e., the space of all possible samples) such that, if a point falls within it, the hypothesis is rejected. The region is, of course, determined by the hypothesis, the nature of the probability distribution of the variables and the degree of assurance or confidence required to decide on rejection.

CRITICAL RESOLVED SHEAR STRESS. The shear stress resolved on a slip plane and in the slip direction that just causes a metal single crystal to undergo slip. Tables 1 and 2 give the critical resolved shear stresses for some typical metals.

The ready occurrence of slip along a crystallographic plane (see **Crystal**) gives rise to the small magnitudes of these stresses which are 10^3 to 10^4 times smaller than the theoretical shear strength of the same crystals. This discrepancy between theoretical and observed shear strengths of single crystals was responsible for the original postulation of dislocations.

Most determinations of the critical resolved shear stress are made using crystals deformed in tension. In this case, the Schmid resolved shear stress equation

$$S_s = S_n \cos \phi \cos \lambda$$

TABLE 1. CRITICAL RESOLVED SHEAR STRESSES FOR FACE-CENTERED CUBIC METALS

Metal	Purity	Slip System	Critical Resolved Shear Stress pounds/square inch
Cu	99.999	{111} ⟨110⟩	92
Ag	99.999	{111} ⟨110⟩	54
Au	99.99	{111} ⟨110⟩	132
Al	99.996	{111} ⟨110⟩	148

TABLE 2. CRITICAL RESOLVED SHEAR STRESS FOR BASAL SLIP

Metal	Purity	Slip System	Critical Resolved Shear Stress pounds/square inch
Zinc	99.999	{0001} ⟨11$\bar{2}$0⟩	26
Cadmium	99.996	{0001} ⟨11$\bar{2}$0⟩	82
Magnesium	99.95	{0001} ⟨11$\bar{2}$0⟩	63

may be used, where S_s is the shear stress resolved on the slip plane and in the slip direction, S_n is the applied tensile stress, and ϕ and λ are the angles between the crystal tensile axis and the slip plane normal and slip direction, respectively. In a given set of previously undeformed crystals of the same composition, tested in tension at the same temperature that slip on the same crystallographic plane, it has been observed that the shear stress, determined with the aid of the above equation, is remarkably constant. It is thus apparent that the critical resolved shear stress is independent of crystal orientation. It does depend on the temperature and increases with decreasing temperature, as may be seen in part (a) of the figure below. The critical resolved shear stress also depends on the composition of the crystal and decreases with increased metal purity, as may be seen in the second figure for the case of silver and copper crystals (part (b) of figure below). Any plastic deformation, because of work hardening effects, will increase the critical resolved shear stress.

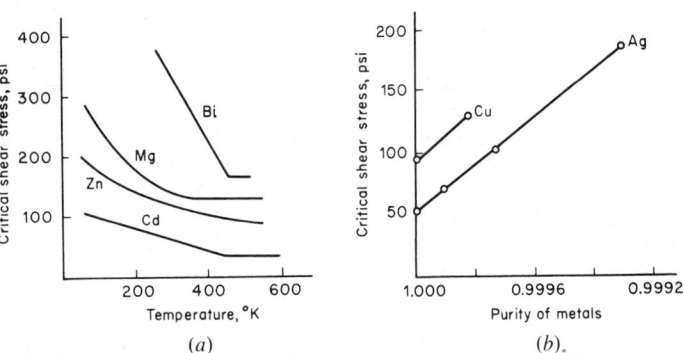

(a) Effect of temperature on the critical shear stress. NOTE: The date on which these curves are based predate that of Table 2. The higher critical stresses in this case correspond to crystals of lower purity; (b) variation of the critical resolved shear stress with purity of the metal.

CRITICAL SOLUTION TEMPERATURE. For two partially miscible liquids, the composition of the two conjugate solutions approach each other with increasing temperature. At the critical solution temperature the two solutions have identical compositions and form one layer.

CRITICAL SPEED. Any body having weight and "springiness" has one or more natural frequencies at which it can vibrate. If forces are applied to the body at any of these natural frequencies a resonance condition will build up, and the amplitude of motion will become quite large. Since a machinery shaft has weight and deflects under load, it has several natural frequencies. It is difficult to balance a shaft perfectly, so that during rotation centrifugal forces are set up within it that act on the shaft at the same frequency as the shaft rotates. If this impressed frequency coincides with a natural frequency of the shaft resonance occurs. This speed of rotation is then known as a critical speed.

Resonance is an undesirable condition as it may cause rubbing of parts and creates high shaft stresses. Consequently the running speed is kept at least 20% away from the critical speed. Since the shaft has a number of natural frequencies it also has a number of critical speeds. For most actual machinery shafts only the first or second critical speed is of importance.

CRITICAL TEMPERATURE. 1. This term is most commonly used to denote the maximum temperature at which a gas (or vapor) may be liquefied by application of pressure alone. Above this temperature the substance exists only as a gas. 2. The critical temperature of a superconducting transition takes place in zero magnetic field.

CRITICAL VELOCITY OF FLOW. If above a certain velocity of flow the nature of the flow changes qualitatively, the velocity is critical for the particular flow system. The criterion is always better expressed as a critical Reynolds number, Mach number, Froude number, or whatever the appropriate nondimensional parameter may be. See also **Turbulent Flow.**

CRITICAL VOLUME. The volume occupied by unit mass, commonly one mole, of a substance at its critical temperature and critical pressure.

CROAKERS (*Osteichthyes*). Of the order *Percomorphi*, suborder *Percoidea*, and family *Sciaenidae*, croakers are named because of the sounds which they produce. Sound is caused voluntarily by strong muscles which are caused to vibrate much as a string instrument. The air bladder acts as a resonating chamber. Although not fully understood, it is believed that because of the correlation of this noise-making with the spawning season and because it also varies from day to night, the croaking is associated with the mating process. There are somewhat over 150 species in the croaker family. Most species prefer shallow water, are usually carnivorous, and habituate both tropical and temperate waters. Several species can adapt to brackish waters. The American freshwater drum (*Aplodinotus grunniens*) is found all the way from Guatemala to Canada. It does not return to the sea. Of considerable interest was the finding some years ago that several species of sciaenids would survive in the salty inland Salton Sea of California. These fishes were introduced from the Gulf of California.

The Atlantic croaker (*Micropogon undulatus*) normally ranges in size from 1 to 4 pounds (0.5 to 1.8 kilograms). It is found from Argentina northward to Massachusetts. The channel bass or redfish (*Sciaenops ocellata*) is not a bass, but is associated with the croaker family. This is quite a large fish, attaining a length of over 50 inches (1.2 meters) and a weight of over 80 pounds (36 kilograms). It is found in the Gulf of Mexico and from Florida northward to Massachusetts. The California white sea bass of the genus *Cynoscion* (*C. nobilis*) and the Atlantic weakfish or common sea trout are also so associated. Occurring in the Gulf of California and weighing up to 225 pounds (102 kilograms), the totuava (*C. macdonaldi*) is the largest member of the family. The spot (*Leiostomus xanthurus*) is found from Texas in the Gulf all the way around to Cape Cod.

Croakers of the genus *Nibea* are well known for their assemblage in schools of millions of fish and for the synchrony of their drumming noises. They are found in Japanese waters. A very few croakers make no noise because they do not have air bladders. These include a number of Atlantic species—the king whiting (*Menticirrhus saxatilis*), the gulf minkfish (*M. focaliger*), as well as the California corbina (*M. undulatus*).

CROCIDOLITE (Blue Asbestos). The mineral crocidolite may be considered as a fibrous variety of the monoclinic amphibole, riebeckite. It is also known as a massive mineral. Its hardness is 4; specific gravity, 3.2–3.3; luster, silky to dull; color, blue or bluish-green. It is found in Austria, France, Bolivia, the Republic of South Africa (the variety known as tiger's-eye); and in the United States, in Massachusetts and Rhode Island. The name crocidolite is derived from the Greek, meaning woof, in reference to its fibrous appearance. See also **Cat's-Eye.**

CROCODILES AND ALLIGATORS. Of the class *Reptilia* (reptiles), subclass *Anapsida*, order *Crocodylia* (crocodiles), and suborder *Eusuchia*, according to classification of Grzimek (1974). Crocodiles are of the family *Crocodilidae*; alligators of the family *Alligatoridae*. The gavials (*Gavialidae*) are closely associated with crocodiles and alligators.

The *Crocodylia*, a uniform group of unmistakable animals, includes the largest modern reptiles. Some of them are said to live for a century, although such an advanced age has not been proved. An American alligator is reported to have lived for 56 years in a zoo; a Nile crocodile has been reported living up to 45 years in captivity. Since crocodilians become much larger in the wild than when confined, it is possible that free-living animals can reach a correspondingly greater age. Crocodilians are characterized by a lizardlike shape and an armored skin, which covers the whole body with large, strong, and partially ossified horny plates.

Apart from one species found only in brackish and seawater, crocodilians live near fresh water shores in the warmer regions of the earth. Although they move about most efficiently in the water, they use various gaits on land—sliding slowly on the belly, stepping along with the legs extended so that the belly does not touch the ground, and even galloping for short distances by moving the fore and hind legs, much as a jackrabbit does when it hops. The forelegs end in five fingers, separated down to the base, while the four toes which terminate the hindlegs are more or less completely connected by webbing, with the three inner toes bearing strong claws. Despite the fact that they are webbed, the feet are not the basis of locomotion in water; rather, the animals swim entirely by serpentine movements of the body or mighty strokes of the laterally flattened, oarlike tail. There is a characteristic comb or crest of tall scales on the tail, double near the base and single from the middle of the tail to its tip. Crocodilians can float in water with the body and tail angled downward, leaving only the nostrils, eyes, and ears above the surface; they evidently do this by suitably distributing the air in the lungs.

Crocodilians are of large to very large size, with lengths up to 7 meters (over 20 feet) or more. In some species, the males grow larger than the females. The skin on the head is firmly fused to the skull; on the neck there are groups of large, sharply ridged bosses, whose number and arrangement can be used in the identification of certain species. The back is covered by thick, rectangular horny plates, partially ossified on the underside. The plates on the ventral surface are smaller, and only among the alligators and in a few crocodiles are they completely ossified. Where dermal ossifications do exist, they are flexibly connected to one another. There are ribs on all the trunk vertebrae, the sacral vertebrae, the first 5 to 10 rail vertebrae, and, in addition and in contrast with all other modern reptiles, even on the two cervical vertebrae. In the abdominal region, there are not only 8 pairs of true ribs, but also 7 or 8 pairs of abdominal ribs or gastralia, lying free in the musculature.

The teeth are situated in deep hollows (alveoli) in the jaws, and usually are rather variable in form and size. The teeth serve only for seizing and holding the prey, not for chewing. Since the thick tongue is firmly fused to the floor of the mouth, it can barely be moved. A long esophagus leads to the rounded, muscular stomach, which in contrast with most other reptiles is distinctly divided into an anterior and a posterior section. Young crocodilians feed mainly on insects, worms, and the smallest fish. As they grow larger, they tend to prefer fish and turtles. Older crocodilians also eat birds and small mammals and frequently larger decayed mammals. Feeding occurs only in the water, but not underwater. After rapid, thorough digestion, the remnants appear as uniformly shaped droppings. The indigestible parts of the prey, such as feathers, are spit out in a mass.

The nostrils lie on the raised tip of the nose; they can be closed by folds of skin. A long nasal passage leads from the nostrils to the choanae, which open far back on the palate. These also can be closed off by a flap of skin in the mouth. Because of this arrangement, crocodilians can lurk under water with their mouths open and still breathe, as long as the nostrils are above the water surface. Since the heart is almost completely divided into four chambers, oxygenated blood and oxygen-poor blood do not mix within it; they do mingle, however, at a perforation near the base of the aortic arches. The arrangement is such that the pulmonary artery receives deoxygenated blood; the right systemic arch (which supplies the head) receives oxygenated blood; the left systemic arch receives mixed blood. This pattern of circulation may well be related to their ability to remain submerged for long periods. Small crocodilians have been known to stay underwater for 45 minutes and the larger ones for over an hour, without breathing.

The eyes have both an upper and lower lid, as well as a semitransparent nictitating membrane which can be drawn over the front of the eye from the inner corner. In contrast with all other reptiles, crocodilians have an external ear—a fold of skin which can be closed over the eardrum.

All crocodilians reproduce by means of white, hard-shelled eggs of about the same size as chicken or goose eggs. The eggs are porous on the surface and weigh between 40 and 90 grams. The females care for the young, occasionally in quite an elaborate manner. When the young are ready to hatch, they break the egg with the egg caruncle, a horny protuberance on the tip of the snout. At first, their voices are squeaky, but as the animals grows there is a transition to a dull roar.

Most scientists divide present-day crocodilians into three families: (1) alligators (*Alligatoridae*), in which the fourth tooth of the lower jaw fits into a laterally closed pit in the upper jaw, while the fourth tooth of the upper jaw is best developed. See Fig. 1. (2) Crocodiles (*Crocodylidae*), in which the fourth tooth of the lower jaw is laid into a notch in the upper jaw, open at the side, so that it is visible when the mouth is closed, while the largest tooth of the upper jaw is the fifth. See Fig. 2. (3) Gavials (*Gavialidae*), a single species in which the snout is greatly lengthened and the teeth are all of the same size and shape.

Alligators. The alligators are New World animals, with the exception of the *Chinese alligator* (*Alligator sinensis*), which achieves a length of about 2 meters (6.4 feet) and occurs in the lower reaches of the Yangtze River. The best known species is the *American alligator* (*A. mississippiensis*), which attains a length up to 6 meters (nearly 20 feet). This species occurs in the southeastern United States. The fingers of the American alligator, as well as the toes, are joined at the base by webbing. For many years, these animals were intensively hunted for their skins, which found many useful purposes. Regulations have been put in place to protect the alligator. Because of past hunting, it became difficult to find a specimen over 3 meters (9.8 feet) long.

Caimans of the alligatorid genus *Caiman* are found from Central America to the central part of South America. They live in the backwaters of rivers or in very slowly flowing waters with muddy bottoms and soft sand banks. There is a ridge on the head running between the eyes like the bridge of a pair of glasses. The bony plates (osteoderms) of the belly armor are particularly well developed. The caimans are dark olive in color. There are two principal species—the *spectacled caiman* (*C. crocodilus*), which achieves a length of about 2.5 meters

Fig. 1. Alligator. (*A. M. Winchester.*)

Fig. 2. Crocodile. (*A. M. Winchester.*)

(8.2 feet) and the *broadnosed caiman* (*C. latirostris*), which is just slightly smaller than the spectacled caiman. A third species is the *black caiman* (*Melanosuchus niger*). The basic color of the black caiman is black, but young animals have yellowish spots and stripes. This species lives in central South America. Since this species, the largest of the caimans, preys on quite large animals that are valued by human hunters, efforts are made to minimize the population of the black caiman. The black caiman attains a length up to 4.7 meters (15.4 feet).

Crocodiles. These animals are found in tropical regions all over the world. The iris is greenish or yellowish; on either side of the upper jaw there are no more than 19 teeth. The majority of crocodilians belong to the genus *Crocodylus*, with 11 species.

The *Orinoco crocodile* (*C. intermedius*) attains a length of some 7.2 meters (23.6 feet) and occurs in the regions of the Orinoco and Amazon rivers in South America. The *American crocodile* (*C. acutus*) is about the same length. The range is large, extending from southern Florida across Central America into northwestern South America and the Antilles. *Morelet's crocodile* (*C. moreleti*) is considerably smaller, attaining a length of about 2.5 meters (8.2 feet). This species occurs in Mexico, Honduras, and Guatemala and is similar to the American crocodile, but is distinguished by a rounded protuberance in front of the eyes. Additionally, there are the *Australian crocodile* (*C. johnsoni*) with a length of about 3 meters (9.8 feet), and the *New Guinean crocodile* (*C. novaeguineae*) with a comparable length and which lives in New Guinea, the Sulu Archipelago, and other Philippine Islands. This species is noted for its long snout. The *salt-water crocodile* (*C. porosus*) attains a length over 7 meters (22.9 feet) and, like the American crocodile, can swim into the open ocean and thus has extended its range over a very large area. It is found from southern India across the Sunda Islands, the Philippines, the Moluccas, New Guinea, the Solomons, and the New Hebrides, as far as northern Australia. The salt-water crocodile lives primarily in coastal areas, both in the sea and in brackish water. Occasionally, animals driven out to sea by the wind make astonishingly long voyages. It has been reported that a salt-water crocodile arrived at the Cocos Islands in the Indian Ocean, after having traveled at least 1100 kilometers (684 miles). This species, like other species of crocodile that consume foods with a high-salt concentration, must be able to dispose of excess salt. This is done by way of the lacrimal glands and the glands associated with the nictitating membranes, which in these crocodiles correspond in both structure and function to the salt glands of sea birds. An identifying characteristic of the salt-water crocodile is the double row of bosses forming ridges on the upper surface of the snout.

Nile Crocodile. This animal (*C. niloticus*) attains a length up to 7 meters (about 23 feet) and once inhabited all of Africa. Since it, like the salt-water crocodile, does not hesitate to go out to sea, it was able to colonize various islands off the coast, and its range includes Madagascar. In the early 1900s, the Nile crocodile was exterminated in what was then called Palestine, and it is now no longer to be found below the second cataract of the Nile, i.e., in Egypt.

Only in a few parts of Africa—for example, at Murchison Falls in Uganda—can one now be certain of an opportunity to observe many large Nile crocodiles at relatively close range. They were so numerous earlier in this century in various parts of Africa, such as Tanzania, that bounties were offered for killing them. Humans have hunted crocodiles because the crocodiles attack domestic animals and occasionally humans, but for a long time this activity had little effect upon the crocodile population. In later years, the value of the skins greatly increased hunting. In recent years, regulations have been installed by some countries in an effort to protect this species.

Crocodilians are important elements of the biota where they live. Usually the large crocodilians spend the night in the water and lie in the sun on land for most of the day; only at midday do they withdraw to the shade or cool themselves briefly in the water. With the onset of darkness, they leave the shore again. Although they are poikilothermic (cold-blooded) animals, they manage in this way to keep the body temperature relative constant—at an average value of 25.6°C, with brief excursions of a few degrees up or down. On land, in the midday heat, crocodilians tend to lie with the mouth wide open. In the absence of sweat glands on the body, this method permits water evaporation from the mucous membranes of the mouth—something like the panting of a dog. Crocodilians are seldom if ever encountered in open water of lakes.

The mouth-opening behavior of crocodiles was mentioned by the Greek historian Herodotus (490–424 B.C.) who reported that a bird, the trochilus, slips between the jaws and picks off leeches there. This has not been proved, but there is an association between certain birds and the crocodiles. The common sandpiper (*Tringa hypoleucos*), which breeds in Europe and spends the winter among the crocodiles, picks off the parasites from their bodies and even runs to meet the crocodiles when they come out of the water. Most modern zoologists do not believe that birds actually go into the open mouths and clean the teeth.

Fully grown crocodiles in South Africa's Kruger National Park kill primarily antelopes, such as the impala, bushbuck, and waterbuck, but they also attack other animals, such as giraffes, buffalo, and young hippopotamuses, wild dogs, porcupines, and lions. Their main food in Kruger Park consists of turtles. It should be mentioned that crocodiles have killed more humans than all of the predatory mammals and poisonous snakes found in Kruger Park. Often crocodiles will consume the remains of dead animals. The teeth of the crocodiles are not suited for tearing apart or chewing up large prey. In the case of a hippo or buffalo which has recently died, the crocodiles can at first only bite off the ears and the tail, for the skin is too firm. For this reason, they often push dead animals into underwater hollows along the banks, so that the skin softens and decay begins.

Several species of crocodile make provisions for their young. Incubation of eggs is not uncommon.

Initially, crocodiles grow rapidly. In the first 7 years, the Nile crocodile increases in length by an average of 26.5 centimeters (10.4 inches) per year. Growth slows as the animal ages. The male animals become sexually mature when at a length of 2.9 to 3.3 meters (9.5 to 10.8 feet). Females do not lay eggs until they are 8 to 12 years old.

Gavialidae. Only one genus of this family of crocodilians remains, the *Indian gavial* (*Gavialis gangeticus*), which attains a length of about 7 meters (22.9 feet). The origin of this species is perhaps to be found among the long-snouted crocodiles of the Cretaceous. The snout is about 3.5 times as long as it is broad at the base.

The Indian gavial is the crocodilian most strictly limited to life in the water. Its legs are quite weak, whereas the oarlike tail is particularly powerful. The gavial lives in the deep, flowing waters in the regions of the Ganges, Mahanadi, and Brahmaptura Rivers of India, as well as in the Koladan River at the mouth of the Maingtha in southeastern Asia. The females, like all crocodilians, lay their eggs on the land, usually on sandbanks.

References

Buffetaut, E.: "The Evolution of the Crocodilians," *Sci. Amer.*, **241**, 4, 130–144 (1979).

Gadow, H.: "Amphibia and Reptiles," Macmillan, London, 1901. A classic reference.

Goodrich, E. S.: "Studies on the Structure and Development of Vertebrates," 2 volumes, Dover, New York, 1958.

Gore, R.: "A Bad Time to be a Crocodile," *National Geographic*, **153**, 1, 91–114 (1978).

Hotton, N., III: "Reptilia," in "The Encyclopedia of the Biological Sciences," (P. Gray, editor), Van Nostrand Reinhold, New York, 1970.

Mertens, R.: "Crocodlia," in "The Encyclopedia of the Biological Sciences," (P. Gray, editor), Van Nostrand Reinhold, New York, 1970.

Romer, A. S.: "Osteology of Reptiles," Univ. of Chicago Press, Chicago, Illinois, 1956.

Scherpner, C.: "The Crocodiles and Alligators," in "Grzimek's Animal Life Encyclopedia," Vol. 6, Van Nostrand Reinhold, New York, 1974.

Wermuth, H.: "Systematik der Rezenten Krokodile," *Mitt. zool. Mus. Berlin*, **29**, 375–514 (1953).

CROCODILIA. An order of large reptiles of long slender build. The skin is armed with bony plates and the long jaws bear many conical teeth set firmly in bony sockets. The animals are chiefly aquatic. They are found only in warmer regions and chiefly in fresh water, although some enter the ocean.

The order includes crocodiles, alligators, caimans, and the gavial.

CROCOITE. The mineral crocoite, lead chromate, corresponds to the formula $PbCrO_4$, and forms prismatic monoclinic crystals, often acicular. It is also found in columnar or granular masses. It has a rather distinct cleavage parallel to the prism, and a less distinct cleavage parallel to the base. It has a conchoidal fracture, is sectile; hardness, 2.5–3; specific gravity, 5.9–6.1; luster, adamantine to vitreous; color, red; streak, orange-yellow, translucent. Crocoite is a secondary mineral believed to be formed by waters containing chromic acid acting upon lead minerals like galena, with which it is associated. It is found in the U.S.S.R., Rumania, Tasmania, Brazil, the Philippines, and Arizona. It is not of commercial importance. The name crocoite is derived from the Greek word for saffron in reference to the color of the powdered mineral.

CROHN'S DISEASE. **Colitis and Other Inflammatory Bowel Diseases.**

CROMWELL CURRENT. A dense ocean current running beneath and in the opposite direction of the South Equatorial Current. It was discovered at a point 150° West and has been traced along the Equator for nearly 3,500 miles (5,600 kilometers), disappearing at the Galapagos Islands. Approximately 700 feet (210 meters) thick and 250 miles (400 kilometers) wide, it has a velocity of approximately 3.5 miles (5.6 kilometers) per hour.

CROOKES TUBE. Sir William Crookes was a pioneer in the study of electric discharge in gases. In the vacuum tubes which he used, and certain forms of which still bear his name, the pressure was reduced to such a point that the bright glow observed at higher pressures practically disappeared. The cathode rays, obstructed by but little residual gas, shot straight across the tube and, impinging upon the opposite wall, caused it to glow with greenish fluorescence. By placing an obstacle, such as a metal plate shaped like a Maltese cross, in the path of the rays, he was able to demonstrate their rectilinear character by the shadow on the fluorescing surface. In one type of tube he interposed a light paddle-wheel of metallic vanes in the path of the rays, and found it driven at high speed as if by a stream of air. It is, however, probable that the greater part of this effect is due to the heating of the bombarded surfaces and that the paddle-wheel really operates as a Crookes radiometer.

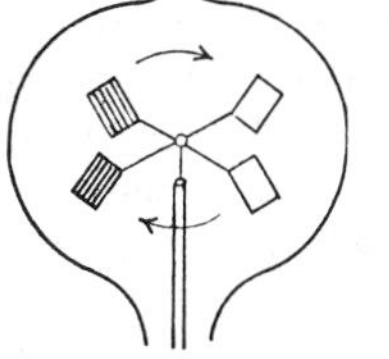

Four-vane radiometer in bulb.

CROP. A thin-walled expanded portion of the alimentary tract (see **Digestive System (Other Life Forms)**) used for the storage of food prior to digestion. Crops are found in many animals, including earthworms, insects, and birds.

CROP (Bird). Poultry.

CROP SURVEILLANCE. Earth Resources Satellites and Geologic Remote Sensors.

CROSS. Plant Breeding.

CROSSBAR SWITCHES. These are matrix-switching devices used to connect any one of a group of inputs to any one of a group of outputs. See accompanying illustration. This is done by energizing two magnets; one selects an input row, and the second selects an intersecting output column. By this action, a set of contacts is closed, connecting input to output. While these switches are used principally in telephone switching, they are adaptable for many testing and control functions. They do the same tasks as those done by rotary switches or step-by-step switches, but their advantage lies in being able to select any input-to-output connection with only two magnet operations.

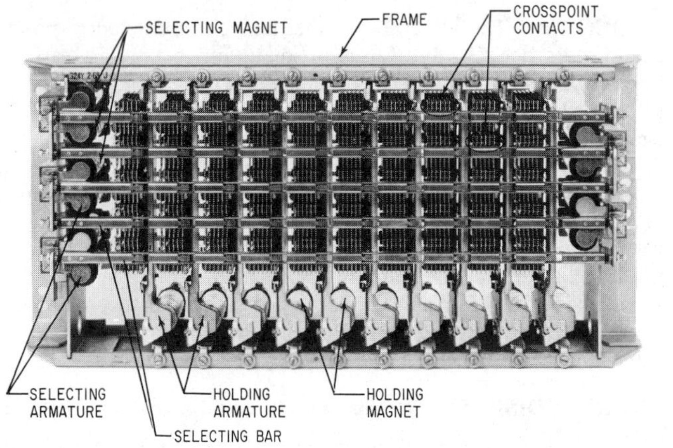

Representative crossbar switch. This particular switch does not fully reflect the effects of miniaturization of this kind of hardware.

Most present crossbar switches operate by partial rotation of two sets of bars mounted at right angles. A selecting bar, in the first set, moves an interposing device, called a selecting finger, between a holding bar in the second set and a group of contact springs. Energizing the holding magnet causes the holding bar to press against the contact springs through the selecting finger. Thus, only the intersection of the two bars will have a group of crosspoint contacts closed for any pair of magnet operations.

Some of the variations on this method of operation are:

(a) The use of draw bars instead of rotating bars.
(b) Mechanical latching to hold the contacts closed without continuous power.
(c) Use of permanent magnets for magnetic latching.

Note that (b) or (c) will require additional electrical operations to release the contacts once they are closed. Even in switches which do not have latching, the selecting magnets are only used while the connection is being set up. Therefore, these magnets are free for use in setting up other connections. The number of simultaneous connections will normally be limited by the number of holding magnets.

In addition to the crosspoint contacts, other contact groups (off-normal contacts) may be supplied which operate whenever a selecting or holding bar is in use.

Crossbar switch operation implies that one side of the contacts in an input row will be electrically in parallel with the similar contacts in that row. Likewise the other side of the contacts will be paralleled along an output column. While this limits the number of crosspoints on a switch that may be used at one time, it allows all inputs to reach all outputs. Paralleling these sets of contacts may be done by either internal or external wiring.

Crosspoints are generally groups of from three to eight normally open contacts. Usual contact metals are palladium and silver with the former preferred for low level or "dry" switching. Off normal contacts vary widely with switch design.

Crossbar switches are made in a variety of matrix sizes in the United States and wider variety abroad. In the United States, most switches are made with ten selecting levels and either 10 or 20 holding levels. Other matrix sizes include 12 × 15 and 12 × 10 arrays. Switches made in some other countries include the foregoing as well as additional configurations, such as 10 × 15, 10 × 52, 9 × 49, 28 × 17, and 30 × 14. Design trends in crossbar switches include the use of wire contact springs, miniaturization, application of the "card release" principle with pretensioned springs, different array sizes, and new methods of latching.

Some of the crossbar switches in use in some systems today have been reduced to one-quarter the size of conventional switches just a few years ago.

CROSSBEDDING. Oblique lamination of certain beds in aeolian or water-laid sediments, caused by current action, is called crossbedding. Crossbedded sediments are found especially in river and stream deltas, alluvial fans and cones, river sand bars and marine sand deposits, and are also characteristic of windblown deposits of all kinds. The different types of crossbedding are useful criteria for helping to determine the physical (including climatic) conditions under which certain types of clastic sediments were deposited. In regions where the formations have been highly deformed, and possibly overturned, crossbedding may also be used by the stratigrapher and structural geologist to determine the original order in which sedimentary strata were laid down.

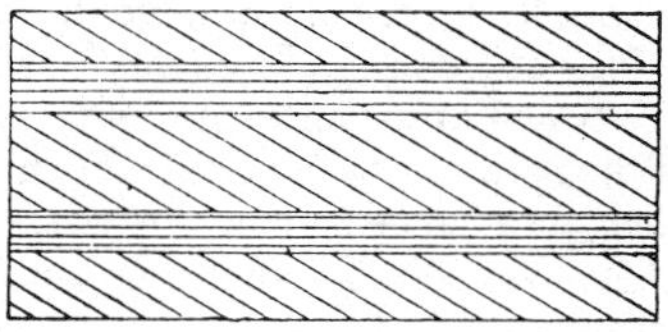

Crossbedding (false bedding). (*"Field Laboratory Manual,"* *Princeton Univ. Press.*)

CROSSCORRELATION DETECTION. A method of detection in which a signal is compared, point-to-point, with an internally generated reference. The output of such a detector is a measure of the degree of similarity of the input and reference signals. The reference signal is constructed in such a way that it is at all times a prediction or best estimate of what the input signal should be at that time.

CROSS EYES. Vision and the Eye.

CROSSLINKING. Rubber (Natural).

CROSS-MATCHING. Blood.

CROSS MODULATION. This is an effect produced in radio receivers which results in the modulation from one carrier being impressed on another carrier. If two modulated carriers are applied to the input of a vacuum tube or transistor which has appreciable curvature in its operating characteristic, there will be numerous frequency components in the output of the tube. However, following selective or tuning circuits will discriminate against many of these but there will be certain ones which will not be eliminated. Some of these represent the sidebands of the modulation of the undesired signal modulated on the desired carrier. Since they occupy the same frequency band as the desired carrier and its normal modulation there is no way in which they may be filtered out, and hence upon detection the audio output will contain both the desired and the undesired audio signals.

CROSS-POLLINATION. Pollination.

CROSS-REACTIVITY (Antigens). Immune System and Immunology.

CROSS SECTION. From its general meaning of a section at right angles to an axis, the term cross section has been extended to mean a measure of the probability of a particular process. It is expressed in units of area, although it is not usually identical with the geometric cross section across which the process occurs. For a collision reaction between nuclear or atomic particles or systems, the cross section is an area such that the number of reactions occurring is equal to the product of the number of target particles or systems multiplied by the number of incident particles which would pass through this area at normal incidence. If n_t is the number of target nuclei or other particles per unit volume (cm^{-3}) of a substance exposed to an incident beam consisting of n_0 particles per unit area and unit time (cm^{-2} sec^{-1}), then the interaction cross section $\sigma = N/n_0 n_t$, where N is the number of reactions of a specified type per unit volume and unit time (cm^{-3} sec^{-1}). The *macroscopic cross section* $\Sigma = \sigma n_t$ is the cross section per unit volume for the process under consideration. Nuclear cross sections include a cross section for fission, a cross section for capture, and a cross section for scattering both elastic and inelastic. Atomic cross sections include the cross section for Compton collision and the cross section for ionization by electron impact. The *total cross section* is the sum of the separate cross sections by which a particle can be removed from a beam. In nuclear processes the customary unit of cross section is the barn.

CROSS-STAFF. In the period prior to the invention of the sextant in the eighteenth century, both the cross-staff and the astrolabe were used by navigators for the purpose of measuring altitude of celestial objects. The astrolabe was the more compact of the two, but the cross-staff was simpler to use, and the results were slightly more accurate, particularly for measuring small altitudes.

The principle and use of the instrument are illustrated in the accompanying figure. A "cross," with a peep sight in its upper end, slides along a rod *EB*. Holding the cross in a vertical position, the observer sights from *E* along the rod toward the horizon at *H*, and slides the cross along until the object under observation appears through the peep sight in the cross. The graduations on the rod indicate directly the value of the angle *HES*, which is the desired apparent altitude. The instrument may also be used to measure the angular distance between any two objects.

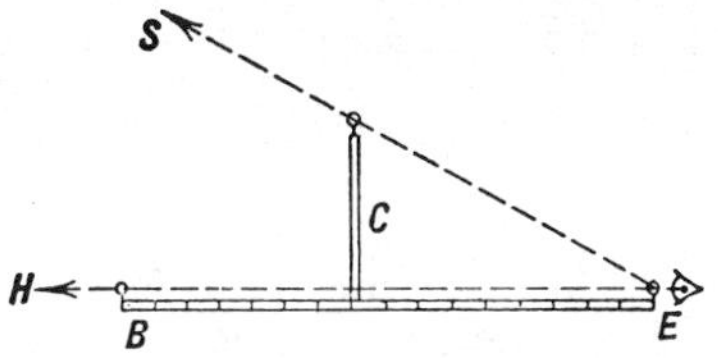

Cross-staff shown schematically.

Within recent years, the instrument has been used to some extent for teaching purposes in elementary courses in astronomy.

See also **Astrolabe; Navigation;** and **Sextant.**

CROSSTALK. This is the undesirable operating condition of a communication system where the electrical effects of one circuit produce an undesired current in another. In the case of a phantom telephone circuit, crosstalk will be troublesome unless the lines with loading coils, terminal equipment, etc., are perfectly balanced electrically. Considerable care must be taken in the manufacture of the loading units in order to minimize crosstalk. See also **Inductive Interference.**

CROTON. Euphorbiaceae.

CROTON BUG (*Insecta, Orthoptera*). A European cockroach, one of the two chief pests of this kind in the United States. Its name is said to be due to its association with water pipes from the Croton reservoir in New York City.

CROUP. Epiglottitis; Larynx.

CROW (*Aves, Passeriformes*). Large birds of the northern hemisphere and Africa. They are black or black and white in color. Together with the ravens, rooks, magpies, jays, and other species they make up the crow family (*Corvidae*).

The importance of the common crow of North America, *Corvus brachyrhynchus*, has been debated. The bird does some good in destroying vermin and insects, but the farmer and conservationist are both against it for its destruction of fruit and grain and the eggs and young of other birds.

Crow.

The crow is omnivorous and gregarious. It is remarkably intelligent as compared with most other birds. Its plumage is soft, dense, and dark. The common North American crow measures about 20 inches (51 centimeters) in length. The tail is composed of 12 feathers. See accompanying sketch. The nest is usually located high in tree tops and is often repaired and used a second or third year. The egg is blue-green with dark markings and of obtuse shape. There are usually five eggs. The female incubates the eggs, but the male assists in feeding the young. The crow lives for many years. These birds prefer large communal rookeries, sometimes inhabited by several tens of thousands of birds. The birds may commute from 30 to 40 miles (48 to 64 kilometers) daily in each direction from the rookery to other points. It is difficult for naturalists to study crows because they are very difficult to approach, using an effective system of sentries to warn of approaching danger. The crow's thieving characteristics are well established.

The fish crow (*C. ossifragus*) is a somewhat smaller bird and is found along tidewater areas. Whereas the common crow eats a variety of foods, from carrion to insects, seeds, fruits, reptilian eggs, and even sometimes small reptiles, the fish crow has a diet of shellfish.

The Jackdaw (*C. monedula*) is a small, gregarious European crow, typically with a gray collar, but also occurring in a wholly black form. These birds are common in Zurich. Both sexes build the nest. Another species of daw occurs in Asia and flocks are usually seen around the temples of the Chalimar Gardens of Kashmir. The term *daw* also is sometimes used to refer to a great-tailed grackle found in the southwestern United States.

CROWN OF THORNS. Euphorbiaceae.

CRT DATA INPUTS/OUTPUTS. Input/Output Devices (Computing System).

CRUCIBLE. A crucible is a vessel made of heat-resistant material and employed to hold a material that in itself is at high temperature, or is to be subjected to high temperature. Crucibles will range in size from the small laboratory types to large ones having a capacity of several tons of molten metal. They are roughly cup- or barrel-shaped, and are made of some material such as clay. In the laboratory, platinum, iron, and porcelain crucibles are used. In the iron and steel industry, clay crucibles have been used, but at the present time most manufacturers employ the graphite crucible, which is made half and half from graphite and fire-clay, well-mixed, molded, and burned to vitrification. The crucible may be used to hold a material being melted or burned, as in some processes for making steel, where the raw material is put in the crucible, which is then set in a hot furnace until the contents are melted. Or a crucible may be used to receive molten metal, which has been produced elsewhere, as from a brass furnace or cupola, in which case it is the means for conveying it from the point of melting to the point of casting, serving thus as the intermediate reservoir between the furnace and the mold. See **Casting.**

CRUCIFERAE. Brassica.

CRUDE OIL. Petroleum.

CRUDE OIL (SNG). Substitute Natural Gas (SNG).

CRUISE CONTROL. Automotive Electronics.

CRUNODE. A point on a curve through which there are two branches of the curve with distinct tangents.

See also **Tangent (Geometry).**

CRUSTACEA. The lobsters, crabs, barnacles, shrimps, and many other species, constituting a class of the phylum *Arthropoda*. A large majority of the 25,000 known species are aquatic.

The Crustaceans are distinguished from other classes of the phylum by the following characteristics: (1) The body is divided into cephalothorax and abdomen. (2) The eyes are compound. (3) Two pairs of antennae are present. (4) Jointed appendages are found on the abdomen on many species. (5) Respiration is usually accomplished by gills.

The principal economic importance of crustaceans is due to the value of some species as food. Lobsters, shrimps, and some of the crabs are caught in large numbers for the market and are regarded as delicacies. Although formerly plentiful, the lobster has been caught in such large numbers on the New England coast of the United States that it now has to be protected by law and brings high prices. The smaller crustaceans are imporant as food for fishes.

Barnacles have long been a nuisance for their part in the fouling of ship bottoms.

The following indicates the complexity of classification of the crustaceans:

Subclass I. *Branchiopoda*. Varying number of body segments with appendages of uniform character, generally foliaceous; abdomen devoid of appendages; carapace usually present.
- Order *Anostraca*. Carapace not developed; eyes stalked. Fairy shrimp.
- Order *Notostraca*. Large dorsal shield-shaped carapace; eyes sessile. *Apus*.
- Order *Conchostraca*. Carapace divided into two valves enclosing entire animal. *Cyzicus*.
- Order *Cladocera*. Bivalved carapace enclosing trunk but not head. Water fleas.

Subclass II. *Ostracoda*. Bivalve carapace, not more than four pairs of appendages on the trunk. *Cypris*.

Sublcass III. *Copepoda*. No carapace, usually five pairs of limbs. *Cyclops* and the parasitic fish-lice.

Subclass IV. *Branchiura*. Parasitic, compound eye and suctorial mouth. *Argulus*.

Subclass V. *Cirripedia*. Sessile as adults, usually six pairs of biramous cirriform appendages in body region; limbless rudimentary abdomen. Calcareous plates support carapace.
- Order *Thoracica*. Non-parasitic. *Lepas*, *Balanus*.
- Order *Acrothoracica*. Boring in shells of Molluscs, fewer than six pairs of trunk appendages. *Alcippe*.
- Order *Ascothoracica*. Parasitic, six pairs of trunk appendages; mouth appendages modified for piercing and sucking. *Petrarca*.
- Order *Apoda*. Parasitic, without mantle or trunk appendages. *Proteolepas*.
- Order *Rhizocephala*. Parasitic, body undergoes extreme degeneration. *Sacculina*.

Subclass VI. *Malacostraca*. Carapace normally present; many appendages on thorax and abdomen.
- Series I. *Leptostraca*. Abdomen with seven segments and telson. Thoracic appendages foliaceous, abdominal biramous. *Nebalia*.
- Series II. *Eumalacostraca*. Six abdominal segments and telson; thoracic appendages leg-like but seldom uniform.
  - Division 1. *Syncarida*. No carapace; first thoracic segment united with head or marked off by a groove.
    - Order *Anaspidacea*. Thoracic appendages with exopodites and lamellar epipodites (gills). *Anaspides*.
  - Division 2. *Peracarida*. Carapace when present leaves at least four of the thoracic segments free.
    - Order *Mysidacea*. Carapace extends over greater part of thorax but fuses dorsally with no more than three segments. *Mysis*.
    - Order *Cumacea*. Carapace coalesces with first three or four thoracic segments and encloses a branchial cavity on each side and forms a rostrum in front. *Diastylis*.
    - Order *Tanaidacea*. Carapace coalesces with first two segments and encloses a branchial cavity. *Apseudes*.
    - Order *Isopoda*. No carapace; oval and flattened; marine, freshwater and terrestrial forms. Pill bugs, woodlice, *Ligia* and others.
    - Order *Amphipoda*. Similar to preceding order but laterally compressed; second and third thoracic appendages nearly always prehensile organ. *Gammarus*.
  - Division 3. *Eucarida*. Carapace forms a cephalothorax.
    - Order *Euphausiacea*. Thoracic limbs do not form maxillipedes. *Euphausia*.
    - Order *Decapoda*. First three thoracic appendages modified as maxillipedes with branchiae usually in series. Prawns, shrimps, lobster, crayfish, hermit crab, crabs.
  - Division 4. *Hoplocarida*. Cephalothorax short, branchiae on abdominal segment.
    - Order *Stomatopoda*. *Squilla*.

CRUSTACEANS (Edible). The crustaceans have been known from early in the earth's history and are considered a highly successful group in their adaptation to changing environmental conditions over the centuries. With their approximately 35,000 species, the crustaceans contain four times as many species as are found among the birds. See also **Crustacea.** Only a relatively few of these species, however, are of importance as food substances—either as food for direct human consumption, or as food or bait for other sea animals, or for use as crop fertilizers. The crustaceans, making up the class *Crustacea* (phylum *Arthropoda*; subphylum *Diantennata*) originated in the sea, and the majority are still marine in character, although some have assumed freshwater and even terrestrial habitats. From a total worldwide marine catch standpoint, crustaceans make up only about 3% of the total, only about one-half of the quantity of mollusks harvested. Marine fishes comprise nearly 90% of the total catch.

The principal classes of crustaceans of interest as food are shrimp (about 62% of the crustacean catch); crabs (28%); lobsters; and crawfish and crayfish. The various edible crustaceans are of the order *Decapoda* (10 legs).

Crabs

The principal crab fisheries of the world are found along the Asian and North American coasts of the central and northern Pacific Ocean; along the shores of North America of the Atlantic Ocean; along the coasts of northern Spain and western France (Bay of Biscay), in the Atlantic south of Ireland, and in the North Sea. Smaller concentrations are found off Chile in South America; and off Brazil in the southwestern Atlantic.

The crab industry of the Pacific coast of North America extends from California northward to Alaska and essentially involves two species: (1) the king crab (*Paralithodes camtschatica*), and (2) the Dungeness crab (*Cancer magister*). Two species of lesser importance are the rock crab (*Cancer* spp.) and the tanner crab (*Chinoecetes* spp.). In Hawaii, a small fishery exists for the Kona crab (*Ranina ranina*). The king crab is confined essentially to Alaskan waters. The Dungeness crab is much more widely distributed and is sometimes referred to as the market crab in California.

The Dungeness crab inhabits shallow waters inshore and estuaries as well as offshore waters on sandy or mud bottoms up to 50 fathoms in depth. The abundance of this species has fluctuated significantly in some areas during the past few decades. Some authorities believe that these deviations arise from natural causes rather than overfishing. Fishery regulations are imposed by various states and generally provide for retention of male crabs above a size at which they have spawned at least once, but size limit is not uniform, ranging from about 16 to 18 centimeters (6.26 to 7 inches) across the carapace. Regulations usually prohibit the retention of female, soft-shell, and undersized male crabs, and provide for escape ports in the pots to enable the smaller females and undersized males to escape. A closed season is applied in some areas to protect the crabs during the molt and soft-shell stages.

In crabs, the oxygen-carrying pigment in the hemolymph is a copper hemocyanin rather than an iron heme compound as in the blood of higher animals. The copper heme pigment is relatively colorless in the hemolymph, but tends to develop an objectionable bluish color after the crab is processed. The bluish color may become especially noticeable in canned crab meat during storage. Because iron and copper tend to promote discoloration, crab meat should be handled and processed without exposure to metals or metal-containing compounds.

Dungeness crab meat is high in protein (18–20%), very low in fat (0.7–1.1%), and has about 90 calories of food energy per 100 grams. The natural sodium level is relatively high, ranging from 153 to 329 milligrams per 100 grams, as compared with an average level of 68 milligrams per 100 grams in marine fish.

King Crab. This species (*Paralithodes camtschatica*) is an 8-legged, spiderlike arthropod covered with spiny projections. Two degenerate legs especially modified for breeding are tucked under the rear carapace margin, thus qualifying it as a decapod. An adult male king crab measures up to 24 centimeters (9.5 inches) across the carapace. Females are considerably smaller. The meat is white with reddish covering and firm in texture. The subtle flavor of king-crab meat makes it popular in salads or as an entree. Meat content represents about 30% of body weight, with most meat being in the legs and shoulders.

King crabs are found in ocean waters from southeastern Alaska to Siberia. Commercial concentrations of king crab have been located in nearly all areas of Alaska, with fisheries conducted inside the continental shelf from southeastern Alaska westward to the Aleutian Islands, in the Bering Sea, and Cook Inlet. Asian fisheries are centered mainly in the Okhotsk Sea and western Kamchatka. Individual migrations of king crab have been recorded at a maximum of about 161 kilometers (100 miles). The norm, however, is dictated according to the habits of the species, which find it moving into shallow areas to spawn in late March and April and returning to deep-water trenches during the summer and early fall. Winter migrations are shoreward or toward offshore shallows where molting and breeding take place in the spring, thus completing the cycle.

A larval king crab is nearly microscopic, physically resembling a shrimp, and free-swimming. Within twelve weeks, characterized by heavy mortalities, the larva metamorphoses into a bottom-dwelling miniature king crab less than 3 millimeters (about $\frac{1}{10}$-inch) long. Young crabs band together in pods, consisting of thousands of individuals, apparently for protection against enemies. Podding, as a function of survival, continues until about the third or fourth years, when the carapaces of both sexes are about 9 centimeters (3.5 inches) long. In the fifth years, the crabs attain sexual maturity. The species may live up to 14 years of age. At maturity, a male king crab may weigh as much as 11.3 kilograms (25 pounds), have an overall leg span of 1.8 meters (6 feet), and contain about 1.8 kilograms (4 pounds) of recoverable meat.

Red Crab. The deepsea red crab (*Geryon Quinquedens*) occurs along the edge of the continental shelf from Nova Scotia to Cuba and specimens are found in the Gulf of Mexico and off the coat of Brazil. The crab is generally found where the water temperature is between 3.3 and 5°C (38–41°F). South of New England, the crab is rarely found in waters less than 170 fathoms deep and the larger concentrations are usually at a depth between 250 and 300 fathoms. Further south along the coast, the crab will stay deeper and specimens in the Gulf of Mexico and off Brazil are caught at about 700 fathoms.

The red crab is about twice the size of the blue crab and grows to a size of 1 kilogram (about 2.25 pounds). Females are more slender and grow to about 0.8 kilogram (1.25 pounds). The body of the crab is squarish and the walking legs are long and slender. On each side of the front edge of the carapace are five short spines or teeth, to which the scientific name *quinquedens* refers. This is not a swimming crab. The color is red or deep-orange. On the average, 24% of the meat is in the claws, 36% in the legs, and 40% in the body. The crab meat has the same color as lobster meat (white and pink), but it is difficult with handpicking to remove all of the shell. Red-crab meat lends itself to pasteurization. Taste panels often rate pasteurized red-crab meat over that of blue-crab meat.

Blue Crab. This species (*Callinectes sapidus*) has supported the oldest crab industry in the United States. Records indicate that blue crab was known in the Chesapeake Bay as early as the 1630s. Most current landings of blue crab are out of Florida, Maryland, North Carolina, and Virginia. Meat yield is about 12% of live weight. The meat is highly perishable because cooking cannot be relied upon to fully sterilize it, and the extensive use of hand picking and packing is a major source of contamination. The great concentration of the industry is found around the Chesapeake Bay.

Stone Crab. This species (*Menippe mercenaria*) is harvested on the Gulf coast, mostly in Florida, with a small product for local consumption in Texas. The crab cannot swim and it lives mostly on shallow flats in high-salinity areas near oyster reefs. It possesses great strength and can crack the shell of a large oyster. Nevertheless, it has a mild disposition and is often taken by hand.

Other minor crabs collected for food usage include the rock crab (*Cancer irroratus*), usually referred to as the sand crab in New England; the Jonah crab (*Cancer borealis*), mainly of regional interest; the tanner crab (*Chionoecetes tanneri*), which is found on the continental shelf of the north Pacific Ocean and valued by the Japanese and Russians; and the Kona crab (*Ranina ranina*), the most important species caught in Hawaiian waters.

Lobsters

The principal lobster fisheries of the world are found along the western shores of the Atlantic Ocean from Labrador south to Florida, and off Brazil (Cape de São Roque); in the Caribbean; off the southern tip of Africa (Cape of Good Hope); in the western Pacific off Japan; in the south Pacific Ocean, east of New Zealand; in the waters east of Tasmania; in the Indian Ocean west of Australia; with smaller lobster fisheries located in the North Sea, in the Atlantic west of Ireland, in the English Channel, in the Bay of Biscay, along the west coast of Africa southward from Cape Blanco, and in the eastern Pacific in waters west of Colombia and Mexico, including Baja California.

The greatest lobster harvest is of three species of true lobsters: (1) the American lobster (*Homarus americanus*), representing about 50% of the total catch; (2) the European lobster (*H. gammarus*), representing about 30%; and (3) the Norwegian lobster (*H. norvegicus*), also sometimes referred to as the Dublin Bay prawn or scampi. The Norwegian lobsters generally are not sold alive becaue they are not hardy. Much of the catch is sold as frozen tail meats.

Along the New England coast, 30 pounds (13.6 kilograms) of lobster usually comprises 20 to 30 lobsters. In contrast, lobsters taken in the offshore fishery on the edge of the continental shelf may weigh from 10 up to 30 pounds (4.5 to 13.6 kilograms). Large lobsters are not as desirable in the live market, but since the giants are filled with succulent meat, an increasing number are being harvested, particularly for processing.

Shrimp

The fisheries for shrimps and prawns may represent only about 1% of the total catch of seafood, but they represent at least 5–7% of the total value. More countries land shrimp than nearly any other kind of marine product and in at least 20 countries, shrimp fishing is a substantial industry. The leading shrimp harvesters are the United States, Mexico, Germany (West), India, Japan, and Pakistan. Other large harvesters include Thailand, Korea, China, Ecuador, Chile, Malaysia, the Netherlands, Norway, Spain, Australia, El Salvador, Greenland, Denmark, Venezuela, Italy, France, Sweden, Colombia, and the United Kingdom. About half the world's supplies of shrimp come from Asian countries. The importance of this has increased considerably in recent years, particularly the harvests from India and Pakistan.

There are over 2000 known species of shrimps (*Natantia*). They inhabit all possible environments between the deep sea and fresh water, but mostly are marine. The body is usually laterally compressed, and a tail fan is present on the relatively long abdomen. Despite this, shrimps spend most of their time on the bottom, walking around on their five pairs of walking legs, or using them to dig in the sediment. The long first antennae, which are held close together, form a tube through which fresh water is brought to the body for respiration. See Fig. 1. Some species have great tolerance for changing environmental conditions. For example, the common shrimp (*Crangon crangon*) does well in the Wattenzee (the Netherlands) in large numbers, despite the great variations in temperature and salinity found in this body of water. The common shrimp feeds at night on other small crustaceans, worms, and mollusks; during the day, it hides in the soft mud. Its life expectancy is 3 years. During this time, a single female can produce up to 20,000 offspring.

Mariculture of pond-reared shrimp has been studied. These investi-

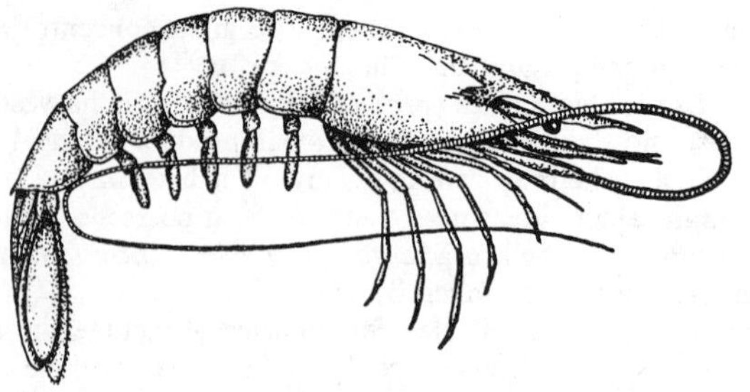

Fig. 1. Shrimp.

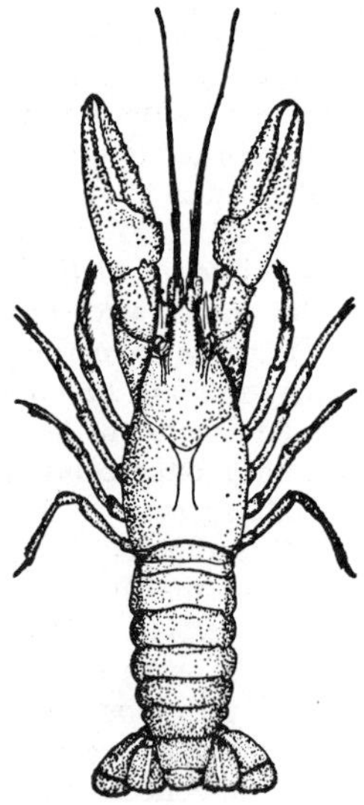

Fig. 2. Crayfish.

gations have included ways to improve and stabilize shrimp production through better pond construction, water management, nutrition, and disease control. Microbial activity is one of the main causes of quality deteoriation of harvested shrimp. It also can be a potential cause of foodborne diseases connected with shrimp.

Crawfish

Crawfish are sometimes referred to as "false lobsters" because of their similar appearance to lobsters.

Spiny Lobsters. This creature is actually a crawfish (suborder *Reptantia*, the crawlers) and can be distinguished from the shrimps (*Na*tantia, the swimmers) by their dorsoventrally flattened bodies and usually powerfully developed pincers on the first pair of walking legs. The abdominal legs are no longer adapted for swimming, but are used by the female as the place to attach her eggs. The spiny lobster (family *Palinuridae*) is recognizable by its antennae, which are longer than the body, and the spiny processes of the carapace. Large, gripping pincers are lacking. Commercial use of spiny lobsters in all parts of the world is attributable to their meaty abdomens.

The European spiny lobster (*Palinurus vulgaris*) has a body length up to 45 centimeters (18 inches) and weighs up to 8 kilograms (17.5 pounds). These figures refer to mature specimens (10–15 years old). This species lives on the rocky coasts of the Atlantic and Mediterranean. It is a valuable food species. It feeds at night on snails, mollusks, and dead animals.

The Cape spiny lobster (*Jasus lalandei*) is a popular item on the menu in South Africa and Australia. The same is true of other *Palinurus* species on the coasts of North and South America. The American species (*Palinurus argus*) has seasonal migrations, during which the animals may walk for more than 100 kilometers (62 miles) along the bottom of the sea and thus be able to discover suitable areas with sufficient food. It is possible that the sounds the animal produces, discovered only relatively recently, aid in regulating these migrations.

One of the largest of the spiny lobsters is the New Zealand species (*Jasus verreauxi*), which can attain a weight of 13.6 kilograms (30 pounds) and a length of about 1 meter (3 feet). Most specimens, however, are considerably smaller.

Spiny lobster fisheries are based mainly in temperate and subtropic countries where the species of the genera *Jasus* and *Panulirus* are the basis for the fisheries.

Crayfish

The crayfish (*Astacidae*) are essentially restricted to the Northern Hemisphere. In the Southern Hemisphere, the ecological niche they occupy is taken over by the parasticids (*Parasticidae*) in South America and Madagascar (but not Africa), and in Australia by the austroastacids. The true crayfish is a freshwater crustacean. It reaches a length of about 15 centimeters (6 inches). See Fig. 2.

The European crayfish (*Astacus astacus*) lives only in clean water and hunts for food at night, feeding on both plant and animal material. Male crayfish can live up to 20 years. In recent years, partially because of pollution, crayfish populations have diminished. This has also been due partly to a fungus disease called "crab plague." The European crayfish population was seriously threatened in 1879 when it was nearly wiped out, and its return has been slow. The American crayfish (*Orconectes limosus*), introduced into Europe in 1890, is immune to the disease and also can survive in quite dirty water. Although it cannot be compared in size with European crayfish, it is now considered the most common inland crustacean used for human consumption.

Antarctic Krill

Although the euphausids (*Euphausiacea*) seem almost shrimplike, they can be distinguished immediately from the true shrimp, which belong to the *Decapoda*, by the end of the tail, which is covered with bristles. The lateral gills are not covered by the carapace, as in the case of the decapods. In addition, the maxillipeds, which are very well developed in the decapods, are lacking. The majority of euphausids are oceanic, inhabiting certain zones in enormous numbers, almost 100 million individuals per shoal. The inhabited areas are characterized by uniform temperatures and salinity, and each species of euphausid is so closely adapted to the environmental conditions in which it is found that it dies, or at least is unable to reproduce, if it drifts into a water zone with different conditions.

The enormous swarms of euphausids are an immense source of readily available food for many inhabitants of the oceanic regions, such as herrings, sardines, and other fishes. Even the giant blue whale and baleen whale sometimes feed exclusively on these tiny animals. The majority of euphausids feed by filtering plant and animal plankton; they make daily migrations up and down with their food source, their movements being dependent upon light intensity. At night, therefore, they are found near the surface and, during the day, at depths of several hundred meters. This can be demonstrated by echolocation, inasmuch as the euphausids form a scatter layer which returns the echo.

The most important members of the euphausids belong to the "krill," as it is called by whalers. *Euphausia superba* inhabits cold seas; *Meganyctiphanes norvegica* lives on the edge of the Gulf Stream in Norway; in addition, members of the genus *Meatoscelis* and the carnivorous genus *Stylocheiron* should be mentioned.

The *Euphausia superba* achieves a length of 4 to 7 centimeters (1.5 to 2.75 inches). Krill is fundamental to the food chain of the Antarctic. Whales, seals, penguins, winged birds, various fishes, squid depend directly or indirectly on krill as a food source. Their existence has been known for many years, but no serious thought toward exploitation of krill commenced until the early 1960s. Krill harvesting and processing was pioneered by the U.S.S.R., followed by the Japanese who produce krill meal and krill protein concentrate. It was in the early 1970s that the U.S.S.R. commenced to market krill butter and krill cheese products. The Japanese have made a type of fish sausage wherein 20% of the usual fish protein is replaced by krill. Products have also been introduced by Germany (West). Countries most active in krill research and harvesting have been the U.S.S.R., Japan, Germany, Norway, Poland, Chile, and Taiwan. The Antarctic treaty signatory nations and the Food and Agriculture Organization (United Nations) have acknowledged the potential of krill protein and have recognized the need for proper management of the resource and harvesting limits.

Authorities have variously estimated the krill stocks from 125 to 200 million metric tons—to as high as 6 billion metric tons. These estimates admittedly have been based upon samplings conducted in relatively small areas. It has been estimated that baleen whales consume some 33 million tons annually; crabeater seals another 100 million tons; fur seals, another 4 million tons; penguins, about 39 million tons. Some experts believe that marine fishes and squid consume another 100–200 million tons annually. These consumption figures, added together, exceed the smallest of the estimates of krill stock. On the other hand, these consumptions represent less than 15% of the higher estimates. Thus, prior to aggressive exploitation of krill to meet some of the protein needs of humans, much more scientific information is required.

References

Dassow, R. W., Jr.: "Crab Industry," in "The Encyclopedia of Marine Resources," (F. E. Firth, editor), Van Nostrand Reinhold, New York, 1969.
El-Sayed, S. Z., and M. A. McWhinnie: "Antarctic Krill—Protein of the Last Frontier," *Oceanus*, **22**, 1, 13–20 (1979).
Everson, I.: "Krill and Squid: Resource Review," in "Biomass," Vol. 2 (S. Z. El-Sayed, editor), Woods Hole Conference on Living Resources of the Southern Ocean, Woods Hole Oceanographic Institution, Woods Hole, Massachusetts, 1977.
George, R. W., and T. G. Kailis: "Lobsters," in "The Encyclopedia of Marine Resources," (F. E. Firth, editor), Van Nostrand Reinhold, New York, 1969.
Grzimek, B.: "Grzimek's Animal Life Encyclopedia," Vol. 1, Van Nostrand Reinhold, New York, 1974.
Idyll, C. P.: "Shrimp Fisheries," in "The Encyclopedia of Marine Resources," (F. E. Firth, editor), Van Nostrand Reinhold, New York, 1969.
Lubimova, T. G., Naumov, A. G., and L. L. Lagunov: "Prospects for the Utilization of Krill and Other Nonconventional Resources of the World Ocean," *J. Fish. Bd. Canada*, **30**, 2196–2201 (1973).
Staff: "Atlas of the Living Resources of the Seas," Food and Agriculture Organization (United Nations), Rome, 1972.
Staff: "Yearbook of Fishery Statistics," Food and Agriculture Organization (United Nations), Rome (Published annually).
Ward, D. R., Nickelson, R., II., and G. Finne: "Relationship between Methylmercury and Total Mercury in Blue Crabs," *J. Food Sci*., **44**, 3, 920 (1979).

CRUST (Ocean). Earthquakes, Seismology, and Plate Tectonics; Ocean; Volcano.

CRUX. Often referred to as the Southern Cross, this constellation consists of four bright stars which form a figure of the approximate proportions of a Latin cross. The best known of the southern constellations, two of the stars are of the first magnitude and two are of the second magnitude. Located between Centaurus and Musca, the constellation lies above the Antarctic Circle and is not seen in northern latitudes.

CRYING (Infant). Sudden Infant Death Syndrome (SIDS).

CRYOEXTRACTION. Vision and the Eye.

CRYOGENIC GYROSCOPE. Gyroscope.

CRYOGENIC PROCESS (Natural Gas). Natural Gas.

CRYOGENIC PROCESS (Turboexpanders). Gas and Expansion Turbines.

CRYOGENIC ROCKET FUELS. Rocket Propellants.

CRYOGENICS. The production and study of phenomena which occur at very low temperatures, i.e., below about 80K. The first step in attaining the required temperature generally involves the liquefaction of a gas or gases. Liquids can exist over a range of temperatures limited by the critical point at the higher end and the triple point at the low-temperature end. It is thus possible to compress a gas to the liquid phase at the critical point and to cool it by boiling under reduced pressure to its triple point. A series of gases having their critical and triple points overlapping can thus be used in a cascade process each being used as the refrigerant for the next in the series. Pictet used this method to liquefy oxygen, using methyl chloride and ethylene as refrigerants. There are, however, no liquids which cover the range from 77K to the critical point of hydrogen, or from 14K to the critical point of helium (5.2K). Thus, liquid hydrogen and helium cannot be produced by the cascade method.

A gas may also be cooled by making it do work in the course of an expansion. When an ideal gas is expanded through an aperture into a constant volume, no work is done, since there are no interactions between the molecules and the molecules themselves occupy no volume. When a nonideal gas is so expanded, however, an amount of internal work ($W = (PV)_{\text{final}} - (PV)_{\text{initial}}$) is done against the intermolecular forces. This work may be positive or negative, resulting in a cooling or heating of the gas. Air is cooled by this Joule-Thomson expansion at room temperature, but hydrogen and helium must be precooled to 90 and 15K, respectively to obtain further cooling upon expansion. Using this method, Kamerlingh Onnes first succeeded in liquefying helium in 1908. Compressed gases may also be made to do external work, for example, by expansion against a movable piston. In this case, the work is always positive and helium may be cooled and liquefied without any precooling by liquid hydrogen. See **Ammonia;** and **Helium.**

With liquid helium readily available in the laboratory, research in the temperature range 5 to 0.8K has become commonplace. By using the isotope of helium ^3He, it is possible to attain temperatures down to about 0.3K since the isotope has a lower boiling point than ^4He. This is about the lowest temperature practically attainable by boiling liquids at reduced pressure. To reach lower temperatures, it is necessary to use magnetic phenomena.

Debye and Giauque pointed out that at 1K the entropy of paramagnetic salts was still fairly large and, moreover, that it was almost all due to nonalignment of magnetic moments and that the entropy of lattice vibrations was very small. If the electron spins are aligned by application of a magnetic field, the entropy of the salt falls to a low value and the heat of magnetization can be extracted isothermally. The salt can then be thermally isolated and demagnetized adiabatically and its temperature will fall. Temperatures in the order of 0.01K can readily be reached by this method and the lower limit for a single demagnetization would appear to be about 10^{-3}K. When demagnetization occurs, the spin system reaches equilibrium temperature in about 10^{-10} seconds. It is found that equilibrium is achieved between the spin temperature and the lattice temperature by spin-orbit coupling in times in the order of a few seconds. Paramagnetic salts have relatively high specific heats at low temperature and hence the cold salt can be used to cool other bodies. However, making good thermal contact with the cooled salt can be difficult.

Kurti, Simon, and Gorter suggested that a further reduction in temperature could be attained if adiabatic demagnetization was performed on nuclear moments rather then the electron spin. The temperature which can be reached in an adiabatic demagnetization is determined by the point at which the entropy of the system in zero external field decreases sharply with decrease in temperature due to the alignment of magnetic moments, i.e., the point at which the interaction energy μh equals kT (h = internal field; μ = magnetic moment). Since the interaction energies of nuclear moments are much smaller than electron spin interactions, much lower temperatures should result. The materials used experimentally were metals cooled to 10^{-2}K by contact with a paramagnetic salt. The thermal isolation of the nuclear spins during demagnetization was achieved naturally by the nuclear-spin-conduction electron relaxation time ($\sim$100 seconds). Thus, while the nuclear spins cooled to between 10^{-5} and 10^{-6}K, the conduction electrons and lattice remained in thermal contact with the cooling salt at 10^{-2}K.

Another method of attaining temperatures below 1K is to take advantage of the fact that the entropy of a superconducting metal is less than that of the metal in its normal state. Quenching of a superconductor by the application of a magnetic field can cause a cooling to about 0.1K. However, since the specific heat of metals is very small at these temperatures, they are not very suitable for cooling other bodies.

Low-Temperature Measurements. Such measurements are usually carried out using secondary thermometers which have been calibrated at certain fixed points previously determined on the absolute scale by a standard instrument. This instrument is generally a constant-volume gas thermometer used at low pressures. When the readings of this instrument are extrapolated to zero pressure, the scale coincides with the thermodynamic scale. In the range of 0.8 to 5.2K, the vapor pressure of ^4He provides the most commonly used secondary scale and it agrees with the absolute scale to within 2 millidegrees over this range. The use of ^3He instead of ^4He increases the coverage to 0.3K.

Resistance thermometers are useful over a wide range. For example, platinum is used from 273 to 15K, and carbon covers the range 20 to 2K and has the advantage of being quite insensitive to magnetic fields. The foregoing are examples of secondary standards where the scales are interpolated between fixed points.

For temperatures below 0.3K, the susceptibility of a paramagnetic salt can be measured and the temperature calculated by extrapolation from Curie's Law $\chi = C/T$. This method gives true values for the temperatures so long as Curie's Law holds. Beyond this region, it is necessary to perform a thermodynamic cycle to determine the relationship between the magnetic temperature T^* and the absolute temperature T. The method of Kurti and Simon is to demagnetize adiabatically from a known temperature on the absolute scale, using a number of different field intensities, and hence to determine the relationship between T^* and the entropy S over the required range of temperature. Measurement of the amount of heat Q necessary to raise the temperature from T_1^* to T_2^* gives the absolute value of the average temperature $T_{1,2}$ from the relationship $\Delta Q = T\Delta S$. The heat is generally supplied to the salt by gamma rays, thus ensuring even heating of the sample, a necessary precaution since the thermal conductivity is poor. The problem of nonlinearity does not arise in the case of nuclear spin demagnetization since, in this case, the susceptibility obeys Curie's Law down to 10^{-7}K.

Cryogenic Phenomena. One of the most interesting phenomena of cryogenics is that of superconductivity, which was also discovered by Onnes. When metals are cooled from room temperature, their resistivities decrease and at low temperature, they attain low values which are fairly independent of temperature. Some metals, however, have a critical temperature below which their resistance goes to zero. Such a metal is known as a superconductor. See also **Superconductors.**

Superfluidity. As liquid helium is cooled below 2.18K, it undergoes a sudden discontinuity of specific heat and a second-order transition to the superfluid state. In this state, the viscosity of the helium becomes a function of the method used to measure it. Measured by an oscillating disk method, the viscosity falls from 23×10^{-6} poise just above the transition to 1×10^{-6} poise at 1.3K. Measured by passage through very fine capillaries, the viscosity is very nearly zero in the superfluid state. Hence, it is postulated that there are two coexisting, noninteracting fluids, one having the properties of non-"superfluid" helium; and the other having virtually zero viscosity and zero entropy. It is interesting to note that the thermal conductivity of superfluid helium is about 2,000 times greater than that of copper. This is the result of the motion of the entropy-free superfluid rather than normal thermal conductivity. See also **Superfluidity.**

Magnetic Properties. There are several fundamental experiments on the magnetic properties of materials which become possible as a result of the low-temperature environment of cryogenics. The first of these was discovered by deHaas and Van Alphen in 1930. They found that at low temperatures, the susceptibility of bismuth single crystals rose and fell periodically as the magnetic field was increased. Later work shows that the periodicity occurs in all metals at low temperatures and is the result of quantization of electron motion perpendicularly to the applied field. This effect was used to determine the Fermi surface of metals.

Radioactive Decay. The method of alignment of nuclear moments has been used to study radioactive decay as a function of nuclear orientation. The aligning field in this case can be either an externally applied magnetic field or an internal crystal field. One of the more striking of these experiments has been the test of the Lee-Yang theory of nonconservation of parity in weak interactions. A third fundamental experiment of interest was the confirmation of London's concept that flux through a superconducting ring is quantized. The ring was a lead tube of 10^{-3} centimeter diameter and it was suspended from a torsion balance. The tube was made superconducting in the presence of longitudinal magnetic fields, and the frozen-in flux was measured and found to be quantized.

Practical Aspects of Cryogenics. In addition to its contributions to fundamental science, the field of cryogenics is bringing about practical applications. One of the most important of these is the superconducting magnet. To achieve magnetic fields in the order of 70 kilogauss by normal methods calls for the expenditure of about a megawatt of power for as long as the field is maintained. The necessity of removing this power by circulation of cooling fluids adds to the inefficiency. When a superconducting coil is used in the persistent current mode, no energy is needed to maintain fields of this magnitude. Alloys of niobium-tin and niobium-zirconium have been found to have the capacity to remain superconducting while they are subjected to high magnetic fields and are carrying the current necessary to produce the field.

There also have been important developments in superconducting circuitry. The basic device in this area has been the *cryotron*, a switch in which the field resulting from current in a superconductor can quench another superconducting superconductor to its normal state. When made in the form of deposited thin films, cryotrons have switching times less than 10^{-8} second.

See also **Josephson Tunnel Junction.** The use of cryogenics in various medical applications is covered under specific entries, such as **Vision and the Eye.**

CRYOGENICS (Resistive). Electric Power Transmission.

CRYOGENIC VACUUM PUMPING. Vacuum Pumps.

CRYOHYDRATE. An eutectic system consisting of a salt and water, having a concentration at which complete fusion or solidification occurs at a definite temperature (eutectic temperature) as if only one substance were present.

CRYOHYDRIC POINT. The eutectic point in cases in which the system contains water.

CRYOLITE. Cryolite, sodium aluminum fluoride, Na_3AlF_6, crystallizes in the monoclinic system but in forms that closely approach cubes and isometric octahedrons. It is usually found massive. Cryolite has an uneven fracture, is brittle; hardness, 2.5; specific gravity, 2.97; luster, vitreous to greasy; color, snow-white but may be colorless, reddish or brownish; translucent to transparent. The only considerable occurrence of cryolite is at Ivigtut, Greenland, where veins of this mineral are associated with granites and gneisses. Small occurrences of cryolite have been noted in the Ilmen Mountains, U.S.S.R., and at Pikes Peak, Colorado. Cryolite has its chief use in the electrolytic production of aluminum, but small amounts are employed in the manufacture of opalescent glass.

The name cryolite is derived from the Greek words meaning frost (ice) and stone in reference to its translucency.

The aluminum industry no longer uses the natural mineral for operation of electrolytic cells except infrequently for start-up operations. The synthetic cryolite used is produced by two main processes: (1) the sodium aluminate from the Bayer process (see also **Bauxite**) is reacted with hydrofluoric acid: $NaAlO_2 + 2NaOH + 6HF \rightarrow 3NaF \cdot AlF_3 + 4H_2O$; or (2) sodium carbonate may be used instead of NaOH:

$$NaAlO_2 + Na_2CO_3 + 6HF \rightarrow 3NaF \cdot AlF_3 + CO_2 + 3H_2O.$$

The quality hydrofluoric acid required for these reactions may be prepared from fluorspar: $CaF_2 + H_2SO_4 \rightarrow 2HF + CaSO_4$. Normally, cryolite is not considered toxic. A continued exposure to finely divided cryolite in the air may lead to *fluorosis*. However, the compound is toxic to insects.

See also terms listed under **Mineralogy.**

CRYOLOGY. That branch of hydrology which is concerned with snow and ice.

CRYOMETER. A low-temperature thermometer.

CRYOPUMP. An exposed surface refrigerated to cryogenic temperature for the purpose of pumping gases in a vacuum chamber by condensing the gas and maintaining the condensate at a temperature such that the equilibrium vapor pressure is equal to or less than the desired ultimate pressure in the chamber. Also referred to as a *cryogenic pump* and not to be confused with a cryogenic fluid pump for circulating cryogenic propellants.

CRYOSCOPE. An instrument for measuring the freezing or solidification point. The Hortvet cryoscope is used for the estimation of added water in milk from the lowering of the freezing point.

CRYOSCOPIC CONSTANT. A quantity calculated to represent the molal depression of the freezing point of a solution, by the relationship

$$K = \frac{RT_0^2}{1000 l_f}$$

in which K is the cryoscopic constant, R is the gas constant, T_0 is the freezing point of the pure solvent, and l_f is the latent heat of fusion per gram. The product of the cryoscopic constant and the molality of the solution gives the actual depression of the freezing point for the range of values for which this relationship applies. Unfortunately, this range is limited to very dilute solutions, usually up to molalities of $\frac{1}{100}$, and must be modified for many solutes. See also **Freezing-Point Depression.**

CRYOSPHERE. Polar Research.

CRYOSURGERY. Vision and the Eye.

CRYOTHALAMOTOMY. Parksinson's Disease.

CRYOTRON. Cryogenics.

CRYPTATE COMPLEXES. Chemical Elements.

CRYPTOCRYSTALLINE. When the texture of a rock is so finely crystalline (that is, made up of such minute crystals) that its crystalline nature is but vaguely revealed even in a thin section by transmitted polarized light, the rock is said to be cryptocrystalline. Among the sedimentary rocks, chert and flint are cryptocrystalline. Lava flows, especially of the acidic type such as felsites and rhyolites, may have a cryptocrystalline ground mass as distinguished from pure obsidian (acidic), or tachylite (basic), which are natural rock glasses.

CRYPTOCRYSTALLINE QUARTZ. Quartz.

CRYPTOCRYSTALLINITY. Mineralogy.

CRYPTOMERIA. Araucarias.

CRYPTOMONADIDA. An order of one-cell animals containing species of constant body form, with flagella.

CRYPTORCHIDISM. Gonads.

CRYSTAL. A macroscopic sample of a solid substance exhibiting some degree of geometrical regularity, or symmetry, or capable of showing these properties after suitable treatment (e.g., cleavage, etching, etc.). Almost all pure elements and compounds are capable of forming crystals.

A perfect crystal is one in which the crystal structure would be that of an ideal space lattice. No such crystals exist, all real crystals containing imperfections which have a strong influence on the physical properties of the crystal.

Structure of Crystals. Early investigators suggested that the regular structure of crystals, embodied in the laws of crystallography, could be explained if they were thought of as built up by the repetition of equal polyhedral cells, fitting together to fill space, each cell representing a characteristic group of particles, perhaps the atoms and molecules of the compound. A rough calculation showed that the spacing of these units in many ionic crystals might be of the same order of magnitude as the wavelengths of x-rays, as deduced from quantum theory. Von Laue suggested, and verified, that diffraction of the x-rays occurs when they are passed through a crystal, suitably oriented. He knew from the density and atomic weights, that the number of atoms in a cubic centimeter of rock salt, for example, is about 4.488 $\times 10^{22}$, and that therefore, if they are equally spaced in all three directions, their distance apart is 2.814×10^{-8} centimeters or 2.814 Å. Certain quantum theory calculations had already indicated that x-rays have wavelengths of this order. It occured to von Laue that if a beam of x-rays were directed upon a crystal and the crystal turned into a suitable position, one might observe interference maxima analogous to those produced with light by a diffraction grating. This proved to be the case, and the result verified beyond question the existence of regular spacings between reflecting planes of some sort, presumably plane arrays of atoms or ions. The kind of pattern obtainable is demonstrated by Fig. 1.

Subsequent analysis of the problem by Bragg resulted in a formula analogous to that for interference of light reflected by thin plates. If the reflecting layers are spaced at equal distances d, and if the wavelength of the incident x-rays is λ, the angle θ between rays and layers necessary for an interference reflection maximum is given by Bragg's law, viz., $\sin\theta = n\lambda/2d$; in which n is an integer. By slowly turning the crystal, the various plane-families are brought into suitable orientations for the production of maxima. The result is a Laue pattern of black spots on the photographic plate placed beyond the crystal to catch the reflections. Another method, developed by Hull and by Debye and Scherrer, secures the necessary angular variation by crushing the crystal to powder and relying upon the fortuitous orientation of the fragments; the pattern in this case being a system of concentric rings, as exemplified by Fig. 2.

While it is very easy, when one knows the structure of the crystal and the wavelength of the rays, to predict the diffraction pattern, it

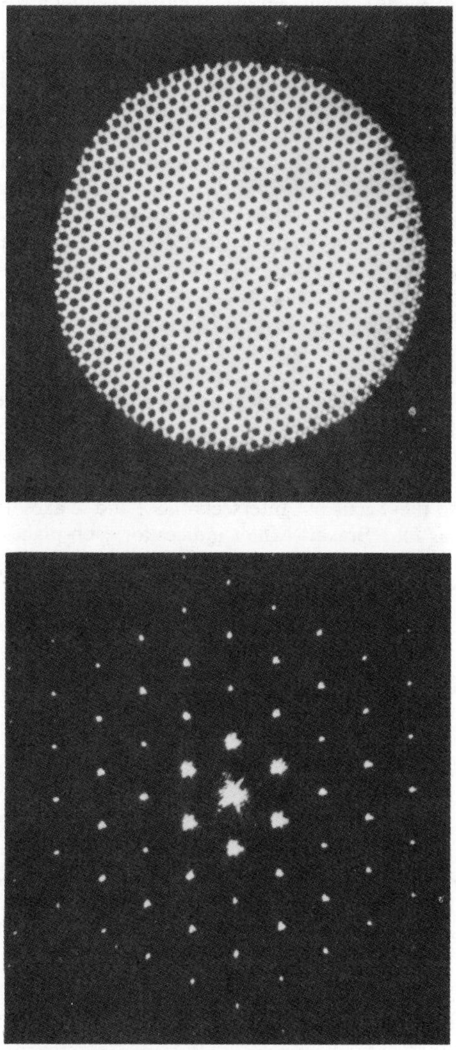

(a) (b)

Fig. 1. Types of x-ray patterns of crystals: (a) Steel balls in a crystalline lattice; (b) Bragg reflections with shape S^2.

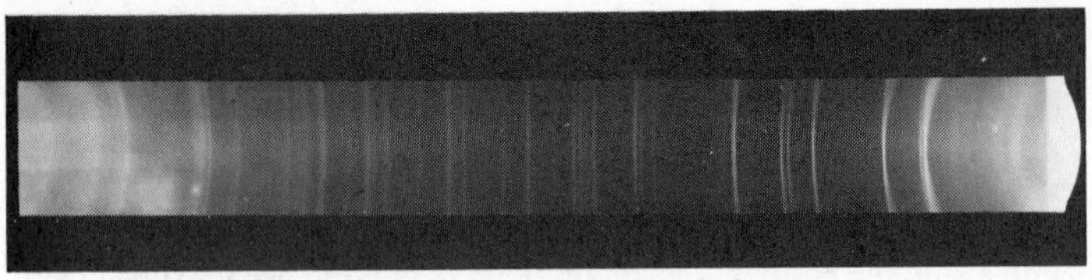

Fig. 2. Concentric ring pattern of crystalline materials.

is quite another matter to deduce the crystal structure in all its details from the observed pattern and the known wavelength. The first step is to determine the spacing of the atomic planes from the Bragg equation, and hence the dimensions of the unit cell. Any special symmetry of the space group of the structure will be apparent from space group extinction. A trial analysis may then solve the structure, or it may be necessary to measure the structure factors and try to find the phases or a Fourier synthesis. Various techniques can be used, such as the F^2 series, the heavy atom, the isomorphous series, anomalous atomic scattering, expansion of the crystal and other methods.

By such methods, the structures of crystals have been determined and all of them can be shown to possess space structures corresponding to one or another of the 14 Bravais lattices. Models of crystal structure are shown in Figs. 3 and 4.

Crystal Systems and Crystallography. The investigative field which deals with the external shapes of crystals and with the geometrical relations between the atomic planes within them is *crystallography*. If a solid crystal is broken, it is found to have separated along certain "cleavage planes" into polyhedral fragments. Even when crushed to powder, the minute grains show this characteristic. Measurements upon the variously shaped pieces reveal that if they were fitted together again, the planes would all be found to belong to one or another plane-family, the members of any one of which are all parallel.

In most crystal systems each of the more prominent crystal faces belongs to one of three plane-families intersecting along what are called the crystal axes. In the hexagonal system there are four. These may be conveniently used as coordinate axes, X, Y, Z, though they are not generally at right angles. Haüy discovered that if the ratio of the intercepts of two crystal planes on one of these axes is a simple fraction, such as $\frac{3}{5}$, the ratios of the intercepts on the other axes are likewise simple. This suggests that the two intercepts on any one axis are multiples of a common unit. The units are, however, generally different for the different axes, bearing to each other ratios called the axial ratios.

It is more convenient to use the reciprocals of the intercepts. For example, a plane might have intercepts equal to 10,000, 15,000, and 6,000 of the respective units. The reciprocals have the ratios 1/10,000:1/15,000:1.6,000, which in lowest terms are 3:2:5. These smallest integers are the Miller indices of the family to which this plane belongs, and the family is thus designated (325). The family (201) is parallel to the Y axis but intersects the X and Z axes. (The hexagonal system has four Bravais-Miller indices for each plane-family.)

If the intercept of any plane has a negative value, that is, if it cuts the axes when extended in a direction opposite to that in the standard arrangement, the fact is shown by a bar over the Miller index, e.g., $(2\bar{2}\bar{2})$. The eight faces of an octahedron are (111), $(1\bar{1}1)$, $(\bar{1}11)$, $(\bar{1}\bar{1}1)$, $(1\bar{1}\bar{1})$, $(11\bar{1})$, $(\bar{1}1\bar{1})$, and $(\bar{1}\bar{1}\bar{1})$, that is, they all belong to the form (111).

Close study of the angles, indices, and axial ratios long since made it clear that every crystalline substance has a structure built upon a space "lattice" characteristic of the substance. It has been established that this is due to the regular arrangement of the atoms, molecules, or ions composing the substance. As shown by the accompanying table, the lattice structures of crystals may be classified into 32 symmetry classes (point groups), which are further divided into seven systems. This topic also is discussed under **Mineralogy.**

Practically all minerals are crystalline, although perfect natural crystals are seldom, if ever, found. Because of the laws of crystallography, however, a crystallographer can usually determine the crystal form of a known species from a fragment of the original crystal, provided that at least two of the crystal faces are visible. Crystalline aggregates are said to be cryptocrystalline when the individual particles are proved

ELEMENTS OF CRYSTAL SYSTEMS

System	Crystallographic Elements	Essential Symmetry	Number of Point Groups
Cubic, or regular	Three axes at right angles: all equal.	4 triad axes; 3 diad, or 3 tetrad axes	5
Tetragonal	Three axes at right angles: two equal.	1 tetrad axis	7
Ortho-rhombic or rhombic	Three axes at right angles: unequal.	3 diad axes, or 1 diad axis and 2 perpendicular planes intersecting in a diad axis	3
Monoclinic	Three axes, one pair not at right angles: unequal.	1 diad axis, or 1 plane	3
Triclinic or anorthic	Three axes not at right angles: unequal.	No axes or planes	2
Hexagonal	Three axes coplanar at 60°: equal. Fourth axis at right angles to other three.	1 hexad axis	7
Rhombohedral or trigonal	Three axes equally inclined, not at right angles: all equal.	1 triad axis	5

Fig. 3. Crystal structure models: (left to right) beta-quartz, rutile, potassium dihydrogen phosphate. (*Bell Laboratories.*)

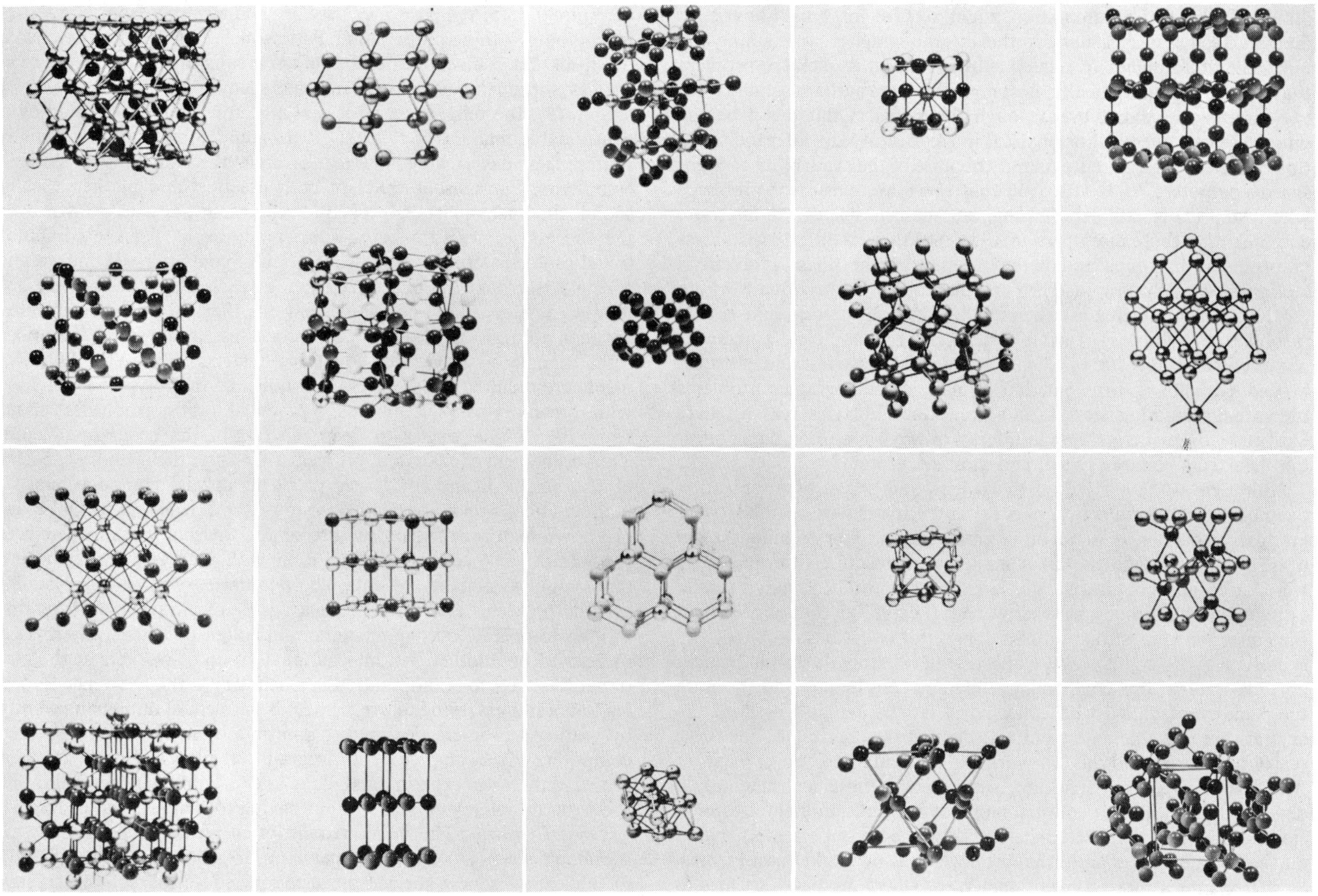

Fig. 4. Crystal structure models: (top row, left to right) cuprite, zincblende, rutile, perovskite, tridymite; (second row) cristobalite, potassium dihydrogen phosphate, diamond, pyrites, arsenic; (third row) cesium chloride, sodium chloride, wurtzite, copper, niccolite; (fourth row) spinel, graphite, beryllium, crabon dioxide, alpha-quartz. (*Bell Laboratories.*)

to have crystalline structure but their crystal faces are exceedingly small or indistinguishable. A mineral is pseudocrystalline if its external form does not correspond with its crystalline structure.

Determination of Crystal Structure. The object of a crystal-structure determination is to ascertain the position of all of the atoms in the unit cell, or translational building block, of a presumed completely ordered three-dimensional structure. In some cases, additional quantities of physical interest, e.g., the amplitudes of thermal motion, may also be derived from the experiment. The processes involved in such crystal-structure determinations may be divided conveniently into (1) collection of the data, (2) solution of the phase relations among the scattered x-rays (phase problem)—determination of a correct trial structure, and (3) refinement of this structure.

The data consist of intensities $I(hkl)$, where h, k, and l (the Miller indices represent a vector triplet which conveniently identifies the beam diffracted from a single crystal. In a typical determination, there may be one to two thousand such $I(hkl)$. The intensity is related to the structure factor $F(hkl)$ by the relation.

$$I(hkl) = KF(hkl)F^*(hkl) \qquad (1)$$

where K is a known relative factor, and where F^* is the complex conjugate of F. The structure factor itself is related to the scattering by the j atoms in the unit cell by the relation,

$$F(hkl) = \sum_j f_j T_j \exp[2\pi i(hx_j + ky_j + lz_j)] \qquad (2)$$

where f_j are the individual atomic scattering factors, T_j are the individual modifications of the scattering as a result of thermal motion, and x_j, y_j, z_j are the fractional positions of atom j along the three crystallographic axes. In a typical determination, j may be between 10 and 60. The *scattering density* $\rho(xyz)$ is derivable from the relation,

$$\rho(xyz) = V^{-1} \sum_{-\infty\, h,k,l}^{\infty} F(hkl) \exp[-2\pi i(hx + ky + lz)] \qquad (3)$$

where V is the volume of the unit cell.

The "phase problem" in crystallography arises because in the usual experiment (Eq. 1) the magnitudes of the complex structure factors are obtained, but not the phases. Yet in order to obtain the scattering density, and hence the positions of the atoms, the phases as well as the magnitudes of the structure factors are necessary (Eq. 3).

Once the phase problem is solved, then the positions of the atoms may be refined by successive structure-factor calculations (Eq. 2) and Fourier summations (Eq. 3) or by a nonlinear least-squares procedure in which one minimizes, for example, $\Sigma w(|F_{obs}| - |F_{calc}|)^2$, with weights w taken in a manner appropriate to the experiment. Such a least-squares refinement procedure presupposes that a suitable calculational model is known.

It is perhaps useful to indicate how the attention of crystallographers to these three steps in the solution of a structure has changed in recent years. In 1954, the time involved in the arduous task of collecting the three-dimensional data—step (1)—needed for the solution of a complex problem was generally short in comparison with the time needed to solve the phase problem—step (2). This time involved in step (2) of course depended (and still depends) upon the complexity of the problem, and on the ingenuity, luck, and perseverance of the investigator, but it was true in many cases that step (2) was the rate-determining step in the entire process. This in part was because little attention was paid to detailed refinements—step (3); in 1954, three-dimensional least-squares refinements of complex structures were out of the question computationally, and even Fourier refinements were rare, for on computing systems advanced for those days (e.g., punched-card tabulators, sorters, and primitive electronic computers), a three-dimensional Fourier summation might require forty man-hours. In fact, in 1954 it was usual for the crystallographer to examine the unit cells of a number of related substances and to pick the problem that was crystallographically most favorable (and perhaps soluble from two-dimensional data), even though this problem might not be the one of greatest chemical or physical interest. Ten years later the situation had changed markedly, mainly because of the availability of high-speed computers. It is still true that there are classes of problems where step (2) is rate-determining, but these problems are far more complex than those attempted in 1954. Yet there is an extensive class of problems in which today the solution of the phase problem is straightforward and rapid. The crystallographer is thus often working on the problem of greatest chemical or physical interest, and is able to obtain a solution in feasible time. Relatively complete refinement of structures is now the rule, since it is a reasonably fast and effortless procedure. Thus it turns out that in many crystallographic problems the rate-determining step is data collection. For this reason, there has been a dramatic increase in interest in ways of making data collection less tedious, more rapid, and more accurate.

Although in the early days the Braggs and others used ionization chambers for the collection of x-ray intensities, these methods were gradually abandoned in favor of photographic film techniques. Up until a few years ago the great majority of structure determinations were based on photographically recorded intensities, usually visually estimated. This process is a slow one: the typical time involved in the collection and estimation of a data set of two thousand intensities is perhaps six to eight weeks. Collection of intensity data from protein crystals is far more challenging and time-consuming, both because the number of data to be collected is far greater and because the crystals are unstable and rapid collection is thus desirable. For these reasons, Harker and his co-workers, particularly Furnas, then at Brooklyn Polytechnic Institute, were among those instrumental in developing scintillation-counter methods for collecting three-dimensional x-ray data. Diffractometers with single-crystal orienters, based on the so-called Eulerian geometry developed by Harker and Furnas, as well as on the more conventional Weissenberg geometry are available commercially and have engendered widespread interest in counter techniques. Data collection by counter techniques, as practiced by most workers, is still an arduous task, since the setting of a number of orientation angles is involved. Program or computer control of such setting operations is an obvious extension. Especially for neutron diffraction studies, such programmed control of diffractometers has been the rule for some time.

Nevertheless, a programmed unit will do only what it was designed to do, whereas a computer can be programmed to perform new tasks or operations as they seem necessary. There are several installations of computer-controlled diffractometers. Counter methods, particularly when semiautomatic or completely automatic, enable more rapid data collection than is possible photographically. What is equally important is that they should also enable more accurate data to be collected. The general level of accuracy of intensities obtained photographically is perhaps 15 to 20%. Such a level has proved sufficient for the solution of conformational or stereochemical problems, but not necessarily for the determination of meaningful descriptions of thermal motion or bonding.

There are two approaches to the solution of the phase problem that have remained in favor. The first is based on the tremendously important discovery of Patterson in the 1930's that the Fourier summation of Eq. 3, with the experimentally known quantities $F^2(hkl)$ replacing $F(hkl)$ leads not to a map of scattering density, but to a map of all interatomic vectors. The second approach involves the use of so-called direct methods developed principally by Karle and Hauptman of the U.S. Naval Research Laboratory. The direct method of phase determination makes use of probability theory to give probable relations between phases of different structure factors.

The *Patterson function* has been the most useful and generally applicable approach to the solution of the phase problem, and over the years a number of ingenious methods of unraveling the Patterson function have been proposed. Many of these methods involve multiple superpositions of parts of the map, or "image-seeking" with known vectors. Such processes are ideally suited to machine computation. Whereas the great increase in the power of x-ray methods of structure determination in the past few years has come simply from our ability to compute a three-dimensional Patterson function, it is reasonable to expect that as machine methods of unraveling the Patterson function are developed, this power will increase many fold.

Step (3), the refinement of crystal structure, continues to enjoy a considerable amount of interest. Reasonably complete refinement is routine these days, owing in large measure to the availability of suitable computers. For reasons that are both practical and mathematically sound, the least-squares approach to refinement has gained favor over the successive structure-factor-Fourier approach. Yet the computational problems often tax this generation of computers. If one assigns a single isotropic thermal parameter to each atom, then there are four parameters, three positional and one thermal, to be determined for each atom. In the least-squares procedure, if one stores the upper right triangle of the normal-equations' matrix, then $\frac{1}{2}N(N + 1)$ elements are required, where N is the number of variables. In a machine with a memory of 32,000 words, a practical limit is reached at about $N = 200$, if one wishes to keep the rest of the program in core. Thus refinement of a 50-atom problem often taxes the memory capacity of the machine, and for larger problems special computational or mathematical tricks are needed. One of these tricks is to make use of known features of the structure or the thermal motion to reduce the number of parameters. But, even with increased numbers and more rapid availability of data, computerized solutions to crystallographic problems still suffer from a degree of uniqueness. Programs written to operate on one machine must often be extensively revised to be used on another. An international group of workers at the National Resource for Computation in Chemistry at the Lawrence Berkeley Laboratory is trying to overcome this problem of interchangeability by writing a program designed to run on any medium or large sized computer. When completed, the program will allow a crystallographer, armed only with experimental data and a computer, to determine the structure of any crystal without having to write special programs.

Crystal Growth. The direct growth of an ideal and perfect crystal is difficult except at very high supersaturations because of the difficulty of nucleating a new surface on a completed surface of the crystal. But, if there is a screw dislocation present, it is not necessary to start a new surface, and growth proceeds in a spiral fashion by the accretion of atoms at the edge of growth steps. The resultant growth spirals have been observed, and it is believed that most crystals grow in this manner. See Fig. 5. But spiral growth is not the only mechanism which enables crystal growth at fast rates. Gilmer, in particular, has employed computer simulation models in studies of crystal growth and has demonstrated that, among other influences, temperature and impurity levels have decisive effects upon growth rates. Again, since different crystal faces have different kinetic properties, the particular crystal plane exposed to growth will also be a partial determinant of growth rate.

In modern technology based upon solid state chemistry and physics, much emphasis is placed upon the availability of elements and com-

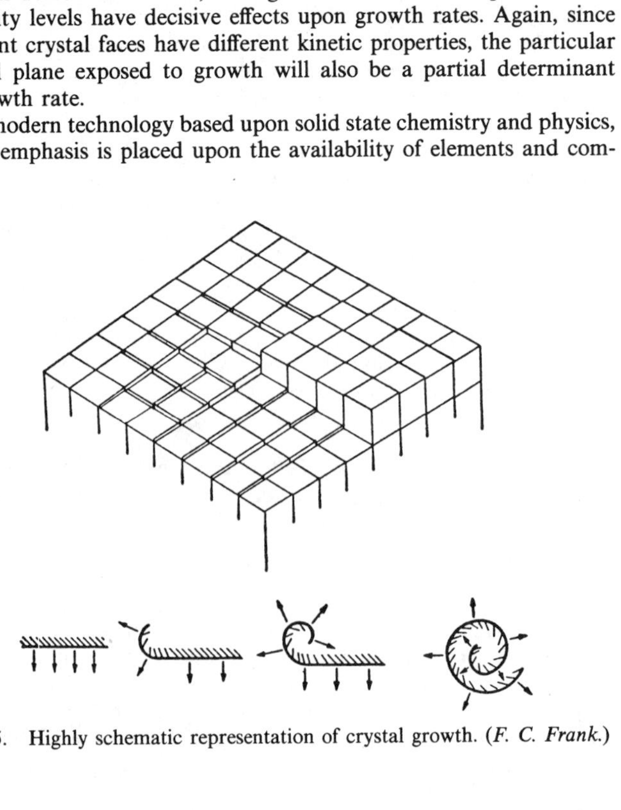

Fig. 5. Highly schematic representation of crystal growth. (*F. C. Frank.*)

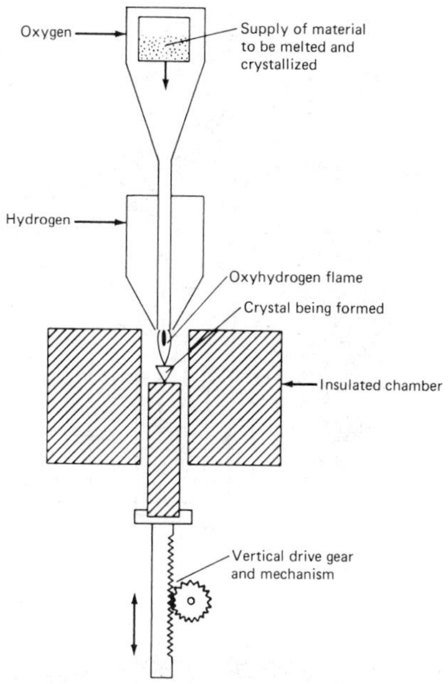

Fig. 6. Verneuil technique for growing crystals.

pounds in single-crystal form. Over the past twenty-five years a highly sophisticated technology has developed in this area. From the relatively simplistic growth of ammonium and potassium dihydrogen phosphates (ADP and KDP) from saturated solutions for reansducer elements, a level has been obtained at which pure metals (Cu, Pb, Al, Ag, Fe etc.), a semi-metals (Si, Ge, As, etc.), and compounds (GaAs, InAs, InSb, InP, etc.) are available and even essential in large single crystal form to the electronics industries. Synthetic gems (rubies, spinels, sapphires, emeralds, and zircons) are single crystals of aluminum, beryllium, and zirconium silicates or oxides with controllled impurity levels of transition elements.

Growth of such single crystals can follow several techniques, with thermodynamic constraints dictating the technique for any particular material: crystallization by cooling a supersaturated solution of a compound in a high-temperature flux; crystallization by dropping powder through an intense flame onto a seed pedestal, known as the Verneuil technique (see Fig. 6); crystallization by pulling a "seed" crystal from the surface of a liquid melt, known as the Czochralski method (see Fig. 7); crystallization by lowering a melt through a small, controlled thermal gradient, known as the Stockbarger technique (see Fig. 8); crystallization by zone-melting, known as the Pfann method (see Fig.

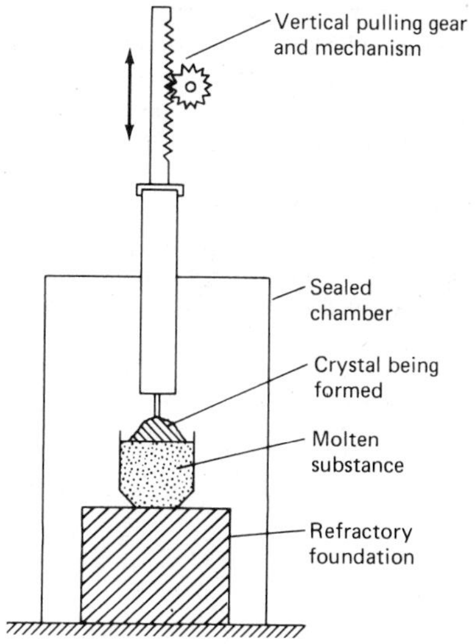

Fig. 7. Czochralski method for growing crystals.

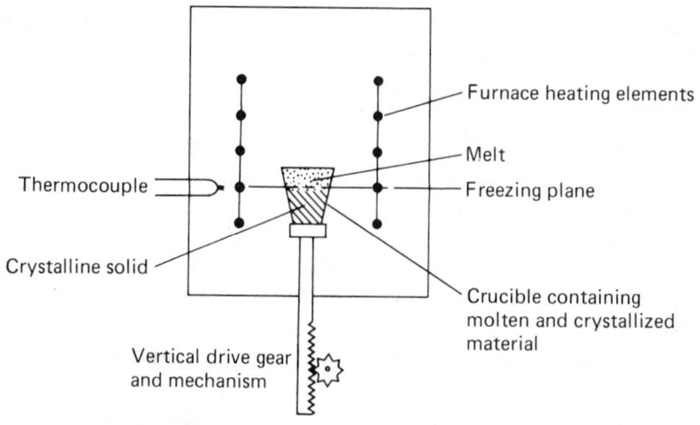

Fig. 8. Stockbarger technique for growing crystals.

9); and crystallization by the vapor-phase approach (see Fig. 10). All of these procedures, however, require three essential ingredients: (1) A good "seed" crystal from which spiral and sometimes oriented growth can occur and develop; (2) highly precise operational conditions—movement of fractions of a millimeter, or temperature variations of 0.5°C per hour; and (3), as previously indicated, materials of a specific impurity level. Given these conditions, single crystals can be grown in large quantities—ranging from the multimillion carat operations of Linde (United States) an Djevaherdijian (Switzerland), using the Verneuil technique, to the multikilo manufacture of single crystal silicon by Texas Instruments Incorporated (United States), employing the Czochralski and zone melt methods, and including a 27-kilogram single crystal of dislocation-free silicon by the Kayex

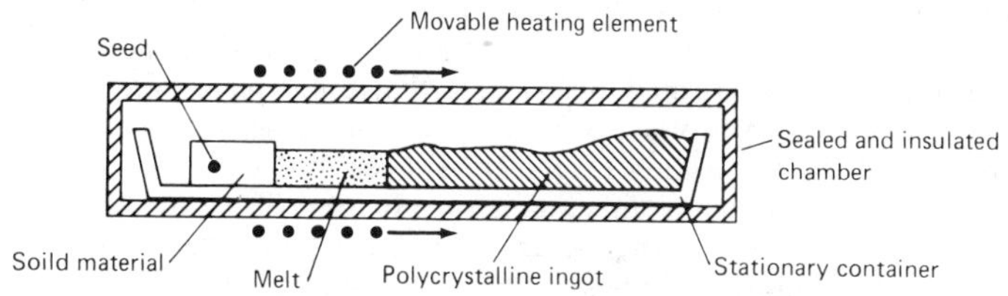

Fig. 9. Zone-melting or Pfann method for growing crystals.

Corporation (United States) grown by a modified Czochralski approach.

See also **Semiconductor.**

Dislocation in Crystalline Solids. This type of imperfection in a crystalline solid is generated as follows: A closed curve is drawn within the solid, and a cut made along any simple surface which has this curve as boundary. The material on one side of this surface is displaced by a fixed amount called the Burgers vector relative to the other side. Any gap or overlap is made good by the addition or removal of material, and the two sides are then rejoined, leaving the strain displacement intact at the moment of rewelding, but afterwards allowing the medium to come to internal equilibrium. If the Burgers vector

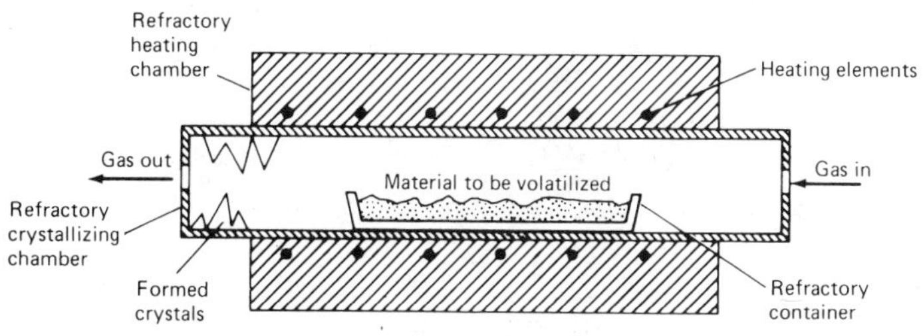

Fig. 10. Crystallization by the vapor-phase approach.

represents a translation vector of the lattice, the weld is invisible, and the dislocation is characterized only by the original curve, or dislocation line and by the Burgers vector.

A dislocation line may only terminate at the surface of the crystal. The energy of a dislocation is largely stored as strain in the surrounding lattice. The important property of a dislocation is its ability to move quite easily through the lattice, and hence to allow the rapid propagation of slip. The general dislocation defined above usually separates into its components, edge and screw dislocations, which may be treated as rather stable entities in the theory. Direct evidence for the existence of dislocations is the observation of dislocation networks in crystals of silver bromide, and, for their motion, from the spiral growth patterns in crystals.

In recent years high resolution electron microscopes have been used to study dislocations and their movements in thin crystals. This is possible in both metals and nonmetals. Other techniques which have been applied successfully are field ion microscopy and x-ray microscopy. Dislocations are important in determining the mechanical and electrical properties of solids, and play an important part in solid state physics. The density of dislocations (i.e., the concentration of dislocation lines), for example, is believed to vary from about 100,000,000 per square centimeter in good natural crystals, through 1,000,000,000 in good artificial crystals up to about 1,000,000,000,000 in cold-worked specimens. These estimates are based on the energy stored by cold work, on x-ray analysis, and on measurements of electrical resistivity. Among the various types of dislocation are the *edge dislocation*, which is defined as having its Burgers vector (line of displacement) normal to the line of the dislocation. An edge dislocation may be thought of as caused by inserting an extra plane of atoms terminating along the line of the dislocation (Fig. 11). For example, if the dislocation were along the Z-axis and its Burgers vector along the X-axis, then one might think of an extra half plane of atoms being inserted at the surface $x = 0$, $y > 0$. Such a dislocation would be of positive sign. An edge dislocation may move easily only parallel to its Burgers vector, i.e., in its slip plane.

The *screw dislocation* has its Burgers vector parallel to the line of the dislocation. In a screw dislocation, the atomic planes are joined together in such a way as to form a spiral staircase, winding round the line of the dislocation (Fig. 12). A screw dislocation is capable of easy movement in any direction normal to itself. The growth spirals formed in crystal growth appear where such dislocations intersect the surface. Edge dislocations of the same sign repel each other along the line between them, but are most stable when arranged vertically above each other. Edge dislocations of opposite signs attract one another, but otherwise prefer to lie so that the line between them makes an angle of 45° with their slip planes. Screw dislocations of opposite sign attract, of like sign repel.

Dislocation Line. The curve separating displaced and undisplaced positions of a crystal, and thus at the center of a dislocation, is termed the dislocation line. Within a distance of one or two lattice constants of the dislocation line the atoms are displaced by an amount more than can be represented fairly as a strain. A screw location may have a substantial hole down the dislocation line, through which impurity atoms may diffuse.

Dislocation Climb. This is a type of dislocation motion, differing fundamentally from slip, that is associated with the edge components of dislocations. In climb, an edge dislocation moves in a direction

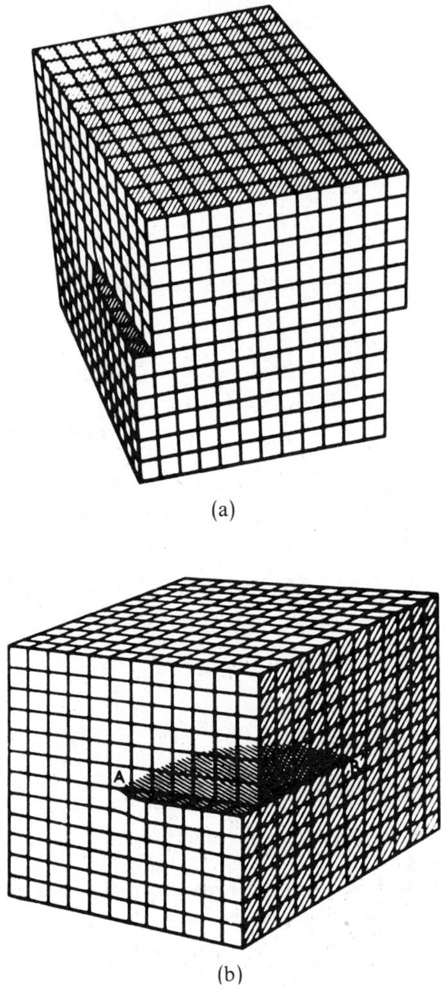

Fig. 12. (a) Simple type of screw dislocation; (b) both screw and edge dislocation along the arc from A to B.

perpendicular to its slip plane as atoms are either added to or taken away from the extra plane of the dislocation. The motion of atoms to and from the dislocation is accomplished by vacancy movements. If an atom in the plane next to the edge jumps out of its position and attaches itself to the extra plane, as indicated in the accompanying figure, a vacancy is created which can then diffuse off into the lattice. Repetition of this process over and over will cause the extra plane to grow in size and, if this occurs, the process is said to be negative climb. On the other hand, if vacancies diffuse up to the extra plane of the dislocation and remove atoms from it, the plane grows smaller in size and positive climb is said to occur.

Dislocation climb only becomes of practical significance at elevated temperatures because of its dependence upon vacancies whose number and mobility depend very strongly on the temperature. Dislocation climb is important in high temperature creep and recovery phenomena. See Fig. 13.

Crystal Slip. This is the process by which a crystal undergoes plastic deformation, as a result of which one atomic plane moves over another. Slip is believed to occur through the movement of dislocations. The total deformation of a given crystal is the sum of many small lateral

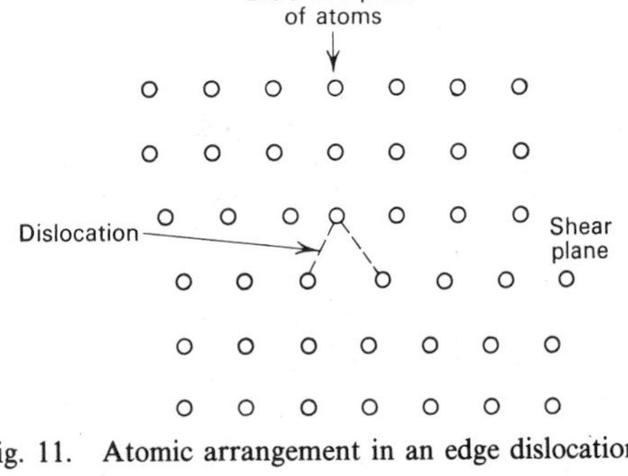

Fig. 11. Atomic arrangement in an edge dislocation.

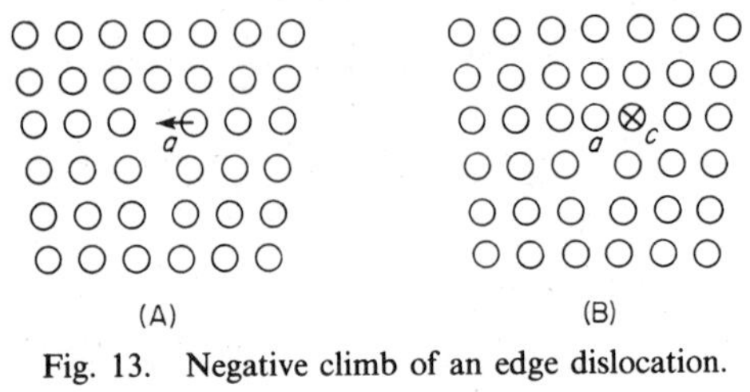

Fig. 13. Negative climb of an edge dislocation.

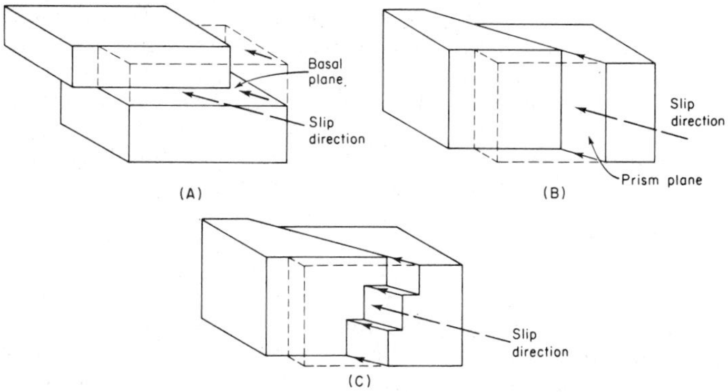

Fig. 14. Schematic representation of cross-slip in a hexagonal metal: (A) slip on basal plane; (B) slip on prism plane; (C) cross-slip on basal and prism planes.

displacements in parallel crystallographic planes of a given family. Moreover, each slip plane becomes more resistant to further deformation than the remaining potential slip planes.

Cross-Slip. This is slip that occurs simultaneously on several slip planes having the same slip direction. See Fig. 14. This type of plastic deformation is normally associated with the movement of screw dislocations. Screw dislocations can move on any slip plane that passes through the dislocation. This is a result of the fact that the slip plane of a dislocation is that plane which contains both the dislocation and its Burgers vector, and the fact that the Burgers vector of a screw dislocation lies parallel to the dislocation itself.

See also **Liquid Crystals;** and **Solid-State Physics.**

References

Bardsley, W., Hurle, D. T. J., and J. B. Mullin: "Crystal Growth," Elsevier-North Holland, New York, 1979.

Buckley, H. E.: "Crystal Growth," Wiley, New York, 1964.

Considine, D. M. (editor): Crystallization and Crystallizers in "Chemical and Process Technology Encyclopedia," pp. 326–336, McGraw-Hill, New York, 1974.

Gilman, J. J. (editor): "The Art and Science of Growing Crystals," Wiley, New York, 1964.

Gilmer, G. H.: "Computer Models of Crystal Growth," *Science*, **208**, 355–363 (1980).

Gruber, Boris: "Theory of Crystal Defects," Academic, New York, 1964.

Hoseman, R.: Diffraction by Matter and Diffraction Gratings in "The Encyclopedia of Physics," (R. M. Besancon, editor), Van Nostrand Reinhold, New York, 1974.

Kroger, F. A.: "Chemistry of Imperfect Crystals," Wiley, New York, 1964.

Lawson, W. D., and S. Nielsen: "Preparation of Single Crystals," Butterworth, London, 1958.

Nowick, A. S. and B. S. Berry: "Anelastic Relaxation Crystalline Solids," Academic, New York, 1971.

Ramachandran, G. N. (editor): "Crystallography and Crystal Perfection," Academic, New York, 1963.

Randolph, A. D. and M. A. Larson: "Theory of Particulate Process—Analysis and Techniques of Continuous Crystallization," Academic, New York, 1971.

Strickland-Constable, R. F.: "Kinetics and Mechanism of Crystallization," Academic, New York, 1967.

Ubbelohde, A. R.: "The Molten State of Matter," Wiley, New York, 1979.

van Gool, W.: "Principles of Defect Chemistry of Crystalline Solids," Academic, New York, 1966.

R. C. Vickery, Hudson Laboratories, Hudson, Florida.

CRYSTAL DETECTOR. A device consisting of a "cat's whisker" bearing on a semiconducting crystal, used in early radio sets and in microwave receivers. It depends for its action on the rectifying properties of the point contact, just as in the point contact transistor.

CRYSTAL (Face-Centered). A type of crystal structure in which atoms occupy the corners and centers of the faces of a cube or other parallelepiped. The face-centered cubic structure is *close-packed.*

CRYSTAL (Ferroelectric). Ferroelectric Effect.

CRYSTAL FIELD THEORY. A theory which was developed in the early 1930's in research on magnetism by Bethe, Van Vleck and others. It applied particularly to the transition metal ions, and is therefore conveniently treated by reference to those ions.

A transition metal ion in a complex or compound is considered to be subject to the electrostatic field of the molecules and ions in its neighborhood, particularly by those constituting its nearest neighbors. In the compounds to which the theory applies, the only nearest neighbors which are important to the theory are either negative ions, or molecules such as NH_3 which have unshared electron pairs and which orient themselves so that the negative end of the electron-pair dipole is directed toward the transition metal ion. The effect of these negative ions or negatively-oriented dipoles (which we shall hereafter call ligands, with the understanding that ligand field theory is a later development of crystal field theory) is to produce a negative field about the positive transition metal ion.

In the absence of this negative field, the *d*-electrons of the central ion have orbitals of equal energy, i.e., degenerate orbitals, but the field of the ligands affects the energies of these orbitals to different degrees. To show how this effect arises, consider the example chosen by Griffith and Orgel. This is the regular octahedral complex MX_6, where M is a metal ion of the first transition series of the elements (see **Periodic Table of the Elements**), and X is a ligand such as H_2O, NH_3 or Cl^-.

The five 3*d* orbitals of M have the forms indicated in the figure, in which the coordinate axes lie along the MX bond directions. It is clear from the figure that the d_{z^2} and $d_{x^2-y^2}$ orbitals have substantial amplitudes in the directions of the ligands but that the orbitals d_{xy}, d_{yz}, and d_{zx} tend to avoid them. Hence the energy of an electron in the d_{z^2} or $d_{x^2-y^2}$ orbitals will be substantially raised by the repulsive field of the ligands, whereas the energy of an electron in the d_{xy}, d_{yz}, or d_{zx} orbitals will be comparatively little affected. Furthermore, it is obvious from symmetry that the degeneracy of the last three orbitals is maintained in the octahedral complex and it can be shown by group theory that the d_{z^2} and the $d_{x^2-y^2}$ orbitals also remain degenerate. Consequently the five *d* orbitals split into a lower group of three and an upper group of two, the two groups being usually designated as t_{2g} and e_g respectively (or sometimes as γ_5 and γ_3, and as $d\epsilon$ and $d\gamma$ respectively).

In tetrahedral complexes it can be shown similarly that the *d* orbitals are again split into groups of three and two, respectively, but now the doubly degenerate orbital is lower. In all other important cases the degeneracy of the *d* orbitals is reduced even further.

In order to understand the electronic structure of an octahedral complex and the optical transitions which it can undergo, it is necessary to appreciate the principles determining the distribution of the *d* electrons among the t_{2g} and e_g orbitals. Let us begin by considering the ground state. Two separate tendencies are at work. The first is the tendency for the electrons to occupy, as far as possible, the orbitals of lowest energy in the ligand field. The second is for the electrons

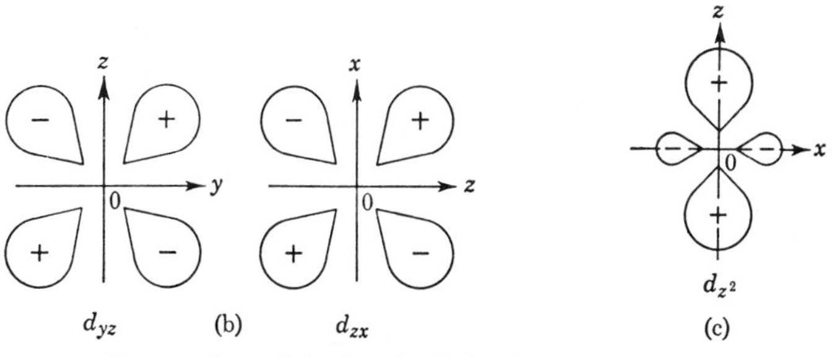

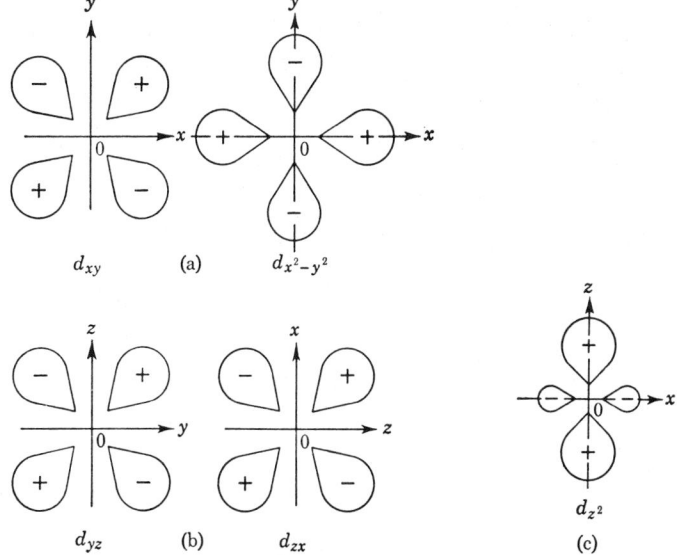

Cross sections of the five *d* orbitals, chosen in real form.

to go into different orbitals with their spins parallel, since this gives a lower electrostatic repulsive energy and a more favorable exchange energy. Let us now see how these ideas apply to an octahedral complex containing n $3d$ electrons.

The ion $[Ti(H_2O)_6]^{3+}$ has one d electron. In the ground state this will obviously occupy one of the t_{2y} orbitals. A transition is possible in which this electron is transferred to one of the e_y orbitals, and this occurs at 20,400 cm^{-1}. (The intensities of such transitions are low—$\epsilon_{max} \approx 10$—as they are symmetry-forbidden.) The converse situation arises in the hydrated copper(II) ion which has nine d electrons. In this ion the vacancy in the d shell is one of the e_y orbitals. This vacancy can be filled by exciting an electron from one fo the t_{2g} orbitals, giving rise to a transition at about 12,500 cm^{-1}. However, the ion $[Cu(H_2O)_6]^{2+}$ is strongly distorted in its ground state so that most of the degeneracy of the t_{2g} and e_g orbitals is removed and there is more than one transition in this region. From these two examples we see that the splitting between the t_{2g} and the e_g orbitals may be quite large, being usually in the range 20–50 kcal $mole^{-1}$.

It is only when we come to consider complexes with several d electrons that complications arise. If there are only two or three d electrons, both the above-mentioned tendencies can be satisfied simultaneously by placing the electrons in different t_{2g} orbitals with their spins parallel. However, when there are more than three d electrons this is no longer possible. If there are 4–7d electrons we then have the choice either of putting as many as possible into the low-energy t_{2g} orbital or distributing them so as to maintain a maximum number of parallel spins. This is illustrated in the accompanying table.

The former choice will be favored if the orbital separation Δ is large, and the latter will be realized if Δ is small. The value of Δ depends primarily on the nature of the ligand and the charge on the ion, and the following generalizations can be made for the first transition series: (1) For hydrated bivalent ions Δ falls in the range 7,500–12,500 cm^{-1}; (2) For hydrated tervalent ions Δ falls in the range 13,500–21,000 cm^{-1}; (3) The common ligands can be arranged in a sequence so that Δ for their complexes with any given metal increases along the sequence. A shortened series is I^-, Br^-, Cl^-, F^-, H_2O, oxalate, pyridine, NH_3, ethylenediamine, NO_2^-, CN^-. Lastly, Δ for the compounds of the second and third series is 40–80% larger than for corresponding compounds of the first series.

With these considerations in mind let us consider in turn the two extreme possibilities, known respectively as the "strong-field" and the "weak-field" case. If Δ is very large the tendency for electrons to go into separate orbitals will be outweighed by their tendency to occupy the t_{2g} orbitals (as against the e_g orbitals) in circumstances where these two tendencies conflict. In the strong-field case, therefore, a complex with up to six d electrons will have all these in t_{2g} orbitals with the maximum number of unpaired spins consistent with the restriction to the t_{2g} orbitals. As examples we may take the hexacyanoferrates(II) and (III) which possess six and five d electrons, respectively, all in t_{2g} orbitals with the maximum number of unpaired spins consistent with the restriction to the t_{2g} orbitals. As examples we may take the hexacyanoferrates(II) and hexacyanoferrates(III) which possess six and five d electrons respectively all in t_{2g} orbitals. The former has no unpaired electrons and the latter one. The next four electrons will then enter the e_g orbitals; the first two will go into different e_g orbitals with their spins parallel as in the octahedral complexes of Ni^{2+}. In Co^{2+} there is just one e_g electron.

The complex $[Co(NH_3)_6]^{3+}$ provides a good example of the strong-field case. The ground state is $(t_{2g})^6$ and the transitions observed at the longest wavelengths involve taking one of these electrons and putting it in an e_g orbital. According to the choice of the orbitals involved the final state may be one of the two triply degenerate states, and the bands associated with the two transitions have been observed for a number of d^6 complexes in each of the three transition series. (The separation between the bands is not in very good agreement with theory, however, so that it is difficult to obtain a reliable value of Δ for such cobalt(III) complexes. The reasons for this are not fully understood at present.)

The weak-field case is that in which the separation Δ is not large enough to overcome the tendency of the d electrons to go into different orbitals with their spins parallel. For example, in the hydrated manganese(II) ion the five d electrons each occupy one of the five d orbitals; this is because the separation Δ is insufficient to break up the highly stable half-filled shell in which all the electron spins are parallel. The same is true of $[Fe(H_2O)_6]^{3+}$ and of the hydrated iron(II) ion, which has six d electrons, the extra one being in one of the t_{2g} orbitals. It may be noted again that only in those complexes containing four, five, six, or seven d electrons is there an important distinction between the strong- and the weak-field cases; if there are one, two, or three d electrons they will necessarily occupy t_{2g} orbitals, while if there are eight or nine the vacancies in the d shell will occur in the e_g orbitals in both the strong- and the weak-field case. (For further discussion, see J. H. Van Vleck, *Phys. Review*, 1932, **41**, 208.)

CRYSTAL FILTER. Filter (Communication System).

CRYSTAL (Griebe and Schiebe Method). Griebe and Schiebe Method.

CRYSTAL HABIT. The external shape of a crystal, which depends on the relative development of the different faces, as well as upon the interfacial angles characteristic of the crystal.

CRYSTAL (Hermann-Manguin Symbols). Hermann-Manguin Symbols.

CRYSTAL (Homometric Pairs). Two crystal structures having the same x-ray diffraction pattern. This is possible because, basically, a

d-ELECTRON ARRANGEMENTS IN OCTAHEDRAL COMPLEXES

Number of d Electrons	Arrangement in Weak Ligand Field		N	Arrangement in Strong Ligand Field		N	Gain in Orbital Energy in Strong Field
	t_{2g}	e_g		t_{2g}	e_g		
1	↑	—	0	↑	—	0	0
2	↑↑	—	1	↑↑	—	1	0
3	↑↑↑	—	3	↑↑↑	—	3	0
4	↑↑↑	↑	6	↑↓↑↑	—	3	Δ
5	↑↑↑	↑↑	10	↑↓↑↓↑	—	4	2Δ
6	↑↓↑↑	↑↑	10	↑↓↑↓↑↓	—	6	2Δ
7	↑↓↑↓↑	↑↑	11	↑↓↑↓↑↓	↑	9	Δ
8	↑↓↑↓↑↓	↑↑	13	↑↓↑↓↑↓	↑↑	13	0
9	↑↓↑↓↑↓	↑↓↑	16	↑↓↑↓↑↓	↑↓↑	16	0

N Number of distinct pairs of electrons with parallel spin.

diffraction pattern depends only on the relative vector distances between the atoms in the lattice, not on their absolute positions in space.

CRYSTAL (Ice). Clouds and Cloud Formation.

CRYSTAL (Ionic). Ionic Crystal.

CRYSTAL (Isomorphous). One of two or more crystals which are similar in crystalline form and in chemical properties and are related in chemical composition in that one or more of the atoms and radicals in one are of similar chemical type to the corresponding atoms or radicals in the other. Usually, one or more of their other atoms or radicals are identical.

CRYSTALLINE CONE. A glassy body between the cornea and the sensory portion of each component of the compound eye in some species of insects.

CRYSTALLINE STYLE. A rod of protein bearing an amylolytic ferment produced continuously by secretion of the cells of a diverticulum or pouch of the intestine in certain bivalve and some gasteropod mollusks. It projects into the stomach, and as it is rotated by cilia it is worn away by contact with a gastric shield and mixed with the food. It is specially concerned with the digestion of carbohydrates and associated with herbivorous duct, ciliary feeding and intracellular digestion of proteins and fats.

CRYSTAL (Liquid). Liquid Crystals.

CRYSTALLIZATION. Crystals are formed (1) from solution, (2) from fusion, and (3) by sublimation.

The formation of crystals from solution, starting with an unsaturated solution, takes place when a solution is evaporated or cooled below the saturation point, except as retarded by supersaturation. Supersaturation is prevented by the addition of seed crystals of the substance. Since, in the case of the majority of soluble substances, solubility increases with increase of temperature cooling below the saturation point favors the formation of crystals. In very few cases, such as sodium sulfate above 32.4°C, calcium sulfate (slightly soluble), calcium hydroxide (slightly soluble), solubility decreases with increase of temperature, and the above statement would not apply. A case of wide scope and great importance is that of crystal formation by precipitation upon mixing two solutions. Actually, this is the same as for substances of greater solubility, since the substance precipitated is first formed in solution and the excess above the saturation point separates as precipitate. As a rule the crystals are larger and more perfect the slower their growth. Conversely, when small crystals are desired, rapid stirring and quick cooling are practical. The smaller the crystals of a given substance, the purer the material generally is. Small crystals may be increased in size by allowing them to stand in the mother liquor before separation.

An instustrial forced-circulation evaporative crystallizer is shown in Fig. 1. Sizes range from 18 inches (~ 46 centimeters) to over 42 feet (12.6 meters) in diameter, with no inherent limit to the size of the vessel. Slurry is moved by the circulation pump through the heat exchanger, where it is subjected to a temperature rise of 2 to 10°F. The heated liquor is discharged tangentially into the body at a point sufficiently far beneath the surface so that the liquor entering tangentially is just at the boiling point for the liquid depth at which it is submerged. As the liquid rotates around the body and rises toward the surface, it starts to boil, and this boiling induces a secondary circulation which creates a spinning toroid of fluid within the body. Depending on the location of tangent inlet with respect to the cone, this toroidal circulation can result in considerable secondary circulation and agitation within the body of the vessel. When properly designed, this type of vessel is capable of producing a smooth boiling action with relatively small amounts of salt being deposited on the walls, while still maintaining a suitable suspension of product crystals within the boiling zone and in the lower part of the vessel. Crystalline materials produced in this type of equipment include sodium carbonate, sodium sulfate, and sodium chloride. Operating cycles of this equipment between washouts to remove salt growth from the walls

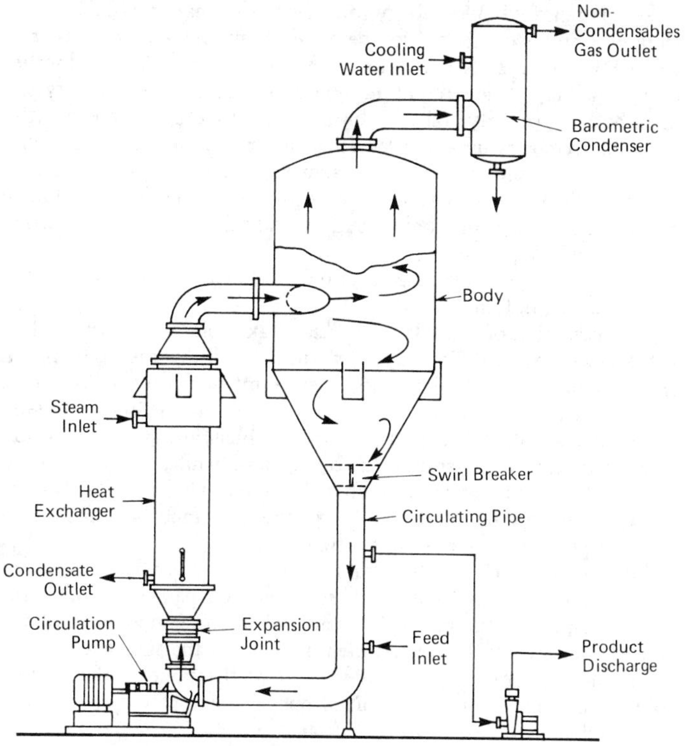

Fig. 1. Forced-circulation evaporative crystallizer.

of the body or from the heat exchangers normally range from 30 to 90 days.

To achieve some control of the number of fine particles within the crystallizer body, and thereby increase the overall particle size, it is necessary to selectively remove the fine particles so that they can be destroyed by the action of heat or dilution. A draft-tube baffle crystallizer of the type shown in Fig. 2 achieves these objectives. A body of growing crystals is suspended by the circulation flowing up the draft tube from the propeller shown close to the bottom of the vessel. From the areas surrounding this body of circulated slurry, a

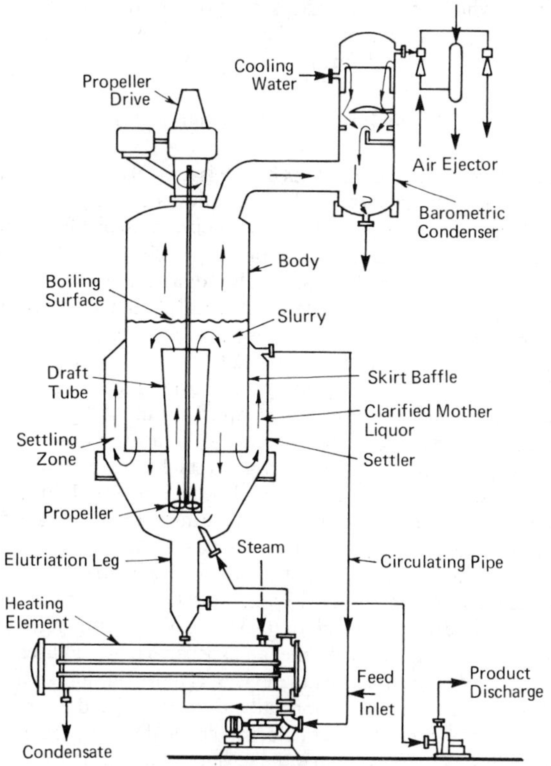

Fig. 2. Draft-tube baffle crystallizer.

stream is removed at relatively low velocity so that gravitational settling will produce a separation between the product-size crystals and relatively fine crystals which are removed with the clarified mother liquor leaving by the circulating pipe. In the draft-tube baffle crystallizer, both the velocity of the liquor in the settling zone and the quantity of liquor removed by the circulating pump are important to insure that the proper end-product size is achieved and that reasonable stability in particle size is obtained. The equipment is used to crystallize potassium chloride, ammonium sulfate, and other relatively fast-growing inorganic salts.

The foregoing descriptions cover only two of several industrial crystallizer configurations.

The formation of crystals from fusion takes place when the melted substance is cooled sufficiently slowly near and below the fusion point. If the cooling is rapid the fusion may result in the formation of an undercooled liquid of rigidity corresponding to a solid. Glasses, whether artificial, such as glass, vitreous enamels, and slags, or natural, such as vitreous rocks and minerals, e.g., obsidianite, are undercooled liquids. Rocks and minerals which have cooled sufficiently slowly from fusion form crystals, for example, granite. An important method of forming pure crystals is zone melting.

The formation of crystals by sublimation takes place when the vapor of a substance is condensed as a solid without passing through the liquid phase in so doing. This occurs when the temperature of the condenser is below that of the melting point of the substance.

The heat of crystallization is in amount the same as the heat of solution of a given substance but of opposite sign.

For references see entry on **Crystal.**

CRYSTALLOBLASTIC. Designating the textures of metamorphic rocks resulting from recrystallization under differential and directed pressures and high viscosity.

CRYSTALLOGRAM. The x-ray diffraction pattern of a crystal, whence the crystal structure may be obtained.

CRYSTALLOGRAPHY. Crystal; X-Ray Analysis.

CRYSTALLOID. Colloid System.

CRYSTAL (Mixed). A crystal consisting of two or more chemical compounds, which may have the same positive radical or the same negative radical, and which, in their pure form, are isomorphous, i.e., have the same crystal form.

CRYSTAL (Optical Mode). Optical Mode.

CRYSTAL OSCILLATOR. This device is a precise mechanical resonator and frequency generator. The need for a stable, accurate, and low-cost frequency generator for the precise control of commercial radio and other higher communication frequencies led to the development of piezoelectric resonators, notably quartz crystals, for a wide variety of applications. The quartz crystal has become highly developed as a frequency standard for timekeeping and as a time-signal generator. See also **Oscillator;** and **Piezoelectric Effect.**

Piezoelectricity was discovered by the Curie brothers in 1880. The term *piezo* is derived from the Greek word meaning "to press." The effect causes a crystal to exhibit electrical polarity when the crystal is subjected to mechanical pressure. Conversely, the crystal is physically deformed when subjected to an electrical potential. Specifically, piezoelectricity is a property of nonconducting solids that have a crystal lattice which does not have a center of symmetry.

Quartz, tourmaline, and rochelle salts and such synthetic crystals as ethylene diamine tartrate (EDT), dipotassium tartrate (DKT), and ammonium dihydrogen phosphate (ADP) have varying suitability as piezoelectric elements. Tourmaline, an expensive material used mainly in hydrostatic pressure-measuring devices, is more durable than quartz and, for a given frequency, normally is more rugged than quartz. Rochelle salt has a greater piezoelectric effect than any other crystal but has the disadvantage of a greater sensitivity to temperature change than quartz. EDT has an advantage over quartz when used in frequency-modulated oscillators because of the wide gap between its resonant and antiresonant frequencies.

All piezoelectric crystals should have a good temperature coefficient, that is, should show as little change in resonant frequency as possible under large variations in temperature. Ideally, the piezoelectric constant of proportionality between the mechanical and electrical variables must be the same for both direct (pressure-to-electricity) and converse effects.

Natural quartz crystal and new synthetic quartz crystal have been found best to meet the properties required of a piezoelectric element. Quartz is hard (7 on Mohs' scale as compared with a diamond, which is 10) and is relatively abundant and stable. Because of the anisotropic structure of quartz, cutting a crystal in different orientations makes it possible to obtain crystals for the widest range of applications, including frequency control, filters, resonators, and electromechanical transducers. Each of these uses depends on the orientation of the crystal cut with respect to the crystallographic axes. As shown in the accompanying figure, the principal axes in quartz are identified as the optic

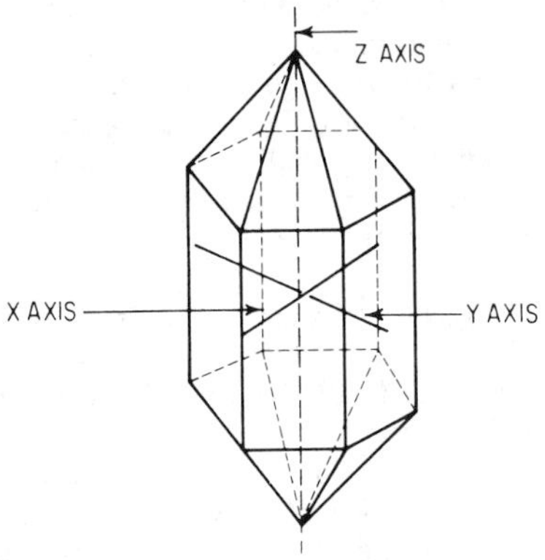

Principal axes in a quartz crystal.

(Z), the mechanical (Y), and the electrical (X) axes. By use of polarized light and x-ray diffraction techniques, the various axes may be properly located, permitting a crystal plate to be cut from the quartz with the performance characteristics desired. After cutting and extensive processing, the crystal plate is mounted at the nodal points of its normal vibration. These points, which also serve as electrical connecting points, allow the crystal to vibrate freely with a minimum of damping. Finally, the mounted crystal is hermetically sealed in a dry inert atmosphere within a crystal holder.

A stability of 1 part in a billion is routinely possible by the use of proportional ovens that can regulate the crystal temperature within $\pm 0.001°C$. By employing multiplying or binary techniques, frequency-generating devices in the range of from less than 1 Hz to more than 600 MHz are available. In the temperature-compensated cyrstal oscillator (TCXO), stability over a wide range of temperatures is obtained by a compensating network that shifts the frequency of the crystal by an amount approximately equal and opposite to the shift of frequency caused by the temperature change. This network eliminates the need for an oven. TCXOs are available with temperature stabilities of ± 0.5 ppm over a temperature range of minus 40 to plus 70°C.

William O. Bennett, John J. Carpenter, Frank Dostal, and E. Van Haaften, Bulova Watch Company, Inc., New York.

CRYSTAL OVEN. A temperature-controlled container for stabilizing the temperature and resonant frequency of a crystal used in a crystal-controlled oscillator.

CRYSTAL PHASES (α-, β-, γ-, ϵ-, η-, etc.). Certain alloy systems may form different crystal structures, according to the relative proportions of the constituents, e.g., Cu-Zn, for which no less than five different phases are known. In many cases, the same crystal structure occurs with quite different constituent metals, so that it is often possible to use the one expression such, for example, as β-phase, to cover a wide variety of compounds all having the same basic structure. This effect is explained by the Hume-Rother rules. Pure substances, as well as alloys, may exhibit more than one crystal structure, depending on temperature and past history, e.g., cobalt, iron, titanium.

CRYSTAL PICKUP. Since piezoelectric crystals produce electrical voltages when subjected to mechanical stresses they offer possibilities for various electromechanical processes. One of these is the phonograph pickup, where the phonograph needle operating in the groove of the record must transmit mechanical motion to something which will convert it into electrical effects so electronic amplifiers may be used. The crystal is one of the most sensitive and at the same time one of the highest fidelity devices for doing this. While many crystals exhibit the piezoelectric effect. Rochelle salt is the most sensitive and is used for pickups, microphones, etc. Among the other crystals used, quartz is noted for its durability. Through a mechanical linkage the motion of the needle is transmitted into mechanical stresses on the cyrstal and hence produces electrical effects which may be amplified and then converted to sound by the loud-speaker.

CRYSTAL (Rotation-Reflection Axis). Rotation-Reflection Axis.

CRYSTAL SEMICONDUCTOR. Semiconductor.

CRYSTAL SIZE CONTROL. Conversion Coatings.

CRYSTAL (Tamm Levels). Tamm Levels.

CRYSTAL (Twinning). Twinning (Crystal).

CTENOID SCALES. Fishes.

CTENOPHORA. Ocean Resources (Living).

CUBE. 1. A solid bounded by six planes, having its face angles all right angles, and its twelve edges equal. 2. The third power of a number or quantity.

CUBE ROOT. Cubic Equation; Square and Square Root.

CUBIC CRYSTAL. Crystal.

CUBIC EQUATION. An algebraic equation of the third degree in one or more variables but usually taken to mean the case of one variable so that its most general form is $a_0x^3 + a_1x^2 + a_2x + a_3 = 0$. Equations for its three roots have been known to mathematicians since the sixteenth century and many different forms of them can be given. All of them are relatively complicated but the following procedure is probably as simple as any.

First substitute $x = y_1 - a_1/3a_0$ to get $a_0y^3 + b_2y + b_3 = 0$ and then convert this into $y^3 + py + q = 0$, where $p = b_2/a_0 = (3a_0a_2 - a_1^2)/3a_0^2$; $q = b_3/a_0 = (2a_1^3\ 27a_0^2a_3 - 9a_0a_1a_2)/27a_0^3$. Its three roots are $y_1 = u + vy_2 = e(u + ev)$; $y_3 = e(eu + v)$, where e is either one of the complex cube roots of unity, $(-1 \pm i\sqrt{3})/2$. In order to calculate u and v let

$$D = \sqrt{\frac{q^2}{4} + \frac{p^3}{27}}$$

then

$$u = \sqrt[3]{\frac{-q}{2} + D}$$

where any one of the three roots may be taken and

$$v = \sqrt[3]{\frac{-q}{2} - D}$$

but the relation $v = -p/3u$ must hold. The solution y_1 is known as Cardan's formula (1515).

The general solution can be somewhat simplified, depending on the nature of the roots.

1. $D^2 > 0$. One root is real, the two others are conjugate imaginary. If the real root of u is chosen, then v is also real and y_1 is the real root of the cubic.

2. $D^2 = 0$. In this case there are three real roots, two of which are equal, and the general equations become $y_1 = 2\sqrt[3]{-q/2}$, $y_2 = y_3 = -y_1/2$.

3. $D^2 < 0$. All roots are real and none are equal. From the definition of D, p, < 0 or $p = -P$, where P is a positive quantity. A trigonometric substitution is now convenient.

$$\cos\phi = \frac{-3\sqrt{3}q}{2P\sqrt{P}}$$

where ϕ can be taken in the first quadrant by proper choice of the sign of $\sqrt{P/3}$. The roots are then

$$y_k = 2\sqrt{\frac{P}{3}}\cos\frac{1}{3}(\phi + 2k\pi),\ k = 0, 1, 2$$

These are the three real roots of $y^3 - Py + q = 0$.

Although straightforward in principle, application of these formulas is quite cumbersome and they are seldom useful in any practical case. Certain equations, combined with graphs and numerical tables, which give approximate roots for the cubic may be found in the literature, but these are also rather complicated. Unless there is some particular reason for doing othewise, one should select a computer aided numerical method (see **Approximate Calculation**) if one needs to solve a cubic equation.

See also terms listed under **Mathematics.**

CUCKOOS AND COUCALS (*Aves, Cuculiformes, Cuculidae*). This family of birds is widespread and many species can be found in all continents except Anarctica. Their length is from 14–70 centimeters ($5\frac{1}{2}$–$27\frac{1}{2}$ inches) and the weight is 25–1000 grams (0.8–35 ounces). The bill curves downward slightly with a protruding hook at the tip of the upper mandible and a deep cleft to the beak. The birds have 13 to 14 cervical vertebrae. The lateral toe is reversed. The main food is insects. Cuckoos and coucals have very different habitats. Most species are unsociable. Of the 128 species, 50 are brood parasites, and the clutch of the remaining species numbers from 2–6 unmarked eggs. In contrast, the eggs of many brood parasites are spotted. Eggs are laid at 2-day intervals, and at intervals of 1 day only in the smallest species. Incubation in the nonparasitic species begins after the first egg is deposited. The young, as they hatch, are rosy-red or dark red and blackish, naked or only very sparsely covered.

Cuckoos and coucals have 7 subfamilies: 1. Cuckoos (*Cuculinae*); 2. *Coccyzinae*; 3. *Phaenicophaeinae*; 4. *Crotophaginae*; 5. Ground Cuckoos (*Geococcyginae*); 6. *Couinae*; 7. Coucals (*Centropodinae*).

Cuckoos usually have inconspicuous colors, such as light gray or light brown to deep red-brown and black. The plumage of many is shiny or shimmering. The basic color is often overshadowed by light or dark transverse bands and less often by longitudinal striping, particularly below, on the wings and the tail. Aside from the large yellow and green areas of the Didric cuckoo, the only vivid colors in cuckoos are the frequently colored bills, the red eyes (mostly in older birds), and the colored or sometimes black, naked areas about the eyes. See accompanying illustration.

The Coucal (*Centropodinae*) reach a length of 36–70 centimeters (14–$27\frac{1}{2}$ inches). The skin is thick and dark; the intermost back talon is almost straight and usually elongated (in other cuckoos it is more curved and short). The wings are short and the legs are long; the bill is powerful. The colors are distributed over large areas, usually black, red-brown ranging to light brown and white, sometimes with white shaft streaks. The birds have a masterful way of using their legs to crawl and climb through the densest grasses and the most tangled foliage and lianas. Their diet is varied: arthropods, such as grasshoppers, ants, centipedes, and scorpions, as well as snails, frogs, lizards, snakes, and small birds. The nests are usually globular and

Yellow-billed cuckoo (*Coccyzus americanus*), a long slender bird, grayish brown above, whitish below, with white marks on the ends of the tail feathers; lower part of beak yellow.

have a side entrance, but also are sometimes hidden under grass, or are bowl-shaped like those of crows. The clutch size is from 2 to 5; the eggs are white and covered with a thick, leatherlike, chalky layer.

There is only 1 genus with 26 species; among them are: 1. The Violet Coucal (*Centropus violaceus*); 2. The Bismarck Coucal (*Centropus ateralbus*); 3. The Pheasant Coucal (*Centropus phasianinus*); 4. *Centropus goliath*; 5. The Common Coucal (*Centropus sinensis*); 6. *Centropus toulou*; 7. The Gabon Coucal (*Centropus anselli*); 8. The Senegal Coucal (*Centropus sensgalensis*); 9. The White-Browed Coucal (*Centropus superciliosus*); 10. *Centropus melanops*. See also **Cuculiformes.**

CUCKOO WASP (*Insecta, Hymenoptera*). Small wasps which lay their eggs in the nests of solitary wasps and bees. Usually metallic blue or green. Family *Chrysididae*.

CUCULIFORMES (*Aves*). This order of terrestrial birds reach a length of 14–70 centimeters ($5\frac{1}{2}$–$27\frac{1}{2}$ inches) and a weight of 25–100 grams (0.8–3.5 ounces). The tail is usually relatively long and has 10 rectrices (a quill in a bird's tail, especially a long feather of use in steering), and 8 in those birds which inhabit the warmer parts of America. The feet are grapplers with the lateral toe either permanently pointing back or reversed. The posterior edge of the sternum has 2 notches on each side; one of them may form a window or else be lacking. The young remain in the nest for a comparatively long time; many leave the nest before they are able to fly. Their area of distribution covers all continents except Antarctica. There are 2 families: 1. Turacos (*Musophagidae*), and 2. Cuckoos and Coucals (*Cuculidae*). There are 40 genera with 146 species.

Turacos differ markedly from cuckoos and coucals. Each family has its diagnostic characteristics and is easily distinguishable. The feathers are shed in such a way that the two neighboring feathers of each one lost persist (saltatory or transilient moulting). In all comparable birds, the wing moulting proceeds from one feather to the next, unless the moult is entirely irregular, or else the feathers are shed simultaneously. Repeatedly, the relationship between the turaco and the gallinaceous birds, particularly the megapodes, the pheasants, and the hoatzins, is emphasized. This may indicate that the cuckoos and their relatives branched off early from the original phylum and that even today they differ little in appearance from the ancestral arboreal birds. See also **Cuckoos and Coucals;** and **Turacos.**

CUCUMBER. Cucurbitaceae.

CUCUMBER TREE. Magnolia Trees.

CUCURBITACEAE. A small family of plants, largely restricted to tropical or warm climates. Most of its 650 species are climbing or trailing herbaceous plants which grow very rapidly. They are mostly annuals. The stems are hollow and in most species abundantly supplied with stiff bristly hairs. The large leaves are borne alternately on the stem, have a distinct, often long, petiole, and show a variety of shapes. The tendrils, which are a conspicuous feature of many members of this family, appear in the axils of the leaves and are interpreted as stems modified greatly. They are very sensitive organs, responding to the lightest touch of any solid substance, and often show a change in the direction of twining in the middle of a single tendril. In many species the nutating or circling movement of the tendril is very rapid. The flowers are axillary, either borne singly or in various types of inflorescence, and are usually yellow or white. The plants are either monoecious or dioecious. The calyx is adnate to the inferior ovary, the corolla is 5-lobed and inserted on the calyx. The stamens are typically five, but show great variation in number through fusions. The inferior ovary is 1- to 3-celled and usually contains many flattened seeds. The latter lack endosperm. The fruit is a variety of berry called a pepo, differing from a berry in that the receptacle enters into the formation of the rind or outer wall. The germination of the seeds of the commonly grown members of this family exhibits one rather striking peculiarity. When the arched hypocotyl emerges from the seedcoats a small peg forms on its lower end. This peg prevents the seedcoats from sticking to the cotyledons, which are withdrawn and carried into the air by the straightening of the arched hypocotyl. Many members of this family are grown in cultivation, as for example squashes, pumpkins, and cucumbers, and certain ornamental species, like *Echinocystis, Momordica,* and some of the gourds.

Cucurbita, pumpkins and squashes. These are rather coarse annual vines having very rough bristly stems, large, long-stalked leaves and axillary (axil) flowers of two kinds. The staminate (stamen) flowers have long stalks while the pistillate flower stalks, or peduncles, are short. Staminate flowers have a rudimentary ovary while pistillate flowers have three staminodia, or vestigial stamens; in both kinds of flowers the 5-lobed yellow corolla is conspicuous. Insect-pollination is usual. It is probable that these plants are native to tropical America, where they have been long culitivated. Cultivation is now widespread.

Pumpkins, *Cucurbita pepo,* are of many varieties, sizes and shapes. Included here are the Field Pumpkin, Sugar Pumpkin, Pie Pumpkin and Mammoth Pumpkin, Fordhook, Scallop (Petty-pans), Crookneck Squashes, and Marrow Squashes. In this group the stems are prickly as a rule, and more or less 5-angled.

Squashes, *Cucurbita maxima,* are plants having cylindrical stems which are hairy rather than bristly. Here are found Hubbard squashes, turban squashes and mammoth squashes, the latter often of immense size and frequent occurrence, sometimes weighing over 100 pounds. *Cucurbita moschata* includes cushaw and cheese types of squashes.

Cucumis, Muskmelons, Cantaloupes and Cucumbers. The plants in this genus are considered to be natives of tropical Asia and Africa and the East Indian Islands, where they have been in cultivation for many centuries. In these plants the tendrils are unbranched, the staminate flowers are borne in small clusters in the axils of the leaves, while the pistillate flowers are solitary. Many more staminate flowers are formed than pistillate, to insure successful pollination, which is almost entirely by insects.

Cucumis melo includes melons of various kinds, among them muskmelons and cantaloupes. Many varieties bear inedible fruits, some of which may be used in making preserves. Few of these are cultivated in American gardens. The fruits have a warted or ribbed skin, but never hairy or spiny. They are probably native to southern Asia.

Cucumis sativus is the cucumber. This is a native of the East Indies. There, and in Asia, cucumbers have been cultivated since earliest times. Many varieties have been developed. Certain varieties, grown under glass, are often seedless. Others are largely grown for pickling. The best pickling fruits are grown in regions having a cool climate. For pickling, the fruits are picked while still young and small. They are first salted in brine, after which they are bottled in vinegar, often with the addition of various spices or other flavorings, such as dill, or mustard, or peppers.

Gherkins, *Cucumis Anguria,* are native to the West Indies. Small cucumbers are also frequently called gherkins.

Citrullus. Watermelons, Citron and Colocynth. These are natives of Asia, Africa and southern Europe. They are coarse trailing vines with branched tendrils and lobed leaves.

Citrullus vulgaris includes the watermelon and citron as varieties.

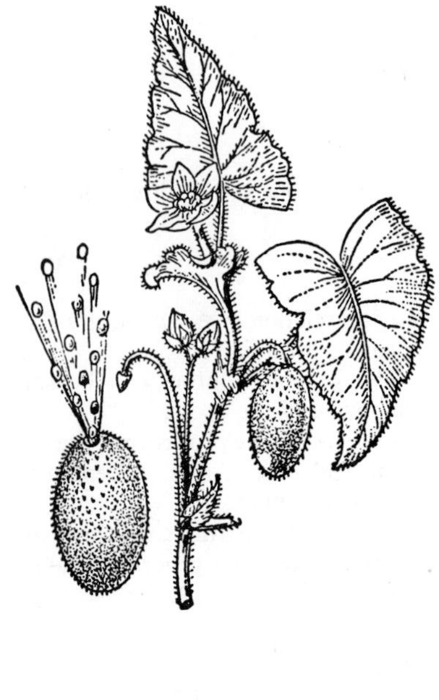

Squirting cucumber. Pressure develops inside the fruit and, as it is detached, a hole is torn through which the seeds are violently discharged with the juice.

It is native to Africa, where it has been cultivated since the time of Egyptian supremacy. Watermelons are grown for the juicy tender flesh. In contrast to them, the flesh of the citron is firm and inedible when raw. It is grown largely for preserving or for pickling. Preserved citron is used in cakes and in the making of certain kinds of bread. Because the juice of the citron is rich in pectin, it is much used in making jellies, especially with fruits naturally lacking in pectin and hence not capable of "jelling."

Many members of this family are grown as ornamentals or for their curious and sometimes useful fruits. Many such are classed as gourds. One of these, *Lagenaria vulgaris*, a native of the Old World tropics, is known as the calabash gourd or bottle gourd, from the shape of its fruit. Excellent flasks are made from the woody pericarp.

Luffa cylindrica, the "bath sponge," is frequently seen in gardens. The vascular tissues of the pericarp from an intricate net which is sometimes used as a sponge. Many species of *Luffa* have edible fruits. Mostly they are natives of the Old World. Another gourd, a native of tropical America, is *Sechium edule*, grown for its edible fruit. Many other species of gourds have curiously ornamental fruits.

Ecballium elaterium, the squirting cucumber, found in Mediterranean regions, has a fruit which when mature is very turgid. When the fruit is broken, the seeds are forcefully ejected by the contraction of the pericarp. From the fruit is obtained a powerful purgative.

Citrullus colocynthis is another Cucurbitaceous plant, the fruit of which yeilds a drug. In this species pulp of the fruit gives colocynth.

Echinocystis lobata, a native American plant, is frequently grown for ornament, and as a vine to cover unsightly places quickly. Its small white flowers are pleasantly fragrant. The staminate flowers are borne in long-stalked many-flowered inflorescences, while the pistillate are borne singly and are very short-stalked. The tendrils of this plant are especially sensitive to touch and move very rapidly.

CUEA. **Ocean; Ocean Resources (Living).**

CULTIVAR. **Hybrid; Plant Breeding.**

CUMACEA. An order of crustaceans made up of a few marine species of small size, highly specialized for living in mud or sand.

CUMIN. **Flavorings.**

CUMMINGTONITE. The mineral cummingtonite is a variety of amphibole which is essentially $(Mg, Fe, Mn)Si_8O_{22}(OH)_2$, the amounts of magnesium and iron varying as they replace one another. Cummingtonite is generally restricted to material containing from 50 to 70% $MgSiO_3$. The name grunerite has been applied to cummingtonite which contains more than 50% of the $FeSiO_3$ molecule. Cummingtonite usually occurs as a dark green to brown fibrous to lamellar mineral. It derives its name from Cummington, Massachusetts. See also **Amphibole.**

CUMULANTS. The cumulants $\kappa_1, \kappa_2 \ldots$ of a probability distribution function are defined by the identity

$$\exp\left[it\kappa_1 + \frac{(it)^2}{2!}\kappa_2 + \frac{(it)^3}{3!}\kappa_3 + \cdots\right] = \left[1 + it\mu_1' + \frac{(it)^2}{2!}\mu_2' + \frac{(it)^3}{3!}\mu_3' + \cdots\right] \equiv \phi(t)$$

where $\mu_1', \mu_2' \ldots$ are the moments of the distribution about the origin and $\phi(t)$ is the characteristic function. $\Psi(t) = \log \phi(t)$ is called the cumulant generating function. κ_1 is equal to the mean μ_1' and the other cumulants can be expressed in terms of the moments about the mean; the first four can be given by $\kappa_1 = \mu_1'$, $\kappa_2 = \mu_2$, $\kappa_3 = \mu_3$, $\kappa_4 = \mu_4 - 3\mu_2^2$.

The cumulants other than the first are invariant under a change of origin. For the normal distribution all cumulants after the second are equal to 0; for the Poisson distribution all the cumulants are equal to μ. $\gamma_1 = \kappa_3/\kappa_2^{3/2}$ and $\gamma_2 = \kappa_4/\kappa_2^2$ provide measures of skewness and kurtosis.

CUMULATIVE EXCITATION. An excited atom, in the metastable state, may receive a further increment of energy by collision, as with an electron, and thus be raised to a still higher energy state. This process by which an atom is raised by collision from one excited state to higher states is known as cumulative excitation. In fact, it is possible for an atom in the metastable state to receive sufficient energy by this process to be ionized and this process is designated as cumulative ionization.

CUMULATIVE PROBABILITY FUNCTION. **Frequency Function.**

CUMULONIMBUS. **Clouds and Cloud Formation.**

CUMULOSE. The term proposed by Merril in 1897 for sediments composed almost, if not entirely, of carbonaceous material, such as peat and lake mucks.

CUMULUS. **Clouds and Cloud Formation.**

CUNNINGHAMIA LANCEOLATA. **Araucarias.**

CUPOLA FURNACE. **Iron Metals, Alloys, and Steels.**

CUPOLA STRUCTURE. In geology, cupola is the term proposed by R. A. Daly in 1911 for a subsidiary dome-like protrusion in the roof of a batholith. Cupolas are supposed to be the reservoirs for the concentrated rising gases for the batholithic magma and may serve as the loci for volcanoes.

CUPRESSACEAE. **Arborvitae; Cedar Trees; Cypress Trees; Juniper Trees.**

CUPRITE. The mineral cuprite, cuprous oxide, Cu_2O, occurs as isometric crystals, usually octahedrons, but may be cubes, dodecahedrons or modified combinations. It also is found as a massive, earthy material. Its fracture is cochoidal to uneven; brittle; hardness, 3.5–4; specific gravity, 6.14; luster, submetallic to earthy; color, red; nearly transparent to nearly opaque. Its streak is shining brownish-red. Cuprite is a secondary mineral resulting doubtless from the oxidation of copper sulfides. It is often found associated with native copper, malachite and azurite.

Cuprite is a fairly common mineral, and of the many localities in which it occurs may be mentioned the Province of Perm, in the U.S.S.R.; Chessy, France; Broken Hill, New South Wales; Corocoro, Bolivia; Andacollo, Chile; Bisbee, Arizona; and Del Norte County, California. Magnificent large transparent red gem crystals, some with a coating of malachite, have been found at Ojunga, S.W. Africa. The name cuprite is derived from the Latin *cuprum*, copper.

CUPRONICKEL. **Copper.**

CUP-TYPE ANEMOMETER. **Wind and Air Velocity Measurements.**

CURARIZATION. **Tetanus.**

CURASSOW (*Aves, Galliformes*). Birds of a few species found in northern South America. They are about the size of turkeys and are arboreal in habit. Excellent as food and sometimes domesticated.

The guan of central and South America is related to the curassows. One species, the chachalaca, *Ortalis vetula*, enters southern Texas. See also **Galliformes.**

CURB PRESS. **Expression (Mechanical).**

CURCULIO (*Insecta, Coleoptera*). The curculios are weevils which damage apple, apricot, cherry, grape, peach, pear, plum, quince, and various other stone fruits. The best known of the curculios is the *plum curculio* (*Conotrachelus nenuphar*), which damages plums, but also has allied specialists for apple, grape, and quince. The adult plum curculio is a small grayish-brown snout beetle, with black and white markings and four prominent dark humps on its back. The larva is a whitish mass, legless, slightly curved, with a brown head and reaching up to $\frac{3}{8}$-inch (9 to 10 millimeters) in length. Both forms of the insect are damaging. The adult feeds on fruit in spring, making crescent-shaped cuts in the fruit in which to lay eggs. The larva makes tunnels in the fruit as it eats its way through its food source. Distribution of the plum curculio is essentially east of the Rocky Mountains in the United States.

Fruit damaged by plum curculio usually falls before it is mature. Apples, for example, drop during May and June. It is highly advantageous to pick up prematurely fallen fruit and destroy it by burning, thus preventing any reinfestation. It is also well to cultivate the soil around all fruit trees affected by this pest during late spring and early summer to destroy the larvae and pupae in their cells in the earth. The adult curculio hibernates under leaves or trash in winter. All trash under which the beetle may find shelter must be destroyed. Overgrown hedges and fences are favorite hiding places for the insect.

Plum curculios are shy and prefer the deep shade when they do their damage. Therefore, trees should be properly pruned to admit the sun. Fertilizers should be used generously but judiciously to otherwise maintain healthy trees.

Natural enemies of the curculio include birds and parasites. Minute wasps (*Trichogramma*) attack the larvae in the fruit; other parasites attack the eggs. Both the adult curculio and larvae are attacked by fungal diseases. Lack of resistance to long and cold winters is also a factor.

Specialist species of the curculio, the habits and damage of which parallel those of the plum curculio, include:

Apple curculio (*Anthonomus quadrigibbus*). This is a grub, white soft, about $\frac{1}{2}$-inch (12 millimeters) in length. Also effects apricot.

Grape curculio (*Craponius inaequalis*). Larva is small, white, and features a brown head. In June and July, the insect infests the grape. The point of entry causes a small black hole in the skin, around which the fruit becomes discolored. The adult grape curculio is a grayish-brown, snout-type beetle, attaining a length of about $\frac{1}{10}$ inch (2.5 millimeters).

Quince curculio (*Contrachelus crataegi*). This insect is a bit larger than the plum curculio and has a different life-cycle. In the fall, the grubs exit the fruits and enter the ground, whereupon they hibernate and transform to adults, energing during the late spring or early summer.

Rhubarb curculio (*Lixus concavus*). This insect is an elongated grub, attaining a length of about $\frac{3}{4}$ inch (18 millimeters), which bores into the crowns and roots of the rhubarb plant.

Walnut curculio. The adult is a beetle, about $\frac{1}{4}$-inch (6 millimeters) long, with a curved snout, and prominent humps and ridges on the wing covers. The larva is a white, legless worm with a brown head and ranging up to $\frac{1}{2}$ inch (12 millimeters) in length. The walnut curculio resembles the pecan weevil. The adult feeds on newly formed nuts and new foliage. Females lay eggs in nuts, causing them to drop prematurely. Distribution is throughout the United States.

CURCUMA. Turmeric.

CURETTAGE. Scraping of an organ, bony cavity, or other portion of the body with a curette or other instrument. Root curettage in the treatment of pyorrhea is an example. The dentist removes the tartar deposits from the roots of teeth with curettes designed to reach between the gum and the tooth. Curettage is sometimes used in combination with corticosteroid therapy for the bridging synechiae seen in Ahserman's syndrome (severe endometritis, as in secondary amenorrhea, either postpartum or postabortion).

CURIE. Units and Standards.

CURIE POINT (or Curie Temperature). Ferromagnetic materials lose their permanent or spontaneous magnetization above a critical temperature (different for different substances). This critical temperature is called the Curie point. Similarly, ferroelectric materials lose their spontaneous polarization above a critical temperature. For some such materials, this temperature is called the "upper Curie point," for there is also a "lower Curie point," below which the ferroelectric property disappears. See also **Ferromagnetism.**

CURIE-WEISS LAW. The transition from ferromagnetic to paramagnetic properties, which occurs in iron and other ferromagnetic substances at the Curie point, is accompanied by a change in the relationship of the magnetic susceptibility to the temperature. P. Curie stated in 1895 that above this point the susceptibility varies inversely as the absolute temperature. But this was found to be not generally true, and was modified in 1907 by P. Weiss to state that the susceptibility of a paramagnetic substance above the Curie point varies inversely as the excess of the temperature above that point. At or below the Curie point, the Curie-Weiss law does not hold.

CURING (Rubber). Rubber (Natural).

CURING (Skins and Hides). Tannin.

CURIUM. Chemical element symbol Cm, at. no. 96, at. wt. 247 (mass number of the most stable isotope), radioactive metal of the Actinide series, also one of the Transuranium elements, mp estimated 1350 ± 50°C. ^{247}Cm has a half-life of 1.64 × 10^7 years. Other long-lived isotopes are ^{245}Cm ($t_{1/2}$ = 9320 years), ^{246}Cm ($t_{1/2}$ = 5480 years), ^{248}Cm ($t_{1/2}$ = 4.7 × 10^5 years), and ^{250}Cm ($t_{1/2}$ = 2 × 10^4 years). Other known isotopes are ^{238}Cm, ^{242}Cm, ^{244}Cm, and ^{249}Cm. Electronic configuration

$$1s^22s^22p^63s^23p^63d^{10}4s^24p^64d^{10}4f^{14}5s^25p^65d^{10}5f^76s^26p^66d^17s^2.$$

Ionic radius: Cm^{3+} 0.98 Å.

First identified in 1944 by G. T. Seaborg, R. A. James and A. Ghiorso, who found ^{242}Cm in the product obtained by bombarding ^{239}Pu with alpha particles of resonance energies. Later L. B. Werner and I. Perlman produced and isolated the same isotope by the action of neutrons upon ^{241}Am.

In experiments the concentration of curium must be kept low in order to avoid the formation of a reducing medium due to the action of the ^{242}Cm alpha particles on H_2O. At a concentration of 10^{-5} molar in curium, and under conditions where americium(III) is oxidized to americium(VI) in the same solution, the curium is not oxidized above the (III) state with ammonium peroxydisulfate.

The solubility properties of curium(III) compounds are in every way similar to those of the other tripositive Actinide elements and the tripositive Lanthanide elements. Thus the fluoride and oxalate are insoluble in acid solution, while the nitrate, halides, sulfate, perchlorate, and sulfide are all soluble.

Solid curium trifluoride has been prepared by drying the fluoride, which precipitates from dilute HNO_3 upon the addition of hydrofluoric acid. Curium trifluoride can be reduced to the metal by heating at 1275°C in a beryllia crucible with barium vapor. The metal is silvery in color and has the properties of an electropositive element in common with the other Actinide elements.

The ion Cm^{3+} is colorless, as are its compounds generally. CmF_3 is hexagonal, Cm_2O_3 is white and CmO_2 is black and hexagonal. $CmCl_3$ is light yellow and hexagonal. Cm^{4+} is known in solution only as the complex fluoride.

In research at the Institute of Radiochemistry, Karlsruhe, West Germany during the early 1970s, investigators prepared alloys of curium with iridium, palladium, platinum, and rhodium. These alloys were prepared by hydrogen reduction of the curium oxide or fluoride in the presence of finely divided noble metals. The reaction is called a *coupled reaction* because the reduction of the metal oxide can be done in the presence of noble metals. The hydrogen must be extremely pure, with an oxygen content of less than 10^{-25} torr.

Curium was first isolated in the form of a pure compound, the hydroxide, of curium-242 (produced by the neutron irradiation of americium-241) by Werner and Perlman at the University of California in the atumn of 1947. Much of the earlier work with curium used the isotopes ^{242}Cm and ^{244}Cm, but the heavier isotopes offer greater advantages mainly because of their longer half-lives. The isotope ^{248}Cm, obtainable in relatively high isotopic purity as the alpha particle decay daughter of ^{252}Cf, is the most practical for chemical studies. See also **Radioactivity.**

References

Asprey, L. B., Ellinger, F. H., Fried, S., and W. H. Zachariasen: "Evidence for Quadrivalent Curium: X-Ray Data on Curium Oxides," *Amer. Chem. Soc. J.*, **77**, 1707–1708 (1955).

Asprey, L. B., Ellinger, F. H., Fried, S., and W. H. Zachariasen: "Evidence for Quadrivalent Curium. II. Curium Tetrafluoride," *Amer. Chem. Soc. J.*, **79**, 5825 (1957).

Kanellakopulos, B., et al.: "The Magnetic Susceptibility of Americium and Curium Metal," *Solid State Commun.*, **17**, 6, 713–715 (1975).

Keller, C., and B. Erdmann: "Preparation and Properties of Transuranium Element–Noble Metal Alloy Phases," *Proc. 1972 Moscow Symp. Chemistry of Transuranium Elements* (1976).

Samhoun, K., and F. David: "Radiopolarography of Am, Cm, Bk, Cf, Es, and Fm," *Proc. 4th International Transplutonium Element Symp.*, Baden Baden, W. Germany, 1975.

Seaborg, G. T.: "The Chemical and Radioactive Properties of the Heavy Elements," *Chem. Eng. News*, **23**, 2190–2193 (1945).

Seaborg, G. T. (editor): "Transuranium Elements," Dowden, Hutchinson & Ross, Stroudsburg, Pennsylvania, 1978.

Stevens, C. M., et al.: "Curium Isotopes 246 and 247 from Pile-Irradiated Plutonium," *Phys. Rev.*, **94**, 4, 974 (1954).

Street, K., Jr., and G. T. Seaborg: "The Separation of Americium and Curium from the Rare Earth Elements," *Amer. Chem. Soc. J.*, **72**, 2790–2792 (1950).

Werner, L. B., and I. Perlman: "First Isolation of Curium," *Amer. Chem. Soc. J.*, **73**, 5215–5217 (1951).

CURL. A vector resulting from the action of the operator del on a vector, **V**. It can be written in Cartesian coordinates in the following forms:

$$\text{curl } \mathbf{V} = \nabla \times \mathbf{V} = \mathbf{i}\left\{\frac{\partial V_z}{\partial y} - \frac{\partial V_y}{\partial z}\right\} + \mathbf{j}\left\{\frac{\partial V_x}{\partial z} - \frac{\partial V_z}{\partial x}\right\} + \mathbf{k}\left\{\frac{\partial V_y}{\partial x} - \frac{\partial V_x}{\partial y}\right\}$$

$$= \begin{vmatrix} \mathbf{i} & \mathbf{j} & \mathbf{k} \\ \partial/\partial x & \partial/\partial y & \partial/\partial z \\ V_x & V_y & V_z \end{vmatrix}$$

The curl of a position vector vanishes; $\mathbf{R} = \mathbf{i}x + \mathbf{j}y + \mathbf{k}z$, $\nabla \times \mathbf{R} = 0$. If the curl of a vector function vanishes everywhere in a certain region, the function is said to be an irrotational vector (or a lamellar vector), in this region. It follows that if **V** is an irrotational vector so that $\nabla \times \mathbf{V} = 0$, then $\mathbf{V} = \nabla\phi$ (**V** is the gradient of ϕ), where ϕ is some scalar function.

European writers often use the word rotation instead of curl and the symbol rot **V**.

CURLEW. **Shorebirds and Gulls.**

CURRENT AMPLIFICATION. 1. Of an amplifier, the ratio of the current produced in the output circuit as a result of the current supplied in the input circuit, to the current supplied to the input circuit. 2. Of a magnetic amplifier control-winding, the ratio of the change in output current to the change in current in the control winding required to produce the output current change. Assuming the change from minimum to maximum output current to be 100%, the nominal current amplification is to be measured over the following range: An output current 20% greater than the minimum to an output current 20% less than the maximum. Current amplification is usually stated for operations of the magnetic amplifier at its rating except for control currents and output currents. The current amplification taken is the minimum that exists for any condition within the rating. 3. Of a multiplier phototube, the ratio of the output current to the photocathode current due to photoelectric emission at constant electrode voltages. Terms output current and photocathode current as here used do not include dark current. This characteristic should be measured at levels of operation that will not cause saturation. 4. Of a transducer, the ratio of the magnitude of the current in a specified load impedance connected to a transducer to the magnitude of the current in the input circuit of the transducer. If the input and/or output current consists of more than one component, such as multifrequency signal or noise, then the particular components used and their weighting are to be specified. By custom, this amplification is often expressed in decibels by multiplying its common logarithm by 20.

CURRENT ATTENUATION. Of a transducer, the ratio of the magnitude of the current in the input circuit of a transducer to the magnitude of the current in a specified load impedance connected to the transducer. If the input and/or output current consist of more than one component, such as multifrequency signal or noise, then the particular components used and their weighting need to be specified. By custom, this attenuation is often expressed in decibels by multiplying its common logarithm by 20.

CURRENT BALANCE. A system of fixed and movable coils of accurately known dimensions so arranged that the force developed between the coils (by the passage of electric current) can be balanced by the force of gravity acting on a known mass. Such an arrangement is used for the absolute determination of the ampere. See also **Electrical Instruments.**

CURRENT CORRECTION ANGLE (Navigation). The angle between the heading of a ship and the course of the ship relative to the earth. In sea navigation, the actual motion of the water relative to the surface of the earth is known as current. The direction in which the water is moving is known as the set, and the speed of motion is called the drift. Three main types of currents are recognized by navigators: general ocean currents, tidal currents, and currents due to wind. General ocean currents are discussed in publications of the United States and other governments, and are plotted on their monthly pilot charts. Likewise, tidal currents are predicted for any particular locality at any time. Currents due to winds are very difficult to determine with any exactness. Experience has shown that a wind that has been blowing steadily for several hours, will produce, in the northern hemisphere, a surface current that sets about 40° to the left. The drift of wind currents is between 1 and 3% of the speed of the wind producing it. Many variables, such as gustiness, steadiness of direction, etc., enter into the determination of wind currents, and only long experience will give anything approaching accurate set and drift. In the article on Dead Reckoning is discussed a so-called "current" that is the "catch-all" for various errors in dead-reckoning navigation.

If a ship is to make good a specified course in a region where a current is known to exist, the so-called current correction angle should be determined to find the proper heading for the ship. The determination of this angle may be made either by graphical or computational methods. Either method is sufficiently accurate, and the graphical method is more commonly used. However, the computational method is fully as rapid as the graphical, and does not require the space and paraphernalia needed for constructing the vector triangle. The graphical method is practically identical with that discussed under **Wind Correction Angle (Navigation).**

The simplicity of the computational method is indicated in the following case: A ship is operating with cruising speed of 15 knots, in a region where a current is known to set 160° with drift of 3 knots. The navigating officer wishes to determine the proper heading for the ship in order that its actual motion relative to the surface of the earth shall be 270° (due west), and he also wishes to know the speed the ship will make good along this course. A freehand sketch is helpful, but by no means necessary. Such a sketch is shown in the accompanying figure. The angle *xec* is known, since the current sets 160° and the course is to be 270°. The side *cs* is known, since the speed of the ship is 15 knots. Then, by means of traverse tables, slide rule, or any form of three-place computing, we have:

$$70^\circ \quad ec = 3.00\text{ k} \quad ex = 1.03\text{ k} \quad xc = 2.82\text{ k}$$

$$esc = 11^\circ \quad cs = 15.00\text{ k} \quad xs = 14.74\text{ k} \quad xc = 2.82\text{ k}$$

$$ex = 13.7k$$

Hence, we have the heading to be used, given by 270° + 11° = 281°, and the predicted speed along the course of 13.7 knots. This

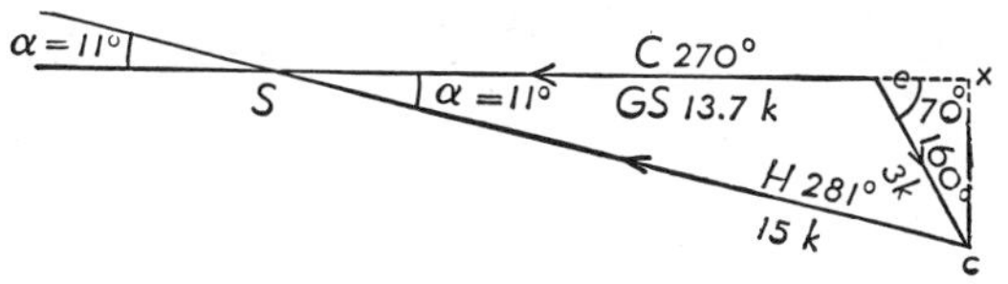

Determination of current correction angle.

work was done in about 2 minutes, using nothing other than a pencil, pad of paper, and the traverse tables.

Any one of the dead-reckoning computers, designed for air navigation, may be used for solving problems of this sort. When using these, care must be taken to remember that they are designed for use with wind, and that wind direction is given as opposite to that in which the air is actually moving; whereas the set of the current is properly given as the direction in which the water is moving.

See also **Course; Dead Reckoning;** and **Navigation.**

CURRENT DENSITY. 1. The limiting value of the time rate of flow of electric charge per unit area (perpendicular to direction of flow) as the area approaches zero. 2. In nuclear physics, a vector such that it component along the normal to a surface equals the net number of particles crossing that surface per unit area and unit time. Commonly referred to simply as current, as in neutron current. 3. By analogy with electric current, a vector representing the time rate of flow of any quantity such as mass or momentum, per unit area in a transport process.

CURRENT (Displacement). Displacement Current.

CURRENT ENERGY (Hydraulic). Ocean Resources (Energy).

CURRENT (Impressed). Impressed Current.

CURRENT (Ocean). Ocean; Ocean Resources (Energy).

CURRENT REGULATOR. Power Sources and Supplies.

CURRENT (RMS). Root-Mean-Square.

CURRENT TELEMETERING. Telemetering (Industrial).

CURRENTS (Ocean). Ocean.

CURRY POWDER. Flavorings.

CURVATURE. A measure of the rate of change of direction of a curve. If τ is the angle made with the *OX*-axis by a point *P* on the curve and *s* is the arc length from some fixed point on the curve to *P*, the curvature at *P* is $\kappa = d\tau/ds$.

The radius of curvature is $\rho = 1/\kappa$. When the equation of the curve is given in one of the equivalent forms: (a) $y = f(x)$; (b) $x = f_1(t)$, $y = f_2(t)$, the parametric form; (c) $F(x, y) = 0$, the corresponding equations for the radius of curvature are: (a) $(1 + y'^2)^{3/2}/y''$; (b) $s_t^2/[x_{tt}^2 + y_{tt}^2 - s_{tt}^2]^{1/2}$; (c)

$$[(F_x^2 + F_y^2)^{3/2}]/(F_{xx}F_y^2 - 2F_{xy}F_xF_y + F_{yy}F_x^2).$$

Equations can also be derived when the curve is described in polar coordinates.

Since any three points determine a circle, choose three arbitrary points P_0, P_1, P_2 on a plane curve and let P_1, P_2 approach P_0 along the curve to obtain a limiting circle for the curve. It is called the circle of curvature at P_0 or the osculating circle. Its center is the center of curvature and its coordinates are $X = x - \rho \sin \phi$; $Y = y + \rho \cos \phi$, where $\tan \phi = y'$.

See also **Circular Curves;** and terms listed under **Mathematics.**

CURVATURE (Measurement of). Spherometer.

CURVATURE OF FIELD (Optics). One of the five geometrical aberrations of lenses. If a system is corrected so that there is no spherical aberration, no coma, and no astigmatism, the images of off axis points will lie on a curved surface called the Petzval surface.

CURVATURE OF LENS (Total). The quantity *K*, defined by the expression

$$K = \frac{1}{r_1} - \frac{1}{r_2}$$

is called the total curvature of a lens. Thus, for a thin lens:

$$\frac{1}{f} = (n-1)\left(\frac{1}{r_1} - \frac{1}{r_2}\right) = (n-1)K$$

CURVE. The locus of a point moving with one degree of freedom; that is, its path is defined with the aid of one parameter. In Euclidean space, this expression would be $x_1 = x_1(t)$, $x_2 = x_2(t)$, . . . , $x_n = x_n(t)$, with $a \le t \le b$. If $x_1(a) = x_1(b)$, $x_2(a) = x_2(b)$, . . . , $x_n(a) = x_n(b)$, the curve is *closed.* If, with this possible exception, distinct values of *t* produce distinct points, the curve is *simple.* If the total length of an inscribed polygonal line approaches a finite limit as the length of each side approaches zero, the curve is *rectifiable.* (See entries following).

CURVE FITTING. It is often of interest to represent a set of experimental data by means of a mathematical equation, which can be used for interpolation or other purposes. If the theoretical relation between the experimental variables is known, the procedure is generally simple but in many cases a purely empirical equation must be assumed. Even where the theoretical equation is known, the latter type may be preferred because it is easier to use. (A typical example is the temperature variation of the heat capacity of a gas. The theoretical equation, exponential in form, is complicated; a polynomial with two or three terms is much more convenient.)

The problem considered here is the following: given a set of numbers $x_1, x_2, x_3 \ldots$ and $y_1, y_2, y_3 \ldots$ it is desired to represent these data by an equation $y = f(x)$. There are thus two parts to the problem (1) to choose an appropriate form for the equation; (2) to evaluate the constants in it.

The first step, usually a graphical one, is a plot of *y* against *x*. If a straight line results, the required equation is $y = mx + l$. If the plot is not a straight line, some change of variable may still give a straight line. Thus if the data fit the equation $y = ax^n$ a plot of log *x* vs. log *y* would be a straight line of slope *n* for $\log y = \log a + n \log x$. When no such transformation reduces the equation to linear form, a polynomial $Y = a + bx + cx^2 + dx^3 \ldots$, where $Y = y$, y^2, log *y*, x/y, etc., should be tried. If the measured values of *x* are in arithmetic progression, as $x = 0$, 0.1, 0.2, 0.3, etc., and the nth differences of *y* are constant, a polynomial with last term of x^n will represent the data exactly. However, the labor involved in calculating the coefficients becomes quite great if *n* is larger than three or four.

Having chosen the appropriate equation, the constants in it must now be evaluated numerically. This can be done: (a) graphically; (b) by the methods of selected points, choosing as many (x_i, y_i) pairs as there are unknown constants and solving the resulting simultaneous equations for the constants; (c) by the method of averages, grouping all of the (x_i, y_i) pairs into a number of sets equal to the number of unknowns, taking their averages, and again solving simultaneous equations; (d) by the method of least squares. Closeness of fit between observed and calculated points usually increases in going from method (a) to (d) but the work involved in computation increases in the same order.

CURVE (Higher Plane). Generally understood to mean one which is not a straight line or a conic section. Thus, if the equation describing the curve is algebraic, $F(x, y) = 0$, the polynomial is of degree greater than two. The curve could also be described by a transcendental equation.

Higher plane curves often receive some attention in elementary calculus courses, as they illustrate many of the principles studied there (see also **Singular Point of a Curve**). The examples discussed in this work include: Archimedes spiral, asteroid, brachistochrone, cardioid, Cartesian oval, catenary, cissoid of Diocles, conchoid of Nicomedes, Cornu spiral, cyclic curves, cycloid, epicycloid, evolute, folium of Descartes, hypocycloid, involute, lemniscate of Bernoulli, limaçon, lituus, logarithmic spiral, oval of Cassini, parabola (cubical and semi-cubical), parabolic spiral, quadratrix of Dinostratus, rose curve, sici spiral, spiral, strophoid, tractrix of Huygens, trisectrix of Maclaurin, trochoid, witch of Agnesi. The exponential, hyperbolic, logarithmic, and trigonometric functions could also be called higher plane curves. See also terms listed under **Mathematics.**

CURVE OF GROWTH (Stellar). Stellar Curve of Growth.

CURVE (Plane). Analytically, a curve is defined by a function of two or more variables but the shape of the curve is described most readily by a graph of the function. Geometrically, a curve can be regarded as the locus of the equation which describes the motion of the generation point.

A curve in two dimensions is called plane; for three dimensions, see **Curve (Space).** In rectangular coordinates, a plane curve can be represented by the equations: (1) $F(x, y) = 0$; (2) $y = f(x)$; (3) $x = f_1(t)$, $y = f_2(t)$. The third form is the parametric equation of the curve. If the parameter t is eliminated from the two equations in (3), the forms (1) or (2) result. If the locus of a curve is described by the end point of a line of variable length, its other extremity being fixed, polar coordinates are generally more convenient than Cartesian coordinates.

Plane curves may be classified in many different ways. Perhaps the simplest, but not the most important mathematically, is by the order, which is the degree of the defining equation in Cartesian coordinates. An algebraic curve of order one is thus a straight line; of order two, a conic section, or one of its degenerate cases, such as a circle. (For algebraic curves of order greater than two and transcendental curves see **Curve (Higher Plane).)**

Analytic geometry is mostly concerned with the simpler types of plane curves but the methods of calculus, differential geometry, and more advanced branches of mathematics are required for a complete study of curves. Plane geometry is the study of certain closed curves like the triangle, quadrilateral, polygon, and circle.

(For further properties of plane curves see **Asymptote; Curvature; Envelope (Mathematics); Evolute; Length of a Curve; Tangent (Geometry).)**

CURVE PLOTTER. Plotter (Curve).

CURVES (Circular). Circular Curves.

CURVE (Single Point of). Singular Point of a Curve.

CURVE (Space). If a curve does not lie in one plane, it is called skew, twisted, or a space curve. It could, for example, be regarded as formed by the intersection of two surfaces. Now, in Cartesian coordinates, a surface is described by an equation $F(x, y, z) = 0$, hence two simultaneous equations in three varaibles $F_1(x, y, z) = 0$ and $F_2(x, y, z) = 0$ will describe a space curve. If one variable, x for example, is regarded as independent, the other two could be expressed as $y = f_1(x)$, $z = f_2(x)$. The first equation can then be interpreted as the projection of the space curve in the XOY-plane and the second equation as the projection of the space curve in the ZOX-plane. Projections on the XOZ- and YOZ-planes could be obtained in a similar way by elimination of y and x, respectively, from the pair of simultaneous equations $F_1 = 0$ and $F_2 = 0$. Finally, a space curve can be represented in terms of a parameter by three equations $x = \phi_1(t)$, $y = \phi_2(t)$, $z = \phi_3(t)$.

The differential properties of space curves lead to relatively complicated formulas in the usual notation of calculus but they are given compactly in vector form. Let **r** be a position vector for a point on the curve and, if s is arc length along the curve, $d\mathbf{r}/ds = \mathbf{t}$ is a unit vector, tangent to the curve. Then, if a prime means differentiation with respect to arc length $\mathbf{t}' = \mathbf{n}/\rho = \kappa\mathbf{n}$, where **n** is a unit vector in the direction of the principal normal to the curve, perpendicular to **t**, and the scalar quantities are ρ, the radius of curvature, and κ, the curvature. Now define another unit vector by the equation $\mathbf{b}' = -\tau\mathbf{n}$, where the scalar τ is called the tortuosity of the curve and **b** is a unit vector in a direction called the binormal. The three unit vectors **t, n, b** constitute a righthanded system of mutually perpendicular vectors. The planes in which they lie are named respectively: (1) normal, **b** and **n**; (2) osculating, **n** and **t**; (3) rectifying, **t** and **b**. The following relations are known as the Serret-Frenet formulas: $\mathbf{t}' = \kappa\mathbf{n}$; $\mathbf{n}' = \tau\mathbf{b} - \kappa\mathbf{t}$; $\mathbf{b}' = -\tau\mathbf{n}$.

The arc length of a space curve is conveniently given in terms of the parametric equation. Between the points t_0 and t_1, the result is

$$s = \int_{t_0}^{t_1} [\phi_1'(t)^2 + \phi_2'(t)^2 + \phi_3'(t)^2]^{1/2}\, dt$$

See also **Surface;** and terms listed under **Mathematics.**

CURVILINEAR ORTHOGONAL COORDINATES. A curvilinear system of coordinates α, β, γ is generated by a system of three mutually orthogonal familes of surfaces α = constant, β = constant, γ = constant. It is assumed that the unit tangent vectors **a, b, c** along the coordinate curves form everywhere a right-handed system, i.e., $\mathbf{a} \times \mathbf{b} = \mathbf{c}$.

An infinitesimal line vector $d\mathbf{l}$ is given by the expression

$$d\mathbf{l} = \mathbf{a}M\, d\alpha + \mathbf{b}N\, d\beta + \mathbf{c}P\, d\gamma$$

A volume element dV is given by

$$dV = M\,N\,P\,d\,d\alpha\, d\beta\, d\gamma$$

CUSHING'S SYNDROME. Adrenal Glands; Pituitary Gland.

CUSPATE FORELAND. A coastal headland of triangular shape, with its apex seaward and its sides concave.

Sketch map illustrating erosion of a headland and development of shore current, beach, spit, bar, cove and stacks, and cuspate foreland. (*W. H. Hobbs.*)

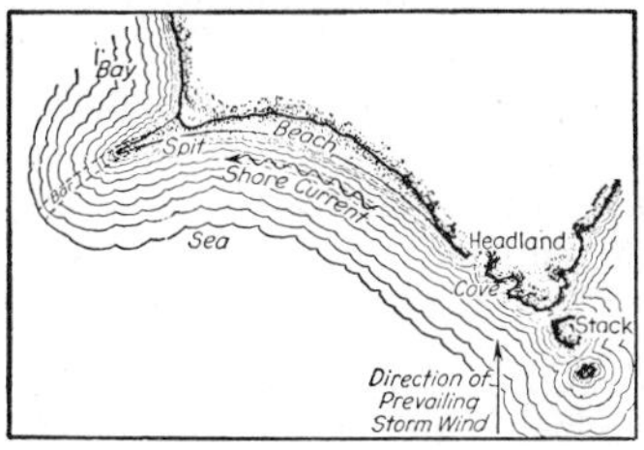

CUSP (Mathematics). A singular point on a curve where there are two coincident tangents. If there is a branch of the curve on each side of the double tangent, the cusp is of the *first kind* (e.g., the semicubical parabola); if the two branches lie on the same side of the double tangent, the cusp is of the *second kind.* If the curve is represented by $f(x, y) = 0$, the condition for a cusp, and also for a point of osculation is

$$\left(\frac{\partial^2 f}{\partial x\, \partial y}\right)^2 - \frac{\partial^2 f}{\partial x^2}\frac{\partial^2 f}{\partial y^2} = 0$$

CUSTARD APPLE. Annona.

CUSUM CHART. An abbreviation for "cumulative sum chart." Such charts are used in quality control. When a series of values x_1, x_2, etc., occur in temporal order, as in the measurement of a variable on components coming off a production line, the cusum chart plots $\Sigma_{i=1}^{k} x_i$ against k and observes the departures from process control when they fall outside permissible sampling limits.

CUTANEOUS TUMORS. Dermatitis and Dermatosis.

CUTICLE. 1. For the use of this term in botany, see **Leaf.** 2. In zoology, cuticle is the outermost layer of the integument. See **Integumentary System.** As applied to the skin of the vertebrates this layer is composed of cells of ectodermal origin but in the invertebrates it indicates a noncellular layer secreted by the underlying cells. A cuticle of the latter type occurs in the parasitic flatworms, the rotifers, the roundworms and annelid worms, and the arthropods. The cuticle of insects is usually called the cuticula.

CUTICULA. 1. The thickened plate at the free end of some epithelial cells. 2. The layer of scales covering a hair. 3. The noncellular layer covering the bodies of some invertebrates. Some entomologists call the outer and inner layers of the chitinous cuticula of insects the epidermis and dermis. The terms epidermis, dermis, cuticle and cuticula are loosely used.

CUTLASSFISHES (*Osteichthyes*). Of the suborder *Trichiuridea,* family *Trichiuridae,* these fishes have an elongated, band-shaped body with a sharply tapering head. See accompanying illustration. The fish is naked or covered with very tiny scales. The mouth opening is broad

Cutlassfish (*Trichirus lepturus*).

and has several large teeth on the jaws and palate. The dorsal fin originates just behind the head runs the length of the body. A finlet may be present. There are 100 to 160 vertebrae. Cutlassfishes are divided into about 25 genera, most with very few species. They are known to have existed since the Lower Oligocene period, and teeth have been found in Eocene layers which resemble present-day *Trichirus* species.

Found in tropical and neighboring seas, cutlassfishes usually are in deeper parts of coastal waters, where they swim rapidly after schooling fishes. The cutlassfish (*Trichirus lepturus*) attains a length up to nearly 5 feet (1.5 meters) and is one of the most widely distributed species. It is encountered in tropical and subtropical parts of the Atlantic, and in the Indian and western Pacific oceans. In the Atlantic Ocean, it follows the warm currents to the coast of England, and occasionally penetrates the Mediterranean. The silver-white body terminates in a thin, almost hair-like tail shaft. Thus, these fishes are sometimes called *hairtails.* In some regions, such as on the Japanese coast (where it moves into shallow water in August and September), the species is extensively fished and regarded as quite flavorful. Usually in reporting total catches, the several species of cutlassfishes are not differentiated.

CUTOFF. 1. A particular point in a cycle, or magnitude of a quantity, at which a mechanism, electric circuit, or other system, cuts off some flow to or from it. Thus, the steam engine cutoff is that percentage of stroke accomplished by the piston, when the inlet valve closes and prevents more steam entering the cycle from the boiler. The cutoff of a Diesel cycle is the fraction of the stroke accomplished when supply of fuel oil to the cylinder is stopped. 2. A technique used in theoretical physics when the theoretical contribution to the value of a physical quantity arising from integration over part of the range of a certain parameter is not to be believed, in particular when such a contribution is infinite. The integral is cut off, usually at some high frequency limit, with the acknowledgment that beyond this limit either the method of approximation, or the theory itself, will have to be modified in the future.

CUTOFF FREQUENCY. 1. Of a transducer, either a theoretical cutoff frequency or an effective cutoff frequency. The latter is the frequency at which the insertion loss of a transducer between specified terminating impedances exceeds, by some specified amount, the loss at some reference point in the transmission band. 2. Of a wave filter, the frequency at which the attenuation begins to increase sharply. In the ideal filter, the attenuation would go to infinity at the cutoff frequency, but in a practical filter, the rise in attenuation is not so abrupt, and never reaches infinity, but does usually go to a very high value. 3. For a given transmission mode in a nondissipative, uniconductor waveguide, the frequency below which the propagation constant is real.

CUT SET. In the theory of linear graphs the concept of a cut set is almost as important as that of a tree. A cut set of a connected graph G is a set of edges such that the deletion of these edges reduces the rank of G by one. Moreover, no proper subset of this set possesses this property.

Clearly, the removal of the cut set of edges must yield an unconnected graph since the number of vertices v is invariant and rank $G = v - 1$. Thus by "cutting" this set of edges the graph is separated into two pieces. One of the pieces can be an isolated vertex. This latter case occurs for example if all the edges incident at a vertex are removed. As a matter of fact, the totality of edges incident at a given vertex is a cut set if and only if the vertex is not a cut vertex. Another and deeper characterization of a cut set is as a minimal set of elements which contains at least one branch from every tree. Again, a single non-circuit element constitutes a cut set whereas a single circuit element does not.

See also **Graph (Mathematics); Tree (Mathematics);** and terms listed under **Mathematics.**

CUT SET (Fundamental). Suppose G is a linear connected graph (see **Graph (Mathematics)**) possessing v vertices and T is one of its trees. Each cut set of G must contain at least one branch of T. Those $v - 1$ cut sets of G which include exactly one branch of T are said to be fundamental with respect to T. The orientation of a fundamental cut set of a directed graph (see **Digraph**) is usually chosen to agree with that of the defining branch.

CUT SET MATRIX. The cut set matrix $Q_a = (q_{ij})$ of a *directed graph G* is defined in the following manner:

a. Q_a has one row for each cut set of the graph and one column for each edge
b. $q_{ij} = 1$ edge j is in cut set i and the orientations agree (see **Cut Set (Oriented)**)
c. $q_{ij} = -1$ if edge j is in cut set i and the orientations disagree
d. $q_{ij} = 0$ if edge j is not in cut set i

The rank of Q_a is $v - 1$, v denoting the number of vertices in G.

The cut set matrix Q_a and vertex matrix A_a are rather intimately related. For example:

1. If G is nonseparable, Q_a contains A_a (with some rows possibly multiplied by -1) as a submatrix.
2. If Q is a cut set matrix of $v - 1$ rows and rank $v - 1$ of a connected directed graph G of v vertices and A the vertex matrix of G,

$$Q = DA$$

where D is nonsingular.
3. Under the same restrictions as in 2, the nonsingular submatrices of Q of order $v - 1$ are in one-to-one correspondence with the trees of G. That is to say, each such submatrix is the *fundamental cut set matrix* of some tree. Conversely, any fundamental cut set matrix appears as a submatrix of Q.

See **Graph (Mathematics); Tree (Mathematics),** and terms listed under **Mathematics.**

CUT SET (Oriented). Let G be a linear connected graph. The removal of a cut set of edges decomposes G into two connected pieces A and B. The cut set is oriented by ordering A and B either as (A, B) or (B, A). Each element of the cut set must have one vertex in A and one vertex in B. Suppose the cut set is oriented as (A, B). Then, an oriented element of the cut set is said to have the same orientation as the cut set if it is directed away from its vertex in A and towards its vertex in B.

See **Graph (Mathematics).**

CUTTLEFISH (*Mollusca, Cephalopoda*). Mollusks, related to the squids but forming a separate family (*Sepiidae*). Used as food in the Oriental region. The shell is the cuttlebone of commerce and the ink sac secretes the pigment sepia. See also **Invertebrate Paleontology.**

CUTTER (Climb Cutting). Climb Cutting.

CUTWORM (*Insecta, Lepidoptera*). The cutworm is the larval form of any of numerous species of moth (usually owlet moths, *Noctuidae*). Or, a cutworm may be described as a caterpillar. Sizes range up to 2 inches (5 centimeters) in length. Most species feed at night and cut plants off at their base, cut leaves from stems, or chew holes in leaves. They most frequently are found curled up in the soil near a freshly damaged plant during the day. The wide variety of cutworms is evidenced by just a few that are illustrated in Fig. 1. Assumption of the daytime curled position of a representative cutworm is shown in Fig. 2. Cutworms are extremely damaging to crops and injurious to maize (corn) in particular. Infestations of cutworms may require replanting a stand of corn. In years past in some areas, an infestation may destroy from 5 to 50% of the corn crop. The army cutworm (not the true armyworm, which is described in another entry by that title) can be devastating to a corn crop. Other crops that can be

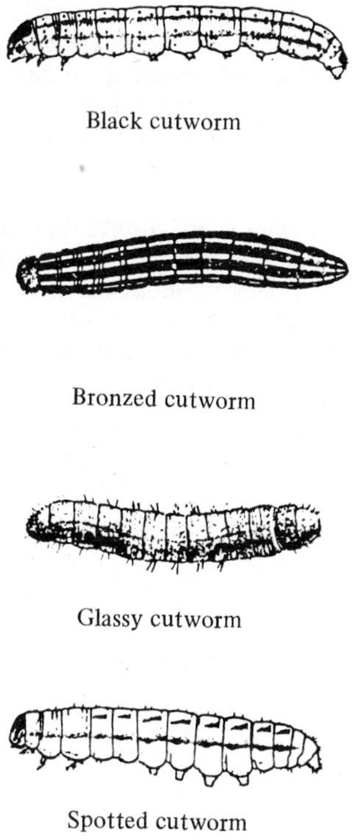

Fig. 1. Larvae of representative cutworms. (*Illinois Natural History Survey.*)

severely affected by cutworms include bean, beet, cabbage, onion, and pea, although very few crops are immune to attack.

The corn cutworm (*Agrotis, Hadena*, etc.) is a soft-bodied caterpillar that feeds on young plants, cutting or segmenting the plant as it proceeds. Preventing measures for control of cutworms includes early plowing of land intended for maize (corn) planting; turning pigs into pasture land to be used later for planting; using a seed-drill to form a line of poisoned bran. Chemicals effective on cutworms include toxaphene, chlordane, and dylox (trichlorfon). The insecticide is applied to the soil surface before planting if cutworms are known to be present. Where damage to young plants is noticed, the chemical should be applied to the soil about the base of the plants. Baits should be applied in late afternoon.

Care should be taken to avoid contamination of edible parts of plants with the insecticide and none of the mentioned chemicals should be used on kale, spinach, turnip, or mustard. At least 21 days should be allowed on cabbage before harvest; and 28 days on collard after the last application of insecticide.

One useful classification of cutworms is: (1) *Solitary, surface cutworms* that feed on a plant just above or just below the soil level. The insect usually moves on to another plant once a sufficient amount of the plant has been consumed to cause it to topple over. This habit emphasizes the great damage that can be done in a very short time. Cutworms that fall into this classification include the black, the bronzed, the clay-backed, and the dingy cutworm. (2) *Climbing cutworms*, which in contrast with the surface cutworm, ascend the plant and attack the buds, leaves, and fruits of vegetable and orchard crops. In this category will be found the variegated and the spotted cutworms. (3) *Army cutworms*, which invade a planted area by the thousands. Practically all vegetation in any given area is consumed by this army of caterpillars, whereupon the insects proceed to a new, untouched field. The rate of damage by such infestations can be total within just a few days. (4) *Subterranean cutworms*, which remain in the soil and consume roots and all underground segments of plants. In this category will be found the pale western and the glassy cutworms.

Fig. 2. Typical curled posture of cutworm resting on top of soil during daytime, usually near a freshly damaged plant.

With most common cutworms, there is only one generation per year. Although there are several variations in life cycle, typically the worms will winter over as small larvae in the soil or under plant residuals and debris. Their feeding commences in the spring and their growth is steady until very late spring or early summer, depending upon locale and species. Then they change to a pupal stage, after which they are adults (moths). About 2 weeks or considerably less for some species are required for the egg stage. Growth from tiny caterpillars only a small fraction of an inch in length up to approximately 2 inches (5 centimeters) in length may require from 2 to 5 months, again depending upon species and locale. At this point the insects burrow into the soil, often to a depth of several inches, where they pupate for several weeks, or in some cases, overwinter. Then the new adults crawl from beneath the surface, aided by tunnels which have been constructed by larvae making their way into the soil. A subclassification in terms of life cycle can be made: (1) *Single-generation* species, most frequently found in the more northern climes. This category of cutworm winters as larvae. It is interesting to note that the insect is confined to one generation per year because the prepupal stage is retarted by relatively high ground temperatures. (2) *Several-generation* species, most frequently found in southern climes. These insects winter over as pupae. Weather conditions greatly affect the numbers of all species of cutworm. Wetness, for example, delays the moths in laying eggs, sometimes preventing this altogether; or flooding of the soil may cause the larvae to surface from within the soil and thus exposing them to their natural enemies, notably parasites.

Some of the more destructive cutworms include the following:

Black cutworm (*Agrotis ypsilon*, Rottemburg). Life cycle varies with location, ranging from a couple of generations per year in the northern states and Canada to perhaps 3 or 4 generations in mid-southern states, such as Kentucky, Tennessee, and Arkansas. The larva ranges from gray to brown in color and has a pale light striping.

Dingy cutworm (*Feltia subgothica*, Haworth). Mainly found in northern climates, produce one generation per year, and winter over as partially grown larvae. The dingy cutworms are resistant to drought and tend to be climbers. Larva is dull brown, featuring a gray dorsal stripe.

Bronzed cutworm (*Nephelodes emmedonia*, Cramer). Essentially a northern species, this insect is injurious to maize (corn), grains, and grasses. There is only one generation per year and the insect winters as a partially developed larva. The larva is a dark bronze color and striped from head to tail by 5 pale brown lines.

Variegated cutworm (*Peridroma saucia*, Hubner or *margaritosa*, Haworth). This cutworm occurs throughout the United States and, in some years, has destroyed many millions of dollars in crops. Preference of the insect is for garden vegetables, vine and tree fruits, and glasshouse plants. From 3 to 4 generations per year are possible. The insect winters over mainly in the form of pupae. Some 60 eggs may be laid at a time, usually on stems and leaves of plants. The larva is of a noticeable yellow color and a W design appears on the eighth abdominal segment.

Spotted cutworm (*Amathes* or *C. nigrum*, Linne). A widely distributed insect throughout Europe, Asia, and North America. Garden vegetable crops are a favorite target of the insect. There are from 2 to 3 generations per year. The insect winters over as a large larva. Eggs are found in groups of about a hundred, mainly on leaves.

Army cutworm (*Chorizagrotis auxiliaris*, Grote). Widely distributed, but well adapted to arid conditions. This insect is a surface feeder and generally operates in very large numbers, as an army, as previously described. Records indicate that over 100,000 acres (40,470 hectares) of wheat were destroyed in Montana in one year by this pest. There is one generation and the insect winters over as a half-developed larva. The eggs are laid upon the soil. Larva is a pale gray-green or perhaps brown. There is a pale black stripe along the back. Although there are no prominent markings, there is a splotchy appearance made up of fine white and brown areas. The skin is granular in texture. See also entry on another species, **Army Worm.**

Pale western cutworm (*Agrotis orthogonia*, Morrison). This cutworm is another major destroyer of crops, notably alfalfa, beet, and small grains in the western United States and Canada. There is one generation per year. Larvae hatch during the winter and early spring, whenever temperatures are sufficiently warm to trigger the development. Their feeding and hence crop destruction is usually complete by early July. The body of the insect is gray with no particular markings. The skin is granular. Eggs are laid in the soil.

Glassy cutworm (*Crymodes, Sidemia devastator*, Brace). Distributed throughout the United States, but not common in the southern states. This insect lives in the soil and like other subterranean species is particularly difficult to diagnose early and to control. A small larva winters over and there is only one generation per year. Similar to a grubworm, the insect's body is a greenish white. The small, somewhat red-colored head has a glassy appearance, hence the name. There are no granulations in the skin.

Yellow-striped armyworm (*Prodenia ornithogalli*, Guenee). Also called the cotton cutworm, this insect occurs in several areas of the United Stated, but is common in the southern states. The insect feeds on a wide variety of crops. Young plants are a special favorite. The larva has a triangular black spot on most segments, coupled with a brilliant orange stripe.

Southern armyworm (*Prodenia eridania*, Cramer). This insect is very common in the southern United States and sometimes is a major pest on vegetable crops. There are four or more generations per year. It is interesting to note that the female moth covers her eggs with whitish hairs. The fully developed larva is gray-to-black and marked with yellow stripes.

CUVIERIAN ORGAN. A defensive organ found in some sea cucumbers (*Holothuroidea*). It is a modified part of the respiratory tree attached to the cloaca, and made up of tubes covered with sticky material. When irritated the animal contracts violently and ejects this structure through the ruptured wall of the cloaca. In the sea water the sticky material forms long adhesive threads which entangle the enemy.

CYANAMIDES. Cyanamide $NC \cdot NH_2$ or $HN:C:NH$ is a white solid, melting point 44°C, boiling point 140°C at 20 mm pressure, transformed at 150°C into cyanuramide, tricyantriamide $(NC \cdot NH_2)_3$. Cyanamide reacts (1) as a base with strong acids forming salts, (2) as an acid forming metallic salts, such as calcium cyanamide $CaCN_2$. Cyanamide is formed (1) by reaction of cyanogen chloride $CN \cdot Cl$ plus ammonia (ammonium chloride also formed), (2) by reaction of thiourea plus lead hydroxide (lead sulfide also formed).

When calcium cyanamide is boiled with water, dicyandiamide $(NC \cdot NH_2)_2$, melting point 207°C is formed (along with calcium hydroxide). Fusion of dicyandiamide with sodium carbonate plus carbon produces sodium cyanide plus ammonia (also some tricyantriamide). Diethylcyanamide $(C_2H_5)_2N \cdot CN$ is a colorless liquid, boiling point 189°C at 748 mm pressure, and when hydrolyzed yields diethylamine $(C_2H_5)_2NH$ plus ammonia plus carbon dioxide. Diphenylcyanamide $C_6H_5N:C:NC_6H_5$ when hydrolyzed yields aniline plus carbon dioxide. Benzylcyanamide $C_6H_5CH_2NH \cdot CN$ is a white solid, melting point 43°C.

CYANIC ACID AND RELATED COMPOUNDS. Cyanic acid, HCNO or HOCN, is a colorless, odorless liquid; soluble in water and in ether; volatile with decomposition when heated; passing at ordinary temperature into a mixture of cyanuric acid, $(HNCO)_3$, and cyamelide, $(CONH)_x$, white solid, which on vaporizing yields cyanic acid; when cyanic acid vapor is rapidly cooled in a freezing mixture, unstable, liquid cyanic acid is obtained, and when the vapor is condensed above 105°C, cyanuric acid

$$(HNCO)_3 \text{ or } CO \begin{array}{c} NH-CO \\ \diagup \qquad \diagdown \\ \qquad \qquad NH \\ \diagdown \qquad \diagup \\ NH-CO \end{array}$$

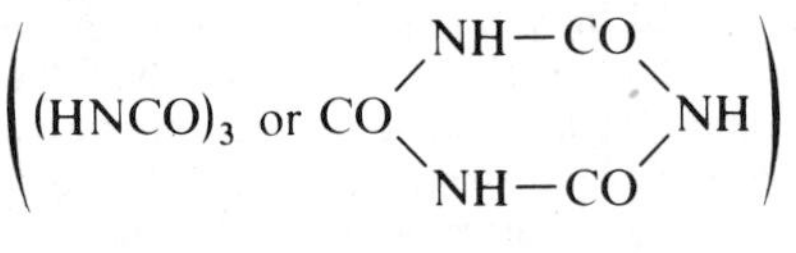

is obtained. Cyamelide dissolves in sulfuric acid unchanged and addition of water causes precipitation of cyamelide; passes into cyanuric acid when warmed with concentrated sulfuric acid, finally into carbon

dioxide plus ammonia; dissolves in sodium hydroxide solution forming sodium cyanate. Sodium cyanate is prepared by heating sodium cyanide and an oxide such as lead monoxide PbO, trilead tetroxide Pb_3O_4, or lead dioxide PbO_2, addition of water and separation of the sodium cyanate solution from the lead oxide by filtration. Sodium cyanate solution upon boiling changes into sodium carbonate plus urea $CO(NH_2)_2$.

Ammonium cyanate, $CNONH_4$, white solid, formed by reaction of sodium cyanate and ammonium sulfate solutions is transformed to urea upon being heated at 100°C. This reaction was carried out in 1828 by Wöhler, and is the first record of a so-called inorganic substance being transformed outside a living organism into a so-called organic substance. The following esters are known:

> Methyl isocyanate, CH_3NCO, boiling point 44°C
> Ethyl cyuanate, C_2H_5OCN, decomposes on heating
> Ethyl isocyanate, C_2H_5NCO, boiling point 60°C
> Phenyl isocyanate, C_6H_5NCO, boiling point 166°C

Ethyl cyanurate 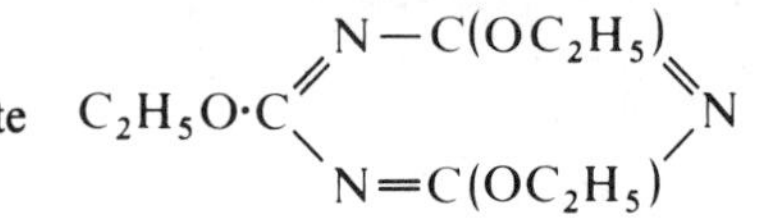
$$C_2H_5O \cdot C \begin{array}{c} N-C(OC_2H_5) \\ \diagup\diagup \qquad \diagdown\diagdown \\ \qquad \qquad N \\ \diagdown \qquad \diagup \\ N=C(OC_2H_5) \end{array}$$

Ethyl isocyanurate 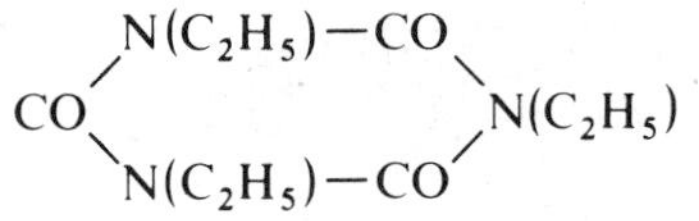
$$CO \begin{array}{c} N(C_2H_5)-CO \\ \diagup \qquad \diagdown \\ \qquad \qquad N(C_2H_5) \\ \diagdown \qquad \diagup \\ N(C_2H_5)-CO \end{array}$$

The extensive use of organic isocyanates in various industrial processes for production of high polymers has brought about tonnage production. Toluene diisocyanate is made by nitrating toluene to the dinitro compound, which is then reduced to the diamine, and treated with phosgene to obtain the diisocyanate:

$$C_6H_5CH_3 \xrightarrow{HNO_3} C_6H_3(NO_2)_2CH_3 \xrightarrow{H}$$
$$C_6H_3(NH_2)_2CH_3 \xrightarrow{COCl_2} C_6H_3(NCO)_2CH_3$$

Toluene diisocyanate is widely used in the manufacture of urethane plastics, particularly the urethane foamed plastics. Another isocyanate, diphenylmethane 4,4'-diisocyanate, is produced by reaction of aniline and formaldehyde, followed by reaction with phosgene:

$$2C_6H_5NH_2 \xrightarrow{HCHO} CH_2(C_6H_4NH_2)_2 \xrightarrow{COCl_2} CH_2(C_6H_4NCO)_2$$

The diphenylmethane 4,4'-diisocyanate is used in the manufacture of solid urethane elastomers (primarily for heavy duty tires) and chemically resistant coatings.

Fulminic acid, HONC, and the fulminates are violently explosive. Utilizing this property, mercuric fulminate $Hg(ONC)_2 \cdot \frac{1}{2}H_2O$, is used as a detonator for other explosives. Mercury fulminate is made by the reaction of ethyl alcohol and mercuric nitrate in excess of nitric acid, from which insoluble mercuric fulminate separates. Silver fulminate, $Ag(ONC)$ is more explosive than mercuric fulminate, and is used in the manufacture of firecrackers. Free fulminic acid may be obtained by reaction of potassium fulminate and excess of ether. It volatilizes with the ether upon distilling, and changes rapidly to metafulminic acid. Related to fulminic acid, is fulminuric acid, $(HONC)_3$, or $NO_2 \cdot CH(CN) \cdot CONH_2$.

CYANOGEN. Cyanogen $(CN)_2$ is a colorless gas of marked characteristic odor, very poisonous, density 1.8 (air equal to 1.0), melting point −28°C, boiling point −20°C, soluble. When passed into water at 0°C, cyanogen forms hydrocyanic acid plus cyanic acid, but at ordinary temperatures the reaction is complex. With sodium hydroxide solution, there is formed with cyanogen sodium cyanide plus sodium cyanate, with dilute sulfuric acid oxamic acid $COOH \cdot CONH_2$, oxalic acid $COOH \cdot COOH$. By reaction with tin and hydrochloric acid, cyanogen is reduced to ethylene diamine $CH_2 \cdot NH_2 \cdot CH_2 \cdot NH_2$. Cyanogen reacts with hydrogen to form hydrocyanic acid, and with metals, e.g., zinc, copper, lead, mercury, silver, to form cyanides. Cyanogen, (1) when burned in air produces a violet flame forming carbon dioxide and nitrogen in the outer part and carbon monoxide and nitrogen in the inner part, (2) when exploded with oxygen produces carbon dioxide or carbon monoxide and nitrogen depending upon the ratio of oxygen to cyanogen (2 volumes oxygen plus 1 volume cyanogen yields 2 volumes carbon dioxide plus 1 volume nitrogen; 1 volume oxygen plus 1 volume cyanogen yields 2 volumes carbon monoxide plus 1 volume nitrogen). The flame spectrum contains characteristic bands in the blue and violet. By means of the electric spark, the electric arc or a red hot tube, cyanogen is decomposed into carbon plus nitrogen. When heated at ordinary pressure at about 300°C, or under 300 atmospheres pressure at about 225°, cyanogen is converted into paracyanogen, a brown powder, also formed when mercuric cyanide is heated. Cyanogen is prepared (1) by reaction of sodium cyanide and copper sulfate solutions, whereby one half the cyanogen is evolved as cyanogen gas and one half remains as cuprous cyanide. From the filtered cuprous cyanide, by treatment with ferric chloride solution, cyanogen is evolved with accompanying formation of ferrous chloride, (2) by heating mercuric cyanide solid, or a mixture of mercuric chloride and sodium cyanide solutions, mercury and mercurous, respectively, being formed, (3) by heating ammonium oxalate $COONH_4 \cdot COONH_4$ with phosphorus pentoxide, water being abstracted. Small amounts of cyanogen are present in blast furnace gas and raw coal gas.

CYANOHYDRINS. The products of the reaction between an aldehyde or a ketone with hydrogen cyanide HCN are termed *cyanohydrins.* Sometimes the compounds are referred to as hydroxycyanides.

$$CH_3CHO + HCN \rightarrow CH_3 \cdot CH(OH) \cdot CN$$

(acetaldehyde) (hydroxyethyl cyanide or aldehyde cyanohydrin)

$$(CH_3)_2CO + HCN \rightarrow (CH_3)_2C(OH) \cdot CN$$

(acetone) (hydroxyisopropyl cyanide or acetone cyanohydrin)

CYANOMETRY. Atmospheric Optical Phenomena.

CYANOSIS. A blue color of the skin and mucous membranes, most marked in the lips, nose, cheeks, ears, hands, and feet. It is due to the presence of abnormally large amounts (in excess of 5% by weight) of reduced hemoglobin (i.e., hemoglobin which has given up its oxygen to the tissues) in the blood. This occurs when there is a failure of aeration of the blood as it passes through the lungs so that insufficient oxygen is picked up by the red cells, or in circulatory failure and stasis of venous blood in peripheral vascular beds. It also occurs when there is abnormal communication between the venous and arterial sides of the circulation as in certain congenital malformations of the heart. Cyanosis is commonly seen in severe heart disease, particularly the congenital type, pneumonia, severe infections, asthma, and emphysema. It is also seen as a result of poisoning with gases, or drugs which interfere with respiration or the rate of absorption of oxygen by the blood. It is also found in wool sorters' disease. For further mention of cyanosis, see specific diseases mentioned above.

CYBERNETICS. This term is derived from the Greek word meaning the science of the steering of ships. It was possibly first used in a broader sense by Ampere to mean the science of the control of society. The word was popularized in the modern sense as the result of the use of the term in a book by Norbert Wiener in 1948 which dealt with control in machines and in living systems. Possibly an acceptable definition of the term as currently used would be "the function of control in machines and animals." There has been a tendency in the United States to emphasize the control aspects, while in Europe, the emphasis has been on the handling of information. The science of control obviously involves three systems of work: (1) closed-loop feedback systems; (2) the manipulation of the information which guides these systems; and (3) processes for filtering out casual disturbances from the information channel. The second of these, obviously, and the third, less obviously, can lead on to a consideration of communication, control, and thought processes in animals. The term tends to be fuzzy because in its overall sense, cybernetics embraces numerous well-established disciplines which already are well covered by their own sets of definitions.

Thus, the concepts of feedback and closed-loop control systems are well advanced and covered in the general field of automatic control and instrumentation, as indeed are adaptive control systems, also sometimes ascribed to cybernetics. As one extends the application of these principles into biology and animal behavior, the interface with biophysics and molecular biology and associated fields is encountered. In recent years, formal dictionaries have greatly narrowed the definition of cybernetics to consider it a comparative study of computers and the animal nervous system in an attempt to understand the nature of the brain. See also **Brain and Nervous System;** and **Control System (Automatic).**

CYBOTAXIS. A condition in which certain liquids, under x-ray examination, give evidence of structure resembling that of crystals. By passing a beam of x-rays through various alcohols and other organic liquids, G. W. Stewart and his collaborators have obtained one, two, or even three diffraction maxima or halos, somewhat like the diffraction rings produced by powdered crystals. These suggest that molecules are temporarily arranged in rows, layers, or stacks like bricks in a pile and that they have one, two, or even three different dimensions or spacings, corresponding, in accordance with Bragg's law, to the different angles of diffraction observed.

A closely related property is exhibited by certain substances known as "liquid crystals," which appear to be intermediate between merely cybotactic liquids and true crystals. In these there appear to be large groups of molecules which, though able to move and turn about, retain their structural arrangement. Such mesomorphic substances manifest even some of the optical properties of crystals, which the former type do not.

See also **Liquid Crystals.**

CYCAD (*Cycadales*). An order of Gymnosperms containing 9 genera and less than 100 species. They first appeared in late Paleozoic times, became a dominant group almost cosmopolitan in distribution in the Mesozoic period. Cycads are now limited to tropical or subtropical regions, often with a very restricted range. Some of the genera are found only in Mexico and the West Indies; others occur only in Australia, Africa and various islands of the Pacific Ocean. Because of their decorative habitat they are frequently grown in cultivation in places outside their natural range. Many are grown as greenhouse plants in temperate regions.

The appearance of the plants is uniform in all genera. Stems are subterranean and tuberous, or above ground and columnar. Columnar stems are usually from 6–10 feet (1.8 to 3 meters) tall, but some species grow much higher. The leaves form a large crown at the top of the stem. They are pinnate except in one Australian genus, *Bowenia,* which has bipinnate leaves which are thick and leathery. The stems are generally thickly clothed in the persistent leaf-bases, which sometimes give an indication of the age of the plants. As determined by the number of leafbases many are several hundred years old. Internally, the stem contains a very large pith and a thick cortex with a narrow cylinder of vascular tissues between them. The primary root is large and extends deep into the ground.

Fig. 1. Cycas, a cycad. A male plant is shown on the left bearing a single, large staminate cone. The female plant, on the right, bears a cluster of megasporophylls, in a loosely arranged cone. (*A. M. Winchester.*)

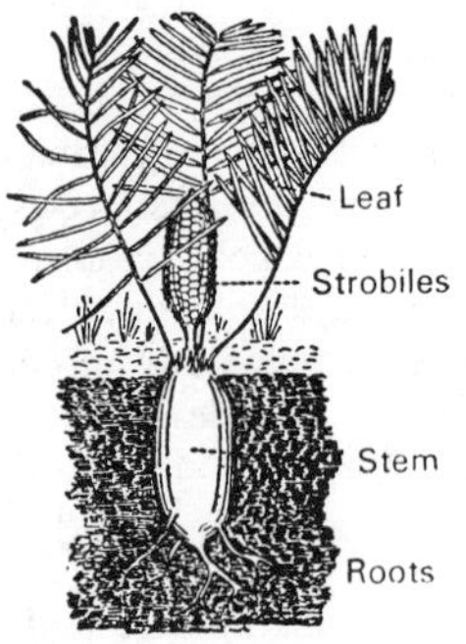

Fig. 2. A cycad. Mature sporophyte of *Zamia* bearing a carpellate strobile. (*Smith, Overton, Gilbert, Denniston, Bryan and Allen, "Textbook of General Botany," The Macmillan Co.*)

All Cycads are dioecious plants. The ovulate strobili, or cones, are usually very large. The different genera have cones which show a very distinct series, ranging from those of *Cycas revoluta* with loosely arranged leaf-like sporophylls to the compact cones of *Zamia.* The large ovules, or megasporangia, are covered by a single thick integument. The male cones are much smaller and always formed of compactly massed sporophylls, each of which bears many sporangia, or pollen sacs. The pollen grains are very numerous and light. Pollination is effected by wind, though insects are frequently seen on the male cones, and may play some part in the pollen transfer.

The pollen grains are caught in a sticky fluid, which covers the micropyle of the ovule. As this sticky substance dries, it shrinks, drawing the pollen grains down through the micropyle. Each pollen grain then puts out a pollen tube which digests its way through the mass of nucellar tissue which surrounds the female gametophyte, and reaches a small chamber which is formed at the micropylar end of the gametophyte and the nucellus. Meanwhile, two sperm cells have been forming in the pollen tube. Each is very large and has a spiral band of cilia wound about its anterior end. Freed from the pollen tube, these sperm pass to the gametophyte, where union of one sperm with the egg occurs. Sperm and egg nucleus presently unite, and the egg is fertilized. The fertilized egg divides to form a mass of cells which is known as the proembryo. At the base of this proembryo is a group of cells which becomes the true embryo. The mature seed of a Cycad has an outer fleshy coat which is variously colored. Inside this there is a hard stony layer which in turn surrounds another layer which is fleshy at first but soon becomes thin and dry. Within is the gametophyte which contains the embryo. Cycad seeds germinate as soon as they are mature.

Apart from their value as decorative plants or as curiosities, few products of importance are obtained from the Cycads. Their leathery leaves remain green for some time after removal from the plant. They are therefore often used on Palm Sunday and for funeral purposes. The seeds of many of them are edible, as is also the central portion of the stem of *Cycas* species. Cycads are sometimes confused with palms.

Cycads probably originated in very early times from some primitive ferns. Even today certain cycads closely resemble ferns. Cycads are "living fossils" which continue to exist in a very restricted range.

CYCLAMATE. Sweeteners.

CYCLE (Computer System). (1) A set of operations that is repeated regularly and in the same sequence. (2) In data processing, the word *cycle* often refers to the storage cycle, namely, the shortest time interval between one fetch (or store) and the next fetch (or store) within the same storage unit. Normally, the storage cycle establishes the pace for the entire computer. In some designs, access to more than one instruction per cycle is possible. Multiple independent storage units and overlapping cycle times also are possible. Storage cycle often is referred to as machine cycle or major cycle. The *clock cycle* of a synchronous computer establishes the commencement of each elementary subtask of the execution routine. This is a much faster cycle, internal to the computer hardware, and is sometimes referred to as the *minor cycle.* See also terms listed under **Data Processing.**

CYCLE (Mechanical). A series of changes executed in orderly sequence, by means of which a mechanism, a working substance, or a system is caused periodically to return to the same initial condition, constitutes a cycle. Many complicated machines or assemblages of machines work in definite cycles. An important form of cycle is the heat engine cycle, in which a series of thermodynamic changes in a working medium periodically return the system to the same thermodynamic level. This working medium may be a gas, as in the Otto and Diesel cycles, or a vapor, as in the steam cycle. See also **Carnot cycle,** for an example of a general ideal cycle. A vapor cycle is so named from the fact that it is conceived as using the same vapor over and over, passing it around what might be thought of as a closed loop of equipment, and subjecting it to various thermodynamic changes by means of which useful mechanical energy is produced from heat. The distinction between vapor and engine cycle should be recognized. An engine cycle considers only the changes occurring within an engine, but a vapor cycle involves, in addition, all changes in the vapor state from the point of leaving the engine until it is again ready to enter it.

CYCLE OF STRESS. The stress variation on a particular plane through a specific point in a body which is subjected to a repeated load is called a cycle of stress. If the stress varies alternately between tension and compression, the variation is known as reversal of stress. The reversal is complete when the alternate stresses are equal in magnitude.

The algebraic difference between the maximum and minimum stresses of a cycle is the range of stress. The endurance limit depends on the range of stress.

CYCLE OILS. Petroleum.

CYCLE STEAL (Computer System). A type of channel operation that has the ability to access main storage for data while the arithmetic unit is performing an operation which does not require a storage access. Where both the channel and the arithmetic unit request a storage access simultaneously, the arithmetic operation typically is delayed until the channel storage request is serviced. Also known as a Direct Storage Access (DSA) or Direct Memory Access (DMA) channel.

CYCLIC ACID. Organic Chemistry.

CYCLIC AMP (Adenosine Monophosphate). Brain and Nervous System.

CYCLIC CURVE. A general class of higher plane curves of which there are many special cases. Consider two circles, one designated as C which is of radius R and fixed in a Cartesian coordinate system; the other, C' of radius r which moves in contact with C. Place the center of C at the coordinate origin, the center of C' on the OX-axis and select a point P on this axis at a distance a from the center of C'. As the latter rolls around the circumference of C the point P will generate a curve given in parametric form by the equations

$$x=(R+r)\cos\phi - a\cos\frac{(R+r)\phi}{r}$$

$$y=(R+r)\sin\phi - a\sin\frac{(R+r)\phi}{r}$$

where the angle ϕ is measured from the coordinate origin. If P is inside or outside the rolling circle ($a \lessgtr r$), the curve is called trochoidal; if $a = r$, so that the point P is on the circumference of the rolling circle, the curve is cycloidal. Furthermore, the two radii can be so related that C' is always on the outside of C and the curves are then epitrochoidal or epicycloidal; if C' rolls around the inside of C the curves are hypotrochoidal or hypocycloidal.

In further special cases, when C' rolls along a plane instead of around another circle, the curve is called a trochoid or a cycloid. Still another special case is the evolute of a circle.

A generalized cyclic curve is produced when the circles are replaced by ellipses or hyperbolas. A pseudocyclic curve involves hyperbolic functions instead of trigonometric functions.

See also **Curve (Higher Plane);** and terms listed under **Mathematics.**

CYCLITIS. **Vision and the Eye.**

CYCLOHEXANE. **Organic Chemistry.**

CYCLOHEXANOL-CYCLOHEXANONE. KA oil is comprised of cyclohexanol, an alcohol, and cyclohexanone, a ketone, and is a principal raw material in the manufacture of nylon 6 and nylon 66 fibers. The KA stands for ketone-alcohol oil. At one time, KA oil was derived principally from phenol, but the majority is presently produced from the oxidation of cyclohexane. The KA oil is converted to adipic acid in the production of nylon 66, but in the production of nylon 6 (polycaprolactam), the KA oil is converted into the monomer, caprolactam.

Cyclohexanol, $CH_2\langle(CH_2CH_2)_2\rangle CHOH$, formula weight, 100.16, mp 23.9°C, bp 160–161°C, sp gr $0.962^{20/4}$, slightly soluble in water, soluble in alcohol and ether. Cyclohexanone, $CH_2\langle CH_2CH_2\rangle_2 CO$, formula weight, 98.14, mp −45°C, bp 155–156°C, sp gr $0.947^{19/4}$, soluble in water, alcohol and ether.

CYCLOID. The path described by a point on a circle as it rolls along a straight line and a special case of a trochoid (see also **Cyclic Curve**). Thus b, the distance of the generating point from the center of the rolling circle equals a, the radius of the circle. Its parametric equations are $x = a(\theta - \sin\theta)$; $y = a(1 - \cos\theta)$ and in Cartesian coordinates $x = a\cos^{-1}(a - y)/a + \sqrt{y(2a - y)}$. There are cusps in the curve, separated by the distance $2\pi a$, every time the point touches the line on which the circle rolls.

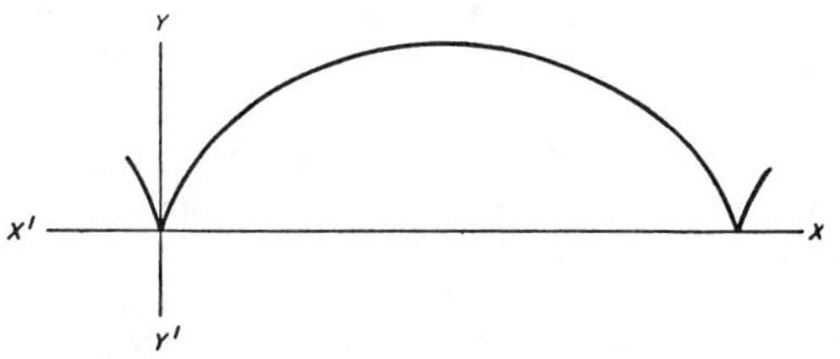

Cycloid.

The teeth of gears are often cut with faces which are arcs of cycloids, so that there is rolling contact when the gears are in mesh.

The inverted arch of the cycloid is a tantochrone for the force of gravity and a brachistochrone, or curve of quickest descent. The evolute of a cycloid is another cycloid. A related curve, sometimes called the companion of the cycloid is $x = a\theta$, $y = a(1 - \cos\theta)$.

See also **Tantochrone; Trochoid;** and terms listed under **Mathematics.**

CYCLOID SCALES. **Fishes.**

CYCLONE. **Atmosphere (Earth); Fronts and Storms.**

CYCLONE-FURNACE (Boiler). **Boiler; Burner.**

CYCLONE SEPARATOR (Steam). **Boiler.**

CYCLONE WAVES. **Atmosphere (Earth).**

CYCLOPROPANE (Anesthesia). **Anesthesia.**

CYCLOSTOMATA. The round-mouthed eels, hag fishes, and lampreys. A class of the phylum Chordata made up of marine and freshwater species resembling slender fishes in form but without hinged jaws, and having a protrusible toothed tongue, single nostril on top of head, and naked glandular skin.

The principal characters of the class are: (1) The notochord is persistent. (2) Cartilaginous neural arches indicate the development of a vertebral column. (3) There are no paired fins. (4) The mouth is a funnel-shaped depression with chitinous (see **Chitin**) teeth. (5) External gill openings are separate.

The two principal subdivisions of the class are listed both as subclasses and as orders in modern classifications.

Subclass *Myxinoidea* (*Hyperotreta*). The hag fishes. Marine species with a poorly developed oral depression. Gill openings far behind head.

Subclass *Petromyzontia* (*Hyperoartia*). Lampreys. Marine and freshwater species with a well-developed oral funnel acting as a sucker. Gill openings immediately behind head.

It is fairly probable that the origins of the jawless fishes can be found in brackish water, i.e., in the intermediate region between salt water and fresh water with its high variations in water temperature, salinity, water flow, depth, and other characteristics—all of which accelerate an adaptation process. Some structural features of the cyclostomes were present primevally and others have been modified as a result of the cyclostome life habits.

The distribution of jawless fishes is limited to the temperate-to-cold waters of the northern and southern hemispheres. The northern species are more highly differentiated from the southern ones than those in the same hemisphere. Water temperature is an important factor in distribution of the sea-living hagfishes. The 10°C boundary is crucial for them. In the cold northern and southern seas hagfishes can penetrate to approximately 100 feet (30 meters), while at the equator they are found at depths of over 3280 feet (1000 meters).

Hagfishes Myxinoidea). These are worm-shaped jawless fishes with nasal opening lying at the fore-end of the body joined with the mouth opening. The species has 4 to 6 barbels on the head; the mouth has 2 rows of protruding teeth and a row of mucous glands are on each side of the stomach. A single fin seam is present and the cartilage gill skeleton is poorly developed. Hagfishes have underdeveloped eyes which are not visible externally. They are pure sea inhabitants, and they lay only a few large eggs which in development lack a larval stage. The hagfish eye is covered by skin and has neither lens, iris, nor muscles, and the corresponding connections to the brain are poorly developed. Experiments have shown that this eye has no particular visual significance; interestingly, light-sensitive organs have been found in the skin, especially in the head and rear regions. A similar condition exists in other boring and hole-inhabiting animals.

Aquarium experiments have shown that common hagfishes become active immediately after some bait has been laid and can find it within 2 minutes. Little is known about distances from which food can be detected, but on the basis of some observations this would amount to 20 to 24 inches (50 to 60 centimeters). Large numbers of hagfishes collect around a good-sized piece of food after a short time. In one case, 123 hagfishes were found at a dead cod; and another time 100 had collected at a bait. Dead animals are quickly devoured by hagfishes. However, there is not a sufficient number of dead fishes or other organisms lying on the ocean floor to serve as the prime food source for larger hagfish populations. Extensive investigations have shown that the main hagfish diet does not consist of carrion, but of organisms inhabiting the ocean floor, such as annelids, echiurid worms (*Echiurus echiurus*), and others. Small snails and mussels less than a millimeter in size have also been found in hagfishes, as well as the remains of shrimp and hermit crabs. Very active crabs almost certainly are consumed only after they have been injured. See also **Hagfishes (Agnatha).**

The North Atlantic hagfish or common hagfish (*Myxine glutinosa*) lives on soft ground. Boring into the floor is accomplished by powerful swimming motions of the rear body portion in a vertical position. The bored-in passages are not coated with mucous, as is the case of some boring fishes.

Lampreys (*subclass Petromyzontia*). These fishes are also eel-shaped, but as adults they have well-developed eyes. The nasal orifice on top of the head has a blind end. There are 7 gill openings and a circular sucking mouth well supplied with teeth is on the lower side of the head. Lampreys lack barbels. They have 2 dorsal fins and 1 caudal fin. Lampreys are free-living inhabitants of salt and fresh water.

Lampreys undergo metamorphosis similar to amphibians and some fishes; they have a juvenile stage, which as a larva is structured differently from the adult form and leads a completely different way of life. In many cases, the larvae even inhabit another environment. Two groups can be differentiated: (1) Migrating forms and, (2) freshwater forms. Members of the former group live as adults in the ocean or in brackish water near the coast, but after a period of time migrate into rivers and streams, where they lay their eggs and where their larvae live until they metamorphose. After metamorphosis, the young larvae migrate again to the coast. Species of the second group spend their lives in fresh water and do not migrate. See also **Lampreys (Agnatha).**

As a result of filter feeding, the growth of lamprey larvae proceeds slowly. The larvae reach a length of less than 1 inch (about 2 centimeters) after 1 year. After 4 to 5 years, the larvae are 4 to 8 inches (10 to 20 centimeters) long.

Well known among the lampreys is the American sea lamprey (*Petromyzon marinus dorsatus*), which has been greatly responsible for reducing the whitefish population in the Great Lakes. See also **Whitefishes.**

References are listed at end of entry on **Fishes.**

CYCLOSTROPHIC FLOW. Gradient Flow.

CYCLOSTROPHIC WIND. Winds and Air Movement.

CYCLOTRON. Particles (Subatomic).

CYCLOTRON (Dating Determinations). Radioactivity and Other Dating Techniques.

CYDIPPIDA. An order of *Ctenophora.*

CYESIS. Pregnancy.

CYGNUS. (the swan.) One of the most striking and interesting constellations of the northern sky, also frequently referred to as the northern cross. Cygnus represents a swan flying with outstretched wings, and legs trailing out behind. Lying, as it does, in one of the most impressive portions of the northern Milky Way, the constellation contains many interesting objects. Its brightest star (α Cygni), known as Deneb, has an intrinsic brightness about 1,000 times that of the sun, and is among the most distant of all of the bright stars. One of the most striking of all of the double stars is β Cygni, whose two bright components, one blue and the other orange, may be easily separated by a small telescope. This star is probably not a true binary, since no physical connection has been determined between the components. Another interesting member of Cygnus is the relatively faint star noted as number 61 on large star maps, the first star to be measured for its distance from the earth. The astronomer Bessel's determination, in 1838, of the stellar parallax of this object opened a field of astronomical research that has done much to solve many problems of general cosmogony. See map accompanying entry on **Constellations;** also **Black Hole; Radio and Radar Astronomy;** and **X-Ray Astronomy.**

CYLINDER (Geometry). A solid bounded by a cylindrical surface and two parallel planes. The terms cylinder and cylindrical surface are often used interchangeably but, as stated, the former is a solid and the latter a surface.

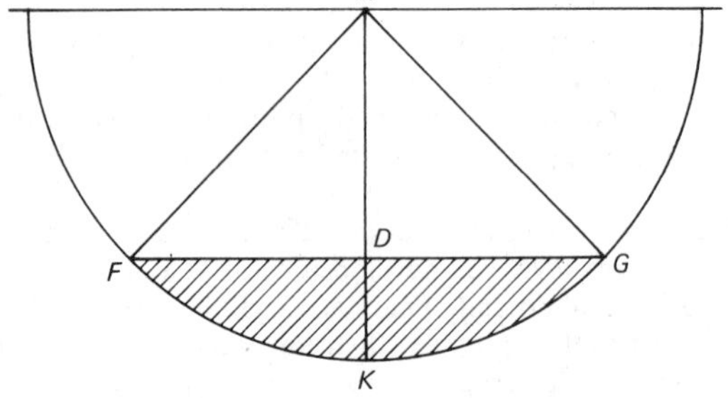

Volume calculations of cylindrical tanks.

Area FGK = Area FMGK − FMG
Area FMGK

$$= \left(\frac{2\angle \text{FMD}}{360}\right) \times \text{Area of Circle.}$$ $\angle$ FMD is determined from its cosine.

Volume of Tank = Area FGK × Length (same units)

The two plane surfaces bounding a cylinder are its bases and the cylindrical surface is its lateral surface. The altitude of a cylinder is the perpendicular distance between its bases. If the elements of a cylinder are perpendicular to its bases, the cylinder is called a right cylinder; otherwise, it is oblique. If the bases are circles, it is a circular cylinder. A cylinder of revolution may be generated by rotating a rectangle about one side as an axis and the result is a right circular cylinder.

The following relations refer to a cylinder of revolution: $S = Ch = 2\pi rh$; $A = 2\pi r(r + h)$; $V = \pi r^2 h$ where the symbols mean: S, lateral area; C, r, circumference and radius of lower base; h, altitude; A, total area; V, volume. For any cylinder, $V = bh$, where b is the area of the base.

Because it is comparatively easy to form, by usual means of manufacture, and because its shape is very well adapted to the resisting of internal bursting pressure, the cylinder is a very common engineering shape. While, of course, anything of cylindrical shape might truly be called a cylinder, it is customary to apply the term to that part which, in conjunction with a closely fitting internal piston will provide an enclosed space the volume of which may be varied by motion of the piston. Expansion of volume of a working medium is the basis of all commercially employed power cycles, and the cylinder and piston have an important place in this field.

Tanks in the form of cylinders also are widely used for storing liquids and gases, while cylindrically-shaped silos are used for storing a variety of bulk solid materials, such as grain and chemical raw materials. Tanks may be erected vertically or horizontally. Calculation of the volume of contents in vertical tanks for various levels simply requires use of the aforementioned formulas. Reference is made to the accompanying diagram for calculation of capacities of horizontal cylindrical tanks.

CYLINDRICAL COORDINATE. A curvilinear coordinate system of right-circular cylindrical surfaces forming families of circles about the origin in the XY-plane of a rectangular Cartesian system (ρ = const.); half-plane from the Z-axis (ϕ = const.); planes parallel to the XY-plane (z = const.). The position of a point in this system is given by (ρ, ϕ, z) where $x = \rho \cos \phi$, $y = \rho \sin \phi$, $z = z$.

More precisely, the system should be called circular cylindrical because elliptic and parabolic cylindrical coordinates are also used.

See also **Coordinate System;** and terms listed under **Mathematics.**

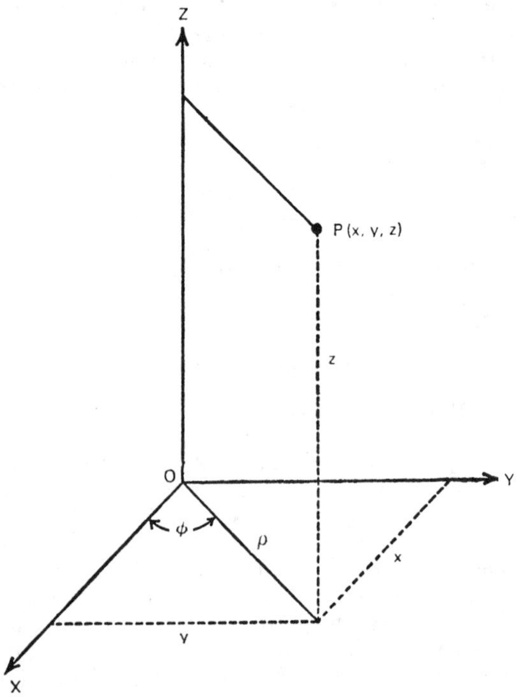

Cylindrical coordinate system.

CYMOPHANE. Chrysoberyl.

CYPRESS TREES. The term *cypress* is not scientifically specific, but as will be noted from the following list of cypress trees, cypresses are found in two families and in various genera, only a few of which are in the genus *Cupressus* (true cypresses). The two families represented are: (1) *Cupressaceae* (cypress family); and (2) *Taxodiaceae* (swamp cypress family). In addition to *Cupressus* under *Cupressaceae*, there is the genus *Chamaecyparis* (false cypresses). A further complication is the fact that several cypresses are called cedars in the commercial world. Thus, some of the cypresses listed below, as noted, are described in the entry on **Cedar Trees.**

Arizona smooth cypress	*Cupressus glabra* (A true cypress)
Bald cypress	*Taxodium distichum* (A swamp cypress)
Classic cypress	*Cupressus sempervirens* (A true cypress)
Hinoki cypress	See Japanese Sawara cypress
Italian cypress	See Classic cypress
Japanese Sawara cypress	*Chamaecyparis obtusa* (A false cypress)
Kashmir cypress	*Cupressus cashmeriana* (A true cypress)
Lawson cypress	*Chamaecyparis lawsonia* (A false cypress, but better known as Port Orford cedar. See entry on **Cedar Trees.**)
Leyland cypress	*Cupressocyparis leylandii* (Closely related to the false cypresses, but of a separate genus known as the Leyland cypresses.)
Mediterranean cypress	See Classic cypress
Mexican cypress (or Montezuma cypress)	*Taxodium mucronatum* (A swamp cypress)
Mexican cypress (or Cedar of Goa)	*Cupressus lusitanica* (A true cypress, but better known as the Cedar of Goa. See entry on **Cedar Trees.**)
Monterey cypress	*Cupressus macrocarpa* (A true cypress)
Nootka cypress	*Chamaecyparis nootkatensis* (A false cypress, but better known as the Alaska cedar. See entry on **Cedar Trees.**)
Pond cypress	*Taxodium ascendens* (A swamp cypress)
Santa Cruz cypress	*Cupressus abramsiana* (A true cypress)
Smooth Arizona cypress	See Arizona smooth cypress
Southern cypress	See Bald cypress
Swamp cypress	See Bald cypress
White cypress	*Chamaecyparis thyoides* (A false cypress, but better known as the Atlantic white cedar. See entry on **Cedar Trees.**)
Yellow cypress	See Nootka cypress

It is interesting to note that in addition to the cypress trees just listed, other genera of *Cupressaceae* (cypress family) include the thujas or arborvitae, the incense cedars, and the junipers; and other genera of *Taxodiaceae* (swamp cypress family) include the dawn redwoods, the giant sequoias, coast redwoods, and a small group known as Pacific cedars. See also **Giant Sequoia;** and **Redwood (Coast).**

The True Cypresses

Probably the best known of the true cypresses is the classic or Mediterranean or Italian cypress, highly regarded for its decor and use in formal landscaping. The tree does very well in the Mediterranean region. It is also found in other parts of the world where it has been introduced and does well provided that climate and soil conditions are appropriate. It is found in the southern United States and along the southern part of the west coast. In northern areas where the classical cypress cannot be used, the Lombardy poplar provides somewhat the same landscaping effect. The classical cypress can attain a height of from 50 to close to 100 feet (15 to 30 meters) and retains its stately, formal and rather narrow columnar shape with a top that resembles a huge rounded pencil point. Of an entirely different contour and skeletal form is the Monterey cypress which thrives right at the edge of the Pacific Ocean along the California coast. It has a rather ancient, windswept appearance, often with proportionately little foliage for its size. However, when protected from high winds and heavy weather, the tree takes on a different appearance, but it is always characterized by multiple stems. When grown in a garden, the tree appears something like a bushy cedar of Lebanon. The Monterey cypress has been introduced into New Zealand in connection with reforestation. The Kashmir cypress is of a blue-green color which attains a maximum height of about 20 feet (6 meters) and is of a weeping contour. It is a difficult tree to raise and hence is quite rare. The Arizona smooth cypress is found in the southwestern United States and hence is well-named. It is pyramidal in shape, somewhat columnar, but not nearly so formal as the classical cypress. The foliage is bluish-gray, with thick upswept branches. True cypresses of the southern hemisphere include the Australian *Callitris* and the Chillean *Fitzroya.* The latter is quite a large tree, native to the Andes. Specimens introduced into favorable and comparable climates in the northern hemisphere have done quite well. The Mexican cypress (or cedar of Goa) is the only true cypress that is referred to as a cedar. The tree was named after Lusitania (Portugal). See Fig. 1.

Fig. 1. Group of broad-topped Monterey Cypress trees in California. (*Photo by A. Gaskill.*)

False Cypresses

Important species include the Lawson, Nootka, and white cypresses, commonly referred to as cedars, and discussed under **Cedar Trees.** The Leyland cypress is a hybrid of the Nootka cypress and the Monterey cypress, and was first developed in Wales about 1925. Since then, the tree has become highly regarded in Britain as an excellent garden tree. The tree is hardy, shapely, dense with foliage and of good color. It is also prized for its fast rate of growth. A 30-foot (9-meter) specimen may be obtained in about 10 years, and is considered capable of attaining a height of about 100 feet (30 meters). The tree usually reaches a height of about 6 feet (1.8 meters) in its first 3 years. The flame-like shape is retained as the tree grows tall.

Swamp Cypresses

The bald cypress is found in the United States from Delaware to Florida and westward through the Mississippi valley and in some midwestern states. The tree is well-named because it prefers swampy environments. It frequently is found standing in water. The height ranges from 80 to 120 feet (24 to 36 meters). Parts of the tree rise out of the water like knees, measuring from 8 to 30 inches (20.3 to 76.2 centimeters) in height. The tree does not develop these knees unless it is growing in a watered area. Roots extend downward from the knees, providing the tree with adequate anchoring when it rises from a muddy, soggy base. The knees per se make interesting souvenirs, as well as providing a material from which to make interesting artistic objects. The knees sometimes are used effectively in flower arrangements. See Fig. 2.

Fig. 2. Cypress swamp near Little Rock, Arkansas. (*U.S. Dept. of Agriculture photo.*)

The bark of the bald cypress is heavy, thick, pale-brown and its broad ridges have a vertical pattern. The branches of the tree are slender; the tips have thick scales. The cone is small, about 1½ inches (3.8 centimeters) in diameter. The wood of the bald cypress is somewhat heavier than white pine, is straight-grained and easily worked. The wood is used in general construction and as an interior finish wood.

The champion cypress trees in the United States as selected by The American Forestry Association are listed in the accompanying table.

For references, see **Trees.**

CYPRINIFORMES. Carp; Bitterling; Fishes.

CYST. 1. A capsule containing semi-solid or fluid material which may be under pressure. The contents of a cyst may be retained normal secretions, e.g., the sebaceous cyst which contains the products of a plugged sebaceous gland; or they may represent a fluid collection associated with a parasitic infection, e.g., echinococcus cysts of the brain, liver or other organs. Many tumors undergo cystic degeneration, especially carcinomas of the ovary, breast, and uterus. Benign cysts occur in the ovary, spleen, lungs, kidney and liver where they are often congenital. Other congenital cysts result from fetal malformations and failures of development, such as the several cysts which occur in the neck. 2. The resting form of an organism in a protective covering. Significant examples in medicine are the cysts of *Endamoeba histolytica* which are found in the stools of patients with amebiasis, and the encysted larva of *Trichinella spiralis* which can be demonstrated in the muscles of individuals who have had trichinosis.

CYSTEINE. Amino Acids.

CYSTICERCOID. A small bladder worm in which the scolex fills the interior.

CYSTIC FIBROSIS. This is a genetic disease, particularly prevalent among Caucasians, that results from dysfunctions of the exocrine glands. The disease occurs in one out of above every 2500 live births. An *exocrine gland* secretes into a cavity or on the body surface, as contrasted with ductless glands which secrete into the tissue fluids and blood. In cystic fibrosis, the exocrine glands produce a thick mucus that clogs the lungs and digestive system, a condition that can be fatal in children. At one time, it was estimated that half of the persons with cystic fibrosis died before reaching the age of 21 years. The availability of antibiotics and improved supportive therapy has increased the number of patients surviving to adulthood. Thus, the disease is no longer exclusively a pediatric problem.

In the late 1970s, researchers at the University of Minnesota proposed that the disorder may be caused by an abnormality in a cellular enzyme, NADH dehydrogenase. Several other groups of investigators have studied the blood serum from cystic fibrosis patients. Data from these studies indicate that cystic fibrosis patients have low concentrations of cholesterol and linoleate (an essential fatty acid). They also show symptoms of vitamin deficiencies (A, K, and E). These substances require lipoprotein for transport to various locations. Other researchers have reported that two major classes of serum lipoprotein are significantly lower in cystic fibrosis patients. These findings may prove important in therapy for the disease.

RECORD CYPRESS TREES IN THE UNITED STATES[1]

Specimen	Circumference[2] (inches)	Circumference[2] (centimeters)	Height (feet)	Height (meters)	Spread (feet)	Spread (meters)	Location
Arizona (Typical or rough) cypress (1977) (*Cupressus arizonica*)	226	574	73	21.9	40	12	Arizona
Cuyamaca cypress (1976) (*Cupressus arizonica*)	70	178	37	11.1	28	8.4	California
Macnab cypress (1972) (*Cupressus macnabiana*)	83	211	55	16.5	38	11.4	California
Mendocino cypress (1976) (*Cupressus goveniana*)	176	474	96	28.8	55	16.5	California
Baker cypress (1976) (*Cupressus bakeri*)	129	328	129	38.7	29	8.7	Oregon
Monterey cypress[3] (1975) (*Cupressus macrocarpa*)	333	846	97	29.1	106	31.8	Oregon
Piute cypress (1976) (*Cupressus arizonica*)	115	292	45	13.5	40	12	California
Santa Cruz cypress (1977) (*Cupressus goveniana*)	180 (at 24 inches)	457 (at 61 centimeters)	75	22.5	60	18	California
Tecate cypress (1976) (*Cupressus guadalupensis*)	88 (at 18 inches)	224 (at 46 centimeters)	47	14.1	38	11.4	California

[1] From the "Social Register of Big Trees," The American Forestry Association (by permission). Dimensions of the swamp cypresses of the genera (*Sequoiadendron* and *Sequoia*) are given in the entries on **Giant Sequoia,** and **Redwood (Coast),** respectively.

[2] At 4.5 feet (1.4 meters).

[3] Introduced.

The principal clinical features presented by cystic fibrosis are pulmonary abnormalities, pancreatic insufficiency, and elevated sweat electrolyte levels. The first two features are attribtued to the production of abnormal mucus. Pulmonary involvement is the most consistent feature of the disorder. The thick mucus secreted in the bronchi block airways and thus impair local defense mechanisms. The usual result is a persistent bronchial infection. Irreversible structural changes occur in the airways. Obstruction by mucus in the gut leads to pancreatic insufficiency, cirrhosis (liver), and intestinal obstructions, among other disturbances of normal function.

The management of cystic fibrosis has traditionally paralleled that for severe and diffuse bronchiectasis. See **Bronchiectasis.**

CYSTINE. Amino Acids.

CYSTINE STONES. Kidney and Urinary Tract.

CYSTITIS. Kidney and Urinary Tract.

CYSTOSCOPY. Kidney and Urinary Tract.

CYTOCHROMES. The cytochrome *c*-cytochrome oxidase system represents the terminal segment of the respiratory chain common to the vast majority of organisms utilizing oxygen as the terminal oxidant in tissue respiration. The complete respiratory chain consists of a number of electron carriers, both protein and nonprotein in nature, organized in a definite sequence within the walls and internal partitions of subcellular organelles known as mitochondria. These structures carry the electrons which come from the substrates being oxidized and eventually react with oxygen. The energy released in several of the many steps of this series of reactions is utilized to make the high-energy compound adenosine triphosphate, a process known as *oxidative phosphorylation.* The high-energy compound is, in turn, employed to drive the many reactions of metabolism which require chemical energy. Every component of the terminal respiratory chain is reduced by the component immediately preceeding it and then reduces the component immediately following it in the chain, itself becoming reoxidized. The *c*-cytochrome oxidase system is common to all vertebrates and invertebrates, plants, as well as numerous microorganisms, and must be distinguished from systems having similar functions, but very different properties, which occur in numerous bacteria.

The cytochromes were first observed by MacMunn as early as 1886. He described their spectral absorption bands in a large variety of organisms and tissues. His discovery was, however, forgotten after a controversy with Hoppe-Seyler had raised doubts as to the validity of some of his conclusions, and it was not until 1925 that Keilin independently, rediscovered the remarkable cytochrome spectrum in the flight muscles of a living insect.

Keilin's observations came at a time when the understanding of tissue respiration had advanced to the point of providing the foundations necessary for the unraveling of the physiological role and chemical nature of cytochromes. The first step had indeed been taken some 40 years earlier by Ehrlich when he found that a variety of animal tissues could transform a mixture of α-napththol and dimethyl-*p*-phenylenediamine to indophenol, in the presence of oxygen. A decade later, the enzyme responsible for this effect had been named indophenol oxidase, and it was shown that its activity was inhibited by cyanide. In the first decades of this century, Warburg, from studies of the catalysis of the oxidation of cysteine by iron-charcoal, considered as a "model" of cellular respiration, concluded that oxygen activation was the all-important process in cellular respiration and that an iron-containing enzyme, the "respiratory enzyme" or *atmungsferment,* is solely responsible for the transport of the oxidizing equivalents of oxygen to the substrates. An opposing view was taken by Thunberg, who had detected a large variety of dehydrogenases in tissues, and by Wieland who used palladium-hydrogen as a "model" of tissue respiration and believed that substrate-specific hydrogen activations were characteristic of all biological oxidation processes, the reaction with oxygen being nonspecific and relatively unimportant.

The controversy as to the respective roles and importance of hydrogen and oxygen activation faded into the background when, following his initial observations, Keilin demonstrated that the four-banded spectrum of cytochrome, observed in a large variety of tissues and organisms, was in fact the spectrum of the ferrous or reduced forms of three distinct cytochromes, cytochrome *a*, cytochrome *b* and cytochrome *c*. Keilin obtained a soluble preparation of cytochrome *c* from baker's yeast, and together with Hartree, in 1938–1939, showed that the indophenol oxidase activity of particulate tissue preparations was simply the result of a nonenzymic reduction of cytochrome *c* by dimethyl-*p*-phenylenediamine, the reduced heme protein being oxidized by indophenol oxidase in the presence of oxygen. Having established the nature of the final steps of tissue respiration, they renamed the enzyme "cytochrome oxidase," since its only function appeared to be the oxidation of cytochrome *c*. There had been no doubt of the overwhelming physiological importance of the system ever since 1934 when Haas found that in a number of tissues the rate of oxygen uptake was identical to that of cytochrome *c* reduction, demonstrating that nearly all of the oxidizing equivalents of oxygen were transmitted by the cytochrome *c*-cytochrome oxidase system.

That the material in tissues reacting directly with oxygen was in fact a heme compound had been shown by the experiments of Warburg and collaborators on the effect of carbon monoxide on tissue respiration, carried out in the late 1920s. Warburg observed that carbon monoxide inhibits the uptake of oxygen by tissues and that this inhibition is reversed in bright light. Using this phenomenon, he succeeded in measuring the absorption spectrum of the carbon monoxide complex of the respiratory enzyme, a spectrum which turned out to be clearly that of a heme compound. Thus, when Keilin and Hartree in 1939 found that in the presence of carbon monoxide, cytochrome *a* showed up as two spectroscopic components, they were able to demonstrate that the new cytochrome, cytochrome a_3 was the substance responsible for the photochemical action spectrum of Warburg. Cytochrome a_3 was thus identified with the respiratory enzyme reacting directly with oxygen, and the system was considered to be composed of three entities, cytochromes *c*, *a*, and a_3, reacting consecutively, like all the other components of the respiratory chain.

Cytochrome *c* consists of a polypeptide chain, from 104 to 108 amino acid residues in length. A single heme prosthetic group is attached by thioether bonds formed between the sulfhydryl side chains of two cysteine residues in the protein and the vinyl side chains of the porphyrin ring as shown by

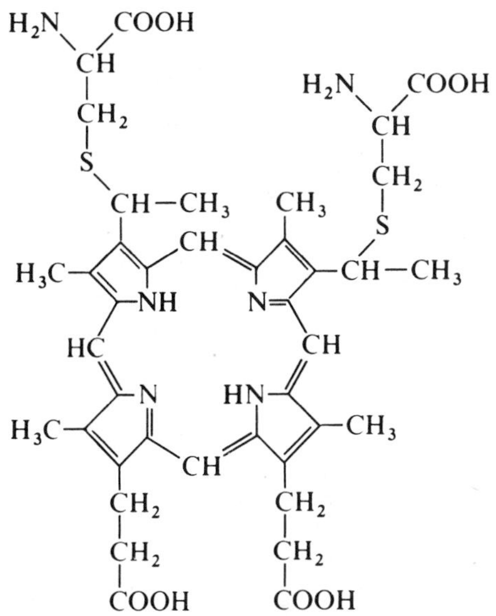

This entry reviewed and updated for the 6th Edition by Ann C. Vickery, Ph.D., University of South Florida, College of Medicine, Tampa, Florida.

CYTOGENIC GLAND. An organ which produces and discharges cells, such as the reproductive glands.

CYTOKINESIS. Cell (Biology).

CYTOLOGY. A branch of biology dealing with the details of cell structure and studies of cell function. Cells may be examined in the living state by use of phase contrast microscopy (see **Microscope**) or with the conventional light microscope if "vital" stains are employed. These selectively stain specific structures in the living cell and render it more discernible. Alternatively, cells killed by chemical or physical fixation may have their components selectively stained for clarification of identity or function.

Knowledge of cytology has been greatly expanded through use of transmission and scanning electron microscopy, which have revealed internal and external details of cells too small to be seen with optical microscopes. An additional very powerful technique in cytology is the use of fluorescence microscopy, in which a specific cell substance such as protein or serum antibody is specifically conjugated with a fluorescent compound, thus rendering the cell substance uniquely visible when examined microscopically under, e.g., mercury lamp excitation.

Ann C. Vickery, Ph.D., University of South Florida, College of Medicine, Tampa, Florida.

CYTOKINESIS. **Cell (Biology).**

CYTOMEGALOVIRUS. **Virus.**

CYTOPLASM. **Cell (Biology).**

CYTOPLASM (Root). **Root (Plant).**

CYTOTOXIC CHEMICALS. Chemical agents that damage cells to which they are applied. They are poisons, to which cells respond with injury, disease, or death. There are multitudes of cytotoxic chemicals; they act by a variety of mechanisms; and they have many different kinds of effects. See also **Carcinogens.**

CYTOTOXINS. **Foodborne Diseases.**

CZOCHRALSKI METHOD. **Crystal.**

D

DAB. **Flatfishes.**

DABCHICK. **Podicipediformes.**

DACITE. The name of a somewhat variable group of extrusive igneous rocks similar to the rhyolites but richer in plagioclase feldspar. Typical dacites are felsitic to porphyritic in texture. Dacites are the extrusive equivalents of the quartz-rich varieties of diorites and are sometimes classified as quartz-bearing andesites. The porphyritic types usually occur toward the center of the thicker dacite flows, dikes and sills, as well as the marginal zones of laccoliths. Dacites are common in the Cordilleran province of North, Central and South America. The term, dacite, was proposed by G. Stache of Austria for lavas in the old Roman province of Dacia.

DACRYDIUM. **Podocarps.**

DACRYOCYSTITIS. Infection of the lacrimal sac of the eye. The lacrimal glands produce tears for lubrication and protection of the eye. When the naso-lacrimal duct becomes obstructed, infection of the tear-producing sac is likely. Dacryocystitis appears most often in infants and in adults over 40 years of age. Secretions which cannot drain through the obstructed duct spill back out through the eye. Medical attention is required to avoid complications, such as infection of the cornea. See also **Vision and the Eye.**

DACTINOMYCIN. **Antibiotic.**

DACTYLOZOOID. A form of polyp found in colonial Hydrozoa which captures prey and brings it to the mouth.

DADDY LONG LEGS. **Crane Fly.**

DAGUERREOTYPE. **Photography and Imagery.**

DAIRY COW. **Bovines.**

D'ALEMBERTIAN. Analogous to the differential operator del in ordinary vector analysis, a four-dimensional operator called quad, and denoted by the symbol $\square$, has components $\partial/\partial x_i$. The contracted second-order derivative of a scalar function ϕ can then be written $\square^2\phi$. It is known as the d'Alembertian operator

$$\square^2 = \frac{\partial^2}{\partial x^2} + \frac{\partial^2}{\partial y^2} + \frac{\partial^2}{\partial z^2} - \frac{1}{c^2}\frac{\partial^2}{\partial t^2}$$

and it is thus similar to the Laplacian operator in three dimensions. The equation $\square^2\phi = 0$ is a wave equation for waves traveling with the velocity of light, c. The d'Alembertian is Lorentz invariant.

D'ALEMBERT PRINCIPLE. The principle, first pointed out by d'Alembert in 1742, that Newton's third law (see **Newton's Laws of Dynamics**) holds for forces acting upon bodies entirely free to move as well as upon fixed bodies in stationary equilibrium. In the former case the "reactions" concerned are due solely to inertia. Thus, in the act of throwing a ball, one pushes upon the ball with a certain force, and the inertia of the ball causes it to push back on the hand with an equal force. The condition of the system during such a process is said to be one of kinetic equilibrium. From this point of view, obviously, any system of bodies must always be in equilibrium, either kinetic or static.

D'ALEMBERT TEST. If

$$\lim_{n\to\infty} |s_{n+1}/s_n| = r$$

and $r < 1$, the series

$$\sum_{n=1}^{\infty} s_n$$

converges; if $r > 1$, it diverges; if $r = 1$, the test fails and the series must be investigated in some other way (see **Convergence**).

This test is often known as Cauchy's convergence test of the second kind but it was published by d'Alembert in 1768, seems to have been known to Waring in 1781, and was not given by Cauchy until 1821.

See also **Cauchy Convergence Test.**

DALLISGRASS. **Grasses.**

DALTONIDE COMPOUNDS. **Chemical Composition.**

DALTON LAW. The law of partial pressures in mixed gases and vapors. If several gases not reacting chemically upon each other are introduced into the same container, the pressure of the resulting mixture is equal to the sum of the pressures which would be observed if each gas were separately enclosed in that container. We may, for example, regard the atmospheric pressure as the sum of a nitrogen pressure, an oxygen pressure, an argon pressure, a carbon dioxide pressure, a water-vapor pressure, etc. The same principle holds for mixtures of the saturated vapors of two or more liquids evaporating in the same closed space, provided one liquid does not dissolve the vapor from the other (as water dissolves ammonia). Like other gas laws, this law is approximately valid only within limits. See also **Combustion.**

DALTON'S LAW (Combustion). **Combustion.**

DAM. Dams are constructed for several purposes, including hydroelectric power generation, navigation, river control and flood prevention, irrigation, water system storage, with the creation of water playgrounds usually being an added dividend. Often, a single dam will serve several purposes. Increasing attention is being given to the total impact of a new dam on the topography and the numerous technical, sociological, and economic factors over the usually wide geographical region affected in some manner by the dam. Dams, over the years, have been constructed usually with one over-riding purpose and, in some instances, have created unexpected impacts of an environmental and sociological nature, sometimes of a positive type; other times of a negative nature. Although accidents involving dams are rare, the public long remembers accidents of this type. A dam collapse, such as that at Vajont, Italy on October 9, 1963, in which there was a loss of 1,800 lives, is not soon forgotten. Consequently, safety in the engineering of dams is of paramount importance and is receiving increasing attention, notably in terms of constructing dams with ample resistance to the effects of earthquakes. Collapse of the Teton Dam (Idaho) in June 1976 is described later in this article.

Like many other engineering achievements which have come under severe scrutiny by environmentalists, the building of dams must be studied carefully, but fairly, not overlooking the tremendous advancements and advantages which have resulted from the generation of low cost power, particularly important to developing nations, and the many thousands of lives that have been saved, not to mention billions of dollars of property damage, because of the effective use of dams in flood prevention projects. Dams also have served in numerous instances to improve upon nature in creating new ecosystems which have contributed to the prevention and expansion of wildlife and natural recreational activities.

Some of the major dams of the world are described in the accompanying table. See also **Hydroelectric Power.** There are scores of additional and what may be considered major dams throughout the world. Over 27 major dams have been built by the Tennessee Valley Authority on the Tennessee River system since 1933. In addition to their use for power generation, these dams make the main stream of the Tennessee River navigable over its 650-mile (1046-kilometer) length from Knoxville to the Ohio River. TVA provides electric power to over two million customers in parts of seven states, as well as large atomic and military installations.

REPRESENTATIVE MAJOR DAMS OF THE WORLD

Name of Dam, River/Basin, Year of Dedication	Height		Crest Length		Volume		Gross Reservoir Capacity		Type of Construction
	(Feet)	(Meters)	(Feet)	(Meters)	(Cubic Yards)	(Cubic Meters)	(Acre-Feet)	(Hectare-Meters)	
ARGENTINA									
El Chocon, Limay, 1974	282	86	7546	2300	17004	13001	17025	2099	E
AUSTRALIA									
Dartmouth, Mitta-Mitta, 1978	591	180	2200	671	18312	14001	5232	645	R
Talbingo, Tamut, 1971	530	162	2300	701	18950	14489	747	92	R
AUSTRIA									
Gepatsch, Faggenbach-Inn, 1965	500	152	1908	582	9810	7501	113	14	R
BRAZIL									
Ilha-Solteira, Paraná-Rio de la Plata, 1973	295	90	20308	6190	29454	22521	27730	3419	EG
Marimbondo, Grande, 1975	315	96	12297	3748	24328	18601	5184	639	E
CANADA									
W. A. C. Bennett, Peace-Mackenzie, 1967	600	183	6700	2042	57203	43737	57006	7029	E
Gardiner, South Saskatchewan, 1966	223	68	16700	5090	85743	65559	8000	986	E
Daniel Johnson, Manicougan-St. Lawrence, 1968	703	214	4311	1314	2950	2256	115000	14180	MA
Mica, Columbia, 1974	**794**	**242**	2600	792	4200	32113	20000	2466	R
COLOMBIA									
Chivor, Bata, 1975	778	237	*919*	*280*	14126	10801	661	82	R
EGYPT									
High Aswan (Sadd-El-Aali), Nile, 1970	364	111	12565	3830	57203	43737	137000	16892	ER
GHANA									
Akosombo-Main, Volta, 1965	463	141	2100	640	10400	7952	120000	14796	R
GREECE									
King Paul (Kremasta), Acheloos, 1965	541	165	1510	460	10686	8171	3850	475	ER
INDIA									
Beas, Beas-Indus, 1975	435	133	6400	1951	45800	35019	6600	814	G
Bhakra, Sutlend-Indus, 1963	742	226	1700	518	5400	4129	8000	986	G
Hirakud, Mahandi, 1956	202	62	15748	4800	25100	19191	6600	814	GE
IRAN									
Reza Shah Kabir, Karoun, 1975	656	200	1247	380	1570	1200	2351	290	A
ITALY									
Alpe Gera, Comor-Adda-Po, 1965	584	178	1710	521	2252	1722	53	7	G
Place Moulin, Buthier-Dora Baltea, 1965	502	153	2181	665	1962	1500	81	10	AG
JAPAN									
Kurobegawa No. 4, Kurobe, 1964	610	186	1603	489	1782	1363	162	20	A
Okutadami, Tadami, 1961	515	157	1575	480	2145	1640	487	60	G
Sakuma, Tenryu, 1956	510	155	963	294	1465	1120	265	33	G
NETHERLANDS									
Afsluitdijk, Zuider Zee, 1932	*62*	*19*	105000	32004	82927	63406	4864	600	E
Brouwershavense Gat, 1972	118	36	20341	6200	35316	27003	466	57	E
Haringvliet, Haringvliet, 1970	79	24	18044	5500	26160	20002	527	65	E
Lauwerszee, Lauwerszee, 1969	75	23	42650	13000	46532	35578	40	5	E
PAKISTAN									
Jari, Jari, 1967	234	71	5700	1737	42400	32419	400	49	E
Mangla, Jhelum, 1967	380	116	11000	3353	85872	65658	5150	635	E
Tarbela, Indus, 1975	486	148	9000	2743	158268	121012	11100	1369	ER
SPAIN									
Almendra, Turmes-Douro, 1970	662	202	1860	567	2188	1673	2148	265	A
SWITZERLAND									
Emosson, Barberine, 1974	590	180	1818	554	1426	1090	184	23	A
Goscheneralp, Goschener, 1960	508	155	1771	540	12230	9351	62	8	E
Grande Dixence, Dixence-Rhône, 1962	935	285	2280	695	7792	5958	325	40	G
Luzzone, Brenno di Luzzone, 1963	682	208	1738	530	1739	1330	71	9	A
Mauvoisin, Drance de Bagnes, 1957	777	237	1706	520	2655	2030	148	18	A
SYRIA									
Tabka, Euphrates, 1975	197	60	14764	4500	60168	46004	11350	1399	E

REPRESENTATIVE MAJOR DAMS OF THE WORLD (*continued*)

NAME OF DAM, RIVER/BASIN, YEAR OF DEDICATION	HEIGHT		CREST LENGTH		VOLUME		GROSS RESERVOIR CAPACITY		TYPE OF CONSTRUCTION
	(Feet)	(Meters)	(Feet)	(Meters)	(Cubic Yards)	(Cubic Meters)	(Acre-Feet)	(Hectare-Meters)	
TURKEY									
Keban, Euphrates (First), 1974	679	207	3881	1183	20900	15980	25110	3096	ERG
UGANDA									
Owens Falls, Lake Victoria-Nile, 1954	100	30	2725	831	—	—	166000	20468	G
UNITED STATES									
Bagdad Tailings, Maroney Gulch, 1973	121	37	2601	793	37304	28523	40	5	E
Castaic, Castaic Cr., 1973	340	104	5200	1585	44000	33642	432	53	E
Cochiti, Rio Grande, 1975	253	77	26891	8196	64631	49417	513	63	E
Copper Cities Tailing 2, Tinhom Wash, 1973	325	99	7598	2316	30003	22940	*4*	*0.5*	E
Cougar, S. Fork McKenzie, 1964	519	158	1600	488	13000	9940	219	27	R
Don Pedro, Tuolumne-San Joaquin, 1971	585	178	1900	579	16760	12815	2030	250	R
Dworshak, N. Fork Clearwater, 1974	717	219	3287	1002	6500	4970	3453	426	G
Esperanza Tailings, Santa Cruz, 1973	121	37	10600	3231	39704	30358	5	0.6	E
Fort Pack, Missouri, 1940	250	76	21026	6409	125612	96043	19133	2359	E
Fort Randall, Missouri, 1956	165	50	10700	3261	50205	38387	5701	703	E
Garrison, Missouri, 1956	203	62	11300	3444	66506	50850	24321	2999	E
Glen Canyon, Colorado, 1964	710	216	1560	475	4901	3747	27000	3329	A
Grand Coulee, Columbia, 1942	550	168	4173	1272	10585	8093	9724	1199	G
Hoover, Colorado, 1936	726	221	1244	379	4400	3364	19755	2436	AG
Hungry Horse, S. Fork Flathead, 1953	564	172	2115	645	3086	2360	3468	428	AG
Ludington, Lake Michigan, 1973	170	52	29301	8931	37703	28828	83	10	E
Navajo, San Juan, 1963	407	124	3648	1112	26841	20523	1709	211	E
New Bullards Bar, North Yuba-Sacramento, 1970	637	194	2200	671	2700	2064	960	118	A
New Cornelia Tailings, Ten Mile Wash, Arizona, 1973	98	30	35600	10851	**274026**	**209520**	20	2.5	E
New Melones, Stanislaus-San Joaquin, 1975	625	191	1600	488	15970	12211	2400	296	R
Oahe, Missouri, 1963	245	75	9300	2835	92008	70349	23591	2909	E
Oroville, Feather-Sacramento, 1968	770	235	6920	2109	78008	59645	3538	436	E
San Luis, San Luis-San Joaquin, 1967	382	116	18600	5669	77666	59383	2039	251	E
Shasta, Sacramento, 1945	602	183	3460	1055	8711	6660	4552	561	AG
Swift, Lewis-Columbia, 1958	610	186	2100	640	15800	12081	756	93	E
Trinity, Trinity-Klamath, 1962	537	164	2600	792	29252	22366	2448	302	E
Tuttle Creek, Big Blue-Missouri, 1962	154	47	7500	2286	22937	17538	413	51	E
Twin Buttes, Conco-Colorado, Texas, 1963	134	41	42463	12493	21442	16395	641	79	E
Twin Buttes Tailings, Santa Cruz, 1973	239	73	11299	3444	38604	29517	209	26	E
Yellowtail, Bighorn-Missouri, 1966	525	160	1480	451	1456	1113	1375	170	A
U.S.S.R.									
Bratsk, Angara, 1964	410	125	16864	5140	18283	13979	**137220**	**16919**	GE
Charvak, Chirchik-Sir Darya, 1970	551	168	2483	757	24983	19102	1620	200	E
Chirkey, Sulak-Caspian Sea, 1975	764	233	1109	338	1602	1225	2252	278	A
Dneprodzerzhinsk, Dnieper, 1964	112	34	118090	35994	28503	21793	1994	246	GE
Irkutsk, Angara, 1956	144	44	8989	2740	16219	12401	37290	4598	GE
Ivankova, Volga-Caspian Sea, 1937	96	29	31398	9570	20207	15450	908	112	GE
Kakhovka, Dnieper, 1955	121	37	5380	1640	46617	35643	14755	1819	GE
Kanev, Dnieper, 1974	82	25	52950	16139	49520	37863	2125	262	E
Kapchagay, Ili, 1970	164	50	1542	470	5078	3883	22813	2813	E
Kiev, Dnieper, 1964	72	22	**177448**	**54086**	57552	44004	3021	372	E
Kremenchug, Dnieper, 1961	108	33	39844	12144	41192	31495	10945	1350	GE
Mingechaur, Kura, 1953	262	80	5085	1550	20400	15598	12970	1599	E
Rybinsk, Volga-Caspian Sea, 1941	98	30	2060	628	3329	2545	20590	2539	GE
Saratov, Volga-Caspian Sea, 1967	131	40	37204	11340	52843	40404	10458	1289	E
Tsimlyansk, Don, 1952	128	39	43411	13232	44323	33889	17715	2184	GE
Vilyui, Vilyui, 1967	246	75	2297	700	3793	2900	29104	3589	ER
Volga-22d Congress, Volga-Caspian Sea, 1958	144	44	13108	3995	33020	25247	27160	3349	ERG
Volga-V.I. Lenin, Volga-Caspian Sea, 1955	148	45	12405	3781	44298	33870	47020	5798	GE
Zeya, Zeya, 1975	369	112	2343	714	3139	2400	55452	6837	G
VENEZUELA									
Guri, Caroni-Orinoco, 1968	348	106	2264	690	4917	3760	14349	1769	GE
ZIMBABWE (Rhodesia)–ZAMBIA									
Kariba, Zambesi, 1959	420	128	2025	617	*1350*	*1032*	130000	16029	A

NOTES: A = Arch; E = Earthfill; G = Gravity; R = Rockfill
Information Source: Bureau of Reclamation, U.S. Department of the Interior.
The largest figure for each category of dimensions is shown in **boldface;** the smallest figure is shown in *italics.*

Dam Classification and Construction

Dams may be classified as follows:

1) Timber dams.
2) Rock-fill dams.
3) Earth dams.
   Plain.
   Core wall.
   Hydraulic fill.
4) Masonry dams.
   Gravity, solid and hollow.
   Arch, single and multiple.

Timber Dams. The timber dam is rarely used because of its short life and the limitation in height to which it may be carried. It is conceivable that in a location where timber is plentiful and cement costly and difficult to transport, and where only a submerged diversion dam is contemplated, the timber dam would be the most economical to construct, even taking due account of its lack of permanence.

Rock-Fill Dams. The rock-fill dam is an embankment of loose rock with either a water-tight upstream face of concrete slabs or timber, or a water-tight core. Where suitable rock is at hand in plentiful amount, a minimum of transport of materials can be realized with this type of dam. Like the earth embankment, it resists damage from earthquake shock very effectively.

Earth Dams. The earth dam, as shown in Fig. 1, is constructed as (a) a simple homogeneous embankment of well-compacted earth, (b) the same, but with a water-tight core or an upstream face pavement, (c) a hydraulic fill in which hydraulic segregation is relied upon to produce a water-tight core.

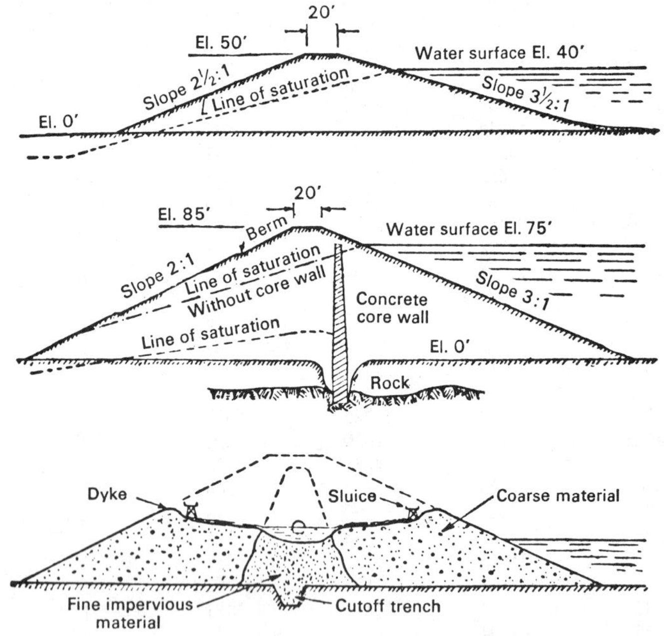

Fig. 1. Earth dams: (top) section through typical homogeneous rolled earth embankment; (middle) section through typical core wall earth dam; (bottom) section through hydraulic fill dam, showing method of construction.

A major dam disaster occurred on June 5, 1976 in southeastern Idaho (44 miles; 71 kilometers north of Idaho Falls) when the earth-fill Teton Dam collapsed just as the water behind it was approaching full reservoir capacity for the first time. Downstream inundation caused 14 deaths and, at a minimum, about $400 million in property damage. The Teton Dam was considered multipurpose—planned to provide irrigation water, flood protection, electric power, and a water-sports recreational area. Site of the dam was in a deep, narrow canyon on the Teton River, which is in the watershed of the Snake River. Height of the dam was about 300 feet (91 meters) and the crest was about 3200 feet (975 meters) long. The reservoir extended some 17 miles (27 kilometers) up the canyon. Engineering of the dam was handled by the Bureau of Reclamation of the U.S. Department of the Interior, a bureau that previously had engineered about 250 earth-fill dams, all of which have performed satisfactorily.

Extensive studies were made of the collapse. The engineers were confronted with a number of unusual problems at the site. Rocks in the canyon walls and beds were highly fractured, thus providing many passages through which water could travel. Some fissures found during excavation of the dam foundation permitted easy passage for a person for up to about 100 feet (30 meters), but such passages did not converged with passages from the other side. Fissures are present at nearly every dam site to some extent and are not considered serious if they do not provide channels for water to reach the downstream face of the dam and thus cause erosion. Corrective measures were taken at the Teton Dam, including extensive cutting of trenches and pumping of grout in an attempt to create an impermeable barrier. Theoretically, the engineering measures taken were sound, but in practice did not suffice.

Major reasons given in the official report (see References) for failure of the dam were: (1) overreliance on a grout curtain that turned out to be imperfect; (2) use of "brittle" and "highly erodible" silts in the core of the dam and the trench fill; (3) selection of a poor geometrical configuration of the trenches; (4) inadequate provisions for collection and safe discharge of seepage or leakage; and (5) insufficient instrumentation to enable construction engineers to be aware of changing conditions in the dam embankment and the canyon walls. It is believed that the dam had probably been eroding for some time before visible signs of failure appeared just hours before the structure collapsed. Most investigatory reports concluded with the observation that the site was so poor that construction there never should have been seriously considered. The Teton Dam collapse initiated an extensive program for improving the inspection procedures for all large dams.

Masonry Dams. Masonry dams are of either the gravity or arch type. Stability is secured in the gravity dam by making it of such a shape and size that it will resist overturning, sliding and crushing at the toe. The dam will not overturn providing the resultant force falls within the base. However, to prevent tension at the upstream face and excessive compression at the downstream face, the dam cross section is usually designed so that the resultant falls within the middle-third at all elevations of the cross section. In the arch dam stability is obtained by a combination of arch and gravity action. If the upstream face is vertical the entire weight of the dam must be carried to the foundation by gravity, while the distribution of the normal hydrostatic pressure between vertical cantilever and arch action will depend upon the stiffness of the dam in a vertical and horizontal direction. When the upstream face is sloped the distribution is more complicated. The normal component of the weight of the arch ring may be taken by arch action, while the normal hydrostatic pressure will be distributed as explained above. Hence, for the gravity type, good impervious foundations are essential, but, for the arch type, firm, reliable support at the abutments (either buttress or canyon side wall) is more important. The most desirable site for an arch dam is a narrow canyon with steep side walls of sound rock. When situated on a suitable site, the gravity dam inspires more confidence in the layman than any other type. It has mass that lends an atmosphere of permanence, stability, and safety. When built upon a carefully explored foundation with stresses calculated from completely evaluated loads, the gravity dam probably represents the art of dam building at its highest point of development. This is an attribute of no mean significance because, due to flood disasters and their tremendous potentialities, fear of flood is a keenly developed human instinct. This factor has led to the adoption of the gravity section in some instances where an arch dam would have been the more economical construction.

Gravity dams are classified as *solid* or *hollow* (Fig. 2a and b). The solid type is the more widely used of the two, although the hollow dam is frequently the more economical to construct. Most forms of hollow dams have been patented by Ambursen and others. The gravity dams can also be classified as *overflow* and *non-overflow*. If the dam is to serve as a spillway section, its downstream face is ordinarily made an ogee curve with the curvature such that there will be no tendency of the water to leave the surface of the concrete, even with maximum water elevation at the crest.

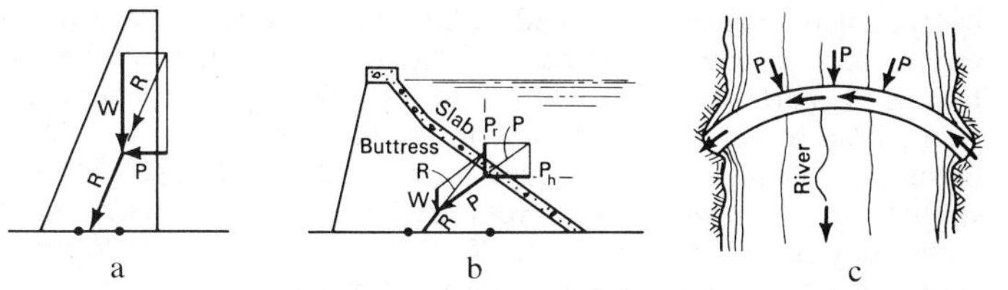

Fig. 2. Comparison of stabilizing forces in dams: (a) solid gravity dam. W is large enough, with respect to P, to incline R sufficiently to fall within the middle third. (b) hollow gravity dam. The slab is inclined enough to produce a vertical water pressure P, which inclines P sufficiently to overcome the effect of a small W, so that R falls within the middle third. (c) arch dam. Water pressure on the up-stream face has the effect of shortening the dam, thereby creating resisting compressive stresses within the dam and tightening it against the abutments.

Two types of single-arch dams (Fig. 2c) are in use; namely, the constant angle and the constant radius dam. The constant radius type employs the same face radius at all elevations of the dam, which means that as the channel grows narrower, as at the bottom, the central angle subtended by the face of the dam becomes smaller. In a constant angle type of dam, this subtended angle is kept a constant and the variation in distance from abutment to abutment at various levels taken care of by varying the radii. The safety of an arch dam is dependent on the strength of the side wall abutments, hence the arch should not only be well seated on the side walls, but the character of the rock in bearing be carefully inspected to determine its ability to take the enormous thrust that will be set up as the water rises. The multiple-arch dam consists of a number of single-arch dams with concrete buttresses as the supporting abutments. The multiple-arch dam does not require as many buttresses as the hollow gravity type. It requires good rock foundation because the buttress loads are heavy.

Special Design Factors. In climates where ice may form on the reservoir surface, the ice expands upon a temperature rise and exerts a force on the top of the dam. Conventionaliy, ice pressure factors of as great as 50,000 pounds per square foot (244,100 kilograms per square meter) have been used as a criterion in dam design for far-northern installations. More recently, Edwin Rose devised a method for calculating these forces, with resulting factors ranging from 2,000 to 10,000 pounds per square foot (9,764 to 48,820 kilograms per square meter). These depend upon the rate of temperature rise and the restraining conditions at the edges of the reservoir. Thus, the former values are now considered quite high from a practical standpoint.

Few areas of the world are free from earthquakes of varying intensity. Inasmuch as earthquakes create both vertical and horizontal accelerations of any object resting upon the surface of the earth, a dam is vulnerable to earthquake damage—if not damage of a major nature, cracks and fissures may develop, the cost of repairing of which can be extremely high, particularly if an important dam must be reduced in or taken completely out of service for a period. Of course, major damage to a dam that may result in collapse or extensive cracks and fissures can add flood damage to an already earthquake damaged region. In seismically active regions, dams have been designed for an acceleration equal to $0.1g$, where g is the acceleration due to gravity. A useful formula for approximating earthquake forces on a dam was developed in 1933 by von Karman:

$$F_e = 0.555\ awh^2$$

where F_e = inertial force of the water on the face of the dam
w = unit weight of water, pounds per cubic foot
a = acceleration due to the earthquake, feet per second²
h = depth of water behind dam, feet

References

Boffey, P. M.: "Teton Dam Collapse," *Science*, **193**, 30–32 (1976).

Boffey, P. M.: "Teton Dam Verdict," *Science*, **195**, 270–272 (1977).

Davis, C., and K. Sorenson: "Handbook of Applied Hydraulics," 3rd edition, McGraw-Hill, New York, 1969.

Parker, A.: "Planning and Estimating Dam Construction," McGraw-Hill, New York, 1971.

Sage, A. P.: "Methodology for Large-scale Systems," McGraw-Hill, New York, 1977.

DAMAGING ULTRAVIOLET RADIATION (DUV). Oxygen.

DAM (Hydraulic Uplift). Uplift (Hydraulic).

DAMINOZIDE PLANT GROWTH MODIFIER. Plant Growth Modification and Regulation.

DAMPED WAVE. This term is ordinarily used to designate electric waves which decrease in amplitude with time. In any oscillatory circuit which contains resistance (and all practical ones do) the energy of the oscillations will be dissipated in resistance losses and the amplitude of the oscillations will gradually decrease unless energy is continually added to the circuit. When energy is added to overcome this dissipation and maintain the amplitude constant, continuous waves result. A capacitor discharging through an inductance will give rise to damped waves and this was the basis of the old spark radio transmitters where the spark gap initiated the discharge and the oscillations continued until all the energy had been dissipated. Since these waves are not as effective for radio transmission as the continuous waves and since they give rise to spurious frequencies they are no longer used for this purpose.

DAMPING. With reference to industrial and scientific instruments, the Scientific Apparatus Makers Association defines damping as the progressive reduction or suppression of the oscillation of a system.

—When the time response to an abrupt stimulus is as fast as possible without overshoot, the response is said to be *critically damped*.

—*Underdamped* when overshoot occurs.

—*Overdamped* when response is slower than critical.

Relative Damping. For an underdamped system, a number expressing the quotient of the actual damping of a second-order linear system or element by its critical damping.

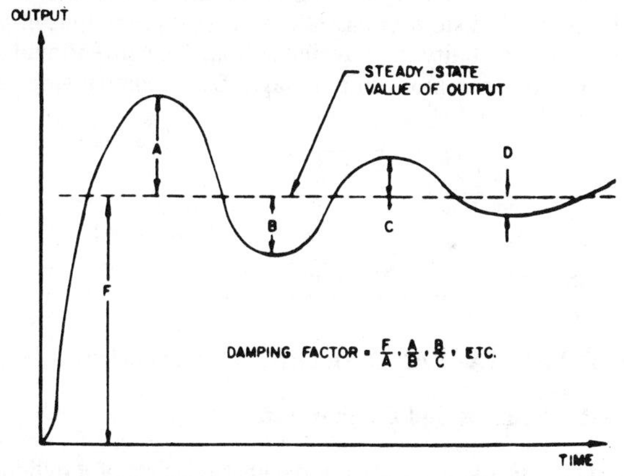

Underdamped response of a system with second-order lag.

For any system whose transfer function includes a quadratic factor: $s^2 + 2z\omega_n s + \omega_n^2$, *relative damping* is the value z, since $z = 1$ for critical damping. Such a factor has a root $-\sigma + j\omega$ in the complex s-plane, from which $z = \sigma/\omega_n = \sigma/\sqrt{\sigma^2 + \omega^2}$

where s = complex variable
ω_n = natural frequency, rad/second
z = damping ratio
σ = real part of the complex variable (s)
$j = \sqrt{-1}$
ω = frequency, rad/second

Damping Factor. For the free oscillation of a second-order linear system, a measure of damping, expressed (without sign) as the quotient of the greater by the lesser of a pair of consecutive swings of the output (in opposite directions) about an ultimate steady-state value. See accompanying figure.

DAMSEL-FLY. Odonata.

DANBURITE. The mineral danburite, $CaB_2Si_2O_8$, calcium-boron silicate, crystallizes in the orthorhombic system in prismatic forms somewhat resembling the mineral topaz. Its fracture is subconchoidal;

brittle; hardness, 7; specific gravity, 2.97–3.02; color, colorless, yellowish-white, yellow, dark wine yellow and brownish-yellow; luster, vitreous to greasy; translucent to transparent. It is found at Danbury, Connecticut, from whence its name was derived. Saint Lawrence County, New York, Switzerland, Japan, and the Malagasy Republic.

DANDELION. ***Compositae*** **or Composite Family.**

DANDRUFF. Seborrhea.

DANIELL CELL. Galvanic Cell.

DARCY. Units and Standards.

DARCY'S LAW. Drainage Systems.

DARIAZ TURBINE. Hydroelectric Power.

DARK ADAPTATION. Vision and the Eye.

DARK CURRENT. A current that flows in photoemissive and photoconductive detectors when there is no radiant flux incident upon the electrodes (total darkness). The dark current may vary considerably with temperature.

DARK DISCHARGE. Discharge (Gaseous).

DARK-FIELD ILLUMINATION. For observing very small particles or very fine lines with a microscope, a condenser is used which sends the light through the object at such angles that it does not pass by transmission into the objective. Small particles or lines serve to diffract the light so that a small particle appears as a bright star against a dark background.

DARLINGTON PAIR. A particular transistor amplifier configuration using two transistors in cascade offering higher input impedance and greater gain stability than available from the conventional cascade connection of two common emitter stages. See accompanying diagram.

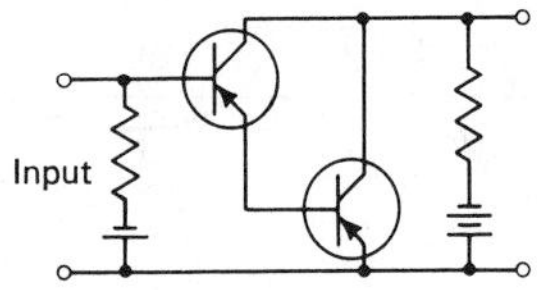

Darlington pair.

D'ARSONVAL GALVANOMETER. Galvanometer.

DARTER. Pelicans and Cormorants.

DASHPOT. In its usual form a dashpot consists of a cylinder with a piston. It may be used to dampen the motion of instruments that are vibrating by attaching the piston to the instrument part and the cylinder to the frame. As the piston moves in the cylinder the fluid, which may be air, oil, or water, squeezes past the piston and offers a resistance to the motion. This resistance is usually considered to be proportional to the relative velocity between the piston and cylinder; or $F = Cv$. The coefficient C of the relationship is primarily a function of the viscosity of the fluid in the cylinder, and the clearance between the piston and cylinder.

If the fit between the piston and cylinder is made very close and the fluid is air, this device can be used to effect a quick, jerking, mechanical motion. This application is used in the valve gear of a Corliss engine. The motion of the piston creates a partial vacuum on one side of the piston that may amount to 10 to 13 psi and produce a pull on the piston rod sufficient to overcome considerable resistance in the linkage. Then the piston is literally dashed back to its original position. Although a spring could be used for the same service it contains greater resiliency or springiness and a tendency for fatigue that is not found in the dashpot.

Dashpots are used in several types of industrial scales. If there were no friction present in a scale or balance, the device would oscillate continuously and would not reach a point of equilibrium where the pointer would come to rest and indicate a reading. Fluid friction originating in a properly designed dashpot possesses the desirable property of decreasing as the velocity of the oscillating parts falls off, so that as the parts are about to come to rest, friction approaches zero. In addition to decreasing oscillation, the dashpot absorbs the shock of sudden loading and, in a large scale, should be located near the supporting levers where the forces are relatively large. The ability of the dashpot to absorb shock effectively decreases as its location approaches the counterbalancing mechanism.

DASSIES. Hyraxes.

DASYURIDS. Marsupialia.

DATA-ACQUISITION COMPUTER. When a digital computer operates in real time to collect physical data, it often is referred to as a data-acquisition computer. The instruction repertoire of the data-acquisition computer typically includes special instruction to facilitate this application. Characteristic of the data-acquisition computer is its ability to attach a wide variety of input or output devices. These include telemetering devices and flying-spot digitizers which are unique to the particular application. The data-acquisition computer also can handle random demands from devices which are asynchronous with respect to other program activity. Although the average data rates may be reasonable, data-acquisition input/output devices may have very high peak data rates. Exemplary of this is a flying-spot digitizer, which may transmit data to a computer at an average rate of 100,000 bytes per second, but with a peak rate of 1,000,000 bytes per second. Generally, external devices transmit data randomly to a data-acquisition computer. Hence, so that the necessary action may be taken by the control program, the data-acquisition computer is designed with an ability to respond very quickly to an external interrupt signal.

See also **Digital Computer; Electronic Data Processing (EDP);** terms listed under **Data Processing.**

DATA BASE. Essentially, an accumulation of information, often well organized and assembled for easy retrieval, as by a data processing system. Data bases often are specialized in terms of information categories, such as business, medical, and science, among others. The term *data bank* is no longer preferred.

DATA FILE (Data Set). In connection with electronic data processing, a data file is a space for a set of related data organized in a specific manner. The phrase also may refer to the *set of data* or *data set.* As an example, an inventory file may consist of 12,000 records of 300 characters apiece, each record representing the current inventory situation for a given part number and containing the part number, part name, automatic reorder point, economic reorder quantity, back order quantity, current quantity in stores, current quantity in work in progress, unit cost, total value, date of last receipt, etc.

As files may be defined in a wide variety of ways on a large variety of media and devices, it has become important to structure them in some ordered way and to try to make them program independent.

Two of the major problems in file management are to insure that the file accessed actually is the correct one and to check if a particular program can be allowed access to this file. As the data processing field has expanded, the number of files available in a given installation has burgeoned—so that currently even rather small users may have hundreds of active files, some of which are confidential and should only be accessible by special programs. Current methods of attaining a degree of program independency are to move a major portion of the file definition from the program to the job control language where the definition may be changed without recompilation of the program. An improved system is to include the file definition material in the initial records of the file (called the file label), to be automatically recovered by the data management system when an access to the file is attempted.

See also terms listed under **Data Processing.**

Thomas J. Harrison, International Business Machines Corporation, Boca Raton, Florida.

DATA MANAGEMENT SYSTEM. Computer Operating Systems; Data File (Data Set).

DATA PLOTTER. **Plotter (Curve).**

DATA PROCESSING. The semi-automatic and automatic handling, manipulation, and computation of information, usually accomplished with the aid of computers and other electronic equipment. Several score topics in this field are described in this volume. The terms are classified and listed in the accompanying table.

DATA (Rejection of). In general, the rejection of any data on grounds of unreliability, bias and so forth. In a narrower sense, statisticians have developed criteria for the removal of observations which look atypical, for example extremely outlying measurements, the presumption being that such observations are either mistakes or, when taken, were disturbed by some transient cause. There are various criteria for the purpose. Most of them depend on a probabilistic argument:

DATA PROCESSING TOPICS DESCRIBED IN THIS VOLUME

General Terms

Computer
Data-Acquisition Computer
EDP Center
Electronic Data Processing
General-Purpose Computer
Scientific Computer
Special-Purpose Computer

Analog Computer Systems

Analog Computer
Slide Rule

Data Acquisition—Input/Output Systems

Aliasing Error (Data Acquisition System)
Analog-to-Digital Converter
Analog Input
Analog Multiplexer
Analog Output
Analog Switch
Aperture Card
Bridge Amplifier
Bright Switch
Carrier Amplifier
Chopper Amplifier
Collator
Common-Mode Rejection Ratio
Common-Mode Voltage

Data Acquisition—Input/Output Systems

Comparator-Amplifier
Data Base
Data File (Data Set)
Differential-Mode Voltage
Digital-to-Analog Converter
Digital Multiplexer
Direct-Coupled Amplifier
Dry-Reed Relay
Encoder
Field-Effect Transistor Switch (FET)
Floating Amplifier
Hollerith
Information Retrieval System
Input/Output Devices
Integrating-Ramp A/D Converter
Keypunch (Computer System)
Light-Coupled Switch
Magnetic Ink Character Recognition
Mercury-Wetted Relay
Microfilm
Microform
Optical Character Recognition (OCR)
Output (Computer System)
Parallel A/D Converter
Parallel-Serial A/D Converter
Pattern Recognition
Peripheral Equipment (Computer System)
Plotter (Curve)
Printer (Data Processing)
Ramp A/D Converter
Readout
Resolution (Computer System)
Sample-and-Hold Amplifier
Scale Factor (Computer System)
Scaling Circuit
Shannon
Shannon Formula
Simulator (Computer System)
Single-Ended Amplifier
Successive-Approximation A/D Converter
Tabulation (Computer)
Terminal (Computer)
Threshold Detector
Voice Recognition
Voltage-to-Frequency A/D Converter

Digital Computers—Hardware/Characteristics

Abacus
Acceleration Time (Computer System)
Access Time (Computer System)
Accumulator (Computer System)
Adder (Computer System)
Address (Computer System)
Address Register
AND (Circuit)
Asynchronous
Bit (Data System)
Block Diagram (Data Processing System)
Buffer (Computer)
Byte
Central Processing Unit (Computer)
Channel (Computer System)
Column (Data Processing)
Core Storage (Computer)
Counter (Computer System)
Cycle (Computer System)
Cycle Steal (Computer System)
Density (Data Processing)
Diagnostics (Computer System)
Digit (Computer System)
Digital Computer
Digital Output
Diode Logic
Diode Transistor Logic
Disk Storage (Computer)
Drum Storage (Computer)
Exclusive OR (Circuit)
Fixed Point Arithmetic (Computer System)
Fixed-Program Computer
Flip-Flop
Gate Circuit
Gate (Computer System)
Half-Adder (Computer System)
Hardware (Computer System)
Hartley
Head (Data Processing)
Housekeeping Operation (Computer System)
Index Register
Instruction Counter (Computer System)
Location (Computer System)
Logic (Computer System)
Logic Diagram (Computer System)
Logical Operation (Computer System)
Magnetic Tape Storage (Computer)
Memory (Computer System)
NAND (Circuit)
NOR (Circuit)
NOT (Circuit)

DATA PROCESSING TOPICS DESCRIBED IN THIS VOLUME (*continued*)

OR (Circuit)
Register (Computer)
Semiconductor Storage (Computer)
Storage (Computer)
Storage-Cycle Time (Computer System)
Synchronous
Turing Machine
Wired-Program Computer
Word (Computer System)

Programs and Programming—Digital Computers

Assembler (Computer System)
Bookkeeping Operation (Computer)
Bootstrap (Computer)
Branch (Computer)
Character (Computer System)
Check (Computer System)
Code (Computer System)
Command (Computer System)
Compiler (Computer System)
Computer Error Procedures
Computer Operating System
Concurrent Operation (Computer System)
Control Algorithm
Conversion Time (Computer System)
Debugging (Computer Program)
Dummy (Computer System)
Emulator (Computer System)
Flag (Computer System)
FORTRAN (Computer Programming)
Instruction (Computer System)
Interpreter
Interrupt (Computer System)
Job (Computer System)
Key (Computer System)
Language (Computer)
Linkage (Computer System)
Loop (Computer System)
Macroassembler
Mask (Computer System)
Microprogram
Module (Computer System)
Multiprogramming (Computer System)

Programs and Programming—Digital Computers

Offline (Computer System)
Online (Computer System)
Operand (Computer System)
Operating System (Computer)
Overlay (Computer System)
Parallel Operation (Computer System)
Patch (Computer Program)
Postprocessor
Program (Computer)
Program Generator (Computer System)
Program Relocation (Computer System)
Programming Flow Chart (Computer)
Quantization
Queue (Computer System)
Real-Time Computing
Record (Computer System)
Reentrant (Computer System)
Resistor Transistor Logic
Restart (Computer System)
Roll In/Roll Out (Computer System)
Routine (Computer System)
Scheduler (Computer System)
Serial Operation (Computer System)
Simulator (Computer System)
Software (Computer System)
Storage Protect (Computer System)
Subroutine (Computer System)
Task (Computer System)
Time Sharing
Time Slicing

if an observation lies more than, say, three times the standard deviation of a whole sample of observations away from the mean, it is unlikely to belong to that sample in the sense of having been obtained under similar conditions.

DATE LINE (International). **International Date Line.**

DATE PALMS. **Asexual Reproduction; Palm Trees.**

DATING (Fossils, Rocks, etc.). **Radioactivity and Other Dating Techniques.**

DATING (Ice Cores). **Polar Research.**

DATOLITE. Datolite, basic calcium boron silicate, $CaBSiO_4(OH)$, occurs in monoclinic crystals of varied habit, mostly short stout prisms, but often in highly modified forms. Datolite reveals no cleavage, its fracture is conchoidal to uneven; brittle; hardness, 5–5.5; specific gravity, 2.9–3.0; luster, vitreous to dull; color, white to gray or may be greenish, yellowish, or brownish. It has a white streak and is transparent to translucent usually, but has been observed opaque.

Datolite is a secondary mineral, being found in veins and cavities associated with zeolites and calcite, particularly in the basic igneous rocks. It has been found in the Harz Mountains, Germany; in the Trentino district, Italy; in Norway and Tasmania. In the United States it has been found in the Triassic traps of the Connecticut River Valley in Massachusetts and Connecticut, and from similar rocks in New Jersey. In Michigan, datolite has been found associated with the copper-bearing rocks of Keweenaw County. This mineral derives its name from the Greek word meaning to divide, in reference to the granular structures of some of the massive varieties.

DAVIDA. **Asteroid.**

DAVISSON-GERMER EXPERIMENT. In 1927, Davisson and Germer conducted a research, the results of which furnished a remarkable confirmation of the basic postulate of wave mechanics. De Broglie had suggested in 1924 that electrons have in some respects the characteristics of waves, and deduced, for the wavelength equivalent to a moving electron, the expression $\lambda = h/mv$, in which m and v are the mass and speed of the electron and h is Planck's constant. If the electron is moving, for example, with a speed corresponding to 65 eV of energy, the corresponding "de Broglie wavelength" is 1.52 angstroms, which is in the x-ray range. This led Davisson and Germer to see whether electrons might be reflected from crystals after the manner of x-rays. They used a single crystal of nickel cut parallel to the (111) planes, and upon varying the electron speed at a fixed angle of incidence, they found not only a distinct "regular" reflection but also a series of diffraction maxima strikingly similar to those obtained with the same crystal for x-rays of varying wavelength. The differences observed were satisfactorily explained as due to the refraction of the nickel for the electron waves.

G. P. Thomson independently reached the same conclusions in 1927. The hypothesis that matter exhibits both corpuscular and wave-like characteristics served as a stimulus for the formal development of quantum mechanics by E. Schrödinger, M. Born, W. Heisenberg, and others. Following this discovery, which eventually led to a Nobel Prize to Davisson and Thomson, electron diffraction was immediately utilized as a tool for the study of the structure of matter.

DAVYDOV SPLITTING. The splitting observed in absorption lines due to excited states in molecular crystals. In the simple case where there are two molecules in the unit cell in which the molecules are differently oriented, a doublet will be observed. The splitting results from the interaction between excited states of molecules within the cell.

DAVY EXPERIMENT. An experiment carried out by Humphry Davy, in 1799, which showed that two pieces of ice or other substance with a low melting-point could be melted by rubbing them together, without any other addition of heat. The experiment helped to disprove the caloric theory of heat.

DAWN REDWOOD. **Redwood (Dawn).**

DAY. **Units and Standards.**

DAYLIGHT FACTOR. The ratio of the daylight illumination at any point in a building to the simultaneous illumination under the open sky.

DAYLIGHT SAVING TIME. **Time.**

D-CELL ADENOMA. **Zollinger-Ellison Syndrome.**

DC-DC CONVERTER. **Power Sources and Supplies.**

DC POWER TRANSMISSION. **Direct-Current Circuits; Electric Power Transmission.**

DC RECTIFIER. **Powrer Sources and Supplies.**

DEACCENTUATOR. A network used in frequency-modulation reception to achieve deemphasis.

DEAD BAND (Instrument). The range through which an input can be varied without initiating a response. Dead band is usually expressed in percent of span. See figure which accompanies entry on **Hysteresis (Instrument).**

DEAD CENTER. In machine tools, the term applies to the stationary center on which work rotates while being machined. In a lathe, the dead center is mounted in the tailstock. In heavy-duty lathes, the pressure on the pivot bearing effected by the center holes and the dead center is often so great that the dead center, or the center hole in the work, may wear excessively, and a rotating dead center, consisting of a center mounted in ball bearings carried in the shank which fits the tailstock sleeve, is often used. In grinding machines both centers are frequently stationary.

An engine is said to be on dead center when the piston is at one end of its stroke and the crank, connecting rod, and piston rod are in alignment. In this position, the pressure does not exert a rotative force on the crank, since it is transmitted directly to the shaft and bearings.

DEAD MEN'S FINGERS. 1. *Porifera.* A branching sponge, *Chalina arbuscula,* found off the Atlantic coast of the United States. Its branches are rounded finger-like projections of white or light gray color. 2. Coelenterate *Anthozoa—Alcyonium digitatum.* Colonial form of various shapes and sizes usually broad-lobed masses with polyps embedded in a fleshy mass attached to stones below tide marks.

DEAD RECKONING. The meaning of this navigation term can be best understood from a consideration of its probable origin. Early navigators deduced their positions at any time by using the distances and directions of motion of their ship since leaving some previously known position. Such positions were determined at frequent intervals and were entered in log books under a column headed by the abbreviation "ded. pos." The calculating or "reckoning" necessary to obtain the entries for this column was known as "ded reckoning," and is now written as dead reckoning.

Dead reckoning is fundamental for the methods of modern navigation, and should be clearly understood. Pilotage, radio navigation, celestial navigation, and other methods for determining positions of a ship or plane all involve the use of dead-reckoning methods. This is particularly true in a case where two lines of position are not observed simultaneously, and the fix must be found by moving one or both of the lines by dead reckoning.

There is a conflict of opinion among modern sea navigators as to just how much material should be included in the dead-reckoning (DR) position. It is certainly true that, originally, only the heading of the ship and the distance run on that heading were included. Some modern sea navigators cling to the original meaning of the term. The effects of current and leeway may be computed independently and the DR position corrected to an estimated position (EP). Some sea navigators include all known or suspected motions of the ship in their dead reckoning, and call the resulting position either DR or EP.

It is an established procedure in air navigation to include the movement of the plane through the air (heading and air distance), and the movement due to the motion of the air itself (wind direction and speed), in determining a DR position. In air navigation, the term estimated position (EP) is used in connection with celestial navigation, and is discussed in the article on that subject.

Before entering upon a discussion of actual methods of dead reckoning, it should be clearly understood that certain errors are bound to appear in the data employed in the calculations. These may be listed under five groups: compass corrections, steering, patent logs or airspeed meters, leeway of a seaborne ship, and predicted current or wind. Under the most favorable conditions, the dead-reckoning position is somewhere within a circle whose center is the DR position and whose radius is between 1 and 2% of the distance run. In conditions of rough sea or turbulent air, the radius may increase to more than 5% of the distance run.

The graphical method of solution is used by practically all air navigators and by a majority of those on the sea. This is essentially the graphical addition of vectors. In the determination of the old-fashioned DR position, the vectors used are total motions within given periods of time. These vectors are drawn to proper scale of distance on a mercator chart, mercator plotting sheet, or small-area plotting sheet, with the tail of the first vector resting on the last-known position, known as the point of departure, and the head of the final vector at the DR or EP position. All vectors must be properly labeled, the direction of each being the heading and the length proportional to the distance run.

The graphical method of solution is illustrated in the following case: At 1200, a ship is in latitude 43° 35′ N and longitude 34° 38′ W. At 1200, the ship is on heading 150°, speed 12 knots; at 1430, heading altered to 125°, speed 12 knots; at 2000, heading altered to 030°, speed 12 knots; at 2400, heading altered to 320°, speed 12 knots. The vector diagram is drawn on a small-area plotting sheet and properly labeled, as shown in Fig. 1. The various distances are obtained for the vectors by simply multiplying the speed by the time. From this diagram, the DR positions for any desired times can be read off directly. For example, the position at 1430 L = 43° 09′.0 N and Lo = 34° 17′.7 W and at 0400 L = 43° 49′.5 N and

Lo = 33° 12′.0 W.

In case a current is predicted for this region, an additional vector may be added to any DR position to obtain the estimated position. For example, if, in the above situation, we assume a current to set 240° with drift 3 knots, vectors (shown dotted in the figure) in direction 240° and lengths 7.5 and 48 miles are added at the 1430 and 0400 DR positions, respectively. This yields a 1430 EP L = 43° 05.3 N and Lo = 34° 26.6 W and 0400 EP L = 43° 25.5 N and Lo = 34° 09.4 W. If leeway is present, it should be treated as an additional compass correction to each heading, as explained in the article on leeway.

The graphical method of solution for dead-reckoning problems requires a considerable amount of paraphernalia, e.g., plotting sheet, parallel rulers or protractor, scale, either pad and pencil or a dead-

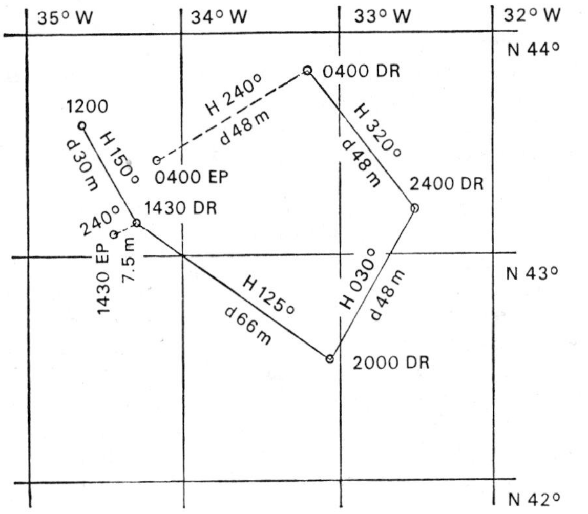

Fig. 1. Vector diagram used in dead reckoning calculations.

reckoning calculator for arithmetic computing, and, most important of all, a flat surface for doing the drawing. On large ships or planes, the navigator has the needed space, and the chart table is frequently equipped with a drafting machine, which simplifies the plotting. On small boats or planes, sufficient space for plotting is difficult to find. Computational methods are just as rapid as those involving plotting and they certainly yield more accurate results. However, it should be emphasized that the advantage of computational methods is only that of convenience in space, for the plotting methods are fully as accurate as the data warrants.

In the computational method, the first step is to solve the right triangle in which the heading vector is the hypotenuse, with the difference of latitude (DL) and the departure (dep) the two legs. The DL is added algebraically to the latitude (L_1) of the point of departure, calling DL+ when north, and − when south. The DR latitude (L_2) is then $L_2 = L_1 + DL$. The departure must be converted from nautical miles to difference in longitude (DLo), in minutes of arc, by the method of middle-latitude sailing, using for the mid-latitude

$$L_m = \frac{L_1 + L_2}{2}$$

The DLo is added algebraically to the longitude (Lo_1) of the point of departure, calling DLo+ when west, and − when east. The DR longitude (Lo_2) is then $Lo_2 = Lo_1 + DLo$. Three-place computing, using slide rule or traverse tables, is sufficient to attain all accuracy warranted by the data. To illustrate the computational method: At 0815 a ship is in latitude 41° 42′ N and Lo = 35° 25′ W. The ship proceeds along heading 243° at 14 knots, and the DR position at 1200 is desired.

L_1	41° 42.0 N	H	243°	Lo_1	35° 25′.0 W
DL	23.8 S	d	52.5 m	DLo	1 02′.5
L_2	41 18.2 N	dep	46.8 W	Lo_2	36 27′.5 W
		L_m	41°.5		
		DLo	62.5 W		

The actual numbers were read directly from traverse tables, and the total time required for solution was 2 minutes. The equipment used was a pencil, scratch pad, and traverse tables.

In more complicated cases, where a number of different headings and distances are used, and in a region where a current is predicted, we set up what is known as a traverse form, in which the columns N, S, E, and W refer to the direction of the individual DL's and departures. At 0400, a ship is on heading 200°, speed 15 knots, and is in latitude 40° 21′.3 N and longitude 124° 34′.6 W. AT 0723, the heading is altered to 300°, speed 15 knots; at 1136, heading altered to 030°, speed 15 knots; at 1314, heading altered to 340°, speed 15 knots. In this region there is a current setting 140° 2 knots. The traverse form used in determining the DR position at 1600 follows:

Traverse Form

H	Time	d	N	S	E	W
200°	3 h 23 m	50.8 m		47.7		17.4
300	4 13	63.3	31.8			54.9
030	1 38	24.5	21.2		12.2	
340	2 46	41.5	39.0			14.2
		Sums	92.0	47.7	12.2	86.5
		Total	44.3			74.3

L_1	40° 21′.3 N	dep	74.3 m W	Lo_1	124° 34′.6 W
DL	44.3 N	L_m	40°.7	DLo	1 37.9 W
L_2	41 05.6 N	DLo	97.9 W	Lo_2	126 12.5 W

To this DR position, the current motion must be added to attain the estimated position. The current motion in 12 hours is 24 miles along 140°, for which DL = 18.4 S, dep = 15.4 m E, and, in mid-latitude, 40°.7 DLo = 20.4 E. Applying these to the DR position, we have EP L = 40° 47′.2 N and Lo = 125° 52.1 W.

Whenever a fix is obtained, a DR position is also obtained. The difference between the two positions is always attributed to "current" by sea navigators, never to errors in their data or in obtaining the DR position. Here the difference in opinion as to whether or not leeway and current should be included in obtaining the DR position introduces confusion. If an old-fashioned DR position (i.e., without current or leeway included) is used, then the difference between fix and DR may be the actual current. If an EP is used, the difference between this position and the fix is a residual current, supplemental to that already used in the determination of the EP. In either case, the difference is called current, with set being equal to the direction from the DR position to the fix, and drift being the distance between the two positions divided by the number of hours elapsed from the point of departure for the dead reckoning. These so-called currents are entered in the ship's log book and later forwarded to the U.S. Coast and Geodetic Survey or other government agencies where statistical discussions yield improvements to previously predicted values.

In air navigation, a slightly different procedure for finding DR position is employed. In the first place, the work is always done by constructing the vector diagram. Before plotting the total distances from the point of departure, a velocity-vector diagram is used to determine the course and ground speed made good along this course. The heading and air speed of the plane are drawn as one vector, and to this is added the vector representing the direction toward which the wind is blowing and the wind speed. The vector sum is a vector giving course and predicted ground speed. Unfortunately, the first man to erect a weather vane balanced it so that the arrow pointed into the wind instead of with it. Still more unfortunately, the weather bureaus have retained this archaic method of publishing wind directions, and air navigators must be particularly careful in constructing their diagrams. After the course and ground speed along the course have been obtained, then a DR position can be obtained by plotting these in the same manner that the sea navigator plots heading and ship's speed.

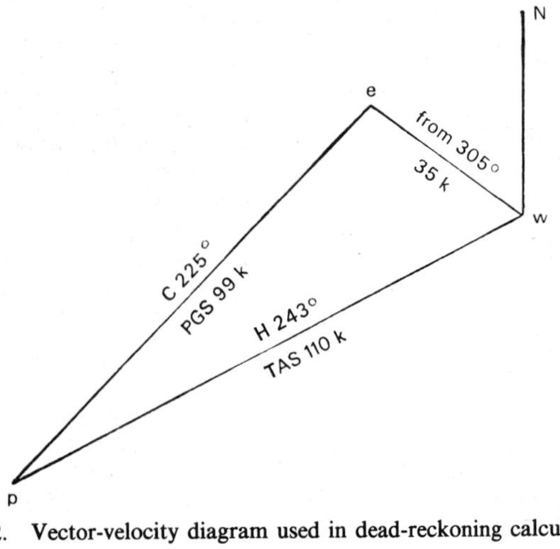

Fig. 2. Vector-velocity diagram used in dead-reckoning calculations.

At 0916, a plane is known to be in latitude 41° 52′ N and longitude 56° 34′ W. The plane is heading 243°, with true air speed 110 knots, and the predicted wind is from 305°, speed 35 knots. The pilot wishes to determine the course and ground speed of the plane and the DR position at 1045. The vector-velocity diagram is shown in Fig. 2. The wind vector, *ew*, must be drawn in the direction in which the wind will move the plane, i.e., wind direction—180°; to this is added the heading air-speed vector, *wp*. The sum, or vector, *ep*, is the course and speed made good. In this case, the course is found to be 225° and the ground speed 99 knots. With these quantities determined, the DR position is found by plotting the distance-vector diagram, shown in Fig. 3. The course 225° is laid off from the 0916 position on a mercator plotting sheet or a small-area plotting sheet, and the distance 99 × 1 h 20 m = 146 m is measured off on the proper scale, giving the 1045 DR position. If the heading is altered, a new velocity-vector diagram must be constructed and a new course and ground speed obtained. Then a distance vector would be added to that already found, and the DR position found at any desired time.

The solution of the velocity-vector diagram is facilitated by the use of any one of a number of dead-reckoning computers.

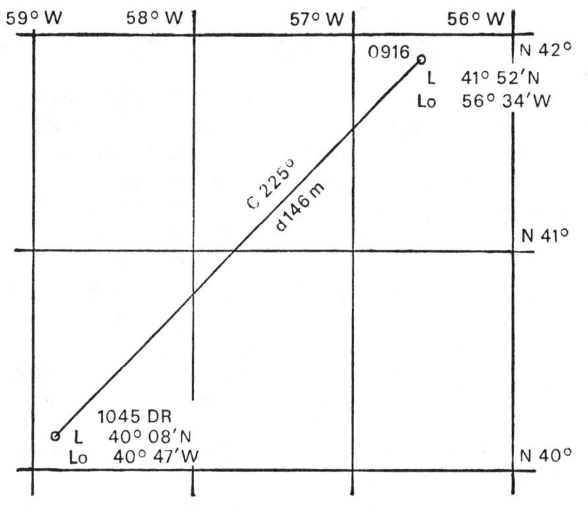

Fig. 3. Distance-vector diagram used in dead-reckoning calculations.

In case a reliable fix is obtained at the same time a DR position is found, and the two do not coincide, the air navigator always assumes that the difference is due to an error in the predicted wind. The line between the point of departure and the fix is known as the track if but one heading is used, or as the track made good if the plane has changed heading since leaving the point of departure. A discussion of this problem will be found under **Air Plot.**

See also **Compass (Navigation); Course; Line of Position; Log (Navigation); Navigation; Pilotage;** and **Plane Sailing.**

DEAD SPOT. 1. A particular geographic location in which radio reception over some band of frequencies is weak or nonexistent. 2. A portion of the receiver tuning range which has poor or no sensitivity, resulting from improper receiver design. 3. Sometimes used with other forms of instrumentation to describe insensitive regions of detection.

DEAD TIME. The interval of time between initiation of an input change or stimulus and the start of the resulting response. See figure that accompanies entry on **Response (Instrument).**

DEAERATION. Because water dissolves, to a greater or less extent, many common gases, it will contain in the natural state a certain amount of dissolved gases, such as oxygen and carbon dioxide. Deaeration is the removal of this dissolved gas. Deaeration at the present is practiced where the gas that water contains would have undesirable effects. Often dissolved oxygen is objectionable because of its corrosive action. This is true in the case of the high-pressure steam boiler, where a small amount of oxygen dissolved in the feedwater may become quite active in attacking the boiler metal under the high pressure and temperature conditions there experienced. Steam boiler operators often treat their boiler feedwater in deaerators to remove this oxygen.

These deaerators are either of the deactivating or the heating type. Deactivating types employ chemical means of deaeration. Deaerating action in a heating-type deaerator is obtained by first reducing the solubility of the gas through heating the water (under pressure); second, reducing the pressure and producing explosive boiling; and third, controlling the agitation of the water subsequent to the second action in a partially evacuated region.

DEAFNESS. Hearing and the Ear.

DEASPHALTING. Petroleum.

DEATH ADDER. Snakes.

DEATH'S HEAD MOTH (*Insecta, Lepidoptera*). A large European sphinx moth, *Acherontia*, whose thorax bears a light mark shaped like a skull.

DEATH WATCH (*Insecta, Coleoptera*). A small beetle of the family *Anobiidae* which burrows in solid wood in buildings. By striking its head against the walls of the burrow it produces a sharp sound which can be heard in a quiet place.

DEBRIDEMENT. The treatment of wounds, especially traumatic, dirty, crushing wounds, by means of excising all injured, contaminated, or devitalized tissue, also, the division of any band of tissue constricting or compressing an organ.

DE BROGLIE ELECTRON THEORY. Electron Theory.

DE BROGLIE WAVELENGTH. A wavelength ascribed to any particle having momentum. For a relativistic particle, the value of this wavelength is given by the expression:

$$\lambda = \frac{h}{mv} = \frac{h}{m_0 v}\sqrt{\frac{1-v^2}{c^2}}$$

where λ is the de Broglie wavelength, h is the Planck constant, m_0 is the rest mass of the particle, v is its velocity, and c is the velocity of light. The observed mass of the particle is m and the momentum is mv.

As an example, the wavelength of an electron moving with a kinetic energy of 1 eV (electron volt) is 1.23×10^{-7} cm, shorter than the wavelength $\lambda = c/v = hc/hv = 1.24 \times 10^{-4}$ cm for a photon with an energy of 1 eV, but longer than the wavelength of a proton moving with a kinetic energy of 1 eV, which is $\lambda = 2.86 \times 10^{-9}$ cm.

See also **Davisson-Germer Experiment.**

DEBUGGING (Computer Program). Computer programs frequently contain errors, especially logical errors, when originally written. These errors are commonly referred to as "bugs" and the process of finding and correcting them is "debugging." Debugging normally accounts for one-third to one-half of the total time and expenditure of program development. Provision must also be made to correct bugs which are found after the program has been released for normal usage. All but the most trivial programs normally are so complex that it is unlikely that the programmer considered all possible paths when originally writing the program, or in the subsequent debugging. As a result, bugs may appear long after the program is released by the programmer for regular production use.

Debugging consists of two phases: (1) "desk debugging," in which the code is carefully checked by hand-checking test cases against the code; and (2) "machine debugging," in which test cases are run on the computer against the program. The last phase is generally much the most efficient due to the large number of test cases which must be run before the program can be regarded as ready for normal usage. If a test case fails, an identification of which portion or subroutine the bug is located in usually can be made with reasonable despatch by looking at the intermediate results. If necessary, special code can be inserted to display the information. However, to find the individual instruction causing the error, a technique called *tracing* is sometimes used.

The tracing technique consists of printing the instruction performed and selected data values of the machine registers whenever the instruction counter is within a range specified by the programmer. The programmer then can see at exactly what instuction the program deviated from its expected path. The drawback is that tracing is extremely time consuming, being limited by printing speed. Tracing normally is done by running the problem program under the control of a special trace program which actually simulates the computer to the problem program. Some computers have special hardware to facilitate tracing.

When programming in higher-level languages, the usual technique is to insert print or output statements at appropriate points in the source program.

Another alternative is to permit the failure to occur, force a halt immediately after the failure, and then print out the contents of storage as a "storage dump." The limits on the usefulness of this technique are the result of the static nature of the dump. In addition, it may be difficult to interpret the dump because of the complexity of large dynamic systems. The dump may be meaningful only to an expert on the system.

See also **Diagnostics (Computer System);** terms listed under **Data Processing.**

Thomas J. Harrison, International Business Machines Corporation, Boca Raton, Florida.

DEBYE-FALKENHAGEN EFFECT. The variation of the conductance of an electrolytic solution with frequency. This effect, which is noted at high frequencies, is also called the dispersion of conductance.

DEBYE-HÜCKEL LIMITING LAW. The departure from ideal behavior in a given solvent is governed by the ionic strength of the medium and the valences of the ions of the electrolyte, but is independent of their chemical nature. For dilute solutions, the logarithm of the mean activity is proportional to the product of the cation valence, anion valence, and square root of ionic strength giving the equation

$$-\log f_{\pm} = Az_{+}z_{-}\sqrt{\mu}$$

See **Electrochemistry.**

DEBYE-SCHERRER-HULL METHOD. A technique of x-ray diffraction, in which a beam of x-rays is directed on a powdered sample of the material (hence the name "powder method") and the diffracted beams received on a photographic plate. Because the powder contains crystals in every orientation, the plate shows a pattern of concentric rings, which is characteristic of the material and may be used to identify it, or to obtain very accurate estimates of the cell dimensions.

DEBYE-SEARS EFFECT. A piezoelectric crystal vibrating in a longitudinal mode in a liquid sets up acoustic waves consisting of regions of compression and regions of rarefaction in the liquid, which alternate at distances of half a wavelength. Hence, if a parallel beam of light shines through such a crystal tank with plate-glass walls, the regions of density and rarefaction act like a plane light diffraction-grating. If the parallel beam from the cell is focused on a single spot when no sound waves are present, first and higher order diffraction-spectra will appear on either side of the zero-order spot when sound waves are present. From the spacings of the diffraction orders, the sound wavelength can be determined, which, together with the frequency, gives the velocity of sound in the liquid. See **Piezoelectric Effect.**

DEBYE THEORY OF SPECIFIC HEAT. The specific heat of solids is attributed to the excitation of thermal vibrations of the lattice, whose spectrum is taken to be similar to that of an elastic continuum, except that it is cut off at a maximum frequency in such a way that the total number of vibrational modes is equal to the total number of degrees of freedom of the lattice.

The Debye temperature is defined by the relation

$$\Theta = \frac{h\nu}{k}$$

where ν is the maximum frequency of the thermal vibrations of the lattice, h is Planck's constant and k is the Boltzmann constant.

DEBYE THEORY (Phonons). Phonons.

DECALESCENCE. Absorption of heat, usually by an alloy without rise of temperature, due to an allotropic transformation.

DECAPODA. 1. The shrimps and prawns, lobsters, crayfishes, and crabs, constituting a large and important order of crustaceans. The thorax is covered by a carapace and bears five pairs of appendages, the first pair chelate grasping structures and the remaining four formed for walking. 2. The cuttlefish, squids and related forms, constituting an order of cephalopods. They possess ten arms, with stalked suckers provided with horny rims, and have a well-developed internal shell.

DECARBOXYLATION COENZYMES. Coenzymes.

DECARBURIZATION. Reduction in carbon content at the surface of steel or cast iron by heating in air or other oxidizing or reducing gases. In heating for hot rolling, forging, or heat treatment, decarburization is usually objectionable, and specially prepared neutral furnace atmospheres may be used to reduce or eliminate it. Molten salt or lead baths are also effective in protecting the surface during heat treatment.

In the case of heat-treated machine parts, surface decarburization is objectionable because it reduces fatigue strength and lowers the wear-resistance of bearing surfaces. Important surfaces of hardened steel parts are often finish-ground, in which case a limited amount of decarburized skin can be removed. Tool steels for cutting tools, punches, chisels, etc., are usually ground sufficiently to remove all decarburization; however, many tools and dies are machined to finish dimensions before hardening and extreme care must be taken to protect the surface.

Decarburization is intentional in the processing of low-carbon sheet steels for electrical applications. In the production of malleable cast iron by annealing white cast iron decarburization is beneficial.

DECAY CONSTANT. The magnitude of some processes diminish in accordance with an exponential function of the time; such as, for example, phosphorescence and radioactive radiations. The falling off or "decay" of such a process may be represented by an equation giving the intensity at the end of a time interval t as $I = I_0e^{-ct}$, in which I_0 is the intensity at the beginning of the time interval and c is the "decay constant." Closely related to c is the half-value period, which is the time required for I to fall to $\frac{1}{2}$ its original value I_0; it is equal to $0.69315/c$. Thus, for example, because the half-value period of radium B is 1,608 seconds, c is $0.69315/1608$ seconds $= 0.000431/$second. This means that approximately 0.000431 of the substance existing at any instant disintegrates during the ensuing second. The reciprocal of c, called the mean life or, under certain circumstances, the relaxation time, represents the time required for I to diminish to $1/e$ or 0.3697 of its original value I_0. It is equal to 1.4427 times the half-value period; for the decay of radium B its value is therefore 2,320 seconds. See accompanying figure.

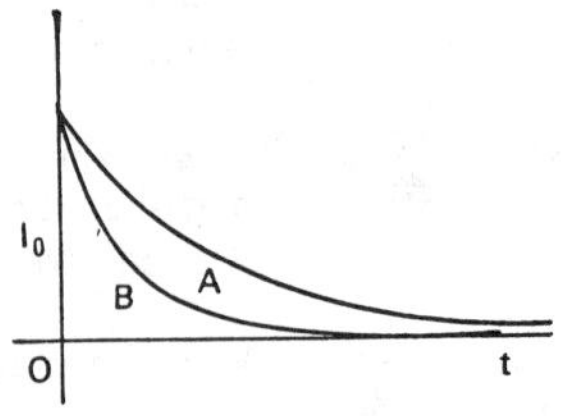

Typical exponential decay curves. The decay constant in *B* is greater than in *A*.

DECAY (Radioactive). Radioactivity.

DECCA NAVIGATION. Navigation.

DECELEBRATION PARAMETER (Universe). Cosmology.

DECIBEL. Units and Standards.

DECIDUOUS FOREST BIOME. **Biome.**

DECIDUOUS PLANTS. Plants which drop their leaves at the end of the growing season.

DECISION FUNCTION. A decision function is a rule of conduct which, at any stage of a sampling investigation, tells the statistician whether to take further observations or whether enough information has been collected, and in the latter case, what decision to make upon it. At each stage beyond the first, the decision function is a function of the preceding observations. Until the development of sequential analysis decision functions were mostly of the simple type, based on a fixed sample size, which enjoined the acceptance or rejection of a hypothesis or set limits to a parameter under estimate. The above definition provides for a sequential situation wherein the investigator may not reach a decision about the hypothesis but proceeds to take further observations.

DECLINATION. The declination of a celestial object is the coordinate in the equatorial system of spherical coordinates measured in the plane of the hour circle through the object from the equator to the object. In case the object is between the equator and the north celestial pole, the declination is said to be north or positive (+); otherwise, the declination is south or negative (−). Declination is ordinarily measured either with a meridian circle or an altazimuth instrument.

The term declination is used by surveyors, and a few others, in place of variation to describe the angle between true and magnetic north. Navigators, both air and sea, always use the term variation for this angle. See also **Celestial Sphere.**

DECODING NETWORK (D/A Converter). **Digital-to-Analog Converter.**

DECOLORIZING AGENT. A substance that removes color by a physical or chemical action. Charcoals, carbon blacks, clays, earths, activated alumina or bauxite, or other materials of highly adsorbent character are used to remove undesirable colors (and often odors) from sugar, vegetable and animal fats and oils, and other substances. In a broad sense, decolorizing agents also embrace bleaches, which usually remove color by chemical reaction.

Activated carbon is one of the most widely used of the adsorbents. It is an amorphous form of carbon characterized by high adsorptivity. The carbon is obtained by the destructive distillation of wood, nut shells, animal bones, or other carbonaceous material. It is "activated" by heating to 800–900°C with steam or carbon dioxide, which results in a porous internal structure. The internal surface area of activated carbon averages about 10,000 square feet (929 square meters) per gram. Numerous uses include applications in the brewing and sugar refining industries.

Diatomaceous earth also finds numerous adsorbant applications in food and chemical processing, not only in decolorizing, but as a filter aid and clarifying agent as well. This is a soft, bulky solid material (88% silica) composed of skeletons of small prehistoric aquatic plants related to algae. See also **Diatoms.** They have intricate geometric forms and expose a great deal of area per unit of weight.

Fuller's earth, also used as an adsorbant, is a porous colloidal aluminum silicate (clay) which has a high natural adsorptive power. See also **Fuller's Earth.**

Silica gel is a regenerative adsorbant consisting of amorphous silica derived from sodium silicate and sulfuric acid. In addition to color adsorbing and bleaching powers, silica gel is used as a dehumidifying and dehydrating agent and as an anticaking agent.

Prior to crystallization in the refining of sugar, bleaching of the syrup is required. This is sometimes effected through treatment of the solution with calcium hypochlorite, usually in the presence of calcium phosphate which serves as a buffer and aids in the final precipitation of calcium from the bleached solution.

The physical properties of representative adsorbents and decolorizers are given in table in entry on **Adsorption Operations.**

DECOMPOSITION (Chemical). A chemical change in which a single chemical substance is broken up into two or more other substances, which differ from each other and from the parent substance in chemical identity. Complete decomposition refers to such a condition of the products that they are not readily decomposed further, e.g., such decomposition products as ammonia and carbon dioxide. *Degradation* refers to gradual decomposition in which the molecule is diminished in size in small steps. See also **Degradation. (Chemical).**

The *heat of decomposition* is the change of heat content when one mole of a compound is decomposed into its elements. This is equal in quantity, but opposite in sign, to the *heat for formation*.

Sensitized decomposition is a chemical decomposition that is brought about by the presence of a second substance which absorbs an exciting radiation. The essential mechanism of the reaction is the excitation of particles of the second substance by the radiation, followed by collisions between these excited particles and molecules to be decomposed. The process proceeds most effectively if the energy difference between the ground state and excited state of the sensitizer is nearly equal to the energy of the decomposition reaction.

Double decomposition is a term used to express the interaction of molecules which exchange one or more of their constituent atoms or radicals.

DECOMPOSITION VOLTAGE. The minimum electromotive force which must be applied to a given solution and given electrodes to produce steady electrolysis. The value of the decomposition voltage is not known precisely because the rise in current-voltage curve does not begin at a sharply defined point. Instead the curve at first rises very slowly with increasing applied voltage; in the neighborhood of the decomposition voltage there is an abrupt change in slope (a sharp bend) and thereafter the curve rises rapidly with applied electromotive force.

DECOUPLING FILTER. In most multistage amplifiers there are certain circuits, such as voltage supplies, common to more than one state. Since these common circuits provide a path through which energy may be fed from the output back into the input of some stages, serious feedback problems would result if something were not done to prevent them. The usual remedy is to insert a decoupling filter in those voltage supply leads which are common to more than one amplifier stage. These filters are frequently resistances in series with the lead and a by-pass capacitor from the device (tube or transistor) side of the resistor to ground. The resistance used must be low enough not to cause a serious loss of voltage and the capacitor should have a reactance which is low compared with the resistance at the lowest frequency for which the circuit is designed. Where the resistance would produce too much dc voltage drop or where it does not give enough filtering action an inductance is sometimes used. For still more effective filtering a second resistance and condenser in cascade may be used.

A practical amplifier employing decoupling filters is shown in accompanying diagram. The signal current i flowing in the collector circuit of the third transistor develops a voltage iZ across the output terminals of the power supply. The voltage developed across the impedance Z will cause additional components of signal current to flow in the base circuits of the second and third transistors. Use of the elements R_1, C_1 and R_2, C_2 causes a reduction of the signal fed back to an amount that does not deteriorate the amplifier performance appreciably. See also **Filter (Communications System).**

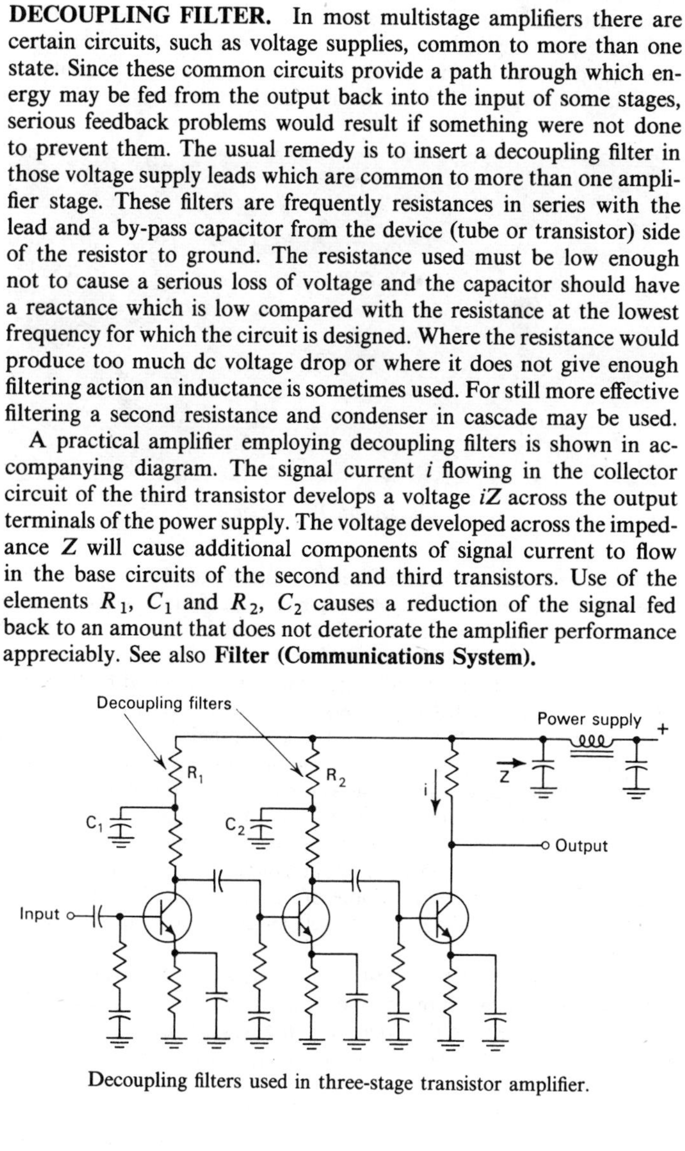

Decoupling filters used in three-stage transistor amplifier.

DECREPITATION. The emission of a crackling sound, commonly by crystals on shattering under the internal stresses resulting from heating.

DECUBITUS ULCER. **Ulcer.**

DEDENDUM. In a gear tooth, the distance from the pitch circle to the root circle or bottom of the tooth space.

DEDIFFERENTIATION. A process of change from a more specialized to a less specialized condition.

DEEP SCATTERING LAYER. During the day zooplankton sink substantially below the ocean's surface to which they rise again in the night. This migration can be detected by echo sounding devices as a "bottom" at mid-depth and is referred to as the "deep scattering layer." The cause of the daily migration is believed to be an aversion to light.

DEEP SEA DRILLING PROJECT (DSDP). **Ocean; Ocean Research Vessels; Ocean Resources (Mineral).**

DEEP VENOUS THROMBI. **Arterial and Venous Disorders.**

DEER (*Mammalia, Artiodactyla*). The deer (*Cervines*) comprise one of the larger groups of the order *Artiodactyla* (even-toed hoofed mammals). The various species of deer represent one of the most widely distributed of mammals throughout the world and appear in most locations with the exception of Australia, New Zealand, and central and southern Africa. These animals, however, have been introduced into Australia and New Zealand. Some species, notably those known as the true deer (*Cervinae*), are widely distributed. Others are found only in certain regions. It is notable, for example, that most of the deer found in South America are species indigenous to that continent and that these species are not found on other continents.

Elk. (*A. M. Winchester.*)

Generalizations are difficult because there are so many varieties. The male deer is called the stag or buck; the female is the hind or doe, and the young are called fawns. The males of most species of deer possess antlers which are shed each year. Antlers project from the animal's frontal bone and their growth is fast, aided by lime and phosphorus in the diet. New antlers are made of a pliable bony material, covered with soft velvety skin and fine hair. The antlers soon branch, patterns varying with different species. The antlers are shed after the mating season. The process is hastened by the animal rubbing the antlers against a tree to accelerate ossification at the base of the antlers. The first antlers grown by a deer are small, but each year the growth becomes larger. The size and quality of the antlers decrease after a buck has passed the prime of his sexual life. The antlers are a major weapon of defense—as well as offense during the mating season.

The deer has short fur, red-brown to gray in color, usually white below. Fawns have white spots on their back and flanks until they are about 6 months old. Upon reaching 1 year, the young deer is graceful, alert, and very fast, attaining speeds as much as 50 miles per hour.

The general organization of the subfamilies making up the *Cervines* is given in the accompanying table.

The musk-deer is a small animal of the Himalayas and stands only about 20 inches high. In this particular species, antlers are lacking and the upper jaw bears a pair of tusks which may project several inches in the males. A related species, *M. sifanicus*, occurs in the northwestern part of the People's Republic of China. The limbs of the musk deer are long, ears are large, the coat is gray-brown, coarse, brittle, and long. The animal takes shelter in thickets of rhododendron, juniper, and birch where available. The diet is comprised of moss, leaves, and grasses. The animal is shy and is mostly active at night. The male musk deer possesses a small sac underneath the stomach which contains a dark brown, viscous substance that at one time was highly valued as a fixative for perfumes; the material is still one of the most effective for this purpose. Because of the number of animals taken for this material, the population has been threatened for a number of years.

The muntjacs are shy, small animals of several species. They also are known as barking deer for the sounds which they can emit. They stand only about 2 feet high and have small two-tined antlers somewhat like those of the American pronghorn. The common Indian species, *C. muntjac*, is called the kakar. Their coloration is any one of several shades of brown.

There are about twelve species of the true deer. They are common and range widely over the earth, with exceptions as mentioned earlier. The Père David's deer is quite rare, numbering only a few hundred, all in zoological gardens or special preserves. It is not known in the wild state, but was discovered by a missionary in China in the mid-1800s in a herd of semidomesticated animals maintained by the Chinese Emperor near Peking. Fallow deer are small and highly preferred for display in parks and zoological gardens. The coloring is orange-red-brown with white spots. As indicated by the accompanying table, it is found only in the Middle East.

Of the Axis deer, the chital is known as the Indian spotted deer, is of medium size, with a reddish color and deep markings of white spots. The male's antlers are three tined. The animal lives in herds in the jungles of India, Sri Lanka, and Southeast Asia. The other axis deer is the hog deer which lives in eastern Asia. The animal is pale brown featuring white spots. It assumes a darker brown coloration during winter.

Several species of the red deer are widely distributed, and it is one of the more common deer in many parts of the world. These deer inhabit woodland areas, but are known to migrate seasonally between mountain pastures in summer and lower valleys in winter. Although the red deer belong to a single genus, there are dozens of races with a vast variety of names applied to them, the names usually bearing some localized significance. With all of these complications, there is about only a dozen species importantly recognized. The so-called real red deer (*Cervus elaphus*) is found in northerly latitudes of Europe, ranging from the British Isles eastward to eastern Siberia. They also are found in North Africa, in the Middle East to Afghanistan, north of the Himalayas to and including the mountainous region of Tien Shan. In Europe and Russia, these deer are sometimes misleadingly termed *wapiti* (to be described later). Special names for these deer, depending upon location, include: maral (Caspian region); hangul (Kashmir); and shou (Tibet).

In North America, the species (*C. canadensis*) is the wapiti, or more commonly termed by Americans as the American elk, or simply the elk. This terminology also is somewhat confusing, particularly to Europeans, because the term *elk* was originally the common name for the moose in the Old World. The American elks or wapitis, whichever term one prefers, once were wide ranging over the northern regions of North America. They are now essentially limited to wilderness or parkland areas in the Rocky Mountains. Where protected, the elk multiplies rapidly and thus can be periodically harvested under official control. The elk competes with the white-tailed deer and mountain sheep for range. The bull elk is known for its bulging or loud, shrill call when challenging another bull. Hunters prize the head and antlers of the elk as trophies. An antler may measure some 6 feet, with six pointed tines per antler. The elk is about 10 feet (3 meters) long, 5 feet (1.5 meters), tall, and weighs up to 1,000 pounds (454 kilograms).

As indicated by the accompanying table, there are several other species of red deer.

The hollow-toothed deer are exclusive to the Western Hemisphere. With the exception of the white-tailed deer and its close relatives, the other species of hollow-toothed deer are found in South America. The white-tailed deer ranges in southern Canada as well as in the eastern and central United States, and southward into Central and South America. Weighing up to 200 pounds (91 kilograms), the white-tailed deer measures up to a length of 40 inches (102 centimeters) and height of just under 3 feet (.9 meters). Exceptional specimens may weigh up to 350 pounds (159 kilograms). This is a valued sporting deer and the flesh is considered quite good. The deer gets its name because the underside of the tail is white as snow.

The mule deer is also common in western Canada and the United States. Its name is derived from its large ears which resemble those of a mule. These animals prefer the forests and are a highly regarded sporting deer where found. They are about the same height as the white-tailed deer, but weigh less, a large buck usually not weighing over 200 pounds (91 kilograms). The key deer is found on the islands off Florida, and essentially is a dwarf white-tailed deer. These animals seldom weigh over 50 pounds (23 kilograms).

The marsh deer, pampas deer, guemals, brockets, and pudus are all inhabitants of South America. The marsh deer is well named because it prefers the swamps and rivers of low country. It is widely hunted for meat and hide. The pampas deer is named for its habitat, the southern savannas and pampas of the southern regions of Argentina, Brazil, Paraguay, and Uruguay. The animal is small and reddish in coloration. It is well known for its speed and jumping ability. Large bucks may weigh up to 250 pounds (113 kilograms). The guemals sometimes are called "horse-camels" and are found in the high Andes. The brockets are delicate, small deer and, in South America, are reminiscent of the duikers of Africa. Their main habitat is the Amazon River basin. Unlike most South American deer which feed at dawn and dusk, the brockets are nocturnal feeders, hiding and

GENERAL ORGANIZATION OF THE DEER

CERVINES

MUSK DEER (*Moschinae*)

MUNTJACS (*Muntiacinae*)

TRUE DEER (*Cervinae*)
- Père David's Deer (*Elaphurus*)
- Fallow Deer (*Dama*)
- Axis Deer (*Axis*)
  - Chital
  - Hog Deer
- Red Deer (*Cervus*)
  - Eastern Red Deer (*C. elaphus*)
    - Maral
    - Hangul
    - Shou
  - American Red Deer (*C. canadensis*) (American Elk or Wapiti)
  - Asian Red Deer (*C. unicolor*)
  - Thorold's Deer (*C. albirostris*)
  - Swamp Deer (*C. duvauceli*)
  - Thamin Deer (*C. eldi*)
  - Sikas Deer (*C. nippon*)
  - Rusas Deer (*C. timoriensis*)
  - Philippine Red Deer (*C. alfredi*)

HOLLOW-TOOTHED DEER (*Odocoileinae*)
- White-tailed Deer (*O. virginianus*)
- Key Deer (*O. v. clavium*)
- Mule Deer (*O. hemionus*)
- Marsh Deer (*Blastocerus*)
- Pampas Deer (*Ozotoceros*)
- Guemals (*Hippocamelus*)
- Brockets (*Mazama*)
- Pudus (*Pudua*)

MOOSE (*Alcinae*)
(Known as the Elk in Europe)

REINDEER (*Rangiferinae*)
- Eurasian Reindeer (*Rangifer tarandus*)
- Caribous (*Rangifer arcticus, etc.*)
  - Barren Ground Caribou
  - Woodland Caribou
  - Mountain Caribou

WATER DEER (*Hydropotinae*)

ROE DEER (*Capreolinae*)

Very Approximate Regional Distribution

Eastern Asia	Middle East	Central Asia	India and South-eastern Asia	Central-Northern Eurasia
Musk Deer Red Deer Water Deer Roe Deer Sikas Hog Deer	Fallow Deer Red Deer	Red Deer Roe Deer	Axis Deer Red Deer Thamin Rusas Muntjacs	Red Deer Moose (Elk) Reindeer Eurasian Reindeer Roe Deer Brockets

Central America	North America	South America	Rare
White-tailed Deer	Red Deer (Elk) White-tailed Deer Moose Caribou Reindeer Mule Deer Key Deer	Marsh Deer Pampas Deer Pudus Guemals Brockets	Père David's Deer Thorold's Deer

sleeping through most of the day. The smallest of all deer is the pudus, about the size of a small terrier dog. They range in the temperature forest zone of the Andes. The small antlers are in the form of spikes. There are records of the pudus being domesticated much as a dog.

The moose is a large deer with a broad muzzle, prehensile upper lip, and high shoulders. The male has broad palmate antlers. The moose is one of the largest of the deer, weighing a half-ton or more. They are some 6 feet (1.8 meters) tall and attain a length of about 9 feet (2.7 meters). Two species occur in North America, one ranging over the northern United States and Canada; the other species is in Alaska. The European species of moose, closely related to the common North American species, is locally called the elk.

As do other reindeer, the Eurasian reindeer has antlers in both sexes, set well back on the head and palmately branched at the tips. These animals are important in Lapland and in other northern parts of the Old World where they have been domesticated. One record indicates that one of these small deer pulled two men in a sled for a total of 16 hours at an average speed of 18 miles (29 kilometers) per hour. The meat and skin are excellent. At one time, the Lapps relied almost entirely on this animal for their livelihood. The milk of these small deer is rich and thick and must be diluted before drinking. From it are made cheese and other products, including an alcoholic drink.

There also have been problems of nomenclature involving reindeer and caribous. The terms essentially can be used interchangeably. The two North American species of caribou and several other species of deer are closely related to the true reindeer. The caribou (*R. arcticus*) is found from Greenland westward to Alaska. These animals travel southward to northern Canadian forests to spend the winters. In the spring, as snow conditions permit, they migrate north. The animal measures about 4 feet (1.2 meters) in height and the males weigh up to 700 pounds (318 kilograms). Although the Eskimos and American Indians did not domesticate caribou as in Lapland, the animal nevertheless has been extremely important to the livelihood of these people. The three species-complexes of caribou include the barren ground caribou, the woodland caribou, and the mountain caribou, their names essentially indicating their habitat. There are wide variations in coloration and other features of these animals. The woodland caribou is one of the largest of the caribous and is of the darkest color. The animal has very keen senses and consequently is difficult to stalk. It is believed that these animals, which are found from Alaska through northern Canada to the northern parts of Maine, originally came from Siberia. At one time, the animals were threatened with extinction, but during the last several years have been protected by law. The mountain caribou is the largest species, rich dark brown in coloration, and is found in British Columbia and Alberta.

The water deer is a small animal found along the margins of the Yangtse Kiang River in the People's Republic of China. The male has long curved tusks in the upper jaw and neither sex has antlers. The species is remarkable among deer, in that it produces from 3 to 6 young at a time.

Roe deer represent a small species, but a rather large population. They prefer a woodland habitat. The antlers of the male reach a little more than a foot in length, and in normal specimens have only three tines. These have been favorite hunting deer for hundreds of years and probably are the best known deer to Europeans. They are found all across Europe, from the British Isles eastward to Siberia and southward to the Caucasus, as well as in central Asia. Pockets of roe deer are found in the woodland areas that separate the highly industrialized communities of Europe.

The mouse deer, so-called because of its deer-like appearance, is not a deer, but is a traguline. See **Tragulines.**

For references, see **Mammalia.**

DEER FLY (*Insecta, Diptera*). Small horse flies with banded wings which are abundant in the eastern woods. In the west the name is applied to snipe flies. All species are annoying to man.

DEER-OXEN. Bovines.

DEFECT (Materials). A term used to include various types of point imperfections in solids, such as vacancies, interstitial atoms, etc., as distinct from extended imperfections such as dislocations. Lattice defects are particularly important in ionic crystals, where they may be created by heating in the vapor of one constituent, thus creating a stoichiometric excess, by bombardment with x-rays and energetic particles, etc. Any crystal must contain a certain equilibrium concentration of defects, as a function of the temperature, purely as a result of thermal agitation (see **Frenkel Defect; Schottky Defect**). Defects are responsible for the diffusion of ions, ionic conductivity, and the complex phenomena related to color centers. The term defect conduction is applied specifically to electric conduction by holes in the valence band of a semiconductor.

A *Frenkel defect* is a lattice vacancy created by removing an ion from its site and placing it at an interstitial position within the lattice. Thus, a Frenkel pair is a vacancy and interstitial. A *Schottky defect* is a lattice vacancy created by removing an ion from its site and placing it on the surface of the crystal. For electric neutrality, the number of cation Schottky defects must equal the number of anion Schottky defects. The number, n, of Schottky defects is given by

$$\frac{n}{N-n} = C_s e^{-W/kT}$$

where there are N lattice points, and W is the energy required to remove an ion from a lattice point, and then add it to the surface. C_s is a numerical factor of the order of 10^3–10^4. This relation may be derived from thermodynamic arguments. It can be shown that the factor C_s includes a vibrational entropy term.

Other basic defects also are recognized. These are *surface*, *electrical defects*, *thermal defects*, and *stacking faults*. Many of these defects are present in stoichiometric crystals, but departure from stoichiometry in compound crystals means that there can be material defects present with excess of either the electropositive or electronegative constituent. In nonstoichiometric solids there will be electrical defects to preserve electrical balance.

DEFECTS (Control of). Statistical Quality Control.

DEFERVESCENCE. A medical term indicating the decline of fever to a normal body temperature.

DEFICIENCY DISEASE. A disease due primarily to lack of certain essential nutritional substances considered essential to normal body function. These include various vitamins, minerals, such as iron, calcium, phosphorus, proteins and amino acids. Diseases of this type common at one time throughout the world were beriberi, scurvy, pellagra, and rickets. During the past century, with a knowledge of their causation and the greater availability of dietary supplements and general dietary knowledge, these diseases no longer are commonplace. See **Dietary Requirements and Trends;** and **Vitamin.** Chemical elements important to the animal body are also described in several entries under the germane element.

However, in the case of protein deficiency, the recognition and importance of such protein–calorie malnutrition (PCM) diseases as kwashiorkor and marasmus have been given proper attention only during the past few decades. These diseases are described in the entry on **Protein.** Protein imbalance in the diet of the infant can also result in PCM-plus syndrome or infantile obesity. This condition is also described in the entry on **Protein.** PCM diseases are encountered most commonly in the underdeveloped countries of Africa, Asia, and Central and South America.

DEFLAGRATING EXPLOSIVE. Explosive.

DEFLATION. In geology, deflation is the action of the wind in removing unconsolidated fine-grained sediments from a land surface. In 1895 dust fell in Missouri which must have come entirely from western Kansas and Nebraska, since the intervening country was covered with ice and snow. Dust from the Sahara has been blown over Germany and England (transported by air 2000 miles); 3218 kilometers).

DEFLECTION ANGLE. In surveying, the angle between a line and the extension of the preceding line is called a deflection angle

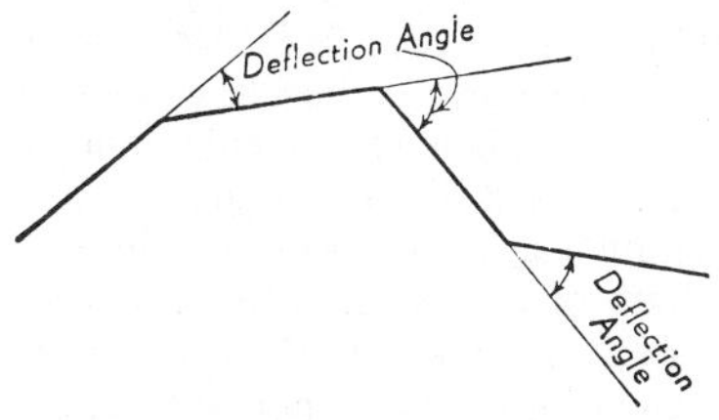

Fig. 1. Representation of deflection angle.

(see Fig. 1). When the survey is a closed traverse the algebraic sum of the deflection angles must equal 360 degrees. Circular curves are frequently laid out by the method of deflection angles, as illustrated in Fig. 2. Before it is possible to lay out points on the curve it is necessary to locate the P.C., P.I., and P.T. and obtain the central angle Δ which is numerically equal to the deflection angle Δ_1. Having the value of Δ_1, the deflection angles d_1, d_2, etc., may be calculated from the assumed chord lengths c_1, c_2, etc. The *transit* is first set up over the P.C. and sighted on the P.I. The deflection angle d_1, which gives the direction of the line from P.C. to x, is then turned off by means of the transit. The chord length c_1 is next measured on the ground by a steel tape which definitely fixes the positions of point x. Point y may be located by turning off the deflection angle d_2, and measuring the chord c_2 from the point x. Other points on the curve may be set in a similar manner. From the geometry of the figure it can be seen that the deflection angle (d_4) to the P.T. is equal to one-half of the central angle.

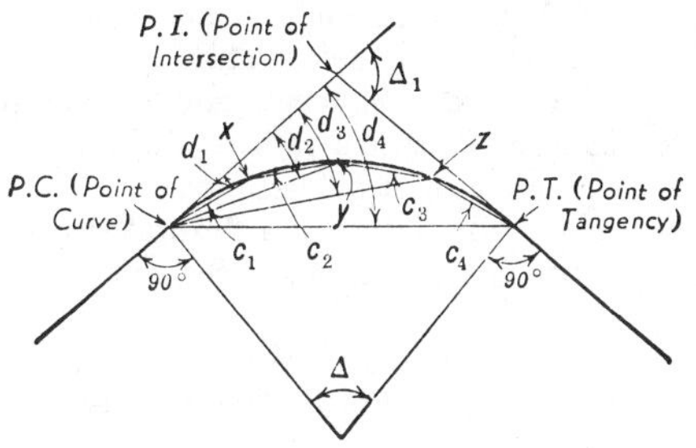

Fig. 2. Application of deflection angles to layout of circular curves.

DEFLECTION OF THE VERTICAL. The angular difference, at any place, between the direction of a plumb line (the vertical) and the perpendicular (the normal) to the reference spheroid. This difference seldom exceeds 30 seconds of arc. Also called station error. When measured at the earth's surface the deflection of the vertical is equal to the angle between the geoid and the reference spheroid.

DEFLECTION SENSITIVITY (CRT). Cathode-Ray Tube.

DEFLECTION (Structural). Loads acting on a structure cause displacement of the structure relative to its original position. This is known as deflection. Deflection is characteristic of all structures since all materials are elastic to a certain extent. Figure 1 represents a beam which has been bent by the action of an external load. The deflections are the ordinates between the original and final positions of the elastic curve. The amount of deflection depends upon the load, stiffness of the material and the dimensions of the beam. The stresses cause the top fibers to compress and the bottom fibers to elongate. Since Hooke's Law states that stress is proportional to strain (deformation) as long as the stress is below the proportional limit, the stress in the beam due to bending will be proportional to the deformation of the fibers. The beam will come to rest or be in a state of equilibrium after the application of a load, when at every section the moment of the internal stresses equals the moment due to the external load.

Deflection is an important element in the design of a load-carrying structure. If strength is the limiting condition for which a beam is designed, the design should be tested for deflection to make certain

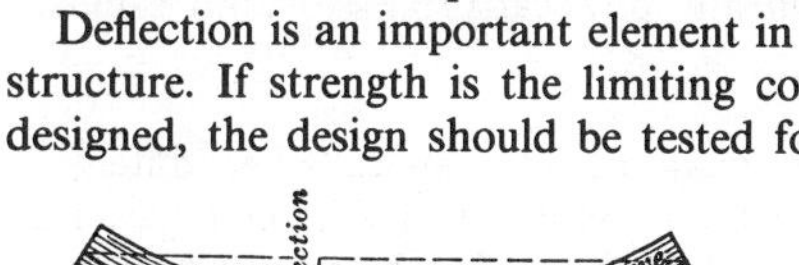

Fig. 1. Deflection of a bent beam.

that the displacements are within allowable limits. Deflection rather than strength often governs the design. This is especially true in the design of beams for buildings, where small deflections, only, are permissible, because of the tendency of the deflection of the beams to crack terrazzo floors, plastered ceilings, etc. Solid rib and trussed bridges are also subject to deflection. Large steel bridges and building trusses are always cambered (see **Camber**) to counteract the effect of deflections. The deflections of beams, rigid frames, arches, trusses and other structures can be obtained by the methods of structural mechanics, both analytical and graphical.

Relative deflections under certain assumed conditions and loadings are determined in order to analyze statically indeterminate structures.

Maxwell's Law of Reciprocal Deflections, which may be applied to any loaded structure, states that the deflection of a point A in an arbitrary direction AC due to any load P applied at B in a direction BD is equal to the deflection of B in the direction BD when the load P is applied at A in the direction AC. (See Fig. 2.)

See also **Beam (Structural)**; and **Elasticity.**

Fig. 2. Application of Maxwell's law of reciprocal deflections.

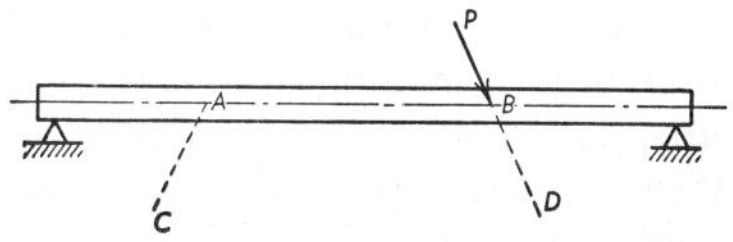

DEFLECTION YOKE. One or more electromagnets placed around the neck of an electron-beam tube for the purpose of producing a magnetic field to deflect one or more electron beams. Also simply termed a *yoke*.

DEFLECTOR. A plate, baffle, or the like that diverts something in its movement or flow, as: (a) a plate that projects into the airstream on the underside of an airfoil to divert the airflow, as into a slot—sometimes distinguished from a spoiler; (b) a cone-like device placed or fastened beneath a rocket launched from the vertical position, to deflect the exhaust gases to the sides; (c) any of several different devices used on jet engines to reverse or divert the exhaust gases; (d) a baffle or the like to deflect and mingle fluids prior to combustion.

DEFOAMING AGENTS. Film breakers or defoaming agents are substances used to reduce foaming caused by proteins, gases, or nitrogenous materials, among others, which may interfere with processing or the desired characteristics of the end-products. Processes particularly prone to foaming conditions include the Kraft process for papermaking, where a very foamy pulp slurry is formed; phosphoric acid production from phosphate rock; beet sugar processing; several fermentation processes; and, in terms of the end-product, latex paints. See also **Paints.**

The terms *defoamer* and *antifoam* or *anti-foaming agent* are frequently used interchangeably. A *defoamer* best describes a substance that kills the foam from above, once it exists. An *antifoaming agent* stops the foam from forming in the first place. Often a defoamer will be a poor antifoaming agent, and vice versa. Defoaming agents, when used in small concentrations, can be quite effective. Where a suitable chemical substance cannot be found, physical means may be required. These may be mechanical, electrical, or thermal in nature. The mechanical devices are fundamentally simple, usually taking form of rotating breaker arms. The presence of a hot surface near a foam tends to destroy the foam. Essentially, a portion of the foam is evaporated, causing the acceleration of its breakdown. It has also been established that electrical discharges tend to weaken or destroy films.

Mechanisms of Defoaming Agents. These agents may operate via a number of mechanisms, but the most common ones appear to be those of entry and/or spreading. The defoamer must first of all be insoluble in the foaming liquid for these mechanisms to function. Second, the surface tension of the defoamer must be as low as possible. The interfacial tension between defoamer and foamer should be low, but not so low that emulsification of the defoamer may occur. Third, the defoamer should be dispersible in the foaming liquid. It was first shown in 1948 that thermodynamically the entry of the defoamer droplet into a bubble surface occurs when the entering coefficient has a positive value. The physics of bubbles is described in entry on **Foam.**

A type of defoamer may consist of a dispersion in oil of fine particles of silica coated with silicone, the silicone surface of the particles causing them to be hydrophobic. The defoaming action of such a formulation can be explained on the basis of the entry mechanism. Hydrophobic particles can act as an emulsifying agent where the defoamer oil constitutes the continuous phase and the foam constitutes the dispersed phase. In experimental trials, it has been found that excessively hydrophobic particles, such as powdered Teflon, do not function as well as silicone-coated particles. An emulsifier particle must be wetted to some extent by the dispersed phase in order to function as an emulsifier.

The more efficient defoaming mechanism of spreading involves transport of underlying liquid so that the liquid is replaced by a film of defoamer which does not support foam. A drop of oleic acid added to water spreads at a velocity of 30 miles (48.2 kilometers) per hour. The mechanical shock to a film by such a defoamer may be considerable. In addition to the foam-destroying aspect, spreading is also of value as a defoamer-dispersion method, particularly in viscous or poorly stirred systems.

Defoaming agents are in three principal categories, but sometimes are used in combination: (1) solubilized surfactants, (2) dispersions of hard particles, and (3) dispersions of soft particles. The fatty acid–fatty alcohol combination in hydrocarbon oil is an example of a solubilized surfactant defoaming formulation. Paraffinic waxes and fatty amides may be used in soft-particle formulations. The most common of the hard-particle formulations is silica or a mineral coated with silicone dispersed in a vehicle. A particle size as small as 0.02 micrometer may be optimal.

Choice of defoaming agents in the food field is somewhat restricted because substances used obviously must be nontoxic and not produce off odors, off colors, or off tastes. Chemical defoaming agents commonly used in food processing include decanoic acid, dimethylpolysiloxane, lauric acid, mineral oil (white), myristic acid, octanoic acid, oleic acid, oxystearin, palmitic acid, petrolatum, petroleum wax (synthetic), silicon dioxide, sorbitan monostearate, and stearic acid. See also **Foam.**

An example from the brewing industry points out the importance of defoaming agents. Advantages of their use include: (1) higher production through increased fermentation capacity—up to 20% more throughout; (2) the lid of the fermentation tank can be left on, reducing oxidation and improving sanitation; (3) lower oxidation rate, which gives a better physical and chemical stability to the beer as the denatured or partially denatured protein levels remain low, and thus reducing turbidity; and (4) less yeast build-up on the sides of the tank, thus reducing cleaning requirements. Inasmuch as foam is a consideration in the final quality of the brewed product, effective foam control during processing can later affect the "head" of the final product, providing for a stable, long-lived, creamy foam. Thus, a defoaming agent for process use must be insoluble in the beer and capable of removal so that it does not detract from the head of the final product.

References

Andres, C.: "Anti-foaming Agent Increases Fermentation Capacity 20%," *Food Processing*, **38**, 5, 58–59 (1977).

Considine, D. M. (editor): "The Encyclopedia of Chemistry," 4th edition, Van Nostrand Reinhold, New York, 1983.

Robinson, J. V., and W. W. Woods: *J. Soc. Chemists*, **67**, 361–365 (1948).

Staff: "Food Chemicals Codex," National Academy of Sciences, Washington, D.C. (Revised periodically).

DEFOLIANT. Agent Orange; Herbicide.

DEFORMATION BANDS. Bands formed inside the crystals of metals by plastic deformation in which the crystal lattice has been rotated into orientations differing from that of the rest of the crystal. See also **Crystal.**

DEFORMATION BONDING. Welding.

DEFORMATION (Continuous, Mathematics). A transformation which shrinks, twists, etc., in any way without tearing. A continuous deformation of an object A into an object B is a continuous mapping $T(\rho)$ of A onto B for which there is a function $f(\rho, t)$ which is defined and continuous (simultaneously in ρ and t) for real numbers t with $0 \leq t \leq 1$ and points ρ of A and for which $F(\rho, 0)$ is the identity mappings of A onto A, $F(\rho, 0) \equiv \rho$, and $F(\rho, 1)$ is identical with $T(\rho)$. With this definition, a circle in the plane can be continuously deformed to a point although a circle around the outer circumference of a torus cannot be deformed continuously into a point, or into one of the small circles around the body of the torus, without leaving the torus; i.e., with all values of $F(\rho, t)$ being points on the torus. It is frequently required that a continuous deformation not bring points together; i.e., that the above function $F(\rho, t)$ be a one-to-one correspondence for each value of t. Then a circle in the plane can be continuously deformed into a disc (a circle and its interior), but not into a cylinder or a sphere. It is said that two mappings T_1 and T_2 of a topological space A into a topographical space B can be continuously deformed into each other if there is a function $F(x, t)$ which has values in B, which is continuous simultaneously in x and t for x in A and $0 \leq t \leq 1$, and for which $F(x, 0) \equiv T_1(x)$ and $F(x, 1) \equiv T_2(x)$ for each x of A. Two mappings are said to be homotopic if they can be continuously deformed into each other. If A is contained in B and T_1 is the identity mapping of A onto A, then T_2 is a continuous deformation (in the above sense) of A into the range of T_2 if T_1 can be continuously deformed into T_2.

See also **Mapping; Topological Space;** and terms listed under **Mathematics.**

DEFORMATION (Materials). The change in the shape of a body which accompanies a stressed condition is called deformation. The total amount of deformation in one direction is the total deformation. Unit deformation is the deformation per unit of length and is commonly called strain. Permanent deformation is known as set. If an axial load is applied to a body, the length and lateral (cross-sectional) dimensions are changed. Poisson's ratio is the ratio of lateral unit deformation to longitudinal unit deformation.

Deformation which is the result of a flexural stress (see **Flexure**) is called bending deformation. Shearing of shear deformation is caused by shearing stress.

DEFORMATION (Plastics). Plastic Deformation.

DEFORMATION (Resilience). Resilience.

DEGASIFICATION. Removal of gas, as applied particularly to the removal of the last traces of gas from wires used in vacuum apparatus, from metals to be plated, and from substances to be used in other specialized applications. Untreated glass always contains water, carbon dioxide, oxygen, and traces of other gases within it and on its surface, and these are ordinarily in a state of equilibrium with the surroundings. When the pressure is reduced, however, the equilibrium is upset, and these gases, being gradually released from solution in and adsorption on the glass, spoil the vacuum. It is usual to drive the gases out of the glass by baking the glass at a temperature of about 350°–500°C, while on the vacuum pump. Degassing of metal is necessary for the same reason as in degassing of glass, but because of the larger quantities of gas present in metals, more complex methods of degassing must be employed. These include baking at high temperature, eddy-current heating, and electron bombardment. See also **Gettering.**

DEGENERACY. In the kinetic theory of gases, a gas which does not obey the ideal gas laws is referred to as a degenerate gas. The greater the deviation of the real gas from the ideal, the greater is its degeneracy.

A *degenerate electron gas* is an electron gas which is far below its Fermi temperature, that is, which must be described by the Fermi distribution. The essential characteristic of this state is that a very large proportion of the electrons completely fill the lower energy levels, and are unable to take part in any physical processes until excited out of these levels.

DEGENERATE STATE. In quantum mechanics, when different states of motion correspond to the same energy level, the states are said to be degenerate. The degeneracy can often be removed by the application of a perturbing field with the effective introduction of a new quantum condition, i.e., the breakup of one eigenvalue into several.

DEGENERATION (Control System). Same as negative feedback. See **Feedback.**

DEGRADATION (Chemical). A gradual decomposition occurring in stages with well-marked intermediate products. For example, the maltose chain loses one carbon atom under certain conditions to produce a sugar with eleven carbon atoms in its skeleton. In fact, the term degradation often means specifically a reduction of the number of carbon atoms in an organic compound, usually an aliphatic compound. Among the specific methods used for this purpose are the Hoffmann reaction, treating an amide with a hypohalite; the conversion of fatty acids to methyl ketones, followed by oxidation; and the Curtius reaction for the conversion of an acid azide to the primary amine.

Kraft Method. A method of reducing the number of carbon atoms in the molecule of an acid, especially a fatty acid by the decomposition of its calcium or barium salt in the presence of a salt of acetic acid, followed by oxidation of the resulting methyl ketone:

$$(RCH_2COO)_2Ca + (CH_3COO)_2Ca \rightarrow 2RCH_2COCH_3 + 2CaCO_3$$
$$\downarrow CrO_3$$
$$2RCOOH + 2OH_3COOH.$$

See also **Decomposition (Chemical).**

DEGRADATION (Energy). The second law of thermodynamics is sometimes referred to as the Law of Degradation of Energy, because of the statement that the entropy of an isolated system is increased by irreversible processes involving energy changes, and that, therefore, the sum of the available energy tends to decrease.

DEGRADATION (Nuclear). In atomic physics, degradation means a loss of energy by collision. Neutron degradation is also called moderation.

DEGRADATION (Oxidative). Antioxidant.

DEGREE CELSIUS. Units and Standards.

DEGREE DAYS. A form of accumulated temperature; a unit employed in heating and air-conditioning calculations. For specifying the nominal heating load in winter, there are as many degree days as there are degrees Fahrenheit difference in temperature between the mean temperature for the day and 65°F (18.3°C) (60°F (15.6°C) in Great Britain). For example, in a month during which the outside temperature averaged 20°F (−6.7°C) for 10 days and 35°F (1.7°C) for 20 days, there are 1050 degree days. This number is found thus: Degree days = 10(65 − 20) + 20(65 − 35) = 1050. One *heating degree day* is given for each degree that the daily mean temperature departs below the base of 65°F (18.3°C). In accumulating degree days over a "heating season," days on which the mean temperature exceeds 65°F (18.3°C) are ignored. The energy requirements for air conditioning or refrigeration are estimated by *cooling degree days*, one given for each degree that the daily mean temperature departs above the base of 75°F (23.9°C). As used by the U.S. Army Corps of Engineers, degree days are computed as departures above and below 32°F (0°C), positive if above, and negative if below. To avoid confusion, it might be well to call this a "freezing degree-day."

The period of time between the highest point and the succeeding lowest point on the time curve of cumulative degree days above and below 32°F (0°C) is called the *freezing season*; while the period of time between the lowest point and the succeeding highest point is called the *thawing season*. The *freezing index* is the number of degree days (above and below 32°F; 0°C) between the highest and lowest points on the cumulative degree-days time curve for one freezing season. The index determined for air temperatures at 4.5 feet (1.3 meters) above the ground is commonly designated as the "air freezing index"; while that determined for temperatures immediately below a surface is known as the "surface freezing index." See also **Climate.**

DEGREE (Electrical). This is $\frac{1}{360}$ of a cycle of alternating current representing one electrical revolution.

DEGREE FAHRENHEIT. Units and Standards.

DEGREE (Geometry). A degree is the unit measured by the central angle subtended by $\frac{1}{360}$ of the arc of a circle. This unit of angular measurement has been extended to a great many practical uses, as, for example, the reading of compass bearings in degrees. A right angle contains 90°; a minute is $\frac{1}{60}$ of a degree, and a second is $\frac{1}{60}$ of a minute.

Degrees also are measured in *radians*. A radian may be defined as an angle which, if placed at the center of a circle, will intercept an arc of a length equal to the radius of the circle. The ratio of the circumference to the radius is identical for all circles. Thus, the radian has a fixed value, that is, it is a constant angle and independent of the size of the circle. The circumference of a circle is 2π times the radius. Thus, the circumference embraces 2π radians and the following relationships apply: 2π radians = 360°; π radians = 180°. One radian equals $180°/\pi$, or 57.29578 degrees. One degree equals 0.0174533 radian.

DEGREE OF SUBSTITUTION. Cellulose Ester Plastics (Organic).

DEGREE RANKINE. Units and Standards.

DEGREES OF FREEDOM. Freedom (Degrees of).

DEGREES OF FREEDOM (Statistics). The number of degrees of freedom in a statistical quantity is the number of independent values necessary to determine it. A sample of n values $x_1, x_2, \ldots, x_n$ has n degrees of freedom but the sum

$$\sum_{i=1}^{n} (x_i - \bar{x})^2$$

is regarded as having $n - 1$ because, for given $\bar{x}$, only $n - 1$ values are assignable at will.

By extension, the number of degrees of freedom of a statistical distribution relates to the degrees of freedom of the distributed statistic. For example, χ^2, being the distribution of the sum

$$\sum_{i=1}^{n} (x_i - \bar{x})^2/\sigma^2$$

has $n - 1$ degrees; and the F-distribution, which concerns the ratio of two independent such quantities has a pair of degrees of freedom, one relating to the numerator and the other to the denominator of the ratio.

DEGREE (Thermal). In essence, the thermal degree represents molecular activity, in that temperature depends upon molecular velocity. There are several temperature scales, each of which is divided into measurement units. The key unit in each scale is a degree. The most common temperature scales are Celsius (Centigrade) and Fahrenheit. Thus, the terms degrees Celsius (°C); and degrees Fahrenheit (°F). See also **Temperature.**

DEHISCENCE. The splitting open of a pod or anther, causing seeds to be ejected with explosive force. See **Seed.**

DEHUMIDIFICATION. A process used in air conditioning and in the process industries in which air or other gases partially saturated with water are subjected (1) to cooling below their dew point, so that part of the water vapor is condensed and thus separated from the gas; (2) to the action of chemical desiccants, either liquid or solid, which adsorb moisture from the gas; or (3) to a combination of both actions. The most commonly used gas in industry is compressed air. This air requires drying and conditioning for trouble-free operation of pneumatic equipment, including tools, instruments, and automatic controllers. In paint manufacture, dry inert gas is used to blanket agitation operations. Dry process gases, such as nitrogen or hydrogen, are used in metal-annealing operations. The manufacture of some transistors requires blanketing with a dry gas during assembly.

Deliquescent dryers (dissolving desiccant types) and refrigeration-type dryers are adequate for most air-comfort applications and some industrial applications. Usually, these systems only partially remove moisture from air and other gases and generally require additional drying equipment. To meet tight drying specifications, a solid, regenerable desiccant which adsorbs the moisture usually is used. Desiccant regeneration usually is accomplished by the application of heat or purging, or a combination of both procedures. Normally, dual drying towers are used for continuous service, allowing one tower to be on-line, while the other tower is being regenerated. Desiccants most frequently used are silica gel, activated alumina, and molecular sieves. Desiccant systems permit efficient dew-point performance in the range of −40 to −100°F (−40 to −73°C) and are available in a wide range of capacities and pressure ratings (from atmospheric pressure up to 340 atmospheres; 5000 psig).

DEHYDRATED FOODS. Antimicrobial Agents (Foods).

DEHYDRATING. Drying (Process).

DEHYDRATION (Chemical). Removal of water from a substance or system or chemical compound, or removal of the elements of water, in correct proportion, from a chemical compound or compounds. The elements of water may be removed from a single molecule or from more than one molecule, as in the dehydration of alcohol, which may yield ethylene by loss of the elements of water from each molecule, or ethyl ether by loss of the elements of water from two molecules, which then join to form a new compound:

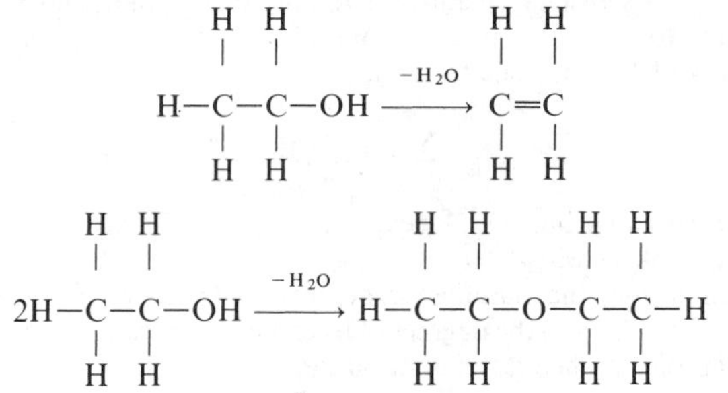

Many reactions known in chemistry under special names, such as neutralization, esterification and etherification are dehydration reactions.

In the food processing field, dehydration is sometimes described as the removal of 95% or more of the water from a food substance, by exposure to thermal energy by various means. The aims of dehydration are reduction in volume of the product, increase in shelf-life, and lower transportation costs, among other factors. There is no clearly defined line of demarcation between drying and dehydrating, the latter sometimes being considered as a supplement of drying. Usually, the direct use of solar energy, as in the drying of raisins, hay, etc., is not lumped in with dehydrating. The term dehydration also is not generally applied to situations where there is a loss of water as the result of evaporation. *Rehydration* or *reconstitution* is the restoration of a dehydrated food product to essentially its original edible condition by the simple addition of water, usually just prior to consumption or further processing. The distinction between the terms drying and dehydrating may be somewhat clarified by the fact that most substances can be dried beyond their capability of restoration. Important food products that are dehydrated include animal feedstuffs, hops, malt, oat, peanut (groundnut), potato, rice, and sweet potato. See **Drying (Process).**

DEHYDRATION (Diarrhea). Diarrhea.

DEHYDRATION (Physiological). A clinical state in which the body does not retain sufficient fluids to maintain the normal functions of blood circulation, excretion of waste products through the kidneys, temperature regulation, and individual cell performance. A common cause of dehydration is excessive loss of fluid through vomiting and diarrhea, frequently accompanying a disease such as intestinal flu (gastroenteritis). Infants exhibit little tolerance for prolonged dehydration and may readily go into coma, followed by death, if the underlying causes of the dehydration are not corrected immediately. Among treatments used to overcome dehydration are (1) use of anti-vomiting preparations, (2) administration of intravenous or subcutaneous fluids, and (3) application of liquids to the colon by enema, inasmuch as the colon readily absorbs fluids. These and other actions, with particular emphasis on clinical measurements as guidelines, comprise a system of body water and electrolyte system management.

The physiology and physical chemistry of body water are described in entries on **Water;** and in the section on "Water Deficiency" in the entry on **Kidney and Urinary Tract.** See also **Heat Exhaustion and Heat Stroke.**

Normally, water enters the body via the gastrointestinal tract and remains in the body for about two weeks. About half of the water is taken in as liquid water; the other half as contained in solid foods. A little over 10% of the water present in the body results from oxidation processes. Where there are no dehydration processes present, water loss normally occurs as (1) liquid loss (60%) by way of urine and feces, about 94% of this via the urine, and (2) vapor loss (40%) by way of the lungs and skin, roughly in equal proportions. Even if all water intake stops, the processes of water loss continue, the body continuing to lose about 2% of its total weight per day as the result of continuing water loss.

DEHYDROCHLORINATION. Organic Chemistry.

DEHYDROGENASE. Coenzymes.

DEHYDROGENATION. A reaction which results in the removal of hydrogen from an organic compound or compounds. This process is brought about in several ways. Simple heating of hydrocarbons to high temperature, as in thermal cracking, causes some dehydrogenation, indicated by the presence of unsaturated compounds and free hydrogen. Catalytic processes often produce commercially-practicable yields of selected dehydrogenated products. The enzyme dehydrogenase is a selective catalyst of this character. There is considerable evidence to indicate that many reactions commonly classed as oxidations, e.g., the oxidation of methanol to formaldehyde are actually dehydrogenations, i.e.,

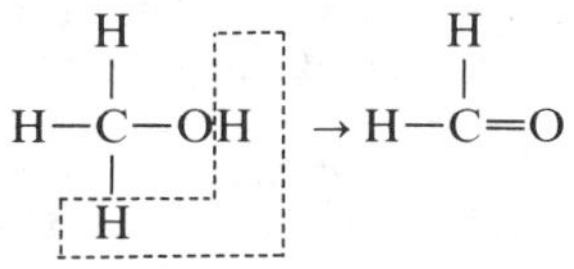

In the chemical process industries, nickel, cobalt, platinum, palladium, and mixtures containing potassium, chromium, copper, aluminum, and other metals are used in very large-scale dehydrogenation processes. For example, acetone (6 billion pounds per year) is made from isopropyl alcohol; styrene (over 2 billion pounds per year) is made from ethylbenzene. The dehydrogenation of *n*-paraffins yields detergent alkylates and *n*-olefins. The catalytic use of rhenium for selective dehydrogenation has increased in recent years. Dehydrogenation is one of the most commonly practiced of the chemical unit processes.

See also **Organic Chemistry.**

DEIMOS. Mars.

DEIONIZATION POTENTIAL. The potential at which conduction in a gas-discharge tube stops, due to the cessation of ionization.

DEIONIZATION TIME. The time required for the grid of a gas-discharge tube to regain control after anode-current interruption. To be exact, the ionization and deionization times of a gas tube should be presented as families of curves relating such factors as condensed-mercury temperature, anode and grid currents, anode and grid voltages, and regulation of the grid current.

DEL. The differential operator used in vector analysis and sometimes also called nabla, since its usual symbol ∇ is thought to resemble an Assyrian harp with the latter name. In Cartesian coordinates it is $(\mathbf{i}\partial/\partial x + \mathbf{j}\partial/\partial y + \mathbf{k}\partial/\partial z)$. When applied to a scalar function it gives the gradient; to vectors, it can give the divergence or the curl.

In the following relations ϕ is a scalar; **U**, **V** are vectors; the asterisk (*) can be either a dot or a cross and, depending on that choice, A, B are either scalars or vectors. $\nabla*(A + B) = \nabla*A + \nabla*B$; $\nabla(\phi A) = \nabla\phi*A + \phi\nabla*A$; $\nabla(\mathbf{U}\cdot\mathbf{V}) = (\mathbf{V}\cdot\nabla)\mathbf{U} + (\mathbf{U}\cdot\nabla)\mathbf{V} + \mathbf{V} \times (\nabla \times \mathbf{U}) + \mathbf{U} \times (\nabla \times \mathbf{V})$; $\nabla\cdot(\mathbf{U} \times \mathbf{V}) = \mathbf{V}\cdot\nabla \times \mathbf{U} - \mathbf{U}\cdot\nabla \times \mathbf{V}$; $\nabla \times (\mathbf{U} \times \mathbf{V}) = (\mathbf{V}\cdot\nabla)\mathbf{U} - \mathbf{V}(\nabla\cdot\mathbf{U}) - (\mathbf{U}\cdot\nabla)\mathbf{V} + \mathbf{U}(\nabla\cdot\mathbf{V})$. If

$$\mathbf{R} = \mathbf{i}x + \mathbf{j}y + \mathbf{k}z,$$

a position vector, $\nabla\cdot\mathbf{R} = 3$; $\nabla \times \mathbf{R} = 0$; $\mathbf{U}\cdot\nabla\mathbf{R} = \mathbf{U}$.

There are six possible combinations where the operator is applied twice, although two of them equal zero identically. They are: (1) $\nabla^2\phi$, the Laplacian; (2) $\nabla^2\mathbf{V}$; (3) $\nabla(\nabla\cdot\mathbf{V})$; (4) $\nabla \times (\nabla \times \mathbf{V}) = \nabla(\nabla\cdot\mathbf{V}) - \nabla\cdot\nabla\mathbf{V}$; (5) $\nabla \times \nabla\phi = 0$; (6) $\nabla\cdot\nabla \times \mathbf{V} = 0$.

DELAY CIRCUIT. When it is desired to delay the output signal of a device or system for a specified time interval with respect to the input signal, this can be accomplished by use of a delay circuit. See **Delay Line.**

DELAYED COKING. Cracking Process.

DELAYED NEUTRON. Neutron; Nuclear Reactor.

DELAY EQUALIZER. An equalizer which is used to correct for delay distortion in transmission systems. Usually a corrective network is employed to make the phase delay (or envelope delay) of a system or circuit essentially constant over a desired range of frequency.

DELAY LINE. In communication and control systems, a delay line may be used to hold a signal for a discrete period and then retransmit the signal with a minimum of distortion. This action can be accomplished in several ways.

In a *magnetostrictive delay line*, electric pulses are converted into acoustical pulses for transmission along a wire and then are reconverted to electric pulses after a delay, depending upon the length of the wire. A simple form of this device consists of two coils, biased with small permanent magnets, which are placed around a strip of magnetostrictive material. A current pulse in the input coil produces a longitudinal stress wave that travels along the wire at about 2×10^5 inches (5.1×10^5 centimeters) per second. When the wave reaches the output coil, it introduces a corresponding pulse. Absorption pads are clamped at the ends of the magnetostrictive strip to eliminate reflected pulses. The device is limited to about 50 microseconds inasmuch as this length of delay occupies a 15-inch (38-centimeter) package, making longer delays rather unwieldy physically. The maximum signal frequency is about 4 MHz and the delay is about 5 microseconds/inch of transmission material.

In a *torsional type magnetostrictive delay line*, the longitudinal stress waves generated in magnetostrictive material are converted to torsional waves in a transmission wire. The torsional pulse induced in the transmission wire travels along in helical fashion at about 1.2×10^5 inches (3×10^5 centimeters) per second. Delays may be as long as 20 milliseconds. Bandpass characteristics are linear over a frequency band equal in width to the center frequency of operation. As an example, a 3-dB bandpass for a line designed to operate at 1 MHz will extend from 0.5 to 1.5 MHz. Lines can be designed to operate at frequencies ranging from 200 kHz to 4 MHz. Principal applications are in electronic memories and recirculating buffers in display systems.

Blocks of glass or quartz also can be used as a medium to transmit acoustical pulses. An electric pulse is converted to an acoustical signal by a transducer on one face of a multifaced block. The signal is totally internally reflected in the block many times before it reaches the receiving transducer on another face. Transducers may be piezoelectric-ceramic or crystalline quartz. Ceramic transducers produce a low loss of about 40 dB at frequencies up to about 10 MHz. Crystalline quartz transducers produce 60- to 80-dB loss, but can be used up to 60 MHz. Longer delays can be generated in glass delay lines and up to 5 milliseconds in fused quartz lines. If ferromagnetic metal is used for the transmission block, the frequency of signals is limited to about 5 or 10 MHz because of absorption of higher frequencies in the metal domains. A glass delay line can be mounted on a circuit board together with integrated circuitry to utilize it as a digital memory. Uses include buffers for computer printers, readouts, and typewriters, in addition to short-time memories for small computers, numerical controls, and electronic calculators.

In connection with pneumatic signals, the delay is about 1 millisecond per foot (0.3 meter) depending upon the propagation velocity of sound waves. Pneumatic delays are used in fluidic circuits to generate time-based functions.

Taped delays can be used in some cases. If a signal is recorded on a moving tape or drum and then read at some distance from the writing head, it will obviously be delayed by the time of transport between reading and writing heads. This approach provides long, easily adjustable delays, but because of the cost and complexity of driving equipment, it is usually available only in special equipment.

Devices used for injecting timed intervals into systems beyond a fraction or few seconds are usually classified as timers rather than delay devices. See also **Timing Controllers.**

DELBRÜCK SCATTERING. Scattering.

DELHI BOIL. Leishmaniasis.

DELIGNIFICATION. Pulp (Wood) Production and Processing.

DELIQUESCENCE. When a substance absorbs moisture upon exposure to the atmosphere, the substance is said to be *deliquescent.* At ordinary temperatures the vapor pressure of water varies as shown in the accompanying table. If the solution of a substance in water has a lower water vapor pressure than that of the atmosphere at the given temperature, water vapor condenses in the solution from the atmosphere until the water vapor pressure of the solution equals the water vapor pressure of the surrounding atmosphere.

VARIATION OF WATER VAPOR PRESSURE WITH TEMPERATURE

Temperature, °C	Water Vapor Pressure, in mm of Mercury	
	At Saturation	At 50% Humidity
0	4.6	2.3
10	9.2	4.6
20	17.5	8.8
30	31.8	15.9
40	55.3	27.7

Substances that are ordinarily deliquescent are sulfuric acid (concentrated), glycerol, calcium chloride crystals, sodium hydroxide (solid), and 100% ethyl alcohol. In an enclosed space, these substances deplete the water vapor present to a definite degree. Other substances are used to accomplish this end by chemical reaction, e.g., phosphorus pentoxide (forming phosphoric acid), and boron trioxide (forming boric acid). Water is absorbed from nonmiscible liquids by addition of such substances as anhydrous sodium sulfate, potassium carbonate, anhydrous calcium chloride, and solid sodium hydroxide. The converse phenomenon is known as *efflorescence.*

See also **Dehumidification;** and **Efflorescence.**

DELIRIUM. A mental disturbance characterized by disorientation, hallucinations, incoherent rambling speech, excitement and usually extreme restlessness. It occurs with high fever, especially in children, or with the toxemia of any severe illness. It is often present in various forms of mental disease and also may result from drugs or abuse of alcohol. Delirium tremens occurs in alcoholics; its chief features are hallucinations, agitation, and tremors of the extremities. It may occur in the chronic alcoholic following a temporary excess or sudden withdrawal of alcohol. It is particularly liable to develop in the alcoholic with pneumonia, in erysipelas, or following trauma. See **Alcoholism.**

DELLINGER EFFECT. The sudden disappearance of sky-wave signals as a result of greatly increased ionization in the ionosphere due to solar storms. The effect may last from 10 minutes to several hours.

DELOCALIZATION ENERGY. Consider the removal of the wall separating two identical boxes containing equal numbers of electrons lowers the zero point energy of the system. In the simplest one-dimensional case, the kinetic energy levels, which were given for each box by the sequence $k^2(h^2/8ml^2)$, become $k'^2(h^2/8m(2l)^2)$ for the united box. k and k' are whole numbers, l is the length of the primitive box, h, the Planck constant and m, the mass of the electron. Each level can be occupied by two electrons. The respective zero point energies, before and after the removal of the wall, are therefore

$$2 \cdot (2^2 + 4^2 + 6^2 + \cdots)(h^2/4m(2l)^2)$$

and

$$(1^2 + 2^2 + 3^2 + 4^2 + \cdots)(h^2/4m(2l)^2)$$

The increased mobility of the π-electrons along a skeleton molecule with conjugated double bonds leads to an analogous lowering of the heat of formation, i.e., to an increased stability of these molecules. This effect is usually called the *delocalization effect*, and the corresponding decrease in energy is called the *delocalization*, or *enlargement energy*.

DELRAC SYSTEM. **Navigation.**

DELTA. The terminal deposit of river-borne sediment in a lake or bay. So called because of its triangular or delta-like ground plan. The cross section or structure of a typical delta is shown in the accompanying sketch. Except in the case of small deltas only the top-set beds can be observed. In the case of Paleozoic, Mesozoic, and Cenozoic deltas it is extremely difficult to distinguish between the top-set, fore-set, and bottom-set beds, and the ultimate determination that a sedimentary formation is of delta origin depends largely on tracing the original source and areal distribution of the sediments, and the presence or absence of marine and terrestrial fossils.

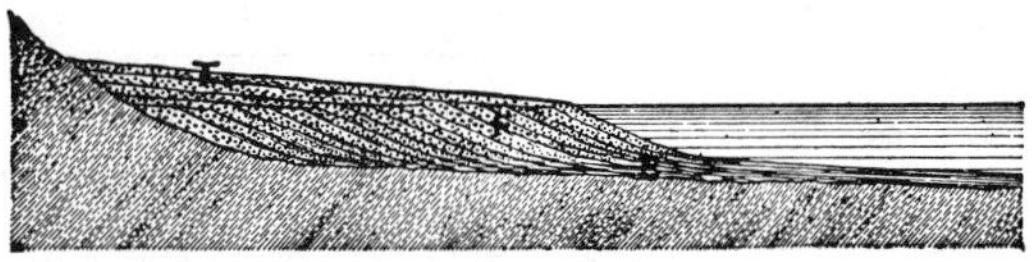

Ideal structure of a delta: (T) top-set beds; (B) bottom-set beds; (F) fore-set beds.

DELTA CONNECTION. The Delta connection is one of the two most frequently used ways of connecting a three-phase alternating-current circuit. The other is the Y connection. A three-phase machine has three coils. These coils have six ends which must, in some way, be connected to the three wires of a three-phase circuit. The Delta connection, as illustrated in the accompanying figure, has the coils connected at three points corresponding to the three-phase circuit. When this is compared with the Y connection, it will become apparent that the line voltage in Delta connection equals the coil voltage, and that the line current in Y connections equals coil current. For a balanced Delta load, the current in each line is $\sqrt{3}$ times the phase (or coil) current. It is also the π configuration of a two-part network.

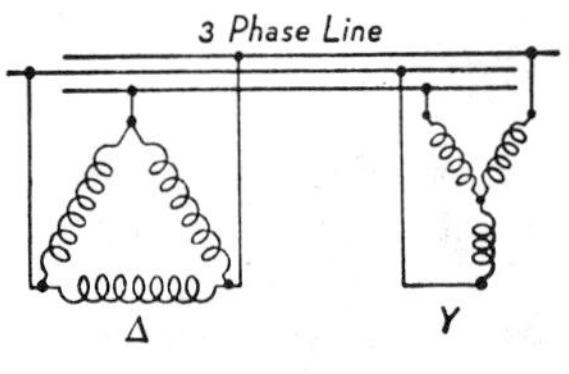

Comparison of delta and Y-connections.

DELTA FUNCTION. Commonly written as $\delta(x)$, it lies outside the usual scope of function theory. It can be defined by the relation

$$\int_a^b \phi(x)\, \delta(x - x_0)\, dx = \phi(x_0)$$

where $\phi(x)$ is continuous and $a \le x_0 \le b$. It can thus be thought of as a function which is zero for every value of x except the origin, where it is infinite in such a way that

$$\int_a^b \delta(x)\, dx = 1, \; a < 0 < b$$

The delta function has played an important role in mathematical physics and applied mathematics since it can be used to represent idealizations such as point charge, or point load, which lead to Green's functions of the differential equations representing various field theories. See also **Dirac Delta Function.**

DELTA-MATCHING TRANSFORMER. A matching network used between half-wave antennas and two-wire transmission lines. The antenna is not cut, as its center and the transmission lines are fanned out before connection to it. The resultant delta-shaped pattern gives the network its name.

DELTA RAY. An electron that is ejected by recoil when a rapidly moving charged particle, e.g., an alpha particle, passes through matter. The use of the term delta ray is sometimes extended to apply not only to electrons, but to any secondary ionizing particles ejected by recoil when a primary ionizing particle passes through matter.

DELTA WAVES (Sleep). **Sleep.**

DEMAGNETIZING FIELD. A body of magnetic material subject to an applied magnetizing force H_a is acted upon by a net magnetizing force.

$$H = H_a - \Delta H$$

where the demagnetizing field ΔH is interpreted as being due to the poles induced on the surface of the body. In the case of a permanent magnet with $H_a = 0$, the demagnetizing field is readily seen to be the field of the magnet itself, which always has such a direction as to oppose the magnetization. The existence of a demagnetizing field is a necessary consequence of the fact that energy is stored in the field external to the magnetized body, and that the sum of this energy and the internal energy must be minimized at equilibrium.

DEMAL SOLUTION. A solution which contains one gram-equivalent of solute per cubic decimeter of solution. It is slightly weaker than a normal solution, in the ratio of the magnitude of the liter to the cubic decimeter.

DEMANTOID. **Garnet.**

DEMEROL. **Pyridine and Derivatives.**

DEMINERALIZATION. **Ion Exchange Resins.**

DEMINERALIZATION (Boiler Water). **Feedwater (Boiler).**

DEMODULATOR. The process by which information is derived from a modulated waveform about the signal imparted to the waveform in modulation is termed *demodulation*—and the devices used to accomplish this function are called *demodulators.*

The *rectifier demodulator,* a device consisting of a diode or diodes through which the amplitude-modulated carrier is passed. The resulting rectified output has an average value proportional to the original modulation. With conditions approximating a perfect rectifier, the linearity of the device is quite good below 10% modulation.

The *envelope demodulator,* a rectifier demodulator whose output is shunted by a capacitance, thus causing the output to be proportional to the peaks of the rectified amplitude-modulated carrier. If low distortion is desired, this process may be used only where the ratio of carrier to the highest modulation frequency is quite large.

The *frequency demodulator,* a device which will produce an output proportional to the variation of the instantaneous frequency of the input voltage. Ideally it will be insensitive to variations in the amplitude of the input wave.

The *product demodulator,* a device whose output is the product of its two inputs, these being the amplitude-modulated carrier and a locally-generated voltage of carrier frequency. This is basically the same device as a product modulator, and with proper filtering, can produce an output proportional to the original modulation.

The *square law demodulator*, a device whose output voltage is proportional to the square of its input voltage. An amplitude-modulated carrier passing through such a device produces an output containing the original modulation signal as well as distortion products. The distortion products increase rapidly as the percent modulation increases.

DE MOIVRE-LAPLACE THEOREM. A theorem which essentially states that the binomial distribution tends to the normal distribution as n, the sample size, tends to infinity. More precisely, if p is the probability of a "success," and n is the number of trials, the distribution of x successes tends to the normal form with mean np and variance $np(1 - p)$.

DE MOIVRE THEOREM. Any power of a complex number in polar form is given by

$$[r(\cos\theta + i\sin\theta)]^n = r^n(\cos n\theta + i\sin n\theta)$$

This formula holds when n is a positive or a negative integer. It also holds when n is fractional, but may then be written in the more general form:

$$[r(\cos\theta + i\sin\theta)]^{-n} = r^{-n}\left[\cos\frac{\theta + k\cdot 360°}{n} + i\sin\frac{\theta + k\cdot 360°}{n}\right]$$

where k takes the values 0, 1, 2, . . . $(n - 1)$ and where r^{-n} denotes the principal nth root of r. The theorem thus gives all of the nth roots of any number.

When $r = 1$, $n = 1$, the result is known as the Euler theorem on the exponential function (see **Exponent**); hence, $(\cos\theta + i\sin\theta) = e^{i\theta}$.

DEMOSPONGIAE. A class of sponges (*Porifera*) of complex structure, including the sponges of commerce and the freshwater sponges.

The members of this class have siliceous spicules which are never 6-rayed, a spongin skeleton, or a combination of spongin and siliceous matter. Some species have no skeleton. The body plan is of the rhagon type.

These orders are recognized:

Order *Myxospongida.* Simple sponges without skeletal structures.
Order *Monaxonida.* Monaxonid spicules only present.
Order *Tetraxonida.* Skeleton siliceous, sometimes with spongin. Freshwater sponges are included in this order with many marine forms.
Order *Keratosa.* Skeleton of spongin fibers. Spicules absent. Commercial sponges belong here.

DENDRIDE (Neuron). Brain and Nervous System.

DENDRITE. A tree-like crystal formed during solidification of metals or alloys. Dendrites generally grow inward from the surface of the mold, extending branches from a central trunk in a manner resembling a fir tree. In alloys, the central portions of a dendritic crystal are richer in higher melting point constituents, while the outer portions consist of lower melting point material which is last to solidify. This form of segregation can usually be eliminated by diffusion during subsequent mechanical working and heat treatment.

The reasons for the branched growth of a crystal into a liquid the temperature of which falls ahead of the solid are easily understood. Whenever a small section of the interface is ahead of the surrounding surface, it will be in contact with liquid metal at a lower temperature. Its growth velocity will be increased relative to the surrounding surface which is in contact with liquid at a higher temperature, and the formation of a spike is only to be expected. Associated with the formation of each spike is the release of a quantity of heat (latent heat of fusion). This heat raises the temperature of the liquid adjacent to any given spike and retards the formation of other similar projections on the general interface in its immediate vicinity. The net result is that a number of spikes of almost equal spacing are formed which grow parallel to each other in the fashion shown in Fig. 1. The *dendritic growth direction* depends upon the crystal structure of the metal.

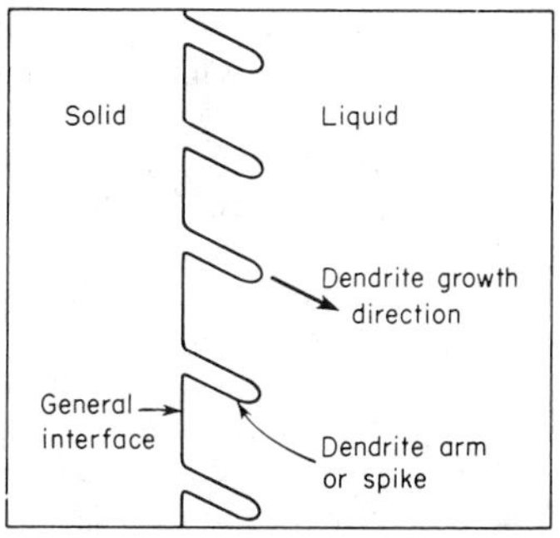

Fig. 1. Schematic representation of the first stage of dendritic growth. A temperature inversion is assumed to exist at the interface i.e., the temperature in the liquid drops in advance of the interface.

The branches, or spikes, in Fig. 1 are primary in nature. Once they have formed growth at the general interface will be slow because here the supercooling is small and the latent heat of fusion associated with the formation of the spikes tends to further decrease its magnitude. At section *bb* in Fig. 2 on the other hand, the average temperature of the liquid is, by definition, lower than at *aa*. However, even on this section at points in the liquid close to the spikes the temperature will be higher than midway between the spikes ($T_A > T_B$) because of the latent heat released at the spikes. There is, therefore, a decreasing temperature gradient not only in front of the primary spikes, but also in directions perpendicular to the primary branches. This temperature gradient is responsible for the formation of secondary branches which form at more or less regular intervals along the primary branches, as shown in Fig. 3.

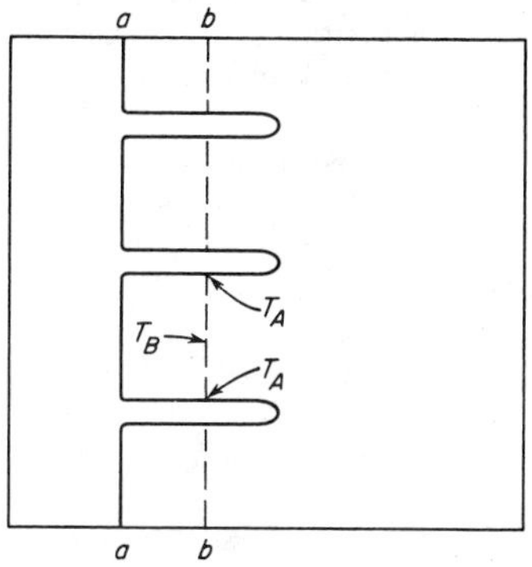

Fig. 2. Secondary dendrite arms form because there is a falling temperature gradient starting at a point close to a primary arm and moving to a point midway between the primary arms.

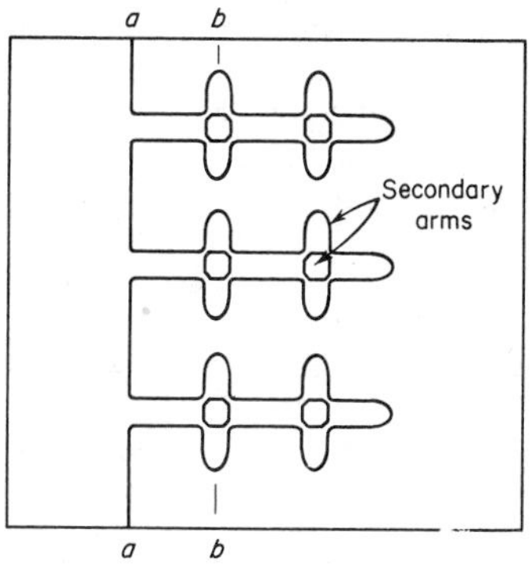

Fig. 3. In a cubic crystal, primary and secondary arms are normal to each other.

DENGUE. A viral infectious disease that occurs in many areas of the world with a warm climate, particularly in Asia and the Pacific region, in West Africa, and in the Caribbean. Major outbreaks have occurred in Jamaica. Occurrence is most common in populated urban areas where humans serve as the primary reservoirs of the virus. The same species of mosquito (*Aedes aegypti*) that transmits yellow fever virus also transmits the dengue virus (Group B arbovirus). These two viruses are practically indistinguishable. In some jungle areas, such as found in Malaysia, it is believed that monkeys also serve as reservoirs for the dengue virus. It is interesting to note that although the same mosquito vector is involved, dengue is common in certain regions where yellow fever is unknown.

There are four serotypes of dengue virus: *Type 1* (commonly encountered in West Africa); *Type 2* and *3* (most commonly found in the Caribbean); and *Type 4* (found in all areas, including Asia and the Pacific region). However, epidemics in Jamaica have mainly been of *Type 1*. Epidemics seem to follow a period of heavy rains, during which time the mosquito breeds vigorously.

Dengue actually takes two forms: (1) *Dengue fever*, the most widely occurring form, usually mild in children, but more serious in adults, although it is rarely fatal without the presence of serious complications. Dengue fever is accompanied by generalized fever, lymphadenopathy, and myalgia. A rash may or may not be present. The incubation period is a few days. Symptoms usually persist for a few days, but in some cases, the recovery period may be much longer. Nonaspirin-type analgesics, such as codeine, can be effective in management of the fever. (2) *Dengue hemorrhagic fever* occurs principally in Southeast Asia and is more serious than dengue fever. The liver is involved, there is hemorrhagic nephritis and often shock. Much greater supportive measures are required than with dengue fever. There must be careful management of fluids, colloids, and electrolytes. Improperly managed, the disease may have a mortality of up to 10%.

Further antimosquito measures are required to reduce the incidence of dengue viral infections.

DENIER. Acetate Fibers.

DENSE. A set *T* of a space *S* is dense in *S* if every point of *S* is a point or limit point of *T*, or in other words, if the closure of *T* is *S*.

DENSE-MEDIA SEPARATION. Classifying (Process); Coal.

DENSIMETER. Any instrument used to determine density.

DENSITOMETER. Photometers.

DENSITY. The density of a substance is its mass per unit volume, usually expressed in grams per cubic centimeter. The specific gravity of the substance is the ratio of its density to that of water, usually at 4°C, or 20°C, or 60°F, in the same units, and is therefore an abstract number independent of units. The density of a body is the ratio of its mass to its volume.

To determine the density of a given substance, it is necessary only to ascertain the volume of a specimen whose mass is known by weighing. This may be obtained from measurements on the dimensions of the specimen, or, in the case of a liquid, by the use of a pycnometer or specific gravity bottle. For solids a more precise method is to measure the buoyant force, upon the specimen, of a liquid of known density in which it is immersed, or by enclosing it in a specific gravity bottle and determining the volume by displacement. The Mohr-Westphal balance is especially designed to give densities of liquids by the buoyant force on a solid sinker of known volume. The hydrometer may also be used for quick determinations of liquid densities. The density of a gas is best obtained by weighing a specimen of it in a large, light bulb of known capacity, concurrently observing the temperature and pressure to which the gas is subjected, much as the pycnometer is used for liquids.

A brief table of densities, all in grams per cubic centimeter, is appended. For gases the densities are at standard temperature and pressure.

Substance	Density	Substance	Density
Air	0.001293	Gold	19.3
Alcohol	0.794	Hydrogen	0.0000899
Aluminum	2.70	Iron	7.86
Carbon dioxide	0.001977	Lead	11.3
Chlorine	0.003214	Mercury	13.55
Copper	8.90	Nitrogen	0.001251
Cork	0.24	Oxygen	0.001429
Gasoline	0.67	Platinum	21.45
Glass	2.4–2.8	Silver	10.5
Glycerine	1.27	Water (4°C)	0.999973

The term density is also applied in length-force-time systems of units, to the weight per unit volume. Other uses of the term density are the blackness of the image on a photographic plate or film and to the ratio of the number of particles or total amount of such a quantity as energy or momentum, carried by or contained in a volume, to that volume. Thus one speaks of energy density, electron density, charge density, etc.

This last usage has given rise to a number of specific applications of the term density. Thus, the *luminous density* is the luminous energy found in a unit volume of space. The *specular density* is the logarithm of the reciprocal of the specular transmittance, and so on. See also **Rectilinear Diameters Law.**

DENSITY (Chemical Elements). Chemical Elements.

DENSITY CURRENT. Ocean.

DENSITY (Data Processing). The number of bits that can be stored per unit of recording medium or the number of integrated circuits fabricated on a single silicon chip or packaged in a single module. With magnetic tape, the density is expressed in terms of the number of bits per inch of track. This also applies to disks, drums, and other mobile magnetic media. In semiconductor storage, density usually refers to the number of bits of storage on a single chip or in a single module. Logic circuit density is normally expressed in terms of circuits per chip.

DENSITY MEASUREMENT. Radioactivity; Specific Gravity.

DENSITY (Rectilinear Diameters). Rectilinear Diameters Law.

DENTAL AMALGAMS. Mercury.

DENTAL CARIES. Caries and Cariology; Fluorine.

DENTINE. Tooth.

DENTITION. The form and arrangement of the teeth in vertebrates. Teeth are so intimately related to the food that they are involved in the fundamental adaptations of the animal. In connection with the study of the many adaptations of teeth, terms have been coined which apply in some cases either to the tooth itself or to the entire dentition, while others apply to the dentition in general.

The primitive form of tooth is apparently that of the sharks, which has a principal sharp flattened point and in some cases fairly prominent lateral points. These teeth are arranged in several rows and are renewed as needed. In other fishes and in the amphibians teeth are also developed in large numbers and in some forms occur elsewhere in the mouth than on the jaws. They are named for the part of the skull with which they are associated, as the vomerine teeth.

In the reptiles and mammals the simpler condition of a row of teeth along each jaw prevails. The teeth may be entirely conical, as in the reptiles, or of various types, as in the mammals, and may be indefinitely renewable or limited to one or two sets. Where teeth are of more than one kind the dentition is said to be heterodont. The forms of teeth include the sharply conical canines, the sharp-edged cutting incisors, and the broad grinders, which include premolars and molars. Renewal is unlimited in the reptiles but in the more highly specialized mammals only one or two sets appear normally. Monophyodont dentition consists of one set and diphyodont includes a temporary set of milk teeth which is replaced by a set of permanent teeth, as in humans.

The numbers of teeth of different kinds are expressed in a dental formula as a distinctive characteristic of mammals. In this formula the teeth of one-half of each jaw are listed in this order: incisors, canines, premolars and molars, and those of the upper jaw are placed above those of the lower jaw. Thus the dental formula of man is 2123/2123 and that of the woodchuck is 1023/1013. The zeros in the latter formula indicate the absence of canines.

The position of the teeth in the jaw is also sometimes indicated by a special term. When placed along the edge of the jawbone the dentition is said to be acrodont, and when placed along the inner margin, pleurodont.

DEODAR CEDAR. Cedar Trees.

DEODORIZING (Edible Oils). Vegetable Oils (Edible).

DEOXIDATION (Steel). Calcium.

DEOXIDIZING AGENT. A compound that has an affinity for oxygen—hence, chemically removes oxygen from many substances. In essence, a deoxidizing agent plays the role that is reverse that of an oxidizing agent. Thus, a deoxidizing agent is a reducing agent. Of course, at one time, oxidation meant simply a combination with oxygen and reduction meant a loss of oxygen. In their broader interpretations, oxidation now refers to the loss of one or more electrons from the outershell of an atom, and the reverse for reduction. Although the broader interpretation also could apply to deoxidation, the term still is interpreted generally as removal of oxygen.

Oxygen frequently is an impurity in various metallurgical processes, particularly melting and refining processes. Deoxidizing agents are commonly used to reduce or remove oxygen from molten metals. Lithium metal, for example, will preferably absorb oxygen (by combination) from copper and copper alloys. Boron carbide also is used as a deoxidizing agent for casting copper. Silicon usually in the form of ferrosilicon, and manganese in the form of ferromanganeses are widely used in the production of steel and iron alloys. Aluminum, titanium, zirconium, and vanadium also play a deoxidizing part in the production of iron alloys. Magnesium is also a powerful deoxidizer and desulfurizer and is used in the production of such metals as beryllium, hafnium, titanium, uranium, yttrium, and zirconium. Zinc is used as a deoxidant in the refinement of silver and gold.

Normally, the process of rock formation is one of oxidation. However, a greenish or yellowish area in a red rock may be developed by reduction (deoxidation) of ferric oxide in the presence of organic materials.

DEOXYRIBONUCLEIC ACID (DNA). A complex sugar–protein polymer of nucleoprotein which contains the genetic code for enzymes in the cell. It occurs as a major component of the genes, which are located on the chromosomes in the cell nucleus. The DNA molecule is a unique and vastly intricate structure; it is comprised of from 3000 to several million nucleotide units arranged in a double helix containing phosphoric acid, 2-deoxyribose, and the nitrogenous bases adenine, guanine, cytosine, and thymine. The spiral consists of two chains of alternating phosphate and deoxyribose units in continuous linkages. The nitrogenous bases project toward the axis of the spiral and are joined to the chains by hydrogen bonds. Adenine always unites with thymine and cytosine with guanine. The complementarity of the bases on the joined chains allows each chain to act as a template for replication of the other when the chains are separated, thus producing two new strands of DNA. The sequence of the bases on the chains varies with the individual, and it is this sequence that governs the genetic code. DNA works in conjunction with ribonucleic acid (RNA).

The foregoing is a highly generalized definition. Within the last few years, considerable new knowledge has been gained concerning DNA and its genetic function. The early studies of genes concentrated on bacterial genes. In these, the bacterial genes are not spread out in pieces. More recent studies in the mid- and late 1970s concentrated on animal viruses, animals, and humans. The genes in these cases, with the possible exception of the histone genes, are found in pieces that are spread out along DNA. Thus, between gene fragments, there are long stretches of DNA, the functions of which are poorly understood as of the present. This discovery of fragmented or spaced out genes in animals raised the question as to whether or not such structures are exceptional or the rule. Subsequent research has indicated that they are the norm.

The discovery, while creating several new and fundamental questions, has provided at least partial answers to some former questions. For example, it has been known for some years that there are large quantities of DNA in the cells of higher organisms, that is, DNA in an amount far in excess of the DNA required if the genes were not in pieces. The spacing out of the genes into fragments accounts for all or part of the excess DNA previously noted.

As of the early 1980s, numerous hypotheses have been formulated by molecular biologists and this fundamental discovery has stimulated a whole new line of research in many laboratories throughout the world. Some scientists have observed that the extra DNA cannot be accounted for simply upon the basis of evolutionary theories. The extra DNA may play a role in controlling gene expression. The complexity and current uncertainty of these hypotheses are beyond the scope of this book at this juncture in the research program. Perhaps the topic will be better clarified at the time of the next edition. Several of the references listed shed further insights.

Studies of DNA in a human cell have suggested that the intricate DNA structure may be as much as 1.2 meters (4 feet) long—and yet it is contained within a nucleus which is less than 0.025 millimeters (1/1000 inch) in diameter. That the DNA structure is in the form of a complex tangle under these conditions is not difficult to imagine. Upon division of the cell, the DNA condenses into strands in duplicate copies, one set for each daughter cell. It follows that the copying must take place while the DNA is in the form of a tangle, but in such a manner to allow easy separation of the duplicate copies. The mechanics and time required for the copying process still remain discrete. As early as 1974, Berezney and Coffey (Johns Hopkins University School of Medicine) estimated from electron micrographs that about 5% of the protein in a nucleus appears to be a rigid skeleton or scaffolding, a structure they termed the *nuclear matrix.* These researchers have proposed that the DNA in the nucleus is attached onto the nuclear matrix at thousands of sites, with the genetic material arrayed in thousands of loops. The loops pass through enzyme complexes located at the matrix sites and in so doing are copied. In the process, the loops are preserved and remain attached to the matrix. It is further postulated that when a cell divides, duplicate copies of loops already joined to a scaffolding are drawn apart to the daughter cells. It has been estimated that genetic material in rat liver cells consists of from 10,000 to 15,000 loops. Thus, the nuclear matrix would have a corresponding number of enzyme complexes.

See also **Genes; Nucleic Acids and Nucleoproteins;** and **Recombinant DNA.**

References

Bauer, W. R., Crick, F. H. C., and J. H. White: "Supercoiled DNA," *Sci. Amer.*, **243**, 118–133 (1980).

Brown, P. O., and N. R. Cozzarelli: "A Sign Inversion Mechanism for Enzymatic Supercoiling of DNA," *Science*, **206**, 1081–1083 (1979).

Cozzarelli, N. R.: "DNA Gyrase and the Supercoiling of DNA," *Science*, **207**, 953–960 (1980).

Griffith, J. D.: "DNA Structure: Evidence from Electron Microscopy," *Science*, **201**, 525–527 (1978).

Kolata, G. B.: "DNA Sequencing: A New Era in Molecular Biology," *Science*, **192**, 645–647 (1976).

Kolata, G. B.: "Genes in Pieces," *Science*, **207**, 392–393 (1980).

Kolber, A. R., and M. Kohiyama (editors): "Mechanism and Regulation of DNA Replication," Plenum, New York, 1977.

Lehman, I. R., and D. G. Uyemura: "DNA Polymerase I: Essential Replication Enzyme," *Science*, **193**, 963–969 (1976).

Maugh, T. H., II: "Phylogeny: Are Methanogens a Third Class of Life?" *Science*, **198**, 812 (1977).

Mitchell, R. M., et al.: "DNA Organization of *Methanobacterium thermoautotrophicum*," *Science*, **204**, 1083–1084 (1979).

Portugal, F. H., and J. S. Cohen: "A Century of DNA," MIT Press, Cambridge, Massachusetts, 1978.

Razin, A., and A. D. Riggs: "DNA Methylation and Gene Function," *Science*, **210**, 604–610 (1980).

Smith, G. P.: "Evolution of Repeated DNA Sequences by Unequal Crossover," *Science*, **191**, 528–535 (1976).

Sparrow, A. H., and A. F. Nauman: "Evolution of Genome Size by DNA Doublings," *Science*, **192**, 524–529 (1976).

Staff: "Undogmatic Toad," in "Science and the Citizen," page 66–69, *Sci. Amer.*, **242**, 4 (1980).

Temin, H. M.: "The DNA Provirus Hypothesis," *Science*, **192**, 1075–1080 (1976).

Timberlake, W. E.: "Low Repetitive DNA Content in *Aspergillus nidulans*," *Science*, **202**, 973–975 (1978).

Yang, R. C. A., and R. Wu: "BK Virus DNA: Complete Nucleotide Sequence of a Human Tumor Virus," *Science*, **206**, 456–459 (1979).

DEOXYRIBOSE. Nucleic Acids and Nucleoproteins.

DEPARTURE (From Nucleate Boiling; DNB). Boiler; Boiling.

DEPARTURE (Navigation). Any course and distance covered by a ship may be resolved into two components at right angles to each other. One of these components will be parallel to a meridian; the other will be along a parallel of latitude, and is known by the term departure. If the lengths of these components are expressed in nautical miles, the component along the meridian may be immediately converted into difference of latitude, because a minute of arc of latitude is practically equal to one nautical mile. The component along the parallel of latitude, expressed in nautical miles and referred to as easting or westing, is known as the departure.

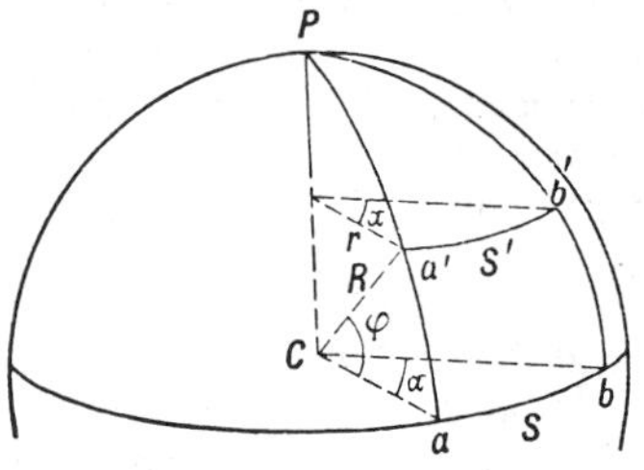

Analysis of components in converting departure into difference of longitude.

For the solution of various problems in dead reckoning and in the sailings, it becomes necessary to convert departure into difference of longitude. The accompanying figure illustrates the problem. Here, we have S', a departure measured along the parallel of latitude ϕ between the meridians $Pa'a$ and $Pb'b$. The arc, S, of the equator represents the difference of longitude corresponding to the departure S'. Planes are passed through S' and S perpendicular to the axis of the earth, PC, and radii are drawn in these planes to include the angles subtended by S' and S. Call R the radius of the earth and r the radius of the arc S'. The angles subtended by S and S' must be equal, and we have at once $S/R = S'/r$. By the definition of latitude (assuming the earth as spherical), we have the plane angle $aCa' = \phi$ and, from the definition of the trigonometric functions, $R/r = \sec \phi$. Therefore, we have $S/S' = R/r = \sec \phi$, or $S = S' \sec \phi$. Since S' is expressed in nautical miles, which are practically equivalent to a minute of arc on the equator, we have at once $S = S' \sec \phi$ as the difference in longitude corresponding to the departure S', the difference in longitude being thus expressed in minutes of arc.

See also **Course; Dead Reckoning;** and **Navigation.**

DEPHLEGMATION. Distillation.

DEPOLARIZATION. This term designates any process of removing polarization. Representative examples include the following:

1. Depolarization of an electric cell, most commonly a dry cell, is commonly effected by a substance added to the cell which prevents the accumulation of reaction products that interfere with the function of the cell. A related purpose may be accomplished by a diaphragm placed in the cell so as to prevent mixing of the reaction products. Such added substances and diaphragms are often called depolarizers, the latter sometimes being called an electrical depolarizer.

2. An optical depolarizer is a device for the resolution of polarized light, to which process the term depolarization may be applied.

3. Electric depolarization or demagnetization may be effected by a depolarization field. When an electric or magnetic field is applied to a macroscopic specimen, the field acting on a given atom contains a contribution due to the charges or poles induced on the surface of the specimen. This field opposes the external applied field, and hence tends to reduce the polarization of the material. See **Demagnetizing Field.**

4. Biological depolarization is a decrease in the membrane potential (i.e., the potential across the cell membrane.)

DEPOLARIZATION (Battery). Battery.

DEPOSIT (Ore). Ore.

DEPRESSANTS (Flotation). Classifying (Process).

DEPRESSION (Meteorology). Atmosphere (Earth).

DEPRESSION (Psychiatric). This is the most common psychiatric disorder and estimates indicate that 2–4% of the general population is affected to some degree by this disorder. As of the early 1980s, the disease ranked first among the need to hospitalize patients for psychiatric reasons. However, it is estimated that only about 8% of the total cases of depression are treated in a hospital environment, a large number of cases not being treated at all. British authorities have stated that among all medical patients, about 25% have symptoms of mild depression. Numerous classifications of depression have been proposed, most of which are of little or no value in establishing a diagnosis. Particularly helpful, however, is the distinction between primary and secondary affective illness, the key to which is the time scale of the depressive event. Where this event occurs before other psychiatric or serious medical diagnosis, the disorder is considered *primary.* The episode of depression is classified as *secondary* when there have been preexisting psychiatric diagnoses, including alcoholism and schizophrenia. Another useful classification is the distinction between a bipolar (also includes episodes of mania) and a unipolar (depression only) condition. Traditionally, all patients who present mania only are still classified as bipolar.

Scientists at the Washington University have developed eight criteria of depression: (1) Loss of appetite and/or weight; (2) disturbance of sleep habits, such as insomnia or hypersomnia; (3) general fatigue, energy loss, and so-called tired feeling; (4) agitated or retarded emotional responses; (5) general loss of interest in otherwise stimulating situations, such as hobbies, social activities, job—often combined with a decrease of sexual drive; (6) a lessened ability to concentrate and think through problems—so-called mixed-up thinking; (7) manifestations of guilt and self-deprecation; and (8) repetitive concerns with death and suicide, including the so-called "Wish I were dead." In analyzing these criteria, the physician or psychiatrist will consider the patient depressed if 5 of the 7 aforementioned conditions are present. The presence of 4 symptoms is considered borderline and requires further patient history and probing. Normally, the physician will not establish a definite diagnosis until the pattern of symptoms has persisted for at least one month. The examiner must be clever in interviewing patients because commonly an individual will deny feeling any depression. Medical practioners and psychiatrists must be sensitive to physical conditions that also may cause some of the symptoms and are quite unrelated to depression. For example, a number of drugs have been associated with depression. These include ACTH, amphetamines, barbiturates, butyrophenones, chlordiazepoxide, diazepam, L-Dopa, methyldopa, oral contraceptives, propanolol, reserpine, spironolactone, and steroids. However, with exception of reserpine and methyldopa, the aforementioned drugs will not produce a syndrome of true depression. It is known, for example, that chloridiazepoxide can produce uncontrolled weeping, but this is not indicative of depression by itself.

Research has indicated that depression results from an alteration of neurochemistry and, in particular, a functional deficiency of two central neurotransmitters—the monoamines (norepinephrine and serotonin). It is hypothesized that these transmitters may be destroyed by various enzymes, notably, monoamine oxidase. However, straightforward explanations at the neuron and even molecular level do not satisfactorily explain the many variations of depression.

Most physicians first attempt to control depression by the use of tricyclic antidepressant drugs. These include amitriptyline, desipramine, doxepin, imipramine, nortriptyline, and protriptyline. Generally these antidepressants are metabolized slowly, with an effective half-life ranging from about 16 hours to over 5 days. It has been found that these drugs possess a "therapeutic window," i.e., they may be ineffective when administered at levels below or above certain dosages. To determine the most satisfactory dosage for a given patient, the drug will be administered in progressively larger doses, considering various side effects. It is estimated that about two-thirds of individuals with depression will respond favorably to the tricyclic drugs. The other one-third of patients may require the use of MAO (monoamine oxidase) inhibitors, such as isocarboxazid, phenelzine, and tranylcypromine. Because MAO inhibitors potentiate the actions of several other drugs, physicians will advise their patients to alert dentists, for example, that they are under MAO therapy. These pharmacologic agents include opiates (meperidine), anesthetics, ganglionic blocking agents, diuretics, and sympathomimetic amines, among others. Certain foods containing tyramine and other aromatic amines also must be avoided. These include fermented cheese, pickled herring, sardines, anchovies, chicken livers, canned or processed meats, pods of broad beans, canned figs, yeast extract, red wine, and beer, among others. Coffee intake, as well as chocolate, sour cream, yogurt, and cottage cheese intake also must be limited. The physician will advise the patient to avoid nose drops, cold remedies, cough syrups, diet pills, and stimulant drugs because of their content of aromatic amines.

Where drug therapy does not succeed, psychiatric counsel is indicated. Electroconvulsive therapy is the most effective somatic treatment of depression and is reserved for unresponsive cases.

The treating physician or psychiatrist will be aware that a number of depressed patients are suicidal. In some cases, the patient will admit to suicidal tendencies when interviewed. In treating such patients, the professional will emphasize self-esteem and self-respect in the patient because self-deprecation is a good indicator of suicidal tendencies.

Through interviewing, the examiner will distinguish true depression from grief and bereavement. Grieving is considered necessary to emotional recovery from personal loss. The examiner will have to sort out the symptoms of grief from those of depression. Somatic symptoms, such as sighing respiration, exhaustion, gastrointestinal symptoms, restlessness, yawning, and choking, often accompanied by feelings of guilt, are indicative of grief and bereavement. Behavior of the bereaved person may include some rather bizarre behavior, such as preoccupation with the image of the deceased, talking with the deceased, a sense of the presence of the deceased, among others. Combined feelings of hostility to some persons and friendliness to others also may be signs of grief. In the normal individual, the period of bereavement may persist up to 3 months. Authorities have suggested three periods of bereavement: (1) searching and protest, (2) disorganization, and (3) reorganization. The second period is dangerous in some persons because suicidal tendencies may develop.

References

Beck, A. T., Kovacs, M., and A. Weissman: "Hoplessness and Suicidal Behavior: An Overview," *JAMA*, **234**, 1144 (1975).

Burrows, G. D., et al.: "Cardiac Effects of Different Tricyclic Antidepressant Drugs," *Br. J. Psychiatry*, **129**, 335 (1976).

Feighner, J. P., et al.: "Diagnostic Criteria for Use in Psychiatric Research," *Arch. Gen. Psychiatry*, **26**, 57 (1972).

Grollman, E. A.: "Talking about Death," 2nd edition, Beacon Press, Boston, 1976.

Lesse, S. (editor): "Masked Depression," Jason Aronson, Inc., New York, 1974.

Moffic, H. S., and E. S. Paykel: "Depression in Medical Inpatients," *Br. J. Psychiatry*, **126**, 346 (1975).

Murphy, G. E.: "The Physician's Responsibility for Suicide," *Ann. Intern. Med.*, **82**, 301 and 305 (1975).

Schneidman, E. S.: "Suicide," in "Comprehensive Textbook of Psychiatry, II," 2nd edition (A. M. Freedman, H. I. Kaplan, and B. J. Sadock, editors), Williams & Wilkins, Baltimore, 1975.

Synder, S. H., and H. I. Yamamura: "Antidepressants and the Muscarinic Acetylcholine Receptor," *Arch. Gen. Psychiatry*, **34**, 236 (1977).

Spiker, D. G., and D. D. Pugh: "Combining Tricyclic and Monoamine Oxidase Inhibitor Antidepressants," *Arch. Gen. Psychiatry*, **33**, 828 (1976).

DEPTH MEASUREMENT (Underwater). Sounding.

DEPTH OF FIELD. In photography, the term depth of field refers to the distance over which satisfactory definition is obtained when the lens is in focus for a certain distance. If, for example, a lens is in focus for an object at a distance of 25 feet and the definition is satisfactory on objects from 20–40 feet, the depth of field extends from 20–40 feet. Depth of field is frequently but incorrectly termed depth of focus, which is the range of image distances corresponding to the range of object distances covered by the depth of field.

The depth of field depends upon: (1) the standard adopted for "satisfactory" definition; (2) the distance of the plane on which the lens is in focus; (3) the focal length of the lens; (4) the relative aperture

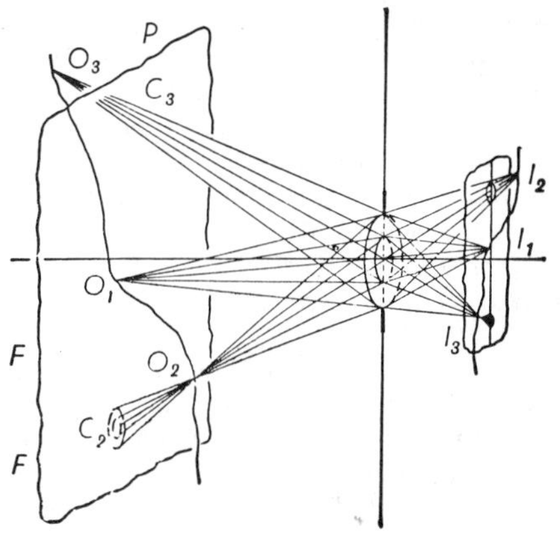

Demonstration of depth of field.

(*f*/number). So far as the last three are concerned, we may note (1) that the depth of field increases with the distance of the plane on which the lens is in focus and (2) that it becomes less as the focal length or (3) as the aperture increases.

Let O_3, O_2, O_1 represent an object, parts of which are at different distances from the lens L. Suppose the lens to be focused on O_1, a point image I_1 will be formed on the focusing screen, or sensitive plate. With the focusing screen, or the sensitive plate, at this point it is clear that the image of O_2 is formed at I_2 *behind* the screen and that of O_3 at I_3 in *front* of the screen. In other words, the position of the image point varies with the distance of the object point and the lens cannot produce a point for point image upon a plane surface, such as the sensitive film, unless the subject itself is a plane. When this is not the case, the image of points nearer or further from the lens is a circular disk rather than a point.

Any disk, however, will appear to the eye as a point if the viewing distance is sufficiently great. At a distance of 10 inches, for example, a circle with a diameter of $\frac{1}{100}$ inch appears as a point to the average eye. On an angular basis this corresponds to about 2 minutes of arc.

Since any unsharpness in any part of the negative is increased when an enlargement is made, the disk, or circle of confusion, in an enlargement is equal to the diameter of the disk in the negative multiplied by the degree (times) of enlargement. Thus, if the largest circle allowable for sharp definition is assumed to be $\frac{1}{100}$ inch and the print is a $5 \times$ enlargement, the maximum diameter of the circle of confusion in the negative is $\frac{1}{100} \times \frac{1}{5}$ or $\frac{1}{500}$ inch and the depth of field should be calculated accordingly.

If it is assumed that the viewing distance is the distance at which the proper perspective is obtained, i.e., a distance equal to the focal length of the lens, then the circle of confusion can be expressed as a fraction of the focal length of the lens. The value generally used is 0.00058 or $\frac{1}{1720}$ of the focal length. If

u = distance focused on,
θ = angular size of circle of confusion ($\frac{1}{1720}$ of the focal length),
τ = effective diameter of lens or the focal length divided by the *f*/number.

then the nearest point sharply defined in front of the plane in focus is

$$\frac{u^2 \tan \theta}{\tau + u \tan \theta}$$

and the greatest distance beyond the plane in focus is

$$\frac{u^2 \tan \theta}{\tau - u \tan \theta}$$

Tables of depth of field are included in most camera manuals and in reference books.

DERIVATIVE (Chemical). A term used in organic chemistry to express the relation between certain known or hypothetical substances and the compound formed from them by simple chemical processes in which the nucleus or skeleton of the parent substance exists. Thus, phenol, aniline, and toluene are said to be derivatives of benzene; and many of the terpenes are derivatives of cymene. Or, when a paraffin such as methane is halogenated, a series of halogen-substitution products may be formed. Depending upon conditions of the reaction, CH_4 plus Cl may yield the monosubstitution produce CH_3Cl (methyl chloride), or the disubstitution product CH_2Cl_2 (methylene dichloride), or the trisubstitution product $CHCl_3$ (chloroform), or the tetrasubstitution product CCl_4 (carbon tetrachloride). In essence, these compounds are derived from methane and thus are methane derivatives as the result of chlorination. Usually the term applies to those compounds where the resulting compound is formed in one step, although a chain of steps may be involved in some cases, depending essentially upon how easy it is to identify the "derivative" with the parent substance. Where a chain of steps is involved, the intervening compounds often are called intermediates rather than derivatives.

DERIVATIVE (Mathematics). The instantaneous rate of change of a function with respect to its independent variable. Let $y = f(x)$ be a given function of its independent variable x, and for an assigned value of x consider a small increase or increment of it, Δx. The dependent variable increases simultaneously by Δy, where $\Delta y = f(x + \Delta x) - f(x)$. The ratio of increments is

$$\frac{\Delta y}{\Delta x} = \frac{f(x+\Delta x) - f(x)}{\Delta x}$$

and, by definition, the limit of the right-hand member when $\Delta x \longrightarrow 0$ is the derivative of $f(x)$, or

$$\frac{dy}{ax} = \lim_{\Delta x \to 0} \frac{f(x+\Delta x) - f(x)}{\Delta x} = \lim_{\Delta x \to 0} \frac{\Delta y}{\Delta x}$$

If the limit exists the function is continuous; if the limit fails to exist at one or more values of x these are said to be singular points for the function.

The process of finding the derivative of a function is differentiation. The usual symbol for a derivative is dy/dx but y', $f'(x)$ and other designations are also used. If the function $y = f(x)$ is plotted in rectangular coordinates, the slope of the curve at the point $x = x_0$ is its derivative, dy/dx, at that point.

The derivative of the derivative of a function is a higher derivative or one of higher order. The derivative of the second order is written in various forms, thus: d^2y/dx^2, $D_x^2 y$, $f''(x)$, f_{xx}, y''. Similar notation is used for derivatives of third or higher order. In general, the nth derivative is the derivative of the $(n - 1)$th derivative and it could be designated as $d^n y/dx$, $D_x^n y$, $f^{(n)}(x)$, or $y^{(n)}$.

Let $u = f(x, y, z, \ldots)$ be a function of several variables. Keep all of the variables constant except one, x for example, and give x an increment Δx. If Δu is the corresponding increment in u, then the limit $\lim_{\Delta x \to 0} \Delta u/\Delta x$ is the partial derivative of u with respect to x. It is denoted by $\partial u/\partial x$, u_x, or, especially in thermodynamics, by $(\partial u/\partial x)y, z, \ldots$ where the subscripts indicate the variables held constant during the differentiation. The partial derivaties $\partial u/\partial y$, $\partial u/\partial z$, etc., are defined similarly.

In general, the partial derivatives are themselves functions of one or more of the variables. They may thus be differentiated again to obtain partial derivatives of higher order:

$$\frac{\partial^2 u}{\partial x^2} = \frac{\partial}{\partial x}\left(\frac{\partial u}{\partial x}\right); \frac{\partial^2 u}{\partial x\, \partial y} = \frac{\partial}{\partial x}\left(\frac{\partial u}{\partial y}\right)$$

$$\frac{\partial^2 u}{\partial y\, \partial x} = \frac{\partial}{\partial y}\left(\frac{\partial u}{\partial x}\right); \frac{\partial^3 u}{\partial x^3} = \frac{\partial}{\partial x}\left(\frac{\partial^2 u}{\partial x^2}\right); \text{etc.}$$

The abbreviated notation u_{xx}, u_{xxx}, etc., is often used.

If the first partial derivative of a function is continuous, $u_{xy} = u_{yx}$. In this case, if each variable is itself a function of a single independent variable t, then one can write

$$\frac{du}{dt} = \frac{\partial u}{\partial x}\frac{dx}{dt} + \frac{\partial u}{\partial y}\frac{dy}{dt} + \frac{\partial u}{\partial z}\frac{dz}{dt} + \cdots$$

which is the total derivative.

A directional derivative is one which gives not only the rate of change of a function but its direction of change. It is most conveniently discussed by the methods of vector analysis.

See also classified index under **Mathematics.**

DERIVATIVE (Rate) CONTROL. In an automatic control system, a control action in which the output is proportional to the rate of change of the input. Derivative control does not exist by itself, but always is in combination with proportional control. The term *anticipatory control* sometimes is used to describe rate or derivative control. The preferred name for this control action is *derivative control* (Process Measurement and Control Section, Scientific Apparatus Makers Association).

Derivative control action is demonstrated by

$$m = K_c\, T_d e' + M$$

where m = manipulated variable
K_c = proportional sensitivity
T_d = derivative time
e = deviation
M = constant
e' = first derivative of e with respect to time de/dt

Curves showing the effects of derivative control action combined with proportional action are given in entry on **Control Action.** See also **Proportional Plus Derivative (Rate) Control.**

DERMAPTERA (*Insecta*). The earwigs. An order of insects made up of species whose forewings, when present, are short leathery wing covers, and whose abdomen bears a pair of appendages like the jaws of a forceps at the tip. The common name is based on the supposition that they enter the ears of human beings. They are plant-eating, insect-eating, and probably in some species scavengers.

DERMATITIS AND DERMATOSIS. *Dermatitis* is an inflammation of the skin and is a term frequently used erroneously as a synonym for *dermatosis,* which means skin disease. Seemingly, there are almost unlimited numbers of disorders which may affect the skin, so that the field in medicine which is concerned with the skin and its disorders and diseases, *dermatology,* is broad and complex.

The symptoms of dermatitis are varied, and include reddening (*erythema*), small blisters, crusting, oozing of fluids, scaliness, cracking or fissuring, and other secondary changes from the normal appearance. The causes of these conditions are many, and include burns, physical irritants, infections, plant and insect poisons, strong chemicals (industrial), nutritional deficiency, disturbances of other parts of the body, and systemic diseases.

Heat, cold, chaffing, and scratching may produce a number of common forms of dermatitis. Sunburn is caused by the ultraviolet rays of the sun. These rays can be filtered out by certain sunscreening agents or by opaque chemicals tinted to skin color. Such compounds are of value for light-sensitive persons and remain effective for about 3 to 4 hours after application. Lotions and creams that produce an artificial tan, or a double tan, are popular. These materials contain dihydroxyacetone, a substance capable of reacting with elements of the skin to bring about tanning within a few hours of application. However, the resulting skin color may have an undesirable yellow overtone. *Dihydroxyacetone confers no protection against actual sunburn.*

Severe Sunburn. This is a more serious condition than once regarded. In a typical sunburn, the initial reddening is followed by the appearance of minute blisters (*vesicles*) which may grow together to form large blisters. The intense itching experienced during this period usually disappears after several days, or at about the time that the outer layer of the skin peels off. Chronic overexposure to the sun may produce more serious changes. Continuous drying of the skin may lead to a thickened, brown or white horny deposit on the outer layers of the skin (keratosis). Injury and formation of cataracts in the eyes may occur. These conditions are prevalent among sailors, ranchers, and others who are constantly exposed to the elements.

The color of the skin is governed largely by the presence of a brown to black pigment called *melanin.* This material is produced by special cells (*melanocytes*) in a complex series of biochemical reactions. The process must begin with the oxidation of tyrosine, an important amino acid, and is catalyzed by tyrosinase, an enzyme found in both human beings and lower animals. How much melanin is produced in this way depends primarily on hereditary factors; under certain conditions, it may be greater than normal, or missing entirely. In some persons, melanin is unevenly distributed, being entirely absent in patches or large areas of skin.

Exposure to sunlight stimulates greater production of melanin in the skin, resulting in a tan if distribution is even, or freckles if it is uneven. This is the mechanism for protecting sensitive skin cells, since the melanin pigment absorbs much of the harmful radiation. Biochemically, the sun's rays affect tyrosinase activity. They remove the inhibitors that normally restrain this enzyme, and hence more melanin is produced. A suntan fades away when the melanin gradually migrates with the epidermal cells toward the surface of the skin and is sloughed off. However, chemical reactions within the skin also destroy some of the pigment.

Skin with a small amount of melanin has a pink color, given to the skin by the blood in the numerous, small, superficial blood vessels (capillaries) which supply it with food and oxygen. See also **Skin.**

There is a growing school of experts who believe that overexposure of the skin to the ultraviolet rays of the sun may cause or contribute to carcinoma.

Exposure to Cold. Chilblain and frostbite are two common dermatoses occurring as the result of overexposure to cold. See also **Chilblain.**

Mechanical Irritants. Skin changes may result from scratching or picking at the skin with fingernails or other objects, or from irritation of the skin by the chafing of clothing. Bedsores on invalids are caused by such mechanical irritation, and can be prevented by proper nursing techniques. Rubbing the skin over a long period of time may cause it to assume a permanent thickened and leathery appearance. When two surfaces of the skin touch each other, such as between the thighs, and cause friction, a resulting inflammation may develop. This condition is known as *intertrigo,* and may be accompanied by cracking, oozing, burning, and itching. Such lesions frequently are complicated by infection, either by yeasts, bacteria, or both, and require attention.

Corns are hard, cone-shaped, thickened areas of the skin which usually appear on the toes as the result of friction or pressure from improperly fitting shoes or socks. The inner portion of the corn is pointed, so that external pressure forces the point of the corn into the underlying tissues with a painful effect. Calluses resemble corns, except that they cover larger areas and have no pointed central core.

Viruses. Many cases of dermatitis are caused by an infectious agent with either infects the skin or invades the body as a whole and causes symptoms of dermatitis. The source of these disorders determines whether the skin itself is the site toward which attention is directed, or whether medical care must be given to the body as a whole.

Fever blisters or cold sores (*herpes simplex*) are among the most common infections of humans. Occurring most frequently in children of the 1-to-5 age group, this condition is caused by a large virus, and the lesions usually appear as an itching group of small blisters on the lips. The base of these blisters may be reddened. At other times, the eruption may occur on the nose, face, ears, genitals, tongue, or any mucous membrane. A fluid exudes from the sores and forms a crust; eventually this flakes off. There occasionally may be a swelling of the lymph nodes in the areas near the sore, but the disease usually disappears spontaneously within a week or two. The virus is thought to be dormant in body tissues, becoming active only in the presence of "trigger mechanisms," such as upper respiratory tract infections, fever, menstrual periods, physical or emotional stress, over-exposure to sunlight, and perhaps the use of certain foods and drugs. Although the infection is seldom severe, herpes simplex of the cornea of the eye, if not properly treated, can result in impaired vision or blindness.

Herpes Zoster. Commonly referred to as *shingles,* this is another relatively common viral disease characterized by the appearance of small patches of blisters the size of a matchhead, on a red base. These appear almost always on only one side of the body. The disease occurs most often in spring and autumn, chiefly in adults. It appears suddenly, preceded by severe pain in the affected area, and sometimes fever. The pain varies greatly in intensity and although the skin symptoms usually subside a few weeks after the initial attack, the pain may last for several months. While shingles is not generally regarded as a dangerous disease, the possibility that it may affect nerves leading to the eyes or other important organs is high.

Molds and Fungi. A large number of skin diseases of varying severity are caused by molds or fungi. Two such diseases are *ringworm* and *athlete's foot.*

Ringworm (*tinea*) is caused by the genera *Microsporum, Trichophyton,* or *Epidermophyton,* all of the fungus family. The disease usually takes the form of one or several raised, round sores on the skin which seem to heal in the center while the edges continue to grow outward. Occasionally, the healed centers become reinfected, and a second ring develops and grows within the original ring. In some types of ringworm, there is no healing of the center and the lesion continues to grow. The sores of ringworm begin as small, slightly raised areas with a reddish color. As they enlarge, they become redder, and often contain one or many blistered areas. There may be a slight itching or burning sensation.

A less common type of ringworm appears most frequently in the crotch or under the arms. Called "jockey-strap itch," dhobie itch, or *tinea cruris,* it does not heal in the center, and may cover large areas of the skin. This type of ringworm lacks the circular appearance of the common disease, but often resembles butterfly wings when the sore spreads over the inner surface of both legs.

Ringworm of the scalp is common among children, but relatively rare in adults. It produces areas of partial baldness, which are usually temporary. Because children are highly susceptible to this type of ringworm, outbreaks sometimes occur in schools.

Normally, the patient with ringworm is treated with griseofulvin, an antibiotic compound which is especially effective against certain fungus infections and is derived from a species of *Penicillium* mold. The drug is usually taken orally for several weeks, but in refractory cases must be taken for 4 months or longer. A few types of ringworm do not respond to this drug.

Ringworm is a highly contagious disease. It can be spread by animals as well as by human beings. Dogs and cats that are not bathed frequently are common sources of human infection. Ringworm may be acquired by direct contact with the infection, or it may spread to other areas of the skin of a single individual. Objects handled by infected individuals also carry the fungus. Occasional sources of infection are the backs and arms of theater chairs, combs, and brushes. Fungi thrive on damp, warm skin, especially in areas such as the crotch, where perspiration is unable to evaporate readily.

Athlete's foot (*tinea pedis*) is a fungus infection. Among primitive peoples unaccustomed to wearing shoes, it is rare. Contrary to general opinion, athlete's foot does not appear to be easily transmitted from one person to another by the use of common showers or shared facilities. Individual susceptibility and foot hygiene appear to be more important. When the skin remains warm and moist for long periods, fungi of the genus *Trichophyton* find optimum conditions to invade the dead outer layer (*stratum corneum*) and begin to grow. Two major types of athlete's foot can be distinguished. In the more common forms, called *intertriginous,* a crack or fissure appears in the skin, usually at the base of the fifth toe or between the fourth and fifth toes. In most cases, there is also a visible mass of loose dead skin clinging between the toes. When this loose skin is removed, the skin beneath appears reddened and shiny. In the second type, *squamous-hyperkeratotic,* the disease commences with a reddening and subsequent scaling and thickening of the skin, usually also between the toes. Sometimes areas with increased amounts of the horn-like material of the skin (*keratin*) are observed and these may resemble calluses. Both types of athlete's foot may spread to cover part or all of the soles. Both feet may be involved, but more frequently attacks occur to a greater extent on one foot than on the other. The hands are rarely affected.

Treatment of serious cases of tinea pedis may require the administration of oral micronized griseofulvin for a period of weeks.

Pityriasis versicolor or *chromophytosis* is a rather common fungus infection in which there appear fawn-colored patches on the trunk and limbs. It is most common among young adults. The colored patches that occur in chromophytosis actually may be lighter than the surrounding skin in dark-complexioned persons.

Actinomycosis is caused by a mold which ordinarily affects the respiratory system. When it infects the skin, it generally involves the mouth, jaw, neck, shoulders, or back. Red swollen areas slowly develop and exude a pus-like material. The involvement of the skin is usually secondary to an infection of the underlying tissues. Actinomycosis is a dangerous condition, but does respond to drugs when it has not become too advanced prior to medical attention. See **Actinomycosis.**

Bacterial Infections. Dermatitis from bacterial infections may be caused by (1) bacteria which are located in the skin itself, (2) bacteria which are distributed in the skin and other parts of the body, or (3) bacteria which are solely in other parts of the body. In many rather common infectious diseases, such as scarlet fever, brucellosis, pneumonia, typhoid fever, rheumatic fever, and meningitis, an eruption or rash may appear on the skin which is not caused by the bacteria, but by secondary effects which the organisms produce. Treatment of patients with these conditions involves the destruction of the bacteria which have invaded the body as a whole. Diseases of this type are discussed elsewhere in this volume. See, for example, **Acne Vulgaris;** and **Sebhorrea.**

The most common of the bacterial skin infections are boils (*furuncles*). These are round, tender, reddened elevations on the skin which contain a central core filled with pus and bacteria, usually staphylococci. See **Boils.**

A number of other pus-forming eruptions of the skin are not uncommon. Barber's itch or *sycosis vulgaris* is a typical example and is caused by an infection with staphylococci. The disease attacks primarily the hair follicles. In this manner, the disease may spread until it eventually affects the entire bearded region. The condition should be distinguished from another form of barber's itch (*tinea barbae*), which affects the lower bearded regions below the jaw, and which is, in reality, ringworm.

Impetigo is a common disease of childhood, but also occurring in adults. It is caused by staphylococci and streptococci. See Impetigo.

Erysipelas or St. Anthony's Fire is a particularly severe streptococcal infection of the skin and subcutaneous tissues. See **Erysipelas.**

Anthrax, the cause of which is *Bacillus anthracis,* is generally contracted from infected animals. See **Anthrax.**

Diphtheria of the skin is rare, and in most cases is caused by infection of some pre-existing wound with the diphtheria organism, *Corynebacterium diphtheriae.* The usual symptoms are a false membrane and gray ulcertation around the swollen edges of the infected area. Such infections, although they are quite small in extent, are almost invariably fatal if not given prompt medical care. Diphtheria immunization is helpful in preventing the occurrence of the disease. See also **Diphtheria.**

Tuberculosis of the skin is more common in Europe than in North America. There are two distinct varieties. In one, the tubercle bacillus (*Mycobacterium tuberculosis*) can be found in the eruptions. This is true tuberculosis of the skin. The other form (*tuberculids*) has many of the characteristics of the true tuberculosis, but the bacillus usually does not occur in the sores. Tuberculin or tuberculin-like substances probably cause this allergic reaction. The treatment of this condition is like the treatment given individuals with tuberculosis. See **Respiratory Systems.**

Gangrene. When deep wounds or lesions become infected with certain types of bacteria, or when the circulation to some tissues is interrupted, *gangrene* may develop. See **Gangrene.**

Infectious Eczemoid Dermatitis. This disease starts with blisters which become infected. The blisters fill with pus, and eventually result in a crusted, oozing, itchy condition. Such symptoms are believed to be caused by a secondary infection on the underlying eczema. In another disorder called *nodular erythema* or *erythema nodosum,* a number of small nodular swellings occur on the shins and other parts of the legs. The reddened swellings, which seem to be below the surface of the skin, are tender for some time, but frequently recede spontaneously. Such a condition usually indicates there is an allergic response of the blood vessels of the skin to some bacterial infection. Causes include streptococcal infection, tuberculosis, rheumatism, septic sore throat, and sensitivity to drugs.

Pityriasis rosea appears to be feebly infectious, although the organism which causes it is unknown. It is usually mild, and is manifested by small salmon-colored patches which eventually coalesce to cause larger pigmented areas. The condition largely affects the trunk and is more prevalent in young adults during spring and summer months. It may disappear spontaneously within a few weeks, but requires medical attention to distinguish it from a number of other more severe skin disorders.

Dermatitis from Plants and Animals. The skin disorders which result from contact with various green plants are largely allergic in nature. A variety of minute animals cause severe skin eruptions by bites, stings, or by burrowing into the skin.

Scabies. This disorder results from infection of the skin by small mites, about $\frac{1}{50}$th of an inch ($\frac{1}{2}$ millimeter) long (*Acarus scabiei*). These mites live on the surface of the skin, but the female burrows into the skin to lay its eggs. During this process the mite may remain under the skin for some time, traveling along and creating an extended tunnel in which the eggs are laid. The young develop in a few days, and then come directly to the surface where they spend their lives until they, in turn, are ready to lay eggs. The typical sign of scabies is the short, winding burrow in the skin which is most often between the fingers and toes. In children, this may be accompanied by tiny blisters on the surface of the skin near the burrow. In this stage of the dermatitis, there is very little itching. Later, the skin may develop an allergy or hypersensitivity to the mite, which causes the severe itching associated with scabies. Consequently, by the time the disease is first noticed, the mite usually has spread over a large portion of the body. Treatment is entirely a matter of ridding the patient of the mites. Underwear and bedclothing must be changed daily for a week or two, or until all eggs are hatched, and care must be taken to avoid infecting other persons. Daily baths and use of a sulfur ointment, benzoate emulsion, gamma benzene heachloride, or crotamiton, frequently discourage any further activity of the mite in the skin. The use of these compounds requires a physician's advice because some individuals cannot tolerate them.

Ticks of varying kinds and sizes may infect the skin and cause severe eruptions. They are picked up frequently in brushy areas or by contact with dogs or other animals that carry them. The female tick attaches itself to the skin by its nose and draws blood for food from the underlying vessels. After several hours, the tick will become filled with blood and drop from the skin. If pulled off by force, the proboscis may be left in the skin and cause an infected sore. The greatest danger from ticks is the possibility that the tick may carry some infectious microorganism and transmit it to the person it feeds upon. Rocky Mountain spotted fever is one such severe infection carried by certain ticks (*Dermacentor andersoni*). See **Rickettsial Diseases.**

Chigger bites result from small mites or bed bugs (*Trombicula irritans*). These mites secrete a keratolytic agent which dissolves the outer layer of skin on which the animal feeds, and causes a red and intensely itching swollen area. The mites are picked up from grasses and brush, and consequently affect the legs and lower portions of the body most often. They accumulate usually underneath garters and belts and in other areas where tight clothing restricts their movement. Chigger bites can be prevented by rubbing wet "sulfur-foam" impregnated material on the extremities before going into chigger-infested areas.

Pediculosis is caused by infestation of the skin by lice (*Pediculidae*). See **Pediculosis.**

Bedbugs. These are small, ororous, wingless bugs (*Cimicidae*) that feed on the blood in a manner similar to that of lice. The reaction to bedbug bites varies among different individuals, but generally consists of small, red punctures surrounded by swollen, inflamed areas. The swellings may be painful. The use of soothing ointments to prevent scratching frequently causes the inflammation to subside in a short time. A building that contains bedbugs requires fumigation in order to destroy them completely, because bedbugs live during the daytime in crevices in the floors and walls and in furniture and are difficult to find. Fumigation may be accomplished with hydrocyanic acid gas (a dangerous poison), or fumes of sulfur, formol, or other vapors. Fumigation should be handled by experts.

Fleas (*Siphonaptera*) live, for the most part, on lower animals, but occasionally infest human beings. They feed on blood, and leave swollen, reddened areas on the skin that may be severe to a sensitive skin. Such diseases as typhus and plague are carried by rat fleas. Infested pets should be treated with various dusting powders and houses can be cleared of fleas by spraying or scrubbing with disinfectants. *Sand fleas* burrow into the skin in order to deposit their eggs. They live in dry sandy soil in warm climates, and most often affect the feet, ankles, and legs. The reddened swelling which they cause may become the size of a pea or larger and is susceptible to secondary infection by bacteria. Strong soap and soothing ointments generally correct this condition after several days. However, the lesion disappears much more rapidly if a physician removes the flea and its "cocoon" from the skin.

Spider and Centipede Bites. These are usually harmless unless they result in secondary infection. Other than immediate pain, they usually produce only an itching, swollen area on the skin surrounded by some degree of reddening. However, a particularly poisonous variety, the *black widow spider* (*Lactrodectus mactans*) may cause severe systemic symptoms that require immediate medical attention.

Hookworm is a small, parasitic worm (*Ancylostoma duodenale* or *Necator americanus*) which is regarded primarily as an intestinal parasite, although it produces characteristic skin symptoms. The dermatosis is caused by an invasion of the skin by the young, or larvae, of the worm, and occurs several months before the general systemic symptoms may be pronounced. The earliest signs are in the soles of the feet, since it is through the soles that the worm originally enters the body from the soil. Small, reddened pimples develop into blisters which may eventually become pus-filled. This disease is serious and requires prompt medical attention. In *sandworm disease*, which is generally caused by hookworms carried by cats or dogs, the larvae may cause extensive winding burrows in the skin. The condition is also referred to as *creeping eruption.*

Swimmers' itch is thought to be caused by a small aquatic worm which invades the skin of persons who swim or wade in contaminated water. Reddened itching pimples or patches develop a day or two after exposure. The disease may be prevented by thoroughly bathing with soap and rubbing the skin with a towel after exposure.

Dermatitis also may be caused by chemicals, malnutrition, and from organic disorders. See also **Allergy.**

In vitamin A deficiency, the skin becomes dry, scaly, and develops small, spiny lesions. Vitamin A is believed to have an important role in maintaining the health of the skin. See **Skin.** Dermatitis caused by a deficiency of vitamin B_2 (riboflavin) is manifest by oily scaling about the ears and nose, and by cracking at the corners of the lips. In *pellagra*, a disease caused by a nicotinic acid deficiency, the dermatitis is found on the parts of the body exposed to sunlight, particularly the face and hands. See also **Niacin.**

Insufficient thyroid function (*myxedema*) is frequently accompanied by a swelling of the skin. The skin also may have a dry and waxy appearance. In a number of other conditions, various fatty and protein materials may infiltrate the skin and give it a lumpy or nodular appearance. In other cases, the skin may become highly elastic (*India rubber skin*), or its attachment to the underlying tissues may become so loose that it hangs in folds from the body.

Porphyria is the name given to a small group of organic diseases (of genetic origin) caused by a metabolic abnormality. In several forms of porphyria, dermatitis may be the only or the most important symptom. The basic organic defect involves the metabolism or *porphyrins*, compounds from which plant and animal respiratory pigments are made. The green chlorophyll of plants is perhaps best known of the porphyrins. In human beings afflicted with porphyria, porphyrins are produced in excess and deposited in the skin and other anatomical areas. In later stages of the disease, porphyrins may alter the mental processes. It has been suggested that the madness of King George III was attributable to porphyria.

Variegate and *cutaneous porphyria* are congenital disorders in which the skin is highly sensitive to trauma on sun-exposed surfaces. *Congenital photosensitive porphyria* is a rare variety of porphyria, in which sensitivity to sunlight is so pronounced that many patients cannot venture outdoors during daylight.

Psoriasis, afflicting 2–3% of the adult white population (much less common among blacks) is of unknown origin. The lesions consist of rounded, reddish, dry, scaly patches covered by grayish-white, micalike scales. These patches may spread and become extensive. The disorder is recurrent, tending to recede during summer in temperate climates. In general, psoriasis is most often seen on the scalp, nails, lower back, knees, and elbows. Heredity is considered to be a factor, although the genetic mechanisms involved are poorly understood. Secondary causative factors may include infection, local trauma, disturbances in body chemistry, and psychosomatic influences. Frequently in psoriasis there are accompanying alterations in the quality of the nails. Ice-pick pits, so-called, and longitudinal ridging on the nails are characteristic. The relationship of psoriasis to arthritis has been known for 150 years. The significance of the association is far from clear and is still controversial. Some authorities now feel that arthropathy in psoriasis represents a variant of rheumatoid arthritis and that there is no justification for recognizing psoriatic arthritis as a separate disease.

Topical therapy for psoriasis include application of coal tar compounds, anthralin, and glucocorticosteroids. Preparations have been available for many years that contain from 2 to 5% of crude coal tar in various bases and may be used in combination with salicylic acid. These compounds usually require application once or twice daily. Coal tar compounds tend to stain the skin and also have a characteristic odor which some patients find objectionable. A synthetic compound, *anthralin* (dithranol) has been available for over 50 years and has been found effective for psoriasis of the scalp and other areas. Apparently, its essential function is inhibition of enzyme metabolism. Topical glucocorticosteroids, although effective, are like the other topical remedies: the skin condition returns when the treatment is halted. Some authorities believe that moderate, sensible exposure of affected areas to sunlight assists in treating psoriasis.

Systemic therapy includes the administration of the antimetabolite *methotrexate*, particularly in cases where the disease may cause severe disability, including serious emotional problems. This drug can produce serious side effects and must be used with discretion. Hydroxyurea also has been used in treating serious cases of psoriasis. A combined therapy (PUVA) or psoralen and longwave ultraviolet light therapy has been used, but results are not permanent. Psoralen is a photosensitizing drug (8-methoxypsoralen) and is given orally, along with frequent periods of exposure to longwave UV radiation.

Malignant Cutaneous Tumors

Not widely appreciated is the fact that skin cancers occur more frequently than cancer at any other site of the body. With exception of melanomas, most skin cancers are amenable to curative therapy. Primary malignant cutaneous tumors may occur in the epidermis, dermis, or subcutaneous tissue. Tumors at other sites of the body also may ultimately metastatize to the skin, in which cases prognosis is frequently guarded.

Sunlight. Despite the apparent misbelief of the fact by millions of avid sun bathers, chronic exposure to sunlight is the most important factor in the development of skin cancer. These cancers are found at a higher rate in outdoor workers, with the greatest incidence among persons with fair complexions and blue eyes.

Premalignant Cutaneous Tumors. A number of skin lesions of a benign nature frequently will develop into malignancies with the passage of time if they are left untreated. The most common of these premalignant skin lesions is *actinic keratosis* (senile or solar keratosis). One or many lesions will occur in areas that are frequently exposed to sunlight, notably the face, forearms, neck, scalp (where not shielded by hair), and the dorsa of the hands. The lesions vary in size from a few millimeters up to a centimeter (0.4 inch) and may be roundish or irregularly shaped. They may be pinkish to red (erythematous) or sometimes of a tan coloration. The lesions may have a smooth or scaly surface. These premalignant lesions, which statistics show will develop into squamous cell carcinoma, can be removed by one of several techniques (excision or curettage and electrodesiccation with local anesthesia; liquid nitrogen cryotherapy). Symptoms of malignancy are enlargement and elevation of the lesions with accompanying inflammation. Topical treatment with 5-fluorouracil has proved successful in many cases. The risk of untreated actinic keratosis hinges on the fact that squamous cell carcinoma, a later consequence, may metastasize and thus affect other organs of the body, leading to death.

Leukoplakia is much like actinic keratosis except that these lesions occur on mucous membranes (lips, other oral mucosa, vulva). The lesions are somewhat elevated, irregularly shaped, sharply bordered white patches. It is believed that these lesions may be aggravated by tobacco juices from pipe smoking and chewing tobacco. Treatment is the same as previously outlined for actinic keratosis.

Bowen's disease also occurs in areas exposed to sunlight. The disease, which is characterized by irregular, scaly, dull red patches, is most commonly seen in males with a fair skin who have had excessive exposure to sunlight. Where lesions appear in normally sun-shielded areas, they may indicate a cancer somewhere else in the body. Exposure to arsenic is another known causative factor of Bowen's disease. For reasons previously given, these lesions also should be excised.

Paget's disease, manifested by sharply defined, red, and scaly lesions, frequently with an accompanying watery discharge, is found mainly in women. Involved is the nipple or areola of the breast. This condition may be secondary migration of mammary carcinoma cells to the skin because almost invariably the lesions are associated with underlying mammary carcinoma. Occasionally these kinds of lesions may occur in other areas of the body and in both sexes, in which instances, the tumors should be excised. The physician will seek out possible underlying malignancies elsewhere, such as eccrine or apocrine gland carcinoma or rectal cancer.

Basal Cell Carcinomas. These may be described as dome-shaped papules or nodules, white to pinkish in coloration, often with an elevated pearly border. In some, scaling, crusting, or central ulceration may be present. Some contain a gelatinous fluid. There is a tendency for these lesions to enlarge with time. Proper identification and diagnosis requires professional expertise. Because these tumors are malignant, they should be excised or otherwise destroyed (desiccation, radiation, etc.) after the diagnosis has been confirmed. A border of normal tissue adjacent to the tumor must be removed to prevent recurrence from invasive strands of tumor cells.

Squamous Cell Carcinomas. These are firm, red nodules which slowly form shallow ulcers surrounded by indurated borders. They are most commonly seen in areas regularly exposed to sunlight—scalp, face, lips, hands. Treatment is similar to methods previously mentioned. *Keratoacanthoma* presents tumors essentially like squamous cell carcinomas. These are more often seen in elderly patients. They also occur in areas that have been generously exposed to sunlight over a period of years. Complete excision is indicated in these cases.

Malignant Melanoma. This is one of the most serious manifestations of skin cancer and leads to the greatest number of fatalities. These tumors usually are seen during and between the fourth and sixth decades of life. White people with light complexions are much more prone to melanoma than persons with dark-pigmented skin and black people. Statistics definitely show a relationship between incidence and exposure to sun. Malignant melanoma is more frequently seen, for example, in the southern half than in the northern half of the United States; in the northern half of Australia than in the southern half. Malignant melanoma is of three forms: (1) *Lentigo malignant melanoma,* with a median age of onset of 70 years. The lesions are various shades of brown and black and the margin of lesion is flat. Lesions are usually found on head and neck and less commonly on hands and legs. Lesions may commence as a freckle-like lesion with an irregular outline and this may grow very slowly over a period of 5 to 15 years. (2) *Superfical spreading melanoma* is the most common type of malignant melanoma, with a median age of onset of 56 years. The lesions are various shades of brown and black; although gray and pinkish-rose colorations are not uncommon. The margins of the lesions are distinctly palpable. The lesions may be found on all body surfaces. However, in women, the lesions usually are found on the lower legs. The lesions begin as small, irregular, brown-pigmented areas with various shades of red, white, and even blue. Growth of the lesions into papules or nodules may require from 1 to 5 years. (3) *Nodular melanoma,* with a median age of onset of about 50 years. The lesions are uniformly bluish-black and sometimes may display a depigmented, irregular halo. The margin of the lesion is palpable. Usually commencing as a papule or nodule with a smooth, scaly, eroded, or ulcerated surface, the lesions grow rapidly (months to about 2 years). Satellite lesions are not infrequent.

Survival rates for the aforementioned three types of melanoma, where treatment is instituted promptly, are highest for Lentigo maligna melanoma and the less potent types of superficial spreading melanomas. This may be as high as 95% (5-year survival rate). In nodular melanoma, where distant mestastases frequently form, the prognosis is considerably poorer. Immunotherapy or chemotherapy, when combined with surgery, may improve the prognosis. Melanomas usually metastasize first through the lymphatic system, with involvement of regional nodes, and then via blood vessels, reaching other areas of the skin, liver, lungs, and brain.

Mycosis fungoides has its origin in the skin, but eventually disseminates to the lymph nodes and internal organs. The disease generally commences with a chronic dermatitis usually involving multiple areas of the body. The condition may not be identified or seriously treated for many years. Then, in middle age, a biopsy may show that a malignant condition has developed. The disease is uncommon. Various forms of chemotherapy, ultraviolet radiation treatments, and other procedures have been used with varying degrees of success.

Sarcomas of the Skin. These may be primary or metastatic and single or multiple. Sarcomas may involve connective tissue, blood vessels, adipose tissue, smooth muscle, or the lymphoid-recticular system. Kaposi's idiopathic hemorrhagic sarcoma is manifested by firm reddish-brown or bluish plaques and nodules on the hands and feet. Most authorities consider Kaposi's sarcoma as a tumor of multicentric origin arising from immature pluripotential vascular cells.

Related topical entries in this encyclopedia include: **Acne Vulgaris; Alopecia; Anaphylaxis; Burn; Callus; Cheilitis; Cuticula; Epidermis; Epithelium; Hair; Hirsutism; Impetigo; Integument/Integumentary System; Pemphigus Vulgaris; Percutaneous; Pruritus; Rash; Sebaceous Cyst; Sebaceous Gland; Seborrhea; Skin; Tactile Organs; Urticaria; Wart;** and **Xerosis.**

References

Adams, R. M.: "Occupational Contact Dermatitis," Lippincott, Philadelphia, 1969.

Andrade, R., et al.: "Cancer of the Skin," Saunders, Philadelphia, 1976.

Cram, D. L.: "Recent Advances in the Pathogenesis and Treatment of Psoriasis," *J. Continuing Ed. Dermatol.* **17**, 25 (1978).

Demis, D. J., Dobson, R. L., and J. McGuire (editors): "Clinical Dermatology," Harper & Row, Hagerstown, Maryland, 1977.

Domonokos, A. N.: "Andrews' Diseases of the Skin," 6th edition, Saunders, Philadelphia, 1971.

Epstein, E., and E. Epstein, Jr. (editors): "Skin Surgery," 4th edition, Charles C. Thomas, Springfield, Illinois, 1976.

Farber, E. M., and M. L. Nall: "The Natural History of Psoriasis in 5600 Patients," *Dermatologica,* **184**, 1 (1974).

Fitzpatrick, T. B., Arndt, K. A., and W. H. Clark, Jr. (editors): "Dermatology in General Medicine," McGraw-Hill, New York, 1979.

Freeman, R. B.: "Carcinogenesis of Skin Neoplasms, Neoplasms of the Skin, and Malignant Melanoma," Year Book Medical Publishers, Chicago, Illinois, 1976.

Lever, W. F.: "Histopathology of the Skin," 5th edition, Lippincott, Philadelphia, 1975.

MacDonald, E. J.: "Epidemiology of Skin Cancer: Neoplasms of the Skin and Malignant Melanoma," Year Book Medical Publishers, Chicago, Illinois, 1976.

Monroe, E. W., and H. E. Jones: "Urticaria: An Updated Review," *Arch. Dermatol.,* **113**, 80 (1977).

Parker, C. W.: "Drug Therapy: Drug Allergy," (in 3 issues), *N. Engl. J. Med.,* **292**, 511 (1975); **292**, 957 (1975); **292**, 957 (1975).

Rippon, J. W.: "Medical Mycology: The Pathogenic Fungi and the Pathogenic Actinomycetes," Saunders, Philadelphia, 1974.

Rook, A., Wilkinson, D. S., and F. J. G. Ebling (editors): "Textbook of Dermatology," F. A. Davis Co., Philadelphia, 1972.

DERMATOMYOSITIS. Myopathy.

DERMESTID (*Insecta, Coleoptera*). Any of the small beetles of the family *Dermestidae,* including the buffalo carpet moth. They damage clothing, woolen articles, and museum specimens.

DERMOLITH. A term proposed by T. A. Jaggar in 1917 for ropy, wrinkled or pahoehoe type of basic lava, such as occurs in the volcanic islands of Hawaii.

DERMOPTERA (*Mammalia*). This order of the *Mammalia* is represented by only one genus, variously called flying lemur, colug, kobugo, kaguan, or kobego. The classification of this gliding mammal has created problems for many years. At one time it was assigned to the bats; at another time to the Insectivores; and at still another to the Primates, by way of the lemurs. Kobegos or flying lemurs are about 18 inches (46 centimeters) in length and have folds of skin along their sides, the front paws and hind feet are fully webbed. Thus, upon leaping, the animals spread their limbs and become furry kites which permit gliding travel for considerable distances. Glides may measure up to 100 feet (30 meters), with only 1 foot (0.3 meter) dropped for every 5 feet (1.5 meters) of essentially horizontal travel. These animals possess several interesting physiological and anatomical qualities, likening them in various ways to other orders of mammals as previously mentioned. One type of kobego is found in Borneo, Java, Malays, Siam, and Sumatra, while a second type occurs in some of the Philippine islands. They diet on leaves, flowers, and fruits. The animals sleep by hanging upside down from a tree branch. Despite their airborne capabilities, the kobegos are unlike bats in most other respects.

DESALINATION. The process of removing dissolved salts, notably sodium chloride, from seawater and brackish waters to yield a potable water for human consumption, process, and irrigation purposes. Water shortages are increasing as the earth's population expands, as per capita requirements for water increase, and as more individuals congregate in areas which have a naturally limited supply of water. Concurrently, water resources are being depleted by declining water tables, by pollution of surface sources, and by the salting-up of ground and river waters. Plants for desalting ocean water and brackish water are becoming economically competitive with the cost of transporting fresh water over long distances. While desalination plants consume large quantities of energy, thus altering the economics some during the last few years, the energy costs for water transporting also have risen.

As of the early 1980s, there are about 510 desalination plants in operation or under construction throughout the world. These plants have a total capacity of some two billion tons of fresh water daily. In addition to direct ocean sources of saline water, a number of the smaller desalination plants produce fresh water at coastal and island locations, on drilling structures, and on shipboard. Logically, many of the larger plants are located in arid regions, such as the Mid-Eastern countries, and in selected locations of high population where demand exceeds supply. Plants in Hong Kong and Rotterdam are respresentative of these situations. See Table 1.

Because of the requirement to minimize production costs, research continues to find more exotic and sophisticated processes that will use low-cost materials and lower-energy inputs. A number of processes are described in Table 2. However, major plants built to date use distillation as the desalting process. A majority of these plants employ multistage flash distillation. There are also several plants that use multi-effect evaporation with falling films, and single or multiple stage vapor compression. A plant in Israel uses the freezing process. Electrodialysis and reverse osmosis (membrane plants) made considerable progress during the 1970s and these processes are a serious consideration when new desalination plants are planned.

TABLE 1. REPRESENTATIVE MAJOR SEAWATER DESALINATION PLANTS

Location	Capacity (Metric Tons/Day)	Process Used
Hong Kong (Lok on Pai)	181690	MSF
United Arab Emirates (Dubai)	113920	MSF
Kuwait (Shuaiba)	95387	MSF
Saudi Arabia (Jeddah)	90846	MSF
Qatar (Ras abu Fontas)	90846	MSF
Kuwait (Doha)	81800	MSF
United Arab Emirates (Abu Dhabi)	81000	MSF
	72700	MSF
	60000	MSF
Saudi Arabis (Riyadh)	46100	MSF
	45500	MSF
Kuwait (Shuaiba)	45400	MSF
Saudi Arabia (Jeddah)	44000	MSF
Israel (Ashdod)	40000	TFH
Saudi Arabia (Riyadh)	39900	MSF
	39900	MSF
United Arab Emirates (Abu Dhabi)	36300	MSF
Sardinia (Porto Torres)	36000	MSF
Netherlands (Rotterdam)	32400	MSF
Netherlands (Terneuzen)	29000	MSF
Italy (Gela)	28800	MSF
Mexico (Rosarita)	28400	MSF

NOTES: MSF = multistage flash; TFH = thin-film horizontal.

Information Source: Office of Water Research and Technology, U.S. Department of the Interior.

As established by the U.S. Public Health Service, water for human consumption should contain no more than 500 parts per million (ppm) of dissolved solids. Seawater contains about 35,000 ppm; brackish water is generally classified as containing 1,000 ppm; the purity of irrigation water varies with the crop and soil conditions, but sometimes is as high as 1,200 ppm.

Multistage Flash Evaporation Process. Seawater is pumped from the ocean, deaerated, treated with scale-control chemicals, and combined with a stream of recycle brine. The saline solution is then pumped through tubes set lengthwise in the top half of horizontal vessels. Vertical baffles (stage dividers) divide each vessel into vacuum cham-

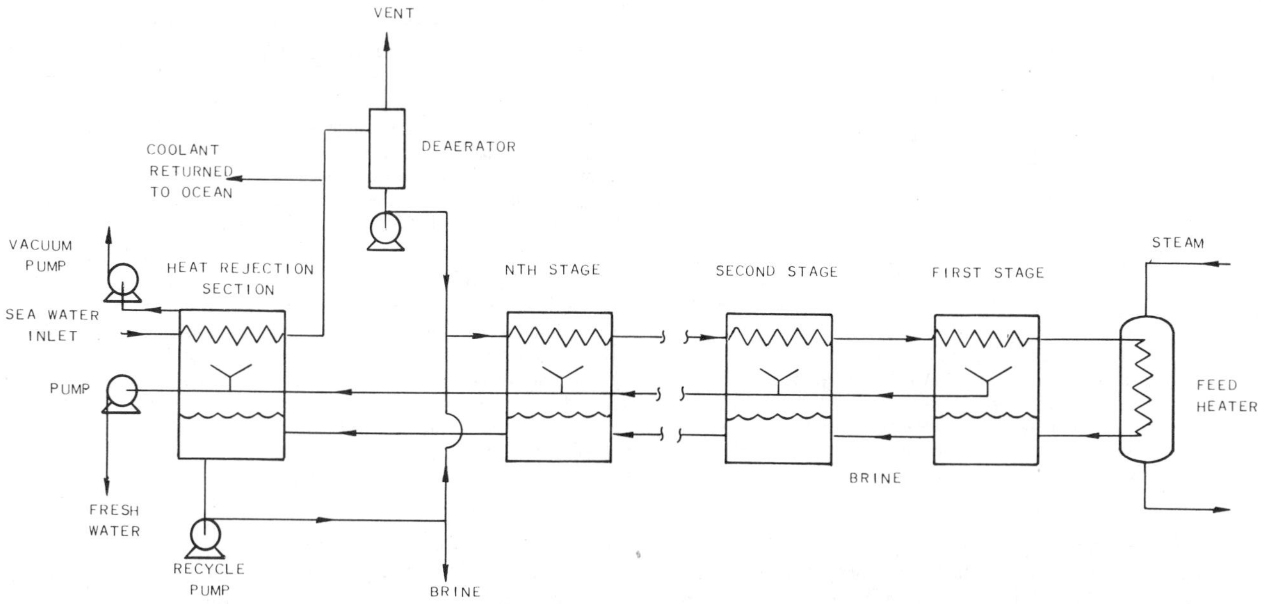

Fig. 1. Multistage flash distillation (MSF) with recycle. (*Fluor Corp.*)

TABLE 2. DESALINATION PROCESSES

MULTISTAGE FLASH DISTILLATION (MSF)

Saline solution is vaporized and pure water is obtained by condensation. The seawater is heated progressively to a maximum of about 250°F (121°C) and then flashed into a number of successive stages, in which operation is under progressively lower pressures. Incoming seawater is heated by condensing vapors. Much experience has been gained from numerous installations. Scale formation tends to limit the top operating temperatures and thus limits efficiency.

MULTIPLE-EFFECT MULTISTAGE FLASH DISTILLATION (MEMS)

Fundamentally, this process is similar to the MSF. The stages, however, are divided among two or more effects in which the brine is recycled. Thus, operation at higher temperature is possible. The process also allows higher blowdown concentrations without scaling. Although efficiency is generally higher than the MSF, capital costs tend to be higher. Top operating temperatures still are limited by scaling and the operation tends to be more complex than the MSF.

VAPOR REHEAT FLASH DISTILLATION

This process is similar to the MSF distillation with the exception that fresh water is recycled and used to condense vapors. Heat is recovered from the fresh water recycle stream by use of heat exchangers. If liquid–liquid heat transfer is used, metallic metal transfer surfaces (as in the distillation process) are eliminated, reducing the scale problem. However, the large volume of liquid handling involved increases the complexity and cost of the process. Capital costs are higher than for the MSF process for similar size plants.

MULTIPLE-EFFECT EVAPORATION (MEE) (Vertical Tube Evaporator Process)

Three or more evaporators are used and steam is used to evaporate portion of the seawater in the first effect. Vapor from the first effect condenses in the second effect and evaporates additional water. This process has been used successfully in numerous industrial evaporation processes for years. The process can operate at higher blowdown concentrations than the MSF. However, more effective antiscaling techniques are required. Initial capital investment is higher than for the MSF, but may be quite competitive for plants exceeding 10 million gpd.

VAPOR COMPRESSION DISTILLATION

The vapor from boiling brine is compressed mechanically, thus increasing the vapor temperature and pressure. The compressed vapor is then fed back into the evaporator to distill more seawater feed. The multiple-effect approaoch can be applied to this concept. Attractive for small size units.

SOLAR DISTILLATION

A large basin is fitted with sloping transparent glass or plastic covers. These serve as condensing surfaces. Saline water fed to the enclosed chamber is heated by solar radiation. Process limited to areas of maximum sunshine. A large land area is required. Advantages include simplicity and minimal fuel costs.

HUMIDIFICATION

Either by solar or fossil fuel energy, the incoming saline water is heated, after which it is sprayed in a tower with dry air flowing countercurrently. The saturated air leaving a first tower is cooled in a second tower where fresh water is condensed and collected. The process is simple and may be competitive in small size units up to about 0.5 million gpd.

SOLVENT EXTRACTION

A solvent is used to remove either the salts from water or water from the salts, followed by separation of solvent and salts, or fresh water by change in temperature of mixture. Appropriate solvents are costly and removal of traces of solvent (possibly toxic or otherwise objectionable) may pose problems. Large volumes of fluids must be handled.

ELECTRODIALYSIS

Electrodes impress an electrical field across a cell through which saline water flows. Cations migrate to the cathode; anions to the anode. The positive ions pass through cation-permeable membranes; negative ions through anion-permeable membranes. Between alternate membranes, the water becomes enriched with or depleted of salts. The energy requirements are in proportion to saline content, but increase rapidly with attainment of high purity. Suited to brackish waters that contain up to 5,000 ppm dissolved solids.

REVERSE OSMOSIS

The saline water is pressurized above its osmotic pressure—then flows over a semipermeable membrane. The membrane is permeable to water, but not to dissolved solids. The fresh water is transported through the membranes. Energy requirements increase less rapidly with increasing saline content than with electrodialysis process. Tailoring of membranes to different saline concentrations is possible. High-pressure equipment is required.

Although high pressures are needed to overcome the osmotic pressure of seawater, membranes have been developed that allow single-pass separation of water containing up to 500 milligrams/kilogram of salts from a 3.5% (weight) seawater solution. Available are cellulose acetate and substituted polyamide membranes both in hollow fine-fiber and sheet configurations.

HYDRATE PROCESS

A gas, such as propane, is mixed with saline water. Insoluble crystals of a solid hydrate are formed. A slurry containing crystals and brine goes to a wash column where crystals are washed and transferred to a decomposition chamber. Here the hydrate crystals are decomposed by altering temperature and/or pressure. The process can operate at near-ambient conditions.

DIRECT FREEZING—VAPOR COMPRESSION

The saline water is sprayed into a vacuum chamber where part of the water is evaporated. This produces a cooling effect, causing formation of ice crystals. A slurry of ice and brine goes to a washer-melter where brine is washed from ice after which ice is melted to yield fresh water. The energy requirements are lower than for distillation and corrosion and scaling problems are minimized. However, large volumes of vapor have to be handled and compressed. High capital costs.

DIRECT FREEZING—SECONDARY REFRIGERANT

A hydrocarbon refrigerant immiscible with water, such as butane, is vaporized in direct contact with the saline water. This produces an ice–brine mixture. The ice crystals are washed to remove the brine. The refrigerant vapor is compressed and condenses as it melts the ice. Some cost advantages over other freezing processes, and energy requirements are lower than for distillation. Removal of hydrocarbon refrigerant from product water poses problem.

ION EXCHANGE

Cation and anion ion exchange resins are used to remove salts. The exhausted resins are regenerated with acids, ammonia, or lime, depending upon the particulars of the process. Basic process well-developed for other industrial uses. Foreign matter can foul ion exchange resins. Pretreatment of some saline waters may be in order. Production costs are about proportional to total dissolved solids removed.

---

bers, each making up a stage of the process. See Fig. 1.

The stages operate at progressively higher temperature over a range of more than 150°F (84°C). Thus, the saline water heats gradually as it passes in the tubes from stage to stage. The water does not boil because it is under pressure.

At the end of the first pass of a round trip through the evaporator, the seawater leaves the highest-temperature stage and enters a brine heater. There its temperature is further increased by steam supplied

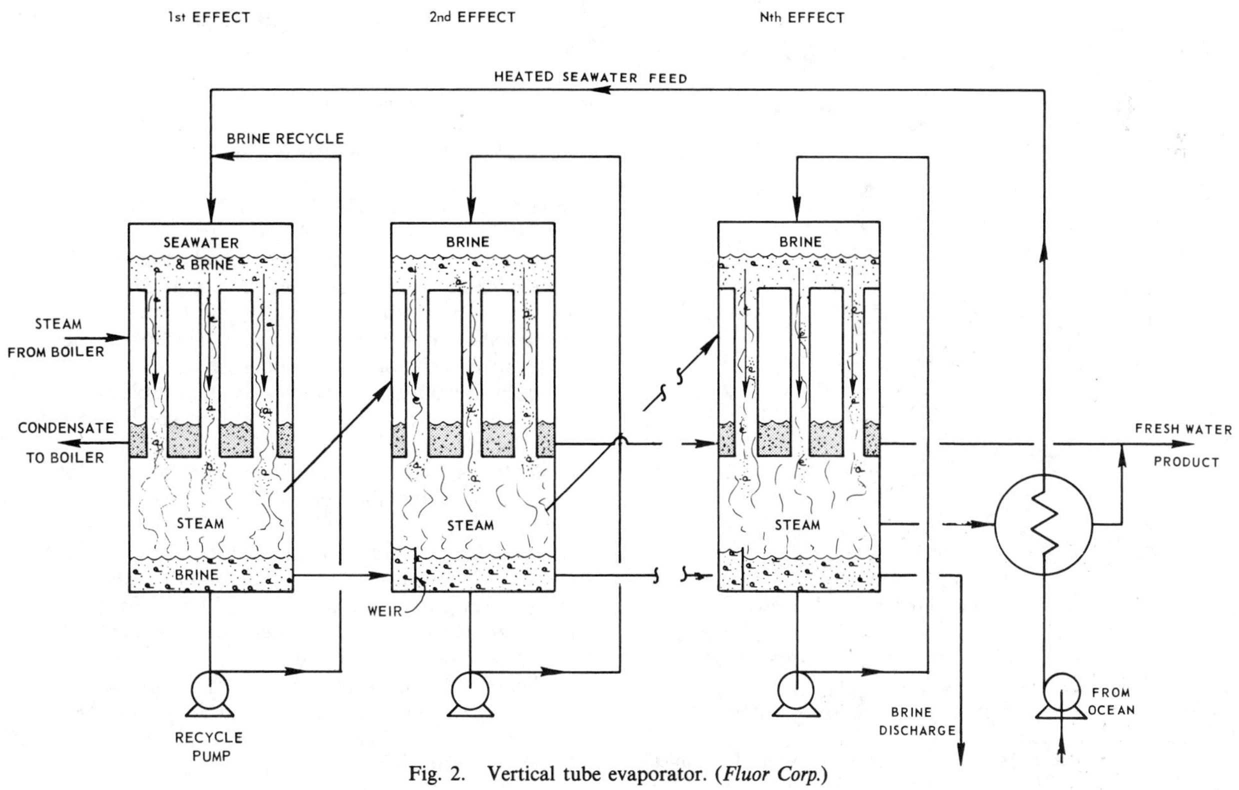

Fig. 2. Vertical tube evaporator. (*Fluor Corp.*)

from either a conventional boiler or from a nuclear steam generator. After this heating, the seawater is introduced to the shell side of the highest temperature, or first stage.

When the brine enters the bottom of this spacious compartment, it suddenly boils (flashes) and releases steam, caused by a reduction of pressure on the brine. The steam rises and contacts the tubes, passing off heat to the seawater in them, and then condenses and drips into troughs beneath the tubes.

The second stage operates at slightly less pressure than the first. This causes the brine and the distilled water in the troughs to flow spontaneously through liquid pressure seals in the stage dividers and into the second chamber. The brine and product water again flashes,

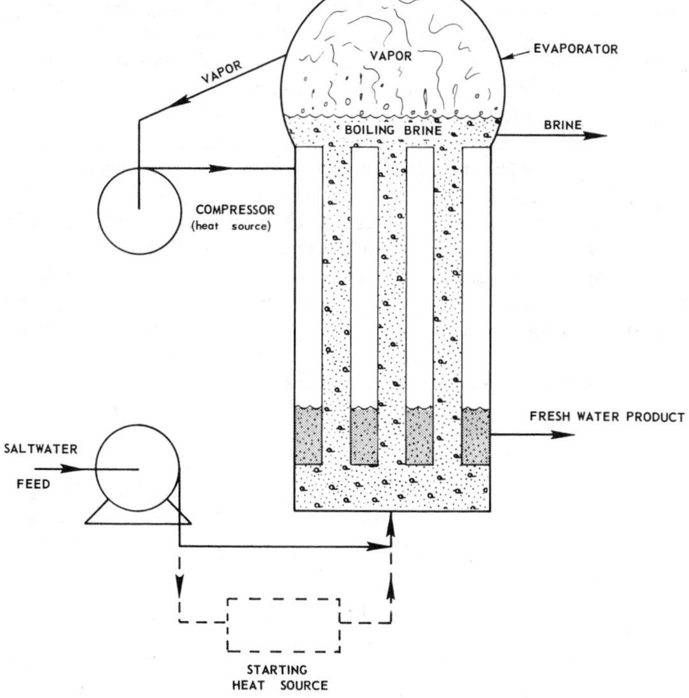

Fig. 3. Elementary vapor compression distillation. (*Fluor Corp.*)

steam condenses on the tubes, and more fresh water is made. The process repeats through the many stages of the system until all the available heat is removed from the brine. Concentrated seawater in excess of that needed for recycling is discharged to the ocean, and very pure fresh water is taken from the troughs as product.

Vertical-Tube Evaporator Process. The seawater is treated and then passed through several preheaters and transferred into the top of a vertical-tube evaporator (first effect). See Fig. 2. Steam from a boiler passes on the outside of the tubes. Heat transferred to the seawater falling inside the tubes causes the steam to condense and the seawater to boil. The condensate collecting in the first effect is ordinarily returned as boiler feedwater. Steam from the boiling seawater is transferred to the second effect, where it is condensed on the outside of the vertical tubes. Some of the brine collecting at the bottom of the first effect is recycled to the top of the first effect. The remainder of the brine is transferred to the bottom of the second effect, where it flashes and releases steam because of a reduction in pressure.

Similarly, condensate from each effect, except the first, flashes in the next effect and produces additional steam. Steam from the second effect is transferred to the next effect, where it heats the brine falling inside the tubes. The process is repeated through additional effects, where more condensate is produced. Each effect operates at a lower pressure than the preceding effect. Steam from the last effect is used to heat incoming seawater. Concentrated brine is discharged to the ocean, and condensate is removed as fresh water product. The vertical-tube evaporator may be combined with other processes.

Vapor-Compression Distillation. This process can be integrated into either of the processes previously described. See Fig. 3. Seawater is pumped into a tubular exchanger within an evaporated chamber. There it is boiled. Steam from the boiling seawater is piped to a compressor, where it pressure and heat content are increased by compression. The compressed vapor then flows outside the tubes in the evaporator. There it is condensed to fresh water, giving up its latent heat to boil the incoming seawater. The method makes use of the fact that the saturation temperature of water increases with pressure.

Solar Distillation. A representative diagram of a simple solar still is shown in Fig. 4. Vaporized seawater condenses on the surface of an air-supported plastic film, collects in troughs, and is removed as product water at selected points. They may be used to recover salt as well as fresh water. Another design configuration is shown in Fig. 5.

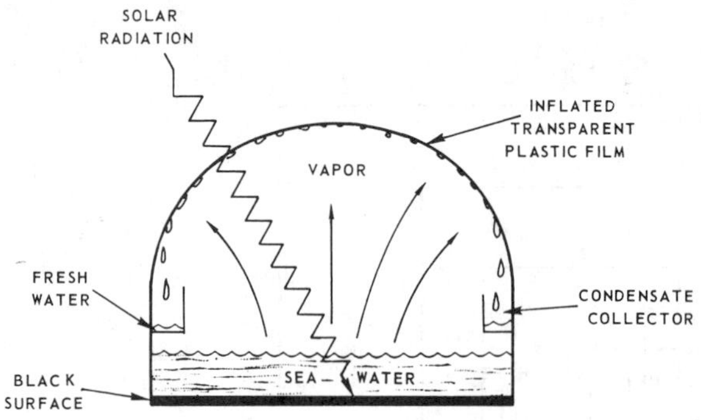

Fig. 4. Simple solar still.

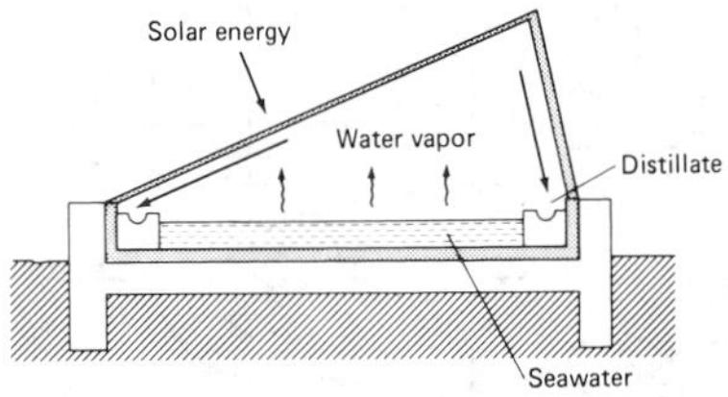

Fig. 5. Solar distillation unit featuring tilted glass panel.

References

McIlhenny, W. F.: "Extraction of Inorganic Materials from Seawater," in "Chemical Oceanography," (J. P. Riley and G. Skirrow, editors), Vol. 4, Chap. 10, Academic Press, London, 1975.

McIlhenny, W. F.: "Ocean Raw Materials," in "Encyclopedia of Chemical Technology," 3rd edition, Wiley, New York, 1981.

Multhaupf, R. P.: "Neptune's Gift—A History of Common Salt," Johns Hopkins Univ. Press, Baltimore, Maryland, 1978.

Spiegler, K. S.: "Principles of Desalination," Academic Press, New York, 1966.

Staff: "Desalting Plants Inventory," Office of Water Research and Technology, U.S. Department of the Interior, Washington, D.C. (Revised periodically).

DESCARTES' LAWS OF REFRACTION. Referring to the incident and refracted rays: they are in the same plane with the normal to the surface, (b) they lie on opposite sides of it, and (c) the sines of their inclinations to it bear a constant ratio to one another, the ratio depending only on the two media involved, not on the angles with the normal. Descartes applied these laws only to ordinary isotropic media.

DESCARTES' RULE OF SIGNS. If in passing from one coefficient of a polynomial equation to the next, there is a change of sign from plus to minus or from minus to plus, this is called a variation of sign; a succession of two like signs, either both plus or both minus, is called a permanence of signs. The number of positive roots of a polynomial equation $P(x) = 0$ with real coefficients is not greater than the number of variations of sign in the polynomial, and the number of negative roots is not greater than the number of variations of sign in the polynomial $P(-x) = 0$.

DESCENDING COLON. Digestive System (Human).

DESCRIBING FUNCTION. The output response of a nonlinear system to a sine wave input will not necessarily be a sine wave. The actual output will depend on the nonlinearity, but will normally have a fundamental mode of the same frequency as the input. The transfer function based on the magnitude ratio and phase angle of the output fundamental versus the input sine wave is defined as the describing function. Application of the describing function allows linear analysis of nonlinear systems.

DESCRIPTIVE GEOMETRY. The theory of graphic representation, invented and developed by Gaspard Monge, 1795. Descriptive geometry deals with the exact representation of objects composed of geometrical forms and of the graphical solution of problems involving the space relationship of these forms.

DESERT. Biome.

DESERTIFICATION STUDIES. Earth Resources Satellites and Geologic Remote Sensors.

DESERTS (Hydrology). Hydrology.

DESICCANT DRYERS. Dehumidification.

DESIGN OF EXPERIMENTS. The science of designing experiments so as to obtain unambiguous results of the highest possible accuracy. Four principles of experimental design may be distinguished.

1. *Replication.* Any treatment must normally be applied to more than one experimental unit. This provides greater accuracy than can be obtained from a single observation, since the experimental errors tend to cancel each other; it also provides a measure of the experimental error derived from the variability between replicates.

2. *Randomization.* The decision as to which experimental unit shall receive which treatment should include a random element. This is designed to ensure that every treatment shall have its fair share of the particularly favorable and the particularly unfavorable experimental units.

3. *Local Control.* Any structure in the properties of the experimental units should be utilized fully. Thus if the units fall into relatively homogeneous groups (neighboring plots in a field, animals from the same litter, etc.) comparisons between treatments should be made as far as possible between units in the same group.

4. *Balance.* If the effect of several different factors is to be tested simultaneously, the experiment should, if possible, be laid out in such a way that the contributions of the factors can be separately distinguished and estimated.

Sir Maurice Kendall, International Statistical Institute, London.

DESMINE. Stilbite.

DESORPTION. The reverse of absorption or adsorption, as in the release of one substance which has been "taken into" another by a physical process, or the release of a substance which has been held in concentrated form upon a surface.

DESULFURIZATION (Steel). Calcium.

DETACHED RETINA. Vision and the Eye.

DETECTION (Radio). The process of separating the intelligence from the carrier upon which it was modulated for transmission. See also **Modulation.** For the detection of amplitude modulated signals some sort of rectification must be used, since the rapid alternations back and forth of the radio-frequency signal are too fast for the loudspeaker to follow, and even if it did follow them the ear could not respond. Among the earliest radio detectors were various crystals such as galena, silicon, and silicon carbide. These materials, when properly mounted, pass current in one direction and not in the other, so by impressing the modulated signal on them they may be made to serve as detectors. The resulting current is a pulsating dc, which varies in magnitude with the amplitude of the modulation on the original carrier. Crystals can be used for ultra-high frequency detection. For home radio and television reception, the most common detectors are semiconductor diodes. The accompanying figure shows a typical diode detector circuit.

The modulated signal is impressed through the coupling transformer

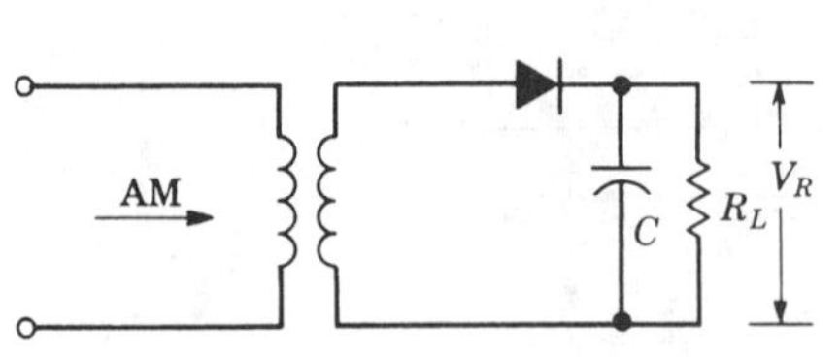

Semiconductor diode detector circuit with filter

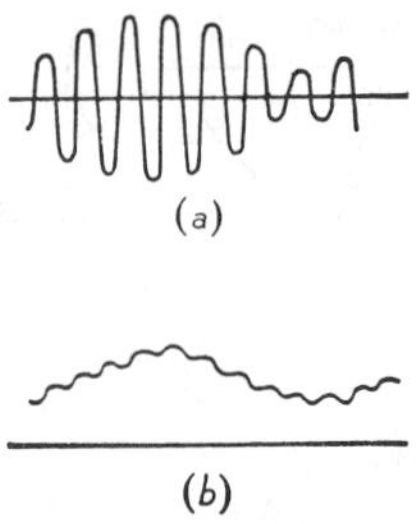

Semiconductor diode detector circuit with filter and load.

from a preceding amplifying circuit. Current flows through the diode and rectifies the signal (see **Rectifiers**). This current charges up the capacitor *C*, which, in turn, discharges through the resistor when the diode is not passing current. The capacitor does not completely discharge, but has a voltage across it similar to that shown in (a), while (b) shows the impressed modulated carrier. It is seen that the voltage across the capacitor follows the modulation closely, and by adding additional capacitance-resistance units, even the slight ripples shown on it may be removed. This voltage may be coupled to the input of a transistor amplifier stage and further amplified before being impressed on the speaker.

In the reception of continuous wave radio signals, a heterodyne detector is often used. For such signals to be audible they must be beat (see **Beat Frequency; Heterodyne**) with another signal to produce an audible frequency, since they have no modulation, and hence, even if detected by the usual means, cannot produce a varying output signal. When the continuous-wave radio signal is mixed in an electronic circuit with a locally generated signal which differs in frequency by some audio amount from it, the resultant output of the device contains various component frequencies. Among these is one equal to the difference between the two frequencies mixed in the input. By proper adjustment of the frequency of the local signal this difference can be made any value desired. A common choice is from 500 to 1000 cycles. This gives a pleasing tone for the output. Such reception is called *heterodyne detection*, a somewhat similar process where the difference is used in the first detector of the superheterodyne receiver where the difference is made fairly high (this is the intermediate frequency and is always well above the audible range). When the heterodyning signal is generated by the detector serving as an oscillator at the same time it is often called *autodyne detection*. (For frequency modulation detection, see **Discriminator.**)

DETECTOR. Any device which indicates the presence of an entity of interest, such as radiant energy, without necessarily yielding quantitative information. The term is essentially similar to *sensor* and *primary element*, namely that part of a measurement system which is exposed to and affected by the variable being measured—as the thermocouple in a temperature-measurement system, or a strain-gage load cell in a force-measurement system. In electronic circuitry, detector denotes that stage of a receiver in which demodulation occurs. In the case of a superheterodyne receiver, this is termed second detector; sometimes termed demodulator.

DETERGENTS. Complete washing or cleansing products which, among other ingredients, contain an organic surface-active compound known as a *surfactant*, the latter possessing marked soil-removal characteristics. Tonnage detergents are synthetic products sometimes referred to as *snydets*. Soaps are alkali salts of long-chain fatty acids and differ significantly in certain important performance properties and are still widely used. In addition to a surfactant, a detergent usually will contain (1) a *builder*, normally an ingredient which will chelate (sequester) or precipitate polyvalent metal ions present in the cleaning solution, particularly calcium and magnesium ions which are present in substantial quantities in hard water supplies; (2) *bleaches*; (3) *corrosion inhibitors*; (4) *sudsing modifiers*; (5) *fluorescent whitening agents* (FWA); (6) *enzymes*; (7) *antiredeposition agents*; and (8) *various additives* to enhance color, scent, and general consumer acceptability.

Surfactants. A surfactant is an organic compound consisting of two parts: (1) a hydrophobic portion, usually including a long hydrocarbon chain; and (2) a hydrophilic portion which renders the compound sufficiently soluble or dispersible in water or another polar solvent. The combined hydrophobic and hydrophilic moieties render the compound surface-active and thus able to concentrate at the interface between a surfactant solution and another phase, such as air, soil, and textile or other substrate to be cleaned.

Surfactants function to penetrate and wet soiled surfaces, to displace, solubilize, or emulsify various soils, notably oils and greases, and to disperse or suspend certain soils in solution to prevent their redeposition. Surfactants usually are classified into: (1) *anionics*, where the hydrophilic portion of the molecule carries a negative charge; (2) *cationics*, where this portion of the molecule carries a positive charge; and (3) *nonionics*, which do not dissociate, but commonly derive their hydrophilic portion from polyhydroxy or polyethoxy structures. Amph"olytic and zwitterionic surfactants are not presently of major commercial importance.

The anionic surfactants are the most commonly used and are linear sodium alkyl benzene sulfonate (LAS), linear alkyl sulfates, and linear alkyl ethoxy sulfates. To obtain improved biodegradability, manufacturers converted a few years ago to linear alkyl chains. Formulas of the major surfactants are:

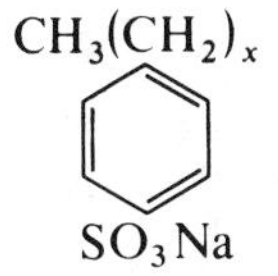

(*x* ranges from 9 to 15)

Linear alkyl benzene sulfonate (LAS)

$CH_3(CH_2)_xOSO_3Na$ (*x* ranges from 9 to 17)

Alkyl sulfate

$CH_3(CH_2)_x(O{-}CH_2{-}CH_2)_yOSO_3Na$ (*x* ranges from 7 to 15) (*y* ranges from 0 to 6)

Alkyl ethoxy sulfate

For automatic dishwashing detergents and laundry detergents, nonionic surfactants also are used, particularly because of their lower sudsing characteristics. Commercially important nonionics include the alkyl ethoxylates, the ethoxylated alkyl phenols, the fatty acid ethanol amides, and complex polymers of ethylene oxide, propylene oxide, and alcohols. Some of these formulas are:

$CH_3(CH_2)_x(O{-}CH_2{-}CH_2)_yOH$ (*x* ranges from 9 to 15) (*y* ranges from 4 to 20)

Alkyl ethoxylates

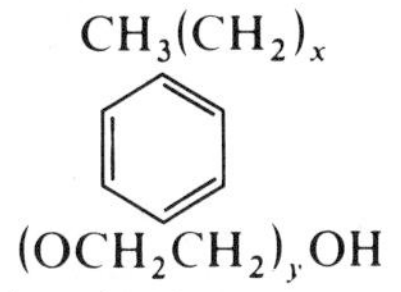

(*x* ranges from 7 to 13) (*y* ranges from 3 to 12)

Ethoxylated alkyl phenols

$CH_3(CH_2)_xCON(CH_2CH_2OH)_y(H)_z$ (*x* ranges from 9 to 17) (*y* ranges from 1 to 2) (*z* is 1 or 0)

Fatty acid ethanol amide

For specialty detergents, such as metal cleaners for electroplating and in connection with accessory laundering products, such as fabric softeners, antistatic, and germicidal preparations, cationic surfactants are used in fairly limited quantities. Tallow trimethylammonium chloride, $CH_3(CH_2)_{13\text{-}17}N^+(CH_3)_3CL^-$, is a representative cationic surfactant used in this manner.

Unlike soaps, synthetic anionic and nonionic surfactants do not form visible insolubles with the calcium and magnesium ions present in hard water. These compounds, along with certain phosphate chelating agents, revolutionized the laundry-product field a number of years ago.

Builders. When a laundry product contains a large quantity of builders it may be termed *heavy-duty* or *built* detergent. By sequestering the calcium and magnesium ions of the water, builders perform the following major functions: (1) polyvalent metal ions are prevented from combining with surfactant (in most cases) to form an adduct which would be less effective than the modified surfactant in cleansing characteristics; (2) polyvalent metal ions are prevented from combining with various soils, such as lipid residues and clays, to form less dispersible residues which adhere tenaciously to the surface to be cleaned; (3) soils removed from a surface are prevented from redepositing back onto the surface because of the dispersing action associated with chelating and charge-distribution effect; (4) additional buffered alkalinity is provided to the wash solution; and (5) destruction and removal of microorganisms is enhanced, particularly important for maintaining commercial sanitation.

The most effective builders appear to be condensed polyphosphates, mainly pentasodium tripolyphosphate (STP) and to a somewhat lesser extent, tetrasodium pyrophosphate. These polyphosphates chelate the polyvalent metal ions to form a soluble complex. Precipitating builders also are used, such as sodium carbonate. Application of this compound is limited because the precipitate formed can deposit on the surface to be cleaned and thus requires special washing procedures. Nonphosphate chelating builders include trisodium nitrilotriacetate (NTA), tetrasodium ethylenediamine tetraacetate (EDTA), and certain other polycarboxylates. Formulas for the condensed polyphosphates are:

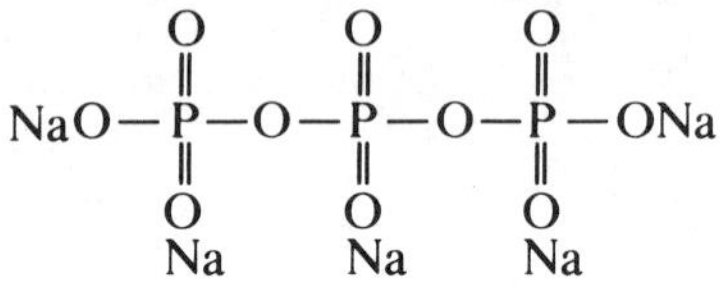

STP

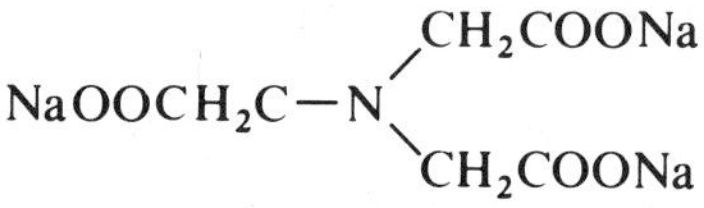

NTA

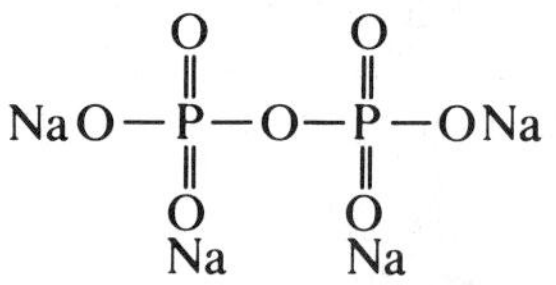

Tetrasodium pyrophosphate

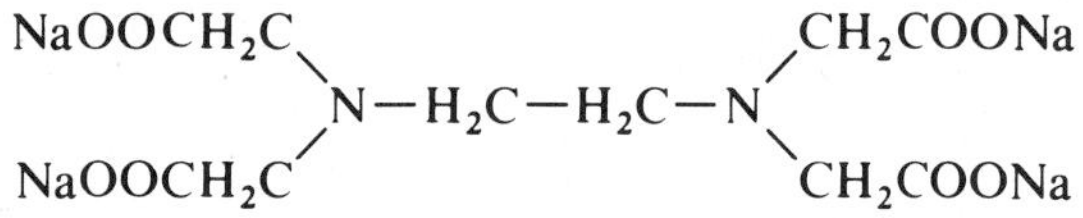

EDTA

Bleaches. Two families of bleaching agents may be used: (1) *hypochlorite* or chlorine-type; and (2) *peroxygen* compounds. The hypochlorite compounds tend to be more powerful in their oxidizing action. Commonly used chlorine-type compounds include potassium dichlorisocyanurate (KDCC), and chlorinated trisodium phosphate. The latter is a physical mixture of NaOCl, H_2O, and Na_3PO_4. Of the peroxygen bleaches, sodium perborate, $NaBO_3 \cdot 4H_2O$ is the most common. Because aqueous bleach solutions have not been found to be sufficiently stable in the presence of other detergent ingredients, it is common practice to include a solid bleaching agent in detergents.

Corrosion Inhibitors. The unmodified alkaline detergent can be corrosive to aluminum, porcelain, and the overglaze on fine china. This type of corrosion is essentially prevented by adding soluble silicates into the detergent mix. The soluble silicates contain varying ratios of SiO_2 and Na_2O.

Sudsing Modifiers. For certain types of detergent products, a sudsing action presents an aesthetic appeal even though it may not add essentially to the functioning of the product. Sudsing can be increased by adding small quantities of anionic surfactants, such as mono- and diethanol amides of C_{10-16} fatty acids. In other instances, sudsing depression is desired. This can be accomplished by the addition simply of the C_{10-16} fatty acids.

Fluorescent Whitening Agents (*FWA*). These agents also are termed brighteners and optical bleaches. They are organic chromophores which absorb incident light in the ultraviolet region and reemit part of the absorbed energy as visible light, usually in the blue region of the visible spectrum. For use in detergents, the chromophore is modified with organic substituents to make it substantive to one or more textile substrates from a laundry-wash solution. Thus, there is enhancement of the brightness and whiteness of the fabrics onto which FWA deposit has been made. The result is that an added portion of incident light is reflected by the fabric. Low levels of sulfonated trazinylstilbenes are used as FWAs in detergents for cellulosic fibers. The trend is toward incorporating a brightening agent into other synthetic fibers during their manufacture.

Enzymes. Low levels of enzymes may be added to detergents and pre-soak products. Proteolytic and anylolytic enzymes attack and loosen soils and stains with protein and carbohydrate substituents (including body soils, numerous food stains, grass stains, blood, and others). The enzymes perform catalytically and quite specifically and thus can be used effectively at low levels, assuring safety for fibers and preservation of textile colors. The essential step needed to make enzymatic action available in textile detergency was the discovery and production of the *B. subtilis* and *B. licheniformis* mutants and their metabolites which remain active under laundering conditions.

Antiredeposition Agents. Substances which tend to prevent redeposition of soils once removed from a fabric include carboxymethyl cellulose and polyvinyl alcohol. The actual mechanisms involved still are under investigation.

Although detergents commonly are supplied as powders, they are available in liquid and paste form. In some instances, the liquid products are essentially the same as the dry detergents without going through the spray-drying operation.

DETERGENT SYSTEM. Colloid System.

DETERMINANT. A square array containing n^2 elements and said to be of order n. Let the element A_{ik} stand in the ith row and kth column of the array, then the determinant in its developed or expanded form is a homogeneous polynomial of the nth degree in these elements.

To evaluate a determinant, form all products, $n!$ in number, by taking one element A_{ik} from each row and column. The subscripts in the products, $i, i', i'', \ldots$ and $k, k', k'', \ldots$ will then include all permutations of the numbers $1, 2, \ldots, n$. Rearrange the subscripts i so that these numbers are in their natural order. The second subscripts k will then require either an even number or an odd number of interchanges to return them also to the natural order $1, 2, \ldots, n$. The value of the determinant is then defined as

$$|A| = \det A = \sum (-1)^h A_{1k_1} A_{2k_2} \cdots A_{nk_n}$$

where the summation is made over all permutations $k_1, k_2, \ldots$ of the subscripts k and h is the number of interchanges needed to restore the natural order.

Another method of evaluation is the Laplace development. Define the minor as any determinant of order $m < n$ obtained by deleting one or more rows and columns of a determinant of order n. If the row and column containing the element A_{ik} is so removed, the resulting determinant is called the complementary minor to A_{ik}. Attach the sign $(-1)^{i+k}$ to the minor and it is called a signed minor or cofactor. With these definitions, the Laplace development can be written as:

$$|A| = \det A = \sum_{i=1}^{n} A_{ik}A^{ik} = \sum_{i=1}^{n} A_{ki}A^{ki}, \qquad k = 1, 2, \ldots n$$

where A^{ik} is the cofactor to A_{ik}.

The following properties are sometimes useful in evaluation of determinants. In each case, the word *column* can always be substituted for the word *row*, or the word *row* for *column*.

1. The value of a determinant is unchanged if rows are changed into columns or if the elements of any row, multiplied or divided by a constant, are added or subtracted from the corresponding elements of another row. Its sign is changed if two rows are interchanged.

2. The value of a determinant is zero if all elements in one row are zero, if two rows are identical, or if all the elements of a row are proportional to those of another row. This property follows from the Laplace development if the i or k in A_{ik} is different from the i or k in A^{ik}.

3. If each element in any row of a determinant is the sum of two or more quantities, the given determinant can be written as the sum of two or more determinants of the same order. Thus, if $A_{ik} = a_{ik} + b_{ik} + c_{ik} + \cdots$ (i fixed), the new determinants will be identical with the old one except for the ith row and this will contain a_{ik} in the first one, b_{ik} in the next one, etc.

4. Determinants may be multiplied together by a rule similar to that used for a matrix product.

5. The partial derivative of a determinant with respect to an element A_{ik} equals its cofactor A^{ik}.

The determinant and its matrix, a closely related array, are of importance in the study of simultaneous equations. However, in any practical case, where the order of a determinant is greater than three or four, the classical methods of expansion (see also **Cramer Rule**) become prohibitively laborious and should seldom be used. Matrix methods are then much more satisfactory.

A similar procedure can also be used to evaluate a determinant. In principle, it depends on the solutions of the following equations; (1) $A_{11}x_1 = 1$; (2) $A_{11}x_2 + A_{12}y_2 = 0$, $A_{21}x_2 + A_{22}y_2 = 1$; (3) $A_{11}x_3 + A_{12}y_3 + A_{13}z_3 = 0$, $A_{21}x_3 + A_{22}y_3 + A_{23}z_3 = 0$, $A_{31}x_3 + A_{32}y_3 + A_{33}z_3 = 1$; etc. Then det $A = 1/(x_1 y_2 z_3 \ldots)$. In practice, the work can be carried on as shown in the following form, which uses a determinant of order 3, as an example:

$$\begin{array}{cccccccc}
 & A_{11}^* & A_{12} & A_{13} & & -A_{12}/A_{11} & -A_{13}/A_{11} & \\
(A) & A_{21} & A_{22} & A_{23} & & 1 & 0 & (B) \\
 & A_{31} & A_{32} & A_{33} & & 0 & 1 & \\
\hline
 & & B_{11}^* & B_{12} & & -B_{12}/B_{11} & & \\
 & & B_{21} & B_{22} & & 1 & & \\
\hline
 & & & C_{11}^* & & & &
\end{array}$$

To calculate the elements B_{ij}, omit the starred row in (A) and multiply rows of (A) by columns of (B). In the same way, $C_{11} = -B_{21}B_{12}/B_{11} + B_{22}$. The value of the determinant is the product of all starred elements, det $A = A_{11}B_{11}C_{11}$. By an obvious extension of the method, determinants of higher order can also be evaluated.

(For properties of some special determinants see **Gram Determinant; Hessian; Jacobian; Secular Determinant;** and **Wronskian.)**

See also terms listed under **Mathematics.**

DETERMINATE STRUCTURE. Any structure in which the reactions and stresses can be found by means of the equations of statics only is a determinate structure. If a sufficient number of such equations cannot be set up from known conditions, the structure is not statically determinate.

DETINNING. Goldschmidt Detinning Process.

DETONATING EXPLOSIVE. Explosive.

DETRITUS. General term for unconsolidated sediments derived from pre-existing rocks by natural agencies. Derived from the Latin word meaning worn.

DEUTERIC. Used by petrologists to describe those alterations in an igneous rock which occur during the later stages of its solidification.

DEUTERIUM. The isotope of hydrogen with mass number 2 is termed deuterium. The symbol D is sometimes used. Using ocean water as a reference, the atomic abundance of deuterium in natural hydrogen is 0.0149%. Deuterium oxide D_2O is known as heavy water and was first identified by Harold C. Urey in 1932. Urey noted a slight shift in the spectrum of deuterium and tritium as compared with protium. The diameter of the electron orbit for deuterium is slightly greater than for ordinary hydrogen, and still greater for tritium. Deuterium and deuterium oxide gained prominence largely because of their excellent properties as moderators in nuclear reactors. See also **Uranium.**

DEUTERIUM-DEUTERIUM REACTION CHAIN. Nuclear Reactor.

DEUTERIUM-TRITIUM REACTION CHAIN. Nuclear Reactor.

DEUTERON. The nucleus of deuterium (heavy hydrogen) is known as deuteron. A particle that contains one proton and one neutron also is termed a deuteron.

DEUTOPLASM. Inert material stored in eggs. Yolk.

DEUTSCHES ELEKTRONEN-SYNCHROTRON. Particles (Subatomic).

DEVELOPING PROCESS. Photography and Imagery.

DEVELOPMENTAL BIOLOGY. The study of how highly specialized tissues and organs form from a series of divisions of a single fertilized egg. This field seeks to answer such questions as to what mechanism commits a cell to a pathway of development that will only become apparent many generations later? How does a given cell "know" that, at some later time, it will be part of the heart instead of the liver of the central nervous system? Researchers in developmental biology are reviewing embryology and growth phenomena in terms of molecular biology. Some specialists in this field suggest that many exciting and unexpected revelations lie ahead. One scientist, for example, has observed that "Contrary to prior beliefs that genes become active during development, it now looks like genes become progressively inactive, suggesting that the driving force behind development is not the increasing activity of genes, but their selective inactivity."

References

Grant, P.: "Biology of Developing Systems," Holt, Rinehart and Winston, New York, 1978.

Kolata, G. B.: "Developmental Biology: Where is it Going?" *Science*, **206**, 315–316 (1979).

Weissbach, H., and S. Pestka (editors): "Molecular Mechanisms of Protein Biosynthesis," Academic, New York, 1977.

DEVIATION. 1. The deviation of x from a is x-a. The absolute value of the deviation of x from a is defined as $|x\text{-}a|$.

2. Light passing through a prism is always deviated away from the diffracting edge if the refractive index of the prism is greater than unity. If i_1, i_2 are angles of incidence and emergence while r_1, r_2 are the corresponding angles of refraction, the deviation of the prism is given by

$$\Delta = i_1 + i_2 - r_1 - r_2$$

If the refracting angle of the prism is $\phi = r_1 + r_2$, then

$$\Delta = i_1 + i_2 - \phi$$

3. For deviation of a compass, see **Compass (Navigation).**

DEVIATION DISTORTION. Distortion in an FM receiver caused by inadequate band-width, inadequate amplitude-modulation rejection, or inadequate discriminator linearity.

DEVIL RAYS. Skates and Rays.

DEVIL'S DARNING NEEDLE. Odonata.

DEVITRIFICATION. The process by which the natural rock glasses, such as obsidian and tachylyte, develop minute but definite minerals, usually quartz and feldspar.

Devitrification also applies to manufactured glasses, denoting crystallization and detected by the appearance of opaque areas. See also **Vitreous State.**

DEVONIAN. The name of a geologic period. Type locality, Devonshire, England. The formations of this period were first studied and described by R. I. Murchison in 1839. The Devonian period began 330 million years ago and lasted for 50 million years. The Devonian formations are well exposed in eastern North America and parts of the North American Cordilleran. In the Appalachian Geosyncline the Devonian is largely represented by an immense thickness of red and brown shales and sandstones of deltaic and estuarine origin, and the transition between the sediments of this system and that of the underlying Silurian is so gradual that the boundary is extremely difficult to locate by physical means alone. In Britain the Devonian is represented by a marine limestone (facies) in the type locality, and a red non-marine sandstone (facies) to the north. This red sandstone (facies) is referred to as the "Old Red" by British geologists. The Scottish "Old Red" contains the famous fossil "fishes" described by Hugh Miller in 1851. These fish include two distinct groups, Ostracoderms and Ganoids, the latter being the supposed ancestors of the Amphibia or first terrestrial vertebrates. The first undoubted evidence

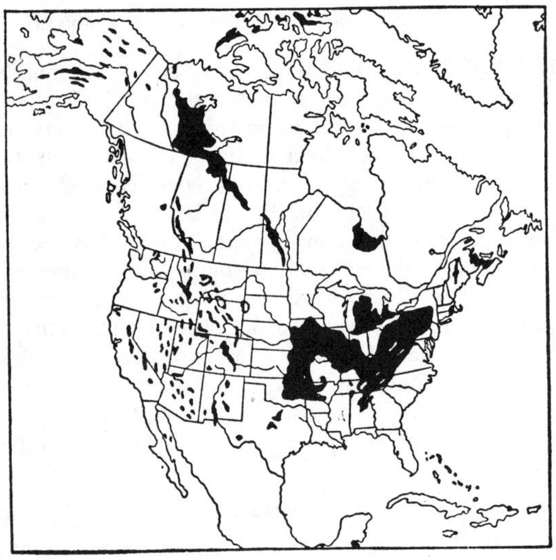

Areas of outcrops (surface distribution) of Devonian, Mississippian, and Pennsylvanian strata in North America.

of terrestrial plants occurs in the Devonian, the late Devonian types being the progenitors of the Carboniferous forms. In England, Scotland, Spitzbergen, western U.S.S.R., and Norway occur great thicknesses of terrestrial, intermontane clastic sediments similar to those found in the Proterozoic. Highly fossiliferous marine sediments, including sandstones, shales and limestones, are particularly well exposed in New York State. Many of the limestone formations contain reefs composed principally of compound corals, Bryozoa and Calcareous Algae. Among the marine invertebrates goniatites and Eurypterids are particularly representative. Other common marine types are corals, Bryozoans (reefs) and Echinoderms, Pelecypods and Trilobites. The spire-bearing Brachiopods (Spirifers), which started in the Silurian, reach their maximum development in genera and species in the Devonian. The only fossil evidence of a terrestrial vertebrate rests upon a footprint (probably that of an amphibian) found in the Upper Devonian of western Pennsylvania. Beginning with the middle and ending with the period, mountain building occurred in the New England States. The principal economic products derived from the American formations are petroleum and natural gas, first exploited in 1859 in western Pennsylvania, New York, Ohio and West Virginia.

DEVONITE. The name given by Johannsen, in 1910, for a variety of porphyritic basalt containing phenocrysts of potassium-rich plagioclase. Type locality Mt. Devon, Massachusetts.

DEW. Precipitation and Hydrometeors.

DEWAR FLASK. A vessel with a double wall, in which the region between the walls has been evacuated, and in which the walls bordering this space have been silvered. With this construction the region within the inner container is very well insulated from the outside. The vessel is commonly used for storage of liquefied gases, and a similar device is used widely for hot or cold beverages. Large Dewar vessels are used for the truck and rail movement of liquefied gases.

DEWBERRY. Rose Family.

DEW CELL. Precipitation and Hydrometeors.

DEW CLAW. Small hoofs of the rudimentary toes, found just above the functional hoofs in some of the even-toed hoofed animals.

DEWLAP. The fold of skin which hangs below the neck in cattle.

DEW POINT. Precipitation and Hydrometeors.

DEW-POINT METER. Hygrometer; Meteorological Instruments.

DEXAMETHASONE. Steroids.

DEXTROROTATORY COMPOUND. Asymmetry (Chemical); Isomerism.

DEXTROSE. Carbohydrates; Starches; Sweeteners.

DEZINCIFICATION. A form of electrolytic corrosion observed in some brasses where the copper-zinc alloy goes into solution with subsequent redeposition of the copper. The small red copper plugs thus formed in the brass are usually porous and of low strength. In recent years, the term dezincification has also been applied in a more general sense to signify any metallic corrosion process that dissolves one of the components from an alloy.

2,4-D HERBICIDE. Herbicide.

DHOBIE ITCH. Fungus.

DIABETES INSIPIDUS. Characterized by the passage of large quantities of urine (polyuria) and usually accompanied by excessive thirst, diabetes insipidus is caused by the absence or insufficiency of vasopressin (antidiuretic hormone). Vasopressin and oxytocin are secreted in the hypothalamus. They are transported along the axons of the neurons and stored in the posterior pituitary. The posterior pituitary is composed of the terminal portions of neurons whose origin is in the hypothalamus. See also **Hypothalamus;** and **Pituitary Gland.**

A mild insufficiency of vasopressin may cause partial diabetes insipidus which may not require treatment unless the condition (polyuria) interferes with sleeping. The disease usually has a sudden onset and the patient will exhibit a strong preference for iced drinks. The diagnosing physician will rule out vasopressin resistance arising from significant renal disease as well as compulsive water drinking (a psychogenic disorder). The most recent diagnostic tool is direct measurement of plasma arginine vasopressin. Where polyuria interferes with sleeping, chlorpropamide, which potentiates the effect of vasopressin on renal concentrating ability, may be prescribed, but only in the confirmed absence of anterior pituitary insufficiency. In more severe cases of diabetes insipidus, vasopressin tannate may be administered intramuscularly every few days. More recently available in some countries is an analogue of vasopressin, known as DDAV. With continued good reports, DDAV may ultimately become the most effective therapy.

DIABETES MELLITUS. This major metabolic disorder is difficult to define because beyond the principal clinical manifestation, *hyperglycemia* (high level of blood glucose), the condition is no longer considered a single disease, but rather a heterogeneous group of diseases, all of which lead to an elevation of blood sugar. Furthermore, diabetes mellitus should be viewed in terms of the numerous other diseases which develop as a consequence of it.

Diabetes mellitus has been known since 1500 B.C. (Ebers papyrus of Egypt). The name *diabetes* was given to the disease by Aretaeus of Cappadocia in the Second Century A.D., who said, "Diabetes is a strange disease that consists in the flesh and and bones running together into the urine." The sweetness of diabetic urine was discovered by an Indian physician in the Sixth Century A.D. Then the word *mellitus*, meaning "honeyed," was first used.[1]

O. Minkowski and Baron Joseph von Mering, physicians of Strasbourg in 1889, removed the pancreases from several dogs in an effort to find out if that gland was essential to life. The researchers were surprised to find that flies were attracted to the canine urine. The urine was analyzed and found high in glucose content. Thus, the first recorded association of the pancreas with glucose metabolism was established.

The search for an antidiabetic substance, presumed to be secreted by the pancreas, was commenced about 1909, and even before such a substance was discovered and identified, it was given the name *insulin.* Early attempts to treat pancreatectomized dogs with crude pancreas extract given orally to the canines proved unsuccessful, and the failure was later explained on the basis that the extract (a protein)

[1] The name *diabetes mellitus* is used in contradistinction to a second type of diabetes called *diabetes insipidus.* Mellitus means honeylike. The name was applied to this type of diabetes because of the sweetness of the urine. Diabetes insipidus, conversely, is a disease characterized by the excessive elimination of urine, but urine that is *insipid* or tasteless. This latter condition is associated with disturbances in a part of the hypothalamic-hypophyseal system and is described under **Diabetes Insipidus.** When the name *diabetes* alone is used, it usually refers to diabetes mellitus.

was destroyed by protein-cleaving enzymes in the gastrointestinal tract. Canadian investigators Banting and Best later extracted insulin from dog pancreas and found that injection of the insulin into diabetic dogs reduced the glucose level of the canine blood quite promptly. This experiment led within a few years to the use of insulin extract in human subjects, a practice which with some improvements continues in use by many diabetics today. The experiment was highly publicized and for awhile it was believed that a "cure" for diabetes mellitus had been achieved. The administration of insulin did ameliorate the primary symptoms of the disease, including death from coma, but several years later it was found that insulin did not cure the many serious long-term complications of the disease.

Insulin is currently derived from cattle and swine. Taking advantage of recombinant DNA gene splicing techniques, research has been conducted to produce a synthetic human virus. (The animal insulin now in use is about 1–2% impure and can cause allergic reactions in some diabetics.) In the procedure, researchers used a complex technique for inserting the genetic code for human insulin into an *E. coli* bacteria strain, a type which is usually found in the human intestine. The synthetic genes were then triggered by the bacteria to produce one of the two protein chains found in human insulin. After being isolated and purified, the two protein chains were combined in the laboratory to create the synthetic human insulin. Because of the newness and concern over this type of technology, commercialization of such products proceeds slowly. See also **Recombinant DNA.**

Types of Diabetes Mellitus

Diabetes mellitus is commonly classified in two categories, essentially differentiating chronology (age of onset) and, to some degree, the type of therapy required:

(1) *Juvenile-Onset Diabetes.* This is insulin-dependent or ketoacidosis-prone diabetes for which insulin is required to prevent ketoacidosis. The disease is usually found in children and adolescents—hence the term juvenile-onset diabetes. In the United States, about 1.5 million people (including 100,000 children) require insulin daily. The rate in black children is only about one-quarter of that in whites, and, in orientals, the rate is even lower.

(2) *Maturity-Onset Diabetes.* Usually first observed by middle-age persons, the disease is essentially controlled by dietary measures and, in some cases, hypoglycemic agents in concert with dietary measures. It is estimated that between 2 and 3 million persons in the United States are in this category, with possibly another 2 million persons who suffer from undiagnosed diabetes mellitus. Reliable statistics are not available. Maturity-onset diabetes may worsen as the result of a number of factors, including lack of patient cooperation, to the point where insulin may be required to control ketoacidosis.

Diabetes-Induced Diseases. As of the early 1980s, diabetes mellitus is considered by many authorities as a major disabling disease and contributing cause of death in the United States and a number of other countries. Diabetes mellitus is the manifestation of a major disturbance in body chemistry, the ultimate consequences of which, over a period of time, are several and serious. Particularly in maturity-onset diabetes, these consequences can be forestalled or even overcome through prudent eating and exercising and precisely following therapeutic regimens. Traditionally, diabetic complications have been ascribed to high glucose levels (blood sugar), the high average of which is sustained over long periods. Although there is important rethinking going on as regards this generality, it nevertheless remains the current majority view and the fundamental therapies used are directed toward using blood glucose levels as the major guideline. Research into the root causes of diabetes mellitus may alter a number of current concepts in the relatively near future.

Diabetes-induced retinopathy is a major cause of poor vision, ultimately leading to blindness. Retinopathy results from deterioration of tiny blood vessels in the eye. See **Vision and the Eye.** In the absence of sufficient natural or administered insulin, certain cells, such as those of fat and muscle, starve for glucose because it cannot enter them unless insulin is present. But the sugar can enter many other cells, such as those of nerve and lens, in the absence of insulin. Because of the resulting high glucose concentrations, the cells are damaged. The manner in which this damage is caused is under intensive investigation. In addition to nerve damage, retinopathy, and cataract formation, diabetes is also associated with damage to small blood vessels. The lining of these vessels (basement membrane) thickens while simultaneously the vessels become more porous. This is believed to be an underlying cause of kidney damage in diabetics. Diabetes is also heavily implicated as a major causative factor in stroke and heart disease.

In addition to diet control and traditional drug therapy, researchers have been investigating more effective and precise controls for blood glucose levels. This immediately suggests an artificial pancreas, toward which considerable work has gone forward. A major problem with this development is that of finding an appropriate sensor for determining blood glucose concentration so that this information can be fed back to the artificial organ, which would then "secrete" insulin at just the right rate. Immunological rejection has markedly slowed progress toward natural pancreas transplants.

Physiology and Biochemistry of Diabetes Mellitus

The Pancreas. The physiology and disorders of the pancreas as related to matters other than diabetes mellitus are described in the entry on **Pancreas.** The pancreas gland contains *acnar cells,* which manufacture digestive enzymes for secretion into the duodenum; and groupings of cells known as the islets of Langerhans, which secrete a variety of hormones, including insulin, into the bloodstream. These islets, first discovered by Paul Langerhans over a century ago, are involved in the melange of metabolic and metabolically induced disorders commonly called diabetes mellitus. It has been estimated from micrographic techniques that there are between one and two million such islets in the pancreas. They are about 200 microns in diameter and make up only about 2% of the mass of the gland. They are generously vascularized. Research to date indicates that there are at least four kinds of cells in each islet. The cells have been differentiated as to function. (1) The hormone *glucagon* is secreted by alpha cells, which constitute about 20% of the islet. (2) The hormone *insulin* is secreted by beta cells, which account for about 75% of the islet. (3) The delta cells, comprising less than 5% of the islet, secrete the hormone *somatostatin.* (4) The PP cells, also comprising less than 5% of the islet, secrete *pancreatic polypeptide hormone.*

Insulin from the beta cells is known to regulate carbohydrate and fat metabolism, especially glucose oxidation. It also stimulates amino acid and glucose transport into cells and protein synthesis. Insulin mediates the uptake of glucose by liver, muscle, and fat cells and the conversion of glucose to glycogen in these cells.

Glucagon from the alpha cells, also known as the HGF factor (hyperglycemic-glycogenolytic factor), is known to increase blood sugar, blood potassium, oxygen consumption, liver glycogenoloysis, gluconeogenesis, and nitrogen and salt excretion, while it decreases liver glycogen, protein formation, gastric juice, and fatty acid synthesis. Glucagon is antagonistic to insulin and reverses the actions effected by insulin.

Somatostatin from the delta cells is known to inhibit the secretion of both insulin and glucagon. The role of pancreatic polypeptide hormone released by the PP cells is still poorly understood.

The sequence of events occurring in the pancreas and associated organs, greatly simplified, proceeds about as follows:

(1) The ingestion of carbohydrates and proteins causes the beta cells to respond directly to the rise in blood glucose and amino acids. This response takes the form of secreting increased insulin. This response is governed both by amount and rate. Where insulin furnished is deficient or too slow, the blood glucose level will remain high for an appreciably longer period of time. In diagnosing diabetes mellitus, physicians employ a glucose-tolerance test as illustrated and explained by the accompanying figure.

(2) The liver is alerted by the higher insulin level—chemically sensing that a supply of fuel (food) is on the way. In a reverse (fasting) situation, low insulin levels signal an inadequate fuel supply.

(3) By way of a process still not fully understood and occurring simultaneously with the build-up of insulin level, chemical signals from the enteric system (in the form of pancreozymin or cholecystokinin) enter the blood, these substances also stimulating the beta cells to increase insulin secretion. There appears to be a quantitative relationship between these signals and the amount of fuel introduced into the system.

(4) Still unproved, but strongly suspected, is the concept that the

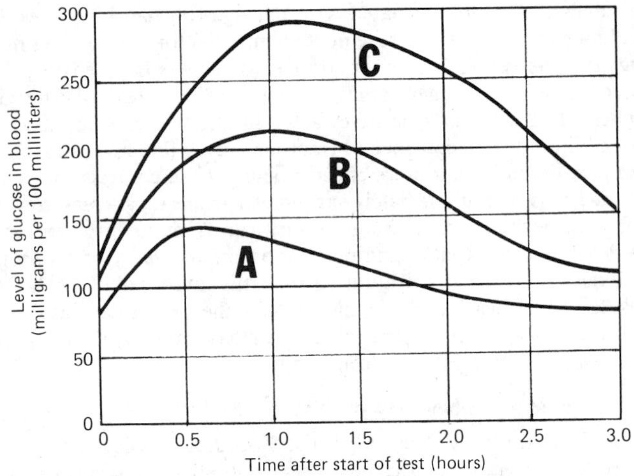

In a glucose-tolerance test, the patient consumes a flavored drink that contains exactly 100 grams of glucose. The blood glucose is measured before the start of the test (time = 0) and at one-hour intervals up to a total of 3 hours. The patient is directed to fast for a period prior to start of the test. Curve A indicates response of a normal, nondiabetic person. Curve B indicates an impaired glucose tolerance, but not quite that of a diabetic. Curve C is typical response for a diabetic, demonstrating how the glucose level, after peaking, remains at an elevated level even after three hours. In normal persons, the glucose level is below 160 milligrams per 100 milliliters after one hour of eating a normal meal (or glucose load given in the test); and a level of 120 mg/100 ml two hours later.

gastrointestinal inhibitory peptide (GIP) and pancreozymin are coded—so as to respectively designate ingestion of carbohydrate and/or protein.

(5) Alpha cells, also responsive to blood glucose and amino acid concentrations, secrete more or less glucagon. A low level of glucose increases glucagon secretion. A high glucose level suppresses it. Where both glucose and amino acid levels increase, the response to glucose predominates, causing glucagon to decrease.

(6) Increased glucagon initiates a response by the liver for increased glycogenolysis. Decreased glucagon deaccelerates the process. See **Liver.**

The liver functions in the "fed" state to consume glucose for the synthesis of glycogen and fat, while using some of it to furnish its own energy needs. In the total process, glucose not only is utilized by muscle for energy and for the synthesis of glycogen, but also is used by adipose tissue for the synthesis and storage of fat. Insulin inhibits the release of free fatty acids from adipose tissue. Thus these acids in the blood are significantly reduced during the "fed" state. Central to the success of these metabolic processes is the optimal maintenance of insulin : glucose : glucagon proportionalities at all times.

Carbohydrate metabolism and fat metabolism are further described in the entries on **Carbohydrates;** and **Lipids,** respectively.

In diabetes mellitus, one of at least three situations may be present: (1) Essentially no natural insulin is available; (2) natural insulin is available, but not in sufficient quantity to handle a normal diet; and (3) according to a relatively recent concept, natural insulin is available, but is not effectively utilized because of malfunctioning or sluggish receptors. Situations (2) and (3) may be present together. Considering the vast number of cells in the body to be served by a markedly reduced number of beta cells, it becomes obvious why increased weight over normal (obesity) antagonizes the disorder. It is also evident why patients are frequently advised to eat less per meal and more often, thus spreading out the demand for insulin during the day rather than concentrating that demand three times per day.

A situation of severe insulin insufficiency, in terms of system responses, marks the fasting (starving) state and the system, recognizing this as an emergency, activates all elements of the system, particularly the liver, to adjust to it. Muscle tissue releases amino acid at an accelerated rate; free fatty acid is released from adipose tissue at an enhanced rate; hepatic (liver) uptake of amino acids and free fatty acids is accelerated; production of glucose and keto acid by the liver is maximized. Thus, although not in need, the body is furnished with fuel in excess. This is reflected by a marked rise in blood glucose levels. In this state, grossly damaging consequences occur if the patient is not immediately treated. High levels of glucose in the urine (glucosuria) and high levels of ketones in the urine (ketouria) initiate polyuria (large quantities of urine), leading to dehydration, hypovolemia (blood loss) and, within a day or two, death. The latter phases of this chain are called *diabetic coma.*

Root Causes of Diabetes Mellitus

Research has been conducted by many investigators in several countries for decades in an effort to understand the root causes and mechanisms of diabetes mellitus. As of the early 1980s, it appears that, like the multifaceted nature of the disorder itself, there are multiple causes, some working alone and others working together. Suspected for many years but only comparatively recently confirmed on the basis of research that still may be characterized as pioneering is the *virus connection* with the disease as well as the role of *autoimmunity.* The role of *genetics* as a predisposing factor has been suspected for many decades and has been accepted in general terms for a number of years in some cases. The details of this connection, however, remain poorly understood, and research in this area continues at a good pace. Much progress has been made since the early 1970s toward improved comprehension of the phenomenon of *insulin resistance.* These studies are oriented toward insulin and glucagon receptors. It is only comparatively recently that receptors have been isolated in the laboratory. While rather simplistic, the antagonistic roles of *obesity* and *lack of exercise,* long associated with the disease, are being increasingly emphasized as drugless therapy. Weight reduction and exercise can achieve strikingly positive results in some patients. These findings are now being better documented as the result of laboratory tests of human subjects. Studies along these lines were commenced at the Yale University School of Medicine in 1979. Other possible causative factors of a biochemical nature are being investigated by researchers. These include the phenomenon of *glycosylation* and the effects of *vitamin D deficiency,* among others. The probability that some *drugs and chemicals* induce diabetes comprises a more recent avenue of research.

Virus-Diabetes Connection. The possible implication of virus infection as a primary cause or triggering event that leads to diabetes mellitus dates back to the early 1900s when a physician in Philadelphia noted that diabetes shortly followed mumps in one of his patients. Over the years, there have been a number of similar reports possibly associating the onset of diabetes closely following an infection (particularly of mumps), but no convincing string of evidence was developed. In fact, some authorities pointed out that if such a connection exists, considering the incidence of mumps and the incidence of diabetes mellitus without such a connection, the combination of circumstances indeed must be rare—and it was further suggested that to support the connection, a special and rare strain of mumps virus must be assumed, or that a relatively few people may have some unusual, possibly genetically determined, response to mumps virus. Researchers in Australia also pointed to a possible connection between rubella (German measles) and diabetes. A more convincing tie between viruses and diabetes came out of research commenced in 1968 at the University of Vermont College of Medicine, where a connection between a variant of encephalomyocarditis (EMC) virus and diabetes in laboratory mice was demonstrated. Examination of the pancreas of infected animals showed damaged beta cells as well as inflammatory white cells in the islets of Langerhans. However, proof that the damage was caused by the virus was not fully developed. In following this avenue of research further, researchers at the National Institutes of Health have more recently demonstrated, through the use of immunofluorescence micrography, that EMC can infect and damage beta cells, primarily by replication of the virus within the cells. Further research has indicated that upon cell damage there is an abnormal release of insulin into the circulation, thus lowering the blood glucose level, but that within a few days the animal's reserve of insulin is depleted to a subnormal level. The animals then demonstrate the symptoms (consumption of excessive amounts of water and food) of juvenile-onset diabetes.

Generalities pertaining to a virus-diabetes connection could not be made at this juncture, however, because it was found that only certain inbred strains of mice developed the disease. After many breeding experiments (crossing and backcrossing), it was concluded that a single gene locus appears to play a major role in determining susceptibility to this particular EMC-induced diabetes. Subsequent research has indicated that the susceptibility to the virus may be a function of the number or type of viral receptors on the surface of the beta cells. During the 1970s, a number of other viruses were tested for a possible diabetes connection, including Coxsackie virus and respiratory-entero-orphan virus, among others. See **Coxsackie Virus.**

The first direct evidence of a connection between a virus infection of a human and diabetes was reported by Notkins and associates (National Institute of Dental Research, National Institutes of Health, and National Naval Medical Center) in 1978. The researchers succeeded for the first time in isolating virus Coxsackie B4 from the pancreas of a 10-year-old boy who had developed a fatal case of diabetes soon after a flu infection. When the same virus was injected into laboratory animals, certain inbred mice developed the same disabling type of diabetes. These findings also strengthened the theory that inherited susceptibility may play a key role, since only certain inbred laboratory animals developed the disease in the course of the experiments.

It is postulated by a number of authorities that viruses may be only one of several fundamental causes of diabetes mellitus and, in fact, may be a minor cause. It is further postulated at this juncture that several and even many viruses may be involved in this connection and, if so, the outlook for a vaccine to prevent diabetes does not appear good in the near future, the situation being somewhat similar to that of the common cold that has numerous causative agents.

Genetics—Heredity. Pyke and associates (King's College Hospital, London) studied 100 pairs of identical twins in genetic research on diabetes mellitus in the early 1970s. Although twins had been studied earlier by other investigators, this was by a substantial margin the largest sample ever undertaken. Among the conclusions drawn was that genetic factors are predominant in maturity-onset diabetes, but that factors, such as virus infection, environmental substances, etc., are required to trigger the disease in a genetically diabetes-prone individual. Conclusions have since been modified. In recent years, considerable attention has been given to determining the relationship between diabetes and the histocompatibility antigens. These antigens are sometimes referred to as the HLA system. In research at the Steno Memorial Hospital (Copenhagen, Denmark), it was found that two HLA systems (B8 and B15) were found in diabetics with an occurrence three times that of nondiabetics—but that this rate applied only to juvenile-onset diabetes and not the maturity-onset type. It was also found that with HLA antigens at the D locus (with more than one high-risk allele present in the same person), the chance of developing juvenile-onset diabetes was increased up to ten times. Later research showed that certain HLA alleles are associated with a decline in the incidence of juvenile-onset diabetes. As pointed out by Notkins (1979), ". . . the high-risk alleles associated with the HLA complex might code for a deficient immune response to agents that preferentially attack beta cells, thereby allowing beta-cell damage and diabetes to result. Conversely, the protective alleles might enhance the host's immune response to such invaders."

Thus, although differing, genetic correlations have been found in both juvenile-onset and maturity-onset diabetes. Considerably more research is required to transform genetic findings into methods of prevention and therapy.

Insulin Resistance. Because of the numerous, sometimes conflicting findings concerning diabetes mellitus and the diseases which it induces, a consensus has developed in recent years that the condition is a rather heterogeneous group of diseases or indeed a multifaceted disease which, in either case, leads to an elevation of the glucose level of the blood. Research, particularly during the last few years, has shown that the pancreas of many maturity-onset diabetics produces ample quantities of insulin. There is, however, a reduced sensitivity of fat and muscle cells to the effects of insulin, a phenomenon sometimes called *insulin resistance.* The mechanics of this phenomenon still are not well understood, but a breakthrough was achieved in 1971 when the insulin receptor from rat liver membranes was first isolated by researchers at Johns Hopkins University School of Medicine.[1] The isolation of hormones is a very tedious and exacting procedure. Some of the details of procedure are aptly described in the Maugh (1976) reference listed. A bit later, other researchers at Georgetown University Medical Center isolated glucagon receptors. This earlier research cleared the way for studying the characteristics of insulin binding. Studies have been conducted in the United States and other countries in which obese mice that exhibited insulin resistance comparable to human maturity-onset diabetes were used. The general reasoning is that if insulin is available and if ways can be found to provide ample well-functioning insulin receptors in cells, then another route to diabetes therapy is available. This avenue of research already has impacted on traditional therapy and will continue to do so as more information is gained and as more physicians become better acquainted with new techniques. The research is still in a comparatively early phase.

Some of the wide diversity of findings include: (1) Although the insulin receptor is better understood, the glucagon receptor is equally important and increasing research is being directed toward it. Where insulin in diabetics may be present in normal, higher, or lower concentrations, as determined by the type of diabetes, it has been established that glucagon is always present in greater than expected levels.

(2) Investigators have found that there are reduced numbers of insulin receptors in insulin-resistant tissues from rodents and humans. One investigator has observed that obese mice bind only 35% as much insulin as do lean mice. Other researchers have found that monocytes from maturity-onset diabetics bind only about 50% as much insulin as monocytes from healthy persons—further, that there are only about 1200 receptors per monocyte from diabetics as compared with 2200 per monocyte from healthy persons.

(3) Some researchers have observed a marked inverse correlation between insulin concentration in the circulating blood and the number of insulin receptors—that is, the greater the insulin concentration in the blood, the fewer the insulin receptors in liver, fat, muscle and blood cells.

(4) Some investigators have found that the level of circulating insulin is increased by a diet high in carbohydrates, but is not accompanied by a reduction in receptor numbers. This condition holds only for an immediate time span after meals—the basal concentration of the carbohydrate-fed mice ultimately returns to a below-normal level.

(5) Researchers have found that large adipocytes from obese rats have about the same number of receptors as smaller adipocytes from lean rats; they also report no difference in glucose transport between the two types of cells, this suggesting no defect in insulin binding in the larger adipocytes. Other researchers do not agree, finding a decrease in number of insulin receptors in large adipocytes. They believe insulin binding is a membrane phenomenon and thus the most important characteristic is the number of receptors per unit area.

(6) Mathematical models of cellular membrane have been constructed, from which it has been suggested that a membrane protein, called an *effector,* must interact before the signal for insulin binding can be transmitted inside the cell.

(7) There is some indication that the increased concentration of circulating insulin may be the cause and the reduced number of receptors the effect. It has been suggested that insulin may directly catalyze the breakdown of insulin receptors. Other researchers observe that the proteolytic activity of insulin is too weak to account for the observed decrease in insulin receptors and thus the effect must be due to a poorly understood complex regulatory mechanism.

And thus the research continues. There is some consensus on a few points. A major cause of continuing diabetes in maturity-onset diabetics is overeating. It is estimated that 80% of such diabetics now being treated are obese. Physicians may stress this more vigorously in the future as the disenchantment with the use of drugs and insulin for diabetic control increases. This ties in with the findings that a principal cause of the disorder is a defect in insuling binding to receptors on the cell surface. There is also growing evidence that exercise can increase insulin sensitivity and insulin cell receptors in healthy subjects and that this finding probably can be transferred to maturity-onset diabetes therapy. In tests at Yale University School of Medicine (1979), six healthy men were tested. It was found that exercise did not increase insulin levels, but it did increase insulin sensitivity by 30% and the number of insulin receptors by 50%.

Protein Glycosylation. As pointed out by Bunn, Gabbay, and Gallop

[1] As early as 1949, researchers at the University of Pennsylvania conducted direct studies of the insulin-receptor interaction by using radioactively labeled insulin.

(1978), protein glycosylation is important in the maintenance of the integrity of plasma membranes and in facilitating the secretion of proteins into the extracellular space. These modifications are usually precisely controlled by enzymes. In contrast, certain proteins may undergo nonenzymatic glycosylation. This appears to be the case with human hemoglobin. Much has been learned concerning the structure and biosynthesis of glycosylated hemoglobins in recent years, including data that have a bearing on certain areas in diabetes research. This includes a more accurate measurement of glucose intolerance, particularly in borderline cases. The new findings are also relevant to studies of the complications of diabetes mellitus. The aforementioned researchers have observed that the organs and tissues most affected by diabetic complications (lens, peripheral nerves, kidney, retina, blood vessels) are not insulin-dependent for glucose penetration and thus achieve high intracellular glucose concentrations during periods of hyperglycemia. These intracellular glucose levels have been shown as causative factors for the formation of some diabetic complications. A particular hemoglobin (A_{Ic}), found in higher concentration in some diabetics, may serve as a useful model of nonenzymatic glycosylation of other proteins that may be involved in the long-term complications of diabetes.

Vitamin D. Researchers at the University of California (Riverside and San Francisco) reported in 1980 that vitamin D deficiency inhibits pancreatic secretion of insulin. The researchers point out that, along with the previously demonstrated presence in the pancreas of a vitamin D-dependent calcium-binding protein and cytosol receptor for the hormonal form of vitamin D (1,25-dihydroxyvitamin D_3), this indicates an important role for vitamin D in the endocrine functioning of the pancreas. Details are given in the Norman et al. (1980) reference.

Drug-Induced Diabetes Mellitus. Just as the connection between viruses and diabetes mellitus at one time seemed remote, with comparatively little research directed along these lines, an association between drugs (as causative factors) and diabetes has not received major attention. It is true that the drug *alloxan* has been known since the 1940s to be capable of destroying beta cells and of inducing diabetes in laboratory animals. Highly selective damage to beta cells can be noted within a few minutes after injection. It was also found in the early 1960s that the drug *streptozotocin* is toxic to beta cells and similarly induces diabetes in laboratory animals. In more recent research, workers at the University of Massachusetts Medical School have shown that controlled multiple doses of the drug can alter beta cells so that they become vulnerable to attack by the animal's immune system. This line of experimentation has also shown that genetic factors are an important factor in susceptibility to the drug.

In 1975, a rodent poison with a molecular structure similar to that of streptozoticin was put on the market in the United States. Accidental injection of the drug caused acute diabetes (requiring insulin therapy) in at least 20 survivors to the exposure. At autopsy, two of the fatalities from such poisoning showed beta cell destruction. A few other drugs and chemicals have been shown to be toxic to beta cells.

References

Bunn, H. F., Gabbay, K. H., and P. M. Gallop: "The Glycosylation of Hemoglobin: Relevance to Diabetes Mellitus," *Science*, **200**, 21–27 (1978).

Coleman, D. L.: "Obesity Genes: Beneficial Effects in Heterozygous Mice," *Science*, **203**, 663–665 (1979).

Cudworth, A. G.: "Type I Diabetes Mellitus," *Diabetologia*, **14**, 5, 281–291 (1978).

Eastman, R. C., and J. S. Flier: "Receptors for Peptide Hormones: New Insights into the Pathopsychology of Disease States in Man," *Ann. Int. Med.*, **86**, 2, 205–219 (1977).

Kolata, G. B.: "Controversy over Study of Diabetes Drugs Continues for Nearly a Decade," *Science*, **203**, 986–990 (1979).

Kolata, G. B.: "The Phenformin Ban: Is the Drug an Imminent Hazard?" *Science*, **203**, 1094–1096 (1979).

Kolata, G. B.: "Blood Sugar and the Complications of Diabetes," *Science*, **203**, 1098–1099 (1979).

Maugh, T. H., II: "Diabetic Retinopathy: New Ways to Prevent Blindness," *Science*, **192**, 539–540 (1976).

Maugh, T. H., II: "Hormone Receptors: New Clues to the Cause of Diabetes," *Science*, **193**, 221–222, 254 (1976).

Maugh, T. H., II: "Virus Isolated from Juvenile Diabetic," *Science*, **204**, 1187 (1979).

Norman, A. W., et al.: "Vitamin D Deficiency Inhibits Pancreatic Secretion of Insulin," *Science*, **209**, 823–825 (1980).

Notkins, A. L.: "The Causes of Diabetes," *Sci. Amer.*, **241**, 5, 62–73 (1979).

Onodera, T., et al.: "Virus-Induced Diabetes Mellitus: Reovirus Infection of Pancreatic Beta Cells in Mice," *Science*, **201**, 529–531 (1978).

Staff: "Report of the National Commission on Diabetes to the Congress of the United States," Department of Health and Human Services, Washington, D.C., (1976).

Ulrich, A., et al.: "Genetic Variation in the Human Insulin Gene," *Science*, **209**, 612–615 (1980).

Varma, S. D., Mizuno, A., and J. H. Kinoshita: "Diabetic Cataracts and Flavonoids," *Science*, **195**, 205–206 (1977).

Yoon, Ji-Won, et al.: "Virus-Induced Diabetes Mellitus: Isolation of a Virus from the Pancrease of a Child with Diabetic Ketoacidosis," *N. Engl. J. Med.*, **300**, 21, 1173 (1979).

DIACETONE ALCOHOL. **Ketones.**

DIAGENESIS. A term proposed by Gumbel in 1888 for the gradual and successive chemical physical changes which take place in sediments previous to or during their consolidation. Diagenesis may also include the numerous processes of lithification but is a useful term only when particularly applied to the more or less contemporaneous chemical alteration of sediments.

DIAGNOSTICS (Computer System). Programs provided to the maintenance engineer or operator to assist in discovering the source of a particular computer system malfunction. Diagnostics generally consist of programs which force extreme conditions (the worst patterns) on the suspected unit with the expectation of exaggerating the symptoms sufficiently for the engineer to readily discriminate between possible faults and to identify the particular fault. In addition, diagnostics may provide assistance in localizing the cause of malfunction to a particular card or component in the system.

There are two basic types of diagnostic programs: (1) online; and (2) offline. Offline diagnostic programs are those which require that there be no other program active in the computer system, sometimes requiring that there be no executive program in the system. Offline diagnostics are typically used for central processing unit (CPU) malfunctions, very obscure and persistent peripheral device errors, or for critically time-dependent testing. For example, it may be suspected or known that a harmonic frequency is contributing to the malfunction. Thus, it may be desirable to drive the unit continuously at various precise frequencies close to the suspected frequency to confirm the diagnosis and then to confirm the cure. Interference from other activities may well make such a test meaningless. Thus, all other activity on the system must cease.

Online diagnostics are used mainly in multiprogramming environment and are vital to the success of real-time systems. The basic concept is that of logically isolating the malfunctioning unit from all problem programs and allowing the diagnostic program to perform any and all functions on the unit. Many of the more common malfunctions can be isolated by such diagnostics, but there are limitations imposed by the interference from other programs which also are using the CPU. See also **Debugging (Computer Program);** and terms listed under **Data Processing.**

Thomas J. Harrison, International Business Machines Corporation, Boca Raton, Florida.

DIAGONALIZATION OF A MATRIX. Given the matrix equation $B = PAQ$ it is said that A and B are equivalent (see **Transformation (Mathematics)**). Given the single matrix A it is often useful to find B, requiring it to have certain properties not possessed by A. The most common case is that B is to be made diagonal, thus simpler in appearance than A. The problem is analogous to that of transforming from one coordinate system to another in order to obtain simpler relations between variables or functions. In the matrix case, diagonalization can be achieved, in general, by specifying the properties of the tranforming matrices P and Q.

Assume first that none of the matrices are singular. Write A in partitioned form as

$$A = \begin{bmatrix} A_{11} & V \\ \tilde{V} & A' \end{bmatrix}$$

where $V = [A_{12}, A_{13}, \ldots, A_{1n}]$, a row vector; $\tilde{V}$ is the corresponding column vector and the remaining elements of A appear in A' which has $(n-1)$ rows and columns. Now choose

$$Q_1 = \begin{bmatrix} 1 & -V/A_{11} \\ 0 & E_{n-1} \end{bmatrix}$$

where E_{n-1} is a unit matrix of order $(n-1)$. It is found by a congruent transformation that

$$\tilde{Q}_1 A Q_1 = \begin{bmatrix} A_{11} & 0 \\ 0 & A'' \end{bmatrix}$$

with A'' of order $(n-1)$. The latter is now subjected to a similar operation with a matrix Q_2, so that it contains two zeros and two non-vanishing elements. Finally, a finite number of operations will reduce A to the desired form

$$\tilde{Q}AQ = D = \text{diag}(\alpha_1, \alpha_2, \alpha_3, \ldots, \alpha_n)$$

where $Q = Q_1 Q_2 Q_3 \cdots Q_n$. The diagonal elements are, explicitly, $\alpha_1 = \Delta_1, \alpha_2 = \Delta_2/\Delta_1, \ldots, \alpha_n = \Delta_n/\Delta_{n-1}$ where Δ_m is a discriminant obtained by omitting all but the first m rows and columns of Δ, the determinant of A.

If one considers a quadratic form, the diagonalization process gives the following result

$$\tilde{x}Ax = \tilde{y}Dy = \alpha_1 y_1^2 + \alpha_2 y_2^2 + \cdots + \alpha_n y_n^2$$

It is equivalent to the change of variable $x = Qy$ and has reduced the form to a sum of squared terms. If desired, one could transform once more to get

$$\tilde{y}Dy = \tilde{z}Ez = z_1^2 + z_2^2 + \cdots + z_n^2$$

where $z_i = \sqrt{\alpha_i}\, y_i$. If the given matrix A is singular with its rank, r less than its order, n the diagonal matrix will have r elements along the diagonal and zero elsewhere.

Properties of the characteristic matrix may also be used for diagonalization. Combining that procedure with the one described here one can simultaneously reduce two quadratic forms to sums of squares. This problem is of some importance in various branches of physics and chemistry, for example, in the mechanical theory of small vibrations as applied to the polyatomic molecule. Suppose $\tilde{x}Ax$ and $\tilde{x}Bx$ are given. Reduce the first form by a congruent transformation, $x = Qy$ to get

$$\tilde{y}\tilde{Q}AQy = \tilde{y}Dy = \tilde{y}[\alpha_i \delta_{ij}]y$$

The second form becomes

$$\tilde{y}\tilde{Q}BQy = \tilde{y}Cy$$

but C is not diagonal. Now substitute $z_i = \sqrt{\alpha_i}\, y_i$ to get $\tilde{y}Dy = \tilde{z}Ez$ and $\tilde{y}Cy = \tilde{z}C'z$, with the α_i combined with C to form C'. Finally, use an orthogonal transformation, $z = Ru$, which produces the desired result

$$\tilde{y}Dy = \tilde{u}\tilde{R}ERu = u_1^2 + u_2^2 + \cdots + u_n^2$$

and

$$\tilde{y}Cy = \tilde{u}\tilde{R}C'Ru = \tilde{u}\Lambda u = \lambda_1 u_1^2 + \lambda_2 u_2^2 + \cdots + \lambda_n u_n^2$$

See also terms listed under **Mathematics.**

DIAGONAL REGRESSION. Regression.

DIAGONAL (Structural). Any inclined web member of a truss other than the end post is known as a diagonal. Under various live load conditions, certain web members of a bridge truss may be subjected to a reversal of stress. Stiff diagonals are web members designed to carry either tension or compression. Tension diagonals are assumed to be incapable of resisting any appreciable amount of compression.

When the diagonals are designed to take tension only (are not stiff) two are required, sloping in opposite directions, in a panel where either one, acting alone, would have its stress reversed. When one of these diagonals is acting the other is out of action for all practical purposes. The one which carries the dead load stress when no other loads are on the bridge is called the main diagonal. The other which comes into play when the main diagonal goes out of action is the counter. See also **Beam (Structural).**

DIAGRAM FACTOR (Steam Cycle). Steam Cycles (Diagram Factor).

DIALLAGE. The mineral term for a calcium-iron pyroxene, similar in chemical composition to diopside but richer in iron oxide. In addition to the typical prismatic cleavage of the pyroxene group diallage has a marked "cleavage" parallel to the vertical pinacoids, known as diallage parting. Diallage is a common constituent of gabbros. The term diallagite was proposed by Cloiseaux in 1845 for rocks particularly rich in diallage. The term diallage is derived from the Greek meaning difference, and referring to the peculiar cleavages of this variety of monoclinic pyroxene. See also **Pyroxene.**

DIALYSIS. The process of separating compounds or materials by the difference in their rates of diffusion through a colloidal semipermeable membrane. Thus, sodium chloride diffuses eleven times as fast as tannin and twenty-one times as fast as albumin. When the process is conducted under the influence of a difference in electrical potential, as from electrodes on opposite sides of the semipermeable membrane, it is called electrodialysis.

An apparatus for carrying out a dialysis usually consists of two chambers separated by a semipermeable membrane of parchment paper latex, animal tissue, or other colloid. In one chamber the solution is placed, and in the other, the pure solvent. Crystalline substances diffuse from the solution through the membrane and into the solvent much more rapidly than amorphous substances, colloids or large molecules.

The use of dialysis in connection with the treatment of kidney diseases is discussed under **Kidney and Urinary Tract.**

DIALYTIC BATTERY. Ocean Resources (Energy).

DIAMAGNETISM. Diamagnetism is the phenomenon in which the magnetization in a substance opposes the magnetizing force which induces it. Diamagnetism is considered to exist in all substances, although in substances exhibiting paramagnetism or ferromagnetism, it is masked by the much greater opposite effect due to the orientation of the magnetic atoms or molecules.

DIAMETER-AREA RATIOS (Of Circles). Circle (Geometry).

DIAMOND. An allotropic form of carbon, diamond occurs in nature and, in comparatively small sizes, is produced synthetically. Diamond crystallizes in the cubic system, is the hardest of known substances (10 on the Mohs scale; 5,500–7,000 on the Knoop scale), specific gravity 3.51–3.521 (20°C), dielectric constant at 10^4 Hz, 16.5, at 10^8 Hz, 5.5, index of refraction 2.417–2.4195. Classically, a diamond crystal may be pictured as a huge polymeric molecule, very tightly packed, with a density about 1.6 times greater than the other allotropic form of carbon, graphite. The normal C–C single bond distances in the atomic lattice of diamond are all 1.54 Å, whereas in graphite the C–C bond distances are 1.42 Å. The tight packing of diamond accounts, of course, for its relatively high density and for its extreme hardness. Graphite, on the other hand, essentially is comprised of two-dimensional molecules, very laminar, and which tend to slide and thus impart lubricity to the substance. A very rough comparison of the structures of these two forms of carbon is portrayed in Fig. 1. Because diamond is rare and beautiful, it is a gem. But, because of its hardness, diamond also is very important industrially in terms of its use in abrasives. The latter aspects of the diamond are described under **Grinding and Polishing Agents.**

The diamond fields of India, now seemingly exhausted, yielded quantities of stones, probably all of those known previous to 1725 when diamonds were discovered in Brazil. That country has produced many fine but mostly small stones.

Diamonds were discovered in 1867 along the Orange River in South Africa, and since then Africa has been preëminent in the production of diamonds; in the seventies and eighties occurred a series of amazing discoveries of diamond fields and stones of extraordinary size. Dia-

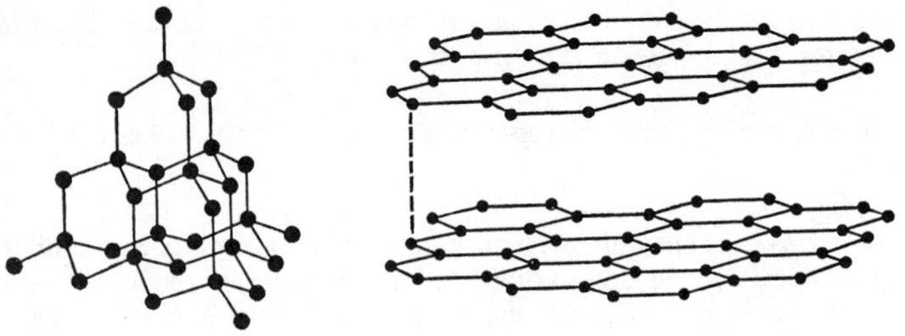

Fig. 1. A gross conceptualization of the space lattices of (a) diamond, and (b) graphite. See the new spatial concept for graphite in the entry on **Carbon.**

monds have also been found in Australia, Borneo, British Guiana, and Arkansas. See also **Carbonado; Kimberlite; Mineralogy.**

Much as the matter has been studied there is no general agreement as to the genesis of the diamond. It is found in alluvial deposits, both unconsolidated and consolidated, indicating the erosion of rocks containing diamonds not only during the present era but also in past geologic time. In Africa diamonds are mined in a dark basic rock of the general nature of peridotite called kimberlite from the town of Kimberley. The kimberlite occurs in vertical "pipes," resembling what once may have been volcanic necks or other types of igneous rock conduits. It is supposed that the diamonds have been formed in and brought to the surface by the magma which was of the general nature of peridotite. Undoubtedly high pressures, and possibly high temperatures as well, are necessary for the development of crystallized carbon in the form of diamonds.

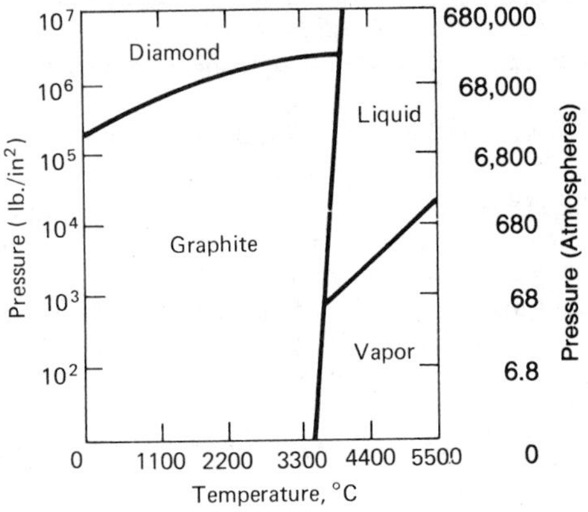

Fig. 2. Phase diagram approximation of carbon, indicating pressure-temperature parameters favoring yield of graphite and diamond. See also the phase diagram in the entry on **Carbon.**

Synthetic Diamonds. Reference to Fig. 2 shows that at high pressures and temperatures the stable form of carbon is diamond, rather than graphite. When graphite is heated and compressed into this region, its layer space lattice undergoes a transformation into the much more closely-knit lattice of diamond. In synthesizing diamond, graphite is introduced in solution in iron, after which high temperatures and pressures are applied to the sample. After the sample has cooled, the iron matrix is dissolved by acid, leaving a diamond residue. The individual pieces in the residue are of industrial grade, containing various impurities. The size usually is on the order of 0.1 mm, suitable for grinding abrasives. Industrial-grade diamonds as large as 1 carat have been produced by the synthetic process, but costs rise proportionately with the size of the diamonds yielded. See also **Gem Stones.** An interesting treatise is "Man-Made Gemstones," by D. Elwell, Wiley, New York, 1980.

Diamond, particularly of gem stone quality, is marketed by the carat. A carat equals about 3.086 grains (troy) or 0.2 grams.

See also **Carbon;** and **Ocean Resources (Mineral).**

DIAMOND-BACK RATTLESNAKE. Snakes.

DIAMOND-BACK WATER SNAKE. Snakes.

"DIAMOND RING" EFFECT. Eclipse.

DIAPAUSE. Boll Weevil.

DIAPHANOMETER. An instrument used to measure the degree of transparency of solids, liquids, or gases.

DIAPHRAGM. 1. The thin layers of muscle which suspend the insect heart within the body. 2. The muscular partition between the thoracic and abdominal cavities of mammals. The mammalian diaphragm is important in respiration. 3. Metal, plastic, and rubber diaphragms find numerous industrial applications in various packings, in pressure gages, as seals, and in diaphragm actuators and diaphragm motor valves. 4. A form of diaphragm is used as a contraceptive device. See **Contraception.**

DIAPHRAGM ACTUATOR AND VALVE. A pneumatic actuator essentially is a pressure-to-stroke transducer. The device translates an air-pressure signal into valve-stem motion. A diaphragm actuator of the type shown in the accompanying figure consists of a flexible

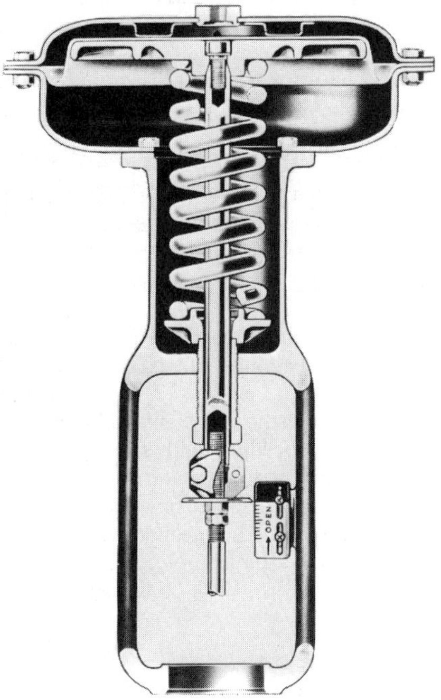

Direct-acting diaphragm actuator. (*Fisher Controls.*)

diaphragm in a pressure-tight housing. The actuator stem connects the diaphragm to the valve stem of the control-valve body. The diaphragm acts as a pressure seal between the upper and lower case. Diaphragms are made from neoprene, buna N, or other suitable elastomers and generally are molded instead of constructed from a flat sheet of material. A properly molded diaphragm provides a nearly constant effective area and gives linearity between loading pressure and travel, makes the actuator easy to assemble and to service, and provides the maximum effective area for a given case size.

In the spring-return diaphragm actuator, the spring opposes the force developed by the loading air pressure acting on the effective area of the diaphragm. This provides a given actuator stem position for a specific value of air-pressure signal received from the controller. Diaphragm actuators can be direct-acting (the type illustrated), or reverse acting. A direct-acting unit strokes down with increasing air pressure; a reverse-acting actuator strokes up with increasing pressure. In the springless type diaphragm actuator, stem position is also proportional to the air-pressure signal, but the opposite side of the diaphragm is pressure-loaded with an independently regulated air supply, or by operation of a double-acting valve positioner. Springless actuators also can be either direct- or reverse-acting. They develop more force than spring-return types, but have the disadvantage of not "failing" in any one position upon unexpected loss of air pressure.

Many thousands of diaphragm motor valves and actuators are installed as parts of important industrial control systems where pneumatic operation of final controlling elements is preferred over electric or hydraulic operation.

See also **Control System (Automatic);** and **Final Controlling Element.**

DIAPHRAGM CELL. Chlorine.

DIAPHRAGM GAS METER. Flow Measurement.

DIAPHRAGM MOTOR VALVE. Valve (Control).

DIAPHRAGM (Muscle). Respiratory System.

DIAPHRAGM (Physiology). The broad sheet of muscular tissue which separates the chest cavity from the abdominal cavity; it contracts with each inspiration and relaxes with each expiration.

DIAPHRAGM SEAL. Seal (Diaphragm).

DIAPHRAGM (Semipermeable). Semipermeable Membrane.

DIARRHEA. Increased frequency of bowel movements, accompanied by a loose consistency of the stools. The condition is particularly dangerous in infants and small children because of the rapidity with which serious dehydration may occur. Diarrhea occurs in connection with a number of human diseases and disorders and can be categorized in terms of principal causes and/or mechanisms: (1) osmotic diarrhea; (2) secretory diarrhea; (3) contact-deficiency diarrhea; and (4) exudative diarrhea. Quantitative data pertaining to the physiology of the normal adult small intestine and colon may be helpful to understanding diarrhea:

Source of Water	Rate (Liters/Day)
Dietary intake of water	2.0
+ saliva	1.0
+ gastric secretions in stomach	2.0
To duodenum	5.0
+ pancreatic secretions	2.0
+ biliary secretions	1.0
To jejunum	8.0
+ jejunum secretions	1.0
	9.0
− jejunum absorption	4.5
To ileum	4.5
− ileum absorption	3.5
To colon	1.0
− colon absorption	0.9
Excreted in stool	0.1

Osmotic Diarrhea. Certain ions, such as divalent magnesium (Mg^{2+}), divalent sulfate (SO_4^{2-}), and trivalent phosphate (PO_4^{3-}), are poorly absorbed by the small intestine. As a consequence, water is retained in the lumen by osmosis. This is exemplified by the effect of strong antacids, such as magnesium hydroxide, used in treating peptic ulcer disease, which can produce a water diarrhea. Saline laxatives also cause osmotic diarrhea. Another frequent cause of osmotic diarrhea is the ingestion of undigestable carbohydrates. If there is insufficient disaccharidase (enzymes) at the intestinal surface membrane, disaccharide (lactose or sucrose) will be retained within the intestinal lumen. The sugars must be converted to monosaccharide before they can be transported. This retention can account for 0.5–2 liters of diarrheal water per day. Also, action of certain bacteria in the small intestine may catabolize 12-carbon sugars to 3-carbon fragments, further enhancing the osmotic effect. See also **Carbohydrates.** With increasing age, commencing in the mid-teens, some individuals become lactase deficient and thus cannot handle the ingestion of milk and dairy products as well as they could when younger. This situation, of course, can be readily alleviated simply by reducing the intake of milk and related products. This cause of diarrhea can easily be confused with the so-called *irritable bowel syndrome,* which is described later. Other offenders in some individuals are legumes which contain oligosaccharides (stachyose, raffinose, etc.) and nondigestible sugar alcohols (mannitol, sorbitol, etc.) which are now widely used as artificial sweeteners. These substances can cause osmotic diarrhea, particularly in children. The administration of lactulase as the therapy of hepatic encephalopathy can cause a watery diarrhea, but this usually is an acceptable side-effect.

Secretory Diarrhea. This condition results from the action of various substances that increase the secretion of water in the bowel system. The enterotoxins produced by certain bacteria (*Escherichia coli, Vibrio cholerae,* etc.), through a complex series of reactions involving intestinal receptors, increase the level of adenosine monophosphate (AMP) and consequently enhance the secretion of sodium, chloride, and water. See **Cholera.** Other microorganisms (viruses and parasites) may cause secretory diarrhea. *Salmonella, Clostridium perfringens,* and staphylococci, among bacteria; viruses (see **Coxsackie Virus; Norwalk Virus);** fungi which causes candidiasis; and parasites which cause giardiasis and amebiasis, among others, also cause varying degrees of diarrhea—either as the principal or one of the major consequences of infection and invasion. See also **Amebiasis.** It is estimated that about half the travelers who complain of diarrhea when in Mexico and a number of other developing countries are infested with *E. coli.* Symptoms are watery diarrhea and severe abdominal cramps. The disease runs a course of 4–5 days. Because of rapid dehydration, this disease in children should receive early attention. Most patients respond well to ampicillin therapy, although in infrequent cases antimicrobial therapy may be indicated. See **Foodborne Diseases.**

The malabsorption of fats may cause a watery secretory diarrhea as the result of irritation of the distal small intestine by long-chain fatty acids. Stools are frequently voluminous and rise to the top of the water in the toilet bowl, causing difficulty in flushing the toilet. Another possible cause of secretory diarrhea stems from the use of certain irritating laxatives, such as cascara and castor oil. It has been reported that sometimes neomycin and para-aminosalicylic acid may antagonize the absorption of fats.

A condition known as *villous adenoma* (leading in about half the cases to carcinoma) is found in the rectum and/or colon and can cause severe secretory diarrhea. A congenital situation in which there is a malfunctioning bicarbonate-chloride mechanism in the ileum may result in excessive secretion of chloride ion and water. Known as *congenital chloridorrhea,* this is a rather uncommon disorder.

Contact-Deficiency Diarrhea. Insufficient contact and mixing action within the bowel system can cause diarrhea. This may result from certain operative procedures, such as resection of small intestine, surgical bypass, or by the presence of a "short circuit" (fistula, etc.) in the small intestine. In some individuals, rapid transit time in the bowel does not permit sufficient time for normal bowel action and results in diarrhea. Sometimes called *spastic colitis, mucus colitis,* or *irritable bowel syndrome,* symptoms include abdominal cramping, urgency to defecate, and loose, mucus covered stools (no blood). This rather poorly understood syndrome accounts for a high percentage of gastrointestinal complaints. Sometimes there is a cyclic nature to the condition, ranging from constipation to diarrhea. Some authorities suggest emotional stress in connection with the disorder, thus accounting for the transient nature of symptoms. Therapy varies considerably and is addressed to particular individual complaints.

Exudative Diarrhea. Principal causes of exudative diarrhea are ulcerative colitis and granulomatous colitis. See **Colitis and Other Inflammatory Bowel Diseases.** The condition also arises in pseudomembranous colitis, and from invasive bacteria, such as **Shigella.** See **Shigellosis.** Shigellosis is also associated with "traveler's diarrhea."

Pseudomembranous Enterocolitis. This complaint has several designations, including antibiotic, diphtheric, necrotizing, postoperative, and staphylococcal enterocolitis, depending upon particular circumstances. Because this is a multifaceted condition, sigmoidoscopy and other laboratory examinations are frequently required for diagnosis. Because the mortality rate of the disease can reach 50–75%, the condition, the symptoms of which are acute diarrhea (containing pus, mucus, and sometimes, fresh blood), high fever (up to 105°F; 40°C), and often dehydration and much less frequently, shock, requires immediate attention. Cardiac arrhythmias may also develop. In recent years, an association has been made between pseudomembranous enterocolitis and certain antibiotics, such as clindamycin and lincomycin; it is also believed that ampicillin, chloramphenicol, and tetracycline may also be implicated in some situations. It has been found that stool examinations of patients where the disorder is antibiotic-induced will show the presence of *Clostridium difficile.* It is considered that the latter may be a direct cause of the disease. The disease runs a course of 1 to 4 weeks. Therapy is usually administration of a broad-spectrum antibiotic.

Other drugs which can cause diarrhea include certain antimetabolites, colchicine, and ethanol, all of which can cause mucosal damage.

References

Dupont, H. L., et al.: "Pathogenesis of *Escherichia coli* Diarrhea," *N. Engl. J. Med.*, **285**, 1 (1971).

Phillips, S. F.: "Diarrhea: View of Pathophysiology," *Gastroenterology*, **63**, 495 (1972).

Schultz, S. G., and P. F. Curran: "Intestinal Absorption of Sodium Chloride and Water," in "Handbook of Physiology," American Physiological Society, Washington, D.C., 1968.

Sleisenger, M. H., and J. S. Fordtran (editors): "Gastrointestinal Disease: Pathophysiology, Diagnosis, Management," Saunders, Philadelphia, 1973.

DIARTHROSES. Joint (Anatomy).

DIASPORE. The mineral diaspore is a hydrous oxide of aluminum corresponding to the formula AlO(OH) occurring in prismatic orthorhombic crystals, usually somewhat flattened, or massive. It displays good cleavage; conchoidal fracture; is brittle; hardness, 6.5–7; specific gravity, 3.3–3.5; luster, vitreous to pearly; color, white, grayish, greenish, yellowish, brownish or colorless; transparent to translucent. Diaspore is found associated with corundum, emery and bauxite, being probably an alteration product of the oxide. It has been made artificially. Diaspore has been found associated with emery in the Ural Mountains, in Asia Minor, in the Island of Naxos, Greece, and in the United States at Chester, Massachusetts. Its name is derived from the Greek word meaning to scatter, because of its decrepitation upon heating.

DIASTEM. A term proposed for a slight hiatus, or loss of record, during the deposition of sediments. As diastems must be contemporaneous with sedimentation they are not to be confused with disconformities.

DIASTOLE. The stage of dilation of the heart or relaxation of the heart muscle. It is during this stage that the chambers of the heart fill with blood.

DIASTROPHISM. A general term for all types and modes of deformation of the earth's crust, including the formation of ocean basins, continents, plateaus, and mountain ranges. Use of the term preferably is limited to large-scale processes. *Tectonism* is a synonym.

DIATHERMAL WALL (also Adiathermal; Diathermaneous). A perfect heat conductor; assuring equality of temperature on both sides of it. Since in a rigorous development of the principles of thermodynamics it is necessary to introduce the concept of a diathermal wall before the concept of heat, it is necessary to adopt the following alternative definition. If two closed systems are placed in contact (i.e., interact) through a diathermal wall, their states cannot be varied independently of one another. If the state of one system is changed, then in general, the state of the other system will also change; the systems are coupled. If the state of one system is described by k properties x_i and the state of the other system is described by l properties y_i, then coupling the two systems with the aid of a diathermal wall implies a functional relationship

$$f(x_1, x_2, \ldots, x_i, \ldots x_k; y_1, y_2, \ldots, y_i, \ldots y_l) = 0$$

The number of degrees of freedom (independent properties) of the combined system is now $k + l - 1$.

DIATHESIS. Congenital predisposition toward any disease.

DIATOM. Diatoms are algae which are very commonly found in both fresh and salt waters. Often they occur in immense numbers, especially in the ocean. The feature which distinguishes them from all other algae is the siliceous wall which encloses them. This is composed of two halves or valves, one of which fits over the other much as a cover fits onto a box. These siliceous walls are often beautifully marked with the finest and most regularly arranged patterns which makes these algae objects of great beauty when seen under a microscope. Because of the regularity of these marks, certain species are used for testing the resolving power of a microscope. Within the wall, the simple protoplast contains several chloroplasts which may contain a brown pigment that masks the chlorophyll present.

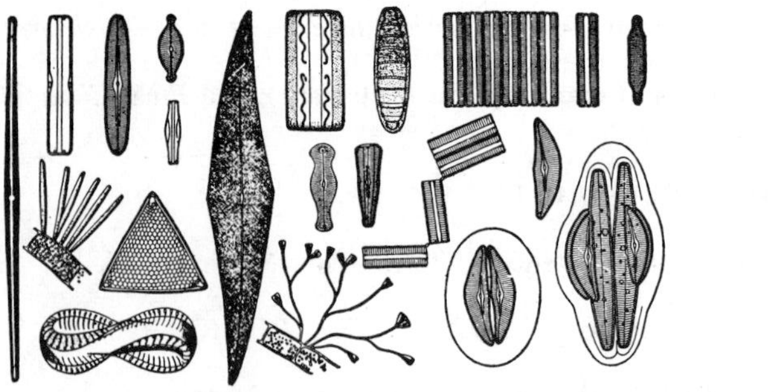

Collection of diatoms showing variation in appearance. (*Kerner, "Natural History of Plants," Blackie & Son.*)

Many diatoms are distinctly unicellular organisms; others stick together to form long chains or are included in a gelatinous sheath which forms extensive branching aggregations. There are two large orders of diatoms, separated according to the shape of the cell. One, the Centrales, comprises those diatoms which are radially symmetrical; the other, the Pennales, those which are not radially symmetrical. Many of the Pennales are bilaterally symmetrical, others irregular. Many members of the Pennales move about in a gliding manner.

Two methods of reproduction are found in diatoms. The more common method is asexual. In this process the protoplast of the cell enlarges, pushing the two valves apart. Then the protoplast divides into two parts, each of which occupies one of the two valves. A new valve is formed over the exposed surface of the protoplast but inside the original valve of the cell. As a consequence the new valve is smaller than the original. As repeated divisions occur the size of the valve gradually decreases. This does not continue indefinitely, however. To bring the cell back to its original size, auxospores are formed. When this happens the protoplast enlarges tremendously and escapes from its walls. It then divides and secretes about itself new walls which have the size of the original cell. Auxospore formation is often a result of sexual fusions. Sexual reproduction is accomplished by the fusion of two diatoms. There are several variations in the method in which this process occurs. Sometimes it is a simple fusion of two protoplasts to form one; in other forms, two amoeboid gametes are formed by each protoplast. Gametes from different protoplasts fuse, forming zygospores. Several other variations are known. Some of the Centrales form small bicilate gametes which fuse.

The diatoms are a very important group of algae. Occurring in immense numbers as they often do, they are the main food substance of many animals, which in turn become food for higher organisms, including man. When a diatom dies and its protoplast disintegrates, the siliceous shell sinks to the bottom of the water. Gradually immense accumulations of diatom valves are formed on the ocean bottom. These may eventually be buried beneath other deposits. They become diatomaceous earth, often forming beds hundreds of feet thick and covering large areas. Diatomaceous earth after removal of organic matter is used as an abrasive and scouring agent, as a filter, and in many other ways.

DIATOMACEOUS EARTH. Decolorizing Agents; Diatoms.

DIATOMACEOUS SILICA INSULATION. Insulation (Thermal).

DIATOMIC GASES. Chemical Elements.

DIATOMITE. A rather dense, chert-like, condensed version of diatomaceous earth. Also defined as indurated diatom ooze. The term frequently is used synonymously with diatomaceous earth. Diatomite is composed principally of the opaline frustules of diatoms. See also **Diatom.**

DIATREME. A general term for volcanic pipes and circular vents, the result of the explosive action of magmatic gases.

DIAZO COMPOUNDS AND DIAZONIUM SALTS. Azo and Diazo Compounds.

DIBENZAMIDE. Imino Compounds.

DIBENZOYL PEROXIDE. Bleaching Agents.

DICED SNAKES. Snakes.

DICHLOROBENZENES. Chlorinated Organics.

DICHOTOMY. A system or method of branching in which the main axis divides into two branches, which may in turn branch in the same manner, as for example the thallus of an alga or an hepatic, or the root or stem of a club moss.

DICHROGRAPH (Circular). Photometers.

DICHROISM. The property of exhibiting two colors, especially of exhibiting one color when viewed in reflected light and another when viewed in transmitted light, as in the case of solutions of chlorophyll. Substances which have this property are termed dichroic. They may also be said to exhibit dichromatism. See also **Cordierite.**

DICHROITE. Cordierite.

DICHROMATISM. This term has two uses in optics, as follows: 1. A type of color blindness in which the eye can distinguish two and only two colors. 2. A substance with two broad, but not equally deep, absorption bands in the visible may appear to have a different color as observed by transmitted light depending on the thickness of the plate of material. Such a material is called dichromatic or dichroic.

DICOTYLEDONS. The larger of the two subclasses of angiosperms is the dicotyledons, containing over 150,000 species. The plants of this subclass have leaves, flowers and seeds distinctly different from those of monocotyledons. The leaves are generally broad and have netted veins; the parts of the flowers occur most frequently in fours or fives or multiples of these numbers. That is, there are 4 or 5 sepals, 4 or 5 petals, 4 or 5 (8 or 10, or more) stamens, and one to many pistils. The embryo of the seed has two cotyledons or seed leaves.

Many important food plants are members of this group; for example, potato, cabbage, carrot, apples, oranges, and peanuts. Other dicotyledons yielding economic products of great value are the various rubber plants, cotton, flax, sugar beets, and soy beans. The number cultivated as ornamental plants is too great to enumerate here.

Dicotyledons vary in size from tiny annuals an inch or less in height to giant trees 350 feet (105 meters) tall. They include herbaceous and woody members, annuals, biennials, and perennials. They are found all over the world, wherever plants can grow. See also **Angiosperms.**

DICROSCOPIC EYEPIECE. An eyepiece for a polariscope or polarizing microscope which gives a comparison view of the same object or field under illumination by the two complementary rays of polarized light.

DICTYONEMA. Invertebrate Paleontology.

DIDELPHIDS. Marsupialia.

DIDYMIUM. Neodymium.

DIDYMIUM GLASS. Didymium glass is a special optical glass which is tinted with mixed oxides of neodymium and praseodymium to give very narrow and sharp absorption bands. One in particular falls at the wavelength of yellow sodium light, so that the glass, which is only faintly tinted to white light, is almost opaque to the yellow sodium light.

DIE. There are several commonly accepted usages of this term. In thread-cutting, a die is a tool resembling a slotted nut, and is used for cutting external threads on screws, bolts, and pipe. In wire drawing, a die is a device used to procure plastic flow of metal; the die is usually, although not necessarily, made with a bell-mouth circular hole, into which the end of the wire is introduced. In the past, hardened steel dies for wire drawing were extensively used; at the present time, either tungsten-carbide or diamond dies are used, since the wear and abrasion on the die are considerably reduced.

DIE CASTING. Die castings are produced by forcing molten metal under pressure into a steel die. The pressure is maintained until solidification is complete. The process is essentially a further development of gravity-feed casting, but the pressure function entails finer detail and better finish. While gravity-feed casting tonnage is greater than that of pressure casting, the latter has a wider field of application and is more important in the quantity production of precision parts. Zinc alloys are generally used for die castings, although aluminum alloys, brass alloys and other non-ferrous metals are used to a considerable extent.

The process of die casting is entirely automatic and requires the following elements: a die-casting machine to hold the molten metal under pressure; a metallic mold or die capable of receiving the molten metal, and designed to permit easy and economical ejection of the solidified product; and a casting alloy that will produce a satisfactory product with suitable physical characteristics.

There are two types of die-casting machines: The first, or air-operated machine, forces the material into the die by high pressure on the surface of the molten metal in a special ladle or goose; and the second, or plunger type machine, forces the material into the die by means of a cylinder and piston which are submerged in the molten metal.

Die-casting dies are constructed in different styles for various production requirements. A single die contains an impression of only one part; a multiple die contains two or more impressions of any one part; a combination die contains one impression only of two or more parts; and a combination-multiple die contains a number of impressions of each of two or more parts. Single dies are comparatively cheap and are used for small-lot production, since they reduce the tool investment to a minimum for any one part. Combination dies, when properly planned, will reduce the total die cost for a given set of castings to a minimum. They are applicable to parts that will always be used in the same quantities and of the same alloy. These parts should be of the same general character and weight. Multiple dies are usually slower to operate than single dies but will give higher production rates for the same labor costs.

Die-casting dies are often vented by permitting air to escape through the clearance in the ejector and core pin bearings. The problem of venting is considerably more important than in sand casting because the mold has no porosity. Sometimes dies are vented by grinding shallow grooves on the parting surfaces of the dies; in other instances plugs with suitable vent grooves are added to the die.

DIE-CASTING ALLOYS. Aluminum; Antimony; Zinc.

DIECIOUS ORGANISMS. Those which produce only one type of reproductive gamete; organisms which are usually classified as either male or female. The opposite of monecious plants which bear both male and female gametes. In botanical usage, however, the word diecious may be applied to trees that bear distinct male and female flowers or cones even though both may be on the same tree. Most higher animals, including all of the arthropods and chordates, are diecious, but most of the higher plants are monecious. Most of the flowers produced by the angiosperms bear both male and female organs.

DIELECTRIC CONSTANT. Oscillometry.

DIELECTRIC-CONSTANT CHEMICAL ANALYZERS. Instruments of this type permit the analysis of certain liquids, solutions, and suspensions—substances of high dielectric constant or loss in the presence of background having a low dielectric constant or loss, such as water in organics, polar molecules in nonpolar solvents, and the detection of different grades of petroleum products in pipelines. Where applicable, the range is from 0.1 to 100%. Because of measurement complexities, the method usually is limited to binary systems. The method is based upon the change of frequency of a tuned circuit (or the required change in tuning of the circuit, or voltage change across the circuit) containing the sample as a dielectric in a capacitative element. Samples must be clean, low in conductance, and nominally temperature-regulated. Radio-frequency measurements with noncontacting electrodes find application where samples are corrosive. The method is similar to oscillometry. See also **Oscillometry.**

DIELECTRIC FILMS. Thin Films.

DIELECTRIC HEATING. The heating of a dielectric material by molecular friction in it as a result of the application of a high-frequency, alternating electric field. Dielectric heating is applicable to nearly all nonconducting materials, such as plastics, wood, and certain liquids. The method is extensively used for the preheating of plastic materials because the materials must be heated uniformly. Prior conventional methods required hours instead of a few minutes with dielectric heating. The method also is used in the printing and dyeing industry, and in the lumber and associated industries for speeding and perfecting the drying of glued joints.

Normally the power required for dielectric heating is provided by some form of oscillator, although the power can be obtained from an amplifier driven by an oscillator. The principal engineering is involved in the design of an oscillator circuit that will provide the proper energy level and in the design of the configuration whereby the energy can be most efficiently imparted to the workpiece.

See also **Microwave Radiation.**

DIELECTRIC THEORY. A dielectric is a material having electrical conductivity low in comparison to that of a metal. It is characterized by its dielectric constant and dielectric loss, both of which are functions of frequency and temperature. The dielectric constant is the ratio of the strength of an electric field in a vacuum to that in the dielectric for the same distribution of charge. It may also be defined and measured as the ratio of the capacitance C of an electrical condenser filled with the dielectric to the capacitance C_0 of the evacuated condenser:

$$\epsilon = C/C_0$$

The increase in the capacitance of the condenser is due to the polarization of the dielectric material by the applied electric field. The terms "specific inductive capacity" or "permittivity" are occasionally used instead of dielectric constant. The constant ϵ appearing in the Coulomb law of force is called the permittivity, but it is also commonly called the dielectric constant. The relative permittivity or dielectric constant is the ratio ϵ/ϵ_0, where ϵ_0 is the permittivity or dielectric constant of free space. In the mks system of units, the dielectric constant of free space is 8.854×10^{-12} farad/m, while in the esu system the relative and the absolute dielectric constants are the same. The relative dielectric constant, which is dimensionless, is the one commonly used. When variation of the dielectric constant with frequency may occur, the symbol is commonly primed. When a condenser is charged with an alternating current, loss may occur because of dissipation of part of the energy as heat. In vector notation, the angle δ between the vector for the amplitude of the charging current and that for the amplitude of the total current is the loss angle, and the loss tangent, or dissipation factor, is

$$\tan \delta = \frac{\text{Loss current}}{\text{Charging current}} = \frac{\epsilon''}{\epsilon'}$$

where ϵ'' is the loss factor, or dielectric loss, of the dielectric in the condenser and ϵ' is the measured dielectric constant of the material.

At low frequencies of the alternating field, the dielectric loss is normally zero and ϵ' is indistinguishable from the dielectric constant ϵ_{dc} measured with a static field. Debye has shown that

$$\frac{\epsilon_{dc} - 1}{\epsilon_{dc} + 2} = \frac{4\pi N_1}{3}\left(\alpha_0 + \frac{\mu^2}{3kT}\right) \tag{1}$$

where N_1 is the number of molecules or ions per cubic centimeter; α_0 is the molecular or ionic polarizability, i.e., the dipole moment induced per molecule or ion by unit electric field (1 esu = 300 volts/cm); μ is the permanent dipole moment possessed by the molecule; k is the molecular gas constant, 1.38×10^{-16}, and T is the absolute temperature. An electric dipole is a pair of electric charges, equal in size, opposite in sign, and very close together. The dipole moment in the product of one of the two charges by the distance between them.

In Equation (1), $\mu^2/3kT$ is the average component in the direction of the field of the permanent dipole moment of the molecule. In order that this average contribution should exist, the molecules must be able to rotate into equilibrium with the field. When the frequency of the alternating electric field used in the measurement is so high that dipolar molecules cannot respond to it, the second term on the right of the above equation decreases to zero and we have what may be termed the optical dielectric constant ϵ_∞, defined by the expression

$$\frac{\epsilon_\infty - 1}{\epsilon_\infty + 2} = \frac{4\pi N_1}{3}\alpha_0 \tag{2}$$

ϵ_∞ differs from n^2, the square of the optical refractive index for visible light, only by the small amount due to infrared absorption and to the small dependence of n on frequency, as given by dispersion formulas. It is usually not a bad approximation to use $\epsilon_\infty = n^2$. The general Maxwell relation $\epsilon' = n^2$ holds when ϵ' and n are measured at the same frequency. The Debye equation may be written in the form

$$\frac{\epsilon_{dc} - 1}{\epsilon_{dc} + 2} - \frac{\epsilon_\infty - 1}{\epsilon_\infty + 2} = \frac{4\pi N_1}{9kT}\mu^2 \tag{3}$$

A much better representation of the dielectric behavior of polar liquids is given by the Onsager equation

$$\frac{\epsilon_{dc} - 1}{\epsilon_{dc} + 2} - \frac{\epsilon_\infty - 1}{\epsilon_\infty + 2} = \frac{3\epsilon_{dc}(\epsilon_\infty + 2)}{(2\epsilon_{dc} + \epsilon_\infty)(\epsilon_{dc} + 2)}\frac{4\pi N_1\mu^2}{9kT} \tag{4}$$

Anomalous dielectric dispersion occurs when the frequency of the field is so high that the molecules do not have time to attain equilibrium with it. One may then use a complex dielectric constant

$$\epsilon^* = \epsilon' - j\epsilon'' \tag{5}$$

where $j = \sqrt{-1}$. Debye's theory of dielectric behavior gives

$$\epsilon^* = \epsilon_\infty + \frac{\epsilon_{dc} - \epsilon_\infty}{1 + j\omega\tau} \tag{6}$$

where ω is the angular frequency (2π times the number of cycles per second) and τ is the dielectric relaxation time. Dielectric relaxation is the decay with time of the polarization when the applied field is removed. The relaxation time is the time in which the polarization is reduced to $1/e$ times its value at the instant the field is removed, e being the natural logarithmic base.

Combination of the two equations for the complex dielectric constant and separation of real and imaginary parts gives

$$\epsilon' = \epsilon_\infty + \frac{\epsilon_{dc} - \epsilon_\infty}{1 + \omega^2\tau^2} \tag{7}$$

$$\epsilon'' = \frac{(\epsilon_{dc} - \epsilon_\infty)\omega\tau}{1 + \omega^2\tau^2} \tag{8}$$

These equations require that the dielectric constant decrease from the static to the optical dielectric constant with increasing frequency, while the dielectric loss changes from zero to a maximum value ϵ''_m and back to zero. These changes are the phenomenon of anomalous dielectric dispersion. From the above equations, it follows that

$$\epsilon''_m = (\epsilon_{dc} - \epsilon_\infty)/2 \tag{9}$$

and that the corresponding values of ω and ϵ' are

$$\omega_m = 1/\tau \qquad (10)$$

and

$$\epsilon''_m = (\epsilon_{dc} + \epsilon_\infty)/2 \qquad (11)$$

The symmetrical loss-frequency curve predicted by this simple theory is commonly observed for simple substances, but its maximum is usually lower and broader because of the existence of more than one relaxation time. Various functions have been proposed to represent the distribution of relaxation times. A convenient representation of dielectric behavior is obtained, according to the method of Cole and Cole, by writing the complex dielectric constant as

$$\epsilon^* = \epsilon_\infty + \frac{\epsilon_{dc} - \epsilon_\infty}{1 + (j\omega\tau_0)^{1-\alpha}} \qquad (12)$$

where τ_0 is the most probable relaxation time and α is an empirical constant with a value between 0 and 1, usually less than 0.2. When the values of ϵ'' are plotted as ordinates against those of ϵ' as abscissas, a semicircular arc is obtained intersecting the abscissa axis at $\epsilon' = \epsilon_\infty$ and $\epsilon' = \epsilon_{dc}$. The center of the circle of which this arc is a part lies below the abscissa axis, and the diameter of the circle drawn through the center from the intersection at ϵ_∞ makes an angle $\alpha\pi/2$ with the abscissa axis. When α is zero, the diameter lies in the abscissa axis, there is but one relaxation time, and the behavior of the material conforms to the simple Debye theory. When, as may arise from intramolecular rotation, a substance has more than one relaxation mechanism, or, when the material is a mixture, the observed loss-frequency curve is the resultant of two or more different curves and, therefore, departs from the simple Debye or Cole-Cole curve.

If the dielectric material is not a perfect dielectric, and has a specific dc conductance k' (ohms^{-1} cm^{-1}), there is an additional dielectric loss

$$\epsilon''_{dc} = \frac{3.6 \times 10^{12}\pi k'}{\omega} \qquad (13)$$

The effective specific conductance is given by

$$k' = \frac{1}{4\pi}\frac{(\epsilon_{dc} - \epsilon_\infty)\omega^2\tau}{1 + \omega^2\tau^2} \qquad (14)$$

It is evident from this equation that k' increases with ω, approaching a limiting value, k_∞, the infinite-frequency conductivity, which is attained when 1 can be neglected in comparison with $\omega^2\tau^2$, so that

$$k_\infty = \frac{\epsilon_{dc} - \epsilon_\infty}{4\pi\tau} \qquad (15)$$

In a heterogeneous material, interfacial polarization may arise from the accumulation of charge at the interfaces between phases. This occurs only when two phases differ considerably from each other in dielectric constant and conductivity. It is usually observed only at very low frequencies, but, if one phase has a much higher conductivity than the other, the effect may increase the measured dielectric constant and loss at frequencies as high as those of the radio region. This so-called Maxwell-Wagner effect depends on the form and distribution of the phases as well as upon their real dielectric constants and conductances. Each type of form and distribution requires special treatment. For a commercial rubber, for example, the observed loss may be

$$\epsilon''(\text{observed}) = \epsilon''_{dc} + \epsilon''(\text{Maxwell-Wagner}) + \epsilon''(\text{Debye}) \qquad (16)$$

DIELS-ALDER REACTION. **Organic Chemistry.**

DIENCEPHALON. A portion of the forebrain of vertebrate animals. The diencephalon is of great importance in the sensory and automatic adjustments of the body. One part, the thalamus, is a major relay station for sensory impulses going to the cerebrum, and it is also concerned with motor coordination. The hypothalamus integrates the various automatic body reactions. For instance, when a dog becomes overheated he begins to pant and his tongue hangs from his mouth. There is also an increased saliva flow which helps to cool the blood flowing through the tongue. If the hypothalamus is injured, however, such reactions do not occur. The dog may become so overheated that he suffers heat prostration, yet his body never undergoes any of these cooling reactions.

DIESEL ENGINE. In a patent dated 1892, Dr. Rudolf Diesel, a German engineer, described an engine to operate on the Carnot cycle. Coal dust was the fuel, and it was to be fed rapidly enough so that isothermal expansion would result. After fuel cut-off, an adiabatic expansion would continue, followed by a compression made isothermal by the injection of water into the cylinder. An adiabatic compression then brought the cycle back to its beginning. A further claim of the patent covered the use of liquid fuels and the spray valve. Early attempts to build this engine resulted in the adoption of a modified cycle which, after much experimentation, was built into a successful working engine.

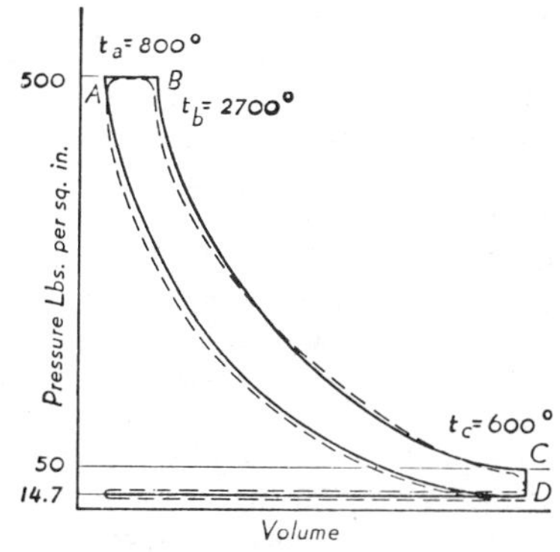

Departure of actual diesel cycle from theoretical on air standard cycle.

Although modified from the inventor's original conception, the modern diesel cycle retains the most important feature, namely that of compression of air to the ignition temperature, followed by timed introduction of fuel. This cycle is shown in the accompanying diagram. The solid line indicates a theoretical cycle, the dotted line shows how a slow-speed actual cycle may depart from the theoretical. Typical temperatures are also indicated. Beginning with point *D* on the cycle, imagine that a cylinder filled with air is closed at the end by a tightly fitting piston. The piston is moved to compress the air without addition or loss of heat through the cylinder walls. As the air is decreased in volume, the pressure rises adiabatically, and it arrives at the condition corresponding to point *A*. The piston is then reversed in direction, and starts to move so as to increase the volume of the air. The air is very hot due to its adiabatic compression. In fact, it is well above ordinary ignition temperatures of petroleum products. As the piston starts to move, carrying the cycle from point *A*, fuel is injected or sprayed into the cylinder just rapidly enough so that its combustion will keep the pressure up while the volume is being increased, at least up to point *B*. At *B* when the outward stroke is partially completed, the fuel is cut off, and the products of combustion expand adiabatically from *B* to *C*, giving work to the piston as they do. At *C* the exhaust valve opens, and the pressure drops to *D*. The line extending horizontally from *D* represents the theoretical exhaust and suction stroke. Adiabatic expansion and compression are not possible in a cylinder which must be well cooled in order to maintain a lubricating oil film. Therefore an actual cycle will not be expected to follow the adiabatic. Another difference between actual and theoretical cases is the composition of the gas within the cylinder. Theoretical studies are made assuming pure air in the cylinder. Actually there is a little burned gas present during the compression, and a great deal of it during the expansion strokes. However, by assuming no friction loss, adiabatic compression and expansion, and air, only, in the cylinder, an expression may be derived for the efficiency of the cycle *ABCD*. Note that the temperatures in the diagram are in degrees F.

The stationary diesel engine is usually a heavily built engine having a piston which reciprocates in a cylinder. A connecting rod is either pinned directly into a trunk type piston, or to a crosshead to which the piston is connected by a piston rod. The connecting rod bears on a crank, which has bearings in the main frame. Added to these parts are the auxiliaries such as valves and valve gear, fuel injection systems, water circulation systems, starting systems, etc. The diesel is more heavily built than the gasoline engine, and is a relatively slow-moving machine. Rotative speeds are commonly 100 to 750 rpm, except that automotive diesels are designed for 2000 rpm and over. The fuel ordinarily employed is a product from crude petroleum.

Fuel is pumped into the cylinder during the first part of the power stroke, correctly metered so that its combustion tends exactly to offset the drop of pressure which would otherwise be experienced. During combustion of the fuel in the diesel engine the pressure remains approximately constant. After a small portion of the power stroke is completed, the fuel is cut off, and the products of combustion do work on the piston expansively.

The principal and important difference between the oil and the gasoline engine resides in the method of ignition. Compression of the air trapped in the cylinder of a diesel engine is employed as its means of ignition. The compression is carried to much higher pressures in the diesel than in the Otto cycle; consequently, the temperature at the end of compression is higher in the diesel cycle. In fact, compresion is carried high enough so that the temperature of the compressed gas exceeds the ignition temperature of the fuel. This is compression ignition. It may require the volume after compression to be only one-fourteenth of that before compression, whereas it is only a fifth or a sixth in the gasoline engine. It is not possible to use compression ignition in the ordinary Otto cycle engine, because an inflammable charge is compressed, and the compression to the ignition temperature would cause spontaneous, uncontrolled, unregulated ignition. However, the charge compressed in the diesel is fresh air—incombustible. This air is compressed until its pressure is over 500 psi (34 atmospheres), and its temperature around 800°F (427°C). Then the oil is injected into the hot compressed air whose temperature is sufficiently high to cause immediate ignition of the spray.

The heart of a diesel engine is the combustion chamber end of its cylinder. The shape of the cylinder head and face of the piston and the design of the nozzle and the injection system must be carefully considered. To obtain the complete combustion necessary to good efficiency, a fuel must be thoroughly mixed with the air charge, so that all particles of it will be burned. There must be good penetration of the oil spray into the highly compressed dense air, and there must be turbulence to insure mixing of the oil spray with the air. These two important characteristics are secured by special design, both of the cylinder head and the spray valve.

In the two-cycle engine there is no valve gear. The absence of this feature is, indeed, the virtue of the two-cycle principle. In the four-cycle engine the exhaust and inlet valves are mechanically operated from a camshaft. Since the diesel engine is commonly rather large, the valves are correspondingly large in girth, and are operated from a massive camshaft. Although the diesel engine is basically a slow-speed type, refinements of design, especially in the combustion chamber, have enabled builders to produce medium- and high-speed compression ignition engines.

Direct-Injection Diesel Engine. Unlike the traditional passenger car diesel engine which employs a precombustion chamber (creates turbulence in air charge and thus assures rapid mixing with the fuel jet in the interest of achieving complete combustion and reduction of smoke emissions), there is no prechamber in the direct-injection system. This system injects fuel into the clearance volume as the piston nears its top position. Used for several years in intermediate-size and heavy-duty diesel engines, direct injection is now under serious consideration for lighter vehicles and is expected to be available during the early 1980s. Because of higher operating temperature, there is an increase in nitrogen oxide (NO_x) emissions.

Considerable opportunity remains for improvement of the standard prechamber diesel engine, including improved combustion chamber design to reduce losses; optimizing the timing and duration of fuel injection; and improved torque characteristics through turbocharging, already incorporated in some designs. Recently available ceramic components may help designers to improve the engine's thermal efficiency.

References

Cross, R. K., Larka, P., and C. G. O'Neill: "Electronic Fuel Injection for Controlled Combustion in Diesel Engines," SAE Paper 810258, Society of Automotive Engineers, Inc., Warrendale, Pennsylvania, 1981.

Heywood, J., and J. Wilkes: "Is There a Better Automobile Engine?" *Technology Review (MIT)*, **83**, 2, 18–28 (1980).

Postulka, A., and K. H. Ies: "Chemical Characterization of Particulates from Diesel Powered Passenger Cars," SAE Paper 810083, Society of Automotive Engineers, Inc., Warrendale, Pennsylvania, 1981.

Ryan, T. W., III, Likos, W. E., and C. A. Moses: "The Use of Hybrid Fuels in a Single-Cylinder Diesel Engine," SAE Paper 801380, Society of Automotive Engineers, Inc., Warrendale, Pennsylvania, 1981.

DIESEL ENGINE (Control). Automotive Electronics.

DIESEL ENGINE (Emissions). Gas and Expansion Turbines.

DIESEL ENGINE (Stratified Charge). Internal Combustion Engine.

DIESEL FUELS. (Petroleum).

DIE SINKING. The process of cutting a recess or cavity in a die for drop forging, press working, or plastic molding.

DIETARY REQUIREMENTS AND TRENDS. Research has been conducted worldwide for many years to determine the nutritional needs of humans and, in more recent years, the relationship between diet and health. Quantitatively the greatest progress has been made in determining the minimally adequate daily requirements for vitamins and minerals. Applicable data for use in the 1980s are given in Tables 1 and 2. Much progress also has been made in relating calorie consumption to human weight for healthy persons, considering sex, height, frame size, and age. A convenient chart, based upon age 25 years, is included here. These data are further refined in terms of age increments and energy needs in Table 3. Incidentally, there is not perfect agreement among authorities pertaining to these data, but there is general consensus. In connection with such data, one must always take into consideration both the occupational and avocational energy requirements of an individual.

The importance of vitamins and minerals is described in considerable detail in a number of entries in this encyclopedia. See entries on the various vitamins. Also see several of the entries on chemical elements which have sections on the biological role of the elements. The relationship of nutrition with disease is also covered in several entries on the cardiovascular system, ulcers, gout, among others. See also **Obesity.**

Dietary Trends. An adequate statistical base for analyzing the trend in human diets is essentially limited to a comparatively few of the industrialized nations. For many years, the thrust of improving nutrition was directed mainly to raising the levels of nutrition in impoverished countries—from gravely subnormal values to minimal acceptable values. This is a task which, as of the early 1980s, is far from complete. Serious nutritional awareness in the advanced countries, with few exceptions, has grown markedly during the past decade or so. See sections on "Kwashiorkor," "Marasmus," "Protein-Calorie Malnutrition (PCM)," and "PCM-Plus" in the entry on **Protein.** Studies have sharpened consumer interest in the importance of scientifically based nutritional principles—not only in terms of the human diet, but for the production of livestock as well.

Dietary Trends in the United States. It was not until about 1910 that the diet patterns of the United States were monitored (U.S. Department of Agriculture) in any formal way. By compiling statistics on food disappearance, subtracting from the quantities of foods grown and processed—the obvious amounts for export, spoilage, etc.—it has been possible to derive estimates of food consumption within the country. Calorie (total) consumption over the years has remained relatively constant. Although subject to some cycling, the proportion of calories derived from protein in the recent diet is roughly equivalent to that of the 1910 diet. During the early part of the century, of course, about half of the protein was derived from grains and other vegetable sources and about half from animal products. In recent years, about 70% of the protein has come from the more costly animal variety. The protein shift is the result of increased consumption of meat, poultry, fish, dairy products, and eggs.

Carbohydrate consumption has declined quite steadily since the early part of the century. This downward trend was also accompanied by a change in the source and nature of the carbohydrates consumed. There were significant drops in the starches. The marked rise in total and refined sugars (in the 1920s) has essentially been maintained,

TABLE 1. RECOMMENDED DAILY DIETARY ALLOWANCES[A]

Person Category	Age (Years)	Weight (Pounds)	Weight (Kilograms)	Height (Inches)	Height (Centimeters)	Protein (Grams)	Fat-Soluble Vitamins: Vitamin A (Micrograms R.E.)[B]	Fat-Soluble Vitamins: Vitamin D (Micrograms)[C]	Fat-Soluble Vitamins: Vitamin E (Micrograms α T.E.)[D]
Infants	0.0–0.5	13	6	24	60	Note 1	420	10	3
	0.5–1.0	20	9	28	71	Note 2	400	10	4
Children	1–3	29	13	35	90	23	400	10	5
	4–6	44	20	44	112	30	500	10	6
	7–10	62	28	52	132	34	700	10	7
Males	11–14	99	45	62	157	45	1000	10	8
	15–18	145	66	69	176	56	1000	10	10
	19–22	154	70	70	177	56	1000	7.5	10
	23–50	154	70	70	178	56	1000	5	10
	51+	154	70	70	178	56	1000	5	10
Females	11–14	101	46	62	157	46	800	10	8
	15–18	120	55	64	163	46	800	10	8
	19–22	120	55	64	163	44	800	7.5	8
	23–50	120	55	64	163	44	800	5	8
	51+	120	55	64	163	44	800	5	8
Pregnant						+30	+200	+5	+2
Lactating						+20	+400	+5	+3

Person Category	Age (Years)	Weight (Pounds)	Weight (Kilograms)	Height (Inches)	Height (Centimeters)	Water-Soluble Vitamins: Vitamin C (Milligrams)	Thiamin (Milligrams)	Riboflavin (Milligrams)	Niacin (Milligrams N.E.)[E]	Vitamin B_6 (Milligrams)	Folacin[F] (Micrograms)	Vitamin B_{12} (Micrograms)
Infants	0.0–0.5	13	6	24	60	35	0.3	0.4	6	0.3	30	0.5[G]
	0.5–1.0	20	9	28	71	35	0.5	0.6	8	0.6	45	1.5
Children	1–3	29	13	35	90	45	0.7	0.8	9	0.9	100	2.0
	4–6	44	20	44	112	45	0.9	1.0	11	1.3	200	2.5
	7–10	62	28	52	132	45	1.2	1.4	16	1.6	300	3.0
Males	11–14	99	45	62	157	50	1.4	1.6	18	1.8	400	3.0
	15–18	145	66	69	176	60	1.4	1.7	18	2.0	400	3.0
	19–22	154	70	70	177	60	1.5	1.7	19	2.2	400	3.0
	23–50	154	70	70	178	60	1.4	1.6	18	2.2	400	3.0
	51+	154	70	70	178	60	1.2	1.4	16	2.2	400	3.0
Females	11–14	101	46	62	157	50	1.1	1.3	15	1.8	400	3.0
	15–18	120	55	64	163	60	1.1	1.3	14	2.0	400	3.0
	19–22	120	55	64	163	60	1.1	1.3	14	2.0	400	3.0
	23–50	120	55	64	163	60	1.0	1.2	13	2.0	400	3.0
	51+	120	55	64	163	60	1.0	1.2	13	2.0	400	3.0
Pregnant						+20	+0.4	+0.3	+2	+0.6	+400	+1.0
Lactating						+40	+0.5	+0.5	+5	+0.5	+100	+1.0

TABLE 1. RECOMMENDED DAILY DIETARY ALLOWANCES[A] *(continued)*

Person Category	Age (Years)	Weight (Pounds)	Weight (Kilograms)	Height (Inches)	Height (Centimeters)	Minerals: Calcium (Milligrams)	Phosphorus (Milligrams)	Magnesium (Milligrams)	Iron (Milligrams)	Zinc (Milligrams)	Iodine (Micrograms)
Infants	0.0–0.5	13	6	24	60	360	240	50	10	3	40
	0.5–10	20	9	28	71	540	360	70	15	5	50
Children	1–3	29	13	35	90	800	800	150	15	10	70
	4–6	44	20	44	112	800	800	200	10	10	90
	7–10	62	28	52	132	800	800	250	10	10	120
Males	11–14	99	45	62	157	1200	1200	350	18	15	150
	15–18	145	66	69	176	1200	1200	400	18	15	150
	19–22	154	70	70	177	800	800	350	10	15	150
	23–50	154	70	70	178	800	800	350	10	15	150
	51+	154	70	70	178	800	800	350	10	15	150
Females	11–14	101	46	62	157	1200	1200	300	18	15	150
	15–18	120	55	64	163	1200	1200	300	18	15	150
	19–22	120	55	64	163	800	800	300	18	15	150
	23–50	120	55	64	163	800	800	300	18	15	150
	51+	120	55	64	163	800	800	300	10	15	150
Pregnant						+400	+400	+150	†	+5	+25
Lactating						+400	+400	+150	†	+10	+50

This information prepared and released by the Food and Nutrition Board, National Academy of Sciences, National Research Council, Washington, D.C. (Revised 1979).

NOTES:

1. Multiply kilograms by 2.2
2. Multiply kilograms by 2.0

A. The allowances are intended to provide for individual variations among most normal persons as they live in the United States under usual environmental stresses. Diets should be based on a variety of common foods in order to provide other nutrients for which human requirements have been less well defined.

B. Retinol equivalents: 1 Retinol equivalent = 1 milligram retinol, or 6 milligrams γ carotene.

C. As cholecalciferol. 10 micrograms cholecalciferol = 400 I. U. vitamin D.

D. α tocopherol equivalents: 1 milligram *d*-α-tocopherol = 1α T.E.

E. 1 NE (niacin equivalent) = 1 milligram niacin, or 60 milligrams of dietary tryptophan.

F. The folacin allowances refer to dietary sources as determined by *Lactobacillus casei* assay after treatment with enzymes ("conjugases") to make polyglutamyl forms of the vitamin available to the test organism.

G. The RDA for vitamin B_{12} in infants is based on average concentration of the vitamin in human milk. The allowances after weaning are based on energy intake (as recommended by the American Academy of Pediatrics) and consideration of other factors, such as intestinal absorption.

† The increased requirement during pregnancy cannot be met by the iron content of habitual American diets nor by the existing iron stores of many women; therefore, the use of 30–60 milligrams of supplemental iron is recommended. Iron needs during lactation are not substantially different from those of nonpregnant women, but continued supplementation of the mother for 2–3 months after parturition is advisable in order to replenish stores depleted by pregnancy.

TABLE 2. ESTIMATED SAFE AND ADEQUATE DAILY DIETARY INTAKES OF ADDITIONAL SELECTED VITAMINS AND MINERALS[A]

		Vitamins		
Person Category	Age (Years)	Vitamin K (Micrograms)	Biotin (Micrograms)	Pantothenic Acid (Milligrams)
Infants	0.0–0.5	12	35	2
	0.5–1	10–20	50	3
Children	1–3	15–30	65	3
and	4–6	20–40	85	3–4
Adolescents	7–10	30–60	120	4–5
	11+	50–100	100–200	4–7
Adults		70–140	100–200	4–7

		Trace Elements[B]					
Person Category	Age (Years)	Copper (Milligrams)	Manganese (Milligrams)	Fluoride (Milligrams)	Chromium (Milligrams)	Selenium (Milligrams)	Molybdenum (Milligrams
Infants	0.0–0.5	0.5–0.7	0.5–0.7	0.1–0.5	0.01–0.04	0.01–0.04	0.03–0.06
	0.5–1	0.7–1.0	0.7–1.0	0.2–1.0	0.02–0.06	0.02–0.06	0.04–0.08
Children	1–3	1.0–1.5	1.0–1.5	0.5–1.5	0.02–0.08	0.02–0.08	0.5–0.1
and	4–6	1.5–2.0	1.5–2.0	1.0–2.5	0.03–0.12	0.03–0.12	0.06–0.15
Adolescents	7–10	1.0–2.5	2.0–3.0	1.5–2.5	0.05–0.2	0.05–0.2	0.1–0.3
	11+	2.0–3.0	2.5–5.0	1.5–2.5	0.05–0.2	0.05–0.2	0.15–0.5
Adults		2.0–3.0	2.5–5.0	1.5–4.0	0.05–0.2	0.05–0.2	0.15–0.5

		Electrolytes		
Person Category	Age (Years)	Sodium (Milligrams)	Potassium (Milligrams)	Chloride (Milligrams)
Infants	0.0–0.5	115–350	350–925	275–700
	0.5–1	250–750	425–1275	400–1200
Children	1–3	325–975	550–1650	500–1500
and	4–6	450–1350	775–2325	700–2100
Adolescents	7–10	600–1800	1000–3000	925–2775
	11+	900–2700	1525–4575	1400–4200
Adults		1100–3300	1875–5625	1700–5100

This information prepared and released by the Food and Nutrition Board, National Academy of Sciences, National Research Council, Washington, D.C. (Revised 1979).

NOTES:

A. Because there is less information on which to base allowances, these figures are not given in the main table of the RDA and are provided here in the form of ranges of recommended intakes.

B. Since the toxic levels for many trace elements may be only several times usual intake, the upper levels for the trace elements given in this table should not be habitually exceeded.

although there was a temporary drop during World War II because of scarcity of supply. The dramatic decline in potato consumption was slowed in the mid-1960s and, during the 1970s, showed a comeback, even if of a mildly cyclic nature. Improvements in the processing of potatoes (dehydrating, freezing, etc.) is largely responsible for the recent upward trend in potato consumption. Over the years, a marked decrease occurred in the consumption of flour and cereal grain products. Together with a decrease in fiber intake, the lowered intake of complex carbohydrates has been of some concern to nutrition and health specialists.

Consumption of oils and fats in the U.S. diet has steadily increased, but with only mild tapering in the early 1980s. Again, there has been a shift in the type of fat consumed. There have been only small changes in saturated fatty acids in the food supply over several decades. Polyunsaturated fatty acids (linoleic acid) increased modestly prior to 1940 and quite noticeably during the last couple of decades. This reflects, of course, the increased consumption of edible oils, margarine, and shortening. Actually, cholesterol intake appears to be only about 10% above that of the diet in the early 1900s. During the last 3 decades, there has been a decline in egg, lard, butter and various dairy product composition, while concurrently there has been a large increase in meat consumption. Thus, the loss of cholesterol intake from one of the changes has been more than made up by the other, but not to a large degree.

It is also interesting to note that vitamin and mineral consumption is at much higher levels than earlier in the century. In 1941, four nutrients (iron, riboflavin, niacin, and thiamine) were added to flour, resulting in a dramatic increase in the per capita availability of these nutrients. The enrichment of cereals has also added nutrients (primarily B-vitamins and iron) to the food supply.

Although calorie consumption has remained essentially constant over the last 70 years, this has been accompanied by a decrease in energy expenditures, thus accounting for substantial increases in obesity of the population.

Dietary Trends in the United Kingdom. As reported by Hollingsworth, against the historical background and analysis of dietary trends, recent British and other medical recommendations about diet (notably coronary heart disease) must be considered. In 1976, the U.K. Department of Health Committee stated, "... that they cannot recommend an increase in the intake of polyunsaturated fatty acids in the diet

TABLE 3. MEAN HEIGHTS AND WEIGHTS AND RECOMMENDED ENERGY INTAKE

Person Category	Age (Years)	Weight		Height		Energy Needs (kcal)	Energy Range (kcal)
		(Pounds)	(Kilograms)	(Inches)	(Centimeters)		
Infants	0.00–0.5	13	6	24	60	Note 1	Note 1
	0.5–1.0	20	9	28	71	Note 2	Note 2
Children	1–3	29	13	35	90	1300	900–1800
	4–6	44	20	44	112	1700	1300–2300
	7–10	62	28	52	132	2400	1650–3300
Males	11–14	99	45	62	157	2700	2000–3700
	15–18	145	66	69	176	2800	2100–3900
	19–22	154	70	70	177	2900	2500–3300
	23–50	154	70	70	178	2700	2300–3100
	51–75	154	70	70	178	2400	2000–2800
	76+	154	70	70	178	2050	1650–2450
Females	11–14	101	46	62	157	2200	1500–3000
	15–18	120	55	64	163	2100	1200–3000
	19–22	120	55	64	163	2100	1700–2500
	23–50	120	55	64	163	2000	1600–2400
	51–75	120	55	64	163	1800	1400–2200
	76+	120	55	64	163	1600	1200–2000
Pregnancy						+300	
Lactation						+500	

The energy allowances for the young adults are for men and women doing light work. The allowances for the two older-age groups represent mean energy needs over these age spans, allowing for a 2% decrease in basal (resting) metabolic rate per decade and a reduction in activity of 200 kcal/day for men and women between 51 and 75 years; 500 kcal for men over 75 years and 400 kcal for women over 75. The customary range of daily energy output shown for adults is based on a variation in energy needs of ±400 kcal at any one age, emphasizing the wide range of energy intakes appropriate for any group of people. Energy allowances for children through age 18 are based on median energy intakes of children these ages followed in longitudinal growth studies. The values are 10th and 90th percentiles of energy intake, to indicate the range of energy consumption among children of these ages.

NOTES:

1. Multiply kilograms by 115. Range (95–145).
2. Multiply kilograms by 105. Range (80–135)

This information prepared and released by the Food and Nutrition Board, National Academy of Sciences, National Research Council, Washington, D.C. (Revised 1979).

as a measure intended to reduce the risk of the development of ischaemic heart disease" on the ground that they did not find convincing, ". . . the available evidence that such a dietary alteration would reduce the risk in the United Kingdom at the present time." However, The Royal College of Physicians Working Party recommended partial substitution of saturated fats by polyunsaturated fats on the ground that ". . . a reduction of saturated fat sufficient to produce a significant fall in plasma cholesterol level is likely to be unacceptable unless there is a measure of substitution by polyunsaturated fats" because we ". . . have become accustomed to a diet rich in fat." Both reports recommended an avoidance or reduction of obesity, by reducing total energy intake. Both reports mention both fats and sugar in this connection, and the Royal College of Physicians' report also mentions alcohol.

Recommendations of the British Nutrition Foundation included: (1) Find a diet that contains less fat than we have been accustomed to eating for some years; (2) correspondingly increase the starchy cereal foods in our diet; (3) find a diet which provides fewer calories although the trend in recent decades has been for greater consumption of "empty calorie" foods, such as fats, sugar and more recently alcohol. With reduced energy requirements because of lower energy expenditure, this implies that a diet of higher nutrient concentration is needed; (4) find a diet which accommodates personal preferences and in no way detracts from the pleasure of eating. And if a change in habits is to be permanent, the alternative diet must be as pleasurable or even more so than the old eating pattern; and (5) find ways of translating nutrition into value for money in view of rapidly rising food prices.

With reference to the foregoing recommendations, it is interesting to note that as early as 1957, Normal Jolliffe and in 1963, Ancel and Margaret Keys advocated the same principles and gave much of the same advice. As a practical application of principles, the British Nutrition Foundation has suggested a plan along the following lines: "The general pattern of a meal is a centre food, a staple food (or filler) and a vegetable or fruit dish. In order to get less fat the proportion of the three fundamental components must be altered. The filler and vegetable or fruit must be increased, and the centre food (meat, cheese, egg, etc.) decreased. Dishes which are examples of this change in relative proportions are pizzas, pasta with Italian sauces, risotto with vegetables and a little fish, curries, Scandinavian salads—in which beans, rice or potato provide a base to which small amounts of tunny fish or smoked sausage are added."

Formal Dietary Recommendations—United States

There has been a strong trend during the past 10–15 years for more central government attention to the people's dietary habits. One of the most controversial (and possibly confusing) recommendations along these lines stemmed from U.S. Congressional interest in nutrition and the issuance of a report "Dietary Goals for the United States," released in February 1977 by the Senate Select Committee on Nutrition and Human Needs. In general terms, these dietary goals call for:

1. A decrease from 42 to 30% of energy (calories) from total fat.
2. A decrease from 16 to 10% of energy from saturated fatty acids.
3. No change in the level of energy from protein (12%).
4. An increase in total carbohydrate consumption to account for 55–60% of energy intake.
5. A two-fold increase in the energy from complex carbohydrates, mainly from grain products and some vegetables.
6. A decrease in sugar from 24% of energy to 15%. This refers to total sugars, including sugars found naturally in foods such as milk and fresh fruit.

The dietary goals also specified about 300 milligrams per day for cholesterol and 3 grams per day for salt. No goal or energy allowance was specified for alcohol, although that provides substantial amounts of energy in many U.S. diets.

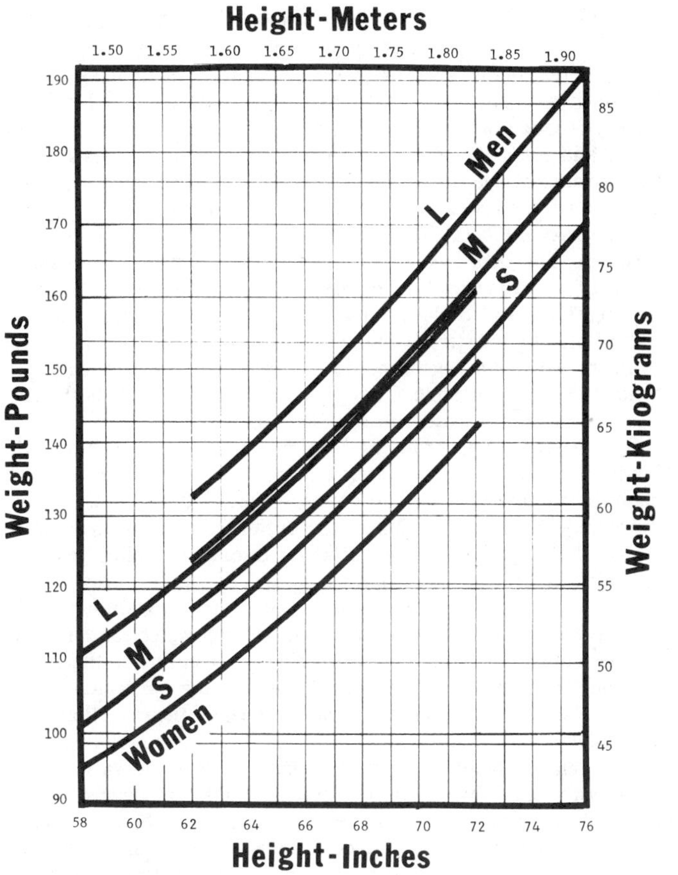

Target weights for healthy individuals (25 years old), considering height and frame size. L = large frame; M = medium frame; S = small frame. Allowances are made for light clothing. (*After Cahill, 1980.*)

The Report translated the foregoing stipulations into lay terms as follows:

1. Eat more fruits and vegetables and whole grains.
2. Eat less meat and more poultry and fish.
3. Cut down on foods high in fat and partly substitute polyunsaturated for saturated fat.
4. Substitute nonfat milk for whole milk.
5. Cut down on eggs, butterfat, and other high cholesterol sources.
6. Cut down on sugar and foods high in sugar.
7. Cut down on salt and foods high in salt.

B. B. Peterkin, in translating the foregoing into terms of a dietary change for an average male (20–54 years of age), summarized the adjustments as follows:

1. Consume two-thirds more grain products.
2. One-fourth more vegetables and fruit.
3. One-eighth more dry legumes and nuts.
4. One-eighth more milk, all of it in form of skim milk.
5. One-half as many eggs.
6. One-half as much refined sugar and sweets, such as syrup, jams, and jellies.
7. One-sixth less fats and oils.
8. One-fifth less meat, poultry, and fish.

Major points of criticism of the controversial proposal include: (1) Few persons in the United States consume diets that are as high in carbohydrate and as low in fat and sugar content as specified; (2) the goals do not take into consideration variation in nutrition requirements of men, women, and children, of different ages; (3) the goals are based upon food disappearance data, probably an inadequate criterion for determining nutritional performance in the past (figures do not agree with actual survey figures from the past).

As of the early 1980s, there appears to be no shortage of advice as regards nutritional recommendations. Recommendations were developed in 1978 by a task force established by the Food and Nutrition Committee of the American Health Foundation. Included were representatives of the American Medical Association and the American Heart Association. The objectives were to develop *Prudent Diet* recommendations aimed at the general population. In general terms, the recommendations of this diet included the following:

1. Caloric intake should be adjusted to achieve and/or maintain desirable body weight, once achieved, generally requires the development of improved dietary habits.
2. Fat intake should be restricted to 35% or preferably less, of calories, and desirably, the intake of saturated fat should be restricted to no more than one-third of the total fat intake both from animal and plant sources.
3. A protein intake of 10 to 15% of total calories is generally adequate.
4. The suggested fat and protein intake implies that 50% or more of the calories should come from carbohydrate, and in order to meet desired micronutrient intakes, this carbohydrate should be obtained largely from cereal and grain products, fruits and vegetables, but it would appear desirable to strive for lower intake levels of simple sugars and alcohol.
5. Care should be exercised that food selection does not compromise the intake of essential nutrients and the selection of a variety of foods can insure an intake of nutrients in accordance with the Recommended Dietary Allowances of the National Academy of Sciences.

Accumulating evidence makes it prudent to reduce intakes of salt, sugar and cholesterol and to increase the consumption of dietary fiber.

Role of Fiber in Diet

Although fiber in the human diet may have been espoused as early as 400 B.C. by Hippocrates who identified bran as a laxative, serious evaluation of the dietary role of fiber, with few exceptions, was not undertaken by food professionals until the early 1970s. Much of the interest in fibers stemmed from reports by D. P. Burkitt (British medical researcher and surgeon), who observed that rural Africans, whose diets are high in fiber-containing foods, have a lower incidence of appendicitis, hemorrhoids, diverticular disease, cardiovascular disease, and cancer of the colon than persons who live in the western, developed nations. Diets in the latter countries are comparatively low in fiber content.

As of the early 1980s, research in this interesting and potentially valuable area of nutrition is quite active, but still in a relatively early stage. Claims of potential benefits from more fiber in the diet in some current instances may prove to be overstated—because 30 or more human diseases and disorders have been mentioned as benefitting in some fashion from increased fiber content. But, research to date has provided a trend of evidence that enhances the role of fiber in human metabolism that goes far beyond the alleviation of constipation, the traditionally accepted role of fiber or "roughage" in the diet.

Definitions of Fiber. A search of the literature reveals inconsistencies in the use of the term *fiber.* The early definition of the Association of Official Analytical Chemists (Washington, D.C.) in the publication "Official Methods of Analysis" (AOAC, 1970) defines *crude fiber as "the residue remaining after treatment with hot sulfuric acid, alkali, and alcohol. It consists primarily of cellulose, lignin, and trace amounts of other polysaccharides."*

This definition appears somewhat narrow in terms of present views of the subject. Based upon the writings of a number of investigators, fiber may be defined as *"that part of plant material in the diet which is resistant to digestion by the secretions of the human gastrointestinal tract—consisting of variable proportions of complex carbohydrates, such as celluloses, hemicelluloses, pentosans, and uronic acids, as well as lignin."*

From an analytical standpoint, because it is difficult to measure the undigested fractions (may be different from one person to the next), a definition of dietary fiber probably should be amended to include, *"the residue remaining after an analytical procedure, such as the Neutral Detergent Fiber method, or the Acid Detergent Fiber method."*

A definition of dietary fiber should include all the components of a food that are not broken down by enzymes in the human digestive tract to produce small molecular compounds which are then absorbed into the blood stream. Thus, dietary fiber includes hemicelluloses, pectic substances, gums, mucilages, as well as certain other carbohydrates in addition to lignin and cellulose. These chemical compounds are found largely in the cell walls of plant tissues.

The term *crude fiber* as traditionally used may represent as little as one-seventh of the total dietary fiber of a given food. It is possible that the term fiber in itself may be somewhat misleading, inasmuch as all components of presently regarded "dietary fiber" are not fibrous in the usual physical sense, while, at the same time, some foods that contain recognizable fibers, such as muscle meats, do not yield undigestible residue.

Burkitt's definition of dietary fiber is "*mostly celluloses and lignin and lignin material, varying in different plants according to type and age. Basically, it passes through the small intestine undigested by our enzymes. A kind of natural and necessary laxative.*"

Wide Variations in Nature of Fibers. As pointed out by the Institute of Food Technologists Expert Panel (1979), the results of feeding "high-fiber" diets differ from one researcher to the next. One explanation for conflicting results may be the result of inadequate analytical methods used for determining fiber data. But even more fundamental is the fact that different fiber components have very different physiological functions and, inasmuch as fiber composition differs with the food source, the physiological effects noted will depend upon the predominant type of fiber present in the experimental diet. Some of these differences are in kind; others are of degree. As an example, the effect of fiber on the level of serum cholesterol has varied widely. Pectin, lignin, guar gum, oat hulls, and barley have been shown to have some cholesterol-lowering effect in human and animal studies, while bran and cellulose have not shown a similar effect.

Digestion of some components of dietary fiber, especially the hemicelluloses, takes place in the colon as a result of bacterial action. White flour, for example, is high in hemicellulose. The volatile fatty acids produced from this soluble fiber in the digestive process attract water from the surrounding tissues by osmosis, and thus may have a cathartic effect. Some fibers, such as bagasse from sugarcane, are very sharp abrasives to the intestinal tract; while others, such as lignin, may actually be constipating. There are such great differences in the physiological effects of the various constituents of dietary fiber that some researchers feel that it is essentially meaningless to consider high-fiber diets in the abstract. They do not deny that needs may exist for components of dietary fiber with specific properties, but rather that these needs may vary with different physiologic states.

Sources of Fibers. A wide variety of foods supply significant amounts of dietary fiber. Fundamental are those foods, such as fruits and vegetables, which provide significant quantities of fiber as the result of their naturally high fiber content. In many other instances, fibrous components (powdered cellulose; rice and soy hulls; soy, corn (maize), rice, wheat brans; coconut residues, citrus byproducts, ground almond skins, groundnut (peanut) hulls, etc.) can be added to processed foods, such as breads and, particularly to fabricated foods which provide excellent opportunities for improving the dietary aspects of many products in this latter category. A number of the fiber-containing components for addition to food products are not strange because for years past some of these materials have been considered additives, but for other purposes, such as thickening and bulking agents.

In both fresh vegetables and fruits, the *total dietary fiber* may appear to be relatively low because of the high water content of these foods. However in terms of solids content, the fiber portion appears substantial in many instances. For example, potato and starchy vegetables furnish appreciable amounts of fiber when consumed in relatively generous quantities. The lignin content of most vegetables is quite low, while that of fruits is greatest in those species which contain lignified seeds (such as strawberry), or lignified cells in the flesh (such as pear). The noncellulose polysaccharides in these foods are usually rich in pectic substances (uronic acids) and in 5-carbon sugars (pentoses).

Consumer interest in fibers has caused several bakeries to develop breads which contain from 6 to 8% crude fiber. This is approximately four times the amount of fiber present in ordinary whole wheat bread. Frequently, powdered foodgrade cellulose will be the fiber component added. Cellulose, of course, has been recognized for years as a food additive, often used as a thickener or to decrease separation of fat and water in other products, but its use in breads is relatively new. When added to bread, it must be listed on the ingredient label. Since cellulose and the water it holds dilute other nutrients in normal bread, "high-fiber" breads are sometimes promoted as "low-calorie" products.

The cereal brans, such as wheat, soy, and corn (maize) bran, can be added to many foods, such as snacks, cookies, bread and other bakery products, and cereal-based foods where the natural whole-grain flavor and color and texture imparted by the brans is consistent with the previous image of these products. Of the ingredients available to date, powdered cellulose is probably the most versatile.

Properties of Fibers. From the standpoint of application in preventive medicine, fiber has two major functional properties: (1) Absorption capacity, and (2) water-binding properties. The cardiovascular diseases apparently relate to the ability of fiber to absorb materials which reduce blood cholesterol. The intestinal diseases appear to be related to fiber's water-binding properties and are influenced by the physical tone of the large intestine. Obesity is indirectly related to both properties, but mostly to factors in the food itself.

The water-binding properties of fibers, such as wheat bran, which binds twice its weight in water, and pectins, which bind more than 5 times, increase the rate and volume of fecal elimination, producing more frequent and softer elimination. It thus relieves intracolonic pressure and the symptoms of diverticular diseases, which develop from chronic constipation. In theory, a high-fiber diet should also be preventive, since it produces better muscle tone in the large intestine.

Constipation. The value of fiber in increasing water content of feces has been mentioned. Fiber tends to effect a transit time in the gastrointestinal tract which is intermediate between being too rapid (diarrhea) and too slow (constipation). Some authorities theorize that the increased volume and softness of the stools, by reducing straining during defecation, is a factor in preventing hemorrhoids and varicose veins.

Diverticulosis. Diverticula are outpouchings that develop in weak areas in the bowel wall. If they are numerous and become inflamed, diverticulitis is the result. Accompanying the condition quite frequently is pain in the lower left side, an alternating diarrhea and constipation, and flatulence. Diverticulitis was essentially unknown prior to the turn of the century, but the incidence of the disorder has increased markedly in industrialized countries. In western countries, as of the early 1980s, it has been estimated that from one-fourth to one-third of the population of older persons may suffer some or all of the symptoms of diverticulitis. Prior to the current recognized treatment with high-fiber diet, it was treated with a low-residue diet, presuming that such a diet would permit healing and cause less irritation to the bowel. See **Diverticulosis and Diverticulitis.**

Cardiovascular Diseases. As previously mentioned, a high-fiber diet may lower the blood cholesterol levels by reducing transit time through the gastrointestinal tract. People on certain high-fiber diets excrete more bile acids, sterols, and fat, implying that the fiber compounds "bind" bile acids and thereby prevent absorption of cholesterol and fat and also the reabsorption of bile acid derived from the body's cholesterol. High serum cholesterol levels have been identified as one of the risk factors in atherosclerosis, although there is disagreement as to whether the actual risk can be reduced by lowering cholesterol levels by means of diet or drugs. Complicating any resolution of the role of dietary fiber in cardiovascular disease are the inconsistent effects produced by dietary fiber from different foods.

Cancer. The hypothesis relating dietary fiber to colon cancer presumes that the slow movement of the feces which occurs with a low-fiber diet allows more time for any carcinogens present in the colon to initiate cancer. Also, the extra water, bile acids, salts, and fat bound by added fibers are assumed to act as solvents to remove a wide variety of chemical factors which may be carcinogenic. A high-fiber diet also may alter the type and number of microorganisms in the colon, which produce compounds convertible to carcinogens. Theories based essentially upon correlations of various population characteristics can, of course, be misleading. As just one example, the incidence of colon cancer in different countries and cultures correlates much better with the consumption of fat in the diet than it does with the consumption of fiber. As of the early 1980s, there has been no proven relationship between bowel transit times and the incidence of colon cancer. Also, there is no proof that constipation leads to cancer, and none that dietary fiber per se has a definable effect on the intestinal flora in humans.

Excessive Fiber in Diet. With emphasis upon the probable beneficial effects of dietary fiber, the question of the consequences of fiber overdosage so to speak is rightfully brought forward. Much less research on this point has been conducted. It has been suggested that too much pectin may cause decreased vitamin B_{12} absorption. This would be an important concern for certain types of vegetarians whose diets are already low in that vitamin and high in fiber. There may also be a significant loss of minerals, particularly zinc, iron, calcium, copper, and magnesium, due to binding of these minerals by phytic acid, present in certain plant-based foods. The high-fiber diets of Africa and India, for example, are associated with a high incidence of such mineral deficiencies and kidney stones, especially in areas where rice is the major calorie source. The rate of stomach cancer in some of these areas is high, which should lead to cautious interpretation of epidemiological studies. Also, fiber, by its sheer bulk, may reduce the total amount of food consumed, thereby resulting in the deficiency of certain nutrients and, possibly in extreme cases, to reducing calories in areas where malnutrition already exists. The Institute of Food Technologists Expert Panel recommends that persons with kidney disease, diabetes, or other diseases should, without question, obtain permission from their physicians prior to consuming bran or making other drastic changes in their diet. Very large amounts of fiber could even cause enlargement and twisting (or "volvulus") of the sigmoid colon and aggravate ulcerative colitis. Both conditions occur in Africa, but rarely in Europe.

References

Burkitt, D. P.: "Some Diseases Characteristic of Modern Western Civilization," *Brit. Med. J.*, **1**, 274 (1973).

Burkitt, D. P., and H. C. Trowell: "Refined Carbohydrate Foods and Disease," Academic, New York, 1975.

Celender, I. M., Shapero, M., and A. E. Sloan: "Dietary Trends and Nutritional Status in the United States," *Food Technology*, **32**, 9, 39–41 (1978).

Hamilton, E. M. N.: "Nutrition: Concepts and Controversies," Food and Nutrition Press, Westport, Connecticut, 1979.

IFT Expert Panel: "Dietary Fiber," *Food Technology*, **33**, 1, 35–39 (1979).

Labuza, T. J.: "Contemporary Nutrition Controversies," Food and Nutrition Press, Westport, Connecticut, 1979.

Mottram, R. F.: "Human Nutrition," 3rd edition, Food and Nutrition Press, Westport, Connecticut, 1979.

National Academy of Sciences: The following books are available from the Office of Publications, National Academy of Sciences, Washington, D.C.:
"Human Vitamin B_6 Requirement"
"Laboratory Indices of Nutrition Status in Pregnancy"
"A Selected Annotated Bibliography on Breast Feeding"
"Folic Acid: Biochemistry and Physiology in Relation to the Human Nutrition Requirement"
"Plant and Animal Products in The U.S. Food System"
"Recommended Dietary Allowances," 9th edition (1980).

Rechcigl, M., Jr.: "Handbook of Nutrition and Foods," CRS Press, Cleveland, Ohio, 1977.

Winick, M.: "The Relationship of Diet, Nutrition, and Health during Various Stages of Life," *Food Technology*, **32**, 9, 42–43 (1978).

DIET (Carbohydrates). **Carbohydrates.**

DIETERICI EQUATION. A form of the equation of state, relating pressure, volume, and temperature of gas, and the gas constant. The Dieterici equation applies a correction to the van der Waals equation to allow for variation in density throughout a gas, due to the higher potential energies of molecules on or near the boundaries. One form of this equation is

$$P = \frac{RT}{V-b} e^{-a/RTV}$$

in which P is the pressure of the gas, T is the absolute temperature, V is the volume, R is the gas constant, e is the natural log base 2.718 . . . , and a and b are constants.

DIETETIC FOODS (Bulking Agents for). **Bodying and Bulking Agents (Foods); Sweeteners.**

DIET (Fishes). **Fishes.**

DIETHYLENE GLYCOL. **Ethylene Glycol.**

DIETHYL PYROCARBONATE. **Antimicrobial Agents (Foods).**

DIET (Protein-Sparing). **Obesity; Starvation.**

DIFFERENCE. The result obtained by subtraction of numbers or other quantities. Also called the remainder.

An especially useful and important type is a finite difference. Its properties are studied in the calculus of finite differences, sometimes called the twin sister of differential and integral calculus. It is applied to problems of interpolation, approximate differentiation and integration, summation of series, and to solutions of difference equations, the analogue of differential equations.

Let $y_0, y_1, y_2, \ldots$ be values of $y = f(x)$, let the corresponding values of the independent variable be $x_0, x_1, x_2, \ldots$ and define a first divided difference as

$$[x_i x_j] = (y_i - y_j)/(x_i - x_j)$$

Second, third, etc., divided differences eventually lead to the nth difference

$$[x_0 x_1 \cdots x_n] = \frac{[x_0 x_1 \cdots x_{n-1}] - [x_1 x_2 \cdots x_n]}{x_0 - x_n}$$

After some manipulation, one obtains

$$y = f(x) = f(x_0) + (x - x_0)[x_0 x_1] + (x - x_0)(x - x_1)[x_0 x_1 x_2] + \cdots + (x - x_0) \cdots (x - x_{n-1})[x_0 x_1 \cdots x_n] + R(x)$$

where $R(x)$ is the remainder. This formula makes it possible to calculate $y = f(x)$ at some value intermediate between tabulated numbers $x_0, x_1, x_2, \ldots$ and $y_0, y_1, y_2, \ldots$ but the procedure is less laborious if the x_k are evenly spaced, so that $x_n - x_0 = nh$, where n is an integer. These quantities could be, for example, numbers taken from a table of logarithms or trigonometric functions. They might also be the result of some experimental measurement. In the latter case, if the measurements were not made at equally spaced values of x, graphical interpolation could be used to get such numbers.

First differences are then defined as

$$\Delta y_0 = y_1 - y_0; \quad \Delta y_1 = y_2 - y_1; \quad \ldots; \quad \Delta y_{n-1} = y_n - y_{n-1}$$

Second differences, third differences, etc., are defined in a similar way and the $(n + 1)$th order differences are

$$\Delta^{n+1} y_0 = \Delta^n y_1 - \Delta^n y_0; \quad \Delta^{n+1} y_1 = \Delta^n y_2 - \Delta^n y_1, \ldots$$

By successive substitution, it is found that

$$\Delta^n y_k = \sum_{r=0}^{n} (-1)^r \binom{n}{k} y_{k+n-r}$$

Quantities of this kind are called diagonal or forward differences.

If $\Delta^m y_k$ is such a diagonal difference of mth order, a horizontal difference is defined as

$$\Delta_m y_{k+m} = \Delta^m y_k \qquad \text{or} \qquad \Delta_m y_n = \Delta^m y_{n-m}$$

Differences of either kind are conveniently displayed in a difference table, which is shown for the case where horizontal differences are used. Other types can be constructed in a similar way:

x	y	$\Delta_1 y$	$\Delta_2 y$	$\Delta_3 y$
x_0	y_0			
x_1	y_1	$\Delta_1 y_1$		
x_2	y_2	$\Delta_1 y_2$	$\Delta_2 y_2$	
x_3	y_3	$\Delta_1 y_3$	$\Delta_2 y_3$	$\Delta_3 y_3$

If h is the interval between equally spaced values of the argument in such a table, constructed from $y = f(x)$, it is often convenient to define

$$\delta f(x) = f(x + h/2) - f(x - h/2)$$

$$\mu f(x) = \tfrac{1}{2} \delta f(x)$$

A central difference, formed by the first of these operators, is related to a finite difference by the equation

$$\delta^m y_{n/2} = \Delta^m y_{(n-m)/2}$$

where m and n are integers, both even or both odd.

For some uses of differences of these types, see **Interpolation;** and terms listed under **Mathematics.**

DIFFERENCE EQUATION. Instead of variables and derivatives, $x, y, y', y'', \ldots$, as in differential equations, a difference equation contains variables and finite differences, Δy, $\Delta^2 y$, etc. It is convenient to change the independent variable from x to s, where $x = hs$, $\Delta x = h$ and then $\Delta s = 1$. With u as the dependent variable, $\Delta u_s = u_{s+1} - u_s$, $\Delta^2 u_s = u_{s+2} - 2u_{s+1} + u_s$, hence the equation may be written in terms of s, u_s, u_{s+1}, etc.

A linear difference equation, the simplest type studied, is

$$\sum_{i=0}^{n} f_i(s)u_i(s) = \phi(s)$$

which is inhomogeneous; homogeneous, if $\phi(s) = 0$. Difference and differential equations are similar in many properties but their solutions are quite unlike. Given an initial value, u_0, one could proceed step-by-step for a first-order case and calculate a table of particular solutions for integral values of s. This could be called a particular discrete solution for it will be a special case of a particular continuous solution containing an arbitrary constant, determined to satisfy the initial conditions. However, this is not the general solution.

Consider $f(s, u_s, w) = 0$, where w is any periodic function with period of unity. Then $f(s + 1, u_{s+1}, w) = 0$ and, by elimination, one obtains $F(s, u_s, u_{s+1}) = 0$, the given difference equation of order one. The solution containing such a periodic function is then the general solution.

By analogy with the case of integral calculus, one may find inverse equations to difference equations. Thus, as simple examples: for a product, $\Delta u_s v_s = v_{s+1}\,\Delta u_s + u_s\,\Delta v_s$; for a power, $\Delta s^{(n)} = ns^{(n-1)}$, etc. With a suitable collection of such formulas, the work of solving a difference equation proceeds in a manner similar to that for a differential equation.

DIFFERENTIABLE TOPOLOGY. Topology.

DIFFERENTIAL DISTILLATION. Distillation.

DIFFERENTIAL ENTROPY (Information). Information Theory.

DIFFERENTIAL EQUATION. An equation involving derivatives or differentials of an unknown function. When partial derivatives occur the equation is a partial differential one, otherwise an ordinary one. The order of the highest derivative occurring is the order of the equation and the highest power of the function or its derivative is the degree. Equations of the first degree are called linear, others are nonlinear. Further classifications will be described later.

Formal methods of solving differential equations were mostly completed by the middle of the eighteenth century and since then the theory of differential equations has been more concerned with existence and validity of solutions, rather than with the actual task of solving the equation. However, applied mathematicians and scientists are more interested in obtaining a solution. For this reason, the material here presented is written from that standpoint.

Given a differential equation, it is hoped that the reader may here find some suggestions as to how it may be solved. The first test is to determine whether it is: (a) a single equation, an ordinary differential equation; (b) a single equation, a partial differential equation; (c) a system of simultaneous differential equations. Proceed then to the appropriate reference, where further classifications will be developed in each case. However, it must be remembered that many integrals cannot be evaluated in terms of known functions and it thus follows that not all differential equations can be so solved. If the classical devices fail, final resort may be made to numerical, graphical, or mechanical methods. Recent developments in high-speed electronic computers have made it possible to complete the solution of many equations, especially nonlinear ones, which would have been difficult, if not impossible, with desk calculating machines. (See entries following.)

DIFFERENTIAL EQUATION (Exact). A differential equation obtained by differentiation of some function $\phi(x, y) = C$, hence, if it is of first order, its form is $P(x, y)\,dx + Q(x, y)\,dy = 0$. Here

$$P = \partial\phi/\partial x; \quad Q = \partial\phi/\partial y$$

and the left-hand side is an exact differential. The necessary and sufficient condition for an exact differential equation is

$$\partial P/\partial y = \partial Q/\partial x$$

which is a special case of the Cauchy-Riemann equation. The solution of such an equation is

$$\int_{x0}^{x} P(x, y)\,dx + \int_{y0}^{y} Q(x, y)\,dy = C$$

where x_0 and y_0 are constants to be chosen as convenient. They are often taken as zero or unity in order to simplify the integration.

In case $P\,dx + Q\,dy = 0$ is not exact, as shown by the fact that

$$\partial P/\partial y \pm \partial Q/\partial x$$

an integrating factor or Euler multiplier μ exists so that $\mu(P\,dx + Q\,dy) = 0$ is exact and can be integrated as explained. An infinity of such factors exists but it is not always easy to find one of them. Some special cases follow, depending mostly on certain forms of the function

$$F(x, y) = \partial P/\partial y - \partial Q/\partial x$$

(a) $F = Qf(x)$, $\mu = \exp \int f(x)\,dx$; (b) $F = -Pg(y)$,

$$\mu = \exp \int g(y)\,dy;$$

(c) if either $Px \pm Qy$ vanishes, but not both, the reciprocal of the non-vanishing quantity is an integrating factor; (d) if neither vanish, the quantity $1/(Px + Qy)$ is an integrating factor if the equation is homogeneous; (e) $yP(x, y)\,dx + xQ(x, y)\,dy = 0$, $\mu = 1/(xP - yQ)$; (f) $F = h(u)(yQ - xP)$, $u = xy$, $\mu = \exp \int h(u)\,du$; (g) $x^2F = -h(v) \times (Px + Qy)$, $v = y/x$, $\mu = \exp \int h(v)\,dv$.

See also **Differential Equation (Integral Solution of)** for some further properties of the integrating factor, especially as applied to equations of higher order.

DIFFERENTIAL EQUATION (Integral Solution of). It is sometimes desirable to obtain the solution of a differential equation, especially one which is linear and of second order, as a definite integral. The theory of the method is closely connected with that of the exact differential equation (see **Differential Equation (Exact)**) and its integrating factor. The concepts can be generalized for equations of order n but will be here illustrated only for $n = 2$.

Let the equation be given in operator form as $L(u) = u'' + p(x)u' + q(x)u = 0$ and suppose that an integrating factor; v exists so that $vL(u)\,dx$ is an exact differential, then the operator adjoint to $L(u)$ is $L(v) = v'' - p(x)v' + [q(x) - p'(x)]v$,

$$vL(u) - uL(v) = (d/dx)[P(u, v)]$$

is the Lagrange identity, and $P = uv[u'/u - v'/v + p(x)]$ is the bilinear concomitant. If the adjoint equation $L(v) = 0$ can be solved, the original equation $L(u) = 0$ is equivalent to the first order equation $P(u, v) = C$, where C is an arbitrary constant. The latter, of course, can always be solved.

An equation identical with its adjoint is said to be self-adjoint. Any differential equation may be put into self-adjoint form if the appropriate factor is introduced.

We now suppose that a solution to a given equation is sought in the form

$$u(x) = \int_a^b K(x, t)v(t)\,dt$$

which means that three properties of the integral must be determined so that $L_x(u) = 0$. These are: (a) the function $K(x, t)$, which is the kernel of the integral; (b) the function $v(t)$; (c) the integration limits, a and b.

The kernel is chosen to satisfy a partial differential equation

$$L_x(K) = M_t(K)$$

where M_t is a linear differential operator in t and $\partial/\partial t$. Operating on the definite integral

$$L_x(u) = \int_a^b M_t\{K(x, t)\}v(t)\, dt$$

and using the Lagrange identity

$$v(t)M_t\{K(x, t)\} - K(x, t)\bar{M}_t\{v, t\} = \frac{\partial}{\partial t} P\{K, v\}$$

where $\bar{M}_t$ is the operator adjoint to M_t, one obtains

$$L_x\{y(u)\} = \int_a K(x, t)\bar{M}_t(v)\, dt + \Big[P\{K, v\}\Big]_a^b$$

If the two terms on the right vanish, the assumed integral satisfies the differential equation. They will vanish if $v(t)$ is a solution of the partial differential equation in the adjoint $\bar{M}_t(v) = 0$ and if the limits of integration are chosen so that the bilinear concomitant vanishes.

See also **Green Function;** and **Integral Transform.**

DIFFERENTIAL EQUATION (Linear). The general linear equation of order n is $p_0(x)y^{(n)} + p_1(x)y^{(n-1)} + \cdots + p_n(x)y = r(x)$. If $D = d/dy$ and $L = p_0D^n + p_1D^{n-1} + \cdots + p_n$, a polynomial in D, the equation may be written symbolically as $L(y) = r(x)$, where D is a differential operator and L is a linear operator of order n. When $r(x) = 0$, the equation is homogeneous but inhomogeneous otherwise.

Methods for solving linear differential equations can be treated best as special cases: (a) $n = 1$, see section (1) which follows; (b) $n > 1$, all p_i = constant, see section (2) of this article; (c) $n \geq 2$, p_i not constant but functions of x, see **Differential Equation of Second Order (Linear).**

(1) *First Order Linear Equation.* The equation, in standard form, can be written as $y' + P(x)y = R(x)$. Consider first the case where $P(x) = 0$ and the variables are seen to be separable (see **Differential Equation (Separable)**) as is also true for the homogeneous case, $R(x) = 0$. In the latter case, the solution becomes

$$y_0 = Ce^{-\phi}, \quad \phi = \int P\, dx$$

a subscript being used on y to distinguish it from the solution of the inhomogeneous case, now to be sought. There are several different procedures, all of which give the same result.

(i) Assume that $y = y_0 v$, where y is the solution of the inhomogeneous equation. Then $y_0 v' = R(x)$, the variables are again separable, hence one more integration yields y.

(ii) Assume that the integration constant C is a parameter depending on x and adjust it to satisfy the inhomogeneous equation. This procedure is called the method of variation of parameters. It was discovered by Lagrange (1736–1813).

(iii) Look for an integrating factor (see **Differential Equation (Exact)**). This is found to be Ce^{ϕ}.

The general solution, found in any of these ways, is

$$y = Ce^{-\phi} + e^{-\phi}\int R(x)e^{\phi}\, dx$$

We note that two integrations are necessary but only one integration constant results since the equation is of first order. Furthermore, if $P(x)$ and $R(x)$ are constants, the methods are still applicable.

An equation reducible to the linear case is Bernoulli's equation.

(2) *Higher Order Linear Equation with Constant Coefficients.* The equation in standard form can be taken as $(D^n + a_1D^{n-1} + \cdots + a_n)y = R(x)$, where the a_i are constants.

Consider first the homogeneous case, where $R(x) = 0$, and form the auxiliary equation

$$(D - r_1)(D - r_2) \cdots (D - r_n)y = 0$$

where the r_i are roots of the algebraic equation $r^n + a_1r^{n-1} + a_2r^{n-2} + \cdots + a_n = 0$. The general solution of the homogeneous equation is then

$$y = \sum_i c_i e^{r_i x}$$

the c_i's being arbitrary constants, provided the roots are all unequal. If one of them, r_g is repeated g times the corresponding factors in the general solution of the differential equation are replaced by $c_g(1 + a_1x + a_2x^2 + \cdots + a_{g-1}x^{g-1})e^{r_g x}$. Imaginary roots, if they occur, have the form $r_\pm = A \pm iB$ and the term in the solution becomes $e^{Ax}(c_+ \cos Bx + c_- \sin Bx)$.

The general solution of the inhomogeneous equation, where $R(x) \neq 0$, equals the solution of the homogeneous equation, as just found and called the complementary function, to which is added some particular integral of the inhomogeneous equation. To find this particular integral, we note that symbolically $\phi(D)y = R(x)$, hence a particular integral is $y = \phi^{-1}(D)R(x)$. In principle, there are two general ways of finding the inverse function, $\phi^{-1}(D)$: (a) it can be factored; (b) it can be written as a sum of partial fractions. There are several special cases, where the labor of the general procedure can be shortened: (1) $R(x) = e^{kx}$, k = constant. Then, if $\phi(D)$ is a polynomial in D, so that $\phi(k) \neq 0$, $\phi^{-1}(D)e^{kx} = e^{kx}/\phi(k)$. (2) $R(x) = e^{kx}f(x)$, $\phi^{-1}(D)e^{kx}f(x) = e^{kx}\phi^{-1}(D + k)f(x)$. (3) $R(x) = \sin ax$ and $\phi(D)$ is an even polynomial in D, write $\phi(D) = F(D^2)$, then $\phi^{-1}(D) \sin ax = \sin ax/F(-a^2)$. Modifications of this case can take care of odd polynomials in D and particular integrals of the form $\phi^{-1}(D)e^{kx}\sin(\cos)ax$.

DIFFERENTIAL EQUATION (Numerical Solution of). When a differential equation cannot be solved analytically, a graphical or numerical method may be tried. Many variations in the latter method have been proposed but they may usually be classified as: (1) Methods of successive approximations or iteration (see **Euler Method for Numerical Solution of a Differential Equation**) by means of polynomials or integral equations (see **Picard Method of Successive Approximations or Iteration**). (2) Expansion in a Taylor series. (3) Runge-Kutta method. (4) Milne method. Each of these procedures has been illustrated by a first order equation. In higher order cases, the equation can be reduced to first order equations by suitable change of variables.

DIFFERENTIAL EQUATION OF FIRST ORDER (Ordinary). The equations considered here are of the type $f(x, y, y') = 0$, where there is only one equation given, not a system of equations; the derivative of order one is an ordinary, not a partial derivative; the degree of the equation could be first, or it could be higher than first.

A geometric interpretation of a first order differential equation is easily given for the equation defining it as an equation of the nth degree in three variables and therefore represents an infinite set of curves, n of which (either real or imaginary) will pass through a given point (x_1, y_1) and the differential equation will therefore determine n slopes at that point.

Having found that a given differential equation is of first order, first consider the simplest case, that also of first degree. It will then fall into one of the following categories, or perhaps can be transformed to fit there. For further information, see the entries indicated: (1) variables are separated or separable (see **Differential Equation (Separable)**); (2) linear (see **Differential Equation (Linear)**); exact, or can be made so by an integrating factor (see **Differential Equation (Exact)**); (4) other methods (see **Differential Equation (Series Solution of)**) and **Differential Equation (Numerical Solution of).**

Now suppose that a given equation is still of first order, but higher than first degree. The general case, of degree n, is $F(x, y, y') = p^n + A_1p^{n-1} + \cdots + A_{n-1}p + A_n = 0$, where $p = dy/dx = y'$ and A_i is a function of x and y. Assuming that the equation can be factored, that is treated as a polynomial in p, its n roots, a_i, which are also functions of x and y, will satisfy the factored form of the equation, which is $(p - a_1)(p - a_2) \cdots (p - a_n) = 0$. Let $f_i(x, y, c_i)$ be a solution of one equation in the product, such as $(p - a_i) = (y' - a_i) = 0$, then it must also satisfy the original differential equation and its general solution will be $f_1(x, y, c)f_2(x, y, c) \cdots f_n(x, y, c) = 0$. It should be noted that only one integration constant c is needed and that the problem of the nth degree equation has been reduced to the solution of n equations of first degree. The difficulties that may arise are thus either algebraic or related to integral calculus.

Particularly simple cases occur if either the dependent or independent variable is missing, so that the equation is either $f(x, p)$ or $f(y, p)$. In either case, solve for p and integrate. If it is easier to solve for x, so that $x = f(p)$, then $1/p = f'(p)\, dp/dy$ and finally, $y =$

$pf'(p)\,dp + C = \phi(p)$. This equation, together with the form $x = f(p)$, can be regarded as a parametric representation of the solution. Elimination of p gives the desired solution to the differential equation. Suitable modifications of the procedure permit one to solve for y instead of x, if this is easier.

Other schemes for higher order equations involve a transformation of variable to convert the equation into one of first degree. A particular simple case of this kind is Clairaut's equation. See **Clairaut Equation.**

DIFFERENTIAL EQUATION OF HIGHER ORDER (Ordinary). The general equation of this type can be solved in finite form in only a few special cases. The most important equation of higher order is the linear one, especially that of second order, for only a few equations in applied mathematics have order higher than two, and most of them are linear. Even restricting one's attention to the linear case, a general method is known only for the one with constant coefficients or those transformable into this type. In most cases, the solution of the linear equation defines a new transcendental function. It can be represented as an infinite series, an infinite continued product, or as a definite integral.

Given a differential equation of this type, $f(x, y, y', y'', \ldots, y^{(n)})$, for solution attempt to fit it into one of the following categories and proceed as indicated.

1. The case, $y^{(n)} = f(x)$. A series of n successive integrations will give the general solution

$$y = \frac{1}{(n-1)!}\int_{x0}^{x}(x-t)^{n-1}f(t)\,dt + C_1(x-x_0)^{n-1} + C_2(x-x_0)^{n-2} + \cdots + C_n$$

where x_0 is chosen for convenience in evaluating the definite integral and can be often taken as zero.

2. The dependent variable is missing, $f(x, y^{(k)}, y^{(k+1)}, \ldots, y^{(n)})$. Let $v = y^{(k)}$ and the given equation is reduced in order from n to $(n-k)$. If this can be integrated to give a solution $v(x)$, then y can be found from $y^{(k)} = v$, which is of type (1). A special and simple example is $f(y^{(n)}, y^{(n-1)}) = 0$, which requires only one integration. Another simple case is $n = 2$. Thus, given $f(x, y', y'') = 0$, let $p = y'$, $p' = y''$ and a first order equation, $g(x, p, p') = 0$ results. Its solution is $p = \phi(x)$, hence $y = \int\phi(x)\,dx + C$.

3. The independent variable is missing, $f(y, y', y'', \ldots, y^{(n)}) = 0$. The order may be reduced by the substitutions $y' = p$, $y'' = pp'$, etc., to give $g(y, p, p', \ldots, p^{(n-1)}) = 0$, which may, or may not, be solvable. The special case, $n = 2$, is simple, for if $f(y, y', y'') = 0$ is given, then the reduced equation is of first order $g(y, p, p') = 0$, with solution $p = \phi(y)$, hence $x = \int dy/\phi(y) + C$.

4. The homogeneous equation. If the equation is homogeneous, of degree k, in $(y, y', y''$, etc., then its form is $y^k f(x, t_1, t_2, \ldots, t_k) = 0$, where $t_1 = y'/y$, $t_2 = y''/y_1$, etc. Let $y = e^{\psi}$, $\psi = \int u\,dx$, and the order of the given equation is reduced from n to $(n-1)$.

A second possibility, somewhat simpler, occurs if the equation is homogeneous in $y, xy', x^2y'', \ldots$, so that $f(y, xy', x^2y'', \ldots, x^ny^{(n)}) = 0$. Change variables $x = e^t$, $y = ze^{nt}$ and the transformed equation is type (3), with independent variable missing.

5. The linear equation. The general linear equation can be written as $p_0(x)y^{(n)} + p_1(x)y^{(n-1)} + \cdots + p_n(x)y = r(x)$, which may be written symbolically as $L(y) = r(x)$, where $L = p_0D^n + p_1D^{n-1} + \cdots + p_{n-1}D + p_n$ is a linear differential operator of order n and $D = d/dy$. If the coefficients $p_k(x)$ are constants, see **Differential Equation (Linear)**; if they are functions of x, and even though L is of order greater than two, see **Differential Equation of Second Order (Linear).** Sometimes, a change of variable will transform the general linear equation into one with constant coefficient. As a typical example, consider

$$\sum_{k=0}^{n} a_k x^k y^{(k)} = f(x)$$

and let $\ln x = t$, $y(x) = Y(t)$. The result is, with

$$D = d/dt,\ \sum_{k=0}^{n} a_k D(D-1)\cdots(D-k+1)Y = f(e^t)$$

which has constant coefficients. This is called Euler's linear equation and also, but apparently incorrectly, Cauchy's equation. Actually, John Bernoulli (see **Bernoulli Equation**) has studied it before 1700, Euler is known to have worked on it about 1740, and Cauchy (1789–1857) only much later. The more general equation

$$\sum a_k(Ax+B)^k y^{(k)} = f(x)$$

is reducible to the Euler equation by the substitution $y(x) = Y(z)$, $z = (Ax + B)$.

6. For other cases, see **Differential Equation (Series Solution of); Differential Equation (Integral Solution of);** and **Differential Equation (Numerical Solution of).**

DIFFERENTIAL EQUATION OF SECOND ORDER (Linear). Equations of this type, with variable coefficients, occur frequently in theoretical physics and chemistry. For ease of discussion, this article is limited to the second order linear equation but the methods may be extended to the nth order case, in some instances.

The general equation is $A_0(x)y'' + A_1(x)y' + A_2(x)y = R(x)$, but the standard form is usually taken as $y'' + P(x)y' + Q(x)y = R(x)$. Let us suppose that a particular integral for the homogeneous case, $R(x) = 0$, can be found. Then by an extension of the methods described under **Differential Equation (Linear),** the general solution of the inhomogeneous equation can be found. As an example of the first of these methods, write the homogeneous equation as $z'' + P(x)z' + Q(x)z = 0$ and assume that z is known, but not $z = 0$. A change of dependent variable $y = zv$ converts the inhomogeneous equation into $zv'' = (2z' + Pz)v' + (z'' + Pz' + Qz)v = R$. But the third term on the left is zero, since it is the corresponding homogeneous equation and the problem has been reduced to a first-order differential equation in v', which is $zv'' + (2z' + Pz)v' = R$. This can presumably be solved (see **Differential Equation of First Order (Ordinary)**), thus, there is no loss in generality if we henceforth restrict the discussion to the linear homogeneous case. There are several possibilities: (1) The coefficients are constants, see **Differential Equation (Linear).** (2) The normal equation. Substitute $y = ve^{-\phi/2}$, where

$$\phi = \int P\,dx$$

and obtain the normal form, $v'' + I(x)v = 0$, where $I(x) = Q - P'/2 - P^2/4$. This, with no term in the first derivative, is sometimes easier to solve than the original equation. Thus, if $I =$ constant, the equation is linear with constant coefficients; if it is a constant divided by x^2, it has become Euler's linear equation (see **Differential Equation of Higher Order (Ordinary).** (3) Integrating Factor. If u is an integrating factor to the given equation, then it satisfies $u'' - Pu' + (Q - P')u = 0$. If this can be solved, the equation in z is linear and of first order. (4) Change of Independent Variable. If $(Q' + 2PQ)/Q^{3/2} = K$, a constant, the new variable

$$w = \int Q^{1/2}\,dx$$

will reduce the equation to case (1) since it has become Euler's equation, see (2). Sometimes, this method will change the equation into another one whose solution is known. Thus, if

$$K\int Q^{1/2}\,dx = 2$$

a change of independent variable will yield Bessel's equation. (5) Riccati Equation. Every second-order linear equation can be converted into a Riccati equation, which is of first order. Unfortunately, this is seldom helpful since the Riccati equation can be solved in finite terms only for a few special cases. (6) Other Methods. Refer to **Differential Equation of Higher Order (Ordinary)** for some special cases of the second order equation. (See also **Differential Equation (Series Solution of); Differential Equation (Integral Solution of)** and finally, if necessary, see **Differential Equation (Numerical Solution of).**)

DIFFERENTIAL EQUATION (Ordinary). An ordinary differential equation, generally taken to mean a single equation with ordinary, not partial, derivatives has as its general solution or primitive an expression containing a number of arbitrary constants equal to the order of the equation. Imposition of boundary conditions specifies the values of the constants and yield a particular solution. Certain special solutions, called singular solutions, occasionally occur which cannot be obtained from the general solution (see **Clairaut equation**).

The primitive of an ordinary differential equation can also be discussed in another way. Consider the equation $f(x, y, c_1, c_2, \ldots, c_n) = 0$, where the c_i are arbitrary and independent constants. Differentiate this equation n times to get $f_x + f_y y' = 0$; $f_{xx} + 2f_{xy}y' + f_{yy}y'^2 + f_y y'' = 0$; $\ldots$, $f_x^{(n)} + \cdots + f_y y^{(n)} = 0$. These equations, $(n + 1)$ in number, can then be used to eliminate the n constants c_i and the final result is the nth order differential equation $F(x, y, y', y'', \ldots, y^{(n)}) = 0$. The process of solving a given differential equation is the reverse of that described; that is, of finding the primitive from which it might have been formed. It should be noted, however, that differential equations are seldom obtained by the procedure just outlined.

Given an ordinary differential equation for solution, classify it according to one of the following and refer to that entry: (a) two variables, first order, see **Differential Equation of First Order (Ordinary)**; (b) two variables, higher than first order, see **Differential Equation of Higher Order, (Ordinary)**; (c) more than two variables, see **Differential Equation (Total)**.

DIFFERENTIAL EQUATION (Partial). Partial Differential Equation.

DIFFERENTIAL EQUATION (Separable). The simplest possible case of a first order differential equation is $y' = \phi(x, y)$, where ϕ can be factored into one of the forms: (a) $f(x)$; (b) $f(x)g(y)$; (c) $g(y)$. In each case, the variables are separated and one integration yields the general solution of the differential equation as:

$$\text{(a)} \quad y = \int f(x)\,dx + C$$

$$\text{(b)} \quad \int dy/g(y) = \int f(x)\,dx + C$$

$$\text{(c)} \quad \int dy/g(y) = x + C$$

Fortunately, many equations of chemistry and physics are of this very simple type and others can be transformed to it as will now be shown.

Given $y' = f(y/x)$, the equation is said to be homogeneous and of the same degree in x and y. Substitute $y = vx$, which results in $v + xv' = f(v)$, the variables are separable, and the solution is

$$\ln x = \int \frac{dv}{f(v) - v} + C$$

Given $y' = f(ax + by + c)$, if either a or b vanish, the variables are immediately separable. If neither vanish, let $v = ax + by + c$, which gives

$$\int \frac{dv}{a + bf(v)} = x + C$$

In the more general case, where

$$y' = f\left(\frac{ax + by + c}{Ax + By + C}\right)$$

provided $(aB - Ab) \neq 0$, let $x = h + u$; $y = k + v(u)$, which gives the homogeneous equation,

$$v' = f\left(\frac{au + bv}{Au + Bv}\right)$$

In case, $(aB - Ab) = 0$, the substitution $v(x) = ax + by + c$ or $v(x) = Ax + By + C$, will immediately separate the variables. Finally, if $y' = y/x + g(x)f(y/x)$, the proper substitution is $y = xu(x)$, which also results in separation.

DIFFERENTIAL EQUATION (Series Solution of). Linear equations, especially of second order, are often solved by assuming an infinite series as a solution

$$y = \sum_{k=0}^{\infty} A_k x^{k+s}$$

Let the given equation be $y'' + P(x)y' + Q(x)y = 0$ and the series be substituted into it. An identity in x results, hence coefficients of each power of x must vanish, and a system of equations in s and the A_i results. That one which comes from the lowest power of x (i.e., $k = 0$) is called the indicial equation and it serves to determine s, which is an exponent of the differential equation.

The indicial equation may be: (1) independent of s, in which case no series solution of the type assumed exists; (2) a polynomial in s, of degree equal to the order of the differential equation; depending on the nature of the roots of the indicial equation, there may then be as many distinct series solutions as the order of the differential equation or fewer; (3) the degree of the polynomial is less than the order of the differential equation and series solutions are again impossible in the usual case.

The remaining equations of the system determine relations between successive coefficients in the series. From them a recursion formula can be found. It is usually a first order difference equation, permitting the calculation of A_{i+1} where A_i is known. In some more complicated cases, a second order difference equation occurs, as in the Mathieu equation. This involves relations between three successive coefficients.

In the study of series solutions, the singular points of the differential equation should be considered. These might be at $x = a_1, a_2, \ldots$ in the general case and the indicial equations and exponents should be determined for each. Solutions near each singular point, as series in powers of $(x - a)$, may or may not be possible as shown by the various exponents. If such a series exists about a singular point $x = a$, it will usually hold over a region up to the next nearest singular point $x = a'$, but it may fail to satisfy the differential equation there. (See also **Analytic Continuation** and **Riemann-Papperitz Equation.**)

DIFFERENTIAL EQUATION (Simultaneous). A simultaneous system of ordinary differential equations might arise in the following way. Suppose two functions were given $f_1(x, y, t, c_1, c_2) = 0$ and $f_2(x, y, t, c_1, c_2) = 0$, where c_1, c_2 are arbitrary constants. Differentiation with respect to t and elimination of the two constants would then result in a pair of simultaneous first order ordinary differential equations $\phi_1(x, x', y, y', t) = 0$, and $\phi_2(x, x', y, y', t) = 0$. Alternatively, suppose $y^{(n)} = F(x, y, y', y'', \ldots, y^{(n-1)})$ were given. Introduce new variables $y_1 = y$, $y_1' = y_2$, $y_2' = y_3$, $\ldots$, $y'_{(n-1)} = y_n$ and these equations, together with $y_n' = F(x, y_1, y_2, \ldots, y_n)$, constitute a set of n simultaneous equations of first order. Thus, by suitable change of variable, a single equation of any order or any system of equations can be reduced to a simultaneous system of first order equations.

In principle, the systems can be treated like algebraic equations if they are linear, considering the unknowns to be the dependent variables and eliminating them one by one. The complete solution will contain as many independent relations between the variables as there are dependent variables. For further details, texts on differential equations must be consulted. Sometimes total differential equations occur as simultaneous systems.

DIFFERENTIAL EQUATION (Total). One containing two or more dependent variables and their differentials or ordinary derivatives with respect to a single independent variable. The latter, however, need not appear explicitly in the equation.

If there are three variables, it has the form $P\,dx + Q\,dy + R\,dz = 0$, where P, Q, R are functions of x, y, z. It may have a general solution $\phi(x, y, z) = C$, where C is a constant, for total differentiation of this result gives

$$\frac{\partial\phi}{\partial x}\,dx + \frac{\partial\phi}{\partial x}\,dy + \frac{\partial\phi}{\partial z}\,dz = 0$$

and, if the partial derivatives have an integrating factor $\mu(\mu = 1$ is included), so that $\mu P = \partial\phi/\partial x$, $\mu Q = \partial\phi/\partial y$, $\mu R = \partial\phi/\partial z$ then a total differential equation of the form indicated has been obtained.

However, it does not follow that every total differential equation has such a solution for there may be no integrating factor. A necessary and sufficient condition for its existence, and hence for a solution of the form assumed, is

$$\begin{vmatrix} P & Q & R \\ \partial/\partial x & \partial/\partial y & \partial/\partial z \\ P & Q & R \end{vmatrix} = 0$$

This is the condition for integrability and, if the determinant does not vanish, the total differential equation is said to be non-integrable, the equation then being known as Pfaff's problem. When there are only two variables, equations of this type are always integrable (see **Differential Equation (Exact)**).

A general method for solution of the integrable case is as follows: (1) integrate assuming some one of the variables to be a constant, but take an arbitrary function of that variable, rather than a constant, as the constant of integration; (2) differentiate the result to get a new differential equation containing the arbitrary function; (3) compare the latter with the original total differential equation and eliminate all quantities except the arbitrary function and its variable; (4) integrate the equation so obtained to get the general solution of the given total equation. Several short-cuts and special cases can often be found.

Simultaneous total equations sometimes occur (see **Differential Equation (Simultaneous)**).

DIFFERENTIAL GAP. Multiposition Controller.

DIFFERENTIAL GEOMETRY. Geometry.

DIFFERENTIAL LEVELING. Differential leveling is a system of surveying whereby the difference in elevation of two remote points is obtained through the use of the surveyor's level and level rod. A chain or tape is not needed. The procedure in differential leveling is illustrated in the figure. BM_1 represents a known bench mark. The elevation of BM_2 is to be found. The rod is held on BM_1 and the level set up so as to take a back sight on the rod. The rodman then advances to a turning point chosen by the instrument operator, and the telescope is swung around for a foresight reading on the rod. The levelman then advances the instrument to a new position, from which he takes a back sight on the rod, which is still at the turning point. This procedure is continued until the rodman reaches the sight of BM_2. The back sight reading, added to the elevation of BM_1, gives the elevation of the level at the first station. The foresight reading, subtracted from the instrument elevation, gives the elevation of the turning point. In this way, by additions and subtractions of back sight and foresight readings, the total difference of elevation between BM_1 and BM_2 is determined. See also **Level (Surveyor's).**

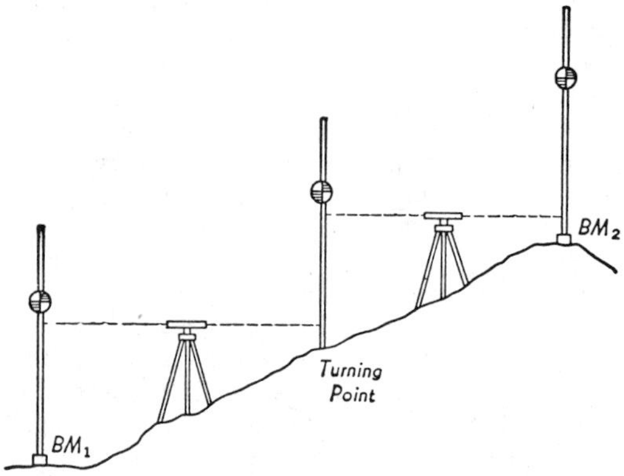

Differential leveling.

The field work which is necessary for determining the elevation of points along a given line such as the center line of a railroad, airport runway, or highway is called profile leveling. Rod readings are taken at regular intervals and also at points of abrupt change of slope. The outline of a vertical section through the center line is called a profile. The profile is obtained by plotting elevations which are the result of profile leveling. Lasers are now used by surveyors in leveling operations, notably in the grading of land for irrigation. See also **Irrigation.**

DIFFERENTIAL MANOMETER. A device for measuring small pressures. The device is best explained by referring to the figure. A U-tube, equipped with an enlarged section (c) at the top of each side, has in it two immiscible liquids, a lower (heavier) liquid *F* and an upper (lighter) liquid *E*. If a pressure difference be set up across *A* and *B*, the liquid *F* will change position giving a head *D* to compensate for the pressure. Because of the enlargements in the upper tubes the top level of liquid *E* changes a negligible amount. The head equivalent to the pressure varies inversely as the difference between the densities of liquids *E* and *F*. By selecting the proper liquids, the difference can be made very small, making for a large head. Thus a pressure which would give only a small reading on an ordinary manometer can be made to produce a large reading on a differential manometer, thus increasing the accuracy of measurement.

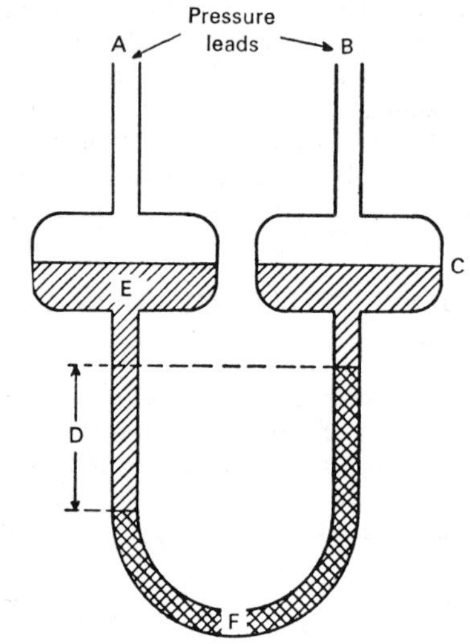

Cross section of a differential manometer.

Differential manometers are widely used in industrial instruments, not only for pressure measurements per se, but in flowmeters, liquid-level meters, and specific gravity meters—where a pressure difference is indicative of a change in another variable. A widely used configuration is in an orifice type flowmeter. Detection of manometer level changes can be accomplished automatically by way of floats, induction coils, and other sensors.

DIFFERENTIAL (Mathematics). If the variable y depends on the single independent variable x, so that $y = f(x)$, their differentials are designated by dy and dx. If $dx \neq 0$, the ratio of the differentials is the derivative of y with respect to x

$$dy/dx = f'(x)$$

Given a function of several independent variables, the total or complete differential is a sum of terms containing partial derivatives as coefficients

$$d\phi(x, y, z, \ldots) = \frac{\partial\phi}{\partial x}dx + \frac{\partial\phi}{\partial y}dy + \frac{\partial\phi}{\partial z}dz + \cdots$$

To emphasize the fact that the function of $d\phi$ has been obtained by differentiation, it is also called an exact or perfect differential. See also **Differential Equation (Exact)**; and **Pfaff Problem.**

DIFFERENTIAL-MODE VOLTAGE. The voltage which appears between two terminals or other points in a circuit, neither of which is necessarily at the system reference potential (usually designated as ground) is termed the differential- (or normal)- mode voltage. In the diagram (next page), $V_A - V_B$ is the differential input voltage. A perfect differential system would indicate only the value of the differential signal voltage V_S. However, in practical installations, unbalances and inaccuracies in the system often result in conversion of some of the common-mode voltage into a differential-mode signal. These causes are described in some detail under **Common-Mode Voltage.**

Thomas J. Harrison, International Business Machines Corporation, Boca Raton, Florida.

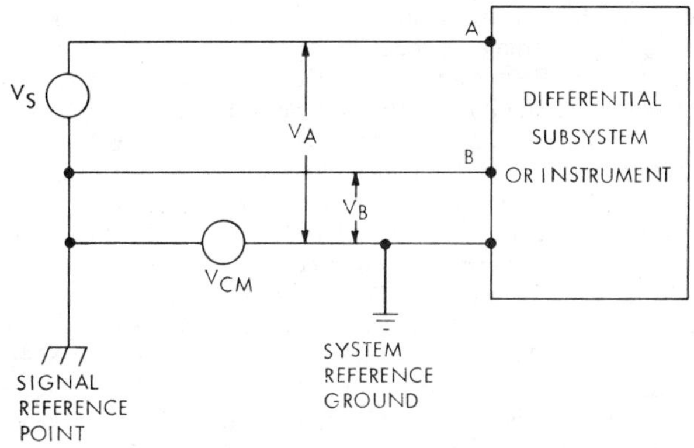

Differential-mode voltage.

DIFFERENTIAL PROCESS. Random Walk.

DIFFERENTIAL PULLEY. Machine (Simple).

DIFFERENTIAL THERMAL ANALYSIS. This analytical method makes possible the detection of thermal transitions accompanying many physical and chemical changes taking place in a substance while it is being heated. Energy changes associated with fusion, vaporization, recrystallization, solid–solid phase changes, oxidation, reduction, dehydration, chemical recombination, and decomposition may be detected and measured and hence such data used as criteria for identifying substances. Suitable instrumentation is used to record exothermic and endothermic reactions as positive and negative deviations from a base line.

DIFFERENTIAL (Vehicle). As a 4-wheel vehicle rounds a corner, the outer wheels travel a greater distance than the inner. The wheels on a wagon are mounted on a dead axle, so that they turn independently of each other. On a live axle, some device which will permit them to revolve at different speeds to compensate for the difference in travel when rounding a curve is necessary. The fundamental operating principle of a simple bevel gear differential is illustrated by the accompanying diagram. The driveshaft has mounted on it pinion *B*, which drives gear *C*. If it were not for the necessity of rounding curves, gear *C* could be rigidly fixed to the live axle *KK*. The differential action is obtained as follows: Gear *C* is not keyed to the axle. The spider *E* is rigidly fastened to the gear and has mounted on it, free to turn, the bevel gear *G*. Gear *G* meshes with gears H_1 and *H*, each of which is keyed to a half of the axle. When traveling straight ahead, gears *G*, H_1, and *H* revolve with the spider, but do not have any motion relative to each other. When rounding a curve, one wheel must travel faster than the other. The difference in rotation of the axle is compensated for by rotation of the differential gear *G* on its pin *F*. Any accelerated motion of one wheel is offset by a retarded motion of the other.

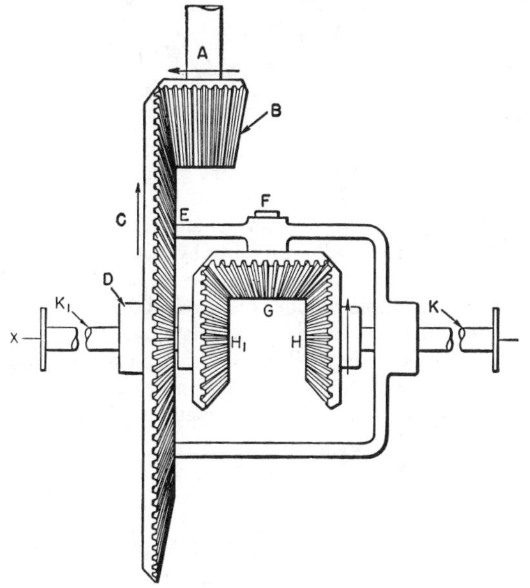

Schematic representation of a bevel gear differential for illustration of operating principle. There are other differential designs, such as limited-slip differentials.

DIFFERENTIATION (Geology). The general process of formation of different types of igneous rocks from a common parent magma. In a broader sense, the term signifies crystallization or recrystallization phenomena that occur in a magma as it cools.

DIFFERENTIATION (Mathematics). Differentiation is the process of finding the derivative of a function with respect to the independent variable. The methods of calculus are used to determine the derivative for certain elementary functions and two or more of the resulting rules are combined to obtain derivatives for more complicated functions. In the following table *y, u, v, w* are functions of the independent variable *x*; *A, B, C* are constants; $y' = dy/dx$; $u' = du/dx$, etc.,

Algebraic Functions

1. $y = u \pm v \pm w \pm \cdots$; $y' = u' \pm v' \pm w' \pm \cdots$; $y = C$; $y' = 0$.
2. $y = Au \pm Bv \pm Cw \pm \cdots$; $y' = Au' \pm Bv' \pm Cw' \pm \cdots$.
3. $y = uv$; $y' = uv' + u'v$; $y = uvw \cdots$; $y' = vwu' + uwv' + uvw' + \cdots$; $y = uv$; $y'/y = u'/u + v'/v$.
4. $y = u/v$; $y' = (vu' - uv')/v^2$; $y = A/v$; $y' = -Av'/v^2$.
5. $y = u^n$; $y' = nu^{n-1}u'$, where *n* is any positive or negative integer or fraction, rational or irrational number but independent of *x*.
6. If *u* is a function of *x* and $\phi(u)$ is a function of *u*, $\phi' = d\phi/dx = (d\phi/du)(du/dx)$.
7. If $x = \phi(y)$ is the inverse function to $y = f(x)$, $d\phi/dy = 1/f'(x)$.

Transcendental Functions

8. $y = e^u$; $y' = e^u u'$; $y = a^u$; $y' = a^u \ln au'$, where *a* is independent of *x*.
9. $y = \ln u$; $y' = u'/u$; $y = \log_a u$; $y' = \dfrac{u'}{u \ln a}$.
10. $y = \sin u$; $y' = \cos uu'$.
11. $y = \cos u$; $y' = -\sin uu'$.
12. $y = \tan u$; $y' = \sec^2 uu'$.
13. $y = \cot u$; $y' = -\csc^2 uu'$.
14. $y = \sec u$; $y' = \sec u \tan uu'$.
15. $y = \csc u$; $y' = -\csc u \cot uu'$.
16. $y = \sin^{-1} u$; $y' = u'/\sqrt{1 - u^2}$.
17. $y = \cos^{-1} u$; $y' = -u'/\sqrt{1 - u^2}$.
18. $y = \tan^{-1} u$; $y' = u'/(1 + u^2)$.
19. $y = \cot^{-1} u$; $y' = -u'/(1 + u^2)$.

For higher-order derivatives, the Leibniz rule is often useful. Partial derivatives are found by methods similar to those described here for ordinary derivatives.

See also **Differentiation (Numerical).**

DIFFERENTIATION (Numerical). A method for calculating the numerical value of the derivative of a function at a given point (x_0, y_0). Graphical or mechanical processes may be used, but more commonly the function is approximated by an interpolation formula which is then differentiated by the rules of calculus. Since the constant term in the interpolation polynomial is lost on differentiation, the series converges more slowly than the original formula. One cannot then hope for highly accurate results in numerical differentiation. Often a carefully made plot on a large scale is more satisfactory than any other method of finding a numerical derivative, for the slope of such a plot can be determined with some precision.

DIFFERENTIATION UNDER THE INTEGRAL SIGN. For the differentiation of a definite integral of a function $f(x, m)$ containing a parameter *m*, when the limits of the integral are constants *a* and *b*,

$$\frac{d}{dm}\int_a^b f(x, m)\,dx = \int_a^b \frac{\partial f}{\partial m}\,dx$$

and when the limits of the integral are u and v, functions of m,

$$\frac{d}{dm}\int_u^v f(x, m)\,dx = \int_u^v \frac{\partial f}{\partial m}\,dx + f(v, m)\frac{dv}{dm} - f(u, m)\frac{du}{dm}$$

DIFFRACTION. In any wave disturbance, the interference pattern resulting from the rays passing through different parts of an opening, or coming from different points around an opaque object, as they unite at each point. Diffraction and interference effects are characteristic of all wave phenomena no matter of what type. They are thus found in electromagnetic waves (light, x-rays, etc.), sound waves, water waves, and in matter waves. Diffraction occurs whenever a wave passes through a restricted aperture, such as a small hole or slit, or around an edge or particle. An example in optics is the case in which light from a point source passes the edge of a postcard and falls upon a white screen; the shadow of the edge is not sharply defined, but deepens to darkness gradually on one side, and is bordered by very narrow alternate bright and dark interference fringes (*diffraction bands*) on the other. (See **Wave Propagation (Huygen's Principle).**) Again, the image of a minute opaque speck under magnification against a bright background is surrounded by concentric diffraction rings. The image of a bright object, such as a star, as formed in the focal plane of a converging lens, is also surrounded by diffraction rings. If two such images are close together the fringe systems will overlap and no matter how much magnification is applied it will never be possible to obtain clear, well-separated, images of the points. The resolving power of an optical instrument may be defined as a measure of the sharpness with which small images very close together may be distinguished. It is directly proportional to the diameter of the objective aperture and inversely proportional to the wavelength of the light. Diffraction thus limits the resolving power, and hence the practicable magnification, of optical instruments.

Diffraction always results in energy being carried into regions that it could not reach if the propagation of the wave were strictly rectilinear. The patterns produced are geometrically similar whenever the ratio of the wavelength to the dimensions of the aperture is the same. Thus a radio wave in the AM broadcast range will be diffracted in passing through a hole 15 meters in diameter to the same extent as blue light passing through a pin-hole of 0.001 millimeter diameter.

For a single slit of width a and light of wavelength λ, falling on the slit at normal incidence, the intensity of light at an angle θ from the normal to the slit is given by

$$I = R_0^2 \frac{\sin^2\left(\dfrac{\pi a \sin\theta}{\lambda}\right)}{\left(\dfrac{\pi a \sin\theta}{\lambda}\right)^2}$$

Fresnel diffraction. The intensity at any point is the resultant of disturbances coming directly to that point from all parts of the exposed wave front. In general, the wave front is spherical or cylindrical, resulting from a source at finite distance, and the point of observation is also at finite distance.

Fraunhofer diffraction phenomena are observed when both the source and the point of observation are effectively at infinite distance from the diffracting object, obstacle, or aperture. This condition is sometimes brought about by passing the light from the source through a collimator before it is diffracted, and then focussing the parallel diffracted rays at the point of observation.

For material particles, according to the de Broglie hypothesis and quantum mechanics, a material particle having a momentum of magnitude p behaves as though it were associated with a wave of wavelength $\lambda = h/p$, where h is the Planck constant. In any physical process in which a particle interacts simultaneously with two or more scattering centers, the waves associated with the scattering from the various centers will undergo interference with one another to produce diffraction analogous to that which would be observed with light of the same wavelength. According to the principles of quantum mechanics, the wave which is associated with the particle is described by the quantum mechanical wave function $\psi(r, t)$ which contains all of the information concerning the state of the particle which one is physically allowed to have. Born showed that $\psi^*(r, t)\psi(r, t)\,d\tau$ should be interpreted as the probability that the particle will be found in the volume element $d\tau$. Thus in regions where the wave functions representing the scattering from the different scattering centers give complete destructive interference, the probability of finding the particle will be zero. In regions where constructive interference occurs, the probability of finding the particle will be enhanced.

The analogy between the diffraction of light, x-rays, etc., on one hand and that of material articles on the other becomes clear if we remember that in the former phenomena it is the probability of finding a photon in a given location that is determined by the square of the absolute value of the wave amplitude. Because it is relatively easy to use electrons or neutrons having wavelengths of the order of one angstrom, electron and neutron diffraction may be used to study crystal structure in a manner very similar to x-ray diffraction. Electrons do not penetrate as deeply into matter as do x-rays, hence electron diffraction reveals structure near the surface; neutrons do penetrate easily and have the advantage that they possess an intrinsic magnetic moment that causes them to interact differently with atoms having different alignments of their magnetic moments.

References

Driscoll, W. G. (editor): "Handbook of Optics," McGraw-Hill, New York, 1978.

Ghatak, A. K.: "An Introduction to Moedern Optics," McGraw-Hill, New York, 1972.

Shannon, R. R., et al.: "New Experimental Data on Atmospheric (Light) Propagation," *Science*, **206**, 1267–1272 (1979).

Smith, F. G., and J. H. Tomson: "Optics" (Manchester Physics Series), Wiley, New York, 1971.

Taylor, C. A.: "Images: A Unified View of Diffraction and Image Formation with All Kinds of Radiation," Crane, Russak & Co., New York, 1979.

DIFFRACTION (Fresnel). Fresnel Diffraction.

DIFFRACTION GRATING. A series of very fine, closely spaced parallel slits, or of very narrow, parallel reflecting surfaces, which, when light is incident upon it at a definite angle, produces a succession of spectra. The complete optical theory is somewhat complicated, but the action of a plane transmission grating may be explained approximately as follows.

A plane, monochromatic light wave W, incident at angle i (see figure), reaches the slits at different times. A lens L receives the waves

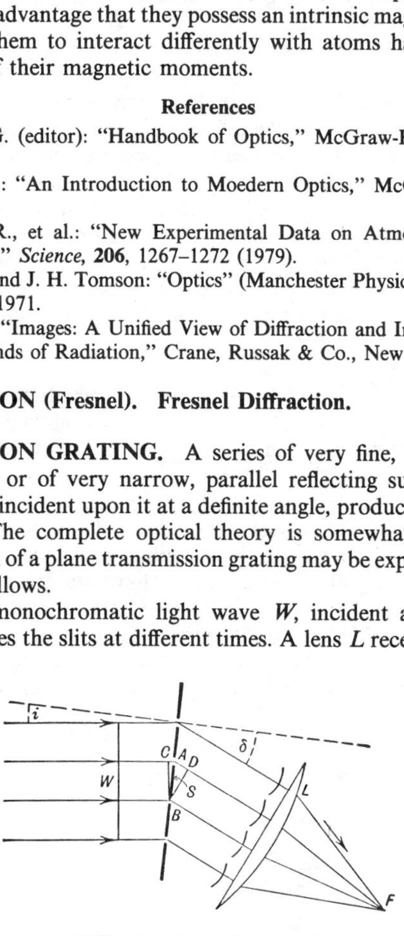

Diffraction by a plane grating.

emerging from any two adjacent slits, A and B (among many others), after they have traveled paths differing by $CA + AD$; that is, by $S \sin i + S \sin \delta$, in which $S = AB$. If the lens is so placed that this path difference is a whole number of wavelengths, $n\lambda$, the successive wave-trains will reach it in the same phase, so that when they are brought to the focus F, they will be in synchronism and will produce a bright image of the distant source. Therefore any angle δ for which this result is possible is subject to the condition

$$S \sin i + S \sin \delta = n\lambda$$

or

$$\sin \delta = \frac{n\lambda}{S} - \sin i$$

Bright images will be produced for those angles δ which correspond to $n = 1, 2, 3, 4, \ldots$; the numbers denote the "orders" of the images.

It is easily shown that for any order the total deviation $(i + \delta)$ is least when $\delta = i$ and therefore when

$$\sin \delta = \frac{n\lambda}{2S}$$

If the incident light is composed of various wavelengths, the corresponding images of any order will appear at different points, since δ varies with λ; and the result is a spectrum. In short, the grating acts as a dispersion piece, and as such is of great value in spectroscopes.

For high dispersion the slits must be very fine and very close together (S small), and for high resolving power (sharpness of spectral lines) the total number of slits must be large. Gratings having several thousand slits to the inch of width are common. They may be made by ruling fine scratches with a diamond point on glass, or, with reflecting gratings, on polished metal. If the rulings are not spaced with absolute regularity, false lines, called ghosts appear in the spectrum.

Rowland was the first to rule reflection gratings on concave metal surfaces. Such gratings eliminate the necessity of the spectroscope collimator or focusing lenses, as they take light direct from the spectroscope slit and form the spectral-line images like a concave mirror. The echelon is another special type of grating.

For a constant angle of incidence, the angular dispersion is given by

$$\frac{d\delta}{d\lambda} = \frac{n}{S \cos \delta}$$

and for small angles from the normal to the grating, cos δ may be replaced by unity. At the point of focus, usually a photographic plate, the so-called normal spectrum will have a constant linear dispersion, often expressed in mm. per angstrom. The resolving power of a grating equals nN, where N is the total number of slits. It, too, is there a constant. For a prism spectroscope both dispersion and resolving power depend on the wavelength of the incident light. The constancy of these two quantities is thus an advantage for a grating instrument.

Typical gratings for the visible and the ultraviolet regions have 6,000–18,000 lines per centimeter, for the infrared, 700–3,000 lines per centimeter. These numbers are roughly equal to the wave number of the light to be dispersed.

For some other special kinds of gratings see **Echelette; Echelle Grating; Echelon;** and **Interferometer.**

DIFFUSE CHARACTERISTICS. Proceeding in all directions, not in any sharply defined path, as in the cases of diffuse reflection, diffuse refraction, and diffuse transmission of radiation.

DIFFUSE NEBULA. Nebula.

DIFFUSER. A diffuser is a passage so shaped that it will change the characteristics of a fluid flow from a certain pressure and velocity to a lower velocity and a higher pressure. The diffusion must be carried out in a well-streamlined passage having smooth interior surfaces, and sides not diverging at so great an angle as to cause the fluid to leave the sides of the diffusing chamber. By reducing the velocity through increasing the cross-sectional area of flow, the pressure may be built up as the velocity head is diminished. Diffusers are applied to centrifugal fans, centrifugal pumps, jet pumps, centrifugal compressors, wind tunnels, and other equipment where it is required to conserve energy by efficiently converting velocity head into pressure.

DIFFUSION. This term denotes the process by which molecules or other particles intermingle as a result of their random thermal motion. The molecules of a gas or of a liquid wander about rapidly, colliding frequently and exchanging kinetic energy, but maintaining a certain aimless progress. If an enclosure contains two gases, the lighter initially above and the heavier below, the gases at once begin to mingle because of their molecular motion. The same is true of a dense solution (as of sugar) and pure water; both the sugar and the water molecules wander across the boundary, so that in the course of time the whole body of liquid attains nearly uniform concentration. The process whereby this is effected is called diffusion. In the case of fluids of different color, its progress may be easily watched.

The rates at which different gases diffuse at a given temperature are inversely proportional to the square roots of their molecular weights. Thus, hydrogen diffuses four times as fast as oxygen. This follows, according to the kinetic theory, from the fact that the molecules of various kinds have the same mean kinetic energy and hence their mean square speeds are in the inverse ratio of their masses. In the case of a solution of non-uniform concentration, the diffusion of the solute from the more to the less concentrated regions takes place in accordance with *Fick's law*, expressed by the equation

$$\frac{dm}{dt} = -DS\frac{dc}{dx}$$

This gives the mass of solute diffused per unit time through a cross-section S, in terms of the concentration gradient dc/dx in the direction x perpendicular to the cross section. D is a constant for the given solute and solvent at a given temperature, and is called the diffusion coefficient. For any one pair of substances, D is found to be proportional to the absolute temperature. It should be stated that these statements apply only to nonelectrolytic solutions.

Diffusion in solids is a phenomenon which occurs rather slowly, but can be observed. Three basic processes may be responsible: (a) direct exchange of atoms on neighboring sites; (b) migration of interstitial atoms; (c) diffusion of vacancies. The first process requires very large energy. The energy to make an interstitial migration is rather large, but many atoms migrate easily. Vacancies are fairly readily formed, and diffuse fairly easily. From the Kirkendall effect it appears that (b) and (c) are the usual processes. The diffusion coefficient is related to the ionic mobility by the Einstein relation.

Another use of the term diffusion is to denote the passage of particles through matter in such circumstances that the probability of scattering is large compared with that of leakage or absorption. It is often limited to phenomena described by a member of the class of differential equations known as diffusion equations.

Diffusion operations are of large importance in chemical and process engineering. Both gaseous and thermal diffusion are used to separate one gas from another. In the case of *gaseous diffusion*, if a binary gaseous mixture at a high pressure is passed over a microporous barrier, a fraction of the gas will diffuse through the barrier into a low-pressure discharge chamber and will be found to be richer in the content of one gas than of the other gas. This is termed *Knudsen diffusion*. The passage of gas mixtures through the barrier is governed by the unequal collision frequency of each molecular species upon the walls of the pores. Fast, so-called light molecules separate from slower, heavier molecules within the barrier. The Oak Ridge, Tennessee plant designed for the enrichment of $^{235}UF_6$ from the naturally occurring uranium hexafluoride that contained 99.3% $^{238}UF_6$ represented the first major application of gaseous diffusion on a large scale. The molecular weight of the hexafluoride of ^{235}U is 349, whereas that of the hexafluoride of ^{238}U is 352. Inasmuch as the rate of diffusion of a gas is inversely proportional to the square root of density, the greatest separation factor for one stage of separation is the square root of 352/349, or 1.0043. Inasmuch as only part of the gas can diffuse through a given barrier, the separation factor is less. Thus, the number of diffusion stages for the Oak Ridge plant was approximately 4,000, requiring a plant that covered several acres of ground. Polymeric barriers also are under study and with scientific improvements, gaseous diffusion may become a widely used means for the recovery of carbon dioxide, helium, and nitrogen from natural gas.

In *thermal diffusion*, a thermal gradient is applied to a homogeneous solution (gas or liquid). This causes a concentration gradient and thus affords a means of separating materials. The logic of thermal diffusion is derived from the kinetic theory of gases and the cage model of liquids. If there is no marked size difference, heavier species tend to concentrate in the cold region. Where materials of identical molecular weight are involved, the larger molecules go to the cold region by virtue of their greater momentum. In the static mode, differential concentration can be established by eliminating convection currents that otherwise would tend to negate the effects of the applied thermal gradient. In the reflux method, hot and cold materials are flowed countercurrently. The reflux usually is provided using the density gradient that results from the imposition of the temperature gradient. Equipment of this latter type usually is referred to as a *thermo-*

gravitational column, or a *Clusius-Dickel column*. Limited applications of thermal diffusion separations include those for concentrating dilute mixtures of isotopic gases. However, equipment costs tend to be high and efficiencies low.

Because chemical processing involves both the mingling and separating of fluids (gases and liquids), an understanding of diffusion processes is fundamental to process design.

References

McGabe, W. L., and J. C. Smith: "Unit Operations in Chemical Engineering," McGraw-Hill, New York, 1967.

Perry, R. H., and C. H. Chilton: "Chemical Engineers' Handbook," McGraw-Hill, New York, 1979.

Slattery, J. C.: "Momentum, Energy and Mass Transfer in Continua," McGraw-Hill, New York, 1972.

Weast, R. C. (editor): "Diffusion Equations and Coefficients," in "Handbook of Chemistry and Physics," 60th edition, CRC Press, Boca Raton, Florida, 1979.

DIFFUSION ANALYSIS. The determination of the relative size or molecular weight of particles by comparing their diffusion rates, or by separating them by differential diffusion methods.

DIFFUSION BONDING. **Welding.**

DIFFUSION (Cell). **Cell (Biology).**

DIFFUSION COEFFICIENT (or Coefficient of Diffusion). A measure of the rate of diffusion of a property, appearing as the factor K in the diffusion equation

$$\frac{\partial q}{\partial t} = K\nabla^2 q$$

where q is the property diffused, and ∇^2 is the Laplacian operator. The diffusivity has dimensions of a length times a velocity; it varies with the property diffused, and for any given property it may be considered a constant or a function of temperature, space, etc., depending on the context.

The most common diffusion process is that for which mass is the property diffused, and hence the differential in the equation can be expressed in terms of concentration change.

DIFFUSION CURRENT. The limiting current which is reached by electrolytic migration of the ions in a solution under the application of a potential difference to the electrodes. As the potential difference is increased the ion current to the electrodes increases rapidly at first but soon reaches a limiting value (the diffusion current value) as the potential difference is increased. If the potential difference is increased still further, a point is ultimately reached at which a new ion species begins to discharge.

The current limit is set by the rate of diffusion (of the ion being discharged) through the depleted layer surrounding the electrode. This diffusion rate is proportional to the ion concentration. For application of this effect, see **Polarographic Analyzers.**

DIFFUSION EQUATION (Einstein). An equation for the mean square displacement of spherical colloidal particles in a gas or liquid, due to Brownian movement. The mean square displacement from their original position after a time τ is

$$x^2 = \frac{RT}{3\pi\eta rN}\tau$$

where R is the gas constant, T is the absolute temperature, r is the radius of the particle, η is the viscosity, N is Avogadro's number. This relationship is only valid for particles of such size that they obey Stokes' resistance law.

DIFFUSION (Fokker-Planck Equation). **Fokker-Planck Equation.**

DIFFUSION (Graham Law). **Graham Law.**

DIFFUSION HYGROMETER. **Hygrometer.**

DIFFUSION LAYER. A layer of solution, actually a double layer, that is in immediate contact with an electrode during electrolysis.

DIFFUSION POTENTIAL. When liquid junctions exist where two electrolytic solutions are in contact, as in the case of two solutions of different concentrations of the same electrolyte, diffusion of ions occurs between the solutions, and the differences in rates of diffusion of different ions set up an electrical double layer, having a difference of potential, known as the diffusion potential or liquid junction potential.

DIFFUSION-TYPE VACUUM PUMP. **Vacuum Pumps.**

DIFLUENCE. The rate at which adjacent flow is diverging along an axis oriented normal to the flow at the point in question; the opposite of confluence. The difluence may be measured by

$$\frac{\partial v_n}{\partial n} \text{ or } V\frac{\partial \psi}{\partial n}$$

where V is the speed of the wind, the n axis is oriented 90 degrees clockwise from the direction of the wind vector, v_n is the wind component in the n direction, and ψ is the wind direction measured in degrees clockwise from a reference direction.

DIGESTER (Process). In the process industries, the term *digester* is used in two principal connections: (1) the digestion of wood chips in the production of pulp prior to the manufacture of paper, and (2) the digestion of sewage sludge in waste-treatment operations. The term also appears in a number of other operations operating under varying conditions and hence a generalized definition is difficult to formulate. In chip digestion (also termed cooking), the chip digester is a large vessel provided with suitable raw-chip feed and cooked-chip discharge ports and equipped with means for heating and maintaining its contents at a specified temperature for a specific time. Batch digesters are vertical, stationary cylindrical pressure vessels into which chips and cooking liquor are charged and in which liquor is constantly moved, either by percolation within the digesters aided by direct addition of steam for heating purposes, or by continual withdrawal of liquor through screened ports and reintroduction of the liquor, after further heating. Modern batch digesters are typically 4,000 to 6,000 cubic feet (113.3 to 170 cubic meters) in volume with pulp capacities of 10 to 20 tons (9 to 18 metric tons).

By contrast in terms of operating parameters, sewage sludge is digested by aerating a lagoon or pond under normal outdoor temperatures, except that below about 40°F (4.5°C), the activity of the bioorganisms which aid in the digestion falls off considerably.

Autoclaves used in the chemical industry also are sometimes referred to as digesters.

DIGESTER (Pulp). **Pulp (Wood) Production and Processing.**

DIGESTIVE ORGAN (Gizzard). **Gizzard.**

DIGESTIVE SYSTEM (Human). The aggregation of organs which are concerned with ingestion, digestion, and elimination. Digestion may be defined as the complex physiological and chemical process for converting foods to forms assimilable by the body cells. Since the food which a person eats must be divided among the billions of cells of the body, it necessarily must be broken down to very small pieces. To accomplish this, the human body has a thorough digestive system which not only grinds the food mechanically, but breaks it up chemically into exceedingly small particles. The circulatory system is then responsible for distribution of the food among the individual body cells.

Digestion is aided by the way food is prepared. Cutting food finely and cooking it may accomplish some of the same steps toward digestion that the body would have to perform. For this reason, proper and adequate food preparation is important for babies as well as for older people whose natural digestive processes cannot take care of all the food they eat.

The body commences its work of digestion as soon as food enters the mouth. The mouth is the front end of a long tube (*alimentary canal*) which extends through the trunk of the body. Foods entering this tube are worked into proper form as they pass along it, and the unusable part of the intake is finally excreted from the other end of the tube, the anus. It is in this sense, that the body can be considered as hollow, since the material in the digestive tract is not truly in the body tissue. Large numbers of bacteria live in certain parts of the digestive tube without disturbing the body in any way because they are outside the body tissues.

The teeth are arranged to grind food into small pieces. Along with the grinding action, the food is mixed thoroughly with saliva in the mouth. The saliva begins at once to digest the food by means of an enzyme in it (*ptyalin*) which breaks down the relatively large starch particles, as found in bread and potatoes, into much smaller pieces. The saliva which is swallowed with the food continues to act on the starch even after the food reaches the stomach. The saliva, in moistening the food, also makes it easier to swallow. When food is swallowed, it takes about 12 seconds for it to reach the stomach. Water and other fluids, however, may reach the stomach in as little time as 1 second, and may pass rapidly through it. Once in the stomach, the food is subjected to the action of the digestive juice which is formed by the stomach lining (*gastric juice*). This juice contains hydrochloric acid and a number of other more complex substances (enzymes) which start changing the food into simpler chemical materials. The most important of these enzymes is *pepsin*, which stimulates the breakdown of proteins into amino acids. The digestive juice oozes from the cells lining the stomach, while at the same time the stomach muscles contract and stretch, causing its shape to change constantly. This churning of the food may break up larger pieces of food, but its most important function is the proper mixing of the food with the digestive juices. Milk is a liquid which would pass immediately through the stomach were it not that the gastric juice contains an enzyme (*rennin*) which stimulates its coagulation or solidification. Consequently, the milk also may be digested in the stomach. Rennin, prepared from the stomachs of farm animals, is used to coagulate milk as the first step in making cheese.

When the food passes into the first 12 inches (0.3 meter) of the intestines (the *duodenum*), it is further broken down by duodenal digestive juices into exceedingly small, submicroscopic particles. The digestive secretions of the intestine are quite alkaline, as contrasted with the highly acidic juices of the stomach. If the food were not thoroughly chewed, the inside of the larger pieces of food could not be reached by the digestive juices of the stomach or intestine. Such pieces of food are not digested and are lost in the feces.

The digestive juices of the intestine come only partially from the cells lining the stomach. Much of the work is done by fluids or digestive juices made in other organs of the body, and carried to the intestine by special tubes or ducts. The bile duct, for example, brings bile from the liver which aids in the digestion of fatty materials. The pancreas is another organ which manufactures digestive juices, and these are sent to the intestine by means of the pancreatic duct. The pancreatic juice is probably the most important of all the digestive juices because it completes the digestive process. It contains starch-digesting enzymes (*amylases*), fat-digesting enzymes (*lipases*), and protein-digesting enzymes (*proteases*). The pancreatic juice also contains the enzymes *trypsin* and *chymotrypsin* which digest proteins, *lactase* which digests lactose (milk sugar), and *steapsin* which digests fats. Actually, there are many other ingredients in the various digestive juices and the biochemistry is quite complex.

Not all the food which is eaten can be digested. The indigestible part continues to move along the length of the intestine. As this mass travels, large amounts of water which have been drunk or have come from the digestive juices are absorbed from it, and it assumes a firmer consistency. The lower part of the intestine is filled with billions of bacteria which, far from being harmful, are extremely valuable. As they grow and reproduce, using the body's waste for food, these bacteria manufacture considerable amounts of vitamins. These vitamins are cast off by the bacteria and are absorbed into the body in pure form. Newborn babies may have a serious vitamin K deficiency during the first few days of life because some time is required for the bacteria to become established in their intestines.

The organs of digestion are not completely mature until the child reaches sexual maturity (puberty). The young child requires a diet consisting largely of mildly seasoned, easily digested foods. The capacity of children's stomachs is small and consequently periods of hunger are more frequent than in adults. Light, in-between-meal snacks are definitely of benefit.

Late premature infants (born 34 to 37 weeks from conception) can swallow, and may be fed from a nursing bottle supplied with a special nipple having a small bulb. Such infants also may be fed by means of a medicine dropper with a piece of rubber tubing attached. The tube is placed in the infant's mouth, and the milk is expressed slowly through the tube. Premature babies who cannot swallow often are fed through a rubber tube (catheter) attached to the barrel of a glass syringe. This method of feeding requires training and skill, for the tube is passed directly into the stomach.

Salivary Glands. Saliva is secreted into the mouth from three paired glands: the *parotid, submaxillary*, and *sublingual* glands. There are also numerous small salivary glands of the cheek and tongue. The parotid lies on the side of the face below and in front of the ear. The salivary secretions of the gland reach the mouth through a duct which runs inward through the fat of the cheek and opens on the inner surface of the cheek at the levels of the crown of the second molar tooth. The saliva secreted by the parotid is thin and watery. Secretion from the sublingual gland is thick and viscid, although the sublingual is the smallest of the main salivary glands. The gland rests immediately below the mucous membrane of the floor of the mouth, beneath the tongue. Its ducts open into the floor of the mouth through small conical elevations (*papillae*), which can be seen by the naked eye.

The submaxillary gland can produce either thick or thin saliva. It can be felt against the inside edge of the lower jaw. A long duct, about 2 inches (5 centimeters) in length, carries the saliva from the submaxillary gland to the floor of the mouth. Secretion of the salivary glands is under control of the nervous system. Stimulation of the body's glands may be by means of nerve impulses or by hormones, the latter form prevailing when rapid response is not essential. Because food remains in the mouth for such a short time, the salivary glands are stimulated by nervous mechanisms. Food placed in the mouth causes a secretion of saliva within 2 to 3 seconds by an unconditioned or inherent reflex. A dry biscuit produces a thin water saliva, while a piece of meat causes a highly viscous saliva which lubricates the meat and enables it to be swallowed easily.

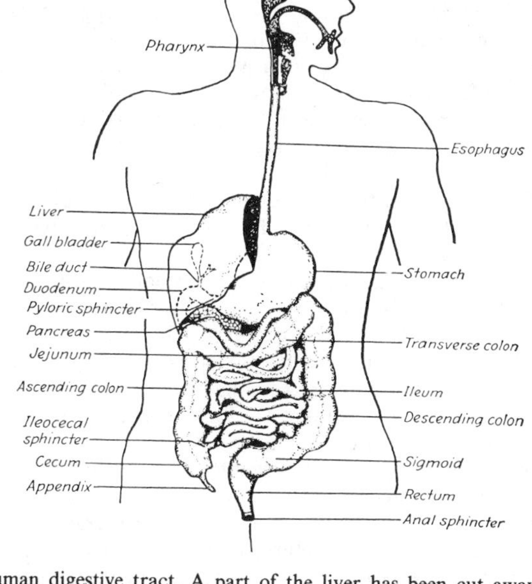

The human digestive tract. A part of the liver has been cut away to show the stomach lying beneath it.

Pharynx and Esophagus. The pharynx is the vertical passage beginning behind the nose and mouth and extending from the base of the skull above to the esophagus below. See accompanying diagram. The pharynx is equipped with three semicircular muscles located one under the other which enable the pharynx to squeeze food down toward the esophagus. One of the most muscular parts of the digestive system is the esophagus. It is a flattened tube passing through the lower part of the neck, the whole length of the chest, and joining the stomach just below the diaphragm. The esophagus is 10 to 12 inches long in the adult. Normally, the entrance into the stomach is kept closed by a muscular contraction in the lower inch or so of the esophagus, which opens as a piece of food approaches, but prevents reflux of acid from the stomach to the esophagus.

The Stomach. This organ is a receptacle in which food accumulates. Some of the earlier processes of digestion take place here, namely, the conversion of food into a viscous fluid. The normal stomach is J-shaped, with a bulge above and to the left of the junction with the esophagus. The shape varies according to whether the person is standing, sitting, or lying down; and according to whether the stomach is full or empty. The stomach lies in the upper left portion of the abdomen, its long axis being nearly horizontal. The stomach narrows to join the small intestine forming a canal (the *pyloric canal*), which has a thick muscular valve (the *pyloric sphincter*). This sphincter remains closed so long as the food in the stomach is solid. The pyloric sphincter relaxes only when the gastric contents have been changed into a semifluid state. If only liquids are taken into the stomach, the pylorus opens and the fluid passes into the small intestine almost immediately. Usually it takes from 3 to 4 hours or somewhat longer before the stomach completely empties a meal into the small intestine. Minute glands in the stomach manufacture hydrochloric acid and certain ferments which break down portions of the food into simpler substances. The muscular coats of the stomach grind and mix the food with the stomach secretions. The physical grinding and crushing is of great importance in normal digestion.

The Bowel. Intestines, the portions of the digestive system from the stomach to the anus, are divided into two main parts, the *small intestine* and the *large intestine.* The small intestine begins at the pyloric sphincter and lies in the abdomen in coiled loops. It is 20 to 22 feet (6 to 6.7 meters) long. Thus, the small intestine occupies the greater portion of the abdominal cavity. It gradually diminishes in size as it extends downward, having a diameter of about 2 inches (5 centimeters) where it joins the stomach, and about 1 inch where it joins the large intestine. The first portion of the small intestine is called the *duodenum*, which is about 8 to 10 inches (20 to 25 centimeters) in length. The duodenum differs from the remainder of the small intestine in that it is fixed to the posterior abdominal wall. The ducts of the liver and pancreas open into the duodenum. The remaining small intestine is divided into the *jejunum* and the *ileum.* The upper 8 feet (2.4 meters) of the small intestine, after the duodenum, is regarded as the jejunum; the lower 12 feet (3.6 meters) as the ileum. The coils of the small intestine are able to move about freely in the abdominal cavity, being connected to the posterior abdominal wall by a fan-shaped sheet of tissue (the *mesentery*), which measures about 20 feet (6 meters) at its free edge and only 5 or 7 inches (12.5 or 17.8 centimeters) at the attachment to the abdomen. The blood vessels, lymph vessels, and nerves serving the intestine lie between the layers of this sheet of tissue.

The piece of food (*bolus*) travels along the small intestine in a series of rushes; the bowel contracts just behind the bolus and relaxes in front of it. This contraction and relaxation occurs in a series of alternating wavelike motions, known as *peristaltic movement.* There are also regular constricting movements of the intestine which occur at a rate of 20 to 30 per minute, kneading the food thoroughly and insuring that the digestive juices are well mixed with it.

The small intestine opens obliquely into the large intestine. A valve (*ileocecal valve*) is located at the junction. The valve permits the passage of the contents of the small intestine into the large intestine, at intervals. It also prevents the return of material into the ileum.

The large intestine begins on the right side of the abdomen just above the rim of the pelvis, and is about 5 feet (1.5 meters) long. Arranged in an inverted horseshoe shape around the small intestine and about 3 inches (7.5 centimeters) in diameter at its commencement, the large intestine gradually narrows to the anus. The large intestine is divided into: the cecum and vermiform appendix; the ascending colon; right flexure of the colon, transverse colon, left flexure, descending colon, sigmoid colon, rectum, and anal canal.

The *cecum* is that portion of the large bowel which hangs below the opening of the ileocecal valve. It is a blind sac to which is attached a worm-like tube, the *vermiform appendix.* The appendix usually is about 3 inches (7.5 centimeters) long, but may be as long as 9 inches (22.9 centimeters) or as short as 1 inch (2.5 centimeters).

The nearly fluid contents of the ileum pass through the ileocecal valve and collect in the cecum. Slowly the contents are forced up into the *ascending colon.* Movement through the large intestine is slow and takes place in periodic rushes. In order to reach the outside of the body, it is necessary for the large intestine or bowel to penetrate the floor of the pelvis. At this juncture, the large bowel is enclosed by two muscles, the internal and external sphincters, which compress the sides of the tube and reduce its cavity to a narrow passage. That part of the large intestine immediately preceding the muscles is termed the *rectum.* The terminal portion of the passage, from the rectum to the external opening or *anus,* is called the *anal canal.* The sphincters remain closed except during defecation. These muscles are voluntary muscles. The waste material of digestion deposited in the rectum is known as *feces* or *fecal matter.* Most of the time the rectum is empty. When the colon becomes full, the fecal matter passes into the rectum. The desire to defecate is a reflex initiated by pressure on the walls of the rectum by the feces.

Absorption of food is effected almost entirely in the small intestine. The large intestine secretes some material and absorbs water, so that the amount of material finally excreted is only about one-third of the weight of material entering. Most of the absorption of fluid takes place in the cecum and ascending colon. The color and odor of the feces are caused by the action of bacteria, which inhabit the large bowel, and by the pigments present in the bile. During starvation, feces continue to be formed from bile, which is emptied into the digestive tract from the liver, and from bacteria and other secretions from the bowel itself.

See also **Achlorhydria; Acidosis; Alimentary Tract; Alkalosis; Amebiasis; Anorexia; Appendicitis; Appendix (Appendix Vermiformis); Basal Metabolism; Colitis and Other Inflammatory Bowel Diseases; Colon; Constipation; Diarrhea; Diverticulosis and Diverticulitis; Foodborne Diseases; Intestinal Protozoa; Obesity;** and **Sigmoidoscopy.** References are listed at the ends of most of the aforementioned entries.

DIGESTIVE SYSTEM (Other Life Forms). One-celled animals and sponges take food into the cell to be transformed by a process of intracellular digestion. While this process persists to a limited extent in coelenterates, flatworms, and mollusks (*Mollusca*), these and all other animals also have some form of digestive system or alimentary tract in which food is retained for extracellular digestion preceding absorption. Secretions are discharged into this tract by the cells of its lining and are mixed with the food. The cavity is lined with endodermal tissue, in many cases supplemented by ectodermal ingrowths at both ends.

The simplest form of digestive system is the enteric cavity of coelenterate polyps. It is little more than a sac with one opening through which food enters and undigested wastes are discharged. In the flatworms a similar condition prevails, but in both the jellyfishes and in some flatworms the cavity is complex, extending throughout the body in a system of canals or branches which distribute the food as well as absorb it. From the roundworms through the remainder of the animal kingdom the system is tubular, opening at one end by the mouth and at the other by the anus.

In the tubular digestive tract, specialization of digestion reaches a maximum. Here, food passes, by the muscular movement of the walls, successively through different regions instead of being mixed indiscriminately, hence each region may subject it to special treatment. The chief regions are those which aid in securing food, simple passages, storage reservoirs, grinding structures, and digestive regions which include chambers and tubular regions which may be long, narrow and coiled with the inner surface thrown into folds or minute finger-like processes (villi) to increase the area of contact. In addition, glandular derivatives of the lining are so highly developed that they become separate organs associated with the tubular tract by slender ducts.

Some of the worms have a very simple tract with a muscular pharynx which aids in securing food and a long simple intestine, in which it is digested. Other animals, including the leeches, insects, and birds, have a crop in which food is stored prior to digestion. The mastax of rotifers and the gizzard of birds are examples of grinding structures.

The mammalian alimentary tract is a good example of regional specialization. The oral cavity with its teeth provides for chewing and some digestion, the pharynx and oesophagus furnish a passage to the stomach, where food is stored and slightly digested, the small intestine completes digestion and absorbs the end-products, the large intestine absorbs water, and the rectum stores the remaining wastes for periodical discharge by way of the anus. Glands associated with this tract are the salivary glands, the liver, and the pancreas.

DIGESTIVE SYSTEM (Ruminants). Bovines (cows, buffalos, oxen), goats, sheep, antelopes, deer, camels, and chevrotains, among others, possess a unique digestive system which enables them to convert many coarse, fibrous plant products, wastes, and byproducts into high-quality feed. These animals can process materials that otherwise would be wasted entirely or returned to the soil. In contrast with the stomach of humans and most other mammals, the corresponding portion of the digestive system of ruminants is relatively large and is divided into four major compartments, as shown in the accompanying figure.

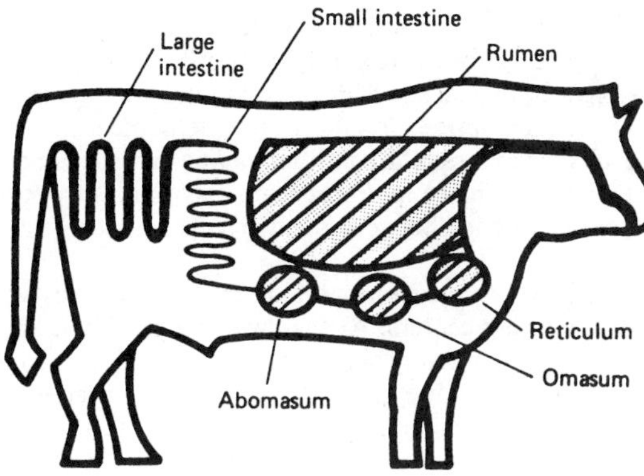

Highly schematic representation of digestive system of a ruminant, showing the four distinct chambers of the stomach: Rumen, reticulem, omasum, and abomasum.

Ruminants are endowed with control devices that provide both the environment required for rapid microbial action and the time needed for bacteria to digest the cellulose, proteins, and other organic constituents in the plant substances eaten. Cellulose is the most abundant chemical component of plants and is the most abundant organic chemical substance on earth. Cellulose, indigestible by humans, is digested by ruminants to the extent of 30–80%. The products, chiefly simple fatty acids, are directly absorbed by the animal during the microbial breakdown of the plant materials. These fatty acids provide the animal with most of its dietary energy.

At the same time, the microorganisms use the chemical fragments produced from the plant material to synthesize microbial tissue containing high-quality proteins and vitamins. The microorganisms synthesize all the essential amino acids needed by humans. The same microorganisms also synthesize other substances essential to the human diet, including all B-complex vitamins and vitamin K.

When a ruminant swallows vegetation or other substances, the material is only partially chewed. The material first goes into the *rumen* (1st stomach or paunch), which performs a number of functions. The rumen chamber, which in a large mature cow may have a capacity of up to 300 pounds (136 kilograms), acts as a storage chamber while the animal continues with its feeding. Between periods of browsing, the animal will ruminate, considered by authorities as a very pleasurable period for the animal. During this period, large pieces of feed material will be regurgitated, an action similar to vomiting, but without any unpleasant sensations to the animal. These larger pieces are more finely ground by the animal by the chewing action of the mouth (chewing the cud). The rumen also preprocesses the food so that bacteria present can assist with a massive breakdown of the food and preparing it for later action by gastric juices. The bacterial action essentially is one of fermentation. Introduction of saliva into the rumen assists in controlling the pH of the rumen. Depending upon the type of feed consumed, there will be a variety of microbial flora—bacteria, yeasts, protozoa—present in the rumen to participate in the biochemical reduction of the feed. During this process, crude fibers are conditioned for further digestive action, essential amino acids and complete proteins are created, and B-complex and K vitamins are produced. During fermentation, significant quantities of gaseous carbon dioxide, methane, and ammonia, and small quantities of hydrogen sulfide and carbon monoxide, are formed. The presence of these gases is relieved by a reflex action of belching, but in some cases, the buildup of gases causes a condition known as *bloat*. This can be a serious condition.

The *reticulum* (2nd stomach or honeycomb) is not fully separated from the rumen. Actually, materials can pass freely from one chamber to the next. The reticulum is named for its folds, forming hexagons (the pattern noted on tripe). A major function of the reticulum is to screen out foreign objects that would not later pass through the remaining stomachs and intestines. Husbandmen sometimes refer to the reticulum as the "hardware" stomach, because it is here that nails, bits of wire, large pebbles, etc., will be captured. Keeping in mind that the process involving the four stomach chambers is continuous, the preprocessed bulk from the rumen (some of which has been regurgitated and returned to the rumen a second time) passes quickly through the reticulum and on to the last two chambers of the stomach. The *omasum* (3rd stomach and also known as "manyplies" because of the presence of many plies or folds of tissue in the organ) functions to reduce the water content of the feedstuffs, as well as to provide additional grinding and squeezing actions on the feed. However, the complete role of the omasum is not fully understood. From the omasum, the material flows into the *abomasum* (4th stomach and sometimes referred to as the "true" stomach). This is called the true stomach because the abomasum acts very much like the stomach of monogastric mammals. Here, digestive juices are admixed with the stomach content and the moisture content is increased. Enzymatic juices in the abomasum aid in protein digestion. However, there is little if any digestion of fat, cellulose, or starch in the abomasum. The latter nutrients are digested and absorbed in the small intestine. As the feed and partially digested materials leave the abomasum, they are highly fluid.

In the last stage of the ruminant's digestion process, the digestive juices are used to break down the microorganisms. During the microbial digestion of the plant materials by the ruminant, the plant proteins are broken down and their nitrogen is released in the form of ammonium. The ammonium is absorbed by the microorganisms and used for producing new proteins. Knowledge of this mechanism has made it possible to increase the efficiency of ruminants in converting low-protein feed, such as cornstalks (maize), by supplementing the diet with synthetic urea, which the microorganisms decompose quickly with the release of ammonium. The artificially supplied ammonium then is absorbed by the microorganisms, like that derived from the plant proteins, to develop new microbial proteins. The cost of nitrogen in the form of urea is relatively low by comparison with the cost of nitrogen in the form of protein.

DIGESTIVE TRACT (Fishes). **Fishes.**

DIGITALIS (*Digitalis purpurea*; *Foxglove*). *Scrophulariaceae.* The foxglove is a biennial often grown as an ornamental plant. The first year of growth produces only the long basal leaves, while in the second year the erect leafy stem 2–5 feet (0.6–1.5 meters) tall is developed. The flowers are borne in a raceme which through the bending of the peduncles or individual flower stalks becomes one-sided. The purple flowers have a five-parted calyx; a tubular bell-shaped corolla obscurely five-lobed, five stamens and a single pistil. They are pollinated mainly by bees. The fruit is a two-celled capsule.

The drug digitalis is prepared mostly from leaves of the second year's growth. These are rather coarse ovate leaves covered with glandular hairs. Decoctions of the leaves have been used in Europe for many years. See **Arrhythmias (Cardiac)**; and **Congestive Heart Failure.**

Digitalis is a valued drug in medicine and is used in certain kinds of heart disease. The chief effects it has on the heart are the regulation of its rate, rhythm, tone, contraction, and conduction of impulses.

As a crop plant, digitalis is grown in England, Germany, and in the United States, especially in Michigan. Propagation is by seeds, which are sown under glass and later transplanted.

DIGITAL COMPUTER. In contrast to an analog computer which operates on continuous data, a digital computer performs a series of discrete operations on numeric or alphabetic data in the solution of a problem. Compared to a calculator in which frequent human operator intervention is required, the computer accomplishes the data processing with little or no intervention by the human user. This entry provides a brief overview of the modern electronic digital computer. For those readers interested in some specific aspect of a digital computer, initial reference to the approximately 200 terms listed under **Data Processing** is suggested. See Fig. 1.

A digital computer may be classified in a variety of ways. One way is to distinguish between *general-purpose* and *special-purpose* computers. The general-purpose computer is designed to satisfy a wide variety of data processing needs with approximately equal speed and efficiency. The special-purpose computer, however, is optimized for the solution of problems associated with a restricted class of applications. With very few exceptions, the special-purpose computer can provide all of the functions of the general-purpose computer, although it may not be as fast or as efficient for some problems. For example, a special-purpose computer may be designed to efficiently handle arrays of data such as those found in a two-dimensional Fourier transform problem. Although this computer could be programmed to perform accounting applications, its facilities would be inefficient for such use. Similarly, the general-purpose computer can be programmed to calculate the two-dimensional Fourier transform, but the solution undoubtedly would take more time.

Within the classification of general-purpose computer, it is common for a manufacturer to provide for several different models which are capable of executing the same program. The difference between models generally is related to the tradeoff between speed of processing and number of circuits (and, therefore, cost). Between computers provided by different manufacturers, however, there usually is little compatibility at the basic instruction level. Programs written in assembler language for one manufacturer's machine will not, in general, run on a computer designed by another manufacturer. However, a high degree of program compatibility is possible if programs are written in a high-level language such as COBOL or FORTRAN.

A second classification is roughly related to the physical size or cost of the computer. In this commonly used terminology, computers are categorized as *microcomputers*, *minicomputers*, and *main-frame* or *large-scale* computers. The prefix *maxi* is sometimes used for the latter classification. Price often is used to distinguish the classes. However, computer prices continue to decline and the functions available for a given price continues to increase. Thus, this means of defining the classes is time dependent.

It is common for a microcomputer design to depend on a single-chip microprocessor as the primary processing element. Thus, computers utilizing such a microprocessor, when combined with some storage and input/output equipment, often are referred to as microcomputers. In addition, the microcomputer generally is physically small and often may be packaged in a desktop enclosure.

To a large degree, but also diminishing with time, the minicomputer is recognized by its package and its variety of possible configurations. The typical minicomputer is mounted in an industry standard 19-inch (~48-centimeter) rack. It is generally offered with a wide variety of separately packaged options and input/output equipment which also are rack-mounted. As a result, the minicomputer has been widely utilized to provide tailored systems for particular applications such as industrial process control. Recently, however, some models of these computers have been marketed in conventional packaging and with limited configuration options. At the same time, some microcomputers are being packaged in a rack. In addition, many current minicomputer designs utilize one or more microprocessors as the central processing unit. Thus, the distinction between the two classes appears to be lessening.

The main-frame or maxicomputer often is associated with large, centralized computing centers. At the lower range of cost and performance, it provides a capability similar to that of the larger minicomputers. At the high end, it may provide computing power hundreds of times greater than that associated with minicomputers, with a correspondingly greater price. Main-frames typically are packaged in cus-

Fig. 1. IBM 4341 computing system. (*IBM Corporation.*)

tomized enclosures with little or no flexibility as to the location of a particular feature. Although many options may be available, the ability to efficiently mix features in an arbitrary manner usually is more limited than in the case of the minicomputer.

It should be clear from this discussion that the class distinctions are imprecise. Across the classes there is a performance range, measurable in terms of instructions executed per second, ranging from about 50,000 instructions per second (ips) to over 5,000,000 ips. Similarly, the price range is from a few thousand dollars to a few million dollars. All of the machines from micro to maxi follow the same basic conceptual design. As a result, the popular use of this classification is, in fact, of little value to the user and tends to be confusing.

It should be noted that the term *computer* and the phrases *electronic data processing (EDP) system* and *automatic data processing (ADP) system* commonly are used as synonyms. Initially, "to compute" was to determine a quantity from other quantities by performing a calculation, an arithmetic operation, or a series of operations. Most data processing operations, however, actually do not require computing in this sense, but rather involve collating, selecting, correlating, etc. information to or from an electronic storage unit. In modern language, these kinds of operations are ascribed to a "computer" whether or not any actual mathematical calculations are being performed on the data.

Elements of a Digital Computer System

Basically, a computing system comprises input and output equipment for physical handling of machine information and instructions, and a central unit which performs the actual electronic processing.

The central unit typically is made up of one or more *central processing units* (*CPUs*) and their associated storage. In larger systems having more than one CPU, the term *central electronic complex* (*CEC*) sometimes is used to describe the set of CPUs. The CPU, in turn, is made up of a control section, an arithmetic and logical unit (ALU), local storage (registers) and, in some larger machines, a input/output channel controller. The control section controls the step-by-step operation of the system by fetching instructions from the storage, interpreting them, and providing the necessary signals to effect control of the ALU and other portions of the system. If a separate channel controller is used, it provides the detailed control of information flow to and from the input/output units. In smaller systems where a separate channel controller is not used, this function is provided by the control section of the CPU. Storage holds both the instructions to be executed, called the *program*, and the data to be processed. These are stored so that they are readily available when needed. Data and instructions also are stored on external storage devices such as magnetic disks and diskettes. The information stored on these devices is read into main storage when needed. (See also **Central Processing Unit (CPU)**; and **Storage (Computer)**).

Actual processing of data is performed in the ALU according to the stored program instructions. Fundamentally, the ALU provides for comparison of two data items, arithmetic operations such as addition and subtraction, and logical operations such as AND and OR. However, most computers provide several hundred different instructions which are combinations of these fundamental operations. For example, a single instruction may be provided which will fetch two operands from storage locations, add them together, and store them in a third storage location. A very important class of instructions provides for altering the sequence of instructions based on the result of a calculation or comparison. For example, when comparing two numbers A and B, if A is less than or equal to B, the computer may "branch" or transfer control to a different set of instruction than if A is greater than B. This provides for handling alternatives in the solution of a problem and is the most significant difference between a computer and a calculator.

Input devices read data to be processed and programming instructions into the main storage unit or, sometimes, into the CPU itself. Common input devices include card readers, keyboards, magnetically encoded document readers, magnetic drum, disk, diskette, and tape units. Output devices transcribe processed data into machine media such as tape, disk, or cards, and to devices providing human-readable information such as cathode ray tube (CRT) displays and printers.

The computer also may accept input from, or provide output to, another computer. When two or more computers are interconnected in this fashion, are in close physical proximity (usually), and cooperatively participate in the solution of a single problem, the aggregate of computers often is referred to as a *multiprocessor*. The interconnection between the computers is by a channel-to-channel link or similar means to allow high speed interchange of information. Also, typically, a single control program directs the operation of the interconnected computers.

A related concept, sometimes called *distributed processing* or *computer networking* also involves interconnected computers. Although precise definitions do not exist, computer networks generally involve two or more computers which are geographically separated and interconnected by serial communication facilities such as telephone lines or dedicated coaxial cable. Information is interchanged between the computers, generally at a rate much slower than in the case of the multiprocessor. The computers in a network generally are under the control of individual, autonomous control programs so that cooperation is effected in a manner similar to a peer group of individuals.

Distributed processing may involve computers interconnected either by slow-speed communication facilities or by high-speed channel-to-channel connections. Distributed processing usually implies, however, that the interconnected processors work cooperatively on the *same* problem (as in the case of the multiprocessor) but not necessarily under control of a single control program; that is, the involved computers are at least semi-autonomous. The reader is cautioned that the distinction between these terms is not clearly defined at the present time, since the concepts of distributed processing and computer networks are just beginning to be widely implemented.

Computer Hardware

The computer *hardware* is the physical equipment which comprises a computer system. This is to be contrasted with the *software*, which is the aggregate of the programs, procedures, and, sometimes, documentation which is necessary to effectively use the hardware. Taken together, the hardware and the software represent the computer system; it is not uncommon, however, to find the term "system" used in reference to only the hardware. The basic structure and cyclical operation of the computer has remained essentially unchanged since the invention of the early electromechanical and electronic computers in the 1940s. The basic sequence of operations is (1) fetch an instruction from storage; (2) decode (interpret) the instruction to determine the operation to be performed and the location of the operand(s); (3) fetch the operands from storage, if necessary; (4) perform the operation on the operand(s); and (5) fetch the next instruction from storage. There often is considerable data manipulation during each of these basic steps. For example, fetching operands from storage may require arithmetic operations to calculate the physical storage location from a relative address provided in the instruction.

The basic speed of the computer is determined by the characteristics of the circuits used to build the hardware. However, and particularly in high performance machines, a number of techniques are used to increase the effective processing speed beyond the limits imposed by circuit speed. One method is to overlap the basic steps outlined above. As one example, unless the location of the next instruction depends on the outcome of the current operation, it is possible to overlap the operation (step 4) and the fetching of the next instruction (step 5). Another technique, called pipelining, utilizes a series of processing subsystems to operate on several instructions concurrently. For example, during the cycle in which the operation specified in instruction n is being executed, a separate processing unit is calculating the effective or physical address need for instruction $n + 1$, while yet a third processing unit is fetching instruction $n + 2$. Occasionally, the instructions being preprocessed ($n + 1$ and $n + 2$) need to be "thrown away" since instruction n causes a branch to an instruction other than $n + 1$. On the average, however, a decrease in total processing time is effected by the concurrent operation.

It is also common to provide for concurrent transfer of information to and from input/output equipment. This is particularly effective since many input/output devices are electromechanical in nature and operate at speeds several orders of magnitude slower than the CPU speed. One technique used in small computers is to utilize *cycle-stealing* or *direct memory access* (*DMA*) between the input/output device and main storage. In this approach, the CPU provides the channel equipment with a starting storage address of the data to be transferred (or, in the case of an input device, the starting location of an input storage area) and the number of data words to be transferred. The channel equipment then independently transfers the information, using ("stealing") storage cycles not used by the CPU, while the CPU executes subsequent program instructions. In larger machines, the channel may have significant functional capability which allows it to do more than merely sustain the information transfer. For example, it may perform error checking on the data or automatically *chain* to the next input/output request upon completion of the current transfer operation.

Data Representation

Binary Notation. Virtually all modern computers utilize the binary number system and binary coded alphanumeric representations for internal data manipulation. The binary number system utilizes 2 as the number base or radix, as compared to 10 used in the familiar decimal number system. Only two digits, 0 and 1, are needed in the binary system, as compared to the ten symbols used in the decimal system. As in the decimal system, the quantity represented by a digit depends on its position in the numeral. Thus, the units position has a value of 1; the next position, a value of 2, the next 4, and so forth. As a specific example, the binary number 11010 represents the quantity $(1 \times 2^4) + (1 \times 2^3) + (0 \times 2^2) + (1 \times 2^1)$ (0×2^0) or 26 (decimal). The binary equivalents of the first 20 decimal integers are:

Decimal	Binary	Decimal	Binary
1	00001	11	01011
2	00010	12	01100
3	00011	13	01101
4	00100	14	01110
5	00101	15	01111
6	00110	16	10000
7	00111	17	10001
8	01000	18	10010
9	01001	19	10011
10	01010	20	10100

Any quantity can be represented in the binary number system, although it requires considerably more digits than decimal notation. Binary representation is used in computers primarily because only two symbols are required and it is easy and economical to build electronic circuits having two states, "on" and "off," corresponding to 1 and 0.

The symbols 0 and 1 are called *bits* (a contraction of the words *bi*nary digi*t*). In some usage, it is common to refer to the 1 as a bit and to the 0 as no bit, even though this is technically incorrect.

Numeric Data Types. Two numeric data types are provided for arithmetic operations on most computers. These are known as *integer* (or *fixed-point*) and *floating point.* Integer data type provides for the representation and manipulation of whole numbers; i.e., . . . −3, −2, −1, 0, 1, 2, 3, ... Integer arithmetic can be used for dealing with fractions (e.g., 0.575 + 1.2) by appropriate scaling of the numbers (i.e., multiplying by factors of 10 in the decimal system to align the decimal point). When utilizing integer arithmetic in this way, the computer programmer is responsible for the explicit scaling of the numbers and for keeping track of the radix point. The advantage of integer arithmetic is that it is faster than floating-point and requires fewer circuits to implement. Its disadvantage is that the range of numbers that can be represented in the typical computer word is quite limited. For example, an 8-bit byte can be used to represent only 0 to 256 (or −128 to +127) and a 32-bit word only represents 0 to 4,294,967,296.

The use of the floating-point data type provides for a greater range of values by representing the quantity as a quantity *and* a scaling factor, in a manner similar to the use of "scientific notation" such as 0.424×10^{12}. In the computer, the fractional part and the exponent representing the scaling factor are stored separately. For example, the numeral might be stored in a computer word in the form:

s	*e*	*s*	*f*

This is interpreted as $\pm 0.f \times 2^{\pm e}$ where *s* represents the sign (±), *e* is the binary exponent, and *f* is the binary fraction. In performing floating-point arithmetic, the computer manipulates the fraction and the exponent separately. Some small computers do not provide hardware for floating-point arithmetic, in which case the programmer is restricted to fixed-point or must rely on programmed subroutines to perform the necessary manipulation.

The precision of the floating point number is determined by the number of bits in the fraction, whereas the span or range is determined primarily by the number of bits in the exponent. Thus, a one byte fraction and a one byte exponent provides representation of quantities from -0.128×2^{128} to $+0.128 \times 2^{127}$.

The advantage of using floating point numbers is the range of values that can be accommodated and the fact that the placement of the radix point is handled by the hardware or software. The disadvantage is that floating-point operations are slower, since the exponent and fraction are manipulated separately.

Character Coding. In addition to numbers, computers must deal with information composed of alphabetic characters and special symbols (e.g., @, #, $, *) and, since the computer circuits are designed to manipulate only 0 and 1, these characters must be coded into a binary notation.

A number of different alphanumeric codes are used. One of the most common is ASCII (pronounced "as-key," and an acronym for American Standard Code for Information Interchange, which is defined by American National Standard X3.4-1977 and its international counterpart ISO 646–1973. This code provides for the representation of 128 alphabetic (upper and lower case), numeric, special symbol, and control characters in a 7-bit byte. Typically, a one-bit parity check is added (see below) and the character is stored as an 8-bit byte in the computer. The following are sample ASCII codes:

Numerals	1	0110001
	2	0110010
	3	0110011
Alphabetic	A	1000001
	a	1100001
	K	1001011
	k	1101011
Special	#	0100011
	(	0101000
	=	0111101
	?	0111111
Control	End of Text (ETX)	0000011
	Bell (BEL)	0000111
	Acknowledge (ACK)	0000110
	Cancel (CAN)	0011000

It is important to note that these codes are indistinguishable from the codes illustrated earlier for numeric quantities. Their interpretation as characters is determined by the intent of the programmer who writes the program for their manipulation. It is possible, and allowable in some computers, for example, to numerically add *A* to 6. Since this usually does not make sense, however, programs usually are designed to detect such operations and, at a minimum, notify the programmer of the use of mixed data types.

Error Detection and Correction. Many codes are designed to assist in ensuring that an error has not occurred in transferring data from one location to another. Such an error might happen if an electrical noise pulse upset a circuit such that it was turned "off," thus converting a 1 into a 0. One of the simplest error-detecting codes, often used internally in computers, is the *parity check.* In this scheme, the value of one bit in the code is selected such that the number of 1s in the coded character is either even (even parity) or odd (odd parity). For example, in even parity, the seven-bit representation 0110101 of a character would be altered by adding a leading 0 to form 00110101, whereas 1010100 would be altered to 11010100 so as to provide an even number of 1 bits, in this case four. If a single bit change error occurs, the number of 1 bits is no longer even and, when tested by the checking circuit, the error would be detected. Note, however, that any even number of erroneous bit reversals is not detectable by a parity check.

More sophisticated error detection codes sometimes are used in portions of computers and in digital communications. These are capable of detecting multiple-bit errors and may be capable of determining the exact error so that it can be corrected automatically. Such codes require additional bits in the representation and, therefore, have a cost in terms of additional circuits and storage capacity.

The Computer Word and Byte. Although single bits of information sometimes are manipulated in a computer, it is common to handle a group of bits in parallel. Through common usage, an 8-bit group usually is the minimum number of bits manipulated in parallel. Although the usage is not technically correct, this 8-bit group usually is called a *byte.* Technically, a byte is any portion of a *word* (see below) and there is a growing tendency to utilize the more specific words octet (8-bit byte), sextet (6-bit byte), etc. to avoid confusion. The octet is a convenient size to represent a character set, since it allows for 256 distinct representations, a number sufficient for the numerals 0–9, upper- and lower-case alphabetic characters, a generous number of special symbols, control codes, and parity.

A computer *word* typically is one or more bytes (usually 8-bit bytes). The internal circuits in the computer usually are designed to transfer or operate on a word, although instructions to allow the individual manipulation of bytes commonly are provided. In addition, the word is the usual amount of information obtained on a single storage access.

Since there is a correlation between the number of bits in the computer word and the number of circuits required in a particular design, and, additionally, there is a correlation between the number of circuits and the classes of computer represented by the so-called micro-, mini-, and maxi-, or main-frame computer, microcomputers typically utilize a one- or two-byte word (8 or 16 bits), minicomputers usually use a two- or four-byte word (16 or 32 bits), and main-frames usually use four- to eight-byte words (32 to 64 bits). However, this is a time-dependent association since, for example, only a few years ago microprocessors used only 4-bit bytes and 32-bit (4-byte) microprocessors are being designed in the early 1980s.

Most instructions are contained in one word. Also, arithmetic and logical operations usually are performed on one-word operands. As previously indicated, however, most computers provide instructions for the manipulation of single bytes. This is particularly useful when manipulating characters coded into bytes. Since the length of the word determines the precision of simple arithmetic operations, the number of bits in the word may not be sufficient for some applications. As a result, many computers provide double-precision arithmetic operations which use two-word operands. These usually require additional computation time, since the implementation merely provides for handling the two words individually. In addition, some instructions are contained in more than one word. This often is caused by the need to include an operand or a physical address in the instruction itself and the number of bits available often is not sufficient to express all of the needed location addresses.

Instruction Representation. A key digital computer concept is that the instructions which control the computer are themselves stored in the computer storage as binary codes. The instruction code is stored in one or more words and is accessed by the control unit when needed. Depending on the particular operation, a variety of information may be contained in the instruction. For example, in an addition operation, the computer must be told where the addend and augend are located and what to do with the resulting sum. For this reason, the format of the stored instruction varies with the operation to be performed. In general, however, an instruction consists of an operation to be performed, one or more addresses indicating the location of the operands, and, possibly, other modifiers or data. Thus, an instruction might be stored in the following format:

OP	R1	R2	R3

where OP represents the Operation Code, and R1, R2, and R3 are three addresses indicating the sources and destinations of operands and results. Using this format, the operation "Add the contents of register 1 to the contents of Register 3 and store the result in Register 2" might be coded as:

0110	0001	0011	0010

Here the Operation Code for ADD is coded as 0110, the address of Register 1 is 0001, the address of Register 2 is 0011, and the address of Register 3 is 0001.

A typical computer is capable of performing several hundred different instructions of this type. Sequences of such instructions direct the detailed operation of the hardware on a step-by-step basis. Determining the exact sequence for the solution of a particular problem is part of the task of the programmer as discussed below.

Although a particular computer may have several hundred instructions, they can be grouped into several major categories:

1. *Arithmetic Instructions.* These include the arithmetic operations, such as addition, subtraction, multiplication, and division. There may be several different ADD instructions, for example, to provide for integer and floating-point data types or to allow addition of quantities stored in particular locations in the computer; separate ADD instructions might provide for addition of quantities located in registers or addition of quantities stored in main storage.
2. *Logical Instructions.* These provide for the logical operations such as AND, OR, and Exclusive OR. Logical operators may act on single bit positions or on whole words or bytes.
3. *Data Movement Instructions.* These provide for the movement of data between various part of the machine. They might include, for example, "Move the contents of storage location *x* into Register *y*;" "Replace the contents of Register *r* with the contents of Register *s*;" or "Move the contents of storage locations *x* through *y* to input/output port *z*."
4. *Control Instructions.* These provide for the control of the sequence of operations. For example, a typical instruction might be "Transfer control to the instruction located in storage location *x*" or "execute instruction *y* if the contents of register 1 is negative, otherwise execute the next sequential instruction."

Basic Hardware Operation

The diagram of Fig. 2 is used to illustrate the basic operation of the computer. Although greatly simplified, the diagram shows the data paths between the functional units found in a typical CPU and its associated main storage. The CPU Control Unit (CCU) consists of circuits which "decode" (interpret) the Operation Code and subsequently provide the timing and actuation signals to control the other functional units in the CPU.

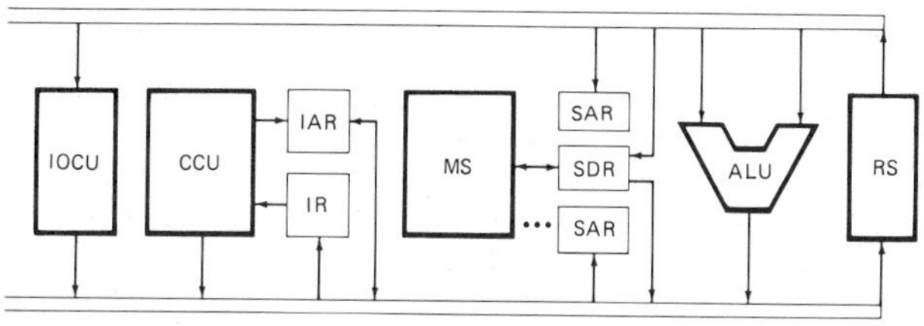

Fig. 2. Hypothetical computer data flow. CCU = central processing unit (CPU) control unit; MS = main storage; SAR = storage address register; SDR = storage data register; ALU = arithmetic-logic unit; IOCU = I/O control unit; RS = register stack; IAR = instruction address register; IR = instruction register.

The Main Storage (MS) is used to store both the program (instructions) and the data on which it operates. Storage is organized as a sequence of locations, typically represented by a byte, which are assigned sequential addresses starting with 1. The amount of storage may vary from as little as a few thousand bytes in a small machine to millions of bytes in a large machine.

The Storage Address Register (SAR) holds the address of the storage location to be activated, either for the purpose of reading the contents of the location or for storing into the location.

The Storage Data Register (SDR) temporarily holds data being read into or out of storage.

The Arithmetic and Logic Unit (ALU) performs the specified arithmetic and logical operations on the data presented at its two inputs. The ALU output is routed to either the register stack, the I/O control unit, or main storage by signals from the CCU.

The Register Stack (RS) is a special-purpose storage unit, consisting of typically 16 to 64 locations that can be used for the temporary storage of data and addresses. It is used in lieu of main storage because it can be accessed much more quickly.

The I/O Control Unit (IOCU) represents the channel or other circuits which provide for the detailed control of input/output units such as video terminals, communication equipment, disk and diskette storage, and data acquisition equipment.

The Instruction Address Register (IAR) contains the location of the instruction currently being executed. It is normally incremented by 1 automatically to access the next sequential instruction.

The Instruction Register (IR) is a temporary storage location in which the current instruction is held during execution.

The basic operational sequence is as follows:

1. Fetch the next instruction from MS.
2. Fetch the operands (if any) from MS or RS.
3. Execute the operation.
4. Fetch the next instruction and repeat.

As a detailed example, consider the following sequence of instructions stored in sequential main storage locations (abbreviations are used to represent the operation codes, but the actual representation in storage is a sequence of binary codes as described earlier):

Storage Location	Instruction	Meaning
1	M,34,8	Move the contents of MS 34 into RS 8
2	M,87,6	Move the contents of MS 87 into RS 6
3	A,8,6,3	Add the contents of RS 8 to the contents of RS 6 and place result in RS 3
4	BCN,3,93,65	If the contents of RS 3 are negative, go to the instruction in MS 93; otherwise, go to the instruction in MS 65

The execution of this program fragment proceeds as follows: The address in the IAR is set to 1, indicating that the instruction in MS location 1 is to be executed. The CCU causes the contents of location 1 to be transferred into the IR where it is decoded by the CCU. The CCU transfers the data address (34) into the SAR and the contents of MS location 34 is transferred into the SDR and then routed to RS location 8. The operation is complete and the CCU increments the IAR by 1, causing it to advance to 2.

Since the IAR is set to 2, the instruction in storage location 2 is transferred into the IR and decoded. The sequence is the same as above except that the data in MS 87 is transferred into RS location 6. At the completion of the transfer, the IAR is incremented to 3.

The instruction from storage location 3 is placed in the IR and decoded. This causes the contents of RS 6 to be routed to one input of the ALU and the contents of RS 8 to be routed to the other input. The two numbers are added together by the ALU and signals from the CCU cause the sum to be stored in RS 3. The IAR is then incremented to 4.

The instruction in MS 4 is transferred to the IR and decoded. This causes the contents of RS 3 to be examined by the CCU to determine whether it is a positive or negative number. Assuming that it is found to be negative, the CCU replaces the contents of the IAR with 65, the address of the next instruction to be executed.

Control of the program has now been transferred to the instructions located at MS 65 and the basic fetch-execute-increment cycle continues. This example is simplified greatly but it is conceptually accurate. In an actual computer, many operations may take place during a single instruction execution. For example, the address of the data may not be explicitly contained in the instruction, so that the CCU may have to fetch data from storage and perform some calculations to determine the actual address of the data. Similarly, a single instruction to send data to a printer may involve a number of data transfers between main storage, registers, and the IOCU, all under control of the CCU.

In this brief entry, it is only possible to present basic concepts and definitions associated with digital computer hardware. Although similar in concept, there are many variations in the hardware design of a computer. Many of the details of these variations and the interconnection of the internal functional elements of the CPU are included as separate entries in this encyclopedia. The reader is referred to these entries or to the references listed at the end of this entry for more comprehensive treatment of specific concepts or hardware elements.

Computer Programming

The computer hardware is controlled by a sequence of instructions which are stored in main storage. The sequence of instructions is called a *computer program* and the act of devising the particular sequence needed to solve a given problem is called *computer programming.* Collectively, the set of programs and associated documentation available for a particular computer is referred to as *software.*

Programming consists of all those activities associated with the production of the actual program. This includes determining the solution to a problem, describing or organizing the solution in such a way that the computer can be used efficiently, coding the solution in a form acceptable to the computer through the use of a programming language, testing the resulting program, and maintaining the program and its documentation over its useful life.

The first phase, problem definition, involves a thorough understanding and description of what the computer is expected to do. This includes understanding what data are available, what errors are likely to be encountered (e.g., a person's name containing a numeral), what data manipulation is necessary, and what results are expected. It is also necessary to understand the characteristics of the data, such as whether certain parameters are positive or negative, the range of numerical inputs, the desired numerical precision, the number of data items to be handled or stored, and the desired format for the results.

Once the problem is defined, a procedure must be devised which provides the desired solution. This typically is done by breaking the problem into a set of subproblems, each of which can be addressed separately. For example, in a payroll calculation, the subproblems might include:

1. Calculating hours worked from starting and ending times.
2. Determining gross wages, including overtime and shift premiums.
3. Calculating income and social security withholding tax.
4. Calculating automatic deductions (e.g., savings bonds or insurance) and arranging for crediting them to the proper account.
5. Calculation of net pay and year-to-date values.

Very often, this decomposition of the problem is facilitated by graphical or other techniques which are useful in organizing the solution. In one popular technique, called *flowcharting,* each subproblem is represented by a block with the relation between blocks illustrated by connecting lines. Blocks having different functions (e.g., input, output, decisions, calculations, etc.) often have distinctive shapes. American National Standard X3.5-1970 describes a set of such symbols and their use. Following the overall decomposition of the problem into subproblems, each of the major steps (blocks in the flowchart) is further decomposed into a more detailed flowchart.

Once the solution is planned, it must be translated into a sequence of computer instructions which represent the program. Although the actual program representation in storage is a series of bits, attempting to write a program in this representation is difficult and error prone, due to the need to manipulate the inconvenient binary codes. As a result, programming languages have been devised to ease the task and reduce the chance for error.

There are many programming languages available for most computers. In general, however, they can be categorized as assembler languages, macroassembler languages, higher-level or procedural languages, and problem-oriented languages. In assembler language, the binary code for each operation code is represented by a mnemonic code. For example, the binary code 01101 meaning ADD REGISTER might be represented by ADR. Similarly, address locations are represented by mnemonics chosen by the programmer; for example, NP may represent Net Pay. The program is coded using these symbols and might appear as follows:

ADR,1,NP
M,34,86
CMP,3,RES

The resultant program, called the *source program*, is used as data for an assembler program which substitutes binary codes for the mnemonic codes and assigns a physical storage address in place of the mnemonic address. The assembler program output is a sequence of binary words (bytes) which can be placed in storage for execution. This binary-coded program is called the *object program*.

It is soon discovered when using assembler language that there are particular sequences of operations which are used frequently. For example, reading a character from a keyboard may require the same 100 assembler language instructions for each character read. Rather than writing this sequence each time it is needed, the whole sequence can be assigned a name such as KEYI, standing for Key Input. Whenever the programmer needs to read a keyboard input, the mnemonic KEYI is inserted in the assembler language program. The sequence of instructions KEYI is called a *macro*. When the program is processed using a macroassembler to produce the object program, the computer inserts the entire instruction sequence represented by KEYI whenever it appears. This technique can reduce significantly the effort needed to code a program.

Procedural languages go a step further and provide even greater function for each statement in the source program. For example, a typical procedural language allows the programmer to write

$$A = B + C / D \times E$$

in order to express a mathematical calculation. Similarly, control statements can be expressed in a form such as:

IF × .GT. y THEN z=x−y ELSE z=y−x

These statements are not executable directly by the computer and must be translated into a sequence of binary-coded instructions. This is done by a program called a *compiler* or an *interpreter* which analyzes each statement and substitutes the necessary sequence of instructions to accomplish the desired result. Each statement in the procedural language source program creates many machine instructions. As a result, the productivity of the programmer is greatly enhanced. A number of high-level procedural languages are in common use. These include, for example, FORTRAN, COBOL, Pascal, LISP, and BASIC. A majority of programming today is done using such languages to express the problem solution.

Problem-oriented languages are similar to procedural languages except that they are designed for the solution of a specific class of problems. They often use a vocabulary which is unique to a particular field. For example, a typical statement in a problem-oriented language for process control applications might be:

AT 1500 HRS CLOSE VALVE V1 WAIT 10 SEC TEST

A source program written in such a language must also be processed by a compiler or interpreter program before it can be executed by the computer.

Once the solution to the problem has been described in a programming language and processed by an assembler, compiler, or interpreter program, it must be tested for correctness before being used. Errors in programs are called "bugs" and the process of finding and correcting them is referred to as "debugging." A variety of techniques are utilized in this process. Most assemblers, compilers, and interpreters include facilities which identify errors associated with the syntax and semantics of the programming language being used and other detectable errors such as an expression involving mixed data types (e.g., attempting to add a character to a number). Errors in the problem solution, of course, cannot be detected by these language processing programs. These are usually detected by visual inspection, sometimes by a second programmer who was not involved in the original writing of the program, and by using test data to determine if the results are as expected. Many errors are associated with the interaction of a particular program with other programs being used concurrently. These are detected when the program is integrated into the total set of programs being used on the computer.

Debugging a program is a difficult task since all possible combinations of data must be considered. It is usually impossible to exhaustively test a program, so it is likely that some bugs will remain undetected until the program has been in use for some time. It is this phenomenon which gives rise to the need for maintenance over the life of the program. See also **Debugging (Computer Program).**

System Programs

Programs can be separated into the two major categories of *application programs* and *system programs*. Application programs are those programs directly involved in the solution of a user's problem and are typically related to the business of scientific purpose of the program. Thus, for example, accounting programs, data base inquiry programs, and airline reservation programs are application programs.

Considerable programming, however, is required just to control the internal operation of the computer, independent of the particular application. The suite of programs used for this purpose often are called "system programs." In general, these provide facilities and services which are useful to the application programmer and serve to free the programmer to concentrate on the details of the application. A major example of this is the so-called "operating system," "monitor," or "executive" program which handles many of the internal details associated with running an application program.

In present computers, it is common for several application programs to be in main storage simultaneously and be executed concurrently. Two terms associated with this type of operation are "time-sharing" and "multiprogramming." In time sharing, each program is executed for a fixed period of time or until the program needs a computer facility which currently is not available. At that time, the program is halted temporarily and another program is executed. After every active program has an opportunity to execute, the computer returns to the original program. Because of the speed of the computer, the individual user often is not aware that their program has been interrupted several times during its execution.

Multiprogramming is a generalization of time-sharing, where the computer facilities are shared among several programs based on the availability of resources, such as input/output devices, or some other criteria. For example, a priority scheme may be used where more important programs are allowed precedence over other programs. Less important programs are not allowed to run until higher-priority tasks are completed.

The scheduling of the programs in a time-sharing or multiprogramming environment is one of the important functions of the operating system. In addition, the operating system often provides detailed control of input/output devices, such as disk files and printers, and security features that, for example, ensure that a program does not access data for which it is not authorized. The computer typically spends more time executing operating system code than it does application programs. As a result, the design of the operating system is crucial to the performance perceived by the user.

In addition to the operating system, a suite of programs referred to as "utility programs" usually are provided. These include the assemblers, compilers, and interpreters needed to process the high-level language source programs. Other utilities are housekeeping programs for rearranging storage, for testing the correct operation of the computer and its peripheral devices, and accounting programs to aid in billing users for their portion of the computer time.

Early History of the Digital Computer

The earliest digital computer was the abacus, invented in the pre-Christian era and still used widely in many parts of the world. Semiautomatic mechanical calculators did not appear until the 17th century, however, when Blaise Pascal, the son of a French tax collector, invented a calculator to assist in adding and subtracting the columns of figures associated with the assessment and collection of taxes. Although never totally successful, the toothed-wheel mechanism invented by Pascal was the basis for the modern mechanical adding machines. Other mechanical calculators followed, including those invented by Leibnitz and Napier, and during the latter half of the 19th Century resulted in business equipment bearing the familiar names of Burroughs and Monroe, to name only two. See Fig. 3; contrast with Figs. 1 and 4.

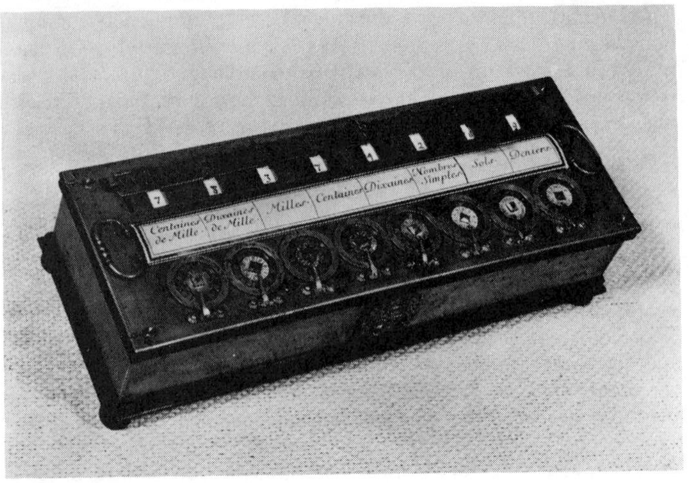

Fig. 3. Pascal's "Machine Aritmétique." (*Furnished courtesy of IBM Corporation.*)

One of the most remarkable developments was that of Charles Babbage who in the 1830s, in England, proposed and partially built his "Difference Engine" to calculate mathematical and astronomical tables. This mechanical device was to be powered by a steam engine and, utilizing the method of successive differences, was to calculate numbers and punch them into copper printing plates. Babbage never succeeded in building his machine, due partially to the lack of adequate mechanical technology, but a scaled-down version was built by Scheutz in Sweden. Babbage then went on in the 1840s to design his "Analytical Engine," which embodied many of the fundamental principles found in the modern digital computer. It consisted of a "mill," which performed the actual calculations; internal storage; and, like the Difference Engine, a mechanism to punch results into copper printing plates. In addition, input instructions and data were to be fed into the machine using punched cards, an idea borrowed from the Jacquard loom of 1799. The machine was never intended to be built, although Babbage's son made several attempts to get support to build it, and the work was forgotten until after the invention of the modern digital computer in the 1940s.

The next significant development was the use of punched card equipment for the 1890 United States census. Dr. Herman Hollerith invented (reinvented?) the punched card as a means of recording data on each citizen and produced equipment for punching and reading the information. The invention allowed the census to be completed in a matter of a few years, rather than the more than ten years originally predicted for hand tabulation methods. Hollerith founded a company which further developed the idea and, during the early 1900s, the use of punched card accounting equipment grew into common use.

Concurrently, there were significant developments in electronics, with the invention of the vacuum tube through the efforts of Edison, DeForest, and Fleming.

These efforts merged and were culminated in the development of the first major all-electronic digital computer by Drs. John Mauchly and J. Prespert Eckert in the Moore School of Electrical Engineering at the University of Pennsylvania, from 1942 to 1946. This machine, called ENIAC (Electronic Numerical Integrator And Calculator) was built as the result of a request by the U.S. Department of War to assist in calculating ballistic tables. It consisted of about 18,000 vacuum tubes, measured 80 × 8 × 3 feet, and consumed 120 kW of power, of which 80 kW was for heating the tube filaments. The machine utilized punched card equipment for input and output and was programmed by means of cards and wired panels. Although powerful for its day, its computing power was significantly less than that found in a small desktop computer of the 1980s.

ENIAC lacked one essential concept found in today's computers, that of the stored program. This was suggested by Dr. John Von

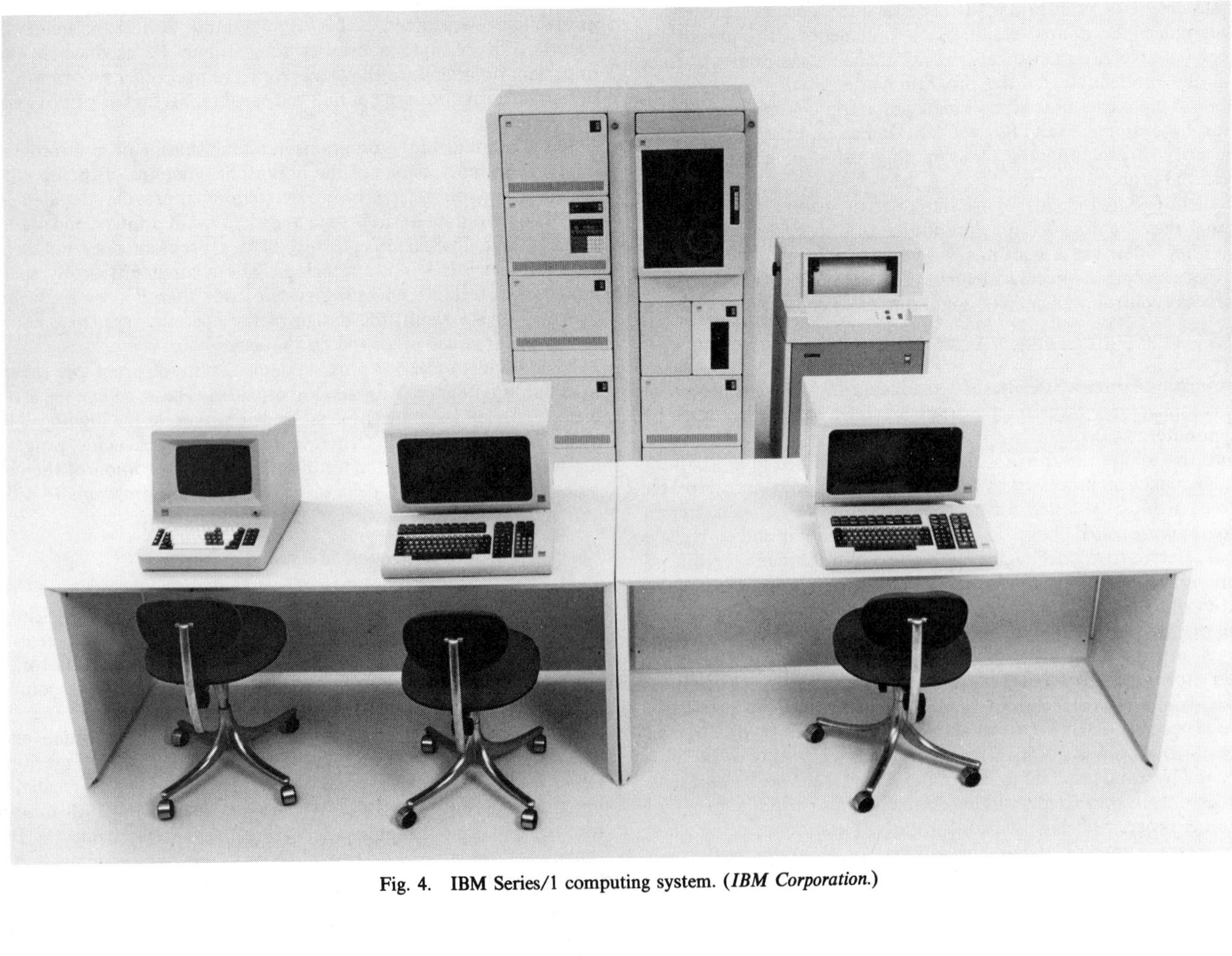

Fig. 4. IBM Series/1 computing system. (*IBM Corporation.*)

Neumann of Princeton University and implemented in BINAC in 1948. The use of the computer storage for the storage of the program provided two advantages which contributed greatly to the flexibility of the machine. The first was that the program instructions were available at electronic speeds, rather than the electromechanical speed dictated by the punched card equipent of ENIAC. Secondly, having the program in storage allowed the program to alter instructions, thereby altering a calculation based on data or programmed controls. Although frowned on today, the ability to alter internally stored instructions was a major factor in realizing the true potential of the electronic computer.

Since these early machines, technology has allowed rapid advances in computers. The invention of the transistor in 1949 by researchers at Bell Telephone Laboratories provided a fast, inexpensive, and low-power device which, when incorporated into computers in the late 1950s, eliminated the bulky, power-consuming vacuum tubes of earlier designs. Progress in transistor technology has been phenomenal and now provides for the fabrication of entire computer processes, many times as powerful as ENIAC, on a tiny chip of silicon about 10 mm square.

This brief history omits thousands of advances and the detailed consideration of the many research scientists and engineers who have contributed to the modern computer. No single contribution or invention "caused" the computer; rather, it was a combination of many technical developments, and the recognition of need on the part of creative businesses that provided the impetus for its past and present development.

References

NOTE: Much more detailed information on various aspects of digital computers can be found in about 200 additional alphabetical entries throughout this volume. Reference to the terms listed under **Data Processing** is suggested.

Goldstine, H. H.: "The Computer from Pascal to von Neumann," Princeton Univ. Press, Princeton, New Jersey, 1972.

Bohl, M.: "Information Processing," 3rd edition, System Research Associates, Chicago, Illinois, 1980.

Harrison, T. J.: "Minicomputers in Industrial Control," Instrument of Society of America, Research Triangle Park, North Carolina, 1978.

Lorin, H. and H. M. Deitel: "Operating Systems," Addison-Wesley Publishing Co., Reading, Massachusetts, 1981.

Aron, J. D.: "The Program Development Process—Part I—The Individual Programmer," Addison-Wesley Publishing Co., Reading, Massachusetts, 1974.

Bell, C. G., Mudge, J. C., and J. E. McNamara: "Computer Engineering," Digital Press, Bedford, Massachusetts, 1978.

Pratt, T. W.: "Programming Languages," Prentice-Hall, Inc., Englewood Cliffs, New Jersey, 1975.

Thomas J. Harrison, IBM Corporation, Boca Raton, Florida.

DIGITAL IMAGING. **Photography and Imagery.**

DIGITAL INPUT/OUTPUT DEVICES. **Input/Output Devices (Computing System).**

DIGITAL MULTIPLEXER. A device that permits sharing a common information path between multiple groups of input or output digital signals. For example, a digital multiplexer can be used to transfer information between the central processing (CPU) of a computer and any one of several digital input or output devices. See also **Input/Output Devices.** A form of digital multiplexer for use in reading information of groups of contact points is shown in the accompanying diagram. To read the status of contacts in group A, a positive signal is transmitted on the group A select line. Should the contact be closed, the positive signal appears on the associated bit line. Bit lines are sampled during the period when the select line is active. The status of the group of contacts is stored in computer memory. Upon completion of the operation, the signal on the select line is removed. When it is desired to read group B, a positive signal is placed on the group B select line and the status of the contacts in group B is placed on the associated bit lines. Diodes in series with the contacts are needed to prevent back circuits; the number of groups that can share the same input bit lines is a function of the back resistance of the diode. See also terms listed under **Data Processing.**

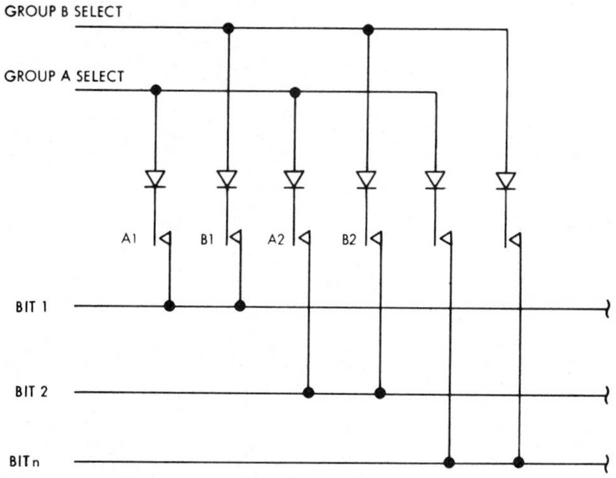

Digital multiplexer.

Thomas J. Harrison, International Business Machines Corporation, Boca Raton, Florida.

DIGITAL OUTPUT. The digital-output information of a computer may be a contact-operate signal or a voltage-level signal. Under the control of the computer program, the digital information is transferred from the central processing unit to a digital-output register. The output of the register may be connected directly to the digital-output signal lines to provide the voltage-level signal. To provide the contact-operate signals, the output of the register is connected to a driver circuit which is connected to the digital output signal lines. When a register position is turned on, the drive circuit provides the current which actuates an external device, such as a relay. The information remains in the register until another computer command to the same group of digital-output points is executed, or the register may be reset as the result of a signal from the external device or after a fixed time interval has elapsed. See also **Input/Output Devices;** terms listed under **Data Processing.**

DIGITAL-TO-ANALOG CONVERTER. Abbreviated D/A converter or DAC. A device for generating an analog voltage or current which is proportional to the value of a digital-input word. Frequently, in systems under the control of a digital computer, analog signals must be generated to cause actuation of analog devices. The latter include positioning mechanisms, such as valve positioners, graphic displays, such as X-Y recorders, oscilloscope displays, and strip-chart recorders. D/A converters also are used apart from computers to provide voltages which are proportional to the settings of input switches.

The majority of D/A converters consist of a D/A converter decoder network, an analog switch for each digital-input bit, a buffer amplifier, and the necessary control logic in a configuration along the lines of Fig. 1. The decoder network is an electronically controllable attenuation network whose attenuation factor is proportional to the position of the input switches. The analog switches, under control of the logic, either connect the reference source to the decoder-network input terminal or ground the terminal. This results in an output voltage that is proportional to the setting of the switches.

The control logic provides digital storage of the input-data word and control of the switch position. Additionally, it may provide decoding of addresses and the digital multiplexing of digital data to one of several D/A converters included in the subsystem. Impedance buffering and current-driving capability is provided by the buffer amplifier. Generally, this is required because the output impedance of the decoder network is relatively high. Driving loads directly from the decoder network can result in loading errors which may be intolerable in a high-accuracy system.

Another D/A converter configuration is shown in Fig. 2. This is comprised of control logic, a decoder network, a reference source, and an operational amplifier.

Decoding Networks. Although an ac decoder network is possible,

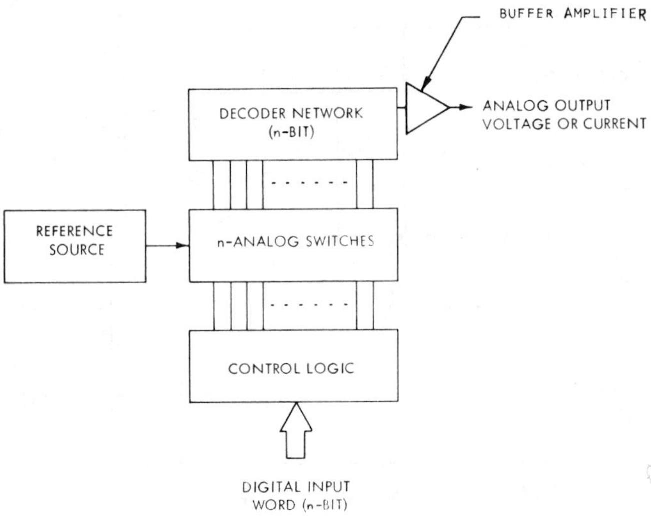

Fig. 1. Representative digital-to-analog converter.

the networks used in D/A converters generally are designed for dc using precision resistors. The two major decoder networks are shown in Figs. 3 and 4. Although the ladder networks shown are n-bit decoders, other codes can be realized by appropriate modifications. As an example, in a weighted-resistor binary-coded-decimal decoder, the configuration is identical, but the resistors are R, 2R, 4R, . . . ; in the ladder decoder, the resistors are R and 2R, but an attenuation network providing an attenuation factor of 0.8 is inserted between each 3-bit section of the ladder decoder to provide the binary-coded-decimal output.

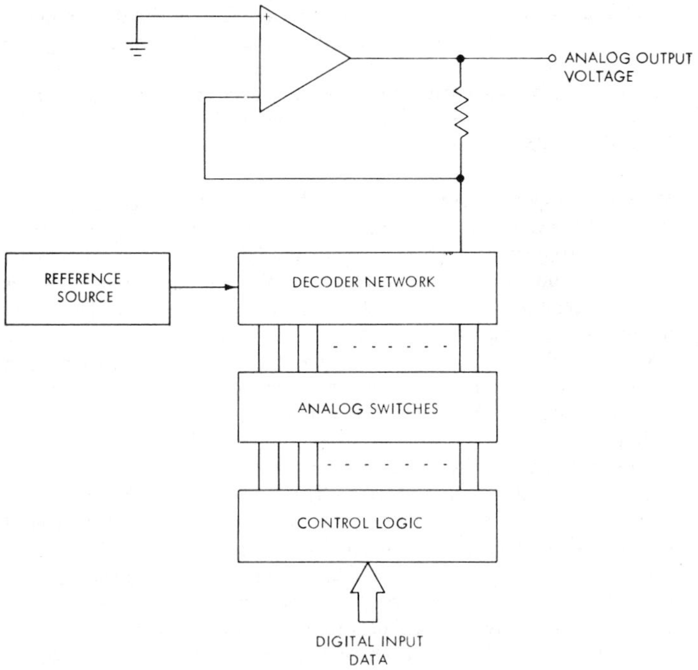

Fig. 2. Operational-amplifier type digital-to-analog converter.

The weighted-resistor network of Fig. 3 is used with a voltage-reference source. Each of the terminals 1, 2, . . . n either is connected to the reference voltage V_r, or it is grounded. If the kth node is connected to V_r and the other $(n-1)$ nodes are grounded, the output voltage V_0 is

$$V_0 = \frac{V_r}{2^n - 1 + G_L/G} \sum_{k=1}^{n} A_k 2^{n-k} \qquad A_k = 1, 0$$

where A_k is an index which is 1 when $k - n$ terminal is connected to the voltage source and 0 when it is grounded. If the conductance $G_L = G$, the expression reduces to

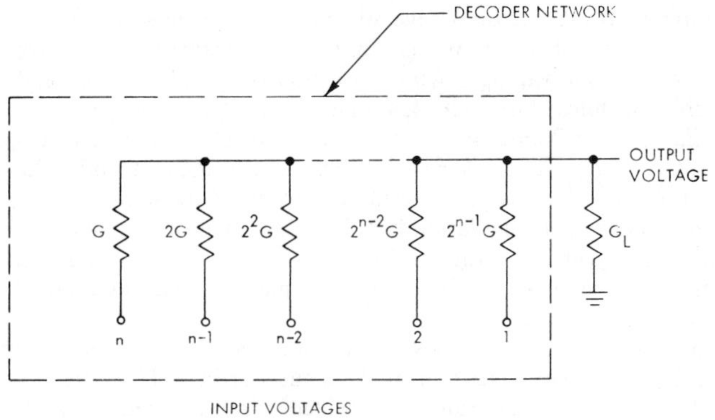

Fig. 3. Weighted-resistor decoder network.

$$V_0 = V_r \sum_{k=1}^{n} A_k 2^{-k} \qquad A_k = 1, 0$$

which shows that the output voltage is binary and is determined by the conditions at the input terminals. The input conductance at the kth terminal when G_L is assumed equal to G is

$$G_{in} = 2^{n-k}(1 - 2^{-k})G$$

and the output conductance of the network is 2^nG, a constant. Although this network is designed for use with a voltage-reference source, a similar network can be designed for use with current-reference sources.

The resistor network of Fig. 4 is commonly known as the R-2R

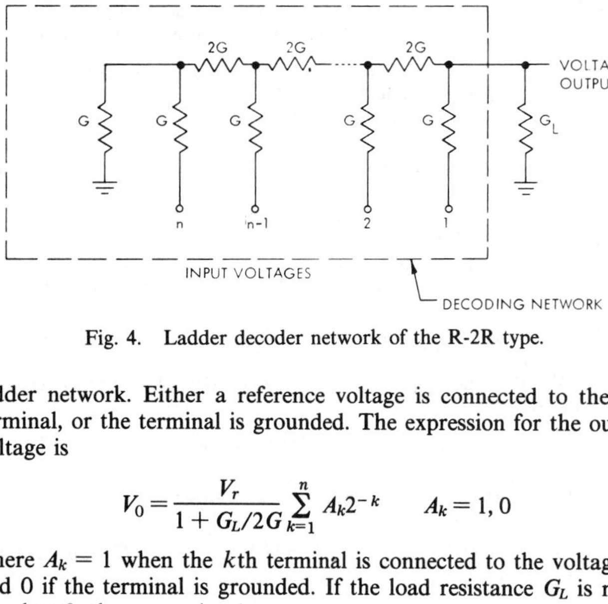

Fig. 4. Ladder decoder network of the R-2R type.

ladder network. Either a reference voltage is connected to the kth terminal, or the terminal is grounded. The expression for the output voltage is

$$V_0 = \frac{V_r}{1 + G_L/2G} \sum_{k=1}^{n} A_k 2^{-k} \qquad A_k = 1, 0$$

where $A_k = 1$ when the kth terminal is connected to the voltage V_r and 0 if the terminal is grounded. If the load resistance G_L is made equal to 0, the expression becomes

$$V_0 = V_r \sum_{k=1}^{n} A_k 2^{-k} \qquad A_k = 1, 0$$

which shows the binary contribution of each input. The output conductance is constant and is equal to $2G$ if $G_L = 0$. The input resistance is also constant and is equal to $2G/3$. A similar network can be used for current sources.

The weighted-resistor network uses a minimum number of resistances. The ratio of the largest to the smallest resistor is 2^{n-1} (or 512 for $n = 10$) as compared with only 2 for the ladder network. The type of resistor used in the network depends primarily on performance limitations and cost. Discrete wire-wound resistors are often used for high-precision networks which have a very low temperature coefficient. Film-type resistors can be used where the temperature-coefficient and precision requirements are less.

See also terms listed under **Data Processing.**

Thomas J. Harrison, International Business Machines Corporation, Boca Raton, Florida.

DIGITAL TRANSMISSION. Telephony.

DIGITATE DRAINAGE. The term applied to the fingerlike pattern of stream valleys. Such a stream pattern usually develops only when the underlying formations are relatively horizontal or if folded, faulted, or metamorphosed, are relatively of equal hardness and solubility.

DIGIT (Computer System). A symbol used to convey a specific quantity either by itself or with other numbers of its set, e.g., 2, 3, 4, and 5 are digits. The base or radix must be specified and each digit's value assigned.

Binary Digit. A number in the binary scale of notation. This digit may be zero (0), or one (1). It may be equivalent to an "on" or an "off" condition, or to a "yes" or "no" condition. Often abbreviated *bit.* See also **Bit (Data System).**

Check Digit. One or more redundant digits carried along with a machine word and used in relation to the other digits in the word as a self-checking or error-detecting code to detect malfunctions of equipment in data transfer operations. See also **Check (Computer System).**

Equivalent Binary Digits. The number of binary positions needed to enumerate the elements of a specific set. In the case of a set with five elements, it will be found that three equivalent binary digits are needed to enumerate the five members of the set: 1, 10, 11, 100, and 101. Where a word consists of three decimal digits and a plus or minus sign, 1999 different combinations are possible. This set would require eleven equivalent binary digits in order to enumerate all of its elements.

Octal Digit. The symbol 0, 1, 2, 3, 4, 5, 6, or 7 used as a digit in the system of notation which uses 8 as the base or radix. See also **Octal.**

Sign Digit. A digit incorporating one to four binary bits which is associated with a data item for the purpose of denoting an algebraic sign. In most binary, word-organized computers, a 1-bit sign is used: 0 = + (plus); and 1 = − (minus). Although not strictly a digit by the above definition, it occupies the first digit position and it is common to consider it as a digit.

Thomas J. Harrison, International Business Machines Corporation, Boca Raton, Florida.

DIGRAPH. A digraph D is a finite set of vertices $\beta_1, \beta_2, \ldots, \beta_v$, together with certain directed lines $\boldsymbol{\beta}_1\boldsymbol{\beta}_2$ (from vertex β_1 to β_2). The possibility that both $\boldsymbol{\beta}_1\boldsymbol{\beta}_2$ and $\boldsymbol{\beta}_2\boldsymbol{\beta}_1$ are lines of the digraph is

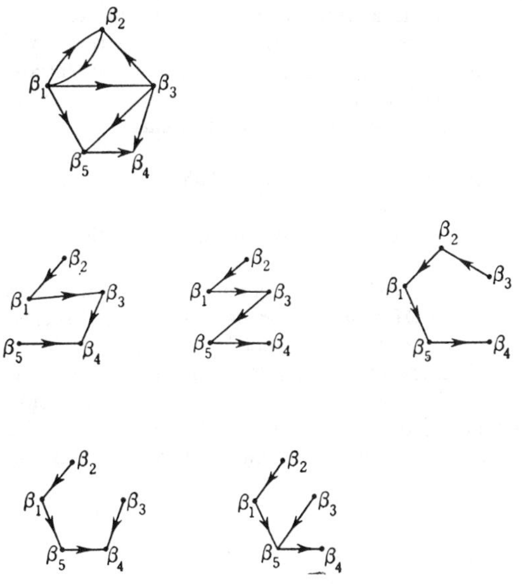

$$M = \begin{vmatrix} 3 & -1 & -1 & 0 & -1 \\ -1 & 1 & 0 & 0 & 0 \\ 0 & -1 & 3 & -1 & -1 \\ 0 & 0 & 0 & 0 & 0 \\ 0 & 0 & 0 & -1 & 1 \end{vmatrix}$$

Digraph and five trees.

not excluded. Thus an oriented graph is a digraph but not conversely.

The adjacency matrix (see **Matrix (Adjacency)**) $A = (a_{ij})$ of a digraph D is defined as follows: $a_{ij} = +1$ if line $\boldsymbol{\beta}_i\boldsymbol{\beta}_j$ is in D and $a_{ij} = 0$ otherwise. It is important to note that A is not necessarily symmetric ($a_{ij} = a_{ji}$) as is the case for the adjacency matrix of a non-oriented graph. A is square and of order v, the number of vertices.

A tree T of D with sink β (a vertex) is

(a) a subgraph of D which contains all the vertices of D;
(b) in T there is a directed path leading to β from every other vertex in D;
(c) every vetex other than β is the initial vertex (see **Vertex**) of exactly one directed line in T;
(d) there is no directed line in T for which β is the initial vertex.

An example of a digraph and its five trees is given in the accompanying figure.

The degree matrix B for D is defined to be that diagonal matrix of order v which is such that the difference $M = B - A$ has every row sum to zero. The matrix tree theorem for digraphs states that the co-factors of the elements in the ith row of M are all equal and each of them gives the number of trees of D with sink β_i. The matrix M associated with the digraph D of the illustrative example above also appears in the figure.

See also **Graph (Mathematics); Tree (Mathematics);** and terms listed under **Mathematics.**

DIHEDRAL ANGLE. Angle.

DIHYDRIC ALCOHOL. Glycol.

DIHYDROCHALCONE. Sweeteners.

DIHYDROXYACETONE. Dermatitis and Dermatosis.

DIHYDROXYPHENYLALANINE (Dopa). Amino Acids.

DIKE (or Dyke). A tabular, intrusive mass of igneous rock which cuts across other igneous rock bodies, such as batholiths; or cuts across the bedding (stratification) of lavas or sedimentary formations. Not to be confused, in the chemical and mineralogical sense, with veins; or, in the structural sense, with sills.

DILATANCY. The property of certain colloidal solutions of becoming solid, or setting, under pressure. Also known as "inverse plasticity" since there is an increase in the resistance to deformation with increase in the rate of shear.

DILATANCY (Rock). Earthquakes, Seismology, and Plate Tectonics.

DILATANT SUBSTANCE. Colloid System.

DILATATION. The increase of volume per unit volume of a continuous material.

DILATION. The fractional increment in volume caused by a deformation or phase change.

DILATION NUMBER. Ratio of the volume of a liquid to the volume of a solid of the same composition at the same temperature.

DILATOMETER. 1. An instrument used to measure small increments in the volume of liquids, as a solid phase separates. 2. An instrument for measuring very small length changes in a solid metal specimen such as occur during thermal expansion or phase transformations.

DILL. Flavorings; Umbelliferae.

DILUTE SOLUTION. Ideal System.

DILUTIONAL HYPONATREMIA. Kidney and Urinary Tract.

DILUVIUM. Derived from the Latin *diluo*, to wash apart, through *diluvialis*, flood. A relatively obsolete geologic term formerly applied to certain water-laid deposits within or bordering the glaciation regions of Europe and North America. Most of the sediments which were previously thought to have been the result of the "flood," and which are now known to be stratified drift, were called diluvium. In Germany, the term corresponds to Pleistocene.

DIMENSIONAL ANALYSIS. Since the quantities represented by the two sides of an equation must have the same dimensions, it is often possible to arrive at the form of an equation connecting physical quantities by a consideration only of the dimensions of the quantities involved, without a detailed theory and without consideration of magnitudes. The equations obtained in this way may be in error by a multiplying constant, which can often be determined empirically. Dimensional analysis has been particularly useful in the development of aerodynamics and hydrodynamics.

DIMENSION MEASUREMENT. Position and Displacement Measurement; Thickness Measurement and Gaging Systems.

DIMETHYLAMINE. Imino Compounds.

DIMETRIC DRAWING. Pictorial Representation.

DIMORPHISM. The occurrence of individuals of two forms in the same species.

In its broadest application dimorphism includes the alternation of forms such as the polyp and medusa of the coelenterates. It is commonly used to indicate less fundamental differences due to the conditions attending the development of essentially similar indiviuals. Thus some of our butterflies differ noticeably in the generation developed during the summer and that which emerges in the spring after passing the winter in an immature stage. This condition is seasonal dimorphism, as also is the occurrence of wet- and dry-season forms in some tropical species. Conspicuous difference between the sexes other than reproductive adaptions is sexual dimorphism. This term is also sometimes applied to the occurrence of two forms in a single sex, like the black and yellow females of the common yellow swallowtail butterfly.

The term is used by mineralogists to describe the phenomenon of certain natural compounds crystallizing in two different forms.

DINGO. Canines.

DINOFLAGELLIDA (*Dinoflagellata*). An order of one-celled animals. Chiefly marine species, whose body is surrounded by an envelope of cellulose, often beautifully figured. *Mastigophora.*

Triggered by the right combination of meteorological and oceanographic conditions, these one-celled organisms can multiply at what has been termed explosive rates to cause the so-called "poisonous tides" or "red tides." Areas of the ocean up to many square miles may be colored yellow, red or olive-green by the presence of these microorganisms. Dinoflagellate toxins resemble the deadly botulinum, blocking nerve impulses that prevent the production of acetylcholine, the substance that acts on the end plate of muscle fiber to make it contract. It has been postulated that in an explosive growth or "bloom" of the dinoflagellate population a biological chain-type reaction may occur—poison from the microorganisms kill fish; the decaying fish increase the supply of nutrients in the seawater; the increased nutrients yield more dinoflagellates, which generate more poison, and so on—until a natural reversal of the conditions which favor the dinoflagellate bloom occurs.

Outbreaks of poisonous tides have occurred along practically all coasts of the United States, but notably Florida, parts of the Gulf Coast, the waters along southern California, and the Gulf of California. Such tides have occurred on the west coast of South America, the west coast of Africa, and the coasts surrounding the tip of India. Less frequently, such tides occur in the English Channel, the Red Sea, and the Gulf of Aden. The southern waters of Japan, the Philippines, Malaysia, northwestern and southeastern Australia, and the north island of New Zealand also have been affected. Shellfish contaminated by such tides have caused numerous deaths, possibly running into the tens of thousands (recurrent epidemics of fish poisoning) although statistics available are not complete or reliable.

There was a particularly bad outbreak in Florida in 1946 at which time some of the coastal water became thick and viscous, killing barnacles, oysters, shrimps, crabs, and turtles. Decaying fish piled along at least 60 miles (96 kilometers) of the beaches. Because the spray from the surf was so irritating, schools and resorts near the shore were closed. There are many outbreaks over the world each year of the red tide, but on the average the condition persists for only a few days at any one location. It is postulated that the outbreaks result from increased concentrations of nutrients that may result from tidal turbulence or by runoff from heavy rains on land. Vitamin B-12 is essential to dinoflagellate metabolism and, in the laboratory, the microorganisms thrive best in diluted seawater that has been enriched with vitamin B-12. This vitamin is produced in large quantities by bacteria and blue-green algae in the soil and salt marshes, from which rains can transport it to the ocean.

Dinoflagellate blooms are frequently followed by blooms of creatures that prey upon the dinoflatellates. Notably, the luminous and predatory dinoflagellate *Noctiluca* appear.

Poisonous tides have been recorded for centuries and it is believed by some that reference in Exodus (7:21) relates to this phenomenon, "And the fish that was in the river died; and the river stank, and the Egyptians could not drink of the water of the river."

DINOSAURS. Fossil Reptiles.

DIODE. A two-electrode device, having an anode and a cathode, which has marked unidirectional characteristics. Many types of diodes are known. See **Electron Tube** for diodes of that kind. There are more importantly crystal and semiconducting diodes. For example:

A *crystal diode* is a diode consisting of a semiconducting material such as a germanium or silicon, as one electrode, and a fine wire "whisker" resting on the semiconductor as the other electrode. Because of its low capacitance, the device finds considerable application as a rectifier or detector of microwave frequencies.

A *semiconductor diode* is a two-electrode semiconductor device having an asymmetrical voltage-current characteristic. A *double-base diode* is a semiconductor diode in which a potential gradient is produced across the base region by the application of a voltage between two electrodes at either end of the base. The correct polarity and magnitude of this voltage causes the diode to exhibit a controllable, negative resistance between one of the base electrodes and the anode. A *junction diode* is a semiconductor diode whose nonsymmetrical volt-ampere characteristics are manifested as the result of the junction found between n-type and p-type semiconductor materials. This junction may be either diffused, grown or alloyed. See **Microelectronics; Semiconductor; Transistor;** and **Zener Diode.**

DIODE (Electron Tube). Electron Tube.

DIODE (Fabrication). Microstructure Fabrication.

DIODE LOGIC. A form of logic circuit using diodes and resistors. The abbreviation DL is sometimes used to refer to this class of circuits. Circuits for performing the function OR and AND are shown in the figure. Considering a positive signal to represent 1, one notes that a positive voltage connected to any one of the inputs *A, B,* or *C* in part (a) of the accompanying figure will result in a positive signal at the output. Thus, the circuit performs the logic OR operation.

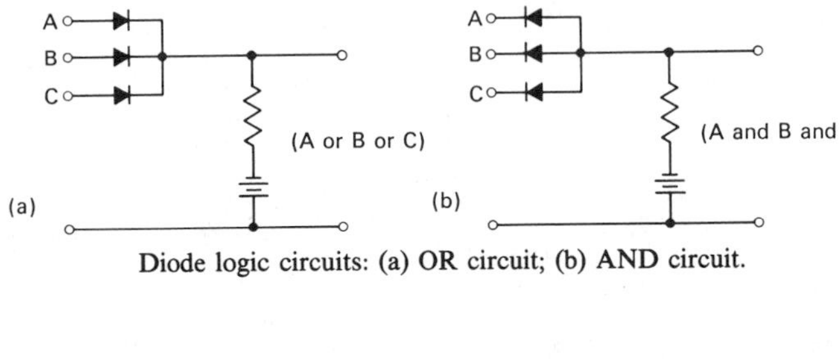

Diode logic circuits: (a) OR circuit; (b) AND circuit.

In (b), the AND operation is accomplished; to obtain a positive (1) output A, B, and C must all be positive (1).

See also **AND (Circuit)**; and **OR (Circuit)**.

DIODE (Rectifier). Rectifiers.

DIODE TRANSISTOR LOGIC. A form of logic circuit using diodes, transistors, and resistors. The abbreviation DTL is sometimes used to refer to this class of circuits. Circuits for performing the logical operations NOT OR (NOR) and NOT AND (NAND) are shown in Fig. 1. The arrangement is an extension of diode logic wherein

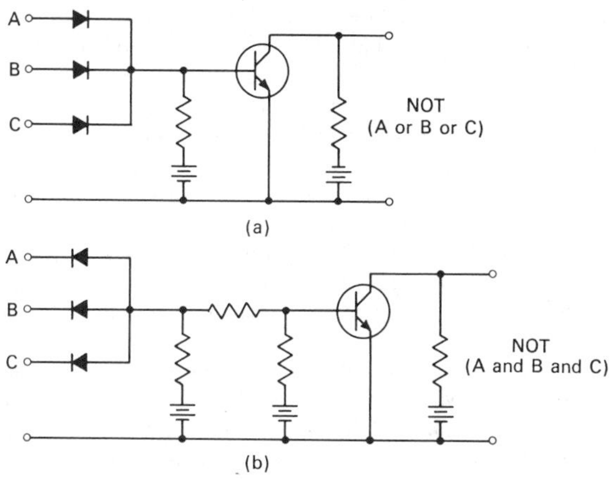

Fig. 1. Diode transistor logic circuits.

the logic operation is effected by diodes and the transistor furnishes an inversion of sign and amplification. A positive signal represents the 1 logic state for the circuits in the figure. In (a) the transistor will be brought into conduction with the collector voltage close to ground potential (0 output, corresponding to NOT 1) when either A or B or C is positive (1 condition). In (b), the transistor will be brought into conduction only when all inputs A, B, and C are in the 1 condition (positive signal). This state will result in a transistor output corresponding to 0 (NOT 1). In both circuits the output of the transistor is positive corresponding to 1, whenever it is cut off. Actual fabrication techniques may combine the function of several of the components shown into a single device. See Fig. 2.

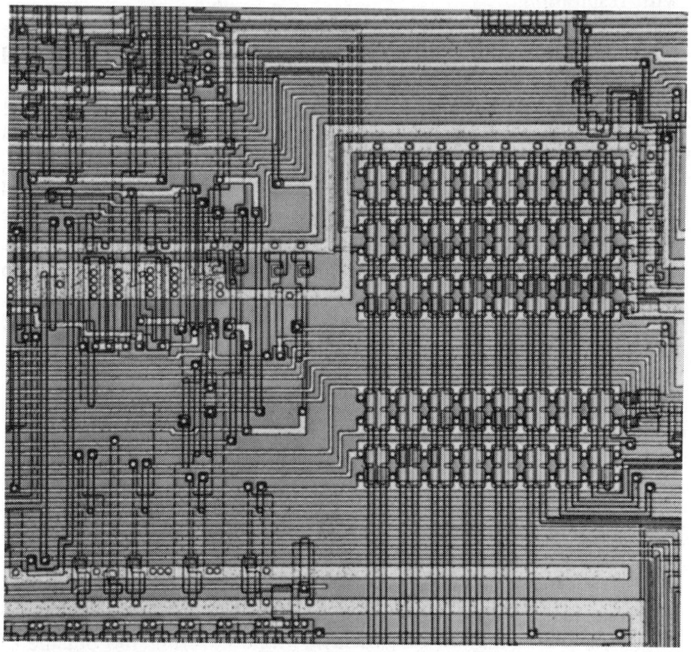

Fig. 2. Photomicrograph of an integrated circuit. Magnification, 100×. (*Polaroid Polacolor ER 4 × 5 Land Film Type 59.*)

See also **AND (Circuit)**; **NAND (Circuit)**; **NOT (Circuit)**; **OR (Circuit)**.

Thomas J. Harrison, International Business Machines Corporation, Boca Raton, Florida.

DIODE (Tunnel). Tunnel Diode.

DIOPHANTINE EQUATION. An equation containing two or more variables and, in general, satisfied by an infinite number of values for each of the unknowns is called indeterminate or Diophantine, after Diophantus of Alexandria (about 320 A.D.). More generally, Diophantine analysis is the study of single equations or systems of them in two or more unknowns, the solutions being required in integers or rational numbers.

In the mid-1600s, a French mathematician (Fermat) noted that it is impossible to satisfy the equation $x^n + y^n = z^n$ when $n < 2$ and the substituted numbers are positive integers. Fermat's proof was not disclosed and, to date, a proof or disproof has not been developed. In recent years Lehmer and Vandiver (University of Texas) have used a computer to test the Fermat theorem in an attempt to prove or disprove it. No single exceptions have been found. In their testing, it was not necessary to test every conceivable combination of numbers, but only to substitute all prime numbers for n. Additional shortcuts were used, that is, n must not divide any of a set of Bernoulli numbers, since otherwise it cannot satisfy the equation. Bernoulli numbers are irregular. For example, the first is 1/6, the third 1/30, the eleventh 691/2,730; the thirteenth 7/6; the seventeenth 43,867/798; and the nineteenth 1,222,277/2,310. It is obvious that numbers further in the series are extremely large. The Lehmer-Vandiver tests included all prime n's up to 4,000, where the Bernoulli numbers are on the order of 10,000 digits long. A computer requires about an hour to test each n under these circumstances. Thus, although computer calculations have not found any exceptions to Fermat's theorem, it remains formally unproved. Needless to say, however, confidence in its truth is high.

See also **Number Theory.**

DIOPSIDE. The mineral diopside is a monoclinic pyroxene corresponding to the chemical formula, $CaMgSi_2O_6$, calcium magnesium silicate. Its crystals, like other pyroxenes, tend to be short stout prisms of square or octagonal cross-section. Compact, granular, lamellar and fibrous varieties are often found. The prismatic cleavage is characteristic, cleavage planes intersecting at angles of 87° and 93°. A basal parting is often noted, but should not be confused with the cleavage. The hardness of diopside is 5–6; specific gravity, 3.2–3.3; uneven fracture tending toward conchoidal; luster, vitreous to dull; sometimes pearly on the base; color, light or dark greens, but may be colorless, gray, yellow or blue, although the latter color is rare.

Diopside is a primary mineral in rocks like diorites, gabbros and the like, but is also found in schists, and, as the result of contact metamorphism, in such rocks as crystalline limestones and dolomites. Diopside is found in association with vesuvianite, garnet, spinel, scapolite, tremolite, tourmaline and similar minerals. It is a rather widespread mineral, important localities being found in the following European countries: Finland, Sweden, Switzerland, Italy; it is found in eastern Siberia near Lake Baikal. In Canada diopside localities are in Lanark and Hastings Counties, Province of Ontario, and in the United States in Lewis and St. Lawrence Counties, New York, and in Maine.

See also **Pyroxene**; and terms listed under **Mineralogy.**

Elmer B. Rowley, Union College, Schenectady, New York.

DIOPTASE. The mineral dioptase is a rather rare copper silicate corresponding to the formula, $CuSiO_2(OH)_2$, occurring in prismatic crystals of the hexagonal system, tri-rhombohedral in form. It may be found in crystalline aggregates or simply massive. Dioptase displays a conchoidal to uneven fracture; hardness, 5; specific gravity, 3.28–3.35; luster, vitreous; color, a beautiful emerald green. It has been found in the U.S.S.R., Congo, Central African Republic, South West Africa, Chile, and in the United States in Arizona. The name is derived from the Greek words meaning through and to see, because cleavage was observed by looking through the crystals.

DIOPTRIC SYSTEM. If an optical system is convergent, it is called dioptric if both focal lengths are positive. If an optical system is divergent, it is called dioptric if both focal lengths are negative. See **Mirrors and Lenses.**

DIORITE. Diorite is a deep-seated igneous rock composed dominantly of sodiaplagioclase feldspar with hornblende, biotite, and (or) augite. Orthoclase may be present in small amounts, also quartz. Any considerable proportion of the latter mineral produces a quartzdiorite. With increasing amount of orthoclase, we have granodiorite, which is generally understood to be a rock intermediate in character between quartz-diorite and granite. If quartz is absent and there are essentially equal amounts of orthoclase and plagioclase the rock is then known as a monzonite from the type locality, Monzoni, in the Tyrol. There are quartz monzonites and, where the deficiency of silica is great enough, nephelite monzonites. Rocks of the latter sort have been reported from the Malagasy Republic. A variety of quartz, diorite, containing both hornblende and biotite, is called tonalite from the Tonale Alps, although the rock found there is more nearly a granodiorite. The word diorite is derived from the Greek meaning distinctive or defining, in contrast to the deceptive dolerites. The diorites are of widespread occurrence.

DIOTOCARDIA. An order of mollusks, *Aspidobranchiata*, mostly marine species.

DIOXIN. Also called dimethoxane (2,3,7,8-tetrachlorodibenzo-*p*-dioxin; 2,6-dimethyl-*m*-dioxan-4-yl-acetate; TCDD), this is a toxic chlorinated hydrocarbon best known as an impurity in the herbicide 2,4,5-T. Some authorities estimate that the compound has a half-life in soil of about one year. Sax ("Dangerous Properties of Industrial Materials," Van Nostrand Reinhold, New York, 1979) reports this as "one of the most toxic materials known. Damaging to guinea pigs at 0.6 part per billion." Experimental laboratory tests show the compound to be a carcinogen. The possible hazard to the atmosphere of dioxin has been a subject of debate among authorities since the late 1970s. For an interesting exchange of viewpoints of this topic, reference is made to letters by Westing and Blair, *Science*, **206**, 1135–1136 (December 7, 1979). For an earlier review, see "Dioxin: Toxicological and Chemical Aspects," Wiley, New York, 1978, edited by F. Cattabeni, A. Cavallaro, and G. Galli. See also "Trace Chemistries of Fire: A Source of Chlorinated Dioxins," by R. R. Bumb, et al., *Science*, **210**, 385–390 (1980).

DIP EQUATOR. Aclinic Line.

DIPHOSPHORYDINE NUCLEOTIDE (DPN). Carbohydrates.

DIPHTHERIA. An acute infection caused by *Corynebacterium diphtheriae* (thin gram-positive rods), first isolated by Loeffler in 1884. The disease is characterized in its early stage by an inflammatory lesion, usually a membranous pharyngitis. Pharyngeal diphtheria is the most frequently occurring form. In addition, the larynx, trachea, nares and middle ear may be involved. A potent exotoxin elaborated by the bacterium produces serious neurologic (myelin degeneration; motor nerve damage), cardiac (myocarditis), and renal (tubulary necrosis) effects. The toxin inhibits protein synthesis. It is the toxin that participates in the formation of the diphtheritic membrane, unique to the disease and a hallmark for diagnosis. The membrane is a gray-to-black, leathery, adherent formation made up of necrotic cellular debris, leukocytes, and bacteria and can cause blockage of the upper airway when it extends into the larynx. Diphtheria infection also takes a cutaneous form, with involvement of the skin as well as of the conjuctiva.

Prior to widespread immunization, particularly of young school children, diphtheria was a major disease and important cause of death. Traditionally, the highest incidence occurred in children under 10 years of age. In more recent years, the disease has been seen in more persons of age 15 or older. The infection is easily spread by way of the respiratory route, can be carried by persons without the disease, and has been a mandatory notifiable disease in the United States and a number of other countries for many years. Among children, the cutaneous form is considered by some authorities as even more contagious than the respiratory form. In 1920, prior to immunization programs, about 200 cases/100,000 population, with 15 deaths/100,000, were reported. The incidence has decreased steadily since then, with about 0.3 cases/100,000 (0.0015 death/100,000) as of the early-1980s. Reported cases in the United States in recent years number less than 100/year, but the total, including unreported cases, may range between 200 and 400. The disease in present times tends to occur in outbreaks, particularly in crowded urban areas, among inadequately immunized children, and unimmunized transient farm workers.

The Schick test is a skin test for determining susceptibility to diphtheria. In one arm, there is an intradermal injection of diphtheria toxin and of purified diphtheria fluid toxoid in the other arm. This test is of value in assessing the level of immunity in contacts of a patient, or in a community where an outbreak has developed. It has been estimated that a minimum of 70–80% of school and pre-school children in urban areas must be immunized in order to prevent community-wide outbreaks of diphtheria. For several years, the common practice has been early immunization of young children, and the administration of diphtheria-tetanus toxoid as a booster every 10 years in older children and adults.

The signs and symptoms of diphtheria appear after an incubation period of 2 to 4 days on the average, but symptoms can occur in 1 day and up to 7 days. Sore throat (in the pharygeal form) and mild fever (100–100.9°F; 37.8–38.3°C) commonly usher in the disease. The previously mentioned membrane may appear early and spread rapidly and may be accomplished by swelling of the neck, resulting in the so-called 'bull neck' in appearance. In the laryngeal form, cough and noisy, difficult respiration occur. The severity of systemic symptoms may be delayed and depend upon the amount of toxin absorbed. In severe pharyngeal cases, prostration will be marked. Nasal diphtheria is localized to the anterior nodes and usually follows a benign course. Cutaneous diphtheria may occur as a secondary infection of a pre-existing wound; or as a superinfection of pre-existing skin lesions (impetigo, infected insect bites, ecthyma, and eczema); or as a primary cutaneous infection, usually commencing as a tender pustule on a lower extremity. The latter condition (sometimes called 'jungle sore') occurs mostly in the tropics.

The exotoxin produces myocarditis in 10–15% of cases and is the usual cause of death in fatal cases. In about 10% of cases, peripheral neuritis occurs within 2 to 6 weeks after onset of the disease. The most dangerous aspect is involvement of the motor nerves associated with respiration.

Therapy may involve administration of equine antiserum. Where there is sensitivity to this serum, desensitization procedures should be attempted.

Diphtheroids. A number of saprophytic organisms (organisms that live on dead organic matter) in the genus *Corynbacterium* and known as diphtheroids frequently occur on the skin, on mucous membranes, and in various environments and, for many years, were not regarded seriously. It has been found that such organisms can cause serious infection, including life-threatening situations, as may be the case in connection with bacterial endocarditis (after cardiac surgery) and in other situations involving surgery, trauma, and immunosuppression, such as reconstructive hip surgery. These conditions usually respond to antibiotic therapy.

DIPHTHEROIDS. Diphtheria.

DIPHYLLOBOTHRIASIS. Foodborne Diseases.

DIPLEXER. Television.

DIPLOBLASTIC. Derived from two embryonic germ layers. The first step in the formation of tissues in the developing multicellular animal is the formation of two layers, one covering the outside of the body and the other lining a cavity within it. These are the ectoderm and endoderm, respectively. The bodies of sponges, coelenterates, and possibly ctenophores develop by the further differentiation of these two layers alone.

DIPLOCOCCUS. **Bacteria.**

DIPLODOCUS. **Fossil Reptiles.**

DIPLOID. The double number of genes or chromosomes (2*N*). In the metazoan animals the somatic or body cells all contain the diploid number of genes and chromosomes—there are two of each kind. For each chromosome of a particular length and shape there is a mate of the same length and shape. The genes on these like chromosomes are also alike in the trait which they influence, but they may differ slightly in the way they effect this trait. The human, for instance has a diploid chromosome number of 46 and the diploid gene number is estimated at about 40,000. The reproductive cells receive only the haploid gene and chromosome number as a result of meiosis. The haploid number is one-half of the diploid (2*N*).

In higher plants most of the tissue again is diploid, but there is a small haploid gametophyte. In some of the lower plants the diploid tissue is very small. See **Alternation of Generations;** and **Cell (Biology).**

DIPLOPIA. **Vision and the Eye.**

DIPLOPODA. The millipedes. A class of the phylum *Arthropoda.* Millipedes are worm-like animals with a head and segmented body. Each segment bears two pairs of legs, excepting the first and last few. The head bears a pair of antennae and in some species a pair of eyes. Most millipedes have a cylindrical body.

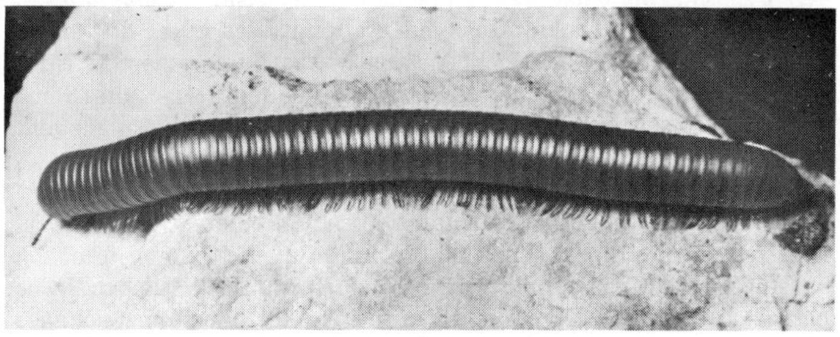

Millipede (*Diplopoda*). (*A. M. Winchester.*)

These animals live in moist places, usually among rubbish on the surface of the ground, and eat decaying organic matter or plant tissues. Some attack roots and are therefore of economic importance.

The class is divided into two orders, *Pselaphognatha* and *Chilognatha.* All common species belong to the latter.

DIP NEEDLE. An instrument consisting essentially of a magnetic needle poised to swing on a horizontal pivot and thus indicate the "dip" or inclination of the earth's magnetic field. Also termed magnetic inclinometer. The zero diameter of the vertical graduated circle must be carefully leveled and adjusted to the magnetic meridian. In order to correct for errors of level, balance, magnetization, and eccentricity, the circle should be reversed north to south, the needle axis should be reversed in its bearings, the magnetization should be reversed, and for each of these positions, both ends of the needle should be read on the circle. A complete observation is thus the mean of sixteen circle readings.

DIPOLE. Two equal electric charges of opposite sign, separated by a small distance, form an *electric dipole.* A circulating current loop, of dimensions small compared to the distance at which it is being observed, can be considered as a *magnetic dipole.* More generally, any given distribution of electric or magnetic charges about a point can be described by means of a series of multipole terms (see **Quadrupole**). The electrostatic potential of the dipole term at a distance r from the point is proportional to $1/r^2$.

DIPOLE (Antenna). **Antenna.**

DIPOLE MOMENT. In the simplest case, let two electric charges $+q$ and $-q$ be separated by the distance **d.** Then the permanent electric dipole moment is the vector $\mathbf{p} = q\mathbf{d}$. More generally, if discrete charges q_i are located at points x_i, y_i, z_i the magnitude of dipole moment is given by $p_\alpha = \Sigma q_i \alpha_i$, $\alpha = x, y, z$. If the charge distribution is continuous, the summations are replaced by integrals. An induced dipole moment can be produced by an electric or magnetic field (see **Magnetism**). Atomic or molecular dipole moments, permanent or induced, are of considerable value in the study of atomic or molecular structure. The magnitude of such moments is usually reported in Debye units. The magnetic dipole moment produced by a current i flowing in a loop area A has magnitude $m = iA$. It is a vector with directional normal to the plane of the loop and sense taken as the direction of progression of a right-handed screw rotating with the current.

DIPOLE RADIATION. A term applied to radiation which occurs as a result of the variation with time of a dipole moment. Dipole radiation may be either electric or magnetic, according to the nature of the dipoles causing it.

DIPOTASSIUM TARTRATE CRYSTAL (DKT). **Piezoelectric Effect.**

DIPROTON. The name given to a combination of two protons in a singular nuclear system. See **Dineutron.**

DIPTERA. Flies, mosquitoes, midges, gnats and other insects. An order characterized by sucking and sometimes piercing mouths and the presence of a single pair of wings. The hind wings are represented by the halteres, slender clubbed appendages, often inconspicuous. A few species lack wings. The metamorphosis is complete and the larvae of many species are known as maggots.

This is one of the largest orders of insects, with about 50,000 described species, and in variety of adaptations it is exceeded by no other. Some species suck the juices of plants, some eat the tissues during larval life, some visit flowers for nectar, some suck blood, some are parasitic inside or outside the bodies of warm-blooded animals, and many are parasitic on other insects or are predacious. Many are scavengers, living on decaying organic matter or on the wastes of animals.

Species of economic importance include some of the plant-feeders, such as the Hessian fly, the blood-sucking horseflies and mosquitoes, and the parasitic bot flies.

Two major suborders of Diptera include *Orthorrhapha* (straight-seamed flies); and *Cyclorrhapha* (circular-seamed flies).

The main families of *Diptera* include:

Anthomyiidae	Root maggots
Asilidae	Robber flies
Blepharoceridae	New-winged midges
Bombyliidae	Bee flies
Braulidae	Bee lice
Cecidomyiidae (also called *Itonididae*)	Gall gnats
Chironomidae	Midges, gnats, and punkies
Chloropidae (also called *Oscinidae*)	Fruit flies, grass stem maggots, eye gnats
Culicidae	Mosquitoes
Dolichopodidae	Long-legged flies
Drosophilidae	Pomace or fruit flies
Ephydridae	Shore flies
Hippoboscidae	Louselike flies
Muscidae (includes *Calliphorinae*)	House flies
Mycetophilidae	Fungus gnats
Oestridae (includes *Gastrophilidae* and *Cuterebridae*)	Bot flies
Ortalidae	Ortalid flies
Phoridae	Humpbacked flies
Psychodidae	Moth flies and sand flies
Rhagionidae (also called *Leptidae*)	Snipe flies

Sarcophagidae	Flesh flies
Simuliidae	Black flies, Buffalo gnats
Stratiomyiidae	Soldier flies
Syrphidae	Flower flies, hover flies, syrphids
Tabanidae	Horseflies, deerflies
Tachinidae	Tachinid flies
Tipulidae	Crane flies (daddy-longlegs)
Trypetidae	Fruit flies

DIQUAT. Pyridine and Derivatives.

DIRAC DELTA FUNCTION. The improper function $\delta(x - x_0)$, so defined that

$$\int f(x)\, \delta(x - x_0)\, dx = f(x_0)$$

for any x. It may be approximated by

$$\delta x = \lim_{h\to\infty} \begin{cases} 0 & ; -h/2 > x \\ 1/h; & -h/2 > x < h/2 \\ 0 & ; \quad x > h/2 \end{cases}$$

The function

$$u(x) = \int_{-\infty}^{\infty} \delta(x)\, dx = \begin{cases} 0; & x < 0 \\ 1; & x > 0 \end{cases}$$

is known as the unit step function. The Dirac delta function is also known simply as the delta function.

DIRAC "HOLE" THEORY. Positron.

DIRAC PARTICLE. Field Theory.

DIRECT ACTING CONTROLLER. A controller in which the absolute value of the output signal increases as the absolute value of the input (measured variable) increases. Compare with **Reverse Acting Controller.**

DIRECT-COUPLED AMPLIFIER. As used in digital data acquisition and instrumentation systems, the term *DC amplifier* has two connotations: (1) a *direct-coupled* amplifier, that is, a low-resistance DC connection is used between each stage and the succeeding stage in the amplifier, and (2) an amplifier with a frequency response that extends to zero frequency (dc). It should be pointed out that although all DC amplifiers (direct-coupled) have a frequency response that extends to dc, not all amplifiers with this type of frequency response are direct-coupled. A carrier amplifier, for example, has a response extending to dc, but it is not direct-coupled.

One type of direct-coupled differential amplifier is shown in the accompanying diagram. Only the first stage of the amplifier is shown in detail because the most important characteristics of an amplifier usually are determined by the first stage of the amplifier. The two or three remaining stages are shown as a single symbol. These stages provide high gain, typically greater than 1,000, and also the output-drive capability of the amplifier. Usually, the conversion from differential input to single-ended output is also effected in these stages.

As shown, the input stage is comprised of a pair of transistors, the bases of which are biased by current sources. A current source also determines the emitter current of the transistor pair. Feedback resistor R_{f1} and emitter resistor R_{e1} determine the gain of the stage. There is a direct coupling of the differential-output signal generated at the collectors of the first-stage transistors and the bases of the transistors in the second stage.

Usually, the gain of the first stage is on the order of 100. Thus, the first stage essentially determines the drift and noise characteristics of the overall amplifier. Because of the manner in which the first-stage is configured, the drift characteristics are determined by the difference in the drifts of the two input transistors. The base-emitter voltage drop is the major contributor to zero-offset temperature coefficient. The overall amplifier drift can be minimized (a few microvolts/°C) by using matched transistors in the differential configuration. Where additional temperature compensation is used in the first stage, a drift of less than 1 μV/°C can be achieved. Noise characteristics are mainly related to the intrinsic noise of the input transistors. Some control over this can be affected by selection of the biasing point. The usual procedure is to select low-noise transistors.

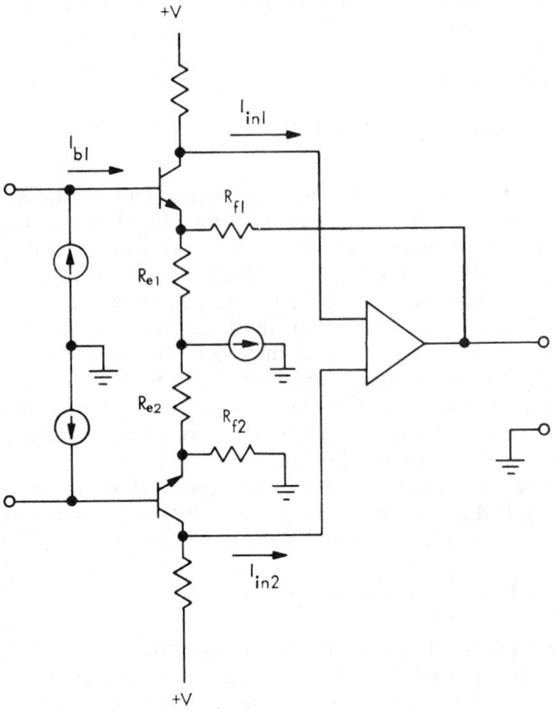

Direct-coupled differential amplifier.

The design of the first stage also affect the common-mode characteristics. Where possible, the configuration should be made symmetrical with respect to ground. Thus, the inclusion of the dummy feedback resistor R_{f2}. Also, the input base-bias sources must be balanced. The balance between the RC constants of the base-to-collector capacitance and the collector resistance determine high-frequency common-mode characteristics. Where a design may be critical, an adjustment of one of these parameters may be required. The common-mode performance of the amplifier also may be improved by common-mode feedback between the first-stage emitter and a common-mode point in a subsequent stage. Feedback is frequently applied to the ground side of the emitter-bias source.

The type of amplifier just described schematically is used extensively in digital-data acquisition and instrumentation systems. Direct-coupled design permits wide bandwidth, typically 0 Hz to frequencies in excess of 50 kHz and at relatively high gains. As compared with some other designs of differential amplifiers, the dc amplifier is relatively inexpensive. For some applications, a disadvantage is the relatively low common-mode voltage capability. The common-mode voltage essentially is limited to 10 to 20 V because of the breakdown characteristics of the input transistors.

See also terms listed under **Data Processing.**

Thomas J. Harrison, International Business Machines Corporation, Boca Raton, Florida.

DIRECT-CURRENT CIRCUITS. Unidirectional current is produced from batteries, from dynamo machinery equipped with commutators, or by means of rectifiers. Direct-current generators are built with their dc magnetic poles on the stator. Armature conductors on the rotor have ac voltages induced in them as they are rotated; the same principle of flux cutting as holds for ac generators. An automatic mechanical switching device, called a commutator, is placed on the shaft. It carries fixed brushes, and with its many insulated copper bars connected to the armature coils, it inverts every other alternation of the voltage to give unidirectional, or dc, voltage at the two armature terminals. It is the commutator that requires the rotor to be the armature so that coils and their switching arrangement always move exactly together. Direct-current generators are generally limited to several thousand kilowatts and their application lies mainly in industrial plants. The disadvantage of dc over the years has been the difficulty of transforming it from low voltage to the high voltage desired for long-distance transmission of electrical power. Difficulties of commutation have prevented generation at high voltages. Silicon-controlled rectifiers and other solid-state devices have been used. In the mid-1950s, development of the high-voltage mercury-arc valve improved the competitive position for dc transmission. Converters based on mechanical switches were tested in England and Sweden in the 1920s and 1930s. As early as 1889, the Swiss engineer Thury developed a system consisting of dc generators and motors connected in series on the dc side. There were installations of this system in Europe over the period of 1890 to 1937, at which time a conversion to mercury-arc valves was made. The fact that dc lines and cables are less expensive than those of 3-phase ac transmission offers a major incentive for continuing to find effective ways for solving dc transmission problems.

Characteristics of some of the more recent high-voltage dc systems are described in the entry on **Electric Power Transmission.**

In a general network composed of resistors and unidirectional electromotive forces, the determination of the currents that flow can be effected by the straightforward application of Kirchhoff's Laws of Networks. Many problems arising in practice either reduce to, or may be simplified materially by reducing portions of a more general network to, combinations of resistors in series or resistors in parallel. As an illustration of the application of this principle, consider the two circuits shown in the accompanying diagram. Application of the Kirchhoff law concerning voltage rises and voltage drops to the circuit at the top yields the result

$$E = R_1I + R_2I + R_3I + R_4I$$

This expression can be written

$$E = (R_1 + R_2 + R_3 + R_4)I$$

or

$$E = RI$$

The latter expression is Ohm's Law which relates the change in potential across the resistance to the current flowing through it. From the expression above, it is concluded that the resistance of a series combination of resistors is equal to the sum of the individual resistances. In

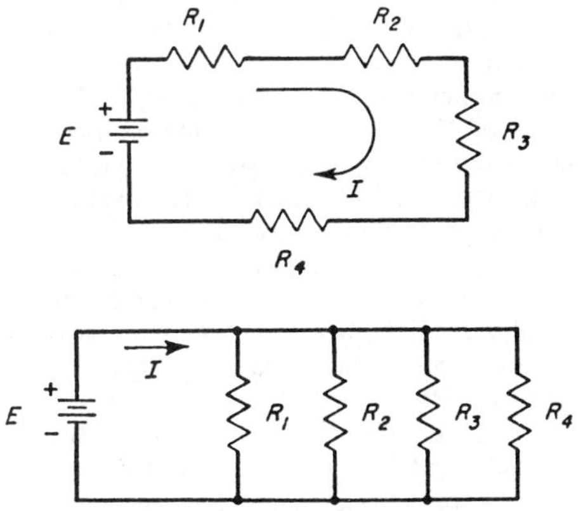

Simple direct-current circuits.

a similar manner, by the application of the Kirchoff current law to the circut at the bottom of the figure, one obtains

$$I = E\left(\frac{1}{R_1} + \frac{1}{R_2} + \frac{1}{R_3} + \frac{1}{R_4}\right)$$

and

$$I = \frac{E}{R}$$

where

$$\frac{1}{R} = \frac{1}{R_1} + \frac{1}{R_2} + \frac{1}{R_3} + \frac{1}{R_4}$$

The reciprocal of resistance is called conductance. By examination of the results just obtained for the parallel connected resistors, one concludes that the reciprocal of the equivalent resistance of a number of resistors connected in parallel is equal to the sum of the reciprocals of the individual resistance values. In terms of the conductance concept, the equivalent conductance of a group of parallel connected resistors is the sum of the individual conductances. The common symbol for conductance is *G*. Complex dc networks consisting of an interconnection of simple resistances and power source elements are normally analyzed by either the mesh or node method. The point at which two or more elements have a common connection is called a *node* and the element itself is called a *branch*. Networks that may be drawn on paper so that no lead must pass over another are referred to as *planar*. A closed path in a network is a *loop*. Loops that do not contain other loops are called *meshes*.

In the *mesh method of analysis*, the circuit is divided up into independent meshes and Kirchhoff's voltage law applied to each mesh. For a three mesh network, the three mesh equation may be written in its following form:

$$R_{11}I_1 - R_{12}I_2 - R_{13}I_3 = E_1$$
$$-R_{21}I_1 + R_{22}I_2 - R_{23}I_3 = E_2$$
$$-R_{31}I_1 - R_{32}I_2 + R_{33}I_3 = E_3$$

where R_{11}, R_{22} and R_{33} are the total resistances in meshes 1, 2, and 3, respectively. $R_{12} = R_{21}$ = total mutual resistance between mesh 1 and 2. $R_{13} = R_{31}$ = total mutual resistance between mesh 1 and 3. $R_{23} = R_{32}$ = total mutual resistance between mesh 2 and 3. E_1, E_2 and E_3 are the three source voltages acting in the three meshes. The equations are solved for I_1, I_2 and I_3, the three mesh currents.

In the *nodal method of analysis* the circuit is divided up into independent nodal pairs and Kirchhoff's current law is applied to all nodes but the reference node. For a three nodal-pair network, one gets

$$G_{11}E_1 - G_{12}E_2 - G_{13}E_3 = I_1$$
$$-G_{21}E_1 + G_{22}E_2 - G_{23}E_3 = I_2$$
$$-G_{31}E_1 - G_{32}E_2 + G_{33}E_3 = I_3$$

where G_{11}, G_{22} and G_{33} are the total conductances connected to nodes 1, 2, and 3, respectively. $G_{12} = G_{21}$ is the total conductance between node 1 and 2. $G_{13} = G_{31}$ is the total conductance between node 1 and 3. $G_{23} = G_{32}$ is the total conductance between node 2 and 3. I_1, I_2 and I_3 are the source currents at node 1, 2, and 3, respectively. Solve for E_1, E_2 and E_3, the three pair voltages between node 1, 2 and 3 and reference node, respectively. The electrical power flowing in a dc circuit is found by multiplying the voltage by the current, the unit of power being watts. Heat which is generated by electrical current flowing through a resistance of *R* ohms for *T* seconds is:

$$\text{Heat} = I^2RT \text{ watt-seconds}$$

The watt-second is a unit of electrical energy so small that 1,055 watt-seconds are required to equal one Btu. See **Electric Circuits; Kirchhoff Laws of Networks.**

DIRECTED GRAPH. Graph (Mathematics).

DIRECT-FIRING SYSTEM. Burner.

DIRECTIONAL ANTENNA. Antenna.

DIRECTION COSINE. Let a set of rectangular coordinate axes in space be chosen and let *L* be any line in space. Through the origin of the coordinate system draw another line *L'*, parallel to the given line *L*. Let α, β, γ be the angles which *L'* makes with the *X*-, *Y*-, *Z*-axes, respectively. These angles are the direction angles of the given line and their cosines $\lambda = \cos\alpha$, $\mu = \cos\beta$, $v = \cos\gamma$ are the direction cosines of the line. When two angles are given, the third can be found, except for sign, by the Pythagorean theorem, for

$$\lambda^2 + \mu^2 + \nu^2 = 1$$

If two rectangular coordinate systems with a common origin are given there will be nine direction cosines, one for each angle between the various pairs of coordinate axes. These cosines, not all of which are independent, can be written as the elements of an orthogonal matrix R and relations between them are conveniently found by matrix or vector methods. If $\mathbf{x}'$ is a vector in one of the coordinate systems and $\mathbf{x}$ is the same vector in the other system, the two vectors are related by the matrix equation $\mathbf{x}' = R\mathbf{x}$.

See als **Angle (Mathematics); Coordinate System;** and terms listed under **Mathematics.**

DIRECTION (Mathematics). The position of one point in space relative to another without reference to the distance between them. Direction may be either three-dimensional or two-dimensional. Direction is not an angle but is often indicated in terms of its angular difference from a reference direction. The *direction angles* of a line are the 3 angles it makes with the positive directions of the coordinate axes. Its direction cosines are the cosines of these angles. Any set of 3 numbers proportional to its direction cosines are called *direction numbers* for the line. A direction at a point on a surface in which the curvature of a normal section is a maximum or a minimum is called a *principal direction.*

DIRECTION OF PROPAGATION. Propagation (Direction of).

DIRECTIVITY (Microphone). Microphone.

DIRECT MEMORY ACCESS (Computer System). Cycle Steal (Computer System).

DIRECTORY-ASSISTANCE COMPUTER SYSTEM. Human Factors Engineering; Voice Recognition and Synthesis.

DIRECT PRODUCT (Of Subgroups). A group G is said to be the direct product of two of its normal subgroups A and B having only the identity in common if every element in G can be represented as the product of an element of A with an element of B, and similarly for the case of more than two subgroups. The group G is said to be completely reducible if it is the direct product of simple groups.

See also items listed under **Mathematics.**

DIRECTRIX. Parabola.

DIRECT STORAGE ACCESS (Computer System). Cycle Steal (Computer System).

DIRICHLET DISCONTINUOUS FACTOR. The even function $f(x)$ defined by

$$f(x) = \begin{cases} 1; & |x| < 1 \\ \tfrac{1}{2}; & |x| = 1 \\ 0; & |x| > 1 \end{cases}$$

It may be expressed as the Fourier integral

$$f(x) = \frac{2}{\pi}\int_0^\infty \cos ux\, du \int_0^1 \cos ut\, dt = \frac{2}{\pi}\int_0^\infty \frac{\sin u \cos ux}{u}\, du$$

which is generally known as the Dirichlet discontinuous factor.

DIRICHLET INTEGRAL. 1. One of the form

$$J_1 = \int_0^a f(x)\frac{\sin kx}{\sin x}\, dx$$

or

$$J_2 = \int_0^a f(x)\frac{\sin kx}{x}\, dx$$

It is the expression of the partial sum of the Fourier series development of $f(x)$. Convergence properties of the integral determine the convergence of the Fourier series. 2. Another integral, also known by the name of Dirichlet, occurs in statistics and statistical mechanics. It is

$$J = \int \cdots \iint x^{r-1}y^{s-1}z^{t-1} \cdots dx\, dy\, dz \ldots$$

in which the variables are given all possible values consistent with the condition that $(x/a)^2 + (y/b)^2 + (z/c)^2 + \cdots$ be not greater than unity. The general integral involves gamma functions. The special case for n variables with $a = b = c = \cdots = R$ gives

$$\left(\frac{R}{2}\right)^n \frac{\pi^{n/2}}{\Gamma(n/2+1)}$$

which is the volume of a hypersphere with radius R in n-dimensional space.

DIRIGIBLES AND AIRSHIPS. Dirigibles, as do other airships, derive their lifting power from aerostatic forces rather than the aerodynamic forces such as those that support the airplane. See **Aerodynamics;** and **Supersonic Aerodynamics.** The lift of an airship is one of buoyancy, and this is derived from the difference between the density of the atmosphere and that of the lifting gas contained by the airship. Hydrogen and helium have been used as lifting gases.

The size of an airship is determined by the volume of lifting gas required to lift the weight of the ship and the load. The lifting power of helium is approximately 58 pounds per 1,000 cubic feet (0.93 kilograms per cubic meter) under standard atmospheric conditions; of hydrogen, 64 pounds for the same volume and conditions (1.03 kilograms per cubic meter). The airship is equipped with a propulsive device and controls so that its attitude, speed, altitude, and course are under control of the pilot. This is in contrast with the free balloon, for which only altitude is controllable, velocity and direction depending essentially upon the wind. See **Balloon.**

The requirements of navigability and aerodynamic efficiency have caused an effective aerodynamic shape to be given the airship. This shape is usually circular in cross section, with a rounded nose and a tapering tail. The ratio of length to diameter, known as the fineness ratio, varies between 5 and 8. In the past, propulsion has been obtained from air propellers driven by internal combustion engines, which are either mounted in nacelles attached to the hull, or are within the hull, and drive propellers by means of shaft and gearing extending through the skin of the hull. The latter location has been considered safe only in helium-filled airships.

There are three principal classifications of airships, i.e., the nonrigid, semirigid, and rigid. The shape of the nonrigid airship is maintained by the pressure of the lifting gas. However, the gas expends and contracts with temperature, and in the contracted state the airship would be limp (the popular name blimp is derived from a wartime designation of this as the B-limp type) were it not for an air-filled ballonet which is built inside the main covering. As the lifting gas expands it forces air out of the ballonet, and when the lifting gas contracts, air scoops fill the ballonet by virtue of the velocity of the airship. The variable volume ballonet can also be used by the pilot to regulate the altitude of the airship by using it to compress or expand the lifting gas, so changing its density and lifting power. The ballonet is emptied for ascent, and refilled for descent. The cabin and engine installations are carried in the car or in nacelles, which are suspended below the envelope.

The rigid airship has a complete metal framework, and its shape is, accordingly, independent of the degree of inflation of the lifting gas cells that are contained within. Fixed equipment, like power plants and living quarters, is rigidly attached to the structural frame. The semirigid airship resembles the nonrigid in that its shape is maintained by gas pressure, but there is a structural keel extending longitudinally from the nose to the tail, with additional structural reinforcement at the nose and at the attachment of the control surfaces. In size, the blimps are the smallest, and the rigid dirigibles the largest, airships.

The framework of the dirigible is made of girders running longitudinally connected by parallel circumferential rings. The circumferential rings must be absolutely rigid, and if the construction is not such that their shape is self-sustaining, they must be braced diametrically. One to three of the longitudinal members at the bottom of the hull are made especially heavy and rigid to form a keel. The keel serves to strengthen and integrate the ship fore and aft, provide main walkways for access to the interior of the ship, and support heavy equipment, such as cabins, engines, and control surfaces.

One of the first dirigible balloons was designed by Henri Giffard. This craft was powered by a 3-horsepower steam engine and in mid-September 1852 flew from the Paris Hippodrome to Trappers, a distance of approximately 17 miles, at a speed of 5 miles (8 kilometers) per hour. Powered by a 3-horsepower Clement engine, the "No 9," a dirigible built by Alberto Santos-Dumont (Brazil) in Paris in the late 1890s, met with some success as a demonstration vehicle. The art of dirigible design was revolutionized by Count Ferdinand von Zeppelin (most of his designs were named after him—*zeppelins*) at the turn of the century. Zeppelin I was 426 feet (130 meters) long and equipped with four Mercedes Daimler 85-horsepower engines. Aluminum was used for the main body structure. Zeppelins were used militarily by the Germans up through World War I and for limited commercial service until 1936, at which time the hydrogen-filled dirigible Hindenburg LZ-129 burst into flames on May 6, 1936 when landing at Lakehurst, New Jersey, in which 25 of the 97 people aboard were killed. In retrospect, considering commercial aviation history, this was not an accident of great dimensions. However, coming at a time when the series of large American-built dirigibles, such as the Shenandoah, Akron, and Macon, had crashed because of weather-imposed accidents, the Hindenburg incident discouraged further dirigible developments. Some French dirigibles also had crashed during that prior period. In fairness to this mode of transportation, it should be observed that the ill-fated LZ-129 prior to its Lakehurst accident had completed ten trips to the United States and eight trips to South America, information that is not usually brought out when the disaster is mentioned. Also, in terms of speculation, it is intriguing to consider what the future of the dirigible might have been had the United States released helium for its use. Dirigible technology at that time was largely concentrated in Germany, of course, and the disaster occurred just a few years prior to the start of World War II—so the fate of dirigible technology was at least, in part, a victim of political considerations.

During the past decade, some interest in attempting to apply the greatly extended know-how gained during the past 40 or 50 years to newly-conceived dirigible or airship concepts has been evidenced, particularly in terms of low-cost mass transportation of goods; and also as a pleasant means of low-altitude, slow-speed touring by air. Meanwhile, the technology is holding on by virture of a very limited number of blimps in use for aerial photography, pleasure cruising, and advertising.

DIROFILARIASIS. **Heartworm Disease (Dirofilariasis).**

DISACCHARIDES. **Carbohydrates.**

DISCHARGE (Coefficient of). Water discharged from an orifice, weir, pipe, etc., theoretically has a velocity which is directly proportional to the square root of the head of water causing flow through the opening. Actually, however, contractions in the stream, surface roughness, and other causes result in the actual velocity being smaller than the theoretical. The theoretical velocity is identical with that of the velocity attained by a freely falling body. It is $\sqrt{2gh}$, wherein g represents the acceleration of gravity, and h the height of fall corresponding to the pressure head creating the discharge. The ratio of the actual velocity to the theoretical is the coefficient of discharge. The coefficient of discharge from circular orifices is affected by the diameter of the orifice, the head, the sharpness of the edge, the velocity of approach to the orifice, and other minor factors. See also **Orifice.**

DISCHARGE (Gaseous). If two electrodes have a gas at low pressure between them and a gradually increasing voltage is applied across them, a series of events takes place as the voltage is raised. First a very small current (microamperes) will flow as the ions and electrons are attracted to the electrodes. These charged particles are present because various radiations (cosmic, etc.) are always present (except in specially shielded enclosures) which ionize the gas molecules. As the voltage is raised, the current finally begins to increase rapidly because the electrons being attracted towards the positive electrode have gained sufficient energy to ionize atoms of the gas and thus generate more carriers of the current. Suddenly the current increases extremely rapidly, and at the same time the voltage across the tube drops. The value of the voltage at which this occurs is the breakdown voltage and the gas has broken into a self-maintaining discharge called a glow. If the current is not limited by circuit resistance it continues to increase, almost instantaneously, while the voltage drops to a low value and the discharge becomes an arc. If the circuit has insufficient resistance the current will reach an enormous value with probable damage to the tube and other circuit elements. The glow discharge is characterized by the ability to pass moderate currents at moderate values of voltage, while the arc will pass very large currents at low values of applied voltage.

Electrodeless Discharge. A current may be maintained in a rarefield gas without the introduction of electrodes into the gas. (1) A tube containing the gas may be placed between external metal plates having a rapidly alternating, high-potential difference. The tube then acts as the dielectric in a condenser, and the gas may become luminous with a discharge across the tube similar to that with internal electrodes. (2) The tube may be surrounded by a helix through which a high-frequency current is passing. In this case, the luminosity takes the form of a ring, inside the tube, coaxial with the turns of the helix. This is due to the alternating electric intensity induced by the current in the helix. The discharge has the characteristics of the positive column in an ordinary discharge tube, except that it forms a closed ring. Striations sometimes appear, in radial planes. If the oscillations in the helix are intense, the ring discharge is confined to the space immediately inside the tube wall; if less so, it extends farther inward. The discharge is facilitated by ultraviolet radiation traversing the gas. Volatile impurities in the tube, such as sulfur or phosphorus, impair the discharge and may stop it. With some gases there is a distinct phosphorescence, called the "after-glow," persisting for some seconds after the helix oscillations cease.

Dark discharge is an electrical discharge in a gas without the production of visible light.

DISCONFORMITY. **Unconformity.**

DISCONTINUITY. A point at which a function is not continuous or not defined. If the point $x = a$ is such that $f(x)$ approaches distinct finite limits as x approaches a from the left or from the right, then a is said to be a *jump discontinuity* of $f(x)$, or $f(x)$ is said to have a jump discontinuity at a. If $f(x)$ can be made continuous at a by giving a suitable definition to $f(a)$, then $f(x)$ is said to have a *removable discontinuity* at a.

DISCRETE CHANNEL. **Information Theory.**

DISCRETE INFORMATION SOURCE. **Information Theory.**

DISCRETE VARIATE. A discrete or discontinuous variable is one for which only a discontinuous set of values can occur. Examples are the number of votes cast in an election, or the number of beta particles emitted by a radioactive substance.

DISCRIMINANT. A relation between the coefficients of a polynomial which is useful in the study of roots and other properties of the function. It is an invariant of the function. In the case of an nth degree polynomial in one variable, the discriminant can be written as a determinant of order $(2n - 1)$ whose elements are simple functions of the coefficients of the polynomial. Vanishing of the discriminant is a condition for equality of its roots. Thus, for a quadratic equation, $ax^2 + bx + c = 0$, its two roots are real and equal if its discriminant, $D = b^2 - 4ac = 0$. See also **Biquadratic Equation; Conic Section; Cubic Equation;** and **Quadratic Equation.**

DISCRIMINANT FUNCTION. **Discriminatory Analysis.**

DISCRIMINATOR. There are two common types of equipment designated by the name discriminator: 1. A circuit used with counters, having the property that only pulses falling between two limits of amplitude (one of which may, however, be 0 or ∞) are recorded. 2. The term discriminator is used widely for a device in which variations in amplitude are derived in response to variations in frequency or phase. A device of this kind is the detector for a frequency modulation receiver. In frequency modulation the intelligence is impressed or modulated upon the carrier by causing variations in frequency. However, our loud-speakers operate upon a change of magnitude of the current through them so some method must be provided in the frequency modulation receiver to convert the frequency variations into amplitude variations. This is the function of the discriminator. It is a special type of balanced rectifier or detector which gives no output at the frequency for which it is tuned (the carrier frequency), but gives an output voltage whose magnitude and polarity are determined by the amount and direction of the frequency deviation of the input signal. As the received signal is varied back and forth in frequency by the modulation, the discriminator gives an output voltage whose magnitude follows these frequency changes. This output voltage may then be amplified and used to drive the loud-speaker. The discriminator is also used with appropriate circuits to tune automatically receivers having automatic frequency control. It is used in certain frequency modulation transmitter circuits as part of the frequency-stabilizing circuit.

DISCRIMINATORY ANALYSIS. Discriminatory analysis is the term used to describe the statistical methods which are brought to bear upon problems of allocating individuals to known classes. Given that an individual may have emanated from one of k populations, the major problem is to allocate it to the correct population with minimum error, usually on the basis of multiple measurements on the individual and a prior set of similar measurements on individuals whose origin is known.

A function of the observations used for the purpose is called a *discriminant function* or *discriminator.*

Discrimination in the exact sense arises when it is known a *priori* that different populations exist. If this is not known, and the problem is to see whether or not the data fall into distinct groups, the question is one of classification. The terms are often confused.

DISCUS. Cichlids.

DISINTEGRATION (Colloidal). Passage of a metal into colloidal solution when the metal is made an electrode under certain conditions.

DISINTEGRATION (Nuclear). A transformation or change involving atomic nuclei. If the disintegration is spontaneous, it is said to be radioactiv; if it results from a collision, it is said to be induced. The rate of disintegration of a radioactive nuclide is measured by a decay constant, which is the probability per unit time, λ, that it will undergo spontaneous transformation. The decay constant is the reciprocal of the mean life of the given system before undergoing transformation. Despite its literal meaning, the term nuclear disintegration refers also to radiative capture, inelastic scattering, beta transformations, and isomeric transitions. (See also **Nuclear Reactor.**) The *nuclear disintegration energy* is the energy evolved, or the negative of the energy absorbed, in a nuclear disintegration; symbol Q. It is equal to the energy equivalence of the sum of the masses of the reactants minus the sum of the masses of the products. (For each reactant or product which is a nucleus, the appropriate mass is that of the corresponding neutral atom.) If the disintegration energy is positive, the disintegration is exoergic; if it is negative, the disintegration is endoergic. Radioactive disintegrations have positive Q-values; nuclear reactions may have values of either sign. The *ground-state nuclear disintegration energy* is the energy evolved when all the reactant nuclei enter, and all the product nuclei end, in their ground states.

DISK DISORDERS. Bone.

DISK ELECTROSCOPE. Electrical Instruments.

DISKETTE (Computer). One form of disk storage utilizes a thin, flexible, plastic disk with one or two magnetic recording surfaces. This diskette, either 8 inches (~20 centimeters) or $5\frac{1}{4}$ inches (~13.3 centimeters) in diameter, is permanently housed in a lubricated protective paper envelope in which it rotates at speeds up to 360 rpm.

Originally conceived as the replacement for the keypunch, the first single-sided, single-density units had a capacity of 256K bytes, or the equivalent of 3200 80-column card images. Currently, a typical 8-inch (~20-centimeter) diskette drive has formatted storage capacity of 1.3M bytes (1.6M bytes unformatted) in a double-density, double-sided configuration, an average access time on the order of 200 msec, and a data-transfer rate of 62.5K bytes per sec.

Diskettes are suited to data-entry, word-processing, and "personal computer" applications, in which data rates are low, response time requirements are modest, and required storage capacities are fairly low.

Synonymous with flexible disk, floppy disk. Also see **Disk Storage (Computer); Storage (Computer);** and terms listed under **Data Processing.**

Thomas J. Harrison, IBM Corporation, Boca Raton, Florida.

DISK (Rupture). Rupture Disk.

DISK STORAGE (Computer). In this form of storage, a magnetic coating on a rotating disk is used. Data are recorded on the surface of the rotating disk by magnetizing the surface in accordance with the pattern of binary data and are read by detecting the magnetic flux. A typical disk organization is shown in the accompanying figure.

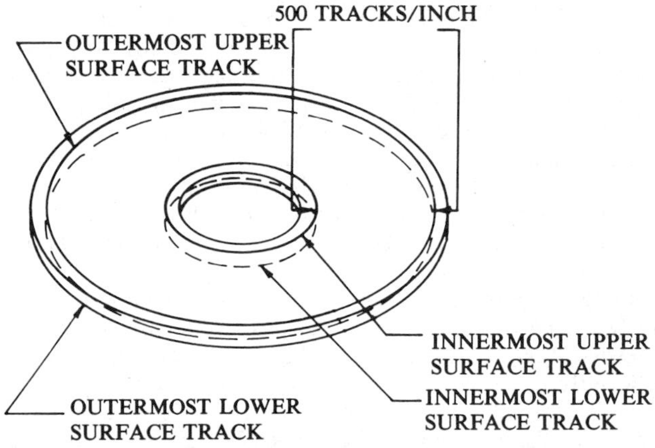

Disk cylinder concept.

Data are recorded on cylindrical tracks on the top and bottom surfaces of the disk. Track density is on the order of 500 tracks per radial inch and data density exceeds 8000 bits per inch (3150 per centimeter) on each track. The disk spins at over 3000 revolutions per minute.

A magnetic head for each of the two data surfaces performs the reading and writing functions. The write portion of the read/write head consists of a coil of wire which magnetizes the magnetic surface of the disk when a current flows through the coil. The read portion of the head consists of a similar coil of wire into which a voltage is induced as it intercepts the magnetic flux on the disk.

The magnetic heads are attached to an access arm which is capable of positioning the head over any one of the cylinders on the disk. In addition, fixed leads may be positioned over tracks on one surface. In order to insure reliable operation, the read/write head must be positioned very close the surface of the disk without actually touching it. One method of accomplishing this is to shape the head surface such that the head "flies" over the surface at a height of about 10 microinches (0.24 micrometer). The disk may also be fitted with nonmovable (fixed) heads positioned over the recorded tracks. This provides minimum access time to recorded information on the track. The resulting characteristics are similar to those of drum storage. See also **Drum Storage (Computer).**

Disk storage is used on a digital computer to store both data and programs. If the disk storage device has replaceable disk packs, the disk packs can be used as a medium of interchange of data with another computer. Although the access time on the disk is slower than either semiconductor or magnetic drum storage (unless fixed heads are used), the cost per bit of information is considerably less. The access arm or carriage is moved forward or backward over the surface of the disk under control of the program in the computer. The data are written on the track in a serial by bit format, i.e., the data are read from the disk a bit at a time and formed into a character or word for transmission to the computer by associated logic circuits.

See also **Storage (Computer)**; and terms listed under **Data Processing.**

Thomas J. Harrison, International Business Machines Corporation, Boca Raton, Florida.

DISK/TABLE FEEDER. Feeder (Volumetric).

DISLOCATION (Bone). Bone.

DISLOCATION (Crystal). Crystal.

DISORDER (Entropy of). Entropy.

DISPERSION. Colloid System.

DISPERSION MATRIX. Dispersion (Statistics).

DISORDER PRESSURE. The contribution to the pressure that arises through the existence of a contribution to the entropy of a liquid through molecular disorder, i.e., from the communal entropy.

DISPERSION (Radiation). The process (or resulting condition) of separating a radiation, a complex sound wave, etc., in accordance with some characteristic such as frequency, wavelength, or energy, into components. The process may be due to diffraction, refraction, or scattering. For example, a prism or diffraction grating disperses white light by sending light of different wavelengths in different directions. Quantitatively, a general measure of such dispersion is the derivative of the deviation with respect to that variable (wavelength, frequency, etc.). Dispersion of a medium is also expressed as the rate of change of refractive index with wavelength (or frequency, etc.).

The dispersion of light may be accomplished through refraction by a prism, diffraction by a grating, or other means. Refractive dispersion is due to the fact that the velocity of light in a given medium, and hence the refractive index, varies with the frequency. In any case, if the deviation Δ produced by the dispersion apparatus is expressible as a function of the wavelength λ, then the measure of the dispersion may be taken as

$$D = \frac{d\Delta}{d\lambda}$$

For example, for a plane diffraction grating at normal incidence, the deviation in the first order is given by $\sin \Delta = \lambda/s$, in which s is the grating space; from which it follows that the dispersion is $D = (s^2 - \lambda^2)^{-1/2}$.

Refractive dispersion is not so simply expressed. For a single refraction at angle of incidence i and with refractive index n, it may be shown that the dispersion is equal to

$$D = \frac{\sin i}{n\sqrt{n^2 - \sin^2 i}} \cdot \frac{dn}{d\lambda}$$

Various attempts have been made to express n as a function of λ, for example, by the Cauchy dispersion equation or by the Sellmeier equation (see **Anomalous Dispersion**).

Different media are commonly compared through some arbitrarily defined "dispersive power." The spectroscopist generally finds it convenient to express dispersion in terms of the number of angstroms corresponding to a distance of 1 millimeter on a photographic plate. Such a linear dispersion will vary with wavelength for a prism spectrograph but will be constant for a grating instrument.

DISPERSION (Sellmeier Equation). Sellmeier Equation.

DISPERSION (Statistics). A measure of dispersion is a quantity indicating the extent to which a set of observations scatter about some central value. Examples are the standard deviation or variance, the mean deviation, and the range. In advanced theory the measure almost always used is the standard deviation or its square, the variance. For a set of variables $x_1, x_2, \ldots, x_n$, the matrix whose element in the ith row and jth column is the covariance of x_i and x_j, is called the *dispersion matrix* or the *variance-covariance matrix.*

DISPERSIVE POWER. If n_1 and n_2 are refractive indexes for wavelengths λ_1 and λ_2 and n is the mean refractive index, the dispersive power is $d = (n_2 - n_1)/(n - 1)$. The refractive indexes n, n_1, n_2 are frequently taken as those of the Fraunhofer D, F, C lines, respectively. The reciprocal of the dispersive power is often called the Abbe number or ν-number.

DISPLACEMENT CURRENT. Consider a capacitor hidden in a "black box" with two terminals. Charging or discharging the capacitor requires a charge flow, or current, in the external connections. Viewed externally, let a current of positive charges flow into one terminal; the equal current of negative charges into the second terminal appears to be a positive current out of the second terminal. It is logically awkward to think of a current into one terminal and out of the other, that doesn't go through the box. Hence, to maintain continuity of current, a "displacement" current is postulated in the capacitor, equal to the "conduction" current in the external connections.

This concept of displacement current was invented by Maxwell to simplify the mathematical equations of electromagnetism; it led to the prediction of electromagnetic waves.

The displacement current is more than a convenient fiction, as is indicated by the fact that the Biot-Savant law holds when the circuit surrounds a displacement current as well as when it surrounds a conduction current. Part of the displacement current can be accounted for as the movement of bound charges within the dielectric, i.e., the creation or reorientation of dipoles. The balance, which is exhibited even in a vacuum, may be better understood as quantum electrodynamics is further developed.

Precisely, the displacement current through a surface is defined as the integral of the normal component of displacement current density over that surface. The displacement current density is the time derivative of the electric induction.

DISPLACEMENT MEASUREMENT. Position and Displacement Measurement.

DISPLACEMENT METERS. Specific Gravity.

DISPLACEMENT SERIES. Activity Series.

DISPOSAL (Radioactive Wastes). Nuclear Reactor.

DISPLAYS (Automotive). Automotive Electronics.

DISSIPATION. This term has three related uses in physics, as follows: 1. The interaction between matter and energy incident upon it, such that the portion of the energy used up in the interaction is no longer available for conversion into useful work. 2. A persistent loss of mechanical energy because of the presence of frictional or friction-like forces. 3. In free oscillatory motion, a persistent loss of mechanical energy due to presence of frictionlike resistance to motion which eventually exhausts the total energy of the system and causes it to come to rest. Such motion is said to be damped.

DISSIPATION FACTOR. The reciprocal of Q, the storage factor.

DISSOCIATIVE REACTION. Amnesia.

DISSONANCE. In acoustics, when two or more musical tones played simultaneously produce an unpleasant effect (due to beats) on the listener, the tones are said to be in dissonance. In optics, dissonance means the formation of maxima and minima by the superposition of two sets of interference fringes from light of two different wavelengths.

DISSYMMETRY OF LIFT. Helicopters and V/STOL Craft.

DISTAL. As used in anatomy, a position distant, or farthest away, from the source of any part.

DISTANCE MEASUREMENT (Astronomy). Cepheids; Period Luminosity Law.

DISTANCE-MEASURING EQUIPMENT. Navigation.

DISTANCE MODULUS. If the apparent magnitude and absolute magnitude of a star, cluster, or galaxy are known, one may calculate the distance to the object in parsecs by means of the relation

$$m - M = -5 \log D - 5$$

When plotting distances, the value $m - M$ (instead of calculating D) is often used, and hence, this quantity is referred to as the distance modulus. See also **Stellar Magnitude.**

DISTANCES (Stellar). Stellar Parallax.

DISTILLATES. Petroleum.

DISTILLATION. This is one of the most important and widely used of the chemical unit operations, both in the laboratory and on a large industrial scale, for separating the components of a liquid mixture. Distillation provides a means for partially vaporizing the mixture and separately recovering the vapor and residue. Consequently, the method is dependent upon the vapor pressures of the components making up the mixture. The vapor pressure of a pure substance is a constant, but varying with temperature. In distillation, the lighter, more volatile components of the original mixture (*distilland*) concentrate in the vapor when heat is applied. Advantage is taken of the fact that the ratios of the component substances in the vapor and liquid phases, except for special situations, are different. The less volatile components concentrate in the liquid residue (*bottoms*). The vapors evolved by distillation are condensed and are termed the *distillate.*

Distillation should be contrasted with evaporation wherein the vapor (frequently water) is not usually condensed (except where it is desired to conserve water). The principal product desired in evaporation is the solid material which remains in the evaporator vessel. Thus distillation would be used to separate two or more miscible liquids, such as glycol and water, whereas evaporation would be used to separate solid sodium chloride from brine.

The effectiveness of separation by distillation is largely determined by differences in the boiling points of the starting components. Where these are widely separated, as in the case of water (100°C) and ethylene glycol (197.6°C), the separation is relatively easy and redistillation is not required. Closer-boiling mixtures, such as the isomers of xylene, are much more difficult to separate by distillation. As the boiling points of the components approach each other, effective separation by distillation becomes more difficult.

Boiling-point diagrams or what also are termed boiling and condensation curves, usually experimentally determined, are useful guides in the design of distillation equipment. Fig. 1 is the phase diagram of a binary system forming a liquid and a vapor phase at constant pressure. Curve I is the boiling curve, which gives the coexistence temperature as a function of liquid composition; and curve II is the condensation curve, which gives the coexistence temperature as a function of the composition of the vapor phase. If the temperature is increased, vaporization begins when the boiling curve is crossed. Inversely, condensation begins when the temperature is decreased below the condensation curve. The use of boiling-point diagrams in still design will be discussed later.

Major Types of Distillation

Distillation can be a batch or a continuous operation. Batch distillation is frequently used in the laboratory for determining the chemical composition of mixed liquids, such as hydrocarbons. In the majority of industrial processes, distillation is continuous. In a batch operation, the charge material is boiled and vapors are removed continuously, condensed, and collected until such point is reached where there is the desired average composition. The separation is not sharp. This type of operation is also termed *simple* or *differential distillation.* In another approach, the mixture may be heated until a definite fraction of the liquid batch is vaporized, during which time the liquid and vapor are kept in intimate contact, i.e., with no vapors being removed. At the prescribed temperature and after the liquid and vapor have had opportunity to reach full equilibrium, the vapor is suddenly withdrawn and condensed. This approach is known as *equilibrium* or *flash distillation.* The method finds wider application in connection with multicomponent systems than with simple binary systems.

In the majority of industrial distillation systems, some of the distillate will be continuously returned to the distillation column. The returned condensate is contacted countercurrently with the rising vapors, thus bringing about an enrichment of the vapor in the more volatile components than otherwise would be accomplished with a single distillation and most often obviates the need for one or more redistillations to obtain the degree of purity desired. This approach is known as *rectification* or *fractional distillation.* The material returned is termed *reflux.* In most rectifying columns, the raw feed to the column is introduced at about the mid-level of the tower or column. The portion of the column above the point of feed is called the *rectifying section;* the portion below, the *stripping section.* Where the feed may be introduced at the top of the column, the entire column then is usually referred to as a *stripping column,* with no reflux used.

Dephlegmation is a means for increasing the efficiency of fractional distillation by forcing the vapors from the still to bubble through shallow layers of condensate in a column or dephelgmator whereby the amount of low-boiling component in the vapor is increased and a substantial portion of the higher-boiling components is retained in the condensate.

Steam distillation is a process whereby compounds which are sparingly soluble in water may be distilled by heating with water or by blowing steam through the mixture. Compounds of relatively high boiling point may be distilled at lower temperatures by this method and thus prevent degradation.

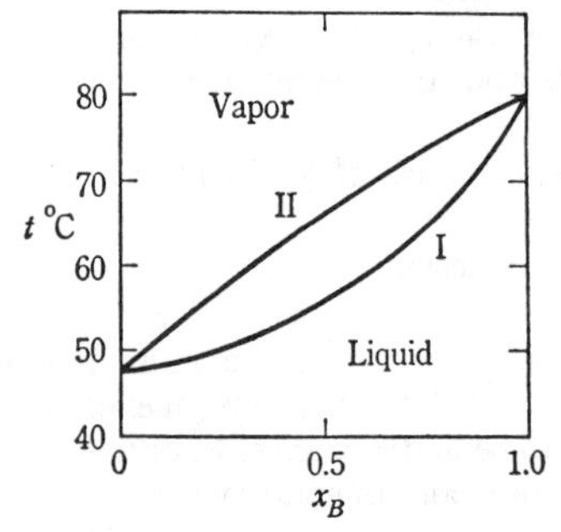

Fig. 1. Temperature-composition of a liquid-vapor system at constant pressure.

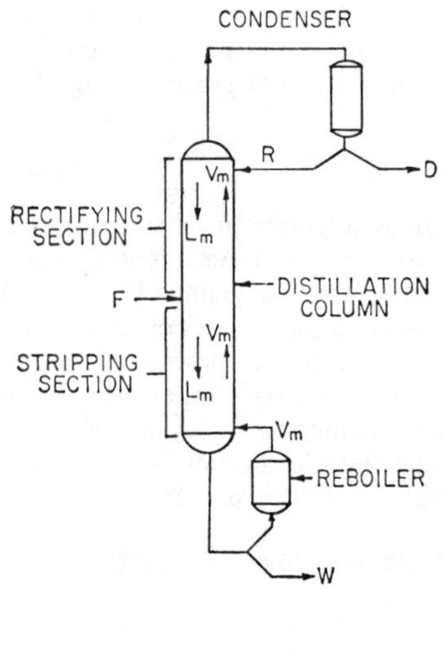

Fig. 2. Typical distillation column.

A representative fractional distillation column of which there are thousands in use in the process industries, notably in the petroleum and petrochemical industries, is shown in Fig. 2. The material balance of the column is:

$$F = W + D$$

$$Fx_F = Wx_F + Dx_D$$

where

F = feed rate, weight-moles/unit of time
W = bottom product, weight-moles/unit of time
D = distillate, weight-moles/unit of time
x_F = mole fraction of low boiler in feed
x_D = mole fraction of low boiler in distillate
x_W = mole fraction of low boiler in bottom product

In this balance, the assumption is made that the molar heat capacities and the latent heats of vaporization of all components are identical. It is also assumed that heat losses from the column and heats of mixing are negligible. With these assumptions, the upward vapor flow and the downward liquid flow in both the rectifying and stripping sections will be invariant within the sections. It is also assumed that accounting for the column heat balance is independent of the compositions of the product streams. Within these qualifications, the internal material balance is:

$$L_n = (1 + b)R$$

$$V_n = D + (1 + b)R$$

$$L_m = L_n + qF$$

$$V_m = L_m - W$$

$$x_W = f\left(\frac{L_m}{V_m}\right)$$

$$x_D = g\left(\frac{L_n}{V_n}\right)$$

where

V_n = vapor rate in rectifying section, weight-moles/unit of time
V_m = vapor rate in stripping section, weight-moles/unit of time
L_n = liquid rate in rectifying section, weight-moles/unit of time
D = distillate rate, weight-moles/unit of time
R = external reflux, weight-moles/unit of time
b = a numerical factor, depending upon the reflux enthalpy or temperature. (It should be noted that b is greater than zero whenever the reflux temperature is below that at the top of the column.)
q = a numerical factor, depending on the feed enthalpy whose value satisfies certain constraints:
$q < 0$, when feed temperature is below feed plate temperature.
$q = 1$, when feed temperature and composition are identical with those of feed plate.
$1 > q > 0$, when feed enters column partially vaporized.
$q = 0$, when feed is fully vaporized and is at saturated temperature.
$q < 0$, when feed is superheated vapor.
f and g are factors which account for several functional relationships which depend upon such column design criteria as the number of plates in column, location of control plates, location of feed plate, and temperature and other conditions specified for control plates.

The foregoing type of material balance is of large value in determining the best form of automatic control to apply to the column in order to maximize yields. This procedure is explained in detail in Reference 1 given at the end of this entry.

Distillation Calculations. In determining the type of packing or trays to be used in a fractionating column, the diameter, height, location of feed, location of reflux return, vapor and liquid rates, and all other specifications for a column to effect a given separation, equilibrium diagrams of the type shown in Fig. 3 are important. Prior to the availability of high-speed computers, distillation column designers depended heavily upon graphical solutions, notably McCabe-Thiele diagrams, named after the early developers of this concept. A typical diagram of this type for a simple binary distillation is shown in Fig. 4. Oversimplifying the method, first an equilibrium curve of the type of Fig. 3 is constructed. Next a 45° diagonal line is drawn. With knowledge of desired final composition, i.e., composition of the liquid received by the top plate from the condenser, X_p, calculate the intercept of an *operating line* with the Y-axis of the chart. This is indicated as #1 on Fig. 4. The term $X_p/O + 1$ is this Y-intercept of the operating line. O is the reflux ratio. Next, the intersection of the diagonal line with the ordinate X_p is marked. This is indicated as #2 on Fig. 4.

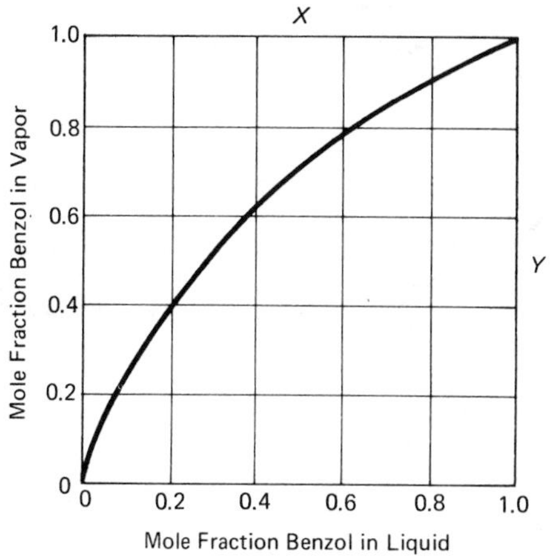

Fig. 3. Equilibrium diagram for the benzol-toluol system.

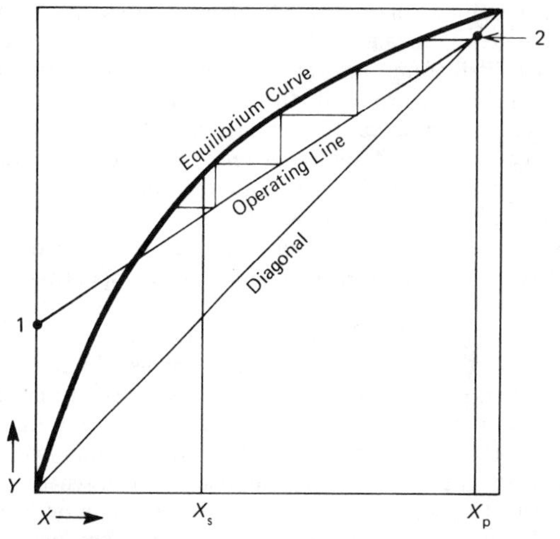

Fig. 4. Representative solution for theoretical plates in rectifying part of a distillation column.

The operating line is then drawn in by joining #1 and #2. Now, commencing at #2, rectangular steps between the operating line and the equilibrium curve are drawn in until it crosses the line $X = X_s$. X_s is the starting composition of the mixture. The number of horizontal steps counted (in this case, five) indicates the number of *theoretical plates* required to accomplish the separation desired. A theoretical plate may be defined as a plate wherein complete equilibrium is reached between the vapor rising from the plate and passing to the plate above—with the liquid leaving the plate and passing to the plate below. An actual plate, of course, will not perform with this efficiency. Thus, the designer, depending upon past experience with plates of certain designs, will include an appropriate margin in specifying the number of actual plates needed.

Azeotropic Systems. An azeotropic system is one wherein two or more components have a constant boiling point at a particular composition. Such mixtures cannot be separated by conventional distillation methods. If the constant boiling point is a minimum, the system is said to exhibit *negative azeotropy*; if it is a maximum, *positive azeotropy.* Consider a mixture of water and alcohol in the presence of the vapor. This system of two phases and two components is divariant. Now

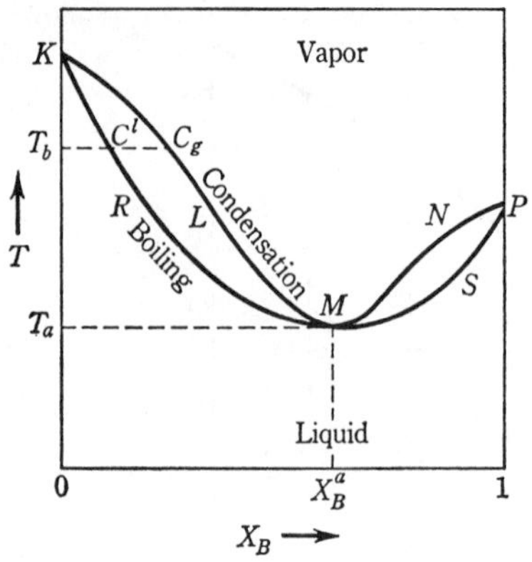

Fig. 5. Boiling-point diagram of an azeotropic system exhibiting negative azeotropy.

choose some fixed pressure and study the composition of the system at equilibrium as a function of temperature. The experimental results are shown schematically in Fig. 5.

The vapor curve *KLMNP* gives the composition of the vapor as a function of the temperature *T*, and the liquid curve *KRMSP* gives the composition of the liquid as a function of temperature. These two curves have a common point *M*, where the curves are tangent. Because of the special properties associated with systems in this state, the point *M* is called an *azeotropic point.* In an azeotropic system, one phase may be transformed to the other at constant temperature, pressure and composition without affecting the equilibrium state. This property justifies the name azeotropy, which means a system which boils unchanged.

Because a number of industrially important liquid mixtures are azeotropic systems, means had to be found whereby they may be separated by distillation. Two approaches, *extractive distillation* and *azeotropic distillation,* are used. In either case, a *separating agent* is added to the column so as to alter favorably the relative volatilities of the feed components. Usually, water or polar organic compounds are found most effective. They increase the liquid-phase non-ideality of one feed component more than another.

In extractive distillation, the agent (sometimes termed solvent) is significantly less volatile than the regular feed components. The agent will be added near the top of the column. The agent behaves as a *heavier-than-heavy* key component. It is also conveniently separated from the product streams. Because the agent usually must be added in fairly substantial amounts, this means that column diameters and heat loads are increased, while plate efficiencies are lowered.

In azeotropic distillation, an agent is selected that will form an azeotrope with one of the feed components. In essence, separation is accomplished between this "new" azeotrope (as an overhead product) and the other feed component as bottoms product. An agent will be selected preferably that will permit easy separation after distillation.

Instrumentation of Distillation Columns. There are numerous ways in which a distillation column can be controlled. Inasmuch as the usual object of distillation is to achieve one or more products of a specified composition, composition control of either the distillate or bottoms product, or both, is one logical approach. However, because of time lags in such a system and because of the difficulty of finding fully applicable composition sensors, other means of control are frequently used.

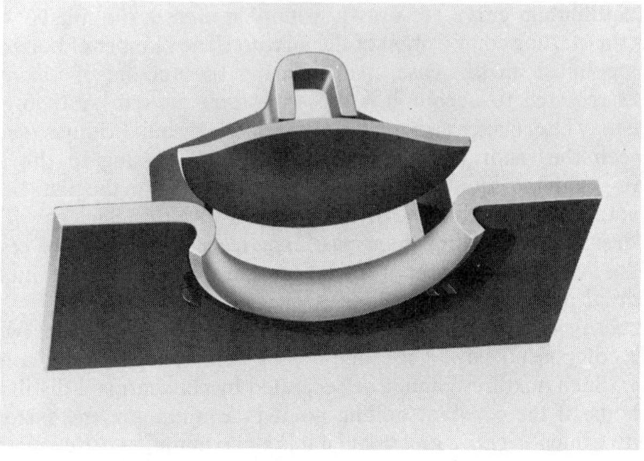

Fig. 6. Cross section of a form of bubble cap. View is from underneath plate or tray upon which cap is mounted.

Fig. 7. Distillation tray incorporating several hundred shallow, tunnel-type caps.

Trays and Packing. Trays with bubble caps or other suitable configurations for enhancing a maximum intermingling of rising vapors with falling liquid in a column are usually used where efficiency and close separations are major considerations. Packed columns, filled with ceramic shapes of various types, such as Berl saddles and Raschig rings, are used primarily where cost and acid-resistance are factors. The sectional view of a form of bubble cap is shown in Fig. 6. A tunnel cap is much more shallow and replete with peripheral performations through which vapor and liquid flow. One column tray incorporating several hundred tunnel caps is shown in Fig. 7. In some very tight separations, several trays separated a few feet apart up the vertical length of the column may be required.

A petroleum crude atmospheric distillation column is illustrated in Fig. 8. This unit separates the crude into three major fractions which then are subjected to later separations: (1) a light straight-run fraction, consisting primarily of C_5 and C_6 hydrocarbons, but also containing any C_4 and lighter gaseous hydrocarbons dissolved in the crude; (2) a naphtha fraction having a nominal boiling range

Fig. 8. View from base upward of atmospheric crude distillation tower.

of 200–400°F (93–204°C); and (3) a light distillate with boiling range of 400–650°F (204–343°C).

See also **Molecular Distillation.**

References

Buckley, P. S.: "Distillation Column Design Using Multivariable Control," *Instrumentation Technology*, **25**, 9, 115–122 (1978); **25**, 10, 49–53 (1978).

Castellano, E. N., McCain, C. A., and F. W. Nobles: "Digital Control of a Distillation System," *Chem. Eng. Progress*, **74**, 4, 56–60 (1978).

Considine, D. M. (editor): "Chemical and Process Technology Encyclopedia," pp. 354–356, McGraw-Hill, New York, 1974.

Danziger, R.: "Distillation Columns with Vapor Recompression," *Chem. Eng. Progress*, **75**, 9, 58–64 (1979).

Hughart, C. L., and K. W. Kominek: "Designing Distillation Units for Controllability," *Instrumentation Technology*, **24**, 5, 71–75 (1977).

Latour, P. R.: "Composition Control of Distillation Columns," *Instrumentation Technology*, **25**, 7, 67–74 (1978).

Lubowicz, R. E., and P. Reich: "High Vacuum Distillation Design," *Chem. Eng. Progress*, **67**, 3, 59–63 (1971).

Perry, R. H., and C. H. Chilton (editors): "Chemical Engineers' Handbook," McGraw-Hill, New York, 1973.

Shinskey, F. G.: "Distillation Control for Productivity and Energy Conservation," McGraw-Hill, New York, 1977.

Wang, J. C.: "Computer Relative Gain Matrices for Better Distillation Control," *Instrumentation Technology*, **27**, 3, 40–44 (1980).

Watkins, R. N.: "Petroleum Refinery Distillation," 2nd edition, Gulf Publishing, Houston, Texas, 1979.

Weisenfelder, A. J., and R. E. Olson: "Solving Recycle Streams in Multicolumn Distillation," *Chem. Eng. Progress*, **76**, 1, 40–43 (1980).

DISTILLATION ANALYSIS. Fractional Distillation Analysis.

DISTILLATION (Crude Oil). Petroleum.

DISTILLATION CURVE (Fuels). Petroleum.

DISTILLATION (Desalination). Desalination.

DISTORTION (Acoustic). A change in waveform. Noise and certain desired changes in waveform, such as those resulting from modulation or detection, are not usually classed as distortion. No acoustic or electroacoustic communication system or receiving apparatus is free from distortion. By this is meant that the waveform of the acoustic output is not strictly similar to that of the acoustic input, so that the quality of the sound is somewhat altered. The human ear is no exception. Vibrations received in the cochlea, with the consequent sensations, do not accurately represent the periodic variations of air pressure in the auditory canal, and we hear subjectively sounds that have no external physical existence. The most important aspect of this auditory distortion is the fact that when two pure musical tones are sounded, which are not of too nearly the same pitch, many persons can distinctly hear other tones in addition to the two actual ones. These subjective sensations are called combination tones, from the fact that they correspond in pitch to tones having frequencies equal to the difference, to the sum, or to other simple combinations of the two actual frequencies.

In the following discussion, tone frequencies are stated in hertz (cycles per second). Thus 528 means 528 hertz.

By far the most conspicuous of these subjective tones is the difference tone. Thus if one plays on a piano the two notes e (333) and c (528), many listeners hear, in addition to these, a note in the key of g (195), in the octave below, which is equal to $528 - 333$. It was formerly supposed that the difference tone was the effect of beats, but Helmholtz detected sum tones and showed, further, that the difference tone sensation is too loud to be attributed to beats.

The cause of distortion in the case of the ear is believed to lie in the fact that the eardrum, or tympanic membrane, does not vibrate symmetrically. Mathematical analysis shows that if the tympanum moves farther inward than outward from its normal position during vibration, the waves passed on from it to the cochlea will contain other frequencies than those making up the original sound, and that in the case of a sound composed of two pure tones of frequencies n_1 and n_2, there will arise also the frequencies $n_1 - n_2$, $n_1 + n_2$, $2(n_1 - n_2)$, etc.

The existence of difference tones is of value in electroacoustic apparatus such as the telephone and amplifier systems, and also in the pipe organ, the deficiencies of which in the lower frequency range are partially offset by the low difference tones that the listener thinks he hears. There is evidence that the "striking note" of certain chime bells is really a difference tone and does not exist among the partials of the complex sounds of these bells. The occurrence of combination tones can hardly fail to be of importance also in the effect on the ear of chords in musical harmony.

DISTORTION (Electromagnetic). An undesired change in waveform. This kind of distortion is one of the major limiting factors in any communications system, being a measure of the failure of the system to reproduce exactly at the receiver the signal which was applied at the transmitter. The amount of this distortion which can be tolerated varies with the different types of communication, and both for electrical and for economic reasons a system is usually not made much better than necessary.

When a sinusoidal signal is passed through a communications system, in general the output signal differs in amplitude and phase from the input signal. For convenience in discussing different kinds of distortion, let the ratio of the output amplitude to input amplitude be designated G and the phase shift between output and input be denoted by θ.

Distortion may be divided into three types: frequency distortion, which is caused by the system and responding to all frequencies equally, that is, G varies with the frequency of the input signal; amplitude distortion, which is caused by the system not responding to amplitude of signal linearly, that is, G varies with the amplitude of the input signal; and finally, phase distortion, which is caused by different frequencies being shifted in phase by varying amounts, that is, θ varies with frequency in such a manner that relative phase relations among the various sinusoidal compbnents of a complex input signal are not maintained. The first two are important in sound communication while all three are important in television.

Frequency distortion is determined by the circuit's ability to respond to a wide frequency band. The average human ear will respond to a range of about 20 to 15,000 Hz, but all of this range is not necessary for most services so the systems are not designed to respond to such a wide range. For telephone service the range from about 250 to 2,800 Hz gives ample articulation and is the uaual band employed. A wider band is not necessary and would greatly increase the cost of apparatus and the difficulties from interference. Radio broadcast service requires a much wider band since it is designed for entertainment and must give a fair reproduction of the original music. Amplitude modulation systems are usually limited to an upper value of about 5,000 Hz to keep the side bands within the allowed channel and to avoid excessive noise pick-up which a wider band would introduce. Frequency modulation, being inherently more noise-free, uses a much wider band, going to the upper limit of audibility. When any system fails to reproduce the applied frequency range linearly it gives frequency distortion, which may or may not be objectionable.

Amplitude distortion usually results from some component saturating when a large signal is applied, frequently the distortion occurring for one polarity of the signal current or voltage and not the other. This has the effect of introducing new frequencies which are harmonics of the original ones.

Phase distortion does not appear to be important for sound since the ear is not sensitive to it, but in picture work it produces very noticeable distortion in the reproduced picture. It is caused by circuit elements offering different impedances at different frequencies and is one of the most difficult forms to eliminate when the frequency range of the sytem is very wide.

For the absence of phase distortion, it is necessary that the curve of phase shift θ vs. frequency be a straight line intersecting the θ axis at some multiple of 180°. If the slope of the curve is not constant, delay distortion is said to occur, whereas if the intercept is not a multiple of 180°, intercept distortion exists. Often compensating networks are introduced in a system to counterbalance certain distortions and give an over-all response free from distortion. The process of correcting frequency and phase distortion is sometimes referred to as gain and phase equalization.

DISTORTION (Harmonic). Harmonic.

DISTORTION (Hysteresis). Hysteresis Distortion.

DISTORTION MEASURE. Information Theory.

DISTORTION (Optical). Geometric Distortion.

DISTORTION (Optics). One of the five geometrical aberrations with spherical surfaces. This aberration is due to the variation in magnification with distance from the axis.

DISTRIBUTED CAPACITANCE (Coil). The capacitance which is inherent in any coil because of the adjacent turns, layers, windings, etc., which are separated by some dielectric material and which have voltage differences between them. The result of this is a capacitance action which lowers the effective inductance of the coil. This capacitance is often considered as lumped and in parallel with the true inductance of the coil.

DISTRIBUTION (Boltzmann's Law). Boltzmann's Distribution Law.

DISTRIBUTION CURVE. The graph of cumulated frequency as ordinate against the variate value as abscissa, namely the graph of the distribution function.

DISTRIBUTION-FREE METHODS. Statistical methods, mostly in the theory of inference, which are independent of the precise form of the parent distribution from which the data emanated.

DISTRIBUTION FUNCTION. The distribution function $F(x)$ of a variate x is the total frequency of members with variate values less than or equal to x. As a general rule the total frequency is taken to be unity, in which case the distribution function is the proportion of members bearing values $\leq x$. Similarly, for p variates $x_1, x_2, \ldots, x_p$ the distribution function $F(x_1, x_2, \ldots, x_p)$ is the frequency of values less than or equal to x_1 for the first variate, x_2 for the second and so on.

DISTRIBUTION (Statistical). The manner in which a set of individuals are classified according to the values of one or more variables, for example the number of men in a given town classified into ranges according to height (a univariate distribution) or the same men classified into a two-way table by height and weight (a bivariate distribution).

Theoretical statisticians recognize a great many distributions expressed in mathematical form, both discontinuous (e.g., binomial distribution, Poisson distribution) and continuous (e.g., normal distribution, exponential distribution, Pearson family of distributions).

DISTRIBUTIVE LAW. If $A(f_1 + f_2 + \cdots) = Af_1 + Af_2 + \cdots$ then the operator A obeys this law when it is applied to a sum of functions. Normally, arithmetic and algebraic operations behave in this way but it is easy to find exceptions to the law. Thus, $\sin(x + y + z + \cdots) \neq \sin x + \sin y + \sin z + \cdots$, which is typical of trigonometric functions.

See also terms listed under **Mathematics.**

DIURESIS. Diuretics; Water.

DIURETICS. These are substances which increase the volume of urine excreted, causing a condition known as *diuresis.* Some natural drugs, such as caffeine, act as diuretics. Some other substances, known as "saline diuretics," when filtered through the renal capsule are incapable of being reabsorbed by the tubules. These substances thus increase the concentration of salts within the tubule so that little of the urine can diffuse back into the blood stream. Some drugs, such as mercurial diuretics, act at the tubules, thus preventing reabsorption. Other drugs, such as digitalis, cause diuresis because of their specific actions on the circulatory system. See also **Congestive Heart Failure;** and **Kidney and Urinary Tract.**

A modern classification of diuretics places them in four categories:

(1) *Thiazides* were discovered during the synthesis of carbonic anhydrase-inhibiting analogues of sulfanilamide. The thiazides inhibit reabsorption of sodium and chloride in the distal convoluted tubules of the kidney. These drugs also increase secretion of potassium in the distal convoluted tubule and collecting ducts and thus may cause a depletion of potassium. The thiazides have relatively few other side effects. Where hypokalemia (deficiency of potassium) is noted, this can be corrected through the use of potassium supplements, usually potassium chloride. Examples of the thiazides include chlorothiazide (Diuril®), hydrochlorothiazide (Hydro-Diruil®), trichlormethiazide (Metahydrin®; Naqua®), and chlorthalidone (Hygroton®). The latter drug differs chemically from the thiazides, but is pharmacologically similar.

(2) *Mercurial diuretics,* which are stronger in their diuretic effects, require parenteral administration. The organic mercurials, such as meralluride (Mercuhydrin®) and mercaptomerin sodium (Thiomerin®) work at the ascending limb of the loop of Henle to inhibit active chloride transport and reabsorption of sodium. Chloride excretion can be extensive, causing hypochloremic alkalosis. Because mercury, a heavy metal, is involved, the mercurials have lost considerable favor.

(3) *Loop diuretics,* such as ethacrynic acid and furosemide, are the strongest diuretics available. These compounds have largely replaced the mercurials. Loop diuretics inhibit tubular reabsorption of sodium chloride in the ascending loop of Henle. These drugs also enhance potassium excretion and thus cause the same kinds of problems as the thiazides. There are a number of side effects, including hearing loss in the case of ethacrynic acid. Because they are potent, the loop diuretics must be prescribed with considerable discretion. The principal examples are furosemide (Lasix®) and ethacrynic acid (Edecrin®).

(4) Frequently *potassium-sparing agents,* although in themselves weak diuretics, are combined with other diuretics. In particular, these agents enhance the actions of the thiazides and loop diuretics, increasing the urinary loss of sodium while decreasing the excretion of potassium. Representative of these agents are spironolactone (Aldactone®) and triamterene (Dyrenium®).

The carbonic anhydrase inhibitors, such as acetazolamide (Diamox®) also decrease the absorption of sodium, bicarbonate, and chloride, but are too weak in their effect to be useful as single agents.

In addition to their use in congestive heart failure and angina pectoris, diuretics are widely prescribed in the treatment of hypertension (high blood pressure), acute respiratory failure, hypercalcemia (in cancer patients), hypertrophic cardiomyopathy, and the syndrome of inappropriate ADH secretion, among others.

See also **Hypertension (High Blood Pressure).**

DIURNAL CHANGES (Meteorology). Atmosphere (Earth).

DIVER. Loon.

DIVERGENCE LOSS (Sound). That part of the transmission loss which is due to the divergence or spreading of sound rays in accordance with the geometry of the system (e.g., spherical waves emitted by a point source).

DIVERGENCE (Mathematics). The scalar product of the differential operator del (∇) and a vector. In Cartesian coordinates

$$\nabla \cdot \mathbf{V} = div\,\mathbf{V} = \frac{\partial V_x}{\partial x} + \frac{\partial V_y}{\partial y} + \frac{\partial V_z}{\partial z}$$

The divergence of a position vector is constant; $\nabla \cdot \mathbf{R} = 3$; $\mathbf{R} = \mathbf{i}x + \mathbf{j}y + \mathbf{k}z$.

If $\mathbf{V}$ represents at each point in space the direction and magnitude of flow of some fluid, such as water or a gas, thermal, or electrical flux, then div $\mathbf{V}$ equals the rate of increase of flow per unit volume.

If the divergence of a vector function of position vanishes everywhere in a certain region, the function is said to be a solenoidal vector in that region. It follows that if $\mathbf{V}$ is a solenoidal vector so that $\nabla \cdot \mathbf{V} = 0$, then $\mathbf{V} = \nabla \times \mathbf{W}$, or $\mathbf{V}$ is the curl of some vector $\mathbf{W}$. See also, for relations sometimes known as the divergence theorem, **Gauss Theorem** and **Green Function.**

DIVERGENCE (Meteorology). Atmosphere (Earth).

DIVERSITY RECEPTION. Fading has been found to vary from place to place at a given time. Thus if a radio signal is received simultaneously at points separated by a few wavelengths' distance it is found that the outputs of the receivers do not all fade together. Diversity reception is a method of utilizing this effect to minimize the fading. Basically such a system consists of 2 or more (3 is quite common) antennae separated by several wavelengths (at least 10 times the wavelength of the received wave is desirable and 3 antennae placed at the vertices of an equilateral triangle give the best positioning) feeding separate radio-frequency receiver channels. The outputs of these channels are then combined to give a single output. By means of automatic volume control circuits, the antenna receiving a non-faded signal supplies most of the output and as the signals at the different antennae fade out and back in, the control system acts to maintain a constant output level. While such a system, because of its complexity, is not suitable for home reception, it is widely used for reception of foreign broadcasts for rebroadcasting. It is also used for transoceanic telephone reception.

DIVERTICULA (Bladder). Kidney and Urinary Tract.

DIVERTICULOSIS AND DIVERTICULITIS. Diverticula are small sacs or pockets (outpouchings) which occur in the colonic mucosa. When diverticula are present, even without symptoms, the disorder is called *diverticulosis.* When inflammation develops, which particularly occurs in the sigmoid colon, the condition is called *diverticulitis.* Diverticula are formed by a spreading of the muscular coat (mucosa and submucosa) of the bowel wall (a type of herniation), causing the bowel to protrude through the wall as a blind pocket. Diverticula may range from as small as 1 to 2 millimeters up to 5 centimeters in diameter. In routine barium enema x-ray examinations, somewhat less than 10% of adults are found to have some colonic diverticula, although the incidence rises to 30–60% among persons over 60 years old. Only a small percentage of persons with diverticula develop the complications of inflammation, i.e., diverticulitis. Nevertheless, 2% of the population of a large country represents a few million cases and thus physicians find diverticulitis, in varying degrees, to be a rather common complaint. Diverticulosis occurs less frequently in countries where people consume less-refined foods that contain appreciable amounts of indigestible fibers, as in Asia and Africa. However, in urban populations of developing countries where dietary habits have altered, the incidence of diverticulosis is increasing. See **Dietary Requirements and Trends.**

In diverticulitis, early symptoms include lower abdominal aching and cramping with pain usually regionalized to the left lower abdomen. Fever is not uncommon. Chills may indicate bacteremia. Watery diarrhea may develop; defecation may be more frequent. Pericolitis may develop, possibly causing painful defecation and, in some cases, complete obstruction. The most common site of inflammation is a narrow-necked diverticulum located in the sigmoid colon. Microperforation may be present. Where there is rebound tenderness, probably indicating a localized peritonitis, a barium enema examination is usually delayed until the acute process has subsided. It is very important that the mass involved in the difficulty be located. There are a number of diseases, particularly in older people, that tend to mimic acute diverticulitis, including Crohn's disease, ulcerative colitis, and pseudomembranous colitis, among others. See **Colitis and Other Inflammatory Bowel Diseases.** In the absence of detailed history, an accurate differential diagnosis is required.

It is estimated that about 70% of patients with acute attacks of diverticulitis can be treated with supportive measures and that they will not experience subsequent attacks. Treatment includes nasogastric suction, fluid and electrolyte replacement, and antibiotic therapy, where the inflammation is severe or where peritonitis is suspected. A broad-spectrum antibiotic, such as ampicillin is frequently the drug of choice. Usually when a patient does not respond to medical supportive measures within 24 to 48 hours, surgical correction is indicated. There are several criteria considered as bases for surgery, including: where barium enema x-ray examinations and colonoscopy do not fully rule out carcinoma of the colon; where attacks of diverticulitis occur frequently; where a fistula is present; where there is chronic infection of the urinary tract coupled with deformity of the bladder; where a giant (measuring from 8 to 25 centimeters) sigmoid diverticulum is present; and where there is persistent low-grade intestinal obstruction. When emergency surgery is required, the mortality rate is about 6%; with elected surgery, about 1 or 2%, with the rate increasing with age of the patient. For some years, it was traditional to perform this surgery in two or three stages. A large number of surgeons today prefer a one-stage procedure.

References

Colcock, B. P.: "Diverticular Disease: Proven Surgical Management," *Clin. Gastroenterol.*, **4**, 199 (1975).

Horner, J. L.: "Natural History of Diverticulosis of the Colon," *Am. J. Dig. Dis.*, **3**, 343 (1958).

Larson, D. M., Masters, S. S., and H. M. Spiro: "Medical and Surgical Therapy in Diverticular Disease: A Comparative Study," *Gastroenterology*, **71**, 734 (1976).

Painter, N. S., and D. P. Burkitt: "Diverticular Disease of the Colon, a 20th Century Problem," *Clin. Gastroenterol.*, **4**, 3 (1975).

Zollinger, R. W., and R. M. Zollinger: "Diverticular Disease of the Colon," *Adv. Surg.*, **5**, 255 (1971).

DIVISION. An operation that is the inverse of multiplication. To divide a number a by a number b is to find a number c, such that $a = bc$. The number a is called the *dividend*, b the *divisor*, and c the *quotient*. The symbols, a/b, $\frac{a}{b}$, and $a \div b$, frequently are used to denote division. Division by zero is excluded from algebra in order that all quotients exist and be unique. The symbols, $b^{-1}a$ and ab^{-1} indicate the products of a and the inverse of b. In algebra, multiplication is commutative and both products equal c. Therefore, dividing a by b is equivalent to multiplying a by b^{-1} (where $b^{-1}b = bb^{-1} = 1$). Multiplication by b^{-1} is meaningful in many applications where $a \div b$, or the other symbols interpreted in the sense of arithmetic or algebra are not well defined (for example, matrices). See also **Synthetic Division.**

Donald R. Hodge, Alexandria, Virginia.

DIVISION (Cell). Cell (Biology).

DKT CRYSTAL. Piezoelectric Effect.

DLVO THEORY. Colloid System.

DNA (Recombinant). Recombinant DNA.

DNA VIRUSES. Virus.

DOBSON SPECTROPHOTOMETER. A photoelectric spectrophotometer which is used in the determination of the ozone content of the atmosphere. The instrument compares the solar energy at two wavelengths in the absorption band of ozone by permitting the two radiations to fall alternately upon a photocell. The stronger radiation is then attenuated by an optical wedge until the photoelectric system of the photometer indicates equality of incident radiation. The ratio of radiation intensity is obtained by this process and the ozone content of the atmosphere is computed from the ratio.

DOCK (Crustacean Damage). Gribble.

DODECAHEDRON. Polyhedron.

DODO. Pigeons and Doves.

DOG BITE. Rabies.

DOGFISHES. Bowfin; Sharks.

DOGWOOD SHRUBS AND TREES. Probably best known for its beautiful trees, the dogwood family (*Cornaceae*) is predominantly made up of shrubs which have toothless leaves, except *Cornus alternifolia*, and staminate and pistillate flowers found on separate plants. The dogwood is found in Europe, Asia, and the United States. The

RECORD DOGWOODS IN THE UNITED STATES[1]

Specimen	Circumference[2] (inches)	Circumference[2] (centimeters)	Height (feet)	Height (meters)	Spread (feet)	Spread (meters)	Location
Alternate leaf dogwood (1972) (*Cornus alternifolia*)	68	173	30	9	50	15	New York
Flowering dogwood (1976) (*Cornus florida*) (Co-Champion)	55	140	55	16.5	56	16.8	Michigan
Flowering dogwood (1971) (*Cornus florida*) (Co-Champion)	61	155	51	15.3	43	12.9	Florida
Flowering dogwood (1969) (*Cornus florida*) (Co-Champion)	76 (at 24 inches)	193 (at 61 centimeters)	36	10.8	40	12	Pennsylvania
Pacific dogwood (1975) (*Cornus nuttallii*)	129	328	50	15	50	15	Oregon
Panicled dogwood (1975) (*Cornus racemosa*)	18	46	38	11.4	24	7.2	Michigan
Red-osier dogwood (1976) (*Cornus stolonifera*)	120	305	17	5.1	18	5.4	Michigan
Roughleaf dogwood (1972) (*Cornus drummondii*)	17	43	31	9.3	20	6	Missouri
Roundleaf dogwood (1975) (*Cornus rugosa*)	132	335	40	12	16	4.8	Michigan
Stiffcornel dogwood (1974) (*Cornus stricta*)	13	33	36	10.8	15	4.5	Florida

[1] From the "Social Register of Big Trees," The American Forestry Association (by permission).
[2] At 4.5 feet (1.4 meters).

C. florida is a small-to-medium size tree, achieving a height of from 20 to 40 feet (6 to 12 meters) and a trunk diameter of from 4 to 6 inches (10 to 15 centimeters). It is an extremely ornamental tree, particularly at springtime. Also known as the flowering dogwood, the *C. florida* has a rather flat top, with essentially horizontal branches. The twig is thin and the foliage is not dense. The leaf is about 4 to 5 inches (10 to 12.7 centimeters) long and clustered at the end of the branch. The flower is large and showy, about 3 to 4 inches (7.6 to 10 centimeters) across. The flowers are four-bracted, creamy white, sometimes tinged with pink. The fruit is a deep scarlet color. The wood is hard, heavy, and close-grained, weighing 46 pounds per cubic foot (737 kilograms per cubic meter). The heart wood has a slight red tinge. The wood has been used where a hard wood of fine quality is desired, but the availability is limited. Applications have included shuttles, skate rollers, golf club heads, and pulleys. The tree is found in woodlands and frequently in large gardens and parks from Maine and Quebec west to Wisconsin and Minnesota and south to Florida and southwest to Texas. It is probably found in the largest numbers in the mountains of North Carolina. The Japanese dogwood (*C. kousa*) is quite similar to the tree just described.

Record dogwood trees in the United States are indicated in the accompanying table. The *C. alternifolia* generally is considerably smaller than the record specimen described in the table. Often it may be a shrub only 6 to 10 feet (1.8 to 3 meters) in height, with a small trunk of perhaps 9 inches (22.9 centimeters) in diameter. The alternating leaves are an exception to the rule of the genus. The flowers of this shrub/tree are white and occur in broad, flat clusters. The fruit is a dark grayish-blue of pea-size. The plant prefers moist thickets and occurs from Quebec westward into Ontario and Minnesota and southward to Alabama and Georgia.

Although the *C. florida* does not do well in Britain, the *C. kousa* appears to be well suited. Both species have performed well in parts of Europe, notably France. The Chinese dogwood (*C. controversa*) is found in Europe and the United States. Although this tree does not feature the large white bracts characteristic of the other species, it does display small flowers and the tree definitely has the general bearing of a dogwood. There are approximately fifty other species of dogwoods.

DOHERTY AMPLIFIER. Amplifier.

DOLDRUMS. Winds and Air Movement.

DOLERITE. The term dolerite, derived from the Greek meaning deceitful, was originally applied to all dark, heavy, fine-grained igneous rocks of doubtful character. It is now used to indicate gabbroid or basaltic types occurring as dikes or sills whose mineralogical composition is plagioclase, feldspar, hornblende or pyroxene or both, olivine and perhaps biotite, magnetite or ilmenite and pyrite. Included in the dolerites are the diabases, which display plagioclase laths in a somewhat radial arrangement, and from this circumstance we have the textural term diabasic which is synonymous with ophitic.

DOLOMITE. The mineral dolomite, the carbonate of calcium and magnesium, corresponds to the formula $CaMg(CO_3)_2$ and closely resembles calcite. Its crystals, rhombohedral in habit, fall in the hexagonal system. Like calcite, it may be massive or granular, some marbles being dolomite rather than calcite. It displays a perfect cleavage parallel to the rhombohedron; subconchoidal fracture, brittle; hardness, 3.5–4; specific gravity, 2.85; luster vitreous to pearly; color varies widely, white, reds, greens, black, browns, yellows or colorless; transparent to translucent. Unlike calcite, dolomite dissolves very slowly if at all in dilute cold hydrochloric acid; powdered dolomite will dissolve in warm acid. This is the common test for the two minerals.

Much dolomite occurs as stratified rocks where it is believed to have been formed by a secondary process, probably by the action of waters charged with magnesium compounds. Dolomite also is found as a vein mineral, as is calcite. Iron or manganese, rarely zinc or cobalt, may replace some of the magnesium. Ankerite is the name given to a mineral whose composition is essentially a calcium-magnesium-iron carbonate. Among the many noted localities for dolomite are Saxony, Switzerland, Italy, France, Spain, Brazil, Mexico; in the United States, Roxbury, Vermont; Lockport, New York; Phoenixville, Pennsylvania; Alexander County, North Carolina; Hancock County, Illinois, and the Joplin District, Missouri, Dolomite was named for Deodat de-Dolomieu, who first described its characteristics. See also **Limestone;** and terms listed under **Mineralogy.**

DOLPHIN. Whales, Dolphins, and Porpoises.

DOLPHINS (*Osteichthyes*). Of the order *Percomorphi*, suborder *Percoidea*, and family *Coryphaenidae*, a dolphin is a marine game fish with a deep head, short snout, and long tapering body. In general form, these fishes appear like the mammalian dolphins. However, the marine dolphin is not to be confused with the air-breathing porpoise also referred to as a dolphin.

There are two species of marine dolphins: (1) *Coryphaena hippurus*, and (2) *Coryphaena equiselis*. *C. hippurus* is the common dolphin which, as an adult, may reach a length of 5 feet (1.5 meters). It is possessed of a long fin and up to 65 rays extending along the back. The color is a beautiful blue and the tail is forked. Adult males may weigh up to nearly 70 pounds (32 kilograms) and have a markedly square-like head. The head of the female is more pleasingly rounded. Females rarely exceed a weight of 35 pounds (16 kilograms). Records of Hawaiian catches indicate an average of about 17 pounds (8 kilograms). Very fast, the dolphin can travel up to 37 miles (60 kilometers) per hour. Dolphins travel singly or in schools and feed on quite a range of fishes. Along the tropical American coasts and the Philippines, the dolphin is not highly regarded as a food fish, whereas in the Hawaiian Islands, *mahi mahi* (*C. hippurus*) is well regarded and is often considered a premium fish. *C. equiselis* is also referred to as the pompano dolphin. The latter is a considerably smaller fish (maximum length of about 30 inches; 76 centimeters) and has fewer rays on the dorsal fin, but otherwise is quite similar to *C. hippurus*.

DOMAIN. 1. A region of space, together with its bounding points, curves, or surfaces. 2. The range of the independent variable for which a function is defined. 3. A region of spontaneous electric or magnetic polarization in a ferroelectric or ferromagnetic crystal. The size and shape of a domain depend on the material and its treatment. The boundary between domains is known as a Bloch wall. See also **Mapping.**

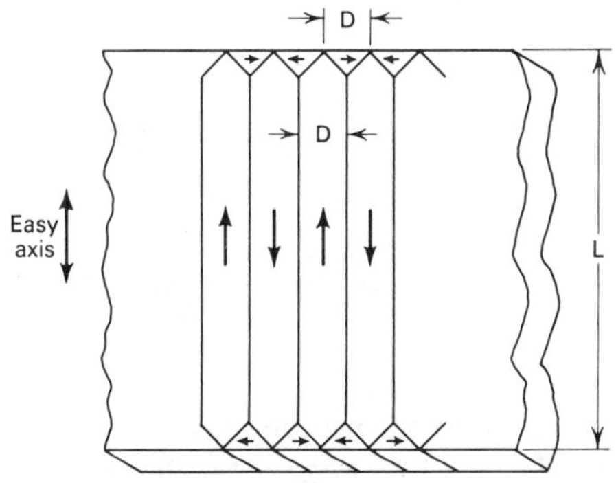

Domain structure. Flux closure domain configuration in a uniaxial crystal. (*Kittel, "Introduction to Solid State Physics," Wiley.*)

DOMAIN STRUCTURE. The theory of the macroscopic behavior of ferromagnetic and ferroelectric crystals depends on their consisting of large numbers of domains, each polarized to saturation but pointing in different directions so as to minimize the energy. An applied field tends to make those domains grow which are already favorably oriented, at the expense of those opposed to the field. Owing to the anistropy energy, domains tend to be oriented along certain directions of easy magnetization, but the detailed arrangement of domains in a crystal is a complicated compromise between the tendency of each domain to be as large as possible and the necessity of creating closed loops of magnetic flux.

DOME. As used by the geologists, this term has several meanings. It is principally applied to mounds of viscous lava, which are squeezed out of volcanoes and solidify without forming lava flows. When portions of the older lavas or ashes are pushed up by the pressure of later lavas, the resulting structure is called a volcanic dome.

DOME CRYSTAL FORM. Mineralogy.

DOMINANCE (Ecology). In ecological communities, certain organisms may exert a much greater influence on the community than the other parts of the community. Those organisms which, in essence, determine the basic nature of a community are referred to as *dominants*. In terrestrial communities, plants are frequently the dominants. It is established, for example, that in a beech-maple community, the amount of shade cast by these trees limits the maturation of shorter plants and the absence of the latter may, in turn, limit the animal species that can survive in the area.

DOMINANCE (Heredity). Heredity.

DOMITE. The term proposed by Von Buch for the trachyte lavas of the volcanic Puy de Dôme district of France. More specifically, trachytes which contain appreciable amounts of oligoclase and hematite.

DONKEY. Horses, Asses, and Zebras.

DONNAN EQUILIBRIUM. If a solution containing a salt with a nondiffusible ion is separated from a solution of an electrolyte containing diffusible ions by a semipermeable membrane, then at equilibrium, an electrostatic difference of potential termed a "membrane potential" will be established between the two solutions separated by the membrane. A difference in osmotic pressure will also be established at equilibrium. Proteins in the presence of simple salts exhibit this phenomenon.

DONOR (Enzyme). Enzyme.

DONOR (Heart). Heart and Circulatory System (Human).

DOOLITTLE METHOD OF SOLVING NORMAL EQUATIONS. Normal equations (1) may be solved readily by the Doolittle technique as simplified by P. Dwyer:

$$\begin{aligned} a_{11}w_1 + a_{21}w_2 + a_{31}w_3 &= a_{41} \\ a_{12}w_1 + a_{22}w_2 + a_{32}w_3 &= a_{42} \\ a_{13}w_1 + a_{23}w_2 + a_{33}w_3 &= a_{43} \end{aligned} \tag{1}$$

where $a_{ij} = a_{ji}$. The solution is given in tabular form, with a check column, and an explanation at the right-hand side.

w_1	w_2	w_3		CHECK	EXPLANATION
a_{11}	a_{21}	a_{31}	a_{41}	a_{51}	$a_{5i} = a_{4i} + a_{3i} + a_{2i} + a_{1i}$, remembering $a_{ij} = a_{ji}$
	a_{22}	a_{32}	a_{42}	a_{52}	
		a_{33}	a_{43}	a_{53}	
a_{11}	a_{21}	a_{31}	a_{41}	a_{51}	$b_{i1} = \frac{a_{i1}}{a_{11}}$
1	b_{21}	b_{31}	b_{41}	b_{51}	$b_{51} = 1 + b_{21} + b_{31} + b_{41}$
	$a_{22.1}$	$a_{32.1}$	$a_{42.1}$	$a_{52.1}$	$a_{i2.1} = a_{i2} - a_{i1}b_{21}$
	1	$b_{32.1}$	$b_{42.1}$	$b_{52.1}$	$b_{i2.1} = \frac{a_{i2.1}}{a_{22.1}}$
					Check = sum of all items in row
		$a_{33.12}$	$a_{43.12}$	$a_{53.12}$	$a_{i3.12} = a_{i3} - a_{i1}b_{31} - a_{i2.1}b_{32.1}$
		1	$b_{43.12}$	$b_{53.12}$	$b_{i3.12} = \frac{a_{i3.12}}{a_{33.12}}$
					Check = sum of all items in row
w_1	w_2	$b_{43.12}$			

In the last row, $w_3 = b_{43.12}$, $w_2 = b_{42.1} - b_{32.1}w_3$, $w_1 = b_{41} - b_{31}w_3 - b_{21}w_2$. Finally, the values of w_1, w_2, w_3 are checked in Equation (1). While we have illustrated the method for three equations, $a_{ij} = a_{ji}$, the method is perfectly general for n equations, $a_{ij} = a_{ji}$.

See also **Equation;** and terms listed under **Mathematics.**

DOPA. Amino Acids.

DOPAMINE. **Brain and Nervous System; Parkinson's Disease.**

DOPING (Semiconductor). **Semiconductor.**

DOPPLER BROADENING. A spreading of radiation frequencies with a resulting broadening of the corresponding spectral line, which takes place when radiating nuclei, atoms or molecules do not all have the same velocity relative to the observer, so that they give rise to different Doppler shifts. For example, since molecules of luminous gases obey the Maxwell distribution law, these effects produce a range of observed frequencies symmetrically distributed about the frequency of the atoms at rest, this range increasing with increasing temperature; and there is a distribution of intensities throughout the broadened line that is determined by the Maxwell distribution of velocities. In absorption spectra, a similar broadening of the lines can result from the motions, relative to the observer, of the absorbing atoms or molecules. In nuclear physics, thermal motions of the nuclei in the material under examination can contribute appreciably to the widths of resonance lines, many of which (for instance, in slow neutron absorption) have natural widths of about 1 eV. See also **Broadening of Spectral Lines.**

DOPPLER EFFECT. In 1842, Christian Doppler predicted that the frequencies of received waves were dependent on the motion of the source or observer *relative to the propagating medium.* His predictions were promptly checked for sound waves by placing the source or observer on one of the newly developed railroad trains. In his original article on the special theory of relativity, Einstein[1] developed the expression for the Doppler shift of light waves which was dependent upon the velocity of the source relative to the observer. From the photon hypothesis for light. Schrödinger[2,3] obtained the same results. Thus, the Doppler effect provides one of the illustrations of the equivalence of the wave and particle descriptions of light.

Classroom demonstrations of the Doppler effect for water waves are made in shallow glass-bottom ripple tanks. Instead of giving the vibrating source a constant velocity, one lets the sheet of water as medium flow continuously by the source.

The circles of Fig. 1 are snapshots of the crests of a water wave

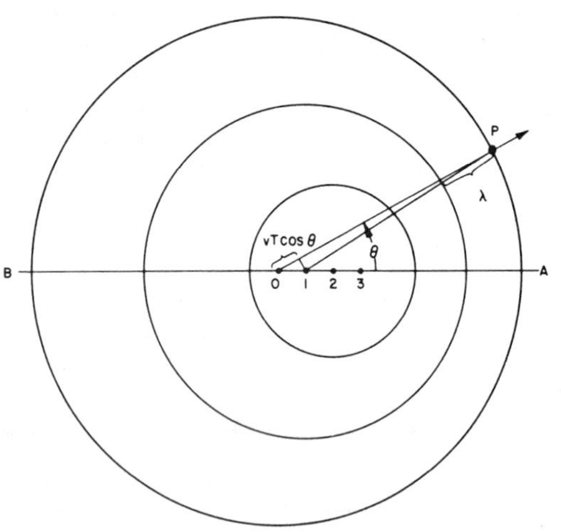

Fig. 1. The source is moving relative to the medium along the line *BA*. The circles represent crest of the wave of an instant. (*Andrews, "Optics of the Electromagnetic Spectrum," Prentice-Hall.*)

or compressions in a sound wave observed when the source is moving at constant velocity v to the right relative to the medium. Points 1 and 2 are positions of the source one and two periods after passing O. The largest circular crest originated at O, the next at 1 and the smallest at 2. A crest is about to leave point 3 at the time the snapshot is taken. If the position P of the observer is a large distance from the source compared to the distance the source moves in one period, then with good approximation we may assume that two successive crests are moving in the same direction as they pass P. If v is the velocity of the source in the direction OA, the source moves a distance vT in one period T. In one period, the source comes closer to P by the amount $vT\cos\theta$, where θ is the angle between the direction of the velocity of the source and the line from the source to the observer at P. Now λ_0 is the wavelength and ν_0 the frequency when the source is at rest; λ is the observed wavelength and ν the observed frequency when the source is in motion. Because of the motion of the source, the wavelength received at P is reduced by $vT\cos\theta$.

$$\lambda = \lambda_0 - vT\cos\theta$$

If c is the velocity of the wave, $T = \lambda_0/c$ and $\lambda = \lambda_0[1 - (v/c)\cos\theta]$, but $\lambda = c/\nu$ and $\lambda_0 = c/\nu_0$. Therefore,

$$\frac{\nu}{\nu_0} = \frac{1}{1 - \frac{v}{c}\cos\theta} \tag{1}$$

when the *source is in motion relative to the medium.*

In Fig. 2 the source is at rest, but the observer at P has a velocity v with respect to the medium. The velocity of the wave relative to the observer is equal to the vector sum of the velocity of the wave relative to the medium and the velocity of the medium relative to the observer. In Fig. 2, c is the velocity of the wave and v the velocity

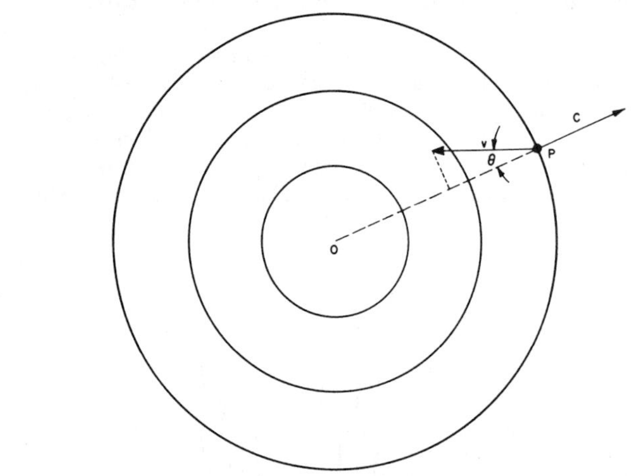

Fig. 2. The source is at rest at point *O* and the observer at point *P* is moving with velocity v relative to the medium. (*Andrews, "Optics of the Electromagnetic Spectrum," Prentice-Hall.*)

of the observer relative to the medium. Let λ_0 be the wavelength, ν_0 the frequency of the source, and ν the frequency received by the moving observer. The radial velocity of the wave relative to the observer is $c + v\cos\theta$ so that

$$\nu\lambda_0 = c + v\cos\theta$$

For an observer at rest $\nu_0 = c/\lambda_0$. Substituting for λ_0, we obtain

$$\frac{\nu}{\nu_0} = 1 + \frac{v}{c}\cos\theta \tag{2}$$

when the *observer is in motion relative to the medium.*

By a postulate of relativity, the velocity of light is the same relative to all observers. The theory of relativity yields the frequency

$$\frac{\nu}{\nu_0} = \frac{1 + \frac{v}{c}\cos\theta_0}{\sqrt{1 - \frac{v^2}{c^2}}} \tag{3}$$

in which $v\cos\theta_0$ is the component of the velocity of the source toward the observer. The angle θ_0 is measured in the source system. If θ is the angle measured in the observers system, then

$$\cos\theta_0 = \frac{\frac{v}{c} - \cos\theta}{\frac{v}{c}\cos\theta - 1} \tag{4}$$

Figure 3 is a graphical plot of ν/ν_0 against v/c for the radial motion in the three cases we have treated. (1) The linear relation is that for

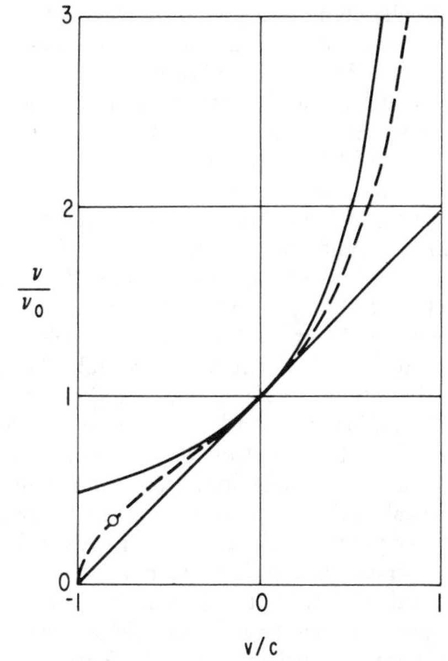

Fig. 3. Graphical plots of the ratio of the observed frequency to the frequency at the source against the ratio of radial velocity to the velocity of the wave for three cases: (1) Sound waves from a moving source; (2) sound waves to a moving receiver; (3) electromagnetic waves. The circle represents the red shift of light received from the most distant galaxies observed.

the observer in motion relative to the medium that propagates sound or other mechanical waves. (2) The other solid curve is for the source of sound in motion. (3) The broken curve represents the Doppler effect for electromagnetic waves such as x-rays, light, and radio waves.

By comparing several spectral lines of elements observed in a star with a laboratory spectrum of the same elements, astronomers use the Doppler effect to measure the radial components of velocity of astronomical bodies toward or away from the earth. Spectra of the edges of the sun's disk are measured to determine the velocities toward and away from the earth. The radial velocities of the principal stars of our galaxy have been recorded. The spectral lines of some of the stars are doublets which periodically come together and separate again indicating that the light comes from two stars revolving about a common center of gravity.

In the expanding universe, the radial velocities of other galaxies away from our galaxy are porportional to their distances from the observer. Thus, the Doppler red shift provides a means of determining the dimensions of the observed universe. In 1964, some of the most intense radio sources were located with high precision by observing these sources when the moon passed in front of them. With this knowledge of position, the same sources were located with a light telescope.[4] The measured red shift was surprisingly high. The sources were not stars as previously thought but the most distant galaxies known. One of them, 3C-9 in the catalogue of radio sources had a red shift $\Delta\lambda/\lambda_0$ equal to 2.0. If this shift were due solely to the Doppler effect, the astronomers had to conclude that the source was moving from us with 80% of the velocity of light. The Doppler frequency and velocity of this source is indicated by a circle on the broken line of Fig. 3.

If a microwave beam is reflected from a moving microwave mirror, such as a person, an automobile or a man-made satellite, the image of the primary source may be considered as another source moving with twice the velocity of the mirror. Since the speed is small compared with the speed of light, the squared terms of Equation (3) may be neglected. Thus

$$\nu = \nu_0\left(1 + \frac{2v}{c}\cos\theta\right)$$

Direct frequency measurements cannot be made to enough significant figures to distinguish ν from ν_0. However, if the two frequencies are combined they give beats or the difference frequency

$$\Delta\nu = \nu_0\frac{2v}{c}\cos\theta$$

Since the beat frequency is proportional to the radial velocity, a frequency meter may be calibrated in miles per hour. The precision of such a speed detector depends upon the frequency of the source being so stable that it varies less than the Doppler frequency shift during the time that the wave travels from the source to the mirror and back. The same phenomena of beats between the direct wave from the source and the wave reflected from a moving mirror may be observed with light. If one of the mirrors of Michelson's interferometer is moved at constant speed, the frequency with which dark bands pass the cross hair is the difference in frequency of the two waves.

The numerator of Equation (3) contains a term for the radial component of velocity. However, the second-order term in the denominator is independent of direction. Thus, as v/c approaches unity, one may expect to detect a *tangential Doppler effect.* Ives and Stilwell[5] have measured the predicted value for the Doppler shift in frequency due to a stream of radiating molecules for which v/c was 10^{-2}. This experiment was a direct proof of time dilatation for the transverse case. In order to separate the tangential from the radial effect, Ives and Stilwell produced a sharply collimated beam of molecules. In order that θ be precisely 90°, they set a mirror accurately normal to the line of observation and altered the line of observation until nearly the same wavelengths were given by direct and reflected light.

See also **Radar; Radio and Radar Astronomy; and Red Shift.**

References

1. Einstein, A.: *Ann, Physik*, **17**, 891 (1905).
2. Schrödinger, E.: *Physik. Z.*, **23**, 301 (1922).
3. Michels, W. C.: *Am. J. Phys.*, **15**, 449 (1947).
4. Greenstein, J. L., and M. Schmidt: *Astrophys. J.*, **140**, 1 (1964).
5. Ives, H. E. and A. R. Stilwell: *J. Opt. Soc. Am.*, **31**, 369 (1941).
6. Gill, T. P.: "An Introduction to the Theory of the Doppler Effect," Academic, New York, 1965.
7. Berkowitz, R. S. (editor): "Modern Radar: Analysis, Evaluation, and System Design," Wiley, New York, 1965.

DOPPLER NAVIGATION SYSTEM. **Navigation.**

DOPPLER RADAR. In a system of this type, use is made of the Doppler shift of an echo caused by the relative motion of target and radar. Differentiation between fixed and moving targets is made possible by observing a shift in frequency. Highly accurate determinations of velocity of moving targets is made possible by measurement of the frequency shift. Both continuous-wave and pulsed Doppler radar configurations are used. See also **Radar.**

DOPPLER SHIFT. The magnitude of the change in the observed frequency of a wave due to the Doppler effect. Expressed in cycles, the magnitude is also termed *Doppler frequency.* See also **Red Shift.**

DOPPLER-TYPE FLOWMETER. **Flow Measurement.**

DORAB (*Osteichthyes*). A member of the order *Isospondyli*, suborder *Clupeidae* (herrings), and family *Chirocentridae*, the dorab is also known as the wolf herring. It is a slender marine fish of Oriental origin, of large size and vicious habits with powerful jaws and formidable dentition. It differs from all other herrings by the presence of a spiral valve in the intenstine. Normally, this is found only in rays and sharks and possibly a limited number of primitive bony fishes. The dorab has much of the outward appearance of a very large herring, but unlike other herring-type fishes, the dorab is an aggressive carnivore. The fish has the typical herring-type knife-edge on the underside of the abdomen. Records indicate the maximum length of the dorab is about 12 feet (3.6 meters), thus several times larger than other herrings. The waters of the central Pacific are its habitat.

DORMOUSE. **Rodentia.**

DORN EFFECT. The production of a potential difference by a powder falling through a liquid, arising from the presence of an electrical double layer around the particles. The double layer is a hypothetical distribution of charge comprising a layer of positive charge and a layer of negative charge very close by. The net charge is usually zero, but fields are produced by the dipole moments of the elements of the double layer.

DORSAL FIN. Fishes.

DORY. John Dories.

DOSIMETER (Radiation). Most commonly, a device worn by persons working around radioactive material to indicate the "dose" of radiation to which they have been exposed. Photographic film badges have long been used because of the sensitivity of photographic emulsions to radiation and the comparatively low cost of this type of detector. The electroscope also has been used for personnel monitoring. In the common form of the instrument, an insulated fiber is supplied with a static charge relative to another electrode. The electrostatic force on the fiber causes it to deflect. The deflection corresponding to a fixed reference voltage is determined as zero dose on the scale. As the electroscope is exposed to radiation, the charge leaks away through the gas and the fiber deflection decreases. The scale of the instrument is calibrated in terms of total radiation dose received. Although primarily used for personnel monitoring, the electroscope also finds application in radiation measurement where the integrated dose at a point location is desired over a long exposure, such as hours or days.

Various chemical dosimeters have been developed based on the ability of ionizing radiation to induce chemical reactions. Among these are the ferrous–ferric dosimeters and the cerous-ceric dosimeters. The latter is used for very high radiation doses. The ferrous dosimeter is one of the most thoroughly studied of the chemical types. This involves the conversion of ferrous ions to ferric ions in an 0.8 normal sulfuric acid solution by radiation-induced oxidation. The detector is most useful for high radiation levels. The yield of the dosimeter usually is expressed in *G* units, which are the number of molecules produced or converted per 100 electron volts absorbed.

Colorimetric dosimeters involving the effect of radiation upon degradation of dye colors also have been used for less accurate determinations of radioactivity. Dysprosium, a rare-earth element, in the form of metal foil and also the oxide, Dy_2O_3, also have been used in dosimeters.

A very helpful reference on this topic is the "Handbook of Radioactivity Measurements Procedures," National Council on Radiation Protection and Measurements," Washington, D.C., 1979.

DOTTEREL. Shorebirds and Gulls.

DOUBLE BOND. Chemical Elements.

DOUBLE BOND (Carbon). Organic Chemistry.

DOUBLE-CURRENT GENERATOR. If a synchronous converter is driven mechanically both ac and dc may be taken from it, the first from the slip ring connections and the second from the commutator connections.

DOUBLE DECOMPOSITION. Decomposition (Chemical).

DOUBLE LAYER (Electric). Dorn Effect.

DOUBLE REFRACTION (Electric and Magnetic). Kerr Effects.

DOUBLE STAR. A pair of stars, both of which, seen from the earth, are so nearly in the same direction that they appear as a single star to the unaided eye, but which may be seen as separate stars through a telescope. Double stars may be either one of two kinds. In cases where the two stars are only apparently close to each other (i.e., lie in approximately the same direction from the earth, but are separated by a great distance in the radial direction), the pair is known as an optical double. In the great majority of cases, however, the stars are actually close enough together to exert strong gravitational attractions on each other, and are in orbital motion relative to each other. Such physically connected stars form what is known as a binary star, and are the chief source of stellar masses. Optical doubles may be distinguished from binaries by observing the pair in a telescope over a period of years. If the distance between the two stars changes progressively over a long period of time, but the position angle remains constant, it may be safely assumed that the motion is due to proper motion alone, and that an optical double is under observation. In the case of a binary, position angle changes progressively, and distance oscillates between a maximum and minimum.

The first recorded discovery of a double star (Zeta Ursae Majoris) was made by Riccioli (1650). The search has continued. During the early part of the present century, the study of double stars and binary stars was central to much astronomical research. At the time when a 36-inch refracting telescope was a highly sophisticated instrument, it was observed by Aitken that "at least 1 in every 18, on the average, of the stars in the northern half of the sky, which are as bright as 9.0 magnitude, is a close double star." Since that time, aided by advanced astronomical equipment and techniques, the estimate has been revised upward. As pointed out by Heintz in the preface of his 1979 book (see list), "Double and multiple stars are the rule in the stellar population, and single stars the minority, as the abundance of binary systems in the space surrounding the sun shows beyond doubt."

Binary star systems are further described in entries on **Binary Stars; Eclipsing Binary; Spectroscopic Binaries;** and **Visual Binaries.** See also **Cosmology.**

References

Batten, A. H.: "Binary and Multiple Systems of Stars," Pergamon, New York, 1973.

Eggleton, P., Mitton, S., and J. Whelan (editors): "Structure and Evolution of Close Binary Systems," Reidel, Boston, 1976.

Heintz, W. D.: "Binary Systems," Reidel, Boston, 1978.

Staff: Gravity and PSR 1913+16," *Astronomy,* **7**, 6, 60–61 (1979).

Staff: "The Mystery of AM CVn," *Astronomy,* **8**, 2, 62 (1980).

Tver, D. F.: "Dictionary of Astronomy, Space, and Atmospheric Phenomena," Van Nostrand Reinhold, New York, 1979.

DOUBLET. 1. Two elements which are shared by two atoms so as to form a nonpolar valence bond. 2. A pair of spectral lines resulting from transitions between a common state and two states which differ only in total angular momentum (*J*), i.e., have identical values of orbital (*L*) and spin (*S*) angular momenta. 3. Two stationary states having common values of (*L*) and (*S*), but different values of (*J*).

DOUBLET LENS. Particularly an achromat having two components. See also **Petzval Surface.**

DOUGH CONDITIONERS. Oxidation and Oxidizing Agents.

DOUGH LEAVENING. Leavening Agents.

DOUGLAS FIR. Fir Trees.

DOVE. Pigeons and Doves.

DOVEKIE. Shorebirds and Gulls.

DOVE PRISM. Prism (Optics).

DOWNHILL TRANSPORT. Cell (Biology).

DOWN'S SYNDROME (Mongoloidism). Caused by chromosomal error, this disorder is found in about one out of every 700 newborn infants. Mothers under 17 and over 40 years of age run a 40% risk of producing a mongoloid child. Affected individuals suffer mental retardation and exhibit characteristic physical features—flattened head, unusual facial contours including flattening at the bridge of the nose, and, in some cases, heart malformations. In the majority of these cases, the error is the presence of three number 21 chromosomes instead of the normal two. This condition is known as *trisomy 21 Down's.* In only about 5% of the cases, the aberration is either a translocation of a piece of chromosome 21 onto another chromosome, or the presence of trisomy 21 or a chromosome 21 translocation in only some tissues of the body. In the past, the statistics have implicated the abnormality to eggs, not sperm. Researchers have recently suggested that the aberration is related to the role of estrogen in reproduction. Eggs are developed only to the prophase stage of cell division until puberty. At that time, the cyclic rise in estrogen triggers one egg, once a month, to ovulate or resume cell division (*miosis*). Thus, in some very young mothers, the normal cycle may not have commenced at the time she became pregnant. In older women, the estrogen concentration decreases, thus decreasing the rate of miosis. It has been suggested by researchers that if miosis is too slow, the spindle does not attach to both ends of the chromosomes by the time the chiasmata (which hold the chromosomes together) have terminalized and thus the dividing cells will have unequal numbers of chromosomes. One cell may lose the 21st chromosome and will not survive. But the extra chromosome may be added to another cell, giving it one extra chromosome. This cell can go on to regular fertilization by the sperm, subsequently resulting in Down's syndrome in the offspring. Quite recently, researchers at the University of Copenhagen found a marked increase in Down's syndrome among children fathered by men 55 years of age or older. Obviously, the explanation offered for the chromosomal aberration evoked by young and older women does not apply in the case of sperm. These new findings are inducing scientists to rethink the basic causes of aberration. Perhaps other, even more important factors enter as causative agents. In the meantime, programs based upon statistical evidence collected since 1876, when mongoloidism was first correlated with older mothers, are proceeding by way of counseling couples that the safest time to create families is when the woman is between 20 and 40 and when the male is between 20 and 45 years old. Counseling is very important because mongoloidism accounts for significant numbers of mentally retarded persons.

DOWNTIME. Production time lost because of major maintenance or process and manufacturing procedural changes. In heavy industries, such as metallurgical operations and typified by the blast furnace operation illustrated, units may be operated continuously for a number of years. Process start-up usually is a time-consuming costly operation. One day of downtime of a large unit can represent hundreds of thousands of dollars in lost production, not only from the unit per se, but also because of this effect on other stages in the manufacturing process. Because of highly excessive corrosion, some chemical processing units may have to be shut down for major maintenance every few months. A production unit in a typical petroleum refinery will be continuously onstream for 1.5 to 2 years. Wherever possible, shutdowns are scheduled long in advance to minimize adverse effects on delivery of products and on inventory control of raw materials and work in process. As much maintenance as possible is done while a unit is in operation. In other industries which are seriously affected by seasonal factors, such as the automotive industry, assembly lines may be down for several weeks, during which time jigs and fixtures for new models are being installed.

Night scene of blast furnace which will operate around the clock for a number of years without downtime.

DOWNWASH. **Aerodynamics; Helicopters and V/STOL Craft.**

DOXYCYCLINE. **Antibiotic.**

DRACO (the dragon). A northern constellation that lies between Ursa Major and Ursa Minor.

DRAFT. 1. The draft of a ship or boat is its depth of flotation; also, the minimum depth of water in which navigation is possible.

2. In foundry practice, in making a pattern which is to be molded in sand, a certain small taper is necessary where the pattern is of such shape as to be rather deeply imbedded in the sand. When the pattern is rapped or jarred loose from the sand, a slight taper, called draft, allows the pattern to pull free during its removal from the mold, without disturbing the side walls.

3. As applied to a gaseous system, *draft* is a pressure-differential that operates to move the gases. *Combustion* requires oxygen—and therefore air. To move this air through the fuel bed and to produce a flow of the gaseous products of combustion out of the furnace, then through the boiler, economizer, etc., requires a difference of pressure equal to that necessary to accelerate the gases to their final velocity, plus friction head losses. This difference of pressure is called draft whether measured above or below atmospheric pressure. The range of pressures required is most easily measured by manometers reading in inches of water.

4. In meteorology, a manometer measures pressures in terms of a displaced column of water. The pressure may be obtained from the manometer reading by employing the factor relating a head of water to the pressure produced at its base. It requires 2.31 feet (0.693 meter) of water to result in a pressure of 1 psi (0.07 atmosphere). The ordinary draft gage is a variation of the U-tube manometer.

5. In heating engineering, requisite draft can be obtained by use of chimneys, fans, steam or air jets, or combinations of these. The chimney is probably the most common, but the least understood of any of them. At one time the chimney was universally used as the sole means for producing a draft and even now it is relied on entirely in many small plants and partially in most of the large ones.

Mechanically draft may be classified as *forced* or *induced*, the former having the combustion air placed under a plenum, the latter referring to gas movement into a region of partial vacuum. With forced draft alone, furnace gases seep outward through cracks in walls, and blow through opened doors and ports. Induced draft alone allows considerable undesirable dilution of the products of combustion unless furnace, casings, ducts, etc., are maintained airtight. A logical compromise is to use both systems in a *balanced draft* adjusted to maintain atmospheric pressure (or a slight vacuum) in the furnace. There, expansion cracks, opened doors, etc., will not cause undesirable gas or air flows.

DRAFT FAN. **Fan.**

DRAFT TUBE. **Hydroelectric Power.**

DRAG. **Aerodynamics; Airplane; Helicopters and V/STOL Craft; Supersonic Aerodynamics.**

DRAG EFFECT. The effect of interionic attraction in reducing the freedom of an ion to move in an electrical field, because of the interference of the ions of opposite charge by which a given ion is surrounded. The drag effect is an essential part of the explanation of the Debye-Hückel theory of the anomalous properties of concentrated solutions of strong electrolytes. See **Electrochemistry.**

DRAG FOLD. In geology, a smaller fold included in major folds. Produced by the shearing stresses set up within a fold by the relative movements of strata parallel to their bedding planes.

DRAG LINE. The drag-line excavator consists of a turntable on wheels or caterpillar treads, which supports the excavating machinery. A long boom is pivoted at its lower end to the turntable, and guyed at its outer end by a rope sheave. Extra long booms are characteristic of the drag-line excavator because the loading of the bucket is by scraping action in contrast to the shoveling action of a grab bucket. By the use of the boom, to which is attached the scraper, with the intermediary of a block and tackle, the operator can place the scraper bucket at a distance of 25–100 feet (7.5–30 meters) from the position of the excavator. The scraper has attached to it a drag line which is wrapped around a powered drum in the cab. As this drag line is reeled in, it drags the scraper bucket and fills it, after which it is hoisted and dumped wherever wished. The drag-line excavator has been built in very large capacities, and is especially suitable for such work as levee building, borrow pit work, etc.

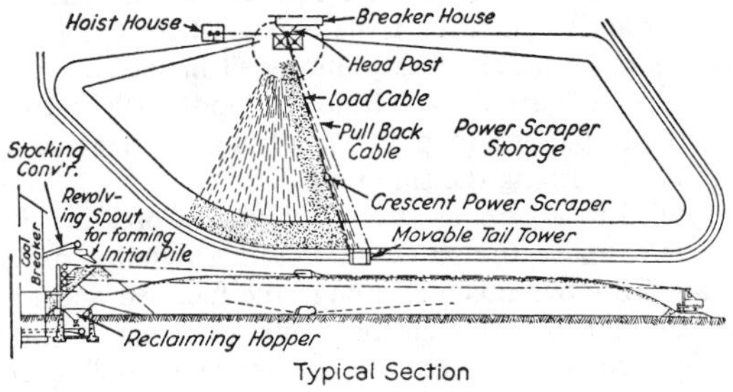

Storage and reclamation by drag scraper.

The drag-line scraper is a means for stocking-out bulk material to storage and reclaiming the same. Drag lines also are used in large coal-mining operations where the coal is near the surface. It is comparatively low in first cost, and adaptable to a storage lot of irregular area. A head post and machinery house is located at the point from which stocking-out begins, and to which reclamation moves. A movable tail tower may be operated at different points along the edge of the storage lot farthest from the head post. Between the tail tower and head post is an endless wire cable, to which is attached a scraper bucket. This passes over the drive sheave in the head tower. The drive may be reversed, and the bucket caused to move out on the stock pile and then be returned, scraping up a full load of loose material as it comes. Typical arrangement of the drag scraper is shown in the accompanying figure.

DRAGONETS (*Osteichthyes*). Of the order *Percomorphi*, suborder *Callionymoidea*, and family *Callionymidae*, dragonets are a rather ugly-appearing fish with a sharp spine of a hook-like nature. The gill opening is very small and located on the upper portion of the head. These fishes are frequently very colorful with slender, flattened bodies. They are found worldwide in both tropical and temperate waters and prefer the bottom. They range from 4 to 8 inches (10 to 20 centimeters) in length. A separate family, the *Draconettidae*, embraces the dragonets of a more primitive nature. These fishes have a somewhat different anatomy, including very broad gill openings. They usually frequent reasonably deep water.

DRAGON FLY (*Insecta, Odonata*). Large insects of powerful flight which eat other insects caught in the air. The immature insect is aquatic and predacious. The order is composed of dragon flies and damsel flies only. All have four slender net-veined wings and long slender bodies, but the dragon flies have fore and hind wings of slightly different shape and hold them extended when at rest. Also called devil's darning needles.

Dragon fly.
(*A. M. Winchester.*)

The dragon fly is considered one of the ancient insects, along with grasshoppers, cockroaches, and crickets. Fossil evidence indicates that some of the early dragon flies measured two feet from wing tip to wing tip. The period of development of the dragon fly ranges from 1 to 4 years, with 10 to 15 larvae stages. The wings appear after the third or fourth molt. Swift, graceful insects that patrol the banks of ponds and lakes, dragon flies are harmless.

DRAINAGE. Hydrology.

DRAINAGE SYSTEMS. Designing an effective drainage system is an important aspect in the civil engineering of numerous structures, notably airport runways, railroad, highways, tunnels, and sport arenas, as well as conventional commercial, industrial, and residential structures. In addition to assuring the convenience, comfort, and safety of nonflooded surfaces, adequate drainage is required to stabilize slopes, prevent mud slides during periods of excessive rainfall, assure dry basements and thus protect an inventory of stored goods, and to prevent the weakening of underlying foundations and ultimate collapse of all or parts of a structure. As water content increases, the strength of soil generally decreases. Attention to drainage is particularly important in areas which may be arid for months, but subject to excessively heavy storm conditions for but brief periods.

Before the best suited drainage system can be proposed, a soil and topographical study of the site is required. The availability of natural channels to provide the most economic means for conveying surface drainage must be ascertained. Careful estimates must be made of the amounts of water that may have to be carried away during the worst possible conditions, such as 50- and 100-year storms. Preliminary studies, while an item of cost which a developer or builder might prefer to put into actual construction rather than planning, frequently will alter plans in a way to conserve total project funding. In some instances, it may be found that provision for adequate draining may be excessively costly and that selection of an alternate site would be the most economical solution. Once construction is commenced and it is found that severe drainage problems are encountered which were not a part of the original plan, drainage system costs may become severalfold what they might have been if part of an original plan.

A useful formula for estimating rainfall runoff is

$$Q = CIA$$

where Q = quantity of runoff, cubic feet/second
C = average runoff coefficient. This is affected by slope, soil, land cover, and time of concentration, and essentially is the percentage of rain that appears as direct runoff
I = intensity of rainfall, inches/hour
A = area under consideration, acres

(Where Q is expressed in cubic meters/second; I in centimeters/hour; and A in hectares, if using coefficients of Table 2, use a multiplying constant (k) of 0.082.)

Where exact land use is known, a precise runoff coefficient can be established. Unfortunately, this is not usually the case, particularly where a variable land use in the lower reaches of the area may not yet be planned or even postulated. Thus, frequently a calculated composite coefficient must be used.

Assumptions made in calculating a composite coefficient include: (1) the maximum rate of runoff considering that a specific rainfall intensity will occur if the duration of the rainfall equals or is greater than the time of concentration. The time of concentration is defined as the time period needed for the water to flow from the far distant point of a drainage basin to a point of flow measurement, (2) the maximum rate of runoff considering a specific rainfall intensity (the duration of which equals or is greater than the time of concentration) will be directly proportional to the rainfall intensity, (3) the peak discharge (per unit area) will decrease as the drainage area increases, and the intensity of rainfall will decrease as its duration increases, and (4) the coefficient of runoff will remain constant for all storms on a given watershed.

Common runoff coefficients are given in Table 1. Some refinements

TABLE 1. COMMON RUNOFF COEFFICIENTS FOR URBAN AREAS

Category of Drainage Area	Runoff Coefficient *C*
Industrial	
Light areas	0.50–0.80
Heavy areas	0.60–0.90
Railroad-yard areas	0.20–0.40
Business	
Downtown areas	0.70–0.95
Neighborhood areas	0.50–0.70
Streets	
Asphaltic	0.70–0.95
Concrete	0.80–0.95
Brick	0.70–0.85
Drives and walks	0.75–0.85
Residential	
Single-family dwelling areas	0.30–0.50
Detached multiunits	0.40–0.60
Attached multiunits	0.60–0.75
Suburban	0.25–0.40
Apartment building areas	0.50–0.70
Parks and cemeteries	0.10–0.25
Playgrounds	0.20–0.35
Unimproved areas	0.10–0.30
Roofs	0.75–0.90
Lawns	
Sandy soil, flat (2%)	0.05–0.10
Sandy soil, average (2–7%)	0.10–0.15
Sandy soil, steep (7%)	0.15–0.20
Heavy soil, flat (2%)	0.13–0.17
Heavy soil, average (2–7%)	0.18–0.22
Heavy soil, steep (7%)	0.25–0.35

SOURCE: Reference 2.

to runoff coefficients are made by special agencies, such as the Los Angeles County Flood Control District which provide runoff coefficients as a function of the soil and area type and of the rainfall intensity for the time of concentration.

The Steel formula is one of the most useful tools for determining the factor I in the foregoing runoff formula. The Steel formula is

$$I = \frac{K}{t + b}$$

where K and b depend on the storm frequency and region of the United States in which the site is located. The United States is broken down into seven regions which take into consideration the 50- and 100-year storm occurrences. Details of these calculating procedures are given in References 1 and 2 at the end of this entry.

In planning a drainage system, the height of the water table at the site is of paramount importance. It is very costly to lower the water table even temporarily during construction. Thus, wherever possible, the new construction should not penetrate the water table. Sometimes it is possible to lower the water table permanently through the use of gravity draining or automatic pumping. These are usually quite costly approaches. In some instances, it may be best to alter the overall development plan and relocate the planned structure to a better suited site.

Necessary to planning a drainage system is a full understanding of the permeability and drainage characteristics of the underlying soil. Permeability is dependent upon the density of the soil, the degree of water saturation, and particle size of the soil. The coefficient of permeability is defined by Darcy's law

$$Q = kiA$$

where Q = rate of flow of water through a soil mass, cubic centimeters/second
i = hydraulic gradient (or total head lost per unit of flow distance, centimeters/centimeter
A = total cross-sectional area of soil through which flow occurs, square centimeter

The Darcy law is useful for estimating gravitational flow, but does not take into consideration the movement of water by capillary action, water held by adsorption, or flow of water by osmosis. For inclusion in the Darcy formula, values of k are given in Table 2. The detailed use of this formula is also presented in Reference 1.

TABLE 2. VALUES OF K AND b FOR STEEL FORMULA

$$I = \frac{K}{t + b}$$

Storm Frequency (Years)	Coefficients	Region of the United States 1	2	3	4	5	6	7
2	K	206	140	106	70	70	68	32
	b	30	21	17	13	16	14	11
5	K	247	190	131	97	81	75	48
	b	29	25	19	16	13	12	12
10	K	300	230	170	111	111	122	60
	b	36	29	23	16	17	23	13
25	K	327	260	230	170	130	155	67
	b	33	32	30	27	17	26	10
50	K	315	350	250	187	187	160	65
	b	28	38	27	24	25	21	8
100	K	367	375	290	220	240	210	77
	b	33	36	31	28	29	26	10

SOURCE: Reference 2. See accompanying map.

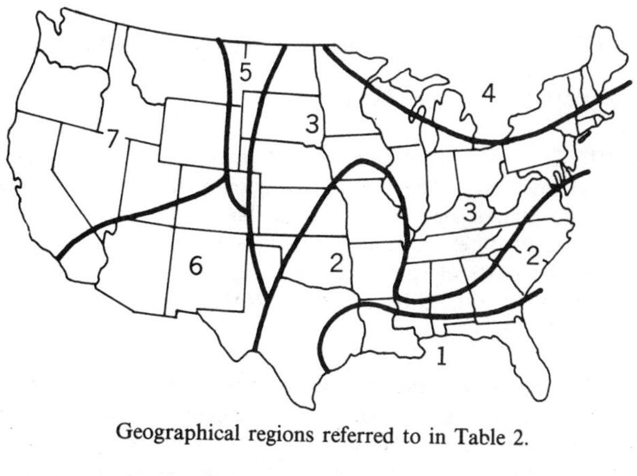

Geographical regions referred to in Table 2.

Numerous drainage designs are available. The simplest and least costly solutions provided they are adequate include the laying of intercepting drains along the contours of the site and, alter construction, covering these areas with heavy mulching and planting to prevent percolation of water downward into the soil. Gravity flow systems are usually best to achieve the permanent stabilization of slopes. If there is artesian pressure present, then vertical wells may be used. In some instances, sand drains or piles are used to compact, stabilize, and drain compressible material. Such piles sometimes are an economical solution for supporting loads on soil with much smaller bearing capacity than sand. In essence, these are cylindrical piles of sand in the ground of 18 to 20 inches (45.7 to 51 centimeters) in diameter and placed from 6 to 10 feet (1.8 to 3 meters) apart. Essentially, the piles perform as wicks for the water squeezed from the soil. They acclerate settlement of the poor material, thus allowing pore water to drain. Once the holes are filled with sand, a drainage blanket and a surcharge are placed over the area. This expedites consolidation of the soil.

Permanent pumped installations become a more costly solution, considering the additional hydraulic engineering required and the years of maintenance of automatic pumping equipment, including standby equipment.

Although electroosmosis also is relatively expensive, it sometimes is the optimal solution for certain conditions. In this system, two electrodes are installed in saturated ground. Direct current is applied and the groundwater normally moves to the cathode. From this point, the water can be pumped out through a wellpoint. Hydrogen ions (positive) are produced by electrolysis of the water. These ions replace the positive ions in fine-grained soils as they move to the cathode. It is the movement of the ions that induces flow of water in the same direction. This method is particularly effective with silts, normally difficult to drain simply by use of open pumping or with wellpoints.

See also **Hydrology.**

References

1. Merritt, F. S., (editor): "Standard Handbook for Civil Engineers," McGraw-Hill, New York, (1968).
2. Chow, V. T.: "Hydrologic Determination of Waterway Areas for the Design of Drainage Structures in Small Drainage Basins," Bul. 426, University of Illinois Engineering Experimental Station, Urbana, Illinois, 1962.
3. Luthin, J. N.: "Drainage Engineering," Wiley, New York, 1965.
4. Dendy, F. E., and G. C. Bolton: "Sediment Yield-Runoff-Drainage Area Relationships in the United States," *J. Soil and Water Cons.*, **31**, 6, 264–266 (1976).
5. Grizzell, R. A.: "Flood Effects on Stream Ecosystems," *J. Soil and Water Cons.*, **31**, 6, 283–285 (1976).
6. Holtan, et al.: "Revised Model of Watershed Hydrology," Tech. Bul. 1518, U.S. Department of Agriculture, Washington, D.C., 1975.
7. Schuhart, A.: "Wetlands Inventory," *Soil Conservation*, **44**, 4, 17–19 (1978).
8. Sutton, J. D.: "Evaluation of Four Completed Small Watershed Projects," Econ. Rept. 271, U.S. Department of Agriculture, Washington, D.C., 1974.
9. Thiel, T. J.: "Soil and Water Conservation Research in the Corn Belt States," Pub. 1283, U.S. Department of Agriculture, Washington, D.C. (Revised periodically).

DRAMAMINE. Hearing and the Ear.

DRAVITE. Magnesium tourmaline, Na > Ca. See **Uvite.**

DRAWBRIDGE. Bridge (Structural).

DRAWING. One use of this term is in reference to an operation performed on metal. The metal is originally in the form either of sheets, solid blanks, or pierced billets. These are formed in suitable dies into many different shapes, such as hollow cylinders, cuplike parts, or solid parts of various shapes. Pipe, wire, various structural shapes, are often formed by drawing. Drawing reduces the thickness of the metal.

DREDGE. An excavating machine for use in river, harbor, or drainage work, is a dredge. A characteristic application of the dredge is in submarine excavation. Dredges are usually floated on water-tight hulls. There are several types, which may be classified according to the way they excavate. The dipper dredge, which is the more common type, is very similar to a locomotive crane, except that it is mounted on a floating hull. Its digging equipment consists of a dipper or grab bucket, which is able to dip as deep as 50 feet below the water surface. This type of dredge may be used in dredging channels or in cutting a wide drainage ditch provided the ditch is large enough to float the dredge. This dredge will dig its own waterway through land, and is particularly useful for drainage work. A dredge where the excavating is done by a number of buckets placed on an endless chain is known as an elevator dredge. A bucket chain is able to elevate the spoil considerably higher than other types. It is frequently used to raise sand and gravel from a stream bed.

The hydraulic dredge digs by suction of the spoil material from the bottom. A pipe is lowered to the area to be dredged. A large power-driven rotary cutter is rigged in front of the entrance end of this pipe. Operation of the cutter breaks up the soft material of the bottom so that it may readily be transported by a current of water. A water pump attached to the upper end creates a powerful flow of water through the pipe which picks up and carries the loose material near the mouth of the pipe. Large, specially designed centrifugal pumps create the flow.

DREIKANTER. Literally a 3-cornered or 3-edged pebble. A term of German origin signifying a pebble that has been sculptured or faceted by natural sandblasting. Such pebbles are usually considered to be proof of the semi-arid or even desert condition under which they have been formed, but they may also be formed in pluvial climates provided that the regolith is composed of porous and shifting sands such as compose glacial sand plains and coastal beaches. They are also called gibbers or glyptoliths.

DREWITE. A term proposed by R. M. Field in 1918 for pure calcium carbonate muds of organic chemical origin. Drewite probably forms the bulk of the fine-grained unfossiliferous limestone from the pre-Cambrian to the present. It was named after G. H. Drew, a pioneer in the study of marine bacteria.

DRIED-FRUIT INSECTS. Wherever dried fruits are produced, whether in the Mediterranean Basin, South Africa, southern Australia, or California, their chief insect pests are the same species. They have been distributed by commerce, probably for several thousand years. Losses caused by insects in dried fruits are difficult to estimate. The loss of weight from insect feeding is usually trivial. The most serious loss is in appearance and quality, which lowers or destroys market value. The presence of insects or any other foreign material in dried fruit obviously is objectionable to consumers. Other losses from insects in dried fruit include the cost of construction and maintenance of facilities for fumigation, the cost of fumigants, and the expense of applying them. Insects are chiefly responsible for the waste involved in culling out damaged dried fruit, and the costs of screening and washing of fruit, general plant sanitation, and cold storage. Special packages designed to resist and exclude insects add to the expense.

Beetles. The beetles associated with dried fruits develop through 4 stages: egg, larva, pupa, and adult. Most of the eggs hatch within 5 days. The eggshells are thin, and the embryos either hatch within 2 or 3 weeks or die. Larval life proceeds through a varying number of molts and abrupt size changes. Feeding is continuous except when the temperature falls below 45°F (7.2°C). Overwintering larvae can feed occasionally on warm days. Beetle pupae develop in cracks in boxes, creases in dried fruits, under bark, or in cells in the soil or ground litter. During the pupal stage, the body structure is completely rearranged, and the resulting adult bears little or no resemblance to the larva from which it developed. The transformation may take only a week or two, or may be prolonged several months by cold weather. As a rule, adult beetles are active feeders. Because some of them live longer in the adult stage than as a larva, they damage dried fruits even more than their larvae do. All of the principal species, except possibly the sawtoothed grain beetle, can fly. They infest commodities by flying or crawling into storage buildings, or by being carried in with dried fruits.

Sawtoothed grain beetle (*Oryzaephilus surinamensis*, Linnaeus). This insect is described in the the entry on **Grain Storage Insects.** In raisins stored for a year or more, the insect can become very abundant. It is not numerous in spring; numbers declining markedly during hot weather.

Merchant grain beetle (*Oryzaephilus mercator*, Fauvel). Similar in habits and appearance to the sawtoothed grain beetle. See Fig. 1. The insect has been found in waste dates, on the ground in the Coachella valley (California) date gardens, and in cull figs in the Fresno

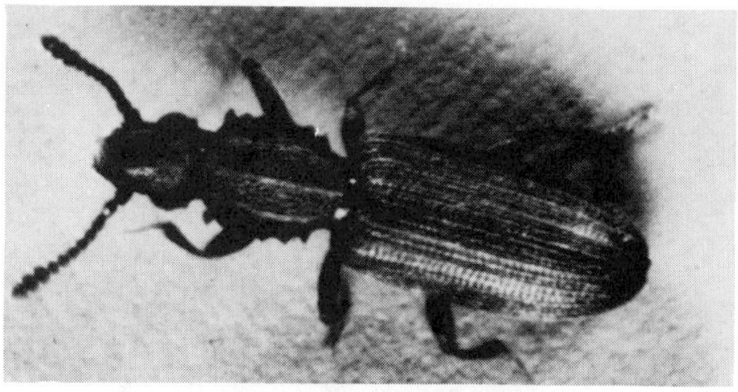

Fig. 1. Merchant grain beetle. (*USDA photo.*)

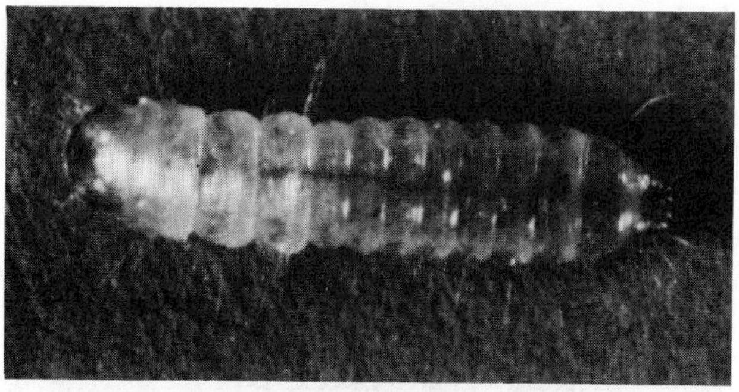

Fig. 3. Larva of hairy fungus beetle. (*USDA photo.*)

district. It does not endure low temperatures well, nor do the females lay as many eggs as the sawtoothed grain beetle.

Small darkling beetles (*Blapstinus* species). These are found in orchards. The most common is *Blapstinus rufipes* (Casey), a dull black, somewhat flattened beetle about $\frac{1}{4}$ inch (6 millimeters) long. These beetles congregate on ripe figs that have fallen to the ground. They may be so numerous as to completely cover a fig. Only fragments of their life history have been recorded. The adults appear in June at about the time the first crop of figs ripen and fall to the ground. The larvae probably develop on plant materials in the soil. As the season advances, they gradually disappear.

These beetles attack tender plants. Young bell pepper plants have been girdled at the soil surface, and as many as 75 adult beetles have been counted around a single seedling. Several species, including *B. rufipes*, have been observed attacking young plants of sugar beet, lima bean, and tomato.

Hairy fungus beetle (*Typhaea stercorea*, Linnaeus). A polished, brown, elongate-oval beetle about $\frac{1}{10}$ inch (2.5 millimeters) long. The body is well covered with short, fine hairs. See Fig. 2. It is a member of the family *Mycetophagidae*, or fungus eaters. It is common in moldy dates lying on moist soil and in moldy raisins. Other foods eaten by this species include stored grain and seeds, tobacco, and cacao. Adults fly in the evening shortly after sunset. On moist, moldy raisins, this beetle develops from egg to adult in 3 weeks. It spends 3 days in the egg, 14 days as a larva, and 4 days as a pupa. See Fig. 3. Newly hatched larvae are unable to develop on clean raisins, but larger larvae can.

Fig. 2. Adult hairy fungus beetle. (*USDA photo.*)

Driedfruit beetle (*Carpophilus hemipterus*, Linnaeus). Adult dried-fruit beetles are about $\frac{1}{8}$ inch (3 millimeters) long and black, with 2 amber-brown spots on each wing cover, one near the tip and a smaller one at the outer margin of the base. See Fig. 4. They are strong fliers, but fly only during daytime when temperature is above 63°F (17.2°C). They may travel as far as 4.5 miles (8.3 kilometers) per day.

Fig. 4. Adult dried-fruit beetle. (*USDA photo.*)

These beetles, of the family *Nitidulidae*, are among the chief pests of ripening and drying figs. Adults and larvae may be found during the winter and at other times in fruit dumps, rotting melons, sticktight pomegranates, dropped peaches, plums, citrus fruits, cull figs, and moist raisins. They do not attack sound fruit, but prefer overripe, fermenting, and rotten fruit. This species thrives in fermenting grape pomace, a winery by product. When raisins are being made, damaged grapes, especially those with bunch rot, attract these beetles to the drying trays. Fruit that is very dry or far advanced in decay ceases to attract them. However, larvae that begin growth in overripe figs, for example, may continue their development after the fruit is fairly dry. Much waste fruit falls to the ground under fruit trees and often squashes and cracks open. In date gardens, where frequent irrigation keeps the soil surface moist, driedfruit beetles abound in fallen waste dates.

Adults feed and larvae develop in a moist, dark environment of yeasty and often moldy pulp. These beetles carry yeasts and mold spores in and on their bodies and inoculate ripening figs with plant diseases. In addition, they carry peach brown rot and bunch rot of grapes. Adults also visit the sap flow of bark wounds in trees that produce a flow of wet and consequently fermentable sap.

Larvae reach a length of $\frac{1}{4}$ inch (6 millimeters) when fully grown. They are white to yellowish. The head and rear end are amber-brown. They are sparsely hairy and have two prominent spinelike projects at the tail end, with two smaller ones in front of them. See Fig. 5.

Fig. 5. Larvae of dried-fruit beetle. (*USDA photo.*)

Pupae are about $\frac{1}{8}$ inch (3 millimeters) long, white or pale yellow, and somewhat spiny. No cocoon is formed. The species usually overwinters as pupae in cells in the soil. Larvae and pupae of nitidulids have been found as deep as 2 feet (0.6 meter) in dry soil in a fig orchard, but most are in the upper 8 inches (3 centimeters). Hundreds of fully grown larvae, pupae, and recently transformed adults may be present under each tree in early spring. New adults emerge late in February and early March. By late April, all have left the soil. A combination of light rainfall and a mild winter favors above-average survival of driedfruit beetles, but a cold, wet hibernation period reduces the population. On warm days in winter, a few adults have been captured in rotary net flight traps.

The species have a short developmental period and a long adult life. At 90°F (32.2°C), the incubation is 1 day, the larval period is 12.4 days, and the pupal period is 5.8 days. The total from egg to adult is 19.2 days, but may be short as 12 days. Mated females may average 103 days of life and males, 146 days. Total egg production may average more than 1000. The record for one female is 2134 eggs, laid over a period of 79 days.

Corn sap beetle (*Carpophilus dimidiatus*, Fabricius). The adult resembles driedfruit beetle, but has no spots and color ranges from brown-yellow through brown-black tinged with red. They are $\frac{1}{16}$ to $\frac{1}{8}$ inch (1.5 to 3 millimeters) long. See Fig. 6. They fly at any time of day. Food habits are similar to those of driedfruit beetle. They are very common in cull grapefruit and cull dates. Corn sap beetles are particularly fond of developing ears of sweet corn that have been previously damaged by insects or birds. Among their host foods are apple pomace, rotten watermelons, stored groundnuts (peanuts) with split hulls, inferior copra, sago flour, coarse bran, and brazil nuts. Periods from egg to adult is about 2 weeks. Adults live about 60 days. Those that overwinter live up to 200 days. Total egg production of one female is about 200.

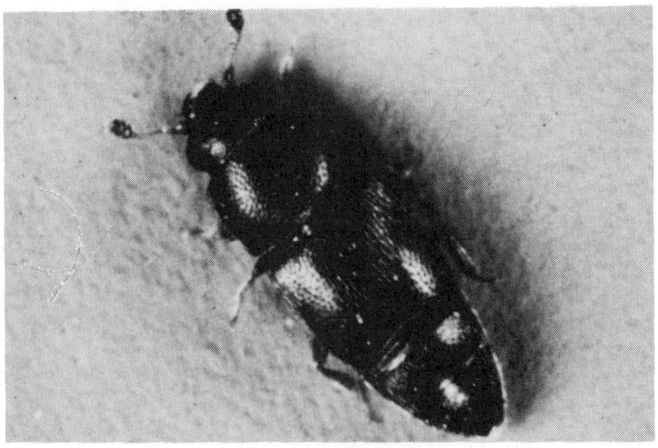

Fig. 6. Adult corn sap beetle. (*USDA photo.*)

Yellow nitidulid (*Haptoncus luteolus*, Erichson). Very similar to driedfruit and corn sap beetle. They are of a blunt shape and yellow-brown. They do not fly during midday, but are numerous in early morning and late afternoon. Large numbers have been trapped in the San Joaquin and Coachella valleys of California.

Pineapple beetle (*Urophorus humeralis*, Fabricius). The adults are shiny black, nearly $\frac{3}{16}$ (4.5 millimeters) long. They fly in daytime and are more numerous in date gardens than in fig. orchards. Pineapple beetles feed in bunches of dates ripening on the palms. They are the dominant species in Hawaiian pineapple fields, where they feed on fermenting trash left from harvesting. In sugarcane fields, they develop in souring cane trash and damage seed pieces of cane underground. In the Coachella valley, they are abundant in waste grapefruit. From egg to adult requires 16.5 days. Females lay an average of about 880 eggs. Adults reared on pieces of pineapple stump lived an average 89 days.

Leadcable borer (*Scobicia declivis*, Le Conte). Cylindrical, dark brown or black beetles from $\frac{1}{8}$ to $\frac{1}{4}$ inch (3 to 6 millimeters) long. See Fig. 7. They develop in dead parts of trees. In central California, they fly in the early evening. These insects rarely enter figs. They sometimes bore holes in the laminated paper covers used to enclose raisin stacks for fumigation. They cause losses in wineries by boring into wine barrels and casks. In the past, the adults have caused considerable trouble by boring holes in lead sheaths of telephone cables, thereby allowing dampness to enter the cable. The insects are attracted to products of fermentation and can be baited with ethyl alcohol traps.

Fig. 7. Adult leadcable borer. (*USDA photo.*)

Date-stone beetle (*Coccotrypes dactyliperda*, Fabricius). A minute cylindrical, shiny, dark-brown beetle. It is a close relative of the bark beetles and is common in seeds of waste dates in the Coachella valley. The insect also infests sweet almonds in the Orient and the hard seeds, known as vegetable ivory, produced by several species of palm in Africa. Buttons made from vegetable ivory are also attacked. Thus another name for the insect is *button beetle.* Other hosts are betel nut, nutmeg, and cinnamon bark.

Moths. None of the adult moths can feed, although they do drink liquids, such as plant nectar. The damage is done while they are larvae.

Raisin moth (*Cadra figulilella*, Gregson). This is a small, gray moth formerly known as *Ephestia figulilella.* Larvae were first noticed in the Fresno, California area in 1928 on muscat raisins. They reached a peak of abundance in 1930. These insects live and develop out-of-doors, although they are often brought into storages with infested commodities. The larvae attack all the usual varieties of drying and dried fruits, fallen figs, and damaged or moldy clusters of grapes and vines. Raisins are attacked until they become unattractively dry. Cottonseed cake, cacao beans, and cashew kernels are among the host foods. Fallen mulberries are important because they are available to the insects early in spring when other food is scarce.

Female raisin moths deposit eggs on all common varieties of drying and dried fruits. From egg to adult requires about 43 days at 83°F (28.3°C). Larvae usually molt 6 times. Larval life is greatly extended during winter. In raisin storage, any larvae that escape fumigation continue to feed, and in the spring they pupate and emerge. In vineyards, most of the larvae pass the winter in cocoons in the upper few inches of soil near the vine trunks and along the wires. In fig orchards, many larvae overwinter in a 6-inch (15-centimeter) band of soil around the tree trunks. The overwintering larvae pupate in the spring. Emergence of adult moths begins in April and reaches a peak in May. There are about three overlapping broods per year. In summer, mated females provided with sufficient water average about 350 eggs, with a record noted of 692 eggs. Most eggs are laid in early evening.

Dusky raisin moth (*Ephestiodes gilvescentella*, Ragonot). Formerly called *Ephestiodes nigrella* (Hulst), this species resembles the raisin moth in size and appearance. They are found in vineyards and in raisin storage areas. There is a new generation about every 2 months.

Indian meal moth (*Plodia interpunctella*, Hübner). This moth also attacks stored grain products and is described in entry on **Grain-Storage Insects.**

Driedfruit moth (*Vitula edmandsae serratilineela*, Ragonot). The adult is a mottled gray and about $\frac{3}{4}$ inch (18 millimeters) long. The moth was first reported in California (Santa Clara County) in 1903. The moth is not abundant, but is frequently collected from stored figs, raisins, and prunes in the Santa Clara and San Joaquin valleys. They were much more numerous in early years when large tonnages of raisins were stored in the San Joaquin valley for up to 4 years without insect abatement measures.

Dried-prune moth (*Aphomia gularis*, Zeller). The larvae are capable of serious damage. When fully grown, they are more than 1 inch (2.5 centimeters) long and produce considerable webbing and coarse excreta. Before spinning a cocoon for pupation, they frequently excavate a shallow depression in the wood of bins or boxes. The insect is now uncommon in California.

Navel orangeworm (*Paramyelois transitella*, Walker).

Flies. The principal species of fly that attacks dried fruits is *Drosophilia.* See entry on **Drosophilia.**

Soldier fly (*Hermetia illucens*, Linnaeus). Black, two-winged insects that resemble some of the four-winged mud-dauber wasps in size, color, and habit. The larvae or maggots are large, brownish, and flattened and have a tough skin. The larvae feed in accumulations of rotten fruit that are dried out, decayed, and black. In New Zealand, they sometimes damage honeybee colonies by feeding on wax, pollen, and honey. These flies overwinter as larvae.

Wasps and Bees. Large numbers of minute *Bracon hebetor* wasps are sometimes seen flying over dried fruit in storage in the fall. They are parasites of storage moth larvae, including the Indian meal moth and the raisin moth. Larger larvae of the moths are paralyzed by the female wasp. Although this wasp kills many moth larvae, its value in suppressing moth infestations has not been determined.

The wasp *Cephalonomia tarsalis* (Ashmead) is small, black, and a common external parasite of the larvae and pupae of the sawtoothed grain beetle. It is only 0.06-inch (1.5 millimeters) long. Winter is passed in the pupal stage. The cocoons of this parasite are frequently seen where sawtoothed beetles are plentiful. However, it is not considered very effective in controlling the beetle.

Fig wasp (*Blastophaga psenes*, Linnaeus). This insect is described in entry on **Chalcid Wasp.**

DRIFT (Geology). Before Louis Agassiz propounded his theory that extensive deposits of sand, gravel, boulder clays, etc., of Northern Europe and North America were the result of the action of great continental ice sheets, it had been suggested that during some period when the land had stood at a lower level, icebergs were swept from the north over the continents and melting, dropped their loads of detritus. Thus some of the material which we now know to be of glacial origin was called drift, because it was believed to have been "drifted" to its place through the agency of ice. This was the theory of Sir Charles Lyell as a substitute for the still older theory of the diluvialists who believed that these glacial deposits were positive evidence of the "deluge." Geologists have retained the term drift to designate all unconsolidated sediments which are determined to be of glacial origin, and still further divide drift into stratified drift and unstratified drift or till.

In Great Britain, the term drift is used for all surficial, unconsolidated rock debris which may be transported from one location and deposited in another. Drift is distinguished from solid bedrock.

In Africa, the term drift is used to indicate a ford or a sudden dip in a road over which water may flow at times. In South Africa, the term is used for a ford in a river.

DRIFT (Instrument). With reference to industrial and scientific instruments, the Scientific Apparatus Makers Association defines drift as a change in the output-input relationship over a period of time; drift (deviation cycle) as the maximum change in output for the same input and operating conditions obtained from a number of test cycles over a specified period of time. The operating conditions may vary between test cycle measurements provided they stay within normal operating conditions of the test device. Typical expression: The maximum drift for any point, at ambient temperature conditions (75 ± 2°F) (23.9 ± 1.1°C), and between tests not exceeding ambient temperature limits of 30 to 130°F (−1.1 to 54°C), was within 0.2% of output span for a period of 30 days.

Point drift is defined as the change in output over a specified period of time for a constant input under specified reference operating conditions. Point drift is frequently determined at more than one input, as for example: at 0%, 50%, and 100% of range. Thus, any drift of zero or span may be calculated. Typical expression: The drift at mid-scale for ambient temperature (70 ± 2°F) (21.1 ± 1.1°C) for a period of 48 hours was within 0.1% of output span.

DRIFT PIN. A tapered steel pin which is used during the fabrication of a member to hold the individual parts together before the fitting-up bolts (bolts which are necessary to draw the parts together subsequent to riveting) are inserted in the rivet holes. Drift pins are also used together with bolts in steel erection to fasten the members to the gusset plates or other connections before the field rivets are driven.

DRILLING (Natural Gas). Natural Gas.

DRIP IRRIGATION. Irrigation.

DRIVE TRAIN CONTROL. Automotive Electronics.

DRIZZLE. Precipitation and Hydrometeors.

DROMEDARY CAMEL. Camels and Llamas.

DROOP (Governor). Governor.

DROP. A small volume of liquid, bounded almost completely by free surfaces. The simplest way to form drops is to allow liquid to flow slowly from the open lower end of a vertical tube of small diameter. When the pendent drop exceeds a certain size it is no longer stable and detaches itself and falls. Drops may also be formed by condensation of a supercooled vapor or by atomization of a larger mass of liquid. The weight of the largest drop that can hang from the end of a tube of radius, a, is nearly

$$mg = 2\pi a\gamma \cos \alpha$$

where γ is the surface tension of the liquid, α is the angle of contact with the tube. This relationship is the basis of a convenient method of measuring surface tension.

DROP FORGING. **Forging; Iron Metals, Alloys, and Steels.**

DROPLET (Atmospheric). **Precipitation and Hydrometeors.**

DROPPING-MERCURY ELECTRODE. **Polarographic Analyzers.**

DROSOMETER. **Precipitation and Hydrometeors.**

DROSOPHILA (*Insecta, Diptera*). Also called the vinegar flies, fruit flies, or pomace flies, species of *Drosophila* are common wherever damaged or overripe fruit and vegetable garbage accumulate. In the mild weather of late summer and early fall, *Drosophila melanogaster* (Meigen) and *D. simulans* (Sturtevant) become abundant in the San Joaquin valley of California, but they do not thrive in the hottest part of summer. In cool weather, *D. pseudobscura* (Frolowa) is more common. More than 50 other species are probably in the area, most of them of minor economic importance. See also entry on **Fruit Fly.**

D. melanogaster, the dominant species in the central valley of California, is a small, clear-winged fly with bright-red eyes and a shining black abdomen, the first 3 segments of which have a yellow band. The body is from $\frac{1}{16}$ to $\frac{3}{32}$ inch (1.5 to 2.5 millimeters) long. These flies are attracted to fermenting fruit waste, melons, piles of peach and apricot pits, damaged grapes, tomatoes, or other fruits. They are common in wineries and in and around tomato and fruit canneries. Cull-fruit dumps on farms and along roadsides swarm with vinegar flies. By leaving rotting fruit and entering figs, the flies inoculate the ripening figs with yeast cells, causing souring. They also contribute to the spread of bunch rot of grapes.

Activity of adult vinegar flies in the field is controlled chiefly by temperature, light intensity, and air movement. They are strong fliers and can fly more than 6 miles per day. They do not move in winds over 5 miles (8 kilometers) per hour. Light intensity of over 150 foot-candles tends to immobilize them. Fly activity may be fairly brisk in a fig tree with heavy foliage. Flies do not congregate in a tree with light foliage because both light and breezes penetrate it. They are active only during the day.

D. melanogaster has the most rapid reproductive rate of any dried-fruit insect. The flies spend only about 24 hours in the egg, 3 days as larva, and 3 days as a pupa, or a total of 7 days. Under some conditions, mature eggs are retained in the body of the female, and such eggs may hatch within 1 hour after they are laid. Adults may lay as many as 2000 eggs, but average about 1000. Females live about 39 days in warm weather, but up to 70 days in cooler weather. Life of the male is about 41 days.

From this fast reproductive cycle and large numbers of offspring produced, it is obvious why the *Drosophila* has been used for innumerable experiments in research and biological laboratories. Much genetic research has centered around these small flies.

DROSOPHILA MELANOGASTER. **Heredity.**

DROSS. The scum that collects on the surface of melted metal as a result of oxidation and the rising to the surface of impurities.

DROUGHT SURVEILLANCE. **Earth Resources Satellites and Geologic Remote Sensors.**

DRUG. A catchall term that embraces chemical substances which are administered to the human body (also livestock, pets, etc.) whether constructively or destructively and whether professionally or amateurishly. Drug as a term has become so all-inclusive in its use that it can be applied to thousands of natural and synthetic substances for scores of specific intentions. The following is an abridged list of categories of agents that variously are called drugs as well as medicines, biologicals, chemicals, pharmaceuticals, remedies, etc. Many of these categories and specific examples given are described under alphabetical entries elsewhere in this volume. See also **Drug Addiction.**

DRUG ADDICTION. A physiological and emotional dependence upon an abnormal use of habit-forming chemicals. Drug addiction is essentially a sociological phenomenon and problem. However, there are many ways in which the basic sciences can contribute toward the alleviation and hopefully ultimate solution of this problem.

Chemists and biologists can assist in the studies of the human reactions to the administration of chemicals to the body by continuing to specify the degree and nature of such reactions and side reactions and thus provide increasingly reliable information for alerting people to what may be expected from the use of such chemicals. Medical specialists can contribute to these evaluations as well as to the development of improved rehabilitation procedures which may be implemented for those people who already have developed a dependence upon chemicals. Notably, the medical profession can continue in its efforts to exert a much tighter control over the use of habit-forming chemicals by prescribing them only when absolutely necessary, particularly at the institutional level, including hospitals and convalescent centers.

The manufacturers of these chemicals, assisted by the scientists in their employ, can exert much more care in the distribution of known habit-forming chemicals and, in particular, by imposing limitations on their production for professional use only. With many habit-forming chemicals, the sociological problem commences with the *overproduction of such materials.*

A large percentage of drug users suffer from some form of personality disorder. Obviously, the continuing, concentrated attention to the unbalanced members of the population by psychiatrists and social scientists must continue. In one study of a group of 1,000 patients admitted to a federal narcotic hospital, only 4% were found to be emotionally stable. The majority of these patients had personality disturbances of the antisocial type, while another large group indicated neurotic disorders. Drug addiction was but one manifestation of their lifelong history of maladjustment. Having failed to solve their emotional conflicts by any other route, they resorted to chemicals to relieve their symptoms of anxiety. Occasionally, drug addiction is seen in association with a major mental illness of which such addiction is but one manifestation.

Some of the oldest historical records contain evidence of the use of various extracts of berries, leaves, beans, roots, fungi, and tree bark for their mental and physiological effects in escaping the drudgery of life. Some of these drugs by comparison with hard narcotics are mild, but nevertheless contribute to a dependence upon chemicals. Other drugs, such as opiates, are treacherous from the beginning because their use produces an overwhelming physical dependency which can only be satisfied by larger and larger doses.

The opium derivatives, which include morphine, codeine, and heroin, are so habit forming that once the craving for them becomes established, the victim's entire emotional life is subjugated to the drugs. Heroin is considered so dangerous that importation into the United States for *any* purpose is banned. Morphine and codeine, however, are used by the medical profession in combating otherwise uncontrollable pain. The problem of addiction arises out of the nature of habit-forming drugs and with the difficulty the human organism has in withstanding them. The addict's dependence upon drugs is twofold. The victim is prompted by a tremendous physical craving, and at the same time the feeling of elation derived from the drug provides temporary support for emotional insecurity. Morphinism is unique among addictions in that sense perception remains clear. Increased resistance to the effect of drugs develops over a period of time so that the addict requires progressively more of the drug to obtain satisfaction.

The cost of heroin is so prohibitive that the recognized means of supporting this habit is to turn to crime. The narcotics user is more interested in rapid relief of craving than in sanitary measures of administering the drug, so frequently the victim employs contaminated syringes for intravenous injections. This often leads to hepatitis (or even malaria) contracted when an unclean needle is passed from one user to another. The user of heroin never can be sure how pure the drug may be, or whether it is admixed with other substances which may be even more lethal than the heroin itself. The rate of deaths from overdosage of heroin reaches its highest point in the overpopulated, low economic areas of large cities.

Sometimes diagnosis is difficult for those skilled in the field and can only be positively made with removal of the patient from all possible sources of supply. Then the characteristic withdrawal symptoms appear, during which the addict cannot conceal his acute physical pain. Among the withdrawal symptoms are *severe cramps in the abdomen and legs, muscular twitching, vomiting,* and *diarrhea.* The patient will be irritable, restless, and unable to relax. Sweating is profuse. Rest and sleep are difficult or impossible to attain.

The drastic treatment required for drug addiction should be attempted only by well-trained personnel with adequate facilities. Withdrawal of the drug may be abrupt, rapid, or gradual. Following withdrawal, a period of psychotherapy and rehabilitation is required. This is an important phase of the treatment, for it has been found that without a minimum of four months of psychotherapy, most patients relapse. The use of methadone and scopolamine in withdrawal procedures remains controversial. While withdrawal from heroin may be eased by this procedure, the end-result is a drug addict, now dependent upon methadone or scopolamine, who now must proceed through still another withdrawal before full rehabilitation is achieved.

Because of the absence of full emotional and mental development, children and youths are most easily drawn into drug habits, unfortunately often by friends and associates. For a very short period, being a member of a drug cult can be a status symbol among the less-gifted, emotionally unstable youths. Warning signals for friends and parents of a drug user include unexplained absences, unaccountable disappearance of valuable objects from the household which may be sold to obtain funds for buying drugs from a "pusher," a withdrawal from some former friends and associates and family members, excessive irritability, and of course where some of the hard narcotics are being used, arms covered with needle punctures.

Addiction to drugs other than heroin follows a somewhat similar pattern, modified only by the peculiarities of each drug. *Cocaine,* which is derived from the leaves of a South American shrub, is taken by sniffing through the nostrils. Unlike morphine, cocaine administration may be skipped for a few days at a time without producing marked discomfort. Cocaine produces a feeling of elation and is used by emotionally unstable and immature persons whose pursuits require a false gaiety or sustained good humor for long periods of time. The pleasurable effects of cocaine are mixed with hallucinations. A commonly reported hallucination is that of insects crawling on the skin. Cocaine sometimes is referred to by users as "snow" and a user is condescendingly called a "snowbird" by the pusher. Cocaine is extremely dangerous because it can produce psychosis, and an overdose can result in death.

Hallucinogens appear to be preferred by the immature person who is in conflict with society. *Marijuana* is a drug of this type and is produced from the dried leaves of Indian hemp. This is the same plant from which *hashish* is derived, but the effects of hashish are five to ten times as potent. Marijuana sometimes is referred to as pot, grass, reefers, joints, and tea. The drug produces a sensation of floating and a gross distortion of time and space perception by interfering with the normal mental processes. Partial evidence indicates that marijuana use is associated with chromosomal changes which can cause birth defects in future generations, a particularly undesirable aspect for youths who ultimately plan to have children of their own.

LSD (lysergic acid diethylamide) is a hallucinogenic agent derived from the fungus, ergot. Anxiety, disorientation, and impaired physical coordination are the usual result of its use. Some users of LSD have met their deaths by leaping in front of autos or out of windows in the mistaken assumption that they were invincible. Hallucinogenic drugs, such as LSD, have been used in research, particularly in connection with mental illness. The rationale is that hallucinogenic drugs produce a "schizophrenic-like" state; the physiological and biochemical mechanisms of the drug action may provide clues to mechanisms involved in schizophrenia; drugs counteracting the drug-induced psychoses may provide a therapeutical approach to schizophrenia. However, to date, the hallucinogens have been of little therapeutic value. It is interesting to observe in this connection that some unstable individuals seeking an escape will make use of a chemical that researchers have used to artificially induce schizophrenia, which is defined as a severe form of mental disease, one of the major psychoses, characterized by loss of contact with reality and by disorganization of the personality.

Bromides are sedatives compounded from the element, bromine. Prescribed for a number of conditions, including convulsive disorders, they also are self-administered as sedatives. The danger from the continued use of bromides lies in the cumulative effects of the drug within the human system. When bromide concentration in the blood rises to a dangerous level, bromide intoxication may result. The symptoms include unusual drowsiness, delirium, and hallucinations, and if not detected the condition may result in death. Bromide intoxication can be readily determined by tests of the blood serum. Upon diagnosis, hospitalization is advisable for elimination of the accumulated bromide from the system. This is done principally by continuous baths and the controlled administration of salt.

Barbiturates are used in sleeping pills and are derived from the organic compound, barbituric acid. These chemicals induce a feeling of relaxation, usually followed by sleep. Barbiturates are sometimes used to provide temporary respite in times of unusual emotion stress. The World Health Organization has stated that barbiturates "must be considered drugs liable to produce addiction."

Amphetamines (benzedrine, dexedrine, and methedrine) sometimes are taken by the emotionally depressed to achieve euphoria and drive, or to find a release from fatigue. Professionally, they are sometimes prescribed in moderate doses for appetite depression in the control of obesity and for relief of mental depression. These drugs have been abused to the extent of preparing them by disturbed, uninformed amateurs in a form for injection directly into the veins. Medical authorities estimate that the life expectancy of a person who "mainlines," i.e., injects methadrine (also called "speed") is approximately five years.

See also **Alkaloids; Cocaine; Codeine; Hallucinogens;** and **Marijuana.**

References

Buerger, A. A., and J. S. Tobis: "Neurophysiologic Aspects of Rehabilitation Medicine," Charles C. Thomas, Springfield, Illinois, 1976.

Broadhurst, P. L.: "Drugs and the Inheritance of Behavior," Plenum, New York, 1978.

Cull, J. G., and R. E. Hardy: "Organization and Administration of Drug Abuse Treatment Programs," Charles C. Thomas, Springfield, Illinois, 1974.

Feit, M. D.: "Management and Administration of Drug and Alcohol Programs," Charles C. Thomas, Springfield, Illinois, 1979.

Hollister, L. E.: "Clinical Use of Psychotherapeutic Drugs," 5th edition, Charles C. Thomas, Springfield, Illinois, 1977.

Iversen, S. D., and L. L. Iversen: "Behavioral Pharmacology," 2nd edition, Oxford Univ. Press, New York, 1980.

Louria, D.: "Overcoming Drugs: A Guide for Parents, Educators and Legislators," McGraw-Hill, New York, 1971.

Petersen, D. M., Whittington, F. J., and B. P. Payne: "Drugs and the Elderly," Charles C. Thomas, Springfield, Illinois, 1979.

Selden, L. S., and L. A. Dykstra: "Psychopharmacology," Van Nostrand Reinhold, New York, 1977.

Vorhees, C. V., Brunner, R. L., and R. E. Butcher: "Psychotropic Drugs as Behavioral Teratogens," *Science,* **205**, 1220–1225 (1979).

Ware, M.: "Operational Handbook for Narcotic Law Enforcement Officers," Charles C. Thomas, Springfield, Illinois, 1975.

DRUGS (Ocean-Derived). Ocean Resources (Mineral).

DRUM (Boiler). Boiler; Feedwater (Boiler).

DRUM-FISH. Croakers.

DRUMLIN. A drumlin is a hill composed of glacial material of unstratified and heterogeneous character usually about 100 feet in height and $\frac{1}{4}$–$\frac{1}{2}$ mile in length, oval in shape, and with its long axis parallel with the general direction of ice movement. Sometimes a mass of bed rock seems to have been the anchor about which the glacial till was deposited. Drumlins are known, however, which contain no bed rock core. Drumloid-shaped hills similar to the smaller glaciated rock features called roches moutonnée are sometimes called rock drumlins.

DRUMSKIN ACTION. The vibration of a wall as a whole under the action of an incident sound wave.

DRUM STORAGE (Computer). A magnetic drum is a metal cylinder which revolves about its axis and on whose surface a magnetic material is deposited which permits the recording of information. The information is recorded on the drum by passing current through a coil of wire in the write head which magnetizes a track on the drum surface in accordance with the pattern of binary data. In order to read the information stored on the drum, a similar coil of wire in the read head is mounted near the drum surface where it intercepts the magnetic flux and transmits the induced voltage to a sense amplifier.

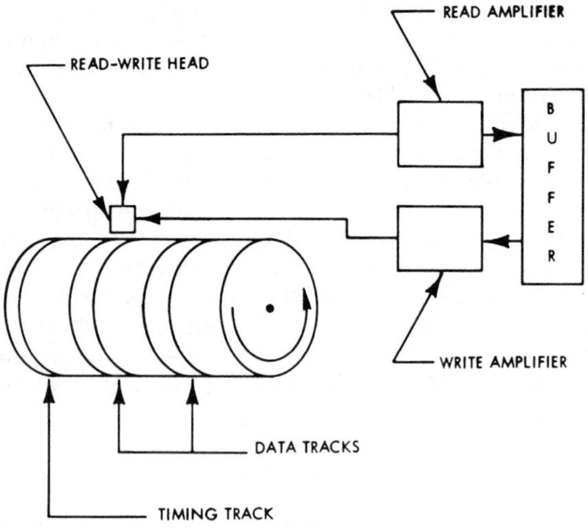

Magnetic drum.

With reference to the accompanying figure, the drum is divided into tracks around the drum circumference and a read/write head is associated with each track. Data may be stored on the drum in serial by bit, serial by digit format, or a parallel by bit, serial by digit format, depending upon the design selected by the manufacturer. A timing track is recorded on the drum to provide the sampling pulses for reading or writing the data and for address determination. The timing track signals increment counters which are used to determine the digit and word locations which are passing under the read/write head at any particular time. Other means for providing timing (e.g., optical) can also be used.

When data are to be written on the drum in a serial by bit, serial by digit format, the drum storage address selects the appropriate write head. When the addressed location passes under the head, the information in the buffer register is gated to the write amplifier which drives current through the write head according to the bit pattern to be written. On a read operation, the selected read head gates the signal to the read amplifier as the addressed data passes under the head.

The average time to read or write a selected address on the drum is one-half the revolution time of the drum. This time may be shortened by placing multiple read or write heads on each track around the periphery of the drum and by providing logic in the drum storage unit to determine which head to select. The number of heads which can be mounted on the drum is a function of the allowable track-to-track density of the magnetic recording material and the physical dimensions of read/write head. By arranging the heads along a helical path, increased track density can be achieved. Another factor which limits the number of read/write heads on the drum storage unit is the cost of the heads and their associated mounting hardware.

Drums make possible large-capacity data storage with the fastest access of any mechanical (moving) storage configuration—with exception of fixed-head disk files, which provide comparable access times. The cost per digit is relatively high because of the number of read/write heads required.

See also **Storage (Computer)**; and terms listed under **Data Processing.**

Thomas J. Harrison, International Business Machines Corporation, Boca Raton, Florida.

DRUPOIDEAE. Rose Family.

DRUSE. A cavity, usually in a sedimentary rock, the walls of which are encrusted with minerals which have been derived, through underground solutions, from the rocks in which the cavities were formed, by solution.

DRY CELL. Battery.

DRY ICE. Clouds and Cloud Formation.

DRY ICE (Fog Clearance). Fog and Fog Clearing.

DRYING MECHANISM (Paint). Paint.

DRYING OILS. Paint.

DRYING (Process). Frequently in the process industries, materials must be dried, i.e., liquid must be removed from a solid or gaseous phase. In most instances, the liquid to be removed is water, although in solvent recovery systems, for example, the liquid may be an organic solvent. See **Dehumidification** for a discussion of the drying of gases. Some materials which must be dried include: (1) *solutions*, colloidal suspensions, and emulsions, such as extracts, milk, blood, waste liquors, rubber latex, and inorganic salt solutions; (2) *slurries*, which are pumpable suspensions, as found in calcium carbonate, bentonite, clay slip, and lead concentrates; (3) *sludges and pastes*, such as centrifuged solids, starch, filter-press cakes, and sedimentation sludges; (4) *powders* that may be relatively free-flowing when wet, but very dusty when dry, including pigments, cement, clay, and centrifuged precipitates; (5) *fibrous solids* and granular and crystalline solids, such as sand, ores, rayons staple, salt crystals, and synthetic rubber; (6) *formed and shapes solids*, such as pottery, rayon cakes, shotgun shells, brick, rayon skeins, lumber, and objects that have been painted or otherwise coated; (7) *sheeted materials*, such as impregnated fabrics, paper, plastic, and fiberboard—in the continuous form; or veneers, wallboards, foam-rubber sheets, and photographic prints—in the individual-piece configuration. Because of this very wide variety of drying requirements, it is obvious that there does not exist what might be termed a universal

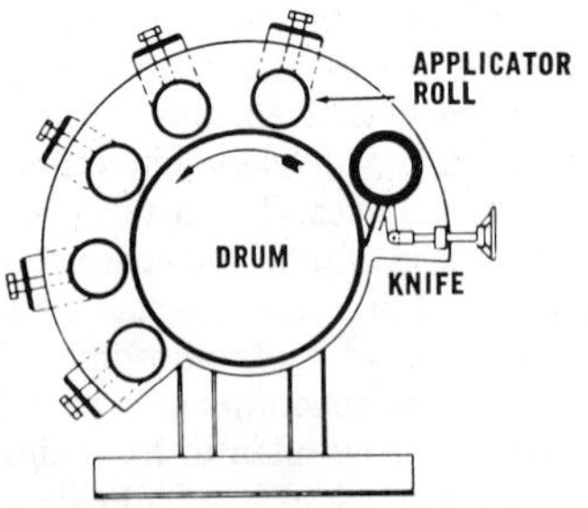

Fig. 1. Single-drum dryer (atmospheric). Dryers of this type may be dip or splash fed (not shown), or, as shown, equipped with applicator rolls. The latter is particularly effective for drying high-viscosity liquids or pasty materials, such as mashed potatoes, applesauce, fruit-starch mixtures, gelatin, dextrine-type adhesives, and various starches. The applicator rolls eliminate void areas, permit drying between successive layers of fresh material and form the product sheet gradually. While single applications may dry to a lacy sheet or flake, the multiple layers generally result in a product of uniform thickness and density with minimum dusting tendencies. (*Buflovak Division, Blaw-Knox Food & Chemical Equipment, Inc.*)

MAJOR TYPES OF PROCESS DRYERS

Continuous Dryers

Direct Heating	Indirect Heating
Tunnel Dryers. Material to be dried is placed on trucks or carts which are moved through a tunnel in which there is a flow of hot gases. Temperature of the tunnel may be zone controlled.	*Drum Dryers.* For materials in liquid and slurry form. One or several drums dip into the liquid or slurry and thus coat the heated drums. The drum temperature is controlled to effect drying during a part of the rotation of the drum from which the dried material is removed by knife prior to dipping one again into the liquid material.
Through-circulation Dryers. Material is supported on a conveying screen that moves continuously. Hot gases from below or above conveyor pass through the material and pick up moisture.	*Cylinder Dryers.* Material in the form of a continuous sheet passes over and around cylinders which rotate and which are heated, usually by steam or hot water.
Rotary Dryers. Material (liquid) is pumped to and showered within a rotating cylinder through which hot gases flow.	*Screw-conveyor Dryers.* A conveyor is housed within a closed, heated housing. This operation may proceed at atmospheric pressure or under vacuum.
Tray Dryers. Material is placed on vibrating trays over or under which hot gases flow.	*Vibrating-tray Dryers.* Similar to the directly heated tray dryer except that the heat is conducted to the trays indirectly (as by electrical heating) rather than by hot gases.
Sheeting Dryers. Material in sheet form passes continuously through a hot chamber. Depending upon product, sheet may be taut (as pinned to a frame), or it may pass through dryer in festoon manner.	*Steam-tube Rotary Dryers.* Material is passed through a long, rotating cylinder. A shell around the cylinder contains steam, hot water, or other heating medium.
Pneumatic Conveyor Dryers. Material is moved in a stream of gas at high-velocity and high temperature and finally collected by a cyclone separator.	

Batch Dryers

Through-circulation Dryers. Material is placed on trays with screen bottoms. Hot gases are blown from below through material.	*Vacuum Rotary Dryers.* Material is subjected to agitation within a stationary, horizontal shell under vacuum. The agitator may be heated to increase drying effectiveness.
Tray and Compartment Dryers. Material is placed on trays which then may be placed on trucks or on permanent shelves within dryer. Hot gases are blown across the trays.	*Agitated-pan Dryers.* Material is placed in covered shallow pans. Pans are jacketed for heating. An agitator stirs the material constantly. This design may be operated at atmospheric pressure or under vacuum.
	Vacuum Tray Dryers. Trays in which material is placed are heated by conduction from supporting shelves. The whole compartment may be under a relatively high vacuum. The material is not agitated.

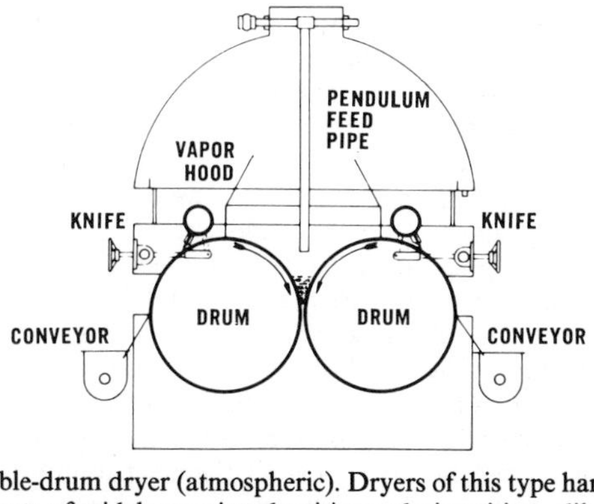

Fig. 2. Double-drum dryer (atmospheric). Dryers of this type handle a variety of food products of widely varying densities and viscosities—dilute solutions, heavy liquids, or pasty materials. A number of products can be dried successfully with this kind of configuration, inasmuch as exposure to temperature above the boiling point is restricted to just a few seconds. The movable drum permits effective control over product film thickness. Feed may be by perforated tube trough, pendulum, or various special configurations. (*Buflovak Division, Blaw-Knox Food & Chemical Equipment, Inc.*)

dryer. Further, universal dryer design criteria are difficult to develop and summarize. See accompanying table.

Basic Concepts of Drying. Two criteria hold for a large number of drying situations, namely, that of breaking the drying process down into two main periods: (1) the *constant-rate period,* the rate of removal of liquid per unit of drying surface is essentially steady, but to qualify the surface of the material must remain fully wet (saturated) during this period; and (2) the *falling-off period,* during which time the rate of drying decreases as it becomes increasingly difficult to move moisture from the capillaries and interstices of the material to the surface where the moisture can be taken up and moved away. With temperature and air flow (if air is the absorbing and moisture conveying medium to be used) well established, the constant-rate period is relatively

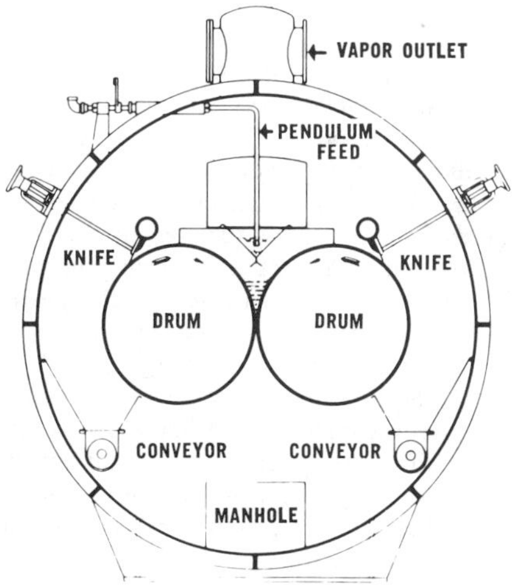

Fig. 3. Single- and double-drum dryer (vacuum). Design configurations like this are applicable whenever products must be dried without exposure to high temperatures or reactive atmospheres. Vacuum operation is particularly suited for vitamin extracts, protein hydrolyzates, soluble coffee, and malt products. Continuous drying without breaking vacuum is achieved by a special conveyance system that utilizes two receivers and air locks. Single-drum feed is usually the pan type with pump and spreading device, or a spray film for materials that are repelled by contact with heated surfaces. Double-drum utilize pendulum or perforated tube feed. Specially designed dispersion devices for feed also can be used. (*Buflovak Division, Blaw-Knox Food & Chemical Equipment, Inc.*)

easy to forecast and to design because this situation is somewhat analogous to that of drying a shallow container of water. Knowing the foregoing conditions as well as the wetted area involved, drying time periods can be predicted. But, in the case of the falling-off period, the dryer designer must be intimately familiar with the mechanism

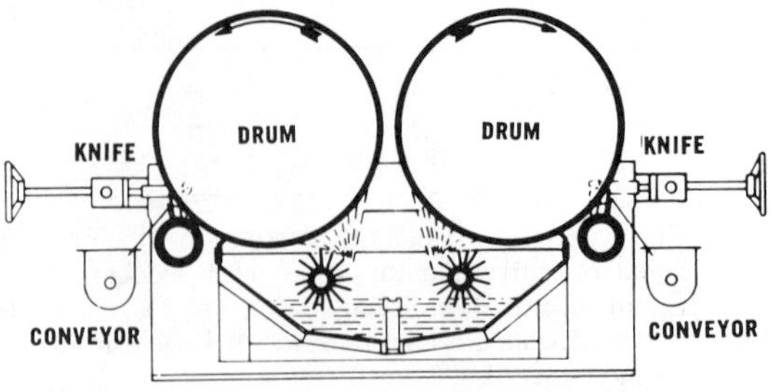

Fig. 4. Twin-drum dryers are designed for handling slurries, corrosive solutions, crystal-bearing or crystal-forming liquids and delicate, heat-sensitive products which may be exposed to high temperatures only for a very limited time. These drums may utilize pendulum or perforated tube for top feed, or splash or spray for bottom feed. Heat-sensitive materials are not in direct contact with the drums. The temperature remains fairly constant and preconcentration is minimized. Cooling and agitation of the material in the pan may also be provided when necessary. (*Buflovak Division, Blaw-Knox Food & Chemical Equipment, Inc.*)

FIRST PASS
IN HOPPER-FEEDER • LEVELER • AIR LOCK • 2 DOWN-THRU ZONES • AIR LOCK • DOFFER • BRUSH

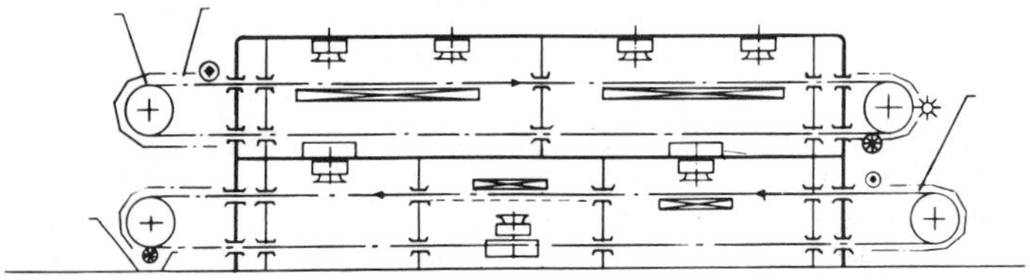

SECOND PASS
COOLING ZONE • UP-THRU & DOWN-THRU ZONES • AIR LOCK • LEVELER
OUT BRUSH • AIR LOCK

Fig. 6. A multi-tier, multi-pass conveyor dryer. Due to space limitations it is sometimes necessary to tier dryers and use gravity or transfer conveyors to carry the product from one tier to the other. The location of the discharge end will vary with the number of tiers. (*The National Drying Machinery Company.*)

whereby moisture is held to the material and the mechanics involved in moving the moisture to the surface where it can be picked up.

Thus, the falling-off period has been broken down into two phases: (1) the *unsaturated-surface drying period*; and (2) the *internal-moisture movement period*. This first period does not impose unusually difficult analysis because essentially the condition is analogous to a partially-filled shallow vessel whose surface is decreasing as drying proceeds. Essentially, this is an extension of the analysis of the constant-rate period, still leaving the internal-movement situation to be predicted. Formulas are available to assist the designer, based upon past experience with specific drying situations and materials. This is a process which is difficult to assess theoretically, and almost always requires experimental runs or comparisons with other at least somewhat similar materials that have been dried successfully. One can generally state that where materials are to be dried to a low-moisture content, the internal-moisture movement period will require the largest portion of the total drying time.

Classification of Dryers. In reviewing the wide variety of drying equipment available, there are a few major classifications that are helpful. There is the distinction between *continuous* and *batch* operations. If a continuous drying operation is desired because it will fit into the overall manufacturing operations best, then this decision will rule out those drying concepts that can be applied only in a batch manner. In some instances, it may turn out that in an otherwise fully continuous manufacturing process, the drying portion will have to be handled by batches simply because, for the particular product, batch drying offers the greatest efficiency. There is also the distinction between *direct heating* and *indirect heating* methods used in dryers. In direct heating, the heat needed is applied by way of immediate contact between the wet material and hot gases. In indirect heating, the heat needed is transferred to the wet material through an intervening medium, commonly pipes or a retaining wall. There are numerous instances, for example, where a product may be too sensitive to withstand exposure to a moving hot gas.

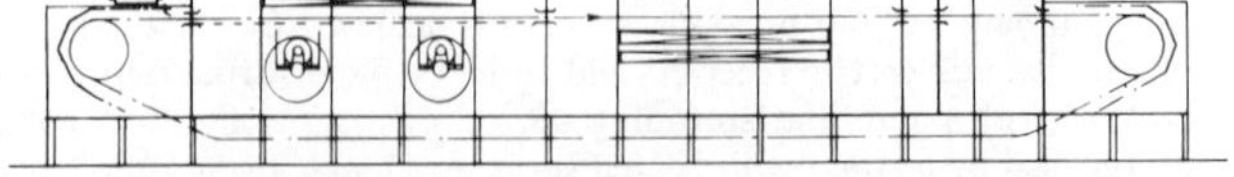

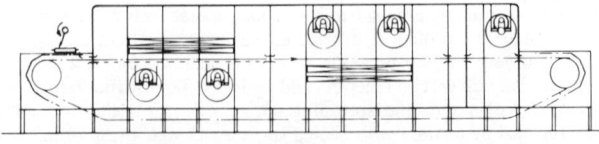

IN CROSS-APRON WIPER-FEEDER • LEVELER • UP-THRU & DOWN-THRU ZONES • AIR LOCK • COOLING ZONE OUT

Fig. 5. Single-stage, single-pass conveyor dryer of convection type commonly used to dry, cool, toast, roast, bake, and heat materials. Such units can be zoned for various product drying temperatures, cooling, humidifying, and conditioning requirements. The air flow through the product can be upward or downward and the heat source can be steam, gas, oil, electric, or waste heat. Dryers with this design configuration are applicable to the processing of breakfast cereals, pet foods, fresh vegetables, fruits, and nuts, among numerous other food substances. (*The National Drying Machinery Company.*)

Since a large majority of drying equipment involves the use of hot gases (usually air), other means of heating can be overlooked. Dielectric heating and freeze-drying, for example, are other means where practical.

Drying equipment also can be made available with means to accelerate the drying process if these means are acceptable to product quality. For example, agitation, stirring, and otherwise keeping the material to be dried in constant motion naturally assists the amount of exposure of surfaces to the drying medium. Numbers of materials, however, cannot be handled in this fashion and requie relatively conservative, still conditions.

Principal design configurations of drum-type dryers are illustrated and described in Figs. 1 through 4; conveyor-type dryers in Figs. 5 through 7.

Dehydration. This operation is sometimes described as the removal of 95% or more of the water from a substance, by exposure to thermal energy by various means. Frequently used in food processing, the aims of dehydration are reduction in volume of the product, increase in shelf life, and lower transportation costs, among other factors. There is no clearly defined line of demarcation between drying and dehydrating, the latter sometimes being considered as a subelement of drying. Usually, the direct use of solar energy, as in the drying of raisins, hay, etc., is not lumped in with dehydrating. The term dehydration also is not generally applied to situations where there is loss of water as the result of evaporation. *Rehydration* or *reconstitution* of a dehydrated product is the act of restoring a product to essentially its original condition by the simple addition of water, usually just prior to use (as consumption of a food product). The distinction between the terms drying and dehydrating may be somewhat clarified by the fact that most substances can be dried beyond their capability of restoration.

In the case of drying foodstuffs, this is a complex combination of coupled heat and mass transfer through natural tissues. The need to take into account the structure of the food, as opposed to considering it as an homogenous solid, is important. In general, the main source of water in a tissue is the cell. Thus, the transport of water to the outside involves migration through the cell and its enveloping structure, through the porous structure of the tissue, and then through the outside boundary layer. To be able to predict drying behavior, it is necessary to establish the extent to which the foregoing transport steps are controlling. There is no general answer because both the cellular structure and the characteristics of the porous-like tissue structure are involved and the roles they play in the transport process may differ widely—simply because tissue properties change from one foodstuff to the next.

The energy requirements for drying have been studied rather intensively in recent years because the cost of a product can be markedly affected by this energy intense operation. Some researchers have observed that drying costs are much lower for air drying and drum drying than for freeze-drying.

A simplified flow diagram for the preparation of dehydrated fruits and vegetables is given in Fig. 8.

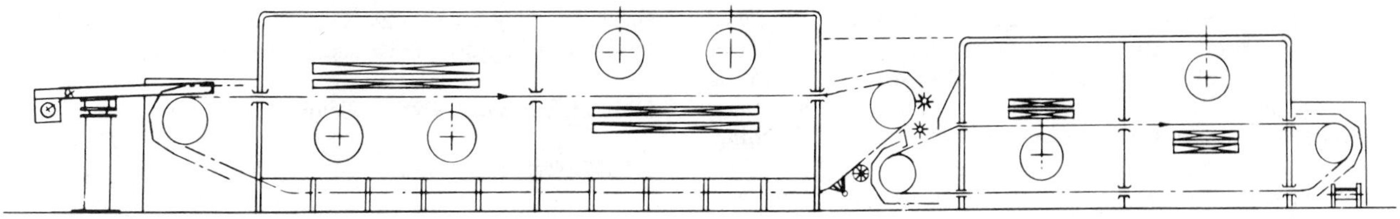

Fig. 7. A multi-stage, 4-zone drying range consisting of a series of single-stage conveyor dryers designed so that the individual stages present a new set of conditions to the product being processed. These stages, in turn, can be zoned to give maximum processing flexibility. Transfer devices between stages reorient the product to provide effective processing uniformity. (*The National Drying Machinery Company.*)

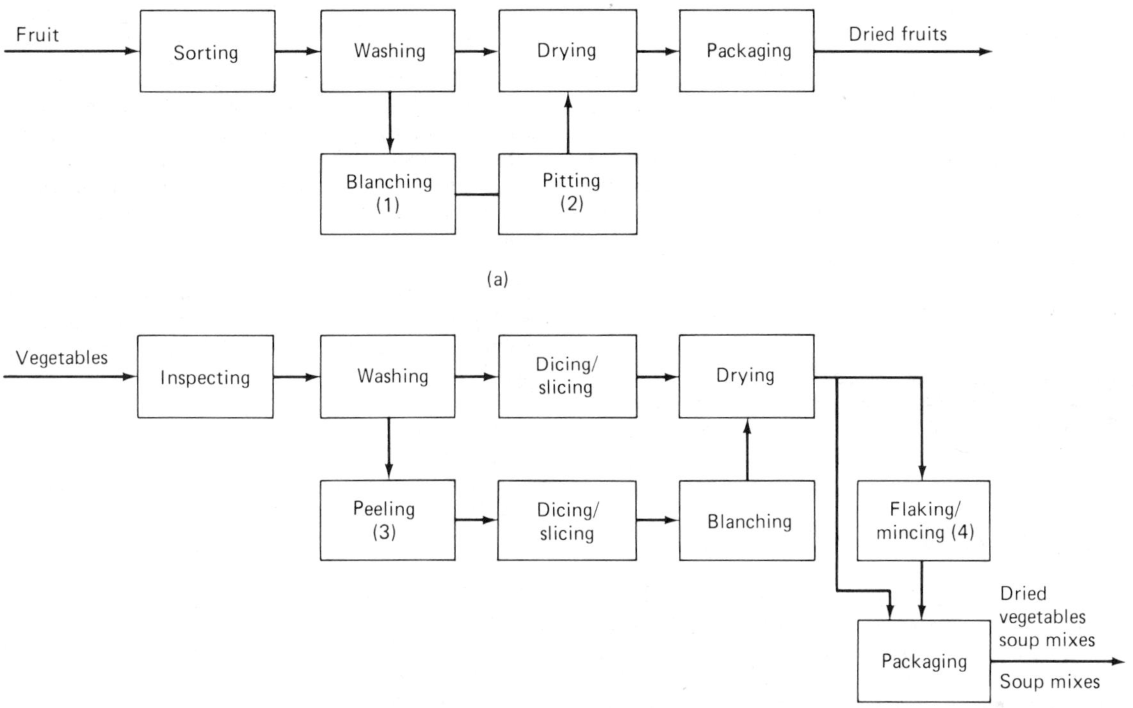

Fig. 8. Simplified flowsheets of operations required in preparation of (a) dried fruit and (b) dried vegetable products. Operations (1) and (2) are required only for certain fruits, such as prunes. Operation (3) is required only for certain vegetables, such as potatoes. Operation (4) is required only for certain end-products.

References

Flink, J. M.: "Energy Analysis in Dehydrations Processes," *Food Technology*, **31**, 3, 77–84 (1977).

Keey, R. B.: "Introduction to Industrial Drying Operations," Pergamon, Elmsford, New York, 1978.

King, C. J.: "Rates of Moisture Sorption and Desorption in Porous Fried Foodstuffs," *Food Technology*, **22**, 4, 165 (1968).

King, C. J., and J. P. Clark: "Water Removal Processes: Drying and Concentration of Foods and Other Materials," Amer. Inst. of Chem. Engr., New York, 1977.

Leed, D. A.: "Five Steps Simplify Selecting a Dryer," *Food Engineering*, **47**, 3, 63–66 (1975).

Mujumdar, A. S. (editor): "Proceedings of the First International Symposium on Drying," Science Press, Princeton, New Jersey, 1978.

Rippen, A. L.: "Energy Conservation in the Food Processing Industry," *J. Milk Food Technol.*, **38**, 11, 715 (1975).

Sapakie, S. F., Mihalik, D. R., and C. H. Hallstrom: "Drying in the Food Industry," *Chem. Eng. Progress*, **75**, 4, 44–49 (1979).

Staff: "Drying Techniques," *Chem. Eng. Prog.*, **75**, 4, 37–60 (1979).

DRY-REED RELAY. An electromechanical, magnetically-actuated, hermetically-sealed switch used mainly for analog multiplexing and for controlling low-power loads (3 to 5 VA). The relay capsule is comprised of overlapped cantilevered contacts which are housed in a hermetically sealed glass capsule. An inert or reducing atmosphere under moderate pressure is contained in the capsule. See accompanying figure. The length of the capsule ranges from less than an inch (2.5 cm) to several inches (8–10 cm), depending upon the power level and application for which the switch is designed. Diameters range from 0.25 to 0.1 in. (0.6–0.25 cm). Nickel–iron alloy contacts, with relatively high permeability and low retentivity, usually are used. Frequently, the contact portion of the reed is electroplated with another metal for the reduction of contact resistance and to increase reliability.

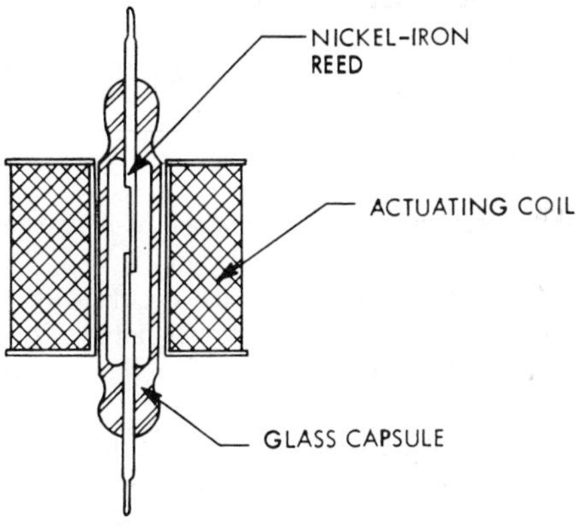

Dry-reed relay.

The switch usually is actuated by a magnetic field that is produced by a coil wrapped around the capsule, or by a magnet located near the capsule. Opening of the switch is accomplished by the restoring spring force of the reeds proper. Through the use of permanent magnets, the reeds can be biased to achieve normally closed, latching, and other contact actions.

The so-called "pick" time of the relay is dependent upon the construction of the capsule and characteristics of the drive signal. Normally, pick times are less than 1 ms (including bounce). Drop times may be as small as a few tens of microseconds. A noise voltage is produced in the reed following closure as the result of magnetostrictive effects. The noise may persist at appreciable levels (greater than 20 μV) for several milliseconds. The reed relay generally is limited to sampling speeds of less than 250 samples/second because of the noise and pick-time characteristics. Typically, relays are specified for a life well in excess of 10^6 operations at full-rated resistive load.

See also **Analog Switch;** and **Mercury-Wetted Relay.**

Thomas J. Harrison, International Business Machines Corporation, Boca Raton, Florida.

***D*²-STATISTIC.** In multivariate analysis, a statistic which measures the "distance" between two populations with identical dispersions. If (α_{ij}) is the dispersion matrix of a p-variate complex, (α^{ij}) is its inverse and δ_i is the difference of means of the ith variate, the distance Δ is defined by

$$\Delta^2 = \sum_{ij=1}^{p} \alpha^{ij}\delta_i\delta_j$$

and for a sample, the definition of the squared distance is similar with estimated means and dispersions on the right. General concept is sometimes referred to as the Mahalanobis' generalized distance.

DUAL GRAPH. Graph (Mathematics).

DUAL PERSONALITY. Amnesia.

DUANE AND HUNT'S LAW. The quantum-energy law for the generation of x-rays. When a cathode particle in an x-ray tube strikes the target, its energy may all be transformed into heat; or it may excite an atom of the metal to emit a quantum of x-radiation. If all the cathode particles have the same energy Ve (voltage times electronic charge), some may give rise to low-frequency quanta and have energy left to heat the target, others may excite quanta of higher frequency and have less energy to spare. But no quantum can be emitted of frequency and energy higher than that corresponding to the original energy of the electrons. This is Duane and Hunt's law. It is expressed by the so-called Planck-Einstein equation. If the highest frequency emitted is $\nu_{\max}$, and h is Planck's constant, the equation may be written $h\nu_{\max} = Ve$. It follows that for a given voltage V, the x-ray spectrum must terminate abruptly at the frequency Ve/h.

DUBIN-JOHNSON DISEASE. Bile.

DUBOIS-TEYMOND FORM. Mean Value Theorems.

DUBOSCQ COLORIMETER. Colorimetry.

DUCHENNE'S DYSTROPHY. Muscular Dystrophy.

DUCK. Poultry; Waterfowl.

DUCK-BILL. Fossil Reptilian Mammals; Monotremata.

DUCTILE FRACTURE. Fracture that occurs with a considerable expenditure of energy and is normally associated with and preceded by extensive plastic deformation.

DUCTILITY. A measure of the ability of a metal to plastically deform without fracturing. Ductility is generally associated with tensile properties or the ability to be cold drawn, as in wire drawing or sheet stamping operations. Percent elongation and reduction of area in the tension test are the usual measure of ductility.

DUCTLESS GLANDS. Gland.

DUGONG. Sea Cows.

DUHEM-MARGULES EQUATION. A relationship (derived from the Duhem Theorem) between the partial vapor pressures of a two-component liquid system and the concentration of the constituents:

$$N_A\left(\frac{\partial \ln p_A}{\partial N_A}\right)_{P,\,T} = N_B\left(\frac{\partial \ln p_B}{\partial N_B}\right)_{P,\,T}$$

or

$$\frac{d \ln p_A}{d \ln N_A} = \frac{d \ln p_B}{d \ln N_B}$$

where N_A = mole fraction of component A, N_B = mole fraction of component B, p_A = partial pressure of component A, the vapor behaving as an ideal gas, p_B = partial pressure of component B, the vapor behaving as an ideal gas, P = total pressure, T = temperature. This relation may only be applied at constant temperature, and if the total pressure remains small enough not to affect the properties of the condensed phase.

DUHEM THEOREM. The state of a thermodynamic system formed by ϕ phases is *completely determined* if one knows: (1) the physicochemical state of each phase, defined, for example, by the intensive variables

$$T, p, x_1^1 \cdots x_c^1, \quad x_1^2 \cdots x_c^\phi \tag{1}$$

where x_i^α is the mole fraction of component i in phase α; (2) the extensive variables for each system. One may take as the independent extensive variables, the masses

$$m^1 \cdots m^\phi \tag{2}$$

of the different phases.

Duhem's theorem states that whatever the number of phases, of components, or of chemical reactions, the equilibrium state of a closed system for which one knows the initial masses is completely determined by *two independent variables.*

The basic difference between the Duhem theorem and the Gibbs phase rule is that the phase rule is concerned only with *intensive* variables while Duhem's theorem is a statement about the *total* number of variables (intensive + extensive) determining the state of a *closed* system. See also **Saurel Theorem.**

DUIKER. Bovines.

DULONG AND PETIT LAW OF SPECIFIC HEATS. It has long been known that the atomic heats of the great majority of elements have nearly the same value at room temperature; in fact, the thermal capacity of a gram atom of most elements is not far from 6 calories per degree. Dulong and Petit expressed this by stating that the specific heats of elements are in inverse proportion to their atomic weights.

That this should be the case for gases, easily follows from the kinetic theory and the principle of equipartition of energy. For example, if the same mean energy per molecule is necessary to raise the temperature of oxygen and of hydrogen 1°, the same is true per atom, and weights of these gases having the same number of atoms will have equal thermal capacities.

For solids the matter is not quite so simple, and the more exacting theories of Einstein, Debye, and others show that the atomic heat should be expected to vary with the temperature. According to Debye, there is a certain characteristic temperature for each crystalline solid at which its atomic heat should equal 5.67 calories per degree. Einstein's theory expresses this temperature as $h\nu_m/k$, in which h is Planck's constant, k is Boltzmann's constant, and ν_m is a frequency characteristic of the atom in question vibrating in the crystal lattice.

DULONG'S FORMULA. Combustion.

DULSE. **Flavorings.**

DUMAS METHOD FOR VAPOR PRESSURE. A small quantity of the liquid the density of whose vapor is required is inserted into a weighed glass bulb which has a neck drawn to a point. The bulb is heated to about 30°C above the boiling-point until the liquid disappears, the vapor being rapidly expelled and carrying the air in the bulb with it. After measuring barometric pressure and temperature of the bath, the bulb is sealed off, cooled, and reweighed. The end of the neck is then broken off under water, when the water completely fills the bulb. The weight of the bulb is again found, full of water, giving its internal volume. From the readings the weight of a known volume of vapor is found, and hence its density. The method is frequently used when a more accurate method than the Victor Meyer method is required. Modifications have been made at lower pressures and higher temperatures.

DUMMY (Computer System). An artificial address, instruction, or record of information that is inserted into a computer system solely to fulfill prescribed conditions, such as to achieve a fixed word length or block length, but without itself affecting machine operations except to permit the machine to perform desired operations. The technique sometimes is used as an aid for debugging a system. See also **Program (Computer).**

DUMORTIERITE. This mineral found within schists and gneisses is valuable for use in the manufacture of high-grade porcelain. It is a borosilicate of aluminum $Al_7(BO)_3(SiO_4)O_3$, crystallizing in the orthorhombic system. Crystals are rare; commonly occurs as fibrous aggregates of blue to pink color, with vitreous luster. Transparent to translucent with hardness of 8.5, and specific gravity of 3.41.

World occurrences include France, the Malagasy Republic, Brazil, and Mexico, with California and Nevada the principal United States localities.

DUNE. An elliptical or crescent-shaped mound of sand which may reach a height of a hundred or more feet. The windward slopes of dunes are gentle, the lee sides are steep. If the dune is crescent-shaped, the convex side will face the direction from which the wind is blowing. Crescent-shaped dunes are called *barchans.* Sand blown up the windward side drops down the lee slope, causing the dunes to migrate slowly. This action can be quite destructive, unless prevented by some

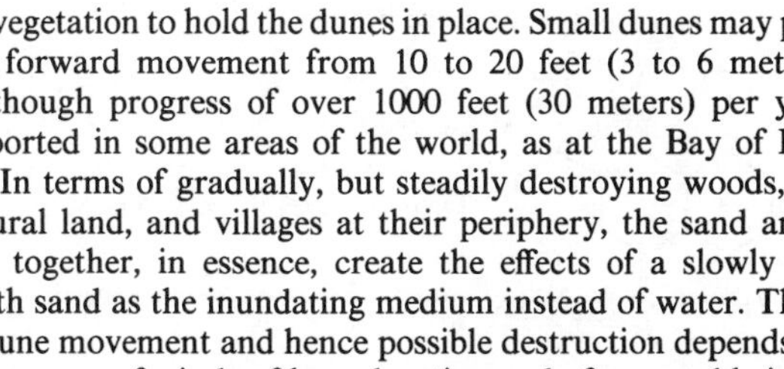

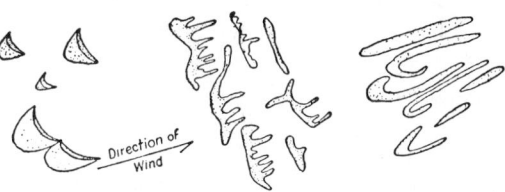

Types of sand dunes. (*Walther and Cornish.*)

form of vegetation to hold the dunes in place. Small dunes may progress in their forward movement from 10 to 20 feet (3 to 6 meters) per year, although progress of over 1000 feet (30 meters) per year has been reported in some areas of the world, as at the Bay of Bicay in France. In terms of gradually, but steadily destroying woods, forests, agricultural land, and villages at their periphery, the sand and wind working together, in essence, create the effects of a slowly moving flood with sand as the inundating medium instead of water. The progress of dune movement and hence possible destruction depends largely on the presence of winds of long duration and of reasonable intensity.

As wind conditions change, so do the dunes. Consequently, there are dunes which reflect conditions of prior geologic and climatic periods, as in the case of the lacustrine dunes found along the southern and southeastern shore of Lake Michigan. Often dunes will be found in areas where former lakes existed. There is also some dune formation along river banks, as for example, the dunes along the eastern banks of the Mississippi, the Missouri, and Rio Grande Rivers in the United States. But generally the inland dunes are found mainly in the large desert areas—the Sahara, Arabian, Gobi deserts and regions of Iran, Turkey, Mongolia, and the United States that are essentially arid. Comparatively few inland dunes are found in the United States, some in California and a few in the sand hill region of western Nebraska.

Oceanic coastal dunes are found in several parts of the world—Brittany, Cornwall, the eastern shoreline of the United States from Cape Cod southward, and the western shoreline, notably along California coasts.

DUNLIN. **Shorebirds and Gulls.**

DUODENAL ULCER. **Ulcer.**

DUODENUM. **Digestive System (Human).**

DUPLEXER. Also known as transmit-receive circuit, a duplexer makes possible the use of the same antenna for both transmission and reception by preventing the flow of damaging amounts of power to the receiver during transmission, without at the same time reducing excessively the input to the receiver during reception.

DUPLEX TELEGRAPH. **Telegraphy.**

DURAIN. The term proposed by Marie Stopes in 1919 for an ingredient of bituminous coal, occurring in bands. In thin, translucent sections durain is seen to be composed and reddish spores in a gray, granular matrix.

DURA MATER. **Brain and Nervous System.**

DURANICKEL. **Nickel.**

DURBIN-WATSON STATISTIC. The usual mode in linear regression, for example

$$y = \sum_{i=1}^{p} \beta_i x_j + \epsilon$$

requires that the residual term ϵ is a random variable and, in particular, that the values from one observation to another are independent. If a set of values of ys and xs is available in temporal order, one problem is to decide whether this assumption is acceptable. The Durbin-Watson statistic, which is dependent on the observed residuals, is designed to test the hypothesis of independence in the true (unknown) residuals.

DURIAN. **Palm Trees.**

DURICRUST. A general term for a hard crust on the surface of, or layer in the upper horizons of, a soil of a semiarid climate. It is formed by the accumulation of soluble minerals deposited by mineral-bearing waters that move upward by capillary action and evaporate during the dry season. It may consist of concentrations of aluminous and ferruginous material (*ferricrete*) upon feldspathic rocks, of siliceous material (*silicrete*) upon argillaceous and arenaceous rocks, or of calcareous material (*caliche* or *calcrete*) upon carbonate rocks. The term sometimes excludes calcareous crusts.

DUST BOWL. **Soil.**

DUST DEVIL. **Fronts and Storms.**

DUST STORM. **Winds and Air Movement; Soil.**

DUTCH ELM DISEASE. **Elm Trees.**

DUTCH MYRTLE. **Bayberry Shrubs and Trees.**

DUTY CYCLE. A term applied to instruments and devices to denote the following characteristics: (1) the time interval occupied by a device on intermittent duty in starting, running, stopping, or idling (an example would be the maximum time that a device can operate continuously), and (2) the ratio of the "on time" interval occupied in operating a device to the total time of one operating cycle (the ratio of the pulse-duration time to the pulse-reception time). An illustration of duty cycle is given in the accompanying diagram.

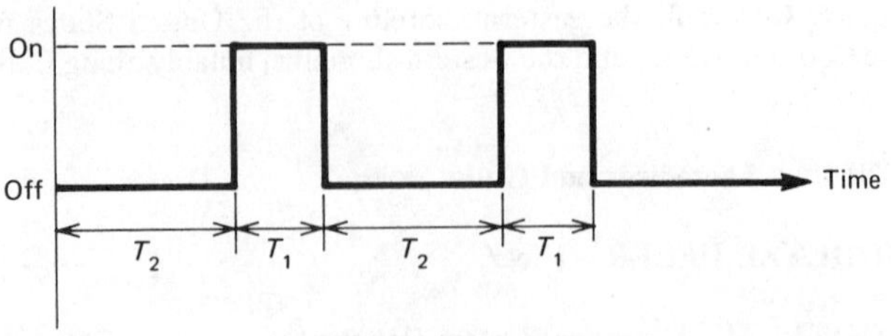

Examples of duty cycle. During period T_1, the device is on. During T_2, the device is off. The duty cycle in percent is $(T_1 T_2) \times 100$.

DWARFISM. Hormones; Pituitary Gland.

DWARF STARS. Giant and Dwarf Stars.

DYADIC. In ordinary vector analysis no meaning is attached to a symbol such as **AB**, Where neither dot nor cross stands between two vectors. Such a quantity is called a dyad and the symbolic sum of two or more dyads is a dyadic polynomial or, more briefly, a dyadic. It is useful in the study of linear vector functions and has applications to theories of rotation and strain, electromagnetism, and optical phenomena in nonisotropic solids.

Consider the vector function

$$\mathbf{r}' = \mathbf{i}(c_1\mathbf{i} \cdot \mathbf{r}) + \mathbf{j}(c_2\mathbf{j} \cdot \mathbf{r}) + \mathbf{k}(c_3\mathbf{k} \cdot \mathbf{r}),$$

which is readily interpreted since the quantities in parentheses are scalars. The same equation, however, can be more simply written as $\mathbf{r}' = (\mathbf{i}c_1\mathbf{i} + \mathbf{j}c_2\mathbf{j} + \mathbf{k}c_3\mathbf{k}) \cdot \mathbf{r}$ or, generally, if $\mathbf{a}_1, \mathbf{a}_2, \mathbf{a}_3, \ldots, \mathbf{b}_1, \mathbf{b}_2, \mathbf{b}_3, \ldots$ are two sets of vectors, as $\mathbf{r}' = (\mathbf{a}_1\mathbf{b}_1 + \mathbf{a}_2\mathbf{b}_2 + \mathbf{a}_3\mathbf{b}_3 + \cdots) \cdot \mathbf{r} = \mathbf{\Phi} \cdot \mathbf{r}$. In the last equation, the direct or scalar product of the dyadic $\mathbf{\Phi}$ and the vector $\mathbf{r}$ yields a new vector $\mathbf{r}'$. The scalar product is not always commutative, for it does not follow that $\mathbf{\Phi} \cdot \mathbf{r} = \mathbf{r} \cdot \mathbf{\Phi}$. A special dyadic is the idemfactor, $\mathbf{I} = \mathbf{ii} + \mathbf{jj} + \mathbf{kk}$, which has the property that $\mathbf{I} \cdot \mathbf{r} = \mathbf{r} \cdot \mathbf{I} = \mathbf{r}$.

Nine fundamental dyads can be formed from unit vectors, **i, j, k.** The nonian form of any dyadic can be written in terms of them and scalar quantities a_{mn} as

$$\mathbf{\Phi} = a_{11}\mathbf{ii} + a_{12}\mathbf{ij} + a_{13}\mathbf{ik} + a_{21}\mathbf{ji} + a_{22}\mathbf{jj} + a_{23}\mathbf{jk} + a_{31}\mathbf{ki} + a_{32}\mathbf{kj} + a_{33}\mathbf{kk}$$

One can continue the algebra of dyadics, defining several kinds of products, reciprocals, conjugates, and other properties. A sophisticated approach to the dyadic regards it as a special kind of tensor with nine components. Each of them is a function of three coordinates and they transform from one coordinate system to another according to rules which can be derived. A dyadic can also be written in matrix form and can be generalized to the concept of triad, tetrad, polyad, and polyadic. See **Tetradic.**

DYAS. The western European term for the Permian system where it is readily divisible into two parts.

DYES (Textile). Special groups within dyestuff molecules lend color to textile substrates. These groups are termed *chromophores* and include: the azo group, —N═N—; the thio group, ═C═S; the nitroso group, —N═O; the carbonyl group, ═C═O; the nitro group, $—NO_2$; and the azoxy group,

—N═N—
O

Organic dystuff molecules also contain *auxochromes*, and these include the amino group, $—NH_2$; the hydroxyl group, —OH; the sulfonic group, $—SO_3H$; and substituted amino groups, $—N(CH_2)_2$ and $—NHCH_3$. Auxochromes increase the solubility of dyestuffs as well as influence the color.

Organic dyestuffs absorb radiation and the color of some dyes derives from the fact that absorption bands fall within the range of radiation visible to the human eye.

There are two convenient ways to classify dyes: (1) by chemical structure; and (2) by dyeing properties. For theoretical purposes and work in the development of new dyes, the chemist prefers the structural classification. The practical dyer and dyeing technologist, however, prefer the classification by dyeing properties. On this latter basis, the main categories are: (1) basic or cationic; (2) acid and premetalized; (3) chrome and mordant; (4) direct and developed direct; and (5) sulfur, azoic, vat, disperse, and reactive dyes. Some examples of these classes of dyes are given in Fig. 1.

Three major steps are involved in fixing dyestuff molecules to fibers: (1) The dyestuff migrates from solution to the fiber interface (substrate) whereupon it is adsorbed on the surface of the fiber; (2) the dye moves or diffuses from the surface into the center of the fiber, whereupon

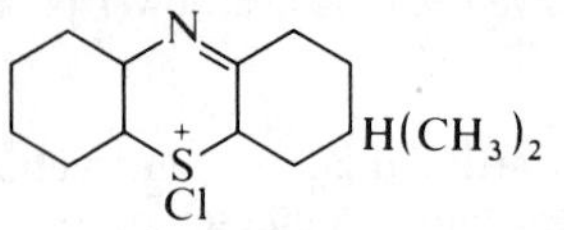

Basic dye: methylene blue

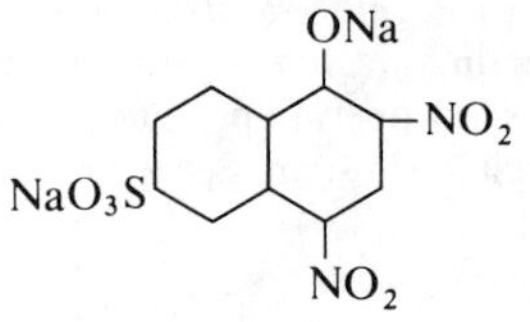

Acid dye: naphthol yellow

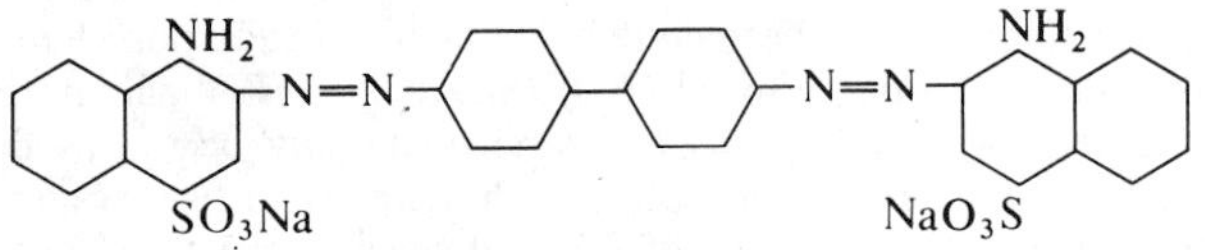

Direct dye: congo red

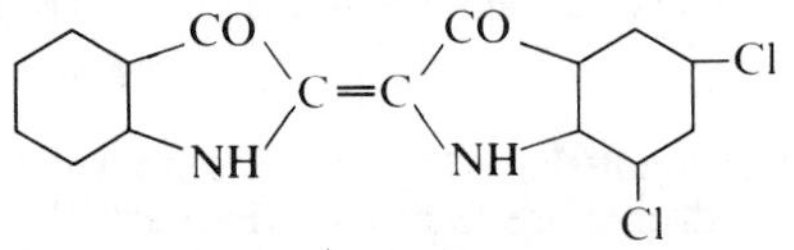

Vat dye: vat blue

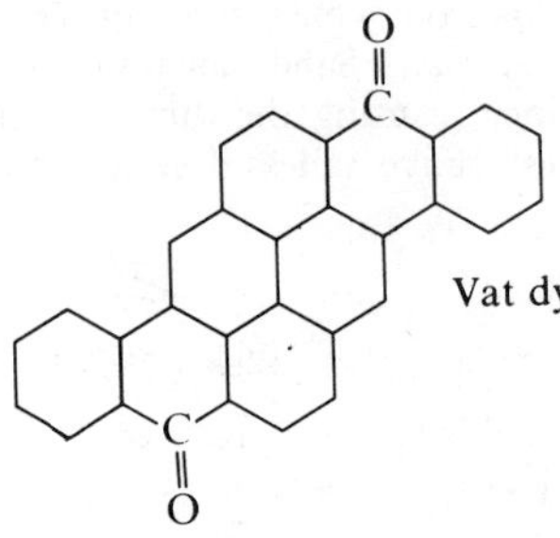

Vat dye: vat blue

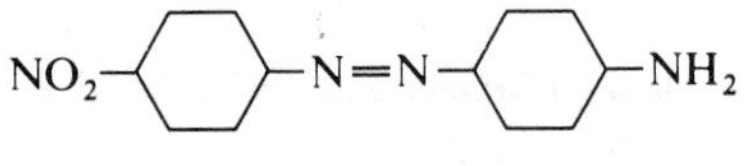

Disperse dye: disperse orange

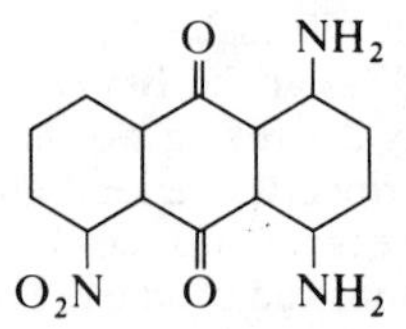

Disperse dye: disperse violet

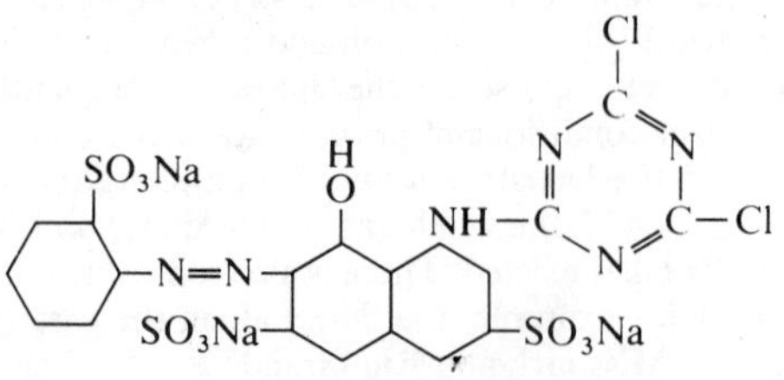

Reactive dye: red

Fig. 1. Representative textile dyes.

it situates around and between fiber molecules; and (3) the actual attachment of the dye molecule to the fiber molecule, utilizing either physical forces, hydrogen bonding, or covalent bonding with the fiber molecule.

Electropotential forces, variations in temperature, and degree of agitation of the solution and substrate affect the movement of the dye molecules from solution to the fiber. Often, the feasibility of employing a specific type of dye on particular fibers is influenced by the relationship between the ionic charge of a fiber and a dyestuff. An elevated temperature helps to break dye micelles into smaller units and this enhances attachment of the dye to the fiber, and subsequently they move among the fiber molecules. Agitation helps to achieve uniformity, but it must not be too great because it can hinder assembly of dye molecules on the fiber substrate.

The process of diffusing the dye into the center of the fiber is related to the pore opening of the fiber and the size of the dye molecule. Phsyical attachment forces are related to the shape and size of the dye molecule, to the planar relation of the dye molecule, and the size of openings between fiber molecules. Dye molecules with a linear and planar shape tend to adsorb to fiber molecules best. Where dye molecules tend to aggregate within the fiber, the molecules will be held well because of their size.

Protein fibers and anionic dyes form ionic or salt linkages, linking with the —NH_3^+ groups in the fiber molecules.

Hydrogen bonding occurs principally in cellulosic fibers. The hydroxyl group of cellulose molecules will form hydrogen bonds with dyes that contain hydroxyl, amino, and azo groups. Critical to hydrogen bonding is the space (expressed in angstrom units) between groups capable of forming the bond.

The most stable bond is a covalent bond which is characteristic of reactive dyes and some acid and cationic dyes.

Dyeing Processes and Machines. The dyeing function occurs in the following major ways: (1) The substrate or fabric is moved through the dye liquor; (2) the dye liquor may be moved through the substrate, a method commonly used in connection with yarns, but also applicable to fabric; and (3) a combination of the two foregoing methods. A dyeing machine should provide adequate movement so that the dye may penetrate the substrate uniformly; the movement must be accomplished so as not to damage the substrate; the equipment must resist corrosive factors involved; generally a uniform temperature must be provided; and facilities should be available so that additional dyes and chemicals may be added without stopping the process. *Leveling* is a term commonly used in dyeing technology and signifies the relative ease with which dye colors the substrate uniformly, evenly, and to the degree desired. See Figs. 2 and 3.

Acid Dyes. Most acid dyes are sulfonic acid salts. These dyes are commonly used for dyeing protein fibers. Acid dyes are quite soluble,

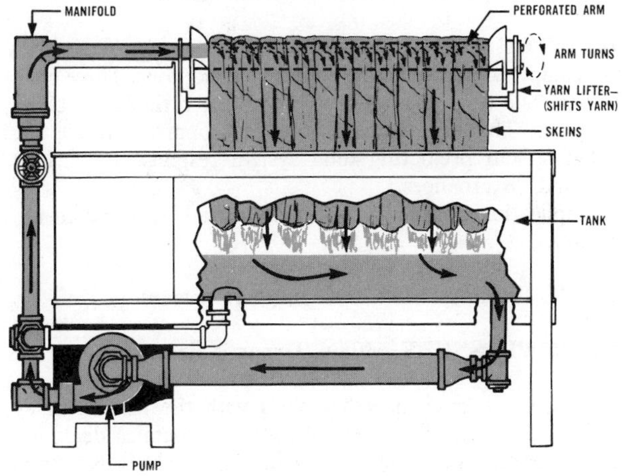

Fig. 2. Cascade machine for skein dyeing. Hanks of yarn are suspended from a perforated stainless steel arm situated above an open tank which contains the dye bath. Dye liquor is continuously pumped from tank through arm which distributes the dye to the skein, returning the liquor to the bath for recycling. The perforated arms are periodically rocked to assure uniform distribution of the dyestuff. (*Allied Chemical Corp.*)

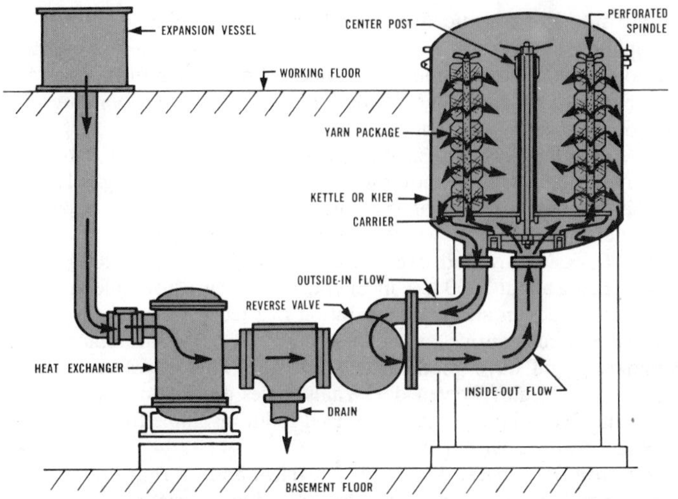

Fig. 3. Package dyeing. Packages of yarn are put on verticle spindles in a portable carrier. Carrier is lowered into dyeing vessel. The dye liquor circulates continuously through the spindle perforations and through the packages. The flow pattern then is reversed. The cycles are repeated until specified shade results. (*Allied Chemical Corp.*)

allowing maximum exhaustion of the dye bath. Acid dyes derive their name because they were originally applied in a mineral or an organic acid bath; and because the anion is the colored portion of the molecule.

Basic or Cationic Dyes. These dyes possess high brilliance and intensity of color. They are not highly soluble in water and may decompose at temperatures exceeding 65°C. Generally, these dyes are not applied along with acid or direct dyes because precipitation may occur. On most fibers, basic dyes have low colorfastness. However, on acrylic fibers, cationic dyes exhibit relatively good colorfastness resulting from some covalent bonding.

Basic or cationic dyes derive their name from the fact that the dye molecules dissociate in water, with the cation being the colored portion of the dye. If anionic sites are present in the fiber, the dye will be attracted to form a covalent bond. Because anionic sites vary in terms of availability with various fibers, the durability of cationic dyes is quite variable.

Direct Dyes. These are water-soluble dyes, easily applied to cellulose fibers, and comprise the largest group of dyes. Sometimes these dyes are called *substantive colors.* The molecules of direct dyes have a tendency to be linear and adsorb onto fiber molecules. Hydrogen bonding and van der Waals forces also contribute to holding the dye onto the substrate. Dyeing is also helped by some aggregation of the dye molecules within the fiber. Frequently, direct dyes possess poor colorfastness and thus require special processing to improve their washability qualities.

Direct dyes are of three principal classes: (1) self-leveling, or Class A; (2) dyes that level with the addition of salt, Class B, and (3) dyes that require addition of salt and careful temperature control, Class C.

Development of the dye by diazotization is probably the most effective means for improving washfastness of direct dyes. For reacting with the free amino group on the dye molecule, developers such as beta-naphthol, *m*-phenylenediamine, phenol, phenylmethylpyrazolone, or resorcine are used. This reaction should occur in situ on the fiber. In this procedure, the substrate is dyed in accordance with the regular direct-dye method. Then, the material is diazotized for about a half-hour in a bath containing about 3% sodium nitrate, and 3–5% sulfuric acid (or 5% hydrochloric acid). Then the dyed substrate is developed for about a half-hour in a cold bath containing one of the aforementioned developers. Developing frequently will change the color because new chromophoric groups may be added to the dye molecule. Thus, pilot runs and experience are mandatory.

Disperse Dyes. These dyes were initially developed for cellulose acetate and at one time were called *acetate dyes.* They are now used with several fibers. Disperse dyes are not soluble in water, but do form a uniform dispersion in a water bath. It is postulated that because of their lack of affinity for water, dispese dyes migrate toward and diffuse into the organic fiber and form a solid solution within the hyrophobic fiber. A more recent postulate states that the dye is transferred from the suspension to the fiber in molecular form. There is also evidence that van der Waals forces and hydrogen bonding also contribute to dye retention.

The colorfastness of disperse dyes varies with the fiber substrate. The dyes can be applied under pressure, necessary for dyeing polyester. In the *thermosol* method of applying disperse dyes, the dye is padded into the substrate, after which it is subjected to dry heat.

Vat Dyes. Natural vat dyes have been used for nearly 4000 years, commencing about 2000 B.C. in Egypt. Vat dyes are insoluble in water, but they reduce in an alkaline solution to a leuco form that is water-soluble and substantive to cellulose fibers. Vat dyes are colorfast to laundering and have good fastness to light. These dyes are held to cellulose molecules by van der Waals forces and hydrogen bonding. Once the dye is applied, it is oxidized to an insoluble form and thus becomes trapped within the fiber.

Reactive Dyes. These dyes form a covalent bond with the fiber molecule. They have excellent washfastness and also resist environmental factors well. Reactive dyes are used mainly on cellulose fibers, but they also have been used successfully with silk, wool, nylon, and acrylic fibers. These dyes contain a chemical group that forms a bond with either a hydroxyl group in the cellulose molecule, or with an amino group in other fibers. Reactive dyes are among the most recently developed dyes and tend to be somewhat more costly than other dyestuffs.

Dyeing of Blended Fabrics. Since each type of fiber reacts somewhat differently to dyestuffs, the dyeing of blends of fibers often presents problems. Often, blends will be dyed in steps, wherein one fiber is dyed and then the fabric is rinsed, after which the second fiber is dyed, and so on. Should the general chemistry and temperature conditions be approximately the same, sometimes two fibers can be dyed in one step.

A brief glossary of dyeing terms not already defined in the text is as follows:

Barré. A dyeing defect showing up as bars or stripes in a fabric.
Blocking. The interference of a dye with the action of another dye in the same bath.
Crocking. The rubbing off of improperly fixed dyestuff.
Dispersing Agent. A detergent or other chemical that accelerates the diffusion of dye molecules through the dye bath.
Exhaustion. The process during which dye molecules leave the liquor and attach to the fiber.
Fastness. The resistance of a dye to loss of color through fading or bleeding.
Grey or Greige Goods. Undyed and unfinished fabric as it comes from the loom or knitting machine.
Merger Number. A number designating yard that is equivalent in its dyeing characteristics.
Migration. The movement or leveling out of dye molecules on goods during the dyeing process.
Pasting. Prewetting dyes or other chemicals and working them into a paste for easier dispersion in the dye bath.
Resisting Agent. A chemical that will repel dye molecules or slow down their rate of attachment to a fiber.
Rope Crease. A dye streak caused by gathering or roping together of a fabric during the dyeing process.
Sampling. The removal of a portion of the goods being dyed to check the color against a standard.
Shading. The deepening of color in one kind of fiber in preference to another as in various union dyeings.
Soaping Off. The removal of residual dyestuff with a soap or detergent.
Sublimation. The loss of dyestuff from a fabric through vaporization.
Union Dyeing. The uniform dyeing of two different fibers in the same yarn or fabric.

See also **Fibers.**

DYNAMICAL ANALOGIES. The formal similarities among the differential equations of electrical, mechanical and acoustical systems which make possible the reduction of mechanical and acoustical systems to electrical networks and the solution of such problems by electrical circuit theory.

DYNAMICAL PARALLAX. An indirect method for the determination of the distances of binary stars. In the determination of orbits of binary stars, the semi-major axis of the relative ellipse is determined in angular units. This length cannot be expressed in any linear units unless the distance of the system is known. With this distance known, the orbit may be solved and the combined masses of the two stars determined in terms of the sun's mass as unity. For the systems thus far solved, the majority of the stars are found to have between $\frac{1}{5}$ and 10 times the mass of the sun. Hence, as a first order of approximation, we may assume that the combined mass for any binary system will not differ greatly from twice the mass of the sun.

The rigorous expression for the so-called harmonic Keplerian Law of Planetary Motion is

$$M_1 + M_2 = \frac{a^3}{P^2} = \frac{\alpha^3}{\pi^3 P^2}$$

where the sum of the masses is in solar units, α is the separation in seconds, π is the parallax, and P is the period in years. Hence, assuming the combined mass of the binary system to be twice that of the sun, and knowing the period of revolution of the system in years, we may find the mean distance between the components of the system in astronomical units. Since, from the orbit, we know the distance between the components in angular units, we can immediately find the angular distance subtended by one astronomical unit at the distance of the star, or, in other words, we can find the stellar parallax of the system.

The parallax thus determined is based on the assumption that the combined mass is twice that of the sun. However, with the parallax thus computed and the apparent magnitudes of the stars known, we can then calculate the absolute magnitude of the system. With this absolute magnitude, we go to the mass-luminosity relation, and get a second approximation to the mass of the system. The process is repeated with this improved mass, and an improved value of stellar parallax obtained. By a sufficient number of approximations, values of the so-called dynamical parallax may be obtained, with an accuracy approaching that of the direct trigonometric determinations.

Several methods have been devised for obtaining dynamical parallaxes for systems that are moving so slowly that orbit determination is impossible. Although such methods do not lead to results of great accuracy, they are valuable for statistical discussions.

See also **Binary Stars; and Kepler's Law of Planetary Motion.**

DYNAMICAL SIMILARITY. Fluid Flow (Dynamical Similarity).

DYNAMICAL VERTICAL. Artificial Horizon (Aircraft).

DYNAMIC BALANCE. Balance (Mechanical).

DYNAMIC CHARACTERISTICS. Important characteristics of a piece of equipment under actual operating conditions. These characteristics define the time response of component variables to changes in input signals. This type of information is important to the designer in analyzing and predicting total system response to various load changes and disturbances.

The application of this type of analysis can be found in all types of industry.

DYNAMIC INSTABILITY. Atmosphere (Earth).

DYNAMIC PRESSURE. Fluid (Dynamic Pressure).

DYNAMICS. The science that deals with the motion of systems of particles under the influence of forces. Dynamics deals with the causes of motion, as opposed to kinematics, which deals with its geometric description, and to statics, which deals with the conditions for lack of motion.

See also **Kinematics; Kinetics; Mechanics; and Rotation (Dynamics).**

Dynamical variables are those variables in terms of which classical mechanics is built up, and which can be given an operational definition. Examples are the coordinates, and the components of velocity, momentum, and angular momentum of particles, and functions of these quantities.

The present state of knowledge recognizes all motion as relative, since no fixed reference is known for finding the absolute motion of any body. The earth, which is frequently used as frame of reference, is rotating about its own axis and is also revolving about the sun. The solar system is a minute part of the Milky Way galaxy that is known to be revolving in space. Beyond this, there is relatively limited knowledge of the nature of motion that exists.

In the field of astronomy, measurements are evaluated in a coordinate system that is located relative to the *fixed* stars. These stars are located at such vast distance from the earth that they appear as points of light that are almost motionless in space. In this frame of reference, the motions of celestial bodies are described with extremely great precision, and the motions of bodies within the solar system can be predicted accurately over periods of hundreds of years.

Some applied areas of dynamics, exemplified by space exploration technology, also require the extreme degree of accuracy that is possible with a celestial frame of reference. In many other areas, this extreme accuracy is not essential and measurements based upon this frame of reference would be tedious and impractical. In such cases, motion may often be adequately described in a coordinate system located relative to the earth.

DYNAMICS (Gases). Gases (Dynamics).

DYNAMICS (Newton's Laws). Newton's Laws of Dynamics.

DYNAMITE. Explosive.

DYNAMO. A dynamo is one of a general class of machines capable of transformation of electrical into mechanical energy, or vice versa. The word is a shortened form of dynamoelectric. A feature of all dynamo machines is the employment of magnetic induction in effecting the transformation. The essential parts of an ordinary dynamo are the armature and the field. One of these is mounted on a rotating shaft, and the other is stationary. Theoretically, the dynamo is perfectly reversible, that is, it may be used either as a generator or a motor. Actually, this is not always possible.

DYNAMOMETER. A dynamometer is an instrument for measuring force, such as a spring balance. Most writers, however, apply the term to certain devices for the measurement of mechanical power. The principal classification is derived from the fact that some types of dynamometers absorb all of the power, which is converted into heat, whereas others transmit the power they receive to some other absorber of power, measuring it during the process. These are called, respectively, absorption and transmission dynamometers.

Absorption Dynamometers. The simplest example is the Prony brake as shown in the accompanying diagram. The mechanical output of an engine, turbine, or motor is called the brake horsepower because one of the most common methods of testing for mechanical output is with the Prony brake. However, the output available at the shaft is called brake horsepower whether measured by brake or not. For example, when an engine drives an electric generator by direct connection, the horsepower available at the coupling between the machines is termed brake horsepower, but would not, under these circumstances, be measured by a brake. In the case of direct connected electric generating sets, the brake horsepower of the prime mover is found by dividing the electrical output from the generator, converted to horsepower, by the efficiency of the generator. The result is the generator input, which is the same as the mechanical output of the prime mover.

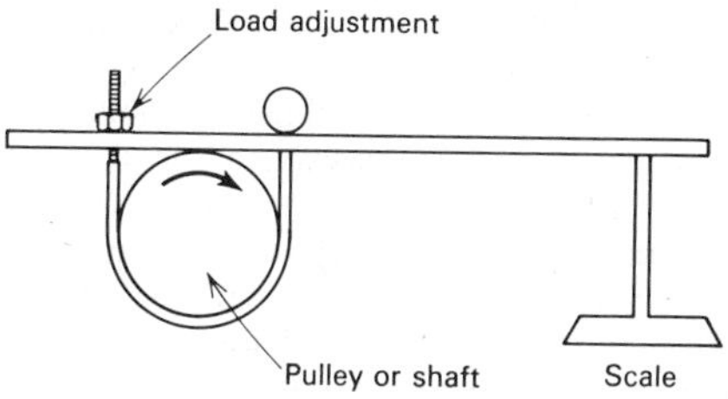

Prony brake.

To measure brake horsepower by a Prony brake, readings of the weight registered on the scales and the speed of rotation of the brake drum are taken. In addition, the measured distance from the center of rotation to the force on the scales must be known. These quantities are substituted in the following formula to obtain brake horsepower:

$$\text{brake horsepower} = \frac{2\pi WRN}{33{,}000}$$

W = net load on the scales, in pounds
R = radial distance in feet to the line of action of the force registered on the scales
N = rotative speed, in revolutions per minute

(Other units are described at end of this entry.)

The W of the above formula is less than the load actually recorded on the scales because the tare weight must be deducted from the scale reading to give net effective weight. The tare weight is an allowance for the fact that the arm of the Prony brake itself gives some reading on the scale due to its unbalanced position, relative to the center of rotation. The tare weight is that weight which, acting at the scales, will give the same moment about the center of rotation that the unbalanced dead weight of the brake would produce.

In the absorption dynamometer class, there are other types which convert the mechanical to heat energy through the medium of mechanical friction. The friction surfaces are variously wooden blocks against metal drums or pulleys, bands with wooden cleats, ropes, or friction-surfaced brake bands. Also there are hydraulic dynamometers which absorb the power by fluid friction. One common arrangement is similar to a centrifugal pump, except that the casing, instead of being rigidly fixed to a bed plate, is freely supported on the propeller shaft. It is restrained from rotating by an attached arm. The restraining moment in the brake arm is measured by platform or spring scales. The energy absorbed appears as a heating of the water in the dynamometer. To prevent it boiling it must be steadily renewed. Thus the energy is carried off in a stream of water entering the dynamometer cool and leaving warm. Air friction has also been set to use in the fan brake absorption dynamometers.

One of the most convenient means for measuring power is to convert it to electrical power (watts). In an electrical dynamometer a generator is slightly modified. The stator is mounted, free to revolve, but restrained from revolving by a brake arm which is attached to it, and to which are fastened weighing scales. The tendency of the casing to rotate with the rotor which is connected to the source of the power is opposed by the brake arm. The force shown on the scales becomes a torque when multiplied by its lever distance from the center of rotation. Since power is torque multiplied by rotative speed, the only other reading necessary from the dynamometer is the speed of the rotor shaft. In all absorption dynamometers the casing is mounted free to revolve under the action of mechanical friction, fluid friction, or magnetic drag. Actual rotation is prevented by the attached brake arm. Power is measured as a torque operating at the rotative speed of the driven shaft.

Transmission Dynamometers. Representative of this class of dynamometers is the torsion type, in which a shaft delivering power is twisted through a small angle by the torque. Such a shaft may be calibrated at rest by measuring the torsional deflection obtained under known torque loadings. This dynamometer has its greatest usefulness where other types are impractical. Measurement of power output from a large marine engine is such a case. Where applications permit breaking into a coupling, an electronic transducer may be used for making both static and dynamic measurements of torque. Units are available for measurement of torque, in increments, from as low as 2.5 to as high as 60,000 inch-pounds.

Range of Dynamometers. Friction (Prony-brake type) dynamometers measure horsepower up to about 200 horsepower and rotating speeds range from 1,000 to 2,000 revolutions per minute. Fluid-friction dynamometers, such as the Froude (ordinary water brake) are useful up to 25,000 horsepower and at revolutions of 10,000 per minute. Of the electric dynamometers, the eddy-current brake type can be used up to 300 horsepower and 6,000 revolutions per minute. The electric generator type has a range up to 30,000 horsepower and speed of 750 to 4,000 revolutions per minute. The torsion-type transmission dynamometer can go up to 50,000 horsepower and speeds of 1,000 to 3,000 revolutions per minute; while the Kenerson transmission dynamometer has a top range of about 100 horsepower at about 1,500 revolutions per minute. Probable errors of dynamometer measurements range from a fraction of a percent up to as much as 5%.

The units of work performed in rotational motion are: (1) In the cgs system, the dyne-centimeter or erg; (2) in the mks system, the newton-meter or joule; (3) in the English system, the foot-pound; (4) in the SI system, the kilowatt (1 kW = 1.341022 horsepower).

DYNAMO THEORY (Earth Magnetism). Earth.

DYNAMOTOR. This is a double armature rotating electrical machine. One of the armatures is wound for low voltage direct current, and serves as a motor armature. The other armature is wound for a high dc voltage and serves as a generator winding. The machine is used for supplying plate voltage to portable radio equipment, being operated from storage batteries on the motor end and supplying the high voltage dc from the generator winding.

DYNATRON OSCILLATOR. Oscillator.

DYNE. Units and Standards.

DYNODE. 1. An additional electrode in a photomultiplier tube, which undergoes secondary emission upon bombardment by photoelectrons, and thus effects amplification. 2. A photomultiplier tube having one or more dynodes.

DYSENTERY. A severe form of diarrhea, characterized by frequent watery stools which contain blood, pus, and mucus. See **Amebiasis** (amoebic dysentery); **Diarrhea;** and **Shigellosis.**

DYSLEXIA. Difficulty in reading or reading disability because of a poor understanding of printed words. Dyslexia may be acquired at any age, if skill in reading is lost because of damage to certain portions of the brain. The term is *loosely* used to describe the condition of a child who is reading two or more grades below his grade level. Approximately one-fifth of all children in the United States have some form of *learning disability,* including *reading disability.**

Slow or poor readers have been divided into several groups, but much confusion exists because different terms are often used to define the same condition. Some forms of dyslexia are caused by obvious conditions, such as prolonged absence from school or brain defects which make reading difficult. All these forms of disability may be called *secondary dyslexias.*

The few children with normal intelligence who have difficulty understanding printed words (*dyslexia*) should not be confused with the larger group of children who learn slowly for various reasons.

About 2–5% of all children have some degree of dyslexia in spite of normal intelligence and absence of obvious causes. Their condition is called *specific primary dyslexia* or *developmental dyslexia.* The underlying basis of primary dyslexia is usually present at birth, although the disturbance is not recognized until several years later when the child is required to learn to read. Primary dyslexia affects boys more often than girls.

Visual Perception. Images of objects are formed in the eyes and the resulting visual impulses are carried to the brain where the complex processes of visual perception take place. Visual perception, difficult to define, means that certain shapes and outlines are recognized as objects. For example, three lines properly arranged may be interpreted as a triangle. Likewise, the parts of the word "d-o-g," by visual perception in the brain, become the word "dog." From intermediate centers of visual perception, impulses travel to higher centers in the brain. Here they are integrated with impulses from other senses. Some of these originate in the ears and are relayed by centers for hearing. Countless other impulses are correlated with previously stored memory of objects and experiences. In the highest centers, finally, abstract concepts of the *meaning* of a letter or a word or a thought are formed.

Visual perception is latent in the new-born child and develops with growth. Visual and other perceptual development is related to the maturity of the child. Visual perception can be aided by training at the proper time and in the proper manner. Simple perceptual training is carried out by the kindergarten teacher. Tests measuring the degree of perceptual development should be used to estimate the reading readiness of the child before the child is given instruction in reading. One must remember that one child's development may be slower or faster than that of other children the same age.

Role of Ocular Dominance. Somewhat like right-handedness or left-handedness, there is a dominance of the right eye or the left eye in many persons. Eye dominance may be on the same side as that of the hand, or it may be on the opposite side. "Crossed dominance" was once considered a cause of dyslexia, but when found, is not considered an accompanying sign and not a cause of dyslexia. Methods to change eye or hand dominance have not helped in the treatment of dyslexia.

Role of the Ophthalmologist. The ophthalmologist is often consulted first in dyslexia. A medical eye examination will be made to discover and correct any disorder of the eyes which might be contributing to the reading difficulty. Since reading disability concerns various bodily and mental functions, it is the ophthalmologist's responsibility with other specialists to arrange an interdisciplinary program to evaluate the child's disability. Diagnosis and treatment involve the ophthalmologist's cooperation with specialists in audiology, psychology, and education as well as pediatricians, psychiatrists, neurologists and other physician specialists. Treatment must be begun as early as possible before secondary frustrations and emotional factors compound the child's reading disability.

Since dyslexia is a complex disability, it is essential to discover those who have no eye problem in order to avoid fruitless expenditure of time and money in nonproductive eye or other exercises.

How each dyslexic child should be taught to read is a problem of educational science. While the ophthalmologist and other medical specialists can offer their expert knowledge in bringing the child's potential to the best level, the actual remedial procedures remain the responsibility of the educator.

Eye defects do not cause specific primary dyslexia since it is based on recognition and not on vision. Eye defects do not cause reversal of letters and other signs of primary dyslexia. Glasses if needed may make it easier for the child to perform his tasks, but glasses are no direct treatment for specific primary dyslexia. In fact glasses, rightly or wrongly prescribed, may create a false sense of security that may delay needed evaluation for the correct approach to treatment.

Recognition of Dyslexia. A medical examination can disclose factors which contribute to *secondary* reading difficulties. Examples are mental retardation, impairment of vision or hearing, and the like. *Specific primary* dyslexia is manifested by difficulty in learning to read despite conventional instruction, adequate intelligence, and sociocultural opportunity. It is dependent upon a deficiency of recognition and comprehension of language symbols. These dyslexic children seem to see and understand in ways different from those regarded as normal. They may reverse letters, the order of letters and words, or may show bizarre spelling. Some cannot recognize sounds. Many have an unreliable visual memory which is often coupled with difficulty of vocal expression.

See also **Vision and the Eye.**

DYSPNEA. Generally, the term applies to an increase in the rate or force associated with respiration. More specifically, dyspnea refers to the discomfort arising from an awareness of the necessity to exert abnormal effort to breathe. The condition may arise from several causes: (1) an interference with normal heart action, (2) some mechanical form of interference with the passage of air in or out of the lungs, (3) an increase in hydrogen ion concentration (lowering of pH) of the blood, as from acidosis, and (4) the overstimulation of sensory nerves associated with pain. Dyspnea occurs, of course, after violent exercise, but in such instances the individual is not unduly alarmed. Persistent occurrences of dyspnea may be symptomatic of heart disease, pneumonia, or emphysema. See also **Congestive Heart Failure.**

* Based upon a position paper prepared by the American Association of Ophthalmology.

DYSPROSIUM. Chemical element, symbol Dy, at. no. 66, at. wt. 162.50, ninth in the Lanthanide Series in the periodic table, mp 1412°C, bp 2567°C, density 8.551 g/cm^3 (20°C). Elemental dysprosium has a close-packed hexagonal crystal structure at 25°C. The pure metallic dysprosium is silver-gray in color and retains its luster at room temperature. Although stable up to approximately 400°C, the metal then oxidizes at a slow rate up to 600°C. Because of its comparative softness, the metal can be worked by conventional equipment to form rod, foil, and ribbon configurations. There are seven natural isotopes ^{156}Dy, ^{158}Dy, and ^{160}Dy through ^{164}Dy. Twelve artificial isotopes have been identified. In terms of abundance, dysprosium is present on the average of 3 ppm in the earth's crust and is ranked 42nd in terms of total abundance, thus making it potentially more available than tin or beryllium. The element first was identified by Lecoq de Boisbaudran in 1886. The metal has a low acute-toxicity rating. Electronic configuration

$$1s^22s^2p^63s^23p^63d^{10}4s^24p^64d^94f^{10}5s^25p^65d^16s^2$$

Ionic radius Dy^{3+} 0.908 Å. Metallic radius 1.775 Å. First ionization potential 5.93 eV; second 11.67 eV. Dysprosium appears to be exclusively trivalent. Other important physical properties of dysprosium are given under **Rare-Earth Elements and Metals.**

Dysprosium occurs in apatite, euxenite, gadolinite, and xenotime. All of these minerals also are processed for their yttrium content. With liquid-liquid organic and solid-resin organic ion-exchange techniques, the separation of dysprosium from yttrium is favorable.

Because dysprosium has a high neutron-absorbing capability, the metal, usually in the form of foil, is used for detecting and measuring nuclear reactions and exposures. Dy_2O_3 also is used in dosimeters. Dysprosium does not emit harmful decay radiations when it is used as a neutron sponge. Further, helium-generated alpha particles which may ultimately crack structural parts for containing nuclear fuel are not generated. Thus, stainless steel containing about 3% Dy_2O_3 sometimes is used in control rod hardware for high-flux-beam reactors. Although Dy_2O_3 catalyzes the polymerization of ethylene and other synthetic reactions, the cost to date has limited these uses. Since dysprosium oxide fluoresces yellow in glass under ultraviolet radiation, it has been considered as an activator for the yellow component of the phosphors used in black-and-white television picture tubes. Investigations are continuing in connection with future uses of dysprosium in thermoelectric, semiconducting, photoelectric materials, and garnet microwave devices.

See references listed at ends of entries on **Chemical Elements;** and **Rare-Earth Elements and Metals.**

NOTE: This 6th Edition entry was revised and updated by K. A. Gschneidner, Jr., Director, and B. Evans, Assistant Chemist, Rare-Earth Information Center, Energy and Mineral Resources Research Institute, Iowa State University, Ames, Iowa. Original 5th Edition entry was prepared by J. G. Cannon, Molycorp, Inc.

DYSTETIC MIXTURE. A mixture of two or more substances in such proportions as to yield the maximum melting point, so that upon altering the proportions the melting point is lowered. Correlative of eutectic mixture.

DYSTROPHIA ADIPOSOGENITALIS. **Pituitary Gland.**

DYSTROPHY (Muscular). **Muscular Dystrophy.**

E

EAGLE (*Aves, Falconiformes*). Large birds of prey with strong hooked beaks, large curved claws, and powerful flight. The American golden eagle, *Aquila chrysaetus*, is an example of the typical members of the group, which also includes the harpy eagles, hawk eagles, harrier eagles, sea eagles, and others. The bald eagle, *Haliaeetus leucocephalus*, national bird of the United States, is one of the sea eagles.

The golden eagle is a land eagle, brownish in color with purple glossy plumage. The head and neck are brown with filtered yellow. The tail is of several shades of brown and is about 3 feet (0.9 meter) in length. Wingspread is about 7 feet (2.1 meters). The legs are fully feathered. The bird ranges from Canada to Mexico, but is relatively rare in the United States. The golden eagle is found to some extent in England and even more so in Scotland. Nests are on high cliffs. Eggs are from 1 to 3 and are spotted. See accompanying photo.

The so-called bald eagle is misnamed because of white feathers which make it appear bald. The size is roughly the same as that of the golden eagle. This bird nests on cliffs or in high tree tops. Preferred dietary items are mainly fish which it finds in nearby lakes and rivers. This and related species of sea eagles are found in numerous locations throughout the world.

The harrier is a moderately large eagle. There are several species, mostly limited to Africa, but one species extends to Asia and Europe. Harrier is also a term used to describe a group of hawks. These hawks are found on all continents. They are slender birds with long wings, and in general are useful as destroyers of reptiles and rodents. The marsh hawk, *Circus hudsonius*, is a North American harrier.

The honey buzzard also is related to the eagles and is named from its habit of robbing the nests of bees and wasps and eating the larvae. One species lives in Europe and Asia and others in the Oriental region.

The harpy is a large, crested eagle and is related to the true buzzards.

Golden eagle. (*John H. Gerard from National Audubon Society.*)

Several species range from Mexico to Patagonia. The harpy is reputed to be the most powerful of the eagles. One species in the Philippines (*Pithecophaga jefferyi*) is known as the monkey-eating eagle. The crowned harpy (*Harpy haliaetus*) is found in southern Brazil.

The kite is a large bird of prey related to the eagle. The bird is found on all continents. Three species, the swallow-tailed kite (*Elanoides forficatus*), white-tailed kite (*Elanus leucurus*), and the Mississippi kite (*Ictinia mississippiensis*) are found in North America. The black kite (*Milvus nigrans*) and the black-winged kite (*Elanus caeruleus*) are found in Europe.

The secretary bird (*Sagittarius serpentarius*) is a remarkable African bird related to the eagles and vultures. The bird is about 4 feet (1.2 meters) tall, with long legs, and is largely terrestrial in habits. It walks and runs very rapidly, and is also a strong flier on the relatively rare occasions when it takes to the air. See also **Falconiformes.**

EAGLE MOUNT. **Reflection Grating.**

EAR. **Hearing and the Ear.**

EARPHONE COUPLER. A cavity of predetermined shape which is used for the testing of earphones. It is provided with a microphone for the measurement of pressures developed in the cavity.

EARTH. Some of the principal dimensions and characteristics of the earth are given in Table 1. Throughout ths encyclopedia, there is much additional information on the earth, as a planet, and pertaining to the life which it supports. Several cross references to major areas of information elsewhere in the volume are given in this immediate text.

In some respects, science is progressing much more rapidly in its accumulation of fundamental data pertaining to the cosmos than pertaining to the earth itself. Beyond the information dealing with the crust of the earth and the atmospheres of the earth, scientific data of a precise nature concerning the inner structures of the earth, such as the mantle and core, remain to be obtained and involve extreme complexity and difficulty. What is known of the inner structure is largely based upon seismic data oceanography, volcanology, and the extrapolation of information collected from other bodies of the solar system. The phenomenon of radioactivity has been very helpful in estimating the composition of the crust. Laboratory techniques remain to be developed that would more accurately duplicate the pressures and temperatures currently estimated of the lower mantle and outer and inner core for evaluation of several geochemical hypotheses.

Much new information pertaining to the earth as a whole and not simply to the oceans per se resulted from the numerous programs of the International Decade of Ocean Exploration (IDOE). These programs are described in considerable detail in the entry on **Ocean.** Studies of lavas and their isotopic ratios have contributed much to an understanding of the inner earth. See **Volcano.** Advances in seismology and plate tectonics have been significant during the past decade. See **Earthquakes, Seismology, and Plate Tectonics.** The exploratory activities of the various earth resources satellites, meterological satellites, and others have given scientific investigations of the earth a perspective that only can be achieved from a position in space. See **Earth Resources Satellites and Geologic Remote Sensors.** See Figs. 1–3.

Data pertaining to the earth as a planet are of relatively recent vintage, considering the long time span over which humankind has used its surface. the rotation and revolution of the earth were not proven until 1725, when Bradley demonstrated the aberration of light.

TABLE 1. SELECTED PARAMETERS AND CHARACTERISTICS OF THE EARTH

FUNDAMENTAL PARAMETERS

Mass	5.975×10^{27} grams (6 sextillion, 588 quintillion short tons)
Volume	$1,083 \times 10^{9}$ km³ (259.8 billion cubic miles)
Surface Area	510.1×10^{6} km² (196,938,800 square miles)
Equatorial Radius	6,378.388 km (3,963.53 miles)
Length of Equator	40,066.59 km (24,901.55 miles)
Length of Meridian	39,999.45 km (24,859.82 miles)
Average Distance from Sun	149,476,000 km (92,900,000 miles)
Average Distance from Moon	384,321 km (238,857 miles)
Length (Degree of Longitude) at equator. (Varies as the cosine of the latitude)	111.296 km (69.171 miles)
Length (Degree of Latitude)	
at equator	110.551 km (68.708 miles)
at poles	111.669 km (69.403 miles)
Magnetic North Pole: (See Fig. 8)	77.3°N, 101.8°W (1980)

AXIS AND ROTATION

Tilt of Earth's Axis away from Perpendicular to Orbit	23°27′
Period of Rotation on its Axis:	
Mean Solar Day	24 hours
Mean Sidereal Day	23 hours, 56 minutes, 4.091 seconds of mean solar time
Period of Revolution about the Sun	
Tropical Year	365 days, 5 hours, 48 minutes, 46 seconds (decreasing at rate of 0.530 second per century)

Variations in Rotation of the Earth:

Secular—A slow secular increase in the length of the day (about 1 millisecond per century) is caused by tidal friction acting as a brake.

Irregular—Believed to be caused by turbulent motion in the core of the earth, the speed of rotation may increase for a short period, such as 5 to 10 years, and then commence decreasing. During a century, the maximum difference from the mean in the length of a day is about 5 milliseconds. Since 1900, the accumulated *difference* (although a compensating effect) has amounted to approximately 40 seconds.

Periodic—There are seasonal variations with periods of 1 year and 6 months. The resulting cumulative effect is that each year the earth is late about 30 milliseconds near June 1 and is ahead that same amount near October. It is believed that the semi-annual effect arises from tidal action of the sun, which distorts the shape of the earth slightly. The maximum seasonal variation in length of the day is about 0.5 millisecond. This annual effect is probably due to the seasonal change in the wind patterns of the Northern and Southern Hemispheres.

COMPOSITION

Average Density	5.517 grams/cm³
Estimated Density of Mantle	3–6 grams/cm³
Estimated Density of Core	10–17 grams/cm³

Occurrence of Elements in Earth's Crust:

Element	%	Element	%
Oxygen	47.	Sodium	2.5
Silicon	28.0	Potassium	2.5
Aluminum	8.0	Titanium	0.4
Iron	4.5	Hydrogen	0.2
Calcium	3.5	Carbon	0.2
Magnesium	2.5	Phosphorus	0.1
Sulfur	0.1	Tin	0.00001
Nickle	0.02	Silver	0.000001
Copper	0.002	All others	<0.48
Lead and Zinc	0.001		

Average Percentage of Metals in Igneous Rocks:

Metal	%	Metal	%
Silicon	27.72	Copper	0.010
Aluminum	8.13	Tungsten	0.005
Iron	5.01	Lithium	0.004
Calcium	3.63	Zinc	0.004
Sodium	2.85	Niobium and Tantalum	0.003
Potassium	2.60		
Titanium	0.63	Hafnium	0.003
Manganese	0.10	Thorium	0.002
Barium	0.05	Lead	<0.002
Chromium	0.037	Cobalt	0.001
Zirconium	0.026	Beryllium	0.001
Nickel	0.020	Strontium	<0.001
Vanadium	0.017	Uranium	<0.001
Rare earths	0.015		

Average Salt Content of Ocean Water 3.5% (wt)

	(Of Total Salts)
Sodium chloride NaCl	77.76%
Magnesium chloride $MgCl_2$	10.88
Magnesium sulfate $MgSO_4$	4.74
Calcium sulfate $CaSO_4$	3.60
Potassium sulfate K_2SO_4	2.46
Magnesium bromide $MgBr_2$	0.22
Calcium carbonate $CaCO_3$	0.34

Composition of the Atmosphere:

	%(vol)
Nitrogen	78.03
Oxygen	20.99
Argon	<0.94
Carbon dioxide	0.03
Hydrogen	0.01
Neon	0.00123
Helium	0.0004
Krypton	0.00005
Xenon	0.000006

ZONAL POINTS AND PARALLELS

		Latitude	
	North Pole	90°00′N	
Frigid Zone			
	Arctic Circle	66°30′N[a]	Northern Hemisphere
North Temperate Zone			
	Tropic of Cancer	23°27′N	
Torrid Zone	Equator	00°00′	
	Tropic of Capricorn	23°27′S	Southern Hemisphere
South Temperate Zone			
	Antarctic Circle	66°30′S[2]	
Frigid Zone			
	South Pole	90°00′S	

[a] Commonly defined as 23°30′ from the respective pole.

CONTINENTS

Dimensions of Continents:

		Distance N to S	Distance E to W
Asia, including islands	44.75×10^{6} km² (17,276,909 mi²)	8,528 km (5,300 mi)	9,654 km (6,000 mi)
Africa	30.28×10^{6} km² (11,688,728 mi²)	8,045 km (5,000 mi)	7,562 km (4,700 mi)
North America including islands	24.26×10^{6} km² (9,368,446 mi²)	8,528 km (5,300 mi)	6,436 km (4,000 mi)

TABLE 1. SELECTED PARAMETERS AND CHARACTERISTICS OF THE EARTH (*continued*)

		Distance N to S	Distance E to W
South America	18.05×10^6 km² (6,970,706 mi²)	7,642 km (4,750 mi)	4,988 km (3,100 mi)
Antarctica	13.38×10^6 km² (5,165,000 mi²)	—	—
Europe	10.22×10^6 km² (3,947,441 mi²)	3,862 km (2,400 mi)	6,275 km (3,900 mi)
Australia	8.53×10^6 km² (3,294,866 mi²)	3,170 km (1,970 mi)	3,862 km (2,400 mi)

Highest Continental Points:

	Elevation above Sea Level	
Asia—Mount Everest, Nepal-Tibet[b]	8.847.7 m	(29,028 ft)
South America—Mount Aconcagua, Argentina	6,959.8 m	(22,834 ft)
North America—Mount McKinley (U.S.A.)	6,193.5 m	(20,320 ft)
Africa—Kibo (Kilimanjaro), Tanzania	5,894.8 m	(19,340 ft)
Europe—Mount El'brus (U.S.S.R.)	5,641.8 m	(18,510 ft)
Antarctica, Vinson Massif	4,138.9 m	(16,860 ft)
Australia, Mount Kosciusko	2,228 m	(7,310 ft)

[b] Based upon Surveyor General of the Republic of India (1954); ±3 m (10 ft) because of snow.

Lowest Continental Points:

	Depth below Sea Level	
Asia—Dead Sea, Israel–Jordan	395.9 m	(1,299 ft)
Africa—Lake Assal, Afars and Issas Terr.	156.1 m	(512 ft)
North America—Death Valley (U.S.A.)	86 m	(282 ft)
South America—Salinas Grandes, Peninsula Valdés, Argentina	39.9 m	(131 ft)
Europe—Caspian Sea (U.S.S.R.)	28 m	(92 ft)
Australia—Lake Eyre	15.9 m	(52 ft)

Highest Temperature:

Azizia, Tripolitania (North Africa), Sept. 13, 1922	57.8°C	(136°F)
In the United States: Death Valley, California, July 10, 1913	56.7°C	(134°F)

Lowest Temperature:

Soviet Antarctic station Vostok, August 24, 1960 (Lower readings unofficially reported in 1982)	−88.3°C	(−126.9°F)
In the United States: Prospect Creek, Alaska, January 23, 1971	−62.2°C	(−80°F)

Most Annual Precipitation:

Mount Waialeale, Hawaii (Island of Kauai)
1,168 cm rainfall/year
460 inches rainfall/year

Economic Areas:

Deserts	12.7×10^6 km²	(4,904,593 mi²)
Steppes	49.77×10^6 km²	(19,217,465 mi²)
Fertile Areas	87×10^{10} km²	(33,588,038 mi²)

OCEANS

	Area	Average Depth
All Oceans	375×10^6 km² (145 million mi²)	—
Pacific Ocean	166×10^6 km² (64.1 million mi²)	4188 meters (13,739 feet)
Atlantic Ocean	86×10^6 km² (33.3 million mi²)	3735 meters (12,257 feet)
Indian Ocean	73×10^6 km² (28.3 million mi²)	3872 meters (12,704 feet)
Southern Ocean (Antarctic region)	36×10^6 km² (13.9 million mi²)	—
Arctic Ocean	14×10^6 km² (5.4 million mi²)	—

Greatest Ocean Depth Measured
11,033 meters (Mariana Trench, Pacific Ocean)
(36,199 feet)

Ranges of Spring Tides in Excess of 5 Meters (16.4 Feet):

Location	Tidal Range Meters	Tidal Range Feet
Burntcoat Head, Nova Scotia (Bay of Fundy)	14.49	47.5
Rance Estuary, France	13.5	44.3
Anchorage, Alaska	9.03	29.6
Liverpool, England	8.27	27.1
St. John, New Brunswick, Canada	7.20	23.6
Dover, England	5.67	18.6
Cherbourg, France	5.49	18.0
Antwerp, Belgium	5.43	17.8
Rangoon, Burma	5.19	17.0
Juneau, Alaska	5.06	16.6
Panama (Pacific Side)	5.01	16.4

MAJOR LAKES

Lake	Area	
Caspian Sea (Europe–Asia)	371,795 km²	(143,550 mi²)
Lake Superior (North America)	210,050 km²	(81,100 mi²)
Lake Huron (North America)	193,730 km²	(74,800 mi²)
Lake Michigan (North America)	175,860 km²	(67,900 mi²)
Lake Erie (North America)	103,600 km²	(40,000 mi²)
Lake Ontario (North America)	90,260 km²	(34,850 mi²)
Lake Victoria (Africa)	69,490 km²	(26,830 mi²)
Aral Sea (Asia)	65,525 km²	(25,300 mi²)
Lake Tanganyika (Africa)	32,895 km²	(12,700 mi²)
Lake Baykal (Asia)	30,510 km²	(11,780 mi²)
Nyasa Lake (Africa)	29,605 km²	(11,430 mi²)

MAJOR RIVERS

River	Flows Into	Length	
Nile	Mediterranean	6,669 km	(4,145 mi)
Amazon	Atlantic Ocean	6,436 km	(4,000 mi)
Ob-Irtysh	Gulf of Ob	5,567 km	(3,460 mi)
Yangtze	East China Sea	5,471 km	(3,400 mi)
Huang	Yellow Sea	4,827 km	(3,000 mi)
Congo	Atlantic Ocean	4,373 km	(2,718 mi)
Amur	Tartar Strait	4,344 km	(2,700 mi)
Lena	Laptey Sea	4,312 km	(2,680 mi)
Mackenzie	Arctic Ocean	4,240 km	(2,635 mi)
Mekong	South China Sea	4,183 km	(2,600 mi)
Niger	Gulf of Guinea	4,183 km	(2,600 mi)
Missouri	Mississippi River	4,076 km	(2,533 mi)
Parana	Rio de la Plata	4,023 km	(2,500 mi)
Mississippi	Gulf of Mexico	3,781 km	(2,350 mi)
Volga	Caspian Sea	3,685 km	(2,290 mi)
Madeira	Amazon River	3,379 km	(2,100 mi)
Yenisey	Kara Sea	3,347 km	(2,080 mi)
Purus	Amazon River	3,218 km	(2,000 mi)

MAJOR FLOODS AND TIDAL WAVES

		Lives Lost
1931	Huang He River, China	3,700,000
1887	Huang He River, China	900,000
1939	Northern China	200,000
1911	Chang Jiang River, China	100,000
1960	Bangladesh (October 10)	6,000
1900	Galveston, Texas	5,000
1960	Bangladesh (October 31)	4,000

MAJOR EARTHQUAKES*

		Lives Lost
1556	Shensi, China	830,000
1976	Tangshan, China (8.2)	655,235
1737	Calcutta, India	300,000

TABLE 1. SELECTED PARAMETERS AND CHARACTERISTICS OF THE EARTH (*continued*)

		Lives Lost
526	Antioch, Syria	250,000
1927	Nan-Shan, China (8.3)	200,000
1730	Hokkaido, Japan	137,000
1290	Chihili, China	100,000
1920	Gansu, China (8.6)	100,000
1923	Tokyo, Japan (8.3)	99,330
1908	Messina, Italy (7.5)	83,000
1667	Shemuka, Caucasia	80,000
1932	Kansu, China (7.6)	70,000
1970	Northern Peru (7.7)	66,794
1268	Silicia, Asia Minor	60,000
1693	Catania, Italy	60,000
1755	Lisbon, Portugal (8.75)	60,000
1783	Calabria, Italy	30,000
856	Corinth, Greece	45,000
1797	Quito, Ecuador	41,000
1755	Northern Persia	40,000
1868	Peru and Ecuador	40,000
1293	Kamakura, Japan	30,000
1531	Lisbon, Portugal	30,000
1783	Calabria, Italy	30,000
1828	Echigo, Japan	30,000
1915	Arezzono, Italy (7.5)	30,000
1935	Quetta, India (7.5)	30,000
1939	Erzincan, Turkey (7.9)	30,000

* Earthquake intensity (Richter Scale) where measured or estimated given in parentheses after location.

MAJOR VOLCANOES

Notable Eruptions (Large Loss of Life and/or Property):

7000 B.C.	Mazama (now Crater Lake, California)
79 A.D.	Mt. Vesuvius, Italy (also 1139, 1631, 1779, 1793, 1872, 1906, 1944)
1815	Tambora, Indsonesia (also 1913)
1902	Santa Maria, Quatemala
1902	Mt. Pelée, Martinique
1912	Katmai, Alaska
1914/1917	Lassen Peak, California
1964	Mt. Agung, Bali
1968	Mt. Batur, Bali
1976	St. Augustine, Alaska (also 1812, 1883, 1902, 1935, 1963–1964)
1976	White Island Volcano, New Zealand
1978	Poas, Costa Rica
1978	Etna, Italy (over 70 eruptions in 2000 years)
1978	Arboukaba, Djibouti
1979	Fuego, Guatemala
1979	Soufrière, St. Vincent
1979	Karkar, Papua New Guinea
1979	Mt. Sinila, Java
1980	Mt. St. Helens, Washington (over 20 eruptions since 1900 B.C.)

Abridged List of Volcanoes (Active and Inactive):

Name of Volcano	Location	Elevation Meters	Elevation Feet
Asia–Oceania			
Klyuchevskaya (1974)	U.S.S.R.	4,850	15,913
Kerintiji (1968)	Sumatra	3,805	12,484
Fujiyama	Japan	3,776	12,390
Rindjani (1966)	Indonesia	3,726	12,225
Semeru (1976)	Java	3,676	12,061
Ichinskaya	U.S.S.R.	3,631	11,913
Koryakskaya (1957)	U.S.S.R.	3,528	11,575
Sundoro (1906)	Java	3,150	10,335
Agung (1964)	Bali	3,142	10,309
Mayon (1978)	Philippines	2,990	9,810
Apo	Philippines	2,953	9,689
Merapi (1976)	Java	2,911	9,551
Tambora (1913)	Indonesia	2,851	9,354
Ruapehu (1975)	New Zealand	2,796	9,174
Balbi	Solomon Islands	2,593	8,508
Asama (1973)	Japan	2,542	8,340
Niigata Yakeyama (1974)	Japan	2,400	7,874
Alaid (1972)	Kuril Islands	2,339	7,674
Ulawun (1973)	New Britain	2,248	7,376
Azuma (1978)	Japan	2,024	6,641
Nasu (1977)	Japan	1,917	6,290
Lamington (1952)	Papua New Guinea	1,830	6,004
Tiatia (1973)	Kuril Islands	1,822	5,978
Batur (1968)	Bali	1,717	5,633
Bagana (1960)	Solomon Islands	1,702	5,584
Keli Mutu (1968)	Indonesia	1,640	5,301
Lokon-Empung (1970)	Celebes	1,579	5,181
Lopevi (1960)	New Hebrides	1,447	4,748
Dukono (1971)	Indonesia	1,087	3,566
Suwanosezima (1979)	Japan	932	3,058
Usu (1978)	Japan	725	2,379
White Island (1979)	New Zealand	321	1,053
Taal (1977)	Philippines	300	984
North America			
Citlaltepec	Mexico	5,676	18,623
Popocatepetl (1920)	Mexico	5,452	17,880
Rainier	Washington (U.S.)	4,395	14,420
Wrangell	Alaska (U.S.)	4,320	14,174
Colima (1975)	Mexico	3,960	12,993
Spurr (1953)	Alaska (U.S.)	3,375	11,073
Baker	Washington (U.S.)	3,316	10,880
Lassen	California (U.S.)	3,186	10,453
Paricutin (1952)	Mexico	3,170	10,401
Shishaldin (1979)	Aleutians (U.S.)	2,858	9,377
St. Helens (1980)	Washington (U.S.)	2,549*	8,364
Katmai (1931)	Alaska (U.S.)	2,285	7,497
Trident (1963)	Alaska (U.S.)	2,070	6,792
Martin (1912)	Alaska (U.S.)	1,830	6,004
Cleveland (1951)	Aleutians (U.S.)	1,730	5,676
Westdahl (1978)	Aleutians (U.S.)	1,532	5,026
Augustine (1976)	Alaska (U.S.)	1,210	3,970
Seguam (1977)	Alaska (U.S.)	1,050	3,445
South America			
Guallatiri (1960)	Chile	6,060	19,883
Cotopaxi (1975)	Ecuador	5,897	19,348
El Misti	Peru	5,825	19,112
Tolima (1943)	Colombia	5,525	18,128
Purace (1977)	Colombia	4,600	15,093
Hudson (1973)	Chile	2,600	8,531
Fernandia (1977)	Galapagos Islands	1,546	5,072
Alcedo (1954)	Galapagos Islands	1,127	3,698
Antarctica			
Erebus (1975)	Ross Island	3,743	12,281
Big Ben (1960)	Heard Island	2,745	9,006
Melbourne	Victoria Island	2,590	8,498
Damley (1956)	South Sandwich Islands	1,100	3,609
Deception Island (1970)	South Shetland Islands	602	1,975
Mid-Pacific			
Mauna Kea	Hawaii (U.S.)	4,206	13,800
Mauna Loa (1978)	Hawaii (U.S.)	4,170	13,682
Haleakala	Hawaii (U.S.)	3,065	10,056
Kilauea	Hawaii (U.S.)	1,222	4,009
Central America			
Tajumulco	Guatemala	4,220	14,502
Tacana	Guatemala	4,092	13,426
Acatenango (1972)	Guatemala	3,976	13,045
Santa Maria (Santiaguito) (1979)	Guatemala	3,772	12,376
Fuego (1979)	Guatemala	3,736	12,258
Irazu (1967)	Costa Rica	3,432	12,245
Poas (1978)	Costa Rica	2,704	8,872
Pacaya (1978)	Guatemala	2,552	8,373
San Miguel (1976)	El Salvador	2,130	6,989
Izaico (1966)	El Salvador	1,965	6,447
Rincon de la Vieja (1968)	Costa Rica	1,806	5,925
El Viejo (San Cristobal)	Nicaragua	1,745	5,725
Arenal (1978)	Costa Rica	1,552	5,092
La Soufrière	Guadeloupe	1,467	4,813

TABLE 1. SELECTED PARAMETERS AND CHARACTERISTICS OF THE EARTH (*continued*)

Name of Volcano	Location	Elevation Meters	Elevation Feet
Pelée	Martinique	1,397	4,584
Soufrière (1979)	St. Vincent	1,178	3,865
Masaya (1978)	Nicaragua	635	2,083
Europe			
Etna (1978)	Italy	3,290	10,794
Vesuvius (1944)	Italy	1,281	4,203
Stromboli (1975)	Italy	926	3,038
Vulcano	Italy	500	1,641
Santorini (1950)	Greece	130	427
Mid-Atlantic Range			
Beerenberg (1970)	Jan Mayen Island	2,277	7,471
Tristan da Cunha (1962)	Tristan da Cunha Island	2,060	6,759
Askja (1961)	Iceland	1,510	4,954
Leirhnukur (1975)	Iceland	650	2,133
Krafla (1978)	Iceland	650	2,133
Helgatell (1973)	Iceland	226	742
Africa			
Kilimanjaro	Tanzania	5,895	19,341
Cameroon	Cameroons	4,070	13,354
Teide (Tenerife) (1909)	Canary Islands	3,713	12,182
Nyirangongo (1977)	Zaire	3,465	11,369
Nyamuragira (1977)	Zaire	3,056	10,027
Ol Doinyo Lengai (1960)	Tanzania	2,886	9,469
Fogo (1951)	Cape Verde Island	2,829	9,282
Piton de la Fournaise (1977)	Reunion	2,631	8,632
Palma (1971)	Canary Islands	2,423	7,950
Karthala (1977)	Comoro Islands	2,361	7,746
Erta-Ale (1973)	Ethiopia	615	2,008

NOTE: Volcanoes listed with no dates within parentheses have been inactive for several decades.

* Prior to May 18, 1980 eruption, elevation was 2,950 meters (9,677 feet).

Several classical proofs were attempted even earlier, using such observations as the variation during the year of the radial velocity of stars and stellar parallax; because of their vey small amplitude, they were not measured until the early part of the nineteenth century. An excellent proof of the earth's rotation is the action of the Foucault pendulum. An analysis of the action of this pendulum shows that it will appear to rotate with a period of one day at the poles, with longer periods at lesser latitudes, and with a period infinitely long at the equator.

Geophysics

Geophysics is the physics of the earth and the space immediately surrounding it and the interactions between the earth and extraterrestrial forces and phenomena. Geophysics consists of a number of interlocking sciences dealing with physical properties of the earth, its interior and atmosphere, its age, motions and paroxysms, and their practical applications. All of these sciences use the methods of physics for measurements and analysis. From observational material, often of an indirect nature, attempts are made to derive abstract models of states and processes through advanced mathematical concepts and, in some cases, through statistical relations.

In essence, geophysics is the study of the earth as a planet—with three basic divisions of activity—the solid-earth, the atmosphere and hydrosphere, and the magnetosphere (solar–terrestrial physics). As a convenience, the term geophysics sometimes has been extended to similar studies of the planets and their satellites. However, for extraterrestrial studies, there are more specific terms, such as selonography which relates to the moon.

Geophysics may be described as an ancient science. In its early stages, it was developed by the Greeks who attempted to determine the shape and size of the earth (Eratosthenes, 275–194 B.C.). Among its most illustrious contributors have been Galileo Galilei (1564–1642); Sir Isaac Newton (1642–1727), who dealt with the motions of the earth and its gravitational field; Karl Friedrich Gauss (1777–1855), who developed the theory of the magnetic field; and Vilhelm Bjerknes (1862–1951), who laid the foundation for the hydrodynamic theories of the atmosphere and the oceans. The roster of distinguished scientists who contributed to this field during the 20th century includes: L.

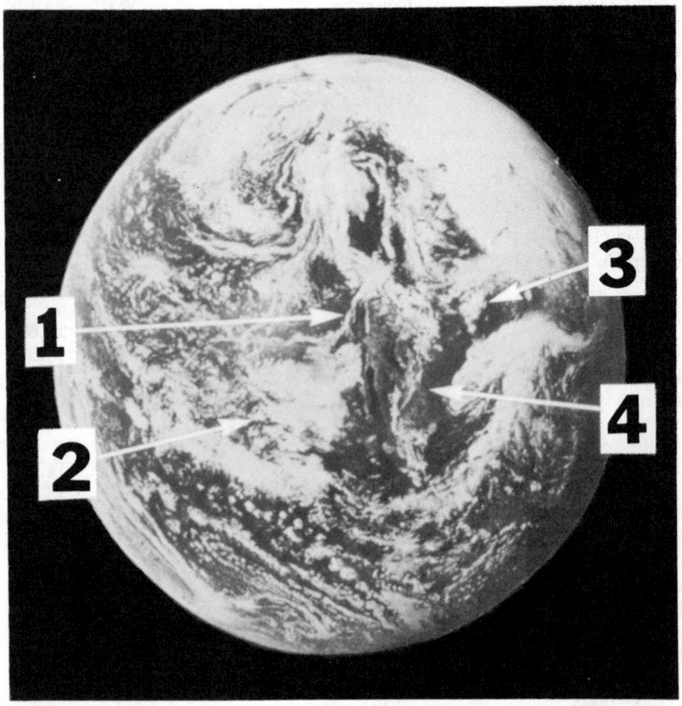

Fig. 1. The Earth as seen from outerspace. Astronauts described the Earth as "a colorful island in the vast sea of blackness." The Northern Hemisphere: (1) United States West Coast; (2) Pacific Ocean; (3) Atlantic Ocean; (4) Gulf of Mexico.

Fig. 2. The Southern Hemisphere as seen from outerspace. The South Polar area is visible under partial cloud cover.

Fig. 3. Small portion of the Earth as viewed from an altitude of 100 miles (161 kilometers). View shows the Hadramaut plateau, the Gulf of Aden, and Somalia.

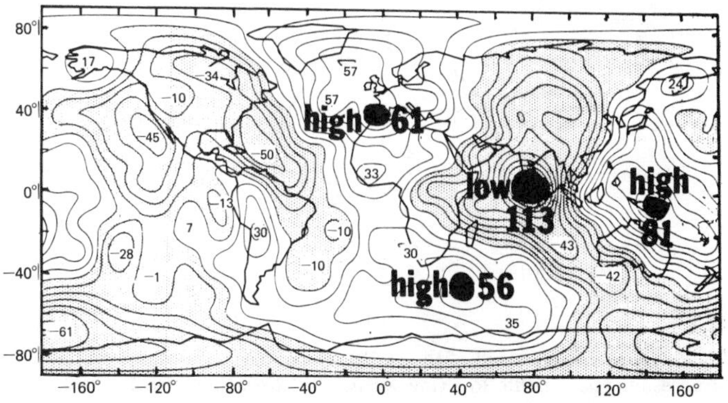

Fig. 4. In 1970, the Smithsonian Institute released what became known as the Smithsonian Standard Earth II. Preparation of the contours shown represented many years of work, using cameras and gravimetric surveys, as well as the first results of measurements by laser tracking of satellites. Over 100,000 photographic observations became part of the data bank. Calculations involved some 200,000 simultaneous equations, solving for about 400 unknowns. These included nearly 300 harmonic coefficients and the station coordinates. This map shows contours of the geoid at 10-meter (32.8 feet) intervals, which are relative to a reference spheroid of flattening 1 part in 198.25. Depressions are shaded areas. The three largest highs and the principal low are indicated. See text. (*After Smithsonian Institution map.*)

Vegard (polar aurora); Sidney Chapman (aeronomy); C. G. Rossby (meteorology); H. U. Sverdrup (oceanography); Sir Harold Jeffreys; F. A. Vening-Meinesz (structure of the earth); and B. Gutenberg and J. B. Macelwane (seismology).

A major series of milestones toward the advancement of scientific knowledge of the earth commenced, on the basis of international cooperation, in the late 19th century, with the First International Polar Year (1882–1883), followed by the Second International Polar Year (1932–1933), fifty years later. For the International Geophysical Year (1957–1959), over 8,000 scientists of 66 nations collaborated and spawned the more recent ventures, including satellite investigations of the earth (Skylab and its predecessors), increased exploration of the Antarctic continent, the International Years of the Quiet Sun (IQSY, 1964–1965), and numerous different and subsequent activities. IDOE (1970–1979) has previously been mentioned.

Geochronology. The study of the age of the earth and its various geological formations. Inadequate earlier methods of sedimentology have been replaced by the use of radioactive decay constants and isotope ratios. This has made it possible to date approximately all major geological eras. For the oldest rocks, the decay of ^{238}U to ^{206}Pb has led to ages of approximately 3×10^9 years. Although authorities do not agree on specific numbers, the length of time that the earth has existed appears to be about 4.5 billion years. For a description of age determination with radioisotopes, see **Radioactivity and Other Dating Techniques.** See also **Evolution.**

Geodesy. Geodesy is principally concerned with the size and shape of the earth and its gravitational field. See Table 2. The first recorded effort to estimate the circumference of the earth (and hence its diameter) dates back to Eratosthenes of Alexandria in the 3rd century B.C. Eratoshenes observed that the sun was overhead at Aswan at midsummer because it shone directly down a well, but in Alexandria, it was 7.2° or 1/50th of a circle away from the vertical. The distance between Aswan and Alexandria was calculated by estimating the time required by a camel to traverse the distance. The calculation was amazingly correct—within 1% of the currently accepted value. This was an example of space geodesy, i.e., using the sun as the reference object, a technique that surfaced again after the launching of the first artificial satellite, Sputnik, in 1957.

The Geoid. Because of the rotation of the earth, its lack of absolute rigidity, crustial mass distribution, and tidal forces, the shape is not perfectly spherical, but approximates that of a triaxial ellipsoid. There is flattening at the poles (polar diameter is shorter than the equatorial). The actual figure of the earth, which is irregular (not just at the poles) is referred to as *the geoid.* See Fig. 4, The largest departure of the actual geoid from the reference geoid (mapped) occurs in a depression of approximately 113 meters (371 feet) south of India. Off New Guinea, the highest hump occurs, about 81 meters (266 feet). There are two other significant humps, one located south of the British Isles and about 61 meters (200 feet) high; and the other hump located south of Madagascar and about 56 meters (184 feet) high.

In constructing the dimensions of the geoid, one is not concerned with the regular physiographic features of the earth—mountains, ocean basis, etc. The concern is the shape of the mean sea-level surface, continued under the land in a logical fashion. This surface is exclusively defined by measuring the variations of the earth's gravitational attraction, both with latitude and longitude. To this must be added the acceleration produced by the earth's rotation. Bomford (1971) defines the geoid as "an equipotential surface of the earth's gravitational potential plus the rotational potential." The geoid is the basic reference shape on which the earth's topography (height above sea level) is superposed.

TABLE 2. GEODETIC PARAMETERS OF THE EARTH

Parameter	Standard Value	Current Estimate and Standard Deviation
Mean sidereal rotation rate, ω	$0.7292115085 \times 10^{-4}$ sec^{-1}	$0.729115085 \times 10^{-4}$ sec^{-1}
Equatorial gravity, γe	9.780490 m/sec^2	$9.780306 \pm .000013$ m/sec^2
Equatorial radius, a	6378388 m	6378160 ± 15 m
Flattening, f	$1/298.25 \pm .03$	$1/298.25 \pm .03$

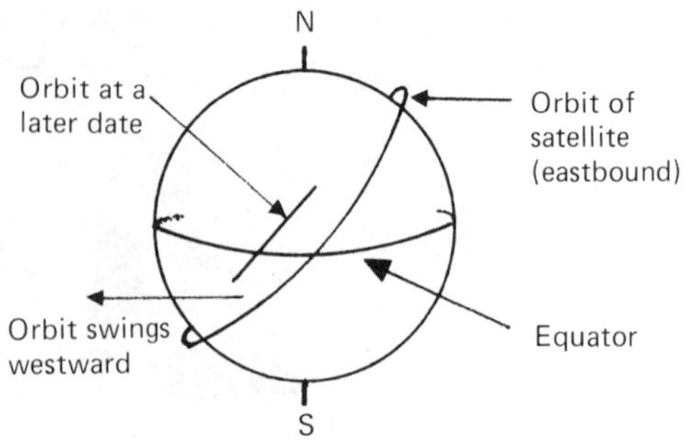

Fig. 5. Gravitational pull of the earth's equatorial bulge causes orbital plane of an eastbound satellite to swing westward. (*After King-Hele.*)

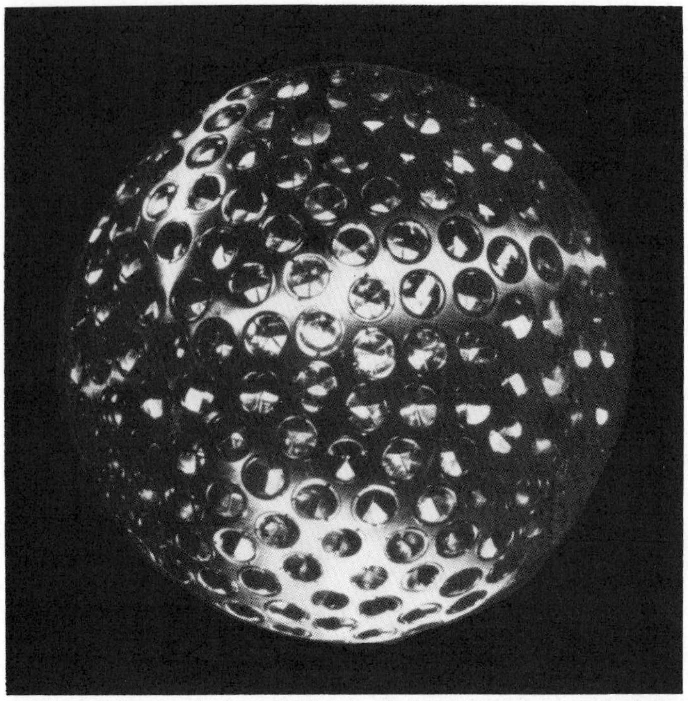

Fig. 7. Laser geodynamic satellite, *Laegeos,* launched in 1976, weighed 400 kilograms (882 pounds), was 0.6 meter (~2 feet) in diameter, carried 426 retroflectors, and established a circular orbit at a height of 6000 kilometers (3728 miles). The satellite was designed for an accuracy of about 2 centimeters when tracked by lasers. (*National Aeronautics and Space Administration.*)

There is a dual relationship between the geoid and orbiting satellites. As pointed out by King-Hele (1976), one effect of the earth's flattening on a satellite orbit is to make the plane of the orbit rotate about the earth's axis in the direction opposite to the satellite's motion, while leaving the orbit inclined at the same angle to the equator. See Fig. 5. As further explained by King-Hele, the most helpful way of analyzing the shape of the geoid is to assume that if it made up of an infinite number of harmonics, the second harmonic defining the flattening, the third harmonic often being called pear-shaped, the fourth harmonic square-shaped, etc. See Fig. 6. Shapes of these kinds would be obtained if one sliced the earth through the poles if only one particular harmonic existed. But, in fact, all harmonics exist. Thus the procedure is one of calculating them separately and then putting them together to yield the final shape.

Laser tracking of satellite has been used for a number of years in making observations for determining the geoid, but these methods have been greatly improved since the mid-1970s. A satellite with corner reflectors was put in orbit in 1965 and followed by a number of like satellites for laser tracking. One method of investigation is to consider the satellite as points in the sky for geometrical triangulation; the other method analyzes the effect of the earth's gravitational attraction on satellite orbits, thus determining how gravity changes over the earth. The earlier satellites were in the *Beacon* and the *Lageos* series. See Fig. 7. The first of the *Magsat* (Magnetic Field Satellite), jointly sponsored by the National Aeronautics and Space Administration (NASA) and the U.S. Geological Survey was launched in October 1979. This was designed as a pole-crossing satellite with a range from 350 to 500 kilometers (217 to 311 miles) with an expected life of about 120 days.

One may ask, why is it so important to accurately determine the shape of the geoid? The effect of the gravitational pull of the earth's equatorial bulge and other irregularities affect the orbit of an artificial satellite. But why were these humps and depressions important prior to the satellite age? For one thing, it has been determined that these humps and depressions tend to migrate with time and undoubtedly are related to migrations of the magnetic poles. (It has not been determined which is the cause and which is the effect.) With greater accuracies of measurement obtainable, as from Doppler radar and laser methods, satellite geodesy is making and will continue to impact all of the earth sciences. This knowledge will assist in making crucial tests of theories about the earth's interior, with some aspects of tectonic plate maps already being reflected in geőid mappings. Further, patterns of convection currents or density irregularities within the earth will have to be meshed with the observed gravitational field. Combined with satellite altimeter measurements, gravitational field determinations will provide an accurate profile of the geoid surface, including earth tides and ocean tides. More knowledge of the polar motion (locus of the point where the earth's axis of rotation cuts the surface) most likely will also be obtained from this line of research. This knowledge may also assist in refining present theories of the earth's magnetism and provide more plausible explanations of how the earth's magnetic field is created in the first place.

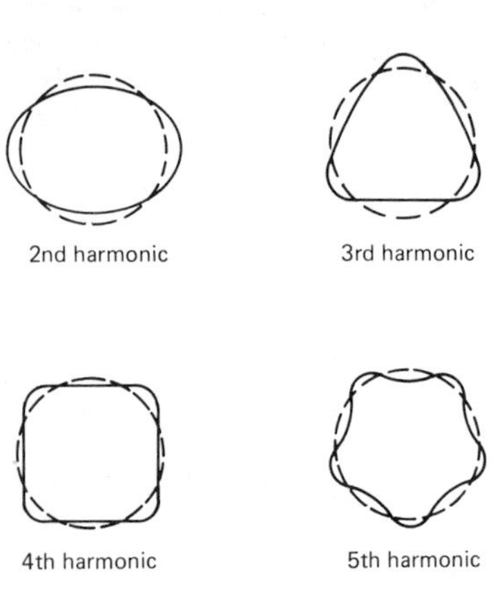

Fig. 6. Harmonics defining the characteristics of the geoid. There are many more harmonics. (*After King-Hele.*)

The measured values of gravity depend on latitude because of the flattening and the variation of the centrifugal force from pole to equator. The normal value of gravity at the earth's surface in centimeters per second per second is represented by

$$\gamma = 978.0516[1 + 0.005291 \sin^2 \phi - 0.0000057 \sin^2 2\phi + 0.0000 \cos^2 \phi \cos 2(\lambda + 6°)]$$

where ϕ and λ are latitude and longitude, respectively.

Gravity measurements have shown that in spite of large mass difference at the surface, the earth is nearly in isostatic equilibrium. Various crustal blocks act as if they floated in a dense subcrustal material. The undulations of the geoid do not exceed 80 meters. Approximate dimensional figures for the earth are: surface area, 510.1×10^6 square kilometers; volume 1.083×10^9 cubic kilometers; average density 5.517 grams per cubic centimeter; mass; 5.975×10^{27} grams; equatorial radius, 6.378 388 kilometers.

The deformation of the solid earth by tidal forces form a specialty. The twice-daily occurring tides are observed by deflections of the vertical or variations of gravity. For the lunar tide, the variations amount to about 0.168 milligal; for the sun, up to 0.075 milligal. (One milligal equals 10 micrometers per second per second.) The maximal elevation of the geoid is 36 centimeters, the largest depression 18 centimeters, for the lunar effect; the total solar tide can rach 25 centimeters. The combined total at new and full moon is 79 centimeters.

There are five principal systems of measurement in geodesy. (1) *Horizontal control* comprises the determination of the horizontal components of position—latitude and longitude—starting from fixed values for a certain point. It includes measurement of distances over the ground by metal tapes or by pulsing or modulating radio or light signals, and measurement of angles about a vertical axis by theodolites. Over the land, the relative horizontal position of points is obtained either by triangulation—a system of overlapping triangles with nearly all angles measured, but only occasional distances measured; or by traverse—a series of measured distances at measured angles with respect to each other; or by trilateration—a system of overlapping triangles with all sides measured. Much of the land area of the world is covered by triangulation, which gives the difference in latitude and longitude between points in the same network with a relative error of about 10^{-5}.

(2) *Vertical control* comprises the determination of heights, which is performed separately from horizontal control because of irregularities in atmosphere refraction. The most accurate method, leveling, measures successive differences of elevation on vertical staffs by horizontal lines of sight taken at intermediate points over short distances (less than 150 meters) balanced so as to minimize differential refraction effect. The datum to which vertical control refers is mean sea level as determined by tide gages. The accuracy is such that the error in difference of elevation between points on the same principal network should be a few tens of centimeters or less.

(3) *Geodetic astronomy* comprises the determination of the direction of the gravity vector and the direction of the north pole at a point on the ground. Astronomic longitude is the angle between the meridian of the gravity vector and the Greenwich meridian and is determined by measuring the time of intersection of a line of sight by a star. In these types of astronomic observation, several stars are normally observed which are selected so as to minimize error due to atmospheric refraction. Astronomic azimuth is determined by the measurement of the horizontal angle between a target and Polaris or other reference star.

(4) *Gravimetry* comprises the determination of intensity of gravitational acceleration. Most gravimetric observations are made differentially, by determining the change, with change in location, of the tension on a spring supporting a constant mass. These measurements are connected through a system of reference stations to a few laboratory determinations of absolute acceleration of gravity. The relative accuracy of gravimetry is about ± 0.001 centimeters per second squared on land and ± 0.005 centimeters per second squared at sea. The principal difficulty in its geodetic application is irregular distribution of observations.

(5) *Satellite tracking* comprises the determination of the directions, ranges, or range rates of earth satellites from ground fixed stations. These observations will be affected both by errors in positions of the station with respect to the earth's center of mass and by perturbations of the orbit by the earth's gravitational field; hence, in conjunction with a suitable dynamical theory for the orbit, they are used to determine the position of tracking stations and the variations of the gravitational field. To minimize refraction effect, directions are determined by photographs of the satellite against the background of fixed stars. Satellites also can be used as elevated targets by simultaneous observations from several ground stations.

The principal practical application of geodesy is to provide a distribution of accurately measured points to which to refer mapping, navigation aids, engineering surveys, geophysical surveys, and so on. The principal scientific interest in geodesy is the indication of the earth's internal structure by the variations in the gravity field.

Geomagnetism. This field of study deals with one of the most important physical characteristics of the earth. Its magnetic field is about 0.3 gauss in strength. The behavior of the entire Earth as a magnet has been known for nearly four centuries, but even today the mechanism that creates and maintains the magnetic field is strictly postulation. Scientists for many years have been seeking a way to explain this phenomenon, which is all the more complicated by the fact that the magnetic north pole migrates, as will be noted from Fig. 8. The concept of a permanent or even semi-permanent lode of magnetic materials was ruled out early—this because it is assumed from experience with such materials that they would lose their permanent magnetism at the high temperatures existing beneath the Earth's crust. Further, it is difficult to believe that such permanently magnetized materials, if they existed in sufficient concentration to create a field, would be able to move as would necessarily be the case in order to explain the migration of the magnetic pole.

There have been several postulations over the years pertaining to the origination and maintenance of the earth's magnetic field. It should be mentioned at the outset that the magnetic field is principally comprised (about 90%) of fields which are generated in some manner within the earth, with an external portion (about 10%) that is caused by an ionospheric ring current. The field can be represented by a dipole, which fluctuates in direction both in time and space. Short-periodic fluctuations are brought about by solar disturbances which cause invasions of protons and electrons into the high atmosphere. These disturbances, which follow broadly solar activity as expressed by sunspots, cause minor and major fluctuations of the earth's magnetic field. The major disturbances, called magnetic storms, can be observed worldwide. The same particle bombardments cause the polar aurora through excitation of oxygen, hydrogen, and nitrogen atoms in the high atmosphere. Much slower variations of the magnetic field are tied to the internal field and possibly connected with convective currents. Paleomagnetic evidence indicates that the magnetic poles are not fixed and that the field has reversed direction several times during the earth's history. In the present era, the magnetic poles and poles of rotation do not coincide.

It has been hypothesized that the migration of the magnetic poles is also an indication of migrations of the inertial axis of the earth. The distribution of the field over the surface of the earth has been mapped in form of the total intensity, the horizontal and vertical intensity, the inclination and declination. The latter is of practical importance for use of magnetic compasses in direction finding. The influence of the earth's magnetic field extends outward into space for many earth radii. It influences the path of invading particles and is responsible for the configuration of the van Allen belts.

The magnetosphere is that region of space in which the geomagnetic field dominates the motion of charged particles. Near the surface of the earth, the geomagnetic field resembles that of a dipole; the best-fit dipole is off center about 440 kilometers and is inclined about 11° to the earth's axis of rotation. Well out into space, the field is severely deformed so that it no longer resembles a dipole field. The most apparent deformation is that due to a plasma of charged particles flowing away from the sun, a flow generally referred to as the solar wind. When the kinetic energy of the directed flow of the solar wind exceeds the energy density of the geomagnetic field, the plasma displaces the field. The diamagnetic properties of the flowing plasma tend to compress and confine the geomagnetic field, at least on the side facing the sun. The side away from the sun may represent an indefinitely extended geomagnetic field. Calculations indicate that the surface of the magnetosphere facing the sun is roughly hemispherical, with dimples over the earth's magnetic poles. Observations by space probes indicate that the distance that the geomagnetic field extends toward the sun is about 10 earth radii.

The magnetosphere is the region of space within which many geophysical phenomena are confined. The van Allen radiation belt apparently extends out to the surface of the magnetosphere on the side facing the sun, although this is not so on the side away from the sun, where the magnetosphere has a tail extending at least as far as the moon's orbit and perhaps much further. Any charged particles from the ionosphere that have sufficient energy to escape from the earth's gravitational field are constrained by the magnetic field to remain in the vicinity of the earth.

The solar plasma that compresses the geomagnetic field and limits the magnetosphere flows away from the sun at a velocity that might be described as hypersonic—the ordered velocity exceeds the average random thermal velocity of particles. Although the medium is so rarefied that collisions are seldom, the particles can interact with one another through the agency of magnetic fields contained within the plasma. As a result, a collisionless shock wave develops in the flow before it reaches the surface of the magnetosphere. Space probes show that the shock front lies about 4 earth radii beyond the surface of the magnetosphere in the direction of the sun.

Generally, the theory of the internal generation of the earth's mag-

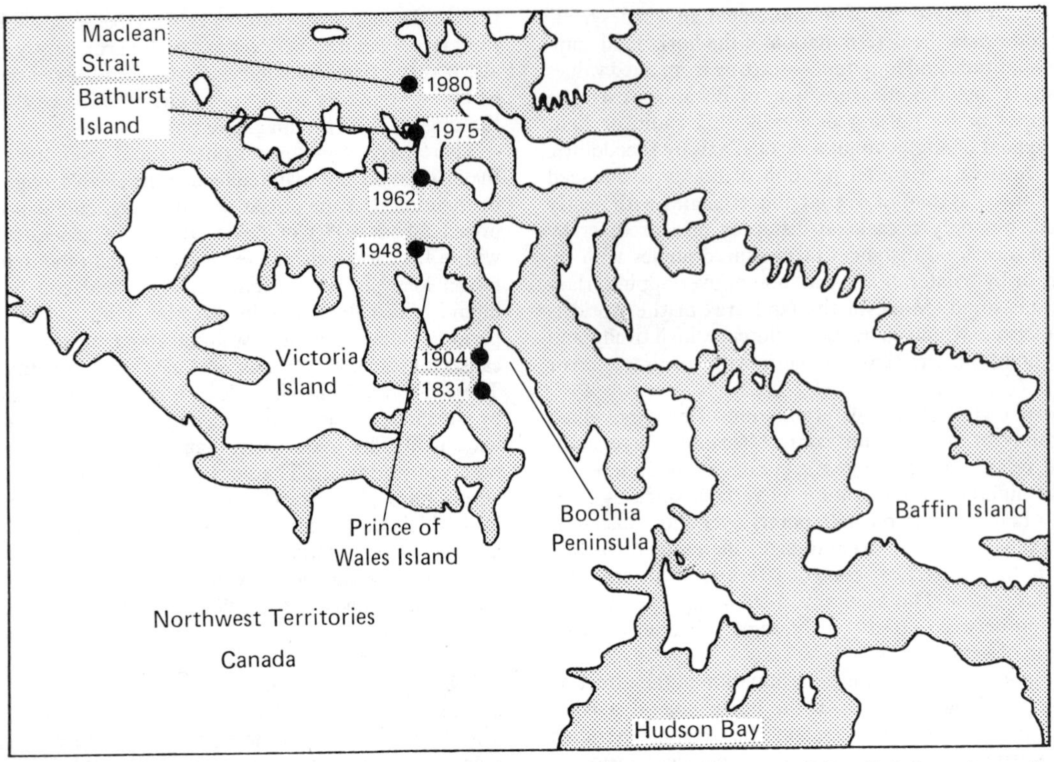

Fig. 8. Migration of the magnetic north pole, a point on the earth's surface about 1400 kilometers (868 miles) from the north geographical pole. The first person to locate the magnetic pole was British explorer, James Ross, in 1831. This point was on the west coast of the Boothia Peninsula. By 1904, Amundsen observed that the pole had moved slightly to the northwest to another point on the peninsula. By 1948, the pole had migrated to the northwestern corner of Prince of Wales Island and, by 1962, had moved to the southern part of Bathurst Island. The magnetic pole continued to move north to another point on Bathurst Island by 1975. As of 1980, the pole was located in the Maclean Strait. The pole has been monitored by Canadian scientists for over 35 years. Recent measurements indicate that the magnetic north pole is moving north by 24 kilometers (14.9 miles) per year and west by 5.7 kilometers (3.5 miles) per year. The mean position of the pole moves about 0.1° per year. Once per 24 hours, the pole makes an elliptical path centered on a mean position for the day. The size of the ellipse reflects disturbances in the geomagnetic field. The editors gratefully acknowledge the following information furnished by Paul H. Serson, Director, Division of Geomagnetism, Earth Physics Branch; Energy, Mines and Resources Canada, 1 Observatory Crescent, Ottawa, Canada.

Date	Latitude	Longitude	Source
1831.4	70.1°N	96.9°W	Ross
1904.5	70.5	96.6	Amundsen
1948.0	73.9	100.9	Dominion Observatory
1962.5	75.1	100.8	Dominion Observatory
1975.0	76.2	100.6	Earth Physics Branch
1980.0	77.3	101.8	Earth Physics Branch

netic field is that a dynamo is set up by the ponderous flow of matter in the earth's core. Flow of materials within the core sets up electric currents, which in turn induce a magnetic field. Since the earth's core cannot be observed firsthand, and because of the temperatures and pressures presumed to exist in the core, only relatively recently has it been possible to approach these conditions in the laboratory.[1] Isolated as it is, what forces energize the fluids in the core and thus set up electrical currents? The core if the earth is estimated to have a radius of about 3500 kilometers (2175 miles), representing about 16.5% of the earth's volume and about 33% of its mass. If all or most of the core is involved in a process of generating electrical currents, this is indeed a large engine by any standard—but particularly so when the magnetic field created is some several hundred times weaker than a toy horseshoe magnet. This is an unusual combination of a huge machine and a comparatively minute output.

Some scientists theorize that the earth's dynamo is self-sustaining. Having once been started with a very small magnetic field, the dynamo will generate its own field without any external supply of magnetism. It is postulated that the earth's dynamo could have been initially activated by the small magnetic field that permeates the entire galaxy (Carrigan and Gubbins, 1979). The dynamo then would take over and generate a much stronger field of its own. Although a self-sustaining dynamo may not require a steady supply of magnetic field, it does require energy to maintain movement of a disk (or metallic liquid core fluids resembling a disk of conducting material. Numerous explanations of this energy supply have been suggested. The Coriolis force is excepted because it acts only perpendicularly to the direction of flow and consequently cannot drive currents against the retarding effects of other forces.

One plausible force is gravitational buoyancy, which through processes of convection, is successfully operative in moving atmospheric currents, ocean currents, and, as one scientists mentioned, even continents. This force causes less dense materials (as in the core) to float or convect upward. Convection, of course, tends to mix fluids and to make them uniform, at which time gravitational buoyancy is no longer operative. Thus, in some way, new materials must be added

[1] As early as 1958, Hughes and McQueen (Los Alamos Scientific Laboratory) achieved pressures up to 75 gigapascals (GPa, equivalent to 0.75 megabar) in the laboratory for the study of pressure–density characteristics of the upper mantle. These researchers found that these characteristics could be closely approximated by a dunite (peridotite in which the mafic mineral is almost entirely olivine with accessory chromite almost always present) of composition $(Mg_{0.9},Fe_{0.1})_2SiO_4$. They also showed that shock-induced polymorphism of silicate minerals could take place on the submicrosecond time scale of shock wave experiments. Much research has taken place since, and continues. Shock wave experiments have provided absolute pressure–density and, more recently, temperature data for minerals at pressures that encompass the assumed entire range present in the earth, i.e., up to 370 GPa or 3.7 Mbar. Shock wave data for the high-pressure phases of these minerals has led to important inferences about the composition of the lower mantle and outer, liquid core of the earth. Explosively driven shock waves were first described in 1955 and stemmed from engineering research on the atomic bomb at Los Alamos.

to the core engine. To test this postulation, Busse and Carrigan (University of California at Los Angeles) set up a rotating plastic sphere in an effort to duplicate the possible fluid motions in the core of the earth. The first model was built in 1973 and research continues.

Geophysicists have suggested a number of mechanisms—gravitational, chemical, thermal—as energy sources to keep the core fluids in motion. It has been suggested that thermal buoyancy could come from radioactivity if the core fluid contained a large enough quantity of radioactive elements. Inasmuch as it is believed that uranium and thorium migrate into the mantle and the crust, with only trace amounts remaining in the core, it has been suggested that perhaps the radioactive isotope present is potassium 40. Another postulation involves the observation that the earth is cooling and hence liberating heat. This is augmented by the latent head of the core liquid as it freezes to form the solid inner core. Perhaps there has been a steady temperature drop of 100 K over some 3 billion years to provide the heat energy needed.

For the reader with a detailed interest in this topic, reference to the Carrigan and Carrigan (1979) paper, as listed at the end of this entry, is suggested.

Meteorology and Aeronomy. These fields are concerned with the physical state and the motions of the atmosphere, which is divided into a number of layers. The lowest is the troposphere with an average thickness of 7 to 8 kilometers in polar regions and 13 kilometers in the equatorial zone. Temperatures decrease to the interface, called tropopause, with the next layer the stratosphere. At the tropopause, polar temperatures average around −55°C; in equatorial regions, −80°C. In the stratosphere, temperatures stay nearly isothermal with height and increase again above 25 kilometers. Above the stratosphere are the mesosphere and ionosphere, and the outermost layer, the exosphere, gradually fades into the plasma continuum between earth and sun. In these higher layers of the atmosphere, complex interactions between the fluxes of electromagnetic radiation of various wavelengths and corpuscular radiation from the sun on one side and the low-density concentrations of atmospheric gases on the other side take place. The particulate radiations are also governed by the earth's magnetic field. Radiations of short wavelength cause a variety of photochemical reactions, the most notable of which is the creation of a layer of ozone acting as an effective absorber of solar ultraviolet and thus causing a warm layer at 30 kilometers in the atmosphere. The upper atmosphere as an absorber of primary cosmic rays shows many interesting nuclear reactions and is an important natural source of radioactive substances, including tritium and carbon 14 which are used as tracers of atmospheric motions and as a criteria of age.

Most manifestations of weather take place in the troposphere. They are governed by the general atmospheric circulation which is stimulated by the differential heating between tropical and polar zones. The resulting motions in the air are subject to the laws of fluid dynamics on a rotating sphere with friction. They are characterized by turbulence of varying time and space scale. Evaporation of water from the ocean and its transformation through the vapor state to droplets and ice crystals, forming clouds and precipitation, are important symptoms of the weather-producing forces. See also **Atmosphere (Earth); Meteorology.**

Geology and Mineralogy

Much of the scientific information pertaining to the earth has been derived from the investigations of geologists and mineralogists over the years. Both geology and mineralogy are old, well established scientific fields, but generally are not included in delineations of geophysics. This illustrates the overlapping of fields of scientific interests that will continue as more and more knowledge of the earth and the cosmos is collected. The more established fields expand their spheres of interest, whereas some of the newer specializations encroach upon the older fields. An interesting, fine distinction between geophysics and geology is noted the "Glossary of Geology" (American Geological Institute). The definition of geology commences, "Geology—study of the planet earth." The definition of geophysics commences, "Geophysics—study of the earth as a planet." Mineralogy, of course, is the study of minerals, their formation, and occurrence. See also **Geology; Mineralogy.**

Geochemistry

Goldschmidt (1954) defined geochemistry as the study of the distribution and amounts of the chemical elements in minerals, ores, rocks, soils, water, and the atmosphere, and the study of the circulation of the elements in nature, on the basis of the properties of their atoms and ions; also, the study of the distribution and abundance of isotopes, including problems of nuclear frequency and stability in the universe. A more succinct definition would be that geochemistry is the study of the chemical constitution of the earth and its chemical changes, either taking place now or having taken place.

Many inferences about the nature and composition of the different zones of the earth are speculative to varying degrees because of the inaccessibility of the interior and because of the differentiated nature of the earth. The overall composition may be deduced from spectroscopic evaluation of solar and stellar radiation, from nuclear chemical and astronomical theories of the origin of the elements and the evolution of the solar system, and from analytical study of the meteorites.

Interior of the Earth

From a practical standpoint, it is urgent that we learn as much as possible about the present and past physical, chemical, and dynamical state of the earth so that we can relate this knowledge to the future impacts of climatic changes, sea level changes, and earthquakes (seismic activity). It is particularly important to learn more about the interior of the earth because, by comparison, the information bank is rich in matters pertaining to the atmosphere.

Geological history of a broad nature has been extended back into time by perhaps 600 million years—this largely resulting from mapping the distribution of various rock formations found on the earth's surface, as well as the study of fossils. It is only within the last several decades that advantage has been taken of various dating techniques, notably through the use of natural isotopes (uranium, thorium, rubidium, potassium, samarium, neodymium, strontium), thus extending portions of our desired knowledge back to 4000 + million years. Field research, geodetic measurements, and mapping were useful in identifying gross trends as they relate to mountain formation and volcanic action—with some indication of continental stability and mobility (early hints of plate tectonics). But a theory to tie these observations together—a geological unfication as it were—was missing. This need has been served, at least in part, very effectively in recent years by the theory of plate tectonics. See also **Earthquakes, Seismology, and Plate Tectonics.**

In recent times, more information is being divulged by the ocean basins than from the continents themselves. It is to be noted that 70% of the earth's crust lies beneath the seas. Plate tectonics theory, while offering a reasonable explanation of several earth phenomena, also has created numerous unanswered questions, but in doing so, has provided the search for further information with a pattern and structure. For example, there is little information and only limited speculation as regards the convective system (differential buoyancy) which causes materials deep in the earth's interior to escape and form crust. What is really poorly understood is the system of counterflow which is always present in a true convective system. It is known that this must extend to depths exceeding 700 kilometers (435 miles) because deep-focus earthquakes demonstrate that the descending slabs penetrate to these kinds of depths. The counterflow system extends below the lithosphere and asthenosphere because the latter are in the uppermost portion of the convective system.

Wetherill and Drake (1980) pose the question as to whether it will ever be possible to infer from geological, geophysical, and geochemical observations the terrestrial conditions and history of the earth all the way to its beginning. Although there is a consensus that only a small amount of this information already has been collected and probably a smaller fraction of this correctly interpreted, improved research methods will yield much more knowledge. However, there is a sense of realism among many investigators to the effect that nature probably has "shredded a lot of paper and erased a lot of tapes" over the eons—removing for all time certain key prior evidence. Within the past decade, for example, much research has gone into the actions of both cold and hot seawater on crustal rock. Deep-sea drilling and manned submersibles are recovering both rocks and seawater that have been altered by exposure to each other. Some scientists have described parts of the ocean bottom as quite leaky, this leakiness allowing constant contact between seawater and crustal rocks, producing significant changes in both. The crust down to 600 meters (1970 feet) shows alteration by cold seawater percolation. Cores obtained by the DSDP project have provided good evidence that the sea floor is permeable. More recently, the vents on the Galápagos Ridge in the Pacific are evidence that seawater sinks into cracks, is heated by hot volcanic rocks, and rises back to the sea, frequently entraining metals and minerals. For more detail, see entry on **Ocean.**

Thus, although there is much to be learned from deep-sea studies, the problems are large. The constitution of the upper mantle beneath the oceans cannot be determined directly because drilling through the entire crust is not yet feasible. Indirect information can be inferred from the velocity of seismic waves reflected or refracted from the mantle, as well as the composition of basaltic magmas whose source region is the upper mantle; and the constitution of ophiolite complexes thought to be uplifted fragments of former oceanic lithosphere. Ultramafic nodules of noncumulate origin, sampled from the mantle by upwelling magmas in oceanic islands, provide additional data. Additional information may be provided by the investigation of blocks of upper mantle uplifted to shallow levels in the oceanic crust and thus accessible to sampling. Bonatti and Hamlyn (1978) describe such an investigation along the Owen Fracture Zone in the western Indian Ocean.

There is growing interest in searching the other planets of the solar system for leads pertaining to the development of the earth. Not many rocks on earth have been found that are older than 2800 million years—one of a few exceptions being the rocks dated 3800 million years old found in western Greenland. The question may be asked, "Why are older rocks so difficult to find?" Volcanic rocks on the moon were easy to find and range from 3300 to 3800 million years in age. These occur on the dark lunar maria. Older lunar rocks and fragments date back to 3900–4000 million years, as found in the brighter highland regions. These findings lead to a plausible interim explanation that the moon must have been hot very early in its history and, if so, the earth may have been even hotter, leading to the conclusion that outgassing (mainly carbon dioxide and steam) was widespread when the earth was very young.

Exploratory observations of Venus have shown that it has an atmosphere mainly of carbon dioxide with an atmospheric pressure at the surface of the planet about 100 times greater than that on earth. Because of the "greenhouse effect," surface temperatures on Venus are about 450°C, as compared with 20°C on earth. Some scientists suggest that Venus was hotter than earth during its early history and thus volatiles escaped from its interior, whereas the earth, being somewhat cooler, held its volatiles in solid or liquid form. These two modes of development would cause a marked difference in the interiors of the two planets. But such generalizations do not explain a number of more recent findings. For example, Venus is comparatively rich in ordinary argon, as well as other noble gases. The much less abundant ^{40}Ar found on earth is attributed to radioactive decay of a potassium isotope, but without an abundance of other noble gases.

General Structure of the Earth

A traditional concept of the earth's general structure is shown diagramatically in Fig. 9. An approximation of various parameters applying to the various layers and regions is given in Table 3.

The earth's crust (lithosphere) is the outer solid geosphere that is separated from the upper mantle by the Mohorovivić discontinuity, commonly referred to as the Moho discontinuity. The lithosphere consists of three shells: (1) a *stratified sedimentary shell,* composed mainly of sedimentary rocks; (2) a *granitic shell,* distributed only beneath the continents and thinning out at the ocean boundaries; and (3) a *basaltic shell,* the structure of which differs whether located under continents or under oceans.

The total volume of the sedimentary shell is equal to 1,050,000,000 cubic kilometers, taking into consideration the consolidation of recent sediments, and 900,000,000 cubic kilometers without volcanic rocks, i.e., about 10% of the volume of the crust and 0.1% the volume of the whole earth. The average thickness of the sedimentary shell is

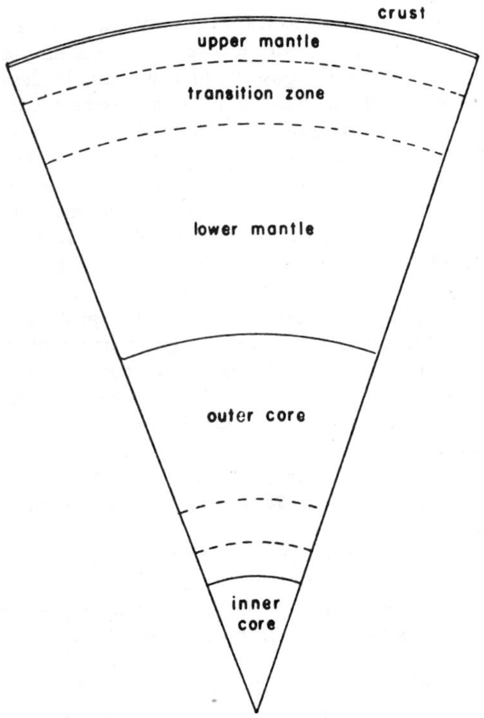

Fig. 9. Generalized section of the earth.

2.0 kilometers; if the area of the shields not covered by sediment is excluded, the average is 2.2 kilometers.

On the contents, about 75% of the volume of all sedimentary rocks is found in geosynclinal areas and only 25% in the platforms, their average thickness being 10 kilometers and 1.8 kilometers, respectively. Clay and shale are the most widespread sedimentary rocks on the continents (42%). Arenaceous, volcanic, and carbonate rocks are approximately equally abundant (20, 19, and 18%). All other rock types, mainly evaporites, comprise about 1%.

The granitic shell is completely restricted to the continents, and its volume and mass are approximately 3,600,000,000 cubic kilometers and 9.8×10^{24} grams, respectively. Acidic granitoids and metamorphic rocks are the main rock types of the granitic shell; the basic and ultrabasic rocks make up less than 15% of the shells volume. These volumetric ratios of rocks lead to the acidic chemical composition of the shell, its typical high content of silica, the concentrations of alkalies $K > Na$, and of the rare elements (uranium, thorium; rare earths; zirconium, niobium, etc.).

The basaltic shell consists of two parts, the continental and the oceanic, differing in structure and apparently in composition. According to one hypothesis, the basaltic shell of the continental crust is formed of strongly metamorphosed rocks of both basic and acid composition, together with a significant portion of magnetic rocks.

The continental crust has considerable thickness and a diversity in composition. The more homogeneous oceanic crust is 86% original oceanic theoleiitic basalts and their metamorphic equivalents. These basalts are characterized by a low content of potassium, rubidium, strontium, barium, phosphorus, uranium, thorium, and zirconium, and high ratios of K/Rb and Na/K, which strongly distinguish them from analogous continental rocks.

The oceanic crust is essentially characterized by the occurrence of ultrabasic rocks, seen in the zones of deep faulting (mid-ocean rift valleys), these rocks being considered outcrops of mantle material.

In summary, about 64% of the whole crustal volume is continental, or 79% when the shelf and subcontinental (quasicratonic) crust are included. The other 21% is oceanic crust. The average thicknesses of the various crustal types decrease from 43.6 kilometers for continental, to 23.7 kilometers for subcontinental, to 7.3 kilometers for oceanic crust. The average thickness of the entire crust amounts to about 20 kilometers.

The chemical composition of the crust as a whole approaches that of intermediate rocks, though it is impossible to find its close analog

TABLE 3. APPROXIMATION OF EARTH'S INTERNAL LAYERING AND VALUES OF SOME PHYSICAL PARAMETERS

Depth (km)		Fraction of Volume[a]	Density (g/cm^3)	Pressure (10^{12} dyn/cm^2)	Gravity (cm/sec^2)	Rigidity (10^{12} dyn/cm^2)	Characteristics of P, S Velocities[b]	Features
0	Crust	0.0155					Complex	Heterogeneous
33	Moho discontinuity							
			3.3	0.01	985	0.6		
	Upper mantle (Region B)	0.1667					Normal gradient	Probably homogeneous
410								
	Upper mantle (Region C)	0.2131					Greater than normal gradients	Transition layer
1000			4.7	0.4	995	1.9		
	Lower mantle (Region D′)						Normal gradients	Probably homogeneous
2700		0.4428						
	Lower mantle (Region D″)						Gradient near zero	Transition layer
			5.7	1.3	1030	3.0		
2900	W-G Discontinuity							
			9.7	1.3	1030	0.0		
	Outer core	0.1516					Normal P gradient	Homogeneous fluid
4980								
	Transition region	0.0028	(12.5)	3.2	(500)	(0.2)	Negative P gradient	Transition layer
5120								
	Inner core	0.0076					Smaller than normal P gradient	Solid
6370[c]			(13.0)	3.7	0	(1.3)		

[a] Volume ratio, crust:mantle:core = 1:51:10.
[b] P = pressure wave; S = shear wave.
[c] The value of 6,370 km depth refers to center of earth.

among them. In rough approximation, the crust's composition can be described as a mixture of the two prevailing types of rocks: (1) granite, and (2) basalt (geosynclinal basalt plus oceanic theoleiite) at a ratio of 2:3. The average chemical composition of the crust changes with depth from the sedimentary shell to the basaltic shell, with a continuous increase in the content of iron, magnesium, and alumina; and a decrease in the amount of combined water. The contents of alkalis (potassium) and silica first increase from the sedimentary shell toward the granitic; and then decrease toward the basaltic shell of the continents and oceans.

The Moho discontinuity is the boundary surface or sharp seismic-velocity discontinuity that separates the earth's crust from the subjacent mantle. The discontinuity marks the level in the earth at which p-wave velocities change abruptly from 6.7–7.2 kilometers per second (in the lower crust) to 7.6–8.6 kilometers per second, or an average of 8.1 kilometers per second (at the top of the upper mantle). The depth of the Moho discontinuity varies from about 5–10 kilometers beneath the ocean floor to about 35 kilometers below the continents. Its depth below some mountain regions may reach 70 kilometers. It is reasoned that the discontinuity represents a chemical change between the basaltic materials above to periodotitic or dunitic materials below, rather than a phase change (basalt to eclogite). The discontinuity should be defined in terms of seismic velocities alone until more fundamental findings are made. The discontinuity is variously estimated to be between 0.2 and 3 kilometers thick. The discontinuity is named after Andrija Mohorovičić (1857–1936), the Croatian seismologist who discovered the phenomenon.

Seismic velocities experience a sharp discontinuity in the interior of the earth. Analysis of data indicates that this discontinuity occurs at a radius of 3,473 kilometers from the center of the earth whose total radius is 6,371 kilometers. The central region is called the core; the outer region beyond the discontinuity is called the mantle. At the core-mantle interface, the density is estimated to increase from 5.5 grams per cubic centimeter to 10.0 grams per cubic centimeter at a pressure of 1.35 mb (1 mb is approximately 1 million atmospheres).

From the total mass of the earth and its moment of inertia in conjunction with the seismic velocities, the deduction can be drawn that the core contains about 1.95×10^{27} grams, constituting about one-third of the mass of the earth, while occupying only one-sixth of its volume. Since no shear waves have been observed in the core, it is generally accepted that the core must be in liquid form, although a small seismic velocity discontinuity in the interior of the core indicates that the central 5% in volume of the core is solid.

It is unlikely that the core is made up predominantly of light elements, since at the estimated density (10 grams per centimeter), the observed seismic velocities would not be in agreement with an empirical but systematic relation between velocity and density. The density of the core and general consideration of the chemical abundance of the elements indicate that the core consists mostly of iron. Analysis of shock compression data reveals that the density of the liquid core

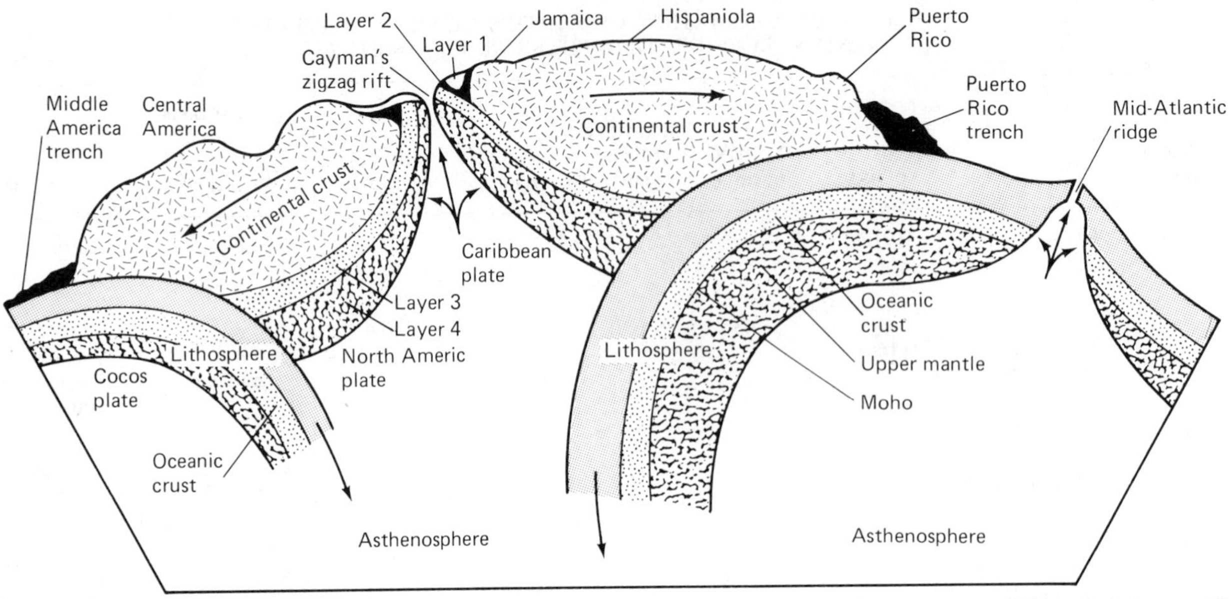

Fig. 10. The Cayman Trough off eastern Central America represents a great tear or shear in the ocean's crust. The trough is some 1500 kilometers (3052 miles) long and about 4 times the depth of the Grand Canyon. It is growing as two plates of the earth's crust (North American Plate and Caribbean Plate) slide past each other, but at a very slow pace. This tectonic action has created very large seismic events, including the February 1976 Guatemalan earthquake that killed some 23,000 people. As the plates separate, magma from the interior slowly rises into the rift valleys, thus continuously building oceanic crust. When plates collide, one edge plunges beneath the other, to be reabsorbed in the interior. Often volcanoes will be found along the overriding edge. Not to scale.

is about 10% less than that of pure iron and that the density at the bottom of the mantle is about 10% greater than that expected of the mantle constituents. These results suggest that some lighter elements are dissolved in the predominantly liquid iron core and that these light elements are there because the mantle material is soluble in the core. The presence of an appreciable amount of impurity in the iron core supports the dynamo theory of the earth's magnetism. This theory requires a fairly high electrical resistance of the core material relative to that of pure iron under atmospheric conditions in order that the eddy currents in the liquid core persist sufficiently long.

Oceanographic investigations during the IDOE programs and since have contributed much new information on seafloor spreading and the formation of ocean ridges and trenches. Such research is exemplified by the current understanding of the nature and mechanics of the Cayman's Zigzag Rift. See Fig. 10. The actual mechanics of crustal formation are poorly understood. Early acoustic sound methods revealed very large zones of deep fracturing, fractures that stretch almost the width of an ocean. These form perpendicular to a ridge wherever one section of the ridge crest is offset horizontally from another. Fracturing also occurs on a very small scale. Such fracturing has been revealed by new instruments, such as side-scanning sonar. Small-scale cracking seems to result from thermal stresses during cooling. Johnson (University of Texas at Dallas) counted the cracks in a DSDP core recovered from 110-million-year-old crust in the Atlantic near Bermuda. See Fig. 11. It was found that the typical crack in the core was about 2 millimeters wide and 150 millimeters long, and that such cracks occurred about every 2 centimeters. Zones of intense cracking were found to occur through a 247-meter (810-foot) core. The end result of all this fracturing appears to be that the crust on and near the ridge crest is very permeable to water, possibly on the order of 20–25%.

Based upon isotope research since 1976, scientists at the California Institute of Technology, in 1979, proposed that the earth's mantle is comprised of two layers, one being a deep layer of unmelted material that has remained unchanged since its creation. This proposal contrasts with the more traditional concepts of a mantle that is a relatively uniform mass of molten material. The new concept is largely based upon the differences in the content of ^{143}Nd in continental lava samples and lava samples from the ocean bottom. See Fig. 12.

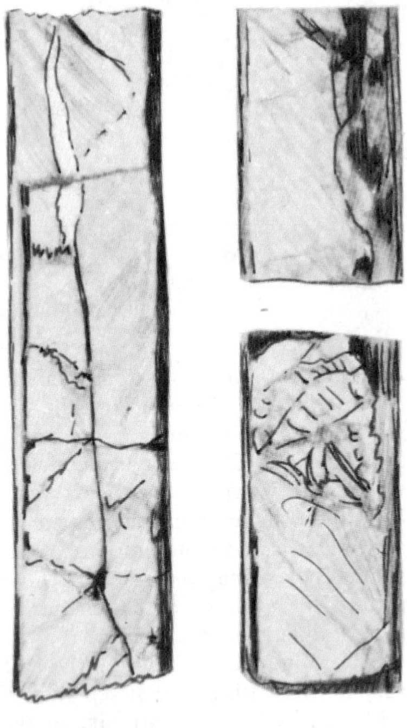

Fig. 11. Cores (8 centimeters in diameter) of fractured crustal rock recovered from 110-million-year-old ocean crust. Natural cracks are filled with carbonates (white) and smectite clays (gray) in the presence of flowing seawater. (*Sketch after Johnson, 1978.*)

The role of volcanoes in the earth system is discussed in some detail in entry on **Volcano.** Studies of the water cycle are described in entry on **Hydrology.**

References

Ahrens, T. J.: "Dynamic Compression of Earth Materials," *Science*, **207**, 1035–1041 (1980).

Ballard, R.: "Window on Earth's Interior," *National Geographic*, **150**, 2, 228–249 (1976).

Bomford, G.: "Geodesy," Oxford Univ. Press, London, 1971.

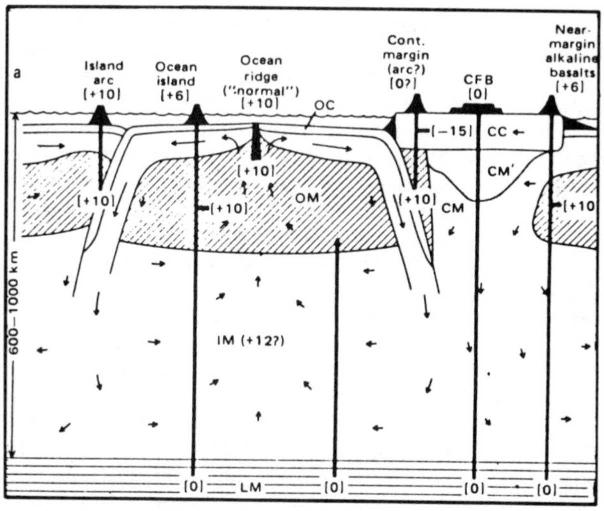

Fig. 12. Model of earth's mantle. LM = unaltered lower mantle, considered to be the source for continental lavas (CFB). Other lavas (ocean ridge normal) arise only from upper mantle (OM and IM) or from mixture of upper and lower material (ocean island). Arrows mark plausible flow pattern. Numbers indicate ratios of isotopes of neodymium. Ocean bottom lava has an excess of ^{143}Nd. In accordance with the conventional "stirred pot" mantle model, all lavas should be enriched in ^{143}Nd. An unmelted primordial layer would preclude whole-mantle convection. (*Proceedings of the National Academy of Sciences*, **76**, 7 and 8.)

Bonatti, E., and P. R. Hamlyn: "Mantle Uplifted Block in the Western Indian Ocean," *Science*, **201**, 249–251 (1978).
Brosche, P., and J. Sundermann (editors): "Tidal Friction and the Earth's Rotation," Springer-Verlag, New York, 1978.
Carrigan, C. R., and D. Gubbins: "The Source of the Earth's Magnetic Field," *Sci. Amer.*, **240**, 2, 118–130 (1979).
Gold, T., and S. Soter: "The Deep-Earth-Gas Hypothesis," *Sci. Amer.*, **242**, 6, 154–161 (1980).
Hadley, D. M., Stewart, G. S., and J. E. Ebel: "Yellowstone: Seismic Evidence for a Chemical Mantle Plume," *Science*, **193**, 1237–1239 (1976).
Hathway, J. C., et al.: "U.S. Geological Survey Core Drilling on the Atlantic Shelf," *Science*, **206**, 515–527 (1979).
Hurlburt, C. S., Jr.: "The Planet We Live On," Time-Life Books, Chicago, Illinois, 1980.
Jacobs, K. A.: "The Earth's Core," Academic, New York, 1975.
Kerr, R. A.: "Seismic Reflection Profiling: A New Look at the Deep Crust," *Science*, **199**, 672–674 (1978).
Kerr, R. A.: "Mantle Geochemistry," *Science*, **203**, 530–532 (1979).
Kerr, R. A.: "Changing Global Sea Levels as a Geologic Index," *Science*, **209**, 483–486 (1980).
King-Hele, D.: "The Shape of the Earth," *Science*, **192**, 1293–1300 (1976).
Kolata, G. B.: "Geodesy: Dealing with an Enormous Computer Task," *Science*, **200**, 421–422 (1978).
McCulloch, M. T., and G. J. Wasserburg: "Sm-Nd and Rb-Sr Chronology of Continental Crust Formation," *Science*, **200**, 1003–1011 (1978).
McElhinny, M. W. (editor): "The Earth," Academic, New York, 1979.
Nakanishi, I.: "Regional Differences in the Phase Velocity and the Quality Factor Q of Mantle Rayleigh Waves," *Science*, **200**, 1379–1380 (1978).
O'Nions, R. K., Hamilton, P. J., and N. M. Evensen: "The Chemical Evolution of the Earth's Mantle," *Sci. Amer.*, **242**, 5, 120–133 (1980).
Pelton, J. R., and R. B. Smith: "Recent Crustal Uplift in Yellowstone National Park," *Science*, **206**, 1179–1182 (1979).
Pitman, W. C.: "Relationship between Sea-Level Change and Stratigraphic Sequence," *Geol. Society of Amer. Bulletin 89*, 1389–1403 (1978).
Press, F., and R. Siever: "Earth," 2nd edition, Freeman, San Francisco, 1979.
Ringwood, A. E.: *Geochem. J.*, **11**, 111 (1978).
Ross, J. V., Ave'Lallemant, H. G., and N. L. Carter: "Activation Volume for Creep in the Upper Mantle," *Science*, **203**, 261–263 (1979).
Smith, R. B., and R. L. Christiansen: "Yellowstone Park as a Window on the Earth's Interior," *Sci. Amer.*, **242**, 2, 104–117 (1980).
Stolz, A., et al.: "Earth Rotation Measured by Lunar Laser Ranging," *Science*, **193**, 997–999 (1976).
Vail, P. R., et al.: "Seismic Stratigraphy and Global Changes of Sea Level," *Seismic Stratigraphy*, Amer. Assn. of Petroleum Geologists Memoir 30, 1980.
Walker, J.: "How to Measure the Size of the Earth with only a Foot Rule or a Stopwatch," *Sci. Amer.*, **240**, 5, 172–182 (1979).
Wetherill, G. W., and C. L. Drake: "The Earth and Planetary Sciences," *Science*, **209**, 96–104 (1980).
Yoshii, T.: "A Detailed Cross-Section of the Deep Seismic Zone beneath Northeastern Honshu," *Tectonophysics*, **55**, 349–360 (1979).

EARTH (Atmosphere). **Atmosphere (Earth).**

EARTH AXIS. Any one of a set of mutually perpendicular reference axes established with the upright axis (the Z-axis) pointing to the center of the earth, used in describing the position or performance of an aircraft or other body in flight. The earth axes may remain fixed or may move with the aircraft or other object.

EARTH CURRENT. A large-scale surge of electric charge within the earth's crust, associated with a disturbance of the ionosphere. Current patterns of quasi-circular form and extending over areas the size of whole continents have been identified and are known to be closely related to solar-induced variations in the extreme upper atmosphere.

EARTH DAMS. **Dams.**

EARTHLIGHT. The illumination of the dark part of the moon's disk produced by sunlight reflected onto the moon from the earth's surface and atmosphere. Also called *earthshine.* Spectroscopic observations reveal that earthlight is relatively richer in blue light than is direct sunlight; this condition results from the fact that an appreciable part of the total earth reflection is backward-scattered light which, in accordance with the Rayleigh law, is relatively rich in the blue and poor in the red. See also **Moon (The).**

EARTHMOVING EQUIPMENT. **Coal.**

EARTH POINT. The point where the forward straight-line projection of a meteor trajectory intersects the surface of the earth.

EARTHQUAKES (Major). **Earth.**

EARTHQUAKES, SEISMOLOGY, AND PLATE TECTONICS. An *earthquake* may be defined as a sudden motion or trembling in the earth caused by the abrupt release of slowly accumulated strain. Synonyms for earthquake include *temblor* (sometimes spelled *tremblor*), *seism, macroseism* (as opposed to microseism), and *shock.* An *earthquake swarm* is a series of minor earthquakes, none of which may be identified as the main shock, occurring in a limited area and time, frequently in the vicinity of a volcano. A *tremor* is a minor earthquake, especially a foreshock or an aftershock. A *foreshock* is a tremor that commonly precedes a larger earthquake or main shock by seconds to weeks (possibly years) and that originates at or near the focus of the larger earthquake. An *aftershock* is an earthquake which follows a larger earthquake or main shock and originates at or near the focus of the larger earthquake. Generally, major earthquakes are followed by a large number of aftershocks, decreasing in frequency with increasing time. Such a series of aftershocks may last many days for small earthquakes or even many months for large earthquakes.

Seismology is the study of earthquakes and, by extension, the investigation of the depths of the earth (as in the case of oil exploration) by way of natural and artificially generated seismic signals. *Seismic* is a term used to describe anything pertaining to an earthquake or earth vibration. *Seismography* is the study of the theory of seismographs. A *seismograph* is an instrument that records vibrations of the earth, particularly of earthquakes or artificially induced energy for the exploration of underlying rock formations and the interior of the earth. The record produced by a seismograph is termed a *seismogram.* The *epicenter* is that point on the earth's surface directly above the focus of an earthquake. A *fault* is a surface or zone of rock fracture along which there has been displacement from a few centimeters to a few kilometers in scale.

Tectonics is that branch of geology dealing with the broad architecture of the upper part of the earth's crust, i.e., the regional assembling of structural or deformational features, a study of their mutual relations, their origin, and their historical evolution. It is closely related to *structural geology,* with which the distinctions are blurred, but tectonics generally deals with the larger, and structural geology with the smaller, features. A *tectonic earthquake* is an earthquake that is due to faulting rather than to volcanic activity. *Plate tectonics* is a theory of tectonics on a global scale, characterized by a relatively small number of large, broad, thick *plates* (blocks composed of areas of the continental and oceanic crust and mantle), each of which "floats" on some viscous underlayer in the mantle and moves more or less independently of the others and grinds against them like ice floes in a river, with much of the dynamic activity concentrated along the periphery of the plates, which are propelled from the rear by sea-floor spreading. The continents form a part of the plates and move with them, like logs frozen in the ice floes. *Sea-floor spreading* is a hypothesis that the oceanic crust is increasing by convective upwelling of magma along the mid-oceanic ridges or world rift system and moving away of the new material at a rate of from 1 to 10 centimeters per year. This movement provides the source of power in *plate tectonics.* This theory is supportive of the continental displacement (drift) hypothesis.

The topics of this entry are very closely associated with topics found in other entries in this encyclopedia. In particular, see entries on **Earth; Ocean;** and **Volcano.** Also consult list of entries to be found in entry on **Geology.**

Properties of an Earthquake

Quite understandably, earthquakes that cause major losses of life and property provoke the greatest attention. However, earthquakes of even greater magnitude may occur in relatively isolated areas which are principally of concern to seismologists.

Three parameters are important in assessing an earthquake: (1) Duration of the earthquake; (2) velocity of the surface movement; and (3) the rate of change of this velocity. In their potential for damage, these three factors are closely related. Some of these relationships are immediately obvious. A very short earthquake of high velocity—only one or two cycles of ground motion—is less damaging than an earthquake causing similar motion for many cycles. An earthquake with high acceleration but low velocity is less damaging than one causing higher velocities. As part of an effort to develop seismic design standards for buildings and structures, these factors have been combined in a map which tentatively characterizes potential earthquake risk in terms of acceleration and velocity throughout the United States. Participants in this development include the American Society of Civil Engineers, the University of California (Seismographic Station, Berkeley), and the Massachusetts Institute of Technology (Department of Civil Engineering). Efforts are sponsored by the National Science Foundation and the U.S. National Bureau of Standards. See Fig. 1.

As early as 1883, the Italian geologist, M. S. De Rossi, and the Swiss naturalist, F. A. Forel developed a scale of one to ten for expressing the intensity of an earthquake. Earthquake intensity may be defined as a measure of the *effects* of an earthquake, notably the effects in terms of people and structures. Earthquake intensity not only will be dependent upon the strength (or magnitude) of the earthquake, but also upon the distance from the epicenter. Intensity also will be markedly affected by local geology, by the numbers and kinds of structures in a given area, as well as the concentration of people within the affected area. Even the time of day may have a large bearing upon the effects, with large numbers of people assembled in factories, schools, offices, etc., during daytime hours. From a scientific standpoint, the *effects* may be of less value than measurements of *magnitude.* However, from an engineering and technological viewpoint, great knowledge has been learned from studying the aftermath of earthquakes, particularly in terms of structures of all types.

The De Rossi-Forel one-to-ten scale was ultimately replaced by the Mercalli scale, devised by Giuseppi Mercalli, an Italian geologist, in 1902. Again, this was an arbitrary scale of earthquake intensity, ranging from I (detectable only instrumentally) to XII (causing almost complete destruction). This scale was later adapted to North American conditions and became known as the *modified Mercalli scale.* This modified scale (MM) was prepared in 1931 by American seismologists, H. O. Wood and F. Neumann and takes into consideration such features as tall buildings, motor cars and trucks, and underground pipes, not included in the earlier, unmodified scale. The scale is described in Table 1.

Richter Scale. Earthquake size as determined by instruments is measured on a logarithmic scale called the Richter *magnitude* scale. In one variation of this scale, the very largest shocks have magnitudes slightly greater than 8.5. Energy in ergs is given empirically by $\log E = 11.4 + 1.5M$, where M is the magnitude. The measurements are based on records made on a standard type of seismograph a distance of 100 kilometers (62 miles) from the epicenter. Usually, seismograms from several different stations contribute to computing the magnitude of an earthquake. In most instances, of course, a station will not be the standard 100 kilometers from the epicenter. Thus, many records must be compared and complex conversion tables help to estimate the final figure. Fortunately, an experienced seismologist, within a few minutes, usually can estimate magnitude with reasonable accuracy from a record made at only one seismographic station. The logarithmic character of the Richter scale is often overlooked by lay people and news reporters. Obviously, an earthquake of 8.0 magnitude is not twice as powerful as one of 4.0 magnitude, but rather it is $10 \times 10 \times 10 \times 10$ times as powerful.

Since most people do not have access to seismographs, but are

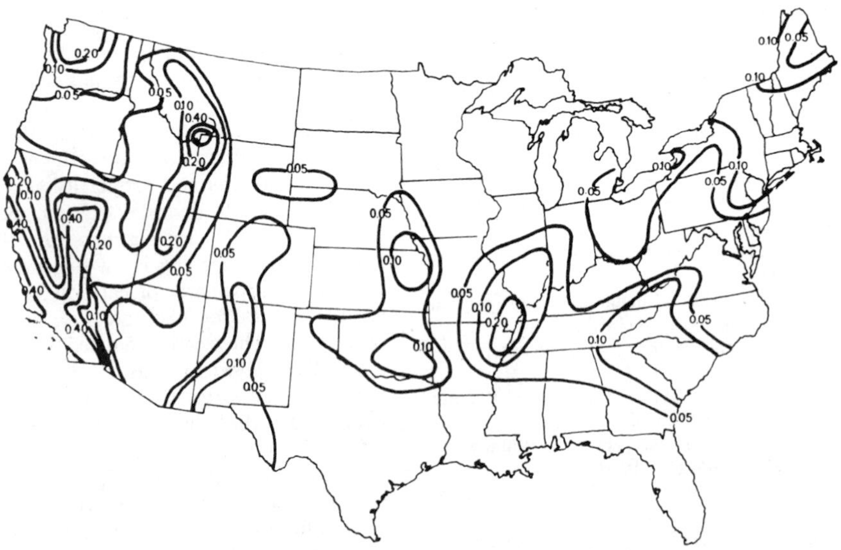

Fig. 1. Effective peak acceleration (EPA) for the lower contiguous United States. Contours represent EPA levels with a nonexceedance probability of between 80 and 95% during a 50-year period. Contours are for values of EPA in units of gravity. EPA is a modification of the peak-recorded instrumental acceleration designed to represent the energy content in the ground motion to which buildings respond. Although the estimate is somewhat more complex, it can be likened to the filtering out of high-frequency spikes of acceleration which structures do not react to because of their considerable inertia. (*American Society of Civil Engineers.*)

TABLE 1. MODIFIED MERCALLI SCALE OF EARTHQUAKE INTENSITIES
(Not to be confused with Richter Scale units)

I. *Not felt by persons, except under particularly favorable circumstances, Dizziness or nausea, however, may be experienced.*
Birds and animals may appear uneasy, disturbed. Trees, structures, liquids, bodies of water may sway gently; doors may swing slowly.

II. *Detected indoors by a few persons, particularly on upper floors of a multistory building; and by sensitive or nervous persons.*
Same manifestations as I, plus hanging objects will swing if delicately suspended.

III. *Detected indoors by several persons, usually as a rapid vibration, but which may not be recognized as an earthquake immediately. Vibration is similar to that from a passing, lightly-loaded truck, or of a heavily-loaded truck from some distance. In some instances, duration is sufficient to be estimated.*
Manifestation of II, plus appreciable movements on upper levels of tall structures. A standing motor car may rock slightly.

IV. *Detected indoors by many persons and outdoors by a few persons. Awakens a few individuals, particularly light sleepers, but experience is not frightening except by persons who have had prior experience with earthquakes. Vibration similar to that from the passing of a heavily loaded truck. Sensation is like that of a heavy body striking the structure, or the falling of heavy objects inside the structure.*
Dishes, windows, and doors rattle; glassware and crockery clink and may crash. Walls and structure frame (particularly of a wooden house) creak, especially if intensity is in the upper range of this class. Hanging objects swing. Liquids in open vessels are disturbed. Parked automobiles rock noticeably.

V. *Detected indoors by practically everyone; and outdoors by most everyone. Slight excitement; some* persons *may run outdoors. Awakens most sleepers. Frightens a few persons.*
Buildings tremble throughout. Dishes and glassware break to some extent. Windows crack in some cases, but not generally. Vases and small or unstable objects frequently overturn. Hanging objects and doors swing generally or considerably. Pictures knock against walls or swing out of position. Doors and shutters open or close abruptly. Pendulum clocks stop or run fast or slow. Small objects move, and furnishings may shift to a slight extent. Small amounts of liquids spill from well-filled open containers. Trees and bushes shake slightly.

VI. *Detected by everyone, indoors and outdoors. Awakens all sleepers. Frightens many people; general excitement, and some persons run outdoors.*
Persons move unsteadily. Trees and bushes shake slightly to moderately. Liquids are set in strong motion. Small bells in churches and schools ring. Poorly built buildings may be damaged. Plaster falls in small amounts. Other plaster cracks somewhat. Many dishes and glasses break, and a few windows break. Knick-knacks, books and pictures fall. Furniture overturns in many instances. Heavy furnishings move.

VII. *Frightens everyone. General alarm, and everyone runs outdoors.*
People find it difficult to stand. Persons driving cars notice shaking. Trees and bushes shake moderately to strongly. Waves form on ponds, lakes, and streams. Water is muddied. Gravel or sand stream banks cave in. Large church bells ring. Suspended objects quiver. Damage is negligible in buildings of good design and construction; slight to moderate in well-built ordinary buildings; considerable in poorly built or badly designed buildings, adobe houses, old walls (especially where laid up without mortar), spires, etc. Plaster and some stucco fall. Many windows and some furniture break. Loosened brickwork and tiles shake down. Weak chimneys break at the roofline. Cornices fall from towers and high buildings. Bricks and stones are dislodged. Heavy furniture overturns. Concrete irrigation ditches are considerably damaged.

VIII. *General fright and alarm approaches panic.*
Persons driving cars are disturbed. Trees shake strongly, and branches and trunks break off (especially palm trees). Sand and mud erupts in small amounts. Flow of springs and wells is temporarily and sometimes permanently changed. Dry wells may renew flow. Damage slight in brick structures built especially to withstand earthquakes; considerable in ordinary substantial buildings, with some partial collapse; heavy in some wooden houses with some tumbling down. Panel walls break away in frame structures. Decayed pilings break off. Walls fall. Solid stone walls crack and break seriously. Wet ground and steep slopes crack to some extent. Chimneys, columns, monuments, and factory stacks and towers twist and fall. Very heavy furniture moves conspicuously or overturns.

IX. *Panic is general.*
Ground cracks conspicuously. Damage is considerable in masonry structures built especially to withstand earthquakes; great in other masonry buildings—some collapse in large part. Some wood frame houses built especially to withstand earthquakes are thrown out of plumb; others are shifted wholly off foundations. Reservoirs are seriously damaged, and underground pipes sometimes break.

X. *Panic is general.*
Ground, especially when loose and wet, cracks up to widths of several inches; fissures up to a yard in width run parallel to canal and stream banks. Landsliding is considerable from river banks and steep coasts. Sand and mud shifts horizontally on beaches and flat land. Water level changes in wells. Water is thrown on banks of canals, lakes, rivers, etc. Dams, dikes, embankments are seriously damaged. Well-built wooden structures and bridges are severely damaged, and some collapse. Dangerous cracks develop in excellent brick walls. Most masonry and frame structures, and their foundations, are destroyed. Railroad rails bend slightly. Pipelines buried in earth tear apart or are crushed endwise. Open cracks and broad wavy folds open in cement pavements and asphalt road surfaces.

XI. *Panic is general.*
Disturbances in ground are many and widespread, varying with the ground material. Broad fissures, earth slumps, and land slips develop in soft, wet ground. Water charged with sand and mud is ejected in large amounts. Sea waves of significant magnitude may develop. Damage is severe to wood frame structures, especially near shock centers; great to dams, dikes and embankments, even at long distances. Few if any masonry structures remain standing. Supporting piers or pillars of large, wellbuilt bridges are wrecked. Wooden bridges that "give" are less affected. Railroad rails bend greatly, and some thrust endwise. Pipelines buried in earth are put completely out of service.

XII. *Panic is general.*
Damage is total, and practically all works of construction are damaged greatly or destroyed. Disturbances in the ground are great and varied, and numerous shearing cracks develop. Landslides, rock falls, and slumps in river banks are numerous and extensive. Large rock masses are wrenched loose and torn off. Fault slips develop in firm rock, and horizontal and vertical offset displacements are noticeable. Water channels, both surface and underground, are disturbed and modified greatly. Lakes are dammed, new waterfalls are produced, rivers are deflected, etc. Surface waves are seen on ground surfaces. Lines of sight and level are distorted. Objects are thrown upward into the air.

interested in earthquakes, there is a relative correlation between magnitude readings of the Richter scale and probable intensity. This correlation is:

Magnitude	Intensity (Probable Effects)
1	Detectable only by instruments.
2	Barely perceptible, even near epicenter.
4.5	Detectable within 20 miles of epicenter. Possible slight damage within a small area.
6	Moderately destructive.
7	A major earthquake.
8	A great earthquake.

As observed by Donovan (1979), in an attempt to provide a more meaningful and precise measure of earthquake size, some seismologists are using a quantity called the *seismic moment.* This represents the product of the change in volumetric strain energy (the change in stress or stress drop multiplied by rigidity), the size of the fault rupture surface, and the actual displacement or throw of the fault. The seismic moment expressed in fundamental physical units becomes an incomprehensibly large number. For example, the seismic moment of the 1971 San Fernando earthquake ($M_L = 6.5$) was 1.4×10^{26} ergs. Research using a magnitude scale based upon seismic moment, developed by Kanamori (California Institute of Technology), has suggested revisions in the relative sizes of some historic earthquakes. The 1906 San Francisco earthquake would be reduced, while the 1960 Chilean earthquake, which may have been the largest event in this century, would be increased. Although seismic moment is more useful for the study of earthquakes, the complications brought about by the need to estimate three parameters have delayed acceptance of this value in the development of standards for earthquake-resistant structures.

In comparing earthquakes, Kanamori first compared the estimated energy release with the magnitude of moderate-sized events. The extrapolation was then made to large events by taking the Gutenberg-Richter relationship between magnitude and energy, and using the energy release computed from the seismic moment. It should be noted that while the new scale provides a better measure of the relative size of events, the recorded magnitudes of events using the Richter scales remains unchanged.

Seismographs. Most modern seismographs are of the inertial type, depending upon measurement of the relative displacement between a point fixed to the earth and a mass loosely coupled to the earth. Other instruments measure relative displacement between two points on earth. There are well over 1000 seismograph stations in the world. However, considering the size of the earth, this is not a large number, particularly in view of the importance of this kind of data. During the past decade, there has been considerable activity underway in the operation of seismographs on the floor of the deep ocean and, during the Apollo Program of the 1960s and early 1970s, seismographs were placed on the moon for seismic investigations of that satellite. There are governmental and private organizations which exchange seismographic information on a worldwide basis. These include the World Wide Seismic Network (WWSN), established in the early 1960s, the Lamont-Doherty Geological Observatory at Columbia University, the Coordinating Committee for Earthquake Prediction (CCEP), the U.S. Geological Survey (Menlo Park, California), the California Institute of Technology, the American Institute of Civil Engineers, the Woods Hole Oceanographic Institution (Woods Hole, Massachusetts), the Scripps Institution of Oceanography (La Jolla, California), and the Japan Meteorological Agency, among others.

The operating principle of a seismograph is shown in Fig. 2. When the ground shakes, the suspended weight, because of its inertia, scarcely moves, but the shaking motion is transmitted to the marker, which leaves a record on the drum. There are, of course, several configuration which utilize this basic operating principle. A representative seismogram is shown in Fig. 3. This is the record of an earthquake of magnitude 6.5 (Aleutian Islands). The record was made at a station some 4,000 kilometers from the epicenter.

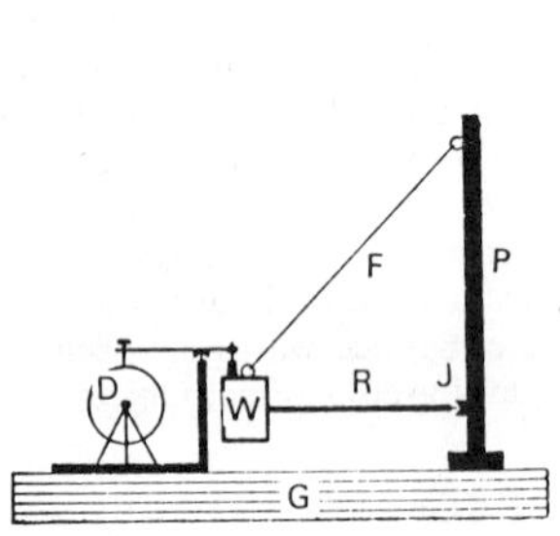

Fig. 2. Principle of a seismograph. *G*, ground; *P*, post set in the ground; *W*, weight; *R*, rigid support contacting the post with a free-moving sharp point at *J*; *F*, flexible wire; *D*, recording drum. Marker extends from *W* to *D*. An improved seismograph will employ an electrical pickup of pendulum movements. The instrument is placed on a concrete base which, in turn, is thoroughly anchored into bedrock so that when the instrument base moves it truly reflects earth movement and not extraneous vibrations.

The basis of seismology is the observation and analysis of elastic energy as it propagates itself through the earth. When mechanical energy is released in a homogeneous earth, it is propagated outward in waves whose fronts are spherical and whose mechanism is alternating compression and rarefaction of the material through which they pass. These waves, called *P* waves, are physically analogous to the sound waves that spread outward from an explosion in air or water. The *P* wave travels at a rate of about 5.6 kilometers per second. It is the first wave to reach the surface. A longitudinal wave, the *P* wave tends to create a "push-pull" effect on rock particles as it passes.

The earth is not generally homogeneous, and though imperfectly elastic, it has rigidity or shear strength which is absent in air or water. This results in a second kind of wave action, in which the material through which the wave passes moves transversely to the direction of wave motion. These shear or *S* waves travel through solid material at a velocity a little more than half that of the *P* waves. The *S* wave causes the earth to move at a right angle to the direction of the wave.

In addition to the *P* and *S* types of waves (called body waves), there is also energy propagation along surfaces or interfaces. At the earth-air interface, waves similar to surface waves in water are propagated. These are called Rayleigh waves and cause, as in water, circular vertical motion of the material through which they pass, in the plane containing their direction of propagation. In solid material, surface waves like *S* waves also occur, the material moving horizontally in a plane transverse to the direction of motion of the wave. These waves are called Love waves (*L* waves). These waves usually are distinguishable only at great distances. *L* waves can cause the swaying of tall buildings as well as slight wave motions in bodies of water at great distances from the epicenter.

If the velocities of the different modes of wave propagation are known, deduction can be made, for example, of the distance between the earthquake and the observation station by measuring the time interval between the arrival of the faster and slower waves.

When earthquake waves move across an interface between earth materials with different elastic properties, their velocity, direction, and phase may be changed, and they will generally give rise to waves of other modes. For example, a *P* wave encountering an interface may give rise to a refracted and reflected *P* wave and a refracted and a reflected *S* wave. A wave may also be refracted even if it does not cross an interface, because of the change in pressure and density of the earth material as the wave penetrates more deeply into the earth.

Waves recorded at an observatory can be analyzed for their directions of travel, times of arrival, and frequency count. When data from different observatories are compared, it can then be deduced what interfaces they have passed through. Such deductions have led to the discovery that the earth is divided into three main layers, the crust, the mantle, and the core, each with its own set of characteristics with respect to the passage of seismic waves and each with its own subdivisions. See also **Earth** (Table 3).

In addition to transient seismic phenomena, such as those due to earthquakes and explosions, there is a continuous background noise in the earth which is measurable with modern seismographs over a wide range of periods. For periods of about one second, ground amplitudes at very quiet locations are less than 1 millimicron. Between

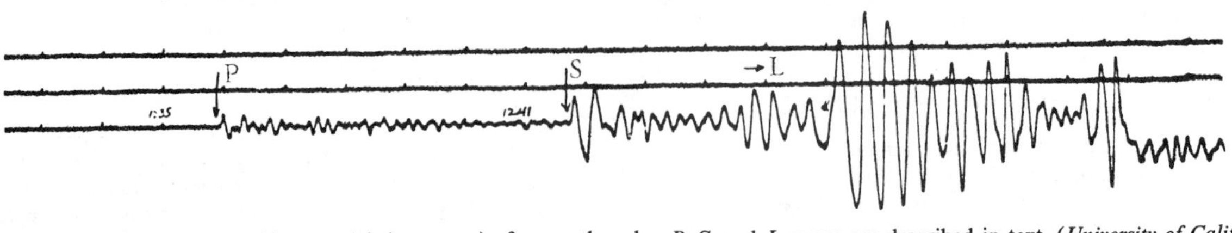

Fig. 3. Representative seismographic record (seismogram) of an earthquake. *P*, *S*, and *L* waves are described in text. (*University of California.*)

periods of 4 and 9 seconds, there is a sharp peak in the noise spectrum. The corresponding waves are called *storm microseisms* and are related, presumably through ocean waves, to meteorological disturbances at sea.

Seismology and Petroleum Exploration. It is interesting to note that, as of the early 1980s, about 95% of the $1 billion worldwide investment in geophysical exploration is made by private firms in the interest of locating petroleum reserves. The results of much of this research are funneled through scientific channels and thus contribute much to an advancing knowledge of seismic phenomena and an understanding of the earth's interior. Knowledge of seismic phenomena accounts for about 90% of the total of all geophysical activity.

Within recent years, it has been found that the traditional two-dimensional seismic prospecting methods are inadequate for solving the three-dimensional problems encountered in petroleum exploration. In three-dimensional seismic methodology, the reflected wave field must be adequately sampled spatially to ensure proper imaging of the subsurface by means of a three-dimensional wave equation. The development of petroleum resources led to the use of seismic exploration techniques. The first field work in the United States began in the mid-1920s. After surface and less expensive means of exploration for gas or oil, seismographic methods may be used. Charges of explosives may be used, or large hydraulic vibrators, mounted on trucks, may be used to induce a shock wave series into the earth. Seismographs measure the reflected or refracted shock waves at carefully spaced locations. Interpretations of data from rock strata more than 6 km below the surface often provide accurate results. Offshore seismic searches are conducted by creating sharp sound wave pulses in the water which travel through the water and deeply into the rock formations below. Unfortunately, seismic information can only indicate subsurface conditions normally favorable to the generation and accumulation of gas or oil—not positively identify their presence. See Table 2.

TABLE 2. REPRESENTATIVE LONGITUDINAL SEISMIC WAVE VELOCITIES

Material	Velocity: Feet per Second	Velocity: Meters per Second
Weathered surface material	500– 2,000	152– 610
Gravel, rubble, sand (dry)	1,500– 3,000	457– 914
Sand (wet)	2,000– 6,000	610–1829
Clay	3,000– 9,000	914–2743
Sandstone	6,000–13,000	1829–3962
Shale	7,000–14,000	2134–4267
Limestone	7,000–20,000	2134–6096
Granite	15,000–19,000	4572–5791
Metamorphic rocks	10,000–23,000	3048–7010
Glacial till (Saskatchewan)	5,000– 7,000	1524–2134
Ice	12,500	3810
Fresh water	4,700– 4,900	1433–1494
Seawater	4,800– 5,000	1463–1524

Data compiled from Jakosky and the Saskatchewan Research Council.

Amplitudes of earthquakes vary from a fraction of a centimeter to several centimeters. The destructive phase of earthquake varies from one minute to but a few seconds. It is estimated that over the earth there are over a million earthquakes each year, the majority of them occurring in regions of recent mountain building. The majority are quite weak. It is estimated that a major earthquake occurs on the earth about once per week. Tremors of the ocean bottom cause seismic seawaves (tsunamis). These seawaves have been known to rise 30 meters (100 feet) or more. When such waves break upon a densely inhabited coast, there is great loss of life and destruction of property. A list of major earthquakes in terms of lives lost is given in Table 1 under **Earth.**

Excluding certain near-surface regions, the velocities of dilational seismic waves range from about 5 kilometers per second in parts of the crust to a maximum of 13.5 kilometers per second at the base of the mantle; corresponding shear wave velocities range from 3 to 8 kilometers per second. The shortest periods of interest in the study of waves from distant earthquakes are of the order of $\frac{1}{3}$ second frequency (frequency = 3 Hz); the longest periods are about 53 minutes (frequency = ~ 1 cycle per hour), and they correspond to a free oscillation of the earth in the fundamental spheroidal mode.

Free oscillations of measurable amplitudes are generated only by the largest earthquakes. The largest earthquakes probably release between 10^{24} and 10^{25} ergs in the form of seismic waves, and the few largest shocks account for most of the energy released in this form. Mean annual release is estimated at 9×10^{24} ergs.

Earthquake Mechanisms. The frequency of occurrence of earthquakes increases by about a factor of 8 or 10 per unit of magnitude *as the magnitude decreases.* On the average, only about 25 shocks with a magnitude of 7 or more occur each year, but it has been estimated that there are at least one million earthquakes per year, most of them quite small. About 5,000 to 10,000 of these are routinely located and studied.

Most earthquakes occur in certain narrow, world-circling belts separated by relatively stable blocks. The circum-Pacific belt is by far the most important, accounting for about 80% of the total activity. Other important features are the trans-Eurasian belt and the mid-ocean belt. The foci, i.e., the points of initiation of the first seismic waves, of most shocks are shallow, i.e., at depths of 60 kilometers (37 miles) or less, but some are as deep as 700 kilometers (435 miles). Most large shocks are followed by a series of smaller aftershocks which occur in the same general region as the main shock. Aftershocks generally decrease in size and in frequency of occurrence with time, but may persist for more than a year following a very large shock. Some earthquakes are preceded by one or a few foreshocks.

Seismic amplitude measurements made by the U.S. Geological Survey in 1978 suggest that foreshocks sometimes may have different focal mechanisms than aftershocks. The ratio of the amplitudes of *P* and *S* waves from the foreshocks and aftershocks of three California earthquakes (Oroville, Galway Lake, and Briones Hills) during the 1975–1977 period showed a characteristic change at the time of the main events. As this ratio is extremely sensitive to small changes in the orientation of the fault plane, a small systematic change in stress or fault configuration in the source region may be inferred. The success of the Chinese in predicting the Haicheng earthquake (February 1975) was based upon the use of long-, intermediate- and short-term precursors. As pointed out by Lindh et al. (1978), the imminent prediction was based partly on the recognition of a swarm of small-to-moderate earthquakes near Haicheng as a potential foreshock sequence. However, foreshocks do not always occur; estimates of their frequency vary widely and they are usually recognizable as foreshocks only in hindsight. To use foreshocks in a predictive mode, one must distinguish them from background seismicity and earthquake swarms.

The most popular model of the earthquake mechanism is based upon the elastic rebound theory,[1] which calls for gradual accumulation of elastic strain in a region prior to the release of the strain energy through rupture or by slippage along a preexisting fault. This model explains many features of a number of earthquakes, but is oversimplified and may not be applicable to deep shocks.

Most of the seismic activity of the earth may be explained by the theory of plate tectonics. In the theoretical model, the earth's surface is made up of a small number of large plates of lithosphere some 100 kilometers (62 miles) in thickness. See Fig. 4. These plates move about with relative velocities of the order of a few centimeters per year. The boundaries of the plates, where plate interactions occur, correspond to the major active seismic belts. The greatest seismic

[1] One consequence of the assumption that the earth is a perfectly elastic body is that elastic moduli and seismic velocities do not depend upon frequency. As observed by Anderson et al. (1977), if the assumption is valid, the moduli determined from seismic body waves can be compared directly with surface wave, free oscillation, and ultrasonic laboratory results and used to compute tidal response. However, "elastic" waves are nondispersive only at very high and very low frequencies, and ideal elastic behavior is only approached at very low temperatures. The absorption of seismic body waves (a wave that travels through the interior of the earth and is not related to any boundary surface) and the decay of free oscillations indicate that the earth is not a perfectly elastic body. It is well known that dispersion must accompany absorption and thus elastic moduli depend upon frequency.

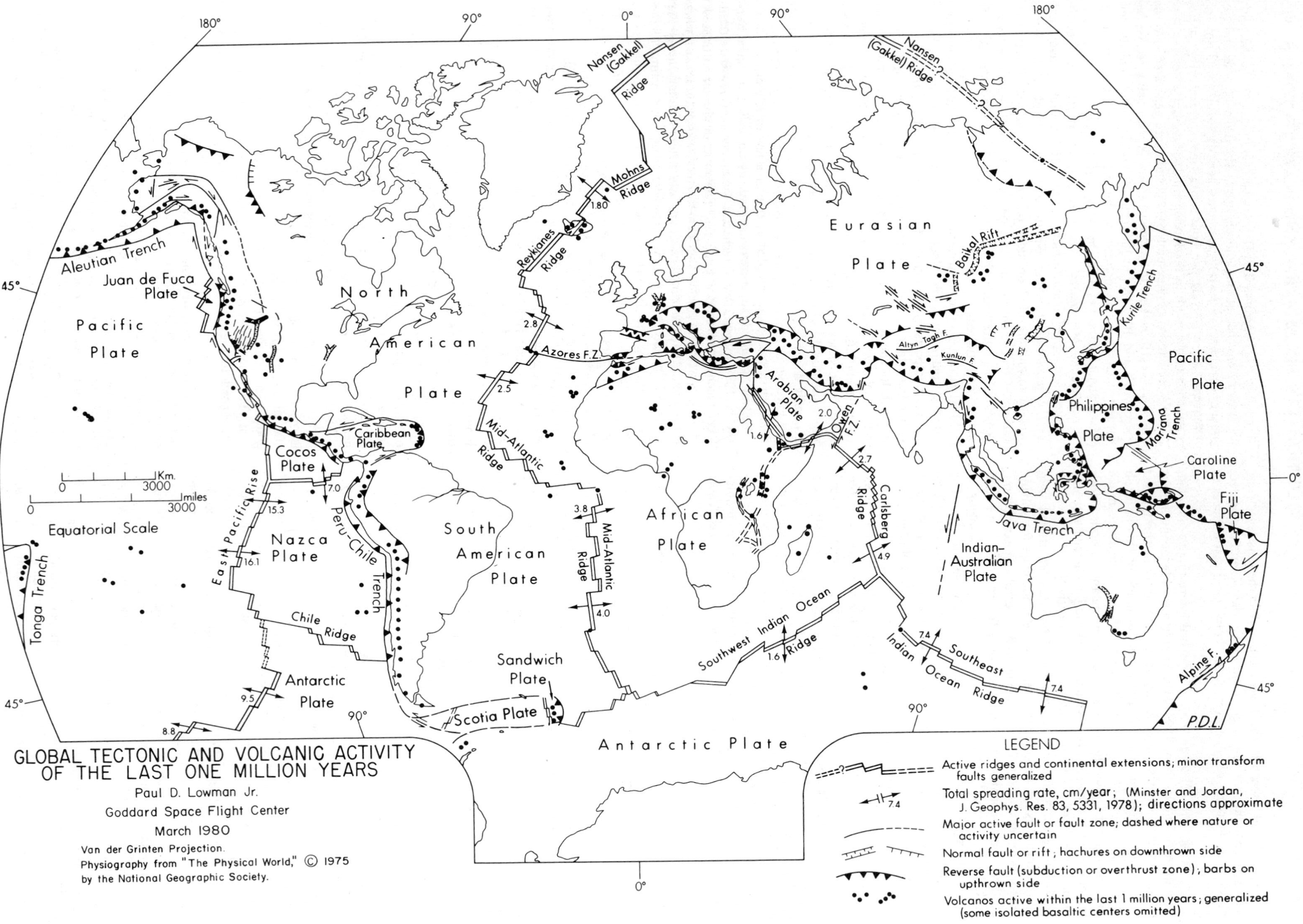

GLOBAL TECTONIC AND VOLCANIC ACTIVITY OF THE LAST ONE MILLION YEARS

Paul D. Lowman Jr.
Goddard Space Flight Center
March 1980

Van der Grinten Projection.
Physiography from "The Physical World," © 1975 by the National Geographic Society.

activity, the largest shocks, and the deepest shocks, occur at places where plates converge (the arcs such as Japan and Tonga), where one plate is thrust beneath another to depths at least as great as the depths of the deepest earthquakes. Where plates diverge (as along the Mid-Atlantic Ridge), or slide past one another (as along the San Andreas Fault in California), seismic activity is shallow, and although substantial, is usually not as great as that of the arcs. The global patterns of the focal mechanisms of earthquakes also fit, in general, the patterns of plate motion. Seismological evidence played an important role in the development of the concept of plate tectonics. See Fig. 5.

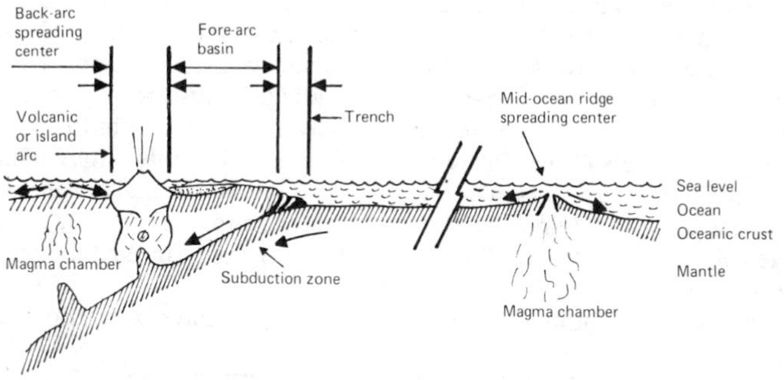

Fig. 5. Important features of plate tectonics. Examples are taken from the Mariana island-arc system. At depths of 90–100 kilometers (56–62 miles), the descending slab commences to melt and magma rises toward the surface, forming volcanic island arcs. The depressed area between the volcanic arc and trench gradually fills with sediments forming the *fore-arc basin.* Other magmas rise to the surface along the Mariana *back-arc basins,* creating a *spreading center* along which new crustal material forms at the surface. *Island arc*: Also known as a volcanic arc, an island arc is usually a chain of islands, e.g., the Aleutians, rising from the deep sea floor and near to the continents. *Trench*: A narrow, elongated depression of the deep-sea floor, with steep sides and oriented parallel to the trend of the continent and between the continental margin and the abyssal hills. Such a trench may be 2 or more kilometers (1.2 miles) or deeper than the surrounding ocean floor, and may be thousands of kilometers long. See also **Earth;** and **Ocean.** (*Sketch suggested by Davin and Gross, 1980.*)

Donovan (1979) observes that most strong-motion earthquake data have been accumulated from events in California and Japan. California and Japan represent two different tectonic regions. The west coast of the United States is not typical of continental margins. The primary seismic feature is the well-known San Andreas fault, with its right lateral strike-slip movement (known as a transform fault), which is usually associated with mid-oceanic spreading ridges where it is of much shorter length and a consequence of the actual three-dimensional nature of the plates moving on the surface of the earth. The San Andreas fault joins the spreading East Pacific Rise and the spreading zones of the Gorda and Juan de Fuca ridges to the north. The realization that the fault was of a transform type was first made by J. Tuzo Wilson (University of Toronto). Wilson recognized that a fault displacement of up to 1000 kilometers (620 miles) required some mechanism that allowed the displacements to occur while satisfying the laws for the conservation of matter. Because of the special nature of the relative movements between the Pacific Plate and the North American Plate, the San Andreas fault may be considered as an extension and a very large example of a mid-oceanic feature.

The majority of continental margin seismic activity stems from direct plate collisions, as explified by the activity along the Chilean coast. The focal depths, which are shallow off the coast, become progressively deeper as activity occurs further inland on the eastern side of the Peru-Chile trench.

In a study by Walker (1976), it was proposed that evidence suggests that similarities in the seismic activity of widely differing tectonic environments may be related to worldwide fluctuating stress fields. The topic of the dynamics of global plate tectonics is discussed under **Volcano.** The evidence is in the form of time series of counts per unit of time of earthquakes with magnitudes greater than a chosen threshold. In the period studied, many of the time series from widely separated regions were found to have activity highs in 1965 and lows

Fig. 4. Principal plates, ridges, and trenches according to the theory of global plate tectonics. The outer surface of the earth, just under the crust and known as the lithosphere, supports a number of rigid plates. These plates may be about 100 kilometers (62 miles) or more in thickness. The continents (about 40 kilometers; 24 miles thick) essentially rest on the larger lithospheric plates. In the case of the western United States, southern California is part of the Pacific Plate, whereas the rest of North America is on the North American Plate. The three main forms of plate boundaries are (1) mid-oceanic ridges, (2) transform faults, and (3) subduction zones. New crust is created along the mid-oceanic ridges by the rise of basaltic magma. This cools and is carried away by sea-floor spreading. The mechanism of sea-floor spreading was conceived as the result of two important discoveries: (1) the mid-oceanic ridge system, and (2) the zebra-stripe pattern of magnetic anomalies which symmetrically flank these ridges. Studies of ocean floor rocks support a presumption that the sea floors are moving away from the mid-oceanic ridges.

Transform faults, along which the mid-oceanic ridges are offset, provde areas where the lithospheric plates can slide past one another. Subduction zones are the sites of destruction of the lithospheric plates. There are places in the ocean trenches where one oceanic plate descends beneath another. Should a second plate have a continental crust, then mountains would be inclined to develop along the continental margin, as exemplified by the Andes. Where two continental crustal plates collide, even greater mountain-building activity is created. Movement of plates so affects the crust that earthquakes, mountain building, and volcano formation become an intimate part of the system. A majority of volcanoes occur in arcs which are parallel to the oceanic trenches and are believed to result from one ocean crustal plate being subducted beneath another plate. Partial melting of the oceanic crust on the descending plate may occur as it passes onto the high-temperature athenosphere. Thus, magma ascends to the surface and forms volcanoes and underlying granitic batholiths. Volcanoes are also situated along the mid-oceanic ridges and, less frequently, over isolated "hot spots" in the athenosphere, exemplified by the Hawaiian volcanoes. See **Volcano.**

in 1966 and 1967. Similar observations had previously been made, using yearly seismic energy totals for differing tectonic regions. Walker suggests that three possible inferences may be drawn from these correlations: (1) Worldwide fluctuations in stress fields produce, or trigger, major earthquakes throughout the world as well as smaller earthquakes along individual rising plate edges; (2) major earthquakes trigger other earthquakes throughout the world, especially along rising plate edges; and (3) minor earthquakes along rising plate edges trigger major earthquakes throughout the world. Walker noted that the times of occurrence and epicentral locations of the triggering earthquakes and the triggered earthquakes included in the study imply stress propagation rates on the order of 10^3 to 10^4 kilometers/year.

Faults. A fault is a great fracture in the crust of the earth along which movement has taken place with the result that the crustal blocks are displaced relative to one another. This movement is in most cases probably intermittent and the actual individual displacements may be very small, but by accumulation may reach tens, hundreds, or rarely thousands of feet (meters). The fractures themselves may often be traced for many miles. The San Andreas fault, horizontal movement along a portion of which caused the earthquake that so severely damaged San Francisco, California, in 1906, has been traced for about 600 miles (965 kilometers).

In describing faults (See Fig. 6), certain terms have been adopted.

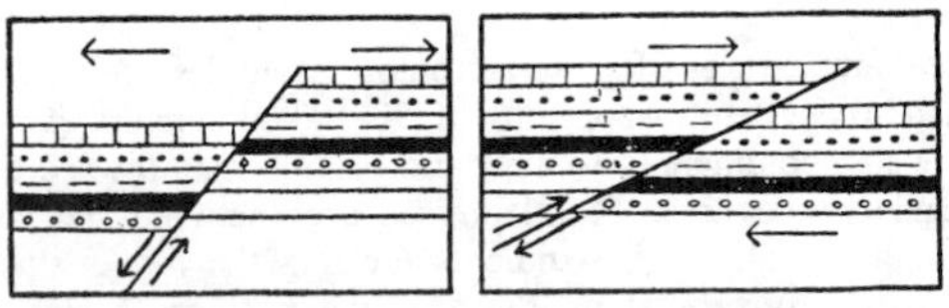

Fig. 6. Structure sections of a single normal fault (left); and simple thrust fault (right).

Those in most common use are given herewith: the fault plane is the plane of the fracture and may be vertical or at some other angle; the angle between the fault plane and the horizontal is called the angle of dip of the fault plane. The angle between the vertical and the fault plane is spoken of as the angle of hade or simply as the hade of the fault. The surfaces of the fault plane are called the walls of the fault. If the fault plane dips, the uppermost wall is called the hanging wall, the lower wall the foot wall. These terms are applied irrespective of whether the fault is normal, with the hanging wall slipping down the dip of the fault plane, or whether it is a reverse fault with the hanging wall apparently pushed up the dip of the fault plane. A normal fault is sometimes spoken of as a gravity fault. The displacement measured along the fault plane is designated as the slip; the displacement measured vertically is called the throw; the displacement measured at right angles to the plane of the involved stratum is called the stratigraphic throw. The amount of horizontal displacement between the ends of a broken stratum measured at right angles to the direction of strike of the fault plane, is called the heave. The visible evidence of the trace of a fault plane at the earth's surface is called the fault trace. The block of the earth's crust which has moved downward, relatively speaking, to the other is called the downthrown block or referred to as the downthrow side of the fault. The other block is called the upthrown block or the upthrow side of the fault. If the strike of the fault plane is essentially at right angles to that of the bedding it is called a dip fault. A strike fault is one in which the movement has been parallel to the strike of the strata involved. A compound fault involving several parallel displacements dipping in the same direction, resulting in a step-like arrangement, is referred to as a step fault. The term graben or trough fault refers to a downthrown area bounded on each side by two or perhaps more faults. A horst is an uplifted area bounded by two or more faults. See Fig. 7.

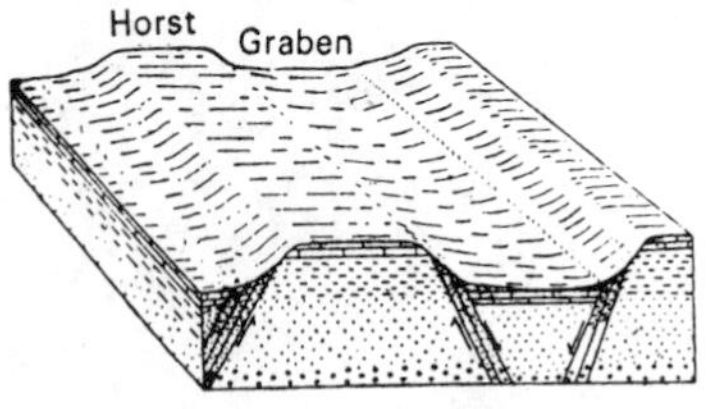

Fig. 7. Block diagram of one type of scenery produced by "normal" or block faulting. It may also be assumed that the graben has been pushed down, and the horst has been pushed up. Neither structure is necessarily entirely the result of tension. Note that the stratigraphic (vertical) order of the formations has not been duplicated or reversed.

Predicting Earthquakes

In the early 1980s, scientists generally are not as optimistic concerning the ability to predict earthquakes (in the short-term) as they were during the 1970s. However, the long-term forecasting of earthquake-prone regions has improved. Researchers were heartened in 1975 when the Chinese successfully predicted the Haicheng earthquake of that year, this prediction probably saving tens of thousands of lives. The methodology used involved, in part, the *dilatancy theory* which requires a detailed model of the *microscopic events* preceding an earthquake. Many observations over a considerable time span are required to fill out the dilatancy model.

Possibly due to too few observations, a subsequent earthquake at Tanshan, China was not effectively predicted in the short-term and reports indicate that over 600,000 people were killed by that quake.

The principal methods (and combinations) used to forecast earthquakes include:

(1) Analysis of historical records—seismic gaps—for forecasting in terms of years and decades.
(2) Groupings of smaller earthquakes—foreshocks.
(3) Dilatancy theory.
(4) Presence of radon gas in groundwater.
(5) Presence of helium in fault outgas.
(6) Electrical conductivity/resistivity of rocks.
(7) Meteorological data and soil tilt.
(8) Animal behavior.

Historical Records—Seismic Gaps. Akin to Mendeleev's search for missing elements to fill out spaces in the Periodic Table of the Elements, seismologists today are making mid- and long-term predictions of earthquakes by identifying regions in which earthquakes have not occurred for a long period, but where theoretically they should have occurred. In other words, seismologists are selecting regions that are "overdue" for seismic action. The Tumaco, Colombia earthquake (December 12, 1979) of magnitude 8, the largest quake in northwestern South America since 1942, was predicted by Kelleher as early as 1972. This event filled a gap in the shallow seismic zone of northern Ecuador and southwestern Colombia, where no sizable earthquake had occurred since a quake of magnitude 8.7 happened in 1906. By carefully studying the region, Kelleher concluded that the direction of region during each prior significant earthquake in the general region was proceeding in a north to northeastern direction toward Tumaco.

In their excellent report of the Tumaco earthquake, Herd et al. (1981) observe that history suggests that another series of large, shallow-focus earthquakes along the Ecuador-Colombia coastline may begin in this century near Esmeraldas. The 1942 Esmeraldas earthquake (magnitude, 7.9), the first of the latest series of northward-progressing earthquakes in the Ecuador-Colombia seismic gap, followed the 1906 earthquake by only 36 years. In the years since 1942, there have been no large-magnitude earthquakes near Esmeraldas. Another shallow-focus earthquake could recur in this new seismic gap at Esmeraldas before the balance of the 1906 seismic gap is filled near Buenaventura.

Another example of the "seismic gap" forecasting technique is the Alaskan earthquake of February 1979, which occurred in a sparsely populated part of the Alaskan coast, some 400 kilometers (248 miles) east of Anchorage. As shown by Fig. 8, the earthquake filled one of at least two seismic gaps, as forecasted as early as 1971. Another large quake in the remainder of the gap, which extends almost to Valdez, is expected within the next decade or so.

Groupings of Smaller Earthquakes—Foreshocks. As of the early 1980s, the search for correlations between the groupings of smaller earthquakes and foreshocks and larger subsequent earthquakes is receiving much attention. The theory is that nondestructive earthquakes and foreshocks indicate a build-up of stress that must eventually be released in the form of a main shock of considerable magnitude. However, in some earthquake-prone regions, such as California, scientists

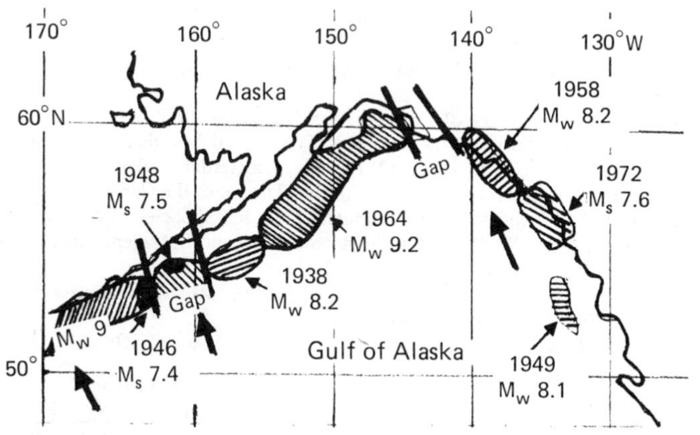

Fig. 8. Two gaps in the Alaska-Aleutian seismic zone. Top gap was noted in 1971. The 1979 earthquake ruptured only the eastern half of the gap. (*After McCann, Lamont-Doherty Geological Observatory.*)

have been frustrated for the last few years by the absence of small quakes from which to obtain data. This area of interest has currently displaced the attention which scientists gave a few years ago to dilatancy-related effects.

Although both events exhibited foreshocks, neither the Izu-Oshima-kinkai earthquake in Japan of 1978 nor the Adak event in the Aleutians could be interpreted with sufficient reliability to warrant a short-term prediction. In retrospect, however, pre-main event data could be used to forecast similar events should precipitating steps be similar, a broad assumption.

Not all earthquakes have foreshocks of at least moderate magnitude. Of 160 earthquakes of magnitude 7 or greater included in a survey by the California Institute of Technology, only half exhibited foreshocks of magnitude 4 or greater. Thus, it is obvious that lesser earthquakes must be included in possible foreshock groupings in an effort to expand the data base and yet remain above the random earth noise level. Clusters of events of magnitude 3 and 4 are now included by some investigators. Also, foreshocks useful to forecasting purposes may be separated by much greater time spans than previously considered. For example, it appears that there was a subtle pattern of seismicity over a ten-year period in the case of the San Fernando, California earthquake of 1971.

Investigators at several institutes have proposed the concept of a "slow earthquake," for which the crustal movement is about one hundred times slower than that of a normal quake. Such a slow quake would not produce higher frequency seismic signals recorded by standard seismographs and thus would go undetected. Nevertheless, such an earthquake could transmit stress from one place to another. In the opinion of one investigator, slow earthquakes may be responsible for a sudden increase of stress to critical levels within only a few days or weeks prior to a final failure. If indeed slow earthquakes do occur, this would greatly complicate the problem of making long-term predictions.

Dilatancy Theory. A precursor that is directly implied by dilatancy is *crustal uplift.* Such uplift may be detected by geodetic or tilt measurements. Where a dilatant zone is sufficiently large as well as shallow, dilatancy could produce a crustal uplift of several centimeters. In the case of an earthquake of magnitude 6 (Odaigahara, Japan, 1960), five tiltmeters at different localities up to 100 kilometers from the epicenter detected precursory movements. Six months prior to the earthquake, rapid tilting in the direction away from the epicenter began to be recorded at two stations, 40 and 100 kilometers from the epicenter. Simultaneously a strainmeter at one of the sites began to show anomalously high rates of extension. These activities continued for 3 months, after which they stopped. However, a station farther to the south began tilting. One month prior to the earthquake, all three stations began tilting in the opposite direction and two additional stations began to tilt rapidly. The data indicate that different stages of dilatancy occurred at different times at different localities. Both the magnitude and the duration of the tilts are explicable on the basis of the dilatancy model. The reverse of tilt just prior to the earthquake indicates some type of short-term precursor, perhaps due to a slight compaction of the dilatant volume."

During the past decade, after intensive research, many researchers have grown wary in their attempts to justify the dilatancy theory as a reliable precursor of major earthquakes. The so-called Palmdale Bulge, located near Los Angeles, has been studied for several years and thus far has yielded no information considered useful for prediction of an earthquake in the area. In fact, some investigators have voiced the opinion that the bulge is not so extensive as once visualized and attribute this to instrumental errors.

Presence of Radon Gas. As early as 1966, the year of the Tashkent earthquake, analysis of waters from a deep well in the hypocentral region indicated a variation in the radon content prior to the earthquake. Radon has a half-life of only 3.8 days, and its lifetime diffusion distance is only several centimeters. Thus, its increased concentration in the Tashkent well prior to the earthquake and its larger aftershocks could have been caused by two means: (1) an increase in the surface area of rock in the epicentral region due to cracking, or (2) by an increase in the flow rate of pore water. Both of these effects are predicted by the dilatancy model. In the case of the Izu-Oshima-kinkai, Japan earthquake of 1978, precursory changes in the radon concentration of groundwater were observed. The distance from the epicenter to a continuous radon-monitoring station at Nakaizu was about 25 kilometers (15.5 miles). A sudden drop and a subsequent increase in the radon concentration recorded 5 days prior to the earthquake were significant. The size of the spikelike change was about 15%. After the earthquake, a remarkable increase in the radon concentration occurred.

In a 1980 paper, five Japanese scientists made the following observations: "Although our understanding of the whole mechanism of radon emission remains unclear, the observed precursory changes must somehow reflect the deformation or damage, or both, to the artesian layer caused by the action of stress release or stress accumulation. Even though we did not predict the occurrence of the Izu-Oshima-kinkai earthquake, our efforts were directed toward obtaining data on reliable premonitory changes through observation, a first step toward earthquake prediction. We conclude that the measurement of the radon concentration in groundwater at carefully chosen wells should offer definitive information on the likelihood of an earthquake."

In California, some radon measurements have shown precursory phenomena and the interpretation of others remains obscure. Peaks of radon emission, lasting several months, centered about two earthquakes of about magnitude 4. Since the spring of 1977, detection systems along the San Jacinto fault, located southeast of Los Angeles, have indicated unusually high levels of radon.

Helium "Spots." Japanese scientists have found that measurements of variations in the degassing rate from a fault can be a good way of inferring the most recent time of displacement along a fault, including displacements, such as creep and stick slippage. Helium is believed to be one of the most ideal elements to measure for this purpose because of its mobility, chemical inertness, and low abundance in the atmosphere. Using helium as a geochemical tracer, the investigators aimed at a qualitative assessment of an active fault related to the occurrence of an earthquake. It should be pointed out that helium surveys already have been attempted for prospecting for natural gases, ores, and active geothermal areas. Application of a similar technique to active fault zones may provide useful information for earthquake prediction. (Wakita et al., 1978).

Electrical Conductivity. A measurement associated with the concept of dilatancy is a significant reduction of the electrical resistivity of rock and soil prior to an earthquake as a result of the influx of water into the dilatant region. In the Garm region of the U.S.S.R., strong correlations were shown some years ago between minima in electrical resistivity and the occurrence of earthquakes. These findings were supported by studies of the electrical resistivity of rocks in the laboratory.

Meteorological Data and Soil Tilt. A popular instrumentation strategy for earthquake prediction is to deploy large numbers of relatively inexpensive tiltmeters, strainmeters, creepmeters, and other geophysical instruments. Most of these instruments are implanted at shallow depths in soil regimes near active faults. In some cases, the sites are within the fault zone. Scientists at the U.S. Geological Survey (Menlo

Park, California), over a two-year period, studied the numerical relation between earthquake activity, temperature, and rainfall, as well as tiltmeter response. Their conclusions were that, if the data contain premonitory earthquake signals, they are buried in local meteorological noise. Separating an earthquake anomaly from the response to surface phenomena becomes more difficult as the earthquake anomaly lead time approaches the rise time of the soil to weather and seasonal variations (Wood and King, 1977).

Animal Behavior. The reports of animal behavior as a precursor of an imminent earthquake are taken seriously by some researchers and not really ruled out by any investigators. The following excerpt from the Haicheng Earthquake Study Delegation (China, 1979) is of interest: "Some instances noted at this time (before the earthquake) were of snakes being found frozen on the road . . . geese flying, chickens refusing to enter their coops, pigs rooting at their fence, cows breaking their halters and escaping, and goats as well as cows being unusually restless. Rats appeared to behave as though drunk. Three well-trained police dogs howled, refused to obey commands, and kept their noses close to the ground as though sniffing." In China, numerous non-scientifically trained persons serve as observers along with scientific coordinators. The content of the foregoing excerpt may be better understood after perusing the Shapley (1976) reference listed.

Most investigations have shown that peculiar animal behavior occurs during or after, rather than before an earthquake. If animal precursors are to be considered seriously, there must be some scientific explanation of the sensing mechanisms possessed by birds, farm animals, pets, etc., that make them more perceptive than humans, or even of scientific instruments. One possible explanation is the superior sensitivity of some animals to perceive very low, negligible (from a human standpoint) vibrations of earthquakes of magnitude 2 and under. These may arrive as very gentle foreshocks of a large quake. Only a very nearby seismometer would pick up such vibrations. If felt by only one instrument in California's dense network of seismometers, for example, the data would not be considered valid. There is also the possibility that animals will detect very low, rumbling booms with a frequency of 50–70 Hz, similar to faint thunder. Humans can detect such noises, but some species, such as pigeons, are far more sensitive to frequencies below 50 Hz. Again, such mild vibrations could arrive ahead of a large shock.

Seismologists have noticed the barking of dogs during the gentle foreshocks of weak quakes as well as the mild aftershocks. Some of these effects were observed during the Willits, California earthquake of 1977. However, an extensive monitoring activity, from which several reports were received, failed to provide convincing documentation of peculiar animal behavior prior to the Coyote Lake, California quake of November 1979. At that time, there were over 1200 volunteer observers in the region.

To date, no details have been worked out (except possibly in China) to include animals in a detection network for providing earthquake warnings sufficiently in advance to be of practical value.

Peculiar Earthquake Phenomena

As mentioned earlier, rumbling noises are definitely associated with some earthquakes and are not a figment of the imagination. At least since the time of Benjamin Franklin, there have been reports of a peculiar light "which illuminated the night sky" during an earthquake. Somewhat parallel to the animal behavior reports, the matter of an earthquake light persists. Thompson (1978) reports that mysterious phenomena, including earthquake light, have been chronicled since ancient times. Geologists noted a brightening of the sky the night before a major earthquake in Turkey (1975). In Japan, sightings and photographs have documented bright night skies before an earthquake. In one modern theory, it is proposed that hydrocarbons convert to methane, which is capable of surviving the high pressures and temperatures deep within the earth. It is further proposed that methane escaping through faults in the land surface catches up small grains of dust or dirt that collide, creating sparks that could ignite the methane and cause the flames and explosions sometimes reported prior to and during earthquakes. It has also been suggested that methane, combining with sulfur to form hydrogen sulfide may also be responsible for the observed uneasiness of animals.

References

Anderson, D. L., et al.: "The Earth as a Seismic Absorption Band," *Science*, **196**, 1104–1106 (1977).

Bolt, B. A.: "Earthquakes," Freeman, San Francisco, 1978.

Bott, M. H. P.: "Sedimentary Basins of Continental Margins and Cratons," *Developments in Geotectonics*, **12**, Elsevier, Amsterdam, 1976.

Bullard, E. C., Everett, J., and A. G. Smith: "The Fit of the Continents around the Atlantic," *Symposium on Continental Drift*, Royal Society, London, 1965.

Charpal, O., et al.: "Rifting, Crustal Attenuation, and Subsidence in the Bay of Biscay," *Nature*, **275**, 706–711 (1978).

Corliss, J. B., et al.: "Submarine Thermal Springs on the Galapagos Rift," *Science*, **203**, 1073–1082 (1979).

Davin, E. M., and M. G. Gross: "Assessing the Seabed," *Oceanus*, **23**, 1, 20–32 (1980).

Donovan, N.: "Earthquake Hazards," *Oceanus*, **22**, 3, 63–79 (1979).

Garwood, N. C., Janos, D. P., and N. Brokaw: "Earthquake-Caused Landslides: A Major Disturbance to Tropical Forests," *Science*, **205**, 997–999 (1979).

Glass, R. I., et al.: "Earthquake Injuries Related to Housing in a Guatemalan Village," *Science*, **197**, 638–643 (1977).

Graebner, R., Wason, C., and H. Meinardus: "Three-Dimensional Methods in Seismic Exploration," *Science*, **211**, 535–540 (1981).

Gray, J., and A. J. Boucot, Editors: "Historical Biogeography, Plate Tectonics, and the Changing Environment," Oregon State Univ. Press, Corvallis, Oregon, 1979.

Hadley, D. M., Stewart, G. S., and J. E. Ebel: "Yellowstone: Seismic Evidence for a Chemical Mantle Plume," *Science*, **193**, 1237–1239 (1976).

Herd, D. G., et al.: "The Great Tumaco, Colombia Earthquake of 12 December 1979," *Science*, **211**, 441–445 (1981).

Jordan, T. H.: "The Deep Structure of the Continents," *Sci. Amer.*, **240**, 1, 92–107 (1979).

Kanamori, H.: "Great Earthquakes at Island Arcs and the Lithosphere," *Tectonophysics*, **12**, 187–198 (1971).

Kerr, R. A.: "Seismic Reflection Profiling: A New Look at the Deep Crust," *Science*, **199**, 672–674 (1978).

Kerr, R. A.: "Earthquakes: Prediction Proving Elusive," *Science*, **200**, 419–421 (1978).

Kerr, R. A.: "Seawater and the Ocean Crust," *Science*, **200**, 1138–1141 (1978).

Kerr, R. A.: "Earthquake Hazards: Real but Uncertain in the East," *Science*, **201**, 1000–1003 (1978).

Kerr, R. A.: "Earthquake Prediction," *Science*, **203**, 860–862 (1979).

Kerr, R. A.: "Another Successful Quake Forecast," *Science*, **203**, 1091 (1979).

Kerr, R. A.: "Prospects for Earthquake Prediction Wane," *Science*, **206**, 542–545 (1979).

Kerr, R. A.: "The Bits and Pieces of Plate Tectonics," *Science*, **207**, 1059–1061 (1980).

Kerr, R. A.: "Quake Prediction by Animals Gaining Respect," *Science*, **208**, 695–696 (1980).

Kerr, R. A.: "How Much is Too Much When the Earth Quakes?" *Science*, **209**, 1004–1007 (1980).

Le Pichon, X., and J. Franchetau: "A Plate-Tectonic Analysis of the Red Sea-Gulf of Aden Area," *Tectonophysics*, **46**, 369–406 (1978).

Lindh, A., Fuis, G., and Mantis, C.: "Seismic Amplitude Measurements Suggest Foreshocks have Different Focal Mechanisms than Aftershocks," *Science*, **201**, 56–58 (1978).

Lloyd, J. J.: "Earthquake Light," *Science*, **193**, 1070 (1976).

Mattill, J. I.: "How Earth Quakes," *Technology Review (MIT)*, **79**, 6, 15 (1977).

McCann, W. R., et al.: "Yakataga Gap, Alaska: Seismic History and Earthquake Potential," *Science*, **207**, 1309–1314 (1980).

Nakanishi, I.: "Regional Differences in the Phase Velocity and the Quality Factor Q of Mantle Rayleigh Waves," *Science*, **200**, 1379–1380 (1978).

Pelton, J. R., and R. B. Smith: "Recent Crustal Uplift in Yellowstone National Park," *Science*, **206**, 1179–1182 (1979).

Pilger, R. H., Jr. (editor): "The Origin of the Gulf of Mexico and the Early Opening of the Central North Atlantic Ocean," Louisiana State Univ. School of Geoscience, Baton Rouge, Louisiana, 1980.

Pitman, W. C., III: "Relationship between Eustacy and Stratigraphic Sequences of Passive Margins," *Geol. Soc. Amer. Bull.*, **89**, 1389–1403 (1978).

Raleigh, C. B., Healy, J. H., and J. D. Bredehoeft: "An Experiment in Earthquake Control," *Science*, **191**, 1230–1237 (1976).

Richter, C. F.: "Earthquake Light in Focus," *Science*, **194**, 259 (1976).

Riecker, R. E. (editor): "Rio Grande Rift: Tectonics and Magnetism," American Geophysical Union, Washington, D.C., 1979.

Shakal, A. F., and M. N. Tosköz: "Earthquake Hazard in New England," *Science*, **195**, 171–173 (1977).

Schlee, J. S., et al.: "The Continental Margins of the Western North Atlantic," *Oceanus*, **22**, 3, 40–47 (1979).

Shapley, D.: "Chinese Earthquake," *Science*, **193**, 656–657 (1976).

Staff: "Past and Future Shocks," *Sci. Amer.*, **239**, 4, 92–94 (1978).

Staff: "Palmdale Bulge," *Science News*, **116**, 24, 404 (1979).
Staff: "Shakier U.S. in 1979," *Science News*, **117**, 9, 136 (1980).
Stukeley, W.: "The Philosophy of Earthquakes," *Natural and Religious Review*, Corbett, London, 1756.
Sykes, L. R., et al.: "Rupture Zones of Great Earthquakes in the Alaska-Aleutian Arc," *Science*, **210**, 1343–1345 (1980).
Takeuchi, A., Nakamura, K., Kobayashi, Y., and K. Hori: "Restoration of Tectonic Stress Field of Cenezoic Central Honshu," *The Earth Monthly*, **1**, *Symposium 6*, 447–452 (1979).
Thompson, R. B.: "Afloat on a Sea of Methane," *Technology Review (MIT)*, **81**, 1, 14–15 (1978).
Wakita, H., et al.: "Helium Spots—Caused by a Diapiric Magma from the Upper Mantle," *Science*, **200**, 430–431 (1978).
Wakita, H., et al.: "Radon Anomaly: A Possible Precursor of the 1978 Izu-Oshima-kinkai Earthquake," *Science*, **207**, 882–883 (1980).
Walker, D. A.: "Yearly Seismic Energy Release: World Totals Versus Ridge System Totals," *Science*, **193**, 886–888 (1976).
Watkins, J. S., Montadert, L., and P. W. Dickerson (editors): "Geological and Geophysical Investigations of Continental Margins," Amer. Assn. of Petrol. Geologists, Tulsa, Oklahoma, 1979.
Watts, A. B., and M. S. Steckler: "Subsidence and Eustasy at the Continental Margin of Eastern Northern America," *Deep Drilling Results in the Atlantic Ocean*, Ewing Series, Vols. 2 and 3, Washington, D.C., 1979.
White, R. S., and D. A. Ross: "Tectonics of the Western Gulf of Oman," *J. Geophysical Research*, **84**, B7, 3479–3489 (1979).
Wilson, M. E., and S. H. Wood: "Tectonic Tilt Rates Derived from Lake-Level Measurements, Salton Sea, California," *Science*, **207**, 183–186 (1980).
Wood, M. D., and N. E. King: "Relation between Earthquakes, Weather, and Soil Tilt," *Science*, **197**, 154–156 (1977).
Yoshii, T.: "A Detailed Cross-Section of the Deep Seismic Zone beneath North-eastern Honshu," *Tectonophysics*, **55**, 349–360 (1979).
Zoback, M. D., and J. C. Roller: "Magnitude of Shear Stress on the San Andreas Fault: Implications of a Stress Measurement Profile at Shallow Depth," *Science*, **206**, 445–447 (1979).
Zoback, M. D.: "Recurrent Intraplate Tectonism in the New Madrid Seismic Zone," *Science*, **109**, 971–976 (1980).

EARTH RATE CORRECTION. A command rate applied to a gyroscope to compensate for the apparent precession of the gyro spin axis with respect to its base caused by the rotation of the earth.

EARTH RESOURCES SATELLITES AND GEOLOGIC REMOTE SENSORS. Water, vegetation, and minerals all have unique signatures in the realm of multispectral light. Within certain limitations, these features will reveal their identity from images produced through the technique of remote sensing. Prior to the formal launching of the first earth resources technology satellites (initially called ERTS, but now named Landsat), the results of earth surveys from high-altitude aircraft greatly aided scientists and conservationists in charting the features of the earth. Earth resources experiments were on the testing agenda of several earlier Gemini and Apollo manned spacecraft flights. Images were made of the earth both in the visible light range and in the infrared and ultraviolet portions of the electromagnetic spectrum. The results were promising and indicated that such data, properly interpreted, could assist in a wide variety of projects—agricultural planning, charting the movement of sea life, monitoring concentrations of air and water pollutants, and furnishing more fundamental knowledge of the earth in terms of geography, cartography, forestry, geology, and hydrology.

Design of the first Landsat (or ERTS-1) satellite was based upon the highly successful Nimbus meteorological satellites which had regularly returned pictures of the earth's weather state since 1964. Landsat-1 was launched in July 1972 for operation in a polar orbit about 914 kilometers (about 570 miles) above the earth. The observatory returns images from two independently functional multispectral sensors. A data collection system on board the observatory also gathers environmental information from earth-based platforms and relays this information to the ground processing facility. Operation of the observatory is controlled by a ground data handling system facility located at the Goddard Space Flight Center of the National Aeronautics and Space Administration. Sensor data are processed into both black and white and color photo images tailored to the needs of users of the information.

Landsat-1 carried a return beam vidicon camera subsystem and a multispectral scanner which furnished an independent view of the earth directly beneath the satellite to data acquisition stations on the ground.[1] See Figs. 1 and 2. Two wideband video tape recorders on board stored up to 30 minutes of picture information for delayed readout. The observatory made 14 revolutions (one every 103 minutes) each day. Ground coverage proceeded westward until global coverage was completed in about 18 days. An active attitude control subsystem maintained the observatory within ±0.7 degree of the local vertical. Two solar panels tracked the sun and provided 500 watts of electrical energy to power the spacecraft.

Three return beam vidicon subsystem cameras viewed the same 100-mile (185-kilometer) square ground scene from the observatory,

Fig. 1. The earth resources technology satellite (Landsat-1) as it appeared in orbit.

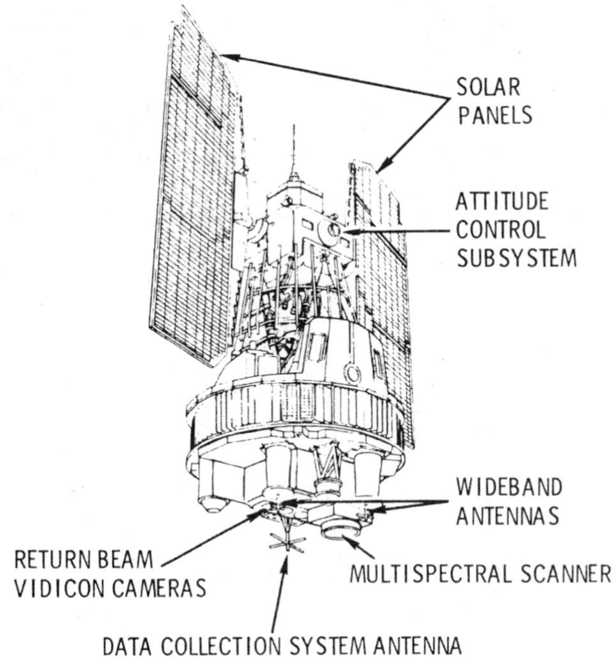

Fig. 2. Principal components of Landsat observatory.

[1] In the interest of providing background to the full Landsat Program, details of Landsat-1 are included.

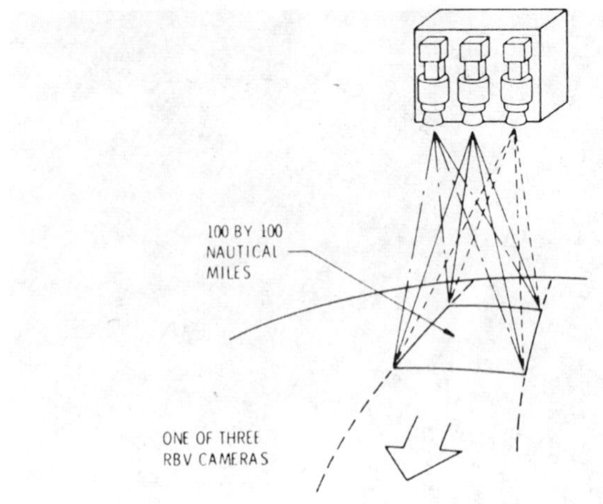

Fig. 3. Return beam vidicon subsystem.

but were sensitive to different spectral bands within the total band from 0.48 to 0.83 micrometers. See Fig. 3. When the cameras were shuttered, images were stored on photosensitive surfaces within each vidicon camera tube, then scanned to produce video outputs. The cameras were scanned in sequence requiring about 3.5 seconds to read out each of the three images. To produce overlapping images of the ground along the direction of satellite motion, the cameras were reshuttered every 25 seconds.

The multispectral scanner subsystem collected data by continually scanning the ground directly beneath the observatory. The width of the strip was identical to the coverage by the return beam vidicon system, i.e., 100 nautical miles. See Fig. 4. Optical energy was sensed by an array of detectors simultaneously in four spectral bands (from 0.5 to 1.1 micrometers). During ground processing, 100 × 100 nautical mile (185 × 185 kilometer) frames were constructed from the continuous strip to correspond with the return beam vidicon information.

The data collection system received data from remote, automatic data collection platforms (located in the United States and Canada) and relayed the information to ground stations via the Landsat-1 spacecraft. Each data collection system platform collected data from as many as 8 sensors which sampled such local environmental conditions as stream flow, snow depth, or soil moisture. Data from any platform was available to users within 24 hours from the time the sensor measurements were made.

The return beam vidicon and multispectral scanner subsystem data were transmitted to the tracking stations via dual wideband (2.2 GHz) S-band data links for recording on magnetic tape. Narrowband telemetry provided for satellite housekeeping, the relay of data collection system data, and such payload-related data as attitude and timing information. Telemetry, tracking, and command subsystems on board the satellite were compatible with tracking stations in NASA's spaceflight tracking and data network.

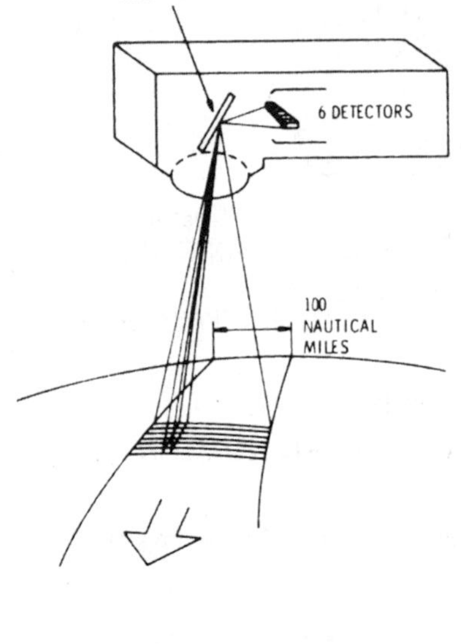

Fig. 4. Multispectral scanner subsystem.

Some of the early achievements of Landsat-1, which provided incentives to expand and improve the Landsat Program, included:

Geology. Nearly 50% additional unmapped faults found branching off the San Andreas fault in California, previously thought to be an accurately mapped seismologically active area.

Hydrology. Shallow substrate water-bearing rocks detected in Nebraska, Illinois, and New York; polluted-water drift charged off the coast of New Jersey; extent of Mississippi River flooding determined.

Cartography. Inaccuracies in map locations of Brazilian lakes and rivers discovered; unmapped lakes located in Iran.

Oceanography. Uncharted reefs detected off Bahama Islands and Australian coast; navigation charts for Bahamas to depths of 4 to 8 meters (13 to 26 feet) produced more accurately than any prior maps.

Agriculture. Area coverage of Texas and Oklahoma categorized as to range and pasture land, forests and water; up to 11 types of crops identified for fields 20 acres (7 hectares) or larger for San Joaquin Valley and Imperial Valley, California.

Urban Development. Dallas-Fort Worth images showed new roads, reservoirs, suburbs, and airports not on maps made three years prior. Essentially the first year of data gathering by the Landsat-1 was devoted by scientists in a variety of fields in learning how to use the data effectively, of literally coping with the immensity of information that can be returned from the satellite and in planning how to make the second satellite. Landsat-2 and others to follow in the series even more effective. Landsat-1 became inoperative in the mid-1970s.

Expanding the Landsat Series. Incorporating improvements and providing expanded coverage of the earth's features, Landsat-2 was launched in 1975, but became of limited functional capability by the late 1970s. In 1978, Landsat-3 was launched. During the early 1980s, Landsat-D (or 4) will be launched.

An expanded view and description of the Landsat-D observatory is given in Fig. 5. As noted in Fig. 6, the telecommunications network will be more complex for Landsat-D than in the case of the prior satellites in the series.

Applications for Landsat Imagery

Landsat data are made available to interested parties on a worldwide basis. In 1977, the U.S. Congress passed a bill requiring the Secretary of the Interior to furnish Landsat information to government and private users, domestic and international, on an equal basis.

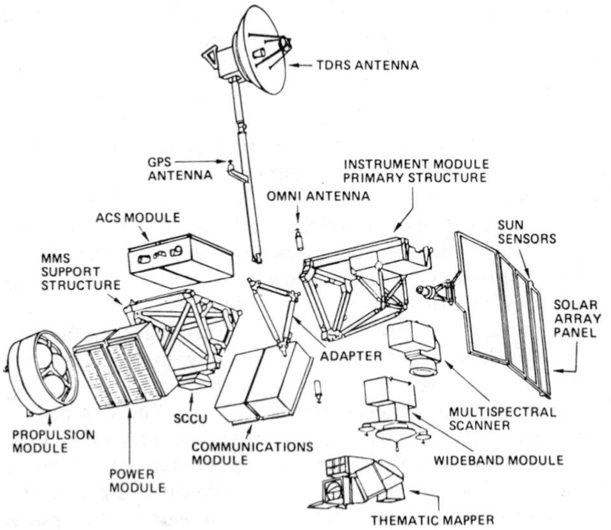

Fig. 5. Expanded view of Landsat-D (or 4) multispectral scanner (MSS) satellite scheduled for launching in the early 1980s. Landsat-D will be launched into a sun-synchronous, polar orbit approximately 700 kilometers (435 miles) high. The satellite will carry two remote sensing instruments: (1) a thematic mapper (TM), an experimental sensor of advanced design to provide scenes with 30-meter (about 100 feet) resolution. It is a 7-channel radiometer. (2) A multispectral scanner (MSS) with 4 channels, 80-meter (262.5 feet) resolution, which is identical to the sensors on Landsats 1 and 2. Landsat-D will have the capacity to generate 800 scenes per day (550 MMS; 250 TM) for all ground stations, as compared with 190 MSS scenes from Landsats 1, 2, and 3. Each scene will cover 34,253 square kilometers (13,225 square miles). (*General Electric.*)

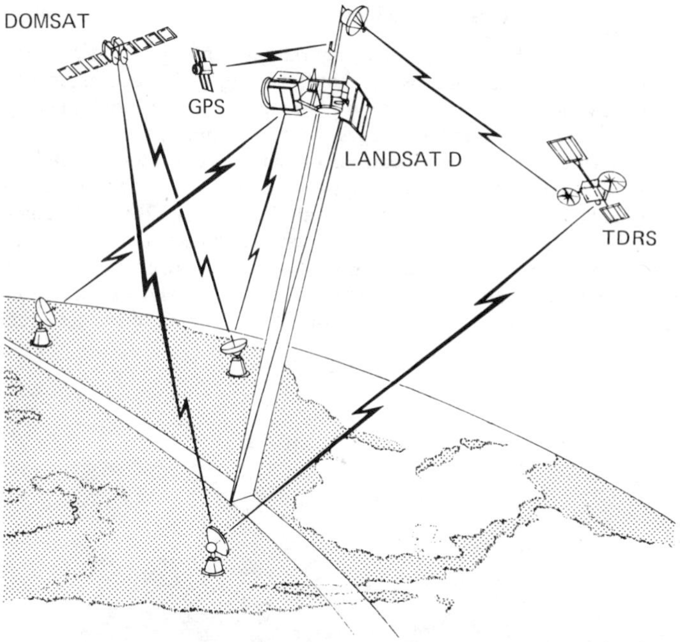

Fig. 6. Spacecraft position in orbit will be determined using the Global Positioning System. Sensors will be programmed from the ground, based upon users' data requests. Data from both sensors will be transmitted from Landsat-D via a Ku-band link to a Tracking and Data Relay Satellite (TDRS) in a geostationary orbit, and then from TDRS to the National Aeronautics and Space Administration's White Sands, New Mexico ground station. The data will then be relayed from White Sands via domestic communications satellite to the Goddard Space Flight Center in Greenbelt, Maryland. Data will also be transmitted directly from the satellite at X-band for TMM and MSS data and S-band for MSS data to Goddard and ground stations in other countries. (*General Electric.*)

Initially, photographs were processed as black-and-white photos, one for each spectral band, with its own information. Later, color composites were prepared by combining processing of black-and-white bands through color filters onto color film. Frequently, they were processed to have the characteristics of infrared photography, a system with which interpreters were familiar because of earlier and some continuing use of infrared photography in conventional aerial missions.

Instrumentation. Because surfaces viewed are frequently of high opacity and have radiation-scattering tendencies, data are generally obtained from the upper micrometer or millimeter layers or zones of the materials. Opacity in the visible and infrared portions of the spectrum results from high absorption coefficients of materials. The presence of water in many of the materials viewed restricts penetration in the microwave range because of high conductivity and dielectric constant. However, by considering changes in surface temperature as induced by diurnal solar heating, a trained analyst sometimes can infer certain additional properties of the imaged materials. But even this technique, where applicable, limits information to a substance depth of ten centimeters or less.

Wavelengths used extend from ultraviolet (0.4 micrometer) to microwaves at 50 centimeters, representing a wavelength range of 10^6. Ultraviolet below 0.4 micrometer is not usually used because of high atmospheric absorption and Rayleigh scattering. Infrared (1 to 3 micrometers) is best suited for determining the composition of observed minerals and rocks. As pointed out by Goetz and Rowan (1981), spectral reflectance of vegetation in the visible and near-infrared differs markedly from that of rock and soils. This spectral region contains intense chlorophyll absorption and the region of high reflectance beyond 0.7 micrometer. Thus, even the small presence of vegetation can alter the spectral signature of rocks and soils. It is interesting to note, however, that the subtle variations in the spectrum of stressed vegetation may effectively outline mineralized areas. Trained analysts also can interpret the distribution of areas of vegetation in terms of soil composition. The mid-infrared region beyond 8 micrometers is valuable for geological mapping because spectral emittance variation can be a basis for discriminating between silicate and nonsilicate rocks. The range from 8 to 14 micrometers is excellent for differentiating types of silicates.

As contrasted with traditional photographic methods, the Landsat multispectral scanner (an optomechanical scanner) permits the creation of digital multispectral images at wavelengths outside the sensitivity of film. The Landsat MSS provides multiple, spatially registered data sets, each taken at different wavelengths. See Fig. 7. The advantages of digital over photographic processing is that transformations applied to the data are not subject to the problems sometimes associated with the processing of film. Digital processing methodologies used include: (1) contrast enhancement; (2) spatial filtering to embrace morphological or structural information; and (3) arithmetic operations, such as ratioing of spectral bands to enhance spectral reflectance differences and to suppress systematic effects, such as topography. Statistical analysis is used to reduce dimensionality in data sets containing many spectral bands or other variables. Sequential ratioing of the MSS bands results in black and white ratio images that can be optically combined to produce a color-ratio composite.

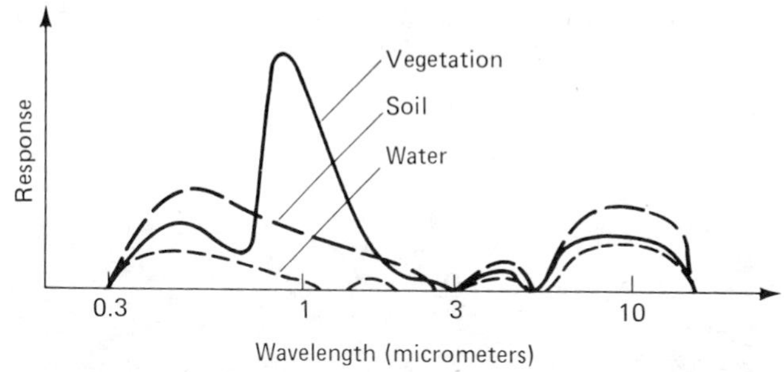

Fig. 7. Light and heat reflected or emitted from the earth will be picked up in selected wavelengths or bands by Landsat-D remote sensing instruments. Each band, from the visible to the far-infrared, will contain specific information about the earth's surface. In essence, the sensors are "tuned" to the wavelengths of vegetation, soil, water, geologic, and other surface materials to collect the spectral data each reflects. (*General Electric.*)

Each image displays 34,253 square kilometers (13,225 square miles) with uniform illumination and the scene composition can be optimized through digital processing of the radiance values.

During the earlier phases of the Landsat program (and continuing), large numbers of Landsat views of the world were made available, not only to scientists in specialized fields, but also for use by science teachers. Landsat views have been very valuable to the teaching of geology, environmental topics, geography, and social and urban studies. A standardized coloring system was developed for these views and the analysis of the views became a challenge to scholars. The color scheme used is:

Red-Magenta—denotes vigorous vegetation, i.e., forests, farmlands (agriculture) near peak growth. Detailed study of different shades of red aid in classification of different types of vegetation as well as the determination of their growth cycle and vigor.

Pink—depicts less densely vegetated areas and/or vegetation in an early stage of growth. Suburban areas around major cities with the green lawns and scattered trees normally show up pink.

White—areas of little or no vegetation but of maximum reflectance. White areas include snow, clouds, deserts, salt flats, ground scarring, fallow fields, and sandy beaches.

Dark Blue to Black—normally identifies water, i.e., rivers, streams, lakes, reservoirs. In the western United States, particularly Idaho and New Mexico, the very dense black features are ancient lava flows rather than water bodies.

Gray to Steel Blue-Gray—indicates towns, cities, urban, populated areas. The colors are produced by the returns from asphalt, concrete, and other anthropogenic features.

Light Blue in Water Areas—denotes either very shallow water or heavily sedimented water.

Light Blue in Areas of Western United States (and similar global areas)—identifies desert shrubland capable of supporting limited grazing activities.

Brown, Tan, and Gold—denotes areas comprised primarily of open woodlands (piñon, juniper, aspen, chaparral) and rangeland suitable for grazing.

Although shown here in black and white, Fig. 8 is a Landsat view of essentially an agricultural area; the principal object of the view is the Salton Sea. In this reproduction, areas that appear red on the original Landsat view are black or nearly black.

As an aid to researchers as well as scholars, grids in the form of transparencies that can be placed over a Landsat image are very helpful in measuring the area of various features. See Fig. 9.

Applications of Landsat Technology

Agriculture. The concept of using remote sensing of agricultural lands and crops was proposed many years ago. The combined use of high-flying aircraft and infrared photography represented an early effort and demonstrated advantages in terms of time and cost as compared with survey crews operating on land. Some uses of satellite surveillance include: (1) early warning of changes affecting production and quality of commodities and renewable resources; (2) specific commodity production forecasts; (3) land use classification and measurements; (4) renewable resources inventory and assessment; (5) land productivity estimates; (6) conservation practices; and (7) pollution detection and impact evaluation, among others. Some of the research

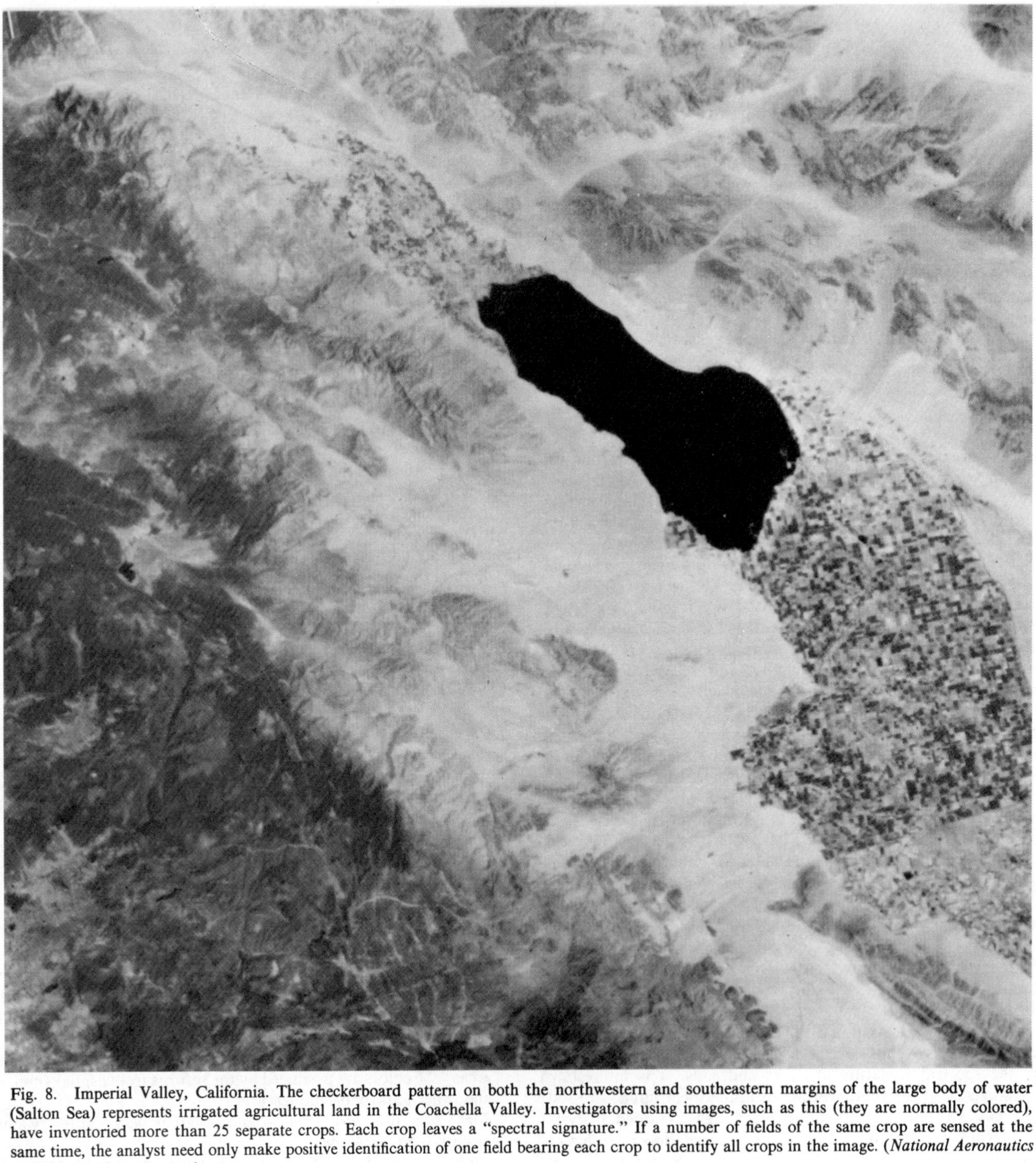

Fig. 8. Imperial Valley, California. The checkerboard pattern on both the northwestern and southeastern margins of the large body of water (Salton Sea) represents irrigated agricultural land in the Coachella Valley. Investigators using images, such as this (they are normally colored), have inventoried more than 25 separate crops. Each crop leaves a "spectral signature." If a number of fields of the same crop are sensed at the same time, the analyst need only make positive identification of one field bearing each crop to identify all crops in the image. (*National Aeronautics and Space Administration.*)

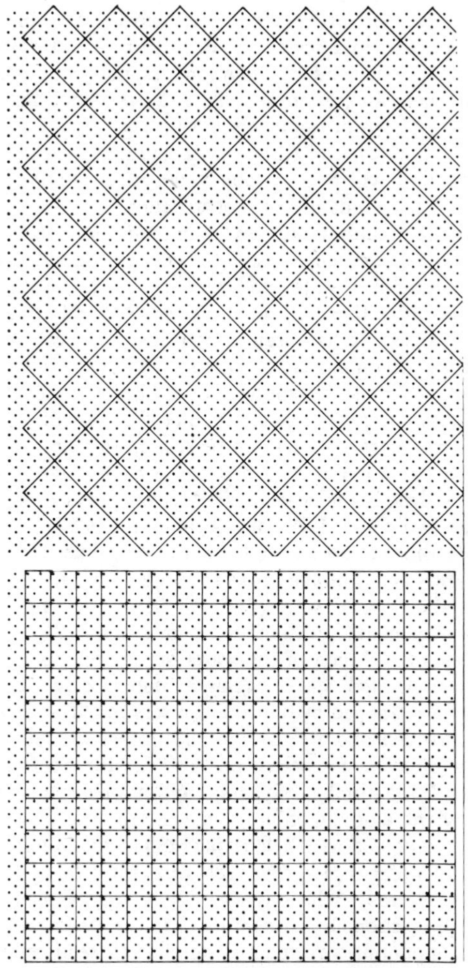

Fig. 9. Transparencies are available with patterns like these which can be overlaid on Landsat images (Scale 1 : 1,000,000) and approximate measurements can easily be made by counting the number of dots contained within the boundaries of interest. When the number of dots is multiplied by 40, the total area of inclusion within a boundary will be given in hectares; or, if multiplied by 98.8 will give the area in acres. Likewise, if the number of dots is multiplied by 0.4, the area will be in square kilometers; or, if multiplied by 0.154, the area will be in square miles. The patterns in this figure have been considerably reduced. (*U.S. Geological Survey, Geography Program.*)

as of the early 1980s is concentrating on the quantitative monitoring of conditions that foretell droughts, and on soil temperature surveillance, including factors that foretell and contribute to grain winterkill. Work on vegetative indexes is being refined, that is, the *green index number* (GIN) which is used to summarize the condition of vegetation within a Landsat scene. With this system, digital data from the Landsat satellite are put into a scale ranging from 0 for bare soil to 15 for healthy green vegetation. Over time, an understanding of the relationship between these readings and actual yields will allow analysts to quantitatively define production possibilities based on the GIN.

In 1974, jointly sponsored by the U.S. Department of Agriculture, the National Oceanic and Atmospheric Administration, and the National Aeronautics and Space Administration, the Large Area Crop Inventory Experiment (LACIE) was established. During the 1974–1978 period, Landsat data were used to estimate wheat production in the United States, Canada, and the U.S.S.R. It also allowed for technology problem definition in Argentina, Australia, Brazil, India, and China. Working with basic units called sample segments—5 × 6 square nautical miles (25,000 acres; 10,000 hectares) which were further broken down into pixels (1.1 acres; 0.44 hectare)—analysts selected training fields from reconstructed pictures. They identified specific crops in these segments and then programmed the computer to pick out other fields that appeared similar. In this way, area estimates were made. Yield models based upon historical weather information were used together with these area estimates to make production forecasts.

The LACIE estimates were mailed each month ahead of the official U.S. Department of Agriculture crop estimate. Results also were compared directly with those obtained on the ground in test areas, or so-called blind spots, of the United States. The LACIE goal—90/90 accuracy at the country level, or estimates within 10% of the final official estimate 90% of the time—was met during the experiment's third phase for the U.S. winter wheat and the U.S.S.R. wheat crops. Researchers reported that one of the most critical problems encountered was distinguishing between spring wheat and spring barley—both grown at about the same time in the northern United States and Canada. Moreover, the resolution of the satellite's multispectral scanner was not capable of separating fallow from wheat in the same areas, which were often strip-cropped in widths of 90 meters (300 feet) or less.

In an improved system, retransmission of processed data via satellite can make it possible to deliver data within 60 hours after collection, as contrasted with 29 days during the experiment described. Assessment and interpretation then could follow within a total of 72 hours after data measurement.

Agricultural land and crop measurement techniques have improved considerably during the past few years. Details cannot be given here, but are included in the "Foods and Food Production Encyclopedia," (D. M. Considine, editor), Van Nostrand Reinhold, New York, 1981.

Soil-Association Mapping. Soil maps ranging in scales from 1 : 15,840 to 1 : 7,000,000 serve such purposes as irrigation and drainage planning, forest management, crop-yield estimates, farm appraisals, and land-use planning. Aerial photography, introduced about 40 years ago, greatly assisted the mapping of soils. Now Landsat can image an 8-million acre scene in one frame, allowing comparisons of soil associations over their entire extent, all at the same instant in time and growth. The four spectral bands of Landsat and the repetitive coverage made possible by the satellite make subtle differences readily apparent and allow vegetative differences (which are usually a function of varying abilities of the different soil associations to produce vegetation) to be used effectively to help separate soil-association landscapes.

Drought and Desertification. Studies of Landsat imagery of the Sahel region of Africa (critically suffering from drought since 1970) revealed a dark polygonal area bounded by straight lines. It resembled areas known to harbor subsurface water. However, the area was some distance away from any obvious subsurface water source. The unusual linearity of the feature led to the conclusion that it was of artificial origin. The area was later visited and it was found to be a carefully managed, fenced-in ranch, differing noticeably in vegetative cover from the uncontrolled grazing areas outside the fence. This ranch demonstrated that uncontrolled animal grazing, characteristic of this region's agricultural practices, may be at least partly responsible for the advance of the Sahara Desert into the Sahel region. These observations led to suggestions to the governments of the affected nations that careful management of grazing areas may lead to reversal of the Sahara's progress and to the reclamation of some of the land already turned into desert.

Snow and Ice Monitoring. In several areas of the world, such as the western United States, snowmelt and glacial icemelt provide a major part of the annual water runoff needed for irrigation, personal and industrial consumption, and hydroelectric power generation. Results from Landsat indicate that snowlines can be observed to within 60 meters (197 feet) under good conditions, that snowcover can be empirically related to water runoff, and that snowcover area can be observed within a few percent. Landsat imagery also has been used by cartographers to meet the need for small-scale mapping by national and international polar scientific projects. It has been used to provide numerous and extensive changes to maps of Antarctica, including particularly the regions around the Ross Ice Shelf, Franklin Island, and various ice tongues. Even though maps of the Arctic are much more accurate than those of the Antarctic, several changes have been incorporated in recent maps of the Arctic region. Accurate identification of lake ice and classification of ice types can provide invaluable information for shipping in the Great Lakes, allowing for more efficient routing of shipping through ice-covered waters and the possible extension of the shipping season. Investigators have reported observing pressure ridges, rotten ice, cracks, leads, ice breakup, and other changes on the Great Lakes and also on lakes in New England. By comparing the 0.6 to 0.7 micrometer (band 5) with the 0.8 to 1.1 micrometer (band 7) observations, investigators have been able to delineate the area of melting ice on a lake.

Several investigators have shown that Landsat imagery can be used as a survey device to determine the abundance and the dimensional characteristics of Antarctic icebergs as a function of general location. With the polar overlap of Landsat imagery, even the prevailing 80% cloud cover yielded sufficient usable imagery for this application. Iceberg heights down to 50 meters (164 feet) have been estimated. Such observations may prove useful later should the concept of towing icebergs to water-starved areas as a source of fresh water become economically practical.

Oil Exploration. Reconnaissance of large sedimentary basins and new exploration provinces provided by Landsat imagery has aided rapid interpretation of large features and quickly focused attention on anomalous areas. Types of specific information gathered by Landsat have included: (1) data on linear features; (2) general lithologic relationships; (3) closed anomalies of various types correlated with known structural features of oil and gas fields; (4) details of structures controlling hydrocarbon accumulation; (5) overall structure of a basin and its major internal structures; and (6) geologic context of the exploration region.

Ore-Deposit Exploration. Ore deposits tend to be controlled during emplacement by fractures in the earth's crust. One group of investigators studied Landsat imagery in detail in central Colorado, well known for its large number of metal-producing mines. Using a snowcovered winter scene of the area, the investigators identified several linears, many of them new discoveries. Some of these linears are nearly straight, but others are curved or arcuate. Most are assumed to be large fractures; the arcuate ones are probably related to underlying intrusions that have cracked the country rock above. The investigators contoured the relative densities of these fractures. Circles 14.5 kilometers (9 miles) in diameter were drawn around the areas of highest fracture density. Of the 10 circles on the Landsat image, 5 coincide with major mining districts, indicating a correlation between extensive fracturing and the localization of minable ore deposits. The remaining 5 circles hopefully will constitute potential targets for exploration. Additional studies have been made of metallic ore deposits in the western Cordillera in Canada and Alaska. Reports of porphyry copper deposits in central Alaska northeast of Fairbanks have been under investigation. On an Landsat image, several of the intrusives have been found along or near a regional linear. The recognition of controlling linears in Landsat imagery can provide an excellent tool in ore-deposit exploration.

In the past, the geologist's basic tool for mineral exploration has been the geologic map. Whenever major improvements in the accuracy and completeness of essential maps can be made, both planning of exploration and eventual field work can be made more efficient. Investigators have shown that Landsat images are potentially capable of increasing the rapidity of geologic mapping and reducing its corresponding costs. An African study has provided a good example. Southwestern Africa is noted for its extensive copper-nickel mineralization. Past mapping has been relatively primitive. Investigators in South Africa received several excellent Landsat images of this mineral-rich area. By piecing together the images to make a mosaic of the entire region, a rather detailed geologic map of the area was produced for the first time.

Flood Area Assessment and Flood-Plain Mapping. Regional Landsat assessments can be valuable in delineating areas of potential flood damage and for focusing relief efforts of local and federal agencies, of mapping plain areas susceptible to flooding, and of locating areas of extensive flooding where additional flood-control works may be necessary. Such matters are of worldwide significance. Some investigators have used image enhancement and transferred these data to base maps at scales up to 1:100,000 to map floods. The flooded areas are easily delineated by standing surface water or by areas of low, near-infrared reflectance due to excessive soil moisture or stressed vegetation. The estimates of flooding over large regions are accurate to within 5%. For better flood-area assessment, additional efforts will use digital products for mapping to obtain the best resolution. The largest scale for this mapping appears to be 1:24,000.

Land Quality. With Landsat data, investigators have mapped strip-mined and reclaimed areas, monitored construction practices, sited new construction, and assessed urban quality. An extensive mapping project of strip-mine and reclamation mapping was completed for areas in Indiana, Ohio, Pennsylvania, and Maryland. Specialists with the U.S. Army Corps of Engineers have prepared vegetation and permafrost maps of portions of Alaska with detail not previously available. Such information will be of much value in evaluating sites for oil pipeline and other kinds of new construction.

Water Quality. Landsat data are being used to measure and monitor water quality in rivers, lakes, estuaries, and near-shore coastal zones. Suspended sediment, a water pollutant, occurs as a result of natural processes as well as those of unnatural processes. The sediment load of near-surface waters has been clearly identified in some cases. In several studies, attempts have been made to monitor lake eutrophication, out of which some innovative interpretive techniques have evolved. Determining average reflectance from small areas of water having uniform characteristics can be related to suspended solid concentration in the water (between 5 and 100 milligrams/liter). High reflectivity of algae in the near-infrared permits detecting algal blooms in both lakes and rivers. Landsat information has been used routinely in making decisions important to the hydrological needs of agricultural and urban areas and national parks in Florida—as, for example, management of the flow of water and its distribution in the southern Florida drainage canal and impoundment system.

Water Resources. Much of the water supply for populations living in the western United States begins as snow. Below the mountain peaks, rivers are trapped by dams and reservoirs. The captured water must be controlled and released in the proper amounts to provide not only for direct water needs, but also for the electrical energy it can generate. For a number of years, Landsat operations have assisted in measuring the amount of snow in western snowpacks. With this information, combined with field data obtained by ground inspections, the volume of water in a spring melt runoff can be anticipated with considerable accuracy. This assists both in energy planning as well as in flood prevention. By attaching a transmitter to a stream gage planted in a mountain stream, it is possible to send the information directly from the stream to Landsat every few hours during the snow-melt season. The satellite transmits the data to the ground station where engineers can enter it into the data bank immediately. When floods do occur, Landsat imagery can be used to show the extent of

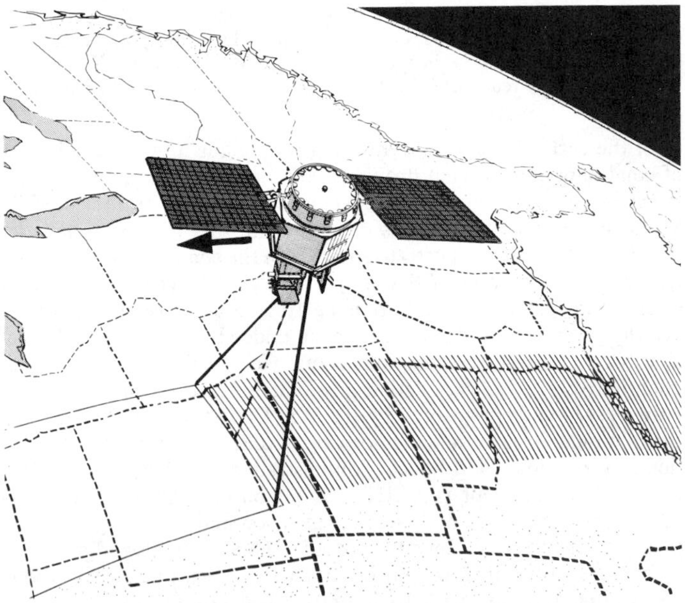

Fig. 10. HCMM (Heat Capacity Mapping Mission), the first spacecraft built to test the feasibility of measuring variations in the earth's temperature was launched on April 26, 1978. Purpose of mission—to produce thermal maps for discrimination of rock types, mineral resources, plant temperatures, soil moisture, snow fields, and water runoff. The experimental satellite travels in a circular, sun-synchronous, 620-kilometer (385-mile) orbit that allows for measuring mid-latitude test areas of the earth's surface for their minimum temperatures and then measuring those same areas for maximum temperatures about 11 hours later. (*National Aeronautics and Space Administration.*)

flooding and to monitor changes in flood plain management. Images can record rivers at normal levels and at flood stages, demonstrating the value of assembling a chronological set of images.

Air Quality and Weather Modification. The ability to detect certain types of air pollution on Landsat-1 imagery was noted by a number of investigators. Air pollution over the Great Lakes had been observed to modify weather. Radiance measurements over water have been used to calculate vertical aerosol burden to about plus-or-minus 10%. In a study of haze in the Los Angeles basin, a correlation was found between radiance and visibility, but the investigator concluded that the radiance values over land were not a particularly sensitive measurement of air pollution. Also, cloud-seeding results were monitored in the Colorado River Basin. The use of analytical models in conjunction with Landsat-1 data was studied by several scientists. The Great Lakes area is being modeled to study the interaction of air pollution and the dynamics of cloud behavior. Consideration is being given to the use of Landsat imagery for verification of models to predict plume and contrail dispersion.

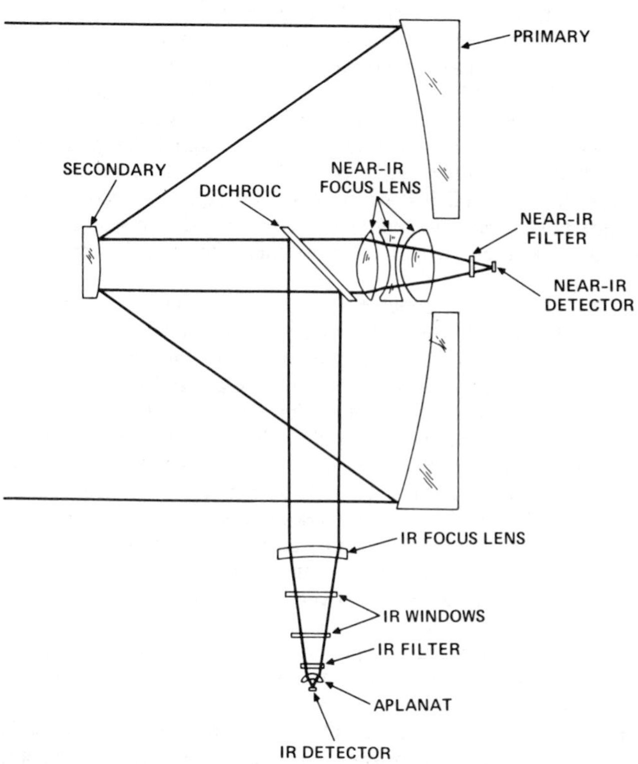

Fig. 11. Optical system of HCMM spacecraft. The instrument module is located on the spacecraft so that the Heat Capacity Mapping Radiometer (HCMR) is earth-pointing. The basic radiometer instrument is a modified spare Surface Composition Mapping Radiometer similar to that flown on Nimbus 5. The HCMR will have a small geometric field of view (less than 1 × 1 milliradians), high radiometric accuracy, and a wide enough swath coverage on the ground so that selected areas can be covered within the 12-hour period corresponding to maximum and minimum temperatures.

The instrument will operate in 2 channels, at approximately 0.5 to 1.1 micrometers (visible and near-infrared) and 10.5 and 12.5 micrometers (infrared), providing measurements of reflected solar energy and equivalent black-body temperatures.

Principal parameters are:

Orbital altitude	620 kilometers (385 miles) nominal.
Angular resolution	0.83 milliradians.
Resolution	500 × 500 meters (1640 × 1640 feet) at nadir.
Scan angle	60 degrees (full angle).
Scan rate	14 revolutions per second.
Scan period	0.07 second per revolution.
Sample rate	One sample per resolution element.
Sampling interval	9.2 microseconds.
Swath width	700 kilometers (434 miles).

Wildlife Resources. Under normal population densities, wildlife species are not resolvable with current Landsat imagery. However, investigators have been successful in identifying and measuring habitat factors, such as vegetation and water, both having direct and indirect influences on wildlife population. In the past, vegetation maps of certain wildlife habitats have been gross generalizations due to the difficulties of access to remote and sometimes hostile terrain. Landsat not only shows remote areas, but does so synoptically. Landsat maps can show between eight and ten vegetation classes, and relatively detailed peripheral boundaries, as compared with former maps of two to four classifications having generalized boundaries. Adding surface water gives a good map of a habitat.

Other Remote Sensing Methods

In addition to Landsat technology, other sensing methods include: (1) airborne radar (X-band, 3 centimeters; L-band, 25 centimeters); (2) orbiting laboratories, such as the downed Skylab, which had a 13-channel multispectral imager; (3) Seasat (equipped with synthetic aperture L-band radar); and (4) the Heat Capacity Mapping Mission (HCMM) which provides day-night thermal and visible images for mapping the thermal inertia of surface materials. See Fig. 10. The optical system of the HCMM is shown in Fig. 11. Investigations for which the HCMM was designed include: (1) urban heating patterns; (2) freeze damage assessment with development of planting date advisories; (3) evapotranspiration rates; (4) soil moisture assessment; (5) thermal mapping for discrimination of geologic units and energy or mineral resource areas and evaluation of thermal modeling and satellite mapping techniques; (6) detection of high potential groundwater pollution; (7) studies of snow hydrology programs; (8) studies of estuarine currents; (9) a topoclimatological and snow-hydrological survey of various countries, such as Switzerland; (10) monitoring of large scale pollution effects in the North Sea, among several others. Scientists from several countries, including Australia, Canada, the European Economic Community, France, Germany (West), Italy, Morocco, Spain, Switzerland, and the United Kingdom, are participating in investigations involving the HCMM.

References

Adrien, P. M., and M. F. Baumgardner: "Landsat, Computers, and Development Projects," *Science*, **198**, 466–470 (1977).

Goetz, A. F. H., and L. C. Rowan: "Geologic Remote Sensing," *Science*, **211**, 781–791 (1981).

Hammond, A. L.: "Remote Sensing," *Science*, **196**, 511–516 (1977).

Idso, S. R., Jackson, R. D., and R. J. Reginato: "Remote-Sensing Crop Yields," *Science*, **196**, 19–25 (1977).

Paul, C. K.: "Satellites and World Food Resources," *Technology Review (MIT)*, **82**, 1, 18–29 (1979).

Sabins, F. F. Jr.: "Remote Sensing; Principles and Interpretation," Freeman, San Francisco, 1978.

Smith, W. L. (editor): "Remote Sensing Applications for Mineral Exploration," Dowden, Hutchinson, and Ross, Stroudsburg, Pennsylvania, 1977.

Staelin, D. H., et al.: "Microwave Spectroscopic Imagery of the Earth," *Science*, **197**, 991–993 (1977).

Tindal, M. A.: "Educator's Guide for Mission to Earth: Landsat Views of the World," Goddard Space Flight Center, National Aeronautics and Space Administration, Greenbelt, Maryland, 1978.

EARTH-SHINE. Moon (The).

EARTH TIDE. The response of the solid Earth to the forces that produce the tides of the sea; semidaily earth tides have a fluctuation of between 7 and 15 centimeters.

EARTHWORK. This term denotes work, the object of which is to alter the surface of the earth to serve some useful constructional purpose. In addition to excavation, building of embankments and trimming of slopes, earthwork also includes the clearing and grubbing of rough land, grading, etc. Excavation of rock and loose rock is usually considered earthwork. Among the most common forms of earthwork are preparation of subgrade for railways and highways, buildings of embankments for hydraulic work and construction of open drainage systems. In the preparation of a roadbed, the original surface of the ground is altered to the required degree by cuts and fills. As far as possible, cut should equal fill, so that the material excavated may be hauled a short distance and used to fill depressions in the proposed roadway. Where the amount of cut is insufficient for filling, the deficiency must be made up by hauling from borrow pits. An excess of cut is deposited on spoil banks.

The vertical dimension of the cut or fill at any station is found by leveling. An egineer's level is set up near the station. The height of the instrument, which is the vertical distance between the line of sight of the level and a reference plane called the datum, is obtained by taking a rod reading on a bench mark or other point of known elevation above the datum. After the height of instrument is determined a rod reading is taken on the ground at the station. This is the ground rod. The difference between the height of instrument and the known grade elevation at the station is the grade rod. The cut or fill is the algebraic difference between the grade rod and the ground rod. The point where the side slope of a cut or fill will intersect the ground surface is marked by a stake called a slope stake. Slope stakes are set before construction is begun.

To measure the amount of cut or fill in earthwork, transverse cross-sections of the cut or fill are measured at regular intervals. These sections are then plotted on paper, and the area computed. Sections are taken close enough so that the volume of earthwork between them is considered that of a prism of length equal to the distance between stations, an area equal to the average of the two sections. This is known as the end area method. A more accurate result is obtained by the use of the prismoidal formula, which states that the volume of the earth equals $\frac{1}{6}(a_1 + 4a_m + a_2)D$, a_1 and a_2 are the end areas, a_m is the mid-section area, D is the length of the section being measured. Earthwork is usually done on a contract basis and paid for on the basis of volume excavated, or volume of fill. A fill of earth usually shrinks 10–20% upon compacting. Rock may be expected to occupy 15–30% greater volume after excavation than before.

If the volume of cut or fill between any two successive stations is given an algebraic sign (+ for cut and − for fill), the algebraic sum of the volumes starting at an initial station, may be represented by a continuous curve called a mass diagram. The abscissa for any point on the curve is its horizontal distance (in units of 100 feet) measured from the initial station; the ordinate is the algebraic sum (in cubic yards) of the volumes up to the point. Plus-ordinates (cut) or plotted above the *x*-axis, and minus-ordinates (fill), below. (See figure.)

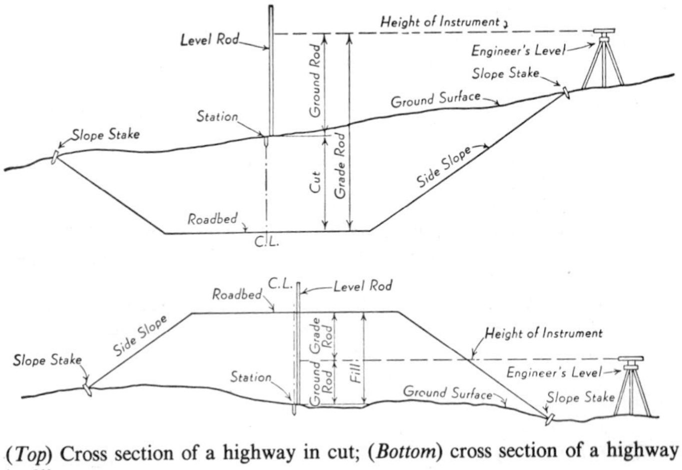

(*Top*) Cross section of a highway in cut; (*Bottom*) cross section of a highway in fill.

EARTHWORM (*Annelida, Oligochaeta*). Terrestrial segmented worms of many species. They burrow in earth containing organic matter on which they live, coming to the surface only in damp cloudy weather and at night. Their activity in loosening and mixing the soil in fields is estimated to be valuable in crop production.

EARWIG. Dermaptera.

EASTERLIES. Winds and Air Movement.

EASTERLY WAVE. Atmosphere (Earth).

EASTERN HEMLOCK. Hemlock Trees.

EASTERN LARCH. Larch Trees.

EASTERN TENT CATERPILLAR (*Insecta, Lepidoptera*). Also known as the apple-tree tent caterpillar (*Malacosoma americanum*, Fabricius), this insect eats the foliage of trees and disfigures them with its nests. The insect is most widespread in the northeastern United States, although it is found throughout the country east of the Rocky Mountains, and in parts of California. Similar species are found in the Rocky Mountains and farther west.

Wild cherry trees and apple trees are most often attacked. Peach, pear, plum, rose, howthorn, and various shade and forest trees are occasionally infested. In spring, their unsightly nests or "tents" are conspicuous on susceptible trees. They are more active in some years than others. They may eat nearly all the leaves of a tree, which weakens the tree, but seldom kills it. Once the caterpillars mature in early summer, they cause no further feeding damage.

One generation of the insect develops in a year. The larvae, or caterpillars, hatch in spring from egg masses, about the time the first leaves are opening. The young caterpillars keep together and spin threads of silken web. After feeding for about two days, they begin to weave their tent in a nearby tree crotch, sometimes joining with caterpillars from other egg masses. As the caterpillars grow, they enlarge the tent until it consists of several layers. In good weather, the caterpillars leave the tent several times per day in search of food, stringing silk after them. In bad weather, they remain between the layers of the tent.

When fully grown, about 6 weeks after hatching, the caterpillar is nearly 2 inches (5 centimeters) long and sparsely hairy. It is black, has a white stripe along the middle of its back, and other white and blue markings. When they are mature, the caterpillars spin cocoons on tree bark, fences, brush, etc. The cocoon is about 1 inch (2.5 centimeters) long and white to yellowish white, depending upon age and presence of caterpillars. Inside the cocoon, the larvae transform to pupae, the resting stage. In early summer, reddish brown moths emerge and the females deposit masses of eggs in bands around twigs. The eggs are covered with a foamy secretion, which dries to a firm, brown covering that looks like an enlargement of the twig. An egg mass usually contains about 200 eggs. The larvae develop inside the eggs, but they do not hatch until spring.

These caterpillar larvae are prey for other insects, toads, and birds. The small wasps (chalcid wasps, etc.) develop as parasites in the eggs, larvae, or pupae. Malathion or methoxychlor applied to the nests and about one foot (0.3 meter) of surrounding area is effective. The spray should be applied before nests are 3 inches (7.5 centimeters) in diameter. The insects are easy to control by hand removal and burning if only a few trees are infested. Wherever possible, wild cherry trees growing near orchards should be removed.

A similar species, the *forest tent caterpillar* (*Malacosoma disstria*), and the *fall webworm* (*Hyphantria cunea*) are sometimes mistaken for the eastern tent caterpillar. The forest species attacks mostly forest trees and seldom fruit trees. The web of the fall webworm can be distinguished from that of the tent caterpillars by the fact that the webworm's nest is made at the tip of a branch instead of at the crotch.

EASTERN WHITE PINE. Pine Trees.

EAST GREENLAND CURRENT. An ocean current flowing south along the east coast of Greenland, carrying water of low salinity and low temperature. The east Greenland current is joined by most of the water of the Irminger current. The greater part of the current continues through Denmark Strait between Iceland and Greenland, but one branch turns to the east and forms a portion of the counter-clockwise circulation in the southern part of the Norwegian Sea. Some of the east Greenland current curves to the right around the tip of Greenland, flowing northward into Davis Strait as the west Greenland current.

EAST PACIFIC RISE. Earthquakes, Seismology, and Plate Tectonics; Ocean; Ocean Resources (Mineral).

EBONY TREE. Persimmon Trees.

EBSTEIN'S ANOMALY. Arrhythmias (Cardiac).

EBULLIOMETER. An instrument, sometimes referred to as an ebullioscope, that measures the property of a substance by noting a deviation from a normal known boiling point. The term applies to apparatus for estimating the percentage of alcohol in a mixture through observation of the boiling point. Beckmann's apparatus for molecular weight determination is an ebullioscope. The ebullioscopic constant is a quantity calculated to represent the molal elevation of the boiling point of a solution by the relationship:

$$K=\frac{RT_0^2}{1000l_e}$$

in which K is the ebullioscopic constant, R is the gas constant, T_0 is the boiling point of the pure solvent, and l_e is the latent heat of evaporation per gram. The product of the ebullioscopic constant and the molality of the solution gives the actual elevation of the boiling point for the range of values for which this relationship applies. Unfortunately, this range is limited to very dilute solutions, not extending to solutions of unit molality.
See also **Beckmann Method;** and **Differential Thermal Analysis.**

EBULLITION. Boiler; Boiling.

ECCENTRIC. The eccentric is a machine element employed to convert rotating to reciprocating motion. Its function is similar to that of the crank. The eccentric is used chiefly for short throws, where it would be undesirable to break the shaft, as is necessary in the case of a crank. It consists of a disk mounted on a shaft in such a way that the geometric center of the disk does not coincide with the center of rotation. The distance between the center of rotation and the geometric center of the eccentric is the throw. This corresponds to the crank-arm distance of an equivalent crank. The eccentric must be used in conjunction with an eccentric strap which surrounds the eccentric, and which transmits the reciprocating motion to an eccentric rod rigidly attached. The eccentric is chiefly used to drive auxiliaries such as valve gear, and where reciprocation of small magnitude is needed. The cam and crank may be employed to provide similar motion.

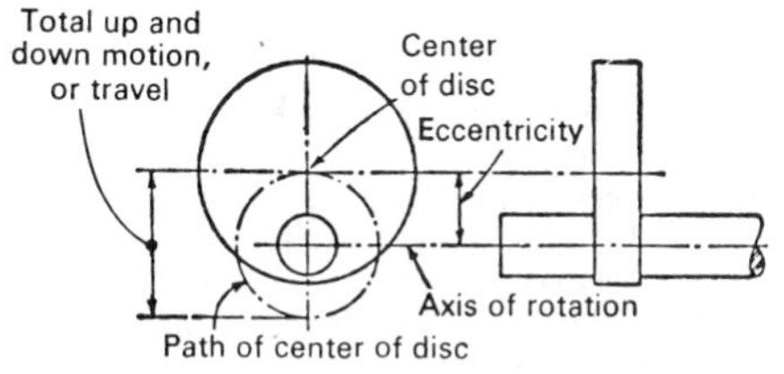

Simple eccentric.

ECCENTRIC CIRCLE. The unmodulated, endless groove provided at the inside of most phonograph records. Its center differs from that of the record, to facilitate the actuation of the trip mechanisms of some types of automatic record changers. The unmodulated sprial groove which leads from the end of the modulated groove portion of a phonographic record to the eccentric circle is termed the *eccentric spiral.*

ECCENTRICITY CORRECTION. Angular Measurements (Eccentricity Correction).

ECDYSIS. The molting or shedding of the cuticula by arthropods. Since the cuticula of these animals is also the rigid skeletal support of the body and is inelastic, it is shed at intervals during growth and a new covering of larger dimensions is formed. In preparation for molting the insect or other arthropod becomes inactive for a time, then by crawling movements crowds forward in the old integument, which splits down the back and allows the animal to emerge. During the resting period preparation is made by the secretion of fluid from the molting glands of the cellular layer and the loosening of the under part of the cuticula. Following the shedding of the old cuticula, a new layer is secreted during a further period of inactivity. All cuticular structures are shed at ecdysis, including the terminal linings of the alimentary tract and of the air tubes if they are present.

The molting of reptiles is sometimes called ecdysis.

ECHELLE GRATING. In studies of high resolution diffraction gratings, it is observed that resolving power at a given wavelength depends only on the ruled width of the grating and the angles of illumination and observation, and not specifically on the number of ruled grooves. The ruling of many inches of a grating with a few tens of thousands of lines per inch is an almost impossible task. An echelle grating has very fine lines ruled much farther apart than is customary. Such a grating has very high resolution, but over only a quite narrow band of wavelengths. Hence it is customary to cross an echelle grating with a second grating (or prism) of lower resolving power, thus producing what is essentially a two-dimensional spectrogram or echellegram.

The resolving power of a diffraction grating depends primarily on the total ruled width and the angles of incidence and refraction. Thus high resolving power should be obtainable with coarse rulings, 100 or so per inch, if the grooves are properly shaped. When the grooves are fixed to reflect light of all colors in a narrow bundle, the resulting echelle grating may have a resolving power equivalent to that of a diffraction grating with 30,000 lines per inch (11,811 per centimeter). It lacks, however, the angular dispersion needed to separate the spectra of consecutive orders; thus this grating is usually crossed with a prism or another grating of lower resolving power.

ECHELETTE. The relative intensities of the light in various orders of a diffraction grating depend on the groove shape of the rulings. With only 1,000–2,000 lines per inch (394–787 per centimeter) of ruled surface and proper groove shape, usually rather broad, 75% or more of the reflected light can be thrown into a single order. Such a grating is called an echelette.

ECHELON. A highly specialized form of diffraction grating, devised by Michelson. It consists of a row of glass plates of exactly equal thickness, packed together to form a miniature stairway of equal risers. The light enters normally to the largest plate at one end (see figure) and emerges at various deviations through the low "risers." It is easily shown that if the thickness of the plates is a, the height of the "risers" b, and the refractive index of the glass n, the equivalent path difference between successive emergent streams for any angle of deviation Δ is $na = a\cos\Delta + b\sin\Delta$; or since Δ is in practice always small, $\cos\Delta = 1$ and $\sin\Delta = \Delta$ (in radians), giving $(n-1)a + b\Delta$. This must be equal to an integral multiple, N, of the wavelength λ for any spectrum line, the deviation of which is therefore:

$$\Delta = N\frac{\lambda}{b}-(n-1)\frac{a}{b}$$

The smallest value N can have (for $\Delta = 0$) is $(n-1)a/\lambda$, which, since a is usually several millimeters and $(n-1)$ is 0.5 or more, is of the order of several thousand. The disposition, viz.,

$$D=\frac{d\Delta}{d\lambda}=\frac{N}{b}-\frac{a}{b}\frac{dn}{d\lambda}$$

is correspondingly large. The echelon is thus especially adapted to the study of the hyperfine structure of spectrum lines. For other instruments of high resolving power, see **Echellette; Echelle Grating;** and **Interferometer.**

Echelon.

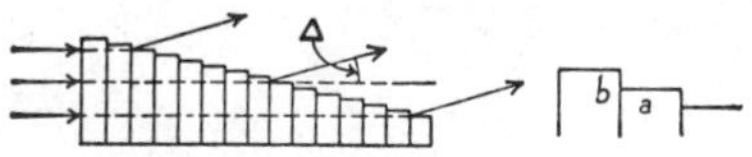

ECHIDNA. Fossil Reptilian Mammals.

ECHINOCOCCOSIS. Hydatid Disease.

ECHINODERMATA. A large division of the animal kingdom including the starfishes, sea cucumbers, brittle stars, sea lilies, sea urchins, and basket stars, all marine animals.

This phylum is characterized by the following structures: (1) The adult is almost perfectly radially symmetrical, although the young are bilateral. (2) The wall of the body contains a hard skeleton in most forms, made up of calcareous bodies called ossicles. (3) The coelom is well developed. (4) A water vascular system is present, consisting of a closed series of tubes opening to the exterior at one point on the body and bearing many delicate sacs, the tube feet or tentacles, which protrude at the surface of the body. (5) There is not special excretory system.

The echinoderms are divided into several classes, which fall into two subphyla:

Subphylum *Eleutherozoa.* Without a stalk.
- Class *Asteroidea.* The starfishes.
- Class *Ophiuroidea.* The brittle stars.
- Class *Echinoidea.* The sea urchins, sand dollars, etc.
- Class *Holothuroidea.* The sea cucumbers.

Subphylum *Pelmatozoa.* With a stalk at least when young.
- Class *Crinoidea.* The feather stars, basket stars, and sea lilies.

See also **Invertebrate Paleontology.**

ECHINODERMS (Fossil). Invertebrate Paleontology.

ECHINOID. Invertebrate Paleontology.

ECHINOIDEA. The sea urchins, keyhole urchins, and sand dollars. A class of the phylum *Echinodermata.*

The members of this class are distinguished by the following characteristics: (1) The body varies from almost globular to thin disks. (2) The tube feet are suckers. (3) The surface bears long spines and pedicellariae. (4) There are no radiating arms. (5) The ossicles are closely associated to form a shell.

Sea urchins live on organic matter of all kinds, including small animals, plant tissues, and waste matter. They are of little economic importance but in some of the Mediterranean countries and to a limited extent in the Orient they are used as food.

The class is divided into the following subclasses:

Subclass I. *Bothriocidaroidea*—extinct.
Subclass II. *Regularia* or *Endocyclica*—periproct encircled by apical system of plates.
Subclass III. *Irregularia* or *Exocyclica*—periproct outside apical system of plates.

Many species flattened. Sand dollars, etc.

Sea urchin (*Echinoidea*). (*A. M. Winchester.*)

ECHO. A wave which has been reflected at one or more points in the transmission medium, or otherwise returned with sufficient magnitude and delay to be perceived in some manner as a wave distinct from that directly transmitted. There are several types of natural echoes, including: 1. The discrete single echo. 2. The discrete multiple echo (a number of successive reflections). 3. The overlapping multiple echo—reverberation. 4. The diffuse echo, due to the scattering of sound by many small objects. 5. The harmonic echo, due to the greater scattering of an overtone than of the fundamental. 6. The musical echo, due to reflection from, or scattering by, a series of objects spaced at uniformly increasing distances from the source. 7. In radar, a general term for the appearance, on a radar indicator, of the radio energy returned from a target. More explicitly, it refers to the energy reflected or scattered back from a target. See also **Sonar.**

ECHOCARDIOGRAPHY. The use of ultrasound in a technique known as *echocardiography* has been available since the late 1960s and has proved of particular value in evaluating such conditions as pericardial effusion, rheumatic disease of the mitral valve, mitral valve prolapse, left atrial tumors, hypertrophic cardiomyopathy, and abnormal dilation of the left ventricle, among other heart problems. See **Heart and Circulatory System (Human).** In heart examinations, the ultrasound beam (narrowly focused) is directed so that it will provide three standard views (left ventricle; mitral valve; aorta and left atrium). The ultrasound reflected from blood-tissue interfaces within the heart and great vessels reveals much concerning the internal anatomy of the heart. The reflected ultrasound radiation is displaced against time, yielding a detailed picture of the motion of the structures of the heart during the cardiac cycle.

In a refinement known as two-sector scanning, a large series of ultrasound beams produces a two-dimensional image of cardiac anatomy. In a device originally developed at the University of Indiana, a transducer that mechanically sweeps through an arc of 30–45° at a rate of 30 times per second is used. Another approach (Duke University; Stanford University) uses an array of transducers to electronically create a wedge of sound. Two-dimensional images appear as slices of the heart. By making two-dimensional images from a number of angles, a three-dimensional image can be created. This has proved particularly advantageous in diagnosing congenital heart defects in which structures of the heart are displaced from their usual positions.

There appear to be no risks or discomfort associated with echocardiography.

References

Feigenbaum, H.: "Echocardiography," 2nd edition, Lea and Febiger, Philadelphia, 1976.
Kolata, G. B.: "Detection of Heart Disease," *Science*, **194**, 1031 (1976).
Luisada, A., Perez, G. L., and P. K. Bhat: "An Atlas of Non-Invasive Techniques," Charles C. Thomas, Springfield, Illinois, 1976.
McDonald, I. G.: "Introduction to Echocardiography," Charles C. Thomas, Springfield, Illinois, 1976.

ECHO SOUNDER. A device used to determine the depth of the sea floor beneath sea level. The device emits a high-pitched sound from the ship's hull. Traveling through sea water at a rate of 4,800 feet (1,440 meters) per second, the sound is echoed back from the sea floor to the ship for detection. Since each leg of the round trip is equal, the elapsed time (in seconds) is divided in two and multiplied by 4,800 to give the total depth in feet (or multiplied by 1,440 to give the total depth in meters). Present models can plot a continuous record of depth as the ship travels.

ECHO CHECK (Computer System). Check (Computer System).

ECHO-RANGING SYSTEM. Sonar.

ECHO-RANGING SYSTEM (Camera). Photography and Imagery.

ECHO SUPPRESSOR. When an electric wave on a line encounters a discontinuity or point at which the impedances do not match (see **Impedance Matching**) some of it is reflected. This reflected wave may return to the sending end of the line with sufficient amplitude to be objectionable. This is especially true in telephone service. While an effort is made to prevent reflections, there are cases where energy is fed back along the line and returns to the sender as an echo. In certain systems two lines (4 wires) are employed for transmission in the two directions and in such systems echo suppressors may be used to suppress the returning wave. One rather simple method of achieving this result is to use a relay to short one line when there is a signal on the other. Thus, if party A is talking and sending a voice signal to B, the voice currents on the line from A to B operate a shorting relay across the line from B to A so party A does not receive his own voice as an echo.

ECLIPSE. A term applied to the obscuration of a celestial body due to the interposition of another body or object. There are fundamentally two kinds of eclipse situations, distinguished by whether (1) the eclipsed object is *self-luminous*; or (2) the eclipsed object normally *shines by reflected light.*

In the first case, an eclipse occurs when an opaque object passes between the luminous body and the observer. The best known eclipse of this type, of course, is an *eclipse of the sun,* as caused by the moon blocking off light from the sun before the light can reach an observer on earth. Depending upon several factors, such an eclipse may be *total,* where to observers within the *shadow path* of the moon all light is blocked off. Or, such an eclipse may be *partial,* where only part of the light is cut off. Details of a solar eclipse are given a bit later.

In the second case, a body that normally shines by reflected light is eclipsed by putting that body in the shadow cast by an opaque body that intervenes in the direct path from the luminous body and the eclipsed body. The best known eclipse of this type is an *eclipse of the moon,* as caused by the earth blocking off light from the sun before the light can reach the moon. Thus, during the period of an eclipse of the moon, an observer on earth will see the moon pass into and out of darkness. Again, depending upon several factors, such an eclipse may be *total,* where the moon is completely within the shadow of the earth, or *partial,* where only part of the sun's light is cut off. Details of a lunar eclipse are given later.

Eclipses of the first kind also apply to instances where the moon may block off the radiation from a star or reflected light from a planet from reaching earth; and also in the case of eclipsing binary stars. See also **Eclipsing Binary.** Where relatively small objects come between earth and sun, as in the case of a planet blocking off radiation from a star, the term *occultation* is usually used instead of eclipse. Occultation of a planet by the moon also may occur. When an apparently very small object intervenes between sun and earth, as in the case of the planet Mercury, the path of the planet across the face of the sun can be observed. The amount of radiation reaching earth from the sun when this occurs is reduced by such a tiny amount that using the term eclipse is hardly appropriate. For situations of this kind, the term *transit* is commonly used.

Eclipses of the sun and moon have been scientifically important as well as very interesting to lay people and, in fact, to some cultures the eclipses are associated with matters of mystique, superstitution, taboos, prophecy, and fear. Mention of eclipses in history is also important to scholars for fixing accurate dates to past events. As an example, reference may be made to an Assyrian table which states; "Insurrection in the city of Assur. In the month of Sivan the sun was eclipsed." This probably also refers to the solar eclipse of June 15, 763 B.C. This is the same eclipse referred to in Amos VII:9: "I will cause the sun to go down at noon, and I will darken the earth in the clear day."

Eclipses, particularly total eclipses of the sun, have been scientifically important in terms of bettering astronomical measurements, the development of solar physics, and proving and disproving various hypotheses and theories, among other scientific objectives. These developments are reviewed briefly a bit later.

Although instruments (see **Sun**) have been developed over the years which essentially duplicate many of the observational advantages of a total eclipse, such as investigating the corona of the sun, large groups of scientists from all over the world continue to travel to areas, sometimes quite inaccessible, to view and carry on special scientific measurements during the period of totality. In recent years, scientists have been joined by growing numbers of lay astronomers who desire the personal experience of witnessing such an event. For example, hundreds of persons, including representatives of the news media, traveled to Kenya's Taita Hills to view the total eclipse of the sun on February 16, 1980.

Geometry of Eclipses. Distances from earth, the relative size or apparent diameters of the participating bodies, and orbital speeds and eccentricities are among the major factors which contribute to positioning the participating bodies at just the right places in the same time frame that result in some kind of eclipse situation. Since all of the foregoing factors relate in a varying manner with time, circumstances for eclipses of the sun and moon occur comparatively infrequently. In only a few years in a given century will circumstances permit up to seven eclipses. In such years, there may be either two of the moon and five of the sun; or three of the moon and four of the sun. Only two years of the 20th century permitted the first of the two foregoing situations—in 1935 and 1982. During this century (1901 through 2000 A.D.), there will have been a total of 375 eclipses, of which 228 will have been solar and 147 lunar eclipses. This is an average of approximately four eclipses per year. Eclipses of the moon, particularly to the layman, may seem more frequent than solar eclipses simply because lunar eclipses are observable over wide areas of earth—in fact, any observer who would be seeing the moon during a given period will witness the moon in eclipse. In contrast, viewing of solar eclipses is limited to a relatively narrow shadow path on earth. Further, the period of totality of a solar eclipse, in the extreme, will not exceed about 7.5 minutes; whereas the moon may be fully darkened up to a period of about 2 hours.

Variations in the sun–earth–moon system not only affect the exact dates on which eclipses will occur, but the type of eclipse (total or partial), the duration of the eclipse, and the locations on earth from which an eclipse can be observed.

With reference to the diagram of Fig. 1 in the entry on **Moon (The),** it will be noted that an eclipse situation can occur only when the moon is at one of its nodes.[1] These nodes occur at those positions when there is a new moon and when there is a full moon. In the idealized diagram of the figure, not allowing for tilt or eccentricity of orbits, these are the points where the moon is most directly between the earth and the sun (new moon); and where the earth is most directly between the moon and the sun (full moon). Total eclipses of the sun occur when there is a new moon; total eclipses of the moon occur when there is a full moon.

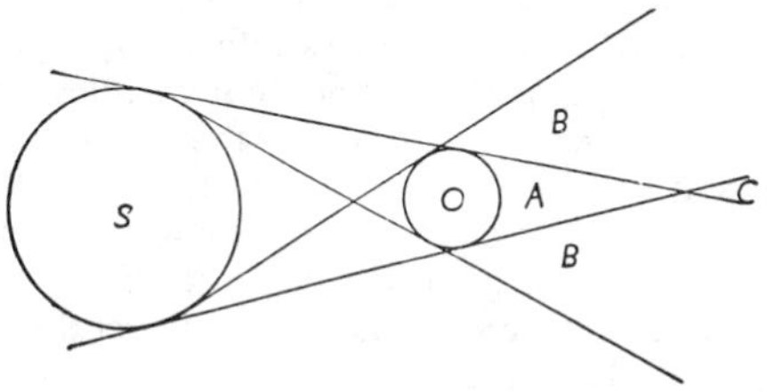

Fig. 1. Formation of shadows for eclipse of sun or moon.

The diagram of Fig. 1 (of the present entry) can be used to illustrate the circumstances of an eclipse of the sun as well as an eclipse of the moon. The opaque body is *O.* The luminous body is *S.*

Eclipses of the Sun. In the case of an eclipse of the sun, consider *S* to be sun; *O* the moon. The moon will cast a shadow toward earth, known as the *umbra,* and represented by area *A.* The umbra is the darkest central part of a shadow. (When a source of light, not a point source, casts a shadow of an object, the shadow consists of two parts—the umbra, just defined, where the region is completely cut off from the light, and the *penumbra* which is partly illuminated by some part of the light. The penumbra is a region of semi-shadow over which the illumination gradually increases from total darkness to full illuniation. From a point within the penumbra, the light source is partially, but not totally occulted by another body.)

[1] Nodes are the points at which the orbit of any satellite crosses the plane of the primary's equator or other fundamental reference plane such as the ecliptic. Movement of these crossing points caused by pertubations is termed the *regression of the nodes.*

Fig. 2. Progress of a total eclipse of the sun at intervals of approximately 10 minutes.

A solar eclipse viewed from the region of the umbra will be a *total eclipse.* Viewed from the regions of the penumbra (*B* areas located on either side of the umbra), the occurrence will be seen as a *partial eclipse* of the sun. For an observer within the umbra, the progress of the solar eclipse will appear as diagramed in Fig. 2. For an observer within the penumbra, the disk of the sun will be viewed with a circular segment cut out, thus with the light reduced, but not completely cut off. The viewing path of the solar eclipse of 1979 is shown in Fig. 3.

In an *annular eclipse* of the sun, the moon obscures the central part of the disk of the sun, but leaves a thin ring of light showing round the circumference. With reference to Fig. 1, this is the situation as viewed from *C.* An annular eclipse occurs when the moon is directly between earth and sun, but when the moon is near its farthest point from earth and where the umbra of the shadow does not quite reach the earth. From a knowledge of the diameters of *S* and *O* and the distance between the two objects, the dimensions of the various parts of the shadow may be calculated. The apparent diameters of sun and moon must be nearly the same to assure a total eclipse. This is not true for annular eclipses.

Because the orbit of the earth–moon system about the sun is eccentric, the length of the umbra cone *A* varies between the following approximate distances: At *aphelion* (point where earth–moon system is farthest from the sun), the length of the embra cone is 236,000 miles (377,800 kilometers); and at *perihelion* (point where earth–moon system is closest to the sun), the length of the umbra cone is 228,000 miles (365,000 kilometers). Also, owing to the eccentricity of the moon's orbit, the distance of the moon from the surface of the earth varies between 221,463 miles (356,334 kilometers) at *perigee* (the point where the moon is nearest to the earth) and 252,710 miles (406,610 kilometers) at *apogee* (the point where the moon is farthest from the earth). Examination of these numbers indicates that with the earth at aphelion and the moon at perigee, the surface of the earth is about 18,200 miles (29,284 kilometers) inside of the apex of the umbra. Under these conditions, the most favorable for an eclipse of the sun, the shadow of the moon on the earth will be a spot about 170 miles (274 kilometers) in diameter and, within this area, a total eclipse of the sun may be observed. Surrounding the spot of totality, there will be a region of about 3000 miles (4827 kilometers) radius, within which the sun will be partially eclipsed. With the earth at perihelion and the moon at apogee, the surface of the earth will be 19,500 miles (31,376 kilometers) beyond the apex of the umbra cone, a condition permitting the observation of an annular eclipse, but where totality is not possible.

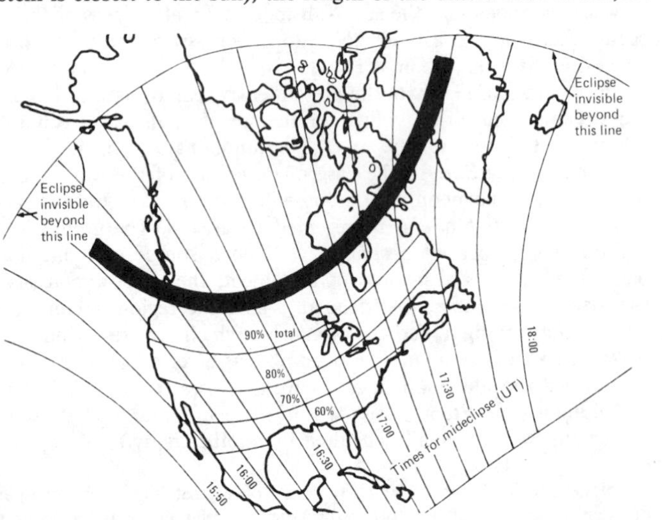

Fig. 3. Viewing path of total solar eclipse of 1979. (*Kitt Peak National Observatory.*)

As the moon revolves about the earth in its orbit, the shadow sweeps along the plane of the moon's orbit with a velocity of about 2100 miles (3379 kilometers) per hour to the eastward. The earth is rotating at such a rate that a point on the equator is moving to the east with a velocity of about 1040 miles (1673 kilometers) per hour. Accordingly, under the most favorable conditions for a solar eclipse (moon at perigee; earth at aphelion; eclipse occurring at noon; observer on equator), the shadow will pass the observer with a speed of $2100 - 1040 = 1060$ miles ($3379 - 1673 = 1706$ kilometers) per hour. The shadow will pass the observer from west to east. The spot will pass the observer in slightly less than 8 minutes. The duration of a partial eclipse, on the other hand, may be several hours.

As stated previously, because the plane of the moon's orbit is inclined to the plane of the ecliptic, an eclipse of either the sun or the moon may occur only when the moon is close to one of the nodes, i.e., close to the plane of the ecliptic, and must also be in conjunction (for an eclipse of the sun), or in opposition (for an eclipse of the moon). Since the earth–moon system revolves about the sun once each year, the line of nodes would pass through the sun twice in each year if the direction of that line were fixed in space. Regression of the moon's nodes causes the line to pass close to the sun three times each year. The period when the line of nodes is close to the sun is known as an *eclipse season.* Two solar eclipses, either total or partial, must occur each year, and five may take place. No lunar eclipse need occur in any year, although three are possible. The minimum number of eclipses in any year is two, both solar.

The sequence of eclipses may be determined by an ancient method known as the *Saros.* The revolution of the moon's nodes is westward at the rate of 19.5° per year. Thus, the sun meets the same node in 346.62 days, the length of an eclipse year. The Babylonians are generally credited for first recognizing the cycle of eclipses. The duration of this cycle is 6585 days (refined to 6585.32 days). This is equivalent to 223 synodic months,[2] and is nearly the same length as 19 eclipse years (6585.78 days). As pointed out by Tver, "After a Saros interval, the sun and moon have returned to nearly the same position relative to each other and to the node, and their distances from earth are nearly the same as before, allowing recurrence of a similar pattern of eclipses." Knowledge of the Saros, as it applied to cycles of lunar eclipses, goes back to about 1000 B.C. The Saros, depending upon the number of intervening leap years, is a period of 18 years, 10.32 days; or 18 years, 11.32 days. The Greeks are accredited with first recognizing the triple Saros cycle of 54 years. This is known as the "exeligmos."

At the end of a Saros cycle, eclipses (both lunar and solar) will recur in the same order and kind (total or partial). However, the essentially identical eclipse will not be observable from the same region on earth—because of the ⅓ day in excess of the 6585 days. During ⅓ day, the earth turns 120° and thus the eclipse will be seen about 120° longitude farther west than before. For example, the February 16, 1980 total eclipse of the sun that was observed in parts of Africa

[2] A synodic month is the interval between successive conjunctions of the moon and sun from new moon to new moon again. This interval is a little more than 29.5 days and varies by more than half a day during the year.

TABLE 1. RECORD OF TOTAL ECLIPSES OF THE SUN SINCE 1962 AND FORECAST THROUGH 1999

Year	Month and Day	Location of Shadow Path	Type of Eclipse	Duration (Minutes)
Saros period beginning in 1962				
1962	February 5	Papua New Guinea, Pacific	Total	4
	July 31	Guyana, central Africa	Annular	4
1963	January 25	South Atlantic	Annular	1
	July 20	Japan, Alaska, Canada	Total	1
1965	May 30	South Pacific	Total	5
	November 23	Northern India, southeast Asia, Borneo and the Pacific	Annular	4
1966	May 20	Northern Africa, the U.S.S.R., and China	Annular	1
	November 12	Pacific, Argentina, South Atlantic	Total	2
1967	November 2	South Atlantic	Total	—
1968	September 22	Arctic, the U.S.S.R.	Total	1
1969	March 18	Indian Ocean, Pacific	Annular	1
	September 11	Northern Pacific, Peru, Brazil	Annular	2
1970	March 7	Pacific, Mexico, Georgia, north Atlantic	Total	3
	August 31	South Pacific	Annular	7
1972	January 16	Antarctica	Annular	—
	July 10	Siberia, Alaska, northern Canada	Total	3
1973	January 4	South Pacific, South Atlantic	Annular	8
	June 30	Guyana, Atlantic, northern Africa	Total	7
	December 24	Pacific, Brazil, Atlantic, Sahara	Annular	12
1974	June 20	Indian Ocean, southwestern Australia	Total	5
1976	April 29	Algeria, Turkey, southern U.S.S.R., Tibet	Annular	7
	October 23	Tanzania, Indian Ocean, southern Australia	Total	5
1977	April 18	Atlantic, South Africa, Indian Ocean	Annular	7
	October 12	Northern Pacific, Venezuela	Total	3
1979	February 26	Northwestern United States, Canada, Greenland	Total	3
	August 22	South Pacific	Annular	7
End of Saros period beginning in 1962. Start of Saros period beginning in 1980				
1980	February 16	Congo, Zaire, Kenya, India, southeast Asia	Total	4
	August 10	Pacific, Brazil, Peru	Annular	3
1981	February 4	South Pacific	Annular	3
	July 31	U.S.S.R., northern Pacific	Total	2
1983	June 11	Indian Ocean, Indonesia, Papua New Guinea	Total	5
	December 4	Mid-Atlantic, Zaire, Somalia	Annular	4
1984	May 30	Mexico, southeastern United States, Algeria	Annular	1
	November 22	Papua New Guinea, South Pacific	Total	2
1985	November 12	South Pacific	Total	—
1986	October 3	North Atlantic, Greenland	Total	1
1987	March 29	Southern Argentina, South Atlantic, Zaire, Somalia	Annular	—
	September 23	U.S.S.R., China, Pacific	Total	4
1988	March 18	Indonesia, Philippines, North Pacific	Total	4
	September 11	Indian Ocean	Annular	7
1990	January 26	Indian Ocean	Annular	—
	July 12	Finland, Siberia, North Pacific	Total	3
1991	January 15	South Pacific	Annular	9
	July 11	Hawaii, Central America, Brazil	Total	7
1992	January 4	Mid-Pacific, California	Annular	12
	June 30	South Atlantic	Total	5
1994	May 10	United States, North Atlantic, Morocco	Annular	7
	November 3	Pacific, middle South America, South Atlantic	Total	4
1995	April 29	Pacific, Peru, Brazil	Annular	7
	October 24	Iran, India, southeast Asia, Pacific	Total	3
1997	March 9	Mongolia, Siberia, Arctic	Total	3
1998	February 26	Pacific, Colombia, North Atlantic	Total	4
1998	August 22	Indonesia, Borneo, Papua New Guinea, South Pacific	Annular	3
1999	February 16	Indian Ocean, northern Australia	Annular	1
	August 11	North Atlantic, central Europe, India	Total	2

TABLE 2. REPRESENTATIVE CIRCUMSTANCES OF TOTAL ECLIPSES OF THE SUN

	1977	1980
Observing Area	Partial phases visible from all 50 of the United States, western and southeastern Canada, Mexico, and Central America. Path of totality commenced in the Pacific Ocean, ran north and parallel to the Hawaiian Island chain, entered South America in the Darien region of Colombia, and ended in Venezuela.	Partial phases visible from all parts of Africa except extreme northern part; all of the Arabian Peninsula; all of India; much of southeast Asia and China. Path of totality passed from the eastern Atlantic to mouth of the Zaire (Congo) River into Tanznia; passed into the Indian Ocean at Malindi, Kenya; touched land again at Ankola, India, and entered Bay of Bengal south of Calcutta; and ended in China near north latitude 26.5° and east longitude 108.5°.
Date	October 12	February 16
Eclipse began	17:47 6	06:16 7
Central eclipse began	18:48 5	07:13 6
Central eclipse at local noon	20:14 4	09:00 7
Central eclipse ended	22:04 9	10:35 9
Eclipse ended	23:05 7	11:32 8
Maximum duration of totality	2 minutes, 37.4 seconds	4 minutes, 12.4 seconds

NOTE: Times given are Greenwich Mean Time. Source of information: U.S. Naval Observatory. For an interesting, informative article on this topic see "Total Eclipses of the Sun," by J. B. Zirker, *Science*, **210**, 1313–1319 (1980).

(including Kenya), in India, and in southeastern Asia, and of a duration of slightly over 4 minutes, was a repeat of the total eclipse of February 5, 1962, of similar duration and observed in Papua New Guinea. After about 3 Saros cycles (54 years), the eclipses are observable in very near the same longitudes from which they were seen 54 years prior. The number of eclipses in 1 Saros is about 70 (41 solar; 29 lunar). Of the solar eclipses, 10 are total, 14 are partial, and 17 are annular. Because of minor deviations not fully accounted for by the Saros cycle, it is estimated that there are longer cycles. For example, it is estimated that a lunar eclipse may repeat itself about 48 or 49 times, over a series lasting about 865 years. A solar eclipse may have from 68 to 75 returns over a cycle lasting some 1260 years.

As early as 1887, Theodor van Oppolzer prepared a comprehensive table of eclipses for the period through the year 2000. See Table 1. As an indication of the characteristics of a total eclipse of the sun, data pertaining to the October 12, 1977 and the February 16, 1980 eclipses are given in Table 2.

The progress of an eclipse of the sun is designated by a series of "contacts." The first contact comes when the edge of the penumbra *B* (Fig. 1) first touches the sun; second contact when the sun first passes into the umbra *A*; third contact when the umbra leaves the sun; and fourth contact when the last edge of penumbra leaves the sun. Accurate recording of the times of these contacts yeilds valuable information regarding the complex motions of the moon.

During the progress of a total solar eclipse, about 1 hour before the totality interval, the moon may be seen gradually commencing to cover the sun's disk. About ¼ hour before totality, a definite diminution of the light can be noticed and the landscape exhibits hues and tones of color that do not quite seem natural. Most birds and animals seem to sense that something unusual is taking place and become restless. As the crescent of light becomes narrower, images of a crescent shape, known as Baily's beads, may be noticed on the ground, particularly where the diminished sunlight is filtered through the leaves of trees and bushes. These light patches are believed to be caused by the penetration of the last rays of sunlight between the irregularities of the moon's surface. Although not always noticed, within 2 to 3 minutes prior to full disappearance of the sun's disk, shadow bands may appear to be moving over the ground. These bands also are present during the same interval after totality and there have been reports of their presence during totality. Although not fully explained, it is believed that these bands are the result of irregular refraction of light in the earth's atmosphere. Darkness falls suddenly and the white corona seems to flash into position. Almost concurrent with the appearance of the corona an intensely bright region, known as the "diamond-ring" effect may be seen.

The full *corona* of a total solar eclipse is shown in Fig. 4. The corona is the outermost part of the sun's atmosphere which becomes visible during a total eclipse of the sun. The shape of the corona changes periodically in sequence with the sunspot cycles and extends outward some 30 solar radii. Estimated temperature of the ionized

Fig. 4. Total solar eclipse of March 7, 1970, showing full corona. (*Sacramento Peak Laboratory.*)

gases in the corona range between 1 and 2 million degrees K. Elements such as iron, nickel, and calcium are contained in the gaseous corona. See entry on **Sun** for more detail. Adjacent to the silhouette of the moon during the interval of totality is the *chromosphere.* This is a layer of the sun's atmosphere that lies just above the *photosphere* or visible surface, and below the corona. The lower chromosphere is mainly neutral hydrogen gas at a temperature of about 7500 degrees K. The upper chromosphere contains ionized hydrogen at temperatures up to about 1 million degrees K. Promiences in the form of bright pink spots which extend outward from the chromosphere usually can be observed.

Considerably dependent upon local climatic conditions, during the period of totality there is usually a noticeable drop in air and ground temperatures. For example, during the February 16, 1980 total eclipse, it was unofficially reported from Ankola, India that the air temperature dropped from 95°F (35°C) to 68°F (20°C) from first contact to totality; while the ground temperature fell from 125°F (51.5°C) to 85°F (about 29°C).

Precautions in Viewing a Solar Eclipse. Kitt Peak National Observatory astronomers caution skywatchers who plan to observe an eclipse of the sun that *partial and permanent eye damage can be caused by looking at the sun, even for only an instant, without adequate protection.* They further observe, "Sunglasses, smoked glass, welder's goggles, photographic neutral-density filters or color film are *not* safe to use, even in combination. They are not dark enough for protection. A solar eclipse is most safely observed by not looking at the sun at all, but instead by watching its image projected on a piece of paper or board.

"To make your own projector, begin with two white sheets of poster board, or heavy paper. Punch a small, round hole in the center of one sheet. This will serve as your 'lens.' Aim the lens at the sun, holding the second board in your other hand. By moving the two boards, you will find that you can project a sharp image of the sun through the lens onto the second board. Information on how to photograph a solar eclipse can be obtained from the Public Information Office, Kitt Peak National Observatory, P. O. Box 26732, Tucson, Arizona 85726."

Total Solar Eclipse Research. Considerable attention of a scientific nature to total eclipses of the sun is generally regarded as having commenced with the eclipse of July 8, 1842. The path of that eclipse passed from Spain through France, italy, Austria, the U.S.S.R., and central Asia. Along the path were located the leading astronomers of Europe. For the first time, large red solar prominences were reliably described. The extent of the corona was measured. Improvements in the examination of the corona and prominences continued during the total eclipse of July 28, 1851. A successful daguerreotype photo was taken in Königsberg, using a telescope with a 2.4-inch (6-centimeter) aperture. Exposure was 1 minute, 24 seconds. The corona and several prominences were well pictured. Scientists were better prepared when the path of the eclipse of July 18, 1860 crossed through Spain, the Mediterranean, and northeastern Africa. Photographic and optical techniques had substantially improved during the 9-year interval between these eclipses and the astronomers also had prior knowledge and experience as a guide to the best exposure time to use.

Spectroscopic techniques were first used in connection with the total eclipse of August 18, 1868. The period of total of 5.5 minutes of this eclipse was relatively long. The bright prominences clearly produced several bright lines, this indicating the gaseous nature of the prominences. Identification was made of the hydrogen lines C and F and of a yellow line, later found to be due to helum and designated the D line. Immediately after the 1868 eclipse, it was erroneously identified as a sodium line. The *flash spectrum* is described under **Sun (The).**

The interest of Janssen was spurred by the 1868 eclipse and he soon found that the bright lines of the prominences could be observed on any clear day without waiting for another eclipse. He soon noted the marked and rapid changes that occur in the prominences. Lockyer came up with the same concept at about the same time, and the achievements of both Janssen and Lockyer were reported at the same meeting of the French Academy.

Observations made of the eclipse of August 7, 1869 showed a bright green line across the continuous spectrum of the corona. For awhile, a new element (*coronium*) was proposed as the source of the line, but spectroscopic analyses of the total eclipses of 1896 and 1898 disproved that.

American astronomers demonstrated serious interest in eclipse-related research in connection with the eclipse of December 22, 1870, when a large group of scientists journeyed to Europe to observe it. Among the findings of the Americans was Young's "discovery" of the reversing layer,[3] but this has since been disproved.

The possible relationship between sun spots and the corona was first investigated during the total eclipse of July 29, 1878, the path of which went diagonally across North America. Langley observed this eclipse from Pike's Peak at an elevation of 14,100 feet (4298 meters). Further evidence of a relationship was noted during the total eclipse of May 17, 1882.

Coupled with a growing sophistication of instrumentation during the later 1800s and up to the present time were discussions of an increasingly complex nature, based upon knowledge gained from and inspired by eclipse studies. Some of these factors are further discussed in entry on **Sun.**

Eclipses of the Moon. Referring back to Fig. 1, for an eclipse of the moon, consider *S* as the sun and *O* as the earth. From the relative dimensions and distances, we find that even under the most unfavorable conditions, i.e., with the earth at perihelion and the moon at apogee, the shadow of the umbra cone of the earth will extend well out beyond the distance of the moon. Hence, there are no annular eclipses of the moon. However, partial eclipses, i.e., eclipses that occur when the moon passes through the earth's shadow far enough off the central line to pass outside the umbra, are quite common. These are known as *penumbral* eclipses. The progress of a total lunar eclipse is shown in Fig. 5. Although not visible in Fig. 5, when the moon is in darkness,

Fig. 5. A lunar eclipse sequence. (*Griffith Observatory.*)

it is not completely dark during eclipse, but is illuminated by a dull-orange light refracted into the umbra by the atmosphere of the earth. For possibly 30 minutes before the moon reaches the earth's shadow, the darkening of the eastern side will proceed quite gradually. Although the edge of the earth's shadow appears sharp to the naked eye, it appears fuzzy with a field glass or telescope. An eclipse of the moon is visible over an entire hemisphere (sometimes a little more) of the earth and thus for persons in any given locality, a lunar eclipse is more common than a solar eclipse, since even though the latter

[3] The now outdated reversing layer concept was considered by Young to be a gaseous layer between the photosphere and the chromosphere of the sun and was regarded as responsible for the Fraunhofer absorption lines in the solar spectrum. The concept proved simplistic, inasmuch as the lines actually arise throughout the photospheric gas.

TABLE 3. LUNAR ECLIPSES THROUGH YEAR 2000

Year	Month and Day	Location Where Moon Is at Zenith	Duration (Minutes) Complete	Total Phase
1982	January 9	Pakistan	214	84
	July 6	Easter Island	224	102
	December 30	Hawaiian Islands	210	66
1983	June 25	Pitcairn Island	130	—
1985	May 4	Mauritius	212	70
	October 28	Bay of Bengal	204	42
1986	April 24	New Hebrides	210	68
	October 17	Arabian Sea	212	74
1987	October 7	Venezuela	22	—
1988	August 27	Samoa	122	—
1989	February 20	Philippines	212	76
	August 17	Central Brazil	220	98
1990	February 9	Southern India	204	46
	August 6	Northeastern Australia	174	—
1991	December 21	Hawaiian Islands	70	—
1992	June 15	Northern Chile	174	—
	December 9	Southern Algeria	212	74
1993	June 4	New Caledonia	220	98
	November 29	Mexico City	206	50
1994	May 25	Southern Brazil	116	—
1995	April 15	Fiji Islands	78	—
1996	April 4	Gulf of Guinea	216	84
	September 27	French Guiana	212	72
1997	March 24	Northwestern Brazil	194	—
	September 16	Maldives	210	66
1999	July 28	Samoa	142	—
2000	January 21	Puerto Rico	214	84
	July 16	Northeastern Australia	224	102

occur more frequently, they are only visible in much more limited areas. On occasions when the sunlight must pass through clouds, the moon's disk may become completely invisible during the period of totality.

Lunar eclipses up to the year 2000 are listed in Table 3.

For references, see entry on **Astronomy;** and **Moon (The).**

ECLIPSING BINARY. When the orbit plane of a binary star lies so nearly in the line of sight of the observer that the components undergo mutual eclipses, the object is known as an eclipsing binary. If the binary is also a spectroscopic binary, and if the parallax of the system is known, we have one of the most valuable specimens for stellar analysis. Eclipsing binaries are variable stars, not because the light of the individual components vary, but because of the eclipses. The most notable of the eclipsing binaries is the star Algol (β Persei).

The light curve of an eclipsing binary is characterized by periods of practically constant light with periodic drops in intensity. In the figure, we have a characteristic light curve of such an object. Here, the eclipse of the larger and brighter primary star by the secondary star is annular, and the eclipse of the secondary by the primary is total.

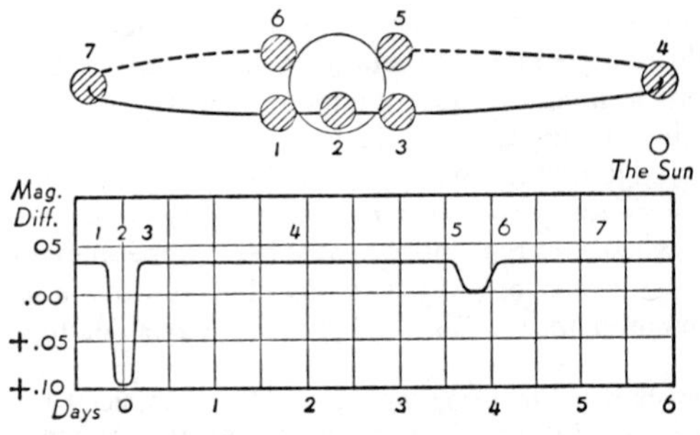

Apparent relative orbit and light curve of the eclipsing binary 1H. Cassiopeiae. (*Curve and orbit determined by Joel Stebbins from his observations with the photoelectric photometer at the University of Illinois.*)

The orbit of an eclipsing binary may be determined from a study of the light curve. In addition to the seven elements of the orbit, it is also possible to determine the relative sizes of the individual stars in terms of the radius of the orbit. In the determination of the orbit of a spectroscopic binary, it is impossible to determine the semimajor axis, a, and the inclination of the orbit plane, i, independently; but a quantity ($a \sin i$) expressed directly in linear units (i.e., miles or kilometers) may be determined. If a star is both an eclipsing and spectroscopic binary, one can determine all seven elements of the orbit, including a and i, in angular units from the light curve, and the quantity ($a \sin i$) in linear units from the spectroscopic data. Hence, one can determine the radius of the orbit in linear units and then get the sizes of the individual stars in linear units. Since the relative masses of the two stars can be obtained from the period, and since the relative sizes are determinable from the combination of the photometric and spectroscopic orbits, then the densities of the individual stars can be found from relative masses and sizes.

See also **Binary Stars; Spectroscopic Binaries;** and **Visual Binaries.**

ECLIPTIC. The great circle cut out on the celestial sphere by the plane containing the orbit of the earth. The ecliptic is the fundamental plane for the system of spherical coordinates in which celestial latitude and longitude are measured. The ecliptic is also the reference plane to which the planes of the orbits of all the members of the solar system are referred.

The plane of the ecliptic is inclined to the plane of the equator by an angle of approximately 23°27′, known as the obliquity of the ecliptic. The two planes intersect in a line known as the line of nodes. The points where this line of nodes intersects the celestial sphere are known as the equinoxes. The apparent motion of the sun in the ecliptic about the earth, due to the actual motion of the earth in its orbit, causes the sun to pass through each one of the equinoxes once each year. The point where the sun crosses the equator from south to north is known as the vernal equinox, and the opposite extremity of the line of nodes is the autumnal equinox. Due to precession, the direction of the line of nodes is continually changing relative to the stars. At present, the vernal equinox is in the constellation of Pisces. It is continually moving along the ecliptic, in a direction contrary to the annual motion of the sun, at such a rate that it will complete one revoltuin of the ecliptic in approximately 26,000 years. See also **Celestial Sphere.**

ECLOGITE. This is a coarse, granular rock composed chiefly of garnet and pyroxene with subordinate amounts of various minerals such as rutile, magnetite, and apatite. Hornblende sometimes is present, replacing the pyroxene, often to the extent that a garnet amphibolite is produced. The origin of eclogites is obscure, they may result in part from the deep-seated metamorphism of gabbronic rocks, but some may have resulted from crystallization of a primary basic magma under conditions of great pressure. They may represent segregations in a highly basic magma analogous to segregations of basic minerals in granites and other common igneous rocks. Seemingly confirmatory evidence of this idea is found in the chunks of ecologite-like material found in kimberlite in the Republic of South Africa.

ECOLOGY. A branch of biology that deals with the relations of organisms and their environment, including their relations with other organisms. It is an interdisciplinarian field, cutting across the life and geophysical sciences. Ecological investigations look into two directions: (1) The nature of environments and the demands which these environments make upon the organisms that inhabit them; and (2) the characteristics of organisms (plant or animal), species, and groups that permit or promote their tolerance of specific environmental conditions. In recent years, particular emphasis has been placed on the studies of groups rather than single species and this has given rise to the term *ecosystem.* Odum defines an ecosystem as "any entity or natural unit that includes living and nonliving parts interacting to produce a stable system in which the exchange of materials between the living and nonliving parts follows circular paths."

Environmental or Habitat Approach. Characteristics of an environment fall into three major categories: (1) *physical*; (2) *chemical*; and (3) *biotic.* In connection with any of these factors, if the presence or absence of a given condition is necessary for sustenance of whatever group, species, etc. that is being considered, such condition is referred to as a *limiting factor.* Physical factors include light, temperature, wind, fire, soil texture, etc. Chemical factors include the composition of the water (pH, dissolved gases and solids, etc.); the composition of the air (pollutants, water vapor, etc.); the composition of the soil (alkalinity, acidity, presence of various elements and compounds); etc. Biotic factors are related to food supply and the presence and behavior of neighboring organisms (predators, parasites, etc.).

Environments can be classified into four major types: (1) *freshwater*; (2) *marine*; (3) *terrestrial*; and (4) *symbiotic.* There are several subdivisions of each.

Freshwater environments are of two principal types: (1) *standing-water* and (2) *running-water habitats.* In the first cateory, there are ponds, lakes, and bogs; in the latter category, streams, rivers, and springs. Limiting factors of significance include temperature, water clarity, concentrations of oxygen and various salts, and evaporation rate in the case of standing waters.

Marine environments, although similar in many respects to freshwater environments, pose special environmental conditions, including depth, salinity, general presence of greater stability, and less, if any, light at great depths. Subdivisions of marine environments include: (1) the *neritic zone*, the relatively shallow waters of the continental shelf; (2) *oceanic region*, the deep waters beyond the continental shelves, which, in turn, is further subdivided into: (a) the *euphotic* zone, the upper portion of the oceanic region where photosynthesis occurs; (b) *bathyal zones*, depths below the euphotic zone but still located on continental slopes; and (c) *abyssal zones*, depths below the euphotic zone in all other parts of the oceans.

Terrestrial environments are divided into nine or more types, called *biomes.* Examples include deserts, tundras, grasslands, savannas, deciduous forests, coniferous forests, tropical forests, and woodlands. These are described under **Biome.** The ocean is also sometimes referred to as a biome.

The environments described thus far have related essentially to nonorganic factors. In the case of the symbiotic environment, the factors of concern are other organisms. There are numerous examples where two or more species may inhabit a given area of close proximity in the absence of predatoriness. Usually, each species present will contribute in some fashion to the well being of others, although what might be called "stand-off" conditions also are found. See also **Symbiosis.**

Communities Approach. Where the interrelationships of a single individual or species with it environment is studied, the term *autecology* may be applied. In contrast, if entire populations are studied as units, the term *synecology* is used. The latter approach is most popular today. The term population is used to describe a group of individuals composed of members of a single species or of several closely associated species that occupy a definite environmental area. Where all of the populations that occupy a given geographic area are studied, the term *biotic community* is used.

The portion of the earth on or in which life exists is termed the *biosphere.* This is a shallow surface region, including the oceans and part of the atmosphere. An ecosystem, as previously defined, can be small and simple or large and complex and, in fact, it is not inaccurate to refer to the entire biosphere as an ecosystem. If life were found in other areas of the cosmos, for example, then it is likely that the first ecosystems to be compared would be those of a scale of the biosphere.

Habitat may be defined as that particular environment in which a population lives. A catfish that likes slow-moving streams and lakes is described as having this particular habitat. Obviously, the variation among habitats considering the tens of thousands of life forms is tremendous, although generalizations can be made. *Niche* is a term used to describe that role played by a population within its community and ecosystem. Numerous factors determine niche—eating habits, predatoriness, etc.

The first branch of ecology to develop beyond the stage of life-history study was the description of vegetation. Its basic method is to study the detailed distribution of vascular plants in terms of communities of various types, the pattern and complexity of which depend largely upon the climate and soil.

The major theoretical contribution of this school is the idea of *seral succession* toward a stable climax. According to this idea, if a new environment is created for terrestrial plants, or an old one drastically changed, the vegetation that first develops on it does not remain unchanged for eternity, but rather alters the environment so that it becomes more suitable for some new and different kind of vegetation and so on. Ultimately, a *climax* vegetation develops which is stable under the prevailing climatic conditions and remains until the climate changes to favor something else, or until some new catastrophe changes the environment drastically once again.

Further study of seral succession showed that vegetation patterns were seldom so simple. The climax community is regarded as a useful concept. Even under constant climatic conditions, some sorts of vegetation are not stable, but undergo local cylical changes. At the other extreme, a few kinds of vegetation replace themselves after disturbance without intervening seral stages. It is clear that the sort of climatic constancy that was implied in the climax idea has not prevailed at least since the beginning of the Pleistocene, and that the climax is best considered an ever-changing end-point toward which vegetation develops rather than a state it actually attains.

Populations. Probably the greatest body of coherent ecological theory has been created by students of populations. The study of field populations has uncovered many interesting phenomena, such as the periodic oscillation of arctic small vertebrate populations and the seasonal changes in abundance of planktonic organisms. These observations have stimulated both laboratory experiments and the development of deductive mathematical theory.

If a small number of organisms is provided with a new unexploited environment, the ensuing population growth curve exhibits a roughly sigmoid form, with an initial phase of very rapid, almost logarithmic increase and a later phase in which the rate of growth gradually declines to zero. Such a population history can be described by a curve of the form:

$$dN/dt = rN\frac{(K - N)}{K} \tag{1}$$

where dN/dt is the instantaneous rate of growth; r the intrinsic rate of natural increase in the absence of crowding; N the population size at any time; and K the maximum population size. A formula of this kind is simple to use and easy to comprehend, and although few populations justify the assumptions on which it is based, it has been used not only for curve-fitting and the description of population growth, but as a point of departure for population theory. The principal developments of importance from it have been the prey-predator equations of Volterra:

$$dN/dt = r_1N_1 - \alpha_1N_1N_2 \tag{2}$$

$$dN_2/dt = \alpha_2N_1N_2 - d_2N_2 \tag{3}$$

and the Gause equations of species interaction:

$$dN_1/dt = r_1N_1\frac{(K_1 - N_1 - \beta N_2)}{K_1} \tag{4}$$

$$dN_2/dt = r_2N_2\frac{(K_2 - N_2 - \beta N_1)}{K_2} \tag{5}$$

In these equations, N_2 is the size of one population and N_1 that of another; α_1 expresses the effect of predatoriness on the prey population, and α_2 expresses its effect on the predator. In the absence of prey, the predator should die at the rate of d_2; K_1 and K_2 are the saturation values of two species grown alone; r_1 and r_2 are the intrinsic rates of natural increase of species N_1 and N_2; and α and β express the inhibitory effects of these species on each other. If $\alpha > K_1/K_2$ and $\beta > K_2/K_1$, then either N_1 or N_2 may win out in competition, the result depending upon the initial concentration of the two species. If $\alpha < K_1/K_2$, and $\beta < K_2/K_1$, the species will coexist. If $\alpha < K_1/K_2$ and $\beta > K_2/K_1$, N_1 will be the sole survivor of competition; and if $\alpha > K_1/K_2$ and $\beta < K_2/K_1$, only N_2 will survive.

Work with models of this sort and with experimental laboratory populations that behave more or less in the ways that the models predict has led to concepts about the *ecological niche*, or the way in which an organism fits into the ecological system of which it forms a part. Some authorities hold that Gause's axiom, which states that two species cannot live indefinitely the same way in the same place under constant conditions, is essentially trivial, while others regard it as one of the great generalizations of ecology. Probably its principal value lies in raising the question of how the niches of apparently similar organisms differ, and so compelling ecologists to examine their material very closely. In this, Gause's axiom is similar to the idea of seral succession, which directs attention to the environmental requirements as well as to the ways in which the environment is changed by them. Both ideas stimulate the acquisition of useful information, although they imply an environmental constancy that is seldom found in nature.

Niche theory has led to renewed interest in the taxonomic diversity of natural communities. MacArthur, assuming that there must be some way in which the niches of organisms in an ecological system did not overlap, developed a model of the relative abundance of the individual species in a natural community. According to this model, the expected abundance of the rth rarest species, where there are n species and m individuals and i is the species rank, is given by

$$\frac{m}{n}\sum_{i=1}^{r}\frac{1}{(n-i+1)} \tag{6}$$

The predictions of the model have been borne out in a number of taxonomically homogeneous cases, but do not seem to hold generally for natural communities.

A different line of approach has been developed by ecologists interested in practical problems of population management, particularly the exploitation of fish populations. The method has been to start with a formula that is essentially a simple equation of continuity:

$$S_2 = S_1 + (A + G) - (M + C) \tag{7}$$

where S_1 and S_2 are the total weight of the population at the beginning and end of the time under consideration; A is recruitment of new individuals to the population; G is growth of the population; M is natural mortality; and C is capture by fishing.

Such an equation is modified as data become available for a specific case and is gradually refined until it is a very powerful predictive tool. The equations become cumbersome in the process, but the mathematics involved is essentially simple, and electronic computation avoids tedium and human errors.

Unfortunately, such models are limited to the populations for which they have been formulated, or for others similar to them in essential respects. There seems little immediate likelihood of the development of simple and accurate general models for the dynamics of natural populations. Many ecologists working with organisms other than fish despair of producing useful models without recourse to stochastic theory. The stochastic approach is appealing, for the processes of population change are certainly not deterministic ones, but such a shift involves a great increase in mathematical complexity. The sophistication necessary to handle stochastic models is rare among ecologists, and it has usually been applied to new methods of constructing a deterministic theory. Cole, for example, used finite difference equations to demonstrate that the reproductive potential of a species depends very greatly upon the timing of reproduction in its life cycle and to point out the critical importance of pieces of information, such as the age at first reproduction, which might otherwise be overlooked in life-history studies. See **Population (Statistics).**

Chemical Aspects. In any community, the interactions of the organisms with each other and with their environment are so complex as to defy complete understanding. One way of asking answerable questions is to restrict attention to the chemical aspects of ecological processes, and to regard organisms as the causes or effects of geochemical processes. This approach has led to an understanding of the quantitative role of organisms in the major chemical cycles, such as the carbon cycle and the nitrogen cycle, and is being strengthened by the use of isotopic methods. The introduction of isotopic tracers into biogeochemical cycles permits the estimation of reservoir sizes and exchange rates and seems likely to lead to new theoretical developments.

Energy Flow. Energy transfer can be considered instead of chemical reactions. See accompanying figure. The principal contribution of the energy flow concept has been to demonstrate that organisms, like most machines, are not efficient converters of energy, so that an organism will incorporate into its body only a small part of the stored chemical energy of its food. This places a severe practical limit on the number of steps that can be maintained in the food chain and has important practical consequences for an expanding human population trying to increase its food supply.

The measurement of energy flow in natural communities presents many difficulties, only some of which are technical in nature. Although the subject is suitable for theoretical development, this has been limited by the lack of methods for dealing with the thermodynamics of open systems. With a better understanding of the thermodynamics of open systems, as being attained by physical chemists, the study of energy flow seems likely to be productive in the future.

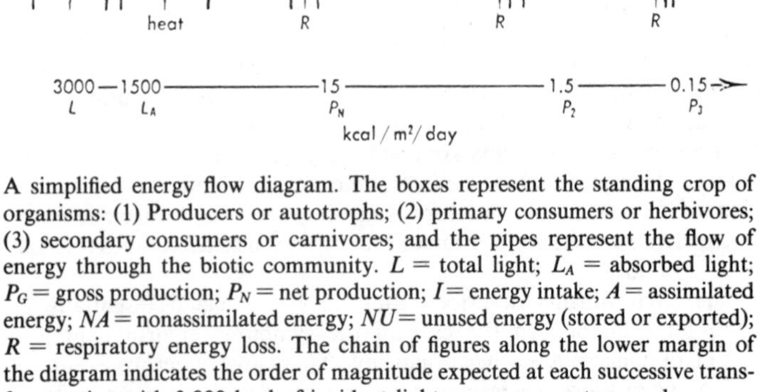

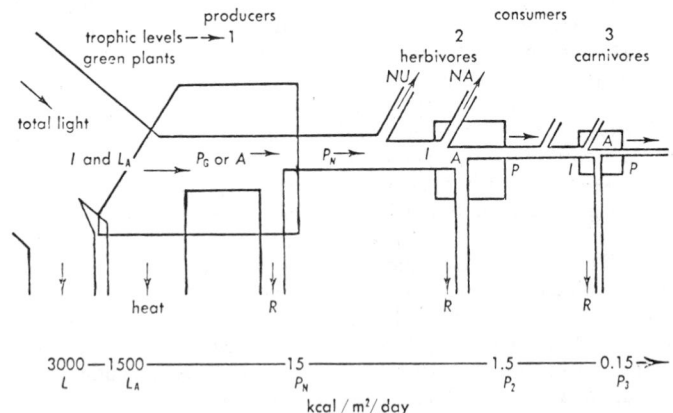

A simplified energy flow diagram. The boxes represent the standing crop of organisms: (1) Producers or autotrophs; (2) primary consumers or herbivores; (3) secondary consumers or carnivores; and the pipes represent the flow of energy through the biotic community. L = total light; L_A = absorbed light; P_G = gross production; P_N = net production; I = energy intake; A = assimilated energy; NA = nonassimilated energy; NU = unused energy (stored or exported); R = respiratory energy loss. The chain of figures along the lower margin of the diagram indicates the order of magnitude expected at each successive transfer, starting with 3,000 kcal of incident light per square meter per day.

Behavior. The experimental analysis of animal behavior patterns has led to a renewed interest among ecologists in the social and psychological aspects of their subject. In particular, studies have focused on territoriality, social stress, hierarchies, and other behavioral mechanisms that control population density, on the transmission of traditional information about nesting and feeding sites, and on feeding behavior as it affects an animal's role in the community. Although some behavioral studies have relevance to species diversity, community stability, and population growth rates, much of this work can be carried on profitably outside an ecological context, and behaviorists seem closer to establishing an independent discipline than most other ecologists.

References

For the interface between ecology and climate and weather, see list of entires under the entry on **Meteorology.**

Brewer, R.: "Principles of Ecology," Saunders, Philadelphia, 1978.

Colinvaux, P. A.: "Introduction to Ecology," Wiley, New York, 1972.

Dickinson, R. E.: "Regional Ecology; The Study of Man's Environment," Wiley, New York, 1970.

Erlich, P. R., Ehrlich, A. H., and J. P. Holdren: "Ecoscience: Population, Resources, Environment," Freeman, San Francisco, 1978.

Grzimek, H. C. B. (editor): "Grzimek's Encyclopedia of Ecology," Van Nostrand Reinhold, New York, 1976.

Hutchinson, G. E.: "An Introduction to Population Ecology," Yale Univ. Press, New Haven, Connecticut, 1979.

Johnson, C. E.: "The Natural World," McGraw-Hill, New York, 1972.

Ketchum, R. M.: "The Secret Life of the Forest," McGraw-Hill, New York, 1970.

MacArthur, R. H.: "Geographical Ecology," Harper and Row, New York, 1972.

May, R. M.: "Stability and Complexity in Model Ecosystems," Princeton Univ. Press, Princeton, New Jersey, 1973.

Meggers, B. J., Ayensu, E. S., and W. D. Duckworth (editors): "Tropical Forest Ecosystems in Africa and South America," Smithsonian Institution Press, Washington, D.C., 1973.

Petrov, M. P.: "Deserts of the World," Wiley, New York, 1977.

Rappaport, D. J., and J. E. Turner: "Economic Models in Ecology," *Science*, **195**, 367–373 (1977).

Staff: "The Global Report to the President," Superintendent of Documents, U.S. Govt. Printing Office, Washington, D.C., 1980.

Treshow, M.: Environment and Plant Response," McGraw-Hill, New York, 1970.

ECOLOGY (Plant Succession). Succession (Plant).

ECOLOGY (Redwood Forest). Redwood (Coast).

ECOLOGY (Sardine). Herring.

ECOLOGY (Society). Society (Ecology).

ECOLOGY (Symbiosis). Symbiosis.

ECONOMIZER. Boiler.

ECOSPHERE. Atmosphere (Earth).

ECOSYSTEM. Ecology.

ECOTONE. The often rather blurred and indefinite boundary between two ecological communities. There is usually some tension between two or more communities present at these boundaries. In some instances, the ecotone may represent a fairly wide strip, sometimes referred to as the transition zone. In other instances, the ecotone may be rather sharply marked, as by a stream, edge of a grassland or forest, water hole, etc. See also **Ecology.**

ECTODERM. Embryo.

ECTOPIC PREGNANCY. Pregnancy.

ECTOPROCTA. Members of this group make up the phylum *Bryozoa.* The included species live in colonies of many forms and are characterized by the retractile tentacles and by the anus lying outside of the circlet of tentacles (lophophore).

The phylum is divided into two orders:

Order *Gymnolaemata.* Lophophore circular. Mouth usually closed by a flap called the operculum. Marine.

Order *Phylactolaemata.* Lophophore horseshoe-shaped or oval. Freshwater species.

ECZEMOID DERMATITIS. Dermatitis and Dermatosis.

EDDINGTON LIMIT. X-Ray Astronomy.

EDDY. 1. By analogy with a molecule, a "glob" of fluid within the fluid mass that has a certain integrity and life history of its own, the activities of the bulk fluid being the net result of the motion of the eddies. The concept is applied, with varying results, to phenomena ranging from the momentary spasms of the wind to storms and anticyclones.

2. Any circulation drawing its energy from a flow of much larger scale, and brought about by pressure irregularities, as in the lee of a solid obstacle.

EDDY CURRENT. A term generally applied to currents set up in a substance by variation of an applied magnetic field. Eddy currents result in power loss and reduction of magnetic flux. Transformer cores and dynamo frames are laminated to break up the iron structure into thin, insulated layers to reduce the effects of eddy currents. Eddy currents also are used to advantage in induction heating and various damping devices.

Use of separately insulated strands of conductors or bundles of wires in some electrical apparatus accomplishes a reduction in eddy currents just as in the case of using the aforementioned thin, insulated sheets. Eddy loss in conductors of circular section, such as wire, is given by

$$P_e = \frac{(\pi r f B_{\max})^2}{4p \times 10^{16}} \text{ (watts/cubic centimeter)}$$

The formula for the loss in sheets is

$$P_e = \frac{(\pi t f B_{\max})^2}{6p \times 10^{16}} \text{ (watts/cubic centimeter)}$$

where r = radius of wire, cm
f = frequency, Hz
$B_{\max}$ = maximum flux density, lines/cm^2
ρ = specific resistance, ohm-cm
t = thickness, cm

See also **Core Loss.**

EDDY CURRENT TESTING. Nondestructive Testing.

EDDY-CURRENT TRANSDUCER. When a nonmagnetic electrically-conductive object is placed in the magnetic field of a coil, the effective inductance of the coil is decreased and its resistance is increased. This results because the field sets up eddy currents in the object that circulate so as to oppose the field that creates them. Moving the object closer to the coil increases the effect. Thus, coil characteristics are a function of spacing between coil and object. This effect is utilized in transducers to sense the position of a specific object or "target."

Various eddy-current transducer systems are available: (1) displacement systems, in which a noncontacting measurement is made of the position of a target, such as a panel being observed for flutter, or a shaft being studied for runout, (2) pressure systems, in which the target is a diaphragm that is displaced by the pressure being measured, (3) accelerometer systems, in which the target is a seismic mass displaced by inertial forces, and (4) differential transformer systems, in which the target is a thin metal sleeve mounted on a ceramic core support that is attached to the mechanism being monitored.

All such systems include electronic circuitry to generate an electrical analog signal related to target position. In some systems, the sensing coil is made part of an impedance bridge circuit so that changes in coil impedance produce an error signal related to target position. In other systems, the coil is part of an oscillator circuit so that the target movements produce frequency changes. Linearity of the output signal may be 1% or less, dependent upon system details. The output voltage may be up to 5 V. The coil excitation frequently may be 1 MHz or higher so that the frequency response of the system can be from 0 to at least 100 kHz. At 1 MHz, the eddy currents are confined near the surface of the target. For example, the "skin depth" in aluminum is 0.0035 inch. Therefore, only a thin target is required and nonconductors can be made into suitable targets by applying, as examples, aluminum foil tape or copper plating.

Operation of eddy-current transducers does not depend upon magnetic properties. Therefore, the devices can avoid problems caused by changes in permeability resulting from temperature changes or stray magnetic fields. Eddy-current transducers are suited to extreme environments and can be fabricated entirely of inorganic and nonmagnetic materials that are highly resistant to the effects of temperature, magnetic fields, and nuclear radiation, including a low cross section of x-rays. The operating principle of the eddy-current transducer is inherently different from the usual magnetic principle, in which coil inductance is increased by an approaching target. These transducers will operate with targets of magnetic materials with reduced, but often with adequate sensitivity.

EDDY VISCOSITY. The turbulent transfer of momentum by eddies, giving rise to an internal fluid friction, in a manner analogous to the action of molecular viscosity in laminar flow (see **Fluid Flow**), but taking place on a much larger scale. If the eddy viscosity is the same at all positions, the equations of motion take the laminar forms but the magnitude of the eddy viscosity must be inferred from other considerations. Values in the atmosphere may be as large as 10^5 cm^2/sec in kinematic units.

EDEMA. Accumulation of excess fluids in the tissues of the body. The condition may rise from several causes. Physiologically the balance of the body fluids between the cells (intracellular fluid), the fluid bathing the cells (extracellular fluid), and the blood plasma is upset. fluid is drawn from the blood into the tissues when there is a higher osmotic pressure in the tissues than in the blood. This higher pressure may be due to an actual increase (e.g., in salt retention due to impaired kidney function) or it may be a relative increase, as in edema associated with low serum proteins in the blood due to nutritional deficiency. Obstruction to venous flow also results in edema, by the mechanical factor of increased pressure in the capillaries. Capillary damage due to infection, bacterial toxins, or to trauma will allow the passage of fluid from the blood into the tissue spaces and produce edema. Exudation of fluid into the extracellular spaces is part of the general process of inflammation and is found at the site of any localized inflammatory reaction or infection.

The common conditions characterized by edema are congestive heart failure, nephritis, varicose veins, cirrhosis, and allergic phenomena such as angioneurotic edema. In congestive failure, the fluid tends to collect in dependent portions, the feet, legs, and over the sacrum. In nephritis these areas plus the loose tissue around the eyes and other easily distensible tissues become edematous. In filariasis, obstruction of lymphatic channels by the parasites results in lymphatic edema, leading to elephantiasis.

See also **Kidney and Urinary Tract.**

EDENTATA (*Mammalia*). This comparatively small order of mammals includes mammals without teeth in the front part of the jaws and with no enamel on the teeth. The feet bear claws. A literal translation of *Edentata* is "without teeth," and thus the term does not seem very appropriate for an order where the majority of animals contained in it do have teeth and, in fact, one of the animals, the giant armadillo, has more teeth than all other mammals with exception of certain species of whales. An approximate classification of *Edentata* includes:

- Anteaters (*Myrmecophagidae*)
  - The Giant Anteater (*Myrmecophaga*)
  - Lesser Anteater (*Tamandua*)
  - The Pigmy Anteater (*Cyclopes*)
    - *Note*: The spiny anteaters are of a different order (*Monotremata*); the scaly anteaters are of the order *Pholidota.*
- Sloths (*Bradypodidae*)
  - The Two-fingered Sloth (*Choloepus*)
  - The Three-fingered Sloth (*Bradypus*)
- Armadillos (*Dasypodidae*)
  - Peludos (*Chaetophractus, . . .*)
  - Giant Armadillo (*Priodontes*)
  - The Cabassou (*Cabassous*)
  - Pebas (*Tolypeutes*)

All of the foregoing creatures are considered to be primitive and of very early origin. Although some huge animals of this type once lived, as, for example, ground sloths, they appear to have become extinct during the 16th century. Some of these animals were as large as elephants. At that time, the ground sloths were found in Patagonia and the environs of the Andes. It is also believed that these animals existed in the southwest of North America and some of the islands of the West Indies.

Specializations of the anteaters include a slender elongate snout, a long sticky tongue which aids in gathering a sufficient number of the small prey, and strong claws for tearing open ant nests. Some of the giant anteaters may attain a length of 6 feet (1.8 meters) from head to tail. One pound of ants may be consumed at a single meal. They are gray-black in color, with long hair and a long bushy tail. One young is produced at birth which rides on the mother's back when very young. See Fig. 1.

The lesser anteater differs considerably from the giant. The ears are large, the muzzle is short, the tail is opossum-like and, although appearing naked, it is covered with dense, hard fur of various colorations. The tail is used in a prehensile fashion. The animal prefers living in trees where it feeds on ants and termites. They have a reputation of being very strong for their size and quite agile when on the ground. They are considered to be excellent defenders of themselves, usually fighting from a sitting position with their strong arms and sharp claws slashing any predators.

The pigmy anteater is short, about 1 foot (0.3 meter) in length, and inhabits the forests of Trinidad and South America. It is of reddish-brown coloration, has small ears, a short snout, and is seldom seen.

Sloths are animals of Central and South America, highly specialized for arboreal life. Their claws are large hooks by which they suspend themselves from branches, and they are so thoroughly adapted to this inverted position that the hair runs from belly to back, opposite to its direction in animals who maintain the usual position. Sloths eat the foliage of cecropia tress by preference.

The sloths are of two forms: The two-fingered (or sometimes called two-toed) and the three-fingered. Actually, there is a need to differentiate between these two types, but the names are extremely misleading because both types have five toes! The so-called two-fingered sloth is about the size of a large domestic cat, covered with very shaggy and coarse fur along the back. The fingers are ideally shaped, hardly requiring flexing for clinging to vines and tree boughs. The animals prefer leaves and fruits. They are known for their durability and ruggedness and can recover from extremely bad wounds and injuries. When provoked, they also can inflict very bad bites and tears. Gener-

Fig. 1. Giant anteater; female with young. (*New York Zoological Society.*)

Fig. 2. Armadillo. (*A. M. Winchester.*)

ally, however, they are considered rather lazy and mindful of their own business.

The three-fingered sloth or *ai* is a smaller animal, with a dense and hard hair coat. They are of a silver-gray coloration and have a bright yellow or cream-colored face. They have an unusual bright yellow and black sunburst type of marking between their shoulder blades. Although the marking may be coincidental, it is interesting to note that the marking is quite similar to the flower clusters of the wild pawpaw tree on which the animals feed. The young, one at a birth, are tiny. Some experimenters believe that the ais can sense color because they refuse to eat artificially-colored cecropia flowers, or any food that is colored by artificial lighting.

Armadillos are burrowing animals with many bony plates in the skin which form a more or less complete armor when the animal rolls up. Several species occur from Argentina northward through South America, and one, the nine-branded armadillo, is found in the southwestern United States. They range in size from the 5-inch (13 centimeters) pichiciago to the 3-foot (0.9 meter) giant armadillo. As with the other animals previously described in the order of *Edentata*, the mother armadillo carries the baby on her back and in case of trouble rolls around it, assuming the shape of a ball wherein the baby is completely hidden. Some of the particular varieties of armadillo include: the peludo (the hairy armadillo of Argentina, *Tatu pilosa*); the tatoquay (the broad-banded armadillo of South America, *Cabassous unicinctus*); the pichi (the pigmy armadillo, *Chalamyphorus*, of Argentina); the peba (the nine-banded armadillo, *Tatusia novemcinctus*, found from southern Texas and New Mexico to Argentina); and the apar (the three-banded armadillo, *Tolypeutes*, of South America). See Fig. 2.

For references, see **Mammalia.**

EDGE. Two distinct points (endpoints) and a line segment joining them. Considering the edge as point-closed (including endpoints) is the conventional procedure although open line segments are also used by some authors. Similarly the insistence upon the distinctness of the endpoints of an edge is motivated by the desire to exclude closed "loops" in which the endpoints coincide.

See **Graph (Mathematics); Tree (Mathematics); Vertex;** and terms listed under **Mathematics.**

EDGE DISLOCATION (Crystal). Crystal.

EDGE EFFECT. In a capacitor comprising two parallel plates, the electric field is normal to the plates except near the edges, where the field lines bulge outward. This edge effect introduces a correction to the capacitance as computed from the parallel field idealization. By giving one plate greater area than the other, and surrounding the smaller plate with an auxiliary guard ring maintained at the potential of the smaller plate, the edge effect is eliminated, and the capacitance between the small plate and the large plate is given by the simple theory.

EDGE SEQUENCE. A subgraph whose edges admit an ordering possessing the following property: Each edge has one vertex in common with the preceding edge and the other vertex in common with the succeeding edge. The words "preceding" and "succeeding" are defined with respect to the ordering imposed on the edges.

EDGE TONES. The tones produced by the splitting of an air-jet by a sharp edge maintained in the jet.

EDGE TRAIN. An edge sequence in which each edge has multiplicity one. A *closed edge train* is one whose terminal vertices coincide. An *open edge train* is one whose terminal vertices do not coincide.

EDGEWORTH SERIES. A form of expansion of a frequency function in terms of derivatives of the normal distribution. For a distribution in standard form (i.e., with zero mean and unit variance), the expansion is

$$f(x) = \left[\exp\left\{\sum_{j=3}^{\infty} k_j \frac{(-D)^j}{j!}\right\}\right]\alpha(x),$$

where k_j is the jth cumulant; D is the operator d/dx and $\alpha(x)$ is the normal function $e^{-1/2x^2}/\sqrt{(2\pi)}$.

See also **Gram-Charlier Series.**

EDP CENTER. An electronic data processing center is a complex comprised of one or more computers, peripheral equipment, and personnel skilled in the operation of computers and associated data processing equipment. Sometimes the distinction is made between an "open shop" and a "closed shop." In an open shop, facilities and assistance are provided to "non-professional" clients in the implementation of their own solutions to their own problems. A closed shop, on the other hand, is limited to use by professional programmers and operators. Clients of closed shops explain their requirements to analysts or programmers on the staff of the data processing center. The user is limited to guiding the problem-solving effort. Operation of the closed shop is essentially similar to the data processing center captively managed by a large industrial firm. The essential elements in the management of an EDP center break down about as follows:

- Computer Center Management
  - Computer operators
  - Data entry equipment
  - General administration and clerical personnel
- Systems and Programming Management
  - Systems planning analysts
  - Application development analysts
  - Application development programmers
  - Maintenance programmers

See list of related terms described in this volume under **Data Processing.**

EDTA (Ethylenediaminetetraacetate). Chelates and Chelation.

EDT CRYSTAL. Piezoelectric Effect.

EEL GRASS (*Zosterna marina*; Najadaceae). Eel grass is a common plant of sandy or muddy ocean shores. The plant has a creeping somewhat fleshy stem which roots freely at the nodes and has short erect branches. Usually it grows in salt water from a foot to over 4 feet (1.2 meters) deep, and is frequently found in tidal pools.

Quantities of eel grass are raked up and dried. It is used for packing, for stuffing for various objects, in the manufacture of certain kinds of wall board and for insulation in walls.

The plants suffer periodically from a certain disease which seriously depletes their numbers. In some regions the natural growth of eel grass has been practically wiped out by this disease.

Freshwater eel grass, *Vallisneria spiralis*, is an entirely different plant, with an interesting method of pollination. It is frequently used in aquaria.

EELS (*Osteichthyes*). Members of the order *Apodes*, eels are elongate slender fishes without pelvic fins and in some species lacking pectoral or median fins.

Eels fall into several classifications: (1) freshwater eels (family *Anguillidae*); (2) parasitic snubnosed eel (family *Simenchelidae*; (3) moray eels (family *Muraenidae*); (4) snake eels (family *Ophichthidae*); (5) snipe eels (family *Nemichthyidae*); (6) deep-sea eels (family *Synaphobranchidae* and family *Serrivomeridae*); (7) conger eels (family *Congridae*); and (8) worm eels (family *Moringuidae*). Also, of the order *Heteromi*, there are deep-sea spiny eels. The latter are not true eels. The so-called electric eel is not an eel. See **Gymnotid Eels (Osteichthyes).**

European Eels. The anguillid eels are found in a number of freshwater areas, such as lakes, pond, rivers, and streams. See accompanying illustration. To spawn, they return to the sea. In the European eel, well known as an edible fish, females do not spawn until they have reached an age of about 12 years (length of about 5 feet; 1.5 meters). The males migrate for spawning much earlier—when they are from 4 to 8 years old (about 20 inches; 51 centimeters long). During this migration, the eels change color—from their normal yellow to silver and, at this time, they also are quite fat and thus most attractive from a food standpoint. Thus, the European fisheries gear their operations to catching the eels during the downstream migration. By instinct, the eels head for the Sargasso Sea in the Bermuda region. They spawn after having traveled thousands of miles, requiring about a year. After spawning, the adults die. The eggs float to the surface from a spawning depth of some 1,500 feet (450 meters). After hatching, the leptocephalous larvae commence the return trip, requiring in this case some 3 years. Usually arriving in European waters in the spring, the young elvers (about 3 inches; 7.5 centimeters long) seek out fresh water. The designers of the Zuider Zee dike did not contemplate the life cycle of the eel and hence millions of the elvers perished until so-called elver runs were created especially for them.

The mystery of eel migration was studied extensively by Danish biologist Johannes Schmidt who, in 1906, proposed a solution to the puzzle. From his investigations, he learned that the American eel (*Anguilla rostrata*) spawns in about the same area as the European eel (*Anguilla anguilla*). The puzzling problem was the difference in time required for the return trip, one year for the American eel; about three years for the European eel. The American eel can be found northward from Brazil to Greenland and Labrador and has penetrated into the central portions of the United States.

There are several species of freshwater eels, but none have been identified in the South Atlantic or eastern Pacific. These latter waters are not sufficiently saline and at the desired temperature at spawning depths.

The blood of the anguillids contains a toxin which affects the nerves and penetrates a wound if exposed during handling and preparation of the eel for food.

Snubnosed Eels. The parasitic snubnosed eeel habitates waters at depths between 4,000 and 5,000 feet (1200 and 1500 meters). Maximum length is about 2 feet (0.6 meter). Found in the waters along the American Atlantic coasts, Japan, and the Azores, the *S. parasiticus* obtains its nourishment by drilling into the bodies of larger fishes.

Moray Eels. There are numerous species of moray eels, of which most attain a length up to 5 feet (1.5 meters), although some have been reported as long as 10 feet (3 meters). Although a number of species are poisonous (with a number of deaths recorded as the result of eating), other species are routinely eaten. The several species of moray eels differ considerably in appearance, due mainly to their rostral development and nostril profile. Some appear as dragons and sea monsters. They prefer temperate or tropical waters and dwelling in a rock or reef environment.

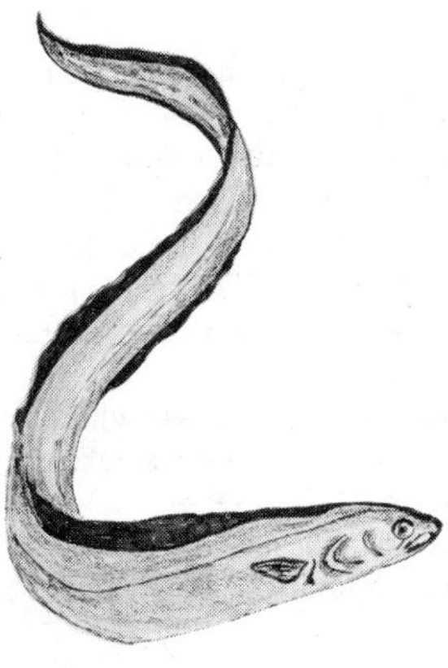

European eel (*Anguilla anguilla*).

Spiny Eels. The deep-sea spiny eels (*Heteromi*) look like eels, but are not related to eels. They are found in very cold waters at considerable depths—in excess of 1000 feet (300 meters) and sometimes reaching a depth of nearly 9000 feet (2700 meters). Some species possess luminous organs along their sides and on their head. They are highly elongated fishes and are seldom seen because of the great depths which they prefer.

Congers. The many genera and species of conger eels (*Congridae*) are distributed in almost all tropical and subtropical seas. The best known of them is the conger eel (*Conger conger*), which reaches a length of about 10 feet (3 meters) and a weight of about 243 pounds (65 kilograms). Distribution is almost worldwide, since it is caught in every ocean except the eastern Pacific. The species prefers rocky coasts, in the crevices of which it hides during the day. It is occasionally caught in brackish water of river mouths. The powerful teeth reflect that this is a predator which is dangerous. Its diet consists of various fishes, crustaceans, and squid. In addition to size, this species is distinguished from the European eel by the longer dorsal fin, which has its base shortly before that of the pectoral fins, and by the scaleless skin.

Not all is understood concerning the spawning of conger eels. Occasionally, females have been captured which have not yet emptied all their contents. Females have also reached maturity in aquariums. After maturity, it is estimated that the highest number of eggs which a female can lay is about 8 million. Pathological changes take place concomitantly with maturation of the gonads, these changes being in the intestinal tract and other organs, including skeleton and teeth. The adults cannot recover from these post-spawning changes and die as a result of them. Eels probably spawn in the open sea at depths of about 8350 feet (2500 meters), but specific spawning grounds have not been found for this group. All other genera and species of this family are considerably smaller than the congers and are almost exclusively deep-sea inhabitants.

Garden Eel. These eels (*Heterocongridae*) are classed with congers by a number of zoologists, while others treat them as an individual family. Garden eels are some 12 to 20 inches (30 to 50 centimeters) long and live in tubes, which extend vertically about 1.5 feet (0.5 meter) deep in loose sand or fine-grained coral sand. It is quite an experience for a diver to encounter such a garden eel "settlement." They often cover 120 square yards (100 square meters) of sandy bottom and the garden eels inhabit the floor at intervals of 8 to 24 inches (20 to 60 centimeters). With a slightly bent fore-end, which protrudes about two-thirds out of the tube, the eels sway to and fro with the head directed against the current, seeking the zooplankton upon which they feed. Such garden eel colonies have only been found at places constantly covered with water, and only where the current is uniform. The garden eels also avoid areas where breakers occur. It is reported that these eels flee from a diver by sliding into the soil when the diver reaches a distance of about 10 feet (3 meters) from them, and only their heads protrude. At a distance of about 3 feet (1 meter), they withdraw completely into their tubes and thus wait for about 5 minutes before shyly looking out again. Most attempts to dig healthy garden eels out of the sand fail, since the animals can dig back into the soil very rapidly. With a poison solution, they can be readily driven out of their tubes. As long as they are uninjured, they swim headfirst with undulating motions over the sandy bottom, in a completely flat position. But after swimming about 3 feet (1 meter) they turn about quickly and rapidly bore tail-first with powerful movements into the sand. Thus, the garden eels are excellent in their maneuvers to avoid capture. Secretion of a mucus substance enables the eels to maneuver in and out of their tubes with ease.

See also **Aquaculture;** and **Fishes.**

EELWORM (*Nematoda*). 1. A group of roundworms of the genus *Heterodera*, parasitic on the roots and underground stems of vascular plants. They cause serious damage to many cultivated crops—potato, sugar beet, beetroot and cereals. Resistant cysts are formed, viable for many years, but the root-knot eelworm causes galls to appear on the roots of tomato, cucumber, and others grown under glass. Entirely satisfactory chemical control has not been achieved as yet, but biological methods using crop rotation have been more successful and economical. 2. The roundworm *Ascaris lumbricoides*, parasitic as adult in the intestine of man and domestic animals, is called an eelworm in some countries.

EFFECTIVE TEMPERATURE (Astrophysics). A measure of the temperature of a star deduced by means of the Stefan-Boltzmann law, from the total energy emitted per unit area. Compare brightness temperature, color temperature. Effective temperature is always less than actual temperature.

EFFICIENCY. The general significance of this term as applied to a device or machine may be expressed as the ratio of output to input of energy or of power. If a dc motor, for example, is operating on 4 amperes at 100 volts (the power input is 400 watts), and if the motor actually delivers only 280 watts of mechanical power, its efficiency at that load is 280 watts ÷ 400 watts, or 70%. In general, the efficiency of a machine varies somewhat with the conditions under which it operates. Usually there is a load for which the efficiency is a maximum. This may be illustrated by a heavy block-and-tackle. For a small load the efficiency would be very low, because of power wasted in bending the ropes; for an excessive load it would again be low, on account of the large friction which would then develop; while for intermediate loads, higher efficiencies would prevail. The concept may be extended to other than purely mechanical systems. Thus the efficiency of an electric lamp may be expressed in candles or lumens of luminous flux (output) per watt of electric power (input); or that of an automobile horn in watts of acoustic power (noise) per watt of electric input. Various types of heat engine exhibit different thermodynamic efficiencies, i.e., the ratio of the work derived in the engine to the heat energy applied to it. See **Thermal Efficiency.** In statistics, the relative efficiency of two consistent estimators t_1, t_2 of a parameter ϕ is defined as the limiting value of the inverse ratio of their variances as the sample size increases. It can be shown that no statistic can have a large-sample variance less than a certain lower bound; estimators whose large-sample variance achieves this lower bound are said to be fully efficient or simply efficient. See also **Machine (Simple).**

EFFICIENCY (Solar). Solar Energy.

EFFLORESCENCE. When a substance evolves moisture upon exposure to the atmosphere, the substance is said to be efflorescent, and the phenomenon is known as efflorescence. At ordinary temperatures, the vapor pressure of water is shown by the accompanying table. If the substance has a higher water vapor pressure than corresponds to that of the atmosphere at the given temperature, water vapor is evolved from the substance until the water vapor pressure of the substance equals the water vapor pressure of the surrounding atmosphere.

Substances that are ordinarily efflorescent are sodium sulfate decahydrate, sodium carbonate decahydrate, magnesium sulfate heptahydrate, and ferrous sulfate heptahydrate. When the saturated solution of a substance in water has a water vapor pressure greater than that of the surrounding atmosphere, evaporation of the water from solution takes place.

See **Deliquescence** for the converse phenomenon.

VAPOR PRESSURE OF WATER

Temperature, °C	Water Vapor Pressure, in mm Mercury	
	At Saturation	At 50% Humidity
0	4.6	2.3
10	9.2	4.6
20	17.5	8.8
30	31.8	15.9
40	55.3	27.7

EFFLUENT STREAM. A stream that flows out of another stream or out of a lake. Also a stream whose upper surface is below the surface of the local ground water table. The term *effluent* is frequently used in the processing field to indicate a gaseous or liquid stream that is discharged after some form of treating or processing.

EFFUSION. Effusion is a general term denoting a process of discharge, that is also used specifically to denote the passage of gas under pressure through a small orifice.

EFFUSIVE. The term applied by geologists to molten material (lava) which has been poured out on the surface of the earth from a vent or fissure, as distinguished from ejected volcanic material (ashes and bombs) and injected magmas (plutonic rocks).

EGG. Poultry; Protein.

EGG (Deposition). Ovipositor.

EGG PLANT. Solanaceae.

EGG (Shell). Shell.

EIDER DUCK. Waterfowl.

EIGENFUNCTION. If a differential or integral equation possesses solutions satisfying the given boundary conditions for only certain values of a parameter λ, such a value is an eigenvalue (proper or characteristic value) and the corresponding solution is the eigenfunction belonging to that eigenvalue. Thus, given the linear operator P, the solutions u to the equation $Pu = \lambda u$ are eigenfunctions of P belonging to the eigenvalue λ. The totality of eigenvalues of any linear operator constitute the complete set, which may be made orthonormal.

Matrix methods are often useful in discussing this subject. If λ is a scalar parameter, $\mathbf{A}$ a square matrix of order n, and $\mathbf{E}$ the unit matrix of the same order, then $\mathbf{K} = [\lambda\mathbf{E} - \mathbf{A}]$ is the characteristic matrix of $\mathbf{A}$. The equation $det\ \mathbf{K} = 0 = \lambda^n = a_1\lambda^{n-1} + \cdots + a_n$, where the a_i are functions of the elements of $\mathbf{A}$, is the characteristic equation of $\mathbf{A}$ and its roots are the eigenvalues, characteristic or latent roots. The trace of $\mathbf{A}$, the sum of its diagonal elements, is the sum of the eigenvalues. Two matrices related by a similarity transformation have the same eigenvalues and hence the same trace.

An eigenfunction is often regarded as a vector in an abstract n-dimensional space; hence, it is often called an eigenvector.

See also terms listed under **Mathematics.**

EIGENVALUE (Proper Value). The concept described by this hybrid word has become extremely important in pure and applied mathematics, and in engineering, physics and chemistry. The German word "*eigen*" means "characteristic," a term already overburdened in mathematical English. The "characteristic values" of a physical system are numbers which describe, for example, the critical frequencies of a suspension bridge or of a rotating shaft, the critical load of a supporting column, or the energy levels of a system in quantum mechanics. In corresponding mathematical language, we shall define the eigenvalues of a square matrix, of the kernel of an integral equation, of a differential equation with boundary conditions, and of an operator or transformation, the last case including all the others.

If $\mathbf{A} = \{a_{ik}\}$ is a square matrix, then the number λ is an eigenvalue (also called a characteristic number, a proper value, a latent root, etc.) of the matrix $\mathbf{A}$ if the determinant

$$\begin{vmatrix} a_{11}-\lambda & a_{12} & \cdots & a_{1n} \\ a_{21} & a_{22}-\lambda & \cdots & a_{2n} \\ \vdots & \vdots & \cdots & \vdots \\ a_{n1} & a_{n2} & \cdots & a_{nn}-\lambda \end{vmatrix}$$

is equal to zero. In other words, an eigenvalue of a matrix is a root of the characteristic equation of the matrix.

As a physical example, consider a suspension bridge which is vibrating with n degrees of freedom about a position of stable equilibrium. Its potential energy at any time will be a function, call it $A(q_1, q_2, \ldots, q_n)$, of its n position coordinates $q_1, q_2, \ldots, q_n$, which is approximated by the quadratic form $\Sigma\ a_{ik}q_iq_k$, the constants a_{ik} being given by $a_{ik} = \frac{1}{2}(\partial^2 A/\partial q_i \partial q_k)$ evaluated at the position of equilibrium. Then, the eigenvalues of the matrix $\{a_{ik}\}$ are the squares of the critical frequencies of the bridge, namely the frequencies at which an impressed force will cause dangerous resonance.

From the theory of linear equations it follows that this definition of the eigenvalues of a matrix can be restated as follows. The constant λ is an eigenvalue of the matrix $\{a_{ik}\}$ if the n homogeneous linear equations in the n unknowns $q_1, q_2, \ldots, q_n$

$$\begin{aligned} a_{11}q_1 + a_{12}q_2 + \cdots + a_{1n}q_n &= \lambda q_1 \\ a_{21}q_1 + a_{22}q_2 + \cdots + a_{2n}q_n &= \lambda q_2 \\ \vdots \quad\quad \vdots \quad\quad \cdots \quad\quad \vdots \quad & \quad \vdots \\ a_{n1}q_1 + a_{n2}q_2 + \cdots + q_{nn}q_n &= \lambda q_n \end{aligned}$$

have a non-trivial solution (i.e., one for which not all the q_i vanish). In other words, regarding the matrix as a linear operator in n-dimensional Euclidean space, the constant λ is an eigenvalue of the operator A which transforms a vector $\mathbf{q} = (q_1, q_2, \ldots, q_n)$ into a vector denoted by $A\mathbf{q}$, if there exists a non-zero vector $\mathbf{q} = (q_1, q_2, \ldots, q_n)$ such that $A\mathbf{q} = \lambda\mathbf{q}$. Such a non-zero vector is called an *eigenvector* (of the operator A) belonging to the eigenvalue λ.

Similarly, let $A(x, y)$ be a continuous function of two variables (the infinite-dimensional analogue of a matrix with rows and columns) defined for $a \leq x, y \leq b$, and let A be the integral operator which transforms a function $\phi(x)$ into a function $f(x)$ according to the formula

$$f(x) = \iint A(x, y)\phi(y)\, dy,$$

where the integration is taken over the square $a \leq x, y \leq b$ and the functions $\phi(x)$ and $f(x) = A\phi(x)$ are infinite-dimensional analogues (in Hilbert space) of the above vectors $\mathbf{q}$ and $A\mathbf{q}$. Then the constant λ is defined to be an eigenvalue of the integral operator A if the integral equation

$$\iint A(x, y)\phi(y)\, dy = \lambda\phi(x),$$

which we may also write in the form $A\phi = \lambda\phi$, has a non-trivial solution $\phi(x)$. The solution $\phi(x)$ is called an eigensolution, or eigenfunction belonging to λ.

The type of eigenvalue problem occurring most frequently in practical applications involves a differential operator. For example, the critical frequencies of a vibrating plate (important in the theory of microphones and elsewhere) are the squares of the eigenvalues of the biharmonic operator discussed below.

In general, let A be any differential operator acting on functions ϕ of any number of variables; e.g., for the clamped plate, $\phi = \phi(x, y)$ and A is the biharmonic operator defined by the formula

$$A\phi = \partial^4\phi/\partial x^4 + 2\partial^4\phi/\partial x^2\, \partial y^2 + \partial^4\phi/\partial y^4.$$

Also, let us consider as admissible only those functions ϕ which satisfy certain boundary conditions; e.g., for the clamped plate, ϕ and its normal derivative must vanish on the boundary. Then the constant λ is an eigenvalue of the operator A if there exists an admissible function ϕ, called an eigenfunction of A, such that $A\phi = \lambda\phi$. Differential eigenvalue problems of this sort are of fundamental importance in the theory of vibrations and buckling and in quantum mechanics.

EIGENVALUES AND EIGENVECTORS OF MATRICES (Methods of Computing). For a given square matrix $\mathbf{A}$ (singular or not), non-null vectors $\mathbf{x}$ exist satisfying $\mathbf{Ax} = \lambda\mathbf{x}$ only when the scalar λ satisfies det $(\lambda\mathbf{I} - \mathbf{A}) = \phi(\lambda) = 0$, called the characteristic equation (cf., the entry **Eigenvalue**) where $\phi(\lambda)$ is a polynomial of degree n, the order of $\mathbf{A}$, called the characteristic polynomial (see **Matrix**). Any root λ of this equation is called an eigenvalue; associated with any root λ there is at least one non-null vector $\mathbf{x}$ called eigenvector belonging to λ. The number of independent eigenvectors belonging to an eigenvalue may equal, but cannot exceed, the multiplicity of λ as a root of the characteristic equation. For normal matrices, including Hermitian, the number equals the multiplicity. For a non-normal matrix, if λ is a root of multiplicity $\nu > 1$, and if fewer than ν independent vectors belong to λ, then there are principal vectors $\mathbf{y} \neq 0$ satisfying $(\lambda\mathbf{I} - \mathbf{A})^\mu\mathbf{y} = 0$ for $1 < \mu \leq \nu$. It is always true that $\phi(\mathbf{A}) = 0$. If there is a polynomial $\psi(\lambda)$ of degree $m < n$ for which $\psi(\mathbf{A}) = 0$, then that polynomial of lowest degree for which this is true is a minimal polynomial, and $\mathbf{A}$ is said to be derogatory; otherwise nonderogatory. The condition for being derogatory is rather stringent and not often satisfied in practice, and the same is true for the presence of principal vectors, but when the conditions are nearly satisfied the corresponding eigenvectors are poorly defined and computational difficulties arise, possibly even insurmountable.

Implicitly or explicitly, to evaluate an eigenvalue requires the solution of an algebraic equation of degree n. The problem is much simpler for a Hermitian matrix, and for these is discussed under that heading. Some of these methods can also be adapted to normal matrices. Here the general case is considered. A method may purport to yield only the characteristic, or possibly the minimal, polynomial, leaving this to be solved by any of the standard methods for solving algebraic equations. Such a method is called direct. Once an eigenvalue is known, an eigenvector belonging to it can be found by solving a system of homogeneous equations. A direct expansion of the determinant is unthinkable when n is at all large.

Since any $n + 1$ vectors in n-space are linearly dependent, if $\mathbf{b} = \mathbf{b}_1 \neq 0$, then in the sequence

$$\mathbf{b}_{i+1} = \mathbf{A}\mathbf{b}_i$$

there is a smallest index $m \leq n$ such that $\mathbf{b}_{m+1}$ is linearly dependent upon those preceding:

$$\mathbf{b}_{m+1} + \beta_1\mathbf{b}_m + \beta_2\, \mathbf{b}_{m-1} + \cdots + \beta_m\mathbf{b}_1 = 0. \tag{1}$$

This represents n equations in $m \leq n$ unknowns, and they are consistent by hypothesis, even if $m < n$. But this is

$$(\mathbf{A}^m + \beta_1\mathbf{A}^{m-1} + \cdots + \beta_m\mathbf{I})\mathbf{b} = 0. \tag{2}$$

Hence, if

$$\psi(\lambda) \equiv \lambda^m + \beta_1\lambda^{m-1} + \cdots + \beta_m, \tag{3}$$

then $\psi(\mathbf{A})\mathbf{b} = 0$ and $\psi(\lambda)$ is either a minimal polynomial or a factor of it, and except for very special choices of $\mathbf{b}$, $\psi(\lambda)$ is a minimal polynomial.

If one forms the matrix

$$\mathbf{B} = (\mathbf{b}_1, \mathbf{b}_2, \ldots, \mathbf{b}_m) \tag{4}$$

whose columns are the $\mathbf{b}_i$, then

$$\mathbf{AB} = \mathbf{BF} \tag{5}$$

where $\mathbf{F}$ is the companion matrix whose form is

$$\mathbf{F} = \begin{pmatrix} 0 & 0 & \cdots & 0 & -\beta_m \\ 1 & 0 & \cdots & 0 & -\beta_{m-1} \\ & & \cdots\cdots & & \\ 0 & 0 & \cdots & 1 & -\beta_1 \end{pmatrix} \tag{6}$$

Its characteristic polynomial is $\psi(\lambda)$. With $\mathbf{b}_1 = \mathbf{e}_1$, the first column of $\mathbf{I}$, this is the method of *Krylov*.

This method is effective for matrices of fairly low order. For matrices of high order the equations to be solved for the β_i tend to become ill-conditioned. A method due to *Hessenberg* yields a matrix $\mathbf{B}$ such that

$$\mathbf{AB} = \mathbf{BG}, \tag{7}$$

where $\mathbf{G}$ is not the companion matrix, but whose form is such that the characteristic polynomial can be expanded directly, and from which it is rather easy to solve for the eigenvectors. To apply this method one selects auxiliary vectors $\mathbf{c}_1, \mathbf{c}_2, \mathbf{c}_3, \ldots$ subject to two mild restrictions but otherwise arbitrary. Ordinarily one can take (with Hessenberg) each $\mathbf{c}_i = \mathbf{e}_i$, a colum of $\mathbf{I}$. One starts again with a vector $\mathbf{b}$, possibly $\mathbf{e}_1$. Now $\mathbf{b}_2$ is a linear combination of $\mathbf{b}_1$ and $\mathbf{Ab}_1$ orthogonal to $\mathbf{c}_1$; $\mathbf{b}_3$ a linear combination of $\mathbf{b}_1$, $\mathbf{b}_2$, and $\mathbf{Ab}_2$ orthogonal to both

$\mathbf{c}_1$ and $\mathbf{c}_2, \ldots$. The restrictions on the $\mathbf{c}_i$ are, first, linear independence, and, second,

$$\mathbf{c}_i^*\mathbf{b}_i \neq 0.$$

With these conditions fulfilled one forms

$$\mathbf{A}\mathbf{b}_1 = \mathbf{b}_1\gamma_{11} + \mathbf{b}_2$$

where

$$\mathbf{c}_1^*\mathbf{A}\mathbf{b}_1 = \mathbf{c}_1^*\mathbf{b}_1\gamma_{11}.$$

Hence γ_{11} is obtained, and therefore $\mathbf{b}_2$. Next

$$\mathbf{A}\mathbf{b}_2 = \mathbf{b}_1\gamma_{12} + \mathbf{b}_2\gamma_{22} + \mathbf{b}_3,$$

where

$$\mathbf{c}_1^*\mathbf{A}\mathbf{b}_2 = \mathbf{c}_1^*\mathbf{b}_1\gamma_{12},$$
$$\mathbf{c}_2^*\mathbf{A}\mathbf{b}_2 = \mathbf{c}_2^*\mathbf{b}_1\gamma_{12} + \mathbf{c}_2^*\mathbf{b}_2\gamma_{22}.$$

From the first of these one obtains γ_{12}, from the second γ_{22}, and finally $\mathbf{b}_3$. The process continues, and $\mathbf{G}$ is seen to have the form

$$\mathbf{G} = \begin{pmatrix} \gamma_{11} & \gamma_{12} & \gamma_{13} & \cdots \\ 1 & \gamma_{22} & \gamma_{23} & \cdots \\ 0 & 1 & \gamma_{33} & \cdots \\ & \cdots & & \end{pmatrix}. \tag{8}$$

It may be observed that when the choice $\mathbf{c}_i = \mathbf{e}_i$ is made, then the first element of $\mathbf{b}_2$ is null as are the first two elements of $\mathbf{b}_3$, the first three of $\mathbf{b}_4, \ldots$.

A further refinement of the method is the biorthogonalization method of *Lanczos*. In this method only $\mathbf{b}_1$ and $\mathbf{c}_1$ are arbitrary. Thereafter, just as $\mathbf{b}_2$ is a linear combination of $\mathbf{b}_1$ and $\mathbf{A}\mathbf{b}_1$ orthogonal to $\mathbf{c}_1$, so is $\mathbf{c}_2$ a linear combination of $\mathbf{c}_1$ and $\mathbf{A}^*\mathbf{c}_1$ orthogonal to $\mathbf{b}_1$, and so for the others. The matrix $\mathbf{G}$ is then tridiagonal, being null everywhere except along, just below, and just above the diagonal. The recursion then never contains more than three terms in any equation.

Another modification is due to Arnoldi, who chooses $\mathbf{b}_i = \mathbf{c}_i$ and requires $\mathbf{B}$ to be an orthogonal matrix. Then (7) is satisfied, but with a matrix $\mathbf{G}$ of the form

$$\mathbf{G} = \begin{pmatrix} \gamma_{11} & \gamma_{12} & \gamma_{13} & \cdots \\ \gamma_{21} & \gamma_{22} & \gamma_{23} & \cdots \\ 0 & \gamma_{32} & \gamma_{33} & \cdots \\ & \cdots\cdots & & \end{pmatrix} \tag{9}$$

similar to (8) but with elements other than ones on the subdiagonal. The development of the characteristic polynomial from $\mathbf{G}$ is almost equally straightforward.

When only one, or few, of the eigenvalues of largest modulus are required a simple iteration scheme is advisable, and is preferred by some even for a complete reduction of rather large matrices. If there is a single proper value exceeding all others in modulus, λ_1, vectors $\mathbf{b}_v$ in the continued sequence

$$\mathbf{b}_{v+1} = \mathbf{A}\mathbf{b}_v$$

approach the eigenvector belonging to λ_1, so that for large v,

$$\mathbf{b}_{v+1} \doteq \lambda_1\mathbf{b}_v.$$

Hence, with any vector $\mathbf{u}$, an approximate value of λ_1 is given by

$$\mathbf{u}^*\mathbf{A}\mathbf{b}_v = \lambda_1\mathbf{u}^*\mathbf{b}_v.$$

Convergence, when it occurs, can be accelerated by application of the delta-square process termwise to the vectors in the sequence.

In the case of a pair of complex roots, convergence does not occur, but in the limit $\mathbf{b}_v$ becomes parallel to the plane of the two eigenvectors. Then if

$$\mu_v = \mathbf{u}^*\mathbf{b}_v$$

the roots of the equation

$$\begin{vmatrix} 1 & \mu_v & \mu_{v+1} \\ \lambda & \mu_{v+1} & \mu_{v+2} \\ \lambda^2 & \mu_{v+2} & \mu_{v+3} \end{vmatrix} = 0$$

approach these roots λ_1 and λ_2 as v increases. In fact, whether or not λ_1 and λ_2 are equal in modulus, but provided only both exceed all others in modulus, the statement is true. Moreover, the roots of

$$\begin{vmatrix} 1 & \mu_v & \mu_{v+1} & \mu_{v+2} \\ \lambda & \mu_{v+1} & \mu_{v+2} & \mu_{v+3} \\ \lambda^2 & \mu_{v+2} & \mu_{v+3} & \mu_{v+4} \\ \lambda^3 & \mu_{v+3} & \mu_{v+4} & \mu_{v+5} \end{vmatrix} = 0$$

approach the three proper values of largest modulus, if such exist (cf. Bernoulli method).

Returning to the two roots, for large v,

$$\mathbf{b}_v \doteq \mathbf{v}_1 + \mathbf{v}_2,$$

$$\mathbf{b}_{v+1} \doteq \lambda_1\mathbf{v}_1 + \lambda_2\mathbf{v}_2,$$

approximately, where $\mathbf{v}_1$ and $\mathbf{v}_2$ are eigenvectors belonging to λ_1 and λ_2. Knowing λ_1 and λ_2, as well as $\mathbf{b}_v$ and $\mathbf{b}_{v+1}$, these equations can be solved for $\mathbf{v}_1$ and $\mathbf{v}_2$.

When any eigenvalue and eigenvector are known it is possible to apply *deflation* as follows: The matrix

$$\mathbf{A} - \lambda_1\mathbf{v}_1\mathbf{u}^*,$$

where $\mathbf{u}$ is any vector satisfying $\mathbf{u}^*\mathbf{v}_1 = 1$ has the same eigenvalues as $\mathbf{A}$ except that instead of λ_1 the new matrix has a zero. Moreover, if $\mathbf{w}$ is an eigenvector of $\mathbf{A} - \lambda_1\mathbf{v}_1\mathbf{u}^*$, belonging to the eigenvalue λ, then $(\mathbf{A} - \lambda_1\mathbf{I})\mathbf{w}$ is a proper vector of $\mathbf{A}$ belonging to the eigenvalue λ. And finally, it is possible to choose $\mathbf{u}^*$ so that every element but the first in the first row of $\mathbf{A} - \lambda_1\mathbf{v}_1\mathbf{u}^*$ is zero. To do this, normalize $\mathbf{v}_1$ to have 1 as the first element (should this be zero, permute rows and corresponding columns of $\mathbf{A}$, and the same elements of $\mathbf{v}_1$ to make the first element non-null), Choose the last $n - 1$ elements of $\mathbf{u}^*$ to be $1/\lambda_1$ times the corresponding elements of the first row of $\mathbf{A}$, and choose the first element of $\mathbf{u}^*$ to satisfy $\mathbf{u}^*\mathbf{v}_1 = 1$. It remains now to consider only a matrix of order $n - 1$, and further reductions are made as further eigenvalues and eigenvectors are found.

In case μ is an approximation to a particular eigenvalue λ_i, closer than to any other λ_j, then $\mu - \lambda_i$ is the numerically largest eigenvalue of $(\mathbf{A} - \mu\mathbf{I})^{-1}$. Hence the solution of the system

$$(\mathbf{A} - \mu\mathbf{I})\mathbf{x}_1 = \mathbf{x}_0$$

for an almost arbitrary $\mathbf{x}_0$ will give a good approximation to the eigenvector belonging to λ_i, and $\mathbf{x}_1^*\mathbf{A}\mathbf{x}_1/\mathbf{x}_1^*\mathbf{x}_1$ will be a better approximation to λ_i. This is *Wielandt's broken iteration*.

Unless it is known a priori that a matrix has principal vectors that are not eigenvectors, and has auxiliary conditions available for determining them, rounding errors will almost inevitably conceal them, but the problem may turn out to be highly unstable. If $\mathbf{A} = \mathbf{P}\Lambda\mathbf{P}^{-1}$ where Λ is diagonal, the numerical stability decreases as $\|\mathbf{P}\|\ \|\mathbf{P}^{-1}\|$ increases in some norm, hence this seems to be an appropriate condition number with respect to this problem. But as $\mathbf{A}$ approaches the nondiagonalizable form, this number becomes infinite, hence at least some of the eigenvalues and eigenvectors become ill defined.

EIKONAL EQUATION. A fundamental equation of wave motion of the form:

$$|\nabla\psi|^2 = n^2(x, y, z)$$

where n is the refractive index of the medium for the waves, ∇ is the vector differential operator, and $\psi(x, y, z)$ is a function called the eikonal, which defines the wave fronts.

EIKONOMETER. A scale, attached to the eyepiece of a microscope, which is seen superimposed on the image, and is used to measure the dimensions of the objects viewed.

EINSTEIN DIFFUSION EQUATION. **Diffusion Equation (Einstein).**

EINSTEIN EMISSION COEFFICIENTS. **Emission Coefficients (Einstein).**

EINSTEIN HEAT CAPACITY EQUATION. **Heat Capacity Equation (Einstein).**

EINSTEINIUM. Chemical element symbol Es, at. no. 99, at. wt. 254 (mass number of the most stable isotope), radioactive metal of the *Actinide* series, also one of the *Transuranium* elements. Both einsteinium and fermium were formed in a thermonuclear explosion which occurred in the South Pacific in 1952. The elements were identified by scientists from the University of California's Radiation Laboratory, the Argonne National Laboratory, and the Los Alamos Scientific Laboratory. It was observed that very heavy uranium isotopes which resulted from the action of the instantaneous neutron flux on uranium (contained in the explosive device) decayed to form Es and Fm. The probable electronic configuration of Es is

$$1s^22s^22p^63s^23p^63d^{10}4s^24p^64d^{10}4f^{14}5s^25p^65d^{10}5f^{11}6s^26p^67s^2.$$

Ionic radius Es^{3+} 0.97 Å.

All known isotopes of einsteinium are radioactive. The first evidence of their existence was obtained by ion-exchange methods applied to coral rocks obtained from Eniwetok Atoll after the thermonuclear explosion. The first pure isotope found was ^{253}Es, produced by prolonged treatment of plutonium-239 with neutrons in the Arco, Idaho, Testing Reactor. The most stable is ^{254}Es, half-life 270 days, and therefore the mass number 254 is carried in the atomic weight table. Others include ^{245}Es–^{246}Es, ^{248}Es–^{252}Es, and ^{253}Es, ^{255}Es.

The ion Es^{3+} is stable. The isotopes of mass numbers 245, 252, 253 and 254 decay by alpha-particle emission; that of mass number 250 by electron capture, those of mass numbers 246, 248, 249, and 251 by both of these processes, while those of mass numbers 255 and 256 emit electrons to form the corresponding fermium isotopes.

Sufficient einsteinium, produced through intense neutron bombardments of plutonium-239 in a reactor, was not available until 1961 to allow separation of a macroscopic amount. Cunningham, Wallmann, L. Phillips, and Gatti worked with submicrogram quantities to separate a small fraction of pure einsteinium-235 compound and measure its magnetic susceptibility. Only a few hundredths of a microgram were available at that time.

References

Cunningham, B. B., Peterson, J. R., Baybarz, R. D., and T. C. Parsons: "The Absorption Spectrum of Es^{3+} in Hydrochloric Acid Solutions," *Inorg. Nucl. Chem. Lett.*, **5**, 519–523 (1967).

Cunningham, B. B., and T. C. Parsons: "Preparation and Determination of the Crystal Structure of Californium and Einsteinium Metals," *Lawrence Berkeley Laboratory Nuclear Chemistry Annual Report*, UCRL-20426, University of California, Berkeley, California, 1970.

Fields, P. R., et al.: "Additional Properties of Isotopes of Elements 99 and 100," *Phys. Rev.*, **94**, 1, 209–210 (1954).

Fujita, D. K., Cunningham, B. B., and T. C. Parsons: "Crystal Structures and Lattice Parameters of Einsteinium Trichloride and Einsteinium Oxychloride," *Inorg. Nucl. Chem. Lett.*, **5**, 4, 307–313 (1969).

Ghiorso, A., et al: "New Elements Einsteinium and Fermium, Atomic Numbers 99 and 100," *Phys. Rev.*, **99**, 3, 1048–1049 (1955).

Seaborg, G. T. (editor): "Transuranium Elements," Dowden, Hutchinson & Ross, Stroudsburg, Pennsylvania, 1978.

Studier, M. M., et al.: "Elements 99 and 100 from Pile-Irradiated Plutonium," *Phys. Rev.*, **93**, 6, 1428 (1954).

Thompson, S. G., Harvey, B. G., Choppin, G. R., and G. T. Seaborg: "Chemical Properties of Elements 99 and 100," *American Chemical Society Journal*, **76**, 6229–6236 (1954).

EINSTEIN LAW (Photochemistry). Stark-Einstein Equation; Stark-Einstein Law.

EINSTEIN SHIFT. According to the relativity theory, when radiation quanta leave a massive source like the sun or a star, they are retarded by the gravitational attraction and hence lose energy. This means that they lose frequency and that the wavelength λ increases. For a star of radius *R* and mass *M*, the fractional increase in wavelength is

$$\frac{\Delta}{\lambda} = \frac{G}{c^2} \cdot \frac{M}{R}$$

in which *G* is the gravitation constant and *c* is the electromagnetic constant (speed of light). The coefficient $G/c^2 = 7.414 \times 10^{-29}$ centimeters per gram. For the sun, $M = 2.3 \times 10^{33}$ grams and $R = 1.394 \times 10^{11}$ centimeters. Then $\Delta\lambda/\lambda = 1.23 \times 10^{-6}$, so that each solar spectrum line should be shifted toward the red by a little over a millionth of its own wavelength.

Measurements of this precision are hardly possible at present. But there are other stars (in particular, the white dwarfs) so massive and so dense that the shift has actually been observed and found to be of the correct order of magnitude. See also **Star.**

EINSTEIN THEORIES. Cosmology; Gravitation; Quantum Mechanics; Relativity and Relativity Theory.

EINSTEIN UNIVERSE. Cosmology; Red Shift.

EINSTEIN X-RAY OBSERVATORY. X-Ray Astronomy.

EJA (*Reptilia, Sauria*). The desert saw viper of Egypt, a vicious poisonous snake.

EJACULATORY DUCT. The portion of the male genital duct between the duct of the seminal vesicle and the urethra.

EJECTOR-TYPE VACUUM PUMP. Vacuum Pumps.

EKMAN SPIRAL (Meteorology). Winds and Air Movement.

EKMAN SPIRAL (Oceanography). As originally applied by Ekman to ocean currents, a graphic representation of the way in which the theoretical wind-driven currents in the surface layers of the sea vary with depth. In an ocean assumed to be homogeneous, infinitely deep, unbounded, and having a constant eddy viscosity, over which a uniform, steady wind blows. Ekman computed with the current induced in the surface layers by the wind have the following characteristics: (1) At the very surface, the water will move at an angle of 45° *cum sole* from the wind direction; (2) in successively deeper layers, the movement will be deflected farther and farther *cum sole* from the wind direction, and the speed will decrease; and (3) a hodograph of the velocity vectors would form a spiral descending into the water and decreasing in amplitude exponentially with depth.

EKYIMAN RELATIONSHIP. Refraction.

ELAND. Bovines.

ELASSER RADIATION CHART. Atmosphere (Earth).

ELASTIC AFTER-EFFECT (or Lag). The time delay which some substances exhibit in returning to original shape after being stressed within their elastic limits. There is some evidence that the magnitude of this time depends on the homogeneity of structure of the substance. For instance, quartz, which has a homogeneous structure, shows almost no lag. Glass, which is a mixture of aggregates, can have a time delay of the order of hours.

ELASTIC CONSTANTS AND MODULI. These constants occur in relationships between the components of strain and stress in a medium which obeys Hooke's law. Under such conditions the strain components may be expressed as functions of the stress components by linear equations of the type:

$$e_{xx} = s_{11}X_x + s_{12}Y_y + s_{13}Z_z + s_{14}Y_z + s_{15}Z_x + s_{16}X_y$$

where e_{xx} is the strain component along *x*-axis, the capitalized quantities are the stress components (the stress component X_x, for example, representing a force applied in the *x*-direction to a unit area of a plane whose normal lies in the *x*-direction) and the *s* terms are called the elastic constants or compliance constants. Also, the stress components may be expressed as functions of the strain components by linear equations of the type:

$$X_x = c_{11}e_{xx} + c_{12}e_{yy} + c_{13}e_{zz} + c_{14}e_{yz} + c_{15}e_{zx} + c_{16}e_{xy}$$

where the *c* terms are called the stiffness constants or elastic moduli.

For the special moduli used in engineering, see **Elasticity.**

ELASTIC CURVE. The curve of the neutral surface of a structural member subjected to loads which cause bending is called the elastic curve. The ordinates between this curve and the original position of the neutral surface represent the deflections due to bending.

ELASTIC DEFORMATION. Resilience.

ELASTIC FLUID. Fluid.

ELASTICITY. The property whereby a body, when deformed, automatically recovers its normal configuration as the deforming forces are removed. Each of its several types is probably due to the action of intermolecular forces which are in equilibrium only for certain configurations.

Deformation or, more briefly, strain is of various kinds; in each case its measure is a certain abstract ratio. For example, the elongation of a rod under tension is expressed as the ratio of the increase in length to the unstretched length. Linear compression is the reverse of elongation. They are both accompanied by a fractional change in diameter, the ratio of which to the elongation is called the Poisson ratio. Shear is a strain involving change of shape, such that an imaginary cube traced in the unstrained material becomes a rhombic prism. The measure of sheer is the tangent of the angle through which the oblique edges have been made to depart from their original perpendicular direction. Volume strain is the ratio of a decrease in volume to the normal volume. Flexure, or bending, and torsion, or twisting, are combinations of these more elementary strains. A straight rod bent into a plane curve undergoes elongation on the convex side and linear compression on the concave side, while there is an intermediate neutral layer which suffers neither.

In tests such as the tensile test it is necessary to differentiate between two basic ways of measuring strain. "Conventional strain" is the increase in length of the gauge section divided by the original gauge length, whereas the "true strain" is the natural logarithm

$$\ln \frac{l}{l_0}$$

where l_0 is the original value and l the instantaneous value of the length of the gage section.

For every strain, there arises, in an elastic substance, a corresponding stress, which represents the tendency of the substance to recover its normal condition. Stress is expressed in units of force per unit area Tensile stress, for example, is the ratio of the force of tension to the normal cross section of the rod subjected to it. Shearing stress is the force tending to push one layer of the material past the adjacent layer, per unit area of the layers. Pressure, expressed in like units, is the stress corresponding to volume compression, etc.

For each type of strain and stress there is a modulus, which is the ratio of the stress to the corresponding strain. In the case of elongation or linear compression, it is commonly called Young's modulus; we also have the bulk modulus and the shear modulus of rigidity. See accompanying table.

In engineering design, Young's modulus is used for tension and compression and the rigidity modulus for shear, as in torsion springs. According to Hooke's law, the stress set up within an elastic body is proportional to the strain to which the body is subjected by the applied load.

Theory of Elasticity. The fundamental quantities in elasticity are second-order tensors, or dyadics: the deformation is represented by the *strain dyadic,* and the internal forces are represented by the *stress dyadic.* The physical constitution of the deformable body determines the relation between the strain dyadic and the stress dyadic, which relation is, in the infinitesimal theory, assumed to be linear and homogeneous. While for anisotropic bodies this relation may involve as much as 21 independent constants, in the case of isotropic bodies, the number of elastic constants is reduced to two.

Let $\mathbf{s}(\mathbf{r})$ be the displacement vector, due to the deformation, of a particle that before the deformation was situated at point P having $\mathbf{r}$ as position vector with respect to some arbitrary origin. A neighboring point Q, whose position vector was $\mathbf{r} + d\mathbf{r}$ before the deformation, will suffer a displacement $\mathbf{s}(\mathbf{r} + d\mathbf{r})$ which will differ from $\mathbf{s}(\mathbf{r})$ by the quantity

$$d\mathbf{s} = d\mathbf{r} \cdot \nabla \mathbf{s}$$

The hypothesis of small deformations means that $d\mathbf{s}$, the change in the displacement vector when we go from P to the neighboring point Q, is very small compared to $d\mathbf{r}$, the position vector of Q relative to P. Consequently, the scalar components of the dyadic $\nabla\mathbf{s}$ are all very small compared to unity. The geometrical meaning of the dyadic $\nabla\mathbf{s}$ is obtained by separating it into its symmetric part $\mathbf{S} = \frac{1}{2}(\nabla\mathbf{s} + \mathbf{s}\nabla)$ and its antisymmetric part $\mathbf{R} = -\frac{1}{2}\mathbf{1} \times (\nabla \times \mathbf{s})$, where $\mathbf{1}$ is the unity dyadic. The antisymmetric part is interpreted as follows: if at some point M the symmetric part vanishes, then we have for the neighborhood of M the relation

$$d\mathbf{s} = d\mathbf{r} \cdot \mathbf{R}_M = \omega_M \times d\mathbf{r}$$

where $\omega_M = \frac{1}{2}(\nabla \times \mathbf{s})_M$ is an infinitesimal vector. This means that the neighborhood of point M undergoes an infinitesimal rigid rotation, without any change in shape or size. Consequently, the deformation is represented by the symmetric part $\mathbf{S}$, which is called the *strain dyadic.*

In a Cartesian orthonormal basis, in which we have $\mathbf{r} = \Sigma_{i=1}^{3} x_i\mathbf{a}_i$, we write $\mathbf{s} = \Sigma_{i=1}^{3} s_i\mathbf{a}_i$, and obtain

$$\mathbf{S} = \sum_{i,\,j=1}^{3} \mathbf{a}_i\mathbf{a}_j S_{ij}$$

where

$$S_{ij} = \frac{1}{2}\left[\frac{\partial}{\partial x_i} s_j + \frac{\partial}{\partial x_j} s_i\right].$$

The diagonal components S_{11}, S_{22}, and S_{33} are the coefficients of linear extension in the directions $\mathbf{a}_1$, $\mathbf{a}_2$, and $\mathbf{a}_3$, respectively, while the non diagonal components $S_{12} = S_{21}$, $S_{13} = S_{31}$, and $S_{23} = S_{32}$ are called shear strains. For instance, $2S_{12}$ is the change in the angle of the dihedron formed by the planes that before the deformation were respectively normal to the directions $\mathbf{a}_1$ and $\mathbf{a}_2$. The shear strains are not essential for the complete representation of a deformation since they can be made to vanish by expressing $\mathbf{S}$ in the basis of its principal axes.

If an infinitesimal element of the body occupies the volume dV before the deformation and the volume dV' after, the relative increase of volume, or volumetric dilatation, is given by

$$\frac{dV' - dV}{dV} = S_{11} + S_{22} + S_{33} = |\mathbf{S}| = \nabla \cdot \mathbf{s}$$

The forces applied to a finite deformable body are either body forces acting on every volume element dV and represented by the notation $dV\,\mathbf{F} = dV\,\rho\mathbf{K}$, where $\mathbf{F}$ is the force per unit volume, $\mathbf{K}$ is the force per unit mass, and ρ is the density, or surface forces acting on every element dS of the bounding surface and represented by $dS\,\mathbf{T}$, where $\mathbf{T}$ is the surface stress, or surface force per unit area. The effect of these applied forces is transmitted throughout the body, so that through any surface element inside the body, there is a force exerted by the matter on one side of the element upon the matter on the other side. Such forces are called internal stresses and are defined as follows: let dS be a surface element completely inside the body, and let us choose arbitrarily the positive sense of the normal $\mathbf{n}$ to this surface element; this defines for dS a positive side, the one containing $\mathbf{n}$, and a negative side. Then $\mathbf{T}_\mathbf{n}$, the stress vector on the positive side of dS is defined as a vector such that $dS\,\mathbf{T}_\mathbf{n}$ is the surface force

NOMINAL MODULUS VALUES OF REPRESENTATIVE METALS
(values in pounds per square inch)

Material	Young's Modulus	Rigidity of Shear Modulus
Magnesium	6.5×10^6	2.4×10^6
Aluminum	10.2×10^6	3.6×10^6
Copper	14.5×10^6	6.1×10^6
Steel	30×10^6	11.6×10^6

on the positive side of dS—i.e., the resultant of all the forces exerted through dS by the matter on the positive side of dS upon the matter on the negative side. In general there is a normal component $\mathbf{T_n \cdot nn}$, which is a pressure or a traction depending upon whether the sign of $\mathbf{T_n \cdot n}$ is negative or positive, and a tangent component $\mathbf{n \times T_n \times n}$ called the shear stress. The value of the stress vector $\mathbf{T_n}$ depends upon the orientation of the normal $\mathbf{n}$, so that we can characterize the state of stress at a point by defining the *stress dyadic* $\mathbf{T}$ through the relation

$$\mathbf{T_n} = \mathbf{n} \cdot \mathbf{T}$$

The mechanical equilibrium conditions applied to an arbitrary volume V, bounded by the closed surface S, and completely inside the deformable body give

$$\int_V dV\mathbf{F} + \int_S dS\,\mathbf{n} \cdot \mathbf{T} = 0$$

and

$$\int_V dV\mathbf{r} \times \mathbf{F} + \int_S dS\,\mathbf{r} \times (\mathbf{n} \cdot \mathbf{T}) = 0$$

By the use of the divergence theorem, the first condition gives the equation

$$\nabla \cdot \mathbf{T} + \mathbf{F} = 0$$

at any point inside the body, and the second condition implies that $\mathbf{T}$ is a symmetric dyadic. On the external surface of the body, we have usually to fulfill the boundary condition

$$\mathbf{n} \cdot \mathbf{T} = \mathbf{T}$$

where $\mathbf{T}$ is the applied external force per unit area. Other boundary conditions can also be met, such that the value of the displacement be prescribed.

For infinitesimal deformations, we assume that the relation between strain and stress is expressed by Hooke's law: the deformation is proportional to the applied force. For isotropic bodies, this linear relation is

$$\mathbf{S} = \frac{1}{E}[(1+\nu)\mathbf{T} - \nu|\mathbf{T}|\mathbf{1}]$$

where E is Young's modulus and ν is Poisson's ratio. These two elastic constants can be defined by considering the stretching of a cylindrical bar by normal traction forces uniformly distributed on the end sections; then we have

$$\text{Young's modulus} = \frac{\text{Normal traction force per unit cross sectional area}}{\text{Relative longitudinal extension}}$$

and

$$\text{Poisson's ratio} = \frac{\text{Relative lateral contraction}}{\text{Relative longitudinal extension}}$$

We can also write

$$\mathbf{T} = 2\mu\mathbf{S} + \lambda|\mathbf{S}|\mathbf{1}$$

where $\mu = E/2(1+\nu)$ and $\lambda = \nu E/(1+\nu)(1-2\nu)$ are Lamé's constants. μ is the rigidity modulus, the only constant necessary when the volumetric dilatation vanishes everywhere.

Substituting the preceding relation into the equilibrium equations, we transform them into

$$2\mu\nabla \cdot \mathbf{S} + \lambda\nabla|\mathbf{S}| + \mathbf{F} = 0 \text{ inside the body}$$

and

$$2\mu\mathbf{n} \cdot \mathbf{S} + \lambda\mathbf{n}|\mathbf{S}| = \mathbf{T} \text{ on the bounding surface.}$$

These vector relations are not sufficient for the complete determination of the symmetric dyadic $\mathbf{S}$. To insure that a solution of the above equations corresponds to a possible displacement vector $\mathbf{s}$, we must be able to integrate the relation

$$\mathbf{S} = \tfrac{1}{2}(\nabla\mathbf{s} + \mathbf{s}\nabla)$$

i.e., from a given expression for $\mathbf{S}$ obtain the value of $\mathbf{s}$. From the vanishing of the curl of a gradient, it is easily seen that this integrability condition, also called the compatibility equation, is

$$\nabla \times \mathbf{S} \times \nabla = 0$$

By elimination of the vector products, we obtain the equivalent form

$$\nabla\nabla \cdot \mathbf{S} + \nabla \cdot \mathbf{S}\nabla - \nabla\nabla|\mathbf{S}| - \nabla \cdot \nabla\mathbf{S} = 0$$

Using the stress-strain relation and the equilibrium conditions, we obtain the Beltrami-Michell form of the compatibility equation:

$$\nabla \cdot \nabla\mathbf{T} + \frac{1}{1+\nu}\nabla\nabla|\mathbf{T}| = -\frac{\nu}{1-\nu}\nabla \cdot \mathbf{F}\mathbf{1} - (\nabla\mathbf{F} + \mathbf{F}\nabla)$$

Finally, by expressing the strain dyadic in terms of the displacement vector, we obtain Navier's form of the equilibrium equations:

$$\mu\nabla \cdot \nabla\mathbf{s} + (\lambda+\mu)\nabla\nabla \cdot \mathbf{s} + \mathbf{F} = 0$$

inside the body and

$$\lambda n\nabla \cdot \mathbf{s} + 2\mu\mathbf{n} \cdot \nabla\mathbf{s} + \mu\mathbf{n} \times (\nabla \times \mathbf{s}) = \mathbf{T}$$

on the bounding surface. Dealing here directly with the displacement vector, there is no need of considering the compatibility equation.

The propagation equation for elastic disturbances is obtained by adding the inertia force to the body force. We get then

$$\mu\nabla \cdot \nabla\mathbf{s} + (\lambda+\mu)\nabla\nabla \cdot \mathbf{s} + \rho\mathbf{K} = \rho\frac{\partial^2}{\partial t^2}\mathbf{s}$$

inside the body. The stress-strain relation and the boundary conditions are not affected, but we generally have to take into account initial conditions.

The energy density u, or energy per unit volume, is given by

$$u = \frac{1}{2}\mathbf{S}:\mathbf{T} + \frac{1}{2}\rho\frac{\partial\mathbf{s}}{\partial t} \cdot \frac{\partial\mathbf{s}}{\partial t}$$

where the first term is potential, or strain energy, and the second term is kinetic energy. The energy flux density vector

$$\mathbf{S} = -\frac{\partial\mathbf{s}}{\partial t} \cdot \mathbf{T}$$

is a vector such that $dS\,\mathbf{n} \cdot \mathbf{S}$ gives the quantity of energy that flows per unit time through the surface element dS in the positive direction of $\mathbf{n}$, the normal to dS. At any point the energy continuity equation

$$\frac{\partial u}{\partial t} + \nabla \cdot \mathbf{S} - \rho\frac{\partial\mathbf{s}}{\partial t} \cdot \mathbf{K} = 0$$

expresses the conservation of mechanical energy.

ELASTIC LIMIT. The maximum unit stress which can be obtained in a structural material without causing a permanent deformation is called the elastic limit.

ELASTICITY (Modulus). Modulus of Elasticity.

ELASTIC REBOUND THEORY. Earthquakes, Seismology, and Plate Tectonics.

ELASTOMERS. Of natural or synthetic origin, an elastomer is a polymer possessing elastic (rubbery) properties. A polymer is a substance consisting of molecules which, are in the most part, multiples of low-molecular-weight units, or monomers. As an example, isoprene (2-methylbutadiene-1,3) is C_5H_8, whereas polyisoprene is $(C_5H_8)_x$, where $x \geq 2$ and normally is from 1,000 to 10,000 for rubbers. Although they differ in composition from natural rubber, many of these high-molecular-weight materials are termed *synthetic rubbers.* See also **Rubber (Natural).**

The serious development of synthetic rubbers commenced in the late 1930s and early 1940s, accelerated by a cutoff of supplies of natural rubber because of political turmoil and war. Synthetic rubbers fall into two major classifications: (1) general-purpose rubbers, but nevertheless the major volume of which is used for tire production; and

(2) specialty rubbers that essentially find little use in tires, but that are important for a number of other categories. Synthetic rubbers have not replaced natural rubber for numerous uses. For large, heavy-duty truck and bus tires, natural rubber tends to run considerably cooler and wears better than a blend of natural and synthetic rubbers. On the other hand, a tire tread made of a blend of styrene-butadiene (SBR) and butadiene rubber (polybutadiene) wears longer than natural rubber in conventional sutomobile usage, where lower temperatures can be maintained.

Styrene-Butadiene (SBR) Rubbers. This series of rubbers includes monomer ratios up to about 50% styrene. The addition of more than 50% styrene makes the materials more like plastic than rubber. The most commonly used SBR rubbers contain about 25% styrene, which is polymerized in emulsion systems at 5–10°C. Most SBR goes into tires, but the type for the tread differs from that of the sidewall or carcass. SBRs for adhesives, shoe soles, and other products also differ. The formulation permits vast varieties of end-products. Among the processing variables that can be manipulated to provide different end-characteristics are temperature, viscosity, use of different emulsifiers and solvents, use of different antioxidants for stabilization, different oils, carbon blacks, and coagulation techniques.

Initial processes for emulsion (E-SBR) called for polymerization at 50°C. However, it was found later that cold processing produced a rubber particularly good for tires. This is sometimes referred to as *cold SBR*. The overall process for making SBR is shown in the accompanying figure. Typical formulations are:

	Parts by Weight	
	Hot SBR	*Cold SBR*
Deionized water	180	200
Fatty acid soap	5	4.5
Styrene	25	28
t-Dodecyl-mercaptan	0.35	0.2
Butadiene	75	72
Potassium persulfate	0.3	
Redox initiator	—	small quantity
Temperature	50°C	5°C
Time	12 hours	8 hours
Conversion	75%	60%

Polymerization of emulsion SBR is started by free radicals generated by the redox system in cold SBR and by persulfate or other initiator in hot SBR. The initiators are not involved in the molecular structure of the polymers. Almost all molecules are terminated by fragments of the chain transfer agent (a mercaptan). Schematically, the molecules are RSM_nH, where RS is the $C_{12}H_{25}S$ part of a dodecyl mercaptan molecule; M is the monomer involved; n is the degree of polymerization, and H is a hydrogen atom formerly attached to the sulfur of a mercaptan. In the case of free-radical-initiated polymerization of butadiene, by itself to form homopolymers or with other monomers for form copolymers, the butadiene will be about 18%; 16% *cis*-1,4; and 66% *trans*-1,4.

Considerable quantities of SBR latex are used in the manufacture of foam rubber, adhesives, fabric treating, and paints. The solids content of latices runs from 50% to 65–70%.

Solution (S-SBR) consists of styrene butadiene copolymers prepared in solution. A wide range of styrene-butadiene ratios and molecular structures is possible. Copolymers with no chemically detectable blocks of polystyrene constitute a distinct class of solution SBRs and are most like styrene-butadiene copolymers made by emulsion processes. Solution SBRs with terminal blocks of polystyrene (S-B-S) have the properties of self-cured elastomers. They are processed like thermoplastics and do not require vulcanization. Lithium alkyls are used as the catalyst.

Stereospecific solution polymerization has been emphasized since the discovery of the complex coordination catalysts that yield polymers of butadiene and isoprene having highly ordered microstructures. The catalysts used are usually mixtures of organometallic and transition-metal compounds. An example of one of these polymers is *cis*-1,4-polybutadiene, the bulk of which is used in tires. However, it must be blended with other materials because of its poor processability and traction.

Butyl Rubber. Known as IIR, butyl rubber is a copolymer of isobutylene and isoprene. The elastomers contain only 0.5–2.5 mole % of isoprene. This is introduced to effect sufficient unsaturation to make the rubber vulcanizable. Polymerizations are usually carried out at low temperature (−80 to 100°C) with methyl chloride as solvent. Anhydrous aluminum chloride and a trace of water serve as catalyst.

Butyl is one of the earlier synthetic rubbers. However, it lost favor after the development of SBR. Butyl rubber is incompatible with natural rubber and is difficult to cure. Chlorobutyl rubbers, a much more recent development and containing up to 1.3% chlorine, apparently do not exhibit these former problems. Major uses for butyl rubber have been inner tubes for tires. The appearance of tubeless tires, however, depressed this market. Butyl rubber makes excellent motor mounts because of its high energy absorption and low rebound. Essentially free of double-bonds, butyl rubber has a high resistance to aging, attractive properties for use in curing bags for tire production and in outside coating materials.

Acrylonitrile-Butadiene Rubbers (NBR). Except for the monomers used, the production of NBRs is quite similar to that described for

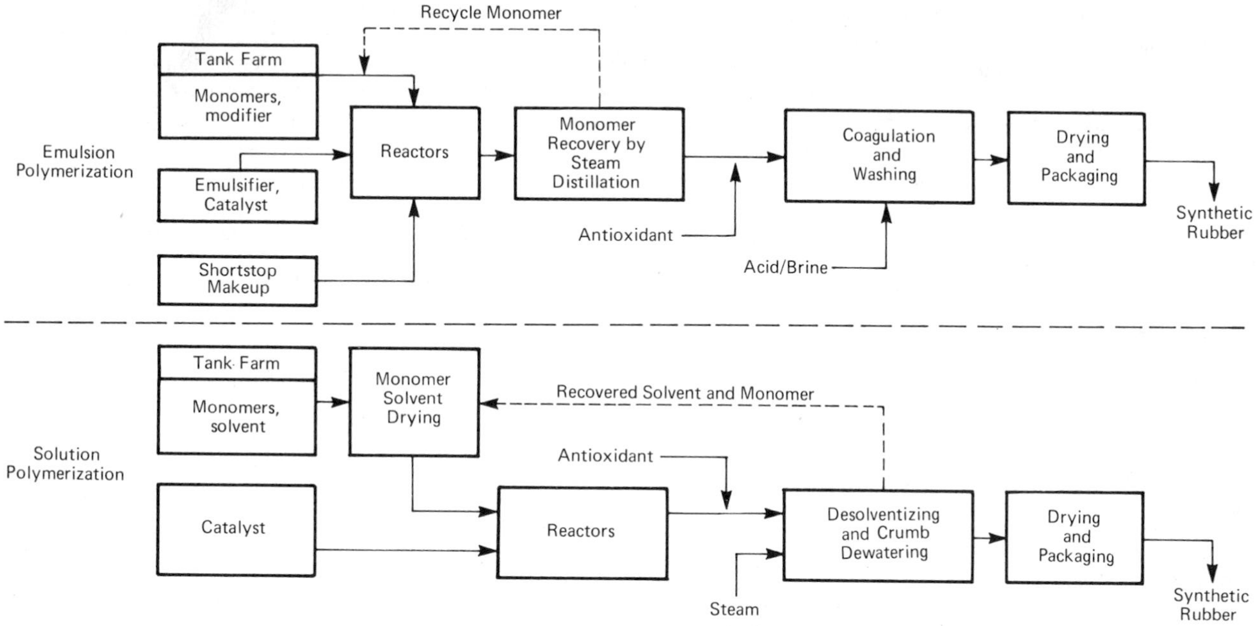

Synthetic rubber production process.

the SBRs. The NBR family is sometimes referred to as the *nitrile rubbers.* The acrylonitrile-butadiene ratios cover a wide range from 15:85 to 50:50. NBRs are noted for their solvent resistance, increasing with the acrylonitrile content. Thus, they are used for gaskets and oil and gasoline hoses, solvent-resistant electrical insulation, and food-wrapping films. Nitrile latices also are used in treating fabrics for dry-cleaning durability. Because the NBRs become quite inflexible (stiff) at low temperatures (actually brittle at about −20°C), they are blended with polyvinyl chloride for some applications.

Neoprene. This family of dry rubbers and latices was introduced in 1932, called *Duprene* by Du Pont at that time. The material is made by the free-radical-initiated polymerization of chloroprene in emulsion systems. As with most synthetic rubbers, a variety of neoprenes is made possible by variation of the polymerization conditions and ingredients. Neoprene is particularly good for its fire-retardant, solvent-resistant, and high-temperature stability properties. The chlorine in each segmer deactivates the adjoining carbon-carbon double-bond, thus making it less sensitive to oxidative attack. Metal oxides, such as zinc oxide and magnesium, serve as curatives rather than sulfur.

Polyurethanes. Polyethylene in solution is treated with chlorine and sulfur dioxide to introduce approximately 1.3% sulfur and 29% chlorine into the polymer. Most of the chlorine is attached directly to the carbon atoms in the backbone of the polymer. The remainder is in the form of sulfuryl chloride groups, $\cdot SO_2Cl$, through which cross-linking occurs in the curing step with metal oxides. The material has good oxidation and ozone resistance and thus overall excellent weather resistance. Calendered stocks are used for lining ditches and ponds, for example.

Thiokol® Rubbers. These are polysulfide rubbers and are prepared by the condensation polymerization of sodium polysulfides with a dichloro (sometimes blended with a trichloro) organic compound. Type A, the first family of rubbers, was made from Na_2S_4 and ethylene dichloride. Thiokols are known for high resistance to organic solvents.

Polyacrylate Elastomers. These are made in emulsion or suspension systems involving the copolymerization of ethyl acrylate with the acrylate esters of higher-molecular-weight alcohols. These materials have excellent solvent-resistant properties and stability at elevated temperatures. A major use is for automatic-transmission gaskets for automobiles.

Silicone Elastomers. These materials have alternating Si and O atoms for a backbone and the members differ mainly in the nature of the organic substituents on the Si atoms and the degree of polymerization. Because of the absence of double-bonds in the backbone, the numerous forms of stereoisomers found in unsaturated hydrocarbon rubbers do not have counterparts in the silicone rubbers. The chemical combination of organic and inorganic materials gives the silicone rubbers useful properties over a wide temperature range (−70 to + 225°C). These rubbers are well known for excellent dielectric stability and high resistance to weathering, oils and chemicals.

Fluoroelastomers. These materials are prepared by emulsion copolymerization of perfluoropropylene and vinylidene fluoride; or of chlorotrifluoroethylene and vinylidene fluoride. Also there are fluorosilicones in this family. Useful at temperatures up to and over 300°C, the fluoroelastomers have excellent resistance to aromatic solvents, acids, and alkalies. They are also among the more costly of the available commercial elastomers.

Ethylene-Propylene Elastomers. Known as EPR, this material is of limited use because it cannot be vulcanized in readily available systems. However, the rubbers are made from low-cost monomers, have good mechanical and elastic properties, and outstanding resistance to ozone, heat, and chemical attack. They remain flexible to very low temperatures (brittle point about −95°C). They are superior to butyl rubber in dynamic resilience.

ELBAITE. **Tourmaline.**

ELDER TREES AND VIBURNUMS. Of the family *Caprifoliaceae* (elder or honeysuckle family), the elder is a small tree, frequently a shrub, which is found in Europe and North America. The box elder (or ash-leaved maple) is a member of the family *Aceraceae* (maple family) and is not directly related to the other elders. Elder trees are of the genus *Sambucus.* These plants are deciduous, considered hardy, with odd, pinnately compounded leaves. There are from 5 to 11 ovate, pointed, finely-toothed leaflets. The flower is small and may be pink or creamy white in color, occurring in flat clusters. The fruit is a small berrylike drupe and may range in color from red, through blues and purples, and black, depending upon the species. There are about 12 species commonly found in North America. The more important include: American elder (*Sambucus canadensis*), which has black-to-purple-to-blue berries, well known as the source of elderberry wine; the blueberry elder (*S. glauca*), which has a blue fruit; the Mexican (*S. Mexicana*) and the velvet elder (*S. velutina*), both of which have black fruits; and the Pacific red elder (*S. callicarpa*), which has a red fruit. There are also some essentially local species, including the blackhead elder (*S. melanocarpa*), and the Florida elder (*S. simpsonii*). See accompanying table for record elder trees.

The American elder is found from Nova Scotia and New Brunswick westward across Canada to Manitoba and south and westward in the United States through Kansas and into Texas and Arizona. The tree tends to become larger as it is found in the west. The elder can be found in the Alleghany Mountains to an elevation of about 3500 feet (1070 meters). It thrives in rich moist soil.

The common elder of Europe, north Africa, and western Asia is the *S. nigra,* which has yellowish flowers and lustrous black fruit. The pithy new wood of this shrub or tree is known for its rather disagreeable odor, which is alleged to even repel flies and other insects.

Closely associated with the elders and also members of the elder family are the viburnums, which also are deciduous, but sometimes evergreen shrubs and trees. These plants have fragrant, clustered white flowers and fruits of fleshy consistency and bright colors. The *Viburnium prunifolium* is commonly referred to as the blackhaw or stagbush. Usually an erect bush, but frequently a tree ranging in height from 10 to close to 30 feet (3 to 9 meters), the plant normally will

RECORD ELDER TREES AND VIBURNUMS IN THE UNITED STATES[1]

Specimen	Circumference[2] (Inches)	Circumference[2] (Centimeters)	Height (Feet)	Height (Meters)	Spread (Feet)	Spread (Meters)	Location
American elder (1975) (*Sambucus canadensis*)	22	56	16	4.8	20	6	Virginia
Hybrid elder (1972) (*Sambucus callicarpa x cerulea*)	39	99	42	12.6	30	9	Oregon
Blueberry elder (1973) (*Sambucus cerulea*)	(at 36 inches) 91	(at 91 centimeters) 231	45	13.5	37	11.1	Oregon
Florida elder (1972) (*Sambucus canadensis*)	34	86	20	6	14	4.2	Florida
Pacific red elder (1976) (*Sambucus callicarpa*)	42	107	25	7.5	21	6.3	Oregon

[1] From the "Social Register of Big Trees," The American Forestry Association (by permission).
[2] At 4.5 feet (1.4 meters).

have a trunk diameter up to 10 inches. The bark is rough and gray-brown and in earlier years was used as a tonic and medicinal. The leaves are dark green, from 1 to 3 inches (2.5 to 7.6 centimeters) in length. The flowers are quite small and white and occur in large clusters approaching 5 inches (12.7 centimeters) in breadth. The fruit is bluish, edible, and sweet. The southern or rusty blackhaw (*V. rufidulum*) is a somewhat smaller species, but quite similar to the stag-bush. Record blackhaws also are listed on the accompanying table.

There are numerous species of the viburnums, just a few of which include: The wayfaring tree (*V. lantana*), very popular in Europe as a hedgerow plant; *V. opulus*, also known in Europe and north Africa as the Guelder rose or cranberry shrub, and particularly well adapted to wet areas. (Incidentally, this shrub is not related to the common cranberry, which is of the family *Ericaceae* (heather family).); the *V. alnifolium*, also known as hobble bush, witch hobble, and wayfaring tree; *V. opulus* var. *americanum*, commonly known as the pimbina, cramp-bark, cranberry tree, or highbush cranberry (these berries are sometimes used as substitutes for common cranberries); *V. pauciflorum*, known as the squashberry bush or squashberry pimbina; *V. acerifolium*, the dockmackie or arrowwood; *V. pubescens*, the downy arrowwood; *V. molle*, the soft-leaved arrowwood; *V. venosum*, similar to *V. molle*; *V. dentatum*, the arrowwood shrub; *V. cassinoides*, the withe-rod Appalachian tea shrub; *V. nudum*, the naked withe-rod; and *V. lentago*, the nannyberry, sheepberry, or wild raisin tree.

ELECTRET. A permanently-polarized piece of dielectric material; the analog of a magnet. Barium titanate ceramics, preferably containing a small percentage of lead titanate, can be polarized by cooling from a temperature above the Curie point in an applied electric field. Electrets are also produced by solidification of mixtures of certain organic waxes in a strong electric field.

ELECTRIC ACCOUNTING MACHINES. Automatic Data Processing (ADP).

ELECTRICAL CONDUCTIVITY. The measure of a material's ability to carry an electric current. An electric conductor is a material which, when placed between terminals having a difference of electrical potential, will readily permit the passage of an electric current. Different materials have different degrees of conductivity, and their effectiveness in this respect is computed as the conductivity. The best conductors are the metals, such as silver, copper, aluminum, platinum, and mercury, but nonmetallic substances such as carbon, saline solutions, and moist earth also are sufficiently conductive so that this property becomes of significance under certain circumstances. By virtue of their cost-conductivity characteristic, copper and aluminum are widely used conductors. They will usually be found as wires or buses. Copper is used more commonly than aluminum, the use of which is preferred for high-voltage transmission lines, where its lighter weight is of definite advantage. Steel as a conductor is inferior to the other two materials mentioned, but its greater strength and resistance to wear have led to its adoption as a conductor of special purposes, such as that of power rail service on electrified railways, and as an inner core of copper or aluminum cables. The resistance of a conductor is its resistivity multiplied by its length and divided by its cross-sectional area.

Commonly Used Electrical Conductors. For practical comparative purposes, the commonly used metals are compared with the International Annealed Copper Standard (IACS). A value of 100% conductivity is assigned to annealed copper. The standard may be expressed in terms of mass resistivity as 0.15328 ohm-grams/square meter, or the resistance of a uniform round wire 1 meter long weighing 1 gram at 20°C. Useful equivalent expressions of the annealed copper standard in terms of various units of mass resistivity and volume resistivity, include:

Value	Units at 20°C
0.15328	ohm-grams/square meter
875.20	ohm-pounds/square mile
1.7241	microohm-centimeters
0.67879	microohm-inches
10.371	ohm-circular mils/foot
0.017241	ohm-square millimeters/meter

CONDUCTORS—ELECTRICAL CONDUCTIVITY AND RESISTIVITY[a]

Material	% IACS (Volume) 20°C	Volume Resistivity (Microohm-Centimeters) 20°C
Annealed copper	100.00	1.7241
Copper alloy (B187-62)	98.40	1.7521
Copper alloy (B188-61)	97.80	1.7629
Copper alloy (B47-64; B116-64)	97.16	1.7745
Copper alloy (B355-65T, nickel coated)	96.00	1.7960
Copper alloy (B246-64, tinned hard)	92.72	1.8595
Copper alloy (B355-65T, nickel coated, soft)	88.0	1.9592
Aluminum	64.94	2.6548
Aluminum alloy (B233-64)	61.5	2.8035
Aluminum alloy (B317-64)	59.0	2.9222
Aluminum alloy (B396-63T)	52.5	3.2839
Copper-clad steel	40	4.3971
Aluminum-clad steel	20	8.4805
Iron (pure)		9.78
Commercial galvanized iron		16–20
Silver	108.4	1.59
Gold		2.3

[a] Representative metals and alloys. There are hundreds of metal combinations used. Frequently, electrical conductivity is not the sole criterion in selecting a conductor. Other factors include density (weight), strength, ductility, and corrosion and abrasion resistance.

All of the foregoing values are equivalent to $\frac{1}{58}$ ohm-square millimeters/meter. Thus volume conductivity can be expressed as 58 mho-square millimeters/meter at 20°C. See accompanying table.

Fundamentals of Electrical Conductivity. The conductivity σ of an isotropic material in a steady direct-current electric field E is defined by

$$j = \sigma E$$

where j is the current density (charge transported per unit time across unit area perpendicular to the current flow). In meters-kilograms-seconds (mks) units, j is measured in amperes per square meter and E in volts per meter, so that σ is in (ohm-meters)$^{-1}$, or mhos per meter. The reciprocal of the conductivity is the electrical resistivity, $\rho = 1/\sigma$. The electrical resistance R of a sample is the ratio of the potential drop across the sample to the total current through the sample. The resistance of a cylindrical sample (with the current flow parallel to the axis of the cylinder) is

$$R = \rho l/A$$

where l is the length of the cylinder and A is its cross-sectional area. If the dimensions are measured in meters and ρ is in mks units, the resistance is given in ohms. The reciprocal of the resistance is the electrical conductance.

Many homogeneous solids and liquids obey Ohm's law for sufficiently small electric fields. Ohm's law states that the current through a sample is proportional to the potential drop across the sample, and thus R, ρ, and σ are independent of the impressed electric field. Samples which do not obey Ohm's law are called *nonlinear*; gases fall into this category, as do many important circuit elements, such as transistors and vacuum tubes. Most metals obey Ohm's law for field up to at least 10^8 volts/meter, though under certain conditions, semiconductors have shown deviations from Ohm's law for fields as low as 10^{-2} volts/meter.

The existence of a finite resistance means that the energy delivered by the electric field to the current carriers is dissipated, being converted to heat (mostly energy of atomic vibrations). The rate of dissipation per unit volume is given by

$$\tfrac{1}{2}\rho j^2 = \tfrac{1}{2}\sigma E^2$$

and is called the Joule heat.

In some solids, the conductivity is *anisotropic*, i.e., the magnitude of j depends upon the direction of E as well as its magnitude, and j and E are not necessarily parallel. If Ohm's law is obeyed, it becomes

$$\mathbf{j} = \sigma E$$

where σ is a second rank tensor. Anisotropic solids include single crystals of materials which do not have cubic crystal structures, and polycrystalline aggregates of such materials in which there exists some preferred orientation such as can be produced by extrusion or cold rolling.

If the applied electric field varies sinusoidally in time (alternating current), the conductivity generally depends upon the applied frequency ν. Appreciable deviations from the dc value may appear for microwave frequencies or greater. In the microwave and optical range, it is common to identify $-\sigma/2\pi\epsilon_0\ \nu$ as the imaginary part of the *dielectric coefficient*, where ϵ_0 is the permittivity of free space $\frac{1}{4}\pi\epsilon_0 = 9 \times 10^9$ newton meter^2/coulomb^2. Thus, σ is closely related to the absorption of electromagnetic energy.

Solids are usually classified as metals, semiconductors, or insulators. Metals are characterized by an increasing resistivity with increasing temperature. Resistivities of metals at room temperature range from about 10^{-8} to 10^{-6} ohm-meters. Behavior at extremely low temperatures is covered under **Superconductors.** Semiconductors are characterized by a decreasing resistivity with increasing temperature (impure semiconductors may show this behavior only at high temperatures). Resistivities of semiconductors at room temperature range from about 10^{-5} to 10^{+7} ohm-meters. Insulators share the same temperature behavior of resistivity as semiconductors, so the difference between the two classes of materials is a simple one of degree only. Generally, materials with room-temperature resistivities greater than 10^7 ohm-centimeters are called insulators. Resistivities as high as 10^{18} ohm-meters have been observed. There are some materials intermediate between metals and semiconductors. For example, the resistivity of manufactured carbon decreases with increasing temperature at low temperatures, and increases at high temperatures. The room-temperature resistivity is also intermediate, varying from 10^{-5} to 10^{-4} ohm-meters, depending upon the conditions of manufacture.

Except in the case of some insulators, the current in solids is carried by electrons. Quantum mechanics has shown that not all the electrons in a solid are free to carry current. The conductivity depends upon the number of free carriers and their ease of motion. The latter factor is expressed by the mobility μ which is defined by

$$v_d = \mu E$$

where v_d is the drift velocity (average velocity of the free carriers produced by the action of the electric field). The drift velocity is usually very much smaller than a typical carrier velocity v (the average of the magnitude of the carrier velocities). The conductivity is then given by

$$\sigma = ne\mu$$

where n is the density of free carriers (in meters^{-3}), e is the magnitude of the electronic charge (1.6×10^{-19} coulomb), and μ is in square meters per volt per second. If more than one group of carriers is present, the total conductivity is the sum of contributions from each group. Quantum mechanics has also shown that the mobility in a pure, perfectly regular crystal would be unlimited. The mobility is limited by the scattering of the carriers by deviations, such as the thermal vibrations of the atoms, impurity atoms, or irregularities in the crystal structure, such as vacancies and dislocations. A simple approximate theory of the mobility allows it to be expressed as

$$\mu = e\tau/4\pi\epsilon_0 m^* = e\lambda/4\pi\epsilon_0 mv$$

where τ is the relaxation time (roughly, the average time between the collisions which a carrier makes with the scattering centers), λ is the mean free path of the carriers, and m^* is the effective mass of the carrier (a concept which comes from the theory of energy bands in solids). If more than one scattering process is involved, the reciprocal of the relaxation time (scattering rate) is approximately the sum of the contributions from each process. At room temperature, mobilities range from very small values (such as 10^{-4} meter^2/volt-second) to 10 meter^2/volt-second; relaxation times range from about 10^{-14} second to about 10^{-12} second, and mean free paths range from about 10^{-9} meter to about 10^{-6} meter. One of the early triumphs of quantum mechanics was the explanation of why the mean free path can be so much larger than the distance between neighboring atoms (of the order of 10^{-10} meter).

In good metals, n is the density of valence electrons, and is thus independent of temperature (except for very small temperature dependence due to the thermal expansion). Because the electrons follow the *Fermi–Dirac* distribution law, the typical velocity v is the velocity at the Fermi surface, which is large (usually about 10^6 meters/second), and independent of temperature. The temperature dependence of the conductivity is that of the relaxation time τ, and the approximate additivity of scattering reates due to different processes results in *Matthiessen's rule* which states that the resistivity is approximately the sum of a temperature dependent part R_T, due to scattering by lattice vibrations, and a part R_I proportional to the concentration of impurities and lattice defects. As the amplitude of the lattice vibrations increases with increasing temperature, the scattering effect and R_T increase. Above the Debye temperature θ (given by $k\theta = h\nu_m$, where k is the Boltzmann's constant and ν_m is the maximum frequency of the lattice vibrations), R_T is proportional to the absolute temperature. At low temperatures, R_T is proportional to the fifth power of the absolute temperature, and at all temperatures it is well approximated by the *Bloch–Gruneisen formula.* At very low temperatures (1 to 18°K), some metals become superconductors and all measurable resistance disappears.

Pure semiconductors at absolute zero temperature have no free electrons. As the temperature is increased, some electrons are excited to current-carrying states (in the *conduction band*). The states that are left unoccupied (in the *valence* band) are also free to carry current, and are called *free holes.* The concentrations of electrons and holes increase very rapidly with temperature, causing the resistivity to decrease. The carrier concentration may be expressed as

$$n = f(T)\exp(-\Delta E/2kT)$$

where T is the absolute temperature, ΔE is the energy gap between the valence and conduction bands, and $f(T)$ is a slowly varying function of temperature. Extra carriers may also be provided by impurity atoms, *donors* contributing electrons and *acceptors* trapping electrons and producing holes. Very small impurity concentrations may make very large changes in resistivity. If most of the free carriers come from impurities, the semiconductor is called *extrinsic*; it is n (negative)-type if electrons predominate, or p (positive)-type if holes predominate. If most of the carriers come from thermal excitation, the concentration of electrons and holes are about equal, and the material is called *intrinsic.* The rate of change with temperature of the mobility of a semiconductor is generally less than the rate of change of the carrier concentration. The mobility is often represented by $\mu = cT^r$, where r varies from about -2.2 to $+1.5$, depending upon the particular material, concentration, and type of defects. The conductivity of intrinsic material can also be expressed as

$$\sigma = g(T)\exp(-\Delta E/2kT)$$

where $g(T)$ is another slowly varying function of T. This equation is often used to analyze experimental data and find ΔE.

Insulators may be thought of as semiconductors with such large energy gaps that n is very small and the resistivity is very high. In addition, the ions in some insulating solids (such as the alkali halides) are free to move and carry a measurable current. The ions move by "hopping" into vacant lattice sites. As they must cross a potential energy barrier, the ionic mobility is proportional to an activation factor $\exp(-\epsilon/kT)$. The resistivities of such solids are of the order of 10^2 to 10^8 ohm-meters at room temperature, but are as low as 10^{-3} to 1 ohm-meter at elevated temperatures.

Another mechanism for electronic conductivity seems to operate in certain oxides and perhaps in some organic semiconductors. In this case, the electrons are localized and move by "hopping" in the same way as the ions in ionic solids.

Because the resistivity depends upon the state of crystalline order, it is used as a tool in studying phase changes, such as solid–liquid, order–disorder, magnetic, and crystal structure transitions. The resistivity of some materials is affected by pressure and mechanical strain.

Many insulators and semiconductors exhibit the phenomenon of photoconductivity, in which the absorption of light produces free carriers and increases the conductivity.

Many conductors show the phenomenon of *magnetoresistance,* which is the increase in resistance when the conductor is placed in a magnetic field. (A very few materials exhibit negative magnetoresistance, which seems to be related to an inhomogeneous structure of the conductor.) The basic cause of the magnetoresistance is the Lorentz force, which causes the electrons to move in curved paths between collisions. Even for isotropic materials, the conductivity and resistivity must be taken as tensors in the presence of the magnetic field, called the *magnetoconductivity tensor* and the *magnetoresistivity tensor.* The off-diagonal elements of the magnetoresistivity tensor are related to the *Hall effect.* The on-diagonal elements for isotropic materials or for the magnetic field parallel to a crystal axis are usually called the longitudinal magnetoresistance (for the current parallel to the magnetic field) and transverse magnetoresistance (for the current perpendicular to the magnetic field). For small values of the magnetic field, the change in resistance is proportional to the square of the magnetic field strength. For large magnetic fields, the resistance may saturate (approach a constant value), continue to increase as the square of the magnetic field strength, or follow a more complex behavior. The magnetoresistance ratio (change in resistance divided by the zero-field resistance) reaches only a few percent for most materials, but at low temperatures for some materials, such as bismuth, approaches 10^6. In general, the transverse effect is larger than the longitudinal one; the effect is larger in high-mobility materials, and is largest for materials with more than one type of charge carrier. The magnetoresistance may be used in conjunction with the Hall effect to deduce the type, density, and mobility of charge carriers, and to obtain information about the Fermi surface.

Electrolytes. The conductance of an electrolyte has the same general definition as that given for any conductor. The same unit is used, the reciprocal ohm (mho). The term conductivity also has the same significance for electrolytes as for solid conductors, being conductance per unit length of path. Another term applied to electrolytes is *equivalent conductance,* which is the product of conductivity and the volume (in cubic centimeters) of solvent containing 1 gram-equivalent weight of solute at a specified concentration, measured when placed between electrodes 1 centimeter apart. The *molar conductance* is defined similarly, except that the weight of solute is 1 gram-molecular weight instead of 1 gram-equivalent weight. The *ionic conductance* is the amount contributed by each characteristic ion to the toal equivalent conductance in infinite dilution. Thus, in the mathematical expression of the law of independent migration of ions:

$$\Lambda_0 = \lambda_+ + \lambda_-$$

in which λ_+ and λ_- are the ionic conductances of cation and anion, respectively, and Λ_0 is the total equivalent conductance of the electrolyte. The *conductance ratio* is the ratio of the equivalent conductance of a given ionic (electrolytic) solution to its equivalent conductance at infinite dilution.

See also **Electrolytic Conductivity Measurements;** and **Superconductors.**

ELECTRICAL CONDUCTORS. Aluminum; Copper; Silver; Superconductors.

ELECTRICAL INSTRUMENTS. The basic electrical analog instrument can be traced to Oersted's discovery (1819) of the relation between current and magnetism. Faraday (1821) learned that a current-carrying conductor would rotate in a magnetic field. Ampere (1821) demonstrated that a current in one conductor attracted or repelled another current-carrying conductor. Sturgeon (1836) wound the current-carrying wire into a coil and suspended it in a magnetic field. Kelvin (1867) placed a soft iron core in the center of the coil, shortening the air gap, increasing the sensitivity, and improving the scale characteristics of the device. D'Arsonval (1881) patented an instrument of this type. Weston (1888) discovered the factors required to produce a permanent magnet system, added soft iron pole pieces, devised current-carrying control springs and produced the first commercial double-pivoted permanent magnet moving coil instruments as such. From Oersted's and Kelvin's work also stem the principles on which polarized iron vane and electrodynamic type instruments evolved through the work of Ayrton and Perry (1881), Bruger (1886), Weston (1889) and many others.

Traceability

Electrical instruments are calibrated in terms of the prime standards of the quantity measured. Such prime standards are established and maintained by the U.S. National Bureau of Standards and several other governmental standards organizations throughout the world. As in the United States, the governmental standards are usually the legal standards for commercial reference and use of the electrical variables, such as amperes, volts, ohms, watts, and so on.

A cornerstone of maintaining accuracy and reliability in the manufacture of numerous electrical instruments is the standard cell, of which one version—the cadmium cell—was developed by Dr. Edward Weston in 1893. Standard cells are described in the entry on **Battery.** Instrument manufacturers periodically send groups of saturated standard cells to the National Bureau of Standards for calibration for use as secondary standards and working standards.

The method of tracing the accuracy of electrical instruments to prime, legal standards is shown by Fig. 1. Extension of current values is accomplished through the use of very low valued resistors; passing the unknown current through them results in a voltage which, in turn, can be compared with the standard cell voltage, and through the use of Ohm's law and the resistance value we can get the value of the current. Such low valued resistors are called shunts, and standard units of this kind are maintained with much care; a few sets of various values even carry calibrations by the National Bureau of Standards.

The direct voltages and currents established at this level are then used to calibrate dc voltmeters, ammeters and ac–dc voltmeters, ammeters, and electrodynamic wattmeters.

Also shown on the chart are ac–dc difference or transfer standards. These are high accuracy instruments such as the thermal and electrodynamic type voltmeters, ammeters and wattmeters which are sent to the National Bureau of Standards for calibration. In this case the Bureau of Standards does not perform an accuracy calibration, but rather determines the difference between the ac calibration at various frequencies and that on direct current. These standards are then used

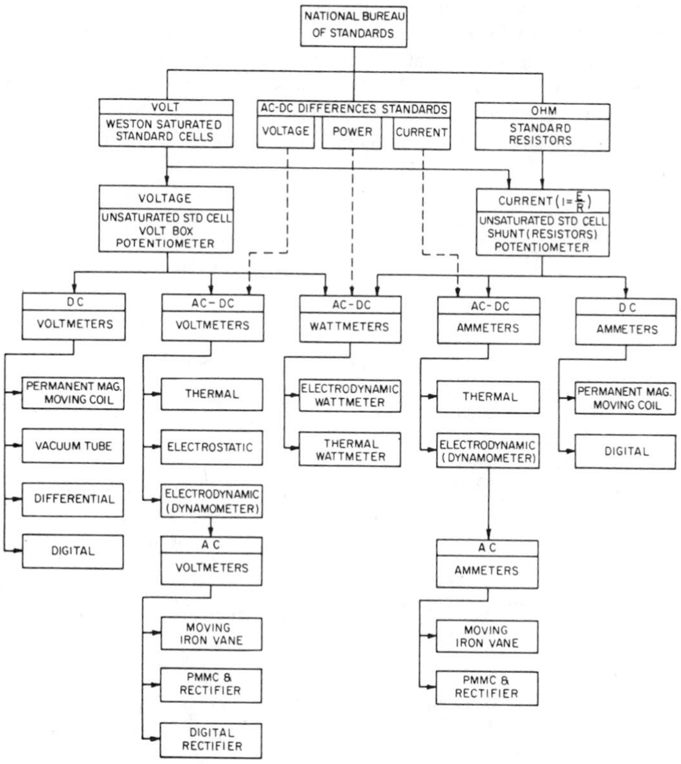

Fig. 1. Pathways for tracing the accuracy of electrical instruments back to the prime, legal standards.

for determining the performance of ac–dc instruments when used on alternating current.

Alternating current voltmeters and ammeters are usually calibrated against electrodynamic voltmeters and ammeters which have been calibrated on direct current and which have a known accuracy on alternating current as determined by the ac–dc difference test.

Resistors used as basic standards in values above 10 ohms are built by the National Bureau of Standards. Carefully heat treated manganin wire is wound on an insulated brass cylinder and cemented in position. Separate connections are provided for input and potential leads. After extended baking to insure dryness the assembly is placed in moisture-free oil and sealed in its brass case. Below 10 ohms the construction is the same except mechanically larger. Such units, properly handled and used, maintain their accuracy within a few parts per million for a long period of time.

Fundamentally, the use of standards of emf and resistance includes, by derivation, the standardization of current through the application of Ohm's law by which the three basic electrical quantities are interconnected. The practical standardization and calibration of electrical measuring instruments is based on these principles.

Polarized Iron-Vane Mechanism

The polarized iron-vane system was the first current measuring mechanism to be reduced to practice. Originally dependent upon the earth's magnetic field to provide the control torque, it was later made both more sensitive and more convenient in use by the use of a permanent magnet. In the form shown in Fig. 2(e), the mechanism came into use in large quantities as an automobile dashboard ammeter. Better suited to mass production in relatively crude form than to greater refinement, this mechanism has lost its position in the field of electrical instruments and has, instead, made a place for itself in the "indicator" field where precision is not demanded.

Moving Iron-Vane Mechanism

In an early magnetic vane mechanism of the suction type the opposing force or restoring torque is provided by gravity instead of the presently used conventional spring. This method was widely used in older instruments. All gravity-controlled instruments had the major disadvantage of being subject to serious position errors.

With reference to Fig. 3, if two similar adjacent iron bars are similarly magnetized, a repelling force is developed between them which tends to move them apart. In the moving iron vane mechanism, this principle is used by fixing one bar in space and pivoting the second so that it will tend to rotate when the magnetizing current flows. A spring attached to the moving vane opposes the motion of the vane and permits the scale to be calibrated in terms of the current flowing.

In the concentric vane mechanism shown in Fig. 4, vanes slip later-

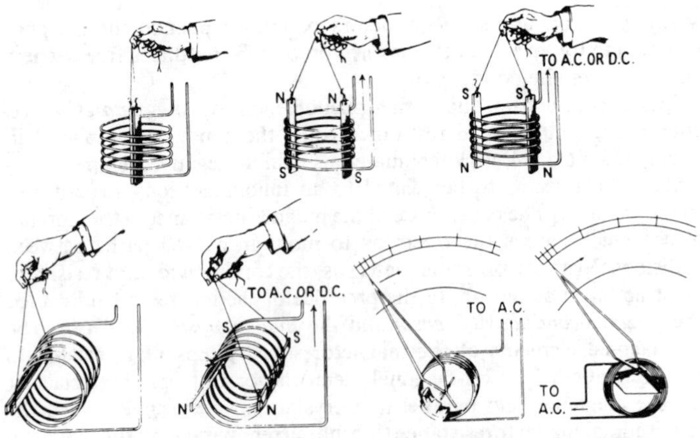

Fig. 3. Principle used in moving iron-vane mechanism.

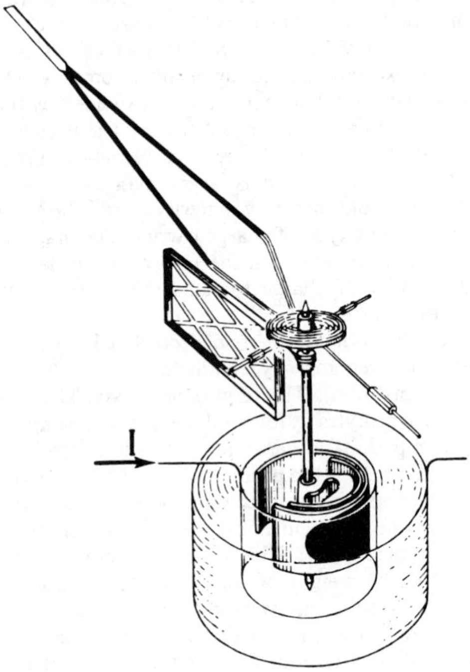

Fig. 4. Concentric vane mechanism.

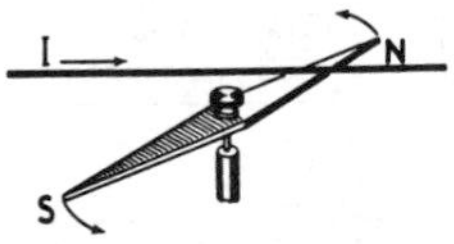

(a) The original electrical indicator—Oersted—1819.

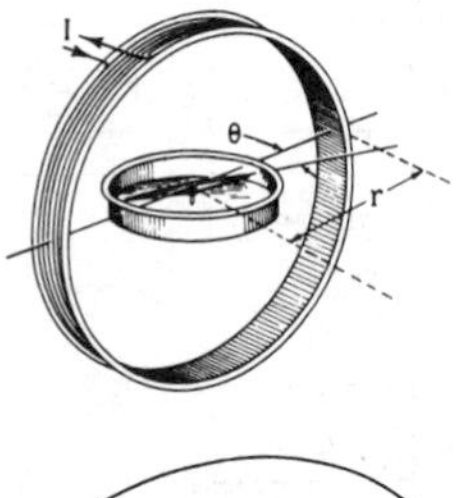

(b) Tangent Galvanometer,

$$I = \frac{10\ Hr}{2\pi N} \tan\theta \text{ amperes.}$$

r = radius of coil.
H = horizontal component of earth's field in gauss.
N = number of turns.

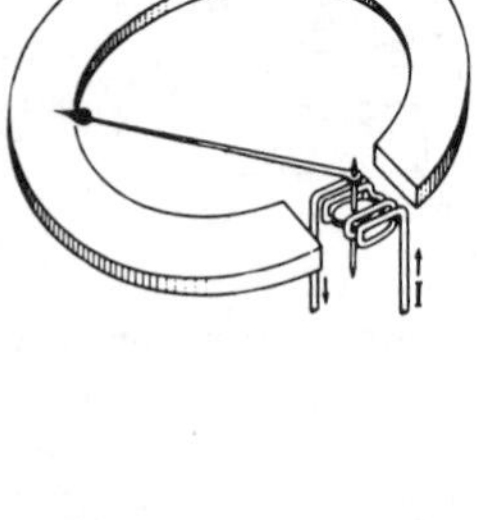

(c) Simple polarized iron vane mechanism. Pointer is driven by an iron vane governed by the resultant field of the permanent magnet and that of the current in the coil.

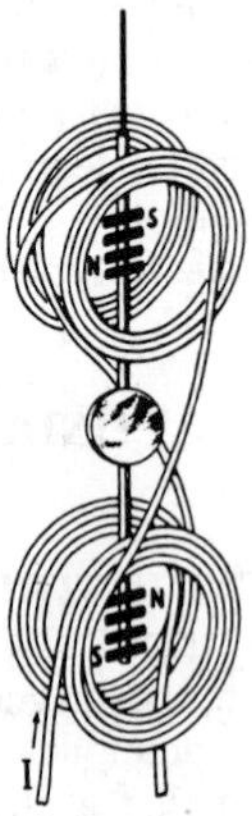

(d) Early type astatic galvanometer —Kelvin—1858. Used on early trans-Atlantic cable circuits. Note lower set of needles has reversed magnetic polarity from upper set, thus reducing the restoring force of the earth's field and increasing the sensitivity.

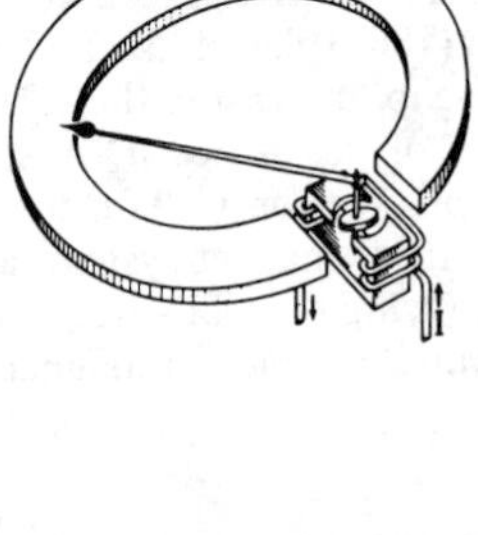

(e) Complete polarized iron vane mechanism. Soft iron core adds to sensitivity, requiring fewer turns of wire and improves scale characteristics. Weston prototype——Model 354 (now obsolete).

Fig. 2. Forms of polarized iron-vane mechanism.

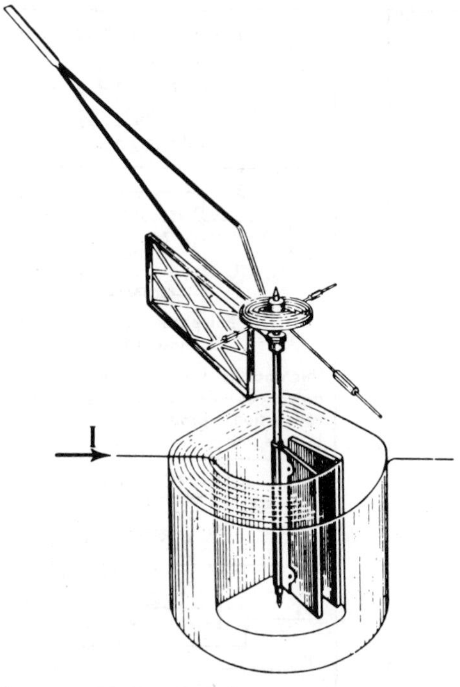

Fig. 5. Radial vane mechanism.

ally under repulsion. This design has only moderate sensitivity and has square-law scale characteristics. The short magnetic vanes result in small direct-current reversal and residual magnetism errors. With this mechanism, it is also possible to shape the vanes to secure special characteristics, thereby opening the scale where needed.

The radial vane mechanism, shown in Fig. 5, opens up like a book under repulsion. Of the polarized iron-vane mechanisms, this is the most sensitive, has the most linear scale, and requires better design and better magnetic vanes for good grades of instruments. An aluminum damping vane, attached to the shaft just below the pointer, rotates in a close-fitting chamber to bring the pointer to rest quickly.

The Electrodynamic Mechanism

Electrodynamic mechanisms are the most fundamental of all of the indicating devices presently used. Form factor variations do not occur because of the complete absence of magnetic materials, such as iron, and the indications are of true RMS values. This mechanism is current sensitive—the pointer moves because of current flowing through turns of wire. It is the most versatile of all of the basic mechanisms since the single-coil movement can be used to indicate current, voltage, or power, ac or dc. Crossed-coil movements can be used for power factor, phase angle, frequency, and capacity measurements. Arrangements of this type are shown in Fig. 8.

An early version of this basic principle was the current balance of Lord Kelvin (1883) and shown in Fig. 6.

Perhaps the most important use of this mechanism is as a transfer instrument between the basic standards of *E*, *I*, and *R*, all of which are defined for direct current only, to alternating current, in which form most of the power of the world is generated, sold, and used. The torque produced in the moving coil of an electrodynamic instrument is proportional to the product of the in-phase components of the currents in the field and moving coils; this mechanism measures products.

The electrodynamic mechanism used in commercially available

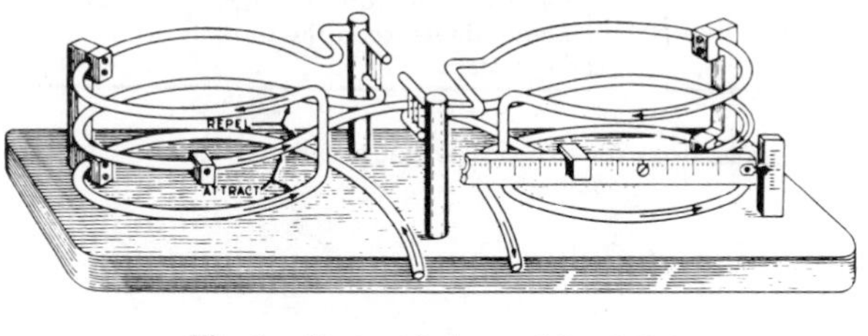

Fig. 6. Current balance of Lord Kelvin.

wattmeters, voltmeters, and ammeters is shown in Fig. 7. With a longer pointer and other minor changes, the mechanism is used in a laboratory standard. With the further modification of crossed-moving coils indicated in Fig. 8, fundamentally the same mechanism is used for measurements of power factor, phase angle, and capacity. In this form or with crossed field coils, this mechanism measures a ratio by balancing two torques.

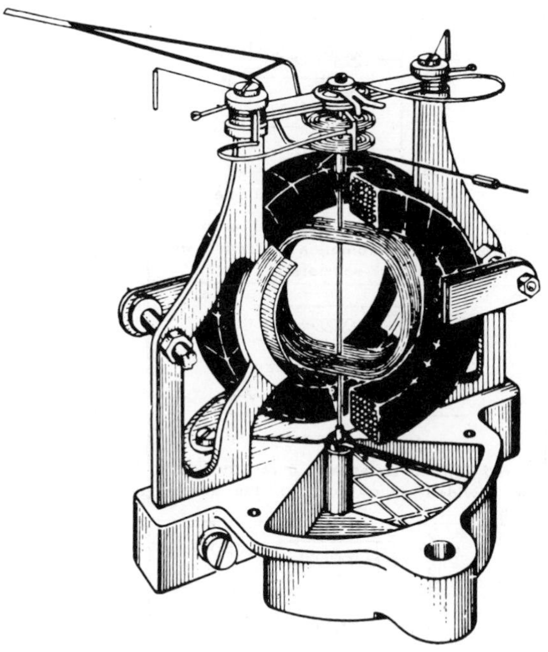

Fig. 7. Electrodynamic mechanism used in some contemporary voltmeters, ammeters, and wattmeters. (*Weston.*)

Damping vanes similar to those used in moving iron instruments (Figs. 4 and 5) rotate in close-fitting chambers and provide proper damping for the pointer motion. If two complete field coil systems are arranged one above the other, each including and acting upon its own moving coil, and if the two moving coils are attached to a common shaft which carries the pointer, the mechanism can be used to measure the total power in a poly-phase ac system. These arrangements are indicated in Fig. 8.

The Electrostatic Mechanism

The electrostatic mechanism resembles a variable condenser. Of all the mechanisms used for electrical indications, it is the only one that measures voltage directly rather than by the effect of current. The torque resulting from the attraction between fixed and moveable plates is a function of the voltage between the plates, the plate area and, inversely, the distance between the plates. For greater sensitivity, this distance must be reduced, clearances permitting, or the plate area (and thus the weight) must be increased. Greater weight, in turn, calls for still greater plate area if a sufficient torque is to be developed to overcome pivot friction in most industrial applications. This limits the use of the instrument to certain special applications, particularly in ac currents of relatively high voltage, where the current taken by other mechanisms would result in erroneous indications. A protective resistor is used in series with the instrument to limit current flow in the event of a short between plates.

A gold leaf electroscope is shown in Fig. 9. Like charges on the ends of the leaf cause a separation of the ends. This is very sensitive as an indicator, but not characterized as an instrument per se. A form of the attracted-disk electroscope as devised by Sir William Snow Harris (1834) is shown in Fig. 10. Guard ring around the disk served to prevent nonuniformity in the electrostatic lines of force. A very large example of the attracted-disk electrometer mounted in a shielded cage and using quartz pillar supports is used at the National Bureau of Standards for voltage standardization up to 300,000 volts. Using this high voltage electrometer, the ratio of transformation of high

SINGLE ELEMENT TYPE

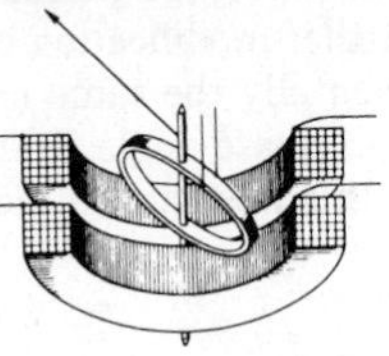

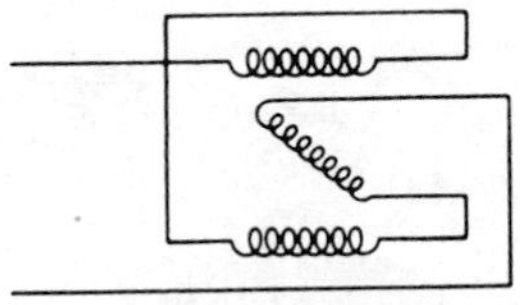

Milliammeter-Voltmeter

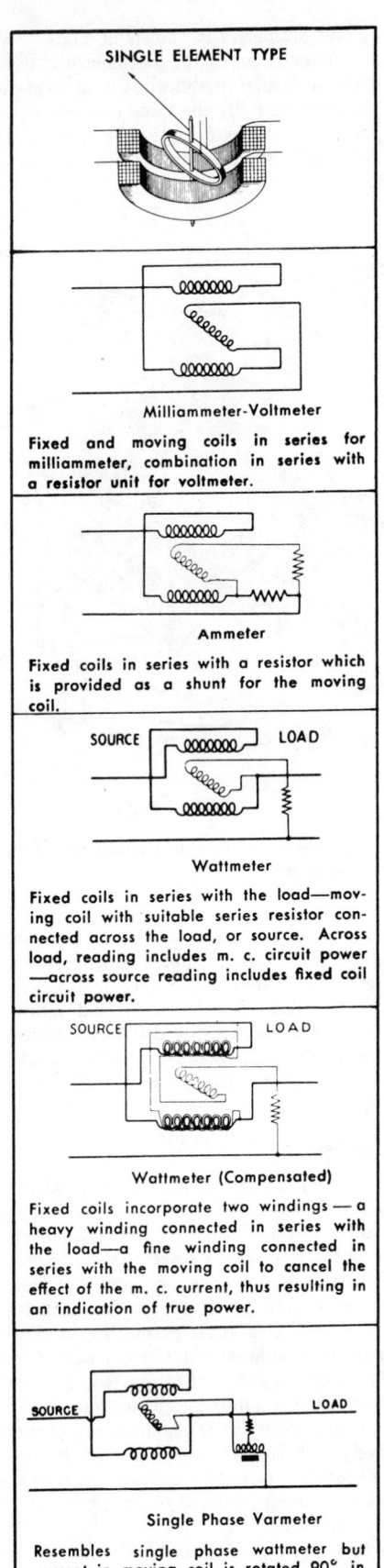

Fixed and moving coils in series for milliammeter, combination in series with a resistor unit for voltmeter.

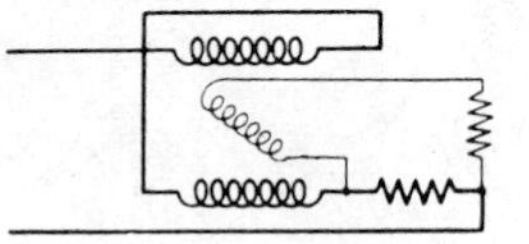

Ammeter

Fixed coils in series with a resistor which is provided as a shunt for the moving coil.

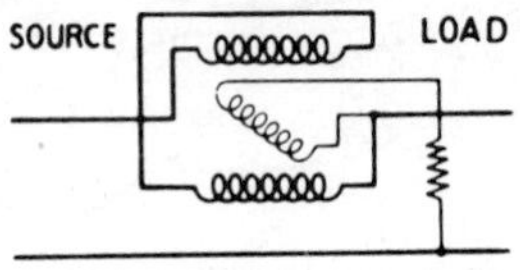

Wattmeter

Fixed coils in series with the load—moving coil with suitable series resistor connected across the load, or source. Across load, reading includes m. c. circuit power—across source reading includes fixed coil circuit power.

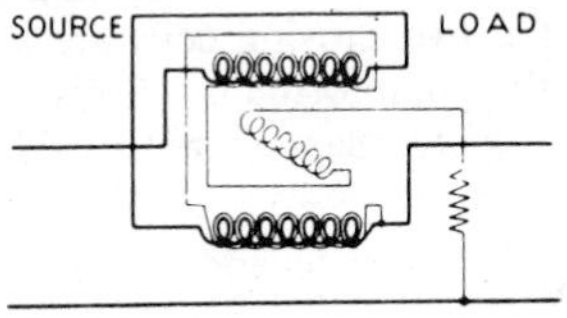

Wattmeter (Compensated)

Fixed coils incorporate two windings — a heavy winding connected in series with the load—a fine winding connected in series with the moving coil to cancel the effect of the m. c. current, thus resulting in an indication of true power.

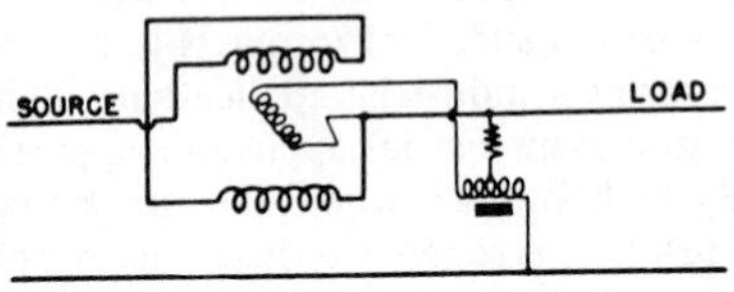

Single Phase Varmeter

Resembles single phase wattmeter but current in moving coil is rotated 90° in time phase.

DOUBLE ELEMENT TYPE

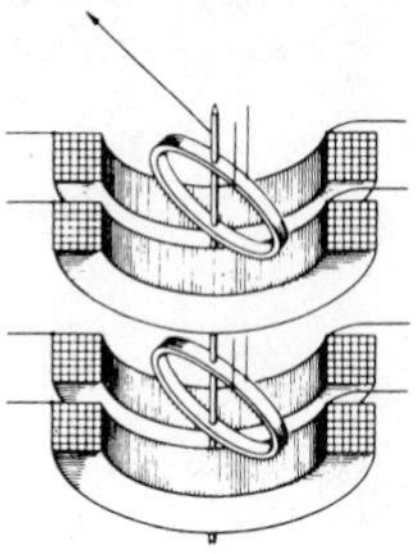

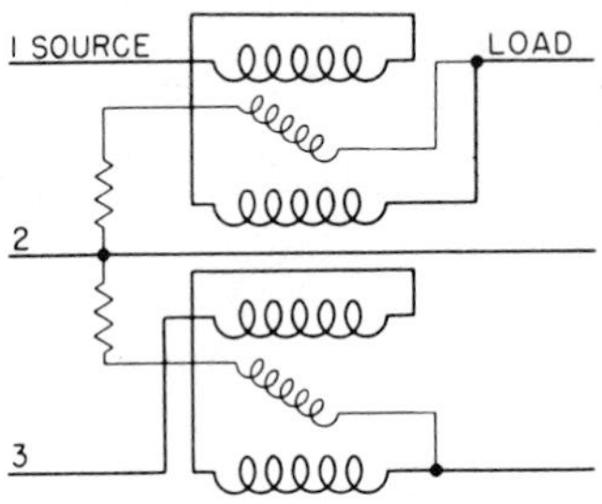

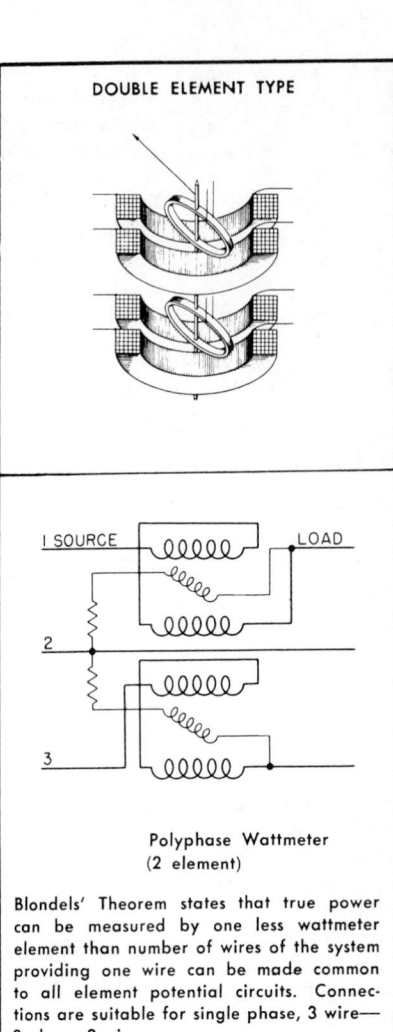

Polyphase Wattmeter
(2 element)

Blondels' Theorem states that true power can be measured by one less wattmeter element than number of wires of the system providing one wire can be made common to all element potential circuits. Connections are suitable for single phase, 3 wire—3 phase, 3 wire.

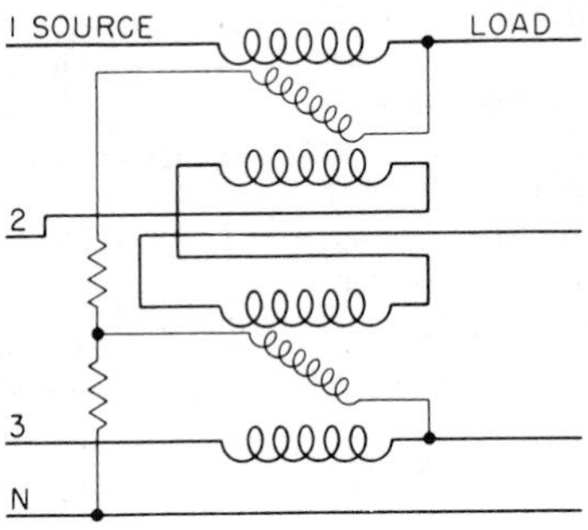

Polyphase Wattmeter
(2½ element)

Above arrangement constitutes a compromise on Blondels' Theorem for measurement of 3 phase, 4 wire system power—will indicate true power providing voltages are balanced. Called 2½ element because of split connection of fixed coils. Note current in line 2 passes thru both upper and lower element, in reverse direction, but is displaced in phase by 60°—its effect is thus half. Summation of two elements equals third element required.

DOUBLE ELEMENT TYPE (Cont.)

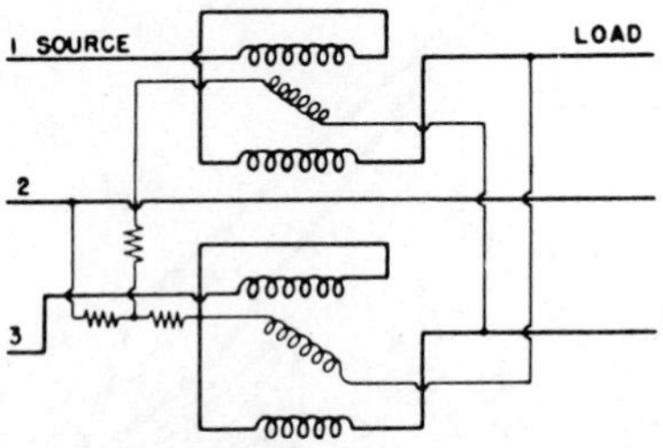

Polyphase Varmeter (2 element)
3 Phase 3 Wire System

This and the 4 wire version below resemble polyphase wattmeters except moving coil connections produce currents 90° rotated in time phase.

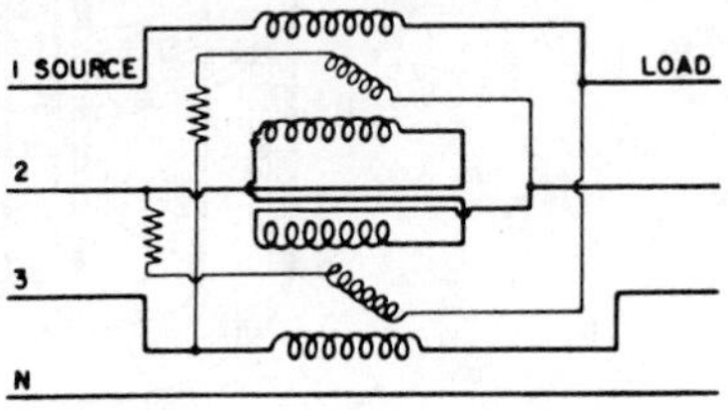

Polyphase Varmeter for 3 Phase
4 Wire System (2½ element type)

CROSSED COIL TYPE

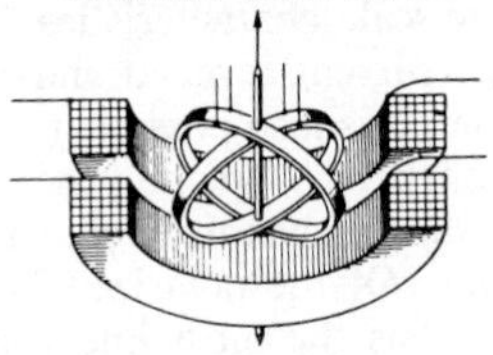

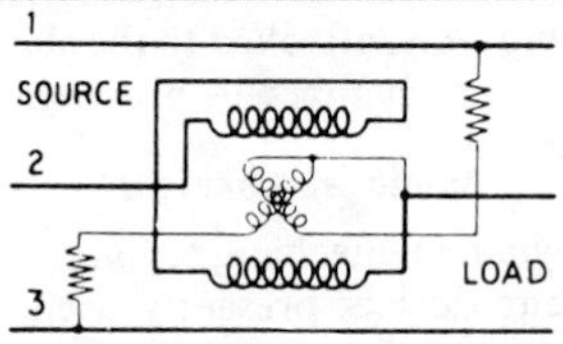

Power Factor Meter (3 phase type)

Crossed moving coils are connected to opposite legs of a 3 phase system—Fixed coils connected in series with line used as common for moving coil connection. Moving system assumes a position depending on power factor of circuit. Correct for balanced loads only.

CROSSED COIL TYPE (Cont.)

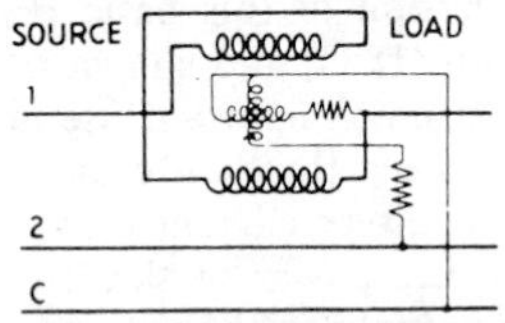

Power Factor Meter (2 phase type)

Essentially similar to 3 phase except pointer is mounted in line with inner coil. Fixed coils not in the common line.
For a single phase Power Factor Meter the double moving coil is again used, with a reactance network to give the required phase difference in the currents in the moving system.

Fig. 8. Physical arrangement of coils in electrodynamic and similar mechanisms for electrical measurements.

voltage potential transformers has been checked for the first time by an independent method.

Fig. 9. Gold leaf electroscope.

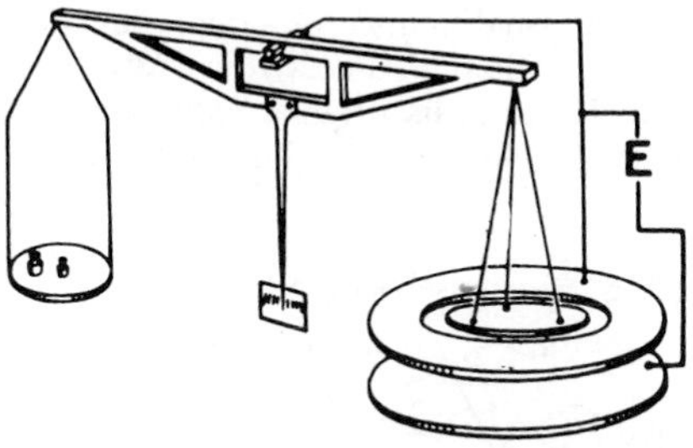

Fig. 10. Attracted-disk electroscope.

Figure 11 illustrates the principle of an early electrostatic mechanism devised by Lord Kelvin in 1887. Moving and fixed plates attract each other causing the moving plate to rotate against the torque of a control spring (not shown). Position of balance thus becomes a measure of

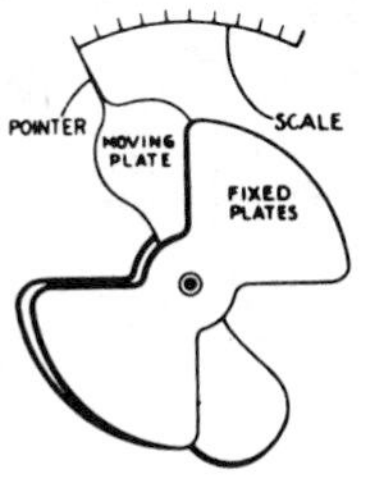

Fig. 11. Early electrostatic mechanism devised by Lord Kelvin.

the potential applied. In contemporary form (Fig. 12), such instruments in $3\frac{1}{2}$-inch (~9 centimeters) panel size are made with multiple sets of plates in ranges from 150 to 3,500 volts.

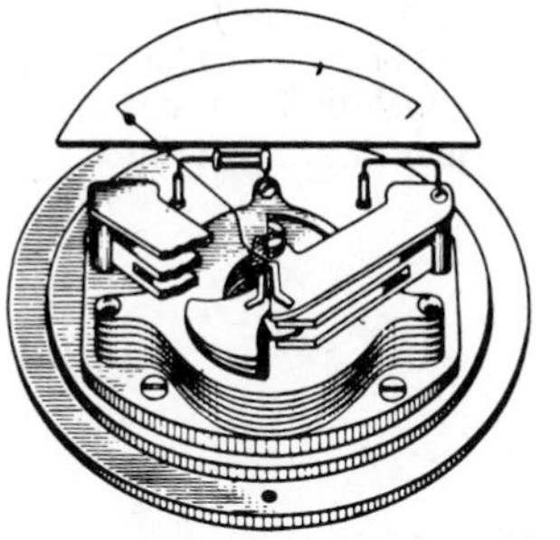

Fig. 12. Contemporary form of moving-and-fixed-plate-type electrostatic instrument.

Permanent-Magnet Moving-Coil Mechanism

The suspension galvanometer is described in entry on **Galvanometer.**

The design shown in Fig. 13 offers the largest magnet in a given space and is used where maximum flux in the air gap is required in order to provide an instrument of lowest possible power consumption. Common practice is to grind one side of an Alnico V block of magnetic material and solder or cement it to soft iron pole pieces to form a "U" shaped magnetic structure.

In recent years, with the advent of Alnico and other improved magnetic materials, it has become feasible to design a magnetic system in which the magnet serves as a core. Such magnets operate at their highest energy product with minimum lengths, thereby making the core magnet mechanism a practical reality. These mechanisms have

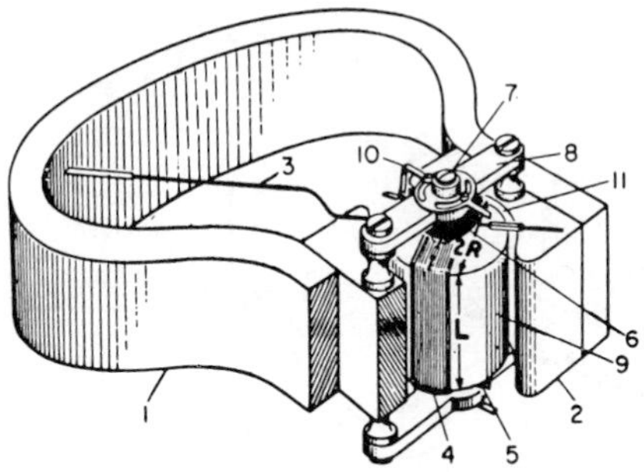

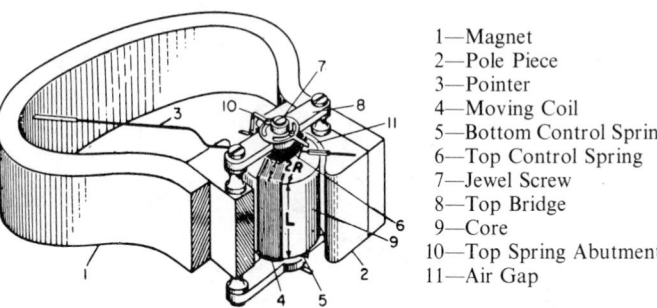

1—Magnet
2—Pole Piece
3—Pointer
4—Moving Coil
5—Bottom Control Spring
6—Top Control Spring
7—Jewel Screw
8—Top Bridge
9—Core
10—Top Spring Abutment
11—Air Gap

Fig. 13. Permanent-magnet moving-coil mechanism with external magnet.

the obvious advantage of being relatively resistant to external magnetic fields, therefore allowing for the elimination of magnetic shunting effects in a panel or the need for magnetic shielding in the form of iron cases. See Fig. 14.

The advantage of the self-shielded feature makes the core magnet mechanism particularly useful for aircraft and aerospace applications, especially where a multiplicity of mechanisms must be mounted in close proximity to each other as exemplified by cross-pointer indicators wherein as many as five mechanisms are mounted in one case to form a unified display. Obviously, the elimination of iron cases and the corresponding weight reduction are of great advantage in aerospace instruments.

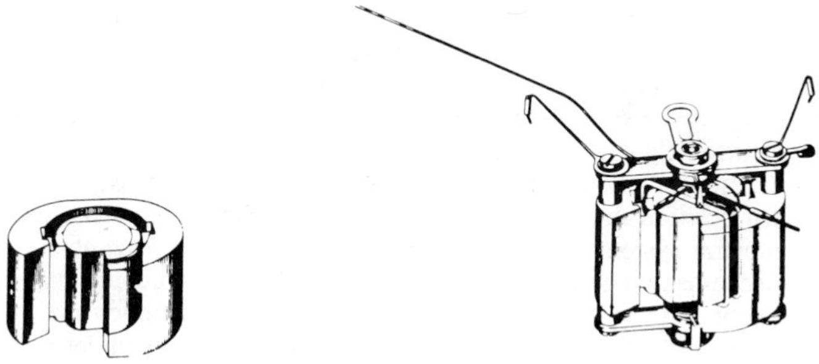

Fig. 14. Core magnet and moving-coil mechanism.

End-Pivoted or Off-Center Coil Type. The end-pivoted permanent-magnet moving-coil mechanism is a variation of the more common center-pivoted type. In this arrangement, the coil rotates in a single air gap, allowing a full-scale deflection as high as 270°. The concept is not new. Meters of this type have been made as far back as 1900, but with low magnetic flux and poor performance. A modern self-shielded version is shown in Fig. 15. This instrument has a scale

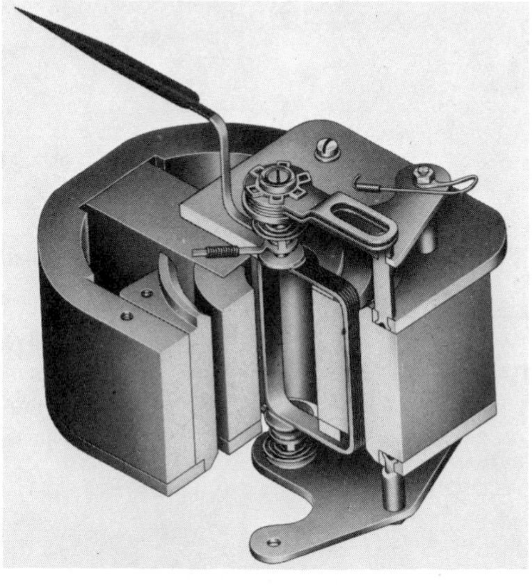

Fig. 15. End-pivoted, self-shielded, permanent-magnet moving-coil mechanism. (*Weston.*)

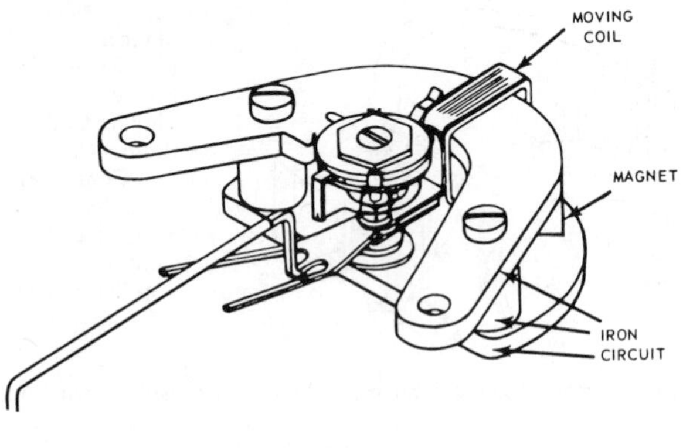

Fig. 16. Typical edgewise meter mechanism.

deflection of 250°. The Alnico magnet extends into the soft iron core, thereby minimizing leakage flux and achieving a high flux density.

The off-center coil type is also used extensively in edgewise panel and aircraft meters. The deflection is usually limited to 60° or less with the magnetic flux concentrated over the smaller angle. An edgewise meter is shown in Fig. 16. In the latter, a flat Alnico magnet is attached to one of two soft iron plates separated by iron bosses to complete the magnetic circuit. The coil rotates over the upper plate and magnet with the lower side of the coil in the active air gap. An important advantage of this construction is that the coil is a counterbalance to the pointer. This is especially important for aircraft mechanisms and other designs with long or heavy pointers.

Taut-Band Instruments. The suspension type mechanism has been known for years. Until recently, the device was utilized for laboratory equipment where high sensitivities were required and available torques were extremely low so that it was desirable to eliminate even the low friction of pivots and jewels. The device had to be used in the upright position since the sag in the very low torque ligaments caused the moving system to come in contact with stationary members of the mechanism in any other position. The taut band instrument enables one to obtain the advantage of the absence of friction of ribbon suspension while also eliminating the sag by placing the ribbons under sufficient tension. This tension is provided by a tension spring so that the instrument can be used in any position. Generally speaking, taut band suspension instruments can be made with higher sensitivities than those using pivots and jewels and can be utilized in virtually every application which is presently served by pivoted instruments. See Fig. 17.

A permanent-magnet moving-coil mechanism is not insensitive to temperature by itself but may be made so by the appropriate use of proper series and shunt resistors of copper and manganin. Magnets and springs decrease in strength and copper increases in resistance with increase in temperature. The changes in the magnet and the copper tend to make the pointer read low on fixed voltage impressed, while the spring change tends to cause the pointer to read high. The effects are not identical, however, with the result that an uncompensated mechanism tends to read low by approximately 0.2% per degree centigrade. For purposes of specification an instrument is considered compensated when the change in accuracy due to 10 degrees change in temperature is not more than one-fourth of the total allowable error.

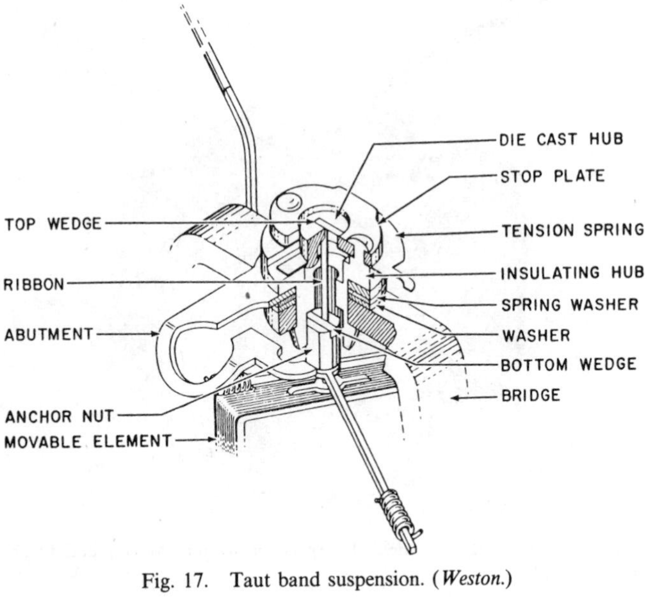

Fig. 17. Taut band suspension. (*Weston.*)

Special Panel Meters. The equivalent scale length of a 7 × 9-inch (17.8 × 22.9 centimeters) pointer type panel meter can be provided by the projected moving-scale instrument in a $2\frac{1}{2} \times 3\frac{1}{8}$-inch (6.4 × 7.9 centimeters) panel space. This is accomplished by mounting a scale of photographic film to the moving coil of a core magnet mechanism in place of the usual pointer. A lamp, lens system and mirror projects this moving scale onto a coated window. The optical system expands the scale 10 times its original value and the portion to be read appears on both sides of a hair line on the coated window. See Fig. 18.

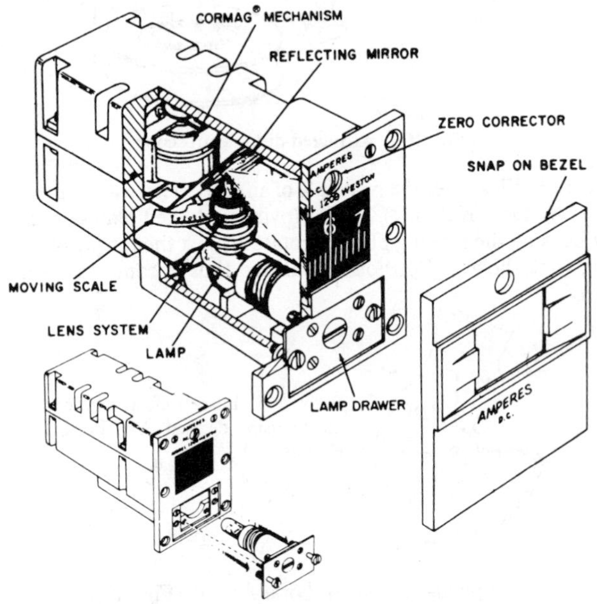

Fig. 18. Projected moving-scale instrument. (*Weston.*)

The application of light beams to panel and seitchboard instruments is further demonstrated by the projected moving pointer instrument (these instruments make use of core magnet or external magnet mechanisms). In these optical type units, the scale is fixed and a pointer of light moves along the scale; scale markings are printed on the

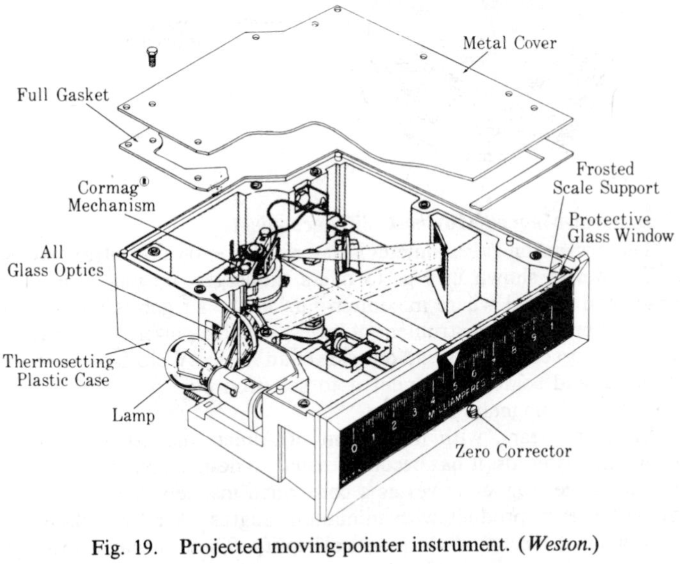

Fig. 19. Projected moving-pointer instrument. (*Weston.*)

front surface of the translucent scale plate which is also the light-diffusing surface. See Fig. 19.

Most of the information in this entry was extracted with permission from "The Instrument Sketch Book," published by Weston Instruments, Div. of Sangamo-Weston, Inc., Newark, New Jersey.

Wattmeters, thermal watt converters, watthour meters, power factor meters, var meters, and instrument transformers are described in the entry on **Electric Power and Energy Measurement.**

ELECTRIC CAR. Records show that 1575 electric automobiles were built in the United States in the year 1900, while only 939 cars equipped with gasoline engines were manufactured. The feasibility of the electric car was established well over eight decades ago. Electric automobiles were particularly popular with women during the early part of this century because of the difficulty in starting piston engines with a crank (prior to the self-starter). Coupled with gross improvements in gasoline-powered engines, the self-starter eclipsed the future of the electric car for many years. By 1930, the electric automobile was largely regarded as a museum piece. The principle of electric propulsion has been applied over the years and to the present, however, in the form of hundreds of thousands of electric trucks, fork lifts, etc. used in numerous materials handling operations. Because these off-road vehicles operate at low speeds and frequently only during one shift out of 24 hours, the standard lead–acid storage battery was adequate.

Emphasis upon the electric car within the past two decades has come from two causes. The first concern, dating back to the 1960s, was that of the growing air pollution caused by gasoline engines. The second concern developed in the 1970s when ultimate shortages in petroleum fuels became apparent. An associated cause, continuing into the present, is overall car operating economy and the realization that for many persons who make numerous short trips per day, as opposed to long cross-country trips, a vehicle with some degradation of performance would be acceptable if there were an accompanying significant cost advantage.

Initial attention was given to the earlier use of lead–acid storage batteries, but for several years, greatest attention has been given to nontraditional battery power sources and to fuel cells. The key objective has been one of finding a high-density power source. This is well illustrated by the case of using lead–acid storage batteries. a 20-gallon (75.7-liter) tank of gasoline can provide about 2.4 million Btus (0.6 million Calories). Lead–acid storage batteries, weighing about the same as that quantity of gasoline, can provide only about 7700 Btus (1940 Calories), or about $2\frac{1}{4}$ kilowatt-hours. The ratio of energy in a tank of gasoline to the energy in the same weight of conventional storage batteries (before they are fully discharged) is more than 300 to 1. However, an automobile engine can convert only about 20% of the energy in the gasoline into driving power, while the electric motor can produce motive effort from nearly all of the electricity delivered by a battery. This reduced the margin to 60 to 1, still a very marked disadvantage for lead–acid batteries as compared with the internal combustion engine.

Prior to briefly describing current battery research, it is in order here to review the lead–acid battery situation further—because there are parallels even though, with the more promising batteries, the situation becomes much more favorable for the battery power source. If it is first assumed that lead–acid batteries are used to power passenger cars, experience indicates that an electric car in city traffic at speeds up to 35 miles (56 kilometers) per hour uses about $\frac{1}{4}$ kilowatt-hour per 10 miles (16 kilometers) traveled. If it is further assumed that the car weighs 3000 pounds (1361 kilograms), half of that weight must be allowed for the battery. The very best lead–acid batteries currently available can deliver 15 watt-hours per pound if discharged over 20 hours. Thus, the battery can deliver, at the maximum, 22.5 kilowatt-hours. However, the battery will be discharged over a 1- to 2-hour period and thus its capacity will be about halved. Thus, the useful available output at most will be 11 kilowatt-hours from a 1500-pound (680-kilogram) battery pack. Inasmuch as the car is assumed to weigh $1\frac{1}{2}$ tons (1361 kilograms), thereby using 0.375 kilowatt-hour per mile, the car will have a maximum range of 30 miles (48 kilometers) at 35 miles (56 kilometers) per hour with a fully charged battery. If, instead of 35 miles (56 kilometers) per hour, the operator elects to travel at 55 miles (88 kilometers) per hour, the range would be cut in half, or approximately 12 to 15 miles (19 to 24 kilometers) per charge.

With no intent to deprecate the potential of the electric car, it is easy to lose sight of the fact that switching to electric cars will not alleviate fully the air pollution or energy shortage problems. Electricity must be generated to recharge the batteries and any marked conversion to electric vehicles would require substantial expansion of electric power generating plants, many of which would probably be fossil fuel fired.

A major advancement that will contribute substantially to the progress of electric cars is the availability of microprocessors for use in controlling the power system. Rheostat controllers can be replaced with more efficient "chopper"-type components. Better advantage can be taken of regenerative braking.

Within the last few years, General Electric and Chrysler, among others, have introduced a zinc–nickel battery with an energy density of 70 watt-hours per kilogram. This is nearly double that of the lead–acid battery (36 watt-hours per kilogram). Electric cars using these batteries may be available on a mass-production basis by the mid-1980s. A zinc–chloride battery, with a theoretical energy density of 830 watt-hours per kilogram has been developed by Gulf and Western Industries. As of early 1981, one European car manufacturer has suggested that an electric car using the zinc–chloride system may be available by the mid-1980s, if not earlier. It has been predicted that such a vehicle would have a range of 150 miles (241 kilometers) at a speed of 55 miles (88 kilometers) per hour. Compared with the zinc–nickel battery, the zinc–chloride system is more complex, requiring storage of the chlorine as a hydrate and thus mandating a cooling unit, circulation pump, and heat exchanger to release chlorine gas from the hydrate as required. Accidental release of chlorine gas must be guarded against.

A number of other batteries have been seriously studied within the last several years and ultimately may show promise for mass-produced electric cars. These power systems include the sodium–sulfur battery, which has a theoretical energy density of 760 watt-hours per kilogram, not much less than the zinc–chloride battery. The sodium–sulfur system uses comparatively low-cost materials, but has a shorter life due mainly to its operating temperature of 300–350°C (572–662°F). Another system (known as the lithium battery) utilizes lithium with sulfur, air, or chlorine. The operating temperature of this battery is even higher than that of the sodium–sulfur system—up to 650°C (1202°F). Many authorities believe that the high-temperature power supply systems will require 10 to 15 additional years of additional development. Reference to the entry on **Fuel Cells** is suggested.

Hybrid Power Systems. The hybrid engine–electric power plant offers another means of operating an electric car—and without tapping the central electric power station. In one system, a small fossil-fueled engine drives a generator on board the vehicle which charges the batteries and drives the vehicle's wheels. The engine can be smaller than required were the electric drive not present. Also, the engine is relieved of some of the transient demands on its performance. Thus, the engine can be optimized for low emissions to a greater extent than if it were the sole driving source.

When the internal combustion engine was first attacked for its emissions problem, a vocal steam power minority claimed that a Rankine cycle engine would be the complete answer to the problem. There was some reason, of course, to believe that the Rankine cycle engine could perhaps be easier to modify for low emissions than the internal combustion engine. This did not prove to be the case in numerous tests. Another Rankine cycle engine problem is fuel economy, some test installations experiencing about three times the fuel usage over a diesel powered bus, for example. Other Rankine cycle engine problems include large weight and size (particularly the heat exchangers for vaporizing and condensing), complexity in the mechanism and controls, cost, water consumption, water freezing, lubrication, and nitrogen oxide emissions.

See also **Automotive Electronics; Battery; Diesel Engine;** and **Internal Combustion Engine.**

ELECTRIC CATFISH. Catfishes.

ELECTRIC CHARGE (Conservation of). Conservation Laws and Symmetry.

ELECTRIC CIRCUITS. Electric charges, at rest and/or in motion, are the fundamental sources of electric and magnetic fields. Broadly speaking, there are three more or less distinct situations involving the spatial distribution of the fields produced by distributions of charge and current. Many important instances arise where these fields are distributed throughout a region of space of vast extent and differ accordingly at separated points in the region. The fields produced by a radio antenna come under this category. On the other hand, important applications are made of devices in which the electric and magnetic fields are confined to a much more limited region of space although the fields still undergo a significant variation in magnitude from point to point throughout the region. A transmission line connecting a television antenna on a roof to its associated receiver is an example of this type of situation. Here the fields are confined to the immediate vicinity of the line conductors but the fields vary significantly as one moves along the line. Finally, however, there are applications utilizing pieces of apparatus in which the electric and magnetic fields are confined to regions of space which are so restricted in spatial extent that one may speak of an electric or magnetic field as having an essentially constant value in the immediate region of the device which is very much greater than at all other points of space. As an example it is noted that if a large number of turns is wound on a small circular core to form an inductance through which a direct current passes, a magnetic field will be created which is concentrated in the immediate vicinity of the coil and vanishes rapidly as one recedes from this location.

It is found, however, that when one deals with changing currents (charges in motion with varying velocity), the spatial dimension alone is not an adequate measure to establish which of the three situations suggested above is involved in a particular application. For sinusoidally varying currents one can define a quantity known as the wavelength which in free space is numerically equal to the velocity of light divided by the frequency of the sinusoidal variation. If spatial extent is measured in wavelengths, then an assemblage of apparatus which is confined to a region which is less than about one tenth of a wavelength in greatest dimension results in a good approximation to the third situation considered above. This possibility, one in which the electric fields and the magnetic fields may be considered to be concentrated in individual pieces of apparatus, is the domain of electric circuits. An electric circuit may be defined as a characterization of an electrical system in terms of the integrated effects of the electric and magnetic fields present in the system. The characterization is an approximation to the actual field problem in which one replaces the actual system by elements having resistance, capacitance, and inductance and by sources of electric potential and electric current. Systems in which the approximation cited is permissible are sometimes called "lumped constant" circuits. Antennas and transmission lines, on the other hand, are often referred to as "distributed constant" circuits. The distinction between these two designations comes from the spatial variation of the electric and magnetic fields as outlined above.

As may be inferred from their definitions, the parameters of resistance, capacitance, and inductance are not necessarily independent of the currents and voltages impressed upon the elements of an electrical system. Whether the resistance of an element is a function of the current through it or not, the relation $e = Ri$ (R is resistance in ohms) is still valid for the connection between the voltage across an element and the current through it. On the other hand, corresponding relations for the coil and the capacitor become more involved if their parameters are dependent on current or voltage. For constant parameters of inductance (L in henrys) and capacitance (C in farads), the voltages and currents in pure inductive and capacitive components assume the form

$$e = L\frac{di}{dt} \quad \text{or} \quad i = \frac{1}{L}\int e\,dt$$

and

$$i = C\frac{de}{dt} \quad \text{or} \quad e = \frac{1}{C}\int i\,dt$$

respectively. If the parameters L and C vary with impressed voltage and current, the above relations are not valid and recourse must be made to the basic law of electromagnetic induction when coils are involved and to the expression for electric current as the rate of change of charge (derivative of charge with respect to time) for circuits containing capacitors.

Whether elements with constant or variable parameters are involved, one may employ Kirchhoff's Laws of Networks to formulate equations representing conditions of equilibrium between the applied voltages and/or currents and the quantities that result. See **Kirchhoff Laws of Networks.** The equilibrium equations may be formulated on the basis of voltages as independent variables and currents as dependent quantities, the system of equations being known as the mesh equations for the circuit. Alternatively, they may be written on the nodal basis, where the sources are current generators and the dependent quantities are nodal voltage differences with respect to a reference node. As an example of the process involved, consider the simple series circuit shown in Fig. 1 and the simple parallel circuit shown in Fig. 2. The

Fig. 1. Simple series circuit.

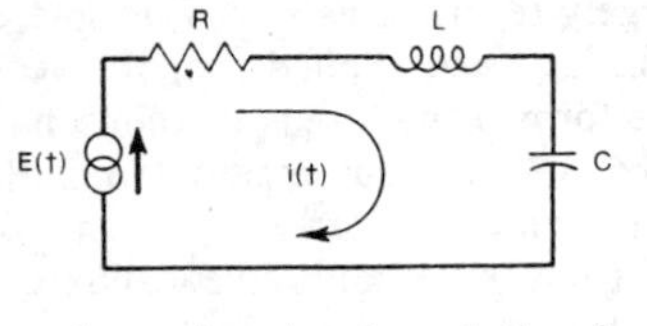

single mesh equation characterizing the series circuit and the single nodal equation describing equilibrium in the parallel circuit are, respectively

$$E(t) = Ri(t) + L\frac{di(t)}{dt} + \frac{1}{C}\int i(t)\,dt$$

and

$$I(t) = \frac{1}{R}e(t) + \frac{1}{L}\int e(t)\,dt + C\frac{de(t)}{dt}$$

The mesh and nodal equations for more complicated circuits have a form similar to these equations with added dependent variable terms and, in general, additional independent variable terms. The equations may be solved by various mathematical means to yield the desired unknown quantities, currents for mesh equations and voltages for nodal equations.

Fig. 2. Simple parallel circuit.

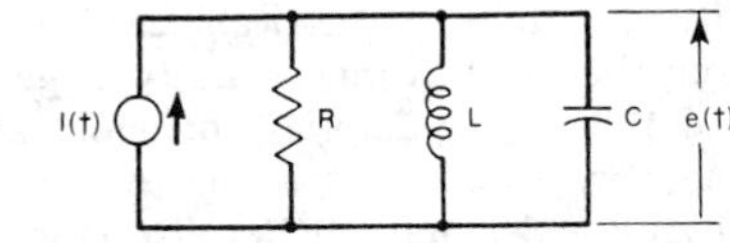

If the applied voltage sources are batteries or alternative sources of electric potential which are constant with time, the solution to the network equations described above will consist of constant terms (including zero as a special case) plus terms which decay exponentially with time. The latter terms are called the transient terms or the transient solution. They correspond to the solution of the homogeneous differential equations. The constant terms represent the steady state solution of the problem. These particular solutions may be obtained by the means that are considered in the section on Direct-Current Circuits where only networks of pure resistances are involved. (In the computation of the steady state response for circuits with constant potential applied, inductors may be replaced by elements of zero resistance and capacitors by resistors of infinite resistance.)

A general form of solution is obtained by assuming exponential applied sources as the form $A\epsilon^{st}$. For dc, $s = 0$; for exponentially increasing or decreasing sources, s is real; for sinusoidal sources, s is imaginary. Thus, for the series circuit shown in Fig. 1 with an exponential source voltage $E\epsilon^{st}$

$$Ri + L\frac{di}{dt} + \frac{1}{C}\int i\,dt = E\epsilon^{st}$$

The transient response is obtained by solving the homogeneous differential equation

$$Ri_t + L\frac{di_t}{dt} + \frac{1}{C}\int i_t\,dt = 0$$

The solution is $i_t = C_1\epsilon^{st} + C_2\epsilon^{s_2 t}$ where

$$s_1 = R/2L + \sqrt{(R/2L)^2 - 1/LC}$$

and $s_2 = R/2L - \sqrt{(R/2L)^2 - 1/LC}$, and C_1 and C_2 initial condition constants. The steady state response is obtained from the particular integral which can be seen directly from the differential equation to be $i_s = E\epsilon^{st}/(R + sL + 1/sC)$. The current $i = i_t + i_s$. Methods of obtaining the steady state solution for sinusoidal sources are given in the section on **Alternating Currents.**

Note that in the exponential expressions in these formulas, the Greek letter epsilon is used to represent the base of natural logarithms (see **e, the Number**). This was done to avoid confusion with the italic *e* that is used for voltage in this entry.

ELECTRIC CLOCK. A clock that employs electric current. Most automobile clocks use current from a battery to automatically rewind a mainspring. In other electric clocks, the mainspring is usually eliminated—the battery current driving the balance wheel through electromechanical contacts. An early line-current system used an auxiliary mainspring that was kept fully wound by the current for use should the line power fail. Later, a synchronous motor controlled by alternating current replaced the balance-wheel system in electric plug-in clocks.

Other types of electric clocks represent developments of the pendulum. See also **Pendulum Clock.** An electromechanically-driven pendulum system, developed in 1843 by Alexander Bain in Scotland, was probably the first electric clock. At a considerably later date, the so-called Western Union clock used two 1.5-volt cells to wind a mainspring that maintained a pendulum while accepting an hourly time-pulse signal by telegraph line to synchronize its time display on the hour with a remote master clock system. The time-pulse signal, in effect, corrected for accumulated errors from all sources.

The later consumer cordless electric clocks use a 1.5-volt dry cell to drive the balance wheel electromagnetically, by means of an electronic circuit. By utilizing a transistorized electronic circuit as a switch to drive the balance wheel equipped with a tiny magnet, these clocks eliminate electromechanical contacts. In another clock, the frequency standard is an electromagnetically-driven 360 Hz tuning fork, powered by a single aspirin-size 1.3-volt mercury-oxide cell. An electronic control circuit is required to drive the tuning fork. The accuracy of this clock averages plus-or-minus two seconds a day. If three tuning fork systems are electrically interconnected, as in a marine navigation clock, the daily deviation in rate drops to less than one second. See also **Tuning Fork.**

In the United States, plug-in electric clocks usually use self-starting shaded-pole synchronous motors, controlled by the 60-cycle alternating line current. Their accuracy, therefore, is dependent upon the line frequency. Throughout the United States, however, line frequency is normally both properly stabilized and corrected several times daily, eliminating accumulated error and permitting an accuracy of up to plus-or-minus four seconds. Where line frequency is not corrected at least once a day, the resulting error displayed by plug-in electric clocks can be far greater. In many locations elsewhere in the world, where line voltage is not normally stabilized, plug-in clocks are less reliable. Where power-line shutoffs are routine or frequent, plug-in clocks can represent a major inconvenience.

William O. Bennett, John J. Carpenter, Frank Dostal, and E. Van Haaften, Bulova Watch Company, Inc., New York.

ELECTRIC CONTROL VALVES. Valve (Control).

ELECTRIC DIPOLE. Dipole.

ELECTRIC EEL. Gymnotid Eels.

ELECTRIC FEEDBACK. In magnetic amplifier terminology, feedback through an electrically conductive network, as differentiated from feedback produced by currents in windings having coupling to the control windings (magnetic feedback).

ELECTRIC FIELD STRENGTH. The magnitude of the electric field vector. This term is sometimes called the electric field intensity, but such use of the word intensity is deprecated in favor of field strength, since intensity connotes power in optics and radiation.

ELECTRIC FIELD VECTOR. At a point in an electric field, the force on a stationary positive charge per unit charge. Under conditions in which the ratio of force to charge is not constant, the field vector is defined as the limit of the ratio as the change approaches zero. This may be measured in newtons per coulomb, in volts per meter or in corresponding units in systems other than the mksa system.

ELECTRIC FLUX DENSITY. At a point, the vector whose magnitude is equal to the charge per unit area which would appear on one face of a thin metal plate introduced in the electric field at the point and so oriented that this charge is a maximum. The vector is normal to the plate from the negative to the positive face. The term electric displacement density or electric displacement is also in use for this term. In an isotropic medium of permittivity ϵ, the flux density is $\mathbf{D} = \epsilon\mathbf{E}$ in rationalized mks units, where $\mathbf{E}$ is the electric field vector.

ELECTRIC GENERATOR SPEED CONTROL. Governor.

ELECTRIC INDUCTION. Induction (Electric/Magnetic).

ELECTRICITY. An isolated atom consists of a small nucleus, itself composed of protons and neutrons, surrounded by a cloud of electrons. The proton and the electron are the ultimate stable particles of electricity. Their charges are equal and opposite, the proton being regarded, by convention, as positive. A normal atom with its full complement of electrons is thus uncharged.

Electrostatic phenomena arise when bodies (or parts of bodies) have an excess of electrons or protons, a state usually produced by transferring electrons, e.g., by means of a battery or by rubbing two dissimilar materials together. Between two positively charged bodies (or two negatively charged ones) there is a repulsive force; between positive and negatively charged ones, an attractive force.

If two bodies at different potentials are connected by a conductor, such as a metal wire in which there are free electrons, the electrons in the wire drift under the influence of the electric field. Such movement of electrical charges gives rise to further phenomena and we speak of an electric current. The current may be one of electrons only, as in a metal, in semiconductors, or in electron tubes. Or the current may be of positive nuclei, as in an isotope separator; or of both positive and negative charges, as in the conduction by ions (atoms that have gained or lost an electron) in liquids or in gaseous electrical discharges.

Electrons flowing in the positive direction give rise, by our convention of signs, to a negative current. In a metal, semiconductor, or conducting liquid, the velocity at which the electrons or ions drift is quite slow, less than one centimeter per second even for current densities in a metal as high as 10^4 amperes per square centimeter. In vacuum devices, such as cathode-ray tubes, the speed of the electrons approaches that of light. If a wire joins two electrostatic charges, the current persists for a short time only, but it may be maintained by means of some source of energy, such as a battery, a generator, a thermocouple, or a solar photoelectric cell.

When a current flows under the influence of a potential difference, the moving charges—electrons in metals, ions in solution—are impeded by collision with the atoms in the conducting metal or liquid. The charges give up to the atoms the energy they acquired as they moved in the electric field, and electrical power is converted into other forms—for instance, into heat in the case of a metal wire. Electric fields are also producted by time-changing magnetic fields, and this principle is extensively exploited, as is motional electromotive force (emf) to generate electric power. Electromotive force and voltage drop are usually regarded as synonymous. When an emf is impressed on a closed metallic circuit, current results.

The electrification of amber by rubbing with wool or fur was observed many centuries ago. Not until the work of Volta, late in the 18th century, was electricity recognized through any but electrostatic phenomena, and investigations on the properties and applications of electric currents were among the most brilliant features of nineteenth-century physics. Even in the 1890s physicists were still asking, "What is electricity?" It had then long been known that an appropriate application of energy will separate electricity into two components, designated as positive and negative; that bodies charged with these components attract each other; and that the energy of separation is yielded upon the reunion of the two components. It remained for J. J. Thomson to recognize the electron, and for the subsequent analysis of atomic structure to identify the proton, and to explain their relations.

Hundreds of entries in this volume deal with electrical fundamentals, equipment, and applications.

ELECTRICITY (Frictional). Frictional Electricity.

ELECTRIC LENGTH. The physical length of a transmission line or its equivalent, corrected for any inhomogeneities that may effect the speed of propagation, and expressed in wavelengths, radians, or degrees.

ELECTRIC LOAD FACTOR. Load Factor (Electric).

ELECTRIC MOTOR. Motor (Electric).

ELECTRIC-NETWORK RECIPROCITY THEOREM. Reciprocity Theorem (Electric-Network).

ELECTRIC POTENTIAL. The electric potential at a point is the work done in moving a unit positive charge from the datum point (sometimes at an infinite distance from the region of interest) to the point in question. The earth's surface has also been used as the datum of potential. From the definition, it can be seen that the potential at the datum is zero. The unit of potential is the volt. The difference of potential between two points is the work done in moving a point charge from the first point to the second. It is equal to the difference in the values of the potentials at the respective points.

ELECTRIC POWER. Electric power is the product of electric current and electromotive force; that is, multiplication of current flowing by voltage forms the basis of the calculation of electric power. In a dc circuit, the current measured in amperes, multiplied by the voltage between wires, is the power in watts. A thousand watts constitute the kilowatt, a larger and more frequently employed unit of electric power.

The voltage and current may not be in phase with each other in an ac circuit and, while the instantaneous power is the product of the instantaneous voltage and current, this out-of-phase relation causes the power to fluctuate between positive and negative values. Hence for the average power (which is usually what is desired) this factor needs to be taken into account in determining electric power is an ac circuit, for it is only that component of the current which is in phase with the voltage that contributes to the average electrical power. The out-of-phase component produces the "wattless power." The power factor measures the fraction of the current that is in phase and available for true power. It is equal to the cosine of the phase difference between voltage and current. In a single-phase ac circuit having current of I amperes, voltage of E volts, and power factor f, the true power is EIf watts. In a balanced three-phase circuit, it is $\sqrt{3}\,EIf$ watts.

ELECTRIC POWER AND ENERGY MEASUREMENT. Over the years, the term *power*, in association with electricity, has tended to lose its true meaning. Thus, power is often found used in nontechnical literature where actually the correct term *energy* should be used. By definition, power is the rate at which energy is transformed or made available and is measured in *watts*. Energy may be defined as the time integral of power, or as the total energy supplied and is measured in *watthours*.

From an economic viewpoint, the most important of all electrical measurements is the measurement of energy. The watthour meter in various forms can be found in nearly every home, factory, highway billboard, and other locations where electrical energy is being purchased. Metering, installation and wiring have been governed by national, industrial, and local codes for so many years that at least in the United States, a particular type of installation is nearly identical everywhere.

For homes or small stores where energy demand is low and fractional horsepower motors are used, the common supply is single-phase, 3-wire. Two voltage levels are available, 120 and 240 volts, depending upon which pair of wires is selected. Electric ranges and heavy-duty home air conditioner motors can take advantage of the high voltage to reduce line currents which reduce losses and permit smaller-size wiring.

Measurement of energy is almost always by means of fixed-installation metering. This provides safety through grounding of the meter enclosure and ease of reading through proper location and mounting. Tamper-proof housings, which are also weatherproof where necessary, are common practice to insure the integrity of readings.

On the other hand, the measurement of power (watts) follows no such set of rigid rules. Very often considerable planning must go into a watt measurement to properly use existing metering or to purchase new equipment so that the test will be valid and results will be within the expected accuracy limits.

Whereas the measurement of energy is almost entirely restricted to 60 Hz, power measurements range from direct current to alternating current, including distorted waves, chopped waves, and missing pulses. A variety of circuits for connecting wattmeters to single-phase and poly-phase systems have been developed over the years. Basic connection diagrams appear in many electrical engineering textbooks. However, the user of a wattmeter is hard pressed to find diagrams covering practical or unusual situations. Wattmeter manufacturers usually offer an instruction book or a bound set of connection diagrams with their instruments. Basic wattmeter connection diagrams appear later in this article.

Power Theory. Since energy is simply the total power over a time period, an understanding of the power equations will provide some background into both power and energy terminology. A direct current circuit under steady-state conditions will produce power, computed as the production of the voltage across the circuit and the current in amperes in the circuit. This will also apply to alternating current circuits as long as instantaneous values of volts and amperes are used. The product of volts and amperes at any instant will give the instantaneous power in watts. However, such a measurement is unusual, difficult to make, and the resulting information is of limited use. Instantaneous power is of interest, of course, in the study of transient phenomena.

Average power in an ac circuit is of far more interest since it is equivalent to dc power and is a measure of mechanical work being done or heat liberated. Wattage or average power has an exact mathematical relation to horsepower or Btus or Calories.

The most basic equation for power, relating voltage, current, power, and the phase angle between the voltage and current, is derived as follows. If both voltage and current are sinusoidal, the average power over a cycle is:

$$P=\frac{1}{T}\int_0^T ei\,dt$$

$$P=\frac{1}{2\pi}\int_0^{2\pi} E_m \sin\theta\cdot I_m \sin(\theta-\phi)\,d\theta$$

where E_m and I_m are maximum values and ϕ is the phase angle by which the current lags behind the voltage.

From $\sin(\theta-\phi)=\sin\theta\cos\phi-\cos\theta\sin\phi$,

$$P=\frac{E_mI_m}{2}\left(\int_0^{2\pi}\sin^2\theta\cos\phi\,d\theta-\int_0^{2\pi}\sin\theta\cos\theta\sin\phi\,d\theta\right)$$

$$P=\frac{E_mI_m}{2\pi}\left[\left(\frac{\theta}{2}-\frac{\sin 2\theta}{4}\right)\Big|_0^{2\pi}\cdot\cos\phi-\left(\frac{1}{2}\sin^2\theta\right)\Big|_0^{2\pi}\sin\phi\right]$$

$$P=\frac{E_mI_m}{2}\cos\phi$$

The RMS values of sinusoidal voltage and current are:

$$E=\frac{E_m}{\sqrt{2}}\quad\text{and}\quad I=\frac{I_m}{\sqrt{2}}$$

Substitution in the previous equation yields:

$$P = EI\cos\phi$$

An immediate concern is what happens to the indications of a wattmeter if the voltage or current or both are not sinusoidal. Since it is possible to synthesize an odd wave shape with higher harmonics of the fundamental frequency, a wattmeter will give correct indications if it is frequency compensated over the span of harmonics. It is to be noted that if a particular harmonic is present in either the current or the voltage but not the other, it does not contribute to the average or active power. Frequency compensation of the wattmeter is still necessary for an accurate measurement.

Wattmeter Construction

Dynamometer. All wattmeters of this type contain a fixed coil (usually divided into two coils), which carries the current, and a moving coil having series resistance connected for voltage, turning within the fixed coil. The torque on the moving system is proportional to the product of the currents in the fixed and moving coils:

$$\text{Torque} = K_1 i_m i_f \frac{dM}{d\theta}$$

where M is the mutual inductance between the two sets of coils and remains constant over the usable scale angle.

The period of the instrument is very long compared with the period of the alternating voltage. Since the instrument movement cannot follow the rapid variations in torque, it will take up a balance position where the driving torque will equal the spring-restoring torque. Deflection represents the average torque:

$$\text{Deflection} = \frac{K_2}{T}\int_0^T ei\,dt$$

which is identical to the average power equation:

$$P = \frac{1}{T}\int_0^T ei\,dt$$

multiplied by a constant.

It follows then that a dynamometer wattmeter is a "true RMS wattmeter" and will take into account the magnitudes of voltage and current as well as the phase angle between them. Furthermore, meter indication will reverse when the flow of power reverses.

Thermal Watt Converter. As with the dynamometer, this type of instrument dates back to the early days of electricity. Heat produced by the voltage and current directly heats thermocouples which are arranged in a network to provide a dc output directly proportional to the wattage input. Figure 1 shows the essential parts of a thermal watt converter, namely, the potential transformer which is connected to monitor voltage, the current transformer which has a double-wound, center-tapped secondary, and the two sets of thermocouples.

The quantity of heat in $R1$ is proportional to $R1(I_{T1})^2$ and in $R2$ is equal to $R2(I_{T2})^2$. A vector diagram can be drawn for the sum and difference of I_p and I_c. From this diagram, equations can be developed which yield the difference in wattage in resistance $R1$ and $R2$ as $4RE1 \cos\phi$. The variable part of this term is $EI\cos\phi$ which is the expression for ac power.

Some early thermal wattmeters used bimetallic elements or liquid-filled thermometers to measure the difference in heat between resistors $R1$ and $R2$. Thermocouples were used in more recent designs. In one design of a thermal wattmeter (Weston), the resistors and the thermocouples are one and the same. They act both as the heating resistors and as the temperature-sensing elements and show a very fast response to power changes. Also because the impedances of the several parts of the circuit are inherently balanced, there is little tendency for interchange of currents or potentials between the ac and the dc portions of the network.

Used mostly by the power industry, especially in totalizing of electric system loads, the watt converter has also found widespread use for measuring the power taken by very large motors where a remote readout is needed. Classed as a true RMS wattmeter, this device will respond to magnitude of voltage and current as well as the phase angle between them. Converter output will also reverse if the flow of power in the system reverses.

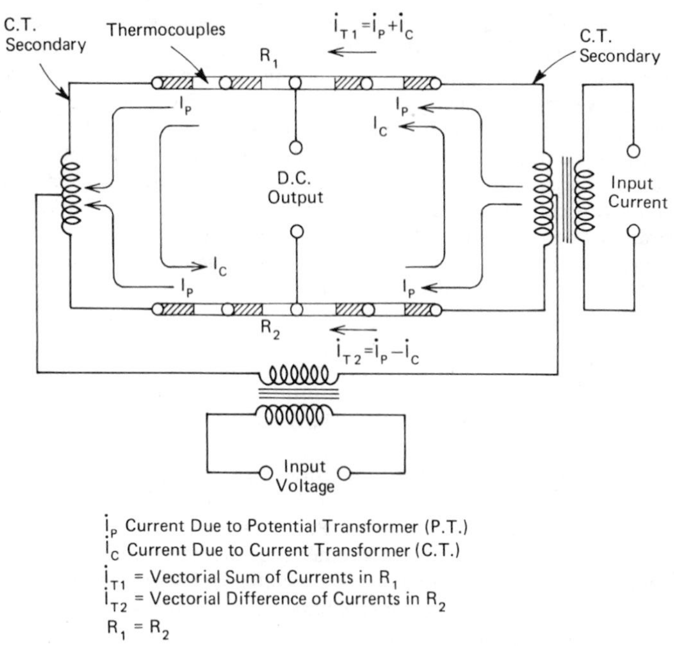

Fig. 1. Basic circuit of single-element thermal watt converter.

Specific Design Considerations. For clarity, it is believed best to comment pertaining to design factors pertaining to a few specific commercial instruments, realizing of course that other designs are also available commercially. Available single-element and 2-element watt converters have a response to 99% in 0.7 second. Due to this rapid response, some protection of the thermocouples is necessary during overload conditions. Converters made by Weston use transformer cores which will saturate at moderate overload and minimize thermocouple burnout or other damage. If repeated overloads occur, the current circuit is usually operated below the 5 ampere normal level. Output is 50 millivolts open circuit per element when connected to a 500-watt load. An available 2-element watt converter has an output of 100 millivolts for 1,000 watts of circuit load. Provision is made for adjusting resistors in the output network so that the output can be reduced to achieve a particular ratio between input wattage and output millivolts.

Since a thermal watt converter has no moving parts to wear, it has been made very rugged by potting the entire circuit in a steel case. Poly-phase wattmeters are easily made by assembling two or more single-element wattmeters and providing for a common output signal. An available 3-element wattmeter was designed for use on military motor-generator sets having 3-phase, 4-wire distribution systems. Such systems can be expected to operate under unbalanced conditions and to correctly read system power, a 3-element wattmeter must be used.

If the internal potential and current transformers are of good quality, the thermal watt converter can be used on frequencies extending to 20,000 Hz. A high-frequency type meter has a working frequency span of 180 to 20,000 Hz. In general, all thermal watt converters use internal transformers which excludes their use on direct currents.

The Low Power Factor Wattmeter

All of the so-called true RMS wattmeters will operate over the full range of power factor from zero to unity. Many low power factor examples can be found in the laboratory, such as motor or transformer testing, core loss tests, and power supply circuits. At zero power factor, a wattmeter will indicate zero watts even with rated voltage and current flowing. It quickly becomes apparent that the major difficulty in making a low power factor measurement is the low indication obtained on the meter. For example, if a wattmeter indicates full-scale with a unity power factor load, it will indicate half-scale with a 50% power factor load for the same level of voltage and current. At 20% power factor, the pointer will move only one-fifth the distance up-scale.

Special wattmeters are available for use on low power factor circuits. They are commonly called 20% power factor meters since the full-scale wattage is equal to the maximum voltage times the maximum current times 0.2. Both the accuracy and readability are improved through the use of this type of instrument. Since the 20% power factor wattmeter is designed to develop five times the torque of a unit power factor type instrument, care must be taken not to apply voltage or current above the maximum values shown on the instrument rating. Large overloads will soon burn out the resistors, fixed coils or moving coils. Small overloads continuously applied will cause deterioration of the overheated insulation. A wattmeter designed for low power factor can be used on higher power factor circuits provided either the voltage or current is sufficiently reduced to keep the pointer on the scale. Likewise, normal unity power factor meters may be used at low circuit power factors provided maximums are not exceeded.

Power Measurement

Power in an ac or dc circuit may be determined indirectly by making appropriate measurements of voltage, current and, where necessary, power factor. Power factor is usually expressed as a decimal value ranging between 0 and 1 and is derived from the cosine of the angle between the voltage and current. Power factor is further designated lead or lag, depending upon whether the current vector is ahead or behind the voltage vector based on counterclockwise rotation of the vectors. When making power calculations, it is not necessary to know if power factor is lead or lag.

Therefore, calculated power is a valid procedure, but with some reservations. Results are accurate only if all quantities of voltage, current, and power factor are correctly measured. The most elusive quantity is power factor. It is rare that a single-phase power factor meter can be used on other than 60 Hz. Even the 3-phase power factor meter has the requirement that the load must be balanced, although some designs have covered several thousand hertz. Modern electronic phase-angle voltmeters give excellent results on good sine waves over a wide frequency span. However, when distorted waves are encountered, results are questionable since many instruments of this type operate on the zero crossing principle. Further, the power factor of a distorted wave has little meaning since by definition it is based on a sine wave. The conclusion is soon reached that the only way to measure power accurately is through the use of a wattmeter.

In recent years, the phase "true RMS wattmeter" has been reserved for the description of the ultimate in wattmeters. This is because some of the types of wattmeters available today will be accurate at only one frequency, must operate over a narrow voltage span, will not operate on dc, or must be worked at a high power factor. Although more descriptive than technically correct, the phrase will probably remain in the literature.

The original, basic, true RMS wattmeter was the dynamometer. Even until recently, this type of instrument was used as the standard wattmeter at the U.S. Bureau of Standards. A dynamometer wattmeter can be calibrated very accurately on dc and then used on ac. It is often the standard used to check other wattmeter devices because it can be made to a high accuracy, is a passive device, and will retain its accuracy for many years.

Probably the most important theorem in electric power measurement is that proposed by Blondel. In essence, it states that to correctly measure total system power, it is permissible to use one less wattmeter than current-carrying conductors. Also, the common point for the potential circuits is the conductor without a wattmeter current connection. The circuit being so measured may be operated at any power factor or condition of current or voltage unbalance. Strict adherence to Blondel's theorem would require the use of three wattmeters or a 3-element type meter to correctly measure a 3-phrase, 4-wire system. Since large commercial systems strive to maintain good voltage balance, a less expensive wattmeter of the $2\frac{1}{2}$-element design may be used and still achieve good accuracy.

Many questions often arise as to the proper wattmeter connections for various types of loads. For example, in a 3-phase, 3-wire circuit, the load can be delta, wye, or some other configuration. The wattmeter is only concerned with the three wires. This leads to a simple pictorial concept for the connection of a wattmeter. Visualize a laboratory bench with an unknown power supply on the left, the connecting wiring across the bench, and an unknown load to the right. Without knowledge of source or load, a true measurement of total system power can be made by following Blondel's theorem. Assume four wires are present and that all may be carrying current. Provide three wattmeters, making the wire without a meter the potential common. It is to be noted that one terminal of the meter current circuit and potential circuit carries an instantaneous polarity marking (usually plus or minus). That means that if (+) of a direct current supply is applied to each of these terminals, the meter will deflect upscale. Likewise if (−) were so applied, the meter will still go upscale. If one terminal is made the opposite polarity, the meter will move downscale. In a multi-meter correction, the (±) current terminals should all be toward the source. Even so, due to load reactance, one wattmeter may produce a reversed indication. Total power then will be the algebraic sum of all meter readings. The measurement has been made without any knowledge of the source or the load. Voltage and current levels must be within the range of the meter to avoid overheating damage.

If more knowledge of the load can be obtained, the immediate benefit would be reduced metering costs. If we still have a 3-phase, 4-wire system, but know that the neutral wire carries no current, then only two wattmeters are needed to give a true reading.

If we further know that both voltage and currents are balanced around the phases, then metering costs can be further reduced by using a single wattmeter and a wye box. When combined with the meter, the two arms of resistance in the wye box form a wye having a neutral point. The wattmeter will then measure a phase power which is a known fraction of the total power. The scales of switchboard meters are usually direct reading, but the indication from a portable wattmeter must be multiplied by the wye box multiplying factor.

If only voltages in this 3-phase, 4-wire example are known to be balanced, then a so-called $2\frac{1}{2}$-element wattmeter is satisfactory.

Adherence to Blondel's theorem will always provide a true power measurement regardless of circuit conditions. A knowledge of the circuit will often point the way to less expensive metering which can provide adequate results.

For the measurement of power in a 2-wire circuit, Figures 2 and 3 show the possible connections. The connection shown in Fig. 2 indicating the wattmeter potential circuit connected on the load side of the current coils is most often used. Readings taken on small wattage loads may easily be corrected for meter loss by opening the load and reading the wattmeter. Although not truly correct under varying loads, this "tare" reading will much improve the accuracy of the measurement.

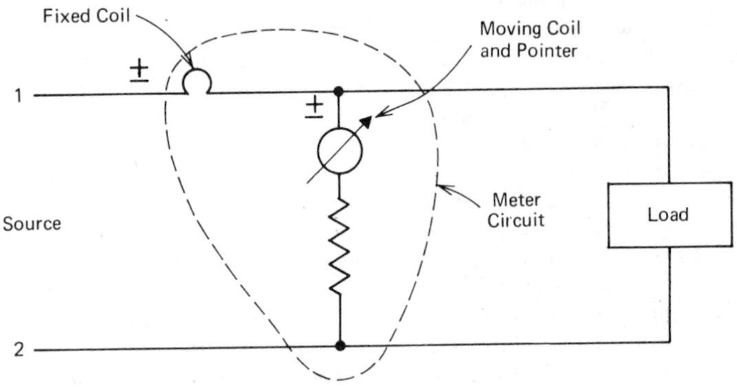

Fig. 2. Single-element wattmeter with potential circuit connected on load side. This connection is most often used.

If the wattmeter current coils are wound for low currents, such as 0.1 ampere, there is sufficient voltage drop across them under load to make the connection shown in Fig. 3 and thus yield a more accurate result.

Figure 4 clearly demonstrates Blondel's theorem of two wattmeters in a 3-wire circuit where the common potential circuit connection is made in the line without a current coil. There are some 2-element wattmeters available which connect the two moving coils into line 2. Such an arrangement is adequate for moderate voltages and accuracy, but the mechanical force set up between the fixed and moving coil due to the electrostatic effect precludes the use of this connection where high accuracy is needed.

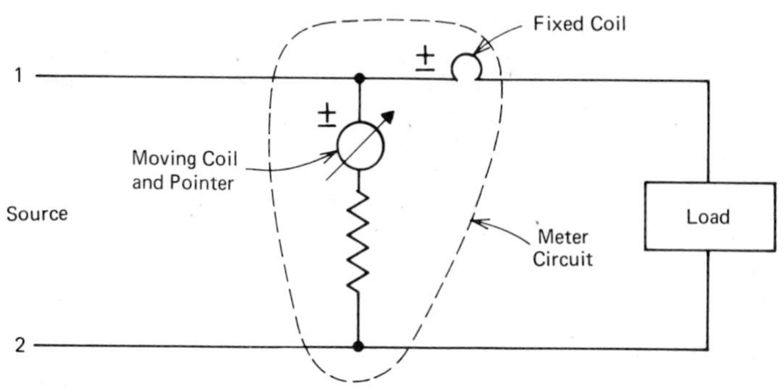

Fig. 3. Single-element wattmeter with potential circuit connected on source side. Usually used when fixed coil has a low current rating and a high voltage drop.

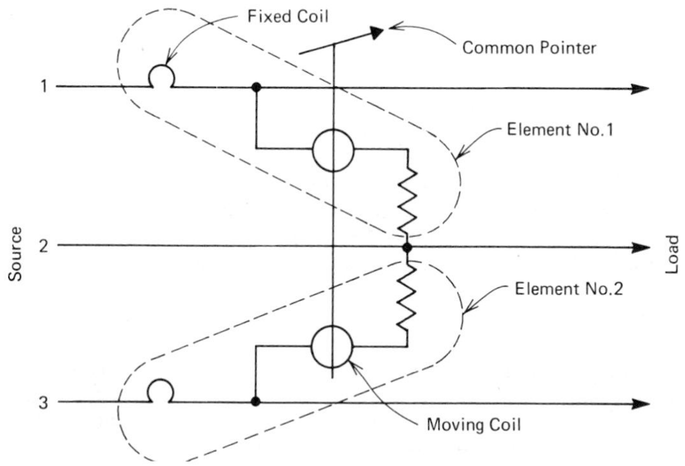

Fig. 4. Two-element wattmeter connected to a 3-phase, 3-wire system.

The $2\frac{1}{2}$-element wattmeter of Fig. 5 will monitor all the current flowing and so is capable of a correct wattage measurement where current unbalance exists. Line to line voltages should be nearly balanced and it is assumed that they are 120° from one another in vector rotation.

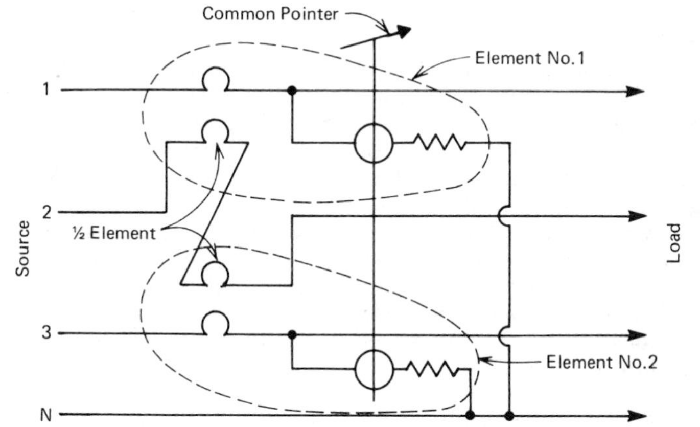

Fig. 5. A $2\frac{1}{2}$-element wattmeter connected to a 3-phase, 4-wire system.

Power Factor

Whenever the voltage and current are not in phase, a third term called power factor must be introduced. The formula is expressed by:

$$\text{Power Factor} = \cos\left(\tan^{-1}\frac{Q}{P}\right)$$

where Q = reactive power, P = active power.

This basic formula may be used on simple, 2-wire circuits as well as a 3-phase system as long as the poly-phase system is balanced in both voltage and current.

When a poly-phase system is unbalanced in any manner, the system power factor will no longer have any specific physical meaning. A numerical ratio can be obtained and is defined as the interval power vector. This is not, in general, equal to the average value of the power factor during the interval.

Power factor is not usually one of the measured quantities when testing low power circuits under laboratory conditions. Commercial users of bulk power very often monitor power factor so as to be able to make adjustments of condenser banks or synchronous motors to keep the power factor as high as possible. This, in turn, usually reduces the cost of the power purchased.

Power Factor Meter. Since the power factor meter can show at a glance the operating condition of a power system, it is most often found on the switchboards of both consumer and supplier of power.

Ratio type movements are commonly found in both single phase and poly-phase power factor meters. The single-phase power factor meter measures the ratio of vars to watts which corresponds to the tangent of the power factor angle when the voltage and current are sinusoidal. A scale can then be drawn for the power factor which is the cosine of the angle.

The poly-phase power factor meter also uses the same basic ratio mechanism as the single-phase meter and is connected 3-phase, 3-wire. This instrument will indicate vector power factor by measuring the angle between line current 2 and line voltage 3–2 and 1–2. The indication is the poly-phase power factor only for balanced voltages and currents when both are sinusoidal. Some years ago, Weston offered a poly-phase power factor meter for use on unbalanced systems. However, due to limited acceptance because of the high cost resulting from the complexity of construction, it was discontinued. It is doubtful that such a meter can be purchased today from any manufacturer.

Single-phase instruments can be scaled in a variety of combinations, such as 0–1 P.F. lag or lead, or 0.3–1–0.3 P.F. Due to the principle of the instrument, not every range combination is possible in the poly-phase power factor meter.

When compared to the other electrical quantities, the *var* is a relatively new term, having been recognized by international agreement in 1930. The letters were taken from volt-ampere-reactive and represent power incapable of producing work. Voltages used in this form of metering are always 90° in vector rotation from that used in wattage measurement. In any alternating current system having sinusoidal voltages and currents operating at other than unity power factor, the real power is less than the volt-ampere-product and is related by the familiar right triangle. See Fig. 6.

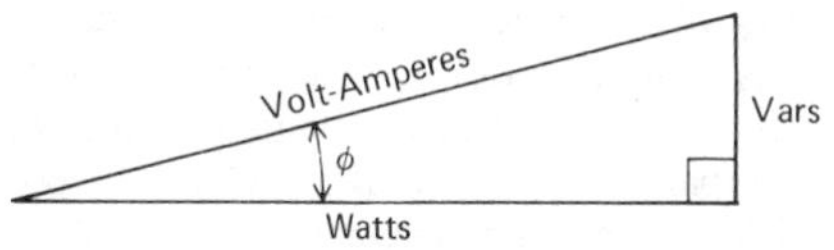

Fig. 6. Real power is less than volt times amperes. ϕ = power factor angle.

Since more iron and copper are required to deliver a given amount of power at a low than at a high power factor, design allowances must be made for any reactive volt-amperes in addition to the designed load power. Line losses are higher and voltage regulation poorer when the power factor is low. Any customer contributing to a low power factor must be expected to pay for this added loss in addition to the actual energy consumed. Var metering is generally used to obtain an estimate of the average power factor of a fluctuating load over a period of time. A recording type meter would be used for this measurement.

$$\text{Var} = EI\sin\phi = EI\cos(90° - \phi)$$

A wattmeter can be converted to a varmeter if the voltage is shifted into quadrature with the line voltage at the load. In single-phase varmeters, this voltage shift is done by means of added reactance. In poly-phase systems, the voltage shift is most easily done by reconnection if all necessary points are available. On 3-phase, 3-wire systems, a var connection may be made to wye-connected resistors forming an artificial neutral. A special poly-phase, phase-shifting autotransformer, also called a potential converter or phasing transformer, is available for var service. When a varmeter is purchased from the manufacturer, the proper scaling, instrument adjustment (watts), and connection diagrams are provided. If it is desired to convert an existing wattmeter to a varmeter, several problems must be avoided (or at least understood).

Varmeter indication can be either up- or down-scale, depending upon whether clockwise or counterclockwise rotation is selected for the new var voltages. Either is correct, but it should be determined that up-scale indication will occur for a leading or lagging power factor. Another more serious problem is that of voltage level. If the new var voltages differ from those used by the basic wattmeter, then, for the same power supplied to the load, the full-scale value and therefore all scale points will be in error. In order to correct this, the meter series resistance would have to be changed (a task for a meter shop with wattmeter calibration equipment).

Varmeters are usually arranged to deflect to the right (up-scale) on leading power factor circuits with phase rotation 1–2–3. Zero center varmeters also have the words IN at the left and OUT on the right. The OUT refers to the flow of power from a source, considered to be located at the left of a diagram, to a load somewhere to the diagram right.

Instrument Transformers

For reasons of personnel safety and good instrument design, it becomes impractical to connect meters directly to circuits having voltages above 1,000 volts and currents above 50 amperes. Furthermore, instruments tend to become inaccurate when directly connected to high voltage because of electrostatic forces that act on the moving system. Shunts are often considered for large ac currents, but have the disadvantages of large power loss at high current ratings and no voltage isolation. Unless a shunt is especially designed for ac use, its inductance, nonuniform split of current in the blocks, skin effect, proximity effect, and nonuniformity of the material all contribute to ac shunt error.

By contrast, potential and current transformers offer a practical and safe means of reducing voltage and current by an exact ratio for instrument use. All instrument transformers have the following basic design objectives: (1) careful attention to an exact primary-to-secondary ratio; and (2) as small a phase angle as possible. Instrument transformers by design have small load (burden) capability because meters do not need large amounts of power and a low burden will enhance the design for best possible accuracy.

In a power station, the instrument transformers which are used for station metering, relay operation, and control services are often several feet tall (a few meters), resulting from a design to withstand very high voltages. For laboratory and shop testing at low distribution voltages, both potential transformers and current transformers are very small in comparison. They weigh perhaps 15 pounds (~7 kilograms) and can be hand carried. Quality potential transformers for laboratory use will support a 25-volt-ampere burden, have a ratio accuracy from 0.1 to 0.5%, and a phase angle of 10 minutes. A core-type design in which the winding surrounds the iron provides the necessary insulation for a high range of 2,300/1,150 or 115 volts. The iron is usually grain oriented silicon steel from 0.012 to 0.025-inch (0.3 to 0.6 millimeter) thick, depending upon the quality of the transformer and frequency range. Current transformers of the toroidal type, using tape-wound, high nickel-iron cores, have phase angles of less than 2 minutes and a ratio error of plus or minus 0.02%. Burden capability is up to 25 VA. Primary-to-secondary insulation is difficult to provide in a toroidal transformer. A rating of 2,500 volts is common for stock transformers with 5,000 working volts pushing the practical limit for this construction in custom designs.

Walter A. Troeger, Weston Instruments, Newark, New Jersey.

ELECTRIC POWER GENERATORS. The majority of electric power produced in the United States is by 3-phase generators. Advantages of 3-phase generators lie in economy of apparatus, lower transmission losses, inherent starting torque for poly-phase motors, and a constant running torque for balanced loading. A generator is built with axial slots for armature coils in a stationary hollow cylindrical iron core called the stator. The windings are placed in the slots so that when carrying current they produce a chosen even number of alternate magnetic poles. The coils over each magnetic pole are grouped in three equal bands to give a 3-phase balanced system of terminal voltages.

An inner rotor has coils which carry direct current to give the same number of alternate magnetic poles as on the stator. Rotor current strength is controlled by a rheostat or voltage from a direct-current generator. Voltages are produced in the stator windings by flux cutting as the rotor magnetic flux sweeps by them, and currents flow when the generator terminals are connected to a 3-phase load impedance. The 3-phase stator line voltages are equal in magnitude and 120 electrical degrees apart in time sequence. So also are the line currents for a balanced 3-phase load. Generator voltages are of the order of 12,000 to 30,000 volts for large machines.

Generator frequency is the product of the pairs of magnetic poles and the speed in revolutions per second. At 60 cycles (60 Hz), a 2-pole generator runs at 3,600 revolutions per minute; a 6-pole generator at 1,200 revolutions per minute. The maximum speed of 3,600 revolutions per minute has been increasingly adopted even for very large machines because high speed means decreased size and weight for a given kilowatt rating and better turbine performance. However, waterwheels and water turbines show best characteristics at much lower speeds, roughly a range of 100 to 600 revolutions per minute. Sixty cycles is the prevailing frequency in the United States. Because of weight and space limitations, 400 cycles is popular in the aircraft industry. Fifty cycles is a common frequency in Europe. See also *Alternator.*

An elementary form of electric generator is shown and described in the accompanying figure.

Direct-current generators are built with their dc magnetic poles on the stator. Armature conductors on the rotor have ac voltages induced in them as they are rotated; the same principle of flux cutting holds as previously described. An automatic mechanical switching device, called a commutator, is placed on the shaft. It carries fixed brushes, and with its many insulated copper bars connected to the armature coils, it inverts every other alternation of the voltage to give unidirectional, or direct-current voltage at the two armature terminals. It is the commutator that requires the rotor to be the armature so that coils and their switching arrangement always move exactly together. Direct-current generators are generally limited to several thousand kilowatts and their application lies mainly in industrial plants of a specialized nature.

Rotating electric generators are driven by a variety of turbines, notably steam, gas, and hydraulic turbines; by gasoline, gas, and diesel engines. These drivers are described under their appropriate alphabetical entries in this volume. The type of driver used is predominately influenced by the type of fuel economically available at a given site as well as local and regional environmental regulations.

Among the major developments within the last few decades are the superconducting turbine generator and the magnetohydrodynamic generator. The application of superconductivity to synchronous machines with rotating field winding was pioneered at the Massachusetts Institute of Technology, commencing in 1967. The development of large superconducting generators for electric utility applications addresses a number of specific needs of the utility industry. By replacing the conventional copper conductor field winding in the rotor of a synchronous generator with a high-capacity superconducting winding that virtually has zero resistance at cryogenic temperatures, important benefits are gained. The most obvious advantage is the elimination of rotor I^2R loss. The resulting reduction in excitation power is accompanied by a similar reduction in rotor ventilation power requirements. Increased power density and the resulting elimination of stator iron at the armature winding are some of the more subtle, but nonetheless beneficial characteristics of the superconducting synchronous machine. See entry on **Superconducting Electric Generators.**

The magnetohydrodynamic generator is one in which a thermally ionized gas is forced at high temperature, pressure, and velocity through a duct situated in a transverse magnetic field. An induced voltage appears in the third mutually perpendicular direction (Hall effect), and this voltage may be tapped by electrodes within the duct.

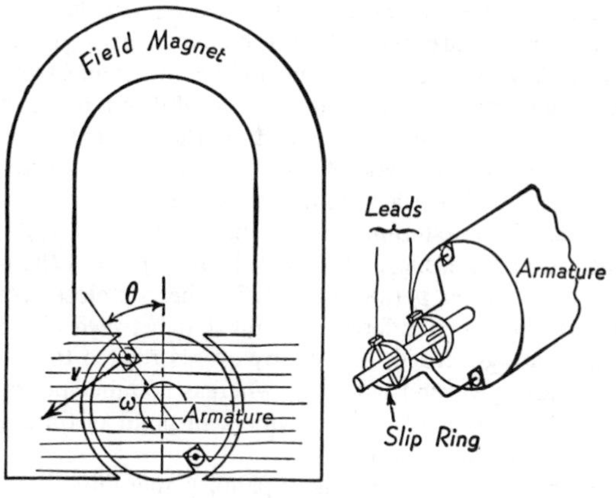

(a)

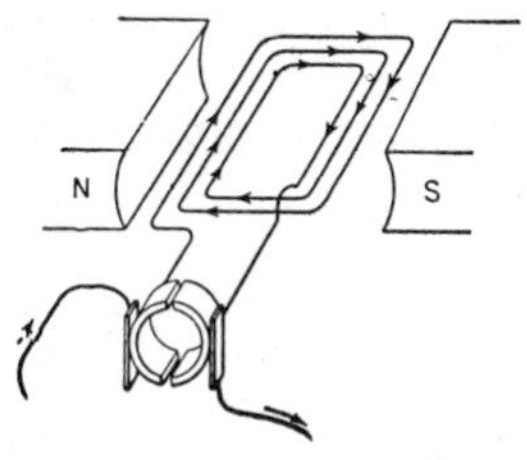

(b)

Elementary electric generator consists of a soft iron core rotating between the poles of a permanent magnet, and having the slots on the surface, in which is embedded a coil of wire. It is apparent that as the coil rotates, carried by the soft-iron armature, it will cut across the flux lines, extending from pole to pole as shown in (a). When this apparatus is connected to stationary leads through the medium of slip rings, and brushes resting thereon, it becomes an elementary alternator. If, instead of slip rings, a split segment, such as that shown in (b), is connected to the ends of the coil, the reverse of the current in the coil will occur when alternate segments of the slip ring (an elementary commutator) are opposite one of the brushes. This gives unidirectional current in the exterior leads, although it would be quite variable with only one coil in the armature.

A uniform and unidirectional current is the result of many single-coil armatures so connected that the resultant current is the sum of several individual outputs. With sufficient overlap, the resulting current will be uniform and unidirectional. When the coil is revolving at a speed of ω radians per minute, at the position indicated, the speed of cutting vertically across flux lines is $v \cos \omega t$. The time is measured from the vertical position of the coil, and angle θ is ωt. When a wire cuts a magnetic field having a flux density represented by b and has a length and velocity represented by l and v the voltage generated is $blv/10^8$ volts, thus the voltage generated any instant t, t being measured from the point of minimum generated voltage, is $blv/10^8 \sin \omega t$.

The dc generator is an ordinary dynamo machine having a multiple-coil winding, the ends of the coils of which are connected to a multiple-segment commutator. The armature is usually rotating, and the field stationary. The field sets up magnetic lines of force, which are cut by the conductors on the revolving armature, giving rise to a generated voltage, which is led through the commutator to a unidirectional external circuit. The iron core is built of laminations of iron insulated from each other by mill scale, or lacquer, or japanning, so that eddy currents which can be generated in the iron core, will be a minimum. The field windings are usually stationary, and the armature rotating; hence low voltage is the usual condition of use of dc generators.

If the exhaust gas from the magnetohydrodynamic generator is used to heat steam for a conventional generator, a larger portion of the thermal spectrum can be utilized and the system efficiency may be raised from the tradition value (about 40%) to possibly 50 or 55%. The art of the magnetohydrodynamic generator is not in the high developmental stage as is the case of the superconducting generator, but active research goes forward. Considerable attention has been given to the potential use of magnetohydrodynamic generator units as infrequently required peaking units. See separate entry on **Magnetohydrodynamic Generator.**

Other related entries in this encyclopedia include **Electric Power and Energy Measurement; Electric Power Production and Requirements; Electric Power System Interconnections;** and **Electric Power Transmission.** Specially designed bulb-type electric generators for use in tidal power facilities are described in entry on **Tidal Energy.**

ELECTRIC POWER (Geothermal). Geothermal Energy.

ELECTRIC POWER (Hydroelectric). Hydroelectric Power.

ELECTRIC POWER PRODUCTION AND REQUIREMENTS. Nine nations of the world generate and consume over three-fourths of the total world electric power. In the United States (1978) this exceeds 408 million kilowatts in terms of electricity generated and a capacity of about 550 million kilowatts, leaving a margin of 25% of total capacity for expansion, handling of shortages during scheduled maintenance, emergency outages, etc. Since this generating capacity is tied together by three major power networks, there is considerable flexibility which enables the importing and exporting of power from one region to the next within a given network. The anticipated peak load requirements by mid-1988 are projected at about 655 million kilowatts, with a total generating capacity of about 810 million kilowatts to allow for ample margin. This projected increase over 1978 is just under 50%. Although there are numerous factors, mainly of a political nature, that can shift the energy sources that will be used to generate power in the late 1980s, current projections indicate that (compared with 1978), coal-fired units will rise from 44.7% to 49.2% of the total; oil-fired units will be down from 16.3% to 12.8% of the total; gas-fired units will be down from 13.5% to 3.4% of the total; nuclear units will increase from 12.8% to 27.3% of the total and hydroelectric units will be down from about 13.3% to 7.5%. In the case of the shrinking percentage of the total, part of this will be due to the lack of construction of new facilities, particularly in the case of hydroelectric capacity; in the case of gas- and oil-fired units, the shrinkage will reflect a reduction of new facilities as well as the conversion of facilities to another energy source, notably coal. Natural precipitation of water also can affect hydroelectric production by several percentage points up or down.

As of the early 1980s Indiana, Ohio, Kentucky, West Virginia, western Pennsylvania, and much of Virginia lead the rest of the country in dependence upon coal; in these states coal accounts for just under 90% of the energy used. The use of coal will continue to expand in most regions, but this is somewhat dependent upon the expansion of nuclear facilities. If no additional restraints and delaying threats further impede the development of nuclear energy, some of the facilities that otherwise might be coal-fired will be nuclear. The greatest impact on gas-fired units will occur in Texas where, as of 1978, nearly 80% of the power plants were gas-fired. By 1988, gas-fired units are expected to account for no more than 20% of the units. In Texas, it is projected that coal will account for a little over 45%, oil about 17%, nuclear about 16%, the remainder being hydroelectric.

In the western states, hydroelectric sources will fall from about 47% of the total (1978) to about 25% in 1988, coal will be up from about 23% to 32%, oil up from about 16% to 18%, gas down from about 11% to 2%, and geothermal energy up from about 0.7% to 3.6%. As pointed out by the Edison Electric Institute, these forecasts are based upon the best information available as of early 1979. See Fig. 1.

Worldwide, electric power production in the United States is followed by that of the U.S.S.R. (about 19.3% of the total), Japan (8.4%), the United Kingdom (7.9%), West Germany (5.9%), Canada (5.2%), France (4.6%), Italy (4%), and Spain (2.1%). It is estimated that close to 20 million kilowatts of electric power are generated in China and slightly less than this figure in India.

Probably more significant in terms of industrial and residential life style is the kilowatt-hours per capita figure. Canada leads the world in this regard, with nearly 10,000 kilowatt-hours produced annually per Canadian citizen. With exception of France, Spain, and Italy, listed above, the ranking for total power production capacity and power generated per capita show no correspondence.

Unfortunately, detailed requirements on electric power use are not readily available for many nations of the world, but there is no doubt that electric power production and consumption reflect the degree of industrialization, including power-consuming mechanization and automation, the degree of residential conveniences (appliances, light-

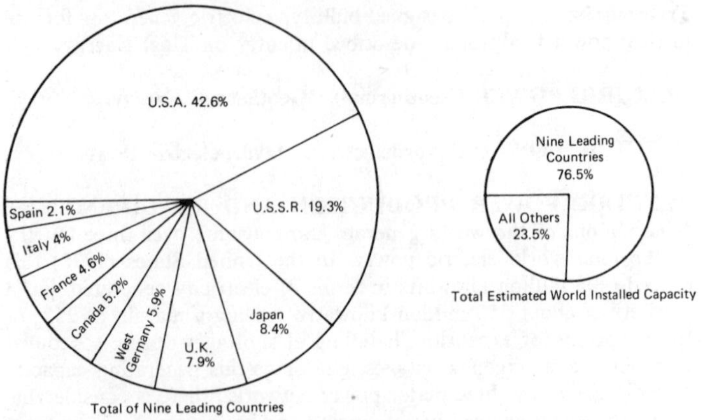

Fig. 1. Distribution of electric generating capacity among nine leading countries.

ing, communications, heating), the size of certain specialized high power consuming industries, such as the mining and metallurgical fields (aluminum, magnesium, steel, etc.–as well as nuclear fuel processing plants), and obviously, to some degree, a wasteful attitude toward the use of electric power.

Annual use of electricity in industry in the United States is estimated at about 46,000 kilowatt-hours per working person. It is interesting to note that no worker, by sheer muscle power, can produce in a day the energy represented by just 1 kilowatt hour of electricity. The average power which a person can exert is estimated at about 35 watts. Thus, a person averaging 250 eight-hour days of manual work per year is estimated to expend energy equivalent to about 67 kilowatt-hours. Thus, a factory worker using some 46,000 kilowatt-hours annually will have the equivalent energy assistance of 683 people helping on the job all year long.

Growth of electric power generation in the United States is shown in Fig. 2.

Definitions of Terms Used in Industry. Terms commonly used in electric generation forecasting include:

Capability. The capability of the power systems is defined as the maximum kilowatt capability of the systems with all power sources available, with no allowances for outages, and (for systems including hydro installations) with sufficient kilowatt-hours to supply the energy requirements of the system. The capability must provide for scheduled maintenance, emergency outages, and system operating requirements, in addition to the estimated load and any unforeseen load. The capability of existing installations is determined by the dependable capacity at the time of the peak load and the energy requirements of the system. Estimated capabilities of those systems served wholly or in part by hydropower sources are determined for median hydro conditions.

The capabilities of all systems normally interconnected are based upon fully coordinated operation. Peak capability refers to the maximum capability, as previously defined, at the approximate time and for the duration of the load referred to, that is, at the time of the December peak load or the summer peak load.

Peak Load. The maximum load encountered on the system during a given period of time. December peak loads represent the maximum load during the winter period. Generally, there is closer coincidence between winter peaks on different systems than between summer peaks. Winter peaks are governed, in part, by events related to the calendar (pre-holiday and year-end activity), whereas summer peaks are governed by local weather conditions which, in point of time, may vary somewhat widely with locations.

Gross Margin. Also known as capability margin, this represents the difference between capability, as previously defined, and peak load. Gross margin, therefore, should be sufficient to provide for scheduled maintenance, emergency outages, and system operation requirements, if practical operating conditions are to be maintained. Any excess in the gross margin over and above the provision for these items is available for unforeseen loads.

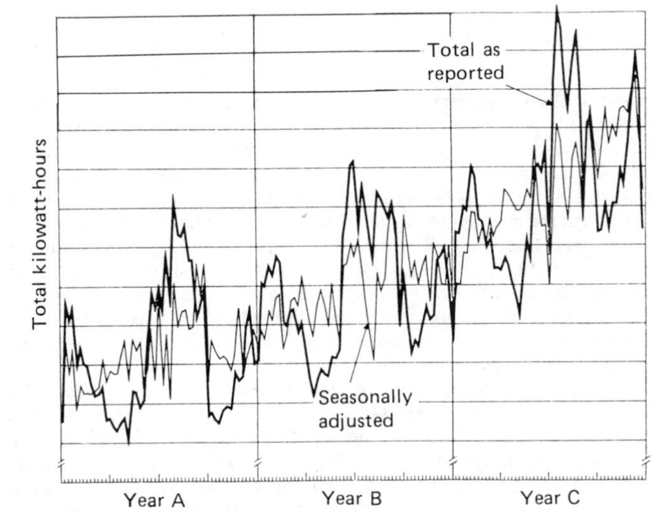

Fig. 3. Patterns of electric output on a weekly basis for three years for the total electric utility of the United States, excluding Alaska and Hawaii.

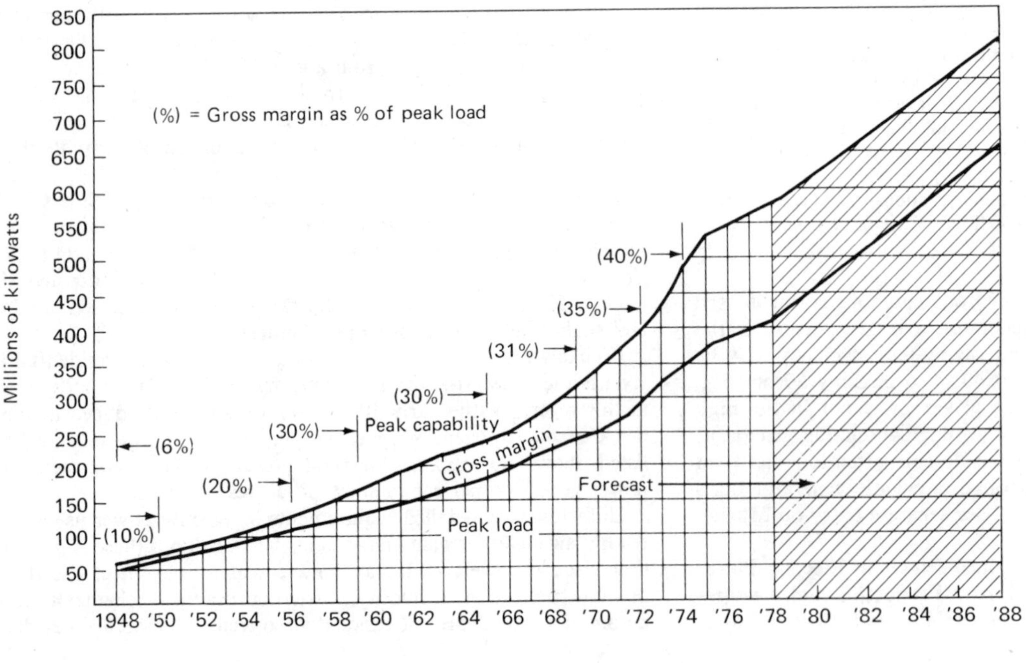

Fig. 2. Historical development of December peak capabilities and gross margins for total electric utility industry of the United States, excluding Alaska and Hawaii.

The variation in weekly power requirements is demonstrated by Fig. 3. This pattern varies to some degree from one region to the next, depending upon local climate, population, extent of industrialization, etc. However, the pattern shown in Fig. 3 is weekly electric output on a totalized basis for the United States. See list of related entries at end of entry on **Energy.**The use of electric power in the United States divides roughly as shown in Fig. 4.

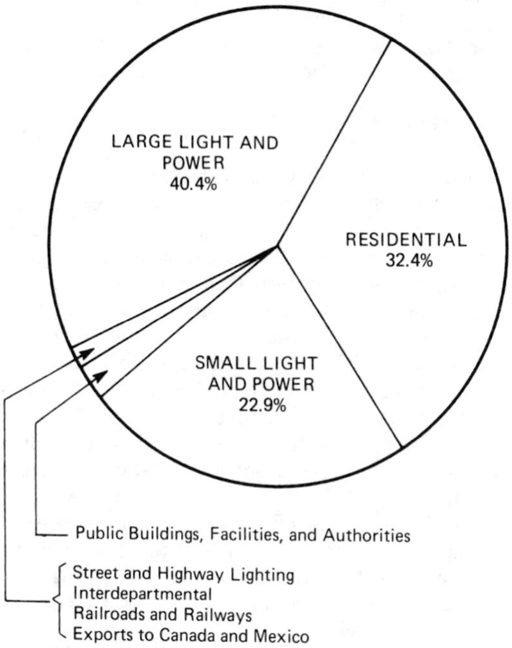

Fig. 4. Distribution of electric power requirements in the United States—average for several years.

References

Staff: "U.S. Energy Outlook—Fuels for Electricity," National Petroleum Council, Washington, D.C. (Revised periodically).

Staff: "Statistical Year Book on the Electric Utility Industry," Edison Electric Institute, New York (Published annually).

Staff: "Semi-Annual Electric Power Survey," Edison Electric Institute, New York (Published semi-annually).

NOTE: Additional statistics and reference material may be obtained from the American Public Power Association, Washington, D.C.; National Rural Electric Cooperative Association, Washington, D.C.: Federal Power Commission, Washington, D.C., National Coal Association, Washington, D.C.; American Gas Association, Arlington, Virginia; National Electrical Manufacturers Association, New York; The Electric Power Research Institute, Palo Alto, California.

ELECTRIC POWER SWITCHING. **Switching (Power).**

ELECTRIC POWER SYSTEM INTERCONNECTIONS. The electrical power requirements or load placed upon a specific power-generating facility may, at times, be greater than or less than the optimal rate in terms of economy and other operating factors. Of course, within the upper and lower limits of capacity of any facility, adjustments in operation can be made, cutting in generating equipment or cutting out units, as may be required. But, it may be more economical for a facility to "buy" power from another generating facility, or depending upon the direction of the demand, to "sell" power to another facility. This can be accomplished smoothly by means of a tie-line, or if large numbers of facilities are so connected, the system may be called an interconnecting network. In the latter situation, not only are individual facilities connected, but whole utility systems to form a vast network encompassing very large geographical areas. In some countries, all or nearly all facilities are so interconnected. In the United States, there are three major power networks, as shown by the map of Fig. 1.

Fundamentally, such networks are established in the interest of maximizing operating economy, but because such networks are usually

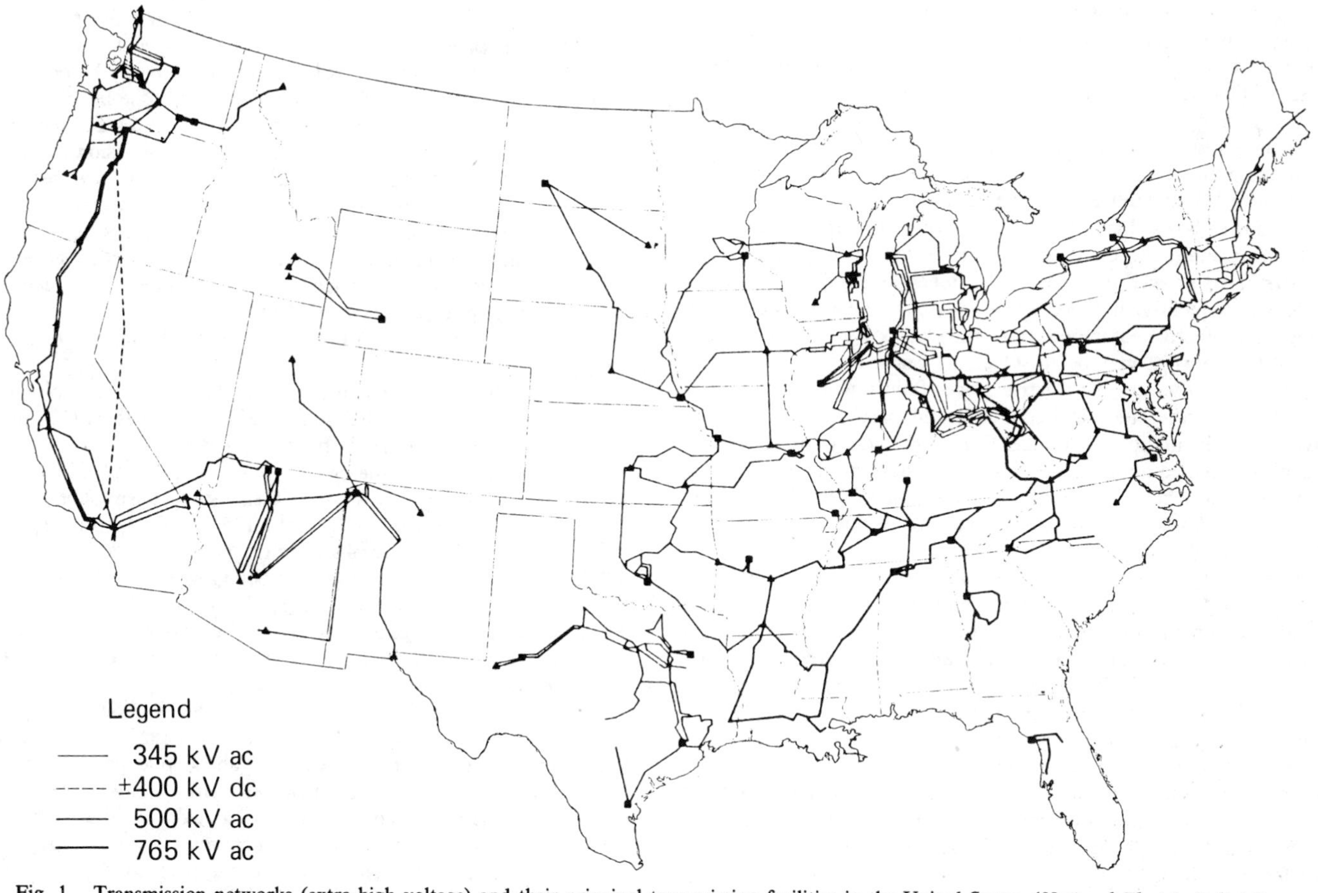

Fig. 1. Transmission networks (extra-high voltage) and their principal transmission facilities in the United States. (*National Electric Reliability Council.*)

publicized at the times of emergencies, the networks are sometimes more associated with emergencies than with effecting economies of operation. Such networks do provide greater reliability of service in the instance of what might be called minor emergencies, such as the purposely scheduled maintenance shutdown of some generating capacity in part of the network, or less frequently, in the case of unscheduled shutdown. T. J. Nagel (1978) describes the situation succintly: "The purpose of interconnections is to expand the scope of the individual power systems so as to enhance both reliability and economy of power supply. In terms of reliability, interconnections provide assistance during generation outages, assist in distributing excessive generation at times of major load outages, and provide support in times of transmission outages. In economic terms, interconnections allow the full exploitation of economies of scale in both generation and transmission. Savings can thus be realized without compromising reliability, through the sharing of risk. Interconnections allow the interchange of power to reduce generation costs."

Interconnections, of course, also add to the abilities of the electric-generating industry (public or private) to accommodate not only for short-term deficiencies in any given region, but for long-term deficiencies as well. During a coal strike in 1978, for example, over a 30-day period, more than three billion kilowatt-hours of electric power was imported into the east-central region of the United States from adjacent regions. Because of the gross capacity of a network, such transfers can be made without overloading any of the major power supply sources and can be accomplished without interruption in service.

As has been shown by very serious blackouts in the United States affecting millions of users over large geographical areas in 1965 and as recently as 1977, interconnection networks are not fully without risk—such blackouts resulting from what may be called a cascading or islanding effect. It has been shown from experience that the larger and more complex the network may be, the greater is the need for sophisticated control equipment, operator training for emergency situations, and meticulous maintenance of all equipment. Equipment failures can cause a domino effect.

From a fundamental standpoint, the great bulk of instantaneous support to make up a generating deficiency in some part of a network is derived from the inertial energy stored in the combined rotating masses of all generating plants in the network. Although such a transfer may be accompanied by a slight, very temporary decline in frequency in the network, this is not a critical factor in making the transfer. The much greater role in sustaining continued service and reliability is the capability of the transmission facilities to handle the resulting shift in power flows. Basically, the flows of power on individual elements of an interconnected network are usually uncontrolled. They are distributed in accordance with the impedances of the circuit elements. Consequently, if these elements are overloaded and disconnected by protective equipment, uncontrolled power failures in some parts of the network may occur. Over the years, and with particular emphasis since the massive electric power failure which blacked out most of the northeastern United States and parts of two Canadian provinces on the night of November 9–10, 1965, many studies, including sophisticated computer assisted programs and simulations of emergency situations, have been made. The National Electric Reliability Council (NERC), formed some years ago, has as one of its objectives the provision of a mechanism for ensuring the coordination of the system planning and operation of all bulk electric power supply facilities in North America. Through this Council, large amounts of information are exchanged between operators. Simulation studies designed to determine the anticipated behavior of an entire interconnected network are made. The NERC is comprised of several regional organizations, as indicated by the map of Fig. 2.

The blackout of 1965 and a subsequent blackout of New York and environs in 1977 were thoroughly investigated and documented. The chain of happenings, including equipment malfunction and network management communications, are beyond the scope of this book, but they make fascinating reading. More detail can be found from the references listed.

Among many areas researched toward the management of power system interconnections in time of crisis is consideration of magnetohydrodynamic generators. See entry on **Magnetohydrodynamic Generator.**

References

Clapp, N. M.: "State of New York Investigation of the New York City Blackout," State of New York, Albany, New York, 1978.

Staff: "System Blackout and System Restoration," First Phase Report 26, Consolidated Edison, New York, August 1977.

Staff: "Seventh Annual Review of Overall Reliability and Adequacy of the North American Bulk Power Systems," National Electric Reliability Council, Princeton, New Jersey, 1977.

Staff: "The Con Edison Power Failure of July 13 and 14, 1977," Final Report, Federal Energy Regulator Commission, Washington, D.C., June 1978.

ELECTRIC POWER (Tidal). Tidal Energy.

ELECTRIC POWER TRANSMISSION. Energy may be transmitted electrically in overhead wires or underground cables as an electronic flow under pressure, the flow being measured in amperes. The voltage is the electric pressure. Electrical energy is proportional to the product of these two quantities. This energy for all practical purposes cannot be transmitted without some losses, the principal one being a resistance loss, which depends upon the current flow and the size of the conductor. In transmitting a given amount of energy, this loss may be reduced by increasing the voltage, with a corresponding decrease in current. Thus, the use of higher voltages increases the efficiency of electric power transmission by decreasing the conductor loss. In addition, there are economic motivations for increasing transmission line voltages. The power that can be transmitted by a line increases with the square of the voltage and decreases directly with the distance. Therefore, for a fixed distance, a doubling of the previous line voltage will permit about four times as much energy transfer. Over the years, the voltages of new transmission overlays have increased progressively.

In 1978, there were approximately 270,000 miles (434,430 kilometers) of AC high voltage transmission lines in the United States ranging from 115 kV to 765 kV, plus another 1,300 miles (2,092 kilometers) of DC up to ±400 kV. This compares to only 215,000 total circuit-miles (345,935 kilometers) in service in 1970. In order to keep pace with the growth in generation capacity, it is speculated that nearly 113,000 more miles (181,817 kilometers) will have to be built by the year 2000. The majority of the overhead lines to be installed will be at voltages greater than 345 kV in order to realize substantial economic benefits. However, the key purposes of the transmission system will continue to be (1) to supply the load centers from new generation, (2) to provide interconnections between individual electric utility systems and pools for achieving more reliable power supply systems, and (3) to allow for economical interchange of energy between systems. The essential economy or savings derived from interties between power systems is that peak loads can be handled with less plant investment than if each region were to install the required capacity to meet its own peaks.

From the inception of transmission, the increase in voltage was based on technology and economics. Early changes in transmission voltage reveal a doubling of the previous level when a change was made. One exception is the single, unique use of 287 kilovolts (kV) from Hoover Dam to Los Angeles. Generally, transmission remained at a level of 230 kV from 1922 until 1953 when the first EHV, 345 kV transmission lines were placed into service. During the mid-1960s, 500-kV-class transmission came into being, with 765 kV coming onto the scene in the early 1970s. Currently, over 1500 circuit-miles of 765 kV lines are in operation in the United States. A historical plot of voltages for the electric utility industry in the United States is shown in Fig. 1.

The majority of the EHV lines through the early 1980s have utilized alternating current. However, because reactive compensation for extra-long high-voltage AC lines is expensive and line losses are very high, HVDC has become an economical alternative. Line losses for DC are approximately 33% lower since DC power, by its nature, is not subject to reactive power losses. The construction of a HVDC transmission line is lower in cost, requiring only two conductors per line as opposed to three per line used in AC three-phase transmission. This allows for lighter towers to be used and consequently narrower

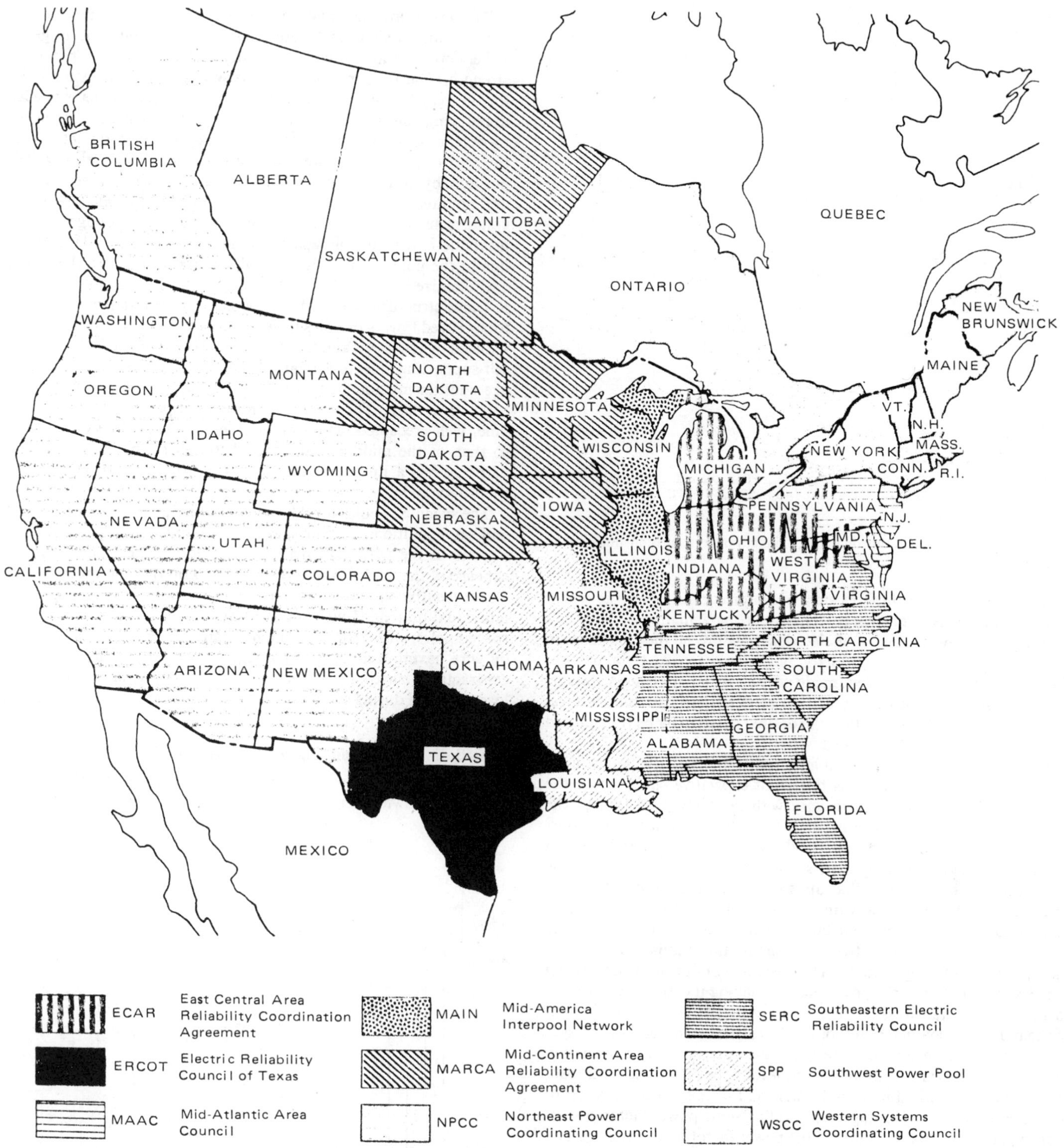

Fig. 2. Nine regional reliability councils that make up the NERC. (*National Electric Reliability Council.*)

rights of way. Less insulation is normally needed on a DC line to deliver the same amount of power. The first modern-day attempt at DC transmission in the United States occurred in 1970 when an 800 kV (±400 kV) line known as the Pacific Northwest Southwest Intertie was placed into operation. This bipolar, overhead line permits an exchange of power at peak load times between two regions 850 miles (1368 kilometers) apart.

Terminal conversion equipment, however, continues to be the major obstacle to widespread use of HVDC. Since a terminal is made up of an AC switchyard plus a valve hall and harmonic filters, size and area requirements are excessive. The development of a solid state thyristor valve, however, has enabled the cost of HVDC to remain stable throughout the 1970s, in addition to simplifying and improving the reliability of the conversion process. Probably the greatest limitation of HVDC is that power can only be transferred between two points. Multiterminal systems of 3 or 4 terminals are possible with the development of a HVDC breaker and the system control. The use of DC transmission will likely be confined to applications where special situations exist. These include the need for: (1) long-distance point-to-point bulk power transfers; (2) asynchronous ties for control of power flow; (3) increased transmission capacity on limited right-of-way; and (4) special stability problems involving remote generation.

As previously mentioned, the energy loss caused by current flowing through the line resistance is not the only loss of energy in power transmission. The long stretches of parallel conductors have a capacitive effect, causing them to draw a current much as a condenser, even though the switches at the far end of the line are open. Furthermore, at high voltages, the air surrounding the conductors becomes

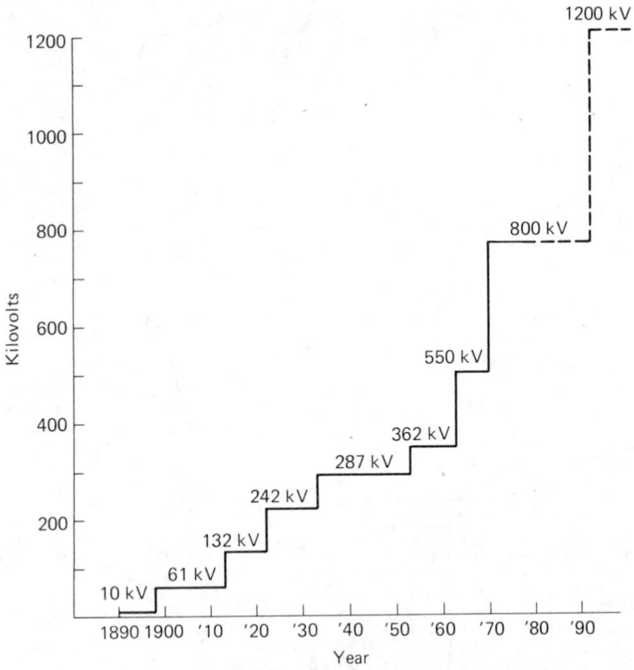

Fig. 1. Record of maximum operating voltages of transmission lines in the United States.

partially ionized, and there exists a brush or corona discharge which represents a leakage of energy. The latter, as a matter of fact, is a limiting factor in the raising of electrical pressure on the transmission line and on the size of conductor, and tends to offset the savings due to the reduction in current made possible by an increase in voltage. Good transmission line design requires proper coordination of voltage, wire size, and line losses, so that the desired power may be transmitted at a minimum total annual cost.

When current flows on a long transmission line, the inductance and resistance of the line, plus the shunt capacitive effect, cause the voltage at the receiving end to vary with the load, even though the voltage is constant at the sending end. This effect is known as transmission line regulation, and can be calculated quite accurately. From the standpoint of the equipment served, it is desirable that this line regulation be controlled so that utilization equipment eventually supplied from step-down transformers may operate within a reasonable voltage range. Regulators and tap-changing transformers are among the means employed to control transmission line regulation. Furthermore, it can be demonstrated that the economy of power transmission varies as the square of the power factor, making it extremely desirable to operate the transmission line at unity power factor. On account of the induction characteristics of most electrical loads, and the capacitive effect of a transmission line itself, the current on a transmission line will tend to vary from lagging to leading or vice versa with load, although customarily it tends to be lagging. For this reason, a synchronous motor has the advantage of providing some power factor correction. Another method to improve the power factor is to connect static capacitors to the line. More recently, a solid state equivalent of the synchronous condenser called a Static VAR Compensator has been developed. The Static VAR Compensator utilizes high-voltage solid state thyristor AC switches to provide continuously variable leading and/or lagging reactive current from standard reactors and capacitors.

The physical exposure of the ordinary transmission line makes it a likely victim of lightning, and no extensive transmission line could be successfully operated without adequate lightning protection. This may consist of lightning arrestors strategically placed, or an overhead grounded guard wire. Direct strokes are usually immediately dissipated by flashover on the insulators, the lightning then finding its way down the pole to the ground. High-frequency induced waves (2000 to 5000 cycles) may not possess flashover potential, and will travel along the line until discharged by an arrestor, or attenuated, i.e., dissipated in resistance loss. Aside from lightning, a line should be protected against overload and short-circuits. This is commonly the function of transformer fuses, substation circuit breakers, or circuit breakers in general.

Turning from electrical to mechanical characteristics of electric power transmission, the material used as a conductor is generally either copper or aluminum, frequently with a steel core for mechanical strength. The former is used in many cases, but the latter is in use for the very high-voltage lines because the larger diameter is effective in reducing corona loss. It also has low relative cost and high strength-to-weight ratio. Except for very low-voltage distribution network lines, the wires are not insulated, and are carried, mechanically held in position, by insulators of porcelain or glass. Sometimes these insulators are mounted rigidly on the cross arm of the poles, but for the higher voltages suspension insulators are used. On high-voltage lines, each insulator is a chain of separate units, so that the voltage from the line to the pole or tower has a uniform gradient across the insulator.

Galvanized steel, self-supporting towers or poles, or wood H- or K-frame structures are used to support overhead transmission conductors. Guyed towers or poles have the least weight, which can be further reduced by the use of aluminum instead of steel. For EHV applications, self-supporting steel towers are most commonly used. The lower voltage subtransmission lines are generally supported by wooden poles using cross arms. The most commonly used pole is of southern yellow pine, well creosoted. The towers or poles should be sufficiently high so that, at the middle of the span between them, the bottom of the wire sag has a minimum clearance over the ground, as specified in the safety code.

The spacing of poles must be chosen with due regard for temperature and sag conditions. Larger sags in the wire between poles give lower stresses in the wire, and permit the use of longer spans, but at the expense of the use of taller poles to maintain the minimum clearance. Consequently, the observer may see extremes of design representing individual designers' viewpoints, varying from relatively low, closely spaced poles, with little sag, to long spans where tall poles are spanned by wire having a much greater sag. At the same time, the effects of contraction caused by lowering of temperature in winter, and the loads suffered during storms, particularly sleet storms, must be guarded against.

Growth in Circuit- and Gigawatt-Miles

Annual additions of transmission in circuit-miles in the United States are shown in Fig. 2. The chart indicates that installations from

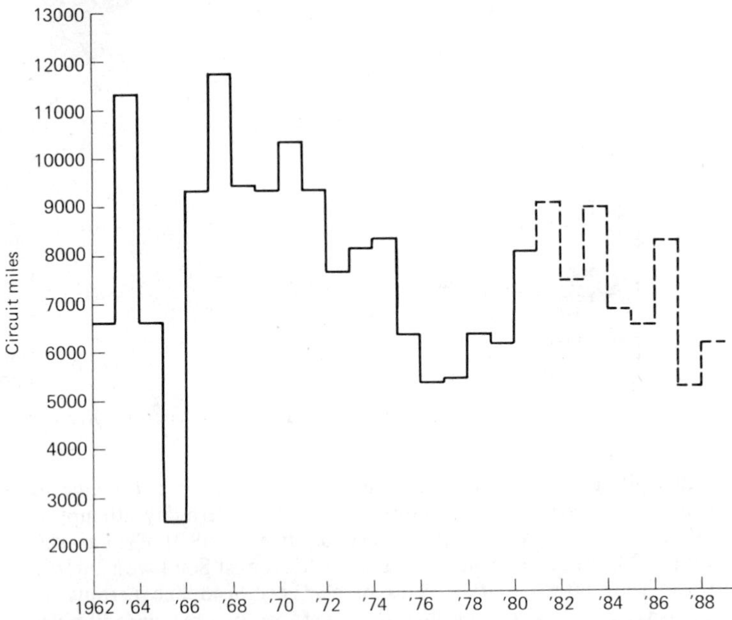

Fig. 2. Annual additions of circuit miles of transmission in the United States (115 kilovolts and above). (1000 circuit-miles = 1609 circuit-kilometers.)

1976 through 1980 were 21% less than those from 1971 through 1975. However, indications are that total annual additions will grow slightly on an annual rate through the year 1988. In order to determine transmission system growth patterns, the parameter "total circuit miles" does not account for the disparity that exists in load capability for

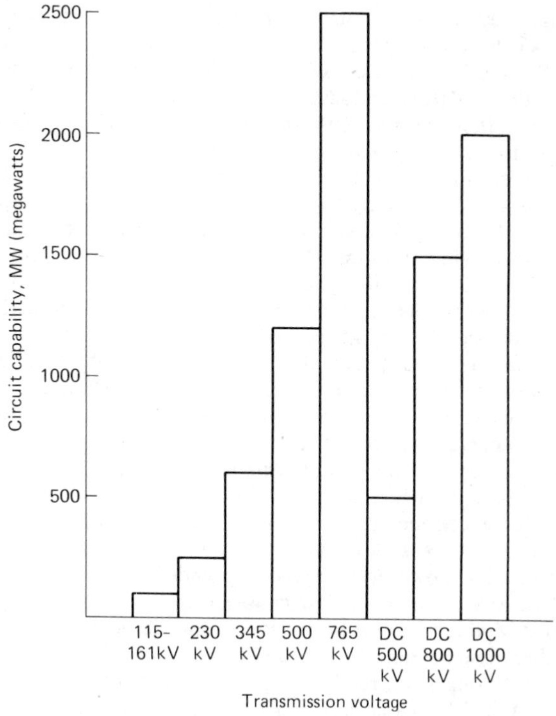

Fig. 3. Typical circuit capabilities.

each of the voltage classes considered. A more meaningful way to measure this growth is in terms of capability of the transmission system to transmit power. Thus, the term gigawatt-mile is an excellent means of expression since it constitutes a measure of transmission capability. The term is analogous to "passenger miles" as used by the airlines, for example, to express the capability to transport people, or of "ton miles" in terms of freight.

Typical load-carrying capacity of circuits operated within the various voltage classes are shown in Fig. 3. Multiplying the circuit capability values by the corresponding circuit-miles, for lines installed and projected, results in the plot of transmission capability shown in Fig. 4. The impact of EHV on the growth pattern of transmission capability is evident from this chart. Until 1960, EHV accounted for less than 10% of the total capacity, while the lower voltage classes were the real workhorses. EHV growth began in 1964, and by 1970 accounted for three-fourths of the doubling in transmission capability that occurred during that period. Capital expenditure postponements by the utility industry on line construction were responsible for the decline shown in the mid-1970s. Although annual additions for 115–230 kV are expected to decline over the next 10 years, EHV is expected to grow at a greater rate for the same period. Voltages higher than 765 kV are not expected to develop commercially until the 1990s. Future DC capacity is projected to be minimal for both 800 kV and 1000 kV voltage levels.

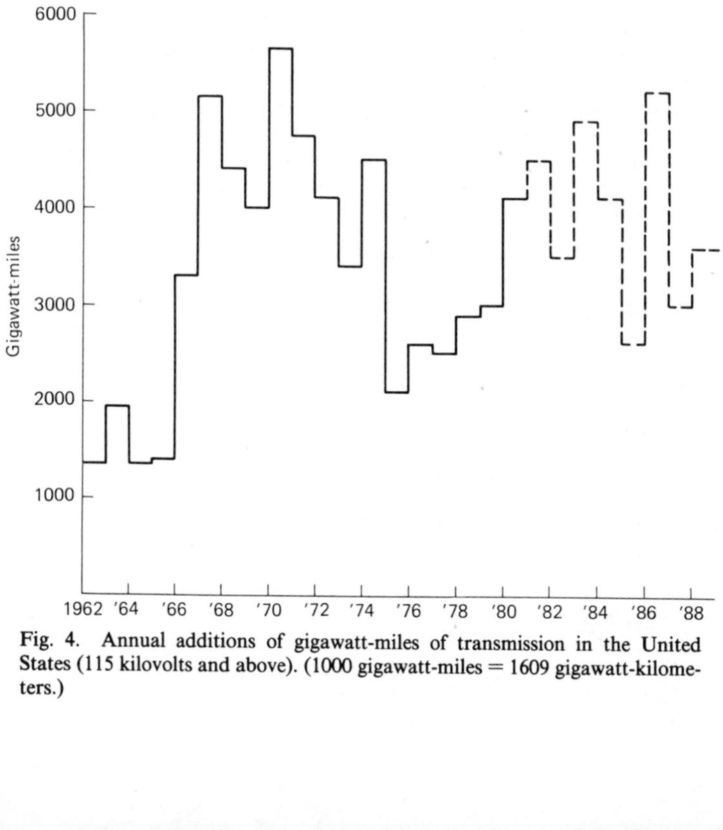

Fig. 4. Annual additions of gigawatt-miles of transmission in the United States (115 kilovolts and above). (1000 gigawatt-miles = 1609 gigawatt-kilometers.)

How does transmission capability growth compare to that of electric power generation? The average generation and circuit-mile additions for various time periods are shown in Fig. 5. In real terms, transmission

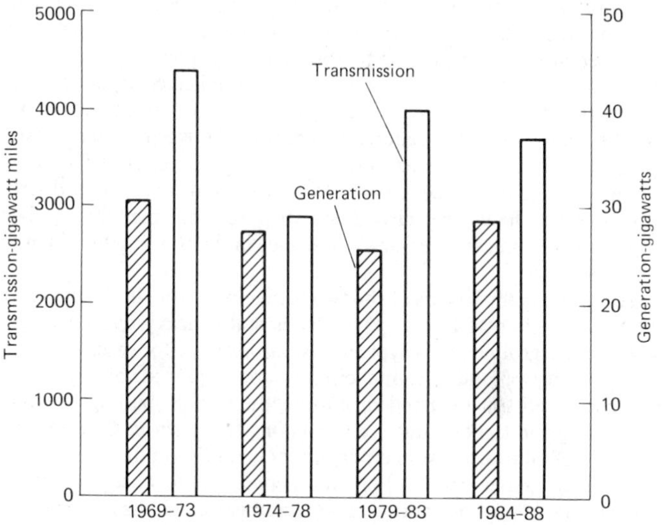

Fig. 5. Average annual additions of generation and transmission capacity in the United States. (1000 gigawatt-miles = 1609 gigawatt-kilometers.)

costs have declined while generation costs have increased in the past decade. For this reason, transmission capability installed has exceeded that of generation since the marginal costs of obtaining capacity and energy by this means is less than for building new generating capacity. The trend of additions to generating capacity in the 1980s is one of little or no growth, so that the trend to more transmission will surely continue. In the long run, however, transmission cannot be a substitute for generation.

Overhead Technical Developments

Current technology and research at various laboratories around the country will greatly affect the future of overhead transmission. Studies now under way are concerned with the design and testing of new insulation types, the evaluation of transmission towers and foundations, conductor motion problems, and the thermal characteristics of conductors and fittings.

While there are economic benefits for three-phase transmission at higher voltages, economic advantages for shifting from very high voltage three-phase transmission to lower-voltage high-phase-order transmission has also been demonstrated. Both six- and twelve-phase systems are being considered developmentally. A prototype of a six-phase line has been built and is currently being tested to obtain data on radio noise and electric field effects. Mechanical testing will be done under various conditions to observe conductor movement. A significant problem to be solved in multiphase transmission (6 or more phases) is relaying and protection.

Reduced rights-of-way and the desire to reduce transmission losses are among the factors that have forced the utility industry to higher-voltage transmission systems. The development of technology for 1200 kV lines and equipment has been in progress in the United States since 1967. Among the problems of ultra-high voltage transmission are radio and audible noise, induction effects, corona and generation of ozone, and mechanical stresses imposed upon the multiple-conductor bundles used. Studies have shown benefits for a single UHV line, compared to two 500 kV lines, when comparing line constructing costs, annual cost of delivered power, and cost of losses. Reliability of a single UHV line is less, however, than two EHV lines, since the single-line probability of outage is greater than that of two lines.

Considerable research and development efforts are underway in the area of DC transmission with concentration on developing more compact, lower-cost terminal converter facilities, increased voltage levels, and the expansion of the market to multi-point projects. Specifically, funded projects include HVDC transmission line research, thyristor valve improvement and uprating, HVDC circuit breaker development, and HVDC/HVAC interaction and system studies.

Underground Technical Developments

Overhead transmission systems are the most logical choice for utilities unless the overhead lines are barred by topography, legislative action, or nonexistent or extremely expensive rights-of-way. Cost has now become a strong factor to consider when planning transmission expansion. The generally accepted ratio of installed costs, underground to overhead, that ranged between 10:1 and 20:1 is no longer used. Studies show that no general comparison between system costs is practical due to the different attributes of each new circuit. The various differences include price and availability of land, population distribution, terrain, load size, circuit length, and local regulations. Local input is becoming an ever larger influence in the transmission planning process.

Although comparisons of the two technologies for rural installations show cost ratios as low as 5:1, cost is still a heavy disadvantage of going underground. In a typical city installation, for example, where the majority of underground circuits are employed, nearly 50% of the circuit cost is incurred in cutting the trench, pulling the cable, backfilling the trench, and reconstructing the surface. Consequently, only 1% of all transmission circuit-miles in service in the United States is installed underground. For this percentage to change, new and improved methods of underground transmission will have to be developed. Research is being sponsored for the development of new cables and more efficient underground installation methods that will result in decreased cost and more reliable systems.

Figure 6 summarizes the new and existing types of underground transmission systems with a time schedule showing their dates of commercial application. A brief discussion of each of these technologies follows.

Pipe. Over 75% of the underground circuits in the United States consist of pipe-type cable systems. This type of system utilizes three oil-impregnated paper-insulated cables enclosed in a single steel pipe with oil under high pressure. This high-pressure, oil-filled (HPOF)

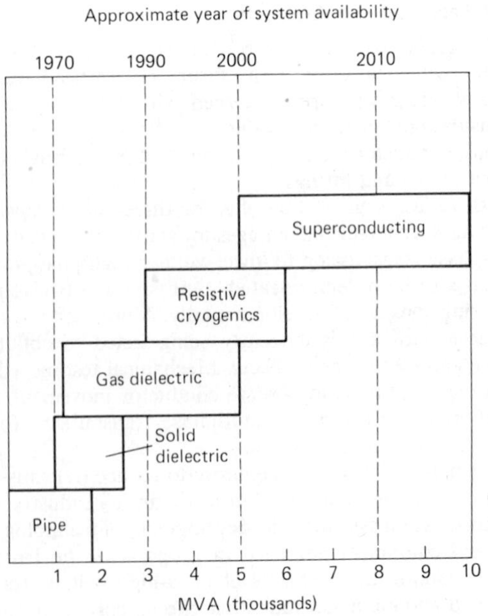

Fig. 6. Comparative capabilities of underground transmission systems.

cable has been in service since the mid-1970s and designs rated up to 500 kV have been successfully tested. Limitations that still exist with HPOF cables include excessive overload temperature rise, high losses in the insulation, and severe charging currents at higher voltages. In addition, they are subject to occasional failures. High-quality semi-synthetic insulation systems and forced-cooling techniques are being investigated, as are rapid techniques for fault location.

Gas Insulated. Both rigid and flexible gas-insulated cables have been developed. In this system, sulfur hexafluoride gas is used for the insulating and cooling medium. Rigid cables of a single-phase design, while capable of transmitting large quantities of power, are only economical for short runs. To overcome this problem, researchers developed a three-conductor, gas-insulated cable. Overall system size is reduced and a 15–25% reduction in cost is possible. Installation is easier and more efficient, because splicing and field welding are reduced considerably. A 345 kV, three-conductor cable has now been installed and will undergo two years of operation and test.

Solid Dielectric. This system normally utilizes cable of the extruded polyethylene type, where a solid insulation of the synthetic material surrounds the conductor. These cables are easier to handle than the previous types because no insulating fluid is involved. The reliability of these cables, however, is heavily dependent on the purity of the dielectric. Any contaminants can cause electrical discharges that result in failure. Methods have been developed, and are now being perfected, to combat these contaminants and enhance cable reliability. Development and testing is currently being carried out for cable designs at 138 kV. A 230 kV cable has also been produced and will be tested after splice and terminal developments have been completed. A 345 kV cable is under development.

Resistive Cryogenics. Since electrical resistance is a function of the conductor temperature, it is possible to reduce the heat losses in a system if operation is at low temperature. The two conductor metals, copper and aluminum, have electrical resistances at liquid nitrogen temperatures (−320°F; −196°C) of one-tenth the magnitudes at normal ambient temperatures. In order for cryogenic systems to be considered efficient, reductions in resistance have to generate savings that match or exceed the cost of refrigeration. Cryogenic systems are currently limited by the high cost and low efficiency of gas expansion-compression refrigerators. Recent attempts to develop a magnetic refrigerator have fallen short of both design goals and what would be needed for a practical refrigerator.

Superconducting. Superconducting systems operate at even lower temperatures, down into the liquid-helium range of −452°F (−266.9°C). Certain metals, such as niobium, lose their electrical resistance at these temperatures with the result that extremely high currents can be transmitted with low losses. Superconducting systems have demonstrated overall efficiency, that is, there is a net gain when savings and costs are compared. A program has begun to manufacture and test a high-current-density prototype Nb_3Ge coaxial transmission cable. Conductors of Nb_3Ge have operated at higher temperatures and current densities than have been possible with any other superconductor. Superconducting systems have the added advantage of higher power-transmitting capability, indicating that further research is likely to be a worthwhile investment.

R. F. Lawrence, Manager, and J. L. Haubert, Engineer, Transmission and Distribution Systems Engineering, Westinghouse Electric Corporation, Pittsburgh, Pennsylvania.

References

Cissna, V. J.: "A-C Power Transmission," in "Standard Handbook for Electrical Engineers," McGraw-Hill, New York, 1978.

Engström, P. G.: "Direct-Current Power Transmission," in "Standard Handbook for Electrical Engineers," McGraw-Hill, New York, 1978.

Geballe, T. H., and J. K. Hulm: "Superconductors in Electric Power Technology," *Sci. Amer.*, **243**, 5, 138–172 (1980).

Hammond, A. L.: "Direct-Current Power Transmission in Brazil/Paraguay," *Science*, **200**, 754 (1978).

Hartline, B. K.: "Power Line Radiation in the Magnetosphere," *Science*, **205**, 1365 (1979).

Kimbark, E. W.: "Direct Current Transmission," Wiley, New York, 1971.

Park, C. G., and R. A. Helliwell: "Magnetospheric Effects of Power Line Radiation," *Science*, **200**, 727–730 (1978).

Thorne, R. M., and B. T. Tsurutani: "Power-Line Harmonic Radiation," *Science*, **204**, 839–840 (1979).

ELECTRIC POWER (Wind Energy). Wind Power.

ELECTRIC RAYS (Torpedinidae). Skates and Rays.

ELECTRIC RESISTANCE. Resistance.

ELECTRIC SHOCK. Experiments with human volunteers carried out by various researchers since 1933 have established that the mean threshold of perception (the first "tingle") experienced from an electric shock is 1.1 milliamperes (mA) for men and 0.7 mA for women. Slightly higher current (called reaction current) provides for the possibility of an unexpected involuntary reaction which might result, for example, in the dropping of a hot pan of grease or a fall from a ladder. In 1967, the American National Standards Institute (ANSI) and the Underwriters Laboratories (UL) were funded to determine reaction current levels. The result of these studies was an ANSI specification in 1970 which limited maximum current leakage levels to 0.5 mA for new, properly operating, 2-wire cord-connected appliances; and 0.75 mA for heavy, movable cord-connected appliances, such as freezers and air conditioners. While this specification predicts that new appliances should never give perceptible shocks, this may not be true after the hardware has aged for a number of years or has been misapplied or exposed to degrading environments. It should be emphasized that a poorly designed or poorly used 120-volt appliance is capable of delivering lethal shocks.

As the intensity of shock current increases, sensations of warm, tingling, and muscular reaction increase and pain develops. Eventually, a current level is encountered at which a person cannot voluntarily "let go" because of muscular spasms and thus the victim is "frozen" to the circuit. This threshold is of great importance because under electrically induced muscular spasms, the musculature of breathing becomes paralyzed. unless the circuit is interrupted, collapse, unconsciousness, and death may follow in a matter of minutes or fractions thereof. Research has shown that this threshold for men is 16mA and 10.5 mA for women. See Fig. 1. In the case of higher-current shocks, resumption of normal breathing may not occur for several minutes after the current is stopped. In these cases, artificial respiration is required to prevent physiological damage. At even higher current levels, heart action becomes seriously impaired or stopped, most frequently by the mechanism of ventricular fibrillation (VF).

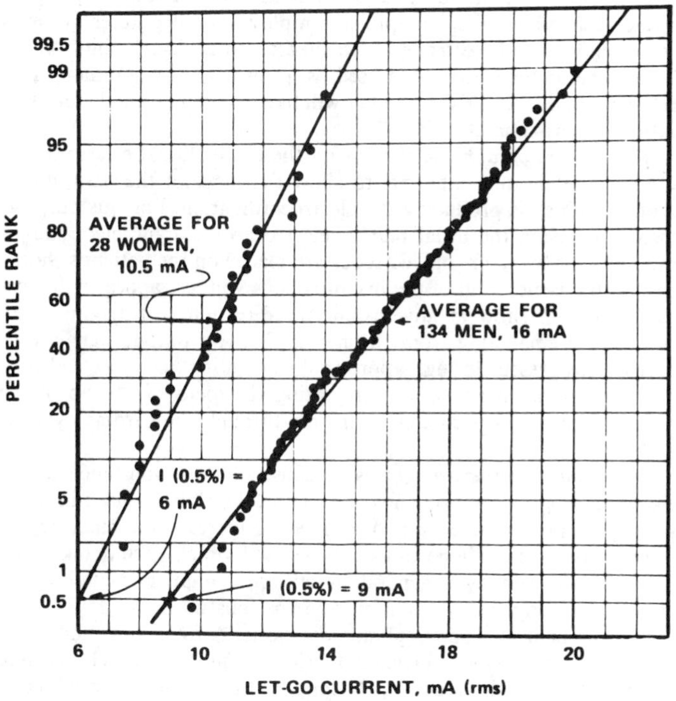

Fig. 1. "Let-go" currents for men follow a normal distribution, as do those for women, using a smaller test sample. Mean values (men and women) are in a ratio of two to three. (*After Dalziel.*)

Ventricular fibrillation, the most common cause of death from electric shock, is the result of desynchronizing the natural electrical impulses which govern the ordered muscular reactions which cause the heart to pump blood. In VF, the heart seems to "quiver" like a bag of jelly without effective pumping. Brain death normally occurs within 4 to 6 minutes when this condition sets in. After the onset of VF, normal heart action seldom resumes spontaneously and unless prompt defibrillation is effected by trained medical personnel, death is inevitable. Defibrillation is accomplished by applying a stronger, shorter duration shock of controlled amplitude and duration to electrodes on the chest. This stops the quivering for an instant and then allows the normal, synchronized heart action to resume. Cardiopulmonary resuscitation technique allows maintenance of circulation by rescuers for many minutes, even for hours with help, until defibrillation can be accomplished. See Fig. 2.

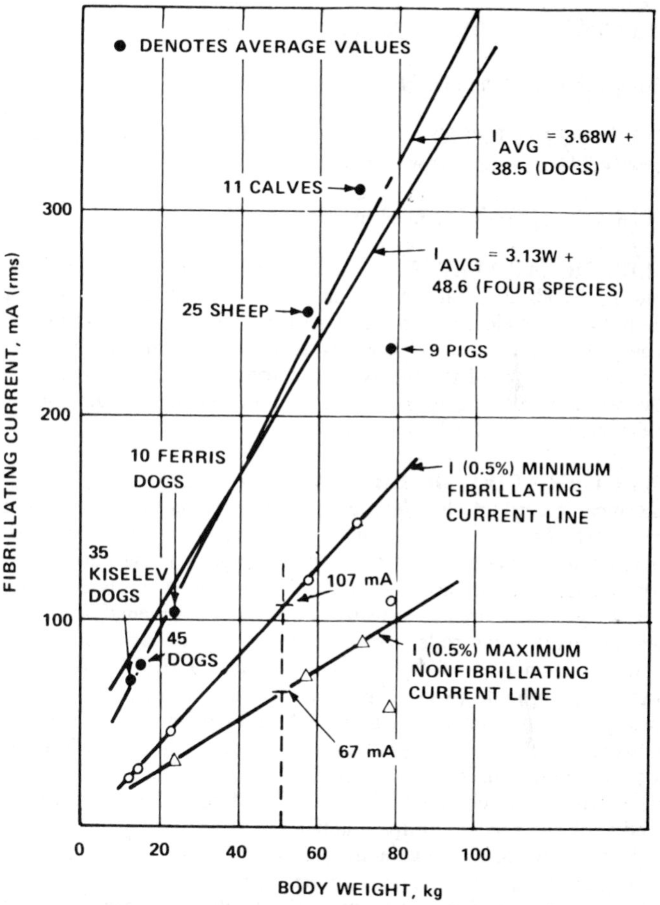

Fig. 2. Fibrillating current versus body weight for animals. These curves combine results of several experiments, including those on calves and pigs. (*GTE Laboratories.*)

Considerable research has indicated that a normal, healthy 50-kilogram (110-pound) adult will not be likely to fibrillate if the current intensity is less than

$$I\,(\text{electrocution threshold}) = \frac{116}{\sqrt{T}}\,\text{mA}$$

where T is in seconds. The relationship to body weight appears to be direct; a 25-kilogram (55-pound) child's electrocution threshold is believed to be one-half the values given by the aforementioned formula. These values have been used in preparation of UL specification 943 which all UL labeled and listed ground fault circuit interrupters (GFCI) must meet. The National Electrical Code, which provides the basis for most building codes and inspections in the United States, requires the exclusive use of UL approved devices.

The 1978 National Electrical Code (NEC) requires the use of GFCIs in all receptacle circuits of swimming pools, bathrooms, garages, and outdoor receptacles for new construction. In addition, all single-phase 120-volt receptacles used at construction sites or for other temporary use are required to have GFCI or specified other safety grounding measures. In many areas and probably in all areas soon, such requirements will also apply to kitchen and laundry receptacles.

Considerably more detail on this topic, including description of GFCIs, can be found in "Your Personal Safety and Ground Fault Circuit Interrupters" by V. C. Oxley, Profile Three 1979, GTE Laboratories, Waltham, Massachusetts. See also **Circuit Breakers.**

References

Dalziel, C. F., and W. R. Lee: "Reevaluation of Lethal Electric Currents," IEEE Industry and General Applications 1GA-4, 5, 467 (Sept.–Oct. 1968).

Dalziel, C. F.: "Electric Shock Hazard," *IEEE Spectrum*, **9**, 2, 41 (1972).

Gordon, A. S., et al.: "Standards for Cardio-Pulmonary Resuscitation," *J. Amer. Med. Soc.*, **227**, 7 (1974).

ELECTRIC SMELTER. **Iron Metals, Alloys, and Steels.**

ELECTRIC WELDING. **Welding.**

ELECTRIDE. **Chemical Elements.**

ELECTROACOUSTICAL RECIPROCITY THEOREM. **Reciprocity Theorem (Electroacoustical).**

ELECTROACOUSTICS. **Acoustics.**

ELECTROCAPILLARITY. The surface tension between two conducting liquids in contact, such as mercury and a dilute acid, is sensibly altered when an electric current passes across the interface. As a result, when the contact is in a capillary tube, the pressure difference on the opposite sides of the meniscus is affected by a current traversing the capillary column, to an extent dependent upon the direction of the current across the boundary.

ELECTROCARDIOGRAPHY. An electrocardiogram (ECG) is a graphical record of the electric potentials produced by activity of the heart. The normal beating of the heart is associated with the production of bioelectric currents in the organ. These currents are not strong, but they are carried to the surface of the body, where they may be measured by sensitive electrical instruments. The heartbeat normally is controlled by a rhythmically occurring electrical discharge that emanates from a spot in the right auricle known as the sino-auricular node (SA). This wave of electrical activity is spread through the heart muscle by a network of fibers known as Purkinje's network. As the wave spreads, the muscle of the heart rhythmically contracts and then gradually relaxes. The contraction, or squeezing of the muscle forces blood from the heart into the blood vessels, producing the needed pumping action.

As the wave of electrical activity spreads across the surface of the heart, the wave produces an electrical field which spreads through the conducting medium of body tissues to the surface of the body. This field can be considered to be produced by an electric-dipole moment which rotates and varies in amplitude in the course of the cardiac cycle. The pattern shown in the diagram was produced by placing electrodes on the left and right arms. The characteristic features of this electrocardiogram are labeled as conventionally described by physicians.

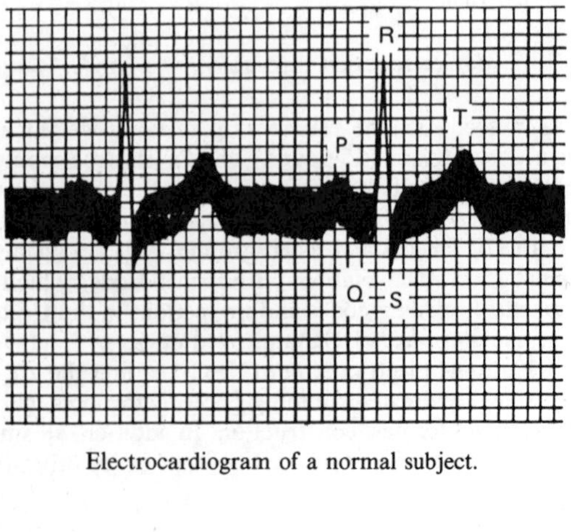

Electrocardiogram of a normal subject.

The key wave is due to electrical activation of the auricles of the heart; the QRS complex occurs at the time that the ventricles (or main pumps) of the heart contract; the T wave is caused by the repolarization or resetting of the electrical system following contraction. The contracting phase of the heart cycle is called *systole,* and the relaxed phase is known as *diastole.* Changes from the normal electrocardiographic pattern can be used to diagnose a heart attack and the progress and recovery from it. The pattern also can be used to diagnose and follow irregular heart rates (*arrhythmias*) and conditions of faulty conduction.

At the surface of the body, the voltage amplitude of a typical QRS complex will vary between 100 microvolts and 2 millivolts. Faithful reproduction of an electrocardiogram can be achieved with a frequency response flat between 0.25 and 100 Hz. There are some pathological situations which produce "splintering" of the QRS complex that can be easily detected only with a frequency response to approximately 1,000 Hz. This is not a common requirement and most recorders do not have a response above 100 Hz. To produce a faithful electrocardiogram, the voltage-amplifier input impedance should be above 0.1 million ohms and preferably above 1 million ohms.

In order to avoid motion artifacts in the record, care must be taken in the choice and application of the electrodes. Chemical action will cause a gradual change in the impedance of the skin-metal interface which slowly will reduce the amplitude of the ECG and, more seriously, will produce an abrupt change in the dc level, as this resistance changes with relative motion between the skin and the electrode. If silver is used as the electrode, chloride ions from the sodium chloride in body perspiration gradually will deposit on the silver reducing the effective surface area and thus increasing electrode resistance. By coating the silver with a silver chloride surface and by further introducing a paste salt bridge between this surface and the skin, the process can be reduced by several orders of magnitude. The skin has a high resistance compared with the subdermal tissues. Thus, for good results, the skin usually is rubbed briskly before application of the electrode paste and the electrodes. Skin resistance typically can vary from 5,000 to 200,000 ohms under varying conditions of perspiration and surface condition—hence the need for a high-impedance input.

Since the body usually is in a 60-Hz electric field and the room in which an electrocardiogram is made is not shielded, the 60-Hz pickup between two points on the body frequently will be one or two orders of magnitude greater than the desired ECG signal. To reduce this effect, ECG amplifiers employ several differential-input stages and a third electrode is attached to an arbitrary point on the body. This latter electrode is used as a "ground" or common-mode reference point. An ECG amplifier will have a common-mode rejection ratio of approximately 3,000 : 1.

Vector Cardiography. A variant on the conventional electrocardiographic presentation attempts to show the locus of the basic dipole moment which is producing the electrocardiogram. For this purpose, instead of using the usual pair of electrocardiographic electrodes, a minimum of two pairs of electrodes are placed on the patient's thorax so that the vector sum of the instantaneous voltages appearing across each pair will show at any instant the direction and magnitude of the voltage emanating from the heart. The two output signals thus derived are passed through appropriate weighting networks and then displayed on an oscilloscope, presenting the locus of the tip of the potential vector as it rotates with time through the cardiac cycle as a closed loop.

Processing Electrocardiograms. Because of the importance of the electrocardiogram as a diagnostic tool and the skill required in interpreting it, means have been developed to increase the efficiency of interpretation. Telephone links can be used which permit taking an electrocardiogram by a general practitioner or at a small hospital and transmitting the information to a specialist in facsimile form for immediate analysis and telephonic diagnosis. A frequency-modulated audio tone generator connected to the transmitting ECG machine can be used. At the receiving telephone, the signal is demodulated and fed into a conventional ECG recorder. Monitors are available which permit the patient to wear a small, battery-operated electrocardiograph which telemeters by FM radio over a relatively short distance.

This permits examination of the patient's heart during exercise and other usual physical and psychologically stressful situations in the patient's life. Magnetic tape recording is also used which permits later analysis of ECGs at high speed, or for spot checking. Digital computer programs also are available for immediate ECG analysis. Because of cost and complexity, this approach involves remote transmission from numerous ECG sources.

Impedance Cardiography. An electrical measure of the mechanical activity of the heart may be made by passing 4 to 5 milliamperes of 100-kHz current vertically through the heart, using electrodes at the neck and wrist. The change in this current as a function of the varying amount of blood in the heart through the cardiac cycle is measured with a pair of electrodes placed along the left and right midaxillary lines at the level of the heart (along the sides of the trunk about 4 inches below the armpits of an adult). Since the blood has appreciably lower resistivity than the body tissues, the impedance between the electrodes of approximately 25 ohms will be seen to decrease during diastole by about 0.1 ohm as blood flows into the heart and to increase sharply during systole as it is forced back out of the current path. Anomalies in the pumping cycle will show up if the blood moves out of the electrical field in an abnormal fashion.

Ballistocardiography. One important measure of heart activity is the force with which the heart ejects blood. If one stands quite still on a spring scale, one will note a small pulsation in the dial reading with each heartbeat. This ballistic recoil of the body as the blood is propelled from the heart can be measured more carefully if the subject lies on a specially constructed bed. Essentially, the bed becomes a platform which can be measured (motion) photoelectrically, or with a linear variable-differential transformer, or with a coil and magnet combination.

For analysis of cardiac output and mechanical heart function, the first- and second-derivative (velocity and acceleration) curves are generated as well as the curve of instantaneous position. These curves can be formed electronically by using high-pass filters, but such filters also increase the effect of noise in the system. To reduce this problem, some ballistocardiographs are made with a velocity pickup made by passing a magnet through a solenoid coil. The voltage appearing across such a coil will be proportional to the velocity with which the magnet is being moved through the coil. It then requires only one differentiation to obtain the acceleration curve and one integration to produce the motion curve, thus markedly improving the signal-to-noise ratio for the system.

Phonocardiography. Electronic stethoscopes have improved the efficiency of the strictly audio-type stethoscope. The latter instrument, of course, has been used for physicians for scores of years to detect heart sounds. An electronic system for heart-sound amplification should have a passband between 10 and 1,000 Hz. For the physician who has become accustomed to a conventional stethoscope, some retraining of the ear is required in switching to an electronic device.

Halter Monitoring. For a number of years, it has been possible to record the ECG of ambulatory patients on magnetic tape continuously over several hours for later playback and interpretation (halter monitoring). Only in recent years, however, has this method come into wide application. Such long-period ECGs are helpful in determining the etiology of dizzy spells or syncope of cardiac origin; studying the nature of palpitations and episodes of tachycardia; and assessing the effectiveness of anti-arrhythmic therapy; among other situations where a comparatively long time span of measurements is required to develop a complete picture.

See **Heart and Circulatory System (Human)** and the list of entries included at the end of that entry.

References

Fletcher, G. F., and J. D. Cantwell: "Exercise and Coronary Heart Disease," Charles C. Thomas, Springfield, Illinois, 1979.

Flowler, N. O.: "Cardiac Diagnosis and Treatment," 2nd edition, Harper and Row, Hagerstown, Maryland, 1976.

Harrison, D. C., Fitzgerald, J. W., and R. A. Winkle: "Ambulatory Electrocardiography for Diagnosis and Treatment of Cardiac Arrhythmias," *N. Engl. J. Med.*, **294**, 373 (1976).

Krohn, L. H.: "High-Resolution Electrocardiography," Charles C. Thomas, Springfield, Illinois, 1976.

Lipman, B. S., Massie, E., and R. E. Kleiger: "Clinical Scalar Electrocardiography." 6th edition, Year Book Medical Publishers, Chicago, Illinois, 1972.

Lister, J. W., et al.: "Arrhythmia Analysis by Intracardiac Electrocardiography," Charles C. Thomas, Springfield, Illinois, 1976.

Littman, D.: "Textbook of Electrocardiography," Harper and Row, Hagerstown, Maryland, 1972.

Parisi, A. F., et al.: "Noninvasive Cardiac Diagnosis," *N. Engl. J. Med.*, **296**, 316 (1977).

Silber, E. N., and L. N. Katz: "Heart Disease," Macmillan, New York, 1975.

ELECTROCHEMICAL MACHINING (ECM). In this method of metalworking, electrical and chemical energy become the cutting edges of the tool. Electrical energy causes a chemical reaction which, in turn, dissolves metal from a workpiece into an electrolytic solution. The ECM tool (cathode) is brought very close to the workpiece (anode). The distance will range from less than 0.001 to 0.010 inch (0.025 to 0.25 millimeter). A low-voltage, high-density direct current passes between them through the electrically conductive electrolyte solution. This solution is pumped through the gap at pressures ranging up to 300 psi (20.4 atmospheros). The solution normally is maintained at about 100–120°F (38–49°C). The current that is passed through the electroyte solution ranges widely. Machines using up to 20,000 amperes have been used. As the current passes from the workpiece to the tool, metallic particles on the surface of the workpiece (ions) are caused to go into solution. These particles are then swept away by the rapidly flowing electrolyte. Some of advantages claimed for ECM include: (1) virtually no tool wear; (2) no burrs are produced; (3) both hard and soft metals can be machined at same rate; (4) usually additional finishing operations are not required; (5) no mechanical stresses are produced in workpiece surface; (6) no thermal effects are produced in the workpiece because of the relatively low temperature of the operation; (7) tough metals often can be removed faster by ECM than conventional methods; (8) good tolerances and repeatability are produced in complex as well as simple shapes; and (9) the process is easily automated. The basics of this process are also applied in deburring, grinding, and polishing operations.

ELECTROCHEMISTRY. That branch of science which deals with the interconversion of chemical and electrical energies, i.e., with chemical changes produced by electricity as in electrolysis or with the production of electricity by chemical action as in electric cells or batteries. The science of electrochemistry began about the turn of the eighteenth century.

Background. In 1796, Alessandro Volta observed that an electric current was produced if unlike metals separated by paper or hide moistened with water or a salt solution were brought into contact. Volta used the sensation of pain to detect the electric current. His observation was similar to that observed ten years earlier by Luigi Galvani who noted that a frog's leg could be made to twitch if copper and iron, attached respectively to a nerve and a muscle, were brought into contact.

In his original design Volta stacked couples of unlike metals one upon another in order to increase the intensity of the current. This arrangement became known as the "voltaic pile." He studied many metallic combinations and was able to arrange the metals in an "electromotive series" in which each metal was positive when connected to the one below it in the series. Volta's pile was the precursor of modern batteries.

In 1800, William Nicholson and Anthony Carlisle decomposed water into hydrogen and oxygen by an electric current supplied by a voltaic pile. Whereas Volta had produced electricity from chemical action these experimenters reversed the process and utilized electricity to produce chemical changes. In 1807, Sir Humphry Davy discovered two new elements, potassium and sodium, by the electrolysis of the respective solid hydroxides, utilizing a voltaic pile as the source of electric power. These electrolytic processes were the forerunners of the many industrial electrolytic processes used today to obtain aluminum, chlorine, hydrogen, or oxygen, for example, or in the electroplating of metals such as silver or chromium.

Since in the interconversion of electrical and chemical energies, electrical energy flows to or from the system in which chemical changes take place, it is essential that the system be, in large part, conducting or consist of electrical conductors. These are of two general types—

electronic and electrolytic—though some materials exhibit both types of conduction. Metals are the most common electronic conductors. Typical electrolytic conductors are molten salts and solutions of acids, bases, and salts.

A current of electricity in an electronic conductor is due to a stream of electrons, particles of subatomic size, and the current causes no net transfer of matter. The flow is, therefore, in a direction contrary to what is conventionally known as the "direction of the current." In electrolytic conductors, the carriers are charged particles of atomic or molecular size called *ions*, and under a potential gradient, a transfer of matter occurs.

An electrolytic solution contains an equivalent quantity of positively and negatively charged ions whereby electroneutrality prevails. Under a potential gradient, the positive and negative ions move in opposite directions with their own characteristic velocities and each accordingly carries a different fraction of the total current through any one solution. Each fraction is referred to as the ionic transference number. Furthermore, the velocity increases with temperature causing a corresponding increase in electrolytic conductivity. This characteristic is opposite to that observed for most electronic conductors which show less conductivity as their temperature is increased.

The concept that charged particles are responsible for the transport of electric charges through electrolytic solutions was accepted early in the history of electrochemistry. The existence of ions was first postulated by Michael Faraday in 1834; he called negative ions "anions" and positive ones "cations." In 1853, Hittorf showed that ions move with different velocities and exist as separate entities and not momentarily as believed by Faraday. In 1887, Svante Arrhenius postulated that solute molecules dissociated spontaneously into *free ions* having no influence on each other. However, it is known that ions are subject to coulombic forces, and only at infinite dilution do ions behave ideally, i.e., independently of other ions in the solution. Ionization is influenced by the nature of the solvent and solute, the ion size, and solute-solvent interaction. The dielectric constant and viscosity of the solvent play dominant roles in conductivity. The higher the dielectric constant, the less are the electrostatic forces between ions and the greater is the conductivity. The higher the viscosity of the solvent, the greater are the frictional forces between ions and solvent molecules and the lower is the electrolytic conductivity.

In 1923, Debye and Hückel presented a theory which took into account the effect of coulombic forces between ions. They introduced the concept of the ion atmosphere, in which at some radial distance r from a central ion, there is, on a time average, an ionic cloud of opposite charge which sets up a potential field whose magnitude depends on the magnitude of r. This interionic attraction leads to two effects on the electrolytic conductivity. Under a potential gradient, an ion moves in a certain direction. However, the ion cloud, being of opposite sign will tend to move in the opposite direction, and because of its attraction for the central ion, will have a retarding effect on the ion velocity and thereby lead to a lowering in the electrolytic conductivity. On the other hand, the central ion will tend to pull the ion cloud with it to a new location. The ion atmosphere will adjust to its new location in time, but not instantaneously, and the delay results in a dissymmetry in the potential field around the ion. This also causes a lowering in the conductance of the solution. These effects become more pronounced as the concentration of the solution is increased; for dilute solutions, below about 0.1 molal, the equivalent conductance decreases with the square root of the concentration. For more concentrated solutions, the relation between conductivity and concentration is much more complex and depends more specifically on individual solute properties.

Interionic attraction in dilute solutions also leads to an effective ionic concentration or activity which is less than the stoichiometric value. The *activity* of an ion species is its thermodynamic concentration, i.e., the ion concentration corrected for the deviation from ideal behavior. For dilute solutions the activity of ions is less than one, for concentrated solutions it may be greater than one. It is the ionic activity that is used in expressing the variation of electrode potentials, and other electrochemical phenomena, with composition.

When electricity passes through a circuit consisting of both types of electrical conductors, a chemical reaction always occurs at their interface. These reactions are electrochemical. When electrons flow from the electrolytic conductor, oxidation occurs at the interface while reduction occurs if electrons flow in the opposite direction. These electronic-electrolytic interfaces are referred to as *electrodes*; those at which oxidation occurs are known as *anodes* and those at which reduction occurs, as *cathodes*. An anode is also defined as that electrode by which "conventional" current enters an electrolytic solution, a cathode as that electrode by which "conventional" current leaves. Positive ions, for example, ions of hydrogen and the metals, are called *cations* while negative ions, for example, acid radicals and ions of nonmetals are called *anions*.

Laws of Electrolysis. In 1833, Faraday enunciated two laws of electrolysis which give the relation between chemical changes and the product of the current and time, i.e., the total charge (coulombs) passed through a solution. These laws are: (1) the amount of chemical change. e.g., chemical decomposition, dissolution, deposition, oxidation, or reduction, produced by an electric current is directly proportional to the quantity of electricity passed through the solution; (2) the amounts of different substances decomposed, dissolved, deposited, oxidized, or reduced are proportional to their chemical equivalent weights. A chemical equivalent weight of an element or a radical is given by the atomic or molecular weight of the element or radical divided by its valence; the valence used depends on the electrochemical reaction involved. The electric charge on an ion is equal to the electronic charge or some integral multiple of it. Accordingly, a univalent negative ion has a charge equal in magnitude and of the same sign as a single electron, and its chemical equivalent weight is equal to its atomic weight, if an element, or to its molecular weight, if a radical. A trivalent ion has $+3$ or -3 electronic charges, depending on whether it is a positive or a negative trivalent ion. For trivalent ions, then, the equivalent weight would be equal to its atomic weight, if an element, or to its molecular weight, if a radical, divided by three.

The quantity of electricity required to produce a gram-equivalent weight of chemical change is known as the *faraday*. A faraday corresponds, then, to an *Avogadro number of charges*. The most accurate determination of the faraday has been made by a silver-perchloric acid coulometer in which the amount of silver electrolytically dissolved in an aqueous solution of perchloric acid is measured. This method gives 96,487 coulombs (or ampere-seconds) per gram-equivalent for the faraday on the unified ^{12}C scale of atomic weights adopted in 1961 by the International Commission on Atomic Weights.

Electrochemical Equivalent. Preferably termed *coulomb equivalent* of an element or radical, this is the weight in grams which is equivalent to 1 coulomb of electricity and is given by the gram-equivalent weight divided by the faraday (96,487 coulombs per gram-equivalent); for example, the electrochemical equivalent of silver is given by 107.870/96,487 or 0.00111797 grams/coulomb where 107.870 is the atomic weight of silver based on the unified ^{12}C scale adopted in 1961. The electrochemical equivalents of other elements may be calculated in like fashion.

In electrolysis and in any electric cell or battery, there is an electromotive force (emf) or voltage across the terminals. This emf is expressed in the practical unit, the volt, which is equal to the electromagnetic unit in the meter-kilogram-second system. In any one cell, the emf is the sum of the potentials of the two electrodes and of any liquid-junction potentials that may be present. Neither of the individual electrode potentials can be evaluated without reference to a chosen reference electrode of assigned value. For this purpose, the hydrogen electrode has been universally adopted and is arbitrarily assigned a zero potential for all temperatures when the hydrogen ion is at unit activity and the hydrogen gas is at atmospheric pressure. A hydrogen electrode consists of a stream of hydrogen gas bubbling over platinized platinum or gold foil and immersed in a solution containing hydrogen ions; the electrochemical reaction is: $\frac{1}{2}H_2(\text{gas}) = H^+ \text{ (solution)} + \epsilon$, where ϵ represents the electron. The potential of the hydrogen electrode, E_H, as a function of hydrogen ion concentration and hydrogen-gas pressure is given by

$$E_H = E_H^0 - (RT/nF)\ln(a_{H^+}/p_{H_2}^{1/2})$$
$$= E_H^0 - (RT/nF)\ln(c_{H^+}f_{H^+}/p_{H_2}^{1/2}),$$

where E_H^0 is the standard quantity assigned a value of zero, R is the gas constant, T the absolute temperature, n the number of equivalents, F the faraday, p_{H_2} the pressure of hydrogen, and a_{H^+}, c_{H^+} and f_{H^+}, respectively, the activity, concentration, and activity coefficient of hydrogen ions. When a_{H^+} and $p_{H_2}^{1/2}$) equal one, $E_H = E_H^0$. For very dilute solutions below 0.01 molal f_{H^+} may be taken as unity without appreciable error.

The standard potentials, E^0, of other electrodes are obtained by direct or indirect comparison with the hydrogen electrode. Values are determined at 25°C. The values for several metals and other elements are given in entry on **Activity Series.** The reducing power of the elements decreased on going down the column from those elements with negative standard electrode potentials to those with positive potentials. These values are for the ions at unit activity, and reversible or thermodynamic values as a function of metal or radical concentration are given by equations similar to the one above. For the general reaction: $M = M^{n+} + n\epsilon$, the potential is given by $E_m = E_M^0 - (RT/nF)\ln a_{Mn^+}$.

In electrolysis, at very low current densities, the potentials of the electrodes approximate in magnitude their reversible values and deviate somewhat from these values because of an IR drop in the solution and possible concentration polarization (the concentration at the electrode surface may differ from that in the bulk of the solution). Also for high current densities, especially for the generation of gases such as hydrogen, oxygen or chlorine, the voltage required exceeds the reversible voltage; the excess voltage is known as overvoltage, or overpotential for a single electrode, and arises from energy barriers at the electrode. Overpotential, in general, increases logarithmically with an increase in current density.

Scope of Electrochemistry. In addition to what has previously been described, it is customary to include under electrochemistry: (1) processes for which the net reaction is physical transfer, e.g., concentration cells; (2) electrokinetic phenomena, e.g., electrophoresis, electro-osmosis, streaming potential; (3) properties of electrolytic solutions if determined by electrochemical or other means, e.g., activity coefficients and hydrogen ion concentration; (4) processes in which electrical energy is first converted to heat which in turn causes a chemical reaction to occur that would not do so spontaneously at ordinary temperature. The first three are frequently considered a portion of physical chemistry, and the last one is a part of electrothermics or electrometallurgy.

The passage of electricity through gases is sometimes included under electrochemistry. However, in electrical discharges in gases, the principles are entirely different from what they are in the electrolysis of electrolytic solutions. Whereas in the latter, ionic dissociation occurs spontaneously as a result of forces between solvent and solute and without the application of an external field, for gases relatively high voltages must be applied to accelerate the electrons from the electrode to a velocity at which they can ionize the gas molecules they strike. In this case, the resulting chemical reaction taking place between ions, free radicals, and molecules occurs in the gas phase and not at the electrodes as in the electrolysis of solutions. Studies of the electrical conduction of gases, accordingly, are generally considered under the physics of gases.

Electrochemistry finds wide application. In addition to industrial electrolytic processes, electroplating, and the manufacture and use of batteries already mentioned, the principles of electrochemistry are used in chemical analysis, e.g., polarography, and electrometric or conductometric titrations; in chemical synthesis, e.g., dyestuffs, fertilizers, plastics, insecticides; in biology and medicine, e.g., electrophoretic separation of proteins, membrane potentials; in metallurgy, e.g., corrosion prevention, electrorefining; and in electricity, e.g., electrolytic rectifiers, electrolytic capacitors.

See also **Battery; Corrosion; Electrolytic Conductivity Measurements; Electrophoresis; pH (Hydrogen Ion Concentration);** and **Polarographic Analyzers.**

References

Delahay, P., and C. Tobias: "Advances in Electrochemistry and Electrochemical Engineering," Volumes 1 through 8, Wiley, New York, 1971.

Heise, G. S., and N. C. Cahoon: "The Primary Battery," Wiley, New York, 1971.

Salamon, M. B. (editor): Physics of Superionic Conductors," Springer-Verlag, New York, 1979.

Yeager, E., and A. J. Salkind (editors): "Techniques of Electrochemistry," Wiley, New York, 1972.

ELECTRODE. In an electric circuit, part of which is composed of other than the usual conductor of copper, or other metal, the terminal connecting the conventional conductor and the conducting substances is an electrode. Examples of electrodes are to be found in the electric cell, where they dip in the electrolyte; the electric furnace, where the electrodes connect the external circuit with the heating arc; and the metallic elements in thermionic tubes and gas-discharge devices, and in semiconductor devices, where they perform one or more of the functions of emitting, collecting or controlling by an electric field the movements of electrons and ions. See also **Graphite.**

ELECTRODE (Battery). Battery.

ELECTRODE CURRENT. Fault Electrode Current.

ELECTRODE (Fuel Cell). Fuel Cells.

ELECTRODELESS DISCHARGE. Discharge (Gaseous).

ELECTRODE POTENTIALS. Activity Series.

ELECTRODIALYSIS. The removal of electrolytes from a colloidal solution by a combination of electrolysis and dialysis. Usually the colloidal solution is placed in a vessel with two dialyzing membranes with pure water in compartments on the other side of the membranes. Two electrodes are inserted in the pure water compartments and an applied emf causes the ions to migrate from the colloidal solution. See also **Desalination;** and **Ocean Resources (Energy).**

ELECTRODYNAMIC MECHANISM. Electrical Instruments.

ELECTRODYNAMOMETER. This term denotes a meter having both a fixed coil and a movable coil, whose deflection depends on the interaction of the magnetic fields produced by the currents of the two coils. The coils may be either in series or in parallel to yield a square-law meter for measurement of voltage or current. Such a meter can be calibrated on dc and used for ac measurements. By connecting one coil across a load, and the other in series with the load, the electrodynamometer becomes a wattmeter.

ELECTROENCEPHALOGRAM. A graphic tracing of the electric impulses of the brain, sometimes abbreviated EEG. The instrument which makes the record is known as an electroencephalograph. The electrical activity of the brain is manifested at the surface of the scalp by small potential changes on the order of 5 to 200 microvolts in a frequency band from 1 to 50 Hz. A sample of so-called brain waves or electroencephalogram is shown by the accompanying diagram.

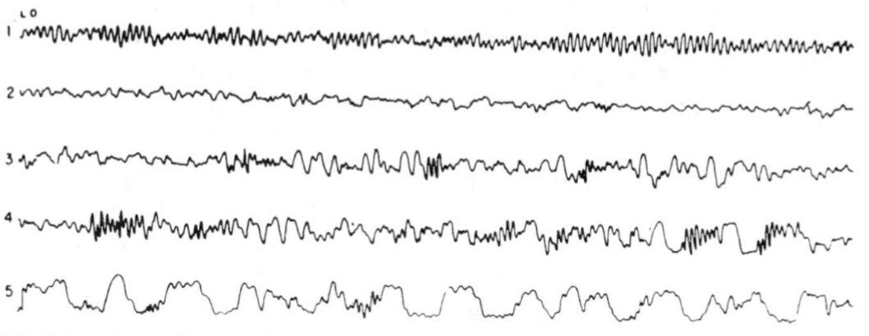

EEG tracings of normal subject during sleep. Numbers shown at left indicate increasing depth of sleep.

When viewed casually, the usual EEG would appear to be simply random noise, but spectral analysis and some direct electroencephalograms show pronounced components at several frequencies. The lowest (1 to 3 Hz) is termed the *delta rhythm.* Next highest are the *theta waves* (4 to 7 Hz). The *alpha rhythm* is the most pronounced and occurs between 8 and 13 Hz. A third pronounced rhythm appears between 13 and 30 Hz and is termed the *beta rhythm.* As shown by the diagram, the alpha rhythm is most pronounced during light sleep, while the delta rhythm appears during deep sleep. One of the uses of the EEG is determination of stages of sleep in sleep research and associated investigations.

In several diseases of the brain, especially epilepsy and cerebral tumors, very slow delta waves of higher-than-average voltage may be seen. Occasionally, as in epileptic seizures, similar waves of very high frequency may occur. In tumors, the appearance of such waves is confined to the area immediately surrounding the tumor; hence electroencephalography has some ancillary value in the localization of cerebral tumors. See **Seizure (Neurological).**

Electroencephalographic techniques also have been used rather extensively in research on sleep and sleep apnea associated with insomnia.

A conventional electroencephalogram is taken by placing a series of 6 to 10 electrodes symmetrically on each side of the head from the front to the back, with a reference electrode on the mastoid bone or some other similar spot. Sometimes the electrodes consist of small pins which puncture the skin to make better contact, but usually the scalp electrode is a small sponge soaked with saline solution and held down with tape. Usually, it is not necessary to shave the hair from the scalp. The usual EEG machine uses a multichannel, direct-writing recorder with 6 to 10 channels and speeds ranging from 2 to 50 millimeters per second, the width per channel typically being about 25 millimeters. Because the EEG amplitude is an order of magnitude smaller than the electrocardiograph amplitude, the criteria for the design of an EEG amplifier apply much more strenuously. See **Electrocardiogram.** Common-mode rejection should be at least 10,000:1, and the recording should be done in a shielded room. In contrast, low-frequency response need not be as good for the EEG as for the ECG; thus polarization of the electrodes is not so critical. When making an EEG recording, it is necessary that the shielded room be relatively free of external stimuli, such as changes in ambient light, sound, or vibration because sudden changes may produce startle-response artifacts.

A technique known as the evoked-cortical-potential makes it possible to see the effect on the EEG of a sensory stimulus. At the same time as the stimulus is initiated (such as a flash of light or burst of sound), an EEG recording is made and stored in a computer memory, usually on a digital basis. Each time a fresh stimulus is applied, the storage is updated, so that after 50 or 100 trials, the digital memory contains an average EEG synchronized with the stimulus. A typical computer memory for this application may contain 2,500 bits; 50 elements in one direction could represent fifty 10-millisecond periods, starting with the onset of the stimulus, while the 50 elements in the orthogonal direction would represent the average amplitude of the EEG for each time period. This technique is useful for research work on sight, sound, touch, and other sensory stimuli and, under some circumstances, in the location of disorders causing blindness or deafness.

Work on the brain and on individual fibers frequently requires the use of microelectrodes to stimulate and detect the responses from very tiny spots. Microelectrodes, consisting of very fine wires surrounded by thin glass insulation, are made down to an active diameter of approximately one micrometer. The impedance of such an electrode can be 10 to 100 million ohms, and the frequency of the signal can have components up to 10 kHz. At the same time, the signal-to-noise ratio can be very low. Thus, the requirements of very high common-mode impedance, very high input impedance, and virtually no input capacitance dictates a special class of amplifier in which the input capacitance is neutralized by feedback. Apparatus is available that allows the simultaneous use of an electrode for producing a stimulus and recording the response. The input and output circuits to the electrode must be carefully isolated to make this possible.

ELECTROENCEPHALOGRAM (Seizure). Seizure (Neurological).

ELECTROENCELPHALOGRAPHY (Sleep Research). Sleep.

ELECTROENDOSMOSIS. Electrophoresis in which the solid is stationary and the water phase is displaced and migrates toward the electrode.

ELECTROFORMING. The electrolytic deposition of metal upon a conducting mold, to make a desired metal object, such as precision tubing or medals. The mold is often of graphite-coated wax, so that it can be removed by melting. See **Electroplating.**

ELECTROGENIC FISHES. Catfishes; Gymnotid Eels; Mormyrids.

ELECTROKINETIC EFFECTS. Movements of particles under the influence of an applied electric field.

ELECTROKINETICS (Filtration). Filtration.

ELECTROKINETIC TRANSDUCER. A transducer that depends for its operation on the dielectric polarization in certain liquids resulting from viscous shearing stress that accompanies flow through porous materials.

ELECTROKINETIC (ZETA) POTENTIAL. The difference in potential between the immovable liquid layer attached to the surface of a solid phase and the movable part of the diffuse layer in the body of the liquid.

ELECTROLUMINESCENCE. Luminescence.

ELECTROLYSIS. Electrochemistry.

ELECTROLYSIS (Hydrogen Fuel). Hydrogen (Fuel).

ELECTROLYSIS-TYPE CHEMICAL ANALYZER. Sometimes referred to as electroplating analyzers, these devices can be used for determining metals and other materials that will plate out on an electrode which is part of an electrolytic cell. Alloys, such as stainless steel, brass, and bronze that contain chromium, copper, lead, iron, nickel, and tin-bearing metals containing bismuth, can be analyzed in this fashion. The sample must be relatively easy to dissolve so that an electrolyte can be formed and it must be sufficiently large to permit plating out a quantity of material that can be accurately weighted. The potential required to effect plating of a specific material should be known in advance. Complex materials usually are identified on a cumulative basis by making stepwise increases in plating potential, with intermittent weighing of the plated electrode.

ELECTROLYTE (Battery). Battery.

ELECTROLYTE (Ionic Mobility). Ionic Mobility.

ELECTROLYTE (Kohlrausch Law). Kohlrausch Law.

ELECTROLYTE SYSTEM (Body). Kidney and Urinary Tract; Potassium and Sodium (In Biological Systems); Water.

ELECTROLYTE (Transference Number). Transference Number.

ELECTROLYTIC CONDUCTIVITY MEASUREMENTS. Although measurements of conductivity have long been used as a means of investigating the properties of electrolytes in solution, such as dissociation, activity, formation of complexes, and hydrolysis, such measurements also provide the basis for instrumentation used in industry to detect the ionic contamination of water and to determine the concentration of simple electrolytic solutions. In this reference, the term electrolytic conductivity has been applied almost exclusively to water solutions of electrolytes in which the mechanism of electrical current transfer is dependent on ions. Solid and fused salts, however, also exhibit electrolytic conductivity.

Electrolytic conductivity (specific conductance) is defined as the electrical conductance of a unit cube of electrolytic solution. It is expressed in the same units as electrical conductivity, i.e., reciprocal ohms per unit length. Most commonly we find:

Mhos/cm, siemens/cm, and siemens/meter

(1 moho/cm = 1 siemens/cm = 100 siemens/meter)

The accompanying table gives the conductivity at 25°C of some common electrolytic solutions at different concentrations, and Fig. 1 shows the shapes of the conductivity-concentration curves for a salt solution and a mineral acid. Typically the conductivity increases to

APPROXIMATE ELECTROLYTIC CONDUCTIVITY OF SOME COMMON ELECTROLYTES IN WATER SOLUTION AT 25°C

% BY WEIGHT	ELECTROLYTE					
	NaCl	NaOH	H_2SO_4	HCl	NH_3	CO_2
	Values in Micromhos/Centimeter					
0.0001	2.2	6.2	8.8	11.7	6.6	1.2
0.0003	6.5	18.4	26.1	35.0	14	1.9
0.001	21.4	61.1	85.6	116	27	3.9
0.003	64	184	251	340	49	6.8
0.01	210	603	805	1,140	84	12
0.03	617	1,780	2,180	3,390	150	20
0.1	1,990	5,820	6,350	11,100	275	39
0.3	5,690	16,900	15,800	32,200	465	55
1.0	17,600	53,200	48,500	103,000	810	
3.0	48,600	144,000	141,000	283,000	1,110	
10.0	140,000	358,000	427,000	709,000	1,120	
30.0	—	292,000	822,000	732,000	210	

a maximum value and then decreases with increasing concentration. Sometimes an additional point of inflection may occur.

Figure 2 shows a family of curves defining the change in conductivity of sodium chloride solutions with temperature. The typical increase in conductivity with temperature is quite apparent. Most solutions exhibit very similar changes. Pure water changes somewhat more with changes in temperature while strong acids and bases change somewhat less. From the foregoing discussion it can be seen that the value of a conductivity measurement is useless without knowledge of the temperature at which the measurement was made.

Electrolytic conductivity is most often measured by placing electrodes in contact with the electrolytic solution which is contained in such a way that the measured electrical conductance between the electrodes can be related to the conductivity of the solution. The conductivity cell, as shown in Fig. 3, most commonly comprises an enclosure made of electrically insulating material such as glass or plastic which serves to hold or isolate a portion of the electrolytic solution and to accommodate the two electrodes. The cell constant of such a device is then used to relate the measured electrical conductance between the electrodes to the actual electrolytic conductivity. Two electrodes 1 centimeter square located on opposite interior faces of a hollow cube 1 centimeter on an edge would have a cell constant of 1/cm, and a measured conductance of 0.005 mhos at 25°C would indicate a conductivity of 0.005 mhos/cm (0.5 siemens/meter) at 25°C.

If the electrical conductance between the electrodes is measured with direct current, the resulting electrolysis and gas evolution interferes with the current passage and changes the composition of the solution. Alternating current greatly reduces these interfering factors, and finds wide use in such measurements. Properly designed conductivity cells operated at appropriate alternating current frequencies obey Ohm's law since the current through the cell is proportional to the applied voltage and the conductivity of the electrolytic solution. Alternating current Wheatstone bridges and conductance meters make up the most widely used instrumentation accepted for electrolytic conductivity measurements. Figure 4 shows a typical ac Wheatstone bridge circuit with R_x, R_s, R_3, and R_4 representing the resistance of the four arms. From the bridge equation, it can be seen that R_3 is proportional to $1/R_x$ which corresponds to the conductance of the conductivity cell. In practice, R_s is usually a temperature sensitive resistive element responsive to changes in solution temperature employed to provide automatic temperature compensation. Changes in solution temperature change R_x and R_s by the same amount, thereby allowing the bridge to remain balanced except for actual changes in solution concentration. Conductivity meters generally apply a constant alternating voltage across the electrodes and respond to the resulting flow of current, which is proportional to the conductivity of the solution. Means of automatic temperature compensation are also included in such circuitry.

It is also possible to measure electrolytic conductivity by means

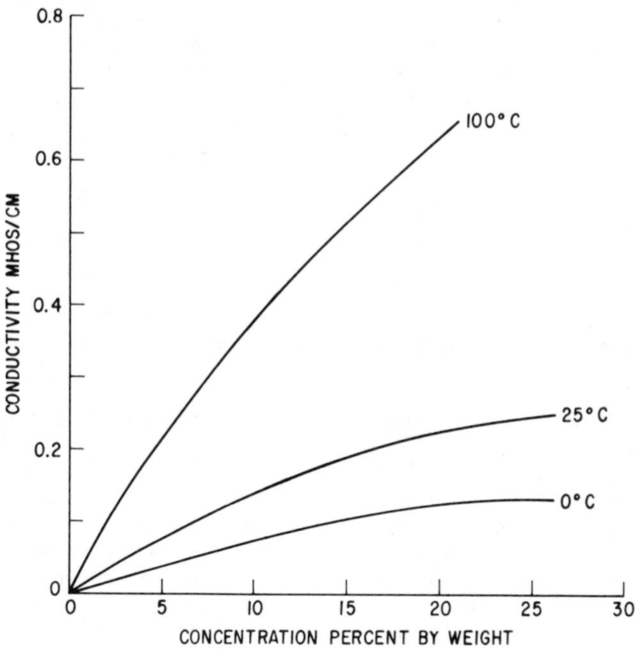

Fig. 2. Temperature dependence of electrolytic conductivity.

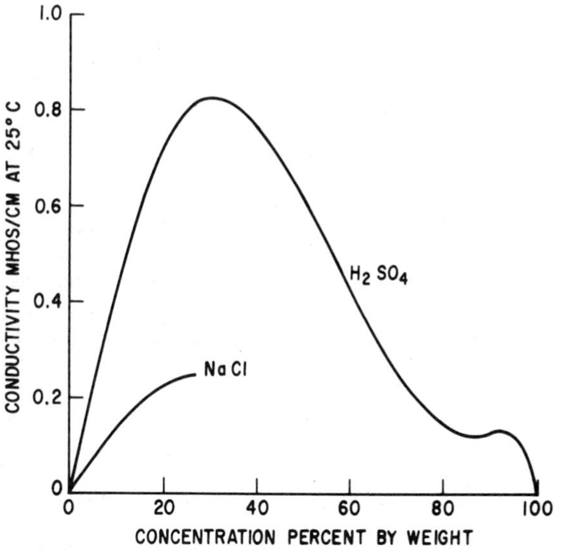

Fig. 1. Conductivity (at 25°C) as a function of concentration.

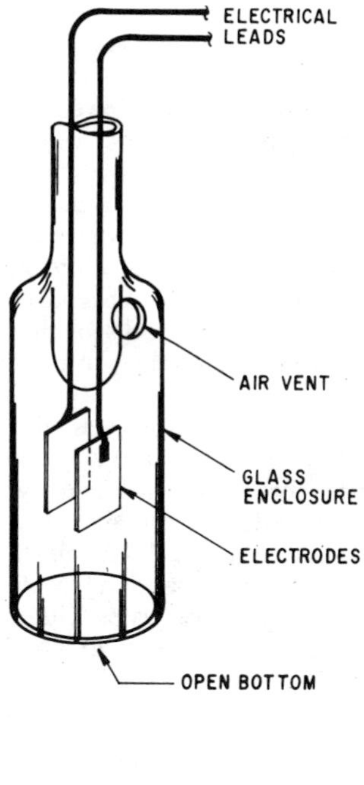

Fig. 3. Dip-type conductivity cell.

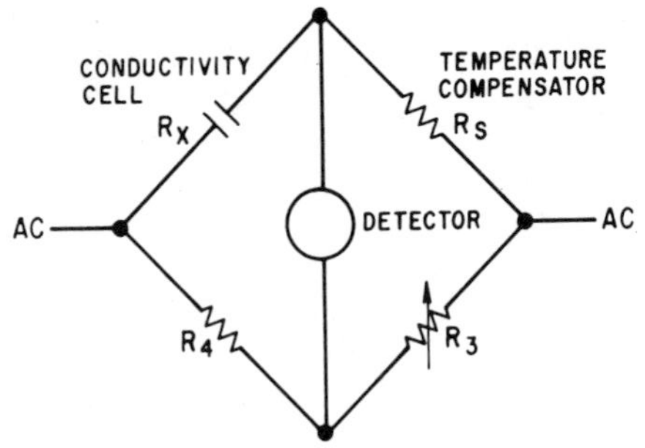

Fig. 4. Wheatstone bridge circuit commonly used to measure electrolytic conductivity.

of electrical induction without the use of contacting electrodes. Such measurements are made by inducing an alternating current in an electrolyte by use of a coil of wire. The magnitude of the induced current is proportional to the conductivity of the electrolyte. In Fig. 5, current is caused to flow in a closed circular path through the electrolyte by a first coil of wire wound on a toroidal core of magnetic material. The magnitude of the current and hence the conductivity is measured by a second similar coil.

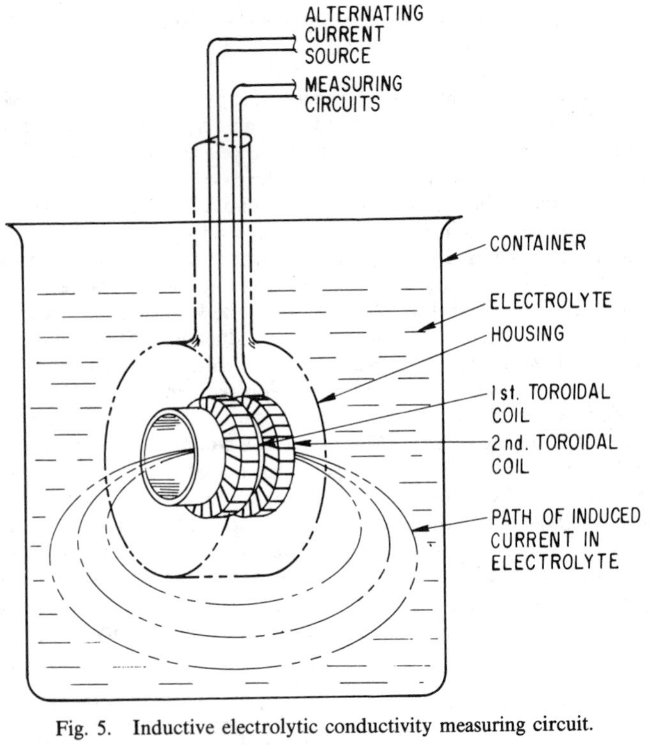

Fig. 5. Inductive electrolytic conductivity measuring circuit.

A typical laboratory-type conductivity cell, which employs two platinized platinum electrodes contained in an open-bottom cylindrical chamber formed from pyrex glass, is shown in Fig. 6. This cell has a cell constant of 0.5/cm and is intended for use in measuring the conductivity of distilled water and other dilute solutions used in the laboratory. This kind of cell is dipped into an open-topped container containing the sample to be measured. Wide use is made in the laboratory of conductivity cells of this type in ascertaining water quality and in screening samples to be titrated or further analyzed by other means.

It is interesting to contrast the glass laboratory cell with the electrodeless cell shown in Fig. 7. This latter cell is fabricated from glass-fiber-reinforced polyester and epoxy plastics. The small hole located near the periphery contains provision for an automatic temperature compensator. This cell contains no solution-contacting electrodes. It is protected by a coat of antifouling paint so that it can be immersed continuously in sea water without accumulation of marine growth. Such cells are used to continuously measure the salinity of estaurine water during site survey work required by governmental authorities

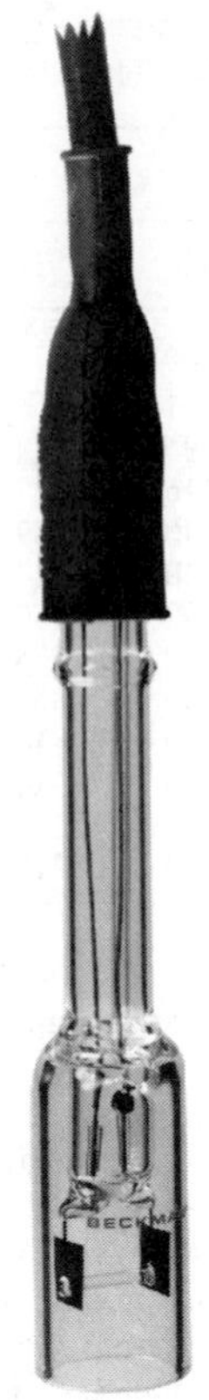

Fig. 6. Laboratory-type conductivity cell. (*Beckman Instruments, Inc.*)

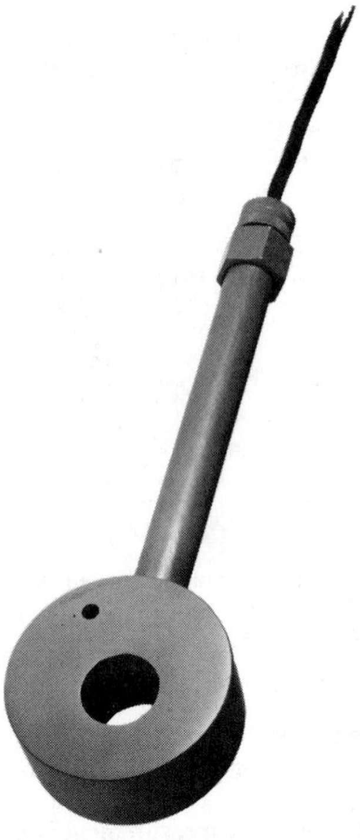

Fig. 7. Electrodeless-type conductivity cell. (*Beckman Instruments, Inc.*)

prior to issuance of approval for construction of various facilities, such as power plants.

References

Fuoss, R. M.: "Conductance-Concentration Function for the Paired Ion Model," *J. Physical Chemistry*, **82**, 22, 2427 (1978).

Janz, G. J.: "Conductance Cell Calibrations: Current Practices," *J. Electrochemical Society*, **124**, 2, 57C (February 1977).

Jones, Grinnell, et al.: A series of seven papers on conductivity measurements published in the *Journal of the American Chemical Society*. (Last of series of these important and fundamental papers was published in the *J. Am. Chem. Soc.*, **57**, 280 (1935)).

MacInnes, D.: "The Principles of Electrochemistry," Dover, New York, 1961.

Pungor, E.: "Oscillometry and Conductometry," Pergamon, Elmsford, New York, 1965.

Elmer, A. Sperry, III, Beckman Instruments, Inc., Cedar Grove, New Jersey

ELECTROLYTIC FLOWING JUNCTION. Flowing Junction (Electrolytic).

ELECTROLYTIC IRON. Iron Metals, Alloys, and Steels.

ELECTROLYTIC TRANSDUCER. A device whose electrolytic resistance is caused to change when exposed to the variable being measured. A simple configuration comprises two electrodes that are separated by a small distance, with the volume between the electrodes occupied by an electrolyte. To be used as a transducer, as in the case of displacement measurement, the distance between the electrodes or the effective cross-sectional area of the electrolyte can be varied. Very little force is required to produce a displacement and the device can be made quite small. The devices are very temperature sensitive, a disadvantage for practical, industrial applications. Also, like other electrolytic cells, the devices are subject to polarization. Because of these limitations, electrolytic transducers are not commonly used.

ELECTROMAGNET. A magnet whose field is produced by an electric current, and which is largely demagnetized upon cessation of the current, is an electromagnet. In order to obtain the strongest field possible, highly permeable soft iron or steel is employed for the core of electromagnets. In an electromagnet the current flows through a solenoid, which is a conductor wound in the form of a helix, and which produces a strong magnetic field coaxial with the helix. The core is placed inside the helix in order to give a magnetic path of the least reluctance. Electromagnets are found in a number of different forms, such as the plain solenoid with cylindrical core, or the horseshoe electromagnet, much used in electric bells, telegraph instruments, and telephones. Very powerful electromagnets are often used to move masses of iron, such as scrap iron, and have the advantage that the loading or unloading of the crane to which the magnet is attached is simply a matter of applying or disconnecting the electric current.

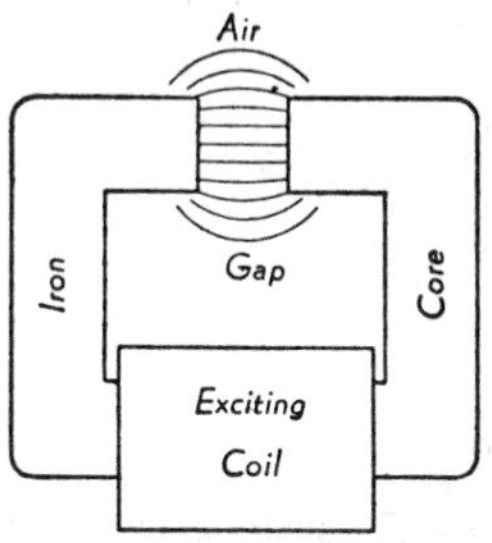

Electromagnet having two poles, one of a large variety of available designs.

For practical calculations, we may write the Bosanquet law $\phi = \mathcal{F}/\mathcal{R}$ for the flux in a magnetic circuit in more useful approximate form:

$$\phi = \frac{nI}{5l/2\pi\mu a} = \frac{nI}{0.796l/\mu a}$$

This refers to a simple, closed magnetic circuit of length l (cm), uniform permeability μ, and uniform cross section a (sq cm), where $\mathcal{F}$ is the magneto motive force ($= nI$) in ampere-turns, and $\mathcal{R}$ is the reluctance, which is equivalent to $0.796l/\mu a$. The denominator is the reluctance of the circuit. If either the permeability or the cross section is not uniform, this reluctance must be separated into parts, each having its own value for l, μ, and a. For example, let us calculate the ampere-turns nI required to produce an induction of 2,200 gausses in the air between the poles of an electromagnet having an iron core of permeability 1,500 and crossection 40 sq cm, the distance around through the iron being 78 cm and the single air gap 6 cm wide. The required flux is 40 sq cm $\times$ 2,200 gausses $=$ 88,000 lines or maxwells. The reluctance of the iron part of the circuit is 0.796 $\times$ 78 cm $\div$ 1,500 $\times$ 40 sq cm $=$ 0.001 reciprocal cm, and that of the air gap, 0.796 $\times$ 6 cm $\div$ 1 $\times$ 40 sq cm $=$ 0.119 reciprocal cm (taking the permeability of air as 1); so that the total reluctance is 0.12 reciprocal cm. That is, 88,000 maxwells $= nI \div 0.12$ reciprocal cm, or $nI =$ 10,560 ampere-turns. The result, though only roughly approximate, is sufficiently accurate for practical purposes. It is to be noted that since the air gap contributes by far the larger part of the reluctance, the flux is very sensitive to variations in its width.

The quantities in this example are representative of operation in the ordinary atmospheric range. Since, however, magnetic flux is, as stated in the formula, directly proportional to the current in the coil, any reduction in the resistance of the wire used should, by increasing the current, increase the flux without changing the size of the electromagnet.

The use of superconductors can significantly increase the effectiveness of an electromagnet. See also **Superconductors.**

ELECTROMAGNETIC EQUILIBRIUM. Equilibrium.

ELECTROMAGNETIC FIELD. Field Theory.

ELECTROMAGNETIC FLOWMETER. Flow Measurement.

ELECTROMAGNETIC HORN. Horn (Electromagnetic).

ELECTROMAGNETIC INDUCTION. Faraday Law of Electromagnetic Induction.

ELECTROMAGNETIC PHENOMENA. The term *electromagnetic* is used to describe the combined electric and magnetic fields that are associated with movements of electrons through conductors; to the combined electrical and magnetic effects exhibited by and used by equipment, apparatus, and instruments; and, in terms of radiation, to describe the radiation that is associated with a periodically varying electric and magnetic field that is traveling at the speed of light, such as light waves, radio waves, x-rays, gamma radiation, and so on.

Electromagnetism. The pioneer discovery of the magnetic effect of the electric current was made by Oersted at Copenhagen in 1820. In experimenting with battery currents, he happened to bring a compass needle near a wire in which there was an electric current, and noted that the needle was deflected. Such a wire is surrounded by a magnetic field so that, to one looking along the wire in the direction from the positive to the negative battery-terminal (the so-called "direction of the current"), the direction of the field, as indicated by the north pole of the compass needle, is clockwise (Ampere's rule).

If the wire carrying the current is placed in a magnetic field perpendicular to its direction, this field reacts with that due to the current in such a way as to give the wire a lateral thrust, perpendicular to both the wire and the field in which it is placed. For a wire of length l carrying current I and placed across a field of intensity B, this lateral force is given by the equation $f = BlI$, which follows from the definition of the ampere. An electric motor is driven by forces thus produced.

If the wire is bent into a circular loop of radius r, still carrying current I, there is produced at its center, perpendicular to the plane of the loop, a magnetic field of intensity $H = I/2r$. This, and the statement in the preceding paragraph, may be shown to be interdependent. If more loops are added, forming a coil of n equal turns close together, the resulting field is n times as great. By winding the n turns along a cylinder, forming a "helix" of radius r and axial length a, one obtains something greatly resembling a bar magnet, the ends of the helix corresponding to the poles. The field intensity at the center of the axis of this helix is

$$H_0 = \frac{nI}{\sqrt{4r^2 + a^2}}$$

The above expressions are all appropriate in the rationalized mksa system.

If we now insert an iron core, we have an electromagnet, and the helix supplies the magnetomotive force nI ampere-turns for a magnetic circuit composed partly of iron and partly of air.

More general calculations of electromagnetic effects are based upon Ampère's law, the Biot-Savart law, and Maxwell's equations.

Electromagnetic Induction. Probably the most noteworthy of the many scientific contributions of the renowned Michael Faraday was his discovery in 1831 of electromagnetic (or more logically, magneto-electric) induction. As exhibited in the usual experimental arrangements, this phenomenon is the setting up, in a circuit, of an electromotive force by reason of the variation of the magnetic flux linked with the circuit; the magnitude of that electromotive force being, as Faraday found, proportional to the rate at which the flux through the circuit, or the "linkage," varies. If the flux linkage with the circuit, in weber-turns, is expressed by $N\phi$ (the actual flux, ϕ, times the number of turns, N), the electromotive force generated by its variation, in volts, is

$$E = N\frac{d\phi}{dt}$$

The electromotive force is positive (counterclockwise) when $d\phi/dt$ is positive, that is, when the flux is increasing, negative when it is decreasing; as viewed by one looking in the direction of the magnetic induction.

Another aspect of the matter is that if a conductor moves through a magnetic field, or if a magnetic field sweeps over a conductor, in such a way that the conductor cuts across the lines of force, the electricity in the conductor experiences forces at right angles to the field and to the (relative) motion. More general still is the Maxwell concept that when magnetic lines of force move sidewise, their movement results in an electric field at right angles to the magnetic lines and to their motion.

Faraday's discovery was almost accidental. Happening to thrust a bar magnet into a coil connected with a galvanometer, he noted a momentary deflection of the needle. If the north pole is thrust downward into the coil, so as to increase the flux linked with the coil, the current will be counterclockwise as viewed from above, and reverses on drawing the magnet out again. See accompanying figure.

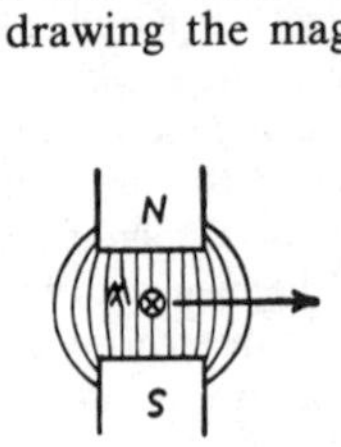

Wire (W) moving to the left across an upward magnetic field has induced in it an emf away from the observer.

The far-reaching consequences of this simple observation can hardly be overestimated. It was the forerunner of the invention of electric generators and alternators, of the original Bell telephone, of the induction coil, of the transformer, of the induction motor, of magnetic damping devices, and of many other electric appliances. It is the basis of Lenz's law and of the Wilson experiment, and the explanation of eddy currents. The volt and the henry are definable in terms of it. The phenomenon is called mutual induction when the variation of current in one circuit causes a variation of magnetic flux through, and hence an electromotive force in, another coupled circuit. It is a curious fact that in a vacuum the linkage through one circuit A due to unit steady current in a neighboring circuit B is equal to the linkage through B due to unit current in A; hence the mutual inductance of two circuits is the same, whichever is the primary circuit. If the circuits are closely coupled and have high self-inductance, and if an alternating electromotive force E be applied to A, the resulting ac induces an electromotive force in B approximately equal to (n_B/n_A) E; in which N_A and N_B represent the numbers of turns in the respective circuits. The principle is utilized in induction coils and in potential transformers.

Electromagnetic Field. Traditionally and in a simplistic way, it may be stated that a wire carrying an electric current is surrounded by a magnetic field whose lines of force are circles with the wire as their axis. This statement implies that the magnetic field is directly traceable to the moving electricity in the wire. This is an oversimplified explanation, however, because each electric "particle" projects into space a radiating field of electric force; and as the particles move along the wire, the lines of force move with them.

Theory of Maxwell. As will be developed later, the Maxwellian theory of the electromagnetic field has been supplanted and refined. Prior to the more detailed description of modern concepts, it may be in order to develop the earlier Maxwell concept. According to the theory of Maxwell, it is the motion of these lines of electric force that sets up the magnetic field transverse to them. More generally, a variable electric field is always accompanied by a magnetic field; and conversely, a variable magnetic field is accompanied by an electric field. The joint interplay of electric and magnetic forces here described is what is called an electromagnetic field, and is considered as having its own objective existence in space apart from any electric charges or magnets with which it may be associated. An essential feature of the theory is that this process, whatever it is, represents a flow of energy at right angles to both electric and magnetic components. The flux density of this energy (corresponding to the intensity of radiation) is represented by what is known as the Poynting vector. Electromagnetic radiation is, on this theory, the propagation of these electric and magnetic stresses through space with the speed of light, somewhat as the much slower waves of elastic stress are propagated through steel. The conditions in an electromagnetic field are expressed mathematically by Maxwell's equations.

When an electric charge is set into motion, it builds about itself an electromagnetic field, and this implies a distribution of energy throughout space. The density of this energy at any point of the field is proportional to the product of the electric and magnetic vector components and the sine of the angle between them (vector product). The total field energy can be obtained by suitable integration, and is greater than that of the purely electric field of a stationary charge. Maxwell's theory treats this excess as kinetic energy, thus endowing the moving charge with an "electromagnetic mass" and an "electromagnetic momentum" inherent in its electrical character.

The Maxwell equations comprise a set of four classic formulae of the electromagnetic theory. They deal with certain vector quantities pertaining to any point of a region under varying electric and magnetic influence. If the point is in empty space, the equations are somewhat simplified; in general, provision must be made for the presence of dielectrics, conductors, or magnetizable bodies. In these equations, **H** is magnetizing force, **B** is magnetic induction, **E** is electric intensity, **D** is electric induction, ρ is electric charge density, **J** is conduction current density, t is time. The "curl" and the "divergence" of a function are well-known operators of vector analysis. The equations, in rationalized mks units, are

$$\text{Curl } \mathbf{H} = \frac{\partial \mathbf{D}}{\partial t} + \mathbf{J} \qquad \text{(I)}$$

$$\text{Curl } \mathbf{E} = -\frac{\partial \mathbf{B}}{\partial t} \qquad \text{(II)}$$

The additional relations

$$\text{Div } \mathbf{B} = 0 \qquad \text{(III)}$$

$$\text{Div } \mathbf{D} = \rho \qquad \text{(IV)}$$

are frequently included as part of Maxwell's system, although they are not independent relations if one assumes the conservation of charge. The last two are also known as the Gauss law. For linear homogeneous isotropic media, $\mathbf{B} = \mu\mathbf{H}$; $\mathbf{D} = \epsilon\mathbf{E}$. The values of μ and ϵ for a vacuum satisfy

$$\mu_v \epsilon_v = 1/c^2$$

where c is the speed of light, μ is permeability and ϵ is permittivity.

Electromagnetic Radiation

The electromagnetic spectrum is the total range of wavelengths or frequencies of electromagnetic radiation. As shown by the accompanying table, this range extends from the longest radio waves to the short cosmic rays. See also **Radio Frequency Allocation.**

Prior to Maxwell's studies of the electromagnetic field in an effort to substitute electric and magnetic forces for elastic forces in the theory of light propagation, light was believed to be a transverse wave motion in an *ether* which behaved like an elastic solid. According to Maxwell's views, it was postulated that light is the result of vibrating electric charges. These set up alternating electric and magnetic fields at right angles to each other and to the direction of propagation, which pass on the energy from one portion of the ether to the next as an electromagnetic wave. (Poynting's theorem states that the rate of this energy transfer is proportional to the product of the electric and magnetic intensities.) The theory was successful in explaining many of the electrical and magnetic properties of light, such as the Faraday effect and the Kerr effect.

Maxwell suggested that it should be possible to produce waves of much longer wavelength than light by causing electricity to oscillate in a conductor. This was a forecast of the radio waves, upon which radio transmission depends, and which exhibit many of the characteristics of light, such as reflection, refraction, diffraction, interference, polarization, etc., but on a gross scale. Other researches showed that the infrared and ultraviolet radiations have these same properties. It was therefore natural to classify them together as the same phenomenon in different frequency ranges. X-rays for a time could not be identified with this group, but their diffraction by crystals finally demonstrated their wave character and they now take their place next to the ultraviolet. Meanwhile the quantum theory put a new aspect on the whole matter, and the gamma rays were added to the radiation family.

Present Concepts of Electromagnetic Theory. The task of electromagnetic theory is to account for the effects of electrical charges in various states of motion. Although historically electromagnetic theory was developed from Coulomb's celebrated law, it is at present more economic to develop it differently. The macroscopic effects are described with remarkable accuracy by the following set of equations (rationalized mks system of units):

$$\mathbf{F} = q\mathbf{E} + q\mathbf{v} \times \mathbf{B} \tag{1}$$

$$\nabla \cdot \mathbf{J} + \frac{\partial \rho}{\partial t} = 0 \tag{2}$$

$$\nabla \times \mathbf{H} = \frac{\partial \mathbf{D}}{\partial t} + \mathbf{J} \tag{3}$$

$$\nabla \times \mathbf{E} = -\frac{\partial \mathbf{B}}{\partial t} \tag{4}$$

$$\mathbf{D} = f_1(\mathbf{E}) \tag{5}$$

$$\mathbf{B} = f_2(\mathbf{H}) \tag{6}$$

$$\mathbf{J} = f_3(\mathbf{E}, \mathbf{H}) \tag{7}$$

provided the functional relationships indicated in Equations (5), (6), and (7) are known explicitly. With these equations and the laws of mechanics, classical electromagnetic theory becomes essentially a branch of applied mathematics.

ELECTROMAGNETIC SPECTRUM SHOWING PRINCIPAL RADIATION CATEGORIES

Frequency Hz	Type of Radiation	Wavelength Cm	
10^{23}			
	Cosmic Rays	10^{-12}	
10^{22}			
		10^{-11}	
10^{21}			
	Gamma Rays	10^{-10}	
10^{20}			
		10^{-9}	
10^{19}			
	X-Rays	10^{-8}	
10^{18}			
		10^{-7}	
10^{17}			
		10^{-6}	
10^{16}	Ultraviolet Radiation		
		10^{-5}	
10^{15}	Visible Light		
		10^{-4}	
10^{14}			
	Infrared Radiation	10^{-3}	
10^{13}			
		10^{-2}	
10^{12}	Submillimeter Waves		
		10^{-1}	
10^{11}			
		1.0	
10^{10}	Microwaves (Radar)		
		10	
10^{9}			UHF
		10^{2}	
10^{8}	Television and FM Radio		VHF
		10^{3}	
10^{7}	Short Wave		HF
		10^{4}	
10^{6}	AM Radio		MF
		10^{5}	
10^{5}			LF
	Maritime Communications	10^{6}	
10^{4}			VLF

Equation (1), sometimes known as the Lorentz force equation, defines the field quantities, **E**, the electric field intensity, and **B**, the magnetic induction, in terms of an observable, the force **F** on a charge q. In Equation (1), **v** is the velocity of the charge relative to the observer. Equation (2) is a statement of the law of conservation of electric charge in terms of the charge density ρ and the total current density **J**. Equation (3) is the differential form of Ampere's law,

$$\oint_{c\ of\ s} \mathbf{H} \cdot d\mathbf{l} = \iint_S \mathbf{J} \cdot d\mathbf{S} = I$$

which relates the magnetic field intensity **H** to the current, including in addition the displacement current density term $\partial\mathbf{D}/\partial t$, which was added by Maxwell to make the law applicable to time-varying fields. The term **J** represents the total current density. Equation (4) is the differential form of Faraday's law of electromagnetic induction. Equations (5), (6), and (7) are functional relationships, for the most part determined experimentally, by means of which the effects of different materials are accounted for. Mathematically, these equations are employed to reduce Equation (3) and (4) to a pair of equations in only two unknowns. In free space, Equations (5), (6) and (7) take their simplest form, respectively, $\mathbf{D} = \epsilon_0\,\mathbf{E}$, $\mathbf{B} = \mu_0\mathbf{H}$, $\mathbf{J} = 0$ (or $J = J_s$, a source current independent of **E** and **H**), where ϵ_0 and μ_0 are constants whose value depends on the system of units (in the mks system $\epsilon_0 = 8.854 \times 10^{-12}$, $\mu_0 = 4\pi \times 10^{-7}$). Since matter itself is a relatively dilute collection of charged particles, it is always theoretically possible to define terms so that the theory is a description of the effects and interactions of charges in free space, with consequently no essential distinction between **D** and **E** or between, **B** and **H**, as indicated above. In practice however effects of materials are usually best handled in another way. Dielectric polarization effects are accounted for by making the **D** vector include the electric dipole moment density **P**, $\mathbf{D} = \epsilon_0\,\mathbf{E} + \mathbf{P}$, and then introducing a material constant, the permittivity ϵ, such that $\mathbf{D} = \epsilon\mathbf{E}$. The relative permittivity of a dielectric material is then equal to one plus the electric susceptibility. Magnetic polarization effects are handled similarly by defining the field vector **B** so that it includes the magnetic dipole moment density **M**, $\mathbf{B} = \mu_0(\mathbf{H} + \mathbf{M})$. The material permeability is then introduced so that it depends upon the magnetic susceptibility analogously, and $\mathbf{B} = \mu\mathbf{H}$. Effects of conductors are represented by a material conductivity σ, such that $\mathbf{J}_c = \sigma\mathbf{E}$. With these simple forms for Equations (5), (6) and (7), Equations (3) and (4) take on the useful form

$$\nabla \times \mathbf{H} = \epsilon\frac{\partial \mathbf{E}}{\partial t} + \sigma\mathbf{E} + \mathbf{J}_1 \tag{8}$$

$$\nabla \times \mathbf{E} = -\mu\frac{\partial \mathbf{H}}{\partial t} \tag{9}$$

provided μ and ϵ are constant in time. The term $\mathbf{J}_1$ here includes currents arising from charges in free space plus any (source) currents which are independent of **E** and **H**. If there are no free charges in the region, $\mathbf{J}_1$ includes only the source currents; these latter are known, so Equations (8) and (9) may be solved for **E** and **H**. Since the equations are partial differential equations, boundary conditions over closed surfaces are required for unique solutions. Boundary conditions on the field quantities, which must hold at any boundary between two regions, may be derived from these equations. The conditions are: across a boundary (a) tangential **E** must be continuous, (b) tangential **H** must be continuous, (c) normal **D** and normal **B** must be continuous. Idealizations of material properties are sometimes helpful. For example, a perfect conductor has no nonstatic fields inside it, and at its surface, tangential **E** and normal **B** are zero, tangential **H** is equal and perpendicular to any surface current density, and normal **D** is equal to any surface charge density.

Two additional equations, especially useful in static problems, may be deduced from Equations (2), (3) and (4):

$$\nabla \cdot \mathbf{D} = \rho \tag{10}$$

$$\nabla \cdot \mathbf{B} = 0 \tag{11}$$

Solutions to the field equations are most readily obtained by imposing a restriction on the time dependence. If the fields are assumed to be independent of time (static), then Equations (3) and (4) or (8) and (9) decouple. One of the equations becomes $\nabla \times \mathbf{E} = 0$. This means that **E** is irrotational and may be represented by a scalar potential function ϕ, $\mathbf{E} = -\nabla\phi$. Combining this with Equation (10) gives the fundamental equation of electrostatics,

$$\nabla^2\phi = -\rho/\epsilon \tag{12}$$

Poisson's equation. This equation for the electrostatic potential is solved by the standard methods of partial differential equations. The boundary conditions on the potential may be found from the boundary conditions on the fields. In practice, it is frequently necessary to solve for the potential and electric field in a restricted region in which the charge density is zero, but the potential at the boundary is held at some particular value(s). The problem then is to solve Laplace's equation, $\nabla^2\phi = 0$, subject to the stated boundary conditions. The standard techniques for solving boundary value problems are employed. However, if the region of interest is partially open, known analytical techniques are sometimes inadequate to solve the problem. In two-dimensional problems of such a difficult type, the method of conformal transformations (conjugate functions) is often helpful.

The main applications of electrostatic theory are in (a) the theory of material properties, (b) the calculation of charged particle trajectories in electron guns, deflection systems, and accelerators (here in conjunction with magnetostatic theory), (c) the calculation of circuit component values, such as capacitance, and (d) the determination of voltage gradients in connection with voltage breakdown problems.

Magnetostatic theory is developed from Equations (11) and (8). Since **B** is divergenceless, it can be represented by the curl of a vector **A**, which is known as the magnetic vector potential. Equation (8) can usually be written in terms of this potential as follows:

$$\nabla^2 \mathbf{A} = -\mu \mathbf{J} \tag{13}$$

Taken one rectangular component at a time, this equation is of the same form as Poisson's equation [Equation (12)] and may be solved in the same way. The boundary conditions on **A** may be found from those on **B** and **H.** In regions with no current, Equation (8) becomes $\nabla \times \mathbf{H} = 0$ so that **H** may be represented by a scalar potential function $\mathbf{H} = -\nabla\phi_m$. In such regions then, in view of Equation (11), the magnetic scalar potential, ϕ_m, must satisfy Laplace's equation

$$\nabla^2\phi_m = 0 \tag{14}$$

provided $\nabla\mu = 0$ in the region. The techniques and solutions of electrostatics are applicable to many magnetostatic problems. Unfortunately, however, in practice many of the systems designed to establish a given magnetic field incorporate ferromagnetic materials. For such materials, the magnetic susceptibility (and hence the permeability) is not independent of the field intensity and the field equations become nonlinear. Present mathematical techniques for handling nonlinear problems are severely limited. Practical magnetostatic problems are, therefore, frequently solved by some approximation. One of the simplest and most useful approximations is a representation by a magnetic circuit. Series and parallel branches of the magnetic circuit may be recognized, and the techniques of linear and nonlinear circuit analysis can be applied to obtain a solution.

Magnetostatic theory is applicable to a myriad of magnetic devices, including deflection systems, motors, generators, relays, magnetic pickup devices, permanent magnets, memories, transducers and coils. To date, the need for particular solutions has frequently arisen before sound analytical methods have been available, so many devices are developed empirically.

Energy is required to establish electric and magnetic fields, and such energy is associated with the fields. The field energy in a given volume may be computed in most cases from a volume integral of one or both of the following energy density expressions $W_e = \frac{1}{2}\epsilon E^2$, $W_m = \frac{1}{2}\mu H^2$, respectively the electrostatic and magnetostatic values.

When the fields are time varying, Equations (8) and (9) are coupled and must be solved simultaneously. Almost invariably, a potential function such as a vector potential or a Hertz potential is introduced. For example, Equation (11) implies that **B** may be replaced by a vector potential such that $\mathbf{B} = \nabla \times \mathbf{A}$. Equation (9) implies the following equation for **E**,

$$\mathbf{E} = -\nabla\phi - \frac{\partial \mathbf{A}}{\partial t} \tag{15}$$

so that **H** and **E** may be replaced in Equation (8), and with the condition on **A**, $\nabla \cdot \mathbf{A} = \mu\epsilon\, \partial\phi/\partial t$, the following equations may be obtained for **A** and ϕ (σ assumed zero here)

$$\nabla^2\mathbf{A} - \mu\epsilon \frac{\partial^2 \mathbf{A}}{\partial t^2} = -\mu\mathbf{J} \tag{16}$$

$$\nabla^2\phi - \mu\epsilon \frac{\partial^2\phi}{\partial t^2} = -\rho/\epsilon \tag{17}$$

That is, both the vector potential **A** and the scalar potential ϕ satisfy a differential equation known as the inhomogeneous wave equation.

Because of their simplicity and practical importance, solutions for those sources and fields which simply oscillate at a single frequency have been studied extensively. In this case, the time is eliminated as an independent variable, as if by a transform operation. (In fact, transform methods are often the best means of obtaining transient field solutions.) In the equations, the time derivatives are replaced by frequency multipliers so that the resulting equations are functions of the space variables only. The vector potential may then be found by standard techniques of partial differential equations and boundary value problems. Having **A**, the field quantity **B** is found from $\mathbf{B} = \nabla \times \mathbf{A}$ and **E** is found from Equation (8). In practice, a theorem which can be derived from the field equations, called the reciprocity theorem, is often helpful. The theorem relates the fields $\mathbf{E}_a$ and $\mathbf{E}_b$ produced respectively by a pair of current distributions $\mathbf{J}_a$ and $\mathbf{J}_b$. The theorem is

$$\iiint \mathbf{E}_a \cdot \mathbf{J}_b\, dv = \iiint \mathbf{E}_b \cdot \mathbf{J}_a\, dv \tag{18}$$

For example, if $\mathbf{J}_b$ is selected to be a point current at point P, directed along x (represented mathematically by a Dirac delta function), then Equation (18), $E_{ax}(P) = \iiint \mathbf{E}_b \cdot \mathbf{J}_a\, dv$, gives a formula for the computation of the field due to $\mathbf{J}_a$ which is equivalent to a superposition integral involving a Greens function.

Perhaps the most fundamental problem of electromagnetic theory is the determination of the fields of a point charge, at rest, in oscillation, or in some general state of motion. For a point charge q, at rest in free space, the solution may be obtained by solving Equation (12) in spherical coordinates. With the point charge at the origin, symmetry conditions may be employed to eliminate the angular variation, and the remaining differential equation in r can be solved subject to Equation (10) to give $\phi_G = (q/4\pi\epsilon_0 r)$ for the potential associated with the point charge. A superposition integral

$$\phi = \iiint \frac{\rho\, dv}{4\pi\epsilon_0 r} \tag{19}$$

may then be employed to find the potentials associated with more complicated distributions. The field of an oscillating dipole, which is equivalent to a point alternating current, is also of great interest. This solution may be obtained from Equation (16) (single frequency version). If the point current is directed along z, the z-component of the vector potential may be found by a procedure similar to that employed for a point charge. The final result is

$$A_{ZG} = \frac{I\,\Delta Z}{4\pi\mu_0 r} \cos\omega(t - \sqrt{\mu_0\epsilon_0}\, r) \tag{20}$$

where $I\,\Delta z$, the current moment, is equal to $q\,\Delta z$, the maximum dipole moment of the oscillating dipole. The factor $(t - \sqrt{\mu_0\epsilon_0}\, r)$ exhibits the time delay required for the effects of the oscillating charges to propagate to distant points. The electric and magnetic fields may be computed from Equation (20) as indicated above. The magnetic field strength produced by an oscillating dipole (point current) is, for example, in the spherical coordinate system (r, θ, φ)

$$H_\varphi = \frac{I\,\Delta Z}{4\pi} \sin\theta \left[\frac{\cos\omega(t - \sqrt{\mu_0\epsilon_0}\, r)}{r_2} - \frac{\omega\sqrt{\mu_0\epsilon_0}}{r} \sin\omega(t - \sqrt{\mu_0\epsilon_0}\, r) \right]$$

This form like Equation (20), shows that the crests and valleys of the field oscillations are propagated in spherical waves at the speed of light $v = (\mu_0\epsilon_0)^{-1/2}$. The solution for a point current may be employed in an integral similar to Equation (19) to find the vector potential of a more complicated distribution of current. Such solutions may also be employed to find the radiation patterns and input impedances of antennas.

The potentials and fields produced by a charge moving in an arbitrary way may also be obtained.

In regions free of source currents and charges, the fields and potential satisfy the homogeneous wave equation [for example Equation (16) with $\mathbf{J} = 0$]. Then one of the simpler solutions which can be obtained is that of the plane electromagnetic wave. With appropriate orientation of the rectangular coordinate system, the solutions show that plane waves may progress along z, with components as follows:

$$E_x = E_0 \cos \omega(t - \sqrt{\mu_0 \epsilon_0}\, z)$$

$$H_y = E_0 \sqrt{\frac{\epsilon_0}{\mu_0}} \cos \omega(t - \sqrt{\mu_0 \epsilon_0}\, z)$$

where E_0 is an arbitrary constant amplitude. Note that **E**, **H** and the direction of propagation are all perpendicular to one another. The Poynting vector, $\mathbf{S} = \mathbf{E} \times \mathbf{H}$, points in the direction of the propagation. Moreover, the power carried through a closed surface by an electromagnetic field may be computed from a surface integral of the Poynting vector.

With single frequency fields in source free regions, both **H** and **E** can be represented by vector potentials, $\mathbf{H}_1 = \nabla \times \mathbf{A}_1$, $\mathbf{E}_2 = \nabla \times \mathbf{A}_2$, and moreover the coordinate systems may be oriented so that $\mathbf{A}_1$ and $\mathbf{A}_2$ each have a single component. In cylindrical systems, this single component is commonly along z. $\mathbf{H}_1$ is then transverse to z (TM) and the set of fields, $\mathbf{E}_1$, $\mathbf{H}_1$, derivable from $\mathbf{A}_1'$, are called TM fields. $\mathbf{E}_2$ is likewise transverse to z and the set of fields, $\mathbf{E}_2$, $\mathbf{H}_2$, derivable from $\mathbf{A}_2$, are called TE fields. This procedure is particularly helpful in problems involving transmission lines and waveguides.

Some of the most interesting and fundamental problems of electromagnetic theory are concerned with the scattering and diffraction of electromagnetic waves. For example, exact solutions are available for the scattering by cylinders and spheres, as well as an infinitely long slit. Approximate solutions are available for many other shapes. The methods are those outlined above, supplemented by generalizations of the principles of Huygens and Babinet.

Another topic of wide interest is the nature of fields in ionized gases or plasmas. The applications range from ionospheric propagation to microwave devices to nuclear apparatus to magnetohydrodynamics to satellite reentry problems. The simplest theory for these effects is developed from Equations (3) and (4) (single frequency version) by separating the ion current term $\mathbf{J}_e = \rho\mathbf{v}$ from **J**, and employing Newton's law to eliminate **v** in favor of **E**, **H** and whatever mechanical constraints are applicable.

Effects peculiar to charges moving with very high velocities have not been included in this discussion. Numerous entries throughout theis encyclopedia are concerned with electromagnetic phenomena. See list of terms listed in entry on **Data Processing** for some of these. See also **Magnetism.**

ELECTROMECHANICAL COUPLING COEFFICIENT. **Coupling (Physics).**

ELECTROMECHANICAL SWITCH. **Analog Switch.**

ELECTROMETER. An instrument for measuring electric charge, usually by mechanical forces exerted on a charged electrode in an electric field. It consists, therefore, of a sensitive voltmeter operating on the principles of electrostatic attraction and repulsion. Thus, if the movement of the goldleaf in an electroscope is observed through a microscope whose ocular is provided with a calibrated scale, the instrument becomes an electrometer, capable of measuring potential differences in millivolts. (Some forms of electrostatic voltmeters operate on the same principle.) If the capacitance of the charged system is known, the rate of movement of the electrometer index may be used to measure the current from the discharging body; ionization currents are often thus measured.

The quadrant electrometer has a thin, oblong, metal plate suspended horizontally in the interior of a flat, circular metal box cut into four quadrants. One pair of opposite quadrants and the suspended strip are connected to the source of potential, the other pair of quadrants is grounded. This causes the strip to turn toward the grounded pair against the torsion of the suspending wire (see figure, *left*). Several electrometers have been designed, depending upon the lateral deflection of a lightly stretched, silvered or platinized quartz fiber in an electric field; they are called string electrometers. The Wulf electrometer employs two such fibers side by side; on being charged, they bulge apart (see figure, *right*). The displacement of the fibers is observed in a micrometer microscope. Some of the special electrometers used for work with cosmic rays are of this type.

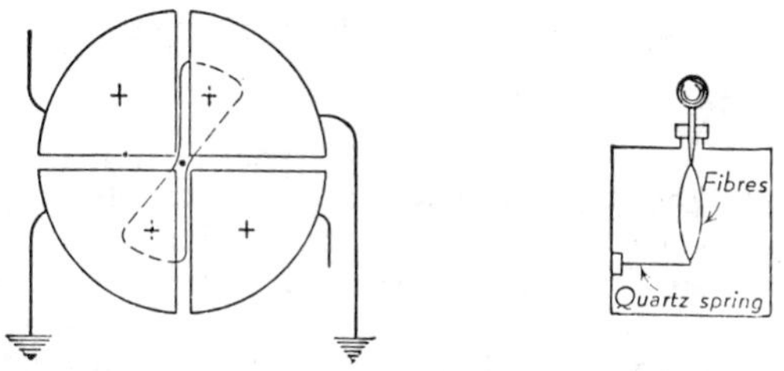

Electrometers: (*Left*) quadrants and strip of quadrant electrometer; (*Right*) quartz fiber electrometer.

Vacuum tube electrometers, which are specially designed amplifiers with high input impedance, have been replaced by quadrant and string electrometers to a large extent.

See also **Electrical Instruments;** and **Electroscope.**

ELECTROMETRIC ANALYTICAL METHODS. **Coulometer; pH (Hydrogen Ion Concentration); Polarographic Analyzers; Titration (Potentiometric).**

ELECTROMOTIVE FORCE. The electric potential difference (emf) between the terminals of any device which is used or may be used as a source of electrical energy, i.e., to supply an electric current. More strictly, the limiting value of that potential difference which is found as the current flowing from the source approaches zero. To avoid ambiguity, the strict sense of the term is often indicated by the use of the qualifying term "open circuit" or "no load."

The open circuit electromotive force of a cell is identical with its reversible potential difference; that of a rotating electrical machine is the potential difference existing across its terminals when the machine is neither receiving nor delivering electric power, i.e., is at the transition point between being a generator and being a motor. When a potential source of electrical energy, such as a capacitor, an inductor, or a rotating machine, is receiving energy from the external circuit, it is said to develop a counter-electromotive force. See also **Kirchoff Laws of Networks.**

Counter electromotive force is the emf generated by a running motor, by virtue of its generator behavior, or by an inductive circuit element through which the current is increasing with time. The total emf in the circuit is the impressed emf minus the counter emf; the current is given by the ratio of this total emf to the resistance in the circuit. The counter-electromotive force is sometimes called the back electromotive force. The *impressed emf* is the open-circuit (no load) emf of a source connected into a network.

Effective or root-mean-square electromotive force is the effective value of an alternating electromotive force and is the square root of the mean value of the squares of the instantaneous values. Where the variation of voltage with time can be expressed by a mathematical function, the value of the effective current can commonly be expressed in terms of the maximum value of the emf. Thus, for a sine wave relationship,

$$E_{eff} = \frac{E_{max}}{\sqrt{2}}$$

ELECTROMOTIVE SERIES. **Activity Series.**

ELECTROMYOGRAPHY. The recording of action potentials from muscles in the living subject by means of a needle electrode inserted through the skin into the muscle and connected to a suitable amplifier and cathode-ray oscillograph. Normal resting muscle shows no changes in potential; contraction gives rise to large numbers of monophasic or diphasic spikes, indicating changes in potential of up to 1 millivolt, each lasting from 5–10 milliseconds. Various diseases and injuries of muscles and their nerves of supply give rise to characteristic and distinct departures from these normal patterns, and hence electromyography is of value in assessing the extent of damage in diseases of the nervous system in which muscle is involved and in estimating the probability, progress and degree of recovery.

ELECTRON. An elementary particle of rest mass

$$m = 9.107 \times 10^{-31}$$

kilogram, a charge of 1.602×10^{-19} coulomb, and a spin quantum number $\frac{1}{2}$. Its charge may be positive or negative, although the term electron is commonly used for the negative particle, which is also called the negatron. The positive electron is called the positron. The electron is a constituent of all matter, thus the normal atom consists of a positively-charged nucleus surrounded by a sufficient number of electrons so that their total charge is equal to the positive charge on the nucleus. The electron also has wave characteristics, with a frequency and a wavelength. According to wave mechanics, an electron traveling with speed v is associated with a "de Broglie wave" train of wavelength $\lambda = h/mv$ (in which m is the electronic mass and h is Planck's constant). See also **Davison-Germer Experiment.**

Bonding Electron. An electron in a molecule which serves to hold two adjacent nuclei together.

Conduction Electron. An electron which plays an important part in electrical or thermal conduction by solids, i.e., by metals or semiconductors, e.g., the electrons in the conduction band, which are free to move under the influence of an electric field.

Electron Donor. 1. When a valence bond between two atoms is that type of covalent linkage in which both the electrons of the duplet are supplied by one atom, then that atom, or portion of the molecule of which it forms a part, is called the electron donor, and the other atom in the linkage is called the electron acceptor. 2. A donor is also an impurity added to a pure semiconductor to increase the number of free electrons.

Electron Duplet. A pair of electrons which is shared by two atoms, and is equivalent to a single, nonpolar chemical bond.

Electron Octet. A group of eight valence electrons which constitutes the most stable configuration of the outermost, or valency, electron-shell of the atom, and hence the form which frequently results from electron transfer or sharing between two atoms in the course of a chemical reaction.

Electron Pair. The negatron and positron that result from the pair-production process or interact to initiate an annihilation process.

Electron Shell. The structure of a neutral atom consists of a positively-charged nucleus with a number of electrons moving about it—the number being such that their total negative charge is equal to the positive charge on the nucleus. These electrons may be assigned to various shells, characterized by different principal quantum numbers.

Electron Spin. The intrinsic angular momentum of an electron, independent of any orbital motion. Spin ($= h/2$) contributes to the total angular momentum of the electron and is quantized. It gives rise to multiplicity in line spectra, which may be characterized by introduction of the spin quantum number.

Electron Transfer. The process of the shifting of an electron from one electrical field to another, as in the formation of an electrovalent bond, in which an electron moving in an orbit about one atom shifts to move in an orbit around the two bonded atoms.

Free Electron. An electron which is not restrained to remain in the immediate neighborhood of an atom or molecule.

Orbital Electron. An electron remaining with a high degree of probability in the immediate neighborhood of a nucleus, where it occupies a quantized orbital.

Photo Electron. An electron ejected from a substance by the action of a single photon of light or other electromagnetic radiation.

Secondary Electron. An electron deriving its motion from a transfer of momentum from primary radiation, which may be either particulate or electromagnetic.

Valence Electron. The electrons in the outermost shell of the structure of an atom. Since these electrons are commonly the means by which the atom enters into chemical combinations—either by giving them up, or by adding others to their shell, or by sharing electrons in this shell—these outermost electrons are called valence electrons.

ELECTRON AFFINITY. 1. Degree of electronegativity, or the extent to which an atom holds valence electrons in its immediate neighborhood, compared to other atoms of the molecule. 2. The work required to remove an electron from a negative ion, and hence to restore the neutrality of an atom or molecule, is called the electron affinity of the atom or molecule.

ELECTRON BEAM. A stream of electrons moving with about the same velocity and in the same direction, so as to form a beam.

ELECTRON BEAM LITHOGRAPHY. Advances in the design and fabrication of electronic devices have resulted in reduced power dissipation and shorter response time. These improvements were made possible by increasing the complexity and sophistication of the circuits and by reducing their size. Early computers had limited capabilities, filled large rooms, and dissipated very large amounts of power. A device of comparable capability, manufactured by the techniques of the early 1980s, can be placed on a single chip of 0.25 square inch (161 square millimeters) or less.

Because the processing cost is nearly independent of the amount of circuitry on a chip, it is economically advantageous to further increase the amount of circuitry. This can be done in either of two ways: (1) increasing the size of the chip, or (2) reducing the size of the circuits themselves. Larger chips exhibit poorer manufacturing yields and less mechanical stability. Hence, increasing the circuit density can best be accomplished by reducing the circuit dimensions. Further size reductions, however, must rely on the development of new techniques, because present optical lithography methods have reached the limit dictated by diffraction and optical aberrations. Additional improvements below a feature size of about 2 micrometers (79 microinches) requires a new fabrication technology.

Electron beam lithography is a recent technology that provides improvements in resolution and edge definition which is capable of reducing the linear dimensions of circuits by at least an order of magnitude, and hence increase the packing density by two orders of magnitude.

Instrumentation for Electron Beam Lithography. Typically an electron beam lithography system accelerates and focuses an intense beam of electrons to draw precise circuit patterns on suitable substances. Electron beam instruments, unlike optical instruments, require a high-vacuum environment for both the electron beam path and the substrate.

An intense beam of electrons from a high brightness source, such as lanthanum hexaboride, is focused to a size of about 0.2 micrometer (8 microinches) or less by electromagnetic lenses. Patterns are generated by moving the beam across the substrate and turning the beam on and off with a blanking system. Two methods are employed to move the beam: (1) small movements are accomplished by deflecting the beam with electromagnetic coils; (2) large movements are made by moving the wafer by a mechanical stage. Positioning precision is achieved in the first case by using high-stability electronics of high numerical precision, while in the second case it is achieved by employing a laser interferometer to monitor the stage motion. Because the beam must move at rates of two to ten million steps per second, a computer is required to control the position. Few computers are capable of the necessary data rates and, therefore, specific custom hardware is usually included in the interface to reduce the required amount of data.

The technique used in writing with an electron beam is similar to that used for tracing figures on cathode ray tubes in computer graphics displays. Two distinct methods have been developed for this purpose: (1) raster scan, and (2) vector scan (sometimes called (*random draw*). Because specific and distinct hardware is required for each, electron beam instruments are designed to use either one method or the other, but not both.

Raster Scan Method. In the raster scan method, the beam is moved across the entire chip using a television-type raster. It is unblanked only where exposure is desired. An instrument designed and built at Bell Labs, EBIS, moves the stage continuously in one direction, *X*, while scanning the beam in an orthogonal direction, *Y*. In this way, a narrow strip is exposed, and the entire wafer is covered by exposing a large number of these strips. As the beam is scanned in the *Y* direction, the hardware must turn it on or off at each point. Currently available computers are not capable of supplying data at the required high rates and, therefore, a large block of information is stored in the hardware and used repeatedly to expose each of many identical areas on the wafer. This method results in fast exposure, but only if many identical chips are to be exposed. The raster scan method is shown in Fig. 1.

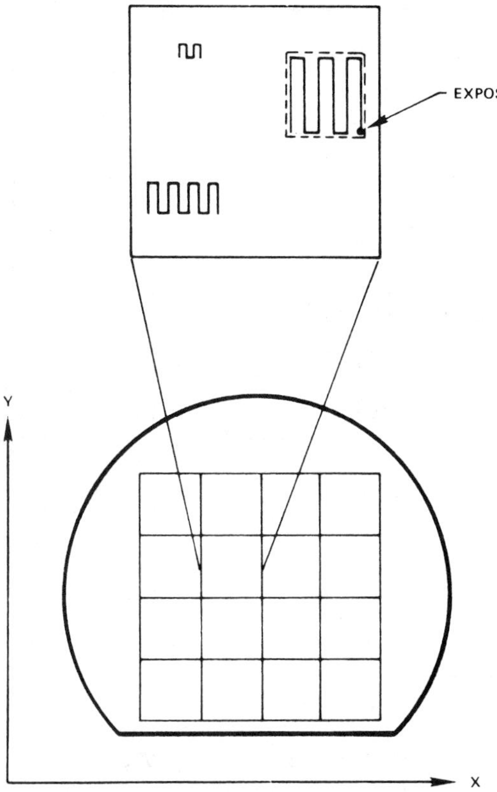

Fig. 1. Exposure of a wafer by the raster scan method.

Advantages of the raster scan method include less stringent requirements on the electronics and on the electron optics. In some cases, it is possible to achieve faster exposures. Deflection of the beam in the *X* direction is limited to that necessary to correct for minor errors in the position of the stage. In the *Y* direction, the beam deflection required is significantly less than that for the vector scan method. At these small beam deflections, field distortion is normally not a problem. Processing times are short if there are many identical chips to be made from a single wafer, but increase as the size of the chip increases.

These advantages are offset by a number of disadvantages. The necessity of "stitching" together many small fields places stringent requirements on the precision of the mechanical stage. The entire wafer must be scanned, and this is time consuming if it is necessary to expose only a small fraction of the area. Furthermore, corrections for the exposure effects of adjacent fields and for wafer distortion are difficult, if indeed possible.

Vector Scan Method. In the vector scan method, the beam is moved only to locations where exposure is desired. The beam is moved over a serpentine path to cover an entire basic element, the portion of the chip which can be exposed by a single computer command. The element is constructed of a series of parallel lines, each line consisting of a row of exposed points. Because the spacing of the points is of the same order as the width of the beam, the entire area is exposed. The maximum point rate is 5 or 10 megahertz, and, because most computers are unable to give commands at this rate, hardware is normally included to move the beam over an entire element. The beam is blanked only for a short time between the tracing of adjacent lines and while moving from one element to another. The vector scan method is shown in Fig. 2.

The vector scan method has the advantage of not using up time moving the beam over areas which are not to be exposed. In addition, it is relatively easy to make corrections for scan field distortions, for the effect of adjacent exposed areas, and for wafer distortion. The necessary size of the scan field, however, is significantly larger than that used in the raster scan method. Consequently, more stringent requirements are placed on the quality of the electron optics, the scan coils, and the amplifier which drives them.

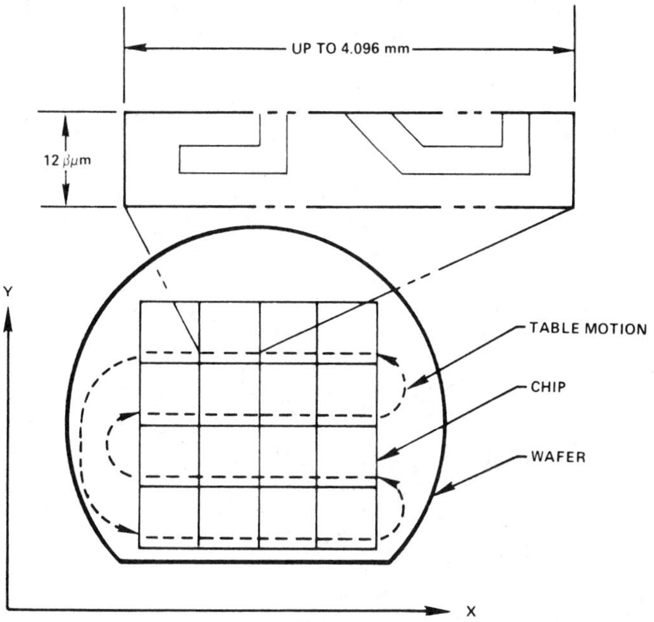

Fig. 2. Exposure of a wafer by the vector scan method.

As an example, an instrument used at the GTE[1] Labs employs a computer[2] to drive an interface which implements the vector scan method. The interface contains a series of electronic counters and digital-to-analog converters which determine the position of the beam. The output from this circuitry is fed to an amplifier which drives the deflection coils. A diagram of the instrument is given in Fig. 3.

Each pattern consists of a group of geometrical elements. The hardware stores the information which defines the coordinates of the lower left corner of each element also its height and its length. The hardware counts the height vector either up or down to expose a single line and decrements the length vector between lines to expose the entire element. The most commonly used element is a rectangle, but the hardware has the additional capability of drawing parallelograms and triangles.

The circuit designer defines the patterns, using another computer, usually one with graphics capabilities. The pattern definition is then placed on a magnetic tape which can be read by the machine's computer. The software in the latter machine then interprets the tape and produces the commands to drive the instrument. This software is capable of accepting control information related to scaling, exposure, and position, either from the magnetic tape or directly from the operator.

The customizing of the system maximizes flexibility in order to facilitate the engineering prototyping of circuits. A single wafer may contain either many identical or many different chips, and exposure and scale may be varied from chip to chip. The reduced number of steps in the process, as compared with optical methods, reduces the time necessary to make a modification and to obtain a finished circuit ready for testing. The result is a shorter turnaround time in the design of new circuits.

[1] GTE Laboratories Incorporated, Waltham, Massachusetts.
[2] Digital Equipment Corporation PDP 11/40 computer.

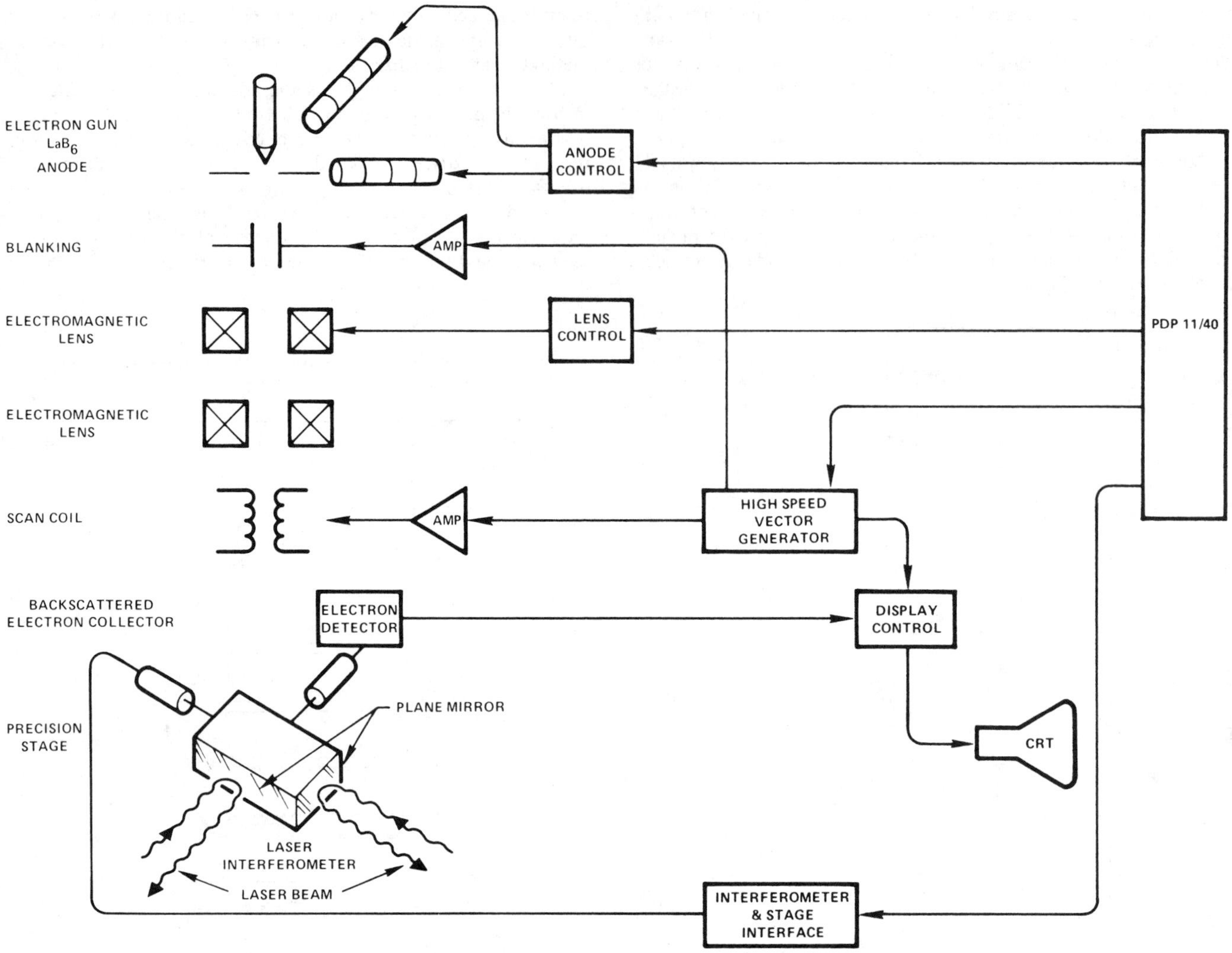

Fig. 3. Electron beam instrument. (*GTE Labs.*)

Fabrication Techniques. Electron beam techniques of fabricating devices are in many ways similar to conventional optical techniques. Processes common to both include growing and removing oxides, depositing metal conductors, and doping the semiconductor. Each process requires that a different pattern be exposed and positioned accurately (registered) relative to previous ones.

Traditional methods of device fabrication define these patterns by a set of masks fabricated by optical techniques, usually using more than one processing step to manufacture each mask. The masks are then used to expose a photosensitive emulsion on the wafer, again by optical techniques. The masks are commonly made of conventional silver halide emulsions on glass plates.

Electron beam techniques make possible two improved methods of fabrication. (1) In the first, the electron beam instrument is used to make a mask, usually of chromium deposited on glass. The masks are then used to expose the resist on the wafer with either X-rays or light. Chromium masks have been manufactured by optical techniques, but those fabricated by electron beam methods make possible improved manufacturing yields even when light is used to expose the wafer. If X-rays are used, it is also possible to reduce the dimensions of the finished circuits.

(2) The second electron beam technique is direct fabrication. Here the wafer itself is coated with an electron-sensitive polymer (resist), which is exposed directly by the electron beam. The wafer is then removed from the instrument, the resist developed, and the wafer processed. Each processing step requires that this sequence be repeated. In some cases, direct fabrication increases the processing time, but the technique has significant advantages, which include flexibility, greater ease of registration, and higher resolution. The devices commonly fabricated by direct electron beam methods have smaller dimensions and, therefore, require greater positioning accuracy. The accuracy is achieved by the use of registration marks consisting of heavy metal deposits or of holes in an oxide layer. The instrument uses electrons scattered from these marks to locate accurately the position of previous patterns. This technique achieves a positioning precision of 0.1 micrometer (4 microinches).

The process used to fabricate registration marks is similar to that used to fabricate the pattern. The device is covered first with a polymeric material and then with a thin aluminum coating which is used to drain charge from the device. Irradiation increases the degree of polymerization in some polymers and it reduces it in others. Chemicals are then used to develop the pattern by removing the unpolymerized material. The remaining polymer is used as a mask to protect areas not to be covered with metal, or not to be doped. Finally, the polymer is dissolved, and any excess metal or dopant is removed. This method is known as the "lift off" technique and is shown in Fig. 4.

The polymeric material used can be either of two general types: (1) In the first, positive resist, the exposed portion is depolymerized and therefore removed by the developer; while (2) in the second type, negative resist, the unexposed area is removed. The electrical charge per unit area (coulombs per square centimeter) that is required to depolymerize or to polymerize the resist is a definition of its sensitivity. The resolution of the resist describes the minimum feature size that can be usefully exposed. Poly(methylmethacrylate), PMMA, is an example of a well-known positive resist with high resolution, but poor sensitivity. Poly(butene-1-sulfone), PBS, and Poly(glycidylmethacrylate), PGMA, are well-known negative resists with higher sensitivity, but less resolution than PMMA. Resolution is degraded by electron scattering from the resist and from the substrate. Because the area exposed by the scattered electrons is much larger than the minimum area of the electron beam, the scattering is the limiting factor in determining the minimum possible feature size.

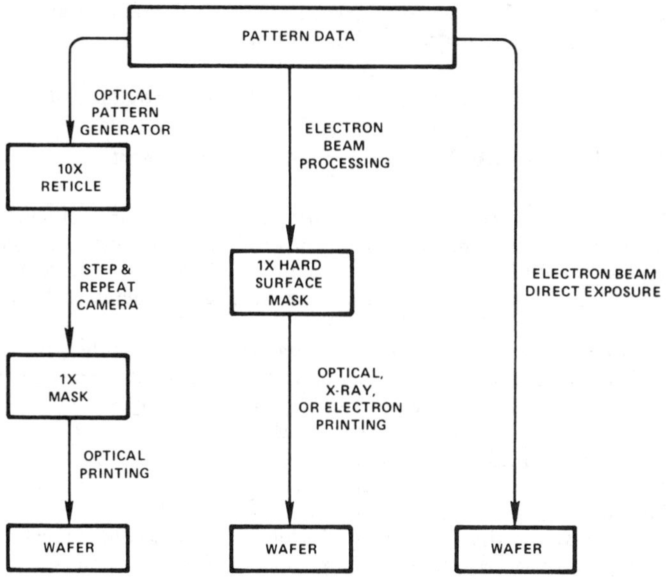

Fig. 4. Comparison of optical and electron beam fabrication.

Because of the complexity of electron beam instruments, a substantial investment in equipment is required. These instruments, however, provide increased flexibility and automated control which are extremely useful in the design and prototyping of circuits. See also **Microstructure Fabrication.**

Adapted from an article by W. D. Jensen, F. B. Gerhard, Jr., and D. M. Koffman, GTE Laboratories, Waltham, Massachusetts for *GTE Profile* magazine.

ELECTRON BEAM WELDING. Welding.

ELECTRON BINDING ENERGY. Binding Energy.

ELECTRON DIFFRACTION. Beams of high-speed electrons exhibit diffraction phenomena analogous to those obtained with light, thus showing the wave-like character of electron beams. Such patterns are useful in the interpretation of the structure of matter.

See also **Electron Microscope;** and **Davisson-Germer Experiment.**

ELECTRONEGATIVITY. This term refers to the relative tendency of an atom to acquire negative charge. This tendency is not precisely defined because no exact theoretical or experimental method of evaluation has been devised. Electronegativity exists because the electronic cloud which surrounds each atomic nucleus is inadequate to block off the nuclear charge completely at the periphery. In other words, although every complete atom is electrically neutral as observable from a distance, a small fraction of the total nuclear positive charge can be detected at any point near the surface of the atom. This "effective nuclear charge" at the surface is relatively insignificant in the absence of low energy vacancies capable of accommodating a foreign electron, as in the atoms of M 8 elements (commonly called the *inert* or *noble gases*). However, wherever a low energy electron vacancy occurs in the outer shell of an atom's electronic cloud, the effective nuclear charge as manifest within that vacancy is of major importance. Indeed, it constitutes the cause and means of chemical bond formation, and largely determines both polarity and strength of the bond. Electronegativity is a measure of the force of this effective nuclear charge within an orbital vacancy at the distance of the atomic radius.

Evaluation of relative electronegativity was first accomplished by Pauling. He considered the energy of a heteronuclear single covalent bond to consist of the geometric mean of the homonuclear single bond energies of the two elements, *supplemented* by an electrostatic or ionic energy resulting from uneven electron sharing in the bond, which he attributed to an electronegativity difference between the two elements. He used the difference between the observed bond energy and the average of the homonuclear energies as a measure of the ionic energy. From such differences for various pairs of atoms he established a relative scale in which the most electronegative element, fluorine, has the value 4.0, and within the same period are O 3.5, N 3.0, C 2.5, B 2.0, Be 1.5, and Li 1.0.

Mulliken defined electronegativity as the average of the "valence state" ionization energy and electron affinity. Gordy suggested that electronegativity is a measure of the electrostatic potential at the surface of an atom, expressed as the effective nuclear charge, $Z_{eff}e$ divided by the radius. Allred and Rochow modified this concept by considering the electrostatic force, $Z_{eff}e^2/r^2$, as the measure of electronegativity. Electronegativities have also been estimated from work functions of metals, from force constants determined by infrared spectroscopy, from nuclear magnetic resonance spectroscopy, and by other methods. When adjusted to the same arbitrary scale, conventionally that established by Pauling, these methods give values in surprisingly good agreement, with only a few minor discrepancies that remain controversial. The principal application of *Pauling scale electronegativities*, which are those almost universally given in textbooks, has been qualitative prediction of bond polarity. That is, a given bond between atoms initially differing in electronegativity is polar with a partial negative charge on the initially more electronegative atom. The degree of polarity increases with increasing electronegativity difference.

Far more valuable applications of electronegativity have been made using a different scale based on the relative compactness of the electronic clouds of atoms. Electronegativities thus derived are approximately a linear function of the square root of the Pauling scale values, and in this sense in generally good agreement. They (see accompanying table) have been used for quantitative estimation of the partial charges

RELATIVE ELECTRONEGATIVITIES OF SOME ELEMENTS
(Relative Compactness Scale)[a]

H	3.55	K	0.42	Rb	0.36	Cs	0.28
Li	0.74	Ca	1.22	Sr	1.06	Ba	0.78
Be	2.39	Zn	3.00	Cd	2.59	Hg	2.93
B	2.93	Ga	3.28	In	2.84	Tl(I)	1.89
				Sn(II)	2.31		
C	3.79	Ge	3.59	Sn(IV)	3.09	Tl(III)	3.02
N	4.49	As	3.90	Sb	3.34	Pb(II)	2.38
O	5.21	Se	4.21	Te	3.59	Pb(IV)	3.08
F	5.75	Br	4.53	I	3.84	Bi	3.16
Na	0.70						
Mg	1.56	Sc	1.30	Y	1.05	La	0.88
Al	2.22	Ti	1.40	Zr	1.10	Hf	1.05
Si	2.84	V	1.60	Nb	1.36	Ta	1.21
P	3.43	Cr	1.88	Mo	1.62	W	1.39
S	4.12	Mn	2.07	Tc	1.80	Re	1.53
Cl	4.93	Fe	2.10	Ru	1.95	Os	1.67
		Co	2.10	Rh	2.10	Ir	1.78
		Ni	2.10	Pd	2.29	Pt	1.91
		Cu	2.60	Ag	2.57	Au	2.57

[a] Values for the transitional elements are tentative estimates only.

on combined atoms, which in themselves permit correlation of a vast quantity of chemical data and interpretations of many chemical phenomena. The partial charges in turn have been applied to the quantitative calculation of bond energy. Furthermore, more recently, a simple quantitative relationship between homonuclear covalent bond energy and electronegativity has been demonstrated. Experimental homonuclear bond energy can be used to calculate electronegativity, or vice versa.

Space does not permit a detailed description of the concepts and methods mentioned here, but an example of the results obtainable may illustrate the principles involved. Silica, SiO_2, consists of silicon atoms initially of 2.84 electronegativity, tetrahedrally surrounded by oxygen atoms, initially of 5.21 electronegativity, each of which bridges two silicon atoms. The principle of electronegativity equalization states that when two or more atoms initially different in electronegativity combine, their electronegativities become equalized to the geometric mean. For SiO_2 the electronegativity of the compound is $(2.84 \times 5.21^2)^{1/3} = 4.26$. The equalization is brought about through the uneven sharing of the bonding electrons. The oxygen being initially more electronegative, attracts more than a half share of the bonding electrons. By spending more than half time more closely associated with the oxygen, the bonding electrons impart a partial negative charge on the oxygen, expanding the cloud through increased repulsions and decreasing the electronegativity. Simultaneously the silicon atoms shrink because of reduced repulsions, and increase in electronegativity because of reduced shielding between nucleus and bonding electrons. The decrease in oxygen electronegativity from 5.21 to 4.26 is 0.95. If oxygen had acquired an electron completely the electronegativity would have dropped by 4.75. The partial charge on oxygen is defined as the ratio 0.95/4.75 = −0.20 (minus because the electronegativity decreased). The silicon is left with a partial positive charge of 0.40.

The silicon-oxygen bond, like all heteronuclear bonds, can be treated *as if* its energy were partly covalent and partly ionic. Instead of ionic energy supplementing the covalent energy, as suggested by Pauling, the ionic energy *substitutes* for a part of the covalent energy. The ionic weighting coefficient is half the charge difference: (0.40 + 0.20)/2 = 0.30. The covalent weighting coefficient, 1.00 − 0.30, is 0.70. For the covalent energy one takes the geometric mean, 59.7, of the homonuclear bond energies of silicon (53.4) and oxygen (66.7, as calculated from the O_2 molecule), multiplies by 0.70, and corrects for bond length by the factor (covalent radius sum)/(observed bond length) = 1.90/1.61. This calculation gives 49.1 kcal/mole of bonds. The ionic energy is simply the conversion factor (to kcal/mole) 332, times the weighting coefficient 0.30, divided by the bond length 1.61, or 61.8 kcal/mole of bonds. The sum, 110.9 kcal, is the Si—O bond energy. Atomization of SiO_2 requires the rupture of four SiO bonds per formula unit; 4 × 110.9 = 443.6 kcal/mole for the atomization energy of SiO_2(c). Subtraction of the atomization energies 108.9 for Si and 119.2 for two O gives −215.5 kcal/mole for the calculated standard heat of formation of SiO_2(c). The experimental value is −217.7.

Electronegativity thus permits a quantitative interpretation of the bonding in SiO_2 or any other compound for which appropriate data are known. The same principles allow a superior alternative to the "ionic" model of nonmolecular solids, and offer high hope of eventually elucidating the thermochemistry of mineral substances in general.

ELECTRON EMISSION. The liberation of electrons from an electrode into the surrounding space. Quantitatively, it is the rate at which electrons are emitted from an electrode.

ELECTRONEUTRALITY. If one describes the properties of electrolytic solutions in terms of ionic species, one has to take account of the fact that the concentrations of all species are not independent because the solution as a whole is neutral.

One generally uses the symbol z_i to denote the charge on an ion measured in units of the charge of a proton (for example, for Na^+, $z = 1$; for La^{3+}, $z = 3$; for PO_4^{3-}, $z = -3$; z is also called the *charge number* of the ion.

If n_i is the number of moles of the ionic species i, the condition of electrical neutrality is

$$\sum_i z_i n_i = 0 \tag{1}$$

Alternatively if one uses the subscript $_+$ to denote positively-charged ions or *cations* and $_-$ to denote negatively charged ions or *anions*, then one may write (1) in the form

$$\sum_+ z_+ n_+ = \sum_- z_- n_- \tag{2}$$

ELECTRON GAS. The term electron gas is used to denote a system of mobile electrons, as, for example, the electrons in a metal which are free to move. In the free electron theory of metals, these electrons move through the metal in the region of nearly uniform positive potential created by the ions of the crystal lattice. This theory when modified by the Pauli exclusion principle, serves to explain many properties of metals, especially the alkali metals. For metals with more complex electronic structure, and semiconductors, the band theory of solids gives a better picture.

ELECTRON GUN. An electrode structure which produces and may control, focus, and deflect an electron beam.

ELECTRONIC A/D CONVERTERS. Analog-to-Digital Converter.

ELECTRONIC COUNTER. Frequency Measurement.

ELECTRONIC DATA PROCESSING. The application of stored-program digital computers to the processing of information. Some typical applications are business, scientific computation, data acquisition, and process control. See also **Data-Acquisition Computer; Digital Computer;** and **Scientific Computer.**

Business Applications. Electronic data processing in business is typically concerned with maintaining files of data. Examples include payroll, inventory control, customer billing and accounting. Characteristically, these applications require large amounts of storage for files and have a high incidence of input and output operations. Historically, this field has been a typical batch type environment, i.e., the data to be processed have been accumulated over a period of time and then are entered into the system at a scheduled time. Telecommunications in recent years has led to the updating of files and processing in real time, i.e., each transaction is processed as it is received from a remote location and the response is sent back to the sending station immediately.

Scientific Applications. Generally, this is concerned with the solution of complex mathematical problems using digital computers. In contrast to business applications, scientific data processing has less input and output of data, but the amount of arithmetic calculations is greater. In general, the problems being solved are not repetitive and a larger amount of machine time is used in compiling new programs than is true in the business environment. See also **Program (Computer).**

Data Acquisition Applications. EDP is typically concerned here with the recording or gathering of physical data in real time. Data acquisition applications generally involve data coming into the system from analog and/or digital input devices, while control of external devices is secondary. A partial list of data acquisition applications would include wind tunnel data collection, missile system checkout, low- and high-energy physics experiments, telemetry data acquisition, electronic component testing, and manufacturing quality control. See also **Real-time Computing.**

Process Control Applications. In this area, EDP plays the role of real time monitoring and controlling of a process by means of a digital computer. Signals from instrument and sensors are scanned and the data are converted to engineering units for either logging or further computations. The calculations commonly performed are for control and for optimization. In control calculations, the digital computer senses the output of a process with appropriate sensors and transducers and calculates the variation from the desired value. If the variation is not within acceptable limits, a new setpoint or control value is calculated and transmitted to the controllers and final control elements (valves, switches, etc.) which control the variables of the process. Optimization calculations are used to maximize the economic value of the output within given constraints. These calculations are much more complex and usually require a simulation model of the entire process which must be periodically verified against actual plant measurements. The optimization program uses economic data and physical plant data to determine optimum operating points for variables in the process.

See also classified index of **Data Processing** terms.

Thomas J. Harrison, International Business Machines Corporation, Boca Raton, Florida.

ELECTRONIC IMAGERY. Photography and Imagery.

ELECTRONIC INTEGRATOR CIRCUITS. Memory Circuit (Analog).

ELECTRONIC MAIL. The concept of mail between people and communities is attributed to the Persians (circa 400 B.C.). The concept of government control of the mail system dates back at least to 1591, as proclaimed by Queen Elizabeth of England (1591). In North America, the Massachusetts colonialists established a postal system in 1639. In modern times, authorities divide mail into several categories, as follows in diminishing order of pieces handled: (1) routine financial documents—bills, statements, and checks, constituting about 40% of total mail handled in the United States; (2) advertising pieces of many types, about 26%; (3) correspondence—personal, business, and government, about 22%; (4) magazines and newspapers, about 11%; and (5) shipment of merchandise by mail, about 1%.

In an electronic mail system, sender input equipment would comprise such items as keyboard terminals, facsimile scanners, and electronic data processors. High-volume mail receivers, such as businesses and governments would require a raster output printer. Mail to households and small businesses would continue to be handled much as it is now except for the long-distance aspects. Here the conventional postal handling service would essentially operate as a high-volume business or government receiver, but providing normal local delivery by footmen. For households and small businesses it is not envisioned that equipment costs will be reduced sufficiently during the next several years to make a total electronic mail system feasible.

As of the 1980s, the technology (except from an economic standpoint) is ready for electronic mail. Consumer acceptance also may provide a hurdle. In principle, private, nonconventional message handling has been practiced for many years—as by railroads, power utilities, and communications companies—that used their own telecommunications facilities. What is new is the extension of this concept, aided by modern electronics and telecommunications equipment, to the general users of mail, in the interest of substantially improving transit time and, later, possibly reducing mail-handling costs. Government regulatory agencies and the availability of communications channels also may be hurdles to overcome. A review of electronic mail can be found in "Electronic Mail" *Science*, **195**, 1160–1164 (1977).

ELECTRONIC MUSIC. **Musical Sound.**

ELECTRONIC RECTIFIERS. **Rectifiers.**

ELECTRONICS. An all-inclusive type of term embracing the study, design, and application of devices whose operation is dependent fully or partially upon the characteristics and behavior of electrons. Since electricity is a fundamental quantity in nature comprising electrons and protons at rest or in motion, any phenomenon or device of an electrical nature would also be covered by the umbrella term, electronics.

The word *electron* was first used in a paper by George J. Stoney in the July 1891 issue of *The Scientific Transactions of the Royal Dublin Society*, entitled, "On the Cause of Double Lines and of Equidistant Satellites in Spectra of Gases." The word *electronics* (Ger. *Elektronik*) was first used to describe the branch of physics now generally called physical electronics. That usage dates back almost to the discovery of the electron in 1897, as witness the names of two early journals in the field, *Jahrbuch der Radioaktivität und Elektronik* (founded in 1904) and *Ion: A Journal of Electronics, Atomistics, Ionology, Radioactivity and Raumchemistry* (1908). In the currently prevalent technological context, the adjective *electronic* and the noun *electronics* (Ger. *technische Elektronik*) date back only to the 1920s.

The discovery of thermionic emission and the utilization of this effect in vacuum tubes set the stage for a new field of technology which, although basically electrical in character, nevertheless differed from the traditional spheres of electrical engineering and the science of electricity and magnetism. Invention of the triode by De Forest in 1907 and the development of new uses for vacuum tubes (or valves), with their particular attributes, notably amplification and oscillation, and with their unique solutions for problems, notably in the communications field, markedly differentiated these devices from prior run-of-the mill electrical and magnetic hardware. Designers of the early 1900s were able to develop practical and important electrical equipment without benefit of formal knowledge of electron theory. In Marconi's experiments of the late 1800s, using magnetic detectors for radio waves, and his dramatic accomplishment of spanning the Atlantic with radiotelegraphy in 1901, there was a definite pointing toward what is now considered modern electronics. This was given further emphasis with the first application of the three-electrode vacuum tube to radio in 1912. The Radio Corporation of America was formed in 1919 to pursue radiotelegraphy developments for the United States, which, at that time, enjoyed cable communications with only two nations in Europe—Great Britain and France. It was at about this time that the ever increasing and tight bond between electronics and communication was formed, which continues to the present and into the foreseeable future.

The first of the "modern" digital computers, the Mark I, was completed in 1944 for the U.S. Navy by Dr. Howard H. Aiken of Harvard University, assisted by International Business Machines Corporation. This was an electromechanical machine controlled by electromagnetic relays, two-position devices developed by Bell Telephone Laboratories as switches for the telephone industry. Other Bell Labs products used in the Mark I were punched paper tape equipment teleprinters and other components developed for communications systems. In today's parlance this machine might be called "electronic," but in its day it did not meet the requirement of incorporating vacuum tubes. However, the ENIAC (Electronic Numerical Integrator and Calculator), also built during the early 1940s by Dr. John Mauchly and Dr. J. Presper Eckert at the Moore School of Electrical Engineering at the University of Pennsylvania in Philadelphia did meet the test, and it was commonly referred to as the first true electronic computer. Instead of electromagnetic relays, some 18,000 flip-flop vacuum tubes originally designed for radar and television were used. And thus the firm and everlasting association between electronics and computing was established.

The close association between electronics and the vacuum tube persisted for many years, until the development of semiconductors. In the 1941 edition of *The Encyclopedia Americana*, the entry on *Electronics* was largely confined to thermionic emission, applications of diodes, grid-controlled thermionic tubes, and triodes as detectors and amplifiers. The definition of electronics given was, "the term applied to a wide variety of applications of the electron theory to engineering practice," which, in retrospect, was an excellent definition for its time.

Specialized publications for the field of electronics commenced in the 1940s. The American Institute of Electrical Engineers and the Institute of Radio Engineers combined in 1962 to form the Institute of Electrical and Electronics Engineers. The new organization was composed of 34 specialized groups, most of them pertaining to some aspect of electronics. In a way, this merger of two professional groups recognized somewhat belatedly the existence of an electronics technology, but also perpetuated some demarcation of electronics from other aspects of electrical technology.

In an excellent, short review of electronics, Pierce (1977) suggested that "electronics has come to mean all electrical devices for communication, information processing, and control." Pierce points out that this would include pre-vacuum-tube devices such as the electric telegraph, which replaced such light-wave communication as semaphore telegraphs, signal flags, and heliographs. It would include the telephone. And, thus, according to this definition, electronics was born nearly 150 years ago with the first electric telegraphs.

In conclusion, the editors of this encyclopedia return to the definition given at the opening of this article—the same definition which was featured in the 5th Edition (1976). It appears that, as the next several decades come and pass, electronics may be the overall umbrella term used for all electrical phenomena and pursuits—because even in current times those electrical phenomena which one might unquestionably label nonelectronic are narrowing, whereas those phenomena and applications where electronics is fully appropriate are widening at a rapid pace.

References

Abelson, P. H., and A. L. Hammond: "The Electronics Revolution," *Science*, **195**, 1087–1091 (1977).

Harper, C. A.: "Handbook of Components for Electronics," McGraw-Hill, New York, 1977.

Landee, R.: "Electronics Designers' Handbook," 2nd edition, McGraw-Hill, New York, 1977.

Linvill, J. G., and C. L. Hogan: "Intellectual and Economic Fuel for the Electronics Revolution," *Science*, **195**, 1107–1113 (1977).

Pierce, J. R.: "Electronics: Past, Present, and Future," *Science*, **195**, 1092–1095 (1977).

ELECTRONICS (Automotive). Automotive Electronics.

ELECTRONICS (Thin-Film). Thin Films.

ELECTRON IMAGE TUBE. Cathode-Ray Tube.

ELECTRON LENS. An electron moving in an inhomogeneous electric or magnetic field in general follows a curved trajectory. It may be shown that the trajectories in certain fields are such that all electrons which pass through a given point subsequently pass through, or close to, a second point, whose location is fixed by the field strengths, and by the position of the first point. The paths of electrons in such a system therefore bear a striking resemblance to the light rays which pass through a lens, and the set of electrodes or of conductors which establish the necessary fields is known as an electron lens. The focal length of the lens is defined in the same way as is the focal length of an optical lens; it may be varied by changing the field strengths. As an example of an electron lens, a slit or hole in a conducting sheet acts to converge or diverge electrons if the electric fields on the two sides of the sheet differ in strength.

ELECTRON LINEAR ACCELERATOR. Particles (Subatomic).

ELECTRON MICROSCOPE. The concepts that eventually led to the development of electron microscopes came out of the discovery of the wave nature of the electron in 1924. The effective wavelength of the electrons varies with accelerating voltage and is less than 1Å: $\lambda = \sqrt{(150/V)}$ Å. This short wavelength makes possible far better resolution and higher magnification in the electron microscope as compared with the optical microscope.

The lenses used in electron microscopes act on the beams of electrons in much the same way that ordinary glass lenses act on beams of light. Most electron microscopes have electromagnetic lenses, although electrostatic lenses also can be used. Application of a uniform axial magnetic field causes the electrons to travel in a spiral path and return to the axis as shown schematically in Fig. 1. Except for the spiral

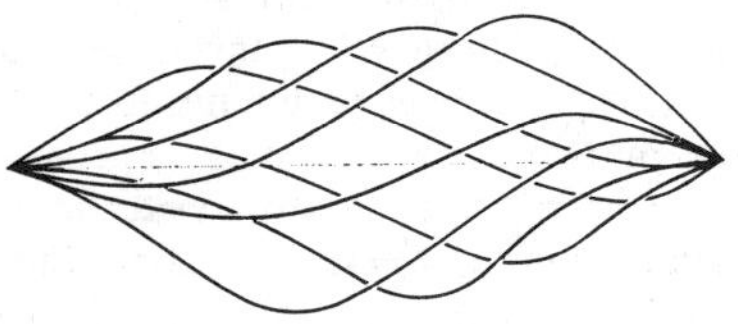

Fig. 1. Courses of electron beams in a homogeneous magnetic field.

part of the motion, this behavior is just like a simple glass lens, and the equations for optical lenses apply, as shown in Fig. 2:

$$\frac{1}{a}+\frac{1}{b}=\frac{1}{f}$$

$$\text{Magnification} = \frac{b}{a}$$

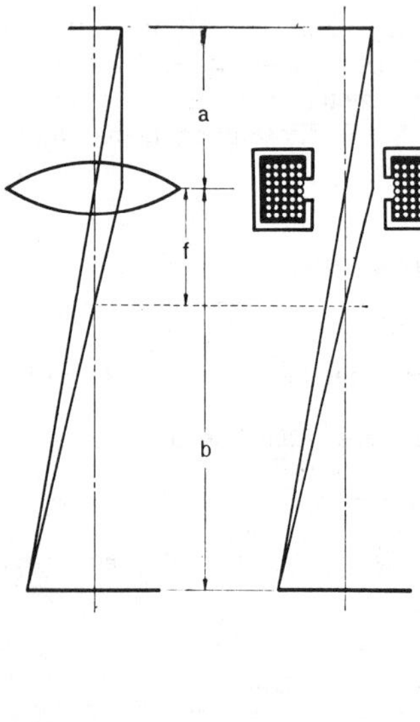

Fig. 2. Lens system used in transmission-type electron microscope. (Left) optical convex lens; (Right) electromagnetic lens.

However, in the electromagnetic lens, the angle of deflection and the focal length depend upon the strength of the magnetic field. This can be controlled by adjusting the current in the coils so that the magnification of the lens can be continuously varied over a broad range. See Fig. 3.

A simple electron microscope, as illustrated in Fig. 4, is operated by producing a beam of electrons from a heated filament, accelerating the beam with a high voltage applied to an anode, and then directing this beam of illuminating electrons onto the specimen. The specimen is in some respects similar to that used for optical microscopy, but because electrons are not so penetrating as light, the specimen must be much thinner (on the order of 1,000Å or less for most materials). For biological materials, these thin sections are produced by ultramicrotomes. Many materials, especially metals, can be thinned chemically or electrochemically. Rough surfaces can be examined by evaporating carbon on the sample and then removing it and using the carbon replica in the microscope.

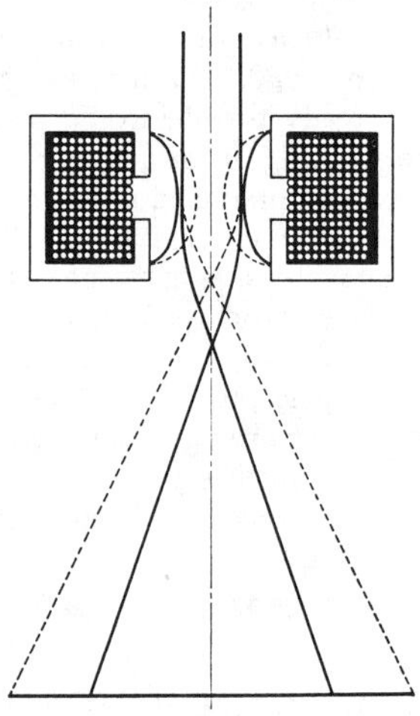

Fig. 3. Relation between magnetic field intensity and magnification in case of large magnetic field intensity (i.e., a large current flowing through the coil); or small magnetic field intensity (i.e., a small current flowing through the coil).

The beam of electrons that passes through the specimen is then magnified by an objective lens and a projector lens and finally strikes either film held in a camera, or a fluorescent screen. The image seen on the fluorescent screen has varying shades of gray that depend upon the distribution of density and thickness of the specimen. This is most

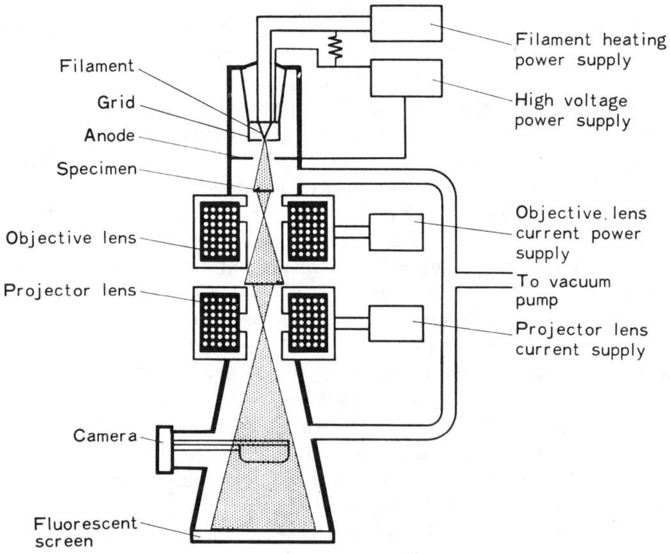

Fig. 4. Construction of a simple electron microscope.

important for biological samples. For crystalline materials, the regular arrays of atoms aligned in critical directions can act like a mirror, diffracting the beam to another direction. This makes it possible to observe imperfections in the atomic arrangement of a material.

More complex electron microscopes use additional lenses, both above and below the specimen. The condenser lenses above the specimen concentrate the electron beam and increase the illumination. The addition of intermediate lenses below the specimen make it possible to go to higher magnification in the final image. Various alignment controls, apertures for the lenses, specimen handling devices, and suitable airlocks and anticontamination traps also are provided.

The central column of the instrument must be maintained as a vacuum because the electrons would be absorbed if any atmospheric gases were present. Typical resolutions obtainable by commercial instruments are on the order of 2 to 5Å. Accelerating voltages from 20,000 to 1 million volts have been used. The higher accelerating voltages are useful for penetrating thick specimens (in some cases, up to 1 micrometer or more).

Scanning-Type Electron Microscope. This type of electron microscope is completely different in principle and application from the conventional transmission-type electron microscope. In the scanning instrument, the surface of a solid sample is bombarded with a fine probe of electrons, generally less than 100Å in diameter. The sample emits secondary electrons that are generated by the action of the primary beam. These secondary electrons are collected and amplified by the instrument. Since the beam strikes only one point on the sample at a time, the beam must be scanned over the sample surface in a raster pattern to generate a picture of the surface sample. The picture is displayed on a cathode ray tube from which it can be photographed.

A block diagram of a scanning-type electron microscope is given in Fig. 5. Major elements of the instrument include the electromagnetic lenses that are used to form the electron probe, the scan coils that sweep the beam over the sample, the detector that collects the secondary electrons, and the amplifying means where the secondary electrons are amplified and fed to the cathode ray tube for display. Since the cathode ray tube is scanned in synchronization with the electron beam, the resulting picture corresponds to the area of the sample being examined.

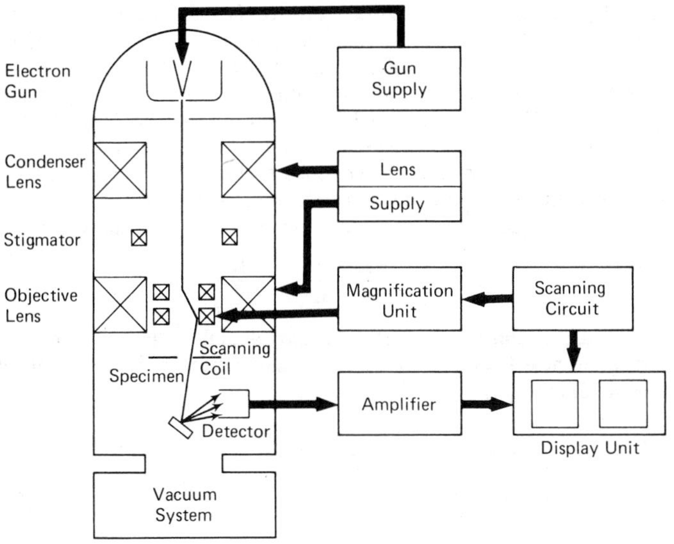

Fig. 5. Simplified diagram of scanning-type electron microscope.

The advantages of the scanning microscope as compared with conventional optical microscopes include superior resolution and depth of field. Resolution of 200Å is obtained with a depth of field several hundred times that of a conventional optical microscope. The scanning microscope also makes possible the display of other kinds of data obtained from the specimen, notably cathodoluminescent photons emitted by fluorescing samples, electrical voltages generated in semiconducting samples by the passage of the electron probe, and characteristic x-rays that can be used to determine sample composition. Also see **X-Ray Analysis.**

The range of magnification of scanning electron microscopes generally is from less than 30 × to about 40,000 ×, limited by the resolution of the instrument. Most specimens can be examined without any special preparation, but nonconducting specimens usually are coated with 100Å to 500Å of metal to conduct away the beam current. A conventional diffusion pump vacuum system is used since a vacuum is required for operation of the electron beam. The image that is formed is easily interpreted as surface topography inasmuch as the illuminating and shadowing effects on the sample are similar to the appearance of large objects as they normally are seen by the unaided eye. The scanning electron microscope has found broad application in the transistor industry to show voltage distributions in such devices. Other uses include industrial quality control and a broad range of industrial and biological research.

ELECTRON MULTIPLICATION. On bombarding certain surfaces by electrons it may happen that each impinging electron expels several electrons from the struck surface. If these electrons are caught in an electric field and driven against another similar surface, each of them may again give rise to several electrons. After several such stages of multiplication, an appreciable pulse may be obtained in this manner, starting with a single electron as beta radiation, or produced photoelectrically by gamma radiation. Quantitatively, the electron multiplication of a device such as a photomultiplier is the ratio of the number of electrons reaching the anode to the number emitted at the cathode.

An electron multiplier is a device for amplifying by a process of electron multiplication. In the accompanying figure, electrons are emitted from the cathode and accelerated to an electrode called a dynode, maintained at about 100 volts positive with respect to the cathode. Secondary electrons knocked out of the dynode are accelerated to a second dynode at a higher potential, and there still produces more secondaries. The cathode, nine dynodes and final anode used in typical multiplier cells are so placed and shaped that the electron beams are efficiently focused at each stage. In typical multiplier cells, the gain per dynode is between four and five. The total gain is $(4)^9$ to $(5)^9$ or about 10^6. In addition to their use in photocells, multipliers are used in radiation detectors.

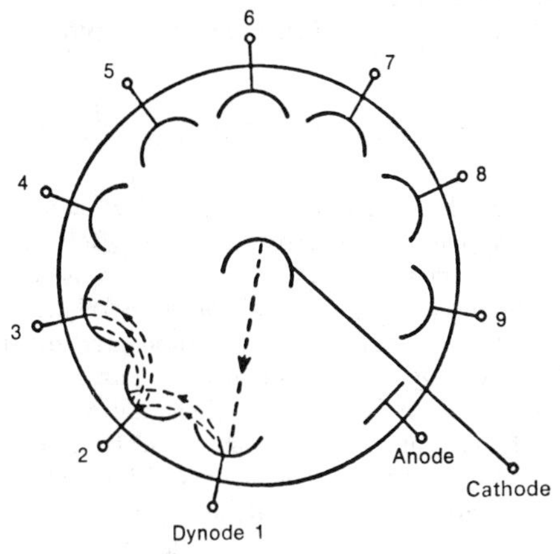

Electron multiplier.

ELECTRON OPTICS. The control of the movement of free electrons by means of electric fields, and their utilization in research investigation of electronic diffraction phenomena in direct analogy to the effect of lenses on light. Electron beams are used to determine the microstructure of crystals and many other materials.

The invention of wave mechanics in 1926 by Heisenberg and Schrödinger introduced a new era in physics. It became apparent that the ideas of classical dynamics could be formally replaced when a stream of particles was considered. According to the well-known De Broglie hypothesis, a wavelength λ can be assigned to any material particle such that $\lambda = h/mv$, where h is Planck's constant, m is the particle mass, and v is the particle velocity.

As one of the many consequences of these ideas, electron optics emerged. Busch demonstrated that the action of a short axially symmetric magnetic field on a beam of electrons was similar to that of a glass lens on light. This analogy has proved useful in the study of

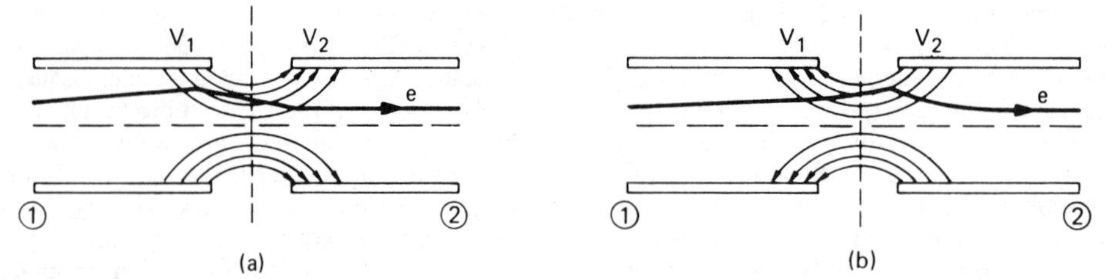

Fig. 1. Electrostatic focusing of electron beam by cylinder lenses 1 and 2. (a) Voltage $V_2 > V_1$; (b) voltage $V_2 < V_1$.

the behavior of electrons in vacuum tubes (valves), magnetrons and klystrons, traveling wave tubes, cathode-ray tubes, and electron microscopes, among other very useful devices. The original concept of the electron microscope was evolved by direct analogy with the light microscope.

An electron which has fallen through a potential V has a kinetic energy $\Gamma = \frac{1}{2}mv^2 = eV$, where e is the electronic charge. Hence $v = (2eV)^{1/2}/m^{1/2}$ and, using the previous equation, the useful formula for expressing electron wavelength in terms of volts becomes:

$$\lambda = \left(\frac{150}{V}\right)^{\frac{1}{2}}$$

where λ is in angstroms.

Electrostatic and magnetic fields control the motion of an electron according to

$$\mathbf{F} = m\frac{d\mathbf{v}}{dt} = e(\mathbf{E} + \mathbf{v} \times \mathbf{B})$$

where $\mathbf{F}$ = force; $\mathbf{E}$ = electrostatic field; $\mathbf{B}$ = magnetic field; $\mathbf{V}$ = electron velocity; and e = electronic charge. (Boldface variables are vectors.) The electron must travel in a vacuum if it is not to be scattered and lose its kinetic energy by collision with relatively massive gas molecules. Nuclear particles, such as protons and neutrons, are each about 1837 times larger in mass than the electron.

Electrostatic Electron Lenses. If in the equation just given, $\mathbf{B} = 0$, then $\mathbf{F} = e\mathbf{E}$. Thus, from the equation, an electron in a field $\mathbf{E}$ experiences a force $\mathbf{F}$ in the direction of the field. Thus in the system shown in Fig. 1 (a), where there is a voltage V_1 on cylinder 1 and V_2 on cylinder 2, and where $V_2 > V_1$, the field and path of an electron are shown. The electron moving from left to right increases its velocity and is deflected toward the axis. After passing the median, the force is away from the axis, but since velocity is increased, the electron spends less time in this part of the field. Therefore, deflection away from the axis is less than it was toward it, and there is a net convergence. If $V_2 < V_1$, as shown in Fig. 1(b), then for an electron beam traveling from left to right the lens will still be convergent because maximum deflection will occur after the electrons have slowed down. All such lenses are convergent for $\mathbf{E} = 0$ on both sides of the lens.

Magnetic Electron Lenses. If, in the foregoing equation, $\mathbf{E} = 0$, then $\mathbf{F} = e\mathbf{v} \times \mathbf{B}$. Then, from the equation, the force $\mathbf{F} = e|\mathbf{v}||\mathbf{B}| \sin \theta\, \boldsymbol{\epsilon}$, where $|\mathbf{v}|$ and $|\mathbf{B}|$ are the numerical values of the vectors, θ is the angle between them, and $\boldsymbol{\epsilon}$ is a unit vector perpendicular to both, indicating the direction of the force. For $\mathbf{F}$ to be greater than 0, $\mathbf{v}$ must be greater than 0. The force constrains the electron to move in a circle of radius $\rho = (\mathbf{v} \sin \theta)/(\mathbf{B}\ e/m)$. At the same time, it moves in a perpendicular direction with velocity $\mathbf{v} \cos \theta$, and therefore, it traces the path of a helix and returns to the axis in a time $T = 2\pi/(\mathbf{B}\ e/m)$ which is independent of both $\mathbf{v}$ and θ. The net effect is that all electrons are focused by $\mathbf{B}$ to produce an image of the source from which they diverge.

The simple magnetic lens consists of a short coil of wire contained in a surrounding shield of magnetic material. A small gap in the case material concentrates the escaping field when the coil is energized. The use of magnetic lenses is shown in entry on **Electron Microscope.** The coils are wound in opposite senses to cancel the spiral distortion of the electron beam which is introduced by the individual lenses.

Electrostatic Deflection. With reference to Fig. 2(a), a voltage V_d exists between plates x_1 and x_2, and an electron enters along the axis with a constant velocity $v = (2\ Ve/m)^{1/2}$; then, because of the field $\mathbf{E}_y$ caused by V_d, a transverse force $\mathbf{F}_y$ causes a motion in this direction to be impinged on the electron, i.e., $m(d^2y/dt^2) = V_d e/d$, from which $dy/dt = (V_d/d)(e/m)t$ and $y = \frac{1}{2}(V_d/d)(e/m)t^2$. The force acts for time $t = \ell/v$; thus deflection $y = \frac{1}{2}(V_d/d)(e/m)\ (\ell/v)^2 = \frac{1}{4}(V_d/d)(\ell^2/V)$. The additional deflection, after $\mathbf{F}$ has ceased to act on the electron, is $y^1 = (V_d/V)(\ell L/2d)$.

Magnetic Deflection. With reference to Fig. 2(b), if B is a sharply defined field, then the deflection on leaving the field is given by $y = \ell^2/2\rho$. Total deflection $D = L \tan \alpha + y$. To a good approximation,

$$D = \frac{\mathbf{B}L\ell}{v} \cdot \frac{e}{m}\left(1 + \frac{1}{2L}\right).$$

The use of controlled electron beams is making possible great advances in the field of microelectronics manufacturing. See also **Electron Beam Lithography.**

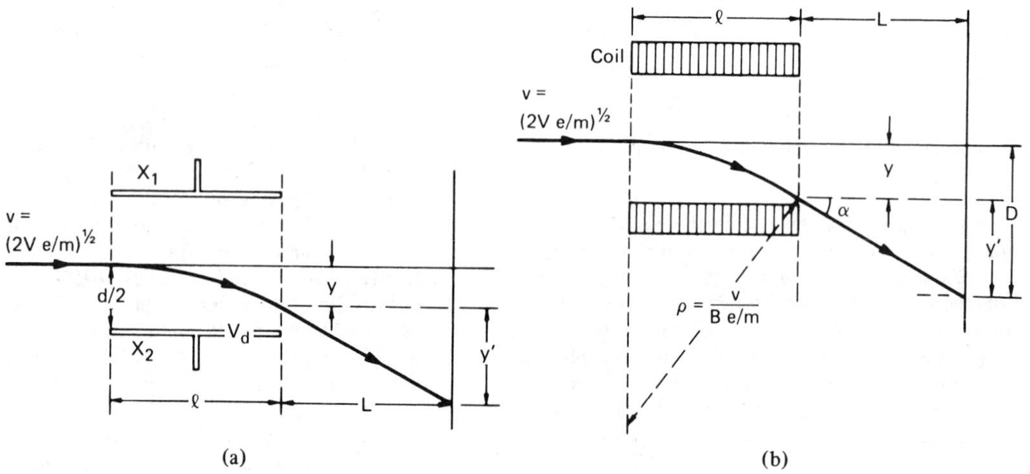

Fig. 2. (a) Deflection of electron beam by electric field between parallel plates X_1X_2. (b) deflection of electron beam by magnetic field. Electron path is an arc of a circle only while it is influenced by field. After leaving the field, the path is linear.

ELECTRON SPIN RESONANCE SPECTROSCOPY. Magnetic Resonance Spectroscopy.

ELECTRON STRUCTURE (Elements). Chemical Elements.

ELECTRON SYNCHROTRON. Particles (Subatomic).

ELECTRON THEORY. In the 1830s, Faraday had tentatively suggested that his experiments in electrochemistry could be interpreted in terms of a small unit of charge attached to ions. This notion of individual "atoms of charge" was somewhat eclipsed, however, by the enormous success of Maxwell's theory of electromagnetism, which was generally interpreted, by 1880, as favoring a view that electrical phenomena were due to continuous charge distributions and motions. G. Johnstone Stoney, in 1874, and Helmholtz, in 1881, had suggested again an atomic interpretation of electricity, but it was not until the brilliant experiments of Perrin, J. J. Thomson, Zeeman, and others in the 1890s that the concept of the electron received firm experimental foundation. Later experiments and theory (Millikan, Bohr, etc.) established the constancy of the electronic charge and interwove the concept of an electron of definite charge and mass into the basic structure of the atom.

The Cathode Ray Controversy. After the discovery of the cathode ray in high-vacuum discharge tubes by Plücker in 1858, there developed, with the experiments of Goldstein, Crookes, Hertz, Lenard, and Schuster, a controversy over the nature of the rays. A predominately German school held that the rays were a peculiar form of electromagnetic rays. The British physicists thought they were negatively charged particles. The controversy provides a classic "case history" of the typical scientific controversy in which two quite different models both explain most, but not all, of the observable facts. For resolution of this controversy, see **Cathode Ray.**

Thomson's Determination of e/m. In 1897 Thomson devised an apparatus in which he could deflect a beam of cathode rays with a magnetic field of induction B and also with an electric field of strength E. If the fields are perpendicular to each other, and to the original path of the beam, and if they occupy the same region, then (with proper polarities and magnitudes of fields) the electric force on the beam can equal the magnetic force, so that the beam hits the same point on a fluorescent screen as when no fields are applied. If e is the charge of a given particle, m its mass, and v its velocity, $v = E/B$. Thus, velocities of typical cathode ray beams could be measured. If the magnetic field is used alone, and the curvature R of the beam is measured, then one can equate centripetal and magnetic field forces $mv^2/R = Bev$, and then deduce $e/m = v/BR$. With v known from the previous experiment, e/m can be calculated. Thomson's early values were not very precise, but later experiments of a similar type gave values close to 1.76×10^{11} coulombs/kg. More recent experiments, drawing on measurements of many kinds, give $e/m = (1.75890 \pm 0.00002) \times 10^{11}$ coulombs/kg.

The Zeeman Effect. In 1896 Zeeman discovered the broadening of spectral lines when a light source was in a strong magnetic field. Experimental refinements by Zeeman and others, and theoretical work by Lorentz and Zeeman, permitted the interpretation of this effect as due to the influence of the magnetic field on oscillating or orbiting negatively charged particles within the light-emitting or absorbing atoms. From the spectroscopic data, the ratio of charge to mass of these hypothetical particles could be shown to be equal to that of cathode rays. The Zeeman effect thus provided the first experimental evidence that the negative particles emitted by atoms when heated (Edison effect) or subject to high fields and/or ionic bombardment (cathode rays) or bombarded by short-wavelength light (photoelectric effect) were, indeed, actual constituents of the atoms and were probably responsible for the emission and absorption of light.

The Charge on the Electron. In the decade following 1897, many different methods were evolved for determination of ionic charges. Some methods depended upon measuring the total charge of a number of ions used as nuclei for cloud droplet formation. Other methods were more indirect—experiments, for example, which, combined with the kinetic theory of gases, could give crude values for Avogadro's number, N. By dividing the Faraday constant (the charge carried in electrolysis by ions formed from one gram-atom of a univalent element) by N, one could determine the average charge per ion. Similarly, the constants in Planck's theory of blackbody radiation, when evaluated experimentally, could provide a numerical value for N, as could certain experiments in radioactivity. All such methods gave values of N of the order of 6×10^{23}, and hence 1.6×10^{-19} coulomb for the ionic charge. None of these methods measured individual charges; strictly speaking, the value for the ionic charge could be thought of only as an average value.

Millikan's experiments with single oil drops, beginning in 1906, provided a method for measuring extremely small charges with precision. He was able to show that the charge on his drops was *always* ne, with $e = 1.60 \times 10^{-19}$ coulomb (modern value) and n a positive or negative integer.

He observed the motions of very small charged oil drops in uniform vertical electric fields. The drops were so small that they moved with constant velocity (except for Brownian fluctuations) for a given force. The force in each case was due to gravity acting on the mass of the drop and to the electric field (if any) acting on the charge, q, on the drop. The charge on a given drop could be changed by shining x-rays upon it. Using Stokes' law, in a form modified to correct for the fact that the drops were *not* large in comparison to the inhomogeneities of the surrounding air, and the velocity of a drop in free (gravitational) fall, Millikan could infer the diameter and mass of a given drop, and then calculate its charge. The charge q always equaled ne. (See Reference 1 or 2 for experimental details.) A few other physicists, in similar experiments, thought they had detected electric charges smaller than Millikan's e, but their experimental techniques were probably faulty.

Millikan's experiment did not prove, of course, that the charge on the cathode ray, beta ray, photoelectric, or Zeeman particle was e. But if we call all such particles electrons, and assume that they have $e/m = 1.76 \times 10^{11}$ coulombs/kg, and $e = 1.60 \times 10^{-19}$ coulomb (and hence $m = 9.1 \times 10^{-31}$ kg), we find that they fit very well into Bohr's theory of the hydrogen atom and successive, more comprehensive atomic theories, into Richardson's equations for thermionic emission, into Fermi's theory of beta decay, and so on. In other words, a whole web of modern theory and experiment defines the electron. (The best current value of $e = (1.60206 \pm 0.00003) \times 10^{-19}$ coulomb.[4]

The Wave Nature of the Electron. In 1924, L. DeBroglie suggested that the behavior of electrons within atoms could be better understood if it were assumed that the motion of an electron depends upon some sort of accompanying wave, the length of which would be h/p (h = Planck's constant and p the momentum of the electron). This suggestion led to the development of quantum mechanics by Schrödinger, Heisenberg, and others. The concept of electron waves provided an explanation for experiments on reflection of electron beams by metallic crystals, carried out from 1921 onward by Davisson and others, and provided an impetus for the experiments of G. P. Thomson on the diffraction of electron beams by thin films.

Other Characteristics of Electrons. In applying quantum mechanics to certain problems in atomic spectroscopy, in 1925 and 1926, Pauli, and Goudsmit and Uhlenbeck found that electrons must possess angular momentum of amount $\pm\frac{1}{2}(h/2\pi)$. Dirac's work on a generalized quantum theory of the electron showed that it possessed a related magnetic dipole moment of magnitude $eh/4\pi mc$. The ratio of the dipole moment to the angular momentum (e/mc) is larger than can be accounted for in classical terms with any homogeneous wholly negative model. The concept of electronic dipole magnetic moment is essential not only in spectroscopy but in theories of ferromagnetism (see **Magnetism**).

One may speak of the "classical radius of the electron," $a = e^2/mc^2$, derived by setting the self-energy of the coulomb field of a charge e contained at a radius a equal to the relativistic rest energy, mc^2 of the electron. This $a = 2.82 \times 10^{-13}$ cm, comfortably smaller than any atom, but larger than the usual estimates of sizes of protons and neutrons.

Positive Electrons. Dirac's paper in 1928 could be interpreted as predicting the existence of electrons that are positive. But until such particles were found experimentally by C. D. Anderson in 1932 in cloud chamber pictures of cosmic ray particle tracks, most physicists preferred other interpretations of Dirac's paper. Positive electrons, or positrons are now known (1) to occur as decay products from certain radioactive isotopes, (2) to be produced (paired with a negative electron) in certain interactions of high-energy gamma rays with intense electric fields near nuclei, and (3) to be the product of certain decays of certain mesons. in principle, positrons could form anti-atoms with nuclei made from anti-protons and anti-neutrons, but in practice almost all positrons produced in the observable universe quickly meet their end by annihilating themselves together with some hapless negative electron. The end product of a positron-electron annihilation is a pair of gamma rays.

ELECTRON TUBE. A device in which electrons are freed from the restraints of a solid conductor, pass across a free space (vacuum or gas at low pressure) and are again collected by a solid conductor, but during this passage in free space are controllable in manners which would be impossible if they had not been temporarily freed. Also known as valves (British), electron tubes were, until the perfection of semiconductor devices in the late 1940s and 1950s, the major components of nearly all electronic circuits and equipment. Although electron tubes continue to be used for certain applications, particularly involving specially-designed tubes, a massive replacement or substitution of transistors and other solid-state devices and approaches for electron tubes has taken place. Whereas early in the period of conversion from tubes to transistors, one would assume that a piece of electronic equipment incorporated electron tube circuits unless otherwise denoted, the assumption now is that electronic equipment circuitry will be solid-state unless otherwise specified.

The remarkable circuit and packing densities now obtainable in microelectronic devices (see **Microelectronics**) have further deemphasized the electron tube. Nevertheless the technology built around electron tubes is classical and merits continued coverage in fundamental scientific literature.

The "revolution" in electronics required a number of years to achieve, recalling that the point-contact transistor was invented by J. Bardeen and W. H. Brattain (Bell Telephone Laboratories) as early as 1948. The earlier semiconductors were quite costly to produce and their mass production under very closely-controlled conditions was difficult to achieve. Good yield and quality control continue to be major problems among semiconductor manufacturers.

An electron tube consists of a heater for kinetic energy excitation of electrons, a cathode which acts as a transfer electrode source of electrons, controlling grid electrodes, and an anode that is maintained electrically positive with respect to the cathode. See Fig. 1. These elements are insulated from each other and enclosed within an evacuated envelope made of either glass, metal, ceramic, or a combination

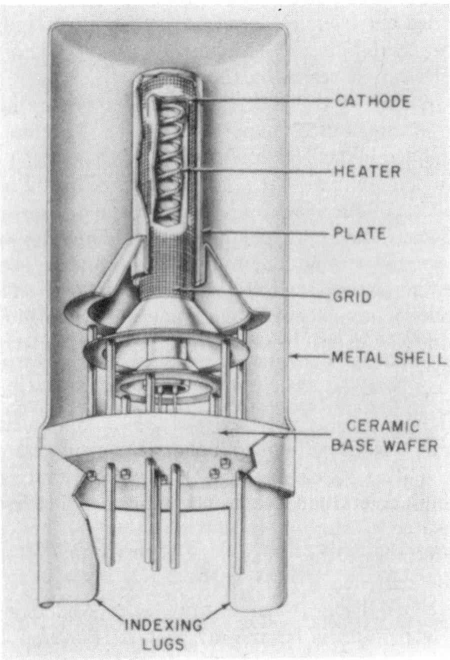

Fig. 1. Principal elements of a triode.

of these materials. A getter is flashed within the tube to absorb any residual gas molecules which could have a harmful effect, electrically and chemically, on the operation of the tube.

When the device has only two electrodes (a cathode and an anode), it is called a *diode.* With the anode maintained electrically positive with respect to the cathode, an electric field results which causes the electrons to move toward the anode. In the external circuit, the electrons flow from the anode through the load impedance and then through the voltage source to the cathode, which acts as a low-work-function transfer medium, and so back to the anode. The work function can be considered to be the total amount of work necessary to free an electron from a solid.

Other electrodes are introduced in some designs between the cathode and the anode in the form of grids. By varying the voltages on these intervening electrodes, it is possible to modify the electric field between the cathode and the anode, and thus to control the current in the external circuit. Tubes having one grid in addition to the cathode and anode are called *triodes.* Tubes with two grids are called *tetrodes*; and tubes with three grids are called *pentodes.* Generally, tubes are labeled in accordance with the total number of active electrodes in a linear arrangement using a common electron stream. Sometimes, two or more sections are enclosed within the same envelope (e.g., a diode–triode or a triode–pentode); these tubes are not referred to in terms of the total multielectrode structure, but they are designated in terms of the respective tube units.

Oxide-coated *filamentary cathodes* can operate at relatively low temperatures of 1,000K because of the low-work-function surface layer. However, they are subject to sputter effects and can also evaporate substantial amounts of material that can deposit onto adjoining electrodes and lead to harmful grid-emission and contact-potential phenomena. Such oxide-coated filaments are high-efficiency emitters at low-wattage inputs and are suited in low-current pulsed high-voltage rectifiers (as for example scanning systems in television receivers).

The indirectly-heated cathode consists of a nickel alloy sleeve coated with alkali earth oxides of barium and strontium, and, inside the sleeve, a heater of alumina-coated tungsten or molybdenum–tungsten alloy wire. The heater wire is in the form of a helical coil or folded strands; it is coated with alumina to insulate the heater wire from the cathode nickel sleeve. In addition, this insulating coating prevents adjoining helix turns or strands of wire from short-circuiting each other. The cathode sleeve is heated by conduction and radiation from the heater. Because the oxide-coated cathode is electrically isolated from the heater, it is called a unipotential cathode, since unlike the filamentary type, there is no voltage drop along its length due to heater current. The indirectly-heated cathode system has two major advantages: (1) the nickel-alloy sleeve acts as a magnetic shield around the heater wire and minimizes the effects of 60-cycle hum from the ac heater supply; and (2) the use of separate heater and cathode circuits permits the design of close-spaced control grids and cathodes which, in turn, result in higher gain amplifier service. Close cathode-to-anode spacing design in rectifier service creates low tube-voltage drops and improved voltage regulation. Thus, electron tube manufacturers have shown preference for the indirectly-heated cathode system.

Amplification Factor. In the conventional tube, the current is controlled both by the voltage applied to the grid and that applied to the plate. Because the grid is closer to the cathode from which the electrons are drawn, a voltage applied to it is more effective in affecting the electron stream than would be the same voltage applied to the plate (it is the passage of these electrons across the cathode-anode space which constitutes the tube current). Amplification factor is a measure of the relative effectiveness of voltages on the two electrodes and is defined as the negative of the ratio of the infinitesimal plate voltage change necessary to counteract a given infinitesimal change in grid voltage in order to keep the plate current constant:

$$\mu = -\left.\frac{\partial e_b}{\partial e_c}\right]_{i_b \text{ constant}}$$

where e_b is the plate voltage, i_b is the plate current, and e_c is the grid voltage.

In multielectrode vacuum tubes an amplification factor may be defined for any two electrodes in a manner similar to that done for the plate and grid of the triode.

Diodes. As previously mentioned, the diode is the simplest form of vacuum tube having a cathode system and anode together with a flashed getter. In well-degassed tube structures, the reducing element content of the cathode nickel alloy creates sufficient active barium sites in the emission oxide matrix so that the unit itself can function as a getter at high operating temperatures. Diodes are used as rectifiers, detectors, dampers, and limiters. These tubes are high-vacuum types in which the internal voltage drop is proportional to the dc load current. When a low constant-voltage drop is desired, mercury vapor tubes are used. The constant-voltage drop of 15 volts is a function of the ionization potential of the mercury vapor since the positively charged mercury ions neutralize the space charge effect. In general, for diodes, when the current demand is less than the temperature-limited value, the current I will vary as the three-halves power of the applied voltage E as given by the following equations. For the plane parallel system: $I = (2.33 \times 10^{-6}E^{3/2})/S^2$ where S is the cathode-to-anode spacing. For the concentric cylindrical system: $I = (14.65 \times †]^{-6}E^{3/2})/(b^2 \times r_a)$ where b is the ratio of the anode radius r_a to the cathode radius r_c. For very low anode voltages, the above approximations are not accurate because the effects of the initial electron velocity and the contact potential result in currents larger than calculated.

Not all electrons leaving the cathode reach the anode. Some electrons return to the cathode by reason of a balance between the initial electron velocity and the image force; i.e., when an electron is emitted from a metal surface and is at some distance from the surface, it induces a charge of equal magnitude but opposite sign in the interior of the metal. As a result, some electrons remain in the space above the cathode and produce a cloud of electrons or "space charge" which repels some electrons back to the cathode and thus impedes the flow of electrons to the anode. The extent of this action and the amount of the space charge are dependent upon the temperature of the cathode, the cathode-to-anode spacing, and the anode potential.

Under fixed temperature conditions, the maximum number of electrons that are emitted is also fixed. The higher the anode potential, however, the lower is the number of electrons remaining in the space charge region; as a result, an increase of the anode voltage results in an increase in current until saturation is reached. Beyond this condition (maximum current at maximum voltage at fixed cathode temperature conditions), additional anode (plate) voltage will only increase the plate current slightly because of the reduction of the work function at the cathode due to the electrostatic effect of the applied field.

The Richardson-Dushman equation yields information pertaining to the work function ϕ of cathode systems and the zero field saturated current I_{s0} with respect to temperature T. The emission equation for a cathode of surface area S has the form $I_{s0} = A_0ST^2e^{-(11610\phi_0/T)}$. The zero field saturated current I_{s0} is extrapolated from a plot of the log I_s (saturated current) as a function of the square root of the applied voltage at the anode V_a. The work function with an external field E is a function of the square root of the external field strength such that $\phi_E = \phi_0 - 3.78 \times 10^{-5}E^{1/2}$.

Triode. When a third electrode, called a grid, is placed between the cathode and the anode, the tube is called a triode. This grid consists of a fine wire wound on two support side rods; the spacing between the turns of the wire is relatively large so that electrons are able to pass from the cathode to the anode. When the tube is used as an amplifier, a varying signal imposed on a negative dc voltage on the grid controls the plate current. As the grid voltage becomes more negative, the plate current decreases while the plate voltage increases back to the original applied potential as a result of the decrease in the voltage drop across the load impedance. In other words, the grid voltage and the plate current are in phase, but the plate voltage is 180° out of phase with the grid voltage. The cathode, grid, and plate of the triode form an *electrostatic system* such that it is possible to equate the plate current I_b to a three-halves power law similar to that for diodes, i.e.,

$$I_b = k\left[\frac{C_{gk}}{C_{pk}} \times E_g + E_b\right]^{3/2}$$

where K is a constant (the perveance) which depends upon the geometry of the tube, and C_{gk} and C_{pk} are the grid-to-cathode and the plate-to-cathode capacitance, respectively. This equation assumes a negligible effect for the initial velocity of electrons from the cathode. Actually, the initial electron velocity creates the space-charge effect so that the cathode is slightly negative (instead of zero) and a potential minimum exists a short distance in front of the cathode. The effect of this shift is to increase the plate current slightly. Because the initial electron velocity distribution depends upon cathode temperature, the plate current will also be influenced by the cathode temperature even though it is limited by the space charge.

The electrical characteristics of triodes are described in terms of three parameters: the amplification factor (μ), the dynamic plate resistance (r_p), and the transconductance (g_m) as given by the following relationships:

$$\mu = -\frac{C_{gk}}{C_{pk}} = -\frac{\partial e_b}{\partial e_g} = -\left.\frac{de_b}{de_g}\right|_{i_b \text{ constant}}$$

$$r_p = \frac{\partial e_b}{\partial i_b} = \left.\frac{de_b}{di_b}\right|_{e_g \text{ constant}}$$

$$g_m = \frac{\partial i_b}{\partial e_b} = \left.\frac{di_b}{de_b}\right|_{e_b \text{ constant}}$$

These relationships are applicable within the region where the family of curves is straight, parallel, and equidistant for equal increments in the parameter. The triode tube can be considered a linear circuit element within small variations, and the quantities μ, r_p, g_m may be used as constants in the analysis of tube performance. A useful approximation for the calculation of plate current I_b is the expression

$$I_b \doteq \frac{2.33 \times 10^{-6}(E_g + E_b/\mu)^{3/2}}{(S_{gk})^2}$$

or

$$I_b \doteq \frac{2.33 \times 10^{-6}\left(E_g + \frac{E_b}{\mu}\right)^{3/2}}{(S_{gp})^2/\mu}$$

where S is the distance between grid-to-cathode and grid-to-plate, respectively and

$$\mu = \frac{g_m E_g}{3/2I_b - g_m E_g}$$

Tetrodes. The effects of interelectrode capacitances between the grid and the plate and between the grid and the cathode sometimes result in coupling between the input and output circuits. Such coupling and impedance mismatching can cause instability and low output performance. The grid-to-plate capacitance can be sufficiently reduced by the introduction of an additional grid electrode, called the screen grid, between the control grid and the plate, and such tubes are known as tetrodes. In practice, the control grid-to-plate capacitance is reduced from several picofarads for a triode to less than 0.01 pF for a screen grid tube when a rf capacitor is connected between the screen grid circuit and the cathode.

The screen grid operates at a positive voltage and supplies an electrostatic force which pulls electrons from the space charge region, through its widely spaced turns of wire, and onto the plate. The plate is shielded from the cathode and exerts insignificant force on the space charge region. The plate current in a tetrode is almost independent of the plate voltage and depends mainly upon the screen grid voltage. With tetrodes, moderately high amplification can be obtained without capacitive feedback from the plate to the control grid. However, for operation in amplifiers where linearity is required between the control grid voltage and the plate current, the transconductance must be constant. To achieve this condition, the tetrode must be operated in the region where the plate voltage is higher than the screen voltage, i.e., where the family of curves is nearly straight, parallel, and equidistant for equal increments.

Tetrodes are not useful for amplification of very large signals because the proximity of the positive screen grid to the plate (at equal or slightly higher positive voltage) permits the capture of secondary electrons emitted from the plate. The capture of the lower-velocity secondary electrons by the screen grid is more pronounced when the plate voltage swings lower than the screen grid voltage. This condition occurs when large signal variations cause an increased voltage drop across the load impedance in the plate circuit.

Pentodes. The effects of secondary emission from the plate in a tetrode configuration are minimized when a fifth electrode is used. See Fig. 2. This fifth electrode (a third grid) is made of widely spaced

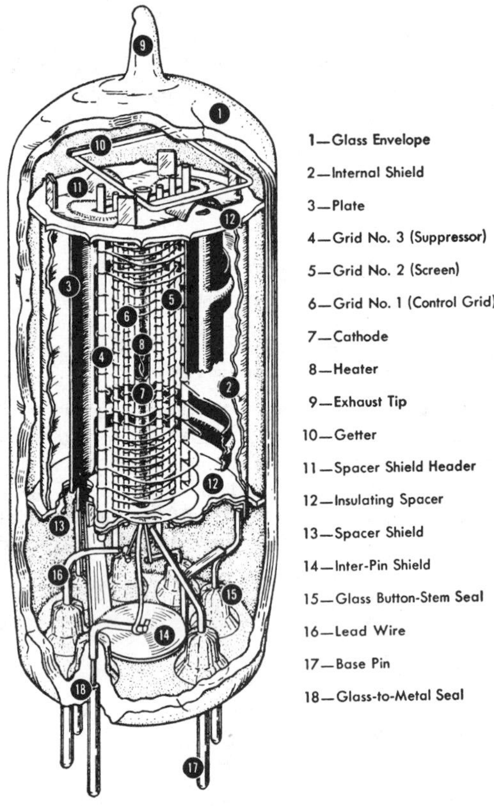

Fig. 2. Principal elements of a pentode.

turns of wire between the screen grid and the plate, and is known as the *suppressor grid*; it is usually connected to the cathode at zero potential. A tube with five electrodes in a linear arrangement, a cathode, three grids, and a plate, is called a pentode. The suppressor grid is negative with respect to the plate and can divert secondary electrons of low velocity back to the plate. Pentodes are capable of high voltage amplification and high power output at low levels of driving voltage on the control grid. When large signal inputs are placed on the control grid, large load impedance voltage drops result in the plate circuit which cause the plate voltage to become momentarily lower than the screen grid voltage without the loss of secondary electrons from the plate.

Instead of a suppressor grid, the interelectrode region between the screen grid and the plate can be virtually used as a space charge region to minimize the loss of secondary electrons from the plate. In practice, a suppressor grid or a beam-confining plate is used to enhance the effect of the electron cloud density. This effect is created primarily by designing the tube electrodes in such a way that the electron stream from the cathode is confined and concentrated in the region between the screen grid and the plate. The space potential is depressed at some point in the interelectrode region in front of the anode to a value below the anode potential to prevent secondary electrons from reaching the screen grid. a tube with this feature of construction is called a *beam-power tube.*

In beam-power tubes, the pitches of the helical turns of the lateral wire of the control grid and the screen grid are made equal and the wires are aligned to confine the passage of electrons into flat beams. A focusing beam plate, or optional use of a wide-spaced suppressor grid electrode, is used to maintain this high electron density. Because of the alignment of the control grid and the screen grid, very few electrons are captured by the screen grid; thus, a beam power tube draws low current in the screen grid circuit. Effective suppressor grid action and low screen grid current permit beam power tubes of moderate size and voltages to operate efficiently at high power outputs with large values of transconductance. The large values of transconductances obtained with beam power tubes make them useful in the design of wide-band amplifiers for television service.

In both tetrodes and pentodes, the control grid is sometimes constructed in a nonuniform manner. The spacing between the adjoining helix turns of the grid lateral wire is made closer at the ends than at the center of the grid. In tubes with such variable pitch grids, the amplification factor decreases when the plate current decreases. Such grid structures also require large negative voltages to produce cut off conditions. These tube types are known as remote cutoff or variable-mu tubes. Tubes having uniform grids are called sharp cutoff tubes. Remote cutoff tubes are used in automatic-volume-control circuits in which the dc component of the control grid voltage is varied in such a manner that the amplification of the tube is made smaller for large signal voltages.

Multifunction Tubes. In addition to the single function multigrid tube such as a tetrode amplifier 6CY5 or a pentode amplifier 6BC5, there are other multi-electrode tubes that perform more than one function, e.g., an oscillator and mixer in a superheterodyne receiver. The term pentagrid converter represents a tube with five grid structures that lie in the electron stream between the cathode and the plate. Two of the grids function as control grids; the remaining grids function as screen or suppressor grids to shield the control grids from the plate, e.g., pentagrid converters 6BE6 and 6SA7 or the 6BY6, a pentagrid amplifier used in color television receivers as a sync separator and sync clipper as part of a gated amplifier circuit.

There are also multi-unit tubes consisting of combinations of diodes, triodes and pentodes in one envelope which use more than one cathode and/or plate. For example, the twin triode 12AX7A used in resistance coupled amplifiers; the 8AW8A in which the triode unit is used in a sync separator circuit and the pentode unit is used as an i.f. or video amplifier; and the triple triode 6EZ8 used in oscillator mixer and AFC circuits in FM receivers.

Microwave. The above mentioned multigrid and multi-unit tubes normally are used in amplification circuits at frequencies up to 500 MHz. However, above this frequency range, the transit time of electrons is short when compared to the voltage cycle and, as a result, the phase angle between the plate current and the plate voltage becomes less than 180°. Accordingly, an increase occurs in the power dissipated at the plate for a given designed power output. At these very high frequencies, electron loading can occur at the control grid because of the short transit time of the electrons from the cathode to the plate. Certain pencil triode tubes are made to operate in the 500-MHz region with regular tube constructional arrangement except that the interelectrode spacing between cathode and grid, and between grid and plate, is made very small—about 1-mil clearance—in order to accommodate the short transit time of electrons.

To avoid the difficulties resulting from the short tansit time of electrons as well as the effects of interelectrode capacitances, microwave tubes have been designed to make use of velocity-modulated beams of electrons. Klystrons, traveling-wave tubes, and magnetrons are special-purpose tubes designed to operate in the microwave region.

See also **Ignitron,** and **Microwave Tubes.**

References

Becker, P. W.: "Design of Systems and Circuits," McGraw-Hill, New York, 1977.

Fink, D. G. (editor): "Standard Handbook for Electrical Engineers," 11th edition, McGraw-Hill, New York, 1979.

Harper, C. A.: "Handbook of Components for Electronics," McGraw-Hill, New York, 1977.

Kloeffler, R. G.: "Electron Tubes," Wiley, New York, 1966.

Landee, R.: "Electronics Designers' Handbook," 2nd edition, McGraw-Hill, New York, 1977.

Seely, S.: "Electron-Tube Circuits," 2nd edition, McGraw-Hill, New York, 1958.

Zbar, P. B.: "Electricity-Electronics Fundamentals," 2nd edition, McGraw-Hill, New York, 1977.

ELECTRON VOLT. This is a convenient unit of energy for calculations in electronics and in connection with ionization or excitation of atoms or molecules. When an electric charge e is transferred from a region where the electric potential is V_1 to one where the potential is V_2, its potential energy changes by an amount equal to $e(V_1 - V_2)$. If the charge e is the electronic charge 1.602×10^{-19} coulomb (as it is when the transferred particle is an electron or a proton), and if the potential difference $V_1 - V_2$ is one volt, the corresponding change in energy is equal to 1.602×10^{-19} joule or 1.602×10^{-12} erg, and is called an electron volt. Thus if a doubly ionized positive oxygen molecule moves in an electric field through a potential drop of 500 volts, it receives $2 \times 500 = 1000$ electron volts or 1.602×10^{-9} erg of energy; and since the mass of the oxygen molecule is about 5.31×10^{-23} grams, this energy would give the molecule, if unimpeded, a speed of about 7.77×10^6 centimeters per second or 48.1 miles per second. The abbreviation for electron volt is eV. In x-rays, nuclear physics, and elementary particle physics, the use of higher energies is encountered and additional abbreviations in common use are keV for 10^3 eV, MeV for 10^6 eV, and GeV (sometimes BeV in the United States) for 10^9 eV.

ELECTROOSMOSIS. The movement of liquid with respect to a fixed solid (e.g., a porous diaphragm or a capillary tube) as a result of an applied electric field. See also **Drainage Systems; Electroendosmosis;** and **Solar Energy.**

ELECTROPHILE. Nucleophile.

ELECTROPHILIC REACTION. The reaction in which an electrophilic reagent attacks a nucleophilic compound. The reagent is taken to be the inorganic substance (in the case of reactions of inorganic and organic substances) or the simpler of two reacting organic compounds. The electron-pair for the bond formed is furnished by the nucleophilic compound. The term *electrophilic* connotes *electron-seeking* and is applied, for example to positively-charged cations, or to reactions brought about by them.

ELECTROPHORESIS. A study of the migration of charged particles, either colloidally dispersed substances or ions, through conducting solutions.

It was observed, many decades ago, that when electricity is passed through a solution containing colloidally dispersed particles, the negatively charged particles move toward the positive electrode, and positively charged particles move in the opposite direction. Moreover, the particles move at differing speeds, depending on such properties as their net electric charge, size and shape, thus making it possible to separate them from a mixture. The charge on a particle may arise from charged atoms or groups of atoms that are part of its structure, from ions which are adsorbed from the liquid medium, and from other causes. It soon became evident that the behavior of collidal particles in an electric field, as compared to that of ions, differed in degree rather than in kind. Although a colloidal particle is much larger than an ion, it may also bear a much greater electrical charge with the result that the velocity in an electric field may be about the same, varying roughly from 0–20 $\times 10^{-4}$ centimeter/second in a potential gradient of 1 volt/centimeter.

To understand the phenomenon of electrophoresis, let us suppose, for simplicity, that a non-conducting particle, spherical in shape, of radius r, and bearing a net charge of Q coulombs, is immersed in a conducting fluid of dielectric constant D, having a viscosity of η poises. Suppose, further, that the particle moves with a velocity of v centimeters/second under the influence of an electrical field having a potential gradient of x volts/centimeter. The force causing the particle to move, namely $Qx \times 10^7$ dynes, is opposed by the frictional resistance offered to its movement by the liquid medium. From Stoke's law, the latter is given by $6\pi\eta rv$. Under steady-state conditions, and by introducing the electrophoretic mobility $u = v/x$, rearrangement yields the expression $u = Q \times 10^7/6\pi\eta r$. It is evident that if the electrophoretic mobility of a particle can be computed, it should be possible to determine Q, the net charge on the particle.

For the micro and moving-boundary techniques, a more rigorous treatment of the problem must take into account such complicating factors as electroosmosis, the actual size and shape of the moving particle, the electrolyte concentration in the solvent medium, and the conductivity of the particle itself. Although the ionographic technique is simple from an experimental standpoint, additional complex factors are introduced due to the presence of the stabilizing agent, namely paper, cellulose acetate, starch, etc. However, it is now possible to introduce suitable corrections for these factors and to arrive at mobility data of sufficient quality to be useful in physical chemical computations.

The most important applications of electrophoresis are to the analysis of naturally occurring mixtures of colloids such as various proteins, lipoproteins, polysaccharides, nucleic acid, carbohydrates, enzymes, hormones and vitamins. Electrophoresis often offers the only available method for the quantitative analysis and recovery of physiologically active substances in a relatively pure state. It provides the most convenient and dependable means of analyzing the protein content of body fluids and tissues, and it constitutes an important tool in most hospital laboratories. The marked differences between normal and pathological serum samples are useful in the diagnosis and understanding of disease. Such changes in the electrophoretic pattern of blood serum are evident in diseases characterized by marked protein abnormalities such as multiple myeloma, nephrosis, obstructive jaundice, liver cirrhosis and various parasitic disorders. Because of the small amount of fluid required, the method is applicable for the study of spinal fluids.

The decade beginning with the year 1950 marked the serious application of this technique. Prior to that time, two methods were generally used for studying the electrophoretic behavior of charged particles in a liquid. In the microscopic method, the migration of particles is observed in a solution contained in a glass tube placed horizontally on the stage of a microscope. The method is suitable for the study of relatively large particles such as bacteria, blood cells or droplets of oil. Its usefulness was extended somewhat by finding that various finely divided inert materials such as tiny spheres of glass, quartz or plastic can, in some instances, be so completely covered with adsorbed protein that they act as if they were large protein particles and respond to an electrical field in terms of the charge on the protein. The method is, today, mainly of historical interest.

In the moving-boundary technique of electrophoresis, the movement of a *mass* of particles is measured, thus obviating the necessity of observing individual particles. The displacement of the particles in an electric field is recorded photographically as the movement of a boundary between a solution of a colloidal electrolyte, such as a protein, and the buffer against which it was dialyzed. The material to be studied is poured into the bottom of a U-tube, and on top of it, in each arm of the U-tube, a buffer solution is carefully layered so as to produce sharp boundaries between the two solutions. Electrodes, inserted in the top of each arm of the tube, are attached to a dc electric source.

In the material under study is a protein, bearing an excess of negative charges on its molecular surface, the boundary will move toward the positive electrode. Since the net electric charge on the protein molecule varies with the acidity of the buffer solution, the charge on the molecule, and hence its velocity, may be varied by varying the acidity of the buffer. As the acidity of the buffer is progressively increased, *i.e.*, as the hydrogen ion concentration is increased or the pH lowered, the velocity of the protein is reduced until a point on the pH scale is reached at which it fails to move (isoelectric point or pI). If the pH of the buffer is further reduced, the protein will acquire a net positive charge and will move toward the negative electrode. The migration velocity of any particular migrant is, of course, directly proportional to the applied voltage gradient.

Other important factors which may affect the observed velocity include the molecular shape and structure of the specific substance under study, the concentration of the buffer solution, or more specifically its ionic strength, the temperature, and electroosmosis. Electroosmosis refers to the constant flow of liquid generally toward the negatively charged electrode.

Through the efforts of many investigators, but especially of Tiselius and co-workers in Sweden, the moving-boundary method was developed into a discriminating and rather accurate technique. Instead of round tubing, the U-tube has a narrow rectangular cross section which provides better cooling of the solutions and improved optical qualities. The U-tube is immersed in a water bath held at the temperature of the maximum density of the solution, approximately 2.8°C for a 0.1N

sodium acetate solution. At this point, a change in temperature produces the least change in density and, therefore, minimum convection. The lowered temperature also increases the resolving power of the apparatus. The apparatus is equipped with a cylindrical lens system to render the boundaries visible as shadows, or "schlieren," from which this system is known as the schlieren method. The method is based on the fact that at a boundary between two transparent materials of different density, the light rays are refracted, thus casting shadows which mark the place of refraction. The instrument produces a photographic diagram in which the abscissa represents the refractive index gradients which can be related to concentration gradients. The areas under the curves are, therefore, proportional to the concentrations of the various components.

The third technique for carrying out electrophoretic separations is known variously as ionography, zone electrophoresis, electrochromatography, etc. Although the rootlets of the technique are discernible in the publications of Lodge dating back to 1886, they withered for all practical purposes until the 1920s when they were revived for a few years by Kendall and co-workers. In 1939, König and Klobusitzky, in Brazil, described a technique they used for the partial separation of a snake venom using paper-stabilized electrolytes. Shortly thereafter, Berraz, in Argentina, described quite a sophisticated apparatus and procedure for the separation of inorganic ions on a narrow strip of filter paper wetted with a conducting buffer solution.

During 1950, several papers on electromigration in paper-stabilized electrolytes were published from a number of countries. From that time on, the number of reports on various applications, modifications and limitations of the technique has grown phenomenally, and this procedure has become one of the important tools in biochemical and clinical chemical research. It lends itself equally well to the study of the electromigration of ionic substances of low molecular weight such as amino acids, peptides, nucleotides and inorganic ions, or of colloidal materials such as proteins and lipo-proteins. Under favorable conditions, substances not only are separated totally from mixtures but may be recovered almost completely. This aspect of the procedure is particularly valuable in work with radioactive materials. Rigid restrictions on the temperature, current and composition of the solutions are largely removed when electrophoresis is carried out in a solution stabilized with a material such as paper, cellulose acetate, starch, polyacrylamide gel or agar. In addition, only minute amounts of material are required, the equipment is relatively simple and inexpensive, and the method can be utilized over a wide range of temperature.

Several methods have been utilized for the quantitative determination of the dyes zones (*e.g.*, protein fractions) on the ionogram. The strip may be cut into sections, the colored material eluted with suitable solvents and its concentration determined in a spectrophotometer fitted with small cuvettes. A key element in the various fractions, *e.g.*, nitrogen, may be determined by the micro Kjeldahl method, and the concentrations of the original components can be computed.

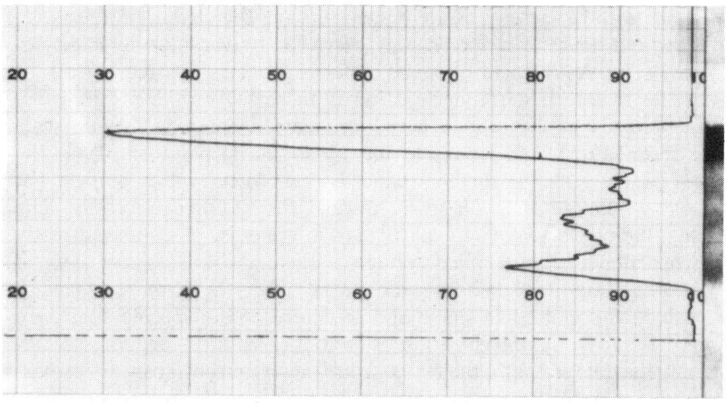

Densitometer tracing of electrophoretically separated fractions of serum proteins from a patient with recurrent acute rheumatic fever. Experiment was carried out under following conditions: Buffer, "Veronal"; ionic strength, 0.05; pH 8.6; potential gradient, 3 volts/centimeter; temperature, 25°C; time, 5 hours; monochromator set at 585 micrometers. Thin line across ionogram from which tracings were made indicates initial point of application of plasma to paper strip. Proteins stained with bromophenol blue. Largest peak represents albumin fraction. Small peaks, in order of increasing size, represent α_1, α_2, β and γ globulin fractions.

The most common method in use today involves direct determination by a transmission densitometer. A motor-driven device moves the strip of paper, cellulose acetate, etc., past the exit slit of a monochromator. The light, after passing through the strip, impringes on a photoelectric cell. The impulse generated is amplified and, after passing through a long converter unit, is fed to a strip chart recorder. Movement of the chart paper on the recorder is synchronized with movement of the ionogram before the exit slit of the monochromator. The areas under the individual peaks of the graph on the chart are proportional to the amount of the component represented by the peak. See accompanying illustration.

ELECTROPHORUS. The simplest of all static machines; devised by Volta in 1816. It consists of a slab of some resinous substance, such as sealing wax or vulcanite, which is negatively charged by rubbing with fur. A metal plate provided with an insulating handle is placed upon the electrified slab. The contact is localized at a few points, so that instead of taking the negative charge off the slab, the metal plate becomes charged by induction, positively on the under side and negatively on the upper. The negative induced charge is now removed by grounding with the finger, and upon being lifted by means of the handle, the plate becomes positively charged all over, often strongly enough to yield bright sparks. Very little of the negative charge on the slab is removed in this process, and it may thus be used over and over to induce an indefinite number of positive charges. The energy is of course furnished by the operator in pulling the metal plate away from the slab. If a slab of glass is used, and rubbed with silk, it becomes positive and the induced charges on the plate are then negative. The instrument is useful in lecture-table demonstrations.

ELECTROPHORUS ELECTRICUS. **Gymnotid Eels.**

ELECTROPHOTOGRAPHIC PRINTING. **Input/Output Devices (Computing System).**

ELECTROPLATING. The coating of an object with a thin layer of some metal through electrolytic deposition. The process is widely used, either for the purpose of rendering a lustrous or noncorrosive finish on some article, or, as in electrotyping, being the principle part of the process. In electroplating, the general object is to employ the article to be plated as the cathode in an electrolytic bath composed of a solution of the salt of the metal being plated. The other terminal, the anode, may be made of the same metal, or it may be some chemically unaffected conductor. A low-voltage current is passed through the solution, which electrolyzes and plates the cathodic articles with the metal to the desired thickness. In this way, table utensils are silver plated, various parts are made weatherproof by cadmium or chromium plating, and a high finish may be imparted through nickel plating. Copper, zinc, and gold are also plated. As the plating proceeds, the strength of the solution must be kept up by the addition of crystals of the plating salt, or a renewal of the anode if it is of the plating metal. A firm bond between the anode and the deposited metal is to be secured when the two metals are of a type which tends to alloy. If this is not the case, some intermediate metal, which will alloy between the base and the plate, is first deposited. For example, in silver plating on steel, the iron would otherwise form a poor bond with the silver, so a thin layer of copper is first deposited on it.

Because of the excellent conducting properties of a metallic salt solution, only a low voltage is required. As this must be dc, the process of electroplating calls for a supply of current from special low-voltage dc generators or rectifiers. The voltage will be of the order of 6 volts or less between anode and cathode.

Some of the solutions used to place various metals are as follows: for silver or gold plating, double cyanide of the metal and potassium; copper plating, copper sulfate; nickel plating, nickel ammonium sulfate. The articles to be plated must be thoroughly and effectively cleaned of all grease and dirt by washing in caustic or acid solutions. While the above is a brief outline of the process of electroplating, in commercial operations there are many troublesome angles which would not be suspected from the foregoing. Irregularity of the plate, poor surface graining, "trees," insufficient bond, and other troubles develop. The overcoming of these requires the use of various expedients, such as careful control of temperature, or the addition of certain colloids and other compounds which have been found effective in preventing formation of defects on the plated articles.

A particular and specialized branch of electroplating is the preparation of plates from printers' type, artists' engravings, etc. This art is known as electrotyping, and is an important phase of electroplating. To make a book plate, for example, melted wax is run over the printers' type as it is locked up in its frame, so that an impression of the type is made in wax. The surface of this wax impression is then made conducting by coating it with graphite. This is then put in an electroplating bath of copper sulfate, and a thin shell of copper built up over the wax mold. The shell can then be separated from the mold and backed up with type metal to give it body. Finally, it is mounted on a wooden block for rigidity.

ELECTROPLATING ANALYZER. Electrolysis-Type Chemical Analyzer.

ELECTROPNEUMATIC CONVERTER. An instrumentation system device that receives electrical signals and converts them to pneumatic signals. These converters are frequently used where it is most effective to measure a variable, such as pressure or temperature, with some form of sensor having an electrical output (thermocouple, strain gage, etc.), but where the most effective means for process control is to use a pneumatically-actuated final controlling element (valve, damper, etc.). Electric-motor actuators, for example, tend to be slower and more costly than diaphragm motor control valves.

Two methods of converting electrical milliampere signals into pneumatic output pressures are commonly used: (1) use of a force-coil motor; and (2) use of a torque motor. Each of these methods uses a nozzle-flapper signal amplifier. The flapper is attached to the coil in the force-coil motor system, and to the armature of the torque motor in the other system. Major advantages of a torque-motor system over a force-motor coil are a much more rugged construction, development of greater force, and less susceptibility to vibration, allowing the transducer to be mounted directly on the final controlling element. The operating principle of an electropneumatic transducer can be gleaned from inspection of the accompanying illustration.

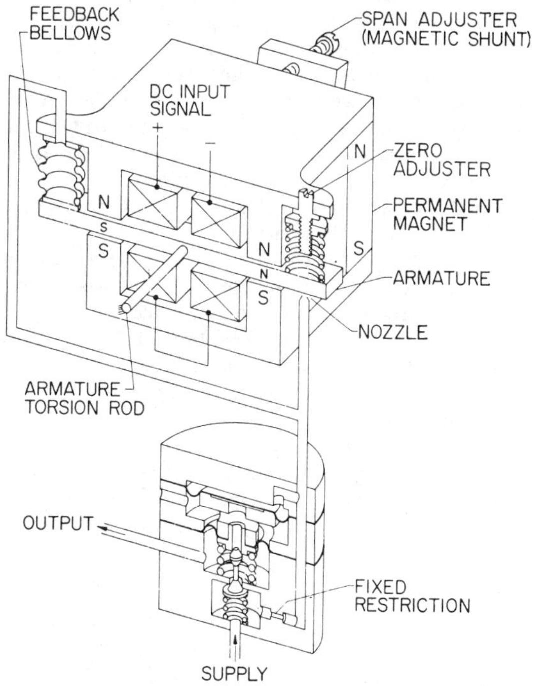

Operating principle of an electropneumatic transducer. (*Fisher Controls.*)

ELECTROPOLISHING. Production of a smooth surface on metals by electrochemical means.

In all electroplating processes, metals (and hydrogen) are deposited on the cathode and dissolved from the anode, except when insoluble anodes are used in which case oxygen is liberated at the anode. Electropolishing is the reverse of electroplating. The work is made the anode and tends to be dissolved. The operating conditions are controlled so that atomic oxygen forms continuously and reacts with the metal surface. Part of this oxygen may be liberated as gas. According to one theory, the high points of the metal surface are most readily oxidized and this oxidized material is thereupon dissolved in the electrolyte or otherwise removed. In any case, selective solution of the high points of the surface tends to give a very smooth finish comparable or superior to a mechanically buffed surface. A wide variety of electrolytes is used. A typical one for stainless steels contains phosphoric acid and butyl alcohol.

All mechanical methods of polishing, including those used for metallographic samples, produce a thin surface layer of work-hardened metal. Electropolishing produces a strain-free surface which is especially suitable for microscopic examination.

An important commercial application of the process is the polishing of stainless steel parts of irregular contour which would be difficult or impossible to buff. Copper and its alloys, Monel metal, aluminum, and many other alloys can be electropolished.

ELECTRORETINOGRAPHY. Vision and the Eye.

ELECTROSCOPE. An instrument for detecting small charges of electricity, or for measuring small voltages, or sometimes, indirectly, very small electric currents, by means of the mechanical forces exerted between electrically-charged bodies. One of the earliest sensitive electroscopes consists of two narrow strips of gold-leaf hanging together in a glass jar. Upon being charged, they stand apart on account of their mutual repulsion. One leaf may be replaced by a stiff strip of brass, so that only the remaining leaf can move. The Wilson electroscope has a single gold-leaf which, on being charged, is attracted by a grounded metal plate tipped at such an angle as to give maximum sensitivity. The Lauritsen electroscope is a rugged, yet sensitive instrument, employing a metallized quartz fiber as the sensitive element. If the movement of the gold-leaf in an electroscope is observed through a microscope whose ocular is provided with a calibrated scale, the instrument becomes an electrometer, capable of measuring potential differences in microvolts. (Some forms of electrostatic voltmeter operate on the same principle.) If the capacitance of the charged system is known, the rate of movement of the electrometer index may be used to measure the current from the discharging body; ionization currents are often thus measured. See also **Electrical Instruments;** and **Electrometer.**

ELECTROSOL. A colloidal solution produced by electrical means, as by passing a spark between metal electrodes in a liquid.

ELECTROSTATIC ACCELEROMETER. Acceleration Measurement.

ELECTROSTATIC ACTUATOR. An apparatus constituting an auxiliary external electrode which permits the application of known electrostatic forces to the diaphragm of a microphone for the purpose of obtaining a primary calibration.

ELECTROSTATIC DEFLECTION. The deflection of an electron beam as a result of its passing through an electrostatic field which has a component perpendicular to the path of the beam.

ELECTROSTATIC DEPOSITION (Paint). Paint.

ELECTROSTATIC GENERATOR. Any apparatus that generates electrostatically a voltage between two terminals. If a sufficiently large electrostatic charge can be accumulated on a particle large enough to be seen visually, it can also be accelerated to measurable speeds by an electrostatic generator. Such acceleration of particles provides a means for laboratory studies of the effects of collisions in space between satellites and micrometeroids. A particle of mass m (in kilograms) carrying a charge q (in coulombs) and accelerated through a potential difference of V volts attains a speed $v = (2Vg/m)^{1/2}$ meters/second. Carbonyl iron spheres one micrometer in diameter, and carrying a charge up to about 3×10^9 volts/meter, have been accelerated to speeds in the 5 to 6 km/sec range in a 2 million volt accelerator. Improvements in particle charging techniques are expected to bring on increases in particle speeds into the hypervelocity range attained by real micrometerooids.

A widely used electrostatic generator is known as a Van de Graaff® generator in which a moving belt is charged by a low-voltage supply, such as a battery, and this charge is deposited onto a hollow, spherically shaped shell, on which a voltage can be developed that is 100 times or more the voltage of the primary supply, depending on the radius of the shell and the electric field at which voltage breakdown occurs. Positive ions or electrons, depending on the sign of the charge on the shell, can then be accelerated from a source inside the shell down an evacuated tube to strike a target at the grounded end of the tube. These charged particles strike the grounded terminal with an energy eV equal to the voltage applied at the source. See also **Particles (Subatomic).**

ELECTROSTATIC LENS. An arrangement of electrodes so disposed that the resulting electric field produce a focusing effect on a beam of charged particles.

ELECTROSTATIC MECHANISM. Electrical Instruments.

ELECTROSTATIC MICROPHONE. Microphone.

ELECTROSTATIC PRECIPITATOR. The concept of the electrostatic precipitator for the removal of particulates from smoke and industrial emissions dates back to the late-1800s and the pioneering work of Frederick Gardner Cottrell. Some of the earliest work in connection with electrical precipitators was directed to copper smelting in the early 1900s. Quoting briefly from Reference 1, "In the first decade of the century there were three well-publicized and classic examples of smelter smoke injury in the United States—at Ducktown, Tennessee, at Salt Lake City, and at Anaconda, Montana. Some indication of the scope of the problem can be seen from a study that was eventually made at the latter and which revealed some startling figures. In a normal day's operation, up and out the stack of that copper smelter went the amazing total of 3,200 metric tons of sulfur dioxide, 200 metric tons of sulfur trioxide, 30 metric tons of arsenic trioxide, 3 metric tons of zinc, and over 2 metric tons each of copper, lead, and antimony trioxide. The marvel is that anything remained, but nothing could more clearly demonstrate that here in full bloom was a cardinal essence of successful invention: a need existed. . . . The emissions from the low stacks of an old plant operated at a neighboring location had killed all vegetation, and losses of livestock by arsenical poisoning had been heavy over the near-lying area. Years after the plant was dismantled, the topsoil of a large area centering at the old site was stripped off, sent through concentrators, and smelted at the new plant with a reported recovery of over $1 million in copper and other metals."

But despite the need and the invention, Cottrell had considerable difficulty over a number of years in gaining acceptance of the electrical precipitator by industry. Today, and for a number of decades past, the electrostatic precipitator has been a major device for combatting air pollution. Since the precipitator functions only against particulates, numerous other items of air pollution control equipment, such as absorbers, scrubbers, and filters, are required and are described elsewhere in this volume.

Operating Principle. The basic operating principle of the electrostatic precipitator is demonstrated by the familiar experiment in which a glass rod is rubbed with a silk cloth; the action gives the rod an electrostatic charge, making it capable of attracting uncharged bits of paper, lint, or cork. In the electrostatic precipitator, it is the collecting surfaces that are grounded, while the charge is created on the particulates to be collected. A representative large electrostatic precipitator is shown in Fig. 1.

The power supply is a transformer-rectifier set which steps up ordinary 220-V ac supply to the high level necessary for precipitator operation, and rectifies it to direct current. The dc voltage is applied to

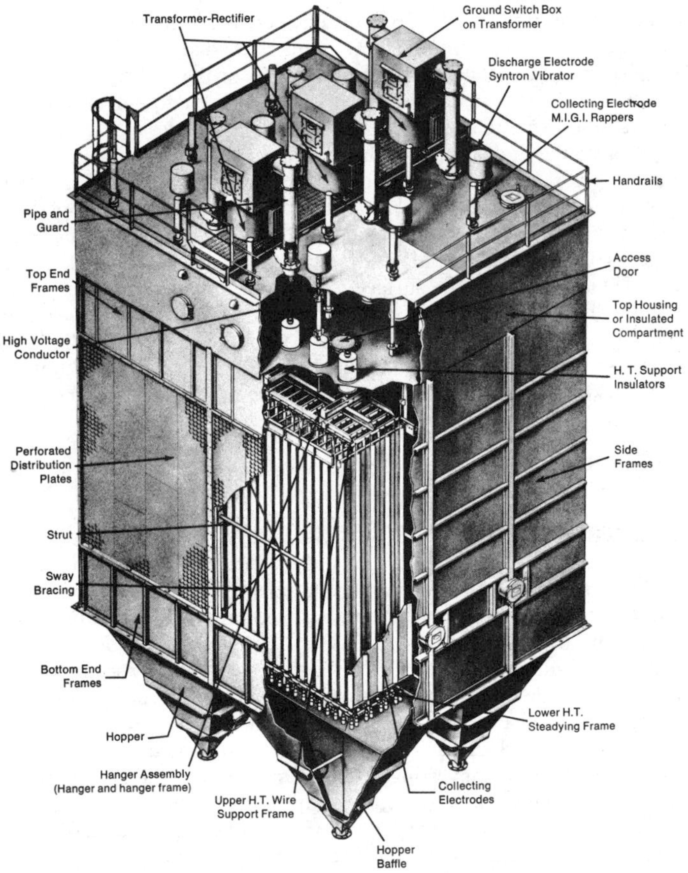

Fig. 1. Industrial-type electrostatic precipitator. (*Research-Cottrell.*)

discharge electrode wires suspended in the gas flow path. See Fig. 2. In the most common industrial type of precipitator, the discharge electrodes hang between rows of collecting electrode plates which form a series of parallel gas flow ducts. The high potential on the discharge electrodes causes a corona discharge, from which electrons migrate out into the gas. These create gas ions, which attach themselves to particulates in the gas and give the particles a charge.

The collecting electrodes are grounded, so that the high potential difference between them and the discharge electrodes creates a powerful electric field through which the gas must flow. According to Coulomb's law, such a field exerts a force on charged particles in the field. In the precipitator, this force moves particles out of the gas stream to the collecting electrodes. In a typical precipitator, the force on a particle 0.5 micrometer in diameter is several thousand times the force of gravity on such a particle.

At the grounded collecting electrodes, the particulates lose their charge. They drain off, if liquid, accumulate until washed off or, more commonly in the case of dry dust, are dislodged by mechanical agitation of the electrodes. In a few applications, the collecting electrodes are vertical pipes, instead of parallel plates, each pipe with a discharge electrode wire hanging down its axis.

The electrostatic precipitator is sensitive to many variable characteristics of the gas stream and the particulates to be collected. When a high collection efficiency is required, even the minor fluctuations that occur in nominally-steady combustion or other processes have an effect

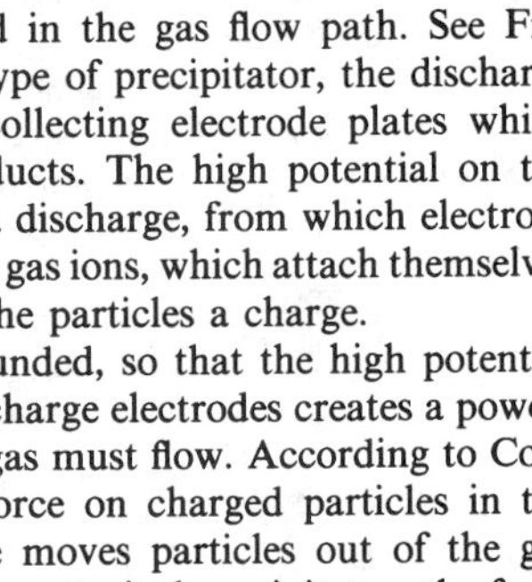

Fig. 2. Arrangement of electrodes in electrostatic precipitator.

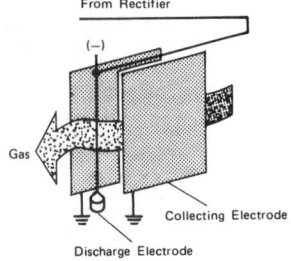

on the voltage level at which the precipitator must operate to achieve constant maximum performance. The voltage applied to discharge electrodes is, therefore, automatically and continuously adjusted by a sensitive feedback control system. The requirement is so severe when extremely high collection efficiency is required, that the power supply and control, and the discharge electrodes, are sectionalized. The voltage in each section follows a control path set by fluctuations in gas and particulate characteristics within that section, independent of conditions existing elsewhere in the precipitator. The rapping intensity (of collecting electrodes to dislodge accumulated particulates) and the frequency of rapping blows are also carefully adjusted to fit the specific application for which the precipitator has been designed; usually they are also sectionalized.

The electrostatic precipitator is a versatile particulate collection device and can be designed for collection of solid particulates, such as fly ash, fluid-bed catalyst fines, or cement kiln dust; or it can collect liquid droplets, such as sulfuric acid mist or tar fumes. The precipitator can operate at below or above atmospheric pressure and can handle dust loadings ranging from less than a grain to more than 100 grains per cubic foot of gas. The units can be designed for operation anywhere within a wide range of gas moisture content levels and for temperatures to 1,500°F (816°C) or higher.

Basic equations which pertain to electrostatic precipitation are given below:

$$\mathrm{Eff} = 1 - \exp(-\mathbf{A}\mathbf{w}/\mathbf{v}) \quad (1)$$

$$\mathbf{w} = \mathbf{E}_0\, \mathbf{E}_p\, \mathbf{a}2/\pi v \quad (2)$$

$$\mathbf{E}_0, \mathbf{E}_p = f(\rho_g, \mathbf{R}) \quad (3)$$

where

Eff = fraction collected, percent, by weight
A = surface area of collecting electrodes
V = volumetric flow rate
w = particle drift velocity or precipitation rate parameter
$\mathbf{E}_0$ = charging field, volts/distance
$\mathbf{E}_p$ = collection field, volts/distance
a = particle radius
η = gas viscosity
ρ_g = gas density
R = particle(s) bulk resistivity

It is of interest to observe that the electrostatic precipitator is independent of the velocity of the main stream flow and, in this respect, unlike all other devices for collecting particulates. The precipitation rate parameter is critically dependent on the field strength (second power), but is dependent only on the first power of the particle radius. Thus, the precipitator is less sensitive to particle size than inertial collectors, although very sensitive to factors which affect maximum voltage at which it can operate. The higher the gas density, the higher the field strength, the higher the particle drift velocity and the higher the efficiency. Field strengths in high-performance electrostatic precipitators usually run in the range of 15,000 volts per inch, or about 60,000 to over 100,000 volts in industrial electrode geometries.

References

Abbott, J. H. and D. C. Drehmel: "Control of Fine Particulate Emissions," *Chem. Eng. Progress,* **72**, 12, 47–51 (1976).

Ardell, M. "Particulate Control for Coal-fired Boilers," *Chem. Eng. Progress,* **75**, 8, 78–82 (1979).

Cameron, F.: "Cottrell," Doubleday, Garden City, New York, 1952.

Hartshorn, W. T.: "Electrostatic Dust Collection from Oil-fired Boilers," *J. Techn. Assn.* of Pulp and Paper Industry, **56**, 6 (June 1973).

Maloney, K. L., et al.: "Low-sulfur Western Coal Use in Existing Small and Intermediate Size Boilers," Environmental Protection Agency, Washington, D.C., July 1978.

Nicholes, G. B., and J. P. Gooch: "Electrostatic Precipitator Performance Model," 5th Quarterly Progress Narrative, Southern Research Institute, Report No. 15, January 1974.

Walker, A. B.: "Characteristics and Electrostatic Collection of Particulate Emissions from Combustion of Low Sulfur Western Coals," 67th Annual Meeting, Air Pollution Control Assn., Denver, Colorado, June 1974.

ELECTROSTATICS. That branch of electromagnetism that deals with the effects of stationary (as opposed to moving) electric charges. The basic law of electrostatics is the Coulomb law. Materials are conveniently classed into conductors and nonconductors. In the former, charges are free to move within the conductor and any charge placed on a conductor will so distribute itself over the surface that the electric field within the conductor is zero and so that the conductor is an equipotential. When an uncharged conductor is placed in an electric field, produced, for example, by a neighboring charged body, a separation of charge on the surface will occur. A charge placed on a nonconductor will remain where it was placed. No material is a perfect nonconductor, but may approximate one very closely. When a nonconductor is placed in an electric field, electric dipoles are induced within it.

Static electric charges may be built up on a body by friction by electrostatic induction, and by other means.

The laws of electrostatics are expressed in terms of electric charges, q, charge densities $\rho = dq/dV$, electric field strengths $\mathbf{E} = \mathbf{F}/q$, and electrostatic potentials $\phi = -\int \mathbf{E} \cdot ds$. The basic law of electrostatics is the Coulomb law of force between charges:

$$\mathbf{F} = \frac{q_1 q_2\, \mathbf{r}}{4\pi\epsilon_0 r^3},$$

with the resulting field due to a single point charge q:

$$\mathbf{E} = \frac{q\mathbf{e}_r}{4\pi\epsilon_0 r^2} \text{ (rationalized mksa units)},$$

where $\mathbf{e}_r$ is a unit vector pointing in the direction of increasing **r**. The field due to a number of discrete charges is the vector sum

$$\mathbf{E} = \frac{1}{4\pi\epsilon_0} \sum_i \frac{q_i(\mathbf{e}_r)_i}{r_i^2},$$

while that due to a continuous distribution of charges is

$$\mathbf{E} = \frac{1}{4\pi\epsilon_0} \int \frac{\rho \mathbf{e}_r\, dV}{r^2}.$$

The potential at some point in space due to a single point charge, referred to an origin of potential at infinity, is $\phi = q/4\pi\epsilon_0 r$, while the potentials respectively due to a number of discrete charges or to a continuous distribution of charges are

$$\phi = \frac{1}{4\pi\epsilon_0} \int \frac{\rho\, dV}{r} \quad \text{and} \quad \phi = \frac{1}{4\pi\epsilon_0} \sum_i \frac{q_i}{r^2},$$

again referred to an origin of potential at infinity. If the potential function ϕ has been determined, then the electric field may be determined from $\mathbf{E} = -\nabla\phi$ where $\nabla\phi$ is the gradient of ϕ. Alternatively, if the field **E** has been determined, the potential ϕ may be obtained from $\phi = \int_\infty^R \mathbf{R} \cdot ds$.

The basic law of electrostatics, the Coulomb law, may alternatively be expressed either as the Gauss law or as the Poisson equation. In the absence of local charges, the Poisson equation reduces to the special case of the Laplace equation.

The electrostatic field is a conservative field, which leads to the fact that is possible to set up a potential function ϕ. This fact may be alternatively stated in terms of the closed line integral $\oint \mathbf{E} \cdot d\mathbf{s} = 0$, or in terms of the curl of $\mathbf{E}$: $\nabla \times \mathbf{E} = 0$.

ELECTROSTATIC SHIELD. **Faraday Shield (Screen).**

ELECTROSTRICTION. The phenomenon wherein some materials experience an elastic strain as the result of an applied electric field, this strain being *independent* of the polarity of the field. In addition to the first-order piezoelectric effect found only in crystals having particular symmetry properties and being linear in **D**, all cyrstals have a second-order electrostrictive effect in which a distortion occurs which is proportional to the square of the electric displacement. The effect is large in ferroelectric crystals, such as barium titanate, and is apparently due to the stresses induced by changing the alignment of the ferroelectric domains upon electrification.

ELECTROSTATIC THEORY. **Electromagnetic Phenomena.**

ELECTROVALENCE. **Chemical Elements.**

ELECTROVISCOUS EFFECT. The change in viscosity of a liquid when placed in a strong electrostatic field. The effect is very small and occurs only in polar liquids.

ELECTRUM. Electrum is a native alloy of gold and silver in which the latter metal may be present in quantities up to 40%. Electrum from the Urals is said to carry 20% copper. The color of electrum is a pale yellow or yellowish-white and the name is derived from the Greek word mentioned in the "Odyssey," meaning a metallic substance consisting of gold alloyed with silver. This same word was also used for the substance amber, doubtless because of the pale yellow color of certain varieties.

ELEMENTARY PARTICLES. Particles (Subatomic).

ELEMENT (Graph). A *circuit element* is an element of a graph G which is contained in some circuit of G. A *noncircuit element* is an element of a graph G which is not contained in a circuit. The removal of a noncircuit element from a connected graph G leaves G unconnected. The removal of a circuit element leaves the connectivity and the number of vertices invariant. An *oriented element* of a graph is an element with an orientation assigned by ordering the vertices of the element. If the vertices β_1 and β_2 of element ϵ_1 are ordered as (β_1, β_2), ϵ_1 is said to be oriented away from β_1 and toward β_2.

See also **Graph (Mathematics)**; and terms listed under **Mathematics.**

ELEMENTS 104, 105, 106, 107. Chemical Elements.

ELEMENTS (Electronegativity). Electronegativity.

ELEPHANT (*Mammalia, Proboscidea*). Like most orders of hoofed animals, excepting the even-toed ungulates, Elephants or Proboscideans (order *Proboscidea*) are dying out and are past their zenith. During the Tertiary and the Ice Age there were many proboscidean species, distributed almost throughout the world. They are now represented by just two genera, each with one species, the last survivors of a large group related to hyraxes and sirenids that developed from primitive ungulates in the Lower Tertiary and have evolved as an independent line since the Eocene (50 million years ago). Although elephants are highly developed animals, they seem, with justification, to be primitive organisms. Their extinct predecessors were among the most distinctive mammals of earlier periods of the earth's history.

The most striking feature of modern elephants, besides their great size, is their trunk, which not only substitutes for a hand as a grasping instrument but is also used for touching objects and for smelling. The trunk develops from the upper lip and the nose, and differs structurally in the two extant elephant species. See Fig. 1. Having arisen from the nose, the trunk of course functions in preceiving odors and in respiration, the latter being particularly evident when elephants swim. However, the trunk's most important function is to obtain food and water. When an elephant drinks, it sucks about a bucketful of water 40 centimeters (16 inches) up its trunk, closes the tip of the trunk with the fingerlike structure at the tip, and squirts the water into its mouth. The trunk is also a formidable weapon. It is likewise

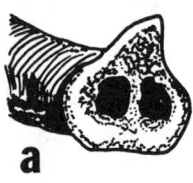

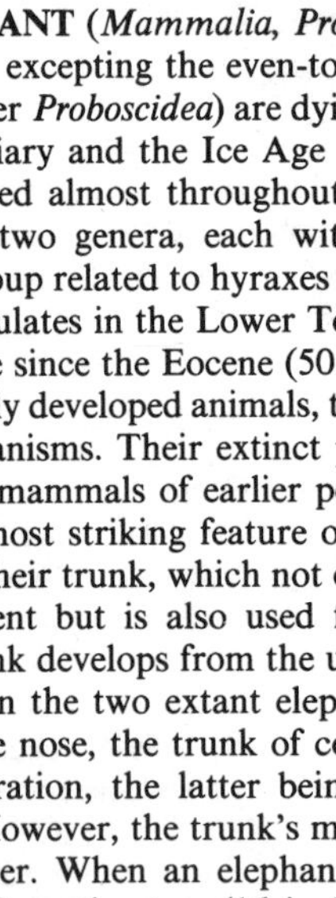

Fig. 1. Tip of the trunk in the (a) Indian elephant (b) African elephant.

a highly sensitive tactile instrument, as we see when an elephant picks up a coin from the floor or tugs great loads about. Large motor-nerve fibers extend up and down the trunk, corresponding to the pyramidal tracts in humans.

The powerful skull bears huge ears, especially the African elephant. The skull bones have spongelike parts, which makes them weigh less than they would if they were solid bone. They are lined with mucous membranes like those found in the nose. The spongelike system reduces the massiveness of the skull in favor of increased mobility, an important compromise since with increasing age the large tusks would become useless if the skull were extremely heavy. The tusks are not elongated canine teeth. They are actually modified upper incisors, and they grow continuously.

Even at birth, elephants have so-called "milk tusks," which can be up to 5 centimeters (2 inches) long. Permanent tusks erupt when the animal is one year old. One third of each tusk is embedded in the upper jaw bones. Since they are incisors, they have a substantial enamel covering, and over a period of years they develop a strong, sharp tip within which the dental cavity is completely filled. Ivory is a mixture of dentine, cartilaginous material, and calcium-salt deposits on the incisors. The other teeth are also quite distinctive structures. Each of the four jaw halves bears six molars, but not simultaneously. No more than two of these molars are present in a jaw half at any one time. Newborn elephant calves possess the first and second molars; the first is as large as a match box, while the second is about the size of a cigarette pack. The molars wear down as a result of chewing action, and since they also rub against the front of the jaw, they break off in lamellae (in layers). The first molar disappears at the age of 3 to 4 years, while the second molar remains for 6 or 7 years. The third molar becomes functional between the age of 3 to 13, the fourth between 6 and 26 years, and the fifth between the age of 16 and 43. The last molar, which is the size of a small brick, begins to erupt when the animal is 33 years old.

Ivory as already mentioned, is largely the dentine of teeth, a material similar to bone but harder and of different minute structure. It is deposited outside of the layer of cells that produce it in the form of small tubules extending toward the outside of the tooth. The chief sources of ivory for commercial purposes are the tusks of various animals. Elephants' tusks have only a little enamel at the tip and are solid ivory except where the pulp cavity invades the base. The tusks of walruses have also been an important source of ivory, although they are inferior to elephant ivory. The material is used extensively for carved ornaments.

The African elephant is the largest of the existing terrestrial animals. The Indian elephant is the third largest, being somewhat smaller than the African Ceratothere or White Rhinoceros. The elephant is characterized by massive structure and by the elongation of the nose and upper lip to form a long prehensile proboscis or trunk. The two upper incisors develop into long tusks in the male and the broad grinding molar teeth grow into position gradually as they are worn off during the life of the animal.

Two species of elephants are recognized: The Indian (*Elephas maximus*); and the African (*Loxodonta africana*). The Indian elephant averages 8 to 9 feet (2.4 to 2.7 meters) in height, but the record is 10 feet, 8 inches (3.2 meters). The African elephant averages about 10 feet (3 meters) in height, with a record of 12 feet, 8 inches (3.8 meters). Adult elephants weight as much as 6 tons (5.4 metric tons). An African elephant may consume up to a half-ton (0.45 metric ton) of grass, twigs, leaves, and fruit in a single day. The African elephant has much larger ears of the two species and the tusks of the African beast are also larger and heavier. See Fig. 2. Numerous extravagant claims have been made from time to time concerning the size of elephant tusks. Reasonable records indicate that the largest tusk from an Indian elephant measured 8 feet, 9 inches (2.6 meters), versus a dimension of 11 feet, 5 inches (3.5 meters) for an African elephant. The tusks weighed 161 and 293 pounds (73 and 133 kilograms), respectively. It is interesting to note that ivory from extinct Mammoths, predecessors of modern elephants, has been dug from the frozen tundra of Siberia and offered to the market.

The life span of the elephant is estimated at about 60 years, although records indicate that some specimens have lived longer. The elephant is fully grown at 25 years, but may breed at 15 to 20 years. The

Fig. 2. Elephants: (above) Indian; (below) African. (*A. M. Winchester.*)

gestation period of the elephant is from 20 to 21 months. The young are weaned at 5 years. The average speed of an elephant is less than 15 miles (24 kilometers) per hour, although it can achieve a charging speed of 17 to 20 miles (27 to 32 kilometers) per hour. The elephant tends to socialize in herds with 100 or more beasts ambling slowly through their habitat. The animal, because of its tremendous capacity for food, must spend a great deal of its time in browsing and foraging. The tusks are used for fighting and can cause terminal destruction to most would-be attackers. The Big Cats are the elephant's main natural threat, particularly in catching and killing baby elephant calves. Over the years, the elephant has steadfastly resisted domestication.

The mother of an elephant calf is not the only individual to care for the infant; other group members join in, their assistance beginning even before the birth of the calf. Elephant calves are sporadically suckled and fed until the end of their second year, but they are capable of taking solid food shortly after birth. However, they are not able to masticate food at this time. Older herd members help them by collecting and cleaning grass, breaking off branches for them, and cutting the food into small pieces. Young elephants even snatch half-chewed food from the mouths of adults. Newborn elephants sleep more frequently and for longer periods than adults. They often interrupt their play fights to lean against each other and sleep. Since elephant calves are curious and playful, they do not pass up any opportunity to observe nearby buffaloes or deer or to chase a hare, a monitor lizard, or another small animal. Their behavior often creates difficulties for the adults.

Elephant mothers cannot continually watch over their younsters, but this problem is overcome by keeping the young in "kindergartens," groups of infants that use the best feeding grounds in an area. The group is guarded by different adults. While most of the adults are off feeding, the "baby-sitter" insures that the young stay together and do not run off from the group. Baby-sitters also guard the young while they sleep and pre-chew their food. The baby-sitter role is apparently an unpopular one and is usually assumed, rather involuntarily, at water holes shortly before the group breaks up. Elephant infants have to be forced out of the water, since they tend to splash about and play with each other interminably when they are in the water. The lead cow usually begins moving off as the other mothers and half-grown animals try to get the youngsters out of the water. However, as the lead cow departs, other potential "baby-sitters" begin leaving the young, and the last cow to be with the infants assumes the role of baby-sitter.

The extremely keen hearing of elephants has long impressed hunters and mahouts, but it was not scientifically studied until 1951. An elephant could distinguish one specific tone from six pairs of pure tones. In one case the various tones were just one note apart from each other. Elephants also learned simple melodies and rhythms that could be recognized independently of the instrument creating the music (i.e., whether the music was performed on violin, piano, organ, or xylophone). More detailed coverage of this topic can be found in an article by Hefner, R. and H. Hefner: "Hearing in the Elephant (*Elephas maximus*)," *Science*, **208**, 518–520, May 2, 1980.

Elephant cows call their young by slapping their ears against the head. When companions meet, they softly "peep" and "rumble." These greeting sounds are apparently understood by mahouts, too. Elephants trumpet when they are surprised by predators or humans, and the shrill tones of a trumpet often initiate either fight or attack by the elephants. When they threaten, they often beat their trunks against the ground, producing a sound like that of automobile tires striking against a hard surface. Vocalizations either originate in the larynx (as in rumbling and roaring sounds) and are then amplified in the air columns of the trunk, or they are created in the trunk itself (trumpeting).

Olfactory signals undoubtedly play an important role in mutual recognition and in communication. When two elephants greet each other each one touches temporal and cheek glands and the mouth and genital region of the other. We do not know much about how important vision is. For a long time it was thought that elephants have poor eyesight, since their eyes are so small in proportion to the huge body. Furthermore, the field of vision is oriented downward and to the side in the normal head position, causing the elephant to have to lift its head to see in front. However, laboratory experiments have shown that the elephant's vision is as good as that of a horse. It is less able to adapt to changing illumination than humans are.

It seems unusual that the world's largest terrestrial animal makes relatively poor use of its food. Their diet consists of grasses, bamboo, roots, bark, wood, and fruits of specific plants. Some of the most popular items are tender bamboo shoots and the leaves. About half of the food swallowed leaves the body undigested. Elephants spend most of the day preparing and eating their food; adults spend 18 to 20 hours daily with this activity! Their short sleep period of just 2 to 4 hours is sufficient for their needs. Elephants, particularly older animals, often sleep standing up. Sleep is usually interrupted at 15 to 30 minute intervals, during which time the animals check the surroundings for danger. Even in deep sleep they become quickly aroused to any disturbances.

The intestines of an adult elephant exceed the length of those in any other mammal: an elephant has 25 meters (82 feet) of small intestine, 1.5 meters (5 feet) of appendix, 6.5 meters (21.3 feet) of large intestine, and 4 meters (15.1 feet) of rectum. An elephant must typically drink 70 to 90 liters (18.5 to 23.7 gallons) of liquid per day. In the wild and in work the elephant's liquid needs are satisfied during the course of several daily baths. Elephants choose their bathing water carefully, since they also drink where they bathe. Of course, an intake of such huge quantities of water means that elephants urinate impressive quantities. Elephants urinate ten to fourteen times every 24 hours. Wild elephants regularly interrupt their feeding and seek water holes or rivers, both for purposes of drinking and for bathing. They frequently slosh about in mud, which helps cool the body off. When the weather is extremely hot they fan themselves with their ears.

Since elephants are huge animals, they have relatively little surface area for heat loss compared to the internal heat-preserving mass. They only breathe 12 times per minute, and the heart beats 40 times per minute. The average body temperature is 39.9°C (104°F). The blood vessels are quite large: arteries leading to the head have a diameter of nearly 2 centimeters (0.8 inches), and the heart weighs about 12 kilograms (26.5 pounds). The great capacity of the circulatory system may be related to the small sleep requirements of elephants. Elephants in their native habitat tolerate cold better than heat, which they will actively avoid. During the day they avoid open sunny spots, instead withdrawing to the shaded jungles.

Hardly any other animal expends so much time and energy caring for its skin, bathing, massaging, and even powdering it (with dust). The thick skin is not nearly as insensitive as it looks, and it requires constant attention. In young elephants the skin is gray to blackish; with increasing age a pink-white appearance develops, beginning at the base and tip of the trunk, along the edges of the ears, the temples, and on the neck. Finally entire sections of skin have this pale color.

Temperature regulation through the sweat glands is apparently ineffective; elephants are unable to work during the midday hours in hot, tropical climates. Asiatic work elephants are usually rested from 10:30 A.M. to 3:30 P.M., since there is too much danger of overheating during these hours. The secretions from the sebaceous glands, at least in captive animals, are apparently of some help in reducing body temperature.

The Indian elephant is tamed for use as a beast of burden and for handling heavy materials, such as timbers. The elephant does not breed freely in captivity, hence wild herds are the source of supply. The animals are trapped in several ways.

Much has been written concerning the lore of the elephant and its behavior. It is well established that the elephant lives in accordance with certain customs. Apparently, sexual dueling is in order among younger males, but only in accordance with strict rules. The elephant is nomadic and apparently follows one of two plans. In the one case, a herd of animals will make a round trip between two locations during the course of a year; or they make what might be termed a "grand circle" tour that may last some ten years. Although traveling in herds is the norm, solitary elephants are sometimes found; or sometimes a very small group of male elephants. The solitary animals may be explained by the presence of illness or crippling, or just plain aging where an older male may not be able to keep the pace. It is known that elephants suffer from arthritis and rheumatism. An elephant may leave the herd for awhile and then return; or temporarily it may join another herd, but apparently the invitation is extended over only a limited time span. Female elephants of all ages generally are part of a large herd. It is well established that other female elephants assist at birth and in the early care of parent and calf.

Much has been written concerning the intelligence and memory of the elephant. The memory is considered exceptional, but not infalible. Like humans, elephants also forget. There is evidence that at least to some degree an elephant understands human language to a point exceeding that of—say a dog obeying a command. A well-trained elephant can take verbal instruction from other than its regular trainer and thus obviously depends upon language sounds and to some degree of complexity. In other words, to some extent, the exact terminology need not be repeated each time or by the same person to achieve a desired reaction from the elephant.

In general terms, the elephant is nomadic, highly social, quite intelligent and, essentially as a result of this intelligence, quite temperamental, quite varied in personality from one specimen to the next, and extremely specialized, thus placing the elephant high if not at the top in the order of mammals in terms of advancement from the primitive state.

Over the years, there have been rumors of pigmy elephants. None have been found. The closest would be some of the Loxodonts found in parts of the former Congo territory in Africa, although these hairy races are far from the size of a pigmy animal.

At one time, the elephant, the rhinoceros, and the hippopotamus were classified in one group known as *Pachydermata* (thick-skinned beasts). These animals all have been reclassified, the elephant now in its own group *Proboscidea.*

For references, see **Mammalia.**

ELEPHANTIASIS. Filiariasis.

ELEVATED RANGE AND SPAN. Range and Span (Instrument).

ELK. Deer.

ELLIPSE. A conic section obtained by a cutting plane parallel to no element of a right-circular conical surface. It is the locus of a point which moves so that the sum of its distances from two foci is a constant. Its eccentricity is less than unity. The standard equation may be taken as $x^2/a^2 + y^2/b^2 = 1$. The curve is a central conic for it is symmetric about both the X- and Y-axes. When placed in its standard position, the center of the ellipse is the coordinate origin, the major axis of length $2a$ is along the X-axis, and the minor axis of length $2b$ is along the Y-axis. The distance from the center to either focus is $\sqrt{a^2 - b^2}$; the eccentricity e is given by $ae = \sqrt{a^2 - b^2}$; the length of the latus rectum is $2b^2/a$; the equations for the directrices are $x = \pm a/e$. The distance from any point on the ellipse to a focus is a focal radius and the sum of two focal radii equals $2a$.

If the semi-major axis equals the semi-minor axis ($a = b$), the ellipse degenerates into a circle.

The polar equation of an ellipse is

$$r = \frac{a(1 - e^2)}{1 - e\cos\theta}$$

and its parametric equations are $x = a\cos\phi$, $y = b\sin\phi$. The equation for the evolute of an ellipse is $X^{2/3} + Y^{2/3} = 1$, where $X = ax/e^2$, $Y = by/e^2$. It is similar in shape to an asteroid and sometimes called that.

Axes of ellipse.

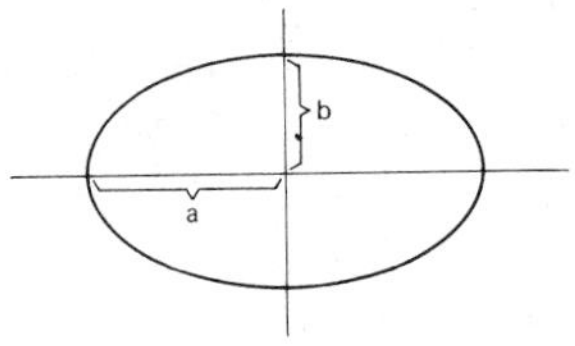

With reference to the accompanying diagram, the area of an ellipse $A = \pi ab$. Circumference $C = 4aE(k)$, where $k = 1 - (b^2/a^2)$ and $E(k)$ is the complete elliptic integral of the first kind. This is an approximation for the circumference $C = 2\pi\sqrt{(a^2 + b^2/2)}$.

See also **Asteroid (Mathematics); Conic Section;** and terms listed under **Mathematics.**

ELLIPSOID. A central quadric surface, given in its standard form with center at the coordinate origin, as

$$\frac{x^2}{a^2} + \frac{y^2}{b^2} + \frac{z^2}{c^2} = 1$$

where a, b, c are the semi-axes. Sections parallel to each of the coordinate planes are ellipses.

If two of the axes become equal, the surface is a spheroid, which can be generated as a surface of revolution. Consider an ellipse in the XZ-plane $x^2/a^2 + z^2/c^2 = 1$, with $a > c$, so that its major axis is along the X-axis of the coordinate system and its minor axis along the Z-axis., There are then two possibilities: (1) rotate the ellipse about its major axis and the result is a prolate spheroid with $a > b = c$; (2) rotate about the minor axis and obtain an oblate spheroid, $a = b > c$. Sections through the surfaces are circles in both cases: when taken parallel to the plane $x = 0$ for the prolate case: parallel to $z = 0$ for the oblate case.

In the final degenerate case, $a = b = c$, the surface is a sphere.

See also **Spheroid; Surface (Of Revolution);** and terms listed under **Mathematics.**

ELLIPSOIDAL COORDINATE. A system based on confocal quadric surfaces. If λ, μ, ν are the three real roots of a cubic equation in a parameter describing such quadrics, they also locate the position of a point in space, for three mutually perpendicular quadric surfaces intersect at the point. If constants are taken so that $a > b > c$, the surfaces are: (1) ellipsoids, $\lambda =$ const., $c^2 > \lambda > -\infty$; hyperboloids of one sheet, $\mu =$ const., $b^2 > \mu > c^2$; (3) hyperboloids of two sheets, $\nu =$ const., $a^2 > \nu > b^2$.

The relation between the rectangular Cartesian coordinates and the ellipsoidal coordinatees of a point are

$$x^2 = \frac{(a^2 - \lambda)(a^2 - \mu)(a^2 - \nu)}{(a^2 - b^2)(a^2 - c^2)}$$

$$y^2 = \frac{(b^2 - \lambda)(b^2 - \mu)(b^2 - \nu)}{(b^2 - a^2)(b^2 - c^2)}$$

$$z^2 = \frac{(c^2 - \lambda)(c^2 - \mu)(c^2 - \nu)}{(a^2 - c^2)(b^2 - c^2)}$$

Since x, y, z occur as squares in these relations they give eight points symmetrically located in the Cartesian system. Some convention must then be adopted for the signs of the ellipsoidal coordinates in order to locate a point uniquely.

See also **Coordinate System;** and terms listed under **Mathematics.**

ELLIPSOMETER. Photometers.

ELLIPTICAL GALAXY. Galaxy.

ELLIPTIC CYLINDRICAL COORDINATE. A degenerate case of ellipsoidal coordinates where the surfaces are: (1) elliptic cylindrical with semi-axes $a = c \cosh u$, $b = c \sinh u$, $u =$ const.; (2) hyperbolic cylindrical with $a = c \cos v$, $b = c \sin v$, $v =$ const.; (3) planes parallel to the XY-plane, $z =$ const. A point in this system has rectangular Cartesian coordinates

$$x = c \cosh u \cos v$$

$$y = c \sinh u \sin v$$

$$z = z$$

and $0 \leq u \leq \infty$; $0 \leq v \leq 2\pi$; $-\infty < z < \infty$.

See also **Coordinate System.**

ELLIPTIC GEOMETRY. Geometry.

ELLIPTIC INTEGRAL. Any integral of the type

$$\int f(x, \sqrt{R})\, dx$$

where f is a rational function of its two arguments and R is a third or fourth degree polynomial in x, with no repeated roots. It may be reduced, by suitable change of variable, to a sum of elementary integrals and one or more of the following types:

$$u_1 = \int_0^x \frac{dt}{\sqrt{(1 - t^2)(1 - k^2t^2)}}$$

$$= \int_0^\phi \frac{dw}{\sqrt{1 - k^2 \sin^2 w}}$$

$$u_2 = \int_0^x \sqrt{\frac{1 - k^2t^2}{1 - t^2}}\, dt = \int_0^\phi \sqrt{1 - k^2 \sin^2 w}\, dw$$

$$u_3 = \int_0^x \frac{dt}{(t^2 - a)\sqrt{(1 - t^2)(1 - k^2t^2)}}$$

$$= \int_0^\phi \frac{dw}{(\sin^2 w - a)\sqrt{1 - k^2 \sin^2 w}}$$

These are incomplete elliptic integrals of the first, second, third kind, respectively. When expressed in terms of $t = \sin w$, they are Legendre's normal forms. The constant k $(0 < k^2 < 1)$ is the modulus and a is an arbitrary constant. If $\phi = \pi/2$, the integrals are called complete.

Series evaluation of the elliptic integrals may be made and numerical tables for them are available. They are called elliptic because they were first studied in order to determine the circumference of the ellipse. Their properties are best studied in terms of their inverse functions. See **Jacobi Elliptic Function; Theta Function; Weierstrass Function;** and terms listed under **Mathematics.**

ELLIS-VAN CREVELD SYNDROME. Pituitary Gland.

ELM TREES. Of the family *Ulmaceae* (elm family), there are close to fifty species of elm trees, notably of Europe and North America. As of the mid-1970s, the story of the elm remains one of grave concern for its future because of the large toll of these trees that has been taken by the Dutch elm disease, a fungus disease carried by a beetle. The plight of the elms has and is continuing to have economic as well as emotional impact. The tree was once a significant source of timber, not on the grand scale of the commercial firs and pines, but for more specialized uses. It is estimated that 70% of the hedgerow trees in the English Midlands are elms. Of the species of elms, there are numerous hybrids, cultivars, and clones, and thus there is a resulting complexity of nomenclature.

The American elm (*Ulmus americana*), also called the white elm, has a natural range from Newfoundland to Florida and westward to the Rocky Mountains. Many of the choice specimens were created from grafting. The height of the tree normally ranges from 50 to 100 feet (15 to 30 meters), with a trunk from 20 to 30 feet (6 to 9 meters) in a circumference, and spread of about 70 to 100 feet (21 to 30 meters). The record elm of this species as selected in 1974 is listed in the accompanying table. About one-third of the way up the tree, the trunk usually divides into a number of very stout branches. The bark is grayish-brown and is furrowed. The leaf is $1\frac{1}{2}$ to 3 inches (3.8 to 7.6 centimeters) in length, alternately spaced, and sharp-pointed. The tree is essentially an ornamental shade trade and widely used for plantings in parks and along streets. Of course, in recent years, the elm disease has caused the removal of large numbers of these trees.

A former champion American elm, located at Dundee, Kentucky. (*Kentucky Division of Forestry.*)

The cedar elm (*U. crassifolia*) is found mainly in the southern states, ranging westward from the Mississippi basin to Texas and parts of Mexico. The tree usually attains a height between 60 to 80 feet (18 to 24 meters) and a trunk diameter in excess of 2 feet (0.6 meter). The record cedar elm is listed in the accompanying table. The tree is characterized by very small leaves, 1 to 2 inches (2.5 to 5 centimeters) in length, considering the other dimensions of the tree. The September elm (*U. serotina*), also known as the southern or red elm, prefers limestone regions and ranges from southern Kentucky westward into Arkansas and southward to the northern parts of Alabama and Georgia. The tree normally attains a height of about 50 to 60 feet (15 to 18 meters) although some specimens become larger, as indicated by

RECORD ELM TREES IN THE UNITED STATES[1]

Specimen	Circumference[2] (inches)	Circumference[2] (centimeters)	Height (feet)	Height (meters)	Spread (feet)	Spread (meters)	Location
American elm (1974) (*Ulmus americana*)	317	805	92	27.6	102	30.6	New York
Cedar elm (1969) (*Ulmus crassifolia*)	98	249	94	28.2	50	15	Texas
Florida elm (1971) (*Ulmus americana var. floridana*)	151	384	68	20.4	78	23.4	Florida
Rock elm (1976) (*Ulmus thomasii*)	134	340	99	29.7	59	17.7	Missouri
September elm (1976) (*Ulmus serotina*)	98	249	71	21.3	70	21	Michigan
Siberian elm (1976)[3] (*Ulmus pumila*)	159	404	120	36	147	44.1	Michigan
Slippery elm (1972) (*Ulmus rubra*)	239	607	90	27	80	24	Pennsylvania
Winged elm (1977) (*Ulmus alata*)	133	338	116	34.8	56	16.8	Florida

[1] From the "Social Register of Big Trees," The American Forestry Association (by permission).
[2] At 4.5 feet (1.4 meters)
[3] Introduced

the accompanying table. The leaves are 2 to 3 inches (5 to 7.6 centimeters) long and are of a narrower contour than found on most other elms. The slippery or red elm (*U. rubra*) has a rough bark, deeply furrowed, scaly, and of a dark-brown color. The leaf surfaces also are rough. The tree normally attains a height of about 70 feet (21 meters). The top is often formed in a broad, irregular manner. This species ranges from the lower Saint Lawrence River area westward to the Dakotas and Nebraska and south and southwestward from western Florida to Texas. The mucilaginous inner bark has been valued as a home medicine of demulcent qualities. The winged elm (*U. alta*), also called Wahoo elm or cork elm, is smaller than most elm trees, with an average height of 40–50 feet (12 to 15 meters). However, the specimen in Georgia, as shown on accompanying table, indicates that the tree can grow to much larger dimensions. The bark is of a light-gray-brown color and is close and fine, with perpendicular ridges. The leaves range from 1 to 2 inches (5 centimeters) in length, usually very narrow with sharp points, and of a deep olive-green color. The foliage usually is reasonably dense. The tree ranges southward from Virginia to western Florida and westward to southern Indiana and Illinois and into Texas.

Other species of elm include: the rock elm (*U. thomasii*) (see table); the Florida elm (*U. americana* var. *floridana*) (see table); the Cornish elm (*U. angustifolia cornubiensis*), a large tree of the British Isles and France; the smooth-leaved elm of Europe and north Africa (*U. carpinifolia*); the Wych or Scotch elm of Europe and north and western Asia (*U. glabra*); the Camperdown, tabletop, Dutch, and Belgian elms, all relatives of *U. glabra*; the Chinese elm (*U. parvifolia*) of northern and central China, Korea, Japan, and Taiwan (disease-resistant); the English elm (*U. procera*); the dwarf or Siberian elm (*U. pumila*); the Jersey or Wheatley elm (*U.* × *sarniensis*); and the Huntingdon or Chinchester elm (*U.* × *vegeta*).

As may have been noted from the foregoing descriptions, the elm did not reach the Pacific west coast of North America. However, in recent years, the disease-resistant Chinese elm has been planted in the western United States. Its branches are long, delicate, and drooping and it makes an excellent park and street tree, especially for certain parts of California.

It is alleged that the Dutch elm disease was imported into North America certainly unintentionally when it was present in a shipment of elm logs. In early America, wood of the American elm and of the rock elm was used extensively for ship building because it does not splinter. Also the wood bends well, but retains its strength, important characteristics for constructing a wooden ship. The fungus carried by the elm-leaf beetle (*Galeruca scanthomelaena*) and the long-horned beetle (*Saperda tridentata*) grows in the new ring of wood, adjacent to the bark. The tree depends upon this for the flow of sap. Once the sap vessels are blocked, a branch dies and this process is repeated until the total tree is lost. Trees of 100 years or more in age can be destroyed in one season. Various insecticides (notably DDT) attack the beetles and the sap-stream also has been injected with the insecticide. Unfortunately the process must be continued and, considering the large number of trees to be treated and the time and money involved, it is not likely that this methodology will save but a relatively small number of the trees.

ELONGATION (Astronomy). Conjunction (Astronomy).

ELONGATION (Poisson's Ratio). Poisson's Ratio.

ELUTION. In general, a process for extracting a solid substance from a mixture of solids by means of a liquid; as in the recovery of a vitamin adsorbed on an adsorbent by means of a solution. Specifically, a process for the recovery of sucrose from molasses. Quicklime in the proportion of 25% of the weight of the molasses is added, the resulting mass is freed from much impurity by percolating (in "elutors") with 35% alcohol and is then decomposed by carbon dioxide which liberates the sucrose. See **Chromatography.**

ELUTRIATION. The separation of solids by the action of water or other liquids: hence, also the washing of a solid by decantation or a related process.

ELUVIUM. General term for unconsolidated, residual sediments.

ELYTRON. The front wing of an insect, modified to form a leathery or rigid wing cover which folds above the body and conceals the hind wings at rest. They usually meet in a straight line down the middle of the back. Elytra are characteristic of the beetles and earwings. They are sometimes short, sometimes cover the entire posterior part of the body, and are sometimes united to form an immovable shield.

EMANATION. Radon.

EMBDEN-MEYERHOF PATHWAY. Carbohydrates.

EMBERIZINAE. Fringillidae.

EMBIIDINA. A small order of rare insects which live in nests and galleries of silk under objects lying on the ground. The few known species are found in warm regions, including the southwestern United States.

EMBOLISM. **Arterial and Venous Disorders; Cerebral Vascular Diseases; Ischemic Heart Disease.**

EMBOSSED GROOVE RECORDING. A method of disk recording which employs a comparative blunt stylus to push aside the material in the modulated groove. No material is removed from the disk. Frequently employed in dictating machines because the disks can be heated and reused, and also because there is no refuse to be cleaned away during the recording operation.

EMBRITTLEMENT. A lowering of the ductility of a metal as a result of physical or chemical changes. Metals may be embrittled under many different conditions. Ordinary steel, wrought iron, and body-centered cubic metals generally, as well as zinc alloys and magnesium alloys suffer a reduction in impact toughness at subnormal temperatures. The effect is only temporary, full recovery of toughness occurring upon return to normal temperatures. Austenitic stainless steels, brasses and bronzes, nickel alloys, aluminum alloys, and lead alloys are not subject to severe embrittlement at low temperatures. Nickel additions to ordinary steels have a favorable effect.

Hydrogen embrittlement of iron and steel may be caused by absorption of atomic hydrogen in electroplating processes or in pickling baths. After such exposure the normal toughness can usually be restored by prolonged aging or a short period of heating at a slightly elevated temperature, as in a steam bath.

Season cracking of heat zinc brasses is a severe form of embrittlement resulting in cracking or distintegration. Somewhat similar forms of stress-corrosion cracking occur in many other metals and alloys. Embrittlement of boiler plate, discussed below, may be considered a special case.

Steels and ingot iron may be embrittled by any heat treatment that deposits films of either oxides or carbides in the grain boundaries.

Caustic embrittlement is the development of brittleness in metals, such as steel or ferrous alloys, upon prolonged exposure to alkaline substances, like caustic soda, in solution. Failures and explosions in boilers and evaporators have been caused by this action. Effective water treatment essentially has eliminated this condition in boilers. See also **Corrosion Embrittlement.**

EMBRYO. The developing individual between the union of the germ cells and the completion of the organs which characterize its body when it becomes a separate organism. The term is difficult to limit because some development occurs after birth or hatching and in some species a considerable period of growth intervenes between the completion of the essential structures of the individual and its assumption of separate life. In the latter stage, the organism is called a fetus if it is a mammal; but this term is not applied to the similar period of birds and reptiles. For the embryo in botany, see **Seed.**

At the moment the sperm cell of the human male meets the ovum of the female and the union results in a fertilized ovum (*zygote*), a new life has begun. But before this new life is transferred from the inner, protected life provided by the mother, the new organism will have acquired an age which may vary from a premature 26 weeks to a postmature 46 weeks.

There is no exact procedure for computing the exact age of a child at birth, because the precise date of ovulation of the mother is not established. Although most authorities agree that ovulation occurs about 14 days prior to the beginning of menstruation, it is recognized that such an estimate is hardly more than an average, and that mature eggs (*ova*) may be liberated either sooner or later. Therefore, it is not known how long the new life inhabits the womb before birth, except in rare cases in which the exact occasion of a fruitful coitus is surely known. But regardless of the length of time spent in the womb by the unborn life, it goes through six preparatory stages before becoming a full-term infant.

The first of these stages is fertilization within one of the Fallopian tubes, which extend from each side of the top of the womb (*uterus*). It is believed that fertilization takes place within the first 24 hours following sexual intercourse. Almost immediately after fertilization of the ovum, the second stage begins. This stage is concerned with the process of cell division. The single-celled zygote becomes a multicelled embryo. The term *embryo* covers the several stages of early development from conception to the ninth or tenth week of life. The early embryo is barely visible without the aid of a microscope; it is considerably smaller than the periods which end the sentence of a typical printed page. The initial series of cell divisions occur as the fertilized egg passes down the Fallopian tube.

Although cell division is still going on in the third stage, when the embryo reaches the womb, the cell cluster has not increased appreciably in size. Up to this time, the embryo is free in the uterus. By the end of the tenth day of development, the fertilized ovum begins to burrow its way into the wall of the uterus. See Fig. 1. This process is known as *implantation*, the fourth stage. It takes about two weeks for the embryo to begin to obtain food from the maternal blood vessels; during this time the developing embryo is probably nourished by the uterine substances it absorbs.

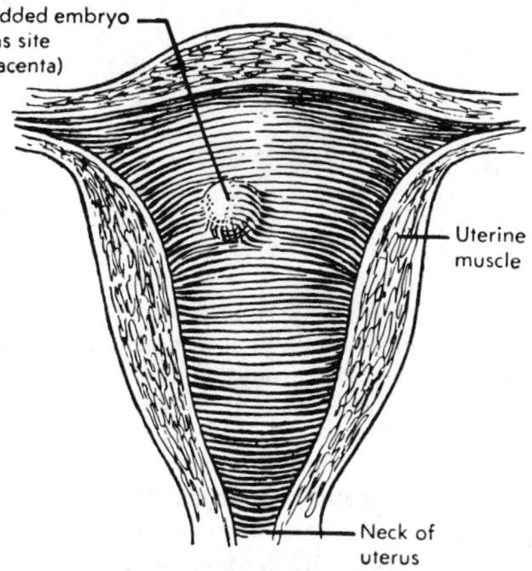

Fig. 1. The human embryo, on about 10th day, becomes embedded in the soft uterine wall. After about two additional weeks, the embryo will derive nourishment through a new placenta which will develop at the site of the attachment.

During the fifth stage the growing new life attains an age of eight to ten weeks, has definitive vital organs, as well as partial ability to balance itself within its fluid environment. When these organs are formed, the individual is called a *fetus*. In the sixth stage of prenatal development, the fetus is prepared for and experiences birth, at which time it is called an *infant*, fully capable of existing as a separate entity in the outer world.

The period of pregnancy is initiated by the union of the sperm and egg. At the moment of fertilization of the egg (*conception*), a new life begins. The period of pregnancy is also referred to as the period of gestation, and its duration from conception to full-term birth varies between 265 and 285 days in normal situations. See also **Pregnancy.**

Early Weeks of Life. During the fourth stage of development, which coincides with the first two or three weeks after conception, the new life is still not much taller than the capital letters upon this page. It can barely be seen and gives little evidence of its presence in the womb. By the third week after conception the embryo indicates its position in the womb by a small elevation. In the weeks since fertilization of the ovum, the weight of the resulting embryo has increased about 10,000 times, its length about 15 times.

Within the first weeks of growth, the outer layers of embryonic cells are undergoing development and providing nourishment; at this time, too, changes are taking place in the thick disc of inner cells (the *blastodisc*) which gives rise to the embryo proper. The uppermost layer of cells of this disc separates from the remainder to form a cavity known as the *amniotic cavity* and the upper layer thus becomes the *amnion*. The cavity remains filled with fluid throughout the prenatal period so that the developing child leads an aquatic existence during the entire period of prenatal life. The amnion is one of the important membranes covering the developing fetus. In this central cell mass, the outer layer of cells on the underside split off to form what is known as the *yolk sac.* The lowermost cells of the blastodisc becomes the *endoderm.* The embryonic structure now appears to be a flattened disc between two hollow sacs (*vesicles*). The upper of these vesicles is the amnion, and the lower is the yolk sac. The relationship of these structures is shown in Fig. 2.

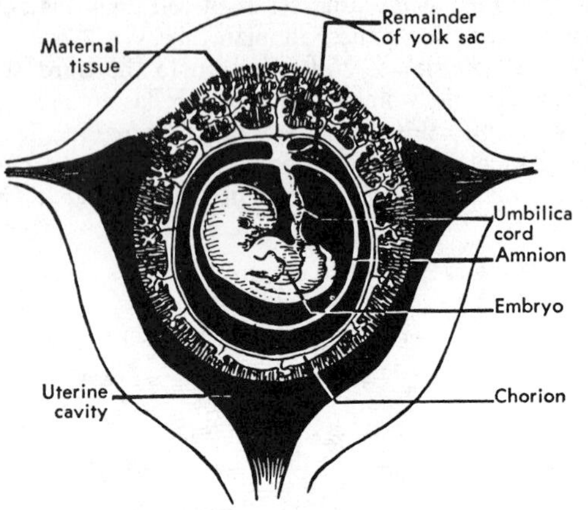

Fig. 2. Protection of developing embryo within the uterus by the chorion and the amnion. The growing child is completely surrounded by amniotic fluid and derives entire nutrition from maternal blood through umbilical cord which is connected to the placenta.

The cells of the embryonic disc segregate to give rise to three main layers of cells which form the definitive organs of the body. The outermost of these cell layers is the *ectoderm,* the middle is the *mesoderm,* and the innermost is the *endoderm.* The middle germ layer, as it is called, rapidly sends out migrant cells which line the entire sac. These mesoderm cells which migrate to the outside of the hollow sphere unite with the external layer to form the *chorion,* a part of which later becomes the *placenta* or the organ by which the fetus obtains food from the mother's blood. Another layer of mesoderm spreads out in a sheetlike fashion to surround the yolk sac. A relatively large amount of mesoderm adheres to the outer membranes to form the body stalk which later will become the *umbilical* cord through which the fetus will receive its nourishment.

Between the second and eighth weeks after conception the three germinal layers differentiate, divide, and combine with each other to lay down the basic body structures, from which the more complicated organism grows. All of the infant body is derived from combinations of the three germinal layers of the embryonic disc. From the mesoderm come supporting structures—muscles, bones, and connective tissues—as well as kidneys, blood and blood vessels, the lymphatic system, and the organs of generation. From the ectoderm are derived the skin and its glands, the hair, the nails, the lens of the eye, the internal and external ear, the mouth and teeth, the mammary glands, the nervous system, and the lower part of the rectum. From the endoderm evolve the respiratory tract, except for the nose; the digestive tract and its glandular outgrowths, including the liver, the pancreas, and the gall bladder; the bladder, and portions of the reproductive organs.

Primitive Streak. During the early part of the third week, the embryonic disc changes shape, so that from the upper surface it appears as an elongated, egg-shaped structure. The cells in the central portion of this disc thicken and form a slight ridge which is known as the *primitive streak.* At one end of this streak, a small knob appears which marks the beginning of the head; in front of this knob the head process forms. The mesoderm is thought by many authorities to give rise to a solid, compact, elongated mass of tissue, the *notochord.* This structure grows forward as well as backward to form the beginning of the backbone. The notochordal tissue continues well into the head region, and on each side of it are formed the bones of the base of the skull.

The primitive streak with its head process divides the embryonic disc into right and left halves. This primitive streak is the first evidence of polarity—cells growing at opposite poles in opposite directions. In the postnatal human animal this polarity of development is clearly evident. An example is the manner in which muscles and tissues grow in opposite directions away from the axis of the spinal cord.

A groove soon courses along the primitive streak, deepens, and presently forms a connecting canal between the amniotic and yolk sac. This is the *neurenteric canal,* forerunner of the *neural* canal. The neutral canal is the forerunner of the entire nervous system, including the brain and the spinal cord.

Having organized the area in which will lie the future head of the embryo, the primitive knot shifts, enlarges into an "end bud," and from this bud the lower half of the body arises.

During the third week of prenatal life the embryo is still a tiny organism the size of a large English pea, and the embryonic disc is about the size of the head of a pin. The body now begins to assume a cylindrical form instead of its previously flattened shape. This change is produced by the edges of the embryonic disc growing downward and enclosing the underlying structures. Concomitantly the underlying endoderm rolls itself into a tube which is to form the digestive system, or as it is properly called, the "gut." The mesoderm gathers itself into a number of small segmentally-arranged bundles called *somites,* which later give rise to the deeper layers of the skin and to the muscles and bones. These somites are formed in rapid succession, so that sometimes they are used as an index to the age of the embryo. See Fig. 3.

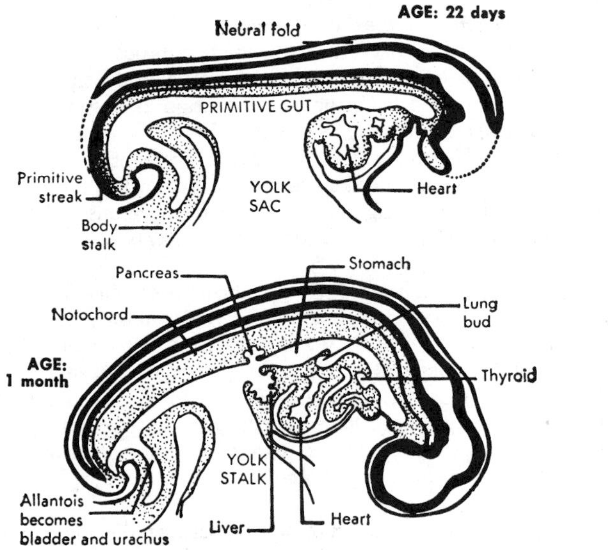

Fig. 3. By end of third week, embryo has commenced to fold over and assumes a more cyclindrical form as contrasted with prior flat form. During this early period, traces of most of the important body structures can first be observed.

Differentiation of Primitive Organ Systems. Near the end of the third week, the embryo has formed the beginnings of most of the important organ systems. The anterior or front end of the neural tube closes and the primitive brain starts forming. On either side of the brain, early in the fourth week, is to be found the first sign of the eyes. These are called the *optic vesicles;* they grow out from the brain and appear as bulges on either side of the early head. Back of or posterior to the future eyes are the *auditory vesicles,* which represent the beginning of the ears.

Toward the end of the fourth week of prenatal life, the original three divisions of the brain called the forebrain, midbrain, and hindbrain now become five divisions. By a symmetrical growth the brain makes a series of bends which cause the head region to curve downward with respect to the remainder of the body. The cranial nerves which later innervate the face begin their formation during this fourth week.

The spinal cord, which is formed from the neural tube, becomes thickened on the underside to develop the primitive nerve cells. Previously, when the margins of the neural tube had formed, there were left behind small clusters of cells. These clusters of cells, by uniting with the upper and lower sides of the neural tube, form the spinal nerves. These nerves grow outward from the spinal cord to innervate the organs of the body as they are being formed.

The heart is formed by the union of two blood vessels underneath the head. The united tube thus formed grows rapidly in length and bends around itself to form the letter "S." Thus by bending back upon itself the *ventricle*, or main pumping organ of the heart, is formed. Rapidly it changes into an incomplete, four-chambered structure. The heart begins to beat during the third week and continues beating throughout the life of the individual.

The lungs develop from endodermal tissue by forming a bud, which later branches into two lung buds. Each lung bud forms two *bronchi*, from which later are formed the *bronchioles* and finally the *air sacs*.

The primitive gut gives rise to some of the glands such as the thyroid, pancreas, thymus, and parathyroid. From the gut, a pouch grows downward and invades the circulatory system to form the liver. The hind part of the digestive system produces a pouch known as the *allantois* which remains relatively undeveloped in human beings during most of embryonic life. The urinary bladder may be considered a remainder of part of the allantois. The limbs first appear during the fourth week as tiny buds from the midside region and from the hind region of the embryo. They later develop into the arms and legs of the fetus.

Early Fetal Development. After the first eight weeks, the embryo has become a fetus; that is, it now roughly resembles the ultimate adult human being. Prior to this time it would have been impossible to determine by observation whether the embryo was that of a human being, pig, goat, dog, or monkey.

During the third month, there is a rapid growth of the fetus so that by the end of the third month the weight has increased eight- to tenfold. The facial features have shown a marked change; the eyes have migrated inward so that they are no longer on the sides of the head. A bulging high forehead, a small slitlike ear, widely separated nostrils, and a large slitlike mouth characterize the earlier part of the third month's development. The upper limbs show sufficient development so that one may readily discern the fingers, wrist, and the forearm. The lower limbs are relatively smaller and less developed. The liver begins to function during this period. The intestine becomes a coiled structure. At the beginning of the third month the internal organs of reproduction have become sufficiently developed to enable one to distinguish between the sexes. The external genitalia, however, are still in the asexual stage, so that externally both sexes appear the same. Some of the bones are beginning to calcify.

See Fig. 4(a) through (f).

The Umbilical Cord. The unborn depends for its nourishment, oxygen supply, and the removal of its waste products, upon the mother's blood supply. From the original multicelled embryo, accessory arrangements have developed apace with the growth of the fetus. These, in conjunction with maternal contributions, provide mechanisms to give the embryo nourishment from the mother. From the larger sac which was contained within the covering membrane and surrounded by ectoderm cells, a two-way cord (umbilical cord) develops. This is the connecting structure between the embryo and the placenta. It is attached to the middle of the fetal abdomen.

The Placenta. The original covering membrane (chorion), in cooperation with certain accommodating cells of the womb, evolves the placenta, which is commonly known as "the afterbirth."

Through the placenta, the bloods of mother and fetus circulate independently, and in entirely separate channels. Maternal blood empties into pockets (*sinuses*) in the placenta, from which food materials are absorbed through the thin walls, to pass into the fetal circulatory system. By a reverse process, waste material is picked up by the maternal blood. In addition to providing oxygen and taking up gaseous and fluid wastes from the fetus, the placenta acts as a digestive area for adjusting foodstuffs in the maternal bloodstream to meet the absorption capabilities of the fetus.

The supportive role of the placenta is essential to fetal health and well-being. Besides keeping the fetus alive, the placenta has the additional function of preparing the uterus and birth canal for the delivery. Several hormones are manufactured by the placenta; these are for use sometimes by the mother and sometimes by the fetus. Should the placenta falter in any of its supportive endeavors, the fetus is in trouble.

If the protective functions of the placenta are impaired, noxious products of the fetal metabolism enter and cause disturbances in the maternal blood. The severe and continued vomiting sometimes experienced in pregnancy may be associated with the incomplete functioning of the placenta. In most cases, the placenta prevents infections from reaching the fetus, although sometimes it fails to protect against such diseases as syphilis, smallpox, and German measles. After the birth of the baby, the placenta is expelled from the uterus.

The Protected Fetus. When the placenta and the umbilical cord have been formed, during the first eight weeks of unborn life, the fetus rests within a closed membrane (amniotic sac) which fills the inside of the womb. This fluid-filled sac absorbs shock, equalizes pressure, prevents the fetus from adhering to its protective enclosure, and provides nourishment. After the first 8 weeks, the primitive, but fast-developing muscular and nervous system of the fetus allows it spontaneous movement. Within the amniotic sac, the fetus has room to rearrange its posture. During the seventh week, the middle vestibule of the ear becomes functionally alive. This development allows the embryo to balance itself. The semicircular canals of the middle ear are structures providing for maintenance of static equilibrium throughout life. At birth of the baby, they are of adult size.

Fourth Month to Birth. During the fourth month, there is considerable development of the abdomen so that the head is less out of proportion to the remainder of the body. Hair begins to appear on the head. During this time, the mother becomes aware of movements of the arms and legs.

During the fifth month the lower abdomen and legs become proportionately larger. The legs and arms show vigorous, active movements during this month. A thin silky hair, which disappears during the succeeding weeks, is deposited over the surface of the body. During the sixth month, the fetus increases in size and the organs in complexity. The embryo is lean, with little fat immediately beneath the skin. The skin is protected by a thick, oily secretion of the external glands. Eyelashes and eyebrows are present, and the eyelids have become separate.

The seventh, eight, ninth, and tenth lunar months are characterized by the maturation of the fetus. There is a layer of fat deposited beneath the skin during the last two months of unborn life. This fat protects and nourishes the infant during its early existence in the external world. During these last months before birth, the organs carry on their functions in much the same manner as they will in the external world. The fetus swallows amniotic fluid which passes through the walls of the stomach and the intestine. The kidneys likewise may function slowly and discharge their contents into the amniotic fluid. Rhythmical movements occur in the intestine and the stomach, but their contents are not emptied into the amniotic fluid. During this period the mother's body is active in the elimination of waste material from the fetal body.

In the ninth lunar month, redness which heretofore has been considerable in the fetal skin, now fades. The body becomes rounded; the nails project. Weight is from five and one-half to six pounds. The fetal infant is complete except for the finishing touches which are accomplished in the tenth and last lunar month before birth.

As the fetal body produces glandular secretions and excretions in preparation for changes to be encountered through birth, the body becomes firm, sturdy, and round. By the time the baby is ready to be born, its many body functions—heartbeat, blood pressure, temperature regulation and, as it is being born, its breathing—have been correlated.

Studies of the Fetus

During the last few decades, there has been a great expansion of the knowledge of the fetus. Problems which have confounded obstetricians for centuries are being analyzed and it is becoming possible to treat the new life as a *patient*, along with the mother. This medical discipline is known as *fetology* and has been made possible mainly through the development of new instrumental exploring and measuring techniques.

Amniocentesis. This is the extraction and analysis of some of the amniotic fluid from the sac surrounding the fetus. Study of this fetal fluid yields clues to many obstetrical problems, for example, the complications that arise in the unborn children of mothers who are Rh-negative, diabetic, or hypertensive. A fairly exact estimate of fetal age can be determined, as well as the sex of the fetus. In amniocentesis, a hollow needle is inserted through the mother's abdomen and the wall of the uterus into the amniotic sac. A small sample of the amniotic

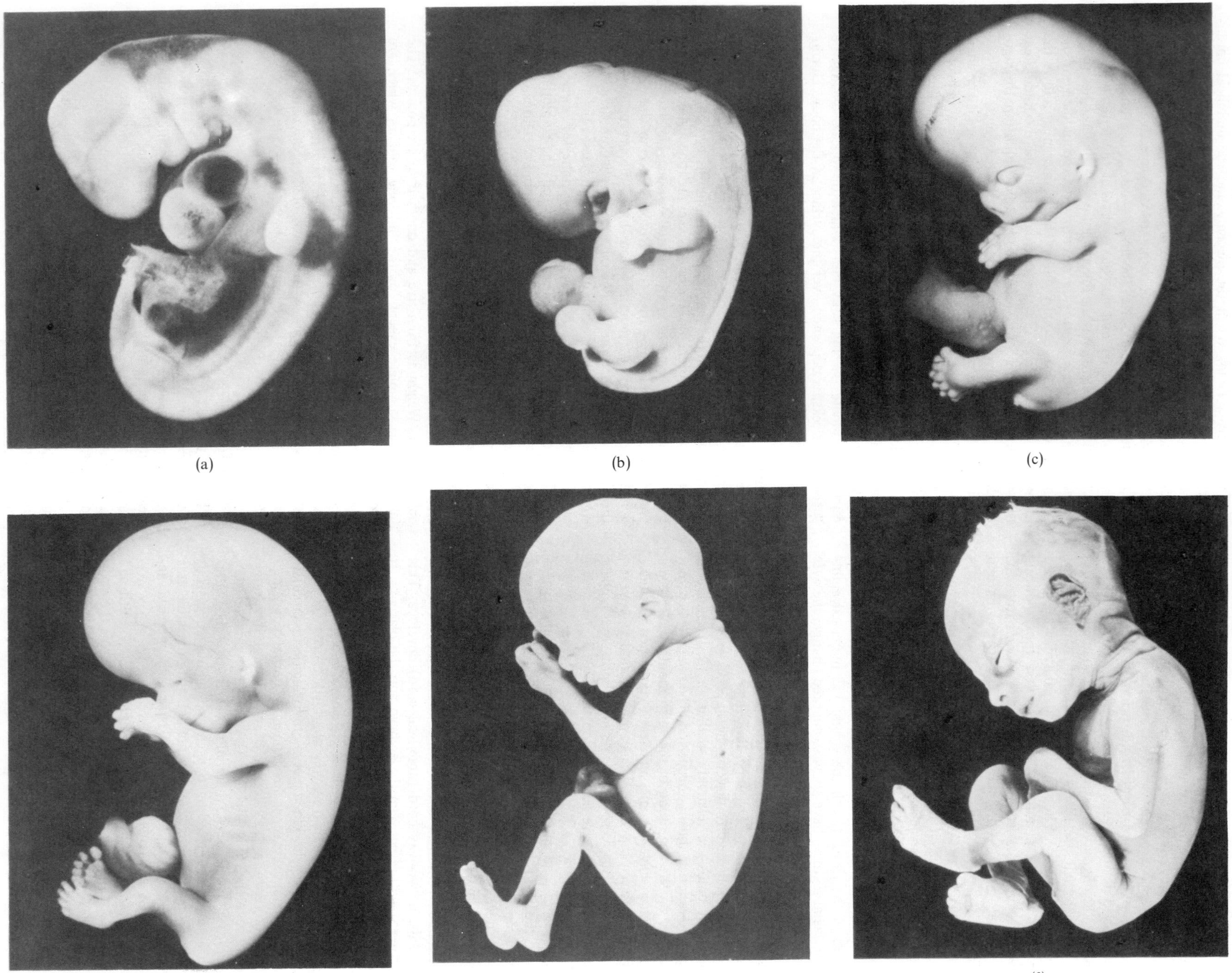

(a) (b) (c)

(d) (e) (f)

Fig. 4. Various stages in development of human embryo and fetus: (a) At 29 days, magnified 9 times; (b) at 37 days, magnified 4 times; (c) at 42 days, magnified 3.5 times; (d) at 56 days, magnified 2 times; (e) at 4 months (now called a fetus), reduced to about $\frac{1}{2}$ actual size (note disproportionate head size); and (f) at 6 months, shown from $\frac{1}{2}$ to $\frac{1}{3}$ actual size (note body is now more proportionate to head). (a) through (d) (*Carnegie Institution, Washington, D.C.*; (*e*) (*f*) *photographed by F. W. Schmidt.*)

fluid is thus obtained. In certain situations, such as Rh incompatibility, a transfusion can be given to the child while in the womb. See also **Blood.** The amniotic fluid contains clues to numerous factors that determine the further development of the fetus and the infant. Because the fluid contains cells from the fetal skin, chromosomal analyses (*karotyping*) can be performed. Potentially dangerous genetic disorders may be diagnosed. Although there is tremendous interest on the part of the obstetrician and, of course, the parents in the state of the fetus, most authorities agree that the procedure should not be undertaken simply out of interest in the sex of the fetus, but rather the technique should be reserved for those instances where family histories or other factors indicate distinct medical advantages to the obstetrician. Although the procedure is generally regarded as safe, it is nevertheless an *invasive procedure.*

Fiberoptic Camera and Endoscope. The *amnioscope* is an illuminated endoscope which can be inserted through the cervix and placed directly against the cervical membrane at any time from the 30th week of pregnancy until delivery. The color of the amniotic fluid will be indicative of whether the fetus is in distress and/or ready for delivery.

The fiberoptic camera is a miniature camera connected to a needle which is inserted into the uterus. Within the needle are fibers that refract light into the lens. Although the instrument yields a picture only one square inch (2.5-centimeters square) in size, it does allow direct observation of the fetus.

Ultrasonography. This is a *noninvasive* procedure and considered by many to be essentially free of risk. The technique can be used to measure the size of the head of the fetus as well as other features. A pulsed beam of ultrasound passing through the fetal head is partially reflected by the skull margins and by the variable density within the brain. A given echo indicates the size of the fetal head. This can be an accurate aid in determination of fetal size and weight. The method can be used after six weeks of pregnancy to visualize the amniotic sac. Progressive ultrasonographic images will indicate the rate of fetal growth. It has been estimated that by the mid-1980s, nearly all infants in the United States, for example, will have been exposed to ultrasound in utero. Because of the growing and widespread use of this technology, some investigators have suggested that, even though investigators have been unable to demonstrate deleterious effects of ultrasound on human chromosomes to date, further research along these lines is warranted. In a 1979 report, one research team indicated that exposure to diagnostic levels of ultrasound increased the frequency of sister chromatid exchange (SCE) as well as in a continuously growing human lymphoblast line. The results confirmed previous findings indicating that ultrasound of diagnostic intensities can affect the DNA of animal cells. Most clinical examinations of the fetus are of short duration and the surrounding maternal tissues attenuate the ultrasonic beam. However, these findings suggest that ultrasound may not be entirely innocuous. (Liebeskind et al., 1979).

Vulnerability of Unborn

Extensive studies have been made of the placenta which serves as the interface between the fetal blood supply and the mother's blood. The placenta anatomically separates the circulatory systems of the fetus and the mother. Exchanges of substances take place across this interface. There are, however, important and sometimes tragic exceptions. In the early 1960s, one of the most dramatic examples of how damaging influences from the outside can reach the fetus via the maternal bloodstream was discovered. In December 1962, a tranquilizing and sleep-inducing drug *thalidomide* was taken off the market and all samples were recalled. The drug had not been approved for use in the United States, but had enjoyed distribution in Germany, the United Kingdom, and a few other European continental countries.

This drug, which had been believed to be safe enough even to be given to babies, was causing a rare malformation in infants of mothers who took the drug during the sixth to eighth week of pregnancy, the period during which the limbs are forming. The most common malformation was *phocomelia* or "*seal limbs,*" in which the arms and legs were often absent and there were seallike "flippers" in their place. Phocomelia was also often accompanied by internal abnormalities, even some affecting the heart. About one-third to one-half of the babies died within a few days of birth.

Since that time, the medical profession and the drug manufacturers have exercised vigilance in this regard and various regulating agencies in different countries have exercised more aggressive controls over drug testing and approval for distribution, with safe administration during pregnancy being a major concern. Only a few examples of offending drugs can be given here. For example, adenine arabinoside, sometimes used to treat acute herpes simplex keratoconjunctivitis and recurrent epithelial keratitis caused by herpes simplex types 1 and 2, should not be given systemically to pregnant women because of its known teratogenicity and toxicity to the embryo (in experimental animals). Anticoagulants, if used during pregnancy, must be administered with extreme discretion. Although it has been shown that heparin does not cross the placenta and thus does not anticoagulate the fetus, it does increase the potential of maternal bleeding. Warfarin anticoagulants, on the other hand, do anticoagulate both the mother and the fetus. Serious problems have been reported as resulting from anticoagulant therapy during pregnancy, including nasal hypoplasia (incomplete development) and stippled epiphyses (parts of bone; see **Bone**), particularly when warfarin was administered during the first trimester of pregnancy. See also **Anticoagulants.** Anticonvulsant drugs have been shown to increase the rate of malformations in offsprings of women who have been on chronic anticonvulsant drug therapy. There are many other examples. Much greater attention is also being given to the effects of chemicals found in the environment.

The effects of ingesting alcohol during pregnancy are mentioned in entry on **Alcoholism.** In addition to those drugs which may be prescribed to nonpregnant women with full justification, but which may be contraindicated during pregnancy, there are the so-called hard drugs (street drugs) which poorly informed persons sometimes take. Drug addiction, of course, is an anathema to the pregnant woman. See **Drug Addiction.** Smoking, a lesser form of addiction, is considered by most authorities in the field as probably harmful in pregnancy over and beyond the effects of heavy smoking on all persons as a precipitating factor of several diseases. The well informed pregnant woman will avoid smoking throughout her pregnancy to assist her body in handling the exceptional demands of that period.

Infections. For many years, rubella (German measles) has been known to be capable of causing birth defects, especially deafness and mental retardation, in children whose mothers had the infection during the first trimester of pregnancy. This usually mild disease can produce devastating effects in the fetus. See **Rubella.**

In addition to the rubella virus, it is now known that other virus infections in the mother may lead to infection in the baby. Depending on the type and severity of the infection, abortion or stillbirth may result, normal development of some of the organs may be prevented (for example, the deafness which often occurs in children of rubella-infected mothers), or the baby may be so infected that its first days or weeks of independent life are an uphill struggle against disease.

Apparently, viruses may infect the fetus at any time from the first few days after conception until immediately before delivery. The incidence of virus infections in fetuses is not known, but it is expected that such infections may account for some otherwise unexplained disorders of the fetus and newborn child.

See also **Congenital Disorders; Embryology; Gonads; In-Vitro Fertilization;** and **Pregnancy.**

References

Annegers, J. F., et al.: "Do Anticonvulsants have a Teratogenic Effect?" *Arch. Neurol.*, **31**, 364 (1974).

Beaconsfield, P., and C. Villee: "Placenta—A Neglected Experimental Animal," Pergamon, Elmsford, New York, 1979.

Beaconsfield, P., Birdwood, G., and R. Beaconsfield: "The Placenta," *Sci. Amer.*, **243**, 2, 94–102 (1980).

Boyd, J. D., and W. J. Hamilton: "The Human Placenta," Macmillan, New York, 1974.

Cadkin, A. V., and M. N. Montew: "Clinical Atlas of Gray Scale Ultrasonography in Obstetrics," Charles C. Thomas, Springfield, Illinois, 1979.

Dinges, D. F., Davis, M. M., and P. Glass: "Fetal Exposure to Narcotics," *Science*, **209**, 619–621 (1980).

Fuchs, F.: "Genetic Amniocentesis," *Sci. Amer.*, **242**, 6, 47–53 (1980).

Garrett, W. J., and D. E. Robinson: "Ultrasound in Clinical Obstetrics," Charles C. Thomas, Springfield, Illinois, 1970.

Gupta, C., Sonawane, B. R., and S. J. Yaffe: "Phenobarbital Exposure in Utero," *Science*, **208**, 508–511 (1980).

Hafez, E. S. E.: "The Mammalian Fetus," Charles C. Thomas, Springfield, Illinois, 1976.
Kolata, G. B.: "Scientists Attack Report that Obstetrical Medications Endanger Children," *Science*, **204**, 391–392 (1979).
Liebeskind, D., et al.: "Sister Chromatid Exchanges in Human Lymphocytes after Exposure to Diagnostic Ultrasound," *Science*, **205**, 1273–1275 (1979).
Marx, J. L.: "The Mating Game: What Happens When Sperm Meets Egg," *Science*, **200**, 1256–1259 (1978).
Marx, J. L.: "Restriction Enzymes: Prenatal Diagnosis of Genetic Diseases," *Science*, **202**, 1068–1069 (1978).
Milunsky, A.: "The Prenatal Diagnosis of Hereditary Disorders," Charles C. Thomas, Springfield, Illinois, 1973.
Milunsky, A.: "Genetic Disorders and the Fetus," Plenum, New York, 1979.
Moghissi, K. S., and E. S. E. Hafez: "The Placenta," Charles C. Thomas, Springfield, Illinois, 1974.
Omenn, G. S.: "Prenatal Diagnosis of Genetic Disorders," *Science*, **200**, 952–958 (1979).
Pavan-Langston, D., Buchanan, R. A., and C. A. Alford (editors): "Adenine Arabinoside: An Antiviral Agent," Raven Press, New York, 1975.
Persaud, T. V. N.: "Prenatal Pathology: Fetal Medicine," Charles C. Thomas, Springfield, Illinois, 1976.
Reinisch, J. M., et al.: "Prenatal Exposure to Prednisone in Humans and Animals Retards Intrauterine Growth," *Science*, **202**, 436–438 (1978).
Scriver, C. R., et al.: "Genetics and Medicine: An Evolving Relationship," *Science*, **200**, 946–952 (1978).
Soloff, M. S., Alexandrova, M., and M. J. Fernstrom: "Oxytocin Receptors: Triggers for Parturition and Lactation," *Science*, **204**, 1313–1314 (1979).
Szabo, A. J., and R. D. Grimaldi: "The Metabolism of the Placenta," *Advances in Metabolic Disorders*, **4**, 185–222 (1970).
Walsh, S., et al.: "The Human Fetal and Neonatal Circulation," Charles C. Thomas, Springfield, Illinois, 1974.

EMBRYOLOGY. The science which deals with the development of the individual from the union of the germ cells to the completion of its bodily structure. Although the term embryo cannot be precisely limited, the science of embryology is concerned with all development prior to birth or hatching.

Development of the fertilized ovum begins with the process of cleavage. Following cleavage a process of gastrulation gives rise to two or three germ layers and from this point the development of specialized tissues and organs goes on by gradual steps, all based on the subdivision and differentiation of many cells.

The processes of change by which germ layers give rise to other structures are varied. In some cases masses of cells grow out in solid protuberances from an existing source. This process is called budding and is exemplified by the appearance of legs and other appendages on the surface of the body. Other structures are developed by the pushing in or out of layers of cells. If the new part pushes in the layer it is said to invaginate, and if it pushes out, to evaginate. Hollow organs may also be formed by the splitting of solid masses and parts may separate by splitting from such masses; either process is delamination. A good example of evagination is the pushing out of a blind sac from the embryonic pharynx of vertebrates to form the respiratory system, and invagination is illustrated by the pushing in of ectoderm to form the stomodaeum which becomes the oral cavity in part. The formation of the vertebrate excretory tubules as solid knots of tissue whose cavities arise by internal splitting is a case of delamination.

The details of development of any species or groups of animals are complex. Vertebrate embryology has been worked out in great detail and is fairly uniform, but the number of invertebrate forms is so great that their embryonic development cannot be concisely summarized. See also **Embryo.**

In the vertebrates, once the germ layers are formed their further development is the formation of organs and tissues and in some species extraembryonic membranes, with the exception of the mesoderm. This layer gives rise to diffuse mesenchyme and its compact portions differentiate into three regions, the dorsal, intermediate, and lateral or ventral mesoderm. The first subdivides into two longitudinal series of metameric masses, the mesodermal somites, flanking the middle line of the body where the notochord lies. This skeletal primordium is independently derived from the same source as the mesoderm. The lateral mesoderm splits to form an outer somatic layer associated with the body wall and an inner splanchnic layer which envelops the viscera. The split forms the coelom or body cavity. From this point the mesoderm, like the other germ layers, gives rise directly to organs of the body. The organs and systems developed from each embryonic tissue are listed under germ layers.

In the field of experimental embryology an effort has been made to learn of the controlling factors in development by subjecting embryos and ova to various unusual conditions. By exposure to chemical stimuli, unusual temperatures, radiation, and the effects of centrifuging, many abnormal results have been recorded. It is evident from these results that development, like the life of the organism, is conditioned by a delicate balance of environmental factors. The response of inherited potentialities to this balance in the development of normal organic structure links embryology very closely with the subject of heredity.

Although the word embryology is most commonly used to refer to the development of the embryo of the vertebrate animals, it also refers to the study of the development of embryos of invertebrate animals and plants.

EMBRYOLOGY (Morphogenic Induction). Morphogenic Induction.

EMBRYONIC DEVELOPMENT. Gastrula.

EMBRYONIC FISSION. The subdivision of a single ovum at some stage in its development into parts which give rise to complete embryos. Polyembryony.

As a result of this process a single egg of many insects (parasitic *Hymenoptera*) and of some rotifers develops into several or many individuals.

EMBRYO (Yolk Sac). Yolk Sac.

EMERALD. This beautiful green variety of the mineral beryl has been known since ancient times and always prized as a gem, both because of its color and relative rarity. It is frequently cloudy or flawed, hence the expression "rare as an emerald without a flaw." The original source of emeralds seems to be the so-called Cleopatra's mines in Egypt, where in a range of low mountains about 15 miles from the Red Sea, they are found in schists. The quality of these emeralds is not high, but there is much evidence of considerable workings in a former period. See also **Beryl.**

Although emeralds are found in the Urals and to some extent elsewhere the most important locality for emerald is at Muso, Colombia, South America, about 75 miles northwest of Bogotá. These mines are believed to be in part at least the source of the emeralds which Cortez and the Spanish conquistadores ruthlessly seized and which were believed for a long time to have come from Peru.

The word emerald is probably derived from the Persian.

EMERGENCY BREATHING. Artificial Respiration.

EMERY. Corundum.

EMESIS. Commonly termed vomiting, this is the expulsion of the stomach contents through the mouth. This is accomplished by reversal of the direction of the normal waves of peristalsis in the gastrointestinal tract. Vomiting may accompany almost any disease, but in particular often accompanies viral gastroenteritis (intestinal flu), intestinal obstructions, kidney infections, intracranial pressures which may result from blood clots, concussions, meningitis, and tumors, and the ingestion of many offensive drugs and poisons. A person also may vomit when exposed to a particularly offensive odor or terribly unpleasant subject, as upon viewing an accident or other tragic event. Usually vomiting is preceded by nausea, an imminent desire to vomit, accompanied frequently be feeling of weakness, faintness, sweating, vertigo, headache, increased pulse rate, and salivation. Numerous drugs are available to encourage as well as to discourage vomiting.

EMETIC. A substance or drug that induces vomiting, either by direct action of the stomach or indirectly by action on the vomiting center in the brain. Examples of emetics include ipecac (emetine derivative), potassium antimony tartrate (tartar emetic), apomorphine, and mustard.

EMETINE. Alkaloids.

EMF. Electromotive Force.

EMF (Measurement). Electrical Instruments.

EMISSARY SKY. Clouds and Cloud Formation.

EMISSION COEFFICIENTS (Einstein). In Einstein's derivation of the Planck radiation formula for a black body three coefficients were introduced which are of importance in the consideration of spectral intensities. Assume two quantum states m and n of a system such that the energy level E_m is higher than E_n; then transition from the upper to the lower state is accompanied by emission of radiation of frequency

$$\nu_{mn} = \frac{E_m - E_n}{h}$$

Assume further that there are a large number of identical systems (atoms or molecules) in equilibrium with black body radiation at a temperature T. Then the rate at which systems pass spontaneously from state m to n, by emission of radiation, is given by

$$-\left(\frac{dN_m}{dt}\right)_1 = A_m^n N_m$$

where A_m^n is known as *Einstein's coefficient of spontaneous emission*, and N_m is the number of systems in state m. But the rate at which systems can pass from the upper to lower state is also dependent upon the density of the radiation, $\rho(\nu_{mn})$, so that there is a second process defined by the relation,

$$-\left(\frac{dN_m}{dt}\right)_2 = B_m^n N_m \rho(\nu_{mn})$$

where B_m^n is known as *Einstein's coefficient of induced emission.*

Furthermore, the system can pass from the lower to the upper state by absorption of radiation, and the rate of this reaction will be given by

$$-\left(\frac{dN_n}{dt}\right)_2 = B_n^m N_n \rho(\nu_{mn})$$

where N_n = number of systems in quantum state n, and B_n^m is known as *Einstein's coefficient of absorption.* See also **Photoelectric Effect.**

EMISSION CONTROLS. Pollution (Air).

EMISSION (Photoelectric). Electron emission from solids or liquids resulting directly from bombardment of their surface by photons. See also **Black Body;** and **Planck Radiation Formula.**

EMISSIONS (Electric Utilities). Pollution (Air).

EMISSIONS (Gas Turbine). Gas and Expansion Turbines.

EMISSIONS (Internal Combustion Engine). Automotive Electronics; Gas and Expansion Turbines.

EMISSIONS (Radioactivity). Chemical Elements.

EMISSIONS (Secondary). Secondary Emission.

EMISSIONS (Vehicular). Automotive Electronics; Catalytic Converter.

EMISSIVE POWER. The emissive power of a body is equal to its emissivity multiplied by the emissive power of a black body at the same temperature. The emissive power of a black body (perfect or complete radiator) is the total radiation from the black body per unit area of radiating surface. See also **Wien Laws.**

EMISSIVITY. The ratio of the radiation emitted by a surface to the radiation emitted by a black body at the same temperature and under similar conditions. The emissivity may be expressed for the total radiation of all wavelengths (total emissivity), for visible light (luminous emissivity) as a function of wavelength (spectral emissivity) or for some very narrow band of wavelengths (monochromatic emissivity). Excepting for luminescent materials, the emissivity does not exceed unity. See accompanying table.

EMISSIVITY (Solar Collector). Solar Energy.

EMITTANCE. The radiant emittance of a source is the power radiated per unit area of the surface. This may be either the radiant emittance per unit range in wavelength, the spectral radiant emittance, or its integral over all wavelengths, the total radiant emittance. If the radiant emittance is evaluated by the standard luminosity function, it is called luminous emittance. For a perfectly diffusing surface, the luminous emittance is equal to π times the intensity luminance.

EMITTANCE (Insulation). Insulation (Thermal).

EMPHYSEMA. Distention of tissues by gas or air within the interstices. Emphysema is one of the most common disabling disorders of the respiratory tract, a condition characterized by overdistention of the lungs with air that cannot be expelled. At the microscopic level, the walls of the tiny alveoli stretch and eventually rupture, reducing the capacity of the lungs to exchange carbon dioxide and water. Chronic bronchitis of many years' duration almost always precedes the development of alveolar overdistension. Emphysema is a progressive disease that is most common in males over 40. Its cause is unknown, but excessive smoking and atmospheric pollution may be factors.

Obstructional emphysema is classified professionally as one of the chronic obstructive lung diseases (COLD). Others in the group include chronic bronchitis, bronchiolitis, cystic fibrosis, small airway disease, and Kartagener's syndrome. In obstructive emphysema, the lungs are sometimes called "floppy lungs," i.e., the lungs lack the important property of elastic recoil. The actual physiological cause of the disease is still poorly understood. Some researchers suggest that destruction of lung tissue may result from the actions of proteolytic enzymes. They have found that the plasma of emphysema patients lack certain substances (such as alpha globulin that is capable of neutralizing various proteolytic enzymes, including elastase) which are present in the plasma of a person without the disease. Some researchers have suggested that cigarette smoke may cause increased destruction of elastin by increasing the release of proteolytic enzymes from various lung cells.

X-rays of the chest of patients with emphysema indicate a number of characteristic features of the disease—hyperinflated chest, a low, flat diaphragm, wide interspaces, and small heart.

Symptoms of emphysema include a long history of cough and the raising of sputum, and shortness of breath. The shortness of breath is noticeable first on exertion and later with walking and other daily activities. Bouts of wheezing are not unusual. Weakness, lethargy, anorexia, and weight loss also may be present. In patients with significant emphysema, the chest is hyperinflated and at times fixed in the inspiratory position. On inspiration, the entire rib cage is lifted and accessory muscles of respiration are used. The diaphragm is flattened and hardly moves.

Although changes in the lungs are irreversible, it is possible to give the emphysema patient considerable relief and to increase the functioning capacity of the lungs. The patient should be encouraged to live a moderately active life, but to avoid any exertion which might increase shortness of breath. To control bronchospasm, patients should make regular use of bronchodilator aerosols. Thick and tenacious bronchial secretions can be thinned with sputum liquefiers, and deliberate coughing will help to bring them up. Exercises should be done to strengthen the abdominal muscles and permit more complete exhalation. Manual compression of the abdomen during expiration will aid in elevating the diaphragm. Elevating the foot of the bed will produce similar results.

Oxygen inhalation is sometimes necessary for relief of shortness of breath, but it must be used with caution. The safest method is with the intermittent positive pressure apparatus which produces adequate ventilation and also removes carbon dioxide. A return to normal ventilatory function cannot be expected in patients with symptomatic emphysema. The normal course of events is relentless progression, and therapy is successful if it maintains the status quo or merely slows the downward trend. A patient with mild to moderate emphysema may live a long and comfortable life, provided all the factors producing bronchospasm and bronchial irritation are controlled. In contrast, the patient with severe emphysema has a greatly reduced life expectancy.

EMPRESS TREE. **Catalpa Trees.**

EMPTY MAGNIFICATION. If the image formed by a telescope or microscope is not magnified enough, some detail of the object which is resolved by the instrument may not be seen by the eye. The detail may be correctly represented in the image formed by the instrument, but its size in that image may be so small that it cannot be resolved by the eye. Magnification which gives an image up to the size such that each detail resolved by the instrument shall be large enough to be resolved by the eye is useful magnification. Any magnification in excess is called empty magnification since it does not reveal any fresh detail in the object.

EMPYEMA. Accumulation of pus in the pleural space, a condition known since the time of Hippocrates. Bacteria may enter the pleural space as the result of bronchopulmonary infections, such as pneumonia, lung abscesses, and bronchiectasis. Empyema may be a complication of thoracic (region between neck and abdomen) surgery. A direct injury to the chest region may be a cause. Rarely a rupture of the esophagus is a causative factor. Traditionally, pneumococcus was the principal infective agent (associated with pneumonia) seen in empyema, occurring in 5–10% of pneumonia cases. Early and effective antibiotic therapy in handling pneumonia cases has greatly diminished this cause. Today, *Staphylococcus aureus* is the major cause. *Pseudomonas aeruginosa, Klebsiella pneumoniae,* and *Escherichia coli* are also seen with some frequency. However, the possible causative agents are many, including anaerobes, *Bacteroides* species, *Fusobacterium,* and *Mycobacterium tuberculosis,* among others.

Present with empyema are fever, dyspnea, chest pain, and cough. With delayed treatment, weight loss usually will occur. Therapy is usually a combination of antibiotic administration and drainage. In acute empyema, thoracentesis or tube thoracostomy may be employed. Surgical drainage may be indicated in chronic empyema. The chronic form of empyema is less frequently seen because of advances in treatment of patients in the acute stage.

The appearance of the fluid from the pleural cavity in empyema varies. It may be watery (*serous*), puslike (*purulent*), or a combination of the two (*seropurulent*). In deep wounds, in addition to the aforementioned procedures, irrigation (*lavage*) with saline solution may be indicated.

EMU. (*Aves, Casuariiformes, Dromaiidae*) The Emus are flightless inhabitants of Australian bush steppes. Three subspecies which lived on coastal islands were exterminated in the last 150 years. Ancestors of today's emus lived in the Upper Pleistocene (50,000–10,000 years ago) in Australia. See accompanying illustration.

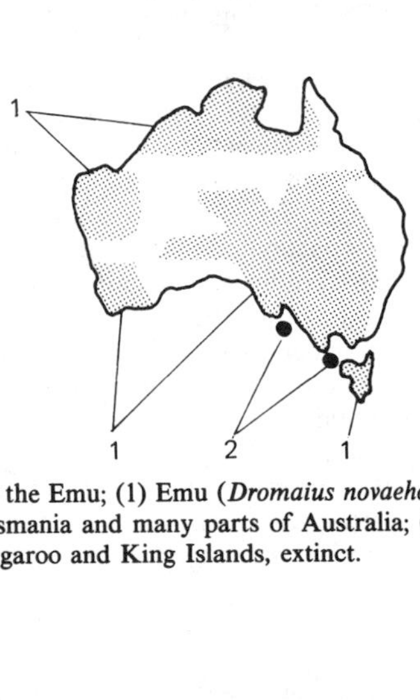

Areas inhabited by the Emu; (1) Emu (*Dromaius novaehollandiae*), Australia, extinguished in Tasmania and many parts of Australia; (2) Black emu (*Dromaius minor*), Kangaroo and King Islands, extinct.

Externally they resemble the rhea. See also **Rhea.** They are about the same size but much more compact and heavier, weighing up to 55 kilograms (121 pounds). The main shaft and secondary shaft (aftershaft) are equal in length so that every feather appears to be double. The wings are small, and are hidden by the rump plumage. There are three toes. The gut and caeca are shorter than in rheas. Their food consists of fruits and seeds. There is no preen gland.

The emu is a fast runner which can reach speeds of up to 50 kilometers (30 miles) per hour. Surprisingly, it also swims well and with endurance.

Incubation and raising young is the male's task, as in the rhea and cassowary. When two emus are paired they stand next to one another with lowered heads and bent necks. They sway their heads from side to side above the ground. Then the female sits down, the male sits down behind her, and he shuffles up and onto her and finally grasps the skin of her nape with his beak. At the same time he utters squeaking or purring sounds, and finally he runs away while the female remains sitting. The nest is a shallow depression located next to a bush. It is simply made with leaves, grass, and bark, and holds 15–25 eggs which come from several females. Incubation takes 25 to 60 days; the great variability is due to pauses during which the male must leave to feed and drink for shorter or longer periods of time. At 2 to 3 years of age, the young are grown and capable of reproduction. See also **Ratites.**

EMULATOR (Computer System). A feature that enables a computer to execute instructions written for a different type of computer. This feature is particularly common where different computers are of similar, but not identical configurations. Emulator implies the use of special logic circuits or microcode to assist in the execution of the instructions of the computer being emulated. The same functions, without the requirement for special circuits or microcode, can be implemented as a special "simulator" program at the expense of storage and performance. See also **Program (Computer).**

EMULSION. **Colloid System; Photography and Imagery.**

ENAMEL. **Paint.**

ENANTIOMORPHIC ALLOTROPY. **Chemical Elements.**

ENANTIOMORPHISM. **Amino Acids.**

ENANTIOTROPY. The property possessed by a substance of existing in two crystal forms, one stable below, and the other stable above, a certain temperature called the transition point.

ENARGITE. A mineral sulfosalt composed of copper, arsenic and antimony, Cu_3AsS_4. Crystallizes in the orthorhombic system. Hardness, 3; specific gravity, 4.451; color, gray to black with metallic luster. From the Greek, meaning distinct cleavage.

ENCEPHALITIS. An infection of brain tissue usually caused by one of several viruses, but which also may be caused less frequently by the ingestion of a drug or toxin, a systemic fungal infection, or by a space-occupying lesion, such as a tumor or subdural hematoma. Postinfectious encephalitis also may occur with mumps, measles, rubella, or chickenpox. The incidence of encephalitis from these infections is no longer common with the exception of rubella, for which no satisfactory vaccine is yet available. The symptoms and course of encephalitis are quite variable, depending upon the causative agent and initial condition of the patient. In encephalitis resulting from some viruses, the symptoms may be a brief illness, so mild that the patient does not seek extra rest; or it may be a grave illness with high fever persisting for several days or weeks. Stupor and weakness of eye muscles are the most notable symptoms in some patients, while less commonly there may be violent delirium, insomnia, and involuntary muscle activity. Muscular rigidity and rhythmical tremor may be seen, as in paralysis. There may be a rapid fatal termination, or the illness may be chronic. Unless the disease occurs during an appropriate viral season for which epidemiologic data are available, diagnosis can be difficult. Specific diagnosis usually requires extensive laboratory examination of blood and cerebrospinal fluid. In the case of enteroviruses, throat and stool cultures may be a useful diagnostic tool. In the case of herpes simplex type 1 encephalitis, a CAT scan will reveal involvement of the temporal, frontal, or parietal lobes. In some cases, open brain biopsy may be indicated for the purpose of cerebral decompression and of ruling out lesions, such as brain abscesses, tumors, or fungal infections.

Simple viral encephalitis usually is self-limiting, running a course of about two weeks. Authorities indicate that less then 50% of the roughly 2000 cases of encephalitis reported in the United States each year are properly diagnosed. The incidence in any given year depends primarily on whether arboviral or enteroviral epidemics occur in any given year. The togaviruses are responsible for most cases of the disease. Other agents, less frequently encountered, include Colorado tick fever virus, varicella-zoster virus, mumps virus, and herpes simplex type 1 virus. Sometimes aseptic meningitis caused by coxsackievirus B presents essentially an encephalitic picture.

Treatment of encephalitis is essentially supportive. In the case of a herpes simplex infection, vaccines may be used if available. Live herpesvirus vaccines are currently being tested. Good experience with the vaccine has been reported in Japan since 1977.

See also **Virus.**

ENCEPHALON. **Brain and Nervous System.**

ENCKE DIVISION. **Saturn.**

ENCODER. A device or subsystem which will accept an input and produce an output in coded form. This definition includes digitallogic configurations which convert a digital input word in one code into an output word in a different code. Examples would include a decimal-to-binary or a decimal-to-octal decoding circuit. Encoder also refers to a device or subsystem which converts an analog quantity into a digital representation by the use of a quantization technique. See also **Quanization.** Electronic and electromechanical analog-to-digital converters, such as shaft-to-digital converters, can be regarded as encoders. There is a trend toward confinement of the term for electromechanical converters, such as the shaft-to-digital encoder. See also **Numerical Control.**

ENDEMIC. A term applied to a disease caused by agents (especially of an infective nature) that are constantly present in a particular human community, leading to a generally higher incidence of the disease in that community than elsewhere. See also **Foodborne Diseases.**

ENDIVE. ***Compositae* or Composite Family.**

ENDOCARDITIS. Inflammation of the endocardium, a thin layer of tissue lining the inner surfaces of the heart. Bacterial endocarditis is a bacterial infection of the endocardium. It accounts for 2% of all organic heart disease. A patient with bacterial endocarditis has an excellent chance for survival with prompt treatment.

Several types of bacteria can cause bacterial endocarditis. It has been known for a long time that bacteria occasionally gain access to the blood vascular system of the body. Usually these invaders are quickly destroyed by the leucocytes, or white cells, of the blood. However, if the bacteria appear in the blood as the result of an infection elsewhere in the body (*septicemia* or blood poisoning), they may be present in very large numbers. Should invading bacteria become attached to the inside of the heart, to one of the valves of the heart, or to the inner wall of one of the major blood vessels, the result is termed bacterial *endocarditis* (affecting the heart); or *endoarteritis* (affecting an artery).

This condition is especially serious because the circulatory tissues are poorly equipped for combating infection. Whereas other tissues of the body may literally wall up an infection so that it can be destroyed by the white cells, the heart and arterial tissues have no such ability to isolate an infection. A large percentage of persons who have bacterial endocarditis have had a previous heart disability. The heart may have some congenital structural defect, or the endocarditis may have resulted from a disease of the heart, such as rheumatic fever. Affected persons usually are young adults, although the disease may attack any age group.

One of the most characteristic signs of bacterial endocarditis is fever, always present in the acute form, but persons with the subacute form may suffer only intermittent fever. The onset of the fever is almost always a result of the presence of free bacteria in the blood stream. The physician may withdraw a sample of blood during a *febrile* period for culture of the organism. The patient also suffers from anemia, which is partly caused by the destruction of red blood cells by the bacteria.

Embolism is also a complication of bacterial endocarditis. Emboli which develop because of the disease may cause *Osler's nodes* in the skin. These are small, raised, reddened areas found most often on the inside of the fingers and toes. They may be somewhat tender, but usually disappear within a few days. Larger and much more painful lumps may appear on the limbs, beneath the skin; usually they remain about a week. Sometimes these are caused by hemorrhage.

When a bacterial embolus lodges within an artery, it may cause a bulging sac from the wall of the artery called a *mycotic aneurysm.* These aneurysms appear in the smaller arteries, such as those that supply the skin. However, they may occur elsewhere.

When emboli become lodged within the blood vessels of the lungs, they produce symptoms similar to those of *hemorrhagic bronchopneumonia.* Emboli affecting the kidneys will cause many of the signs and symptoms of kidney malfunction, but rarely cause fatal *nephritis.* An embolus lodging in the brain may result in widespread damage to nervous tissue by cutting off the blood supply to nerve centers. Probably because of toxins manufactured by the bacteria, the smaller blood vessels (the capillaries) often become unusually fragile. The rupture of the walls of these tiny vessels causes a hemorrhage; the resulting symptoms depend upon the location of the capillaries affected. When capillaries in the skin are affected by the toxins, numerous small, purplish spots appear in the skin. They may be seen almost anywhere in the skin or mucous membrane. When they appear under the nails, the spots often resemble splinters. There may be capillary ruptures on the surface of internal organs, notably the heart and kidneys. In addition to these signs, the spleen usually becomes enlarged and may feel tender to the touch.

All individuals suffering from bacterial endocarditis may not exhibit all of the foregoing symptoms and signs. An individual having an infection of the right side of the heart might well exhibit signs in the lungs, since they are supplied with blood by the right side of the heart. Conversely, an individual who has an infection of the left side of the heart, the aorta, or the mitral valve, will be more likely to have systemic symptoms—emboli in the skin and organs, kidney involvement, enlargement of the spleen, and aneurysms.

In almost all cases, the infection can be controlled by one more of the various antibiotic drugs. However, the dosages must be large and prolonged to insure that the drugs destroy the bacteria. The usual period for antibiotic administration is about one month. In most cases, blood tests will show that the bacteria are resistant to one or more types of antibiotic, so that the treatment may be even longer.

Kidney malfunction caused by bacterial endocarditis may be permanent, restricting the patient to reduced activity.

The individual with chronic heart disease should discuss this with the dentist or surgeon before undergoing tooth extraction, or simple ear, nose, or throat operations. These procedures may be especially dangerous, since bacteria from a throat infection or tooth abscess enter the blood stream in large numbers and, consequently, infect damaged areas of the heart.

See also list of entries at the end of the entry on **Heart and Circulatory System (Human).**

ENDOCRINE SYSTEM. The endocrine system is made up of several organs situated in various parts of the body, all of which are characterized by their ability to produce active chemical substances called *hormones.* The organs of the endocrine system are called *glands of internal secretion* because they secrete the products of their activities directly into the blood stream for distribution throughout the body. *Endocrinology* is the science of the ductless glands and an *endocrinologist* is one who specializes in the science of the ductless glands and their function.

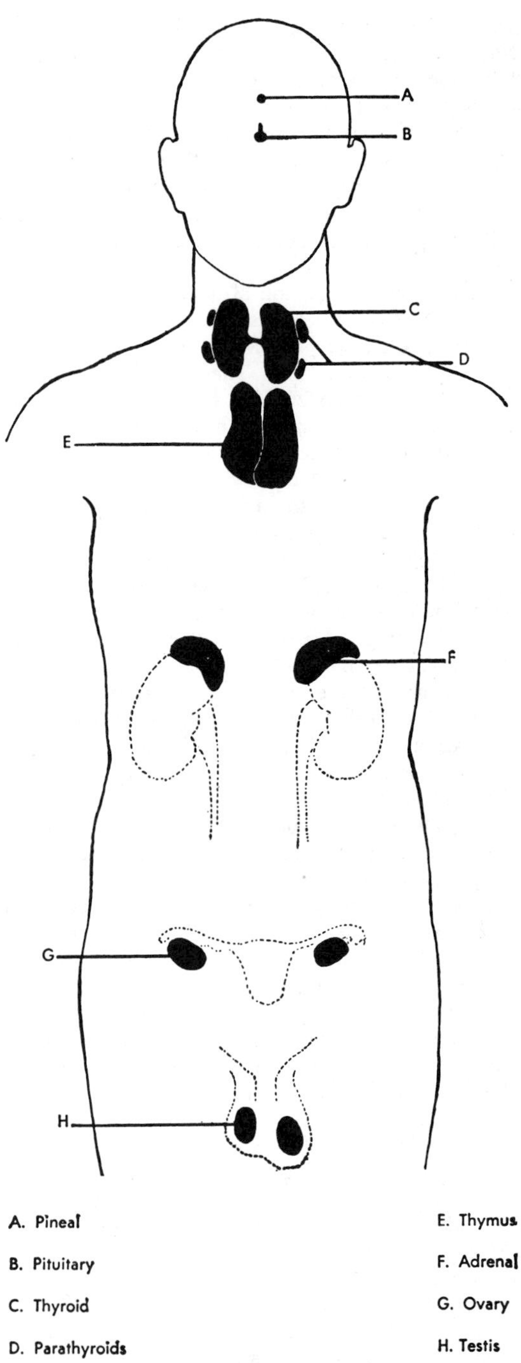

Distribution of the major endocrine glands in the human body.

As indicated by the accompanying diagram, the endocrine glands are: (1) the *pituitary*, located at the base of the brain, which manufactures hormones that act as growth-stimulators of children, assist in the process of sexual maturation, and assist in the coordination of the activity of other endocrine glands, is thus sometimes referred to as the "master gland"—therefore, disorders of the pituitary can have grave significance; (2) the *hypothalamus*, which is part of the diencephalon, the central portion of the brain, and which is concerned with the regulation of many autonomic functions, including body temperature, sleep, behavior, appetite, and emotional response; (3) the *thyroid*, which is located in the neck, and which regulates the general speed of most of the chemical reactions in the cells of the body, regulating the calcium level in the blood, and playing a role in growth, development, and metabolism; (4) the *parathyroids*, located adjacent to the thyroid gland in the lower front portion of the neck and which maintain normal concentrations of calcium in many of the body tissues; (5) the *gonads* (*testes* in the male; *ovaries* in the female), which are the fundamental organs of reproduction and which produce gonadal hormones (respectively male and female hormones) that set secondary sex characteristics; (6) the *adrenals*, sometimes referred to in humans as the *suprarenal glands*, located near the top of the kidneys, and which secrete numerous hormones, notably *adrenalin* (epinephrine), which increases blood pressure, speeds up the respiration and the heartbeat, augments the amount of sugar in the blood, and in stressful situations, gives the individual a feeling of added strength and aggressiveness; (7) the *thymus*, which lies just above the heart, and which plays a central role in establishing the immunological capacities of the body, producing a hormone which is responsible for the production of cells with the capacity to make antibodies and reject foreign elements; (8) a part of the *pancreas*, notably the *islets of Langerhans* (see **Diabetes Mellitus**) which secrete insulin, glucagon, and gastrin, important in the regulation of sugar levels in the blood; and (9) sometimes classified as part of the endocrine system, the *pineal gland*, a small gland attached to the posterior part of the brain, behind and above the third ventricle, the functions of which remain somewhat obscure, but which secretes a hormone known as melatonin, an agent affecting pigment cells. Additionally, the *placenta* (the spongy structure in the uterus through which the fetus is nourished) also secretes hormones and thus is closely related to the endocrine system, and functional during pregnancy.

Some endocrine glands also perform other activities. For example, the pancreas produces a digestive secretion which is passed into the small intestine through a system of ducts.

Hormones fall into two general categories: (1) *messengers*, and (2) *managers*—to use a simple analogy. Most of the endocrine glands produce one or more of each type. The messengers are those hormones which act upon tissues and organs outside of the endocrine system to speed or slow their normal functions. The managers also carry messages, but always within the endocrine system. They are responsible for the fact that some endocrine actions occur constantly, some occur periodically, some during certain years only, and some occur only once during the life of the individual.

One of the principal functions of the nervous system is to correlate all the parts of the body so that they may function harmoniously as a unit. In achieving this harmony, the nervous system is supplemented by the endocrine system; the two systems mutually affect each other. Mental strain, fear, anger, or any emotional state affects the activities of the endocrine glands. A well-known action is the accelerated production of adrenalin during a state of rage or fear. More subtle changes occur constantly in the endocrine system, elicited by some form of nervous stimulation. In some instances, the changes in endocrine activity are reflected in perceptible acts; in others, they remain imperceptible. The crying spells, hot flashes, and irritability sometimes seen in women during premenstrual tension or menopause illustrate this point.

The hypothalamus is part of both the nervous system and the endocrine system. The nerves connecting it to the pituitary as well as to the centers of neurosecretion permit a direction and indirect influence by the nervous system upon its endocrine counterpart. This influence is all the greater because the pituitary, in turn, has a definite influence over other endocrine glands. The hypothalamus also produces hormones which stimulate and which inhibit the pituitary gland. See **Brain and Nervous System.**

See also **Addison's Disease; Adrenal Glands; Diabetes Insipidus; Gland; Gonads; Hormones; Parathyroid Glands; Pineal Gland; Pituitary Gland; Thymus Gland;** and **Thyroid Gland.**

For references see the lists at the ends of the aforementioned entries.

ENDODERM. Embryo.

ENDOGENIC. A term used by geologists to denote processes originating within the earth.

ENDOGENOUS. Originating and developing from within—without influence of external factors and stimuli. In medical usage, a disorder, frequently of a biochemical nature, which arises "naturally" from and within a cell or organ in the absence of external contributing factors. Such malfunctions may arise from genetic abnormalities that are present with an individual entity and require no stimuli to generate symptoms of their presence with the exception of the passage of time.

ENDOMETRIOSIS. Gonads.

ENDOMORPHISM. That phase of contact metamorphism which takes place in the intrusive magma rather than in the walls of the rock mass which it invades.

ENDOPHRAGMAL SKELETON. An internal framework found in some crustaceans. It is made up of apodemes derived from the exoskeleton.

ENDOPLASTIC RETICULUM. Cell (Biology).

ENDOPTERYGOTA. Insects whose wings are concealed beneath the integument of the larva during development. Such insects have complete metamorphosis; hence, this term is synonymous with Holometabola.

ENDORPHINS. Brain and Nervous System.

ENDOSCOPE. An instrument equipped with a lighting and lens system used for visual examination of the interior of a body organ or cavity.

ENDOSMOSIS. A type of osmosis in which the solvent dialyzes into the system. Exosmosis is the reverse process. The two processes may be illustrated by the conditions in the living cell; when the plasma is hypertonic, solvent passes from the cell into the plasma (exosmosis); when the plasma is hypotonic the solvent passes from the plasma into the cell (endosmosis).

The movement of the liquid relative to colloidal particles under an applied electrical field is termed electroendosmosis.

See also **Ocean Resources (Energy).**

ENDOSTYLE. A groove lying in the median line of the ventral wall of the pharynx in the tunicates, lancelets (*Amphioxus*), and larval lampreys (*Cyclostomata*). It is ciliated (see **Cilium**) and secretes mucus. Food particles swept into the pharynx are caught by the mucus and carried forward to peripharyngeal grooves which run around the pharynx to a dorsal median hyperbranchial groove. In this groove the food is carried to the intestine.

The thyroid gland of vertebrates evolved from the endostyle.

ENDOTHELIUM. The delicate lining of the organs of circulation. It is one cell in thickness and is continuous throughout the closed passages with the exception of the sinusoids. The walls of capillaries are made up of little more than the endothelium.

END POINT (Chemical Reaction). Chemical Reaction Rate.

END POINT (Distillation). Petroleum.

END-STAGE RENAL DISEASE (ESRD). Kidney and Urinary Tract.

ENEMA. Constipation.

ENERGY. In most contemporary texts and those of the last several decades, energy generally has been defined simply as "the ability or capacity to do work." This is a broadening of the earlier definition in terms of Newtonian mechanics which was "a property of moving masses."

The concept of energy is central to thermodynamics, quantitative chemistry, and electromagnetism. Consider Einstein's mass-energy equation, $E = mc^2$ for the interconversion of mass and energy, where E = energy in ergs; m = mass in grams; and c is the velocity of light in centimeters per second. Or, Planck's equation, which expresses the fundamental law of quantum theory, stating that the energy transfers associated with radiation are made up of definite quanta of energy proportional to the frequency of the radiation: $E = h\nu$, where E = the value of the quantum units of energy; ν = the frequency of radiation; and h is the elementary quantum of action, more commonly known as Planck's constant (6.6256×10^{-27} erg-second—the proportionality factor that, when multiplied by the frequency of a photon, gives the energy of the photon).

Although the fundamental definition of energy can be brief, it immediately calls for an explanation of work, and of power. In the strict physical sense, work is performed only when a force is exerted on a body while the body moves at the same time in such a way that the force has a component in the direction of motion. The amount of work done during motion from point "a" to point "b" can be expressed by

$$W = \int_a^b F \cos\theta \, ds$$

where F is the total force exerted and θ is the angle between the direction of F and the direction of the elemental displacement, ds. In the cgs system, the unit of work is the *dyne-centimeter* or *erg*; in the mks system, the *newton-meter* or *joule*; and in the English system, the *foot-pound.*

In rotational motion, the definition just given can be exactly applied, but it is often convenient to express the force as a torque and the motion as an angular displacement. The work done will be

$$W = \int_a^b \tau \cos\theta \, d\omega$$

where in this case, θ is always the angle between the torque τ, expressed as a vector quantity and the elemental angular motion $d\omega$, also expressed as a vector. The units of work performed in angular motion will, of course, be the same as in the case of linear motion. Notice that the definition of work involves no time element.

Power is defined as the rate at which work is performed. The average power accomplished by an agent during a given period of time is equal to the total work performed by the agent during the period, divided by the length of the time interval. The instantaneous power can be expressed simply as

$$P = dW/dt$$

In the cgs system, power has the units of *ergs per second*; in the mks system, units of *joules per second* (or *watts*); and in the English system, units of *foot-pounds per second.* A common engineering unit is the *horsepower,* defined as 550 foot-pounds per second; or 33,000 foot-pounds per minute. The SI unit of power is the *watt.* 1 watt = 1 joule per second. (1 joule is the work done by 1 newton acting through a distance of 1 meter.) 1 joule = 1 watt-second = 10^7 ergs = 10^7 dyne-centimeters. The SI unit of force is the newton. (1 newton = 10^5 dynes). See also entry on **Units and Standards.**

Now, returning to the basic definition of energy as the capacity for performing work. This definition may be better understood when stated as: "The energy is that which diminishes when work is done by an amount equal to the work so done." The units of energy are identical with the units of work previously given.

Energy can exist in a variety of forms, some more recognizable as being capable of performing work than others. Forms in which the energy is not dependent upon mechanical motion are generally referred to as forms of *potential energy.* The most common example in this category is gravitational potential energy. A body near the earth's surface undergoes a change in potential energy when it is changed in elevation, the amount being equal to the product of the weight of the body and the change in elevation.

Potential energy also may be stored in an elastic body, such as a spring or a container of compressed gas. It may exist in the form of chemical potential energy, as measured by the amount of energy made available when given substances react chemically. Potential energy also exists in the nuclei of atoms and can be released by certain nuclear rearrangements.

Kinetic energy is the energy associated with mechanical motion of bodies. It is quantitatively equal to $\frac{1}{2}mv^2$, where m is the mass of a body moving with velocity v. In the case of rotational motion, the kinetic energy is more easily calculated, using the expression $\frac{1}{2}I\omega^2$, where I is the moment of inertia of the body about its axis of rotation and ω is the angular velocity. Kinetic energy, like all forms of energy, is a scalar quantity (having magnitude but not direction). In a system made up of an assembly of particles, such as a given volume of gas, the total kinetic energy is equal to the sum of the kinetic energies of all the molecules contained in the volume. Calculation of the energy of such systems is very successfully treated theoretically on the basis of statistical averages.

Within a given system, energy may be transformed back and forth from one form to another, without changing the total energy of the system. A simple example is the pendulum, in which the energy is periodically converted from gravitational potential energy to kinetic energy and then back to gravitational potential energy. A similar situation, but on a submicroscopic scale, occurs in solid materials where the atoms are vibrating under the effect of interatomic rather than gravitational forces. As the temperature of a solid increases, the energy associated with the vibration of the atoms increases.

The example just given illustrates how, on a macroscopic scale, heat can be considered a form of energy. Regardless of the material involved, any amount of heat absorbed or released may be quantitatively expressed as an amount of energy. A *gram-calorie* of heat is equivalent to 4.19 joules, and in the English system, a *British thermal unit* (Btu) is equivalent to 778 foot-pounds.

Potential energy is also present in electric and magnetic fields. The energy available in a region of electric field is equal to $E^2/8\pi$ per unit volume, where E is the electric field strength. Within a given volume, the total energy represented by the electric field is the integral of $E^2/8\pi$ over the volume. Similarly, the energy represented by a magnetic field may be independently calculated by integrating $H^2/8\pi$ over any given volume, where H represents the magnetic field strength. In the case of an electrically charged capacitor, the total energy in the electric field, and hence in the capacitor, can be shown to be $\frac{1}{2}CV^2$. Here C is the capacitance and V the electric potential to which the capacitor is charged. Similarly the total energy in the magnetic field associated with an inductor carrying an electric current is $\frac{1}{2}LI^2$, where L is the inductance and I is the current.

Electromagnetic radiation is a combination of rapidly alternating electric and magnetic fields. Energy is associated with these fields and is exchanged between the electric and magnetic forms. This energy in a quantum of electromagnetic radiation, such as light or gamma radiation, can be expressed in different ways, but is commonly expressed as $E = h\nu$, as previously mentioned.

For particulate radiation or any very rapidly moving mass, the expression previously given for the kinetic energy, $\frac{1}{2}mv^2$, is not accurate when the velocity approaches that of the velocity of light. The theory of relativity requires a correction be made, and the exact kinetic energy, T, may be calculated in terms of the mass, m_0, of light in vacuum, c, as follows:

$$T = m_0c^2\left[\left(1 - \frac{v^2}{c^2}\right)^{-1/2} - 1\right]$$

Notice that this formula may also be written:

$$T = (m - m_0)c^2$$

where m is the variable quantity $m_0(1 - (v^2/c^2))^{-1/2}$. This quantity represents the mass of the body, reducing to m_0 when v is zero, and approaching infinity as v approaches the speed of light.

This example illustrates another result of the theory of relativity, namely, the equivalence of mass and energy. Rewriting the last equation,

$$m = m_0 + \frac{T}{c^2}$$

The mass is seen to increase linearly with the kinetic energy of the body, the proportionality factor being c^2. It should be noted that even the rest mass, m_0, represents an amount of energy equal to m_0c^2. The total energy of a body of mass, m, can be generally given as:

$$E = mc^2 \quad \text{or} \quad E = m_0c^2 + T$$

In dealing with radiation, whether particulate or electromagnetic, it is customary to express energy in terms of electron volts. An electron volt is equal to the amount of work done when an electron moves through an electric field produced by a potential difference of one volt. One electron volt is equivalent to 1.60×10^{-12} erg. When charged particles, such as electrons or protons, are given kinetic energy by an accelerator, their kinetic energy is stated in terms of electron volts (eV), million electron volts (MeV), or billion electron volts (BeV).

A basic principle of physics known as the conservation of energy requires that within any closed system, the total energy must remain constant. Energy can be changed from one form to another; but the total, so long as no energy is added to or lost from the system, must be constant. In the case of the swinging pendulum, decreases in kinetic energy reappear as increases in potential energy and vice versa. Eventually, of course, the pendulum will stop due to the effect of frictional forces. At that time, all of the kinetic energy and gravitational potential energy will have been converted to heat.

In another example involving a radioactive atom, the total energy represented by the atom and the emitted radiation must be constant. If a gamma ray is emitted, the rest mass of the atom will be decreased by an amount equivalent to the sum of the energy of the gamma ray and the recoil kinetic energy of the atom, which will be very small. If a beta ray is emitted, the rest mass of the atom will be decreased by an amount equivalent to the sum of the rest mass of the emitted electron, the kinetic energy of the electron, and the recoil kinetic energy of the atom.

Entropy. In the mathematical treatment of thermodynamic processes there occurs very often a quantity, now relating energy to absolute temperature, now associated with the probability of a given distribution of momentum among molecules, and again expressing the degree in which the energy of a system has cased to be *available energy*. Its mathematical form suggests that these are all aspects of a single physical magnitude. Application of the second law of thermodynamics leads to the conclusion that if any physical system is left to itself and allowed to distribute its energy in its own way, it always does so in a manner such that this quantity, called entropy, increases; while at the same time, the available energy of the system diminishes. This has led to the observation of a so-called "order of merit" for the various forms of energy:

FORM OF ENERGY	ENTROPY PER UNIT ENERGY
Gravitation	0
Energy of rotation	0
Energy of orbital motion	0
Nuclear reactions	10^{-6}
Internal heat of stars	10^{-3}
Sunlight	1
Chemical reactions	1–10
Terrestrial waste heat	10–100
Cosmic microwave radiation	10^4

With reference to the foregoing listing, the energy usually flows from higher levels to lower levels—in a direction such that the entropy increases. Thus, cosmic microwave background radiation is defined as the ultimate heat sink, i.e., it represents the ultimate in energy degradation with no lower form in which to be converted.

As pointed out by Dyson (see reference listed), the universe evolved by the gravitational contraction of objects of all sizes, from clusters of galaxies to planets. In considering that thermodynamics appears to favor the degradation of gravitational energy to other forms, why is it that after an estimated 10 billion years since cosmic evolution, gravitational energy remains the predominant form? Dyson explains this in terms of a series of phenomena which he calls "hangups." These are, in essence: (1) The cosmos is large to the extreme; distances between objects are tremendously long; the average density is extremely low. Thus, matter cannot collapse gravitationally in a time shorter than the "free fall time." In relating free fall time (t) with density (d) with the formula, $Gdt^2 = 1$, where G is the constant in Newton's law of gravitation, it is apparent that the free fall time is extremely long. Dyson estimates the time at about 100 billion years. But, since the density of our own galaxy is estimated at one million times that of the universe, more than the size hangup is required to preserve the galaxy. (2) An extended object cannot collapse gravitationally if it is spinning rapidly. The object assumes a stationary orbit revolving about the inner parts instead of collapsing. Thus, the earth has not collapsed into the sun. Other examples of the spin hangup at work include galaxies, planetary systems, double stars, and the rings of Saturn. (3) Hydrogen "burns" to form helium when it is heated and compressed. But, this thermonuclear burning releases energy, which opposes any further compression. Thus, a star with a lot of hydrogen cannot collapse gravitationally beyond a certain point until the hydrogen is burned up. It is estimated that the sun has been "stuck" on this thermonuclear hangup for 4.5 billion years and will need another 5 billion years to burn hydrogen before its gravitational contraction can be resumed. (4) Dyson points out that whereas a thermonuclear bomb is made mainly of heavy hydrogen, the sun contains ordinary hydrogen with only a trace of the heavy hydrogen isotopes. Whereas heavy hydrogen can burn explosively by strong nuclear reactions, ordinary hydrogen can react with itself only by the weak-interaction process. This proton-proton reaction proceeds about 10^{18} times more slowly than a strong nuclear reaction at same density and temperature. Dyson attributes at least three fortunate circumstances to the weak-interaction hangup: (a) without it, there would not have been a long-lived and stable sun; (b) the ocean would constitute an excellent thermonuclear high explosive; (c) hydrogen has survived rather than having been consumed in the initial, hot, dense phase of the evolution of the universe. (4) Because the transport of energy from the hot interior of the earth to the surface requires billions of years, the earth remains geologically active, these processes deriving their energy from the original gravitational condensation of the earth estimated as some 4 billion years ago. See also **Geothermal Energy.** (5) Dyson concludes the list of hangups with a special surface tension hangup, accounting for the survival of fissionable uranium and thorium nuclei in the earth's crust. They contain a high positive charge and excessive electrostatic energy such that they are ready to explode when triggered. However, before this can happen their surface must be stretched into a nonspherical shape. This process is opposed by an extremely powerful force of surface tension, estimated at about 10^{18} times stronger than that of a drop of water. Thus, it is estimated that fewer than one in a million of the earth's uranium nuclei fission spontaneously.

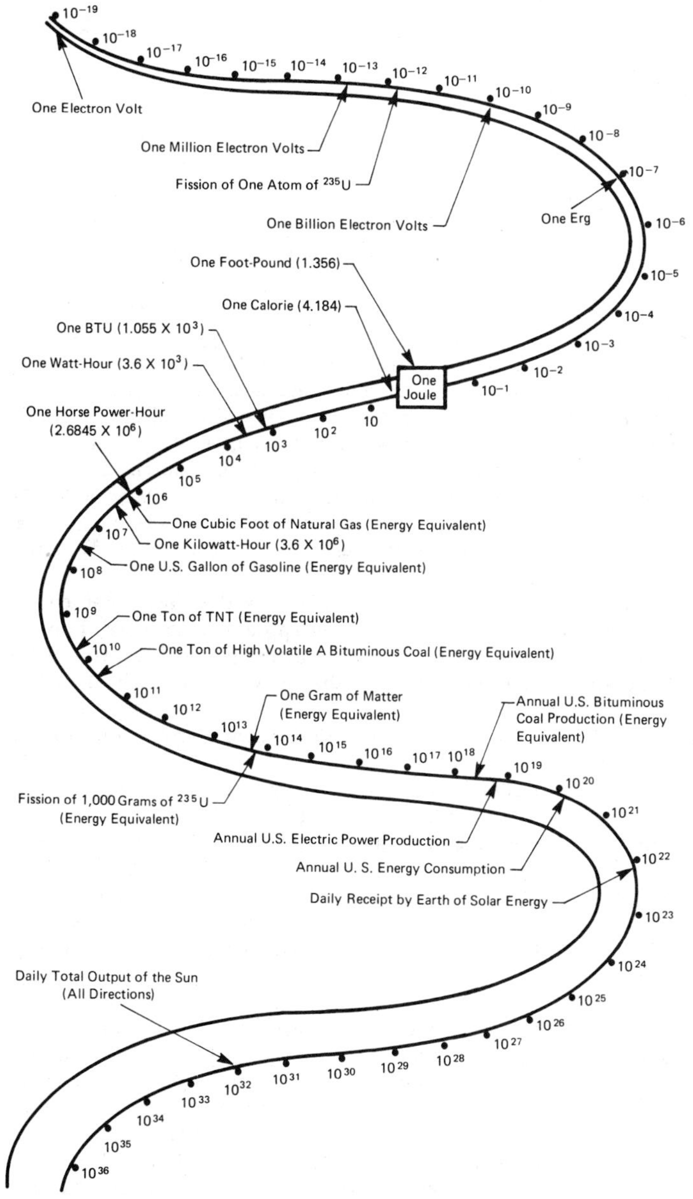

Spectrum of various energy quantities. (SOURCE: *Omnibix U.S.A.*)

Energy Technology

Numerous entries in this volume are devoted to descriptions of energy technology—coal technology, gas technology, nuclear energy technology, petroleum technology, shale oil, tar sands, geothermal technology, solar technology, wind energy, etc. Specific entry titles are listed at the end of this entry, as well as other entries relating to various aspects of energy.

Breadth in the "Packaging" of Energy. In the accompanying diagram, a packet of energy of 1 joule (1 newton-meter) is represented by the box in the upper center. Various energy packets, ranging from 1 electron volt (10^{-19} joule) to the daily energy output of the sun (total—in all directions) of 10^{32} joules are indicated.

Perfecting contemporary energy resources and power generation and consumption and developing presently nonconventional energy resources and systems possibly pose the greatest challenge to scientists and technologists in the last quarter of this century. In particular, scientists and technologists must be encouraged to work better together in an effective and realistic fashion to create constructive solutions to the problem of the energy/environment interface.

See also the following entries in this book: **Boiler; Boiling; Burner; Carburetor; Catalytic Converter; Coal; Combustion; Diesel Engine; Electric Power; Electric Power Generators; Electric Power Production and Requirements; Electric Power Transmission; Feedwater (Boiler); Fuel; Fuel Cells; Gas and Expansion Turbines; Geothermal Energy; Hydroelectric Power; Hydrogen (Fuel); Internal Combustion Engine; Kinetic Energy; Natural Gas; Nuclear Reactor; Oil Shale; Petroleum; Potential Energy; Solar Energy; Steam Engine; Substitute Natural Gas (SNG); Supercharger; Switching (Power); Tar Sands; Thermoelectric Cooling; Tidal Energy; Turbine (Steam).** For waste energy, see **Biomass and Wastes as Energy Sources.**

References

Asbury, J. G., Geise, R. F., and R. O. Mueller: "Electric Heat," *Technology Review (MIT)*, **82**, 3, 32–46 (1980).

Ashworth, J.: "Renewable Energy for the World's Poor," *Technology Review (MIT)*, **82**, 2, 42–49 (1979).

Berg, C. A.: "Process Innovation and Changes in Industrial Energy Use," *Science*, **199**, 608–614 (1978).

Bodansky, D.: "Electricity Generation Choices for the Near Term," American Public Power Association National Conference, Seattle, Washington, June 1979.

Considine, D. M. (editor): "Energy Technology Handbook," McGraw-Hill, New York, 1976.

Diennes, L., and T. Shabad: "The Soviet Energy System," Winston, Washington, D.C., 1979.

Dyson, F. J.: "Energy in the Universe," *Sci. Amer.*, **224**, 3, 50–59 (1971).

Funk, G. L.: "Automation Turns Energy Conservation Theory into Reality," *Chemical Engineering Progress*, **76**, 4, 46–52 (1980).

Grey, J., Sutton, G. W., and M. Zlotnick: "Fuel Conservation and Applied Research," *Science*, **200**, 135–142 (1978).

Harte, J., and M. El-Gasseir: "Energy and Water," *Science*, **199**, 623–634 (1978).

Hayes, E. T.: Energy Resources Available to the United States, 1985–2000," *Science*, **203**, 233–239 (1979).

Hirst, E., and B. Hannon: "Effects of Energy Conservation in Residential and Commercial Buildings, *Science*, **205**, 656–661 (1979).

Homes, J. (editor): "Energy, Environment, Productivity," National Science Foundation, Washington, D.C., 1973.

Hoogendoorn, C. J., and N. H. Afgan (editors): "Energy Conservation in Heating, Cooling, and Ventilating Buildings," 2 volumes, Hemisphere Publishing, Washington, D.C., 1978.

Inhaber, H.: "Risk with Energy from Conventional and Nonconventional Sources," *Science*, **203**, 718–723 (1979).

Kalhammer, F. R.: "Energy-Storage Systems," *Scientific American*, **241**, 6, 56–65 (1979).
Karkheck, J., Powell, J., and E. Beardsworth: "Prospects for District Heating in the United States," *Science*, **195**, 948–955 (1977).
Martin, W. F., and F. J. P. Pinto: "Energy for the Third World," *Technology Review (MIT)*, **80**, 7, 48–56 (1978).
Paisson, B. O., et al.: "Biomass as a Source of Chemical Feedstocks," *Science, 213*, 513–571 (1981).
Pound, R. V.: "Radiant Heat for Energy Conservation," *Science*, **208**, 494–495 (1980).
Rebnett, J. D.: "Engineering Approaches to Energy Conservation," *Chemical Engineering Progress*, **74**, 5, 43–88 (1978).
Reynolds, J. Z.: "Power Plant Cooling Systems: Policy Alternatives," *Science*, **207**, 367–372 (1980).
Shinskey, F. G.: "Energy Conservation through (Automatic) Control," Academic, New York, 1977.
Staff: "Energy in Transition—1975–2010," Freeman, San Francisco, 1980.
Staff: "U.S. Energy Outlook—Fuels for Electricity," National Petroleum Council, Washington, D.C. (Revised periodically).
Staff: "Statistical Year Book on the Electric Utility Industry," Edison Electric Institute, New York (Published annually).
Staff: "Semi-Annual Electric Power Survey," Edison Electric Institute, New York (Published semi-annually).
Staff: NOTE: There is almost a continuous stream of publications emanating from the following organizations: American Public Power Association, Washington, D.C.; National Rural Electric Cooperative Association, Washington, D.C.: Federal Power Commission, Washington, D.C.; National Coal Association, Washington, D.C.; American Gas Association, Arlington, Virginia; American Petroleum Institute, Washington, D.C., National Electrical Manufacturers Association, New York; The Electric Power Research Institute, Palo Alto, California.
Tybach, L., and L. J. P. Miffler (editors): "Geothermal Systems," Wiley-Interscience, New York, 1981.
Ward, G. M., Sutherland, T. M., and J. M. Sutherland: "Animals as an Energy Source in Third World Agriculture," *Science*, **208**, 570–574 (1980).
Zaleski, C. P. L.: "Energy Choices for the Next 15 Years: A View from Europe," *Science*, **203**, 849–851 (1979).
Zvolner, W. M.: "Conserving Energy through Computerized Building Automation," *In-Tech*, **26**, 3, 35–38 (1979).

ENERGY BALANCE (Planet). Heat Balance (Planet).

ENERGY BANDS (Semiconductor). Semiconductor.

ENERGY CONSERVATION. Conservation Laws and Symmetry; Insulation (Thermal).

ENERGY CONVERSION. Battery; Biomass and Wastes as Energy Sources; Coal; Fuel Cell Geothermal Energy; Natural Gas; Solar Energy; Substitute Natural Gas (SNG); Tidal Energy; Wind Power.

ENERGY DENSITY (Battery). Battery.

ENERGY (Earthquake). Earthquakes, Seismology, and Plate Tectonics.

ENERGY (FREE). Free Energy.

ENERGY (Fuel Cell). Fuel Cells.

ENERGY (Geothermal). Geothermal Energy.

ENERGY (Impact). Impact.

ENERGY (Kinetic). Kinetic Energy.

ENERGY LEVEL. A stationary state of energy of any physical system. The existence of many stable, or quasi-stable, states, in which the energy of the system stays constant for some reasonable length of time, is an essential characteristic of quantum-mechanical systems, and is the basis of large areas of modern physics.

1. The motions of electrons within an atom may be described by various orbitals, which are wave functions corresponding to various quantized orbits. The characteristic spectra of atoms are emitted when an electron shifts from an orbital of higher energy to one of lower energy. Such transitions are shown on an energy level diagram. Figure

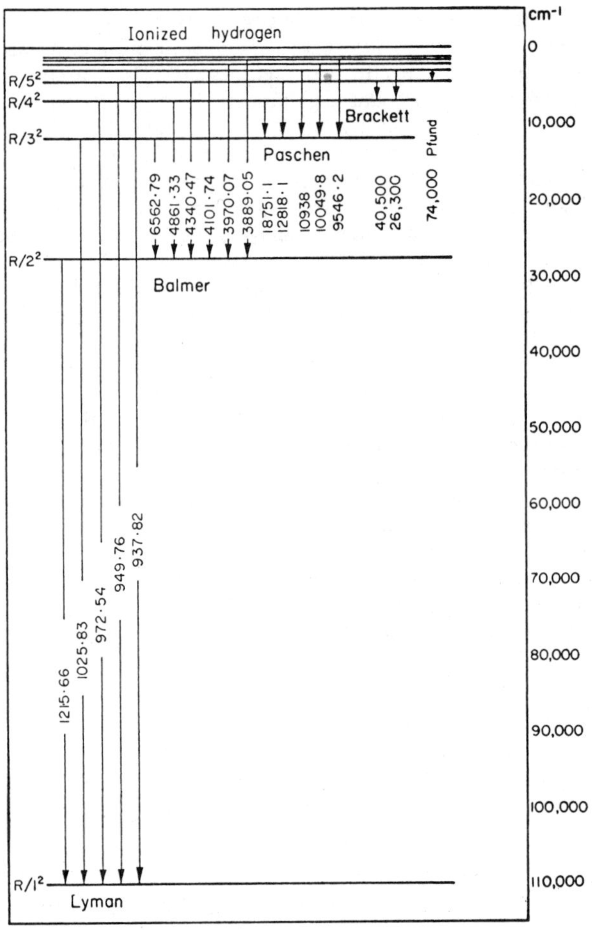

Fig. 1. Energy levels of the hydrogen atom.

1 shows the energy level diagram for hydrogen, on which the various spectral series are designated. The atomic energy level diagrams for other elements having only one electron in their outer "shells" (that is, those orbitals in an atom which are designated by the highest principal quantum number possessed by any of the electrons in that atom) are quite similar to those of hydrogen, as is apparent in Fig. 2, which is the atomic energy level diagram for lithium.

On the other hand, the atomic energy diagram for elements having more than one electron in their outer "shells" is more complex, as is seen from the energy level diagram for calcium (Fig. 3), which has two electrons in its outer "shell."

2. The fine structure of x-ray lines may be described by an energy level diagram that is closely similar in appearance to those of optical spectra (Fig. 4). However, the selection rules permit fewer transitions of the kind that produce x-rays (that is, between "shells") so that the x-ray energy levels (and the x-ray spectra they represent) are simpler than the atomic energy levels (and their corresponding optical spectra).

3. Even more complex optical and x-ray energy diagrams are found in heavy elements, which have multiplet levels (that is, groups of spectral lines very close in frequency and their corresponding levels). The transition elements exhibit particularly high multiplicities (e.g., chromium and iron have them as high as seven and manganese as high as eight). Moreover, in magnetic and electric fields the Zeeman and Stark effects cause splitting of the energy levels and further complexity.

4. As might be expected, the molecular energy level diagrams are far more complex than the atomic ones. Figure 5 shows some of the transitions that may occur between only three levels of one molecular energy state in the ultraviolet band system of the nitrogen molecule.

5. Just as electronic transitions of the orbital electrons of the atom are represented by energy level diagrams, so are the nuclear changes in radioactivity. Figure 6 shows the decay process of Pb-210, in which this lead isotope emits an electron (and a neutrino) to produce a bismuth-210 nucleus in an excited state. Emission of a gamma ray

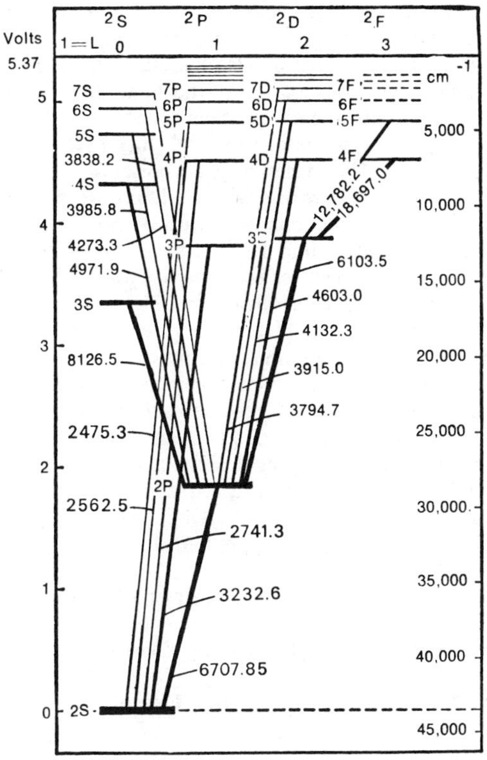

Fig. 2. Energy level diagram of the lithium atom. (*After Grotian.*)

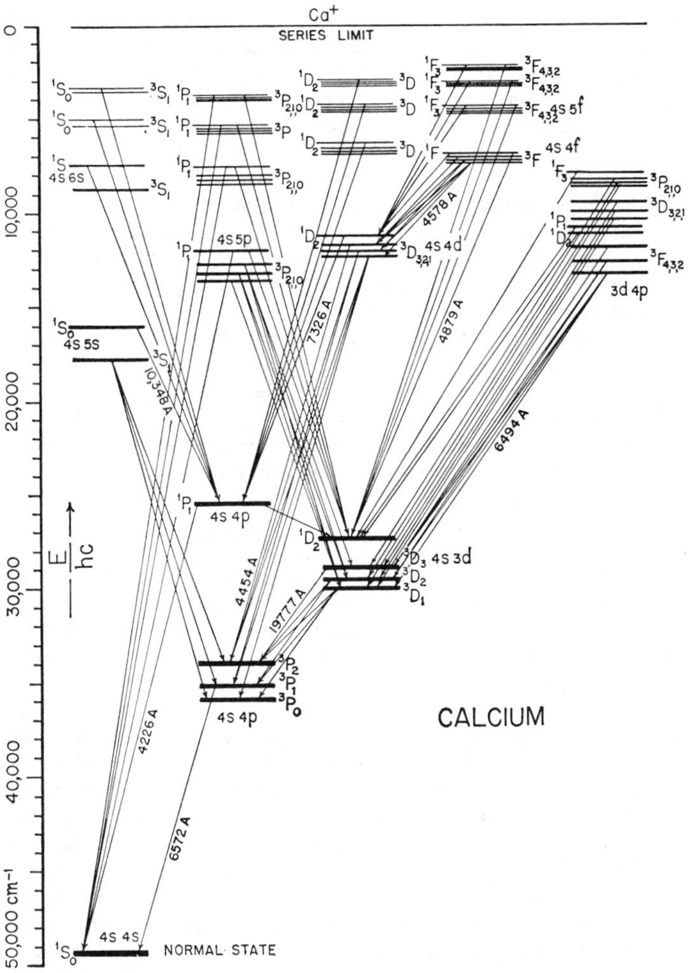

Fig. 3. Energy level diagram for neutral calcium atoms, showing electron configurations and some of the more prominent spectral transitions.

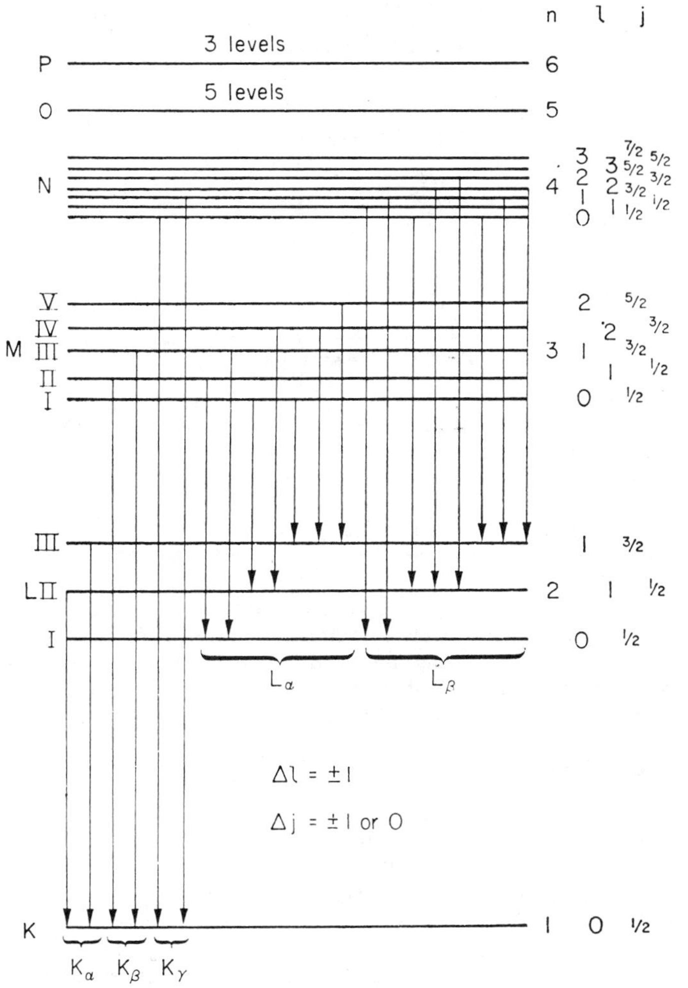

Fig. 4. Fine structure of x-ray energy levels.

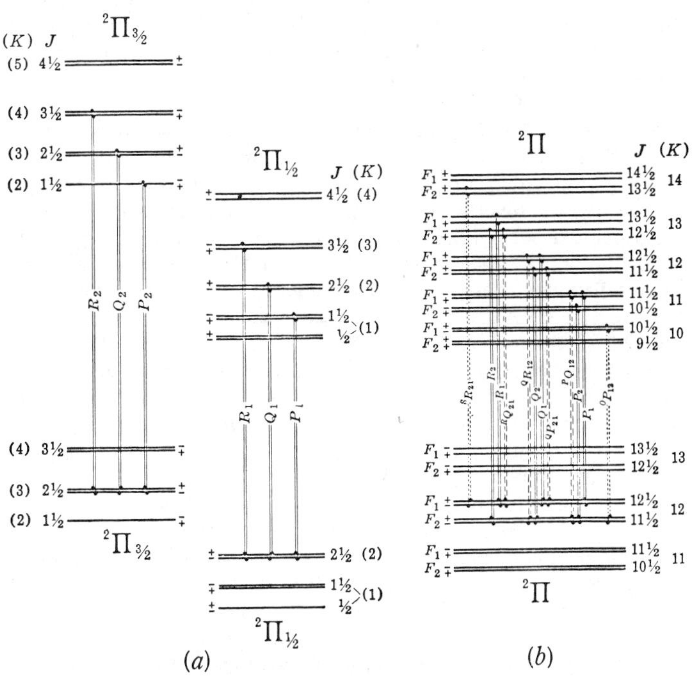

Fig. 5. Energy level diagrams for ${}^2\Pi$–${}^2\Pi$ bands: (*a*) in Hund's case, (a) [${}^2\Pi$(a)–${}^2\Pi$(a)]; (*b*) in Hund's case, (b) [${}^2\Pi$(b)–${}^2\Pi$(b)]. Only one line of each branch is given. In the designation of the branches, components of the same Λ-type doublet are not distinguished. In (*a*), the ${}^2\Pi_{1/2}$ and ${}^2\Pi_{3/2}$ levels form the $F_1(J)$ and $F_2(J)$ series, respectively. The dotted branches in (*b*) do not appear when both ${}^2\Pi$ states belong strictly to Hund's case (*b*).

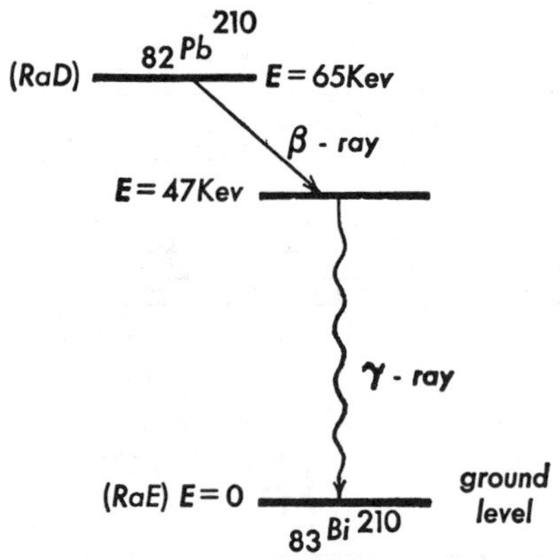

Fig. 6. Energy-level diagram illustrating beta-ray emission followed by gamma-ray emission.

effects deexcitation, yielding a bismuth-210 nucleus at ground level. The latter in turn emits another electron to become polonium-210 (not shown in figure).

6. Of course, energy level diagrams are not limited to single atoms, molecules, or nuclei. If a block of stone, or any other object, is represented at different heights, the change in potential energy may be shown on a diagram. Moreover, a system of entities may also be represented on an energy level diagram.

ENERGY LEVELS. Selection Rules (Energy Levels).

ENERGY LOSS. Combustion.

ENERGY MANAGEMENT. This term has become popular during the past decade. Essentially it is the common sense approach to energy usage, aided by instruments, computers, and controls, to eliminate waste and penalties that may be involved in the use of certain kinds of energy at certain times of the 24-hour period. Generally, the term has been used in connection with heating, ventilating, and air conditioning (HVAC) system for public and private buildings. Depending upon the age of a building and upon the degree of sophistication desired, the initial and operating costs of an energy management system range widely. Within recent years, telephone utilities have been encouraging the use of existing wiring within a building to avoid extensive new conduiting for carrying signals from sensors to computers and controls. The increasing availability, as of the early 1980s, of microprocessor-based systems is providing further impetus for computerized building automation, through modularity for incremental growth and for meeting the diverse but downscale needs of smaller buildings.

As pointed out by Zvolner (Johnson Controls, Inc.), many of the opportunities for economy in building automation result from flexibility in scheduling energy-intensive equipment to meet operating loads. For example HVAC systems are typically sized to handle peak conditions and are consequently run at partial load most of the time; likewise, process units, such as water heaters or charge tanks, have inherent storage capacities and therefore may be interrupted for short periods without affecting service. The opportunities for economy are evident in lowering demand charges in the billing structures of electric, steam, and other utility companies.

Demand control systems are usually interfaced directly with the utility metering equipment, to obtain signals indicating the outset of each billing interval and the rate of energy consumption. Some systems predict demand level at the ends of the periods, based upon rates of consumption and elapsed time. Loads are shed when the projected value exceeds an upper limit and restored when the estimate reaches a lower boundary. Another approach is to establish an ideal rate, usually a linear function with an offset to prevent control action too early in the interval. The system sheds or restores loads depending upon the departure of the measured consumption from the target curve. Once the decision is made to shed or restore load, the computer selects units to be deactivated or energized. The criteria normally considered include capacity to be removed or added, duty cycles of equipment, present and projected needs for resources, and lengths of time that devices have been out of service.

Building automation systems also can start or stop equipment to minimize energy consumption. Heat and lights in work areas can be operated in accordance with predetermined occupancy schedules. Likewise, computer systems can control duty cycling when HVAC load is below peak. The cycles may be prescheduled or determined algorithmically in accord with space conditions, to achieve the maximum energy conservation consistent with desired comfort levels.

Most large air conditioning systems have means for switching between outside and return air for economy—with the dry bulb temperature of the outdoor air used as the controlling factor. Thermodynamic considerations indicate that better energy efficiency can be achieved if enthalpy as well as temperature is considered.

The energy management concept can be applied to any kind of energy-consuming operation. It represents the comingling of excellent advance planning which takes full advantage of past historical records, the use of instruments to sense critical variables, and the programming of computers to make decisions, based upon sensor inputs, to effect control through various actions—valving, switching, etc.—all in the interest of optimizing energy supply and demand.

ENERGY (Nuclear). Nuclear Reactor; Sun (The).

ENERGY (Nuclear Binding). Binding Energy.

ENERGY (Nuclear Fission). Nuclear Fission.

ENERGY (Potential). Potential Energy.

ENERGY (Solar). Solar Energy.

ENERGY STATE TERMS. Terms designating the discrete energy states of a particle in a system. Thus the energy states of an atom are called $S, P, D, F, \ldots$ terms, respectively, corresponding to the values 0, 1, 2, 3, . . . of L, the resultant angular momentum quantum number of the atom. The energy states of a molecule are called $\mathbf{\Sigma}$, $\mathbf{\Pi}$, $\mathbf{\Delta}$, $\mathbf{\Phi} \ldots$ terms, respectively, corresponding to the values 0, 2, 3, . . . of Λ, the electronic orbital angular momentum (about internuclear axis) quantum number.

The letters indicating the value of L are usually preceded by a superscript denoting the multiplicity and followed by a subscript denoting the total angular momentum quantum number J. In addition, the principal quantum number is often written as a coefficient. Energy state terms and their transitions are shown in energy level diagrams.

Magnetic Energy State. A magnetic dipole of a moment μ in a magnetic field of flux density B has an energy that depends on orientation, $E = -\mu B \cos\theta$, or in vector notation $E = -\boldsymbol{\mu}\cdot\mathbf{B}$, in mksa units ($-\boldsymbol{\mu}\cdot\mathbf{H}$ in emu). In atomic and nuclear systems the orientation of $\boldsymbol{\mu}$ relative to $\mathbf{B}$ is quantized, only certain values of $\cos\theta$ being allowed. Transitions between these allowed magnetic energy states may take place with the emission or absorption of electromagnetic (magnetic dipole) radiation of frequency given by the Bohr condition:

$$\nu = \Delta E/h, \quad \text{or} \quad \omega = \Delta E/h$$

Particles such as electrons, protons, nuclei, etc., have intrinsic magnetic moments $\boldsymbol{\mu} = egh\mathbf{I}/2M$, where $\mathbf{I}h$ is the spin angular momentum, g the g-factor of the particle, and M is the mass of the electron or the mass of the proton. The magnetic energy states are thus given by $E = -heg\mathbf{I}\cdot\mathbf{B}/2M = -hegmB/2M$. Here m is the magnetic quantum number, which can take on the values $-I, -(I-1), \ldots (I-1), I$ where I is the spin quantum number ($\frac{1}{2}$ for electrons and protons). The energy is also often written $E = -\mu_B gmB$ or $-\mu_N gmB$ where μ_B and μ_N are the Bohr magneton and nuclear magneton respectively. The magnetic quantum number can only change by ± 1 as a result of the emission of radiation, so that there is only one emission or absorption frequency $\omega = egB/2M$.

Negative Energy State. 1. Any bound state, in which the sum of the kinetic energy and the potential energy, the latter reckoned relative to zero at infinity, is less than zero. The existence of such states is essential for the stability of any system that is not surrounded by a region of positive potential energy, such as the Coulomb barrier.

2. A consequence of the Dirac electron theory is that there exist electron states of *negative* total energy (including both rest mass energy and kinetic energy). Electrons in such states of negative energy are unobservable, only electrons of positive total energy being observable. The allowed positive and negative states are shown in the diagram (only $E > m_0c^2$ and $E < -m_0c^2$ are allowed in a field-free region). If a γ-ray photon of energy greater than $2m_0c^2$ (where m_0 is the rest mass energy of the electron) is absorbed by an electron of negative energy, it will be lifted into a positive energy state and will become observable. The positron is identified with the hole that is left behind.

ENERGY STORAGE SYSTEMS. Battery; Fuel Cells; Hydroelectric Power; Natural Gas; Petroleum; Solar Energy; Tidal Energy.

ENERGY (Thermonuclear). Nuclear Reactor; Sun (The).

ENERGY (Tidal). Tidal Energy.

ENERGY TRANSFER (Biological). Biological Energy Transfer.

ENERGY UNITS. Units and Standards.

ENGELMANN SPRUCE. Spruce Trees.

ENKEPHALINS. Brain and Nervous System.

ENGINE. In common usage, the term engine is used widely for devices which produce motion. In stricter technical sense, an engine is said to transform energy, especially heat energy, into mechanical work. Among the prime movers, those in which the power originates in a piston and cylinder are classed as engines, while those with purely rotative motion are known as turbines.

ENGINE (Air). Gas and Expansion Turbines.

ENGINE (Diagram Factor). Steam Cycles (Diagram Factor).

ENGINE (Four-Cycle). Four-Cycle Engine.

ENGINE (Internal Combustion). Automotive Electronics; Internal Combustion Engine.

ENGINE (Reaction). Reaction Engine.

ENGINE (Rocket). Rocket Propellants.

ENGINE SPEED CONTROL. Governor.

ENGINE (Steam). Steam Engine.

ENGINE (Thermal Efficiency). Thermal Efficiency.

ENGINE VIBRATION MEASUREMENT. Torsional Vibration Measurement.

ENGLER VISCOSIMETER. Viscosity.

ENHANCER (Flavor). Flavor Enhancers and Potentiators.

ENRICHMENT. 1. Also "secondary enrichment." The term applied by students of ore deposits to the natural processes by which the lower levels of an ore deposit are enriched at the expense of the upper levels, or the original protore. Particularly applied to lodes in which the sulfide ores have been concentrated by the leaching of the upper levels of the vein and redeposition below the groundwater table. Important ore minerals belonging to this type are chalcocite and argentite.

2. Any process which changes the isotopic ratio; in reference to uranium, it is a process which increases the ratio of ^{235}U to ^{238}U in uranium by separation of isotopes. Enrichment processes include thermal diffusion, gaseous barrier diffusion, and centrifugal and electromagnetic separation. Laser technology is also used.

See **Uranium.**

ENRICHMENT (Fuel). Nuclear Reactor.

ENSEMBLE. A collection of similar systems considered in statistical mechanics. Ensembles were introduced by Gibbs, and their importance lies in the fact that the average behavior of a system in an ensemble can often be used to predict the behavior of an actual physical system. Usually, all systems in an ensemble are supposed to have the same number of constituent particles. Such ensembles are called *petit ensembles.* Examples of petit ensembles which are used extensively are the *microcanonical* and *macrocanonical ensembles.* In a *microcanonical ensemble*, the variation in energy (or other independent variable) of all the systems lies within an infinitesimal range. Over this range, the assembly is in statistical equilibrium. In a *macrocanonical ensemble*, there is present a collection of identical microcanonical ensembles. In both of them, and any other *canonical ensemble*, the distribution in energy of the system is given by the Boltzmann factor. In a *cooperative ensemble*, the interactions between the systems composing the ensemble are not negligible. The state of a given system is largely determined by the states of the neighboring systems, while in an *ideal ensemble*, such as a perfect gas or ideal solution, these interactions can be neglected.

The *density of an ensemble* is written as the quantity ρ defined in such a way that $\rho\, d\Omega$ is the fraction of systems in an ensemble which have values of the momenta and position coordinates of all the particles in the system corresponding to a point in gamma-space within the extension in phase $d\Omega$. For grand ensembles one must suitably alter this definition, *grand ensembles* being defined by Gibbs as ensembles which do not have the same numbers of particles. The most often used grand ensemble is the *grand canonical ensemble*, which has a density ρ defined by

$$\rho = e^{-q+\nu n-\beta\epsilon}$$

where q is a (normalizing) constant, $\beta = 1/kT$ (k, Boltzmann's constant; T, absolute temperature), ϵ, the energy of the system, $\nu = \beta g$ (g, the partial thermal potential), and n the number of particles of the system.

ENSIGN FLY (*Insecta, Hymenoptera*). Small parasitic insects whose abdomen is elevated on a slender stalk above the thorax. It has been likened to a flag and gives the common name to the group. The ensign flies make up the family *Evaniidae.* In all species whose habits are known, the larvae are parasitic in the eggs of cockroaches.

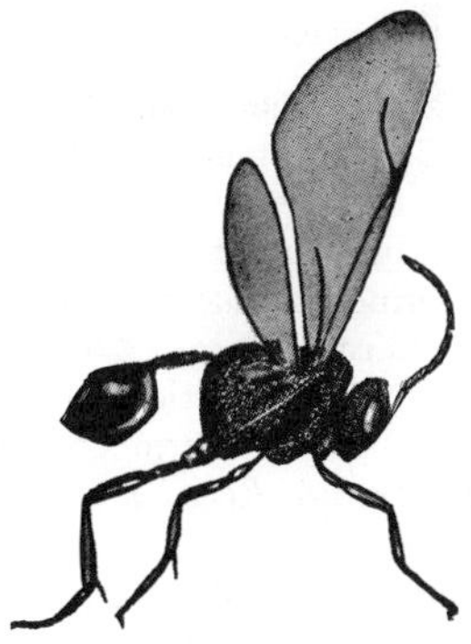

Ensign fly.

ENSTATITE. The mineral enstatite is an orthorhombic pyroxene, rarely in distinct crystals, usually found as fibrous or lamellar masses or perhaps compact. It has one easy cleavage parallel to the prism; brittle with uneven fracture; hardness 5–6; specific gravity 3.2–3.4; luster pearly to vitreous, sometimes somewhat metallic in bronzite, a variety of enstatite carrying up to 15% ferrous oxide. FeO. Color grayish to greenish or yellowish-white, green and brown. Chemically, enstatite is a silicate of magnesium, $MgSiO_3$. It occurs in igneous rocks which are high in magnesium content, like gabbros, diorites, and pyroxenites, and less commonly in metamorphic rocks. Meteorites of both the stony and metallic types have been shown to contain enstatite. It has been found at many places in Europe, Czechoslovakia, Austria, Bavaira, Germany, Norway, and the Republic of South Africa. In the United States it occurs in Putnam and St. Lawrence Counties, New York; Lancaster County, Pennsylvania; Jackson County, North Carolina, and near Baltimore, Maryland. The name enstatite is derived from the Greek word meaning *opponent*, in reference to its refractory nature; it is almost infusible. See also **Pyroxene.**

ENTERIC CAVITY. The digestive cavity, *enteron.* This cavity forms by the splitting or invagination of the inner germ layer early in embryonic (see **Embryo**) development and persists as a sac with one opening to the exterior in the coelenterates and flatworms. In this form it is also called the archenteron.

In animals with a tubular alimentary tract the enteric cavity becomes the primitive gut. Its endodermal lining becomes the glandular digestive tissue and gives rise to large glandular masses in some species, and in the terrestrial vertebrates also produces the respiratory system.

ENTERIC FEVER. Foodborne Diseases.

ENTERITIS. Any inflammation of the small intestine, usually accompanied by fever, pain in the abdomen, diarrhea and other constitutional symptoms. See **Diarrhea.**

ENTEROCOELA. Animals whose bodies contain no other cavity than that used for digestion. Coelenterates, ctenophores, flatworms, nemerteans, roundworms and rotifers.

ENTEROVIRUSES. Virus.

ENTEROZOA. Animals which have a digestive cavity, the enteric cavity. All animals but the protozoans and sponges.

ENTHALPY. The *enthalpy, H* or *heat content,* of a substance is a thermodynamic property defined as the *internal energy, E,* plus the product of the pressure, *P,* times the *volume, V,* of the substance

$$H = E + PV \tag{1}$$

The enthalpy is an extensive state function; its value depends only on the state and the amount of the substance and not on its previous history. It has the units of energy and it is usually expressed in calories (or kilocalories).

For a process at *constant pressure* ($\Delta P = 0$), in which the only work performed is the mechanical pressure-volume work ($P\,\Delta V$), the *change in enthalpy,* ΔH, is equal to the heat adsorbed by the system, q (hence the name heat content):

$$\Delta H = \Delta E + P\,\Delta V = q \tag{2}$$

This relation is a direct consequence of the definition of enthalpy by Equation (1) and of the mathematical statement of the first law of thermodynamics, namely that the change in internal energy, ΔE, is equal to the heat adsorbed minus the work done ($q - P\,\Delta V$). It is clear that this thermodynamic relation does not define absolute values of enthalpy or internal energy. Changes in enthalpy, however, are readily measured by calorimetric techniques, and the relative enthalpy values are sufficient for all thermochemical calculations.

Enthalpy-Temperature Relation and Heat Capacity. When heat is adsorbed by a substance, under conditions such that no chemical reaction or state transition occur and only pressure-volume work is done, the temperature, *T,* rises and the ratio of the heat adsorbed, over the differential temperature increase, is by definition the heat capacity. For a process at constant pressure (following Equation (2)), this ratio is equal to the partial derivative of the enthalpy, and it is called the *heat capacity at constant pressure,* C_p, (usually in calories/degree-mole):

$$\left(\frac{\partial H}{\partial T}\right)_p = C_p \tag{3}$$

The temperature dependence of *H* for a substance remaining in the same physical state can be expressed as a function of C_p by integration of Equation 3.

If a substance undergoes a transformation from one physical state to another, such as a polymorphic transition, the fusion or sublimation of a solid, or the vaporization of a liquid, the heat adsorbed by the substance during the transformation is defined as the *latent heat of transformation* (transition, fusion, sublimation or vaporization). It is euqal to the enthalpy change of the process, which is the difference between the enthalpy of the substance in the two states at the temperature of the transformation. For the purpose of thermochemical calculations, it is usually reported as a molar quantity with the units of calories (or kilocalories) per mole (or gram formula weight). The symbol *L* or ΔH, with a subscript *t,f* (or *m*), *s,* and *v* is commonly used and the value is usually given at the equilibrium temperature of the transformation under atmospheric pressure, or at 25°C. For a substance undergoing one phase transformation, with a latent heat Δh_t at a temperature T_t, the enthalpy change between two temperatures, T_1 and T_2, such that $T_1 < T_t < T_2$, is given by

$$H_T - H_T = \int_{T_1}^{T_t} C_p'\,dT + \Delta h_t + \int_{T_t}^{T_2} C_p''\,dT \tag{4}$$

where C_p' and C_p'' are the heat capacities of the substance in the two different physical states. For several successive transformations, additional terms are added. The accompanying diagram illustrates the temperature dependence of enthalpy and heat capacity.

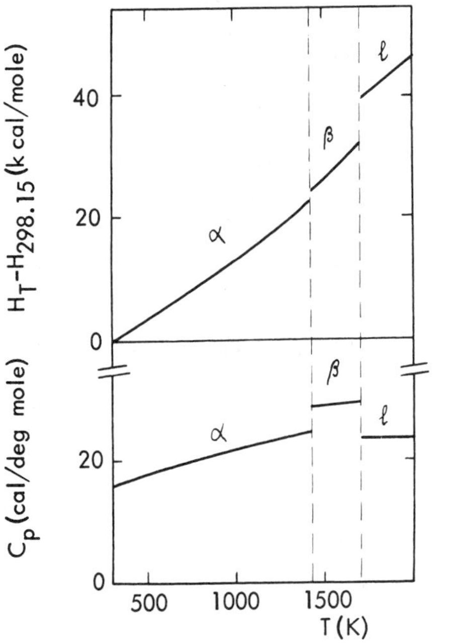

Example of the temperature dependence of enthalpy relative to 25°C, $H_T - H_{298.15}$, and heat capacity, C_p. The data are for fluorite, CaF_2 (*K. K. Kelley, 1960*). The discontinuities in the lines correspond to the α to β transition (1424 K) and the fusion (1691 K).

Very precise measurements of the heat capacity of liquids and solids can be obtained by calorimetric techniques at relatively low temperatures (below 200°C) and they can be extrapolated down to the absolute zero of temperature (−273.15°C) by reliable theoretical expressions. In that temperature range, heat capacity data are usually very accurate and enthalpy values are obtained by integration (Equation (4)). The most reliable method for determining high-temperature enthalpies and heat capacities is the dropping method (or method of mixtures) which consists of dropping the substance under investigation from a furnace at a known temperature into a calorimeter at room temperature. This method determines directly the change in enthalpy (or heat content) of the substance between the temperature of the furnace and that of the calorimeter. Heat capacities are obtained by differentiation (Equation (3)). The measurement of heat capacity of gases is usually more difficult, and their thermodynamic properties can be more accurately calculated by methods of statistical mechanics based upon energy level of gas molecules obtained from spectroscopic data, or upon the knowledge of the molecular configuration and the vibration frequencies of the molecules.

Molar enthalpy data for elements and inorganic compounds above room temperature are usually tabulated in the form of the heat content above a reference temperature, usually 298.15°K =25°C. They are represented by: $H_T - H_{298.15}$ in calories/mole. The data are correlated over a range of temperature by empirical equations such as a series of powers of the absolute temperature or such as the following expression adopted by K. K. Kelley (1960) for his extensive compilation of data on inorganic compounds:

$$H_T - H_{298.15} = aT + bT^2 + c/T + d \quad (5)$$

where T is the absolute temperature (°K) and a, b, c, d are constants determined from experimental data. The corresponding equation for heat capacity is:

$$C_p = a + 2bT - c/T^2 \quad (6)$$

Standard Enthalpy of Formation. For the convenience of tabulation and computation of thermodynamic data, it is essential to present them in a commonly accepted form relative to a single standard state of reference. At all temperatures, the *standard state* for a *pure liquid or solid* is the *condensed phase under a pressure of 1 atmosphere.* The standard state for *a gas* is the *hypothetical ideal gas at unit fugacity* (equivalent to a "perfect gas" state), in which state the enthalpy is that of the real gas at the same temperature when the pressure approaches zero. Values of thermodynamic quantities for standard-state conditions are identified by a superscript ^{0}, and H^0, for instance, is the enthalpy change of a reaction when reactants and products are in the standard state.

The *standard enthalpy of formation,* ΔHf^0 (also represented by ΔH_f^0 or simply H_f^0), of a substance at a given temperature is by definition, the enthalpy change when 1 mole of the substance in its standard state is formed, isothermally, at the indicated temperature from the elements, each in its standard state. Usual units are kilocalories/mole. *For all elements* in their *stable form at* 25°*C* (298.15°K), the *enthalpy of formation is zero.* If solid substances have more than one crystalline form, the most stable one is taken as the standard state, and the others have slightly different enthalpies. This convention about zero enthalpy is arbitrary but universally accepted, and it may be compared to the arbitrary choice of zero for terrestrial altitudes. The combination of enthalpies of formation, enthalpies of transition, and heat capacities makes possible the calculation of the enthalpy of a substance, in a given state at a given temperature, relative to a commonly accepted reference.

Enthalpy calculation for mixtures is more complex than for pure substances and its discussion is beyond the scope of the present article. *Aqueous solutions* (q.v.), however, are very important from a geochemical point of view, and reliable data are usually available. The *enthalpy of solution* is the enthalpy change resulting from the dissolution of a substance; it is a function of the solute concentration and its values are reported accordingly in the literature. For a *solute in aqueous solution,* a *standard state* is defined as the *hypothetical ideal solution of unit molality* (1 mole of solute per 1000 grams of water). In this state, the partial molal enthalpy and heat capacity of the solute are the same as in the infinitely dilute real solution. Although it is impossible to prepare a solution of only one ionic species (since the system must remain neutral), it is convenient to apportion the enthalpy (and other thermodynamic properties) between the various ions. This apportionment is not unique and an additional convention has to be made, namely that the *standard enthalpy of hydrogen* ion in aqueous solution (aq) at unit activity, ΔHf^0 for H^+ (aq), *is zero.* The properties of a neutral electrolyte in the standard state are equal to the algebraic sum of the values corresponding to the individual ions.

Heat of Reaction and Gibbs Free Energy Change. For a chemical reaction, at constant pressure with only pressure-volume work performed, the heat adsorbed by the process, q, is equal to the enthalpy change, ΔH, or the sum of enthalpies of the products of the reaction minus the sum of enthalpies of the reactants (taking into account the amount of each).

$$q = \Delta H = \Sigma H_{\text{products}} - \Sigma H_{\text{reactants}} \quad (7)$$

Those heat effects can be easily calculated when the enthalpies of formation and the enthalpy-temperature relations are available for the substances considered. Usually, the *heat of reaction* is defined as the heat evolved by the process, and it is equal to the enthalpy change but opposite in sign, while heats of fusion or vaporization always refer to the heat adsorbed, and for heats of solution the usage varies. In order to avoid any confusion, it is recommended to express heat effects of chemical process by reporting the enthalpy change, ΔH.

Early chemists thought that the heat of reaction, $-\Delta H$, should be a measure of the "*chemical affinity*" of a reaction. With the introduction of the concept of *netropy* (q.v.) and the application of the second law of thermodynamics to chemical equilibria, it is easily shown that the true measure of chemical affinity and the driving force for a reaction occurring at constant temperature and pressure is $-\Delta G$, where ΔG represents the change in thermodynamic state function, G, called *Gibbs free energy* or *free enthalpy,* and defined as the enthalpy, H, minus the entropy, S, times the temperature, $T(G = H - TS)$. For a chemical reaction at constant pressure and temperature:

$$\Delta G = \Delta H - T\Delta S \quad (8)$$

and the Gibbs free energy change can be obtained by calculating the enthalpy and entropy change and applying Equation (8). The criterion for *a spontaneous chemical reaction* is that ΔG be *negative,* and a chemical equilibrium corresponds to the condition $\Delta G = 0$.

Conversely, if the Gibbs free energy change is known as a function of temperature at constant pressure, the enthalpy change can be obtained by a relation which is an alternate form of the *Gibbs-Helmholtz equation,* and which can be derived from Equation (8).

$$\left(\frac{\partial(\Delta g/T)}{\partial(1/T)}\right)_P = \Delta H \quad (9)$$

This means that ΔH is the slope of the line representing $\Delta G/T$ versus $1/T$ at constant pressure.

For references, see entries on **Heat Transfer;** and **Thermodynamics.**

ENTHALPY EQUATION. **Thermodynamics.**

ENTHALPY (Steam). **Boiler.**

ENTISOLS. **Soil.**

ENTNER-DOUDOROFF PATHWAY. **Carbohydrates.**

ENTOMOLOGY. The science that deals with all facts pertaining to insects. Because of the large number of insect species and their frequent economic importance, the principal divisions of the science have been systematic entomology and economic entomology. Classification is difficult and intricate, and demands the constant service of specialists. The economic field is of such importance, especially to agriculture, that the national and state governments maintain organizations for the scientific study of insects which also assist in their control. See **Insect Control and Insecticides.**

Although entomologists may specialize in insect morphology or physiology, work of this type is less extensive than more practical studies and is of more general biological interest; hence, it takes its place largely in the subsidiary sciences of general biology.

ENTOMOPHILOUS PLANTS. **Pollination.**

ENTOMOSTRACA. A division of the class *Crustacea* formerly used to include all but the subclass *Malacostraca.*

ENTOPROCTA. A group of animals formerly classed with the *Bryozoa* but now placed in a separate phylum.

ENTRANCE SLIT. A narrow slit in an opaque screen through which light enters a spectrometer. The spectrum formed is the image of this slit in each wavelength of light present. A narrow slit is necessary for good resolution to avoid great overlapping of these images. However, the smaller the entrance slit, the less radiation enters the spectrometer. Hence a slit width must be used which is a compromise between the resolution desired and the necessary light intensity for proper observation or detection.

ENTRENCHED MEANDER. Also termed *incised meander,* a river valley which has a distinctly meandering old-age pattern (longitudinal profile), and a V-shaped or canyon-shaped, youthful transverse profile.

The meandering course of the river is inherited from the time when it flowed at, or close to, base level, that is on a relatively flat surface. Subsequent uplift of the region quickens the flow of the stream, hence its downcutting power, without necessarily altering its inherited meandering course. Entrenched meanders are therefore physiographic evidence of the rejuvenation of the erosive power of an old-age stream. Entrenched meanders may suggest the first stage in a new cycle of erosion, but usually imply an interruption in the normal cycle due to relatively sudden uplift of the region before the entire area has been reduced to a peneplain.

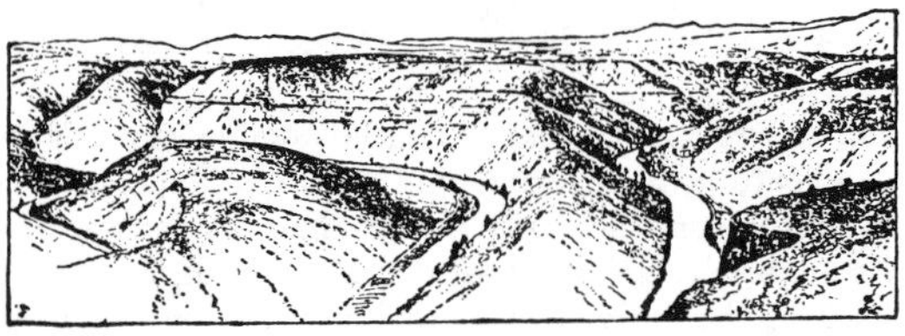

A rejuvenated region showing entrenched meanders. Yakima Canyon, Washington. (*G. O. Smith.*)

ENTROPY. 1. In the mathematical treatment of thermodynamic processes there occurs very often a quantity, now relating energy to absolute temperature, now associated with the probability of a given distribution of momentum among molecules, and again expressing the degree in which the energy of a system has ceased to be available energy. Its mathematical form suggests that these are all aspects of a single physical magnitude. Application of the second law of thermodynamics leads to the conclusion that if any physical system is left to itself and allowed to distribute its energy in its own way, it always does so in a manner such that this quantity, called "entropy," increases; while at the same time the available energy of the system diminishes. This law applies to the universe as a whole, hence the proposition that the total entropy increases as time goes on. An interesting conclusion as to entropy in the vicinity of absolute zero is expressed by the Nernst heat theorem; viz, that all physical and chemical changes in this region take place at constant entropy. Any process during which there is no change of entropy is said to be "isentropic." This is true, for example, of an adiabatic process in which there is no dissipation of energy, i.e., one which is also a reversible process. In thermodynamic discussions entropy is commonly classed, along with temperature, pressure, and volume, as one of the variables defining the state of a body, and is often graphed as such on thermodynamic diagrams.

2. In information theory, entropy is a measure of the uncertainty of our knowledge.

3. In thermodynamics, entropy is defined by the equation

$$dS = dQ/T$$

where dS is an infinitesimal change in the entropy of a system, dQ is the infinitesimal amount of heat that enters the system, and T is the absolute temperature.

In statistical mechanics, entropy is

$$k \log_e P + \text{constant}$$

where k is Boltzmann's constant, and P is the statistical probability of the state considered.

Standard Entropy. The total entropy of a substance in a state defined as standard. Thus, the standard states of a solid or a liquid are regarded as those of the pure solid or the pure liquid, respectively, and at a stated temperature. The standard state of a gas is at 1 atmosphere pressure and specified temperature, and its standard entropy is the change of entropy accompanying its expansion to zero pressure, or its compression from zero pressure to 1 atmosphere. The standard entropy of an ion is defined in a solution of unit activity, by assuming that the standard entropy of the hydrogen ion is zero.

Entropy of Disorder. That part of the entropy of a substance that is due to a disordered arrangement of the particles as opposed to a similar but ordered arrangement. The most clear-cut example is the order-disorder transition in binary alloys, in which virtually the whole entropy change is of this kind. The entropy change on fusion of a solid is largely due to entropy of disorder.

See also **Energy.**

ENTROPY (Kinetics). Kinetic Theory.

ENTROPY (Of Information). Information Theory.

ENTROPY OF VAPORIZATION. Hildebrand Rule.

ENTROPY (Thermodynamic Law). Thermodynamics.

ENTRY CORRIDOR. Depth of the region between two trajectories which define the design limits of a space vehicle which will enter a planetary atmosphere.

ENVELOPE DEMODULATOR. Demodulator.

ENVELOPE (Mathematics). The equation of a curve usually contains one or more constants, in addition to the dependent and independent variables. If the constants are regarded as variable parameters, a family of curves is generated as these parameters are assigned a series of different value. In case only one such parameter is involved, the equation for the family of curves can be written as $f(x, y, t) = 0$. If another curve, or group of curves, is a common tangent to the given family, it is said to be the envelope and it can be found by solving the two equations.

$$f(x, y, t) = 0; \quad \partial f/\partial t = f_t(x, y, t) = 0$$

to obtain x and y as a function of the parameter t.

The procedure can be generalized for families of curves which depend on two or more parameters.

See also **Circular Curves; Tangent** (Geometry); and terms listed under **Mathematics.**

ENVIRONMENT. The assemblage of material factors and conditions surrounding the living organism and its component parts.

Environment includes both external and internal factors. In the external environment inanimate objects and the forces associated with them constitute the physical environment, and the living things and their derivatives with which the animal may be associated constitute the organic environment. Within its body it maintains an organization which constitutes an internal environment to which all of its parts respond directly, whether or not they also have external contacts.

See also **Ambient Conditions; Atmosphere (Earth); Biome; Ecology; Pollution (Air); Supersonic Aerodynamics; Water Pollution;** as well as entries on various planets; and section on "Radioactive Wastes" in entry on **Nuclear Reactor.**

ENVIRONMENTAL CONTROLS (Air). Pollution (Air).

ENVIRONMENTAL WASTES (Radioactive). Nuclear Reactor.

ENVIRONMENT (Controlled). Frequently, in industry and research facilities, it is necessary to provide a planned and carefully controlled environment in which to conduct manufacturing, assembly, inspection, and test operations as well as scientific investigations. Clean rooms, temperature-controlled rooms, sound rooms, and dry rooms are terms that are often applied to special types of controlled environments. Some environmental systems are designed to control a single condition; in other cases, many variables are controlled. A clean room may require only dust particle control and a comfortable temperature for the personnel who work in the room. In the case of a room for calibrating dimensional standards and for measuring precision parts, very close temperature and humidity also will be required. Controlled environments, of course, also are required in hospitals and medical facilities, in horiticultural establishments, in museums, and numerous other industrial and commercial structures. There is a fine division between what may be termed a controlled environment and what may be inferred from the term *air conditioning.* Essentially a controlled environment differs from an air-conditioned space on two counts: (1) a controlled environment may include the measurement and very care-

Environmentally-controlled room for the measurement and inspection of precision parts and assemblies. Dust particle, temperature, relative humidity, static pressure, and air velocity are carefully controlled.

ful control of many environmental factors other than temperature and humidity which are the primary concerns of air conditioning, and (2) a controlled environment implies much greater precision in control and usually for reasons other than the comfort and well being of the occupants of a space. Further, a controlled environment may not even involve air, but rather may be concerned with the control of an aqueous environment, or of nonatmospheric factors, such as protection from shock, vibration, change of position, and so on.

Space does not permit a delineation of the many types of controlled environments encountered in industry and research. Because of the tight control requirements, an example is included of environmental control for a metrology standards laboratory and for so-called "clean rooms."

The International Organization for Standardization adopted 20°C (68°F) as the standard reference temperature for length measurement in 1951. Environmental systems for linear measurements provide temperature control at 20°C ± 0.03°C, maintain humidity between 40 and 45% (R.H.), and remove 99.97% of dust particles above 0.3 micrometer in size at the filter. Outside of the metrology laboratory, but in dimensional inspection rooms concerned with production parts, the temperature is controlled at 20°C ± 0.60°C.

Modular environmental enclosures are available in package form for on-site assembly in existing or new buildings. Each is a complete, prefabricated package containing wall and roof panels, assembly hardware, environmental control system, and numerous optional features. The modular design permits a room to be enlarged or changed to meet new requirements or to be disassembled and moved to another location in event of a change in plant layout. One modular design uses 4-foot wide aluminum clad insulated wall and roof panels. The panel has a rigid polystyrene core of insulation between sheet aluminum and a panel-wide air duct. The inside portion of the room is painted hardboard or porcelain enamel on steel. The conditioned air flows into the room draft-free through thousands of tiny holes in the ceiling and is drawn off at floor level and returned to the overhead plenum chamber upward through the panel-wide air ducts.

Environmental rooms usually have air lock entry chambers. An air shower is included where critical control of duct contamination is required. A blast of air for a measured length of time is designed to remove foreign particles from an individual's outer clothing, usually a smock or uniform.

Modular room sizes range from 12 × 12 × 12 feet (approximately a 3.5-meter cube) to 18 × 40 × 12 feet (approximately 5.5 × 12 × 3.5 meters). Ceiling height is frequently 12 feet (3.5 meters). Room features include wide service doors, material pass-through airlocks, flush mounted lighting fixtures, and outside facilities for servicing temperature, humidity, and dust controls. A facility of this type is shown in accompanying illustration.

Generally, clean rooms using either vertical or horizontal laminar air flow are designed to control contamination and other variables as follows:

Particle Count—not to exceed 100 particles per cubic foot (approximately 3500 particles per cubic meter) of air of a size of 0.5 micrometer and larger.

Temperature—72°F ± 1°F (22.2°C ± 0.6°C) under static conditions in a horizontal plane 1 foot (0.3 meter) above the floor.

Relative Humidity—50% maximum. This depends upon use.

Air Velocity—60 feet (18 meters)/minute (minimum) to 100 feet (30 meters)/minute (maximum), down to the floor in a straight-line flow.

Make-up Air—15% minimum of total air used for air conditioning.

Positive Static Pressure—0.05 to 0.10 inch (1.3 to 2.5 millimeters) of water.

ENVIRONMENT (Ecological Types). Ecology.

ENZOOTIC. Term describing any disease of animals whose incidence and distribution resemble those of endemic disease in man.

ENZYMATIC BROWNING. Antioxidant.

ENZYME. Collectively, the term *enzymes* refers to a group of proteins which catalyze a variety of chemical reactions. Over a thousand enzymes have been identified. In 1964, the International Union of Biochemistry officially named and listed nearly 900 enzymes. Since that time, progress has continued toward developing a consistent and standardized procedure for naming the enzymes. The existence of enzymes has been known since the early 1600s, mainly from the observation of their role in digestion and fermentation processes used to make alcohols and other allied products. Only in comparatively later years were the simpler enzymes isolated. Urease was produced in crystalline form in 1926. Other enzymes that were later isolated include amylase, carboxy-peptidase, chymopapain, papain, pepsin, and starch phosphorylase. See also **Enzyme Preparations.**

Enzyme complexes are generated by living cells. They function as catalysts in reactions that involve the metabolism of living organisms and thus play vital roles at practically all levels of food involvement—production, processing, and consuming, whether by fish, bird, insect, primate, etc. Enzymes and the reactions in which they participate thus are ever present during the entirety of the food chain—from start to finish. Investigations in botany, pursuits of agronomy, studies of nutrition, inquiries into plant and animal pathology, and the numerous other aspects of the sciences that are involved in life processes, when probed in depth, ultimately encounter the vital roles played by enzymes.

Expanding the knowledge of the life processes depends critically upon furthering the investigation of enzymes, their constitution, structure, synthesis, and behavior. Common properties of enzymes include: (1) their predominant, established role as catalysts for several types of chemical reactions, often providing the means of effecting chemical conversions otherwise difficult and at lower rates of energy expenditure; (2) their structure which suggests that most enzymes are simple or conjugated proteins; (3) their relatively high sensitivity to environmental conditions, including temperature and pH, and the presence of certain organic and inorganic materials; and (4) their origin from living cells. The environmental tolerance of enzymes closely parallels other substances associated with life processes. They tolerate a relatively narrow temperature span with denaturation (deactivation) occurring at temperatures generally above 50°C (122°F) and greatly reduced activity often occurs well above the freezing point of water. Enzymes have a low tolerance to a pH below 4, a minimal to no tolerance of certain organic solvents such as alcohol and acetone, and destruction by numerous organic and inorganic substances.

Unlike most inorganic catalysts, enzymes are very specific for the reactions which they catalyze. As a case in point, an acid catalyst will yield glucose, fructose, and galactose in the hydrolysis of raffinose (a trisaccharide). But, diastase will yield melibose and fructose; emulsin will yield sucrose and galactose. The glucosidic linkages are hydrolyzed at about equal rates with an acid catalyst, whereas the enzyme catalysts act on just one kind of linkage even though the difference in linkages is small. Whereas acids may catalyze numerous compounds, including amides, acetals, and esters, a given enzyme will confine its actions to a very specific compound or related group. Because of this behavior, mixtures of enzymes often can be effective.

In addition to the very large role that enzymes play in life processes and medicine and in industrial fermentation and related processes, enzymes are finding a growing role in industrial products, such as detergents, where enzymes tend to break down proteins to water-soluble proteoses or peptones. Obviously for purposes of this type, enzymes must be selected that can remain active at relatively high pH values (8.5 to 9.5) and remain stable for long periods of storage.

The great number of reactions catalyzed by the enzymes in living organisms can be indicated by mentioning some of the major types. They include all the oxidation processes by which the organism obtains its energy—mechanical and thermal; the hydrolysis processes by which food carbohydrates, proteins, and fats are broken down into simpler molecules capable of direct oxidation or of use by the organism in constructing its own structure; and all the detoxification reactions by which many harmful substances that may be absorbed by the organism, as well as its normal waste products, are converted into forms suitable for excretion.

Activators. Most enzymes can function only with the assistance of certain other substances. These are broadly designated as *activators*, and are commonly grouped into two classes. The first is that of the nonspecific activators, which take no part in the reaction and appear to act by their effect upon the enzyme itself. The most important of these are the metallic ions K^+, Na^+, Rb^+, Cs^+, Mg^+, Ca^{2+}, Al^{3+}, Zn^{2+}, Cd^{2+}, Cr^{2+}, Mn^{2+}, Fe^{2+}, Co^{2+}, Ni^{2+}, Cu^{2+}. The second class of activators, mentioned earlier in this entry, are organic molecules, which enter into the reaction itself, often as carriers of a particular group. These substances are the group discussed in the entry on **Coenzymes.** In general, they are regenerated in their original form by other processes, so that they are not strictly substrates. On the other hand, the nicotinamideadenine nucleotides, which act as hydrogen carriers for various oxidoreductase reactions, may well be regarded as substrates.

A *substrate* may be defined as a substance modified by the action of an enzyme, or by the growing upon it of microorganisms. A *coenzyme* is a low-molecular-weight organic substance which can attach itself and thus supplement specific proteins to form active enzyme systems.

Inhibitors and Primers. Two other adjunct substances are the *inhibitors*, which retard or block enzyme action; and the *primers*, which enhance, or in some cases, are essential to it. An example is the priming of polyribonucleotide phosphotransferase by short ribonucleotide polymers.

Structure of Enzymes. The knowledge of the structure of enzymes is growing at a rapid rate, but much research remains before a high confidence level can be established pertaining to even the fundamentals of certain basic reactions.

One of the intensively investigated enzymes is ribonuclease (trivial name), which has the systematic name, polyribonucleoside 2-olionucleotide-transferase. This enzyme transfers a phosphate group from one position to another within a polynucleotide, forming a cyclic compound, and the pancreatic form of this enzyme can also catalyze the transfer of the phosphate group to water, which is a step in the depolymerization of RNA.

The molecular weight of this enzyme was found (by analysis of its constituent amino acids) to be 13,700. It was found to consist of a single polypeptide chain, internally cross-linked by four cystine residues, as evidenced by the failure of any drop in molecular weight to accompany the oxidation of all four cystines to cysteic acid, and also by the occurrence of only one terminal $—NH_2$ group and one terminal —COOH group.

From this point, the primary structure was fully determined. The *primary structure* of an enzyme, or other protein, is the number, length, and composition of the polypeptide chains, the linear arrangement of their amino acids, and the number and position of the cross-links between chains. (The geometrical configuration of the molecule, which is usually a three-dimensional coiled and folded structure, and the side chains and their interactions were not determined.)

To determine the primary structure, after oxidation of the disulfide bridges between the cysteine molecules, the enzyme was cleaved into a series of linear polypeptides by the enzymatic action of trypsin. The fragments were separated by chromatography, and their individual amino acids were split off by acid hydrolysis. Then by repeating this process with another enzyme, a different series of polypeptide fragments were obtained, because the chains split at different points. By studying the overlaps of the two series, the fragments could be arranged in linear order. Combining the peptide structure so determined with the amino acids found gave the provisional primary structure of the enzyme shown in the accompanying figure.

The final step in complete elucidation of the three-dimensional structure of ribonuclease was made by researchers at the Roswell Park Memorial Institute, Buffalo, New York. This group, headed by Dr. David Harker, employed x-ray diffraction techniques. Roughly 500,000 diffraction points were recorded and the data so obtained was fed to a computer.

The elucidation of the structure of ribonuclease follows that of lycozyme by a group at London's Royal Institute headed by Dr. David C. Phillips, and that of the other protein, myoglobin, for which Dr. Max F. Perutz and Dr. John C. Kendrew of Cambridge University received the Nobel Prize in chemistry in 1962.

Note in this structure of enzyme that specific points are marked as those at which other enzymes act. This feature of "active sites" is characteristic of enzyme behavior. In the case of the enzymes trypsin or chymotrypsin, the active sites for peptide or ester hydrolysis contain the functional groups of two histidine residues and of a serine residue. The ester enters the active region, forming temporary bonds with the enzyme at that point, the —OR group of the ester becoming bonded to a hydrogen atom of the enzyme. Then the bond between the hydrogen atom and the enzyme breaks, releasing the alcohol of the ester. As a next step, H_2O adds from the solution to the complex, and by another bond rupture, the acid part of the ester is released, leaving the enzyme in its original condition. The overall reaction is a simple hydrolysis of the ester,

$$R'COOR + H_2O \rightarrow ROH + R'COOH$$

but a large number of steps may be involved.

The determination of sites of active centers is effected not only by splitting of enzymes, but also by treating them with temporary or permanent inhibitors, and determining their points of attachment. Other methods of studying enzymes are by means of enzyme induction and enzyme repression.

Enzyme Induction. An example of *enzyme induction* is the growth of the bacteria *Escherichia coli* in a suitable culture medium. If no beta-galactoside is added to the medium, the bacteria form scarcely any of the enzyme that hydrolyzes that sugar. The addition of the sugar to the medium increases the production of the enzyme by the cell by as much as 10,000 times. On the other hand, the same bacteria will produce the enzyme tryptophan synthase only if tryptophan is absent from the culture medium. These observations are useful, not only in interpreting enzyme action, but also in determining its relationship to genetics, for in this case, the genes determining the ability to synthesize both enzymes are present on the chromosome map of the organism.

There are methods used to study enzymes other than those of chemical instrumental analysis, such as chromatography, that have already been mentioned. Many enzymes can be crystallized, and their structure investigated by x-ray or electron diffraction methods. Studies of the kinetics of enzyme-catalyzed reactions often yield useful data, much of this work being based on the Michaelis-Menten treatment. Basic to this approach is the concept that the action of enzymes depends upon the formation by the enzyme and substrate molecules of a complex, which has a definite, though transient, existence, and then decomposes into the products of the reaction. Note that this point of view was the basis of the discussion of the specificity of the active sites discussed above.

A simple enzyme reaction may thus be written as

$$E + S \leftrightarrows ES \rightarrow \text{products}$$

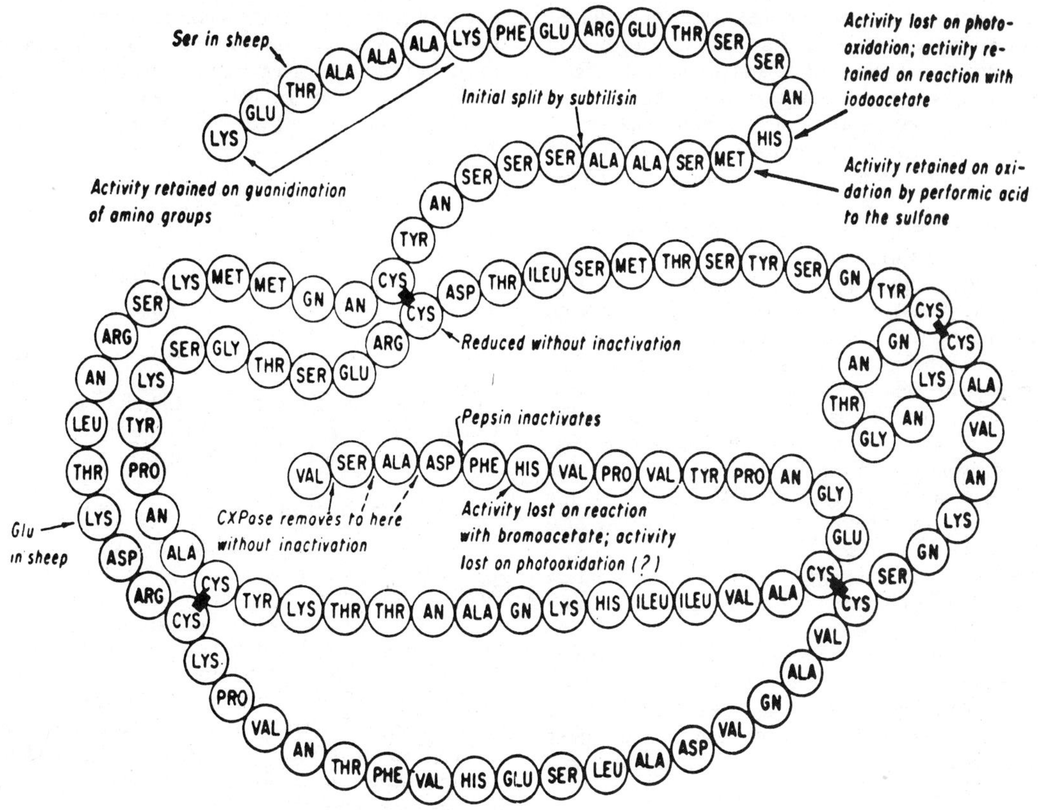

Ribonuclease. Abbreviations used: alanine (ALA), adenine (AN), arginine (ARG), aspartic acid (ASP), cysteine (CYS), cytosine (CN), glutamic acid (GLU), glycine (GLY), guanine (GN), histidine (HIS), isoleucine (ILEU), leucine (LEU), lysine (LYS), methionine (MET), phenylalanine (PHE), proline (PRO), serine (SER), threonine (THR), thymine (TN), tryptophan (TRY), tyrosine (TYR), uracil (UN), and valine (VAL).

In the Michaelis-Menten treatment, this equation can be regarded as the result of the three processes:

rate of formation of $ES = k_1[E][S]$
rate of decomposition of ES into products $= k_2[ES]$
rate of decomposition of ES into original reactants $= k_3[ES]$

where the terms in square brackets denote concentrations of E, S, and ES, and the k's are rate constants. By representing the ratio $(k_2 + k_3)/k_1$ by K_m, the *Michaelis constant*, we can obtain a form of the Michaelis equation

$$\frac{[E_t]}{v} = \frac{K_m}{k_3 + [S]} + \frac{1}{k_3}$$

where E_t is the total concentration of enzyme (as distinguished from $[E]$, the concentration of free enzyme), and v is the velocity of the reaction. By plotting $[E_t]/v$ against $1/[S]$, we can obtain, at the axis-intercepts, values of $1/K_m$ and $1/v$.

This approach has been successfully extended to the more complex enzymatic reactions involving inhibitors, activators, and even multiple reactions in which the successive action of more than one enzyme is involved.

The observation follows that in addition to the specificity of the enzymes with respect to reaction type and to structure of the substrate, the action is also confined to a single configuration of the substrate. If the molecular structure of the substrate is unsymmetrical (asymmetric) and therefore two compounds exist—one the mirror image of the other—a specific enzyme will act upon only one of the stereoisomers. This specificity is undoubtedly due to the fact that the region of interaction on the enzyme is also asymmetric and exists in only one form or configuration. Racemic mixtures or certain substrates may sometimes be separated by making use of the fact that enzymic action will affect only one of the two forms.

Enzyme Repression. Repression as applied to biochemical reactions is a process of feedback control whereby a cell limits its production of the substances produced within it. An example that has been investigated shows the nature and mechanism by which this limitation is effected. It has been found that production of the amino acid L-isoleucine by cells of the bacterium *Escherichia coli* is repressed in the presence of an excess of the product. This excess is obtained experimentally by adding the substance to the culture medium in which the bacterium is grown. A form of the L-isoleucine is used that has been labeled with a radioactive isotope, so that the mechanism of the repression can be followed.

By this means, it has been found that the excess of L-isoleucine has two distinct effects—one that is relatively slow, and another that is rapid. The slower effect is to repress production by the cell of all the enzymes required to catalyze the series of biochemical reactions in the metabolic pathway by which the cell synthesizes L-isoleucine. The fast effect is to inhibit production of the enzyme for the first reaction in the series. This enzyme is L-threonine deaminase, which removes the amino group from L-threonine as a preliminary step to its oxidation and reamination in order to produce L-isoleucine from it.

The indpendent existence of these two effects was demonstrated by the discovery among mutations of *E. coli*, a mutation that exhibited only one of the effects, and of another mutation that exhibited only the other, leading to the conclusion that two distinct genes are involved in the control system.

An even more striking instance of feedback control is found in the synthesis of DNA (see **Nucleic Acids and Nucleotides**). As pointed out in that entry, normal DNA is composed of the nucleotides deoxyguanosine, deoxycytidine, deoxyadenosine, and thymidine, and the amounts of the first and second of these are the same, as are those of the third and fourth. Obviously, close control is required of the amounts of these nucleotides that are synthesized by the cell, if they are to be made in the quantities required for DNA synthesis. Evidence has been found that the enzyme carbamoylphosphate: L-aspartate carbamoyl transferase, which catalyzes the reaction between aspartic acid and carbamoyl phosphate (which has deoxycytidine triphosphate, CTP, as its final product), is inhibited by an excess of the CTP, and is also initiated (or activated) by an excess of deoxyadenosine triphosphate, which requires an equal amount of CTP to react with it in forming DNA. There are thus both positive and negative feedback controls on the synthesis of the enzymes that catalyze the synthesis of the nucleotides. Since all enzymes are proteins, the mechanism of this control is believed to be that suggested for protein synthesis in the entry on **Nucleic Acids and Nucleotides.** See also **Recombinant DNA.**

An aspect of the control of enzymatic action that is related to the effect of initiation (or activation) just discussed is the effect of *induction,* which can readily be illustrated experimentally. Many years ago, it was found that the yeast, *Saccharomyces ludwigii,* although able to ferment many sugars, was ineffective on lactose (milk sugar), because it did not synthesize the necessary enzyme, lactase (β-D-galactoside galactohydrolase). However, if this yeast was grown for several generations on a medium containing lactose, it acquired the ability to make lactase, and its subsequent generations retained that ability. In the years since this discovery, so many instances of induction have been discovered that they are regularly cited in discussions of the properties of those enzymes for which they are known, as are also the repressing and blocking substances.

Experimental evidence is accumulating to lead to an explanation of the precise genetic and molecular mechanism of the action of both repression and induction.

Since all enzymes are proteins, and since, as explained in the entry on nucleic acids, proteins are synthesized by a process of replication, starting from the DNA of the genes, it is considered that there are "repressor" and "inductor" molecules, which, respectively, either block or play a necessary part in that process for the particular enzyme.

See also **Bile; Carbohydrates; Detergents; Starch.**

As shown by the accompanying table, enzymes are classified into six groups: (1) oxidoreductases, (2) transferases, (3) hydrolases, (4) lyases, (5) isomerases, and (6) ligases or synthetases. The main group to which an enzyme belongs is indicated by the first figure of the code number. The second figure indicates the subclass; for the oxidoreductases, it shows the type of group in the *donors* which undergoes oxidation; for the transferases, it indicates the nature of the group which is transferred; for the hydrolases, it shows the type of bond hydrolyzed; for the lyases, the type of link which is broken between the group removed and the remainder; for the isomerases, the type of isomerization involved; and for ligases, the type of bond formed.

The third figure of the code number, indicating the sub-sub class, shows for the oxidoreductases the type of acceptor involved; for the transferases and hydrolases, it shows more precisely the type of group transferred or bond hydrolyzed; for the lyases, it shows the nature of the group removed; for the isomerases, it indicates in more detail the nature of the isomerization; and for the ligases, it shows the nature of the substance formed. Thus, an enzyme number, commonly indicated by the prefix EC, provides fairly detailed information about a specific enzyme.

Categories of Reactions. The comparative simplicity of the classification scheme bears testimony to the underlying unity of enzymatic catalysis.

Oxidoreductases. The overall reaction catalyzed by the oxidoreductases can be written as hydrogen transfer, and these enzymes might be considered to be merely one section of the transferases. The oxidoreductases are classified separately because of their large number and because of their great biological importance in bringing about the main energy-yielding reactions of living tissues.

Transferases. The main groups of transferases are concerned with the transfer of *one-carbon* groups, acyl groups, glycosyl residues, amino- and other nitrogen-containing groups, phosphate, and sulfate. Oxidoreductases and transferases together represent about half or more of the enzymes presently recognized. A general reaction for both oxidoreductases and transferases can be written:

$$AX + B \rightleftharpoons A + BX$$

Hydrolases. These enzymes include esterases, glycosidases, peptidases, deaminases, and enzymes which hydrolyze acid anhydrides (such as the pyrophosphate group in adenosinetriphosphate). Many hydrolases have been shown to be able, under appropriate conditions, to catalyze transfer reactions; a high concentration of acceptor is usually necessary, since there is competition between the added acceptor and water for the group transferred. The detailed mechanism in these cases probably involves transfer of a part of the substrate onto a group on the enzyme, with subsequent transfer to an acceptor or hydrolysis, e.g., for a hydrolase acting on a substrate AB to produce AOH and BH:

$$EH + AB \rightarrow E - A + BH$$

and

$$E - A + X \rightarrow E + AX$$

or

$$E - A + H_2O \rightarrow EH + AOH$$

These hyrolases, if not all, can therefore be regarded as transferases which include H_2O among their possible acceptors. Under normal conditions, in aqueous solution, hydrolysis will be the dominant reaction.

Lyases. Enzymes in this grouping catalyze reactions of the type:

$$AX - BY \rightleftharpoons A = B + X - Y$$

Molecules, such as H_2O, H_2S, NH_3, or aldehydes, are added across the double bond of a second unsaturated molecule. Decarboxylases, such as those acting on amino acids can be regarded as lyases (carboxy-lyases), assuming CO_2 and not H_2CO_3 to be the immediate product of decarboxylation. Over one hundred lyases are known.

Isomerases. These include enzymes which bring about reactions similar to those in several other groups, but distinguished in that the reaction takes place entirely within one molecule, which is not cleaved, so that the overall reaction is

$$A \rightleftharpoons B$$

Thus, there are intramolecular oxidoreductases (e.g., ketolisomerases), intramolecular transferases (e.g., phosphomutases), and intramolecular lyases. About fifty isomerases are known.

Ligases. These enzymes catalyze reactions which are more complex than those of the other groups and must involve at least two separate stages in the reaction. The overall result is the synthesis of a molecule from two components with a coupled breakdown of adenosine triphosphate, or some other nucleoside triphosphate. In general, this may be written:

$$X + Y + ATP \rightarrow XY + AMP$$
$$+ \text{Pyrophosphate (or ADP + Phosphate)}$$

These enzymes, of which about fifty are known, are of great importance in the conservation of chemical energy within the cell and in the coupling of synthetic processes with energy-yielding breakdown reactions.

References

Baker, T. S., Eisenberg, D., and F. Eiserling: "Ribulose Bisphosphate Carboxylase: A Two-layered, Square-shaped Molecule of Symmetry 422," *Science*, **196**, 293–295 (1977).

Brattsen, L. B., Wilkinson C. F., and T. Eisner: "Herbivore–Plant Interactions; Mixed-Function Oxidases and Secondary Plant Substances," *Science*, **196**, 1349–1352 (1977).

Cornforth, J. W.: "Asymmetry and Enzyme Action," *Science*, **193**, 121–125 (1976).

Dixon, N. E., et al.: "Metal Ions in Enzymes using Ammonia or Amides," *Science*, **191**, 1144–1150 (1976).

Eisenberg, D.: "Enzyme Structure Control," Academic, New York, 1970.

Gray, J. C.: "Enzyme Catalysed Reactions," Van Nostrand Reinhold, New York, 1971.

Gutfreund, H.: "Enzymes: Physical Principles," Wiley, New York, 1972.

Jakoby, W. B.: "Enzyme Purification and Related Techniques," Academic, New York, 1971.

Meister, A.: "On the Enzymology of Amino Acid Transport," *Science*, **180**, 33–39 (1973).

Ory, R. L., and A. J. St. Angelo (editors): "Enzymes in Food and Beverage Processing," American Chemical Society, Washington, D.C., 1977.

Perlman, G., and L. Lorand: "Proteolytic Enzymes," Academic, New York, 1970.

Plowman, K. M.: "Enzyme Kinetics," McGraw-Hill, New York, 1972.

CLASSIFICATION OF ENZYMES

1. OXIDOREDUCTASES
   - 1.1 *Acting on the* CH—OH *group of donors*
     - 1.1.1 With NAD or NADP as acceptor
     - 1.1.2 With cytochrome as an acceptor
     - 1.1.3 With O_2 as acceptor
     - 1.1.99 With other acceptors
   - 1.2 *Acting on the aldehyde or keto group of donors*
     - 1.2.1 With NAD or NADP as acceptor
     - 1.2.2 With a cytochrome as an acceptor
     - 1.2.3 With O_2 as acceptor
     - 1.2.4 With lipoate as acceptor
     - 1.2.99 With other acceptors
   - 1.3 *Acting on the* CH—CH *group of donors*
     - 1.3.1 With NAD or NADP as acceptor
     - 1.3.2 With a cytochrome as an acceptor
     - 1.3.3 With O_2 as acceptor
     - 1.3.99 With other acceptors
   - 1.4 *Acting on the* $CH—NH_2$ *groups of donors*
     - 1.4.1 With NAD or NADP as acceptor
     - 1.4.3 With O_2 as acceptor
   - 1.5 *Acting on the* C—NH *group of donors*
     - 1.5.1 With NAD or NADP as acceptor
     - 1.5.3 With O_2 as acceptor
   - 1.6 *Acting on reduced* NAD *or* NADP *as donor*
     - 1.6.1 With NAD or NADP as acceptor
     - 1.6.2 With a cytochrome as an acceptor
     - 1.6.4 With a disulfide compound as acceptor
     - 1.6.5 With a quinone or related compound as acceptor
     - 1.6.6 With a nitrogenous group as acceptor
     - 1.6.99 With other acceptors
   - 1.7 *Acting on other nitrogenous compounds as donors*
     - 1.7.3 With O_2 as acceptors
     - 1.7.99 With other acceptors
   - 1.8 *Acting on sulfur groups of donors*
     - 1.8.1 With NAD or NADP as acceptor
     - 1.8.3 With O_2 as acceptor
     - 1.8.4 With a disulfide compound as acceptor
     - 1.8.5 With a quinone or related compound as acceptor
     - 1.8.6 With a nitrogenous group as acceptor
   - 1.9 *Acting on heme groups of donors*
     - 1.9.3 With O_2 as acceptor
     - 1.9.6 With a nitrogenous group as acceptor
   - 1.10 *Acting on diphenols and related substances as donors*
     - 1.10.3 With O_2 as acceptor
   - 1.11 *Acting on* H_2O_2 *as acceptor*
   - 1.12 *Acting on hydrogen as donor*
   - 1.13 *Acting on single donors with incorporation of oxygen (oxygenases)*
   - 1.14 *Acting on paired donors with incorporation of oxygen into one donor (hydroxylases)*
     - 1.14.1 Using reduced NAD or NADP as one donor
     - 1.14.2 Using ascorbate as one donor
     - 1.14.3 Using reduced pteridine as one donor
2. TRANSFERASES
   - 2.1 *Transferring one-carbon groups*
     - 2.1.1 Methyltransferases
     - 2.1.2 Hydroxymethyl-, formyl-, and related transferases
     - 2.1.3 Carboxyl- and carbamoyltransferases
     - 2.1.4 Amidinotransferases
   - 2.2 *Transferring aldehydic or ketonic residues*
   - 2.3 *Acyltransferases*
     - 2.3.1 Acyltransferases
     - 2.3.2 Aminoacyltransferases
   - 2.4 *Glycosyltransferases*
     - 2.4.1 Hexosyltransferases
     - 2.4.2 Pentosyltransferases
   - 2.5 *Transferring alkyl or related groups*
   - 2.6 *Transferring nitrogenous groups*
     - 2.6.1 Aminotransferases
     - 2.6.3 Oximinotransferases
   - 2.7 *Transferring phosphorus-containing groups*
     - 2.7.1 Phosphotransferases with an alcohol group as acceptor
     - 2.7.2 Phosphotransferases with a carboxyl group as acceptor
     - 2.7.3 Phosphotransferases with a nitrogenous group as acceptor
     - 2.7.4 Phosphotransferases with a phospho-group as acceptor
     - 2.7.5 Phosphotransferases, apparently intramolecular
     - 2.7.6 Pyrophosphotransferases
     - 2.7.7 Nucleotidyltransferases
     - 2.7.8 Transferases for other substituted phospho-groups
   - 2.8 *Transferring sulfur-containing groups*
     - 2.8.1 Sulfurtransferases
     - 2.8.2 Sulfotransferases
     - 2.8.3 CoA-transferases
3. HYDROLASES
   - 3.1 *Acting on ester bonds*
     - 3.1.1 Carboxylic ester hydrolases
     - 3.1.2 Thiolester hydrolases
     - 3.1.3 Phosphoric monoester hydrolases
     - 3.1.4 Phosphoric diester hydrolases
     - 3.1.5 Triphosphoric monoester hydrolases
     - 3.1.6 Sulfuric ester hydrolases
   - 3.2 *Acting on glycosyl compounds*
     - 3.2.1 Glycoside hydrolases
     - 3.2.2 Hydrolyzing N-glycosyl compounds
     - 3.2.3 Hydrolyzing S-glycosyl compounds
   - 3.3 *Acting on ether bonds*
     - 3.3.1 Thioether hydrolases
   - 3.4 *Acting on peptide bonds (peptide hydrolases)*
     - 3.4.1 α-Aminoacyl-peptide hydrolases
     - 3.4.2 Peptidyl-amino acid hydrolases
     - 3.4.3 Dipeptide hydrolases
     - 3.4.4 Peptidyl-peptide hydrolases
   - 3.5 *Acting on* C—N *bonds other than peptide bonds*
     - 3.5.1 In linear amides
     - 3.5.2 In cyclic amides
     - 3.5.3 In linear amidines
     - 3.5.4 In cyclic amidines
     - 3.5.5 In cyanides
     - 3.5.99 In other compounds
   - 3.6 *Acting on acid-anhydride bonds*
     - 3.6.1 In phosphoryl-containing anhdrides
   - 3.7 *Acting on* C—C *bonds*
     - 3.7.1 In ketonic substances
   - 3.8 *Acting on halide bonds*
     - 3.8.1 In C-halide compounds
     - 3.8.2 In P-halide compounds
   - 3.9 *Acting on* P—N *bonds*
4. LYASES
   - 4.1 *Carbon-carbon lyases*
     - 4.1.1 Carboxy-lyases
     - 4.1.2 Aldehyde-lyases
     - 4.1.3 Ketoacid-lyases
   - 4.2 *Carbon-oxygen lyases*
     - 4.2.1 Hydro-lyases
     - 4.2.99 Other carbon-oxygen lyases
   - 4.3 *Carbon-nitrogen lyases*
     - 4.3.1 Ammonia-lyases
     - 4.3.2 Amidine-lyases
   - 4.4 *Carbon-sulfur lyases*
   - 4.5 *Carbon-halide lyases*
   - 4.99 *Other lyases*
5. ISOMERASES
   - 5.1 *Racemases and epimerases*
     - 5.1.1 Acting on amino acids and derivatives
     - 5.1.2 Acting on hydroxyacids and derivatives
     - 5.1.3 Acting on carbohydrates and derivatives
     - 5.1.99 Acting on other compounds
   - 5.2 *Cis-trans isomerases*
   - 5.3 *Intramolecular oxidoreductases*
     - 5.3.1 Interconverting aldoses and ketoses
     - 5.3.2 Interconverting keto- and enol-groups
     - 5.3.3 Transposing C=C bonds
   - 5.4 *Intramolecular transferases*
     - 5.4.1 Transferring acyl groups
     - 5.4.2 Transferring phosphoryl groups
     - 5.4.99 Transferring other groups
   - 5.5 *Intramolecular lyases*
   - 5.99 *Other isomerases*
6. LIGASES OR SYNTHETASES
   - 6.1 *Forming* C—O *bonds*
     - 6.1.1 Aminoacid-RNA ligases
   - 6.2 *Forming* C—S *bonds*
     - 6.2.1 Acid-thiol ligases
   - 6.3 *Forming* C—N *bonds*
     - 6.3.1 Acid-ammonia ligases (amide synthetases)
     - 6.3.2 Acid-amino acid ligases (peptide synthetases)
     - 6.3.3 Cyclo-ligases
     - 6.3.4 Other C—N ligases
     - 6.3.5 C—N ligases with glutamine as N-donor
   - 6.4 *Forming* C—C *bonds*

Segal, H. L.: "Enzymatic Interconversion of Active and Inactive Forms of Enzymes," *Science*, **180**, 25–32 (1973).
Wang, D. C., et al.: "Fermentation and Enzyme Technology," Wiley, New York, 1979.

ENZYME PREPARATIONS. During the past several years, a number of commercially prepared enzyme preparations have been available to processors, notably for use in the food industry. These preparations fall into three basic categories: (1) Animal-derived preparations; (2) plant-derived preparations; and (3) microbially derived preparations.

Fruit juices, jams, and jellies, corn (maize) syrups and sweeteners, structured protein foods, and tenderized meats are exemplary of products, the quality of which has been improved through the use of enzyme preparations. Principal areas of development in food-grade enzyme research have been toward upgrading quality and byproduct utilization, higher rates and levels of extractions, synthetic food development, sweetener development, improving flavor of foods, and the stabilization of food quality and nutrition. Enzyme preparations also are used in the detergent field.

Animal-derived enzyme preparations include catalase (bovine liver), lipase, pepsin, rennet, and trypsin. Plant-derived preparations include bromelain, cellulase, ficin, malt, papain, and pectinase. Microbially derived preparations include amylases, carbohydrase, catalase, glucose oxidase, lipase, protease, and zymase.

ENZYME PROTEINS. Brain and Nervous System.

ENZYME (Restriction). Recombinant DNA.

EOCENE. A subdivision of the Tertiary of the geologic time-scale. Type locality, near Paris, France. Term first proposed by Lyell in 1832. The Eocene began approximately 60 million years ago and lasted for about 20–30 million years. The greatest thickness of formations of this period occur in Wyoming. The principal areas of deposition in the United States are: (1) the unconsolidated marine gravels, glauconite sands and clays which overlap the Cretaceous marine sediments of the Atlantic Border; (2) the marine limestones, terrestrial sandstones and lignites of the Gulf Coast; (3) the marine sediments of the Pacific Coast; (4) the terrestrial intermontane deposits of the Western interior. The plants of this period suggest worldwide warm climate. The fossil plants include many of the modern genera, such as the beeches, dogwoods, walnuts, maples and elms. Fossil vertebrate skeletons show that the mammals are now dominant, although many of the existing orders of reptiles and birds also lived at this time. The mammalian fauna may be divided into two principal groups: (1) the archaic types which did not survive the Eocene; (2) the progenitors of the modern mammals, including the ancestors of the camels, pigs, horses, rats, and primitive monkeys. The principal surviving archaic forms are the creodonts (primitive flesh-eaters), uintatheria (hippopotamus-like forms), and zeuglodons (marine mammals). The mineral resources of this period are described under Tertiary.

EOHIPPUS. Fossil Reptilian Mammals.

EOLIAN DEPOSITS. Sediments and sedimentary rocks which are largely, if not entirely, composed of wind-blown material. Desert sands are typical aeolian sediments, characterized by relatively uniform, well-rounded particles whose surfaces are usually covered with microscopic pits due to their mutual bombardment during transportation. This pitting gives each sand grain a frosted appearance. Wind-blown sediments frequently show characteristic cross-bedding, ripple marks (miniature dunes) and wind-faceted pebbles (glyptoliths). Further evidence of their origin is the absence of fossils. Aeolian deposits are usually largely composed of quartz sand. An important fine-grained wind-blown deposit is loess. Extensive desert deposits are also composed of gypsum, salt, etc.

EON (Geology). Geologic Time Scale.

EOPHYTIC. A paleobotanic division of geologic time. This term signifies the time during which algae were abundant. See also **Paleobotany.**

EOSINOPHILIA. Bronchial Asthma.

EOSINOPHILIC GRANULOMA. Bone.

EOSINOPHILS. Blood.

EÖTVÖS BALANCE. Torsion Balance.

EÖTVÖS-RAMSEY-SHIELDS LAW. Surface Tension.

EPEIROGENY. A term signifying broad and relatively widespread or continental uplife as compared with mountain building or Orogeny.

EPHEDRINE. Alkaloids.

EPHEMERIS TIME (ET). Equation of Time; Time.

EPHEMEROPTERA. The mayflies, also known locally as shad flies, salmonflies and June bugs. The adults are sluggish insects with slender filaments at the caudal end of the body and large triangular front wings. The hind legs are much smaller, in some species rudimentary. The immature insect is aquatic and in most species feeds on decaying vegetable matter. It may live for several years, while the adult stage lasts only a few days.

May fly (*Ephemeroptera*). (*A. M. Winchester.*)

Mayflies of one species emerge as adults in large numbers within a short period and are sometimes very abundant near favorable bodies of water. They fly at twilight and can sometimes be seen in large gray clouds at a distance of more than a mile over the islands of Lake Erie, where they are especially abundant. In the cities bordering the lake they are attracted to lights and their dead bodies are sometimes swept up in bushels after a heavy flight. Under such conditions they are a nuisance but not a serious pest. Their value as food for fishes more than offsets what little harm they do.

EPHYRA. A saucer-shaped jellyfish larva with a deeply notched margin which is produced by the segmentation of the primary scyphistoma larva and develops directly into the adult jellyfish.

EPICARDIUM. 1. The thin covering of the vertebrate heart, continuous with the lining of the pericardial cavity. 2. Outgrowth from the branchial sac in many ascidians, which takes part in budding.

EPICENTER. Earthquakes, Seismology, and Plate Tectonics.

EPICHLOROHYDRIN. A highly reactive and industrial important compound with the structural formula

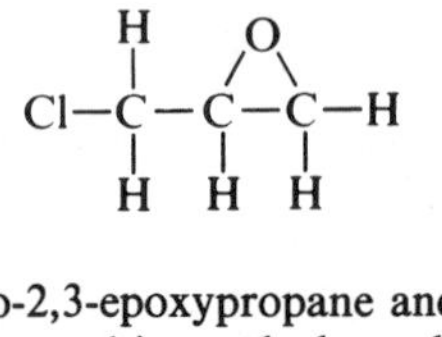

It is also called 1-chloro-2,3-epoxypropane and is classified as an organic epoxide. The compound is a colorless, clear, mobile liquid with an odor something like chloroform. Molecular weight, 92.53; freezing point, −57.1°C; boiling point, 116.07°C; density 1.1750 g/cm^3 at 25°C with reference to water at 4°C. Solubility is 6.53 g/100 g of water.

Epichlorohydrin is made by the chlorohydrination of allyl chloride, in which 1,2-dichlorohydrin and 1,3-dichlorohydrin are produced as intermediates.

One of the most common epoxy resins is produced by the reaction between epichlorohydrin and bisphenol A. See also **Epoxy Resins.** The compound also is used in the production of epichlorohydrin-based rubbers which have good aging, high resiliency, and flexibility at low temperatures, advantage of which is taken in automotive and aircraft parts, seals, gaskets, hose, belting, wire, and cable jackets. These rubbers also have good resistance to solvents, fuels, oils, and ozone. A number of wet-strength resins for use in the paper industry also are derived from epichlorohydrin, including (a) epichlorohydrin-modified polyamides; and (b) the addition of epichlorohydrin to high-molecular-weight polyalkylene polyamines. The advantages of these resins is that no alum or acid medium is required for incorporating the resin into the cellulose pulp. During the drying process, the resin cross-links and thus yields a paper with permanent wet-strength properties. Ion-exchange resins also can be prepared by reacting epichlorohydrin with ethylene diamine or a similar amine. The resulting material is a stable, water-insoluble anion-exchange resin.

In addition to its use in the production of epoxy resins epichlorohydrin is used in large quantities in the manufacture of glycerin. Other uses include textile applications where it is used to modify the carboxy groups of wool, thus increasing durability and improving moth resistance; in the synthesis of antistatic agents, wrinkle-resistant agents, and coating sizings. Effective against the larvae of certain insects, the compound is used in control chemicals for agriculture where permitted.

EPICONTINENTAL SEAS (or Epicontinental Marginal Seas). Shallow bodies of water deeper than continental shelves and having somewhat greater relief. Generally somewhat greater than 600 feet (180 meters) in depth. See **Continental Shelves.**

EPICYCLIC GEAR TRAIN. Combinations of gears having a motion resulting from rotation about an axis which, in itself, is in rotation, are known as epicyclic trains. A simple epicyclic gear train, consisting of three gears and an arm, is shown in the figure. Mechanism of

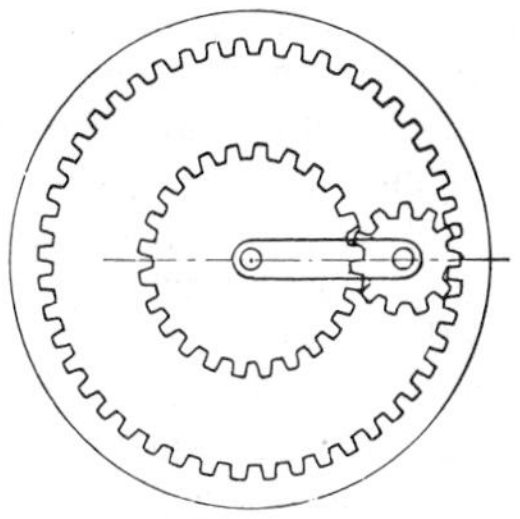

Epicyclic gear train.

this nature is sometimes used for speed reducers. The ratios of speed of the driven and driving elements are found by the following simple rule: consider, first, the gears locked and the entire mechanism turned one revolution; then the arm locked and the fixed gear turned one revolution in the opposite direction to the first step. The algebraic sum of these two separate motions will give the absolute number of turns of any gear, and from this the speed ratio may be found.

EPICYCLOID. A higher plane curve, which is a special case of a cyclic curve. A circle of radius r rolls around the outside of a fixed circle of radius R and a point on the circumference of the moving circle traces out the curve. Its equation in parametric form is

$$x = (R + r)\cos\phi - r\cos\frac{(R+r)\phi}{r}$$

$$y = (R + r)\sin\phi - r\sin\frac{(R+r)\phi}{r}$$

The curve consists of a set of congruent arches. The first one will occur at some value of ϕ between 0 and $2\pi r/R$ and the kth one between $2\pi r(k-1)/R$ and $2\pi kr/R$. It will not repeat itself, however, unless $2\pi kr/R$ is a multiple of 2π for k, an integer. This means

Epicycloid.

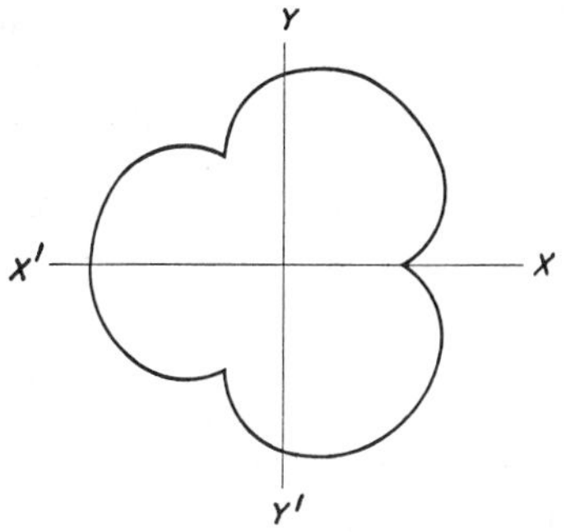

that $r/R = m/n$, where m/n is a rational fraction. Thus $k = n$, there are n arches to the curve, an equal number of cusps, and it winds around the fixed circle m times.

The special case of $r = R$ is called a cardioid. If the generating point for the epicycloid is at a distance $a \neq r$, the curve is an epitrochoid.

If an epicycloid rolls on a straight line, the locus of the center of the fixed circle is an ellipse, which is called the roulette of the epicycloid. The evolute of an epicycloid is a similar epicycloid.

See also **Cardioid;** and terms listed under **Mathematics.**

EPIDEMIC ENCEPHALITIS. Encephalitis.

EPIDEMIC TYPHUS. Rickettsial Diseases.

EPIDEMIOLOGY. That branch of medical science which is concerned with the study of disease as it appears in its natural surroundings, and as it affects a community of people rather than a single individual. Epidemiology is largely concerned with infectious diseases; it has a statistical as well as an experimental side. The statistical includes the gathering of facts about the incidence, mortality rate, relation of climate, age, sex, race and many other factors to the appearance of a given disease. From these data, valuable information which is important in controlling epidemics of disease is obtained. In experimental epidemiology, the investigator may produce epidemics in laboratory animals for the study of certain problems, or he may go out in the field and study epidemics in man by laboratory means. Such work is important in elucidating the cause, the modes of transmission, and the insect vectors possibly concerned, as well as other facts of fundamental importance in public health. Numerous governments throughout the world maintain statistical and data processing centers which provide periodic reports on many aspects of diseases. Most states and provinces maintain their own records and also report and exchange information with a central government office, such as the Center for Disease Control of the U.S. Public Health Service, located in Atlanta, Georgia.

EPIDEMIOLOGY (Heart Attack). Ischemic Heart Disease.

EPIDERMIS. In insects, the outer layer of the noncellular cuticula; here the cellular layer is called the hypodermis. In other invertebrates the cellular layer covering the body is called the epidermis. In vertebrates, the epidermis is the outer cellular layer of skin. The human epidermis consists of four layers, from without inward as follows: (1) a layer of horny flattened cells, (2) a layer of transparent cells, (3) layers of granular cells, (4) a layer of rounded pigmented cells. The outermost layer of cells on the younger parts of plants is also called the epidermis. See also **Dermatitis and Dermatosis.**

EPIDERMIS (Leaf). Leaf.

EPIDIORITE. A term applied to gabbros, dolerites, and diabases, the augite of which has been partly altered to hornblende, thus approaching a diorite in mineral composition. The term is derived from the Greek, meaning upon, plus diorite.

EPIDOTE. This mineral is a hydrous silicate of calcium, aluminum, and iron with the formula, $Ca_2(Al, Fe)_3Si_3O_{12}(OH)$. The ratio of aluminum to iron ranges from 6:1 to 3:2. Epidote is found in prismatic monoclinic crystals, which may be acicular to fibrous. Fine granular and compact masses are common. The mineral displays one good cleavage, an uneven fracture; is brittle; hardness, 6–7; specific gravity, 3.25–3.5; luster, vitreous to resinous; typical color, pistachio green, but may be yellowish- to brownish-green, sometimes red, yellow, gray, white or colorless. Colorless to grayish streak; transparent to opaque. The characteristic color of ordinary epidote makes it usually an easily identified mineral.

It occurs commonly in metamorphic rocks as gneisses and schists; however, it seems probable that under certain conditions it may appear as a primary mineral, for example in granitic rocks. The Urals, Austria, Switzerland, Italy, France and Norway are known for their occurrences of fine epidote crystals. In the United States epidote has been found in excellent specimens at Franconia and Warren, New Hampshire; Huntington, Massachusetts; Willimantic and Haddam, Connecticut; Chaffee County, Colorado, and Riverside County, California. The word epidote is derived from the Greek. The name pistacite, from the Greek word meaning the pistachio nut, has been occasionally applied to this mineral. It has been used as a gemstone but is in little demand for this purpose.

See also terms listed under **Mineralogy.**

EPIGENETIC. A term used by petrologists to denote physical and chemical changes, particularly in igneous and sedimentary rocks, which are clearly secondary to (later in time) the conditions under which the rock originated. This term is commonly used by the students of ore deposits to designate minerals formed after the enclosing wall rocks, in contrast to those minerals formed contemporaneously with the wall rocks. The latter minerals are said to be syngenetic. In the case of the sedimentary rocks, the term is used to describe textures, structures and mineral aggregates, of nonmetamorphic origin, which have originated during the postlithification history of the formation. Thus flint, chert, and concretions, may be described as being either epigenetic or syngenetic.

EPIGLOTTITIS. The pharynx is a common passageway of the respiratory and digestive system. See **Pharynx.** A valvelike structure at the base of the tongue, the *epiglottis*, projects backward over the larynx during swallowing, and thereby prevents food from entering the larynx. Acute *epiglottitis*, an infection of this valve, is one of the most rapidly progressive and sometimes more lethal than any of the other infections of the upper respiratory tract—particularly among children between ages of 2 and 8 years. The condition starts suddenly, with very severe sore throat and fever, leading to dysphagia (difficulty in swallowing). There may be retention of secretions and drooling. Respiratory obstruction may occur within a few hours. The condition may occur in adults, but progresses more slowly because of the larger size of the airway in an adult.

Acute epiglottitis is considered a medical emergency. Inasmuch as examination of the pharynx may not produce an accurate diagnosis and because tongue depressors may elicit spasms, an immediate lateral-view x-ray of the neck should be made. Frequently, swelling of the epiglottis will be indicated. Where the x-ray does not satisfactorily confirm the condition, indirect laryngoscopy may be undertaken. Tracheostomy is the usual means taken to provide a restoration of the airway. Nasotracheal intubation is also used. To avoid delays, direct laryngoscopy may be undertaken. An important element of diagnosis is that of distinguishing epiglottitis from croup caused by viral laryngitis or other laryngeal conditions, and other possible causes of airway obstruction. In the case of croup, the epiglottis will be normal or only mildly inflamed.

The principal cause of acute epiglottitis is *H. influenzae* Type B. However, other pathogens may be involved. They include pneumococci, streptococci, and staphylococci. Physicians stress the need for immediate action in these cases, the initial therapy usually consisting of high-dose intravenous chloramphenicol (but considering its side effects) and penicillinase-resistant penicillin. Steroids may be administered to reduce edema and a mist tent may be a helpful supportive measure to assist breathing. In about half of the cases of this disease, nasotracheal intubation or tracheostomy is required.

EPILEPSY. Seizure (Neurological).

EPINEPHRINE. Alkaloids; Adrenal Glands; Brain and Nervous System; Endocrine System; Hormones.

EPIPELAGIC ZONE. The uppermost and very shallow layer of water in the oceans into which sufficient light passes to enable phytoplankton to convert the available carbon dioxide into food by means of photosynthesis. This zone, rarely more than 700 feet (210 meters) in depth, is the habitat of the major part of the sea's living matter.

EPIPHARYNX. A fold on the inner surface of the upper lip (labrum) of the insect mouth.

EPIPHRAGM. A tough membrane of calcified mucus with which some snails close the aperture of the shell during periods of drought.

EPIPHYSIS. An area of cartilage near the ends of long bones which ossifies separately and later becomes part of the long bone; bone growth in length takes place in this area until ossification halts growth. Also called *epiphyseal cartilage.* See also **Bone.**

EPIPHYTES. A striking feature of tropical forests is the abundance of plants which grow attached to other plants. These attached plants are called epiphytes, which means plants growing on other plants. They are found both on the main trunk and on the branches, often far above the ground. In many cases the epiphyte grows on the under side of a branch to which it is firmly fastened by its roots. Epiphytes gain nothing but support, and a more favorable position of growth because of better light conditions and other environmental factors; they do not obtain any nutrients from the supporting plants, as parasites would. Particularly noteworthy among epiphytes are many ferns, aroids and especially orchids. Frequently, as in the orchids, the roots of the plants are so modified as to absorb water directly from the atmosphere.

In the strictest meaning of the word, many lower plants including algae, fungi, lichens and mosses are epiphytes, since they are often found growing on other plants, not only in tropical regions but in temperate regions as well.

EPISTASIS. The inhibition by a gene at one locus on a chromosome of the expression of a gene at other locus. This form of suppression should be distinguished from simple dominance, which is associated with members of allelic pairs. The gene which does the inhibiting is said to be epistatic to the gene which is inhibited. The gene which is suppressed is said to be hypostatic. For instance, the gene for albinism is epistatic to the gene for black coat color in guinea pigs. A guinea pig may be homozygous for the dominant gene for black coat, but if he is also homozygous for the recessive gene for albinism, the coat is white. This is known as recessive epistasis, since two genes for albinism must be present. In dominant epistasis, only one gene is necessary to cause the inhibition. In man there is a dominant gene for dwarfism (Chondrodystropic dwarfism) and a single gene for this condition can be epistatic to the genes for normal growth.

EPISTAXIS. Hemorrhage from the nose. Nosebleed.

EPITAXIAL FILM. Microelectronics; Microstructure Fabrication; Semiconductor; Telephony.

EPITAXY. Oriented intergrowth between two solid phases. The surface of one crystal provides, through its lattice structure, preferred positions for the deposition of the second crystal.

EPITHELIUM. Tissue which covers surfaces and lines hollow organs, and the derivatives of these tissues, whether solid or hollow. All epithelial tissues are made up of closely associated cells with very little intercellular material and most of them have one surface free and the other connected with an underlying tissue.

Epithelia are classified according to the form of cells, the number of layers, and the embryonic origin and location. See accompanying illustration. In flat, pavement, or squamous epithelium the cells are much thinner than their diameter. Cuboidal epithelium is made up of cells approximately as thick as their width. They are not strictly cuboidal but are polyhedral prisms. Columnar epithelium contains cells that are much higher than their diameter. Glandular epithelia are usually of the two latter forms. Any of these forms of cells may occur in more than one layer as a stratified epithelium, but flat epithelia are more often stratified and thicker cells usually form a single layer, or a simple epithelium. In some cases, the epithelium is made up of cells of several forms in two or three layers, which change with movements of the part. Such a tissue is called traditional. Others appear to have several layers of cells but all are attached to the underlying tissue, rising to various heights. This is a pseudo-stratified epithelium.

According to origin and position, two kinds of epithelia derived from the mesoderm are recognized. Of these, endothelium lines the circulatory organs and mesothelium lines the body cavity. Other special types of epithelium are derived from each of the three germ layers. The free surfaces of some bear cilia.

The cells of many epithelia produce special secretions, and in some cases, these glandular layers are highly developed to form massive structures known as glands.

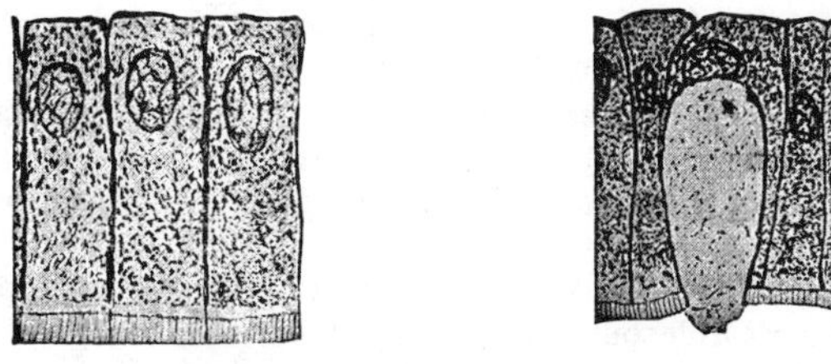

(a)   (b)

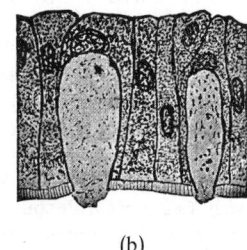

Forms of epithelium: (a) Simple columnar epithelium from intestinal lining; (b) goblet cells from epithelium lining large intestine. (*Kimber and Gray, "Textbook of Anatomy and Physiology," Macmillan.*)

Most epithelia rest on a thin basement membrane or *membrana propria* derived from the connective tissues.

EPITHERMAL NEUTRON.   Neutron.

EPITOKE.   The posterior sexual portion of the body in certain annelids, the Heteronereis. In one of the Pacific species this part of the worm breaks off when mature and rises to the surface. In this stage it is the palolo worm of the Pacific Islands. The anterior asexual portion (atoke) remains in the burrow and produces another epitoke.

EPIZOOTIC.   Term descriptive of any disease in animals whose incidence and distribution resemble those of an epidemic in man.

EPOCH.   Geologic Time Scale.

EPOXIDES (Foods).   Antimicrobial Agents (Foods).

EPOXY PAINTS.   Paint.

EPOXY RESINS.   A family of thermosetting resins known for their excellent mechanical and electrical properties, dimensional stability, resistance to high temperatures and numerous chemicals, and for their strong adhesion to glass, metal, fibers, and numerous other materials. Structurally, the epoxy groups are three-membered rings with one oxygen and two carbon atoms. The most common epoxy resins are made by reacting epichlorohydrin with a polyhydroxy compound, such as bisphenol A, in the presence of a catalyst. Epoxy resins produced in this fashion are known as diglycidyl ethers of bisphenol A (bis-A). The structural formula is:

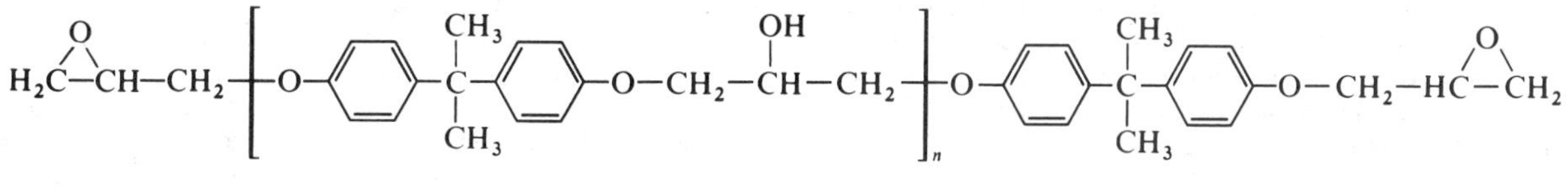

By changing the ratio of epichlorohydrin to bis-A, resins range from low-viscosity liquids to high-melting solids. The structure shown represents a solid epoxy novolak resin. The epoxy phenol novolak resins are the most important. Basically they are novolak resins whose phenolic hydroxyl groups have been converted to glycidyl ethers. Epoxidized novolaks are used principally in solid single-stage molding compounds and high-temperature laminating systems.

The liquid cycloaliphatic epoxies normally are produced by the peracetic acid expoxidation of cyclic olefins where the epoxide groups are attached directly to the cycloaliphatic ring. These materials have better weatherability, arc and tracking resistance and dielectric strength over conventional epoxies.

Usually the major makers of resins, hardeners and other chemicals for epoxy systems do not supply finished compounds. The compounding is done by specialized firms and by some large epoxy users. The epoxy resins per se are not finished products, but are reactive chemicals to be combined with other chemicals to yield systems capable of conversion to a predetermined thermoset structure.

Epoxy resins are cured by cross-linking agents known as hardeners or by catalysts which promote self-polymerization. Some of the cross-linking agents used include the primary and secondary aliphatic polyamines, such as diethylenetriamine, triethylenetetramine, tetraethylenepentamine, diethylaminopropylamine, and piperazines. Most of these materials are liquids of moderately low viscosity that can be blended with the resins at room temperature.

Because of excellent electrical properties, epoxies are used in casting, potting, and encapsulation of electrical/electronic parts. Advantage is taken of the low shrinkage of the epoxies, with absence of cracking or separation of the resins from the parts during cure. Encapsulated parts vary from miniature coils and switches that may weigh but a few grams to large motors and insulators that may weigh several pounds.

Epoxy resins also find use in making chemical-loaded molecular sieves, adhesives, and protective coatings. Epoxy-based adhesives are widely used for bonding dissimilar materials like plastics and metal,

Glass and other fibers coated with epoxy resin are filament-wound into containers of high strength. (*Union Carbide Corp.*)

$$H_2C-CH-CH_2 \left[O- \underset{}{\bigcirc} -\overset{CH_3}{\underset{CH_3}{C}} - \bigcirc -O-CH_2-\overset{OH}{CH}-CH_2 \right]_n O- \bigcirc -\overset{CH_3}{\underset{CH_3}{C}} - \bigcirc -O-CH_2-HC-CH_2$$

wood and metal, and ceramics and rubber. Minimum pressure is required to obtain a satisfactory bond. For such applications, epoxy adhesives are available as one- or two-part systems. The one-part systems require curing at elevated temperatures. The two-part systems can be cured at room temperature, but both have better resultant properties if cured under heat. Epoxy resin-based coatings provide outstanding chemical resistance, toughness, flexibility, and adhesion to most substrates.

Epoxy resins are used in the chemical industry because of their excellent resistance to attack by many corrosive chemicals. One application is a protective coating for container, pipe, and tank liners, floors, and walls. High-pressure vessels are made by filament winding with cycloalphatic epoxies. See accompanying illustration. Corrosion resisting pipe and fittings are also produced by filament winding.

EPSOMITE. Epsomite is normally found as an efflorescence on mine and cave walls. It belongs to the orthorhombic crystal system, being a hydrous sulfate of magnesium, $MgSO_4 \cdot 7H_2O$, with vitreous to earthy luster and colorless to white, of transparent/translucent quality. Hardness of 2–2.5, and specific gravity of 1.68. Very bitter to the taste. Found with the soluble salt lake deposits in Stassfurt, Germany, and in limestone caves in Kentucky, Tennessee and Indiana; also in several California and Colorado abandoned mines.

EPSOM SALTS. **Magnesium.**

EPSTEIN-BARR VIRUSES. **Virus.**

EQUAL-AREA MAP. A map so drawn that a square mile in one portion of the map is equal in size to a square mile in any other portion. This is obtained by changing the scales along the meridians and parallels in inverse proportion to each other.

Equal-area maps covering the whole globe, such as the Sanson-Flamsteed sinusoidal projection, are approximately elliptical. In Lambert's azimuthal equal-area projection, the parallels of latitude come closer together as the equator is approached (see **Lambert Projection**). Any map from the center of the whole world or of a hemisphere necessarily distorts the shape of a region far from the center of the map, but such maps are useful for some climatological studies in which the correct representation of area is important. For limited parts of the globe, equal-area projections are quite practical. (Cf. **Conformal Map.** See **Map Projections.**)

EQUAL-ENERGY SOURCE. A light source for which the time rate of emission of energy per unit of wavelength is constant throughout the spectrum.

EQUALIZER (Network). It is very difficult to construct apparatus and particularly lines for communication circuits which will pass all the necessary frequencies equally. Often it is necessary to insert networks whose function is merely to restore the original relation of the various frequencies. These are called equalizers. Usually they are loss circuits which will cause more loss in those frequencies which had been attenuated less and the opposite for those attenuated the most. Thus the equalizer shows a frequency response which is the inverse of the system it is intended to equalize, so the result of connecting it in the circuit is to restore the over-all response to a flat characteristic. Equalizers may be used to correct either frequency distortion, phase distortion, or both. The term equalizer is also used for the series of connections sometimes made in dc machines to insure equal current distribution in the conductors.

EQUAL LOUDNESS CONTOURS. **Acoustics.**

EQUATION. An equality involving two or more numbers or functions. Equations are classified by terms describing the functions in them or they are given the name of a mathematician who discovered or studied the equation. Depending on the functions, the equation could be algebraic or transcendental, with many subclasses in both of these types. (For further information about equations, see **Differential Equation; Difference Equation; Diophantine Equation; Integral Equation;** and **Polynomial.**)

A plane curve is usually described by a single equation in two variables representing rectangular or polar coordinates. Sometimes it is preferable to represent the curve by parametric equations expressing the coordinates separately in terms of a third variable, the parameter. Parametric equations of surfaces and of curves in space are also useful.

The equation of a given locus is the equation which is satisfied by the coordinates of all points of this locus, and of only such points.

When it is of interest to find the solution of an equation, approximate methods are often useful. See **Equation (Numerical Solution of).**

EQUATION (Numerical Solution of). Given equations in one or more variables, it is often required to determine the unknown quantities numerically from a sufficient set of known quantities. Provided graphical methods are not used, the following are cases which sometimes occur.

1. The roots of a polynomial. The quadratic equation in one variable is readily solved by a formula. The cubic and biquadratic equations can also be solved in this way but the method is so tedious in any practical case that other procedures are preferred. See, for example, **Graeffe Method; Iteration (Method of); Newton's Formula for Interpolation; Regula Falsi (Method of).** The reverse problem also occurs: given a set of points (x_i, y_i) to be represented by a polynomial, the coefficients of the equation are to be found. (For appropriate procedures, see **Curve Fitting.**)

2. Systems of simultaneous equations. An explicit solution is possible with Cramer's rule but this involves the evaluation of several determinants, a laborious task. If there are more than three equations other methods are more satisfactory.

3. Transcendental equations. These can also be solved by methods similar to those used for roots of polynomials.

See also terms listed under **Mathematics.**

EQUATION OF CONTINUITY. **Continuity (Equation).**

EQUATION OF STATE. Also called *characteristic equation,* a relation, empirical or derived, between thermodynamic properties of a substance or system. The equation of state must be single-valued in terms of its variables. This is a direct consequence of the concept of state.

There exist systems, namely systems which undergo processes involving hysteresis (plastic deformation or ferromagnetism, for example) for which no equation of state can be indicated. Although the laws of thermodynamics may apply to such systems, the rigorous results of classical thermodynamics are not applicable because the science of thermodynamics is developed on the assumption of the existence of the single-valued function.

In the realm of classical thermodynamics, equations of state are assumed given. They can be derived from first principles only by the methods of statistical mechanics and quantum mechanics. These rely on the adoption of suitable molecular models for substances, and so far no universal, generally applicable model has been discovered even for narrow classes of substances such as gases.

It is an experimental fact that every thermodynamic system possesses a definite number n of independent properties which determine its state. Consequently, an equation of state is a relation between n properties (mutually independent) chosen (otherwise arbitrarily) as the independent properties $(x_1, x_2 \cdots x_n)$ of the system and one more property, the dependent property y. Hence the equation of state is a function of the form

$$F(y, x_1, x_2, \ldots x_n) = 0$$

The simplest thermodynamic systems possess two independent properties, consequently the simplest equation of state is written in terms of three variables. When it is written in terms of pressure p, volume V, and absolute temperature T, it is called the p-V-T relation for the system or the thermal equation of state. When one of the caloric-thermodynamic properties (better called caloric properties, because p, V, T are thermodynamic properties also), such as enthalpy, entropy, Gibbs function, or work function (Helmholtz function) are given, the equation is called a thermodynamic equation, or better, a caloric equation of state, although the latter is not a commonly accepted designation.

Even in the case of a simple system, one equation of state, e.g., the equation $f(p, V, T) = 0$, does not necessarily determine the form of all the other equations of state. This is connected with the fact that the derivation of the other equations of state may involve the integration of partial derivatives which leads to the appearance of whole functions in the integration "constant." An equation from which other equations of state can be derived by differentiation only, is called a *fundamental equation* (of state). In the case of a pure substance in a specified phase, the p, V, T relation does not constitute a fundamental equation with respect to the properties U, H, S, G, A, or their derived properties C_p, C_v, γ, etc. Consequently, it is possible to have two or more substances whose p, V, T relations are identical but whose specific heats, for example, are different.

In the case of continuous systems, for which the state changes from point to point, for example, a flow field of a viscous fluid, it is assumed that at every point, the equation of state is the same as for a homogeneous system and does not involve the gradients of the thermodynamic properties. Hence, such systems can only be studied with the aid of thermodynamics if local departures from equilibrium are small (near-equilibrium processes), i.e., if the gradients of the thermodynamic properties are not too great.

An equation of state must necessarily involve a finite (even if very large) number of independent variables. The particular variables which are chosen as independent is immaterial, on condition that they are mutually independent, and that their number is appropriate to the physical nature of the system.

Equations of states of various types of systems are numerous. The Curie equation is the equation of state of a paramagnetic solid. The Beattie and Bridgeman equation, Berthelot equation, Clausius equation, Dieterice equation, Keyes equation, and van der Waals equation are other examples in this category.

It should be noted that equations of state for systems which consist of several components, rather than a single substance, can be written by introducing the variables $N_1, N_2, \ldots N_c$, which are the respective mole numbers of the components present.

EQUATION OF TIME. The interval by which the true Sun is ahead or behind the mean Sun. The Sun's apparent motion in the sky varies throughout the year because the earth's speed in its elliptical orbit varies slightly from perihelion to aphelion, while the mean Sun is assumed (by definition) to travel with a constant speed equal to the average speed of the true Sun. The interval never exceeds 17 minutes.

EQUATOR. A plane perpendicular to the axis of rotation of the earth and passing through the center of the earth will intersect both the surface of the earth and the celestial sphere in great circles. These great circles are known as the terrestrial and celestial equators. Some navigators refer to the celestial equator as the equinoctial. See also **Thermal Equator.**

EQUATOR (Geomagnetic). Geomagnetic Equator.

EQUATORIAL AIR. Air Mass; Atmosphere (Earth).

EQUATORIAL COORDINATES (Astronomy). Equatorial coordinates are a system of spherical coordinates, in which the origin may be the eye of the observer (in which we have the apparent system of coordinates), the center of the earth (geocentric system), the center of the sun (heliocentric system), or the center of the Milky Way (galactocentric system). The fundamental line in the heliocentric system is the line joining the poles of rotation of the earth, which cuts the celestial sphere in its poles of rotation. The plane perpendicular to the fundamental line through the origin is the celestial equator. The fundamental direction in the plane may be either the point of intersection of the local meridian with the celestial equator, which is above the horizon, or the vernal equinox.

To locate an object in this system of coordinates, a plane is passed through the object and the line joining the poles of rotation, and this plane cuts out a great circle, known as an hour circle, on the celestial sphere perpendicular to the plane of the equator. The declination of an object is the angular distance of the object north (+) or south (−) of the celestial equator measured in the plane of the hour circle through the object. The hour angle of the object is the angular distance, measured in the plane of the equator, from the point of intersection of the meridian above the horizon to the point of intersection of the hour circle through the object, in the direction of apparent rotation (west) of the celestial sphere. The right ascension of the object is the angular distance, measured in the plane of the equator from the vernal equinox to the point of intersection of the hour circle in a direction (east) contrary to the direction of apparent rotation of the celestial sphere. For convenience, both right ascension and hour angle are frequently expressed in units of hour, minutes and seconds of time, rather than the more common angular notation of degrees, minutes, and seconds of arc.

Due to the fact that the local meridian remains fixed as the celestial sphere apparently rotates, the hour angle of an object is continually changing. Since both the vernal equinox and the hour circle rotate with the celestial sphere, both the right ascension and declination of the object remain fixed as the sphere rotates. However, both right ascension and declination change slowly due to precession and nutation. In tabulating these coordinates in star catalogues, the values are given for the position of the equinox for some particular date, and the corrections necessary to reduce the positions to the present date must be applied.

See also **Celestial Sphere.**

EQUATORIAL COUNTERCURRENT. A current moving eastwards and caused by the return of lighter water piled up on the inner margins by the North Equatorial Current and South Equatorial Current on the western side of the ocean basin. This countercurrent lies in a band of a calm air, the doldrums (see **Winds and Air Movement**).

EQUATORIAL EASTERLIES. Winds and Air Movement.

EQUATORIAL TELESCOPE. A telescope so mounted that it may be moved parallel to the equatorial coordinates of hour angle and declination; sometimes called simply an equatorial. In this form of mounting, one axis, known as the polar axis, is parallel to the axis of rotation of the earth, and the other axis, known as the declination axis about which the telescope may be rotated, is perpendicular to the polar axis. From this arrangement of the axes, the telescope as it is rotated about the declination axis, must move in a plane perpendicular to the equator, and hence, parallel to hour circle or in the direction of declination. The rotation of the declination axis about the polar axis, with the telescope remaining fixed in declination, will cause the telescope to move parallel to the equator, and hence, in the coordinate of hour angle.

The majority of equatorials are carried on a single pier. The difficulty with this form of mounting is that the telescope will frequently run into the pier when the hour angle is close to zero. To avoid this, several other methods of supporting the polar axis have been devised. Perhaps the most common is the so-called English mounting, in which the two ends of the polar axis are supported on separate piers, with the telescope free to pass through zero hour angle.

The most difficult adjustment of the equatorial is to get the polar axis strictly parallel to the axis of rotation of the earth. When this has been accomplished, the instrument is by far the most convenient of all forms of mounting. If the instrument is rotated about the polar axis from east to west at exactly the same rate that the earth is rotating about its axis from west to east, the telescope lens will remain fixed relative to objects on the celestial sphere. Hence, once the instrument is set on a star, clockwork may be devised to keep the telescope "following."

See also **Equatorial Coordinates;** and **Telescope.**

EQUATORIAL VORTEX. Atmosphere (Earth).

EQUATORIAL WAVE. Atmosphere (Earth).

EQUILIBRIA (Forces). Statics (Graphical).

EQUILIBRIUM. In the elementary sense of the macroscopic (visible to the naked eye) system, equilibrium is obtained if the system does not tend to undergo any further change of its own accord.

Mechanical and Electromagnetic Systems. Equilibrium in mechanical and/or electromagnetic systems is reached when the vectorial summation of generalized forces applied to the system is equal to zero. In any potential field, that is, gravitational or electric vector potential, force can be expressed as gradient of potential (magnetic force however, is a curl of a vector potential). The potential energy therefore has an extremum at the equilibrium configuration. For example, a system such as a mass suspended by a string against the gravitational force (or its weight) is at mechanical equilibrium if the tensile force in the string is equal to the weight of the mass it supports. The d'Alembert principle further states that the condition for equilibrium of a system is that the virtual work of the applied forces vanishes.

Thermodynamic Systems. When a hot body and a cold body are brought into physical contact, they tend to achieve the same warmth after a long time. These two bodies are then said to be a thermal equilibrium with each other. The zeroth law of thermodynamics (R. H. Fowler) states that two bodies individually at equilibrium with a third are at equilibrium with each other. This led to the comparison of the states of thermal equilibrium of two bodies in terms of a third body called a thermometer. The temperature scale is a measure of state of thermal equilibrium, and two systems at thermal equilibrium must have the same temperature.

Generalization of equilibrium consideration by the second law of thermodynamics specifies that the state of thermodynamic equilibrium of a system is characterized by the attainment of the maximum of its entropy. Thermodynamic coordinates are defined in terms of equilibrium states.

Equilibrium between two phases of a system is reached when there is no net transfer of mass or energy between the phases. Phase equilibrium is determined by the equality of the Gibbs functions (also called free enthalpy, free energy, or chemical potential) of the phases in addition to equality of their temperatures and stresses (such as pressure and/or field intensities—intensive properties). Equilibrium of first-order phase change requires continuity of slope or first derivative of the Gibbs function with respect to an intensive property and is generalized as the Clapeyron relation. Second- and higher-order phase changes are given by the condition of continuity of curvature or second derivative of the Gibbs function and so on.

Chemical or nuclear equilibrium of a reactive system is reached when there is no net transfer of mass and/or energy between the components of a system. At chemical or nuclear equilibrium, the Gibbs function of the reactants and the products must be equal according to stoichiometric proportions, in addition to uniformity in temperature and stresses. Chemical equilibrium is summarized in the form of the Law of Mass Action. The trend for the displacement from an equilibrium state is specified by LeChâtelier's principle.

Thermodynamic equilibrium is reached when the condition of mechanical, electromagnetic, thermal, phase, and chemical and nuclear equilibrium is reached.

Stability of Equilibrium. A process or change of state carried out on a system such that it is always near a state of equilibrium is called a quasi-stationary equilibrium. This requires that the process be carried out slowly. If a mechanical system is initially at the equilibrium position with zero initial velocity, then the system will continue at equilibrium indefinitely. An equilibrium position is said to be stable if a small disturbance of the system from equilibrium results only in small, bounded motion about the rest position. The equilibrium is unstable if an infinitesimal displacement produces unbounded motion. In the gravitational field, a marble at rest in the bottom of a bowl is in stable equilibrium, but an egg standing on its end is in unstable equilibrium. When motion can occur about an equilibrium position without disturbing the equilibrium, the system is in neutral (or labile, or indifferent) equilibrium, an example being a marble resting on a perfectly flat plane normal to the direction of gravity. It is readily seen that stable equilibrium is the case when the extremum of potential is a minimum.

When dealing with general thermodynamic systems, the fact that entropy tends to a maximum in the trend toward equilibrium of a natural process generalizes the above mechanical consideration with respect to stability. An equilibrium state can be characterized as a stable equilibrium when the entropy is a maximum; neutral equilibrium when displacement from one equilibrium state to another does not involve changing entropy; and unstable equilibrium when entropy is a minimum. Any slight disturbance from an unstable equilibrium state of a system will lead to transition to another state of equilibrium.

Statistical Equilibrium. In the microscopic sense, that is, treating systems in terms of elemental particles such as molecules, atoms, and other material or quasi-particles (such as photons in radiation, phonons in solids and liquids, rotons in liquids), equilibrium states are recognized as the most probable states. An equilibrium state of a system is therefore defined in terms of most probable distributions of its elements among microscopic states which may be defined in terms of energy states. In this sense, statistical equilibrium is a condition for macroscopic equilibrium and an equilibrium state of a system is one of its extremal states. In the methods of statistical mechanics, the probability of distribution is expressed in terms of the density of distributions in the phase space. Based on the Liouville theorem, if a system is in statistical equilibrium, the number of the elements in a given state must be constant in time; which is to say that the density of distribution at a given location in phase space does not change with time. For an isolated system, the distribution is represented by a microcanonical ensemble. At equilibrium, no phase point can cross over a surface of constant energy, and the density of distribution is preserved. In this case individual molecules of a system can be represented by phase points. Any part of an isolated system in statistical equilibrium can be represented by a canonical ensemble. A subsystem of a large system in thermal equilibrium also behaves like the average system of a canonical ensemble. A system and a constant temperature bath together can be considered as an isolated system. A phase point in a canonical ensemble can represent a large number of molecules, thus accounting for strong interactions. A canonical ensemble is characterized by its temperature and is therefore pertinent to the concept of thermal equilibrium. When applied to equilibrium of systems involving mass exchange, such as a chemical system, we have a "particle bath" in addition to a constant temperature bath. The pertinent representation for equilibrium including mass exchange as well as energy exchange is known as a grand canonical ensemble, which accounts for the chemical potentials of its elements.

When applied to a system with a large number of elements, the distributions are measured by thermodynamic probability (W); the most probable distribution is such that W is a maximum. This optimal principle is consistent with the condition of maximum entropy (S) cited under **Entropy.** The Boltzmann hypothesis states that $S = k \ln W$, where k is the Boltzmann constant.

Depending on the specifications of W, namely, those of Maxwell-Boltzmann (for low concentration of distinguishable particles, weak interaction and high temperature, such as a dilute perfect gas), Fermi-Dirac (for elemental particles with antisymmetric wave functions at high concentrations of indistinguishable particles and low temperatures, such as electrons in metal), or Einstein-Bose (for elemental particles with symmetric wave functions, such as He^4 at high concentration of indistinguishable particles and low temperature), equilibrium distributions take different forms. The Maxwellian speed distribution in a dilute perfect gas is a distribution based on Maxwell-Boltzmann statistics.

As a consequence of molecular considerations, when two systems are connected for transfer of mass without significant transfer of energy, such as two containers at different temperatures connected by a capillary tube, we have the relation of thermal transpiration.

Trend toward Equilibrium. The mechanism by which equilibrium is attained can only be visualized in terms of microscopic theories. In the kinetic sense, equilibrium is reached in a gas when collisions among molecules redistribute the velocities (or kinetic energies) of each molecule until a Maxwellian distribution is reached for the whole bulk. In the case of the trend toward equilibrium for two solid bodies brought into physical contact, we visualize the transfer of energy by means of free electrons and phonons (lattice vibrations).

The Boltzmann H-theorem generalizes the condition that with a state of a system represented by its distribution function f, a quantity H, defined as the statistical average of $\ln f$, approaches a minimum when equilibrium is reached. This conforms with the Boltzmann hypothesis of distribution in the above in that $S = -kH$ accounts for equilibrium as a consequence of collisions which change the distribution toward that of equilibrium conditions.

Consideration of perturbation from an equilibrium state leads to methods for dealing with rate processes and methods of irreversible thermodynamics in general.

Fluctuation from Equilibrium. A necessary consequence of the random nature of elemental particles in a body is that the property of such a body is not at every instant equal to its average value but fluctuates about this average. A precise meaning of equilibrium can only be attained from consideration of the nature of such fluctuations. In the above, we have repeatedly considered a "large" number of particles. It is important to know how large a number is "large." When considering fluctuation of energy from an average value in an isolated system, the ratio of the two is given to be proportional to $1/\sqrt{N}$, where N is the total number of elements in the system. This is also the magnitude of the fluctuation of number of particles in a system involving transformation of phases and chemical and nuclear species. An equilibrium state is one at which the longtime mean magnitude of fluctuation from the average state is independent of time and this magnitude has reached a minimum value.

Large perturbation from a given state of fluctuation leads to a relaxation process toward a state of equilibrium. The relaxation time, for instance, measures the deviation from quasistationary equilibrium of a process which is carried out at a finite rate.

EQUILIBRIUM (Atmosphere). Atmosphere (Earth).

EQUILIBRIUM (Biological). Biological Equilibrium.

EQUILIBRIUM (Chemical). Chemical Equilibrium.

EQUILIBRIUM DIAGRAM. A diagram showing the phase fields of an alloy system under the conditions of complete equilibrium using as coordinates the temperature, the compositions in terms of the components, and the pressure. The most frequently used equilibrium diagrams in metallurgy are drawn with the pressure considered constant. See iron-carbon diagram under **Iron Metals, Alloys, and Steels.** See also **Distillation.**

EQUILIBRIUM DIAGRAMS (Distillation). Distillation.

EQUILIBRIUM DISTILLATION. Distillation.

EQUILIBRIUM SENSING (Human). Hearing and the Ear.

EQUILIBRIUM (Three-Phase). Three-Phase Equilibrium.

EQUILIBRIUM VOLATILITY (Gasoline). Petroleum.

EQUINES. Horses, Asses, and Zebras.

EQUINOX. The line of intersection of the plane of the earth's equator with the plane of the ecliptic (the line of nodes of the earth) intersects the celestial sphere in two diametrically opposite points known as the equinoxes. As seen from the earth, the sun apparently passes through each of the equinoxes once each year, passing through the vernal equinox on approximately March 21st and through the autumnal equinox on approximately September 22nd.

The great circle passing about the celestial sphere through the equinoxes and the pole of the ecliptic is known as the *equinoctial colure.*

The *autumnal equinox,* for either hemisphere, is the equinox at which the sun "retreats" into the opposite hemisphere. In northern latitudes, the time of this occurrence is approximately September 22nd. The *vernal equinox,* for either hemisphere, is the equinox at which the sun approaches from the opposite hemisphere. In northern latitudes, this occurs approximately on March 21st. See also **Celestial Sphere.**

EQUISIGNAL RADIO-RANGE BEACON. Am aircraft-guidance radio beacon that transmits two distinctive signals which may be received with equal strength only in certain sectors called equisignal sectors.

EQUIVALENCE PRINCIPLE. It is always possible at a point in space-time to transform to a (in general accelerated) coordinate system such that the effects of gravity will disappear over a differential region in the neighborhood of the point. As a particular case, if there are two observers, one uniformly accelerated with acceleration g and not in a gravitational field, the other not accelerated but held in a uniform gravitational field g, the results of mechanical and optical experiments performed by the two observers will be identical. See **Gravitation;** and **Relativity and Relativity Theory.**

EQUIVALENCE THEOREM. The field in a source-free region bounded by a surface could be produced by a distribution of electric and magnetic currents on that surface that would be equivalent, for points inside the surface, to the actual external sources.

EQUIVALENT BINARY DIGITS. Digit (Computer System).

EQUIVALENT CIRCUIT. This term is applied to an electrical circuit which is electrically equivalent to another circuit, or sometimes, to a mechanical device. Equivalent circuits of mechanical systems or electromechanical systems such as loudspeakers enable the designer to apply methods of circuit analysis and often obtain a solution easily which would be very difficult if not impossible otherwise. The equivalent circuit method is used extensively in the analysis of communication circuits, particularly those involving vacuum tubes and transistors. These circuits do not yield the static currents and voltages (those existing in the absence of applied signals) but their use provides a method of computing the response to alternating voltages or dc changes, provided both are of small magnitude. Both of these are usually the ones of interest. Similarly, many other types of electrical circuits may be simplified in terms of equivalent circuits, sometimes giving all the necessary solutions, sometimes giving solutions for limited conditions, but, in most cases, greatly decreasing the labor involved in analyzing the circuit or equipment.

EQUIVALENT ELECTRONS. For an atom, electrons in the same orbital (whereby they have the same principal quantum number and the same azimuthal quantum number). For a molecule, electrons having the same quantum numbers, apart from spin, and the same symmetry g or u.

EQUIVALENT WIDTH (Stellar Spectra). Spectral Equivalent Width (Stellar).

EQUIVOCATION. A measure of the average ambiguity of a received signal. It is the residual remaining uncertainty when the received signal has been interpreted.

ERA (Geology). Geologic Time Scale.

ERBIUM. Chemical element symbol Er, at. no. 68, at. wt. 167.26, eleventh in the Lanthanide Series in the periodic table, mp 1529°C, bp 2868°C, density 9.066 g/cm³ (20°C). Elemental erbium has a close-packed hexagonal crystal structure at 25°C. The pure metallic erbium is silver-gray in color and retains its luster at room temperature, not affected by moisture or normal atmospheric gases. Large pieces of the metal do not oxidize readily even when heated. Fine chips and powder, however, will ignite and burn. Because of its comparative softness, the metal can be worked by conventional equipment. The metal should be annealed after size-reduction. There are six natural isotopes ^{162}Er, ^{164}Er, ^{166}Er through ^{168}Er and ^{170}Er. Twelve artificial isotopes have been prepared. The natural isotopes are not radioactive. In terms of abundance, erbium is present on the average of 2.8 ppm in the earth's crust, making its potential availability about equal with uranium. The element was first identified by C. G. Mosander in 1843. The thermal-neutron-absorption cross section of erbium is 160 borns per atom, relatively high and tenth among the natural elements. The metal has a low acute-toxicity rating. Electronic configuration

$$1s^2 2s^2 2p^6 3s^2 3p^6 3d^{10} 4s^2 4p^6 4d^{10} 4f^{11} 5s^2 5p^6 5d^1 6s^2.$$

First ionization potential 6.10 eV; second 11.93 eV. Ionic radius Er^{3+} 0.881 Å. Metallic radius 1.758 Å. Other important physical properties of erbium are given under **Rare-Earth Elements and Metals.**

Erbium occurs in certain types of apatites, xenotime, and gadolinite.

These minerals also are processed for their yttrium content as well as for other heavy *Lanthanide* elements. With liquid-liquid organic and solid-resin organic ion-exchange techniques, the separation of erbium from the other elements is favorable.

Because of the metal's high thermal-neutron-absorption cross section, it has been of much interest in terms of use in nuclear reactor hardware. When an erbium-activated phosphor is coated onto a gallium-arsenide diode, the latter emits infrared radiation, which is converted to visible light by the phosphor. Through variation of the energizing power and by use of a combination of rare-earth-activated phosphors, the primary colors of light can be produced. Thus, erbium holds promise for use in display panels and color-television picture tubes. An erbium hydride-hydrogen system at a fixed temperature creates an extreme vacuum and when used for comparative purposes makes it possible to measure vacuums in the range of 10^{-4} to 10^{-11} torr with much precision. The system has been used for the calibration of ionization gauges used for very high vacuums as found in outer space. Erbium is in an early stage of investigation for application to lasers, semiconductor devices, garnet microwave devices, ferrite bubble devices, and catalysts.

See references listed at ends of entries on **Chemical Elements;** and **Rare-Earth Elements and Metals.**

NOTE: This 6th Edition entry was revised and updated by K. A. Gschneidner, Jr., Director, and B. Evans, Assistant Chemist, Rare-Earth Information Center, Energy and Mineral Resources Research Institute, Iowa State University, Ames, Iowa. Original 5th Edition entry was prepared by J. G. Cannon, Molycorp, Inc.

ERECTILE TISSUE. Gonads.

ERECTOR. A vehicle used to support a rocket for transportation and for placing the rocket in an upright position within a gantry.

ERG. Units and Standards.

ERGODICITY. Generally, this word denotes a property of certain systems which develop through time according to probabilistic laws. Under certain circumstances a system will tend in probability to a limiting form which is independent of the initial position from which it started. This is the *ergodic property.* A stationary stochastic process (x_t) may be regarded as the set of all realizations possible under the process. Each such realization may have a mean m_r. If the process itself has a mean $E(x_t) = \mu$, the ergodic theorem of Birkhoff and Khintchine states that m_r exists for almost all realizations. If, in addition, $m_r = \mu$ for almost all realizations the process is said to be ergodic. In this sense ergodicity may be regarded as a form of the law of large numbers applied to stationary processes.

ERGOSTEROL. Photochemistry and Photolysis.

ERGOT. A fungus which grows upon and replaces the grain of rye. Ergot in the human body contracts smaller arteries and smooth muscle fibers, especially that of the uterus. Ergotism is a diseased condition of humans and domestic animals resulting from eating grasses or grain infected with ergot fungus; or from excessive uses of the drug made from the fungus. Ergotamine, one of the alkaloids of ergot, has been used in the treatment of migraine.

ERGOTAMINE TARTRATE. Headache.

ERICACEAE. Heather Shrubs and Trees.

ERIDANUS. A very long constellation that extends from the equator to the southern horizon. The constellation contains the star Achernar.

ERINACIDS. Moles and Shrews.

ERLANGEN PROGRAMM. Geometry.

ERMINE. Mustelines.

EROSION (Geology). A fundamental process of geology which brings about alterations in rocks and other major features of the physical earth. Some of the results of erosion are beneficial, as exemplified by the gradual disintegration, transformation, and transportation of certain kinds of rocks to form soil. On the other hand, erosive processes can leach out or carry away excellent soil, and thus destroy agricultural productivity, sometimes over wide areas, by the action of wind and water. For soil erosion, see **Soil.**

In a general sense, erosion refers to the reduction of the land surface toward sea level by the various agencies of weathering, stream action, glacial action, wind action, chemical action, and other forces. A majority of these forces occur naturally and essentially are beyond control by way of human intervention. On the other hand, erosion also results from unplanned (in terms of long-term erosion) by various projects which may create or accelerate already existing natural erosive processes. Alteration of the course of streams and certain types of agricultural methods can cause serious erosion problems.

The surface of the earth, subject to both natural and artificial forces, is constantly undergoing an overall process that may be termed *weathering.* Erosion is a major result of this process. Erosion processes may be classified into two major groups: (1) *chemical erosion*; and (2) *mechanical erosion.* The actions of erosive agents, such as air, wind, and water, may act both in a chemical and mechanical fashion. See also **Glacier.**

Chemical Erosion. The principal agents of chemical erosion are water and air. Both agents attack the content of the original rocks and thus bring about changes in chemical composition with accompanying physical changes. In falling through the atmosphere, rain water (*meteoric water*) picks up small amounts of oxygen, carbon dioxide, and other gases (notably in areas where air pollutants are present). The addition of these ingredients to otherwise pure water increases the solvent power of the water. Water in soaking downward through the soil reaches rocks which often contain numerous cracks and fissures, thus affecting the rock to cause a considerable area of exposed surface. Further, the solvent power of the water may be increased as it acquires new chemical substances from the rocks themselves. For example, water attacking pyrite will pick up sulfur and through a complex process ultimately convert this to sulfuric acid, which is highly corrosive. Particularly, the combined action of oxygen in the air with moisture from rain can become highly corrosive, as witness the immediate formation of rust upon exposed iron (ferrous) materials. Water plus carbon dioxide can yield corrosive carbonic acid. Thus, in processes of oxidation and carbonation, the chemical weathering of rocks in both erosive and corrosive. Water alone through the process of hydration can cause decomposition of some rocks. For example, two parts of hematite, Fe_2O_3, will unite with three parts of water to form limonite (iron rust), $2Fe_2O_3 \cdot 3H_2O$.

The erosive results of groundwater are principally chemical in nature. Although the corrosive solutions may move slowly, over periods of time extensive changes in rock structures can be brought about, particularly the creation of a large thickness of mantle rock. Thus, enormous caverns may be created, such as the Carlsbad Cave (New Mexico) or the Mammoth Cave (Kentucky). Also, rich ore deposits may result from such actions, as witness the great copper deposits found in the western United States or the iron deposits of the Lake Superior region. The analysis of ground waters testifies to their prior erosive performance. In descending order of content, groundwaters will be found to contain (1) calcium carbonate and calcium sulfate; (2) colloidal silica; (3) sodium carbonate, sodium sulfate, and sodium chloride; (4) magnesium carbonate; and (5) potassium carbonate—all materials derived from prior contacts with rocks and soil. To a lesser extent, ocean water also works in a similar fashion along the coasts.

The mechanical effects of groundwater become significant when the volume and velocity of the water reach a point where the flow can pick up and transport solid particles of rock. Sometimes this process follows or is concurrent with the aforementioned chemical actions of groundwater. The lower portion of Mammoth Cave, exhibiting gravels, sands, and muds, provides evidence of the extensive action of groundwater. The principal results of the mechanical action of groundwater which may be observed from the surface include soil creep, landslides, and rock streams. The underlying slippery nature of shales and clays when wet assist the process of alteration. There are rock streams in the Rocky Mountains of the United States a mile or more in length. So-called "walking mountains" (as reported in China) are the result of such phenomena on an extremely large scale. See **Cave.**

The mechanical actions of the oceans are far more important than their chemical actions in terms of erosion. These include the actions of waves, currents, and tides. A *terrace* or *wave-cut beach* is formed by the action of waves cutting into a land surface of moderate relief. A *sea-cliff* may result where relatively high land is undercut by waves. The chalk cliffs of England and France are examples. Where particularly soft, vulnerable rock structures are encountered by wave action, *sea caves* may be formed. The Blue Grotto (Naples) is a notable example. Sometimes spouting caves are formed. See **Blowhole.** Where rocks contain vertical joints, these may be eroded at a greater rate, producing a cliff with deep indentations. Where the eroded rock may become separated from the land mass, it may form a *chimney* or *stack.* See also **Chimney Rock.** Where there is a rock connection remaining at a higher elevation, a *natural bridge* will be the result.

Waves of the oceans also may form deposits along the shore, built from the deposition of materials by the mechanical work of the waves. Such terms as *wave-built terraces, barrier beaches, spits, hooks, bars,* and *tombolos* are used to describe these resulting features. See **Barrier Beach.**

The wind also serves as a mechanical agent for picking up, carrying, and later depositing solid materials. Numerous erosive features result. Wind-worn pebbles (glyptoliths) give evidence of wind erosion. These are also referred to as dreikanters. See **Dreikanter.** The ravages of wind removal and transportation of soil are only too evident in certain dust bowl areas, generally the result of poor agricultural methods.

When material is moved by wind and ultimately released, two dominant types of deposits occur: (1) *loess*; and (2) *dunes.* See **Dune.** Loess is composed of dust and silt, usually comprised of quartz grains, ranging in size from about 0.1 millimeter, and clay. Loess is characterized by a buff-to-yellow color and great porosity. Loess may be observed along both sides of the Mississippi River (Louisiana and Mississippi, north Illinois and Iowa); also the Missouri River and other tributaries in the Mississippi Valley. Loess occurs extensively in central Europe and in Tibet, Mongolia, and China. Thicknesses of loess deposits in China up to 300 feet (91 meters) are reported. The deposits in Europe are usually quite thin; those in the United States range from 10 to 20 feet (3 to 6 meters) in thickness.

The particle size of sand in sand dunes ranges in diameter from 0.05 millimeter up to that of coarse gravel. Most dunes are made up of quartz, but gypsum dunes are found in New Mexico. Dunes of calcareous oolites are found in the Bahamas and Bermuda; rather low dunes of dry clay occur in Montana.

Although of a seasonal nature, snow drifts are an aspect of the dune phenomenon.

Dunes range from a few feet to over 400 feet (120 meters) in height. They may cover a few square feet (a fraction of a square meter) to up to several square miles (kilometers). The dunes in desert regions commonly are 200 to 400 feet (60 to 122 meters) in height. Usually the axis of a dune is at right angles to the prevailing wind. Nearly all sand dunes migrate to a measurable extent unless they are covered with vegetation. Migrating dunes will advance upon forests, farms, highways. Dunes buried a number of cities of ancient times (Babylonian, Chaldean). Migration of dunes can be slowed or stopped through the planting of surface grasses and shrubs of a type that will withstand the prevailing climate.

The carrying of volcanic dust is a special form of wind deposit. Such deposits may occur long distances from their sources, as witness the volcanic dust deposits in southern Nebraska and north-central Kansas.

Cycle of Erosion. This is a term used to generally describe the work of rivers and streams—erosional, transportational, and depositional. The erosional work of rivers and streams sculptures the surface of the earth into a variety of forms. The river erosion pattern of any region will depend upon the climate, the relative hardness and solubility of the formations, the structure, and the degree to which the erosive process has completed its work, with or without interruptions caused by diastrophism. The stream pattern of a region is not only indicative of the structural control, but also of the stage in its erosional history. The ideal complete cycle of erosion begins with uplift of a region with low altitude and ends with reduction of the uplifted region to a peneplain. See accompanying diagram.

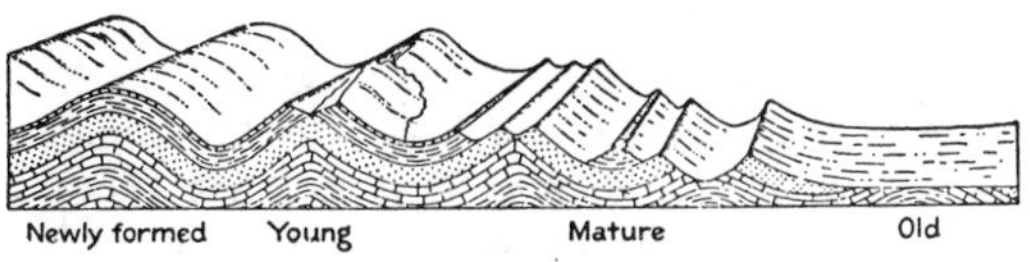

Successive stages in the normal cycle of erosion in a region of folded rocks.

Soil Conservation through Erosion Control. Modern programs of erosion control have been organized on a large scale in the United States, Australia, New Zealand, and in several countries in Africa, south of the Sahara. In nearly all countries where agriculture is important there has been recognition of the problem of erosion.

The cultivation of irrigated rice in oriental countries reduces erosion to a minimum on nearly all of the land actually used for rice. In China and India, much of the land between the irrigated paddy fields has lost its upper layers of soil to sheet erosion and is scored by deep gullies. This kind of problem has been alleviated in Japan by intensive reforestation and by other erosion-control measures on land used for unirrigated crops.

Erosion in many European countries has been kept at a low level for several generations through enlightened land-use programs and intensive cultivation and fertilization; but much of the upland soil bordering the Mediterranean Sea has been lost to erosion. Fertile floodplains around the Mediterranean have been badly damaged by deposition of coarse debris washed from cultivated uplands.

Modern soil conservationists recommend using the land in ways that will produce the greatest income consistent with the least loss of soil. They endeavor to protect sloping lands by planting soil-conserving crops, by contour cultivation, by alternation of strips of close-growing crops with clean-cultivated ones, and by using graded terraces and grass-covered waterways to control the rate of flow or runoff of water. Shallow gullies may be filled by plowing, and deep gullies may be controlled by check dams or vegetative cover. However, terraces and dams may be harmful if not well maintained. The large volumes of water collected behind terraces and check dams sometimes breaks through, sweeping away large volumes of soil, and scours out new gullies. Ditches that were dug to drain wet lowland soils for farming have provided lower base levels for water flow, and as a consequence gullies have eaten their way back from the ditches high into the adjacent uplands. Many examples of this exist in eastern Nebraska and western Iowa.

The bad effects of wind erosion can be ameliorated on cultivated land by planting strips of close-growing crops in alternation with strips of sod or fallow land arranged at right angles to the prevailing wind. Stubble mulch in semiarid wheat land is highly effective. List-furrowing across the wind is effective in some areas, and shelterbelts of trees and shrubs have been helpful to a limited extent. Wind erosion of mulch and peat beds can be greatly reduced by planting wind-breaks at right angles to the prevailing wind.

References

Baver, L. D., Gardner, W. H., and W. R. Gardner: "Soil Physics," 4th edition, Wiley, New York, 1972.

Bowles, J. E.: "Engineering Properties of Soils and Their Measurement," McGraw-Hill, New York, 1970.

Dasmann, R. F.: "Environmental Conservation," 3rd edition, Wiley, New York, 1972.

England, J., and R. S. Bradley: "Past Glacial Activity in the Canadian High Arctic," *Science*, **200**, 265–270 (1978).

Foth, H. D., and L. M. Turnk: "Fundamentals of Soil Science," 5th edition, Wiley, New York, 1972.

Hillel, D.: "Soil and Water: Physical Principles and Processes," Academic, New York, 1971.

Stuiver, M., Heusser, C. J., and I. C. Yang: "North American Glacial History Extended to 75,000 Years Ago," *Science*, **200**, 16–21 (1978).

Sugden, D. E., and B. S. John: "Glaciers and Landscape: A Geomorphological Approach," Arnold, London, 1976.

EROSION (Soil). Soil.

ERRANTIA. A division of marine segmented worms including chiefly free-swimming species. Sometimes ranked as an order and sometimes as a division of the order *Polychaeta.*

ERRATIC. In geology, an ice-carried boulder or block, sometimes weighing many tons, which because of its lack of similarity to the bedrock or formation on which it rests, and the peculiarity of its position, must have been transported to its present resting place by a glacier or an iceberg. When erratics occur in sufficient quantity to form relatively pronounced topographic features these are called moraines. When erratics of similar or identical lithology show a well-defined lineal distribution from the parent outcrop they are called boulder trains.

ERROR. The word error may be used in two ways: 1. In general, it is a mistake in the colloquial sense, for example, an error in copying, an error of reference, or an error of interpretation. In statistics, the word error is used to denote the difference between an occurring value and its "true" or "expected" value. Specific types of errors and error functions follow:

Error, Approximation. In general, an error due to approximation in numerical calculations as distinct; e.g., an error of observation. More particularly, a rounding error.

Error, Experimental. In general, any error in an experiment whether due to stochastic variation or bias. More specifically, the expression is used to denote the essential probabilistic variation to be expected under repetition of the experiment, not actual mistakes in design or avoidable imperfections in technique. It is the aim of good experimental design to provide valid measures of the experimental error in the more restricted sense.

Error Band. In estimation or prediction, the estimated or predicted value is bracketed by a range of values (determined by standard errors, confidence-intervals or similar methods) within which the value may be supposed to lie with a certain probability. This is called the error band.

Error Function. The definite integral, also called the Gauss error function,

$$\operatorname{erf}(t) = \frac{2}{\sqrt{\pi}} \int_0^t e^{-y^2}\, dy$$

When the results of a series of measurements are described about an average by a Gaussian curve, erf(ha) is the probability that the error of a single measurement lies between $\pm a$, where h is the precision index.

Values of the integral, as functions of t, have been tabulated. Sometimes the more general function

$$E_n(t) = n! \int_0^t e^{-y^n}\, dy$$

is discussed and $E_2(t)$ is called the error integral. See also **Fresnel Integral.**

Error in Equations. An equation in variables or variates may be inexact, either because the equation is not a complete representation of the situation (as in a demand-supply equation which omits other factors such as income or employment) or because it is disturbed by extraneous sources of variation (as in an autoregression equation). These departures from the relationship expressed by the equation are known as errors in the equation; as distinct from effects such as observational errors in the variables themselves.

Error of Estimation. In general, the difference between an estimated value and the true value. More specifically, in regression analysis where the regression equation is used to estimate the "dependent" from given values of the "independent" variates, the difference between the estimated and the observed value of the dependent variate.

Error of First Kind. If, as the result of a statistical test, a statistical hypothesis is rejected when it ought to be accepted, i.e., when it is true, then an error is committed. This class of error is termed an error of the first kind and is fundamental to the theory of testing statistical hypotheses associated with the names of Neyman and (E.S.) Pearson. The frequency of errors of the first kind can be controlled by an appropriate selection of the regions of acceptance and rejection; that is to say, by choice of appropriate critical regions it is possible to ensure that the probability of committing an error of the first kind is an assignable constant.

Error of Measurement. Much of the routine work in any experimental research in the physical sciences is concerned with the eliminating, minimizing, or compensating for observational errors. By the error of any measurement is meant the result of the individual measurement minus the true value of the quantity measured. It may thus be either positive or negative. Errors of measurement may be broadly classified into two types: instrumental errors and personal errors. An instrumental error is any error in measurement which results from the properties of the instruments used in the measurement. Instrumental errors may be divided into scale errors, which result from improper calibration of the instrument, and reproducibility errors, which result from the failure of the instrument to give the same indication whenever it is subject to the same input signal. The latter type may be treated as accidental errors, the former may not. A personal error is any error which results from the tendency of an observer to misread an instrument, e.g., to read consistently high or low. Personal errors are not distributed in the same manner as accidental errors, and their magnitudes may be estimated only by the comparison of observations made by different observers. An error of sampling is the discrepancy between the estimate derived from a sample and the true value. The occurrence of errors in this sense is inherent in the incompleteness of sample coverage; they are not "mistakes" in the ordinary sense of the word.

Error of Observation. An error arising from imperfections in the method of observing a quantity, whether due to instrumental or to human factors.

Error of Second Kind. If, as the result of a test, a statistical hypothesis is accepted when it is false, i.e., when it should have been rejected, then an error has been made. This class of error is termed an error of the second kind and, like errors of the first kind it is fundamental to the Neyman-Pearson theory of testing statistical hypothesis. Unlike the error of the first kind, however, it is not, in general, controlled by the simple process of selecting regions of acceptance and rejection. The customary procedure in choosing tests of hypotheses is to fix the magnitude of the first kind of error and, with this restriction, to minimize the second kind of error.

Error Signal. See **Signal (Instrument).**

Error Variance. The variance of an error component. Thus, if the generating model of a set of data consists of certain systematic components together with a stochastic component, the variance of the latter is in the error variance. The expression can also be understood in a wider sense, as the variance of error in repetitions of an experimental situation, whether the "error" is due to sampling effects or not. It makes for clarity if expressions such as "error variance" are eschewed in favor of "residual variance" but the use of the former type of wording is very widespread.

Sir Maurice Kendall, International Statistical Institute, London.

ERROR (Aliasing). Aliasing Error (Data Acquisition System).

ERROR (Common-Mode). Common-Mode Voltage.

ERROR (Computer Program). Debugging (Computer Program); Diagnostics (Computer System).

ERROR CORRECTION CODES. Information Theory.

ERROR (Instrument Burden). Burden (Instrument).

ERROR (Multiplexing). Analog Multiplexer.

ERROR SIGNAL. Signal (Instrument).

ERROR THEORY (Gerontology). Gerontology.

ERUPTIVE. This term has the same general geological meaning as effusive, but is sometimes used in the much more general sense as synonymous with igneous.

ERYSIPELAS. An acute inflammation of the skin (*superficial cellulitis*) due almost always to infection by Group A streptococci. In the newborn, Group B streptococci may cause the disease. The lesion is red, indurated, edematous, and spreads peripherally. Usually the bridge of the nose and cheeks are involved, but other areas of the face and head (ears) may be the site of the lesion. Fever is present from the onset. The disease in most cases is self-limiting. With the use of penicillin therapy, improvement may be achieved within one or two days, but the lesion may not subside for several days.

ERYTHRITE. A mineral of the composition, $Co_3(AsO_4)_2 \cdot 8H_2O$, isomorphous with annabergite. The color ranges from rose to crimson. The mineral sometimes contains nickel and occurs in monoclinic crystals, in earthy forms (as a weathering product of cobalt ores) in the oxidized portions of the veins, or in globular and reniform masses. The mineral sometimes is referred to as *erythrine*, *cobalt bloom*, and *peachblossom ore*.

ERYTHROBLASTOSIS FETALIS. Kernicterus.

ERYTHROCYTES. The red corpuscles of the blood. See also **Blood.**

ERYTHROMYCIN. Antibiotic.

ERYTHROPOIESIS. Anemias.

ESCAPE CHARACTER (Computer System). Character (Computer System).

ESCAPEMENT. Spring Clock.

ESCAPE ROCKET. A small rocket engine attached to the leading end of an escape tower, which may be used to provide additional thrust to the capsule to obtain separation of the capsule from the booster vehicle in an emergency in a manned spacecraft.

ESCAPE TOWER. A trestle tower placed on top of a space capsule, which during liftoff connects the capsule to the escape rocket. The escape tower is of such length as to protect the capsule from the heat of the escape rocket in case the rocket is used to separate the capsule from the booster vehicle during ascent. The tower is ultimately separated from the capsule if ascent is normal.

ESCARGOT. Mollusks; Snail.

ESCARPMENT. A term used by physiographers and structural geologists to denote a line of steep slopes or cliffs.

ESCLANGON EFFECT. The deviation of a ray of reflected light due to motion of the mirror in a direction oblique to its surface.

ESKER. Certain long, often winding, ridges formed of stratified sands and gravels, which occur within the glaciated regions of Europe and North America are called eskers. These pronounced topographic features are frequently several miles in length and because of their peculiar and uniform shape somewhat resemble railway embankments. Eskers represent the deposits of glacial streams which flowed within and under glaciers. After the retaining ice walls melted away the stream deposits remained as long winding ridges. Synonyms for esker include os, eschar, serpent kame, and Indian ridge.

ESOCIDS. Pike.

ESOPHAGEAL GLAND. Glands associated with pouches on the sides of the esophagus in earthworms. They secrete calcium carbonate. Also called calciferous glands.

ESOPHAGUS. The tubelike passage that connects the lower end of the pharynx (throat) with the stomach. This tube is about 10 inches (25 centimeters) long. It lies in front of the spine as it descends through the chest and passes through the diaphragm just before it enters the stomach. The walls of the esophagus are made up of circular and straight muscle fibers which allow for wavelike contractions to move food downward toward the stomach. The inner surface contains many glands which secrete mucus for lubrication of the walls. Difficulty of swallowing is a symptom associated with a number of esophageal disorders and is called *dysphagia*. Diagnostic aids sometimes used in identifying esophageal disorders include barium contrast x-rays, esophagoscopy (using a fiberoptic instrument which permits examination of the tube), and an acid perfusion test, the latter being particularly helpful in the diagnosis of esophagitis. A weak hydrochloric acid solution is administered through a nasoesophageal tube alternately with a saline solution. The presence of pain in response to hydrochloric acid during the test cycle is an excellent indicator of esophagitis and helps the physician differentiate this condition from the similar pain of coronary artery disease.

The principal disorders of the esophagus are as follows. (1) *Chronic peptic esophagitis* is quite common and usually responds well to intermittent antacid therapy. Onset of the condition usually commences an hour or so after meals. There may be regurgitation of small amounts of gastric contents into the mouth. The typical symptom is commonly referred to as *heartburn*. Unresponsive cases may require surgery. This condition is not always associated with a hiatus hernia. (2) *Hiatus hernia* is caused by a portion of the stomach protruding through the hiatus of the diaphragm and into the thoracic cavity. Barium x-ray examination and esophagoscopy are usually required to affirm the disorder. Because the symptoms of chronic peptic esophagitis and hiatus hernia are so similar, a physician may commence with antacid therapy. If the condition persists, cimetidine therapy may be initiated. In the absence of success, surgical repair of the hernia is indicated, along with restoration of the gastroesophageal junction. (3) *Carcinoma of the esophagus*, to date, has proved to be a very high risk situation. Because of late diagnosis, only about 20% of patients are surgical candidates. The surgery is technically difficult, and may be accompanied by radiation therapy. The five-year survival rate after surgical therapy alone has been estimated at about 10%. Dysphagia is an early symptom of this condition. Other disorders of the esophagus include: (4) lower esophageal ring; (5) diffuse esophageal spasm; (6) achalasia; (7) collagen vascular disease; and (8) esophageal diverticulum.

See also **Digestive System (Human).**

ESPUNDIA. Leishmaniasis.

ESSENTIAL AMINO ACIDS. Amino Acids.

ESSENTIAL HYPERTENSION. Hypertension (High Blood Pressure).

ESTERIFICATION. Cellulose Ester Plastics (Organic); Esters.

ESTERS. The compound resulting from the reaction of an alcohol with an acid is termed an *ester*. The reaction is termed *esterification* and is accompanied by the yield of H_2O along with the ester. The reaction is highly reversible and hydrolysis will occur in the reverse direction when H_2O remains present. The formation of ethyl nitrate from ethyl alcohol and nitric acid typifies a simple esterification: $C_2H_5OH + HNO_3 \rightleftharpoons C_2H_5NO_3 + H_2O$.

Under normal conditions, esterification occurs slowly and inasmuch as the reaction is fully reversible, an equilibrium is reached which tends to withhold completion of the reaction in either direction.

The esterification reaction can be speeded up by the use of a catalyst. Such a catalyst is hydrogen ion, as from HCl or H_2SO_4. Side reactions may occur, HCl furnishing some organic chloride, and H_2SO_4 causing dehydration of the alcohol. Phosphoric acid generally avoids both these results. Salts that hydrolyze to furnish hydrogen ions are also used, e.g., zinc chloride, aluminum sulfate, ferric chloride, sometimes by the addition of acid to these salts.

The equilibrium point can be displaced to produce more ester by increasing the relative amounts of either the acid or the alcohol as desired.

A complicating factor of considerable significance when recovery of the ester is to be made by distillation is the existence of 2-component (binary) azeotropes of constant boiling points with (1) ester and H_2O and (2) ester and alcohol, and 3-component (ternary) azeotropes with

AZEOTROPES ENCOUNTERED IN ESTERIFICATION PROCESSES

Ester [Alcohol]	bp °C		Azeotrope bp °C	Ester %	Other Component %
Ethyl acetate	77.1	a	70.4	93.9	6.1 water
[Ethyl alcohol	78.3]	b	71.8	69.2	30.8 alcohol
		c	70.3	83.2	7.8 water 9.0 alcohol
n-Propyl acetate	101.6	a	82.4	86.0	14.0 water
[*n*-Propyl alcohol	97.2]	b	94.7	49.0	51.0 alcohol
		c	82.2	59.5	21.0 water 19.5 alcohol
iso-Propyl acetate	91.0	a	77.4	93.8	6.2 water
[*iso*-Propyl alcohol	82.5]	b	81.3	40.0	60.0 alcohol
		c			
n-Butyl acetate	126.2	a	90.2	71.3	28.7 water
[*n*-Butyl alcohol	117.8]	b	117.2	53.0	47.0 alcohol
		c	89.4	35.3	37.3 water 27.4 alcohol
n-Amyl acetate	148.8	a	95.2	59.0	41.0 water
[*n*-Amyl alcohol	137.8]	b			
		c	94.8	10.5	56.2 water 33.3 alcohol

The ternary azeotrope has the lowest boiling point and in distillation, as long as the three components are present, it distills off first, then the binary azeotrope distills off as long as its components are present. Since esters are not very soluble in water, water can be added to the distillate, and then water plus alcohol are removed. Upon redistillation the ester soon distills off pure.

Many variations in details are used in applying the principles of esterification.

(3) ester and H_2O and alcohol, selected from the accompanying table.

Esters are high-tonnage chemicals. Among the more important esters are normal, secondary, and isobutyl acetates, ethyl acetate, normal, secondary, and isoamyl acetates, and methyl acetate. These acetates are used primarily in the lacquer industry. Cellulose nitrate is used in the plastics, lacquer, and explosives industries; cellulose acetate in the plastics and lacquer industries; glyceryl trinitrate in the explosives industry; while cellulose xanthate (viscose) is an important synthetic product for textiles. In the specialized field of plasticizers for the plastics and lacquer industries, numerous synthetic esters are used; as examples, butyl stearate, diamyl phthalate, dibutyl oxalate, dibutyl phthalate (also for smokeless powder), dibutyl sebacate, dibutyl tartrate, diethylene glycol monostearate, diethylene glycol distearate, diethyl phthalate, dimethyl phthalate, diphenyl phthalate, glyceryl tripropionate, isobutyl phthalate, tributyl borate, tributyl citrate, tributyl phosphate, tricresyl phosphate, triethylene glycol dihexoate, triethylene glycol dioctoate, triethyl citrate, triethyl phosphate, triphenyl phosphate. Methyl methacrylate ester is an important plastic. Ethyl silicate is used to cover concrete, brick and stone with a coating of silicic acid to resist water penetration. Diocytl phthalate is used as a plasticizer in cable and wire insulation, and dimethyl phthalate as an insect repellent.

See also **Organic Chemistry.**

ESTIMATED POSITION (EP). Navigation.

ESTIMATION (Theory of). Given a sample from a population whose specification involves one or more parameters, it is necessary to form estimates of the parameters. Usually, many different estimates of a given parameter can be derived, and the theory of estimation is concerned with the properties of these different estimates. Together with the estimate itself, it is useful to provide some idea of its precision and this is commonly done by specifying an interval which is intended to contain the true value of the parameter (see **Confidence Interval;** and **Fiducial Inference**).

A basic result in the theory of estimation states that, under general conditions, a consistent estimator has a sampling variance in large samples which is not less than a certain lower bound. Statistics whose variance attains this lower bound are said to be efficient. A still more valuable property, which applies for all sample sizes, is that of sufficiency; a sufficient statistic is one which summarizes all the information on the parameter that is contained in the sample. R. A. Fisher has shown that the method of maximum likelihood leads to efficient estimators, and to sufficient estimators where these latter exist. The method of minimum chi-square also leads to efficient estimates in large samples, but in finite samples may fail due to the occurrence of small expected frequencies.

ESTIVO-AUTUMNAL MALARIA. Malaria.

ESTROGEN. Gonads; Hormones; Steroids.

ESTROGEN (Infertility). Infertility.

ESTUARY. The wide mouth of a river, or arm of the sea, where the tide meets the river current, or flows and ebbs. It may also be defined as "a body of water in which the river water mixes with and measurably dilutes seawater" (Ketchum, 1951). These definitions do not overlap completely, because a lagoon connected with the sea may also be affected by the tide.

Some scientists prefer to describe the environment in terms of the salinity of the water (saline, brackish, or fresh), but saline water is not restricted to marginal marine areas. Such a description does not consider, therefore, the most characteristic aspect of the estuarine environment—that it is a region of steep and variable gradients in the environmental conditions (Fig. 1).

If the physiography of estuaries is a sole consideration, they can be defined as "bodies of water bordered by and partly cut off from the ocean by land masses that were originally shaped by nonmarine agencies. They are usually perpendicular to the coast line and most of them occupy the drowned mouths of stream valleys and are, there-

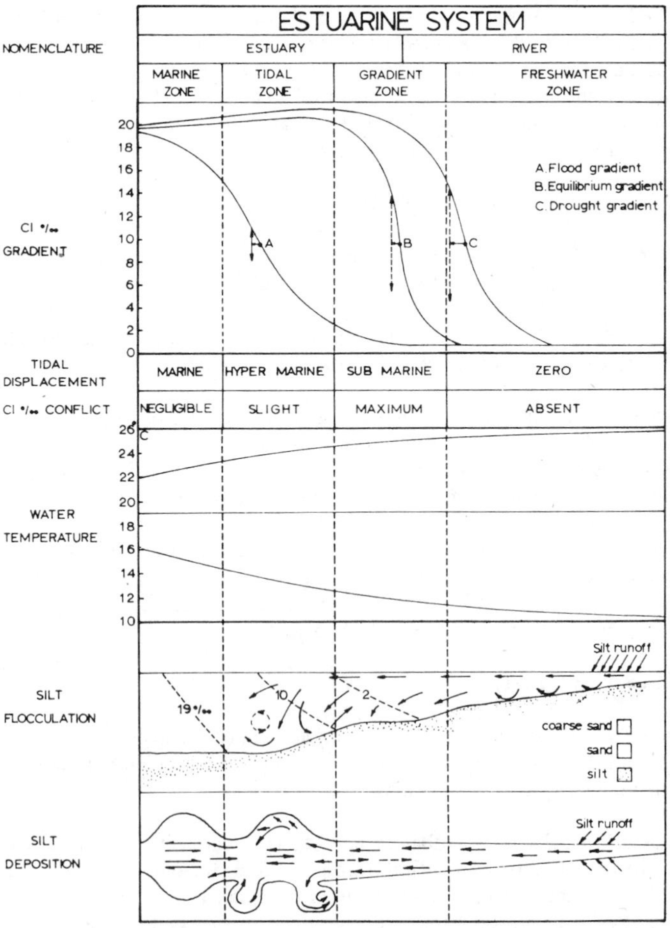

Fig. 1. Some zonal features of a composite Australian estuarine system.

fore, usually considered as evidence of submergence" (Emery and Stevenson, 1957).

Classification of Estuaries. There is no system which is universally used to classify estuaries. A broad classification separates normal (or positive) estuaries, in which freshwater inflow exceeds evaporation, from inverse (hypersaline) estuaries in which evaporation exceeds freshwater inflow. Neutral estuaries are those in which neither evaporation nor river discharge dominates.

Estuaries along most coastlines have been formed partly by the subsidence of the land mass and partly by the rise in sea level. These embayments are usually elongate indentures of the coastline with rivers flowing in from the landward ends. Deep estuaries are known as *rias.* In eastern North America most estuaries are shallow with irregular, or dendritic, shore lines and are normal estuaries.

Along the Gulf Coast of the United States, marine processes have built a series of barrier islands parallel to the coastline. Most of the islands extend across the mouths of estuaries, forming a lagoon and decreasing the width of the estuarine entrance to the open sea.

The exchange of water, in such cases, between the estuary and the open sea is modified by the intervening lagoon in which evaporation may exceed freshwater inflow. The waters in the estuary, then, have salinities higher than normal as a result of the exchange with the lagoonal water.

Water Characteristics and Circulation. The important feature in an estuary is the intermixing of seawater with the freshwater from land drainage. This interaction usually produces a variation, both horizontal and vertical, in the salinity of estuarine waters. In normal estuaries, salinities range from nearly zero at the river's mouth, to approximately 30% at the seaward extremity. In addition, there is generally an increase in salinity with depth.

An inverse estuary also has greater salinities at depth, but the highest salinities are at the head of the embayment rather than at the mouth. There may be a difference of several parts per thousand between the salinity at the head and that of normal seawater (Fig. 2).

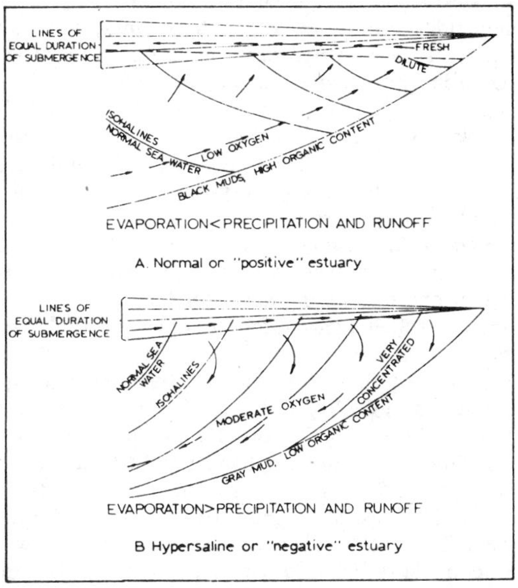

Fig. 2. Schematic sections of the two basic types of estuaries.

Temperature. The water in estuaries, and especially that overlying the tidal flats, is relatively thin, so it follows the variations in temperature of the atmosphere more closely than does the water of the open sea. The water is much colder in winter and warmer in summer than is the sea. The diurnal variation is also greater than in the sea.

There are pronounced variations in water temperature with depth. During the winter, the water is cold and nearly isothermal at all depths. In some instances, in response to cold weather, the surface water may become a degree or so colder than the deeper water. During the summer, solar radiation and minor wind mixing produce a high temperature at the surface with less change at depth. The difference between surface and bottom temperatures may also be influenced by warm or cold river water flowing over the dense seawater.

Where evaporation exceeds river inflow, the summer surface water may become so saline as to sink to the floor of the estuary and flow out of the entrance beneath the incoming seawater. The temperature of the deep water may then be higher than that at the surface.

Circulation. Water movements in estuaries result mainly from the interaction of tides, river flow, and wind. The tides and river flow are usually the dominant factors. In Gulf Coast estuaries, where the tide range is small and river discharge is at times negligible, wind-induced currents are most important.

Stable estuaries. In normal estuaries, the distribution of temperature and salinity, and the circulation pattern, are controlled almost exclusively by the tide range and the river inflow. Tidal currents tend to produce turbulent mixing of the river water and the seawater. However, the low-density freshwater above the seawater results in a stable vertical stratification which resists mixing. As a consequence, the relative magnitudes of the river discharge and the tidal flow are significant in controlling the physical structure of the water in the estuary.

Where the river flow is large in relation to the tidal exchange, the seawater enters the estuary as a salt-water "wedge" along the bottom. However, there is frictional drag between the overlying freshwater, the salt-water wedge, and the bottom. The relative velocities of the seaward-flowing freshwater and the intruding salt-water wedge control the magnitude of the friction factor. Thus, the actual position of the wedge is closely dependent on the volume of the river flow. When the volume of river discharge is great, the wedge extends only a short distance into the estuary and, of course, vice versa (Pritchard, 1952).

The salinity of the salt-water wedge remains similar to that of the open sea because there is only minor mixing with the seaward-flowing freshwater. However, at the interface between the two types of water, waves form and sometimes intrude into the surface water. Thus, the salt content in the upper layers increases slightly as the water moves seaward. Even so, throughout the estuary, a sharp salinity gradient exists between the two water layers.

The loss of salt water from the wedge to the upper layer is compensated by a flow of water from the sea (Fig. 2). The exchange from below takes place all along the upper interface of the wedge. As a result, there is a flow directed upstream at all positions within the wedge. The landward-moving water in the salt wedge is minor, however, and of the two, the seaward flow of surface water is far greater.

Partly mixed estuaries. Where tidal movements are great as compared to the volume of river discharge, mixing between the seawater and freshwater is sufficient to destroy sharp interfaces. The salt wedge, in such cases, does not exist as an identifiable feature, but a transition layer of definitely increasing salinity does occur. In such an estuary, however, the salinity in both the upper and lower layers decreases toward the head of the estuary.

The chief cause of currents in estuaries in which the waters are partly mixed is the tide. As in the stratified estuary, there is a net water movement superimposed on the tidal currents—a net seaward flow at the surface and a net flow toward the head in the deeper layers. These water motions are not as well defined as in a stratified estuary, and there is no sharp current interface. The flow from the deeper layers toward the surface decreases toward the head of the embayment (Fig. 2). The volume rate of seaward flow increases, therefore, toward the mouth (Pritchard, 1952).

Mixed estuaries. In wholly mixed estuaries, the movements induced by the tide are far greater than those produced by the river inflow. The waters are completely mixed and are isohaline from the surface to the bottom. At all depths, the salinity decreases from the mouth to the head.

In such estuaries, the outward flowing water is deflected to the right, in the northern hemisphere, because of earth rotation. Thus, in wide estuaries, the salinity is less on the right side (looking toward the estuarine mouth) than on the left. A net seaward flow exists along the right side and a net landward flow on the left. Water also moves laterally across the estuary from the left to the right side resulting in horizontal mixing.

In narrow, well-mixed estuaries, mixing induced by tidal action may be great enough to eliminate any lateral salinity gradient. There is a net seaward flow in all waters and the only difference in salinity is the normal decrease toward the head.

Estuaries bordered by lagoons. Along coastal regions where barrier islands extend across the mouths of estuaries, the water bodies are usually so shallow that mixing by winds is sufficient to produce homogeneous water. Tidal currents are only significant through the inlets between the barrier islands. The total volume of water which flows in and out is relatively small. As a consequence, the rise and fall of the tide and tidal currents are minor within the estuary, and the most significant currents are from wind action.

There is, necessarily, a net flow of water out of these shallow estuaries sufficient to remove the water added by freshwater discharge. The large cross-sectional area of the estuary and the dampening effect of the coastal lagoon reduce the flow so that it is normally not directly measurable. The net motion may be completely modified by wind action to the extent that high water and constant, net inflow may occur during times when prevailing winds blow up the estuary. Strong winds blowing from the land reverse this effect and result in extremely low water levels in the estuary and the extrusion of estuarine waters many miles to sea.

Seiches in estuaries. In some of the larger estuaries, the periodic flooding by the tides is supplemented by seiches, long stationary waves. The simplest seiche is one whose node is at the mouth of the estuary and the antinode near the head. The period of the seiche is controlled by the length and depth of the body of water, and where its natural period nearly coincides with that of the tide, as at the Bay of Fundy, a great fluctuation in sea level occurs (about 15 meters). In most estuaries, the seiche is only a few centimeters and is obscured by the much greater tidal amplitude.

Waves. Wind waves are small in estuaries because of the short fetch and the shallow water. They usually cause little erosion although when the tide is high, waves may stir up the muddy sediment on tidal flats. Waves may transport some sand and, because the largest waves come across the widest part of the estuary, they form sand spits pointing upstream in tidal channels.

A tidal bore (a wave of translation) is common in narrow estuaries and tidal channels. As a result of the shape of the entrance and bottom friction, the flooding tide is held back for a time until the water finally rushes up the channel as a steep wall of water. Bores may be from a few centimeters to several meters in height and move at velocities as great as 10 knots. The character of a tidal bore is determined, in part, by the river discharge which must be sufficient to hold back the tide for a period of time.

Estuarine sediments. The inorganic sediments of estuaries are derived from inflowing rivers bordering sea cliffs, the sea floor outside the estuary, and the reworked deposits of tidal flats and marshes along the shores. Regardless of the source, much reworking of sediment occurs within estuaries. Erosion, too, is evident from the migration of tidal channels and the muddy color of the water when no river inflow is taking place. Some estuaries have entrances narrow enough so that tidal currents scour the bottom locally, leaving rocky or gravelly bottoms. The prevailing condition, however, must be one of deposition, and the average rate of deposition is greater than that of the open sea.

References

Armstrong, R.: "White Oak River Dying," *Tideland News*, Swansboro, North Carolina, February 20, 1980.

Clark, J., and S. McCreary: "Prospects for Coastal Resource Conservation in the 1980s," *Oceanus*, **23**, 4, 22–31 (1981).

Emergy, K. O., and R. E. Stevenson: "Estuaries and Lagoons," *Geol. Soc. Am. Mem.*, **67**, 1, 673–750 (1957).

Ketchum, B. H.: "The Flushing of Tidal Estuaries," *Sewage Ind. Wastes*, **23**, 2, 198–209 (1951).

Lauff, G. H. (editor): "Estuaries," American Association for the Advancement of Science, Publ. 83, Washington, D.C., 1967.

ETHANE. C_2H_6, formula weight 30.07, colorless, odorless gas, mp −172°C, bp 88.6°C, sp gr 1.05 (air = 1.0), practically insoluble in H_2O, moderately soluble in alcohol. The compound burns when ignited in air with a pale faintly luminous flame; forms an explosive mixture with air over a moderate range. With excess air, products of combustion are CO_2 and H_2O. Ethane is among the chemically less reactive organic substances. However, ethane reacts with chlorine and bromine to form substitution compounds. Ethane occurs, usually in small amounts, in natural gas. The fuel value of ethane is high, 1,730 Btu per cubic foot. Ethane may be prepared by reaction of magnesium ethyl iodide in anhydrous ether (Grignard's reagent) with H_2O or alcohols. Ethyl iodide, bromide, or chloride are preferably made by reaction with ethyl alcohol and the appropriate phosphorus halide. Important ethane derivatives, by successive oxidation, are ethyl alcohol, acetaldehyde, and acetic acid.

ETHANE RECOVERY. **Gas and Expansion Turbines.**

ETHANOL. **Ethyl Alcohol.**

ETHANOLAMINES. There are three ethanolamines, all hydroxyamines, and all high-tonnage industrial chemicals. Production is about 300 mil pounds annually.

Monoethanolamine, $NH_2CH_2CH_2OH$, industrial symbole (MEA)
Formula weight 61.08, mp 10.5°C, bp 171°C, sp gr 1.018

Diethanolamine, $NH(CH_2CH_2OH)_2$, industrial symbol (DEA)
Formula weight 105.14, mp 28.0°C, bp 270°C, sp gr 1.019

Triethanolamine, $N(CH_2CH_2OH)_3$, industrial symbol (TEA)
Formula weight 149.19, mp 21.2°C, bp 360°C, sp gr 1.126

Mono- and triethanolamine are miscible with H_2O or alcohol in all proportions and are only slightly soluble in ether. Diethanolamine will dissolve in H_2O up to 96.4% at 20°C, is very soluble in alcohol, and only slightly soluble in ether.

Wurtz first reported the ethanolamines in 1860, but they were not used commercially on any scale until the late 1920s. All of the compounds are clear, viscous liquids at standard conditions and white crystalline solids when frozen. They have a relatively low toxicity. Industrially, the compounds are important (1) because they form numerous derivatives, notably with fatty acids, soaps, esters, amides, and esteramides; and (2) for their exceptional ability for scrubbing acidic compounds out of gases. Monoethanolamine, for example, will effectively move H_2S from hydrocarbon gases. The compounds also remove CO_2 from process streams and, where desired, the CO_2 may easily be recovered by heating the absorptive solutions. The soaps of the ethanolamines are extensively used in textile treating agents, in shampoos, and emulsifiers. The fatty acid amides of diethanolamine are applied as builders in heavy-duty detergents, particularly those in which alkylaryl sulfonates are the surfactant ingredients. The use of triethanolamine in photographic developing baths promotes fine grain structure in the film when developed. Ethanolamine also is used as a humectant and plasticizing agent for textiles, glues, and leather coatings; and as a softening agent for numerous materials. Morpholine is an important derivative.

In early processes, the ethanolamines were prepared by reacting ethylene chlorohydrin $ClCH_2 \cdot CH_2OH$ with NH_3. Current processes react ethylene oxide $\langle(CH_2)_2\rangle O$ with NH_3, usually in aqueous solution. The ratio of mono-, di-, and triethanolamines varies in accordance with the amount of NH_3 present. This is controlled by the quantities of MEA and DEA recycled. Higher NH_3-ethylene oxide ratios favor high DEA and TEA yields, whereas lower ratios are used where maximum production of MEA is desired. The reaction is noncatalytic. The pressure is moderate, just sufficient to prevent vaporization of components in the reactor. The bulk of the H_2O produced in the reaction is removed by subsequent evaporation. The dehydrated ethanolamines then proceed to a further drying column, after which they are separated in a series of fractionating columns, not difficult because of the comparatively wide separation of their boiling points.

e (The Number). A transcendental number, used as the base of the system of natural or Napierian logarithms. It is defined by

$$e = \lim_{n\to\infty}(1 + 1/n)^n$$

or by

$$e = \lim_{x\to 0}(1 + x)^{1/x}$$

It is represented by the infinite series

$$e = 1 + \frac{1}{1!} + \frac{1}{2!} + \frac{1}{3!} + \frac{1}{4!} + \cdots + \frac{1}{n!} + \cdots$$

and it equals 2.71828, approximately. In this book, the number *e* is written in italic form, except in equations involving electrical quantities, where the roman e is used to avoid confusion.

See also **Logarithm;** and terms listed under **Mathematics.**

ETHEPHON PLANT GROWTH MODIFIER. Plant Growth Modification and Regulation.

ETHER. Anesthesia.

ETHER (Chlorinated). Chlorinated Organics.

ETHERS. The homologous series of ethers has the formula C_nH_{2n+2}. Structurally, the ethers have an oxygen linkage between two radicals (R—O—R′). R and R′ may be the same as in dimethyl ether CH_3—O—CH_3; or they may differ as in ethylisopropyl ether C_2H_5—O—C_3H_7. The latter may be referred to as a *mixed ether.* Mixed ethers frequently are made from mixed alcohols. Where R and R′ are alkyls, the ether may be called an *alkyl ether* or an *alphyl oxide.* They may be considered to be derivatives of the monohydric alcohols. Each ether is isomeric with a saturated alcohol. Both diethyl ether and butyl alcohol are $C_4H_{10}O$. Also, there are many isomeric ethers, starting with $C_4H_{10}O$. Methylpropyl ether and diethyl ether are isomeric. Where compounds such as these have the same general formula, are members of the same family, and differ only by the alkyl group present, they are termed *metameric.*

Since they are similar structurally to the alcohols, phenols also form ethers. An example of an aromatic ether is methylphenyl ether (anisole) C_6H_5—O—CH_3. There are few ethers where both R and R′ are aryls. The structure of thioethers is similar to the other ethers, but with a sulfur atom in the link instead of an oxygen atom, as R—S—R′. Examples of thioethers include diethyl sulfide C_2H_5—S—C_2H_5 and methylethyl sulfide CH_3—S—C_2H_5, which is a mixed thioether.

The properties of the ethers may be summarized by: (1) with the exception of dimethyl ether which is a gas, the ethers are volatile, mobile, inflammable liquids that are lighter than H_2O; (2) they are relatively inert chemically, not being acted on by alkali metals or alkalis and not reacting with dilute acids; (3) they form substitution products when reacted with chlorine and bromine; and (4) they are decomposed when heated with strong acids, yielding esters:

$$\underset{\text{(diethyl ether)}}{(C_2H_5)_2O} + 2H_2SO_4 \rightarrow \underset{\text{(ethyl hydrogen sulfate)}}{2C_2H_5\cdot HSO_4} + H_2O$$

$$\underset{\text{(methylethyl ether)}}{CH_3\cdot O\cdot C_2H_5} + 2HBr \rightarrow \underset{\text{(methyl bromide)}}{CH_3Br} + \underset{\text{(ethyl bromide)}}{C_2H_5Br} + H_2O$$

Ether. $(C_2H_5)_2O$, formula weight 74.12, mp − 116.3°C, bp 34.6°C, sp gr 0.708. Probably the best known of the ethers, diethyl ether, commonly called simply *ether,* is slightly soluble in H_2O (1 volume in 10 volumes H_2O) and is miscible with alcohol in all proportions. Ether dissolves iodine and many organic substances, e.g., oils and fats, waxes, resins, and alkaloids and hence is widely used as a solvent for these substances in the preparation of numerous products, including explosives and collodion. Ether explodes in oxygen in the presence of a flame or spark, yielding H_2O and CO_2. When heated with an acid, such as H_2SO_4, ether yields ethyl alcohol. With phosphorus halides, ethyl halide (2 moles) is formed, Ether reacts with HNO_3 to form ethyl oxide.

Although still used medically, at one time ether was the major anesthetic, for which it must be scrupulously pure. In addition to various side effects which may result from the use of ether as an anesthetic, it is a definite hazard in the operating room because of its explosive properties, particularly in enriched oxygen atmospheres.

See also **Organic Chemistry.**

ETHIOPIAN REALM. Biome.

ETHMOIDAL SINUS. Sinus.

ETHMOIDITIS. Sinusitis.

ETHOXYLATED ALKYL PHENOLS. Detergents.

ETHNOBOTANY. The study of the intererelations of primitive man and plants.

ETHYL ALCOHOL. C_2H_5OH, formula weight 46.07, colorless liquid with mild characteristic odor, mp − 114.1°C, bp 78.32°C, sp gr 0.789. Also known as *ethanol,* the compound is miscible in all proportions with H_2O or ether. When ignited, ethyl alcohol burns in air with a pale blue, transparent flame, producing H_2O and CO_2. The vapor forms an explosive mixture with air and is used in some internal combustion engines under compression as a fuel. See also **Fuel.** Such mixtures are frequently referred to as *gasohol.*

Anhydrous ethyl alcohol is made from the constant boiling mixture with H_2O (95.6% ethyl alcohol by weight)—(1) by heating with a substance such as calcium oxide, which reacts with H_2O and not with alcohol, and then distilling, or (2) by distilling with a volatile liquid, such as benzene (bp 79.6°C) which forms a constant low-boiling mixture with H_2O and alcohol (bp 64.9°C), so that H_2O is removed from the main portion of the alcohol; after which alcohol plus benzene distills over (bp 78.5°C). Anhydrous ethyl alcohol is required for certain purposes as a solvent and reagent and fuel applications.

Commercially, ethyl alcohol is marketed by the proof gallon, 200 proof on the scale representing pure alcohol (100%). When the term *alcohol* alone is used, it refers to a liquid that ranges from 188 to 192 proof (94% to 96% ethyl alcohol). When the terms *grain alcohol, high-purity* alcohol, or *pure* ethyl alcohol are used, these usually refer to a liquid that is 190 proof. In most countries, beverage alcohol is highly taxed and to make the product available for nonbeverage purposes, denaturants will be added. Denaturants include methyl alcohol, pyridine, benzene, kerosene, pine oil, mixtures of primary and secondary aliphatic higher alcohols, and hydrogenated organic compounds. Thousands of nonbeverage industrial and commercial products, notably food extracts, toiletries, pharmaceuticals, solvents, and cleaning products, contain denatured ethyl alcohol.

Worldwide, ethyl alcohol is the basis for a huge alcoholic beverage industry, offering a wide range of products wherein the alcoholic content varies from a few to over 50% (100 proof). Industrially, ethyl alcohol is very important high-tonnage raw and intermediate material for numerous processes, and is used extensively in solvents, antiseptics, antifreeze compounds, and fuels.

Production. Natural fermentation is the oldest process for making ethyl alcohol and still constitutes the principal means for creating the alcoholic content of beverages. Except in connection with other alcohol-containing products, industrial producers of ethyl alcohol use processes other than fermentation. For fermentation, almost any agricultural raw material with a carbohydrate content in the form of sugars or starches that are easily converted to sugars can be used. Once the raw materials are in the form of sugars, yeast enzymes are added to commence natural fermentation. Traditionally, in the United States, industrial alcohol prepared by fermentation has used blackstrap molasses, which contains up to 50% sugars and can be easily fermented. The starting mash is prepared by diluting the molasses with H_2O to bring the sugar content down to about 15% (weight). The mash is slightly acidified, after which invertase (enzyme to convert sucrose) and zymase (enzyme to convert glucose and fructose) are added. The products are ethyl alcohol and CO_2. Yeast activity is sustained by the addition of nutrients. With careful control of temperature and acidity, the fermentation process can be completed in about two days. The resulting mash (beer) usually contains about 12% ethyl alcohol which is recovered from the beer by distillation. See **Fermentation.**

In modern industrial ethyl alcohol plants, the compound is produced in two principal ways: (1) by *direct hydration of ethylene,* or (2) by *indirect hydration of ethylene.* In the direct hydration process, H_2O is added to ethylene in the vapor phase in the presence of a catalyst: $CH_2{:}CH_2 + H_2O \rightleftharpoons CH_3CH_2OH$. A supported acid catalyst usually is used. Important factors affecting the conversion include temperature, pressure, the $H_2O/CH_2{:}CH_2$ ratio, and the purity of the ethylene. Further, some by-products are formed by other reactions taking place, a primary side reaction being the dehydration of ethyl alcohol into diethyl ether: $2C_2H_5OH \rightleftharpoons (C_2H_5)_2O + H_2O$. To overcome these problems, a large recycle volume of unconverted ethylene usually is required. The process usually consists of a reaction section in which crude ethyl alcohol is formed, a purification section with a product of 95% (volume) ethyl alcohol, and a dehydration section which produces high-purity ethyl alcohol free of H_2O. For many industrial uses, the 95%-purity product from the purification section suffices.

In the indirect hydration process, ethylene first is absorbed in concentrated H_2SO_4 to form mono- and diethyl sulfates: $CH_2{:}CH_2 + H_2SO_4 \rightarrow CH_3CH_2OSO_3H$; and

$$2CH_2{:}CH_2 + H_2SO_4 \rightarrow (CH_3CH_2)_2SO_4.$$

The ethyl sulfates then are hydrolyzed to ethyl alcohol: $CH_3CH_2OSO_3H + H_2O \rightarrow CH_3CH_2OH + H_2SO_4$; and

$$(CH_3CH_2)_2SO_4 + 2H_2O \rightarrow 2CH_3CH_2OH + H_2SO_4.$$

Remaining steps in the process include recovery and purification of the crude ethyl alcohol and reconcentration of the dilute H_2SO_4. The crude ethyl alcohol is steam-stripped from the dilute acid solution, followed by distillation for purification.

Azeotropes. The physical properties of ethyl alcohol are influenced by the hydroxyl group that imparts hydrogen-bonding characteristics and polarity to the substance that are analogous to water. Ethyl alcohol displays a highly nonideal behavior in numerous solutions, forming several azeotropes. The list of binary azeotropes of ethanol is long, including acetonitrile, benzene, carbon disulfide, chloroform, ethyl acetate, hexane, toluene, and water. See also **Azeotropic System.**

Chemistry. Ethyl alcohol reacts (1) with sodium metal, forming sodium ethoxide C_2H_5ONa plus hydrogen gas, (2) with phosphorus chloride, bromide, iodide, forming ethyl chloride, bromide, iodide, respectively, (3) with H_2SO_4 concentrated, forming at 100°C ethyl hydrogen sulfate $C_2H_5OSO_2OH$, at 140°C diethyl ether $(C_2H_5)_2O$, at 200°C ethylene $CH_2{:}CH_2$, (4) with organic acids, warmed in the presence of H_2SO_4, forming esters, e.g., ethyl acetate $CH_3COOC_2H_5$, ethyl benzoate $C_2H_5COOC_2H_5$ (see various individual acids), (5) with magnesium methyl iodide in anhydrous ether (Grignard's solution), forming methane as in the case of primary alcohols, (6) with calcium chloride to form a solid addition compound $4C_2H_5OH \cdot CaCl_2$, which is decomposed by H_2O, (7) with oxygen, using sodium dichromate solution and H_2SO_4, to form acetaldehyde (and acetic acid), using air, in the presence of acetic bacteria, to form vinegar (dilute acetic acid along with the substances present in the alcohol used, e.g., wine, cider), (8) with HNO_3 (a) concentrated, free from nitrogen tetroxide, to form ethyl nitrate, (b) dilute to form glycollic acid, (c) concentrated acid containing nitrogen tetroxide (fuming HNO_3) explosive reaction, (9) with chlorine (or bromine) to form chloral CCl_3CHO (or bromal).

See **Organic Chemistry.**

References

Batter, T. R., and E. R. Wilke: "A Study of the Fermentation of Xylose to Ethanol by *Fusarium Oxysporum*," NTIS, LBL-635 (June 1977).

Benson, W. R.: "Biomass Potential from Agricultural Production," in "Proc. Biomass—A Cash Crop of the Future?," Midwest Research Inst. and Battelle Columbus Laboratories, Kansas City, Missour, March 1977.

Sutton, O. C., et al.: "Ethanol from Agricultural Residues," *Chem. Eng. Progress*, **75**, 12, 52–58 (1979).

ETHYL CELLULOSE. A versatile thermoplastic cellulose ether that is compatible with a wide variety of solvent systems, resins, oils, and plasticizers. This versatility permits a wide diversity of end-product properties. It is an excellent film former from a wide range of neat, lacquer, or dispersion formulations. Molded ethyl cellulose has excellent toughness, flexibility, and shock resistance. Useful temperature range is from about −40 to +100°C. In the preparation of ethyl cellulose, wood pulp or cotton linters with a high alpha-cellulose content are reacted with ethyl chloride and sodium hydroxide. Structural formulas of cellulose and ethyl cellulose with complete (54.9%) ethoxyl substitution are:

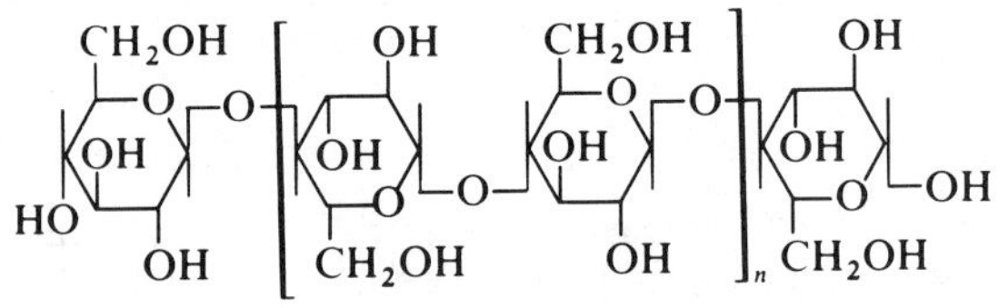

Cellulose, $n > 500$

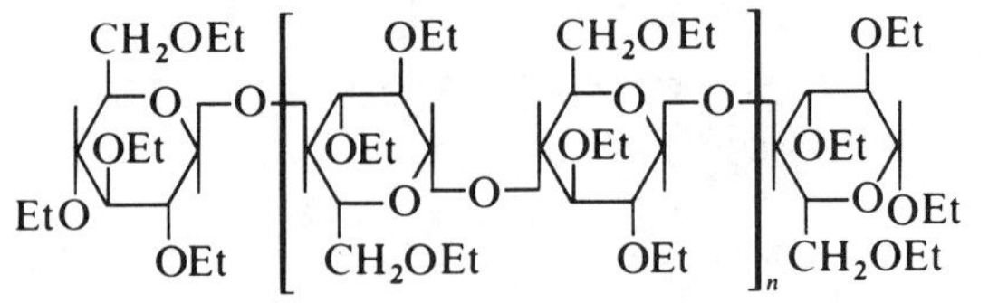

Tri-*O*-ethyl cellulose, $n = 50$–150; Et = CH_3CH_2

The natural color of ethyl cellulose is colorless to light amber, but it can be formulated into a wide range of transparent, translucent, and opaque colors. The material should be dried before molding because it is slightly hygroscopic. Compression molding temperatures range from 121–200°C and pressures from 500–5,000 psi. Injection molding temperatures range from 175–260°C and pressures from 8,000–32,000 psi. Strong acids decompose the material, but weak acids and strong alkalis have only a slight effect. Weak alkalis do not attack the material. Ethyl cellulose is soluble in a large number of organic solvents.

Among the commercial uses for ethyl cellulose are strippable coating for metal parts, paper coatings, and in medicinal tablets. An interesting application is coating for bowling pins. Ethyl cellulose sheeting is tough, flexible, and transparent, yet sufficiently rigid to withstand rough handling.

ETHYL CHLORIDE. Chlorinated Organics.

ETHYLENE. C_2H_4, formula weight 28.03, colorless gas with slight odor, normal bp −103.7°C, critical pressure of 49.98 atmospheres, and critical temperature of 9.5°C, density 1.26 grams per liter (0°C and 760 mm), sp gr 0.97 (air = 1.0), very slightly soluble in H_2O, slightly soluble in alcohol, soluble in alcohol. Ethylene burns when ignited in air with a luminous flame. The presence of ethylene in coal gas is chiefly responsible for the luminosity of the latter gas. Ethylene forms an explosive mixture with air and has a high fuel value, 1,615 Btu per cubic foot.

Even though there are few direct end-uses for ethylene, it is probably the most important petrochemical feedstock, both in terms of quantities used and economic value. Ethylene is the feedstock for ethylene oxide, ethylbenzene, ethyl chloride, ethylene dichloride, ethyl alcohol, and polyethylene, most of which, in turn, are used to produced hundreds of other end-products. Most ethylene is produced by steam cracking of ethane, or propane.

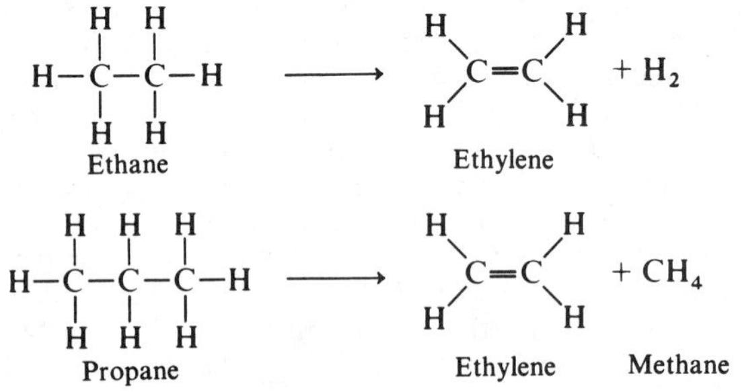

Ethylene also may be produced from other paraffinic or naphthenic hydrocarbons. The reactions are highly endothermic (34,400 kcal/kg mole of ethane cracked at approximately 900°C) and proceed in the direction indicated at temperature exceeding approximately 620°C without a catalyst.

Ethylene is of importance as a petrochemical feedstock because of its great versatility in reacting to form several chemical intermediates. The double-bond provides reactivity; the compound also has the ability to homopolymerize and copolymerize with other monomers. Some of the important reactions involving ethylene include:

Chlorination

$$CH_2{=}CH_2 + HCl \xrightarrow[\text{cat.}]{\text{acidic}} CH_3{-}CH_2Cl$$

Ethyl chloride

Oxidation

$$CH_2{=}CH_2 + 1/2O_2 \longrightarrow$$

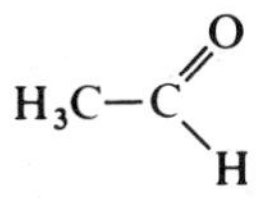

Acetaldehyde

Hydration

$$CH_2{=}CH_2 + H_2O \longrightarrow C_2H_5OH$$

Ethyl alcohol

Oxychlorination

$$CH_2{=}CH_2 + 2HCl + 1/2O_2 \longrightarrow C_2H_4Cl_2 + H_2O$$

Ethylene dichloride

Ethylene dichloride is used for the production of vinyl chloride.

Alkylation

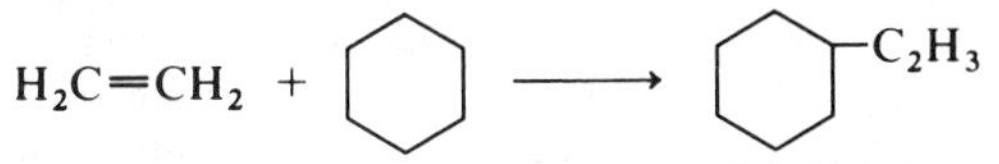

Ethyl benzene

Ethyl benzene is used for the production of styrene.

Polymerization

$$nCH_2{=}CH_2 \longrightarrow (\text{—}CH_2\text{—}CH_2)_n$$

High- and low-density polyethylene

Ethylene is also oxidized in large quantities to ethylene oxide:

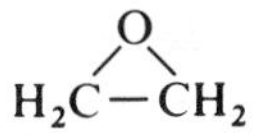

At one time, ethylene was produced by the dehydration of ethyl alcohol over alumina.

Almost any naphthenic or paraffinic hydrocarbon heavier then methane can be steam-cracked to yield ethylene. The preferred feedstock in the United States has been ethane and/or propane recovered from natural gas, or from the volatile fractions of petroleum. However, because of long term uncertainties pertaining to natural gas, many producers have been turning to heavier petroleum fractions, such as gas oils, as feedstocks. The consumption of ethylene throughout the free world is estimated to be about 40×10^9 pounds per year.

Ethylene reacts (1) with the halogens to form substitution halides; (2) with hypochlorous and hypobromous acid to form ethylene chlorohydrin or ethylene bromohydrin, respectively; (3) with hydrogen iodide or bromide (not chloride) to form ethyl iodide or ethyl bromide; (4) with hydrogen, in the presence of a catalyst, e.g., finely divided nickel at 150°C, to form ethane; (5) with concentrated sulfuric acid at 160°C to form ethyl hydrogen sulfate; and (6) with potassium permanganate to form ethylene glycol, although glycol is preferably made from ethylene dichloride or chlorohydrin.

In addition to its uses in the preparation of intermediates for a large variety of petrochemical reactions, ethylene is used as an anesthetic, as a fuel with oxygen for high-temperature flames, and as a coloring and ripening agent for citrus fruits and tomatoes. Ethylene chlorohydrin is used as an agent for decreasing the dormant period of seeds. See also **Polyethylene.**

ETHYLENEDIAMINE TARTRATE CRYSTAL (EDT). Piezoelectric Effect.

ETHYLENE DIBROMIDE. Bromine.

ETHYLENE DICHLORIDE. Chlorinated Organics.

ETHYLENE GLYCOL. This compound, $HOCH_2CH_2OH$, is traditionally associated with its use as a permanent-type antifreeze for internal-combustion engine cooling systems. However, since the early-1960s, large tonnages of ethylene glycol have been used in the production of polyesters for fibers, films, and coatings. The compound also finds important uses in hydraulic fluids, in the manufacture of low-freezing-point explosives, glycol ethers, and deicing solutions. Di- and triethylene glycols are important coproducts usually produced in the manufacture of ethylene glycol. Diethylene glycol, $HOCH_2CH_2OCH_2CH_2OH$, is used in the production of unsaturated polyester resins and polyester polyols for polyurethane-resin manufacture, as well as in the textile industry as a conditioning agent and lubricant for numerous synthetic and natural fibers. It is also used as an extraction solvent in petroleum processing, as a desiccant in natural gas processing, and in the manufacture of some plasticizers and surfactants. Triethylene glycol,

$$HOCH_2CH_2OCH_2CH_2OCH_2CH_2OH,$$

finds principal use in the dehydration of natural gas and as a humectant.

In one process for the manufacture of the aforementioned glycols, ethylene oxide is formed by direct oxidation of ethylene with oxygen over a silver catalyst. After purification, the stabilized ethylene oxide is mixed with a large excess of water, preheated, and fed to an ethylene oxide reactor. Here the ethylene oxide and water react under high temperature and high pressure to form principally ethylene glycol, with the other aforementioned glycols as coproducts. The crude glycols are dehydrated and then recovered individually as highly pure overhead streams from a series of vacuum-operated purification columns.

Principal properties of the glycols are summarized in the accompanying table.

PROPERTIES OF ETHYLENE, DIETHYLENE, AND TRIETHYLENE GLYCOLS

	Ethylene Glycol	Diethylene Glycol	Triethylene Glycol
Molecular weight	62.07	106.12	150.17
Boiling point (760 mm Hg)	197.6°C	245°C	287.4°C
Vapor pressure (20°C)	0.06 mm Hg	< 0.01 mm Hg	< 0.01 mm Hg
Specific gravity (20°C/20°C)	1.1155	1.1184	1.1254
Freezing point in air (760 mm Hg)	−13°C	83°C	99°C
Water solubility	- - - - - - - -	- complete -	- - - - - - - -

SOURCE: Glycols, Shell Chemical Company Bull. IC: 67–58.

ETHYLENE OXIDE. $((CH_2)_2)O$, formula weight 44.05, liquid, mp −111.3°C, bp 13.5°C, sp gr 0.887. The compound is miscible in all proportions with H_2O or alcohol and is very soluble in ether. Ethylene oxide is slowly decomposed by H_2O at standard conditions, converting into glycol $CH_2OH \cdot CH_2OH$. Ethylene oxide is a very high-tonnage chemical, approaching nearly 4 billion pounds annually. In terms of consumption (1) 60% for manufacture of ethylene glycol, the latter being an antifreeze compound as well as a raw material for production of polyethylene terephthalate used in the manufacture of polyester fibers, (2) 12% for preparation of surfactants, (3) 8% for the manufacture of ethanolamines, (4) 10% for production of ethylene glycols which are used in plasticizers, solvents, and lubricants, and (5) 10% for making glycol ethers which are used as jet-fuel additives and solvents. See also **Rocket Propellants;** and **Antimicrobial Agents (Foods).**

Direct oxidation of ethylene in the presence of a silver catalyst is the predominant large-scale process used: $CH_2{:}CH_2 + \frac{1}{2}O_2 \rightarrow ((CH_2)_2)O$. The yield is aproximately 70% of the theoretical. For maximum yield, very careful temperature control is required, the yield dropping as the temperature climbs. The side reaction: $CH_2{:}CH_2 \rightarrow 3CO_2 + 2H_2O$ is the main factor for reducing yield. Thus far, silver has proved to be the most effective catalyst. Several compounds have been investigated that can inhibit the side reaction and also be compatible with the catalyst. These compounds have included ethylene dichloride, ethylene dibromide, alcohol, amines, and organometallic compounds, but their success has been limited. Plants have been designed to use either air or pure oxygen for oxidation. Selection presents an interesting study in economics because (1) where air is used, a purge reactor and associated purge absorber are required (not required by the O_2 process), and (2) where O_2 is used, both a CO_2 removal system and an O_2-making facility are required. The trend is toward the oxygen system with the ethylene oxide plant located near an air-separation plant.

ETHYLENE OXIDE (Propellant). Rocket Propellants.

ETHYLENE (Plant Hormone). Plant Growth Modification and Regulation.

ETHYLENE-PROPYLENE ELASTOMERS. Elastomers.

ETHYLENE-VINYL ACETATE COPOLYMERS. Known as EVA copolymers, these materials are polyolefins which can be processed like other thermoplastics, but which approach rubbery materials in softness and elasticity. The resins meet regulatory requirements for use in direct contact with food in food-processing machinery and in packaging applications. They are used in a number of applications to replace plasticized polyvinyl chloride and rubber. EVA copolymers require no curing or plasticizer. Parts made from EVA have little or no odor. Their elasticity is permanent. The copolymers can be injection-, blow-, compression-, transfer- and rotationally molded or extruded into film, sheeting, pipe, and profiles. EVA copolymers offer advantages over polyvinyl chloride and rubber in that they have good clarity and gloss, stress-crack resistance, good barrier properties, low-temperature flexibility and toughness, good adhesive properties, and good resistance to ultraviolet radiation. Their main limitation is a comparatively low resistance to heat and solvents. The resins soften at a temperature of about 70°C. EVA copolymers are not attacked by alcohols, glycols, or weak organic acids. However, to a varying degree, the materials are attacked by chlorinated hydrocarbons, straight-chain paraffinic solvents, and benzene and its derivatives.

ETHYLPROPYLACETIC ACID. Synthesis (Chemical).

ETIDOCAINE HYDROCHLORIDE. Anesthesia.

ETIOLATION. This is the effect of darkness on a living plant. It is a matter of common observation that plants grown in dark contain little or no chlorophyll and so are nearly white. Green plants placed in darkness lose their chlorophyll. Eventually, when the food reserves are exhausted, the plant dies.

Effect of darkness on the elongation of stems. The bean seedlings on the left were grown in the light; those on the right were grown in darkness. (*Winchester, "Modern Biological Principles," 2nd 1971, D. Van Nostrand Co.*)

Besides the lack of chlorophyll, plants grown in the dark have other characteristics. In dicotyledons, the internodes of the stem become excessively elongated and very slender. The leaves fail to expand normally. In monocotyledons, the leaves become very long and usually very narrow, but the stem shows little change. Etiolated plants never bear flowers, unless the flower buds are well developed before the plants are darkened.

Internally, the tissues are soft and lack strength, the cells being very large and having thin walls. Very little differentiation occurs, the conducting tissues being very much reduced. Leaves which form in darkness show very little of the structure characterizing a normal green leaf, but are almost entirely composed of loosely arranged parenchymatous cells.

See also **Plant Growth Modification and Regulation.**

ETIOLOGY. Knowledge of the cause of any disease or abnormal condition.

EUCALYPTUS TREES. Of the family *Myrtiaceae* (myrtle family), eucalpytus or gum trees are of the genus *Eucalyptus* and are native to Australasia. The tallest trees in Australia are eucalyptus and the *E. regnans* is regarded as the tallest of the nonconifers anywhere in the world. Eucalyptus trees, notably the blue gum (*E. globulus*), were introduced into California in the 1880s, whereupon it immediately thrived. This same species was introduced into Ethiopia in the late 1800s, where it also has fared well and much accepted because of a general lack of timber in that area. Lines of eucalptus trees on the borders of citrus orchards in California are a common sight. It also has been reported that eucalyptus trees have been successful in reducing the occurrence of mosquitoes in marshy areas, the trees assisting in dewatering wet, boggy soil. There are some 600 species of eucalyptus, many of which are quite localized. Thus, generalization are difficult. Some of the more important species include:

Blue gum	*E. globulus*
E. caesia	
Cider gum	*E. gunnii*
Ghost gum	*E. papuana*
E. leucoxylon "Rosea"	
Lemon-scented gum	*E. citriodora*
E. miniata	
E. obliqua	
Red-flowering gum	*E. ficifolia*
E. regnans	
Red-flowering gum	*E. filifolia*
River red gum	*E. camaldulensis*
Snow gum	*E. niphophila*
Spinning gum	*E. perriniana*
Tasmanian snow gum	*E. coccifera*
Woodward's gum	*E. woodwardii*

The red-flowering gum is exceedingly showy when in bloom. Not a tall tree, this species seldom exceeds 50 feet (15 meters) in height. In season, masses of bright-red blossoms in large and heavy clusters nearly cover the tree. The clusters are from 6 to 10 inches (15.2 to 25.4 centimeters) across. Where trees are planted along pathways, the massive blossoms create clearing problems.

The blue gum has a blue-gray bark, quite smooth. The tree may exceed 300 feet (90 meters) in height. The leaf is dark green, 6 to 12 inches (15.2 to 30.4 centimeters) long and about $1\frac{1}{2}$ inches (3.8 centimeters) broad. Growth is rapid and long periods of drought can be withstood. It will survive temperatures as low as 20°F (4.4°C) for short periods. Although the best known of the species introduced into California, there are at least 80 species of eucalyptus growing in the state. Other species include: The white gum or manna gum (*E. viminalis*) with a gray-white bark and height of from 50 to 60 feet (15 to 18 meters). The leaf is from 4 to 8 inches (10 to 20.3 centimeters) in length. The tree grows fast and may have a girth of up to 3 feet (0.9 meters) within a 12-year period. The tree can tolerate poor soil. The dollar-leaf eucalyptus (*E. pulverulenta*) has blue-green leaves, essentially round in shape, and from 2 to 3 inches (5 to 7.6 centimeters) across. The tree grows to about 30 feet (9 meters) in height. The stalk is narrow, ranging up to about $2\frac{1}{2}$ inches (6.4 centimeters). The ornamental foliage is frequently used by florists and artists.

The Tasmanian snow gum is planted in England. The snow gum can withstand the climes of Mount Kosciusko (New South Wales) up to an altitude of 7000 feet (2100 meters).

Eucalyptus oil, an essential oil, is derived from the leaves of several species of these trees. The oil is used in various medicinal and household preparations.

EUCLASE. The mineral euclase is a silicate of beryllium and aluminum corresponding to the formula $BeAlSiO_4(OH)$, which crystallizes in the monoclinic system. It has a perfect prismatic cleavage; hardness, 7.5; specific gravity, 3.1; luster, vitreous; is colorless to sea-green or blue. It has been used to a very slight extent for jewelry as its transparent crystals somewhat resemble the aquamarine. Euclase occurs in the Minas Geraes region, Brazil, associated with topaz and beryl, and also in the Ural Mountains, where it is found in gold-bearing sands. The name euclase is derived from the Greek, meaning easiness and fracture, in reference to its easily cleaved crystals.

EUCLIDEAN GEOMETRY. **Geometry.**

EUCLIDEAN SPACE. A generalization of the algebraic, geometrical and topological properties of the line, the plane, and three-dimensional space to n dimensions.

A Euclidean n-space is an n-dimensional linear space provided with a scalar product. A concrete realization of Euclidean n-space is the set of n-tuples $(\lambda_1, \lambda_2, \ldots, \lambda_n)$ of real numbers with the scalar product of $(\lambda_1, \lambda_2, \ldots, \lambda_n)$ and $(\mu_1, \mu_2, \ldots, \mu_n)$ defined as $\lambda_1\mu_1 + \lambda_2\mu_2 + \cdots + \lambda_n\mu_n$.

In mathematical models for physical systems, it is often necessary to consider spaces having more than three dimensions; for example, six dimensions are needed to locate two particles in Euclidean three space.

EUCLID'S ALGORITHM. **Algorithm.**

EUCRYPTITE. **Lithium (For Thermonuclear Fusion Reactors).**

EUDIOMETER. A graduated tube closed at one end in one form of which two platinum wires are sealed so that a spark may be passed through the contents of the tube; used to measure the volume changes in the combustion of gases.

EUGLENOIDIDA. An order of 1-celled animals. *Mastigophora.*

EUHEDRAL CRYSTALS. **Mineralogy.**

EUKARYOTIC CELLS. **Cell (Biology).**

EULAMELLIBRANCHIATA. In some classifications, an order of bivalve mollusks containing the oysters.

EULER ANGLE. One of three parameters describing the orientation of a rigid body relative to a Cartesian coordinate system (x, y, z) fixed in space. Suppose another coordinate system (x', y', z') is fixed in the body. Then the two systems may be made coincident by three successive rotations, applied in the appropriate order, and the three angles are the Euler angles.

See also **Coordinate System.**

EULER EQUATION. The condition that the integral

$$\int_{x1}^{x2} I(x, y, y')\,dx$$

have a stationary value is

$$\frac{\partial I}{\partial y} - \frac{d}{dx}\frac{\partial I}{\partial y'} = 0$$

the latter being known as the Euler equation in the calculus of variations. A solution $y = f(x)$ satisfying this equation, if it exists, is an extremal and is a maximizing or a minimizing curve. (Note that $y' = dy/dx$.)

There are other equations also known by the name of Euler (see **Differential Equation of Higher Order (Ordinary); Euler Equation; Exponent;** and **Partial Differential Equation.**

EULER FORMULA FOR COLUMNS. **Column (Structural).**

EULERIAN COORDINATES. Any system of coordinates in which properties of a fluid are assigned to points in space at each given time, without attempt to identify individual fluid parcels from one time to the next. Since most observations in meteorology are made locally at specified time intervals, an Eulerian system is usually, though by no means always, more convenient. A sequence of synoptic charts is an Eulerian representation of the data.

EULER-MACLAURIN FORMULA. An equation for evaluating a definite integral. If $f(x)$ is known explicitly and its derivatives are finite at both limits of the integral, or if the derivatives may be determined numerically, the formula may be used. It can be written as

$$\int_a^b f(x)\,dx = h\left[\frac{y_0}{2} + y_1 + y_2 + \cdots + \frac{y_n}{2}\right] - \sum_{\text{odd } k} \frac{h^{k+1}}{(k+1)!} B_{k+1}[y_n^{(k)} - y_0^{(k)}]$$

where y_0 and y_n are values of $f(x)$ at $x = a$ and $x = b$, respectively; $y_1, y_2, \ldots$ are intermediate values at equally spaced intervals h of the independent variable; B_k are the Bernoulli numbers; $y_n^{(k)}$ and $y_0^{(k)}$ are the kth derivatives at $x = b$ and $x = a$, respectively. The formula may also be used to evaluate the finite sum indicated by the first term on the right-hand side of the equation, provided the integration can be performed.

Since the Euler-Maclaurin formula is an asymptotic series, hence not convergent, it should be used with care.

See also terms listed under **Mathematics.**

EULER-MASCHERONI CONSTANT. A number, also often called simply Euler's constant, which occurs in one definition of the gamma function. It can be defined by several equivalent infinite integrals, one example being

$$C = \int_\infty^0 e^{-t} \ln t\,dt$$

Its numerical value is 0.577215665 . . . but the quantity $\gamma = 1.781072\ldots$, defined by $\ln \gamma = C$ is sometimes defined as the Euler Mascheroni constant.

See also **Gamma Function;** and terms listed under **Mathematics.**

EULER METHOD FOR NUMERICAL SOLUTION OF A DIFFERENTIAL EQUATION. An iteration method for $y' = f(x, y)$, where initial values x_0, y_0, y'_0 are known. If the interval between successive values of the independent variable is small enough to write $\Delta x = h$, then an approximate solution of the differential equation at $x_1 = x_0 + h$ is ${}^1y_1 = y_0 + hy'_0$. An approximation to y' at x_1 is $y'_1 = f(x, {}^1y_1)$, which gives an improved value of y_1, which is ${}^2y_1 = y_0 + (h/2)(y'_1 + y'_0)$. The process is then repeated. The method just described is actually a modification of that proposed by Euler. However, both it and the original Euler method are slow in converging and give results of low accuracy. While they are suitable for starting the solution of a differential equation, more accurate methods are preferred for continuing the solution.

EULER'S FIRST INTEGRAL. **Beta Function.**

EULER'S SECOND INTEGRAL. **Gamma Function.**

EULER THEOREM ON A HOMOGENEOUS FUNCTION. A function $f(x, y, z)$ is homogeneous and of order n, if

$$f(kx, ky, kz) = k^n f(x, y, z)$$

where k is a constant. Moreover, such a function, if it has a derivative, satisfies the partial differential equation

$$xf_x + yf_y + zf_z = nf(x, y, z)$$

A harmonic function is a typical example. The theorem can be generalized to any number of variables.

(For Euler's theorem on the exponential function, see **Exponent.**)

EUNOMIA. Asteroid.

EUPHAUSIACEA. A small order of marine crustaceans.

EUPHORBIACEAE. Also known as the spurge family, this is a large family of usually cactus-like trees and shrubs which often yield a milky juice or latex. They occur in many tropical and temperate regions. There are a few herbaceous species, especially in cooler regions. Many of the tropical species are interesting xerophytes, plants capable of enduring the driest climates. Often these have a habit very similar to that of species of cactus, with which they may easily be confused, especially if not in flower. However, nearly all members of the spurge family contain a milky juice which exudes from them when the surface is cut or broken. This milky juice, or latex, will readily distinguish them from cacti, which lack latex. Leaves, when present, are usually alternate and have stipules. In many species, such as the frequently cultivated *Euphorbia splendens*, or "crown of thorns," the leaves soon drop off, leaving a spine-covered stem. In one genus, *Phyllanthus*, leaves are frequently reduced to minute scales, and the stem flattened and green; in these the small pinkish flowers are borne around the edge of the flattened stem. In some species of *Euphorbia* the leaves near the top of the stem become brilliantly colored, as in *Euphorbia pulcherrima*, the poinsettia, where they are bright red. Such leaves, surrounding the inconspicuous flower masses, are often mistaken for parts of the flower.

The inflorescence, in members of this family, is often very complex. In many species the individual flowers are crowded together in such a way as collectively to resemble a single large flower. The flowers are unisexual. The plants are either monoecious or dioecious. In many cases the flowers entirely lack both calyx and corolla, in others a calyx is present, but no corolla, while in some both calyx and corolla are present. They are regular flowers with the perianth commonly 5-parted. The number of stamens varies from one to many; in many cases they are variously united; in some, as in the castor bean, they are branched. The ovary is usually 3-celled, with one or two ovules in each cell. The fruit is a capsule, which when mature often opens with considerable force, throwing the seeds out, often to considerable distances. The seeds have an abundant endosperm and a caruncle.

Among the members of this family are some plants of great economic importance. Many others are poisonous plants.

Hevea brasiliensis, the Para rubber plant, is perhaps the most valuable member. Species of *Ricinus* supply castor oil. Species of *Manihot*, a South American genus, yield cassava or mandioc, a starchy foodstuff, prepared from the large roots. *Manihot esculenta*, for instance, has long been cultivated in Brazil. It is a large, somewhat bushy herb with long-petioled leaves, the smooth blades of which are deeply cleft into 3–7 lobes. The roots, which have the appearance of sweet potatoes, are eaten in much the same way as sweet potatoes. Grated, they yield a starchy product used like bread. From the roots tapioca may be prepared. The poisonous principle which is present in many species of *Manihot* is removed by squeezing or destroyed by heating. From other species of this genus may be obtained, by tapping, a milky juice which is a source of rubber. See also **Rubber (Natural).**

Hura crepitans, the sand box tree, is another member of the family of some slight commercial value. The plant is a fairly large tree the stem of which is covered with short, sharp spines, and bears long-petioled toothed leaves. The fruit is composed of numerous hard carpels which, when mature, explode violently, throwing the seeds out forcibly. These fruits, about 3 inches (7.5 centimeters) in diameter, were formerly gathered and wired to prevent bursting. When dry they were used as containers for the fine sand which was then used to blot ink—hence the common name of the tree. The wood of the tree is used locally, but rarely exported. The milky juice of the tree is very poisonous. This juice, mixed with meal or similar substances, and thrown into the waters of a stream or lake, stupefies the fish present therein, so that they may be readily captured. The poison does not render the fish unfit for human consumption.

Several species of *Croton* yield important purgative drugs. *Croton eluteria* gives Cascarilla bark, used as a tonic.

Jatropha curcas, a small shrub or tree, bears egg-shaped green fruits which contain a high percentage of an odorless oil called "curcas" oil, used in making paints, as a lubricant, and in soapmaking. From the leaves of this tree, natives of the Philippine Islands prepare a fish poison used in much the same way as that obtained from *Hura*.

Aleurites are tropical trees bearing small many-seeded fruits extremely rich in oil. *Aleurites triloba*, the candlenut of the orient, produces a fruit extensively used for food and for light. *Aleurites cordata*, a native of China, is the "varnish-tree." *Aleurites fordii* yields tung oil.

EUPHORIA. A feeling of well-being which is not necessarily justified by the physical condition of the patient. It is seen in certain mental disturbances, manic depressive psychoses, paresis, and, temporarily, as the result of administration of certain drugs.

EUPHOTIC ZONE. The layer of a body of water which receives ample sunlight for the photosynthetic processes of plants. The depth of this layer varies with the water's extinction coefficient, the angle of incidence of the sunlight, length of a day, and cloudiness; but it is usually 80 meters or more. The depth of compensation is the lower boundary of the euphotic zone. See also **Ecology.**

EUROPA. Jupiter.

EUROPEAN CORN BORER (*Insecta, Lepidoptera*). A small moth, *Pyrausta nubilalis*, whose larva bores in the stems of plants, especially Indian corn. The species is closely related to certain North American moths and was first noticed as a pest in the eastern part of the United States about 1920. Since then it has spread westward.

EUROPEAN EEL. Eels.

EUROPEAN LARCH. Larch Trees.

EUROPEAN MOUNTAIN ASH. Ash Trees.

EUROPEAN SILVER FIR. Fir Trees.

EUROPIUM. Chemical element symbol Eu, at. no. 63, at. wt. 151.96, sixth in the Lanthanide Series in the periodic table, mp 822°C, bp 1529°C, density 5.245 g/cm^3 (20°C). Elemental europium has a body-centered cubic crystal structure at 25°C. The pure metallic europium is silver-gray in color under vacuum, but oxidizes readily in air and must be handled in an inert atmosphere. Europium is very soft as compared with the other rare-earth metals. Two stable isotopes of the element occur naturally ^{151}Eu and ^{153}Eu. Upon absorption of thermal neutrons, ^{151}Eu forms ^{152}Eu with a half-life of 13 years; ^{153}Eu forms ^{154}Eu with a half-life of 16 years. The latter further decays to ^{155}Eu with a half-life of 1.7 years. In terms of abundance, europium is present on the average of 1.2 ppm in the earth's crust, making its potential availability greater than antimony, bismuth, or cadmium. The element was first identified by Sir William Crookes in 1889. Europium dissolves readily in dilute mineral acids and reacts with H_2O at room temperature. The metal is not known to be toxic but because of its high reactivity in air, great care in handling is mandatory. Electronic configuration

$$1s^2 2s^2 2p^6 3s^2 2p^6 3d^{10} 4s^2 4p^6 4d^{10} 4f^6 5s^2 5p^6 5d^1 6s^1$$

Ionic radius Eu^{2+} 1.09 Å, Eu^{3+} 0.950 Å. Metallic radius 1.995 Å. First ionization potential 5.67 eV; second 11.25 eV. Oxidation potential $Eu^{2+} \rightarrow Eu^{3+} + e^-$, 0.43 V. Other important physical properties of europium are given under **Rare-Earth Elements and Metals.**

Europium occurs in the rare-earth fluocarbonate mineral bastanite, mainly found in southern California. The mineral contains between 0.09 and 0.11% Eu_2O_3. Other minerals, such as xenotime and monazite, also contain europium compounds and sometimes are used as sources of the element.

Because of the desirable nuclear properties of the element, europium has received serious consideration for the construction of nuclear reactor hardware. Earlier commercial unavailability of the element, however, favored the use of other materials. Some small reactors have been constructed in which europium molybdate has been the major control-rod component. With much increased availability of the metal in recent years, the prospects of further usage of europium in reactor design are good. A europium-activated yttrium orthovanadate $Eu:YVO_4$ has shown promise as a red phosphor for commercial television. An increase of 40% in light output has been claimed. With this system, the average color television set would require about $\frac{1}{2}$ g of Eu_2O_3 and 6 g of Y_2O_3. The stimulus resulting from this discovery resulted in the development of other new phosphors involving europium in various host matrices. These new materials have been used in high-intensity mercury-vapor lamps, general-purpose fluorescent lamps, x-ray screens, charged-particle detectors, and neutron scintillators. In some optically-read memory systems, ferromagnetic europium chalcogenides (sulfides, selenides, and tellurides) have been used. Other electronic and semiconductor uses of europium are under serious investigation.

See references listed at ends of entries on **Chemical Elements;** and **Rare-Earth Elements and Metals.**

NOTE: This 6th Edition entry was revised and updated by K. A. Gschneidner, Jr., Director, and B. Evans, Assistant Chemist, Rare-Earth Information Center, Energy and Mineral Resources Research Institute, Iowa State University, Ames, Iowa. Original 5th Edition entry was prepared by J. G. Cannon, Molycorp, Inc.

EURYPTERIDS. Invertebrate Paleontology.

EUSTACHIAN TUBE. A slender canal between the pharynx and the middle ear of vertebrates. It permits the equalization of pressure on the two surfaces of the eardrum. See also **Hearing and the Ear.**

EUTAXIC. A term proposed by Keyes in 1901 for obviously stratified sedimentary ore deposits as contrasted with those which are unstratified. The latter he designated as ataxic.

EUTECTIC. An eutectic reaction is a reversible isothermal transformation in which, during cooling, a single liquid phase is transformed into two or more solid phases, the number of solid phases being equal to the number of components. In a given alloy system, at a fixed pressure, all phases will have fixed compositions during the isothermal transformation. The temperatures at which the freezing occurs is known as the eutectic temperature, while the composition of the liquid phase is called the eutectic composition. On a temperature-composition binary phase diagram, the eutectic point is determined by the eutectic composition and the eutectic temperature. In general, an alloy of the eutectic composition freezes at a minimum temperature. For this reason, eutectic compositions, or compositions close to the eutectic, are frequently used in low melting point solders.

By a similar usage in petrology, a eutectic is a discrete mixture of two or more minerals, in definite proportions, which have simultaneously crystallized from the mutual solution of their constituents. The eutectic point is the lowest temperature at any given pressure at which the above physical-chemical process may take place. The eutectic ratio is the ratio by weight of two minerals which originate by the above process.

EUTECTOID. This is a phase transformation analogous to an eutectic where a single solid phase, instead of a liquid phase, is transformed into two or more different solid phases. The number of solid phases in the resulting eutectoid structure is equal to the number of components in the system. Under very slow cooling, the eutectoid transformation should occur at the eutectoid temperature. However, due to the sluggishness of solid state transformations, there is usually some hysteresis with the transformation temperature depressed on cooling and raised on heating. Under equilibrium conditions, the compositions of the various phases are fixed in an eutectoid reaction just as they are in an eutectic transformation.

The best known eutectoid reaction is that which occurs in steel where the austenite phase, stable at high temperatures, transforms into the eutectoid structure known as pearlite. In this transformation, the austenite phase, containing 0.8% carbon in solid solution, transforms to a mixture of ferrite (nearly pure body-centered cubic iron) and iron-carbide (Fe_3C). At atmospheric pressure, the equilibrium temperature for this reaction is 723°C. This temperature is the eutectoid temperature.

In binary alloy systems, a eutectoid alloy is a mechanical mixture of two phases which form simultaneously from a solid solution when it cools through the eutectoid temperature. Alloys leaner or richer in one of the metals undergo transformation from the solid solution phase over a range of temperatures beginning above and ending at the eutectoid temperature. The structure of such alloys will consist of primary particles of one of the stable phases in addition to the eutectoid, for example ferrite and pearlite in low-carbon steel. See also **Iron Metals, Alloys, and Steels.**

EUTROPHICATION. Limnology.

EVA COPOLYMERS. Ethylene-Vinyl Acetate Copolymers.

EVACUATED-TUBE COLLECTOR. Solar Energy.

EVAPORATION. The evaporation of a liquid consists in the escape from the main body of the liquid of those molecules which, in their thermal agitation, are moving with a sufficient speed to break through the surface tension; that is, whose kinetic energy exceeds the work function of cohesion at the surface. Since only a small proportion of the molecules are at any instant located near enough to the surface and are moving in the proper direction to escape, the rate of the evaporation is limited. It is easy to see why it proceeds more rapidly with higher temperature, and why liquids of low surface tension are relatively volatile. Also, as the faster moving molecules emerge, those left behind have less average energy, and the temperature of the liquid is thereby lowered. If the evaporation takes place in a closed vessel, the escaping molecules accumulate as a vapor above the liquid. Many of them return to the liquid, such returns being more frequent, the greater the density and pressure of the vapor. Presently the processes of escape and return come to equilibrium; the vapor is then said to be "saturated," its density and pressure no longer increase, and the cooling effect ceases. Even a warm breeze cools the skin because it removes the evaporating perspiration and prevents saturation.

Evaporation is a major chemical engineering unit operation for bringing about separations of liquids and solids and, in particular, to recover the solute (such as a dissolved salt) from the solvent (frequently water). Usually, the main object of the separation is the solute. The pulp and paper industry is a large user of evaporation equipment. In pulp mills, after the digestion system, the pulp is leached with water and the chemical solids are dissolved out almost completely by a pulp-washing system. The recovered liquid from these operations is fed to an evaporator, generally at about 15% total dissolved solids content. The evaporator removes much of the water and in so doing concentrates the liquid to 55–65% total dissolved solids, whereupon the solution then can be further processed in a chemical recovery furnace. Other types of pulp processing also involve chemical-containing solutions which must be evaporated for recovery of valuable chemicals. Evaporation also is used extensively in the production of table and industrial salt (sodium chloride) as well as other salts, in caustic-chlorine production, in the phosphate industry, and in food processing. Evaporators can be large structures as the illustrations indicate.

Evaporation is in principle the same operation as plain distillation, with the modifications in practice that (1) the vapor may or may not be recovered, (2) the residue in the evaporator may or may not contain solids, and (3) vacuum evaporation is frequently used in a single compartment or in multiple stages with each successive stage operated at an increasing vacuum utilizing the heat of condensation of the vapor from the preceding stage. In multiple stage evaporators there is a saving in the cost of heat and an increased expenditure for apparatus. Vacuum evaporation is frequently utilized to lower the temperature to which a substance is subjected and thus avoid decomposition by passing a current of warm dry air over the substance. Combined high-vacuum and very low temperature evaporation or drying is practiced in the final removal of water vapor from frozen penicil-

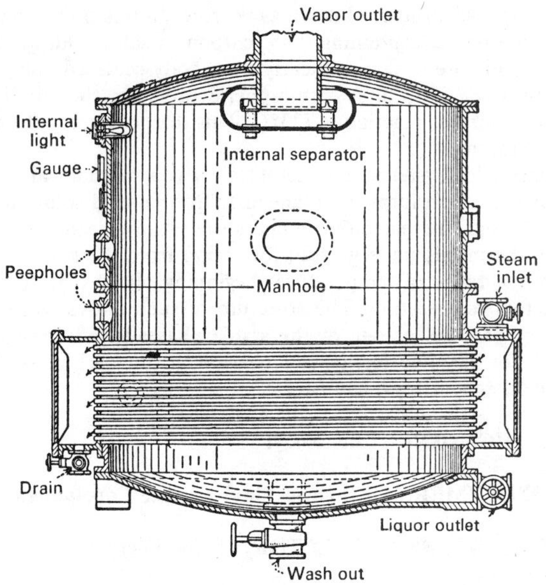

Fig. 1. Horizontal-tube evaporator.

lin, due to the heat-sensitive nature of this material. Water vapor passes from the place of higher concentration, that is, the substance, to the place of lower concentration, that is, the air, and is thus removed from the substance. If oxygen of the air reacts with the substance, an inert gas such as nitrogen may be substituted for air.

An evaporator system may be single effect (Fig. 1), in which the steam is produced from one evaporator, or multiple effect, in which the steam is produced from several evaporators in series. In a multiple effect system the vapor from one evaporator becomes the heating steam in the succeeding. Unusual conditions met in industrial or steam heating plants may require so large a fraction of make-up as to warrant double, triple, or quadruple effect evaporators. The central generating station ordinarily employs single effect and rarely requires more than a double effect system. The ratio vapor produced/steam used is about 0.8 for the single effect, 1.5 for the double effect, and 2.5 for the triple effect system. See Fig. 2. Evaporator feed is sometimes preheated to increase evaporator capacity.

Fig. 2. The sixth-effect evaporator body of a multiple-effect evaporation system.

Evaporators are classed as film, flash, or submerged-tube types. The first and last are steam-tube types; in the former the raw water trickles over the hot tubes, in the latter the tubes are entirely surrounded by the water being evaporated. The flash type produces steam by dropping the pressure on water at the saturation temperature. The excess heat flashes part of the water into steam, then the remainder is drawn off, reheated, and again flashed.

References

Charm, S. F.: "Fundamentals of Food Engineering," 3rd Edition, AVI, Westport, Connecticut, 1978.

Rosenblad, A. E.: "Three Evaporator Flow Diagrams that Adapt to Current Energy Costs and Emission Requirements," Rosenblad Corp., Princeton, New Jersey, 1975.

Schwartzberg, H. G.: "Energy Requirements for Liquid Food Concentration," *Food Technology*, **31**, 3, 67–76 (1977).

Staff: "Evaporation Technology," *Chem. Eng. Progress*, **72**, 4, 41–71 (1976).

EVAPORATION (Cloud). Clouds and Cloud Formation.

EVAPORATION (Desalination). Desalination.

EVAPORATION (Hydrology). Hydrology.

EVAPORATION (Oceans). Atmosphere-Ocean Interface.

EVAPORATION PAN. Precipitation and Hydrometeors.

EVAPORATIVE CRYSTALLIZER. Crystallization.

EVAPOTRANSPIROMETER. Precipitation and Hydrometeors.

EVECTION. A perturbation of the moon in its orbit due to the attraction of the sun. This results in an increase in the eccentricity of the moon's orbit when the sun passes the moon's line of apsides and a decrease when perpendicular to it. Evection amounts to 1 degree 15 minutes in the moon's longitude at maximum.

EVEN-EVEN NUCLEI. Atomic nuclei that contain an even number of protons and an even number of neutrons.

EVENT. A happening, represented by a point (x, y, z, t) in the space-time continuum. It is a fundamental assumption of the theory of relativity that all physical measurements reduce to observations of relations between events.

EVERSION THEORY. Gerontology.

EVOKED-CORTICAL POTENTIAL. Electroencephalogram.

EVOLUTE. The evolute of a given curve is the locus of its centers of curvature. See also **Involute.**

EVOLUTION. (1) The development of an organism toward perfect or complete adaptation to environmental conditions to which it has been exposed with the passage of time; (2) the theory that life on earth developed gradually from one or several simple organisms (appropriate molecules) to more complex organisms. Sometimes called *organic evolution.* The term *evolutionary biology* is also used.

In contemplating the great diversity and vast numbers of species of life on earth, during post-Renaissance times, Linnaeus (1707–1778) and many of his contemporaries ascribed these to a concept of special creation—"there are just so many species as in the beginning the Infinite Being created."

In his *Philosophie Zoologique*, the French philosopher, Jean Baptiste de Lamarck, observed in 1809, "the existence in organisms of a built-in drive toward perfection; the capacity of organisms to become adapted to 'circumstances' [environment in modern terminology]; the frequent occurrence of spontaneous generation; and the inheritance of acquired characters, or traits."

The fourth of the foregoing observations was shown to be inaccurate, but about a half-century later, Darwin (1859) accepted the concept, i.e., "assumed that the use or disuse of a structure by one generation would be reflected in the next generation." Some other theorists who followed Lamarck and Darwin also accepted the validity of Lamarck's fourth observation and it remained for the German biologist, August Weissman (1834–1914), to stress the impossibility (or at least the improbability) of that observation. This constituted the first of several modifications of early hypotheses and the later theory, progressing from Lamarckism to Darwinism, to neo-Darwinism, and to the more recent synthetic theory and to the current conceptual developments.

Seldom stressed is the fact that although the theory of evolution is some 120 years old (Darwinism) or 170 years old (Lamarckism), the theory, in terms of research required to round out its full development and implications, is still in its infancy. Compared with the life sciences information bank of the early 1980s, the early theoretical endeavors were conducted in a scientific vacuum. Thus, the emergence of advanced genetics, cytology, molecular biology and numerous other sciences relative to evolutionary biology have impacted and will continue to impact on those scholars who pursue the theory in their efforts to construct a continuum of events that led from a lifeless earth to living organisms and systems as we know them today.

The difficult tasks facing investigators in this field today are typified, in part, by the observations of one authority on the concept of chemical evolution of life: "The evolution of the genetic machinery is the step for which there are no laboratory models; hence one can speculate endlessly, unfettered by inconvenient facts. The complex genetic apparatus in present-day organisms is so universal that one has few clues as to what the apparatus may have looked like in its most primitive form." (R. E. Dickerson, 1978)

As of the early 1980s, there appears to be a trend toward more effective comingling of various scientific disciplines in an effort to weave a tighter and more coherent network of information concerning the theory of evolution. These include, among many others, the improved integration of findings of paleontologists, archeologists, anthropologists—and chemists and physicists who have developed improved age-estimating and dating techniques. As pointed out by Woodruff (1980) in a review of Stanley's book (see reference list): "Paleontology is currently undergoing an exciting rejuvenation, and Stanley and his fellow paleobiologists (as they are now called) have introduced some scientific rigor into a traditionally descriptive field. Now, in place of inspired speculation, we see attempts to test hypotheses derived from theoretical population ecology against the extensive fossil record. . . . Evolutionary biologists can no longer ignore the fossil record on the ground that it is imperfect."

Possibly the most important conference on evolutionary biology held since the 1940s convened in Chicago in the fall of 1980. As reported (Lewin, 1980), the principal issues discussed at the meeting were (1) the tempo of evolution; (2) the mode of evolutionary change; and (3) the constraints on the physical form of new organisms. Dominating the field of evolutionary biology for several decades, the Modern Synthesis concept of evolution (so named by Julian Huxley in 1942) was reexamined in the light of many intervening years of progress in the biological sciences. In essence, this concept assumes that the pace of evolutionary change is slow, that the direction of evolutionary change is governed by natural selection involving small variations, and that the variants that survive are those that are environmentally superior. The proceedings are well encapsulated by Lewin. As one scientist at the conference observed, "I hope that this meeting will lead to a reapproachement. I hope it will set the basis for a reconstruction of ideas."

In a practical sense, disregarding the challenge and fascination of the theory, the primary usefulness to date of the theory of evolution has been its role as a unifying concept for the biological sciences as we know them today—not dissimilar to the role played by the earlier (Stahl, 1600–1734) phlogiston theory (later supplanted by the laws of thermodynamics) which brought a degree of order to physics.

In essence, the various theories of cosmogony (origin of galaxies, stars, planets, etc.) furnish the prelude to the theory of evolution. In commenting on the relative chores of the cosmogonist and the evolutionary biologist, Mayr (1978) observed: "For one thing, it (biological evolution) is more complicated than cosmic evolution, and the living systems that are its products are far more complex than any nonliving system." Over the future years of continuing investigation and conceptualization, the efforts of these two fields nevertheless require coordination and tight information transfer because the theory of evolution is time sensitive even if in terms of billions of years—because the best estimated age of the earth, of which only the last portion has embraced environments suitable to nurture and sustain life, must encompass all of the events described by the evolutionary biologist.

The many activities which investigators have undertaken in the past and are undertaking today to further mold and refine the theory of evolution are too numerous, complex, and detailed to report here. Several references are listed at the end of this entry for further reading.

Related topics include **Cell (Biology); Fossil Amphibians; Fossil Birds; Fossil Fishes; Fossil Reptiles; Fossil Reptilian Mammals; Genes;** and **Invertebrate Paleontology.**

References

Ayala, F. J.: "The Mechanics of Evolution," *Sci. Amer.*, **239**, 3, 56–59 (1978).

Dickerson, R. E.: "Chemical Evolution and the Origin of Life," *Sci. Amer.* **239**, 3, 70–86 (1978).

House, M. R. (editor): "The Origin of Major Invertebrate Groups," Academic, New York, 1979.

Lewin, R.: "Evolutionary Theory Under Fire," *Science*, **210**, 883–887 (1980).

Lewontin, R. C.: "Adaptation," *Sci. Amer.*, **239**, 3, 212–230 (1978).

Lillegraven, J. A., Kielan-Jaworowska, Z., and W. A. Clemens (editors): "Mesozoic Mammals," Univ. of California Press, Berkeley, California, 1980.

May, R. M.: "The Evolution of Ecological Systems," *Sci. Amer.*, **239**, 3, 194–206 (1978).

Margulis, L.: Origin of Eukaryotic Cells," Yale Univ. Press, New Haven, Connecticut, 1970.

Mayr, E.: "Evolution," *Sci. Amer.*, **239**, 3, 46–55 (1978).

Raup, D. M., and S. M. Stanley: Principles of Paleontology," Freeman, San Francisco, 1978.

Schopf, J. W.: "Biostratigraphic Usefulness of Stromatolitic Precambrian Microbiotas: A Preliminary Analysis," *Precambrian Research*, **5**, 2, 143–173 (1977).

Schopf, J. W.: "The Evolution of the Earliest Cells," *Sci. Amer.*, **239**, 3, 110–138 (1978).

Simpson, G. G.: "Splendid Isolation," Yale University Press, New Haven, Connecticut, 1980.

Smith, J. M.: "The Evolution of Behavior," *Sci. Amer.*, **239**, 3, 176–192 (1978).

Stanley, S. M.: "Macroevolution," Freeman, San Francisco, 1979.

Szalay, F. S., and E. Delson: "Evolutionary History of the Primates," Academic, New York, 1979.

Valentine, J. W., and E. M. Moores: "Plate Tectonics and the History of Life in the Oceans," *Sci. Amer.*, **230**, 4, 80–89 (1974).

Valentine, J. W.: "The Evolution of Multicellular Plants and Animals," *Sci. Amer.*, **239**, 3, 140–158 (1978).

Washburn, S. L.: "The Evolution of Man," *Sci. Amer.*, **239**, 3, 194–208 (1978).

Williams, G. C.: "Adaptation and Natural Selection: A Critique of Some Current Evolutionary Thought," Princeton Univ. Press, Princeton, New Jersey, 1966.

Woodruff, D. S.: *Science*, **208**, 716–717 (1980).

EVOLUTION (Mathematics). Square and Square Root.

EWING'S SARCOMA. Bone.

EXACERBATION. When used in a medical description, exacerbation means an intensification (possibly arising from irritation, aggravation of underlying causes) of pain and other symptoms of a disease or disorder that, in chronic situations, may occur intermittently and rise above a general background of tenderness, discomfort, etc.

EXCAVATION. Earthwork.

EXCESS AIR. Burner; Combustion; Gas and Expansion Turbines.

EXCHANGE DEGENERACY. An exchange process which does not entail a change in value or configuration. For example, by the Heitler-London theory, the essential reason for the strong attraction (or repulsion), of the two H-atoms in the H_2 molecule is the exchange degeneracy, i.e., the fact that for very large internuclear distance, by exchange of the two electrons of the two atoms a configuration results that is indistinguishable from the original configuration. Therefore, as they approach, there arises an interaction between them which may be treated mathematically as electron exchange.

EXCHANGE ENERGY. A specifically quantum-mechanical effect which has no classical analog. It is due to the interaction between two systems that arises, or could arise from the continuous exchange of a particle between them.

Suppose, for example, that two electrons are in states that allow them to come close together. Then, because they are indistinguishable particles, one could not tell the difference if they exchanged states. That is, one must combine with the original description (i.e., wave-function) a function in which the electrons have actually changed places. it can easily be shown that two such combined states are possible—the symmetric and antisymmetric combinations—and that in each of these the energy is significantly different from that of the original state. Exchange energy is the origin of covalent bonding, of ferromagnetism and antiferromagnetism, probably of nuclear forces (where exchange energy could arise by exchange of π-mesons between nucleons, giving rise to an effective potential which involves an operator which exchanges the spins, isotopic spins and/or positions of the particles) and of numerous other physical phenomena.

EXCHANGE FORCES. Nonclassical, quantum mechanical forces, that arise from the phenomena of exchange and that account for the exchange energy. The binding of a hydrogen molecule and the covalent bonding in molecules can be looked upon loosely as due to the exchange of electrons among the atoms. The Coulomb force between charged particles can be looked upon (in quantum electrodynamics) as due to the exchange of photons. The nuclear force between nucleons, can be looked upon as due to the exchange of charged or neutral π-mesons.

EXCHANGE (Particle). 1. A quantum mechanical concept based on the idea of identical particles, which are particles having the same intrinsic properties, such as rest mass, spin and charge. For example, suppose that two electrons are in states that allow them to come close together. Then, because they are indistinguishable particles, one could not tell the difference if they exchanged states. Thus the wave function of the system must be such that an exchange of the electrons leaves the magnitude of the wave function unchanged, except possibly for sign, i.e., the wave function must be either symmetric or antisymmetric to an exchange of the two particles. Particles whose total wave function (including both space and spin coordinates) is symmetric under an exchange operator obey the Bose-Einstein statistics. Particles whose total wave function is antisymmetric obey the Fermi-Dirac statistics.

2. Exchange is also used more specifically as the exchange of one particle between two others, as in the exchange of the single electron between the two identical protons in the hydrogen molecular ion, or the exchange of a meson between two nucleons.

Some concepts have been altered during recent years. Check entry on **Particles (Subatomic).**

EXCITATION. This term has three common uses in physics and engineering: 1. Addition of energy to a system, whereby it is transferred from its ground state to a state of higher energy, called an excited state. 2. The field excitation of dynamo machines, meaning the current or voltage of the field circuit. 3. In vacuum-tube and transistor circuits, the input signal of any stage is commonly called the excitation. Thus in a radio receiver, the signal picked up by the antenna supplies the excitation for the first state, the output of the first supplies the excitation for the next, and so on.

EXCITATION CURVE. In nuclear physics, a graphical relationship between the energy of the incident particles or photons, and the relative yield of a specified nuclear reaction.

EXCITATION ENERGY. Level Width (Excitation Energy).

EXCITER. This term has four common uses in engineering: 1. In antenna terminology, the portion of a transmitting array of the type which includes a reflector, which is directly connected with the source of power. 2. In transmitters, the oscillator which supplies the carrier or subcarrier frequency voltage to drive the stages which ultimately lead to the final power output stage. In FM systems this unit includes all the frequency generating, modulating and frequency multiplying circuits of the transmitter. 3. In photoelectric reproduction of film, the lamp which supplies a light source of constant amplitude. 4. A generator used to supply the field currents of a larger direct current generator or of an alternator.

EXCITER (Vibration). Vibration Exciter.

EXCITING CURRENT (Transformer). This is the current which supplies the core losses and magnetizing current for a transformer. When the transformer is supplying a load this excitation current is one component of the total input, the other being the component which balances the load current. The current establishes the magnetic field in the core, furnishing energy for the no-load power losses in the core. Also called *magnetizing current.*

EXCLUSIVE OR CIRCUIT. A logical element which has the properties that if either of the inputs is a binary 1, then the output is a binary 1. If both the inputs are a binary 1 or 0, the output is a binary 0. In terms of Boolean algebra, this function is represented as $F = AB' + BA'$, where the prime denotes the **NOT** function. With reference to the transistor exclusive **OR** circuit shown in the accompanying diagram, the output is positive when either transistor

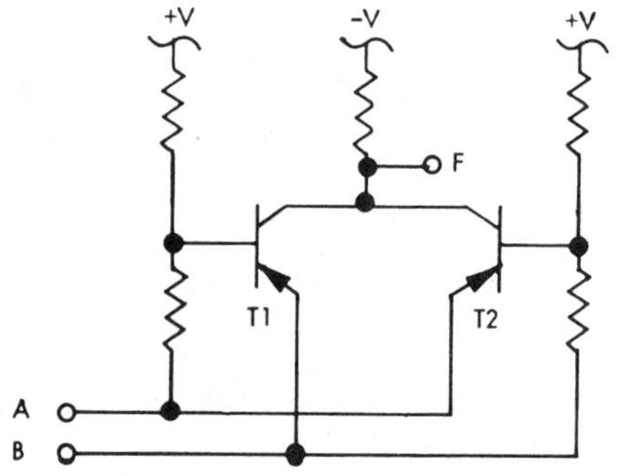

Exclusive OR circuit.

is in saturation. When input A is positive and B is negative, transistor T_2 is in saturation. When B is positive and A is negative, transistor T_1 is in saturation. When A and B are either both positive or both negative, then both transistors are cut off and the output F is negative.

Although shown as discrete devices in the diagram, fabrication using large-scale integrated circuit technology may utilize other circuit and device configurations.

Thomas J. Harrison, International Business Machines Corporation, Boca Raton, Florida.

EXCRETION. The removal of the waste products resulting from the chemical transformation of materials in the body.

The oxidation of materials derived from foods for the release of energy may produce carbon dioxide and water whether the compound oxidized is a protein, a carbohydrate, or a fat, but since proteins contain nitrogen and other elements in addition to carbon, hydrogen, and oxygen, they also give rise to more complex waste products. The chief nitrogenous wastes of animals are urea and uric acid. The elimination of all of these compounds and other substances of like derivation is excretion.

Many small animals, including both protozoans (*Protozoa*) and more complex forms, apparently discharge these wastes from the surface of the body generally, while in others a special excretory system occurs. Even in those forms which have a complex excretory system, any moist surface directly or indirectly exposed to the medium surrounding the animal is favorable for the diffusion of materials into or out of the body, and so may carry on excretion. The wastes passed out in this manner are largely carbon dioxide and water, although the discharge of water may also take out dissolved solids. Thus the lungs of a terrestrial vertebrate eliminate carbon dioxide and the sweat glands of the skin of some animals discharge water with other materials, including nitrogenous wastes, in solution. By far the greater part of the complex wastes is eliminated by the excretory system.

In complex animals other organs than those directly involved in the elimination of wastes may play an important intermediary role. The circulatory system of the vertebrate, for example, transports all wastes from the tissues where they are formed to organs which act upon them and finally to the centers which remove them from the body. The liver removes some substances, including complex organic compounds resulting from the destruction of old red blood cells, and discharges them in the bile by way of the intestine. It also transforms ammonia and amino acids, resulting from the oxidation of proteins, into urea which is returned to the blood to be removed by the kidneys.

See also **Urine.**

EXCRETORY SYSTEM. An organic system whose principal or only function is the removal of complex wastes from the animal body. The excretory system of humans is discussed under **Kidney and Urinary Track.**

Some animals lack an organized excretory system, discharging wastes from the surface of the body generally, but others, even among the one-celled animals, have special excretory structures. The contractile vacuoles of protozoans are supposed to carry out this function.

In the flatworms a special excretory system based on the flame cell appears. Flame cells are large and hollow, with a group of cilia projecting into the cavity whose movement drives out the liquid discharged by the cell. The cavity of each flame cell joins a small duct and these ducts converge to form larger ducts which ultimately open at the surface of the body. Flame cells emptying by ducts into a vesicle connected with the caudal end of the alimentary tract also occur in rotifers.

Roundworms have two slender excretory canals along the sides of the body which unite to empty by a single pore near the anterior end.

In the segmented worms, the body cavity becomes involved in excretion. Two forms of tubes, the coelomoducts and nephridia, open from the coelom to the exterior in these worms. These organs are segmentally arranged ciliated tubes with a funnel-shaped inner end and a minute opening externally. They are variously associated in the excretory organs of different species and in some do not open into the coelom but are provided with cells much like flame cells which are called solenocytes.

Arthropods of different classes have special excretory structures, including the coxal glands of scorpions, said to be derived from coelomoducts, and the Malpighian tubules of insects. The latter are slender tubules opening into the alimentary tract at the caudal end of the stomach and blind at their other end.

The occurrence of solenocytes in the lancelets of the phylum *Chordata* is unusual in this phylum, since in the true vertebrates a pair of kidneys are the chief excretory structures. They are developed from intermediate mesoderm. The excretory unit in these organs is a minute tubule which has in its primitive form a ciliated funnel leading from the coelom. At their lateral ends the series of tubules unite to form a duct which grows back to empty into the cloaca. The tubule is associated with a knot of blood vessels near the coelomic opening (the nephrostome). In a more advanced stage of development excretory tubules lack the nephrostome and have the wall expanded to form Bowman's capsule, embracing the knot of blood vessels which is called a glomerulus. This unit, known as a renal corpuscle, is found in the kidneys of most vertebrates. The tubule leading from it is also specialized for the removal of wastes from the blood.

Three pairs of kidneys are found in different vertebrates and appear in succession in the embryos of the higher classes, the reptiles, birds, and mammals. In cyclostomes and embryos of fishes and amphibians the kidneys are pronephroi, lying well forward in the body and made up of tubules of the primitive type. The pronephroi are vestigial in embryonic reptiles, birds and mammals. Functional kidneys in these embryos and in the adults of cyclostomes, fishes and amphibia are the mesonephroi, lying behind the pronephroi and made up of closed tubules. As they develop, these tubules connect with the duct formed by the pronephroi; this duct is then called the mesonephric or Wolffian duct. In adult reptiles, birds and mammals the mesonephroi are replaced by the metanephroi, lying still farther back. Their tubules develop in a mass of tissue surrounding a blind diverticulum of the mesonephric duct.

The connection of the excretory ducts with the cloaca persists in many vertebrates but in the true mammals this passage splits to form a dorsal rectum and a ventral urogenital sinus which receives the Wolffian duct. An expanded reservoir, the urinary bladder, developed ventrally in connection with the cloaca, ultimately receives the ducts of the metanephroi, while the remainder of the mesonephric duct persists in the male as the main duct of the testis. The relations of all these parts differ greatly greatly in animals of different groups. In all animals with metanephroi the ducts leading from the kidneys are called the ureters and a separate duct from the urinary bladder to the exterior is the urethra.

EXCRETORY SYSTEM (Human). Kidney and Urinary Tract.

EXHAUSTER (Burner). Burner.

EXHAUSTION THEORY. Gerontology.

EXHAUSTS (Automotive). Automotive Electronics; Carburetor; Catalytic Converter; Internal Combustion Engine; Petroleum.

EXIT SLIT. A narrow opening in an opaque screen; when a spectrum is produced upon the screen, by a spectrometer, the slit passes only a small portion of the spectrum, which is then focused onto a detector.

EXOCRINE GLANDS. Cystic Fibrosis; Gland.

EXOCYCLOIDA. An order of sea urchins. *Echinoidea.*

EXOGENETIC. A general term designating all surficial, or near surficial, geologic processes such as: erosion, deposition, and secondary enrichment of ore bodies. The term is not particularly applicable to volcanism.

EXOPHTHALMIC GOITER. Thyroid Gland.

EXOPHTHALMOS. Thyroid Gland.

EXOPTERYGOTA. Insects whose wings appear in the immature stages as external flaps. These insects have gradual or incomplete metamorphosis, hence the term embraces the *Paurometabola* and *Hemimetabola* of some writers.

EXOSKELETON SYSTEM. Integument; Integumentary System; Skeletal System.

EXOSMOSIS. An osmotic process by which a diffusible substance passes from the inner or closed, to the outer parts of a system, as in the loss of substances from a portion of a plant root to water in the surrounding soil.

EXPANDER (Signal). That part of the communication circuit designed to expand the volume range back to the original value as it is normally compressed for transmission. Weak signals are attenuated and strong signals are amplified.

EXPANSION JOINT. Metals constituting pipes have the property possessed by all materials of expanding with increase of temperature. Were they constrained to a fixed length, a reaction equivalent to the force required to compress the pipe through a deformation equal to the prevented expansion would be set up. For all but very short steam lines this force is too large to incorporate in the piping system. The same force would be present, theoretically, in the short line, but the supports would have enough elasticity to take the small expansion. In long lines the expansion is permitted by the use of suitable joints and bends. See accompanying figure.

Both packed and packless expansion joints are used for saturated

Expansion loops that carry geothermal steam from wells to electric generating station more than 1.5 miles (2.4 kilometers) distant. This installation is at *The Geyers*, located in Sonoma County, California about 90 miles (145 kilometers) northeast of San Francisco. (*Pacific Gas and Electric Co.*)

steam at pressures up to 250 psi (17 atmospheres). High temperature has a deteriorating effect on packing; however, packed joints have been designed for high temperature by protecting the packing by air-cooled sleeves. Expansion joints take up expansion at one point by allowing relative motion of the two sections of pipe connected by the joint. Usually one pipe end is anchored by a rigid connection to the body of the joint but occasionally the double slip joint in which both pipe ends are free to move in the joint is used.

When expansion is to be taken by the flexibility of the pipe itself various forms of pipe bends are used. This way of caring for expansion is free of the temperature-pressure limitations of the expansion joints and also of any maintenance work such as the repacking of joints. Consequently, it has been the standard for boiler and turbine leads and for long runs of high-pressure piping of all sorts. Its principal drawbacks are the added friction losses, the expense of fabrication (most bends are special jobs), and the space required.

EXPANSION TURBINE. **Gas and Expansion Turbines.**

EXPANSION WAVE. **Aerodynamics; Supersonic Aerodynamics.**

EXPECTED VALUE. The expected value of a function of variate values is its mean value in repeated sampling. If $F(x)$ is the cumulative distribution function of a variate x, the expected value of a quantity t depending on x is

$$\int_{-\infty}^{\infty} t\, dF(x)$$

It is in fact the average or mean value of t over the distribution of x.

EXPERIMENTS (Statistical). **Design of Experiments.**

EXPLORATION (Coal). **Coal.**

EXPLORATION (Mineral). **Earth Resources Satellites and Geologic Remote Sensors.**

EXPLORATION (Natural Gas). **Natural Gas.**

EXPLOSIVE. In the conventional sense, a solid, gas, or liquid material which, when triggered, will release a great amount of heat and pressure by way of a very rapid, self-sustaining exothermic decomposition. This entry does not describe nuclear explosives.

There are two principal classes of explosives: (1) *deflagrating explosives* whose burning processes are rather slow—with progressive reaction rates and buildup of pressure that create a heaving action; and (2) *detonating explosives*, which are characterized by very rapid chemical reactions, thus causing tremendously high pressure and brisance (shattering action). In the latter, detonation waves may obtain a velocity in excess of 20,000 feet per second. The decomposition of cellulose nitrate used in propellants typifies the deflagrating type: $C_{24}H_{30}N_{10}O_{40} \rightarrow 5N_2 + 10H_2 + 5H_2O + 11CO_2 + 13CO$. The decomposition of nitroglycerine typifies the detonating type: $4C_3H_5(ONO_2)_3 \rightarrow 12CO_2 + 10H_2O + 6N_2 + O_2$.

Black powder, using KNO_3 or $NaNO_3$, charcoal and sulfur was probably the first explosive developed and is attributed either to Chinese or Egyptian ingenuity. The time of first use occurred before the birth of Christ. This is a deflagrating explosive and was adapted for blasting purposes as early as the 1600s.

Black powder (gunpowder) consists of an intimate mixture of finely divided solids, 75% potassium nitrate, 15% carbon, 10% sulfur. Powders for sporting guns contain a slightly larger percentage of potassium nitrate (75 to 78%, smaller percentage of carbon (15 to 12%), and a variation in sulfur from 9 to 12%. Mining or blasting powders, where large volumes of gas are desired, may have 14 to 21% carbon and 13 to 18% sulfur. When ignited, potassium nitrate supplies oxygen for the combustion of explosives, of carbon to carbon dioxide and of sulfur to sulfur dioxide. One gram of powder yields 250 to 300 milliliters of gas measured at 0°C and 760 mm pressure. The heat evolved per gram is 500 to 700 calories, and the temperature of the explosion is estimated at 2,700°C.

Among the earliest high explosives were *mercury fulminate*, $HgC_2N_2O_2$, developed late in the seventeenth century, and *nitrostarch*, $C_{12}H_5(ONO_2)_{30}$, discovered by Braconnot in 1832 and still used as a sensitizing ingredient in modern commercial explosives. *Nitrocotton* was produced in 1838 by Dumas and Pelouse by treating cotton and paper with nitric acid, in the same way that Braconnot treated starch with HNO_3. *Nitroglycerin*, $C_3H_5(ONO_2)_3$, was first made by Sobrero. Early nitroglycerin formulations were highly dangerous and caused numerous accidents. In 1867, Nobel found that nitroglycerin could be rendered safe by absorbing it in a porous material, such as kieselguhr, or diatomaceous earth. After formulation of this first *dynamite*, Nobel introduced $NaNO_3$ and later NH_4NO_3 into his dynamite formulations. Nobel, in 1875, while experimenting with cellulose tetranitrate, mixed collodion with nitroglycerine, resulting in the development of blasting gelatin. The development of the blasting cap used with a safety fuse allowed for safe, positive initiation of dynamite.

Wilbrand, in 1863, first prepared *trinitrotoluene* (TNT), $C_6H_2(CH_3)(NO_2)_3$. The material was not manufactured in production quantities until about 1900. The German military recognized the advantages of TNT as a replacement for *picric acid* which they had used earlier. TNT was used extensively during World War I and became a standard military explosive.

Tollens, in 1891, prepared *pentaerythritol tetranitrate* (PETN), $C(CH_2NO_3)_4$, but this compound was not commercially available until after World War I. Commercial production had to await a lowering in the cost of formaldehyde and acetaldehyde used in its production.

Henning, in 1899, discovered *cyclotrimethylenetrinitramine* (cyclonite-RDX), but its potential was not realized until about 1920. RDX was used extensively during World War II as a component of numerous cyclotols, plastic explosives, and bursting charges.

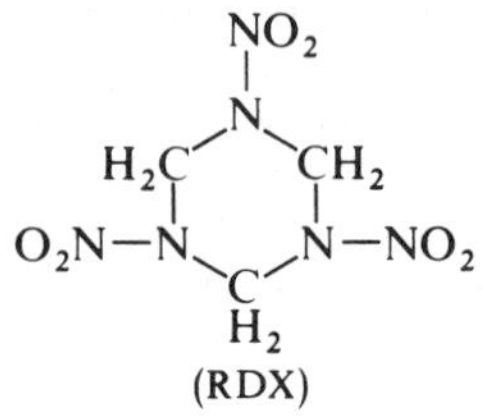

(RDX)

Although ammonium nitrate, NH_4NO_3, was known to have explosive qualities as early as 1659, it was not used much in explosive formulations until about 1867. At that time, it was used by Nobel to replace a portion of nitroglycerin in dynamite. Because of critical toluene shortages during World War I, NH_4NO_3 was used as a way of conserving TNT. Mixtures containing 80% NH_4NO_3 and 20% TNT; or 50–50 mixes were used. These became known as 80:20 or 50:50 *amatols* and were used as military explosives for shells and bombs. In World War II, particularly by the Axis powers, amatols were used, again to conserve TNT. The Texas City, Texas disaster of 1947 (explosion of ship loaded with NH_4NO_3) reemphasized the potential of this substance for explosives. Later, Robert Akre used a combination of prilled NH_4NO_3 and carbon black (94:6 mix) in making an explosive for blasting in open-pit strip mines. The substance was patented under the name *Akremite.* Later experiments included mixing liquid hydrocarbons to replace the carbon black. This resulted in ANFO explosives. It was found that diesel fuel oil can be mixed with NH_4NO_3 with a consistent quality. Commencing in the late 1950s, ANFO explosives became widely accepted.

Unfortunately, the water resistance of ANFO is low and numerous experiments in attempting to dry-package it were not markedly successful. This shortcoming of ANFO led Cook, Farnum, and others to develop *slurry explosives.* These materals are comprised of oxidizers, such as NH_4NO_3 and $NaNO_3$, fuels, such as coals, oils, aluminum, and other carbonaceous materials, sensitizers, such as TNT, nitrostarch and smokeless powder, and water—all mixed with a gelling agent to form a thick, viscous explosive having excellent water-resistant properties. These explosives are made as cartridged units or are mixed at the site in bulk and then pumped into place.

Quarry blast utilizing high explosives initiated with electrical delay devices (millisecond delay electric blasting caps). (*Trojan-U.S. Powder Div., Commercial Solvents Corp.*)

The largest consumers of commercial explosives are the mining, construction, and seismic-prospecting industries. Explosives are also used in agriculture for blasting stumps, setting posts, breaking up hardpan, clearing land, digging wells, and blasting drainage and irrigation ditches. Explosive technology also is applied by industry in the forming, cladding, bonding, hardening, and welding of metals. On a small scale, explosive-actuated devices are used in valves, switches, and relays, as well as for cutting, punching, riveting, and fastening metals.

See accompanying illustration of a typical quarry blast using a millisecond delay electric blasting cap.

References

Johannson, C. H., and P. A. Persson: "Detonics of High Explosives," Academic, New York, 1970.

Langefors, U., and B. Kihlstrom: "Modern Technique in Rock Blasting," 2nd edition, Wiley, New York, 1967.

Meyer, R.: "Explosives," Verlag Chemie International, New York, 1977.

White, J. E.: "Seismic Waves," McGraw-Hill, New York, 1965.

EXPLOSIVE WELDING. Welding.

EXPONENT. Also called index, it is the power to which an algebraic expression is raised. It can be positive or negative, integral or fractional. If m and n are positive integers and a, b are numbers or functions, the following are some of the properties of the exponent;

$$a^m \times a^n = a^{(m+n)}$$

$$a^m / a^n = a^{(m-n)}$$

$$a^{-m} = 1/a^m$$

$$(a^m)^n = a^{mn}$$

$$a^{1/m} = \sqrt[m]{a}$$

$$\sqrt[m]{\sqrt[n]{a}} = \sqrt[mn]{a}$$

$$a^{m/n} = \sqrt[n]{a^m}$$

$$(ab)^m = a^m b^m$$

$$(a/b)^m = a^m / b^m$$

$$\sqrt[m]{ab} = \sqrt[m]{a}\,\sqrt[m]{b}$$

$$\sqrt{a/b} = \sqrt[m]{a}/\sqrt[m]{b}$$

An exponential function is transcendental, an example being $y = ab^x$, where a, b are constants and x is the independent variable. If $a = 1$, it is the inverse of the logarithmic function. Some properties of its curve are: (a) it is not symmetric to either the X- or Y-axis or to the origin; (b) it intersects the Y-axis at $y = 1$ but its asymptote is the X-axis; (c) no finite value of x makes y infinite. If $b < 1$, $y \rightarrow 0$ as $x \rightarrow \infty$ and decreases continuously as x increases; if $b > 1$, $y \rightarrow 0$ as $x \rightarrow -\infty$ and increases continuously with x; if $b = 1$, the curve becomes the straight line $y = 1$.

The most convenient value to choose for b is the transcendental number e. Its curve is similar to the more general one described for the number b but its slope is different.

An exponential equation contains one or more exponential functions. It can often be solved by taking the logarithm of both sides and solving the resulting algebraic equation.

Euler's theorem on the exponential function is $\cos x \pm i \sin x = e^{\pm ix}$. See also **De Moivrė Theorem;** and terms listed under **Mathematics.**

EXPONENTIAL DISTRIBUTION. A distribution of the form

$$dF = \frac{1}{\sigma} \exp\left(-\frac{x-m}{\sigma}\right) dx, \quad m \leq x \leq \infty$$

The parameter σ is the standard deviation of the distribution and is equal to the distance of the mean from the start.

EXPONENTIAL INTEGRAL. Integral (Logarithmic).

EXPONENTIAL SMOOTHING. A method used in forecasting a variable from values of that variable occurring at previous points of time. It relies on a weighted average of those previous values, and on the reasonable assumption that values in the more remote past are of less influence on the present, the weights attached to the values diminish for the less recent values. Such a forecast at time t of a variable u_t, for example, might be

$$u_t(\text{forecast}) = (1-\beta) \sum_{i=1}^{\infty} \beta^{j-1} u_{t-j}$$

The coefficients diminish according to the exponent of the coefficient β; hence the name of the procedure. More elaborate forms of the same basic method are also known as exponential smoothing.

EXPOSURE CONTROL (Camera). Photography and Imagery.

EXPRESSION (Mechanical). The separation of liquids from solids by compressively squeezing certain liquid-containing substances, such as separating oils from vegetable seeds and nuts. Equipment is designed to permit the liquids to be removed while still retaining the solids between the compressing surfaces. Expression equipment takes several configurations. In the *plate press*, the material to be expressed, such as fruit or seeds, is wrapped in special plate cloths. These then are placed between a series of hydraulically operated plates, which act upon the materials much as a vise. The pressure usually is applied in stages, with the maximum pressure applied close to the end of a 20-to-45-minute total cycle. The *box press* is similar with the exception that shallow boxes enclose the pressed cake on two sides, thus simplifying the folding of the press cloths. The *cage press* consists of a cylinder, finely perforated, with a hydraulically-operated ram. Essentially, the material to be expressed is squeezed at one end of the cylinder by the ram. The expressed oil flows through the perforations. Several pressings (strokes of the piston or platen) are usually required. In the *pot press*, the cage is replaced by a series of short, superimposed steam-heated pots. Usually a series of pots is used in each press, the bottom of each pot serving as the ram for the pot below. The *curb press* also is similar to the cage press, but operates at lower pressures with fewer drainage channels. The *screw* press consists of a continuous screw or worm that rotates within a cyclinder housing lined with perforated plates. A powerful squeezing action results from the taper of the screw. In the *V-disk press*, two conical disks face each other in a suitable casing. As the disks rotate, they converge from a point of maximum gap (point of feed) to a point of minimum gap (point of discharge). Designs with hydraulic systems permit the disks to oscillate during operation to maintain constant pressure. The *roll press* operates on the principle of the old-fashioned clothes wringer, with two to three rollers through which the feed is passed, the expressed liquids collecting in a trough below.

Operating characteristics of the various forms of expression equipment are summarized in the accompanying table.

EXPRESSION EQUIPMENT

Type of Press and Applications	Oil, Fat, Liquids in Feed	Cake Discharge
Plate. Vegetable seeds, nuts, olives, fruit, notably flaxseed	30–35%	5–10% oil
Box. Vegetable seeds, nuts, notably peanuts and cottonseed	30–35%	5–10% oil
Cage. Most oil seeds and nuts. Almost any oil material, notably copra and castor beans	35–50%	5–10% oil
Pot. Fats, such as cocoa butter not liquid at room temperature	30–50%	6–10% fat
Screw (Low-pressure type). Wastewater slurries, wood and wood pulp; food and beverage products	90–97%	25–50% solid
Screw (High-pressure type). Vegetable seeds and nuts; rubber; rendered materials	30–35%	3–5% oil
V-disk. High-polymer resins, spent grains, wood and pulp products, food and starch products, wastewater slurry	85–97%	25–55% solid
Roll. Alkali cellulose, sugar cane, wood and pulp products	92–98%	30–55% solid

EXSECANT. Trigonometric Function.

EXSOLUTION. Mineralogy.

EXSTROPHY (Bladder). Kidney and Urinary Tract.

EXTENSIVE AIR SHOWER. Cosmic Rays.

EXTENSOMETER. A device for determining small linear dimensional changes caused by the application of a stress. Thus, an extensometer is used to determine the changes in length of the gage section of a tensile specimen during a tensile test.

EXTENSOR REFLEXES. Brain and Nervous System.

EXTERNAL HUMAN FERTILIZATION. In-Vitro Fertilization.

EXTERNAL STORAGE (Computer). Storage (Computer).

EXTEROCEPTORS. Brain and Nervous System.

EXTINCTION COEFFICIENT. 1. A measure of the space rate of diminution, or extinction, of any transmitted light; thus, it is the attenuation coefficient applied to visible radiation. The extinction coefficient is identified in a form of Bouguer's law (or Beer's law):

$$dI = -\sigma I\, dx$$

or

$$I = I_0 e^{-\sigma x}$$

where I is the illuminance (luminous flux density) at the selected point in space, I_0 is the illuminance at the light source, and x is the distance from the source.

When so used, the extinction coefficient equals the sum of the medium's absorption coefficient and scattering coefficient, each computed as a weighted average over all wavelengths in the visible spectrum. So long as scattering effects are primary, as in the lower atmosphere, the value of the extinction coefficient is a function of the particle size of atmospheric suspensoids. It varies in order of magnitude from 10 km^{-1} with very low visibility to 0.01 km^{-1} in very clear air.

The extinction coefficient is related to the transmission coefficient τ as follows:

$$\tau = e^{-\sigma}$$

2. In oceanography, the extinction coefficient is a measure of the attenuation of downward-directed radiation in the sea. The coefficient K is defined by

$$K = 2.303 \log \frac{I_{\lambda 1}}{I_{\lambda 2}}$$

where $I_{\lambda 1}$ is the intensity of radiation of a given wavelength λ on a horizontal surface and $I_{\lambda 2}$ is the intensity on a horizontal surface 1 meter deeper. K varies with wavelength, with the nature of the scattering particles, and with the presence of dissolved colored substances.

EXTRACELLULAR FLUID (ECF). Chloride; Kidney and Urinary Tract; Water.

EXTRACTION (Liquid–Liquid). Sometimes referred to as solvent extraction, this operation is effected by treating a mixture of different substances with a selective liquid solvent. At least one of the components of the mixture must be immiscible or partly miscible with the treating solvent so that at least two phases can be formed over the entire range of operating conditions. To be effective, one or more of the components must be dissolved from the mixture by the solvent

TABLE 1. LIQUID–LIQUID EXTRACTION USING EQUAL VOLUMES IN SINGLE EXTRACTION

Concentration Ratio $\frac{B}{A}$	Volume Ratio $\frac{B}{A}$	Amount of C in B	Amount of C in A	Fraction of C in B	Fraction of C in A
10	1	$10 \times 1 = 10$	$1 \times 1 = 1$	$\frac{10}{11} = 0.91$	$\frac{1}{11} = 0.09$

TABLE 2. EFFECT OF SUCCESSIVELY SMALLER PORTIONS ON LIQUID–LIQUID EXTRACTION

	Concentration Ratio $\frac{B}{A}$	Volume Ratio $\frac{B}{A}$	Amount of C in B	Amount of C in A	Fraction of C in B	Fraction of C in A
First extraction	10	0.5	$10 \times 0.5 = 5$	$1 \times 1 = 1$	$\frac{5}{6} = 0.83$	$\frac{1}{6} = 0.17$
Second extraction	10	0.5	0.17×0.83	0.03	0.14	0.03
Combined					0.97	0.03

preferential to the other components present. The solvent-rich phase that contains the preferentially dissolved component is termed the *extract layer*. The residual phase formed by the undissolved component (or diluent) and usually containing some solvent is called the *raffinate layer*. Either layer may be at the top or bottom of the separating vessel, depending upon relative densities. Other forms of solvent extraction include leaching, washing, and precipitative extraction (*salting out*).

Liquid–liquid extraction finds application in separating the components of condensed mixtures where vaporization methods, such as distillation or evaporation, may be impractical. This condition may arise because the substances to be separated may have comparable volatilities, are relatively nonvolatile, are heat-sensitive, or have one component present in very small concentration.

Liquid–liquid extraction finds wide application throughout the processing industries, including the manufacture of toluene, uranium vanadium, amino acids, coal-tar products, lube-oil refining, protein processing, and solvent refining of coal and oil shales.

Solvent extraction also is an important laboratory operation as in the recovery of oils from oil-bearing material. The material is placed in a porous container and subjected to treatment with solvent. The solvent containing some dissolved material passes through the porous membrane, leaving the undissolved residue in the container. The principle of countercurrent extraction may be utilized in consecutive containers, or the solvent may be vaporized from the solution, condensed onto the material and, by means of a syphon in the apparatus, withdrawn periodically to the solution compartment below, as in the Soxhlet type of apparatus. When a third substance is of different solubility in two nonmiscible liquids, this substance may be separated from the solution of lower concentration by shaking with the preferential solvent, and then separating the two liquid layers. The desired substance may be recovered from the solution by evaporation of the solvent. The effectiveness of separation is increased by the use of a given amount of extracting solvent in successively smaller portions rather than by a single extraction with the total amount.

Example: upon shaking 1 volume of liquid A plus 1 volume of liquid B, assume a concentration ratio of 1 (concentration in B)/10 (concentration in A) of the third substance C. See Table 1. Upon shaking 1 volume of liquid A plus $\frac{1}{2}$ volume of liquid B and, after separation, shaking 1 volume of liquid A (containing the residue of C) plus $\frac{1}{2}$ volume of liquid B, the results are indicated by Table 2.

A single equal-volume extraction would, therefore, remove 91% of C from A, whereas a double half-volume extraction would remove 97%.

EXTRACTION PROCESS (Coal). Coal.

EXTRACTIVE DISTILLATION. Distillation.

EXTRACTIVE METALLURGY. That phase of metallurgy dealing with the removal of metals from minerals. Methods are discussed under individual metals.

EXTRADURAL HEMATOMA. Brain (Injury).

EXTRAEMBRYONIC MEMBRANES. A series of structures developed in connection with the embryos of vertebrates (reptiles, birds and mammals), but not as parts of the body itself. They relate the embryo to its environment in several ways. These membranes are the allantois, amnion, chorion, serosa, and yolk sac.

EXTRAORDINARY INDEX. The refractive index for the extraordinary ray in a crystal showing double refraction, measured perpendicular to the optic axis (in which direction its value differs most from the index for the ordinary component). If an unpolarized ray of light strikes the surface of calcite or other crystal normally showing double refraction it will be divided into two transmitted rays. One ray, the ordinary, will not be bent, while the other ray, the extraordinary, will be bent on entering the crystal. Rotation of the crystal about the entrant ray causes the extraordinary ray to rotate about the ordinary ray.

EXTRATROPICAL CYCLONE. Fronts and Storms.

EXTRUSION. A majority of stock plastic shapes (bars, cylinders, special cross sections) are made in this way. Thermoplastic materials are heated in a plasticizing cylinder and by means of a rotating screw are forced through a die to provide the desired cross section. A variation of the process is used for extruding coatings of soft plastic materials over other materials. Almost any profile can be imparted to the product, but of course, variations in profile are limited to two dimensions. The tooling costs for extrusion are low compared with injection molding. Thickness of the material can be controlled quite precisely. Production rates are high.

Certain metals, including aluminum, and various rubbers are also extruded.

Extrusion is also combined with mixing in some applications. In the type of device shown in the accompanying diagram, the material

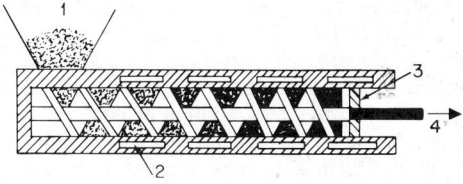

Mixer-type extruder. (1) Charge stock; (2) heating or cooling chambers in extruder jacket; (3) die; (4) extruded product.

is fed as a dry solid, is fluxed in the barrel to form a paste, and then resolidified at the discharge. Action in the barrel is one of shearing, rubbing, and kneading. One continuous screw or two screws rotate in the closely-fitting barrel. The work of the screw is augmented by forcing the material through breaker screens and around breaker disks just before the material is forced out through the exit nozzle. Such machines are often used for extruding soft chemical and food mixes which do not require fluxing, as well as for the extrusion of hard plastics, some of which must be fluxed at temperatures above 400°F (204°C). Wires can be covered and shapes of intricate cross section can be produced. Plastic resins also are blended in extruders to form pellets for later press and injection molding.

EXTRUSIVE. A property of an igneous rock that has been ejected onto the earth's surface. Lava flows and detrital material, such as volcanic ash, are extrusives. See also **Mineralogy.**

EXUDATIVE DIARRHEA. Diarrhea.

EXUMBRELLA. The upper or convex surface of the body of a jellyfish or other medusa.

EYE BAR. A heat-treated tension member formed from a single piece of steel. The finished eye bar consists of a body having a rectangular cross section and two circular heads containing holes for pins, which are used to connect the eye bar when it forms part of a structure.

In the fabrication of an eye bar, the ends of a steel plate, of the correct cross-sectional area and length, are heated and upset to form the heads. The heads are next rolled to remove any unevenness resulting from the upsetting operation. While the ends are still hot, holes are punched out which are smaller in diameter than the finished pin holes. The bars are then subjected to special heat treatment which produces a high tensile strength. After cooling, the pin holes are bored to exact size simultaneously.

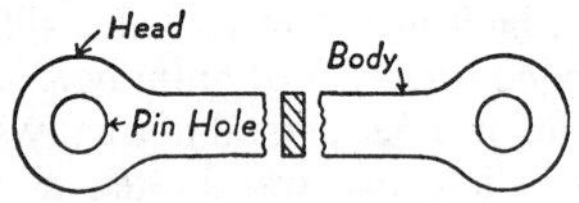

Eye bar.

Eye bars make excellent tension members since the heat treatment enables them to carry higher tensile loads than the ordinary built-up steel members. As the eye bar is a very slender member, it cannot be used where there is a possibility that it will have to carry compressive stress. These members are used in the cable anchorages of suspension bridges. Eye bar chains are sometimes used instead of wire cable for suspension bridges. In the past, eye bars were extensively used for tension members of truss bridges. The use of riveted joints instead of pin-connected joints made eye bars obsolete for such use.

EYE (Human). Vision and the Eye.

EYELID. A fold of skin which can be drawn over the eye in vertebrates above the fishes. Three eyelids are the maximum. These are an upper and lower lid and a third eyelid or nictitating membrane which passes between the others and the eyeball from the inner to the outer margin of the eye. The eyelids contain glands whose secretion lubricates the apposed surfaces of the lids and eyeball, and in the mammals bears a row of stiff hairs, the cilia or eyelashes.

In humans, *hordeolum* (sty) is a common infection of one or more of the small glands of the eyelids, usually caused by staphylococci. Children are especially susceptible. A sty begins as a small, reddened area on the margin of the lid. Pain is almost always present and is directly related to the amount of swelling. In severe cases, the entire eyelid is swollen. A few days after its appearance, the sty develops a yellow center, caused by the formation of pus, and usually erupts a few days later. A single sty may not require medical attention unless it is quite painful. When a number of sties appear, or when they recur often, general health and diet should be evaluated.

Chalazion is a swelling or enlargement of one of the oil glands of the eyelid. This is caused by obstruction of the gland's duct. The skin moves loosely over the swelling. The physician may prescribe topical medication, but, if this fails, a simple surgical procedure is performed which removes the mass, leaving no visible scar.

Blepharitis is a relatively common condition in which the margins of both eyelids become red and inflamed. Blepharitis can be caused by bacterial infection or it may be an extension of seborrheic dermatitis, involving the scalp, eyebrows, and, at times, the ears. Blepharitis may occur only as redness with slight crusting, or it may cause itching, burning, and edema of the eyelids, lacrimation, and hypersensitivity to light. The lids often become stuck together overnight from the accumulation of dried secretions.

Ptosis is a condition in which one or both upper eyelids droop. This is caused by the failure of the levator muscles of the eyelid to operate properly. The abnormality may be congenital or acquired. Congenital ptosis, when severe, may be treated by surgical alteration of the involved muscles.

Edema of the eyelids usually results from allergies to eyedrops, drugs, or cosmetics. Trichinosis also may produce eyelid edema. See **Trichinosis.**

See also **Vision and the Eye.**

EYEPIECE. Also known as the *ocular*, the lens, or system of lenses, closest to the eye in an optical instrument such as a telescope or a microscope. The eyepiece is usually a magnifying device used for the purpose of detailed examination of the real image formed by the objective of the instrument. It is usually designed to act as a collimator to the light from the objective, so that the light from each point of the image formed by the objective emerges in parallel or nearly parallel rays. Hence, in using a telescope or microscope in proper adjustment, the eye should be focused as though looking at a distant object.

The simplest type of eyepiece is either a simple convex or concave lens, of relatively short focus, so placed as to serve as a magnifier for the image formed by the objective. Because of the spherical and chromatic aberrations of the simple lens of short focus, a combination of lens is usually employed as an eyepiece. The two most common types of compound eyepieces are the Huygens (Fig. 1) and the Ramsden (Fig. 2). In these eyepieces, the lens *F* is known as the field lens and the lens *E* as the eye lens. The Huygens eyepiece is placed slightly inside the focus of the objective, and the field lens of the eyepiece forms a real image *R* in the plane *S*, from which the rays emerge parallel from *E*. In the Ramsden type, the field and eye lenses combine to render the light from the real image *R*, formed in the plane *S* by the objective, parallel upon emergence from the eyepiece. Since the Huygens type eyepiece is placed inside the principal focus of the objective, a reticle or filar micrometer cannot be used, although a reticle may be placed inside the eyepiece itself in the plane *S*. The Ramsden type, on the other hand, is focused directly upon the plane of the real image from the objective, and a reticle or micrometer may be placed in this plane. Eyepieces of the Ramsden type, which are simple magnifiers focused upon the real image from the objective, are known as positive eyepieces; while eyepieces placed inside the principal focus of the objective, as in the case of the Huygens type, are known as negative eyepieces. (See entries on individual optical instruments for further discussion of positive and negative eyepieces.)

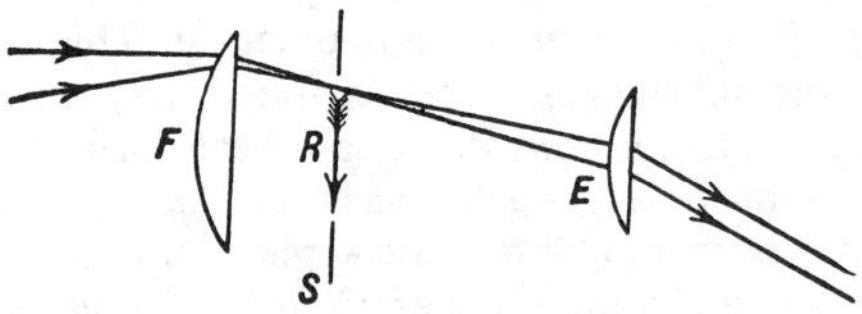

Fig. 1. Huygens eyepiece.

Both the positive and negative eyepieces give a view of the image from the objective in the same orientation as that image is formed. This means that the observer will see the image of a distant object inverted. Although this is no disadvantage in microscopes and in astronomical telescopes, it is intolerable in a telescope or field glass to be used for observation of distant terrestrial objects. The simple concave lens, as used in the so-called Galilean telescope or opera glass, gives an erect image of a distant object. To avoid the aberrations of the simple concave lens, various "erecting systems" are used in terrestrial telescopes. Some of these erecting systems employ prisms, as in the case of binoculars, or complicated systems of lenses.

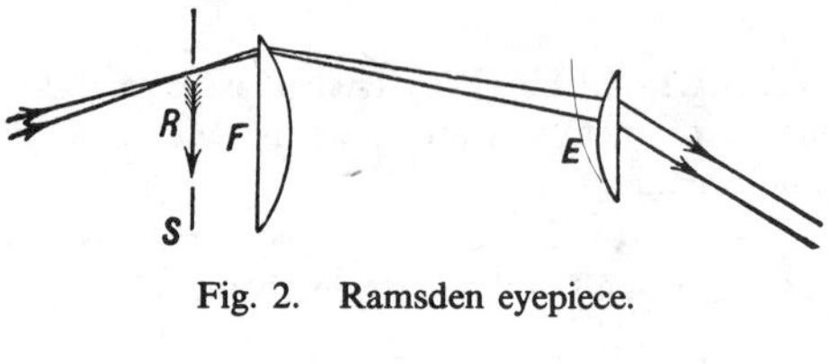

Fig. 2. Ramsden eyepiece.

EYEPIECE (Microscope). Microscope.

EYE (Vertebrate). A sensory organ which is stimulated by light, particularly an organ whose stimulation results in the formation of a mental image of the objects from which the light is reflected or radiated.

Details of the human eye are described under **Vision and the Eye.**

Although most eyes enable the animal to form visual images, that is to see in the usual sense of the word, some light-sensitive organs are capable only of perceiving light and the direction from which it comes. Pigment spots in some of the 1-celled animals are supposed to be light-sensitive and in flatworms and a few insects the eyes are formed of a group of sensitive cells partially isolated by pigment. The structure of these eyes shows no possibility of their forming images.

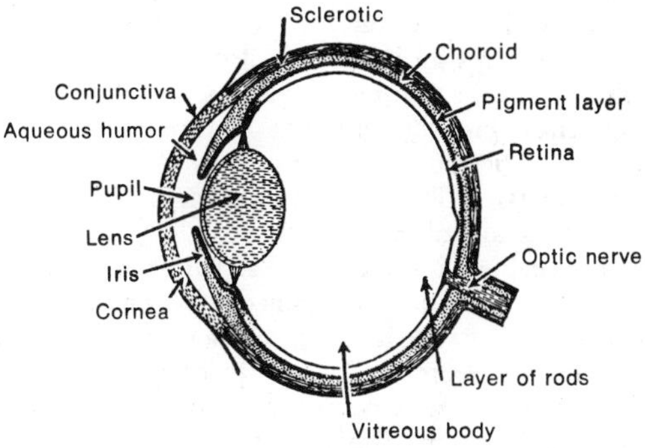

The vertebrate eye.

Eyes of reasonable complexity are found in some of the segmented worms and three types of complex eyes occur in the phyla *Mollusca, Arthropoda,* and *Chordata.* These three forms of eyes have been extensively studied and are known in detail.

Arthropod eyes are of two kinds, simple and compound, of which one or both may occur in a single individual. The sensory end organ in both forms is the retinula, a group of visual cells surrounding a central optical rod or rhabdom. In many simple eyes a portion of the cuticula is thickened to form a biconvex lens opposite to a group of retinulae. Compound eyes are made up of many ommatidia, each consisting of a similar lens forming a facet of the cornea of the entire eye, and an underlying retinula, with intervening crystalline cells and, in some species, other structures.

Both mollusks and vertebrates, including, of course, humans, have camera eyes, although their development and structure differ. All camera eyes have a lens suspended before a chamber lined with a sensory layer, the retina. In front of the lens is another chamber and in front of that a transparent cornea which acts as a lens in terrestrial animals. The eye is insulated by a heavily pigmented layer which surrounds it except where the lens is suspended and extends in front of the lens as the iris. The iris, activated by muscles, controls the size of its central opening, the pupil, through which light enters the eye. Light passing through the lens is focused on the retina in a sharp image and the varied stimuli acting on nerve endings result in a definite mental picture. Such eyes are provided with muscles which direct them toward objects to be observed. They also have muscular focusing devices which move the lens in relation to the retina or vice versa, or control the curvature of the lens as in the human eye.

The action of the different kinds of eyes results in different kinds of vision.

For vision in fishes, see **Fishes;** in snakes, see **Snakes.**

F

FABRICATION. Fabrication is the action of constructing or forming a structure composed of a number of separate elements which must be joined together in one way or another, according to a definite plan. The common methods of engineering fabrication include fusion methods, such as welding; adhesion, exemplified by gluing and soldering; and pinned connections, illustrated by bolting, riveting, and doweling. Fabrication will also include those operations necessary upon the several elements in order to fit them for assembly; also such trimming, polishing, or adjusting operations as will put the complete structure in its final shape. Included in this category are operations like drilling, shearing, milling, polishing and plating.

FABRIC (Geology). The term proposed by Cross, Iddings, Pirrson, and Washington in 1902 for the shapes and arrangement of crystals in an igneous rock. Best defined as the arrangement of the constitute constituents in a rock, i.e., flow fabric in lava, stratification or bedding in sedimentary rocks; foliation in metamorphic rocks.

FACE-CENTERED CRYSTAL. Crystal (Face-Centered).

FACIAL PARALYSIS. Bell's Palsy.

FACIES. In geology, the sum total of the inorganic and organic characteristics of a sedimentary formation. Obviously, different facies of a sedimentary formation (sedimentary time unit) are of the same age; but similar sedimentary facies may represent different formations. Fossils may be useful in determining the age of a facies, provided

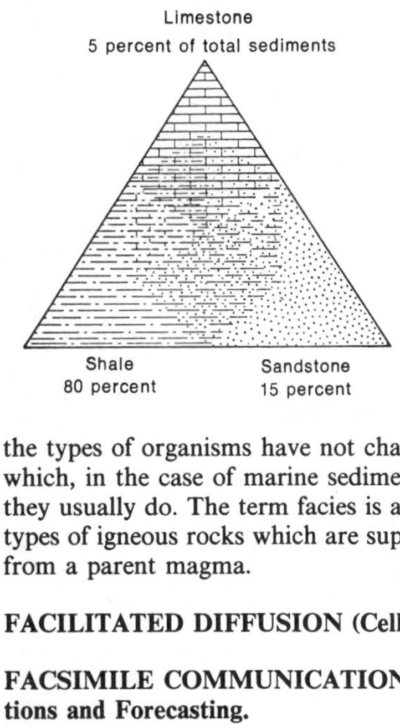

The relative abundance of the three principal types of sedimentary rocks and their intergradations or facies. (*Field, "Outline," Barnes & Noble, Inc.*)

the types of organisms have not changed with the change in habitat, which, in the case of marine sediments (such as limestones or shale) they usually do. The term facies is also used to designate gradational types of igneous rocks which are supposed to have been differentiated from a parent magma.

FACILITATED DIFFUSION (Cell). Cell (Biology).

FACSIMILE COMMUNICATION (Weather). Weather Observations and Forecasting.

FACSIMILE TRANSMISSION. Also known by such names as phototelegraphy, radiophoto, telephoto, telephotography, and wirephoto, facsimile transmission is the communication of a photograph, map, or other fixed graphic material by electrical means over a usually fairly long distance. The information is converted into signal waves and entered into the transmission system which may be wire or radio. At a remote receiving point, there is a facsimile receiver to reconvert the signal information into a copy of the original material.

The graphic information to be transmitted is broken into a sequence of elemental parts by scanning. This may be done by mounting the picture on a revolving drum and projecting a very small beam of light on or through it. The light is reflected (or transmitted) to a phototube, the light and hence the phototube output being proportional to the picture density. The light is moved along the picture (parallel to the axis of the drum) at such a rate that it displaces axially its own width for each revolution of the drum. Thus the spot of light progressively covers ever spot on the picture. The output of the phototube is an electrical breakdown of the picture, and is then modified for transmission. For radio and land lines this means modulation upon a suitable carrier. There are several methods of doing this but these are merely details of the system.

For submarine cable transmission the signal is amplified by dc amplifiers and fed directly to the cable since these cables cannot handle high frequencies. At the receiving end the modulated signal is demodulated and fed to the recorder. This varies with different systems but two types are in wide use. One depends upon a variable light on a photographic paper or film. The light is varied in different ways, one being to use a specially constructed gas-filled lamp whose intensity varies with the signal, others use light valves, but in each case the result is a spot of light whose intensity varies with the picture focused on the paper. There are some direct-developing papers, but others require a development process after removal from the receiver.

The other method of recording is to utilize a paper sensitized to electrical current passage and to pass the received signal current (after detection, of course) through it. Regardless of the method of recording it is necessary for the receiver to be synchronized with the transmitter. Sometimes this is accomplished by depending upon the stability of the power systems and the use of synchronous motor drives; other methods involve the transmission of synchronizing pulses periodically.

FACTOR. The most common use of this term is the mathematical one. If $P(x)$ is a polynomial in one variable of degree n and x_1 is a zero of the polynomial so that $P(x_1) = 0$, then $(x - x_1)$ is one of its factors. In symbols, $P(x) = a_0 + a_1 x + a_2x^2 + \cdots + a_nx^n$; $P(x) = (x - x_1)Q(x)$, where $Q(x)$ is of degree $(n - 1)$. Now $P(x_1) = 0$, hence identically

$$P(x) \equiv P(x) - P(x_1) \equiv a_1(x - x_1) + a_2(x^2 - x_1^2 + \cdots + a_n(x^n - x_1^n)$$

and each term on the right has $(x - x_1)$ as a factor.

Continuing in this way $Q(x)$ can also be reduced to a factor and a term of degree $(n - 2)$. The fundamental theorem of algebra or factor theorem states that there exists one and only one set of constants x_i so that $P(x) = (x - x_1)(x - x_2) \cdots (x - x_n) = 0$. The constants are the roots of $P(x)$.

To factor a polynomial means to find two or more polynomials whose product is the given polynomial.

In statistics, in addition to the above mathematical usage, the term factor is used in three other senses: (1) to denote a quantity under examination in an experiment as a possible cause of variation, e.g., in a "factorial" experiment, (2) (adapted from psychology) in multivariate analysis, to denote a function of the observed variates, usually linear, which may be regarded as part of those variates; and hence as a "factor" of the variation, (3) to denote a constituent item in an average or index-number.

FACTOR ANALYSIS. Suppose we have a multivariate sample of values $(x_1, x_2, \ldots x_k)$. We may be prepared to assume an underlying set of variables $z_1, z_2, \ldots z_p$, with $p < k$ such that each x is a linear function of the z's together with a part specific to itself,

$$x_i = a_{i1}z_1 + a_{i2}z_2 + \cdots + a_{ip}z_p + s_i$$

Factor analysis is the name given to the techniques for estimating the various parameters in a model of this kind. Many such techniques exist.

The z's are known as common factors, the s's as specific factors. If one of the z's contributes to all the x's, it is called a general factor, while one that contributes to some but not to others is called a group factor. The a's are called the factor loadings or saturations; a common factor with both positive and negative loadings is called bipolar. The proportion of the variance of a particular x accounted for by the common factors is called the communality of that x.

FACTORIAL. If n is a positive integer, factorial n means the product $1 \cdot 2 \cdot 3 \ldots n$. It may be denoted by the symbols $\underline{n|}$ or $n!$, but the latter form is generally preferred and the former seldom seen, except in older books. By convention, $0! = 1$. Generalization of such a product to include negative numbers or those which are not integers results in the gamma function.

See also **Binomial Series; Stirling Number.**

FACTORIAL EXPERIMENT. A factorial experiment is one whose treatments are made up of combinations of the variants of several factors. The factor variants may be qualitative—e.g., crop varieties—or quantitative—e.g., amounts of fertilizer applied—but in either case are referred to as levels. Referring to the observations for convenience as yields, any contrast between the mean yields for different levels of one factor, averaged over all levels of the other factors in the experiment, is said to belong to the main effect of the factor. The mean yields for all combinations of the levels of two different factors, averaged over all levels of the other factors, can be set out in a two-way table; a contrast between the entries in this table which is orthogonal to all the main effect contrasts is said to belong to the first order interaction between the two factors. Interactions of higher order are similarly defined.

The underlying notion behind the analysis into main effects and interactions is that of additivity of the effects of the different factors. If the effects are perfectly additive, all interactions are zero; the presence of first or higher order interactions discloses the presence of less or more complicated departures from additivity. In many fields of application, interactions involving 3 or more factors are often small enough to be neglected.

One of the most useful types of factorial experiment is that in which each factor has the same numbers of levels. An experiment with s factors each at p levels is called an s^p experiment, and each replicate will contain s^p treatments. In practice, p is often 2 or 3, but if several factors are to be investigated s^p may be large and heterogeneity of the experimental material may lead to loss of accuracy. To avoid this, each replicate may be laid down in a number of blocks in such a way that the contrasts between blocks coincide with treatment contrasts that are of little interest, such as the higher order interactions. This device is known as confounding.

If the true values of the high order interactions are zero, their estimates will provide estimates of experimental error. An experiment with many factors can then be carried out in a single replication, a suitable estimate of error still being available. Extending this notion, it is possible to carry out an experiment using only a suitably chosen fraction $1/g$ of all the possible treatment combinations. In this case it is found that the contrasts that can be estimated are sums of groups of g main effect and interaction contrasts; each member of such a group is called an alias of all the others. If it can be arranged that all the aliases of the contrasts that are of interest are high order interactions, the interpretation of the results remains reasonably unambiguous. Experiments with single or fractional replication enable a large number of factors to be investigated in a single experiment without the amount of experimental material reaching unmanageable proportions.

FACTOR OF SAFETY. The factor of safety is a number expressing the relation between the utmost endurance of a structural part, or of a complete structure, to the maximum actual demand that may be expected ever to be made upon it. But the factor of safety is not merely some ratio to allow for inaccuracies, lack of knowledge, or absence of confidence. Indeed, it has a very definite rational basis which becomes more apparent as the conditions which govern the factor of safety become known. Factor of safety has many different forms, a few of which will be given below.

It includes a combination of the allowances necessary to be made in the use of practical data, including an allowance for the lack of precision with which certain stress conditions can be ascertained.

If the engineer could definitely specify the usage and the care to which his product would be put, a large element of the so-called factor of safety would not be necessary. It must include allowance for unavoidable shocks or jars which might be expected during the working life of the structure. The material used may not be homogeneous in character, or uniform in all deliveries. Then there is always the desire to be well on the safe side when, due to failure of some part, life will be endangered. Usually, the factor of safety is taken as the ratio of the ultimate strength claimed or accepted for the material, to the working stress used for design, and presumably reached under maximum design loading. When failure is measured by excessive deformation, the factor of safety might well be based on the elastic limit. Likewise, for elements subjected to repeated reversals of stress, the endurance limit should replace ultimate strength. In most old, well-organized fields of design, professional and standardizing societies have undertaken to specify working stresses for the commonly used materials, thus indirectly covering the factor of safety.

Factors of safety do not always bear this name; for example, the safety of a masonry dam against overturning is contained in a computed ratio of the overturning moment due to water pressure, divided by the stabilizing moment of the masonry weight. Safety in aircraft design is contained in a carefully and scientifically determined "load factor" made up in accordance with certain rules promulgated by a governmental bureau.

FACULAE. Sun (The).

FACULTATIVE ANAEROBES. Bacteria.

FADEOMETER. This term designates an instrument for determining the resistance to fading of substances and materials upon exposure to the action of radiant energy (commonly artificial sunlight or ultraviolet light) under controlled conditions.

FADING (Communications). The fading of radio signals is inherent in the transmission of such signals and at best can only be partially compensated for in the receiver by automatic volume control circuits, diversity reception, etc. The compensation may often be made entirely satisfactory if the fading is of the simplest type, but if it is selective, compensation is not always satisfactory. Radio waves going out from the transmitter travel along various paths to the receiver, some of the waves travel along the ground, others are reflected from the ionosphere. In the broadcast band fading is usually caused by signals which have been reflected from the ionosphere combining vectorially with signals which have traveled along the earth (these are called respectively sky wave and ground wave). The sky wave does not return to the earth near the transmitter so there is no fading in this region, and at great distances from the transmitter the ground wave has died out so again there is no fading due to this cause. In the intermediate region both waves may be present and if the phase of the two signals is such that they cancel fading results. Since the ionosphere is continually changing, the phase of the reflected sky wave may cause cancellation at one instant and addition of the signals at the next. Different frequencies travel somewhat different paths in the ionosphere so the time to reach the receiver is different for the different sideband frequencies. Thus one frequency may reach the receiver to add to the ground wave, while another may cancel. This produces what is known as selective fading and the output of the receiver is badly distorted. It should be realized that this is an effect of the transmission and not a characteristic of a given receiver. Both types of fading may be produced by two sky waves which have traveled different paths from the transmitter to the receiver. This is the cause of fading at the very high frequencies where the ground wave does not get far enough from the transmitter to cause any trouble.

FAGACEAE. Beech Trees; Chestnut Trees; Oak Trees.

FAHRENHEIT DEGREE. Units and Standards.

FAHRENHEIT TEMPERATURE SCALE. Temperature.

FAILURE (Metal Fatigue). Fatigue (Metals).

FAILURE (Structural). The inability of a structure or a structural member to perform its proper function causes a condition known as failure. This condition may be the result of sudden fracture as in the case of brittle materials or the excessive deformation of ductile materials. Another cause of failure is a lack of equilibrium between the external loads and resisting forces such as exists in structures which fail by sliding or overturning. Structures also may fail because of poor design. See also **Factor of Safety.**

FAINTING. Syncope.

FAIRING. An object is said to be faired if it is constructed to streamline shape, or has attached to it supplementary bodies which cause it to assume a shape of some degree of excellence of streamlining. The term has its major usefulness in aircraft nomenclature where many instances of fairing of parts in the exposed wind streams are present. Fairing of exposed struts, wheels, cabins, etc., does much to reduce wind drag and increase performance, although retraction of the landing gear eliminates the need of fairing or streamlining. Sometimes the part is actually built in a streamline shape, and sometimes the streamline shape is obtained by enclosing the part in a streamlined case, or by attaching to it a shaped piece of some light material, such as balsa wood. The best faired object in the subsonic region is one whose shape approaches that of a tear drop, having a ratio of length to width of approximately 3.5. For speeds above sonic, shapes other than the tear drop are being developed for better drag characteristics at high speeds.

FALCIPARUM MALARIA. Malaria.

FALCON (*Aves, Falconiformes*). Large birds of prey closely related to the hawks and eagles and like them in appearance. They are found throughout the world. One species of the Old World is called the windhover, *Falco tinnunculus*, or, in common with other species, kestrel. Another is the merlin, *F. aesalon*. The peregrine falcon, *F. peregrinus*, has been widely used for catching game and other birds.

Several species of falcons and merlins occur in North America, among them the duck hawk, *F. anatum*; pigeon hawk, *F. columbarius*; and the little sparrow hawk, *F. sparverius*. See also **Falconiformes.**

FALCONIFORMES (*Aves*). The ability to hunt prey is most highly developed in this order. These birds were also called birds of prey, which, however, is not a specific name since many birds of other orders also hunt for living prey. They are particularly distinguished by the "weapons" they use to overcome their prey, namely the short, hooked beak, with the upper mandible strongly curved, and in particular the strong feet with long toes, and the highly developed sharp claws which are an excellent tool for grasping. By far the majority of this order are adapted to capturing prey, which may be insects, amphibia, reptiles, small birds, or small mammals. A few species prefer a vegetarian diet.

Falconiformes (raptors) are easily distinguished from the members of other orders of birds. The body is strong, compact, and wide-breasted; the large head is generally rounded, and only rarely elongated. The neck is normally short and strong, and only seldom is it long. The rump is short, the breast and limb muscles are strong, and the short, hooked beak is laterally compressed. The cutting edges of the upper mandible project like scissors over those of the lower mandible. The foot is short, strong, with long toes, and an outer toe which in some species can be rotated (to either the front or the rear). The claws are more or less bent, and when strongly bent they are pointed and form grasping tools for seizing prey.

There are 4 families: 1. The New World Vultures (*Cathartidae*); 2. the Secretary Birds (*Sagittariidae*); 3. Goshawks (*Accipitridae*); and 4. Falcons and related forms (*Falconidae*), with a total of 291 species. The distribution is worldwide.

In view of the extensive changes which the natural environment has undergone at the hands of man, we have today a quite different understanding of the role of predators in their various habitats. From their relationship with other animals and in some cases, plants, it can be seen that they are not just "robbers," as used to be assumed, but they play rather a definite role in the balance of nature. They remove animal corpses and feed to a great extent on sick or weak animals, thus ensuring that the population of their prey animals remains healthly and able to compete.

This applies particularly to the members of the two large families of the goshawks (*Accipitridae*) and the falcons (*Falconidae*). According to their abilities when seizing and killing prey, they may be divided into two functional groups: 1. the grasping killers, having a beak for tearing, hooking and cutting; and 2. the grasping holders, having a beak for tearing, hooking, and biting. The first group includes most goshawks, the second, the pigmy and true falcons. The two groups are distinguishable not only on the basis of their beak structure, but also according to the structure and use of their feet, which have become converted into grasping tools.

The feet of grasping killers, as a rule, have short toes; only bird and bat hunters are an exception. The hind toe and the inner front toe have stronger or particularly strongly developed claws. The center claws and outer toes are clearly weaker, even among those birds that prey on small animals. These features are particularly evident on the feet of eagles and the goshawk. The feet of grasping holders, on the other hand, have claws which show no such differences in length. Only the claw of the hind toe is a little larger in these birds. The beaks of the grasping killers have sharp cutting edges on the sides of the upper mandible, as well as a tearing hook. With such beaks they can cut into the tough skin even of larger animals. The beaks of the grasping holders, on the other hand, serve to split open the back of the prey's head, whereby the animal is held with the foot. In the falcons, the upper mandible has a "tooth" behind the hook which fits into an indentation of the lower mandible.

When prey has been seized and killed by the foot or the bite of the beak into the back of the head, it is plucked more or less carefully if it is a bird or mammal; from insects the coarse pieces of chitin, such as the wing covers of beetles are removed. Nevertheless, much indigestible material, such as feathers, hairs, or pieces of chitin, are consumed with the morsels of food. The food is predigested in the crop by gastric juice which is "pumped up," and is then squeezed into the stomach, which dissolves everything that is digestible. Feathers and hair are gathered into clumps and are regurgitated through the beak as "pellets," generally after 16–18 hours.

All raptors of the functional group of grasping killers, with tearing, hooking, and cutting beaks, build their own nests. Nests may be on the ground, but are usually found on trees. They may be newly built, or the birds may utilize old nests of other raptors or even of crows and ravens as a base. In contrast, the grasping holders with tearing, hooking, and biting beaks use ready-made platforms, namely ledges on cliffs or old nests in trees.

The territories occupied by individual pairs of raptors during the breeding season are divided into the hunting area and the nest area. This division, as well as the size of the hunting and nest areas, offer a good insight into the function of the two. The hunting area of hunters of very small animals is small, becoming larger with the size of the prey of the raptor species. Some, like the migrants among the raptors, hunt over several continents. Others, like the goshawks and sparrow hawk, do not hunt near the nests during the breeding season, and only actions directly related to the nest and the young are performed in it, such as the preparation and distribution of the prey. Hunting is clearly separated from care of the eggs and young. In species which often breed socially, hunting and nesting areas cannot be separated.

The role of predators in the population of other inhabitants of their environments has nothing at all to do with our concepts of usefulness or harmfulness; rather the predator has an extraordinary significance with the maintenance of the balance of nature. See also **Caracara; Condor; Eagle; Falcon; Hawk;** and **Vulture.**

FALLAWAY SECTION. A section of a space vehicle that is cast off and separates from the vehicle during flight, especially such a section that falls back to the earth.

FALLOPIAN TUBES. Gonads.

FALLOPIAN TUBES (Blocked). Infertility.

FALLOPIAN TUBES (Infected). Salpingitis.

FALLOUT (Radioactive). The term *fallout* generally has been used to refer to particulate matter that is thrown into the atmosphere by a nuclear process of short time duration. Primary examples are nuclear weapon debris and effluents from a nuclear reactor excursion. The name fallout is applied both to matter that is aloft and to matter that has been deposited on the surface of the earth. Depending on the conditions of formation, this material ranges in texture from an aerosol to granules of considerable size. The acrodynamic principles governing its deposition are the same as for any other material of comparable physical nature that is thrown into the air, such as volcanic ash or particles from chimneys. Therefore, many of the principles learned in studies of fallout from nuclear weapons can be applied to studies of other particulate pollution in the atmosphere.

The topographic distribution of fallout is divided into three categories called (1) local (or close-in); (2) tropospheric (or intermediate); and (3) stratospheric (or worldwide) fallout. No distinct boundaries exist between these categories. The distinction between local and tropospheric fallout is a function of distance from source to point of deposit, while the primary distinction between tropospheric and stratospheric fallout is the place of injection of the debris into the atmosphere, above or below the tropopause. Whether the radioactive debris from a nuclear weapon becomes tropospheric or stratospheric fallout depends on yield, height, and latitude of burst (the height of the tropopause is a function of latitude).

Because air acts as a viscous medium, a drag force is developed to oppose the gravitational force that acts on airborne particulate matter. This makes the velocity of fall dependent on particle size. The larger particles (diameters greater than about 20 micrometers) have a higher rate of settling and create local fallout. Smaller particles injected below the tropopause are carried by prevailing winds over large regions of the surface of the earth and create the tropospheric fallout. Tropospheric fallout particles larger than about 0.1 micrometer diameter continually mix through the circulating air mass that is in contact with the surface of the earth and gradually settle to the ground, or are washed down by rain or snow. Many smaller particles form nuclei for raindrops. Parts of the tropospheric fallout may remain in the atmosphere for a month or more, long enough to circle the earth several times. The mean residence time above the tropopause of stratospheric fallout is from 5 to 30 months, during which time it completely encircles the earth. It gradually returns through the tropopause, primarily in certain regions where mixing between the two layers is more probable.

The exact characteristics of the radiation associated with fallout depend on the nature of the nuclear processes from which its radioactivity originates. Generally, these radioactive nuclides are fission products formed from the fissioning of uranium or plutonium, but, under appropriate circumstances, considerable quantities of radioactivity can be formed through nuclear reactions induced by neutrons that are produced by the weapon or reactor. The radiation problems associated with local fallout are usually those of high-intensity gamma-ray radiation fields resulting from the relatively large quantities of radioactive material that fall back to earth within a few tens of miles from the point of origin. The important radioactive materials consist in this case of short-lived fission products and neutron-induced radioactive nuclides. The hazards of worldwide fallout come more from the problems of the long-lived radionuclides, such as ^{134}Cs, ^{137}Cs, and ^{90}Sr, that can enter the human food chain and ultimately be absorbed by the body.

For a nuclear weapon burst in air, all materials in the fireball are vaporized. Condensation of fission products and other bomb materials is then governed by the saturation vapor pressures of the most abundant constituents. Primary debris can combine with naturally-occurring aerosols, and almost all of the fallout becomes tropospheric or stratospheric. If the weapon detonation takes place within a few hundred feet of (either above or below) a land or water surface, large quantities of surface materials are drawn up or thrown into the air above the place of detonation. Condensation of radioactive nuclides in this material then leads to considerable quantities of local fallout, but some of the radioactivity still goes into tropospheric and stratospheric fallout. If the burst occurs sufficiently far underground, the surface is not broken and no fallout results.

C. Sharp Cook, Professor of Physics, The University of Texas at El Paso, El Paso, Texas.

FALLOW DEER. Deer.

FALL WIND. Winds and Air Movement.

FALSE ACACIA. Acacia Trees.

FALSE CIRRUS. Clouds and Cloud Formation.

FALSE CLEAVAGE. This is also called strain-slip cleavage by the British geologists. It differs from the typical slaty cleavage in that it is obviously associated with incipient foliation of metamorphic rocks.

FALSE COLOR IMAGES. Photography and Imagery.

FALSE CORAL SNAKE. Snakes.

FALSE CYPRESS. Cypress Trees.

FALTUNG. Convolution.

FALX CEREBRI. Brain and Nervous System.

FAMILY (Organisms). Taxonomy.

FAMOUS (Project). Ocean.

FAN. When large volumes of gas, usually air, must be moved with practically no compression (pressure rise of 1 psi or less), the device used is termed a fan (as contrasted with compressor). Centrifugal fans fulfill a variety of needs, such as ventilating, heating, combustion draft, and drying applications.

In contrast to the high-pressure centrifugal compressor, the fan has narrow blades and very little compression occurring in the blading. However, the blade action on the gas is to increase its speed, thus requiring a diffusion to gain pressure. This diffusion is accomplished in a scroll case surrounding the wheel and comprising the casing of the fan. Diffuser guide vanes may sometimes be inserted in the scroll case to improve efficiency by reducing turbulence. Simple radial balding is sometimes used on account of its cheapness and simplicity, but most fans have blading that is curved. Figure 1 shows the appearance of wheels incorporating in the one case forwardly, and in the other, backwardly curved blading. Vector diagrams show that for

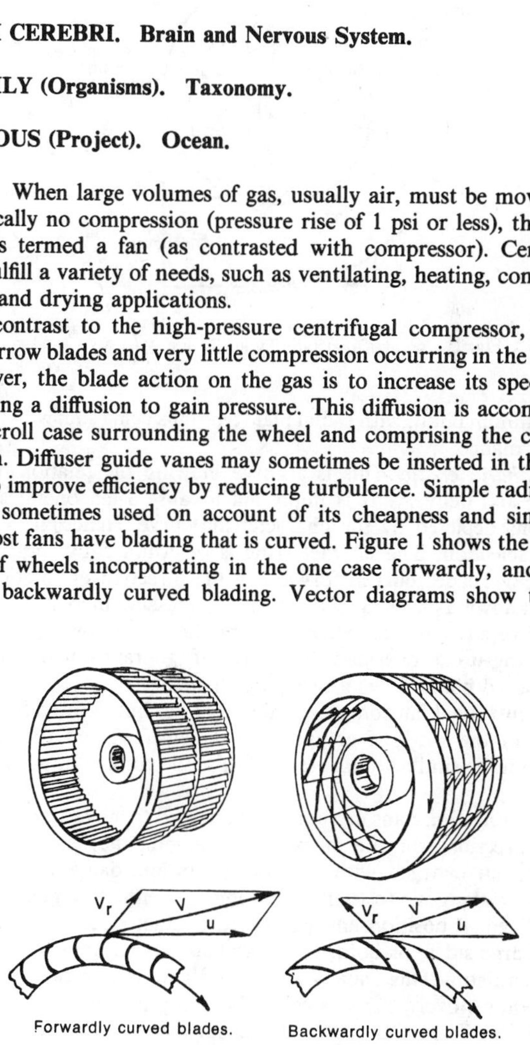

Fig. 1. Fan wheels, showing blading.

the same relative velocity of the gas leaving the wheel and the same wheel speed, the absolute velocity is greater with forwardly curved blades. Achievement of efficient diffusion is therefore more important with forwardly curved blades and pressure increase is greater. Conversely for the same pressure increase, forwardly curved blading may operate at lower rim speeds. However, backwardly curved blading can be built so that somewhere near the best operating point (maximum efficiency) pressure rise diminishes more rapidly than volume increase and thereby induces a self-limiting feature in power consumption that is desirable in many applications. Power demand continues to increase with discharge in forwardly curved blading until well past the best operating point.

Since the pressure increments are small, an excellent approximation which simplifies the energy relation is to assume a constant volume flow through the fan. Given the weight of gas flowing per minute, W, and the draft, D (expressed in feet of air), the power imparted to the air is:

$$\text{air horsepower} = \frac{WD}{33{,}000}$$

Given the volume delivered per minute, V (in cubic feet), and dynamic pressure, P (pounds per square foot), power is also:

$$\text{air horspower} = \frac{PV}{33{,}000}$$

The mechanical efficiency of a fan is the ratio of one of the above theoretical powers to the required drive power. Multivane centrifugal fans (Fig. 2) will usually exhibit an efficiency of from 70 to 80% at

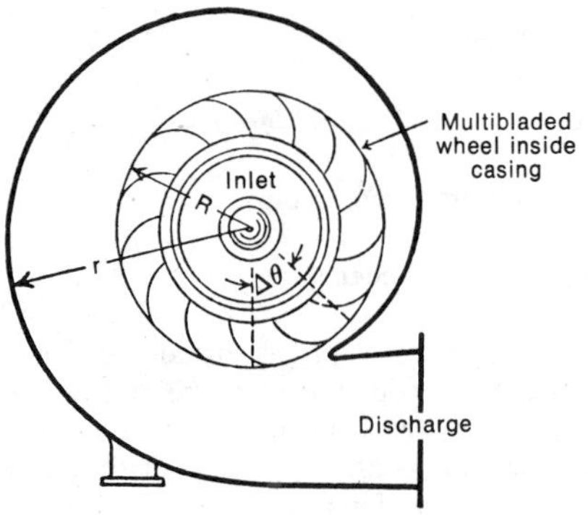

Fig. 2. Centrifugal fan. Typically $r = R\left(1 + \left(\frac{\theta°}{360}\right)\right)$.

their optimum point, with radial plate type fans being somewhat poorer in performance.

Fan Characteristic. This is a curve showing the relation between pressure and delivery. It is important because a fan operates at the conditions depicted by the characteristic curve. Hence, it is a basis for fan selection. The characteristic is determined by the shape of the blades. Blades curved forward in the direction of rotation have what is known as a rising characteristic—pressure increases with volume delivered. This characteristic is productive of low tip speed but fans having it can overload their drives if ignorantly handled. Backward curved blades have a drooping characteristic.

The centrifugal fan compresses the air or gas but slightly. In modern fan theory the work of compression is neglected and the action is assumed to be similar to a reversed hydraulic turbine or a centrifugal pump.

Ventilating Fan. Fans in this service are used to move air in predetermined directions for the purpose of changing the air in buildings, removing air charged with offensive odors and dangerous contaminants, furnishing air to tunnels and mines. Ventilating fans are commonly fixed in position and connected to duct systems on the inlet or discharge sides, or both. Propeller fans have a limited use in this field. Ventilating fans usually are centrifugal because the application often requires overcoming considerable static pressure.

Draft Fan. These fans are used to supplement or supersede chimney action in the production of draft—hence find applications in large furnace, boiler, and other combustion systems, and frequently in cooling tower applications. In application, fans may be used as forced draft or induced draft fans. In a forced-draft application, the fan pushes the air past a finned air cooler, for example; the induced-draft fan pulls the air past the cooler. Draft fans are designated as plate (paddle wheel), multivane, or propeller type, the latter seldom used for this type of service. The plate fan is employed to some extent, but the multivane centrifugal fan is the most common type. Backwardly curved blade wheels are usually selected for forced draft service, because of the high speed, suitable for direct motor drive, the self-limiting power demand (a necessary feature when two or more fans are operated in parallel), and high static efficiency. Induced draft fans handle hot chimney gas. Forwardly curved blades which develop a given draft at lower speeds than those with backward curvature are frequently chosen for this service, since the speeds and the centrifugal stresses in the wheels will be least.

Propeller Fan. These fans operate with air flow parallel to the axis of rotation of the fan. Generally, these fans may be grouped into (1) fans with thin sheet blades of metal or composition material or even flexible material and (2) fans with blades of airfoil section. Examples of the first class are the table fans, small ventilating fans, ceiling fans, unit heater fans, and radiator cooling fans. With the exception of certain lowspeed ceiling fans, these are characterized by high rotative speed and low efficiency. The blades are often stamped in a single piece from a sheet of metal, then twisted slightly and mounted on the motor shaft. These fans are employed to move air but are not satisfactory if the air is to be forced against any pressure increment. The second class of propeller fans reflects a more scientific application of the axial flow principle. Whereas the first class moves the air largely by impulse against the face of the fan blades, the airfoil sections of the latter class perform in accordance with airfoil theory of lift. While more expensive, they are also more efficient, although the latter advantage is gained in sizes more suitable to industrial than domestic usage. The axial flow fan can move large volumes against light static pressures and is frequently more compact and more readily applied than the centrifugal type.

FANGLOMERATE. The term proposed by Lawson in 1913 for a conglomerate composed of the coarser clastic sediments deposited at the head of alluvial fans.

FANNING BEAM. A radiant energy beam, as a radar beam, which sweeps back and forth over a limited arc.

FARAD. **Units and Standards.**

FARADAY CELLS. **Polarimetry.**

FARADAY DARK SPACE. This term denotes the nonluminous region between the negative glow and the positive column in a gas-discharge tube (Crookes tube) at moderate pressure.

FARADAY DISK MACHINE. This device consists essentially of a copper disk, rotated so that it "cuts" the flux between the poles of an electromagnet, serves as a low voltage dc generator by virtue of the induced, radial electromotive force. One output brush contacts the axle; the other, the rim beyond the magnet pole-gap. It is a form of the homopolar generator.

FARADAY EFFECT. **Magneto-Optical Rotation.**

FARADAY FERRITE CIRCULATOR. **Circulator (Microwave).**

FARADAY LAW OF ELECTROMAGENTIC INDUCTION. The electromotive force induced in a circuit is

$$\mathcal{E} = \oint_l E \cos\theta \, dl = -\frac{d\phi_m}{dt}$$

where $\oint_l E \cos\theta \, dl$ is the line integral of the electric field intensity around a closed path and $d\phi_m/dt$ is the rate of change with time of the magnetic flux through the area enclosed by the path. The electromotive force is given in volts and the magnetic flux, in webers.

FARADAY SHIELD (Screen). An electrostatic shield made of a series of parallel wires connected at one end. The common point is grounded. Electromagnetic waves are not influenced by the screen.

FARADAY'S LAW (Magnetic Flowmeter). Flow Measurement.

FARM CROP SURVEILLANCE. Earth Resources Satellites and Geologic Remote Sensors.

FAR POINT OF THE EYE. This term designates the nearest point on which the eye is focused when fully relaxed.

FARSIGHTEDNESS. Vision and the Eye.

FASCIA. 1. Layers of connective tissue composed largely of regularly arranged fibers. They cover muscles. 2. Marks in the form of bands.

FAST NEUTRON. Neutron.

FAST REACTOR. Nuclear Reactor.

FAT BODY. 1. A large mass of fatty tissue found in insects. It serves for the storage of food during larval life, since it is much smaller in the adult, and is apparently a reservoir for nitrogenous wastes since it contains deposits of uric acid. 2. A mass of fatty tissue located near the gonads in Amphibia.

FATIGUE (Corrosion). Corrosion Fatigue.

FATIGUE (Metals). Failure of metal parts by progressive cracking caused by repeated application of stress. Most fatigue failures start at the surface where discontinuities in section such as square shoulders, screw threads, or even tool marks cause a high concentration of stress.

Internal discontinuities may also start a fatigue crack, the most notable example being "transverse fissures" in rails which are believed to originate in areas within the rail section known as "flakes," a defect originating during the cooling period after hot rolling. Once a minute crack is started anywhere in the section, the root of the crack becomes the seat of high stress concentration upon subsequent applications of tensile stress, thus the crack will spread until the section is too weak to carry the load and the remaining portion will fracture suddenly.

The portion of the section which failed progressively will be worn quite smooth due to the rubbing action of successive stress applications (e.g., alternate tension and compression in a rotating member loaded as a beam), while the suddenly fractured portion will have the usual crystalline appearance which is characteristic of fractures in heat-treated steels. For this or other reasons fatigue failures have wrongly been blamed on "crystallization" of the metal.

All metals are crystalline and no alteration in the size or shape of the grains or crystals takes place in service during or before fatigue failure. (Exceptions might be made in the case of lead and other alloys which recrystallize when cold worked at room temperature.)

The fatigue strength, also called endurance limit, is the maximum stress which can be applied repeatedly without failure. In the case of steel, tests are run at a given maximum stress to 10,000,000 reversals or cycles of stress unless failure occurs earlier. It has been found that failures do not occur in steels after a successful run of this duration (4 days at 1700 rpm or less than 1 day at 10,000 rpm). In the case of aluminum alloys and certain other non-ferrous metals, fatigue, failures have occurred after much longer runs, hence tests are sometimes made to 500,000,000 cycles. The materials do not have a true endurance limit and the number of reversals of stress are stated in reporting the fatigue strength.

The most common test is a rotating beam type in which a carefully machined and polished sample is loaded as a beam while rotating in anti-friction bearings. As any point in the periphery rotates from top to bottom to top position the stress changes from maximum compression to maximum tension and back to maximum compression. From 4 to 8 or more individual tests at various maximum stress levels may be required to determine the endurance limit.

The endurance limit for smooth test specimens run in normal atmospheres at room temperature is an ideal or limiting value. In the case of steels not hardened, or heat treated to moderate hardnesses, the smooth specimen endurance limit is approximately one-half of the tensile strength. The endurance limit-tensile strength ratio is less than one-half for many other materials.

The presence of stress raisers, particularly at the surface, will lower the endurance limit. Notch sensitivity can be evaluated as the ratio of the endurance limit of a standardized notched specimen to that of a smooth specimen. Fatigue tests run in a corrosive medium, either gaseous or liquid, generally give much lower endurance limits than tests run in normal atmosphere. Corrosion-fatigue is responsible for many service failures of shafts and other stressed parts of pumps, engines, or processing equipment operating in corrosive media. Protective coatings are sometimes used to improve service life. Nitriding of alloy steel parts has proved very effective.

In order to guard against fatigue failures in critical parts such as connecting rods, they should be fabricated from high-quality steels using designs that avoid regions of stress concentration such as sharp fillets and engraved part numbers. They should be finished over all, avoiding tool or grinding marks. As a further aid in obtaining high fatigue strength such parts are being surface peened by a shot blasting process which work-hardens the surface and sets up compressive stresses in the surface layers. This raises the endurance limit, apparently by reducing the maximum tensile stresses which can be developed at the surface in normal operation.

FAT (Metabolism). Lipidoses.

FATS AND OILS (Diet). Dietary Requirements and Trends.

FATTY ACID ETHANOLAMIDE. Detergents.

FATTY ACIDS. Carboxylic Acids; Chlorinated Organics; Vegetable Oils (Edible).

FAULT. Earthquakes, Seismology, and Plate Tectonics.

FAULT ELECTRODE CURRENT (or Surge Electrode Current). The peak current that flows through an electrode under fault conditions, such as arc-backs and load short-circuits.

FAULT-LOCATING BRIDGE. Bridge Circuits (Electrical).

FAUNA. The animal population of a region.

FAYALITE. Olivine.

FEATHER. Birds.

FEATHERBACKS (*Osteichthyes*). Of the order *Isospondyli* and family *Notopteridae*, the featherback is of peculiar form and found in Africa and the Oriental region. They are characterized by a very long anal fin, extending nearly the full length of the fish on its undersurface to the tip of the tail, thus masking the presence of a tail fin. The dorsal fin is slender, delicate, featherlike. Hence the name of this fish. The rippling action of the long anal fin provides the propulsion and the feathery dorsal fin serves as the rudder. Featherbacks are a freshwater fish and comprise four species. They are highly regarded as food fishes. *Notopterus chitala*, the largest of the species, attains a length of about 3 feet (0.9 meter) and is found in the waters of India and Australia. Spawning of featherbacks in eastern nations has been encouraged because of their demand as a food fish. A somewhat smaller species (*N. notopterus*) occurs in the same geographic regions. The species *Notopterus afer* and the so-called false featherfin (*Xenomystis nigri*) both are found in West Africa. The latter is a small fish of about 6 inches (15 centimeters) in length, and sometimes is selected for aquariums.

FEATHERING (Aircraft). An object of flat plate shape in a fluid stream has maximum resistance to relative motion if its largest area is placed in an attitude perpendicular to the fluid stream, and minimum when the smallest area is so placed. Conditions occasionally arise when it is desirable to have a maximum resistance at one point of a cycle of events, and minimum at another. For example, during the cycle of a rowing stroke, the blade of the oar should have maximum resistance to motion in the water, whereas during the return stroke it should have minimum air resistance. Certain experimental types of lifting planes or airfoils have been built involving this same action. Some helicopters use the feathering action, often in conjunction with flapping of the blades, to equalize the lift on advancing and retreating blades. The act of first presenting the surface of maximum resistance, followed by the surface of minimum resistance on a return stroke, is known as feathering. Controllable pitch propellers have been equipped with hub mechanisms and controls to permit the pilot to feather the blades when the engine was either throttled or inoperative. This action assists control and performance of multi-engined aircraft having one or more engines inoperative; it also allows a single-engined airplane to reach higher terminal diving speeds. Feathering the propeller will keep a damaged engine from "windmilling" and increasing the damage; moreover, the resistance of a feathered propeller is far less than that of a windmilling propeller. Feathering is also applied to the change in the angle of setting of the blades of a helicopter either to secure control or to equalize lift on the advancing and retreating sides. See also **Airplane; Helicopters and VTOL Craft.**

FEBRILE. Characterized by fever.

FECHNER COLORS. Visual sensations of color induced by intermittent achromatic stimuli (white light).

FECHNER FRACTION. If the eye can just distinguish an object whose brightness differs by an amount dB from a large field of brightness B, the contrast sensitivity may be measured by dB/B, sometimes called the Fechner fraction.

FEEDBACK. Systems in general have a direction of signal flow which can be called the principal transmission path. The direction from input to output of any amplifying component is an example of this idea. If in addition to the primary transmission path, there exists one or more paths which allow signals to flow in the opposite direction and which are joined to the primary path at each end, then a feedback system exists. Feedback is the transfer of energy from the output to the input of a system, or from one part of a system to another in a direction opposite to the main flow of energy. This concept is illustrated in Fig. 1, which gives the block diagram of a feedback amplifier with

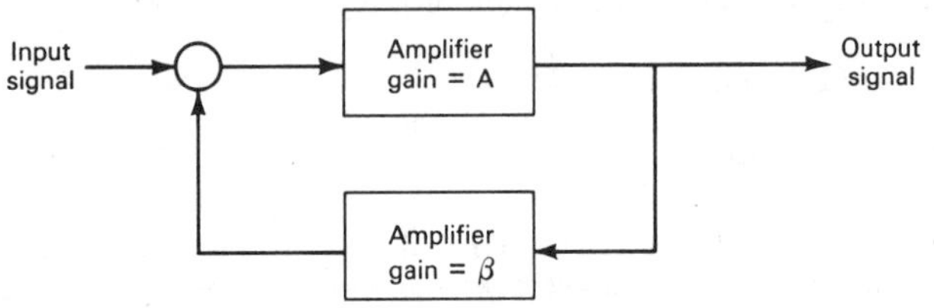

Fig. 1. Single-loop feedback system.

a main transmission path having a gain A and a transmission path from output back to the input which has a gain β.

Figure 1 illustrates the essential elements of a feedback system. They may be considered to be an input, an output, a transmission path which develops some measure of the output signal, a device which permits the comparison of the input signal and measure of the output signal in order to develop a so-called error signal, and a device or devices in which the error signal is converted to the desired output signal. The feedback system, when viewed in these terms, attempts to make the measure of the output equal to the input signal. Servomechanisms and regulators are typical examples of feedback systems.

It is evident from Fig. 1 that a complete transmission path or feedback loop is formed by the combination of the forward and backward paths essential for feedback to exist. A system having only one feedback loop is called a single-loop system, whereas systems having more than one loop are called multiple-loop systems. Figure 2 shows a block diagram of one such system.

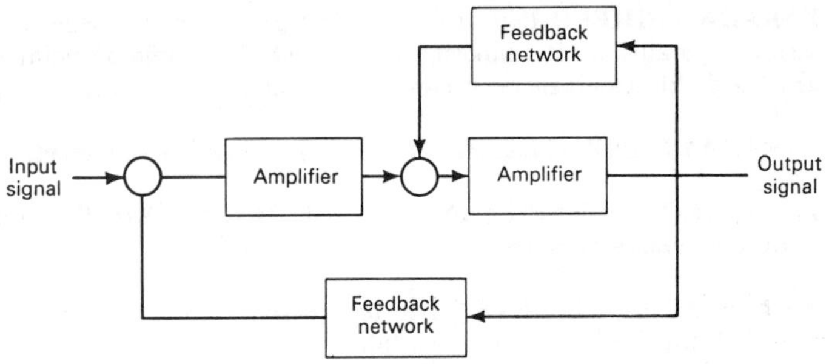

Fig. 2. Multiple-loop feedback system.

In practice, there are many systems that may be classified as feedback systems due to their dependence of one quantity upon another occurring at some later point in the causatory chain. In general, however, the more important class of feedback systems are those in which the feedback loop contains an active element as shown in Figs. 1 and 2. It is to such systems that feedback theory is mainly applied. It is also only such systems which are capable of sustained oscillations or instability, a property which distinguishes them from purely passive systems or systems including active elements which are not included in feedback loops.

This last statement is, however, only applicable to what might be called engineering systems. Feedback theory has been applied with some measure of success to other fields, notably economics and ecology. It is known that economic systems are liable to sustained oscillations (the so-called booms and slumps) as also are, for instance, the population densities of certain interacting animal species (e.g., lynxes and foxes). The active elements in these systems, however, are often difficult to assess.

The main reasons for introducing feedback into a system are as follows: (1) to modify the gain-frequency characteristic of a system; this includes, for instance, modifying the degree of stability of the system, modifying its cutoff frequency, making the gain more uniform over the working band of frequencies, etc.; (2) to minimize the effects of nonlinearities of a system; (3) to minimize the effects of the variation of system parameters; (4) to minimize the output noise level of a system; (5) to modify, particularly in feedback amplifiers, the input and output impedances of the amplifier; (6) to limit the amplitude of one or more system quantities. e.g., torque in an electromechanical control system.

Associated with any feedback loop is the gain of the transmission system which constitutes the loop. This gain is called the *loop gain.* The negative value of the gain is spoken of as the *return ratio.* In the system considered in the first figure, the loop gain is $A\beta$ and the return ratio is $-A\beta$. A quantity of fundamental importance in feedback systems is one called the *return difference,* which is defined as unity plus the return ratio. For the system considered earlier, the return difference is $1 - A\beta$. For single-loop systems, of which Fig. 1 is representative, the return difference $1 - A\beta$ profoundly affects the system characteristics. The gain of the complete system relative to the gain of the amplifier (gain A) will be increased if $|1 - A\beta| < 1$. The feedback is then spoken of as positive or regenerative. On the other hand, the system gain will be less than A if $|1 - A\beta| < 1$. This condition is referred to as one of inverse feedback (negative or degenerative).

When negative feedback is employed, the system with feedback is found to possess improved characteristics relative to the nonfeedback version in direct proportion to the magnitude of the return difference. Stability of gain change as a result of system parameters, reduction of harmonic distortion and extraneous signals introduced in output stages, and improved frequency response (lower distortion) are among the benefits resulting from the use of negative feedback. Since the overall system gain is inversely proportional to the value of return difference attained, it is seen that the benefits are obtained at the price of a reduction in gain. In many practical applications, the potential benefits far outweigh their cost.

There are situations where the use of positive feedback or regeneration is advantageous. This is particularly true in amplifier applications where the gain of an amplifier can be increased so that the output signal obtained with a prescribed constant input signal is far in excess of that obtainable from the same amplifier without the feedback. The Q of the tuned circuit used in an amplifier stage may be increased materially by the use of regenerative feedback. If the amount of positive feedback oscillator, a device which furnishes an output signal without need for an input.

Inasmuch as feedback amplifiers and feedback oscillators have the same structural form, it is important to have a basis for determining the behavior in this condition of a specific structure. Whether a given feedback amplifier acts as an oscillator or as a well-behaved amplifier is a function of the transient response of the device. To act as an oscillator it must have transient response terms, which increase with time when power is first applied. For an amplifier to be classed as stable, on the other hand, the initial transients must decay with time. One simple test for stability against oscillation is known as the Nyquist criterion. The test may be used not only for amplifiers, but for any single- or multiple-loop feedback system. In its simplest form, which applies to single-loop systems only, it states: Plot the imaginary component of the complex number representing the loop gain $A\beta$ frequency (0 to ∞). If the resulting curve encloses or passes through the point -1, the system will oscillate. If the point is not closed, the system will be stable. This criterion is easy to employ, using laboratory measurements, and it finds considerable practical application.

FEEDBACK AMPLIFIER. Amplifier.

FEEDBACK CONTROL. A basic form of automatic control action in which a measured variable is compared with its desired value to produce an actuating error signal which is acted upon in such a way as to reduce the magnitude of the error. See accompanying diagram of a simple feedback control system. See also **Feedback.**

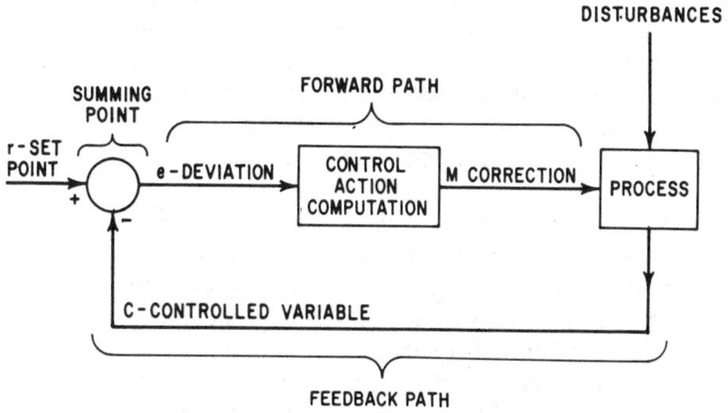

Fundamental elements of feedback control. r = set point; c = controlled variable; e = error or deviation; M = manipulated variable.

A feedback loop or *closed loop* is a signal path which includes a forward path, a feedback path, and a summing point, and which forms a closed circuit. Because of the recirculation or feedback, the control action computation, which determines the corrective action required to maintain the measured variable at the setpoint, is critical to avoid cycling and instability. Where the feedback is positive, i.e., in phase with the deviation, the error is magnified and the process cannot be controlled. Only when negative feedback is present can the deviation be reduced. A reversal of phase must occur at the summing point in order that the effect of e (the control input) is opposite to that of c (the controlled variable). All feedback controllers produce a phase shift of 180° to give a negative feedback.

Feedback control is fundamental to closed-loop controllers. There are several forms of control action which embrace this principle. See **Control Action.**

By contrast, an open loop or *open-loop control* may be defined as a single path without feedback. For example, a process or machine that is preprogrammed to function on a time basis and does not take into consideration continuous measurements of the end results as a criterion for adjusting the control system is open loop. In such a system, no information is fed back to alter the action of the controller.

FEEDBACK CONTROL (Automotive). Automotive Electronics.

FEEDBACK (Governor). Governor.

FEEDBACK (Hormonal System). Endocrine System; Hormones.

FEEDBACK SIGNAL. Signal (Instrument).

FEEDER (Gravimetric). Frequently in the form of belt scales, the function of a gravimetric feeder is to continuously weigh material as it leaves a hopper, chute, or bin while in transport to a process. The gravimetric feeder delivers a desired amount of material within a given time period and, in this sense, are solid mass flowmeters, the setpoint (desired rate) of which can be adjusted by the operator. Some situations require adding a small quantity of additive material to an uncontrolled flow of bulk material. A belt scale may be used to measure the "wild flow" stream whose rate signal feeds the ratio input circuit of a set-point controller which incorporates a ratio adjustment that establishes the ratio of additive to wild flow. An increase or decrease in the wild flow rate produces a corresponding change in the additive flow rate of the feeder, thereby maintaining the correct proportion. More than one additive may be involved in such a system. The wild flow may be of a solid or a liquid; the additive materials may be solids or liquids.

The use of gravimetric feeders in a multiunit proportioning system involving the simultaneous blending of three bulk materials is shown in Fig. 1. The system incorporates a master control that permits in-

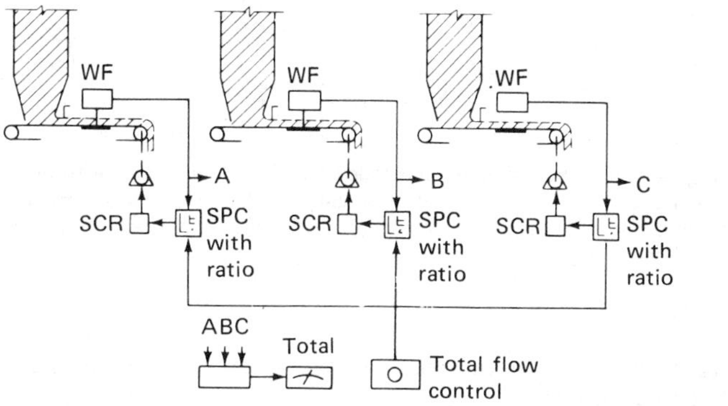

Fig. 1. Multiunit proportioning system involving the simultaneous blending of three bulk materials. WF = weigh feeder; SPC = set point controller; SCR = silicon-controlled rectifier drive.

creasing or decreasing the total flow rate of the system. This may be accomplished in several ways: (1) A manually adjusted power supply that provides a master reference signal to feed the set-point controllers for each feeder; (2) a demand signal generated by a primary control loop designed to maintain a process variable that is affected by the total feed from the system; or (3) a total flow control loop that maintains a uniform total flow by summing the individual feed rates. The last mentioned provides feedback to a total flow set-point

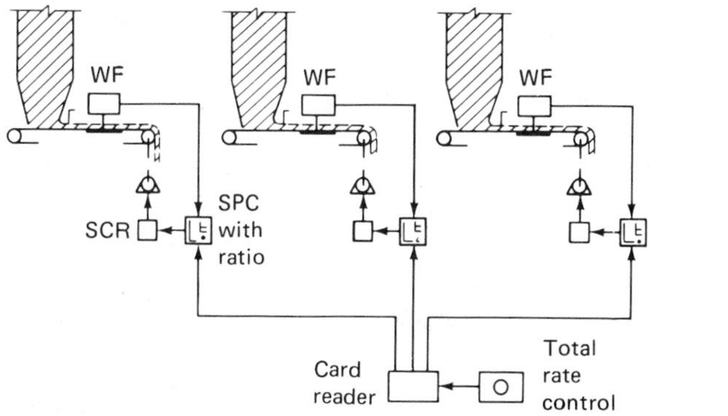

Fig. 2. Multiunit proportioning system with card reader. WF = weigh feeder; SPC = set-point controller; SCR = silicon-controlled rectifier drive.

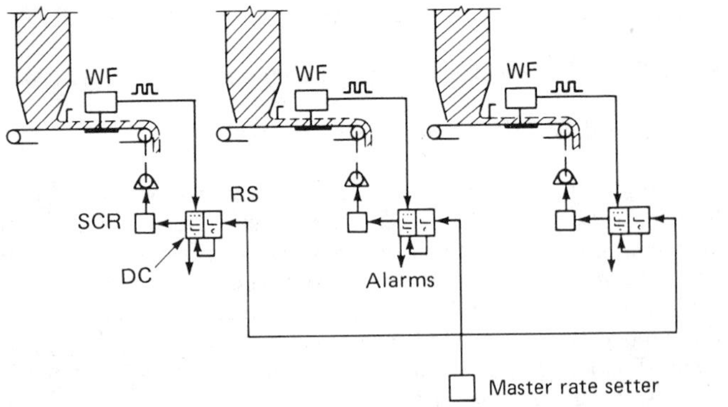

Fig. 3. Multiunit digitally controlled bulk blending system with memory. WF = weigh feeder; RS = ratio station; SCR = silicon controlled rectifier drive; DC = digital control.

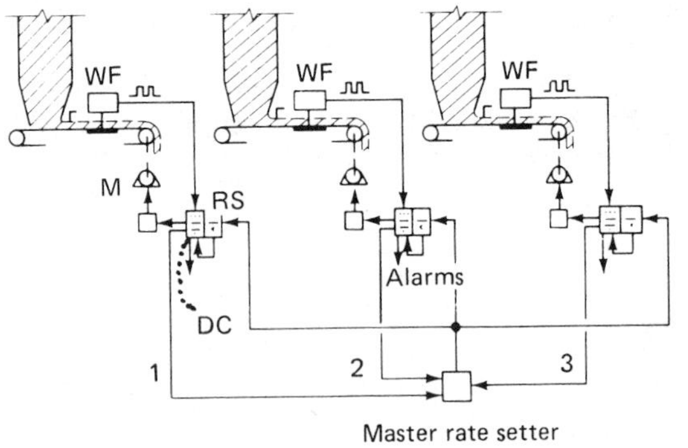

Fig. 4. Multiunit digitally controlled bulk blending system with automatic pacing feature to accommodate a lagging feeder. WF = weigh feeder; DC = digital controller; SCR = silicon controlled rectifier drive; RS = ratio station; 1,2,3 = spacing signals.

controller which, in turn, provides the demand signal to the individual feeder controllers.

In the system shown in Fig. 2, card readers are used. This situation involves a large number of formulations for custom-blending applications. Prepunched cards establish the percent of each ingredient in the blend. The outputs from the card reader are cascaded into each of the set-point controllers as the set-point signal for the individual feeder. This system eliminates possible operator error in adjusting the controller set points.

A multiunit system with digital control memory is shown in Fig. 3. Digital blending systems provide the most accurate means for automatic, continuous control flow of bulk materials and are dependent entirely on the accuracy of the digital pulses transmitted from the individual weigh feeders—with a control accuracy of plus or minus one pulse. In the system of Fig. 3, the output signal from the weigh feeder is a series of pulses whose value is a specific increment of weight (such as 2 pounds, 3 kilograms, etc.). The pulse frequency (pulses per minute) is directly proportional to the flow rate. Thus, 200 pulses per minute, with a pulse value of 10·weight units may equal 2000 pounds or kilograms/minute. These weighted pulses are compared with a set-point frequency established by a ratio station which receives digital pulses from a master rate setter, the latter establishing the total flow rate of the system.

Automatic pacing provides the solution to systems where one or more materials may lag the remainder of the system at start-up because the material characteristics may produce a resistance to flow from the storage bins. The system of Fig. 4 is identical with that of Fig. 3, except that the controllers are equipped with a special pacing circuit that detects an underfeed condition and provides a control signal feedback to the master rate setter to reduce the master reference signal, thus slowing down the rest of the feeds and permitting the lagging feeder to "keep pace" and thus prevent blends that are off specification.

Belt scales are applied and installed on practically any size, length, and shape of conveyor—from 1-foot (0.3-meter) wide belts handling as low as 1 pound (0.45 kilogram) per minute at a minimum belt speed of 1 foot (0.3 meter) per minute to a 10-foot (3-meter) wide belt handling as high as 20,000 short tons (18,000 metric tons) per hour at speeds up to 1000 feet (300 meters) per minute.

The basic design and applications of a conventional belt conveyor scale are shown in Fig. 5. A weigh platform (or weigh bridge), on which are mounted standard conveyor idlers, is used to sense the weight of the material passing over it. A belt-speed pickup system measures the speed of the conveyor belt. Weight and speed signals are transmitted to a multiplier whose product or true rate output signal is integrated and displayed on a continuous totalizer. The rate, of course, is equal to weight times speed. The basic components used to sense the weight and speed may be mechanical, electrical, pneumatic, or hydraulic.

Of a different design principle is the loss-in-weight feeder shown in Fig. 6. The decreasing weight of the material in the weigh hopper is continuously sensed and compared with a diminishing programmed setpoint which is preestablished to satisfy the flow rate desired to process. The deviation between these two signals is measured and converted for corrective action to the feeder whose speed is adjusted accordingly to maintain the programmed set point—hence the desired rate of flow. When the material in the weigh hopper reaches a selected low-level point, the hopper is quickly refilled to a selected higher level, thereby permitting uninterrupted flow rate to process.

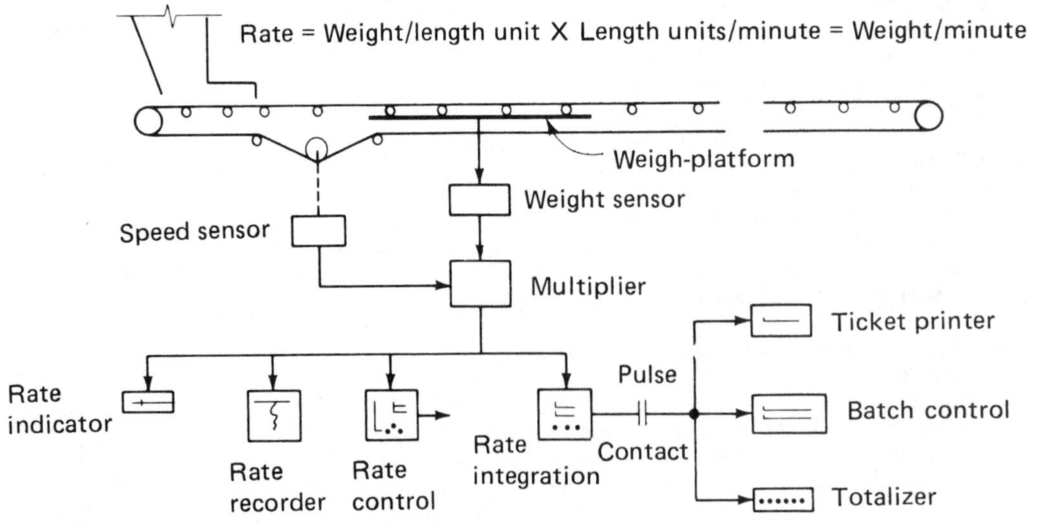

Fig. 5. Basic belt conveyor scale showing variety of display and control instrumentation options.

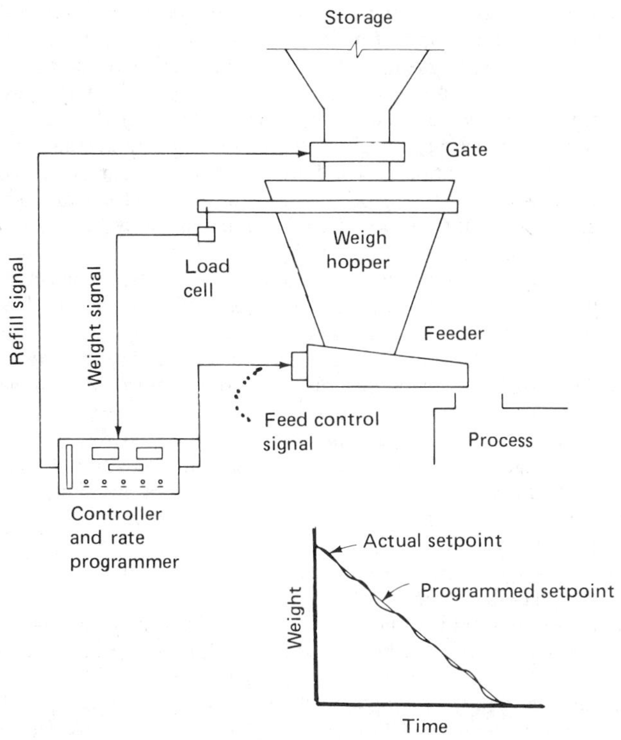

Fig. 6. Loss-in-weight type gravimetric feeder.

FEEDER (Parts). A mechanical device used to orient parts of a wide variety of shapes, sizes, and materials (metal, wood, glass, plastic, rubber, and ceramic) for feeding to processing, manufacturing, assembly, and packaging operations. The parts may range from 0.02 inch to 12 inches (0.5 millimeter to 30.5 centimeters) in length. The device can be driven electromagnetically or pneumatically. An electromagnetic unit is shown in the accompanying illustration. Devices of this type are very important in automation systems.

The bowl, which may vary from 3 to 48 inches (7.5 to 1.22 centimeters) in diameter, is constructed with a spirally-inclined track around its inside perimeter on which the parts are moved by electromagnetic vibration. While in motion, the parts encounter various selective devices which assure their proper orientation for discharge. The device creates a vibratory feeding at a steady 3600 vibrations per minute from half-wave rectified 60 Hz ac. At high frequency, the incremental part movement of approximately 0.03 inch (0.8 millimeter) cannot be detected. Instead, it appears that the parts are flowing like a fluid. Independent rheostat-adjustable vertical and horizontal amplitude automatically match the feed angle with feed rate for optimum parts handling.

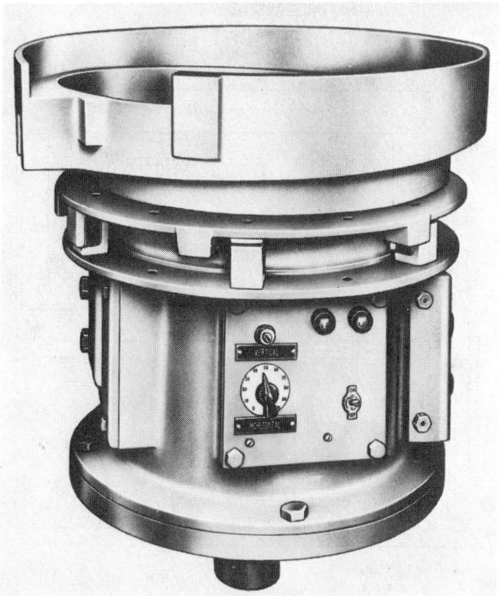

Electromagnetic vibratory parts feeder. (*Syntron.*)

FEEDER (Vibratory). A mechanical device used to provide controlled bulk material flow from storage to processes, or to and from various processes. Generally, these feeders can be driven hydraulically, electromechanically, pneumatically, or electromagnetically.

The electromagnetic vibratory feeder is basically a two-mass spring-connected system. One of the masses is the trough in which the bulk material flows; the other mass is the base or reaction mass. The two masses are connected through a set of leaf springs which allow the two masses to vibrate relative to each other. The electromagnetic feeder creates vibratory feeding at a steady 3,600 vibrations per minute from a half-wave rectified 60-cycle alternating current. The feeders have no rotating parts, bearings, eccentrics, or sliding joints which eliminate the need for lubrication and minimize wear.

In order for the two-mass driven system to take advantage of the natural magnification factor, the system must have its natural vibration frequency close to the operating frequency. This is referred to as subresonant tuning, that is, operating below the resonant frequency. A subresonant feeder can operate favorably regardless of high headloads on the feeder that may be caused by large hopper openings, wide heavy troughs, or the need to convey material a considerable distance from the hopper to the end of the trough. In addition to headload, the damping effect of the bulk material being handled must be considered in feeder design and selection. The damping effect of the material is a direct measure of the energy that is absorbed by the material moving from the hopper along the vibrating trough. Headloads, when considered alone, tend to cause a feeder to increase in amplitude under the effects of this load. The damping effect tends to decrease the amplitude. In practice, as proved by tests, the two effects tend to offset each other in a subresonant-tuned feeder. The general result—a feeder system that operates at reasonably constant amplitude regardless of material effects. Operating details are shown in Fig. 1.

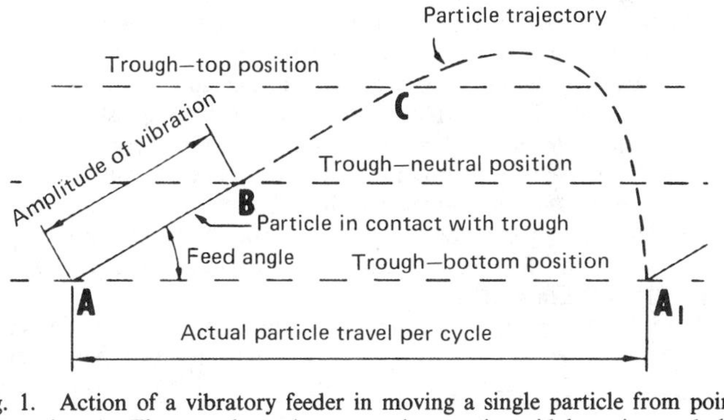

Fig. 1. Action of a vibratory feeder in moving a single particle from point A to point A_1. The trough motion approximates sinusoidal motion and the particle is in contact with the trough surface from the lowest point to approximately the midpoint of the stroke (about one-fourth of a cycle from point A to point B). At this point, the particle has been accelerated to its maximum velocity and leaves the trough surface on a free flight trajectory (trough is decelerating from point B to point C) rejoining the trough surface at point A_1, the lowest point of the stroke—which completes one cycle.

A heavy-duty feeder is shown in Fig. 2. Feeders are available with capacities ranging from 1,250 pounds to 2,000 tons (567 kilograms to 1800 metric tons) per hour. Feeders having dual, twin, or dual twin electromagnetic drives are available with maximum capacities up to 3,500 tons (3150 metric tons) per hour.

Vibratory feeders are available with a number of controls: (1) *Basic manual control*—providing manual adjustment of feed-rate control; (2) *constant feed rate*—where the flow of material through the feeder is held constant regardless of load change on the feeder or variations in line voltage; (3) *two-rate automatic scale control*—particularly useful for bagging and batch weighing applications, external scale controls automatically provide primary fast rate of material flow, secondary

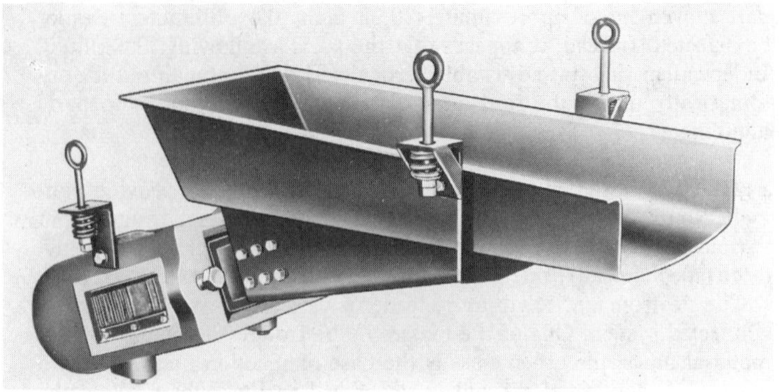

Fig. 2. Heavy-duty electromagnetic vibratory feeder.

dribble flow, and shut-off at a pre-selected weight; (4) *heated liner control*—a temperature-controlled heated liner plate in the trough prevents sticking of certain substances and build-up of high-moisture content materials; (5) *remote control*—the ability to adjust the feeder at extended distances from the feeder; (6) *pneumatic signal control*—a pneumatic positioner adjusts the setting of the feed rate control, sometimes desirable where the other process controls are pneumatic rather than electrical; (7) *solid state dc signal control*—designed to vary the feed rate directly or inversely in proportion to the value of an incoming dc signal which at minimum must be 1.5 V at 5 mA: (8) *solid state proportional control*—which permits individual material flow settings on multiple vibratory feeders. This is particularly useful for blending operations. The gross tonnage can be varied by a master control which will maintain the same ratio of materials flowing on each feeder; and (9) *solid state load monitoring control*—where the flow of material from a vibratory feeder is varied to maintain a selected load level on an ac motor driven crusher (or similar processing equipment) by monitoring the motor load.

FEEDER (Volumetric). A device for metering a particulate solid by volume. Functionally, volumetric feeders for solids are analogous to volumetric flowmeters for fluids and are important elements of many process control installations. Basic characteristics of a volumetric feeding device are: (1) a determinable geometry that, in essence, becomes the basic unit of flow measurement, and (2) a controlled rate of transfer of material from the feeder to the process. These characteristics are illustrated by the rotary lock feeder shown in Fig. 1. Here, the geometry is fixed for a given size unit by cross-sectional area *A*. This is the area bounded by two adjacent vanes or lobes of the rotor, the housing, and the width of the rotor. The rate of transfer is determined by the rotor speed, which may be fixed or variable.

The following factors are important in specifying a feeder for a particular application:

Particle Size—may determine the minimum dimensions and geometry of the feeder, such as the pitch of a helix feeder which must be adequate to allow material to enter and fill the helix. Particle size affects flow characteristics. Fine particles (micrometer size) as found in titanium dioxide pigment tend to exhibit high cohesive and adhesive values, requiring agitation of the material and special materials of construction to obviate build-up of material within the feeder.

Bulk Density—(Weight/Unit Volume) for most applications must be reasonably constant because the process requirement is for uniform feeding by weight. An exception would be a reduction mill where uniform volume of feedstock provides uniformity of load. Factors contributing to bulk density variation include modes of conveying or bin loading which may create aeration and segregation; and compaction due to extended storage under a head of material, or due to excessive vibration. Thus, upstream material handling must be consistent to insure feeder linearity, reproducibility, and repeatability.

Adhesion—often *Teflon* and polished surfaces are used to prevent sticking of material to contact surfaces.

Cohesion—the tendency for particles of a material to adhere one to the other can cause bridges (self-supporting arches within the material), particularly when the material is under pressure. This causes flow interruptions. Normally, agitation or vibration is provided to maintain flow.

Abrasion—a problem solved by use of special materials of construction or replaceable wear elements.

Moisture—normally is not a major problem where only surface moisture or moisture inherent to the material, such as that in a wood particle, are present. A small percentage of moisture in finely powdered material (2–5%) may adversely affect flow properties and feeder performance. Hygroscopic or deliquescent materials may modify within a feeder, causing loss of feed. A totally enclosed design, possible with a gas purge, may be required.

Segregation—often occurs where materials consist of a range of particle sizes and of different materials, thus upsetting uniform feeding. Correct hopper loading, eradication of dead space, and isolation from vibration may assist in maintaining material uniformity.

Degradation—special designs may be required to handle materials of a friable nature. The volumetric belt feeder is inherently best suited for such materials.

Control Factors—ease and sensitivity of control must be provided, such as by the advantages of a 10-turn versus a 1-turn potentiometer. Backlash in transmission drives must be minimized and taken into consideration where it exists, especially in connection with feeders that are manipulated by remote automatic controllers. Where mechanical transmissions are used involving hydraulic or oil lubrication, initial heating of the oil will result in a change of viscosity and may affect transmission speed. Voltage variation in a dc motor or vibrator may result in changes of transfer rate. In some cases, voltage and temperature compensation may be indicated.

Other Factors—include fire and explosion prevention, operator and

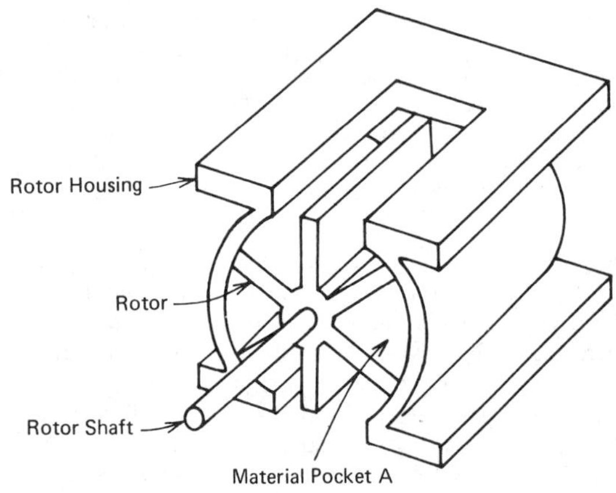

Fig. 1. Rotary lock feeder. The design is used widely for feeding to pressurized or vacuum systems and for feeding aerated or readily aerated materials since it provides a basic seal. The vanes normally incorporate a sealing strip at the top.

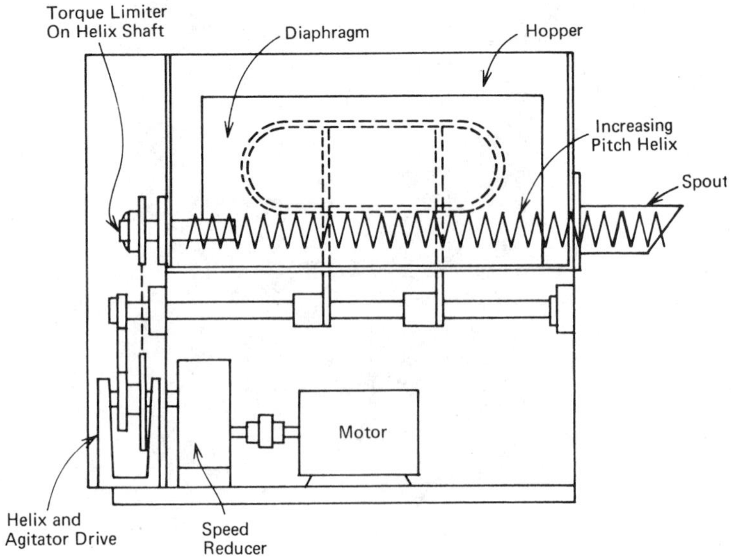

Fig. 2. Helix feeder.

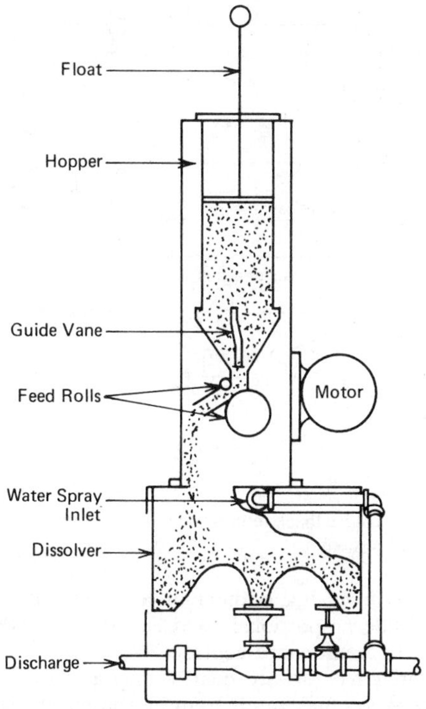

Fig. 3. Roll-type volumetric feeder with dissolver.

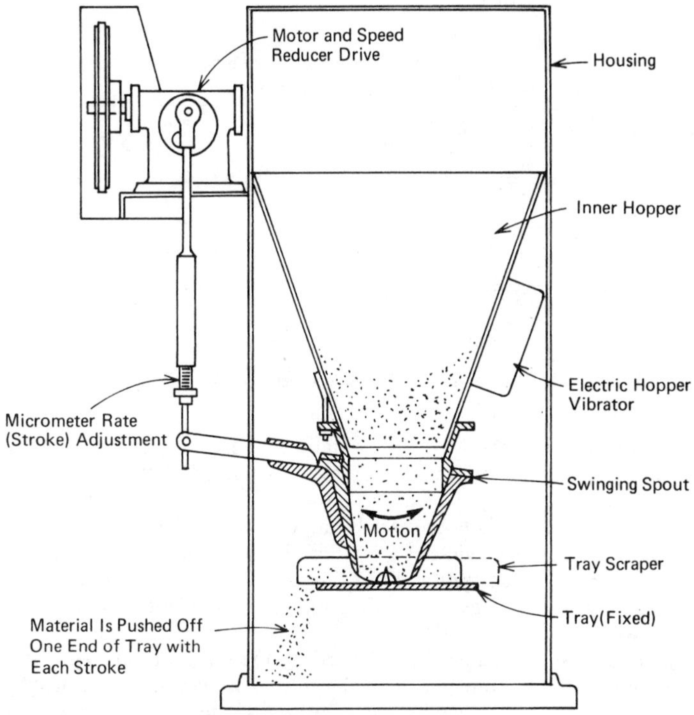

Fig. 5. Typical tray feeder.

maintenance skills, and both upstream and downstream conditions that may affect flow of material to and from the feeder.

Major Types of Feeders

In addition to the rotary lock feeder shown in Fig. 1, there are several other major types of volumetric feeders.

Helix Feeder. This design, including the auger and screw feeder, depends on filling and emptying the helix. With reference to Fig. 2, the metering geometry is determined by the bore of the discharge tube and by the helix pitch. Inherently, there is possibility of slip within the material in an open-type helix operating at high revolutions per minute. Thus, the feeder may be nonlinear. Vibrators may be used to insure filling of the helix and where this is constant in character, nonlinearity may be evident at low rates, but with good repeatability. Because of pitches on the helix, a feeder of this type evidences a pulsating discharge.

Roll-type Volumetric Feeder with Dissolver. As indicated by Fig. 3, the feeder geometry is described by the distance between rolls and by the width of the rolls. The rate is determined by the speed of rotation. The design is inherently sound for providing good distribution of material, and is particularly suited where material must be wetted, as with slurries and solutions.

Disk/Table Feeder. As shown by Fig. 4, the feeder geometry is determined by the cross-sectional area of the groove. The rate is determined by disk speed. The design is well suited for low-rate feeding of fine particle materials. The table-type feeder, a basic design variant, maintains a fixed layer of material between two fixed radii of the table. The rate is determined by entering a discharge plow into the

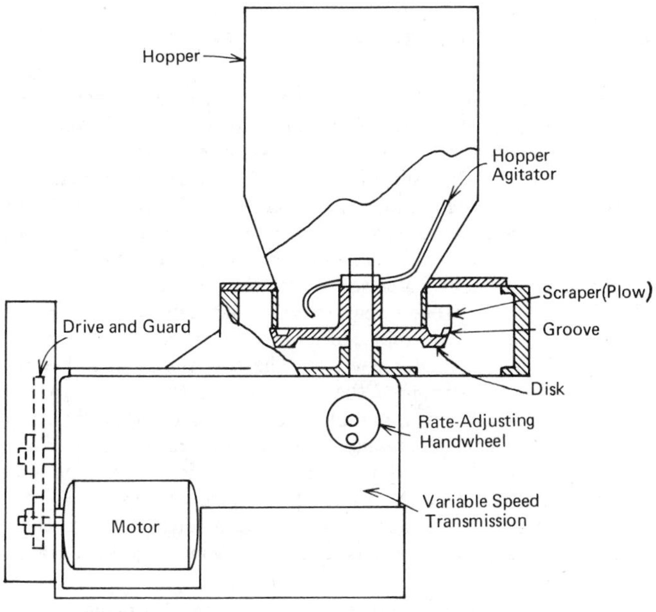

Fig. 4. Disk/table feeder.

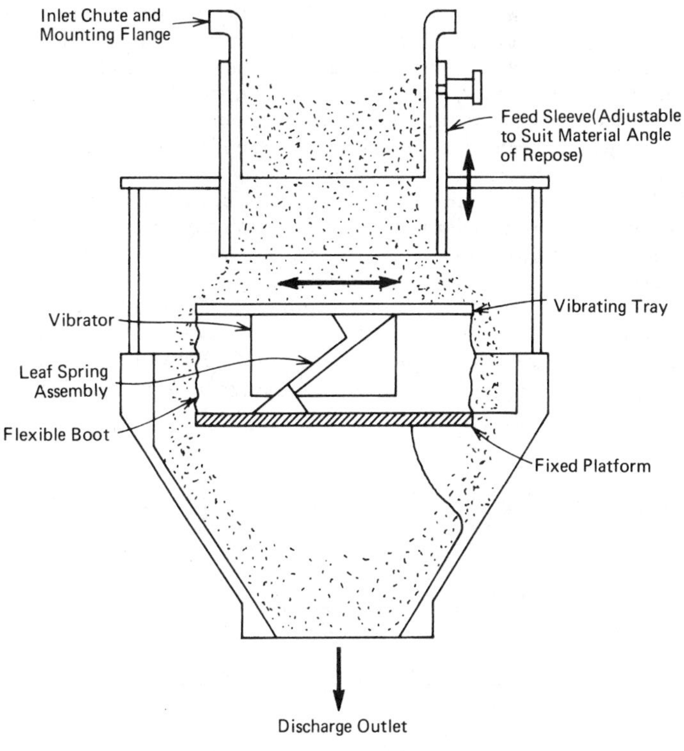

Fig. 6. Vibrating-tray feeder.

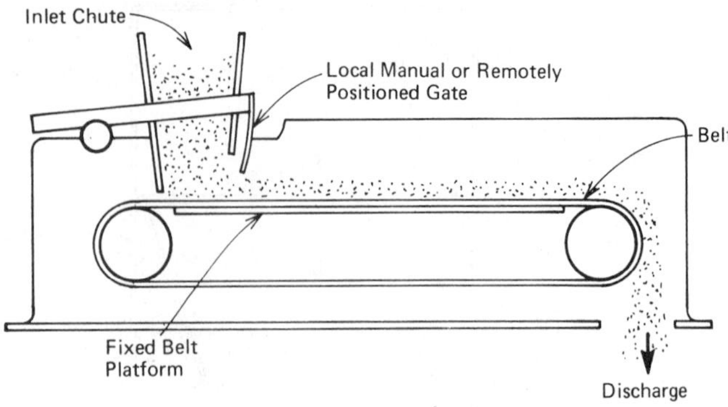

Fig. 7. Volumetric belt feeder.

material on the table, which normally rotates at a constant speed. Units of this type are available up to diameters of 20 feet (6 meters). Large sizes not only provide a controlled feed rate, but effectively constitute a bin or hopper discharging device as well.

Tray Feeder. As evidenced by Fig. 5, the geometry is described by the width of the tray and the distance between the tray and the oscillating nozzle. The rate is determined by oscillation of the inlet throat.

Vibrating-tray Feeder. As indicated by Fig. 6, the geometry is described by the circumference of the inlet section and by the height of the inlet section above the tray. The rate is determined by vibration of the tray. See also **Feeder (Vibratory).**

Volumetric Belt Feeder. As illustrated by Fig. 7, the feeder geometry is determined by the gate height and by the gate width. The rate is determined by belt speed, by gate height, or by a combination of the two factors. Belt feeders often are applied to handling friable materials inasmuch as the material is transported without pressure. These feeders also are well suited to materials having cohesive properties, such as detergents, since little or no pressure is generated which may cause formation of agglomerates.

Norman J. P. Taylor, BIF, a Unit of General Signal Corporation, Rexdale, Ontario, Canada.

FEEDFORWARD CONTROL. An automatic control action in which information concerning one or more conditions that can disturb the controlled variable is converted into corrective action to minimize deviations of the controlled variable. Feedforward control action can be combined with other types of control action to anticipate and minimize deviations of the controlled variable. Predictions used in feedforward control normally are obtained by computations with relation to a mathematical model and information obtained experimentally from measurements of the process to which application of feedforward control is contemplated.

Feedback control solves the control problem by a method of trial and error, whereas feedforward control requires a much greater understanding and quantitative knowledge of the process in order that a mathematical model can be established.

There are instances in connection with very difficult automatic control situations where feedforward control will succeed where feedback control has proved inadequate. In particular, feedforward control may be well adapted to processes which are subject to frequent load changes

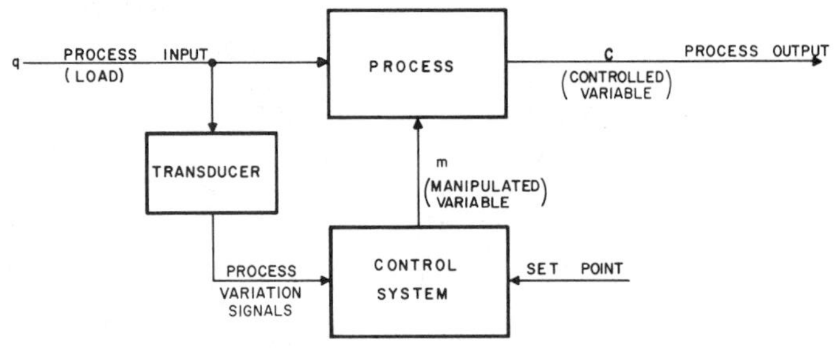

Fig. 1. Feedforward control system.

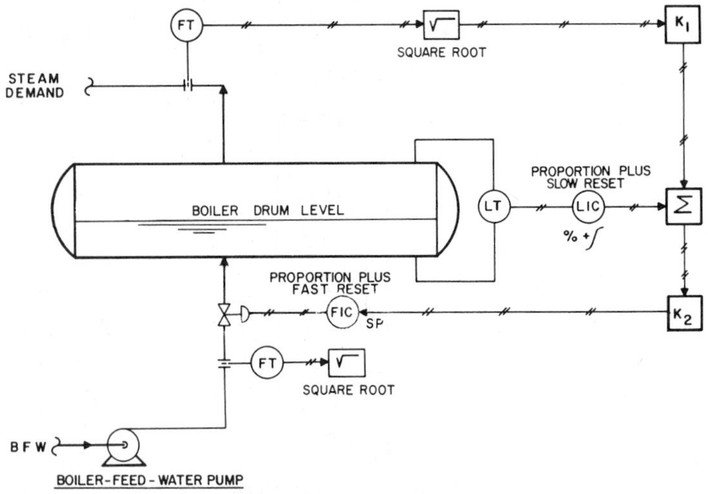

Fig. 2. Level-control feedback loop is used to reset feedforward loop in steam-generation control system.

and to processes involving optimal or adaptive control. Processes which require wide proportional band and long integral time often are candidates for feedforward control.

A representative feedforward control loop is illustrated in Fig. 1. In cases where errors may result from measurements, computations, and other factors in establishing the mathematical model, a remedial measure is to apply manual or automatic feedback control so as to correct for accumulated errors. A solution of this type is shown in Fig. 2. Here a level control feedback loop is used to reset the feedforward loop. A block diagram of the system is shown in Fig. 3. A non-self-regulating process is not controllable by feedforward action alone. A properly designed feedforward system will not affect the stability of feedback controllers involved.

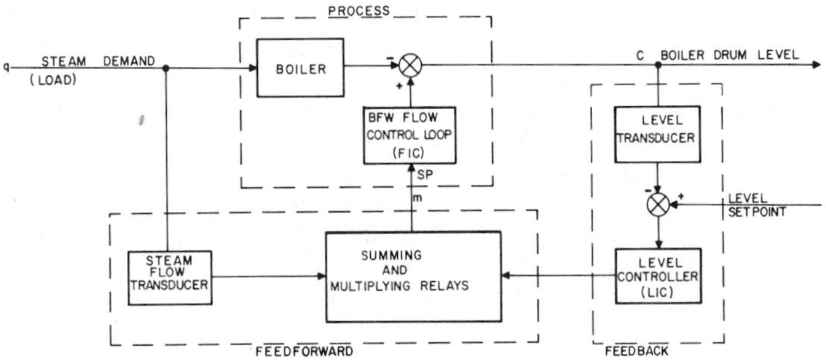

Fig. 3. Block diagram of steam-generation control system shown in Fig. 2.

FEEDWATER (Boiler). No boiler can operate efficiently or dependably if its heat-transfer surfaces are allowed to foul with scale or if corrosion is permitted to occur.* Thus, water treatment must include conditioning of: (1) Raw-water supply; (2) condensate returns from process steam or turbines; and (3) boiler water. Proper conditioning of these waters will provide freedom from deposits on internal surfaces, absence of corrosion of internal surfaces, and prevention of foaming and carryover of boiler-water solids into the steam.

Steam lost due to process requirements, blowdown, or leakage out of the system has to be replaced; the replacement water added to the system is termed makeup water. The condensate, together with the makeup water, comprise the feedwater to the boiler. In some plants, only a small percentage of condensate is returned; in others, almost all of the steam generated is recovered as condensate. Feedwater enters the boiler and is evaporated into steam, leaving behind solids to concentrate in the boiler water. If the concentration of solids in the boiler water exceeds certain limits, the quality of steam can be impaired by carryover. Also, boiler-water solids may settle out on

* Basic information for this entry from "Steam—Its Generation and Use," copyrighted by The Babcock & Wilcox Co., New York (39th Edition, 1978).

the boiler surfaces as sludge. The concentration of solids in the boiler water can be controlled by removing a portion of the water, either intermittently or continuously. This bleeding of a portion of the boiler water from the drum is termed blowdown.

In Universal-Pressure boilers, there are no drums to concentrate the boiler-water salts and impurities, and blowdown is not utilized. Purification takes place by continuously passing all or part of the condensate through demineralizers in a process called condensate polishing.

Raw Water Treatment. The type and amount of dissolved and suspended matter contained in raw water varies, of course, with the source, such as lake, river, or well, and notably from one geographical area to the next. The major dissolved materials in water are silica, iron, calcium, magnesium, and sodium compounds. Metallic constituents occur in various combinations with bicarbonate, carbonate, sulfate, and chloride radicals. In solution, the metal ions carry a positive charge (cations) and the bicarbonate, carbonate, sulfate, and chloride ions are negatively charged (anions). Scaling occurs when calcium or magnesium compounds in the water ("water hardness") precipitate and adhere to boiler internal surfaces. These hardness compounds become less soluble as temperature increases, causing them to separate from solution. Scaling causes damage to heat-transfer surfaces by decreasing the heat-exchange capability. The result is overheating of tubes, followed by failure and equipment damage.

Porous deposits will allow concentration of boiler-water solids. This concentration, particularly if strong alkalies are present, will result in severe corrosion of the tube surfaces. External treatment of water is required when the amount of one or more of the feedwater impurities is too high to be tolerated in the boiler system. Generally, the first step in water processing involves coagulation and filtration of suspended material.

Sodium-Cycle Softening. Also called sodium zeolite softening, this process utilizes resin materials that have the property of exchanging the hardness constituents, calcium and magnesium, for sodium. The process continues until the sodium ions become depleted or, conversely, the resin capacity to adsorb the calcium and magnesium no longer exists. When this occurs, the resin is said to be exhausted, and is regenerated by passing a solution of salt (sodium chloride) through it. Water, after passing through the zeolite process, contains as much bicarbonate, sulfate, and chloride as the raw water. Only the calcium and magnesium have been exchanged for sodium ions. There is no reduction in the overall amount of dissolved solids; neither is there a reduction of alkalinity content. When it is necessary to reduce the amount of dissolved solids, zeolite must be coupled with other methods, such as hot lime zeolite softening.

Hot Lime Zeolite Softening. In this process, hydrated lime is used to react with the bicarbonate alkalinity of the raw water. The precipitate is calcium carbonate and is filtered from the solution. To reduce silica, the natural magnesium of the raw supply can be precipitated as magnesium hydroxide, which acts as a natural absorbent for silica. These reactions are carried out in a vat or tank that is located just ahead of the zeolite softener tank. The effluent from this tank is filtered and then introduced into the zeolite softener. There is always some residual hardness leakage from the hot-process softener to be removed in the final zeolite process. The hot-lime process operates at about 104°C. At this temperature, the potential for the exchange of sodium for hardness ions is greater than at ambient temperature. A system of this type is shown in Fig. 1.

Hot Lime Zeolite—Split Stream Softening. Many raw waters softened by the first two processes would contain more sodium bicarbonate than is acceptable for boiler feedwater purposes. Sodium bicarbonate will decompose in the boiler water to give caustic soda. Caustic soda in high concentrations is corrosive and promotes foaming. The American Boiler Manufacturers Association (ABMA) has adopted the standard that the alkalinity content should not exceed 20% of the total solids of the boiler water. Split stream softening provides a means for reducing the alkalinity content.

This requires a second zeolite tank that has a zeolite resin in the hydrogen form in addition to the usual tank with the resin in the sodium form. The two tanks are operated in parallel. In one tank, calcium and magnesium ions are replaced by hydrogen ions. The effluent from this tank with the resin in the hydrogen form is on the acid side and has a lower total-solids content. The total flow can be proportioned between the two tanks to produce an effluent with any desired alkalinity as well as excellent hardness removal. When the hydrogen resin is exhausted, it is regenerated with acid.

Demineralization and Evaporation. At boiler drum pressures over 1,000 psi (68 atmospheres), demineralization or evaporation of the makeup water is generally desirable. A water that closely approaches theoretical chemical purity can be obtained by either of these processes.

Evaporation as a source of purified water does not involve ion exchange. It is actually a distillation process, consisting of evaporation, leaving most of the solids behind, and recondensation of the purified water. While evaporated water is quite satisfactory, economics generally favors demineralization.

Demineralization, like the zeolite process, involves ion exchange. The metal ions are replaced with hydrogen ions. In addition, the salt anions are replaced by hydroxide ions by means of a specially-prepared resin saturated with hydroxide ions. The two types of resins can be located in a separate tank. In this system, the two tanks are operated in series in a cation-anion sequence. The anion resin is regenerated after exhaustion with a solution of sodium hydroxide. The cation resin is regenerated with an acid, either hydrochloric or sulfuric acid. Some leakage of cations occurs in a cation exchanger, resulting in leakage of alkalinity from the anion exhanger.

In another arrangement, known as the mixed-bed demineralizer, the two types of resins are mixed together in a single tank. In this system, cation and anion exchanges take place virtually simulta-

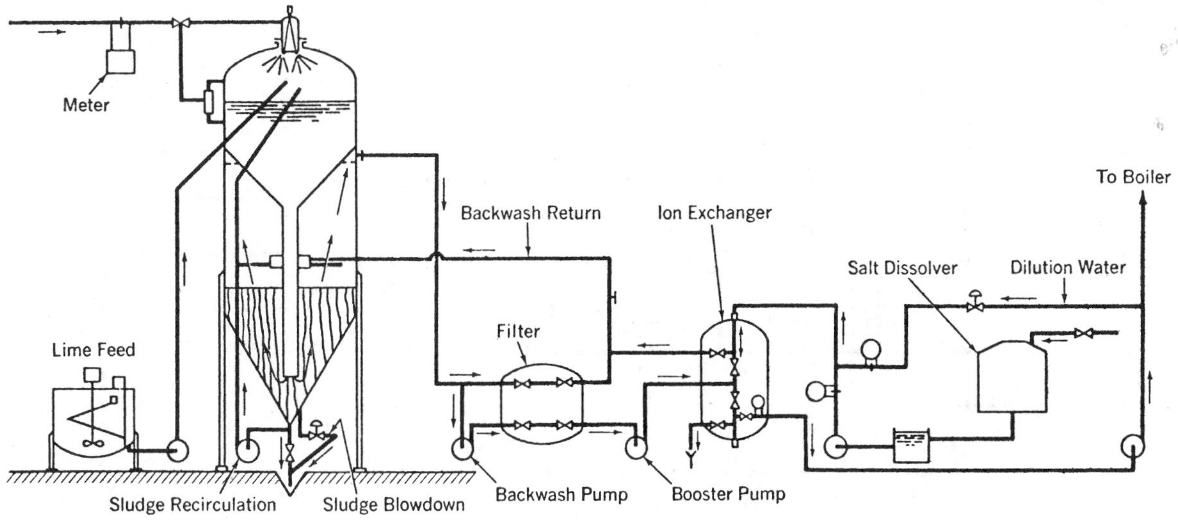

Fig. 1. Flow sheet of typical hot lime zeolite softening process. ("*Betz Handbook of Industrial Water Conditioning.*")

neously, resulting in a single irreversible reaction that goes to completion. Regeneration is possible in a mixed bed because the two resins can be hydraulically separated into distinct beds. The cation resin is approximately twice as dense as the anion resin. Resins can be regenerated in place or sluiced to external tanks for this purpose.

The raw water for drum-type boilers operating above 2,000 psi (136 atmospheres) drum pressure and for once-through units should be prepared by passing water through a mixed-bed demineralizer as a final step before adding to the cycle. The effluent from demineralization is approximately neutral. With nearly all salts removed, the problem of chemical control of the boiler water is minimized.

Treatment of Condensate. In most cases, condensate does not require treatment prior to reuse. Makeup water is added directly to the condensate to form boiler feedwater. In some cases, however, especially where steam is used in industrial processes, the steam condensate is contaminated by corrosion products or by the in-leakage of cooling water or substances used in the process. Hence steps must be taken to reduce corrosion or to remove the undesirable substances before recycling the condensate to the boiler feedwater.

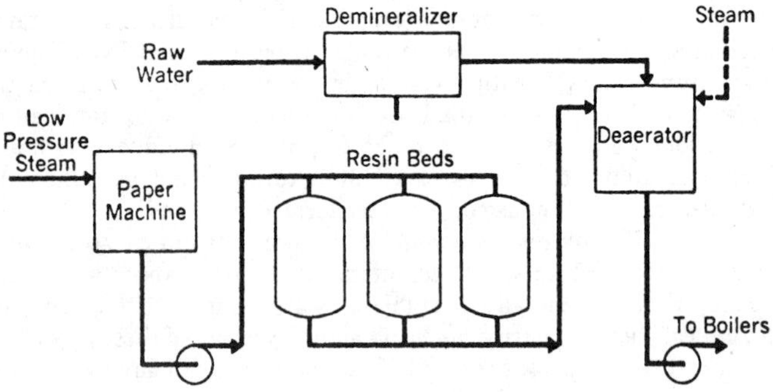

Fig. 2. Condensate purification system used in paper mill. (*Cochrane Div., Crane Co.*)

The presence of acidic gases in steam makes the condensate acidic with consequent corrosion of metal surfaces. In such cases, the corrosion rate can be reduced by feeding to the boiler water various chemicals that produce alkaline gases in the steam. The addition of neutralizing and filming amines to boiler water or condensate is one approach. Neutralizing amines (ammonia, morpholine, cyclohexylamine, and hydrazine) and filming amines (octadecylamine acetate) are volatile alkalizers that distill with the steam and neutralize acids in the condensate. Hydrazine, which is also an excellent oxygen scavenger, is included with the volatile alkalizers. It decomposes in the boiler, forming ammonia, hydrogen, and nitrogen. The ammonia provides pH control in the condensate.

Many types of contaminants can be introduced to condensate by various industrial processes. They include liquids, such as oil and hydrocarbons, as well as many kinds of dissolved and suspended materials. Each installation must be studied for potential sources of contamination. A condensate purification system used in a papermill boiler cycle is shown in Fig. 2. The resin beds not only remove dissolved impurities by ion exchange, but also serve as filters to remove suspended solids. It is necessary to backwash and regenerate these resin beds periodically.

Condensate-Polishing Systems. A condensate-polishing system is a requisite to maintain the purity required for satisfactory operation of once-through boilers. High-pressure drum-type boilers (over 2,000 psi; 136 atmospheres) can and do operate satisfactorily without condensate polishing. However, many operators recognize the benefits of condensate polishing in high-pressure plants, including: (1) Improved turbine capability and efficiency; (2) shorter unit startup time; (3) protection from the effects of condenser leakage; and (4) longer intervals between acid cleanings. See Fig. 3.

Two types of condensate-polishing systems are available. Deep-bed demineralizers operate at flow rates of 40 to 60 gallons/minute/square foot (1624–2441 liters/minute/square meter) of bed cross section. This type requires external regeneration facilities. The deep-bed system has a higher initial cost with possible lower operating costs, especially during initial unit startup. Its greater capacity for removing ionized solids permits continued operation with small amounts of condenser leakage. The deep-bed system is usually operated with the cation resin in the hydrogen form, but the ammonium form can also be used.

The cartridge-tubular type uses smaller amounts of disposable resins, eliminating the need for regeneration. One system of this type uses cation resin in the ammonium form. Because of the many considerations involved, an evaluation of the alternate types should be made.

Feedwater Equipment. The quality of feedwater recommended for various conditions is given in the accompanying table. The pre-boiler equipment, consisting of feedwater heaters, feed pumps, and feed lines,

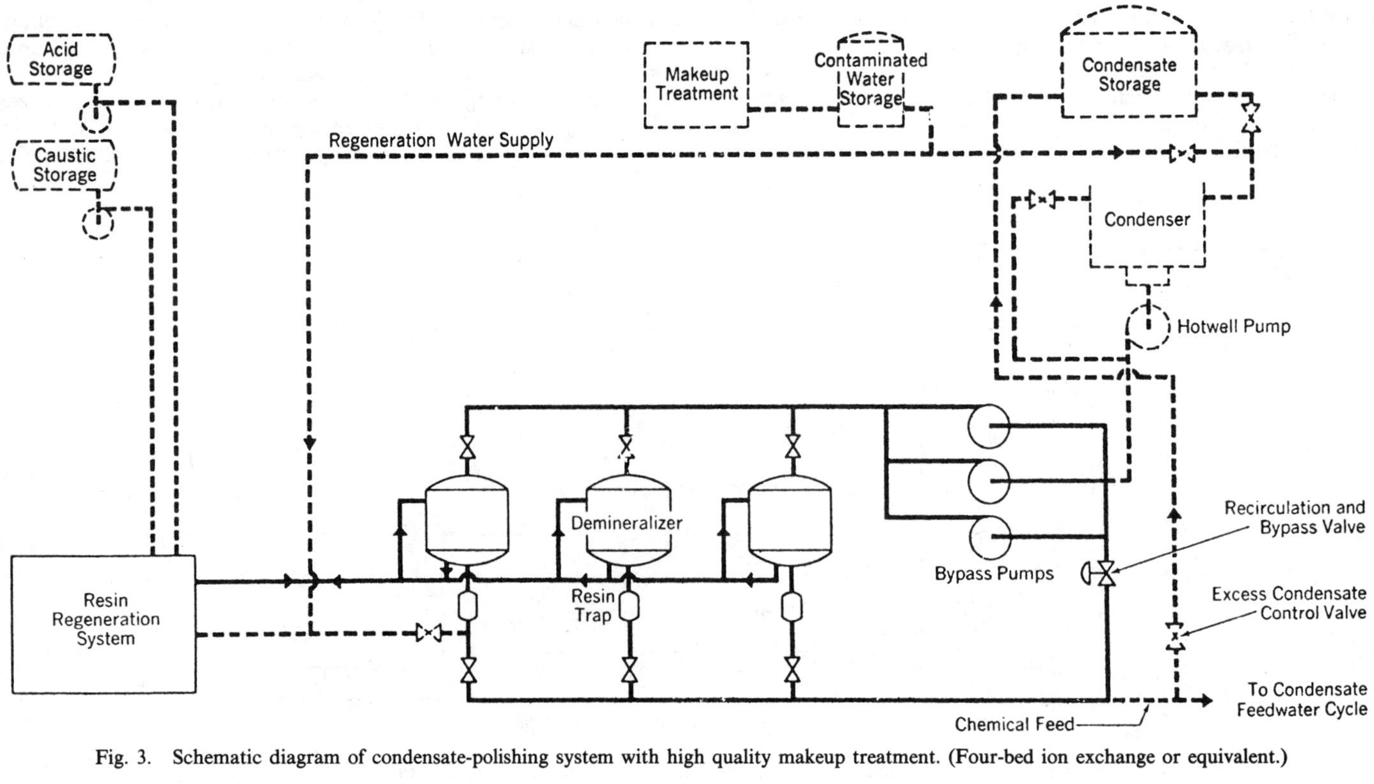

Fig. 3. Schematic diagram of condensate-polishing system with high quality makeup treatment. (Four-bed ion exchange or equivalent.)

RECOMMENDED LIMITS OF SOLIDS IN BOILER FEEDWATER

Drum Pressure	Below 600 psi	600 to 1,000 psi	1,000 to 2,000 psi	Over 2,000 psi
Total solids, ppm			0.15	0.05
Total hardness (as ppm $CaCO_3$)	0	0	0	0
Iron, ppm	0.1	0.05	0.01	0.01
Copper, ppm	0.05	0.03	0.005	0.002
Oxygen, ppm	0.007	0.007	0.007	0.007
pH	8.0–9.5	8.0–9.5	8.5–9.5	8.5–9.5
Organic	0	0	0	0

may be constructed of a variety of materials, including copper, copper alloys, carbon steel, and phosphor bronzes. To reduce corrosion, the makeup and condensate must be at the proper pH level and free of gases, such as carbon dioxide and oxygen. The optimum pH level is that which introduced the least amount of iron and copper corrosion products into the boiler cycle. Although the pH level must be established for each installation, it generally ranges between 8.0 and 9.5.

Oxygen Control. The presence of gases, particularly oxygen, leads to corrosion of the boiler and cycle equipment. This type of attack will occur in an operating boiler as well as in an improperly stored idle boiler. The consequent effect of dissolved oxygen in feedwater is pitting of the internal surfaces. This is most prevalent in the economizer, the steam drum, and the supply tubes. The pitting may be general or selective. In either case, if allowed to proceed unchecked, the reliability of the unit and useful service life will be adversely affected.

The most logical approcah is to expel gases from the system. The usual method is by use of a deaerating heater. This equipment must be kept in prime operating order over the complete load range. If the deaerator operates under vacuum at low loads, the entrance of air must be prevented. Oxygen concentrations at the deaerator outlet should be consistently less than 0.007 ppm. As a further assurance against the destructive effect of dissolved oxygen, a residual quantity of an oxygen-scavenging compound should be maintained in the system.

Most operators use sodium sulfite for chemical scavenging of oxygen. The amount of sulfite that can be safely carried decreases as pressure increases. See Fig. 4. At the high temperatures associated with the higher pressures, sulfite decomposes into acidic gases that can cause increased corrosion. Consequently, sulfite should not be used at pressures greater than 1,800 psi (122 atmospheres). On boilers having spray attemperation, the sodium sulfite should be added after the

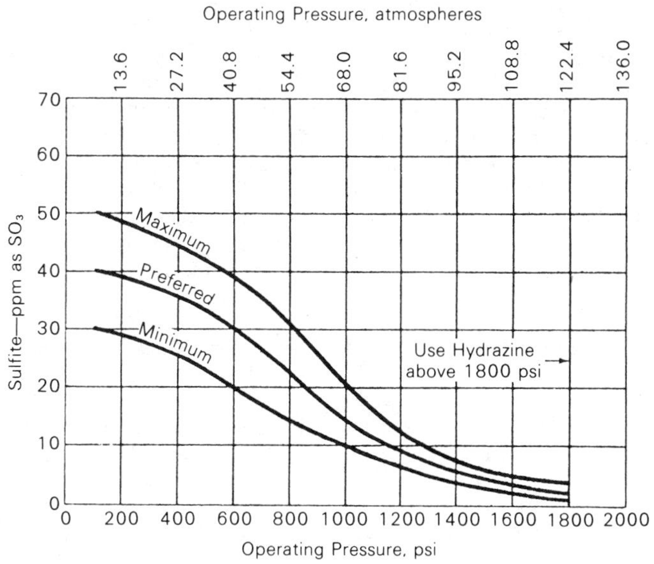

Fig. 4. Usually recommended sulfite residual in boiler water.

attemperator take-off point. About 8 ppm of sodium sulfite is required to remove 1 ppm of oxygen.

Hydrazine is an alternate scavenger, offering two principal advantages: (1) The decomposition and dissolved-oxygen reaction products of hydrazine are volatile. Consequently, they do not increase the dissolved-solids content of the boiler water, nor do they cause corrosion where steam is condensed. (2) Experience has shown that condensate pH will usually stabilize in the range of 8.5–9.5, if a 0.06 ppm hydrazine residual is maintained at the boiler inlet. This eliminates the need for pH treatment of the condensate-feedwater.

Direct Treatment of Boiler Water. Chemicals can be added directly for internal treatment of boiler water to prevent scale formations caused by hardness constituents and to provide pH control. Internal treatments must be used with extreme care to prevent buildup of total solids in the boiler water and carryover to the steam.

There are four methods of internal treatment in common use on natural-circulation drum-type boilers: (1) Phosphate-hydroxide (conventional treatment); (2) coordinated phosphate; (3) chelant; and (4) volatile. Methods 1 and 2 are intended to control the boiler water pH and to precipitate the calcium and magnesium compounds as a flocculent sludge, so that they can be removed in the boiler blowdown rather than being deposited on heat-transfer surfaces. Method 1 maintains an excess of hydroxide alkalinity. Method 3 involves the addition of a complex metal-chelant compound, such as ethylenediamine-tetracetic acid (Na_4EDTA) or nitrilotriacetic acid (NTA). In Method 4, no solid chemicals are added to the boiler or pre-boiler cycle. The pH of the boiler water and condensate cycle is controlled by adding a volatile amine. In Methods 1 through 3, either sulfite (up to 1,800 psi; 122 atmospheres) or hydrazine can be used as the oxygen scavenger. Above 1,800 psi (122 atmospheres) or with the volatile treatment (Method 4), hydrazine is used.

Phosphate-Hydroxide (Conventional-Treatment Method). This is the most prevalent method of treatment for industrial boilers operating below 1,000 psi (68 atmospheres). It involves the addition of phosphate and caustic to the boiler water. Caustic is added in sufficient quantity to maintain a pH of 10.5 to 11.2. A boiler treated with phosphate and caustic is less sensitive to upsets than with other methods of feedwater control. The primary purpose of phosphate addition is to precipitate the hardness constituents. The calcium reacts with phosphate (at proper pH) to precipitate calcium phosphate as calcium hydroxyapatite, $Ca_{10}(PO_4)_6(OH)_2$. This is a flocculent precipitate that tends to be less adherent to the boiler surfaces than simple tricalcium phosphate, which is precipitated below 10.2 pH. Caustic reacts with magnesium to form magnesium hydroxide or brucite, $Mg(OH)_2$. This precipitate is formed in preference to magnesium phosphate at a pH above 10.5 as it is less adherent.

FEHLING'S SOLUTION. This is an alkaline solution of copper hydroxide and sodium or potassium tartrate in sodium hydroxide used either as a mild oxidizing agent or as a test for easily oxidizable groups such as aldehyde groups. The formation of cuprous oxide is a positive test, its color red but often yellow at first. Glucose reacts with Fehling's solution to form cuprous oxide.

FELDSPAR. Feldspar is the name of a group which includes the most important of the rock-forming minerals, making up perhaps as much as 60% of the earth's crust.

This group of minerals consists of three silicates: a potassium-aluminum silicate, a sodium-aluminum silicate, and a calcium-aluminum silicate ($KAlSi_3O_8$, $NaAlSi_3O_8$, and $CaAl_2Si_2O_8$) and their isomorphous mixtures.

The various members of the feldspar group show many characteristics in common. Crystallizing in the monoclinic and triclinic systems, they show similarity of crystal habit, cleavage and other physical properties as well as similar chemical relationships.

Orthoclase, $KAlSi_3O_8$, derives its name from the Greek words meaning right or straight, and fracture, because its two cleavages are at right angles to each other. It crystallizes in the monoclinic system and its crystals are usually prismatic; it occurs also in coarsely cleavable masses. Hardness, 6; specific gravity, 2.56–2.58; luster, vitreous to pearly; colorless to white, gray, yellow or red, rarely green. Twin crystals not uncommon.

Orthoclase is a common constituent of many igneous rocks and is often found in huge masses in pegmatite veins. Localities for orthoclase are so numerous as to prohibit a complete list. Adularia (from Adular) is essentially a pure potassium silicate; when pearly and opalescent it is called moonstone and frequently used for jewelry. These opalescent varieties are known to be an intergrowth of orthoclase and albite. A glassy kind of orthoclase, sanidine, is found in the trachytes of the Drachenfels, Germany. Beautiful moonstones come from Ceylon and Switzerland; in the United States, from California and Virginia.

Orthoclase is found in the New England pegmatites, in New York, Pennsylvania, Virginia, North Carolina, Arkansas, Texas, Colorado, California, and elsewhere. Its commercial use is in the manufacture of porcelain and as a constituent in scouring powders.

Microcline, $KAlSi_3O_8$, is chemically the same as orthoclase, but belongs to the triclinic system, the prism angle being slightly less than a right angle (89°30′), hence the name microcline from the Greek meaning small, and to slope. Microcline is like orthoclase in all physical properties and can be distinguished from it surely only by optical examination. Under the polarizing microscope microcline displays a minute multiple twinning which results in a grating-like structure that is unmistakable. It is probable that much orthoclase would, upon proper examination, prove to be microcline. Amazon stone or amazonite is a beautiful green microcline occurring in the Ilmen Mountains in the Urals, Italy, Norway, the Malagasy Republic; and in the United States, in the Pikes Peak region, Colorado, Virginia, North Carolina, and sparingly in the pegmatites of New England.

The name amazon stone is derived from the application of this term to some green mineral found by the Spaniards among the aborigines of the Amazon Valley in South America. As no microcline is known to occur in the region there must have been some confusion with another green-colored substance.

A soda microcline, anorthoclase, is known, which is probably an isomorphous mixture of $KAlSi_3O_8$ and $NaAlSi_3O_8$, the sodium-aluminum silicate being in the greater proportion. The soda feldspar albite, $NaAlSi_3O_8$ and the calcium feldspar anorthite, $CaAl_2Si_2O_8$ form an isomorphous series from pure albite at one end to pure anorthite at the other, the two molecules appearing to be completely miscible one with the other. The members of this series are spoken of as the soda-lime (or lime-soda) feldspars, and as a group are called the plagioclase feldspars from the Greek meaning *oblique* and *fracture*, referring to the two cleavages at an angle that differs slightly from a right angle. Nearly always present are the striations, fine parallel lines, resulting from minute multiple twinning, which, never seen on orthoclase or microcline, are therefore an important diagnostic feature.

More or less arbitrarily, four intermediate plagioclase feldspars are recognized between albite and anorthite; these are listed below together with the approximate percentage of each molecule present.

	% of $NaAlSi_3O_8$	% of $CaAl_2Si_2O_8$
Albite	100 to 90	0 to 10
Oligoclase	90 to 70	10 to 30
Andesine	70 to 50	30 to 50
Labradorite	50 to 30	50 to 70
Bytownite	30 to 10	70 to 90
Anorthite	10 to 0	90 to 100

Albite is so called from the Latin, *albus*, in reference to its usual pure white color. It is a sodium aluminum silicate corresponding to the formula $NaAlSi_3O_8$. It crystallizes in the triclinic system commonly in tabular crystals. Twinning is very common, thin twinning lamellae producing a series of fine striations on certain crystal faces. There are two good cleavages at an angle of 86°24′ to each other. Hardness, 6; specific gravity, 2.62; luster, vitreous to pearly. It may be colorless to white or gray and transparent to opaque.

Albite is a relatively common and important rock-making mineral associated with the more acid rock types and in pegmatite dikes, often with rarer minerals like tourmaline and beryl. There are many famous localities in Europe in the Swiss and Austrian Alps, the Urals, the Harz Mountains, in Italy, France, and Norway. Brazil has yielded fine specimens. In the United States, notable localities are Paris and Auburn, Maine; Chesterfield, Mass.; Haddam, Connecticut; Amelia County, Virginia; and the Pikes Peak region of Colorado. It is used in the ceramic industries and also in the manufacture of artificial teeth.

Anorthite was named by Rose in 1823 from the Greek meaning oblique, referring to its triclinic crystallization. The physical properties are essentially the same as for albite, except that the specific gravity of anorthite is somewhat greater, 2.74–2.76. Anorthite is characteristic of the basic igneous rocks such as gabbro and basalt. Anorthite is found in the lavas of Vesuvius and Monte Somma, Italy; in Finland; Japan; and in the United States, in Sussex County, New Jersey.

The intermediate members of the plagioclase group are all very similar and with the exception of certain labradorites, cannot be distinguished from each other ordinarily save by optical means. Oligoclase is a common mineral in such rocks as granites, syenites, diorites, their extrusive equivalents and many gneisses. It is a frequent associate of orthoclase. The word oligoclase is derived from the Greek meaning little, and fracture, in reference to the fact that its cleavage angle differs slightly from 90°. Sunstone is mainly oligoclase (sometimes albite) spangled with flakes of hematite.

Andesine is a characteristic mineral of rocks such as diorites which contain a moderate amount of silica and related extrusives, such as andesites. Because of its occurrence in these latter, andesine derives its name from them as well as from the Andes Mountains.

Labradorite is the characteristic feldspar of the more basic rock types like diorite, gabbro, andesite or basalt and it is usually associated with some one of the pyroxenes or amphiboles. Labradorite frequently shows a beautiful play of iridescent colors due to minute inclusions of another mineral. However, the labradorescent phenomenon has not been fully determined. The classic locality for this mineral is of course Labrador, whence its name. It is a constituent there of the rock anorthosite and is found in the anorthosites of the Provinces of Quebec and Ontario, and in the Adirondack region in New York State.

Bytownite, named from Bytown, the former name for Ottawa, Canada, is a rare mineral occasionally found in the more basic rocks.

The feldspars crystallize from the magma in both extrusive and intrusive rocks; they occur as contact minerals, in veins and are developed in many sorts of metamorphic rocks, e.g., albite schists. They may also be found as mechanical deposits in various sedimentary rocks. See also terms listed under **Mineralogy.**

Elmer B. Rowley, Union College, Schenectady, New York.

FELDSPATHOIDS. **Nepheline.**

FELINES. **Cats.**

FELSITE. Felsites are defined by American geologists as dense, fine-grained, light-colored rocks rich in silica, hence classified with the rhyolites, from which some of them have been formed by devitrification. Felsites may occur as intrusive dikes but in general are found as extrusive rocks. They frequently occur interbedded with volcanic ash, tuff or breccia. According to American usage any light-colored lava whose ground mass or matrix is so fine-grained that the individual minerals cannot be distinguished by the naked eye (macroscopically) may be roughly classified as felsite, hence the prevalence of the term felsitic texture. When felsites show phenocrysts they are called felsite porphyries. The term felsite was first applied by Gerhard in 1814 to the fine ground mass (matrix) of porphyries, and is therefore one of the oldest, commonly used, petrological terms.

FEMALE HORMONES. **Endocrine System; Gonads; Hormones; Infertility; Pituitary Gland.**

FEMIC. This term is used by petrologists to designate the more common ferromagnesian minerals such as pyroxene and olivine. Rocks which are relatively rich in femic minerals are said to be urafic.

FEMINITY. **Gonads; Hormones.**

FEMORAL HERNIA. **Hernia.**

FEMUR. **Bone; Skeleton.**

FENESTRATED MEMBRANE. **Arterial and Venous Disorders.**

FENESTRA VESTULI. **Hearing and the Ear.**

FENNEC. **Canines.**

FENNEL. **Umbelliferae.**

FERBERITE. **Wolframite.**

FERGUSONITE. A mineral multiple oxide containing niobium (columbium), tantalum, and titanium, corresponding to formula $YNbO_4$. Essentially an oxide, or niobate-tantalate of yttrium with varying amounts of erbium, cerium, and iron. Crystallizes in the tetragonal system. Hardness, 5.5–6.5; specific gravity, 5.6–5.8; color, variable. Named after Robert Ferguson (1799–1865), Scotland.

FERMAT PRINCIPLE. This is a law of optics recognized nearly 300 years ago by Fermat. It states that when light proceeds by any path from a point A to another point B, the time required in its passage is either a minimum or a maximum as compared to other, arbitrarily chosen, adjacent paths. If the light is reflected from A to B by a plane surface, or is refracted at a plane surface on its way from A to B, the time is a minimum. For a curved reflecting surface, the time is a minimum if the surface has less curvature than the "aplanatic" surface osculating with it at the same point (i.e., the surface which gives rise to no spherical aberration); and this holds true also for a curved refracting surface. In these cases the law is known as the "principle of least time." But if the reflecting or refracting surface has greater curvature than the aplanatic surface at the same point, the time for the actual path is a maximum; that is, if the light could be made to follow a path through any other point of the curved surface than the one it actually passes through, it would do so in less time. For all points on a given aplanatic surface, the time is the same, and the light, if unobstructed, actually does follow paths through all of them.

FERMAT THEOREM. **Diophantine Equation; Number Theory.**

FERMENTATION. A form of respiration which requires no oxygen. There is an incomplete breakdown of food; carbon dioxide and other products, such as alcohol, are formed. The word is commonly used to refer to the conversion of sugars (and sugars derived from starch) into ethyl alcohol by the enzymes of yeast.

The process of fermentation has been used from prehistoric times in the preparation of foods and beverages, but the causative agents of fermentation were not recognized until the middle of the nineteenth century. The end-products resulting from the natural fermentation of glucose, namely, alcohol and carbon dioxide, were identified by Gay-Lussac in 1810, but it was thought that this process resulted from contact catalysis and the decay of animal or vegetable materials. This explanation was refuted by the work of Pasteur (1857) on the lactic acid fermentation. In the course of this investigation, Pasteur determined that fermentation was caused by living cells, that different microbial species caused different fermentations, that the nitrogenous materials present served only to support the growth of the cells, that lactic acid was produced when cells (removed from the fermentation mixture) were added to a sugar solution, and that the natural fermentation yielded both alcohol and lactic acid, but that the amount of each could be altered by changes in pH. In later studies, Pasteur showed that the conversion of glucose to alcohol, $C_6H_{12}O_6 \rightarrow 2CO_2 + 2C_2H_5OH$, was caused by yeast cells growing under anaerobic conditions, thus leading to the definition that fermentation was "life without air." A more modern definition of fermentation would be those energy-yielding reactions in which organic compounds act as both oxidizable substrates and oxidizing agents. Anaerobic reactions in which inorganic compounds are utilized as electron acceptors may be termed "anaerobic respirations," whereas reactions in which oxygen serves as a terminal electron acceptor are respirations.

Almost any organic compound may be fermented provided it is neither too oxidized nor too reduced, since it must function as both electron donor and electron acceptor. In some fermentations, a compound is degraded via a series of reactions in which intermediates in the sequence act as electron donors and acceptors; in others, one molecule of the substrate may be oxidized while another molecule is reduced, or two different organic compounds may be degraded after a coupled oxidation-reduction reaction. These fermentations provide energy required for the growth of a variety of cells. In addition, many microorganisms can carry out, in appropriate conditions, a number of fermentative reactions (e.g., oxidations, reductions, cleavages) which do not yield useful energy, or do not yield sufficient energy for growth.

In view of the great variety of different compounds which may be fermented and the enzymatic capabilities of different microorganisms, it is not surprising that numerous compounds important in industry (e.g., ethyl alcohol, butyl alcohol, acetone, 2,3-butylene glycol), in the production, preservation, and seasoning of food (e.g., lactic, citric, and glutamic acids), and in medicine (e.g., vitamins are extracted from the yeast carrying out the alcoholic fermentation) may be produced most cheaply through microbial fermentations. In addition, fermentations continue to be important in the production of foods (e.g., the lactic and propionic acid fermentations in the making of cheeses), beverages (e.g., the alcoholic fermentations in the making of wine and beer), and in the leavening of breads (by the carbon dioxide produced in the equation previously given). It should be pointed out that fermentation, while sometimes requiring the least expensive processing equipment to handle, and often fairly low-cost raw materials, is not always the most economic route. At one time, nearly all industrial alcohol (ethyl) was produced via fermentation. Currently, most ethanol is prepared by the direct hydration (vapor phase, catalytic) addition of water to ethylene; or by the indirect hydration of ethanol via the sulfation-hydrolysis process. See **Ethyl Alcohol.** However, with conservation measures possibly affecting the availability of hydrocarbons for chemicals (ethylene is produced by thermally cracking hydrocarbon feedstocks), interest in fermentation of agricultural feedstocks may return.

In addition to alcoholic fermentation, there are hundreds of other types of fermentative processes. Some of these include:

Amolytic fermentation—the fermentation of starch, but specifically it is an incomplete fermentation of starch in which simple sugars are not produced.

Butyric Fermentation—in which butyric acid is produced. The organisms producing this type of fermentation are mainly anaerobic like *Clostridium butyricum.* Some organisms, such as *Clostridium tetani,* the organism causing tetanus, and *Clostridium botulinum,* the organism causing botulism, also produce this type of fermentation.

Lactic Fermentation—in which lactic acid is produced. This is an important fermentation for the preservation of food. *Lactobacillus bulgaricus, L. casei,* and *Streptococcus lactis* are used for the manufacture of dairy products, such as sour cream. *Lactobacillus plantarum* is used in the preservation of certain vegetables, such as the production of pickles and kraut.

Controlled Oxidative Fermentations—by which a number of industrial chemicals are produced. *Citromyces,* for example, can be used for the production of citric acid from sugar. *Aspergillus niger* will yield oxalic acid by partial oxidative fermentation, but if the mold is permitted to remain in contact with the acid, it will convert it to carbon dioxide.

Some sugars such as glucose may be completely oxidized to carbon dioxide by certain bacteria, most molds, and some yeasts. Such microorganisms produce complete oxidation by fermentation. Many bacteria and yeasts are able to produce a gassy fermentation. The gaseous end product in the fermentation of vegetable products with *Leuconostoc mesenteroides* and *Lactobacillus brevis* is carbon dioxide. The gaseous end products of the coliform group are carbon dioxide and hydrogen.

Ropy Fermentation—which causes the spoilage of foods. Ropy milk is caused by *Aerobacter aerogenes, Lactobacillus bulgaricus, L. casei,* and *Alcaligenes viscosus.* Ropy bread is caused by members of the *Bacillus mesentericus* group which are identical with or are strains of *B. subtilis* or *B. pumilus.* Rope in maple syrup is produced by *A. aerogenes.*

A characteristic cultural reaction of *Clostridium perfringens* and many other clostridia is known as a *stormy fermentation.* When the organism is inoculated into milk the lactose is fermented and the casein is coagulated.

The anaerobic respiration which takes place in the muscles of higher animals, when insufficient oxygen is available for a complete breakdown of the food, is also called fermentation. Lactic acid and carbon dioxide are the products of this type of fermentation.

References

Bodyflet, F. W., and H. Yang: "Utilization of Cheese Whey for Wine Production," Proceedings of 70th Annual Meeting of American Dairy Science Association, Manhattan, Kansas, 1975.

Bungay, H. R.: Inexpensive Computers Aid Continuous Fermentation," *Chem. Eng. Progress,* **76,** 4, 53–54 (1980).

Humphrey, A. E., and E. K. Pye: "Enzyme Technology, Past, Present, and Future," Proceedings of the 1st RANN Symposium on Productivity, National Science, Foundation, Washington, D.C., 1973.

Humphrey, A. E.: "Fermentation Technology," *Chem. Eng. Progress,* **73,** 5, 85–91 (1977).

Kunkee, R. E.: "Malo-lactic Fermentation," *Adv. Appl. Microbiol.,* **7,** 235 (1967).

Pirt, S. J., "Principles of Microbe and Cell Cultivation," Halsted Press, New York, 1975.

Webb, A. D. (editor): "Chemistry of Winemaking," Advances in Chemistry Series, Vol. 137, American Chemical Society, Washington, D.C., 1974.

FERMENTATION (Amino Acid). **Amino Acids.**

FERMENTATION (Carbohydrate). **Carbohydrates.**

FERMENTATION YEASTS. **Yeasts and Molds.**

FERMI. The name used by some scientists to designate a length of 10^{-15} meter = 10^{-13} centimeter, a length equal approximately to the radius of a nucleon.

FERMILAB. **Particles (Subatomic).**

FERMI LEVEL. **Fermi Surface; Solids (Band Theory).**

FERMIONS. Those elementary particles for which there is antisymmetry under intra-pair production. They obey Fermi-Dirac statistics. Fermionic hadrons are called *baryons;* other hadrons are *mesons.* See also **Baryons;** and **Mesons.** The experimentally observed fermions are the leptons and baryons, each of which has a spin angular momentum of $\sqrt{3}\ h/4\pi$, where h is Planck's constant, which is equivalent to stating that each fermion has a spin quantum number of magnitude $\frac{1}{2}$. See **Particles (Subatomic).**

FERMI RESONANCE. In polyatomic molecules, two vibrational levels belonging to different vibrations (or combinations of vibrations) may happen to have nearly the same energy, and therefore be accidentally degenerate. As was recognized by Fermi in the case of CO_2 such a "resonance" leads to a perturbation of the energy levels that is very similar to the vibrational perturbations of diatomic molecules.

FERMI SELECTION RULES. A set of selection rules for beta decay which state that an allowed transition between parent and daughter states must have no change of spin quantum number and no change of parity.

FERMI SURFACE. Of a metal, semi-metal, or semiconductor that surface in momentum space which separates the energy states which are filled with free or quasi-free electrons from those which are unfilled. Such a surface exists simply because the electrons obey Fermi-Dirac statistics. It is a surface of constant energy and is sometimes called the Fermi level.

If one considers an elementary model of a metal consisting of a lattice of fixed positive ions immersed in a sea of conduction electrons which are free to move through the lattice, every direction of electron motion will be equally probable. Since the electrons fill the available quantized energy states starting with the lowest, a three-dimensional picture in momentum coordinates will show a spherical distribution of electron momenta and, hence, will yield a spherical Fermi surface. In this model, no account has been taken of the interaction between the fixed positive ions and the electrons. The only restriction on the movement or "freedom" of the electrons is the physical confines of the metal itself.

A short derivation, starting with the Schrödinger equation, shows that the total energy of an electron (and thus also its kinetic energy) is given by

$$E = \hbar^2 k^2/2m = p^2/2m$$

where $\hbar$ is Planck's constant divided by 2π, k is the magnitude of the electron wave vector, m is the mass of the electron, and p is its momentum. A plot of E against k is then a parabola, as shown in Fig. 1(a). The Cartesian components of those values of k which are

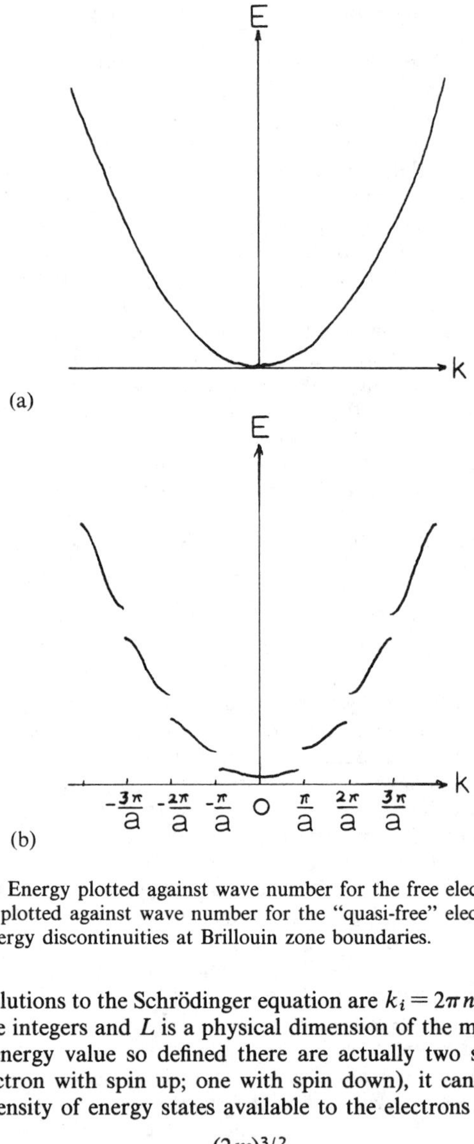

Fig. 1. (a) Energy plotted against wave number for the free electron model. (b) Energy plotted against wave number for the "quasi-free" electron model, showing energy discontinuities at Brillouin zone boundaries.

possible solutions to the Schrödinger equation are $k_i = 2\pi n_i/L$, where the n_i's are integers and L is a physical dimension of the metal. Since for each energy value so defined there are actually two states (one for an electron with spin up; one with spin down), it can be shown that the density of energy states available to the electrons is

$$g(E) = \frac{(2m)^{3/2}}{2\pi^2 \hbar^3} E^{1/2}$$

where $g(E)\ dE$ is the number of states in the energy range E to $E + dE$. Then $n(E)$, the number of electrons per unit volume occupying energy states in this energy range, is

$$n(E)\ dE = g(E) f(E)\ dE$$

where $f(E) = \{\exp[(E - E_f)/bT] + 1\}^{-1}$, a function characteristic of particles which obey Fermi-Dirac statistics. In this expression, T is the absolute temperature, b is Boltzmann's constant, and E_f is a parameter depending on the number of electrons involved. This turns out to be the Fermi energy. E_f can be evaluated by integrating $n(E)\ dE$ from $E = 0$ to $E = \infty$ and recognizing that the integral is equal to N, the total number of electrons per unit volume. The result (at $T = 0$ K) is

$$E_f = \frac{\pi^2 h^2}{2m}\left(\frac{3N}{\pi}\right)^{2/3}$$

At $T = 0$ K, for $E < E_f$, $f(E) = 1$, while for $E > E_f$, $f(E) = 0$. Physically, this means that the probability of a state below the Fermi level being occupied is one; whereas for states with $E > E_f$, the occupancy probability drops abruptly to zero. For temperatures greater than absolute zero, the occupancy probability drops smoothly from 1 to 0 in a range of energy of width approximately equal to bT. This shell of partially filled states gives rise to the following definition: *The Fermi level is the energy level at which the probability of a state being filled is just equal to one half.*

A numerical evaluation of the Fermi energy for a simple metal having one or two conduction electrons per atom yields a value of approximately 10^{-11} erg, or a few electron volts. The equivalent temperature, E_f/b, is several tens of thousands of degrees Kelvin. Thus, except in extraordinary circumstances, when dealing with metals, $bT \ll E_f$; i.e., the energy range of partially filled states is small, and the Fermi surface is well defined by the foregoing statement. It must be noted, however, that this is not necessarily true for semiconductors where the number of free electrons per unit volume may be very much smaller.

The foregoing description provides a qualitative look at the physics of metals and, under some circumstances, semi-metals and semi-conductors. A more detailed analysis requires that the effects of the ions in the lattice be recognized. This can be accomplished by introducing the periodic potential due to the lattice through which the electrons must move. Then the electrons are no longer "free," but, depending on the strength and character of the potentials and the approximations used in solving the Schrödinger equation, act as "quasi-free" particles. Another approach is the "tight-binding approximation"; occasionally a combination of the two approaches is used. In any case, introduction of lattice effects changes the characteristics of the model; the total energy and kinetic energy of an electron are no longer equivalent. The periodic lattice can be described in terms of Brillouin zones, each of which is large enough (in momentum space) to accommodate two electrons per atom. The Brillouin zone boundaries appear to the electrons as Bragg reflection planes or energy discontinuities, resulting in an energy versus wave number plot as shown in Fig. 1(b).

For many metals, the "nearly free" electron description corresponds quite closely to the physical situation. The Fermi surface remains nearly spherical in shape. However, it may now be intersected by several Brillouin zone boundaries which break the surface into a number of separate sheets. It becomes useful to describe the Fermi surface in terms not only of zones or sheets filled with electrons, but also of zones or sheets of holes, that is, momentum space volumes which are empty of electrons. A conceptually simple method of constructing these successive sheets, often also referred to as "first zone," "second zone," and so on was demonstrated by Harrison. An example of such construction is shown in Fig. 2.

This construction works out quite well, for example, for aluminum which has three valence electrons per atom. Experiments and more elegant theoretical calculations show that the fourth zone is totally

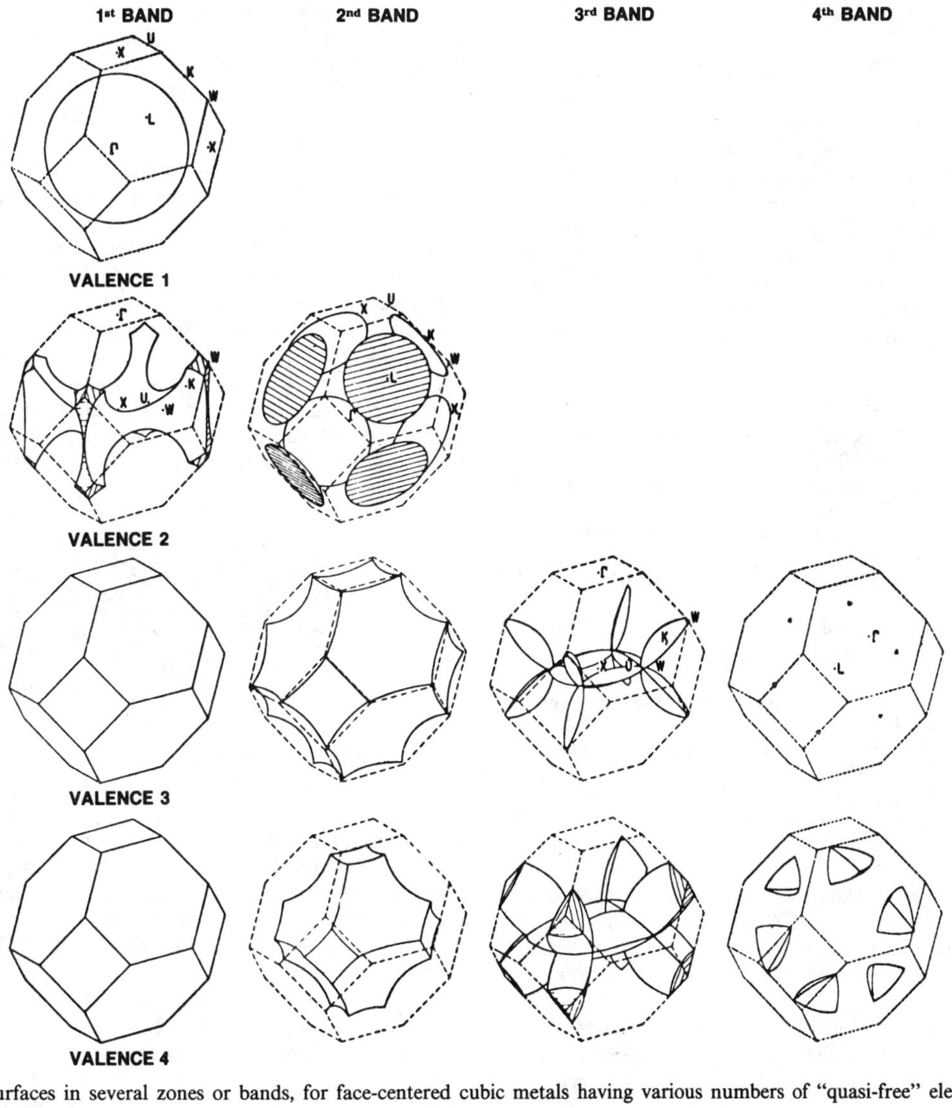

Fig. 2. Fermi surfaces in several zones or bands, for face-centered cubic metals having various numbers of "quasi-free" electrons per atom, as constructed by Harrison.

unoccupied and that the third zone is not multiple-connected in the manner shown.

See also **Semiconductors;** and **Solid-State Physics.**

FERMIUM. Chemical element symbol Fm, at. no. 100, at. wt. 257 (mass number of the most stable isotope), radioactive metal of the Actinide series, also one of the Transuranium elements. Both fermium and einsteinium were formed in a thermonuclear explosion which occurred in the South Pacific in 1952. The elements were identified by scientists from the University of California's Radiation Laboratory, the Argonne National Laboratory, and the Los Alamos Scientific Laboratory. It was observed that very heavy uranium isotopes which resulted from the action of the instantaneous neutron flux on uranium (contained in the explosive device) decayed to form Es and Fm. The probable electronic configuration of Fm $1s^2 2s^2 2p^6 3s^2 3p^6 3d^{10} 4s^2 4p^6 4d^{10} 4f^{14} 5s^2 5p^6 5d^{10} 5f^{12} 6s^2 6p^6 7s^2$. Ionic radius 0.97Å.

All known isotopes of fermium are radioactive. They include isotopes of mass numbers 248($t_{1/2}$ = 0.6m.), 249($t_{1/2}$ = 150s.), 250($t_{1/2}$ = 0.5h.), 251($t_{1/2}$ = 7h.), 252($t_{1/2}$ = 23h.), 253($t_{1/2}$ = 4.5d), 254($t_{1/2}$ = 3.24h.), 255($t_{1/2}$ = 22h.), 256($t_{1/2}$ = 160m.), and an isotope of mass number 257 which has a half-life of about 10 days. The mass number of fermium given in atomic weight tables is 253, the mass number of the isotope of longest half-life (except possibly for the last one cited). The first three decay by alpha-particle emission only; the others, through ^{255}Fm, by that and other modes, including spontaneous fission for ^{254}Fm and ^{255}Fm. Isotopes of 256 and higher decay by this mode only. Fermium isotopes are made (1) by heavy iron bombardment of uranium and plutonium isotopes, (2) by the action of alpha particles on californium isotopes, and (3) from several of the other transuranium elements by multiple neutron capture—or in the case of the heaviest isotopes, from einsteinium by single neutron capture.

The ion Fm^{3+} is stable, and fermium is probably exclusively trivalent.

The discovery of fermium (also einsteinium) was not the result of very carefully planned experiments, as in the cases of the other transuranium elements, but fermium and einsteinium were found in the debris of an atomic weapon test in the Pacific in November 1952. Researchers, using the Oak Ridge High Flux Isotope Reactor (HFIR) which produced 3.2-hour ^{254}Fm, determined the magnetic moment of the atomic ground state of the neutral fermium atom with a modified atomic beam magnetic resonance apparatus. In essence, this observation represented that of a macroscopic property of the metallic 0-valent state of fermium.

References

Diamond, H., et al.: "Heavy Isotope Abundances in Mike Thermonuclear Device," *Phys. Rev.* **119**, 6, 2000–2004 (1960).

Fields, P. R., et al.: "Additional Properties of Isotopes of Elements 99 and 100," *Phys. Rev.*. **94**, 1, 209–210 (1954).

Ghiorso, A., et al.: "New Elements Einsteinium and Fermium, Atomic Numbers 99 and 100," *Phys. Rev.*, **99**, 3, 1048–1049 (1955).

Goodman, L. S., et al.: "g_J Value for the Atomic Ground State of Fermium," *Phys. Rev.*, **4**, 2, 473–475 (1971).

Hulet, E. K., Hoff, R. W., Evans, J. E., and R. W. Lougheed: "79-Day Fermium Isotope of Mass 257," *Phys. Rev. Lett.*, **13**, 10, 343–345 (1964).

John, W., Hulet, E. K., Lougheed, R. W., and J. J. Wesolowski: "Symmetric Fision Observed in Thermal-Neutron-Induced and Spontaneous Fission of ^{257}Fm," *Phys. Rev. Lett.*, **27**, 1, 45–48 (1971).

Seaborg, G. T. (editor): "Transuranium Elements," Dowden, Hutchinson & Ross, Stroudsburg, Pennsylvania, 1978.

Thompson, S. G., Harvey, B. G., Choppin, G. R., and G. T. Seaborg: "Chemical Properties of Elements 99 and 100," *Amer. Chem. Soc. J.*, **76**, 6229–6236 (1954).

FERNS. Members of a relatively small division of the plant kingdom, the ferns (*Filicales*) comprise about 4,500 species of plants, few of which are very large. But in past times, and especially in the Carboniferous period, plants of this division were more numberous and many of them of very large size.

At the present time, the ferns are found in nearly all parts of the world, especially in regions where there is plenty of moisture. They are particularly numerous in the tropics. There also the largest of the ferns, called tree ferns (Fig. 1), are found. These tree ferns have

Fig. 1. Fern fronds typical of the finely divided leaves found in the ferns. (*Winchester, "Modern Biological Principles," 2nd edition, D. Van Nostrand.*)

an erect usually unbranched trunk bearing at its top a crown of much dissected leaves of large size. Tree ferns are from 10–30 feet in height; a few may reach a height of 50 feet (15 meters).

In nearly all ferns the stem is a slender structure. In many species it grows underground as a creeping horizontal rhizome; in other species it is shorter and erect, but seldom rises much above the surface of the ground. From the stem numerous fine wiry roots extend into the ground. The leaves of ferns having creeping rhizomes are borne singly at the nodes; those of erect-stemmed ferns are borne in a group which forms a close crown. In many ferns the leaf, often called a frond, is pinnately compound, and the individual pinnae themselves compound. Other ferns have entire leaves. Young leaves are circinately coiled, the tip of the leaf being in the center of a tight coil, which unrolls from the base upward. Often these young leaves are covered with a mass of dense brown hairs. On the lower surface of the frond reproductive bodies are formed. In many ferns these are found on all pinnae; in others they occur only on special pinnae which are commonly very much modified in size. These reproductive bodies are commonly borne in compact groups called sori, which are often covered by a protective structure, the indusium. The reproductive bodies are stalked sporangia containing spores, which are freed by the action of certain special cells of the sporangium. (Fig. 2.) These

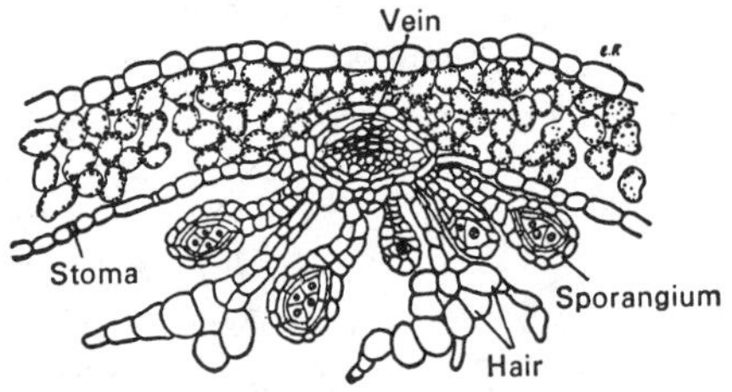

Fig. 2. Section through leaf and sorus of *Polypodium* showing sporangia in various stages of development.

cells have thick inner walls and in many ferns form a distinct row called the annulus. When the atmosphere is dry the cells of the annulus of a mature sporangium lose water and gradually contract. As a result, a considerable strain is exerted by the thin outer wall of these cells, so that the annulus is pulled backwards, a break occurring in a group of thin-walled cells known as the stomium. Finally the tension becomes too great and the annulus snaps back violently, catapulting the spores out of the sporangium (Fig. 3.)

These spores, carried by air currents to a region where moisture is sufficient, germinate. They do not, however, form a new fern plant. Instead they develop a small delicate green plant called the prothallium

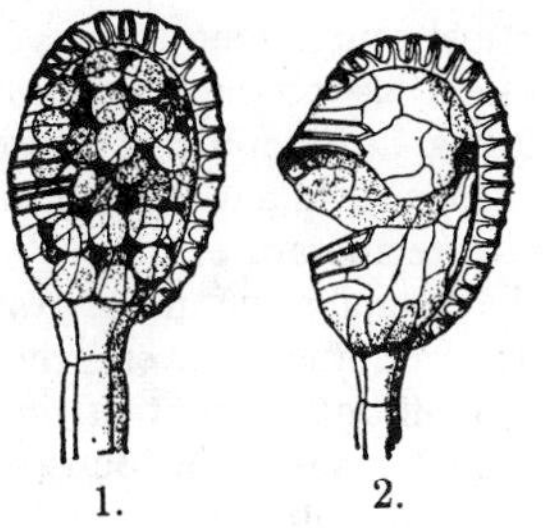

Fig. 3. Fern sporangia: (1) unopened sporangium filled with spores; (2) the empty sporangium after the annulus has returned to its first position.

(Fig. 4). In many ferns this is a heart-shaped body one cell thick. In others it is a branched object resembling certain species of algae, and in still other ferns it is a small tuberous body. From the lower surface of the prothallium numerous short rhizoids grow down and

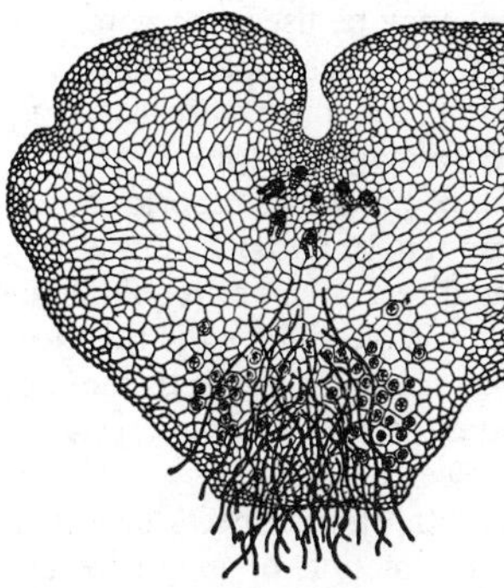

Fig. 4. Fern prothallium. View of the under side showing archegonia near the apical notch and antheridia among the rhizoids near the base. (*Sinnott, "Principles and Problems,"* *McGraw-Hill.*)

anchor it firmly in the substratum. On the lower surface also, reproductive bodies are formed. These consist of two kinds, commonly found on the same prothallium. One, usually in the basal portion of the prothallium is the antheridium (Fig. 5a). An antheridium is a small

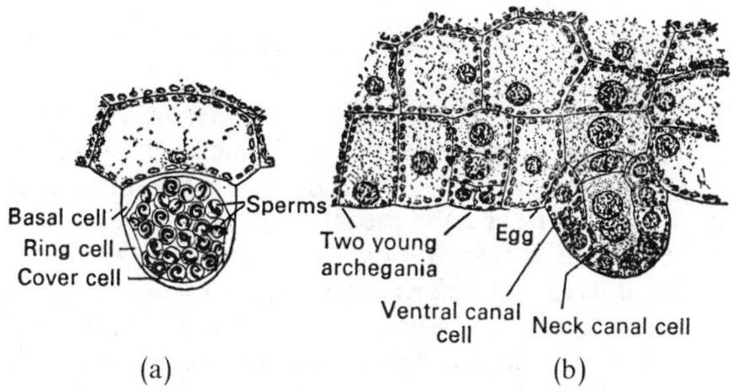

Fig. 5. (a) mature antheridium of fern; (b) two young archegonia of fern and one mature one. (*Chamberlain, "Elements of Plant Science,"* McGraw-Hill.)

multicellular organ in which are formed many small sperms. Fern sperms are spirally coiled cells each having a group of cilia at the tip. The female sex organ is the archegonium (Fig. 5b). These are commonly found near the notch of the prothallium. Each archegonium is a small organ consisting of a basal layer of cells surrounding the single large egg cell and a short tube surrounding a row of cells known as the neck cells. These break down, forming the neck canal, through which the sperm swim to unite with the egg. The fertilized egg or zygote immediately starts dividing and gives rise to a new fern plant. The prothallium of the fern is the gametophyte generation. It is green and very much smaller than the sporophyte, of which it is entirely independent (Fig. 6). Water is absolutely necessary for the prothallium to give rise to a new sporophyte (Fig. 7).

The ferns are of minor commercial importance. Christmas fern, *Polystichum acrostichoides*, has thick evergreen leaves which are often used for decorative purposes. This fern grows wild in open woods of the north temperate region. Another fern, the maidenhair, *Adiantum*

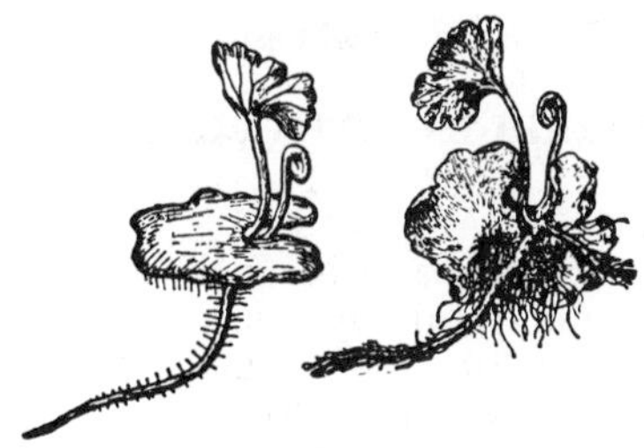

Fig. 6. Fern embryo sporophyte still attached to its parent (the gametophyte), but differentiated into its parts and making its own food: (left) as seen from above; (right) as seen from below.

pedatum, and related species, is frequently grown as a decorative plant because of its delicate fronds. However, the fronds wilt too quickly to be of much use if cut from the plant. Species of *Osmunda*, including the Cinnamon fern, the interrupted fern, and the royal fern, are often planted in shady places for ornament. From these ferns is obtained a coarse fiber used as a potting substance on which to grow epiphytic orchids. The commonly used Boston fern, *Nephrolepsis exaltata bostoniensis*, was derived from the tropical sword fern (*N. exaltata*). Other common decorative ferns include the New York fern, the lady fern, and evergreen wood fern.

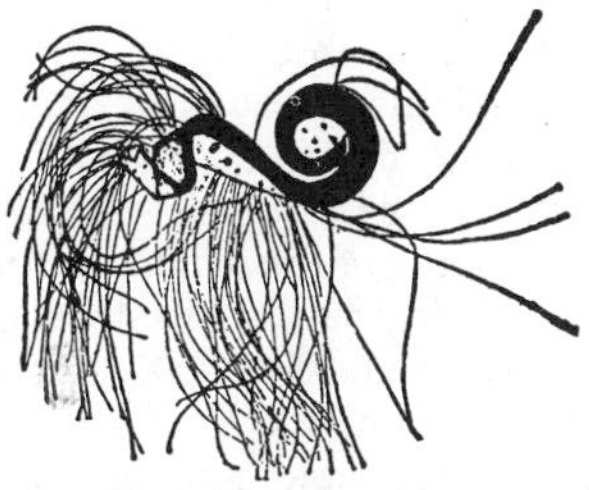

Fig. 7. A single fern microgamete, killed and stained so that the parts of the cell are visible. (*Steil.*)

Sometimes planted as a curiosity. *Camptosorus rhizophyllus*, the walking fern (Fig. 8) gets its name because the tips of the fronds bend down to the ground and take root. New plants are formed at these points. This is a special form of vegetative reproduction. Several ferns propagate themselves by forming small bulbils on the surface of the fronds. Eventually these drop off and take root, giving rise to new plants. *Cystopteris bulbifera* is a fern propagating in this way. One of the commonest and best known ferns is the common brake, *Pteris aquilina*, which grows on dry hillsides and in open woods.

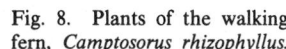

Fig. 8. Plants of the walking fern, *Camptosorus rhizophyllus.*

FERREDOXIN. Nitrogen.

FERREED. A high-speed switching device with relay-like mechanical contacts used in electronic telephone switching. It consists of magnetic reeds with contacts placed in a magnetic field produced by a material having semi-permanent magnetic properties. A field coil is placed around the semi-permanent magnetic material to excite it. A pulse of current through the coil permanently magnetizes the material and the relay closes. A second pulse of current in the opposite direction demagnetizes the material and the relay opens.

FERREL LAW. Atmosphere (Earth).

FERRET. **Mustelines.**

FERRET-BADGER. **Mustelines.**

FERRIMAGNETISM. A type of magnetism, macroscopically similar to ferromagnetism, but microscopically more like antiferromagnetism in that the magnetic moments of neighboring ions tend to align antiparallel. These moments are, however, of different magnitudes, and hence may still have quite a large resultant magnetization. Ferrimagnetism is observed in the ferrites having the general formula $MeFe_2O_4$, where Me denotes a divalent anion, usually of an alkaline earth metal.

FERRITE. The existence of ceramic magnetic materials capable of combining the resistivity of a good insulator (10^{12} ohm-cm) with high permeability was announced by Snoek (1946). Shortly thereafter, Néel (1948) introduced the term ferrimagnetism to describe the novel magnetic properties of these materials. A simple ferrite is composed of two interpenetrating ferromagnetic sublattices with magnetizations $M_a(T)$ and $M_b(T)$ which decrease with increasing temperature and vanish at the Curie point, T_c. In a ferromagnetic material, the resulting saturation magnetization M would be $M_a + M_b$. See also **Ferromagnetism.** However, in a ferrite, strong antiferromagnetic interaction between sublattices results in antiparallel alignment, and $M = M_a - M_b$. In general, $M_a(T) \neq M_b(T)$, and the material behaves in most respects like a ferromagnet, exhibiting domains, a hysteresis loop, and saturation of the magnetization at relatively low applied magnetic fields. Practical values for saturation magnetization and Curie temperature range from 250–5,000 oersteds (19,984–397,887 ampere turns/meter) and from 100 to 600°C.

Ferrites resemble ceramic materials in production processes and physical properties. The high resistance of ferrites makes eddy-current losses extremely low at high frequencies. The direct current resistivities correspond to those of semiconductors, being at least one million times those of metals. Magnetic permeabilities may be as high as 5,000 and dielectric constants in excess of 100,000. Ferrites provide design advantages over strip and powder cores for filter cores, deflection transformers, and yokes and in antenna rods, pulse transformers, delay lines, waveguide elements, and a number of other electronic components.

Ferrimagnetic materials have spinel, garnet, and hexagonal structures. A typical spinel ferrite is $NiFe_2O_4$. Other ferrites may be obtained by substituting magnetic (cobalt, nickel, manganese) or nonmagnetic (aluminum, zinc, copper) ions for some of the nickel or iron ions, e.g., $Ni_{1-y}Co_yAl_xFe_{2-x}O_4$, where x and y may be varied to modify M and T_c. Yttrium iron garnet (YIG), $Y_3Fe_5O_{12}$, is the classical ferrimagnetic garnet which combines very low magnetic loss with high resistivity. Substitution of magnetic rare-earth ions (gadolinium, ytterbium, holmium, etc.) for Y and of nonmagnetic ions (gallium, aluminum) for some of the Fe ions leads to many different compositions with a wide range of M and magnetic loss. The rare-earth ions form a third magnetic sublattice with attendant magnetization M_c antiparallel to the resultant magnetization $M_{a,b}$ of the two Fe sublattices. Since M_c and $M_{a,b}$ exhibit different variations with temperature, the net magnetization may vanish twice, at F_c and at an intermediate temperature called the compensation point, T_{comp}, where $M_c = M_{a,b}$.

A typical hexagonal ferrite is $BaFe_{12}O_{19}$. Again, other magnetic ions, such as manganese, cobalt, and nickel may be introduced to produce wide variations in M and T_C. Hexagonal ferrites are characterized by large aniostropy fields with an axis of symmetry which may be either a direction of hard (planar ferrites) or easy (uniaxial ferrites) magnetization.

To distinguish among major fields of applications, ferrites can be separated into five groups: Soft, square-loop, hard, microwave, and single-crystal ferrites.

Soft ferrites have a slender S-shaped hysteresis loop with low remanence and low coercive force permitting easy magnetization and demagnetization with little magnetic loss. These ferrites are uniquely suited to low-loss inductor and transformer cores for radio, television, and carrier telephony.

Square-loop ferrites are materials exhibiting an almost rectangular hysteresis loop with two distinct states of remanence and with a coercive force of a few oersteds. All practical square-loop ferrites have a spinel structure. The Mg-Mn(Zn) system has retained an important position in computer memory applications. Lithium-nickel ferrites and more complex systems containing Li, Mn, and Al have shown stability and fast switching over a wide range of temperatures.

Hard ferrites are characterized by hexagonal structure, a hysteresis loop enclosing a large area, and a coercive force of several thousand oersteds. These ferrites can store a significant amount of magnetic energy and have been used as permanent magnets in loudspeakers, small motors, generators, and measuring instruments.

Microwave ferrites have garnet, spinel, or hexagonal crystal structure and very low electric and magnetic loss factors. In general, the required M increases with the frequency f of applications. Substituted and pure garnets, Mg-Mn-Al ferrites and Mg-Mn ferrites are used at the lower part of the microwave spectrum where $M = 200$ to 3000 gauss is adequate. In the millimeter wave region, $f = 30$ to 100 GHz, Ni-Zn ferrites ($M = 5000$ gauss) and hexagonal ferrites of various compositions may be used. Microwave ferrite devices, such as isolators, circulators, switches, phase shifters, limiters, parametric amplifiers, and harmonic generators are based upon interactions of rf signals with the ferrite magnetization.

Single-crystal ferrites of practical importance are rare-earth garnets grown in a flux of molten lead oxide. Some of these are optically transparent, permitting direct observation of magnetic domains. Interactions of infrared and visible light with the electron spins is called the magneto-optic effect. It permits electronic modulation of a beam of light which propagates through a single-crystal garnet. These devices are also of interest in laser technology.

Single-crystal, rare-earth garnet sheets have been grown on a substrate with a preferred direction of magnetization perpendicular to the plate. In these plates, tiny round magnetic domains called "bubbles" can be formed by an applied magnetic field. These bubbles can be propagated, erased, and manipulated to perform binary functions in computers, including logic, memory, counting, and switching. See also **Bubble Domain; Rare-Earth Elements and Metals.**

FERRITE CIRCULATOR. **Circulator (Microwave).**

FERRITIC STEELS. **Chromium; Iron Metals, Alloys, and Steels.**

FERROELECTRIC CRYSTALS. **Ultrasonics.**

FERROELECTRIC EFFECT. The phenomenon whereby certain crystals may exhibit a spontaneous dipole moment (which is called ferroelectric by analogy with ferromagnetic—exhibiting a permanent magnetic moment). The effect in the most typical case, barium titanate, seems to be due to a polarization catastrophe, in which the local electric fields due to the polarization itself increase faster than the elastic restoring forces on the ions in the crystal, thereby leading to an asymmetrical shift in ionic positions, and hence to a permanent dipole moment. Ferroelectric crystals often show several Curie points, domain structure and hysteresis, much as do ferromagnetic crystals.

FERROELECTRIC MATERIALS. The dielectric analogs of ferromagnetic materials. Their uses parallel those of ferromagnetic materials in such applications as magnetostrictive transducers, magnetic amplifiers, and magnetic information storage devices. Rochelle salt was the first ferroelectric material to be discovered and the barium titanate ceramics are materials of this type.

FERROMAGNETISM. The property of certain materials that gives them relative permeabilities noticeably exceeding unity, in practice from 1.1 to 10^6. Such materials generally exhibit hysteresis, hence can be used for permanent magnets. Ferromagnetism is an extreme case of paramagnetism, and results from the spontaneous alignment of the electron magnetic moments associated with spin even in the absence of an externally applied field.

Ferromagnetism exemplifies cooperative phenomena in solids. It is characterized by a spontaneous macroscopic magnetization M (magnetic moment per unit volume) in the absence of an applied magnetic field at temperatures below a critical value, known as the Curie temperature, T_C. This property is exhibited by the transition metals, iron, cobalt, and nickel; by the rare-earth metals, such as gadolinium, terbium, dysprosium, holmium, erbium, and thulium; and by a variety of alloys, compounds, and solid solutions involving the transition, rare-earth, and actinide elements. See also **Rare-Earth Elements and Metals.** Curie temperatures range from a fraction of a degree to hundreds of degrees Kelvin.

The apparent permeability of a magnetic material at microwave frequencies is affected (in the presence of a transverse, steady field) by the precession of electron orbits in the atoms. If the microwave frequency equals the precession frequency, resonance occurs and the apparent permeability reaches a sharp maximum. The resonance frequency depends upon the strength of the transverse field. Thus, a thin film of a ferromagnetic substance placed in a static magnetic field H is found to be capable of absorbing from an oscillating field whose magnetic vector is perpendicular to H at a frequency given by

$$\omega = \left(\frac{ge}{2mc}\right)(BH)^{1/2}$$

where b is the magnetic induction associated with H, e and m are the charge and mass of the electron, c is the velocity of light, and g is very near to 2, the Landé factor for free electrons.

The exchange interaction between electrons in neighboring atoms can be shown to depend on the relative orientations of the electronic spins. If it should turn out that parallel spins are favored, there is a strong tendency for all the spins in the lattice to become aligned, the transition to the ordered state corresponding to the Curie point. The concept of localized spins (e.g., *d*-electrons in the transition metals) is confirmed by neutron diffraction, but the theory is incomplete at the stage of calculating the actual magnitude and sign of the interaction.

FERROMANGANESE. Manganese.

FERRONICKEL. Nickel.

FERROSILICON ALLOYS. Silicon.

FERROVANADIUM. Vanadium.

FERRY-PORTER LAW. This law states that the critical fusion frequency (the frequency at which a flicker disappears) for the fovea of the eye is proportional to the logarithm of the luminance.

FERTILE MATERIAL. Material which is not in itself fissionable, but which, in a nuclear reactor, may be transformed into fissionable material through a nuclear transformation; e.g.:

$$^{232}\mathrm{Th} + n \longrightarrow {}^{233}\mathrm{Th} \xrightarrow[23\ \mathrm{min}]{\beta-} {}^{233}\mathrm{Pa} \xrightarrow[27.4\ \mathrm{days}]{\beta-} {}^{233}\mathrm{U}$$

the ^{233}U is fissionable.

FERTILITY. Infertility.

FERTILIZATION (Animal). Embryo; Gonads; Hormones.

FERTILIZATION (Plant). Flower; Pollination.

FERTILIZATION (Spoiled Soil). Revegetation.

FERTILIZER. A substance, but often a combination of substances, of organic composition, natural and/or manufactured, in solid or liquid slurry forms (in some cases in the gaseous phase) made available to plants to promote normal, healthy, and often vigorous growth. Most frequently added to soils, fertilizers also are applied directly to plant parts above ground (foliar sprays), in nutrient fluids as furnished in hydroponic systems, and by irrigation systems.

Unless poisoned or severely leached, some soils still may retain some of the nutrients required by growing plants and may support plants of a weak, straggly nature, with submarginal yields and poor quality for a number of years. Some soils may be generally poor, that is, they originally did not contain any of the primary plant nutrients in adequate concentration; or they may be poor soils because they have an imbalance of nutrients. It should be stressed that the proper chemical nutrients must be present along with suitable physical properties of the soil if highest yields and quality are to be achieved. A given soil may be classified as generally rich and yet lack the needed concentration of only one or two micronutrients. Further, considering the soil–plant system, soils should be customized to the needs of specific crops; or, in reverse, certain crops should not be planted on soils that are severely out of balance with their needs.

Because of the intensity of cultivation in many regions of the world, notably among the larger producers of food and fiber crops, where there is repetitive use of the same tracts of land year after year, the uniformity of nutrient content, as may have been present in the virgin soil, cannot be assumed. In fact, experience has shown that the same soil class may vary widely in nutrient content from one field to the next. The ability of soils to provide plant nutrition is a reflection of how the soils have been artificially treated (fertilized and conditioned) over the remote and recent past and what kinds of crops have been grown on them. With increasing use of control chemicals, soils require increasing frequency of testing and analysis, not only for required plant nutrients, but also for the possible presence of chemicals that tend to accumulate rather than dissipate downward to deeper ground levels or to the groundwater. Some substances may not be biodegradeable and hence may be present from one growing season to the next.

Particularly during the last few decades, astute growers have learned to depend upon reliable scientific analysis of their soils, augmented by tissue analyses of plant parts, and sometimes total plant analysis, as a basis for planning an annual fertilizing program. Perhaps not sufficiently stressed are problems that can arise from overfertilization as well as from underfertilization. Crops vary immensely in their ability to tolerate deficiencies and excesses of nutrients. Over-fertilization also can cause serious pollution problems. Over-fertilization also adds a needless element to the cost of food production.

Basic Fertilizer Functions. Major reasons for adding fertilizer to soils and plants include: (1) replenishment of chemical elements that have been reduced or exhausted by the soils to the crops previously grown or leached from the soils as the result of poor tillage practices, overirrigation, natural flooding, and, in some cases, adding nutrients that are naturally deficient in a given type of soil; and (2) customizing the nutrient content of soils to particular growing objectives.

Fertilization, in combination with irrigation, has been responsible for converting vast semiarid and arid lands with lean soils into land useful for production of a number of important food and fiber crops. This effective combination is particularly important to many of the developing countries that have large holdings of land in these categories.

Although essentially self-evident, the fact that crops exhaust the soil of key chemical ingredients is made even clearer by examination of Table 1. In the analysis, nitrogen is lost and does not appear in the ash, but other methods for analyzing for nitrogen confirm its significant content in crops. Carbon, oxygen, and hydrogen, of course, are made available to crops by way of water and the atmosphere. See Table 1.

Principal Categories of Nutrients. The nutrients required by plants fall into three categories: (1) *Primary nutrients or elements*—nitrogen, phosphorus, and potassium, because they are generally required by plants in larger amounts and are often present in more limited amounts in soil; (2) *secondary nutrients or elements*—calcium, magnesium, and sulfur, because they generally are not so limited in soils and they are required in smaller amounts; and (3) *micronutrient elements*, of which there are several and which are required in very small amounts. See Table 2.

Nitrogen Requirements

Although the requirements for nitrogen are well understood today, this was far from self-evident just a few centuries ago. See Fig. 1. It seems reasonably safe to postulate that the early growers of plants practiced rudimentary forms of fertilization without understanding what they were doing beyond "something taken from the soil must be returned." The latter observation in itself was rather profound for ancient peoples. It remained for the awakenings of chemistry in the 1600s and notably by the early scientists of the late 1700s and

TABLE 1. CONSTITUENTS OF ASH OF NORMAL CROPS

Crop and Part	Pounds of Constituent Removed per Acre of Ground (Kilograms per Hectare given in parentheses)								
	Silica	Potash	Soda	Magnesia	Lime	Ferric Oxide	Chloride	Sulfate	Phosphate
Grain	15	14	7	2	8	1	0	0	36
	(16.8)	(15.7)	(7.8)	(2.2)	(9)	(1.1)	(0)	(0)	(40.3)
Straw	233	33	1	28	12	6	4	13	11
	(261)	(37)	(1.1)	(31.4)	(13.4)	(6.7)	(4.5)	(14.6)	(12.3)
Roots	27	143	17	46	18	4	12	46	26
	(30.2)	(160)	(19)	(51.5)	(20.2)	(4.5)	(13.4)	(5.5)	(29.1)
Tops	3	89	17	72	10	3	50	39	29
	(3.4)	(99.7)	(19)	(80.6)	(11.2)	(3.4)	(56)	(43.7)	(32.8)
Hay	78	38	12	45	7	1	4	9	15
	(87.4)	(42.6)	(13.4)	(50.4)	(7.8)	(1.1)	(4.5)	(10.1)	(16.8)

TABLE 2. PRINCIPAL CATEGORIES OF PLANT NUTRIENTS

Primary Nutrients or Elements	Secondary Nutrients or Elements
Nitrogen (N)	Calcium (Ca)
Phosphorus (P)	Magnesium (Mg)
Potassium (K)	Sulfur (S)

Micronutrients

Element	General Range in Soils: Pounds/Acre	Kilograms/Hectare
Boron (B)	20–200	22.4–224
Manganese (Mn)	100–10,000	112–11,200
Zinc (Zn)	10–600	11.2–672
Copper (Cu)	2–400	2.2–448
Iron (Fe)	10,000–200,000	11,200–224,000
Molybdenum (Mo)	1–7	1.1–7.8

early 1800s to commence investigations of the technical links between plants and their nutrients.

Chilean saltpeter, $NaNO_3$, was the first of the chemical nitrogenous fertilizers. Ammonium sulfate, $(NH_4)_2SO_4$, made available as a byproduct of coal-gas produced in large quantities prior to wide use of natural gas, was also an early fertilizer, but followed Chilean saltpeter by several years. Ammonium sulfate is still an important source of nitrogen, but no longer holds the lead role in most parts of the world.

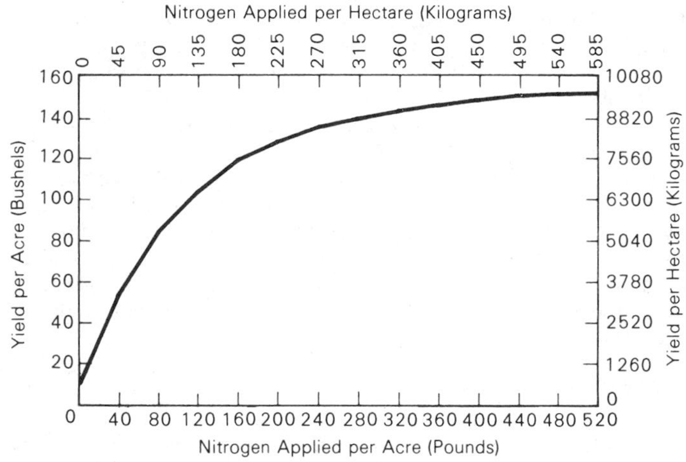

Fig. 1. Effect of nitrogen fertilizer application on crop of irrigated corn (maize) in state of Washington. It should be pointed out that the very high yields were obtained under experimental, carefully controlled conditions. Although much higher yields are obtained in some regions, the average yield in the United States is 87.3 bushels/acre (5489 kilograms/hectare). The world average is 45 bushels/acre (2829 kilograms/hectare).

In the early 1900s, attempts to fix atmospheric nitrogen in a compound that could be applied directly to soil were made. During that period, nitrogen fixation was a much-discussed aspect of industrial chemistry, much as synthetic fuels are a timely topic today. For some years, calcium cyanamide, $CaCN_2$, was produced in the United States, largely as the result, at that time, of the newfound hydroelectric energy of the Tennessee Valley region. A large facility was located at Muscle Shoals, Alabama. The last plant for making $CaCN_2$ in the United States was closed in June 1971. In Norway, also because of low-cost hydroelectric energy once available, production of calcium nitrate by way of first producing nitric acid was pioneered.

The first breakthrough in the large-scale synthesis of ammonia resulted from the work of Fritz Haber (Germany, 1913), who found that ammonia could be produced by the direct combination of nitrogen and hydrogen in the presence of a catalyst at a relatively high temperature and very high pressure. See also **Ammonia.** During the interim, much improved processes for producing synthetic ammonia have been designed.

For many years, the bulk of nitrogen fertilizers has been based upon ammonia, as synthesized. There has been an increasing trend in some major agricultural regions and countries to favor the use of liquid nitrogen for some crops. These products include anhydrous ammonia, aqua ammonia, ammonium salts in solution, and numerous combinations, such as formulations also containing phosphate salts. Because of the relatively easy solubility of these materials, some ecologists have expressed concern over so-called *nitrogen runoff.* Fortunately, however, most soils will fix ammonium ions quite rapidly, thus reducing a pollution threat.

It is interesting to observe that the compound urea, NH_2CONH_2, was discovered (in urine) as early as 1773 and first synthesized by Wöhler in 1828. See also **Urea.** Although there was an obvious connection between urea and life processes, little if any thought was given to its use as a fertilizer until after World War I, when the German firm BASF (Bosch) developed a process for synthesizing urea from carbon dioxide and ammonia. Ureaform fertilizers, now widely used, are the result of reacting urea and formaldehyde to provide a form of controlled-release nitrogen. Urea also can be coated with sulfur (itself required by some soils) to reduce the rate of solution. And slowly soluble compounds, such as isobutylidene diurea, can be made, but because of high cost, these are marketed only to horticulturists rather than growers of commercial crops.

Although not chemical fertilizers (artificially produced), the importance of natural nitrogeneous materials as returned to the soil in various degrees of *organic farming* are of extreme importance and markedly reduce the total demand for chemical nitrogen fertilizers. Significant nitrogen needs are met by returning *crop residues* to the soil. There are some negative aspects to this practice as well, however, because residues may contain insects, fungi, bacteria, and weed seeds, which generally must be controlled chemically. *Green manure* is a crop purposely grown for plowing under to enrich the soil. Animal manure also is an important contributor to the soil, not only in terms of nitrogen, but phosphate and potash as well. Also, certain crops, notably legumes, create more nitrogen for the soil than they use—

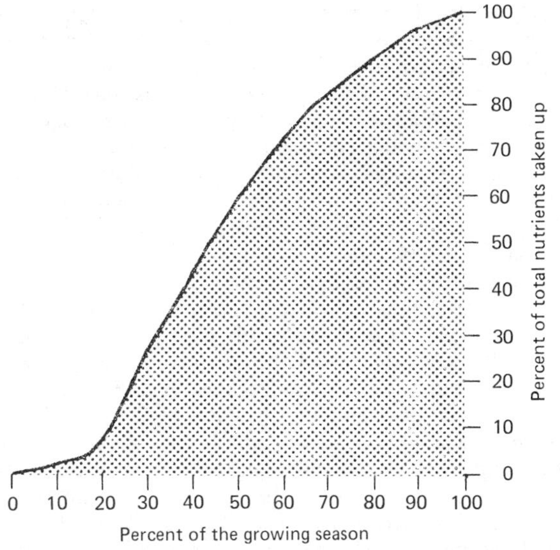

Fig. 2. Composite nutrient uptake chart representing an average of many plants. Each plant has its own specific nutrient uptake curves. (*Source*: *Leitch reference listed.*)

this as a result of bacteria that inhabit the root zones of these plants. Collectively, these kinds of materials are sometimes called *organic fertilizer.* (See Fig. 2)

Other chemical fertilizers used as sources of nitrogen include ammonium nitrate, which contains approximately 35% nitrogen by weight. Calcium ammonium nitrate contains between 20.5 and 28% nitrogen by weight, depending upon the amount of calcium carbonate added. Ammonium sulfate nitrate contains from 26 to 28% nitrogen by weight. See also **Nitrogen** and specific nitrogen compounds described in this book.

Phosphorus Requirements

Deficiencies of available phosphorus in soils are a major cause of limited crop production. Phosphorus deficiency is regarded by some authorities as the most critical mineral deficiency in grazing livestock.

Liebig and other investigators in the mid-1800s indicated that the much earlier noted fertilizing properties of bones were derived mainly from their phosphate content and that the treatment of bones with sulfuric acid increased their effectiveness in soils. With these advantages proclaimed, a large market for what was then called "chemical manure" resulted in a shortage of bones throughout the European agricultural community. Fortunately, a number of years later, several guano deposits were located in Peru and quite a bit later, phosphate minerals were located in Florida and other parts of the world. Guano fertilizers today are used mainly for special horticultural products, whereas commercial chemical phosphate fertilizers are derived from phosphate rock.

In formulating phosphate fertilizers, solubility is of particular concern. Good solubility assures availability of phosphorus to the plants. For example, in water and in alkaline and neutral soils, the apatites, which are $Ca_5(PO_4)_3R$ (where R is usually but not always fluorine), are quite insoluble and hence of little value to the soil. Tricalcium phosphate is also quite insoluble. On the other hand, these compounds are moderately soluble in acid soils and thus can be used with discretion. Somewhat in contrast, dicalcium phosphate, $CaHPO_4$, is quite soluble in acid soils and only moderately soluble in water and alkaline and neutral soils. Fortunately, monocalcium phosphate $Ca(H_2PO_4)_2$, is soluble in water and all moist soils. The presence of iron and aluminum phosphates also has an effect on total solubility, these compounds being insoluble in water, but soluble in weak acids. Some countries require total water solubility and thus monocalcium and ammonium phosphates must be used. In other areas, slight solubility in water or appreciable solubility in weak acids is adequate to meet regulations. Some humic soils can assimilate ground phosphate rock even without chemical treatment. Commonly in phosphate fertilizer manufacture, phosphorus is recovered from phosphate rock. See also **Phosphorus;** and **Phosphoric Acid.**

Nomenclature. Because the ammonium phosphate fertilizers contain two of the primary plant nutrients, it is pertinent at this point to briefly comment on the manner in which these fertilizers are named. A series of three numbers, separated by dashes, is used to indicate the primary nutrient content of fertilizer mixtures. In order, from left to right, the numbers show the percentage of nitrogen, phosphoric oxide, and potash:

5 = 10 = 5

5% Total Nitrogen (N)	10% Available Phosphoric Oxide (P_2O_5)	5% Soluble Potash (K_2O)

Single Superphosphate is produced in large quantities and is the oldest of the water-soluble phosphates. The material contains about 20% P_2O_5 by weight. Single superphosphate is made by reacting ground phosphate rock with 70% sulfuric acid. This reaction results in a solid mass of monocalcium phosphate and gypsum. The fluorine and silicon evolved are removed by water scrubbing.

Wet-Process Orthophosphoric Acid contains 30–54% P_2O_5 by weight. When sulfuric acid is added to phosphate rocks in a proportion greater than needed to make single superphosphate, orthophosphoric acid, H_3PO_4, is produced. This acid is used as an intermediate in preparing other phosphates.

Triple Superphosphate is made by acidulating phosphate rock with phosphoric acid. The concentrated triple superphosphate produced is essentially monocalcium phosphate containing very little gypsum. The principle use of triple superphosphate is in mixed fertilizers to make P_2O_5 available in water-soluble form.

Ammonium Phosphates. Although several ammonium phosphates can be prepared, only the mono- and the di-compounds are produced for fertilizer use. In some processes, anhydrous ammonia is reacted with phosphoric acid, with the resultant slurry converted to solid form by drying. The ratio of ammonia to phosphoric acid can be varied between 1 and 2, and consequently several product grades can be made.

Nitrophosphates. Phosphate rock is readily dissolved in nitric acid to yield a mixture of calcium nitrate, phosphoric acid, and monocalcium phosphate. When calcium nitrate is converted to solid form, the material is highly hygroscopic and thus not desirable for packaging and storing. This is overcome by forming calcium nitrate tetrahydrate crystals through chilling and later removal by filtering or centrifuging. Ammoniation of the mother liquor produces a mixture of ammonium phosphate, dicalcium phosphate, and ammonium nitrate. These can be concentrated and prilled or granulated. Thus, in terms of the traditional fertilizer nomenclature, that is, $\%N = \%P_2O_5 = \%K_2O$, numerous product grades are possible according to raw-material ratios and process conditions. By adding potash to 20–20–0, a formulation of 15–15–15 can be obtained, as one of numerous examples.

Nonorthophosphates. If 54% orthophosphoric acid is dehydrated to remove the remaining free water, a pyro acid is yielded. Continued heating removes more water and various insoluble compounds are formed. Evaporation by submerged combustion or under vacuum yields a eutectic between ortho and pyro acids with a P_2O_5 content of about 72% by weight. This superphosphoric acid can be ammoniated to yield liquid ammonium polyphosphate (APP) fertilizers, such as 10-34-0. Because of their strong sequestering properties, these liquids keep impurities in solution, as well as the salts of micronutrient metals which are insoluble derivatives of orthophosphoric acid.

APP fertilizers also are produced by reacting ammonia directly with wet-process phosphoric acid and dissolving the melt in ammonia solution to produce 10-34-0. By adding potash and a bit of clay, nitrogen-phosphorus-potassium suspensions, such as 13-13-13, are possible. If the melt is granulated, a solid with proportions 12-57-0 will result. Thus, by adding urea and potash, a wide variety of mixes (28-28-0, 19-19-19, and so on) are possible.

Potassium Requirements

Researchers in the early 1800s found that potassium is an essential plant nutrient although the details of its function were unknown. The potassium in wood ashes was first used in Europe, later to be replaced by sylvite, KCl, and carnallite, $KCl \cdot MgCl_2 \cdot 6H_2O$, found in deposits in Germany. During the interim, billions of tons of these materials have been located. A number of sites are continuously mined. Large reserves are found in the U.S.S.R., Canada, and east and west Germany. Significant deposits also are found in Israel and Jordan (Dead Sea region), Spain, France, the United Kingdom, and the United States. Deposits are as water-soluble salts, such as sylvite, and minerals with vaying content of magnesium.

The physiological role of potassium in life processes is described in entry on **Potassium and Sodium (In Biological Systems).** Potassium is a usual, but variable constituent of most soils. However, available potassium may be depleted through loss arising from leaching by rain water, flooding, and overirrigation, or through loss to crops continuously planted on a given tract. Thus, potassium is categorized as a primary nutrient. Potassium differs from most other essential constituents of plant cells in that it is not built into the cell as part of an organic compound, but rather it is an ion from a soluble inorganic or organic salt. Potassium ions may chelate with cellular constituents, such as polyphosphates. The ion is of the correct size to fit into the water lattice adsorbed to the proteins in the cell. In general, potassium ions are attracted to protein or other colloidal or structural units having a negative charge. Mucopolysaccharides within the cell, on the cell surfaces and of the intercellular structures, are of particular importance in holding potassium. Active centers or other configurational features of the proteins in the cell may be affected or altered by the potassium held by electrostatic or covalent binding. There are several enzyme systems which are activated by potassium.

In plants, the meristematic tissues in general are particularly rich in potassium, as are other metabolically active regions, such as buds, young leaves, and root tips. Potassium deficiency may produce both gross and microscopic changes in the structure of plants. Effects of deficiency include leaf damage, high or low water content of leaves, decreased photosynthesis, disturbed carbohydrate metabolism, and low protein content, among other abnormalities. The importance of potassium is also reflected by livestock who consume plants and feedstuffs prepared from plant materials.

Potassium fertilizers are prepared from the potassium minerals previously mentioned. The ores are crushed, beneficiated, crystallized, and dried to commercial *potash* or *muriate* (KCl), in various grades and particles containing from 60 to 62% KCl. Relatively small amounts of other potassium salts are used as fertilizers. Some vegetables and tobacco are adversely affected by high chloride concentrations and some growers prefer to use potassium sulfate, K_2SO_4, or potassium nitrate, KNO_3.

Potash is frequently applied to the soil along with salts containing nitrogen and/or phosphoric oxide, P_2O_5, in amounts varied for different soil and crop requirements. One method used is that of combining crushed muriate with moist nitrogen- and P_2O_5-containing compounds, followed by granulation of the mixture. Another method is that of dry-blending materials, such as urea, diammonium phosphate, and potash, and applying the mixture to the soil. The total water solubility of potash results in full initial K_2O availability in moist soils. However, in clays, when rainfall or irrigation is limited, excessive chloride build-up can be harmful.

Calcium Requirements

The role of calcium in biosystems is described in entry on **Calcium (In Biological Systems).** Fortunately, as the fifth most abundant element in the earth's crust, calcium is usually available to plants in abundance. Nevertheless large amounts of calcium are added to soils by virture of the use of lime (calcium oxide, CaO) as means to adjust the pH of soils. Among many reasons why soil pH is so important are the effects of acidic soils on the availability of manganese, a micronutrient, and also of aluminum, more recently identified as toxic to some crops. The use of dolomitic lime, $CaMg(CO_3)_2$, also is an effective way to correct magnesium deficiencies. Agricultural lime is not water soluble and, therefore, it will not correct soil acidity immediately after application.

Sulfur Requirements

The role of sulfur in biosystems is described in entry on **Sulfur (In Biological Systems).** Sulfur in some form is required by all living organisms. Among important sulfur-containing compounds are the amino acids cysteine, cystine, and methionine; the vitamins thiamine and biotin; and certain complex lipids, such as sulfatides, among others. In the chain from soils to plants to animals, including humans, inorganic sulfur (sulfate ion, SO_4^{-2}) is taken up by plants and converted within the plant to organic compounds (sulfur amino acids). The most important feature of sulfur in the food chain is that plants use inorganic sulfur compounds to make the aforementioned amino acids, whereas animals use the sulfur amino acids for their own processess and excrete inorganic sulfur compounds.

Although sulfur is a widespread element in the earth's crust, ranking as the 14th element in abundance, it is obviously less abundant than calcium and, further, it is not so evenly distributed. Thus some soils show sulfur deficiency. The trend toward high-analysis fertilizers without sulfur can create a need for more deliberate use of sulfur fertilizers.

Sulfur is present in ammonium sulfate, used as a fertilizer, but ammonium sulfate is only one of many fertilizers that may be used. Elemental sulfur and sulfur-containing compounds are also used for controlling various plant pests. The soils also pick up sulfur from air pollution. Nevertheless, soil analysis should be made to provide an accurate diagnosis of whether or not a particular soil may require additional sulfur.

Magnesium Requirements

The role of magnesium in biosystems is described in entry on **Magnesium (In Biological Systems).** Magnesium is generally abundant in the earth's crust, ranking 8th in abundance. Nevertheless, magnesium deficiencies are quite common. Magnesium deficiency is a fairly common cause of poor crop yields, especially among crops produced on sandy soils. Accumulation of magnesium from the soil by plants is strongly affected by the species of plant. Legumes usually contain more magnesium than grasses, tomatoes, corn (maize), regardless of the level of magnesium in the soil. A high level of available potassium in the soil interferes with the uptake of magnesium by plants, and thus magnesium deficiency can occur even though there are adequate amounts of the element in the soil. The role of magnesium highlights the systems aspects of agricultural management, exemplified by need to maintain a proper magnesium intake from pasture and from feedstuffs—because when animals are fed diets primarily of grains, a proper balance among magnesium, calcium, and phosphorus must be maintained to minimize danger of urinary calculi. Magnesium deficiency among cattle (grass tetany or grass staggers) is observed most frequently when animals are first grazed on lush grass or wheat pastures, indirectly indicating the relative low uptake of magnesium by certain crops.

Magnesium deficiences are easily corrected by applying magnesium minerals, such as kieserite or dolomite.

Micronutrients

The principal micronutrients and their deficiencies in soils of the United States are shown in Table 3. Even though the traditional micronutrients may be required only in minute quantities, deficiencies can lead to diseased crops and stunted livestock. See also entries on **Boron; Copper (In Biological Systems; Iron; Manganese; Molybdenum (In Biological Systems);** and **Zinc (In Biological Systems).**

Much more detail on all aspects of fertilizers, including worldwide consumption, methods of application, etc., can be found in the "Foods and Food Production Encyclopedia," (D. M. Considine, editor), Van Nostrand Reinhold, New York, 1981.

References

Bress, D. F., and M. W. Packbier: "Major Urea Plants," *Chem. Eng. Progress*, **73**, 5, 80–84 (1977).

Considine, D. M. (editor): "Foods and Food Production Encyclopedia," Van Nostrand Reinhold, New York, 1982.

Considine, D. M. (editor): "The Encyclopedia of Chemistry," 4th Edition, Van Nostrand Reinhold, New York, 1983.

Leitch, D.: "Matching Fertilizer Application to Crop Requirements," Chevron Chemical Company, San Francisco, California (Revised periodically).

Sheldrick, W. F.: "The Fertilizer Industry in Developing Countries," *Chem. Eng. Progress*, **75**, 10, 21–27 (1979).

Slack, A. V., and G. R. James (editors): "Ammonia, Part IV (Fertilizer Science and Technology Series), Vol. 2, Marcel Dekker, New York, 1979.

TABLE 3. STATUS OF MICRONUTRIENT DEFICIENCIES IN SOILS OF THE UNITED STATES
(Alaska and Hawaii not included)

State	Boron			Copper			Iron			Manganese			Molybdenum			Zinc		
	ND	Mod	Sev	ND	Mod	Sev	ND	Mod	Sev	ND	Mod	Sev	ND	Mod	Sev	ND	Mod	Sev
Alabama		x		x			x				x		x				x	
Arizona	x				x			x			x		x				x	
Arkansas		x		x			x			x			x			x		
California			x			x			x		x			x				x
Colorado	x			x				x		x			x			x		
Connecticut			x	x			x			x				x		x		
Delaware			x	x			x				x				x	x		
Florida			x			x			x			x		x				x
Georgia		x		x			x			x				x			x	
Idaho		x		x				x		x				x			x	
Illinois		x		x			x				x			x		x		
Indiana		x			x			x				x		x		x		
Iowa		x		x				x		x			x			x		
Kansas		x		x					x	x				x			x	
Kentucky		x		x			x			x			x				x	
Louisiana		x		x				x		x			x				x	
Maine			x	x			x			x			x			x		
Maryland			x	x			x			x				x		x		
Massachusetts			x	x			x			x				x		x		
Michigan			x			x	x					x			x		x	
Minnesota		x			x				x		x		x				x	
Mississippi		x		x				x		x			x				x	
Montana			x	x				x			x		x				x	
Nebraska		x		x					x	x				x				x
Nevada	x			x				x		x			x			x		
New Hampshire		x		x			x			x			x			x		
New Jersey			x		x			x				x		x			x	
New Mexico	x			x					x	x			x				x	
New York			x		x		x					x		x			x	
North Carolina			x		x			x				x	x				x	
North Dakota	x			x				x		x			x				x	
Ohio			x	x			x					x	x			x		
Oklahoma		x			x				x	x			x				x	
Oregon			x		x			x			x			x			x	
Pennsylvania			x	x			x				x			x		x		
Rhode Island		x		x			x				x		x			x		
South Carolina			x	x			x				x		x				x	
South Dakota	x			x				x			x		x				x	
Texas		x		x				x			x		x				x	
Utah	x			x				x				x		x				x
Vermont		x		x			x			x			x			x		
Virginia			x		x			x				x	x				x	
Washington		x		x			x			x				x				x
Wisconsin			x			x	x				x			x				x
Wyoming	x			x			x			x			x			x		

Source: University of Wisconsin

ND = no deficiency.
Mod = moderate deficiency.
Sev = severe deficiency.

Staff: "Fertilizer Situation," U.S. Department of Agriculture, Washington, D.C. (Issued several times each year).

Staff: "Annual Fertilizer Review," Food and Agriculture Organization (United Nations), Rome (Issued annually).

Staff: "Production Yearbook," Food and Agriculture Organization (United Nations), Rome (Issued annually).

Swenson, R. J.: "The Need for Soil Testing," Chevron Chemical Company, San Francisco, California (Revised periodically).

Trivedi, R. N., Chari, K. S., and V. Pachaiyappen: "Pollution Control in Fertilizer Industry," Fertiliser Assn. of India, New Delhi, 1979.

FESCUE. **Grasses.**

FESSENDEN OSCILLATOR. **Oscillator.**

FETCH. **Atmosphere-Ocean Interface.**

FETOLOGY. **Embyro.**

FETUS. **Amniocentesis; Embryo.**

FEVER. The elevation of body temperature above its normal range. The normal range varies slightly with the individual and with the time of day. The basic normal temperature is considered to be 98.6°C (37°C). Although a slightly lower body temperature during morning hours, as compared with later in the day, is normal, any swing of over one Fahrenheit degree during a 24-hour period is considered abnormal. Rectal temperatures usually are more reliable than those of the mouth or armpit. During diseases and disorders, body temperature may rise several degrees without permanent damage to organs and vessels. However, a rise above approximately 108°F (42°C) indicates very rapid metabolic rates and, as the temperature increases further, these rates are extremely difficult to reverse, absolutely necessary of course if the body temperature is to return to normal. Body temperature rise to about 112 to 114°F (44.5 to 45°C) usually is fatal, not only because of permanent damage to the neuronal cells of the brain, but because of the virtual impossibility of reversing the individual's metabolic rates at such high temperatures.

Fever occurs most commonly in infections, but may accompany a variety of other ailments. The type of fever often is characteristic for the disease, and may be described according to the constancy with which the elevated temperature is maintained. Thus, a continuous fever is one that is maintained at a fairly constant level for several days. When there are moderate fluctuations, the temperature is said to be remittent. If the temperature approaches or reaches normal during some part of the day, but rises considerably at other times, the fever is said to be intermittent. Rises in temperature may be gradual, or very sudden following chilly sensations or shaking chills. Fever may terminate slowly by lysis or suddenly by crisis.

The mechanism of fever is best appreciated when one realizes that the normal temperature regulation of the human body is extremely reliable and amazingly precise. It is interesting to note, for example, that a healthy individual (in the nude so that effects of clothing are eliminated) can be exposed to dry air temperatures as low as 50°F (10°C) or as high as 165°F (74°C) for many hours and still not alter the internal body temperature by over a degree or two. This regulation is possible because of the several mechanisms available to the body to control heat generation and heat loss and the unusual sensory arrangements that control these mechanisms. It is believed that fever results from an abnormal production of proteins which find their way into body fluids during the process of certain fever-producing diseases. These proteins react upon the hypothalamus. Available mechanisms for increasing body temperature include: (1) *vaso-constriction,* thus decreasing the flow of blood to the skin and consequently diminishing the transfer of heat from the body surface, (2) *increased metabolism* by releasing hormones to cells which step up the metabolic rate, causing increased generation of heat, and (3) *induced shivering,* termed *chill,* which always is an indication that body temperature is rising. Shivering stimulates muscular tone and hence promotes the quantity of muscularly-produced heat within the body. In the reverse process, during the lowering of a fever temperature, vasodilation commences, sweating starts, and there is decreased muscle tone. Sweating, by promoting evaporative cooling, is a very effective body coolant. With the wide use of antibiotics, the true course of a fever may not be observed. In a normal fever situation, the appearance of sweating after a long period of fever is termed the crisis, indicating that the body temperature is commencing to lower toward normal, and consequently the bacteria, virus, or other cause of the fever has been overcome.

Fever of Undetermined Origin (FUO). A low-grade fever that persists (weeks to months) in a patient without obvious cause is one of the most baffling challenges facing the physician. Frequently, such situations are not brought to the attention of a physician for a period of time during which the patient is not unduly concerned. In 1961, Petersdorf and Beeson made a marked contribution toward studying such cases by placing FUO into three categories: (1) The duration of fever is over 3 weeks; this criterion eliminates the usual febrile (fever-producing) illnesses and the fever (pyrexia) sometimes present in the post-operative patient. (2) The fever may periodically exceed 101°F (38.3°C); this criterion rules out persons, frequently young women, who "naturally" have mild hypothermia of 99.1–110.4°F (37.3–38°C). (3) The nature of the condition is truly obscure, that is, still unresolved after a diagnostic effort extending over several days. Statistics show that, in the long run, most cases of FUO are atypical manifestations of common diseases rather than rare kinds of disorders and diseases. Research has shown a host of such atypical situations, adding to the difficulty of the problem. Fever generally has daily patterns and specific diseases may present specific alterations of these patterns, thus some clues. A part of the problem thus arises from the prior administration of antipyretics, such as salicylates, which can convert a FUO to a remittant pattern with altered daily peaks. A small number of patients with FUO are not diagnosed satisfactorily, but in such cases the fever usually abates over a period of several months. It may be hypothesized that, in such cases, certain temperature-regulating mechanisms of the body remain unknown or are very poorly understood and that these mechanisms may malfunction for periods of time without serious consequences.

Relatively common diseases which are prone to cause FUO as an atypical symptom in a minority of patients include: (1) Systemic infections, such as military tuberculosis, infective endocarditis, bacteremia resulting from chronic meningococcemia, brucellosis, listerosis; (2) localized infections and abscesses, such as hepatic infections (liver; e.g., cholangitis), intraperitoneal infections, intra-abdominal abscesses, and urinary tract infections; (3) uncommonly seen diseases, such as psittacosis, toxoplasmosis, Q fever, and mycotic infections; (4) neoplasms, frequently associated with Hodgkin's disease, myeloma, and chronic lymphocytic leukemia; (5) granulomatous diseases, such as sarcoidosis; (6) inflammatory bowel disease; (7) alcoholic hepatitis and cirrhosis; (8) pulmonary emboli; (9) drug-induced fever; and (10) quite uncommonly, "false" elevation in temperature, usually as the result of a psychiatric disorder. This condition is sometimes called *factitious fever.*

References

Aduan, R. P., et al.: "Facetious Fever and Self-induced Infection," *Ann. Intern. Med.,* **90,** 230 (1979).

Deller, J. J., Jr., and P. K. Russell: "An Analysis of Fevers of Unknown Origin in American Soldiers in Vietnam," *Ann. Intern. Med.,* **66,** 1129 (1967).

Jacoby, G. A., and M. N. Swartz: "Fever of Undetermined Origin," *N. Engl. J. Med.,* **289,** 1407 (1973).

Petersdorf, R. G., and P. B. Beeson: "Fever of Unexplained Origin," *Medicine (Baltimore),* **40,** 1 (1961).

Simon, H. B., and S. M. Wolff: "Granulomatous Hepatitis and Prolonged Fever of Unknown Origin," *Medicine (Baltimore),* **52,** 1 (1973).

Simon, H. B., and G. H. Daniels: "Hormonal Hyperthermia: Endocrinologic Causes of Fever," *Am. J. Med.,* **66,** 257 (1979).

FEVER BLISTER. **Dermatitis and Dermatosis; Virus.**

FEYNMAN POSITRON CONCEPT. **Positron.**

FIBER GLASS. Glass in fibrous form. The material generally has properties similar to the glass from which it is made except that the tensile strength may be increased up to over 100 times that of the base glass. History records the use of strands of glass for decorating vases by the early Egyptians. The famed Venetian craftsmen had a limited knowledge of drawing glass fibers, but it was not until the 1930s and 1940s that glass producers perfected a way to make fibers commercially.

Two forms of glass fibers are produced; a staple or short-length fiber or monofilament, and continuous strand composed of many-monofilaments bonded together in a threadlike form. The continuous strands are often chopped into short lengths, ranging from $\frac{1}{8}$-inch to 2 inches (3 millimeters to 5 centimeters) or longer, and this product is referred to as *chopped strand.* Staple fibers are used for thermal and acoustical insulation. Continuous strands are used for yarn (as in fabrics), tire cord, and plastic reinforcement. Both thermoset and thermoplastic resins are reinforced by chopped strand. Some varieties of chopped strand are also converted to monofilament paper, which is utilized in roofing shingles, flooring materials, and other products. Products using continuous filaments have excellent tensile strength (as high as 400,000 pounds per square inch; 2759 megapascals) compared with organic fiber strengths of less than 150,000 psi (1034 megapascals).

The base glass is made by heating raw materials, such as silica sand, limestone, dolomite, clay, boric acid, soda ash, and other minor ingredients, in a high-temperature furnace. See also **Glass.** Typical glass-fiber compositions are given in Table 1. Fiber made from electrical-grade glass E is used most commonly for yarn, tire cord, and plastic reinforcement because of its high strength and electrical properties. Specialty glasses, although low in volume of production, fill important needs. S glass is a superior-strength glass primarily for defense applications, such as missile cases. C glass is more chemically resistant than E and is used for battery separator plates and chemical filters. Alkali glass A is used to some extent in the production of plastic reinforcement products.

Continuous-Fiber Products. In the "direct-melt" process, shown in the accompanying figure, raw materials are fed to a tank furnace to

TABLE 1. FORMULATIONS FOR TYPICAL FIBER GLASS TYPES.

Ingredient	Type of Glass, wt%				
	E	Insulating	A	S	C
SiO_2	54	63	73	64	65
Al_2O_3	14	5	1	24	4
MgO	4	2	2	10	3
CaO	19	6	10	—	14
R_2O	0.5	16	14	—	8
B_2O_3	8	7	—	0–2	6
Fe_2O_3	0.3				
F_2	0.2	1			

NOTE: R = rare-earth element.

convert the mixture to glass. The glass flows to forehearths, which have platinum-alloy bushings or spinnerettes in the bottom. The bushings contain many holes, or orifices, each of which supplies a small stream of molten glass from which monofilaments are drawn. Mechanical attenuation, which produces a forming package, is accomplished by attaching the fibers to a rotating drum which turns up to 20,000 peripheral feet (6,000 meters) per minute.

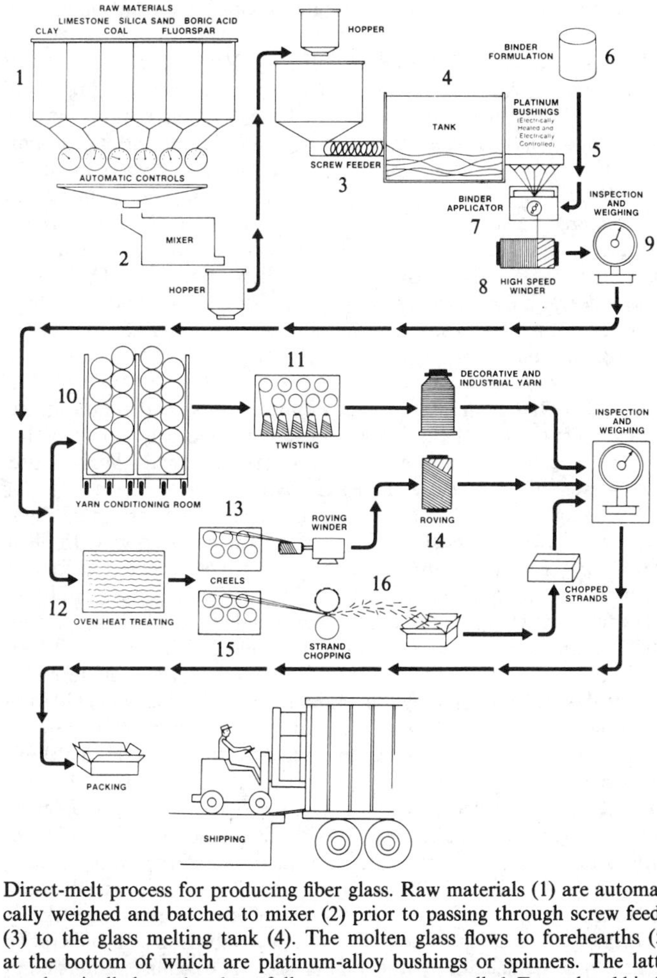

Direct-melt process for producing fiber glass. Raw materials (1) are automatically weighed and batched to mixer (2) prior to passing through screw feeder (3) to the glass melting tank (4). The molten glass flows to forehearths (5), at the bottom of which are platinum-alloy bushings or spinners. The latter are electrically heated and carefully temperature-controlled. Formulated binder material (6) is applied to the newly formed filaments (7) prior to high-speed winding (8). After weighing and inspecting (9), the wound multi-filament (in the form of a strand) follows one of three paths in accordance with desired end product. In making decorative and industrial yarn, the packages are placed in a conditioning room (10) prior to twisting (11). For the production of roving and chopped strand, the material from inspection operation (9) passes to an oven (12), where the filaments are heat-treated. This is followed by creeling (13) and roving winding (14) for production of roving. Following creeling (15), chopped strands (16) may also be made. There are several additional weighing and inspecting stations. (*PPG Industries.*)

The "marble-melt" process consists of producing 1-inch (2.5-centimeter) marbles by a separate tank furnace. The marbles are then fed to a bushing unit, which is heated by electrical resistance. From this point, the process is identical to the direct-melt process.

Sizing. Because of the basic character of glass, the filaments are somewhat fragile and tend to abrade each other in close contact. A protective coating or sizing is necessary for the production, processing, and end use of all continuous-fiber glass products. Generally, a fiberglass sizing or binder for textile or reinforcement products may contain (1) a film former, generally resinous in nature, that forms a strand or thread from grouped monofilaments; (2) a lubricant to aid in processing and end use of the fiber-glass product; and (3) additives to accomplish specified purposes, e.g., providing antistatic characteristics.

Sizings for plastic reinforcement also will have a coupling agent, such as a chrome complex, a silane, or combination of these two, to assure an interfacial bond between the glass surface and the resin matrix. Yarns for weaving normally have an oil-starch sizing. These coatings are applied before winding of the forming package.

Filaments ranging in number from 20 to 2000 are then gathered together as a thread or strand before winding. As shown by Table 2, the filaments are available in many diameters and are letter-designated. In continuous-fiber products, filaments range from designation B to U.

TABLE 2. CODING SYSTEM FOR FIBER GLASS DIAMETERS.

Designation	Diameter $\times 10^{-5}$ in.
AAA	<3.0
AA	3.0– 5.9
A	6.0– 9.9
B	10.0– 14.9
C	15.0– 19.9
DE	23.0– 27.9
G	35.0– 39.9
H	40.0– 44.9
K	50.0– 54.9
P	70.0– 74.9
Q	75.0– 80.0
R	81.0– 85.0
S	86.0– 90.0
T	91.0– 95.0
U	96.0–100.0

NOTE: To convert to microns (micrometers), multiply by 25.4×10^{-2}.

Continuous-filament products are designated by a letter-number system which specifies properties important to end users. For example, listing a strand as ECK67.5 (200) 630 indicates that the material is made from E glass and is a C continuous fiber of K diameter; the strand contains 200 monofilaments and has a yield of 67.5×100, or 6,750 yards per pound. (In the metric system, yield is expressed in Tex, or grams per kilometer. The number 486,235 divided by yards per pound is equal to Tex. Hence the yield in this case would be 72 Tex). This product is coated at the bushing with 630 binder, an oil-starch type making it suitable for weaving into fabric.

Forming packages composed of wound strands normally are not supplied to industrial users without further processing. Strands are twisted and plied before being woven into fabric. A plied yarn, for example, is coated with a latex binder before being used as a tire-cord reinforcement.

End products made from continuous strand include fire-resistant curtains, reinforced tires and transmission belts, and many reinforced plastic items, such as boats, auto bodies, corrosion-proof pipe, roofing panels, and missile cases.

Staple-Fiber Products. Monofilament, short-length fibers are used for thermal and acoustical insulation, filtration, and cushioning. These products are made in basically three ways: In the high-temperature blast-jet process, 30-mil-diameter fibers or rods first are produced by a bushing-type process. The coarse primary rods then are filamentized by a high-temperature, high-velocity blast burner. The blown mass of filaments is collected on a conveyor belt and bonded together by an inert thermosetting resin in a manner which creates many tiny air spaces throughout the material. The bonding process may be modified to produce flexible blankets, rigid board, or special molded shapes, such as pipe insulation. Coatings, facings, or jackets usually are applied for reflective, vapor-barrier, or decorative purposes.

In another process (replacing the high-temperature blast-jet process in some areas), a stream of molten glass is directed onto a rapidly rotating wheel which contains holes in its periphery. Centrifugal force directs glass through each hole to create fibers. A third process involves conversion of 2-to-6-inch (5-to 15-centimeter) chopped strand and textile-type yarns into separate and random monofilaments by a garnetting machine.

Properties of Staple Fibers. Fiber diameters range from AAA to G (Table 2), with the largest production volume in the range C–G. Thermal conductivity of glass-fiber products is influenced by fiber diameter, density or compactness of the fiber mass, and temperature conditions. Generally, thermal conductivity ranges between 0.20 and 0.80 (Btu)(in.)/(hr)(ft^2)(°F). In metric units, this is: 0.029 and 0.115 watt/meter-Kelvin (W/m·K).

Temperature of applications ranges between near absolute zero to 593°C (1100°F), or to the softening point of the glass. Unbonded mat is used at extreme temperature conditions, whereas standard bonded insulation covers the range of −40 to 232°C (−40 to 450°F). Fiber-glass acoustical products are particularly good energy absorbers at the frequency levels of 500–2000 Hz. Fiber diameter, density, and method of mounting control the absorbing characteristics.

L. Dow Moore, PPG Industries, Pittsburgh, Pennsylvania.

FIBERGLASS INSULATION. Insulation (Thermal).

FIBERIZATION. Pulp (Wood) Production and Processing.

FIBER (Kapok). Silk Cotton Trees.

FIBER (Ramie). Ramie Fibers.

FIBER-REINFORCED COMPOSITES. An early prototype of these materials can be said to be reinforced concrete. It resembles them in that it consists of linear elements of one material (steel) in a matrix of another (concrete) and owes its effectiveness to the fact that these two materials form a bond quite readily and that the steel contributes its high tensile strength to the composite, since concrete alone is stronger in compression than tension.

While the steel is added to reinforce concrete largely in the form of rods or bars, many of the newer types of composite materials, both those under investigation and those which have come into use, contain linear additives far smaller in cross section, that is, in the form of fibers, often in quite small cross section area so that they are known as "whiskers." Among the most widely used are the glass-fiber reinforced plastics which have the further advantage of relatively low cost. Like glass fibers, those of boron, tungsten, aluminum oxide, silicon carbide, carbon, beryllium, and boron nitride show very high moduli of elasticity and have come into some use as reinforcing materials. They are of particular interest, of course, in applications where a very high strength-to-weight ratio is desired, such as in helicopter rotor blades, aircraft structures, re-entry vehicles and gas turbine engines. A major problem that arises in this technology is that of obtaining effective bonding between the fiber and the matrix, and accordingly a number of methods of combining the two have been developed, which also determine the alignment and spacing of the fibers. The three major methods are:

1. *In-situ fiber-reinforced composites* are produced by mixing selected powders and controlling the mechanical working so as to elongate the strengthening phase.

2. *Honeycomb composites* contain continuous fibers or filaments of the strengthening phase, in which each filament is surrounded by a continuous matrix of the ductile phase.

3. Chopped fibers in a powder metal matrix, in which fibers averaging up to $\frac{1}{2}$ inch long are dispersed in a suitable metal powder matrix and aligned after the blending operation.

There are a number of methods of producing fibers for incorporation in a matrix. They include (1) drawing from the melt, as applied particularly to glass fibers, (2) extrusion, which has been used with aluminum oxide, (3) ejection through a spinnerette at high temperature, which has been used for silica and silica-carbon combinations, (4) pyrolysis, which has been used with various plastic fibers to produce from them carbon and graphite filaments and (5) vapor deposition, which is especially effective for the production of high-strength single-crystal fibers.

FIBERS. Long, thin, threadlike, strong, and flexible, fibers, both natural and synthetic, are the fundamental structural components of yarn, thread, string, rope, paper, woven, and matted goods. Fibers usually exhibit considerable elasticity, the ability to return to their original dimension without permanent stretching. Most fibers tend to interlock or mechanically bond with other fibers, forming fiber matrices. Natural fibers are usually quite *un*uniform; for example, cotton staple ranges from $\frac{1}{2}$ to $2\frac{1}{2}$ inches (1.3 to 6.4 centimeters) in length, with a diameter of about 1/1,000 inch (0.025 millimeter). Some synthetic fibers are made in the form of very thin filaments and thus may be quite uniform, with fiber length controlled and tailored for specific applications. Many materials required for the production of synthetic fibers are derived from petroleum and, along with synthetic plastics and resins, the synthetic fiber industry contributed importantly to the great growth of the chemical and petrochemical industries. To some extent, the essential raw materials for synthetic fibers are threatened because of competition for the same raw materials that are consumed for power in terms of fuels—both petroleum and natural gas.

The use of fibers dates back to antiquity. Very early uses and still found among primitive peoples are *tying* applications, using easily obtainable fibers from plants. Fibers have been used for centuries for making rough cordage, huts, and rope suspension bridges. Broomcorn and broomroot fibers have long been used for what might be termed *brush* applications. Straw, bamboo, rattan, and palm leaves were among the early *plaiting* and *rough-weaving fibers* for use in furniture making and basketry and, of course, are still extensively used in various parts of the world for these purposes. The use of various reeds, husks, and grasses as *filling fibers* is very old, but large tonnages of such fibers still are used for packing materials, upholstery padding, and like applications.

The use of fibers in nonwoven sheetlike products is quite old, although there has been a resurgence of interest in so-called nonwovens during recent years. Felts, paper, and, more recently, some of the nonwoven disposable garments and products, notably for hospital use, are representative of the use of fibers for what might be termed *matting* and webbing applications. But by far the most important use of fibers, and the application most often visualized in connection with their use, is for the manufacture of knitted and woven textile fabrics and products. Spinning of the fibers to make yarn increases the utility of fibers for uses which far exceeded the imagination of those persons who first applied fibers in their cruder forms.

Although fibers can be classified in numerous ways, in terms of present-day technology, they are fundamentally classified as (1) natural fibers, and (2) synthetic fibers. The principal natural fibers are cotton, wool, and to a much lesser extent, silk, flax, and mohair. Synthetic fibers have made inroads into the use of all natural fibers, but the greatest impact has occurred in connection with the latter three fibers. Cotton continues to be a major textile fiber, measured in terms of billions of pounds used per year. Cotton is one of the most versatile of all fibers and blends well with synthetics. This is also true of wool, but to a somewhat lesser extent. See also **Cotton; Flax;** and **Silk.**

Synthetic Fibers. Introduced in 1910 as a substitute for silk, rayon was the first artificial or synthetic fiber. Rayon, of course, differs completely in chemical constitution from silk. Rayon typifies most reconstituted or synthetic fibers which perform almost as well and, in a number of respects, far better than their natural "counterparts." Some of the more recently developed synthetic fibers have little if any resemblance to naturally available fibers and thus entirely new types of end-products with previously unobtainable end-qualities are available.

It is interesting to note that some authorities define a synthetic fiber as a "noncellulosic fiber of synthetic origin," a definition which excludes rayon and acetate. Other authorities, however, include rayon and acetate, along with nylons, polyesters, acrylics, and others, in

GENERIC DESIGNATIONS AND DEFINITIONS OF SYNTHETIC FIBERS.

NOTE: Unless otherwise noted, moisture regain is stated for a relative humidity of 65% at 70°F (21.1°C). RT = regular tenacity; HT = high tenacity; IT = intermediate tenacity. * indicates variation of this property with particular brand of fiber.

ACETATE FIBERS

A manufactured fiber in which the fiber-forming substance is cellulose acetate. Where not less than 92% of the hydroxyl groups are acetylated, the term triacetate may be used as a generic description of the fiber. A portion of the molecule may appear as:

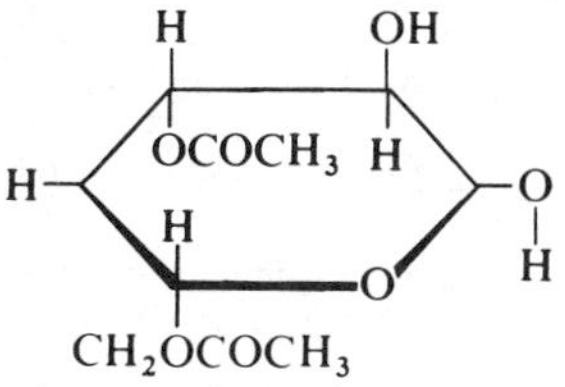

(acetate)

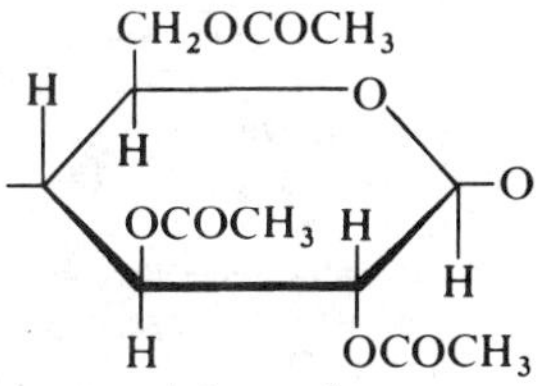

(triacetate)

Specific gravity: 1.3–1.32
Moisture regain: 3.2% (triacetate); 6.3–6.5% (acetate)
Tensile strength: 18–22 × 10^3 psi (124–152 MPa) (triacetate)
20–24 × 10^3 psi (138–166 MPa) (acetate)
Excellent to impervious to aging. Good resistance to mildew discoloration and sunlight (acetate), although there may be some loss of strength from long exposure to sunlight. Triacetate has poor resistance to sunlight. Fair resistance to abrasion.
Attacked by strong oxidizing agents. Resists common solvents, normal hypochlorite, peroxide bleaching. Dissolves or swells in acetone, ketones, trichloroethylene, concentrated and glacial acetic acid, and methylene chloride.
Triacetate does not stick when ironed at cotton setting temperature, 450°F (232°C). Melts at 572°F (301°C). Acetate sticks at 350–375°F (177–191°C). Softens at 400–445°F (204–230°C). Melts at 500°F (260°C).
Dyes: Dispersed and developed dyes are commonly used. Acid dyes are used for printing. Solution dyed available.
Types available:
Triacetate (filament)
Acetate (filament and staple)

ACRYLIC FIBERS

A manufactured fiber in which the fiber-forming substance is any longchain synthetic polymer composed of at least 85% by weight of acrylonitrile units:

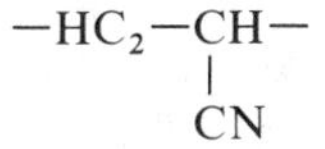

A portion of the molecule may appear as:

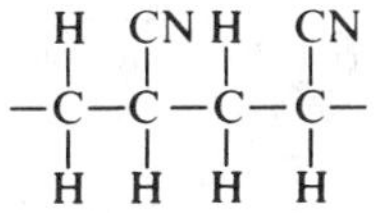

Specific gravity: 1.16–1.18
Moisture regain: 1.0–2.5%
Tensile strength: 30–54 × 10^3 psi (207–373 MPa)
Excellent resistance to mildew and aging. Good resistance to sunlight and abrasion.
Good resistance to bleaches and common solvents.
Generally good resistance to mineral acids and weak alkalis.*
Safe ironing temperature up to 300°F (150°C).*
Does not support combustion.
Dyes: Disperse and cationic.*
Principal brands:
Acrilan®, Monsanto (staple)
Creslan®, American Cyanamid (staple and tow)
Orlon®, DuPont (staple and tow)
Zefran®, Badische (staple)

AREMID FIBERS

A manufactured fiber in which the fiber-forming substance is a longchain synthetic polyamide in which at least 85% of the amide linkages are attached directly to two aromatic rings. Amide linkage: $-\underset{\underset{O}{\|}}{C}-NH-$
Specific gravity: 1.38–1.44
Moisture regain: 4.5–7% (at 55% RH)
Tensile strength: 90–400 × 10^3 psi (621–2760 MPa)
Excellent resistance to mildew and aging. Prolonged exposure to sunlight causes deterioration, but fibers are self-screening. Good abrasion resistance.*
Some are degraded by bleaching; others are not affected. No degradation in solvents, except slight loss of strength from exposure to sodium chlorite.*
Unaffected by most acids, except some strength loss after long exposure to hydrochloric, hydrobromic, nitric, and sulfuric acid. Generally good resistance to alkalis.*
Difficult to ignite—does not propagate flame—does not melt. Decomposition temperature is from 700 to 930°F (371 to 499°C).*
Dyes: Industrial yarn is nondyeable. Staple is dyeable with cationic dyes.
Principal brands:
Kevlar®, DuPont (filament)
Nomex®, DuPont (staple, tow and filament)

FLUOROCARBON FIBERS

Fiber formed of longchain carbon molecules whose available bonds are saturated with fluorine. A portion of the molecule may appear as:

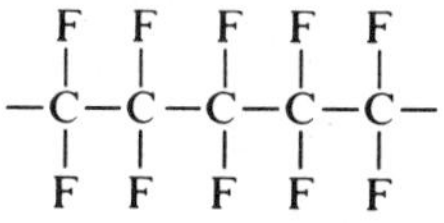

Specific gravity: 0.8–2.2
Moisture regain: 0
Tensile strength: 25–115 × 10^3 psi (173–794 MPa)
Good to excellent resistance to mildew, aging, sunlight, and abrasion.
Essentially inert to bleaches and solvents except for alkali metals at high temperature and/or pressure. Fluorine gas and chlorine trifluoride react with fibers at high pressures and temperatures.*
Essentially inert to acids and alkalis.
Very heat resistant. Usually can be safely handled from −350 to +550°F (−212 to +288°C).* Melts between 550 and 620°F (288 and 327°C).*
Dyes: Some cannot be dyed; others can be pigmented and dyed with selected solvent system.
Principal brands:
Gore-Tex®, W. L. Gore (expanded PTFE staple, filament, tow, and slit film-RT)
Teflon®, DuPont (TFE multifilament, staple, tow and flock; FEP monofilament)

GLASS FIBERS

A manufactured fiber in which the fiber-forming substance is glass.
Specific gravity: 2.48–2.69
Moisture regain: None
Tensile strength: 313–700 × 10^3 psi (2160–4830 MPa)
Not attacked by mildew, although binder may be affected by it. Excellent resistance to aging and sunlight.
Unaffected by bleaches and solvents.
Resists most acids and alkalis.
Nonburning. Generally holds 75% tensility up to 650°F (343°C). Softens between 1560 and 1778°F (843 and 970°C). Melts at 2720°F (1493°C).
Dyes: Resin-bonded pigment systems. Vat, acid, or chrome dyes will tint.
Available from numerous manufacturers.

MODACRYLIC FIBERS

A manufactured fiber in which, when not qualified as rubber or anidex, the fiber-forming substance is any longchain synthetic polymer composed of less than 85%, but at least 35% by weight of acrylonitrile units:
A portion of the molecule may appear as:

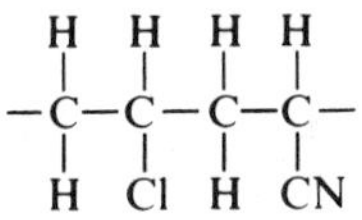

Specific gravity: 1.35–1.37
Moisture regain: 2.5–3.0%
Tensile strength: 29–47 × 10^3 psi (200–324 MPa)

Good to excellent resistance to mildew, aging, and sunlight. Good resistance to abrasion.
Good resistance to bleaches, dry-cleaning fluids, and most common solvents. Dissolves in warm acetone and acrylic-type solvents.
Resistant to most acids and good resistance to weak alkalis; some discoloration may result.*
Boiling-water shrinkage, about 1%. Good resistance to shrinkage in dry heat, with about 5% shrinkage at 390°F (200°C). Pressure and heat at 300°F+ (150°C+) may cause stiffening and discoloration.*
Does not support combustion.
Dyes: Neutral-premetalized, cationic (basic), and disperse.*
Principal brands:
Monsanto (staple)
Verel®, Eastman (staple-regular)

NYLON FIBERS

A manufactured fiber in which the fiber-forming substance is a longchain synthetic polyamide in which less than 85% of the amide linkages are attached to two aromatic rings. Amide linkage: —C(=O)—NH—

A portion of the Nylon 6,6 molecule, based upon hexamethylene diamine and adipic acid may appear as:

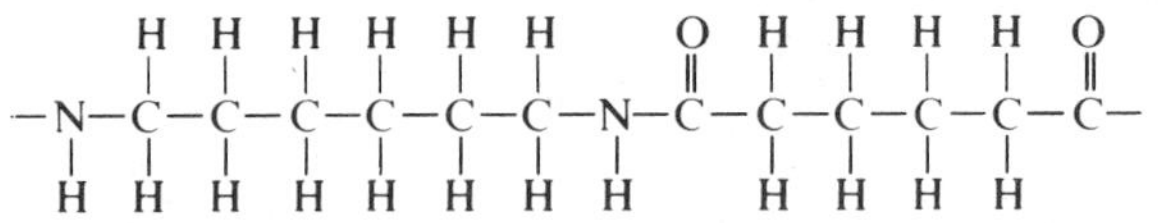

A portion of the Nylon 6 molecule, based upon caprolactam, may appear as:

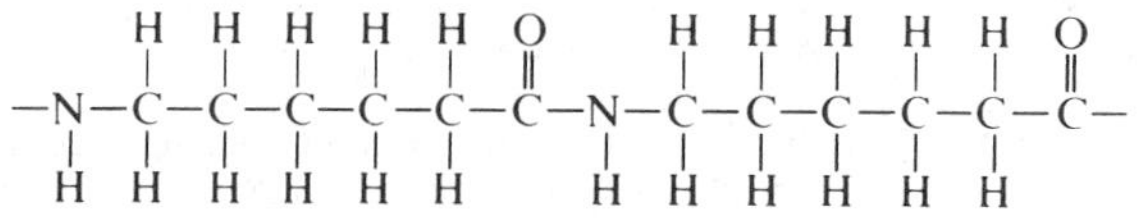

Specific gravity: 1.03–1.14 (most = 1.14)
Moisture regain: 2.8–5%*
Tensile strength: 40–134 × 10^3 psi (276–925 MPa)*
Excellent resistance to mildew, aging, and good-to-excellent resistance to abrasion. Prolonged exposure to sunlight causes some deterioration.
Excellent resistance to bleaches and other oxidizing agents. Generally insoluble in most organic solvents except some phenolic compounds.
Strong oxidizing agents and mineral acids may cause degradation of some brands. However, generally unaffected by most mineral acids, except when hot. Dissolves with partial decomposition in concentrated solutions of hydrochloric, sulfuric, and nitric acids.* Substantially inert in alkalis.
Sticking temperature is about 445°F (229°C).* Melts between 480 and 525°F (249 and 274°C).* Some yellow slightly if held at 300°F (150°C) for several hours. Decomposes between 600 and 730°F (316 and 388°C).*
Dyes: Has marked affinity for all types of dyestuffs, including pigment, direct, acid, premetalized acid, disperse, chrome, and vat colors, including complex types.*
Principal brands:
Nylon 6, DuPont and others (staple, monofilament and filament-RT and -HT; staple and tow)
Nylon 6,6, DuPont and others (staple and tow; monofilament and filament-RT; and filament-HT)
Qiana®, Du Pont (filament-RT)

OLEFIN FIBERS

A manufactured fiber in which the fiber-forming substance is any long-chain synthetic polymer composed of at least 85% by weight of ethylene, propylene, or other olefin units. A portion of the molecule may appear as:

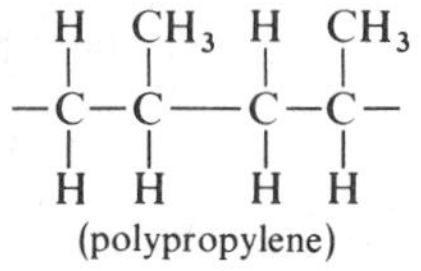

(polypropylene)

Specific gravity: 0.9–0.96
Moisture regain: Negligible (polyethylene); 0.01–0.1% (polypropylene)
Tensile strength: 11–90 × 10^3 psi (76–621 MPa)*
Not attacked by mildew. Good to excellent resistance to sunlight, abrasion, and aging.
Resistant to bleaches and most solvents, but some swelling in chlorinated hydrocarbons at room temperature and dissolves at 160°F (71°C) and higher.*
Excellent resistance to acids and alkalis, with exception of oxidizing agents, such as chlorosulfonic acid and concentrated nitric acid.
Softens at 225–235°F (107–113°C) (polyethylene); at 285–330°F (141–166°C) (polypropylene).
Melts at 230–250°F (110–121°C) (polyethylene); at 320–350°F (160–177°C) (polypropylene).
Dyes: Traditionally fibers are pigmented during manufacture, but some can be dyed with disperse, acid, and chelating dyes and certain vats, sulfurs, and azoics.
Types available:
Polyethylene (monofilament—conventional low density; and high density)
Polypropylene (staple and tow-isotactic; mono- and multifilament-isotactic)

POLYESTER FIBERS

A manufactured fiber in which the fiber-forming substance is any long-chain synthetic polymer composed of at least 85% by weight of an ester of a substituted aromatic carboxylic acid, including but not restricted to substituted terephthalate units:

$$p(\text{—R—O—}\underset{\|}{\text{C}}\text{—}C_6H_4\text{—}\overset{}{\underset{\underset{\text{O}}{\|}}{\text{C}}}\text{—O—})$$

and parasubstituted hydroxybenzoate units:

$$p(\text{—R—O—}C_6H_4\text{—C—O—})$$

A portion of the molecule may appear as

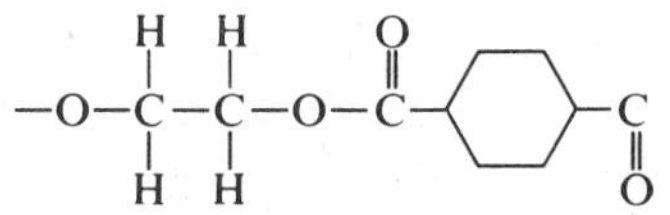

Specific gravity: 1.34–1.39 (most = 1.38)
Moisture regain: 0.4%
Tensile strength: 33–165 × 10^3 psi (228–1139 MPa)*
Good to excellent resistance to mildew and sunlight, although prolonged exposure to full sunlight degrades some brands (strength loss). Abrasion resistance ranges from good to excellent.
Good to excellent resistance to bleaches, soaps, synthetic detergents, dry-cleaning agents, sea water, and perspiration. May be soluble in some phenolic compounds.
Sticking temperature is 440–445°F (227–230°C). Melts between 480–500°F (249–260°C).*
Good resistance to most mineral acids. Dissolves with partial decomposition in concentrated sulfuric acid. Good resistance to weak alkalis. Moderate resistance to strong alkalis.*
Dyes: Disperse, azoic, and cationic dyes.*
Principal brands:
Avlin®, Avtex (filament-RT)
Dacron®, DuPont (staple and tow; partially oriented filament; filament-RT; filament-HT)
Encron®, American Enka (staple-RT; Filament-RT; producer-textured filament)
Fortrel®, Fiber Industries Div. Celanese (staple, RT AND HT; filament, RT and HT)
Kodel®, Eastman (staple, RT, IT, and HT; filament-RT)
Monsanto®, (staple-RT; partially oriented filament; producer-textured filament)
Trevira®, Hoechst (staple, partially oriented filament; filament-HT)
A.C.E.®, Allied Chemical (filament-HT)
S-3, S-3H®, IRC (filament-HT)

RAYON FIBERS

A manufactured fiber composed of regenerated cellulose, as well as manufactured fibers composed of regenerated cellulose in which substituents

GENERIC DESIGNATIONS AND DEFINITIONS OF SYNTHETIC FIBERS (*continued*)

have replaced not more than 15% of the hydrogens of the hydroxyl groups. A portion of the molecule may appear as:

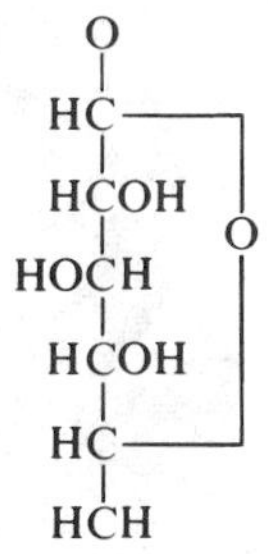

Specific gravity: 1.46–1.54
Moisture regain: 11–13%
Tensile strength: 28–66 × 10^3 psi (193–455 MPa)*
Attacked by mildew. Resistant or stable to aging.* Mostly good resistance to sunlight and abrasion, but long exposure may yellow some intermediate rayons.*
Not affected by solvents. Insoluble in common organic solvents. Some attacked by strong oxidizing agents, but generally not damaged by hypochlorite or peroxide.*
Most rayons behave to acids much as cotton. Hot dilute or cold concentrated acids cause disintegration of fibers. Strong alkaline solutions cause swelling and reduce strength.*
Fibers do not melt, but may lose strength above 300°F (150°C) and decompose between 350 and 464°F (177 and 240°C).*
Dyes: Direct, vat, fiber-reactive, sulfur and pigment.*
Principal types and brands:
Cuprammonium, available from several sources (filament).
Viscose, available from several sources (filament and staple-RT and IT)
Avril, Avril Prima, Avril III®, Avtex® (staple-HT)
Zantrel 700®, American Enka (staple)

SPANDEX FIBERS

A manufactured fiber in which the fiber-forming substance is a longchain synthetic polymer comprised of at least 85% of a segmented polyurethane. A portion of the molecule may appear as

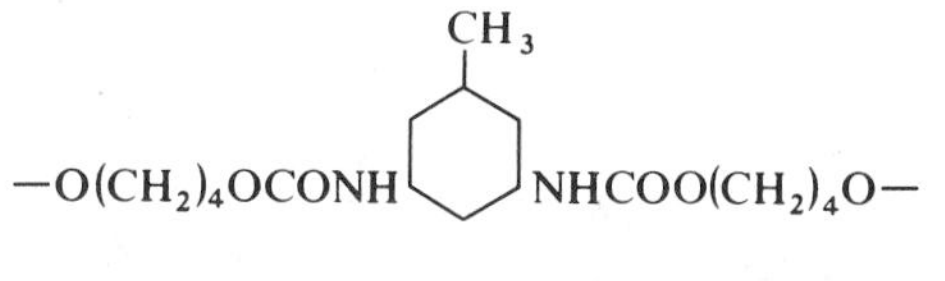

Specific gravity: 1.2
Moisture regain: <1.0–1.3
Tensile strength: 11–15 × 10^3 psi (76–104 MPa)
Good to excellent resistance to mildew, aging, sunlight, and abrasion.
Good resistance to deterioration by bleaches, but some discolored slightly by hypochlorite bleaches. Resistant to solvents, including dry-cleaning fluids, and oils, except glycols.*
Good resistance to mild acids and alkalis, but may be degraded by strong acids and alkalis at high temperatures. Some are slightly yellowed by dilute hydrochloric and sulfuric acids.*
Sticking point from 347 to 420°F (75 to 216°C).* Melts at 511–518°F (267–269°C).*
Dyes: Good affinity for most classes of dyes, but disperse, acid, and premetalized dyes are generally preferred.*
Principal brands:
Glospan/Cleerspan®, Globe (multifilament)
Lycra®, DuPont (coalesced monofilament)
Numa®, Ameliotex® (multifilament)

FOR COMPARISON, THE FOLLOWING NATURAL FIBERS ARE INCLUDED

COTTON

General formula, $(C_6H_{10}O_5)_x$. Chemical composition is cellulose.
Specific gravity: 1.54
Moisture regain: 7.0–8.5%
Tensile strength: 60–120 × 10^3 psi (414–828 MPa)
Cotton fibers are quite resistant to thermal degradation. After about 5 hours at 250°F (121°C), material yellows. Decomposes above 300°F (150°C).
Cotton is attacked by cold concentrated acids and by hot dilute acids. Alkalis cause mercerization, but without danger. Cotton is quite resistant to most solvents.
Cotton fabrics have an excellent hand, good abrasion resistance, excellent pilling resistance, excellent stability to repeated launderings (if preshrunk), fair sunlight resistance, excellent colorfastness, good wash and wear performance (if resin-treated), and good wrinkle resistance (if resin-treated).
Safe ironing temperature: 425°F (219°C)
Dyes used: Direct, vat, azoic, basic, mordant, pigment, sulfur, and fiber-reactive.

WOOL

General formula, $(C_{42}H_{157}O_{15}N_5S)_x$. Chemical composition is keratin.
Specific gravity: 1.32
Moisture regain: 11–17%
Tensile strength: 17–29 × 10^3 psi (117–200 MPa)
Wool shows marked effects of thermal degradation above 212°F (100°C). Scorches at 400°F (204°C); chars at 570°F (299°C).
Wool is destroyed by hot sulfuric acid; otherwise it is quite resistant to acids. Strong alkalis destroy the material; it is attacked by weak alkalis. Wool is quite resistant to most solvents.
Wool fabrics have an excellent hand, fair abrasion resistance (but good in carpets), good pilling resistance (pills form, but tend to break off), poor stability to repeated launderings, good sunlight resistance, good colorfastness, poor wash and wear performance, and good wrinkle resistance.
Safe ironing temperature: 300°F (149°C)
Dyes used: Acid, milling, chrome, mordant, vat, and indigo.

SILK

Silk is comprised essentially of the protein (fibroin).
Silk fabrics have an excellent hand, fair abrasion resistance, good pilling resistance, good stability to repeated launderings, poor sunlight resistance, good colorfastness, poor wash and wear performance, and good wrinkle resistance.
Safe iron temperature: 300°F (149°C)

FLAX

A bast fiber.
Flax fabrics (linen) have an excellent hand, fair abrasion resistance, fair pilling resistance, good stability to repeated launderings, fair sunlight resistance, excellent color fastness, very poor wash and wear performance, and poor wrinkle resistance.
Safe ironing temperature: 450°F (232°C).

the full spectrum of synthetic fibers. The reasoning behind the fine distinction is that, with cellulose-derived synthetics, one commences with a naturally fibrous material and grossly modifies it, whereas with most other synthetics, the starting materials are strictly chemicals that bear no relationship whatever to a fibrous structure, many of the starting ingredients actually being in the gaseous or liquid phase.

In classifying synthetic fibers, there is also a narrow, twilight zone between fibers and elastomers. There are elastomers with fiberlike qualities; and vice versa. For example, spandex is a fiber with rubber-like qualities. See also **Elastomers.**

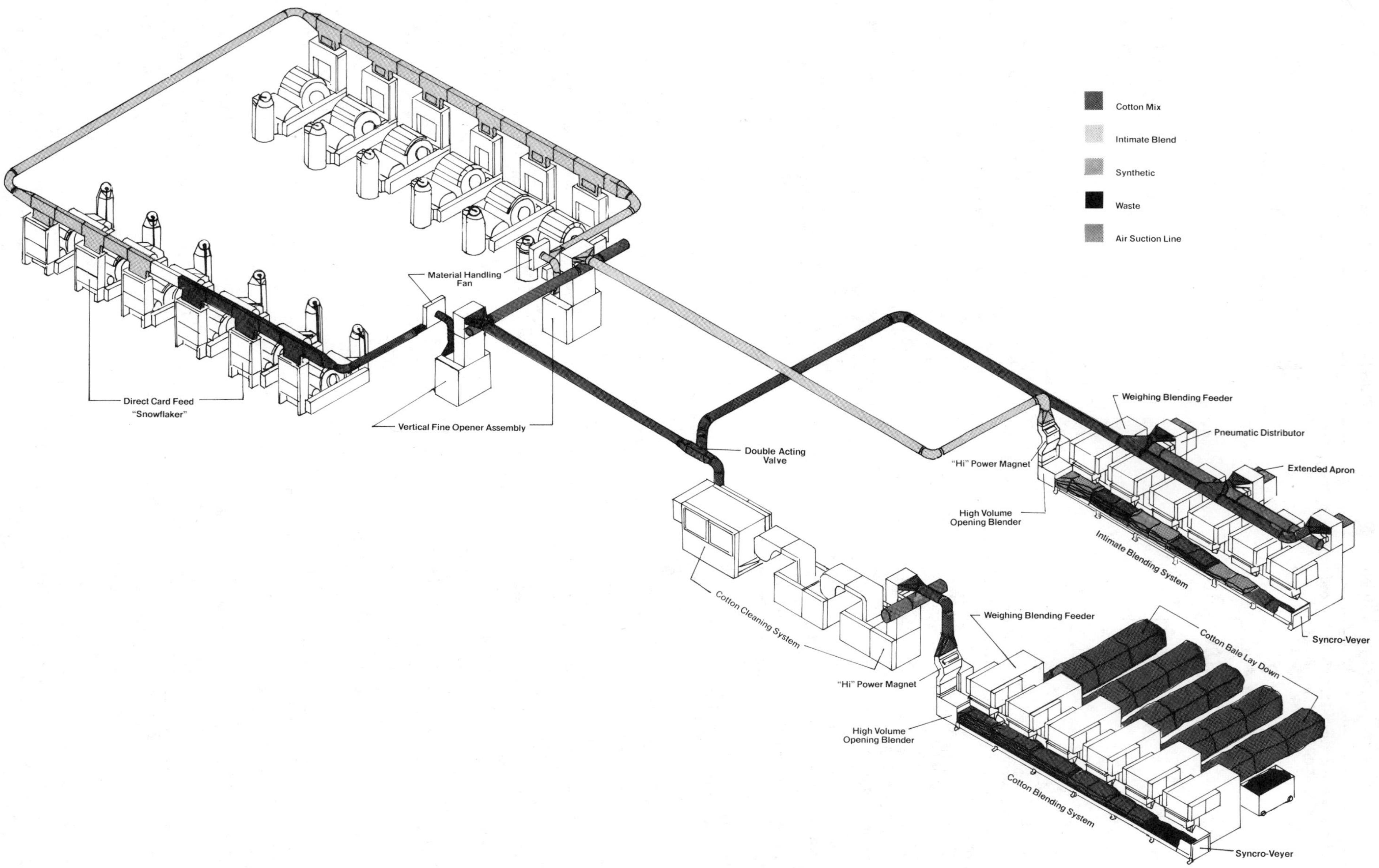

A system for blending (weigh-pan blending and direct card feed) cotton and synthetic fiber substances. (*Fiber Controls®, a Total Systems Corporation, Gastonia, North Carolina. The system is manufactured under one or more of the following U.S. patents*: 2,995,783; 3,111,857; 3,132,709; 3,142,348; 3,196,967; 3,225,848; 3,308,786; 3,355,773; 3,421,659; 3,649,082; 3,671,078; 3,700,091; 3,901,555; and RE-25609).

Particularly in the period between the late 1930s and late 1950s, there was a vigorous development of new and different synthetic fibers, largely stemming from the research efforts of competing firms. Much of the technology was well guarded and there was an overwhelming tendency to give all new fibers tradenames rather than generic designations. This created much confusion, particularly in the marketplace. Some of the former tradenames were lost because they were widely and variously used and ultimately became the generic term for a class of fibers. As a tool for obtaining clarification, and for protecting buyers of various synthetic fibers, the U.S. Congress passed the Textile Fiber Products Identification Act which is administered by the U.S. Federal Trade Commission. The principal synthetic fibers as so defined are listed in the accompanying table. For more complete definitions, refer to ASTM Standards on Textile Materials.

Many textile products combine the advantageous physical and chemical properties of two or more fibers (synthetic or natural). Careful preblending operations are usually required. The accompanying flowsheet illustrates a system for blending cotton with synthetics.

See also **Acetate Fibers; Acrylamide Polymers; Acrylic Fibers; Cotton; Elastomers; Fiber-Reinforced Composites; Flax; Mohair; Polyester Fibers; Polymerization; Silk;** and **Wool.**

References

Staff: "Properties of the Manmade Fibers," *Textile Industries*, Atlanta, Georgia (Published annually).
Staff: "Manmade Fiber Chart," *Textile World*, McGraw-Hill, Atlanta, Georgia (Published annually).
Staff: "ASTM Standards on Textile Materials," American Society for Testing and Materials, Philadelphia, Pennsylvania (Revised periodically).

FIBERS (Dietary). Dietary Requirements and Trends.

FIBONACCI NUMBERS. Medieval investigator, Leonardo ("Fibonacci") da Pisa proposed a sequence of numbers, commencing with 1, adding 1 to itself, to yield 2, and thereafter adding the last two numbers in the series to yield the next number in the series. Thus, 1, 1, 2, 3, 5, 8, 13, 21, 34, 55, 89 ... etc. Although an apparent curiosity, Fibonacci numbers bear a surprising relationship to botany and classical art. Examples of Fibonacci numbers in nature include the spirals of tiny florets found in the core of daisy blossoms. There are sets of clockwise and counterclockwise spirals. Each set has a predetermined number of spirals. In daisies, there are 21 clockwise and 34 counterclockwise, the 21:34 ratio being made up of two adjacent Fibonacci numbers. Opposing spirals in pine-cone scales bear a 5:8 ratio. The leaves of several trees and the bumps on pineapples bear a 8:13 ratio. After the number 3 in the Fibonacci series, adjacent numbers have a ratio of approximately 1:1.6 ($\frac{21}{13} = 1.6154$; $\frac{34}{21} = 1.6190$; $\frac{55}{34} = 1.6176$ ... etc.). This is essentially the value of the Golden Ratio or Golden Section (1:1.618) which occurs in circles, decagons and pentagons. The best known example of the Golden Ratio is the Golden Rectangle, the sides of which bear this ratio. See accompanying diagram. The Parthenon at Athens is patterned in accordance with the Golden Rectangle, but it is doubtful that the builders at that time were consciously aware of the Golden Ratio. See also **Number Theory.**

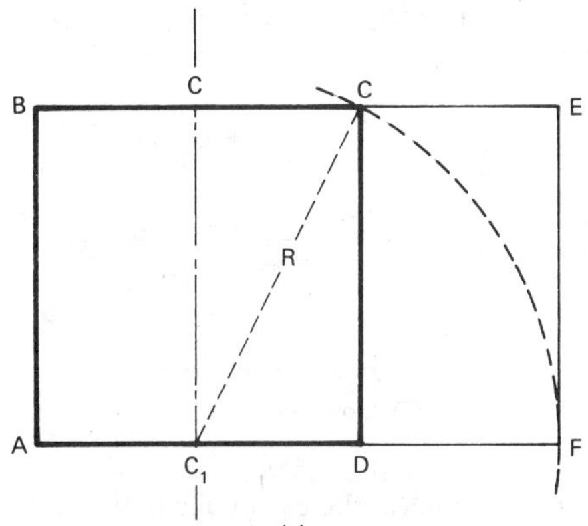

(a)

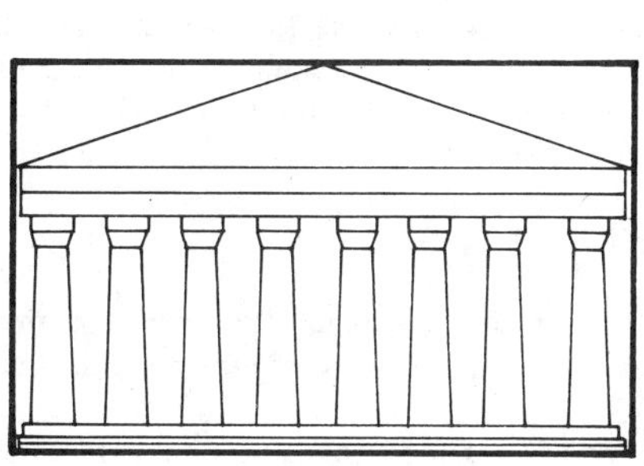

(b)

Golden rectangle. (a) To construct a golden rectangle, commence with square *ABCD*. Divide square into two equal halves on either side of center line CC_1. Using circle radius *R*, extend portion of circle to intersect extension of *AD* at point *F*. Resulting rectangle *ABEF* is a golden rectangle with $AB:AF = 1:1.6$. Rectangle *CDEF* also is a golden rectangle. (b) The façade of the Parthenon at Athens fits almost perfectly inside a golden rectangle.

FIBRIL. A minute threadlike structure in a cell, also the smaller components of the intercellular white fibers of connective tissue.

Fibrils occur near the surface of smooth muscle cells and connective tissue cells (border or myoglia fibrils and fibroglaia fibrils, respectively) and in fully differentiated muscle and nerve cells (myofibrils and neurofibrils, respectively).

FIBRILLAR PROTEINS. Contractility and Contractile Proteins.

FIBRIN. Blood.

FIBRINOGEN. Protein.

FIBROADENOMA. Breast.

FIBROIDS. Gonads.

FIBROMA. A benign tumor of fibrous or connective tissue. Fibroid tumors may occur in various parts of the body but are most often found in the uterus, where, in combination to a varying degree with proliferated muscle fibers, they constitute the fibromyoma which may grow to great size, produce pressure symptoms, and abnormal bleeding. Uterine fibroids are the most common cause for removal of the uterus.

FIDELITY (Communications). The degree with which a system, or a portion of a system, accurately reproduces at its output the essential characteristics of the signal which is impressed upon its input. The term is usually applied to an amplifier or communications circuit to indicate its ability to reproduce the original signal without distortion. A high-fidelity audio amplifier will amplify all audio frequencies equally, whereas a poor-quality amplifier will badly attenuate certain frequencies, usually the low and high frequencies, while passing the middle audio range. Many radio receivers have fidelity controls. These are in the intermediate frequency circuits and have the effect of broadening the pass band of such circuits. Most of the loss of the high frequencies in the radio is due to the various selective circuits cutting out the higher sideband frequencies, and, since these represent the higher audio frequencies, the high frequencies are missing in the output. The use of fidelity controls allows the selective circuits to be broadened for receiving high-fidelity local programs where there is not likely to be adjacent station interference, yet allow the circuits to be returned to the selective condition for tuning in weaker stations which cannot override interference.

FIDUCIAL INFERENCE. A type of statistical inference introduced by Sir Ronald Fisher. It is accepted as valid by some authorities but rejected by rather more. The basic idea may be illustrated as follows: let $f(x, \theta)$ be a frequency distribution of a statistic x, dependent on a parameter θ. In probabilistic terms, this would be thought of as a distribution of x which was determined when θ was fixed, x varying from sample to sample. Fiducial theory regards it as giving, for fixed x, a set of admissible values of θ, usually a range within which the true value may be supposed to lie.

FIELD. The term generally connotes associated and surrounding phenomena and finds numerous uses in science. 1. In mathematics, a set of elements (see later separate entry). 2. A field of force. 3. A radiation field. 4. A sound field. 5. A field of view. 6. In electric motors and generators, the field is the part of the machine that furnishes the magnetic flux that reacts with the armature to produce the desired machine action. The field may be the fixed part of the machine or, as is usually the case in synchronous motors and generators, the rotating part of the machine. 7. The electromagnetic energy radiated from an antenna system of a radio transmitter. 8. In a cathode-ray tube, one set of scanning lines making up a part of the final picture. 9. An electric field. 10. A magnetic field. 11. A gravitational field. 12. An assigned area in a computer record to be marked with information.

FIELD COIL. The coil used to provide the magnetizing force in motors, generators and electrodynamic loudspeakers.

FIELD (Conservative). A field for which the work done in moving an isolated particle of unit mass around a closed path is zero:

$$\int \mathbf{F} \cdot d\mathbf{r} = 0$$

where **F** is the force acting on the particle and $d\mathbf{r}$ is an infinitesimal vector displacement along the path.

FIELD-EFFECT TRANSISTOR SWITCH (FET). An FET switch generally is a switch for controlling analog signals and is comprised of one or more field-effect transistors. In analog-switching and multiplexing applications the metal-oxide-semiconductor FET (MOS FET) is used. The absence of an offset voltage make the FET suitable for switching low-level signals. Further, the high input impedance provides excellent isolation between signal path and drive voltage.

The FET can be used for switching at rates in excess of 100,000 samples per second even though it is not as fast as bipolar switches. A relatively high "on" resistance is the main disadvantage of the FET. "On" resistance ranges from 50 to 200Ω. However, some devices are obtainable with less than 5Ω "on" resistance.

There is no inherent offset voltage because there is no *pn* junction in the signal path as there is, for example, in the bipolar transistor. The FET is a majority carrier device in which the conductance of a conducting channel between two electrodes, the source and the drain, is controlled by a signal applied to a third electrode, the gate. A field established between the channel and the gate, which is insulated from the channel by a thin layer of material (usually silicon dioxide), controls the width of the conducting channel, hence resistance, in the MOS FET. Thus, there is no *pn* junction in the signal path from source to drain. Consequently, there is no offset voltage as in the case of a bipolar transistor.

The FET is a voltage-controlled device—hence has a very high input impedance. The input impedance in the MOS FET is the capacitor formed by the gate and channel, separated by the insulating layer. Thus, the input impedance is very high, usually in excess of $10^9 \Omega$. Leakage currents are very low. These are determined in the MOS FET by the dielectric properties of the insulating layer. The high input impedance and low leakage thus provide excellent isolation between drive signal and the signal being switched.

See also **Analog Switch;** and terms listed under **Data Processing.**

Thomas J. Harrison, International Business Machines Corporation, Boca Raton, Florida.

FIELD (Electromagnetic). Electromagnetic Phenomena.

FIELD (Free). A field (wave or potential) in a homogeneous, isotropic medium free from boundaries. In practice it is a field in which the effects of the boundaries are negligible over the region of interest. In acoustics, the actual pressure impinging on an object (e.g., electro-acoustic transducer) placed in an otherwise free sound field will differ from the pressure which would exist at that point with the object removed, unless the acoustic impedance of the object matches the acoustic impedance of the medium.

FIELD FREQUENCY. Frequency.

FIELD (Gravitational). Gravitation.

FIELD INTENSITY. The field strength of a magnetic and/or electric field at any point. In radio the field intensity at the receiving antenna determines the signal which will be induced in the antenna and hence is an important consideration in the design of the transmitter. By proper choice of the power and frequency of the transmitter and type of transmitting antenna considerable control may be exercised over the field intensity at the receiving point.

FIELD LENS. The lens of a compound eyepiece farthest from the eye, the other lens being called the eye lens.

FIELD (Mathematics). A set of elements for which addition and multiplication are defined (and lead to elements in the set) such that: both operations are commutative and associative, and multiplication is distributive over addition; also there are two elements, conveniently called 0 and 1, such that $0 + a = 1a = a$ for all a in the set; finally, subtraction and division are always possible and unique (except that division by zero is not possible); that is, for every a there exists a unique x such that $a + x = 0$ and a unique y (unless $x = 0$) such that $ay = 1$. Well-known examples of fields are: the set of real numbers, of complex number, of rational numbers, of residue classes, modulo a prime, etc.

FIELD-OF-PLANE MIRROR. The words applied to an optical system apply to a plane mirror. The exit pupil is the pupil of the observer's eye, the entrance pupil is the behind-the-mirror virtual image of the observer's eye pupil. The field stop is the edge of the mirror. The region in front of the mirror is the object field, and the region behind the mirror is the image field.

FIELD OF VIEW. Of an optical instrument, the field of view is the angle subtended at the eye by the largest object, the whole of which can just be seen. This angle in the case of a simple telescope consisting of two converging lenses, is $a_2/(f_1 + f_2)$, with reference to the accompanying figure, where a_2 is the aperture of the eye lens

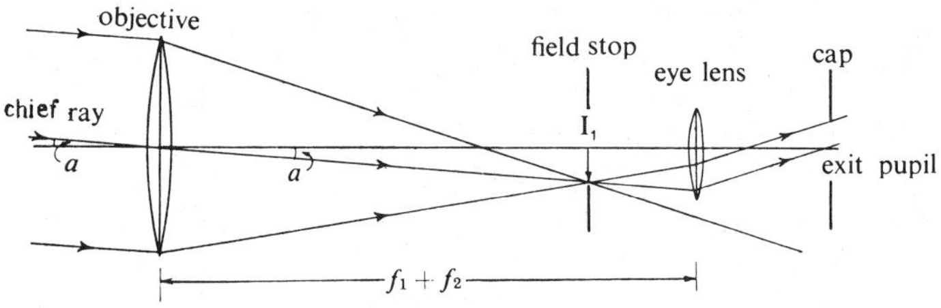

Simple telescope.

and f_1 and f_2 are the focal lengths of the objective and eye lens, respectively. If f_1 is doubled, thereby doubling the magnifying power, the field of view is nearly halved (f_2 is small compared with f_1). The falling off in brightness of the edge of the image, owing to some rays from the objective failing to strike the eye lens, is called vignetting, and is avoided or kept to a reasonable value by placing a circular opening, called the field stop, of suitable size where the image formed by the objective is situated.

FIELD PRESSURE. Altimetry.

FIELD STOP. An opening, usually circular, in an opaque screen which determines the field of view of an optical instrument.

FIELD STRENGTH. The strength of a field is its magnitude. Thus an electric field has a strength (volts per meter) at a given point, as well as a direction; its strength is, therefore, the magnitude of its electric field vector, **E.**

FIELD THEORY. The description of the physical world has evolved profoundly through the ages, at times because of, at other times being the cause of, sweeping changes in our philosophical, mathematical, and experimental knowledge. Greek geometry concerned itself essentially with properties of "objects as such," a triangle or a cube being studied, for example, without any thought of their spatial environment; the Ptolemaic system enhanced this view into a clockmaker's dream, where celestial bodies parade around the earth, rigidly driven in circular motions. Only with Descartes' analytical geometry did objects become "portions of space" and the properties of space itself the main object of study; with Galileo and Newton, a correct science of dynamics was born, which permits the prevision of an amazing number of mechanical phenomena in that space from a few first principles.

Field theory studies the phenomena of the physical world as due to interactions which propagate through space; the "geometrical emptiness," which is the space of mathematics, becomes the medium into and through which actions take place or, even more drastically, a structure which is itself determined by the properties of matter, as in general relativity.

Suppose two bodies interact in space, e.g., the sun and the earth with Newton's law, or two electric charges with Coulomb's law. Two pictures of this situation are equally possible and correct. One is that this interaction cannot be conceived if *both* bodies are not there and that we should study primarily its effects without looking for a detailed mechanism for its propagation from one body to the other; this is the description of "action at a distance," in which *forces* are the main concepts and space is a vacuum into which bodies follow trajectories determined by the forces acting upon them. The other picture consists in imagining that each body, whether alone or not, modifies the structure of the space which surrounds it, geometrically or because in each point of that space there is now potentially a force, which becomes active if another body occupies that point, but should be conceived as existing there in any case; the main objective is here to study how these "fields of force" are created in space by material objects and how they propagate; this is the point of view of "action with contact," which finds its full development in field theory.

Mathematically, a field is characterized by assigning to each point of space a quantity which is *intrinsically* associated with it; a temperature, for instance, or a velocity, or a tensor or a spinor of arbitrary rank. "Intrinsic" means that if we change our frame of observation, this quantity does *not* change; supposing, e.g., that our field is that of the velocities at a given instant of all the points of a moving fluid, if we rotate our coordinate system we shall observe different values for the components of those velocities, just because we, not the velocities, have changed position. It is therefore essential that, together with the specification of the field quantities, their transformation laws also be assigned under changes of the reference frame; these laws are indicated by the description of the field quantity as a "scalar" (which does not change), a "vector" (which changes with the same law as the coordinates), a "tensor," etc.; the complete specification of all such possible laws is a standard chapter of group theory.

Physically, we have to account for the creation or the existence of the field, by describing the field quantities as generated by "sources," such as positive or negative charges for the electromagnetic field, or the sources and sinks of hydrodynamics. Moreover, we have to describe in which way the values of the field quantities change when the point at which they are considered, or the time, is changed. In the absence of discontinuities, for instance in vacuum, one expects these values to differ by infinitesimal amounts if the corresponding points are infinitesimally close, in some way which is typical of the field considered; in other words, that the rates of change of the field quantities with respect to the space coordinates and time be connected by relations which specify both how these changes can occur compatibly with the geometrical properties of space, and how they are related to the sources. Group theory determines all the possible forms which are permissible for these relations, which take the name of *field equations*; each field theory is characterized by a special set of field equations, which are clearly *partial differential equations*.

Relativity and quantum theory have played a great role in the development of field theory; we shall briefly discuss, later, their influence both in the explanation of new physical phenomena and in the mathematical formulation of the theory.

Field theory has taken an entirely new shape with the so-called second quantization, which has led to several modern developments, of which some embody faithfully the concepts outlined thus far and others instead represent new views in natural philosophy; this is still a matter of controversy at present, and it is yet unpredictable whether a reasonably lasting description of nature will come out of such attempts or whether a new drastic turn in human thought will be necessary before we can hope to understand the fundamental laws of physics. Be that as it may, the ideas and the computational techniques of field theory have proved already of invaluable help in the description of many phenomena, from particle physics to superconductivity.

It is convenient to examine first the theories in which the field quantities are ordinary functions of space and time points, regardless of whether they have a direct physical meaning (as with the velocities of hydrodynamics and the electromagnetic forces) or not (as with the wave function which obeys a Schrödinger equation). This comprises of course most of classical and modern physics; mathematics permits again, however, a tremendous conceptual simplification. In the study of continuous media or fields one is, most often, interested in one of the following classes of phenomena:

1. Phenomena which consist of the propagation of some action; the medium itself is not transported from one place to another; typical is the propagation of waves, whether they be seismic, fluid or electromagnetic;

2. Phenomena in which there is transport or diffusion of a quantity in a medium: of heat in a wall, of solute in a solvent, of neutrons in a pile;

3. Equilibrium phenomena: deformations of strained elastic bodies, electro- or magnetostatic fields as determined by charges and boundaries.

Each class is ruled by essentially one type of equation. Let $\Delta = \partial^2/\partial x^2 + \partial^2/\partial y^2 + \partial^2/\partial z^2$ denote the Laplace operator; $\phi = \phi(x, y, z, t)$ the field quantity; F some function of x, y, z, t, ϕ and, at most, of the first-order derivatives of ϕ; v a velocity; and D a diffusion constant. The corresponding equations can be brought into the standard forms:

$$\Delta\phi - \frac{1}{v^2}\frac{\partial^2\phi}{\partial t^2} = F \text{ (hyperbolic partial differential)} \tag{1}$$

$$\Delta\phi - \frac{1}{D}\frac{\partial\phi}{\partial t} = F \text{ (parabolic partial differential)} \tag{2}$$

$$\Delta\phi = F \text{ (elliptic partial differential)} \tag{3}$$

If the field quantity has more than one component, one may deduce for each of its components an equation which is essentially of the same type, although it may be difficult or impossible to obtain an independent equation for each component.

This classification of physical phenomena according to the type of equation to which their study can be reduced is of the greatest importance: Equations (1), (2) and (3) are called in fact "the equations of mathematical physics"; more specifically, Equation (1) is also called the wave equation, Equation (2) the heat equation, and Equation (3) the Laplace or potential equation. The study of their mathematical properties gives complete information on all the physical phenomena which they describe.

The equations of quantum mechanics can also be brought, at least formally, into the form of Equation (1) or (2); the intervention of complex quantities modifies the situation somewhat, in a way which we cannot discuss here.

Electromagnetic phenomena fall typically into the category of Equation (1): each of the components of the electric field $\mathbf{E} \equiv (E_x, E_y, E_z)$ and of the magnetic field $\mathbf{H} \equiv (H_x, H_y, H_z)$ satisfies, in vacuum, Equation (1), with $F = 0$; the connections between $\mathbf{E}$ and $\mathbf{H}$ are given by the Maxwell equations, which characterize completely the theory, and lead in vacuum to the result just mentioned. When $\mathbf{E}$ or $\mathbf{H}$ does not vary with time, Equation (1) reduces to Equation (3), thus yielding electro- or magnetostatics.

The electromagnetic field, i.e., the vectors $\mathbf{E}$ and $\mathbf{H}$, generated by a distribution of moving charges or currents confined within a limited volume has a part which becomes dominant at a large distance from that volume, because it decreases only with the inverse of that distance (instead of the inverse-square law of static fields); this part constitutes the *radiation* field, which is responsible for the transmission of energy and signals (the radiated energy is, of course, supplied by the mechanism which drives the generating charges or currents). This is easy to understand: the energy radiated through a large sphere around the source is proportional to the area of the sphere times the square of *E*; it vanishes therefore with increasing radius for all but the radiative component, for which it stays constant: energy is actually removed from the source and radiated away to all distances. The study of radiation is a most important part of the theory, both macroscopically (telecommunications, radar) and microscopically (atoms, nuclei, elementary particles).

Relativity and quantum mechanics have extended and modified profoundly the classical picture presented so far. The very concepts of space and time change with special relativity: events which are simultaneous for an observer are not such when seen by another observer in uniform motion with respect to the first, because time and space are mixed together by the Lorentz transformations which relate the reference frames associated with the two observers. As a consequence, the laws of nature can retain their universal validity only if they are formulated in the same form by any such observer, i.e., if their form is not altered by a Lorentz transformation—technically speaking, if they are "Lorentz covariant." This requirement becomes a stringent dogma; it suffices to determine, with the help of group theory, the possible equations for any conceivable relativistic field theory; it is of invaluable help, when computations are made, in checking or correcting them.

The nonrelativistic Schrödinger equation for the wave function of a particle is, but for the appearance of complex quantities, of the type (2) described before: this is not acceptable in a relativistic world, because time and space are not treated alike. One needs either an equation which contains only second-order derivatives, or one with only first-order derivatives; for a free particle, this leads either to the Klein-Gordon equation, which is of type (1), or to the Dirac equation, which contains linearly only the first-order derivatives of the wave function, but has a mathematical structure which necessarily assigns special physical properties to the particles described by it. It was one of the greatest triumphs theoretical physics ever witnessed, to discover that such properties are actually displayed by all particles which obey the Dirac equation: spin, and the existence for each Dirac particle of a corresponding *antiparticle*, i.e., a particle having the opposite mechanical and electrical properties.

The requirement of relativistic covariance has thus led to fundamental physical discoveries; for each particle obeying the Dirac equation, the corresponding antiparticle has been experimentally found in nature; electron and positron, proton and antiproton, neutron and antineutron, etc. What is more, the theory predicts that a particle-antiparticle pair can be created in a collision phenomenon, if sufficient energy is available, or can annihilate itself, giving away its energy in the form of electromagnetic radiation or other particles.

The classical theory allowed only for the electromagnetic radiation emitted by moving charges or currents; the creation or absorption of particles in collision phenomena, as well as the creation or annihilation of pairs, were outside its scope and possibilities. A new formulation of the theory was needed, which could account consistently for all such phenomena, handling situations in which particles can be created and destroyed in any numbers. The formalism devised for this purpose is that of quantum field theory.

The basic idea is to describe each type of particle by means of a field which is not any more an ordinary numerical function of space and time, but an "operator," i.e., a quantity which changes the number of particles existing in any given state of the system. If the field operator is known, one can then evaluate the probability of a given state (so many particles, with determined energies and momenta) changing into an equally determined, different state. If the particles do not interact among themselves or with other particles, no change is possible; if there is interaction, the field operator has a structure which can cause such transitions. The field equations appear to be essentially the same as those of the classical Maxwell, Klein-Gordon, Dirac theories, etc.; their structure is however fundamentally different, because they now must be equivalent to infinite sets of ordinary equations, which couple states with different and ever-increasing numbers of particles.

Fields which are associated with particles obeying Gose-Einstein statistics (of which any number can be found in any given state) have radically different mathematical properties from fields associated with particles obeying Fermi-Dirac statistics (of which at most one can be found in any given state); examples of the first are photons (the massless neutral quanta of the electromagnetic field), pions (massive particles, with or without electric charge, which are believed to be responsible for nuclear forces), etc.; examples of the second are electrons and positrons, protons and antiprotons, etc.

The passage from numerical fields to operator fields is called "second quantization"; quantum field theory deals with operator fields.

Striking successes have been met with this approach. From a quantitative point of view, they are confined mostly to electrodynamics, where very small deviations from the values predicted by the nonquantized theory, which were observed in the measurement of the magnetic moment of the electron and in the so-called Lamb shift, were accounted for with amazing accuracy by quantum field theory. Qualitatively, the new conceptual framework has proved extremely useful in understanding elementary phenomena, especially with the help of the diagrams devised by R. P. Feynman, which give a simple intuitive picture of collision and radiation processes involving elementary particles. Very little has been achieved quantitatively, though, for theories other than electrodynamics, because of the tremendous mathematical difficulties which arise as soon as the simplest approximation techniques are not applicable because the interaction is too strong; nevertheless, these ideas have proved greatly helpful in many ways, in combination with general principles of symmetry, Lorentz invariance and causality.

A beautiful consequence of this conception, which assumes that particles are the quanta of a field (as photons were recognized by Einstein to be the quanta of the electromagnetic field) was the discovery of H. Yukawa, that whenever such quanta have a mass different from zero, the force they create between two bodies which interact by exchanging such quanta with each other must be an exponentially decreasing function of distance; this force can become of a coulombian type only if the mass of the quanta vanishes. Thus, the Coulomb force can be explained as due to the exchange of photons among electric charges, the nuclear forces (which have typically short ranges) as due to the exchange of massive particles among nucleons. Exchanges of this nature are not observable in the laboratory, because this runs against Heisenberg's indeterminancy principle; if enough energy is supplied, however, such quanta can actually break loose and do appear as the particles created in collision processes.

The mathematical difficulties encountered in quantum field theory are many, and there is as yet lack of agreement as to the best way to circumvent some of them. Besides mathematical complexity, which prevents all but the simplest calculations, there are many unsolved problems of mathematical rigor and apparent inconsistencies which can be removed only by delicate analyses. Typical of the latter is the fact that unsophisticated calculations give infinite values for masses and charges of interacting particles, and a painstaking analysis is required to retrieve from them the significant physical values; this is the so-called renormalization procedure, which copes with infinities which partly are already present in the classical theory (such as the infinite electromagnetic contribution to the mass of a point-like charged particle, when computed from Maxwell's equations) and partly originate from the new formalism (which permits, for instance, pair creation).

It is not yet certain whether such difficulties are due to the lack of adequate mathematical techniques or are the expression of a fundamental inadequacy of the theory to describe ultimate laws of nature. For this reason, while, on the one hand, the attention of some theoreticians has been directed to perfecting the mathematical foundations of quantum field theory (giving rise to axiomatic field theory and to more rigorous methods of obtaining and studying the quantum field equations, etc.); on the other hand, most physicists have been trying new avenues, such as the S-matrix theory, dispersion relations, the so-called Regge poles, etc.; these approaches have certainly led to very useful results, but they leave altogether at least as many doubts as hopes.

Whatever may be the future prospects of field theory as the correct means for describing the fundamental laws of nature, its tremendous usefulness in providing a conceptual framework, in inspiring new ideas, and in suggesting computational techniques has been overwhelmingly demonstrated in the last decades. It has now found a new, very fertile ground of application in the study of systems containing a very large number of particles, where it has already provided a reasonably good quantitative understanding of superconductivity and superfluidity, and promises many other results of interest in the study of solids and liquids.

References

NOTE: See also references listed at ends of entries on **Particles (Elementary); Quantum Mechanics;** and **Relativity and Relativity Theory.**

Caianiello, E. R.: "Lectures on Field Theory," Academic, New York, 1961.

Hagan, C. R., Guralnik, G., and V. S. Mathur (editors): "Proceedings of the International Conference on Particles and Fields," Wiley, New York (Various dates).

Heisenberg, W.: "Introduction to the Unified Field Theory of Elementary Particles," Wiley, New York, 1967.

Henley, E., and W. Thirring: "Elementary Quantum Field Theory, McGraw-Hill, New York, 1962.

Katz, A.: "Classical Mechanics; Quantum Mechanics; Field Theory," Academic, New York, 1965.

Lurie, D.: "Particles and Fields," Wiley, New York, 1968.

FIGHTER AIRCRAFT. **Airplane; Helicopters and V/STOL Craft; Supersonic Aerodynamics.**

FIG TREES. **Mulberry Family.**

FIGURE OF MERIT. 1. General term for various graphical relationships which summarize certain desirable features of amplifying devices. For example, the figure of merit for a magnetic amplifier has been defined as the ratio of power amplification of a given control winding to the response time of the magnetic amplifier, under specified control circuit conditions. 2. The current required to produce one device deflection (usually one millimeter on a scale at a distance of one meter) of a galvanometer. 3. Any quantity which expresses quantiatively the performance of a measuring device.

FILAMENT. The resistive element (1) in a common electric lamp and (2) in a thermionic tube through which current is passed to provide the temperature required for thermionic emission. The surface of the filament may supply the emission, or the filament may be employed as a heater for an indirectly heated cathode.

The term also is used in astronomy in connection with prominences. In the textile field, particularly in connection with synthetic fibers, individual strands of extruded nylon, rayon, and so on are often referred to as filaments. Often, many filaments are twisted together to form yarn.

FILARIASIS. A parasitic disease of the tropics caused by various species of *Filaria*, a genus of nematode or thread worms chiefly *Wuchereria Bancrofti, W. Malayi, Onchocerca volvulus* and Loa loa. The parasites are commonly found in tropical countries but cases of infection have been reported in the Southern states. Both anopheline and culicine mosquitoes serve as intermediate hosts of *Wuchereria.* For *Onchocerca volvulus* the vector is a black fly of the family *Simulidae*, and for Loa loa, the Tabanid fly, *Chrysops.* The adult forms develop in the human body after larval forms are transmitted by the bite of the infected insect. The incubation period is usually one year or less.

Symptoms depend on the degree of infestation and the parts of the body involved. The parasites show a predilection for the lymphatic structures. High fever, lymphangitis and swelling of the lymph nodes are common.

After repeated attacks permanent thickening of tissue occurs accompanied at times by great swelling of the affected parts. The swelling is caused by blockage of the lymph channels by the parasites obstructing the flow of lymph.

When the legs are involved the condition is called elephantiasis, as the legs become tremendously swollen with thickened, fissured skin. The external genitals, especially the scrotum, may be similarly involved. Blindness is a serious complication of onchocerciasis; Loa loa produces localized tumors known as Calabar swellings.

The drug most commonly used in the treatment of the disease is an antifilarial agent, diethylcarbamazine.

FILAR MICROMETER. An instrument for measuring small distances in the field of an eyepiece. The filar micrometer consists fundamentally of two parallel wires, one being fixed, and the other capable of motion in the direction perpendicular to its length by means of an accurately cut screw. The pitch of the screw is carefully determined for various temperature conditions, and the head of the screw is graduated so that whole revolutions and fractions thereof may be read.

For astronomical purposes, the filar micrometer is somewhat modified. In the accompanying figure, AB is a plate of brass carrying two wires, H and F, which are accurately perpendicular to each other. This plate may be rotated, and the index I sweeps over a circle graduated in degrees, with a vernier reading fractions of a degree. A second plate CD may be moved over the surface of the first plate by means of the accurately calibrated screw S with the graduated head P. This plate carries on its lower surface, so that it will be practically in the same plane as F, a wire M, which is set accurately parallel to F.

Diagram of filar micrometer.

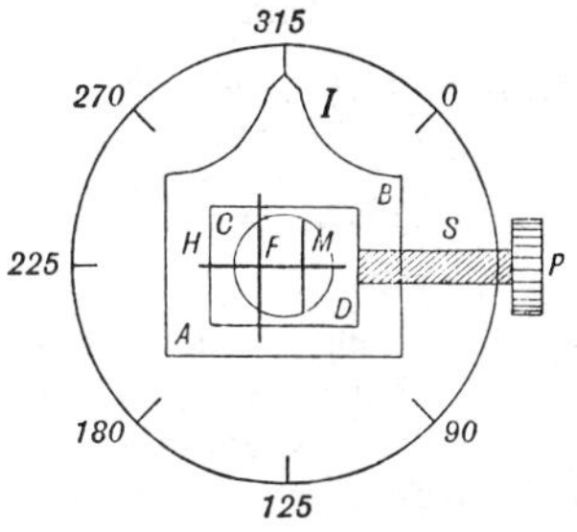

This instrument is so mounted that F and M are in the focal plane of the objective of a telescope, with the optic axis of the instrument passing through the center of the opening in AB. When so mounted, the filar micrometer is one of the most valuable instruments for measurement of small angular distance.

For the study of double stars, the first adjustment of the instrument is to point the telescope at a star, preferably near the equator, and rotate the instrument until the star will move, due to the rotation of the earth, along the wire H. H, being now parallel to the equator, the reading (R_1) of the index is taken. Next, the wires M and F are placed in coincidence, and the head reading (D_0) is taken. The telescope is then directed at the double star under investigation and is rotated until H passes through both stars; then the reading (R_2) of I is taken. With one component of the pair of stars being held on F, the screw is turned until M passes through the other star. The reading (D_1) is taken of the head, also taking into account the number of whole revolutions in the process. $R_2 - R_1$ is defined as the position angle of the double star, and $D_1 - D_0$ (when converted into angular units) is the distance.

The filar micrometer may also be used to determine the position of one astronomical body relative to another close object whose spherical coordinates are known. For this purpose, the wire H is first held at setting (R_1), (i.e., parallel to the celestial equator); the wire F is held on one of the two objects; and the wire M is set on the other. The distance thus measured is parallel to the equator (i.e., is proportional to difference in right ascension of the two objects). The instrument is then rotated through 90°, and the difference in declination may be measured.

See also **Celestial Sphere.**

FILE. A cutting tool for smoothing surfaces, breaking corners, removing burrs, and sharpening tools. The cut of a file denotes the degree of coarseness and the character of the teeth. Single-cut files have single rows of parallel teeth extending the length of the file; double-cut files have two parallel rows of teeth crossing each other; rasps have individual, or disconnected, teeth, as shown in the figure.

A mill file is rectangular in section, and is single-cut and tapered in both width and thickness. A flat file is like a mill file, but is double-cut. A hand file is double-cut, of parallel width and tapered thickness, and has a safe edge (one which is smooth and has no teeth). Half-round files are usually double-cut, with one flat and and one curved surface. Round files are of circular, tapered section, and may be single- or double-cut. Three-square and handsaw files are tapered and have a triangular cross section; they are used for filing to sharp corners, and for saw filing and sharpening.

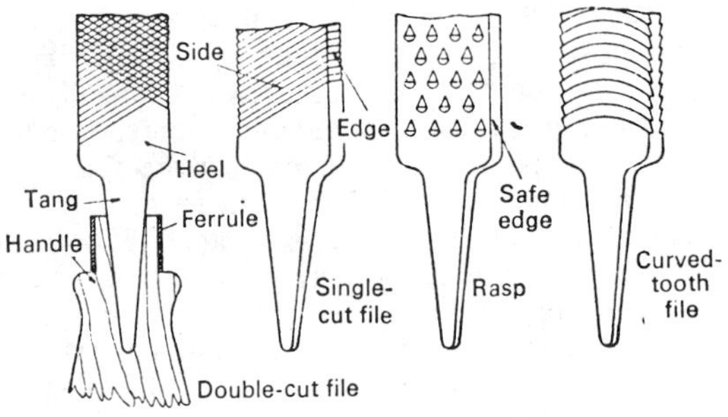

Types of files.

FILICALES. Ferns.

FILLED BAND. An energy band in which every level is occupied by an electron. See also **Energy Band.**

FILLER. Paint.

FILLET. Fillet is the term employed to describe a concave section of a body which is used to reinforce a re-entrant angle formed by the intersection of two plane surfaces. It is purposely incorporated in patterns from which castings are to be made, since lines of stress radiating from sharp edges and corners are set up in the casting during cooling. These will be prevented, and the casting made measurably stronger, if the outside edges are rounded, and the inside filleted. Filleting is also employed to improve the appearance of a corner, or hide a crack, and for the purpose of replacing an angularity with a smoothly rounded surface. The junction of two parts, such as plates at right angles to each other, is often made with a weld of the fillet type, in which a small fillet of welding metal is laid down in the angle created by the intersection of the surfaces of the plates.

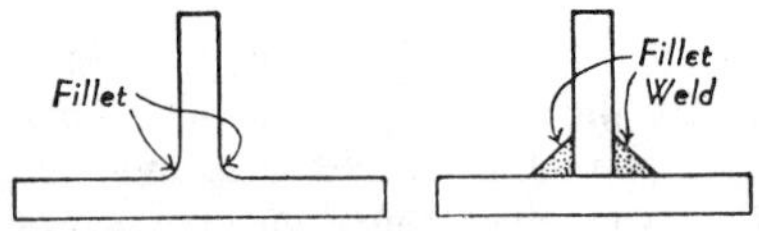

Types of fillets: (*Left*) casting; (*Right*) weld.

FILM (Boiling). Boiler; Boiling.

FILM BREAKER. Defoaming Agents.

FILM (Bubble). Foam.

FILM (Electronic). Microelectronics; Microstructure Fabrication; Semiconductor; Telephony.

FILM (Half-Silvered Surface). Half-Silvered Surface.

FILM (Photographic). Photography and Imagery.

FILM (Structure). In its most general usage, film means any thin sheet of material used for covering, coating or wrapping, or any thin layer that enters into the structure, usually on or near the surface, of a substance or object. The term film also denotes the monomolecular layer which is formed on the surface of a solution or at an interface between two immiscible liquids. The adsorption is of such a nature that the free surface energy is a minimum. Insoluble and non-volatile substances placed on the surface of a liquid such as water may also under certain conditions spread out on the water surface to give a monomolecular film. Adsorbed films of gases or liquids are also formed on solids such as mica, sodium chloride, glass and metals. In some cases, such as the adsorption of vapors on solids at relatively high pressures, it seems that the films may be thicker than one molecular layer and may attain thickness of three or four molecules.

Condensed Film. A surface film in which the molecules are closely packed and steeply oriented to the surface. The molecular packing approaches that observed in the crystalline state.

Expanded Film. A state of film intermediate in area and other properties between gaseous and condensed films.

Gaseous Film. A film in which molecules move about independently on the surface and their lateral adhesion for each other is very small. At low surface pressures (π) and large area (A), a gaseous film obeys the relation $\pi A = kT$. At higher pressures an equation of the form $(\pi A - A_0) = xkT$ holds, where x is a constant.

Liquid-Expanded Film. This film occupies a much larger area than a condensed film, but is still a coherent film. It can form a separate phase from a gaseous film with which it is in equilibrium, and obeys the relation

$$(\pi - \pi_0)(A - A_0) = C$$

where π is the surface pressure, A the surface area, and A_0 the co-area of the molecule.

FILM THICKNESS AND MEASUREMENT. Radioactivity.

FILM (Thin). Thin Films.

FILTER (Communications System). A type of frequency discriminating network designed to select or pass certain bands of frequencies with low attenuation and cause very high attenuation to other frequencies. Filters may be classified according to their characteristics (e.g., low-pass, high-pass, bandpass, band-elimination) or according to their circuits (e.g., ladder, lattice, π and T). A low-pass filter passes all frequencies below a certain value, known as the cutoff frequency, with very little attenuation and then produces high attenuation for all above this value. A high-pass filter passes all above the cutoff frequency and a band-pass filter will pass a band or bands of frequencies and produce attenuation for all frequencies outside these pass regions. A band-elimination filter offers high attenuation to all frequencies in a certain band and low attenuation to all other frequencies.

Filters are made, normally, of combinations of series and shunt elements which are as pure reactance as can be economically attained, the purer the reactance the better the filter. Figure 1 shows the basic or prototype circuits for low-, high-, and bandpass ladder type filters. For more perfect action several sections may be connected in tandem, producing a ladder appearance. These sections may be connected as T or π circuits as shown. While these simple circuits are adequate for some purposes most filters are composed of several sections, each introducing a specific characteristic. A discussion of the low-pass T connected filter will serve to illustrate the usual practice. The prototype shown in Fig. 1 will cause little attenuation up to the cutoff frequency but neither will it cause a high attenuation just above the cut-off, the attenuation increasing very gradually. To overcome this difficulty a section known as a derived section is connected in series with the prototype. This derived section, shown in Fig. 2, gives a very high attenuation at the resonant frequency of the shunt branch. If this reasonant frequency is near the cutoff frequency the attenuation will rise rapidly at cutoff. Unfortunately, this section does not keep a high attentuation as the frequency is raised so it must be used with the prototype to give high attentuation at all frequencies above the cutoff. Often several derived sections, each having a different resonant frequency for its shunt branch, are used. The input and output impedance characteristics of these sections are not very satisfactory so a matching section is connected on each end to correct this. All of these various sections are designed to match one another and to have the same cut-off so the resulting filter, called a composite filter, will have a low attenuation pass band with no irregularities and then a high attenuation for all frequencies above this band. Other types of filters are built up in a similar manner. Quartz crystals are equivalent to resistance, capacitance and inductance networks having a very high **Q**

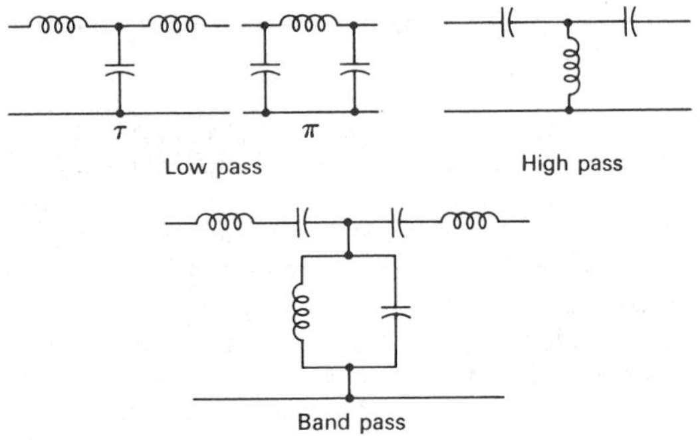

Fig. 1. Typical filter circuits.

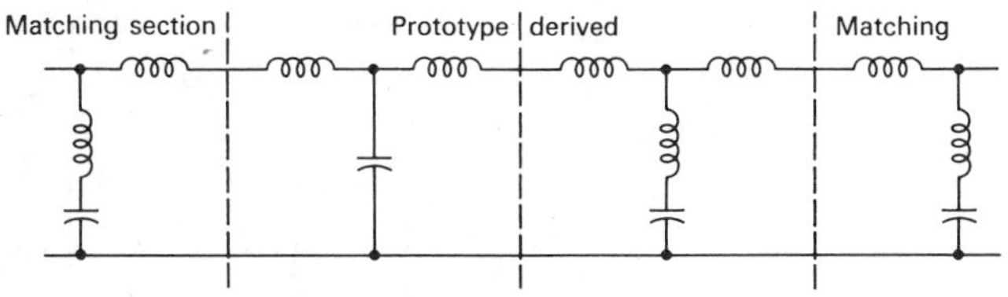

Fig. 2. Composite low-pass filter.

so are ideal for filter components. Through the use of these crystals telephone carrier filters may be made to very close limits and thus many more carrier channels may be superimposed on the lines. These filters usually use a lattice type connection. See Fig. 3. A brief outline of the filter arrangement for a telephone line having a telegraph channel, a voice frequency channel and several carrier channels will serve to illustrate the use of the several filters discussed here. See Fig. 4. The telegraph signals are below the voice frequencies, and above the voice frequencies are several carrier channels. Each of these bands represents a separate communication which must be routed along the proper path at the terminals, all going over the same pair

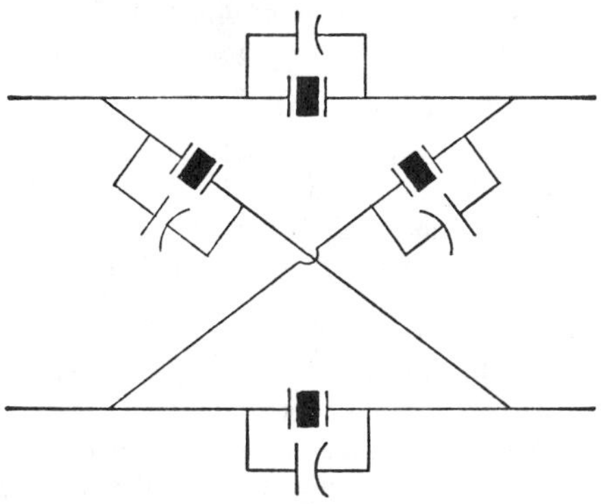

Fig. 3. Lattice-connected crystal filter indicating crystal elements.

of line wires between stations. A high-pass filter connected across the line with its output going to the carrier equipment will allow the carrier frequencies to pass to this equipment and will block the voice and telegraph frequencies which are below its cutoff. A low-pass filter connected across the line at the same point will pass these lower frequencies and block the carrier frequencies. Then if the output of this low-pass filter feeds a high-pass filter with cutoff at the lower voice frequency the voice currents may be passed on to the phone circuits and the telegraph signals blocked. Similarly, a low-pass filter may be used to pass the telegraph and block the telephone currents. The carrier currents which were routed to the carrier circuits by the first high-pass filter are then routed to their respective channels for demodulating by band-pass filters, each designed to pass one channel.

The filters used in the power supplies of various electron tube circuits usually consist of series inductance and shunt capacity, being really low-pass filters which cut out all the ac components of the rectified voltage and current. The band-pass filters used in intermediate frequency amplifiers are double-tuned, inductively coupled circuits which pass a band very narrow in proportion to the mid frequency.

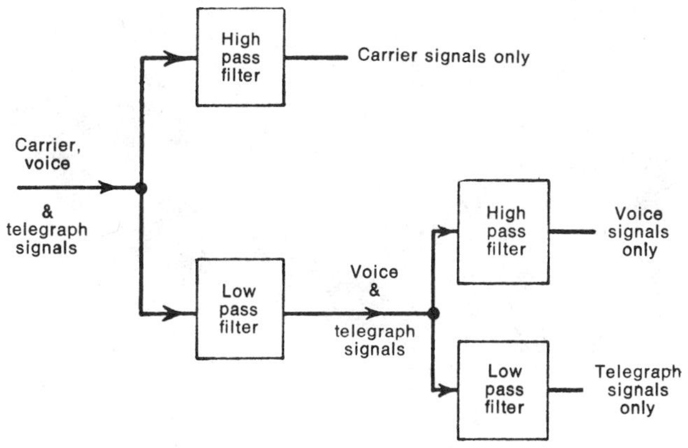

Fig. 4. Filter connection in multichannel telephone line.

Decoupling Filter. In most multistage amplifiers, there are certain circuits, such as voltage supplies, common to more than one stage. Since these common circuits provide a path through which energy may be fed from the output back into the input of some stages, serious feedback problems would result if something were not done to prevent them. The usual remedy is to insert a decoupling filter in those voltage supply leads which are common to more than one amplifier stage. These filters are frequently resistances in series with the lead and a by-pass capacitor from the device (tube or transistor) side of the resistor to ground. The resistance used must be low enough not to cause a serious loss of voltage and the capacitor should have a reactance which is low compared with the resistance at the lowest frequency for which the circuit is designed. Where the resistance would produce too much dc voltage drop or where it does not give enough filtering action an inductance is sometimes used. For still more effective filtering a second resistance and condenser in cascade may be used.

Decoupling filters are used in the three-stage transistor amplifier shown in Fig. 5. The voltage developed across the impedance Z will

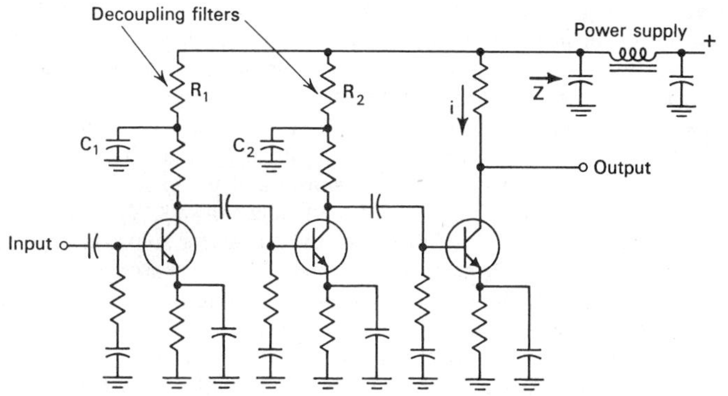

Fig. 5. Decoupling filters used in three-stage transistor amplifier.

cause additional components of signal current to flow in the base circuits of the second and third transistors. Use of the elements R_1, C_1 and R_2, C_2 causes a reduction of the signal fed back to an amount that does not deteriorate the amplifier performance unduly.

Paralleled-Resonator Filter. A bandpass filter in which output and input waveguides are coupled together through several resonators. The resonators are resonant, coaxial lines with a center tuning-screw for alignment purposes. The coaxial resonators are coupled to the input and output guides by means of screws, which distort the field in the main guides. The phase of the coupling depends upon whether the screw is inserted into the bottom or the top of the waveguide. The coaxial resonators are placed a guide half-wavelength apart, so that they are effectively in parallel. The ends of the main guides are short-circuited a guide quarter-wavelength past the last resonator, so that the resonators are at a high-voltage position along the main guides.

Resistance-Capacitance Filter. A low-pass filter circuit sometimes used to lower the ripple content of a rectifier power supply. In its simplest form it consists of a series resistor, followed by a shunt capacitor.

T-Section Filter. A wave filter, one section of which can be shown schematically as given in Fig. 6. If $Z_1 = Z_3$, the filter is said to be symmetrical.

PI-Section Filter. A wave filter, one section of which can be shown schematically as given in Fig. 7 (overleaf). If $Z_1 = Z_3$, the filter is said to be symmetrical.

Fig. 6. T-section filter.

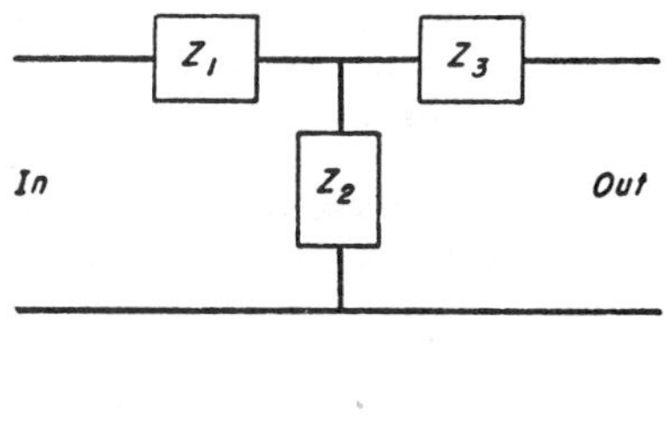

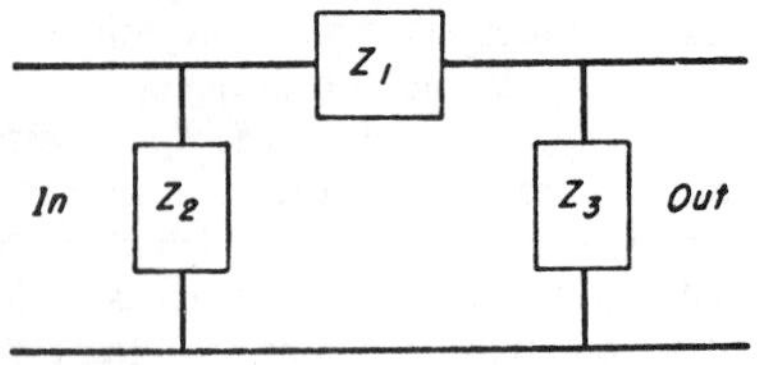

Fig. 7. Pi-section filter.

FILTER (Interference). Interference Filter.

FILTRATION. A very common requirement of several industries, such as chemical and biologicals manufacturing, food processing, ore processing, and water and waste treatment, is the separation of solids that are suspended in liquids. Filtration is a principal means for effecting such separation. Other means include centrifuging, clarifying, and sedimentation, described elsewhere in this volume.

In filtration, the suspension containing the solids is caused to pass through a porous medium. Numerous filtering media are used, including paper, cloth, and wire cloth. Filtration may be conducted under positive pressure or vacuum.

Electrokinetics and Filtration. Process engineers in many industries have long regarded filtration as a very efficient and relatively low-cost means for product recovery, clarification, stabilization, and sterilization. For many decades, a primary component of many filters has been asbestos. See also **Asbestos.** Occupational hazards associated with asbestos were reported as early as 1964. The negative findings on asbestos stimulated the search for substitute filtering media. However, this procedure has been proceeding slowly and led process engineers to restudy the fundamentals of filtration mechanisms. With a better understanding of these fundamentals, the search for substitute media may be accelerated. Prior to recent years, most research leading to development of models of the filtration process concentrated essentially on physical mechanisms and provided little, if any, attention to the electrochemical phenomena that are involved. Several researchers during the past decade have demonstrated a resemblance of filtering to coagulation and have noted that electrochemical phenomena can be controlling factors.

Because asbestos has been such a successful medium, many researchers decided to restudy the mechanical and electrochemical characteristics of this medium. Yada's research in 1967 demonstrated that chrysotile fibers (asbestos) have (1) a hollow cylindrical form with an average outer diameter of 50 to 80 $\times 10^{-10}$ meter (1 angstrom = 10^{-10} meter); (2) that all such tubes are not simple cylinders, but that some may be spirally wound layers; and (3) that the distance between spiral layers may range between 4 and 7 $\times 10^{-10}$ meter. In short, chrysotile fibers possess an exceptionally large surface area per unit weight. See also **Chrysotile.** Also, in 1968, Riddick showed that asbestos has an isoelectric or zero point charge at a pH of 8.3 and thus has a positive charge when in neutral, aqueous solution. It is to be noted that, with few exceptions, the majority of natural substances are negatively charged under these conditions. This observation led to the conclusion that a suitable substitute for asbestos fibers should have an isoelectric point about pH 7 when in neutral, aqueous solutions.

Wnek (1974) described the operation of a filter bed in this way: "Various physical transport mechanisms convey the particles to the surface of the medium. If the medium and particles are of opposite charge, electrokinetic attraction will deposit particles on the bed. If the medium and particle charges are of the same sign, repulsion will occur and deposition will be hindered, if not prevented. As the particles accumulate, the charge on the medium decreases, diminishing its ability to remove particles. Loss of efficiency starts at the top of the medium bed, and eventually electrokinetic removal of particles ceases, although deposited particles may have sufficiently reduced the pore size of the medium to allow further particle retention by straining. As the top of the medium bed becomes saturated, the lower parts remove more and more particles, until they too become saturated. At this point, if the medium bed has not become mechanically plugged, electrokinetic breakthrough will occur, i.e., charged particles will pass through."

Recent research has been directed toward chemically treating otherwise suitable filter media (sand, perlite, diatomite, etc.) so as to impart a positive charge (as found on chrysotile fibrils) instead of the negative charge of untreated materials. One example of this modification research is shown in Fig. 1.

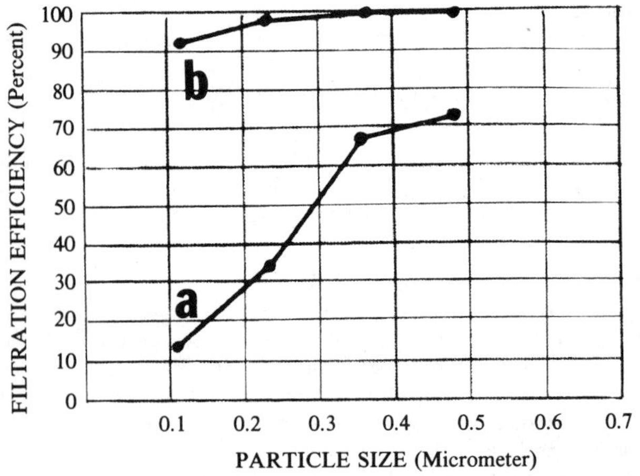

Fig. 1. Effect on filtration efficiency of modification of surface charge of diatomite: (a) untreated diatomite; (b) diatomite treated with melamine-formaldehyde colloid. (*After Fiore/Babineau.*)

Filtration Equipment Configurations

Filters for processing operations may be broadly classified into two distinct groups: (1) continuous, and (2) intermittent. There are many operating schemes used and only a few of the principal designs can be described here. Additional types are listed in the accompanying table.

Rotary Drum Filters. This design is probably the most versatile and widely used continuous filter in the process industries. The rotary drum filter makes it possible to concentrate slurry solids to dry (moist) cakes, to wash solubles from such cakes when needed, and to produce a clarified effluent. The continuous drum filter was first offered commercially in the United States by E. L. Oliver (1908). The basic rotary drum filter is shown in Fig. 2. The principle of operation and fundamental elements of the rotary drum filter are illustrated in Fig. 3.

A horizontal drum is partially submerged in a vat that contains the slurry to be filtered. A vacuum is applied through a central valve on the drum shaft to individual compartments or sections that provide support and drainage for the filter medium. The filter cake is formed while the sections are immersed. When the sections emerge (because of continuous rotation of the drum), additional dewatering takes place as air passes through the cake, thus displacing a significant portion of the mother liquor. Before final dewatering, wash water may be applied to remove any remaining soluble solids. Discharge of the dewatered cake is effected by cutting off the vacuum and applying a reverse air blow. As the cake separates from the filter cloth, a scraper blade deflects it whereupon it is dropped to a conveyor or discharge trough below.

Means for applying vacuum during cake-forming, washing, and de-

Fig. 2. Basic rotary drum filter. (*Dorr-Oliver Inc.*)

CLASSIFICATION OF INDUSTRIAL FILTERS

Class and Type of Filter	Representative Applications
CONTINUOUS FILTERS	
ROTARY DRUM FILTERS	
Basic Design	Sewage sludge, titanium, lime mud, dyes, citrates, catalyst, flue dust, cane mud, clay, red mud, electri-furnace dust, zinc residue, calcium carbonate, peanut butter
String Filter	Starch, gluten, carbonates, antibiotics
Precoat Filter	Juices, wines, antibiotics, petroleum
Cell-less Filter	Adipic acid, fibers, coal, ore, carbonation mud
Traveling-Medium Filter	Sewage and paper-mill sludges, gluten, flue dust, starch, and various chemicals
Top-Feed Filter	Salt and similar crystalline materials
Internal Feed Filter	Metallurgical concentrates, iron ore
Totally-Enclosed Filter	Petroleum dewaxing, solvent slurries, catalyst, hazardous materials
Pulp Filter	Fiber washing, thickening and recovery (kraft, sulfite and other pulps)
ROTARY DISK FILTERS	Metallurgical slurries (taconite, copper, lead), coal, cement, fiber recovery (paper-mill save-all), cement, paper sludge
HORIZONTAL FILTERS	
Rotating-Pan	Rapid-settling solids (sand, coal, salt cake, gypsum), pulp fibers
Tilting-Pan	Gypsum, phosphoric acid manufacture
Traveling-Belt	Medium and coarse solids, fibers
FILTER-THICKENER	
Filter Press (Plate and frame)	Pigments, catalysts (continuous reactors)
Rotating-Disk	Metal hydroxides, coal refuse
Leaf Filter	First carbonation juice (beet sugar)
Tube Filter	White liquor slurry (recausticizing)
ULTRAFILTERS	
Membrane Types	Electrocoating paints, enzymes, proteins, fine solids
INTERMITTENT FILTERS	
LEAF FILTERS	
Stationary Leaf Filter	Red mud, contact clays, pigments, nickel catalysts
Rotating-Leaf Filter	Pharmaceuticals, clay and zinc slurries
Traveling-Medium Filter	Machine coolants, cutting oils, sludges
Filter Press (Multiple plate and frame)	Pigments, dyes, carbon clay, various chemicals
Tube Filter	Chemical wastes
Nutsche Filter	Chemicals, wastes (small volumes)
DEEP-BED FILTERS	
Sand or Coal Beds	Process water, process liquors
PRESSURE FILTERS	
Cartridge Filters	Numerous industrial chemicals, fluids, oils, water, paint

watering during one portion of the cycle and for cutting off the vacuum and causing an air blow during cake discharge is effected by an automatic valve.

The drum usually revolves slowly, requiring from 2 to 10 mintues per revolution. In some applications, however, as in the case of free-draining wood pulp, the filter may revolve at 4 to 5 revolutions per minute. A level controller maintains a continuous supply of slurry in the vat. The filter is permitted to operate continuously until the filter cloth fills up with solids to the point where efficiency is affected. The time for a continuous run, of course, varies considerably with the type of slurry involved.

Drums are furnished up to about 14 feet (14.2 meters) in diameter

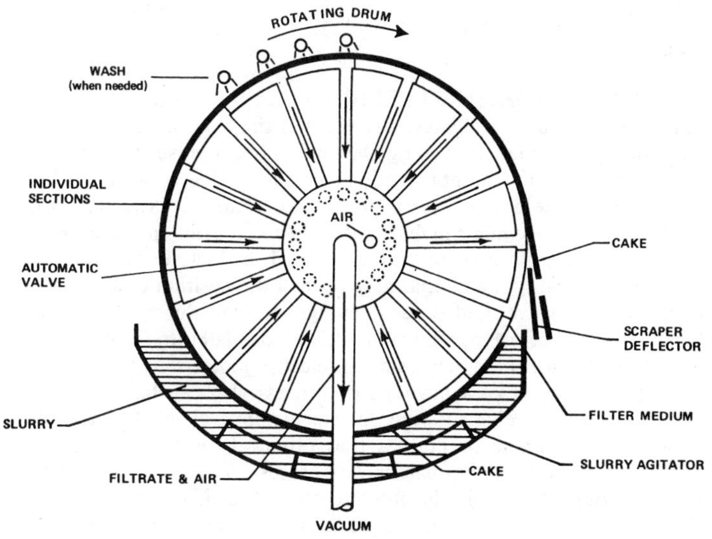

Fig. 3. Configuration of rotary drum filter.

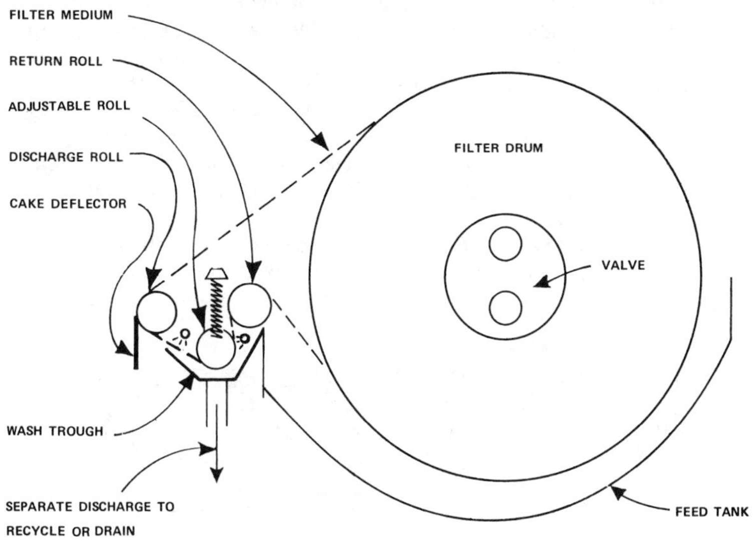

Fig. 4. Overall design configuration of rotary belt filter.

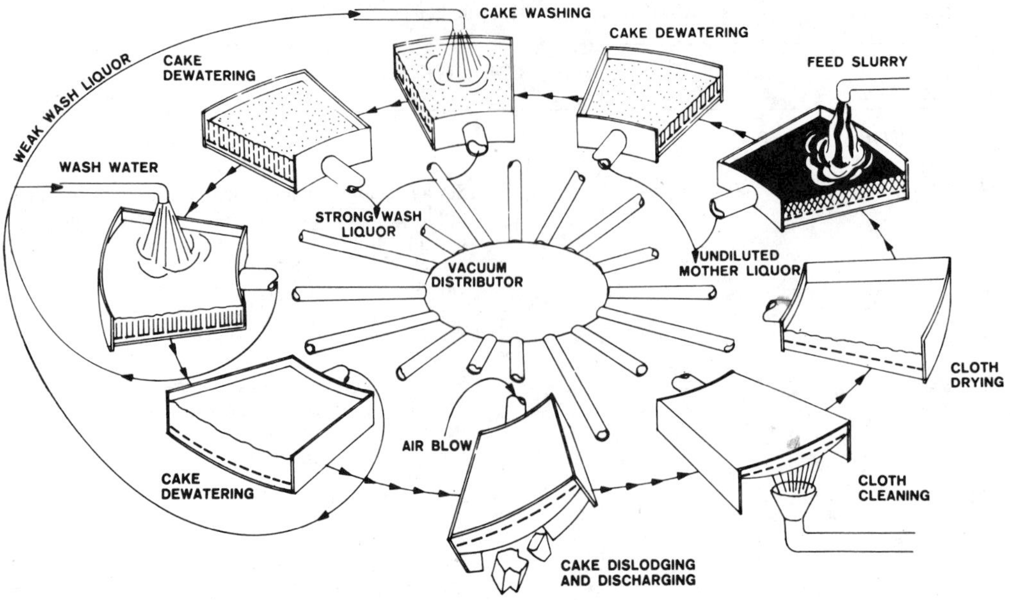

Fig. 5. Operating principle of tilting-pan filter (*Bird Machine Co.*)

and up to 34 feet (10.2 meters) or more in length. Filtration area ranges from 10 to 1,500 square feet (0.9 to 140 square meters). Larger units are used on wood-pulp and special metallurgical applications, particularly in uranium, gold, and copper ore processing.

String Filter. This is a modification of the basic rotary drum filter. A series of endless strings from 0.5 to 1 inch (12 to 25 millimeters) apart are wrapped around the drum and two external rolls. The strings lift the cake off the drum at about 60 degrees beyond top center. As the cake and embedded strings arrive at the discharge roll, the strings separate from the cake since their direction is reversed. The cake proceeds to discharge by gravity.

Belt Filters. These designs were introduced in the mid-1950s with the objective of lengthening the life of the filter-medium through the use of external washing on a continuous basis. Improvement of clarity of filtrate was also an objective. The overall design configuration is shown in Fig. 4. Depending upon application and effectiveness of continuous washing, the filter medium will perform well over a period ranging from 6 weeks to 6 months, or even longer.

Internal-feed drum filters accomplish filtration on the inside of the drum surface. With the feed on the inside, there is no need for an external slurry vat as with the conventional rotary drum. A pool of slurry is maintained in the bottom of the drum to a depth of several inches. When the formed cake approaches top center, the vacuum is cut off and a reverse, pulsating blow is applied to discharge the cakes to a chute or conveyor below.

Tilting-Pan Filters. Principally used for filtering gypsum in phosphoric acid production, the tilting-pan filter is obtainable in sizes up to about 1,800 square feet of filtration surface. Although complex, the large units are quite practical and operate generally without excessive maintenance. They effect a three-stage countercurrent washing with minimum dilution of the filtrate. The cake is discharged by gravity. Washing after each cycle minimizes blinding of the filter medium and lengthens the life of the cloth. The principle of operation is illustrated in Fig. 5.

Horizontal Traveling Belt Filter. This design is a combination of a filter medium and a drainage belt which travels over a series of fixed vacuum boxes. The design is shown in Fig. 6. Feed is supplied by a distributor. The slurry is contained by side dams along the filter belt until dewatering and washing are completed. The cake discharge is by gravity. The filter medium, on the return side, is washed by sprays applied to both sides.

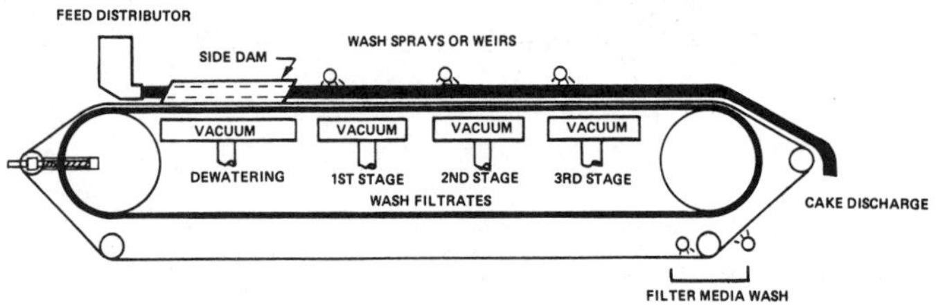

Fig. 6. Horizontal traveling belt filter.

See also **Ultrafiltration.**

References

Fiore, J. V., and R. A. Babineau: "Filtration," *Food Technology*, **33**, 4, 67–72 (1979).

Kirjassoff, D., Pinto, S., and C. Hoffman: "Ultrafiltration of Waste Latex Solutions," *Chem. Eng. Progress*, **76**, 2, 58–61 (1980).

Miles, H. V.: "Filtration," in the "Chemical and Process Technology Encyclopedia," (D. M. Considine, editor), McGraw-Hill, New York, 1974.

Perry, R., and C. Chilton: "Chemical Engineers' Handbook," 5th edition, McGraw-Hill, New York, 1973.

Severeson, S. D., et al.: "The Economics of Fabric Filters and Precipitators," *Chem. Eng. Progress*, **76**, 1, 68–73 (1980).

Smith, C. V.: "Electrokinetic Phenomena in Particulate Removal from Water by Rapid Sand Filtration," Ph.D. thesis, Johns Hopkins University, Baltimore, Maryland, 1966.

Staff: "Filtration and Separation," *Chem. Eng. Progress*, **73**, 4, 57–91 (1977).

Wnek, W.: "Electrokinetic and Chemical Aspects of Water Filtration," *Filt. Separ.*, **11**, 237 (1974).

Yada, K.: "Study of Chrysotile Asbestos by a High Resolution Electron Microscope," *Acta Cryst.*, **23**, 704 (1967).

Yao, Y. M., Habibian, M. T., and C. R. O'Melia: "Water and Waste Water Filtration: Concepts and Applications," *Environ. Sci. Tech.*, **5**, 11, 1105 (1971).

FIN. Fishes.

FINAL CONTROLLING ELEMENT. That element of a control system which directly changes the value of the manipulated variable. Commonly, the final controlling element will be a valve in a pipeline, as for example, the automatically controlled valve in a steam line that throttles the flow of steam to a heat exchanger. Other forms of final controlling elements include dampers and louvers that, for example, throttle the flow of air or gas through a duct. Pneumatic and hydraulic power cylinders that may control the position of a machine element, slide valve, and other openings, orifices, ducts, etc., which affect the value of the manipulated variable also fall into the classification of final controlling elements. An electric motor that finally positions some element of a machine or process also can be considered a final controlling element. The final controlling element causes the action on a machine or process that ultimately is fed back to the controller by way of the machine or process. Whereas the primary or measuring element is the interface with the machine or process on the measurement side of an automatic control system, the final controlling element is that interface on the control side of the system.

The official definition (ANSI C85.1) of a final controlling element is "That forward controlling element which directly changes the value of the manipulated variable."

See also **Control System (Automatic)**; and **Valve (Control).**

FINAL MASS. The mass of a rocket after its propellants are consumed.

FINAL VALUE THEOREM. Laplace Transform.

FIN ATTENUATOR. Attenuator.

FINCH (*Aves, Passeriformes*). Seed-eating birds of many species, found chiefly in the northern hemisphere but to a limited extent in Africa and South America. They are small to moderately large and have a strong beak, usually conical and in some species very large.

In addition to the species whose names indicate their relation with the group, such as the greenfinches and the chaffinches, the finches include birds with distinctive names. The goldfinches, grosbeaks, cardinals, bramblings, siskins, linnets, redpolls, sparrows, canaries, and crossbills belong here. See also entry on **Passeriformes.**

Some species not described elsewhere in this volume include: The American goldfinches (*Astraggalinus tristis*), brightly yellow colored, with top of head, wings, and tail black, and with two white bands on the wings. The female is more drab, mostly olive. The male loses much of its coloration during wintertime. The goldfinch is about 5 inches (13 centimeters) long. A European species (*Carduelis carduelis*) is even more brilliantly colored, having a bright-red face. A goldfinch is sketched in the accompanying diagram.

Goldfinch (*Astragalinus tristis*).

The Siskin is a quietly colored finch, numerous species of which are found throughout the temperate part of the northern hemisphere as well as South America. The pine siskin of North America, *Spinus pinus*, also called the pine finch, is finely streaked with brownish coloration on a lighter background: There are a few yellow marks on wings and tail.

The woodpecker finch (*Camarhynchus pallidus*) is the only bird known that uses a tool for obtaining food. The bird has been observed to visit holes made in trees by woodpeckers. By using a thorn picked up from the ground, the bird probes the holes, finding likely insects which stick to the thorn. This is a means of extending and making the beak of the bird more functional.

FINENESS RATIO. 1. The ratio of the length of a body to its maximum diameter, or, sometimes, to some equivalent dimension—said especially of a body such as an airship hull or rocket.

2. A term analogous to the thickness ratio of an airfoil, and often the inverse of that ratio. For example, the cross section of a streamlined strut or structural member may have a symmetrical airfoil shape whose fineness ratio would be defined by the ratio of the chord length to the maximum thickness, whereas the thickness ratio would be the inverse.

Airship shapes are characterized by fineness ratio, which usually varies from 5 to 8.

See also **Supersonic Aerodynamics.**

FINE STRUCTURE. In atomic spectra, the occurrence of a spectral line as a doublet, triplet, etc., due to the interaction or coupling between the orbital angular momentum and the spin angular momentum of the electrons in the emitting atoms. Fine structure is also exhibited by spectra of particles. See also **Atomic Spectra.**

FINGER LAKE. A glacial U-Valley or rock bowl which forms the basin for a fresh-water lake. Because of the character and origin of the basin, finger lakes are relatively long and narrow. Type locality, the Finger Lake region of New York State.

FINGERPRINT. Bertillon System.

FINITE GRAPH. Graph (Mathematics).

FINNAN HADDIE. Codfishes.

FIORD. A fiord is a glacially overdeepened valley, usually narrow and steep-sided, extending below sea level and occupied therefore by salt water. Typical fiords are to be found in Alaska and Norway; their depths, sometimes as much as 4000 feet (1219 meters), indicate that they are glaciated valleys which have been invaded by the sea after the disappearance of the glaciers. The word fiord is a variant of the Norwegian term for these features, *fjord.* The long fiord-like bays of the New England coast line are sometimes referred to as fiards.

FIREBALL. A term used in astronomy to designate those meteors which are large enough to be apparently brighter than the planet Jupiter. They frequently leave a trail which may be visible for several minutes. Not infrequently a distinct sound is heard either during, or shortly after, the observation of a fireball. See also **Bolide.**

FIRE BRAT (*Insecta, Thysanura*). A wingless insect, *Thermobia domestica*, clothed with silky scales and bearing long antennae and three slender filaments at the opposite end of the body. Related to the silverfish but found in warm places; economic importance and control similar. The fire brat feeds on book binding, wallpaper paste, crackers, and other starch containing substances. The insect is very active in its movements, is approximately one-half inch in length when mature, and prefers damp locations, such as basements.

FIRE-BRICK. Fire-brick is a type of brick capable of withstanding high temperatures and is used to line flues, stacks, furnaces, etc. Good resistance to heat flow is not to be secured simultaneously with refractoriness—indeed, the most refractory bricks generally have the highest thermal conductivities. Where necessary, insulation is added to minimize heat leaks. It is important for the refractory brick to be satisfactory on a number of points in addition to refractoriness, for resistance to melting is only one of several requirements to be met. Among these might be cited resistance to erosion by ash-laden gases, and to the fluxing action of molten slag. A good refractory should not spall badly under rapid temperature changes. The structural strength of fire-brick should hold up well as its temperature approaches the fusion temperature.

Modern installations often impose furnace conditions so severe that refractories other than fire-clay are needed. High aluminum and silicon carbide refractories are typical of these. The heat conductivities of the super-refractories are larger than those of fire-clay brick, and such construction should be backed up with high temperature insulation. Silicon carbide blocks are the most refractory and have the quality of resisting clinker adhesion better than ordinary fire-brick. Their fusion temperature is about 4000°F (2204°C).

Clay fuses at from 2800 to 3200°F, (1538 to 1760°C) the upper limit being for flint clay and the lower for the plastic form which, due to its cementing qualities, is especially valuable in fire-brick manufacture. Red brick is not suitable for refractory service, nor is insulating brick. There are several fire-clay furnace cements on the market that are adaptable to monolithic lining. The standard size of fire-brick and insulating brick is 9 inches by $4\frac{1}{2}$ inches by $2\frac{1}{2}$ inches (22.9 by 11.4 by 6.4 centimeters).

FIRECLAY. This term is chiefly used by British geologists to designate the leached clays, rich in silica and alumina and low in alkalies and lime, which lie directly beneath coal beds. These clays are of economic importance because they are refractory, and do not melt when heated to high temperatures.

FIREFLY (*Insecta, Coleoptera*). Harmless, soft-bodied beetles with a luminous organ in the abdomen. The flashing of these insects apparently enables them to find mates. The insects commonly are called lightning bugs. The wings are blue and lay flat to the body when at rest. The head is of orange coloration with two dark spots. The energy

Firefly. (*A. M. Winchester.*)

system for the lighting or flashing is of extremely high efficiency and involves enzyme reactions with oxygen. It has been found that the insect regulates the flashes of light by controlling the amount of air admitted to its bioluminescent organ. Details of the energy system will be found under **Bioluminescence.**

FIREMOUTH. Cichlids.

FIRE POINT. Petroleum.

FIRE-RETARDANT PAINTS. Paint.

FIREWALL. Industrial or commercial buildings, standing adjacent, with common division wall, may be required to have special attention given to this wall, as to apertures, thickness and material, so as to prevent ignition or transmission of conflagration from one building to another. Such a wall would be a firewall.

Aircraft having the engine mounted in the nose of the fuselage are required to have a firewall. This consists of a bulkhead of sheet steel, dividing the engine compartment from the remainder of the fuselage.

FIRMING AGENTS (Foods). Certain foodstuffs, such as apples, potatoes, and beans, tend to be rather fragile when subjected to processing operations prior to packaging (canning, freezing, etc.) and, if not treated in some way to retain their natural firmness to a relatively high degree, the mouthfeel of the final product will be disappointing (mushiness versus slight chewiness or crispness). Certain chemical substances can be added prior to or during processing to protect and retain natural firmness. These substances include: aluminum potassium sulfate, aluminum sodium sulfate, aluminum sulfate, calcium carbonate, calcium chloride, calcium citrate, calcium gluconate, calcium hydroxide, calcium lactobionate, calcium phosphate (monobasic), calcium sulfate, and magnesium chloride, among others.

Researchers have found that calcium lactate, in particular, can be an effective agent for preserving the firmness of apple slices during processing and prior to canning or freezing. Studies have shown that calcium salts participate in firming the tissues of various fruits and vegetables by forming calcium pectates. Calcium citrate is quite useful for firming peppers, potatoes, tomatoes, lima beans, and snap beans. Manufacturers of pet foods have found that from 1 to 2.5% monoglyceride contributes to the firming of pet foods, as well as aiding in the prevention of fat separation.

FIRST AID (Artificial Respiration). Artificial Respiration.

FIRST AID (Cardiac Massage). Cardiac Massage.

FIRST ORDER REACTION. Chemical Reaction Rate.

FIRST ORDER SYSTEM. A system whose performance characteristics are presented in the form of a first order differential equation

$$r(t) = k_1 \frac{dC(t)}{dt} + k_2 C(t)$$

where r = system input
C = system output
t = time
k_1 and k_2 are coefficients

If both coefficients are constants, the Laplace transform should also be first order:

$$\frac{C(s)}{R(s)} = \frac{1}{k_1 S + k_2}$$

For simplicity, all initial conditions were assumed zero. The accompanying figure demonstrates graphically the time response of a first-order system which is subject to a step input, r_0.

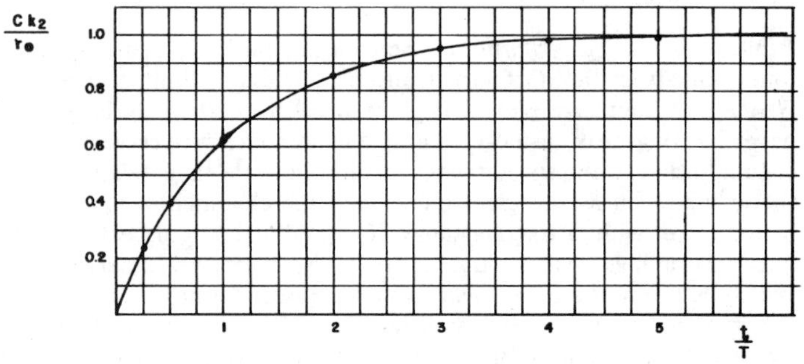

Time response of a first-order system subjected to a step input. System time constant, $T = k_1/k_2$.

FIR TREES. Members of the family *Pinaceae* (pine family), these trees are of the genus *Abies* with the notable exception of the Douglas fir, which is of the genus *Pseudotsuga.* The term *fir* is used rather loosely and some dictionary definitions refer to the firs collectively as cone-bearing evergreens similar to the pine(s). Thus, botanists often use the term *true fir* to distinguish the true firs (also called *silver firs*) from other trees which may have the term fir associated in some way with their name. A case in point, shown in the following list, is the Chinese fir, which is not a member of the pine family, but rather it is of the *Taxodiaceae* (swamp cypress family). It should be pointed out that the Douglas fir is not a true fir as previously described. The Douglas fir, a truly magnificent tree, has the rather unattractive title of "false hemlock," the literal translation from the Latin *Pseudotsuga.* However, the full name, *Pseudotsuga menziesii,* honors the founder of the tree, explorer Menzies, who reported it in 1791. The fir trees are, of course, conifers. The general description of conifers is given in the entry on **Conifers.**

Some of the important species of fir trees include:

Algerian fir	Abies *numidica*
Alpine fir	*A. lasiocarpa*
Arizona cork fir	*A. arizonica*
Balsam of Gilead	See Balsam fir
Balsam fir	*A. balsamea*
Bristlecone fir	See Santa Lucia fir (Not to be confused with the bristlecone pine)
California red fir	*A. magnifica*
Caucasian fir	*A. nordmanniana*
Chinese fir	*Cunninghamia lanceolata* (Member of *Taxodiaceae* family)
Cilician fir	*A. cilicica*
Colorado white fir	*A. concolor*
Common silver fir	*A. alba*
Douglas fir	*Pseudotsuga menziesii*
East Himalayan fir	*A. spectabilis*
Flaky fir	*A. squamata*
Formosan silver fir	*A. kawakamii*
Forrest's fir	*A. delavayi forrestii*
Fraser's fir	*A. fraseri*
Grand fir (or Giant fir)	*A. grandis*
Greek fir	*A. cephalonica*
Japanese fir	*A. firma*
Korean fir	*A. koreana*
Low's silver fir	*A. concolor iowiana*
Manchurian fir	*A. holophylla*
Maries' fir	*A. mariesii*
Nikko fir	*A. homolepis*

Noble fir	*A. procera*
Pacific silver fir	*A. amabilis*
Red fir	See Pacific silver fir
Red silver fir	See Pacific silver fir
Sacred fir	*A. religiosa*
Santa Lucia fir	*A. bracteata*
Szechwan fir	*A. sutchuenensis*
Shasta fir	See California red fir
Siberian silver fir	*A. sibirica*
Sicilian red fir	*A. nebrodensis*
Spanish fir	*A. pinsapo*
Veitch fir	*A. veitchii*
West Himalayan fir	*A. pindrow*
White fir	See Colorado white fir

Several of the champion fir trees in the United States, as selected by The American Forestry Association for inclusion in its "Social Register of Big Trees," are listed in Table 1. See also accompanying figure.

TABLE 1. REPRESENTATIVE RECORD FIR TREES IN THE UNITED STATES[1]

Specimen	Circumference[2]		Height		Spread		Location
	(inches)	(centimeters)	(feet)	(meters)	(feet)	(meters)	
Balsam fir (1962) (*Abies balsamea*)	84	213	116	34.8	33	9.9	Michigan
Bigcone fir (1973) (*Pseudotsuga macrocarpa*)	264 (at 84 inches)	671 (at 213 centimeters)	145	43.5	85	25.5	California
Bristlecone fir (1976) (*Abies bracteata*)	162	411	182	54.6	38	11.4	California
California red fir (1972) (*Abies magnifica var. magnifica*)	320	813	180	54	48	14.4	California
Coast (typical) fir (1972) (*Pseudotsuga menziesii*)	545	1384	221	66.3	61	18.3	Washington
Corkbark fir (1972) (*Abies lasiocarpa var. arizonica*)	157	399	95	28.5	33	9.9	New Mexico
Fraser fir (1972) (*Abies fraseri*)	98	249	83	24.9	48	14.4	North Carolina
Grand fir (1972) (*Abies grandis*)	248	630	231	69.3	46	13.8	Washington
Noble fir (1972) (*Abies procera*)	340	864	278	83.4	47	14.1	Washington
Pacific Silver fir (1972) (*Abies amabilis*)	259	658	206	61.8	38	11.4	Washington
Rocky Mountain fir (1976) (*Pseudotsuga menziesii var. glauca*) (Co-Champion)	211	536	156	46.8	44	13.2	Orgeon
Rocky Mountain fir (1976) (*Pseudotsuga menziesii var. glauca*) (Co-Champion)	236	599	130	39	—	—	Idaho
Shasta red fir (1971) (*Abies magnifica var. shastensis*)	325	826	176	52.8	41	12.3	California
Subalpine fir (1972) (*Abies lasiocarpa var. lasiocarpa*)	253	643	129	38.7	22	6.6	Washington
White fir (1972) (*Abies concolor*)	335	851	192	57.6	35	10.5	California

[1] From the "Social Register of Big Trees," The American Forestry Association (by permission).
[2] At 4.5 feet (1.4 meters).

TABLE 2. ENGINEERING DATA ON WOOD FROM VARIOUS FIR TREES

Common Name for Species	Green Condition			Air-Dried (12% Moisture)		Maximum Crushing Strength (Parallel to Grain)		Maximum Tensile Strength (Perpendicular to Grain)	
	Moisture Content (Percent)	Weight/ Cu. Foot (Pounds)	Weight/ Cu. Meter (Kilograms)	Weight/ Cu. Foot (Pounds)	Weight/ Cu. Meter (Kilograms)	(Psi)	(MPa)	(Psi)	(MPa)
Douglas fir, coast type	38	38	609	34	545	3860	26.6	300	2.1
Douglas fir, Rocky Mountain-type	38	35	561	30	481	3000	20.7	350	2.4
Balsam fir	117	45	721	25	401	2400	16.6	180	1.3
Commercial white fir	111	46	737	27	433	2830	19.5	310	2.1

SOURCE: U.S. Forest Products Laboratory.

Noble fir (U.S. National Champion) located at Gifford Pinchot National Forest, Washington. (*U.S. Forest Service photo.*) For dimensions, see Table 1.

Highlights of the distribution of fir trees in North America include: The balsam fir is the only native fir tree of the northeastern United States and Canada. It is found from Newfoundland and Labrador west to Hudson Bay, Michigan, and Minnesota, south to the high Alleghanies and into Virginia. It is found at high altitudes in the White mountains of New Hampshire. The closely related Fraser's fir is found in the southern Alleghanies and in the mountains of Virginia, North Carolina, and Tennessee, commonly near the mountain summits, infrequently occurring below 4,000 feet (1,200 meters). The red fir and noble fir prefer the middle altitudes and are found in the Sierra and Cascade ranges of the west, northward from southern California into Canada. A variety of the red fir, the Shasta fir, is found in the area around Mount Shasta. The Colorado white fir ranges widely and is found from New Mexico northward into Oregon. The Alpine fir occurs on the Pacific slopes, ranging from New Mexico northward to the Pacific Northwest and into southeastern Alaska. The grand fir prefers lowlands or moist mountain slopes and ranges from the northern coast of California into southern British Columbia. The white fir is found on mountain slopes from Arizona and New Mexico northward through Nevada, Utah, the Cascade range of California, and into southern Oregon. The Pacific silver fir (sometimes simply called silver fir or *Amabilis* fir) prefers the country north of the Cascade range into Washington, British Columbia, and southern Alaska. The noble fir is also a Pacific mountain fir, ranging from the Cascade Mountains northward into Washington and Oregon. The Douglas fir is found in the Rocky Mountains and ranges north-westward to British Columbia and the Pacific Coast. The Rocky Mountain form has the varietal name of *Glauca*. Other forms range from Mexico up the coast and into Canada.

Probably the second tree in Europe in terms of height attained is the common European silver fir (*A. alba*). Other European firs include the Greek fir and the Spanish fir, the latter occurring along the Mediterranean. The Caucasian firs occur in the Caucasus mountains, which interestingly are at the same latitude as northern California where so many of the firs thrive. Well known firs of Japan include the Nikko and the Veitch, the latter occurring in part in the region of Mount Fuji. Maries' fir is also of Japan. Chinese fir trees include Forrest's fir and the Manchurian fir. Fir trees of the Himalayan region include both the East and West Himalayan firs. The Chinese "fir" (*Cunninghamia lanceolata*) is a major timber-producing tree in China, outdistanced only by bamboo. Details of the numerous species of firs, of course, cannot be detailed here. As representatives, however, the balsam fir and the Douglas fir are described briefly.

The balsam fir is a large stately tree with horizontal branches. The leaf is persistent and the cones erect. The fragrance is slightly tangy. The Canadian balsam produces turpentine. The wood is soft, but not considered of a high commercial grade. The weight is about 26 pounds per cubic foot (416.5 kilograms per cubic meter); specific gravity of 0.983 to 0.997. The tree is widely used as a Christmas tree. Souvenir pillows are often stuffed with balsam needles. They are also used for packing boxes and for paper pulp.

Authorities consider the Douglas fir as one of the greatest of all trees. The oldest tree of this species still living is estimated at 1,300 years of age. This tree was discovered in 1825 by David Douglas, a visiting botanist from Scotland, on a trip to North America. The Douglas fir grows thick and lush at an altitude of about 8,000 feet (2,400 meters). The tree does not grow at altitudes higher than 10,000 feet (3,000 meters). It does well on thin mountain soil. The tree root system is shallow, the average roots going down only about four feet. The tree requires full sun. In very hard rains, the tree is sometimes uprooted.

The larger Douglas firs average about 80 to 100 feet (24 to 30 meters) in height. See Table 1. Usually the spread is about 55 to 60 feet (16.5 to 18 meters). In some specimens, the tree may rise 100 feet (30 meters) before the first branching is visible. The California Forestry Department has stated that "The Douglas fir supplies the world market with more products for use than any other species." Some of the logs cut for lumber are 10 feet (3 meters) in thickness and a large truck can handle only three at a time.

The bark, containing tannin, is a dark gray-brown color, resinous, and from 5 to 9 inches (12.7 to 22.9 centimeters) thick. The bark provides effective protection from insects, animals, and fire. The bark ridges are so deep that it is possible for a person to bury a fist in the crevice. The wood is soft, easy to work, light, strong, and firm. Large sheets of knot-free wood are obtained by sawing the base of older trees. Some of the many uses for this wood include heavy construction work, building of homes and factories, mine props, telephone poles, bridge building, and ship docks. Strong, large decorative doors for homes and buildings are made from it on a large scale.

Tannin is recovered from the bark. Crushed bark is also used in the manufacture of acoustical tile and flooring.

Some engineering constants of the commercial wood from fir trees are given in Table 2. For references, see **Tree.**

FISHER. **Mustelines.**

FISHER'S *z* DISTRIBUTION. Fisher's *z* distribution was derived by R. A. Fisher in 1924:

$$P_z\,dz = \frac{2 n_1^{n_1/2} n_2^{n_2/2}}{B\left(\frac{n_1}{2}, \frac{n_2}{2}\right)} \frac{e^{n_1 z}}{(n_1 e^{2z} + n_2)^{(n_1+n_2)/2}}\,dz, \qquad -\infty < z < \infty$$

$z = \frac{1}{2} \log s_1^2/s_2^2$, $B(n_1/2, n_2/2)$ is the Beta function, s_1^2 is an estimated variance with n_1 degrees of freedom and s_2^2 is an independent estimated variance with n_2 degrees of freedom. The *z* distribution is used in the analysis of variance. The *z* distribution becomes the normal curve in case $n_1 = 1$, $n_2 = \infty$ after the transformation $z = \frac{1}{2} \log \chi^2$, $0 \leq \chi^2 < \infty$; it becomes the χ^2 distribution after the transformation $e^{2z} = \chi^2$, $n_1 = n$, $n_2 = \infty$, $0 \leq \chi^2 < \infty$; and becomes Student's *t* distribution after the transformation $z = \frac{1}{2} \log t^2$, $n_1 = 1$, $n_2 = n$, $0 \leq t < \infty$. In the limit as n_1 and $n_2 \rightarrow \infty$ in any manner whatever, the *z* distribution approaches normality with asymptotic mean $\frac{1}{2}(1/n_2) - (1/n_1)$ and asymptotic variance $\frac{1}{2}(n_1 + 1/n_1^2) + (n_2 + 1/n_2^2)$.

The *z* distribution has been extensively tabulated at the significance levels.

FISHES. There are two classes of fishes: (1) Cartilage fishes (Class *Chondrichthyes*); and (2) bony fishes (Class *Osteichthyes*). Both of these classes are of the phylum *Chordata.*

The *Chondrichthyes,* sometimes also called elasmobranchs, include the sharks, rays, and skates. They are among the most generalized of the vertebrates that have complete vertebrae, movable jaws, and paired appendages. The term is derived from *chondros* (Greek for cartilage), and *ichthys* (fish).

The *Osteichthyes,* sometimes also referred to as *Pisces,* are considered the true fishes. They are cold-blooded vertebrates with a bony skeleton, jaws, four pairs of gills in a common cavity, skin with scales, and normally paired fins. The term is derived from *osteon* (Greek for bone; and ichthys, fish).

In general the fishes are characterized by: (1) the skin is kept moist by glandular secretions and in many species is covered with scales; (2) the appendages are in the form of fins, used in swimming; (3) the gill slits persist in the walls of the pharnyx and serve as the seat of the respiratory gills; and (4) the heart has only two principal chambers; an atrium or auricle and a ventricle.

It is estimated that there are about 25,000 species of fishes, found in approximately 36 orders, and some 400 families. In addition to the conventional anatomical and physiological bases (taxonomy) for classifying the fishes as shown in the accompanying tables, there are several other ways to classify or characterize the fishes which are helpful to understanding these animals. Much of this information is included in specific entries in this volume for well over 100 species. The species separately described are indicated by an asterisk in Tables 1 and 2.

In learning about a particular species, it is helpful to know as much as possible concerning: (1) the size—length, girth, and weight—from birth through adulthood, in terms of averages as well as records of maxima; (2) the longevity, ranging from days to many years; (3) performance parameters and temperament factors, such as speed and mobility, voracity versus docility, aggressiveness versus a desire to hide (burrowing, etc.); (4) environmental needs or preferences which, in essence, determine habitat (or vice versa), including (a) fresh waters versus saline marine or brackish waters; (b) shallow, mid-depth, or deep waters; (c) clean versus turbid waters; (d) rushing, fast-moving waters versus quiet waters; (e) open seas, lakes, and rivers versus secluded coral reefs, rocks, gravel or muddy bottoms; (f) cold, temperate, or tropical waters; (5) diet and eating habits—carnivorous, herbivorous, omnivorous—and daily and seasonal feeding patterns; (6) reproduction and spawning habits—bearing of live young, ovipositors, egg layers—and migration patterns tied to spawning needs; (7) aside from spawning needs, other migratory habits and the general range of distances normally covered (entire life in a coral reef, pond, or river versus roving the open seas for hundreds, even thousands of miles); (8) relationships with other fishes of same or other species (schooling versus singular or small groups), as part of a community over and beyond seeking other fishes for food, or conversely, serving as prey (examples of symbiosis, such as the relationship between the remora and the shark, the role of cleaner fishes, relation of cucumber fish with sea cucumber, etc.); (9) unique characteristics or features, such as lack of red blood cells in the ice fishes, the true flying ability of the South American hatchet fishes, etc.); and (10) very important and of much interest, of course, is the interface between the various fishes and people—fishes as sources of food, oils, chemicals, and other useful items of commerce; as means for maintaining natural balances and preservation of desirable environments; and in the reactions of fishes to people-generated environments, in terms of vulnerability or hardiness to numerous kinds of pollution, etc.

Considering the foregoing, partial list of characterization factors and realizing that the parameters of each factor are in most cases very wide and that examples of species of fishes are found which fit

TABLE 1. CLASSIFICATION OF THE CHONDRICHTHYES

(Cartilage Fishes)

(* indicates separate entry in this volume)

Order	Family	Example(s)
Selachii		Sharks*
	Chlamydoselachidae	Frilled Shark
	Hexanchidae	Sixgill and Sevengill Cow-sharks
	Carchariidae	Sand Sharks
	Scapanorhynchidae	Goblin Sharks
	Isuridae	Mackerel Sharks
	Alopiidae	Thresher Sharks
	Cetorhinidae	Basking Shark
	Rhincodontidae	Whale Shark
	Orectolobidae	Carpet and Nurse Sharks
	Scyliorhinidae	Catsharks
	Pseudotriakidae	False Catsharks
	Triakidae	Smooth Dogfishes
	Carcharhinidae	Requiem Sharks
	Sphyrnidae	Hammerhead Sharks
	Heterodontidae	Hornsharks
	Pristiophoridae	Saw Sharks
	Squalidae	Spiny Dogfishes
	Dalatiidae	Spineless Dogfishes
	Echinorhinidae	Alligator Dogfish
	Squatinidae	Angel Shark*
Batoidei		Skates and Rays*
	Torpedinidae	Electric Rays
	Rhinobatidae	Guitarfishes
	Pristidae	Sawfishes
	Rajidae	Skates
	Dasyatidae	Stingrays, Whiprays, Butterfly Rays, and Round Rays
	Myliobatidae	Eagle Rays, Bat Rays, and Cow-Nosed Rays
	Mobulidae	Devil Rays
Chimaeroids		Chimaeroids*
Subclass *Holocephali*	*Chimaeridae*	Short-Nosed Chimaeras or Ratfishes
	Rhinochimaeridae	Long-Nosed Chimaeras
	Callorhinchidae	Plow-Nosed or Elephant Chimaeras

TABLE 2. CLASSIFICATION OF THE OSTEICHTHYES (Bony Fishes)
(* indicates separate entry in this volume)

ORDER	FAMILY	EXAMPLE(S)
Cladistia	*Polypteridae*	Bichirs*
Chondrostei	*Acipenseridae*	Sturgeons*
	Polyodontidae	Paddlefishes*
Protospondyli	*Amiidae*	Bowfin*
Ginglymodi	*Lepisosteidae*	Gars*
Isospondyli	*Elopidae*	Tarpon*
	Albulidae	Bonefish or Ladyfish
	Alepocephalidae	Slickhead Fishes*
	Clupeidae	Herring* Sardines*
	Dorosomidae	Gizzard Shad*
	Dussumieriidae	Round Herring
	Chirocentridae	Dorab* (also Wolf Herring)
	Engraulidae	Anchovy*
	Chanidae	Milkfish*
	Salmonidae	Salmon*
		Trout*
	Coregonidae	Whitefishes*
	Thymallidae	Grayling*
	Osmeridae	Smelts*
		Capelin*
	Galaxiidae	Galaxiids*
	Gonostomatidae	Brisstlemouths*
	Sternoptychidae	Hatchet Fishes*
	Chauliodontidae	Viper Fishes*
	Osteoglossidae	Bony Tongues*
	Pantodontidae	Butterfly Fish*
	Hiodontidae	Mooneye
	Notopteridae	Featherbacks*
	Mormyridae	Mormyrids*
	Gonorhynchidae	Sandfish (also Beaked Salmon)
Haplomi	*Esocidae*	Pike* (includes Pickerel and Muskellunge)
	Umbridae	Mud Minnows
	Dalliidae	Blackfish*
Iniomi		Iniomous Fishes*
	Synodidae	Lizard Fishes
	Aulopidae	Thread-Sail Fishes
	Chlorophthalmidae	Greeneyes
	Ipnopidae	Grid-Eye Fishes
	Bathypteroidae	Spider Fishes
	Myctophidae	Lantern Fishes
	Harpodontidae	Bombay Duck
	Paralepididae	Barracudinas
	Scopelarchidae	Pearleyes
	Evermannellidae	Saber-Tooth Fishes
	Alepisauridae	Lancet Fishes
	Omosudidae	Hammerjaw
	Anotopteridae	Javelin Fish
Giganturoidea	*Giganturoidea*	Deep-Sea Giganturid Fishes
Lyomeri		Deep-Sea Gulper Eels
Ostariophysi	*Characidae*	Characids* (includes Piranha)
	Gymnotidae	Gymnotid Eels* (includes Electric Eel and Knifefishes)
Group: *Cypriniformes*		Bitterling*
	Cyprinidae	Carp* (includes Minnows)
	Catostomidae	Suckers*
	Cobitidae	Loaches*
	Gyrinocheilidae	Gyrinocheilids
	Homalopteridae	Hillstream Fishes
Suborder: *Siluroidea*		Catfishes*
	Doradidae	Doradid Catfishes
	Callichthyidae	Callichthyid Catfishes
	Loricariidae	Plecostomus*
	Aspredinidae	Banjo Catfishes
	Ariidae	Ariid Marine Catfishes
	Plotosidae	Plotosid Marine Catfishes
	Clariidae	Clariid Catfishes
	Siluridae	Silurid Catfishes
	Pimelodidae	Pimelodid Catfishes
	Bagridae	Bagrid Catfishes

(*continued*)

TABLE 2 (*continued*)

Order	Family	Example(s)
	Trichomycteridae	Parasitic Catfishes
	Ictaluridae	North American Catfishes
	Schilbeidae	Schilbeid Catfishes
	Mochocidae	Upside-Down Catfishes
	Malapteruridae	Electric Catfish
Apodes		Eels*
	Anguillidae	Fresh Water Eels
	Simenchelidae	Parasitic Snubnosed Eel
	Muraenidae	Moray Eels
	Ophichthidae	Snake Eels
	Nemichthyidae	Snipe Eels
	Synaphobranchidae and Serrivomeridae	Deep-Sea Eels
	Congridae	Conger Eels
	Moringuidae	Worm Eels
Heteromi		Deep-Sea Spiny Eels
Synentognathi		
	Belonidae	Needlefishes
	Hemiramphidae	Halfbeaks
	Scomberesocidae	Sauries
	Exocoetidae	Flying Fishes*
Microcyprini		
	Amblyopsidae	Blind Fish*
	Cyprinodontidae	Egg-Laying Topminnows
	Poeciliidae	Viviparous Topminnows*
	Goodeidae	Goodeids
	Anablepidae	Four-Eyed Fishes*
	Jenynsiidae	Jenynsiids
	Adrianichthyidae	Andrianichthyids
Salmopercae		
	Percopsidae	Troutperch
	Aphredoderidae	Pirateperch
Solenichthys		Tube-Mouthed Fishes
	Syngnathidae	Pipefishes*
		Seahorses*
	Solenostomidae	Ghost Pipefishes
	Indostomidae	Indostomids
	Centriscidae	Shrimpfishes
	Macrorhamphosidae	Snipefishes
	Fistulariidae	Cornetfish
	Aulostomidae	Trumpetfishes
Anacanthini	*Gadidae*	Codfishes*
	Merlucciidae	**Hake***
	Coryphaenoididae	Deep-Sea Rattails or Grenadiers
Allotriognathi		
	Lampridae	Opah*
Berycomorphi	*Holocentridae*	Squirrelfishes* (also Soldierfishes)
	Berycidae	Alfonsinos
	Anomalopidae	Lantern-Eye Fishes
	Polymixiidae	Barbudos
	Monocentridae	Pinecone Fishes
Suborder: *Xenoberyces*		Gibber, Prickle, Bigscale, and Whalefishes
Zeomorphi		John Dories*
	Antigoniidae	Boarfishes
	Grammicolepidae	Grammicolepids
Percomorphi		Perchlike Fishes
Suborder: *Percoidea*		
	Serranidae	Bass* (includes Groupers)
	Theraponidae	Tigerfishes
	Kuhliidae	Aholeholes
	Centrarchidae	Sunfishes*
	Priacanthidae	Catalufas or Bigeyes
	Apogonidae	Cardinal Fishes
	Percidae	Perches and Darters* (includes Walleyes)
		Redfish*
	Malacanthidae	Blanquillos
	Pomatomidae	Bluefish

(*continued*)

TABLE 2 (*continued*)

Order	Family	Example(s)
	Rachycentridae	Cobia*
	Carangidae	Carangids* (includes Cavallas, Jacks, Pompanos) Scad*
	Coryphaenidae	Dolphins*
	Centropomidae	Robalos (Snook and Glassfish)
	Lutianidae	Snappers*
	Nemipteridae	Nemipterids
	Lobotidae	Tripletails
	Leiognathidae	Slipmouths
	Gerridae	Mojarras
	Pomadasyidae	Grunts*
	Sciaenidae	Croakers*
	Mullidae	Goatfishes or Surmullets
	Lethrinidae	Lethrinids
	Sparidae	Porgies* (includes Sea Breams)
	Monodactylidae	Fingerfishes
	Toxotidae	Archerfishes
	Kyphosidae	Rudderfishes
	Girellidae	Nibblers
	Platacidae	Batfishes
	Ephippidae	Spadefishes
	Scatophagidae	Scats
	Chaetodontidae	Angelfishes* (includes Butterfly Fishes)
	Nandidae	Leaf Fishes and Nandids
	Cichlidae	Cichlids*
	Cirrhitidae	Hawkfishes
	Embiotocidae	Surfperches
	Pomacentridae	Damselfishes
	Labridae	Wrasses*
	Scaridae	Parrotfishes*
	Trichodontidae	Sandfishes
	Opisthognathidae	Jawfishes
	Trachinidae	Weeverfishes*
	Trichonotidae	Sand Divers
	Kraemeriidae	Sand Lances
	Uranoscopidae	Electric Stargazers
	Dactyloscopidae	Sand Stargazers
	Chaenichthyidae	Ice Fishes*
Suborder: *Acanthuroidea*		
	Zanclidae	Moorish Idol
	Acanthuridae	Surgeonfishes*
Suborder: *Siganoidea*	*Siganidae*	Rabbitfishes*
Suborder: *Trichiuroidea*		
	Trichiuridae	Cutlass Fishes and Hairtails*
	Gempylidae	Deep-Sea Snake Mackerels or Escolars
Suborder: *Scombroidea*		
	Scombridae	Mackerels*
		Tunas*
	Istiophoridae	Billfishes* (includes Marlins, Sailfishes, and Spearfishes)
	Xiphiidae	Swordfish (described under Billfishes)
	Luvaridae	Louvar
Suborder: *Gobioidea*	*Eleotridae*	Sleepers
	Gobiidae	Gobies*
	Taenioididae	Eel Gobies
	Rhyacichthyidae	Loach Goby
	Callionymidae	Dragonets*
Suborder: *Blennioidea*		Blennies*
	Blenniidae	Scaleless Blennies
	Clinidae	Scaled Blennies or Klipfishes
	Stichaeidae	Pricklebacks
	Pholidae	Gunnels
	Anarhichadidae	Wolf Fishes and Wolf Eels
	Zoarcidae	Eelpouts
Suborder: *Ophidioidea*		
	Brotulidae	Brotulids
	Ophidiidae	Cusk Eels
	Carapidae	Cucumber and Pearl Fishes

(*continued*)

TABLE 2 (*continued*)

Order	Family	Example(s)
Suborder: *Stromateoidea*		Butterfishes*
Suborder: *Anabantoidea*	*Anabantidae*	Labyrinth Fishes*
Suborder: *Channoidea*	*Channidae*	Snakeheads
Suborder: *Mugiloidea*		
	Sphyraenidae	Barracuda*
	Atherinidae	Silversides*
	Mugilidae	Mullets*
Suborder: *Phallostethoidea*	*Phallostethidae*	Phallostethids
Suborder: *Polynemoidea*	*Polynemidae*	Threadfins
Scleroparei		
	Scorpaenidae	Scorpionfishes and Rockfishes
	Triglidae	Sea Robins*
	Peristediidae	Armored Sea Robins
	Cottidae	Sculpins*
	Rhamphocottidae	Grunt Sculpin
	Agonidae	Sea Poachers and Alligator Fishes
	Liparidae	Snailfishes
	Cyclopteridae	Lumpsuckers
Suborder: *Dactylopteroidea*	*Dactylopteridae*	Gurnards*
Thoracostei		Sticklebacks and Tubenose
	Aulorhynchidae	Tubenose
Hypostomides	*Pegasidae*	Sea Moths*
Heterosomata		Flatfishes*
	Bothidae	Left-Eyed Flounders
	Pleuronectidae	Right-Eyed Flounders
	Soleidae	Soles
	Cynoglossidae	Tongue Soles
Discocephali	*Echeneidae*	Remoras* (or Suckerfishes)
Plectognathi		
	Balistidae	Triggerfishes
	Tetraodontidae	Puffers*
	Canthigasteridae	Sharp-Nosed Puffers
	Diodontidae	Porcupine Fishes* (including Burrfishes)
	Ostraciontidae	Trunkfishes
	Molidae	Ocean Sunfishes
Malacichthys	*Icosteidae*	Ragfishes
Xenoptergyii	*Gobiesocidae*	Clingfishes
Haplodoci	*Batrachoididae*	Toadfishes and Midshipmen
Pediculata		Anglerfishes*
	Lophiidae	Goosefishes or Monkfishes
	Antennariidae	Frogfishes
	Ogocephalidae	Batfishes
Suborder: *Ceratioidea*		Deep-Sea Anglers
Opisthomi	*Mastacembelidae*	Spiny Eels
Synbranchii	*Synbrachidae*	Swamp Eels
Actinistia		Coelacanths*
Dipneusti		Lungfishes*

the extremes as well as the full spectrum between these extremes, it becomes obvious that generalizations are difficult to put down. And, of course, it is not surprising to find so many different kinds of fishes.

Cartilage Fishes

Unlike the bony fishes, the skeleton of the cartilage fishes, as their name implies, is composed fully of cartilage, i.e., gristle. There are two postulates pertaining to the development of these fishes—one to the effect that the cartilage fishes are more primitive forms from which bony fishes later developed; a more recent and popular view is that the cartilage fishes are degenerate forms. The cartilage fishes (chondrichthyes), although of a relatively limited number of orders and families (Table 1) as compared with the bony fishes, embrace a wide variety of sizes, shapes, and other characteristics. A few of the representative forms are shown in Fig. 1. The whale shark is the largest of

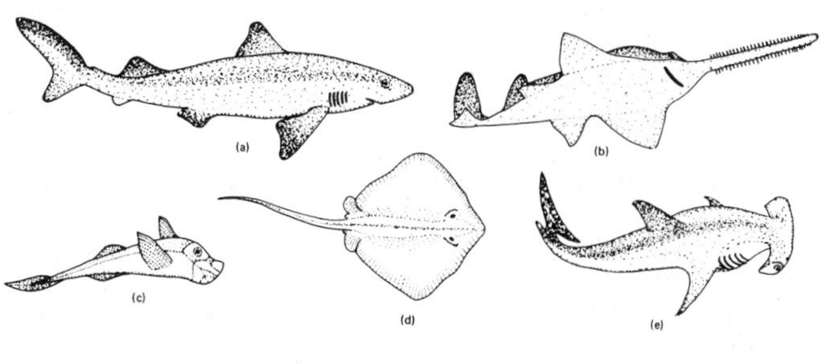

Fig. 1. Representative species of the chondrichthyes (cartilage fishes): (a) Shark; (b) sawfish; (c) ratfish; (d) sting ray; and (e) hammerhead shark. Examples are not to scale.

the chondrichthyes and, with exception of the whale, is the largest living vertebrate animal.

The skeleton of chondrichthyes is divided into two portions: (1) the axial skeleton which is comprised of the vertebral column and the skull; and (2) the appendicular skeleton composed of the fins and the pectoral and pelvic girdles which support the fins. The chondrichthyes are characterized by a spiral fold of mucus membrane within the intestine through which food moves in a corkscrew fashion. This is known as the spiral valve and may have a dozen or more "screw-threads." The valve assists in slowing the passage of food through the digestive system, allowing more time for absorption. These fishes have large livers, a gall bladder, pancreas, and spleen. As with all true fishes, only venous blood is pumped through the heart. Oxygenation occurs in the capillaries of the gills. Respiration is by means of the gills. The major differences between the chondrichthyes and the bony fishes are the cartilage construction of the skeleton and the absence of a number of organs and features, including air bladder, true scales, bones, and lungs.

The Bony Fishes

As will be noted by comparison of Tables 1 and 2, there are many more orders, families, and species of the bony fishes (osteichthyes) than the cartilage fishes. There is an extremely wide range of sizes, shapes, and characteristics and considerably more structural and physiological variation than displayed by the chondrichthyes. Unlike the chondrichthyes, the bony fishes have thousands of freshwater as well as marine or saltwater species. The external features of a representative bony fish are shown in Fig. 2.

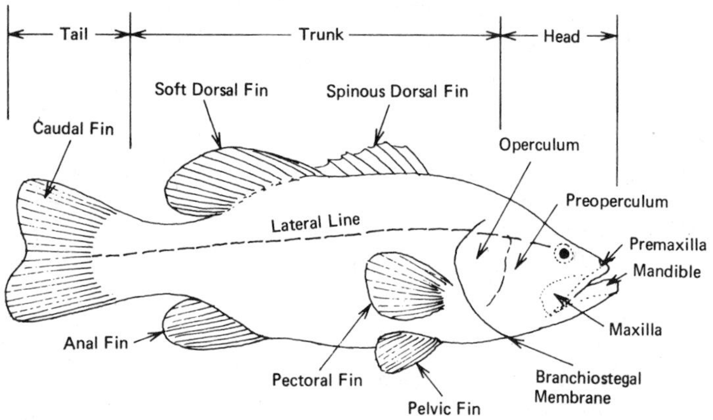

Fig. 2. Principal featurs of a "typical" bony fish. In this case, a largemouth black bass.

Some of the terms used to describe various organs and features of the bony fishes (some terms also applying to the chondrichthyes (include:

(*Adipose Fin*)—A fleshy fin without supporting spines, occurring behind the dorsal fin in some fishes.

(*Air bladder*)—A pouch found in some fishes, derived from the gut and filled with a mixture of gases, chiefly oxygen and nitrogen. It regulates the buoyancy of the body. Also called gas bladder; swimming bladder.

(*Barbel*)—A long, thin, tactile, and projecting "whisker" found extending from the head of some fishes.

(*Branchial Arteries*)—Blood vessels which lead both to and from the gills in fishes. Those which lead into the gills are the afferent branchials; those which lead out are the efferent branchials.

(*Canine Teeth*)—Proportionately large, sharp teeth, often conically shaped.

(*Caudal*)—With reference to the tail. The caudal fin is the tail fin.

(*Caudal Peduncle*)—That portion of the body of a fish located just forward of the tail fin.

(*Dorsal Fin*)—Located on the back of the fish—the main fin on the back.

(*Fin*)—A broad thin appendage, primarily an organ of stabilization, steering, and locomotion. The two prinicpal kinds of fins are the median and the paired. Median fins extend from the body along the median line of the dorsal surface, around the caudal end as the tail fin or tail, and forward as far as the vent. Paired fins include a pectoral pair attached to the pectoral girdle of the skeleton and the pelvic pair attached to the pelvic girdle. All fins are supported by bony (osteichthyes) fin rays or by cartilaginous (chondrichthyes) fin rays.

The median fin is broken up in most fishes to form separate dorsal fins and a ventral anal fin in addition to the caudal fin or tail. A caudal division of the dorsal portion is sometimes developed into the fleshy adipose fin. The tail fin may be heterocercal or homocercal. The former type has the end of the vertebral column extending into or toward its dorsal margin; while the latter is evenly developed above and below the skeletal axis.

Paired fins are variously modified to form sensory lobes or supporting structures.

(*Gill*)—The respiratory organ of aquatic animals. See also **Gill.**

(*Gular Plate*)—Only present in some fishes and located under the throat, a hard plate protecting the under portion of the throat.

(*Lateral Line*)—A longitudinal line, often quite apparent, on each side of the body of a fish. Contains vibration and pressure sensing organs.

(*Lateral Line Organs*)—Sensory organs located on the head and in a line (along the lateral line). They are unlike the sensory organs of terrestrial vertebrates, but are assumed to perceive vibrations of low frequency.

(*Mandible*)—Lower jaw.

(*Maxilla*)—The largest of the bones forming the upper jaw.

(*Milt*)—The sperm of fishes.

(*Operculum*)—Covering over the gills.

(*Oviparous*)—Egg-laying.

(*Ovoviviparous*)—Production and hatching of eggs within the female [and that hatch within the female] with the young being born alive.

(*Parasphenoid*)—A dermal bone supporting the roof of the mouth in fishes.

(*Pectoral Fins*)—A pair of fins which are attached to the shoulder girdle.

(*Pelvic Fins*)—The pair of hind fins. Also called ventral fins.

(*Photophore*)—A light-producing organ.

(*Premaxilla*)—Bones of the upper jaw (front).

(*Preoperculum*)—Anterior cheek bone.

(*Pseudobranchia*)—A small respiratory structure on the inner surface of the operculum in certain fishes; spiracular or vestigal hyoidean gill, serving as a supplementary gill.

(*Roe*)—Fish eggs.

(*Rostrum*)—A snout. The term is applied particularly to the elongated snouts of certain fishes.

(*Vent*)—External opening of the intestine. Anus.

(*Viviparous*)—Young are born alive.

Some Characteristics of the Fishes

Although fishes are found in practically all unpolluted waters of the earth ranging from the warm water of the tropics to the very cold waters of the arctic and antarctic, most species prefer a relatively narrow span of water temperature. Most species are sensitive to temperature changes and usually cannot tolerate rapid fluctuations of over 12 to 15°F (~ 7–8°C).

Water Balance. This is critical to fishes. In the case of salt water fishes, the body juices are less saline than the outside water and the fishes are consequently in the everpresent danger of dehydration. Saltwater fishes must drink copious amounts of water to compensate for that lost through the gills and skin. Some accumulated salt is passed through the digestive tract and excreted. In some fishes, special gill cells aid in returning the water back to the sea. In contrast, the body juices of the freshwater fish are more saline than its water environment. Thus, freshwater fishes are in the everpresent danger of swelling and thus they do not drink. The water coming in through the gills and skin passes through the kidneys and is carried away as waste. Freshwater fishes generally produce copious quantities of urine.

Locomotion. A fish swims by pushing water aside. This is done by creating a wiggling motion through the use of the head, tail, and to some degree, other fins, depending upon the species of fish. While difficult to detect, the movements, forward or backward, are essentially snakelike. The various disk-shaped fishes and triggerfishes make important use of their pectoral fins in achieving motions. In most other fishes, the fins serve various steering and guidance functions. Stabilizing actions are the principal functions of the dorsal and pectoral fins, the pectoral in particular being used for balancing and turning. Pelvic fins usually serve as stabilizers. The tail fin usually performs a number of functions, including propulsion, steering, and stabilizing. In fast-swimming, streamlined fishes, the dorsal and anal fins may fold back, essentially retracting into grooves, thus becoming flush with the body. However, for every generalization of the type just described, there are exceptions. Skates and rays, for example, obtain most of their locomotion by way of much enlarged side fins making them appear as though flying through the water.

The shapes of representative chondrichthyes were given in Fig. 1. Representative shapes of osteichthyes are given in Fig. 3.

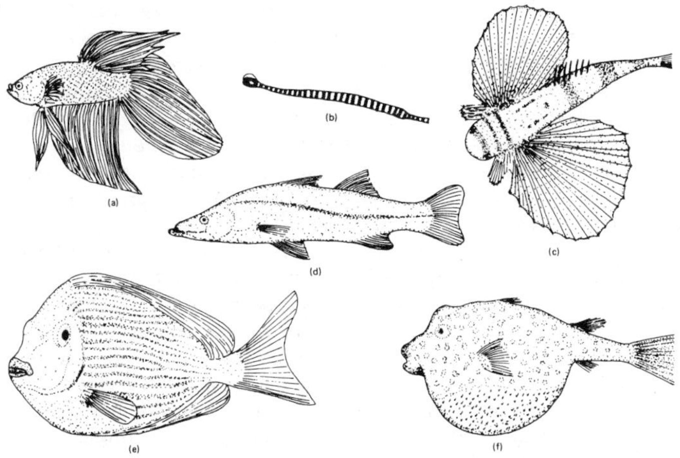

Fig. 3. Representative species of the osteichthyes (bony fishes); (a) Siamese fighting fish (fresh water); (b) banded pipefish (marine); (c) flying gurnard (marine); (d) snook (marine, occasionally fresh water); (e) striped surgeonfish (marine); and (f) southern puffer (marine) shown after self-inflation with air. Examples are not to scale.

Buoyancy. This is accomplished in most fishes by means of a gas bladder. This is a gland-lined airtight sac in the gut of the fish. The glands remove gases from the bloodstream of the fish and direct the gases to the bladder. By regulating the amount of gas in the bladder, the fish may select whatever depth may be desired. Regulation usually is quite rapid and operable from the surface to a depth of about 1,200 feet (360 meters). If a hooked fish is brought to the surface at too rapid a rate, the balloon may burst, causing the stomach to be forced out of the mouth. The gas or air bladder as it is commonly called varies in size in proportion to the fish. Where the bladder is small in comparison with the size of the fish, vertical adjustments to pressure usually are accomplished more readily. Some fishes do not have air bladders (for example, mackerels) and, consequently, to maintain a given level in the water, these fishes must constantly move, lest they sink.

Protection. This is achieved by a flexible armor in the form of scales. These vary in shape and construction. As do the fingernails in humans grow as the person grows in size, so do the scales of a fish. Although the chondrichthyes are not formally recognized as having skin with scales, they do have primitive placoid scales, which may be described as toothlike, but of very small dimensions so that the result is that a shark, for example, has an outer surface something like sandpaper. Primitive fishes, such as the garpike, have ganoid scales. Each scale is shaped like a diamond and is attached to adjacent scales by joints. Ganoin, a material which imparts a glossy ivory-like appearance to the fish, is the outside coating for these scales. Thus the scales are well held together on the under and outer surfaces.

Ctenoid and cycloid scales are the most commonly found types among the bony fishes. Both scales are roughly circular in shape with a fine design appearing something like the annual-ring patterns in the cross section of a tree trunk. However, one side of the "circle" is more like a bump. In the ctenoid scales, this bump has a comblike edge; in the cycloid scales, the bump is smooth. Thin, lightweight, and flexible, these scales are arranged in neat, overlapping rows.

Vision. The seeing apparatus of most fishes parallels human vision in many respects. It is interesting to note, however, that surface-feeding fishes must allow for refraction of light as it bends at the air-water interface. An important difference in the structure of eyes in most fishes is the result of very limited light under water and the absence of rapidly changing amounts of light. Hence, fishes require no eyelids and they do not require a sophisticated iris to regulate the light passing into the eye. Fish eyes are spherical and essentially rigid. The lens is filled with water. Consequently light passing from outside is not refracted at the lens interface. With this spherical structure, theoretically a fish should be able to focus well, but this is debatable. Fish have monocular vision, that is, objects on the right are sensed by the opposite side of the brain and conversely with objects at the left. With eyes on each side of the head, more than one direction can be noted at any one time. See also **Four-Eyed Fishes.** Studies indicate that most fishes have rudimentary color conception, but it is not believed that, in the fish's world of largely subdued greenish-blue tones, color plays an important role, but this requires further study. Sportsmen certainly believe that color is important in the hooking of various fishes, such as trout. It is known, of course, that fishes do recognize patterning, if not color. The example is often given of the vertically-striped pilot fish, less than 12 inches (0.3 meter) in length, that can safely accompany a shark. It is generally assumed that the shark may recognize the pilot fish by its stripes. Color per chance may play a role in warning fishes of certain highly-colored poisonous fishes. Coloration certainly is used to advantage by many fishes in their ability to blend in with their backgrounds, changing color as well as patterning.

The vision process of fishes is quite complex and for greater detail, the reader is referred to the Davison (1969); Endler (1980); Levine (1979, 1980); McFarland (1975); and Munz (1977) references listed at the end of this entry.

Sound. It is only within the past decade or two that concerted efforts have yielded scientific information on how fishes make sounds and how they detect sounds. Certain fundamentals have been known for a long time, but the intricate mechanisms involved have been poorly understood. At one time, authorities disagreed on the importance and indeed the existence of acoustic systems in many species of fishes, but it has been established that all or most fishes have a hearing sense, far more refined in some species than in others.

It is well understood that considerable background noise (the noise created by swimming and, in particular, by thousands of companions in schooling fishes; as well as the noise from wave action near the surface) must play a major role in the hearing process, in particular how this background noise is sorted out from the noise of predators. It has been shown that many species cannot respond to sounds above about a few hundred cycles per second (Hz). However, clupeoid and ostariophysine fishes can respond to higher frequencies.

Sound can be presented to fishes in two fundamental ways: (1) a sinusoidal change in pressure, as that which can be measured by a hydrophone; and (2) a back-and-forth motion of the water, expressed as particle displacement or particle velocity, which is more difficult to detect, requiring sensors in three planes. Pressure and particle motion are related to each other and to the distance of the sound source. Particle motion can give information about the direction of the sound source; pressure cannot.

As pointed out by Blaxter (1980), particle displacement is sensed in some species by mechanoreceptors, located in the acousticolateralis system which comprises the inner ear and the lateral line. In both the ear (labyrinth) and the lateral line, the basic receptor is the neuromast organ, consisting of a group of sensory cells, each with a bundle of ciliary hairs embedded in a gelatinous cupula. The neuromasts have directional qualities. Many species of teleost have a swimbladder that responds to sound pressure by pulsating in sympathy with the passing compressions and rarefactions. The sound pressure will then be reradiated as displacements, thus creating a secondary near-field within the fish that can stimulate the displacement receptors.

For a more detailed review of this topic as of the early-1980s, the paper by Blaxter (1980) is suggested. Further interesting papers include Hawkins (1973); Hoar and Randall (1971); Schuijf and Hawkings (1976); and Travolga et al. (1981), listed at the end of this entry.

Nervous System. Most fishes have large numbers of nerve organs in their skin, as well as barbels which can be used as feelers. Hence, they are equipped with a sense of touch. The fish nervous system consists of brain, spinal cord and nerves. The elongated brain has five segments: Forebrain, diencephalon, midbrain, cerebellum, and brain stem. The spinal cord proceeds from the brain stem down to the tail. It is a round cord which is enclosed by the vertebral column. Twelve nerve pairs (the "cranial nerves") are sent from the brain to the sensory organs of the head and to the head musculature. The spinal cord transmits nerve impulses from the brain and, on the basis of the number of vertebrae, sends signals to an equal number of spinal nerves with a ventral motor root and a dorsal sensory root. The latter are associated with the sympathetic nervous system. This nerve system runs in the form of two fine nerve strands underneath and along the vertebral column. In cartilaginous fishes, the forebrain is highly developed and the halves are not completely divided. All other fishes have clearly divided brain halves, which in lungfishes are large, and in teleost fishes are small and supplied with olfactory lobes. The brain weighs about 1:1300 of the body weight (as in a gar, for example). The brain cortex is, in contrast to higher vertebrates, convoluted to just a small degree.

One of the best developed senses in the fish is that of smell. A shark's ability to detect blood from long distances is well established. With the exception of the lungfishes, fishes do not have taste buds in their mouths. However, such organs could be located elsewhere. From the gulping eating habits of most fishes, however, it would appear that taste is not a consideration.

The pain threshold is very high in fishes, with no physiological component—as presently considered by most investigators. Fish have no equivalent to the human cortex where nerve impulses are integrated. The performance of injured fishes bear out these observations.

Musculature. The muscle comprises a large part of the fish body. Its arrangement on the skeleton permits the very flexible fish movements. On the head is the facial musculature of the jaw and gill arches. The chewing muscles, the muscle groups of the gill cover, and the mandibular and gill arch muscles together form an independent muscle group proceeding from the sides. The strong lateral trunk muscles for propulsion of the fish are developed from the vertebrae and extend from the back of the head to the root of the caudal fin and are symmetrical on both sides of the vertebral column. They consist of numerous muscle segments, one behind the other, which are separated by fine connective tissue partitions; they also consist of a dorsal part and a ventral portion, between which lies a wall of connective tissue. Corresponding to the arrangement of the axial skeleton, each part falls into small muscle packets (*myomeres*). The musculature is held to the vertebral column by connective tissue partitions (*myocommata*). The fin musculature develops from split off trunk musculature. Here the depressors of the fin rays, which fold them together, are differentiated from the lifters which spread the fins out.

Respiration. Breathing occurs by internal gills which lie between the pharynx and body wall in pockets and take up oxygen dissolved in the water. The gills are thin layers of skin with blood vessels, which are free in sharks and protected by a gill cover in teleost fishes. Water enters the mouth and is expelled through the gills. It thereby runs over or through the gills, the surface of which is permeated by blood vessels. The blood takes up oxygen and releases carbon dioxide. The surface area of the gills of a carp measures about 0.5 square meter (775 square inches). The embryos of sharks and lungfishes have external gills like those of amphibians. When the oxygen supply is low, fishes snap at the surface for air to enrich their supply. Very sensitive trout die when the oxygen content declines to 1.5 cubic centimeters/liter; and the less sensitive carp do not perish until oxygen has dropped to 0.5 cubic centimeter/liter.

Blood Circulation. The heart in fishes lies behind the gill arches and in front of the pectoral girdle in a pericardium with a partition between it and the visceral cavity. The heart has one atrium and one ventricle separated by valves. See Fig. 4. In sharks, rays, and sturgeons, a muscular element (the *conus ateriosus*) lies between the chambers and the aorta, the inside of which has from two to eight halfmoon-shaped valves. Only a remnant of this structure is present in teleost fishes. In its place, there is an enlargement of the aortic branch with a *bulbus ateriosus* which prevents a backflow of blood pumped through the heart. Lungfishes have an atrium which is divided

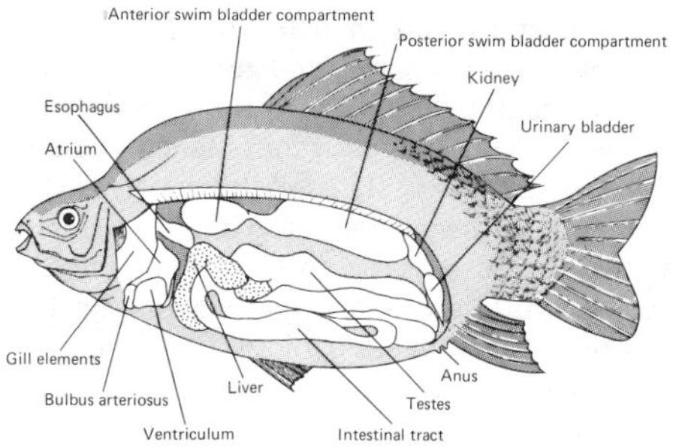

Fig. 4. Internal organs of a representative fish. Carp (*Cypriniformes*).

into two parts by a partition, thus separating oxygen-poor (*venous*) blood (from the liver) from oxygen-rich (*arterial*) blood out of the lung. The heart pumps the blood through the *aorta ascendens* to the gills, and blood replenished with oxygen passes from the gills to the *aorta descendens*, which supplies all the organs with blood. Venous blood circulates back to the heart from the *sinus venosus*.

Fishes also have a lymphatic system which empties into the veins. A few glands which secrete internally empty their contents into the blood circulating through them. Among these, the thyroid is responsible for metabolism; the thymus for growth; the pancreas for producing digestive juices and control of blood sugar levels; the pituitary gland for fat and carbohydrate metabolism, propagation of coloration material, blood pressure, stimulation of the smooth muscles, and other functions. Other significant organs include the gonads, which influence sexual activity; the adrenals, which influence breathing rate as well as heartbeat; the sympathetic nervous system and vascular muscles.

Digestive Tract. In contrast to the lampreys with their round mouths and rasping tongues, fishes have a jawed mouth adapted for swallowing food. It leads to the pharynx and from there into the muscular esophagus. The esophagus leads into the stomach without any striking transition, and from there the small intestine extends all the way to the anus. A short large intestine completes the system and has an external opening. The teeth are highly variable in shape and number; they correspond to the placoid sacles of cartilaginous fishes. They lie on the jaw edges, on the palate and pharynx bones, and sometimes on the hyoid bone and gill arches. In predatory fishes, they are pointed or conical, in other fishes wide and smooth or chisel-shaped. Species which feed on minute crustaceans (for example pipefishes) are toothless. Plankton feeders (e.g., salmon) have elongated thorny processes on the gill arches, which serve as sieves and retain the plankton. Carp have teeth only on the lower pharyngeal bone.

In herring and perch, the esophagus leads to an enlarged muscular stomach with a blind end which acts as a gizzard. The gut is winding and has a mucous membrane inside arranged longitudinally. In sharks, sturgeons, and lungfishes, the gut convolution system merges into a single screw-shaped spiral fold which increases the surface. There are no salivary glands. The liver is large and rich in fat. Sturgeons and many teleost fishes have several hundred short glandular blind tubes in the beginning of the mid-intestine. It is assumed that they increase the absorptive area. The anus is always located ventrally, but in genera *Gymnotus* and *Fierasfer* in the throat region. Sharks and lungfishes have cloacae; thus the intestine culminates with the urinary and genital tracts in a hollow cavity which leads to the outside through the anus.

Genital Organs. With the exception of a few hermaphroditic forms, most species are bisexual and have paired genital organs which lie on both sides of the vertebral column in the visceral cavity. In a few cases, they are united to form a single sack. In sharks, lungfishes, and ganoid fishes, the Müllerian duct, a part of the original ureter, takes up eggs released from the ovary and passes them to the outside. A part of the mesonephron serves as the spermatic duct. In salmon and eels, the eggs pass into the visceral cavity through a break in the ovarian wall and are passed out through paired genital pores behind the anus. In most teleost fishes, the testes and ovaries connect immediately with the spermatic ducts and oviducts, respectively, which together with the urinal tract terminate at the genital papilla behind the anus.

Male sharks and rays as well as a few teleost fishes have copulatory organs in the form of long cartilaginous appendages on the ventral fins. Other fishes giving birth to live young (e.g., toothed carps) have copulatory organs formed from the anal fin. In the European bitterling female, the genital pore is extended to form a 2-inch (5-centimeter) long egg-laying tube during spawning.

Fish hobbyists do not generally use the sex organs—or primarily sexual characteristics—to recognize and differentiate the sexes, but instead rely on other derived differences between males and females, the secondary sexual characteristics. These include color differences or spawning colors of males, their spawning eruptions, and the increased size of particular fins during spawning. Occasionally, the sexes are distinctively different; for example, in the deepsea fish genus *Ceratias*, the dwarf-sized males hang firmly to the females with their mouths, parasitize the females for nutrition and, in essence, are only sperm-creating creatures for the females. In black bass (*Centropristis*), a sexual transformation exists; females which have already undergone a breeding period become transformed to males. Parthenogenesis (development of young without fertilization) has been confirmed in the carp *Carassius auratus gibelio*.

Spawning. See separate entry in this volume on **Spawning.**

Feeding Characteristics. There is much evidence among the fishes of adaptation to means for gaining food and for handling it, once contacted. Many tools of attack have been developed. The sawfish has an extremely elongated snout which is equipped with from 15 to 30 pairs of teeth to form a crude, but effective saw which the fish flails through the water, randomly maiming prospective morsels, returning to eat the pieces at will. A paddlefish has an enormous mouth, swimming with it wide open. The greatly elongated paddle attached to the upper portion of the head is not generally accredited with assistance in food gathering, but when finally investigated it may well prove that the paddle has this major function. Some mormyrids have lower jaws shaped like trowels which they use for "shoveling" when scrounging muddy bottoms for insects and worms. Some fishes, such as the suckermouth catfish, are equipped with powerful sucking mechanisms for obtaining food. The electrogenic fishes, such as the electric catfish or electric eel, use their voltage-generating batteries both for detecting the presence of food and for defense. The interesting angler fish dangles a lure hanging from a "pole" in front of its mouth to attract small fishes which can quickly be gulped or sucked into the mouth. The archerfish propels pellets of water to hit likely prospective food items. Experiments have shown that the fish can hit a target with reasonable accuracy up to a distance of three or more feet (0.9+ meter).

Lateral Line. Known as the *linea lateralis*, the lateral line detects water pressure which acts on the swimming or motionless fish. This extends from the head to the tail on both sides of the body and maintains contact with the water via its numerous pores. If the fish approaches an obstacle or a larger fish, the lateral line senses the resulting pressure change, and furthermore determines the direction and strength of currents so that those fishes which when spawning must swim upcurrent can find even the smallest streams. At night, or in muddy water, the lateral line protects the fish from bumping into objects, since these objects reflect the waves generated by the fish. Involuntary reflex equilibrium movements stimulated by the inner ear and lateral line are carried out by the paired fins and lateral musculature of the trunk and enable the fish to maintain its normal position in the water.

Velocity. The velocity of fishes is highly variable. Some carp can swin 7.5 miles (12 kilometers) per hour; barbels reach 11.2 miles (18 kilometers) per hour; while pike attain 15.5 miles (25 kilometers), and trout 21.7 miles (35 kilometers) per hour. Among marine fishes, tuna can swim about 13.7 miles (22 kilometers) per hour; sharks about 22.4 miles (36 kilometers) per hour; and swordfish up to 60 miles (90 kilometers) per hour.

Coloration. Fish coloration stems from chromatophores located along the border of the epidermis and the subcutis, and also from fatty dyes (red, orange, or yellow lipophores, which are carotenoids dissolved in fat). Color cells are under central nervous system control and when stimulated either cause color to appear or eliminate coloration already present. Changing coloration enables fishes to adapt quickly to their surroundings. Skates and flatfishes not only assume the color of the floor, but also the sandy or gravel texture by forming imitative spots on the upper surface of their bodies. See also **Flatfishes.** Other environmental factors, such as higher water temperature, oxygen deficiency, or behavioral factors like fear, aggression, and similar stimuli can cause color changes. An aggressive male stickleback can change its appearance at short intervals. A color change appears in many fishes (such as salmon) when they are ready for spawning. The male develops a very colorful display pattern, and the epidermis develops wart-like spawning eruptions—as in salmon, chondrostomes (*Chondrostoma*), and European roaches.

The epidermis gives off a silvery mass consisting of fragments of decomposing material of the protein guanine; this forms an effective protective coloration because a predator fish on the bottom sees a prey fish swimming above only as part of the glittering water surface. Bottom dwellers and deepsea fishes generally are dark-colored. Coloration is lacking entirely in blind cavefishes, and the most colorful species are those of tropical inland waters and lakes.

Luminosity. Some sharks, e.g., the spiny dogfishes (*Squalidae*), emit a greenish phosphorescent light created by small light organs which lie throughout the skin. The deepsea scaly dragonfishes (*Stomiatoidei*) have light organs consisting of a convex or concave lens which rests in one of the dermal vesicles of the epidermis. The walls of this vesicle are formed by glandular cells which form the luminous material. If the walls contain black pigment, they act like the concave reflector of a dark lantern; in other cases, the outer skin layer lies over the upper surface of the lens and has about the same effect as the shutter of a camera. The luminous organs are arranged in two rows on both sides of the body. In addition, there are luminous organs on the head and jaws, and beneath or just behind the eyes.

Fishes of genus *Anomalops* from the Indian Ocean have a large luminous organ beneath the eye which sits on a movable flap which can be turned inward and fits in a cavity under the eyes. Lanternfishes (*Myctophidae*) have luminous organs on their stomachs and fins which have a jewel-like glitter. The *Galantheathauma*, caught in 1962 by the *Galathea* Expedition, has a light organ inside its mouth on the foregum; any prey attracted by the light has merely to swim inside the mouth, which then closes. The toadfish *Porichthys* from California coastal waters has about 340 light organs on head and body. These bright white spots consist of lens, gland, reflector and dye layer and they are so bright that a newspaper can be read 10 inches (25 centimeters) from them with no other light source present. The light comes from luminous bacteria or from the chemical reaction in mucus secreted by the gland cells. See also **Anglerfishes.** Lighting serves many functions to fishes so equipped, including species recognition.

Schooling. A school is a group in which all individuals are headed in the same direction, are uniformly spaced in close rank, and are swimming at the same speed. This is a general, but not exclusive definition. A logical analysis of schooling behavior begins with the recognition that regimented behavior of individuals is essential to the unity of the school. The school remains a unit when the individuals are coordinated with one another, either collectively or unilaterally. Schooling behavior is not defined by the spatial orientation of individuals, but rather, the spatial orientation is defined by the occupation of the school; that is, whether it is stationary, feeding, or swimming off to some other locality. The independent variable is the intact unit and the variables of spacing, direction, and speed of individuals are dependent upon the occupation of the school.

Schooling behavior changes with age in some species. The young of the black bullhead (*Cottus gobio*), for example, form dense stationary schools with disoriented individuals in contact with one another, or moving schools of regularly spaced individuals. The intensity of schooling declines in older fish. In most species, the schooling response persists once it has developed among the juveniles, but juveniles rarely coexist in schools of larger fish because they occupy exclusive habitats as, for example, the occupancy of shallow marginal waters by juvenile minnows or suckers, or the surface as opposed to depths by juvenile North Sea herring. In a dynamic sense, juveniles may be segregated because they cannot swim fast enough to stay with the larger fish, or because the fish establish dominance hierarchies and drive away others of different sizes. The segregation of schools according to size, and thus of age, of the fish continues throughout the life span of the species and the age-exclusiveness of schools is most clearly defined in the first few years, when the age–length correlations are most distinct. This does not mean that the individuals within a school are of uniform length—they may vary by a factor of 2 or 3, depending on the mean. However, the mean length between schools differs significantly.

Some changes in schooling behavior are related with either short-term or long-term environmental variations, which are generally interpreted as sources of physiological stress. In the laboratory, increased stress, as caused by low temperatures, elevated carbon dioxide or chlorine content, or stimulants, such as strychnine or caffeine, may intensify the schooling response, as do noises, sudden motion of objects, a strange object in the aquarium, or novel surroundings, which might be created merely by transferring the fish to a different aquarium. The Schreckstoff (fright substance) released by injury to the skin in certain fish, notably the freshwater ostariophysids, initiates a flight reaction that is rapidly transmitted as a visual sign and causes fish to form tight schools or to seek cover, depending on species. In the absence of Schreckstoff, a flight reaction is transmitted by visual means. A reduction in stress as might be represented by the monotony of conditions in the labratory aquarium induces a relaxation in the regimentation of schools, to different degrees in different species, and the response may vary with the size of the aquarium.

In nature, schools are generally spindle-shaped, and if only two or three fish are involved, they usually resort to following one another. Compact pods or balls are frequently seen, and a common formation is the mill, in which the school rotates in one spot like a large wheel on a fixed axis. The significance of such formations is generally unknown, although some are thought to represent reproductive or protective behavior and others may represent a response to physical conditions. Many species become quiescent during the cold seasons, and some, such as the bullheads, are known to form dense congregations in specific areas, suggestive of denning among snakes. The overall form of a school may depend upon the number of individuals comprising it. Changes in the structure of a large school of black mullet (*Mugiloidei*) were correlated, along the axis of progression, with metabolic reduction of dissolved oxygen. This is an interesting example of an aquatic group establishing its own limiting conditions.

There is no dichotomy between schooling and nonschooling fishes. The two categories lie at opposite ends of a continuous scale expressing time devoted to schooling behavior. Occupying the top of the scale would be species of whitefish, herring, sardine, mackerel, and tuna, and, at the bottom, solitary predators, such as trout, pike, and barracuda, or demersal or seclusive species, such as the sculpins and flatfishes.

A more detailed analysis of schooling motivation and behavior is beyond the scope of this volume. Much research has been done; much remains. Further reference to the John (1969) and the Burgess-Shaw (1979) sources listed at the end of this entry is suggested. In addition to knowledge of schooling for strictly scientific purposes, this information also is very useful in improving the catches of various fisheries throughout the world.

Depth Preferences of Various Species. There is a considerable bank of scientific knowledge as regards the depth preferences of various orders and suborders of the fishes and, in fact, broad classifications have been made. There is less knowledge as regards those factors which bring about a differentiation of habitat in terms of depth; on how some species can vertically migrate considerable distances, while others remain constantly in certain depth zones. Some research into this topic has been conducted by the Scripps Institution of Oceanography (La Jolla, California) as well as other similar institutions. Some interesting findings are reported in the Siebenaller-Somero (1978) reference listed at end of this entry.

As pointed out by Bruchhausen et al. (1979), until quite recently there was much speculation as regards the existence of life in the relatively deep culdesacs beneath the large Antarctic ice shelves. Earlier evidence was confined to various specimens collected through natural cracks in the shelf ice; or fishes collected near the leading edge of the shelf. In 1978, a research group lowered baited traps and a camera through the Ross Ice Shelf, Antarctica, to a depth of some 1960 feet (597 meters) below sea level. This experiment revealed the presence of fish, many amphipods, and at least one isopod. Details of this experiment are reported in *Science*, **203**, 449–450 (1979).

Other Entries on Fishes in this Book. For convenience, the various fishes and related topics described in separate alphabetical entries in this encyclopedia are listed below:

Alewife
Anchovy and Anchoveta
Angelfishes
Angel Shark
Anglerfishes
Aquaculture
Barracuda
Bass
Bichirs
Billfishes
Bitterling
Blackfish
Blackhorse
Bleak
Blennies
Blind-Fish
Bluegill
Bony-Tail
Bony Tongues
Bowfin
Bristlemouths
Butterfishes
Capelin
Carangids
Carp
Catfishes
Characids
Chimaeroids
Cichlids
Cobia
Codfishes
Coelacanths
Croakers
Cutlassfishes
Cyclostomata
Dolphins
Dorab
Dragonets
Eels
Featherbacks
Fisheries
Flatfishes
Flying Fishes
Fossil Fishes
Four-Eyed Fishes
Galaxiids
Gars
Gill
Gizzard Shad
Gobies
Grayling
Grunts
Gurnards
Gymnotid Eels
Hagfishes (*Agnatha*)
Hake
Hatchet Fishes
Herring
Ice Fishes
Ichthyology
Iniomous Fishes
John Dories
Labyrinth Fishes
Lampreys (*Agnatha*)
Loaches
Lungfishes
Mackerels
Menhaden
Milkfish
Mormyrids
Mullets
Opah
Paddlefishes
Parrotfishes
Perches and Darters
Pike
Pipefishes
Plecostomus
Porcupine Fishes
Porgies
Puffers
Rabbitfishes
Redfish
Remoras
Salmon
Sardines
Scad
Sculpins
Seahorses
Sea Moths
Sea Robins
Sharks
Silversides
Skates and Rays
Slickhead Fishes
Smelts
Snappers
Spawning
Squirrelfishes
Stone Roller
Sturgeons
Suckers
Sunfishes
Surgeonfishes
Tarpon
Teleostica
Trout
Tunas
Viper Fishes
Vivaparous Topminnows
Weeverfishes
Whitefishes
Wrasses

References

Atema, J.: "Smelling and Tasting Underwater," *Oceanus*, **23**, 3, 4–18 (1980).
Blaxter, J. H. S.: "Fish Hearing," *Oceanus*, **23**, 3, 27–33 (1980).
Bond, C. E.: "Biology of Fishes," Saunders, Philadelphia, 1979.
Budelmann, Bernd-Ulrich: "Equilibrium and Orientation in Cephalopods," *Oceanus*, **23**, 3, 34–43 (1980).
Burgess, J. W., and E. Shaw: "Development and Ecology of Fish Schooling," *Oceanus*, **22**, 2, 11–17 (1979).
Davis, H. S.: "Culture and Diseases of Game Fishes," University of California Press, Berkeley, California, 1977.
Davison, H.: "The Eye," Vol. 1, Academic, New York, 1969.
Endler, J. A.: "Natural Selection on Color Patterns in Poecilia reticulata," *Evolution*, **34**, 1, 76–91 (1980).
Grobecker, D. B., and T. W. Pietsch: "High-Speed Cinematographic Evidence for Ultrafast Feeding in Antennariid Anglerfishes," *Science*, **205**, 1161–1162 (1979).
Halver, J. E.: "Fish Nutrition," Academic, New York, 1973.
Hartline, B. K.: "Coastal Upwelling: Physical Factors Feed Fish," *Science*, **208**, 38–40 (1980).
Hawkins, A. D.: "The Sensitivity of Fish to Sounds," *Oceanogr. Mar. Biol. Ann. Rev.*, **11**, 291–340 (1973).
Hoar, W. S., and D. J. Randall (editors): "Fish Physiology," Vol. 5, Academic, New York, 1971.
John, K. R.: "Schooling Behavior," in "The Encyclopedia of Marine Resources," (F. E. Firth, editor), Van Nostrand Reinhold, New York, 1969.
Levine, J. S., and E. F. MacNichol, Jr.: "Visual Pigments in Teleost Fishes," *Sensory Processes*, **3**, 95–131 (1979).
Levine, J. S.: "Vision Underwater," *Oceanus*, **23**, 3, 19026 (1980).
McFarland, W. N., and F. W. Munz: "The Evolution of Photopic Vision Systems in Fishes," Part 3, *Visual Research*, **15**, 1071–1080 (1975).

Moller, P.: "Electroperception (Fishes)," *Oceanus*, **23**, 3, 44–54 (1980).
Munz, F. W., and W. N. McFarland: "Evolutionary Adaptions of Fishes to the Photic Environment," in "Handbook of Sensory Physiology," (F. Crescitelli, editor), Vol. 8, No. 5, Springer-Verlag, New York, 1977.
Nelson, J. S.: "Fishes of the World," Wiley, New York, 1976.
Nikolsky, G. V.: "Ecology of Fishes," Academic, New York, 1963.
Nixon, M., and J. B. Messenger (editors): "The Biology of Cephalopods," Academic, New York, 1977.
Ryan, P. R.: "Geomagnetic Guidance in Bacteria, and Sharks, Skates, and Rays," *Oceanus*, **23**, 3, 55–60 (1980).
Schuijf, A., and A. D. Hawkins (editors): "Sound Reception in Fish," Elsevier, Amsterdam, 1976.
Siebenaller, J., and G. N. Somero: "Pressure-Adaptive Differences in Lactate Dehydrogenases of Congeneric Fishes Living at Different Depths," *Science*, **201**, 255–257 (1978).
Sinderman, C. J.: "The Principal Diseases of Marine Fish and Shellfish," Academic, New York, 1970.
Tavolga, W. N., Popper, A. N., and R. R. Fay (editors): "Hearing and Sound Communication in Fishes," Springer-Verlag, New York, 1981.
Wilcox, R. S.: "Ripple Communication (Fishes)," *Oceanus*, **23**, 3, 61–68 (1980).
Zahl, P.: "The Four-Eyed Fish Sees All," *National Geographic*, **153**, 3, 390–394 (1978).
Zahl, P.: "Dragons of the Deep," *National Geographic*, **153**, 6, 838–845 (1978).

FISHES (Fossil). Fossil Fish.

FISHES (Respiratory Organs). Gill.

FISHES (Skeletal System). Skeletal System.

FISHES (Sound Production). Sound Production (Physiology).

FISHES (Water Metabolism). Water.

FISH FARMING. Aquaculture.

FISH FLY (*Insecta, Neuroptera*). Species related to the corydalus. The larvae are aquatic and the adults are found near water. The insect is related to the alder fly. The larvae are sometimes used as fish bait.

FISHING (Ocean). Ocean.

FISH LOUSE (*Crustacea, Copepoda*). Minute marine and fresh-water animals parasitic on fishes.

FISH MEALS, OILS, AND PROTEIN CONCENTRATES. For many years, various species of fish, such as menhaden, tuna, groundfish, herring, sardine (at one time prior to recent periods of scarcity), and miscellaneous so-called industrial fish (monkfish, sculpin, sea robin, squirrel hake, shark, and ray) have been used as a source of nutritional meals and oils for use in feeding livestock, including poultry. When the concept of high-energy diets for livestock became popular in North America in the early 1950s, the feed industry became increasingly aware of various alternative sources in the formulation of balanced feeds. During interim years, the consumption of fish source materials has been essentially one of economics, balancing the costs of fish meals against those of other sources. The base cost of fish has risen largely because of increased labor and overhead costs and thus the use of fish sources tends to cycle with the economy.

Menhaden is the most important species for fish meal. Although tuna are not caught for the primary purpose of reduction, enormous numbers of them are canned and considerable wastes result from processing. The material thus becomes a raw material for reduction. Similarly, the wastes from various groundfish (alewife, salmon, haddock, ocean perch, whiting, cod, pollock) are used. The herring caught in New England waters is mainly used for canning, whereas the Alaska herring is reduced to meal. Industrial fish, previously mentioned, are mainly used for canned or otherwise preserved pet foods, but some tonnage is used for reduction. Sardines, no longer in generous supply, are principally canned, but wastes are sold for reduction.

In the United States, two principal methods of reduction are used: (1) Wet rendering, in which the oil is removed before the fish material is dried; and (2) dry rendering in which the oil is removed after drying. Wet rendering is most commonly used and is particularly well adapted to the rapid production of meal and oil from oily fish. In addition to the meal and oil produced, condensed solubles may also result from this method of processing. Dry rendering is well adapted to production on a small scale from fishery materials of low oil content, such as fillet waste from haddock and cod. Continuous dry reduction is also used with shrimp and crab scrap.

Wet Rendering. The principal steps of the wet rendering process are: (1) cooking, wherein the oil and water in the fish can be separated from the solid protein easily and economically in subsequent pressing operations. Overcooking and undercooking the fish results in an unsatisfactory product from the pressing operation, and thus cooking must be carefully temperature and pressure controlled. (2) Pressing (frequent screw-type press) squeezes both oil and water from the fish so that the resulting materials have a low oil content and are thus economical to dry. The product is called presscake. (3) Centrifuging is used instead of the formerly used settling tanks for recovery of the oil from the liquid portion. Two centrifuges are usually used—first a sludger, which handles liquor containing oil, water, and some suspended solids; then fresh hot water is added to the emulsion for processing in a second centrifuge, known as the oil purifier, where the last traces of solids and water are removed. The solubles are stored and sold separately or a portion of them may be added back to the presscake, the resultant product, after drying, being called full meal in contrast to regular meal, which has no added solubles. (4) Drying in direct-heat driers, steamtube driers, or air-lift driers is required to reduce the moisture content of meal down to about 9% to prevent spoilage and to make the product easier to handle and more economical to ship. (5) Deodorizing of the mositure-laden gases resulting from the removal of the fine meal particles is required because of objectionable odors present. Generally, the odor components of the gases must be removed before they may be discharged into the atmosphere. (6) The dried product, called scrap or unground fish meal must be cured. In general, fish oils are highly reactive, being characterized by a high degree of unsaturation, permitting easy combination of the oil with oxygen of the air, releasing considerable heat. This can result in charring and fire. To prevent this, some operators add antioxidants, usually to the dried scrap, but occasionally to the presscake before it is dried.

Dry Rendering. Although many variations in this method exist, the fishery material usually is loaded into a large, steam-jacketed, cylindrical drier. Inside the drier is a rotating scraper, which brings all material into quick contact with the hot inside wall, yet prevents the material from sticking. The drying is done either under vacuum or at atmospheric pressure. The oil is separated from the dried scrap by batch pressing in hydraulic presses. No product other than oil is recovered from this operation. After the oil has been expelled, the remaining solid material is ground into meal, called whole meal, or left unground as cake.

Fish Protein Concentrate. Sometimes referred to as FPC, this material has been defined as a stable product suitable for human consumption, prepared from whole fish or other aquatic animals or parts thereof. Protein concentration is increased by the removal of water and, in certain cases, of oil, bones, and other materials. The traditionally dried or otherwise processed fish meals do not fall within this definition. Many millions of dollars have been invested in research and pilot plants directed toward the production of fish protein concentrates during the past few decades. The objective has been the development of an effective protein source, particularly for the less affluent countries.

The use of minced fish as a basis for food products, such as *kamoboko*, a traditional fish paste, is a practice of long standing in Japan. Modern methods have been applied to this ancient art, and a major industry has arisen in Japan during the past few decades, producing a varied array of food products from minced fish.

Fish Protein Derivatives. Basically, these derivatives are prepared by reacting the myofibrillar protein with the acid anhydride under slightly alkaline conditions. The reacted proteins are then precipitated from solution with hydrochloric acid and extracted with hot azeotropic isopropanol to remove residual lipids. The acetylated proteins are then neutralized with sodium hydroxide to solubilize the derivatives, after which they are dried.

References

Martin, R. E. (editor): "Second Technical Seminar on Mechanical Recovery and Utilization of Fish Flesh," National Fisheries Institution, Washington, D.C., 1974.

Spinelli, J., et al.: "Expanded Uses for Fish Protein from Underutilized Species," *Food Technology*, **31**, 5, 184–187 (1977).

Stone, H.: "Novel Proteins—An Alternate Food Source," *Food Prod. Dev.*, **10**, 64 (1976).

FISH POISONING (Foods). Foodborne Diseases.

FISH PROTEIN. Protein.

FISSION (Biology). A process of reproduction by the subdivision of the parent body into two or more approximately equal parts which become independent individuals.

Fission is a common form of reproduction in the one-celled animals. Division of the cell into two parts, known as binary fission, is accomplished by mitosis. In very simple species, such as *Amoeba*, reproduction is no more than cell division, but in species with constant body form, each half of the cell differs from the other and reproduction is not complete until it has undergone a reorganization with the development of all structures characteristic of its kinds. Some protozoans, notably parasitic species, go through a process of subdivision into a number of parts simultaneously. This process is multiple fission or sporulation. Fissions occurs in multicellular animals of simple structure, including polyps and flatworms, by a gradual reorganization accompanied by the constriction and breaking of the body. Bacteria, as well as a number of the one-celled algae, also reproduce by fission.

FISSION (Nuclear). Nuclear Fission.

FISSION REACTOR. Nuclear Reactor.

FISSURE IN ANO. A break or crack in the skin just internal to the margin of the anus. it is exceedingly common condition, one that usually complicates hemorrhoids and causes excruciating pain. The crack become infected, is raw and tender. Simple early cases can be treated locally with success. Many causes cannot be cured without surgery.

FISTULA. An abnormal channel or communication leading from a body organ. (1) Anal fistula may result from infection of the anus and is usually the result of infective invasion of the numerous tiny glands or crypts, which abound in the tissues adjacent to the anus. In addition to the production of a fissure, hemorrhoid, or abscess, a fistula may be created. This happens if the infection spreads through the wall of the anus, causing an abscess in the tissues around the anus. This abscess may burst through the skin around the anus or back into the rectum. In either case, the abscess cavity has two openings—the original site of entry of infection and the point where it burst through. Unless treated, a chronic discharging fistula may be formed. General treatment is to remove both the gland in which the infection originated and the infected tract once the infection is under control. (2) Bladder fistula is described under **Kidney and Urinary Tract.** (3) Vascular fistulas are abnormal communications between vessels and defects in the septa of the heart. These cause shunts in circulation, causing large amounts of blood to be returned to one or both ventricles. In this situation, the output of one or both ventricles is increased in proportion to the size of the shunt. Specifically named vascular fistulas are: *Arteriovenous fistula* (both ventricles); *patent ductus arteriosus* (left ventricle); *atrial septal defect* (right ventricle); and *ventricular septal defect* (both ventricles).

FITCH. Mustelines.

FIT (Convulsive). Convulsion; Seizure (Neurological).

FITTIG REACTION. The formation of aromatic hydrocarbons from aryl or aryl and alkyl bromides by the use of sodium, e.g., bromobenzene plus ethyl bromide plus sodium forms ethylbenzene plus sodium bromide $C_6H_5Br + C_2H_5Br + 2Na \rightarrow C_6H_5 \cdot C_2H_5 + 2NaBr$.

FITZGERALD-LORENTZ HYPOTHESIS. Sometimes referred to as the contraction hypothesis, this was proposed by Fitzgerald in 1893 to account for the null result of the Michelson-Morley experiment, that in moving through the ether with velocity v a body contracted in the direction of such motion in the ratio

$$\left(1 - \frac{v^2}{c^2}\right)^{1/2} : 1$$

This was applied in 1895 to the Lorentz theory of the electron. The special relativity theory (see **Relativity and Relativity Theory**) yields the same contraction, but ascribes it to the relative motion of the body and the observer rather than to the motion of the body through the ether.

FIXED BEAM. Beam (Structural).

FIXED BED. In processing terminology, a fixed-bed installation (usually a reactor) requires that materials in the solid phase that are to be reacted with gases and vapors remain in a fixed location. In other words, the flow in such equipment is that of the materials in the gaseous or vapor phase. The solid materials require careful preparation to permit a maximum of surface to be exposed to the gases which pass through them and to avoid the occurrence of channels through which the gases would pass without contacting the bulk of the solids. Beds of this type are often used in connection with various catalytic operations and are used, for example, in the production of benzene, in catalytic reforming, hydrocracking, hydrotreating, vinyl chloride monomer production, and in ion-exchange operations. A later reaction development, but also one that is several decades old, is the fluid-bed reactor in which the solids to be reacted are essentially "fluidized" and intermix with other solids or with gases in a rapidly-moving turbulent stream—in contrast with the solids remaining in a fixed bed. Whether or not a fixed bed is used instead of a fluidized bed is determined by numerous factors, notably time and cost. Fixed-bed reactors tend to be simpler and less costly, but often require cyclic operation because of the need to replenish the solids (catalysts, etc.). Thus to achieve continuous production, two or more beds are required (one on stream; the other regenerating). See also **Fluidization.**

FIXED OILS. These are fats, compounds of glycerin and various complex fatty acids. Fixed oils are often called the nonvolatile oils, in distinction to the essential or volatile oils, which are readily vaporized by heat. It is characteristic of the fixed oils to leave a spot when dropped on paper. Many of them remain liquid at common room temperatures, others are solid at such temperatures. Solid forms are usually called fats, a purely arbitrary distinction since slight changes in temperature will cause many of them to change from liquid to solid or vice versa.

Fixed oils, especially those of economic importance, are largely obtained from the seeds of plants. They have a high energy value, so form a valuable food if they prove palatable.

Various methods are employed to obtain the oil from the vegetable tissues. Quite commonly, the seeds containing the oil are subjected to great pressures in hydraulic presses. This may be done without heating, but is frequently facilitated by heating the seeds, the oil being then hot-pressed instead of cold-pressed. More recently, the screw press has come into use. This press has a rotating screw which presses the ground seeds under very high pressure through a cagelike cylinder. The oil is squeezed through openings in the cylinder walls, while the seed meal is discharged through an opening in the end of the cylinder. This procedure is advantageous in that it is continuous, and the machine need not be stopped for loading. A third method of obtaining the oil is by means of solvents. Following expression of the oil from the plant tissues, various methods of refining, decolorizing and deodorizing are employed.

Fixed oils are usually classified into three groups, drying, semidrying, and nondrying oils. Often a fourth group is made of those which are usually seen in solid form, the vegetable fats, although they differ but little otherwise from the other groups.

Drying oils are those which on exposure to air form a tough elastic film. Linseed oil from flax seeds is one of the most important and is largely used in making paints and varnishes. Tung oil, obtained from the fruits of *Aleurites fordii*, is a valuable oil much used in the manufacture of waterproof varnishes and quick-drying enamels. The tree, a native of China and Japan, has been introduced into Florida. Other drying oils are nut oil, from walnut seeds, poppy seed oil, hemp seed oil, and sunflower oil, the latter largely a product of the U.S.S.R.

Nondrying oils are those which remain permanently greasy or sticky, becoming rancid after a time. Among these oils the most important are olive oil, castor oil from the seeds of the castor bean plant, rape seed oil, peanut oil, almond oil, used medicinally, and tea seed oil.

Semidrying oils are intermediate in nature. The principal semidrying oils are cotton-seed oil, soybean oil, corn or maize oil and sesame oil. The latter is obtained from the seeds of *Sesamun indicum*, a member of the *Pedaliaceae*, cultivated in India, China and Japan, where the oil is much used as a food oil and for cooking.

Nondrying oils which are ordinarily solid are palm and palm-kernel oil, coconut oil, and cocoa butter. Another interesting oil of this group is macassar oil, obtained from the seeds of *Schleichera trijuga*, one of the *Sapindaceae*, occurring in tropical Asia. The oil was formerly much used as a potential "hair restorer," necessitating the use of removable covers, or antimacassars, on the backs of upholstered chairs. The same tree also yields a useful timber.

FIXED-PITCHED PROPELLER. **Airplane.**

FIXED-POINT ARITHMETIC. 1. A method of calculation in which operations take place in an invariant manner, and in which the computer does not consider the location of the radix point. This is illustrated by desk calculators or slide rules, with which the operator must keep track of the decimal point. Similarly with many automatic computers, in which the location of the radix point is the programmer's responsibility. See **Floating-Point Arithmetic.**

2. A type of arithmetic in which the operands and results of all arithmetic operations must be properly scaled so as to have a magnitude between certain fixed values.

See also terms listed under **Mathematics.**

FIXED POINT ARITHMETIC (Computer System). A method of storing numeric data in a computer such that the data are all stored in integer form (or all in fractional form) and the user postulates a radix point between a certain pair of digits. Consider a computer whose basic arithmetic is in decimal and in which each computer word consists of seven decimal digits in interger form. If it is desired to add 2.796512 to 4.873214, the data are stored in the computer as 2796512 and 4873214, the sum of which is 7669726. It is recalled that a decimal point between digits 1 and 2 has been postulated. The result, therefore, represents 7.669726. Input and output conversion routines often are provided for convenience. These routines can add or delete the radix point in the external representation and align the data as required internally.

Inasmuch as the arithmetic operations usually are the basic arithmetic operations of the computer, fixed-point operations are fast and thus preferred over floating-point operations. It is important, of course, that the magnitude of the numbers be much better known than for floating-point numbers, since the absolute magnitude is limited by word size and the availability of double-length operations. For many applications, fixed-point calculations are practical and do increase speed.

See also **Floating-Point Arithmetic.**

FIXED-POINT SYSTEM. **Point.**

FIXED-PROGRAM COMPUTER. A computer in which the sequence of instructions are permanently stored or wired in, and perform automatically and are not subject to change either by the computer or the programer except by rewiring or changing the storage input. See also **Wired-Program Computer.**

FIXING PROCESS (Photo). **Photography and Imagery.**

FIX (Navigation). The position of a ship as determined by the intersection of two or more lines of position. Since lines of position are determined in all of the observational types of navigation, pilotage, celestial, or radio, it is obvious that a fix may be defined as any position of a ship determined by observational means.

All observational methods are subject to unavoidable errors of observation; therefore, a line of position is not strictly a line, but is a band of constant width, as in the case of a line determined by celestial navigation, or of width that increases from the point of observation (e.g., in pilotage or radio navigation). Hence, a fix is not strictly a point, but is a polygon, whose area depends upon the accuracy of the observational material employed in determining the "line." With two lines of position, the area will be a quadrilateral, and the area will be a minimum when the lines are at right angles to each other. Three lines of position will seldom, in practice, intersect in a point, but will bound a triangle. The standard symbol for indicating the position of a fix on a geographic plot is a triangle with a dot at the center. This symbol is used no mattter how many lines are employed in determining the position, the point being the point of intersection of two lines, or, when more than two lines are used, the center of the polygon.

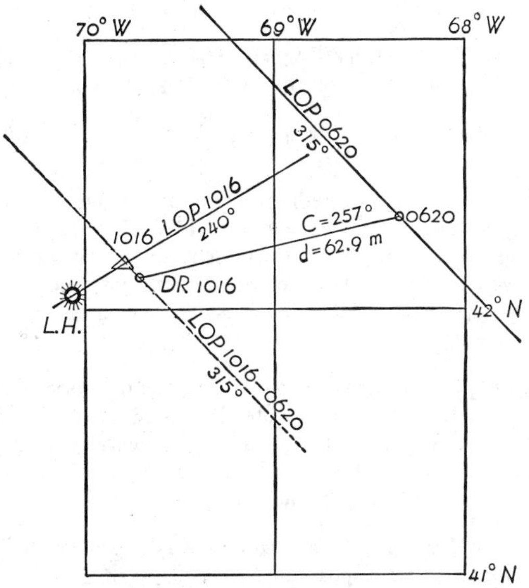

Scale diagram on mercator plotting sheet.

When the observations for the lines of position can be taken simultaneously, a fix can be immediately determined for the instant at which the lines were observed. Simultaneity of observation is seldom possible, and the question of movement of the ship between observations must be carefully considered. It is safe to say that, if the ship does not move more than 2 miles (3.2 kilometers) between observations, the fix may be considered as the intersection of the lines as directly plotted, with the time of the fix recorded as the average of the times of observation.

If the ship has moved appreciably (more than 2 miles; 3.2 kilometers) between observations, the method of running fix must be employed. To obtain a running fix, the selected time of the fix is usually the time of observation of some one of the lines, and the other lines are advanced or retarded to that line by standard dead-reckoning methods. In certain cases, e.g., when a noon position is desired and the only observations available are one before noon and another after noon, both lines are moved, one forward and the other backward, to the desired time. Before considering a specific case, it should be remembered that, by definition, a line of position is a line on which a ship is situated. For the purpose of moving the line, any conveniently located position may be selected, advanced, or retarded by dead reckoning, and a line drawn parallel to the observed line through the DR point will be the advanced line. The use of parallel rulers is convenient, but not essential, for this purpose. If the line of position is appreciably curved, due to the fact that the center of the circle of position is relatively close to the observer, the position of the center must be found, advanced by dead reckoning, and a new circle drawn, using the radius of the original circle.

The following situation illustrates the method for obtaining the running fix of a line of position obtained by celestial navigation and by pilotage:

At 0620, a ship is in L (latitude) = 42° 20′.8 N & Lo (longitude) = 68° 20′.1 W, and is proceeding at 16 knots on course 257°. At 0620, a line of position is obtained by celestial navigation, with the line running in the direction 135°–315°. At 1016, a lighthouse in L = N 42° 02′.4 & Lo = 70° 03′.7 W is sighted on bearing 240°. The 1016 position of the ship is desired.

In the figure, the problem is plotted on a mercator plotting sheet, with all lines labeled in accordance with U.S. Navy procedure. Through the 0620 position, the line of position (LOP) is drawn in the 135–315°. The course line is drawn in direction 257°, and the 1016 DR position is plotted on this line 62.9 miles (101.2 kilometers) (the distance run at 16 knots between 0620 and 1016) from the 0620 position. Through this DR position, a line is drawn parallel to the 0620 LOP, thus giving the advanced line of position. Through the lighthouse, a line is drawn in the 240°–060° direction, and this is the 1016 LOP. The 1016 fix is determined to be at L = 42° 09′.6 N & Lo = 69° 46′.5 W. See also **Course; Dead Reckoning; Navigation;** and **Pilotage (or Piloting).**

FIZEAU EXPERIMENT (Light). This experiment provided confirmation of the conclusion that the speed of light increases by an amount

$$v\left(1 - \frac{1}{n^2}\right)$$

when moving through a medium, of refractive index n, which itself is moving with velocity v. A beam of light was divided into two parts which were sent in opposite directions through two tubes filled with flowing water, and interference fringes were observed as a function of the velocity of the water. The result is a consequence of special relativity theory.

FLAG (Computer System). A bit or character of information attached to a character or word to indicate the boundary of a field. Also an indicator used frequently to tell some later part of a program that some condition occurred earlier; or an indicator used to identify the members of several sets which are intermixed.

FLAGELLATES. This is a large group of organisms, of particular interest because of the position they occupy in the organic world. There are both plant (described here) and animal (*Mastigophora*) flagellates.

They are usually one-celled organisms, of extremely complex structure. Many of them have no cell wall of cellulose, lack the green pigment, chlorophyll, and are definitely animals. Others possess a distinct wall of cellulose and have chloroplasts, and are set off as plants. The separation of these two groups is not sharp, however, and some of the plant members are obviously very closely related to very similar animal forms. Therefore it is impossible to stress the differences which are used as a basis for classification. Characteristic of the flagellates is the flagellum, a long lash-like extension from the protoplast. The vibrations of the flagellum propel the organism through the water, in which they usually occur.

FLAGELLUM. A long slender hair-like projection from a cell which may be used for a variety of purposes. In many plant and animal sperms it serves as a means of locomotion. It also serves this purpose in the flagellated bacteria and in the flagellated protozoans. In the sponges, the flagella of many of the cells lining the pores keep a current of water passing into the body. A flagellum differs from a cilium primarily in length and number. Cilia are much shorter and usually are present in greater quantity.

The electron microscope shows that there is a group of mitochondria at the base of each flagellum. The flagellum is made of an outer membrane, beneath which is a cable of nine protein fibers in a circle and with two fibers in the center.

FLAME CUTTING. Cutting of ferrous metals by oxidation, using a stream of oxygen from a blowpipe or torch. The metal is preheated to a bright red, approximately 1500°F (816°C) by fuel gas jets in the cutting torch. The stream of oxygen is then applied through a central jet. Once oxidation of iron to Fe_3O_4 begins, the heat of the reaction plays a large part in the continuation of the process. Approximately 30% of the molten metal is removed without actual oxidation by the mechanical washing action of the stream of gas and burnt metal. The preheating gases are oxy-acetylene, hydrogen, natural gas, city gas, etc.

The oxygen lance is a form of flame cutting in which oxygen, supplied through an iron pipe, is the only agent used. It is used for heavy-duty cutting.

Very heavy sections may be cut with the blowtorch. Close dimensional tolerances can be maintained, using machine operated torches, thus flame cutting has become a production tool as well as a means of salvaging scrap. Underwater flame cutting is possible at depths of 135 feet (40.5 meters) or more using special practice.

FLAME HARDENING. Surface hardening of steel or cast iron by heating a thin surface layer to the hardening temperature with an oxy-acetylene flame, followed by rapid cooling. Depending on the nature of the part to be hardened, either the torch system or the work itself may be moved. Cylindrical parts are rotated before a stationary flame. An air jet or liquid spray following the torch is used to quench-harden the surface. The relatively cool metal in the interior hastens cooling of the surface by conduction. The depth of flame hardening may be less than $\frac{1}{16}$ inch to about $\frac{1}{4}$ inch (1.6–6 millimeters), depending on the thickness of the section and the service requirements. Distortion is generally less than in parts hardened by general heating and quenching.

Since no hardening agent such as carbon or nitrogen is added to the surface of the steel by this process, only steels having sufficient carbon to harden readily upon quenching are used for flame hardening. The most desirable range is 0.35–0.70% carbon. The hardening treatment is followed by a low-temperature tempering treatment to relieve quenching strains. Typical applications of flame hardening are gear teeth, cams, bearing surfaces, rail ends, crankshafts, and many other machine parts and tools.

FLAMELESS VAPOR EXPLOSION. Natural Gas.

FLAME PHOTOMETRY AND SPECTROMETRY. The basic principle of flame emission spectrometry rests on the fact that salts of metals, when introduced under carefully controlled conditions into a suitable flame, are vaporized and excited to emit radiations that are characteristic for each element. Correlation of the emission intensity with the concentration of that element forms the basis of quantitative evaluation.

The determinations of sodium and potassium constitute the majority of published applications. However, the flame is a suitable emission source for at least 45 elements, which may be grouped as follows:

1. *Elements determined*: aluminum, barium, boron, calcium, cesium, chromium, copper, iron, lead, lithium, magnesium, manganese, potassium, rubidium, sodium, strontium.

2. *Elements determined but sometimes overlooked*: antimony, arsenic, bismuth, cadmium, cobalt, gallium, indium, lanthanum, nickel, palladium, rare earths (except cerium), rhodium, ruthenium, scandium, sliver, tellurium, thallium, tin, and yttrium.

3. *Elements with distinctive but less sensitive flame spectra*: beryllium, germanium, gold, mercury, molybdenum, niobium, rhenium, selenium, silicon, titanium, tungsten.

4. *Elements determined by indirect means*: bromine, chlorine, fluorine, iodine (although bromine, chlorine and fluorine can be determined by their metallic halide spectra), phosphorus, and silicon.

Materials in which these elements are determined by flame spectrometry include water, glasses, cement, soils, fertilizers, plant materials, biological fluids and tissues, petroleum products and metallurgical products.

The *flame spectrometer*, used in emission spectrometry, consists of (1) the pressure regulators and flow meters for the fuel gases; (2) the atomizing device; (3) the flame source; (4) the optical system; (5) appropriate photosensitive detectors; and (6) the electrical circuit for measuring or recording the intensity of the radiation. Depending upon the use intended, the instrument may be a relatively simple assemblage of interference filters and a photo-detector, i.e., a flame photometer, or it may be an elaborate prism or grating monochromator, i.e., a flame spectrometer such as the instrument illustrated.

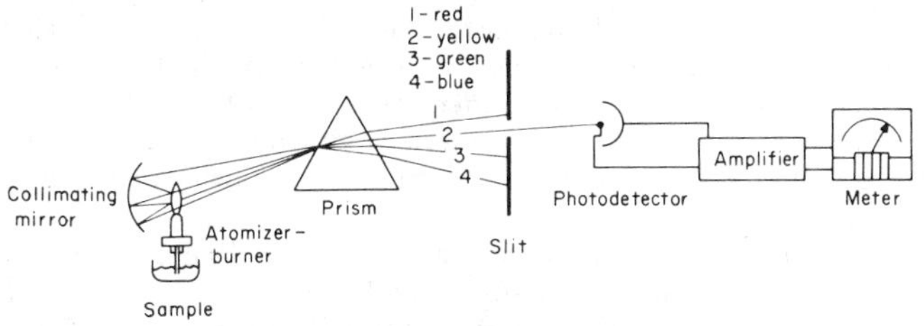

Schematic diagram of a flame spectrometer.

Virtually all flame spectrometers rely on atomization to deliver a steady flow of solution to the flame. The solution is drawn through a capillary positioned either concentric with or at right angles to the annulus or capillary from which the aspirating gas (oxygen or air under pressure) enters. At the tip of the solution capillary, the liquid is sheared off and dispersed into droplets by the blast of oxygen or air.

The best isolation of radiant energy can be achieved with flame spectrometers that incorporate either a prism or grating monochromator, those with prisms having variable gauged entrance and exit slits. Both these spectrometers provide a continuous selection of wavelengths with resolving power sufficient to separate completely most of the easily excited emission lines, and afford freedom from scattered radiation sufficient to minimize interferences. Fused silica or quartz optical components are necessary to permit measurements in the ultraviolet portion of the spectrum below 350 nanometers. See also **Analysis (Chemical); Atomic Spectroscopy; Photometers;** and **Spectro Instruments.**

FLAME-RETARDING AGENTS. A material used as a coating on or a component of a combustible product to raise its ignition point. The protection provided is usually only partial, and most materials so treated will burn when exposed to sufficiently high temperatures. The three principal types of agents are: (1) *nondurable,* consisting of water-soluble inorganic salts, which are easily removed by washing or accidental exposure to water; (2) *semi-durable* (removed by repeated laundering or dry-cleaning); and (3) *durable* (not affected by laundering or dry-cleaning). The latter types include or have included in the past organic compounds of bromine and chlorine, and insoluble metal salts. Antimony trioxide, tricresyl phosphate and other phosphate esters, chlorendic acid, etc., are effective, as well as cellulose-reactive agents. Zinc carbonate in high volume concentration will render a rubber or plastic compound self-extinguishing.

In 1972, flammability standards for children's sleepwear were established in the United States. In an effort to confer flame-resistant properties to the fabrics used, manufacturers began to use a number of chemical additives, notably organic halogens or phosphate esters, or both. One of the most widely used was *tris*-(2,3-dibromopropyl)phosphate, commonly called tris-BP. Other closely associated compounds were used. At a considerably later date, some researchers found that tris-BP and related compounds were carcinogenic, among other negative qualities. There is much room for further research into finding effective flame retardants that do not have adverse side effects. A few years ago, Blum and Ames prepared an interesting summary of flame-retardant additives as possible cancer hazards (*Science,* **195,** 17–23 (1977)).

FLAME TEMPERATURE. Combustion.

FLAMINGO (*Aves, Phoenicopteri*). Large wading birds of several species found in the warm regions of the world with the exception of Australia. They have very long legs and neck and a broad beak bent sharply downward at the middle. Red or rosy shades are characteristic in their plumage. Some authorities regard the flamingo as related ancestrally to ducks, geese, and swans. Only in comparatively recent years has the flamingo been placed in its own class, the *Phoenicopteriformes.* Flamingos are considered among the most beautiful of all birds—graceful, friendly, but gregarious. They range in length up to 6½ feet (2 meters) and may be as much as 5 feet (1.5 meters) in height. Although these birds essentially are mute, they do make a chattering noise with their beak, which at times can become quite loud. In flight, the neck stretches forward and the legs slant backward to aid in their streamlining. Like some other water birds, the flamingo has a filtering mechanism as part of its bill. The bill is boxlike and can be used in the manner of a scoop. Nests are constructed of dirt and mud in the form of mounds from 12 to 16 inches (30–41 centimeters) in diameter. One or two chalk-white eggs are incubated by both parents. They require from 30 to 32 days to hatch. The chick is downy and able to run around almost immediately after hatching. When sleeping, the flamingo rests on one foot, drawing the other up into the feathers, with the knuckle part sticking out far behind. The flamingo is found from the Bahamas to South America and the Galapagos Islands. Some domesticated flocks are found in Florida. The birds also are found in the high Andes and in parts of France and Spain. The birds often fly in formation. See also **Phoenicopteri.**

FLANDERS STONE. Graphite.

FLANGE. A rim or projection extending completely around the object which is flanged. Thus, a flange is distinguished from an ear, which is a similar projection, but which extends only a small portion of the circumference. Flanges are employed for a great many different purposes, among which is the juncture of adjacent shafts by flanged couplings, the flanges providing area through which connecting bolts may be passed. Flanged wheels are commonly used to maintain the position of a wheel and axle group upon parallel rails; pipe flanges, for the connection of pipes which are not to be welded, or that do not have threaded connections.

Flanges are usually used for pipe sizes larger than 2 inches where disassembly is required. A variety of types and facings is available. Although there is a large amount of metal in a flange, careful and precise machining is required only on the facing. Flanged joints do not require severe diametral tolerances on the pipe. Meticulous alignment before assembly of flat-face and raised-face flanges usually is not required. Assembly and dissassembly of flanged piping requires smaller wrenches than for equivalent-size screwed pipe assembly. Flanged-end pipe is furnished in a limited selection of metals. Where the required prefabricated flanged pipe is not obtainable, flanges are attached to the pipe by various kinds of joints. Flanged end fittings and valves are obtainable in most sizes and of most pipe metals.

Welding-neck flanges make possible joints that are as strong as the pipe, both for static and cyclic loading.

A loose flange pipe joint illustrates the general principle of flange construction. See accompanying diagram.

Loose flange (Van Stone) pipe joint.

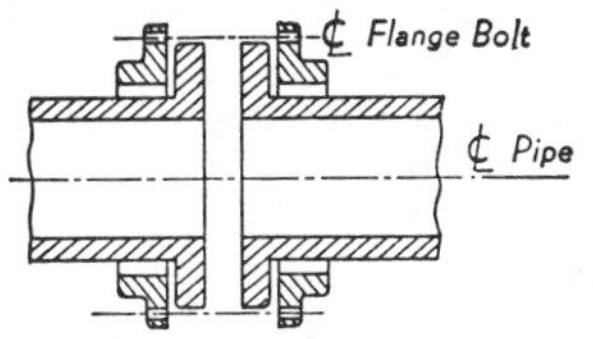

Some pressure vessels also are commonly furnished with flanged connections. Pumps, steam traps, filters, strainers, pipe fittings are commonly available with flange ends, particularly in larger sizes.

FLAP ATTENUATOR. Attenuator.

FLAPPER-NOZZLE SYSTEM. Pneumatic Controller; Transmission (Pneumatic).

FLAPS. Aerodynamics.

FLARE STARS. These are low-mass main-sequence stars that show erratic changes in their brightness. Changes are sudden, short-lived (minutes), and are unpredictable. Generally, these changes are attributed to local flares on the surface of the stars and have been considered to be similar to large solar flares. The brightness change may range from a few hundredths of a magnitude to 2 magnitudes or more in the visible spectrum. Accompanying this brightness change there has been a similar detectable burst in radio noise from the stars (of the order of 8×10^{-5} watts per square meter per cycle per second). Even the smallest flares observed appear to be more violent than the strongest solar flares by a factor of 10 or more. Best known of the flare stars are red dwarf (or UV Ceti) flare stars. All are also characterized by Hα emission, indicative of strong chromospheres. The nearest flare star is estimated to be about 8.6 light-years (2.6 parsecs) distant from earth. This puts the star at a distance about 0.5 million times as far as the sun is distant from the earth.

A number of these stars show rotationally modulated light variations. These stars, the BY Draconis stars, appear to be members of close binary systems with periods of a few days. Their light variations can be explained well by assuming their surface to be covered with large numbers of sunspots, forming large active regions. The processes occurring in the complex magnetic fields characteristic of such regions could thus also help to explain the observed flaring activity and active chromospheres. They are related to the RS CVn stars which also display enhanced, solar-like, activity.

Steven N. Shore, Assistant Professor of Astronomy, Warner and Swasey Observatory, Case Western Reserve University, Cleveland, Ohio.

FLASHBACK VOLTAGE. The peak inverse voltage at which ionization occurs in a gas-tube.

FLASH DISTILLATION. Distillation.

FLASH DISTILLATION (Desalination). Desalination.

FLASH FLOOD. Fronts and Storms.

FLASHING (Structural). Flashing is a method of sealing joints on buildings, especially around roofs, chimneys, gutters, and valleys, in order to render them water-tight. The method consists of using strips, or shingles, or flashing material which are worked into the normal roof surface and turned over the joint. Where the flashing turns up, as along a brick wall, it is necessary to counter-flash, that is to let a strip of flashing material into the brickwork and bend it down over the other flashing. A sloped shingle roof flashed against a brick wall requires flashing shingles which are worked under the top course of the regular shingles and turned up along the bricks. Corresponding counter-flashing let into the brick is bent down over these flashing shingles. A joint between the gutter and cornice is made weatherproof by flashing. Usually the flashing extends from under the lowermost course of shingles and is turned down over the edge of the gutter. The most common flashing materials are tin-coated sheet iron and copper. Lead and zinc are used to a limited extent.

FLASHING (Thermal). Liquids may exist with thermal stability at high temperatures provided they are subjected to sufficiently high pressure. Water, for example, may be heated to about 700°F (371°C) without boiling if under a pressure of 3,200 psi (217.7 atmospheres). It is true of liquids in general that the lower the pressure on them the lower the boiling temperature and the lower the heat contained in the "saturated" liquid. Thus high-temperature liquids when passed from a region of pressure sufficient for stability into a low-pressure region are not able to contain all the heat originally possessed as heat of fluid, and will be spontaneously partially evaporated by the surplus. This violent readjustment to thermal equilibrium is called "flashing," and is a common occurrence, having many uses and occasionally creating hazards. For example, the destructiveness of a boiler explosion arises mainly from the violence of flashing action since the water originally contained in a ruptured boiler drum at 600 psi (40.8 atmospheres) pressure suffers an almost instantaneous *four hundred fold* expansion in volume.

FLASHOMETER. A device for studying the time-intensity distribution of flashes of light.

FLASH POINT. The lowest temperature at which an oil will volatilize to yield sufficient vapor to form with air an inflammable gaseous mixture, demonstrable through the production of a flash on contact with a small open flame. The flash point occurs at a temperature lower than the burning point, which is the lowest temperature at which the production of combustible gas occurs rapidly enough to support a steady flame. It is also to be noted that the flash point is the temperature of formation, under the test conditions, of the lower explosive mixture of the substance tested, with air. (The higher explosive mixture is the *maximum* concentration of vapor, with air, which will sustain combustion.)

The flash point is an important characteristic of oils used for various purposes, such as lubrication, because a low flash point indicates the presence or absence of undesirable lower-boiling compounds. On the other hand, the flash temperature is less important then the burning temperature in determining the fire risk of an oil. The flash point is tested experimentally by heating the oil under certain specified conditions in a cup. A thermometer is suspended in the oil so that the temperature may be read during the test. Periodically an open test flame is introduced through an opening in the cover to detect the slight explosive puff which follows when the flash point has been reached.

FLASH POINT (Fuel). Petroleum.

FLASH SPECTRUM (Solar). Sun (The).

FLATFISHES (*Osteichthyes*). In commercial fishery statistics, a number of fishes, essentially flat in physical form, are lumped together and reported as flatfishes. These include turbot, flounders, halibuts, soles, and plaice. Some of these species are among the most desirable of the fishes for eating fresh or for freeze-preserving.

Flounders. Of the order *Pleuronectiformes*, flounders are distributed in all seas, primarily in warm and temperate zones. A few species are also found in the Arctic and on the borders of Antarctic waters. Some even penetrate fresh water of rivers. Almost all flounders are shelf inhabitants (i.e., shallow seas), and only a few are found at greater depths (down to about 4900 feet; 1500 meters). The size of adult flounders varies from just a few centimeters (i.e., little sole) to several meters. Thus, the halibut, which can attain the length of swordfish and tuna, is exceeded in size only by the whale shark, basking sharks, and the arapaima.

The chief characteristic common to all flounders is the greatly flattened, broadened body. Unlike all other vertebrates, it is not dorsoventrally oriented (i.e., from back to belly), as in rays, but laterally. This has evolved within the group as an adaptation to bottom dwelling. Flounders may have evolved from species which were similar to present-day perch. When flounder larvae hatch, they look like the larvae of all other fishes. During this planktonic stage, the flounder larvae drift about aimlessly. As they continue growing, they swim in the normal pattern for some time, until metamorphosis begins in the open water, whereby the larva is transformed to a very small flounder-shaped fish. The most striking aspect of this metamorphosis is the migration of one of the eyes from one side of the body to the other; finally, both eyes lie on the same side. This becomes the new upper side of the body, while the eyeless (blind) side becomes the underside. This eye migration occurs shortly before the fishes begin their bottom-dwelling life. There are numerous other physiological points of interest concerning the flounder, but these are beyond the scope of this volume. There are both righteyed and lefteyed flounders. See Grzimek reference listed at end of entry on **Fishes.**

As indicated by Fig. 1, some flatfishes have an uncanny ability to alter their color and pattern to match that of their background. Experimentally, it has been found that this ability is lost when the fishes are blinded and thus this pigmentation accommodation must result from visual stimuli.

Witch Flounder. This fish (*Glyptocephalus cynoglossus*) is an important North American species in the inshore gill-net fishery which takes place from fishing communities on the eastern coast of Newfoundland. Typically, the nets are emptied 24 to 48 hours after setting, and the catch immediately returned to modern freezing plants to be iced and processed. The fishery presently takes place during the months

(a)

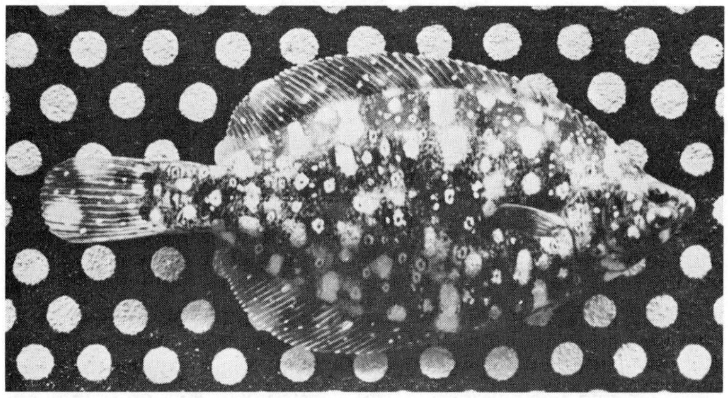

(b)

Fig. 1. The flounder can change pattern and coloration to blend with its surroundings: (a) Pebble-like markings; (b) polka-dotted pattern. (*A. M. Winchester.*)

of June to October, from close inshore to as far as 50 miles (80 kilometers) from the coast. Traditionally, the market has regarded witch flounder as a premium flatfish with prices paid accordingly. A study of the chemical and sensory changes during storage of witch flounder was made by Shaw et al. in 1977. See list of references at end of entry on **Fishes.**

Turbot. This is a lefteye flounder (*Scophthalmidae*). See Fig. 2. These fishes inhabit the northeastern Atlantic Ocean from Iceland and the Scandanavian coast to the Mediterranean and the Baltic Seas and the Gulf of Bothnia. Its rather thick body is quite wide, with practically a round circumference and a short tail shaft. The eye-side of the body has many bony hooks which develop from modified scales. Turbot, like soles, can change their background color with such success that they are virtual disappearing magicians. The species grows very large. Large turbot are many years old. Sexual maturity appears in the fifth year, at a length of about 1 foot (30 centimeters). Due to intense fishing, very few turbot attain a length of more than double that just mentioned. Most of them are much smaller than that when caught and many are either caught or destroyed prior to spawning.

Fig. 2. Turbot (*Scophthalmus maximus*).

Turbot was greatly valued in antiquity. Turbot is mentioned in Roman literature as a delightful food. Turbot feed on fishes, such as soles and haddock, and occasionally sand lances, pipefishes, and oceanic gobies. There are several closely related fishes that are of some commercial, often of regional, importance, including the *Black Sea turbot* (*Scophthalmus maeoticus*); the much smaller turbot *S. auosus*, the closest North American relative to turbot; the *megrim* (*Lepidorhombus whiffiagonis*), a turbot distributed on Europe's western coast from Scandinavia to the Iberian Peninsula and off Iceland; the *Norwegian topknot* (*Phrynorhombus norvegicus*), which inhabits the European coasts from Murmansk to southwestern England and is also found off Iceland.

Species of lefteye flounders (*Bothidae*) are numerous. *Scald fishes* (*Arnoglossus*) are widely distributed on northwest European coasts and in the Indo-Pacific ocean. The *Arnoglossus laterna* inhabits the waters from the Oslo Fjord and the British Isles, and across the Mediterranean into the Sea of Marmara. It is much more slender than turbot species. The *scale fish* (*Bothus*) is distributed primarily along the Atlantic coast of the United States. It is also found westward and southward to the Azores and Angola, respectively. The *Engyproscopon grandisquama* is distributed from the coasts of eastern and southern Africa, Indonesia, Japan, and Australia, and has unusually large scales. This is a very popular eating fish in Japan. A peculiar scald fish (*Pelecanichthys crumenalis*) frequents the Hawaiian Islands, named because its protruding lower jaw is reminiscent of a pelican.

The *summer flounder* (*Paralichthys dentatus*) is one of the best-known species. It is found along the Atlantic coast of North America. Summer flounder (actually a turbot) is a valuable commercial species and exhibits exceptional color transformation. During warm seasons of the year, summer flounder occur in large schools, moving into coastal waters, where they are found on sandy and muddy bottoms. Some even penetrate the fresh water of rivers.

Other valuable commercial species from the same genus include the *California halibut* (*Paralichthys californicus*), from the California coast; *P. olivaceus* from Japan; and *P. microps*, from the coast of Chile.

Halibut. This fish (*Hippoglossus hippoglossus hippoglossus*) is the largest member of the flounder family, reaching lengths up to 7.5 feet (2.3 meters). See Fig. 3. In rare cases, halibut grow to be giants.

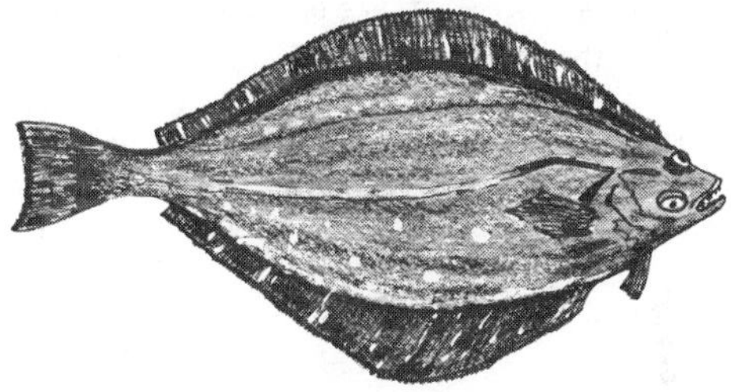

Fig. 3. Halibut (*Hippoglossus hippoglossus hippoglossus*).

In 1884, near Hammerfest, Norway, a halibut weighing 528 pounds (240 kilograms) was caught. One specimen caught in 1935 on Iceland's north coast was 12 feet (3.7 meters) long; 17.7 inches (45 centimeters) thick; and weighed 585 pounds (266 kilograms). It was so large that when brought to market there were few bidders for it. The largest halibut ever recorded was 15.4 feet (4.7 meters) long and weighed 726 pounds (330 kilograms). The average length and weight of those which are presently brought to fish markets are much smaller. The far-flung distribution of halibut extends from Spitzbergen, Iceland, and the Murmansk coast, south to the Bay of Biscay, and in northwestern Atlantic Ocean from southwestern Greenland to Newfoundland.

Not a true arctic species, halibut exists so far north only because it stays warm in the Gulf Stream. This powerful fish is rather elongated, with a tapered head and curved caudal fin edges. The eye-side of the body has a uniform gray-brown to dark olive-brown hue. The underside is white, and the powerful jaws have sharp teeth. While young halibut occur in shallow water, between 115 and 230 feet (35 to 70 meters), the adults are found in depths of nearly 3000 to 3280 feet (700 to 1000 meters). They usually live there above sandy or gravel bottom, very rarely above mud or rocky floors. During summer, halibut prefer the banks of moderately deep to shallow water along the coasts. In winter, they retreat to deeper water.

Halibut carry on extensive migrations between these banks and away from them. They spawn from late December to April, at a water temperature of about 43 to 44.5°F (6 to 7°C). Large females lay up to 3.5 million eggs. They depart from the spawning grounds, which are in the Lofoten-Finnmark region. They migrate northward to Bear Island and into the White Sea, where their feeding grounds are located. Iceland is an important spawning region between February and April. Halibut live for about 25 years. Sexual maturity is attained after 8 to 10 years. Spawning occurs at considerable depths. Eggs have been found between Iceland and the Faroe islands, floating at depths between 1970 and 2625 feet (600 and 800 meters) in water that was 3280 and 6560 feet (1000 to 2000 meters) deep. The halibut diet is quite diverse, consisting chiefly of various fishes, including cod, haddock, scorpion fishes, poachers, grenadiers, herring, sand lances, several flounder species, skates, wolffishes, and mackerel. Lobsters, large mussels, squid, and echinoderms are also eaten. One marine bird species, the razor-billed auk, has been found in the stomach of a halibut. According to some researchers, halibut can stun or kill small cod by striking them with the tail. Halibut are primarily prey to seals and the Greenland shark. The liver of halibut, which when cooked produces high-quality cod liver oil, is rich in vitamins.

The *Pacific halibut* (*Hippoglossus hippoglossus stenolepis*) is a subspecies of the Atlantic relative and is found in the North Pacific Ocean from the Bering Sea and Alaska to the Sea of Okhotsk and the California coast. It was a favorite catch of the American Indians prior to colonization of North America.

The *Greenland halibut* (*Reinhardtius hippoglossoides*) is a dark, rust-colored (upper side) fish. Distribution of this much smaller species extends across the deeper arctic waters of the northeastern Atlantic Ocean as far east as Novaya Zemlya, around Iceland, and southward to the Norwegian coast. In the northwestern Atlantic Ocean, it occurs from western Greenland southward to the Newfoundland banks. The chief catch areas are near the coasts and in fjords. Greenland halibut feeds primarily on fishes, such as cod, small perches, and on shrimp. Greenland halibut is eaten in large quantities by saddleback seals and belugas.

Plaice. This fish (*Pleuronectes platessa*) is one of the small-mouthed species, all of which are placed in the subfamily *Pleuronectinae.* Plaice is the best-known and most marketed flounder species in some regions. Distribution is on the Atlantic coastal waters from the White Sea and Iceland, and the North Sea and Baltic Sea, to the Gulf of Cadiz in southern Portugal. It is less commonly found in the western Mediterranean. Before heavy fishing started, plaice sometimes reached a length of nearly 3 feet (1 meter). See Fig. 4. In present times of heavy fishing,

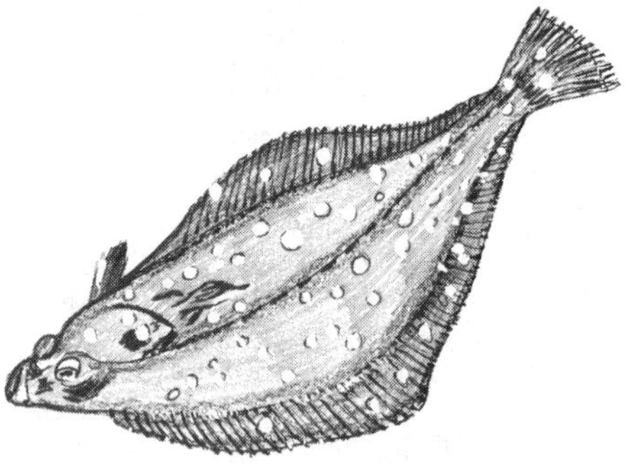

Fig. 4. Plaice (*Pleuronectes platessa*).

a plaice that is about 2 feet (61 centimeters) long is considered fairly large. In the North Sea, females attain sexual maturity in their fifth to sixth year. Males generally mature 1 year earlier. Most plaice in the Barents Sea, however, do not mature for 9 to 13 years. In the North Sea, the spawning season occurs from January to March. Large groups of eggs are found in a great depression in the ocean off the Netherlands. One female can lay from 50,000 to 0.5 million eggs in a single spawning period, depending upon her size. Development of the eggs lasts 10 to 20 days, depending upon temperature. The young migrate landward into shallow-water regions. They spend the early part of their life in the tidal zone near beaches. They do not return to deeper water until they have reached a good length. In spite of their limited mobility, plaice form schools and undertake extensive spawning and feeding migrations. The most important component of their varied diet is small mussels. The plaice break mussel shells with their large, powerful throat teeth, a feature characteristic of this species. In the western Baltic Sea, plaice may crossbreed with flounders or dabs.

The *Pacific plaice* (*Pleuronectes pallasi*) is distributed in the northwestern Pacific Ocean and the waters around Kamchatka. It is an important commercial species.

The *dab* (*Limanda limanda*) differs from the plaice, in that it has a rough upper-body surface. Coloration varies from pale-yellow and brownish to a dirty greenish hue. The dab is distributed on the western European coasts of Iceland and from the White Sea to the Bay of Biscay, also penetrating the Baltic Sea and the Gulf of Finland. The dab is most prevalent in the southern North Sea. The species feeds mostly on hermit crabs, isopods, amphipods, echinoderms, mussels, and various worms.

Sole. This fish is of the suborder *Soleoidei* and is another important flounder. See Fig. 5. Soles not only swim, but also crawl along the

Fig. 5. Common sole (*Solea solea*).

floor. Soles are generally nocturnal. During the day they are deeply buried in mud. With few exceptions, soles are distributed in equatorial tropical and subtropical equatorial waters. A number of species are found in fresh water. The sole (*Solea solea*) is distributed from the Mediterranean and northwestern Africa coasts northward to Trondheim, Norway. It attains a length up to nearly 20 inches (51 centimeters).

Other Flounders. The *smoothback flounder* (*Liopsetta glacialis*) is found around the North Pole along the arctic coasts. It is characterized by a somewhat elongated body and a rather thick, smooth, scaleless zone between the eyes. The species inhabits water which is very near the freezing point. It is important to regional natives.

The *American winter flounder* (*Pseudopleuronectes americanus*) inhabits the Atlantic coast of North America from Labrador to Chesapeake Bay. This commercially important species is occasionally found in the brackish water of river mouths and, rarely, in the fresh water of rivers. Unlike almost all other flounders, its eggs do not float, but adhere to the floor.

The *North Atlantic flounder* (*Platichthys flesus*) is a well-known flounder species and is found on the coast from the White Sea to the western Mediterranean and Black Sea, as well as far into rivers in fresh water. There are varieties which spend years in rivers. Prior to pollution, one of this species was found in the Thames, upriver from London.

The *starry flounder* (*Platichthys stellatus*) contributes in an important way to the flounder catch off the western coast of the United States. The starry flounder also ascends rivers, far into fresh water.

See also **Fishes.** References are included at the end of that entry.

FLAT-PLATE AREA (Equivalent). That area of an imaginary flat plate, normal to an air stream, whose drag would be the same as that of an actual body at the same air speed is the equivalent flat-plate area of the body. The coefficient of drag of a flat plate is known to vary with the aspect ratio and the area, but it is frequently assumed as 1.28. However, as the area is an imaginary one at all events, the drag coefficient may as well be assumed equal to 1. It is seen that equivalent flat-plate area might have two values, f and f', which are related thus: $f = 1.28f'$.

If D = drag in pounds of a body in air stream of dynamic pressure q,

$$f = \frac{D}{q}$$

$$f' = \frac{D}{1.28q}$$

See **Aerodynamics.**

FLAT-PLATE COLLECTOR. **Solar Energy.**

FLATTENING OF THE EARTH. The ratio of the difference between the equatorial radius (major semiaxis) and the polar radius (minor semiaxis) of the earth to the equatorial radius. Also called *compression.* The flattening of the earth is the ellipticity of the spheroid and equals the ellipticity of the ellipse forming a meridional section of the spheroid. If a and b represent the major and minor semiaxes of the spheroid, and f is the flattening of the earth,

$$f = (a - b)/a$$

The magnitude of the flattening is sometimes expressed by stating the numerical value of the reciprocal of the flattening, $a/(a - b)$. See also **Earth.**

FLATULENCE. The accumulation of excessive gas in the stomach or intestine. Causes of stomach gas include: (1) excessive belching which takes in more air than is expelled, (2) consumption of inordinate amounts of carbonated beverages, and (3) swallowing air while eating. Intestinal gas may result from (4) bacterial action on ingested foods, and (5) air that is swallowed and forced into the intestinal tract by peristalsis. Gas also forms in the intestine after abdominal surgery, resulting from a short term paralysis of the intestinal tract making it difficult for the patient to pass off the gas naturally. A rectal tube can be used to relieve gas pains of the latter origin. Particular foods, such as raw vegetables, including tomatoes and beans, often will produce excessive gas in the intestine.

A number of preparations are available for the treatment and relief of flatulence. Sometimes various formulations combine antacids with antiflatulents. Generally, the active ingredients of antiflatulents fall into the following categories, which frequently are used in various combinations in a given preparation: (1) Various digestive enzymes, such as amylase, lipase, protease, cellulase, among others. Pancreatin from beef pancreas is a source of pepsin, providing multiple digestive enzymes. (2) Simethicone. (3) Bile constituents. (4) Antacids, such as aluminum and magnesium hydroxide. (5) Sedatives, such as the belladonna alkaloids hyoscyamine sulfate, atropine sulfate, and scopolamine hydrobromide. These compounds are claimed useful in the relief of variously named conditions, such as dyspepsia, distension, fullness, gastric hyperacidity, mucus-entrapped air or "gas," gastritis, hiatal hernia, peptic ulcer, peptic esophagitis, and common heartburn (pyrosis). Some compounds are not compatible with certain antibiotics, such as the tetracyclines.

Injectible preparations, such as dexpanthenol, are sometimes used immediately after abdominal surgery to minimize the possibility of paralytic ileus.

See also **Antacids.**

FLATWORM (Fluke). **Fluke.**

FLATWORMS. **Platyhelminthes; Turbellaria.**

FLAVIN ADENINE DINUCLEOTIDE (FAD). **Coenzymes; Vitamins.**

FLAVIN MONONUCLEOTIDE (FMN). **Coenzymes.**

FLAVONOIDS. A group of aromatic, oxygen-containing heterocyclic pigments widely distributed among higher plants. They constitute most of the yellow, red, and blue colors in flowers and fruits. (Most of the other pigments are carotenoids. See also **Carotenoids.**) The flavonoids include the catechins, leucoanthocyanidins and flavonones, flavanols, flavones, the anthocyanins, and the flavonols. See also **Anthocyanins.**

FLAVOR ENHANCERS AND POTENTIATORS. A *flavor enhancer* is a substance which when present in a food accentuates the taste of the food without contributing any flavor of its own. This is reminiscent of the role of a catalyst in a chemical reaction which promotes a reaction without chemically participating in the reaction. Although not usually regarded as a flavor enhancer, common salt, if not used excessively, enhances the taste of food substances. Salt does not fully meet the definition of an enhancer, however, because the salt is detectable as salt unless added in very minute quantities.

Monosodium glutamate for many years has been the best known and most widely used of the flavor enhancers. MSG is normally effective in concentrations of a relatively few parts per thousand, but far less powerful than the more recently developed flavor potentiators. Like enhancers, *potentiators* do not add any taste of their own to food substances, but intensify the taste response to the flavorings already present in the food. Because a potentiator is more powerful, smaller quantities of the substances are required than in the case of the enhancers.

The chronology of MSG commenced centuries ago when certain seaweeds were used in the Far East to improve the flavor of soups and certain other foods. It was not until 1908, however, that the curiosity of K. Ikeda (University of Tokyo) caused him to study the seaweed *Laminaria japonica,* traditionally used by Japanese cooks to enhance food flavoring. After much research on the seaweed, MSG was isolated and identified as an excellent flavor enhancer, particularly for high-protein foods. As an aside, it is interesting to note that Ritthausen in Germany had isolated glutamic acid as early as 1866 and his associates had prepared the sodium salt of the acid, namely monosodium glutamate. But the path of research in Germany was targeted in other directions and the flavor enhancing qualities of MSG were left to Ikeda to determine.

The Japanese throughout the first half of this century produced glutamic acid by extraction from natural materials, a slow, costly method. But, as the demand for MSG grew, cost tended to be a secondary factor. It was not until 1956 that Japanese microbiologists succeeded in developing the first industrial production of L-glutamic acid by means of fermentation. The problem of producing glutamic acid, as well as a number of other important amino acids, by fermentation was the lack of suitable strains of microorganisms for starting the cultures.

Initially, the Japanese researchers were successful in isolating microbial strains from natural sources that possessed good abilities to excrete and accumulate a large amount of the amino acid in the cultural broth, but only under very carefully controlled conditions. After much experimentation, a large-scale MSG production was achieved by the fermentation route. Sugar beets are now the most common raw material used.

Although listed as a GRAS substance (generally regarded as safe) for many years, questions concerning the safe usage of MSG have arisen from time to time and, as of the early 1980s, MSG still remains somewhat controversial. It is known that overconsumption of MSG can produce an illness, usually of just a few hours duration, in some persons. This is commonly referred to as the "Chinese Restaurant Syndrome" and is described in the entry on **Foodborne Diseases.** The possible seriousness of any deleterious effects of MSG tend to be countered by the many years the substance has been used by thousands of food processors and millions of chefs and home food preparers. As of 1979, the topic was well summarized by Krueger (see references).

The 5′-Nucleotides. Also dating back many years in the Far East is the knowledge that bonita tuna possesses a substance that very effectively enhances the flavor of foods. However, it was not until 1913 that S. Kodama (University of Tokyo) commenced a serious investigation directed toward identifying and isolating the substance from tuna. Initially, Kodama believed that the substance was the histidine salt of 5′-inosinic acid, but later found that the substance was actually 5′-inosinic acid itself. This nucleotide was found to be many more times as effective as MSG. Further research has shown that these nucleotides are present in many natural foods, including numerous species of fishes, beef, pork, and chicken.

The nucleotides, in addition to their effectiveness at much lower concentration, have been found to be superior to MSG for certain types of foods over and beyond those of a high-protein nature. It also has been observed that the nucleotides tend to create a sense of increased viscosity, providing more body, for example, to soups. Japanese manufacturers are now producing a series of the nucleotides by the enzymatic hydrolysis of ribonucleic acid. As of the early 1980s, these compounds are enjoying a high volume of production and consumption. It should be pointed out that the nucleotides are frequently used along with MSG. Some researchers point out that while the nucleotides and MSG have a lot in common, there is a considerable difference in their application. The nucleotides are up to 100 times more effective than MSG on a weight basis, and whereas MSG has been a favorite of processors for the enhancement of "meaty flavor," the range of the nucleotides is broader, modifying salty or sweet flavors and suppressing many undesirable flavors. The nucleotides are not a replacement for MSG. The substances do have a synergistic effect when used together. Generally 1 gram of nucleotide used with 50 grams of MSG will have the same flavor intensifying result as 100 grams of MSG alone.

Because the market is so large, research continues at a good pace in seeking other potentiators. Established since the early 1940s, *maltol* is effectively used in foods that are high in carbohydrates, such as beverages, jams, and gelatins. Claims of reducing sugar content by 15% in products using maltol have been made. Other potentiators used or proposed include dioctyl sodium sulfosuccinate, *N,N'*-di-*o*-tolylethylenediamine, and cyclamic acid.

References

Furia, T. E., and N. Bellanca: "Fenaroli's Handbook of Flavor Ingredients," CRC Press, Boca Raton, Florida, 1971.

Krueger, J.: "MSG: One of the Food Industry's Most Studied Ingredients," *Processed Prepared Foods*, **148**, 1, 128–140 (1979).

Staff: "An Introduction to Nucleotide Seasonings," Ajinomoto, Tokyo, 1980.

Staff: "Ribonucleotide Flavor Enhancers," Takeda, Tokyo, 1980.

FLAVORINGS. Substances added to various food substances to impart or intensify flavor account for a significant amount of chemical food additives consumption. The *flavor* of a food substance is the combined sensation of *taste* and *odor* as perceived by the eater/drinker of that substance. Although the components (*flavorings*) are present in food substances, the full aspects of flavor require intimate contact between substance and consumer. The odors emanating from a bakery tend to be richer and more pleasant than the bread itself; the flavor of coffee seldom attains the richness of aroma that one perceives in the vicinity of a coffee roasting plant. Flavor is a unique combination of nerve impulses on the brain centers as the result of actions upon receptors located on the tongue and in the lining of the nose and is thus the result of interaction between the food substance and the consumer.

In terms of total flavor sensation, many authorities agree that odor is usually more important than taste. Experience, of course, demonstrates the marked reduction of flavor sensation when the nasal passages are partially blocked, as in the case of a common cold. In such instances, the layperson may refer to the "flat taste" of the food. In actuality, the taste buds are functioning normally; it is the odor component of flavor that is missing.

The odor component of flavor is made up of at least two vectors. Sniffing a substance without contact with the tongue provides a partial indication of odor, that is, molecular vapors or gases pass directly to the olfactory sensors in the nose via the nasal cavities. This vector might be called the absolute external odor or fundamental odor of a substance. This vector is dependent upon the vapor pressure (volatility) of the food substance itself. The other vector of odor is what researchers call internal odor because the molecules reach the olfactory sensors by way of the pharynx, a flattened tubular passage that connects the back of the mouth with the nasal cavities. In the mouth, the food substance is wetted by saliva, altering not only the vapor pressure of the flavoring agents present, but sometimes exposing more and different flavorings, thus affecting flavor intensity and quality. It is well known, of course, that exceedingly dry substances tend to be odorless or nearly so. The odor of a polished metallic surface, for example, is difficult for most persons to detect. The addition of only modest amounts of moisture to most dry substances significantly increases their fundamental odor, as the result of increasing vapor pressure and by activating all flavoring substances present. The effect of moisture on odor is dramatically illustrated by the dog at the fireside and the dog that has just come in out of the rain.

Technology of Flavorings

To remark that the overall topic of flavorings is intricate and complex is indeed an understatement. Far from exhaustive in its coverage, for example, the excellent "Food Chemicals Codex" describes well over 300 flavoring agents. "Fenaroli's Handbook of Flavor Ingredients" describes in detail nearly 200 natural flavorings and nearly 750 synthetic flavorings—and neither of these publications attempts to fully cover the field.

Natural Flavorings. These substances come from a number of plant sources—bushes, herbs, shrubs, trees, weeds—specific parts of which are used as flavoring sources. These include arils, balsams, barks, beans, berries, blossoms, branches, buds, bulbs, calyxes, capsules, catkins, cones, exudates, flowering tops, flowers, fronds, fruits, gums, hips, husks, juices, kernels, leaves, needles, nuts, oils, oleoresins, peels, pits, pulps, resins, rhizomes, rinds, roots, seeds, shoots, stalks, stigmas, stolons, thallus, twigs, wood, and wood sawdust—as well as some entire plants that are ground up as a united flavoring source. The culture of plants as flavoring resources represents an impressive segment of world food production.

Natural flavorings are prepared for commerce in various ways. Adding to the complexities of raw natural flavorings are factors of timing and maturity. Some plant parts are only suitable when green (unripe); others must be fully ripe or nearly so. The timing of harvest can be critical. For example, jasmine flowers must be harvested before dawn. The roots of the orris plant must be aged two years before they are ready for the commercial market. Several of the natural flavorings are also obtainable in two or more quality or grade classes. For example, there are at least four classes of crude camphor oil, in addition to the true or distilled camphor oil.

Only a relatively few animal sources of fundamental flavoring substances are used. There is the musk deer, *Moschus moschiferus* L., found in the Himalayan highlands. The reddish-brown secretion of the male is the odorous principle identified as 3-methylcyclopentadecanonone-one, the principal use of which is in perfumery as a fixative, but which has been reported as an additive in certain food products. There is the civet, *Viverra civetta* Schreber, a cat that lives in Africa and southeastern Asia. The glandular secretion is of main interest in perfumery as a fixative, but it has been reported in foodstuffs at low levels (about 4 parts per million). There is the beaver of genus *Castor* of the northern climes of Alaska, Canada, and Siberia, the dried and ground glandular secretions of which are also used in perfumery, but also in chewing gum up to concentrations of 400 ppm. The flavoring additive is known as castoreum. And there is beeswax, a crude yellow wax that represents a secondary secretion of the honeybee. In addition to use as a modifier in perfumery, the substance is used up to levels of 5 ppm to enhance the flavor and textural qualities of honey. There always has been a close link between the technology of flavorings for the food field and of fragrances used in perfumes, cosmetics, and related products.

In order of decreasing value per kilogram (approximate $/kilogram, 1980, is given in parentheses), some of the important natural essential oils, used for both their taste and odor, are: Rose oil (attar of roses) ($1501); neroli (orange flower oil) ($987); orris oil ($942); sandalwood oil ($110); vetiver oil ($67); cassia oil ($67); geranium oil ($60); ylang ylang (cananga oil) ($53); peppermint oil (mentha piperita) ($45); cedar leaf oil ($31); bergamot oil ($31); caraway oil ($26); palmarosa oil ($25); origanum oil ($24); lavender oil (spike oil) ($24); thyme oil ($22); lime oil ($20); patchouli oil ($17); nutmeg oil ($17); onion and garlic oil ($16); pineneedle oil ($14); lemon oil ($13); pine oil ($13); anise oil ($12); lignaloe (Boise de rose oil) ($10); rosemary oil ($10); pettigrain oil ($9); cornmint oil (*Mentha arvensis*) ($9); lemongrass oil ($6); cinnamon oil ($6); clove oil ($5); bitter almond oil ($4); citronella oil ($4); sassafras oil ($4); eucalyptus oil ($3); camphor oil ($3); cedarwood oil ($2); grapefruit oil ($1); and orange oil ($1).

Synthetic Flavorings. Although the synthetic flavorings have supplemented and replaced the natural flavorings to a substantial degree because of cost and technical reasons, this has not relegated natural flavorings to a minor role. Synthetic flavorings are made from chemical raw materials, such as petrochemicals, *via the route of organic synthesis.* Major unit operations involved include addition, condensation, cycli-

SYNTHETIC EQUIVALENTS (APPROXIMATIONS OR ANALOGUES) OF VARIOUS NATURAL FLAVORS

Almond	Tolualdehyde (*o, m, p*)
Apple	Allyl butyrate, cyclohexylvalerate, isovalerate, propionate; Benzyl isovalerate; Butyl isovalerate, valerate; Cinnamyl formate, isobutyrate, isovalerate; Citronellyl isovalerate; Cyclohexyl acetate, butyrate, isovalerate; Ethyl isovalerate, valerate; Isopropyl acetate, valerate; 2-Methylallyl butyrate; Methyl butyrate; Terpenyl isovalerate.
Apricot	Allyl butyrate, cyclohexylcaproate, cyclohexylvalerate, propionate; Amyl phenylacetate; Benzyl formate, propionate; Butyl propionate; Cinnamic acid; Citronellyl acetate; gamma-Decalactone; gamma-Dodecalactone; Ethyl cinnamate; Geranyl butyrate, isobutyrate, isovalerate, Heptyl acetate, propionate; Methyl ionone; Phenylethyl dimethyl carbinol; Phenylpropyl alcohol; Propyl cinnamate; Santalyl acetate; Tetrahydrofurfuryl propionate; gamma-Undecalactone.
Banana	Cyclohexyl acetate, butyrate, propionate; Ethyl valerate.
Butter	Diacetyl.
Caramel-Butterscotch	Maltol.
Caraway	D-Carvone.
Cheese	Hexanoic acid; Isovaleric acid (rancid).
Cherry	Allyl benzoate; Anisyl butyrate, propionate; Cyclohexyl cinnamate, formate; Methyl anthranilate; Rhodinyl formate; Tetrahydrogeraniol; Tolualdehyde (*o, m, p*).
Chocolate	Tetrahydrofurfuryl propionate.
Cinnamon	Cinnamaldehyde; Alpha-methylcinnamaldehyde.
Citrus	Decanal dimethyl acetal.
Cloves	Methyl cinnamate; isoeugenol.
Cocoa	Neryl butyrate; Phenylpropyl cinnamate.
Coconut	Allyl undecylate; Ethyl undecylate; Methyl undecyl ketone; gamma-Nonalactone; gamma-Octalactone.
Cognac	Allyl pelargonate; Cyclohexyl caproate.
Cola	2-Ethyl-3-furylacrolein.
Currant	Cyclohexyl butyrate; Guaiol acetate (black currant); Linalyl acetate, isobutyrate, propionate (black currant); Methyl ionone; Methyl propionate (black currant).
Fatty	Decanal; Ethyl nonanoate; Heptyl alcohol; Lauryl alcohol, aldehyde; Nonanal; Octanal; 1-Octanol; Undecanal; 10-Undecenal.
Flowery	Anisyl alcohol; Benzyl acetate, phenylacetate; Cinnamic acid; Cinnamyl acetate; Citronellyl formate; Cresyl acetate; Decanal; Dimethyl benzyl carbinol; Dimethyl benzyl carbinyl acetate; Ethyl anthranilate; Geranyl acetate; Hydroxycitronellal dimethyl acetate; Linalool; Linalyl acetate; Methyl benzoate; Penethyl acetate; 2-Phenylpropionaldehyde; 3-Phenylpropionaldehyde.
Flowery/Fruity	Anisyl acetate; Cinnamyl isovalerate; Citronellyl; Ethyl laurate, octanoate; Geranyl butyrate, propionate; Nonyl acetate.
Fruity	Benzyl propionate; Butyl acetate; Cinnamyl anthranilate, formate; Citronellyl acetate, butyrate, isobutyrate; propionate; Delta-decalactone; Diethyl malonate; Dimethylbenzyl carbinyl acetate; Delta-dodecalactone; Ethyl *p*-anisate, benzoate, butyrate, heptanoate, hexanoate, maltol, nonanoate; Isoamyl butyrate, hexanoate, isovalerate, cinnamate; Linalyl isobutyrate, propionate; Maltol; Methyl benzoate, cinnamate; 2-Methylundecanal; Nerolidol; Octanol; Octyl formate; Phenethyl isobutyrate, isovalerate; gamma-Undecalactone.
Grape	Allyl salicylate; Cinnamyl anthranilate; Guaiol acetate; Isobutyl anthranilate; Isovalerophenone; Octyl isobutyrate; Phenylpropyl acetate, ether.
Grapefruit	Styralyl acetate.
Green Leaves	Allyl anthranilate
Hawthorne	Acetanisole; *p*-Methoxybenzaldehyde.
Heliotrope	Piperonal
Honey	Allyl phenoxyacetate, phenylacetate; Benzyl cinnamate; Carvacryl acetate; Cinnamyl butyrate; *p*-Cresyl acetate; *p*-Cresyl ethyl ether; *m*-Cresyl phenylacetate; *p*-Cresyl phenylacetate; Cyclohexyl phenylacetate; Ethyl phenoxyacetate, phenylacetate; Guaiol phenylacetate; Isobutyl phenylacetate; Linalyl butyrate; Methyl phenylacetate; Phenethyl acetate, butyrate, phenylacetate; Phenylacetic acid; Propyl phenylacetate; Santalyl phenylacetate.
Lemon	Citral; Citronellal.
Licorice	Methylcyclopenteneolone.
Maple	Methylcyclopenteneolone.
Melon	Cinnamaldehyde; Ethyl hexadienoate; Methyl amyl ketone; Octyl butyrate.
Menthol	3-*p*-Methanol.
Mushroom	Hexyl furan carboxylate.
Mustard	Allyl formate.
Orange	Linalyl anthranilate.
Peach	Allyl cyclohexylcaproate, cyclohexylvalerate, undecylate; Amyl phenylacetate; Anisyl alcohol, butyrate; L-Citronellol; Cyclhexyl caproate, cinnamate; gamma-Dodecalactone; Ethyl cinnamate; Isopropyl benzyl carbinol; Methyl methylanthranilate; Methyl nonyl ketone; Methyl octine carbonate; gamma-Octalactone; Oxyl acetate; Phenethyl alcohol, isovalerate, salicylate; Phenylallyl alcohol; Phenylpropyl isobutyrate; Propyl cinnamate; Rhodinyl formate; gamma-Undecalactone.
Pear	Benzyl butyrate; 2-Ethylbutyl acetate; Ethyl heptylate; Hexyl acetate; Hexyl furan carboxylate; Isoamyl acetate; 2-Methylallyl caproate; Methylheptenone; Propyl acetate.
Pineapple	Allyl caproate, cyclohexylacetate, cyclohexylbutyrate, cyclohexylpropionate, 2-nonylenate, phenoxyacetate; Benzyl formate; Bornyl acetate; *n*-Butyl acetate; Butyl isobutyrate; Cinnamyl acetate; Decanal dimethyl acetal; Ethyl butyrate, hexadienoate, phenoxyacetate; Hexyl butyrate; 2-Methylallyl caproate; Methyl beta-methylpropionate; Methyl undecylate; Propyl isobutyrate.
Plum	Butyl formate; Citronellyl butyrate, formate, propionate; gamma-Decalactone; Guaiol butyrate; Heptyl formate; Hexyl formate; Isoamyl formate; Isopropyl formate, propionate; Linalool; Neryl propionate; Phenethyl formate (green plum); Phenethyl isobutyrate (green plum); Phenylallyl alcohol; Phenylpropyl butyrate; Propyl formate; Terpenyl butyrate.
Raspberry	Benzyl salicylate; alpha-Ionone; Isobutyl cinnamate; Methyl ionone; Neryl acetate; Santalol.
Rose	Phenethyl alcohol; Phenethyl dimethyl carbinyl isovalerate; Rhodinol.
Rum	Ethyl formate; isobutyl formate.
Sassafrass	*p*-Propyl anisole.
Spearmint	1-Carvone.
"Spice"	Eugenol; Isoeugenyl acetate; 3-Phenylpropyl acetate.
Strawberry	Anisyl formate; Benzyl isobutyrate; Cuminic alcohol; Ethyl methylphenylglycidate; Ethyl phenylglycidate; Isoamyl salicylate; Isobutyl anthranilate; Methylacetophenone; Methyl cinnamate; Methyl naphthyl ketone; Nerolin; Neryl isobutyrate; Phenylglycidate
Vanilla	Propenyl guaiethol; Vanillydene acetone.
Violet	alpha-Ionone; beta-Ionone; Methyl-2-octynoate.
Walnut	gamma-Octalactone.
Wine	Ethyl acetate; heptylate.
Wintergreen	Allyl salicylate; Methyl salicylate.

zation, dehydrogenation, esterification, hydrogenation, oxidation, pyrolysis, reduction, and saponification. An example of this approach is cinnamaldehyde ethylene glycol acetate, which possesses a soft, warm, spicy odor reminiscent of cinnamon. This is prepared by reacting cinnamaldehyde with ethylene glycol. Cinnamaldehyde is prepared by the condensation of benzaldehyde with acetaldehyde in the presence of sodium or calcium hydroxide.

Synthetic flavorings are also produced *via the route of preparation from natural raw materials* that have a close relationship with the end-product. An example of this approach is synthetic eucalyptol, which commences with purified and concentrated eucalyptol obtained by fractionation of the essential oil.

Although a number of synthetics represent essentially identical replacements of natural flavorings, in many instances the synthetics are *approximations* of the natural materials, or represent entirely new flavorings which have significantly broadened the spectrum of available flavorings and flavor sensations. The flavorist has many more substances to call on and many more options than were available a few decades ago. Frequently, there are cost advantages for synthetics, but cost is not always the predominating factor. The flavorist determines the most suited natural and/or synthetic flavoring, with physical, chemical, and organoleptic qualities considered along with cost.

The synthetic equivalents (approximations or analogues) of the common natural flavors are listed in the accompanying table, which gives only an abridged listing.

The flavoring field is further complicated by the very large number of flavors that are available from natural and/or synthetic sources and by the many hundreds of flavors of slightly different emphasis that can be prepared from the material available. Even with much progress in the technology used by the flavorist, there remains a large element of art (experience) along with the science. Sophisticated instruments, such as gas chromatographs and mass spectrometers, among others, are available to assist the professional flavorist, but in developing flavorings for new products or in exerting flavor quality control over mass-produced food products, the need frequently arises for a human panel of sniffers and tasters, professionally termed *sensory evaluation panels* or monitors. See also **Sensory Evaluation.**

Among numerous factors to be considered by the flavorist are flavor yield (total flavor sensation per unit of flavoring substance) and the so-called harmony of flavoring when two or more substances are used. Combinations of flavoring substances can produce synergistic effects, both of a positive and negative nature. The use of artificial sweeteners versus natural sugars usually alters the required flavorings quite markedly. The inclusion of a flavor potentiator also must be considered as part of the total flavoring profile. Incidentally, the flavorist uses the term *compound* to describe an intimate mixture of two or more flavoring substances (possibly some natural; some synthetic), as contrasted with the traditional definition of chemical compound. Commonly, a flavoring compound will be an esoteric, proprietary blend of substances. Traditional strawberry flavoring for soft drinks may include amyl acetate, amyl butyrate, butyl isovalerate, ethyl acetoacetate, ethyl butyrate, ethyl caproate, and ethylfuran carbonate—where the beverage is to be sweetened with natural sugar. On the other hand, if synthetic sweetener is used, the formulation will be quite different: aldehyde C_{18}, diethyl acetal, geranil, beta-ionone, maltol, neroli essential oil, octanyl dimethyl acetal, phenethyl alcohol, terpineol, and vanillin. The multiplicity of ingredients to obtain a given flavor is also illustrated by the flavoring substances used in an imitation rose flavor: aldehyde C_8, citral, citronellol, geraniol, linalool, phenethyl alcohol, and rhodinol. Many scores of additional examples of this nature are given and excellently described in the Furia (1971) reference listed.

The flavorist uses terms reminiscent of sound harmonics to describe the roles of certain ingredients in a flavor compound. The *top note* indicates the flavor first perceived by the food monitor or consumer, and is usually a flavoring substance with a volatility relatively higher than the other flavor components present. The *main note* (sometimes called *middle note*) is the predominating flavor of the food substance, coming on strong, so to speak, immediately after the top note is sensed. The *bottom note* (sometimes called *undertone*) is composed of flavors that are perceived slightly later in the cycle of tasting or smelling a food substance. An optimal blend of the various notes results in what is sometimes called a *full-bodied flavor*, important, for example, in coffee.

Physiological Aspects of Flavors

Receptor cells especially sensitive to chemicals are found in virtually all animals. By convention, those receptors normally excited by contact with chemicals in liquid phase at relatively high concentrations are termed taste or *gustatory receptors*, although the distinctions between taste and smell are not critical at cellular or molecular levels.

Physiologists use the term *papillae* to identify so-called taste buds on the surface of the tongue. There are several thousand papillae on the human tongue. Because the tongue takes so much abuse when a person is eating a variety, often very coarse and rough foods, some authorities believe that the papillae are constantly being renewed. Small fibers, almost like tiny hairs, extend from these cells to the surface of the tongue. Some investigators have described the papillae as being of three general shapes: those that look like tiny mushrooms; those that appear like miniature hills with moats around them; and the tiny threads and cones. Observable differences in shape tend to reinforce earlier theories which embrace the use of differently structured organisms to sense different taste categories, notably the traditional four basic taste sensations—sweet, sour, bitter, salty. These concepts have been difficult to refine or confirm.

Chemical aspects of taste receptor functions can be studied by recording the patterns of electrical potentials in receptor cells while the cells are being stimulated with pure chemicals of known structures and properties. Since the mid-1950s, when this method was first successfully applied to single taste receptor cells, using receptors on the mouth parts of a fly, many earlier theories of taste stimulation have been revised.

Intracellular recordings from taste cells of rat and hamster show that even primary receptor cells are sensitive to three or four of the so-called basic taste modalities. Consequently, it is generally held that a variety of different receptor *sites* commonly exist on the receptor membrane of any one receptor cell. Biochemical characterization of events at receptor sites has progressed in analyzing electrolyte and carbohydrate stimulation.

Electrolyte stimulation is chiefly a function of monovalent cations in all animals which have been studied. Consequently, the receptor sites are thought to be anionic. The pH relationships of stimulation also indicate that strongly acidic (e.g., PO_4^{2-} or SO_4^{2-}) receptor groups are involved. Calculations of free energy changes of the reaction between salt and receptor site give values between 0 and -1 kcal/mole; and low ΔF values suggest that the reaction involves only weak physical forces. The reaction occurs extremely rapidly, since typical nerve impulses can be recorded within 1 millisecond after stimulating electrolytes are applied. In blow-flies, $0.004M$ NaCl, which produces 1 impulse per second, represents the threshold for behavior response. These thresholds appear to be somewhat higher in humans.

No one type of receptor site or reaction can account for the extreme structural specificities observed. A curious assortment of molecules can elicit "sweet" sensations. Early studies on a variety of organisms demonstrated that ring structures and D-isomers were more stimulating in polyol compounds than straight-chain and L-isomers. Thus, inositol (Fig. 1), with its ring structure, was found to stimulate. The straight-chain polyhydric alcohols sorbitol, dulcitol, (Fig. 2), and mannitol did not stimulate. Possession of an alpha-D-glucopyrano side linkage was found to generally increase the stimulating capacity of sugars. Maltose, with a 1,4-linkage; turanose, with a 1,3-linkage; and the nonreducing sugars stimulate. Lactose, with its 1,4-linkage, and melibiose, with a 1,6-linkage, both lack the alpha link and are relatively nonstimulating.

Conformation, as well as configuration, is important in determining the stimulating power of sugar molecules. Glucose, which exists in solution almost entirely in an aldopyranose "chair" formation has derivatives of both 1C and C1 conformations (Figs. 3 and 4). Those of the C1 type are considerably the more stimulating. The hydroxyl groups attached to C3 and C4, inclined 19 degrees above and 19 degrees below the adjacent plane of the molecule, appear to be necessary for the critical linkage at the receptor site. Lack of effects by metabolic inhibitors (azide, fluoride, iodoacetate, etc.) or of temperature effects upon the initial excitatory process, suggests that this step depends upon specific physical rather than chemical reactions.

The nature of other polyol receptor sites and the molecular basis for genetic and species differences in taste capabilities remain largely unknown. Saccharin (*o*-sulfobenzime, Fig. 5) exemplifies both puzzles, since its molecule does not fit any known sugar receptor site, yet it is confused with sugar stimuli by humans and other primates, but probably not by nonprimate animals. The substitution of other groups for one hydrogen (dotted lines in Fig. 5) renders saccharin tasteless. The genetic basis of taste has been studied with phenylthiocarbamide (PTC). A strong bitter taste of PTC depends upon the chemical components indicated by dotted lines in Fig. 6, and upon possession of a dominant "taster" gene in humans. Curiously, a small change in the molecule (Fig. 7) yields a product 250 to 300 times sweeter than sugar.

Taste receptors for water have been reported to occur on mouth parts of mammals and invertebrates. Specialized amino acid and amine receptors are found on the legs of many anthropods. The mechanisms by which adequate stimuli initiate nerve impulses in these cells offers a rich field for further investigation. Some stimuli, especially longchain hydrocarbons, are known to act in an opposing manner, i.e., by decreasing, rather than increasing, the output of receptor impulses. Their effects resemble the actions of narcotics. Some authorities suggest that taste sensation, as ultimately perceived, probably results from a complex coded pattern of augmented or depressed frequencies of nerve impulses, originating in the different cells of a heterogeneous population of taste receptors.

Sense of Smell. For many years, physiologists have explained that the sense of smell is located in the mucus lining inside the nose. Traditionally, it has been observed that a person with a "dry" nose has little if any sense of smell. Molecules characteristic of certain flavorings must be moistened by the mucus before they can be detected. Traditionally, it also has been observed that these nerve cells (estimated to be a million or more) tend to become blocked (refuse further transmission of signals) upon prolonged exposure to any given odor. This blocking phenomenon can occur within just a few minutes after exposure to certain, usually powerful odors. When all is sorted out concerning the mechanics of tasting and smelling, it is highly likely that operationally the cells in the nose lining and the papillae of the tongue will be highly similar, if not identical—simply because many other interrelationships have been shown to be similar. Of course, over the years, some researchers have approached the phenomena of taste and odor separately. Some of the odor detection theories proposed have included: (1) the *vibrational theory* (Demerdach; Dyson; Wright); (2) the *stereochemical theory* (Amoore, Johnston, Naves); (3) the *theory of interfacial adsorption* (Beck; Davies); and the *profile functional group theory* (Beets).

The aforementioned theories are concerned with the size and shape of odorant molecules, but differ in certain underlying concepts. For example, accommodating for functional groups, electron donor–acceptor characteristics, as well as the sorptive nature of odorants on sensor sites. The vibration theory largely concentrates on the far-infrared and Raman spectral characteristics of odoriferous substances. The remaining theories concentrate on structural and behavior characteristics of odorant molecules, stressing direct interactions physically, chemically, and biologically with the olfactory sensor system.

The vibrational theory stresses so-called osmic frequencies (Wright) as setting up resonances in the sensory organs. In commenting on this concept, Dravnieks observed that spectra are codes describing molecular structures and shapes with emphasis on the distribution of atomic masses, distances, and bond polarities. Questions regarding corresponding intramolecular vibrations as direct factors in odor discrimination are independent of the validity of the spectral code.

In connection with the stereochemical theory, Amoore, although emphasizing the importance of molecular shape and size, also gives important emphasis to such chemical factors as electrophilic–nucleophilic characteristics, rotational moment, and the presence of functional groups. In the theory, the functional groups play a more important role in small molecules than in larger ones. Size and shape similarities were analyzed and demonstrated by preparing silhouettes patterned from three-dimensional molecular models.

Critics of the Amoore theory point out that electrophysiological data do not support the concept that sensors are sensitive to size and shape. Other authorities, however, point out that molecular size and shape, operating with a reactive character, most likely are odor relevant, but that this is not explained satisfactorily by the Amoore theory. Another objection to the theory is that equal weight is given to positive and negative differences in odorant molecules. Further, it is observed that the theory does not account for odor blocking or fatiguing (*anosmia*).

The profile functional group theory is a relatively complex two-step concept. First, it is visualized that the functional groups of the odorant molecule interact with the receptor site, thus causing a given orientation of the molecule. This, in turn, determines the final odor-relevant profile. Any similarity in profile at the receptor site causes similar signal transmissions to the brain. Because some odorant molecules will react more strongly at receptor sites than others, stronger signals will be transmitted.

The interfacial adsorption theory proposes that the orientation of odorant molecules is dependent upon their behavior at the hydrophilic–hydrophobic interface, taking into account interaction with the mucus and adjacent olfactory membrane.

Considerably more detail on flavorings can be found in the "Foods and Food Production Encyclopedia" (D. M. Considine, editor), Van Nostrand Reinhold, New York, 1982.

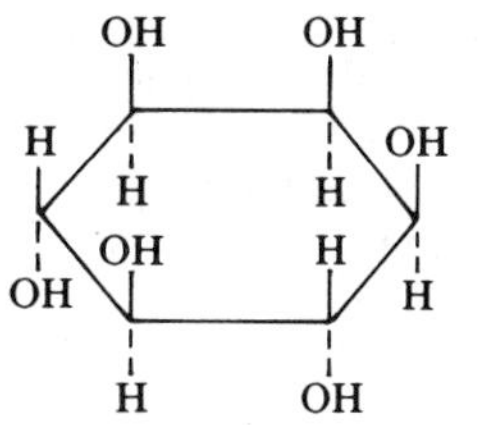

1

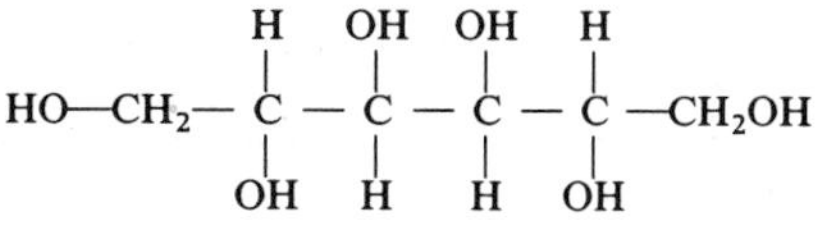

2

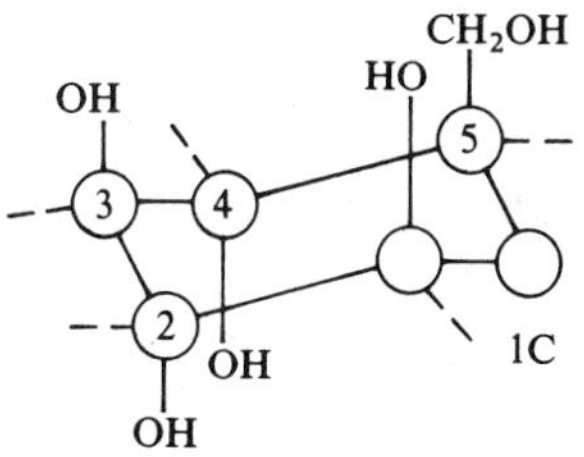

3

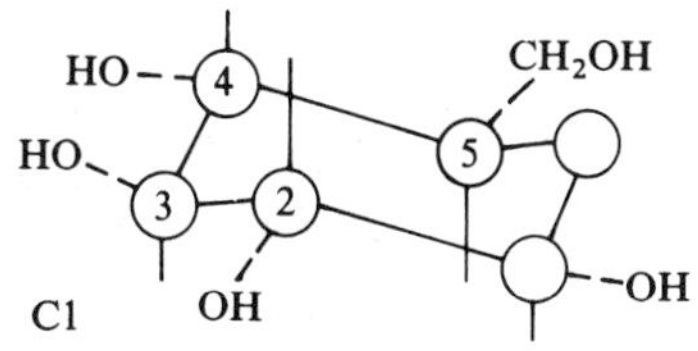

4

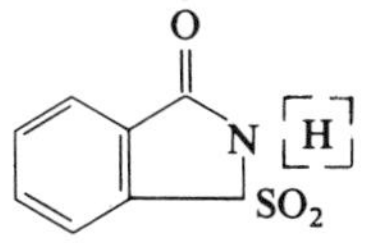

5

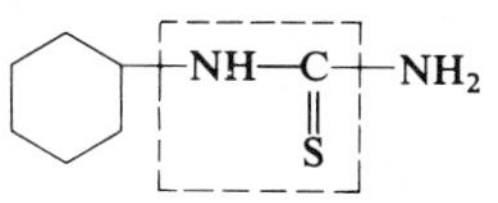

6

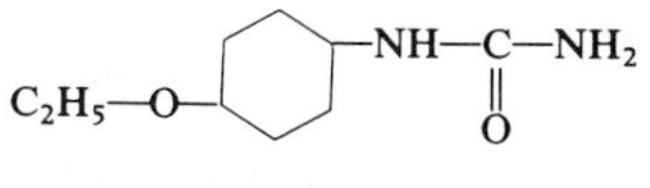

7

Figs. 1 through 7. Molecules illustrating relationships between structure and effects on taste receptors. See explanation in text. (*After Hodgson.*)

References

Amoore, J. E., Johnston, J. W., Jr., and M. Rubin: "The Stereochemical Theory of Odor," *Sci. Amer.* (February 1964).

Arvidson, K., and U. Friberg: "Human Taste: Response and Taste Bud Number in Fungiform Papillae," *Science*, **209**, 807–808 (1980).

Beck, L. H.: "A Quantitative Theory of the Olfactory Threshold Based upon

the Amount of the Sensory Cell Covered by an Adsorbed Film," *Proc. New York Academy of Sciences*, **116**, 448 (1964).
Beets, M. G. J.: "Structure and Odor," in "Molecular Structure and Organoleptic Quality," Soc. of Chem. Ind., London, 1961.
Beets, M. G. J.: "Odor and Molecular Constitution," *Amer. Perf.*, **76**, 54 (1961).
Beets, M. G. J.: "Some Aspects on the Problems of Odor," *Parf. Cosm. Savons*, **5**, 4 (1962).
Davies, J. T., and F. H. Taylor: "Olfactory Thresholds: A Test of a New Theory," *Perfum. Essent. Oil Rec.*, **46**, 1, 15 (1955).
Demerdache, A., and R. H. Wright: "Low-Frequency Molecular Vibration in Relation to Odor," in "Olfaction and Taste" (T. Hayaski, editor), Vol. 2, Pergamon, Elmsford, New York, 1979.
Dethmers, A. E.: "Utilizing Sensory Evaluation to Determine Product Shelf Life," *Food Technology*, **33**, 9, 40–42 (1979).
Dorland, W. E., and J. A. Rogers, Jr.: "The Fragrance and Flavor Industry," Dorland, Mendham, New Jersey, 1977.
Dravnieks, A.: "Comparison of Theories on Relations between Odor Parameters and Other Properties of Odorants," NATO Advanced Study Institute on Odor Theories and Odor Measurements, Robert College, Blebes, Istanbul, Turkey, 1966.
Dyson, G. M.: "Raman Effect and Concept of Odor," *Perfum. Essent. Oil Rec.*, **28**, 13 (1937).
Furia, T. E., and N. Bellanca: "Fenaroli's Handbook of Flavor Ingredients," CRC Press, Boca Raton, Florida, 1971.
Furia, T. E.: "Handbook of Food Additives," 2nd edition, CRC Press, Boca Raton, Florida, 1972.
Heath, H. B.: "Flavor Technology," AVI, Westport, Connecticut, 1978.
Hodgson, E. S.: "Taste Receptors," in "The Encyclopedia of Biochemistry," (R. J. Williams and E. M. Lansford, Jr., editors), Van Nostrand Reinhold, New York, 1967.
Johnson, J. W., and A. Sandoval: "Organoleptic Qualities and the Stereochemical Theory of Olfaction," *Proc. Sci. Sect. Toilet Goods Assoc.*, **34** (1960).
Katz, M. H., and G. Jacobson: "Flavor Chemistry," Symposium, 39th Annual Institute of Food Technologists, Saint Louis, Missouri, 1979.
Kazeniac, S. J.: "Flavor Trends in New Foods," *Food Technology*, **31**, 1, 26–28 (1977).
Naves, Y. R.: "The Relationship between the Stereochemistry and Odorous Properties of Organic Substances," in "Molecular Structure and Organoleptic Quality," Soc. of Chem. Ind., London, 1957.
Staff: "Food Chemicals Codex," National Academy of Sciences, Washington, D.C. (Revised periodically).
Staff: "Essential Oils," Foreign Agriculture Circular, U.S. Department of Agriculture, Washington, D.C. (Issued periodically).
Wolfe, K.: "Use of Reference Standards for Sensory Evaluation of Product Quality," *Food Technology*, **33**, 9, 43–44 (1979).

FLAX (*Linum usitatissimum*; *Linaceae*). Flax is the name given to the pericycle fibers of *Linum usitatissimum*, a plant native to Europe. They were probably the first vegetable fibers to be used by man. Picture-writings at Thebes not only show the plant, but also give details of the processes used in making cloth from the fibers. Egyptians, Greeks, ancient Hebrews, and Romans knew the fiber and used it. Mummy cloths are often of linen.

The flax plant is a slender annual attaining 4 feet in height and branching slightly. It has small lanceolate leaves and clear blue flowers. When mature it bears seed capsules containing ten seeds each and about $\frac{1}{4}$ inch (6.3 millimeters) in diameter. Successful cultivation demands an abundance of potassium and phosphorus in the soil and plenty of moisture. The plant is cultivated not only for the fibers, but also for oil. The best fibers are obtained from plants grown in cool regions, while the best oil is derived from plants grown in tropical countries like India.

For preparing flax, the plants are pulled or cut before they are mature, and stripped of all leaves and seed capsules. The denuded stems are then tied in small bunches and immersed either in stagnant or slowly running soft water, where they are left for several days. During this time, the stems are attacked by bacteria, which bring about fermentation, causing a breaking down of the woody tissues and a partial separation of cells due to the action of the bacterial enzymes on the pectic substances binding the cells together. This process is called "retting." Sometimes the flax stems are spread on the ground in a thin layer and left exposed to the action of dew and sunshine for a few weeks. The same result is obtained, the process being now called "dew-retting." The retted stems are removed from the water, washed and cleaned of as much non-fibrous material as possible. This process is known as "Scutching." "Hackling" follows, and is a sort of combing which removes any remaining non-fibrous material. The fibers thus obtained are in reality bundles of cells which occurred as pericycle fibers in the stem. Good fibers vary from 12–36 inches (0.3–0.9 meter) in length, while many shorter ones are obtained. Short and tangled fibers are called "tow." The fibers vary in color from yellowish to dirty gray, largely depending on the attention paid to the retting process. They are soft and flexible, capable of division into smaller bundles of fewer cells. They are very strong, each cell possessing a uniformly thick wall which surrounds the very slender central cavity or lumen.

The principal use of the fibers is in the manufacture of thread requiring great strength, such as shoe thread, bookbinding thread, fish line and fish-net twine, and also in the making of fine cloth such as table-linen and handkerchief linen. All cloth made from flax fibers is called linen. Flax fibers conduct heat much more rapidly than cotton and so cloth made from them is cooler and much favored in tropical countries.

In addition to the fibers, flax plants yield a valuable oil, called linseed oil. The seeds are about 40% oil. In making this, the seeds are crushed by machinery, heated to 74°C and treated with naphtha, which extracts the oil, or the oil is removed in a continuous screw press. This oil, a drying oil, is used in the manufacture of paints, varnishes, and patent leather, as well as in making linoleum and oil-cloth.

The oil cake left after the oil is pressed from the seeds is used directly as a cattle food or is ground up into oil meal and used for the same purpose.

New Zealand Flax (*Phormium tenax*; *Liliaceae*). The fibers of this flax occur as schlerenchyma sheaths surrounding the vascular bundles in the long, straight, rather stiff leaves of the plant. The plant, a native of New Zealand, has been introduced into Australia and other countries. In the United States, it is cultivated to some extent, often as an ornamental plant. The leaves, from 4–8 feet (1.2–2.4 meters) long and up to 8 inches (20 centimeters) wide, may be 20% fibrous material. To obtain the fiber the leaves are cut off and scraped to remove much of the non-fibrous material. After this the fibers are combed out and cleaned. They are very white, soft, flexible, lustrous and tough. They may be 5 feet (1.5 meters) long. The principal use of New Zealand flax is for binder twine, baling rope, and cordage, often in combination with sisal or other fibers. A fine cloth resembling linen duck can be woven from it.

FLEA (*Insecta, Siphonaptera*). Small insects with transversely compressed bodies, sucking mouths, and no wings. See accompanying illustration. They live as bloodsucking ectoparasites in the adult stage

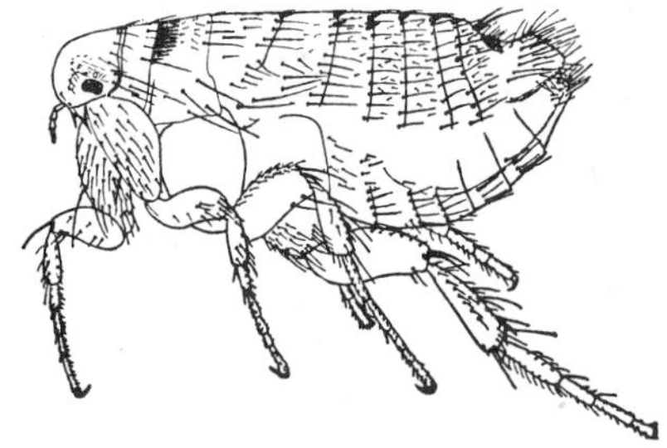

Flea.

on the bodies of mammals and less frequently on birds. They progress chiefly by leaping. They have well developed legs and possess great jumping ability (13 inches; 32.5 centimeters by some species in a single leap). The adults have short stout antennae and their mandibles are long and saw-toothed. The larvae have biting mouths and eat fragments of organic matter in the debris about the sleeping places of the hosts. The larvae are not parasitic. Stages in the development of the flea are egg, larva, pupa, and adult. The eggs are tiny, with a yellow coloration. An average flea is about $\frac{1}{8}$-inch (3 millimeters) long. Although fleas principally affect humans, dogs, cats, and rats, they also can be serious pests to poultry and other animals about the farm. In addition to their severe annoyance, fleas are carriers of serious diseases, the best known of which is the plague. Specific kinds of fleas inclue:

Human flea (*Pulex irritans*). Frequently bites around the legs and there may be 3 or 4 bites in a line. As contrasted with ticks and lice, fleas move from one host to the next. This flea also is found on hogs and commonly breeds in hog houses, as well as on dogs, cats, goats, domestic rats and some wild animals, such as skunks, coyotes, and badgers. This species is most often found in the Mississippi valley, in Texas, and westward to the Pacific Coast.

Dog flea (*Ctenocephalides canis*) and *cat flea* (*C. Felis*) are probably the most widespread of all fleas and will attack other animals and humans.

Northern rat flea (*Nosopsyllus* or *Ceratopsyllus fasciatus*, Bosc). This insect occurs in the northern United States and it is estimated that fleas of this species predominate on rats in that region (about 70%).

Sticktight flea (*Echidnophaga gallinacea*). This insect infests poultry and occasionally annoys humans and pets in the southern United States. Young chickens and other poultry are sometimes killed by heavy infestations.

Oriental rat flea (*Xenopsylla cheopis*). Although widely distributed throughout the United States, the insect is most abundant in southern California and the southern United States. In these regions, about half the fleas found on rats are of this species. The rat flea is capable of transmitting murine or endemic typhus from rats to humans. This disease, known as *Rickettsia typhi*, differs from Old World typhus. A very large percentage of the cases of this disease reported occur in the regions where the oriental rat flea is present. This species also serves as an intermediate for communicating dog tapeworm (*Dipylidium caninum*).

Chigoe flea (*Tunga* or *Dermatophilus penetrans*, Linne). A tiny, red-to-brown flea found in the West Indies, Mexico, and some of the southern United States. This insect, about $\frac{1}{25}$-inch (1 millimeter) in length, attacks humans, hogs, and domestic animals. Their habits and life cycle parallel those of other fleas, depending upon animal blood for their existence. They develop in the soil or in filthy debris. When abundant, this flea can be a severe pest to pigs and dairy animals. In humans, it tends to penetrate the body by way of the skin between toes and under fingernails, and the wound can be quite painful. The sore produced frequently becomes infected. The body of the flea must be removed with sterile instruments. When the female chigoe flea is full of eggs, she may be as large as a small pea. The chigoe flea must not be confused with the chigger, another small and annoying pest. See **Chigger.**

Control Measures. Malathion, methoxychlor, rotenone, or pyrethrum (pyrethrins) are used for killing fleas on dogs and cats. For destroying fleas in dairy barns and poultry houses, malathion is often used.

FLEA-BEETLE (*Insecta, Coleoptra*). These insects are found on several crops and tend to be specialists. The *apple flea-beetle* (*Graptodera foliacea*) is tiny, about $\frac{1}{5}$-inch (5 millimeters) in length or less, of a brassy-green color, and is a leaf feeder. The species *Halticini* attacks beet, cabbage, horseradish, potato, and tomato. This beetle is small and of the jumping type. It attacks the leaves of the plant, producing large numbers of holes and ultimately destroying the leaves. The *grapevine flea-beetle* (*Graptodera chalybea*) is about $\frac{1}{4}$-inch (6 millimeters) in length, is of a blue metallic color, and feeds on buds and tender shoots of the vine in the spring. Treatment of these beetles is usually the application of chemicals to kill the grubs. Mechanical means of shaking the vine can also be effective, particularly for comparatively small plantings.

FLEA BITE. Dermatitis and Dermatosis.

FLETCHER-MUNSON CURVES. Loudness Level; Musical Sound.

FLEXION REFLEXES. Brain and Nervous System.

FLEXURE. Flexure is a term which is used to denote the curved or bent state of a loaded beam. A horizontally located beam, transversely loaded with vertically directed load, offers an example of load-carrying ability derived through flexure. In flexure, an elastic structural material undergoes a deflection sufficient to set up in its material stresses which will support the load. Deflection under load is an essential and necessary part of the process of load carrying by a beam, for until the deflection has occurred, there are set up in the beam no resisting forces. Thus if an unloaded beam is perfectly straight and horizontal, it must assume a slightly curved position if any external load is supported by it. The only way in which a loaded beam could be straight would be to have had an initial deflection in a direction opposite to the loading.

The so-called ordinary, or common, flexure theory establishes a relation between the fiber stresses at any point in a beam and the bending moment causing these stresses. This theory is based on two fundamental assumptions. The first assumption is that a cross-section which was a plane before bending remains a plane after bending. This implies that the unit deformations are proportional to the distance from the neutral axis. The second assumption is that the fiber stresses are proportional to the deformations resulting from these stresses. If a tension member is subjected to an axial load in a testing machine it will be found that, for stresses below the proportional limit, the ratio of the unit stress to the unit deformation is a constant called the modulus of elasticity. This would also be true if the test specimen were a short compression member. In order to reconcile the second assumption it must be further assumed that the fibers act similar to test specimens and that the modulus of elasticity is the same for tension and compression.

It will now be shown how deflection and load-carrying ability are interrelated in a beam (see figure). First, it will be assumed that the

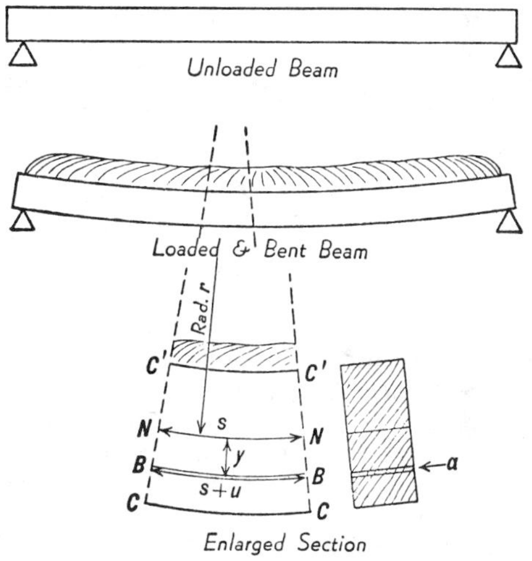

Beam unloaded and in flexure.

structural material is elastic, that is, within the elastic limit the stress is proportional to the strain inducing it, and that it is a homogeneous material. The results produced by materials not exactly meeting these specifications are usually in good accord with the theory based on these assumptions.

Assume that there is a beam of rectangular cross-section mounted horizontally between simple supports. If one were further to assume this material is weightless, the axis of the beam would be absolutely horizontal. Next a gravity load is placed on the beam, resulting in a certain deflection which sets up resisting couples within the beam, enabling it to carry the load. It must be evident that after a static condition is reached, the external bending moment thus imposed on the beam must, at any point, be balanced by an internal moment arising out of the stresses in the material of the beam. Next consider the enlarged section of the bent beam. If the section is taken sufficiently small, it can be assumed to be bent in the arc of a circle whose radius is r. The upper fibers, i.e., $C'C'$, are naturally compressed or shortened in length, and the lower, CC, are stretched. At some intermediate plane, NN, there must exist an unstretched fiber whose length is the same as it possessed in the unloaded state. This axis of no strain is called the neutral axis. If its length is s, then the length of a typical fiber such as BB, located at a distance y from the neutral axis, is $s + u$, in which u represents the stretch, u/s is the percentage stretch, or strain, of the material. From the geometry of the figure it is apparent that the strain $u/s = y/r$. Since stress is proportional to strain, the factor of proportionality being the modulus of elasticity E, it follows that the stress on $BB = f = Ey/r$. Referring to the cross section of this small element of the beam, the end area of fiber BB is taken as a. The stress f acting on this area produces an elementary internal force of af. Above the neutral axis there are similarly produced forces, but oppositely directed. The sum of all these longitudinal forces is, of course, zero, since the beam is static; however, at any cross section they produce, *in toto*, a moment around the neutral axis which is exactly equal to the external bending moment at that section. For example, the moment of the force acting on the fiber BB is afy about the neutral axis. The total moment, then, is the $\Sigma\ afy$ about the neutral axis.

Substituting Ey/r for f the total moment equals

$$\frac{E}{r}\int ay^2 = \frac{E}{r} I$$

The last step shows how the moment of inertia enters into the flexure formula. Since r is not a convenient quantity to work with, a substitution of f/y is made for E/r, resulting in the common flexure formula:

$$f = \frac{My}{I}$$

In this formula, M is the bending moment at a section where the moment of inertia is I, and f is the unit stress at y distance from the neutral axis.

It is readily shown that the neutral axis is coincident with the centroid of the cross section of the beam. From the above, we extract the following equation:

$$af = \frac{E}{r} ay$$

$$\int af = \frac{E}{r}\int ay$$

$\int af$ is the total force within the beam parallel to the neutral axis, and is zero, as explained above, but this results also in $\int ay$ being equal to zero, which can be true only when the distance y is a moment arm around the center of gravity of the cross-sectional area.

The flexure formula is valid as long as the stresses are within the proportional limit and if the neutral axis is a principal axis. For oblique loading, a separate stress is found with respect to each of the two principal axes. These stresses are then combined. In the derivation of this formula it is assumed that the horizontal stresses are the only internal forces which resist the external bending. As a matter of fact, the true maximum tensile or compressive unit stress, called a principal stress, is the resultant of the bending and the shearing stress acting at the point. But, as has been previously stated, the stresses which are obtained by the flexure formula are reasonably correct for ordinary design purposes.

FLEXURE (Section Modulus). Section Modulus.

FLICKER. Woodpeckers and Toucans.

FLICKER SENSITIVITY (Eye). Ferry-Porter Law.

FLIGHT. Aerodynamics; Airplane; Balloon; Dirigibles and Airships; Helicopters and V/STOL Craft; Supersonic Aerodynamics.

FLIGHT (Birds). Birds.

FLIGHT DATA RECORDER. Regulatory agencies of many nations require airline aircraft to be equipped with a flight data recorder and a cockpit voice recorder. These devices record information that may be vital to the analysis of unusual and unexpected flight circumstances. Hence, these recorders must be protected to withstand severe environmental conditions, as may be encountered in accident survival. The flight data recorder shown in the accompanying illustration is contained within a protective enclosure of heavy steel with successive layers of insulating and ablative materials. The housing is designed to exceed impacts of 100 g's and a fire of 2,000°F (1093°C) temperature for 30 minutes. The tape portion is designed to pass a 48-hour seawater immersion test. The housing also provides shielding against accidental erasure through short circuits or other random external electric current, producing a magnetic field.

Flight data recorder.

Variables recorded include altitude from −1,000 feet to +50,000 feet (−305 meters to +15,420 meters); air speed; heading recording, operated from the aircraft compass system through a synchro repeater recording to an accuracy of ±2°, with a system electrically independent of the recorder, i.e., powered from the same source as the compass to eliminate errors; vertical acceleration, with signal obtained from remotely mounted vertical accelerometers with recording accuracy of ±0.2 g; and a timing system which records at one-minute intervals on the edge of the recording medium. The accuracy of the timing system is not dependent on the tape speed or primary power frequency. Additional features include a channel to record aircraft pitch from the vertical gyro and circuitry to record marker beacon signals.

The recording medium used is high-tensile-strength stainless metal foil upon which the record is made by engraving a line approximately .001 inch (0.025 millimeter) in width. The foil can withstand severe accident environments without need for thermal, mechanical, or magnetic protection. The method of engraving prevents the record from appearing on the other side of the foil so that simultaneous recording on both sides of the foil may be made; or the tape may be turned over and used. The tape is contained in an easily replaceable magazine.

A separate cockpit voice recorder is of similar overall construction and records all voice communications between crew members on the flight deck; all voice communications transmitted or received by radio; and all voice communications on the aircraft intercom system. The recorder will preserve the last 30 minutes of the crew's voice communications to aid in accident investigation. The record amplifier, bias oscillator/mixer and monitor are packaged in subassemblies which plug into the main chassis. Assembly circuit boards are mounted on rugged structure, with point-to-point wiring (in contrast to printed wiring) used throughout. The magnetic tape used is stored in a quickly removable magazine which fits within the protective enclosure and requires no threading or adjustment when installed. The tape is stored within the magazine as a continuous loop; no spools or reels are used. An associated microphone monitor system is mounted overhead in the cockpit to pick up crew members' voices. Circuitry is provided with the recorder to erase tape by operating a cockpit control upon termination of an uneventful flight.

FLIGHT ENVELOPE. Also called a *V-n* or a *V-g* diagram, indicates the relationship of limit load factors with corresponding aircraft speeds for ready reference and use by the aircraft stress analyst. (See figure.)

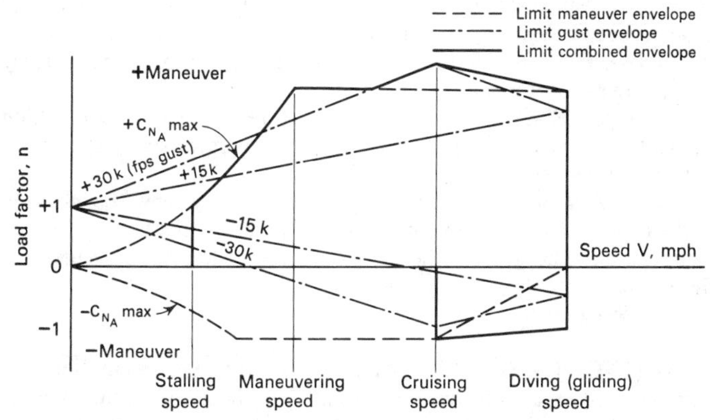

Typical *V-n* diagram (flight envelope).

The envelope represents in empirical form, data similar to those that might be obtained from accelerometer records made during definitive flight tests.

FLIGHT PATH. The flight path of any aircraft is the path in space described by the center of gravity of the aircraft.

FLINT. Flint is a rock composed essentially of a crypto-crystalline form of silica. It is very dense and tough, breaking with a conchoidal fracture; colors, usually dark grays, blues, or browns, often black. It occurs chiefly as nodules and masses in chalks and limestones. Flint is particularly interesting because it was used by primitive man for making instruments (artifacts) for thousands of years before he learned to use bone and metal. Flint still remained an essential mineral resource for making fire, including the flint locks on guns, until the close of the eighteenth century. From the dawn of civilization the best flint has come from Belgium and the coastal chalks of the English Channel and the Paris Basin. See also **Chert.**

FLINTY CRUSH-ROCK. A term proposed by the Scottish geologist Clough for the almost structureless, flinty portion of mylonite, an extreme product of dynamic metamorphism.

FLIP-FLOP. A bistable device, the output of which assumes one of two stable states, depending upon the state of the most recently applied input signal. Also known as a toggle. The flip-flop in computer systems is used for storing information. A direct-coupled flip-flop or latch, constructed of AND and OR logical elements, is shown in Fig. 1. The output is fed back to the input. A set pulse initially turns on the output, and the output then provides its own input even though the set pulse is removed. The circuit remains "latched" in this condition until a reset pulse breaks the feedback loop.

A gated transistor flip-flop is shown in Fig. 2. When the flip-flop is in the "off" condition, transistor T_4 is "off" and the "off" output F is at +V. This represents a binary 1 state. Transistor T_1 is "on" and "on" output E is at 0 V. This represents a binary 0 state. Transistor T_3 is in a state of low conduction, and its emitter voltage is not sufficient to forward-bias T_4. Transistor T_2, however, is in a state of heavy conduction and its emitter voltage forward-biases T_1 and thus holds it "on." To change the output state of the flip-flop, a conditioning gate of 0 V is applied to input C while the signal on input D is still positive. Thus, when the set pulse at input D goes to 0 V, a negative shift appears at the emitter of T_3. T_3 instantly goes into heavy conduction. The reduced collector voltage causes T_2 to go into a state of low conduction. The voltage at the emitter of T_2 is reduced and biases T_1 "off." When the ac transient has receded, the T_3 emitter voltage

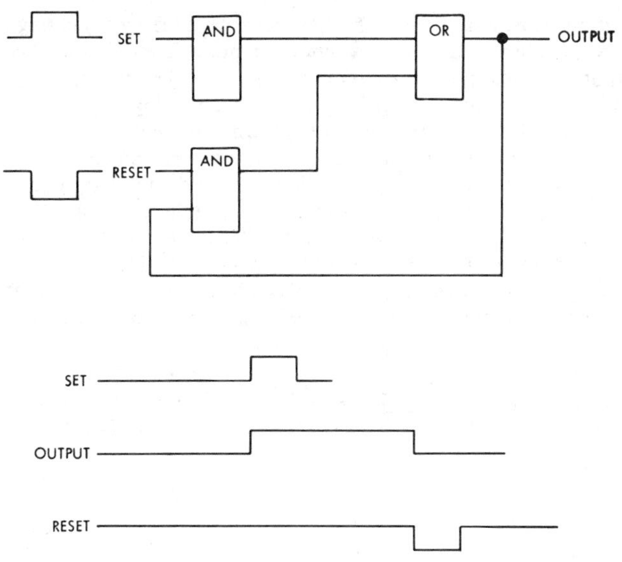

Fig. 1. Direct-coupled flip-flop or latch.

holds T_4 "on." The flip-flop can be turned "off" by applying gate and pulse inputs to points A and B, respectively.

A flip-flop of this type may be connected for binary operation by connecting A to F and C to E, and by applying the input pulse to both B and D. Since the gate must be applied before the pulse arrives, the output alternately changes state whenever a pulse is applied. The circuit may be used to build a shift register by connecting the "off" output of the previous stage to gate C, connecting the "on" output of the previous stage to gate A, and applying a pulse to inputs B and D.

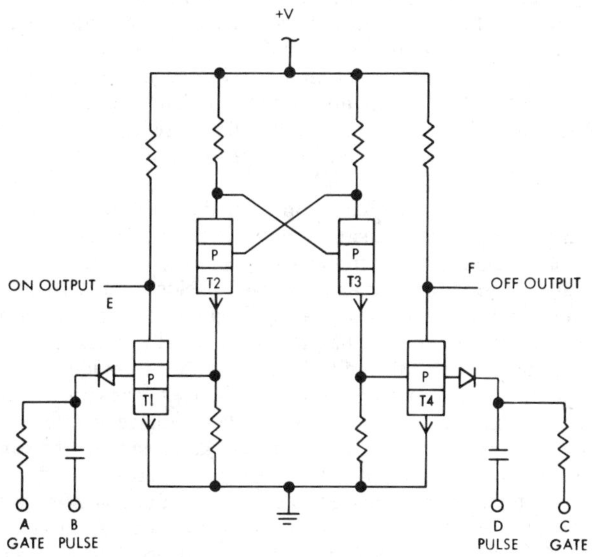

Fig. 2. Gated direct-coupled flip-flop.

See also terms listed under **Data Processing.**

Thomas J. Harrison, International Business Machines Corporation, Boca Raton, Florida.

FLIP-FLOP (Fluidic). Fluidics.

FLIPPER. An appendage of the aquatic mammals in which the digits are enveloped by continuous tissue so that the entire structure forms a flat paddle for swimming. Flippers occur in seals, manatees, whales, and related forms.

FLOATING AMPLIFIER. Also known as an isolated amplifier, the design does not require that the input and output signals be referred to the same signal reference point (ground). Generally, differential-input amplifiers meet this definition. However, the term *floating* normally excludes an amplifier where the input reference point and the output reference point are common, that is, either through a signal conductor or a power supply. Specifically, floating amplifier refers to an amplifier which includes a four-terminal coupling device, such as light-coupled signal-transmission element or a transformer.

The most common floating amplifier used in digital-data acquisition and instrumentation systems is the carrier amplifier. See also **Carrier Amplifier.** The input signal is coupled to the output amplifier, demodulated, and filtered to produce an output signal. As shown in the accompanying diagram, there may be feedback from the output to the input by means of a similar modulator-demodulator combination.

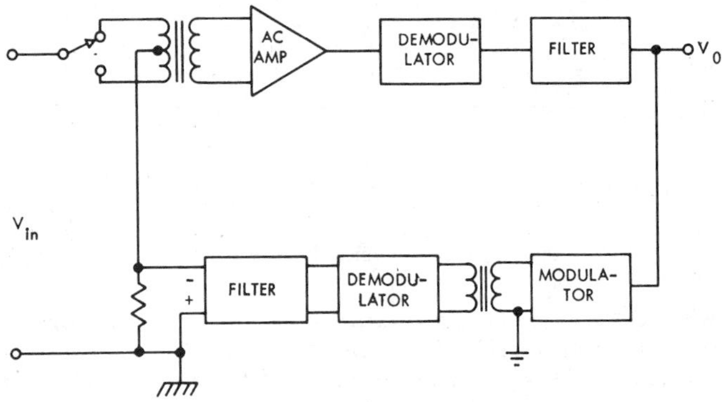

Floating-carrier amplifier with feedback.

The four-terminal isolation property of the transformer provides complete isolation of the amplifier input and output, in that the signal reference point (ground) for the input may be completely independent of the output-signal reference point. Thus, even though the preamplifier and postamplifier may be single-ended, the amplifier performs as a differential amplifier. The breakdown-voltage limitation of the transformer determines the maximum difference in the ground potential between the input and output circuits. Normally, this is quite high (hundreds of volts). Thus, this amplifier design is particularly well adapted to uses that require the amplification of signals in the presence of high common-mode voltages. Several commercial designs are available.

The achievement of a high common-mode rejection ratio is made difficult in the design of these amplifiers because of the unbalanced nature of the preamplifier and unbalances in the coupling transformer. Typically used for these designs are well-shielded transformers with low interwinding capacitances. Also, if optimum performance is to be achieved, the guard-shielding technique usually is needed.

As compared with other differential-amplifier techniques, the isolated carrier amplifier is relatively expensive. A primary advantage is the input/output isolation. A major disadvantage is a reasonably narrow band-width (normally less than 10 kHz), imposed by carrier-frequency limitations and by transformer characteristics. If the guard shield is not correctly connected to the source of common-mode voltage, the ac common-mode rejection ratio frequency is markedly degraded. In summary, where optimum performance is to be achieved, the design is restricted to three-wire systems.

See also terms listed under **Data Processing.**

Thomas J. Harrison, International Business Machines Corporation, Boca Raton, Florida.

FLOATING CONTROLLER. An automatic controller in which the control action provides a predetermined relation between the deviation and the rate of travel of the final controlling element. The final controlling element moves relatively slow toward either one or the other of its two extreme positions, depending on whether the controlled variable is above or below the setpoint. The output of the controller can remain at any value in its operating range when the actuating error signal is zero and constant. Hence the output is said to float. Types of floating control action include:

Single Speed Floating Control. The final controlling element moves at a single rate regardless of the amount of deviation. A single-contact on-off controller can produce single speed floating action if used with a slow-running reversible electric motor driven valve. When the temperature, for example, is at or above the setpoint, the valve will run toward its closed position at a single speed. When the temperature is below the contact setting, the valve will run toward its open position at the same single speed.

Usually a *neutral zone* is used in connection with single speed floating control. When the value of the controlled variable is in the neutral zone between contact setting, no contact is made and the valve remains motionless. It is possible for this form of control to produce a nearly exact correlation for any load condition.

Multispeed Floating Control. In this system, the final controlling element is moved at two or more rates, each rate corresponding to a definite range of values of deviation.

Proportional Speed Floating Control. In this system, the position of the final controlling element is changed at a rate that is proportional to deviation. The greater the deviation, the faster the movement of the valve.

See also **Control Action.**

FLOATING-POINT ARITHMETIC. A method of calculation that automatically accounts for the location of the radix point. This is usually accomplished by handling the number as a signed mantissa times the radix raised to an integral exponent; e.g., the decimal number $+88.3$ might be written as $+.883 \times 10^2$; the binary number $-.0001$ as $-.11 \times 2^{-2}$. See **Fixed-Point Arithmetic.**

FLOATING-POINT SYSTEM. Point.

FLOAT (Swallow). Swallow Float.

FLOCCULATION. Water Pollution.

FLOCCULI. Sun (The).

FLOCCUS. Clouds and Cloud Formation.

FLOODING (Irrigation). Irrigation.

FLOODLIGHTING. Illumination.

FLOOD PLAIN. The relatively level surface, or surfaces, within a river valley, caused by the depositional work of the river, especially during flood or high-water.

FLOOD PROJECTION. A method of facsimile pick-up which utilizes a completely illuminated subject. Scanning is achieved by an aperture which moves between the subject and light-sensitive, pick-up device.

FLOOD PROTECTION. Hydroelectric Power.

FLOOD RECONAISSANCE. Earth Resources Satellites and Geologic Remote Sensors.

FLOODS (Major). Earth.

FLOOR BEAM. A beam which is the direct support of the floor load of a building and transfers this load to the adjacent girders or columns is a floor beam or floor joist. These beams are commonly of steel, reinforced concrete, or wood. However, prestressed, precast concrete is gaining acceptance.

The term floor beam is also used to designate the transverse beams of the floor system of a bridge which transmit load to the longitudinal girders or trusses.

FLOPPY DISK. Diskette.

FLOPPY VALVE SYNDROME. Heart and Circulatory System (Human).

FLOQUET THEOREM. Mathieu Equation.

FLORIDA CURRENT. All the northward moving water from the Straits of Florida to a point off Cape Hatteras where the current ceases to follow the continental slope. It is one of the swiftest of ocean currents (flowing at a rate of 2 to 5 knots).

FLOTATION. Classifying (Process).

FLOTATION (Coal). Coal.

FLOUNDER. Flatfishes.

FLOUR (Wheat). Wheat.

FLOWAGE (Rock). This term, as used by geologists, signifies the internal movement of clays and rocks when stressed beyond the elastic limit. Important types of flowage are: granulation and foliation, although the latter is primarily a physical-chemical rather than a purely mechanical process, as in the other types of flowage.

FLOW CHART. A graphic representation of the major steps of work in process. The illustrative symbols may represent documents, machines, or actions taken during the process. The area of concentration is on where or who does what rather than how it is to be done.

FLOW CHART (Computer Programming). Programming Flow Chart (Computer).

FLOW (Coefficient of Discharge). Discharge (Coefficient of).

FLOW CONTROL (Bulk Solids). Feeder (Vibratory); Feeder (Volumetric); Weighing.

FLOW DIAGRAM. A graphical representation of a sequence of operations. In machine computation, a diagram with labeled boxes, arrows, etc., showing the logical pattern of a problem, but not ordinarily including machine commands.

FLOWER. That part of a plant which is involved in the sexual reproductive process of angiosperms. The formation of the flower is preliminary to the production of fruit with its seeds.

Flowers may be borne at the tip of a stem or a branch, in which case they are said to be terminal flowers. Or they may be borne in the axils of leaf primordia and called axillary. In very many plants the structure which subtends the flower does not develop into a leaf like the other leaves of the plant. Instead, it may remain very small and inconspicuous, or it may grow larger, but be of shape quite unlike that of an ordinary leaf. These structures which subtend flowers are called bracts. In a few plants the bracts are large and brilliantly colored, so that at times they are much more showy than are the flowers. The scarlet bracts of the *Poinsettia* and the white or pink bracts of the flowering dogwood are examples. See Fig. 1. In many monocotyledons, as Palms and Arums, there is a single large bract which subtends and often more or less surrounds the flowers. Bracts of this kind are called spathes. The striped "pulpit" of the Jack-in-the-Pulpit and the white bract of the calla lily are well-known examples. The bract may surround and protect the flower in the bud.

Flowers may be borne singly or they may be associated in a cluster which is known as an inflorescence. See Fig. 2. A single flower is called a solitary flower, and the stalk which supports it is a peduncle. The stem which supports an inflorescence is also a peduncle, while the individual flowers of the inflorescence are supported on pedicels. When there is a distinct axis extending through an inflorescence it is called a rachis. The arrangement of the flowers of an inflorescence varies in different groups. A common, and primitive, form of inflorescence is the raceme, in which the floral shoot grows at the apex and bears many pedicels, each ending in a single flower. The first flowers to open are those at the base of the raceme. If, in an inflorescence of this sort, each branch is a raceme bearing several flowers, it is called a panicle, or a compound raceme. If the flowers of the raceme are borne directly on the main axis, the inflorescence becomes a spike. A secondary spike, common in grasses, is a spikelet. A catkin is a spike which droops. A corymb is a modified raceme in which the lower pedicels of the inflorescence grow faster than the upper ones, so forming a more or less flat-topped cluster. An umbel differs from a corymb in that it has no central rachis, all the pedicels of the inflorescence rising from a common point. More commonly umbels are compound, each of the main stalks of the inflorescence bearing an umbel at its tip. The inflorescence of the onion is an umbel, that of wild carrot a compound umbel. The inflorescence of the composite family is a head, which may be considered as an umbel in which the flowers are all sessile, without pedicels, on the apex of the stem. A cyme is an entirely different type of inflorescence. In the cyme the first flower to open is at the tip of the cluster. Below it on the stem are a number

Fig. 1. Blossom of the flowering dogwood tree.

Fig. 2. The fragrant inflorescence of *Viburnum carlesi*, a flowering shrub, is sometimes called the pink snowball. (*Roche*)

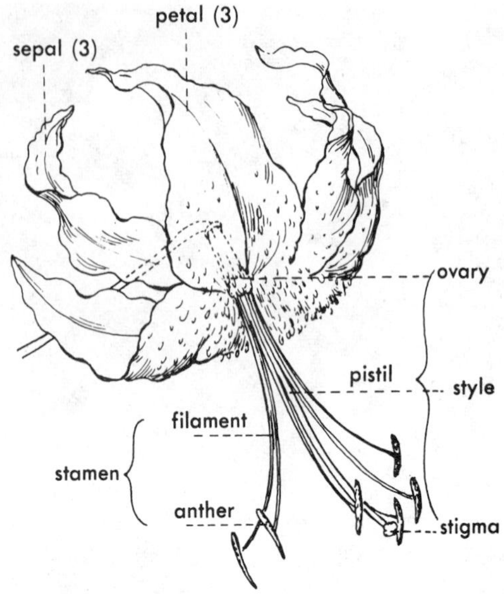

Fig. 3. Reproductive parts of a lily flower.

of bracts. From the axils of these bracts branches develop and also end in a flower. This successive branching may be many times repeated, but always the flower terminates the stem and opens. Combinations of these types of inflorescences are found in many plants. A spadix is an inflorescence of the spike form with elongated axis, sessile flowers, and enveloping leaf, the spathe.

A flower consists of an axis, called the receptacle or torus, and, attached thereto, the pistils, the stamens, the petals, and the sepals. See Figs. 3 and 4. The pistils and stamens are the essential organs of the flower, the petals and sepals are accessory organs. Any flower which has all four organs is a complete flower, while that which lacks one or more is incomplete. If the organs missing be either stamens or pistils, the flower is imperfect or unisexual. A perfect or bisexual flower has both sets of essential organs. If only the stamens are present the flower is staminate; if only pistils, it is pistillate. If the two kinds of flowers, staminate and pistillate, are found on the same plant, that species of plant is monoecious. When the two unisexual flowers are found on different plants, the species is dioecious. Infrequently flowers are borne which lack both stamens and pistils, and are sterile. When the flower lacks sepals and petals, it is a naked flower.

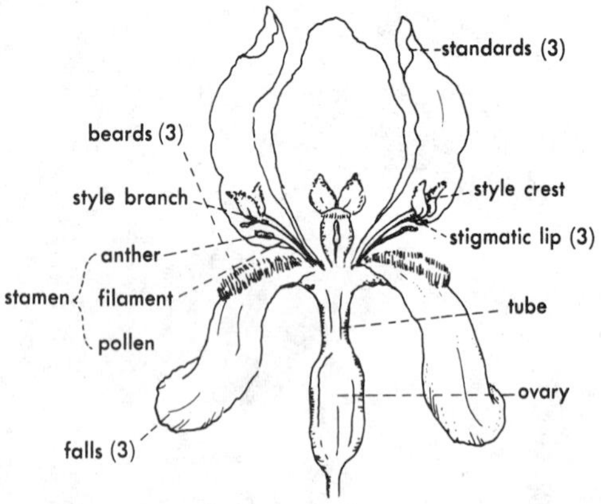

Fig. 4. Reproductive parts of an iris flower.

Flowers may also be distinguished as regular and irregular. Regular flowers are those in which all the members of any set of organs are alike, forming a flower which is radially symmetrical or actinomorphic. Often the organs of one or more sets are not alike, forming an irregular flower. An irregular flower may be bilaterally symmetrical or zygomorphic, one half being a mirror image of the other.

The accessory floral organs, the sepals and petals together, constitute the perianth. In the complete flower, the perianth is composed of a whorl of sepals which is called the calyx and inside the calyx a whorl of petals called the corolla. The sepals are usually green and small. The function of the calyx seems to be to protect the other parts of the flower before the flower bud opens. The petals are usually thin and bright colored or white. The term corolla is applied to all the petals together. The corolla appears in a wide variety of shapes and colors. Its function seems to be to attract animals, especially insects, to the flower and so bring about pollination. The number of sepals and petals is constant for each species of plant. In the flowers of monocotyledons there are usually three of each, while in dicotyledons it varies from four to many, with five a very common number. As a further attraction to insects, many flowers possess special glands called nectaries which secrete a sweet fluid, nectar. Usually these nectaries are situated at the bases of the petals, though they may be found in many other places in the flower. Nectaries known as extrafloral nectaries are found on petioles of leaves or on the stipules.

The stamens, or microsporophylls, taken together consitute the androecium. Usually a stamen consists of two parts, a stalk or filament and an anther. The filament may be short and stout, or more commonly long and slender, raising the anthers well above the base of the flower. The anther when first formed is an undifferentiated mass of cells. As it develops, four groups of cells become set off from the surrounding cells. In these masses, which usually appear as linear strands, certain cells undergo meiosis and become microspores. The sac which contains them is therefore a microsporangium. A microspore develops into a pollen grain. The anther sac, or sporangium, when mature usually opens by two longitudinal slits, or by special pores, and frees the pollen grains. The number of stamens in a flower varies from one to many. Often there are vestigal stamens, or staminodia, present in the flower; in some plants these are large and brightly colored, in others they are small and inconspicuous.

The pistil is the central organ of the flower. See Fig. 5. A single

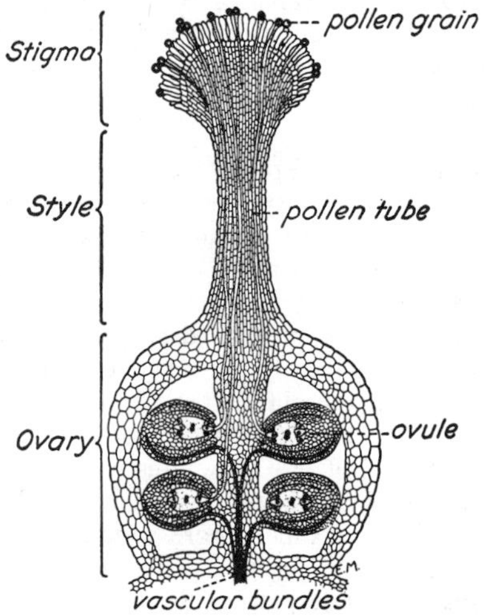

Fig. 5. Longitudinal section of flower pistil showing pollen tubes growing through the style and entering the ovules. Diagrammatic.

pistil or several pistils, which may be separate or partly, or even completely, united, is called a gynoecium. A pistil is composed of one or more modified leaves called carpels or megasporophylls. When there are two or more somewhat united carpels, the pistil is called compound. A pistil is composed of a basal ovary (ovulary), a terminal stigma, and usually an intermediate style, which is often long and slender. The stigma is a receptive organ, the surface of which is often either sticky or hairy. It is to the surface of the stigma that the pollen grains are carried when pollination takes place.

The style may be very much elongated to lift the stigma above the other parts of the flower and so increase the probability of pollination. The ovary has one or more cavities, or loculi. In these are located the ovules, which will become the seeds. The ovules, are attached to the wall of the ovary or to a central column by a small stalk called the funiculus, through which the developing ovule receives nourishment. That region of the ovary to which the ovules are attached

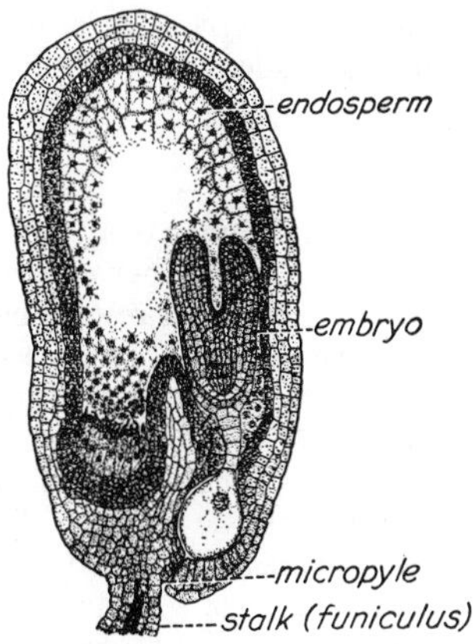

Fig. 6. Ovule of shepherd's purse, containing an embryo and endosperm. (*Chamberlain, "Elements of Plant Science," McGraw-Hill.*)

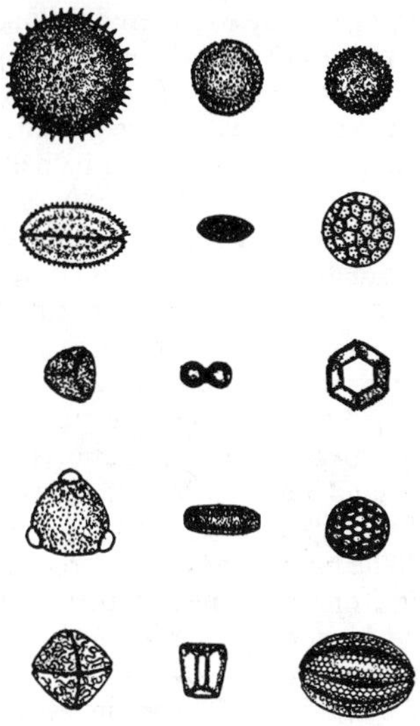

Fig. 8. Pollen grains of various kinds of plants. (*Pope, "Botanical Gazette."*)

is called the placenta. The number of ovules in a single loculus varies from one to many.

Each ovule first appears as a minute rounded projection on the wall of the ovary or the columella. See Fig. 6. In the early period of its development this projection consists of an undifferentiated mass of cells known as nucellar tissue. One or two layers of cells, known as the integuments, rise from the base of the projection and finally almost completely surround the nucellar tissue. A minute opening through the integuments, called the micropyle, is left connecting the cavity of the ovary to the surface of the nucellar tissue. Within the nucellar tissue a very important series of cell divisions has taken place. While there are many variations of the process as it occurs in different species, the process is essentially as follows. Within the mass of nucellar tissue a single cell has become differentiated from all the others by its larger size and denser cytoplasmic content. This is the megaspore mother cell. It divides twice in rapid succession to form a row of four cells. These two rapidly succeeding cell divisions take place in such a way that the four cells have the haploid or reduced number of chromosomes. The process is known as meiosis. Three of the four cells disintegrate and are lost. The fourth, or megaspore, is usually the one nearest the micropyle. It enlarges greatly, while by three successive divisions its nucleus divides to form eight nuclei, all contained within the wall of the very much enlarged female gametophyte, commonly called the embryo sac. The arrangement of these eight nuclei is quite uniform. In most plants there are four of them at each end of the embryo sac. See Fig. 7. One from each end moves to the center

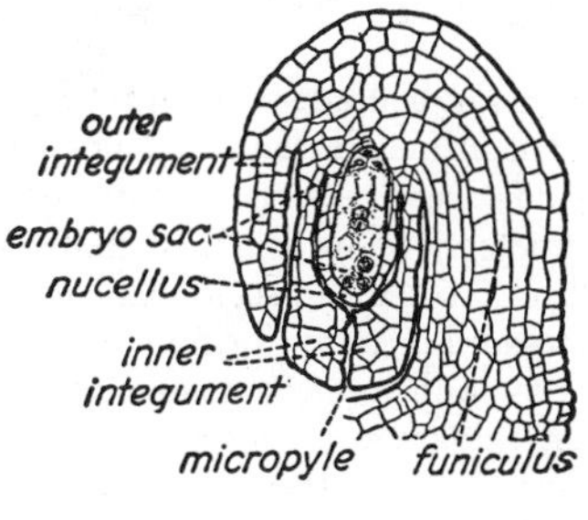

Fig. 7. Mature megagametophyte of lily within an ovule. (*"Textbook of General Biology and Botany," Holman and Robbins, John Wiley.*)

of the embryo sac, where they form an intimate association and eventually fuse. These two are the polar nuclei. Of the three nuclei which remain at the micropylar end of the embryo sac, one becomes larger than the other two. This is the egg nucleus or megagamete; the other two form cells which are called the synergids. The three nuclei at the opposite end of the embryo sac form cells called the antipodals. The mature embryo sac is thus a seven-celled body, with seven nuclei.

The pollen grain is carried to the stigma by various agents. The pollen grains of different species of plants are very characteristically shaped, and are often strikingly beautiful because of the many ridges or protuberances with which the outer wall is marked. See Fig. 8. At first a pollen grain contains a single nucleus. This nucleus divides before leaving the anther and gives rise to two nuclei, one of them called the tube nucleus and the other the generative nucleus. The pollen grain germinates when it reaches the stigma, putting out a slender pollen tube which grows down through the tissues of the style and into the ovary. In the pollen tube the generative nucleus divides to produce two sperm nuclei. There it grows towards an ovule, which it enters through the micropyle. The pollen tube continues to grow until its tip reaches the embryo sac. Into this the two sperm nuclei are discharged. One of them fuses with the egg nucleus of the embryo sac, while the other passes to the polar nuclei and fuses with them. The nucleus which is formed by the fusion of these three nuclei is called the primary endosperm nucleus; it contains three times the haploid chromosome complement. The act of fusion of the male nucleus with the egg nucleus is called fertilization. From the endosperm nucleus there is formed by repeated division and subsequent wall formation a mass of triploid tissue known as the endosperm, which surrounds the developing embryo. The endosperm nourishes the embryo during the early stages of its growth. In many plants such as the bean the endosperm is not formed at all or entirely absorbed by the developing embryo, while the seed is still immature; in others the endosperm forms a considerable part of the mature seed. Seeds with endosperm are known as albuminous seeds; those without endosperm as exalbuminous seeds.

The act of fertilization causes the immediate growth of the fertilized egg. In most plants a series of cell divisions takes place, forming a short filament of cells which is called the proembryo. The appearance of the proembryo varies in different plants. The terminal cell of the proembryo becomes by repeated cell divisions a spherical mass of cells which is the beginning of the true embryo. The embryo grows rapidly and becomes differentiated into three regions, a primitive root, or radicle, a primitive shoot, and cotyledons. This embryo is surrounded by the tissues of the nucellus and the integuments, which have grown larger to form seed coats coincident with the growth of the embryo. The mature ovule becomes the seed, and the ovary which contains it becomes the fruit.

See also **Compositae or Composite Family.**

FLOWING JUNCTION (Electrolytic). A method of setting up a junction between two electrolytes which has important advantages for use in electrochemical measurements, especially of electrode potential. In its simplest form, this arrangement consists of an upward current of the heavier electrolyte meeting a downward current of the other at a point where a horizontal outlet tube joins the vertical tube through which the liquids enter.

FLOW MEASUREMENT (Fluids). The need for instruments to measure the amount of liquids and gases flowing through closed conduits or open channels (as encountered in irrigation water systems, for example) is widespread. The technology of flow measurement is one of the oldest branches of instrumentation. The basic principles used in many modern flowmeters were first pioneered well over a century ago. In industry, flow measurements serve two primary purposes: (1) As a means for accounting fluid commodities that, unlike solid materials, cannot be counted piece by piece or conveniently weighed; and (2) as the basis for controlling processes and manufacturing operations. For establishing materials and energy balances, the quantities of water, steam, compressed air, and other relatively low-cost commodities as well as a vast variety of chemicals must be accounted for—as received, while in process, and as shipped. Liquids and gases, of course, can be handled in what might be termed "packaged form" as in the cases of tank cars, tankers, pressurized cylinders, and so on and quantities thus determined directly by gravimetric or volumetric measurements. But, during the processing, fluids are not handled in this manner, but flow from one point to another in pipes. Also, aside from overseas transfers, most fluids of commerce are transported from source to consumer by way of pipelines—crude petroleum, refined products, natural gas, water, and so on. Essentially, therefore, flow measurement is addressed to the problem of determining quantities and rates of flow of fluids as they pass through closed conduits of various forms.

The range of flow-measurement applications is extremely wide, varying from cryogenic temperatures up to as high as 3,000°F (1,650°C) and higher in some instances. Flowmeters are used to measure flows at pressures as low as 10^{-1} millimeters of mercury to as high as 15,000 psig (1,021 atmospheres). With advances in high-pressure technology, special flowmeters have been developed to operate at pressures in excess of 100,000 psig (6,804 atmospheres) and more. The nature of fluids measured varies from very clean, noncorrosive substances, such as pure water or air, to highly corrosive gases, such as chlorine, or liquids, such as sulfuric acid, and from light, nonviscous gases of dense, highly viscous materials, some of which will only flow at elevated temperatures. Some fluids, such as slurries, contain high concentrations of suspended materials that will settle out rapidly unless kept in a turbulent condition.

A major factor contributing to wide diversity in flowmeter design is that of the magnitude of the flow to be measured—varying from perhaps a few cubic centimeters per hour, as encountered in medical and other laboratory research, to several millions of gallons per day, as found in large water and sewage systems.

It is not usual, therefore, to find that there are literally scores of flowmeter designs, each of which is well adapted to specific use conditions, but worthless for others. There is no universal flowmeter. The foregoing complexities also make it difficult to classify flowmeters in a crisp, scientifically attractive manner. One reasonably satisfactory classification divides flowmeters into: (1) Discrete quantity types which, in essence, as they measure flow divide the flowing fluid into specific volumes and then count the number of these volumes that have passed during a measured time interval; and (2) inferential flowmeters which *indirectly* determine flow rate rather than actual flow quantity. With rate-type meters, the total quantity of fluid that has passed a particular point is established by integration which, of course, fully takes into consideration the varying rates of flow. Even in stabilized conditions, instantaneous flow rates will vary and where this occurs in large conduits, momentary ups and downs in flow rate, unless continuously inputted to the integration system, can cause reading errors in terms of many tons per day.

Classification of Flowmeters. From a practical standpoint, a convenient breakdown of flowmeter types (by operating principle) is as follows: (1) *Head flowmeters,* which operate by measurement of the pressure differential or head across a suitable restriction to flow, such as an orifice, placed in the pipeline; (2) *variable-area flowmeters,* which operate on a variation of the operating principle of the head meter; (3) *magnetic flowmeters,* the operating principle of which is based upon Faraday's law of electromagnetic induction; (4) *positive-displacement flowmeters,* which measure flow directly in quantity terms by subdividing the flowing fluid into discrete volumetric units; (5) *velocity-type flowmeters,* which convert a measurement of velocity of flow into terms of flow rate; and (6) *open-channel flowmeters,* such as weirs and flumes, which essentially translate a level measurement into terms of flow rate. Important to some uses, but much less utilized flowmeters include acoustic, thermal, and radioisotope tracer flowmeters.

In the case of most flowmeters, the fluid being measured comes into intimate contact with the metering device per se and, in fact, several of the most commonly used methods require the presence of an object, such as an orifice, within the pipe. Where the pipeline must be broken in some fashion, the meter is sometimes called "intrusive" or wettable. In relatively few designs, as in the instances of some types of acoustic or ultrasonic flowmeters, flow is sensed from outside the pipe, and such meters may be called non-intrusive or non-wettable.

Head Flowmeters

Head-type flow meters are probably the most common of all flowmeters. In this type of meter the rate of flow, in quantity per unit time, is determined from the pressure drop or head across a restriction in the flow path. Common examples of this type of meter include orifice plates, venturi tubes, and flow nozzles. See **Orifice.** This type of meter operates on the principle of energy conversion between static pressure and velocity. The velocity increase that results from a restriction in a line will have associated with it a decrease in static pressure. Thus, the flow rate may be determined from a measurement of the pressure drop or head, as it is commonly called. A sketch of a venturi meter together with a normal static pressure distribution, is shown in Fig. 1. A simple U-tube manometer is also shown to indicate one method

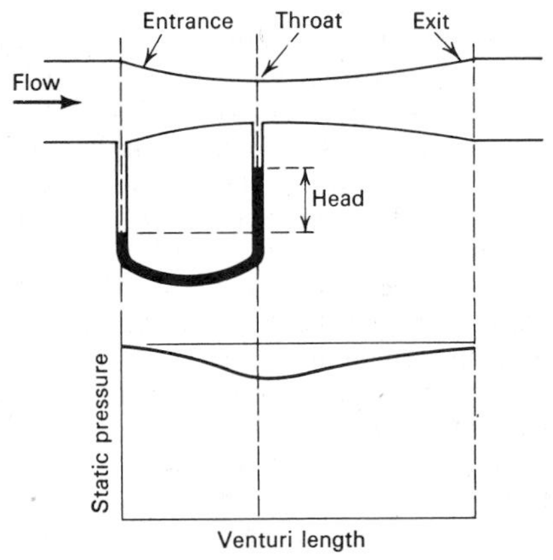

Fig. 1. Typical venturi meter and static pressure distribution.

for measuring the head or pressure differential. Recording and controlling equipment may be actuated by directed measurements of the pressure differential. Although theoretical considerations will yield results within a few percent of actual flow rates, it is common practice to calibrate meters or to use experimentally determined correction factors. This is particularly true in the case of gas flow where the energy of the compressed fluid must be considered. References to detailed studies on this subject may be found in most standard works on hydraulics and fluid flow.

Orifice plates available for installation in conduit range from $\frac{1}{2}$-inch to 72 inches (1.3 centimeter to 183 centimeters) and will normally have a range from 0 to 10 inches (25.4 centimeters) of water up to 1 to 400 inches (2.5 to 1016 centimeters) of water. In commercial meters, accuracy ranges from ±0.25 to 0.5% of fullscale. Advantages of orifice plates as the differential-pressure creators for flow measurement include: (1) Low cost; (2) available in numerous materials; (3) wide range of pipe sizes; (4) characteristics are well known and predictable with years of applicational experience; and (5) can be used with simple types of differential-pressure devices. Some of the limitations of orifice plates include: (1) Relatively high permanent pressure loss; (2) tendency to clog—not useful for slurries or entrained particles; (3) the flow range-ability for a given plate is limited to about 3:1; (4) square-root characteristics; (5) straightening vanes upstream often needed; (6) connecting piping problems encountered (freezing condensation, etc.) unless orifice is integral with a transmitter; (7) characteristics tend to change with time of use due to erosion, corrosion, and sealing; and (8) accuracy is dependent upon care during installation.

A *flow nozzle,* often used instead of an orifice plate, consists of a bell-shaped approach section of eliptical profile attached to a cylindrical throat tangent to the ellipse. The smoothness of the approach to tangency, the length of the cylindrical portion of the nozzle, and the locations of the pressure taps have an important bearing on the discharge coefficient. Flow nozzles are normally available in pipe sizes from 3 to 24 inches (7.6 to 61 centimeters) (diameter). Advantages claimed for flow nozzles include: (1) Under proper installation conditions, permanent pressure loss can be a bit lower than the orifice plate; (2) available in numerous materials; (3) can be used with simple types of differential-pressure devices; (4) can handle fluids containing solids that settle; and (5) widely acceptance performance for high-pressure/temperature steam flow. Limitations include: (1) Higher cost than orifice plate; and (2) limited to moderate pipe sizes, with a top of about 48-inch (122-centimeter) diameter in special cases.

The *venturi tube*, as shown in Fig. 1, creates a much smaller permanent pressure loss than the orifice plate and hence is attractive for numerous installations. Other advantages include: (1) Handles suspended solids; (2) available in very large pipe sizes—widely used for large flows, as in water and sewage plants; (3) available in numerous materials of construction; (4) characteristics well established with years of applicational experience; and (5) best accuracy for head flowmeter differential producers. Its limitations include high cost and unavailability generally for pipe diameters below 6 inches (15 centimeters).

The *pitot tube*, described in entry on **Pitot Tube**, also is used for making flow measurements in closed conduits, with a general range of pipe diameters from 3 to 48 inches. However, the device is of relatively minor importance in commercial flow measurement. It is an effective tool for laboratory use or for spot checks, but its tendency to plug when the flowing fluid contains small amounts of solid matter, its velocity-range limitations, and its sensitivity to abnormal velocity-distribution effects limit its commercial usefulness.

Variable-Area Flowmeters

In general, area-type flowmeters, operate on the same principle as the head type meter. The basic difference is determined on the basis of what variable is maintained constant. In the head-type meter the area of the flow restriction is constant and flow rate is determined from variations in pressure differential across the restriction. In area-type meters the area of the restriction is varied to give a relatively constant pressure differential. The flow rate is thus a function of the area measurement. Common-area meters include rotameters and piston-type meters. A sketch of a rotameter is shown in Fig. 2. In this

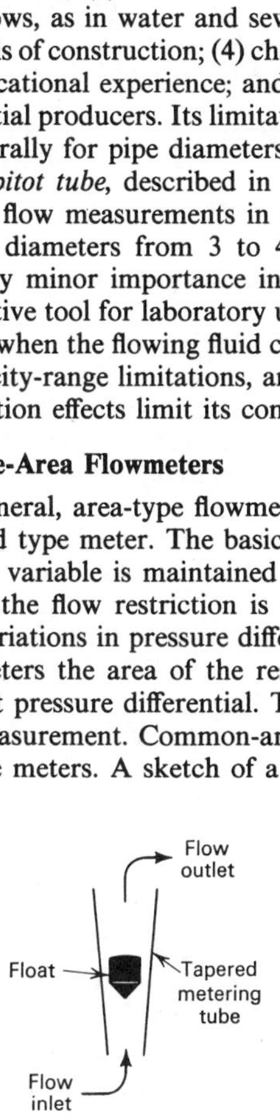

Fig. 2. Principle of rotameter.

meter, the float weight is balanced against the upward force of the fluid. Changes in flow rate result in upward force changes, with the float seeking an equilibrium position in the tapered metering tube. Although rotameters are very often made of glass for visual indication of flow rate, devices are available to sense float position for recording and controlling.

In addition to the rotameter, there are two other major designs of variable-area meters: (1) *Orifice and tapered plug meters.* A meter in this category is equipped with a fixed orifice mounted in an upright chamber. The float has a tapered body with the small end at the bottom and is allowed to move vertically through the orifice. The fluid flow causes the float to seek the equilibrium position. (2) *Piston-type meters.* In these meters, the piston is accurately fitted inside a sleeve and is lifted by fluid pressure until sufficient port area in the sleeve is uncovered to permit the passage of flow. The flow is indicated by the position of the piston.

Tapered-tube rotameters are generally available for pipe sizes ranging from $\frac{1}{2}$ inch to 4 inches (1.3 centimeters to 10 centimeters) in diameter, although they can be obtained as well in considerably smaller sizes essentially for laboratory applications. Some of the advantages of rotameters include: (1) Easily equipped with magnetic, electronic, induction, or mercury-switch alarms, or transmitters; (2) viscosity-immune bobs are available; (3) several rotameters mounted side-by-side provide convenient flow comparisons; (4) relatively low cost; (5) handle a wide variety of corrosives; (6) can be installed without regard to fittings or straight pipe preceding or following the meter; and (7) exceptionally effective for small flows. Limitations include: (1) Must be mounted vertically; (2) limited to relatively small pipe sizes and capacities unless mounted in a by-pass; and (3) relatively low temperature limits.

In 1979, a rotameter of an interesting new format was announced. This instrument determines flow rate by electronically observing the effect of flow on the frequency of a magnetic ball that oscillates vertically in a glass tube. A switch at the top of the vertical flow tube energizes an electromagnet which causes the ball to snap upward through the flowing medium. When the ball again descends to its lower position, it interrupts a light beam, this causing another pulse to be transmitted to the electromagnet, thus repeating the cycle. When there is zero flow, the ball oscillates vertically at 20 Hz. When the flow increases, the ball settles more slowly, requiring longer travel time to reach the light beam. At maximum flow conditions, the oscillating frequency is reduced to about 3 Hz. Accuracy and repeatability of $\pm 2\%$ have been reported. For adaptation to horizontal operation, the meter is equipped with electromagnets and light sources at either end of the flow tube. As reported by Bailey (1980), "Inside the tube are 2 balls, one magnetic, the other nonmagnetic. The magnetic ball functions to rapidly return the nonmagnetic ball to an upstream point after one measuring cycle is completed. The nonmagnetic ball 'floats' in the fluid stream, interrupting first one light beam, then the other as a stream motion tracing element."

Magnetic Flowmeter

Originally, the magnetic flowmeter was developed to measure the volume flow rate of electrically-conductive fluids. The meter is particularly effective in connection with fluids which present difficult handling problems, such as corrosive acids, rayon viscose, sewage, rock-and-acid slurries, sand-and-water slurries, paper pulp stock, rosin size, detergents, tomato pulp, and beer. The meter is mainly applicable to liquids which have a conductivity of one micromho per centimeter or greater. The accuracy is $\pm 5\%$ fullscale for systems; the flow range is very wide—from 0.01 to 150,000 gallons (0.04 liter to 5,678 hectoliters) per hour; the size range is from 0.1 to 96 inches (2.5 millimeters to 2.44 meters) (diameter); and temperature range maximum is approximately 350°F ($\sim$ 180°C). The principal advantages of the magnetic flowmeter include: (1) No obstruction in the pipe; (2) pressure drop equals that from the same length of straight pipe; (3) flow can be measured in either direction; (4) use for particularly difficult fluids; and (5) requires no metering runs or straightening vanes.

A magnetic flowmeter transmitter is shown in Fig. 3. Operation

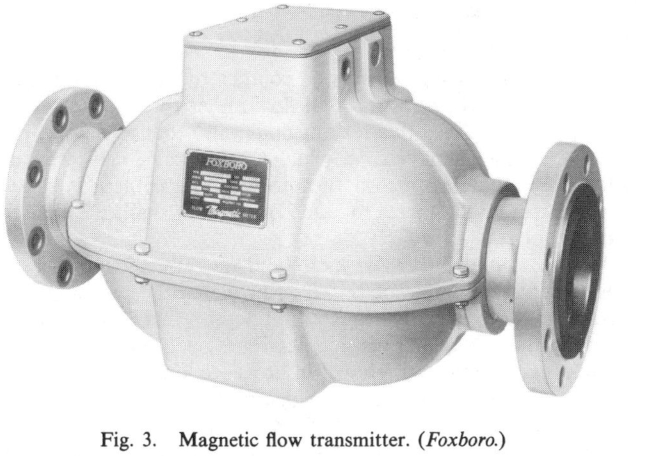

Fig. 3. Magnetic flow transmitter. (*Foxboro.*)

of the meter is based on Faraday's law of electromagnetic induction which states that the voltage E induced in a conductor of length d moving through a magnetic field h is proportional to the velocity v of the conductor,

$$E = Chdv$$

where C is a dimensional constant.

The voltage is generated in a direction that is mutally perpendicular to both the velocity of the conductor and the magnetic field. A graphic

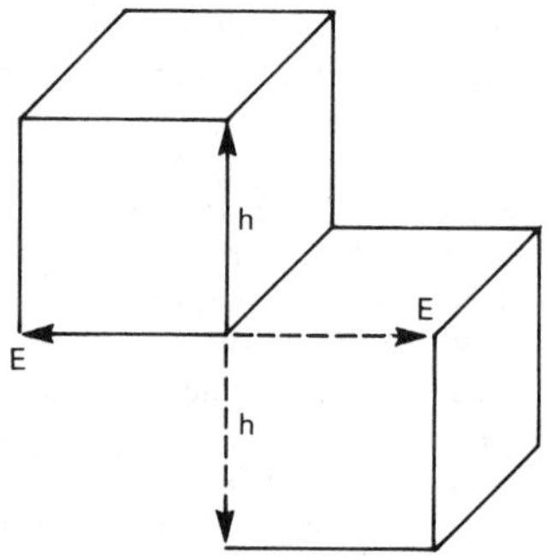

Fig. 4. Graphic representation of formula applicable to a magnetic flowmeter.

representation of the foregoing formula is given in Fig. 4. The magnetic field *h* alternates at power line frequency. Hence, the generated voltage *E* alternates at the same frequency. Special magnetic flowmeters are available which use dc fields. The latter meters are used to measure highly conductive liquid metals, such as sodium, where performance of the ac design is not satisfactory. A cross section of the magnetic flowmeter is shown in Fig. 5. The fluid to be measured is the moving

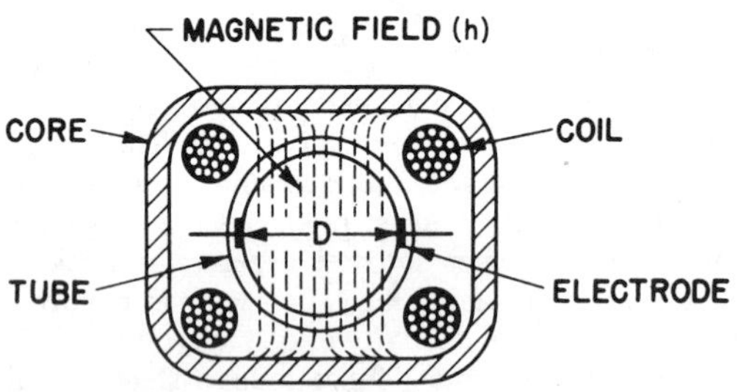

Fig. 5. Cross section of magnetic flowmeter transmitter.

conductor. It passes through the tube and the alternating magnetic field *h*. This action generates a voltage that is proportional to the average velocity of the fluid, assuming a fixed value for *h* and for the diameter of the tube *d*. In actuality, the velocity of the fluid is not usually the same at all points across the tube, but has varying flow profiles—from laminar to turbulent—for different flow conditions. The flowmeter generates a voltage that is proportional to the average fluid velocity of the flow profile at the plane of the electrodes.

Positive-Displacement Flowmeters

Positive-displacement meters, in contrast to variable-head and area meters, are used to measure total flow rather than flow rate. In this type of meter, a fixed quantity of fluid will produce a given motion or rotation to a piston or vane, which through some registering device will then indicate or record the quantity. Since there is a positive displacement of the meter for each quantity of fluid, the time factor is not involved.

Nutating-Disk Meter. The household water meter is probably the most common example of this type of meter. The accuracy of this type of meter over a wide range of flow rates makes it ideally suited to applications involving varying flow rates over long periods of time. A nutating-disk meter is shown in cross section in Fig. 6. Although there are design differences from one manufacturer to the next, the basic operation of the meter is as shown in Fig. 7. Each cycle (complete movement) of the measuring disk displaces a fixed volume of liquid. The only moving part in the measuring chamber is the disk. The liquid enters through the inlet port and fills the spaces above and below the disk, which fits closely and precisely in the measuring chamber. As liquid moves into the chamber, the disk is moved in a nutating path. The motion of the disk is controlled by a cam which keeps the lower face in contact with the bottom of the measuring chamber on one side, while the upper face of the disk is in contact with the top of the chamber on the opposite side. Meters of this type are generally available for pipe diameters from ½ inch to 2 inches (1.3 centimeters to 5.1 centimeters) with a flow rate ranging from two gallons (7.6

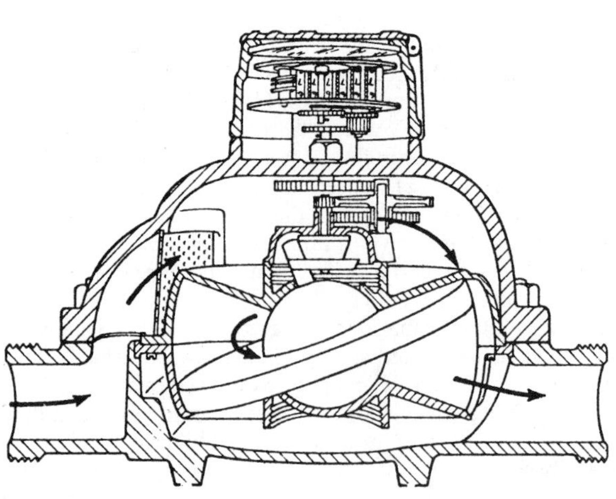

Fig. 6. Sectional view of typical nutating-disk meter.

liters) per minute minimum up to about 160 gallons (606 liters) per minute.

Oscillating-Piston Meter. The operating principle of the oscillating-piston meter is similar to that of the nutating-disk meter. The important contrsst is that the mechanical motion takes place in one plane (no wobble). The principle is illustrated in Fig. 8. Pipe sizes range from ½ inch up to 2 inches (1.3 centimeters to 5.1 centimeters). Flow rates range from 0.75 up to 160 gallons (2.8 liters to 606 liters) per minute. The meter has good accuracy, especially for low flow rates. It is easy to install, maintain, and is readily adaptable to automatic liquid-batching systems. It possesses limited power in terms of driving mechanical accessories.

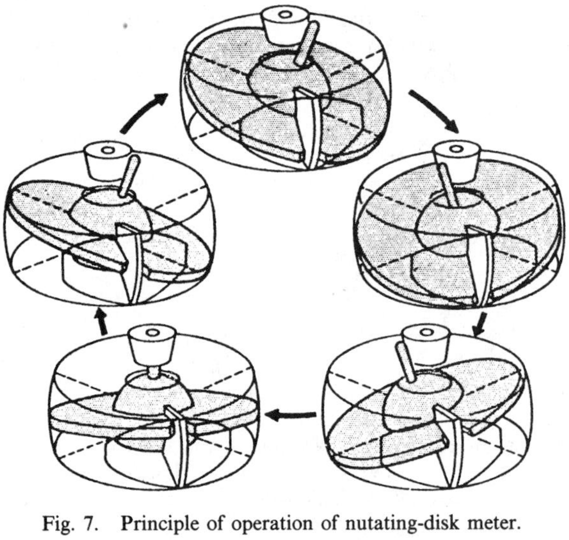

Fig. 7. Principle of operation of nutating-disk meter.

Rotating Meters. The principal positive-displacement meters of the rotating type are the lobed-impeller flowmeter and the sliding-vane and retracting-vane rotary meters. In the *lobed-impeller meter* (Fig. 9) two rotors revolve with fixed relative position inside a cylindrical housing. The measuring chamber is formed by the wall of the cylinder and the surface of one-half of one rotor. When the rotor is in a vertical position, a definite volume of fluid is contained in the measuring compartment. As the impeller turns, owing to the slight pressure differential between inlet and outlet ports, the measured volume is discharged through the bottom of the meter. This action occurs four times for a full revolution, the impeller rotating in opposite directions and at a speed proportional to the volume of fluid flow. These meters are available in pipe sizes from 1½ to 24 inches (3.8 to 61 centimeters) (diameter) for handling from 8 gallons (30 liters) per minute up to 25,000 barrels (39,743 hectoliters) per hour, with an accuracy of about ±0.20%. Advantages include performance at relatively high temperatures (400°F; ~ 205°C) and pressures up to 1,200 psi (81.6 atmospheres). The meters have a low pressure loss, are available in numerous materials of construction, applicable to gases and a wide range

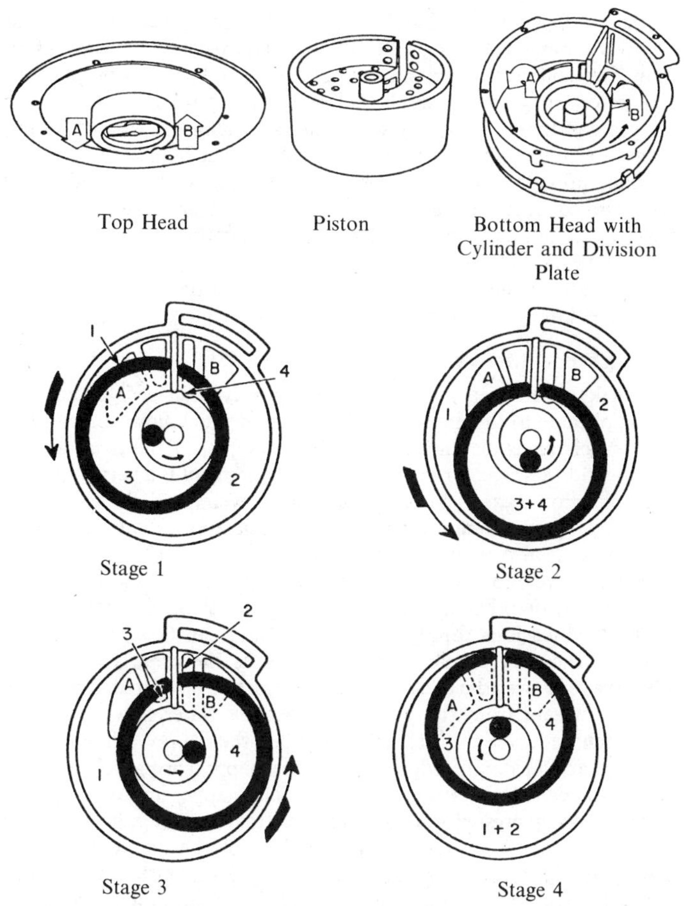

Fig. 8. Principle of operation of oscillating-piston meter. (*Rockwell.*)
Stage 1. Spaces 1 and 3 are receiving water from the inlet port A, and spaces 2 and 4 are discharging through the outlet port B.
Stage 2. The piston has advanced and space 1, in connection with the inlet port, has enlarged, and space 2, in connection with the outlet port, has decreased; while space 3 and 4, which have combined, are about to move into position to discharge through the outlet port.
Stage 3. Space 1 is still admitting fluid from the inlet port and space 3 is just opening up again to the inlet port; while spaces 2 and 4 are discharging through the outlet port.
Stage 4. Fluid is being received into space 3 and discharged from space 4; while spaces 1 and 2 have combined and are about to begin discharging as piston moves forward again to occupy position as shown in Stage 1.

of light to viscous liquids, including asphalt. Limitations include susceptibility to damage from entrained vapors, tendency to be bulky and heavy in larger sizes, relatively high cost, and slip at low flow rates (they are best used at high rates).

As illustrated in Fig. 10, in the *sliding-vane rotary flowmeter,* vanes are moved radially as cam followers to form the measuring chamber. In the *retracting-vane version* of Fig. 11, the vanes are articulated. The sliding-vane type contains a cylindrical rotor that revolves on ball bearings around a central shaft and stationary cam. As fluid flows against an extended blade, the resulting rotation of the rotor and action of the cam cause the blades to act as cam followers, creating measuring chambers that measure fluid throughput. Capillary action of the metered fluid seals the blades to form measuring cavities. In the retracting vane design, fluid entering the meter is deflected downward against the extended blade, causing rotation of the measuring

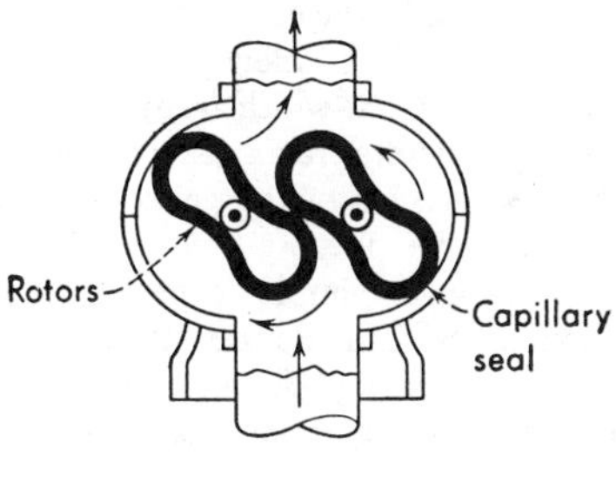

Fig. 9. Lobed-impeller flowmeter.

Fig. 10. Sliding-vane rotary meter.

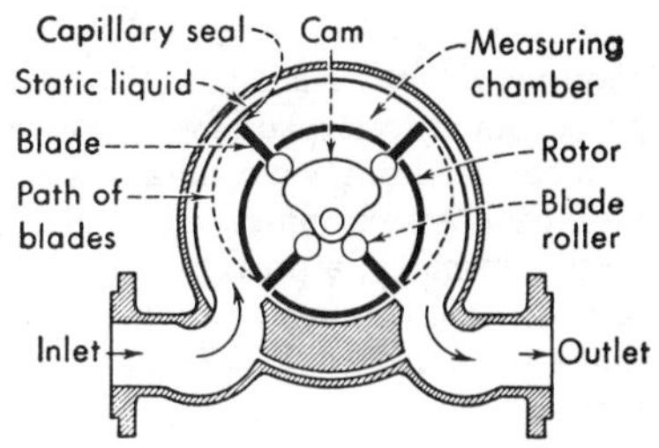

element. Sealing of the metering chamber is accomplished by incorporating features of the piston-ring seal. The vanes or blades, coated with a resilient material, are spring-loaded to provide the wiping action of a piston ring. Both of the vane-type rotary meters are useful up to temperature of 400°F (± 205°C) and pressures up to 500 psi (34 atmospheres) (sliding-vane type) and 900 psi (61 atmospheres) (retracting-vane type). Sliding-vane meters are available in pipe sizes up to 16 inches (41 centimeters); retracting-vene meters are limited to maximum size of 4-inch (10-centimeter) pipes.

Fig. 11. Retracting-vane rotary meter.

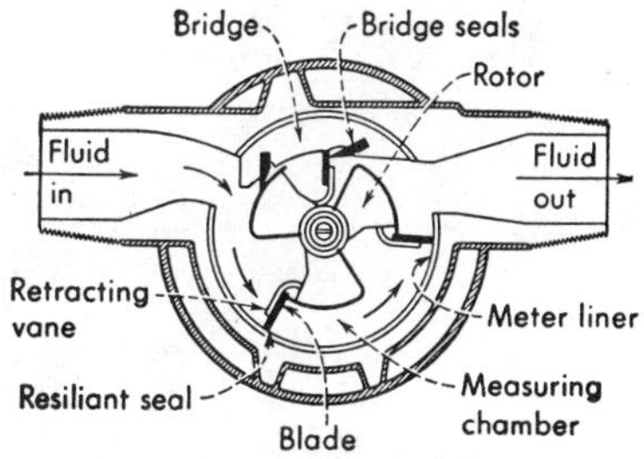

Bellows Gas Meter. The type of meter shown in Fig. 12 is widely used in both commercial and domestic gas service. It is comprised of four measuring compartments which operate simultaneously. Some of the compartments are filling while others are emptying, but all are always conforming to set conditions, thereby insuring a uniform delivery of gas. The number of times each measuring chamber is filled and emptied is registered, thus indicating the total volume in cubic feet on the index. The register or meter is operated from a crank

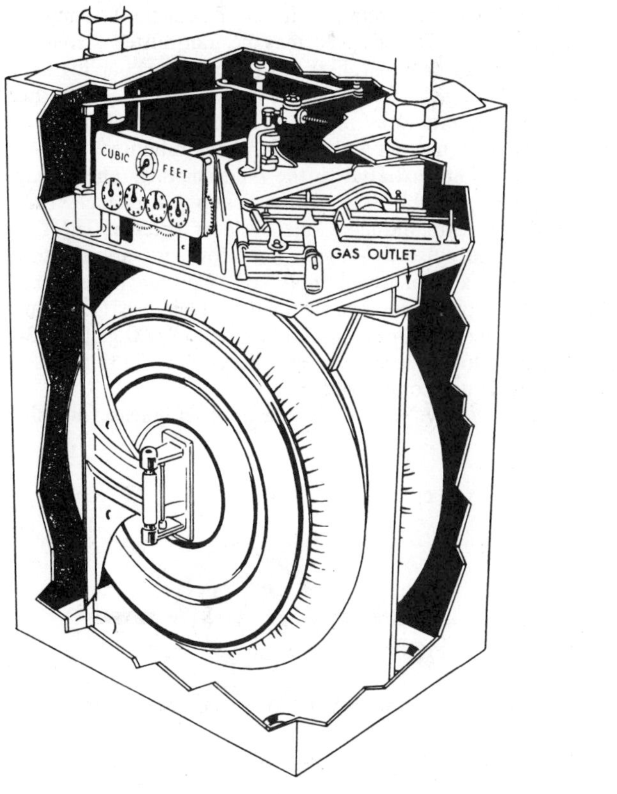

Fig. 12. Bellows type gas meter. (*American Gas Association.*)

that is rotated by movement of the diaphragms. Synthetic rubber diaphragms are usually used to insure that displacement is directly proportional to stroke. Two valves of the D-slide type are actuated by a central crank via a suitable linkage. Motion of the meter mechanism cannot occur unless a pressure differential (hence gas flow) takes place. Normally, a pressure differential of 0.10-inch (2.5 millimeters) of water will commence actuation of the meter. Maximum capacities range from 150 to 17,000 cubic feet (4.2 to 481 cubic meters) per hour.

Velocity-Type Flowmeters

Turbine Meter. In this device, a turbine wheel (rotor) is set in the path of a fluid stream. The flowing fluid impinges on the turbine blades, imparting a force to the blade surface and setting the rotor into motion. When a steady rotational speed has been reached, the speed is proportional to fluid velocity. In the type of design shown in Fig. 13, a multiblade rotor is mounted at right angles to the axis

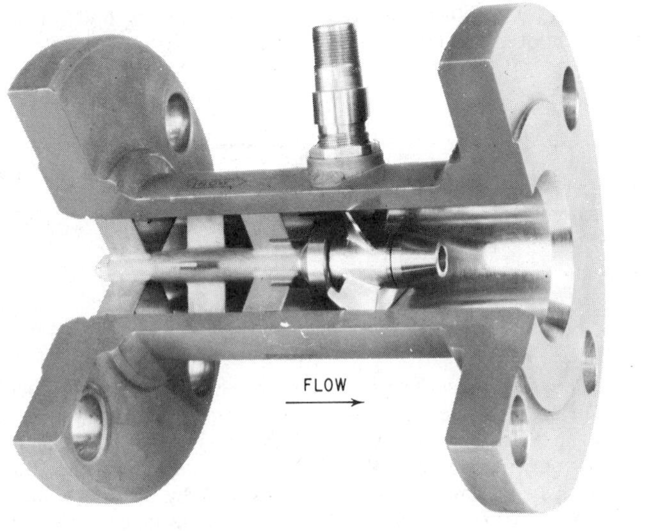

Fig. 13. Turbine flowmeter with interchangeable shaft-support section and flanged housing. (*Foxboro.*)

of the flowing fluid. The rotor is supported by ball or sleeve bearings on a shaft which is retained in the flowmeter housing by a shaft-support section. Rotor speed may be transmitted through the meter housing by a mechanical shaft, with a magnetic coupling to an external shaft, or through a suitable gland in the housing. In another method, a signal is generated by means of a magnetic pickup coil, consisting of a permanent magnet with coil windings, mounted in close proximity to the rotor, but external to the fluid channel. The passage of each rotor blade past the base of the pickup coil changes the total flux through the coil, inducing one cycle of voltage. The frequency of the generated pulses is a measure of flow rate, and the total number of pulses is a measure of total flow. Turbine meters are generally available in pipe sizes from $\frac{1}{2}$ inch to 12 inches (1.3 centimeters to 30 centimeters) (diameter), handling up to 9,000 gallons (341 hectoliters) per minute in larger size at fullscale flow. Turbine meters require viscosity-change compensation.

Vortex Velocity Meter. This type of meter consists of a short section

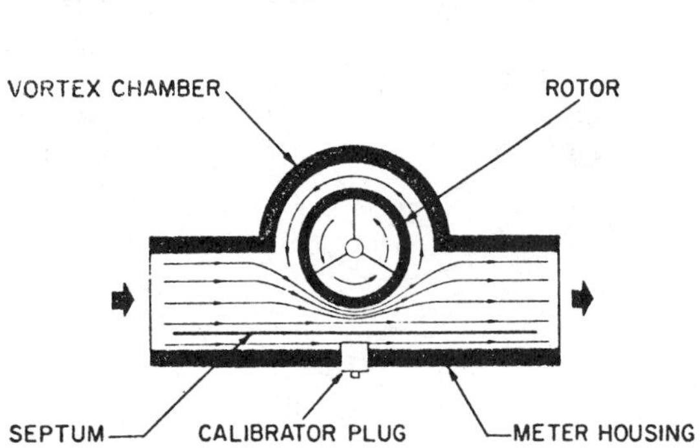

Fig. 14. Principle of vortex velocity flowmeter.

of pipe with an offset chamber in which a machined and balanced rotor is mounted. See Fig. 14. The rotor is fed by a portion of the main flow so that the total number of rotor revolutions is proportional to the total gas flow through the meter. This principle compensates for gas density changes over wide ranges. Standard meter sizes range from 2- to 6-inch (5- to 15-centimeter) pipe diameters with maximum working pressure up to 1,500 psi (102 atmospheres). Once calibrated, the claimed accuracy is $\pm 0.1\%$. Such meters are available with a variety of readouts and secondary instrumentation.

A remarkable feature of the vortex meter is that the same meter can be used on liquid or gas, and the calibration remains the same for either fluid—within normal accuracy tolerances.

Open-Channel Meters

The flow-measurement principles previously described generally are not applicable to the measurement of large-volume flows encountered in open channels, as encountered, for example, in the measurement of water and sewage. Irrigation water measurement is a very important example. For these types of applications, weirs and flumes are the principal measurement means used. In certain applications, the Kennison nozzle is also used.

Weirs. A weir consists of a partition or bulkhead of timber, concrete, sheet metal, or other fabrication, having in its top edge an opening of fixed dimensions through which a stream can flow. The opening is called the weir notch; its bottom edge is the crest; the depth of water passing over the crest is the head (must be measured at a definite point upstream from the weir); the sheet of water flowing through the notch and over the weir crest is called the nappe. When the weir has a sharp upstream edge so that the nappe springs clear of the crest, it is called a sharp crested weir. The nappe, immediately after leaving a sharp crested weir, suffers a contraction along the horizontal crest called crest contraction. If the sides of the notch have sharp upstream edges, the nappe also is contracted in width, and the weir is said to have end contractions. With sufficient side and bottom clearance dimensions of the notch, the nappe suffers maximum crest and end contractions, and the weir is said to be fully contracted. Of the various types of weirs, the three most commonly used are described briefly as follows:

(*V-Notch Weir*)—This device is particularly recommended for metering flows less than one cubic foot (0.03 cubic meter) per second (equivalent to 0.65 million gallons (24,603 hectoliters) per day) and is suitable for measuring slowly-changing flows up to 10 cubic feet per second. Extensive experiments have been made to determine the calibration data for V-notch weirs with included angles of 60° and 90°; two acceptable formulas are given in Fig. 15.

(*Rectangular Weir*)—This device is capable of high-capacity metering and is simple and inexpensive to construct. To assure complete contraction of the nappe, the side and bottom clearance dimensions of the notch must equal or exceed those shown in Fig. 15. The traditional formula for discharge of water through a suppressed rectangular weir is the Francis formula as indicated in Fig. 15.

(*Cipolletti Weir*)—This weir, also shown in Fig. 15, is similar to the rectangular weir except for sloping sides (1 horizontal to 4 vertical) of the notch. The design has the advantage of a simplified discharge formula which is more convenient to handle.

Other weirs include the hyperbolic (Sutro), broad crested, and round crested weirs.

Parshall Flumes. The Parshall venturi-type flume consists of a converging upstream section; a downward sloping throat; and an upward sloping, diverging downstream section. It is usually made of reinforced concrete, but may be of wood and, because of size, usually constructed on the site. Stainless steel and fiber glass reinforced plastic liners have been used for metering corrosive solutions. Surfaces are true planes, finished smooth with close adherence to specified dimensions. Parshall flumes have been constructed in sizes with throat widths ranging from 3 inches (7.5 centimeters) to 40 feet (12 meters) for measuring flows up to many million gallons per day. Some idea of the general shape of a Parshall flume can be gleaned from Fig. 16.

Kennison Nozzle. This device is an unusually simple device for measuring flows through partially-filled pipes. The nozzle successfully copes with low flows, wide flow ranges, and liquids containing suspended solids and debris. Because of its accuracy, nonclogging design,

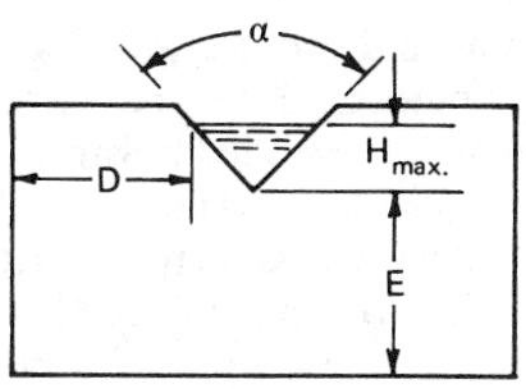

(a) V-Notch Weir

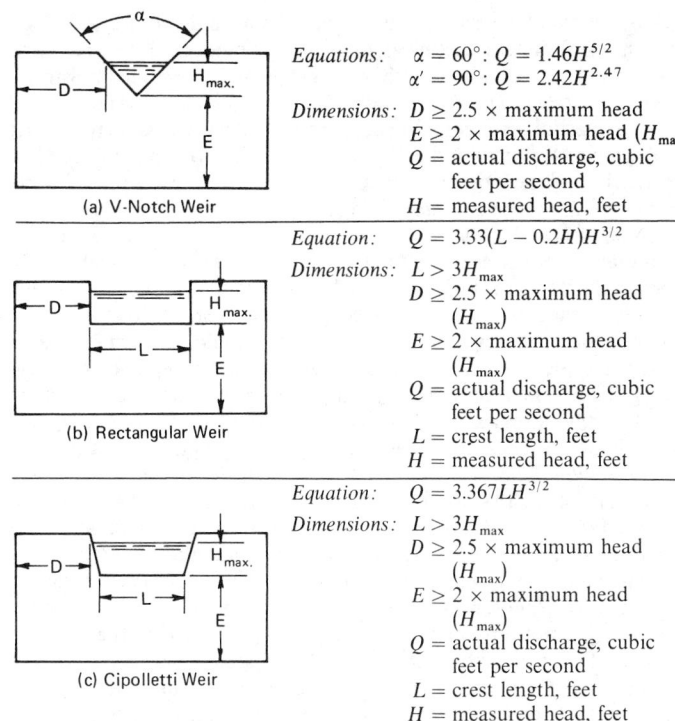

Equations: $\alpha = 60°$: $Q = 1.46H^{5/2}$
$\alpha' = 90°$: $Q = 2.42H^{2.47}$

Dimensions: $D \geq 2.5 \times$ maximum head
$E \geq 2 \times$ maximum head (H_{max})
Q = actual discharge, cubic feet per second
H = measured head, feet

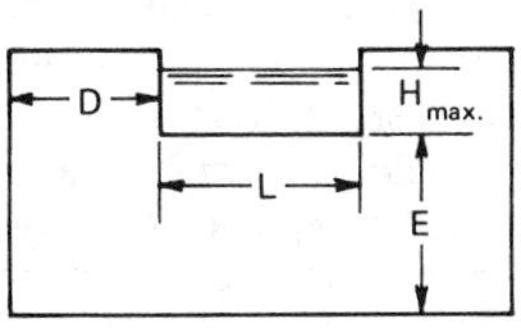

(b) Rectangular Weir

Equation: $Q = 3.33(L - 0.2H)H^{3/2}$

Dimensions: $L > 3H_{max}$
$D \geq 2.5 \times$ maximum head (H_{max})
$E \geq 2 \times$ maximum head (H_{max})
Q = actual discharge, cubic feet per second
L = crest length, feet
H = measured head, feet

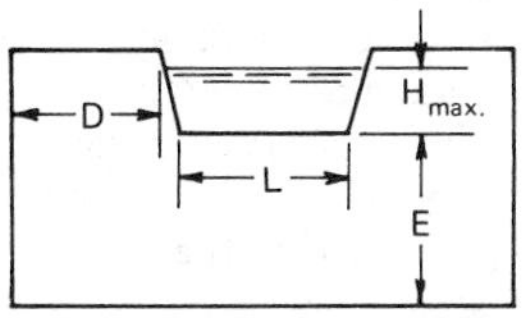

(c) Cipolletti Weir

Equation: $Q = 3.367LH^{3/2}$

Dimensions: $L > 3H_{max}$
$D \geq 2.5 \times$ maximum head (H_{max})
$E \geq 2 \times$ maximum head (H_{max})
Q = actual discharge, cubic feet per second
L = crest length, feet
H = measured head, feet

Fig. 15. Principal types of weirs.

and desirable head-versus-flow characteristics, the Kennison nozzle is well suited for the measurement of raw sewage, raw and digested sludge, final effluent, and trade wastes. Capacities range from 0.13 to 20 million gallons (4921 to 757,000 hectoliters) per day. The unobstructed flow path (with a self-scouring action) prevents clogging by debris.

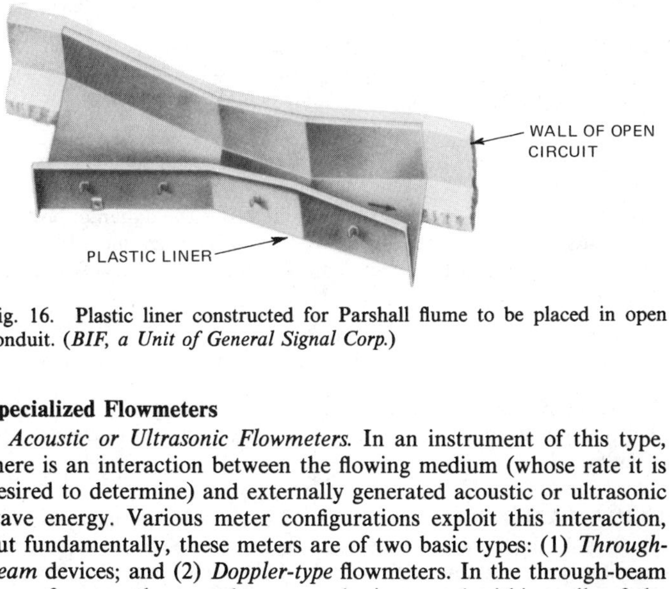

Fig. 16. Plastic liner constructed for Parshall flume to be placed in open conduit. (*BIF, a Unit of General Signal Corp.*)

Specialized Flowmeters

Acoustic or Ultrasonic Flowmeters. In an instrument of this type, there is an interaction between the flowing medium (whose rate it is desired to determine) and externally generated acoustic or ultrasonic wave energy. Various meter configurations exploit this interaction, but fundamentally, these meters are of two basic types: (1) *Through-beam* devices; and (2) *Doppler-type* flowmeters. In the through-beam type of meter, the transducers can be immersed within wells of the pipe, so that energy may be projected into the stream. The transducers are installed on an angle to the flow. Acoustic wave trains are transmitted upstream and downstream from crystals driven by wave generators. The crystals must be able to transmit and receive. This may be accomplished by using sets of crystals, or by alternating the transmitting and receiving function through electronic switching. The frequencies of the acoustic wave trains at the receiving crystals depend on the acoustic velocity in the flowing fluid, whether the signal is sent upstream or downstream, and the angle of transmission. The upstream and downstream frequencies are $f_u = (v + V\cos\alpha)2l$ and $f_d = (v - V\cos\alpha)2l$, where v is acoustic velocity, V is fluid velocity, α is the angle of transmission (measured with reference to fluid flow direction), and l is the distance between crystals. The difference $f_u - f_d$ equals $V\cos\alpha/l$. Since α and l are fixed by instrument configuration, the fluid velocity is proportional to $f_u - f_d$. Mass flow may be obtained by computation from density and velocity measurements.

Since the acoustic wave trains can be transmitted through walls of a test section, the test section can be a straight section of pipe in which the flowing fluid does not contact the sensing elements. This feature minimizes pressure drop and fulfills the stringent sanitary specifications of most medical and food processing applications. Although a wide variety of fluids can be measured, slurries containing solids or air bubbles must be avoided because of their influence on sound transmission.

In the *Doppler* or *reflecting* configuration, a requirement is that the moving fluid contain "objects" from which the externally generated radiation can be reflected. Thus, this principle is applicable to fluids containing undissolved solids, particles, bubbles, etc. These are referred to as "scattering sites." Lynnworth (1979) describes a number of other ultrasonic flowmeter configurations which, as of the early 1980s, remain in an early developmental stage.

Thermal Flowmeter. In meters of this type, heat in measured quantity is added to the flowing stream upstream of the measurement. The cooling effect (heat dissipation) of the flowing stream becomes a measure of flow. Either the temperature rise of the stream at a point downstream or the amount of energy required to maintain a heated element at a constant temperature can be measured.

Simon (1977) reported a slow-flow velocity sensor based upon the hot-wire anemometer principle. Instead of measuring the temperature rise in a wire, this meter measures the temperature rise in the fluid. In an experimental design (NASA Langley Research Center, Hampton, Virginia), a simple light dimmer controls power to the heat source. A sensor is placed in an air stream, such as a wind tunnel, air conditioning duct, or ventilation system, and temperature differentials at each channel's inlet and outlet are recorded. The sensor can detect flows between 0.2 and 20 feet (0.06 and 6 meters) per second and, with slight modifications, can be used to detect liquid flows. Channel velocity is calculated from the Reynolds number and is proportional to the free velocity of the mainstream. The constant of proportionality is determined by calibration in a known flow.

Radioisotope Tracer Flowmeter. Two principal techniques are used for measuring flow utilizing radioisotopes. In the *peak timing* method, a gamma emitter, such as ^{60}Co or ^{124}Sb is injected quickly at a point close to the section of the pipe in which the velocity is to be determined. The time of passage of the peak of the tracer wave is determined by using two detectors (counters or scintillation detectors), located at a known distance apart and external to the pipe. The detectors usually are connected to a rate meter and recorder. The time interval for the fluid to travel between the two detectors is calculated from the distance on the recorder chart between the two radiation-level peaks; or it is determined by having the arrival of the activity at the first detector start a timer which is stopped by the arrival of the radioactive material at the second detector. The measured time interval is divided by the distance between the two observation points to calculate linear flow rate. The principle of the *dilution method* is similar to the thermal flowmeter previously described and involves introducing a known concentration of radioisotope material at an upstream point and measuring the dilution at a point downstream. Thorough mixing between the two points is mandatory to success of this approach.

Laser Doppler Flowmeter. Fluid flow can be determined by measuring the doppler shift in laser radiation scattered from particles in the moving fluid stream. No sensor or transducer is necessary in the moving fluid stream. The laser radiation focal point can be moved across the flow tube to measure velocity profiles. Fluid flows from 0.01 to 5,000 inches (0.25 millimeter to 12,700 centimeters) per second have been measured in this manner. Contaminants, such as smoke, may have to be added to gases to provide scattering centers for the laser beam.

References

Bailey, S. J.: "Flowmeters," *Control Engineering*, **25**, 1, 49 (1978).
Bailey, S. J.: "Oscillating Ball Device Measures Ultra-low Flows," *Control Engineering*, **26**, 5, 75 (1979).
Bailey, S. J.: "Flowmeters—A Review," *Control Engineering*, **27**, 4, 75 (1980).

Chow, V. T. (editor): "Handbook of Applied Hydrology," McGraw-Hill, New York, 1964.
Emerson, L. P.: "Head Flowmeters," in "Process Instruments and Controls Handbook," (D. M. Considine, editor), McGraw-Hill, New York, 1974.
Fees, C. E.: "Variable Area Flowmeters," in "Process Instruments and Controls Handbook," (D. M. Considine, editor), McGraw-Hill, New York, 1974.
Irwin, L. K.: "Flow Measurement in Open Channels and Closed Conduits," Volumes 1 and 2, National Bureau of Standards, Washington, D.C., 1977.
Lomas, D. J.: "Selecting the Right Flowmeter," Part I, *Instrumentation Technology*, **24**, 5, 55–62 (1977).
Lomas, D. J.: "Selecting the Right Flowmeter," Part II, *Instrumentation Technology*, **24**, 6, 71–77 (1977).
Lynnworth, L. C.: "A Checklist of Ultrasonic Flowmeters," *In-Tech*, **26**, 11, 62–63 (1979).
Miller, R. W.: "Principles and Practice of Flow Meter Engineering," 10th edition, McGraw-Hill, New York, 1981.
Morris, H. M.: "Ultrasonic Flowmeters—A Review," *Control Engineering*, **26**, 8, 41 (1979).
Simon, W. E.: "Slow-flow Velocity Sensor," *Instrumentation Technology*, **24**, 11, 75 (1977).
Staff: "Orifice Meter Constants, Handbook E-2," American Meter Company, Philadelphia, Pennsylvania (Updated periodically).
Staff: "Fluid Meters: Their Theory and Application," American Society of Mechanical Engineers, New York (Updated periodically).
Staff: "British Standard Code on Flow Measurement," British Standards Institute Technical Committee on Temperature Measurement, Gas Flow Measurement, and Pressure Measurement, London, England (Updated periodically).
Staff: "Handbook for Gas Measurement," Southern California Meter Association, Los Angeles, California (Updated periodically).
Walker, R. J.: "Instruments to Measure Flow for Water and Wastewater Treatment," *In-Tech*, **27**, 6, 49–53 (1980).

FLOWMETER (Fluidic). Fluidics.

FLOWMETER (Laser Doppler). Laser.

FLOW STRESS. The instantaneous true tensile stress required to cause continued plastic deformation at a given value of the strain.

FLOW STRUCTURE. A type of banding in effusive igneous rocks (lavas) due to the alignment of minerals, inclusions or gas cavities during the movement of the still molten but viscous material. It is not to be confused with foliation.

FLUCTUATION. In thermodynamics, one deals with matter in bulk, and hence, usually considers uniform systems. However, all matter is built up of atoms, and its atomistic nature will produce fluctuations which can be studied by statistical means.

For most physical quantities, a system consisting of a large number of particles will show a Gaussian or normal distribution, and the relative fluctuations of a quantity G will usually be given by an equation of the type

$$\frac{\langle G^2\rangle - \langle G\rangle^2}{\langle G\rangle^2} = \frac{1}{\langle G\rangle} \qquad (1)$$

where the $\langle\,\rangle$ indicate average values. As most $\langle G\rangle$ will be proportional to the number of particles in the system, the relative fluctuations will usually be negligibly small.

There are, however, cases where fluctuations become experimentally observable. The first case is that of Brownian movement, where small particles in suspension will undergo fluctuations in the uniform pressure and hence show a random walk phenomenon. The second case is where Equation (1) breaks down. This will happen near a critical temperature when, for instance, critical opalescence occurs.

If we are dealing with systems of bosons or fermions, the equation must be slightly changed, but the conclusion that the left-hand side is usually very small for systems consisting of many particles remains valid.

FLUCTUATION NOISE. Random noise which has a uniform energy versus frequency distribution. Examples are thermal noise and shot noise.

FLUID. A state of matter in which only a uniform isotropic pressure can be supported without indefinite distortion; so, a gas or a liquid. The distinction between highly viscous liquids and solids is a difficult one, the same material acting as an ordinary liquid under some circumstances and as a solid under others. Fluids may be described in various ways. A *perfect fluid* is frictionless offering no resistance to flow except through inertial reaction. A *homogeneous fluid* has the same properties at all points. An *iostropic fluid* has local properties that are independent of rotation of the axis of reference along which those properties are measured. An *incompressible fluid* is a fluid whose density is substantially unaffected by change of pressure. The behavior of a real fluid is similar to that of an incompressible fluid only if the pressure variations in the flow are small compared with the bulk modulus of elasticity. For a fluid in motion in a gravitational field with velocities of order v, it is necessary that both v and $\sqrt{gh}$ should be small compared with the velocity of sound in the fluid. (h is the depth of the fluid and g the acceleration due to gravity.) An *elastic fluid* is a fluid for which elastic stresses and hydrostatic pressures are large compared with viscous stresses. A *viscous fluid* has an appreciable fluid friction. A *Newtonian fluid* is a viscous fluid in which the viscous stresses are a multiple of the rate of strain. The contact of proportionality is the fluid *viscosity*. A *Maxwellian fluid* is a viscous fluid in which the stress-strain relationship includes the relaxation effect (which takes a measurable time) of the relaxation of the elastic stresses set up by a sudden deformation. A *thixotropic fluid* is a fluid whose viscosity is a function not only of the shearing stress, but also of the previous history of motion within the fluid. The viscosity usually decreases with the length of time the fluid has been in motion. Such systems commonly are concentrated solutions of substances of high molecular weight, or colloidal suspensions. See also **Viscosity.**

FLUID (Constant-Stress Layer). In the boundary-layer of a fluid flowing over a solid wall, the shear stress varies with distance from the wall but it may be considered nearly constant within a small fraction of the layer thickness. The concept is of particular importance in turbulent flow where it leads to a theoretical derivation of the "law of the wall," the logarithmic distribution of mean velocity. The constant stress layer is the best-known example of the equilibrium flows near a wall.

FLUID (Dynamic Pressure). The pressure necessary to accelerate a fluid from rest to a speed of V is the dynamic pressure equivalent to that speed. If ρ is taken as mass density of the fluid,

$$\text{dynamic pressure, } q = \frac{\rho V^2}{2}$$

This conception is useful in aerodynamics as the dynamic pressure represents the unit air pressure acting on a surface increment in atmospheric air moving with velocity V over the surface. By Bernoulli's theorem,

$$p = p_0 - \frac{\rho v^2}{2} = p_0 - q$$

The vacuum caused by air in motion over a surface is greatest when this imaginary dynamic pressure is greatest since q represents the vacuum.

FLUID DYNAMICS. The study of the motion of matter in the gas, liquid, plastic or plasma state. When restricted to flow of incompressible (i.e., constant density) fluids, the term *hydrodynamics* is used. When dealing with electrically conducting fluids with magnet fields present, the term *magneto-fluid dynamics* is used. When dealing with practical problems of air flow past airplane wings, through ventilating equipment, etc., the term *aerodynamics* is used. See also **Aerodynamics.**

Basically two fundamental approaches are used: (1) continuum or field dynamics and (2) kinetic theory and nonequilibrium statistical mechanics. The study of fluids tends to be quite complex.

Continuum Dynamics. In this approach, fluid properties, such as velocity, density, pressure, temperature, viscosity, conductivity, among others, are assumed to be physically meaningful functions of three spatial variables x_1, x_2, and x_3, and time t. Nonlinear partial differential equations are set up to relate these variables. Such equations have no general solutions even for the most restrictive boundary conditions.

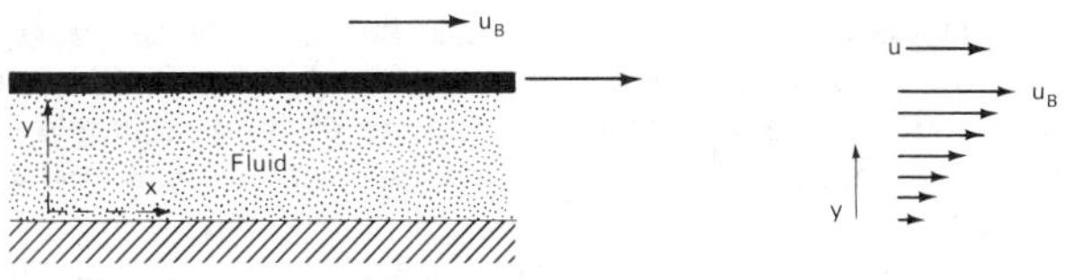

Fig. 1. Couette flow.

But solutions are carried out for very idealized flows. Couette flow is one of these. See Fig. 1.

The flow is between parallel plates, lower plate at $y = 0$ at rest, upper plate y_B moving with constant speed u_B in the x direction. Stress throughout the fluid is constant, given by $P_{xy} = \mu(du/dy) = \mu(u_B/y_B)$. This is pure shear flow and experimentally is often considered to define and measure the viscosity coefficient μ, assumed constant for the homogeneous fluid. The velocity profile appearing at the right in Fig. 1 shows by velocity arrows of different length at the various positions of y how the velocity varies with position. Steady flow (no dependence of any quantity on time), constant pressure, constant density, and laminar flow are additional assumptions for Couette flow. The flow is realized experimentally by confining the fluid in the narrow annulus between rotating concentric cylinders of nearly equal radius; the cylinders rotate at different speeds.

In Fig. 2, the special flow (Poiseuille flow) is in a pipe of uniform

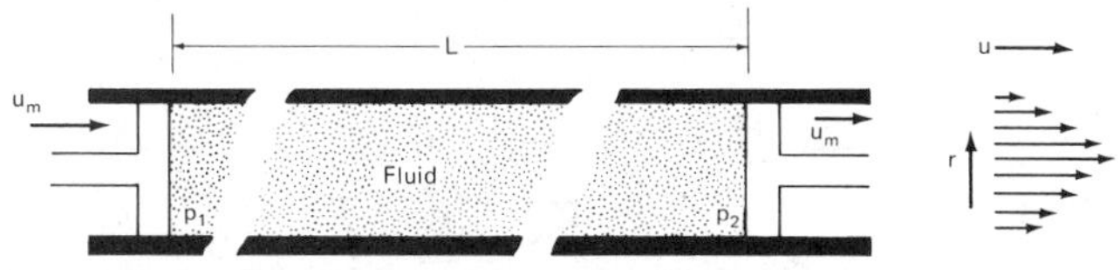

Fig. 2. Poiseuille flow.

cross section, pressure is assumed to be constant across each cross section but to vary linearly with distance x along the axis of the pipe so that $dp/dx = (p_1 - p_2)/L$. Pistons driving the flow are assumed to be infinitely far away, so that the flow velocity, parallel to pipe axis, has the same dependence upon y and z for all x. The velocity profile is parabolic in both the two-dimensional case (infinite parallel plates) and in the circular cross-section case. Mean flow velocity u_m and viscosity coefficient are assumed constant; the flow is assumed steady and laminar. For a circular cross-section pipe of radius a, at any distance r from the center, $u = 2u_m\,(1 - r^2/a^2)$, and the volume passing a cross section per second is $Q = \pi a^2 u_m = \pi a^4(p_1 - p_2)/8\mu L$. Since these formulas do not apply near pipe entrances, discretion in applying them to pipes of finite length is necessary even when the flow is steady and laminar. See also **Fluid Flow; Laminar Flow;** and **Turbulent Flow.**

Other examples of idealized solutions are one-dimensional flow of an ideal gas through a normal shock wave, flow of an ideal gas without viscosity through a pipe of slowly changing cross section (wind tunnel), and one-dimensional finite waves in an ideal gas. Numerous other solutions involve making whatever approximations and assumptions necessary to obtain descriptions of observed flows.

Kinetic Theory. In the kinetic theory and nonequilibrium statistical mechanics, fluid properties are associated with averages of properties of microscopic entities. Density, for example, is the average number of molecules per unit volume times the mass per molecule. While much of the molecular theory in fluid dynamics aims to interpret processes already adequately described by the continuum approach, additional properties and processes are presented. The distribution of molecular velocities (i.e., how many molecules have each particular velocity), time-dependent adjustments of internal molecular motions, and momentum and energy transfer processes at boundaries are examples.

When motion of the fluid consists of only small fluctuations about a state of near-rest, the continuum equations are linearized by neglecting nonlinear terms and become the equations of acoustics. A large variety of fluid motions are described as sound waves; when the small-motion or acoustic description can be used, the principle of *superposition* is valid. This powerful principle allows addition of simple simultaneous motions to represent a more complex motion, such as the sound reaching the audience from the instruments of a symphony orchestra. The superposition principle does not apply to large-scale (nonacoustical) motions, and the subject of fluid dynamics (in distinction from acoustics) treats nonlinear flows, i.e., those which cannot be described as superpositions of other flows. See also **Superposition (Principle of).**

Since sound waves travel with a speed relative to the fluid, waves moving in a moving field can sometimes be carried off in a direction opposite to the direction of sound travel. The flow where this occurs is called *supersonic*; the flow speed is greater than the sound speed at the spot where the flow is supersonic. Supersonic flow occurs around high-speed vehicles and missiles, and in pipes when high pressure gas escapes through a nozzle into a region of sufficiently lower pressure. A steady supersonic flow always must pass through a *shock front* to slow down to subsonic flow again.

The continuum description of flow fails to describe nearly all actual flow because actual flows when looked at carefully are *turbulent.* Turbulent flows have violent and erratic fluctuations of velocity and pressure which are not associated with any corresponding fluctuations of the boundaries containing or driving the fluid. Turbulence is generally considered to be the manifestation of the nonlinear nature of the fundamental equations. Under certain conditions, nonturbulent or *laminar* flow exists. A common example is cigarette smoke rising from a cigarette held at rest; near the cigarette, the stream is smooth and straight, or laminar, and further up the flow breaks into turbulance. See also **Reynolds Number.**

References

March, N. H., and M. P. Tost: "Atomic Dynamics in Liquids," Wiley, New York, 1977.

Hansen, J. P., and I. R. McDonald: "Theory of Simple Liquids," Academic, New York, 1976.

Stephan, K., and K. Lucas: "Viscosity of Dense Fluids," Plenum, New York, 1979.

FLUID FLOW. Just as there are many types of fluids, so there are, partly as a result, many types of fluid flow. *Uniform flow* is steady in time, or the same at all points in space. *Steady flow* is flow of which the velocity at a point fixed with respect to a fixed system of coordinates is independent of time. Many common types of flow can be made steady by a suitable choice of coordinates. *Rotational flow* has appreciable vorticity, and cannot be described mathematically by a velocity potential function. *Turbulent flow* is flow in which the fluid velocity at a fixed point fluctuates with time in a nearly random way. The motion is essentially rotational, and is characterized by rates of momentum and mass transfer considerably larger than in the corresponding laminar flow. *Laminar flow* is flow in which the mass of fluid may be considered as advancing in separate laminae (sheets) with simple shear existing at the surface of contact of laminae should there be any difference in mean speed of the separate laminae. If turbulence exists, its effect is confined to a lamina, and there is not exchange of momentum between laminae. *Streamline flow* is flow in which fluid particles move along the streamlines. This motion is characteristic of viscous flow at low Reynolds numbers or of inviscid, irrotational flow. *Secondary flow* is a less rigorously defined term than many of the foregoing types of flow. The flow in pipes and channels is frequently found to possess components at right angles to the axis. These components which take the form of diffuse vortices with axes parallel to the main flow form the secondary flow. Three types may be mentioned: 1. Secondary flow in curved pipes or channels, being a motion outwards near the flow center and inwards near the walls. 2. Secondary flow in straight pipes and channels of non-circular section, being a motion along the walls toward corners or places of large curvature and from there to the center of the flow. This only occurs in turbulent flow. 3. Secondary flow in pulsating flow. This is due to second order effects and is particularly striking with ultrasonic waves.

For many types of flow, calculations are complex, and where they can be made at all, require the methods of tensor analysis of the stresses and strains involved. One important relationship that is widely useful in the systematic study of fluid flow in the equation of continuity.

A great simplification of the fluid calculations can be effected by assuming that the fluid is perfect, homogeneous, totally incompressible, not viscous, and that therefore its properties are not affected by changes in temperature or pressure. While such an ideal fluid does not exist, its properties are often approached closely enough by real fluids so that calculations based upon it are often useful in practice.

Thus, elementary hydraulics always includes Bernoulli's law, and it is repeated here as being of great importance to the subject of fluid flow.

Bernoulli's law is

$$\frac{p}{w}+\frac{v^2}{2g}+z=\text{a constant}$$

and the symbols are defined as follows:

$p=$ the static pressure in pounds per square feet
$w=$ the specific weight of fluid in pounds per cubic feet
$v=$ the velocity in feet per second
$g=$ the gravitational acceleration in feet per second per second
$z=$ the potential, or "elevation," head in feet

Bernoulli's law states that in steady flow, the total head is a constant at any point and equal to the sum of the pressure head, p/w, the velocity head, $v^2/2g$, and the potential head (z). Since there is actually a loss of head between any two points due to friction, the difference between the total heads at any two points must equal the friction head when the flow is steady.

In this equation, $v^2/2g$ represents the velocity head, a pressure which could be recovered by the efficient reduction of the velocity in a conduit of expanding cross section. When this velocity head is multiplied by the specific weight of the fluid, it is reduced dimensionally to the unit of pressure, pounds per square feet, and may be designated the dynamic pressure, in counter distinction to the static pressure p. In the case of an expansionable fluid, such as gas, the total energy at two points in the flow must be the same, that is, the heat energy plus the kinetic energy of motion must be constant. It is necessary to invoke this law of continuity of energy in dealing with the flow of gases or vapors through nozzles. See also **Viscosity.**

FLUID FLOW (Boundary Layer). Motion of a fluid of low viscosity, such as air or water, around a stationary body or through a stationary conduit possesses the free velocity of an ideal fluid everywhere except in an extremely thin layer immediately next to the stationary body. Many of the phenomena of fluid flow may be studied and analyzed without consideration of this boundary layer, but, thin as it may be (usually a few thousandths of an inch), its internal mechanics must be understood and evaluated in certain of the phenomena of fluid motion. Some of the more important of these are:

1. The magnitude of the maximum lift coefficient of the airfoils.
2. Profile drag of airfoils.
3. The drag of bluff bodies.
4. The large variations of drag coefficient at critical Reynolds number for laminar-turbulent transition.
5. The transfer of heat through suface films.

Many of the phenomena of the boundary layer are explainable on the basis of the theory advanced by Prandtl at the University of Göttingen laboratory nearly half a century ago. In the same flow-research group were others, like Blasius, who broadened and experimentally confirmed the original hypotheses.

An elementary understanding of the effect of fluid viscosity will be had by considering a two-dimensional flow along the upper surface of a very thin flat-plate, as shown in Fig. 1. The thickness of the boundary layer, greatly exaggerated, is y_v; the free stream velocity is V, the variable velocity in the boundary layer is u. A basic assumption of the theory is that a fluid layer of infinitesimal thickness resting

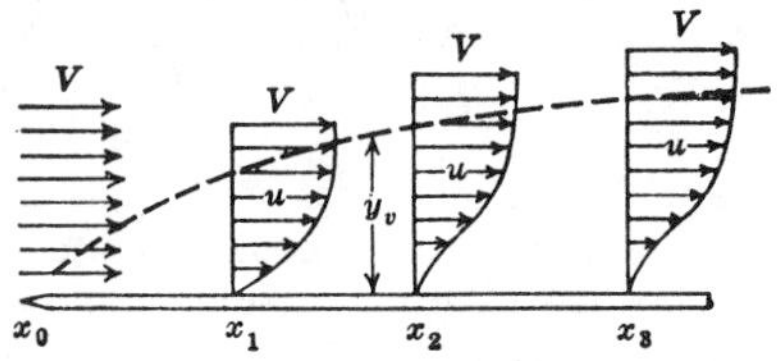

Fig. 1. Velocity profiles in the boundary layer. Dotted line is hypothetical upper edge of the boundary layer.

against the plate "sticks" to it so shearing force of the next fluid layer on the stationary layer determines the skin friction. Assuming that the boundary layer consists of lamina of fluid sliding on each other, the velocities of these lamina increase with y until, at the edge of the boundary layer, $u = V$. A series of boundary layer velocity profiles for stations x_1, x_2, x_3 are drawn to enable the reader to visualize the effect of friction on the momentum in the boundary layer and the thickening of it due to lower average u.

Note also the variation of the profile near the surface of the plate. Skin friction has steadily decelerated the individual fluid particles. The profile at x_3 indicates that the lower portion has come to rest. This is known as the stagnation point. Air-flow phenomena in this region are important in many ways, especially when there is an intended rising downstream pressure gradient, as in diffuser tubes or over the surface of airfoils.

Fluid flow in a divergent tube is illustrated in Fig. 2. On the lower

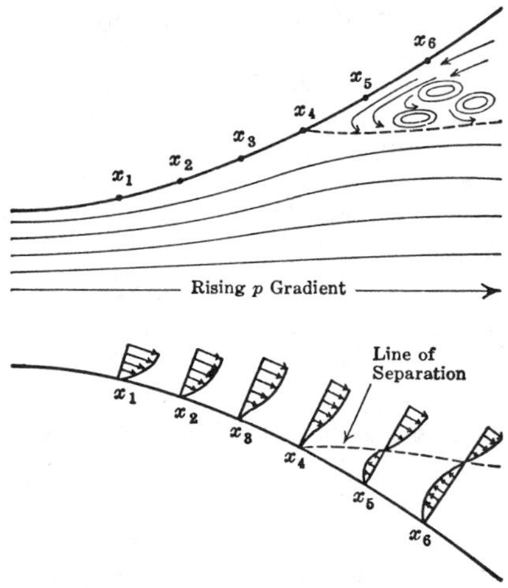

Fig. 2. Effect of boundary layer viscosity of flow on a diffuser.

profile, greatly enlarged velocity profiles are shown for the boundary layer at stations $x_1, x_2, \ldots x_6$. The x_4 profile indicates a stagnation point. Since the pressure gradient of the diffuser has a pressure at x_5 above that at x_4, a reverse flow is produced toward the stagnation point. Streamlines drawn in the upper half of the tube show what happens to the fluid flow. The region of the surface of separation between the reverse flow along the wall and the forward flow is unstable and breaks up into random vortices. Kinetic energy is irreversibly transferred to heat and the diffusion fails to produce the expected pressure gradient. Had the divergence of the conduit been sufficiently small, the pressure gradient would have been lowered per unit length and turbulence in free stream or boundary layer would have delayed the stagnation point.

For the airfoil with burbled or partly burbled air flow (Fig. 3), a

Fig. 3. Partial burble on airfoil at high angle of attack: (A) Stagnation point; (B) line of separation.

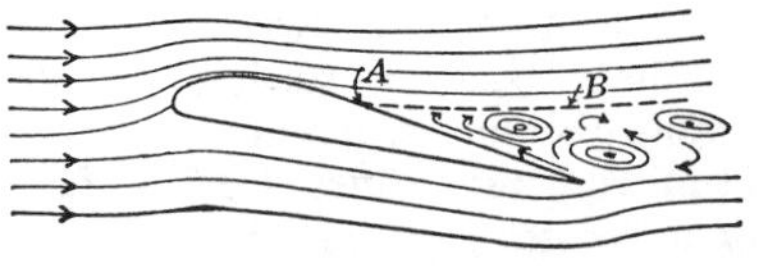

similar explanation exists. Streamline flow over the upper surface of the airfoil increases in velocity as angle of attack (and circulation Γ) increases. However, the surface friction in the boundary layer is decelerating the air particles next to the surface and, with high adverse pressure gradient (existing at high angle of attack) opposing the kinetic energy of the boundary layer, the velocity profiles ultimately show a stagnation point toward the rear of the airfoil. The flow separates from the surface and vortices form a turbulent wake in place of the streamline wake previously existing. Once the separation surface moves onto an airfoil, minor increases of angle of attack bring it rapidly forward. High circulation strength is no longer needed to fulfill the requirement of unity of upper and lower flows at the trailing edge (Kutta's hypothesis) so Γ decreases sharply, and with it the lift. Boundary layer theory accounts in this manner for the maximum lift coefficient. At the same time the extended turbulent wake sharply increases the profile drag coefficient.

Now return to a view of the nature of flow in the boundary layer. It has been called laminar, and so it is for values of the Reynolds number below a *critical* value. But for years, beginning about the time of Osborne Reynolds' experiments and revelations in the field of fluid flow, it has been known that the laminar property disappears and the flow suddenly becomes turbulent when the critical Vl/ν is reached. Usually flow starts over a surface as laminar but after passing over a suitable length the boundary layer becomes turbulent, with a thin laminar sublayer thought to exist because of damping of normal turbulent components at the surface (Fig. 4).

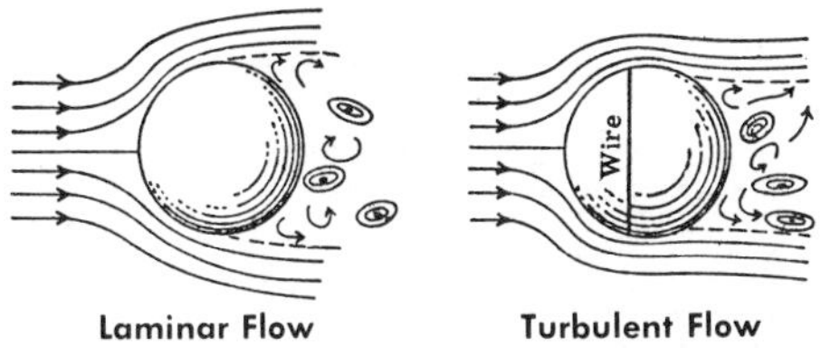

Fig. 4. Boundary layer turbulence reduces width of low-pressure wake.

This transition has profound effects in all fluid dynamics, and certainly so in aerodynamics. The velocity profile in the boundary layer becomes fuller near the surface on account of the higher average kinetic energy of the layer created by turbulent energy exchange from layer to layer. The effective viscosity is therefore larger in turbulent than laminar flow, the turbulent boundary layer thickens more rapidly downstream, the skin friction increases.

The importance to aerodynamics is the beneficial effect of turbulence on the wake existing in the rear of bluff bodies. Even thin airfoils become bluff bodies at high angles of attack. A turbulent boundary layer has more kinetic energy than a laminar one. This carries the air farther toward the rear of the surface before a stagnation point is reached and so reduced the width of the wake and that part of the profile drag. This reduction of drag can be, and usually is, much larger than the increase of skin friction so turbulence has a good effect on both profile drag and maximum lift coefficient of an airfoil. The critical value of Reynolds number lies between 200,000 and 2,000,000, being affected by the initial turbulence existing in the air prior to meeting the surface. Decreasing initial turbulence increases the critical Reynolds number. The effect of turbulence on the boundary layer is strikingly displayed by observing the drag wake behind a smooth sphere mounted in an airstream whose velocity is just under that required to produce breakdown of the laminar boundary layer. An artificial roughness is provided in the form of a fine wire or thread encircling the sphere in the laminar flow. The boundary layer, of course, becomes turbulent downstream from this irregularity resulting in a delayed stagnation point and narrower wake.

See also **Aerodynamics.**

FLUID FLOW (Conservation Laws). The fundamental conservation laws of physics can be used to obtain the basic equations of fluid motion, the equations of continuity (mass conservation), of flow (momentum conservation), of energy (first law of thermodynamics). In addition, conservation over the whole flow system imposes constraints on the flow that can be very useful. The best known of these is the *Kármán momentum integral* for nearly unidirectional mean flow,

$$\frac{d}{dx}\int \rho\nu(\nu_1 - \nu)\,dy - \frac{dP}{dx}\int \frac{(\nu_1 - \nu)}{\nu_1}\,dy = [\tau]$$

equating the momentum flux by flow to the total forces applied by pressure gradients and boundary stresses. See also **Bernoulli's Law.**

FLUID FLOW (Divergence of). The rate of decrease of the fluid density ρ is given by the equation of continuity,

$$\frac{\partial\rho}{\partial t} + \text{div}(\rho\mathbf{u}) = 0$$

where $\mathbf{u}$ is the flow velocity. A divergence of the mass flow $\rho\mathbf{u}$ implies that the density is decreasing.

FLUID FLOW (Dynamical Similarity). Two geometrically similar fluid flows are dynamically similar if the flow field of one may be transformed into the flow field of the other by the same change of length and velocity scales that was necessary to make the boundary conditions identical. If the equations of motion of the flow are made nondimensional by expressing velocities and lengths as fractions of these scales, these equations contain a number of nondimensional coefficients that determine the character of the flow. The general condition for dynamical similarity is that all these coefficients should be the same for the two flows. For geometrically similar flows, the conditions to be observed are generally expressed as the Reynolds number, the Prandtl number, the Grashof (or Rayleigh) number, the Mach number, and the Froude number. See also **Froude Number; Grashof Number; Mach Number; Prandtl Number;** and **Rayleigh Number.**

FLUID FRICTION. The flow of any actual fluid must of necessity be attended by the presence of friction, due to the physical nature of fluids, none of which meets the requirements of the ideal fluid, as mentioned in fluid flow. A great deal of time and attention have been devoted to the study of the properties of a flowing fluid. The frictional effects present in the flow of liquids have been rationalized much more thoroughly than for vapors and gases. However, for all three the friction is found to depend upon the nature of the fluid itself, its viscosity, and upon the conduit which contains it. On account of the different molecular arrangement of liquids, vapors, and gases, the study of friction of fluids has become a specialized study of each of these three.

Fluid flow rarely follows the commonly accepted idea of streamlines, since the velocities necessary for viscous flow of this nature are almost always lower than those found expedient to employ. Most flows are turbulent in nature, and become turbulent at a definite velocity, the value of which was studied by Reynolds, and is incorporated in the well-known Reynolds Number. A general thermodynamic equation of energy of a fluid under flow conditions would be as follows:

Gain in kinetic energy + gain in potential energy + net work received + energy liberated by any chemical change = change in heat content between two states.

In the case of a liquid, this equation can be considerably simplified:—in fact, it becomes Bernoulli's well-known equation—but in the case of compressible fluids, which may also undergo some change of form, such as condensation or compression, the longer equation applies. Most practical problems in fluid friction arise in connection with the flow of fluid through pipes.

FLUIDICS. The technology of sensing, controlling, and information processing with devices that use a fluid medium and whose operation is based solely upon the interaction between fluid streams. The particular function of each device, none of which have moving parts, is dependent upon the geometric shape of the device.

Fluidics is considered to have commenced in 1959 with the work of B. M. Horton, R. E. Bowles, and R. Warren, scientists at the Diamond Ordnance Fuze Laboratories (U.S. Army), Washington, D.C. Some of the principles employed in fluidics involve earlier discoveries by scientists, such as Henri Coanda, a Rumanian engineer who, in 1926, identified the wall-attachment phenomenon for fluid jets—a principle which is now known as the Coanda effect.

Fluidic principles have been applied in the construction of amplifiers, oscillators, computing and logic elements; analog controllers; flowmeters; temperature proximity and dimensional gaging sensors; process control valves; and level controllers.

Characteristics. The major characteristics of fluidic devices include:

(1) *No moving parts*—The absence of moving parts provides fluidic devices with a potential for high reliability.

(2) *Use any fluid*—Almost any gas, liquid, or slurry may be utilized in fluidic devices.

(3) *High amplification*—A high-energy stream is controlled by a low-energy stream. Gains as high as 1,000 or more have been achieved.

(4) *No practical size limitation*—Fluidic devices can be built in prac-

tically any size. There is no reasonable limit, for example, as to how large a process control valve can be. Conversely, miniaturized elements have fluid passageways on the order of 0.010 inch (0.254 millimeter).

(5) *Wide choice of construction materials*—Almost any material that can be cast, molded, machined, formed, or etched can be used to construct a fluidic device.

(6) *High response speeds*—Oscillators have been made to operate at frequencies of 10,000 Hz. The flow in a large 4-inch (10.2-centimeter) liquid-diverting valve can be switched completely in 0.1 second.

(7) *No shock or water hammer*—Even though switching occurs at high speed, there is no shock or water hammer involved when a liquid fluidic diverting valve is switched.

(8) *Unaffected by environment*—Since fluidic devices can be built of nearly any material, the devices can operate under extremes of temperature, vibration, and radiation. Also, operation is unaffected by electricity and magnetism.

Principle of Operation. The simplest form of fluidic amplifier (Fig. 1) uses cross-directed fluid jets. The main jet can be deflected in proportion to the strength of a cross-directed control jet. If receiving apertures are placed downstream, the deflection manifests itself as a relative change in flow and pressure at the apertures. This type of amplifier works entirely on the momentum-exchange principle. However, because so much energy is lost due to turbulence, such a device is quite inefficient.

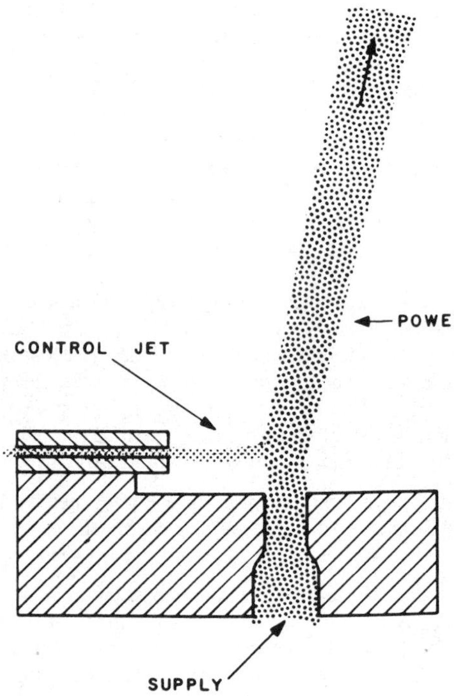

Fig. 1. The basic fluidic amplifier.

An additional principle employed in most fluidic devices is the Coanda effect, sometimes referred to as the wall-attachment principle. To illustrate this principle, assume a device having a cross section as shown in Fig. 2 and plates top and bottom so that the downstream end and left side are open. An end view is shown in the figure. Assume this device is operating open to atmosphere and that compressed air is supplied to the power nozzle. At the moment when supply air is turned on, the main jet tends to project in a straight line [Fig. 2(a)]. Note that the jet, being rectangular in cross section, seals against the top and bottom plates and has the effect of a barrier or movable wall running parallel to the side wall. Whenever a jet flows into a body of stagnant fluid, it entrains some of the surrounding stagnant fluid and starts it into motion. In this case, as the ambient air is entrained and ejected along both sides of the main jet, replenishing air continuously moves into this region. Along the open wall, the replenishing air moves in, unimpeded, and the average pressure along this side is essentially atmospheric. However, along the right side of the jet, the replenishing air must flow down through the restricted opening between the wall and the jet boundary. The average pressure on the wall side, therefore, will be somewhat below atmospheric.

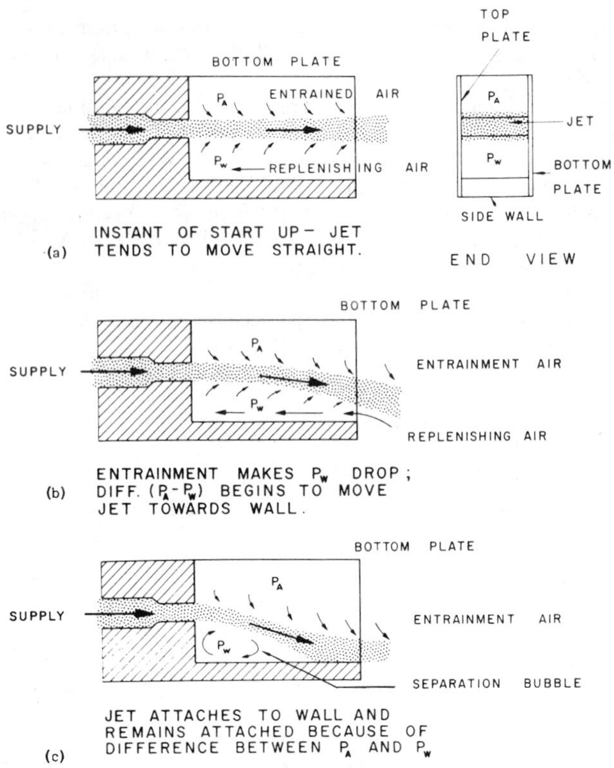

Fig. 2. Demonstration of Coanda effect.

The resultant differential in pressure across the two sides of the jet causes the jet to move closer to the wall, as indicated in Fig. 2(b). This, in turn, further restricts the passage down through which replenishing air must move, making the pressure along the wall decrease further, while the differential across the jet correspondingly increases.

This action is regenerative and continues until it terminates with the jet attached to the wall, as shown in Fig. 2(c). In this condition, a vortex forms in the region between the inside boundary of the jet and the wall. This region is also referred to as a separation bubble. The pressure in this region would be at a high vacuum in the example given, and the jet would remain attached to the wall because of the differential pressure impressed upon it.

To detach the jet from the wall, it would be necessary to supply sufficient replenishing air to the separation bubble. This could be done by supplying the air through a control port, as will be shown in succeeding examples.

The main jet and the ambient fluid in this example need not have been air. These principles apply even if the jet and ambient fluids are different from air. Likewise, the jet and ambient fluids can be different from each other. In fluidic control valves, for example, it is common to control a liquid jet by means of ambient air. The Coanda effect, therefore, is the basis for the "memory" function obtainable in these devices; and the positive feedback which it provides results in high amplification and much more efficient operation.

Bistable Amplifier. Figure 3(a) shows a cross section of a bistable amplifier, or flip-flop, with the main jet attached to the right outlet and both control ports closed to atmosphere. This unit would be made in a sandwich form with top and bottom plates for sealing the jet. All openings would be rectangular in cross section. In order to switch the jet, the right control port is opened, allowing replenishing air to enter. Immediately, the point of attachment of the jet begins to move outward [Fig. 3(b)], and the jet shifts to the left until it attaches to the left wall, as shown in Fig. 3(c). Once it is attached to the left wall, the right control port can be closed again and the main jet will remain locked to the left wall because of the Coanda effect. This construction is the basis for a two-position diverting valve or a bistable logic element with memory.

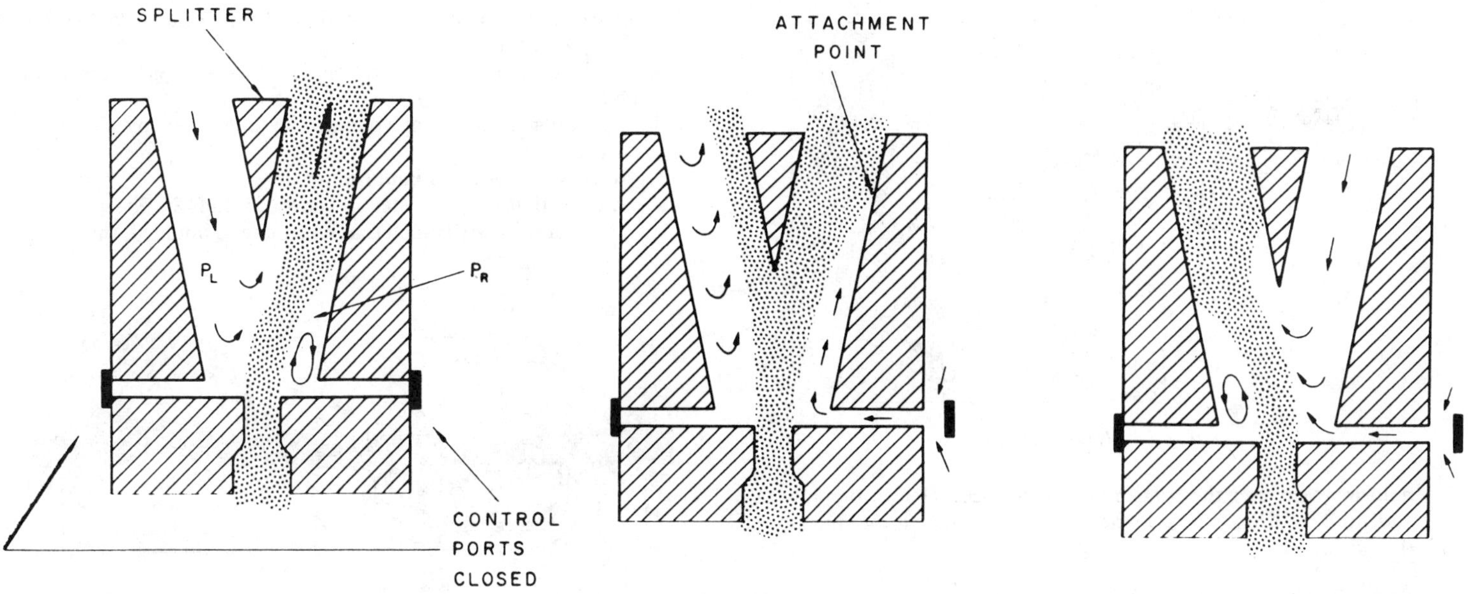

(a) Bi-stable amplifier flip-flop locked to right outlet

(b) Right control port opened, begins diversion of power jet

(c) Diversion completed

Fig. 3. Actions of the fluidic bistable amplifier.

Proportional Amplifier. To make a proportional amplifier, the divider of the unit shown in Fig. 3 would be moved closer to the nozzle. This would allow the operation to stabilize with the flows at the two outlets at any proportion to each other. Such a unit would be controlled by varying the relative opening of the two control ports.

Oscillator. In the oscillator shown in Fig. 4, the instant the power jet switches to the right outlet, ambient air begins to flow counterclockwise through the circular feedback path. This airflow is preceded by propagation of a shock wave which moves at the speed of sound. When the shock wave front collides with the main jet, sufficient energy is transferred to cause the main jet to be diverted to the left outlet. The identical action then occurs when the succeeding shock wave travels down the right outlet and clockwise through the circular feedback path. The frequency of oscillation in this type of unit is a function of the length of the feedback path. As previously mentioned, oscillators of this type have been made to operate at frequencies of 10,000 Hz.

Low frequency oscillators can be made by a different design in which part of the power jet flow through the outlets is alternately collected and fed back to a control port. The feedback flow causes the separation bubble to increase in size, shifting the point of attachment outward, until the main jet switches.

Pulse Counter. This device, also known as a flip-flop or an alternator, is basic for counting and binary computation. In a pulse counter, each time an input pulse is received, the main jet flow switches to an alternate outlet. In Fig. 5, when the main jet is switched to the right outlet, a slight priming draft is set up, running counterclockwise through the heart-shaped feedback path. When a pulse signal appears, the pulse signal flow follows the path of least resistance and joins the priming draft flow. This, in turn, switches the main jet from the right to the left outlet. As long as the pulse signal is maintained, it remains locked to the right wall of the feedback path by the Coanda effect. When the pulse signal is removed, the main jet remains locked to the left outlet also by the Coanda effect. Once the main jet is switched to the left outlet, a priming draft develops in a clockwise direction, thus setting up the unit for the next pulse.

Vortex Amplifier. In this device, shown in Fig. 6, if the upper control port is open, the power jet flow is diverted radially toward the outlet (dashed line) and the resistance to flow is minimal. If the lower control port is opened and the upper port closed, the power jet is aimed tangentially toward the cylinder wall and the fluid resists moving through the outlet because of the law of conservation of momentum. The flow, however, is forced to spiral inward toward the outlet and as the radius decreases, the velocity increases, shear stresses in the

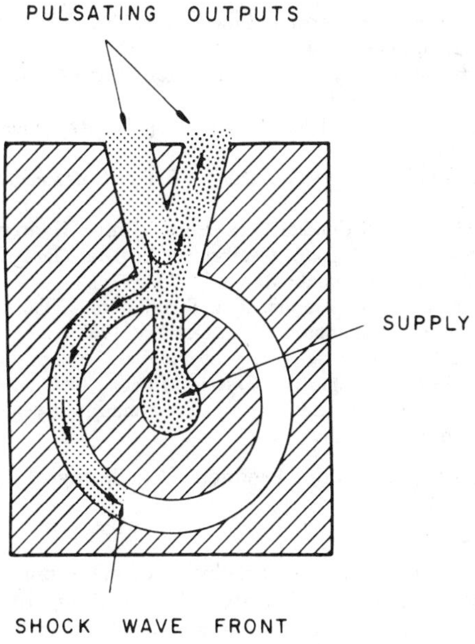

Fig. 4. Fluidic oscillator.

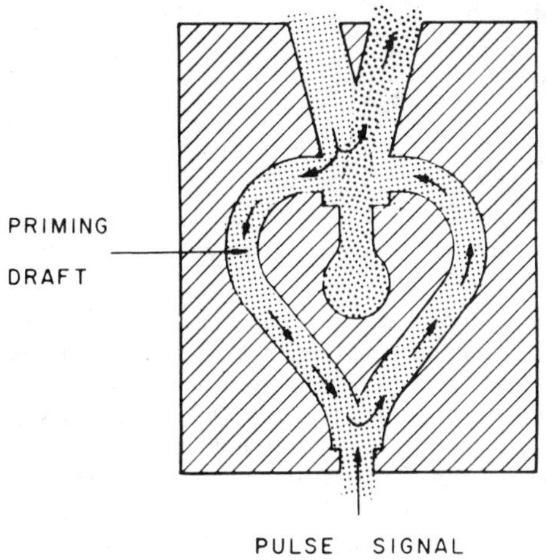

Fig. 5. Fluidic pulse counter.

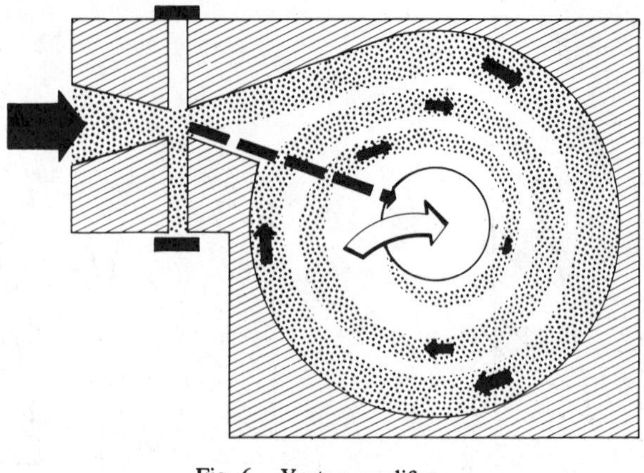

Fig. 6. Vortex amplifier.

fluid increase, and the pressure drop increases. Hence, the flow rate and pressure drop can be made to vary by changing the direction of the power jet.

Applications for Fluidic Devices

In the early 1960s, much attention was directed toward fluidic logic systems, but interest waned in favor of electronic logic as extraordinary advancements were made through integrated circuit technology. Nevertheless, fluidic logic systems and proximity sensors have been applied in control of machine tools and sequential manufacturing operations. Pulse counters are used in shift registers and counters. In areas other than logic, oscillators have been employed as temperature sensors and to provide timing functions. A fluidic oscillating spray nozzle is standard on the windshield washer system of several makes of automobiles. Among consumer products is a pulsating shower head. Major areas of development and reduction to practice have been in the field of process control valves and flow metering.

Fluidic Control Valves. These are available as two-position diverting and proportional diverting valves. They are similar in construction and operation to the bistable amplifier shown in Fig. 3. They can be applied to control gas, liquid, or slurry streams. The low-energy controlling fluid can be atmospheric air, gas, or liquid.

Figure 7 shows a 4-inch (10.2-centimeter), two-position liquid-diverting valve. It operates by opening control ports to atmosphere. The control ports can be as small as $\frac{9}{64}$ inch (3.6 millimeters). All the flow can be made to switch from one outlet to the other in $\frac{1}{10}$ second. And the switchover occurs with no shock or water hammer.

Fig. 7. A 4-inch (10.2-centimeter), two-position liquid-diverting valve.

The most common application for fluidic valves has been in controlling the flow of cooling water to the radiators on diesel-electric locomotives.

Fluidic Flowmeter. A fluidic oscillator forms the basis of a unique, in-line flowmeter which has the following advantages:

1. Linear output with frequency directly proportional to flow rate.
2. Rangeability up to 30:1.
3. Unaffected by shock, vibration, or field ambient temperature changes.
4. Calibration in terms of volume flow unaffected by changes in fluid density.
5. No moving parts; no impulse lines.

A cross section of the meter is shown in Fig. 8. The geometric shape of the meter body is such that when flow is initiated, the flowing stream attaches to one of the side walls by Coanda effect. A portion of the main flow is diverted through a feedback passage to a control port. The feedback flow increases the size of the separation bubble. This peels the main flow stream away from the wall until it diverts and locks on the opposite wall, where the feedback action is similar. The frequency of the self-induced oscillation is a function of the feedback flow rate, which, in turn, is directly proportional to the flow rate of the mainstream.

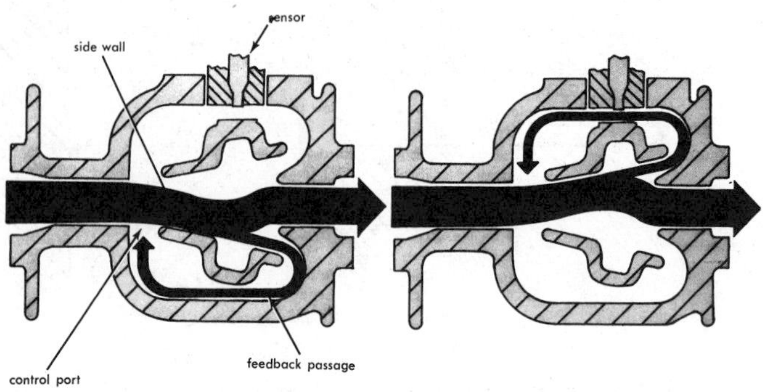

Fig. 8. Cross-sectional views of fluidic flowmeter.

A heated thermistor sensor is located in one of the feedback passages. When the mainstream is diverted toward the sensor, the feedback flow, and hence the carry-away heat, are a maximum. When the mainstream is diverted to the opposite wall, the fluid in the area around the sensor is stagnant. The fluctuation in voltage required across the sensor thus is synchronized with the oscillating frequency of the meter. Integrating these electrical pulses gives a signal proportional to the instantaneous rate of flow. The pulses can also be totalized to give an accurate reading of total flow.

References

"Advances in Fluidics," 1967 American Society of Mechanical Engineers Fluidics Symposium, Fluid Amplifier Associates, Ann Arbor, Michigan.
"Fluid Jet Control Devices," American Society of Mechanical Engineers Symposium, New York, 1962.
"Fluidic State of the Art Symposium Proceedings," Harry Diamond Laboratories, Washington, D.C., 1974.
Humphrey E. F. and D. H. Taramoto: "Fluidics," Fluid Amplifier Associates, Ann Arbor, Michigan, 1967.
Kirshner, J. K.: "Fluid Amplifiers," Fluid Amplifier Associates, Ann Arbor, Michigan, 1967.
Kirshner, J. K., and Katz: "Design Theory of Fluidic Components," Academic, New York, 1976.
"Proceedings of Cranfield Fluidics Conference," British Hydromechanics Research Association, 2nd Conf., Cambridge, England (January 3–5, 1967); 4th Conf., Coventry, England (March 17–20, 1970).
"Proceedings of the Fluid Amplification Symposium," Diamond Ordnance Fuze Laboratories, Washington, D.C., October, 1967.
"Proceedings of the 20th Anniversary of Fluidics," Symposium, American Society of Mechanical Engineers, Chicago, Illinois, 1980.
Schaedel: "Fluidic Elements and Networks" (German), Vieweg & Son, Braunschweig and Wiesbaden, 1979.

C. L. Mamzic, Moore Products Co., Spring House, Pennsylvania.

FLUIDITY. The property of a substance that expresses its ability to flow, as contrasted with viscosity, which is the resistance to flow. Fluidity is a measure of the rate at which a fluid is deformed by a shearing stress, and is mathematically the reciprocal of the viscosity.

FLUIDIZATION. This term is used in engineering to denote the preparation of a solid material having a particle size and other properties such that it may be handled, in many respects, like a fluid; and the technology of handling it under such conditions. One of the early developments in this field was pulverized coal and its development received great impetus in its use in catalytic processes in the petroleum industry, and it has since been extended far more widely. One great advantage of having a solid catalyst that flows like a liquid is that its particle size can be much smaller than possible in the old type of solid-bed catalyst, where entrainment in the passing fluid was objectionable. This smaller particle size greatly increases the surface area (which increases as the inverse square of the particle radius) and therefore increases the effectiveness of the catalyst. Another advantage of fluidization is greater ease and hence lower cost of handling, which has permitted the use of catalysts in processes that require their frequent reactivation.

Other applications of fluidization have been made to such materials as sodium chloride (table salt), soda ash, sodium phosphate, sodium sulfate, starch, talc, magnesium oxide, dry clay, boric acid, hydrated lime, and various high polymers in powdered or "bead" form. Fluidization is especially effective in loading and unloading materials from railroad cars and trucks, as well as in moving them about within the plant.

FLUID MECHANICS. The study of the mechanical properties of fluids, including hydrostatics, hydraulics, hydrodynamics and gas dynamics (compressible flow).

FLUID SEAL. Fluids are occasionally required to be held under pressure in enclosed regions through the walls of which one or more moving shafts must protrude. Unless a seal is provided the fluid will leak through the clearance, left for mechanical reasons between shaft and container, if its pressure exceeds that of the surroundings, or it will be diluted by inflow if at a lower pressure. Such seals are not difficult to provide if the shaft has no motion other than rotation or axial reciprocation. Common examples of such seals are (1) the "stuffing boxes" of double-acting reciprocating compressors, pumps, and engines where the piston rod passes through the cylinder, and (2) the stuffing boxes and "glands" of centrifugal pumps and steam turbines where rotating shafts protrude from the casing. Piston rings, although not described here, could be called fluid seals, as could also the method of sealing off fuel gas in large telescoping gas holders with a liquid. There are several applications of the floating inverted bell as a gas holder using a liquid seal.

A stuffing box is a recess in the wall surrounding the point of exit of the shaft, arranged to receive a soft and pliant packing such as treated hemp, or leather, which is compressed in the box and pressed firmly against the shaft by the pressure exerted against it from a *gland*. The gland is tightened against the packing by screw threads until leakage is reduced to a negligible amount but not enough to produce seizure or excessive frictional heating of the shaft.

Stuffing boxes are used primarily to seal against leakage around

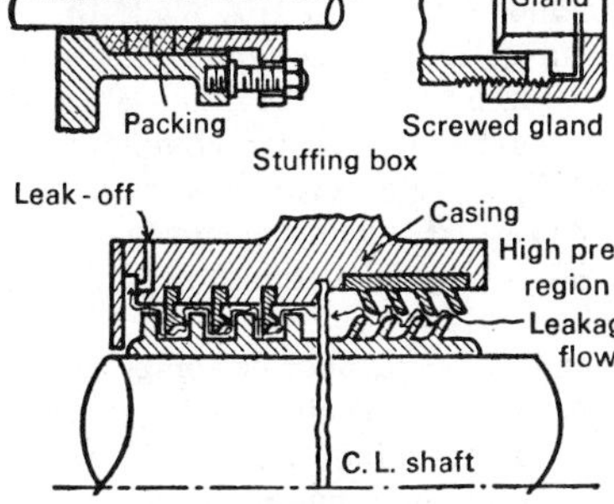

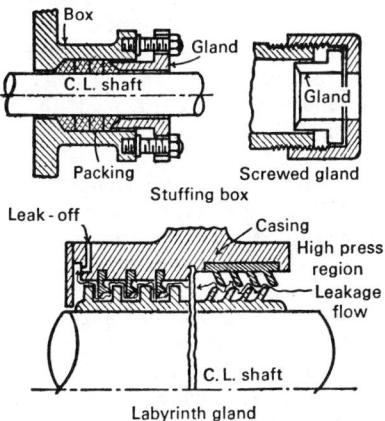

Types of fluid seals: (*Above*) screwed-gland stuffing box; (*below*) laybrinth gland.

small shafts. Leakage glands are preferred for large diameter shafting—or whenever no mechanical contact is wanted in the seal. A typical example is the *labyrinth gland* of the steam turbine. Here the steam is allowed to flow through the clearance space but is given so many turns that the fluid friction and vortices resulting from numerous changes of direction account for the pressure difference across the gland with only a small flow of the fluid through it.

The diagram shows an outflow of leakage steam. When a vacuum exists inside the casing there would be an air inflow through this gland. To prevent this the gland may be designed for admission of high-pressure steam or water to its midpoint, from which there is two-way leakage flow to the edges. Thus water or condensable steam replaces air inflow.

FLUKE. 1. A parasitic flatworm. *Trematoda.* These worms attack many species of animals, passing through a complicated life cycle involving two or three hosts. The liver-fluke of the sheep, which spends part of its life in the body of a snail and becomes adult in the liver of the sheep, is a well-known example. A number of species attack man, particularly in the Orient and in warmer countries. They fall into four groups, the liver, lung, blood and intestinal flukes, depending on the part of the body in which they thrive. 2. The broad horizontal lobes of the whale's tail. 3. Fish, *Platichthys flesus* and *Paralichthys dentatus*, belonging to the flatfish family.

FLUME. An open channel for conveying water for some special purpose, such as water power, washing, etc., is a flume. Flumes are frequently constructed of lumber having the boards placed in the direction parallel to the flow, these often being planed on the wetted side. However, flumes are also constructed of concrete, brick, etc. The flow in a flume is created by the slope of the bottom, releasing a certain amount of energy of position, which is converted into energy represented in the friction between the water and the flume, provided the flow in the flume is uniform.

Specially-designed flumes are also used in connection with the measurement of flow in open channels. *Parshall* flumes have been constructed in sizes with throat widths ranging from 3 inches to 40 feet (7.6 centimeters to 12 meters) for measuring flows up to 1,500 million gallons (56.8 million hectoliters) per day. The flume operates essentially on the principle of the weir wherein the height of the level of the flowing medium in the device is a measure of the flowing volume. See also **Flow Measurement.**

FLUORESCENCE. This term has three common usages: 1. The process of emission of electromagnetic radiation by a substance as a consequence of the absorption of energy from radiation, which may be either electromagnetic or particulate, provided that the emission continues only as long as the stimulus producing it is maintained. That is, fluorescence is a luminescence which ceases within about 10^{-8} second after excitation stops; this period of time being the lifetime of an atomic state for a normal allowed transition. 2. The term fluorescence may also be applied to the radiation emitted, as well as to the emission process. 3. In x-ray terminology, the term fluorescence may be used in the more specific sense (than given in the general definition above) to denote the characteristic x-rays emitted as a result of the absorption of x-rays of higher frequency.

FLUORESCENCE YIELD. The probability that an atom whose electronic structure has been excited will emit an x-ray photon, in the first transition, rather than an Auger electron. The value of the fluorescence yield lies between 0 and 1, characteristic of the particular state of excitation of an atom of a particular element. The K-shell fluorescence yield increases with increasing atomic number, and is the sum of the $K \rightarrow L_{II}$, $K \rightarrow L_{III}$ . . . transitions.

FLUORESCENT SCREEN. A plate coated with a material readily fluorescent. It is used to observe certain patterns or other properties of invisible radiations, such as x-rays, from the fluorescent radiations emitted by the screen. It is also used to form the visible image in cathode-ray tubes as used in oscilloscopes and television tubes.

The distinction between fluorescent and phosphorescent screens is frequently not clearly defined. Fluorescent screens are properly those with only a very short glow-period after the exciting radiation (or electron beam) is extinguished.

FLUORINE. Chemical element symbol F, at. no. 9, at. wt. 18.9984, periodic table group 7a (halogens), mp −219.62°C, bp −188.1°C, density 1.696 g/1 (gas at 0°C), 1.108 g/cm^3 (liquid at bp). Fluorine is a pale yellow gas, poisonous, very reactive, combines with most other elements in the dark, except it does not combine readily with oxygen. Critical pressure is 55 atm; critical temperature is −129.2°C. First identified by Scheele in 1771, but not isolated until 1886 by Moissan who electrolyzed fused potassium hydrogen fluoride in a platinum apparatus. Fluorine is a high-tonnage chemical, used mainly in the production of fluorides, in the synthesis of fluorocarbons, and as an oxidizer for rocket fuel.

First ionization potential 17.42 eV; second, 34.6 eV; third, 58.02 eV; fourth, 84.88 eV; fifth, 113.0 eV; sixth, 152.9 eV. Oxidation potential $F^- \rightarrow \frac{1}{2}F_2 + e^-$, −2.85 V; $2F^- + H_2O \rightarrow F_2O + 2H^+ + 4e^-$, 2.1 V; $HF \rightarrow \frac{1}{2}F_2 + H^+ + e^-$, 3.03 V. Other important physical characteristics of fluorine are given under **Chemical Elements.**

Production: Because fluorine is the most reactive element and one of the strongest oxidizing agents known, its preparation caused difficulties for many years. The requirements for fluorine for separating ^{235}U from ^{238}U during the development of the atomic bomb accelerated research on finding improved production methods. Much research went into the development of compounds and materials that would resist the actions of fluorine for use in diffusion plants. Materials finally selected and in use in modern fluorine production plants include (1) a cathode integral with a mild-steel cell body, (2) a carbon anode, (3) a steel cell head, including a Monel skirt, and (4) a Monel screen diaphragm. The skirt is required to prevent admixture of the fluorine gas formed with the hydrogen gas also formed. The electrolyte consists of a fused mixture of potassium fluoride and hydrofluoric acid. The overall reaction: $2HF \rightarrow H_2 + F_2$. Hydrogen is liberated at the cathode; fluorine at the anode. The potassium fluoride is required because hydrofluoric acid of the purity required does not conduct an electric current.

Fluorine causes both chemical and thermal burns and, unfortunately, they may not be detected immediately, depending upon the concentration. Further, upon contact with the skin, fluorine gas reacts with water in the skin to form hydrofluoric acid, an excellent solvent for protein. Personnel directly involved in the handling of fluorine must be equipped with gauntlet gloves, neoprene rubber apron, chemical goggles and when in atmospheres above the TLV, gas masks with cannisters must be used. In emergency situations where there is a very high concentration of fluorine, the area must be evacuated, directing personnel upwind.

All containers, processing equipment, and piping to be used in fluorine service first must be passivated before use and thereafter designated for fluorine service. These requirements result from the severe oxidizing characteristics of fluorine gas. Passivation removes any easily oxidized materials, such as paint, pipe dopes, metal oxides, grease, and metal filings. During the procedure, a metal fluoride film will form on metal surfaces, thus minimizing further corrosion of the metal by fluorine.

Special permits are required to ship fluorine. Generally, fluorine is transported as a nonliquefied compressed gas in seamless steel or nickel cylinders. Upon receipt, multijacketed dewars frequently are used to contain the product.

Chemistry and Compounds: Fluorine exhibits in common with the other halogen elements a marked readiness to form singly charged negative ions, as would be expected from the fact that these atoms need only one electron to acquire an inert gas configuration. However, the electron affinity of fluorine (3.74 eV) is not the highest of the four common halogens, but is less than that of chlorine and bromine (4.02 eV and 3.78 eV respectively). The greater reactivity of fluorine in aqueous solution is due to the fact that its lower electron affinity is more than offset by its lower energy of dissociation (38 kcal against 58 kcal for Cl_2 and 46 kcal for Br_2) and the higher energy of hydration (122 kcal for F^- against 89 kcal for Cl^- and 81 kcal for Br^-). The overall result is to give the system $F^- \rightarrow \frac{1}{2}F_2 + e^-$, the largest negative oxidation potential (−2.85 V) of any simple ion to its element.

The reactions of fluorine have, in general, high temperature coefficients. At low temperatures its reactivity with hydrogen is very slight, but becomes rapid and even violent at higher temperatures and in the presence of impurities. Fluorine reacts with all metals, the vigor of the reaction and the composition of the resulting fluoride depending upon the temperature and the reactivity of the metal. Sulfur, silicon, carbon, and antimony ignite in fluorine; cesium, rubidium, and potassium form trifluorides, which, however, do not contain trivalent cations, while the noble metals react only at very high temperatures. Unlike the three alkali metals mentioned, however, most metals do not form fluorides in exceptional oxidation states, and, in fact, many elements form oxyanions of higher valence than they do fluorocomplexes. Thus manganese forms two fluorides, MnF_2 and MnF_3, and its fluorocomplex ions of highest valence are MnF_6^{2-} and MnF_5^-; chromium forms four stable fluorides, CrF_2, CrF_3, and CrF_4, and CrF_5 (and possibly CrF_6).

Four binary compounds of fluorine and oxygen have been reported, O_4F_2, O_3F_2, O_2F_2, and OF_2. The polyoxygen difluorides are produced from the elements by action of the silent electric discharge at low pressures, lower temperature favoring higher oxygen content. Dioxygen difluoride is a yellow to orange solid, melting at about −160°C. It decomposes rapidly at temperatures above −25°C. The others are even less stable. Oxygen difluoride, OF_2, is prepared by passing fluorine rapidly through weak NaOH solutions or by electrolysis of liquid HF solutions of H_2O or other oxygen compounds. It has an O—F bond distance of 1.4Å, F—F, 2.22Å, and FOF angle, about 105°. It reacts with metals to form fluorides and oxygen, with other halides to form fluorides, oxygen, and the other halogens, and with other compounds usually to yield fluorides.

Hydrogen fluoride is the most stable of the hydrogen halides (heat of formation 64 kcal). Its bond moment shows the compound to have a marked covalent character. In the pure state, liquid hydrogen fluoride is slightly more conducting than pure water, and in the anhydrous state it reacts only with the alkali metals, alkaline earth metals (excluding beryllium and magnesium) and with thallium. Liquid HF is an extremely strong acid, having a Hammett acidity function of 10.2 (compared to H_2SO_4 11.3), HCl, hydrobromic and hydriodic acids being essentially unionized in it, although the dielectric constant is comparable to that of water. However, in 0.1 N aqueous solution, hydrogen fluoride is only about 15% ionized. A correlative property is the extensive polymerization through hydrogen bonding, various polymers being present. In the vapor state, the degree of polymerization depends upon temperature and pressure, varying from mostly monomer to linear hexamer or even higher polymers, with the ring hexamer being particularly favored. The liquid likewise contains monomer and polymers, the average degree of polymerization being three or four. The units of the polymers undergo very rapid exchange. In aqueous solution there is a strong tendency for fluoride ion to associate with hydrogen fluoride molecules, forming the symmetrical HF_2^- ion. It reacts with metal oxides and hydroxides to produce the fluorides, and with metal ions to produce complex ions, or their salts. It reacts with phosphorus pentoxide to produce complex fluorides, oxyfluorides, and, on continued action, mono-, di- or hexafluorophosphates.

Due to the high oxidation potential of fluorine, and the small size of the fluoride ion, the element enters into many compounds with the other halogens. Diatomic compounds of this type include ClF, BrF, and IF (the latter two having been identified but not isolated in a pure state), tetratomic compounds include ClF_3, BrF_3, and IF_3 hexatomic ones include BrF_5, and IF_5, while the octatomic type is limited to one member, IF_7. These compounds are discussed under the entry for the halogen forming the donor atom in the compound.

Fluorine forms polyhalide anionic complexes including $IFBr^-$, $IFCl_3^-$, BrF_4^-, and IF_6^-, which occur in salts of the higher alkalies, as well as cations such as BrF_2^+, which occurs in such salts as BrF_2SbF_6, BrF_2AuF_4, BrF_2SO_3F, etc. The difluoroiodate ion, $IO_2F_2^-$, is also known in such salts as KIO_2F_2.

Organic compounds: Organic fluorine compounds are made by reaction of the corresponding alkane chloro-compounds with silver fluoride, mercurous fluoride, antimony trifluoride, titanium tetrafluoride, and the arene fluoro-compounds by the diazo-reaction using hydrogen fluoride, and otherwise. The effect of the continued replacement of hydrogen atoms by fluorine atoms is an initial increase in reactivity, followed by a reversal of this effect, so that the highly substituted compounds are relatively inert. See also **Fluorocarbon.**

Biological Aspects of Fluorine. Fluorides are not required for plant growth, but in animals, including humans, low levels of fluorides have been shown to have beneficial effects on teeth and on bone structure. Growth increases in experimental animals have been reported when low levels of fluorides have been added to purified diets. However, fluoride substances show toxicity in both animals and plants when encountered in fumes and dusts from industrial facilities as well as natural emissions from the eruption of volcanoes. Abnormally high levels of fluoride in water also have caused fluorine toxicity in animals and mottled teeth in humans.

Fluorides do not usually move from the soil to plants and on to livestock feedstuffs and human foodstuffs in amounts that are toxic. Injury to plants from fluoride in the soil has been noted on soils that are too acid for the satisfactory growth of most plants. On limed soils or soils with sufficient calcium for optimum growth, any fluorine added to the soil reacts with the calcium and other soil constituents to form insoluble compounds, which are not taken up by the plants. Rock phosphate and some kinds of superphosphate fertilizers contain large amounts of calcium fluoride, but the fluorine content of the plants grown on soils that have been heavily fertilized with these phosphates is not appreciably increased. Tea and some other members of the *Theaceae* family are the only plants that take up very much fluorine from the soil.

While the soil-to-plant segment of the food chain contains some built-in safeguards against fluorine toxicity, this toxicity occurs as a result of the deposition of airborne fumes and dusts on the above-ground parts of plants, followed by the consumption of these contaminated plants by animals, including humans. Also, fluorine toxicity has been caused by direct inhalation of the fumes and dusts, or by drinking water with abnormally high fluorine levels. If the fumes and dusts are mixed into the soil, they will be inactivated and will not find their way into the food chain in toxic amounts.

The safeguards against toxicity provided by the chemistry of fluorine in soils make it unlikely that applying fluorine-containing compounds to soils will be a useful way to insure that plants will contain sufficient fluorine to prevent dental caries. However, tea and mechanically deboned meats may contribute to these needs. When increased fluoride intake is desirable, carefully-controlled direct additions to drinking water, to dentrifrices, or to specific foods are more promising than adding fluorides to soils that produce food crops.

Fluoridation. In a broad sense, this term would signify the addition of fluorine to a substance much as chlorination means the addition of chlorine. In a more specific, but commonly used sense, fluoridation means the addition of very small amounts of a fluoride-containing compound to water supplies for the purpose of preventing dental caries. It has been shown over a number of years that the introduction of about 1 part per million (ppm) of fluoride to drinking water will reduce the incidence of tooth decay in children by as much as 60% as compared with similar groups of children who consume nonfluoridated water. Because of the striking nature of these findings, a few cities in the United States commenced experimental treatment of water supplies during the mid-1940s. As of the early 1980s, it is estimated that close to 100 million persons in the United States are now supplied with fluoridated water.

The commonly used compound is sodium fluoride or sodium silicofluoride in a dry crystalline or powdered form. Hydrofluosilicic (flusilicic) acid is also used in liquid form. Inasmuch as concentrations of fluoride in excess of 1.5 ppm may cause mottling of tooth enamel, it is mandatory to exercise very careful control to maintain the desired 1.0 ppm dosage. Water supply samples are frequently tested by municipal authorities. Fluoride concentrations may be determined by colorimetric or electrometric methods, and the latter can be adapted to continuous reporting and controlling.

The concept of fluoridation has created numerous controversies among the populace, a situation that occurs frequently when decisions to install fluoridation systems for the first time are under consideration. Until the mid-1970s, it was believed that such practice was unquestionably safe and that arguments against fluoridation were essentially emotionally motivated. However, some second thoughts are now being taken, particularly with reference to possible reactions of fluorine with certain pollutants now found in raw water supplies that once were not present.

Feedstuffs. Excessive amounts of fluoride in the soil can cause tooth and bone damage in livestock. Parts of Arkansas, California, South Carolina, and Texas have soils abnormally high in fluorine content. In serious situations, diarrhea and emaciation will be exhibited by the livestock. The effects depend upon the fluorine source and species of livestock. Exceptionally high fluoride levels can be encountered near smelters where pollution safeguards have not been installed or are ineffectively maintained. As compared with other livestock, pigs can tolerate much more fluorine (up to nearly 300 ppm of fluorine derived from rock phosphates).

Fluorine in Tea. The majority of foods found in the average diet contain 0.2–0.3 ppm or less fluorine in the food as consumed. Tea and seafoods are notable exceptions (McClure, 1949). Different values are reported for fluorine content of various teas by different investigators (Wang et al., 1949; Fabre and de Campos, 1950; de Campos, 1950; Zimmerman et al., 1957; Quentin et al., 1960; Okada and Furuya, 1969; Cook, 1970; Venkateswarlu and Sita, 1971). The fluorine content of tea depends upon the origin of the plant, the type of soil and fertilizer, the age of the leaves, and the time of harvesting (Garber, 1962).

In 1978, investigators at the University of Teheran (Iran) undertook a study to find out the fluorine content of teas consumed in Iran and to evaluate the potentiality of tea as a contributor of fluorine. Tea is an important item in the Iranian diet and drunk mostly by laborators and peasants; furthermore, diluted infused tea is used as a supplement in between breast feedings of infants. The investigators concluded that, considering the optimal intake of fluorine of 1 milligram per day suggested for protection from dental caries, the drinking of tea in Iran provides about half of this amount without considering the fluorine content of water and other sources.

Fluorine Content of Mechanically Deboned Beef and Pork. One question that has arisen from time to time in connection with mechanically deboned meat (MDM) is its possible fluoride content because some microscopic bone particles may be present in the product. Investigators Kruggel and Field, Division of Biochemistry and Division of Animal Science, University of Wyoming, made a study of this in 1977. Samples were collected from regions where high levels of fluoride occurring in the water and vegetation have been reported. Higher magnesium, iron, and fluoride contents were found in beef MDM from the western and midwestern regions of the United States when compared with the southern region. Higher iron and fluoride contents were found in beef MDM than in pork MDM. One conclusion drawn was that the consumption of fluoride from MDM and other foods combined would be far below the 20–80 milligrams or more of fluoride that must be consumed daily to produce toxicity (Food and Nutrition Board, 1974). Mottling of teeth in children has been observed at fluoride concentrations in the diet and drinking water of 2–8 ppm. A frankfurter containing 10% MDM would contain about 1.7 ppm fluoride. Since the daily fluoride intake in many areas of the United States is not sufficient to afford optimal protection against dental cavities (Food and Nutrition Board, 1974), products which contain MDM may be of value in furnishing needed fluoride and in reducing the incidence of tooth decay (Kruggel/Field, 1977).

Fluorine in Marine Sponge Halichondria moorei. It is well known that many marine organisms accumulate the halogens iodine, bromine, and chlorine. However, reports of fluorine accumulation have been rare. Thus the report of findings by Gregson et al. (1979) that the marine sponge *Halichondria moorei* has a fluorine content of 10% of the total dry weight is of interest. In this species, the fluorine occurs as potassium fluorosilicate, which is known to be a powerful anti-inflammatory agent. It is of interest that closely related sponge varieties of the same habitat contain little if any fluorine—and, further, that the habitat is free of fluorine except for the small amount naturally present in seawater.

References

Allaway, W. H.: "The Effect of Soils and Fertilizers on Human and Animal Nutrition," Agriculture Information Bulletin 378, Cornell University Agricultural Experiment Station and U.S. Department of Agriculture, Washington, D.C., 1975.

de Campos, P.: "Fluorine Content of Tea Cultivated in Sao Paulo," in *Chem. Abstr.*, **44**, 22, 11498, 1952.
Cook, H. A.: "Fluoride Intake through Tea by British Children," *Fluoride Quart. Rept.* **3**, 12 (1970).
Fabre, R., and P. de Campos: "Distribution of Fluorine in Plants—Tea Leaves," *Ann. Pharm. Franc.*, **8**, 391 (1950).
Food and Nutrition Board: "Effects of Fluoride in Animals," National Academy of Sciences, Washington, D.C., 1974.
Food and Nutrition Board: "Recommended Diary Allowance," National Academy of Sciences, Washington, D.C., 1979.
Garber, K.: "Plants and Fluorine" (in German), *Qualitas Plant. Mater. Vegetabiles*, **9**, 33 (in *Chem. Abstr.*, **52**, 2, 2589, 1962).
Gregson, R. P., et al.: "Fluorine is a Major Constituent of the Marine Sponge *Halichondria moorei*," *Science*, **206**, 1108–1109 (1979).
Kruggel, W. G., and R. A. Field: "Fluoride Content of Mechanically Deboned Beef and Pork from Commercial Sources in Different Geographical Areas," *J. Food Sci.*, **42**, 1, 190–192 (1977).
McClure, F. J.: "Fluoride in Foods," *U. S. Public Health Report*, **64**, 1061 (1949).
Okada, F., and K. Furuya: "Fluoride Content in Tea" (in Japanese), *Chago Gijutsu Kenkyu 37*, 32 (in *Chem.* Abstr., **71**, 23, 245, 1966).
O'Keeffe, M., and J. O. Bovin: "Solid Electrolyte Behavior of $NaMgF_3$: Geophysical Implications," *Science*, **206**, 599–600 (1979).
Quentin, K. E., et al.: "Analytical Determination of Small Amounts of Fluorine in Foods and Waters. 4. Fluorine Studies in Foods" (in German), *Chem. Abstr.* **54**, 6, 5967, 1960.
Venkateswarlu, A., and P. Sita: "A New Approach to the Microdetermination of Fluoride Adsorption-Diffusion Technique," *Anal. Chemi.* **43**, 78 (1971).
Wang, T. H., et al.: "Fluorine Content of Fukein Teas," *Food Research*, **14**, 98 (1949).
Zimmerman, P. W., et al.: "Fluorine in Food with Special Reference to Tea" (in German), *Chem. Abstr.*, **52**, 1, 610, 1958.

FLUORITE. Fluorite is a calcium fluoride mineral CaF_2 crystallizing in the isometric system, often in superb cubic crystals. Twinned crystals are common, usually as cubic penetration twins. It is found in many diverse geological environments, from vein material associated with metallic ores, especially lead and silver, to sedimentary formations associated with celestite, gypsum, dolomite, and calcite, as a component mineral in high-temperature pneumatolytic deposits with cassiterite, topaz and tourmaline, and in pegmatites. Exceptional crystals are found in Alpine type veins on quartz crystals from Switzerland. Also occurs as massive compact to granular aggregates. Possesses perfect 4-directional cleavage planes, with uneven to splintery fracture. It is a brittle mineral with a hardness of 4 and a specific gravity of 3.180. Vitreous luster when crystallized, dull to glimmering in massive material. Colorless when pure, but shades of blue, green, yellow, brown, white, rarely rose-red and pink, are known, including intermediate color graduations of each type. Certain colored crystals appear blue by reflected light, green by transmitted light. This phenomenon may be a product of heat, ultraviolet light, pressure, or exposure to radiation, as from x-rays. Varying color zones are commonly observed in areas parallel to the crystal faces. Massive varieties may also exhibit parallel zones of varying color.

Phosphorescence is not uncommon when certain fluorites are exposed to sunlight, ultraviolet rays, or are heated. Vivid fluorescence is a common attribute of many fluorites, with blue to violet fluorescence predominant. The word fluorescence is derived from the mineral name, fluorite, owing to its strong fluorescent character.

Certain dark blue fluorite from Bavaria known as *antozonite* contains free fluorine and calcium, which when released either by grinding or exposure to cathode rays produce a distinctive odor, caused by the reaction of the fluorine with water.

Fluorite is a ubiquitous mineral and is so widespread in its occurrence that only the most noteworthy can be mentioned. The English localities at Cumberland, Durham, and Weardale are world famous. Exceptionally beautiful banded material of blue fibrous character from Derbyshire, known as Blue-John, has been much used for decorative carved pieces, such as vases and other ornamental objects. Norway has produced exceptional specimens from the famous Kongsberg silver veins, as well as yttrium-rich fluorite from northern Norway associated with rare-earth minerals. Fine material has been obtained from the Transvaal in the Republic of South Africa, Tasmania, and Australia. Large quantities of fluorite is mined in Mexico at Guadalcazar and Guanajuato.

Notable United States localities include Hardin and Pope Counties in Illinois, and also adjacent Kentucky areas where it is intimately associated with calcite, barite, quartz, with minor galena and sphalerite in sedimentary rock veins. Large deep green masses yielding exceptional cleavage octahedrons were obtained from Westmoreland, New Hampshire. Macomb, New York produced large sea-green crystallized cubes. Various sedimentary formations in Ohio have yielded fine brown crystals associated with celestite. Many occurrences are known throughout Colorado and Idaho. Optical-quality crystals have been obtained from Madoc, Ontario, Canada in association with barite and calcite; also in British Columbia near Grand Forks, and at several localities in Mexico.

Fluorite is highly valued as a flux in the manufacture of steel; also as a raw material for hydrofluoric acid. When of optical quality, the mineral is used for lens and prisms in scientific instruments. See also terms listed under **Mineralogy.**

Elmer B. Rowley, Union College, Schenectady, New York.

FLUOROCARBON. A number of organic compounds analogous to hydrocarbons, in which the hydrogen atoms have been replaced by fluorine. The term is loosely used to include fluorocarbons that contain chlorine; these should properly be called chlorofluorocarbons or fluorocarbon chlorides, since it is these which are thought to deplete the ozone layer of the upper atmosphere. Fluorocarbons are chemically inert, nonflammable, and stable to heat up to 260–316°C. They are denser and more volatile than the corresponding hydrocarbons, and have low refractive indices, low dielectric constants, low solubilities, low surface tensions, and viscosities comparable to hydrocarbons. Some are compressed gases; others are liquids. These compounds were once used extensively in aerosol packages. They are used as refrigerants, solvents, blowing agents, fire extinguishers, lubricants and hydraulic fluids—as components of complete systems.

Fluorocarbon polymers include polytetrafluoroethylene, polymers of chlorotrifluoroethylene, fluorinated ethylene–propylene polymers, polyvinylidene fluoride, and hexafluoropropylene, among others. These are thermoplastic substances, resistant to chemicals and oxidation; noncombustible; with broad useful temperature range (up to 285°C); with high dielectric constant; resistant to moisture, weathering, ozone, and ultraviolet radiation. Their structure comprises a straight backbone of carbon atoms symmetrically surrounded by fluorine atoms. These materials are available as powders and dispersions for further processing, as films, sheets, tubes, rods, tapes, and fibers. They find use in high-temperature wire and cable insulation, other electrical equipment, chemical processing equipment, coatings for cooking utensils, piping, gaskets. Among the fluorocarbon polymers are a number of fluoroelastomers. These polymers are amorphous, thermally stable, noncombustible, have low glass transition temperature (−77°C), and are generally resistant to attack by solvents and chemicals.

FLUOROMETERS. In fluorescence analysis, the amount of light emitted characteristically under suitable excitation is used as a measure of the concentration of the responsible material under observation. Thus, the method is closely related to colorimetric or spectrophotometric analysis, in which the amount of light absorbed characteristically is used to measure the concentration of the dissolved species.

The main advantage of fluorescence methods is their high sensitivity, about one part in 10^8, in many determinations both inorganic and organic. This is two or three orders of magnitude better than absorption methods, where the sensitivity is limited by the necessity of detecting a very small fractional decrease in the light transmitted by the solution.

In fluorescence, the situation is inherently more favorable. Inasmuch as zero concentration corresponds to darkness (neglecting reagent blanks), and the sensitivity depends on detecting the first faint emission of light as the concentration is increased, advantage can be taken of highly sensitive detectors, such as photomultipliers, and high-intensity ultraviolet sources for excitation. Combining these with sophisticated electronic and optical techniques has led to a remarkable achievement; under favorable conditions, it is possible to detect Rhodamine 5DGN down to the extremely low concentration of one part in 10^{12} in an instrument designed for tracing ocean currents with a fluorescent dye marker.

The use of fluorescent methods requires that the substance to be determined is fluorescent under suitable irradiation, or can be made so by a chemical reaction. Among organic substances, fluorescence is shown mainly by aromatic compounds (including such hydrocarbons as benzene, naphthalene, anthracene, and their derivatives) rather than the aliphatic series. Among the metal ions, only a few show intrinsic fluorescence, such as uranium and thallium, but many others can be determined fluorometrically by adding a specific reagent which reacts with the metal to form a fluorescent complex. Thus aluminum is complexed with the dye Pontachrome BBR, beryllium with morin, and zirconium with flavenol.

Various sources are used for exciting fluorescence, including proprietary lamps, mercury vapor lamps, and the xenon arc lamp. A tungsten lamp may be used for substances having a strong excitation band above 450 mμ. Both the desired excitation band and emission band may be isolated by means of interference filters. However, since the desired excitation band is usually in the ultraviolet, the tungsten lamp is not in general use, but may have specific applications. It gives a band spectrum and does not have the sharp line limitation of the mercury vapor lamps.

One proprietary lamp is similar to the ordinary fluorescent lighting tube, but contains a phosphor which emits an abundance of radiation in the 350 to 360 mμ region of the spectrum. These lamps are usually of 5 or 15 watts and operate with a simple starter and ballast. The phosphor emits visible light, which must be excluded by a filter. Mercury vapor lamps provide the only practical type of metallic arc used in fluorometry. These are designed to operate at high pressure or low pressure. The high-pressure type was made with a mercury arc at a pressure of about 8 atmospheres surrounded by a protective envelope.

If the emission is in the visible spectrum it may be estimated by visual comparison with standards. In any range of the spectrum, the intensity of the emission may be measured with a phototube, a barrier layer cell, or with a photographic plate and densitometer. By far the most common procedure is to use a phototube or an electron multiplier phototube attached to a microphotometer or a recorder.

Fluorometers are made, containing the lamps and measuring devices just discussed, along with filters and other components. The accompanying illustration shows the functional elements of a fluorometer.

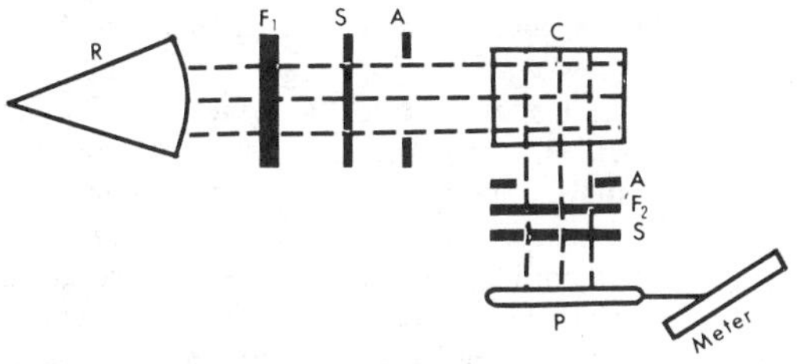

Block diagram of a simple 90°-axis fluorometer, top view: *R*, radiation source; F_1 and F_2, filters; *S*, shutters; *A*, apertures or slits; *C*, sample container.

A fluorometer constructed with two monochromators is called a *spectrofluorometer.* With the spectrofluorometer, two types of information can be obtained easily: the wavelength of best excitation and the wavelength of the strongest emission. Two curves are generally plotted on the recorder for each fluorescing material: an excitation curve and an emission curve. The *excitation spectrum* is a plot of the wavelength of the exciting source against the intensity of the emission. The excitation wavelength producing the greatest intensity of emission would seem to be best exciting wavelength. However, this statement is true only for the particular light source and grating.

See also **Analysis (Chemical); Photometers;** and **Spectro Instruments.**

FLUOROSCOPE. A device that consists of a fluorescent screen, often mounted on one wall of a dark box having a hooded opening opposite to it into which an observer may look. When x-rays or other exciting radiations fall on the screen, it glows brightly. The fluoroscope is generally used to observe x-ray shadows cast by objects and, therefore, serves as a convenient means of making preliminary x-ray examinations, both for medical and nonmedical applications. Although cardiac fluoroscopy is obsolete for routine cardiac evaluation in most cases today, the method remains the best for detection of intracardiac calcification. Fluoroscopy is used to investigate aortic stenosis (valvular heart disease). Fluoroscopy is often superior to chest x-rays because the method will usually reveal aortic valvular calcification, whereas plain chest films usually will not. Because of radiation exposure, the use of fluoroscopy is quite limited today. Some years ago, the principle was employed in the fitting of shoes, an application no longer permitted in most countries.

FLUTTER. 1. Oscillation of definite period but unstable character set up in any part of an aircraft by a momentary disturbance, and maintained by a combination of aerodynamic, inertial, and elastic characteristics of the member itself (example, buffeting). Flutter and vibration are treated theoretically and experimentally in the field of aeroelasticity.

2. In communication practice, (a) distortion due to variations in loss resulting from the simultaneous transmission of a signal at another frequency, or (b) a similar effect due to phase distortion. (c) In recording and reproducing, the deviations in reproduced sounds from their original frequencies, which result in general from irregular motion during recording, duplication or reproduction.

FLUTTER RATE. The number of frequency-excursions in cycles per second, in a tone which is frequency-modulated by flutter. Each cyclical variation is a complete cycle of deviation, for example, from maximum-frequency to minimum-frequency and back to maximum-frequency at the rate indicated. If the overall flutter is the resultant of several components having different repetition rates, the rates and magnitudes of the individual components are of primary importance.

FLUVIAL. Derived from the Latin, meaning river, and used by geologists and physiographers to denote a river as the agent, i.e., fluvial sediments.

FLUX GUIDE. In induction-heating usage, the term flux guide denotes magnetic material to guide electromagnetic flux in desired paths. The guides may be used either to direct flux to preferred locations or to prevent the flux from spreading beyond definite regions.

FLUX (Luminous). The time rate of flow of luminous energy or the radiant flux evaluated by means of the standard luminosity function.

FLUX (Physics). 1. A quantity proportional to the surface integral of the normal (perpendicular) force field intensity over a given area.

$$\text{Flux} = K\int_S F_N\,dS$$

where F_N is the normal component of a field (e.g., gravitational, electric, magnetic), and K is the constant of proportionality between the field and the flux density (permittivity, permeability, etc). 2. A term which denotes the volume or mass of fluid or particles transferred across a given area perpendicular to the direction of flow in a given time.

There are many specific applications in physics of the term flux. For electromagnetic radiation, it signifies the energy per unit time, or the power passing through a surface. For photons or particles, flux is the number per unit time passing through a surface. In nuclear physics, flux is used by many people to mean the product nv, where n is the number of particles per unit volume and v is their mean velocity. However, the ICRU in 1962 adopted the term flux density.

FLUX REFRACTION. When a ferromagnetic body composed of two pieces of different magnetic permeability is placed in a magnetic field, or when a dielectric composed of two adjacent portions of different dielectric constant is placed in an electric field, the lines of magnetic induction in the former case, and the lines of electric displacement in the latter, if oblique to the interface, abruptly change their direction. The phenomenon is thus somewhat analogous to the refraction of light. But the law is different. Whereas in the case of light, the ratio of the sines of the angles of incidence and refraction is constant, in the case of flux refraction it is the ratio of the tangents of the angles that is constant.

For an electric current flowing across a boundary between two conductors of different electrical resistivity, there is a refraction of the lines of flow, likewise obeying the tangent law.

FLUX (Slag). A material added to the contents of a smelting furnace or a cupola for the purpose of purging the metal of impurities, and of rendering the slag more liquid. The flux most commonly used in iron and steel furnaces is limestone, which is charged in the proper proportions with the iron and fuel. The slag is a liquid mixture of ash, flux, and other impurities.

FLUX (Solder). A material which by its chemical action facilitates the soldering and brazing of metals. Such a flux applied to a metallic surface cleans it and renders it receptive to amalgamation with the solder or brazing metal. Some fluxes are rosin, for soldering tin; muriatic acid, for galvanized iron and other zinc surface; and borax for brazing.

FLY. A 2-winged insect belonging to the order *Diptera.* Also commonly applied with some qualifying word to many flying insects with membranous wings, such as May fly, dragon fly, stone fly, and caddis fly. These four examples belong to as many different orders.

Flies are probably responsible for the transmission of more infectious diseases than any other insect. True flies have only one pair of wings.

FLY-BACK. 1. The shorter of the two intervals of time which comprise a sawtooth wave. 2. The retrace motion of an electron-beam, as, for example, in a picture tube.

FLYING FISHES (*Osteichthyes*). Of the family *Exocoetidae*, the flying fishes have the ability to jump and sail through the air. As described under the entry on **Characids** (*Ostariophysi*), the South American flying hatchet fish is probably the only fish that has true flying ability, that is in applying motive power while in the air and thus an ability to extend its jump. This is accomplished in the flying hatchet fish by movement of the pectoral fins during flight. In a lesser degree, this may also be true of the African freshwater butterfly fish. The other known flying fishes, although capable of spectacular "flights," are essentially jumpers and gliders. The *Cypselurus californicus* is the largest known of the flying fishes. It ranges from southern California down to Baja California. It has four wings in an 18-inch (46-centimeter) form. Of the flying fishes, flights up to 150 feet (45 meters) and of 3 seconds duration have been noted. Exceptional flights up to 13 seconds, with an average speed of about 35 miles (56 kilometers) per hour, also have been recorded. A well known 2-wing form is the 10-inch (25-centimeter) *Exocoetus volitans,* found worldwide in tropical waters. See also **Gurnards.**

FLYWHEEL. Prior to the serious concerns over future energy sources, the flywheel was normally envisioned in its conventional use as a way to steady the speed of rotating equipment. The ability of a flywheel to provide remarkable assistance to speed control of a rotating machine lies in its capacity to absorb and release energy with small variations in speed. This is also the quality of a flywheel that can make it useful as an energy-storage medium in certain kinds of energy systems.

Since the kinetic energy contained by a rotating flywheel is $\frac{1}{2}I\omega^2$, I being the moment of inertia of the mass about the center of rotation, ω being the angular velocity in radian units, energy will be absorbed when ω changes slightly only upon the condition that I be a large quantity. Consequently, flywheels are characterized by large moments of inertia. Mass alone is no criterion for the "flywheel effect," however, because the moment of inertia is involved as well as the disposition of the mass. Thus, the shape of the body becomes important. In practice, not only are flywheels massive, but also the mass is placed *as far as practical from the center of rotation*, as provided, for example, by a heavy rim. The flywheel effect, then, is obtained whenever a large mass having a large moment of inertia about its center of rotation is constructed. The large moment of inertia about the center of rotation steadies the rotational motion in the face of uneven power impulses through the process of absorbing or releasing kinetic energy by slight changes of angular rotation.

In the hydraulic turbine field, the flywheel effect has a very special meaning. The magnitude of the flywheel effect of a hydraulic turbine is determined by the weight of its rotating element, multiplied by the square of the radius of gyration of the same.

Energy Storage. A major problem in attempts to use so-called unconventional sources of energy is the severe limitations imposed by the need to store energy from a cyclical source (direct heat radiation from the sun, mechanical energy from the sun in the form of wind, etc.). Central electric power-generating stations, which are now tied together by way of vast interconnecting networks in many countries, and which also have oversize units to handle peak loads, are to be contrasted with smaller wind-generating facilities which, for example, located in remote regions or developing countries, cannot utilize a grid for the transfer of energy. Thus, the possible use of flywheel technology is being considered along with batteries and other less conventional energy storage approaches.

The theoretical limit placed upon the amount of kinetic energy which a flywheel can store is the strength or usable stress of the wheel. In other words, what is the safe rotating speed beyond which there is a risk of the wheel disintegrating (flying apart)? As observed by Millner (*Technology Review* (*MIT*), **82**, 2, 33–40, (1979), considering the present state of the art, "The cost of storage, say per kilowatt-hour, is proportional to the strength-to-weight ratio of the material and its cost per pound. High strength per unit weight is important only in weight-limited applications, such as vehicles; for low-cost stationary systems, expensive materials such as titanium, steel, and aluminum, from which flywheel rotors traditionally have been made, are precluded. Rotors made of these metals cost over $200 per kilowatt-hour stored."

Researchers at the M.I.T. Lincoln Laboratory indicate that, considering a material capable of storing 10 to 20 watt-hours per pound (22 to 44 watt-hours per kilogram) would have to weigh from 1 to 2 tons (0.9 to 1.8 metric tons). It is envisioned that such a rotor would be from 3 to 4 feet (0.9 to 1.2 meters) in diameter and about 3 feet (1 meter) in thickness. This could be appropriate for a solar- or wind-powered residence, for example, that consumes some 25 kilowatt-hours of energy per day and needs 1 day of storage. For a large, commercial installation, such as a school, apartment building, or shopping center, the rotor would have to be about ten times heavier and perhaps have a diameter of about 6 feet (1.8 meters). For comparison, a utility substation with the requirement to store ten times the energy of a commercial situation would have to weigh some 20 tons (18 metric tons) and perhaps be 12 feet (3.6 meters) in diameter. In terms of safety, such wheels could be installed in pits in the ground which would absorb the energetic impact of a disintegrating rotor.

Some materials (*anisotropic*) have their strength principally in one direction and experiments have shown that they can outperform solid steel, for example, as material for a rotor. Examples of anisotropic materials that may be appropriate for flywheels include high-quality steel wire and coated glass fibers. New materials along these lines are being developed. A "bare-filament" flywheel made from such materials has been likened to a spool of yarn. The fragmentation patterns of these structures are also less dangerous, as they have a tendency to disintegrate more slowly (unwinding, so to speak).

To take energy out of storage from a rotating flywheel, conversion to electricity, using the rotor as part of an electric generator, is the immediately obvious means. Conventional motor–generator design requires either brushes or slip rings to transmit current between stator and rotor. Investigators have been working on a brushless motor–generator contained inside a vacuum chamber. Numerous configurations have potential. One of these is a special circuit called a *cycloconverter.* As described, this device is made up of a group of electronic switches that are placed between the flywheel output phases and the load. In a programmable way of manipulating the switches in accordance with a timing sequence, high frequency output of a flywheel's motor–generator can be made to approximately duplicate a desired output voltage.

Another important aspect that must be studied and perfected in instances where a flywheel is expected to store energy for comparatively long periods (days or weeks) is the need for a minimum of drag (as from air) on the rotor and very minimal friction in the bearings.

Bearing specifications for the task are quite demanding. Considering the residential-size flywheel previously described, the bearings must support a ton or two of mass and revolve at from 10,000 to 15,000 revolutions per minute. The bearings also must be able to tolerate some conditions of imbalance. Researchers have not found ball, roller, and needle bearings very successful under these conditions. Magnetic bearings, which remove physical contact between rotor and stator and are used successfully in the aerospace field, appear to have considerable potential for energy-storing flywheels. Researchers have indicated that about ten pounds (4.5 kilograms) of magnets can support the weight of the rotor. A servosystem can be used to stabilize the rotor position. Particularly when installed in an evacuated system, the life expectancy of such bearings is expressed in terms of several years.

References

Cormack, A., and W. Schmill: "Integrated Power/Attitude Control System Study," Rockwell International Corp., NADA CR-2383, National Aeronautics and Space Administration, Washington, D.C., 1974.

Mallon, B., and R. Kuhn: "Bibliography for Flywheel Energy Storage Systems," D.O.E./STOR, UCRL-52637, 1979.

Millner, A. R.: "Flywheel Components for Satellite Applications," M.I.T. Lincoln Laboratory, Tech. Note 1978–4, 1978.

Millner, A. R.: "Flywheels for Energy Storage," *Technology Review* (*MIT*), **82**, 2, 32–40 (1979).

Staff: "Economic and Technical Feasibility Study for Energy Storage Flywheels," Rockwell International Corp., ERDA-76-65, UC-94B, 1975.

Strnat, K (editor): "Proceedings of the Third International Workshop on Rare Earth Cobalt Permanent Magnets and Their Applications," Univ. of Dayton School of Engineering, Dayton, Ohio, 1978.

FOAM. A tightly packed aggregation of gas bubbles, separated from each other by thin films of liquid. If foams were not so common, their existence would cause some wonderment. None of the obvious properties of a liquid would lead one to suppose that thin liquid films could sustain themselves for any appreciable time against the effect of gravity. The existence and stability of a foam depend, in fact, on a surface layer of solute molecules, which form a structure quite different from that of the underlying liquid inside the interbubble film.

At the surface of a liquid, molecules are in a state of dynamic equilibrium, in which the net attractive forces exerted by the bulk of the fluid cause molecules to move out of the surface; this motion is counterbalanced by ordinary diffusion back into the diluted surface layer. The equilibrium results in the surface layer being constantly less dense than the bulk fluid, which creates a state of tension at the surface. The tension can be somewhat relieved by adsorption of foreign molecules either out of the bulk solution, or out of the vapor phase. Soluble substances that have a strong tendency to concentrate in the surface layer are collectively known as surface-active agents; examples are soap, synthetic detergents, and proteins. The excess concentration of solute at the surface reduces the surface tension of water. The general relation, in the form of a differential equation, was first deduced thermodynamically by Gibbs. It is called the Gibbs adsorption theorem, and is

$$\mu = -\frac{c}{RT}\frac{d\gamma}{dc}$$

where μ is the excess concentration at the surface, c is the bulk concentration, and $d\gamma/dc$ is the change of surface tension with concentration of solute.

An excess of solute at the surface, as measured by $+\mu$, can be termed *positive adsorption* to distinguish it from an excess of solvent at the surface ($-\mu$), or *negative adsorption.* According to the Gibbs equation, positive adsorption and the lowering of surface tension always appear simultaneously. When a fresh liquid surface is newly created, however, and before the excess solute molecules have had time to diffuse to the surface, the surface tension must remain high.

Lord Rayleigh showed, by means of a vibrating jet experiment, that about 5 milliseconds are required for the surface tension of a fresh surface to reach equilibrium. During this time, the tension continuously declines from a high initial value of about 70 dynes/centimeter to a final equilibrium value of about 35 dynes/centimeter. The cause of the stability of a foam film resides in this effect. Should, for any reason, the equilibrium surface layer be disturbed, fresh surface is created and the tension immediately increases; a difference in surface tension cannot, however, be sustained for long because of the mobility of the liquid, which flows in response to the higher tension toward the area in which it has appeared. The first reponse of the liquid is no doubt just at the surface, but a considerable quantity of the bulk liquid is dragged along to the area of high tension. The following simple experiment illustrates the effect. Pour a layer of water on to a thin metal plate, and touch the underside of the plate with a piece of ice. A high surface tension is created in the cold water just above the ice, and the motion of the surrounding liquid toward the colder area is noted immediately.

On a foam film, the stress that creates regions of higher surface tension is always present. The liquid film is flat at one place and curved convexly at another, where the liquid accumulates in the interstices between the bubbles. The convex curvature creates a capillary force that sucks liquid out of the connected foam films (Laplace effects), so that internal liquid flows constantly from the flatter to the more curved parts of the films. As the liquid flows, the films are stretched, new surface of higher tension is created, and a counter-flow across the surface is generated to restore the thinned-out parts of the films (Marangoni effect). In this way, the foam films are in a constant state of flow and counter-flow, one effect creating the conditions for its reversal by the other. Pure liquids do not foam because of the absence of a Marangoni effect. See accompanying figure.

Dynamic equilibrium in a stable foam film. The Marangoni effect reverses the destructive action of the Laplace effect.

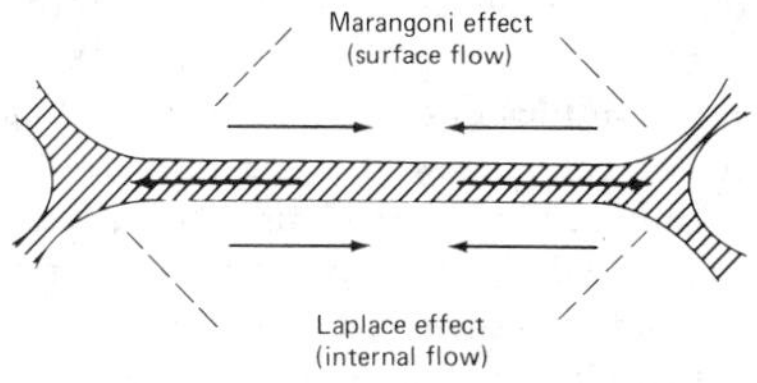

The Marangoni effect maintains the stability of the foam films even against other disruptive actions, such as hydrodynamic drainage, that, like the Laplace effect, cause stretching of the films. To the Marangoni effect can be traced all the resilient ability of foam films for elastic recovery after external mechanical shock. Foam films sometimes have this property to a remarkable degree. Lead shot, cork balls, mercury drops, and jets of water can be dropped through some foam films without causing rupture. In the fragility and brittleness of aged foams are seen the effects of impaired resilience, probably due to the extreme depletion of solution from old films by prolonged drainage.

While the primary stabilizing factor in foam is the resilience of the film, provided by the Marangoni effect, in special cases additional surface-layer phenomena are significant. These include gelatinous surface layers and low gas permeability. Such effects can add significantly to the stability of the foam, resulting in such relatively stable structures as meringue, whipped cream, fire-fighting foams, and shaving foams.

When a foam is first produced—as by bubbling a gas through a liquid, each bubble is a little sphere, separated from its neighbors by thick liquid partitions. But soon after the foam is formed, a large amount of liquid drains away by gravity and the spheres of gas become closer together. At this stage, there is a passage of gas from one bubble to another, through the curved liquid convexities that separate them.

Bubble Pressure. The pressure within a bubble of gas in liquid is greater than the pressure in the surrounding liquid by $2\gamma/R$, where γ is the surface tension and R is the bubble radius. If, as in a soap bubble, a bubble of gas is separated from the surrounding gas by a thin film of liquid, the pressure difference is twice this value.

In a foam, where γ is the same for every bubble, the pressure inside each bubble is inversely proportional to its size, i.e., the gas inside the smaller bubbles is at a higher pressure than the gas inside the larger bubbles. When, through drainage, the bubble wall becomes thin enough to be permeable, the gas in the smaller bubbles diffuses into adjacent larger bubbles to equalize the pressure. This spontaneous process increases the average bubble size without any coalescence of bubbles taking place by film rupture. The final, stable equilibrium product is a fragile, honeycomb structure, in which the separating films have plane surfaces. At this stage of the life-history of a foam, it is particularly vulnerable to external mechanical or thermal shocks, or air-borne contamination. Sir James Dewar kept plane soap films in a horizontal position inside a closed bottle for several months. In the open air, the bubbles would have ruptured almost instantly. An interesting treatise is "Science of Soap Films and Soap Bubbles," by C. Isenberg, Tieto Ltd., Clevedon, Avon, England, 1979.

When comparing the foam stability of one solution with that of another, it is necessary to attend to the means by which foam is made, inasmuch as foams of quite different characteristics can be obtained from the same solution by different treatments. The area of interfacial surface, and the mechanical efficiency with which it is created, are the determining factors. The smaller the bubble, the more persistent the foam; and more foam can be produced if excessive agitation is avoided.

Different physical properties of foams are utilized in various industrial applications. In ore flotation, advantage is taken of the presence of an air–liquid interface and of the buoyancy of the bubble. Finely divided solid particles that are not wetted by aqueous solutions serve as the partitions between bubbles, and so rise with the foam. Thus, hydrophobic sulfide particles can be separated from hydrophilic silica. In fire-fighting foams, use is made of the ability of a foam to retain a noncombustible gas (such as carbon dioxide), thus preventing air from reaching the fire. Foams in which the liquid phase is solidified are useful because of their insulating property and the low density conferred by the gaseous phase. Solid foams can be flexible, as in sponges, because the foam cells have all been ruptured in the course of production. Cellulose sponges, foam rubber, and polyurethane foam are examples of this type. See also **Foamed Plastics.** Solid foams with closed cells are also made to form rigid structures.

In many industrial processes, excessive foaming of liquid causes waste and delays. Excessive foaming is particularly troublesome in the paper industry, the refining of beet sugar, the manufacture and use of glue, and numerous other examples are found in the food processing industry. The foaming can often be inhibited by the addition of an insoluble liquid that is able to spread spontaneously, by virtue of surface-tension forces, over the surface of the foam films even as they are being formed. The spreading of the insoluble droplet is so violent, and the spreading liquid drags along with it so much of the underlying film, that a hole is gouged in the film, which is thus destroyed.

Defoaming agents may be classified as solubilized surfactants, as dispersions of hard particles, and as dispersions of soft praticles. The classifications more often than not overlap. In all cases, a liquid non-aqueous vehicle is present, even where the defoamer is represented as a solid formulation. Water may also be present, particularly in emulsified silicone formulations.

NOTE: This is part of a longer article, much of which has been contributed by S. Ross, in 'The Encyclopedia of Chemistry,' 4th edition (D. M. Considine, editor-in-chief), Van Nostrand Reinhold, New York, 1983.

FOAMED PLASTICS. The great variety of uses for foamed rubber has in some measure brought about, and in some measure been accompanied, by an even greater development of foamed plastics. In fact, one of the most striking features of this development in plastics has been the wide variety of processes, and the consequent great range of products.

Foamed plastics range in density anywhere from $\frac{1}{10}$ to 65 pounds per cubic foot (1.6 to 1041 kilograms per cubic meter); they range in consistency from rigid materials suitable for structural use to flexible substances for soft cushions; they range in cellular formation from the open- or interconnecting-cell type to the closed- or unicell type. Their electrical, thermal, mechanical and chemical properties show a similar variation. Many types of foamed plastics may be produced "on the job"; this procedure often offers important production or construction advantages.

Many methods have been developed for the manufacture of foamed plastics, and void-containing plastics in general. For convenient discussion, they may be classified in three groups: (1) Methods for adding gas to the plastic mass during processing; (2) methods for producing gas in the plastic mass during processing; and (3) methods for forming a plastic mass from granules, and thus obtaining a cellular structure.

1. An obvious method of forming a foamed plastic is to whip air into the plastic mass before it sets. This method is used for ureaformaldehyde and polyvinyl-formaldehyde plastic foams. A related method is to introduce air (or some other gas or volatile solvent) into the plastic mass, and then to form pores of the desired size by expanding the gas bubbles by application of heat, or reduced pressure.

2. Methods for producing gas chemically during the reaction or reactions that produce the plastic are as almost varied as these reactions themselves. For example, a polyester resin and an aromatic diisocyanate react to form a resin prepolymer, which then reacts with water to form a urethane polymer (plastic). Since carbon dioxide gas is also formed in this reaction (which is a condensation), its presence causes the urethane resin to be cellular. Another example is a condensation which eliminates water from the reacting molecules. This process is used in making phenolformaldehyde resin foams, where the reaction is so exothermic (yields so much heat), that the water is produced as steam bubbles that expand the plastic. However, some reactions for forming plastics do not yield gaseous by-products; in such cases gases may be produced by adding to the materials a "blowing agent," that is, a substance which decomposes to form a gas, at a temperature below the gel temperature of the plastic. Substances so used include dinitroso compounds, such as dinitroso pentamethylenetetramine, and hydrazides, such as benzene sulfonyl hydrazide. These compounds evolve nitrogen gas, but inorganic blowing agents, such as bicarbonates which evolve carbon dioxide, have also been used. This technique is applied with many kinds of plastics, including polyethylene, silicone, epoxy, and vinyl resins.

3. An example of the granular type of foam-forming plastics is the polystyrene product furnished as small granules for this use. They expand on heating, and fuse together to form rigid unicellular materials.

FOCAL COLLIMATOR. A type of collimator consisting of an objective lens at one end of a tube, and a pair of cross hairs placed accurately in its focal plane at the other end.

FOCAL LENGTH. The distance from the object or image principal point to its corresponding focal point is the object or image focal length.

FOCAL POINT. The object and image points in an optical system conjugate to infinity.

FOCOMETER. An instrument for measuring the focal length of an optical system.

FOCUSING. The process of controlling the convergence and divergence of a beam of particles or radiations.

FOCUSING COLLISION. The focusing phenomenon when a bundle of energy passes down a row of atoms following on the collision of an energetic particle with a lattice atom in a crystal. This phenomenon is important in estimates of energy losses when considering radiation damage in crystals.

FOG AND FOG CLEARING. Fog is a hydrometeor, a visible aggregate of minute droplets or ice crystals suspended in the atmosphere near the earth's surface, the result of condensation and consequent formation of water droplets or ice crystals in the atmosphere. Fog differs from cloud only in that the base of fog is at the earth's surface, while clouds are above the surface. It is easily distinguished from haze by its appreciable dampness and gray color, and it reduces visibility to a greater extent than mist. Fogs of all types occur when the temperature and dew point of the air becomes identical (or nearly so), provided that sufficient condensation nuclei are available. When this identity of temperature occurs through the cooling of the air to its dew point, fogs are produced.

According to international definition, fog reduces visibility to below 1 kilometer (0.62 mile). In United States weather observing practice, fog that hides less than 0.6 of the sky is called *ground fog.* At one time, a mixture of smoke and fog was termed *smog.* However, the meaning of smog has broadened in recent years to include numerous combinations of air pollutants, with or without fog present. See also **Pollution (Air).**

Temperature Structure in Fog. All fogs, except steam fog, are capped by a layer of air whose temperature is greater than the temperature of the fog. This inversion of temperature sometimes reaches magnitudes of 50°F (~ 28°C). Inversions of temperature thermally isolate the foggy air from the warmer, usually much drier and clear air above. Fog usually will not clear unless the inversion is destroyed or swept away. Over warm ocean waters and in the subtropics and tropics, inversions of temperature near the earth surface are rare. For this reason, fogs are also uncommon over that zone of the earth (between 30°N and 30°S).

Supercooled fog is fog having a temperature less than 0°C. It consists of small droplets at temperatures less than freezing which can exist as liquid down to approximately −40°C. Below this temperature, ice crystals tend to form automatically and the fog changes to ice fog.

Fog Seeding for Thinning and Dispersion. Supercooled fog, when "seeded" with dry ice (solid carbon dioxide) particles or with vaporizing liquid propane gas, will change to ice crystals which fall to the ground and clear the fog. Dry ice particles falling through supercooled fog leave a trail of very small ice crystals which become nuclei upon which the water vapor of the supercooled droplets is deposited. The ice nuclei grow and the droplets disappear. Much of the ice crystal particles fall to the ground, thus clearing the fog. Vaporizing liquid propane causes excessive cooling which, in turn, stimulates the creation and growth of ice nuclei. These ice particles act in the same manner as in the case of dry ice seeding, and the fog is cleared.

Commercial airports have been cleared of supercooled fog for many years by dry ice seeding as a standard practice. Military air fields have been cleared by use of both propane and dry ice methods.

Warm fog is fog wherein the air temperature is above freezing. The aforementioned methods of seeding do not function in warm fog. Unfortunately, 95% of all fogs are warm fogs that blanket airports, highways, and harbors. Warm fogs have been seeded with a variety of chemicals and agents in an attempt to clear, or at least to thin them. None has proven more than slightly successful. Ordinary table salt has proven to be the most useful, but it is corrosive and ecologically unacceptable. A variety of organic and inorganic chemicals in various combinations, with and without electrical charges, has been tried, but with only slight success. A very fine spray of water droplets that are highly charged electrically when injected into warm fog has some promise.

Helicopter downwash has been successful in a very limited number of cases. To be successful, the fog must be very shallow and the air above the fog both dry and warm. Exhaust plumes from jet engines have had some success in a number of European airports, notably at Orly (Paris).

Basically, however, the problem of warm fog clearing remains unsolved.

Specific Types of Fog

In addition to the types of fog already described, specific types include:

Advection fogs owe their existence to the flow of air from one type of surface to another. Surface temperature contrast between two adjacent regions is necessary for their formation. The usual type of advection fog is formed when relatively warm and moist air drifts over much colder land or water surfaces. Over land, examples of this type are found when moist air drifts over snow-covered areas; over water, when moist warm air drifts over currents of very cold water. The latter happens with southerly or easterly winds blowing from the Gulf Stream over the Labrador Current. Coastal and lake advection fog forms when warm and moist air flows offshore onto cold water (summer), or when warm moist air flows onshore over cold or snow-covered land (winter).

Frontal fogs, associated with fronts, can occur when (a) warm and cold air masses are mixed in a frontal zone, (b) air is suddenly cooled over moist ground, or (c) precipitation falls into cold stable air and raises the dew-point temperature. In the latter precipitation-type fog, the greater the temperature difference between the relatively warm rain (or snow) and the colder air layer, the more rapidly will the fog develop.

Radiation fog is a major type produced over a land area when radiational cooling reduces the air temperature to or below its dew point. Thus, a strict radiation fog is a nighttime occurrence, although it may begin to form by evening twilight and often does not dissipate until after sunrise. It forms over land and not over water because water surfaces do not appreciably change their temperature during hours of darkness.

Steam fog (also called *sea smoke, arctic fog,* or *sea mist*) is most commonly formed when very cold air drifts across relatively warm water. No matter what the nature of the vapor source (warm water, industrial combustion exhaust, breath), its equilibrium vapor pressure is greater than that which corresponds to the colder air; thus, the water vapor, upon becoming mixed with and cooled by the cold air, rapidly condenses.

Upslope fog is formed when air flows upward over rising terrain and is, consequently, adiabatically cooled to or below its dew point. It will form only in air that is convectively stable, never in air that is unstable, because instability permits the formation of cumulus clouds and vertical currents.

See also **Clouds and Cloud Formation; Precipitation and Hydrometeors; Visibility;** and other entries listed under **Meteorology.** For references, see entries on **Climate;** and **Meteorology.**

Peter E. Kraght, Certified Consulting Meteorologist, Mabank, Texas.

FOG TRACKS. Linear regions of condensation, produced in air or other gases that are supersaturated with water vapor, by the passage of electrified particles. Fog tracks are useful in following the courses and collisions of such particles.

FOKKER-PLANCK EQUATION. An equation originally occurring in the theory of diffusion when drift is taken into account. It may be written in the form

$$\frac{\partial v(x, t)}{\partial t} = -2c\frac{\partial v(x, t)}{\partial x} + D\frac{\partial^2 v(x, t)}{\partial x^2}$$

where $v(x, t)$ is the probability density for displacement x at time t, D is the diffusion coefficient and c represents drift. The equation occurs in the theory of stochastic processes as a limiting case of random walk or additive processes.

FOLIC ACID. Sometimes referred to as the antianemia factor or folacin and earlier called vitamin B_c, vitamin M, and the *L. casei* factor, the chemical name for folic acid is pteroylglutamic acid. Most animals require folic acid. The substance is synthesized by bacteria in some vertebrates, including human, rat, dog, pig, and rabbit. Exogenous sources are required by most other vertebrates and invertebrates. In ruminants, synthesis of folic acid occurs in the rumen, but some researchers believe that newborn lambs require a dietary supplement. The most common manifestation of a deficiency in livestock is development of a characteristic macrocytic, hyperchromic anemia (also called megaloblastic anemia). Bone marrow changes, red cells are large and immature, usually with an accompanying reduction of white cell numbers. Folic acid deficiency in poultry retards growth. Other disorders include glossitis, diarrhea, gastrointestinal lesions, intestinal malabsorption, and sprue.

In 1931, Wills demonstrated a factor from yeast active in treating anemia. In 1938, Day et al. found yeast or liver extracts active in treating anemia in monkeys. Hogan and Parrot, in 1939, showed how anemia in chicks could be prevented by using liver extract. The *L. casei* growth factor was isolated from liver and yeast by Snell and Peterson in 1940. Hutchings et al., in 1941, found the *L. casei* factor also essential for chicks. Also, in 1941, Mitchell, Snell, and Williams isolated bacterial (*S. lactis* R.) growth factor similar to *L. casei* factor from yeast and named the substance folic acid. Stokstad, in 1943, reported *L. casei* factor from liver more active than from yeast; and provided evidence of multiple factors. Pteroylmonoglutamic acid was finally isolated, the structure proved, and the substance synthesized by Angier et al. in 1946. Commercial production of folic acid is either by extraction from yeast or liver, or by synthesis wherein 2,3-dibromopropanol, 2,4,5-triamino-6-hydroxypyrimidine, and para-aminobenzoyl glutamic acid are reacted.

Folic acid and derivatives are involved biologically in the synthesis of nucleic acid, coenzyme in purine–pyrimidine metabolism, serine–glycine conversion, differentiation of embryonic nervous system, one-carbon transfer mechanisms, metabolism of tyrosine and histidine, formation of active formate and methionine, and the synthesis of choline. Antagonists of folic acid include aminopterin (4-amino-pteroylglutamic acid), methotrexate (amethopterin), pyrimethamine, and 4-amino-pteroylaspartic acid. Synergists include biotin, pantothenic acid, niacin, vitamins B_1, B_2, B_6, B_{12} C, and E, somatotrophin (growth hormone), and testosterone.

Folic acid coenzymes are derivatives of tetrahydrofolic acid. See also **Coenzyme.** Structurally, these are:

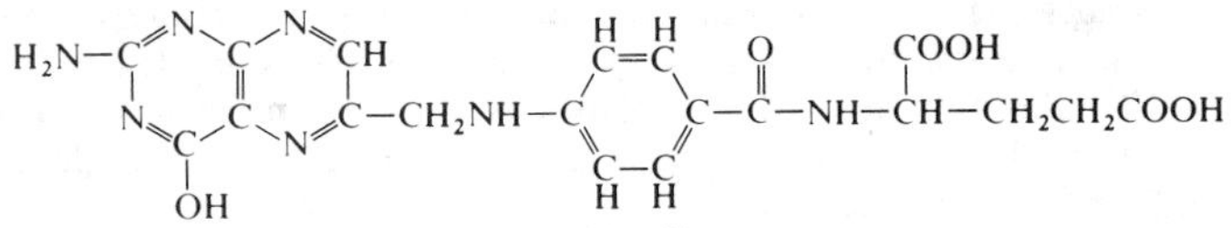

Folic acid

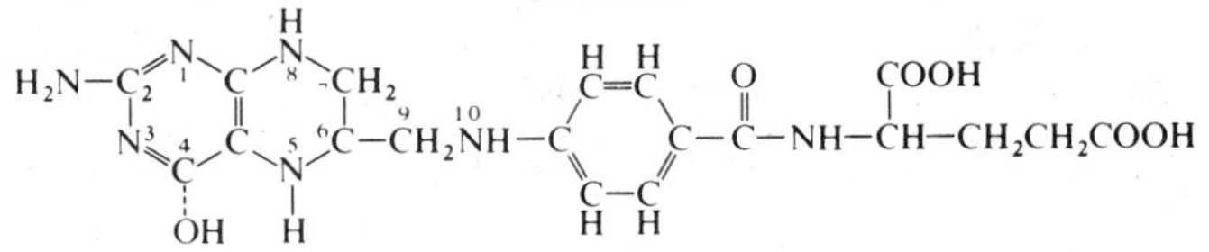

Tetrahydrofolic acid

One-carbon fragments in various oxidation states are: (1) formyl (—CHO); (2) hydroxymethyl (—CH_2OH): and (3) methuyl (—CH_3). The coenzyme forms of folic acid have one of these groups attached to either the 5-N or 10-N of tetrahydrofolic acid. One folic acid coenzyme, methyltetrahydrofolate (CH_3—FH_4) transfers in methyl group to homocysteine to yield methionine, in a reaction which also requires a vitamin B_{12} coenzyme:

$$\underset{\text{Homocysteine}}{HS-CH_2CH_2\overset{\displaystyle NH_2}{\overset{|}{C}}HCOOH} + CH_3-FH_4 \xrightarrow{B_{12}\text{ coenzyme}} \underset{\text{Methionine}}{CH_3-S-CH_2CH_2\overset{\displaystyle NH_2}{\overset{|}{C}}HCOOH} + FH_4$$

As pointed out by Chen and Cooper (Division of Food Science and Nutrition, California State University, Northridge, California), the group of compounds denoted by the term *folacin* is a heterogeneous group of derivatives with a similar basic structure and biological function. Folic acid is the basic structural unit in these compounds. Other monoglutamate folates are formed when the pteridine moiety of this basic molecule is reduced or substituted in the 5-N or 10-N position. In addition, all of these monoglutamate folates may be transformed into polyglutamates of various length by the addition of glutamic acid residues to the basic molecule.

Distribution and Sources. In the biosynthesis of folic acid, para-aminobenzoic acid, glutamic acid, and some substances yet to be identified serve as precursors. Para-aminobenzoylglutamic acid is an intermediate in the synthesis. In plants, folic acid is produced within the leaves, seeds, and cereal germ. Production also occurs in algae, fungi, and bacteria, as in the intestines of the species previously mentioned. The liver is the primary storage site.

Natural sources of folic acid include the following:

High folic acid content (90–300 micrograms/100 grams)
Asparagus, dry beans (lentils, limas, navy), liver (beef, chicken, lamb, pork), spinach, wheat bran, yeast
Medium folic acid content
Beef kidney
Low folic acid content
Most fruits, nuts, vegetables, grains, and dairy products.

Bioavailability of Folic Acid. Factors which cause a decrease in bioavailability include (1) high urinary excretion; (2) destruction by certain intestinal bacteria; (3) increased urinary excretion caused by vitamin C; (4) presence of sulfonamides which block intestinal synthesis; and (5) a decrease in absorption mechanisms. Increase in bioavailability can be provided by stimulating intestinal bacterial synthesis in certain species. No toxicity due to folic acid has been reported in humans.

Some of the unusual features of folic acid noted by investigators include: (1) Folic acid antagonists used in cancer therapy with temporary remissions; (2) folic acid occurs in chromosomes; (3) folic acid is distributed throughout cells; (4) needed for mitotic step metaphase to anaphase; (5) antibody formation decreased in folic acid deficiency; (6) choline-sparing effects; (7) analgesic in humans—pain threshold is increased; (8) antisulfonamide effects; (9) enterohepatic circulation of folate; (10) synthesized by psittacosis virus; (11) concentrated in spinal fluid.

Thermal Destruction of Folacin. Chen and Cooper (1979) have pointed out that the existence of folacin in numerous forms coupled with the lack of differentiation between forms has led to complications in characterizing the substance and in establishing meaningful tables for folacin content of foods. Improved methods have come into use for determining the specific forms of folacin present in foods (Chan et al., 1973). The value derived from separation and quantification of folates has been limited due to lack of information concerning the stability characteristics of various folate forms. These forms have been found to vary over a wide range in their thermal stability (Cooper et al., 1978; Paine-Wilson and Chen, 1979). The presence of oxygen has been implicated as a possible factor in degradation of folates during heating. Heat processing of canned and sealed foods has resulted in markedly higher folacin retention than found in home cooking methods (Huskisson and Retief, 1970). Rolls and Porter (1973) found that the percentage of folacin destroyed during heat processing of milk is affected by the level of residual oxygen in milk. OBroin et at. (1975) found that ascorbic acid has a protective effect on folates, possibly due to its action as a reducing agent. Studies of the effect of temperature on folacin have been less thoroughly documented. Thus Chen and Cooper (1978) undertook a study of the effects of temperature and presence of oxygen and reducing agent on the thermal stability of two naturally occurring labile folates. Their conclusions—loss of 5-methyltetrahydrofolic acid in aqueous solution followed first-order kinetics with a low Arrhenius activation energy of 9.5 kcal/mole. Tetrahydrofolic acid was extremely labile during heating and the stability of both forms of folate was increased in the presence of ascorbate and nitrogen atmosphere.

See also **Anemias.**

References

Brown, G. M.: "Folic Acid" in "The Encyclopedia of Biochemistry" (R. J. Williams and E. M. Lansford, Jr., editors), Van Nostrand Reinhold, New York, 1967.

Chan, C., Shin, Y. S., and E. L. R. Stokstad: "Studies of Folic Acid Compounds in Nature. 3: Folic Acid Compounds in Cabbage," *Can. J. Biochem.*, **51**, 1617 (1973).

Chen, T. S. and R. G. Cooper: "Thermal Destruction of Folacin: Effect of Ascorbic Acid, Oxygen and Temperature," *J. Food Sci.*, **44**, 3, 713–716 (1979).

Cooper, R. G., Chen, T. S., and M. A. King: "Thermal Destruction of Folacin in Microwave and Conventional Heating," *J. Amer. Dietet. Ass.*, **73**, 406 (1978).

Jaenicke, L.: "Folic Acid Coenzymes," in "The Encyclopedia of Biochemistry," (R. J. Williams and E. M. Lansford, Jr., editors), Van Nostrand Reinhold, New York, 1967.

Huskisson, Y. J., and F. P. Retief: "Folate Content of Foods," *South Afr. Med. J.*, **44**, 12, 362 (1970).

Kutsky, R. J.: "Handbook of Vitamins and Hormones," Van Nostrand Reinhold, New York, 1973.

O'Broin, J. D., et al.: "Nutritional Stability of Various Naturally Occurring Monoglutamate Derivatives of Folic Acid," *Amer. J. Clin. Nutr.*, **28**, 5, 438 (1975).

Paine-Wilson, B., and T. S. Chen: "Thermal Destruction of Folacin: Effect of pH and Buffer Ions," *J. Food Sci.*, **44**, 717–722 (1979).

FOLIUM OF DESCARTES. A higher plane curve, which is a special case of a cissoid, defined by the equation

$$x^2 + y^3 = axy$$

or in parametric form by $(1 + t^3)x = at$; $(1 + t^2$. The coordinate origin is a node for the curve and the coordinate axes are the two corresponding tangents. The asymptote to its two infinite branches is $3(x + y) + a = 0$.

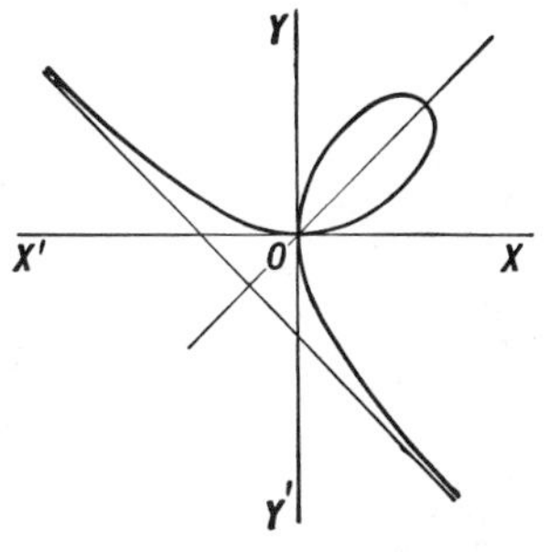

Folium of Descartes.

See also **Curve (Higher Plane)**; and terms listed under **Mathematics.**

FOLLICLE. In zoology, a glandular cavity or sac. The small spherical chambers in the thyroid gland, the hair follicle, the fluid-filled cavity in the vertebrate ovary in which the ovum develops, and the chambers of the arthropod testis are examples of the application of the term.

FOLLICLE CELL. Cells of the ovarian follicle which nourish the ovum during its development. They occur in the Graafian follicles of vertebrates and in the open follicles of invertebrates. In the insects, specialized cells of similar function contained within the follicle are termed nurse cells.

FOMALHAUT (α Piscis Austrini). The southernmost first magnitude star visible from the latitude of New York. Formalhaut rises almost simultaneously with Capella and is above the southern horizon during the evening in the autumn months. It was one of the royal stars of astrology, ruling over the southern sky. Ranking eighteenth in apparent brightness among the stars, Fomalhaut has a true brightness value of 14 as compared with unity for the sun. Fomalhaut is a white, spectral type A star and is located in the constallation Pisces Austrinus south of the ecliptic. Estimated distance from the earth is 23 light years. See also **Constellations.**

FONTANELLE. Any one of the places in an infant's skull that have not become bony at birth. These places are usually situated at each of the angles of the two parietal bones. By the age of 14 months, the anterior fontanelle is, from the functional standpoint, closed, and is the last to become so. Fontanelles also occur in the skulls of other vertebrates, in their immature form.

FOODBORNE DISEASES. In most countries with well-developed health care systems, foodborne diseases fall within the province of public health authorities. Organizational structure varies from one country to the next, but usually the enforcement of regulations and collection of statistics is accomplished at the local level. Superimposed over city, county, and state or provincial agencies will be various national organizations—as, in the United States, the Public Health Service, the Center for Disease Control, and the Food and Drug Administration. The general absence of catastrophic outbreaks of foodborne diseases in many countries is testimony of the workability of the system and of the progress that has been made over the years. Control over food processors has been quite effective. It is not surprising, then, that a significant part of the foodborne disease problem remaining arises from actions taken by persons who are less experienced and inadequately aware of foodborne pathogens and poisons—as, for example, the home canner or the gracious host who inadvertently serves slowly warmed-over turkey at a community picnic that has been delayed several days because of inclement weather.

It is logical that the first interest in foods as means for transmitting communicable diseases would be concentrated on the main killers of earlier times. In the 1850s, it was found that water and milk were responsible for the spread of cholera and thyphoid fever. Although they could not grow the organisms in the laboratory at that time, investigators did develop epidemiological evidence. Early interest concentrated on the association of human sewage with drinking water and foods, and other sources of the microorganisms were overlooked for many years. As comparatively recently as the 1920s, one public health pioneer insisted that animals were the prime reservoir of salmonellae (except for *S. typhi* and most of the paratyphoid bacilli). Today, it is recognized that there is a high rate of excretion of salmonellae among poultry, swine, and cattle, that not all abattoirs can control the spread of infection from gut contents to meat, and that cross contamination from raw to cooked foods in food-preparation areas is responsible for a large proportion of foodborne salmonellosis. Thus, it turns out that animal sewage is far more important than human in search for the source of an infective agent. Commercial interests militated against the confirmation that the heedless preparation of animal feeds from contaminated raw materials resulted in salmonella excretion in poultry and livestock.

During the past 50 years, the number of bacterial agents found to be implicated in foodborne diseases has increased markedly. As well as organisms of the Salmonella and Shigella groups, *Staphylococcus aureus, Clostridium perfringens, C. welchii, Bacillus cereus, Vibrio parahaemolyticus, Escherichia coli*, and certain streptococci, among others, have been responsible for outbreaks of food poisoning.

Classification of Foodborne Diseases

(1) *Poisonings* are caused by consuming toxicants which are found in tissues of certain plants and animals, metabolic products (toxins) formed and excreted by microorganisms (bacteria, fungi, algae) while they multiply in foods; or poisonous substances which may be intentionally added (but not later removed) or incidentally added to foods as a result of production, processing, transporting, and storing.

(2) *Infections* are caused by the entrance of pathogenic microorganisms into the body and the reaction of body tissues to their presence, or to the toxins which they generate within the body. Intestinal infections may be manifested by *in vivo* enterotoxin production or mucosal penetration. After mucosal penetration, the organisms multiply in the mucosa or pass into other tissues. See Fig. 1.

Diseases transmitted by foods can be placed into seven broad causative categories: (1) bacterial diseases, (2) viral and rickettsial diseases, (3) parasitic diseases, (4) fungal diseases, (5) plant toxicants and toxins, (6) toxic animals, and (7) poisonous chemicals, including radionuclides. Examples of principal diseases in these categories are given in Table 1. Although the listings of this table are helpful from the standpoint of classifying sources, the list tends to give equal weight to all diseases and does not convey seriousness, or frequency of occurrence. Several of the diseases occur rarely and, in some instances, are highly regionalized.

In those countries with extensive communications networks and where there is advanced epidemiology and microbiological attention given to foodborne diseases, even one case of certain diseases can cause considerable excitement, particularly precipitating dramatic emphasis in the public press when recalls of certain processed foods may be involved. Worldwide reporting, however, is very poor. Even in some of the advanced countries, only a percentage of foodborne diseases is reported and thus many do not become part of formal statistics. For example, the Committee on Salmonella of the National Research Council (United States) estimates that about two million cases of salmonellosis occur in the country each year, but that only an average of about 20,000 isolations are made each year.

For the United States, over a period of three years in the 1970s, frequency of occurrence by major categories of foodborne diseases are shown in Table 2.

Epidemiology. Common terms used in the science of epidemiology (the study of the pattern of disease and the factors that cause the disease) include:

(1) *Pattern*—of a disease is the composite of the relationships of time, place, and person within a group of cases. Time refers to onset of illness; place refers to residence, geography, and food sources; and

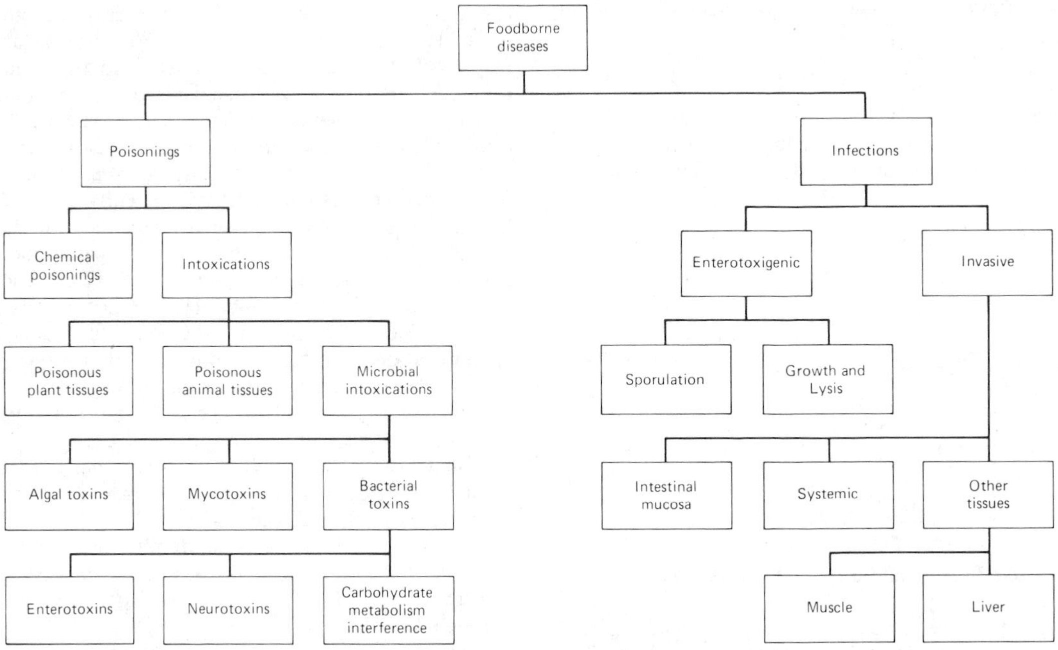

Fig. 1. Classification of foodborne diseases by basic sources and routes of development. (*From Bryan, 1976.*)

TABLE 1. BROAD CATEGORIZATION OF FOODBORNE DISEASES

BACTERIAL DISEASES

Salmonelloses
  Salmonellosis
  Enteric fever (typhoid fever)
  Enteric fever (paratyphoid fever)
Staphylococcal Intoxication
Botulism
Clostridium perfringens (*C. welchii*)
  Type A Disease
  Type C–Enteritis Necroticans
Bacillus cereus
  Gastroenteritis
Vibrio parahaemolyticus
  Infection
Arizona hinshawii
  Arizona infection
Pseudomonas cocovenenans (toxins)
  Bongkrek poisoning
Proteus spp. (histaminelike substances)
  Scombroid poisoning (Scombrotoxism; Saurine poisoning; Fish poisoning)

Note: Bacterial diseases usually transmitted by other means, but sometimes foodborne, include: Shigellosis; enteropathogenic *Escherichia coli* infection; beta hemolytic streptococcal infections, such as scarlet fever, septic sore throat; yersiniosis (*Yersinia enterocolitica* infection or pseudotuberculosis); cholera; brucellosis; tuberculosis; diphtheria; tularemia; anthrax; Haverhill fever.

Bacterial diseases in which proof of transmission by foods is inconclusive include: Enterococcal infection; Klebsiella food infection; enterobacter infection; *Pseudomonas aeruginosa* infection; *Bacillus subtilis* infection; listeriosis, among others.

VIRAL AND RICKETTSIAL DISEASES

Hepatitis A (Infectious hepatitis)
Poliomyelitis
Bolivian hemorrhagic fever
Russian spring-summer encephalitis (Diphasic milk fever)

TABLE 1. BROAD CATEGORIZATION OF FOODBORNE DISEASES (cont.)

PARASITIC DISEASES (Helminthic)

Trichinosis
Taeniasis

Note: Other parasitic diseases always or usually transmitted by foods, but less commonly encountered, include: cysticercosis; diphyllobothriasis; spargnosis; angiostrongyliasis; anisakiasis; fasciolopsiasis; echinostomiasis; heterophyid infection; opisthorchiasis; metagonimiasis; fascioliasis; paragonimiasis; dicrocoeliosis; hymenolepiasis diminuta; gnathostomiasis; intestinal myiasis.

Parasitic diseases which are usually transmitted by other means, but which may be foodborne, include: Amebiasis (amebic dysentery); ascariasis; trichuriasis; *Capillaria hepatica* infection; echinococcoses hydatidosis; alveolor hydatid disease; balantidiasis (balantidal dysentery); giardiasis; coccidiosis (Isospora infection); dientamoeba infection; toxoplasmosis; sarcosporidiosis.

FUNGAL DISEASES

Mycotoxicoses reported in humans
  Alimentary toxic aleukia (ATA)
  Urov diseases (Kaschin-Beck disease)
  "Drunken-bread" poisoning
  Akakabi-byo (Red mold disease)
  Ergotism (Saint Anthony's fire)
  Epidemic polyurea
  Toxic moldy rice disease
  Muco-mycotoxic disease
  Aflatoxicosis

Note: Mycotoxicoses of animals that may also potentially affect humans include: Strachybotryotoxicosis; ochratoxicosis; aspergillus toxicosis; moldy corn toxicosis; facial eczema.

Mushroom-associated diseases
  Mushroom poisoning–Cell destruction
  Mushroom poisoning–Neurological effects
  Mushroom poisoning–Enteritis type
  Mushroom-alcohol intolerance
Mycotic infections
  Phycomycosis

TABLE 1. BROAD CATEGORIZATION OF FOODBORNE DISEASES (cont.)

PLANT TOXICANT AND TOXIN DISEASES

Alkaloids
Jimson weed and nightshade poisoning; senecio poisoning; hemlock poisoning; epidemic dropsy; manchineel poisoning; laburnum poisoning; solanine poisoning; green hellebore poisoning; delphinium and monkshood poisoning; yew poisoning; daffodil bulb poisoning; jessamine poisoning; colchicine poisoning; nicotine poisoning.

Glycosides
Cyanide poisoning; goiter; baneberry poisoning; buckeye poisoning; oleander, lily-of-the-valley, and black hellebore poisoning; pokeweed, corn cockle, and finger cherry poisoning; tung nut poisoning.

Toxalbumins
Castor bean and jequirity poisoning; favism.

Resins
Water hemlock poisoning; mountain laurel poisoning

Other Toxicants, Toxins, and Allergens
Milk sickness; cocculus poisoning; ackee poisoning; lathyrism; oxalate poisoning; mistletoe poisoning; nutmeg poisoning; *Leucaena glauca* poisoning; djenkol poisoning; carotenemia; esophageal cancer.

Note: There are many other plants and substances that can cause toxic and allergenic illness, including common house and garden plants.

TOXIC ANIMALS (Disease Sources)

Fish
Ciguatera poisoning; moray eel poisoning; file fish poisoning; tetraodon or puffer fish poisoning; scombroid poisoning; clupeoid poisoning; elastobranch and chonrichytes poisoning; chimaeroid poisoning; cyclostome poisoning; gempylid poisoning; hallucinogenic fish poisoning; ichthyohepatoxism; freshwater fish poisoning; Haff or Yuksov disease; Minamata disease.

Shellfish
Paralytic shellfish poisoning; oyster poisoning; callistin shellfish poisoning; abalone poisoning; whelk poisoning.

Other Animals
Cephalopod poisoning; sea urchin poisoning; sea anemone poisoning; sea cucumber poisoning; horseshoe crab poisoning (mimi poisoning); turtle poisoning; hypervitaminosis A; porpoise poisoning; toxic quail poisoning.

POISONOUS CHEMICALS (Disease Sources)

Metallic Containers
Zinc, cadmium, antimony, copper, lead, tin poisoning.

Intentional Additives
Nitrite poisoning; niacin poisoning; triorthocresyl phosphate poisoning; diphenylhydatoin intoxication; Oriental (Chinese) restaurant syndrome (excessive monosodium glutamate); potassium bromate poisoning; beer drinkers' cardiomyopathy (cobalt acetate); margarine disease; phenolphthalein poisoning.

Incidental and Accidental Food Additives
Organic phosphorus poisonings (from insecticides); chlorinated hydrocarbon poisoning (from insecticides); carbamate poisoning (from insecticides); fluoride poisoning (from rodenticides); sodium monofluoroacetate poisoning (from rodenticides); thallium poisoning (from rodenticides); warfarin poisoning (from rodenticides); arsenic poisoning (from insecticides and herbicides); phosphide poisoning (from rodenticides and matches); barium poisoning (from rodenticides); nicotine sulfate poisoning (from insecticides and tobacco products); alkyl-mercury poisoning (from fungicides) epoxy resin poisoning (from contaminated grains); chromium poisoning (from vending machines); calcium chloride poisoning (contaminated popsicles); cyanide poisoning (from fumigants); lye poisoning (from household chemicals); methyl

TABLE 1. BROAD CATEGORIZATION OF FOODBORNE DISEASES (cont.)

POISONOUS CHEMICALS (Disease Sources) (cont.)

alcohol poisoning (paint solvents and bootleg whiskey); chronic cadmium poisoning (Itai Itai or ouch ouch disease (mining wastes deposited in rice paddies); yusho (rice oil disease—from salad oil); selenium poisoning (home-grown foods with high concentrations of selenium, such as monkey coconut).

Allergens or Enzyme Deficiencies
Gastrointestinal food allergies
Allergens react with antibody and form histamine or histamine-like substances; (Depends upon sensitivity of individual—not general)
Milk (protein constituents)
Milk products
Egg (whites)
Cereals (wheat, buckwheat, corn, rice, rye, oats)
Fish and seafoods
Meats
Nuts
Spices
Vegetables (celery, string beans, lima beans, tomatoes)
Fruits (oranges, strawberries, bananas, lemons, watermelon)
Preservatives

Disaccharide intolerance
Lactose, sucrose, or isomaltose; (High incidence in blacks and orientals)
Milk and other foods containing disaccharides

Food-drug combinations
Amine poisoning
Tyramine in cheese can be degraded to p-hydroxyphenylacetic acid by monoamine oxidase inhibitors in certain tranquilizers, causing hypertensive attacks.
Mushroom-alcohol intolerance (Specifically inky cap mushroom)
Listed under fungal diseases. Disulfiramlike (antabuse) constituents of mushroom interfere with normal metabolism of alcohol. Attacks can occur even if alcohol is consumed 48 hours after eating mushrooms.

Radionuclides
Food contamination may arise from fallout, reactor plant accidents, radioactive wastes, naturally occurring radioactive substances.
Strontium^{89} (milk)
Strontium^{90} (green leafy vegetables, milk, milk products)
Iodine^{131} (milk)
Cesium^{137} (green leafy vegetables, milk, milk products, meat, shellfish, fish)
Phosphorus^{32} (green leafy vegetables)
Barium^{140} (milk)
Ruthenium^{106} (laverbread made from seaweed components)

Source: Bryan (1976) reference listed.

persons refers to age, sex, occupation, ethnic group, social attributes, and food history of persons involved.

(2) *Causal Factors*—are the agent, reservoir, vector, vehicle, host, and their interrelationships, as well as conditions that permit the agent to survive and/or multiply. These factors constitute the chain of infection for bacterial foodborne diseases, and the chain of actions in connection with other foodborne disease forms. In an epidemiologic investigation, the outbreak must be described in terms of its distribution and each of the causal factors must be determined.

(3) *Epidemic*—the occurrence of a group of illnesses of similar nature in a community or region, clearly in excess of normal expectancy and derived from a common or propagated source. The minimum number of cases that indicates an epidemic will vary with the infectious agent, with the size and type of population exposed, with previous experience or lack of exposure to the diseases, and with the time

TABLE 2. PRINCIPAL FOODBORNE DISEASE OUTBREAKS AND CASES
(Confirmed) (United States–Cumulative for 3-year Period in Mid-1970s)

Etiology and Specific Disease	Outbreaks (Number)	Outbreaks (Percent of All Diseases)	Cases (Number)	Cases (Percent of All Diseases)
BACTERIAL	337	64.3	16,852	92.1
Salmonella	101	19.3	8241	45.0
Staphylococcus	114	21.8	4770	26.0
Clostridium perfringens	37	7.1	1791	9.8
Shigella	12	2.3	898	4.9
Group A *Streptococcus*	1	<0.2	325	1.8
Yersinia enterocolitica	1	<0.2	286	1.5
Vibrio parahaemolyticus	2	0.3	222	1.2
Bacillus cereus	6	1.1	119	0.7
Clostridium botulinum	58	11.1	91	0.5
Suspect Group D *Streptococcus*	3	0.5	88	0.5
Arizona hinshawii	1	<0.2	15	<0.1
Vibrio cholerae	1	<0.2	6	<0.1
CHEMICAL	128	24.4	586	3.2
Ciguatoxin	51	9.7	237	1.3
Heavy metals	14	2.7	133	0.7
Miscellaneous chemicals	19	3.6	116	0.6
Scombrotoxin	18	3.4	47	0.3
Monodosium glutamate	7	1.3	20	0.1
Mushroom poison	12	2.3	15	<0.1
Paralytic shellfish poison	5	1.0	15	<0.1
Puffer fish tetrodotoxin	1	<0.2	2	<0.1
Neurotoxic shellfish poison	1	<0.2	1	<0.1
VIRAL	12	2.3	572	3.1
Hepatitis A	11	2.1	492	2.7
Echo, Type 4 (Enteric cytopathogenic human orphan viruses)	1	<0.2	80	0.4
PARASITIC	47	9.0	294	1.6
Trichinella spiralis	42	>8.0	278	1.5
Entamoeba histolytica	1	<0.2	9	<0.1
Toxoplasma gondii	1	<0.2	4	<0.1
Anisakis spp.	2	<0.3	2	<0.1
Diphyllobothrium latum	1	<0.2	1	<0.1
TOTAL of all confirmed foodborne diseases	524		18,304	
AVERAGE per year (over 3-year span)	175		6101	

Data base: Bryan (1976) reference listed.

and place of occurrence. Thus, an epidemic is relative to the usual frequency of the disease in the same area, among specified populations at the same season of the year. A single case of a communicable disease (as botulism or typhoid fever) long absent from a population or a single case that is the first invasion by a disease (as *Vibrio parahaemolyticus* infection) not previously recognized in an area is also to be considered epidemic. The word epidemic must be defined in contradistinction to the word endemic, which is the habitual presence of a disease or infectious agent within a given geographic area or the usual prevalence of a given disease within such an area. An obvious example of endemic and epidemic can be seen from the graph of pneumonia-influenze deaths (used as example, although not foodborne) over a span of some 3 years. See Fig. 2.

(4) *Epidemic curve*—is a graphical representation of cases according to the distribution of the time of onsets of cases. It is usually constructed as a histogram (see Fig. 3), but sometimes as a frequency polygon. In both types of graphs, the frequencies or number of cases are plotted on the ordinate axis and the time of onset of illness along the abscissa. The duration of each interval may be in hours, intervals of a few hours, etc. The interval selected depends upon the disease in question and the span of time over which cases occurred. A histogram represents each case or group of cases by a block in a unit of time. If more than one case occurs in a time interval, a block representing each additional case is stacked on the initial block. There are no vertical or horizontal spaces between blocks as in bar graphs.

(5) *Common-source or point epidemic*—an epidemic that is spread by a vehicle, such as foods, milk, water, or fomites (inanimate objects—dishes, utensils, table tops, etc. that may be contaminated) shared by the victims. In a common-source outbreak, when many persons are exposed simultaneously, the relative uniformity of the incubation period for a specific disease results in a single cluster of cases in time. Cases occur rapidly after the first onset, reach a peak (*point*), and then decline. The duration of the epidemic will be within the range of the incubation period of the disease. The histogram of Fig. 3 indicates a common-source situation of a salmonellosis outbreak.

(6) *Common-source, single-event epidemic*—is typified by onsets of cases that follow a single situation, as when a group of people eat the same contaminated food (*common source*) at the same place within a short period of time—as during a particular meal (*single event*). Distribution of cases will be stretched out in time when all persons have not eaten the contaminated food at the same time. Such a situation may be more accurately called a *common-source, multiple-event epidemic.*

(7) *Propagated epidemic*—an epidemic that is transmitted from a human or animal reservoir by direct or indirect contact with a host. Reservoirs may be in the incubation period of a disease—they may be mild or missed cases, convalescents, or healthy *carriers.* For transmission to occur, the population must have enough susceptibles to give the infection a chance of spreading. As the disease spreads, those infected become immune, and the supply of susceptibles is depleted to a point at which spread tapers off and finally ceases. The rapidity with which contact-spread epidemics reach a peak and the extent of their duration depend upon infectivity of the agent, length of incubation period, initial proportion of susceptibles in the population, and degree of crowding and intimacy of contact. In zoonoses (animal parasites), the change in the proportion of susceptibles among animals and the distribution of the animals are factors that affect time distribution of cases. In vectorborne disease outbreaks, conditions that favor the development of a vector and the length of time that is required for the pathogens to develop in a vector will alter the time distribution of cases. A typical contact-spread epidemic is shown in Fig. 4. Data on 2 outbreaks of infectious hepatitis are shown in Fig. 5. One set of data defines a common-source outbreak in a city, the other set defines a propagated outbreak in the surrounding county in which the city is located.

(8) *Incubation period*—the time between exposure of a susceptible person to an infection or toxic agent and appearance of the first sign or symptom of the disease. In foodborne outbreaks, it is the time between *ingestion* of contaminated or toxic food and the first symptom. The incubation period may be long or short, depending upon the peculiarities of each specific host-parasite relationship and the ease with which the infecting organism or toxin finds access to the tissue in which its primary multiplication or toxic reaction takes place. When a chemical which is an irritant to the gastrointestinal tract is ingested in sufficient quantity, vomiting follows in a short time, usually within an hour. Infectious hepatitis, which affects the liver, may have an incubation period of 15 to 30 days. The duration of the incubation period is an important characteristic of each disease. During an outbreak investigation, the incubation period cannot be determined until the responsible meal or time of ingestion of the incriminated food has been identified.

(9) *Vector*—a means for transferring organisms from place to place. It is sometimes difficult to determine whether vectors are sources of

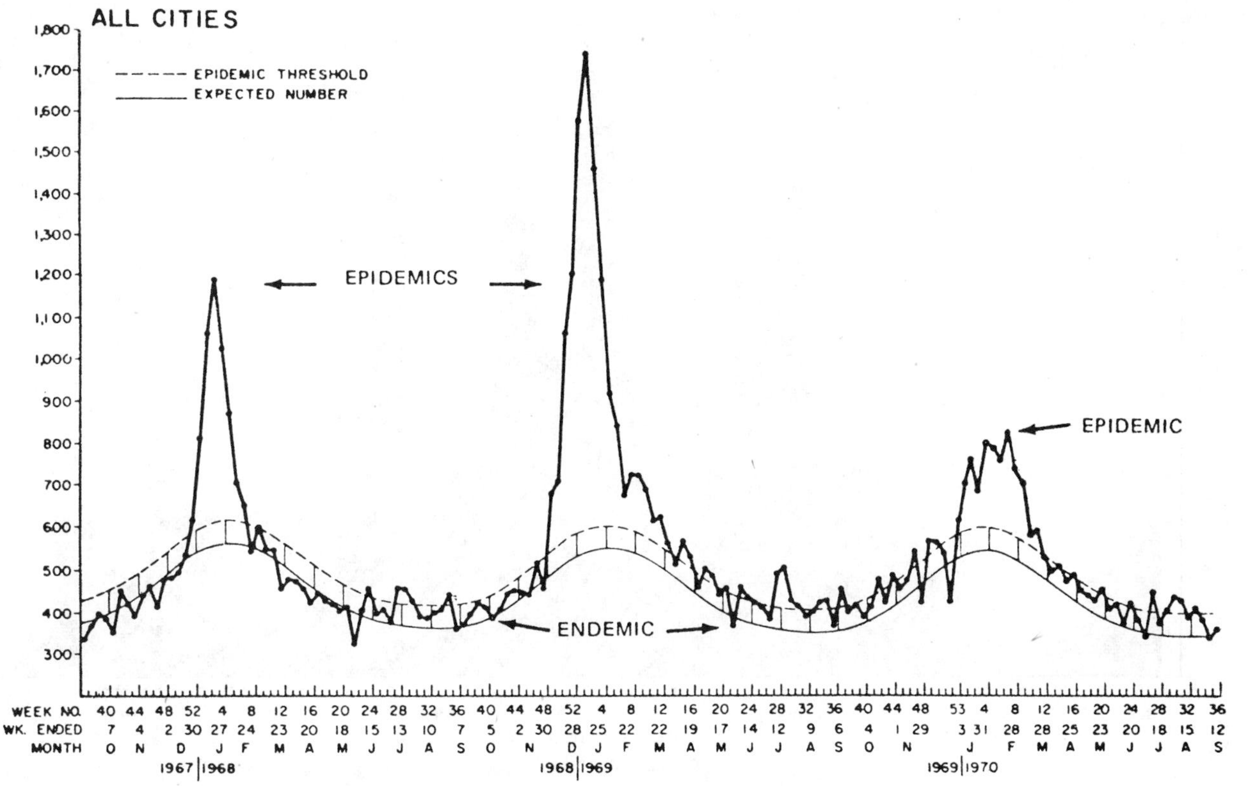

Fig. 2. Illustration of epidemic and endemic. The graph illustrates how an epidemic can be recognized. Epidemics are seen as a significant excess of cases (deaths in this example) over that which is expected on the basis of accumulated experience. The solid endemic line shows the seasonal pattern of the deaths, while the upper, broken line contains the upper level of variations below which 95% of expected observations should fall. When the observed numbers of deaths exceed the broken line for more than 2 successive weeks, an epidemic is indicated. Epidemics are observed during 3 successive years. From past experience of the occurrence of a disease in a community, epidemics can be forecasted. In this illustration, a forecast can be made that epidemics of death caused by pneumonia and influenza will occur again in winter and have higher peaks in the uneven numbered years. Generally, this technique can be used in connection with foodborne diseases, but involving a whole new set of causal factors for each disease being considered. (*Public Health Service, U.S. Department of Health and Human Services*).

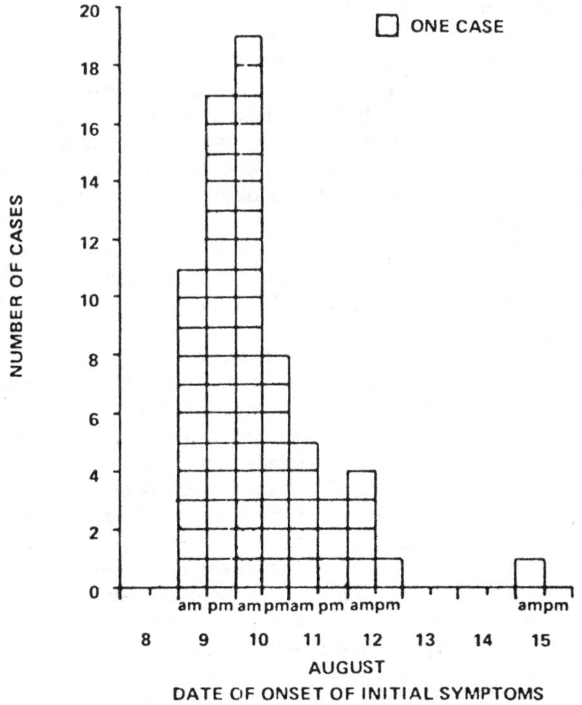

Fig. 3. Histogram (epidemic curve) of a common-source outbreak of salmonellosis. (*Modified from Goldman, 1970.*)

contamination, or whether they pick up organisms from their environment. They do reflect the contamination of their environment and may transfer such contamination. If the environment of a plant, for example, is contaminated with a particular pathogen, finished products almost inevitably become contaminated.

(10) *Outbreak*—in the United States, an outbreak of foodborne disease is a situation in which two or more people experience a similar illness, usually gastrointestinal, after ingestion of a common food and where epidemiologic analysis implicates the food as the source of the illness. However, only one case of botulism or chemical poisoning constitutes an outbreak. Outbreaks are divided into two categories:

(a) *Laboratory-confirmed outbreak*—in which laboratory evidence of a specific etiologic agent is obtained and specified criteria are met.

(b) *Outbreak of undetermined etiology*—in which epidemiologic evidence implicates a food source, but adequate laboratory confirmation is not obtained. These outbreaks are further subdivided into four subgroups by incubation period of the illness:

Less than 1 hour—probable chemical causative
From 1 to 7 hours—probably *Staphylococcus* infection
From 8 to 14 hours—probable *Clostridium perfringens* infection
More than 14 hours—various other infectious agents.

Much of the information required by public health authorities in developing foodborne disease patterns stems from personal interviews with persons who have suffered (or are still suffering) the illness and other people who appear to possess immediate knowledge of suspected places (household, restaurant, picnic, etc.), where food was probably ingested and where it may have been contaminated prior to ingestion.

Important Foodborne Diseases

Salmonellosis. Salmonellosis is a foodborne disease that is caused by salmonellae (bacteria) other than *Salmonella typhi* and *Salmonella*

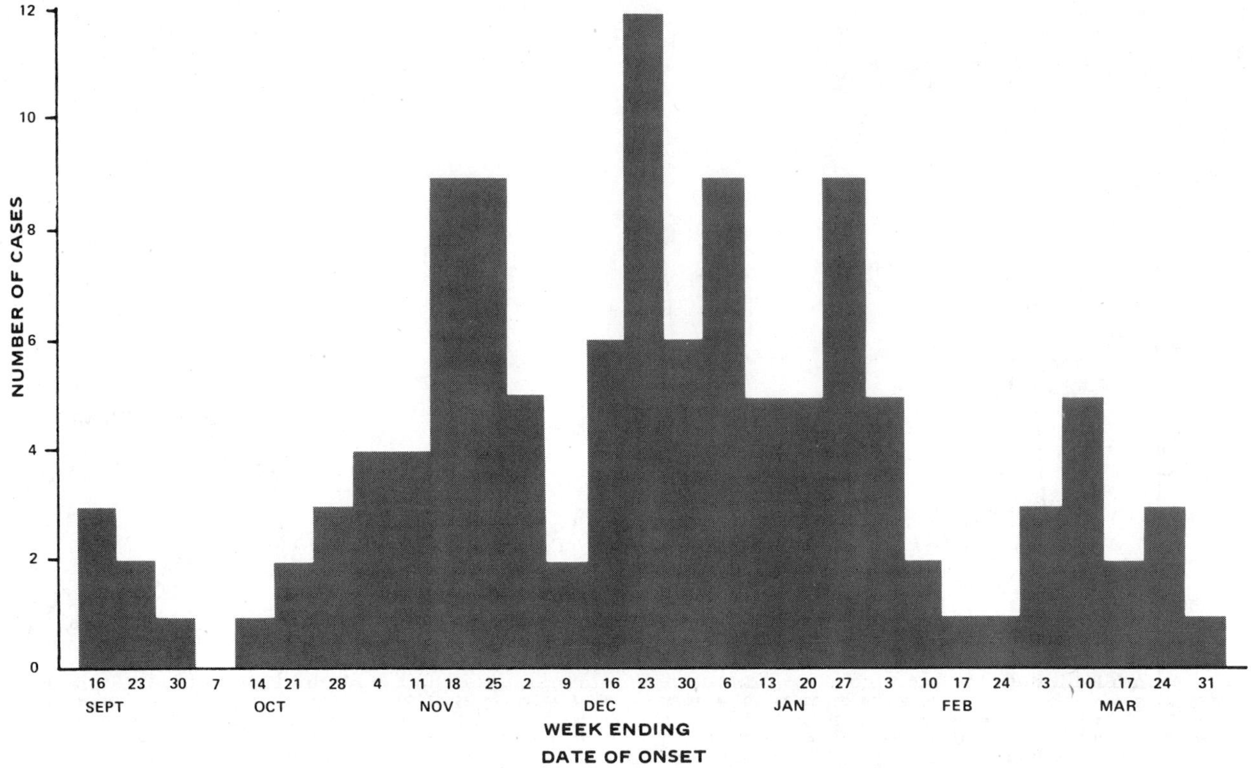

Fig. 4. Epidemic curve of a propagated outbreak of infectious hepatitis. (*Adapted from Davis and Hanlon.*)

paratyphi. Salmonellae are gram-negative, nonsporeforming (mostly) motile rod, aerobic, facultatively anaerobic microorganisms. There are over 1600 known serotypes, but only about 50 of these occur commonly. Species involved include *Salmonella choleraesuis* and *S.*

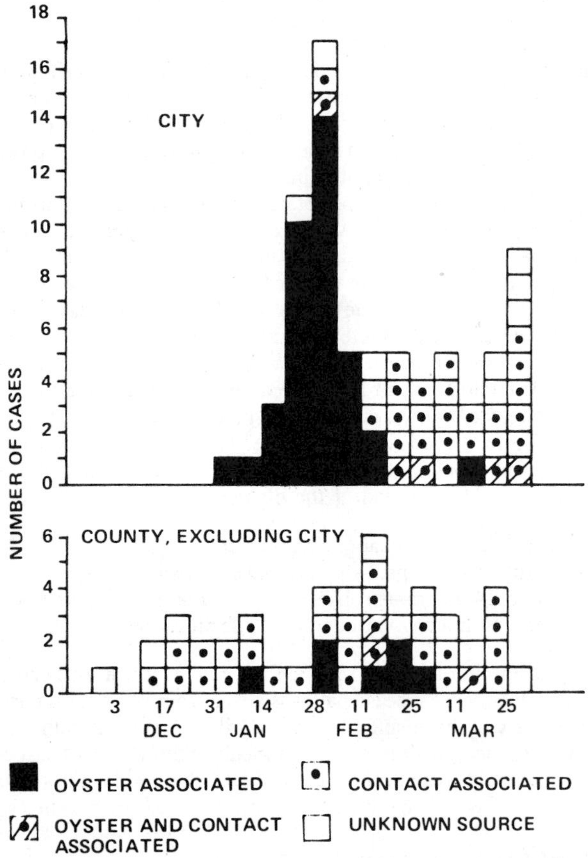

Fig. 5. Epidemic curves of a common-source outbreak with secondary infections and a propagated outbreak of infectious hepatitis. (*Modified from Mason and McLean.*)

enteritidis. Organisms possess O (somatic) and 2 phases of II (flagellar) antigens.

Incubation Period and Symptoms. Commonly, the incubation period ranges from 12 to 36 hours after food ingestion, although it may be as short as 5 and as long as 72 hours. Symptoms are one or more of the following: Diarrhea, abdominal pain, chills, fever, vomiting, dehydration, prostration, anorexia, headache, malaise. Duration of the disease may be several days. Enteritis or focal infection may also occur.

Most Likely Food Sources. Meats, poultry, eggs and their products. Other incriminating foods include: Coconut, cottonseed protein, chocolate candy, dry milk, smoked fish, and yeast.

Typical Sources and Reservoirs of Etiologic Agent. Feces of infected domestic or wild animals and humans.

Susceptibility and Carrier State. Most susceptible persons are infants, the elderly, and malnourished and those with concomitant diseases. However, the disease can affect persons of any age or state of health.

Preventive and Control Measures. Foods should be chilled rapidly in small quantities; foods should be cooked thoroughly. Egg products and milk must be pasteurized. Cross-contamination from raw to cooked foods must be avoided. Strict sanitary measures on the farm and in processing plants must be taken. Feed ingredients for livestock and poultry should be heat-treated. Very importantly, food and feeds must be protected from all excreta.

Other Salmonellae. Typhoid fever or enteric fever can be transmitted by food, *Salmonella typhi* is similar to other salmonellae, but is adapted to the human host. It possesses VI (capsular) antigens as well as O and H antigens. Water is also commonly implicated in outbreaks of typhus. High-protein foods, raw salads, milk, shellfish are the usual categories of foods involved. These are largely foods that have been handled, and then eaten without further heat treatment. *Paratyphoid fever,* also sometimes called enteric fever, can be transmitted by foods. *Salmonella enteritidis* is also similar to other salmonellae, but is more or less adapted to the human host. This microorganism causes a blood stream infection. The disease is of a milder and shorter duration (1 to 3 weeks) than typhoid fever. Foods typically involved are milk, shellfish, raw salads, and eggs.

See also **Typhoid Fever.**

Staphylococcal Intoxication. Also termed *staphyloenterotoxicosis,* staphylococcal intoxication is caused by *Staphylococcus aureus.* The etiologic agent is enterotoxin A, B, C, D, E, or F of *S. aureus*. This microorganism is gram-positive, nonsporeforming, nonmotile cocci occurring in irregular grapelike groupings. Other properties of the microorganism include—aerobic; facultatively anaerobic; coagulase-positive; ferments mannitol; grows well in 10% salt media; produces lipase and hemolysin; often produces orange or yellow pigments and heat-stable enterotoxin; resistant to antibiotics.

Incubation Period and Symptoms. Commonly, the incubation period ranges from 2 to 4 hours, although it may be as short as 1 and as long as 7 hours after food ingestion. Symptoms are one or more of the following: Sudden onset of nausea, salivation, vomiting, retching; diarrhea; abdominal cramps; dehydration; sweating; weakness; prostration. Fever usually does not occur. Disease is usually of a short duration of about 1 or 2 days.

Most Likely Food Sources. Cooked ham; meat products; poultry and dressings; sauces and gravies; cream-filled pastry; potato, ham, poultry, and fish salads; milk; cheese; hollandaise sauce; bread pudding; and many high-protein leftover foods.

Typical Sources and Reservoirs of Etiologic Agent. Nose and throat discharges; hands and skin, particularly from infected cuts, wounds, and burns; boils, pimples, acne; feces. The anterior nares of humans are the primary reservoir. In cows and ewes, it is the mastitic udder. In poultry, the arthritic and bruised tissue. In most cases, where the disease develops, it will be from food that has been contaminated after cooking or processing.

Preventive and Control Measures. Very similar to the general measures described under "Salmonellosis." In particular, ill persons suffering colds, infected cuts, and diarrhea, should not be permitted to be in the food-preparation area. Thorough cooking, reheating, and pasteurization destroy the microorganisms, but unfortunately, they do not destroy any toxin that may be present. Wherever possible, foods should be prepared on the same day they are served. As previously mentioned, freezing does not fully control the microorganism, with outbreaks occurring from time to time after thawing frozen cream-filled pastries.

Treatment of the illness consists of fluid and electrolyte repletion. Death rarely occurs, except in infants or in elderly or debilitated persons.

Botulism. Botulism is a foodborne disease that is caused by the toxins A, B, E, or F of *Clostridium botulinum*. Toxins C and D cause botulism in animals. Type G, found relatively recently, has not, to date, caused any human cases of the illness. The microorganism is a gram-positive, sporeforming, motile rod. Neurotoxins are formed which interfere with acetylcholine at peripheral nerve endings. The spores are among the most heat-resistant of those found in connection with foodborne diseases. The toxins are heat labile. Fatalities have occurred in cases in which victims have tasted only a very small amount of the toxin-containing food. Despite the potent aspects of botulinus toxin, improvements in respiratory care have helped to reduce the mortality in cases of botulism from 60 to 20%. There is also available a trivalent antitoxin, which when administered promptly, is very effective in most cases in modifying the course of the disease. Cases of botulism also can result from infection of wounds by *C. botulinum*, but the great majority of cases result from ingestion of contaminated food.

Cases and outbreaks of botulism occur in many parts of the world and in all regions of the United States, but the disease is most frequently found in the western United States, notably in Alaska, California, Colorado, Oregon, and Washington. Cases in Alaska, in several instances, have been attributed to toxins released from raw fish that has been aged in plastic bags, from seal meat that has been stored in oil, and from smoked salmon that has been wrapped in seal skins.

Incubation Period and Symptoms. Commonly, the incubation period ranges from 12 hours to 36 hours after food ingestion, although it may be as short as 2 hours and as long as 6 days. Symptoms are one or more of the following: Nausea, vomiting, abdominal pain, and diarrhea—signs which may appear early. Headache, vertigo or dizziness, lassitude, double vision, disphagia, dyspnea, ataxia, dry mouth, weakness, constipation, respiratory distress, respiratory paralysis may develop. Partial paralysis may persist for 6 to 8 months. Sensorium is usually clear. When the disease is fatal, death usually occurs within 3 to 10 days.

Of paramount importance is the maintenance of ventilation of the patient. The treating physicians will check for impending respiratory failure by taking serial measurements of vital capacity; actual respiratory failure can be recognized by changes in arterial blood gas levels. As lifesaving interventions, endotracheal intubation, eventual tracheostomy, and use of mechanical ventilators are available. The use of guanidine to increase acetylcholine release from nerve terminals remains controversial.

Infant Botulism. Botulism in infants was first recognized in 1976 and has appeared with surprising frequency since that time. Some authorities now believe that infant botulism may be incriminated in connection with some cases of *sudden infant death.*

Some authorities believe that infant botulism may not be caused by ingestion of toxin, but rather by the colonization of the intestine by *C. botulinum*, followed by absorption of the toxin. Honey has been implicated in about one-third of the cases and, consequently, some physicians recommend that honey not be fed to infants under 1 year of age. Onset of the disease (between 5 and 20 weeks) is characterized by constipation, followed by cranial neuropathy which produces muscle weakness. Infants displaying such symptoms have sometimes been described as "floppy." Sudden death may result from hypoventilation and apnea.

Twenty-four-hour consultative and laboratory services are maintained by the Center for Disease Control (CDC) in Atlanta, Georgia, from which polyvalent and monovalent botulinal antitoxins also are obtainable.

Most Likely Food Sources. The sources particularly pertinent to Alaska have been described. Foods generally implicated are improperly canned, low-acid foods (green beans, corn [maize], beets, asparagus, chili peppers, mushrooms, spinach, figs, olives, tuna). Smoked fish and fermented foods also have been implicated.

Preventive and Control Measures. Cans should be heated at high temperatures under pressure for sufficient time. Home-canned foods should be cooked thoroughly, by boiling and stirring for a considerable time. In addition to adequate refrigeration, addition of acidic substances and salt can be helpful. The availability of antitoxins has been mentioned.

Clostridium Perfringens. During sporulation in the gut, *Clostridium perfringens* (*welchii*) type A microorganisms release a proteinous enterotoxin. However, large numbers of vegetative cells must be ingested. This microorganism is a gram-positive, sporeforming, nonmotile, encapsulated short rod. It is anaerobic and produces lecithinase. There are both heat-resistant (some survive boiling for 1 to 5 hours) and heat-sensitive spores. Heat shock encourages spores to germinate. There are approximately 82 known serotypes.

Incubation Period and Symptoms. Commonly, the incubation period ranges from 8 to 24 hours, with a median of 12 hours. Symptoms include one or more of the following: Acute abdominal pain, diarrhea, occasional dehydration and prostration. Nausea, vomiting, fever, and chills are rare. The disease is of short duration (1 day or less).

Although *C. perfringens* food poisoning is essentially confined to incidences of gastroenteritis, the microorganism when present in severe body wounds can have serious consequences. For infection to occur, a lowered oxidation-reduction potential, which allows spore germination and bacterial multiplication, is required. Tissue necrosis, impairment of blood supply, and introduction of foreign bodies provide such an environment. Major determinants of pathogenicity for invasive clostridia are their potent exotoxins. Once bacterial multiplication and toxin production occur at a site of injury, rapid invasion and destruction of healthy tissue ensue.

Most Likely Food Sources. Cooked meat and poultry that has remained at room temperature for several hours or slowly cooled. (See Fig. 6.) Gravy, stew, and meat pies. Illness from *Clostridum perfringens* in meat products is a major concern in the food industry. Long-time, low-temperature (LTLT) cookery for beef or other meat in the processing industry, food service establishments, and the home creates potential for substantial growth of *C. perfringens* during cooking.

Preventive and Control Measures. Foods should be chilled rapidly in small quantities. Excellent personal hygiene by all personnel (including household) should be practiced. Hot foods should be held at 140°F (60°C) or higher. Meats marketed as cured meats should be processed under rigid restrictions. Sewage should be disposed in a sanitary man-

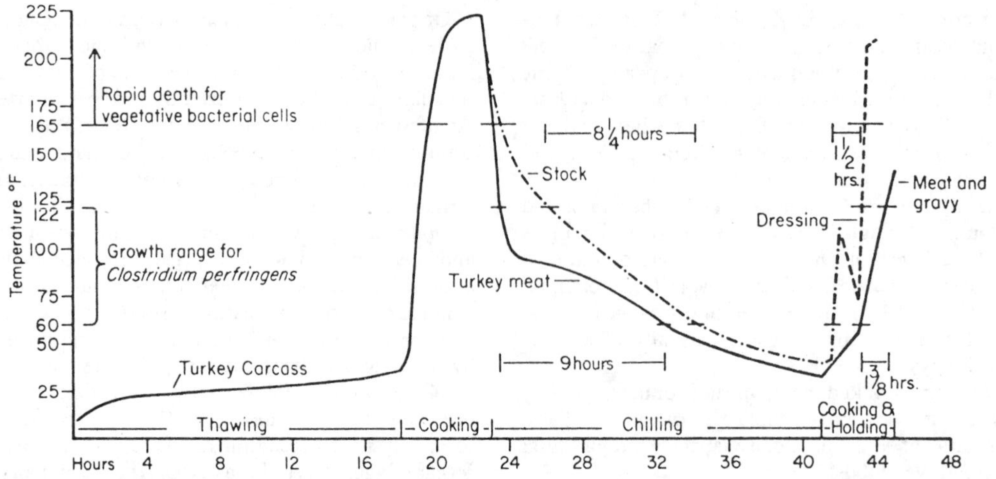

Fig. 6. Time-temperature relationship of a turkey preparation in a school lunch kitchen. In this case, *C. perfringens* could multiply during the 7 to 8 hours in which the meat and stock were in a refrigerator. The organisms present in the meat and gravy would have survived the reheating process. This particular chart was constructed as part of a time-temperature study of the thawing, cooking, chilling, and reheating practices that were used for turkeys which were inplicated in a foodborne disease outbreak. (*From Bryan et al., 1971.*)

ner. Thorough cooking will destroy vegetative cells, but not the heat-resistant spores. When reheating leftover foods, they should reach a minimum temperature of 165°F (74°C).

Clostridium perfringens Type C. This is a gram-positive, sporeforming, nonmotile rod, anaerobic microorganism. This microorganism was formerly called type F. Lecithinase and necrotoxin are produced. The strains differ in minor antigens. The incubation period ranges from 6 hours to 6 days, but is usually about 24 hours. Symptoms include diarrhea, prolonged abdominal pain, gangrene of small intestine, shock, toxemia. Case fatality rate of this type is about 40%.

Most likely food sources include pork, other meats, and fish. Preventive measures include eating a balanced diet. Foods should be chilled rapidly in small quantities. As with type A, thorough cooking will destroy vegetative cells, but not the heat-resistant spores. Leftovers should be reheated to 160°F (71°C) and hot foods should be held at temperatures greater than 140°F (60°C).

Type *C Clostridium perfringens* disease is sometimes called *enteritis necroticans* or *Pig-Bel.*

Bacillus Cereus Gastroenteritis. The etiologic agent of this foodborne disease is exoenterotoxin of *B. cereus.* The microorganism is a gram-positive, sporeforming, motile rod, frequently in the form of chains. Lecithinase is produced.

Incubation Period and Symptoms. Commonly, the incubation period ranges from 8 to 16 hours, but may be as short as 1.5 to 5 hours. Symptoms include one or more of the following: Nausea, abdominal cramps, watery diarrhea, some vomiting. Where there is a short incubation period, nausea and vomiting predominate. In many ways, the disease is similar to staphylococcal intoxication. The illness is usually of short duration (1 day or less).

Most Likely Food Sources. Custards, cereal products, puddings, sauces, meat loaf, fried rice. Preventive and control measures are essentially the same as those described for *Clostridium perfringens* type A illness.

Vibrio Parahaemolyticus Infection. This foodborne illness is produced by *Vibrio parahaemolyticus* microorganisms. The microorganism is gram-negative, straight or curved motile rod and is aerobic and facultatively anaerobic. The microorganism resists 7%, but not 10% sodium chloride media. Possesses O, K., and H antigens.

Incubation Period and Symptoms. Most commonly, the incubation period is 12 hours, although it may be as short as 2 hours and as long as 48 hours. Symptoms include one or more of the following: Abdominal pain, diarrhea (watery stools containing blood and mucus), usually nausea and vomiting, mild fever, chills, headache, prostration. Recovery usually occurs within 2 to 5 days.

Most Likely Food Sources. Raw foods of marine origin. Saltwater fish, shellfish, crustaceans, and various fish products. Cucumbers and other salty foods have been implicated. Reservoirs of the microorganism are sea water and marine life. A large percentage of the illnesses have been reported in Japan, particularly in warmer months. Occurrence is less in the United States and other parts of the world.

Preventive and Control Measures. Foods must be cooked thoroughly. Chilling should be rapid and in small quantities. Cross contamination from saltwater fish must be avoided. Seawater should not be used for rinsing foods to be eaten raw, or for cleaning.

Shigellosis or Bacillary Dysentery. This disease is usually transmitted by interpersonal spread. However, common-source foodborne outbreaks are reported from time to time. The etiologic agents include *Shigella sonnei, S. flexneri, S. dysenteriae,* and *S. boydii.* Shigellae are host specific to humans and thus food substances can be contaminated only by contact with the feces of infected humans. Foods most likely to be contaminated include moist, mixed foods; milk; beans; potato, tuna, shrimp, turkey, and macaroni salads; apple cider; and poi. The incubation period ranges from 1 to 7 days, but usually is less than 4 days. Symptoms are quite variable, ranging from mild to severe, involving abdominal cramps, fever, chills, diarrhea, headache, tenesmus, lassitude, prostration, nausea, and dehydration.

Yersiniosis. This sometimes foodborne disease, also known as *pseudo-tuberculosis,* is caused by *Yersinia pseudo-tuberculosis,* or *Y. enterocolitica,* which are gram-negative, motile rods. Coccoid forms predominate in young cultures. Microorganisms are aerobic, facultatively anaerobic. The incubation period is from 24 to 36 hours or longer. Likely food sources include pork and other meats, raw milk, unpasteurized ingredients of milk products, such as chocolate, and any other contaminated raw or leftover food. Sources and reservoirs include the urine and feces of infected animals, frequently rodents, dogs, and pigs. The microorganisms also are found in soil, dust, and water. Control measures include cooking foods thoroughly and protecting food supplies from contamination by rodents.

Relatively uncommon, yersiniosis occurred in outbreak proportions in central New York (near Utica) in the late 1970s. The ailment affected 218 children and, because the disease mimics appendicitis, 13 needless appendectomies were performed. The disease was first traced to school eating facilities and finally narrowed down to the milk. Upon investigation, it was found that unpasteurized chocolate flavoring had been added to the milk after the milk per se had been pasteurized. Prior to this incident, only about 100 cases had previously been recorded. The first epidemic occurred in the United States in 1973 among four rural North Carolina families, living with poor sanitary facilities. Contaminated milk caused that outbreak.

Foodborne Viral and Rickettsial Diseases. Epidemiological evidence has confirmed the foodborne transmission of several viral diseases, including hepatitis A, poliomyelitis, Bolivian hemorrhagic fever, Russian spring-summer encephalitis, and Q fever. There are also several viral diseases which could possibly be transmitted by foods, but where solid proof is lacking, and these include summer grippe (Coxsackie group A viruses), epidemic myalgia (pleurodynia), echo virus infections (enteric cytopathogenic human orphan viruses), hepatitis B (serum hepatitis), and nonbacterial gastroenteritis (possibly caused by the Norwalk agent). See **Coxsackie Virus; Norwalk Virus.**

Hepatitis A (*Infectious Hepatitis*). Caused by hepatitis virus A, this illness has an incubation period of from 10 to 50 days, but is about 30 days on the average. This is a systemic infection characterized by constitutional and gastrointestinal manifestations and by injury to the liver. Fever, malaise, lassitude, anorexia, nausea, abdominal discomfort, bile in urine, and jaundice all are symptoms. Severity usually increases with the age of the patient. Duration of the illness ranges from a few weeks to several months.

Suspect food sources include shellfish, milk, orange juice, potato salad, frozen strawberries, glazed doughnuts, whipped cream cakes, and a variety of sandwiches. Sources and reservoirs of the virus are the feces, urine, and blood of infected human cases and persons incubating or convalescing from the disease. Although foodborne, the main routes of transmission are person-to-person contact and drinking water. Good preventive measures include cooking foods thoroughly to inactive virus and to avoid pollution of shellfish growing areas. Drinking water must be fully treated, including chlorination. Known cases of hepatitis A should be isolated for 7 to 10 days after jaundice, and all equipment used for parenteral injections should be thoroughly sterilized. Give gamma globulin to contacts.

Charting of hepatitis outbreaks are given earlier in this entry. See Fig. 4.

General Study of Viruses Associated with Food Supply. In 1977, Kostenbader and Cliver (Food Research Institute, Department of Food Microbiology and Toxicology—Department of Bacteriology and World Health Organization Collaborating Centre on Food Virology—University of Wisconsin, Madison, Wisconsin) conducted a study on the association of viruses with food processing and distribution, with emphasis on the food processing plant. Seven plants were selected for examination and the details are well reported in the listed reference. The researchers concluded, in part: No viruses were found in plants processing vegetable products, nor in three of those working with animal products. Incoming swine at one slaughter plant frequently harbored viruses in their intestines; these were apparently not infectious for humans and were not seen in the final product. An agent too elusive to be characterized was found in frozen ground beef patties from another plant; this agent did not appear to be infectious for humans. A further survey of sanitary sewers at nine plants showed that the incidence of intestinal virus infections in processing personnel was below detectable limits. No virus was detected in 60 samples (2 of each of 6 foods from 5 markets). Research for virus-food connections continues.

Toxic Fishes and Shellfish. These diseases include scombroid poisoning, ciguatera poisoning, moray eel poisoning, ichthyohemotoxism (fish serum toxin), file fish poisoning, tetradon or puffer fish poisoning, clupeoid poisoning, elasmobranch and chondrichthyes poisoning, chimaeroid poisoning, cyclostome poisoning, gempylid poisoning, hallucinogenic fish poisoning, ichthyohepatoxism, freshwater fish poisoning, Haff or Yuksov disease, Minamata disease, paralytic shellfish poisoning, oyster poisoning, callistin shellfish poisoning, abalone poisoning, whelk poisoning, horseshoe crab poisoning, turtle poisoning, hypervitaminosis A, porpoise poisoning, and toxic quail poisoning. It will be noted from the names of these diseases that they are, in many instances, quite specific to a given order or species of animal. Also, that the maritime animals predominate. A few of these diseases are described here in some detail.

Ciguatera Poisoning (*Ichthyosarco toxism*). The exact nature of the etiologic agent of this foodborne disease (one aspect of *seafood poisoning*) has not been definitely established. Some authorities believe that the toxin is accumulated through the food chain in bottom-dwelling large fish (barracuda, red snapper, amberjack, and grouper), which are usually caught near ocean reefs, predominantly in the region between latitudes 35°N and 35°S. Eleven orders, 57 families and over 400 species of marine animals have been incriminated in connection with this poisoning. Some authorities believe that any marine fish may be potential transvector of ciguatoxin. The poison is concentrated in fish gonads, liver, intestines, and muscles. Good prevention is to avoid eating these parts of the suspect fishes. Eating of unusually large reef fishes should be avoided. There is no reliable method of detecting poisonous fishes by their appearance. Neither frying, baking, boiling, broiling, stewing, steaming, drying, salting, nor other ordinary cooking method destroys ciguatoxin.

The first signs of poisoning are usually observed within 3 to 5 hours after food ingestion, but may be as long as 24 hours. Symptoms begin with nausea, vomiting, abdominal cramps, and diarrhea. Paresthesias involving the lips, tongue, mouth, pharynx, and extremities are a prominent feature of the disease. Other symptoms include myalgias, arthralgias, blurred vision, photophobia, transient blindness, and cranial nerve palsies. In severe cases, bradycardia, hypotension, and respiratory paralysis appear early.

Symptoms usually subside after a few days, but weakness and sensory disturbances can persist for quite some time in rare cases. Fatalities are rare. There is no specific treatment for the disease. Where respiratory paralysis develops, mechanical ventilation may be indicated.

Scombroid Poisoning (*Scombrotoxism; Saurine poisoning*). This foodborne disease comprises another aspect of seafood poisoning. The etiologic agents are histaminelike substances and possibly saurine (*Proteus* spp.). Histidine in flesh is broken down by action of *Proteus* spp. or other organisms. It is thermostable and can withstand boiling for at least 1 hour. Histamine is a capillary dilator. Symptoms of the poisoning are evident from a few minutes after ingestion to about 1 hour. There is intense headache, dizziness, nausea, vomiting, metallic or peppery taste, diarrhea, facial swelling and flushing, epigastric pain, throbbing of carotid and temporal vessels, rapid and weak pulse, burning of throat, thirst, difficulty in swallowing, edema, and itching of skin. Although an illness of great discomfort, recovery usually occurs within 12 hours.

Likely sources of this poison are scombroid fishes, such as tuna, bonito, mackerel, and skipjack. Good prevention requires refrigeration of fish immediately after they are killed. The fish should be consumed promptly, if not immediately refrigerated.

Upon noting first symptoms, and if vomiting and diarrhea have not already occurred, ipecac and cathartics can be used initially. Antihistamines and bronchodilators may provide symptomatic relief. Of several hundred reported cases of scombroid poisoning, no fatalities have occurred.

In February 1973, scombroid fish poisoning occurred in 232 persons who had eaten from either of two lots of commercially canned tuna. Cases occurred in four states (United States), with no reported hospitalizations or deaths. Patients became ill about 45 minutes after eating the fish. Symptoms lasted about 8 hours. Contaminated fish contained histamine levels of 68 to 280 milligrams/100 grams of fish muscle. This was the first reported outbreak associated with a commercially canned fish product. Numerous outbreaks have occurred in Japan where fresh fish is a substantial part of the diet. One Japanese report described 14 outbreaks involving 1215 persons within a 4-year period. Some cases of so-called "tuna fish allergy" probably represent scombroid fish poisoning.

Paralytic Shellfish Poisoning (*Dinoflagellate poisoning*). This foodborne disease is still another aspect of seafood poisoning. The etiologic agent is saxitoxin, a neurotoxin that blocks neuromuscular junctions. It is an alkaloid and relatively heat-stable. It is water-soluble. During red tides, cell counts of plankton blooms may reach from 20 to 40 million per milliliter. Onset of the illness usually occurs within less than 1 hour after food ingestion. Symptoms include tingling or burning and numbness around lips, finger tips and extremities. In severe cases, muscle paralysis develops and leads to dysphonia, dysphagia, and ventilatory impairment. Death caused by respiratory paralysis may occur within the first 12 hours.

The gastrointestinal tract must be emptied of any unabsorbed toxin, with care taken to avoid aspiration. Measures designed to support the cardiovascular and respiratory system should be commenced quickly. Use of a mechanical ventilator may be required.

Oyster Poisoning (*Asari or Venerupin poisoning*). The etiologic agent is venerupin (asaritoxin), which is a heat-stable toxin. Toxicity remains even after 1 hour of boiling. The toxin is organotropic, affecting mainly the liver. Symptoms may be evident within 6 hours, or as long as 7 days after food ingestion. However, onset of the illness usually is apparent within 24 to 48 hours. Symptoms include anorexia, abdominal pain, nausea, vomiting constipation, headache, malaise, nervousness, halitosis, bleeding of mucus membrane of nose, mouth, and gums, delirium. There is no paralysis. Case fatality rate is as high as 33%.

Food sources are oysters and clams (asari), including *Crassostrea gigae, Dosinea japonica,* and *Tapes semidecussata.* Prevention is essentially control of shellfish harvesting practices.

Abalone Poisoning. The etiologic agent is described as a photodynamic principle that causes photosensitization. The substance is stable to boiling, freezing, and salting. Timing of onset of disease is dependent upon exposure to sunlight. Where accompanied by exposure to sunlight, onset is sudden, with burning and stinging sensation over entire body, prickling sensation, itching, erythema, edema, skin ulceration on parts of body exposed to light. The food source is the Japanese abalone (*Haliotis discus,* or *H. sieboldi*). Best prevention is to make certain that there are no viscera of abalone in food served. It has been suggested that a seaweed of the genus *Desmarestia* may be the source of the toxin.

Tetraodon or Puffer Fish Poisoning. There are about 90 toxic species of puffer fish (fugu, blowfish, globefish, porcupine fish, molas, burrfish, balloonfish, and toadfish). The etiologic agent is tetrodotoxin (tetraodontoxin). This is a neurotoxin (paralysis of central nervous system and peripheral nerves). The toxin is stable to boiling except in alkaline solution. Toxin is water-soluble. The toxin mainly attacks nerve endings by blocking movements of all monovalent cations. Onset of illness usually occurs within 10 to 45 minutes after ingestion, although it may be delayed up to 3 hours. Symptoms including tingling or prickly sensation of fingers and toes, malaise, dizziness, pallor; numbness of lips, tongue, extremities; ataxia, nausea, vomiting, diarrhea, epigastric pain, dryness of skin, subcutaneous hemorrhages and desquamation; eyes fixed, reflexes lost, respiratory distress; muscular twitching, tremor, incoordination, muscular paralysis, intense cyanosis. Case fatality rate is close to 60%.

Best prevention is complete avoidance of eating puffers. The sale of puffers in Japan is governed by regulation. Puffer cooks and restaurants must be licensed. Proof of experience in preparing puffers is required.

Poisons of Fungal Origin. Possibly, the most publicized of the foodborne diseases of fungal origin is that caused by the mycotoxins present in certain varieties of mushroom. There are however, a number of mycotoxicosis diseases emanating from other fungal sources. Reported in some detail here is mushroom poisoning and aflatoxicosis.

Mushroom Poisoning. Some authorities believe that enthusiasm for so-called organic foods and experimentation with natural hallucinogens have probably contributed to the increased incidence of serious and fatal mushroom poisoning that has occurred in the United States during the 1960s and 1970s. Warnings ad infinitum pertaining to the danger of foraging for mushrooms have not proved entirely adequate to prevent tragic cases of poisoning by wild mushrooms. These cases especially occur in the fall when the mushrooms, which are in the reproductive part of the fungus, can be easily harvested. Experience has indicated a wide range of susceptibility to mushroom toxins among individuals. Also, the severity of the poisoning may depend upon the season, the degree of maturation of the mushrooms, and logically the number of mushrooms ingested.

The etiologic agents are cytotoxins, the most important of which is amatoxin produced by mushrooms belonging to the genera *Amanita* and *Galerina.* These include those mushrooms commonly referred to as death angel, death cup, destroying angel. The so-called "false morel" mushrooms of the genus *Helvella* are also implicated. Case fatality rate from poisonings by *Amanita* and *Galerina* approach 50%, while with *Helvella* the fatality rate is considerably lower. It must be emphasized that just one or two mushrooms ingested can cause death.

Mushroom toxins are a protoplastic poison that disrupts integrity of cellular membranes. The amatoxin molecule is a complex protein, a cyclic octapeptide (Fig. 7) that, upon ingestion, inhibits ribonucleic acid polymerase II in the victim's cells, thus interfering with protein synthesis. In turn, this results in damage to membranes of cell walls, of organelles, and of nuclei. Lethal hepatic and renal tubule destruction may occur after ingestion of but one cap containing amatoxin. Authorities do not agree as regards the toxicity of phalloidine, a similar substance with a 7-membered cyclopeptide ring structure, which is produced by *Amanita phalloides.*

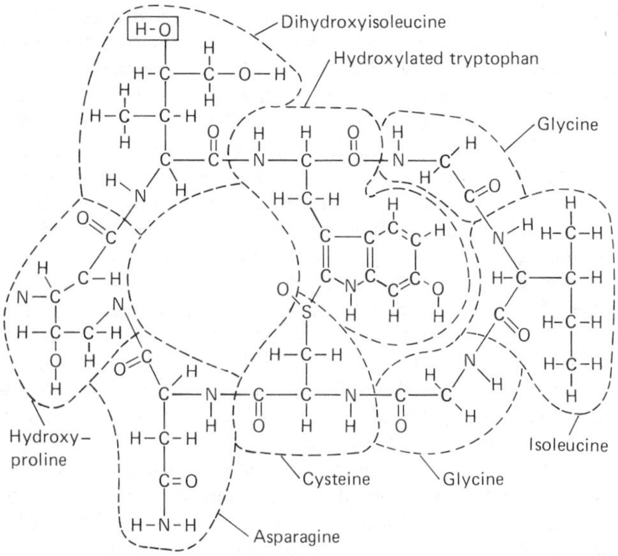

Fig. 7. Amatoxins appear to be made up of eight amino acids (within the dotted lines). These are joined by peptide bonds and thus a continuous cyclopeptide ring is formed. It will be noted that a sulfur atom serves to link the side chains of two of the amino acids and thus a bridge across the ring is created. The OH group contained within the box (for illustrative purposes here) is apparently critical to this molecule's toxicity. The specific toxin illustrated here is alpha-amanitin.

Early symptoms include abdominal pain, followed by severe vomiting, diarrhea, and fever. Dehydration, hypovolemia, and electrolyte loss may ensue, and hematuria and blood-streaked diarrhea also may occur. After a day or two, the gastrointestinal symptoms abate and the patient may appear to improve. However, by the third or fourth day, hepatic and renal failure become evident. Jaundice, hypoglycemia, oliguria, bleeding, delirium, and coma supervene and, in cases this advanced, 40 to 90% of the patients die. Laboratory confirmation of the presence of amatoxin can be obtained by performing thin-layer chromatographic analysis of samples of mushrooms or of vomitus, gastrointestinal aspirates, or stool specimens.

Supportive measures used by treating physicians will include the correction of fluid and electrolyte depletion, important during the initial phase of treatment. Hemodialysis has been used in acute renal failure.

In a number of instances of amatoxin poisoning, thioctic acid has been effective in preventing death. Widely used in Europe, the drug is available in the United States on an experimental basis. The principal side effect associated with thioctic acid has been hypoglycemia. Inquiries regarding thioctic acid should be directed to the National Institute of Health, Bethesda, Maryland.

Toxin ingestion from some mushrooms (or at varying times) may be neurotoxins that are less dangerous than mushroom cytotoxins. Muscarine is one of these neurotoxins. Onset of the illness is rapid. Symptoms include salivation, perspiration, peripheral vasolidation, lacrimation, bradycardia, nausea, vomiting, abdominal cramps, copious watery diarrhea, slow, irregular pulse, pupil constriction, asthmatic breathing, cardiac or respiratory failure (rare). Sensorium is ordinarily clear. Other neurotoxins, with somewhat similar effects, include ibotenic acid, muscimol, uscazone, tricholomic acid, and psilocin. The latter produces psychotomimetic manifestations, causing elevated mood, laughter, hallucinations.

Mushroom-alcohol intolerance is an interesting aspect of certain mushroom reactions. This occurs when alcohol is consumed (ranging from within 24 to 48 hours) after ingesting inky cap mushrooms (*Coprinus attramentarius*). This also occurs in connection with the species *Clitocybe claviceps* when ingested along with, or before beer or sake. Apparently, the disulfiramlike (antabuse) constituents of mushrooms interfere with normal metabolism of alcohol. Normally onset of the illness appears within $\frac{1}{2}$ to 2 hours. Symptoms include flushing (purplish-red face), metallic taste, paresthesia of extremities, palpitation, dyspnea, hyperventilation, tachycardia, feeling of swelling hands, nausea, and vomiting. Attacks usually have a duration from 30 minutes to several hours. Prevention is the abstention from alcohol for several days after consuming either of these varieties of mushrooms, and particularly avoiding overheating *C. atramentarius.* Some authorities suggest avoidance of the latter species altogether.

Aflatoxicosis (*Aflatoxin Poisoning*). The etiologic agents of this foodborne illness are Aflatoxin B_1, B_2, G_1, and G_2 from the *Aspergillus flavus-oryzae* group. These fungi are found worldwide and grow on practically any substrate. They are carcinogenic to rats, ducks, and trout and some authorities implicate them in involvement with human cancers. These toxins are heat-stable. Suspect food sources include cottonseed meal, Brazil nuts, palm kernels, groundnuts (peanuts), corn (maize), other cereals, and animal feeds. Preventive measures include control of moisture during storage of aforementioned materials, prevention of damage during harvesting, insect control, fungicidal mold control, and physical removal of contaminated products, such as groundnuts.

Onset of the illness may require a few weeks. Symptoms include low-grade fever, jaundice, ascites, and edema of feet. Fatty infiltration and cirrhosis of liver also may occur. The appearance of aflatoxin is usually random and unpredictable. For example, in Arizona in 1978, about 17,000 tons of cottonseed were stored in a facility that caught fire. The fire was extinguished with water. Subsequent rains and warm weather encouraged the *Aspergillus flavus* mold which produced the aflatoxin. Later tests of milk in grocery stores showed aflatoxin above the 0.5 part per billion (U.S. Food and Drug Administration guidelines) and dairies in the region reacted by dumping some 200,000 gallons of milk. Cases in children with kwashiorkor in India and Africa have been reported.

Poisons of Parasitic Origin. Parasitic diseases always or usually transmitted by foods include trichinosis, diphyllobothriasis, and anisakiasis. Amebiasis (amoebic dysentery) is also food and water related. See **Amebiasis.** Toxoplasmosis is also food related. See **Toxoplasmosis.**

Trichinosis (*Trichinelliasis*). The etiologic agent is *Trichinella spiralis,* threadlike roundworms (nematodes). The larva excyst in duodenum, females invade mucosa of small intestine, larvae travel via blood and lymph, encyst in muscle, the incubation period varies from 4 to 28 days, but usually requires about 9 days. In the first stage of the disease (intestinal invasion), symptoms include nausea, vomiting, diarrhea, abdominal pain. In the second stage (muscle penetration), there may be an irregular, but persistent fever, edema of the eyes, profuse sweating, muscular pain, thirst, chills, skin lesions, weakness, prostration, labored breathing. In the third stage (tissue repair), there may be generalized toxemia, myocarditis. There will be a high eosinophil blood count.

Because of muscular involvement, there are difficulties with speech, swallowing, and chewing, and usually swelling of the upper eyelids, a characteristic of the disease. Mild cases usually run their course in about 2 weeks; severe cases about 6 weeks. Seldom does any permanent muscle damage result. Contaminated food sources may include pork, bear meat, walrus flesh, and dog meat. Best prevention is by careful control over pork production, making certain that garbage is not fed to pigs. Thorough cooking of all pork also is an excellent safefuard.

Diphyllobothriasis. The etiologic agent is *Diphyllobothrium latum* (broad or fish tapeworm). The flatworm (cestode) attaches to mucosa of small intestine. Length may be 30 feet (9 meters) or more. Symptoms are often trivial or absent. Nausea, vomiting, weakness, dizziness, diarrhea or constipation and anemia may occur in serious cases. Deficiency of vitamin B_{12} may occur because of competition of worm for this vitamin. Treatment with one of several antihelminths, such as oleoresin aspidium or quinacrine, is usually successful.

Most common food sources are raw or partly cooked or inadequately pickled freshwater fish, such as pike and pickerel.

Anisakiasis. The etiologic agent is *Anisakis* spp., a roundworm (nematode). The adult worm lives in the intestine of fish-eating sea mammals. Larvae are found in herring. Outbreak reports have been made in the Netherlands, Japan, the United Kingdom, and the United States. Herring is the likely contaminated food source, particular if raw or partially cooked and pickled or smoked. Best prevention is thorough cooking of herring, or, freshly caught fish can be frozen at −4°F (20°C) and held for 24 hours; or the fish can be preserved with high concentrations of sodium chloride and held for 10 days.

References

Adams D. M., Jr., and H. M. Wehr: "Hazardous Anaerobes and Mycotoxins," *Inst. Food Technol. Symp.*, Saint Louis, Missouri, 1979.

Bryan, F. L.: "Diseases Transmitted by Foods," U.S. Dept. of Health, Center for Disease Control, Atlanta, Georgia, 1976.

Duncan, C. L., and D. B. Rowley: "*Clostridium perfringens* Foodborne Illness," *Inst. Food Technol. Symp.*, Saint Louis, Missouri, 1979.

Fields, M. L., Zamora, A. F., and M. Bradsher: "Microbiological Analysis of Home-Canned Tomatoes and Green Beans," *J. Food Sci.*, **42**, 4, 931–934 (1977).

Hughes, J. M., and M. H. Merson: "Fish and Shellfish Poisoning," *N. Engl. J. Med.*, **295**, 1117 (1976).

Hultin, H. O., and M. Milner (editors): "Postharvest Biology and Biotechnology," Food and Nutrition Press, Westport, Connecticut, 1979.

IAMFES: "Procedures to Investigate Foodborne Illness," Intl. Assoc. Milk, Food, and Environ. Sanitarians, Inc., Ames, Iowa, 1976.

Jemmali, M. (editor): "Mycotoxins in Foodstuffs," Pergamon, Elmsford, New York, 1976.

Kostander, K. D., Jr., and D. O. Cliver: "Quest for Viruses Associated with Our Food Supply," *J. Food Sci.*, **42**, 5, 1253–1257 (1977).

Merson, M. H., et al: "Scombroid Fish Poisoning," *J. Amer., Med. Assoc.*, **228**, 10, 1268–1269 (1974).

Miller, M. W.: "Yeasts in Food Spoilage," *Food Technology*, **33**, 1, 76–80 (1979).

Minor, T. E., and E. H. Marth: "Staphylococci and Their Significance in Foods," Elsevier, Amsterdam, 1976.

Woodicka, V. O.: "Food Ingredient Safety Criteria," *Food Technology*, **31**, 1, 84–88 (1977).

FOOD CHAIN. The series of events which convert solar energy, initially by way of photosynthesis, to food for plants and animals. The basic work is accomplished by leaves in the case of terrestrial plants, and by plankton in the sea. Then, more complex creatures consume plants and/or smaller (usually) organisms and so on in a chain of events involving ever increasing sizes and appetites of the consumers and ever decreasing amounts of food available in the chain. Studies have shown that for a human to gain 1 pound (kilogram) as the result of consuming seafood, the sea must generate 1000 pounds (kilograms) of living matter. At the base of the food pyramid or first link in the chain are the trillions of plankton which create the 1000 pounds (kilograms) of basic chemical food. Then, a variety of crustaceans, smaller fishes, and creates reduce this 1000 pounds (kilograms) to 100 pounds (kilograms). Then, the game and food fishes in their feeding reduce the 100 pounds (kilograms) to 10 pounds (kilograms). To gain 1 pound (kilogram) of weight, the human must consume about 10 pounds (kilograms) of fish. This may be stated more accurately by indicating that the 10 pounds (kilograms) of the fish may sustain 1 pound (kilogram) of weight in the human, these latter factors depending, of course, upon a widely ranging metabolism among individuals.

FOOTING. A footing is a foundation structure used to distribute wall or column loads to the bearing soil or to the piling. There are two general classifications of column footings. The isolated spread footing (see figure) supports one column, and the combined footing supports two. Footings must be spread wider than the base of a column or wall for a number of reasons. They must give stability by reducing the unit pressure below that at which there might be local settlement along one side of the foundation. They must be thick enough to resist the punching shear of the column, and they must not flare so rapidly as to cause them to be weak in bending. Footing for walls, columns, etc. may be very conveniently and economically made of reinforced concrete. Except for very heavy loads, the footing slab is usually square

Sectional view of spread footing.

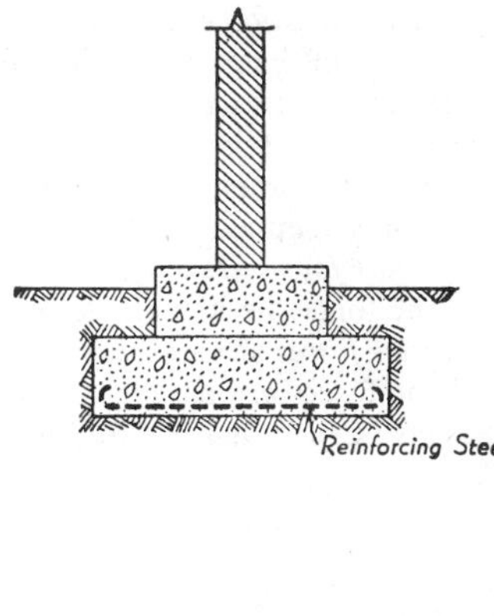

or rectangular and of constant thickness. Design is usually made by semirational rules and formulae since the distribution of stress in such a structural unit is very complex. For very heavy column loads the footing is usually made in the form of a truncated pyramid or preferably in a series of steps giving the "stepped footing." The latter type, illustrated in the figure, predominates, due to the ease with which forms may be constructed. In case the soil is so weak as to require an extremely wide footing, a steel I-beam grillage is incorporated in the footing.

FOOT (Physiology). 1. The ventral protuberance of the body of a mollusk, usually an organ of locomotion. 2. The terminal portion of a jointed appendage which comes into contact with the supporting surface. In quadrupedal vertebrates the term is applied to both fore and hind appendages and in bipedal forms to the latter only.

FORAMEN. An opening, especially in bone, through which other structures pass, such as nerves and blood vessels.

FORB SEEDING. Revegetation.

FORBUSH DECREASE. Cosmic Rays.

FORCE. Three general uses of the term force are:

1. In dynamics, the physical agent which causes a change of momentum, measured by the time rate of change of momentum. If the speeds involved are low compared with that of light, a force may be defined as proportional to the mass m of a body and to the acceleration $\mathbf{a}$ of the body which is produced by the force. Thus $\mathbf{f} = km\mathbf{a}$, where k is a constant for a given system of units, and has the dimensionless value unity in length-mass-time systems or $1/g$ in length-force-time systems, g being the acceleration due to gravity. Force is a vector quantity, requiring both a magnitude and a direction for its complete specification.

2. In statics, the physical agent which produces an elastic strain in a body. Static forces are equated to dynamic forces by the method of allowing a weight to produce a strain and then by allowing the same weight to fall under the action of gravity.

3. From its initial conception, which was purely mechanical as expressed in (1) and (2), above, the term force extends to denote loosely any operating agency, such as coercive force, electromotive force, and magnetomotive force.

Furthermore, there are a number of compound terms denoting various special forces:

An *impressed force* is any external force acting on a particle in a dynamical system. The resultant force on each such particle can always be resolved into a resultant external impressed force and a resultant internal constraint force. Thus in the case of a particle suspended by a string, the weight is the impressed force and the tension in the string is the constaint force.

A *restoring force* is the elastic force which acts on a particle or portion of a mechanical system when displaced from equilibrium and whose direction is such as to return the system to equilibrium. In simple cases, the restoring force is linear, i.e., proportional to the first power of the distance. For some physical systems, however, the restoring force may be proportional to the second or higher power of the distance.

A *tangential force* is a force associated with a wheel or disk always acting perpendicular to the radius, e.g., frictional force between rolling wheel and a surface, or frictional force between a belt and a pulley wheel.

An *attractive force* is a force acting on a particle such that the acceleration of the particle is in the direction of the agency responsible for the force. This agency can be another single particle, a collection of particles or a region acting as the source of a field.

Concurrent forces are a system of forces such that there exists a single point common to all the lines of action of the forces.

Coplanar forces are a system of forces such that the lines of action of all the forces lie in a single plane.

Parallel forces are a system of forces such that the lines of action of all the forces are parallel to each other.

See also **Centrifugal Force; Centripetal Force; Coriolis Effect; Electric Power Generators; Electromotive Force; Electrostatics; Machine (Simple); Magnetism; Mass-Energy Equivalance; Rheology;** and **Van der Waals Forces.**

FORCE-BALANCE ACCELEROMETER. Acceleration Measurement.

FORCE-BALANCE TRANSDUCER. A transducer in which the output from the sensing member is amplified and fed back to an element which causes the force-summing member to return to its rest position.

There are scores of examples of where force-balance transducers are applied in industrial measurements. The use of a force-balance transmitter to measure a liquid level in a pressureized tank is shown by the accompanying diagram. As the head of liquid varies, the pressure on one side of the differential pressure sensor bellows changes. This pressure is rapidly equalized by instrument air, and the greater or lesser pressure required is registered by the pressure gage, calibrated in this case in units of liquid level.

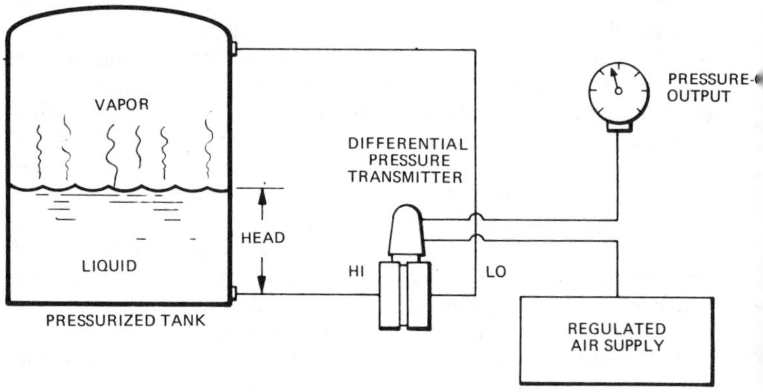

Force-balance transmitter used in level measurement of liquid in a pressurized tank.

FORCE EQUILIBRIA. Statics (Graphical).

FORCE GAGE (Bourdon). Bourdon Tube.

FORCE (Torque). Torque.

FORCE (Traction). Traction.

FORCE UNITS. Units and Standards.

FORCE VARIABLE. Variable (Process).

FORECASTING (Climate). Climate.

FORECASTING (Earthquake). Earthquakes, Seismology, and Plate Tectonics.

FORECASTING (Hurricane). Hurricane Prediction.

FORECASTING (Weather). Weather Observations and Forecasting.

FORESHOCK. Earthquakes, Seismology, and Plate Tectonics.

FORESKIN. The prepuce, a prolongation of the skin of the shaft of the penis or clitoris, covering the head of either of these organs. It is this fold of skin that is removed by circumcision.

FOREST BUFFALO. Bovines.

FOREST (Lumbering). Wood.

FOREST (Nitrate Loss). Nitrogen.

FOREST (Redwood). Redwood (Coast).

FOREST SURVEILLANCE. Earth Resources Satellites and Geologic Remote Sensors.

FORGING. The shaping of metal by hammering or cogging. Only the malleable metals may be worked successfully. While most metals may be shaped in the cold state, the application of heat increases plasticity and permits greater deformation with any given expenditure of energy.

In usual forging practice, heated billets or stock are used. Extremely large billets or ingots are heated in soaking pits which are usually fired with gas or oil fuel and which often are designed to utilize regenerative heating principles. Billets or blooms of smaller size are heated in open muffle furnaces fired with gas or oil.

Stock forged by hand is generally small and this may be heated in a forge. A forge is a hearth made from fire-clay, firebrick, or tamped sand so arranged that a fire of coal, coke, or charcoal may be built upon it. Air is supplied from a blower through tuyères, or openings, suitably placed. A hood above the forge removes the products of combustion from the forge room. In using a forge, the piece to be worked is buried in the hot coals of the thick carbonaceous fuel.

Care must be exercised in the heating of all metals for forging. Metals that are heated too long at too high a temperature oxidize rapidly, forming an excessive amount of scale which not only wastes metal but also prevents the production of smooth surfaces. Temperatures may be measured with thermocouples or with optical pyrometers and standard practices may be developed to give uniform results for any forgeable metal. Some metals are susceptible to carburization, decarburization, attack by sulfur gases, and intergranular oxidation. Furnace atmospheres during heating may be controlled to prevent any of the before-mentioned conditions. High sulfur fuel is a common source of sulfur contamination.

Many metals and alloys forge best in a given temperature range and those that may be worked only in a very narrow range of temperature are gemerally considered the most difficult to forge because this characteristic requires frequent reheating if the total deformation is great.

In addition to fabricating an identified shape, the operation of forging improves the quality of metals. The coarse crystals of metal resulting from solidification in an ingot mold are kneaded and refined. Blow holes and layers of slag are consolidated and usually welded together. This results in a more ductile and stronger product than cast metal with much greater resistance to shock and to fatigue stresses. Hammer forging imparts a higher degree of refinement on the surface of the work while pressed forgings exhibit better average quality throughout thick pieces.

Upsetting is the process of increasing the cross section of stock at the expense of its length. Swaging or drawing-down operations increase length of stock at the expense of reduction in cross-section. Setting down is a localized swaging operation. Bending processes may be classified as angular or as curvilinear. Punching is an operation carried out to remove a slug of metal through shearing. Cutting-out is the process of cutting large holes by progressively using a hot chisel over a hole in the swage block. Saddening is a series of light shaping blows useful in preparing some metals for heavy forging operations.

Die blocks for power forging are made from carefully manufactured alloy steels. Such blocks must have the ability to withstand severe strains imposed by high pressures on hard metals. Long life with a minimum of impression wear is a prime requisite.

Power hammers are used for unit-production forging operations on work that cannot be feasibly hand-hammered, and also on comparatively small work where labor costs may be reduced by their use. There are two general types of hammers, steam or air hammers and trip and helve hammers. In the steam hammer, stream pressure is exerted on both faces of a vertically reciprocating piston, which is connected to the ram by a piston rod, to raise the ram and also to aid in striking the blow. The steam pressure on the downward stroke imparts additional velocity to the falling weight. The ram will therefore strike a blow whose full rating will be about double that of a gravity ram.

A trip hammer has a vertically reciprocating ram that is actuated by a toggle connection driven by a rotating shaft at the top of the hammer. The shaft is driven by a cone clutch which in turn is driven by a second shaft and pulley. The speed of the ram and the resultant effect of the blow are determined by the varying speed of the shaft. Trip hammers are built in sizes from 15 to 500 pounds.

Helve hammers are made in the same sizes as trip hammers and are used for similar classifications of work. The helve hammer consists of a horizontal wooden helve, pivoted at one end with a hammer at the other, and a cam or eccentric between the pivot and the hammer. The cam raises the hammer which falls and strikes a blow by the force of gravity. The action of a helve hammer is essentially that of a hand sledge since the wooden helve is somewhat elastic.

Light hammer blows are comparatively superficial in effect, while a heavy more slowly delivered blow penetrates and influences the material structure to a much greater extent. This effect is of particular importance in large forgings. Therefore, steam-actuated hydraulic presses are employed. Hydraulic presses for forging purposes range in size from 200 to 15,000 tons.

Drop forging is the process of shaping hot metal by forcing it into die cavities by the application of sudden blows. There are two principal forms of drop hammers that are used for drop-forging operations—the steam hammer and the board drop. The operation of the steam hammer has been described previously; the board drop has a hammer or ram fastened to maple boards whose upper ends pass between two rotating rolls. When the rolls are moved towards each other, they lift the board. At the top of the stroke the rolls automatically separate and release the board, permitting the ram to drop and strike the blow. Most board drop hammers are equipped with a device for changing the length of fall and consequently the force of the blow. Forging presses, in which the vertical ram is actuated by a pitman operated by an eccentric, are used for impact extrusion, hot forging, and hot and cold coining operations. These presses are available in capacities up to 2,000 tons.

Forging rolls are often used for the breakdown operations preliminary to drop forging, and for reducing short thick sections to long slender sections. Their action is similar to that of rolling mills, but the forging rolls make use of only a portion of a revolution to reduce the stock; the remainder of the roll has a blank clearance space for the ends of the stock that are to be left full size. The operator stands at the back or emerging side of the rolls, and when the clearance space appears, he places the work into this space. When the reducing portion of the rolls comes in contact with the work, it reduces the stock and ejects it from the rolls towards the operator. At the next open portion of a revolution of the rolls, the stock is again inserted for a second reducing operation. By this method only as much of the length of stock is reduced in area as is required. Forging rolls contain a number of grooves, depending upon the number of passes required.

Machine forging, as distinguished from drop forging, is an upsetting or heading process applied to forgings made from bar stock. A forging machine consists essentially of three dies: a movable die opposed to a stationary die (the two are used for gripping the bar stock) and a third or header die which moves in a plane parallel to the parting surface between the clamping dies, and handles the major portion of the work of forging. The die operation is automatic and is controlled by a treadle-operated clutch. The operator inserts the bar, steps on the treadle and the movable die closes, gripping the bar. The header die moves forward to perform its operation and returns to starting position, and the movable die opens, completing the cycle. The operator then moves the bar to the next die cavity and again steps on the treadle to begin the next cycle.

See also **Iron Metals, Alloys, and Steels.**

FORM. A homogeneous polynomial in $2n$ variables $x_1, x_2, \ldots, x_n, y_1, y_2, \ldots, y_n$ is a bilinear form. If **A** is a matrix; $\tilde{\mathbf{x}}$, a row vector; **y**, a column vector, then the polynomial may be written in matric form

$$A(x, y) = \tilde{\mathbf{x}}\mathbf{A}\mathbf{y} = \sum_{i,\, j}^{n} A_{ij}x_i y_j$$

Since **A** is square and **y** a column, **Ay** is also a column. Then **x**, a row, times **Ay**, a column, gives a (1×1) matrix or a scalar as indicated by the last term on the right in the preceding equation.

Suppose that new variables are introduced by the nonsingular transformations $\mathbf{x} = \mathbf{Px}'$, $\mathbf{x} = \mathbf{Qy}'$, then $'A(x, y) = \mathbf{x}'\mathbf{PAQy}' = \mathbf{x}'\mathbf{By}' = A'(x', y')$. In the special case where $\mathbf{P} = \mathbf{Q}^{-1}$, then $\mathbf{x} = \mathbf{Q}^{-1}\mathbf{x}'$, $\mathbf{y} = \mathbf{Qy}'$ and **x, y** are called contragredient variables since their transformations are opposite. The variables are cogredient if $\mathbf{P} = \mathbf{Q}$.

The quadratic form results when $x_i = y_i$. The coefficient of x_ix_j $(i \neq j)$ is then $(A_{ij} + A_{ji})$ but the value of the sum will remain unchanged if the elements of **A** are taken as $(A_{ij} + A_{ji})/2$ for every i, j. This means that **A** is a symmetric matrix. When the value of a form is positive for all real values of the variables, **A** is positive definite; if it is positive or zero, semidefinite.

If the variables are complex and the matrix is Hermitian, the form is also called Hermitian

$$H(x, x) = \mathrm{E}H_{ij}x^*_iX_j = \mathbf{x}^\dagger\mathbf{Hx}; \quad H_{ij} = H^*_{ji}$$

where the asterisk denotes the complex conjugate and the dagger the associate matrix. In spite of the appearance of complex elements, the Hermitian form is real.

See also **Polynomial;** and terms listed under **Mathematics.**

FORMALDEHYDE. HCHO, formula weight 30.03, colorless gas with pungent odor, mp −92°C, bp −21°C, sp gr 0.815 (at −20°C). The gas is very soluble in H_2O, alcohol, and ether. Formaldehyde usually is produced and marketed as a 37% (weight) solution in water. From 3 to 15% methyl alcohol normally is added as a stabilizer to prevent paraformaldehyde formation. The commercial trend is to furnish a more concentrated product (up to 50% HCHO by weight) which contain as little as 0.5 to 1% methyl alcohol. The addition of special stabilizing agents and storage at elevated temperatures reduces the formation of paraformaldehyde.

Polymerized formaldehyde (trioxane) is a ring compound of anhydrous formaldehyde with the formula $(HCHO)_3$. See also **Acetal Resins.** Trioxane is a colorless crystalline solid with a pleasant odor, mp 62°C, bp 115°C, sp gr 1.17. This compound is used as a tanning agent and solvent and as a source of dry HCHO gas. Because trioxane ignites readily at 113°C and burns with an odorless, hot flame, it has been furnished in tablet form as a replacement for solidified alcohol in portable heating applications.

Paraformaldehyde $(CH_2O)_x$, sometimes called paraform, is an amorphous white powder and may be used for applications where an aqueous solution of HCHO may not be desirable. It finds use as an antiseptic and as a catalyst and hardener for certain synthetic resins. Formaldehyde gas also may be dissolved in methyl or butyl alcohol for applications where H_2O is undesirable.

Formaldehyde is a high-tonnage chemical and, in addition to the uses already mentioned, finds wide application in the manufacture of urea-formaldehyde resins (growing annually at a rate of about 6%), melamine-formaldehyde resins (growth rate of about 5%), and acetal resins (growth rate of about 10%). Formaldehyde also is used in the production of pentaerythritol which, in turn, is used in the manufacture of lubricant additives, resin esters, pentaerythritol tetranitrate, and alkyd resins. Formaldehyde also is required in the manufacture of hexamethylene tetramine, a compound important in explosives manufacture and as a resin-curing agent. Other uses for formaldehyde include the production of ethylene glycol, acrylic esters, urea-formaldehyde fertilizers, textile-treating agents, tetrahydrofuran for elastomeric fibers, trimethylol propane for urethanes, and as a solvent for synthetic and natural resins. The growing need for nitrilotriacetic acid (NTA) and isoprene, for which formaldehyde is required, will account for additional tonnage production in the near future.

Production. All major commercial processes for making formaldehyde initially yield an aqueous solution of HCHO. In over 90% of the installations, methyl alcohol is the chargestock. Other feedstocks uncommonly used include methane, hydrocarbon gases, and dimethyl ether. Those processes commencing with methyl alcohol are of two types: (1) the *silver-catalyzed* process in which formaldehyde results from a combination dehydrogenation–oxidation reaction: $CH_3OH \rightarrow HCHO + H_2 + \frac{1}{2}O_2 \rightarrow H_2O$. The first part of the two-step reaction is endothermic, the second part is exothermic; and (2) the *oxide-catalyzed* process in which methyl alcohol is directly oxidized: $CH_3OH + \frac{1}{2}O_2 \rightarrow HCHO + H_2O$. The latter is an exothermic reaction. A process representing the silver technology is shown in the accompanying flowsheet. The process essentially consists of two sections: (1) the synthesis portion which yields products containing from 3 to 15% methyl alcohol, and (2) the distillation portion which is required only where the methyl alcohol content of the final product must be low. In the oxide-catalyzed process, catalysts used generally are mixtures of oxides of molybdenum, iron, and vanadium.

Formaldehyde reacts with many chemicals in a marked manner, (1) with ammonio-silver nitrate (Tollen's solution), to form metallic silver, either as a black precipitate or as an adherent mirror film on glass, (2) with alkaline cupric solution (Fehling's solution), to form cuprous oxide, red to yellow precipitate, (3) with rosaniline (fuchsine, magenta) which has been decolorized by sulfurous acid (Schiff's solution), the pink color of rosaniline is restored, (4) with NaOH, yields methyl alcohol plus sodium formate, (5) with NH_4OH, when evaporated, yields hexamethylene tetramine "urotropine" $(CH_2)_6N_4$, white solid, mp 263°C, (6) with sodium or hydrogen peroxide in sodium hydroxide, yields sodium formate, (7) with manganese dioxide and H_2SO_4, forms methyl, dimethoxymethane $CH_2(OCH_3)_2$, colorless liquid, bp 42°C.

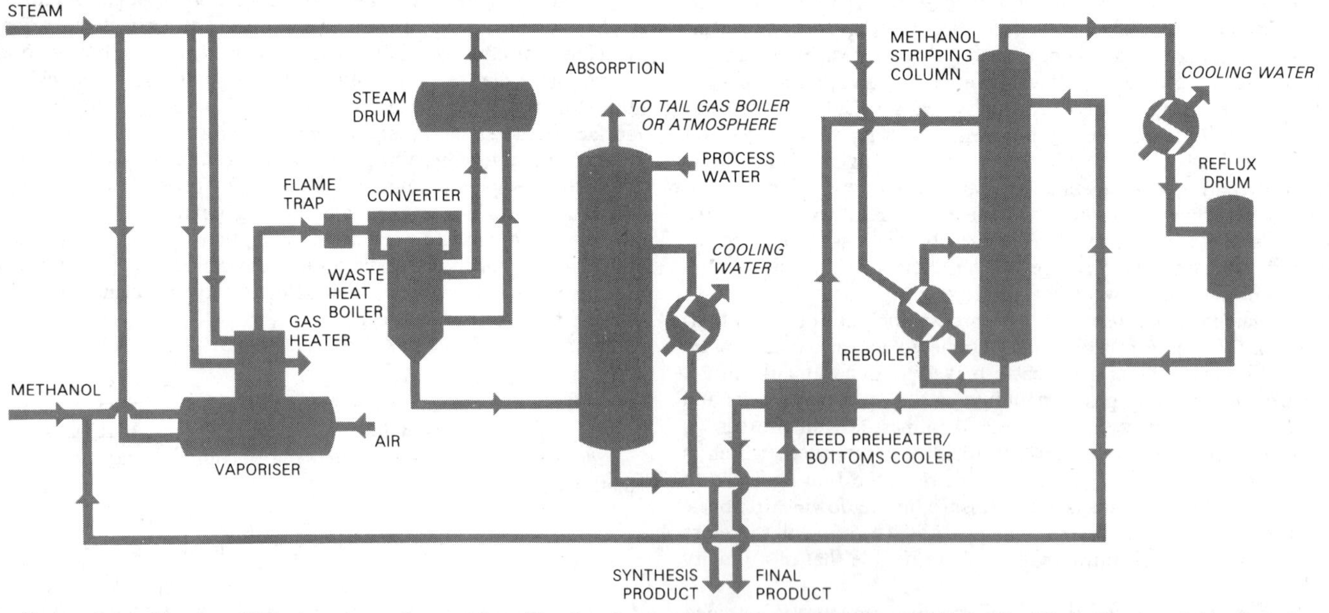

Process for making formaldehyde, using a silver catalyst. (Developed and operated by Imperial Chemical Industries Limited and available through Davy McKee (Oil & Chemicals) Ltd., London, England. Commercial plants are operating in the United Kingdom, the United States, Scandinavia, Australia, New Zealand, and South Africa.)

Formaldehyde gas, when cooled under certain conditions, yields trioxymethylene, metaformaldehyde $(CH_2O)_3$; formaldehyde solution, when evaporated, upon standing, or upon being subjected to low temperatures, yields paraformaldehyde $(CH_2O)_x$, white solid, from which formaldehyde is regenerated upon heating; dilute formaldehyde, in the presence of calcium hydroxide solution, yields a mixture of sugars called formose from which fructose $C_6H_{12}O_6$ has been prepared, suggesting the intermediate formation in nature of formaldehyde in the photosynthetic process of the conversion of carbon dioxide to sugars. Formaldehyde stands chemically between methyl alcohol on the one hand—to which it can be reduced—and formic acid on the other hand—to which it can be oxidized.

Formaldehyde is commonly detected by the Schiff test (above), and confirmed by the formation of a dimethyl derivative with a mp of 189°C.

D. G. Sleeman, Davy McKee (Oil & Chemicals) Ltd., London, England.

FORMAL SOLUTION. Molar Concentration.

FORMANT (Speech). Voice Recognition and Synthesis.

FORMATION (Ecology). A group of associations that exist together as a result of their closely similar life pattern, habits, and climatic requirements. See also **Biome.**

FORMATION (Geology). A distinct lithologic unit which may be used in geologic mapping. The term formation is usually confined to bedded or stratified rocks, including lava flows and volcanic ejectamenta.

FORMATIVE. An archeological term to designate the cultural stage following the Archaic. The Formative stage is characterized by the development of agriculture and a settled population.

FORM FACTOR. 1. A quantity used to describe the shape of a periodic signal. It is defined as the ratio of the root-mean-square (effective) value of the signal to its average value. For signals having no constant (dc) component, the half-period average is the value used in evaluating the ratio. The form factor is smaller for a signal with a flat top than for one which is peaked. The factor has a value of 1.111 for a sine wave and 1.000 for a square wave. 2. A factor introduced into a theory, usually by physical and nonrigorous arguments, to allow consequences of the theory to be computed without contributions from values of a parameter for which the theory is not applicable.

FORMIC ACID. HCOOH, formula weight 46.03, colorless liquid, mp 8.6°C, bp 100.8°C, sp gr 1.220. Sometimes referred to as methanoic acid or hydrogen carboxylic acid, this compound is miscible with H_2O, alcohol, or ether in all proportions. Formic acid occurs in some living organisms, such as nettles, ants, and caterpillars. Commercially, the compound is obtained from the black liquor of sulfite paper mills where it is present as sodium formate. In the laboratory, it may be prepared by reacting oxalic acid and glycerol at about 110°C, or by reacting solid lead formate with H_2S gas at about 100°C, yielding anhydrous formic acid. In the textile and leather industries, formic acid is used as a reducing agent, particularly in connection with the chrome dyeing of wool. Formic acid, in a reaction with glycerol at 220°C, is a source of allyl alcohol. Miscellaneous uses for formic acid have included brewing (acts as a fermentation assistant), electroplating for pH and redox adjustment, food preservative, gemicide, and coagulant for rubber latex. The compound also is useful in the preparation of metallic formates and esters. The most important ester is methyl formate $HCOOCH_3$, a solvent for acrylic resins and cellulose esters as well as an intermediate for certain organic syntheses.

Formic acid solution reacts (1) with hydroxides, oxides, carbonates, to form formates, e.g., sodium formate, calcium formate, and with alcohols to form esters, (2) with silver of ammonio-silver nitrate to form metallic silver, (3) with ferric formate solution, upon heating, to form red precipitate of basic ferric formate, (4) with mercuric chloride solution to form mercurous chloride, white precipitate, (5) with permanganate (in the presence of dilute H_2SO_4) to form CO_2 and manganous salt solution. Formic acid causes painful wounds when it comes in contact with the skin. At 160°C, formic acid yields CO_2 plus H_2. When sodium formate is heated in vacuum at 300°C, H_2 and sodium oxalate are formed. With concentrated H_2SO_4 heated, sodium formate, or other formate, or formic acid, yields carbon monoxide gas plus water. Sodium formate is made by heating NaOH and carbon monoxide under pressure at 210°C.

FORMULA (Chemical). Chemical Composition; Chemical Formula.

FORMULA TRANSLATION (Computer). FORTRAN (Computer System).

FORSTERITE. Olivine.

FORTRAN (Computer Programming). An acronym standing for FORmula TRANslation. It is a programming language designed for problems which can be expressed in algebraic notation, allowing for exponentiation, subscripting, and other mathematical functions. The FORTRAN compiler is a routine for a given machine which accepts a program written in FORTRAN source language and produces a machine language routine object program.

FORTRAN was introduced by IBM in 1957 after development by a working group headed by John W. Backus. It was the first computer language to be used widely for solving numerical problems and was the first to become an American National Standard. Numerous enhancements have been added to the language. The current version is described by American National Standard ANSI X3.9–1978. Realtime extensions are described in ANSI/ISA Sol.1–1976 and ANSI/ISA S61.2–1978.

FORTRAN Source Program. When a problem is coded in FORTRAN, the programmer produces a FORTRAN source program consisting of a set of statements. A given statement may express an algebraic equation to be solved, or a logic decision to be performed. There are five types of statements in FORTRAN: (1) *arithmetic statements* which express a series of arithmetic operations to be performed in algebraic notation; (2) *input/output statements* which allow the programmer to control the transfer of data between the central processing unit (CPU) and the outside; (3) *control statements* which allow the programmer to govern the sequence in which the statements are to be executed (control may be exercised on a logical-decision basis); (4) *specification statements* which allow the programmer to define the nature of the data being manipulated; and (5) *subprogram statements* which allow the user to define the subroutines which may be called at any point within the program.

FORTRAN Compiler. A FORTRAN compiler is a program for a specific computer which accepts as input the FORTRAN source program and translates it into a machine-language object program. The latter then may be loaded into the computer. Some compilers yield a very rapid translation, but the object code produced is inefficient in terms of core requirements and execution speed. Other compilers require more time for the translation, but the object code is much more efficient in both core and execution time required. In a process-control situation where the user is coding control algorithms in FORTRAN, normally these programs are compiled only once, but executed many times. Thus, one of the primary considerations when evaluating a digital computer and its associated FORTRAN compiler for a control application is the efficiency of the machine-language object code generated by the compiler.

See also terms listed under **Data Processing.**

Thomas J. Harrison, International Business Machines Corporation, Boca Raton, Florida.

FOSSAS. Viverrines.

FOSSIL AMPHIBIA. The first known skeletons of amphibia occur in formations of Devonian age. These earliest known amphibia are called *Stegocephalia,* because their heads are covered or "roofed" with thick dermal bones. In some species, both the back and stomach were covered with bony plates or scales. Thus the earliest known amphibia

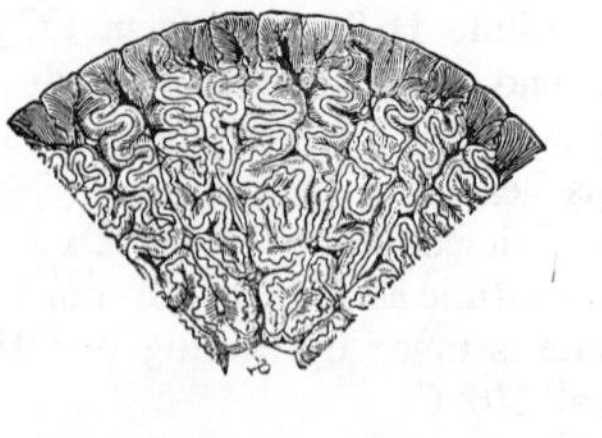

Fig. 1. Transverse section of a labyrinthodont tooth. (*After Owen.*)

(*Stegocephalia*) were armored as compared with most of their modern descendants. Other particularly significant anatomical features are (1) a third, or pineal, eye, (2) presence of a ring of plates around the two lateral eyes, (3) the arrangement of the dermal plates in the head, (4) conical teeth, showing an infolding of the dentine and enamel. This type of tooth structure is particularly characteristic of one group called *Labyrinthodonts* (see Fig. 1 and Fig. 2). All the aforementioned features suggest ancestral relationship to the ganoid fishes. The fossil record of the amphibia is the poorest of all the terrestrial vertebrates. They are therefore not particularly valuable as index fossils, their place being taken by the Mesozoic reptiles and Cenozoic mammals.

Fig. 2. A Pennsylvanian amphibian (Labyrinthodont), *Eryops.* The creature attained a length of 6 to 8 feet (1.8 to 2.4 meters). (*American Museum of Natural History.*)

FOSSIL BIRDS. Probably one of the most famous fossils in the world is *Archaeopteryx*, or the flying reptile which had feathers (Fig. 1). Fossil birds are rare, therefore it is all the more remarkable that this "missing link" between the reptiles and the birds should have been discovered as a natural lithograph in the fine-grained lithographic limestones of Upper Jurassic age at Solenhofen, Bavaria. Only two specimens of this genus are known. *Archaeopteryx* was about the size of a crow, with the combined reptilian and bird-like claws upon each of the three fingers terminating the wings, and lateral arrangement of the feathers in the tail. There have been several theories as to the origin of *Archaeopteryx.* One is that it evolved from a small carnivorous bipedal Triassic dinosaur. Toward the close of the Mesozoic the birds evolved rapidly. Fossils from the Cretaceous show both flying and diving forms, such as *Ichthyornis,* shown in Fig. 2. The birds did not lose their teeth until the Tertiary. It is also interesting to note that one of the largest known birds that ever lived, *Dinornis maximus* from New Zealand, stood 12 feet (3.6 meters) high and has only recently become extinct. The nearest living relative of the ancestral bird (*Archaeopteryx*) appears to be the Hoactzin of the Amazon Valley. The embryonic history of this bird repeats many of the adult features of *Archaeopteryx,* and the young Hoactzin still retains claws on its wings, but these claws disappear in the adult stage. As has been previ-

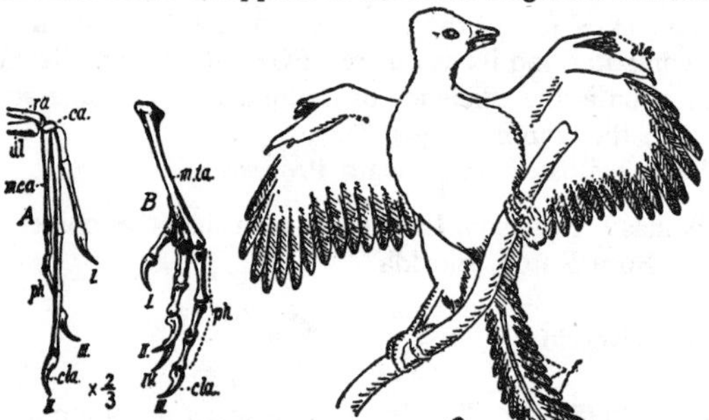

Fig. 1. The earliest known bird, (*Archeopteryx macurura*), from the Jurassic: (A) Right hand; (B) right foot; (C) restoration, modified after Pycraft. (*Shimer, "Introduction to the Study of Fossils," The Macmillan Co.*)

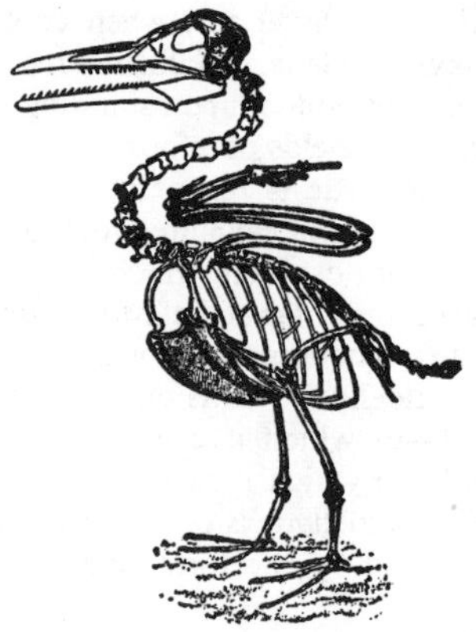

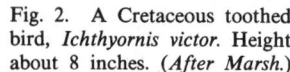

Fig. 2. A Cretaceous toothed bird, *Ichthyornis victor.* Height about 8 inches. (*After Marsh.*)

ously implied, birds do not make good index fossils primarily because of their rarity as fossils even in the later periods of the earth's history.

Reports on the finding of bird fossils in the Dry Mesa quarry of eastern Colorado may require a further evaluation of the claims of *Archacopteryx* as being the earliest bird. A brief summary of the new fossils may be found in *Science,* **199**, 284 (1978).

FOSSIL FISHES. The earliest fish may be represented by a "fish-scale" in the Cambrian. Problematical fish remains have also been found in the Ordovician, and small fin spines in the lower Silurian. A famous upper Silurian occurrence is the Ludlow bone bed of England. Probably none of the aforementioned bones and scales belonged to true fishes but to the so-called "bony-skinned fishes" or Ostracoderms, which are particularly characteristic of the Devonian. See Fig. 1. The Ostracoderms became extinct at the close of the Devonian, and there is no direct ancestral connection between them and either the sharks or the true fishes, although it is probable that all the types mentioned had a common ancestor. The Ostracoderms were armored with placoid bony plates or scales. They had no interior skeleton, and certain types such as Pteraspis and Cephalispis were probably adapted to bottom-feeding in relatively quite marine waters. Other

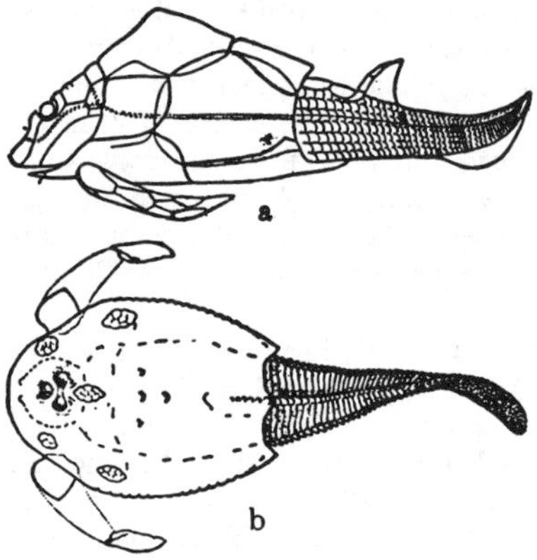

Fig. 1. Devonian ostracoderms: (a) *Pterichthys testudinarius,* restored (*Dean, after Woodward*); (b) *Tremataspia* (*restored, after Patten*).

types such as Bothriolepis and Pterichthys probably frequented fresh-water ponds and lakes. The earliest known sharks occur in the late Silurian or Devonian, and the true fishes (Pisces, or bony fishes) complete one phase of their development during the upper Paleozoic with a decline during the early Mesozoic. See Fig. 2. The Cenozoic has seen the gradual increase of a new phase with rapid and continuous expansion in diversity of form and adaptability during the late Tertiary

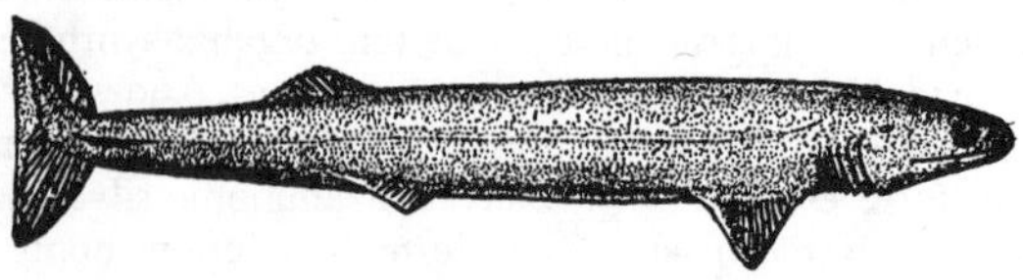

Fig. 2. A Paleozoic (early Mississippian) Selachian or shark, *Cladoselache fyleri.* (*Restored by Dean.*)

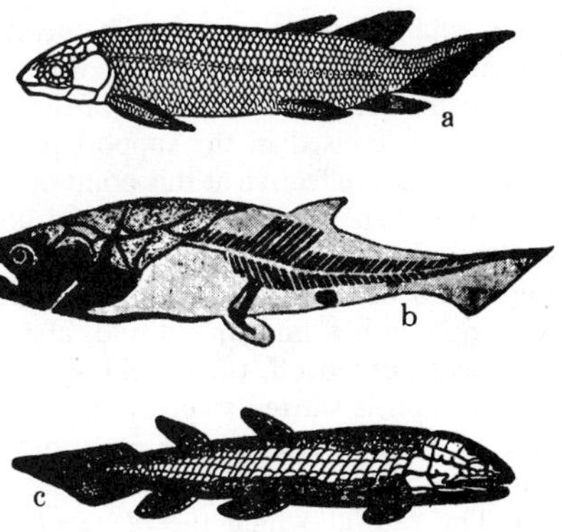

Fig. 3. Devonian fishes: (a) dipnoan; *Dipterus valenciennesi* (*restored by Traquair*); (b) arthrodiran, *Coccosteus decipient* (*restored by Woodward*); (c) ganoid, *Osteolepsis* (*restored by Nicholson*).

and Quaternary. One of the most interesting groups of the fossil fishes are the Dipnoi, or lung fishes. See Fig. 3. The ganoids first appear in the Devonian and are probably closely related to the progenitors of the first terrestrial vertebrates. The structure of the pectoral fins foreshadows a primitive limb and foot. The arrangement of the bones in the head is somewhat similar to that in the earliest known Amphibia (Stegocephalians) also, a peculiar ring of bony plates around the eye, and the conical teeth of infolded dentine and enamel are similar to those of the earliest known amphibians.

FOSSIL-FUEL-FIRED BOILERS. **Boiler; Burner; Feedwater (Boiler).**

FOSSIL FUELS (Pollution). **Pollution (Air).**

FOSSIL INVERTEBRATES. **Invertebrate Paleontology.**

FOSSIL REPTILES. The first indication of reptiles is in the Pennsylvanian. It is believed that most of these reptiles were sluggish creatures, probably not much more active than their amphibian progenitors, and equally adapted to both water and land. There is indication that some forms, however, such as the Pelycosaurs, or sail-backed lizards, had a great dorsal "fin," and thus appear to have been highly specialized. Studies indicate that most of the forms were lizard-like in appearance. The Cotylosauria were probably the most primitive of Paleozoic reptiles, showing Stegocephalian characteristics. The Theriodontia is another interesting group of reptiles which apparently lived during the Permian and Triassic periods. Some fossil skeletons of these creatures have been found, but only in South Africa. Fossils indicate that these creatures were reptilian, their teeth show a differentiation into incisors, canines, and molars.

The reptiles reached their all-time development apparently during the Mesozoic, thus this era has been termed the age of reptiles. See accompanying figure. Among the Mesozoic reptiles, the great group of the Dinosaurs command particular attention because they appeared to range in size from the smallest bird-like creatures to huge forms, both carnivorous and herbivorous, rivaling the mightiest creatures that ever lived on earth. The adaptation of these creatures to changes in environment is considered to have been quite extensive.

Studies indicate that the dinosaurs became extinct at the close of the Mesozoic. Modern reptiles, such as lizards, snakes, and turtles, appear to be but remnants of earlier reptiles.

FOSSIL REPTILIAN MAMMALS. It is believed that the rise of the reptilian mammals began as early as the Triassic, as disclosed by small lower jaws and teeth. These animals have been classified as multituberculates, so named because specimens show that the teeth have coned or tuberculated surfaces. Possible living descendents of this group include the spiny ant-eater, echidna (duck bill), and the duck-billed mole, *Ornythorynchus*, found in Australia. Small mammalian-type jaws similar to those of insectivores have been discovered in the Cretaceous. It is theorized that the Mesozoic creatures were probably egg-laying, relatively small in size and unable to usurp the

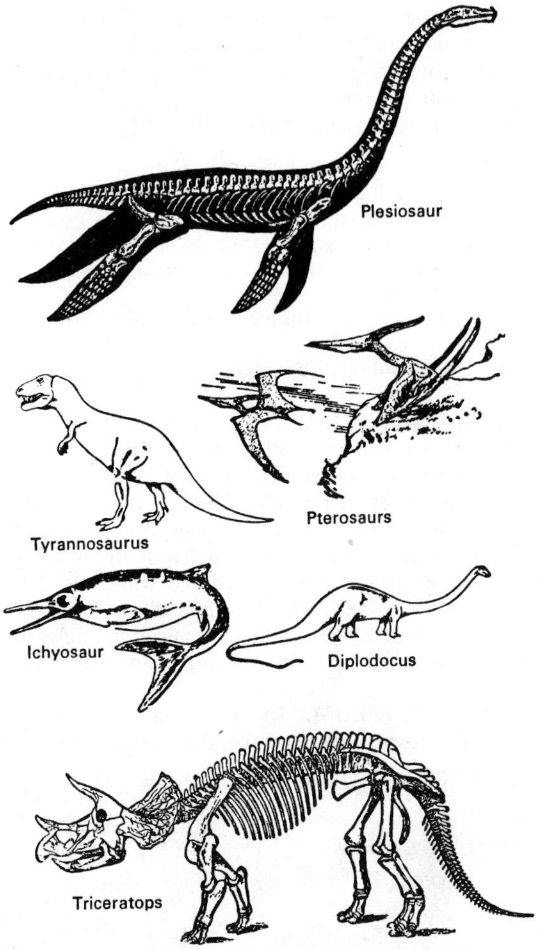

Artist's conception of prehistoric reptiles. (*Field, "Outline of Geology," Barnes & Noble.*)

place of the great host of dinosaurs which had become adapted to all types of environment. Following the extinction of the dinosaurs at the close of the Mesozoic, it is believed that several mammalian types evolved shortly after the start of the Cenozoic, and especially from the Oligocene and thereafter.

The reptile-like mammals of the Mesozoic were probably followed by archaic mammalian types of the Paleocene. It is believed that nearly all of the forms were small, primitive, with long and heavy tails, short limbs, and five digits on each foot. Studies indicate that the principal flesh-eaters were the creodonts. See accompanying diagram. Of present mammals, only the insectivores and marsupials appear to resemble the credonts. Eohippus, a diminutive horse, cursorial rhinoceroses, the creodonts, and several other forms also were probably present during the Mesozoic. One was a large-hoofed giant herbivore or hippopotamus-like creature called the uintatherium. By the Oligo-

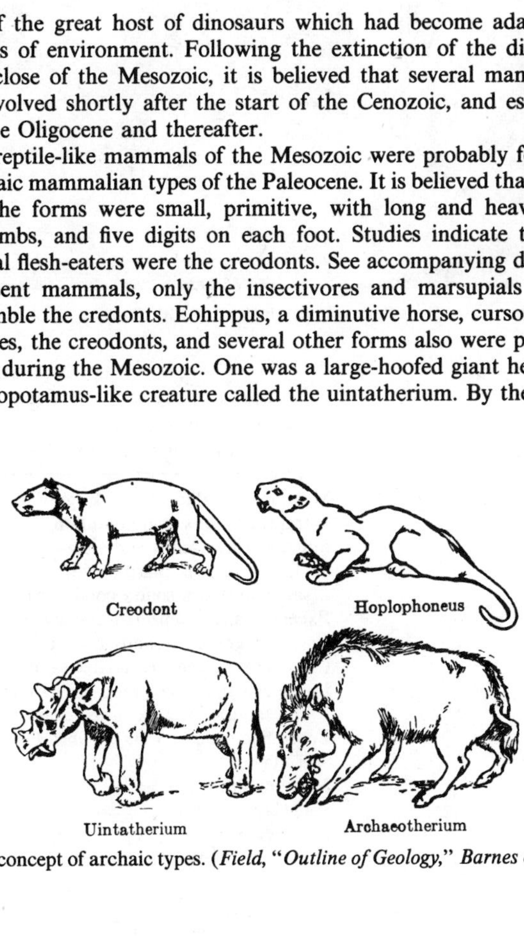

Artist's concept of archaic types. (*Field, "Outline of Geology," Barnes & Noble.*)

cene, a large number of forms appeared to taking to grazing on the open plains. Studies indicate that the true carnivores probably replaced the creodonts, including the Hoplophoneus, which resembled saber-toothed cats. The remainder of the Cenozoic apparently was a period of greater adaptation to both water and land.

References

Desmond, A. J.: "The Hot-Blooded Dinosaurs," Dial, New York, 1976.
Gartner, S., and J. P. McGuirk: "Terminal Cretaceous Extinction Scenario for a Catastrophe," *Science*, **206**, 1272–1276 (1979).
Ostrom, J.: "A New Look at Dinosaurs," *National Geographic*, **154**, 2, 152–185 (1978).
Staff: "Mass Extinction: More Theories," *Science News*, **116**, 21, 356 (1979).

FOSSILS. Paleontology; Taphonomy.

FOUCAULT KNIFE-EDGE TEST. A concave mirror or a lens may be tested by placing a pinhole-source of light at twice the focal length behind the lens or at the center of curvature of the mirror (or as near the center as possible). The eye placed at the image of the pinhole should see the entire lens or mirror illuminated. A knife-edge moved across the image immediately in front of the eye will determine where the image is located and will show defects in the lens or mirror by irregular darkening of the image. This is the customary test used by amateur telescope-mirror makers. A diffraction grating may be tested in a similar manner using a slit instead of a pinhole.

FOUCAULT PENDULUM. In 1851, Foucault performed his celebrated pendulum experiment at Paris, designed to give physical proof that the earth is in rotation about an axis. The pendulum, consisting of a very large iron ball suspended by a steel wire over 200 feet (60 meters) long, was suspended from the center of the dome of the Pantheon. Great care was exercised in the support for the wire so that no external forces should be effective at this point other than a vertical force to prevent the system from falling. On the floor, immediately under the pendulum, a layer of fine sand was placed so that the direction of the swing could be observed. The pendulum was started by drawing it to one side with a fine thread and, after the system was at rest, the thread was burned off, thus avoiding any lateral motion.

Foucault pendulum. (*R. W. Porter.*)

After such a pendulum is started swinging in one plane, it is soon observed that the plane of swing is apparently deviating slowly (in the clockwise direction in the northern hemisphere and the opposite in the southern). The rate at which the plane of swing deviates is equal to 15° per sidereal hour multiplied by the sine of the latitude. Thus at the pole it would make a complete rotation in one sidereal day, while at the equator it would not rotate at all. At Paris (latitude 48° 50′ N) the rate of deviation is about 11° 18′ per hour.

Foucault reasoned quite correctly that, in accordance with Newton's Laws of Dynamics, the direction in space of the plane of swing should not change unless the pendulum was acted upon by some external force other than that of gravitation and the counteracting force parallel to the direction of gravitation at the support. That the direction of the plane does apparently change can be accounted for only on the hypothesis that the earth is in rotation. Foucault's demonstration attracted wide scientific and popular attention and was accepted as a conclusive proof that the earth does rotate upon an axis, a fact which was not universally accepted at that time.

FOUCAULT ROTATING MIRROR. Foucault arranged a light-source, lens, rotating mirror and distant mirror. While light was traveling from the rotating mirror to the distant mirror (20 meters) and back, the rotating mirror turned slightly so that after the second reflection from the rotating mirror the reflected beam was slightly displaced (2′ 40″) from its original path. From these data and the angular speed of the rotating mirror, Foucault computed a speed of light as 2.98×10^{10} centimeters/second. The faintness of the returned beam prohibited use of a greater light-path.

FOUNDATIONS. The structural foundation is that part of a structure which transmits the loads of the superstructure to the supporting material. Foundations used in modern construction are of concrete, usually reinforced concrete. Some common types of foundation structures are isolated spread footings and combined footings (see **Footing**) and mat, or raft, foundations, in which all columns unload on a heavy, continuous slab. It is generally essential that the settlement of a structure be kept to a minimum, and it is particularly essential that any settlement be uniform for all parts of the structure. Nonuniform settlement can result in important structural damage. These settlement requirements can be fulfilled by proportioning the foundation structures in such a way that the soil pressures are not excessive and at the same time are of about equal intensity, under the various footings

ALLOWABLE BEARING VALUES ON SOILS

Type of Soil and Condition	Pounds per Square Foot	Kilograms per Square Meter
Massive crystalline bedrock (granite, gneiss, traprock—in sound condition)	100	488
Foliated rock (schist and slate—in sound condition)	40	195
Sedimentary rock (hard shales, siltstones, sandstones—in sound condition)	15	73
Exceptionally compacted gravels or sands	10	49
Compact gravel or sand-gravel mixtures	6	29
Loose gravel; compact coarse sand	4	20
Loose coarse sand or sand-gravel mixtures; compact fine sand, or wet, confined coarse sand	3	15
Loose fine sand or wet, confined fine sand	2	10
Stiff clay	4	20
Medium-stiff clay	2	10
Soft clay	1	5

SOURCE: New York State Building Construction Code.

or the mat. If poor soil is found at a relatively shallow depth below a foundation, it may be necessary to drive piles through the poor stratum to better soil or to rock. The foundation structures are then supported by the piles. If the material under the foundation is structurally sound rock, having a bearing value within safe limits, there will be no appreciable settlement of the structures; but there is bound to be settlement in structures whose supporting medium is earth, since it is a compressible material. See accompanying table.

The machine foundation performs more than a simple bearing function. It must:

1. Distribute the weight of the machine, the machine bedplate, and its own weight over a safe subsoil area. If heavy unbalanced vertical kinetic forces are produced by the machine, they should be added to the dead weight and the bearing power of the soil must be well in excess of these vertical forces.

2. Provide sufficient mass to absorb machine vibration. Satisfactory foundation weight for this factor is not readily calculable. The accompanying table is given to provide an indication of minimum weights.

3. Be rigid enough to prevent undue deflection of any part of the machine bedplate.

FOUR-COLOR MAP THEOREM. The fact that very complex maps of geographical regions, such as nations of a continent, states or provinces of a nation, or counties of a state, etc., require a maximum of four colors to provide full differentiation between the numerous area segments has puzzled mathematicians for many years. Many attempts have been made to show that a fifth color may be required in some cases and, intuitively, this would seem to be the case. But no map has been made to date which defies the Four-Color Map Theorem.

With the assistance of about 1,200 hours of high-speed computer time, which in effect amounted to coloring some 2,000 maps in as many as 200,000 ways, researchers Appel and Haken (University of Illinois), in 1976, proved the theorem. However, the method of proof used is uncongenial to some mathematicians.

Demonstration of the Four-Color Map Theorem. In place of colors in this diagram, shading, diagonal crosshatching, horizontal rules, and white are used for differentiation. If one commences a map of the counties of Ohio with Cuyahoga County, all counties along and south of Lake Erie can be differentiated with only three "colors" until Ashland and Mahoning counties are reached, at which time a fourth "color" must be used. The three original "colors" continue to suffice until Tuscarawas County is reached. If one were to complete the map of the 188 counties of Ohio, it would be found that no more than the four "colors" would be required to provide distinct differentiation between all counties. In hundreds of years of mapmaking, for area differentiation alone, cartographers never have found need for a fifth color.

FOUR-CYCLE ENGINE. An abbreviated expression for four-stroke cycle. The four-stroke cycle is one upon which either the Otto or Diesel types of internal combustion engines may operate, since it describes, not thermodynamic aspects of a cycle, but rather the sequence by which the cylinder is charged and exhausted. The four-stroke cycle is described as follows. Beginning with a suction or induction stroke, the cylinder is filled with a fresh charge by the outward motion of the piston. Next, on the return motion, this charge is trapped in the cylinder by closure of all valves leading to and from the cylinder, and is thereby compressed. On the next outward stroke, the power stroke, the fuel is burned or exploded to the accompaniment of energy liberation, a great deal of which is made usefully available on the power stroke. The final, or fourth, stroke is a return stroke, or exhaust stroke, during which the contents of the cylinder are exhausted through a port opened by an exhaust valve. Thus the four strokes are suction, compression, power and exhaust. During the suction stroke, an inlet valve is open; during the exhaust stroke, an exhaust valve is open. The advantage of the four-cycle principle is that it gives a full stroke for induction of the fresh charge, and another full stroke for scavenging of the burned gas. In this way it promotes high volumetric efficiency. A disadvantage of the four-cycle principle is the intermittent delivery of power. This contributes to making the four-cycle engine bulky in comparison with the two-cycle, but by employing multicylindered engines a steady flow of power may be secured through overlapping of power strokes. See also entry on **Automotive Electronics.**

FOURDRINIER MACHINE. Papermaking and Finishing.

FOUR-EYED FISHES (*Osteichthyes*). Of the order *Microcyprini*, family *Anablepidae*, the well-named four-eyed fishes have divided eyes which, in essence, provide them with four eyes. The eyes are horizontally partitioned such that the corneas are divided and there are separate retinas in the rear of the eyes. Normally, the four-eyed fishes swim with the upper halves of their eyes protruding. Any object above the water line is focused on the lower retina; any object below the water line is focused on the upper retina. The lens system for air vision differs from that for underwater vision. This is accommodated in the *Anableps* by utilizing an oval-shaped lens which is thicker in the portion required for underwater vision. One other fish has a divided eye—the small blenny *Dialommus fuscus*, found in the Galapagos Islands. But, in the blenny the division is vertical rather than horizontally. Surprisingly, studies to date have not indicated that the *Anableps*' dual-vision system assists it in foraging for insects flying above the water. The fish feeds underwater. Therefore, it is believed that the dual system provides protection from predators both from above and from within the water. The average adult length of the fish is from 6 to 8 inches. *Anableps* is found in northern South America and northward through Central America to southern Mexico.

FOURIER ANALYZER. Spectrum Analysis.

FOURIER SERIES. A single-valued function, continuous except possibly for a finite number of finite discontinuities in the interval $-\pi$ to π, and with only a finite number of maxima or minima in that interval, may be represented by a Fourier series

$$f(x) = \sum_{n=1}^{\infty} a_n \sin nx + \frac{b_0}{2} + \sum_{n=1}^{\infty} b_n \cos nx$$

where the coefficients are given by

$$\pi a_n = \int_{-\pi}^{\pi} f(t) \sin nt \, dt$$

$$\pi b_n = \int_{-\pi}^{\pi} f(t) \cos nt \, dt$$

A change of variable may be made so that the interval extends from 0 to n, 0 to $2n$, L to $-L$, etc. The series may be generalized to permit the expansion of a function of several variables. Fourier analysis is the process of representing a function in a Fourier series.

If the range of the variable is $-\infty < x < \infty$, the Fourier integral is

$$2\pi f(x) = \int_{-\infty}^{\infty} f(z)e^{iy(x-z)}\,dy\,dz$$

or, in a more common form, if $f(x)$ is real the exponential function may be replaced by cos $y(x - z)$. (See also **Integral Transform; Dirichlet Discontinuous Factor;** and terms listed under **Mathematics.**)

FOURIER THEOREM (Sound). Musical Sound.

FOVEA. The central portion of the retina of the eye where vision is most distinct. The fovea of a normal eye subtends an angle of about 2°.

FOWL. Poultry; Waterfowl.

FOWLERITE. Rhodonite.

FOX. Canines.

FOX-BATS. Bats.

FOX BITE. Rabies.

FOXGLOVE. Scrophulariaceae.

FOXTAIL. Grasses.

FRACTION. The quotient of two integers. If the quotient is less than one, the fraction is proper; otherwise, improper. The set of all possible fractions constitutes the rational numbers.

A fractional function is a rational function but it is not a polynomial. However, it can be expressed as a quotient of two polynomials.

A fractional equation can often be solved by converting it to an equality between two polynomials, the process known as clearing of fractions.

A continued fraction is an infinite sequence $\{s_n\}$ with members formed from the sequences $a_1, a_2, \ldots$ and $b_0, b_1, b_2, \ldots$ according to the following directions. The term s_n is found by replacing the denominator of s_{n-1} by $(b_{n-1} + a_n/b_n)$. When written in the conventional way such a fraction is rather awkward for it would read

$$b_0 + \cfrac{a_1}{b_1 + \cfrac{a_2}{b_2 + \cfrac{a_3}{b_3 + \cdots}}}$$

It can be simplified by writing it on one line with the plus sign properly placed, as

$$b_0 + \frac{a_1}{b_1} + \frac{a_2}{b_2} + \frac{a_3}{b_3} + \cdots$$

or, in a two-line form

$$\begin{pmatrix} a_1, a_2, \ldots \\ b_0, b_1, b_2, \ldots \end{pmatrix}$$

Some modifications of these notations also appear.

When there are a finite number of terms, the fraction is called terminating; if there are an infinite number of terms, nonterminating. Convergence behavior of such fractions can be studied by generalization of the methods used for series.

Sometimes continued fractions are suitable as solutions of a linear second-order differential equation, which will be assumed given in the form

$$y = A_0(x)y' + B_1(x)y''$$

If this equation is differentiated n times the result is

$$y^{(n)} = A_n y^{(n+1)} + B_{n+1} y^{(n+2)}$$

where

$$A_n = (A_{n-1} + B_n')/(1 - A_{n-1}')$$

and

$$B_{n+1} = B_n/(1 - A_{n-1}')$$

The ratio y/y' is a continued fraction and it is the reciprocal of the logarithmic derivative of the solution to the differential equation. The result is found to be

$$y/y' = A_0 + \frac{B_1}{A_1} + \frac{B_2}{A_2} + \cdots$$

The method does not seem to be readily extended to differential equations of higher order. Its most familiar application in mathematical physics is to the Mathieu equation.

When a given rational fraction is resolved into a sum of simpler fractions the individual terms in the sum are called partial fractions. The process of resolving a fraction in this way is the inverse process to that of reducing to a common denominator. If the fraction is of the form $f(x)/g(x)$, where the degree of the numerator is less than that of the denominator, $g(x)$ may be written as

$$g(x) = a_0(x - x_1)^{r_1}(x - x_2)^{r_3} \cdots (x^2 + 2b_1x + c_1)^{s_1}(x^2 + 2b_2x + c_2)^{s_2} + \cdots$$

Here a_i is one of the real roots of $g(x) = 0$, it being repeated r_i times. The quadratic expressions arise from complex roots of the polynomial, each occurring s_i times. Each linear factor in $g(x)$ gives rise to terms of the form

$$A_1/(x - a) + A_2/(x - a)^2 + \cdots + A_r/(x - a)^r$$

where the A_i are constant factors to be determined, a is any one of the a_i roots and r corresponds to r_i. Similarly, if $Q(x) = (x^2 + 2bx + c)$, the quadratic factors give terms of the form

$$(B_1 + C_1x)/Q + (B_2 + C_2x)/Q^2 + \cdots + (B_s + C_sx)/Q^s$$

The undetermined coefficients A_i, B_i, C_i in these fractions may be obtained by clearing of fractions and equating coefficients of like powers of x on both sides of the equality, which is an identity in x. Other devices are often possible, such as substituting special values of x. An integral may sometimes be evaluated by converting the integrand into a sum of partial fractions.

See also **Mathieu Equation;** and terms listed under **Mathematics.**

FRACTIONAL DISTILLATION ANALYSIS. The composition of mixtures can be determined by physically separating the components or groups of components by fractional distillation. The components can be identified by measurements of boiling point, refractive index, color, thermal conductivity, and other instrumental means; or by chemical analysis. To ascertain proportions in a mixture, the volumes of various constituents collected during distillation are collected and measured. Generally, this is a batch type laboratory tool and unsuited for on-line use. This form of direct analysis can be used for any mixture of gases or liquids, the components of which (1) differ in boiling point sufficiently to enable separation, (2) are not thermally decomposed, and (3) do not form a solid phase at the conditions of distillation. Laboratory apparatus of this type often is highly automated. The method is particularly useful for the analysis of hydrocarbons, refrigerants, organic solvents, and oils. Some analyses which were almost exclusively effected in this manner at one time are now handled by chromatographs. See also **Distillation.**

FRACTOGRAPHY. The study of fracture surfaces, particularly in metals primarily employing photographic techniques.

FRACTURE. Bone.

FRACTURE (Brittle). Brittle Fracture.

FRACTURE (Mineral). Mineralogy.

FRACTURE STRESS. The maximum principal true stress at the instant of fracture.

FRACTUS. Clouds and Cloud Formation.

FRAGARIA. Rose Family.

FRAMED STRUCTURE. A framed structure is one in which a group of structural members (tension, compression or flexural members) are joined together in the form of triangles or rectangles in order to support given loads and distribute them to the supports in a definite manner. Space frames, rigid frames and trusses are all framed structures. See also **Space Frame.**

FRAME FREQUENCY. Frequency.

FRAME (Rigid). Rigid Frame.

FRAME (Sampling Theory). Especially in sample surveys, the frame is the explicit display of the population from which the sample is to be chosen; for example, census records for sampling human beings or detailed maps for the selection of farms.

FRANCIS FORMULA (Weir). Flow Measurement.

FRANCIS TURBINE. Hydroelectric Power.

FRANCIUM. Chemical element symbol Fr, at. no. 87, at. wt. 223 (mass number of the most stable isotope), periodic table group 1a, mp 26–28°C, bp 676–678°C, density 2.4 g/cm^3. To date, 22 isotopes of francium, with mass numbers ranging from 203 to 224, have been identified. All are radioactive. See also **Radioactivity.** Although Mendeleev visualized that element 87 would occupy the bottom position among the alkali metals in his periodic classification, the discovery of francium did not occur until 1939 when it was confirmed by Marguerite Perey, a collaborator of Marie Curie. Many earlier attempts had been made, including the observations of J. A. Cranston in 1913, the search for the element in radioactive ores by O. Hahn and G. Hevesy in 1926, and efforts by M. C. Guében in 1932. Also, earlier investigations by S. J. Meyer, V. Hess, and F. Paneth in 1914, when they were studying the emissions by ^{227}Ac, contributed to the network of information that finally led to the firm identification of francium. See also **Chemical Elements.**

Studies indicate that no further isotopes of francium with half-lives longer than those already known should exist. Thus, among the first 101 elements of the periodic chart, francium is the most unstable. The short half-lives of the isotopes explain the difficulties in isolating and confirming the element and of learning more of the detailed chemistry of the element.

The isotopes of francium that occur in nature are found in thorium and uranium ores, in which they are continually formed by disintegration chains. These start with ^{232}Th ($4n$ family), ^{237}Np ($4n+1$ family), and ^{235}U ($4n+3$ family). It is estimated that 1 ton of natural uranium contains about 3.8×10^{-3} g of ^{223}Fr and 10^{-17} g of ^{221}Fr. The separation of these isotopes from natural sources involves long and complex chemical procedures. Like the other alkaline elements, francium remains in solution when other elements are precipitated as carbonates, hydroxides, fluorides, chromates, sulfates, and sulfides. However, at a pH of 9, francium can be extracted from solution by nitrobenzene in the presence of sodium tetraphenylborate. Extraction from a very dilute sodium solution can be effected with dipicrylamine in nitrobenzene solution, the separation being much easier to effect than in the case of cesium. Francium can be separated from rubidium and cesium by chromatography on cation-exchange resins or on mineral exchangers.

The heavy isotopes of francium can be formed by irradiation of uranium or thorium by protons of high energy; the lighter isotopes can be obtained by nuclear reactions induced in gold, tellurium, or lead targets by heavy ions.

Because of the difficulties in obtaining any significant quantities of francium, use of the element is confined to scientific investigations. ^{223}Fr is used for the measurement of ^{227}Ac. Studies have shown that francium fixes itself in induced sarcomas in rats. Because of the short half-lives of ^{223}Fr and ^{212}Fr, which would cause no radiation risk to organisms, the property could become useful for the early diagnosis of certain kinds of cancers.

See list of references at end of entry on **Chemical Elements.**

FRANCK-CONDON PRINCIPLE. This principle states that an electronic transition in a molecule or crystal takes place so rapidly that the positions and velocities of the nuclei are virtually unchanged in the process. Therefore, only transitions satisfying these conditions occur with reasonable probability. The unchanged positions of the nuclei can be represented by vertical lines between energy levels within the potential energy curves. Unchanged velocities occur only if transitions are between those parts of an energy level for which the momentum of the nucleus is essentially unchanged.

FRANCOLIN. Partridge.

FRANKINCENSE. Boswellia Tree.

FRANKLINITE. The mineral franklinite is a zinc-iron-manganese mineral whose formula may be written (Zn, Mn^{2+}, Fe^{2+})(Fe^{3+}, Mn^{3+})$_2$O$_4$, but the composition varies considerably, in respect to the amounts of the several metals that may be present. Its isometric crystals have an octahedral habit; it may be coarse or finely granular or compact. It shows a parting parallel to the octahedron; fracture, uneven; brittle; hardness, 5.5–6.5; specific gravity, 5–5.2; luster, usually metallic, occasionally dull; color, black; streak, brown to black; opaque; may be slightly magnetic. Only in one place in the world does franklinite occur in quantity: at Franklin Furnace, New Jersey, from whence it was named. Here there are two bodies of this mineral, which is used as a zinc ore, about 3 miles distant from each other. The franklinite is found in pre-Cambrian limestones that are associated with gneisses believed to be of igneous origin and responsible for the mineralization. Associated minerals are willemite, zinc, silicate, and zincite, zinc oxide, manganoan calcite.

FRANK-READ SOURCE. A mechanism for the continuous generation of dislocations, and hence allowing for plastic deformation of crystals. A segment of dislocation line anchored at its ends can be pulled out into an arc which may then fold back round the anchored ends until it meets itself, whereupon a closed dislocation loop is detached, whilst the original line is reformed. This process may evidently be repeated indefinitely, the successive dislocation loops moving out to the surface of the crystal.

FRASER'S FIR. Fir Trees.

FRATERNAL TWINS. Gonads.

FRAUNHOFER DIFFRACTION. Diffraction.

FRAUNHOFER LINES. The dark lines constituting the absorption spectrum exhibited by sunlight are frequently called the Fraunhofer lines. There are thousands of these lines, of which Fraunhofer, early in the nineteenth century, first observed the most prominent. To these particular lines he assigned letters for reference purposes. Some of these lines, together with their origin and approximate wavelengths, are listed as follows:

A	Terrestrial oxygen	7594 Å	(extreme red)
B	Terrestrial oxygen	6867 Å	(red)
C	Hydrogen	6563 Å	(red)
D_1	Sodium	5896 Å	(yellow)
D_2	Sodium	5890 Å	(yellow)
E	Iron	5270 Å	(green)
F	Hydrogen	4861 Å	(blue)
G	Iron and Calcium (group)	4308 Å	(violet)
H	Calcium	3968 Å	(extreme violet)

The lines of solar origin are due to absorption by gases and vapors in the solar atmosphere. See also **Sun.** The radiation from the hot core of solar gases is characterized by a continuous spectrum. In passing through the cooler gases of the sun's outer atmosphere, radiation will be absorbed at those frequencies which match frequencies of atomic transitions in the cool gas. These blocked out lines in the continuous spectrum show up as dark lines against the background and are called Fraunhofer lines. The rare gas helium was first discovered as a consequence of these observations.

FRAUNHOFER REGION. In antenna terminology, that region of the field in which the energy flow from an antenna proceeds essentially as though coming from a point source located in the vicinity of the antenna.

FREDHOLM EQUATION. A special type of integral equation, the one of second kind being

$$\phi(x) = f(x) + \lambda \int_a^b K(x, z)\phi(z)\,dz$$

where a, b, λ are constants; the kernel, $K(x, z)$ and $f(x)$ are given; $\phi(x)$ is sought as a function of x, the independent variable. If the left-hand side of the equation equals zero, it becomes of the first kind. Either could also be homogeneous, i.e., $f(x) = 0$.

Fredholm has also given a method for solving integral equations. It is based on his observation that the solution of a system of n linear, algebraic equations applies to the integral equation case as n becomes infinite. The details of the method are lengthy, although straightforward. The solution is similar to that in the Liouville-Neumann series where the resolvent is the ratio of two convergent series in the parameter λ. It generally converges but there are exceptional cases.

See also **Liouville-Neumann Series;** and terms listed under **Mathematics.**

FREE ATMOSPHERE. Winds and Air Movement.

FREE CONVECTION. Grashof Number.

FREEDOM (Degrees of). 1. The number of variables which must be fixed before the state of a system may be defined according to the phase rule. The relationship between the number of degrees of freedom (F), the components (C), and the phases (P) of a system is expressed by the formula,

$$F = C + 2 - P$$

Thus a pure gas has two degrees of freedom. At any temperature its volume and pressure are variable, but if one of these is fixed then the other is automatically determined for, at any given temperature and pressure, each pure gas assumes one, and only one, volume at equilibrium.

2. The number of independent coordinates necessary for the unique determination of the position of every particle in a dynamical system is the number of degrees of freedom. Each degree of freedom is represented by a coordinate which can vary with time independently of all the rest. Thus a single particle which may move anywhere in three-dimensional space has three degrees of freedom. A particle constrained to move on a surface has two degrees of freedom. A system composed of three particles has 9 degrees of freedom; for it takes nine independent coordinates to specify the positions of the particles in space, and their arrangement may therefore be changed in nine different ways. A single rigid body, on the other hand, has 6 degrees of freedom, since it may have motions of translation in three coordinate directions and it may also rotate about any one of the three coordinate axes through its center of mass. Any actual motion of the body is in general made up of all six, its linear motion being the resultant of three linear components and its rotation the resultant of three angular components. Each molecule of a diatomic gas has 7 degrees of freedom; viz., the six just mentioned for the molecule as a whole (regarded as a rigid body), and, in addition, one corresponding to the possible vibration of the two atoms toward and from each other. If the body is not rigid, the number of degrees of freedom may be virtually infinite. It should, however, be added that, because of restrictions imposed by the quantum theory, not all of the possible degrees of freedom can in general be expected to participate in changes of molecular energy.

3. The number of degrees of freedom in a statistical quantity is the number of independent values necessary to determine it. A sample of n values $x_1, x_2, \ldots, x_n$ has n degrees of freedom but the sum

$$\sum_{i=1}^{n} (x_i - \bar{x})^2$$

is regarded as having $n - 1$ because, for given $\bar{x}$, only $n - 1$ values are assignable at will.

By extension, the number of degrees of freedom of a statistical distribution relates to the degrees of freedom of the distributed statistic. For example, χ^2, being the distribution of the sum

$$\sum_{i=1}^{n} (x_i - \bar{x})^2/\sigma^2$$

has $n - 1$ degrees; and the F-distribution, which concerns the ratio of two independent such quantities has a pair of degrees of freedom, one relating to the numerator and the other to the denominator of the ratio.

FREE ELECTRON. Electron.

FREE ELECTRON THEORY OF METALS. The most characteristic property of a metal is its electrical conductivity. It was early recognized (by Drude) that this could be explained if the electrons in the metal were relatively free to move. A model of a metal as a gas of free electrons, moving in the region of nearly uniform positive potential created by the ions of the crystal lattice, although satisfactory in some respects, led to serious difficulties, until it was pointed out by Sommerfeld that the electrons must obey the Pauli exclusion principle and hence constitute a highly degenerate Fermi-Dirac gas. On this basis, one can calculate the electronic specific heat, magnetic susceptibility, Hall coefficient, Wiedemann-Franz ratio, thermionic emission, and other physical properties.

The reason for the success of this simple theory is that in certain elements, notably the alkali metals, the outer electrons are only very loosely bound to the remainder of the atom, and when the atoms are brought together to make up the crystal lattice these electrons can jump from ion to ion as if they were free. To understand the behavior of the more complex metals, and of semiconductors, it was necessary to introduce the further complications of the band theory of solids.

See also **Electron; Solids (Band Theory);** and **Solid-State Physics.**

FREE ENERGY. There are two quantities to which this term has been applied. 1. The Gibbs free energy, which is also called the Gibbs function and the thermodynamic potential, is most generally understood when the term free energy is used without qualification. It is defined by the equation

$$G = U - TS + pV$$

where U is the internal energy, T, the absolute temperature, S, the entropy, p, the pressure, V, the volume, and G, the Gibbs free energy (the letter F is also used to denote this quantity).

2. The Helmholtz free energy, which was called the psi function, and which is perhaps most commonly known as the work function. It is defined by the equation

$$A = U - TS$$

where U is the internal energy, T, the absolute temperature, S, the entropy and A, the Helmholtz free energy. (The letter ψ or the letter F are sometimes used instead of A for this quantity.) The decrease in A is equal to the maximum work done on the system in a constant-temperature, reversible change. In terms of the partition function, $A = -RT \ln Z$. Like the Gibbs free energy, the Helmholtz free energy is a thermodynamic potential, although the latter term is commonly used to refer specifically to the Gibbs free energy.

FREE ENERGY CHANGE. The change in the Gibbs free energy for a chemical reaction, defined as

$$\Delta G = \sum_{r=1}^{n} \nu_r g_r$$

where g_r is the molar Gibbs free energy of the rth component in the pure state, under the same conditions of temperature and pressure as those in which the reaction takes place. ν_r is the stoichiometric coefficient of the rth component.

This quantity ΔG is equal to the *maximum net* work available (i.e., work, other than work of expansion, in a reversible process) for a given change in state under constant temperature and pressure.

FREE FALL. 1. The fall or drop of a body, such as a rocket, not guided, not under thrust, and not retarded by a parachute or other braking device. 2. The free and unhampered motion of a body along a Keplerian trajectory, in which the force of gravity is counterbalanced by the force of inertia.

FREE GYROSCOPE. **Gyroscope.**

FREE MACHINING. Many alloys are prepared in free machining variants or forms which require less power for machining, give better surface finishes, and are less wearing on the tools. These alloys have incorporated in them small particles of another metal or some compound that act as stress raisers which cause chips to break off easily during machining. Sulfur, selenium, lead, and bismuth are among the elements added or left in the alloys for this purpose.

"FREE" PENDULUM CLOCK. **Pendulum Clock.**

FREE RADICAL. Although free radicals have been defined as highly reactive groups of atoms containing unpaired electrons, this definition is imprecise as it would include ions, such as those of the lanthanide and actinide series, which not only possess such unpaired electrons, but also—because of this—exhibit the color, magnetic and other characteristics of "free radicals." The term is in general reserved for short-lived alkyl radicals possessing a magnetically noncompensated electron, or somewhat longer-lived, larger organic molecules of the aryl-alkyl type similarly possessing unpaired electrons in their valence shell.

A few very reactive inorganic radicals, such as $NO\cdot$, $ClO_2\cdot$, or $NO_2\cdot$, may also be construed as "free," but knowledge of the chemistry of free radicals is best exemplified by unsaturated organic fragments, such as $CH_3\cdot$, $C_6H_5\cdot$, among others. These free radical fragments are formed by breaking one or more bonds in a stable molecule by photolysis, electrolysis, pyrolysis, and some other processes. The first free radical to be synthesized was that of triphenylmethyl by Moses Gomberg in 1900. Paneth and Hofeditz, in 1926, found free radicals in the pyrolysis of lead tetramethyl, and Rice subsequently demonstrated their existence in the breakdown products of many organic compounds.

Because of the affinity of their unpaired electrons, free radicals have short lives, tend to dimerize and thus lose their reactivity. Because of their generally short half-lives (1–100 milliseconds), detection and identification of these entities is essentially through spectrophotometric methods. However, in solid systems, free radicals can be trapped for appreciable lengths of time and at least one of these, 2,2-diphenyl-1-picrylhydrazyl, has such a long half-life that it is sold as such for the photometric determination of tocopherol.

Free radicals formed by photolysis play significant roles in the chemistry of the earth's atmosphere, not the least of which is the destruction of ultraviolet-protective ozone by fluorocarbons from anthropogenic sources. But the most important reactions involving free radicals as intermediates in organic chemistry are polymerizations, such as may develop from a free radical and ethylene. These may also be ensured through use of substances known to produce free radicals, e.g., benzoyl peroxide, of which the benzoyl free radical can initiate a chain reaction which may continue indefinitely; or, it may lose CO_2 itself to give a new free radical, phenyl, which is itself capable of polymerization.

A free radical chain reaction proceeds through a succession of free radicals. In the photochemical chlorination of an alkane, the initiating step is the homolytic fission of chlorine molecules to produce chloroalkane molecules and chlorine free radicals. These two reactions constitute the propagating step. However, the chlorine free radicals may also combine to form chlorine molecules or react with the alkane free radicals to form chloroalkane molecules. Both of these reactions constitute terminating steps of the chain reaction. It should be noted, however, that the foregoing sequence cannot take place in the dark. Exposure to light allows the series of reactions then to proceed rather violently.

See also **Cancer Research.**

R. C. Vickery, Hudson Laboratories, Hudson, Florida.

EDITOR'S NOTE: The biological roles of free radicals are described in a series of books: Pryor, W. A. (editor): "Free Radicals in Biology," Vol. 4 (latest), Academic, New York, 1980.

FREE-RADICAL MECHANISM (Oxidation). **Antitoxidant.**

FREE RADICAL REACTIONS. **Free Radical; Organic Chemistry.**

FREE RADICAL ROCKET FUEL COMPONENTS. **Rocket Propellants.**

FREE ROTATION. The power of two structural entities joined by a linkage to occupy any position relative to each other and relative to their planes of symmetry. For example, in diphenyl C_6H_5-C_6H_5 the two planar benzene rings may occupy any position relative to each other from parallelism to perpendicularity of the planes of their rings. However, the *o*-, *o'*-dinitrodiphenic acids can be separated into optical isomers, showing that the position of the substituents cannot be planar, i.e., that the free rotation of the rings has undergone steric hindrance.

FREE SURFACE. A phase boundary between two fluids. In hydrostatic equilibrium such a surface must coincide with a gravitational equipotential (neglecting the effects of surface tension), and its movements about the equilibrium position are considered in the theory of surface waves.

FREE TURBULENT FLOW. A free turbulent flow is one not confined by solid or rigid boundaries and so free to spread into the ambient non-turbulent fluid by a process of turbulent entrainment. Examples are wakes, jets and rising plumes of hot air. The boundary-layer may be considered as a free turbulent flow on one side only. The importance of the classification is that free turbulent flows show considerable homogeneity of the turbulent motion, in contrast to wall turbulent flow.

FREE VECTOR. **Affine Tensors and Free Vectors.**

FREE VOLUME. A liquid differs from a solid having the same type of packing of the molecule in having a certain additional volume, the free volume, which provides the necessary looseness in the structure to permit free movement of the molecules. The concept of free volume is used in several theories of the liquid state.

FREEZE. **Climate.**

FREEZE-CONCENTRATING. In lieu of evaporation and other means for concentrating liquid substances, notably foods, freeze-concentrating may be used, particularly where it is desirable to retain volatile constituents, as in the instances of increasing the alcohol content of wines and to prepare and preserve flavor, as in the case of orange juice or coffee extract concentrates. From an energy standpoint, the energy required by freeze-concentrating is considerably less than that used by evaporative systems.

There are difficulties associated with the achievement of high concentrations through freezing. The liquid viscosity may increase so much as concentration increases and freezing point drops that difficulty in handling the ice–concentrate mixture and of separating the concentrate from the ice will arise. Improvements in recent years have been made by the use of ripening-induced growth of large ice crystals through sacrificial melting of small, easily-formed subcritical ice crystals; and the development of wash columns which provide efficient solute recovery from the ice–concentrate mixture. Even then, it is estimated that 35–50% dissolved solids represents the maximum concentration that can be practically achieved by freeze-concentration. By comparison, the liquid food concentrations achievable using two other attractive concentrating methods are even less: 20–35% by reverse osmosis; and 20–30% by ultrafiltration. However, these latter processes, when considered in conjunction with (prior to) evaporation may be attractive from an energy-expenditure standpoint. A proprietary freeze-concentration system is depicted in accompanying diagram.

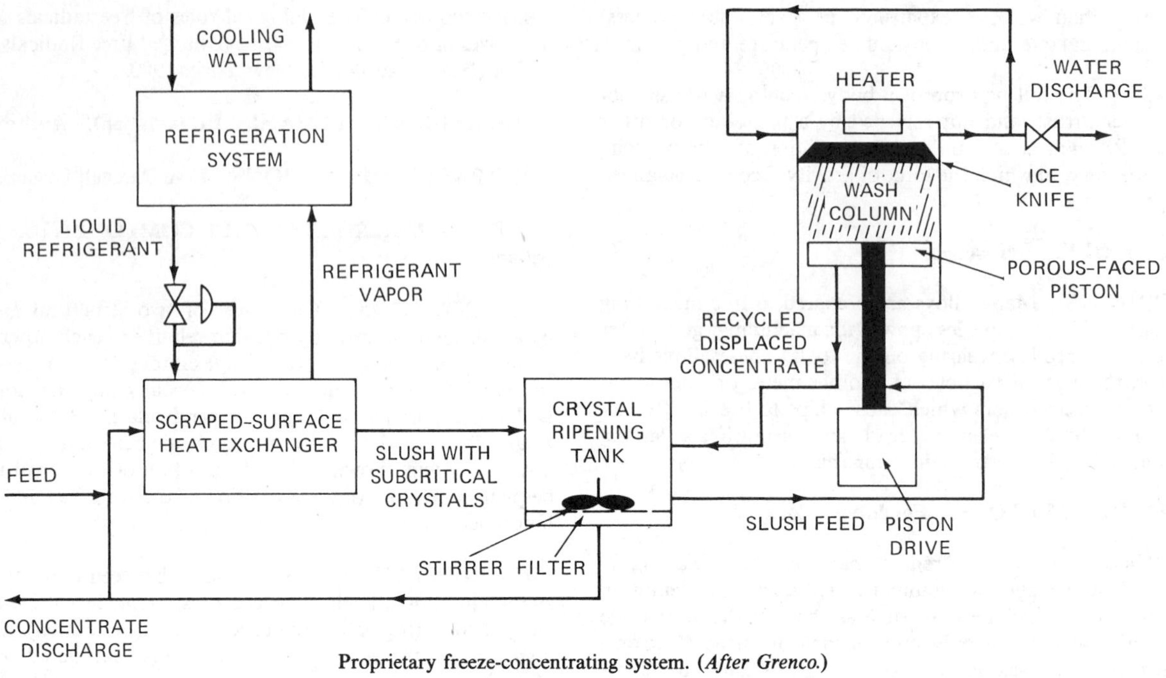

Proprietary freeze-concentrating system. (*After Grenco.*)

References

Leninger, H. A., and W. A. Beverloo: "Food Process Engineering," Reidel, Boston, 1975.

Schwartzberg, H. G.: "Energy Requirements for Liquid Food Concentration," *Food Technology*, **32**, 3, 67–76 (1977).

Tjijssen, H. A. C.: "Freeze Concentration," in "Advances in Preconcentration and Dehydration of Foods," (A. Spicer, editor), Wiley, New York, 1974.

FREEZE-DRYING. A process for removing moisture from a wet material by bringing the material to the solid state and subsequently subliming it. This process is used for drying and preserving a number of products, notably food products, including instant coffee, vegetables, fruit juices, and meats. The needs of the food industry, coupled with those of pharmaceutical manufacturers, accelerated research into this process several years ago and commercial applications of freeze-drying are now commonplace in these industries.

The wet material in the form of a wet solid or in the form of a suspension or solution is frozen under vacuum or at atmospheric pressure, followed by transforming the ice into vapor and removing it. In the usual case, the dried material remaining will be a spongy mass of about the same size and shape as the original frozen mass and frequently will be found to have excellent stability, convenient reconstitution when placed in cold water, and will maintain flavor and texture sometimes indistinguishable from the original materials. These properties differ markedly with various materials. Some products are much better adapted to the process than others.

Usually materials to be freeze-dried are complex mixtures of water and several other substances. When such materials are cooled below 32°F (0°C), pure ice crystals will separate out first. With further cooling, the mass will become rigid as the result of formation of eutectics. (A eutectic is that particular mixture out of a possible combination of two or more mixtures of materials that has the lowest melting point.) See Fig. 1. Most food products and biologicals solidify completely at a temperature in the range of −5 to −100°F (−15 to −73°C). At solidification of the entire mass, all of the free water has been transformed into ice. Only a small quantity of the original water, the bound water, remains fixed in the internal structure of the material.

The quality of the finished product as well as the rate of drying will be affected by the size, shape, and size distribution of the ice crystals which form during freezing. These properties also will be affected by the homogeneity of the frozen mass. Thus, freezing must be effected under carefully controlled conditions (time, pressure, and temperature). Large ice crystals result from slow freezing rates. These may be injurious to certain substances. On the other hand, too-rapid freezing results in small ice crystals, which may cause undesirable color and texture changes.

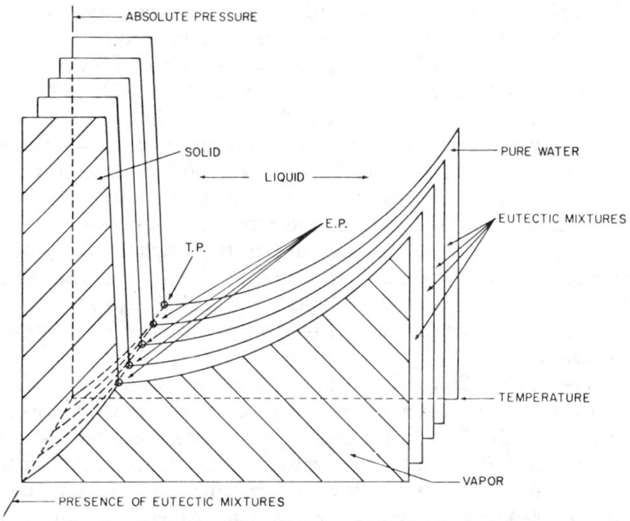

Fig. 1. Eutectic phase diagram in freeze-drying process.

Sublimation (*or Primary Drying*). For the sublimation phase of the process, the frozen material usually is subjected to a vacuum of about 4.6 millimeters of mercury. The ice-crystal sublimation process can be regarded as comprised of two basic processes: (1) Heat transfer, and (2) mass transfer. In essence, heat is furnished to the ice crystals to sublime them; the generated water vapor resulting is transferred out of the sublimation interface. Thus, it is evident that sublimation will be rate-limited by both resistances to heat and mass transfer as they occur within the material.

As the sublimation interface recedes in the material (See Fig. 2), the dry layer presents a resistance to the flow of water vapor and a pressure difference must exist between the ice interface and the surface of the dry layer. A large pressure difference will facilitate high mass-transfer rates; however, the maximum allowable sublimation temperature at which no melting will occur and the cost of the vacuum equipment restrict this driving force to a limited range. In practice, the maximum allowable temperature and corresponding pressure at the sublimation interface is in the range of +15° to −40°F (−9.4° to −40°C) and 2000 to 100 micrometers of mercury, respectively.

The rate of heat input to the frozen material is a function of the operating-vacuum method of heat transfer and the properties of the dried product. The operating vacuum determines the pressure difference and, in turn, the rate of mass transfer, which must be in balance with the rate of heat input. Otherwise, either melting will occur at

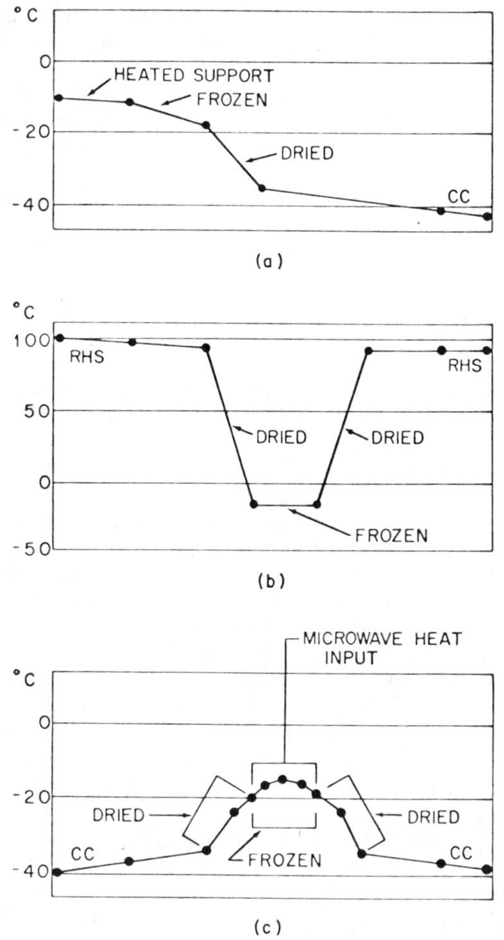

Fig. 2. Heat-input methods for freeze-drying processes: (a) Conduction, (b) radiation, (c) microwave. CC = cold condenser; RHS = radiant-heat device.

the sublimation interface and the purpose of freeze-drying will be defeated; or the sublimation temperature will decrease and the cost of processing will increase.

The heat required for sublimation (1200 Btu per pound of ice; 664 kilogram-calories per kilogram of ice) can be supplied by conduction, radiation, electric resistance, microwave, or infrared heating. Three methods of heat input that have been investigated extensively are shown in Fig. 2. Depending on the method of heat transfer, the temperature gradient between the sublimation interface and the heat source is limited by the maximum temperature which can be tolerated on the surface of the dry layer or frozen mass. For radiation, the dry layer should not be heated to the point where charring or decomposition occur. For conduction, melting of the frozen mass in contact with the heating element should be avoided.

In most commercial applications, conditions are such that the rate of sublimation is controlled by heat transfer. The development of techniques for improving the heat-input rate is the objective of many investigations.

Desorption (or Secondary Drying). Upon completion of sublimation of the ice crystals, final dehydration is carried out to remove the bound water which did not crystallize out during freezing and is bound by adsorption phenomena to the dried product. The product temperature is increase to 80–120°F (27–49°C), and under high vacuum, the bound water and oxygen are removed from the dried product. The rate of desorption is considerably slower than sublimation. Although the bound water is only 5–10% of the total water in many substances, the secondary drying may require up to 35% of the total drying time.

Drying Rates. Drying a frozen material proceeds initially at a constant rate with rapid evolution of water vapor. As the sublimation interface recedes within the product, water-vapor evolution decreases. This is the start of the falling-rate period. When only bound water remains within the cellular structure of the product, the desorption period begins. During the constant-rate period, the sublimation rate can be expressed in terms of the heat of sublimation of ice and the heat-rate equation:

$$\text{Rate of sublimation} = \frac{UA\Delta T}{\Delta H_{\text{ice}}}$$

The overall heat-transfer coefficient U depends upon the properties of the dry product and the method of heat transfer. The heat-transfer rate A is influenced by the mechanical design of the heating elements and the conditioning of the frozen mass. The temperature gradient ΔT is limited by the maximum allowable temperatures at the sublimation interface and dry-layer surface. In the constant-rate period, the first one-half to two-thirds of the drying cycle, about 80% of the water is removed.

Processes and Equipment

In addition to the three fundamental operations just described, the freeze-drying process involves several other operations necessary to achieve an economically feasible system for large-scale production. The general commercial process comprises: (1) Preparation of the material; (2) freezing; (3) conditioning of the frozen mass; (4) drying, that is, sublimation and desorption; and (5) conditioning the product. See Fig. 3.

Preparation of the Material. It is not always economically practical to subject a product in its original state to freeze-drying. One or more operations may be required to prepare the product. Wet solids, such as fruits and meats, are usually ground or sliced to facilitate

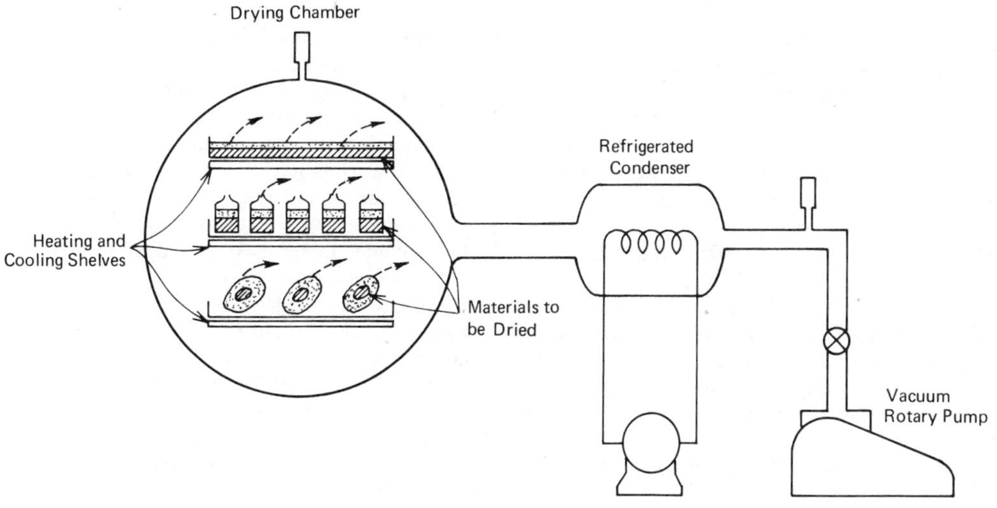

Fig. 3. Schematic diagram of freeze-drying process.

drying by increasing the surface and reducing the thickness. Coffee extract and fruit juices are preconcentrated in order to minimize the water to be removed by sublimation and, in turn, to reduce the processing cost or to ensure that the final product will not be fragile.

Freezing. (1) *Vacuum cooling*: The material freezes itself by the evaporation of water as it is subjected quickly to high vacuum. (2) *Direct contact*: The material is immersed in a cold liquid or in a stream of cold air or inert gases. (3) *Indirect contact*: The material is frozen on cold surfaces.

Freezing the material is accomplished in the vacuum chamber where drying takes place in a separate piece of equipment, or if the frozen mass requires some conditioning prior to drying.

Vacuum cooling is normally accomplished in the drying chamber. Its advantage is that large quantities of water are removed rapidly and no prior refrigeration is required. On the other hand, volatile flavor components are removed which may affect the quality of the final product. Also, removing water from the outer layer prior to complete freezing makes the cellular structure of certain materials collapse, and subsequent drying and reconstitution may be inhibited. Therefore, this freezing technique has limited application.

In direct-contact freezing, wet solid materials are placed in cold chambers and sprayed directly or immersed in cold air, inert gases, or liquid referigerants. Solutions or slurries may be frozen by spraying them in a cold stream of gas or liquid. Poor control of the rate of freezing limits these techniques to cases where quick freezing is desirable. Indirect-contact freezing is generally carried out in trays placed on refrigerated shelves inside the vacuum chamber.

A combination of indirect- and direct-contact equipment is commonly used. Trays containing the material to be dried may be placed on refrigerated shelves in a cold chamber and blasted with cold air, or inert gas or freezing belts may be used. Freezing belts in cold rooms are excellent for continuous operations. By controlling the temperature of the surrounding air and the refrigerant temperature along the length of the belt, good control of the freezing rate can be maintained.

Conditioning of the Frozen Mass. Materials frozen in a bulky or block form, meat, and solutions frozen on a belt or trays require further processing before drying. Granulation or slicing of such materials increases the available surface and minimizes the resistance to heat and mass transfer during drying. Standard size-reduction devices operating in cold chambers at about −50°F (−46°C) are sometimes used.

Substances which do not freeze into a rigid solid at low temperatures, such as fruit juices, may be subjected to a devitrification treatment to avoid the soft-glass structure detrimental to optimum drying.

Drying. The conditioned frozen material to be dried is placed in a vacuum chamber, where sublimation and desorption of water occur. As soon as the chamber has been evacuated and the optimum vacuum has been reached (0.5 to 0.05 millimeters of mercury), heating is applied so that the ice sublimes. For large-scale production of food products, the combination of conduction and radiation that results from circulating a hot fluid through coils or plates has proved quite satisfactory for heating.

A number of vacuum-chamber designs have been used. For large installations, custom engineering and fabrication are dictated to assure optimum performance for a given product. The three vacuum chambers commonly used may be classified as batch, semicontinuous, and continuous.

Batch units are frequently cylindrical shelf-type driers equipped with heating and cooling coils or plates. Designs for better heat input, that is, spikes or expanded metal sheet that penetrate the frozen mass or movable heated shelves compressing the frozen materials, have met with some success.

Semicontinuous units are usually long, cylindrical tunnels. Trays with the frozen material are continuously conveyed through a series of heated zones. Interlocks are used in both ends for proper vacuum control. The frozen material is heated along the length of the tunnel, with each zone maintained at a different temperature, and is removed as a fully dried product.

Continuous units are designed to move the frozen material continuously on the heated surface and transfer it through a series of zones in which the temperature and vacuum are maintained at different levels. The frozen material is fed via interlocks in the chamber. Vibrators or other mechanical means are used to maintain the product in continuous motion.

During the constant-rate drying period, the temperature of the heat source (radiation) is from 200–300°F (93–149°C) for many food products. Thus, high heat-input rates are achieved. This temperature is reduced to 125° to 150°F (51° to 66°C) in the falling-rate and desorption periods to avoid charring and decomposition of the dried products.

The drying process is discontinued when the residual moisture content is sufficiently low to ensure good preservation of the specific product. This may range from 1 t 3%.

Water Removal. The three common methods for removing the water vapor are condensers, direct- and indirect-contact; desiccants, such as calcium chloride and zeolites; and vacuum pumps. The indirect-contact refrigerated condenser offers a good arrangement for large-scale operations. The condensing surface can be located in the drying chamber, or in a separate chamber. The water vapor condenses and forms an ice layer on the cold surface and subsequently is removed by intermittent melting or scraping.

Vacuum Pumps. The function of these pumps is to evacuate the drying chamber quickly without allowing the prefrozen material to melt—and thereafter to reduce the pressure progressively to the desired vacuum and maintain it at this level by removing the noncondensable gases. The vacuum equipment can be either an oil-sealed rotary vacuum pump, or a multistage stream-ejector system.

Conditioning of the Product. The high porosity and low moisture content of the freeze-dried product require that the vacuum be broken and packaging be done under a dried inert-gas blanket, in many cases, to prevent oxidation during storage and maintain the low moisture content. Carbon dioxide or nitrogen are commonly used.

Variety of Problems. Each type of food product presents specific problems that affect the design and operation of freeze-drying processes. For some products, such as freeze-dried coffee (annual production of 50 million pounds; 22.5 million kilograms), the process has reached a mature and sophisticated stage. In this process, coffee extract with 20–25% solids is the raw material. Major steps in the process include: (1) Clarification of the extract; (2) freeze-concentration of the extract to 30–40% solids; (3) freezing extract to a completely frozen mass at −13 to −45°F (−25 to −43°C); (4) granulation of the frozen mass; (5) sublimation of the ice at a vacuum of approximately 200 micrometers of mercury absolute; and (6) drying the final product to a moisture content of 1–3%. For batch driers, the overall drying cycle is on the order of 6–8 hours.

Intensive research continues in the interest of improving the freeze-drying process for certain foods and for expanding the use of this process to a broader spectrum of food products. The Sharma reference (listed at end of this entry) provides a good summary of work in this area being done in connection with freeze-dried meat patties. In another reference, Schmidt describes a comparison of vacuum and atmospheric freeze-drying of carrots. As pointed out by James M. Flink (Massachusetts Institute of Technology), "Specific information on the costs of producing processed foods is generally not available in the scientific literature, it presumably being considered proprietary information by those having the best data, the food processing industry." In the Flink reference, based upon available data, a cost comparison is made of three food-preserving processes—canned, frozen, freeze-dried, and freeze-dried compressed.

Increasing attention in the late 1970s is being given to leafy vegetables, such as spinach. The large surface area of spinach seems well adapted for freeze-drying and compression. Foods produced by freeze-drying have less weight and a preserved flavor and structure, but the volume, in terms of packaging, transportation, and storage, is not changed. In an effort to alleviate this problem, different methods of compressing the freeze-dried products have been developed to eliminate most of the void spaces. Many fruits and vegetables have been compressed and then reconstituted to a normal appearance and texture. Most of the research has been directed toward vegetables. When properly preconditioned, freeze-dried foods can be compressed with little or no fragmentation. Freeze-dried foods have an average bulk density of 0.3 gram per cubic centimeter. With existing technology, it may be possible to compress most foods to a bulk density of 0.9 gram per cubic centimeter without interfering with reconstitution. This concept has been studied in some depth at Texas A & M University. See Wisakowsky reference at end of this entry.

Microwave Freeze-drying. Because microwave energy penetrates very well into ice, it would appear to offer an excellent solution to the heat-transfer problems of conventional freeze-drying. Because the microwave process has a much shorter drying cycle, decreased equipment capacity can be an economic result. Researchers have found that microwave energy utilization efficiencies range from 65% to 90% over much of the drying cycle. But microwave heating is dependent upon relatively high-cost electrical energy. Regardless of the economics, a major problem remaining with microwave heating is the melt-back of the frozen core and/or overheating in the dried layer. This situation occurs when microwave energy is put into the food faster than the sublimation-diffusion process can remove it, causing the pressure at the ice interface to rise above the triple point, usually resulting in melting of the ice. When melt-back occurs, the microwaves couple selectively into the water rather than the ice, thus causing intense local heating, accelerated melting, and a "runaway" condition (first reported by Gunn in 1967). Also for certain products, such as meat, a maximum allowable dried layer temperature of 140°F (60°C) should be observed during the process so that thermal degradation of the dried product can be prevented. This requires a matching of the electric heating rate with the mass flow rate in order to optimize the process. In early 1977, researchers at the University of Waterloo (Ontario, Canada) prepared a mathematical analysis of this problem (see Ang references at end of this entry).

References

Ang, T. K., Ford, J. D., and D. C. T. Pei: "Microwave Freeze-drying of Food—A Theoretical Investigation," *Int. J. Heat Mass Transfer*, **20**, 517 (1977).

Ang, T. K., Ford, J. D., and D. C. T. Pei: "Optimal Modes of Operation for Microwave Freeze-drying of Food," *J. Food Sci.*, **43**, 2, 648–649 (1978).

Flink, J. M.: "Energy Analysis in Dehydration Processes," *Food Technology*, **37**, 3, 77–84 (1977).

Flink, J. J.: "A Simplified Cost Comparison of Freeze-Dried Food with Its Canned and Frozen Counterparts," *Food Technology*, **31**, 4, 50–56 (1977).

Gejl-Hanse, F., and J. M. Flink: "Freeze-Dried Carbohydrate Containing Oil-in-Water Emulsions: Microstructure and Fat Distribution," *J. Food Sci.*, **42**, 4, 1049–1055 (1977).

Gunn, R. D.: Ph.D. Thesis, University of California, Berkeley, California, 1967.

Morgan, J. P., and D. F. Farkas: "Some Effects of Selected Plasticizing Agents on the Reversible Compression of Freeze-dried Cooked Beef Cubes," *J. Food Sci.*, **43**, 4, 1133–1136 (1978).

Noyes, R.: "Freeze-Drying of Foods and Biologicals," Noyes Development Corporation, Park Ridge, New Jersey, 1968.

Pintauro, N.: "Soluble Coffee Manufacturing Processes," Noyes Development Corporation, Park Ridge, New Jersey, 1969.

Porter, W. L., Levasseur, L. A., and A. S. Henick: "Evaluation of Some Natural and Synthetic Phenolic Antioxidants in Linoleic Acid Monolayers on Silica" (related to rancidity of whole tissue, freeze-dried foods), *J. Food Sci.*, **42**, 1533–1535 (1977).

Schmidt, F. W., Chen, Y. S., Kirby-Smith, M., and J. H. MacNeil: "Low Temperature Air Drying of Carrot Cubes," *J. Food Sci.*, **42**, 5, 1294–1298 (1977).

Sharma, S. C., and E. Seltzer: "Development of Procedures to Minimize Mechanical Damage in Freeze-Dried Meat Patties," *J. Food Sci.*, **42**, 5, 1336–1343 (1977).

Sivetz, M.: "Overview: Freeze-Drying Coffee," *World Coffee and Tea* (August 1971).

Steinhart, J. S., and C. E. Steinhart: "Energy Use in the U.S. Food System," *Science*, **184**, 397 (1974).

Stone, H.: "Food Products and Processes," Stanford Research Institute, Menlo Park, California, 1972.

Tuomy, J. M.: "Freeze-Drying of Foods for Armed Services," U.S. Army Food Laboratory Technical Report 70-43-FL, Natick, Massachusetts, February 1970.

Van Pelt, W. H.: "Dutch Process Cuts Cost of Freeze Concentration," *Food Engineering*, **47**, 11, 77–79 (1975).

Wisakowsky, E. E., Burns, E. E., and M. C. Smith: "Factors Affecting the Quality of Freeze-dried Compressed Spinach," *J. Food Sci.*, **42**, 3, 782–789 (1977).

Zarkarian, J. A., and C. J. King: "Asymmetry in Freeze-Drying," *J. Food Sci.*, **43**, 3, 992–997 (1978); "Acceleration of Limited Freeze-Drying in Conventional Dryers," ibid., pages 998–1001.

FREEZE-PRESERVING. Knowledge of the fact that food substances remain edible for longer periods of time when cooled probably dates back to antiquity, centuries before the process for making ice was developed. The latter led to cold storage, a practice that persisted for several decades and which, of course, remains useful for a number of products in current times and in certain regions. Cold storage was first limited by the minimum achievable temperature dictated by the melting point of ice. Chemicals to depress the freezing point were an additional step toward cold-preservation. Generally accredited with the initial breakthrough from cold-storage practices to present freezing technology was the step to quick-frozen foods, pioneered by Clarence Birdseye, among others, in the late 1920s. Consumers began to accept the fact that fresh, high-quality food when frozen quickly and when retained at a temperature of about −17.8°C (0°F) was a good substitute for fresh produce out of season. Frozen and stored in this way, these foods represented a marked improvement over the earlier available, slowly cooled food products.

Freeze-preserving is an across-the-board operation in the food industry of countries with advanced technology. There are relatively few foods—vegetables, fruits, fish, poultry, and meat—that cannot be frozen with reasonable success.

Fundamentals of Freezing. In food materials, water is the major component.* Thus, when foods are cooled below 0°C, ice formation occurs, starting at a temperature between 0 and −3°C (32 and 26.6°F), which depends upon the molar concentration of soluble cell components. As the temperature is progressively reduced, more and more water is turned into ice and the latent heat of ice formation adds to the sensible heat involved in cooling both ice and the unfrozen portion. This leads to large variations in heat capacities while thermal conductivities also change considerably, mainly because the thermal conductivity coefficient of ice is nearly four times greater than that of water. For most biological materials, the largest part of the freezing process takes place in a temperature interval between −1 and −8°C (30.2 and 17.6°F), while the largest variations of heat capacity occur between −1 and −3°C (30.2 and 26.6°F). Only at temperatures ranging from −20 to −40°C (−4 to −40°F) and below, there is no more measurable change with temperature in the amount of ice present, and the remaining water, if any, can be considered as non-freezable. However, for practical purposes, a lower limit to the phase-change interval can be defined on the basis of a ratio of ice to total water content of, say, 90%. This choice, in addition to providing an easily applicable criterion, allows one to approximate heat capacity and thermal conductivity curves, above and below the phase-change zone, by means of constant values. These techniques are described by Bonacina et al. (1974).

Rebellato et al. (1978) have developed a finite element analysis approach to freezing processes in foodstuffs and apply this method to the computation of temperature distributions in foodstuffs of irregular shape during freezing in an air-blast tunnel. A number of other researchers, including Cleland and Earle (1977, 1979), also have been tackling the very important and mathematically complex problem of predicting the freezing times of various foodstuffs—fundamental to the design and operation of freezing equipment and to the scheduling of production. Cleland and Earle carried out a set of experiments to determine the freezing time of cylindrical and spherical blocks of a widely used food analogue material (Karlsruhe test substance, a defined 23% methylcellulose gel the thermal properties of which closely model those of real food materials (Riedel, 1960).

Wide Range of Freezing Configurations

The food processor has several options available when selecting the best freezing format for a given set of product characteristics and marketing objectives. Methods can be classified in several ways, as for example the media used to contact and extract heat from the food substance—air or other gases, liquids, or mechanical contact.

Air-blast systems commonly take the form of large rooms, tunnels, or cells. In a room, the air velocity may be low or range up to 1500 feet (457 meters) per minute. The temperature for air-blast freezing usually ranges from −29 to −40°C (−20 to −40°F). Blast freezing requires longer than other available methods and product quality can-

* Paragraph summary prepared by L. Rebellato, S. Del Giudice, and G. Comini (see reference list).

not be assured unless efficiently insulated. Insulation can be improved by using cold storage doors and air curtains. Where insulation is inadequate, frosting will occur on the coils, lowering the refrigerating capabilities. Air-blast freezing also can be effected in a tunnel on a fluidized bed, or on a belt. For individually quick frozen (IQF) products, fluidized-bed, air-blast systems are frequently used. An advantage of the fluidized bed is that it keeps the product in motion and separated during the freezing process. As pointed out by Shyette (1979), small products are moved along through a tunnel on a belt while they are frozen. The process is fast and accepted as a standard freezing method for IQF products.

Plate freezers generally are limited in application to prepackaged products. In this system, the food substance is placed in direct contact with refrigerated metal plates (usually steel or aluminum). Cooling coils are located within the interior of the metal plates. The required contact refrigeration time ranges from about 30 to 90 minutes, depending upon size and nature of food substance.

Liquid-immersion freezing has grown in acceptance during the past few years. This system requires placing the product in a bath of cooling liquid, which must be nontoxic, noncorrosive, have a low freezing point, low viscosity, and high thermal conductivity. Wrapping of the product is required in many cases. Salt solutions and propylene glycol are frequently used. Advantages over air-blast freezing include operational energy savings that result from high heat-transfer coefficients and high heat capacities.

Aqueous freezants, composed of a single solute species, such as an inorganic salt, acid, or base, or an organic compound, such as sucrose, have been proposed and used for freezing fruits, vegetables, and fish. Freezants containing more than one solute also have been suggested, including sodium chloride with minor amounts of calcium chloride or potassium chloride added to water or sea water for freezing fish (Holston and Pottinger, 1954), glucose with sucrose (Bartlett, 1941), glycerin with ethyl alcohol (Bland, 1936), and glucose with sodium chloride (Butlet and Slavin, 1959). As recently as 1973, a sodium chloride or sodium chloride–sugar mixture was utilized in a prototype "hydrofreeze" aqueous system developed by Marco (Seattle). Flavor impact of aqueous freezants was viewed as an important factor in limiting commercialization of aqueous freezing methods. More recently, Cipolletti et al. (1977) have investigated freezant composition as a means of minimizing flavor impact, as well as considering operational characteristics, such as freezing point and viscosity.

In considering a ternary system (15% sodium chloride, 15% ethanol, and 70% water), among other observations, Cipolletti et al. noted that: (1) Freezing times to 0°F (−17.8°C) as fast as 2 minutes were achievable for carrots and peas; (2) photomicrographs showed greatly reduced cell damage as compared with air-blast frozen samples; (3) sodium chloride uptake in products after freezing varied from a minimum of 0.89% for peas to a maximum of 2.06% for carrots; (4) ethanol uptake in products after freezing was small (0.05–0.27%); and (5) no significant organoleptic preference differences were indicated for peas, snap beans, and corn, frozen either by immersion with the salt–ethanol–water medium (with or without blotting) or by air blast. Preference for air-blast frozen carrot samples was indicated by a panel because of the absence of salt. Panels evaluating mixed vegetable samples containing carrots, peas, beans, and corn indicated no exclusive preference between immersion-frozen and air-blast frozen vegetables.

Cryogenic freezing has gained wide acceptance during recent years. Some of the advantages of this method over blast freezing are immediately obvious from the accompanying table. Because cryogenic freezing is very fast and accomplished at extremely low temperatures (down to −196°C; −320°F), less dehydration occurs. Problems of cell damage, caused by sharp ice crystals formed during slower freezing processes, are largely overcome with short freezing times. Also, the sooner a product is deeply frozen, the sooner will be the halting of bacterial and enzyme degradation. (See observations of Kraft et al. later in this entry.)

For cryogenic freezing, nitrogen is used in several forms—as a shower of liquid droplets, as a liquid bath for direct immersion, or as a cold gas. Carbon dioxide is used as a liquid or in solid "snow" form. When used in a tunnel for IQF applications, liquid carbon dioxide can freeze products at a temperature from −62 to −78°C (−80 to 109°F). Fluorocarbons and halocarbons also have been used in conjunction with tunnel and spiral-type freezers that are used in IQF methods. The advantages of rapid freezing by cryogenic systems, including reduced tissue damage and improved quality, have been summarized many times, including reports by Meryman (1956) and Sills (1969).

Because of large amounts of ground beef used in frozen meat items, particularly as hamburger for fast-food chains (Harr and Minard, 1977), numerous studies have been conducted on the microbiology of the product and its ingredients (Chestnut et al., 1977). Centralized processing with cryogenic freezing of ground beef may provide better sanitation in handling than traditional processing methods. When Duitschaever et al. (1977) examined frozen ground beef patties from retail markets, the investigators found that the microbiological quality was better than that of fresh ground beef. But Kraft et al. (1979) observed that cryogenic freezing is a commonly accepted method for preserving bacterial cells, and any system involving cryogenic food freezing must consider effects on microorganisms in the foods. Thus, Kraft et al. undertook a study designed to evaluate effects of meat composition and freezing by conventional or cryogenic methods on microbial numbers and types in beef patties. Beef patties of different fat content or containing soy protein were frozen by liquid nitrogen, liquid carbon dioxide, or mechanical freezing and stored at −18°C (−27.8°F) for 5 months. Cryogenic freezing produced significantly greater reduction in total viable mesophils and psychotrophs than did mechanical freezing. Patties containing 30% fat provided greater survival than observed with 20% fat or added soy protein. After frozen storage, predominant flora consisted of species of *Moraxella Acinetobacter* (42%) and *Pseudomonas* (32%). The former group was favored by mechanical freezing, but a great proportion of *Pseudomonas* spp. survived cryogenic freezing than mechanical freezing.

In connection with means for chilling poultry, Arafa and Chen

COMPARATIVE PERFORMANCE OF FREEZING METHODS FOR SELECTED SUBSTANCES

Commodity	Cryogenic Freezing		Blast-Freezing
	Liquid Freon	Liquid Nitrogen	
Strawberry			
Freezing time	3 minutes	5 minutes	900 minutes
Temperature after freeze	−25°C (−13°F)	−28°C (−18°F)	−20°C (−4°F)
Percent weight loss	0.0	1.4	2.7
Mushroom			
Freezing time	3 minutes	5 minutes	180 minutes
Temperature after freeze	−30°C (−22°F)	−26°C (−15°F)	−20°C (−4°F)
Percent weight loss	0.0	1.9	2.5
Beef			
Freezing time	4 minutes	8 minutes	180 minutes
Temperature after freeze	−28°C (−18°F)	−50°C (−58°F)	−20°C (−4°F)
Percent weight loss	0.1	1.4	1.3

SOURCE: "Freezing Equipment Influence on Weight Losses," Sture Astrom, Helsingborg, Sweden.

(1978) found chilling of broilers by liquid nitrogen exposure resulted in lower Warner-Bratzler shear values and longer sarcomere lengths, indicating more tender meat when compared with immersion-chilled broilers. Liquid nitrogen chilling also resulted in higher cooking yields. Experienced panelists showed a preference for the liquid-nitrogen chilled product. It was also observed that liquid-nitrogen chilling of poultry resulted in a product with a longer shelf-life when compared with that of immersion-chilled broilers.

References

Arafa, A. S., and T. C. Chen: "Liquid Nitrogen Exposure as an Alternative Means of Chilling Poultry," *J. Food Sci.*, **43**, 3, 1036–1037 (1978).

Bartlett, L. H.: "Quick Freezing of Foodstuffs," U.S. Patent 2,418,745 (1941).

Bland, O. S. H.: "Improvements Relating to the Preservation of Perishable Articles by Freezing," B. P. 454,200 (1936).

Bonacina, C., et al.: "On the Estimation of Thermophysical Properties in Nonlinear Heat-conduction Problems with Special Reference to Phase Change," *Int. J. Heat Mass Transfer*, **17**, 861 (1974).

Butler, C., and J. W. Slavin: "Fishery Products," ASRE Data Book: Refrigeration Applications, Vol. 1, No. 1, page 1406 (1959).

Chestnut, C. M., et al.: "Bacteriological Quality of Ingredients Used in Ground Beef Manufacture," *J. Food Sci.*, **44**, 213 (1977).

Ciobanu, A., et al.: "Cooling Technology in the Food Industry," International Scholarly Book Services, Forest Grove, Oregon, 1976.

Cipolletti, J. C., Robertson, G. H., and D. F. Farkas: "Freezing of Vegetables by Direct Contact with Aqueous Solutions of Ethanol and Sodium Chloride," *J. Food Sci.*, **42**, 4, 911–916 (1977).

Cleland, A. C., and R. L. Earle: "A Comparison of Analytical and Numerical Methods of Predicting the Freezing Times of Foods," *J. Food Sci.*, **42**, 5, 1390–1395 (1977).

Cleland, A. C., and R. L. Earle: "A Comparison of Methods for Predicting the Freezing Times of Cylindrical and Spherical Foodstuffs," *J. Food Sci.*, **44**, 4, 958–963 (1979).

Duitschaever, C. L., et al.: "Bacteriological Evaluation of Retail Ground Beef, Frozen Beef Patties, and Cooked Hamburger," *J. Food Protection*, **40**, 378 (1977).

Harr, J. W., and M. E. Minard: "Cryogenic Meat Processing—Cost and Quality," *Proceedings of Meat Industry Research Conf.*, page 25, Chicago, Illinois, 1977.

Kraft, A. A., et al.: "Effect of Composition and Method of Freezing on Microbial Flora of Ground Beef Patties," *J. Food Sci.*, **44**, 2, 350–353 (1979).

Meryman, H. T.: "Mechanics of Freezing in Living cells and Tissues," *Science*, **124**, 515 (1956).

Rebellato, L., et al.: "Finite Element Analysis of Freezing Processes in Foodstuffs," *J. Food Sci.*, **43**, 1, 239–243 (1978).

Riedel, L.: "Eine Prüfsubstanz für Gefrierversuche," *Kaltechnik*, **12**, 222 (1960).

Shyette, B.: "Options Open to Freezers," *Processed Prepared Food*, **148**, 4, 46–48 (1979).

FREEZING CURVE. A plot of the temperature (or of any quantity which is a function of temperature) against time, obtained as a substance is allowed to cool from above its freezing point to below that temperature. Freezing curves are used in the calibration of thermocouples and other temperature measuring devices.

FREEZING POINT. The temperature at which a liquid solidifies under any given set of conditions. It may or may not be the same as one of the following: (a) The melting point, or the temperature at which a solid substance changes from solid to liquid form; (b) the "true freezing point," or the temperature at which the liquid and solid forms of a substance exist in equilibrium at a given pressure, usually one standard atmosphere; (c) the *ice point*, or the temperature at which a mixture of air-saturated pure water and pure ice may exist in equilibrium at a pressure of one standard atmosphere. The freezing point is not an "equilibrium" property of a substance; it applies to the liquid phase only. It is somewhat dependent upon the "purity" of the liquid, the volume and shape of the liquid mass, the availability of freezing nuclei and the pressure acting upon the liquid.

The freezing point of a solution becomes proportionately lower with an increasing amount of dissolved matter. Therefore, since natural water almost invariably contains some solutes, its freezing point usually is found to be slightly below 0°C. For example, bulk samples of normal seawater freeze at about −1.9°C, or 28.6°F. The *maximum freezing point* is that temperature for a particular composition of a two-component or multi-component liquid system at which the freezing point is higher than that for any other composition or for the pure components.

The *minimum freezing point* is that temperature for a particular composition of a two-component or multi-component liquid system at which the freezing point is lower than that for any other composition or for the pure components.

FREEZING-POINT DEPRESSION. The freezing point of a solution is, in general, lower than that of the pure solvent and the depression is proportional to the active mass of the solute. For dilute (ideal) solutions

$$\Delta T = Km$$

where ΔT is the lowering of the freezing point, K, the *cryoscopic constant* for the given solvent, and m, the molality of the solution.

There are several methods for measuring this depression: In the *Beckmann Method* the freezing point of pure solvent and that of solution is measured by a special type of thermometer, the "Beckmann thermometer." The solvent or solution is contained in a double-walled glass apparatus and placed in a freezing mixture not more than 5°C below the freezing point of the solvent. By rapid stirring when the liquid has supercooled about $\frac{1}{2}$°, crystallization is induced and the temperature rises to the freezing point.

In the *equlibrium method* a relatively large amount of solvent crystals are allowed to form and the system allowed to come to equilibrium. The temperature is recorded and some of the solution withdrawn and analyzed.

See also **Cryoscopic Constant.**

FRENKEL DEFECT. A lattice vacancy created by removing an ion from its site and placing it at an interstitial position within the lattice. Thus a Frenkel pair is a vacancy and interstitial.

FREQUENCY. 1. In general, the number of repetitions of a periodic process per unit time, i.e., the inverse of the periodic time. Cycles per second is commonly used for the expression of many frequency phenomena. The abbreviation for cycles per second, cps, has been replaced by hertz, abbreviated Hz, as the SI unit of frequency. Thus, 60 cps equals 60 Hz and so on.

Frequency is defined mathematically for a periodic quantity in which time is the independent variable, as the number of periods occurring in unit time. If a periodic quantity, y, is a function of the time, t, such that

$$y = f(t) = A_0 + A_1 \sin(\omega t + a_1) + A_2 \sin(2\omega t + a_2) + \cdots$$

then the frequency is $\omega/2\pi$. Unless otherwise specified, the unit is the cycle per second (hertz).

2. In electricity, frequency is the number of complete alternations per second of an alternating current. Sixty cycles per second is the standard electrical frequency for alternating current generation in the United States and many nations throughout the world. In some areas, 25- and 50-Hz systems still persist. In an alternator, the number of alternations per second of the output is the speed, in revolutions per second, multiplied by half of the number of poles. The number of poles in alternators is usually 2 or 4 for steam turbine-driven alternators, or 24, 26, 28, 30, 36, 48 or 60 in the case of engine-driven alternators.

3. In acoustics, the frequency represents the number of sound waves passing any point of the sound field per second.

4. In the case of light or other electromagnetic radiation, frequency may be expressed in this same way, but is usually so enormous (500 million million cycles per second for yellow light—10^{12} Hz or 10 terahertz) that wavelengths or wave numbers (reciprocal of wavelength measured in centimeters) are ordinarily used instead. Radio frequencies are commonly given in thousands of cycles per second (formerly kilocycles; now kilohertz, kHz); millions of cycles per second (formerly megacycles; now megahertz, MHz); or thousands of megacycles per second (formerly kilomegacycles; now gigahertz, GHz).

There are also various compound terms denoting restricted uses of the term frequency, such as:

Audio Frequency, which is a frequency corresponding to a normally audible sound wave.

Infrasonic Frequency, which is a frequency lying below the audio-frequency range.

Ultrasonic Frequency, which is a frequency lying above the audio-frequency range.

Instantaneous Frequency, which may be defined mathematically as the time rate of change of the angle of a wave which is a function of time.

Basic Frequency, which is that particular frequency which is arbitrarily chosen as most important of a wave (oscillatory quantity) having sinusoidal components with different frequencies.

Fundamental Frequency, which may have various meanings: (1) The lowest possible frequency of vibration of a system characterized by normal modes of vibration (for example, a vibrating string or organ pipe). Otherwise stated as the greatest common divisor of the component frequencies of a periodic wave or quantity. (3) Of a periodic quantity, the frequency of a sinusoidal quantity which has the same frequency as the periodic quantity.

Natural Frequency, which is the period of free oscillation of a body; or the applied frequency at which a coil exhibits electrical resonance; or the lowest frequency at which there is a standing wave on an antenna.

Angular Frequency, which is the frequency expressed in radians per second. It is equal to the frequency in cycles per second multiplied by 2π.

Center Frequency, which is (1) the average frequency of the emitted wave when modulated by a sinusoidal signal; (2) the frequency of the emitted wave without modulation.

Frame Frequency, which denotes, in television, the number of times per second that the frame is scanned.

Field Frequency, which denotes, in television, the product of frame frequency and the number of fields per frame.

Line Frequency, which denotes, in television, the number of times per second that a fixed vertical line in the picture is crossed in one direction by the scanning spot. Scanning during vertical return intervals is counted.

FREQUENCY ALLOCATION. **Radio Communications.**

FREQUENCY BAND. A term usually applied to a group of closely related frequencies, but common usage applies it in two slightly different senses. The first of these is synonymous with channel, meaning the band of frequencies associated with a carrier under modulation. The second usage means a group of different carrier frequencies all designated for the same purpose. In this sense there may be many frequency channels within the given band. Thus the standard broadcast band contains many broadcast channels, each having its carrier spaced 10 kilohertz from the next. See also **Radio Frequency Allocation.**

FREQUENCY CHANGER. **Power Sources and Supplies.**

FREQUENCY DEMODULATOR. **Demodulator.**

FREQUENCY DEVIATION. 1. In amplitude-modulated or continuous wave transmission, the amount by which the carrier frequency varies from its assigned value. 2. In frequency modulation, the peak-difference between the instantaneous frequency of the modulated wave and the carrier frequency.

FREQUENCY DISTORTION. **Distortion (Electromagnetic).**

FREQUENCY DISTRIBUTION. A specification of the way in which the frequencies of members of a population are distributed according to the values of the variates which they exhibit. For observed data the distribution is usually specified in tabular form, with some grouping for continuous variates. A conceptual distribution is usually specified by a frequency function or a distribution function.

FREQUENCY DIVIDER. Often in high-frequency measurements it is desirable to have available frequencies which are submultiples of some given frequency, i.e., the given frequency divided by a whole number. These subfrequencies may be obtained by the use of a harmonic generator which, in this application, is called a frequency divider. Oscillators have a strong tendency to synchronize with an injected frequency which is not too different from their normal value. In certain types this tendency is so strong that a harmonic of the oscillator will synchronize with an injected frequency. It is this type which serves as a divider, the frequency to be divided is fed into the circuit of the dividing oscillator, causing the oscillator frequency to change so the proper harmonic is locked in with the injected signal. This, of course, means that the oscillator frequency is fixed at a submultiple value of the original frequency. The various harmonics of the synchronized oscillator then gives a series of frequencies which might be said to be divisions of the original frequency. The multivibrator oscillator can be used. See also **Pulse Generator.**

FREQUENCY-DIVISION MULTIPLEX. The process or device in which each modulating wave modulates a separate subcarrier and the subcarriers are spaced in frequency. Frequency division permits the transmission of two or more signals over a common path by using different frequency bands for the transmission of the intelligence of each message signal.

FREQUENCY DOMAIN ANALYSIS. **Signal Generator.**

FREQUENCY DOUBLER. The oscillator of a radio transmitter can be operated at a lower frequency than the output frequency of the transmitter. This has several advantages. Among them may be mentioned less reflected effect on the oscillator and hence more stable operation, elimination of neutralization, stronger crystals than if they were ground for the higher output frequency, and in the case of frequency modulation a proportionate increase in the degree of modulation. Doubling is used in frequency modulated transmitters since the frequency deviation is doubled each time the frequency is doubled. In some transmitters of this type there may be a dozen or more doubler stages for this purpose.

FREQUENCY DRIFT. A term for the gradual change of carrier frequency which a poorly designed radio transmitter may undergo with time. It is usually caused by temperature effects in the oscillator. The term is also applied to the drift of the local oscillator frequency used in a superheterodyne receiver.

FREQUENCY (Electric) MEASUREMENT. Electric frequency may be defined as the number of periodic swings or cycles of voltage or current in a unit of time. Measurement can be accomplished by: (1) counting the number of cycles in a given time, (2) measuring the time in which a given number of cycles occurs, (3) balancing an impedance bridge in which the impedances of the arms are known functions of frequency, (4) tuning electric circuits to a condition of resonance, (5) resonating tuned reeds magnetically, and (6) using a deflection-type frequency meter.

Counting requires electronic means, including sources of pulse trains, pulse counters, and electronic switches (gates). Very wide ranges of frequency up to many millions of hertz are measurable in this manner. Time measurements require similar instrumentation. Aforementioned methods (3) and (4) require manual adjustment of capacitors or other bridge elements and the observation of a balance or resonance detector in a bridge circuit. Tuned reeds are limited to a fairly narrow range of electromagnetically produced mechanical oscillations. Instruments based upon this method usually extend little beyond the common power frequencies. The usual deflection type frequence meter employs a rotatable iron vane whose position is determined by the superimposition of two stationary magnetic fields whose resultant changes its orientation as a function of frequency. These fields are produced by two coils, one of which is connected into a resistive circuit; the other into an inductive circuit. As the frequency increases, the current in the latter decreases. Such instruments are widely used in electric power installations and are calibrated for the range of frequencies encountered in this field.

Oscilloscope. Comparison of frequencies with an oscilloscope involves the generation of Lissajous figures on the face of a cathode-ray tube. A source of known frequencies is connected to one set of deflection plates; the unknown frequency is connected to the other set. If one frequency is a harmonic or integral multiple of the other,

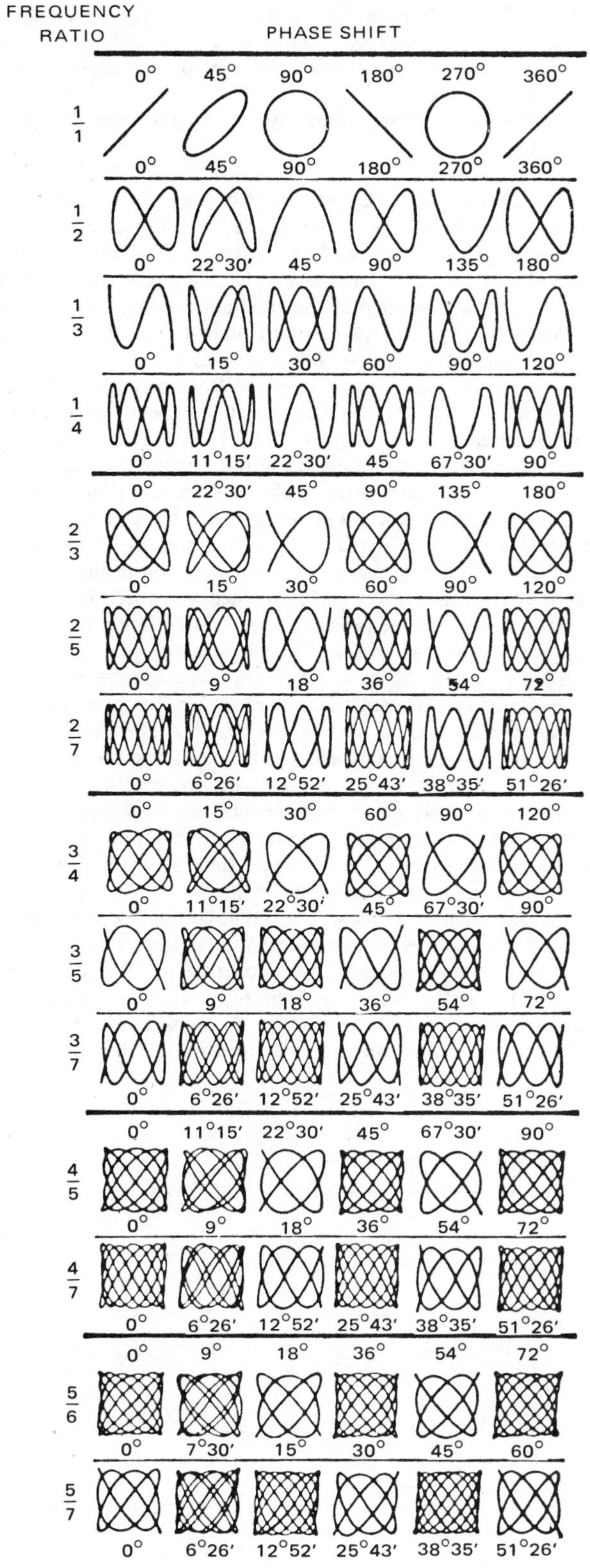

Fig. 1. Frequency patterns (Lissajous figures) for most commonly encountered frequency ratios. The generation of Lissajous figures on a cathode-ray tube is a common method of frequency comparison. A source of known frequencies is connected to one set of deflection plates and the unknown frequency is connected to the other set. Where one frequency is a harmonic or integral multiple of the other, a stable and recognizable pattern is displayed. This diagram shows patterns where the equipment is arranged such that the unknown frequency is connected to the vertical input; the known frequency to the horizontal input. As the phase between two applied signals is varied, the Lissajous pattern shifts. A specific unknown frequency can be measured by the Lissajous method if an interpolation oscillator is used with a frequency standard to provide continuous adjustment of the standard frequency. (*Tektronix, Inc.*)

a stable and recognizable pattern will be displayed. Lissajous figures for the most frequently used frequency ratios are shown in Fig. 1.

Electronic Meters. The input signal is amplified and converted to a square wave which charges a capacitor through a diode. A second diode discharges the capacitor through the meter circuit. The current through the meter is proportional to the rate of the charging pulses. Hence it is proportional to the frequency of the input signal. In another configuration, the input waveform drives a trigger circuit which generates a series of negative trigger pulses. These actuate a constant-current source and a linear-timing circuit. The output of the latter turns off the current source, thus determining the width of the current pulse delivered to the meter circuit. One such stable pulse is passed through the meter circuit for each input cycle. The meter averages the pulses it receives and presents an indication that is proportional to the average frequency.

Heterodyne Meter. This instrument is essentially a frequency-calibrated stable oscillator, a mixer, and an indicator. An amplifier for driving the beat indicator is also included. The output signal may be used to drive external indicators, such as headphones, oscilloscopes, or recorders. With the unknown frequency applied to the mixer, the oscillator usually is tuned for a zero beat, which is indicated by a minimum or zero output from the mixer. The oscillator frequency then will be the same as, or a subharmonic of, the frequency of the applied signal. Heterodyne techniques also are used when the oscillator is tuned for a difference frequency in a specified range of frequencies instead of for a zero beat. The difference frequency then is determined by a separate frequency-measuring circuit to obtain the desired information. By using oscillator harmonics to zero beat against the unknown, the basic range of the instrument may be extended by 50 to 100 times the highest fundamental frequency available from the oscillator. Numerous heterodyne frequency meters also include crystal calibrators which are used to check accuracy at various points on the oscillator frequency-control dial. A heterodyne frequency meter with crystal calibrator can provide measurement accuracy of up to 1 part in 10^6 or better.

Electronic Counter. This instrument comprises a time-base generator, a signal gate, and decade-counting units. Frequency is measured by counting the number of input cycles over a precisely controlled period of time. The time-base generator develops control signals which are applied to the signal gate. When the first or *start signal* is received, the signal gate opens to pass input pulses from the unknown frequency source to the decade-counting units. When the second or *stop signal* is received, the signal gate closes to prevent further input pulses from reaching the decade-counting units. Totalization of input pulses by the decade-counting units during the interval when the gate is open is a measure of the input-signal frequency. Frequency measurement accuracy of an electronic counter is plus-or-minus one count, plus or minus the accuracy of the time-base generator. In one standard arrangement, frequencies up to 500 MHz or more are measured with a 10-MHz counter and a frequency converter. With a transfer oscillator or other harmonic generator-mixer arrangements, the range may be extended to at least 40 kMHz. At lower frequencies, the accuracy of measurement is the basic accuracy of the electronic counter. Inherent characteristics of transfer oscillators limit the measurement accuracy at higher frequencies to about 1 part in 10^7.

Slotted-Line Measurement. An electronic counter and transfer oscillator can be used for very accurate measurements of frequency in the uhf and microwave bands. For some measurements where an accuracy of the order of plus-or-minus 0.5% is sufficient, the frequency may be determined with a slotted line or slotted section by observing the standing-wave pattern. See Fig. 2.

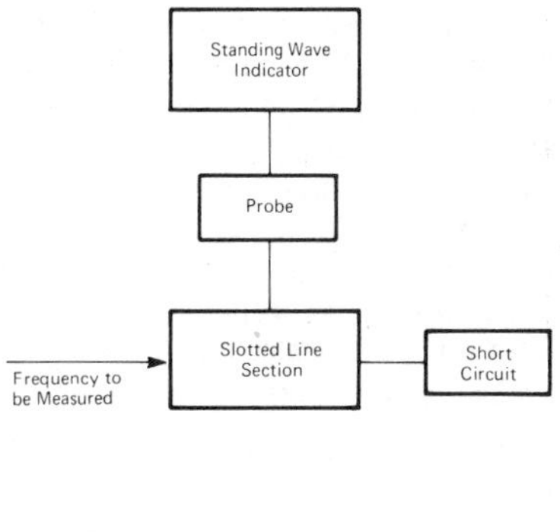

Fig. 2. Frequency measurement by slotted-line method.

Lumped-Constant Wavemeter. These instruments are used up to frequencies of 1,200 MHz. A crystal detector is coupled to a resonant LC circuit. For frequencies up to 100 MHz, a fixed inductor and a variable capacitor are used to form the resonant circuit. Plug-in coils of various inductance values can be used to cover the frequency range. The capacitor dial is calibrated to read frequency directly. When the wavemeter coil is in the presence of the field of the unknown signal, the circuit is tuned to resonate at the same frequency as the unknown signal. The output of the crystal detector is indicated by the meter circuit. In wavemeters operating above 100 MHz, a butterfly resonant circuit is used. This element varies both the inductance and capacitance simultaneously. The accuracy of a lumped-constant wavemeter depends upon the Q of the resonant circuit. In general, accuracies between $\frac{1}{4}$ and 1% are obtained. See Fig. 3.

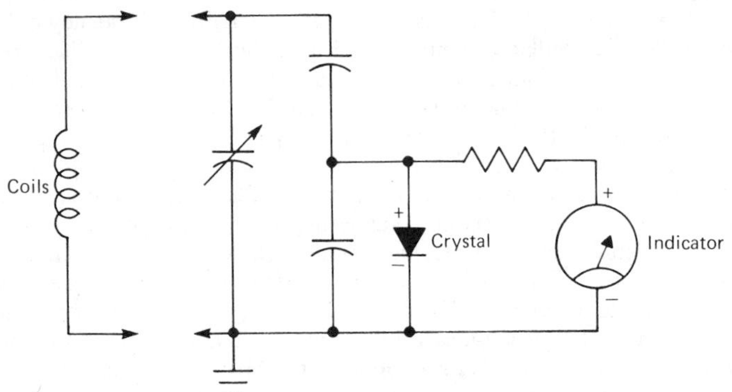

Fig. 3. Lumped-constant wavemeter.

Cavity-Type Wavemeter. These instruments are applicable for determining frequency in a waveguide system. Typically, a cylindrical cavity and piston are used. As the position of the wavemeter piston is varied, the distributed inductance and capacitance of the cavity, and hence the resonant frequency, are changed. In a reaction-type wavemeter, there is a drop in the transmitted power when the piston is adjusted for resonance because power is absorbed in the tuned cavity. Thus, a wavemeter in a waveguide system should be detuned so that maximum power will be transmitted to the load. Accuracy of a cavity-type wavemeter depends upon the selectivity or Q of the resonant cavity. Commercial wavemeters usually have an accuracy of from 0.1 to 0.01%.

FREQUENCY FUNCTION. An expression giving the frequency of a variate-value x as a function of x; or, for continuous variate, the frequency in an elemental range dx. Unless the contrary is specified the total frequency is taken to be unity, so that the frequency function represents the proportion of variate-values x. From a more sophisticated standpoint the frequency function is most conveniently regarded as the derivative of the distribution function. The generalization to more than one variate is immediate.

If the distribution is regarded as defining the probabilities of occurrence of the values of x, the frequency function is sometimes called the *probability density function* and the distribution function itself is called the *cumulative probability function.*

FREQUENCY MODULATION. Modulation.

FREQUENCY MONITOR. A frequency monitor is used to give a continuous indication of any departure of a radio transmitter's frequency from its assigned value. This is usually done by comparing the station frequency with that of a crystal controlled oscillator in the monitor. The monitor frequency and the transmitter frequency are heterodyned and the beat frequency measured. Some monitors are adjusted to give zero beat frequency when the station is on the correct value, others give a difference frequency of some convenient value such as 1,000 Hz. The beat frequency is fed to a circuit feeding an indicating instrument which shows zero for correct transmitter frequency and deflects to the right or left for high or low transmitter frequency. The scale is calibrated in cycles and the instrument indicates cycles off frequency.

FREQUENCY POLYGON. A graph of the frequency distribution formed by graphing the class frequencies as ordinates and the class marks as abscissas and then joining these points by straight lines. A histogram is to be preferred.

See also **Frequency Distribution; Histogram;** and terms listed under **Mathematics.**

FREQUENCY RESPONSE. By the frequency response of a system or component is meant the variation of the gain and phase shift plotted as a function of the frequency of the applied (sinusoidal) signal. This information is normally presented graphically as a Bode or Nyquist plot. Figure 1 shows a Bode plot of a component which has the characteristic curve of a single time constant or first-order lag. A Nyquist plot of the same component is shown on Fig. 2.

As can be seen in Fig. 1, the gain versus frequency record is plotted on log-log coordinates, and the phase versus frequency is on a semilog basis. This method has found its greatest reception in the field of process control. The gain which is the ratio of the change in output to input signal magnitude is often referred to as the *magnitude ratio.* In some cases, a semilog plot will be used with the gain or magnitude ratio plotted in terms of decibels. Since the gain in dB (decibels) is equal to 20 × log (gain), this is merely a variation in method.

In some cases, attenuation versus frequency is used. By definition, attenuation equals the reciprocal of gain. The choice of log-log or semilog plot remains as before. The phase curve is generally given as a semilog plot with phase angle or lag versus frequency.

The choice of scales results from fundamental concepts involving system design and analysis, as well as practical considerations. The wide range of frequencies covered by most systems, for example, requires the log scale in order to graph the information adequately. In actual design work, many components can be adequately represented by a single time constant, or by a more complex transfer function which includes two or more single time constant forms. Thus, to get the component response it may be necessary to combine several single time constant responses. The same combination is necessary when component curves are combined to give the overall system response. To combine responses, gains must be multiplied and phase angles added as is the normal procedure with vector quantities. By plotting gain on a log scale or the log of the gain on a linear scale, multiplication can be accomplished by an addition of distances mea-

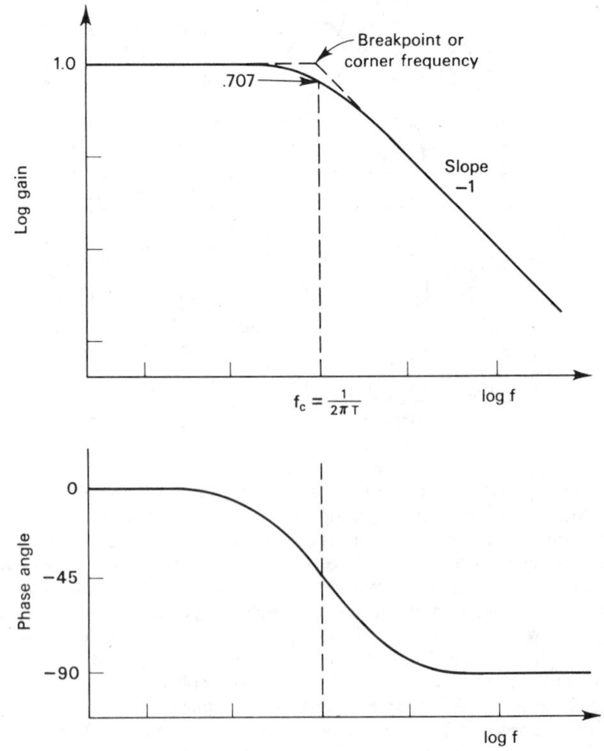

Fig. 1. Bode plot of single time constant.

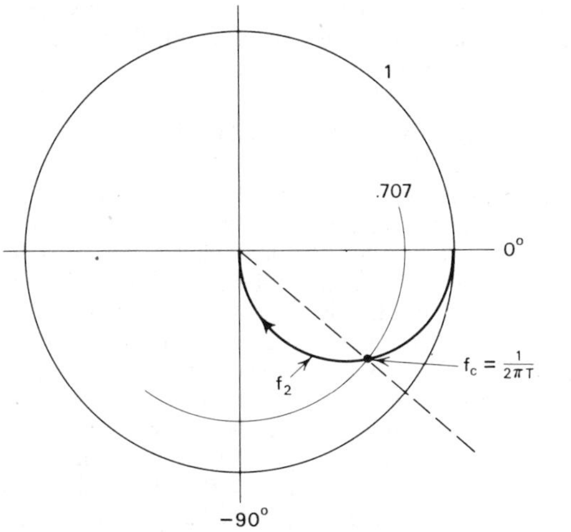

Fig. 2. Nyquist plot of a single time constant.

sured from the unity reference. The same process of addition of distances may be accomplished for phase angles by using a linear plot.

The ease with which system responses may be obtained in this manner is even more evident when the characteristics of the single time constant response of Fig. 1 are analyzed. The shape of the gain and phase curves for a single time constant are constant or fixed. Their relative position on the graph is determined by the zero frequency or steady-state gain, and the value of the constant T as shown in Fig. 1. The breakpoint or corner frequency occurs at a frequency equal to $\frac{1}{2}\pi$ T. This point is also established by the high and low frequency asymptotes. At this point the actual gain is equal to 0.707 times the zero frequency gain and the phase angle is $-45°$. If the zero frequency gain is some value other than 1, the gain curve is merely shifted up or down on the gain scale. The phase curve does not change.

The same properties are demonstrated on the Nyquist polar plot shown on Fig. 2. This method of presenting frequency response data has found its greatest reception in analysis and design of electronic amplifiers and servomechanisms. In this method of presentation, frequency values must be noted on the plot as shown on Fig. 2. Both the Bode and Nyquist plots may be used for system analysis and design. Since both methods present the same information, the choice of method will depend on objectives and preference of the investigator.

Although it is in many cases possible to predict the frequency response of a component or system using analytical techniques, it is often necessary to obtain or confirm results using experimental techniques. Procedures for experimental testing in most cases depend on the type of system under study. Common to all tests, however, will be a sine wave signal generator and at least two channels of recording equipment to record system input and output. The gain and phase angle at each frequency are determined from these records.

Frequency response data are commonly applied to amplifiers, microphones, loudspeakers, control components, and complete systems where dynamic considerations are important. Valuable information, concerning system dynamics can be obtained using this technique. For example, the frequency response of an audio amplifier usually drops off markedly at the low-frequency end and at the high-frequency end, the exact point depending upon the circuit and the values of various components used. The response of a high-fidelity amplifier (one which would reproduce without appreciable frequency discrimination) should cover most of the audio band, i.e., from about 20 to 15,000 Hz. However, this is much better than necessary except for the most refined work, and a range of 30 to 10,000 Hz is usually considered high fidelity. Modern microphones have responses which will cover this latter range, whereas some will extend over the entire audio band. Loudspeakers require the use of dual speakers or special resonant chambers, horns, etc., to cover this range.

The frequency response of video amplifiers used for television purposes extends without appreciable loss from 30 Hz to upwards of 3 MHz.

See also **Transfer Function.**

FREQUENCY RESPONSE MEASUREMENTS. Signal Generator.

FREQUENCY-SHIFT KEYING. Sometimes abbreviated FSK, that form of frequency modulation in which the modulating wave shifts the output frequency between predetermined values, and the output wave is coherent with no phase discontinuity.

FREQUENCY SHIFT THEOREM. Laplace Transform.

FREQUENCY SWING. In frequency modulation, the peak difference between the maximum and the minimum values of the instantaneous frequency.

FREQUENCY SYNTHESIZER. Signal Generator.

FREQUENCY TELEMETERING. Telemetering (Industrial).

FREQUENCY TRANSLATION. Often in communication circuits, it is desirable to transfer the group of frequencies in a particular frequency channel to another location in the frequency spectrum. The transfer process is frequency translation and is performed by beating the frequencies to be moved with the fixed frequency. The sum or difference of the original frequencies and the beating frequency then gives the same channel at another point in the spectrum (actually two new channels are created, one having sum frequencies and one containing the difference frequencies). By the proper choice of the beating frequency the new location can be fixed as desired. The process is used extensively in carrier telephony. It is also used in the doubling process of some frequency modulation systems, since the necessary doubling would raise the final frequency too high if it were not translated back to a lower value. Translation differs from doubling in that it does not bear a harmonic relationship with the original frequency and does not change the absolute width of the channel.

FREQUENCY UNITS. Units and Standards.

FRESNEL-ARAGON LAW. 1. Two rays of light polarized in the same plane interfere in the same manner as ordinary light. 2. Two rays polarized at right angles do not interfere. 3. Two rays polarized at right angles from ordinary light and brought into the same plane of polarization do not interfere in the ordinary sense. 4. Two rays polarized at right angles (obtained from plane polarized light) interfere when brought into the same plane of polarization.

FRESNEL DIFFRACTION. Two uses of this term are: 1. The radiation field transmitted through an aperture in an absorbing screen at distances large compared to the wavelength, and to the dimensions of the aperture, and yet small enough to require consideration of the effect of the phase-differences between secondary wavelets, even along the normal to the screen. (See **Fraunhofer Lines.**) 2. The diffraction effect obtained when either the source of the radiation or the observing screen or both are at a finite distance from the diffracting aperture or obstacle, that is, the wave fronts are spherical rather than plane as in the case of Fraunhofer diffraction. See also **Cornu Spiral; Diffraction.**

FRESNEL INTEGRAL. If the error function, for real values of the variable t, is separated into real and imaginary parts

$$(1 + i)^{-1}\,\mathrm{erf}(t) = C(t) - iS(t)$$

the resulting real functions are the Fresnel integrals

$$C(t) = \int_0^t \cos\theta x^2\,dx; \quad S(t) = \int_0^t \sin\theta x^2\,dx$$

where $\theta = \pi/2$. See also **Cornu Spiral.**

FRESNEL MIRROR. Two mirrors are inclined to each other at a small angle ϕ. A monochromatic light source in front of them will produce two virtual light sources and a suitably placed screen will show consecutive interference fringes at a linear distance apart *b*, as in the Young interference experiment. The wavelength of the incident light is given by $\lambda = 2b\phi$, provided that the incident light is made parallel by a lens between the source and the mirrors.

FRESNEL REGION. In antenna terminology, the region between the antenna and the Fraunhofer region. If the antenna has a well-defined aperture D in a given aspect, the Fresnel region in that aspect is commonly taken to extend a distance $2D^2/\lambda$ in that aspect, λ being the wavelength.

FRESNEL ZONE. Any one of the array of concentric surfaces in space between transmitter and receiver (or between radar antenna and target) over which the increase in distance over the straight line path is equal to some multiple of one-half wavelength. Also called *half-period zone.* Outside of rather unusual multipath transmission of radio energy in the free atmosphere, Fresnel zones are of importance primarily in studying the interference lobes produced by the interaction of a direct and a surface-reflected wave. Thus, for a given path, reflected radio energy arriving at the receiver from any point along any of the surface Fresnel zones will be some multiple of 180° out of phase with the direct wave, thereby producing destructive or constructive interference as the multiple is odd or even, respectively.

FRIABILITY. A term used in medicine to describe the appearance of blood and/or bleeding and particularly applicable to examinations of the gastrointestinal tract. After a surface has been swabbed clean, any presence of minor bleeding can be observed as, for example, by a sigmoidoscope. See also **Sigmoidoscopy.**

FRICTION (Air). Aerodynamics; Supersonic Aerodynamics.

FRICTIONAL ELECTRICITY. This familiar phenomenon is technically known as triboelectrification. When two dissimilar substances are rubbed together, they become oppositely electrified; and if either is an insulator, it retains a charge. For example, if glass is rubbed with silk, the glass becomes positive and the silk negative. Careful experiments make it appear probable that this is a type of contact potential difference, and that the friction serves only to bring about surface contact over a larger area. Accurately ground and polished disks of steel and glass, when pressed firmly together to ensure close contact and then separated, show the same effect, the glass again being positive. Various experimenters have arranged lists of substances in such order that when any two are pressed or rubbed together and then separated, the one higher in the list becomes positive with respect to the other; but the data have often been conflicting. Coehn concluded from his experiments that the potential difference of the charges developed by two contacting dielectrics is proportional to the difference between their dielectric constants, the one having the greater constant being positive.

FRICTION (Fluid). Fluid Friction.

FRICTION GEARING. A nonpositive form of power transmission for rotating shafts. In the usual forms, friction gears may consist of cylindrical wheels, for transmitting power between parallel-axis shafts, and beveled wheels, for transmitting power between shafts whose axes will intersect if produced.

If friction wheels are assumed to operate without slip, the surface speed of both wheels must be equal. The velocity ratio of a pair of wheels is therefore inversely proportional to their diameters, or, if a 4-inch wheel rotating at 300 rpm drives a 5-inch wheel, the larger wheel will rotate at 240 rpm. In friction wheel drives, the driven wheel should be made of the harder material for if slip occurs, the softer driving wheel will wear uniformly about its periphery, and will not cut grooves into the surface of the driven wheel.

The power H that may be transmitted by cylindrical friction wheels is given by

$$H = fFV/33{,}000$$

where f is the coefficient of friction between the wheel surfaces, F the contact pressure, and V the peripheral velocity, in feet per minute. Restrictions imposed by the ability of the friction wheel material to withstand crushing, and by the very heavy loads that the contact pressures impose on the shaft bearings, limit the use of friction gearing to light load transmission.

FRICTION LAYER. Winds and Air Movement.

FRICTION (Mechanical). The chief causes of friction are the interlocking of the minute irregularities on the rubbing surfaces, adhesion between the surfaces, and the identation of the softer by the harder body. Friction between solid bodies may be classified as sliding and rolling. The laws of sliding friction were investigated by Coulomb, who found that, approximately and within limits, (1) the friction between two surfaces is slightly greater just before motion begins than when the surfaces are in steady relative motion; (2) the friction is proportional to the force pressing the surfaces together; (3) it is independent of the area of contact, and (except at start) of the speed of relative motion. The constant ratio of the friction to the force pressing the surfaces together is called the coefficient of friction, some typical values of which are as follows:

Dry wood on dry wood	0.35
Leather on metal	0.55
Iron on stone	0.50
Wood on stone	0.40
Stone on stone or brick	0.65
Well-oiled metals	0.05

By means of such coefficients, it is possible to calculate what the friction will be between two bodies, as a wooden sill on a stone foundation, when the force pressing them together is given.

The angle at which a plane surface must be inclined for a solid block to slide steadily down it is the angle of friction; its tangent is the coefficient of friction between plane and block. Lubrication greatly reduces the coefficient by separating the solid surfaces. Rolling friction, due to the indentation of the surfaces in rolling contact, is much less than sliding friction, as illustrated by the use of ball-bearings. The viscosity of liquids and gases is sometimes called "internal friction."

FRICTION (Rolling). Rolling Friction.

FRICTION (Skin). Skin Friction.

FRICTION (Traction). Traction.

FRICTION WELDING. Welding.

FRICTION (Wind-Earth). Atmosphere (Earth); Winds and Air Movement.

FRIEDEL-CRAFTS REACTION. Aluminum chloride anhydrous, introduced by Friedel and Crafts, is used as reagent, generally in CS_2 solution to avoid rise in temperature, for the preparation of (1) arylalkyl hydrocarbons, (2) di- and triphenylmethane and derivatives, and (3) aryl-alkyl and diaryl ketones. Other chlorides, such as those of zinc, iron(III), and tin(IV), are often effective in certain cases.

1. Aryl-alkyl hydrocarbons. The reaction takes place between benzene or its homologues and the alkyl haloid, thus:

$$\underset{\text{Benzene}}{C_6H_5 \cdot H} + \underset{\text{Methyl chloride}}{Cl \cdot CH_3} \longrightarrow \underset{\text{Toluene}}{C_6H_5 \cdot CH_3} + \underset{\text{Hydrogen chloride gas evolved}}{HCl}$$

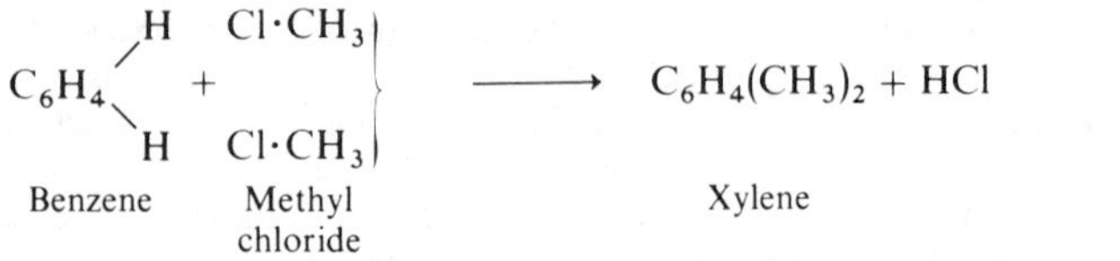

2. Di- and triphenylmethane, derivatives. The reaction takes place between benzene and benzyl haloid or methylene haloid in the case of diphenylmethane, and between benzene and benzal haloid or chloroform in the case of triphenylmethane, thus:

$$\underset{\text{Benzyl chloride}}{C_6H_5CH_2Cl} + \underset{\text{Benzene}}{HC_6H_5} \longrightarrow \underset{\text{Diphenylmethane}}{C_6H_5CH_2C_6H_5} + HCl$$

$$\underset{\text{Methylene chloride}}{H_2CCl_2} + \underset{\text{Benzene}}{\left.\begin{array}{l}HC_6H_5\\HC_6H_5\end{array}\right\}} \longrightarrow \underset{\text{Diphenylmethane}}{C_6H_5CH_2C_6H_5} + \begin{array}{l}HCl\\HCl\end{array}$$

$$\underset{\text{Benzal chloride}}{C_6H_5CHCl_2} + \underset{\text{Benzene}}{\left.\begin{array}{l}HC_6H_5\\HC_6H_5\end{array}\right\}} \longrightarrow \underset{\text{Triphenylmethane}}{C_6H_5CH\begin{array}{l}C_6H_5\\C_6H_5\end{array}} \begin{array}{c}HCl\\+\\HCl\end{array}$$

$$\underset{\text{Chloroform}}{HCCl_3} + \underset{\text{Benzene}}{\left.\begin{array}{l}HC_6H_5\\HC_6H_5\\HC_6H_5\end{array}\right\}} \longrightarrow \underset{\text{Triphenylmethane}}{C_6H_5CH\begin{array}{l}C_6H_5\\C_6H_5\end{array}} \begin{array}{c}HCl\\+HCl\\HCl\end{array}$$

3. Ketones. The reaction takes place between benzene and paraffin or benzenoid acyl haloid thus:

$$\underset{\text{Acetyl chloride}}{CH_3COCl} + \underset{\text{Benzene}}{HC_6H_5} \longrightarrow \underset{\text{Acetophenone}}{C_6H_5COCH_3} + HCl$$

$$\underset{\text{Benzoyl chloride}}{C_6H_5COCl} + \underset{\text{Benzene}}{HC_6H_5} \longrightarrow \underset{\text{Benzophenone}}{C_6H_5COC_6H_5} + HCl$$

The keto-group occupies the position para to alkyl already present. Two acyl groups have been placed in mesitylene to form diacetylmesitylene:

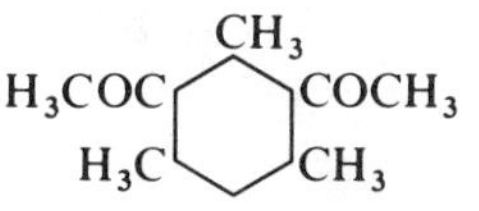

Summarizing: benzene or its homologues plus paraffin-substituted haloid in the presence of aluminum chloride anhydrous react with the elimination of hydrogen chloride. In several cases an intermediate compound of the reactants with aluminum chloride has been identified.

Other reactions involving aluminum chloride anhydrous are:

1. Xylene plus benzene to yield toluene, and the reverse, namely, toluene to yield xylene plus benzene. Boiling temperature.

2. Benzene, toluene and homologs chlorinated by reaction with chlorine gas.

3. Benzene sulfinated by reaction with SO_2. Benzene sulfinic acid $C_6H_5 \cdot SOOH$ formed.

FRIEDLÄNDER'S BACILLI. Respiratory System.

FRIGATE BIRD. Pelicans and Cormorants.

FRIGATE MACKERELS. Mackerels.

FRIGORIE. Refrigeration Cycle.

FRILLED SHARK. Sharks.

FRINGILLIDAE. A family of several hundred species of higher song birds. These birds usually have 12 tail feathers with 9 primaries. Most of them are gregarious, eat seeds, berries, fruits, and they migrate. They are normally found in warmer climates. The plumage ranges from dull and drab to brilliant and bright colors, often dependent upon the season of the year. The males tend to be much the more colorful. There are numerous well-known songbirds in this family. Some sing while in flight during migration.

The size and complexity of the family has required a breakdown into four subfamilies: (1) *Richmondeninae*, the cardinals, grosbeaks, dickcissels, and saltators; (2) *Emberizinae*, the ground finches, New World sparrows, Old World finches, and buntings; (3) *Geospizinae*, the Galapagos and Cocos Island finches; and (4) *Carduelinae*, the crossbills, northern grosbeaks, siskins, and canaries.

See also **Passeriformes.**

FRITILLARY (*Insecta, Lepidoptera*). A butterfly of the genus *Argynnis.* Most species are red-brown with black markings and are spotted beneath with silver.

FROG HOPPER (*Insecta, Homoptera*). Small jumping insects (*Cercopidae*) whose form faintly resembles that of the frogs. The immature insect sucks the sap of a plant and secretes about itself a protective frothy mass, hence they are also called spittle insects or spittle bugs.

FROGS AND TOADS. Of the class *Amphibia* (amphibians), subclass *Anuromorpha*, order *Anura* (anurans), according to the classification by Grzimek (1972). This order contains six suborders with seventeen families, 250 genera, and 2500 species. The terms *frog* and *toad* are general ones, based on superficial appearances, and do not indicate any phylogenetic relationships. Various anuran families contain species which look like tree toads, but are not related to them; toadlike species which are unrelated to toads; and species which look very much like true (ranid) frogs, but which are not even in the family *Ranidae.*

Membership in a particular anuran family cannot be determined by external appearance, body size, skin characteristics, pupil shape, or finger and toe form. Three different families contain very round, bulky anurans which look like they have been inflated. Some of them are so highly arboreal that they are almost never on the ground; others spend most of their lives burrowing beneath the earth; and still others never voluntarily leave water. Anurans from various families have adapted to rain forests, prairies, mountain streams, or to quiet ponds. In different families, highly specialized means of caring for the developing young have been developed independently, thus carrying the eggs on the back, or adhering egg mass to branches above a pool are found in different species. The latter adaptation permits the hatching larvae to fall into the water, where they will continue their development. Clearly, external appearance and life habits are not criteria for systematically arranging anuran species. The shape of the vertebrae is an especially important characteristic for systematic arrangement of the anurans. It is used to differentiate the six suborders. See accompanying table.

CLASSIFICATION OF ANURANS BY SHAPE OF THE VERTEBRAE

Amphicoela—The vertebrae are amphicoelous (biconcave), and the intermediate vertebral element is undivided, a primitive feature. There are two families: the leiopelmatids (*Leiopelmatidae*) and the ascaphids (*Ascaphidae*).

Aglossa—The vertebrae are opisthocoelous (concave only in the rear). No tongue. There is one family: *Pipidae.*

Opisthocoela—The vertebrae are opisthocoelous. Tongue is present. There are two families: the discoglossids (*Discoglossidae*) and the ophrynids (*Rhinophrynidae*).

Anomocoela—The vertebrae are either procoelous (concave in front) or amphicoelous, with free intermediate vertebral elements. There are two families: the spadefoot toads (*Pelobatidae*) and the pelodytids (*Pelodytidae*).

Diplasiocoela (the true frogs or ranids)—The presacral vertebrae are amphicoelous, while all the others are procoelous. This combination is termed diplasiocoelous. This suborder also contains families in which all the vertebrae are procoelous. There are four true frog families: True frogs (*Ranidae*), rhacophorids (*Rhacophoridae*), narrow-mouthed toads (*Microhylidae*), and phrynomerids (*Phrynomeridae*).

Procoela—These species have a uniformly procoelous vertebral column, occasionally with free intermediate elements. There are six families: Pseudids (*Pseudidae*), toads (*Bufonidae*), atelopodids (*Atelopodidae*), tree frogs (*Hylidae*), leptodactylids (*Leptodactylidae*), and centrolenids (*Centrolenidae*).

Only the most primitive frogs have ribs; they should not be confused with the often long processes of the vertebrae. The higher anurans, such as true frogs and toads, lack ribs. This enables one to touch the side of a toad and determine from the feel of its stomach whether or not it has just eaten well. Other features used in ordering anurans are found in the skeleton (such as structure of the pectoral girdle) and arrangement of muscles. Dentition can vary greatly among anurans, even within a single family. There are frogs with teeth in the upper jaw, and others without any teeth. Only one genus (*Amphignathodon*) has teeth in both jaws. The structure of the hands and feet are such a great reflection of the life habits of individual species that they can be used as disguishing characteristics. Fully developed or vestigial webbing may or may not be found between the toes or between both fingers and toes. The tips of the fingers and toes may be tapered or widened to form a T-shape, and they may or may not have adhering pads on their undersides.

For a few larger anuran groups, certain secondary sexual characteristics are important distinguishing features. Male true toads, for example have rudimentary ovaries. These have served as a classic illustration of the fact that in vertebrates, development into one sex rests in part on the suppression of the other sex. If the testicles of a male toad are removed, the Bidder's Organ eventually forms a functioning ovary.

In most anuran species, the female is somewhat or considerably larger than the male; she may be up to twice his size. Females generally call more softly and have a smaller repertoire of sounds than males; they may even be entirely unable to produce sound. Anurans are the first vertebrate order with a middle ear and an eardrum; they have a larynx and vocal cords which are acted upon by air coming from the lungs into the mouth. With few exceptions, all anuran males can amplify their sounds by using vocal sacs located in the floor of the mouth cavity. These sacs are resonators which can greatly magnify the volume of the sound the frogs produce. In some species the mating call can be heard up to one kilometer (0.6 mile) away.

Frogs do not have armor, but only a relatively thin layer of skin. Their body fluids evaporate directly through their skin, and a frog which is kept away from moisture can lose half of its weight through evaporation in a few hours. The humidity borderline at which a frog will still live depends upon the kind of adaptations the individual species has made to its environment. Spadefoot toads (*Scaphiopus*) of North America inhabit arid regions and spend almost their entire lives underground in order not to become dehydrated. These toads will survive a loss of up to about 49% of their body weight through evaporation. Terrestrial and arboreal species can usually tolerate losses of between 36 and 44% of body weigh. In contrast, the pig frog (*Rana grylio*), which is aquatic, cannot withstand a loss of more than 31% moisture. A hungry frog will not eat if it is too dry. Courting frogs will not mate, and even copulating frogs will separate if they enter dry conditions.

Yet in spite of their critical dependence upon temperature and humidity, the anurans have occupied a large variety of ecological niches; they are found in water, semi-deserts, on the equator and in the far north, on coasts and in mountains up to 3000 meters (9843 feet). They spawn in ice-cold water (0.5°C) and in hot springs where the temperature is 34°C. This occupation of various niches can only succeed if the individual species has developed the necessary adaptations for such living conditions. One species which has adapted to one habitat will perish if it is placed in another. The European ranid (*Rana temporaria*) lives as far north as the Arctic Circle and spawns in Alpine lakes which are still partially iced over. If the weather is too warm in the spring, it cannot lay eggs properly; the tadpoles of this species cannot tolerate too much warmth. The southern border of the distribution of this species is determined by the warmer climate. Toads also require a cold spring in order to have mature eggs and to initiate mating behavior. The winter cold is a kind of pacemaker by which the creatures measure the season. In contrast, the panther toad (*Bufo regularis*) from the Congo region of Africa measures its mating behavior by the rainy period. In the prairies of North America, rainfall is minimal. The spadefoot toads (*Scaphiopus*) found that they could not maintain such a short spawning season as would be necessary if their mating were regulated by rainfall. They must be prepared to mate throughout the year. If there is a strong rainfall, the spadefoot toads appear at the surface of the ground, collecting in puddles, and laying their eggs the same night. Frogs in equatorial regions where rainfall occurs regularly do not have a restricted breeding season.

Most anurans are twilight and nocturnal creatures; they leave their hidden recesses at dusk, based on a specific brightness level. Thus, European toads first appear when it is dark enough that the human eye can no longer perceive individual details. Other species come out somewhat earlier and a few may be active during the day. The tree frog first enters the water in the evening, but it does not completely avoid light. During the day it sleeps directly in the sunlight just above the water. While most adult anurans avoid sunlight, the young and freshly metamorphosed toads and frogs show an opposite behavior. Young British toads, European toads, frogs, and the young of several North American species leave their breeding waters by sunlight, and do not assume their nocturnal life for several days.

Amphibians have evolved from fresh water ancestors and they are adapted to the ionic concentration and osmotic pressure occurring in fresh water. Their eggs also require fresh water, whether in pools, puddles, or other spawning sites, in order to develop normally. The eggs of most anuran species perish if the water has the slightest amount of salt in it. However, there are a few species which can tolerate low salt concentrations. The Philippine frog (*Rana cancrivora*) from Manila Bay can tolerate a salinity level of about 2.6%.

The effect of air pressure fluctuations on the behavior of frogs has been established, but this is not well understood. It is possible that frogs utilize atmospheric pressure as an indication of ensuing unsettled conditions. Cold air masses, for example are often associated with high pressure zones and it is possible that the air pressure communicates a coming temperature drop to the frogs.

In seeking food, anurans react to moving objects and thus are heavily dependent upon their sense of vision. It has been proven that they also use their sense of smell (olfaction); if they are fed mealworm larvae in a terrarium, they will also react to the ground on which these larvae were raised. Movement of prey is highly important to anurans. Most anurans, excepting the South American neotropical toad (*Bufo marinus*) will not even react to their favorite food if it does not move. The European frog (*Rana esculenta*) has bulging eyes which give it a field of vision which is almost 360 degrees. See accompanying figure. If the frog has spotted a mealworm, it turns its head (or, if necessary, its body) so that it is facing directly at the prey. Then it extends its tongue toward the prey; the worm sticks to the tongue, and the tongue is then rolled back inside the mouth. The mealworm is swallowed without chewing. Bugs, earthworms, young mice, and other larger prey are seized with the jaws; sometimes the hands may help to push the resisting large prey into the mouth. The European toad pulls earthworms between its fingers, ridding them of the largest dirt particles. After the prey is swallowed, the toad removes dirt from the edges of its mouth, using its hands. Excepting specialists which feed just on termites, anurans are not very selective within the framework of their prey configuration scheme. However, *Rana temporaria* prefers snails over ants and bugs, while the European toad prefers insects. Tree frogs also prefer flies over worms. Frogs and toads can learn from bad experiences and they will avoid wasps and bees. From the frog's viewpoint, all living organisms around it fall into three groups—anything smaller can be eaten; creatures of the same size are usually to be clasped during the breeding season, or less often, fought; and creatures that are considerably larger than the frog should be avoided. However, nearly all frogs will not flee from a larger creature unless the latter shows movement. This explains the "cannibalism" which is often seen in anurans. The frog will feed on freshly metamorphosed young which appear at the surface of the water, or any of its own young which correspond to the prey configuration scheme.

Frog. (*A. M. Winchester.*)

When several species occur at the same breeding site, occasional interspecific matings occur, in spite of the presence of various isolating mechanisms. This can be caused when a female of one species can not differentiate the call of a male of another species from that of an appropriate mate. Also, a male may clasp and copulate with a female of another species. Unlike mating, which is relatively uniform throughout the anuran families, egg laying is not carried out in any one way. In fact, not all anurans lay eggs; *Nectophrynoides* toads bear fully developed young. In other species the male usually takes part in spawning, fertilizing the eggs when they are released from the female's cloaca. One notable exception to this is the tailed frog (*Ascaphus truei*) in which the female lays eggs which have been internally fertilized months before. Other females assume a specific position to signal to the male, sitting on the female's back, that she is about to lay eggs. This position causes the male to release his sperm.

There is a great variety in the number of eggs laid and the size of the eggs, as well as the time interval in which eggs are laid and the instinctive behavioral patterns associated with egg laying. Initially, the freshly hatched larvae hang firmly onto the egg envelopes and nearby plant parts, using the adhering fibers which are near the future mouth; they remain motionless there for a fairly long time. The little motion they do initiate is caused by cilia on their skin. During this stage of development, the larvae breathe through external gills. However, a membrane soon grows over the gills, leaving just a single hole on the left side of the body as a means of releasing water from the body. The eyes become functional; the mouth and anal openings break through; and the tail develops considerably and has a high crest. The jaws become hardened and form a beaklike structure, and small rows of horny ridges form around the mouth opening. The number of continous rows of these horny ridges above and below the "beak" is an important characteristic used in identifying species membership of many tadpoles.

The larvae finally develop into free-swimming tadpoles, which feed on algae, plant parts, and bits of animal matter. Respiratory water is taken in through the mouth and released through the gill opening; the water moves through a filter apparatus which enables the tadpole to take up bits of food floating in the water. Toward the end of the larval development period, the hind legs gradually develop, while the fore legs do not appear from the gill pockets until just before completion of metamorphosis. Meanwhile the lungs have developed and the tadpoles frequently come to the surface to breathe. The internal gills and the tail degenerate and the horny jaws are cast off; the narrow mouth opening of the tadpole widens to form the broad frog mouth. The metamorphosing tadpole pushes partway on land, and the intestine becomes shorter, an adjustment made to the change from a plant to an animal diet. However, during this initial period on land the tadpole does not feed at all. When the tail has become reduced to just a stump, the small (15 millimeter) as in the case of the *R. temporaria* frog, or the larger *R. esculenta*, leaves the water for the first time. Metamorphosis is complete.

Anurans molt at regular intervals. They undergo a series of ritualized movements in removing their old skin, pushing it off in one piece and eating it all. A few hours before the onset of molting, a fluid is secreted by the mucus gland beneath the outermost dead horny skin layer, pushing it away from the new layer of skin. Then, numerous perforations form along certain seams, and the skin divides along these seams.

Anurans in deep rest in winter or when asleep in dry seasons are difficult to arouse, even with fairly strong stimulation. However, the short periods of sleep, lasting just a few hours, may be ended immediately if an enemy or desirable prey appears. Like birds and mammals, anurans usually have specific sleeping locations and also assume a characteristic sleep posture. Anurans have some behavioral patterns which are not seen so frequently and which may only be elicited in conflict situations. Frogs will yawn, but this is not related to being sleepy. They apparently yawn when frustrated.

References

Heusser, H. R.: "Frogs and Toads," "Lower Anurans," and "Higher Anurans," in "Grzimek's Animal Life Encyclopedia," Vol. 5, Van Nostrand Reinhold, New York, 1972.

Goin, C. J.: "Amphibia," in "The Encyclopedia of the Biological Sciences," (P. Gray, editor), Van Nostrand Reinhold, New York, 1970.

Inger, R. F.: "The Development of a Phylogeny of Frogs," *Evolution*, **21**, 369–384 (1955).

McCoy, C. J.: "Anura," in "The Encyclopedia of the Biological Sciences," (P. Gray, editor), Van Nostrand Reinhold, New York, 1970.

Orton, G. L.: "The Bearing of Larval Evolution on Some Problems of Frog Classification," *Syst. Zool.*, **4**, 105–119 (1955).

Tihen, J. A.: "Evolutionary Trends in Frogs," *Amer. Zool.*, **5**, 309–318 (1965).

FRÖHLICH'S SYNDROME. **Pituitary Gland.**

FRONTAL BONE. Bones of the vertebrate skull lying between and behind the eyes, usually paired but in most human skulls united to form the single large bone of the forehead.

FRONT FOCAL LENGTH. The distance from the primary focal point to the front vertex of a lens. Its reciprocal is called the neutralizing power of the lens.

FRONTAL FOG. **Fog and Fog Clearing.**

FRONTAL LOBE. **Brain and Nervous System.**

FRONTAL WAVE. **Atmosphere (Earth).**

FRONTS AND STORMS. In meteorology, generally, a front is the interface or transition zone between two air masses of different density. Since the temperature distribution is the most important regulator of atmospheric density, a front almost invariably separates air masses of different temperature. Along with the basic density criterion and the common temperature criterion, many other features may distinguish a front, such as pressure troughs, a change in wind direction, a moisture discontinuity, and certain characteristic cloud and precipitation forms.

The term front is used ambiguously for: (a) *frontal zone*, the three-dimensional zone or layer or large horizontal density gradient, bounded by (b) *frontal surfaces*, across which the horizontal density gradient is discontinuous (frontal surface usually refers specifically to the warmer side of the frontal zone); and (c) *surface front*, the line of intersection of a frontal surface or frontal zone with the earth's surface or, less frequently, with a specified constant-pressure surface. The formation or intensification of a frontal surface is called *frontogenesis*. The disintegration or weakening of a frontal surface is called *frontolysis*.

The equilibrium slope of a front is given by the relation:

$$\tan \alpha = \frac{2\omega \sin \phi}{g} T_m \left[\frac{v_2 - v_1}{T_2 - T_1} \right]$$

where $\tan \alpha$ = tangent of α, the angle of slope of the frontal surface to sea level
ω = angular velocity of the earth
ϕ = latitude angle
T_m = mean temperature of the air masses
T_1, T_2 = temperature of each air mass, respectively
v_1, v_2 = wind velocity parallel to the frontal surface in each air mass, respectively

From this relation, it is apparent that steep equilibrium frontal slopes result from high mean temperatures, high latitudes, large differences in the velocity components, and small differences in the temperatures between the two masses. The slope of a front is always such that cold air underlies warm.

The world's major front in terms of air mass contrast and susceptibility to cyclonic disturbance is the *polar front*. According to the polar-front theory, this is a semi-permanent, semi-continuous front separating air masses of tropical and polar origin. The semi-continuous, semi-permanent *arctic front* separates the deep, cold arctic air and the shallower, basically less cold polar air of the northern latitudes; it is comparable to the *antarctic front*, which separates the antarctic air and the polar air of the southern oceans.

Cold Fronts. When cold air on one side of a front is replacing warm air horizontally, the front is called a *cold front*. At the approach, passage and recession of a cold front, the following phenomena commonly occur in order:

(1) *Sky and weather*: Increasing cloudiness of altocumulus, cumulus, or stratocumulus; overcast with heavy showers; clearing to scattered cumulus.

(2) *Temperature and wind*: Wind from southerly or westerly direction (northerly and westerly in the Southern Hemisphere), with temperature steady or rising slightly; wind shifting sometimes with violent gusts and high velocity from the southwesterly to the northwesterly (northwesterly to southwesterly in the Southern Hemisphere); and

temperature starting to drop; wind fresh from the new direction and temperature dropping steadily.

(3) *Pressure*: Barometer falling prior to wind shift, then rising rapidly after shift and continuing to rise.

Warm Fronts. When warm air on one side of a front is replacing cold air horizontally and overriding the cold air, the front is called a *warm front.* Over the average warm frontal surface, a wide band of clouds and precipitation is present. Precipitation fog is also likely to form over a considerable area. At the approach, passage, and recession of a warm front, the following phenomena usually occur in order:

(1) *Sky and weather*: Increasing cloudiness, first cirrus, then cirrostratus, altostratus, rain, or snow; stratus, nimbostratus, scud, sometimes fog for a considerable period of time; clearing to scattered or broken stratocumulus and cumulus.

(2) *Temperature and wind*: Wind from some easterly quadrant, and temperature steady and relatively low; wind-shift at front passage from easterly quadrant to southerly quadrant (easterly quadrant to northerly quadrant in the Southern Hemisphere) and considerable temperature rise; more or less steady southerly winds (northerly in the Southern Hemisphere) with much higher temperatures.

(3) *Pressure*: Barometer falling until wind shifts, then steady or falling very slightly.

Other Characteristics of Fronts. Cold-front slopes are greater than, and warm-front slopes are less than, the equilibrium slope. Cold fronts move with relatively steady velocity, although subject to some acceleration. Warm frontal movement is commonly erratic, but generally slower than cold fronts.

Fronts are further distinguished in the following ways: When warm air is ascending the frontal surface up to high altitudes, the front is an *anafront*; when descending the frontal surface (usually at a cold front), a *katafront.* An *upper front* is present in the upper air (i.e., generally, above 850 millibars) but does not extend to the ground. A *secondary front* is one that may form within a baroclinic cold air mass that is, itself, separated from a warm air mass by a primary frontal system. A *quasi-stationary front* (commonly called a *stationary front*) is one that is stationary or nearly so, moving at a speed of less than about five knots. This front has approximately the equilibrium slope. In synoptic chart analysis, a quasi-stationary front is one that has not moved appreciably from its position on the last previous synoptic chart (three or six hours before).

An *occluded front* is a composite of two fronts, formed as a cold front overtakes a warm front or a quasi-stationary front. A cold occlusion results when the coldest air is behind the cold front. In this case, the cold front undercuts the warm front and, at the earth's surface, coldest air replaces less-cold air. A warm occlusion is formed when the coldest air lies ahead of the warm front, and when the cold front is forced aloft at the warm-front surface. A neutral occlusion results when there is no appreciable temperature difference between the cold air masses of the cold and warm fronts. In this case, frontal characteristics at the earth's surface consist mainly of a pressure trough, a wind-shift line, and a band of cloudiness and precipitation.

The extent to which frontal theory is to be modified and the nature of the modifications are, as yet, very controversial questions. For example, a front known as the *intertropical front* (or *equatorial* or *tropical front*) is presumed to exist within the equatorial trough separating the air of the Northern and Southern Hemispheres. It has been generally agreed that this front cannot be explained in the same terms as the fronts of higher latitudes.

Extratropical Cyclone

Cyclonic-scale storms that are not tropical cyclones are referred to as extratropical cyclones. During development, youth, and maturity, these storms are composed of definite parts including warm and cold fronts and warm and cold sectors. Temperate zone cyclones usually involve two air masses. That part occupied by cold air is known as the cold sector; this constitutes more than one-half and often practically all the cyclone area. The smaller part occupied by the warmer air mass is known as the warm sector. Beyond maturity, the extratropical cyclones approach vortex structure. See **Atmosphere (Earth).**

The most frequent form of extratropical cyclone is the wave cyclone; nearly all storms of the temperature zone are of this type. Wave cyclones form and move along a front, and are named for the wave-like deformation of the front produced by the circulation about the cyclone center.

Wave Cyclone

Cyclones that develop in the temperate zone are waves on frontal surfaces in contrast with tropical cyclones in which fronts are not prominent features. Cyclones develop more frequently along a polar front than any other place, but they also appear on the intertropical front and on other minor fronts. It is on the polar front, however, that wave cyclones develop into major storms and climax their lives as large-scale vortices. With reference to Fig. 1, wave cyclone development occurs in the following sequence:

(1) A fairly prominent front exists between two air masses of some density contrast. The frontal slope is approximately in equilibrium.

(2) A perturbation of wave form, with the major motion nearly horizontal rather than vertical (as in gravity waves), develops on the frontal surface. One part of the front is bent toward the cold air as a warm front, and an adjacent part is bent from equilibrium toward the warm air as a cold front. A deep indentation into the cold sector (cold air mass) occurs with the crest of the wave at the maximum point of indentation.

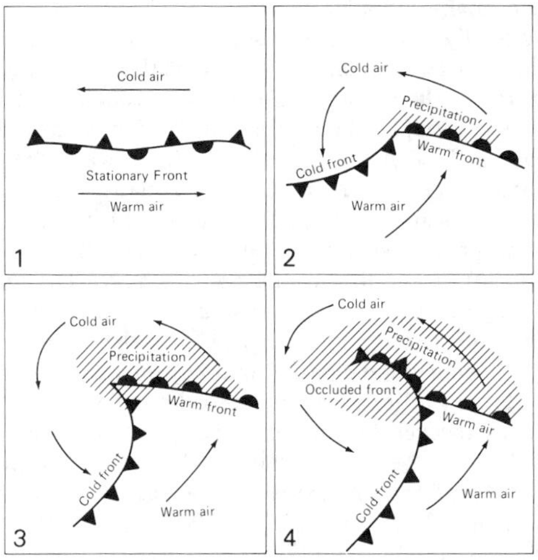

Fig. 1. History and development of an extratropical cyclone, forming an occlusion.

(3) The wave moves along the frontal zone, but while it moves it also begins to occlude, i.e., the warm sector begins to close because the cold front moves more rapidly than the warm front. Pressure falls about the center, but most rapidly in the direction toward which the crest of the wave is moving.

(4) As the cold front overtakes the warm front, occlusion continues and the warm-sector air is squeezed aloft between the frontal surfaces.

After the fronts have occluded and the warm sector is gone, only an approximate vortex remains, which slowly loses intensity. Typical warm and cold frontal weather are associated with both fronts, respectively. In the fall, winter, and spring months of the year, most of the rain and snow falling over the United States and Canada is due to wave cyclones. Movement of the cyclones is generally easterly with a north or south component, but it should be remembered that wave cyclones tend to move along the frontal zone in which they developed. Rate of movement varies from less than 100 miles (161 kilometers) per day to as much as 1,000 miles (1,609 kilometers); on the average, about 400–500 miles (644–805 kilometers) in summer and 600–700 miles (965–1126 kilometers) in the winter. Velocity of the wave crest (center of the cyclone) is greatest during the early stages of the storm and normally decreases as it approaches maturity. Wave cyclones often occur in a series with a definite wavelength, thus having a uniform spacing. Under the influence of a well-established major circulation pattern, the paths of cyclones can be predicted and the general aspect of weather estimated for several days. With transient general circulation patterns, however, the paths of cyclones become quite variable and accurate prediction from one group of cyclones to the next becomes next to impossible.

The history and development of an extratropical cyclone in the Southern Hemisphere is illustrated in Fig. 2.

Instability Line

The general term for any nonfrontal line or band of convective activity in the atmosphere is *instability line.* This includes the developing, mature, and dissipating stages. Earlier, *squall line,* instead of instability line, was the overall term used. In current practical usage, the term squall line may be applied to the mature stage of the convective activity (consisting of a line of active thunderstorms), while the term instability line may refer only to the less active phases.

Instability lines are usually hundreds of miles long (not necessarily continuous), 10 to 50 miles (16 to 80 kilometers) wide, and are most often formed in the warm sectors of wave cyclones. Unlike true fronts, they are transitory in character, ordinarily developing to maximum intensity in less than 12 hours and then dissipating in about the same time. Maximum intensity is usually attained in late afternoon.

Instability lines are of two types: (1) cold-front squall lines, composed of numerous thunderstorms, squalls, or showers, occur at the cold front due to the cold air behind the front underrunning and lifting the warm air ahead of the front. (2) Prefrontal squall lines originate from a combination of several factors: (a) a tongue of moist air in advance of and parallel to a cold front; (b) strong southerly winds at low levels ahead of the cold front, carrying warm air northward at a rapid rate; and westerly winds at high levels, carrying comparatively cold air over the northward-moving warm air, thus creating instability; (c) lifting of the air well in advance of the cold front, because of its fluid properties, and because of convergence in the warm sector of a cyclone. These conditions are most often satisfied in the warm sector of a temperate-zone cyclone just ahead of the cold front.

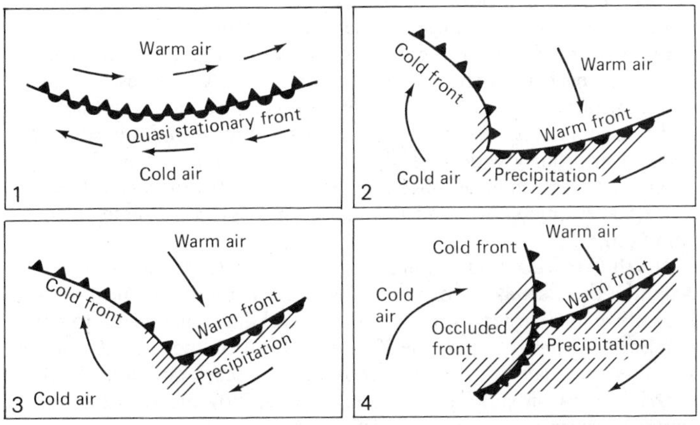

Fig. 2. History and development of an extratropical cyclone in the Southern Hemisphere.

Prefrontal instability lines may begin with a few scattered showers or thunderstorms, but develop rapidly into a solid line of showers or thunderstorms located from 50–200 (80–322 kilometers) miles ahead of the cold front. Both prefrontal and cold-front instability lines, once developed, travel perpendicular to their leading edge and with the cold front. Prefrontal instability lines are usually the more violent and destructive of the two, often being accompanied by tornadoes, hail storms, gusty surface winds, and severe turbulence.

Law of Storms

Historically, the general statement of (a) the manner in which the winds of a cyclone rotate about the cyclone's center; and (b) the way that the entire disturbance moves over the earth's surface is known as the Law of Storms. The formulation of this "law" was due largely to the investigations of Brandes (1826), Dove (1828), and Redfield (1831). This knowledge of the general behavior of storms led to the issuance of rules for seamen, instructing them in means of navigation to avoid the dangers of storms at sea.

Storms of the Middle Latitudes

Major storms of the middle latitudes are associated with one or more of three features:

(1) A large and vigorous area and center of low pressure.

(2) An abnormally large pressure gradient, i.e., large difference in air pressure horizontally over a large region.

(3) A vigorous cold front accompanied by strong winds, a substantial shift in wind direction and precipitation usually coming from cumulonimbus clouds.

Tornadoes and violent thunderstorms, although often devastating, are small in size insofar as the affected area.

Low-Pressure Storms. The speed of the wind blowing around a center of low pressure can, theoretically, increase without limit because the forces associated with the pressure gradient and the centripetal acceleration are opposing each other. However, friction along the earth's surface causes an inflow of air toward the eye of the storm which thereby places a limit on the speeds that, in reality, are attained. Convergent flow also can contribute to the inflow of air. Notwithstanding, very strong winds do occur in middle-latitude cyclonic storms and these winds extend over a large area.

In the North American region (United States and Canada), the most frequent track for middle-latitude storms of low pressure areas is along the east coast of the continent. Storms begin in the general area of the coastal waters of the Carolinas. Then they grow to cover an ever-increasing area, and move toward the northeast with increasing wind speeds and decreasing pressure at the eye of the storm. These storms usually reach their maximum intensity in the area of eastern Canada, or over associated coastal waters. Wind speeds of 75 knots are not uncommon and winds of over 50 knots frequently cover a very large area. The precipitation shield (snow in winter; rain in summer; mixed rain and snow in spring and fall) may extend to a radius of 1,000 miles (1609 kilometers) from the eye of the storm. These storms are most frequent in winter and spring when they cause snowfall upwards of 2 feet (60.9 centimeters). (1 knot = 1.8532 kilometers/hour)

A second region of middle-latitude storms associated with low-pressure centers, especially in winter, is the west coast—from northern California, along British Columbia, to Alaska. Rain in the lower coastal plains and snow in the mountains is frequently equivalent to 2 to 3 inches (5.1 to 7.6 centimeters) of rainfall. Winds along the coastal areas reach a velocity of 100 knots in gusts.

A third region where middle-latitude low-pressure storms develop, but less frequently, is the flat plains area between the Rocky Mountains and the Appalachians. Storms in this region are primarily winter and spring phenomena. They can cause widespread paralysis of travel because of deep snow and large snow drifts. These storms usually begin in the southern states and move northeast or north-northeast into Canada where they eventually dissipate.

The energy required to feed these large low-pressure storms is derived mainly from the contrast of temperature in the tropical air mass and the polar airmass that lie adjacent to each other across a very pronounced frontal surface. In addition, energy is dervied from the condensation of large amounts of precipitation. Convective instability in the tropical air mass also contributes. Initially, the middle-latitude storm of low pressure begins as a minor disturbance on a pronounced front separating the polar and tropical air masses. When the distribution of potential energy that is needed to drive the storm, the location of the jet stream, and the advection of vorticity into the storm area are all in a favorable state, the initial disturbance grows rapidly and within the span of a day a very large and violent storm may be in progress. The storm usually reaches peak strength at the end of the second day, after which it decays slowly.

Storms from Abnormally Large Pressure Differences. Strong winds not associated with low-pressure systems can be considered as middle-latitude storms. These strong winds arise from the presence of an abnormally large pressure difference, horizontally, over relatively short distances. Actually the sea level pressure can be above normal at the same time an abnormally large pressure gradient exists in that region. Excessive winds always accompany large pressure gradients and these winds may be unidirectional, i.e., not curved as they are about a low-pressure eye. Areas of abnormally large pressure gradients and strong winds frequently are elongated, being relatively narrow, but long downwind. This shape can create a large fetch over open water and, in this case, onshore winds pile up water in coastal areas far inland from the position of usual high tide. This condition of strong corridors of onshore winds is relatively common on the west coasts of North America and Europe; and on the North Sea.

A *blizzard* is a severe weather condition characterized by low temperatures, very strong winds that carry either or both falling and drifting snow. Blizzard conditions are considered as present when the wind exceeds 30 knots, the temperature is well below normal, and visibility is reduced to $\frac{1}{8}$ mile ($\frac{1}{5}$ kilometer). A severe blizzard condition exists when the winds exceed 40 knots, the temperature is below 10°F (−12.2°C), and visibility is essentially zero in driving and drifting snow.

Vigorous Cold Fronts. These also are storms of the middle latitudes although they are of short duration at any one location. Along a vigorous cold front, there are strong winds in both air masses, a sharp shift in wind direction along and during the frontal passage. A large temperature drop occurs. Thunderstorms, heavy showers, and a zone of substantial precipitation frequently extend for 500 to 1,000 miles (805 to 1,609 kilometers) along the front. Duration of storm conditions is from less than an hour to a half-day. Damage inflicted by a vigorous cold front is caused mainly by the strong gusty winds and the sudden wind shift. Hail in thunderstorms, although not common, does occur.

Flash Flooding. This is also a storm condition in the middle latitudes. In the tropics and subtropics, tropical storms usually are accompanied by heavy rain and rainfall in excess of 1 inch (2.5 centimeter) is more or less expected. This is not the case in the middle latitudes where most rain-producing conditions generate less than 1 inch (2.5 centimeter) of rainfall. Flash flooding occurs when rain-producing cloud systems cause rainfall rates well above normal and the rain continues to fall at these excessive rates for several hours to a half-day or more. Occasionally, excessive rainfall continues for several days. The total ranifall, which is far above normal over relatively small areas, cannot be accommodated by runoff and absorption of the ground. Consequently, water accumulates rapidly, causing flooding. The suddenness of rising waters accounts for the term *flash flood.*

Flash flooding usually occurs under stationary or very slow moving clusters of showers and thunderstorms, or is caused by stationary or slowly-moving low-pressure storms with an unabated inflow of air of high water content.

Storms of the Tropics

Storms of the tropics are usually devoid of fronts that are so characteristic of middle- and northern-latitude storms. Tropical storms derive their energy from convective instability and condensation of copious quantities of water that produces abundant rain. Lesser tropical storms do not have a closed circulation, as do the larger storms. The lesser storms tend to be compact in area and move westward in both the Northern and Southern Hemispheres. Because they occur primarily between 30°N and 30°S, the Coriolis force is a minor factor in their winds. Instead, the winds are related to the pressure gradient and centripetal acceleration as in a large vortex.

The generally-accepted classification of tropical cyclones includes: (1) *Tropical disturbances,* which are a common phenomenon in the tropics and subtropics. Winds are not strong; rain may be heavy; there is no closed circulation about a storm eye. (2) *Tropical depressions,* which are storms with winds up to about 30 knots. There is usually a minor eye and closed circulation. Rain is usually heavy. (3) *Tropical storms,* which have winds between 30 knots and about 60 knots. There is a recognizable center, i.e., an eye about which there is a closed circulation. Rain is heavy to torrential. (4) *Hurricanes and typhoons,* which are large and violent storms with winds in excess of 65 knots. Winds sometimes reach 175 knots. There is a pronounced eye and the circulation is a large whirl about the eye. Rain is torrential.

Tropical cyclones begin over warm ocean waters. They travel westward as long as they stay nearer than 30° to the equator in either the Northern or Southern Hemisphere. After they pass 30°, they often begin a poleward drift and, if they reach 40°, they begin to move toward the east. The season and region of origination are varied.

Some general observations pertaining to storms in the tropics include:

(1) Over the tropical portion of the eastern North Atlantic, the period of incubation is August and September.

(2) In the Caribbean and Gulf of Mexico regions, the season is June through November.

(3) Off the west coast of Mexico, the storms are present from June through November.

(4) Over the tropical portions of the western North Pacific, the season goes from May through December.

(5) The tropical portion of the western South Pacific spawns storms from December through April.

(6) The storm season of the south Indian Ocean is from November through April.

(7) In the north Indian Ocean, cyclones are present from April through December.

Hurricane. A hurricane is a severe tropical storm in the North Atlantic Ocean, Caribbean Sea, Gulf of Mexico, and in the Eastern North Pacific of the west coast of Mexico. By international agreement, tropical cyclones are classified as hurricanes if they have winds of 65 knots or higher. A *typhoon* is distinguished from a hurricane only in that it occurs in the western Pacific Ocean.

Surface pressure in a hurricane is very low at the center of eye of the storm, but rises rapidly outward toward the periphery. Because of the large pressure gradient, winds are of high velocity, blowing counterclockwise in the Northern Hemisphere and clockwise south of the equator. Velocities of 100 mph (160 kph) are relatively common near the storm's center, and 50–75 mph (80–121 kph), over a wide area can be expected. It is probable that velocities of 175 mph (282 kph) and perhaps more, have occurred in some storms. Mountainous waves and confused seas result from such winds. The quadrant of the storm where the velocity of wind and the velocity of the storm are additive is the quadrant where the greatest wind velocity is present, usually the north or west quadrant in the northern hemisphere and the south or east quadrant in the southern hemisphere. Only in the center of the storm (the eye) is there little or no wind, although the sea remains confused. In the dangerous quadrant, a hurricane moving toward a coast line drives before it a rising sea known as the storm wave, or hurricane wave, which often lifts ocen water to great heights, along coastal areas. Great depth of clouds and torrential rain normally occur over a fairly wide area about the storm center. Rainfall locally amounts to as much as 40–50 inches (100–120 centimeters); a 20-inch (50-centimeter) rainfall during the passage of a single storm is common.

North American hurricanes originate near the equator over the Atlantic, the Caribbean, or Gulf of Mexico. From the point of origin, they move in a westerly direction for several days, then begin to curve northward. After they pass the 30th latitude north, they usually assume a path somewhat east of north and begin to accelerate considerably. Hurricane season is the period from June to November, but an occasional hurricane may occur out of season.

As of the early 1980s, in the United States, National Weather Service forecasters issue warnings for hurricanes and severe thunderstorms. Tornado warnings are described later. A Hurricane Watch means that an existing hurricane poses a threat to coastal and inland communities in the area specified by the Watch. A Hurricane Warning means hurricane force winds and/or dangerously high water and exceptionally high waves are expected in a specific coastal area within 24 hours.

Tropical Cyclone (Indian Ocean Area). At maturity, this is one of the most intense and feared storms of the world; winds exceeding 175 knots (200 mph; 322 kph) have been measured, and its rains are torrential. Fully mature tropical cyclones range in size from 60 to well over 1,000 miles (97–1,609 kilometers) in diameter. The surface winds spiral inward cyclonically, becoming more nearly circular near the center. The wind field pattern is that of a circularly symmetric spiral added to a straight current in the direction of propagation of the cyclone. The winds do not converge toward a point, but become, ultimately, roughly tangent to a circle bounding the eye of the storm.

Willy-Willy is a term sometimes used for a severe tropical cyclone in the Australian area.

More Localized Storm Phenomena

Thus far in this summary, wide-area type storm conditions have been described. However, as mentioned, these larger storms are often accompanied by varied, more localized storm situations.

Thunderstorm. In general, a local storm invariably produced by a cumulonimbus cloud and always accompanied by lightning and thunder, usually by strong gusts of wind, heavy rain, sometimes by hail, and occasionally by tornadoes. A strong convective updraft is a distinguishing feature of a thunderstorm in its early phases, while a strong downdraft in a column of precipitation marks its dissipating stages. It is usually of short duration, seldom over 2 hours for any one storm.

Thunderstorms are a consequence of atmospheric instability, and constitute, loosely, an overturning of air layers in order to achieve a more stable density stratification. They originate as a result of the occurrence of three factors, the absence of any one of which will not permit their formation: (1) sufficient number of vertical perturbations in the air to cause parcels of air to start upward; (2) sufficiently unstable air to permit the vertical perturbations to rise with accelerated velocity up to great enough heights to ensure penetration of the freezing level; (3) sufficient moisture to ensure the development of cloud droplets and, eventually, snow and rain in the rising parcels of air.

The first requirement may be fulfilled by mechanical lifting of air over terrain or up a frontal surface, or by heating of the surface air over warm ground or water. The second requirement may be fulfilled by one of four processes in the atmosphere: (1) heating from below, i.e., along the ground or the water; (2) cooling from above, i.e., advection of cold air at high altitude; (3) simultaneous heating from below and cooling from above; (4) differential cooling of a layer of air in such a manner that the top grows colder than the bottom.

Necessary moisture is not always present in the atmosphere to assure thunderstorm development, even though the other two requirements are met. Moisture content of the air must be such that saturation can be caused by the two other requirements, and its distribution must be such that lower levels reach saturation first. Because moisture is distributed horizontally by large-scale air currents in tongues and islands in the atmosphere, the distribution of thunderstorms often assumes these patterns.

Over land, heating from below normally occurs during daylight hours; over water, whenever cold air flows from cold continents over warm water, or from cold water over warmer water. Thermal or air-mass thunderstorms of this type are at a maximum over land in the late afternoon, and are restricted almost entirely to the spring and summer of the year. Thermal or air-mass thunderstorms occur over oceans at any time of day or night and generally are at a maximum during winter and spring. Air-mass thunderstorms are isolated and distributed at random.

Cooling from above by nocturnal radiation from the moist air takes place over ocean areas where the surface water temperature is maintained virtually constant. This creates instability, and results in early morning thunderstorms at sea.

Cooling aloft may also occur by advection of cold air over a layer of warm air. Thunderstorms of this type are usually high-level, with bases one or two miles above the earth. They occur over both land and water. In North America, they are common to the region just east of the Rocky Mountains during summer.

On some occasions, both heating from below and cooling from above occur simultaneously. Resulting thunderstorms may lie as low as a few thousand feet or as high as several miles. Heating from below may arise from any of the processes that raise the temperature of the lower levels of the atmosphere, but high-level cooling is due to advection of cold air.

Differential cooling of an air layer occurs when an entire layer of air is lifted. If the air remains unsaturated, the top will cool slightly more rapidly than the base. If the layer remains unsaturated at its top but becomes saturated at its base, lifting will cool the top rapidly but the base only slowly, causing a rapid trend toward instability. Lifting occurs along cold and warm fronts, along mountains, and at a relatively slow rate in air that is undergoing convergence.

Lightning. The electrical condition of the earth's surface and of the atmosphere in stormy weather is quite different from its normal, fair weather state. Over a level stretch of country in fine weather, there is distributed a negative surface charge estimated at about 0.00027 electrostatic units per square centimeter, or 0.0014 coulomb per square mile. Above this, the electric potential of the atmosphere increases with elevation at the rate of about 100 volts per meter, the upper atmosphere being, apparently, positively charged. The earth, the atmosphere, and the ionosphere thus form a vast condenser, through the dielectric of which there is constant leakage because of ionization. That which maintains the charges against this leakage is not well understood.

In a rapidly developing cumulonimbus cloud, the top of the cloud becomes positively charged and the area near the freezing level becomes negative. Electrification does not occur to any great extent until ice is present in one of its many forms found in thunderstorms (mostly hail). At a temperature just less than freezing, ice particles coated with liquid water set up a fairly strong dipole, with the negative charge inside. Differential falling rates and updrafts strip the water shell off the ice pellet, leaving the ice negative and the small droplets, which are carried upward, positive. The cloud thus acts as a huge static machine, with drops as carriers, which operates until the electric stress becomes so great that it causes a discharge of lightning between the charged surfaces of the same cloud, or between two clouds, or between a cloud and the induced charge on the earth under it. These activities, of course, greatly modify the distribution of charge and potential in the surrounding area, changes that can be detected by electrometers suitably placed.

It has been estimated that over the entire earth the frequency of lightning averages about 100 flashes every second, and that this rate of discharge represents something like 4,000,000,000 kilowatts of continuous power. The flashes are often very long, sometimes several miles, and have been estimated to be from 4–6 inches (10–15 centimeters) in diameter. Often several flashes, each of very short duration, follow in quick succession over nearly but not quite the same path, thus producing the illusion of forking. Because light travels at the rate of 186,000 miles ($\sim$300,000 kilometers) per second, and sound at about 1,100 feet (375 meters) per second, lightning flashes are seen and vanish long before the rumble of thunder is audible, unless discharge occurs near the observation point. "Sheet lightning," so called, is merely the reflection or scattering of light from distant flashes by clouds.

Thunder is the term used to describe the sound emitted by rapidly expanding gases along the channel of a lightning discharge. Some three-fourths of the electrical energy of a lightning discharge is expended, via ion–molecule collisions, in heating the atmospheric gases in and immediately around the luminous channel. In a few tens of microseconds, the channel rises to a local temperature of the order of 10,000°C, with the result that a violent quasi-cylindrical pressure wave is sent out, followed by a succession of rarefactions and compressions induced by the inherent elasticity of the air. These are heard as thunder.

Thunder is seldom heard at points farther than about 15 miles (21 kilometers) from the lightning discharge, with 25 miles (40 kilometers) an approximate upper limit, and 10 miles a fairly typical value of the range of audibility. At such distances, thunder has a characteristic rumbling sound of very low pitch, due to the strong attenuation of the high-frequency components of the original sound. The rumbling results chiefly from the varying arrival times of the sound waves emitted by portions of the sinuous lightning channel, and secondarily from echoing and from the multiplicity of strokes that results when a lightning flash is made up of a series of electrical discharges. Inasmuch as lightning travels at the speed of light and thunder at the speed of sound, the time interval between visible lightning and audible thunder can measure the distance to the point where the discharge occurred. The speed of sound averages just a little more than 1,000 feet (305 meters) per second in such situations and a stop watch activated at the instant lightning is seen will measure the seconds until the thunder is heard. Multiplying the seconds by 1,000 gives the distance in thousands of feet, or roughly in $\frac{1}{5}$ mile ($\frac{1}{3}$ kilometer).

Tornado. A tornado is a violently rotating column of air, pendant from a cumulonimbus cloud, and nearly always observable as an inverted cloud cone or tube, popularly called a "funnel cloud." On a local scale, the tornado is the most destructive of all atmospheric phenomena. Its vortex, commonly several hundred yards (or meters) in diameter, whirls usually cyclonically with wind speeds estimated

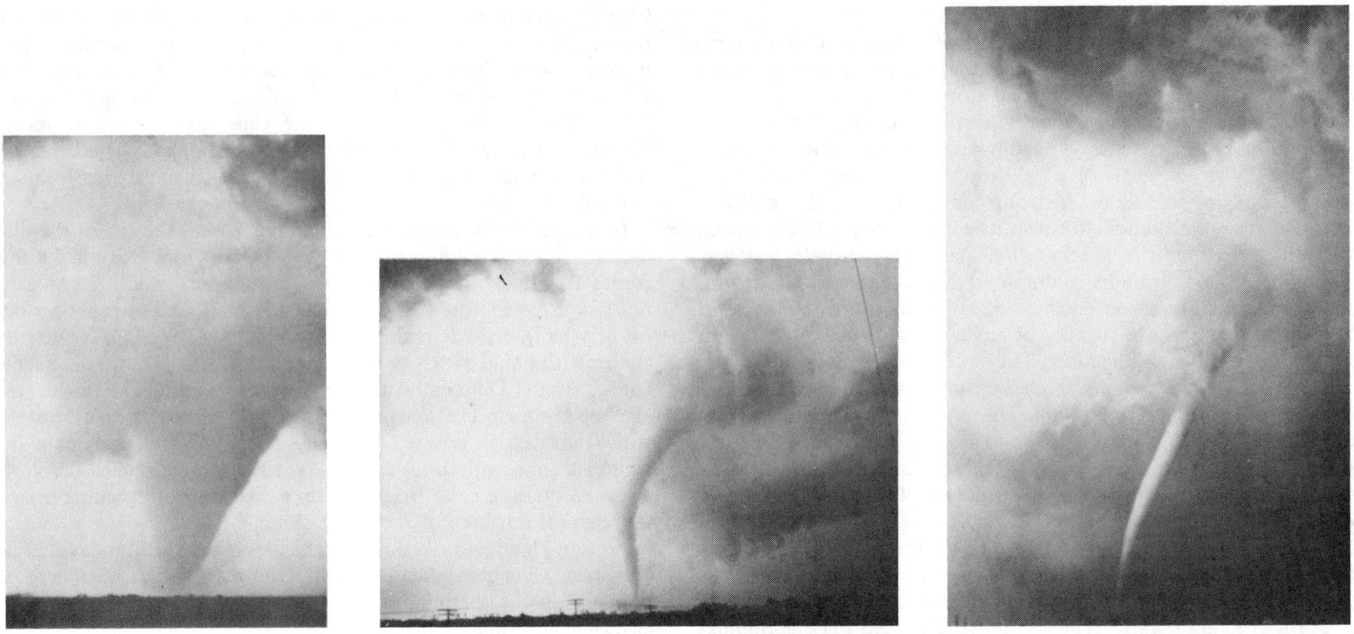

(a) (b) (c)

Fig. 3. A major tornado struck the small farming community of Union City, Oklahoma on May 24, 1973 and subsequently was much studied by meteorologists. The tornado was on the ground 26 minutes, attaining a maximum width (at cloud base) of nearly 600 meters (1968 feet). Even though the funnel narrowed toward the ground, the width of the damage path consistently equalled funnel width at cloud base. The tornado life cycle consisted of four distinct parts: the *organizing stage* (visible funnel intermittently touching ground with continuous damage path), the *mature stage* (tornado at largest size), the *shrinking stage* (entire funnel decreasing to thin column), and the *decaying stage* (fragmented, contoured funnel). Even in its final stages, the tornado retained its destructiveness.

According to Joseph H. Golden (U.S. National Oceanic and Atmospheric Administration) and Daniel Purcell, the life cycle of this type of tornado, in many respects, resembles the typical Florida Keys waterspouts. Both commence with surface evidence of vortex existence before a visible funnel cloud has descended a significant distance toward the surface. Approaching the mature stage, the tornado and waterspout exhibit spiral inflow characteristics with a distinct boundary between warm, moist air and cool, dry air. The cooler air mass from a nearby precipitation area apparently cuts off the flow of warm, moist air into the tornado's circulation, leading to vortex decay. The visible funnel becomes thin, increasingly tilted and distorted as it dissipates. Major differences between this type of tornado and waterspout appear to be vortex and parent cloud scales and, to a lesser extent, vortex lifetimes and intensities. Both vortices may evolve rapidly through their respective life cycles without evolving through every stage. This tornado was described in much detail in *Monthly Weather Review*, **106**, 1, 3–11 (1978), published by the American Meteorological Society.

(a) *Early part of mature stage.* During mature stage, stronger rotation is apparent around the edge of the wall cloud. A "feeder band" of low-hanging fractocumulus clouds spiraled into the upper portion of the tornado funnel from the northeast.

(b) Maximum tornado size was attained about 4 minutes after the start of the mature stage, with a funnel width of 590 meters (1936 feet) at cloud base and 155 meters (509 feet) at 150 meters (492 feet) above ground. The tornado's shape was that of a broad, truncated, inverted cone—a characteristic of the mature tornado stage. The total mature stage lasted approximately 8 minutes.

(c) *Shrinking stage.* A large bend developed in the middle of the funnel as the tornado turned sharply to the southeast. As the tornado passed through the middle of Union City, it shrank rapidly. Whereas the upper part of the funnel was 290 meters (951 feet) at the northwest edge of town, it measured only 60 meters (197 feet) in width at the same height 3.5 minutes later. The damage path likewise narrowed from 220 meters (722 feet) to 160 meters (525 feet). Some people who took shelter in town insisted that there were two separate tornadoes. They believed the tornado they saw approaching town before taking shelter was entirely different from the one they saw heading away from town when they came out.

The entire lifetime of the Union City tornado spanned 26 minutes, the longest stage being the organizing stage (10 minutes), followed by the mature stage (8 to 8.5 minutes), the shrinking stage (5.5 minutes), and the decaying stages (2 minutes). All stages were marked by significant differences in funnel width, funnel shape, damage path width, and damage characteristics.

(*National Oceanic and Atmospheric Administration, Boulder, Colorado.*)

at 100 and rarely up to 300 knots per hour. Its general direction of travel is governed by the motion of the parent cloud, which is usually toward the northeast.* Very low pressure prevails inside a tornado because of its great rotational winds. When it passes over a structure, however, the structure virtually blows up because of the strong winds impinging on it, which also blow in through its windows before the lowest pressure occurs. Thus, the often made analogy of a tornado and a huge vacuum sweeper is not accurate. Tornadoes occur on all continents, but are most common in Australia and the southern and midwestern parts of the United States. See Figures 3 and 4. The term tornado is also used to describe a violent squall or whirlwind of small extent that occurs during summer months along the west coast of Africa. The *FPP Scale* is used by professional meteorologists to classify specific tornadoes. See accompanying table.

During the past 40 years, tornadoes reported have evidenced a significant gain, due partly to more severe weather conditions and primarily because of better meteorological instrumentation, communications, and thorough reporting. In the 1940s, there were 155 tornadoes reported for an average year, with resulting lives lost of 180 per year. In terms of tornadoes reported, the figure trebled during the 1950s, with 280 tornadoes reported for an average year, but with a decrease of deaths reported to 141 lives per year. In the 1960s, the number of tornadoes reported continued to increase, with 683 tornadoes reported in an average year. Fatalities from tornadoes for an average year in the 1960s, however, dropped to 94 deaths. Tornadoes reported continued to increase during the 1970s, with an average of 861 tornadoes reported per year, the fatality rate remaining essentially the same—at about 100 deaths per year. As with other natural disasters, the extent or intensity of tornadoes is frequently reported in terms of lives lost, or of property and crop damage rather than in terms of scientific parameters.

Weather warning systems have been instrumental in lowering the loss of life to tornadoes. In the 1950s, there were 0.294 reported deaths per tornado; this figure fell to 0.138 during the 1960s; and fell further to 0.116 in the 1970s. The practice, as of the early 1980s, of the National Weather Service forecasters in the United States is to issue a Tornado Watch for a specific area where it is reasonably possible that tornadoes may occur during the valid time of the watch. A Watch

* In North America.

Fig. 4. The thunderstorms of the High Plains near the east slope of the Rockies are frequent hail producers, but they are not well known for their tornadoes. However, on August 14, 1977, a mini-outbreak of three tornadoes occurred near Bennett, Colorado. View of one of these tornadoes is shown here. (*Colorado State Patrol.*)

serves the function of alerting people to watch for tornado activity, make whatever preparations may be indicated, and to listen further for a Tornado Warning. A Tornado Warning means that a tornado has been sighted or indicated by radar. In recent years, there has been increasing use of doppler radar to discover and track tornadoes. See also **Radar.**

In an average year in the United States, several tornadoes will cause sufficient loss of life and property and crop damage to be reported nationwide by the mass communications media. The extent of tornado activity varies considerably from one year to the next. Since the early 1900s, the greatest number of tornadoes in the United States occurred in 1973, with a total of 1109 tornadoes reported.

Highly disastrous tornadoes reported have included: (1) March 18, 1925, 689 lives were lost in a series of tornadoes in Illinois, Indiana, and Missouri; (2) March 21, 1932, 268 lives lost in a series of tornadoes occurring in Alabama; (3) April 5, 1936, 216 fatalities were reported from a tornado in Tupelo, Mississippi; (4) April 9, 1947, 169 lives were lost in a series of tornadoes in Kansas, Oklahoma, and Texas; (5) March 21, 1952, a series of tornadoes in Arkansas, Missouri, and Tennessee took 271 lives; (6) April 11, 1965, 271 deaths were reported from a series of tornadoes in Illinois, Indiana, Michigan, and Wisconsin; (7) February 21, 1971, 110 lives were lost in tornado activity in the Mississippi Delta region; (8) April 3 and 4, 1974, in a 2-day period of tornado activity, 350 lives were lost in Alabama, Georgia, Kentucky, Ohio, and Tennessee, the majority of deaths occurring in or near Xenia, Ohio; (9) April 11, 1979, 50 or more lives were lost to a series of tornadoes in Oklahoma and Texas.

Reporting of tornado activity from other parts of the world is less frequent and statistics are often less reliable. A very costly tornado occurred in southeast Bangladesh on April 1, 1977, taking an estimated 600 lives. A tornado of April 16, 1978 caused over 500 deaths in Orissa, India.

Waterspout. This is a funnel-shaped tornado cloud at sea or over water of lakes or rivers. It is the equivalent of a tornado, but does not usually reach the same violence; it is comparable to a dust devil over land.

Dust Devil. A small, but vigorous whirlwind over a sandy area, which picks up dust and sand, swirling the particles upward. The average height is about 600 feet, but a few dust devils have been observed up to several thousand feet.

Tropical Showers. The main difference between tropical showers in cumulonimbus clouds and thunderstorms is the absence of ice in the tropical showers. Electrical charging, lightning, and thunder are substantially less and often completely absent. The mechanism for heavy rainfall in tropical showers is the virtual exponential growth of the larger water droplets as they begin their descent under the influence of gravity. The large droplets, therefore, precipitate out as rain and the small droplets remain to form the upper part of the

FPP TORNADO SCALE

	F Maximum Windspeed Per Hour		*P'* Path Length		*P''* Path Width	
Scale	Miles	Kilometers	Miles	Kilometers	Yards	Meters
—	< 40	< 64	< 0.4	< 0.6	< 6	< 5.5
0	40–72	64–116	0.4–0.9	0.6–1.4	6–17	5.5–15.5
1	73–112	118–180	1.0–3.1	1.5–5.0	18–55	16.4–50.1
2	113–157	182–253	3.2–9.9	5.1–15.9	56–175	51.0–159.3
					Miles	Kilometers
3	158–206	254–332	10–31	16–50	0.1–0.3	0.2–0.5
4	207–260	333–418	32–99	52–159	0.4–0.9	0.6–1.4
5	261–318	420–512	100–315	161–507	1.0–3.1	1.6–5.0
6	—	—	—	—	3.2–9.9	5.1–15.9
7	—	—	—	—	10.0–31.6	16.0–50.8

NOTE: The FPP Tornado Scale was devised initially by Fujita and Pearson for classifying tornadoes based upon their intensity (F), path length (P'), and path width (P''). It can also be used to describe downbursts. Because of anticipated wider path width of downbursts, the path-width scale was extended up to scale 7. These data first appeared in connection with the Satellite and Mesometeorology Research Project, Department of the Geophysical Sciences, University of Chicago, Chicago, Illinois. Later data were given in the "Manual of Downburst Identification for Project NIMROD," T. Theodore Fujita (May 1978). More background will be found in: "Tornadoes Around the World" by Fujita in *Weatherwise*, **26**, 2, 56 (April 1973).

cumulonimbus cloud. Airborne radar does not paint any significant echo of the upper parts of tropical showers under which torrential rain may be falling.

Squall. By common nautical definition, a squall is a severe local storm considered as a whole, i.e., winds, cloud mass, and precipitation (if any), thunder, and lightning. In more official terminology, a squall is a strong wind characterized by a sudden onset, a duration of the order of minutes, and a rather sudden decrease in speed. In United States observational practice, a squall is reported only if a wind speed of 16 knots or higher is sustained for at least 2 minutes, thereby distinguishing it from a gust.

A *line squall* occurs along an instability line, i.e., a nonfrontal line or narrow band of active thunderstorms. A *white squall* appears suddenly in tropical or subtropical waters; it is so called because the usual squall cloud is absent, and thus the only warning of its approach is the whiteness of a line of broken water or whitecaps. A *black squall* is accompanied by dark clouds and generally by heavy rain. The *willywaw* of the Straits of Magellan, the *sumatra* of the Malacca Strait, and the *abrolhos* of the Abrolhos Island off Brazil are violent squalls that occur seasonally.

See **Atmosphere (Earth); Precipitation and Hydrometeors;** and other entries listed under **Meteorology.** See list of references at ends of entries on **Climate;** and **Meteorology.**

The assistance of Dr. Joseph Golden, National Oceanic and Atmospheric Administration, Boulder, Colorado, in reviewing the text on tornadoes is hereby acknowledged.

Peter E. Kraght, Certified Consulting Meteorologist, Mabank, Texas.

FROST. Precipitation and Hydrometeors.

FROST-POINT HYGROMETER. Hygrometer.

FROTHING AGENTS. Classifying (Process).

FROUDE NUMBER. The flow of an inviscid, incompressible fluid in two geometrically similar flow systems is dynamically similar if the Froude number of the two systems is the same. The number is defined as $V/(gl)^{1/2}$, where V is the velocity scale and l the length scale of the system considered, and g is the acceleration due to gravity. This criterion is only relevant if a free surface is present. Naval architects use an equivalent parameter, the *speed-length ratio,* defined as the quotient of the speed of the ship (or ship model) in knots divided by the square root of the length in feet.

FROZEN PACK. Freeze-Preserving.

FRUCTOSE. Carbohydrates; Sweeteners.

FRUIT. As commonly used in the food industry, the word *fruit* signifies tree fruits—apple, apricot, cherry, peach, pear, plum of deciduous trees; or the important family of citrus fruits; or the bushberries—blackberry, raspberry, and strawberry, among others.

Botanically, a fruit is the ripened ovary of the flower, with or without other associated parts and, thus, this definition greatly broadens the number of food commodities that fall under the umbrella of fruits. There are many kinds of fruits. Usually, they are separated into two classes—dry fruits and fleshy fruits.

Growth of Ovary into a Fruit. The fruit begins its existence as the ovary of the flower. After pollination and fertilization have occurred, embryos begin to develop in one or many ovules inside the ovary. As this growth continues, the ovule gradually becomes a seed, and the ovary wall or *pericarp* may grow larger or thicker, may store relatively large amounts of food, or may undergo other changes. Eventually the seeds reach maturity, and about the same time, the fruit ripens. The final form of the fruit is characteristic of the particular species of plant.

Three layers of cell tissue are sometimes recognizable as the ovary matures. The outermost layer is the *exocarp,* which is usually a thin layer, often an epidermis only one cell thick. The innermost layer is the *endocarp.* Between these layers is the *mesocarp,* in which the vascular tissues ordinarily occur. The relative thickness and appearance of these layers vary greatly in different fruits.

During the growth of the ovary into a fruit, other flower parts or adjacent stem tissue may also change and become an integral part of the fruit. In a strawberry, for example, the red pulp is not the ovary, but a very much enlarged and modified stem tip, the receptacle of the flower. A large part of the pineapple is stem, not ovary.

Basic Forms of Fruit

Berry. A true berry consists of a fleshy fruit, derived entirely from the ovary of a flower and its contents. Usually many seeds are embedded in the flesh. Common examples are tomato, grape, gooseberry, and currant. See Fig. 1.

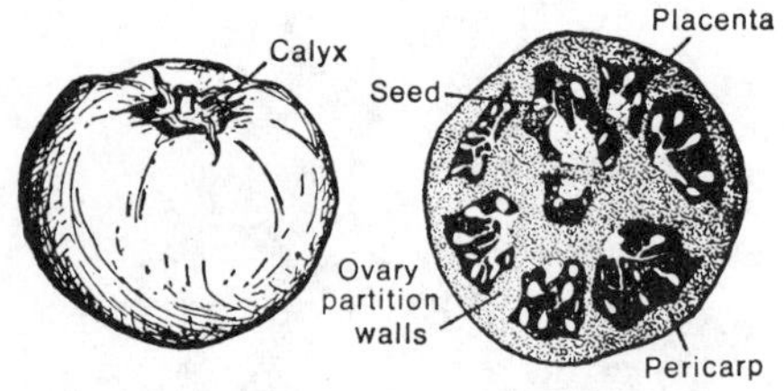

Fig. 1. A berry-type fruit, the tomato (*Lycopersicon esculentum*). *Left:* surface view; *Right:* cross section.

Hesperidium. The hesperidium is a berrylike fruit which is represented by citrus fruits (orange, lemon, grapefruit, mandarin). It differs from a true berry in having a leathery rind of ovary tissue containing oil ducts, and many membranous, juice-filled sacs in place of solid flesh. See Fig. 2.

Fig. 2. A hesperidium, the orange (*Citrus sinensis*) shown in cross section.

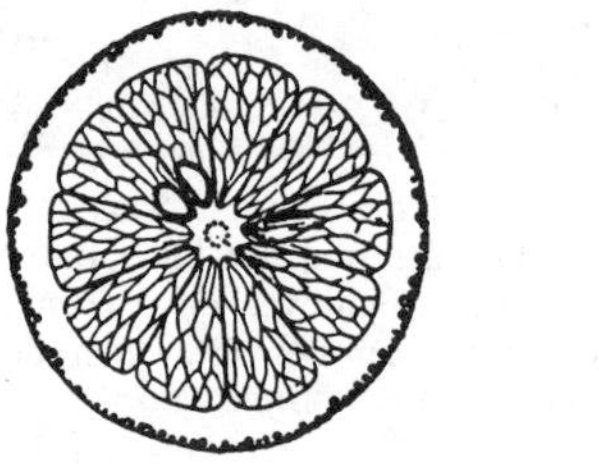

Pepo. A pepo is represented by the cucumber, squash, and pumpkin. These fruits resemble berries to a certain extent. The hard outer covering originates from the receptacle of the flower. (In the case of the hesperidium, the rind arises from ovary tissue). See Fig. 3.

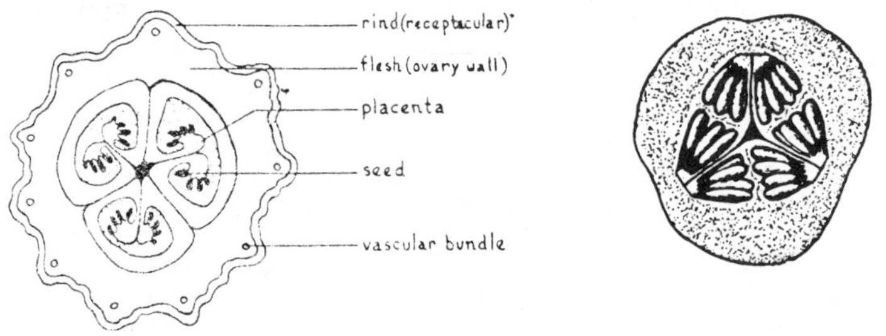

Fig. 3. A pepo, the cucumber (*Cucumis sativus*). Cross sectional views.

Drupe. A fleshy fruit with a thin, edible, outer skin derived from the ovary is called a drupe. A layer of edible flesh of varying thickness

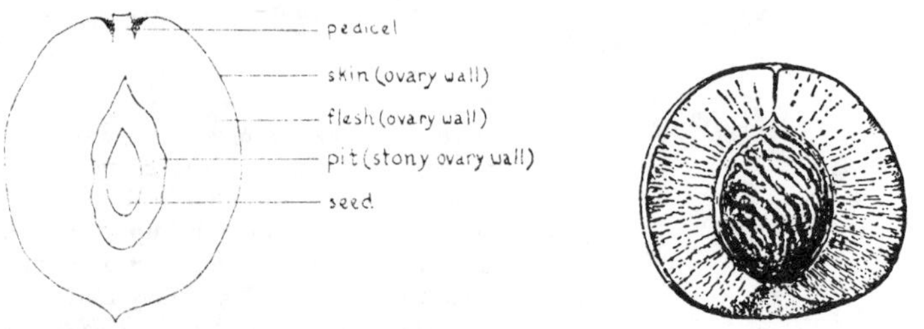

Fig. 4. A drupe, the peach (*Prunus persica*). Cross sectional views.

lies beneath the skin. Within this is the stone or pit, which is actually a hard inner wall of the ovary. Enclosed within the pit is the seed. The cherry, peach, and plum are typical drupes. They are also called *stone fruits.* The raspberry consists of a cluster of small, individual drupes, or druplets. Botanically speaking, the raspberry is *not* a berry. See Fig. 4.

Aggregate. An aggregate fruit is one which is formed from numerous carpels of one flower. The fruit, therefore, consists of a cluster of small, individual fruitlets. Examples are blackberry, raspberry, and strawberry. The fruitlets of the blackberry and raspberry are actually small drupes. In the strawberry, the seedlike achenes are fruitlets, embedded in a fleshy, edible floral receptacle. See Fig. 5.

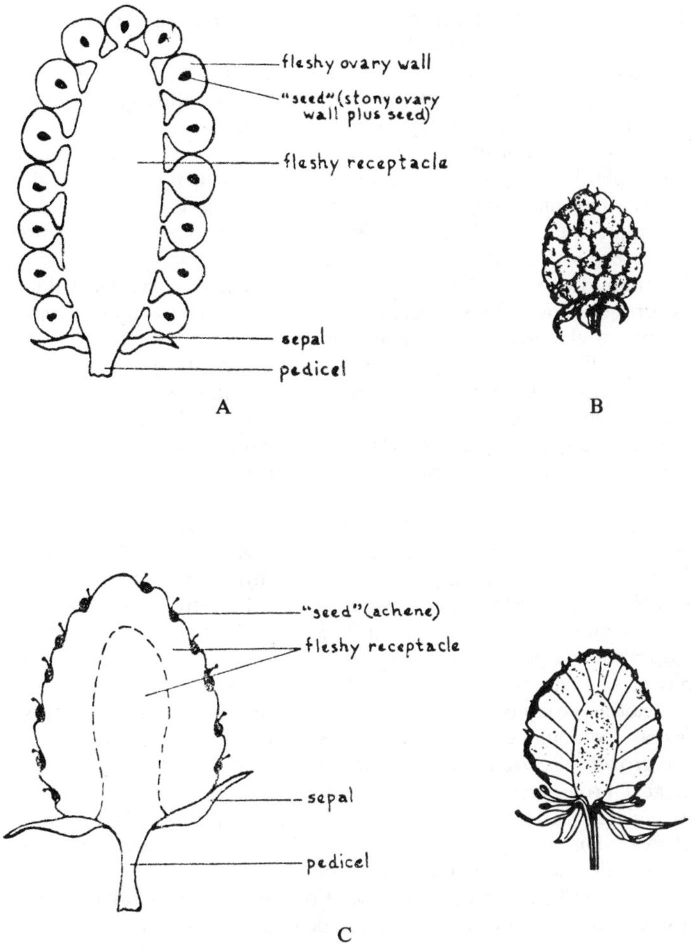

Fig. 5. Aggregate fruits. (A) Blackberry (*Rubus*), (B) raspberry (*Rubus*), (C) Strawberry (*Fragaria*). (*After Kessler.*)

Multiple Fruit. A multiple fruit is formed from individual ovaries of several flowers. Fruits of mulberry, fig, and pineapple constitute common examples. In the pineapple, portions of the flower stalk, sepals, petals, and ovaries of many flowers are fleshy and edible, and all are so tightly compressed together that they appear fused to each other.

Pome. The pomes are fleshy fruits consisting of a thin skin and outer zone of edible flesh. Common examples are the apple and pear. The fleshy portion beneath the skin is ovary tissue. The core in the center consists of a number of seed-containing, leathery little compartments called carpels. These are derived from the inner ovary wall. See Fig. 6.

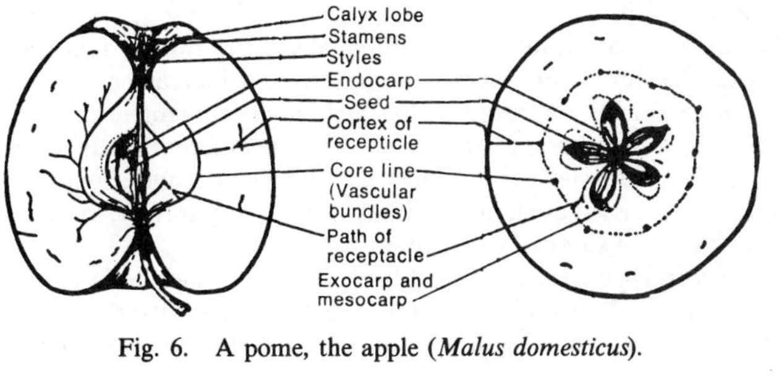

Fig. 6. A pome, the apple (*Malus domesticus*).

Legume. The main characteristic of a legume is the shell-like pod containing a number of relatively large seeds. Peas and lima beans are typical legumes. The pod which has developed from a single ovary dries out as it matures, splits into two halves, and releases the seeds. See Fig. 7.

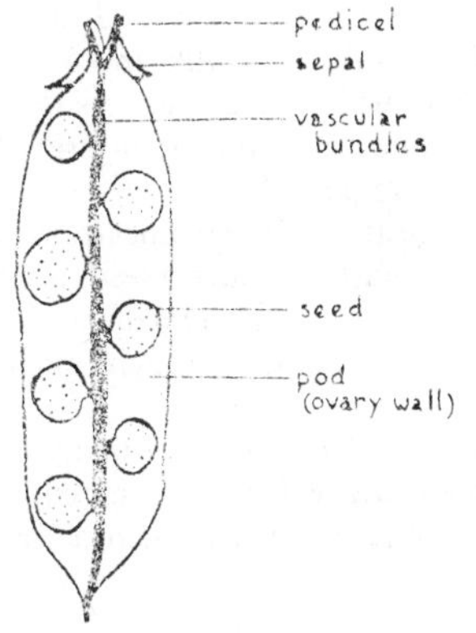

Fig. 7. A legume, an opened pea pod (*Pisum*). (*After Kessler.*)

Capsule. A capsule is somewhat like a legume, but differs in that it consists of more than one seed compartment and splits along more than two lines when ripe. The fruit of okra is a familiar example.

Caryopsis. The kernel of sweet corn is a kind of fruit called a *caryopsis.* The more or less horny outer coat is the ovary wall. This is firmly attached to the seed coat of a single seed. The remaining portions (endosperm and embryo) comprise the seed in this case.

Nut. A nut is defined as a hard, dry, single-seeded fruit, partly or entirely enclosed in a husk, which remains with the fruit as it ripens. Common examples are chestnuts and filberts. Although the term "nut" is popularly applied to many hardshelled fruits that may be stored dry, many of these are *not* true nuts. The groundnut (peanut), for example, is not a nut, but it is a legume. The almond is actually the pit of a drupelike fruit and the Brazil nut is actually a seed. See also **Nut.**

Dry Fruits. The dry fruits are separated into *dehiscent* fruits, those which split open when ripe, and indehiscent fruits, which do not do so. Common dehiscent fruits are the legune, the follicle, and the capsule; dry indehiscent fruits are the achene, the caryopsis or grain, the samara, and the nut. A follicle is similar to a legume, but splits along one side only. Milkweed pods are follicles. The fruits of the columbine and larkspur are also follicles. A capsule is a dehiscent fruit which develops from a compound ovary. The fruit of a lily or an iris is a capsule. The achene is a single-seed indehiscent fruit which when mature has the seed from the ovary wall except at the point of attachment. Fruits of the buttercup are achenes; also the fruits of the strawberry, which are the small hard bodies borne on the surface of the "berry." The achene of the Compositae family differs from the others in having the calyx tube coalesced with the ovary wall. A caryopsis or grain is very similar to an achene but has the seed coat fused with the pericarp so that the seed cannot be removed from the ovary wall. The fruits of all cereal grasses are caryopses. The samara is an indehiscent fruit which has a wing. The fruits of the maple and elm trees are samaras.

Dissemination of Seeds. Often the structure of the fruit is directly correlated with the dissemination of the seeds. The principal agencies for fruit dispersal are the wind, water, and animals. Many fruits are provided with wings, thin, bladelike structures which enable the fruits to drift slowly downward and generally away from the parent plants. The fruits of the maple are provided with wings. The fruits of the dandelion and thistle are familiar to all, though usually they are called seeds. In them, a group of slender hairs forms a parachute which enables the fruit to drift far away from its parent plant into new region. Sometimes, the fruit lacks any special structures which will aid in its dissemination, but is itself very easily blown about. Many fruits, especially those of plants which grow along the shores of streams, are carried about by the water. The large fruits of the coconut are often carried far from the parent plant in this manner.

Animals are an important means of dispersal of fruits and seeds. There are two ways in which fruit may be thus carried. Very commonly, hooks or barbs are formed on the surface of the fruit and are easily caught on the fur of a passing animal, and stay there until some mechanical action breaks off the hooks or barbed bristles and allows the fruit to fall. Some fruits are covered with a sticky coating which causes them to adhere to the coats of passing animals, to be rubbed off later at some other location. Still other "seeds" have a soft, often tasty, outer wall, or edible mesocarp, and are eaten by animals. The hard inner wall of the fruit resists action of the animal's digestive juices so that the seed passes uninjured through the digestive tract and is voided in a region often far distant from the place where it was eaten. Partial digestion of the endocarp may even aid the liberation of the seed for germination. Tomato seeds are particularly resistant to attack by the digestive juices of humans and often will be capable of germination even after passing through extensive sewage treating processes.

The explosive splitting of the walls of many fruits ejects the seeds violently, hurling them away from the parent plant. In some rare species, seeds may be hurled by as much as 50 yards (45 meters). See Fig. 8.

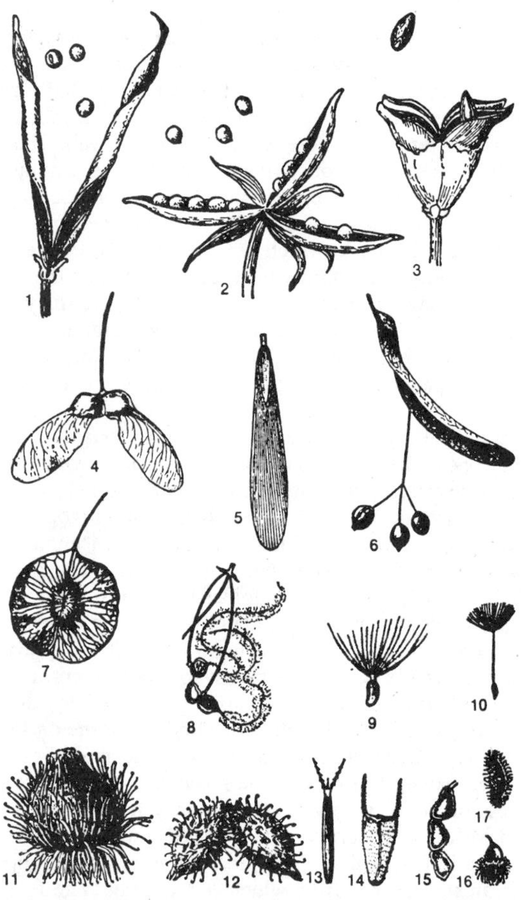

Fig. 8. Dry fruits with devices for dispersal of their seeds. (1–3), by sudden dehiscence; (4–10), by wind; (11–17), by animals. Items shown are: (1) wild bean, (2) violet, (3) witch hazel, (4) maple, (5) ash, (6) basswood, (7) elm, (8) *Clematis*, (9) thistle, (10) dandelion, (11) burdock, (12) cocklebur, (13) Spanish needle, (14) beggar's trick, (15) beggar's lice, (16) agrimony, (17) carrot.

Carbohydrate Patterns of Fruits

Soluble carbohydrates are synthesized in the chloroplasts of green plants and those not utilized immediately in respiration are translocated to other parts of the plant. These translocated carbohydrates may be utilized in respiration and growth, or may be stored as reserve foods. In many plants, the most conspicuous site of stored foods is to be found in the fruit.

All fruits undergo four stages of development: (1) Following fertilization, a fruit grows by cell division. (2) There follows a period of cell enlargement, during which time sap-filled vacuoles are formed. Sugars accumulate in the vacuoles; the cytoplasm which, up to this stage, consisted chiefly of proteins, now contains starch. When a fruit has attained full growth, it may be considered *mature*, but not necessarily *ripe*. (3) Ripening ensues during the third stage of development, during which period substances responsible for flavor and aroma are formed, acidity is reduced, sugars increase, and a certain amount of softening occurs. Softening is the result of conversion of pectin substances in the cell wall from the insoluble to the soluble form. (4) The fourth stage of development, called *senescence*, begins when ripening is essentially complete.

Fruits can be divided into two groups: (1) Those with a starch reserve; and (2) those without a starch reserve, although there are some fruits that fall midway between these classes.

Fruits with a Starch Reserve. Typical of fruits with a *starch reserve* are apple, banana, and pear. It has been shown, for example, that invert sugar and sucrose increase throughout the growing period of the apple fruit, but starch reaches its maximum when ripening processes begin. During the course of ripening, therefore, the starch is hydrolyzed to sugar. During the early stages of ripening, the soluble pectin substances also develop. The sugars in a ripe apple consist mainly of glucose, fructose, and sucrose.

Fruits like the apple, pear, and banana, with their carbohydrate reserve, can be harvested in the *mature green* and permitted to finish their ripening process during storage. Other fruits, like citrus, raspberries, cherries, among others, do not develop a carbohydrate reserve and must, therefore, be ripened on the tree if they are to ripen at all.

Rather pronounced carbohydrate transformations take place in the banana during ripening. It has been shown that when the fruit changes from the green to the ripe stage, total carbohydrates drop from 26.6% to 19%, soluble carbohydrates increase from 1.3% to 17%, and insoluble carbohydrates decrease from 25.3% to 2%. In general, reducing sugars show a gradual increase during post-harvest ripening. The behavior of nonreducing sugars is determined by the variety of fruit.

Another group of substances, known as *tannins*, is sometimes classified as compound carbohydrates. These substances accumulate during the growth of certain fruits, accounting for astringency in the unripe stage. Unripe persimmons, olives, bananas, and dates are characterized by high tannin content. This is true to a lesser extent of certain varieties of pear. During the ripening of these fruits, astringency is reduced as tannins are converted into insoluble forms.

Fruits Without a Starch Reserve. Fruits which do not accumulate a large carbohydrate reserve are typified by citrus fruits, blackberries and raspberries, cherries, peaches, plums, strawberries, and others. During ripening on the tree or bush, these fruits show an increase in sugars and a decrease in acids. Following harvest, fruits without a starch reserve may develop a characteristic color, soften (in some types), and lose a slight amount of acid through respiration, but they will not show any increase in sugar. A good variety of orange has been shown to contain 10.6% soluble solids (mainly sugars) and 0.85% acids when acceptable to consumers.

Several exceptions or variations from the general rule can be found in this second group. Lemons, for example, do not undergo the same changes during ripening as those in oranges and grapefruit. The lemon fruit, during growth and maturation, does not increase in sugar. Free acids in the juice increase during ripening and predominate over sugars in the ripe fruit.

In the avocado, total sugar content decreases during maturation. With the loss of sugar, there is a concomitant increase in oil. Dates are unique not only because of the high sugar content in ripe fruits, but also because different varieties accumulate different kinds of sugar. *Barhee*, for example, accumulates mostly glucose and fructose and is, therefore, classed as an invert-sugar variety. *Deglet Noor*, in contrast, contains mainly sucrose when ripe.

Vegetable-Type Fruits. Although a popular distinction is made between fruits and vegetables, technically there is no valid distinction provided that the commodity in question meets the basic definition of fruit as previously given. Thus, if the edible portion of the plant is a leaf, petiole, stem, or root, it is definitely a vegetable. From the popular standpoint, a fruit is more frequently eaten raw as a dessert, and it possesses a characteristic aroma and flavor due to the presence of various organic esters. A vegetable is ordinarily eaten cooked, or when raw, as a salad or relish. It is the product of a herbaceous plant, rarely of a shrub or tree.

Fruits borne on succulent vines, if of economic importance, may be called vegetables, although botanically they may be true fruits. In this category are included tomato, muskmelon, watermelon, and pumpkin.

Mature green tomatoes contain a very slight amount of starch which disappears upon ripening. Reducing sugars increase with ripening, but only traces of sucrose have been found in these fruits in various stages of ripening.

Muskmelons, honeydew and casaba melons undergo an increase in total solids, total sugar and sucrose during ripening and a decrease in invert sugar. The same general changes occur in the watermelon. Both pumpkin and squash differ from related species in that immature fruits contain as much sugar as ripe ones.

Carbohydrate Accumulation in Seeds. Seeds may be grouped into three categories: (1) Those in which carbohydrates represent the main food reserve. The cereal grains are typical of this group. (2) Seeds that accumulate large quantities of proteins. Many of the legumes (peas, beans, etc.) are in this group. (3) Seeds in which large quantities of oil are stored. Sunflower, almond, macadamia, and castor bean seeds belong to this group.

Starch and hemicellulose predominate in seeds in which large quantities of carbohydrates are stored. During the early stages of development of this type of seed, there occurs a gradual increase in sugar up to a maximum. Subsequently, sugars decrease and starch and other polysaccharides increase. Some seeds are used as foods when the seeds are still rather succulent. Sweet corn is an example of this type. Four stages in the development of sweet corn have been described: (1) The *pre-milk stage*, in which the exudate (when skin is broken) is opalescent. The ratio of sugar to starch at this stage is about 1.9. (2) Next is the *milk stage* when the sugar/starch ratio is about 0.750. (3) In the *early dough stage*, the sugar/starch ratio is about 0.2. (4) In the *final dough stage*, the sugar/starch ratio is about 0.15.

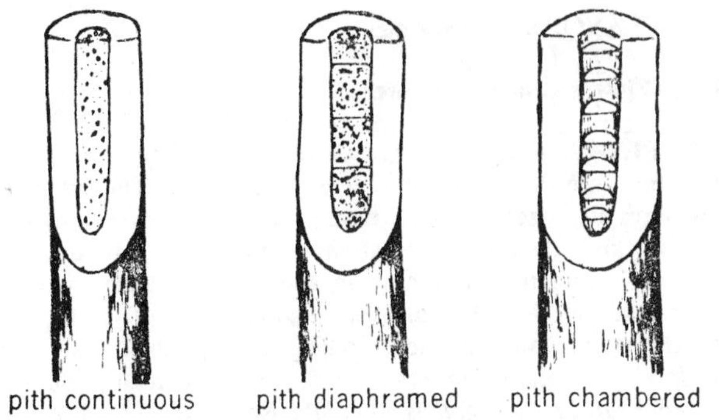

Fig. 10. Pith is the soft central part of a twig. Where the pith is continuous, it is a solid, homogeneous material, not divided into compartments. In diaphragmed pith, cross membranes of denser material extend across the pith. In chambered pith, the central portion of the twig is divided into empty horizontal chambers by cross partitions. (*University of Georgia College of Agriculture.*)

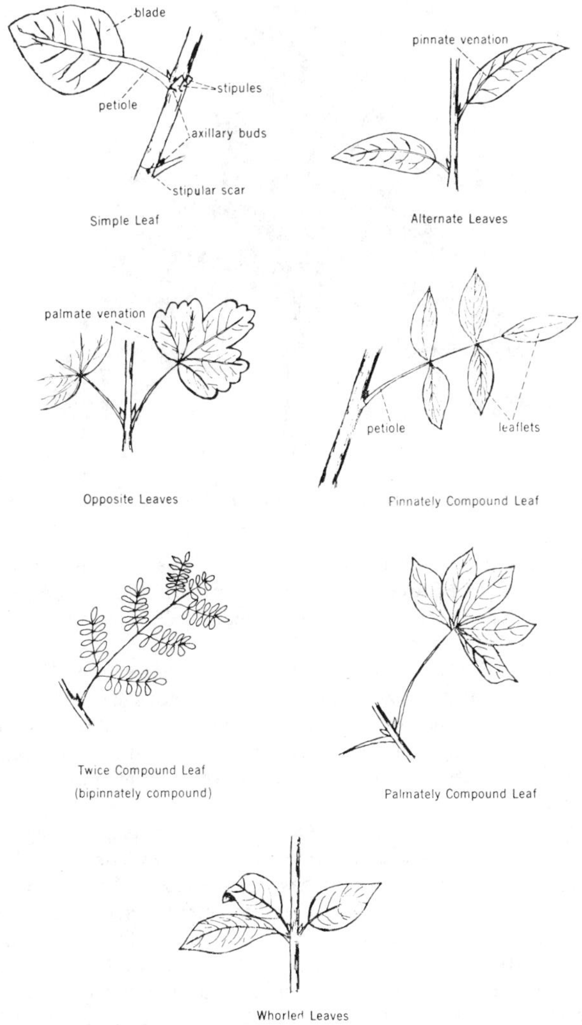

Fig. 9. Various configurations of leaves. (*University of Georgia College of Agriculture.*)

Although some seeds, like peas, are rich in protein, they also accumulate carbohydrates, although to a lesser extent than do the carbohydrate-rich seeds, and the changes from sugar to polysaccharides proceed in the same order during ripening.

Seeds of the coconut possess two rather different features—their large size and the presence of liquid endosperm (milk) during the maturing stages. Ripening stages of coconut have been divided into three parts: (1) Before the formation of the endosperm, when invert sugar and amino acids accumulate in the milk; (2) when the loss of water from the nut takes place and sucrose appears in the milk; and (3) when a sudden rise in the oil content of the endosperm and a loss of nutrients in the milk occur.

Some seeds have the carbohydrate reserves stored as hemicellulose in the tertiary, much thickened cell walls of the endosperm of the cotyledons, instead of in the interior of the cells as in the case with stored starch. The most striking seed of this kind is the seed of the ivory nut palm (*Phytelephas macrocarpa*) from South America. Seeds of the date palm also store carbohydrate in the form of hemicelluloses. Hydrolysis of this seed yields glucose, fructose, mannose, galactose, arabinose, and xylose.

Tree Characteristics

There is a wide variation in the basic growth habits of the various fruit-producing trees and plants. Some of the terms used in describing the botanical features of trees are illustrated in Figs. 9, 10, and 11.

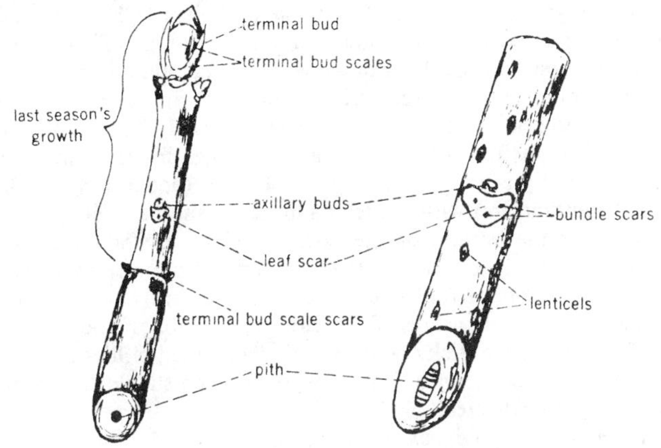

Fig. 11. Illustrations of several terms used in describing trees. (*University of Georgia College of Agriculture.*)

FRUIT FLAVORS. **Flavorings.**

FRUIT (Flower Ovary). **Flower.**

FRUIT FLY (*Insecta, Diptera*). Of the family *Trypetidae*, the fruit fly can be quite damaging to several fruit crops. The adult female usually lays her eggs in plant tissue. Larvae of several species are borers, working their way into stems, mining into leaves, and, most damaging, boring into and moving about in the fleshy portions of fruits and vegetables. They also produce galls. The fruit fly tends to specialize, as indicated by the following descriptions:

Apple Maggot (*Rhagoletis promnella*). Described in entry on **Maggot.**

Cherry Fruit Fly (*Rhagoletis cingulata*, Loew). See Fig. 1. A primary

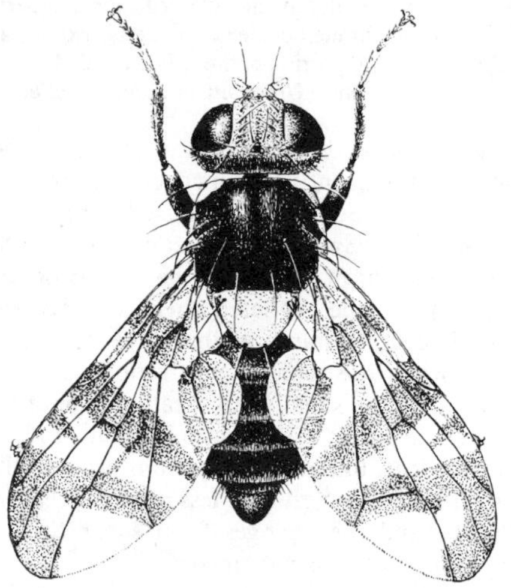

Fig. 1. Cherry fruit fly. (*USDA diagram.*)

cause of deformed and wormy cherries. Often, one side of the fruit will be decayed and shrunken, while the other side is healthy. The maggots of this fly are yellow-white and footless and range up to $\frac{1}{4}$ inch (6 millimeters) in length. The head is pointed and is used for boring into fleshy fruit. Sometimes the very small maggots are difficult to find, but traces of their burrows will be apparent. Sometimes the damage is not fully detected until the fruit is processed. In addition to the cherry, this insect damages pear and plum. A closely related species is the *black cherry fruit fly* (*R. fausta*, Osten Sacken) whose primary target is sour cherry. These insects winter in the pupa stage. The pupa is brown and shaped something like a capsule. Time in this stage is long, ranging up to 150 days. The adults usually emerge when the outside temperature has risen to 40°F (4.5°C), but this habit varies considerably, depending upon locale. The adult fly leaves the soil in late spring. The females puncture cherry leaves and fruits with their ovipositor, each female laying nearly 400 eggs over a laying period of about 25 days. Chemical sprays and trapping are used as control measures.

After larvae development is completed within the cherries, larvae drop to the ground and change to pupae in the soil. Cherry varieties that are harvested early are likely to contain larvae. Therefore, to control the pest, infested fruit must be destroyed and an area around the tree should be cultivated. All cherries that appear damaged on the tree should be picked and immediately burned. Traps reduce the number of adult flies before they lay their eggs in the fruit. Traps may be made by coating a small piece of wood (about 6 × 8 inches; 15 × 20 centimeters) with a sticky substance, such as Tanglefoot or TacTrap. At the bottom of the board, a small jar or bottle filled with ammonium carbonate is attached. A few holes are punched in the jar lid so that the fumes of the bait can get into the air. Several of these traps are suspended from the lower limbs of the tree. For full effectiveness, the sticky board should be cleaned of flies and other debris at periodic intervals. The sticky substance should be renewed periodically as indicated.

Currant Fruit Fly (*Epochra canadensis*).

Mediterranean Fruit Fly (*Ceratitis capitata*, Wiedeman). See Fig. 2. The insect occurs widely in Bermuda, Hawaii, and in most subtropi-

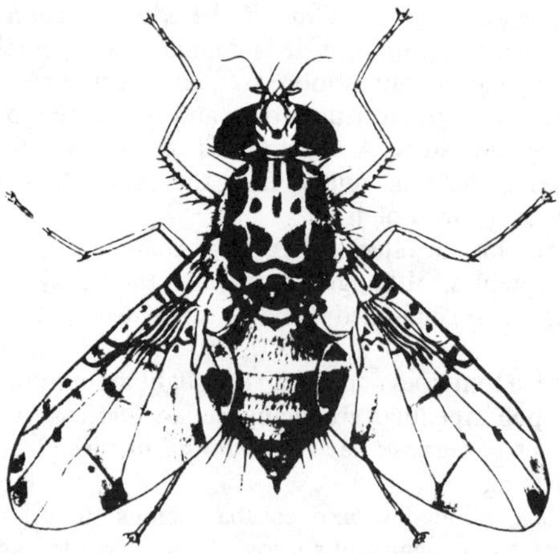

Fig. 2. Mediterranean fruit fly. (*USDA diagram.*)

cal countries. There have been several invasions of the fly into Florida. Several million dollars were invested in clearing an area of 10 million acres after an invasion of the fly in 1929. The fly had been found in scattered locations within this area. Later, in 1956, additional large funds were invested in eradicating the insect from 28 counties in Florida. The fly attacks citrus and deciduous fruits, notably apple, grapefruit, nectarine, orange, peach, pear, plum, quince, as well as coffee. An invasion of several regions of California by the Mediterranean fruit fly occurred during the summer of 1981. See also **Drosophila.**

Mexican Fruit Fly (*Anastrepha ludens*, Loew). See Fig. 3. This insect

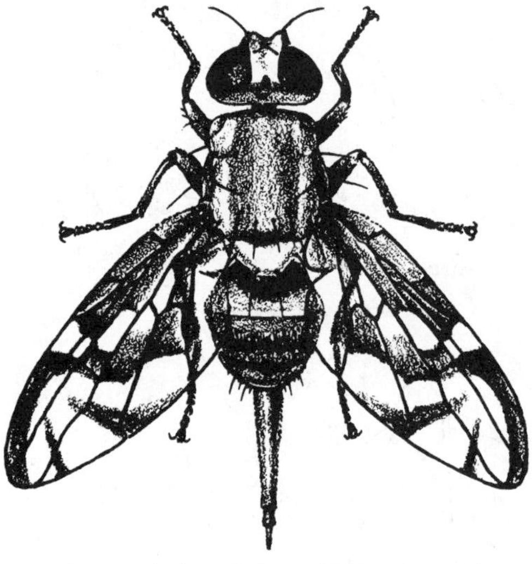

Fig. 3. Mexican fruit fly. (*USDA diagram.*)

is principally a pest on citrus and mango. The fly is an important economic pest in Mexico and Central America. From time to time, members of this species have been trapped in California.

Olive Fruit Fly (*Dacus oleae*).

Oriental Fruit Fly (*Dacus dorsalis*, Hendel). See Fig. 4. This insect injures all varieties of citrus and most deciduous fruits, as well as avocado, banana, melon, tomato, and many other plants. In 1946, the oriental fruit fly was introduced into Hawaii from the Marianas Islands. The fly is also found in Burma, India, the Philippines, and Taiwan.

Walnut Husk Fly or Maggot (*Rhagoletis completa*, Cresson). The adult fly is pale yellow with brown eyes, stiff brown hairs on abdomen, and with transparent wings that have dark stripes. The larva is white or pale beige and up to $\frac{1}{2}$ inch (12 millimeters) in length. The maggot feeds in the husk of maturing nuts and reduces the quality of the kernels. Distribution is throughout the United States. Natural circumstances often assist the nut grower by causing the fly to emerge early or late in the season. The female cannot lay its eggs successfully until the walnut husk becomes soft. During July and August, unsuccessful egg-laying attempts have been observed on eastern black walnuts that

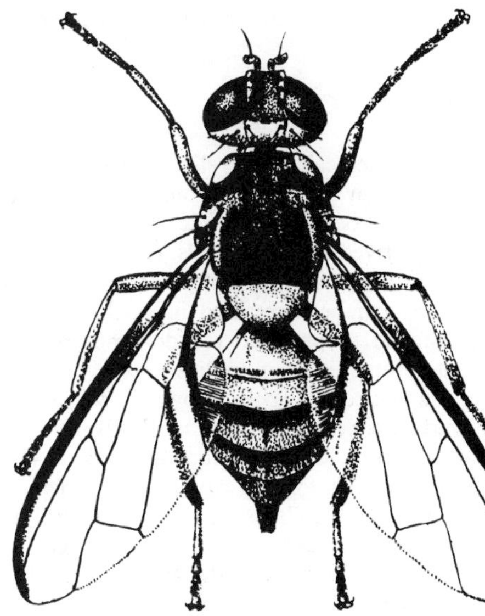

Fig. 4. Oriental fruit fly. (*USDA diagram.*)

were sound and without blemish. However, eggs have been found in mid-August in husks that have naturally or mechanically induced abrasions.

Husk maggot problems on the eastern black walnut can be greatly reduced by selecting late-maturing varieties. Larvae which hatch from eggs laid after September 20 in the latitude of Maryland are no problem because they will not mature.

See also **Drosophila.**

FRUIT FLY (Heredity Studies). Heredity.

FRUITS. Antimicrobial Agents (Foods).

FRUITS (Hypobaric Storage). Hypobaric (Controlled-Atmosphere) Systems.

FRUIT TREES. Citrus Trees; Rose Family; Trees.

FRUITWORM (*Insecta*). The *western raspberry fruitworm* (*Byturus bakeri*, Barber) and the *eastern raspberry fruitworm* (*Byturus rubi*, Barber) are among the most destructive insects on loganberry and raspberry. Both the adult and larva forms are destructive. The adult beetles range from $\frac{1}{8}$ to $\frac{1}{6}$ inch (3 to 4 millimeters) in length and are of a light-brown color. The beetles eat buds, blossoms, and leaves. Eggs are laid on the blossoms and small fruits just getting started. Shortly thereafter, thin whitish grubs are hatched. They are about $\frac{1}{3}$ inch (8 millimeters) in length. The grubs bore into the fruit, damaging it. When fully grown, they drop to the ground, where they pupate until early spring. Rotenone is an effective control chemical.

The *gooseberry fruitworm* (*Zophodia convolutella*, Hübner) eats the pulp of the fruit, causing the berries to dry up, to prematurely color, and become brownish in appearance. Rotenone is an effective control chemical.

Apple fruitworms, sometimes called green fruitworms, are green to green-white caterpillars that look like climbing cutworms. They are up to 1.25 inches (about 3 centimeters) long. The larvae have white or yellow striping on each side. These worms eat leaves and make large holes in fruit. Generally, they are found in the northern United States. Green fruitworms do most of their damage to young fruit in May. However, some fruitworms continue to damage fruit until mid-June. During the first week of June, most of the caterpillars attain their full growth. At this time, they burrow into the soil beneath the trees to a depth up to 3 inches (7.5 centimeters) and construct an earthen cell. About mid-September, the moths emerge and go into hibernation in sheltered nooks. Some pupae, however, do not become moths until early the following spring. When the insect reaches the caterpillar stage, it can be handpicked from small trees and destroyed. Because of its large size, the caterpillar is easily identified. To prevent further damage to the fruit, the ground under the trees should be thoroughly cultivated to a depth of about 4 inches (10 centimeters). This destroys the caterpillars in their earthen cells. Cultivation reduces the number of adult moths that will emerge and lay eggs in subsequent generations.

Cherry fruitworms are found in the northwestern United States. The adult is a small grayish-black moth with a wingspan of about $\frac{1}{4}$ inch (6 millimeters). The larva is a whitish-pink worm with a black head and up to $\frac{3}{8}$ inch (9 millimeters) long. The larva bores into fruit and feeds on pulp, causing rough, brownish areas in the pulp and on the skin. The cherry fruitworm winters as a fully-grown larva in a silken cocoon, tunneled inside the pruned stub of a dead twig, under bark, or debris on the ground. The larva pupates in May and the adult emerges about 1 month later.

Controlling the pest involves pruning away all dead branches and twigs and burning them to kill overwintering larvae. Cleaning up bark and debris on the ground also reduces populations in the spring. Damaged and prematurely dropped fruit should be picked up and destroyed. Placing bands around the tree, as in the case of controlling the codling moth, also is effective. Whenever silken cocoons are found, they should be destroyed immediately. See also **Codling Moth.**

FUCHS THEOREM. If a second-order linear differential equation has regular singular points at ∞ and at x_k, $k = 1, 2, \ldots, n$, and no other singularities, its general form is

$$y'' + \frac{p_{n-1}(x)y'}{F(x)} + \frac{p_{2(n-1)}(x)y}{F^2(x)} = 0$$

where $F(x) = (x - x_1)(x - x_2) \cdots (x - x_n)$ and $p_i(x)$ is a polynomial in x of degree $\leq i$. Such an equation is said to be of Fuchsian type. The theorem may be extended to equations of any order.

The general Fuchs theorem is concerned with linear differential equations, the determination of a fundamental set of solutions relative to each of its singular points, and the behavior of these solutions. See also **Riemann-Papperitz Equation.**

FUEL. In the conventional sense, a fuel is a material or combination of materials which, when burned with air, produces heat. This heat, in turn, can be used in numerous ways—as in the conversion of water to steam. The steam, in turn, can be used in many ways—as in a steam turbine to produce electricity. Fuels also are burned for the purpose of obtaining explosive or mechanical energy—as in an internal combustion engine where heat per se is an inevitable, but undesired by-product. The term *fuel* is also used in connection with nuclear reactions—as the material, such as uranium and plutonium isotopes, which undergoes fission and, in so doing, yields heat energy. Fuel also appears in the term *fuel cell*, in which chemical reactions other than what may be considered as conventional combustion are carried out to yield electrical energy.

Conventional fuels may be solids (coal, coke, wood, etc.); liquids (fuel oil, gasoline, alcohol, etc.); or gases (natural gas, synthetic gases, hydrogen, etc.). Where natural fuels are derived from geochemical processes in the earth over long periods of time, as coalification in the production of peat and various grades of coal from prehistoric vegetation, the term *fossil fuel* is applied. The principal fuels in this very large and important class of fuels are coal, petroleum, and natural gas. Where fuels are produced by chemical means, often by synthesis from other materials, the term chemical fuel may be applied, as in the instances, for example, of alcohol (either from natural fermentation or synthesis), hydrogen (from chemical or electrochemical reactions, including the electrolysis of water), and various synthesis gases (water gas, producer gas, town gas, etc.). Rocket fuels generally are chemical-type fuels.

Some of the desired properties of fuels, depending upon particular applications, include heat content, that is, Btus or calories released upon combustion per unit weight or volume of the fuel. Energy density is particularly important where a fuel is used in some form of vehicle where the fuel must be carried and thus part of the fuel is expended simply to transport itself. Cleanliness of burning is of major concern, not only in terms of pollutants that may be produced as the result of combustion, but also in terms of additional costs of equipment that may be required to process and handle the by-products, such as flue gas and ash. Natural gas, in this respect, is close to an ideal fuel for many applications because it is clean-burning with no ash and when burned under proper conditions, produces essentially harmless carbon dioxide and water. Economics and availability, however, frequently dictate that a fuel of more inferior burning qualities, such as coal, be used instead of natural gas. The trend of many years to switch from coal, to fuel oil, to natural gas in large central station installations has been reversed in recent years as the result of growing shortages of natural gas and petroleum fuels (reflected in prices)—with a reawakened appreciation for the readily available coals (in the United States and some other industrial nations). There is also much growing emphasis upon what might be termed nonconventional fuel sources, such as oil shales, tar sands, and, of course, on the liquefaction and conversion of coal into more attractive fuel products, such as high-Btu pipeline gas, as well as liquid petroleum-like fuels derived from new raw materials and new processes.

Fuels are extensively covered in this volume. In particular, reference to entries on **Coal, Fuel Cells, Hydrogen (Fuel), Natural Gas, Oil Shale, Petroleum, Substitute Natural Gas (SNG), Tar Sands** and **Waste Conversion to Energy**—is suggested. Reference is also suggested to topics that relate closely to fuels, in that they tend to lessen the burden on or substitute for conventional fuel uses. These topics include **Battery, Geothermal Energy, Hydroelectric Power, Nuclear Reactor,** and **Solar Energy.** For analysis of fuels, see **Calorimetry;** and **Combustion.**

Although, more detailed information is given in specific fuel entries throughout this volume, the accompanying table is helpful in making comparisons of heating values of various fuels.

HEATING VALUE OF VARIOUS FUEL SUBSTANCES[a]

Gases	Heating Value: Btu/cubic foot at 60°F and 30 inches mercury pressure	Heating Value: Kilogram-calories/gram-molecular weight
Acetylene	1,455	312
Butane	3,200	680
Carbon monoxide	317	67.1
Ethane	1,730	368
Ethylene	1,615	332
Hydrogen	319	68.4
Methane	995	211
Natural gas	975–1,180	207–253
Substitute natural gas (SNG		
Pipeline quality (high-Btu)	950–1,050	202–224
Low-Btu quality	400–600	85–128

Liquids	Kilogram-calories/gram	Kilogram-calories/gram-molecular weight
Benzene	10.0	782
Ethyl alcohol	7.13	328
N-Heptane	11.5	1,150
N-Hexane	11.5	990
Methyl alcohol	5.34	171
N-Octane	11.4	1,303
N-Pentane	11.6	883
N-Propyl alcohol	8.00	481
Toluene	10.2	934

Solids	Kilogram-calories/gram
Carbon (Amorphous to CO_2)	8.08 (97.0 kg-cal per gram-molecular weight CO_2)
Carbon (Amorphous to CO)	2.49 (29.9 kg-cal per gram-molecular weight CO)
Cellulose	4.21

[a] Where applicable, higher heating value (water condensed as formed in combustion) figures are given.

For references, see lists of references that accompany specific entries on various fuels.

FUEL CARBURETION. Carburetor.

FUEL CELLS. The fuel cell is an electrochemical device which directly combines hydrogen and oxygen from air to produce electricity and water. With prior processing, a wide range of fuels, including natural gas and coal-derived synthetic fuels, can be converted to electric power. The basic process is highly efficient, pollution-free, and since single fuel cells can be assembled in stacks of varying sizes, systems can be designed to produce a wide range of output levels and thus satisfy numerous kinds of applications.

The concept of the fuel cell has been known for well over a century, but was not utilized in a practical sense until advent of the various space-exploration programs. Cells developed by Bacon (1960) in the late 1930s provided the basis for the Apollo power plant and led to a sophisticated technology of cells using alkaline electrolyte and pure hydrogen and oxygen as reactants.

The adaptation of this technology to the seemingly endless possibilities in terrestrial applications proved to be difficult because operation on hydrogen of lesser purity and air rather than oxygen presented many problems. These shortcomings of the early alkaline cell were overcome by new cell types of which the phosphoric acid cell is the most advanced.

Operating Principles of Fuel Cells

As an energy conversion device, the fuel cell is distinguished from a conventional battery by the fact that the electrodes are invariable and catalytically active. Current is generated by reaction on the electrode surfaces which are in contact with an electrolyte. As a rule, fuel and oxidant are supplied as required by the current load; water is continuously removed.

Single Cell. Under load, the voltage of one individual fuel cell element is less than one volt. Therefore, the assembly of many cells, connected in series as a stack, is necessary. Each individual cell contains the elements needed for feeding reactants to the electrode surface, and removal of water from the cell, as shown schematically in Fig. 1 for a hydrogen-air cell with acid electrolyte.

The electrode reactions are comprised of the oxidation of hydrogen on the anode (the negative electrode) to hydrated protons with the release of electrons; and on the cathode the reaction of oxygen with protons to form water vapor with the consumption of electrons. Electrons flow from the anode through the external load to the cathode and the circuit is closed by ionic current transport through the electrolyte. In an acid cell, the current is carried by protons.

Reactants in this cell need not be pure. Hydrogen may be extracted

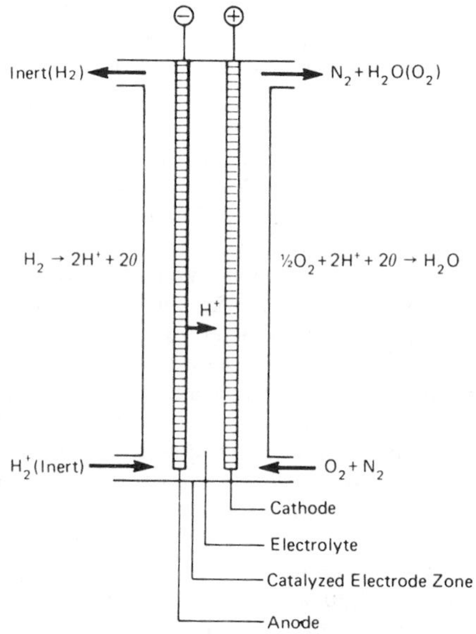

Fig. 1. Principles of operation of the hydrogen-air cell with acid electrolyte. Product water is removed by the flowing air.

from fuel mixtures and oxygen from air. Since product moisture is formed in an acid cell on the cathode, the air depleted in oxygen can be used for water removal if the cell is operated at a sufficiently high temperature to vaporize the water as it is formed.

The electrode has a central function in cell operation. In its catalyzed layer, it provides a large number of sites where gases and electrolyte can react. By virtue of a porous configuration, fast reactant transport and removal of inerts and product moisture is possible. The electrode also provides a path for current to flow to the terminals and serves to contain the electrolyte. The latter not only provides ionic conduction, but also assures separation of the reactants.

Cell Voltage. The cell voltage and the free energy of the underlying reaction are defined by

$$U=\frac{\Delta F}{nF}$$

where U= theoretical cell voltage; n= number of electrons transferred in the reaction; and F= Faraday constant.

Since $\Delta F = \Delta H - T\Delta S$, it follows that, depending upon the value of ΔS, the electrical energy to be derived from the cell, can be larger or smaller than the energy ΔH obtained by direct combustion of the fuel. ΔH= reaction enthalpy for the current-generating reaction; ΔS= entropy change; and T= absolute temperature.

For the hydrogen-oxygen couple, the corresponding theoretical cell voltage at 25°C is 1.23 volt if liquid coater is formed; or 1.18 volt if the product water is vaporized.

The thermodynamically possible conversion efficiency, however, is only partly realized in a practical fuel cell. Two basic losses are encountered: (1) the ohmic loss and (2) the electrode polarization, that is, the deviation of the actual from the thermodynamic electrode potential. The polarization is the result of the irreversibility of the electrode process, that is, the activation polarization and the voltage loss which develops from concentration gradients of the reactants. This leads to the current-voltage characteristics as shown in Fig. 2.

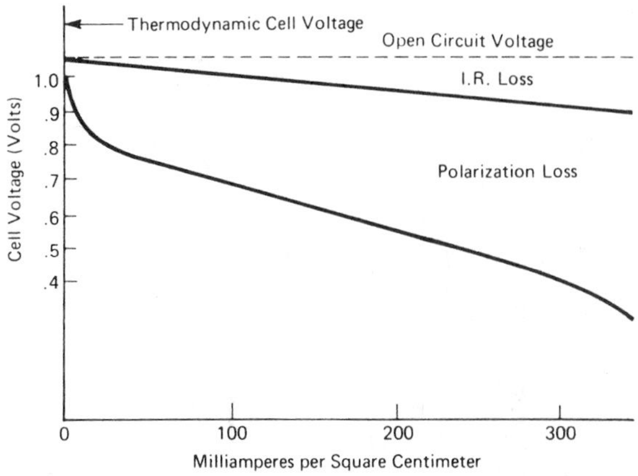

Fig. 2. Current-voltage characteristic of hydrogen-air cell with phosphoric acid electrolyte. Operating temperature is 125°C (257°F).

Cell Technology

Fuel cells currently in use (early 1980s) or in an advanced state of development are based upon the electrochemical systems summarized in the accompanying table. The selection resulting from electrochemical as well as systems and cost considerations includes cells with aqueous and fused-salt electrolyte and high-temperature cells in which ionic transport is provided by oxygen mobility in the solid state. The alkaline cell with platinum-activated electrodes, a highly efficient cell, but sensitive to reactant impurities, is used in spacecraft power plants. The remaining cells are considered for commercial power generation and are air-breathing.

Commercial cells operate at elevated temperatures and, in larger power plants, also at elevated pressures. This leads to an improvement in cell performance and also provides reject heat at a useful temperature level.

In molten carbonate cells and solid-electrolyte cells, the temperature of the reject heat is sufficiently high to permit integration of these cells with coal gassification systems (Dawes/Peterson, 1981).

Historically, the fuel cells listed in the table were not the first to be perfected. The first practical fuel cell was built in connection with the National Aeronautics and Space Administration's Gemini Program by the General Electric Company and relied on an unconventional electrolyte, namely, a solid polymer electrolyte membrane. Referred to as an "ion exchange membrane," it consisted of a lacelike organic structure, with anionic groups firmly bonded and hydrogen ions loosely held in the polymer chain. This provided sufficient mobility for ionic transport. A main advantage of the ion-exchange membrane was the elimination of the need for electrolyte containment because of its well-defined boundaries. Separate electrodes were not required since a layer of platinum black bonded to the surface served that function.

Phosphoric Acid Matrix Cell

Among air-breathing cells, those with phosphoric acid electrolyte are the most advanced. The matrix-type cell construction is used. In this configuration, a limited amount of electrolyte is trapped in a microporous structure by capillary forces. As a result, thin, highly porous, and comparatively low-cost electrodes can be employed inasmuch as the electrodes are not required to contain the electrolyte. In practice, for electrolyte absorption, a thin layer of Teflon-bonded silicon carbide powder is directly applied to the electrode surface. The electrodes are made up of 0.3 millimeter-thick Teflon-impregnated

MAJOR ELECTROCHEMICAL SYSTEMS FOR FUEL CELLS

Electrolyte	Current Transport	Operating Temperature		Electrode Catalyst	Reactants		Operating Pressure	
		°C	°F		Fuel	Oxidant	PSIG	Atmospheres
Aqueous potassium hydroxide (KOH)	OH^-	20–90	68–194	Nickel, silver, or platinum metals	Hydrogen	Oxygen	15	1.02
Concentrated phosphoric acid (H_3PO_4)	H^+	190	374	Platinum metals	Impure hydrogen*	Air	Atmospheric to 120	8.16
Fused alkali carbonate	CO_3^{2-}	600–800	1112–1472	Nickel, silver	Impure hydrogen*	Air	Elevated	
Stabilized zirconium oxide	O^{2-}	700–1000	1292–1832	Base metal oxides	Impure hydrogen*	Air		

* Hydrogen-containing mixtures, such as steam reformate.

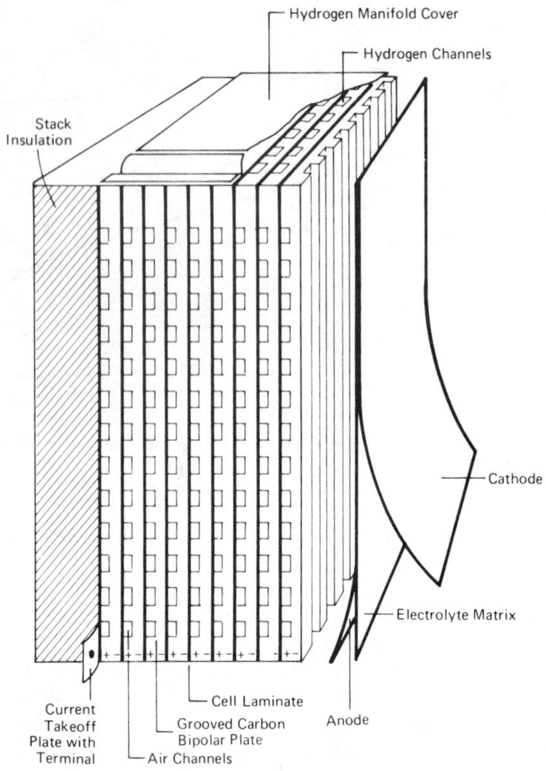

Fig. 3. Stack section of phosphoric acid matrix cell. Operation and design of cell are exceedingly simple. Product water is removed with air.

carbon fiber sheets, which are activated on one surface with a layer of platinum dispersed on carbon black. See Fig. 3.

Single cells are sandwiched between carbon plates with a suitable pattern for current collection and grooved for reactant distribution. The direction of flow channels for air and hydrogen fuel are perpendicular to each other. Air is used to remove the product water as it is generated and may also serve to remove reject heat (Adlhart, 1972).

In large cells, in order to minimize temperature gradients, heat is removed through cooling plates or coils located in the stack either by recirculation of liquid coolants (Johnson/Kaufman, 1981) or by generation of steam.

Fuel Sources for Fuel Cells

The reactants converted in the fuel cell to electric power are hydrogen and oxygen from air. Hydrogen, however, is not the primary fuel source in many modern cells. It is obtained from selected fossil fuels to be substituted eventually by coal-derived synthetic fuels. These fuels are converted by suitable processing, such as steam reforming, into a hydrogen-containing gas stream from which hydrogen is extracted in the fuel cell for power generation. See also **Coal Conversion Processes.**

As of the early 1980s, the preferred fuel for major applications is methane derived from natural gas—substitute natural gas (Rigney, 1981). See also **Substitute Natural Gas (SNG).** The cost is favorable, a distribution system is in place, and the availability in quantity appears assured for the remainder of the century.

Alternative fuels under consideration are light distillates, coal gas, and fuel-grade methanol. The latter is a potentially coal-derived liquid fuel. Methanol can be steam reformed at low temperatures and is considered partly for this reason as well as for its adaptability to smaller, transportable fuel-cell power plants (Smith, 1981).

Fuel Cell Power Plants and Their Applications

The developments of fuel cells first undertaken for the space program were directed to the advantages of high energy density and the advantage of cryogenically stored reactants (hydrogen and oxygen). The cell used in the Space Shuttle is considerably advanced over the earlier cells used in the *Gemini* and *Apollo* programs and weighs only 0.1 that of the earlier cells per kilowatt of power generated (Gitlow, 1980).

Energy density is not as much an overriding consideration in potential terrestrial uses as it is in space programs. In the case of ground-based installations, the inherent conversion efficiency, with the resulting fuel savings, most likely will be the primary objective.

As indicated by Fig. 4, thermal efficiencies of fuel cells are expected

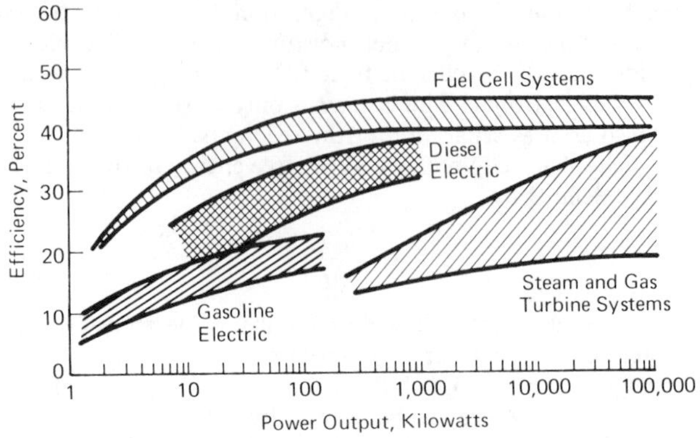

Fig. 4. Thermal efficiency of fossil fuel-operated fuel-cell power plants compares favorably with conventional means of energy conversion. The efficiency is reduced in small units mainly because of losses in the fuel processing. (*United Technology Corp.*)

to be superior to conventional means of energy conversion. Thus, an efficiency of 38% (based upon higher heating value) is projected for initial fuel-cell systems, with a potential for over 50% efficiency in the future. Moreover, the efficiency is essentially constant over a wide range of loads and, to a degree, independent of power plant size. Among other appealing features is the low level of emissions: fuel-cell power plants are expected to produce only a small fraction of the pollutants released by conventional fossil-fueled power plants. The potential for modular design also is distinctly appealing. This feature should allow mass production of units and an ability to add power generating capacity quickly when required.

In contrast with the comparatively simple hydrogen-oxygen dc power plant used in spacecraft, the commercial fuel-cell system, operating on carbonaceous fuel and air, entails a much higher degree of technical complexity. As shown by Fig. 5, a commercial system is comprised of three subsystems: (1) the reformer section; (2) the fuel-cell power section; and (3) in most instances, an inverter to generate utility-grade alternating current. To achieve high fuel conversion efficiency, close thermal integration of the fuel cell and the fuel-processing section is required. In particular, part of the reject heat in the fuel cell is used to generate steam for the reforming of the fuel.

The status of major development programs in the early 1980s is described in the following paragraphs. Currently, the development relies exclusively on cells with phosphoric acid electrolyte. This is the only cell that at present is considered to be sufficiently developed for large-scale commercialization during the remainder of this decade.

On-Site Fuel Cell Program. This program is supported by some gas utilities and the federal energy program in the United States.

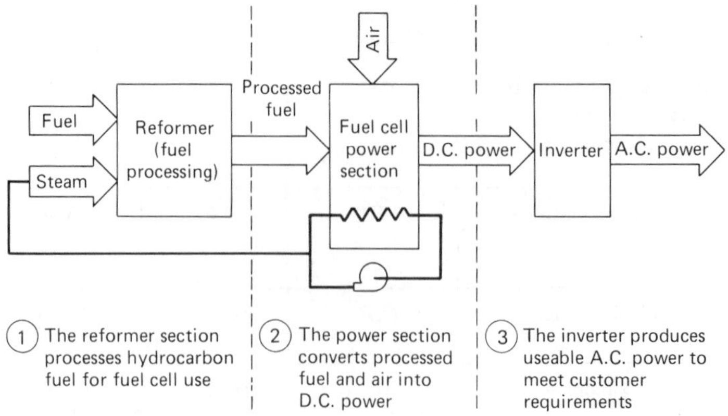

Fig. 5. Three principal subsystems required to convert hydrocarbon fuel to electric power.

The goal is development of systems which will supply electric power and heat for multifamily residential, commercial or industrial buildings (Sperberg/Woods, 1981). Fueled with natural gas and rated at 40 to 250 kilowatts, these systems are expected to benefit from the high electrical efficiency of the fuel cell as well as the high overall efficiency that results from cogeneration of heat and electricity at the point of end use.

The On-Site Program is the successor to an effort (called TARGET) launched in 1967 by a group of gas utilities. Field testing of 40-kilowatt units commenced in 1982 and commercial service is planned for the mid-1980s (Grevstad, 1981). The technical characteristics of field test units include:

(*Electric Power Rating*)—40-kilowatt grade, 3-phase alternating current grid connectable or capable of isolated operation.

(*Thermal Energy*)—Hot water supply at 160°F (71°C).

(*Fuel Efficiency*)—Overall fuel efficiency (including electrical and thermal energy), up to 80% at full electrical loads. Electrical generation efficiency (based on lower heating value), 40% at one-half to full electrical loads.

(*Environmental Impact*)—Low air-pollution emissions (based upon federal standards for the cleanest fossil-fueled, central electric generating stations): less than 0.1% of the particulates standard; less than 0.1% of the sulfur oxides standard; less than 10% of the nitric oxides standard; and negligible smoke and low noise generation (60 decibels at 15 feet; 4.5 meters).

Megawatt Demonstration Facilities

As of the early 1980s, two demonstration projects are in progress. One of these is at a Consolidated Edison Company site in New York; the other is in Tokyo. The fuel-cell power plants are rated at 4.8 megawatts and are scheduled to commence operation by 1984 or earlier. Their operation will allow the electric utility industry to assess the advantages of fuel cells for their specific needs. Efficiency, environmental compatability, and modular character are expected to make fuel cells attractive for dispersed peaking and load following in environmentally constrained urban and suburban zones. The possibility to recover reject heat is another important consideration. The 4.8-megawatt demonstration module in New York has been designed to use natural gas or petroleum distillate having a sulfur content of less than 500 parts per million and an end boiling point below 450°F (232°C). The electrical conversion efficiency of the system is expected to be 38%, based upon the higher heating value of the fuel.

Mobile Fuel-Cell Systems

The interest in smaller, transportable fuel-cell systems is longstanding (Kurpit/Gillis, 1974), with traction applications being the major potential use area. The preference for fuel cells over conventional modes of power generation is not exclusively based upon the presumed improved fuel efficiency. The preference also includes the expected superior general operational capabilities, such as increased reliability, lower operational and maintenance requirements, silent operation, and reduced emissions. Several major development programs are in progress (Huff, 1981), including an effort by the U.S. Department of Defense. The latter program is well advanced and is expected to lead to field testing of a 1.5 kilowatt system in 1983. See Fig. 6. The U.S. Army program focuses primarily on the application of phosphoric acid fuel cell technology for the development of tactical power units to replace the 1.5-, 3-, and 5-kilowatt standard engine generators. These fuel-cell power plants use methanol which is steam-reformed at low temperatures.

Fig. 6. Methanol fuel-cell system (1.5 kilowatt). Device uses air-cooled phosphoric acid cell type. (*U.S. Army photograph.*)

References

Adlhart, O. J.: "The Air-Cooled Matrix Type Phosphoric Acid Cell," in "From Electrocatalysis to Fuel Cells," Univ. Washington Press, Seattle, Washington, 1972.

Bacon, F. T.: "High Pressure Hydrogen-Oxygen Fuel Cells," in "Fuel Cells" (F. J. Young, editor), Van Nostrand Reinhold, New York, 1960.

Berger, C.: "Handbook of Fuel Cell Technology," Prentice-Hall, Englewood Cliffs, New Jersey, 1968.

Bockris, J. O'M., and T. Srinivasan: "Fuel Cells: Their Electrochemistry," McGraw-Hill, New York, 1969.

Breiter, M. W.: "Electrochemical Processes in Fuel Cells," Springer-Verlag, Berlin, 1969.

Dawes, M. H., and J. R. Peterson: "Carbonate Fuel Cell Technology Progress," Intersociety Energy Conversion Engineering Conference Proceedings, Atlanta, Georgia, 1981.

Gitlow, B.: "Strip Cell Test and Evaluation Program," Final Report, Contract NAS3-20042, National Aeronautics and Space Administration, Washington, D.C., 1980.

Grevstad, P. E.: "40 Kilowatt On-Site/Integrated Energy System," National Fuel Cell Seminar, Norfolk, Virginia, 1981.

Huff, J. R.: "Progress in Fuel Cells for Transportation," National Fuel Cell Seminar, Norfolk, Virginia, 1981.

Johnson, G. K., and A. Kaufman: "Phosphoric Acid Technology Progress," National Fuel Cell Seminar, Norfolk, Virginia, 1981.

Kurpit, S. S., and E. A. Gillis: "1.5 Kilowatt Indirect Methanol-Air Fuel Cell Power Plant," Engineering Foundation Conference on Methanol Fuel, New England College, Henniker, New Hampshire, 1974.

Liebhafsky, H. A., and E. J. Cairns: "Fuel Cells and Fuel Batteries: A Guide to Their Research and Development," Wiley, New York, 1968.

Rigney, D. M.: "Fuel Cell Power Plants—A User's Group Evaluation," National Fuel Cell Seminar, Norfolk, Virginia, 1981.

Smith, G. A.: "Methanol-Based Fuel Cell Systems," National Fuel Cell Seminar, Norfolk, Virginia, 1981.

Sperberg, R. T., and R. R. Woods: "Cogeneration with Fuel Cells," Intersociety Energy Conversion Engineering Conference Proceedings," Atlanta, Georgia, 1981.

Williams, K. R.: "Introduction to Fuel Cells," Elsevier, Amsterdam, 1966.

Otto J. Adlhart, Senior Research Associate, Engelhard Industries Division, Engelhard Minerals and Chemicals Corporation, Iselin, New Jersey.

FUEL (Coal). Coal.

FUEL CONVERSION FACTOR. Nuclear Reactor.

FUEL CYCLE. Nuclear Reactor.

FUEL ECONOMY. Automotive Electronics.

FUEL (Electric Power Generation). Electric Power Production and Requirements.

FUEL (Gas Turbine). Gas and Expansion Turbines.

FUEL (Hydrogen). Hydrogen (Fuel).

FUEL INJECTION. Automotive Electronics; Diesel Engine.

FUEL (Natural Gas). Natural Gas.

FUEL (Nuclear). Nuclear Reactor.

FUEL OIL. **Petroleum.**

FUEL-OXYGEN-SCRAP STEEL PROCESS. **Iron Metals, Alloys, and Steels.**

FUEL (Petroleum). **Petroleum.**

FUEL POISON. **Nuclear Reactor.**

FUELS (Rocket and Missile). **Rocket Propellants.**

FUEL (Substitute Natural Gas). **Substitute Natural Gas (SNG).**

FUEL (Tar Sands). **Tar Sands.**

FUGACITY. Only perfect gases obey exactly the ideal gas law, which is the basis for the derivation of many other equations for the properties of gases. Therefore we cannot substitute the measured pressure of real gases for the p term in such equations without more or less inaccuracy. Since, however, calculations are simplified by using ideal equations for real gases, the quantity fugacity is defined as the equivalent pressure of a real gas for which the ideal gas equations are valid, so that by tabulating calculated values of fugacity corresponding to measured pressures for real gases, we can use the relatively simple equations derived for real gases.

For example, the chemical potential for a mixture of ideal gases can be written in the form

$$\mu_i = \mu_i^*(T) + RT\ln p_i \tag{1}$$

where p_i is the partial pressure of component i. By analogy one may write for a mixture of real gases

$$\mu_i = \mu_i^*(T) + RT\ln p_i^* \tag{2}$$

where $\mu_i^*(T)$ is the same function as for the ideal gas, while all the effects of molecular interactions (that is, of the departure of the real gas from ideality) are included in the p_i^*. This function $p_i^*(T, p, n_1, \cdots n_c)$ is called the fugacity of component i. This definition, due to G. N. Lewis, permits the preservation for real gases of the general form of the equations for ideal gases, with the fugacities replacing partial pressures.

In the lower pressure limit, p_i^* reduces to p_i.

FULGURITE. A vertical, sometimes branching tube of fused quartzitic sand formed from the intense heat developed when the sand is struck by lightning.

FULL ANNEALING. A term applied to a heat treatment of steels in which the steel is heated into the austenite range and slowly cooled back to room temperature.

FULLER'S EARTH. A fine-grained earthy substance similar to clay, both in appearance and composition, but lacking the usual plasticity, possessing a higher water content, and usually high in magnesia. The material consists mainly of hydrated aluminum silicates, such as the clay minerals, montmorillonite and palygorskite. Generally, it is believed that fuller's earth was formed as a residual deposit as the result of decomposition of rock in place, perhaps by the devitrification of volcanic glass. The color of the material ranges from light brown through yellow and white to light and dark green. Fuller's earth is used for decolorizing oils, degreasing raw wool, and as a natural bleaching agent.

FULMAR. **Petrels and Albatrosses.**

FULMINATE OF MERCURY. **Mercury.**

FULMINATING. In medical usage, a sudden, often unexpected, surge in the progress of an infective process, proceeding from acute to superacute—usually of a serious nature and requiring immediate and special attention.

FULMINIC ACID. **Cyanic Acid and Related Compounds.**

FUMARIC ACID. **Isomerism.**

FUMAROLE. Derived from the Latin *fumus*, smoke, the term fumarole is applied to openings in the earth's crust, often in the neighborhood of volcanoes, which emit steam and gases such as carbon dioxide, hydrochloric acid, and hydrogen sulfide. A special name, solfatara, from the Italian *solfo*, sulfur, is given to fumaroles that emit sulfurous exhalations. Perhaps the greatest area of fumarole activity is the famous Valley of Ten Thousand Smokes, adjacent to Katmai Volcano, Alaska.

FUME. A suspension of fine solid or liquid particles (0.2 to 1 micrometer in diameter) in a gas. Technically, fumes are collodial systems formed from chemical reactions, such as combustion, distillation, subliimation, calcination, and condensation.

FUMING NITRIC ACID. **Rocket Propellants.**

FUNCTION. A mathematical expression describing the relation between variables; the function taking on a definite value, or values, when special values are assigned to certain other quantities, called the arguments, or independent variables of the function. If there is one independent variable, the dependent variable y may be determined explicitly by the equation $y = f(x)$ or implicitly by $f(x, y) = 0$. If there are several independent variables, the forms are $y = f(x_1, x_2, \ldots, x_n)$ or $f(x_1, x_2, \ldots, x_n, y) = 0$.

The precise definition of a function, namely as a set of ordered pairs, the first element of each pair being an argument of the function and the second its corresponding value, was first introduced by Dirichlet. A set-theoretical definition is simply a many-one relation, that is, a relation which to any element in its domain relates exactly one element in its range.

A function may be classified in many other different ways but the principal categories are: 1. Continuous or discontinuous. 2. Single-valued, if the dependent variable is uniquely determined when a number is assigned to the independent variable as $y = 2x$; multivalued (see also **Singular Point of a Function**) if two or more values of y can result when x is fixed, for example, $x^2 + y^2 = 4$. 3. Algebraic, if the variables involve only algebraic operations (see **Albegra,** where subclasses such as rational, irrational, power, polynomial, fractional, and radical functions are defined). 4. Transcendental, if it is not algebraic (see **Transcendental** for further definitions as well as those of its main subclasses: exponential, logarithmic, trigonometric, elliptic, etc.). 5. Real or complex, depending on the nature of the variable (see **Complex Variable**).

A function is said to be even if $f(x) = f(-x)$ or odd if $f(x) = -f(-x)$. Typical examples are x^2, $\cos x$ and x^3, $\sin x$. Other names often given to functions with special properties are: harmonic, homogeneous, integral, inverse, linear, orthogonal, periodic.

The zero of a function is a value of the argument for which the function vanishes. Thus a zero of $f(x)$ is a root of the equation $f(x) = 0$.

A functional is a function whose argument or independent variable is a curve or surface, hence a corresponding function. It may also be described as a function of a function.

Many functions are named for mathematicians. Some well-known examples of this type are listed under **Generating Function.**

FUNCTIONAL ANALYSIS. A branch of mathematics concerned with the study of linear topological space, their conjugate spaces, and the continuous operators between such spaces.

During the latter part of the nineteenth century Volterra and Fredholm succeeded in solving certain classes of integral equations. The subsequent development of their ideas by Hilbert, Riesz, and others led to a more careful examination of the topological linear spaces over which these equations were defined. In addition, the successful development of integration theory by Lebesgue and others provided numerous concrete examples of topological linear spaces. One further source of stimulation was the seminars of Hadamard in which the term "functional analysis" was introduced—a functional is a function defined on a space of functions.

Functional analysis has played an important role in much of the development of analysis over the last several decades. It provided the abstract framework and underpinning for the study of linear partial differential equations, harmonic analysis, distribution theory, and the foundations of quantum mechanics.

See also **Topology;** and terms listed under **Mathematics.**

FUNCTIONAL GROUP ANALYSIS. Analysis (Organic Chemical).

FUNCTIONAL GROUPS (Organic). Organic Chemistry.

FUNCTIONAL RESERVES. The ability of the body to accomplish additional muscular or other activity and useful work beyond the normal level of activity of an individual.

FUNCTION (Singular Point of). Singular Point of a Function.

FUNDAMENTAL FREQUENCY. Frequency.

FUNDUS. The base of an organ; the part farthest removed from the opening of the organ.

FUNGAL ARTHRITIS. Arthritis (Infectious).

FUNGICIDE (Pyridine). Pyridine and Derivatives.

FUNGUS. Any of a group of thallophytic plants (phylum *Thallophyta*) mainly characterized by an absence of chlorophyll. Examples of fungus include mushrooms, toadstools, smuts, rusts, molds, and mildews. There are well over 70,000 species, with widely diverse habits and characteristics. Thus generalizations are difficult. Among the fungi are numerous microscopic unicellular forms, as well as plants of elaborate structure and considerable size. Fungi grow in almost every habitat where organic sutstances exist and external conditions are suitable. Numerous species are found in water, either fresh or saline. Others are adapted to life on land, or in the ground. During the short summers of the Arctic, certain species appear. Fungi are particularly abundant in the tropics because warm, humid climates tend to favor the existence of numerous species.

Over 40 species of fungi produce diseases in humans and other mammals. There are those fungi which attack only the hair, skin, and nails; and there are those species that invade deeper tissues of major internal organs to produce serious systemic diseases. Many more species produce a variety of diseases among plants, the various species tending to specialize in attacking certain kinds of plants and certain parts of plants. Crop damage from fungus infections runs into the many millions of dollars annually, which damage, of course would be many times greater if effective controls were not applied.

Nature of Fungi. Except for the absence of chlorophyll, the structure of fungi resembles that of the algae, the other main division of the thallophytes. The vegetative body of a fungus, except for the unicellular forms, is always composed of slender branching threads, or hyphae, making up what is known as *mycelium.* Mycelia in many cases are colorless, but may contain pigments of every color. Each hypha is ordinarily composed of a row of cells, each containing one (or more) minute nuclei. In most species, cross walls are rarely formed, the hypha being coenocytic. Even the largest, most complex fungi are composed entirely of tangled masses of hyphae, which may be loosely aggregated or so densely packed as to form a hard body suggestive of woody structure, as for example, in the bracket fungi. See also **Bracket Fungi.**

In their reproductive processes, fungi are quite similar to algae. both are asexual. Sexual reproduction occurs in the life histories of these plants, which also often show very distinctly an alternation of vegetative growth and reproductive activity. As may be expected in so diversified a group of plants, a considerable variety of reproductive processes occurs.

Among the lower forms, many of which occur in water, asexual reproduction is accomplished by means of zoöspores. The zoöspores are formed in sporangia from which they escape at maturity. After a period of motility, each zoöspore settles down, loses its cilia and at once gives rise to a new plant. In the nonaquatic fungi asexual reproduction ordinarily occurs by means of nonmotile spores, called conidia, which have a rigid cell-wall. These conidia, often produced in immense numbers, are carried about by air currents, sometimes to great distances, and on reaching a favorable habitat, germinate to form a new plant. The methods of sexual reproduction found in fungi are extremely varied, and can best be considered under the different groups. Sexual reproduction usually occurs in a distinct body, the sporophore, which forms in many fungi a very conspicuous part of the life cycle of a fungus. This is frequently the only part recognized by the ordinary observer. In each kind of fungus the sporophore assumes a very definite and distinct form. In the cup fungi the sporophore is frequently a saucer- or cup-shaped structure. Other types are found in the familiar mushroom; in the puff-balls; and in the bird's-nest fungus, the sporophore here having many small somewhat spherical objects contained in an open cup-like body. In all of these, spores are formed, often in unbelievable numbers; a common puff-ball contains millions of them. The spores are borne about in the air currents, and germinate when brought to a favorable environment. It is obvious that many spores must fail to reach such a favorable spot, otherwise the world would be overrun with fungi.

Classes of Fungi. The true fungi are separated into three classes: (1) the Phycomycetes, in which the mycelium is nonseptate and coenocytic; (2) the Ascomycetes, characterized by having spores borne in special sacs or asci; and (3) the Basidomycetes, distinguished by the basidium, a spore-bearing cell which typically bears externally four spores (in some cases more or less). In addition to these three classes, there is another group known as the Imperfects, or Fungi Imperfecti, which contains those forms of plants in which the sexual or perfect stage is not known and which, therefore, cannot be assigned to one of the three aforementioned classes.

Phycomycetes. This group of fungi is so diverse in habit as to suggest polyphyletic origin, quite probably from several different groups of green algae, to which many of them show remarkable similarity. Simpler members of the Phycomycetes consist of but a single cell, while other species have a well-developed branching mycelium, always composed of hyphae possessing no cross-walls. Many species grow in water and are known as Water-molds. Others grow out of water. Among the latter are some of the common destructive parasites.

Also known as the lower true fungi, there are some 1500 species of the Phycomycetes. A number of these species are parasitic on crop plants.

Ascomycetes. The majority of the some 40,000 species making up this group of fungi are small, often minute, while a relatively few species attain heights of 3 to 4 inches (7.5 to 10 centimeters), with a diameter of 1 to 2 inches (2.5 to 5 centimeters). Occasional individuals are even larger. All are characterized by the ascus, or spore-sac, commonly an elongate cylindrical body containing eight spores. Much more detail on the physical nature of these fungi is given in entry on **Ascomycetes.**

Many members of the Ascomycetes are of major economic importance because of their destructive parasitic habit and damage to food and fiber crops. Among the Ascomycetes are also many beneficial species, such as truffles and morels, and, of course, the many yeasts which are basic to fermentation processes. These factors are also described in the entry on **Ascomycetes.**

Basidiomycetes. Most of the fungi commonly observed are members of this group of fungi, which includes toadstools, mushrooms, puffballs, and many other forms. The characteristic feature which distinguishes them from other fungi is the basidium, typically a club-shaped structure bearing four spores at its apex. The Basidiomycetes are discussed in considerable detail in entry on **Basidiomycetes.**

Rust Fungi. The rust fungi (*Uredinales*) are parasitic basidiomycetes which owe their popular name to the reddish color of the spore masses in some of the commonest species. Because many of the thousand species found in North America attack important cultivated and wild plants, they are of great economic importance. The group possesses a remarkable variety of spore types. Many species have five kinds of spores. It is notable that any given species has a very limited range of host plants.

One of the best-known species of Rusts is the common Wheat Rust, *Puccinia graminis,* which has long been known. Five spore forms are included in its complex life-history.

On wheat plants, and on various grasses, there appear during the summer on the stem and leaves reddish spots which on examination are found to contain large numbers of one-celled spores which are called urediniospores. See Fig. 1. These are scattered by the wind

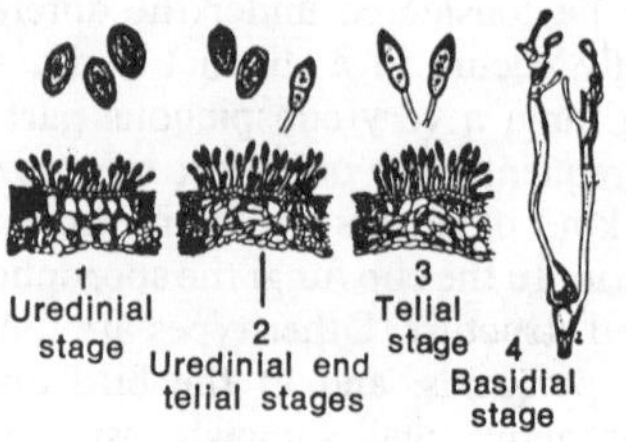

Fig. 1. *Puccinia graminis* in wheat leaves: (1) uredinial stage; (2) uredinial stage changing to telial stage; (3) telial stage; (4) basidiospore germinating and forming basisia and basidiospores. (*Tulasne.*)

and reinfect wheat plants continuously during the growing season. Near the end of the growing season, when the host plant is maturing, a new form of spore appears either with the urediniospores or in separate pustules. These spores are two-celled, of dark color, and have a very thick wall. They are called teliospores, or winter spores, and are able to survive the winter season independent of any host plant, for they can neither attack one nor parasitize it.

In the spring each cell of the teliospore puts out a short tube which becomes four-celled. From each of the four cells a small spore is formed: this is the basidiospore and the four-celled tube is therefore a basidium. The one-celled basidiospores are carried by air-currents to suitable host plants, which in this case are not wheat plants, but barberry plants. In contact with the young leaves of the latter plant the basidiospore develops a short tube which penetrates the leaf epidermis and forms a mycelium within the leaf. After a time this mycelium gives rise to a new type of spore which appears on the upper surface of the barberry leaf in small pustules called pycnia or spermagonia. See Fig. 2. These contain masses of small hyphae, from the tips of

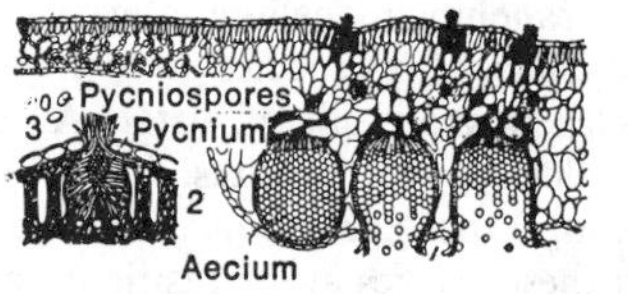

Fig. 2. 1. Section of barberry leaf infected with *Puccinia graminis* and producing pyenia and aecia; (2) section through pyenium; (3) three pycniospores.

which are cut off minute one-celled bodies called pycniospores or spermatia, which seem incapable of reinfecting the host plant.

Soon after the formation of the pycnia there appear on the under side of the leaf clusters of orange-colored cups. These are the aecia or cluster cups, in which are formed chains of tightly packed aeciospores. These spores, when released, cannot reinfect barberry plants but must be carried to wheat plants before they can grow. On the wheat plant each aeciospore puts out a short germ tube which penetrates the tissue of the leaf or stem within which it forms an extensive mycelium, from which the urediniospores and later teliospores are formed.

So it is apparent that for the completion of its life-history *Puccinia graminis* must have two very different host plants. Many rusts show this character of requiring alternate hosts, and are called heteroecious. Other rusts complete their life cycle on a single host: they are said to be autoecious.

Another rust of great economic importance, especially in the northern United States and Canada, is the white pine blister rust, *Cronartium ribicola*, which, like wheat rust, has two alternate hosts, white pine having the pycnia and aecia, and currants and gooseberries the uredinia and telia stages.

In many rusts one or more of the spore forms may be entirely lacking. For example, in *Gymnosporangium juniperi-virginianae* there is no uredinial stage, while in the common hollyhock rust, *Puccinia malvacearum*, pycnia, aecia, and uredinia are all lacking, only the teliospors and basidiospores being formed.

Smut Fungi. The smut fungi (*Ustilaginales*) are also parasitic basidiomycetes, so named because of the conspicuous masses of sooty black spores which they form externally on the host plant. Infection by smuts is seldom fatal to the host plant but does seriously reduce its size and may even prevent seed formation completely. The fungus grows as a septate mycelium which penetrates between the cells of the host plant; into these cells it sends haustoria, which obtain nourishment therefrom. Presently this septate mycelium gives rise to immense numbers of spores. Often the presence of the mycelium causes the host tissue to enlarge tremendously, producing irregular tumor-like growths. These are particularly conspicuous in Corn Smut. The spores are thick-walled unicellular objects capable of surviving for some time under unfavorable conditions. On germinating, each spore develops a short germ tube, or promycelium, which becomes from 1–4 cells long. Each of these cells produces a spore. The promycelium becomes a basidium; and the spores, basidio-spores. These spores are capable of infecting new host plants, producing therein a mycelium. Conjugation between cells of the mycelium occurs so that each cell comes to have two nuclei. As growth continues, the two nuclei of any cell divide simultaneously so that every cell continues to have two nuclei. When spore formation occurs, the two nuclei fuse, dividing again when the spore germinates. Many variations of this process are found; in many smuts the promycelium buds off from its apex many cells. Often fusion between two of these cells occurs immediately, even before they are separated from the promycelium.

Because of the rapidity with which smuts may spread and the great reduction of seed production which their presence may cause, smuts are of great economic importance. One species, Corn Smut, *Ustilago zeae*, causes the loss of millions of bushels of corn yearly. Oat smut, *Ustilago avenae*, may cause a 30% reduction in yield, while other smuts are equally important. Control of the parasites may be obtained by rotating crops; but due to the resistant nature of the spores, at least three years should elapse before replanting an infected field to the same crop. Other methods of control consist of soaking seeds in various solutions, such as formaldehyde solution in water, or dusting infected seeds with copper compounds.

Important Plant Diseases Caused by Fungi

As is true of insects, nematodes, bacteria, viruses, and other injurious pests of crop plants, some of the fungi function on a rather broad spectrum, attacking several plants, whereas other species of fungi specialize and confine their attacks on one or just a few plants. This also applies to method of attack, some species of fungi injuring numerous portions of a plant, while others confining their injury only to steams or leaves and roots, etc. With severe infections, of course, regardless of the point of attack, the end result is that the plant withers and dies, usually within a short period.

An abridged list of important fungus diseases of crops would include:

Alfalfa. Crown wart of alfalfa disease. First noted in Ecuador in 1895 and later confirmed in Europe and the United States in the early 1900s. First reported in California in 1909. The disease is caused by the organism *Physoderma alfalfae* Karling. The disease occurs mostly in the western states.

Apple. Bitter rot of apple disease. This has been known since the early 1800s, when it was recognized in Europe. Causative organism is *Glomerella cingulata* Spauld. & Schrenk. Generally found in the United States east of the Rocky Mountains and south of latitude 40°N. *Apple rust* occurs widely in Europe and in North America east of the Rocky Mountains. Causative organism is *Gymonsporangium juniperi-virginianae* Schw. It was noted in the early 1800s. *Apple scab* was first noted in Sweden in 1819 and in Germany a few years later. It was noted in the United States in 1834 and in the British Isles in 1845. However, it did not reach Australia until 1862. Causative organism *Venturia inaequalis* Wing.

Banana. A vascular disease known as *banana wilt* is caused by *Fusarium oxysporum* f. *cubense* Snyder & Hansen.

Barley. Loose smut of barley is caused by *Ustilago nuda* Rostr. and is one of three smuts that affect barley.

Bean. A major disease of dry and snap beans (*Phaseolus vulgaris* L.), or bean anthracnose, was first described in Germany in 1875. The disease also affects several other types of bean and is found essentially worldwide. Evidence of the disease is shown in Fig. 3. Rust fungus also occurs on beans. These fungus diseases are most common during cool, moist summers. The fungus is carried on seeds and lives in soil and on remains of diseased plants. In parts of the world, planting of anthracnose-free seeds has proved quite successful. Crop rotation also is a good cultural practice. Since the late 1800s, resistance to

Fig. 3. Snap beans grown in the southern United States that were fieldgraded and shipped to northern markets without refrigeration, during which time they were destroyed by anthracnose. (*USDA photo.*)

the disease has been noted to vary considerably from one variety of bean to the next. Genetically developed resistance has proved interesting and complex and, to date, has been valuable in developing disease-resistant seeds.

Celery. This plant is subject to both early and late blight and to Fusarium yellows (*F. oxysporum* f. *apii* Snyder Hansen). Late blight of celery was first recorded in Italy in about 1890, and shortly thereafter in Denmark, Germany, and the eastern United States. It was not reported in the British Isles until about 1910. The seed-borne disease is now reported worldwide. If not controlled, the disease can have serious economic consequences. The causal organism is *Septoria apiicola.* Evidence of the disease on celery leaves is shown in Fig. 4; on stalks in Fig. 5. Effective control chemicals include a spray or dust

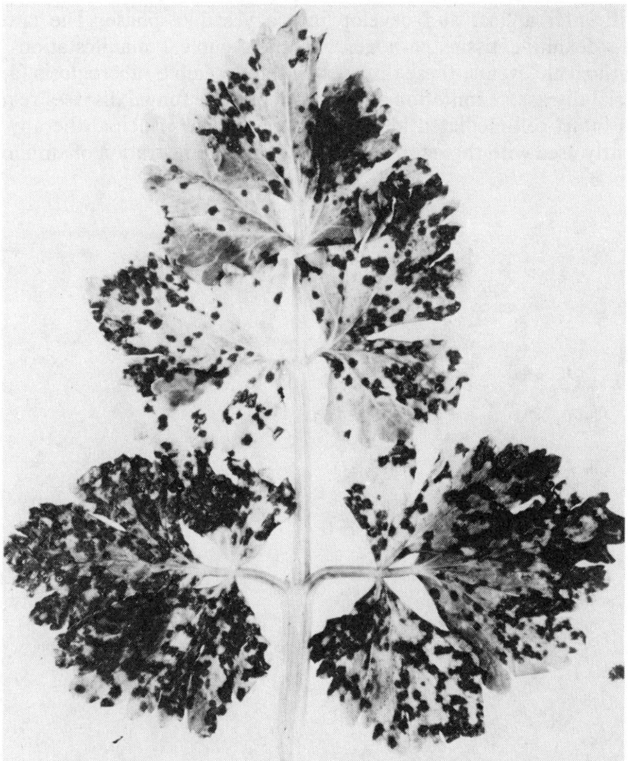

Fig. 4. Celery plant showing extensive evidence of late blight disease. (*USDA photo.*)

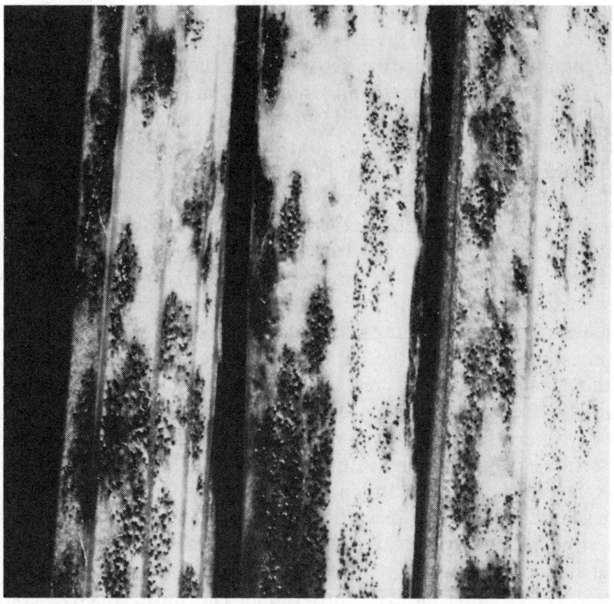

Fig. 5. Stalks of celery showing effects of late blight disease. (*USDA photo.*)

containing a fixed copper fungicide or zineb, or ziram. Crop rotation is effective.

Cereals. Ergot of grains and grasses, caused by *Claviceps purpurea* (Fr.) Tul., is a serious economic disease and occurs worldwide. The fungus affects the flowering parts of the affected plants. Rye is particularly severely affected. See Fig. 6. Infected flowers produce ergot sclerotia rather than the normal kernels, thus reducing yields. The sclerotia

Fig. 6. Ergot present in the head of rye. The purple-black fruiting bodies replace grain in the rye head, destroying the plant as a food crop. (*USDA photo.*)

contain alkaloids which subsequently can be quite injurious if the affected grain is fed to livestock. See also *Ergot.* The most effective control is use of ergot-free seed as well as rotation of susceptible grain crops with legumes and other resistant plants, thus reducing the population of overwintering organisms. Other fungus diseases of cereals include head blight (scab), seedling blight, powdery mildew, and black stem rust caused by *Puccinia graminis* Pers., among others.

An interesting interrelationship between barberry and wheat rust has been known since 1804. Experiments during the early 1800s proved that barberry hosts the organism and that it is easily transferred from barberry to cereal plants. Eradication of barberry in some regions has greatly alleviated the problem.

Chestnut. A major fungal disease of chestnut is *Endothia* canker or chestnut blight. In the early 1900s, this disease destroyed most of the American chestnut trees in the Appalachian ranges of the eastern United States. The disease was first noted in New York City in 1904 and, within a few years, was noted several hundred miles distant in western Pennsylvania and in northern Georgia. Discovery of the disease in China in 1913 and a bit later in Japan indicates that the disease may have been brought to North America on nursery stock from the Orient. In the early 1930s, the disease was reported in Italy, with further findings of it in various parts of Europe as far east as the U.S.S.R. and southeast into India. The most effective controls found by chestnut producers is the use of the resistant Asiatic varieties of the tree. A concerted research program continues toward the development of resistant species that are more suitable to American climate.

Citrus. One of the most destructive of fungus diseases to citrus, as well as stonefruits, is caused by *Armillaria mellae* Quel. This is an important economic disease of the Pacific coast citrus crops in the United States. The disease, first identified in 1873, invades the root system of the citrus tree.

Further descriptions of specific fungus infections of plants are beyond the scope of this book. However, more of a panorama of the widespread nature of fungus diseases can be gleaned from Figs. 7 through 10.

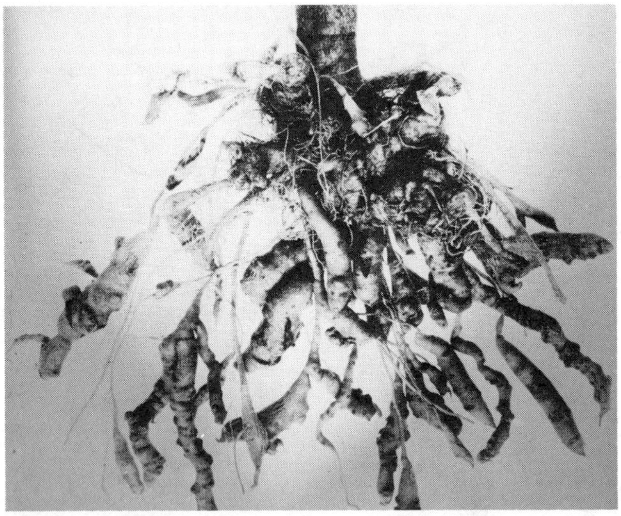

Fig. 7. Malformed, spindlelike roots of cabbage plant affected by clubroot fungus disease. (*USDA photo.*)

It is interesting to note that the *late blight of potato,* caused by the fungus *Phytophthora infestans* Dby., has been known in Europe since the 1500s, having been introduced there from South America. The disease was first reported in the United States about 1830. There was a severe epidemic of the disease in Europe in 1845, when the disease was a major cause of the *Irish potato famine.*

Human Diseases Caused by Fungi

In addition to causing diseases and discomforts of an annoying nature, such as athlete's foot, jockey-strap itch, and ringworm, certain species of fungi are involved in human diseases of a more serious nature.

In North and South America and Europe, the principal mycotic (fungus) infections are blastomycosis, coccidioidomycosis, histoplasmosis, and to a lesser extent sporotrichosis. These diseases are described in separate alphabetical entries in this book. Actinomycosis is sometimes reported as a fungus disease, but the causative agent really is a gram-positive bacteria. These bacteria have been reported erroneously because they have a characteristic filamentous, branching shape which resembles the hyphae of fungus organisms. See also **Actinomycosis.**

Fig. 8. Extensive evidence of downy mildew on young grape berries. (*Cornell University, Department of Plant Pathology.*)

Mycotic infections have a number of common points. They usually occur in fairly well-defined geographic regions, they usually occur in the soil, and they are easily aerosolized and thus easily spread by air currents and contracted by inhalation (not always). The fungus is dimorphic, existing in its natural habitat as a mold (mycelium) which bears infectious spores. Upon airborne dissemination, the spores will enter a host and develop into a yeastlike phase. The latter is the definitive tissue pathogen. In their clinical manifestations and pathogenic events, these diseases closely resemble tuberculosis (a bacterial disease). Limitation and control of these fungal diseases requires an intact-cell-mediated immune response. The principal therapy currently used with these fungus infections is administration of amphotericin B.

Fig. 9. Appearance of onion smudge on onions otherwise ready for market. (*USDA photo.*)

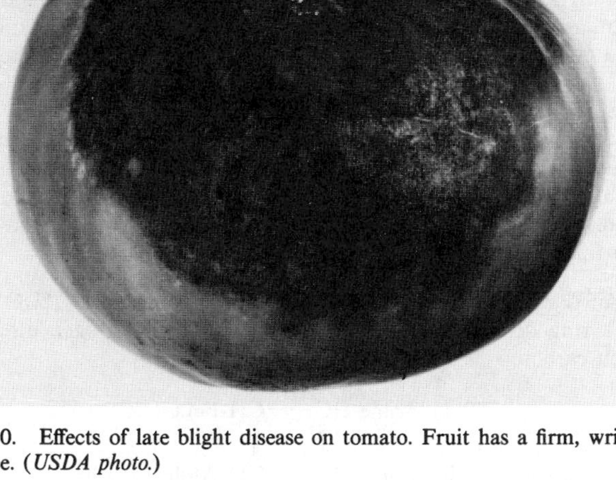

Fig. 10. Effects of late blight disease on tomato. Fruit has a firm, wrinkled surface. (*USDA photo.*)

The involvement of fungus microorganisms in a number of other diseases is described in connection with entries on several specific diseases in this book, as in the case of arthritis (fungal arthritis), etc. In particular, consult the entry on **Dermatitis and Dermatosis.**

The poisonous nature of many fungi has received wide publicity, and probably accounts for the popular aversion to this group of plants. The toxic substances present in the fungus are products of its metabolism, not substances absorbed from without. It may be that these substances are some sort of waste products accumulating in the cells. Some people have suggested that they are a means of protection against animals which might otherwise eat the plant. However, many animals eat with impunity fungi which are violently toxic to man, so it seems difficult to maintain this explanation. The toxic substances present in fungi are various. Closely related species may contain quite different poisons; a single species may have more than one poison. The effect of the poison on the human body varies. One group of poisons is taken into the body some hours before its effects become evident. Then abdominal cramps and nausea develop; vomiting occurs; thirst arises and diarrhea. These symptoms continue for hours, usually (but not always) ending in death. This is the type of poisoning caused by *Amanita phalloides. Amanita muscaria* is less violent in its action. The poison of this fungus acts on the nerve centers, causing lack of coordination, illusions and delirium, as well as gastric disturbances. *Amanita muscaria* is only very rarely fatal. This fungus is used by native tribes of northeastern Siberia as a stimulant. In addition to the *Amanitas,* many other fungi are of poisonous nature. It also seems that the effect varies among different people. What one finds edible may be definitely toxic to another. This renders even more difficult the problem of satisfactorily determining harmful species.

The saprophytic forms of fungi attack foodstuffs, causing complete spoilage; attack fabrics which they mildew and so ruin; and attack and cause the destruction of timbers. While on the surface these processes appear negative and destructive, it should be stressed that some forms of destruction in nature are necessary as means of preparation for new growth. Were all rotting prevented, the accumulation of dead material would soon become so great as to hinder and stop life. Only the lower plants, notably fungi and bacteria, are able to break down complex matter to forms in which it is again available for higher organisms.

On the positive side for fungi is the important role played by them in the development of various antibiotics. See **Antibiotic.** Also, many edible forms of fungi are available. Of these, many are species which grow wild. To distinguish those species which are edible from those which are harmful is an ever present problem. To style the edible species mushrooms and reject the others as toadstools does not solve the problem, since it first becomes necessary to define the terms "toadstool" and "mushroom." And there is no obvious distinction. The only safe rule to follow is that of total abstinence from any doubtful species until one is absolutely certain that it is safe. It is thus only natural that man has turned to the cultivation of fungi. Of all the edible species known, only a few have been successfully cultivated. Of these, only one is commercially important, the mushroom, *Agaricus campestris.* (See **Agarics.**) Mushrooms are much fancied for their palatability and fine flavor. It is interesting to note that one group of tropical ants feeds largely on fungus plants, which they grow in their nests in an advanced state of cultivation.

References

Bennett, J. E.: "Chemotherapy of Systemic Mycoses," Parts 1 and 2, *N. England Jrnl. Med.,* **290**:30, 320 (1974).

Emmons, C. W., et al.: "Medical Mycology," 3rd edition, Lea and Febiger, Philadelphia, 1977.

Bourke, P. M. A.: "Emergence of Potato Blight—1843–1846, *Nature,* **203,** 805–808 (1964).

Halisky, P. M.: "Physiological Specialization and Genetics of the Smut Fungi," *Bot. Rev.,* **31,** 114–150 (1965).

Pappagianis, D.: "Coccidioidomycosis. Infectious Diseases," 2nd edition, Charles C. Thomas, Springfield, Illinois, 1958.

Shaw, M.: "The Physiology and Host-parasite Relations of the Rusts," *Ann. Rev. Phytopathology,* **1,** 259–294 (1963).

Walker, J. C.: "Plant Pathology," 3rd edition, McGraw-Hill, New York, 1969.

Westcott, C.: "Westcott's Plant Disease Handbook," 4th edition, Van Nostrand Reinhold, New York, 1980.

FUNGUS DISEASES (Human). Blastomycosis; Coccidioidomycosis; Dermatitis and Dermatosis; Fungus; Histoplasmosis; Sporotrichosis.

FUNGUS GNAT (*Insecta, Diptera*). Small two-winged flies whose larvae live on fungi and decaying vegetation. Family *Mycetophilidae.*

In general terms, a fungus gnat might be described as a mosquito-like fly. The larvae are creamy white or gray and feed on fungus when maggots.

The species (*Sciaridae*) or dark-winged fungus gnat is sometimes seen migrating in larvae form over the ground in an inch-deep snake-like line when hunting leaf mold or other food. The eyes of the dark-winged gnat are extremely close together. The species (*Sciara tritici*) attacks roots of wheat and mushroom beds. Sometimes members of the species (*Mycetophilidae*) will bore in potato tubers, causing a scab. The female of this species is wingless. The species (*Arachnocampa luminosa*) are found in great numbers in some of the caves of New Zealand. Their bioluminescent light is vividly seen in the dark caves. These insects spin webs that hang down for catching insects. In Europe and North America, the species (*Ceroplatus*) is found in large numbers. These insects also are luminescent and build webs on which they crawl. Most are predacious. Generally, the fungus gnats are from $\frac{1}{8}$ to $\frac{1}{4}$ inch in length when adults and prefer swampy, damp places.

FUNGUS (Lichen). Lichen.

FUNGUS (Rust). Rust Fungus.

FUNGUS (Smuts). Smuts.

FUNICULAR POLYGONS AND CATENARIES. If a closed loop of cord or rope is pulled at several points by forces in various directions, it forms a figure, plane or otherwise, known as a funicular polygon. The external forces acting on the loop at the vertices are, for equilibrium, subject to the same conditions as a set of noncurrent forces acting on a rigid body; while the three forces concurrent at each vertex, including the tensions in the loop itself, may be represented by an equilibrium triangle, and the several triangles, fitted together to form the equilibrium polygon for the external forces (Fig. 1). Figure

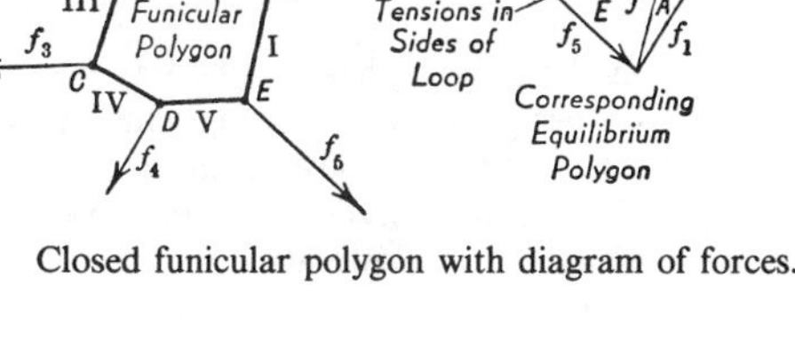

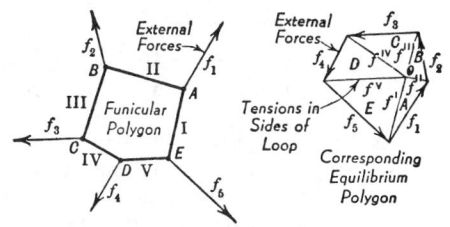

Fig. 1. Closed funicular polygon with diagram of forces.

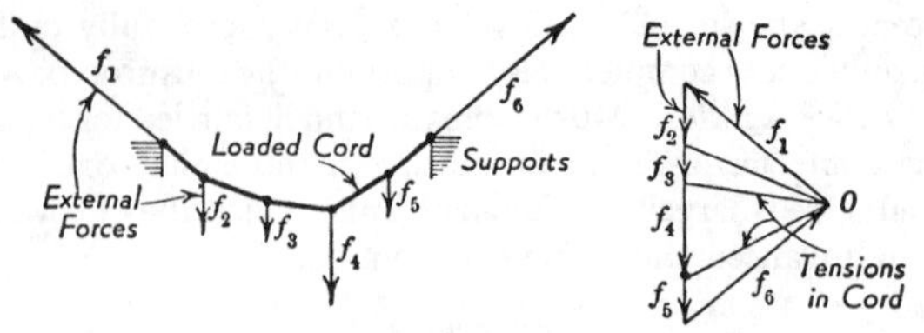

Fig. 2. Suspended cable with unequal loads.

2 gives the corresponding analysis for an open cord supported at the ends and loaded by weights hung vertically from it. In Fig. 3 the weights are equal, have equal horizontal spacing, and are hung close together. The form of the cord in this case approximates a parabola. This condition practically obtains with the cables of a suspension bridge. The point O in each figure is located by drawing from the extremities of any side of the external-face polygon lines parallel to the sides adjacent to the corresponding vertex of the funicular polygon.

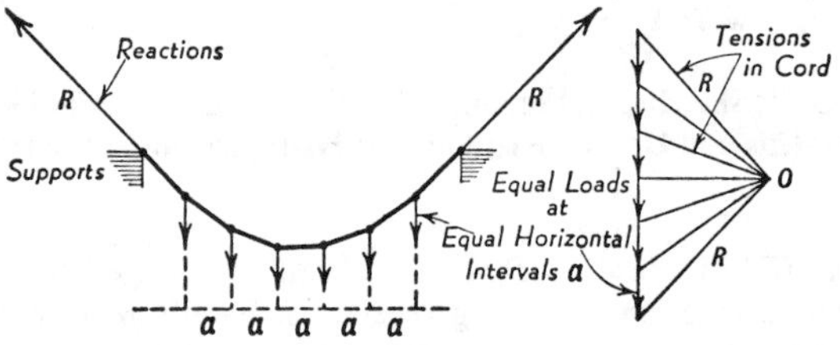

Fig. 3. Suspended cable with equal loads, as in a suspension bridge.

If a suspended cord is loaded uniformly along its length (not horizontally), as by its own weight, it assumes the form of a catenary. The equation of this curve may be written

$$y=\frac{a}{2}(e^{x/a}+e^{-x/a})=a\cosh\frac{x}{a}$$

The hyperbolic cosine form is convenient for numerical computations; a represents the Y-intercept of the curve. It is an interesting property of a catenary cable that if at any point it is hung over a pulley, and enough cable cut off to reach down to the X-axis (Fig. 4), the weight

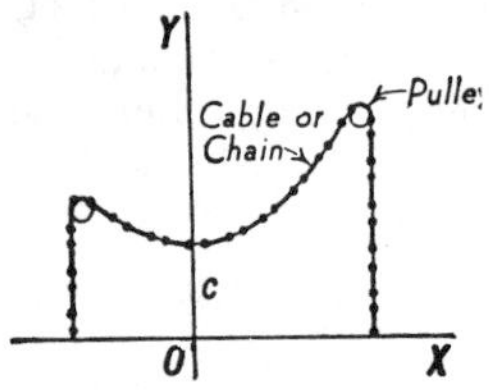

Fig. 4. Tension of chain in catenary balances weight of chain hanging down to X-axis.

of this portion will just sustain the tension on the other side of the pulley. The funicular-polygon theory is sometimes useful in solving equilibrium problems involving nonconcurrent forces. (See **Statics.**)

See also **Catenary; Statics;** and terms listed under **Mathematics.**

FUNICULUS. A cord. Specifically: 1. A structure attaching the alimentary tract of the bryozoans to the body wall. 2. A slender segmented part of the antenna of some insects, just before its terminal segment. 3. A bundle of nerve fibers in its sheath. 4. Tracts of nerve fibers in the central nervous system of vertebrates. 5. The stalk which attaches the seed to the placenta of the fruit.

FUNNEL CLOUD. Fronts and Storms.

FUR. The fine soft hair of many mammals. Also the pelt, the skin of the animal bearing the fur, especially when made up to be worn as a scarf.

Fur shares with feathers the highest position as a protection against cold. It owes this property to the insulating value of the layer of air held among the fine hairs that compose it. In many animals the vestiture consists of a thick wooly under layer interspersed with longer and heavier hairs which form a smooth surface.

The best furs are those of animals which live in the colder latitudes and especially the semi-aquatic species. The sea otter of the northern Pacific produces a most durable fur, and mink and muskrat are examples of higher-priced and lower-priced furs. Fox and skunk are among the commercially important terrestrial species. All of these animals should be taken in the winter to furnish durable pelts. Summer fur is not only thinner but separates readily from the skin.

The preparation of fur for the market now involves so many processes, such as plucking, shearing, and dyeing, that only an expert can judge skins dependably. Trade names add to the confusion and give false dignity to many furs of very modest worth. Many of the seals on the market, for example, are clipped and dyed muskrat or rabbit fur.

FURANE AND RELATED COMPOUNDS. Furane C_4H_4O contains a ring of 1 oxygen and 4 carbons, with 1 hydrogen attached to each carbon:

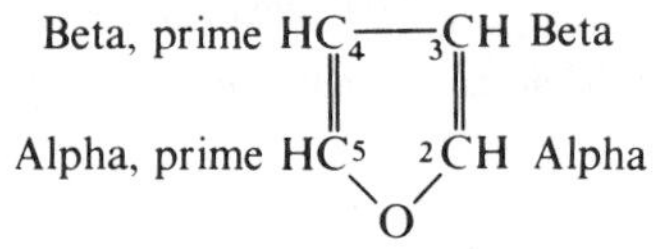

Furane is a colorless liquid, boiling point 32°C, insoluble in water, soluble in alcohol or ether. Furane vapor produces a green coloration on pine wood moistened with hydrochloric acid. Furane may be made from mucic acid, $COOH(CHOH)_4COOH$, by dry distillation into pyromucic acid, $C_4H_3O \cdot COOH$, and then heating the latter under pressure at 270°C. Furane derivatives are known, namely, methyl, primary alcohol, aldehyde, carboxylic acid, in which the group attachment is at carbon number 2:

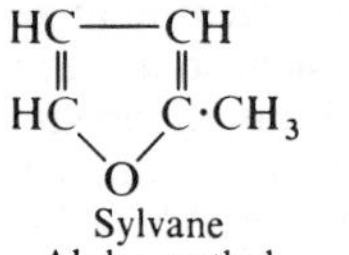

Sylvane
Alpha-methyl furane;
boiling point 65°C

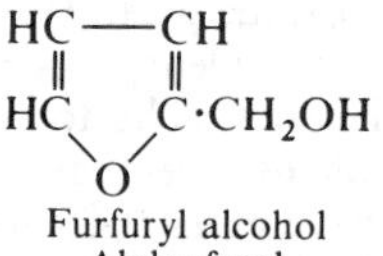

Furfuryl alcohol
Alpha-furyl carbinol;
boiling point 170°C (750 mm)

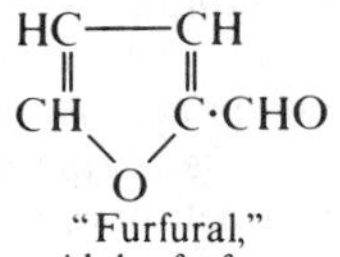

"Furfural,"
Alpha furfuraldehyde;
boiling point 160°C (740 mm)

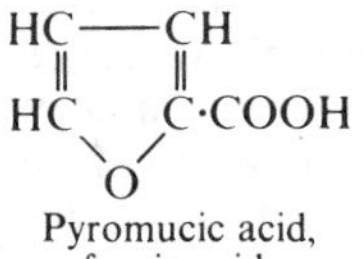

Pyromucic acid,
furoic acid,
furane-alpha-carboxylic acid;
melting point 133°C
boiling point 230°C

See **Furfuraldehyde.**

Coumarone is benzofurane C_8H_6O or $C_6H_4CH{:}CH \cdot O$ or

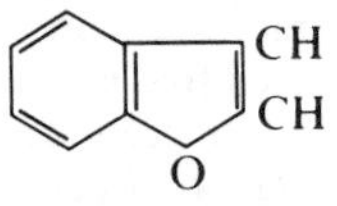

colorless liquid, boiling point 173°C, and diphenylene oxide is dibenzofurane $C_{12}H_8O$ or $C_6H_4 \cdot C_6H_4 \cdot O$ or

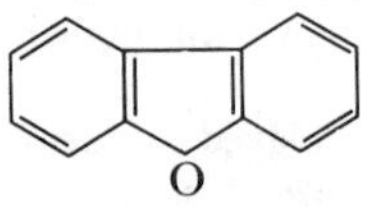

white solid, melting point 81°C, boiling point 288°C.

Coumarin is 1,2-benzopyrone $C_9H_6O_2$ or $C_6H_4OCOCH{:}CH$ or

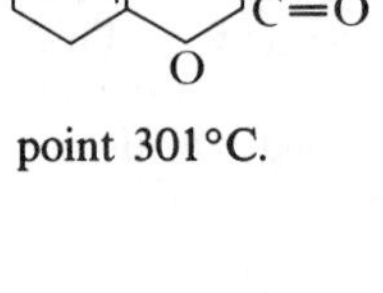

white solid, melting point 67–8°C, boiling point 301°C.

Gamma-pyrone, $C_5H_4O:O(4)$, is a gamma-ketone (4) containing a ring of 1 oxygen and 5 carbons with 1 hydrogen attached to each of 4 carbons, namely, 2,3,5,6.

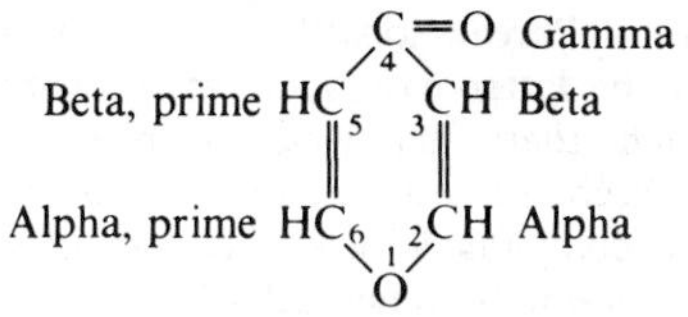

Gamma-pyrone is a colorless liquid, melting point 32°C boiling point 218°C.

Pyrone derivatives are known, e.g.,

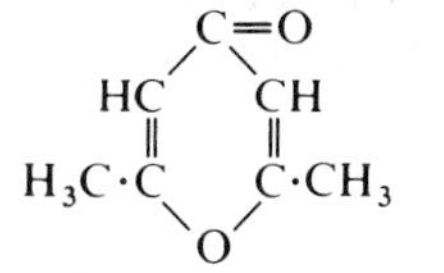

Alpha, alpha prime dimethyl-gamma-pyrone

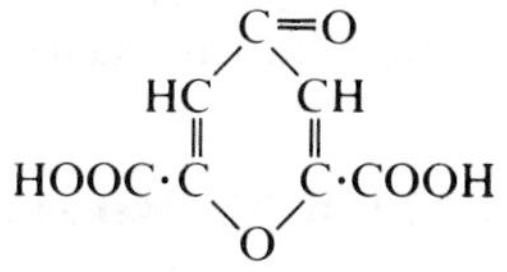

Chelidonic acid gamma-pyrone-alpha, alpha-prime-dicarboxylic acid

Chromone is benzo-pyrone $C_9H_6O_2$, a white solid, melting point 59°C

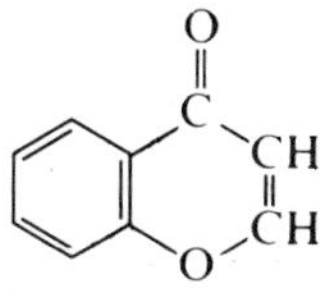

and chromane is colorless liquid, boiling point 214°C, 750 mm.

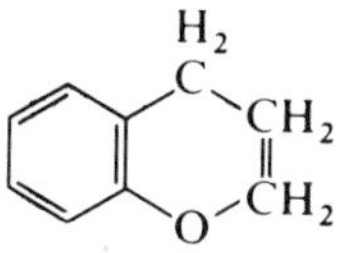

Flavone is phenyl chromone: 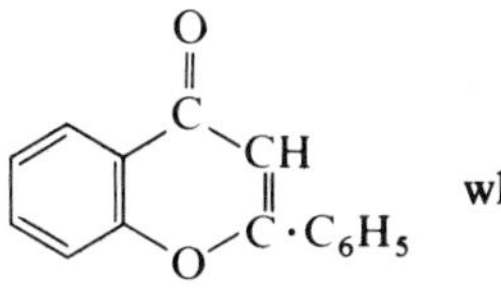 white solid melting point 97°C.

Xanthone is dibenzo-pyrone, $C_{13}H_8O_2$ or 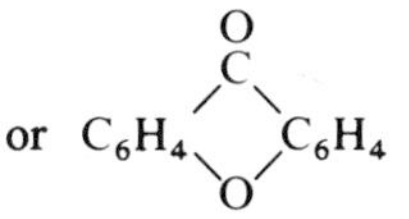 or

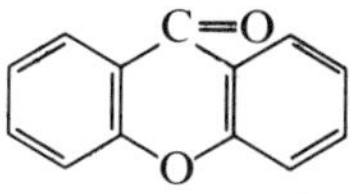 white solid, melting point 174°C, boiling point 351°C, and xanthene is 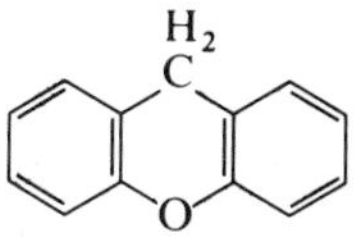 white solid, melting point 100°C, boiling point 315°C. From chromone and xanthone a number of yellow dyes are made, which dyes also occur in nature. Such dyes are chrysin, fisetin, buteolin, morin, quercetin, rhamnetin.

Where oxygen of furane is replaced by sulfur, thiophene is the compound, and of coumarone, benzothiophene; and where oxygen of furane is replaced by nitrogen (group=NH), pyrrole, and of coumarone, indole.

FURFURALDEHYDE. 2-$C_4H_3O \cdot CHO$, formula weight 192.16, colorless, odorous (pungent, almond-like) liquid aldehyde, mp −38.7°C, bp 161.7°C, sp gr 1.159. Also known as 2-furaldehyde or 2-furancarboxaldehyde, this compound becomes brown in color when in contact with air. Furfural is modestly soluble in H_2O (up to 8% by weight at 20°C) and is miscible in all proportions with alcohol and ether. At atmospheric pressure, a mixture of furfural and H_2O (65%) forms a minimum-boiling azeotrope when a distillation temperature of 97.9°C is reached.

Aside from a darkening in color, furfural is relatively stable thermally and does not exhibit changes in physical properties after prolonged heating up to 230°C. The reactions of furfural are typical of those of the aromatic aldehydes, although some complex side reactions occur because of the reactive ring. Furfural yields acetals, condenses with active methylene compounds, reacts with Grignard reagents, and provides a bisulfite complex. Upon reduction, furfural yields furfuryl alcohol; upon oxidation, it yields furoic acid. It can be decarbonylated to furan.

Furfural is obtained commercially by treating pentosan-rich agricultural residues (corncobs, oat hulls, cottonseed hulls, bagasse, rice hulls) with a dilute acid and removing the furfural by steam distillation. Major industrial uses of furfuraldehyde include (1) the production of furans and tetrahydrofurans where the compound is an intermediate, (2) the solvent refining of petroleum and rosin products, (3) the solvent binding of bonded phenolic products, and (4) the extractive distillation of butadiene from other C_4 hydrocarbons.

When pentoses, e.g., arabinose, xylose, are heated with dilute HCl, furfuraldehyde is formed, recognizable by deep red coloration with phloroglucinol, or by the formation, with phenylhydrazine, of furfuraldehyde phenylhydrazone $C_4H_3O \cdot CH:NNHC_6H_5$, solid, mp 97°C.

FURNACE (Boiler). Boiler; Burner.

FURNACE (Glass Melting). Glass.

FURROW IRRIGATION. Irrigation.

FURUNCLE. Boils.

FUSED QUARTZ. Glass.

FUSEE. Spring Clock.

FUSE (Electric). A common protective or circuit-breaking device for low-voltage electric circuits. It is an over-current protector, and since the current must first heat the metal, there is a time delay in fuse "blowing" that is inversely proportional to the current. This characteristic is called "inverse time element." The ordinary fuse consists of a calibrated length of conductor whose resistance is so chosen that when a certain current flow through it is exceeded, it fails to lose by radiation enough of the resistance heat to keep its temperature below melting. The fuse is enclosed in a protective case which forms the contact points to connect it into its circuit.

Although still used in connection with various electrical and electronic appliances, residential and building circuits presently are protected by circuit breakers. See **Circuit Breaker.**

FUSIBLE ALLOYS. Low-Temperature Alloys.

FUSION (Heat of). Very simple experiments show that the fusion of a given mass of any crystalline substance requires a definite quantity of heat. The quantity required per unit mass, without any change of temperature, is called the heat of fusion of the substance. It may be measured by means of a calorimeter. The fused substance is introduced into the calorimeter at a temperature somewhat above its melting point and allowed to cool, the heat evolved being measured. At the melting point it ceases to cool for a time, but continues to give out heat as it solidifies; and when all congealed, it begins to cool again. At this stage the process is terminated; and the total heat evolved, with corrections for the cooling before and after solidification calculated from the known specific heats, gives the heat of fusion. For ice the value is about 79.71 calories per gram.

FUSION (Nuclear). The combination of two light nuclei to form a heavier nucleus, with the release of the difference of the nuclear binding energy of the products and the sum of the binding energies of the two light nuclei. Examples are:

$$^2H + ^2H \rightarrow ^3He + n + 3.27\ \text{MeV}$$

$$^2H + ^3H \rightarrow ^4He + n + 17.59\ \text{MeV}$$

$$^2H + ^6Li \rightarrow ^8Be \rightarrow 2\,^4He + 22.37\ \text{MeV}$$

Fusion reactions can take place only if the reacting nuclei possess sufficiently high energies to overcome their mutual Coulomb repulsion and to approach within the range of nuclear forces, hence they are favored by high temperatures. See also **Nuclear Reactors.**

FUSION (Nuclear) REACTORS. Lithium (For Thermonuclear Fusion Reactors); Nuclear Reactor.

FUSION (Phase Change). A change from the solid to the liquid phase of matter. In crystalline bodies, and, as has now become understood, also in many other solids not exhibiting well-defined crystal structure, the atoms are held in positions of stable equilibrium by intermolecular forces. They of course move with thermal agitation, but their movements are oscillatory and do not carry them outside a limited range of distance from their equilibrium positions. Stable equilibrium may, however, become unstable when the system is disturbed beyond a certain limit. Thus if a solid body is sufficiently heated, the molecules break loose from their stable configuration and wander about or diffuse among each other. When this condition has become general, the body exhibits the characteristics of a liquid, and we say it has undergone fusion. In some cases, such as ice, the change is quite abrupt, the substance having a well-defined melting point; in others, like glass or pitch, it is gradual. The difference is probably due to the more uniform potential energy of the atoms in the former case, so that they all "break loose" at the same stage of thermal agitation; while in the latter case some atoms require more energy to dislodge them than others. In any case the process requires a supply of energy which is recognized as the heat of fusion. With most substances, fusion is accompanied by an increase in volume; but with some, like ice, the volume becomes definitely less.

Fusion, as an order-disorder transition, is the concept that fusion of a crystalline solid is essentially a change from the almost perfectly ordered solid state to a disordered liquid state. The vacant spaces in the crystal lattice correspond to the other component in the binary alloys which undergo order-disorder transition in the pure form. Evidence from x-ray diffraction measurements indicates that short-range order is retained during fusion but long-range order is lost.

FUSION WELDING. Welding.

G

GABA (Gamma-Aminobutyric Acid). Brain and Nervous System.

GABBRO. Gabbro is a deep-seated and often very coarse-grained igneous rock composed of plagioclase feldspar, usually labradorite or bytownite and monoclinic pyroxene, with occasionally as accessories olivine (when it is then called olivine gabbro), biotite, magnetite, ilmenite, and hornblende. Norite is a variety of gabbro, carrying orthorhombic pyroxene, usually hypersthene instead of the monoclinic sort. Troctolite is essentially olivine and plagioclase. Quartz gabbros are known and have probably been derived from magmas somewhat oversaturated with silica. On the other hand, essexites represent gabbros whose parent magma doubtless had an insufficiency of silica resulting in the formation of nephelite. Gabbros are frequently rich in sulfides that may be of commercial value, a notable occurrrence of which is at Sudbury, Canada. Here a norite carrying chalcopyrite and nickeliferous pyrrhotite forms the most important deposits of nickel known. Gold, silver and platinum are also recovered from this ore.

GABOON VIPER. Snakes.

GADIDAE. Codfishes.

GADOLINIUM. Chemical element symbol Gd, at. no. 64, at. wt. 157.25, seventh in the Lanthanide series in the periodic table, mp. 1,312°C, bp 3,273°C, density 7.901 g/cm^3 (20°C). Elemental gadolinium has a close-packed hexagonal crystal structure at 25°C. The pure metallic gadolinium is silver-gray in color, slow to tarnish in normal atmospheres. The metal is soft, malleable, and easy to fabricate with normal tools provided that processing temperatures are maintained below 150°C. The turnings and chips of gadolinium are mildly pyrophoric and care must be exercised in their handling. There are seven natural isotopes of gadolinium: ^{152}Gd, ^{154}Gd through ^{158}Gd, and ^{160}Gd. Eleven artificial isotopes have been prepared. The natural isotopes are not radioactive. In terms of abundance, gadolinium is present on the average of 5.4 ppm in the earth's crust, making it potentially more available than tantalum, tin, or tungsten. The element was first identified by J. C. G. Marignac in 1880. The natural isotopic mixture of gadolinium has the greatest thermal-neutron-absorption cross section of all elements, 40,000 barns. This is approximately 10 times greater than the next two elements, samarium (5,800 barns) and europium (4,300 barns). However, gadolinium is limited to nuclear applications mainly as a start-up and shutdown material because only two of the natural isotopes ^{155}Gd and ^{157}Gd behave in this manner. These are separated by isotopes which do not so react—hence, no chain relationship exists. ^{155}Gd and ^{157}Gd make up 31% of the total weight of elemental gadolinium. The metal has a low acute-toxicity rating. Electronic configuration

$$1s^22s^22p^63s^23p^63d^{10}4s^24p^64d^{10}4f^75s^25p^65d^16s^2.$$

Ionic radius Gd^{3+} 0.938 Å. Metallic radius 1.801 Å. First ionization potential 6.16 eV; second 12.1 eV.

Other important physical properties of gadolinium are given under **Rare-Earth Elements and Metals.**

Gadolinium reacts vigorously with dilute mineral acids, but is practically inert to strong bases and boiling H_2O. Gadolinium is an active reducing agent for metals, including iron, chromium, manganese, tin, lead, and zinc. The major sources of gadolinium are xenotime, monazite, gadolinite, and residues from uranium mining.

Although the nuclear properties of the element are attractive, gadolinium has enjoyed rather limited applications in reactor technology. An important discovery in the 1960s showed that gadolinium iron garnets (called GIGs) $Gd_6Fe_5O_{12}$ possess a crystalline structure which finds useful application in microwave frequency control, circulators, isolators, and bandpass filters in electronic circuitry. Gadolinium oxide also is used as the host matrix in the red phosphor for color television picture tubes, where it is activated by europium. Gadolinium oxysulfide Gd_2O_2S is used as an x-ray image intensifier making possible less x-ray dosage for medical explorations. Along with yttrium and lanthanum activated by cerium, gadolinium is used in a phosphor for single-gun beam-indexing flying-spot scanning cathode ray tubes. Gadolinium also provides magnetic properties when alloyed with cobalt, cerium, iron, and copper ($Co_{3.5}CuFe_{0.5}Ce$) in permanent magnets, imparting a desirable negative temperature coefficient of magnetic saturation. A glass with magnetic properties (5% wt Gd_2O_3) has been produced. Gadolinium metal and several of its salts are under consideration for use in a magnetic heat pump device.

See references listed at ends of entries on **Chemical Elements;** and **Rare-Earth Elements and Metals.**

NOTE: This 6th Edition entry was revised and updated by K. A. Gschneidner, Jr., Director, and B. Evans, Assistant Chemist, Rare-Earth Information Center, Energy and Mineral Resources Research Institute, Iowa State University, Ames, Iowa. Original 5th Edition entry was prepared by J. G. Cannon, Molycorp, Inc.

GADWELL. Waterfowl.

GAGE BLOCKS. Measurement (Manufacturing).

GAGE (Device). An instrument or device for measuring or comparing some physical characteristics, such as size, pressure, temperature, force, water level, and surface quality. As contrasted with sophisticated recording and controlling instruments, gages are frequently manually read and often hand-applied, as in the case of the gages used in the machining and metalworking field. There are instances, however, where the term *gage* is applied to costly, complex instruments, as in the vacuum-measurement field. Gaging also can be fully automated as in the application of pneumatic and electrical gages for the continuous "go-no/go" inspection of parts. No fixed rules have been established for guidance in use of the term.

GAGE INVARIANCE. Conservation Laws and Symmetry.

GAGE LINE. A gage line marks the limits of any standard distance used repeatedly. Structural shapes are punched or drilled on lines called gage lines. The gage lines may be varied to suit the details so long as the minimum required edge distance and clearance for punching and drilling are maintained. In some fabricating shops, holes are made with multiple punches or drills.

GAHNITE—ZINC SPINEL. The mineral gahnite is isometric with an octahedral habit but may appear as dodecahedrons or modified cubes. Chemically it is zinc aluminate corresponding to the formula $ZnAl_2O_4$. There is a tendency for cleavage parallel to the octahedron, fracture varies from conchoidal to uneven; brittle; hardness 7.5–8; specific gravity 4.6; luster, vitreous; color ranges from dark green through various shades of greenish- or bluish-black, yellowish-black or grayish, subtransparent to almost opaque. Gahnite is found in association with other zinc minerals at several European localities, notably in Bavaria and Sweden. In the United States it is found at Franklin and Sterling Hill, New Jersey; at Rowe, Massachusetts and in Maryland, North Carolina, Georgia and Colorado. Gahnite was named in honor of the Swedish chemist, J. G. Gahn.

GAIN (Antenna). Antenna.

GAIN BANDWIDTH PRODUCT. The gain bandwidth product is equal to the product of amplification of an amplifier stage at mid-band, multiplied by the bandwidth of the amplifier. The *bandwidth* is defined as the difference Δf between the two frequencies at which the power output is a specified fraction, usually one-half, of the mid-band (resonance) value.

GAIN (Magnitude Ratio). With reference to industrial and scientific instruments, the Scientific Apparatus Makers Association defines gain for a linear system or element as the ratio of the magnitude (amplitude) of a steady-state sinusoidal output relative to the causal input; the length of a phasor from the origin to a point of the transfer locus in a complex plane.

The quantity may be separated into two factors: (1) a proportional amplification often denoted as K which is frequency-independent, and associated with a dimensioned scale factor relating to the units of input and output; and (2) a dimensionless factor often denoted as $G(j\omega)$ which is frequency-dependent. Frequency, conditions of operation, and conditions of measurement must be specified. A loop gain characteristic is a plot of log gain versus log frequency. In nonlinear systems, gains are often amplitude-dependent.

Closed Loop Gain. The gain of a closed loop system, expressed as the ratio of the output change to the input change at a specified frequency.

Derivative Action Gain (*Rate Gain*). The ratio of maximum gain resulting from proportional plus derivative control action to the gain due to proportional control action alone. See **Derivative (Rate) Control.**

Dynamic Gain. The magnitude ratio of the steady-state amplitude of the output signal from an element or system to the amplitude of the input signal to that element or system, for a sinusoidal signal. It may be expressed as a ratio, or in decibels as 20 times the $\log_{10}$ of that ratio for a specified frequency.

Loop Gain. The ratio of the change in the return signal to the change in its corresponding error signal at a specified frequency. The gain of the loop elements is frequently measured by opening of the loop, with appropriate terminations. The gain so measured is often called the open loop gain.

Proportional Gain. The ratio of the change in output due to proportional control action to the change in input. Illustration: $Y = \pm PX$, where $P =$ proportional gain; $X =$ input transform; $Y =$ output transform. See also **Proportional Control.**

Static Gain. The value of the gain approached as a limit as frequency approaches zero.

GAIN (Transmission). 1. A general term used to denote an increase in signal power in transmission from one point to another; usually expressed in decibels and used to denote transducer gain. 2. The ratio of the output of a transducer to the input, even when these quantities are not measured in terms of power. Thus reference is made to the voltage gain or current gain of an amplifier.

GAL. Units and Standards.

GALACTIC COORDINATES. A system of spherical coordinates having as its fundamental plane that of the Milky Way, i.e., the plane of the galaxy. The pole of this system, for the epoch of 1950, is located at $12^h\ 49^m .0$ right ascension and $+27°\ 24'\ .0$ declination. The angle of inclination with the equatorial coordinate system is $62°\ 36'\ .0$, and, in 1950, the intersection of these two systems occurred at $18^h\ 49^m .0$. The zero point for latitude and longitude in the galactic coordinate system is determined by the center of the Milky Way. In 1950, this center point was given as $17^h\ 42^m .4$ right ascension and $-28°\ 55'$ declination. This basic coordinate system has been revised several times due to observations from radio astronomy, and may be revised again.

GALACTOSEMIA. A disease caused by an inborn error of carbohydrate metabolism. Mental retardation is a clinical feature of the disease. The normal conversion of galactose, a sugar found in milk, is prevented by the absence of the enzyme galactose-1-phosphate uridyl transferase. Removal of milk, the only food source of galactose, from the diet in infancy prevents development of the condition. In the treatment of older patients by diet changes, all symptoms of the disease disappear except intellectual impairment. Success in treating galactosemia by dietary means has spurred the search for other inborn errors of metabolism among the mentally retarded.

GALAGO. Lemur.

GALÂPAGOS RISE. Ocean; Ocean Resources (Energy); Ocean Resources (Living); Ocean Resources (Mineral).

GALAXIDS (*Osteichthyes*). Of the order *Isospondyli*, family *Galaxiidae*, the galaxiids are scaleless, elongated fishes, quite small, seldom exceeding 6 inches in length. The *Galaxias alepidotus* (New Zealand species) was first identified in the late 1700s. It is an exception among the galaxiids in that it can attain a length up to 12 inches (30 centimeters) with some specimens recorded up to about 23 inches (58 centimeters). Even this long variety weighs but about 3 pounds (1.4 kilograms). The *Neochanna apoda* (New Zealand brown mudfish) is well known for its absence of ventral fins and is reminiscent of various lungfishes which can survive for many weeks in dried mud. Adult mudfish attain a length of about 6 inches (15 centimeters). The *Galaxias attenuatus* (whitebait) also occurs in New Zealand as well as Australia. Of interest is the fact that this species is the only galaxiid that is found in two or more locations, probably explained by its ability to tolerate fresh and brackish water, with a preference for salt water when an adult.

The fact that galaxiids are not found in the Northern Hemisphere has puzzled naturalists for many years. The matter is even more puzzling because habitats of practically the exact nature preferred by the galaxiids in the Southern Hemisphere are also found in many locations in the Northern Hemisphere.

GALAXY. In general terms, a galaxy is a system of stars, nebulae, and interstellar gas and dust, the statistics of which are essentially beyond realistic human comprehension. The earth and solar system represent but miniscule components of a great spiral galaxy familiarly known as the Milky Way. This galaxy is some 100,000 light-years (~30,000 parsecs) across (linear diameter)[1] and is estimated to contain over 100 billion stars. The earth and solar system are located in a relatively unpopulated part of the galaxy in a position well away from the center. This is a fortuitous position for observing much of the galaxy. The optical telescope reveals only a small portion of the Milky Way because regions of it are obscured by great dust clouds and other interfering phenomena. With the initiation of radio astronomy in 1932, a technique by which electromagnetic radiation at radio, rather than visible, wavelengths is measured, it became possible to study radio-emitting stars, radio galaxies, and hydrogen in space, thus circumventing the limitations of the optical telescope. More recently, investigations that measure radiation in other portions of the electromagnetic spectrum have furthered the knowledge of the galaxies. See also **Gamma Ray Astronomy; Infrared Astronomy; Milky Way; Radio and Radar Astronomy; Ultraviolet Astronomy;** and **X-Ray Astronomy.**

Galaxy Configurations. Traditionally, galaxies have been divided into four classes, a system proposed by Hubble in 1925. These are: (1) *Elliptical galaxies* (E); (2) *spiral galaxies* (S); (3) *barred spiral galaxies* (SB); and (4) *irregular galaxies* (I). Research, largely undertaken during the 1970s and early 1980s, has shown that galaxies are of greater complexity and much more dynamic than previously considered. With the advent of radio astronomy, galaxies were further classified into: (a) a *weak emitter*, or *ordinary galaxy*, such as the Milky Way, which typically radiates 10^{38} erg/second in radio waves, com-

[1] Many authorities acknowledge that the diameter is probably even greater than this figure, but available optical data have not been sufficient to ascertain the details of structure and composition, let alone exact dimensions. It still has not been established whether the Milky Way is a Type Sb or Sc spiral galaxy. See also **Milky Way.**

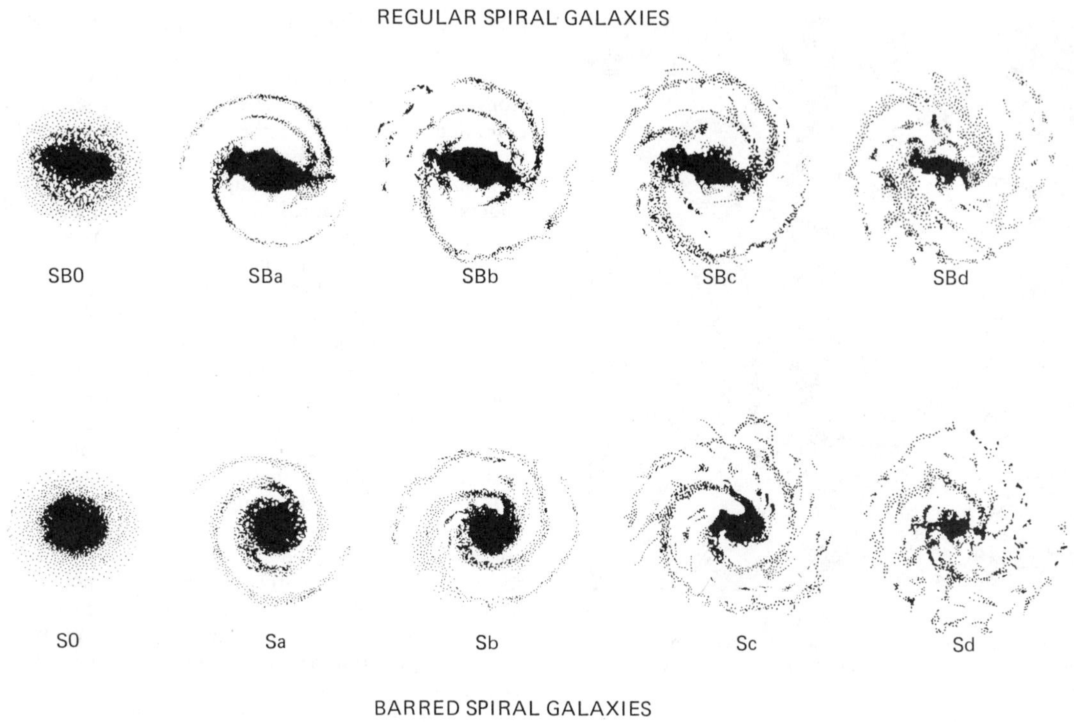

Fig. 1. Generalized and schematic configurations of principal types of galaxies as proposed by Hubble (1925).

pared with 10^{44} erg/second in the opitcal region; and (b) a strong emitter, or *radio galaxy*, which may produce up to 10^{45} erg/second in the radio range alone. In both types of galaxy, the radio continuum is accounted for on the synchrotron theory. See **Synchrotron Radiation.** Further refinements in classification are described a bit later.

Diagrams of various classes of galaxies closely following Hubble's early proposal are given in Fig. 1. Radio astronomy observations to date suggest that the Milky Way is of the Sb or Sc class. The shape-predominant method of classifying galaxies derived logically from the photographic evidence provided by optical equipment. A variety of galaxy forms is also shown in Figs. 2 and 3.

Elliptical Galaxies. When viewed through the telescope, this class of galaxy appears as an elliptical disk. No spiral arms are apparent. Per unit volume of space, this is the most abundant type of galaxy and ranges from the most massive to the least massive of the galaxies. Typically, a disk galaxy incorporates few or no young stars and no gas or dust. Elliptical galaxies apparently have a smooth structure—with a smooth center extending out to a diffuse, irregularly defined edge. Although not fully understood, the elliptical galaxies appear to differ one from the other principally in their ellipticity—from round (Type E0) to a 3 : 1 axis ratio (Type E7). In Fig. 2, a cluster of galaxies in Virgo is shown. Thousands of galaxies form a rich, loose, and irregular cluster which appears to have no central concentration. The two large objects in the right-hand portion of the view are the elliptical galaxies M84 and M86. The giant M49 elliptical galaxy NGC 4472 (Type E1), shown in Fig. 4, is a nearly circular elliptical galaxy. Typical of the elliptical galaxy, this object shows no evidence of recent star formation and little or no gas between the stars. Some 60 million light-years (18.4 million parsecs) away from earth, the mass of M49 is approximately 10^{12} times that of the sun and from 5 to 10 times more massive than the Milky Way.

Fig. 2. Group of clusters in Virgo, including M84 (NGC 4374) and M86. Thousands of galaxies form a rich, loose irregular cluster which appears to have no central concentration. Many subcondensations of galaxies are seen in it. This type of cluster comprises most types of galaxies. (*Kitt Peak National Observatory.*)

Fig. 3. Known as Stephan's Quartet, this group of galaxies (NGC 7317; 7318A; 7318B; 7319; 7320) is located in Pegasus. Four of these galaxies have a redshift of about 6000 kilometers/second, while one (the largest) has a redshift of only 800 kilometers/second. (*Kitt Peak National Observatory.*)

Fig. 4. The nearly-circular elliptical galaxy M49 (NGC 4472), Type E1, located in Virgo. Such galaxies have nearly no dust or gas between their stars, and show no evidence of recent star formation. (*Kitt Peak National Observatory.*)

Galaxy M87 (NGC 4486), a type E0 elliptical galaxy, was the first extragalactic x-ray source to be found by rocket astronomy. Optically, M87 is a large elliptical galaxy characterized by an extended jet, which emerges from the nucleus to a distance of about 5000 light-years (1500 parsecs). The jet is clearly visible in the lower, right-hand inset of Fig. 5. The bluish light of the jet is highly polarized, indicating synchroton radiation. Radio astronomers discovered an intense compact core only 4 light-months in diameter, which may be the origin of the x-radiation (inverse Compton interactions between relativistic electrons and the high density of radio photons in the nuclear region). Other features of the radio image include an extended halo and a fan jet. The fan jet may suggest that gas clouds are expelled from a compact rotating body. One observer estimates that repeated releases from the central body may furnish a few million solar masses to the various jet forms. It is postulated that to sustain a reservoir of material and energy, the gas may be constantly accreting onto a large rotating mass in the nuclear region. It is further postulated that the origin of the gas could be planetary nebulas separated from old-population red giants. If one compares this situation with our own galaxy, the rate of evolution of planetary nebulas in M87 should be about 30 per year.

More recently, the jet has been explained as having been ejected from the nucleus of the galaxy in one or a series of explosions about a million years prior to the state that is presently being observed from earth. X-ray emission provides evidence of explosive activity, commonly encountered in elliptical galaxies as well as in other poorly understood astronomical objects, such as quasars. Energy release in such an explosion may be at a rate that is a trillion times that of the sun, a phenomenon that puzzles modern astronomers.

Early *Uhuru* satellite observations revealed that an x-ray emitting cloud about 1 million light-years (over 300,000 parsecs) across envelops the galaxy. At a later date, analysis of the x-ray spectrum was made by the British *Ariel 5* satellite and also by the NASA OSO-8 satellite. These data, which show the presence of highly ionized iron, indicated that the x-ray radiation emits from a diffuse gas at a temperature of 30 million degrees Kelvin. These findings indicated that the x-ray radiation arises from a thermal and not a nonthermal source.

Investigators, in attempting to explain the presence of this cloud, currently offer three postulates. In one, the gas is considered as forming continuously so that the cloud is constantly replenished; in another, the galaxy as it is currently observed is at a point in its developmental history such that there has not been sufficient time for the gas to dissipate; in the third, some force is confining the gas to the galaxy. A number of researchers tenatively accept the last of these postulates, assuming the holding force to be gravity. Phenomena of this kind are further described in entries on **Black Hole; Cosmology;** and **Quasars.**

In an early study of bright galaxies, observers concluded that all normal spiral and irregular galaxies are probably weak radio sources. In contrast, the strong emitters tend to be elliptical galaxies, often distinguished by peculiar optical phenomena, and some observers believe that the elliptical galaxies constitute the bulk of the 1000+ discrete sources cataloged. Early studies indicated that these sources have power-law spectra, but more recent and reliable data indicate the presence of curvature in many spectra (as displayed in a log frequency–log flux diagram, where a power law is a straight line). The value of observations below a wavelength of 4 centimeters in defining the spectral shape was demonstrated. Later, interferometric studies enabled the dividing of the resolved sources into three groups on the basis of brightness distribution: simple, double, and core-halo sources.

Fig. 5. Galaxy M87 (NGC 4486), Type E0, located in Virgo. The poorly understood jet extending from the galaxy (see photo inset) is a strong radio source. (*Kitt Peak National Observatory.*)

Fig. 6. Spiral galaxy (M31; NGC 224) in Andromeda. Elliptical companion galaxy (M32; NGC 221) appears just above the central region; another elliptical companion (NGC 205) appears below to the left. (*Hale Observatories.*)

Spiral Galaxies. Probably the most familiar and some of the most optically observed of the galaxies are the spiral galaxies, of which the Milky Way is one (Type Sb or Sc). The Milky Way is among the larger of the regular (not barred) spiral galaxies, as is also the Andromeda Galaxy (M31; NGC 224), which is one of the closest spiral systems and one of the most easily visible from earth. Distance from earth approximates 2.2 million light-years (0.7 million parsecs). See Fig. 6. Companion elliptical galaxies are also shown in the view. See also **X-Ray Astronomy.**

The spiral arms of these galaxies contain abundant dust, gas, and newly formed, bright, massive, hot, bluish stars, which often occur in clusters. The central regions have little gas and dust and are dominated by old, red giant stars. Generally, spiral galaxies have the outlines of flattish, lens-shaped disks with a maximum thickness at the center equal to approximately 10–15% of the diameter. Mass calculations and brightness observations indicate that spiral galaxies may contain from 1 billion to 100 billion or more individual stars. With reference to the diagram of Fig. 1, the Type Sb and Sc spiral galaxies are among the most frequently studied and photographed. See Fig. 7(a)–(f).

Among the observable spiral galaxies, there are relatively few that provide a view of the edge of the disk. Some of the galaxies which can be observed in a reasonably edge-on position are shown in Fig. 8(a)–(d).

Barred Spiral Galaxies. From observational data to date, it appears that a small minority of spiral galaxies have a bright bar that slices across the nucleus. The two arms begin at the ends of the bar and wind outward. Contrast this with the normal spiral galaxies, which have a central region (nucleus) to which a number of spiral arms appear to be attached. See Fig. 1 and Fig. 7(f).

Type SO Galaxies. It will be noted from Fig. 1 that Types SO and SBO galaxies differ considerably from the other spiral galaxies

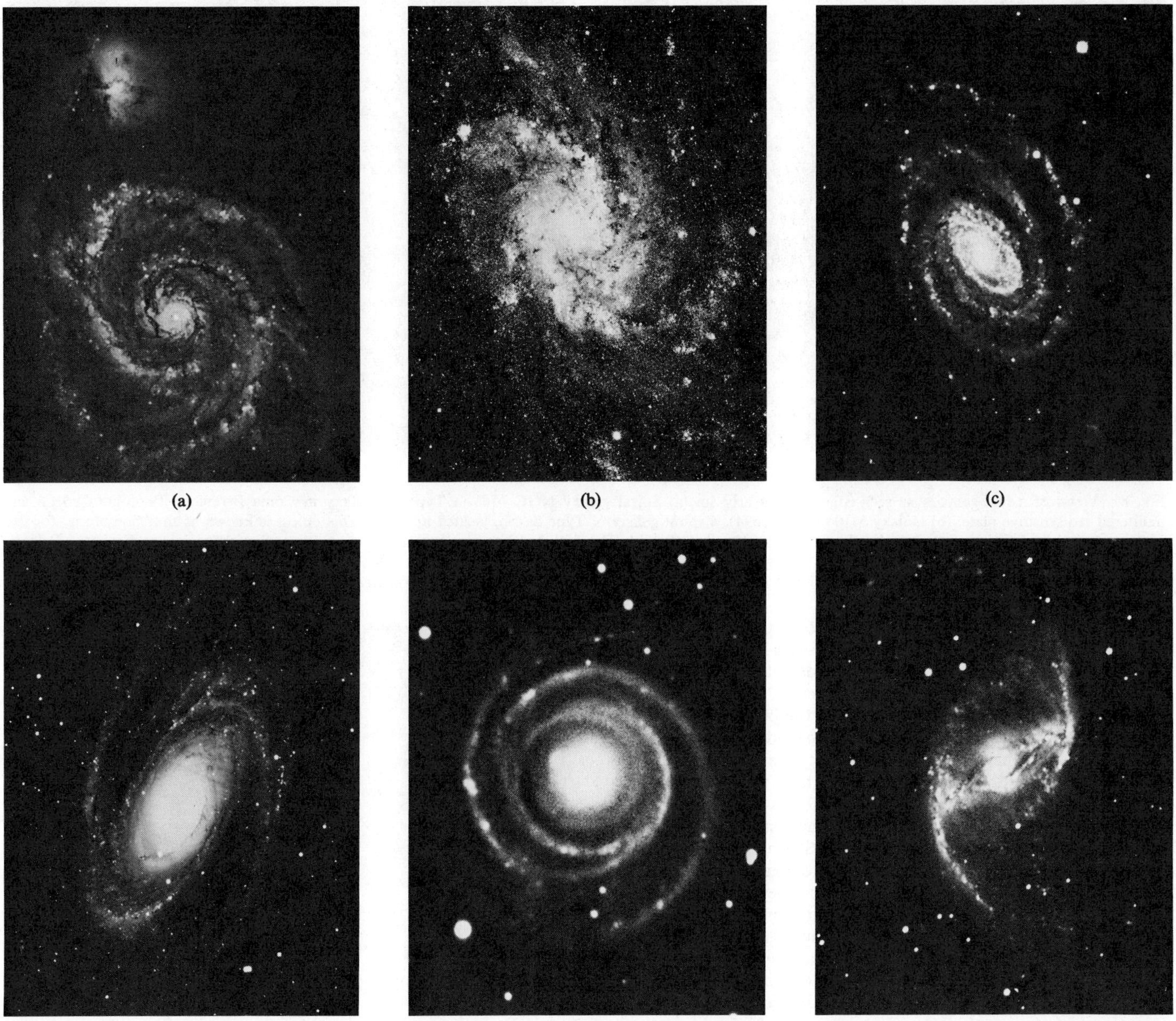

Fig. 7. Representative spiral galaxies: (a) Whirlpool galaxy (M51; NGC 5194) and its satellite galaxy (NGC 5195) in Canes Venatici. Note sharp, bright nucleus. The companion is classified as an irregular galaxy. (b) Galaxy M33 (NGC 598) in Triangulum. This is one of the nearest of the spiral galaxies to the Milky Way, some 2.3 million light-years (~0.7 million parsecs) distant. It is a Type Sc galaxy. (c) Galaxy NGC 5364, a Type Sc galaxy located in Canes Venatici. (d) Galaxy M81 (NGC 3031), a Type Sb galaxy located in Ursa Major. (e) Galaxy NGC 4622, a Type Sb galaxy located in Centaurus. This is a member of the Centaurus cluster of galaxies. Within its remarkably smooth and thin spiral arms there are millions of bright young stars. Distance from earth is about 200 million light-years (~61 million parsecs). (f) Galaxy NGC 1530, a Type SBb galaxy, located in Camelopardalis. Note that this is a barred-type spiral—with a barlike, elongated center as compared with the more circular or elliptical centers. (*Sources of illustrations*: (*a*) *Lick Observatory*; (*b, c, d,* and *f*) *Kitt Peak National Observatory*; (*e*) *Cerro Tololo Inter-American Observatory.*)

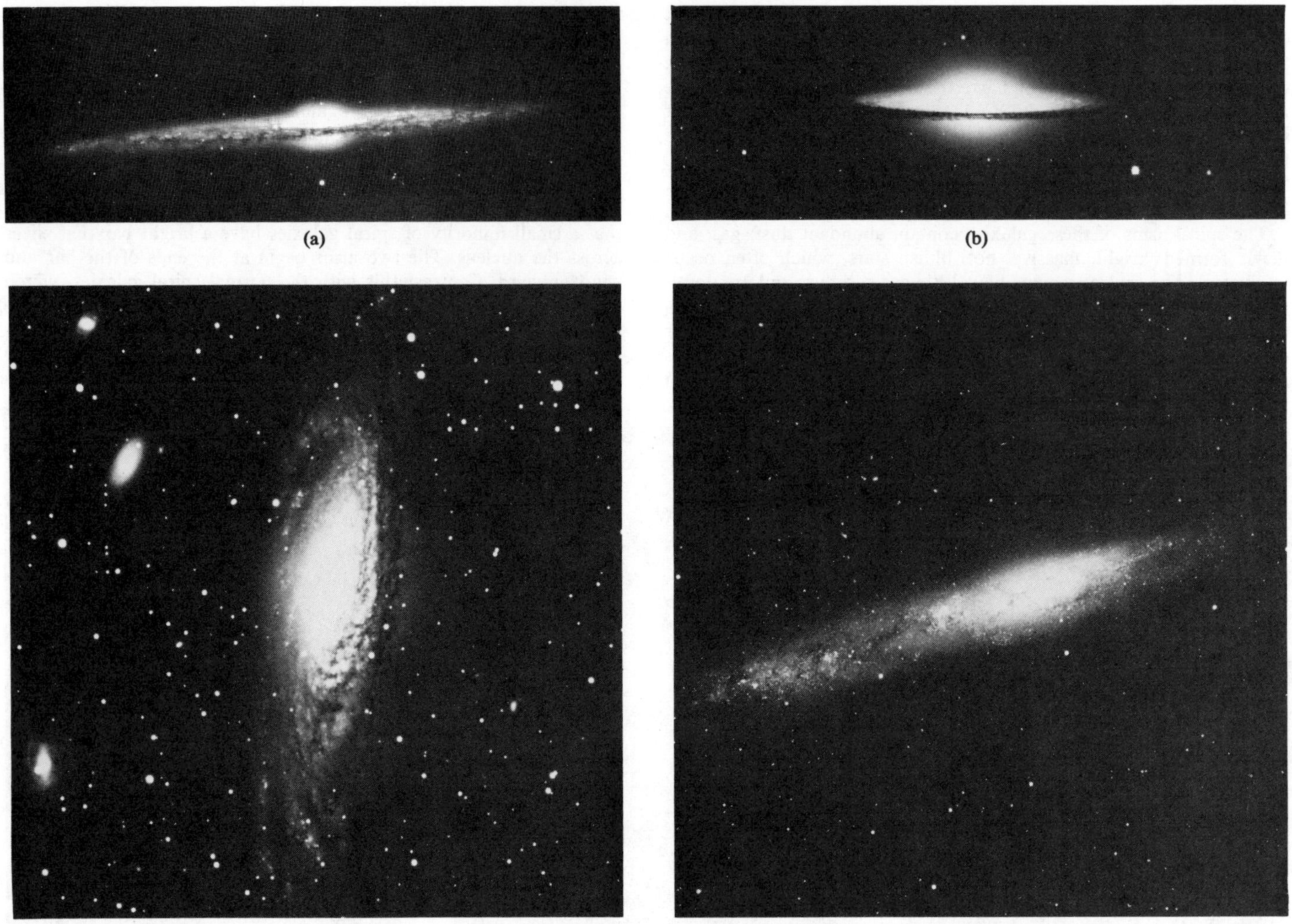
(a) (b) (c) (d)

Fig. 8. Views of a few galaxies as seen edge-on or nearly so: (a) Spiral galaxy NGC 4565, a Type SB galaxy in Coma Berenices. Photographed on an unfiltered red-sensitive plate. (b) Galaxy M104 (NGC 4594), a spiral galaxy of Type Sa/Sb, located in Virgo. This object is known as the "Sombrero" for its edge-on appearance. This galaxy is inclined only about 6° to line of sight. The dark band across the galaxy's center is composed of dust and gas. (c) Galaxy NGC 7331, a spiral galaxy of Type Sb, located in Pegasus. (d) Galaxy NGC 55, a spiral galaxy of the SBm (barred) type and located in Sculptor. (*Sources of Illustrations*: (*a*) *Hale Observatories*; (*b* and *c*) *Kitt Peak National Observatory*; (*d*) *Cerro Tololo Inter-American Observatory*.)

depicted. The two morphologically distinct parts of a disk galaxy are: (1) the *central bulge*, which in many cases is roughly spheroidal; and (2) a comparatively thin *disk* that extends outward. Observations show a great variation in the characteristics of these two parts or subsystems. The relative size and extent of the bulge vary from one galaxy to the next. In some galaxies, the bulge predominates; in others, the disk. Research has shown that the bulge in most disk galaxies is completely or essentially devoid of young stars, with star-formation occuring in the disk. The SO galaxies differ in that the disks are smooth and lack young stars and star-forming complexes. As pointed out by Strom and Strom (1979), the disks of SO galaxies lack evidence of the gas needed for future star formation. SO galaxies are common in large galactic clusters, whereas the other types of spiral galaxies, such as the Milky Way, tend to be located in regions that are relatively unpopulated by galaxies. Within the proper environment, some authorities propose that a spiral galaxy can become a smooth disk without spiral arms, and that this is most likely to occur in regions with large clusters of galaxies, rather than in isolated, widely separated galaxies which are not part of a rich cluster. See Fig. 9.

Ring Galaxies. A galaxy of this type has a prominent, bright ring surrounding the center. In some cases, this center is faint; in others, bright. See Fig. 10. It has been postulated that the ring galaxies may be the result of collisions between pairs of galaxies. Once formed, the configurations appear to be stable.

Irregular and Peculiar Galaxies. The observed galaxies which do not fit well into established criteria (principally shape) are usually termed *irregular galaxies.* Some authorities have broken the irregular class into two categories—the Magellanic Cloud type, and all others. Irregular galaxies (with Q and B stars and emission nebulae) are designated Irr I; those which cannot be resolved into stars are designated Irr II. The closest of the irregular galaxies to earth are the two large, cloudlike objects in the southern sky (the Magellanic Clouds), which are actually companion galaxies to the Milky Way galaxy. See Fig. 11. The larger of these clouds has sometimes been called a barred spiral (Type SBm) with one arm. The largest gaseous nebula in the Large Magellanic Cloud is called 30 Doradus. See also **Nebula.**

Fig. 9. Galaxy NGC 4753, a Type S0 galaxy, located in Virgo. The underlying galaxy is nearly elliptical, but the dust lanes are peculiar in that they do not appear to occur in spiral arms. (*Kitt Peak National Observatory.*)

The two Magellanic Cloud galaxies are rather irregular in shape and are considerably smaller than the Milky Way. Their distance is

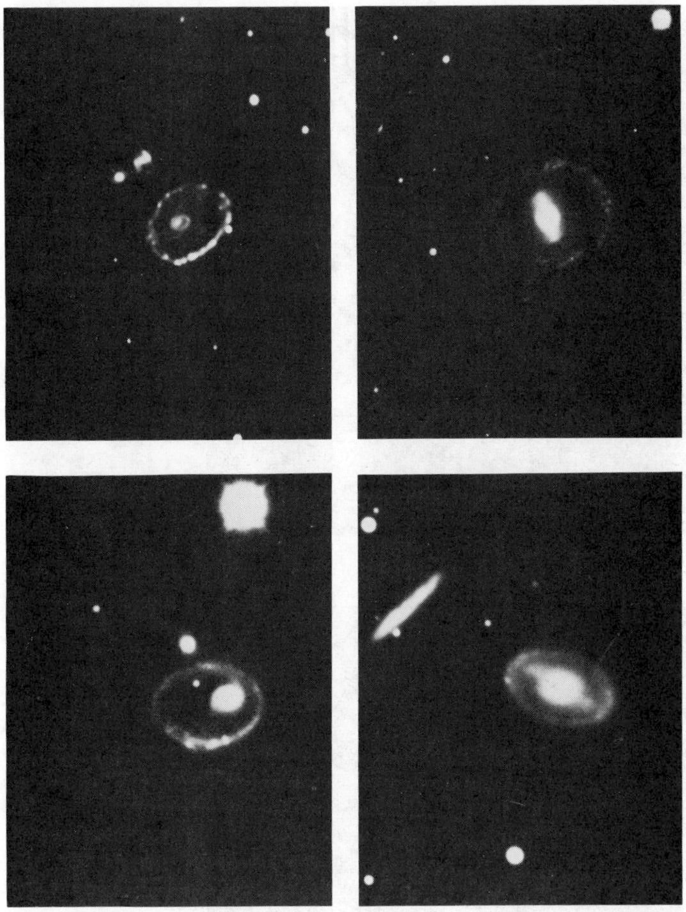

Fig. 10. Four ring galaxies. (*Kitt Peak National Observatory; Cerro Tololo Inter-American Observatory.*)

on the order of 800,000 light-years (~245,000 parsecs) from earth. It is estimated that these galaxies contain on the order of 10 billion stars each. The Magellanic Clouds primarily contain Population I type stars, with lots of gas and dust, although they also exhibit such Population II type objects as globular clusters and cluster-type variables stars.

Examples of other irregular and peculiar galaxies are shown in Fig. 12.

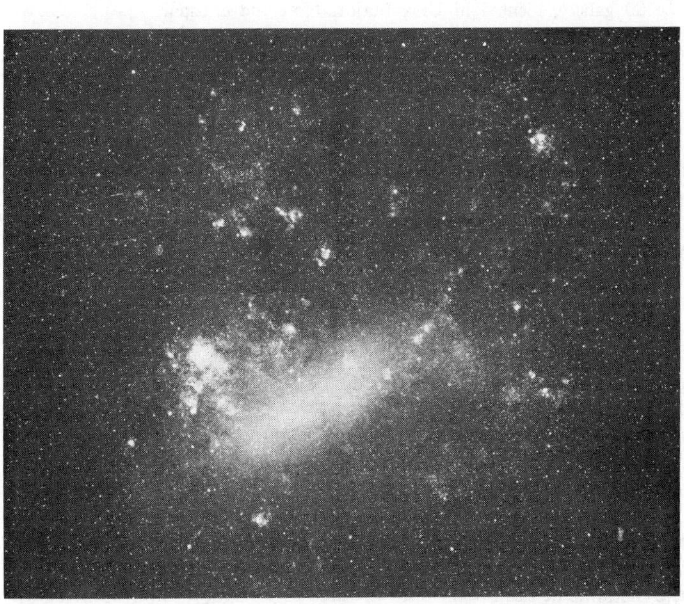

Fig. 11. Large Magellanic Cloud photographed in Hα light. (*Photographed by Karl G. Heinze with the Mt. Wilson 10-inch reflector at the Lamont-Hussey Observatory, Bloemfontain, South Africa*)

Seyfert Galaxies. In general terms, a Seyfert galaxy is any galaxy that has a very bright nucleus showing a high excitation spectrum with broad emission lines. More specifically, a Seyfert galaxy has a small nucleus, often bluish in color and emitting radio energy. Some authorities believe that they are related to quasars and may have explosive activity proceeding in their centers. Two Seyfert galaxies are shown in Fig. 13(a) and (b).

A small percentage of all galaxies are Seyfert galaxies. These objects may contain from 10^9 to 10^{10} stars within a diameter of about 1000 light-years (~300 parsecs). High-velocity gas clouds, hot gas, and nonthermal processes are indicated by strong, widened optical emission lines and a polarized continuum.

Radio Galaxies. Any galaxy, including Seyfert galaxies, that emits measurable amounts of radio radiation may be called a *radio galaxy.* Numbers of these have been identified in recent years. They fall into both the normal and peculiar classes of galaxy. A normal radio galaxy is not necessarily normal in its optical and other properties; rather, its radio emission is considered normal. A peculiar radio galaxy may emit hundreds to millions of times the radio emission a of normal radio galaxy. Some galaxies are peculiar in terms of both radio and optical characteristics. Frequently, these are single galaxies that show evidence of explosive activity in their centers. A jet extending from the nucleus, as shown in Fig. 5, is indicative of instability. Radio galaxies that appear to involve two or more interacting or colliding galaxies also have strong radio emissions. The term, "violently active galaxy" has been introduced into the literature in recent years for those objects with strong emissions in the radio and sometimes the x-ray spectrum.

Clusters of Galaxies. It is not unusual for many physical phenomena in nature to occur in clusters. This is indeed the case with galaxies. Even prior to the accumulation of much knowledge of galaxies, early investigators recognized the predisposition of nebulae to collect in bunches, so to speak. In the late 1800s, over 11,000 "nebular objects" were mapped for the "New General Catalogue," published by J. L. E. Dreyer. In 1921, C. V. L. Charlier published the sky map shown in Fig. 14 from these cataloged objects. Most of the previously listed nebulae were found to be galaxies. The equator of the map corresponds with the central plane of the Milky Way. Because this plane is obscured by dust, few galaxies are shown in the central plane. Later knowledge indicated that clusters of galaxies are much more evenly distributed, as contrasted with the polar concentrations of the map. However, the map portrayal is significant in the manner in which it emphasizes the general clustering of galaxies, rather than uniform spacing.

As early as 1935, Shapley cataloged 25 clusters of galaxies, suggesting that clustering was related to the evolutionary processes of the Universe. Clusters of galaxies have already been shown in Figs. 2 and 3. Clusters in Perseus and in Coma Berenices are shown in Fig. 15(a and b). By definition, a cluster of galaxies is a group of associated galaxies, usually within 10–100 galaxy diameters of each other.

Globular clusters of stars do not exhibit the usual features of galaxies and are believed to be much older objects. These clusters are described in the entry on **Star.**

Black Hole Hypothesis. In recent years, particularly during the 1970s and early 1980s, scientists have gathered much new data pertaining to galaxies and notably the violently active galaxies. Analysis of these data has led to much speculation and hypothesizing as regards the energy source for such galaxies. Some astronomers are of the opinion that black holes are at the center of galaxies. As early as the late 1960s, a few astronomers, in their study of quasars, suggested that a black hole model appeared to be the only reasonable approach. Some scientists currently observe that the most plausible model of the mechanism that can be constructed within the framework of conventional physics is the black hole. Other possibilities include a dense cluster of stars in the core of active galaxies, these stars providing the energy required as the result of frequent supernova reactions, or a *spinar* or *magnetoid* (an object that has not fully collapsed to the characteristic diameter of a black hole). Some observers suggest that the spinar or magnetoid concept may be correct, but only as a "way station" leading to a black hole. This discussion has also led to mention of a "new physics," which in itself has a fuzzy definition, but indicates the need for further developments and knowledge in other related areas of physics—these then to be applied to the galaxy energy machine

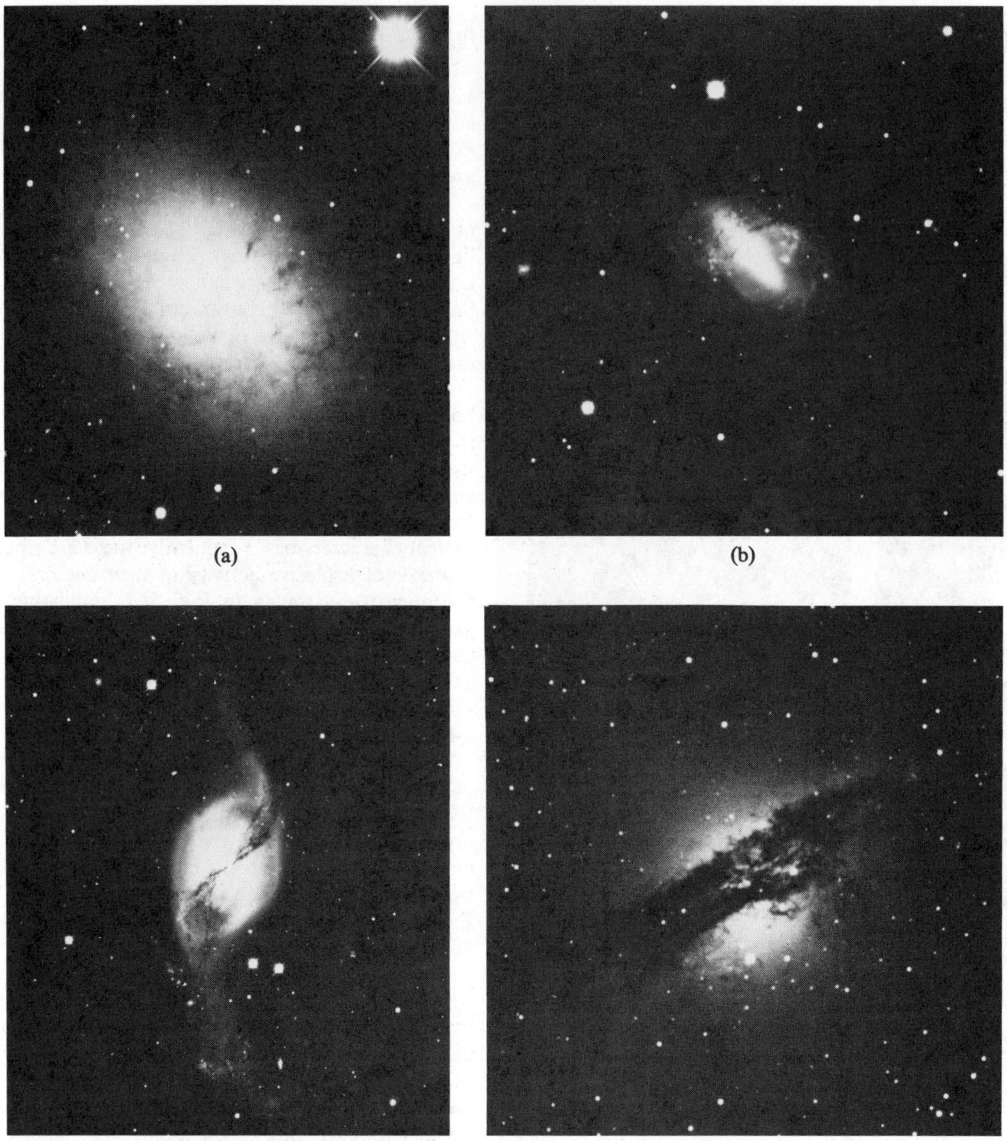

Fig. 12. Irregular and peculiar galaxies: (a) Irregular (Type II) galaxy (NGC 3077), located in Ursa Major. Note that the dust lanes do not follow the usual pattern. (b) A peculiar Type S0 galaxy (NGC 2685), located in Ursa Major. Note that there are two axes of symmetry. (c) Another peculiar Type S0 galaxy, located in Ursa Major. Note the unusual absorption features. (d) A Type E0 elliptical galaxy located in Centaurus. This is a strong radio source and is the nearest known violent galaxy. It is also an x-ray source. (*Sources of illustrations*: (*a through c*) *Kitt Peak National Observatory*; (*d*) *Cerro Tololo Inter-American Observatory.*)

problem. This general topic is described in further detail in entry on **Black Hole.**

Local Group of Galaxies. Approximately 20 of the nearest galaxies, which appear to form a cluster, are sometimes referred to as the Local Group. However, most of the mass is contained in the Milky Way and the Andromeda galaxies. In terms of increasing distance from earth, the local Group galaxies include: Milky Way (Earth is a part of this); Large and Small Magellanic Clouds; Ursa Minor System; Draco System; Sculptor System; Fornax System; Leo I System; Leo II System; NGC 6822; NGC 185; NGC 147; IC 1613; M31; M32; NGC 205; and M33.

Very Early Galaxies. Accepting the speed of light throughout the cosmos as a finite value, it follows that light reaching earth from distant objects enables researchers to look back into the history of the universe. Some scientists consider that the universe was created ~16.5 billion years ago. The extent to which this backward look can be made depends upon the ability of observational instruments used and, in principle, it should be possible with superpowerful and sensitive instruments to observe galaxies "in the making"—thus revealing the whole gamut of steps that occur in galaxy formation. If, as is generally thought, most galaxies follow the same developmental pattern, observations of "primeval galaxies" should provide information concerning the formation of the Milky Way (estimated to be some 13 billion years old) and other Local Group galaxies.

As of the early 1980s, the instruments available are not capable of separating primeval galaxies from other distant, faint, or diffuse objects. The next scheduled improvement in observational capability will be a very powerful telescope in space. As of the early 1980s, plans call for putting this instrument in earth orbit during the mid-1980s. Thus, by the end of the present decade, it is hoped that the new instrument (resolution 10 times greater and faint object detection capability 100 times greater than early 1980 instruments) will reveal some of the primeval galaxies in early stages of development.

As explained in the entry on **Red Shift,** the wavelength of radiation received from distant objects increases as the velocity of recession of the object increases. Thus, red shift can be used as a measure of the distance of the object and, in turn, of its age. With present equipment, it has been determined that galaxies of about the same age as the Milky Way (distance of ~20 million light-years; ~6.1 million parsecs) recede with a red shift of 0.1% (velocity of recession = 0.001).

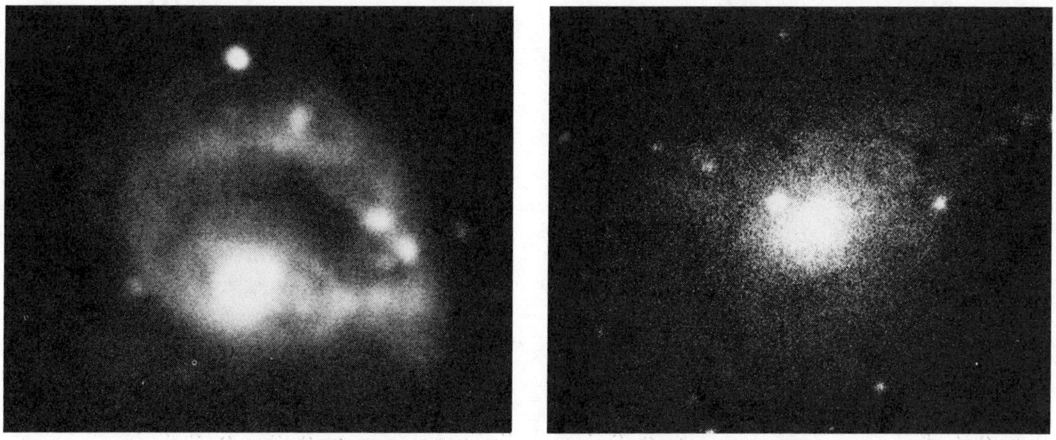

(a) (b)

Fig. 13. Seyfert galaxies: (a) Distorted ring galaxy that has a violently active Seyfert nucleus. (b) A peculiar galaxy (NGC 1275), located in Perseus and known to astronomers as Perseus A. This galaxy is called a Seyfert galaxy because it has large amounts of hot plasma in it. It is a strong x-ray source. (*Kitt Peak National Observatory.*)

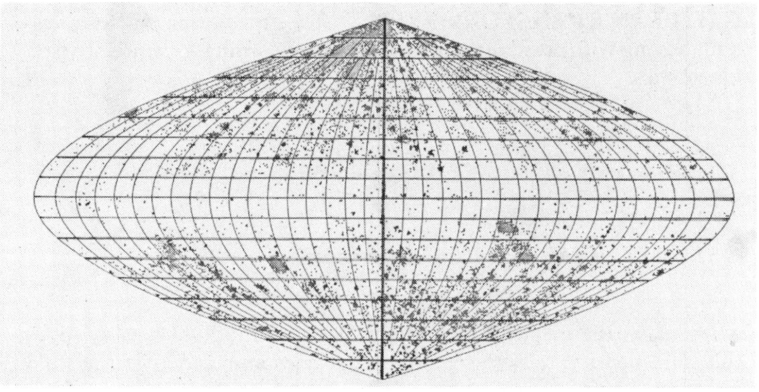

Fig. 14. Many years prior to much detailed knowledge of galaxies, this map portraying clustering of galaxies was prepared by Charlier (1921).

Some quasars (believed to occur at the center of otherwise normal galaxies) show a red shift of 0.1, but quasars are more numerous at a red shift in the vicinity of 2.5. Some scientists reason, therefore, that a primeval galaxy will have a red shift greater than 2.5. Meier and Sunyaev (1979), by extrapolating backward from the present universe to the time when galaxies were touching one another, estimate that an upper limit on the red shift of a galaxy in formation occurs at a value of approximately 100. Considering the state of the art, it should be mentioned that the most distant galaxy detected to date has a red shift of only 0.75.

Analyses of the spectra of observed distant objects aids much in their identification. Normal galaxies do not have emission lines, but do exhibit red shifts, whereas stars have emission lines, but do not exhibit red shifts. It could be postulated that a primeval galaxy with a significant population of bright, massive, hot stars, which would cause the gas between them to ionize, would produce a spectrum with emission lines. As pointed out by Meier and Sunyaev, primeval galaxies could be mistaken for quasars because of their strong emission lines and large red shifts; this introduces the speculation that, if primeval galaxies generally incorporate stars brighter than normal galaxies, perhaps at least some of them could be within the detection range of even present-day instruments and simply not recognized as such. In anticipation of the problem of sorting out possible spectral data from primeval galaxies, a number of scientists have developed models (admittedly relatively unrefined at this juncture) to predict what the spectra of a primeval galaxy may be, stressing the differences between the spectrum of the galaxy and that of a quasar. Further information on galaxy formation is given in the entry on **Cosmology.**

References

Abell, G. O.: "Clustering of Galaxies," *Ann. Rev. Astron. Astrophys.*, **3**. 1965.

Bahcall, N. A.: "Clusters of Galaxies," *Ann. Rev. Astron. Astrophys.*, **15**, 505–540 (1977).

(a) (b)

Fig. 15. Examples of clusters of galaxies: (a) Two clusters in Perseus: *h* and χ Persei. (b) Large clusters of galaxies in Coma Berenices. This huge cluster contains more than 1000 galaxies, each a large system of stars in itself. Such regular-type clusters generally include a large number of S0 and E galaxies, and are often sources of x-ray radiation. (*Kitt Peak National Observatory.*)

Becklin, E. E., et al.: "Infrared Observations of the Galactic Center, I: Nature of the Compact Sources," *Astrophys. J.*, **219**, 1, 121–128 (1978).
Berkhuijsen, E. M., and R. Wielebinski (editors): "Structure and Properties of Nearby Galaxies," Reidel, Boston, 1978.
Bok, B. J., and P. F. Bok: "The Milky Way," Harvard Univ. Press, Cambridge, Mass., 1981.
Burton, W. B. (editor): "The Large-Scale Characteristics of the Galaxy," in "Proceedings of the International Astronomical Union: Symposium 84," Reidel, Boston, 1979.
Darling, D.: "What Makes a Spiral Galaxy?" *Astronomy*, **7**, 7, 6–20 (1979).
Fall, S. M., and D. Lynden-Bell (editors): "The Structure and Evolution of Normal Galaxies," Cambridge Univ. Press, New York, 1981.
Geballe, T. R.: "The Central Parsec of the Galaxy," *Sci. Amer.*, **241**, 1, 60–70 (1979).
Giacconi, R.: "The Einstein X-Ray Observatory," *Sci. Amer.*, **242**, 2, 80–102 (1980).
Gordon, M. A., and W. B. Burton: "Carbon Monoxide in the Galaxy," *Sci. Amer.*, **240**, 5, 54–67 (1979).
Gorenstein, P., and W. Tucker: "Rich Clusters of Galaxies," *Sci. Amer.*, **239**, 5, 110–128 (1978).
Gott, J. R., III: "Recent Theories of Galaxy Formation," *Rev. Astron. Astrophys.*, **15**, 235–266 (1977).
Hodge, P. W.: "The Andromeda Galaxy," *Sci. Amer.*, **244**, 2, 92–101 (1981).
Larson, R. B.: "The Formation of Galaxies," in "Galaxies: Sixth Advanced Course of the Swiss Society of Astronomy and Astrophysics" (L. Martinet and M. Mayor, editors), Geneva Observatory (Switzerland), 1976.
Larson, R. B.: "The Origin of Galaxies," *Amer. Scient.*, **65**, 188–196 (1977).
Longair, M. S., and R. A. Sunyaev: "Young Galaxies, Quasars and the Cosmological Evolution of Extragalactic Radio Sources," in "Radio Astronomy and Cosmology: International Astronomical Union Symposium No. 74" (D. L. Jauncey, editor), Reidel, Boston, 1977.
Longair, M. S., and J. Einasto, (editors): "The Large Scale Structure of the Universe," Reidel, Boston, 1978.
Meier, D. L.: "The Optical Appearance of Model Primeval Galaxies," *Astrophys. J.*, **207**, 2, 343–350 (1976).
Meier, D. L., and R. A. Sunyaev: "Primeval Galaxies," *Sci. Amer.*, **241**, 5, 130–144 (1979).
Metz, W. D.: "Violently Active Galaxies: The Search for the Energy Machine," *Science*, **201**, 700–702 (1978).
Oemler, A.: "The Structure of Elliptical and cD Galaxies," *The Astrophys. J.*, **209**, 3, Part 1, 693–703 (November 1, 1976).
Oort, J. H.: "The Galactic Center," *Ann. Rev. Astron. Astrophys.*, **15**, 295–362 (1977).
Peebles, P. J. E., and S. Brand: "A Million Galaxies: Computer Photo-Map of the Galaxies Brighter than 19th Magnitude Visible from Earth's Northern Hemisphere," *CoEvolution Quarterly*, Sausalito, California (1978).
Penzias, A. A.: "Nuclear Processing and Isotopes in the Galaxy," *Science*, **208**, 663–669 (1980).
Rieke, G. H., Telesco, C. M., and D. A. Harper: "The Infrared Emission of the Galactic Center," *Astrophys. J.*, **220**, 2, 556–567 (1978).
Sandage, A., Sandage M., and J. Kristian (editors): "Galaxies and the Universe: Volume 9, Stars and Stellar Systems," University of Chicago Press, Chicago, Illinois, 1975.
Shu, F. H.: "Spiral Structure, Dust Clouds, and Star Formation," *Amer. Scient.*, **240**, 4, 72–82 (1979).
Toomre, A.: "Theories of Spiral Structure," *Ann. Rev. Astron. Astrophys.*, **15**, 437–478 (1977).
Tver, D. F.: "Dictionary of Astronomy, Space, and Atmospheric Phenomena," Van Nostrand Reinhold, New York, 1979.

GALE. **Winds and Air Movement.**

GALENA. The mineral galena, lead sulfide, PbS, crystallizes in the isometric system, usually in cubes or cube-octahedron combinations, less frequently in octahedrons. It is often found in cleavable masses, but may be granular or fibrous. The highly perfect cubic cleavage is an important characteristic of this mineral: it may, however, sometimes show an octahedral parting. Its hardness is 2.5; specific gravity, 7.58; luster, metallic; color, lead gray; streak, grayish-black; opaque. Galena is the most important ore of lead and in addition often carries values of silver; it is then known as argentiferous galena. It occasionally is actually mined as a silver ore. Sometimes galena contains small amounts of zinc, cadmium, antimony, bismuth, and copper as sulfides.

Galena is a very common and widely spread mineral, it occurs in veins and beds in various rocks, both crystalline and sedimentary. Some of these deposits are doubtless replacements, others seem to show a close connection with intrusive igneous rocks. Of the many European localities, the classics are Freiberg, Saxony, and the silver mines of the Harz Mountains. This mineral has been found in the lavas of Vesuvius, in Italy, and fine specimens came from Cornwall and Cumberland, England. Australia, South America, Chile, and Peru produce galena. In the United States, Missouri, Illinois, Iowa, and Wisconsin contain large and important galena deposits. In Colorado and Idaho it has been mined for its silver content. Galena is usually associated with sphalerite, smithsonite, and at Phoenixville, Pennsylvania, with beautiful pyromorphite crystals. The name is derived from the Latin *galena*, a term which was applied both to the lead ore and slag from refining.

See also terms listed under **Mineralogy.**

Elmer B. Rowley, Union College, Schenectady, New York.

GALIDINES. **Viverrines.**

GALILEAN SATELLITES. **Jupiter.**

GALILEAN TELESCOPE. A form of telescope that has a divergent lens for ocular and in which no real image is formed. The field of view is small, but the whole telescope is shorter than conventional telescopes of comparable power. Commonly used in opera glasses. See also **Telescope.**

GALILEAN TRANSFORMATION. The transformation to a system moving with constant relative velocity according to nonrelativistic kinematics:

$$\begin{aligned} dx' &= dx - v_x\,dt \\ dy' &= dy - v_y\,dt \\ dz' &= dz - v_z\,dt \\ dt' &= dt \end{aligned}$$

GALILEO NUMBER. This is defined as

$$N_{Ga} = d_p^3 \rho_f(\rho_s - \rho_f) g/\mu^2,$$

where d_p is the mean particle diameter; ρ_f is the fluid density; ρ_s is the solid density; g is the gravitational constant; and μ is the fluid viscosity.

GALLBLADDER AND BILIARY TRACT DISEASES. The gallbladder is a pear-shaped organ (see figure) situated on the underside of the liver on the right side of the body just below the ribs. This organ serves as a reservoir for the bile and by means of the cystic duct it communicates with the common duct through which the bile secreted by the liver passes to the duodenum. The gallbladder is about 3 inches (7.5 centimeters) in length and 1–$1\frac{1}{4}$ inches (2.5–3 centimeters) in diameter. It holds about $1\frac{1}{2}$ ounces (29.5 milliliters) of bile. When fatty substances are ingested, the normal gallbladder empties the stored, concentrated bile into the common duct. Upon passing into

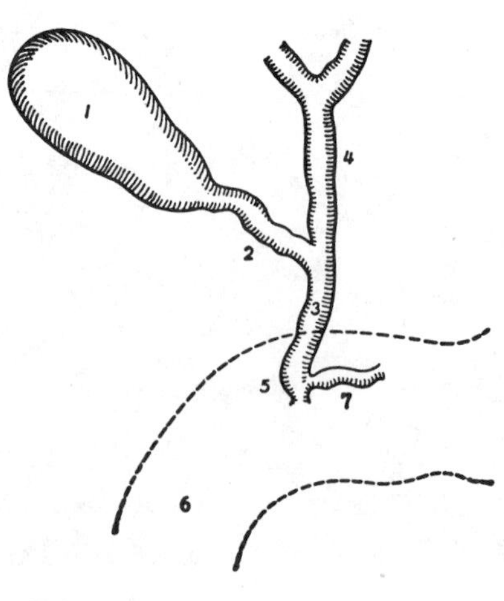

Biliary tract: (1) gallbladder; (2) cystic duct; (3) common bile duct; (4) hepatic duct; (5) opening of bile duct into the duodenum; (6) duodenum; (7) duct from the pancreas. The biliary system is sometimes referred to as the biliary tree.

the duodenum, the bile participates in a very important way in the digestion of food, notably fats. The characteristics and role of bile in the process of digestion are described in considerable detail in the entry on **Bile.**

Principal diseases of the gallbladder and biliary tract are: (1) *cholelithiasis* (presence of stones in the gallbladder); (2) *cholecystitis* (inflammation of the gallbladder due to obstruction of the cystic duct); (3) *choledocholithiasis* (stone lodged in the common bile duct after passing from the gallbladder and through the cystic duct); any of these three diseases may be acute or chronic; (4) *chronic cholangitis* (chronic inflammation in the hepatic biliary tree); and (5) *idiopathic hyperbilirubinemia* (defect in bilirubin transport). Gilbert's syndrome (most common), the Crigler-Najjar syndrome, and the Dubin-Johnson syndrome are examples of idiopathic hyperbilirubinemia and are described in the entry on **Bile.**

The formation of gallstones derives from physicochemical changes that occur in or produce a change in the composition of the bile. Although the root causes of these changes remain poorly understood, a pathway has been described to demonstrate the alteration in bile required to produce cholesterol gallstones, but to date this theoretical approach has not led to effective ways to prevent gallstone formation and subsequent diseases. Gallstones are rather commonly found in otherwise healthy persons, particularly between ages of 55 and 65 years, where their occurrence is found in 10% of the males and about 20% of the females. It is estimated that 15 million persons in the United States alone have gallstones. Of these people, about 300,000 undergo surgery for gallstone removal each year. The occurrence is higher among persons with Crohn's disease (exceeds 20%). See **Colitis and Other Inflammatory Bowel Diseases.** There is no hard evidence to the effect that gallbladder and biliary tract diseases are related to heredity. There has been an interesting finding, however, to the effect that Pima and Chippewa Indians have a much higher occurrence of gallstones. For example, it is estimated that 70% of Pima women over 25 years old have cholelithiasis, although it may be asymptomatic.

Although cholesterol is present in normal bile only to the extent of about 5%, it is the major cause of gallstones because of its insolubility in water. Precipitation of cholesterol occurs unless it is maintained in solution by the action of bile salts. The somewhat complex physical chemistry of bile is described under **Bile.** From 85 to 90% of gallbladder stones seen in patients in the United States and Europe are predominantly composed of cholesterol, which has formed on a nidus of cholesterol. Stones may range in size from a few millimeters to one or more centimeters in diameter. In contrast, the gallstones found in patients in the Orient are bilirubinate stones, which are formed by an entirely different process. Bilirubinate stones, believed to be caused by deconjugation of bilirubin diglucuronide by the action of β-glucuronidase from the microorganism *Escherichia coli*, are uniformly associated with *E. coli* infections.

Gallstones may be present in a person for many years without symptoms and may not be discovered except by a routine abdominal x-ray made to explore some other complaint. The presence of gallstones, even when asymptomatic, poses a threat because of the risk of their causing acute cholecystitis. However, knowledge of the presence of stones by no means suggests surgical removal, particularly in persons beyond middle age. Elective surgery (*cholecystectomy*) is frequently suggested for asymptomatic patients under 60 years of age. The overall mortality for persons under 60 years of age is about 0.4%, where it ranges from 1 to 4% in persons over that age. The long-term risks of not operating, in terms of ultimate development of acute or chronic disease, is unknown and apparently differs much from one individual to the next. These mortality statistics reflect the practice in large medical centers; percentages may be greater in smaller, less well equipped hospitals.

In recent years, drug therapy has been available as an alternative to surgery and can be used with some patients. The objective is to eliminate the supersaturation of the bile with cholesterol, keeping the latter in solution and thus unavailable for the formation of stones. Further, the therapy is directed toward dissolving stones already present, thus eradicating the problem fully. Chenodeoxycholic acid has proved to be effective, although in about 10% of patients a choleretic diarrhea is produced to the extent that the therapy is impractical. The exact manner in which chenodeoxycholic acid functions in the biliary tract is poorly understood, but it has been demonstrated that, over a period of about 12 months of administration of the drug, small radiolucent stones (under 5 millimeters in diameter) can be fully or partially dissolved. Up to two years of treatment may be required to dissolve stones that are 5 to 10 millimeters in diameter. The physician must make certain that the gallbladder has the ability to concentrate therapeutic bile acid, otherwise stone dissolution will not occur. A gallbladder that, as the result of chronic cholecystitis, is essentially nonfunctioning is not a candidate for this type of therapy. The earlier chenodeoxycholic acid therapy is commenced after discovery of stones, the greater are the chances of its success.

Certain fears were initially expressed as regards this type of therapy, one of these being the possible elevation of serum cholesterol with corresponding increase in the risk of atherogenesis. Concern was also expressed as regards possible liver damage. To date, there is no evidence that either condition may result. However, chenodeoxycholic acid and its analogue ursodeoxycholic acid are not approved for general use in some countries, drug regulatory officials believing that additional history is required. One of the overall objections to this therapy is the relatively long period over which it must be used to effect ultimate dissolution of the stones.

Because the process of formation of gallstones, beyond a comprehension of the general principles of biliary cholesterol saturation, has not been thoroughly delineated, approaches to the prevention of gallstone formation are relatively rudimentary. It has been established that prolonged fasting is favorable to biliary cholesterol saturation. Thus, patients who are shown to be prone to stone formation should avoid overnight fasting regimens except when absolutely necessary for other reasons. Some researchers also have shown that bran in the diet lowers biliary cholesterol saturation. Suspected, but not confirmed, are estrogens and clofibrate as possible augmenters of cholesterol saturation.

At one time, it was believed that various dyspeptic symptoms (flatulence, heartburn, fat food intolerance) were early symptoms of cholecystitis, but it has been found that these symptoms tend to occur in the normal population to about the same degree. *Acute cholecystitis* is manifested by very severe acute abdominal pain. The pain is frequently characterized by undulations, i.e., by rising to a very marked intensity for a few seconds, followed by a relatively few minutes of subsidence before the intensity returns. However, a general level of pain may persist between the peak intensities. Some authorities consider the waxing and waning nature of the pain as an essential characteristic of biliary tract disease. Description of the pain varies from one patient to the next—from excruciating to a deep ache or cramp. The pain and associated sensations motivate most persons to immediately seek medical attention. In addition to pain, there is loss of appetite, mild to rather severe nausea and vomiting, and fever in the range of 100 to 102°F (38 to 39°C).

Diagnosis to differentiate acute cholecystitis (cystic duct obstruction) from *emphysematous cholecystitis* (gas forming in gallbladder from bacteria, such as *Clostridium perfringens*, other clostridia, *Escherichia coli*, and anaerobic streptococci), and other acute intra-abdominal processes, such as acute appendicitis, pancreatitis, and severe acute viral hepatitis, includes abdominal x-rays, intravenous cholangiography, and abdominal ultrasonography. Some physicians regard the latter noninvasive technique very highly and consider them to be about as reliable as oral cholecystography. Although there are some disadvantages in the use of narcotics, a drug such as meperidine (Demerol®) may be given for extreme pain. In the absence of evidence of sepsis or localized infection, antibiotics usually are not administered. However, if subsidence of the attack has not occurred within a period of several days, antibiotics may be given as a precautionary measure.

For persons who have had an acute attack of cholecystitis, surgery is usually suggested, particularly for patients under 60 years of age. For older patients and those with other complications, such as diabetes, a more conservative therapy may be elected, such as the use of chenodeoxycholic acid as previously mentioned.

References

Admirand, W.: "The Pathogenesis of Gallstones," in "Gastrointestinal Disease: Pathophysiology, Diagnosis, Management," (M. H. Sleisenger and J. S. Fordtran, editors), Saunders, Philadelphia, 1975.

Bartrum, R. J., Harte, C. C., and S. R. Foote: "Ultrasonic and Radiographic Cholecystography," *N. Engl. J. Med.*, **296**, 538 (1977).
Bennion, L. J., and S. M. Grundy: "Risk Factors for the Development of Cholelithiasis in Man," *N. Engl. J. Med.*, **299**, 1221 (1978).
Thistle, J. L., et al.: "Chemotherapy for Gallstones Dissolution: I. Efficacy and Safety," *JAMA*, **239**, 1041 (1978).
Warren, K. W., and E. G. C. Tan: "Diseases of the Gallbladder and Bile Ducts," in "Diseases of the Liver," (L. Schiff, editor), 4th edition, Lippincott, Philadelphia, 1975.
Watts, J. M., Jablonski, P., and J. Toouli: "The Effect of Added Bran to the Diet on the Saturation of Bile in People without Gallstones," *Am. J. Surg.*, **135**, 321 (1978).

GALL (Botany). Abnormal outgrowths in plants caused by plant or animal parasites or induced by certain chemicals, which attack various parts of the plant. While no part of the plant is immune, galls most frequently occur in those regions composed of actively growing cells, such as leaves, or the cortical tissue of the stem, or young roots. The irritation caused by the parasite may cause numerous cell divisions which result in a tremendous increase in the affected tissues.

The organisms which cause gall formation are many. Nematode worms often enter the roots of plants and cause the formation of irregular tumorous growth. These same organisms often infect the larger brown algae and cause hypertrophies, or at least gall-like malformations. Many parasitic fungi cause galls to form in the tissues which they attack. Galls occur in the leaves and stems of blueberry and cranberry bushes, due to fungus infection by *Exobasidium vaccinii*, a basidiomycete. The hyphae of the fungus penetrate the cells of the host, which enlarge tremendously in consequence. All chlorophyll in these enlarged cells is destroyed, and a red pigment forms, causing the galls to appear very conspicuous. Several species of *Taphrina*, a fungus of the ascomycete group, cause galls in the leaves of many plants. Those caused by *Taphrina aurea* in the leaves and fruits of poplar trees are especially common. Many rusts also cause gall formation.

Possibly the most striking and best known galls are caused by insects. A gall-producing insect lays its eggs in the tissues of the plant. Apparently as a result of the irritations caused by the young larvae, the surrounding cells become greatly enlarged, and the gall is formed. The galls caused by each species of insect have a very characteristic shape. The leaves and stems of rose bushes, for example, are frequently infected. One insect causes a smoothly spherical gall to form; the gall produced by another is similarly shaped but studded with stiff spines; while a third causes the formation of a dense growth of matted, branched hairs, forming a structure an inch or more in diameter. Within, there may be a single insect larvae, or many, feeding on the loose parenchymatous inner tissues of the gall and protected from enemies by the firm outer layers. Often the young buds of willow twigs are parasitized, causing bud galls to form. As the bud grows older, the internodes enlarge tremendously in diameter but elongate very little, so that a gigantic bud is formed.

The leaves of oak trees are very commonly parasitized by gall-forming organisms, both fungus and insect. Considerable value attaches to these galls, because of the large accumulation of tannin occurring in the developing gall.

GALLERY FOREST. **Biome.**

GALL GNAT (*Insecta, Diptera*). Small 2-winged flies of many species constituting the family *Cecidomyiidae*. Most are plant feeders as larvae and produce galls on the plants that they attack. Others are predacious or scavengers.

GALLIFORMES (*Aves*). This order of gallinaceous birds (ancestors of modern poultry) are mostly medium to large in size, with only a few small species. The length is 12–235 centimeters (5–92½ inches), and the weight is 45–11,000 grams (1½ ounces to 24 pounds); in domesticated forms the weight reaches 22,500 grams (49½ pounds). There are 10 primaries; the outer secondaries are generally very short. The feathers often have a well developed aftershaft. Generally downy feathers are found only on the pterylae. There are no powder downs, but the preen gland is present. The males of many species are often very colorful, with widespread iridescent colors. The females generally have a protective coloration. They have very strong breast muscles which enable them to fly up quickly (except for the hoatzins). They are predominantly ground birds with strong feet. They have a strong beak, and almost always, a roomy, distensible crop which acts as a food reservoir. There is a very strong gizzard between whose grinding surfaces, with the help of small stones swallowed for this purpose, grains and green food are ground up. They generally have a long caecum for cellulose digestion. There is a gall bladder in all species.

There are two suborders: 1. *Galli*, including the families Mound Builders, Curassows, and Pheasants and pheasantlike birds; and 2. *Opisthocomi*, with the crested fowl (the hoatzin) as the sole species. There are 94 genera and 263 species in total. They are distributed over most of the world, in semideserts, steppes, savannahs, forests, and cultivated country, and mountains up to far above the tree line (6,000 meters; 19,686 feet). All gallinaceous birds like to bathe in dust or sand, but not in water.

The *Galli* are of importance to man, for they include four widely distributed domestic birds, including the domestic chicken. See also **Poultry.** The great majority of *Galli* can reproduce when one year old. Most species lay many eggs. In the European partridge, up to 26 eggs have been found in one clutch. Incubation is performed almost without exception by the hen alone. Mound-building birds do not incubate at all. Newly hatched chicks have a dense, protectively colored down plumage and are soon able to feed themselves. They can fly in the first few weeks, sometimes even in the first few days. The wings of young of the true *Galli* are, however, still incomplete, having only seven short primaries. They lack secondaries. This "first wing" is much smaller than that of adults but suffices for the chicks' flight. With the increase of the bird's weight, the primaries and secondaries which were lacking grow. The inner primaries, which are too short, are replaced by longer ones. The replacements fit in with the outer primaries of later growth, which from the start are about the final length, and so are not necessarily replaced.

All true *Galli* not only have a "first wing" of short duration, but they also have a smaller, still incomplete "first tail" in many species. Its surface area is in accordance with the needs of the chick during the first weeks. As adults, many species moult the tail from the inside towards the outside (centrifugally). Others moult from the outside towards the inside (centripetally). Still others begin the moult in each half of the tail with a feather which lies between the central one and the outermost one.

All other *Galliformes* (excepting the hoatzin) are united in the large family *Phasianidae*. The size and weight are quite variable, ranging from only 45 grams (1½ ounces; Chinese painted quail) to 22.5 kilograms (49½ pounds; domestic turkey). There are primitive species and highly specialized ones, as well as many intermediate ones. In species which have remained primitive, males and females both have a uniform, camouflaging plumage. In highly specialized species, males have bright plumage colors, decorative ornaments, excessively large decorative feathers, and colorful distensible structures on the head and neck. These decorative feathers are important in courtship display.

They are ground dwellers. Their food consists mainly of vegetation, grains, berries, roots, conifer needles, etc., but many insects and other small animals are also eaten. The union of the sexes is extraordinarily variable, including monogamy, polygamy, or virtually no bond at all. As with birds in general, the more complicated the male's decoration and courtship behavior, the less is its participation in the rearing of offspring. Their nests are built on the ground and rarely, with the exception of the tragopans, in trees. In most cases only females incubate. The young are precocial. They are distributed over most of the world, but are absent on many islands. In America they are represented by grouse, one tribe or group of *Perdicinae*, the toothed quails, and the turkeys.

There are nine subfamilies (grouse, tragopans, pheasants, turkeys, argus pheasants, and peafowl). Altogether there are 75 genera with 204 species. See also **Poultry.**

GALLIUM. Chemical element symbol Ga, at. no. 31, at. wt. 69.72, periodic table group 3a, mp 29.78°C, bp 2403 ± 0.5°C, density 5.90 (solid at 20°C), 6.095 (liquid at 29.8°C), 5.445 (liquid at 1100°C). Elemental gallium has a one-face-centered orthorhombic crystal structure. Among the elements, gallium (like mercury) is liquid at ordinary temperatures. Gallium is a white, tough metal, but so soft that it can be cut with a knife. A freshly exposed surface soon oxidizes superficially to a bluish-gray color. When heated about 500°C, the metal burns in air. Gallium is only slightly affected by H_2O at room temperature, but reacts vigorously in boiling H_2O. The metal is only slowly attacked by concentrated acids, but does dissolve readily in aqua regia. The two stable isotopes of gallium are ^{69}Ga and ^{71}Ga. The eight radioactive isotopes include ^{64}Ga through ^{68}Ga, ^{70}Ga, ^{72}Ga, and ^{73}Ga. All have a relatively short half-life, the longest, ^{67}Ga with a half-life of 78 hours. See also **Radioactivity.** Gallium was one of the elements predicted by Mendeleev from his early periodic arrangement of the chemical elements. The element first was identified by Francois Lecoz de Boisbaudran in 1875 from observations in a spectroscopic study of zinc blende. In terms of abundance, gallium ranks 31st among the elements, with about 15 ppm in the earth's crust.

First ionization potential 6.00 eV; second, 20.43 eV; third, 30.6 eV. Oxidation potentials $Ga \rightarrow Ga^{3+} + 3e^-$, -0.52 V; $Ga + 4OH^- \rightarrow H_2GaO_3^- + H_2O + 3e^-$, 1.22 V.

Other important physical characteristics of gallium are given under **Chemical Elements.**

Gallium occurs in very small amount in zinc blende, magnetite, pyrite, bauxite, and kaolin of certain localities. A few parts per million is present in Oklahoma zinc ores. The recovery of gallium from zinc flue dust is effected by solution of the dust in excess of HCl, addition of potassium chlorate, and distillation to remove germanium. When the residue is converted into sulfate, fractional electrolysis of the slightly acid solution removes zinc, and the gallium is obtained almost free from indium. The only known deposit of gallite, $CuGaS_2$, is in southwest Africa. The mineral contains about 1% gallium. The most important commercial source of gallium is bauxite which contains up to 0.01% gallium. The metal is recovered from the sodium aluminate used in the extraction of aluminum from bauxite. In one process, calcium hydroxide is mixed with the sodium aluminate solution. At this juncture the ratio of gallium to aluminum is about 1 to 3,000. By precipitating and filtering out calcium aluminate, a gallium-rich solution remains. The filtrate then is agitated with CO_2 which precipitates more aluminum out as aluminum hydroxide. At this point, the enriched gallate-in-caustic solution contains approximately 0.2 grams of gallium per liter. This solution is used as an electrolyte in a mercury cathode cell. The gallium amalgamates with the mercury. It is dissolved out of the mercury with boiling NaOH in the presence of iron which serves as a catalyst. At this point, the concentration is approximately 80 g of gallium per liter. The process is repeated several times, after which the gallium concentrate is electrolyzed, using a stainless steel cathode on which the gallium plates out. The gallium is easily removed from the cathode by raising the temperature above the melting point. For highly-pure metal, subsequent purification processes are required, including (1) crystallization as monocrystals, (2) chemical treatment with acids or oxygen at high temperatures, or (3) repeat resolution in pure boiling NaOH and reelectrolyzing. A metal of 99.99999% purity thus can be obtained.

Uses: The availability of gallium in very high purity is important to its use as a semiconductor in various electronic devices, such as diodes, laser diodes, and electroluminescent diodes. The compound usually used in these applications is gallium arsenide GaAs which is prepared by reacting hydrogen and arsenic vapor with gallium oxide Ga_2O_3 (prepared from very pure metal) at a temperature of about 600°C. Properties of the GaAs so produced include: intrinsic electron concentration, 10^7; energy gap, 1.38 eV at 20°C; electron mobility, 8,800 cm^2/V-s.

Gallium arsenide also is used in solar batteries. Gallium metal is used as an activator in luminous paints and phosphors, as well as in arc rectifiers, dental amalgams, as a sealant in vacuum systems, in transistors, and in some organic syntheses. Because the metal expands upon solidifying (3.1%), it should not be stored in fragile containers. Although potentially useful in high-temperature thermometers because of its liquidity over a wide temperature range, these applications have been limited, partially because of the high cost of the element.

Chemistry and Compounds: Gallium metal is quite corrosive to most other metals because of the rapidity with which it diffuses into the crystal lattices of metals. For example, only a very small amount of gallium in contact with an aluminum plate or sheet will result in immediate embrittlement as the result of the diffusion of gallium through the grain boundaries separating them. Gallium readily forms alloys with most metals over 600°C, including barium, copper, gold, iron, lead, lithium, magnesium, manganese, nickel, platinum, silver, sodium, titanium, vanadium, zirconium, and zinc. The few metals that tend to resist attack by gallium are molybdenum, niobium, tantalum, and tungsten.

Gallium trihalides include the trifluoride, tribromide, triiodide, and the trichloride. The trichloride is readily formed by heating the metal with chlorine or HCl, is soluble in ether, and like aluminum chloride, is effective as a catalyst in various organic reactions. Both the trichloride and the tribromide are dimeric in the vapor state. Other known trivalent gallium compounds are the sesquisulfide, sesquisulfate (which forms double salts analogous to the alums), trinitrate, nitride, sesquioxide (which is polymorphic like alumina), and trihydroxide, which is, however, of variable composition, and which forms salts, the gallates, in alkaline solution.

Known gallium(II) compounds include the sulfide, selenide, telluride, dichloride, and dibromide. The last two are unstable, reacting vigorously with water to give hydrogen, and also undergoing oxidation, or disproportionation to the metal and the gallium(III) compound. They also are diamagnetic and their structure is $Ga^+[GaX_4]^-$.

Simple gallium(I) compounds are also unstable, but Ga^+ may be stabilized in the presence of large anions, e.g., in $Ga[AlCl_4]$. The sulfur and selenium compounds Ga_2S and Ga_2Se have been shown to exist, but the oxide is uncertain.

Triethylgallium and trimethylgallium have been prepared, but are extremely reactive, even with air and H_2O. Like aluminum and indium, gallium forms a number of chelated oxy compounds, almost all of which are of 6-coordinate type. They include the stable crystalline inner complexes of which the β-diketones coordinate in the proportion of 3 molecules of diketone per atom of gallium. Trioxalato as well as dioxalato salts are known, and compounds such as 8-quinolinol and substituted 8-quinolinols form trimolecular chelate rings involving nitrogen and donor oxygen.

Gallium, like boron, forms a dimeric hydride, Ga_2H_6, from which a series of tetrahydrogallates, containing the GaH_4^- ion, is derived.

Gallium and most of its compounds are not highly toxic. For rats and rabbits, the LD_{100} has been established at approximately 100 mg of gallium per kilogram.

See list of references at end of entry on **Chemical Elements.**

GALLIUM ARSENIDE SOLAR CELL. Solar Energy.

GALLIUM (Laser). Telephony.

GALLON. Units and Standards.

GALLSTONES. Gallbladder and Biliary Tract Diseases.

GALL WASP (*Insecta, Hymenoptera*). A minute insect whose attack on plants produces galls. They are of many species, making up the subfamily *Cynipinae.*

GALTON BOARD. Probability.

GALVANIC ACTION (Corrosion). Corrosion.

GALVANIC CELL. Also known as a voltaic cell, an electrolytic cell that produces electric energy by electrochemical action. Although a battery may comprise only one cell, there may be several cells making up a battery and thus cell and battery are not fully synonymous. See also **Battery.**

There are many ways in which a voltage difference can be produced in an electrochemical cell. The simplest cell, thermodynamically, is the "concentration cell" in which electrolyte or electrode materials are incorporated into half-cells in differing concentrations; a half-cell is a system involving an electrolyte and a single electrode. When the half-cells are connected, the free energy change accompanying the transfer of one substance from high to low concentration results in the liberation of electrical energy. The gravity cell is a type of two-electrolyte cell in which the separation between the two ionic solutions is maintained by means of gravity. An example is the Daniell cell in which a cupric sulfate solution in contact with a copper electrode is below a zinc sulfate solution in contact with a zinc electrode. The difference in specific gravity of the solutions prevents, or at least retards, mixing. The Daniell cell also belongs to the classification of displacement cells in which the essential chemical reaction is the ionization and entry into solution of atoms of one element, and the discharge and deposition from solution of the ions of another. Concentration cells, although interesting theoretically, are not important commercially.

The majority of economically important cells consists of two dissimilar electrodes of metal or metal compounds, immersed in an aqueous solution of an acid, base, or in some cases a salt. The negative of a fresh cell is typically in the metallic state, while the positive is usually an oxide, or occasionally, a salt of the metal. During discharge, the negative electrode is oxidized as electrons leave it via the external circuit, and the positive is reduced. Since by definition an anode is an oxidation electrode, in the literature the negative is generally called the "anode" and the positive the "cathode." This conforms to accepted electrochemical terminology, although it is the cause of some confusion.

Although galvanic cells theoretically might look more attractive than heat engines as sources of electric power, since the energy changes are not subject to the limitations of the Carnot cycle, the cost comparisons of delivered power to date do not work out that way. In fact, on a kilowatt hour basis, the cost spread is 70 to 100 : 1 for a rechargeable battery, and many hundreds to 1 for primary cells, such as flashlight cells. This is because of inefficiencies in electrochemical operation, high material costs, high cost of the tightly controlled production operations necessary, etc. Galvanic cells have grown in importance because of the strength of other needs, such as that for a portable supply of power, for power at a place far distant from the prime power source, for a reserve or emergency source, etc. There are also needs for a source of pure direct current or for a stable reference voltage which can be provided by galvanic cells. Since the early-1970s, much interest has been regenerated in the use of galvanic cell (battery power) for small electric automobiles in an effort to reduce emissions from internal-combustion engines.

Fuel cells, although operating as galvanic cells at the electrodes, are in a separate class in that they provide direct, single-site conversion of original raw materials into electrical power, obviating the boiler-turbine-transmission-rectifier chain that precedes the production and use of ordinary batteries. See also **Fuel Cells.**

Corrosion also results from the action of oftentimes numerous galvanic cells where dissimilar metals and electrolyte (as from excessive moisture, humidity, acidic atmospheric ingredients, etc.) provide all of the electrochemical necessities for a transfer of material that causes metals to gradually "waste away" and weaken various structures. See also **Corrosion.**

GALVANIZING. The application of a layer of zinc to the surface of iron and steel for protection from corrosion is known as galvanizing. Galvanizing is one of the most effective means of preventing rusting because of its low cost and ease of application compared with other metallic coatings.

In the hot dip process, the sheets or other articles to be coated must be free from scale, dirt, grease, etc., and are usually prepared by pickling and washing before immersion in molten zinc commercially known as spelter. Articles fabricated from iron and steel sheets and wire are hand-dipped. Sheets and wire are handled mechanically.

An increasing proportion of sheet-metal products is being coated as sheet or strip before fabrication. This requires a tightly adhering coating to prevent peeling during stamping or forming operations. In order to obtain good adherence in hot-dipped coatings special processing is necessary, especially with the heavier weights of coating which give longer protection. For lighter weight coatings a duplex bath consisting of a layer of molten lead under the molten zinc is often used. The steel sheet passes through the lead, which does not adhere, and up through the zinc. The time in which the steel is in contact with the spelter is greatly reduced and consequently less zinc is deposited.

Some galvanized sheets are annealed after dipping in order to form a coating consisting entirely of iron-zinc compounds, a process which tends to increase resistance to peeling. For lightweight sheet, electrodeposition of the zinc is common practice. See also **Zinc.**

GALVANOMETER. An instrument for measuring electric currents, usually by means of their magnetic effect. Observations are made by noting the deflection produced by the reactive torque exerted between an electric current and a magnet. Galvanometers may be divided broadly into two classes, according to whether the coil is stationary and the magnet turns, or vice versa.

Perhaps the most highly developed of the first type is the Kelvin astatic galvanometer. This has two magnets equally magnetized but antiparallel mounted on the same suspension, one above the other, and each magnet is surrounded by a coil. The two coils are joined in series and are oppositely wound, so that a current through them will turn their respective magnets in the same direction. The earth's uniform field has no effect upon such an astatic pair of magnets; but there is a large control magnet, placed above the pair, against whose field the current turns the suspended system. The movement is observed by the usual mirror-and-scale or optical lever device. Galvanometers of this type are now used essentially for teaching and demonstration.

Among galvanometers of the second type, that of d'Arsonval is best known. (See figure.) The magnet in this instrument is a fixed,

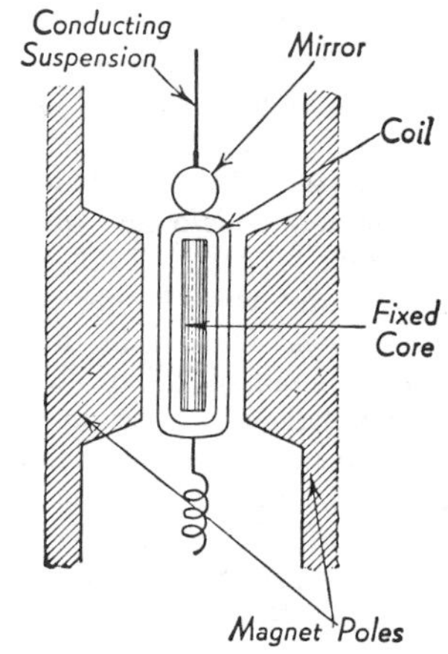

Essential parts of a d'Arsonval galvanometer.

permanent magnet of the horseshoe or double-horseshoe form, with a light, rectangular coil suspended in the strong field between its poles, the suspension carrying the feeble current. The current causes the coil to turn in the field. Often a fixed iron core is supported inside the movable coil to concentrate the field.

If these galvanometers are undamped, they will give a "throw" when a charge of electricity is sent through them, and the charge can be thereby measured. Such an instrument, with a heavy coil, called a ballistic galvanometer, is useful in capacitance measurements. The oscillations may be damped by shunting.

There are also string galvanometers, in which a straight, slender wire carrying the current is thrust to one side by a magnetic field; and vibration galvanometers, in which the strong vibrates in synchronism with the alternating current traversing it.

Many of these instruments are classics and principally of historical interest today. See also **Electrical Instruments.**

GAMBLING ODDS. Game Theory.

GAMETE. A sexual reproductive cell or germ cell which normally unites with another to produce a new individual. In humans and other mammals, gametes are produced in the gonads—the ovaries and testes—where meiosis takes place. By this process, the number of chromosomes is reduced from the diploid of somatic cells to the haploid number. The diploid number of chromosomes is restored at fertilization when the egg and sperm fuse to form the zygote.

The gametes of some primitive organisms are of one form; these are single cells that swim about in the water. Such organisms are said to be isogamous and the germ cells are called isogametes. In most species, however, only the male gametes retain the power of locomotion. The female gametes are larger inert cells and the organisms are called heterogamous. The male cells of these species are known as sperms or spermatozoa and the female cells as ova or eggs. Because of its smaller size the male gamete is also known as a microgamete, and the female gamete as a megagamete.

The union of two unlike gametes (heterogametes) is called heterogamy. The cell which is formed by the union of two gametes is called the zygote; from it a new plant or animal develops. The cells or organs in which gametes are formed in plants are called gametangia. In heterogamous plants, the gametangia containing sperms are called antheridia; those containing eggs, either oögonia or archegonia. In animals, the gamete-producing bodies are known as gonads. The male gonads are the testes and the female gonads are the ovaries.

The development of two forms of gametes permits both the freedom of movement necessary to bring the two cells together for fertilization and the storage of the protoplasm and food necessary for the development of any body of reasonable size and complexity to a stage in which it can secure more materials for itself. By the delegation of one function to each kind of cell neither is subject to harmful restriction.

In most species of animals, the sperm is a minute cell with a slender flagellum or tail whose undulating movements propel it through the water or the seminal fluid. The main part of the sperm is the head, which contains an apical body and the nucleus. Behind the head is a neck, or a middle piece of more complex structure, from which the sperm aster involved in the fertilization process sometimes develops. The sperms of some worms and arthropods lack the flagellum although many bear processes of other kinds. They are much less motile than flagellate sperms but they are said to move slowly by amoeboid action or by means of their processes.

Ova are more compact cells, often spherical in form. They contain abundant cytoplasm and in many species an enormous amount of food material (yolk, deutoplasm), as in the egg of a bird. Here the yolk is the egg cell or ovum proper, but the living protoplasm is a tiny mass at some point on its periphery. Ova may also have special envelopes such as the albumen or white of the bird's egg, the shell membrane, and the shell. In nearly all mammals the ovum has little yolk and becomes separated except at one point from the surrounding layers by a large space filled with fluid. The whole structure is known as a Graafian follicle. Insect eggs are enclosed by a shell called the chorion which is often beautifully sculptured and strangely shaped. Where such coverings occur, a minute opening, the micropyle, sometimes provides an entrance for the sperm.

This entry reviewed and updated for 6th Edition by Ann C. Vickery, Ph.D., University of South Florida, College of Medicine, Tampa, Florida.

GAME THEORY. The theory of games is a mathematical theory, founded by J. Von Neumann, dealing with optimal behavior in situations involving conflict of interest. It is so called because a mathematical theory requires a precise statement of the possible courses of action available to all "players" involved, and of the outcome from any combination of actions. This situation arises naturally in parlor games, but military or economic problems that can be formulated precisely enough are amenable to the theory.

The theory is based on utility theory. Roughly speaking, this asserts that one can analyze all games as if they were played for money and each "player" (or more generally each team) were trying to maximize the *average* value of his net monetary gain from the game. (But the currency in which the game is played may be an abstract one known as "utiles" that is not linearly related to any real currency.) The *sum* of the gains to all players may always be zero, in which case the game is a "zero-sum" game. Otherwise, it is a "non-zero-sum" game.

Another important concept is that of a "pure strategy." This is a precise statement of what the player will do in any conceivable situation. In practice, one often decides one's course of action as the situation develops, but in principle, one could make all decisions in advance and choose a complete strategy. When all players have determined their strategies, the outcome is determined, either as a sure thing or as a probability distribution, and the average net gain to each player is determined. One can therefore (in principle) construct a table giving the average net gain, known as the "pay-off" to each player, given each player's strategy. Such a table is known as a "pay-off table."

The concept of a strategy has been generalized to allow a player to choose between his pure strategies at random but with prescribed probabilities. So a strategy may involve using the ith pure strategy with probability p_i. Such a strategy is called a mixed strategy if more than one p_i is nonzero.

The general theory of games centers on the theory of two-person zero-sum games. This deals with situations involving just two players, say A and B, whose interests are diametrically opposed. The main theorem of the theory of games is that any such game in which both players have only a finite number of possible pure strategies has a unique value v in the following sense. There exists at least one (pure or mixed) strategy for A such that he gains at least v on the average whatever B does, and furthermore there exists at least one (pure or mixed) strategy for B such that he loses at most v on the average whatever A does. In these circumstances a strategy for A that guarantees v (on the average) regardless of B's actions is called "optimal," since any serious attempt to gain more can be regarded as irrational on the grounds that it relies on B acting against his own best interests. Similarly a strategy for B that guarantees losing at most v (on the average) is "optimal." (If $v = 0$ the game is said to be "fair.")

Sometimes both players have optimal pure strategies, in which case the game is said to have a "saddle point." If the number of pure strategies is not too large, a saddle point can easily be determined by inspecting the pay-off table. If there is none, the value of the game and the optimal strategies can be determined by linear programming: the variables are the probabilities that A uses each pure strategy, the constraints are that these probabilities are nonnegative and sum to 1, and that the expected pay-off $\geq v$ against each of B's pure strategies. The objective function to be maximized is v. The dual linear programming problem then determines B's optimal strategy.

A mathematically trivial but conceptually important consequence of the main theorem is that a player's chances of achieving the value of the game are in no way jeopardized if he announces the optimal mixed strategy he is using to his opponent in advance. On the other hand, he must certainly not announce which particular component of a mixed strategy he will use. One can think of a mixed strategy as a mathematically precise description of a policy of occasional bluffing. The theory provides optimal bluffing policies for two-person zero-sum games.

There are great computational problems in studying games with a large or even infinite number of pure strategies available to both players, but some work has been done on them. But extensions of the theory require further concepts, a few of which are outlined below.

A theory of statistical decision making has been developed round the concept of a game between the statistician and nature. Many people think it is unduly pessimistic for the statistician to behave as if nature is doing its very best to frustrate him; but the use of game-theoretic terminology has helped to clarify some of the issues involved.

There is no uniquely obvious generalization of the concepts of the value of a game and the associated optimal strategies when the game is not zero-sum, or when there are more than two players. An "equilibrium point solution" has been defined as a set of (pure or mixed) strategies for each player such that no one can gain by changing his strategy if the other players persist with their present strategies.

Other work on the general theory is based on the idea that the players will group themselves into coalitions. The outcome to a coalition can then be determined on the assumption that those outside the coalition will combine to oppose it. The problem is then reduced to studying which coalitions will form, and what payments should be made to each member of the coalition to prevent him from being attracted by a counter-offer from another potential coalition.

Any game has a "Shapley value" for each player, which has desirable mathematical properties and can be derived as follows. Imagine that a coalition of all the players is formed in a random order, starting with a single player and adding one player at a time. Each player is then assigned the advantage gained by the coalition as a result of his joining it. The average value of this advantage for all orders of formation of the coalition is then the Shapley value.

In spite of the conceptual problems with arbitrary games, it has been shown that for plausible models of real economic situations in terms of games with large numbers of players, all the conflicting concepts reduce to the classical economic theory of free competition as the number of players increases and the importance of any individual player decreases.

GAMETOPHYTE. One of the two generations which alternate with each other in the life-history of many plants is called the gametophyte generation. It is the generation in which the gametes or sexual cells are formed. The cells of plants in this generation have the reduced or haploid number of chromosomes, which is half the number found in cells of plants of the diploid sporophyte generation.

The plant or part of a plant which forms this haploid phase of the plant life cycle is known as the gametophyte. In the ferns and some relatives, the gametophyte is a small plant separate from the sporophyte and independent of it. In the seed plants, the gametophyte is very small and parasitic on the sporophyte. In liverworts, the gametophyte is the dominant generation and the sporophyte is smaller.

Gametogenesis is the formation of gametes. In animals, this is usually accompanied by meiosis.

GAMMA-AMINOBUTYRIC ACID (GABA). Brain and Nervous System.

GAMMA CAMERA. Radioactivity.

GAMMA DISTRIBUTION. Pearson Distributions.

GAMMA FUNCTION. The infinite integral, sometimes called Euler's second integral:

$$\Gamma(z) = \int_0^\infty e^{-t} t^{z-1}\, dt$$

It converges for all positive, real values of z. Its properties include:

$$\Gamma(z+1) = z\Gamma(z)$$
$$\Gamma(z)\Gamma(1-z) = \pi \csc \pi z$$
$$\Gamma(\tfrac{1}{2}) = \sqrt{\pi}$$

when $z = n$, a positive integer, $\Gamma(n) = (n-1)!$, hence this is often called the factorial function.

The Weierstrass definition of the function is

$$1/\Gamma(z) = ze^{Cz} \prod_{n=1}^{\infty} (1 + z/n)e^{-z/n}$$

where C is the Euler-Mascheroni constant:

$$C = \lim_{n\to\infty} (1 + \tfrac{1}{2} + \cdots + 1/n - \ln n)$$
$$= 0.577215\ldots$$

Another definition is that of Euler:

$$\Gamma(z) = \lim_{n\to\infty} \frac{(n-1)!}{z(z+1)(z+2)\cdots(z+n-1)}\, n^z$$

See also **Beta Function,** which is Euler's first integral, and **Stirling Formula.**

Sir Maurice Kendall, International Statistical Institute, London.

GAMMA GLOBULIN. The fraction of the protein globulins of the blood plasma in which are found the antibodies. The administration of gamma globulins may be effective in the preventing of poliomyelitis, measles, and heptatitis A and sometimes as an second choice in connection with hepatitis B. It is most commonly used in connection with infections following exposure to hepatitis A virus. Public health authorities usually suggest that persons in contact with hepatitis A patients and who have not been previously infected should be treated with gamma globulin. Where the source of hepatitis is identified as water, food, or other materials with which the public has been in contact, very early use of gamma globulin can reduce development of the disease within a community. Unfortunately, however, in such situations the persons exposed may already be in the incubation period, making the administration of gamma globulin ineffective. Gamma globulin is also used prophylactically in connection with persons who plan to travel to tropical and other regions where heptatis A virus may be contacted. Gamma globulin is also useful as a preventive measure for persons who handle primates in laboratories, zoos, etc. Such preventive measures must be taken at 4-to-6-month intervals to ensure protection. See also **Immunology and Immunization;** and **Virus.**

GAMMA RADIATION. A photon, or quantum of electromagnetic radiation, that is emitted when an atomic nucleus undergoes a transition from one of its excited energy levels to a lower level. The name *gamma ray* was applied in the earlier years of radioactivity investigations, while the exact nature of these radiations was still a mystery. Gamma-ray energies range from 10^4 to 10^7 eV. They are often emitted as a part of a nuclear reaction, when an atomic nucleus is left in an excited state, or during an isomeric transition. Gamma rays also can be emitted following alpha-particle decay, beta-particle decay, or orbital electron capture, if the daughter nuclide is left in an excited state.

In the strictest sense, the term gamma ray is applicable only to photons produced as a result of transitions in atomic nuclei. However,

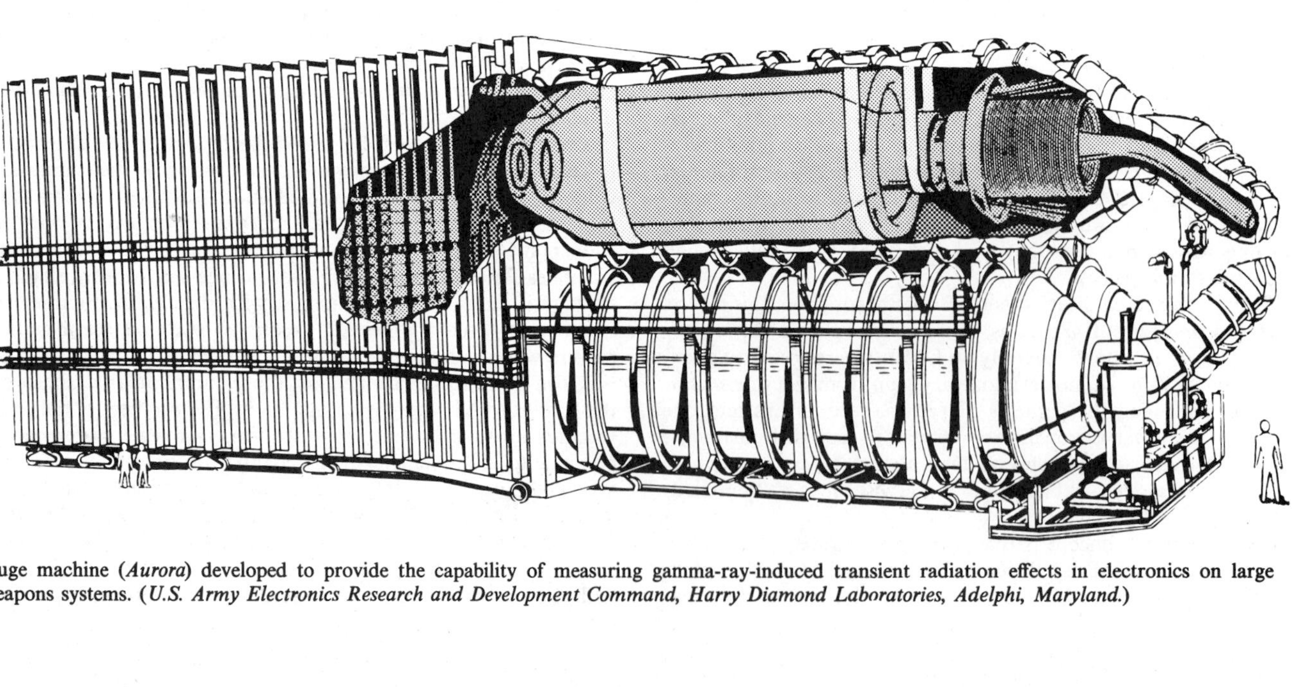

Huge machine (*Aurora*) developed to provide the capability of measuring gamma-ray-induced transient radiation effects in electronics on large weapons systems. (*U.S. Army Electronics Research and Development Command, Harry Diamond Laboratories, Adelphi, Maryland.*)

the term is also sometimes used to denote bremsstrahlung radiation produced when the high energy electrons in the beam of an electron accelerator, such as an electrostatic generator, a betatron, a synchrotron, or a linear accelerator, strike the target of that accelerator.

Gamma rays carry away the full energy of the transition with which they are associated. As a result, if detecting systems are used that are capable of absorbing the full energy of the gamma ray, a spectrum of gamma-ray numbers as a function of energy shows a series of distinct peaks, each associated with an individual gamma-ray transition. On the other hand, the discrete energy characteristics of gamma rays are more difficult to observe if the detecting system separates the effects of different types of gamma-ray interactions with matter, such as the Compton, photoelectric, and pair-production interactions. Under certain circumstances, a transition that would normally be expected to emit a gamma ray may sometimes release its energy through an internal conversion process.

A huge machine for inducing and measuring gamma-ray radiation effects in the electronics of large weapons systems is shown in the accompanying figure.

See also **Particles (Subatomic); and Radioactivity.**

GAMMA-RAY ASTRONOMY. The development of spacecraft capable of reaching out where incoming gamma rays are not too greatly attenuated made it practical to design a gamma-ray telescope. In the general configuration, an incoming gamma ray strikes a "sandwich" of sodium iodide and cesium-iodide material, in which the pair production process occurs, whereby the gamma-ray photon is converted into a positron and a negatron, provided that its energy exceeds the energy equivalent of their total mass. Cesium is used because its high nuclear charge increases the probability of the process. The positrons and negatrons enter a Cerenkov detector, which is viewed by photomultiplier tubes. The sodium-iodide-cesium-iodide layer is also viewed by such tubes; both units are connected to a circuit which registers a count only when both the layers and the Cerenkov detector record an event. As a further guard against spurious counts (those arising from particles or radiation other than gamma rays), the system is contained in a case, which passes gamma rays without reaction, but which gives a signal when charged particles are encountered. Thus, such coincidence events are not counted.

Gamma ray bursts were first noted in mid-1967 by several of the Vela satellites. These satellites were designed to monitor the Nuclear Test Ban Treaty. The bursts of radiation were characteristic of nuclear debris and the observers might have presumed that a hydrogen bomb had been set off in space. For obvious reasons of security, the observations were not made available to scientists without security clearances until 6 years later. Study of the records clearly indicate that the bursts were not coming from sources within the solar system. A number of other experiments in space, including Apollo 16, further confirmed the existence of such gamma ray bursts. The bursts are characterized by a short starting pulse, from 0.1 to 4 seconds in length, followed by one or more pulses, with the event entirely over within a minute.

In regular gamma ray astronomy, radiation with an energy of 50 MeV is investigated. The reported bursts occur between 0.1 and 1.2 MeV. At the lower end of the energy range, it is difficult to distinguish between gamma rays and x-rays. The bursts are sufficiently intense to allay any questions that the signal is simply a fluctuation of background noise.

As of the end of 1973, 23 such events had been recorded. Although there is some question pertaining to the locations of the bursts, it is known that they do not come from the earth, sun, or planets. With one exception, the directions do not correspond to any of the closest stars. In the late-1960s. Colgate suggested that supernovas could be the source of sharp gamma ray pulses. However, supernovas do not occur within our galaxy with sufficient frequency to account for the bursts. Six rather elaborate theories were proposed—a modified theory for sources in extragalactic supernovas, four theories pertaining to sources in various types of stars in the galaxy, and another relating to the possible breakup of relativistic "beebees" within the solar system. The topic is explored in more detail by Metz (*Science*, **182**, 1234–1237 (1973).

Because gamma rays from interstellar space cannot be reliably measured from locations on the earth's surface—they are lost amid the confusion of gamma rays created in the atmosphere by cosmic ray bombardment—useful data must be gathered from high-altitude balloons (earlier studies) and orbiting satellites. Three satellites, launched during the 1970s, collected much gamma ray information. These included the Second Small Astronomy Satellite (SAS-2), launched in 1972 and shown in accompanying diagram; the COS-B satellite, launched in 1975 by the European Space Agency; and the High Energy Astronomy Observatory (HEAO-1). These satellites furnished fundamental data which will be exploited by the forthcoming Gamma Ray Observatory (GRO) planned for launch by the Space Shuttle sometime during the mid-1980s. The proposed satellite will orbit the Earth at a distance of 250–300 miles (~400–480 kilometers).

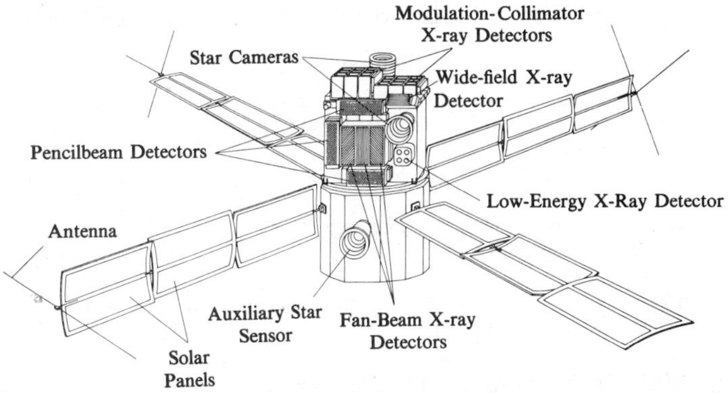

The SAS-2 satellite (NASA) which has detected strong gamma radiation emanating from the plane of the Milky Way. The satellite also has found that weak gamma radiation is widespread throughout space. (*NASA diagram.*)

Resulting from the most energetic processes in the universe (supernovae, pulsars, black holes and sources still not fully understood), high-energy gamma rays make possible the observation of regions previously obscured from Earth. Among the most interesting of observations from the satellites has been the very bright gamma ray emissions from the plane of the Milky Way. It is postulated that these emissions arise from areas of high matter or cosmic ray density.

The Crab Nebula, of which the fastest radio pulsar known is also a part, was one of the first point sources of gamma rays observed. Knowledge from gamma ray studies is being correlated with other advanced observing techniques. Data from SAS-2 indicate that there is diffuse gamma radiation coming from throughout the sky. Much remains to be understood about this phenomenon. With the combination of radio and gamma ray observations, some of the most active regions of star formation and cosmic ray production have been identified. These regions appear to be located about midway between the Sun and the center of our galaxy. Some scientists have called a ring about the center of our galaxy a "cosmic ray bagel." This ring is located some 15–20,000 light years from the galactic center. This distribution of cosmic rays correlates well with the region of highest concentration of supernova remnants and pulsars—the latter believed also to have resulted from supernovae. Thus, cosmic rays are no longer considered to be mainly extragalactic in orgin, but rather they are generated within the galaxy as well.

The COS-B satellite, with a highly directional gamma ray detector, has revealed a number of new gamma ray sources in the Gemini–Taurus, Perseus, and Cygnus regions. The Italian astronomer G. F. Bignami has commented, "The most intriguing feature about these new sources is certainly that they cannot yet be correlated to any known celestial object, in spite of various searches having been carried out. It is, of course, possible that one is dealing with known gamma ray emitters like pulsars which are for some reason so far undiscovered in other bands, but the exciting possibility also exists that we are slowly coming up with representatives of a totally new class of celestial entities—a situation reminiscent of the black holes discovery from the evidence of the first galactic x-ray sources."

A few scientists have observed that gamma ray detection may make it possible to distinguish matter from antimatter for the first time. (Antimatter has been defined as the electrical inverse of normal matter, with positively charged electrons circling a negatively charged nucleus. Minute amounts of antimatter purportedly have been produced in particle accelerators.) Some scientists have postulated that possibly entire galaxies may be made of antimatter. Cramer (University of Washington) has observed, in commenting on the proposition that the circular polarization of gamma rays emitted by antimatter stars may be different from that of matter stars, "It is as if the rifle barrels that shoot radiation to us from matter and antimatter have opposite riflings, and spin their bullets in opposite directions." Cramer and Braithwaite (University of Texas) have observed that the beta decay process which is part of the nuclear fusion reactions of a star may provide the means for distinguishing matter stars from antimatter stars. It is presumed that when these particles slow in their movement through space they will produce gamma radiation of opposite polarization. Cramer has proposed a "gamma ray circular polarimeter" for use in future satellite observatories. These would detect the "twist" of gamma radiation that emanates from various novae or supernovae.

References

Chupp, E. L.: "Gamma-Ray Astronomy," Reidel, Boston, 1976.

Hartline, B. K.: "Bursts of Gamma Rays Baffle Astronomers," *Science*, **207**, 858–859 (1980).

Leventhal, M., and C. J. MacCallum: "Gamma-Ray-Line Astronomy," *Sci. Amer.*, **243**, 1, 62–70 (1980).

Meredith, D.: "Gamma Sky," *Technology Review (MIT)*, **80**, 3, 20 (1978).

Staff: "Electromagnetic Radiation of Celestial Processes," *Sci. American*, **241**, 2, 74–80 (1979).

Trombka, J., et al.: "Gamma-Ray Astrophysics: A New Look at the Universe," *Science*, **202**, 933–938 (1978).

GAMMA RAY BURSTS. Cosmic Rays.

GAMMA-RAY SPECTROSCOPY. Gamma rays of concern here originate in the nucleus of radioactive isotopes, i.e., chemical elements whose nuclei are unstable and emit radiation as they decay to stable states. Such radioactive isotope disintegration follows rules that are always the same for the same nucleus. These rules can be set down in a so-called decay scheme. An example is shown in Fig. 1 for the

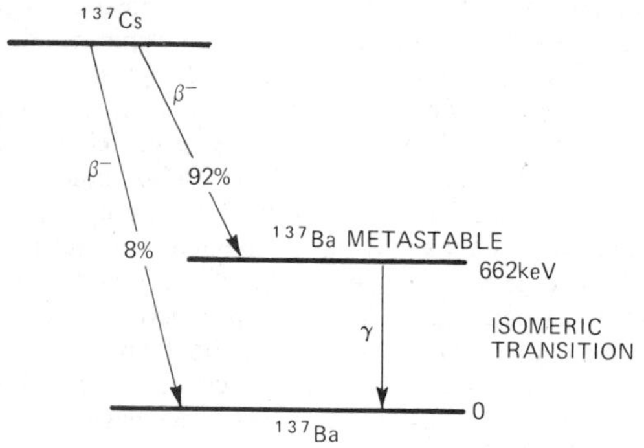

Fig. 1. Decay scheme for ^{137}Cs.

case of the radioisotope ^{137}Cs (cesium-137). The basic decay scheme shown indicates that cesium-137 decays into ^{137}Ba (barium-137) by emitting beta particles (electrons). Eight percent of the cesium nuclei decay directly into barium-137 nuclei; then about 2.5 minutes later, the excited nuclei decay to the lowest energy or ground state by emitting gamma rays having an energy level of 662 keV. Some heavy nuclei emit alpha particles. An alpha particle is a ^{4}He (helium-4) nucleus (two protons and two neutrons). The cesium-137 isotope, with a nucleus containing a total of 137 neutrons and protons, disintegrates with a half-life of 30 years. Since the number of nuclei is halved, the amount of radiation (intensity) is halved. With existing electronic systems, half-lives between 10^{-10} second and 10^{10} years can be measured.

Like most natural events, radioactive decay is not a uniform function. Consequently, the term *half-life* is meant to describe the value that would result if an infinite number of half-life measurements were made and the average calculated. Individual decays, however, follow a Poisson distribution, i.e., the standard deviation is equal to the square root of the number of observed decay events. This fact enables the experimenter to calculate the probable accuracy of his result, assuming no instrumentation inaccuracy.

Gamma Ray Detection. Gamma rays are high energy electromagnetic radiation with very short wavelengths (10^{-18} to 10^{-11} cm). They penetrate matter deeply—on the average much more deeply than do alpha and beta rays, which are charged particles. It is their deep penetration that makes gamma rays useful in the laboratory and industry, in much the same way as x-rays. X-rays originate from shell transitions by orbital electrons, whereas gamma rays originate in the nucleus. Gamma rays usually are detected by observing effects that they produce in matter and when they encounter an atom. Important among these effects are: (1) the photoelectric effect; and (2) the Compton effect. The photoelectric effect occurs when the gamma ray strikes one of the orbital electrons of the atom, transferring its energy to the electron. This process produces a free electron and an ionized atom. The Compton effect arises in the case where the gamma ray strikes an orbital electron without imparting all of its energy to the electron. The electron is detached from the atom but receives only part of the gamma energy. The remaining energy persists as a scattered gamma ray with lower energy than the initial ray. This scattered ray may further collide with one or more other atoms, freeing other electrons. These types of interactions occur variously in nuclear radiation detectors. In each detector type, some observable reaction results, and in one manner or another produces an electrical output charge suitable as an input for an electronic measuring system.

Gamma Ray Spectra. Measurements of gamma radiation are chiefly made in two ways: (1) a record is made of the number of counts as a function of energy, in which case a gamma ray spectrum is obtained; and (2) time relations are observed, in which case several types of information may be desired. A gamma spectrum, as measured by an ideal system, might appear as in Fig. 2 which is the ideal spectrum

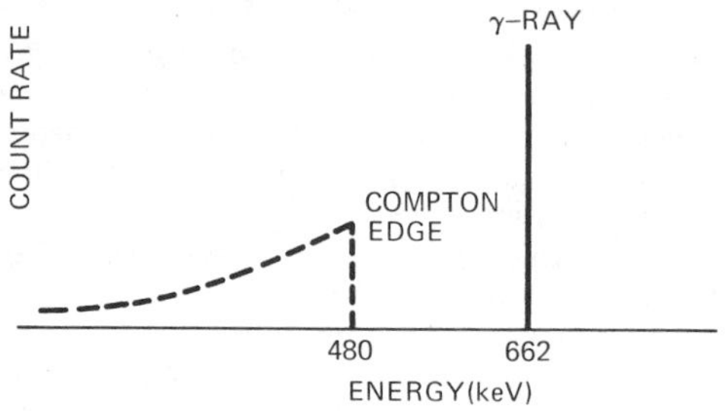

Fig. 2. Ideal gamma spectrum of ^{137}Cs.

of the cesium-137 gamma radiation phenomena discussed earlier. In this spectrum, a large peak appears at 662 keV—caused by the gamma energy radiated when the metastable barium-137 nucleus returns to its ground state. There is also a continuum representing the energies imparted to Compton-scattered electrons. In practice, the spectra measured are not so well defined. See Fig. 3. Most noticeable is that the peaks of the spectrum are broadened to a greater or lesser extent by the characteristics of the devices used to detect gamma rays. Relating to this broadening as a measure of system quality, is its "resolution." This is a function both of the detector and of the associated circuitry. Resolution commonly is defined as the ratio of the full width at half the maximum height of the peak (FWHM) to the energy of the center of the peak. Thus, resolution indicates how well the detector can separate or resolve two different energy peaks. Typical resolutions for common gamma ray detectors range from about 10% to a few tenths of a percent. Also evident in Fig. 3 is a backscatter peak, which results because a large number of gamma rays squarely strike

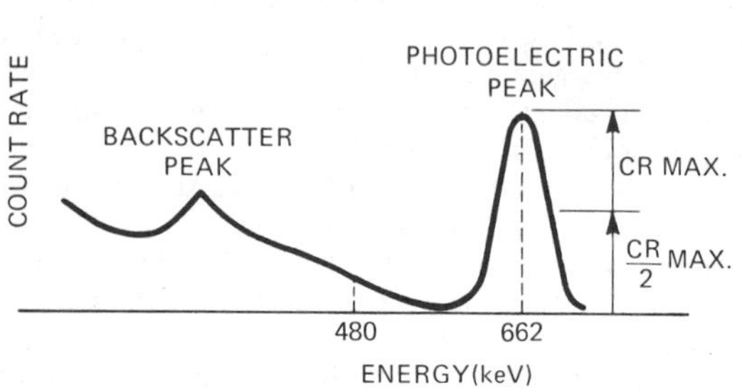

Fig. 3. Typical gamma spectrum of ^{137}Cs as obtained instrumentally.

matter between the source and the detector, losing much of their energy before detection.

Energy Measurements. The measurements usually made in gamma ray work fall into two broad groups: (1) those made of the energy of the radiation; and (2) those made of its timing relative to another event. In addition, counting without regard to energy (often called gross counting) is also done to measure the intensity of the radiation. See Fig. 4. Intensity is measured in terms of counts/minute (or second).

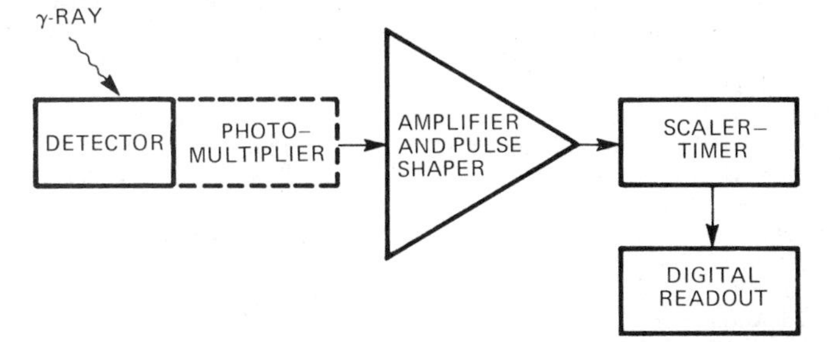

Fig. 4. Gross counting measures radiation intensity of gamma ray regardless of energy.

Time Measurements. The second general class of measurements is one in which the time of occurrence of the gamma ray relative to a reference event is of interest to the experimenter. Such situations occur when one gamma radiation is known to occur a specific interval of time after a trigger event.

Detectors. Commonly used detectors include scintillation, semiconductor, and gas proportional detectors. The scintillation detector often is preferred where high efficiency is more important than resolution—efficiency defined as a measure of the probability that an incident gamma ray will interact with the material in the detector. Semiconductor types are used increasingly, particularly where high resolution is required.

Signal Processing. The signal from the detector is a relatively short current pulse; the time integral of this current impulse is a charge proportional to the energy of the absorbed radiation. The preamplifiers and amplifiers which follow these detectors convert this impulse of charge into a voltage pulse whose height (peak amplitude) is proportional to energy. Thus, signal processing prepares the charge from the detector for the final step, pulse height analysis. In the case of a timing measurement, signal processing prepares the charge signal for use with a timing pick-off (time discriminator). See also **Radioactivity.**

Tracy S. Storer, Hewlett-Packard, Palo Alto, California.

GAMMA SPACE. Phase space of $2fN$ dimensions, the coordinates being f generalized coordinates and f generalized momenta for each of the N particles of the system, each particle having f degrees of freedom. It is the phase space of the whole gas and was called Γ-space by Ehrenfest to distinguish it from the phase space of one molecule (μ-space).

GAMOW-TELLER SELECTION RULES. A set of selection rules for beta decay which state that an allowed transition between parent and daughter states must have no change of parity but can have a spin-quantum-number change of either 0 or 1, except that no $0 \rightarrow 0$ transitions are allowed. See also **Beta Decay.**

GANGLIA. Brain and Nervous System.

GANGLION. In zoology, a ganglion is a small mass of nervous tissue isolated from the central system but containing cell bodies as well as fibers. Many ganglia bear special names. The brain of many invertebrates, for example, is also called the cerebral ganglion, and the more numerous centers of the molluscan nervous system bear names, such as the visceral ganglia and the pedal ganglia. The dorsal root of each nerve arising from the vertebrate spinal cord bears a spinal ganglion and the sympathetic system contains numerous ganglia.

In medicine, a ganglion is a tense globular cystic swelling usually on the back of the wrist or hand, communicating with one of the tendon sheaths or nearby joints. It is formed by a synovial membrane, and is filled with a thick gelatinous fluid. The nature of a ganglion is obscure. It represents either a degeneration in the involved synovial tissue, or simply a herniation of this tissue. The treatment is by mechanical rupturing, or by surgical excision.

GANGLIOSIDES. Identified by Kleng in 1935, the gangliosides are a family of acidic glycolipids that are characterized by the presence of sialic acid. The compounds bear a strong negative charge and are unusual in that they contain both hydrophobic and hydrophilic regions. These compounds are membrane components. Plasma cell membranes are rich with gangliosides. It has been suggested that gangliosides participate in the transmission of membrane-mediated information in living systems. As described by Fishman and Brady, "the carbohydrate portion of gangliosides is made up of molecules of sialic acid, hexoses, and *N*-acetylated hexosamines. The hydrophobic moiety is called ceramide, and it consists of a long-chain fatty acid linked through an amide bond to the nitrogen atom on carbon 2 (C-2) of the amino alcohol, sphingosine. Oligosaccharides are linked through a glycosidic bond to C-1 of the sphingosine portion of ceramide." Svennerholm (1963) suggested the configuration given by the accompanying diagram. The role of gangliosides is still rather discrete, but Fishman and Brady (1976) have studied in some detail the interaction of cholera toxin with ganglioside-deficient cells, as well as their interaction of cholera toxin with ganglioside-deficient cells, as well as their interaction with glycoprotein hormones and their effect on the action of these hormones.

References

Fishman, P. H., and R. O. Brady: "Biosynthesis and Function of Gangliosides," *Science*, **194**, 906–915 (1976).
Klenk, E.: *Z. Physiol. Chem.*, **235**, 24 (1935).
Svennerholm, L.: *J. Neurochem.*, **10**, 613 (1963).

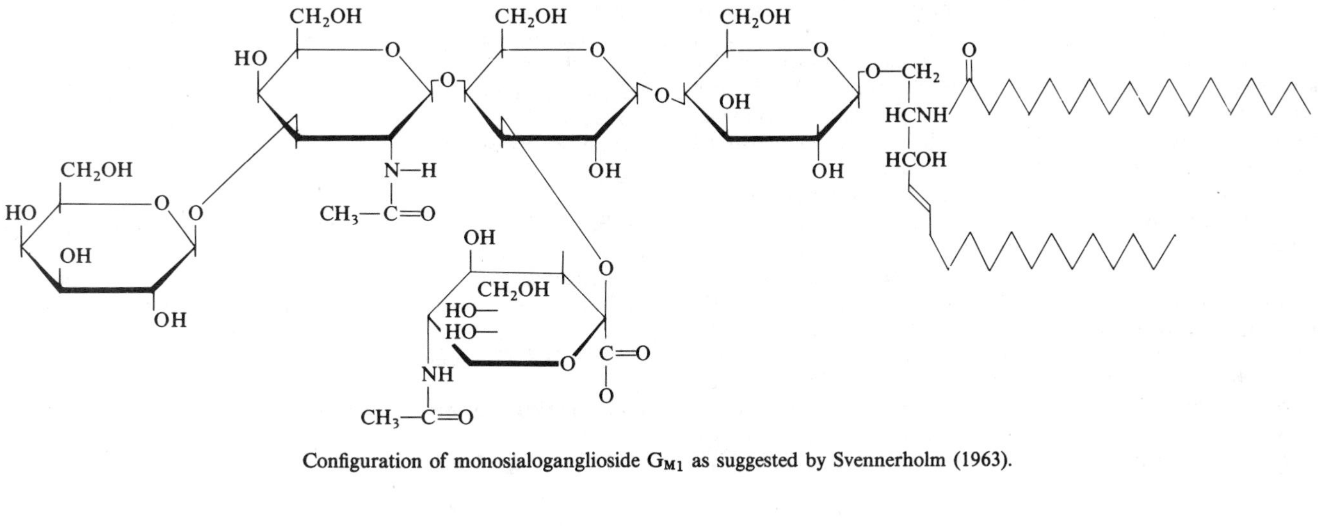

Configuration of monosialoganglioside G_{M1} as suggested by Svennerholm (1963).

GANGLIOSIDOSIS. Lipidoses.

GANGRENE. The death of localized tissue, frequently involving the extremities—fingers, arms, toes, feet, and, in some instances, the ears, nose, and cheeks. Depending upon the cause, however, gangrene may occur in several parts of the body, including lungs, colon, among others. Gangrene may result from physical causes, where in some manner the circulation of blood is stopped or greatly imparied to certain organs. Thus, in cases of injury, severe crushing of tissues may destroy their viability by interfering with the circulation. Inflammation of an area may be so intense as to shut off circulation by strangulation of blood vessels. Circulation may be impaired by thrombosis or clotting. Gangrene may result from arrest of circulation, however produced, as is seen in various diseases causing obstruction of arteries or veins. Examples are severe hardening of the arteries (arteriosclerosis). Chemical and physical agents, including corrosives such as phenol, or prolonged exposure to heat or cold (frostbite) may cause local death of tissue. Nearly all forms of gangrene are accompanied by some kind of infection which spreads the condition. In serious cases, amputation of extremities or parts thereof may be indicated.

In cases of *gas gangrene*, there is infection of tissues around a wound by certain anaerobic bacteria, commonly *Clostridium perfringens*, which grow best deep in tissues away from air. The infection is necrotic and rapidly spreading, usually accompanied by massive edema, gaseous infiltration, and discoloration of tissues. The organisms liberate a toxin which destroys tissue, particularly muscle, and they produce gas by fermenting muscle sugars. Much of the early information on gas gangrene, also referred to as *clostridial myositis*, was obtained during World War I. Many of the war wounds were infected with gas-producing organisms, often from contamination either directly or indirectly with fecal matter contained in the soil. In peacetime, gas gangrene may be precipitated by extensive industrial and transportation injuries, surgery on biliary tract or colon, infarcted bowel (incarcerated hernia), and arterial disease, among other causes. In rare instances, it may result from intramuscular injection, as of epinephrine.

Abscess and gangrene of the lung may be secondary complications of more severe cases of pneumonia. The signs of lung abscess include fever, sweating, and the production of a thin, brown, puslike, foul-smelling sputum.

Progressive bacterial synergistic gangrene may occur around a colostomy or ileostomy opening, in proximity to a chronic ulcer, and sometimes in association with the use of wire stay sutures in surgery. Lesions produced are painful and are surrounded by a rim of gangrenous skin. Causation is usually mixed bacteria, with microaerophilic streptocci, such as *Streptococcus Aureus*, or gram-negative bacilli, such as *Proteus*, being implicated. The condition will spread without treatment. Antibiotics, particularly penicillin G, are effective in arresting the process. However, the condition is most frequently controlled by combining wide surgical excission with parenteral antibiotics.

Anaerobic (*Clostridial*) *cellulitis*, in itself relatively benign gas-forming infection of the skin and subcutaneous tissues without involvement of muscle or toxemia, may predispose the patient to *streptococcal gangrene*, also called *necrotizing fascitis*. In such situations, gangrene reaches the subcutaneous tissue with necrosis of the overlying skin. Again, it is usually the extremities that are involved. The region of involvement may take on a dusky blue coloration. Bullae (blisters) which exude a reddish-black fluid are present. Bursting of the bullae is followed by extensive cutaneous gangrene. Treatment is by removal of necrotic tissue, combined with antibiotic therapy for the streptococcal infection. There is a short incubation of cases of gas gangrene (clostridial myositis), ranging from 8 to 72 hours.

For many years, polyvalent (*C. perfringens*, *C. septicum*, and *C. novyi*) equine antitoxin has been used in the treatment of clostridial myonecrosis. However, because the efficacy of the antitoxin was not proved conclusively, there is no longer production of the antitoxin for clinical use. Surgical debridement of affected tissue is commonly practiced. Hyperbaric oxygen therapy (100% oxygen at 3 atmospheres of pressure over periods of about 2 hours is frequently used in conjunction with debridement procedures and antibiotic therapy. Inasmuch as the infective agents are anaerobic, the presence of concentrated oxygen slows or even stops spread of the infection. The mortality rate in cases of gas gangrene ranges from 15 to 30%, but is as high as 50% when the abdominal wall is involved. Supportive measures include blood transfusions, plasma infusions, and electrolyte replacement to counteract any anemia, hypovolemia, and shock involved. The action of various antimicrobial agents against anaerobic bacteria is described in the Sutter (1976) reference listed.

References

Altemeier, W. A., and W. D. Fullen: "Prevention and Treatment of Gas Gangrene," *J. Amer. Med. Assn.*, **217**, 806 (1971).

Gorbach, S. L., Thadepalli, H., and J. Norsen: "Anaerobic Microorganisms in Intraabdominal Infections," in "Anaerobic Bacteria: Role in Disease" (R. M. DeHaan and V. R. Dowell, Jr., editors), Charles C. Thomas, Springfield, Illinois, 1974.

Holdeman, L. V., Cato, E. P., and W. E.C. Moore: "Current Classification of Clinically Important Anaerobes," in "Anaerobic Bacteria: Role in Disease" (R. M. DeHaan and V. R. Dowell, Jr., editors), Charles C. Thomas, Springfield Illinois, 1974.

MacLennan, J. D.: "The Histotoxic Clostridial Infections of Man," Monograph, *Bacteriol. Rev.*, **26**, 177 (1962).

Sutter, V. L., and S. M. Finegold: "Susceptibility of Anaerobic Bacteria to 23 Antimicrobial Agents," *Antimicrob. Agents Chemother.*, **10**, 736 (1976).

Winstein, L., and M. A. Barza: "Gas Gangrene," *New Engl. J. Med.*, **289**, 1129 (1973).

GANGUE. The essentially valueless mineral aggregates or rock of an ore.

GANISTER ROCK. This term was originally applied to a siliceous underclay occurring in certain coal beds in the north of England. Now it is often applied to highly siliceous, fine-grained rocks used for refractory purposes or to a mixture of ground quartz and fireclay used for furnace linings.

GANNET. Pelicans and Cormorants.

GANOID SCALES. Fishes.

GANTRY. A frame structure that spans over something, as an elevated platform that runs astride a work area, supported by wheels on each side; short for gantry crane or gantry scaffold.

Gantry Crane. A large crane mounted on a platform that usually runs back and forth on parallel tracks astride the work area. Often shortened to gantry.

Gantry Scaffold. A massive scaffolding structure mounted on a bridge or platform supported by a pair of towers or trestles that normally run back and forth on parallel tracks, often used to assemble and service a large rocket as the rocket rests on its launching pad. Often shortened to gantry.

GANYMEDE. Jupiter.

GAP. In geology, a gap is an opening through a ridge connecting the valleys or lowlands on either side. Gaps may be formed by a river which earlier in the cycle of erosion was able to cut its way through the hard rocks now making up the ridge. If the stream is still flowing through this opening, it is spoken of as a water gap; if the stream has disappeared because of its diversion or for other reasons, it is then spoken of as a wind gap.

An electric gap is the distance separating two electrodes between which a spark or arc is caused to pass. A magnetic gap is the distance across an air gap separating two parts of a magnetic circuit. The clearance between pole pieces and rotor of dynamo machinery is such a gap.

GAP LENGTH. In longitudinal magnetic recording, the physical distance between adjacent surfaces of the poles of a magnetic head. The effective gap length is usually greater than the physical length and can be experimentally determined in some cases.

GARDEN EEL. Eels.

GAREFOWL. Shorebirds and Gulls.

GARGANEY. Waterfowl.

GARNET. The name garnet is now applied to a group of very important minerals crystallizing in the isometric system and showing the same habitat of dodecahedrons and trapezohedrons. Garnets belong to the nesosilicate group of silicate minerals and conform to the general formula $A_3B_2(SiO_4)_3$. The elements represented by A and B, respectively, may include calcium, magnesium, manganese, and ferrous iron; aluminum, ferric iron, chromium or titanium. While garnets show no cleavage, a dodecahedral parting is rarely noted; fracture conchoidal to uneven; some varieties very tough and valuable for abrasive purposes and for polishing eyeglass lenses. The hardness of garnet varies between the different varieties from 6.5 to 7.5, and the specific gravity from 3.4 to 4.3. Luster, vitreous to resinous; colors, red, yellow, brown, black, green, or colorless; transparent to opaque. The word garnet is derived from the Latin granatus, a grain.

In general, six varieties of garnet are recognized, based on their chemical composition: grossularite (which is also called hessonite and cinnamon-stone); pyrope; almandine or carbuncle; spessartine; uvarovite; and andradite. Grossularite is a calcium-aluminium garnet which corresponds to the formula $Ca_3Al_2(SiO_4)_3$; the calcium may; however, be in part replaced by ferrous iron and the aluminum by ferric iron. The name grossularite is derived from the botanical name for the gooseberry, *grossularia*, in reference to the green garnet of this composition found in Siberia. Other shades are the well-known cinnamon brown, reds, and yellows. Because of its inferior hardness to zircon, which mineral the yellow crystals resemble, they have been termed hessonite, from the Greek meaning inferior. Curiously, in the gem-bearing gravels of Ceylon, both zircon and hessonite are found and indiscriminately called hyacinth. This term, from the Greek, was apparently a general term used by Pliny for the transparent varieties of corundum; later it was used for yellow zircons.

Grossularite is found in crystalline limestones with vesuvianite, diopside, wollastonite and wernerite. Among the many localities are the Urals, Italy, Switzerland, Mexico, and, in the United States, Maine and New Hampshire. Fine specimens are obtained from the Jeffrey Mine, Asbestos, Quebec, Canada.

Pyrope, sometimes called Cape ruby, is ruby-red in color and chemically a magnesium aluminum silicate with the formula $(Mg, Fe)_3Al_2(SiO)_3$; the magnesium may be replaced in part by calcium and ferrous iron. The color of pyrope varies from deep red to almost black. The transparent pyropes are used as gems, but some have a slight tinge of yellow. The name pyrope is derived from the Greek word meaning *fire-like*. A sub-variety of pyrope from Macon County, North Carolina, is of a violet-red shade and has been called rhodolite, from the Greek meaning *a rose*. In chemical composition it may be considered as essentially an isomorphous mixture of pyrope and almandine, in the proportion of two molecules of pyrope to one molecule of almandine. Pyrope is found at Teplitz and Aussig, Bohemia; in the Kimberley diamond mines in the Republic of South Africa; in Australia and elsewhere. In the United States, important localities are in Arizona, New Mexico, and Utah.

Almandine is the modern gem the carbuncle, although in Pliny's time this term was used for almost any red stone. The term carbuncle is derived from the Latin *carbunculus*, meaning a little spark. The name almandine is a corruption of Alabanda, a locality in Asia Minor where, in ancient times, these red stones were cut. Chemically alamandine is an iron-aluminum garnet corresponding to the formula $Fe_3Al_2(SiO_4)_3$. The deep red transparent stones are often called precious garnet and used for gems. Almandine occurs in metamorphic rocks like mica schists usually associated with typically metamorphic minerals such as staurolite, kyanite, and andalusite. Good gem material comes from India and Brazil. Almandine is also found in Australia, Alaska, Africa, Norway, Sweden, the Malagasy Republic, and Japan. In the United States almandine with 11.48% MgO pyrope content is found in the gneisses of the Adirondack region of New York, sometimes of very large size, in New England, and elsewhere.

Spessartine is manganese aluminum garnet, $Mn_3Al_2(SiO_4)_3$. The name of this mineral is derived from Spessart in Bavaria, a well-known European locality. Spessartine of a beautiful orange-yellow comes from the Malagasy Republic. Violet-red spessartine has occured in rhyolites in Colorado and Maine. Uvarovite is a calcium chromium silicate the formula being $Ca_3Cr_2(SiO_4)_3$. It is a rather rare garnet, bright green in color, usually in small crystals associated with chromite in serpentines, sometimes in crystalline limestones or schists. It is found in the Urals, the Republic of South Africa, Canada, and, in the United States, in California and Pennsylvania. Andradite, calcium-iron garnet, $Ca_3Fe_2(SiO_4)_3$, is of variable composition and may be red, yellow, brown, green, or black, or of intermediate shades. The subvarieties topazolite, yellow or green, demantoid, green, and melanite, a black sort, are recognized. Andradite is found both in deep-seated igneous rocks like syenite as well as in serpentines, schists, and crystalline limestones. Demantoid has been called the "emerald of the Urals" from its occurrence there. Varieties of andradite are found in many localities in Europe: Italy, Switzerland, Norway, and Saxony. In the United States it is found at Franklin, New Jersey; Magnet Cove, Arkansas; and elsewhere.

See also terms listed under **Mineralogy.**

Elmer B. Rowley, Union College, Schenectady, New York.

GARNET (Gadolinium-Iron). Gadolinium.

GARNET (Synthetic). Yag and Yig.

GARNIERITE. This mineral occurs at amorphous masses, presumably as a product of secondary alteration of nickel-bearing peridotites. It is a hydrous silicate of nickel and magnesium, $(Ni, Mg)_3Si_2O_5(OH)_4$. Hardness is 2–3; specific gravity 2.2–2.8, and characterized by its apple green color with dull-to-earthy luster. An important nickel-ore mineral is found with chromite and serpentine in New Caledonia. Additional localities include the Republic of South Africa, the U.S.S.R., the Malagasy Republic, and Oregon and North Carolina in the United States.

GARS (*Osteichthyes*). Of the order *Ginglymodi*, there are approximately eight species, all of which have what might be termed a "crocodilian" appearance. They are heavily armored with ganoin scales, usually in the form of diamonds or rhomboids. These are flat plates with no interlocking as found in conventional fish scales. They move slowly under normal circumstances, but are capable of very fast movements when striking for food. Much as a crocodile, the gar is a slasher, with rapid sidewise movements in its efforts to tear away at its food. Gars are well known for stealing bait from the fisherman's hook. They prefer shallow areas with lots of underwater vegetation and thus it is not surprising to find that one of their natural habitats is in the Florida Everglades. Seminole Indians eat smaller gars.

Gars, like crocodiles, have ball-and-socket joints, unlike most other fishes which have concave vertebrae. The longnose gar (*Lepisosteus osseus*) is found in waters eastward from the Mississippi basin. This species is easily identified by its very long jaws and by length of head and location of eyes which are large. As with other gars, this species prefers salt or brackish water, although it will survive for several years in fresh water. Alligator gars, of which there are a couple of species, definitely prefer fresh water and cannot survive for long periods in salt water. The largest of the gars, the tropical gar (*Lepisosteus tristoechus*) can attain a length of from 10 to 12 feet (3 to 3.6 meters) and is eaten in parts of Mexico. The scales also can be used in ornamental jewelry.

Gars have not been found west of the Rocky Mountains, but are found mostly in the eastern United States, up into southeastern Canada and as far south as Costa Rica. See accompanying view of a long-nosed gar.

Long-nosed gar (*Lepisosteus osseus*). (*A. M. Winchester.*)

GARTER SNAKE. **Snakes.**

GAS. 1. A state of matter, in which the molecules move freely and consequently the entire mass tends to expand indefinitely, occupying the total volume of any vessel into which it is introduced. Gases follow, within considerable degree of fidelity, certain laws relating their conditions of pressure, volume, and temperature. Gases mix freely with each other, and they can be liquefied. 2. The term is sometimes used as distinct from vapor, particularly to indicate a substance having a critical temperature below room temperature.

The fundamental gas laws are described elsewhere in this volume. In particular, see **Equation of State.**

An inert gas is a gas that does not react chemically. The rare gases of the atmosphere were long considered to be completely inert. Also known as noble gases, these included argon, helium, krypton, neon, radon, and xenon. Definite compounds of radon and xenon, for example, have been identified in recent years, but generally their identification as being inert is well justified. Some gases are termed permanent gases, including oxygen, nitrogen, and hydrogen, which require low temperatures and, in practice, high pressures for their liquefaction. The term arises from the fact that in the early years of scientific investigation of these materials, long before the conditions of liquefaction were obtainable, it was believed that these gases could not be liquefied under any circumstances, and hence termed permanent gases.

The laws pertaining to the forces of gas pressure and to the flow of gases are based ultimately upon the kinetic theory, but certain principles can be stated without analyzing their origin to that extent. To a first approximation, the ideal gas law, or the Boyle-Charles law, represents the dynamics of gases at rest. At a given temperature, the pressure of a body of gas varies inversely as its volume, and hence directly as its density (Boyle law); and at a fixed volume, the pressure is a linear function of the temperature, varying at the same rate ($\frac{1}{273}$ per centigrade degree) for all gases (Charles law). But dynamic processes in a gas are complicated by the fact that change in volume is, in general, accompanied by change in temperature, so that simple dynamics is overshadowed by thermodynamics. It was for this reason, for example, that the correct formula for the speed of sound in air proved, for a time, elusive. A gas is highly compressible, and this property affords ready opportunity for the energy of mechanical impulses, which would be merely transmitted by a noncompressible fluid, to be transformed into heat, or for the gas to use its thermal energy to create impulses of its own. The same circumstance complicates the effect of gravity. The atmosphere is not an ocean of uniform density and definite depth; its pressure and density are logarithmic functions of the altitude. The forces associated with moving gases form the subject-matter of aerodynamics.

See also **Atmosphere (Earth).**

GASAHOL. **Biomass and Wastes as Energy Sources.**

GAS AMPLIFICATION. The two common uses of this term are: 1. For a gas-filled counter of ionizing radiations, such as a proportional counter or a Geiger counter, the gas amplification is the ratio of the total number of ion pairs produced in the counter during a single ionizing event to the number of primary ion pairs produced by the radiation that initiates the event. The formula $V = Ane/C$ expresses the size of the pulse V in volts appearing across the counter electrodes in terms of the gas amplification A, the number of electrons formed in the initial ionizing event n, the electron charge in microcoulombs e, and the distributed capacity of the central wire system in microfarads C. 2. In a gas phototube, the ratio of radiant or luminous sensitivities, with and without ionization of the contained gas, is the gas amplification factor.

GAS ANALYZERS (Combustion-Type). The concentration of combustible gases must be determined and controlled in manufacturing operations and other industrial situations for several reasons, including: (1) safety—to avoid explosions by maintaining concentrations well below the lower explosive limit; also to avoid the toxic effects of most combustible gases on operating personnel, (2) efficiency—to maintain optimum concentrations for combustion and other chemical reactions where such gases may be used, and (3) detection of faulty operating equipment and procedures. In combustion-type analyzers, the very quality one is seeking (combustibility) is used as the basis of instrumentation.

The most commonly used method employs a self-heated "hot wire" detector, usually platinum. The wire also serves as a combustion catalyst. Where the combustible gas to be measured also contains air, the mixture simply is fed to a "hot wire" detector whereupon combustion occurs. A temperature sensor, such as a thermocouple, may detect the temperature rise and this, in turn, is a measure of the concentration of the gas. More frequently, the electrical resistance of the "hot wire" itself is measured as the means for detecting temperature rise, much as occurs in a typical electrical resistance thermometer. Where the sample does not contain an excess of oxygen, then air or oxygen must be added to the sample line in carefully controlled quantities,

HEATS OF COMBUSTION OF TYPICAL COMBUSTIBLE GASES[a]

		Heat of Combustion *Hc* at 25°C and Constant Pressure to Form					
		H_2O (gas) and CO_2 (gas)			H_2O (liq) and CO_2 (gas)		
Gas	Formula	kcal/mole	cal/g	Btu/lb	kcal/mole	cal/g	Btu/lb
Hydrogen	H_2	57.7979	28,669.6	51,571.4	68.3174	33,887.6	60,957.7
Carbon monoxide	CO				67.6361	2,414.7	4,343.6
Methane	CH_4	191.759	11,953.6	21,502	212.798	13,265.1	23,861
Ethane	C_2H_6	341.261	11,349.6	20,416	372.820	12,399.2	22,304
Propane	C_3H_8	488.527	11,079.2	19,929	530.605	12,033.5	21,646
n-Butane	C_4H_{10}	635.384	10,932.3	19,665	687.982	11,837.3	21,293
Isobutane	C_4H_{10}	633.744	10,904.1	19,614	686.342	11,809.1	21,242
n-Pentane	C_5H_{12}	782.04	10,839.7	19,499	845.16	11,714.6	21,072
Isopentane	C_5H_{12}	780.12	10,813.1	19,451	843.24	11,688.0	21,025
Neopentane	C_5H_{12}	777.37	10,775.0	19,382	840.49	11,649.8	20,956
n-Hexane	C_6H_{14}	928.93	10,780.0	19,391	1,002.57	11,634.5	20,928
n-Heptane	C_7H_{16}	1,075.85	10,737.2	19,314	1,160.01	11,577.2	20,825
n-Octane	C_8H_{18}	1,222.77	10,705.0	19,256	1,317.45	11,533.9	20,747
n-Nonane	C_9H_{20}	1,369.70	10,680.0	19,211	1,474.90	11,500.2	20,687
n-Decane	$C_{10}H_{22}$	1,516.63	10,659.7	19,175	1,632.34	11,473.0	20,638
Benzene	C_6H_6	757.52	9,698.4	17,446	789.08	10,102.4	18,172
Toluene	C_7H_8	901.50	9,784.7	17,601	943.58	10,241.4	18,422
Ethylene	C_2H_4	316.195	11,271.7	20,276	337.234	12,021.7	21,625
Acetylene	C_2H_2	300.096	11,526.2	20,734	310.615	11,930.2	21,460

[a] Values for additional gases and vapors may be obtained from the National Bureau of Standards, Washington, D.C. and the American Petroleum Institute, New York.

but added well in excess of combustion requirements so that the reaction occurring within the detector will be limited only by the amount of combustible gases or vapors present. Wheatstone bridge circuitry usually is used in these instruments.

The quantity of heat released is related to the concentration of combustibles by reference to a table of heats of combustion. See accompanying table. It is important to note that an analyzer of this type is nonspecific, that is, the instrument is not capable of differentiating between different compositions of combustibles. Inasmuch as the output of the instrument is a function of the rate of combustion and heat of reaction, such analyzers frequently are calibrated in terms of *percent combustibles expressed as percent hydrogen.* However, where it is known in advance that a specific combustible will predominate in the gas stream, the instrument may be calibrated specifically in terms of that component.

Where a bridge circuit is used, a reference detector is required. The reference gas may be air, or the sample gas also may be used if the catalytic characteristics of the "hot wire" are poisoned or destroyed purposely. The latter method has the advantage of compensating for thermal conductivity changes that may occur in the sample as the result of changing sample compositions.

In another type of combustibles analyzer, the sample gas is burned in a small pilot flame, the temperature of which is detected by a thermocouple. The presence of combustibles in the supply of gas to the pilot causes the flame temperature to increase proportionally with concentration. This method is preferred where substances may be present in the gas stream that may poison the catalytic properties of the other form of detector.

Combustible-type gas analyzers are obtainable in combination with oxygen analyzers. In portable form, this combination of instruments is used for testing various types of combustion processes.

GAS ANALYZERS (Thermal-Conductivity Type). Different gases vary considerably in their ability to conduct heat. These variations make it possible to determine the concentrations of a number of gases commonly encountered in laboratory research and industrial processes. Although the relationship between thermal conductivity and gas composition has been investigated widely and so reported in the literature, in general it is not practical or profitable to make detailed calculations of thermal conductivity in designing and applying this type of instrument in gas analysis work. Data available on the thermal conductivity of gases normally is reliable only to within ±5% and, therefore, such calculations usually are confined to obtaining a broad estimate of likely sensitivity of an instrument over a limited range of composition. Further, thermal-conductivity gas analyzers normally are confined to determinations of binary gas mixtures. The method is nonspecific and nonabsolute and thus depends upon empirical calibration. Because the method is so simple, reliable, relatively fast, and convenient to adapt to continuous recording and control, however, this is one of the most widely used gas analysis methods.

The hot-wire gas analysis cell was introduced by Koepsal in 1908 and the principle of the hot wire (in various forms) remains the key approach to thermal-conductivity gas analysis. A typical cell is comprised of an electrically conductive, elongated sensing element that is mounted coaxially inside a cylindrical chamber which contains the gas. By passage of an electric current through the element, the cell is maintained at a temperature considerably higher than the cell walls. The equilibrium temperature is reached when all thermal losses from the wire are equalized by electric power input to the element. If the element is made of a material with a suitable temperature coefficient of resistance, it may serve the dual role of heat source and sensor of the equilibrium temperature. The difference of temperature between the element and the cell walls, reflected by the temperature rise of the element at equilibrium, is a function of electric power input and combined rate of heat loss from the wire by gaseous conduction, convection, radiation, and conduction through the solid supports of the element. Proper cell design and geometry makes it possible to maximize the heat loss due to gaseous conduction. Thus, a rise in the temperature of the element at constant electric power input is inversely related to the thermal conductivity of the gas within the cell.

Normally, a Wheatstone bridge is used to measure the resistance change of the sensing element. The electric current required to energize the bridge also is used to heat the wire. A single hot-wire cell is impractical because of the delicate sensitivity of such an arrangement to changes in ambient temperature and bridge-supply voltage. Commonly, two cells are used in adjacent arms of the bridge. A reference gas is contained in one of these cells. Thus, the bridge responds to the difference in temperature rise of the two cells and consequently depends only upon the difference in thermal conductivities of the sample gas and the reference gas.

While thermal-conductivity gas analyzers are widely used directly for on-line process measurements, they also find wide application in gas chromatographs for determining gas concentration after chromatographic separations. For the quantitative analysis of a binary gas mixture, a useful sensitivity of 1% of full-scale or better is obtainable. The full-scale range varies with the gas mixture and is indicated for several binary mixtures in Table 1. The practical limits of the method are given in Table 2.

The variation of thermal conductivity of binary mixtures does not always follow a simple linear law. Water vapor in air and ammonia in air are examples of nonlinear cases.

TABLE 1. PRACTICAL RANGE OF THERMAL-CONDUCTIVITY METHOD TO BINARY GAS MIXTURES

Mixture	Practical Full-scale Range
Air–carbon dioxide	0–5.3% air in CO_2 0–7.3% CO_2 in air
Air–sulfur dioxide	0–1% air in SO_2 0–3% SO_2 in air
Air–oxygen	0–40% air in O_2 0–38% O_2 in air
Air–helium	0–2.4% air in He 0–0.4% He in air
Nitrogen–carbon dioxide	0–5% N_2 in CO_2 0–7% CO_2 in N_2
Nitrogen–hydrogen	0–2.3% N_2 in H_2 0–0.3% H_2 in N_2
Nitrogen–oxygen	0–55% N_2 in O_2 0–52% O_2 in N_2
Nitrogen–argon	0–5% N_2 in Ar 0–7% Ar in N_2
Hydrogen–helium	0–10% H_2 in He 0–12% He in H_2
Carbon dioxide–oxygen	0–6.4% CO_2 in O_2 0–4.4% O_2 in CO_2

TABLE 2. REPRESENTATIVE APPLICATIONS OF THERMAL-CONDUCTIVITY METHOD

Mixture	Appropriate Comparison Gas
H_2 in CO_2	H_2, CO_2, or $H_2 + CO_2$
H_2 in O_2	O_2, air, or H_2
H_2 in N_2	H_2, N_2, or air
H_2 in Cl_2	H_2 or Cl_2
H_2 in air	H_2 or air
H_2 in CH_4	H_2, CH_4, or $H_2 + CH_4$
H_2 in water gas ($H_2 + CO$)	H_2, or $H_2 + N_2$
Ne in air	Air
He in air, N_2, or O_2	He, air, H_2, or O_2
Cl_2 in air	Air
HCl in air	Air
Acetone in air	Air
O_2 in enriched air	Air
NH_3 in air	Air
SO_2 in air or N_2	Air or N_2
Water vapor in air, N_2, or O_2	Air, N_2, or O_2
Ar in N_2, air, or O_2	N_2, air, or O_2
CO_2 in air, N_2, or flue gas	Air
Benzol in air[a]	Air or N_2

[a] Requires pretreatment by combustion, converting benzol to CO_2 and H_2O.

GAS AND EXPANSION TURBINES. Fundamentally, the gas turbine operates on the concept of the Brayton or Joule cycle (constant-pressure cycle) which was originally used to describe the operation of an air engine, a compressor and a combustion chamber. In the air engine, air entered the compressor wherein the pressure was increased. Fuel burning in the combustion chamber raised the temperature of the compressed air under constant-pressure conditions. The resulting high-temperature gases were then introduced to the engine where they expanded and performed work. The excess work of the engine over that required to compress the air was available for operating other devices, such as a generator. The cycle is illustrated in Fig. 1, with the following equations applying:

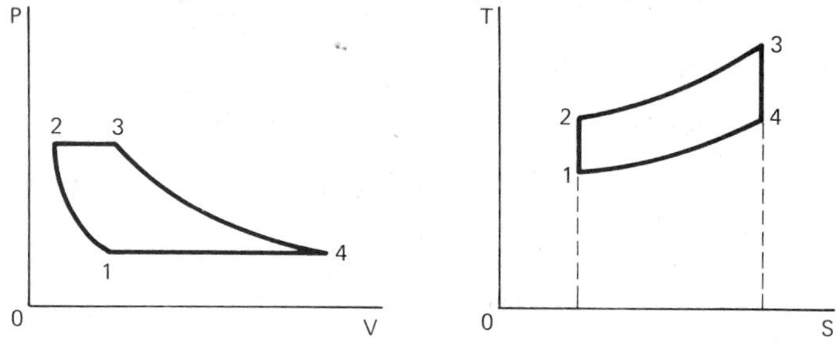

Fig. 1. Brayton or Joule cycle.

$$V_3/V_2 = V_4/V_1 = T_3/T_2 = T_4/T_1$$

where V = total volume; $T = t + 459.69$ = absolute temperature = deg R

$$\frac{T_2}{T_1} = \frac{T_3}{T_4} = \left(\frac{V_1}{V_2}\right)^{k-1} = \left(\frac{V_4}{V_3}\right)^{k-1} = \left(\frac{p_2}{p_1}\right)^{k-1/k}$$

where $k = c_p/c_v$; c_p = specific heat at constant pressure; c_v = specific heat at constant volume; p = absolute pressure, pounds per square foot (1 pound/square foot = 47.88 Pascals = 4.88 kilograms/square meter);

$$(W) = Jmc_p(T_3 - T_2 - T_4 + T_1)$$

where W = external work performed on surroundings during change of state, foot-pounds; J = mechanical equivalent of heat = 778.26 foot-pounds per Btu = 4.1861 joules per cal; m = mass of substance under consideration, lb_m;

$$\text{Efficiency} = (W)/JQ_{23} = 1 - (T_1/T_2)$$

where Q = quantity of heat absorbed by the system from the surroundings, Btu.

In the gas turbine, the air compressor and engine of the foregoing scheme are replaced by an axial flow compressor and gas turbine. Although the turbine is only part of the whole assembly, in modern terminology, the complete assembly is commonly referred to simply as a gas turbine. Air is compressed in the compressor after which it enters a combustion chamber where the temperature is increased while the pressure remains constant. The resulting high-temperature air then enters the turbine, thereby performing work.

Gas turbines usually are rated according to power output (sea level and 80°F; 26.7°C). Some European designs are rated at 60°F (15.6°C). The power output and efficiency are larger for those fuels which produce larger volumes of products of combustion, inasmuch as the compressor does not do any work on additional volume. Gas turbines are classified by the physical arrangements of the component parts, and categories include: (1) single-shaft; (2) two-shaft; (3) regenerative (heat exchanger is used to recover exhaust losses and heat air to the combustor(s); (4) intercooled (heat removed between compressors); and (5) reheat (heat added between turbines). Various configurations of gas-turbine systems are shown in Figs. 2 through 8.

Efficiency. The overall efficiency of a gas turbine is a function of the compressor and turbine efficiencies, ambient air temperature, nozzle inlet temperature, and the type of cycle used. The compressor and turbine are designed for high efficiency. The first-stage gas temperature establishes material and stress conditions for the first set of rotating blades. To the gas temperature at these blades is added the

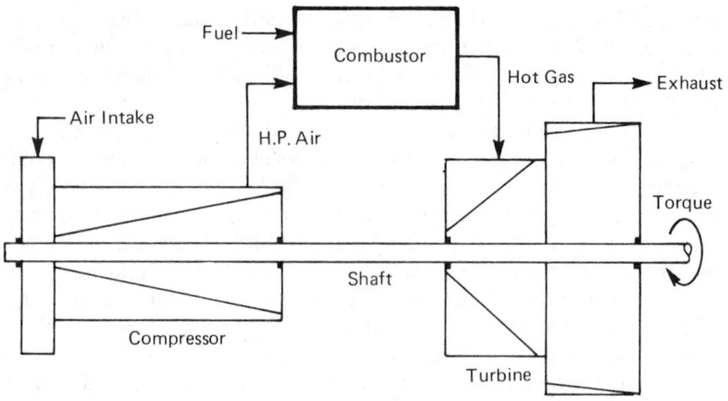

Fig. 2. Gas-turbine configuration exhibiting basic Brayton or Joule cycle.

temperature drop across the first-stage nozzles to determine the inlet temperature of the turbine. This may vary from 704 to 816°C for industrial turbines and usually will be higher for aviation gas turbines. The higher values are usually used in impulse turbines.

In a simple-cycle turbine, there is (for each turbine inlet temperature) an optimum pressure ratio producing the highest possible efficiency. The efficiency and optimum pressure ratio increases with increasing turbine inlet temperatures. These pressure ratios vary from 4 (at 704°C) up to 6 (at 816°C).

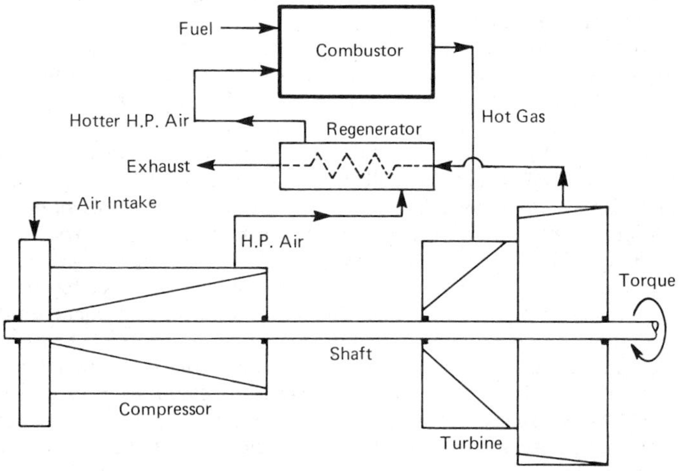

Fig. 3. Gas-turbine configuration with regeneration.

Regenerative cycles favor lower pressure ratios which result in low compressor discharge temperatures, thus allowing greater recovery of heat from the turbine exhaust gases. High-ratio regenerative plants use intercoolers in the compressor circuit to lower the compressor discharge air temperature.

Although any type of efficient compressor can be used, such as positive displacement (Lysholm), centrifugal, and axial flow, most industrial gas turbines use axial-flow compressors. The turbine may

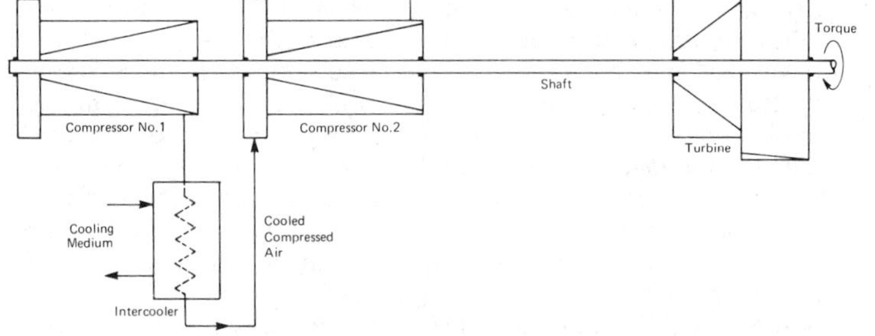

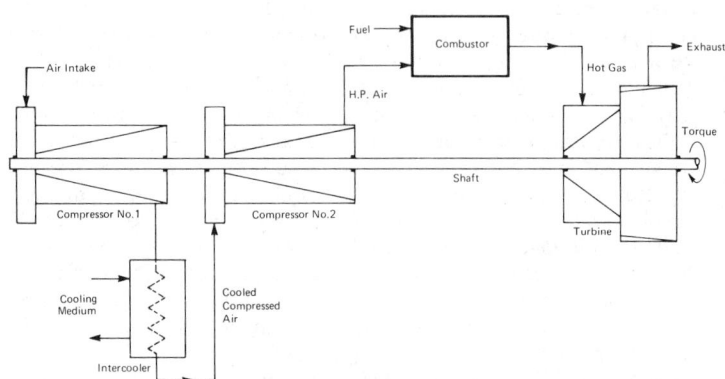

Fig. 4. Gas-turbine configuration with intercooling.

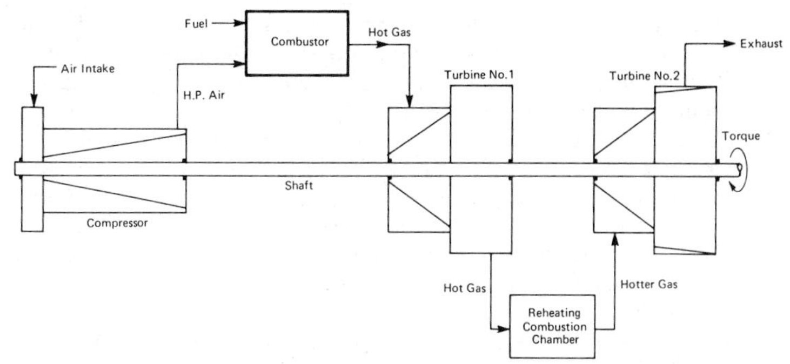

Fig. 5. Gas-turbine configuration with reheating.

have impulse or reaction blading. To minimize losses, air from the compressor discharge flows through the combustor directly into the turbine nozzle. Throttle valves are not used because the resulting pressure drop decreases overall efficiency.

A gas turbine has a large amount of excess air. The combustor is designed with an inner portion burning only part of the air to achieve

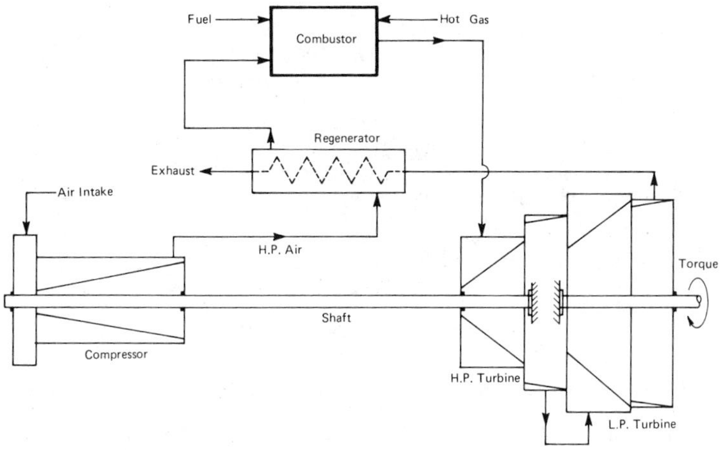

Fig. 6. Regenerative-cycle gas turbine. Two-shaft arrangement with separate power turbines in series.

high combustion temperatures and efficiency. Products of combustion are effectively mixed with the remainder of the air to minimize temperature stratification. Each turbine may have one large combustor or several smaller combustors operating in parallel.

Open- and Closed-Cycle Types. Most gas-turbine installations are of the open-cycle type, using atmospheric air as the working medium

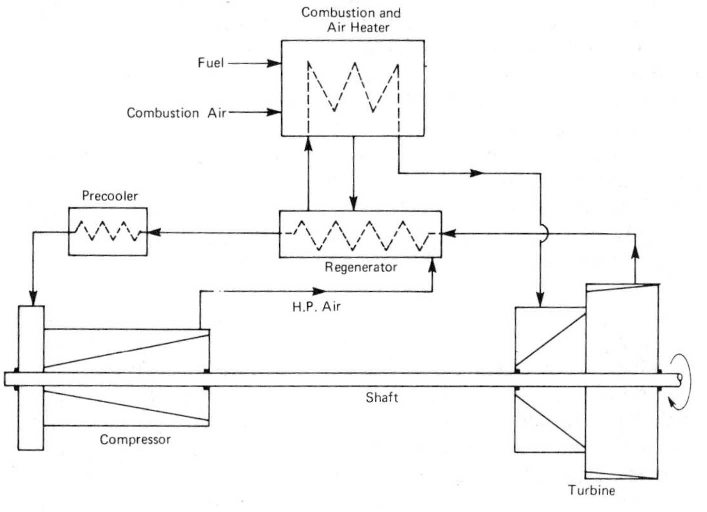

Fig. 7. Closed-cycle gas turbine.

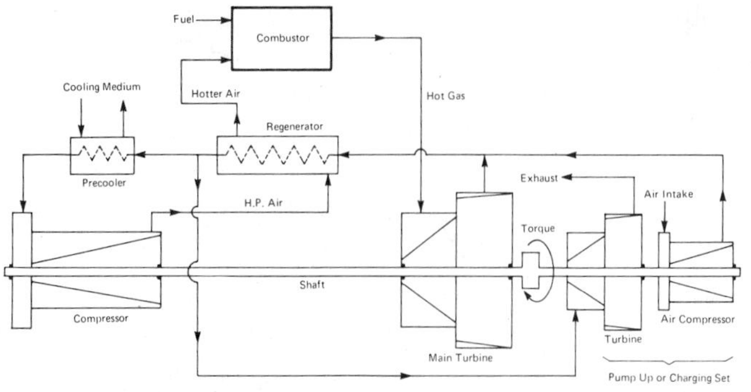

Fig. 8. Semiclosed internally-fired gas-turbine cycle.

and burning relatively clean fuels. Where dirty fuels are used, it is possible to locate the burner in the gas-turbine discharge, using a heat exchanger to heat the air discharged by the compressor. In closed-cycle installations (Figs. 7 and 8), it may be desirable to use other gases, inasmuch as efficiency increases as the specific heat ratio (c_p/c_v) decreases. Optimum plant efficiency occurs at increasingly higher pressure ratios with decreasing values of (c_p/c_v). However, for convenience, most closed systems use air.

Closed systems can provide a high plant efficiency over a power range from 25 to 100% by varying the turbine exhaust and compressor inlet pressure from atmospheric to about 60 psig. These installations require costly heaters, located between compressor discharge and turbine inlet, and large coolers, located between the turbine exhaust and the compressor suction. Usually, combustion of a fuel provides the heat source, and cooling water the cooling medium.

Overloads. Even if only temporary in nature, a large overload can cause a single-shaft gas turbine to shut down, inasmuch as its fuel input is limited by the inlet overtemperature protective system. If the torque requirements of the driven machine do not decrease sufficiently with speed reduction, then the gas turbine will continue to slow down. This results in higher exhaust temperatures. The exhaust temperature control system will either shut off the fuel valve, or further reduce fuel input, causing the turbine to decrease its speed and finally shut down. Carefully matching the load characteristics of the driven equipment with those of the driver can prevent such occurrences.

Single-Shaft Gas Turbines. The wide acceptance of the single-shaft turbine arises from its low cost and compactness in terms of power output per cubic foot of machinery space. Disadvantages include a relatively low operating speed range and sensitivity to atmospheric temperature. The low operating speed range arises from: (1) the quantity of air flow induced by the compressor is proportional to its speed; and (2) the back pressure produced by the turbine nozzles is proportional to air flow. At low speeds, the turbine power is decreased by low air flows and secondarily by the effect of low pressures on allowable inlet air temperatures. At low flows, the decreased pressure at the turbine inlet may require a reduction of turbine inlet temperature to maintain the exhaust temperature within design limitations. This results in a further reduction in power. In most applications, it is necessary to unload the turbine during startup.

Two-Shaft Gas Turbines. A wider operating speed range is provided by the more costly two-shaft machine which consists of a high-pressure turbine driving the air compressor and a low-pressure turbine on a separate shaft to provide output power. See Fig. 9. A variable-area nozzle can be used in the low-pressure turbine to increase the operating speed range. Change in the fuel input to the high-pressure turbine causes the speed and quality of air flow to change. The low-pressure turbine power output is changed by varying the quantity of air flow and the nozzle area of the power turbine.

Air/Temperature Relationships. The air flow to a gas turbine is inversely proportional to the absolute air temperature at the compressor inlet. Inasmuch as the compressor discharge pressure is set by the turbine nozzles (proportional to flow), this results in decreased turbine power output during hot weather, and increased power during cold weather. In hot, dry areas, hot incoming air can be cooled by evaporation using water injection. In locations where the summer season is short, it may be possible to obtain rated power by increasing

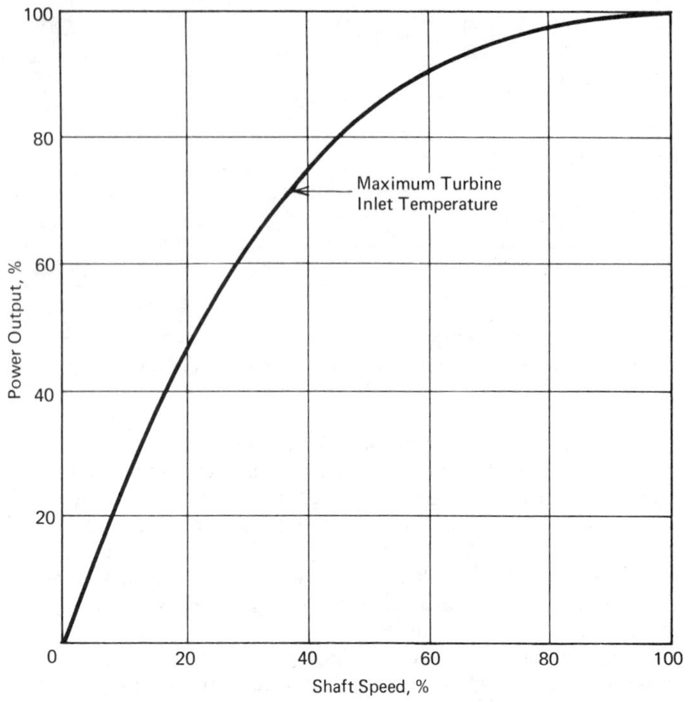

Fig. 9. Operating range for various speeds and loads for two-shaft gas turbine.

the turbine temperature for a short period without appreciably shortening the life of the equipment. In extremely cold temperatures, high air pressures will exist at the turbine inlet and should be considered in the design of the gas turbine.

Since the power required by the compressor is approximately twice as great as the shaft output, a 1% change in compressor efficiency will result in a 2% change in shaft power. A 1% change in turbine efficiency will produce a 3% change in shaft power. Therefore, it is important that all losses be minimized and that sufficiently large inlet and exhaust piping or ducts be used.

Startup of Gas Turbines. A gas turbine is started by bringing it up to starting speed by application of external power (electric, air, gas) and maintaining this speed for several minutes in order to purge the casing. Some machines require that the casing or rotor be heated slowly by burning a nominal amount of fuel in the combustors for several minutes. The turbine inlet temperature is then increased rapidly to a value above the design temperature, thus producing sufficient power in the turbine to bring the set up to full speed. Some installations will require a blowoff valve to prevent surging during startup. The starting power requirements of an unloaded gas turbine will range between 5 and 10% of the full-load speed. Two-shaft turbines will require slightly more starting power than single-shaft machines. By opening the nozzles of the low-pressure turbine, the load is not driven during startup.

Fuels. A wide variety of fuels can be used in gas turbines. The major fuel requirements are that: (1) the fuel does not form ashes which will deposit on the blades and interfere with operation; (2) the fuel does not contain dust which will erode the blades; and (3) the fuel does not contain uninhibited vanadium. Commonly used fuels include natural and refinery gas, blast-furnace gas, fuel oils (including heavy residuals), and the growing application of gas turbines in cycles involving gases derived from coal and other previously nontraditional sources.

The simple cycle-gas turbine is relatively inefficient with almost all of its losses in the hot exhaust gases. When exhaust gases can be used in a boiler or for process heating, the combination of turbine and heat-recovery apparatus results in a high-efficiency plant. Integration of the gas turbine with process requirements also can result in high efficiency.

Gas Turbine Standby Generator Sets

Over the last few years, with growing awareness of the possibility of more frequent utility power outages, thousands of gas turbine standby generator sets have been installed in industrial plants, hospitals, apartment and office buildings, hotels, motels, schools, and other public facilities. Further, to insure the continued reliable operation of communications systems and computer centers in the event of brownouts and blackouts, gas turbine standby generator sets have been widely accepted as emergency substitutes. Generator sets also find extensive application in offshore exploration and drilling activity. Mobile generator sets are used on construction projects, in military and other emergency operations. Such sets have been used to furnish portable power for pumps in connection with fighting forest fires, for example.

Popular standby electric generator sets are designed for 225, 900, and 2,800 kilowatts. Although gas-turbine sets have a slightly higher initial cost, they offer favorable long-term economy when compared with reciprocating engine generator sets. The cross section of a gas turbine (*Solar* engine) used in standby generator sets is shown in Fig. 10. This is a single-shaft industrial gas turbine engine with antifric-

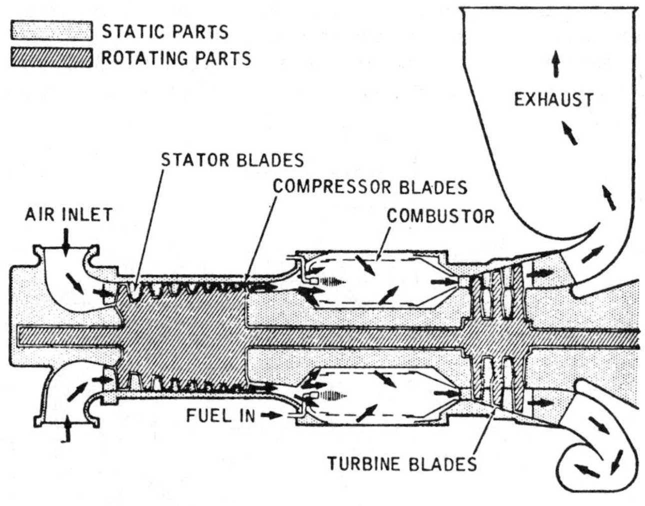

Fig. 10. Cross section of Solar gas turbine. (*Solar International Turbines.*)

tion bearings and integral speed-reducing gear-box to 1,800 revolutions per minute. The nominal engine rating is 1,200 horsepower, based on standard conditions of 15°C at sea level, with zero inlet and exhaust losses. A flexible drive coupling connects the output shaft of the gearbox to the generator shaft. A cutaway view of the turbine is shown in Fig. 11, and of the gear reduction drive in Fig. 12.

An 8-stage axial compressor, fully annular combustion chamber, and 3-stage axial flow turbine are arranged with a straight-through flow path. Advantages of this configuration include minimization of flow losses and a cylindrical structure of considerable rigidity and stability. The accessory drive gearbox with starter mounting pad is located behind the air inlet. The air inlet housing, compressor case, compressor diffuser, combustion chamber case, turbine case, exhaust diffuser, and turbine output housing are interconnected to form the main structural shell of the engine. The reduction drive assembly is comprised of a 2-stage planetary gear train which reduces the turbine output shaft speed from 22,300 revolutions per minute to 6,000 revolutions per minute in the 1st stage and to 1,800 (1,500 for 50 Hz sets) revolutions per minute in the 2nd stage. The primary stage consists of the turbine-connected sun gear, stationary planets, and a rotating ring gear. The 2nd stage is a true planetary system, comprised of a sun gear, rotating planet gears, and a stationary ring gear. The final output is transmitted through the rotating gear carrier to the output shaft.

Air is drawn into the air inlet and compressed by the 8-stage axial flow compressor. The compressed air is directed into the combustion chamber in a steady flow. Within the annular combustion chamber, fuel is injected into the pressurized air. During the engine starting cycle, the fuel/air mixture is ignited by a high-voltage spark from the ignitor plug and continuous burning is maintained as long as there is adequate flow of pressurized air and fuel. The hot, pressurized gas from the combustion chamber expands through the turbine, dropping in pressure and temperature as it drives the turbine. In this

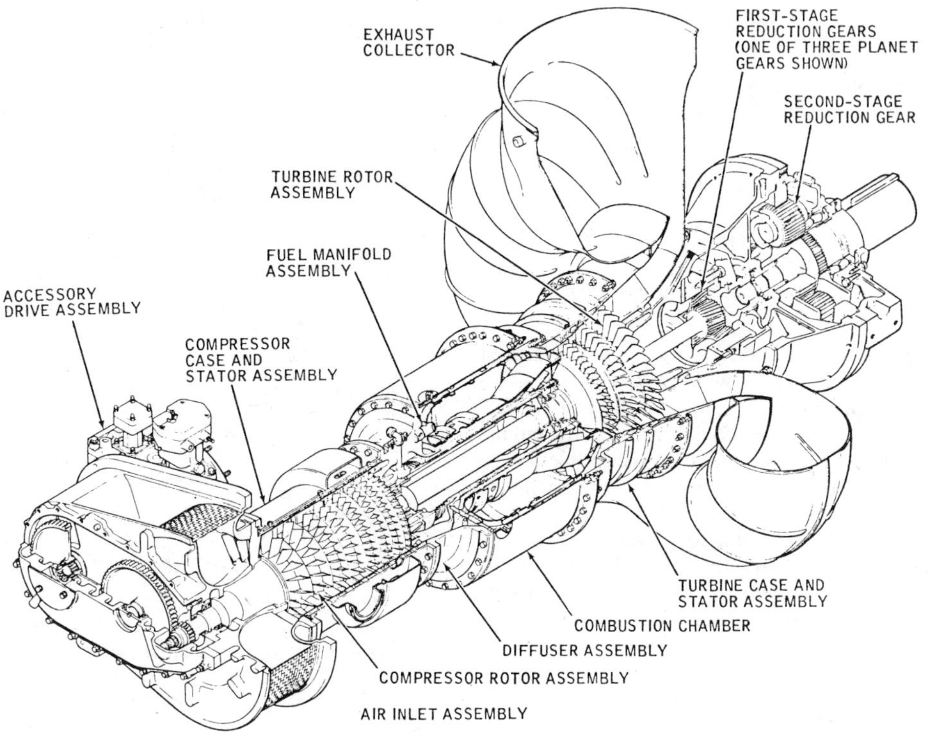

Fig. 11. Solar single-shaft gas turbine. (*Solar International Turbines.*)

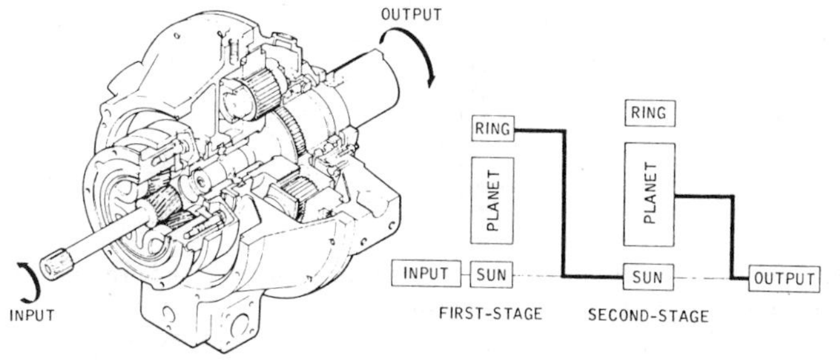

Fig. 12. Reduction drive used with turbine illustrated in Fig. 11.

way, the energy of the fuel is transformed into the rotating power of the turbine output shaft.

The turbine drives the compressor and the external load (in this case, a generator) through a common shaft. During the starting cycle, the starter supplies the driving power for the gas turbine up to approximately 15% rated speed, at which time the gas turbine "lights off" and accelerates to 100% of rated speed. The starter remains engaged after light-off and assists in the acceleration from 15 to 60% speed. Approximately 20% of the available air mass is required for combustion; the remainder is used to cool the combustion chamber and reduce the temperature of the combustion gases to less than 843°C at the inlet to the first turbine stage. The exhaust temperature varies with the generator load and has a maximum value of ~502°C. Exhaust products have a 17 to 19% oxygen content.

Natural Gas Compressor Sets

There has been a rapidly growing trend since about mid-1969 to apply gas-turbine-powered centrifugal compressors in the oil and gas industry. The sets usually consist of a two-shaft, axial-flow gas turbine driving a centrifugal gas compressor. Such systems are used for a variety of purposes in the natural gas industry, both onshore and offshore. They are available with from 1 to 8 stages to meet various flow and pressure conditions. The systems can compress gas at ratios of over 5 to 1 for a wide range of flows. Tandem units, with a turbine driving two or three compressors, can achieve ratios up to 35 to 1. Systems are operating at an altitude of more than 7,000 feet (2134 meters) in New Mexico where they are subject to temperatures ranging from below −17.8°C to over 37.8°C. Systems also are operating in desert areas (Libya, Iran, West Pakistan, etc.), in jungle climates (South America, etc.), and on drilling platforms in various seas of the world. There is nearly a 10 to 1 weight reduction in the gas-turbine centrifugal compressor approach as compared with reciprocating compressor units. In one example, for comparable performance, a gas turbine system will weigh about 33,500 pounds (16,344 kilograms) as compared with a reciprocating compressor unit that will weigh more than 300,000 pounds (136,080 kilograms). The gas turbine centrifugal compressor sets are used in connection with gas transmission, gathering, production, boosting, and refrigeration.

The cross section of a two-shaft gas turbine engine (*Solar*) engine is shown in Fig. 13. The gas turbine engine has an output-load shaft (power turbine) which is mechanically independent of the gas producer section and permits a wide range of speeds at full power. The power transfer diagram is that shown in Fig. 14. This design contributes to better part-load fuel economy. In this engine configuration, called a two-shaft (or split-shaft) design, the first two turbine stages drive the engine compressor only; gases leaving the gas producer turbine flow through the free power turbine, forming a fluid coupling. The remaining energy is absorbed by the power turbine and is transferred to the output shaft. The speed of the gas producer is closely associated with the power level of the engine. Consequently, a speed governor on this shaft is used to control the power level. The power turbine runs at a speed which is dependent only upon the load and the engine is protected against overspeed in the event the load is removed.

The gas turbine engine requires, for stoichiometric combustion, ap-

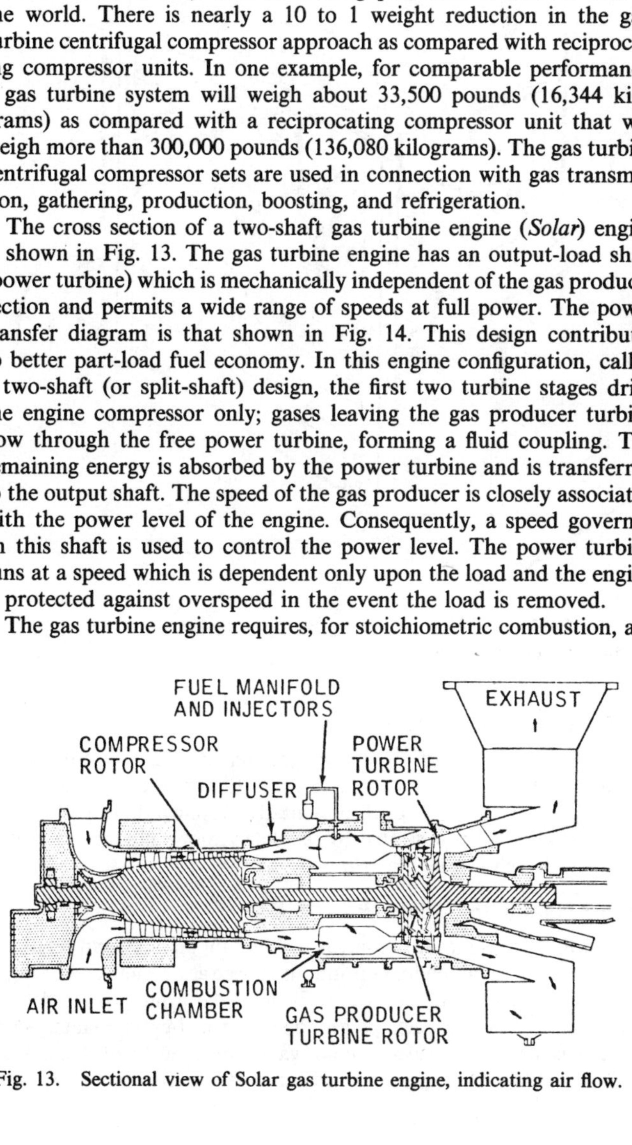

Fig. 13. Sectional view of Solar gas turbine engine, indicating air flow.

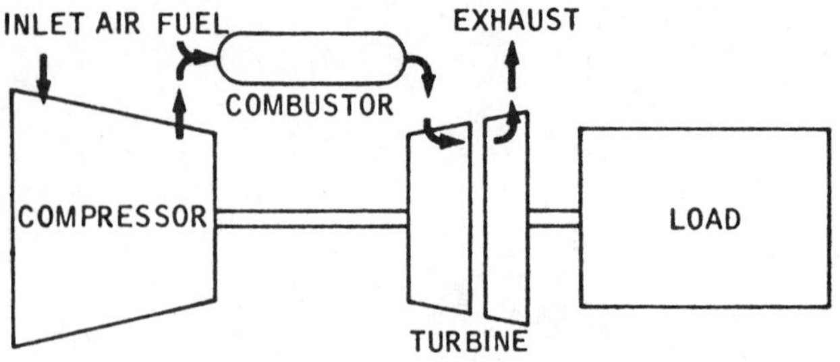

Fig. 14. Power transfer in two-shaft turbine engine.

proximately one-fourth the total air handled. The excess air is used to cool the combustion chamber and mixes with the products of combustion to reduce the gas temperature at the inlet to the first turbine stage. The cooling employed at the combustion chamber of the engine keeps metal temperatures in the turbine section at acceptable levels. Cooling is accomplished by air, thus eliminating the requirement for cooling water.

Recuperator. The addition of a recuperator to a system of this type will lower fuel consumption by about 20%, or from 9,800 Btu (2470 Calories) per horsepower-hour to 7,800 Btu (1966 Calories) per horsepower-hour. A system with recuperator installed is shown in Fig. 15.

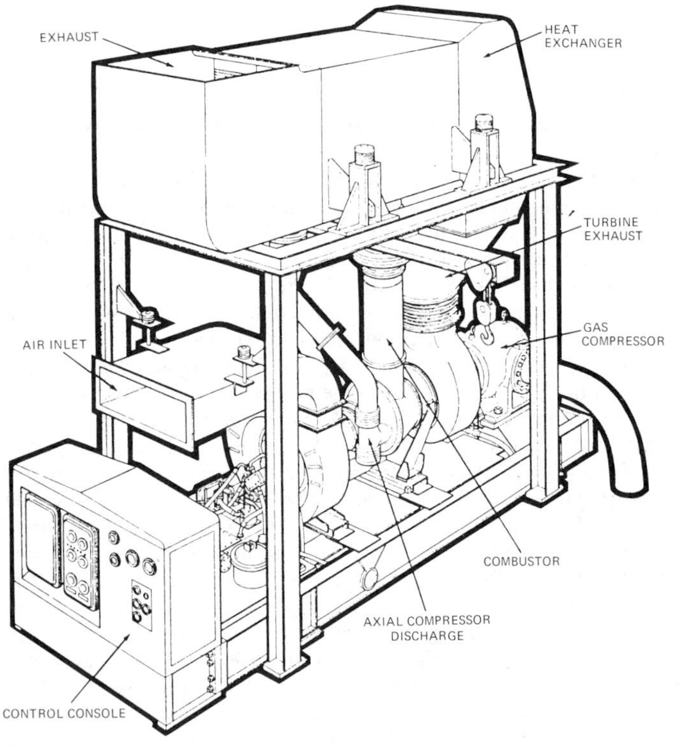

Fig. 15. Solar gas turbine natural gas compressor set equipped with retrofitted combustor and recuperator.

Gas Turbines in Offshore Activity

The compactness and relatively light weight of gas turbine power packages has been particularly attractive to offshore installations in the petroleum and gas field. Representative application of gas turbine power installations is indicated by the diagram of a platform on Cook Inlet, Alaska, shown in Fig. 16.

Gas Turbines in the Utility Power Field

In addition to the wide use of gas turbines for standby and relatively low-kilowatt production for continuous service as previously described, gas turbines have made marked inroads in the utility power field as greater emphasis has been placed on combined-cycle plants. Some of the reasons for the acceptance of gas turbines in this field include: (1) high availability of gas turbines can be achieved when these units

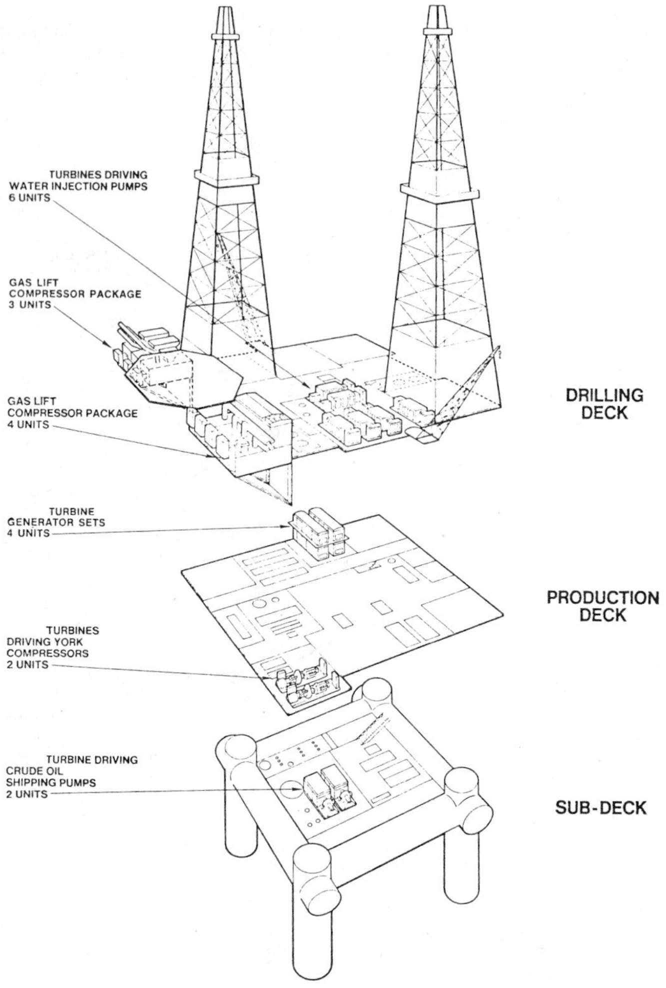

Fig. 16. Use of gas turbine-powered systems on drilling platform on Cook Inlet, Alaska. (*Solar Turbines International.*)

are operated at a relatively constant load for extended periods and when they are properly maintained. Outstanding performance of this equipment in pipeline service has been proved. (2) In some cases, it is more economical to order a combined-cycle plant than to invest capital in pollution control equipment for an older station that might not withstand the rigors of cycling operation; (3) environmental impact of combined-cycle systems should be less than that of peaking/cycling units because of reduced cooling water requirements and the utilization of higher quality fuels; and (4) construction costs of combined-cycle plants are relatively low because most components are assembled at the factory and shipped in modules to the plant site. A major problem remaining for preengineering combined-cycle systems is the reduction of nitrogen oxides from gas turbines.

Combined-Cycle Plants

Over the past decade, electrical energy shortages have resulted in the use of many peaking units at operating levels in excess of their optimum economic levels.* Those peaking units are employed to supply the intermediate load band, along with former base-load units that have had to be shifted to intermediate service. The latter are usually older units that lack modern control systems and, therefore, lack the control flexibility that intermediate generation requires. In short, the intermediate load band often is not being supplied as efficiently and as flexibly as it could be.

In addition, large interconnecting ties require increasingly close control of area power generation to meet the security requirements of the grid, i.e., to keep the system from breaking up. Such control can best be provided by reserve power supplied by peaking and interme-

* Basic data for several following paragraphs were furnished by Westinghouse.

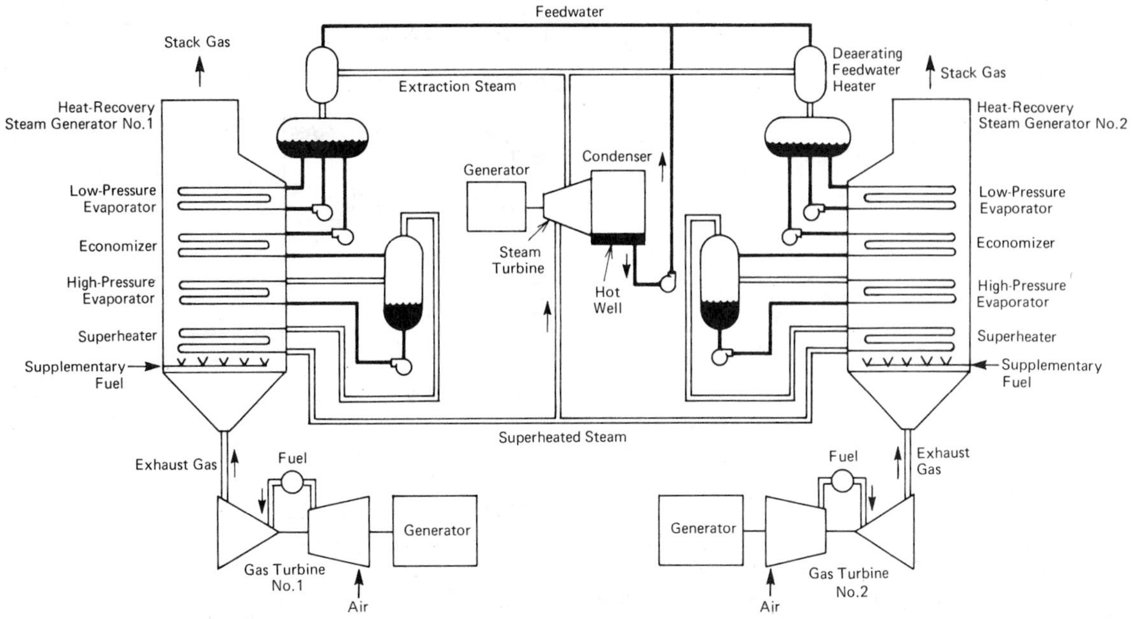

Fig. 17. Combined-cycle power plant. Exhaust energy from two gas turbine-generator units is used to develop steam for a conventional steam-turbine generator unit. Plant burns either gas or oil. Because the steam cycle provides only about one-half the plant output, the need for cooling water is approximately one-half that of a conventional all-steam power plant of same size. Each gas turbine exhausts into its own heat-recovery steam generator, where additional fuel is burned. The hot gases generate steam, which is supplied to a common header at 1200 psi (81.6 atmospheres) and 510°C. From there it goes to the single-cycliner axial-exhaust steam turbine. All three generators are hydrogen cooled. (*Westinghouse.*)

diate plants that are able to start up rapidly and alter their power outputs rapidly in response to the grid's dynamic needs.

Well-engineered combined-cycle plants can meet those needs for increased efficiency and flexibility because they blend the best features of peaking and base-load generation by combining one steam and two gas turbines. Gas turbines are fast in starting up and in responding to changes in power demands, but they are relatively inefficient by themselves; steam turbines are slower in startup and response, but are more efficient. A good engineering design will combine them in such a way as to use waste heat from the gas turbines to generate steam for the steam turbine. See system diagram of Fig. 17.

The result is a plant that has an excellent heat ratc when the steam turbine is operated with one or both gas turbines, and yet has a fast startup and control flexibility. In the plant shown, heat rate is 8,000 Btu (2,016 Calories)/kilowatt-hour, with the steam turbine and both gas turbines operating. For comparison, the heat rates of typical modern steam base-load plants and gas-turbine peaking plants are, respectively, about 9,000 and 12,000 Btu (2,268–3,024 Calories)/kilowatt-hour. The gas turbines can be operated independently of the steam turbine, and of each other, for flexible plant operation. Startup of the entire plant from hot standby requires about one-fourth the time required by typical steam plants.

Thus, although a plant of this type may be designed primarily for intermediate generation, it can provide service ranging from peaking to base-load generation as needed and as influenced by the relative cost of fuels.

Plant operation includes both transient and steady-state modes, which are defined in the control room by the state of bistable or multistable switches, breakers, or valves, and by the state of closed-loop controls, such as on/off switches. The transients are long-term modes, such as acceleration and the building up of steam pressure. Steady-state operating modes are:

(1) *Hot Standby, Ready to Start.* Steam-cycle components are kept hot by electric heaters. The plant can reach full load from this mode in approximately 1 hour.

(2) *Running Mode.* The turbine-generator units selected for operation are running at 3,600 revolutions per minute and all associated auxiliaries are in normal operation. Excitation has been applied to the generators and all equipment is ready for synchronization with the line. The running turbines are operating primarily on speed control, and the afterburners (which are auxiliary firing units in the heat-recovery steam generators) are on. The steam turbine bypass valve acts as a back-pressure control, opening and closing as needed to maintain constant pressure in the main steam header.

(3) *Power Generation, Gas Turbines Only.* When the steam turbine is not needed for power generation or is shut down for routine maintenance, the gas turbines can be used alone to generate power. Startup to base load takes about 35 minutes. The steam generated can be bypassed to the condenser, or the heat-recovery steam generators can be drained and vented.

(4) *Power Generation, Combined Cycle.* Ordinarily, the two gas turbines are synchronized first and then the steam turbine. The steam temperature required to roll the steam turbine (determined by the rotor temperature) is attained by loading the gas turbines and/or firing the afterburners. It is also possible to start the plant with one gas turbine and the steam turbine and then bring the second gas turbine on line later.

Regardless of the plant operating configuration (the combination of turbine-generator units selected by the operator), the control system offers a choice of four control operating levels. From highest to lowest level, they are: (1) plant coordinated control; (2) operator automatic control; (3) operator analog control; and (4) manual control. It is not necessary for all generating units to be operated at the same level of control.

Emission Aspects of Gas Turbines

To get the best performance from heat engines, engineers are concerned with obtaining high combustion efficiency.* It is unfortunate that the easiest way to achieve high efficiency is to operate at maximum combustion temperatures. While this produces low levels of unburned hydrocarbons and carbon monoxide emissions, it also results in the highest levels of nitrogen oxide (NO_x) emissions. As shown by Fig. 18, the generation of NO_x is directly related to high combustion temperatures. There are several ways by which combustion temperatures can be reduced. However, the most effective way is to burn fuel at lean mixtures. Combustion temperatures are related to air/fuel ratios with peak temperatures occurring at the ideal air/fuel ratio of about 15 pounds of air to 1 pound of fuel. From Fig. 19, it will be noted that only by combustion at high air/fuel ratios (lean mixtures) will emissions of all components be at a reduced level. It is evident that the key to controlling combustion emissions is the development of

* Basic data for several following paragraphs were furnished by Solar Turbines International.

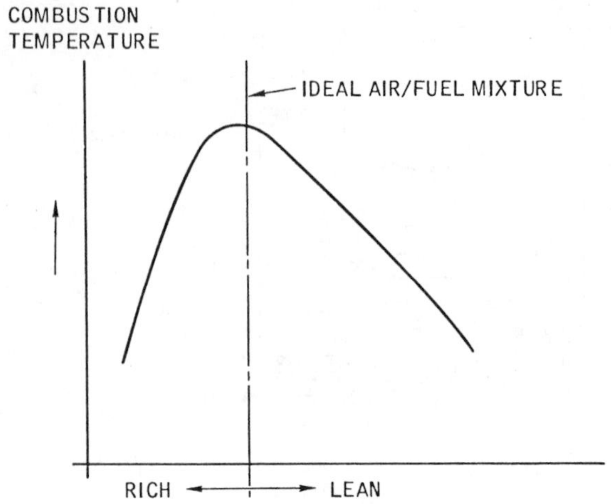

Fig. 18. Temperature mixture relationships.

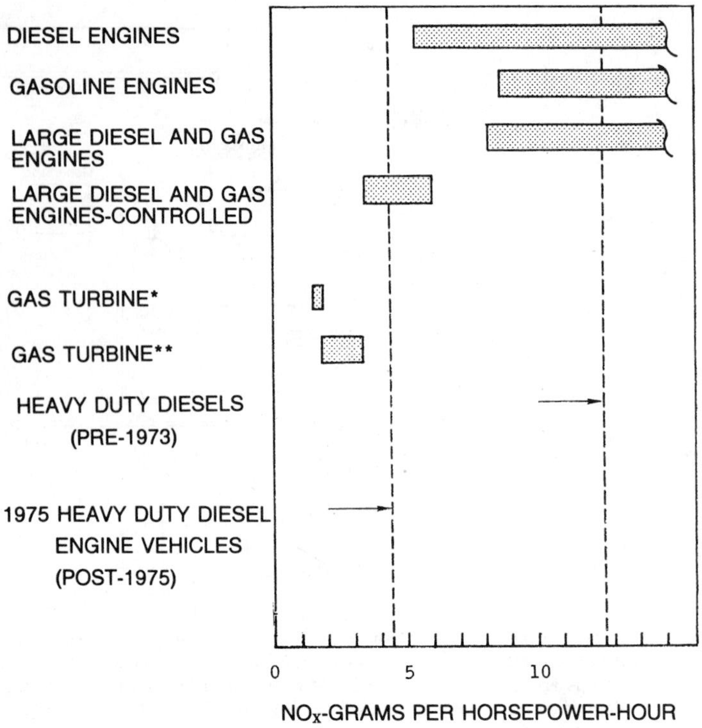

Fig. 20. Interassociation of emissions and engine type. *Essentially uncontrolled emissions. **Controlled emissions.

successful methods of efficiently burning fuel at very lean mixtures, i.e., low combustion temperatures. This requires modification of existing combustion systems. Relatively speaking, such physical modifications are less difficult with the gas turbine than with the piston engine.

The relative levels of NO_x emission for gasoline engines, diesel engines, large gas engines, and several gas turbines are shown in Fig. 20. The NO_x data on piston engines have been taken from publications and reports from various sources and represent a fairly wide cross section of designs. The fact that engine design characteristics, such as rated revolutions per minute, fuel injection method, turbocharged or unturbocharged, etc. have an effect on NO_x levels accounts for the wide range of NO_x levels in the piston engine. It is evident from Fig. 20 that gas turbines, even without an attempt to control their emissions, have lower levels of NO_x than controlled large gas engines. Although data are not included here, this is also true of carbon monoxide and unburned hydrocarbons.

There are two factors inherent to the conventional gas turbine that account for low emissions of the three major pollutants:

(1) Combustion temperatures in gas turbines are low since they use about 3 times the air needed to burn the fuel consumed. Typically, gas turbines have an available air/fuel ratio of 50 to 60:1, which results in an oxygen-rich environment. In comparison, the gasoline engine (spark ignition) and diesel engines (including gas engines) require an operating air/fuel ratio of about 15:1 and 25:1, respectively. The air/fuel ratio operating ranges of piston engines and gas turbines and the effect on NO_x formation are shown in Fig. 21.

(2) Since the gas turbine is a continuous-flow machine, the time available for combustion of the fuel and the attainment of thermal equilibrium is considerably greater than in a piston engine.

A significant difference influencing the control of exhaust pollutants from piston engines and gas turbine engines lies in their mechanical concept. With the piston engine, all of the thermodynamic processes (compression of air, combustion of fuel, and expansion of gases to produce power) take place within the confines of a single element, the cylinder. The gas turbine, on the other hand, achieves its compression, combustion, and expansion in three separate and distinct components, which permits a great deal of latitude in the design of each component. This is of special significance since, with the gas turbine, there is freedom to design pollutant-reducing combustion systems techniques that cannot be used in the piston engine. The turbine combustor

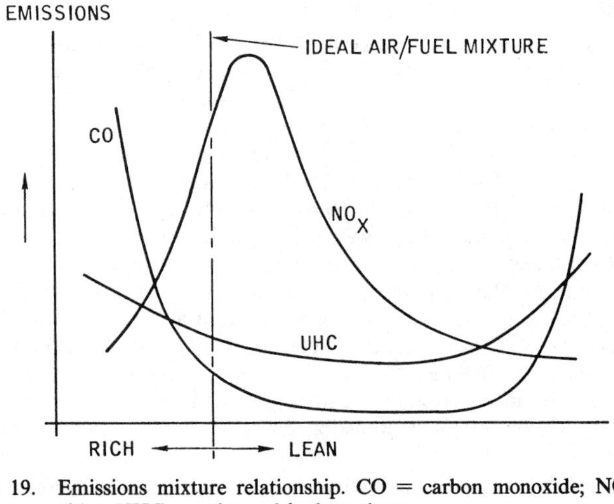

Fig. 19. Emissions mixture relationship. CO = carbon monoxide; NO_x = nitrogen oxides; UHC = unburned hydrocarbons.

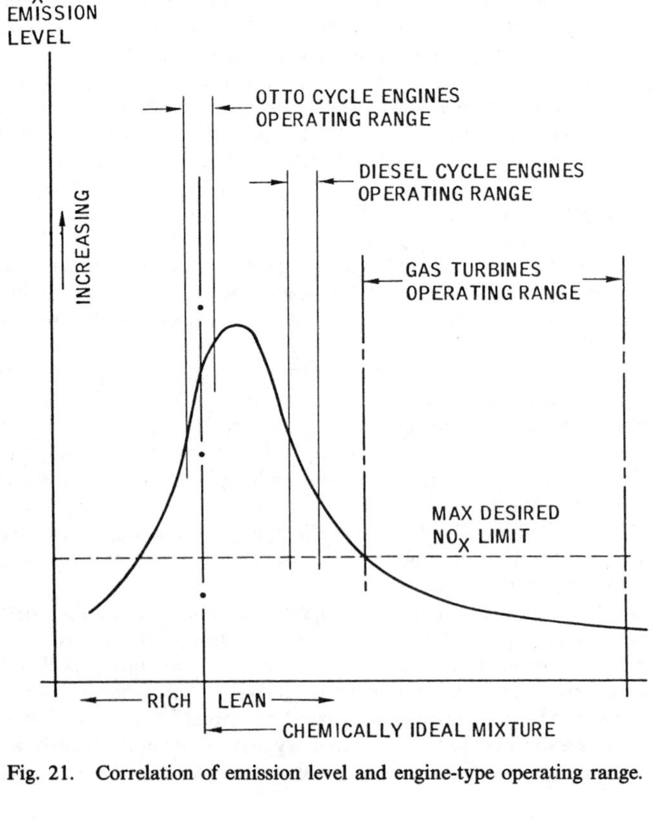

Fig. 21. Correlation of emission level and engine-type operating range.

can be varied in size and shape and can utilize various arrangements for injecting the fuel, premixing the air, and controlling combustion temperatures (air/fuel ratio) to minimize the generation of NO_x, CO, and unburned hydrocarbons. Further, these techniques generally can be applied without penalty to engine performance, power, or fuel consumption.

The piston engine does not have such flexibility. Within the more or less rigid confines of the cylinder combustion chamber, there is little latitude for design innovations. The problem is aggravated by the very short time available to burn fuel within the cylinder, a characteristic unavoidable with the intermittent, explosion-like combustion of the piston-engine. The short combustion time interval makes it essential that the Otto-cycle (spark ignition) engine operate closely to a chemically ideal ratio of air and fuel, about 15:1, if the engine is to run smoothly and efficiently. The explosion of this mixture generates pressures that push combustion temperatures to a peak of about ~2,593°C, a condition that produces nearly maximum NO_x. Although diesel engines (including natural gas engines) operate at somewhat higher air/fuel ratios of 20 to 25:1, they also produce relatively high levels of NO_x emissions because they operate at even higher combustion pressures, which lead to high combustion temperatures.

Although the piston and cylinder arrangement of the piston engine offers limited opportunities to reduce emissions through design innovation, the combustion process itself can be modified or influenced to reduce the levels of emissions. Some of the techniques are:

1. Retard the ignition timing.
2. Operate at the leanest possible air/fuel mixture (sacrificing efficiency).
3. Reduce compression ratio.
4. Controlled burning of fuel (stratified charge).
5. Dilute combustion mixture by recirculating exhaust gases.
6. Increase the operating temperature of the cylinder and head.

It is significant to note that experience to date indicates that to make substantial reductions to NO_x, more than one of the techniques is required. Whatever the combination, the objective is to lower peak combustion temperatures and, therefore, cylinder gas pressure. In the long run, this penalizes performance in power and/or fuel consumption. The bar entitled, "Large Diesel and Gas Engines—Controlled" in Fig. 20 illustrates the level to which NO_x emissions from large engines of this type can be controlled—accompanied by varying degrees of power and fuel consumption penalties.

Another approach to lowering emissions from piston engines is that of treating the exhaust gases before they are ejected into the atmosphere. This consists of adding an afterburner (thermal reactor) to complete the combustion of CO and unburned hydrocarbons and a catalytic reactor to break down NO_x. In order for the catalyst to reduce a nitric oxide by a significant amount, the exhaust must contain a high level of carbon monoxide. This, in turn, requires that the piston engine be operated with a rich fuel/air mixture which unavoidably increases the fuel consumption. See also **Catalytic Converter (Internal Combustion Engine).**

Emission Techniques for Gas Turbines. Research indicates that combustion temperatures in gas turbines must be held within a range of 1,204 to 1,705°C to hold all emissions to an acceptable level. The gas turbine combustor is relatively insensitive to shape and size in regard to its ability to operate efficiently. This gives the designer much greater freedom to modify and arrange the combustor so that it can operate at low combustion temperatures than is available to the designer of the piston engine. Burning fuel efficiently at lean mixtures and low temperatures requires that:

1. The fuel be thoroughly evaporated and uniformly mixed with the air before burning.
2. Burning the mixture uniformly and as rapidly as possible.
3. Additional (secondary) air be added downstream of the combustion zone to assure complete burning of CO and unburned hydrocarbons.
4. The combustor liner wall temperature to be high enough to eliminate fuel quenching, reducing the tendency to form CO, unburned hydrocarbons, and smoke.

A comparison of the traditional combustion design concept with that for achieving lower emissions is given in Fig. 22. In the traditional

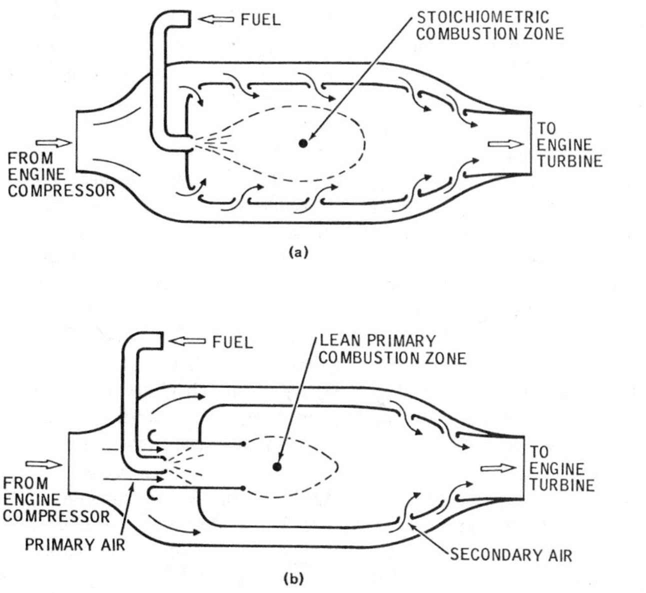

Fig. 22. Gas turbine combustion design principles: (a) traditional; and (b) low-emissions design approaches.

combustor design (a), the primary concern is stable combustion over all engine operating conditions with good combustion efficiency. This is accomplished by establishing a central combustion zone that will burn at near-stoichiometric conditions. This approach accomplishes stability and efficiency objectives, but the resulting high combustion temperatures produce NO_x levels that are higher than attained otherwise.

By modifying the combustor, as shown in (b), the combustion process can be controlled to a lower temperature. An abundance of primary air is premixed with the fuel to produce a homogeneous mixture of air and fuel vapor. This lean mixture, which ranges between a 30:1 air/fuel ratio at full engine load to 45:1 at no load, is burned in the primary combustion zone to maintain a combustion temperature within the range of 1,649° to 1,204°C. Secondary air brought in through the combustor liner downstream from the combustion zone completes the burning process, controls the combustion liner metal temperature, and dilutes the combustion gas to a proper operating temperature level.

Expansion Turbines

An expansion turbine converts the energy of a gas or vapor stream into mechanical work as the gas or vapor expands through the turbine. The expansion process occurs rapidly and the heat transferred to or from the gas is usually very small. Consequently, in accordance with the first law of thermodynamics, the internal energy of the gas decreases as work is done and the resultant temperature of the gas may be quite low, thus giving the expander the ability to act as a refrigerator as well as a work-producing device. As a result, turboexpanders have been widely used in the cryogenic field to produce the refrigeration needed for the separation and liquefaction of gases. By common usage, the terms *turboexpanders* and *expansion turbines* specifically exclude steam turbines and combustion gas turbines.

Turboexpanders may be classed into two broad categories: (1) axial-flow; and (2) radial-flow. Axial-flow turbines are those in which the gas flow is essentially parallel to the axis or shaft of the turbine. Turbines of this type resemble a conventional steam turbine and may be single-stage or multistage with impulse or reaction blading, or combination of impulse and reaction blading. Turbines of this type are not usually used for producing low temperatures, but are basically power-recovery devices and find application where flow rates, inlet temperatures, or total energy drops are quite high. Radial-flow turbines are those in which the gas flow is essentially at right angles to the turbine shaft. Flow may be radially inward or outward, but commercially available turbines are usually the radial-inward-flow type. Radial-flow turbines are usually single-stage and have combination impulse-reaction blades and a rotor that resembles a centrifugal-pump impeller. The gas is jetted tangentially into the outer periphery of the rotor and flows radially inward to the "eye," from which the gas is jetted backward by the angle of the blades so that it leaves the rotor without spin and flows axially away.

These latter machines usually have an efficiency of from 75 to 88%, usually operate at very low temperature, operate often on small or moderate streams, dictating a comparatively high rotating speed, and incorporate effective shaft seals to conserve the process stream. Commonly established operating limitations for turboexpanders are an enthalpy drop of 40 to 50 Btu (10–12.6 Calories)/pound/stage of expansion, and a rotor-tip speed of 1,000 feet (300 meters) per second. Commercial turboexpanders are available up to 2,500 psig inlet pressure and inlet temperatures of over 538°C. The permissible liquid production in the expanding stream varies with discharge pressure; it may be as high as 20% (weight) in the discharge, provided the turboexpander has been specifically designed to handle liquids.

Power Recovery. A potential application for the turboexpander exists whenever a large flow of gas is reduced from a high pressure to some lower pressure, or when high-temperature process streams (waste heat) are available at moderate pressures. When such conditions exist, they should be examined to determine if the use of a turboexpander is justified. In such cases, a turbine can be used to drive a pump, compressor, or electric generator, thus recovering a large portion of the otherwise wasted energy. In applications of this type, careful consideration should be given to the temperature drop which will occur in the expander. Sometimes it may be necessary to heat or dry the inlet gas to avoid low exhaust temperatures, or the formation of liquids.

Refrigeration. Turboexpanders used as components of refrigeration systems offer many possibilities to the designer of refrigeration cycles. They may be used in closed cycles with a pure gas, such as nitrogen, which is alternately compressed and expanded to provide the required refrigeration through a heat exchanger. Various types of open cycles also can be devised so that the process stream to be cooled passes through the expander, thus eliminating the need for the low-temperature heat exchanger. Liquid products can be produced directly from the turboexpander in this manner provided that the expander is specifically designed for this type of service.

The first turboexpander designs took advantage of a "free" pressure drop that was available at particular locations. Since then, the technique has been refined and enlarged to embrace practically every situation encountered in extracting hydrocarbons from a mixed gas stream; even where "free" pressure drop is not available. Designs were improved through better utilization of construction materials, design of the control system by simulation of operations in a computer, and the development of interlocking instrumentation.

The foregoing refinements enabled the utilization of the turboexpander economically in plants where full pressure restoration is required. The turboexpander system also has been used for recovery of propane alone and, in some cases, has been found to be more economical than the oil absorption process. Other turboexpander processes include dehydration and dew point control of wellhead gas streams, utilizing pressure reduction of 2,000 to 10,000 psig, (136 to 680 atmospheres) and nitrogen rejection together with heavier hydrocarbon recovery.

Some inherent advantages of turboexpander systems include: (1) the final cryogenic operating temperature is obtained from the turboexpander and not achieved by costly low-level external refrigeration; (2) product separation pressure is set to give the most desirable equilibrium conditions; (3) plants are compact and inherently simple; (4) capital investment and operating costs are usually low, as much as 40% savings over conventional ethane recovery method; and (5) maintenance requirements are low.

With reference to Fig. 23, the turboexpander system operates as follows: feed gas is first dehydrated and sometimes alcohol is injected at strategic points to protect further against the formation of ice or hydrates. Next, the feed is chilled by heat exchange with the residue gas. Condensed liquids are then separated and the vapors delivered to the expander. A direct-connected compressor recovers the expander energy, boosting the pressure of the condensate stripper overhead gas. Residue gas can be further compressed to any delivery pressure. Alternatively, the compressor can be used to boost the pressure of the feed gas to obtain additional refrigeration. By using a turboexpander to remove energy from the gas, refrigeration is materially increased

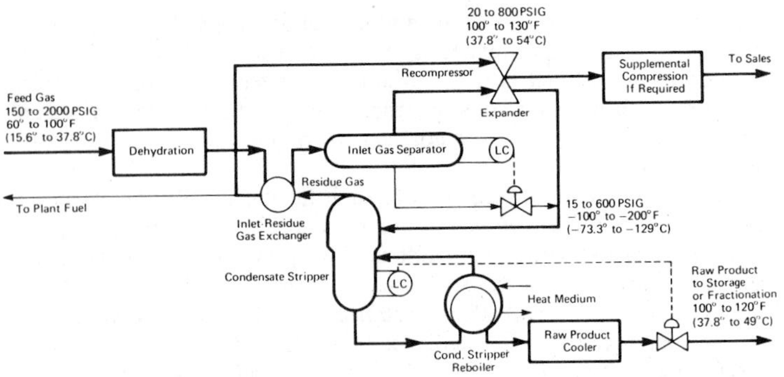

Fig. 23. Expander system for gas processing. LC = Level controller. (*Fluor Corp.*)

and the temperature is lowered below those conditions obtainable from simple adiabatic expansion.

Liquids condensed at the expander outlet and in the feed chilling step are fed to a fractionation system. The bottoms from fractionation represent the desired product mixture, which can have an ethane content equivalent to 90% or more of the ethane in the feed gas. Virtually 100% of propane and heavier hydrocarbons in the feed gas can be recovered. Flexibility in product composition is obtained either by adjusting the expander outlet pressure, or by stripping undesirable components overhead in the condensate stripper. Pressure difference across the expander is the principal energy source. Thus, minimal amounts of electric power and fuel are required for pumps, dehydrator regeneration heat, and stripper reboiler heat.

Because of the small size of turboexpander rotating elements, the forces and stresses at high speed (7,500 to 45,000 revolutions per minute) are equal to or less than those encountered in lower-speed rotating machinery. Any part of a turboexpander usually can be replaced in about 3 hours downtime.

GAS BLADDER. **Fishes.**

GAS BURNER. **Burner.**

GAS CALORIMETER. **Calorimetry.**

GAS CHROMATOGRAPHY. **Chromatography.**

GAS CONSTANT. The constant of proportionality R in the equation of state of a perfect gas $pv = RT$, when referring to one grammolecule of gas. R has the value of 1.985 calories per mole degree (C°).

GAS-COOLED FAST BREEDER REACTOR. **Nuclear Reactor.**

GAS-COOLED REACTOR. **Nuclear Reactor.**

GAS DENSITY (Measurement). **Specific Gravity.**

GAS DISCHARGE. A conduction current in a gas due to ionization. The discharge is self-maintaining if the source of the ionization, such as an external electric field, is sufficient to cause creation of the necessary supply of ions as by collision between molecules and electrons. Breakdown may be due to an external ionizing source, but once the discharge is initiated it continues unaided.

A non-self-maintaining discharge is due to ionization of the gas from an external source other than the applied voltage. Also known as a field-intensified discharge or a Townsend discharge.

GAS EFFLUENT TREATMENT. **Pollution (Air).**

GAS ENGINES (Emissions). **Gas and Expansion Turbines.**

GASEOUS DIFFUSION. **Diffusion; Graham Law.**

GASEOUS DISCHARGE. **Discharge (Gaseous).**

GASES (Ionization). Ionized Gases.

GAS-FIRED BOILER. Boiler; Burner; Feedwater (Boiler).

GAS FOCUSING. A method of concentrating an electron beam by the action of ionized gas. The positive ions produced in the gas compensate for the negative space charge of the electrons.

GAS GANGRENE. Gangrene.

GAS GIANTS (Planets). Jupiter; Neptune; Saturn; Uranus.

GAS (Gibbs Paradox). Gibbs Paradox.

GASIFICATION PROCESSES (Coal). Coal.

GASIFICATION PROCESSES (SNG). Substitute Natural Gas (SNG).

GAS (Ideal). Ideal Gas Law.

GAS-INSULATED CABLES. Electric Power Transmission.

GAS (Intestinal and Stomach). Flatulence.

GAS (Joule-Thomson Effect). Joule-Thomson Effect.

GASKET. The gasket is a layer of packing material firmly held between contact surfaces on two pieces whose joint is to be sealed with the gasket. Gaskets are made in many different shapes to suit the shapes of the various mating pieces. The simplest type of gasket is that used to seal the joint between two circular flanges, as might be used in making a joint in piping. The gasket is made of a thin sheet of material, satisfactory for the service, having through it a hole corresponding to the internal diameter of the pipe. After being inserted between the flanges, flange bolts draw the latter tightly together, compressing the gasket material until it tightly seals the joint. Some gaskets are toroidal in shape, but most are flat. The services they perform range from that of sealing against leakage of the liquids in water lines to rendering gas-tight such high-temperature joints as those in engine exhaust manifolds.

To prevent gaskets from blowing out, male-and-female or tongue-and-groove facings may be required so that hard gaskets can be seated adequately.

Compressed asbestos-sheet gaskets are usually used for service up to 300 pounds (136 kilograms) pressure and 400°C. For higher-pressure and higher-temperature service (up to 590°C), metal-asbestos spiral-wound type gaskets are often used. Such gaskets also find wide application for high-pressure steam service, but they are usually used with a smooth-flange finish. When used with raised facing, a solid metallic ring on the outside (to limit gasket compression) is supplied with spiral-wound types.

GAS (Knudsen Flow). Knudsen Flow.

GAS LASER. Laser.

GAS METER. Flow Measurement.

GAS (Natural). Natural Gas.

GAS OIL. Petroleum.

GASOLINE. Petroleum.

GASOLINE (Carburetion). Carburetor.

GAS (Perfect). Perfect Gas.

GAS PROCESSING. Coal; Gas and Expansion Turbines; Natural Gas; Substitute Natural Gas (SNG).

GASSER. Natural Gas.

GAS (Substitute). Substitute Natural Gas (SNG).

GASTEROPODA (or *Gastropoda*). The snails, slugs, and allied forms, constituting a class of the phylum *Mollusca.* This group includes a large number of marine and freshwater species and many that are terrestrial.

The chief structural characteristics of the gasteropods are these: (1) most species have a dorsal visceral hump which is often spirally twisted. (2) A head is present, bearing eyes and tentacles. (3) The mouth is provided with a toothed organ called the radula. (4) The foot is usually a broad creeping organ. (5) Respiratory ctenidia lie in the mantle cavity of some species, and in others, the walls of the cavity are the respiratory organ. (6) In many species, a shell, conical or spirally coiled, encloses the visceral hump.

Gasteropods are of relatively little economic importance. Snails are eaten in Europe and the abalones of the Pacific Coast are also used as food. The shell of the abalones furnishes beautifully iridescent mother-of-pearl for costume jewelry and there is an extensive traffic in the shells of many species among collectors.

The group is classified as follows:

- Subclass *Streptoneura.* Usually with a shell closed by a horny shield, the operculum, when the animal is retracted.
  - Order *Diotocardia* (*Aspidobranchiata*). Abalones, limpets, and other marine species. A few freshwater forms.
  - Order *Monotocardia* (*Pectinibranchiata*). Whelk, periwinkle, and many other marine forms and a few freshwater species.
- Subclass *Opisthobranchiata.* Shell small and internal, sometimes lacking.
  - Order *Tectibranchiata.* Sea hare, sea butterflies or pteropods with the foot expanded into wing-like lobes. All marine.
  - Order *Nudibranchiata.* Marine species without shells. Often with complex dorsal processes. Sea lemon; nudibranchs.
- Subclass *Pulmonata.* Shell usually present but without an operculum. Mantle cavity sometimes the only respiratory organ. Mostly freshwater and terrestrial species, a few marine.
  - Order *Basommatophora.* Eyes at the bases of the posterior tentacles. Many common snails.
  - Order *Stylommatophora.* Eyes at the tips of the posterior tentacles. Common snails and slugs. (See also **Invertebrate Paleontology.**)

GASTEROPODA (Fossils). Invertebrate Paleontology.

GAS THERMOMETER. Thermometer (Filled-System).

GASTRECTOMY. Ulcer.

GASTRIC JUICE. Bile; Digestive System (Human).

GASTRIC ULCER. Ulcer.

GASTRIN. Diabetes Mellitus; Hormones.

GASTRITIS. An inflammation of the lining membrane of the stomach, occurring in an acute or chronic form. The various gastric juices, such as enzymes, pepsin, hydrochloric acid, rennin, and lipase, as well as a heavy, protective mucus, are secreted by the stomach lining. In gastritis, these functions are disturbed and the digestive process is impeded. Gastritis may result from the ingestion of poisonous corrosives. Toxic substances associated with certain infections also may initiate or aggravate gastritis. Therapy depends upon the type and source of the gastritis. Differential diagnosis is frequently required to determine the exact causative factor. Disturbance of the stomach lining in gastritis may range from tiny hemorrhagic areas to ulceration. Duration of therapy may range from several days to several weeks.

GASTROENTERITIS. An inflammation of the gastrointestinal tract which may be caused by a number of factors. One of the most frequent causes is foodborne disease, notably salmonellosis. See **Foodborne Diseases.** A number of viruses, collectively called enteroviruses, such as the echoviruses, the rotoviruses, and specific related agents, such as the Norwalk agent, can cause gastroenteritis of varying severity and time span. See also **Coxsackie Virus;** and **Norwalk Virus.** *Bacillus cereus* can cause outbreaks of self-limited gastroenteritis that lasts 12 to 15 hours. The disease has been associated with the ingestion of a number of foods, particularly fried rice, sauces, and meat, which contain enterotoxins as the result of inadequate refrigeration. One of the major symptoms of gastroenteritis is diarrhea. See **Diarrhea.**

GASTROENTEROLOGY. The study of the diseases and functions of the stomach and intestines. A gastroenterologist is a specialist in the diagnosis and treatment of the disorders of the stomach and intestines.

GASTROENTEROSTOMY. Ulcer.

GASTROINTESTINAL CANCER. Cancer and Oncology.

GASTROINTESTINAL TRACT. Digestive System (Human).

GASTROPID. Invertebrate Paleontology.

GASTROPODA. Gasteropoda.

GASTROSCOPE. A device for examining the interior of the stomach by direct visualization. A form of endoscope. See also **Endoscope;** and **Ulcer.**

GASTROTRICHA. A group of minute animals found in fresh and salt water on the bottom and among the debris accumulated there. They move chiefly by means of cilia and have cement glands whose secretion attaches them temporarily to supports. They have a tubular alimentary tract and an excretory system consisting of two tubules with flame cells. The group is ranked by some writers as a class in the same phylum as the rotifers and by others as of uncertain relationship.

Two orders are recognized: *Macrodasyoidea,* made up of marine species with numerous cement glands and *Chaetonotoidea,* made up of marine and freshwater species with a single pair of cement glands at the caudal end of the body, or none.

GASTROZOOID. A form of individual in hydrozoan colonies whose function is to digest food for the colony.

GASTRULA. The stage in embryonic development in which the initial differentiation of tissues is evident. The gastrula is typically a sac whose wall is composed of the two germ layers, an outer ectoderm and an inner endoderm. The cavity lined by the endoderm is the archenteron and the opening to the exterior is the blastopore.

The gastrula is formed from the blastula by the process of gastrulation. Typically the wall of the spherical blastula caves in on one side and the invagination progresses until this side is in contact with the opposite wall. In some coelenterates, however, the two germ layers appear as a solid mass of endoderm surrounded by a layer of ectoderm, and the archenteron forms by the splitting of the inner mass. In animals with abundant yolk, modifications also appear. In birds, for example, the stage approximating the blastula is a disk of cells on the surface of the yolk and the endoderm may be formed by the folding under of this layer at one point on the margin or by a more diffuse process of polyinvagination, recently discovered. Later the folded edge undergoes a concrescent growth until it doubles on itself and fuses to form the primitive streak, equivalent to a closed blastopore.

The mesodermal layer also appears in the gastrula of triploblastic animals. Its formation is extremely variable but it usually grows out from the indeterminate zone about the blastopore where ectoderm and endoderm join.

GAS TURBINE ENGINE (Automobile). Internal Combustion Engine.

GAS WELDING. Welding.

GATE CIRCUIT. A circuit which amplifies or passes a signal only in the presence of an appropriate synchronizing or "gating" pulse which "opens" the gate. Also used to refer to the various logic functions and circuits used to realize computer designs, such as the AND, OR, NOT, NOR, and NAND. See also **AND (Circuit); Gate (Computer System); OR (Circuit); NOT (Circuit); NOR (Circuit);** and **NAND (Circuit).**

Transistor transmission gate.

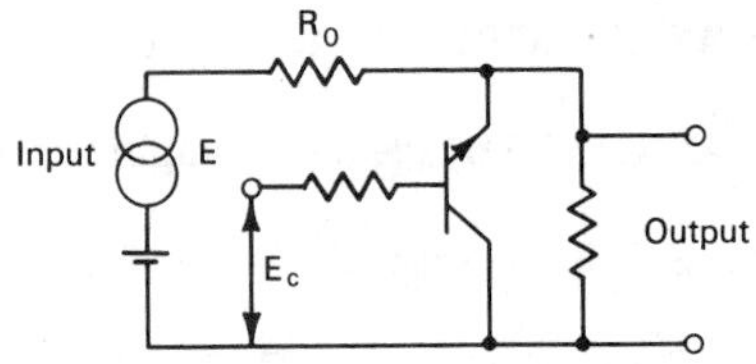

GATE (Computer System). A circuit having a binary output which is fully determined by the binary state of its input signals, such as in the AND and OR gate circuits. Also, a signal which permits an AND circuit to pass a signal. Usually the gate signal is of longer duration than the signal to make certain that coincidence occurs. In conditioning the set pulse of a flip-flop, for example, the gate must precede the set signal in order that the negative shift will be recognized by the transistor. See also **Flip-Flop.**

GATE-TURNOFF SWITCH. An electronic device (GTO) that operates like a silicon-controlled rectifier with exception that the high current conduction state can be interrupted by a negative pulse applied to the gate electrode. Used in dc switching applications.

GATING. 1. The process of selecting those portions of a wave which exist during one or more selected time-intervals, or which have magnitudes between selected limits. 2. The function or operation of a saturable reactor or magnetic amplifier which causes it, during the first portion of the conducting alteration of the ac supply voltage to block substantially all of the supply voltage from the load, and during latter portion allows substantially all of the supply voltage to appear across the load, is called gating or gating action. The "gate" is said to be virtually closed before firing and substantially open after firing.

GAUCHER'S DISEASE. Lipidoses.

GAUGE THEORIES. An excellent and brief description is given by Hung and Quigg (1980): "At the base of the unification of interactions (particles) is the idea of gauge invariance, which draws its name from some early investigation by Weyl (1951) into a possible connection between scale changes and the laws of electromagnetism. Weyl's specific attempt to deduce electromagnetism from a symmetry principle—invariance under a change of length scale at every position of space-time independently—ran afoul of quantum mechanics, but the general strategy and the name have survived. Indeed, gauge theories constructed to embody various symmetry principles are now believed to provide the correct quantum descriptions of the strong, weak, and electromagnetic interactions. The simplest example of a gauge theory is electromagnetism itself." More details can be found in *Science,* **210**, 1208–1209 (1980). The Weyl reference cited is "Space-Time-Matter," Chap. 4, sect. 35, page 282, translation published by Dover, New York (1951). Reference to "A Unified Theory of Elementary Particles and Forces," by H. Georgi, *Sci. Amer.,* **244**, 4 (1981) is also suggested. See also **Conservation Laws and Symmetry;** and **Particles (Subatomic).**

GAUSS. Units and Standards.

GAUSS CONFORMAL. Lambert Projection.

GAUSS EYEPIECE. A Ramsden eyepiece that has a thin plate of glass placed between the two lenses at 45° to the optical axis. It is useful in setting a telescope normal to a plane-reflecting surface. See also **Telescope.**

GAUSS HYPERGEOMETRIC EQUATION. A canonical form of the Riemann-Papperitz equation, where the regular singular points have been shifted to $x = 0, 1, \infty$. Its form is $x(1-x)y'' + [c - (a + b + 1)x]' - aby = 0$, in which a, b, c are parameters related to the exponents of the equation. Its general solution around the point $x = 0$ and convergent for $|x| < 1$, provided c is not a positive or negative integer or zero, is

$$y = AF(a, b, c; x) + Bx^{1-c}(1 + a - c, 1 + b - c, 2 - c; x)$$

In this solution, $F(x)$ is the Gauss hypergeometric function, an infinite series, and A, B are constants of integration.

The series can be written as

$$F(a, b, c; x) = \frac{\Gamma(c)}{\Gamma(a)\Gamma(b)} \sum_{n=0}^{\infty} \frac{\Gamma(a+n)\Gamma(b+n)}{\Gamma(c+n)n!} x^n$$

where Γ is the gamma function. The coefficient of the general term x^k is

$$\frac{a(a+1)(a+2)\cdots(a+k-1)b(b+1)\cdots(b+k-1)}{c(c+1)\cdots(c+k-)k!}$$

Many well-known functions, including the Legendre and other polynomials, are compactly represented by the hypergeometric series. Thus $F(1, b, b; x)$ is the ordinary geometric series and $F(a, b, b; x)$ is the binomial expansion of $(1-x)^{-a}$.

If one lets $b \rightarrow \infty$ in the hypergeometric equation, the singular points at 1 and ∞ merge by confluence to give an irregular singular point at ∞. The result, called the confluent hypergeometric equation (also the Pochhammer-Barnes equation), is

$$xy'' + (c - x)y' - ay = 0$$

Its general solution is

$$y = AF(a, c; x) + Bx^{1-c}F(1 + a - c, 2 - c; x)$$

where

$$F(a, c; x) = \frac{\Gamma(c)}{\Gamma(a)} \sum_{n=0}^{\infty} \frac{\Gamma(a+n)x^n}{\Gamma(c+n)n!}$$

is the confluent hypergeometric series. The notation ${}_1F_1$ is often used for this series and ${}_2F_1$ for the Gauss hypergeometric series. The first subscript indicates the number of factorial terms in the numerator and the second subscript the number of such terms in the denominator. The more general series ${}_nF_m$ has also been studied.

Related equations, obtained from this by change of variable, are those of Hermite, Weber, and Whittaker. Any differential equation with three regular singular points or less can be expressed as a special case of the Gauss equation or its confluent form.

See also **Gamma Function; Gegenbauer Function; Series;** and terms listed under **Mathematics.**

GAUSSIAN DISTRIBUTION. Normal Distribution.

GAUSSIAN NOISE. Noise.

GAUSS-MARKOFF THEOREM. A theorem in statistics to the general effect that the best estimator of a parameter from a population, among the class of estimators which are linear in the sample values, is obtained by the method of least squares. "Best" in this sense means that the estimator is unbiased and has minimal variance.

GAUSSMETER. Magnetometer.

GAUSS METHOD FOR QUADRATURES. The integral to be evaluated

$$y = \int_a^b f(x)\,dx$$

is replaced by

$$\int_a^b \phi(x)\,dx = A_0y_0 + A_1y_1 + \cdots + A_ny_n$$

as in the Newton-Cotes method, and the difference between the two integrals is minimized by fixing the values of $2(n+1)$ quantities; $(n + 1)$ values each of the coefficients A_k and the dependent variable y_k. The latter do not result from equally spaced intervals in x, the independent variable, as is the case for most quadrature formulas. The $2(n+1)$ parameters are determined by solution of simultaneous equations, which in turn depend on the real roots of the Legendre polynomials. Numerical tables for these parameters are available.

This procedure is more accurate than any of the other numerical integration methods but it is somewhat more awkward to use. Note also that if y is a physical quantity, measured when some other quantity, x (the temperature, for instance) is kept constant, the Gauss method specifies what values of x must be chosen.

See also **Quadrature.**

GAUSS PRINCIPLE OF LEAST CONSTRAINT. Motion (Gauss Principle of Least Constraint).

GAUSS THEOREM. A relation between multiple integrals which in Cartesian coordinates is

$$\int_\tau \left(\frac{\partial u}{\partial x} + \frac{\partial v}{\partial y} + \frac{\partial w}{\partial z}\right) dx\,dy\,dz = \int_S (\lambda u + \mu v + \nu w)\,dS$$

The quantities u, v, w are functions of x, y, z having continuous first derivatives within a volume τ and they approach their values on the bounding surface continuously. The outward normal to the surface has direction cosines λ, μ, ν.

A vector form of the theorem, often known as the divergence theorem, is

$$\int_\tau \nabla \cdot \mathbf{V}\,dt = \int_S \mathbf{V} \cdot d\mathbf{S}$$

where the vector **V** has components (u, v, w). However, the first form of the theorem holds even when u, v, w are not components of a vector.

A physical interpretation of the vector equation may be made, for if **V** represents the flux density of an incompressible fluid, $\nabla \cdot \mathbf{V}$ is the amount of fluid which flows from a volume dt per second. The volume integral is thus the total loss of fluid, which must equal the rate of flow across all boundaries of the volume, and that equals the surface integral.

This theorem is also called the Green lemma or theorem. See **Green Function.**

GAVIAL. Crocodiles and Alligators.

GAVIIFORMES (*Aves*). Large birds with submarinelike swimming habits, much larger than most ducks and with shorter necks than geese. They are powerful swimmers with short legs, webbed feet and characteristic strong, sharp beaks. See accompanying illustration. They have a thick neck, heavy body, black head, and are fast in flight, at times achieving a speed up to 60 miles (97 kilometers) per hour. In flight the outline is hunch-backed and gangly, with a slight downward sweep to the neck and the big feet projecting beyond the tail. They build large and bulky nests close to water, mounding together grass and weeds. There are usually two olive green eggs with black spots. The incubation period is 28 days. The young take to the water almost immediately after hatching. See also **Loon.**

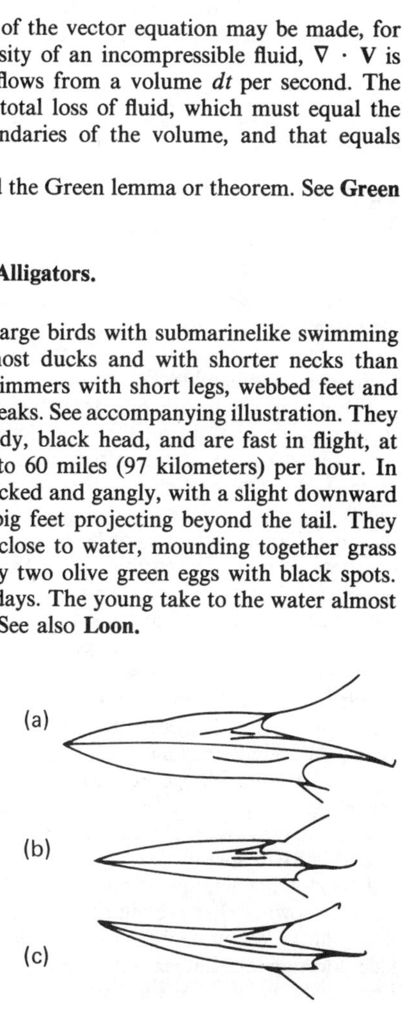

Gaviiformes. Beaks of Loons; (a) Common Loon (*Gavia immer*); (b) Pacific Loon (*Gavia arctica pacifica*); (c) Red-Throated Loon (*Gavia stellata*).

GAYAL. **Bovines.**

GAZELLE. **Antelope.**

GAZELLE-GOATS. **Goats and Sheep.**

GEARING. The use of toothed wheels for transmitting rotary or reciprocating motion from one machine element to another. The principal forms of toothed gearing are spur gearing, helical gearing, bevel gearing, and worm gearing.

GEARING (Friction). **Friction Gearing.**

GEARING (Helical). **Helical Gearing.**

GEARING (Spur). **Spur Gearing.**

GEARS AND GEAR TRAINS. **Machine (Simple).**

GEAR TRAIN. Two or more gears, transmitting motion from one shaft to another, constitute a gear train. If spur, bevel, or worm gears are used, the velocity ratio is inversely proportional to the numbers of teeth in the gears. A pair of spur gears, directly connected, result in a reversal of direction; if the driving gear drives an intermediate idler, which in turn drives the driven gear, the only effect of the idler is to cause the driven gear to rotate in the same direction as the driver. (The same effect can also be obtained by using an internal gear and a pinion.) If a two-gear idler, or compound gear, in which both idlers are fastened either to the idler shaft or to each other, is used, and where the driver engages one of the compound gears, and the driven gear the other compound gear, the velocity ratio is equal to the product of the two trains. The back gearing of a lathe or a milling machine is a familiar example of a compound gear train; a gear attached to the driving pulley drives a large gear mounted on the back gear shaft; a small gear on this shaft in turn drives the spindle gear. See also **Epicyclic Gear Train.**

GEAR TRAIN (Epicyclic). **Epicyclic Gear Train.**

GEAR (Worm). **Worm Gearing.**

GECKO. Of the class *Reptilia* (reptiles), subclass *Lepidosauria*, order *Squamata* (scaly reptiles), suborder *Sauria* (lizards), infraorder *Gekkota*, according to classification of Grzimek (1972). The infraorder *Gekkota* comprises the families *Gekkonidae* (geckos), *Pytopodidae* (snake lizards), and *Dibamidae.* Even though the geckos do not resemble these other families externally, certain anatomical characteristics indicate a close relationship to the snakelike pytopodids and the almost worm-shaped dibamids.

The geckos are lizards of extraordinary diverse form; furthermore, the group is quite old on the evolutionary scale, as can be inferred from the structure of the vertebrae, the presence throughout their lifetime of vestiges of the notochord, the shape of the hyoid bone, the fleshy tongue, and peculiarities of the scales. In the course of their development, the geckos in the subtropics and tropics conquered a variety of habitats—so that today they are found from the desert to the rain forest—in each case suitably modified in form. The geckos are small animals—at most about 40 centimeters (15.7 inches) long. The body is flattened. The large eyes are covered by a transparent scale, and noctural species have a slit pupil. The feet are specially constructed, the fingers and toes often bearing broad clinging lamellae on the underside. In contrast with other lizards, the geckos are quite vocal; the sounds they make range from quiet chirping and squeaking to loud barking. The geckos are the only living reptiles which really make extensive use of their voices, and in this respect they stand comparison with amphibians, birds, and mammals. There are 83 genera and about 670 species.

It is not uncommon to find geckos hanging head down on walls or ceilings when they are hunting insects. In managing these acrobatic feats, the geckos employ their toes, which are broadened on the underside, forming lamellate cushions. Here there are countless microscopic hook cells which, like the bristles of a brush, engage in the tiniest irregularities of a surface. This enables the geckos to run even on vertical surfaces. It was thought at one time that the lamellae exerted a sort of suction or even secreted a sticky substance, but investigation has proved these concepts to be incorrect. On a surface polished to a very high degree, even a gecko cannot adhere and slips like a person walking on ice. The normal mechanism of release and reattachment of the hook cells proceeds so rapidly that one cannot follow it by eye.

Many geckos have a striking ability to change color. Usually, they are lighter by day and darker by night. When one picks up a common gecko (not a simple feat, because of its agility) the skin feels soft and velvety. Like that of all reptiles, it is covered with scales, but their edges lie flush with one another, and do not overlap as in the other lizards and snakes. The gecko molts at intervals; as a rule the skin first breaks open at the head and is stripped off toward the back. Most geckos eat the shed skin, at least in part. If one tries to catch a common gecko, its tail may suddenly break off (autotomy). The breaks tend to occur at specialized places in the bodies of the tail vertebrae, and are brought about by muscular contraction. The cast-off parts regenerates, but cannot be autotomized again, since it now is supported by an unsegmented rod of cartilage without preformed break points. Discarding the tail affords the gecko a certain protection, for the violently jerking cast-off piece can distract a predator and give the gecko a chance to flee. The loss of the tail occurs so frequently among geckos that one often has trouble finding animals with the original tail.

Like most geckos, the common gecko (*Tarentola mauritanica*) is active in the twilight and night. Geckos prefer spiders, beetles, butterflies, millipedes, crickets, and cockroaches, although larger species, such as the Caledonian gecko (*Rhacodactylus leachianus*) also takes young lizards, mice, and small birds. The Madagascar gecko (*Phelsuma* spp.) is active during daylight and prefers plants, particularly fruit. The Japanese gecko (*Gekko japonicus*), which becomes very tame in a terrarium, readily accepts fruit and candy. The *Gehyra mutilata* has been called the "sugar lizard" because of its preference for sweet, fermenting substances.

Many geckos have become established in the vicinity of human dwellings and often are specifically adapted to coexistence with humans. The common gecko is frequently encountered on the rough stone walls of Mediterranean houses, and even within the house. Having become accustomed to the presence of humans, the nocturnal geckos have to some extent lost their fear of people. As a result, geckos living on the coasts or in ports sometimes "stow away" on ships. Thus transported, several species of geckos have greatly expanded their ranges to quite remote parts of the earth. The common gecko has moved from northern Africa to the port cities of southern France and to the Canary Islands, and in some cases will be found in the South Pacific.

Considerably more detail on various species of gecko will be found in Chapter 7, "The Geckos and Their Relatives," in "Grzimek's Animal Life Encyclopedia," Vol. 6, Van Nostrand Reinhold, New York, 1972.

GEE NAVIGATION. **Navigation.**

GEGENBAUER FUNCTION. A solution of the differential equation

$$(x^2 - 1)y'' + (2n + 1)xy' - a(a + 2n)y = 0$$

For integral values of a, the solution becomes the Gegenbauer polynomial. This equation is a special case of the Gauss hypergeometric equation.

GEGENSCHEIN. A slight increase in intensity of the zodiacal light at a point on the eclipse 3° west of the antisolar point. The gegenschein appears as a soft glow against the sky, oval in shape, a few degrees wide and 10–15° in length. It is so faint that it cannot be observed on a night when there is any moon or when the patch falls in the vicinity of the Milky Way. A dust tail of the earth under radiation pressure would explain the 3° lag, and an ordinary photometric function would explain the photometric properties.

GEIGER COUNTER. Also called a Geiger-Müller or G-M counter, the name Geiger counter is now rather commonly applied to a gas-

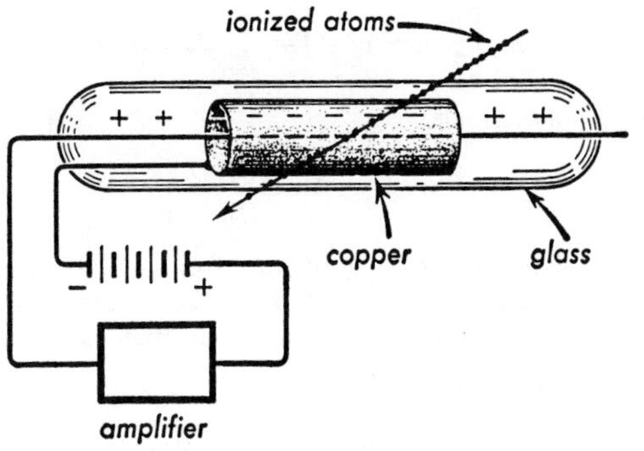

Fig. 1. Geiger counter.

filled detector of ionizing radiations of the general design indicated in Fig. 1. When operating in the Geiger region the tube produces an output voltage pulse of approximately constant magnitude for each ionizing event that takes place within the cylindrical electrode. The development of this output pulse depends on the production of an avalanche of ionization along the central wire electrode, possible only if the central wire is of sufficiently small diameter (typically less than 0.010 inch) that a very high field gradient exists in the immediate vicinity of the wire. This high field gradient causes electrons attracted toward the central electrode to attain sufficiently large kinetic energies that they can ionize additional atoms of the gas inside the tube. Additional electrons, probably produced by photons emitted when some of the ion-electron pairs recombine, are attracted toward the central electrode, and produce complete ionization of the entire region immediately surrounding the central wire in about 10 microseconds. Because of their low mobility, the positive ions produced near the central wire build up as a sheath to destroy the high voltage gradient and render the tube inactive. This action results in a pulse of approximately constant magnitude for each ionizing event.

After the Geiger counter discharges and produces a pulse, it remains inoperative for a period of time called the *dead time.* This is the time required for the positive-ion sheath to move out from the wire to a position where the electric field can recover so that another avalanche can form. The *resolving time* of the counter is larger than the dead time and is determined by the point at which the pulse size becomes large enough to again trigger the electronic equipment. See Fig. 2. The *recovery time,* larger still than the resolving time, is

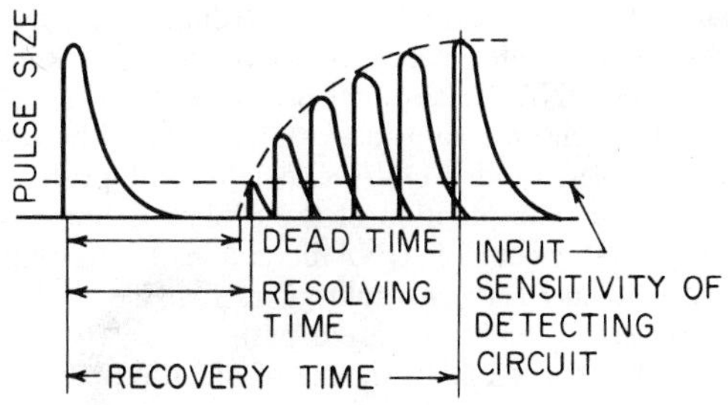

Fig. 2. Dead time and subsequent gradual recovery of pulse size in a Geiger counter.

that point where the pulse again gains its original amplitude. All these factors determine the speed at which a counter can operate without losing a large number of counts. The dead time and recovery time are of the order of 100 to 200 microseconds for the typical Geiger counter.

A large number of different quenching vapors may be used for filling Geiger counters. Amyl acetate, ether, and alcohol have had wide use. Halogen gas has been used as the quench vapor. The halogen molecule does not dissociate as does the polyatomic molecule. In actual practice, the useful life of an organic quenched counter is of the order of 10^8 counts whereas that of a halogen quenched counter may be 10^{10} counts or more. Halogen counters, unlike organic counters, are not damaged when subjected to voltages above the plateau region.

Geiger counters are made in a variety of shapes other than the most common shape of the cylindrical shell and axial wire. Cleanliness can not be overstressed in counter construction. The counter should be washed with a detergent or alcohol, followed with several rinses of distilled water and then dried by heating. The central wire should have no sharp projections and be free of dust and lint. A typical counter of 1-inch (2.54 centimeter) diameter shell and 0,001-inch (0.025 millimeter) wire, filled with ethyl alcohol to a pressure of 1 cm of Hg and argon to a pressure of 9 cm of Hg, will operate at approximately 1,000 V. The pulse size varies with the counter dimensions and the voltage above the Geiger threshold. See Fig. 3.

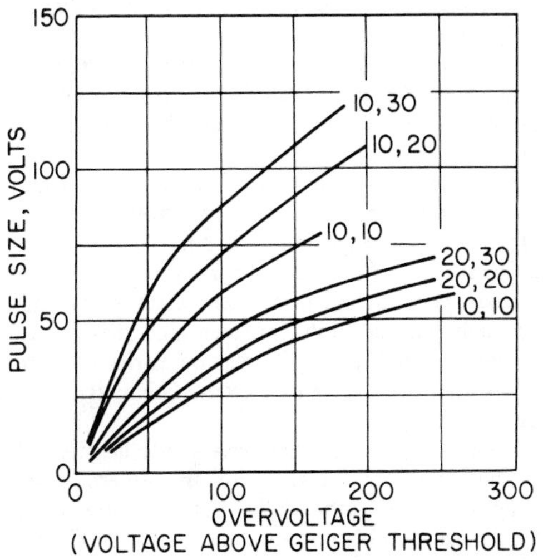

Fig. 3. Pulse size as a function of counter dimension. Tube dimensions are indicated as cathode radius (millimeters) and wire radius (thousandths of a millimeter).

GEIKIELITE. Magnesium analogue of ilmenite.

GEL AND GELATION. **Celloid System.**

GELATIN EMULSION. **Photography and Imagery.**

GEMINI (the twins). A constellation, marking the third sign of the zodiac, which has been recognized as a pair of twins from remote antiquity. The twins have not always been human, however; the Egyptians considered them as a pair of kids, and the Arabians as a pair of peacocks. By far the most familiar names for the two bright stars of this constellation are the names of the warrior brothers, Castor and Pollux, sons of Jupiter and Leda. Both these stars are interesting objects as seen through a 3-inch telescope, Castor being a fine binary and Pollux being a multiple star having at least six components. There is also a fine star cluster in this constellation, which can easily be seen with a field glass and can be detected with the unaided eye on a clear moonless night. (See map accompanying entry on **Constellations.**)

GEMMULE. Reproductive bodies of sponges derived from asexual internal buds, enclosed in protective capsules which enable them to withstand severe conditions of temperature and drought.

GEM STONES. A gem stone is a mineral substance which because of its beauty or rarity is in demand for ornamental purposes, chiefly personal adornment. The origin of such use for what we now call gem minerals is lost in the dim vistas of early human history. Ancient records describe the various gem stones, and archeologists find them in their investigations of bygone peoples. When we look at a collection of minerals with their bright colors and varying degrees of transparency or light-reflecting power, we cannot doubt that primitive man was much attracted by them and valued them greatly. We may imagine, too, that the occasionally found crystals with their regular geometric forms were more highly prized than broken fragments of the same minerals. Later they learned to polish them. Apparently the oldest form into which stones were shaped is that known as *en cabochon,* a French term derived from the Latin word for head and referring to its rounded shape. The forms were either hemispherical or hemiellipsoidal. The Emperor Nero is supposed to have had a large emerald cut en cabochon, and, indeed, for several centuries after his time this seems to have been the only sort of cutting employed. The supposedly accidental discovery in 1475 that diamonds would mutually scratch each other began the era of modern gem cutting. Previously it had been believed that diamonds were so hard that they could not be artificially shaped. At first, however, little progress was made in fashioning gems other than polishing a number of facets without any definite arrangement.

We owe to Vicenzio Peruzzi, a Venetian, the credit for devising the so-called "brilliant cut," the style of the modern diamond cutting, which, except for certain refinements due to a more thorough understanding of the behavior of minerals toward light, remains the same as in Peruzzi's day. At the present time, transparent stones of all sorts are usually "brilliant cut," whereas translucent or opaque are cut en cabochon.

Since time immemorial, dealers in gems have used as the unit of weight the carat, undoubtedly introduced from the east. The word is derived from the Greek meaning a small horn, referring to the pods of the locust tree, *Ceratonia siliqua*, a common Mediterranean tree whose seeds were said to have been taken as the unit of weight in buying and selling gems. In the nineteenth century the actual weight of the carat differed slightly in different countries of Europe, from a little under to somewhat over $\frac{1}{5}$ of a gram. The metric carat is exactly $\frac{1}{5}$ of a gram.

In recent years, synthetic stones have made large inroads in both the jewelry and industrial fields. Although numerous gem materials have been manufactured synthetically, only a few are cut as gemstones for the jewelry trade, notably corundum, spinel, emerald, rutile, garnet, sphene, and strontium titanate. Synthetic diamond for industrial use is produced by subjecting a carbonaceous material to very high temperature and pressure. One of the processes in use for creating synthetic crystals of corundum and spinel was developed by Auguste V. L. Verneuil (1856–1913), a French mineralogist and chemist. The starting composition is an alumina powder which then is melted in an oxyhydrogen flame, whereupon a series of drops are formed that build up the boules of the synthetic gems. See also **Beryl, Corundum, Diamond.**

GENERAL CIRCULATION MODEL (GCM). Climate.

GENERALIZED RIEMANN HYPOTHESIS. Number Theory.

GENERAL-PURPOSE COMPUTER. As contrasted with a special-purpose computer which is optimized toward solving particular programs, the general-purpose computer is a stored-program digital computer capable of solving a large variety of problems. Even with this distinction, however, computers in the general-purpose category are available in configurations which make them more suitable for certain classes of applications than others. Thus, there are commercial, scientific, process-control, and data-acquisition computers. See also **Computer;** terms listed under **Data Processing.**

GENERAL RELATIVITY THEORY. Relativity and Relativity Theory.

GENERATING FUNCTION. A method of representing a function in terms of another function containing one or more variables. These generating functions give the function to be generated as coefficients involving the new variable parametrically. Each of the polynomials, cited as examples, is the solution of a differential equation generally known by the name of the mathematician indicated.

1. Legendre polynomials

$$(1-2xy+y^2)^{-1/2} = \sum_{n=0}^{\infty} P_n(x)y^n$$

2. Associated Legendre polynomials

$$\frac{(2m)!(1-x^2)^{m/2}y^m}{2^m m!(1-2xy+y^2)^{m+1/2}} = \sum_{n=m}^{\infty} P_n^m(x)y^n$$

3. Bessel function of integral order

$$\exp\left[\frac{z}{2}(u-1/u)\right] = \sum_{n=0}^{\infty} J_n(x)u^n$$

4. Hermite polynomials

$$\exp[x^2-(z-x)^2] = \sum_{n=0}^{\infty} \frac{H_n(x)z^n}{n!}$$

The Hermite polynomials find applications in statistics in the Gram-Charlier Type A series and are also useful in certain applications in physics to problems of heat and quantum mechanics.

5. Laguerre polynomials

$$(1-z)^{-1}\exp\left(\frac{-xz}{1-z}\right] = \sum_{n=0}^{\infty} \frac{L_n(x)z^n}{n!}$$

6. Associated Laguerre polynomials

$$(-1)^k(1-z)^{-1}\left(\frac{z}{1-z}\right)^k \exp\left(\frac{-xz}{1-z}\right) = \sum_{n=k}^{\infty} \frac{L_n^k(x)z^n}{n!}$$

7. Chebyshev polynomials

$$\frac{1-xy}{1-2xy+y^2} = \sum_{n=0}^{\infty} T_n(x)y^n$$

Sir Maurice Kendall, International Statistical Institute, London.

GENES. Genes are the physical units of heredity. Precise definitions for them have changed considerably over the years as more has been learned about the chemical nature of the genetic material and its function.

Currently, genes may be defined as segments of genetic material which determine the sequence of amino acids in specific polypeptides, such that there is a one-to-one relation between gene and polypeptide. This definition applies at least to those genes called *structural* genes because they determine the primary structure of proteins. Other kinds of genes may exist as discussed in the following.

Structural Genes. So far as known, structural genes in all organisms are composed of nucleic acids. In the RNA viruses, the genes are RNA (ribonucleic acid) only, but in all other organisms, the DNA viruses and the cellular forms which all possess both DNA (deoxyribonucleic acid) and RNA, the gene material is either known to be DNA, or assumed to be for good reason.

The genes of viruses and bacteria appear to consist of nucleic acid unaccompanied by closely bound protein. Ordinarily this naked nucleic acid is in the two-stranded condition; exceptions are known among both the RNA and DNA viruses some of which possess single-stranded genetic material. In those organisms with true nuclei, the genetic material is always double-stranded DNA associated with protein ordinarily of the histone type. The function of the protein is not considered to be genetic. It probably controls DNA in its role of determining protein structure. Also it may serve to hold genes together and attached to the chromosomes of which they are a part.

Structural genes carry out their role of dictating protein structure by producing a messenger RNA (mRNA) which is a single strand of RNA containing nucleotide bases complementary to one of the strands of the double-stranded DNA of the gene from which it is copied or "transcribed." The evidence is that the same DNA strand of a gene is always transcribed into mRNA. In this way, only one kind of mRNA is made for each gene. In the transcription process the C, T, A and G bases of the DNA determine G, A, U and C, respectively, in the mRNA strand. Transcription effectively constitutes *gene action.* By definition, if a gene is not actively forming mRNA, it is inactive or "turned off."

Each kind of gene is different from every other gene in its DNA sequence. Hence, as many different kinds of mRNA are formed as there are different genes in the organism.

Genes in eukaryotic cells are often not colinear with their products. Instead genes contain intervening sequences of DNA (*introns*) which result in a gene that is much longer than required for the simple coding of amino acid sequence. An enzymatic reaction, gene splicing, is required for the expression of the genes. That is, the entire gene, including introns, is transcribed as a long mRNA precursor. The intervening sequences are clipped out and the ends rejoined to yield the mRNA with the correct coding sequence for the gene product. After their formation, the mRNA strands attach to ribosomes in the cytoplasm, and the process of protein biosynthesis commences. The significant point to be emphasized here is that the sequence of nucleotide bases, of the "genetic code," in a particular gene is reflected in a specific sequence of amino acids in the polypeptide produced through the protein synthetic mechanism.

The one-to-one relation between gene and polypeptide is a more accurate statement of the situation than the earlier one gene-one enzyme hypothesis. It is now known that a number of proteins are constituted in their functional state of subunits which are polypeptides. When subunits are all identical, the one gene–one protein statement holds with certain exceptions. However, proteins such as vertebrate lactic acid dehydrogenase (LDH) and hemoglobin are known to be made up of different subunits. For example, the dominant adult hemoglobin in man contains both α and β polypeptides as subunits. These have somewhat different amino acid sequences, and each has been shown to be under the control of a different gene. The genes are not even on the same chromosome. A similar situation has been found for LDH which may be made up of at least two different subunits, each one again under the control of a separate gene.

A term which is currently used by many synonymously with structural gene is *cistron*. Its original definition was based on complementation tests. If two chromosomes bearing the same kinds of genes (homologous chromosomes) are introduced into the same cell, "product interactions" may be observed between the genes of the same type, *i.e.*, genes which control the same kind of polypeptide. If two genes of the same type are mutant, but mutant at different sites, they may *complement* and produce a protein which has an activity comparable to the nonmutant even though each mutation alone or together on the same chromosome, can produce only a mutant, inactive protein. Those mutants which do not complement with the production of an active protein are said to have mutational sites within the same cistron.

Mutation of Genes. A change in the base sequence of the DNA constituting a gene results in an inherited alteration in the code and is called a gene mutation. Changes in base sequence may conceivably result from (1) the deletion or addition of one or more nucleotide pairs in the DNA chain, (2) changes in one or more bases along the chain, or (3) inversion of a segment of the chain.

Good evidence for the occurrence of the first type of mutation exists at least in bacteriophage of the T series which infect *Escherichia coli.* The deletion or addition of a single base pair into a DNA chain of a gene should be expected to cause considerable difficulties in the translation of the code in the derived mRNA, into an amino acid sequence. For example, if the mRNA of the nonmutant strain has the sequence:

$$\overline{\text{GCU}}\ \overline{\text{AAU}}\ \overline{\text{GAA}}\ \overline{\text{UUU}}\ \overline{\text{AAA}}\ \overline{\text{CAU}}\ \overline{\cdots}$$

which is read in triplets from left to right to give a particular sequence of amino acids, say ala · asp NH^2 · glu · phe · lys · his, a deletion of a base in the mutant would change the "reading frame" starting at the point of deletion. Thus, the sequence

$$\overline{\text{GCU}}\ \overline{\text{AAU}}\ \overline{\text{G}\underset{\text{A}}{\downarrow}\text{AU}}\ \overline{\text{UUA}}\ \overline{\text{AAC}}\ \overline{\text{AU}\cdots}$$

would produce the sequence ala · asp NH_2 · asp · leu · asp NH_2 as one possibility. A similar result would be expected from a duplication, by shifting the reading frame. Such mutations as these should be expected to produce "nonsense" sequences of amino acids after the point of change and are referred to as frame shift mutations.

Mutations which are the result of the simple changing of bases, say G $\leftrightarrows$ A, C $\leftrightarrows$ T, or G $\leftrightarrows$ C, or A $\leftrightarrows$ T should obviously cause changes in a single triplet rather than a whole sequence. As a result only a single amino acid change should occur in the polypeptide, if but a single base is changed. Many mutant proteins from a variety of organisms are now known which have but a single amino acid change from the nonmutant, and are therefore presumably the result of a single, or adjacent changes within a single triplet. The nonoccurrence of single mutations causing the substitution of two adjacent amino acids within a chain is evidence that the genetic code is not overlapping. As might be expected, a number of different amino acids may be substituted for the nonmutant acid, but the number of substitutions has been found to be limited for any particular amino acid. This also has connotations for the nature of the genetic code.

Gene mutations of other types such as inversions probably occur in addition to the two discussed above, but techniques have yet to be devised to analyze them.

For the present it is enough to say that a mutation may occur at any point within a gene. Theoretically there should be as many "mutational sites" within a gene as there are nucleotide pairs.

Gene instability, the sudden occurrence of high mutability of a normally stable gene, has been described at many different loci in maize, the galactose region of *Escherichia coli,* and in the white locus of *Drosophila.* Studies of the molecular basis for this instability in bacteria have identified transposable elements (*transposons*) as the agents responsible. A transposable element is a segment of DNA capable of transposition intact from one position in the genome to another. In addition to promoting their own transposition, transposable elements can also promote inversion, deletion, and transposition of adjacent chromosomal DNA sequences, resulting in increased occurrence of mutations. Indeed transposable elements have now been found adjacent to many mutant genes in bacteria. Such transposable elements are also thought to occur in eukaryotic cells.

Gene Recombination. When two homologous (*i.e.*, bearing the same kinds of genes) chromosomes are paired in synapsis (as in early meiosis in the nucleated organisms) recombination may occur. Recombination is the exchange, usually equal, of segments of chromosomes. Thus, if a chromosome marked:

A b C D e f g H i J

recombines with one marked:

A b c d E f G h I j

between d and e, the recombinant products will be A b C D E f G h I j and A b c d e f g H i J. This natural process presumably occurs in all organisms both *between* or *within* genes. First, it provides a powerful tool for establishing that the genes are ordered linearly on the chromosome, and in what order, and second, it allows one to establish that there exists a colinearity between the genetic material of a gene and the polypeptide it produces. This has been done by mapping a number of mutational sites for a gene that determines one of the polypeptides (protein A) forming the enzyme tryptophan synthetase in *Escherichia coli.* Each mutant produces a modified protein A which can be shown to differ from the wild type by a single amino acid substitution. When the order of the mutant sites on the *coli* chromosome was compared to the order of amino acids within protein A affected by the mutations, it was found that they were the same. This fundamental finding could only have been possible with the use of a recombination analysis.

Controlling Genes. Genes which do not carry codes for the synthesis of proteins which constitute the enzymes, structural components, etc., of the cell almost certainly exist. These genes may produce proteins, but the proteins presumably act by the regulation of the activity of the structural genes, turning them on and off according to circumstances within the cell.

Examples of such genes are found in *Echerichia coli.* These, termed *regulator genes,* presumably produce substances, possibly proteins, which prevent or repress structural genes from synthesizing mRNA unless other substances, the inducers, are present to inhibit the repressor substances. Alternatively, repressor substances from other types of regulator genes are active in repression only when certain substances activate the repressor substances. The reason for the existence of these genes would seem to be for the regulation of metabolism by preventing the overproduction of enzymes when their substrates are not present, or of end products such as amino acids. In the latter case, the end product is usually considered to be the substance which activates the repressor produced by the regulator.

Genetic Code. Genetic information stored in the genes, as a linear sequence of the bases (A, C, G, and T) in deoxyribonucleic acid molecules, is transcribed into a complementary base sequence (U, G, C, and A, respectively) in the messenger RNA molecules; this "coded message" contained in the mRNA, as a linear sequence or 4-letter "language," is "translated" in the process of protein biosynthesis into a linear sequence of the 20 amino acids within the protein polypeptide chain synthesized. Each nuclotide triplet or "code word" consisting of one of the 64 possible triplet combinations of U, G, C, and A nucleotides in a messenger RNA molecule may specify one particular amino acid for incorporation into the polypeptide chain. It appears that certain amino acids may be specified by more than one of the 64 nucleotide triplets; in this respect, the genetic code is said to be "degenerate." A few particular triplet "words" may have special functions, such as to signal polypeptide-chain initiation, or chain termination. The first identification of a particular triplet as the code word for a particular amino acid was the discovery that the sequence UUU (in the form of polyuridylate) appears to be the "code word" specifying incorporation of phenylalanine into a polypeptide, in a cell-free, *in vitro* system containing ribosoimes and other required components.

Evidence that a nucleotide *triplet* (and not some smaller or larger run of nucleotides) is the "code word" for incorporation of a specific amino acid has come from studies of the fine structure of genes or DNA of a bacteriophage (virus). Many tentative formulations of a "code dictionary" of messenger RNA triplets, with the corresponding amino acid specified by each triplet, have been proposed, on the basis of both experimental results (primarily those of the Nirenberg group and of the Ochoa group) and theoretical considerations. The exact determination of the genetic code, or pattern of correspondence between each possible nucleotide triplet of mRNA and the amino acid specified by that triplet for incorporation into proteins, has been an active field since about 1961.

Biochemical Genetics. Beadle defined biochemical genetics as "the branch of biology which seeks to define hereditary units in terms of the chemistry of their structures and of their functions." It has emerged into a rapidly expanding field through the fusion of many of the facts, concepts, ideas, and techniques of two classically independent branches of science until it encompasses many of the central problems of living systems. The fusion of genetics and biochemistry came about because of the specific pressure to explain biological phenomena on a molecular level, typifying the trend in modern biology commenced some years ago.

The recognition of the relation between genes and enzymes that grew out of the work leading to the "one-gene-one-enzyme" hypothesis stimulated the quest for recognition and elucidation of the biochemical nature of genetic material. By 1940, genetic and cytogenic studies clearly pointed to the chromosomes of higher organisms as the structures within which genes were to be found. In 1936, the microspectrophotometric observations of Casperson had made it clear that chromosomes contained large amounts of nucleic acid and refinements of the Feulgen staining technique made it clear that this nucleic acid was primarily deoxyribonucleic acid (DNA). However, extraction from isolated nuclei and extension of Casperson's microspectrophotometric techniques indicated that the DNA probably existed primarily as a nucleoprotein complex in the living cells. The question, then, was what was the real genetic material? Genes might be made of DNA, protein, or a complex of both as mucleoproteins. The situation was complicated by the lack of knowledge of the chemical structure of DNA. It had been widely assumed during the 1930s that DNA might consist of a repetitive series of tetranucleotides, each of which contained adenylic, thymidylic, guanylic, and cytidylic nucleotides. If this were the case, all DNA would be identical or at least so similar that it was difficult to imagine how it could contain the information required for it to be the genetic material. It thus seemed likely that DNA might function merely as a structural element of chromosomes.

In the early 1940s evidence was presented from several sources on the chemical structure of DNA that made the repetitive tetranucleotide model untenable and made it seem likely that there might be a very large number of different kinds of DNA. The idea that DNA itself might be the genetic material received its greatest support from the work of Avery, MacLeod, and McCarty who succeeded in showing that the transforming factor in pneumococcus was almost certainly DNA. Evidence accumulated from many sources, until, by 1952, it seemed likely that DNA was the primary genetic material for living systems ranging from virus to mammals. In addition, during the same period, evidence was accumulating on the chemical structure of DNA that made it seem increasingly probable that it had the required chemical properties. Chargaff pointed out the basic relation between the purine and pyrimidine bases (adenine equals thymine, and guanine equals cytosine) in DNAs from a wide variety of sources and with very divergent over-all base composition. Watson and Crick used this relation with x-ray diffraction data of the type provided by Wilkins to work out the helical double-stranded structure of DNA in which a highly specific hydrogen bonding occurs between the two chains of the polymer. This specific hydrogen bonding between adenine and thymine or guanine and cytosine accounted for the Chargaff rules of DNA base composition, and provided a structure that seemed to provide the required genetic stability.

As DNA began to be recognized as the primary genetic material, the central problem changed from what to how: how is the information contained in DNA and how is it transmitted into proteins. In 1956, Ingram provided some insight into this problem by demonstrating that sickle-cell hemoglobin isolated from humans carrying the mutant gene for sickle-cell anemia differed from normal hemoglobin by only one amino acid.

Work from many sources led to the recognition that genes control the sequence of amino acids in polypeptides and that most mutations cause specific changes in this amino acid sequence. These changes may affect the properties of the protein formed from the mutant polypeptide, such as enzymatic activity, solubility or heat stability. This led Benzer to refine and restate the "one-gene-one-enzyme" hypothesis as "one-gene-one-polypeptide chain."

The emerging concept held that DNA, the genetic material, contained the information necessary to position specific amino acids into sequences in a related polypeptide and that this information was somehow encoded into the sequence of bases of the DNA. It was soon discovered that a ribonucleic acid (RNA) intermediate, messenger RNA, was involved and that the code consisted of a sequence of three nucleotides for each amino acid. See **Cancer Research; Clone;** and **Recombinant DNA.**

Reviewed and updated for 6th Edition by Ann C. Vickery, Ph.D., University of South Florida, College of Medicine, Tampa, Florida.

References

Arber, W.: "Promotion and Limitation of Genetic Exchange," *Science*, **205**, 361–365 (1979).

Bostock, C. J., and A. T. Sumner: "The Eukaryotic Chromosome," North-Holland, Amsterdam, 1978.

Bregliano, J. C., et al.: "Hybrid Dysgenesis in *Drosophila melanogaster*," *Science*, **207**, 606–611 (1980).

Brill, W. J.: "Agricultural Microbiology," *Sci. Amer.*, **245**, 3, 198–215 (1981).

Caplan, A. I., and C. P. Ordahl: "Irreversible Gene Repression Model for Control of Development," *Science*, **201**, 120–130 (1978).

Carlberg, D. M.: "Essentials of Bacterial and Viral Genetics," Charles C. Thomas, Springfield, Illinois, 1976.

Cohen, S. N., and J. A. Shapiro: "Transposable Genetic Elements," *Sci. Amer.*, **242**, 2, 40–49 (1979).

Crow, J. F.: "Genes That Violate Mendel's Rules," *Sci. Amer.*, **240**, 134–146 (1979).

Davidson, E. H.: "Gene Activity in Early Development," 2nd edition, Academic, New York, 1977.

Davidson, E. H., and R. J. Britten: "Regulation of Gene Expression: Possible Role of Repetitive Sequences," *Science*, **204**, 1052–1059 (1979).

Dawkins, R.: "The Selfish Gene," Oxford Univ. Press, New York, 1976.

De Robertis, E. M., and J. B. Gurdon: "Gene Transplantation and the Analysis of Development," *Sci. Amer.*, **241**, 6, 74–81 (1979).

Dubochet, J., and M. Noll: "Nucleosome Arcs and Helices," *Science*, **202**, 280–286 (1978).

Gateff, E.: "Malignant Neoplasms of Genetic Origin in *Drosophila melanogaster*," *Science*, **200**, 1448–1459 (1978).

Gedda, L., and G. Brenci: "Chronogenetics," Charles C. Thomas, Springfield, Illinois, 1978.

Goodenough, U. W.: "Genetics," 2nd edition, Holt, Rinehart and Winston, New York, 1978.

Hopwood, A.: "The Genetic Programming of Industrial Microorganisms," *Sci. Amer.*, **245**, 3, 90–122 (1981).

Khorana, H. G.: "Total Synthesis of a Gene," *Science*, **203**, 614–625 (1979).

Kolata, G. B.: "Overlapping Genes: More than Anomalies?" *Science*, **196**, 1187–1188 (1977).

Lucchesi, J. C.: "Gene Dosage Compensation and the Evolution of Sex Chromosomes," *Science*, **202**, 711–716 (1978).

Mange, A. P., and E. J. Mange: "Genetics: Human Aspects," Saunders, Philadelphia, 1980.
McKusick, V. A., and F. H. Ruddle: "The Status of the Gene Map of the Human Chromosomes," *Science*, **196**, 390–405 (1977).
Miller, J. H., and W. S. Reznikoff (editors): "Gene Regulation," Cold Spring Harbor Laboratory, Cold Spring Harbor, New York, 1976.
Morton, N. E., and C. S. Chung: "Genetic Epidemiology," Academic, New York, 1978.
Nagl, W.: "Endopolyploidy and Polyteny in Differentiation and Evolution," North-Holland, Amsterdam, 1978.
Nathans, D.: "Restriction Endonucleases, Simian Virus 40, and the New Genetics," *Science*, **206**, 903–909 (1979).
Omenn, G. S.: "Prenatal Diagnosis of Genetic Disorders," *Science*, **200**, 952–958 (1978).
Rosenberg, H., Singer, M., and M. Rosenberg: "Highly Reiterated Sequences of SIMIANSIMIANSIMIANSIMIANSIMIAN," *Science*, **200**, 394–402 (1978).
Schull, W. J., and R. Chakraborty (editors): "Human Genetics," Academic, New York, 1979.
Schwartz, R. M., and M. O. Dayhoff: "Origins of Prokaryotes, Eukarotes, Mitochondria, and Chloroplasts," *Science*, **199**, 395–403 (1978).
Scriver, C. R., et al.: "Genetics and Medicine: An Evolving Relationship," *Science*, **200**, 946–952 (1978).
Staff: "Mendel was No Mendelian," in "Science and the Citizen," *Sci. Amer.*, **241**, 4, 76–78 (1979).
Vogel, F., and A. G. Motulsky: "Human Genetics," Springer-Verlag, New York, 1979.
Wright, S.: "Evolution and the Genetics of Populations," Vol. 3, Univ. Chicago Press, Chicago, Illinois, 1977.
Yunis, J. J. (editor): "New Chromosomal Syndromes," Academic, New York, 1977.
Yunis, J. J. (editor): "Molecular Structure of Human Chromosomes," Academic, New York, 1977.

GENERATOR SETS. Gas and Expansion Turbines.

GENERATOR SPEED CONTROL. Governor.

GENES. Cell (Biology).

GENES (Plant). Plant Breeding.

GENET. Viverrines.

GENETIC CODE. Cell (Biology).

GENETIC CONTROL OF INSECTS. Insecticide and Pesticide Technology.

GENETIC ENGINEERING. This term began appearing in the literature during the late 1970s. The term, as of the early 1980s, suggests the purposeful manipulation (engineering) of life processes, particularly those techniques which arose from the concepts of recombinant DNA, first reported in this book in the 5th edition (1976). This current, 6th edition (1982) follows the less dramatic terms of genetics and microbiology in accordance with its conservative policy of adhering to well-established nomenclature. Admittedly, as of the early 1980s, there is a competitive race among many scientists in this field, the solid results of which can best be reported for the permanent literature a few years hence. One respected publication in the general scientific field groups a number of endeavors under the umbrella of genetic engineering, such as industrial microbiology and industrial microorganisms, including the genetic programming of industrial microorganisms and the microbiological production of foods, beverages, pharmaceuticals, industrial chemicals, and the application of microbiology in agriculture. These topics are described in some detail in *Sci. Amer.*, **245**, 2 (1981) under the overall title of industrial microbiology.

Many topics currently considered by some reporters as genetic engineering, but not so identified, are covered in this encyclopedia. Fundamentals are described in such entries as **Biochemical Individuality; Cell (Biology); Clone; Genetics; Recombinant DNA;** and several others. The application of genetic knowledge to agriculture, for example, is described in such entries as **Insecticide and Pesticide Technology; Nitrogen;** and **Plant Breeding.** Industrial microbiology, although in a new dress, is, in reality, a rather old field and is applied to many of the products, such as foods, pharmaceuticals, and other substances based upon the processing of materials derived from various life forms. During the next five to ten years, genetic engineering, which involves genetic programming, is expected to become an established high technology.

GENETIC RESOURCES (Plant). Plant Breeding.

GENETICS. A branch of biology dealing with heredity; a study which seeks to explain the similarities and differences that exist between parents and their offspring. Gregor Mendel is sometimes called the father of modern genetics. Through crosses of garden peas, Mendel worked out the basic principles of inheritance which apply to all forms of life. The methods of study used by geneticists may be divided roughly into five basic branches. (1) Experimental breeding of plants or animals, which enables geneticists to deduce the methods of inheritance of specific characteristics. Statistical methods are used extensively in analyzing the results obtained. (2) Pedigree collection and analysis are often used in studies of human genetics where experimental breeding is not possible. Pedigree charts may be prepared showing the inheritance of specific traits in all of the members of a family line which can be traced. (3) Cytology (cytogenetics) is a study of the chromosomes and other cellular structures that play a part in heredity. This study has come to include the nature of the gene itself and how it reproduces. Human cytology has made great strides recently with the improved methods of preparing smears of cells to show the chromosomes. (4) Biochemical study of the genes has given geneticists much information on the way in which genes act. How the genes transmit messages to parts of the cytoplasm which, in turn, produce the cell proteins according to the message has been investigated thoroughly. Also, through an analysis of gene action, biochemical geneticists, working with such diverse organisms as molds, bacteria, and human beings, have been able to trace the course of the breakdown of particular amino acids in the cells and to learn of abnormalities which arise when a gene fails to produce a particular enzyme. (5) Population genetics deals with the distribution of genes in various populations. Human population geneticists have traced population migration and the intermixing of races through an analysis of the frequency of the various blood antigens. See also **Cell (Biology).**

For references see list under **Heredity.**

This entry reviewed and updated for 6th Edition by Ann C. Vickery, Ph.D., University of South Florida, College of Medicine, Tampa, Florida.

GENETICS (Biochemical). Biochemical Individuality.

GENETICS (Fishes). Blind-Fish; Fisheries.

GENETICS (Nitrogen Fixation). Nitrogen.

GENEVA RULES (Organic Chemistry). Organic Chemistry.

GENITALIA. Gonads.

GENITAL TRACT DISEASES. Gonorrhea and Gonococcemia; Herpes Simplex Virus Diseases; Lymphogranuloma Venereum; Syphilis.

GENTAMICIN. Antibiotic.

GENOTYPE. The actual gene constitution of an organism as opposed to the phenotype or visible expression arising from those genes. The gene which causes albinism in humans is recessive and is represented by *a*. The dominant allele of this gene is *A*. Since each person is diploid with respect to his or her genes, there are three possible genotypes of this particular gene locus: *AA*, *Aa*, and *aa*. The resultant albinism or normal pigment production would be the phenotype produced. See also **Cell (Biology).**

GENUS. Taxonomy.

GEOCENTRIC COORDINATES (Astronomy). Any system of coordinates on the celestial sphere that uses for its origin, or reference point, the center of the earth. Practically all coordinates published in an ephemeris or almanac are geocentric in character. See also **Celestial Sphere.**

GEOCENTRIC PARALLAX. The origin of the apparent systems of spherical coordinates is a point on the surface of the earth, whereas the origin of the geocentric systems is at the center of the earth. For obvious reasons, all observations must be taken in the apparent system. For the solution of most problems, geocentric coordinates are desired. The transfer from one system to the other is made by applying a correction for geocentric parallax.

In the figure, C is the center of the earth, of radius R, and O is the position of an observer on the surface. OC is the direction of gravity at O; OH the direction of the astronomical horizon; and CH a parallel direction drawn through the center of the earth. S and S' represent two positions of an object at distance ρ from the center of the earth, S being the position when the object is on the horizon. At S'', the object has an apparent altitude h' and a geocentric altitude h. P is defined as the horizontal parallax of the object and is the angle subtended at the object by the radius of the earth. For rigor, the quantity usually defined is the mean equatorial horizontal parallax, which is the angle subtended by an equatorial radius of the earth at the object, when the object is on the horizon, and at its mean or average distance from the earth. The equatorial horizontal parallax is tabulated in Ephemerides for all members of the solar system for selected dates. Inspection of the figure indicates that $\sin P = R/\rho$.

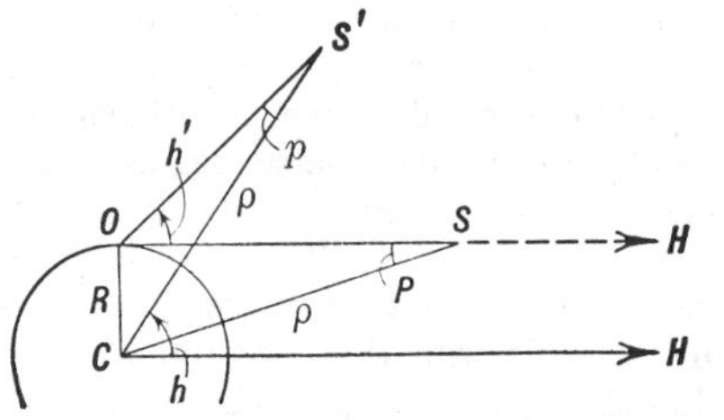

Geocentric parallax in altitude.

The geocentric altitude h is greater than the apparent altitude h' by the angle p, which is defined as the geocentric parallax in altitude. In the oblique plane triangle COS', we have

$$\frac{R}{\rho} = \frac{\sin p}{\cos h'}$$

but we have already seen that $R/\rho = \sin P$, whence,

$$\sin P \cos h' = \sin p$$

Now both P and p are such small angles that, without sensible errors for most problems, except those dealing with the moon, we have $p = P \cos h'$, giving the geocentric parallax in altitude in terms of the equatorial horizontal parallax and the apparent altitude of the object. For objects outside the solar system, the value of P is far too small to be appreciable in even the most refined observations.

If other spherical coordinates than altitude are to be used, the geocentric parallax in altitude may be transformed to the desired quantities by solution of the astronomical triangle or other triangles on the celestial sphere.

See also **Astronomical Triangle; and Celestial Sphere.**

GEOCHEMISTRY. Earth.

GEOCHRONOLOGY. Geologic Time Scale.

GEOCORONA. Atmosphere (Earth).

GEOCRONITE. A mineral sulfide of lead, antimony and arsenic, Pb_5SbAsS_8. Crystallizes in the monoclinic system. Hardness, 2.5; specific gravity, 6.4±; color, gray to blue with metallic luster; opaque.

GEODE. A hollow concretion or nodule whose inside walls are lined with crystals, commonly of quartz or calcite.

GEODESIC. That curve on a surface connecting two fixed points which has an extreme length (maximal or minimal). In three-dimensional Euclidean geometry, the geodesic is clearly a straight line; if the path is constrained to the two-dimensional surface of a sphere, it is a segment of a great circle. In the non-Euclidean geometries appropriate to the general relativity theory, the geodesic is the path followed by a particle upon which no electromagnetic forces act. Such a path is the straightest path in a four-dimensional space-time continuum.

Geodesic Parallels on a Surface. Consider a singly infinite family of geodesics. The singly infinite family of curves on the surface which cut these orthogonally at each point are called geodesic parallels. The distance between two geodesic parallels measured on the surface along any geodesic of the family is the same and is called the *geodesic distance* between the two geodesic parallels.

GEODESY. Earth.

GEODETIC ASTRONOMY. Earth.

GEODETIC COORDINATES. Quantities which define the position of a point on the spheroid of reference with respect to the planes of the geodetic equator and of a reference meridian.

GEODETIC DATUM. A datum consisting of five quantities, the latitude, longitude and elevation above the reference spheroid of an initial point, a line from this point, and two constants which define the reference spheroid. Azimuth or orientation of the line, given the longitude, is determined by astronomic observations. Alternatively, the datum may be considered as three rectangular coordinates fixing the origin of a coordinate system whose orientation is determined by the fixed stars, and the reference speheroid is an arbitrary coordinate surface of an orbiting ellipsoidal coordinate system.

GEODIMETER. An electronic-optical device that measures ground distances precisely by electronic timing and phase comparison of modulated light waves that travel from a master unit to a reflector and return to a light-sensitive detector where an electric current is established. It is frequently used at night and is effective with first-order accuracy up to a distance of 5–40 kilometers (3–25 miles). The ultimate precision of the geodimeter over that of the telluorometer is roughly by a factor of 3. A *tellurometer* is a rugged, lightweight portable electroic device that measures ground distances by determining the velocity of a phase-modulated, continuous microwave radio signal transmitted between two instruments operating alternately as a master station and a remote station. The instrument has a range up to 65 kilometers (35–40 miles).

GEODUCK (*Mollusca, Lamellibranchiata*). A giant clam found on the Pacific coast of North America. It attains a weight of more than 6 pounds (2.7 kilograms) and is edible.

GEOGALE. Moles and Shrews.

GEOGRAPHIC COORDINATES. Geographic coordinates provide a method for determining the position of a point on the surface of the earth by means of a system of spherical coordinates. Because of the fact that the earth is not a sphere, but is, in reality, an oblate spheroid, technically, the system of coordinates cannot be strictly spherical. The geographic method of representation of the position of points on a spherical earth by means of latitude and longitude was first applied by Ptolemy in the construction of his atlas of the world during the second century of the Christian era. Also called *geographical coordinates* or *terrestrial coordinates.*

GEOGRAPHIC HORIZON. Horizon (Geographic).

GEOGRAPHOS. Asteroid.

GEOGRAPHY. Literally, the study, description and mapping of the surface phenomena of the earth or other planets without, necessarily, a consideration of the origin of the phenomena. See **Earth.**

GEOID. The particular geopotential surface that most nearly coincides with the mean level of the oceans of the earth. For mapping purposes, it is customary to use an ellipsoid of revolution as an adequate and convenient approximation to the geoid. The dimensions and orientation of the assumed ellipsoid may represent an attempt to find the ellipsoid that most nearly fits the geoid as a whole, or they may represent an attempt to fit only a particular part of the geoid without regard to the remainder of it. When mention is made of the dimensions of the earth, the reference is usually to the dimensions of the ellipsoid most nearly representing the geoid as a whole. See **Earth.**

GEOLOGIC DATA. **Earth Resources Satellites and Geologic Remote Sensors.**

GEOLOGIC TIME SCALE. The geologic time scale (Table 1) combines the traditional classical time classification of rocks long used by geologists with numerical time boundaries based on radiometric age measurements.

This time scale is sometimes referred to as the "absolute" time scale, but more appropriate are the terms "geochronologic" or "radiometric" time scale to distinguish it from the older relative time classification. In the latter, the principles of stratigraphic and faunal successions have played predominant roles, and three major time divisions, the Cenozoic, Mesozoic, and Paleozoic eras, were named for recent, middle, and ancient life, respectively. This classification with subdivisions into periods and epochs was well worked out prior to 1850. See also **Radioactivity and Other Dating Techniques.**

Early speculations about the time rates of geologic processes led to attempts to place limits in years on subdivisions of geologic time as well as on the age of the earth. Various methods were tried to calculate intervals of geologic time based on the rates of deposition, erosion, development of life, accumulation of salt in the ocean, and so forth. In all these calculations various assumptions had to be made, commonly on the most meager information. It is not surprising, therefore, that time intervals calculated by different individuals varied greatly.

Many prominent geologists, physicists, and chemists have concerned themselves with the problem of measuring geologic time with the objective of attaining some quantitative values for the intervals represented in the geological classification. One of the foremost is Arthur Holmes whose papers on the subject span half a century (1911–1960). His outstanding work in this field has won him wide recognition. In 1956 Holmes was awarded the Penrose Medal of the Geological Society of America, and in 1964 he shared the prize of the G. Unger Vetlesen Foundation at Columbia University for his contributions to the geologic time scale. A special volume, "The Phanerozoic Time Scale," was dedicated to Holmes by the Geological Society of London, and Table 1 is adapted from a time scale that resulted from the Holmes Symposium.

Phanerozoic Time Scale. In 1913, in a small volume entitled, "The Age of the Earth," Holmes outlined how age determinations based on the principles of radioactive decay, in conjunction with geological data on the maximum known thicknesses of rocks assigned to the various geological periods, might be used to construct a quantitative time scale. The ratios of the daughter products, helium and lead, to the parent uranium, were used to calculate these early radioactivity ages. This approach was used by Jospeph Barrell in a monumental paper "Rhythms and the Measurements of Geologic Time" in 1917, in which he presented a time scale (Table 2). Holmes' first extended time scale for the Phanerozoic was published in 1933 (Table 2). European scientists who made important contributions up to this time include, among many others, such illustrious names as Charles Lyell, Charles Darwin, Archibald Geikie, Lord Kelvin, Lord Rayleigh, Lord Rutherford, W. J. Sollas, John Joly, and A. de Lapparent. In addition to Barrell, Americans who made significant contributions were chemists such as B. B. Boltwood and F. W. Clarke, and geologists including G. F. Becker, J. D. Dana, G. K. Gilbert, Charles Schuchert, and C. D. Walcott.

The Barrell and Holmes time scales were based on U-Pb age calculations based on chemical determinations. In 1933, F. W. Aston showed by mass spectrographic analyses that lead is composed of a number of isotopes whose abundance ratios he determined in several samples of common lead. This work was followed in 1939–1941 by papers by A. O. Nier. HIs U-Pb age calculations based on mass spectrometric measurements heralded modern geochronology. The precise isotopic measurements of Aston and Nier provided Holmes with the information he needed for his 1947 geologic time scale (Table 2).

TABLE 1. PHANEROZOIC TIME SCALE

ERA	PERIOD	EPOCH	BEGINNING OF INTERVAL (MILLION YEARS)
	QUATERNARY	PLEISTOCENE	1.5-2
CENOZOIC	TERTIARY	PLIOCENE	7
		MIOCENE	26
		OLIGOCENE	37-38
		EOCENE	53-54
		PALEOCENE	65
MESOZOIC	CRETACEOUS	Upper	100
		Lower	136
	JURASSIC	Upper	162
		MIDDLE	172
		Lower	190-195
	TRIASSIC	Upper	205
		Middle	215
		Lower	225
PALEOZOIC	PERMIAN	Upper	240
		Lower	280
	CARBONIFEROUS: PENNSYLVANIAN		325
	CARBONIFEROUS: MISSISSIPPIAN		345
	DEVONIAN	Upper	359
		Middle	370
		Lower	395
	SILURIAN		430-440
	ORDOVICIAN	Upper	445
		Lower	500
	CAMBRIAN	Upper	515
		MIDDLE	540
		Lower	570
	PRECAMBRIAN		

TABLE 2. VERSIONS OF THE POST-PRECAMBRIAN TIME SCALE (Millions of Years)

Geologic Division	Barrell (1917)	Holmes (1933)	Holmes (1947)	U.S.S.R. (1960)	Kulp (1961)
Pleistocene	1–1.5	1	1	—	1
Pliocene	7–9	15	12–15	10	13
Miocene	19–23	32	26–32	25	25
Oligocene	35–39	42	38–47	—	36
Eocene	55–65	60	58–68	70[a]	63[a]
Cretaceous	120–150	128	127–140	140	135
Jurassic	155–195	158	152–167	185	181
Triassic	190–240	192	182–196	225	230
Permian	215–280	220	203–220	270	280
Carboniferous	300–370	285	255–275	320	345
Devonian	350–420	350	313–318	400	405
Silurian	390–460	375	350	420	425
Ordovician	480–590	440	430	480	500
Cambrian	550–700	510	510	570	600

Paleocene.

Since World War II, great progress has been made in geochronology with the introduction of the K-Ar and Rb-Sr techniques for age determinations and with the application of U-Th-Pb isotopic age determinations to minerals such as zircon.

New data from a number of geochronology laboratories were used in Kulp's (1961) time scale. Table 2 also includes the geochronologic scale compiled by the Commission on Absolute-Age Determination of Geologic Formations of the U.S.S.R. Academy of Sciences.

Precambrian Time Scale. Although geologists surmised that a great deal of time is represented in the Precambrian rocks, the immensity of this interval was not fully comprehended or accepted until isotopic age determinations became available. The age of the earth is now commonly taken as 4.55 billion (10^9) years, the age obtained in 1956 by C. C. Patterson by comparing the abundance ratios of lead isotopes for meteorites with terrestrial lead. Many isotopic ages in the range from 2,500 to 3,600 million years have been determined on mineral and rock samples from the Precambrian shield areas of the Americas, Africa, Australia, and Eurasia.

The lack of fossils in the Precambrian rocks and the metamorphic changes which they have undergone have made extremely difficult the task of deciphering the stratigraphic succession. Locally, as for example, in the Lake Superior region, the succession and a classification of Precambrian rocks have been worked out, but there is no universally accepted classification. Similarly, the radiometric time scale for the Precambrian succession is still in an elementary form compared to that for the Phanerozoic. The metamorphic processes which have affected in varying degree the minerals of the Precambrian rocks also affected the parent-daughter nuclide ratios. Isotopic ages, therefore, are difficult to interpret in areas of complex metamorphic history and reflect metamorphic events rather than the time of first emplacement or first crystallization. Most of the progress that has been made in Precambrian geochronology has come through the dating of major periods of orogeny. See Table 3.

The development of an ordered stratigraphic succession with the use of fossils to correlate rocks in widely separated areas is one of the remarkable achievements of the geological profession. Paleontological and stratigraphic methods remain the most applicable and reliable for correlation in Phanerozoic rocks. Isotopic age measurements now provide a long-needed method for deciphering the succession of Precambrian rocks. In addition, radioactivity age measurements make possible a quantitative approach to the study of geologic history and processes.

TABLE 3. PRECAMBRIAN TIME SCALES

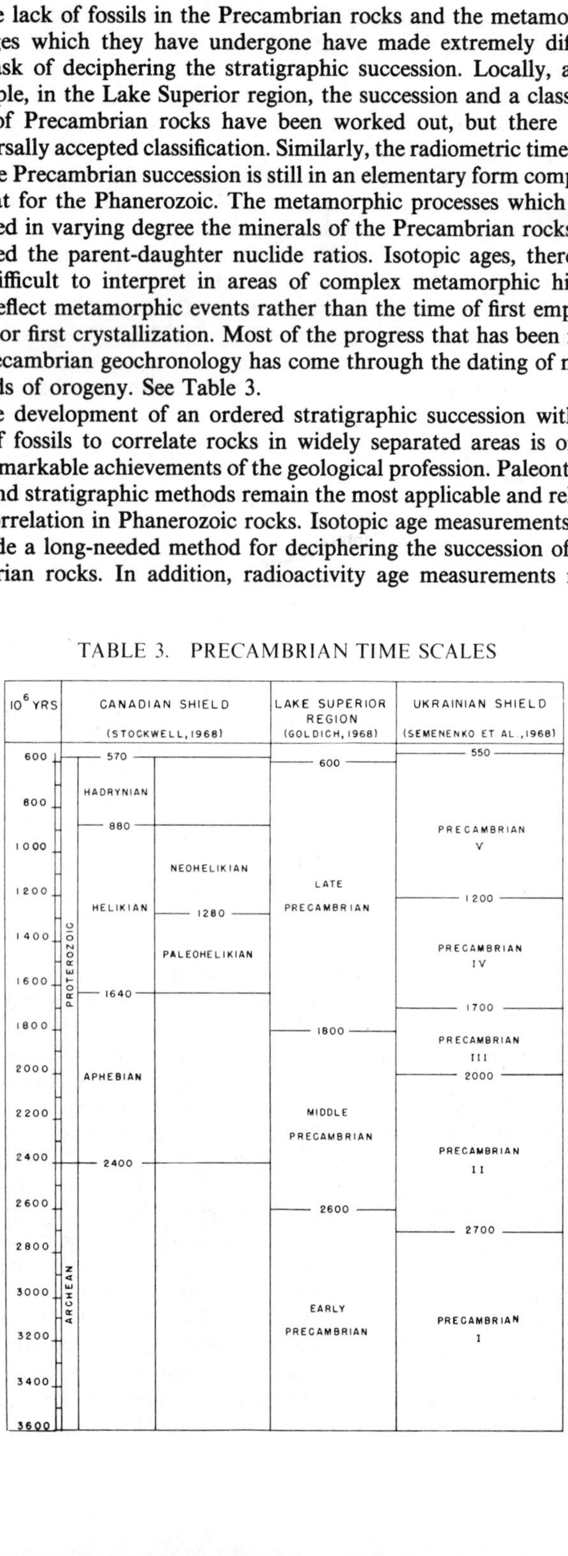

Terms. Some terms used in the field of *geochronology* (the study of the earth and other components of the cosmos with relation to the passage of time) include time periods which are designated by various terms (units) which have a relationship with each other, but which are not of precise or consistent span, as say the units of time, temperature, pressure, etc. used in other measurements. This relationship is shown in a relative way by the following:

Eon—A very large part or grand division of geologic time; the longest of the geologic time units. Sometimes defined as one billion (10^9) years.

Era—An era includes two or more *periods*, during which rocks of the corresponding erathem were formed.

Period—A subdivision of an era, during which the rocks of the corresponding system were formed.

Epoch—A subdivision of a period, during which the rocks of the corresponding series were formed.

Age—A geologic time-unit shorter than an epoch and longer than a subage, during which the rocks of the corresponding stage were formed.

Subage—A rarely used term, shorter than age, during which the rocks of the corresponding substage were formed.

References

Barrell, J.: "Rhythms and the Measurements of Geologic Time," *Geol. Soc. Amer. Bull*, **28**, 745–904 (1917).

Kulp, J. L.: "Geologic Time Scale," *Science*, **133**, 1105–1114 (1961).

McElhinny, M. W. (editor): "The Earth," Academic, New York, 1979.

Press, F., and R. Siever: "Earth," 2nd edition, Freeman, San Francisco, 1979.

Staff: "The Phanerozoic Time-Scale," Symposium of Geological Society of London, *Quart. J. Geol. Soc. London*, **120s** (1964).

Stockwell, C. H.: "Geochronology of Stratified Rocks of the Canadian Shield," *Can. J. Earth Sci.*, **5**, 693–698 (1968).

Wetherill, G. W., and C. L. Drake: "The Earth and Planetary Sciences," *Science*, **209**, 96–104 (1980).

GEOLOGY. As defined by the American Geological Institute in its excellent "Glossary of Geology," geology is . . . "The study of the planet Earth. It is concerned with the origin of the planet, the material and morphology of the Earth, and its history and the processes that acted (and act) upon it to affect its historic and present forms. In the pursuit of that knowledge, the science considers the physical forces that influenced, and continue to influence, change; the chemistry of its consituent materials; the record and age of its past as revealed by the organic remains that are preserved in the layers of its crust or by interpretation of relic morphology and environment. Clues to the origin of the Earth are sought through the study of extraterrestrial bodies and their atmospheres that may reflect an earlier stage of this planet, or whose history may share the events and forces that created the Earth. All of the knowledge obtained through the study of the planet is placed at the service of man, to discover useful materials within the Earth; to identify stable environments for the support of his constructed arts and utilities; and to provide him with a foreknowledge of dangers associated with the mobile being."

There are several hundred entries relating directly or indirectly to geology in this encyclopedia. Some of the more important entries are:

Ataxic
Atoll
Attrition (Geology)
Augen-Gneiss
Aureole (Geology)
Authigenous
Autochthonous
Autoclastic
Avalanche (Geology)
Avulsion

Bad Lands
Baraboo
Barchan
Barranco
Barrier Beach
Barrier Reef
Basal Conglomerate
Basalt
Base Level
Basic Rock
Basin
Batholith
Bayou
Bedding
Bedrock
Bentonite
Berm (Beach)
Berm (Geology)
Binary Granite
Bioherm
Biostrome
Bitumen
Blowhole
Bog
Bog Lake
Bole
Bolson
Bore (Tidal)
Boring (Soil)
Boss
Bostonite
Boulder
Boulder Clay
Breccia
Bronze Age
Buhrstone
Bysmalith

Caldera
Caliche (Nitrate)
Caliche (Soil)
Cambrian Period
Camptonite
Canadian
Canga
Canyon
Capillary Fringe
Carbuncle (Geology)
Cataclastic
Cataclysm
Catawberite
Cauldron-Subsidence
Cave
Cementation
Cenophytic
Cenozoic
Charnockite
Chattermark
Chert
Chimney Rock
Chlorite Schist
Chonolith
Cienega
Cipolin
Cirque
Clarain
Classic
Clastic Rock
Cleavage (Geology)
Coal Balls
Cokite
Columnar Structure
Comagnetic
Concretion
Cone-in-Cone Structure
Confluent
Conglomerate
Connate Water
Consequent Streams
Copper Age
Coral Reef
Corrasion
Consanguinity
Country Rock
Creep (Geology)
Cretaceous Period
Crevasse
Cross-Bedding
Cryology
Cryptocrystalline
Crystalloblastic
Cumulose
Cupola Structure
Cuspate Foreland

Dacite
Deflation
Delta
Dermolith
Detritus
Deuteric
Devitrification
Devonian
Devonite
Diagenesis
Diastem
Diastrophism
Diatom
Diatreme
Differentiation (Geology)
Digitate Drainage
Dike
Diluvium
Diorite
Dolerite
Dome
Domite
Drag Fold
Drainage Systems
Dreikanter
Drewite
Drift (Geology)
Drumlin
Druse
Dune
Durain
Dyas

Earth
Earth Current
Earthquakes, Seismology, and Plate Tectonics
Earth Resources Satellites and Geologic Remote Sensors
Earth Tide
Eclogite
Effusive
Eluvium
Endogenic
Endomorphism
Entrenched Meander
Eocene
Eolian Deposits
Eophytic
Epeirogeny
Epidiorite
Epigenetic
Erosion (Geology)
Erratic
Eruptive
Escarpment
Esker
Estuary
Eutaxic
Exogenetic
Extrusive

Fabric (Geology)
Facies
False Cleavage
Fanglomerate
Felsite
Femic
Finger Lake
Fiord
Fireclay
Flattening of the Earth
Flinty Crush-Rock
Flood Plain
Flowage (Rock)
Flow Structure
Fluvial
Formation (Geology)
Formative
Fulgurite
Fumarole

Gabbro
Ganister Rock
Gap
Geodimeter
Geologic Time Scale
Geomorphology
Geoscience
Geosyncline
Geotectocline
Geyser
Geyserite
Glacial Deposits
Glacier
Glomerate
Gneiss
Gossan
Gouge (Geology)
Graded Bedding
Gradient (Geology)
Granite
Granitoid
Granulite
Graywacke
Greenstone
Greisen
Grit
Ground Moraine
Groundwater

Hälleflinta
Hanging Valley
Hardpan
Hogback
Holocrystalline
Homecline
Hoodoo
Hornblendite
Hornfels
Hydatogenesis
Hydrology
Hydrosphere
Hypabyssal
Hypogene
Igneous Rock
Imbricate Structure
Incompetent
Influent
Iron Age
Isocline (Geology)
Isogeotherm
Isogonic Line
Isomagnetic
Isopach

Jaspilite
Joint (Geology)
Jurassic Period

Kame

Laterite
Lava
Limnology
Liquation
Lithic Stage
Lithification
Lithology
Loess

Maar
Magma
Mesa
Mesozoic
Metamorphic Rocks
Mississippian Period
Mountain
Muskeg

New World

Obsidian Volcanic Glass
Old World
Oligocene
Ophitic Texture
Ordovician
Orogeny
Outcrop
Overfold
Overlap
Overthrust
Ox Bow Lake

Paleocene
Paleontology
Paleophytic
Paleozoic
Palynology
Patina
Pedology
Pegmatite
Pele's Hair
Peneplain
Pennsylvania Period
Perched
Perlite
Permian Period
Petrogenesis
Petrology
Phenoclast
Phenocryst
Phreatic
Physiography
Pillow Lava
Pino
Plateau
Playa
Pliocene
Pneumatolysis
Polar Research
Porphyry
Post-Classic
Pot Hole
Proterozoic
Pumice
Pyroclastic

Quaquaversal
Quartzite
Quartz Porphyry
Qaternary
Quicksand

Rhyolite
Ripple Mark
Roches Moutonnées
Rock
Rock Flour

Saltation
Sandstone
Sandstone Dike
Savannah
Schist
Scoria
Septarian Structure
Seriate Fabric
Shale
Silification
Sill
Siltstone
Silurian Period
Skarn
Slate
Soil
Stack
Stalactite
Steppe
Steptoe
Stone Age
Strath
Sratigraphy
Striation
Stromatolith
Structural Geology
Stylolite
Superimposed River Valley
Superposition (Law of)
Swamp
Syenite
Syncline

Tachylyte
Taconite
Talus
Tephra
Terminal Moraine
Tertiary
Thermal Gradient
Till
Tiltmeter
Time-Stratigraphic Unit
Tombolo
Trachyte
Trap
Trellis Drainage
Triassic Period
Tsunami
Tuff
Tundra

Ultrabasic
Unconformity (Geological)

Variolite
Vein (Geology)
Ventifact
Volcano

Water Resources
Watershed
Weathering
Whinstone

Xenocryst
Xenolith
Xenomorphic
Xenotopic

Yardang
Yazoo Stream
Yosemite
Youth (Geological)

Zigzag Ridge
Zingg Pebble Classification
Zone

For the geological aspects of energy, see also **Coal; Fuel; Geothermal Energy; Hydroelectric Power; Natural Gas; Oil Shale; Petroleum; Tar Sands;** and **Tidal Energy.**

GEOLOGY (Coal). Coal.

GEOLOGY (Geothermal Energy). Geothermal Energy.

GEOLOGY (Lunar). Moon (The).

GEOLOGY (Natural Gas). Natural Gas.

GEOLOGY (Petroleum). Petroleum.

GEOLOGY (Planetary). See specific planets.

GEOLOGY (Sensors). Earth Resources Satellites and Geologic Remote Sensors.

GEOLOGY (Structural). **Structural Geology.**

GEOMAGNETIC DISTURBANCE. **Aurora and Airglow.**

GEOMAGNETIC EQUATOR. The terrestrial great circle everywhere 90° from the geomagnetic poles. Geomagnetic equator should not be confused with magnetic equator, the line connecting all points of zero magnetic dip. See also **Aclinic Line.**

GEOMAGNETIC LATITUDE. Angular distance from the geomagnetic equator, measured northward or southward through 90° and labeled N or S to indicate the direction of measurement. Geomagnetic latitude should not be confused with magnetic latitude, the magnetic dip. Phenomena closely related to the earth's magnetic field are often plotted according to geomagnetic latitude rather than geographic latitude.

GEOMAGNETIC POLE. Either of two antipodal points marking the intersection of the earth's surface with the extended axis of a dipole assumed to be located at the center of the earth and approximating the source of the actual magnetic field of the earth. That pole in the Northern Hemisphere (latitude, 78½°N; longitude, 69°W) is designated north geomagnetic pole, and that pole in the Southern Hemisphere (latitude, 78½°S, longitude, 111°E) is designated south geomagnetic pole. The great circle midway between these poles is called geomagnetic equator. The expression geomagnetic pole should not be confused with magnetic pole, which relates to the actual magnetic field of the earth.

GEOMAGNETISM. **Earth.**

GEOMETRICAL ACOUSTICS. **Acoustics.**

GEOMETRICAL OPTICS. This branch of physics treats light as if it were actually composed of "rays" diverging in various directions from the source and abruptly bent by refraction or turned back by reflection into paths determined by well-known laws. The idea that light travels in straight lines is here uppermost, while its wave character and other physical aspects are disregarded. Thus the image of a point *A*, if "real," is simply another point *B* through which the rays diverging from *A* ultimately pass after the several reflections or refractions produced by the mirrors, lenses, etc., of the optical system. If the image *B* is "virtual," the rays appear to be diverging from it, but only because their direction has been so changed that if produced backward the lines along which they now travel would intersect at *B*. A real image of a lamp may easily be formed by a reading glass; a virtual image, by a plane mirror.

The chief advantage of this mode of visualizing the behavior of light is the simplicity with which problems may be solved by geometrical constructions. The same formulae which are deduced by the methods of geometrical optics may be arrived at, but often with much more labor, by treating light as composed of waves and studying the changes of wave front.

GEOMETRICAL SYMMETRIES. **Conservation Laws and Symmetry.**

GEOMETRIC DISTORTION. Any aberration which causes the reproduced image to be geometrically dissimilar to the perspective plane-projection of the object.

GEOMETRIC MEAN. **Average.**

GEOMETRIC PROGRESSION. **Progression.**

GEOMETRIC VARIABLES. **Variable (Process).**

GEOMETRY. A comprehensive branch of mathematics which is concerned with the properties, measurement, and relations between lines, angles, surfaces, and solids. Classically, the methods of Euclid, who probably lived from 330–275 B.C., were used. These were based on a number of definitions, five postulates, and nine general axioms. The definitions, which were accepted without proof, included statements on point, line, solid, proposition, hypothesis, theorem, etc. The axioms were also accepted without proof, the following being typical: if equals are added to or subtracted from equals the sums or remainders are equal; the whole is equal to the sum of all its parts and greater than any of its parts. Among the postulates, a construction admitted to be possible, typical examples are: a straight line can be drawn from one point to another and can be produced indefinitely; a circumference can be described from any point as a center and with any given radius.

Plane geometry is mostly the study of angles, triangles, polygons, circles, and other figures which can be drawn with ruler and compass; solid geometry involves figures in three dimensions, such as planes, spheres, cubes, polyhedra. Trigonometry is a specialized geometry of the triangle.

Until the nineteenth century, the geometry of Euclid was unquestioned, even though mathematicians had always been unable to prove his fifth postulate. In its classical statement, this postulate takes the form: "If a straight line falling on two straight lines makes the interior angles on the same side less than two right angles, the two straight lines, if produced indefinitely, meet on that side on which are the angles less than two right angles."

An equivalent, and shorter, statement of this postulate is: "Through a point outside a line only one line can be drawn parallel to the given line."

In the nineteenth century, the conclusion was reached that a logical system of geometry could be constructed without use of the fifth postulate and consistent systems were constructed denying it. In one of these it was assumed that through any point there are two or more parallel lines which do not intersect a given line in the plane. This system, which was developed by C. F. Gauss (1777–1855), Wolfgang and John Balyai, and N. I. Lobachevsky (1793–1856) was named *Lobachevskian* or (later) *hyperbolic geometry*. It leads to the conclusion that the sum of the three angles in a triangle is less than two right angles. This system uses, directly or indirectly, all of Euclid's axioms and postulates except the fifth postulate, and all of his theorems which do not depend upon it.

Another non-Euclidean geometry was developed by G. Riemann (1826–1866) and was named *Riemannian* or (later) *elliptic geometry*. It replaced Euclid's fifth postulate by one denying the existence of *any* parallel lines. Therefore, unlike hyperbolic geometry, it rejected a number of Euclid's first 28 propositions, such as the sixteenth and its consequences. (The sixteenth proposition of Book I of Euclid is stated as: "In any triangle, if one of the sides is produced, the exterior angle is greater than either of the interior and opposite angles.") Riemann rejected the infinitude of the line.

In 1872 at the University of Erlangen, Felix Klein presented the so-called Erlangen Programm embodying a definition of geometry which would embrace, as subgeometries of projective geometry, the various non-Euclidean geometries as well as Euclidean geometry itself. This definition of Klein's was as follows:

"A geometry is the study of those properties of a set *S* which remain invariant when the elements of *S* are subjected to the transformations of some transformation group." A transformation *T* of a set *A* onto a set *B* is defined as a one-to-one correspondence between the elements of *A* and those of *B*.

To formulate a geometry by this definition, we need only choose a fundamental element (e.g., a point, line or circle), a set (or space) *S* of these elements (e.g., a plane or a spherical surface of points, a plane of lines, or a pencil of circles), and a group of transformations to which the fundamental elements are to be subjected. Then the definitions and theorems of the geometry consist of the properties invariant under the group of transformations.

A further development that followed the discovery of non-Euclidean geometry was the formulation of precise sets of axioms for Euclidean and non-Euclidean geometries. Axiom sets for the former include those of Pasch (1882), Peano (1889), Hilbert (1899), Veblen (1904), Forder (1927), Birkhoff (1932), Robinson (1940) and Levi (1960). Their common purpose was to place the entire structure of Euclidean geometry upon the simplest possible foundation, that is, to choose a minimum number of undefined elements and relations, and a set of axioms concerning them, with the property that all of Euclidean geometry can be deduced logically from them without any further appeal to intuition.

There are several specialized branches of classical geometry or mathematical techniques related to it. *Analytical geometry*, developed by the French mathematician and philosopher, Rene Descartes (1596–1650) and Pierre Fermat (1601–1665), another Frenchman, is an application of algebraic results in geometry. In two dimensions, considerable attention is given to conic sections; in three dimensions, to the quadric surfaces. It is also called *coordinate geometry.*

Descriptive and *projective geometries* developed from the interest of both painters and mathematicians in the problem of describing three-dimensional figures on a plane. Prior to the time of the Renaissance artists in general had been satisfied with symbolic representations of persons and objects but subsequently they became increasingly desirous of greater realism in their work, Albrecht Dürer, the German painter and engraver (1471–1528), is thought by some mathematical historians to be the inventor of descriptive geometry. Somewhat later, the French mathematician Gaspard Monge (1746–1818) placed the subject on a firm mathematical basis. As indicated before, it is the graphical description of objects in three dimensions and the mathematical technique of mechanical drawing.

A more generalized development of geometric figures is *projective geometry.* Its founders include Gaspard Desargues, a French engineer (1593–1662); Blaise Pascal, French geometer and philosopher (1623–1662); Jean Victor Poncelet, French mathematician and general in the armies of Napoleon (1788–1867). Although projective geometry, like descriptive geometry, uses projection and section it is different from the latter. One of its important objects is the study of properties invariant under projection and section.

Differential geometry is essentially the application of differential and integral calculus to the study of curves and surfaces. Methods of tensor calculus are frequently used and it is the chief mathematical apparatus of relativity theory. Its developers include Gauss and Riemann.

Still other geometries follow readily from Klein's definition given earlier in this entry. Let us take a more specific statement of that definition. "Let S be a set of points P such that a unique point P corresponds to an ordered pair of real numbers (x, y) and let S' be a set of points $P'(x', y')$. Let a one-to-one correspondence between points P and P' be given by the transformations $x' = f(x, y)$; $y' = g(x, y)$. If a set of these transformations form a group, then the properties of the associated geometry are determined by the functions f and g. For example, if f and g are linear functions, we have *affine geometry, similarity geometry*, etc., depending upon the particular group of linear functions formed. If the group of transformations consists of the rational functions f and g such that the inverse transformations are also rational and if those functions are continuous, the associated geometry is called *algebraic geometry.*

Topology is a study of one-to-one bicontinuous transformations. The usual description of topology as rubber-steel geometry is sufficiently suggestive for two-dimensional space only, but the topological spaces of most importance in applied (or pure) mathematics have an infinite number of dimensions. See also **Linear Topological Space** and **Topology.** Numerous geometrical objects, shapes, and bodies are described in detail in this volume. Consult the terms listed under **Mathematics.**

GEOMORPHOLOGY. This term has replaced the earlier term physiography to denote the full scientific interpretation of the origin of topographic features, or the purely physical attributes of scenery. This relatively distinct department of the earth sciences includes the study of the origin of all topographic features in terms of process or processes of erosion and in their relation to geologic structure.

GEOPHYSICS. Earth.

GEOPOTENTIAL. The potential energy of a unit mass relative to sea level, numerically equal to the work that would be done in lifting the unit mass from sea level to the height at which the mass is located; commonly expressed in terms of dynamic height or geopotential height. Unit geopotential is equal to the potential of unit mass lifted a unit distance in a force field of unit strength. Distance upward can be measured in terms of differences in geopotential.

The geopotential Φ at height z is given mathematically by the expression

$$\Phi = \int_0^z g\,dz$$

where g is the acceleration of gravity. Distance upward can be measured in terms of differences in geopotential.

GEOPOTENTIAL HEIGHT. Atmosphere (Earth).

GEOPOTENTIAL SURFACE. Atmosphere (Earth).

GEOPRESSURE ENERGY. Petroleum.

GEOSCIENCE. The collective disciplines of the geological sciences and thus the term is synonymous with geology.

GEOSECS (Project). Ocean.

GEOSPIZINAE. Fringillidae.

GEOSTROPHIC FLOW. Gradient Flow.

GEOSTROPHIC WIND. Winds and Air Movement.

GEOSYNCLINE. Dana's definition of a geosyncline (1873) is a depression which has been produced by lateral compression and which is filled with sediments. Although Dana, in his original definition, suggested that subordinate ridges might be formed in the bottom of the geosyncline during its formation, it remained for Emile Haug, in his "*Traité de Geologie*," to emphasize these ridges (geanticlines) in relation to the tectonics of the Alps. According to L. W. Collet, "A geosyncline is situated between two continental masses and is destined to be filled with sediments, some of which are derived from the geanticlines which develop in it." According to R. M. Field, "A geosyncline originates in a continental block as a great trough, the locus for the accumulation of marine and terrestrial sediments, which are derived from concomitant geanticlines formed in or on the margins of the geosyncline." The geophysical and geological study of the great island arcs, such as the East Indies and West Indies, strongly intimates that the foredeeps in front of the arcs represent geosynclines which have not been filled with sediments while they were being formed. The pronounced deficiency of gravity associated with these foredeeps suggests great down buckle of the crustal or continental type of rocks called Sial into the more basic subcrustal couch called Sima. (See **Earthquakes, Seismology, and Plate Tectonics.**)

GEOTECTOCLINE. Term proposed by H. H. Hess and R. M. Field (1938) for the deformed prism of sediments in (of) the geosyncline. (See **Earthquakes, Seismology, and Plate Tectonics.**)

GEOTHERMAL ENERGY. In the usual sense, geothermal energy is regarded as useful energy that can be extracted from naturally occurring steam and hot water found in the volcanic and young orogenic zones of the earth. Surface manifestations include hot springs, fumaroles, steam vents, and geysers. Such regions may exist without surface manifestation and astute geologists can forecast with some reliability where test bores may be made. Frequently these areas will be found close or relatively close to those areas where natural manifestations are present.

Until the last decade, geothermal energy sources were considered almost exclusively in terms of the kinds of natural phenomena just mentioned. These are the geothermal resources, such as Larderello (Italy), Wairakei (New Zealand), Geysers (California), and Reykjavik (Iceland), which have been successfully exploited for a number of years and whose outputs generally have been expanded in recent years. These regions are characterized by a unique combination of geologic and hydrologic features which brings a supply of water close to rock magma and which is capable of generating very large quantities of steam, hot water, or both. The specific characteristics of any given source range rather widely and thus generalizations are difficult to make.

These natural geothermal regions lie in the earthquake and volcano belts along the earth's crustal plates. If reference is made to Fig. 4 in the entry on **Earthquakes, Seismology, and Plate Tectonics,** it will be noted that known areas of geothermal activity lie in these belts, notably along the west coasts of North and South America as far north as Alaska; then around the western Pacific to locations, such as Kamchatka, Japan, the Philippines, Indonesia, and along into southern Asia and into southern Europe. These are regions where seismic and volcanic activity is relatively current, with numerous major and minor seismic events having occurred along these belts during the past few decades.

In these belt-situated areas, magma works close to the surface of the earth. The areas are characterized by crustal weakness. In these areas of crustal weakness, the normal geothermal gradient may be exceeded by a factor of ten or more.[1] At present drilling depths, even along these plate boundaries, there are only a relatively few locations where geothermal energy is sufficiently close to the surface have been located and/or exploited. It is also true, of course, that seismic activity is far from uniform along the plate boundaries, another factor which makes generalization difficult. It is suspected, however, that with much greater drilling depths, numerous additional geothermal energy sources could be located and exploited.

Within the past decade mainly, another category of geothermal energy source has been seriously considered by a number of researchers and long-range energy planners. This source is described as *hot dry rock* geothermal technology and considers nearly all of the rocks which underly the earth's surface, inasmuch as there is a geothermal gradient essentially universally present. This technology is described later in this entry. It is a recent technology and in its very early stages of investigation and development, compared with the conventional geothermal energy sources. In the long term, it could offer an extremely large and valuable source of energy.

Although considerable progress has been made in increasing the energy derived from geothermal sources over the last decade (as at Geysers in California), it must be stressed that, placed against the total backdrop of energy needs of the United States and other countries where geothermal energy is being exploited, the percentage of the total energy requirements furnished by geothermal sources is quite small. Thus, a lot of progress can be made in the geothermal energy field and still go essentially unnoticed. Geothermal energy is relatively costly to exploit and, depending upon location, it also poses a number of environmental problems. In an energy-short world, a situation which may continue for a number of decades until energy technology can catch up with needs, obviously every effort should be made to continue to place emphasis on geothermal energy. These observations are made here to provide a sense of realism because geothermal energy, along with wind energy, various forms of solar energy capture, and ocean thermal energy, is all too frequently mentioned as one of the energy resources that is in place "for the asking" so to speak. As will be clear from scanning the remainder of this entry, the effective exploitation of geothermal energy requires major capital expenditures backed up by a relatively sophisticated technology.

Background. Expanding upon the prior definition of thermal gradient, the rate of heat conduction outward from the interior of the earth to the surface is estimated to average about 1.5 calories per centimeter per second. One estimate indicates that, over a one-year period, this flux to the total surface of the earth amounts to over 10^{20} calories. Heat stored in rocks beneath the United States alone (to a depth of 10 kilometers) has been estimated to be on the order of 6×10^{24} calories. Other estimates are given later in this entry. In terms of current technology, however, these large numbers are not as exciting as they appear. It has been observed that, over the next few decades, to be practically retrievable, heat from the earth must be concentrated in geothermal reservoirs, as previously mentioned, where the energy has accumulated and been in storage over long periods of time through geological processes.

Record of Utilization. The use of geothermal energy for purposes other than the heating of bathing pools began in Italy in the late eighteenth and early nineteenth centuries near the present site of the Larderello field. Larderello, and the more recently-exploited area, Mt. Amiata, are located on the west side of Italy, not far from Pisa. Steam from fumaroles and shallow bore holes was first used to aid the extraction of boric acid from the hot pools. That industry persisted for many years. In 1904, as a result of a dispute with the local electric utility, Prince Piero Conti, owner of the fields, decided to connect a generator to a steam engine driven by the natural steam. The success of that operation led to the installation of the first geothermal power plant, with a capacity of 250 kilowatts, installed in 1913. Increasing exploitation led to the installed capacity (1980) of approximately 385 megawatts.

Development of the Larderello field has been characterized by a lot of innovation, along with multipurpose utilization. Early developments were directed toward combining electric power production with the extraction of boron and other chemicals in the geothermal fluids. By using heat exchangers, a clean fluid could be used in the turbines, but as the value of the chemicals declined and as turbines were improved in construction to resist corrosion and abrasion, plants using the intermediate heat exchangers were replaced by direct-intake turbines. The direct intake turbines could be constructed at lower costs and, because there were no losses at the heat exchangers, more power could be produced per unit of steam.

Another innovation practiced in Italy has been the installation of relatively small (1.5 to 5 megawatt) back-pressure turbines that exhaust directly to the atmosphere. These are frequently used on individual wells very early in the development of new fields. Some of the advantages of using back-pressure turbines include: (1) they will handle steam containing large quantities of noncondensable gases, such as carbon dioxide, which sometimes exceed 30% (weight) of gases in a newly-opened field. Thus, gas that has become concentrated over a long period of time in the upper part of the reservoir is released and the ratio of noncondensable gases to steam is improved to the point where it can be used in conventional condensing turbines. (2) Another advantage is that reservoir temperature-pressure-volume relationships can be determined by production testing, and reservoir life predictions can be made prior to commitment of funding for more extensive developments. Revenues obtained from the sale of electricity during the testing period, sometimes extending over 2 to 3 years, can make a significant return of exploration costs. In recent years, other geothermal reservoirs have been discovered south of Larderello.

New Zealand. As early as 1932, scientists in New Zealand commenced investigation of thermal manifestations, such as hot springs and geysers, on North Island. It was not until 1948, however, that serious study began to appraise the geothermal resources, with the target of building a geothermal power station. By 1953, it was shown that the Wairakei area showed sufficient steam for construction of a power plant. The first Wairakei station was completed in 1958, and a second in 1963. New Zealand is well endowed with geothermal resources. A thermal area extends over a belt about 250 kilometers long and up to about 50 kilometers wide across the North Island between the central group of volcanic mountains (Mount Ruapehu, Mount Ngauruhoe, and Mount Tongariro) and the White Island volcano in the Bay of Plenty. See Fig. 1. Within this area is to be found a diversity of thermal activity—geysers, fumaroles, hot springs, and pools of boiling mud. Wairakei is one of several active areas where it is known that aquifers containing water up to and exceeding a temperature of 300°C exist. A view of the Wairakei Valley from the air is shown in Fig. 2.

Some 60 bores supply steam to the power station. Half of them are high-pressure bores producing steam at about 180 psi (12.2 atmospheres); others have intermediate pressure at about 80 psi (5.4 atmospheres). Well over 100 bores have been drilled, including those required for exploration.

Any extensive exploitation of underground water usually results

[1] According to Smithsonian tables, the rate of variation of temperature in soil and rock from the surface of the earth down to depths of the order of kilometers is, on the average, about +10°C per kilometer. The thermal gradient varies greatly from place to place, depending on the geological history of the region, the thickness and strength of the crustal rocks, the conductivity of the upper rocks, and, in some regions, the radioactivity of underlying rocks. In terms of exploitation, the thermal gradient is of the utmost importance because it is directly related to drilling depth.

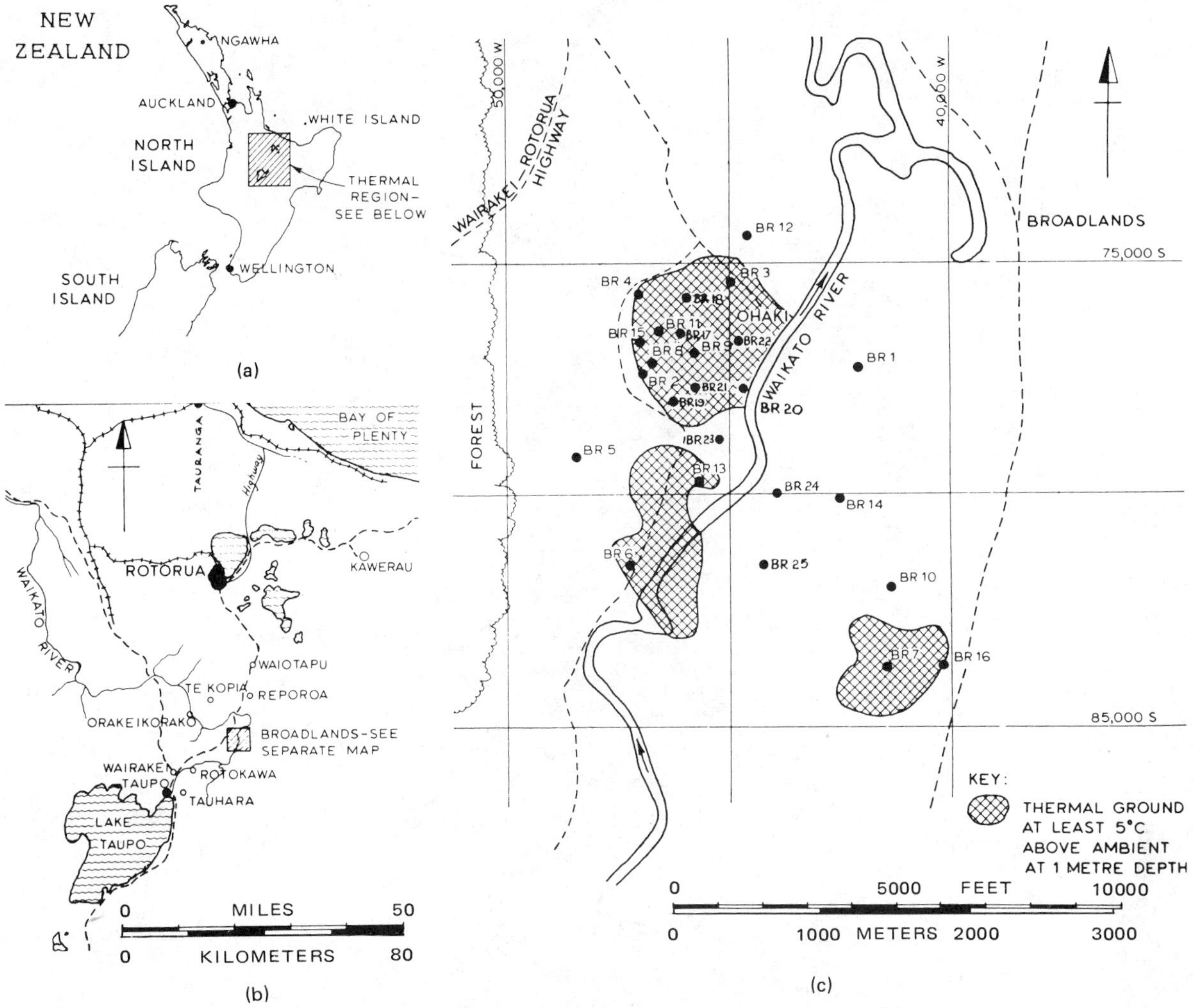

Fig. 1. Geothermal energy in New Zealand: (a) North and South Island, showing thermal region; (b) thermal area; (c) Broadlands area. (*Ministry of Works and Development, Wellington North, New Zealand.*)

in a slow decline in output from all bores, and those at Wairakei are no exception. Output is still tending to fall off, but power generation is not lessened. When pressure at the well head is lowered, a greater mass of steam can be obtained. This, together with the use of hot water, has enabled power station output to be maintained. Some high-pressure bores with a very low level of output have been reduced as far as intermediate pressure and connected to the intermediate pressure system.

Scientists have built up a picture of underground conditions at Wairakei. It shows a vast area of hot fissured rock, thousands of feet deep and several miles wide, filled with water and the products of a million years or so of intense volcanic activity. Over a large part of this area, water slowly seeps down to the bottom, where it is heated. The source of this water is thought to be rain.

Thus, there is a large hot water reservoir which is likely to be long lasting, but continued scientific vigilance and further intensive investigation will be needed. There is no evidence of direct interaction between widely-spaced bores. Bores fed directly by large fissures do not interact even if fairly close together, but those depending mostly one permeable ground may do so. Tests of two bores 90 feet (27 meters) apart in porous ground have shown that they react to each other slightly, the output of one being about 10% higher with the other closed, but another two bores that are 60 feet (18 meters) apart and penetrate fissured formation do not affect each other. There is direct communication between the bottoms of yet other adjacent bores, but no reduction in output has been observed. General effects observed so far over the whole steam field are a fall in pressure and temperature at depth and subsidence of the ground surface over an area of about $1\frac{1}{2}$ square miles (3.9 square kilometers). General indications, in spite of substantial decline in both pressure and temperature, are that full-scale production can continue for many years.

Hot water makes up about 80% (weight) of output from bores and is an important source of generating capacity although most of it is run to waste. Water at high temperature and high pressure boils and produces steam when the pressure is reduced. This process takes place in the steam bores and also in well separators. As hot water leaves the separator under pressure, it discharges to the silencer through a controlling orifice in the bypass pipe. See Figs. 3 and 4. As the pressure falls to near-atmospheric, large volumes of steam are generated. The water is led to waste and all the steam billows out from the top of the towers. This accounts for the waste steam which puzzles many visitors and is the most noticeable feature of the whole steam field. What appears to be a great waste of energy, however, is in reality steam which cannot be used. When the pressure of hot water at the high-pressure wells is lowered suddenly, steam for the intermediate pressure system can be evolved. This presents a method of using some of the hot water.

The total length of the main steam lines at Wairakei is more than 12 miles (19 kilometers) mostly from 20 to 30 inches (51 to 76 centimeters) in diameter. There are many additional miles of branch lines.

The work at and near Wairakei has touched only a small part of New Zealand's geothermal resource, which is estimated to approach 2000 megawatts, of which only about 10% of that capacity is exploited for electricity generation. In addition to the generation of electricity at Wairakei, geothermal energy is used industrially at a pulp and paper mill at Kawerau, and has been used for many years for domestic and small-scale commercial and industrial use in the City of Rotorua, and other parts of the thermal region.

During recent years, exploration drilling has been carried out in several other geothermal fields. One of these, Broadlands, has been drilled up to a proven 150 megawatts. The other areas are Orakeikorako, Reporoa, Rotokawa, Tauhara, Te Kopia, Waiotapu, and Nga-

Fig. 2. Wairakei Valley from the air.

Fig. 3. A typical wellhead set-up using a bottom outlet cyclone. The twin tower silencer to the left provides complete control over the water, as well as reducing noise to an acceptable level. The well is located in the foreground and is connected to the separator by the large sweep bend. The steam and water are separated by simple centrifugal action, the steam flowing from the cyclone, to the ball check vessel located at the left of the cyclone and thence to the steam mains through the branch line running out to the right of the photo. The tangential water outlet is connected to the water drum, and thence to the silencer.

Fig. 4. A double flash unit. A single intermediate pressure separator is used, but because of the higher specific volume of steam at lower pressures, two intermediate low-pressure separators are required. These are the two taller vessels on the left. The two squat vessels are the water drums. Also, two silencers are necessary to cope with the amount of water finally discharged to waste.

Fig. 5. Well 20 Broadlands discharging 400 metric tons per hour at an enthalpy of 305 kcal per kilogram. Two silencers are necessary to provide adequate control of the water.

wha. With the exception of Ngawha in the extreme north, these areas all lie within the thermal region of the North Island.

All New Zealand geothermal fields so far drilled are classified as hot water fields. Formation temperatures and consequently the enthalpy of discharge vary from field to field. That at Wairakei is about 260°C; the highest temperature so far measured is 307°C at Broadlands. See Fig. 5. Generally, the chemistry of the fields is similar, although there are differences in detail. A common feature is a low total dissolved solid content of about 4,000 parts per million. This eases a number of problems in utilization found in other fields throughout the world.

Utilization of future fields probably will be for electric power generation, although other possibilities cannot be overlooked. Future developments will be quite different as compared with Wairakei. Current work, both in New Zealand and in other countries, on techniques, such as reinjection, chemical recovery, two-phase transmission (steam and water), and the binary cycle, will have marked effects on the appearance and efficiency of future schemes.

The Geysers, California. In the United States, the major geothermal development is at The Geysers, about 90 miles (145 kilometers) north of San Francisco. This development commenced in 1960 with a 12,500 kilowatt generating plant. Installed capacity is now 665 megawatts, which makes it the largest geothermal development in the world as of the early 1980s. In Mexico, just south of the California border, a 150 megawatt geothermal plant is powered by hot water rather than steam. The hot water is derived from a thermal zone that extends northward under the Imperial Valley and the Salton Sea. Until the technology for exploitation of hot dry rock regions, as mentioned later, is developed, the Salton Sea region appears to be the only other geothermal region in the United States that is attractive from an electricity-generating standpoint. There are numerous other zones, however, where sufficient energy may be derived for local heating purposes (such as exploited in Iceland).

The geological situation at The Geysers is envisioned about as shown in Fig. 6. About 20 miles (32 kilometers) below the crust of the earth, a molten mass or magma is still in the process of cooling. In some places, earth tremors of the early Cenozoic era have caused fissures to open and the magma to come quite close to the surface. This process can cause active volcanoes and, where there is surface water, hot springs and geysers. The hot magma is also responsible for steam vents, like those found at The Geysers. The steam thrown off by cooling magma is called magmatic steam. Where surface water seeps down into porous rock heated by magma, the steam formed is called

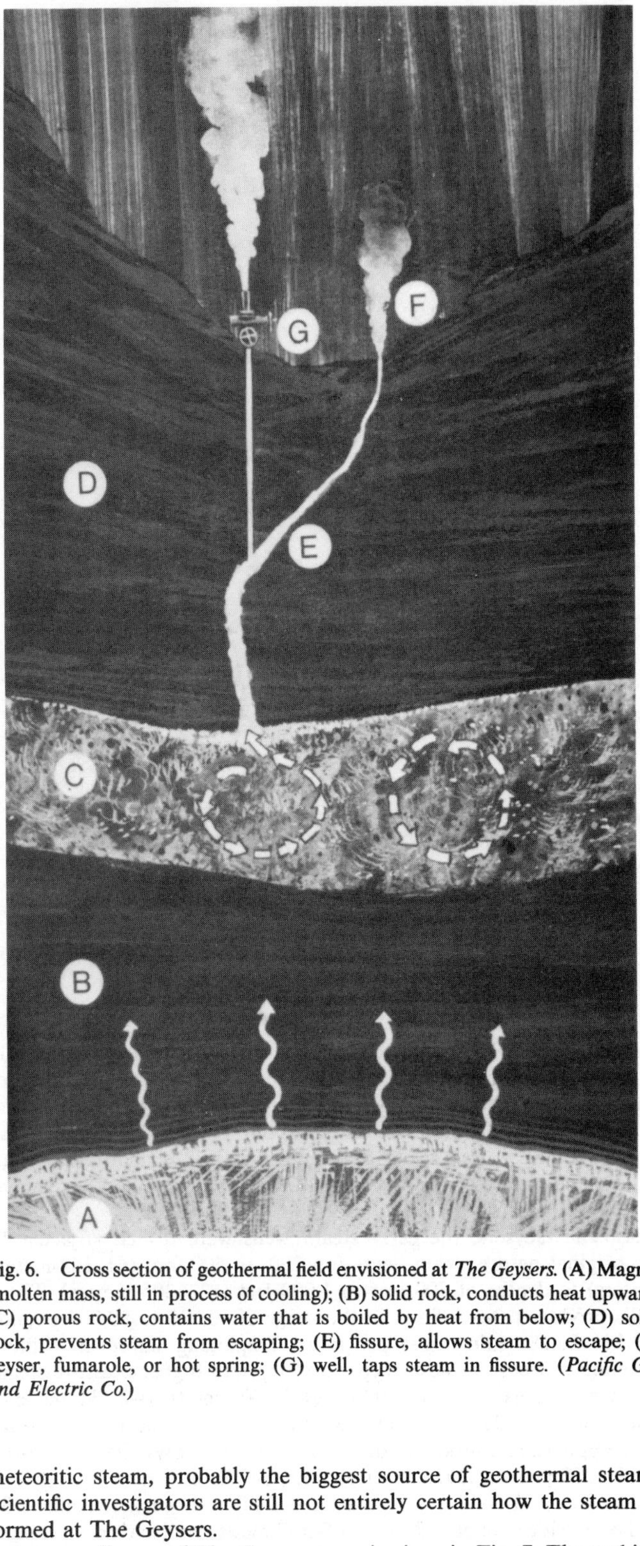

Fig. 6. Cross section of geothermal field envisioned at *The Geysers.* (A) Magma (molten mass, still in process of cooling); (B) solid rock, conducts heat upward; (C) porous rock, contains water that is boiled by heat from below; (D) solid rock, prevents steam from escaping; (E) fissure, allows steam to escape; (F) geyser, fumarole, or hot spring; (G) well, taps steam in fissure. (*Pacific Gas and Electric Co.*)

meteoritic steam, probably the biggest source of geothermal steam. Scientific investigators are still not entirely certain how the steam is formed at The Geysers.

An overall map of The Geysers area is given in Fig. 7. The turbine generator used for Unit 1 had originally been installed in a conventional steam plant at Sacramento, California in 1924. This geothermal unit commenced operation in 1960 with a capacity of 11,000 kilowatts. Unit 2, which started up in 1963, is in the same building and has a rating of 13,000 kilowatts, making a total capacity of 35,000 kilowatts for that location. These two units are pictured in Fig. 8. Unit 3 with a capacity of 27,000 kilowatts was commissioned in 1967, followed by Unit 4 with an identical capacity, which was commissioned in 1968. Units 3 and 4 are located adjacent to each other about 1.5 miles (2.4 kilometers) northwest of the first two units. Unit siting is determined mainly by the location of the steam wells. Units 5 and 6, each with a capacity of 53,000 kilowatts, were put into operation in 1971. Units 7 and 8, also 53,000 kilowatts capacity each, were

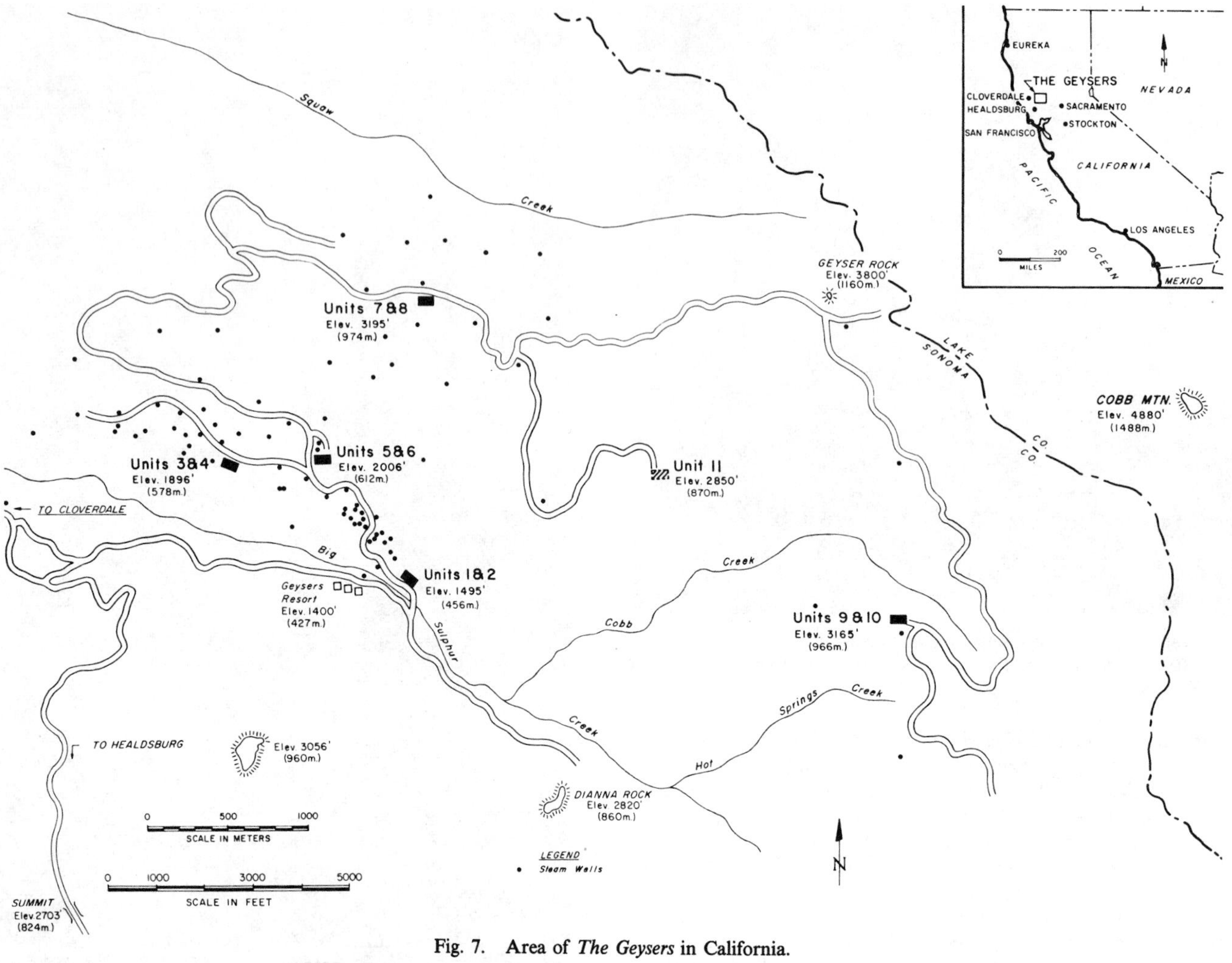

Fig. 7. Area of *The Geysers* in California.

completed in 1972. See Fig. 9. Similarly, Units 9 and 10 (53,000 kilowatts capacity each) were installed in 1973. Unit 11 (106,000 kilowatts) was installed in 1974; and Unit 12 (also 106,000 kilowatts) went into operation in 1975, bringing the total of the total area as of end-1975 at 608,000 kilowatts. By the early 1980s, additional installations and improvements brought the capacity up to the current 665,000 kilowatts.

At The Geysers, the early steam wells were drilled adjacent to the original nature steam vents (on 200- to 500-foot (61–152-meter) centers) to depths of 400 feet to 1,000 feet (122–305 meters). These wells produced steam flows in the range of 40,000 to 80,000 pounds (18,144–36,288 kilograms) per hour. Employing improved drilling techniques, wells are now deeper and tap into higher-pressure steam zones at depths between 2,000 and 7,000 feet (610–2,134 meters). Many of these deep wells are far removed from the natural steam outcroppings, and produce considerably greater flows. One was tested at 380,000 pounds (172,368 kilograms) per hour.

The steam supplying a typical 53,000-kilowatt capacity unit is 36 inches outside diameter, $\frac{3}{8}$-inch (9.5-millimeter) wall carbon steel pipe. This would typically be connected to about seven producing wells. Centrifugal steam separators are installed in the steam pipes to remove any particulate matter and moisture. The steam contains about 1% noncondensable gases in the following approximate amounts: carbon dioxide, 0.79%; ammonia, 0.07%; methane, 0.05%; hydrogen sulfide, 0.05%; nitrogen and argon, 0.03%; and hydrogen, 0.01%.

The steam also contains powder-like dust which deposits out in protected areas of the turbines. This dust builds up on the inside of the turbine blade shrouds in the first two stages. In lower stages, the buildup appears to be washed away by water in the steam. This shroud buildup has caused blade and shroud failures. Earlier units have had heavier-duty replacement blades and shrouds installed to mitigate the problem. A turbine water-wash program also may improve the situation.

Hydrogen sulfide in the air causes serious problems in the electrical equipment because it is corrosive to copper, copper alloys, and silver. Tin alloy coatings have been found to resist corrosion effectively although they have not been satisfactory on current-carrying contact surfaces. Aluminum seems to be particularly impervious to attack, as are stainless steel and some of the precious metals. Platinum inserts or plating appear to be a good solution to the problem with contacts. Protective relays are particularly vulnerable to attack and special relays constructed with noncorrosive materials must be used. Also in the newer units, the relays, communication equipment switchgear, and generator excitation cubicle are placed in a clean-room environment. The multilevel room is maintained at slightly positive pressure with clean air from activated carbon filters.

Where two units are housed in one building, they share the same high-voltage transmission line, and have a common 480-volt station service bus. Any electrical faults that occur beyond either of the two generator breakers requires that both units be tripped and, in addition, that the oil circuit breakers be opened.

The power cycle for all units is essentially similar. Steam from the wells is introduced into the turbines which exhausts to direct-contact condensers located directly below the turbine. The combined condensed steam and cooling water are pumped by two condensate pumps to the cooling water tower. The turbine back pressure on all units is about 4 inches (100 millimeters) of mercury absolute. Cooled water from the tower basin is returned to the condenser by gravity and the vacuum head developed by the condenser. Since the cooling tower evaporation rate is less than the turbine steam flow, an excess of water is developed in the cycle. This flow is dependent upon the dry-bulb temperature and relative humidity, but there is a surplus under all operating conditions. For several years, this excess water from the units has been returned to the wells for reinjection into the steam reservoir. The reinjection method was tried initially with some concern over the effect that it might have in quenching the

Fig. 8. Units 1 and 2 of Pacific Gas and Electric Company's geothermal power plant in Sonoma County, California. The electric generating facility is shown at right of center of photo.

producing steam wells. However, it has proven successful. It is believed that reinjection can extend the productive life of the steam reservoirs since it is felt that there may be more heat in the reservoir than there is vapor to extract it. Two-stage steam-jet ejectors are used to purge the noncondensable gases from the turbine condenser. The condensers for these ejectors are also of the direct-contact design.

The steam turbines are fabricated largely of the manufacturers' standard materials for low-pressure, low-temperature service. Blades and nozzles are typically of 11–13% chrome steel. Carbon steel is used for the turbine casings. Austenitic stainless steel inserts are provided in the casings opposite the rotating blades to prevent moisture erosion of the casing.

Iceland. Geothermal energy for space heating is of great importance in some countries, such as Iceland, where about 50% of the population enjoys such heating for its homes. Plans have been made to extend this type of heating to 70% of the Icelandic population within a few years. The geothermal fluids for such applications usually come from geothermal reservoirs at temperatures ranging from 60° to as high as 150°C (140–302°F). Thermal fluids within this temperature range occur at economically acceptable depths in Iceland and some other parts of the world.

Although geothermal space heating may serve a single house in the rural area, the most usual approach in Iceland is one of district heating services which serve whole population centers. As a rule, space heating by geothermal energy causes minimal pollution problems, inasmuch as there is no smoke and the warm effluents are distributed widely to the sewage system. In many areas where such systems have been installed, the cost of energy provided is very low when compared with fossil fuels. A depreciation time for equipment of from 20 to 30 years is usually used for economic evaluations.

Most distribution systems for hot water for space heating are single-pipe systems, which involve the discharge of the water to the sewage system after use. The distribution temperature of the water is preferably in the 80 to 90°C (176 to 194°F) range and will cool down to around 40°C (104°F) upon us. The supply mains to the distribution system will ordinarily discharge into storage tanks which help in taking care of daily fluctuations in hot water load. See Fig. 10. Booster pumping is usually necessary in order to maintain sufficient pressure in the distribution system. The distribution network in towns is installed underground in the streets. Street mains larger than 3 inches in diameter may be placed in concrete channels and are insulated by rockwool or aerated concrete. The channels are embedded in a hard core, together with concrete drainpipes. Minimum inclination of these channels is kept at 5%. At street junctions, the channels may meet in concrete chambers, where valves, fastening bolts, and expansion joints are placed. These chambers are ventilated and either drained from the bottom, or if that is not possible, they will have a pump pit. Smaller street mains and house connections from street mains may be insulated with polyurethane foam insulation.

A district heating system must be tailored to the local climate. The most important characteristics in this regard is the variation in daily outside temperature over the year. Since every heating arrangement for houses must ultimately have a capacity to provide comfort on the coldest day, obviously there must exist some overcapacity most of the time.

The ultimate cost of geothermal energy for space heating in such systems is usually nearly proportional to the maximum capacity required. Therefore, various approaches are used in order to increase the annual load factor. The latter term is defined as the ratio of total energy used to the basic design capacity. Some of the methods used include:

1. The system is designed for an outside temperature somewhat

Fig. 9. Units 7 and 8 at Pacific Gas and Electric Company's geothermal power generating facility.

Fig. 10. Hot water reservoirs serving the Reykjavík geothermal heating system. (*Photo by Mats Wibe Lund.*)

higher than that of the coldest day of the year, assuming the need for boosting from other sources for a few days each year.

2. The system may include a fossil-fuel booster which is intended for raising the temperature of the water during the coldest spells.
3. The system may include a local geothermal underground reservoir, where deep well pumps are installed in the drillholes. This arrangement may yield increased production for a limited time by pumping at a draw-down of the water level.

Generally, central heating systems are used for houses. The hot water is usually admitted directly to these systems and discharged to sewage after use. Hot domestic water for faucets is also supplied directly. Inferential water meters with a magnetic coupling between the flow sensor and register mechanism are frequently used. The maximum flow of hot water is also controlled by sealed maximum-flow regulators. Sometimes only maximum-flow regulators are used. When direct supply is not advisable, as in the case of water with high mineral content that would cause much scaling, heat exchangers may be used between the hot water and the water circulating in the central heating system.

In addition to use for house heating, most public buildings in Iceland are geothermally heated. A geothermally-heated swimming pool in Reykjavík is shown in Fig. 11.

Agricultural and Related Applications. Geothermal energy for heating greenhouses is important in Iceland and some other countries. Since the temperature of the heat source will vary greatly from one location to the next, as well as variations in heating requirements, the surface area of the radiator system (often consisting of bare pipes) must be carefully tailored to local conditions. Heating fluid temperature exceeding 100°C is rarely practical. Small greenhouses may take advantage of heat in the effluent from ordinary space-heating systems. The most important crops of heated greenhouses of this type include cut flowers, tomatoes, cucumbers, and seedlings of many varieties. Animal husbandry, fish farming, and hatching stations also frequently take advantage of available geothermal hot water.

Process Heating. Since geothermal energy resources exist in a number of countries, it is of interest to point out some common factors which affect the viability of exploitations. Since the applications for heating in cold climates and the generation of electrical power are obvious uses and receiving considerable attention, it may be well to concentrate on the possibilities for process heating uses in any climate. Perhaps the three most important questions are: (1) What products may utilize the heat in geothermal fluids? (2) What are the potential savings or advantages as compared with competitive energy approaches? and (3) If there is a logistic disadvantage involved in site location, can this be offset by the lower-cost energy?

Because present technology is largely tailored to the use of fossil fuels, no conclusive answers can be sought directly from present engineering and economic practice. It may be helpful, however, to begin a search by studying the conventional processes which use fossil fuel-generated steam. And there also will be found cases where geothermal fluids may be used with an advantage even for cases where no steam is used in the conventional processing of today. Some examples include: (1) the use of indirect heating in a process instead of direct contact heating—for example, a steam-tube dryer instead of a direct-fired dryer; (2) there may exist a choice of several processes for any one specific objective. One process may permit the use of geothermal energy with a great advantage, while another may not require any heat, but entail other high-cost categories; and (3) the availability of geother-

Fig. 11. Geothermally-heated swimming pool in Reykjavík, Iceland. (*Photo by Mats Wibe Lund.*)

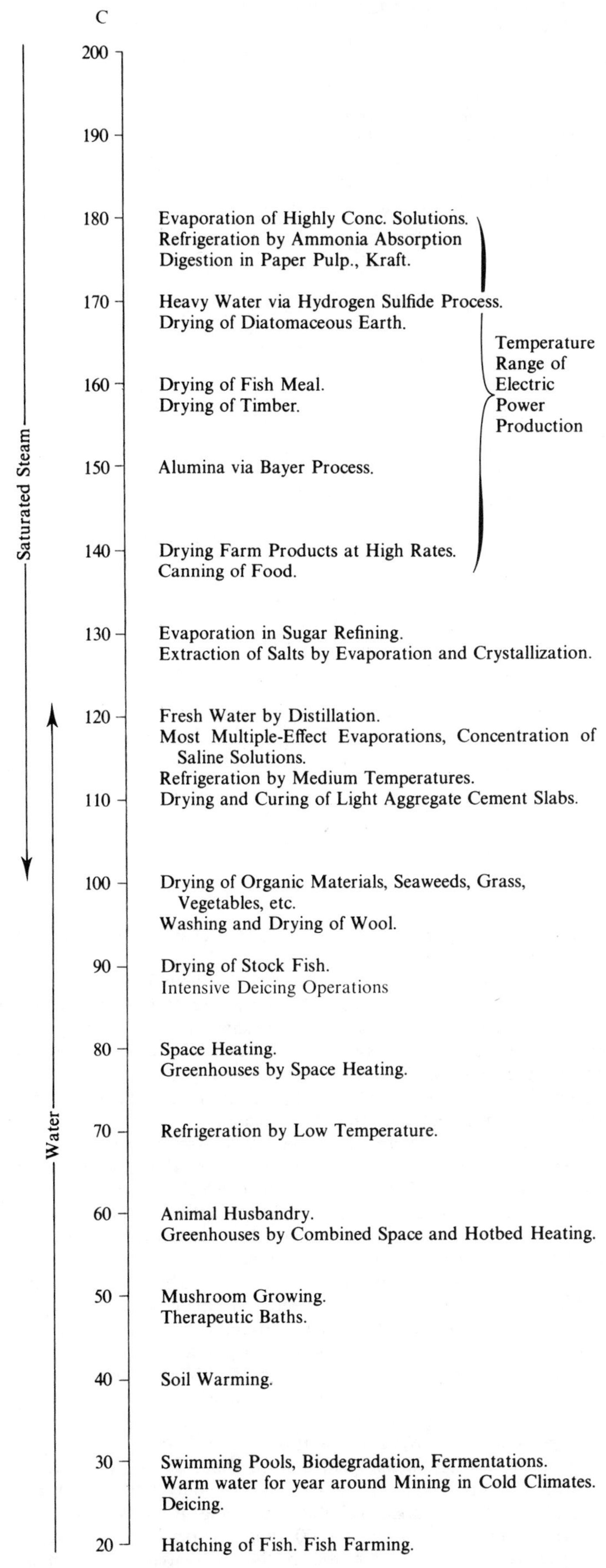

Fig. 12. Applications versus temperature range of geothermal water and steam.

TABLE 1. EXAMPLES OF PROCESS DESIGN FEATURES FOR GEOTHERMAL STEAM AND WATER

	Geothermal Steam		Geothermal Water	
Operation	Type	Examples	Type	Examples
Drying	Indirect heating	Steam tube dryers Drum dryers	Indirect heating	Multideck conveyor dryer
Evaporation	Primary heat exchangers accessible	Forced circulation evaporators	Counter-current heaters	Preheaters
Distillation	Steam distillation	General equipment	—	—
Refrigeration	Freezing	Ammonia absorption	Comfort cooling	Lithium bromide absorption
Deicing	—	—	Direct application Indirect heating	Dredging and pavement deicing

mal energy may call for a completely new process. General applicational areas are given in Fig. 12 and in Table 1. The steam requirements and steam per unit product value for several products of the chemical and process industries are given in Table 2.

The economic importance of geothermal energy in a specific process may be judged by the share it has in the value of a product. This often can be roughly evaluated in terms of the steam or the amount of fossil fuel which would otherwise be required. The effect of a different design, and hence different investment, also enters into the calculation. Numerous cases are known where the equivalent share of thermal energy may be from 5 to 20% of the value of a product. Examples of existing and planned application of geothermal energy for process uses are given in Table 3. An industrial plant for drying seaweed, located in northern Iceland, is shown in Fig. 13.

The major industrial plants currently in operation have amply demonstrated that geothermal energy is a versatile source of energy. There are examples where process heating, space heating, and electric power production have been integrated into the same overall system. Because there is a large variation in geothermal energy sources, optimal utilization of that energy can be achieved only through individual analysis of each source. There are, however, a few generalizations: generalizations:

1. When electric power is the main objective, there generally are ample opportunities for use of waste heat, at least in those plants using wet steam. In such instances, geothermal water may be rejected at elevated temperatures, which subsequently can be used for space heating, fresh water production, and some industrial applications.
2. When space heating is the main objective, secondary electric power generation is possible in some cases. There are numerous secondary applications (greenhouses, soil warming, heating of swimming pools, etc.).
3. When process heating is the main objective, depending upon the geothermal source, some generation of needed electrical power may be possible and, as in the other cases, there usually is ample opportunity for secondary heating applications.

Geological Aspects of Geothermal Systems

Within the last few years, a new concept of the outer few hundred kilometers of the earth has developed. This concept is embodied in the plate tectonic model of the earth. This concept is described in some detail in the encyclopedia entry on **Earthquakes and Seismology.** It is conceptualized that the surface of the earth, including the sea floor, is divided into several rigid plates which are moving relative to each other. The plates are composed of lithosphere which includes oceanic or continental crust or both, veneering and combined with

TABLE 2. CONSUMPTION OF STEAM AND STEAM USED PER PRODUCT VALUE IN SOME ESTABLISHED FUEL-BASED PROCESSES

Product and Process	Steam Requirements Kilograms Steam/ Kilogram Product	Steam per Unit Product Value
Heavy water by hydrogen sulfide process	10.000	151
Ascorbic acid	250	45
Viscose rayon	(70)	42
Lactose	40	130
Acetic acid from wood via Suida process	35	159
Ethyl alcohol from sulfite liquor	22	142
Ethyl alcohol from wood waste	19	123
Ethylene glycol via chlorohydrin	13	45
Casein	13	10
Ethylene oxide	11	33
Basic magnesium carbonate	9	37
35% hydrogen peroxide	9	23
85% hydrogen peroxide from 35% H_2O_2	$4\frac{3}{4}$	—
Solid caustic soda via diaphragm cells	8	121
Acetic acid from wood via solvent extraction	$7\frac{1}{2}$	34
Alumina via Bayer process	(7)	106
Ethyl alcohol from molasses	7	45
Beet sugar	$5\frac{3}{4}$	26
Sodium chlorate	$5\frac{1}{2}$	28
Kraft pulp	$4\frac{1}{5}$	32
Dissolving pulp	$4\frac{1}{5}$	—
Sulfite pulp	$3\frac{1}{2}$	26
Aluminium sulfate	$3\frac{1}{2}$	79
Synthetic ethyl alcohol from ethylene	3	20
Calcium hypochloride, high-strength	$3\frac{1}{3}$	50
Acetic acid from wood via Othmer process	$2\frac{3}{4}$	13
Ammonium chloride	$2\frac{3}{4}$	21
Boric acid	$2\frac{1}{4}$	20
Soda ash via Solvay process	2	60
Cotton seed oil	2	9
Natural sodium sulfate	$1\frac{4}{5}$	54
Cane sugar refining	$1\frac{2}{3}$	8
Ammonium nitrate from ammonia	$1\frac{1}{2}$	20
Ammonium sulfate	$\frac{1}{6}$	5

the uppermost part of the mantle. Oceanic lithosphere is from 75 to 100 kilometers thick, while continental lithosphere is about 150 kilometers thick. Beneath the lithosphere lies the athenosphere.

The composition of the athenosphere is not known, but seismic data indicate that it is a zone of partial melting (upper portion) with several probable density transitions in the lower portion. Along the belt of oceanic ridges, the plates are moving apart at a rate of a few centimeters per year, causing gaps. New mantle material (*magma*) fills these gaps. In the direction of plate motion away from the ridges, plates must converge, one plate sinking or subducting beneath the other. Deep oceanic trenches form at these boundaries. Beyond the trenches, volcanic arcs are produced. These are accompanied by shallow to deep seismicity. Such boundaries are typified by Japan, Indonesia, Kamchatcka, the Aleutian peninsulas, and the Andes of South America. Where plates are converging, both having a veneer of continental crust, the crust is less dense and cannot sink. Thrust faulting, folding, and thickening of the crust marks these boundaries. Examples are the Himalayan and Alpine mountain regions. At plate boundaries where neither spreading nor subduction is occurring, the plates slide past each other along great fractures which are called transform faults. The San Andreas fault system is a prime example as it connects the East Pacific Ridge which enters the Gulf of California to the Gorda Ridge lying off the Oregon–California coast and marks the boundary between the *American Plate* and the *Pacific Plate.* Spreading of the plates is largely confined to the ocean ridges which lie in the deep ocean. The Red Sea and the Gulf of California rifts probably developed only within the last few million years and deep ocean is yet to be attained. The great East African Rift is unusual in that separation is occurring within the continental lithosphere. Continued spreading of this rift may eventually split the African continent and produce new ocean floor. The driving mechanism for plate motion is not understood, but appears to be associated with convective movement of the mantle. *The energy is supplied,* whatever the mechanism, *by the internal heat of the earth.*

It is along the spreading and converging plate boundaries that abnormal terrestrial heat flow occurs. Mass transfer of heat by magmas generated from the mantle brings heat to shallower levels of the crust. From these heat sources, geothermal systems are developed. All of the prospective high-enthalpy geothermal areas of the world are found within the belts of geologically young volcanism and crustal deformation produced by moving lithospheric plates.

Fundamental Geothermal Systems. These systems develop in the upper few kilometers of the earth's crust from a source of heat at some greater depth. The geothermal fluid which contains dissolved minerals and salts is heated and becomes less dense. Where the overlying rock is permeable, a convection cell or system is created. For containment, a cover of impervious rock must overlie the system, thus preventing escape of the fluid to the surface. The thermal gradient is high in the covering rock and decreases rapidly within the upper part of the geothermal system where convection becomes pronounced. The temperature then varies little with depth and is called the base temperature. This portion of the system constitutes the reservoir. Leaks from the reservoir to the surface are manifested by steam vents, hot springs, geysers, and fumaroles.

Vapor-Dominated Systems. In this type of system, saturated to slightly superheated steam (temperature about 250°C; pressures of 30 to 35 bars) is produced. The reservoir generally consists of highly fractured or porous rocks. Well flows may range from a few thousand kilograms per hour to over 250,000 kilograms per hour from depths ranging from 1,000 to 2,500 meters. Noncondensable gases in the steam may range from considerably less than 1% of the steam to 5% or more. Noncondensable gas content may be much higher initially, but diminishes with production and indicates past accumulation in the reservoir.

TABLE 3. EXAMPLES OF EXISTING AND PLANNED APPLICATIONS OF GEOTHERMAL ENERGY FOR PROCESS USE

Product	Country	Applications	Form of Geothermal Energy
Pulp and paper	New Zealand	Evaporating, digesting, drying	Primary and secondary steam
Timber drying and seasoning	New Zealand, Iceland	Drying, seasoning	Steam, hot water
Diatomite processing	Iceland	Drying, heating, deicing	Steam
Hay drying[a]	Iceland	Drying	Hot water
Seaweed drying	Iceland	Drying	Hot water
Washing of wool	Iceland, USSR	Heating and drying	Steam
Curing and drying of building material	Iceland	Heating and drying	Steam, hot water
Stock fish drying	Iceland	Drying	Hot water
Salt recovery from seawater	Japan	Evaporation	Steam
Salts from geothermal[a] brine	Iceland	Evaporation	Steam
Boric acid recovery	Italy	Evaporation	Steam
Brewing and distillation	Japan	Heating and evaporation	Steam

[a] Planned.

The hydrostatic pressure, abnormally low, in these reservoirs indicates they are sealed from groundwater infiltration. It is believed that they developed from high-temperature, liquid-dominated systems which seal their cooler margins through time by precipitation of dissolved material, mainly silica. Further slow escape of water forms a steam space and a deep liquid phase, probably a very hot brine. Heat is received from a source beneath the system, probably a magmatic intrusion.

The steam fields at The Geysers, California, Larderello, Italy, and Matsukawa, Japan are typical examples of the vapor-dominated system. Reservoir characteristics are similar for all. The Geysers reservoir rocks are indurated, highly fractured graywacke sandstone and volcanic rocks. Porous limestone and dolomite are the reservoir rocks of the Larderello region, and fractured volcanic rocks serve as the reservoir at Matsukawa.

Liquid-Dominated Systems. These may be conveniently divided into two types: one having high enthalpy fluids above 200 calories per gram and one having low enthalpy fluids below this point. This division tends to separate fluids useful for generating electric power from those most useful for other purposes.

An important physical difference between the liquid and the vapor-dominated systems is the fact that the reservoir pressures in the liquid systems are near hydrostatic pressures, or around 0.1 bar per meter of depth. So at depths of 1,000 to 2,500 meters pressures are 100 to 250 bars, contrasting to the 30 to 35 bars in the vapor-dominated system.

High-enthalpy systems contain waters with dissolved solids ranging from around 2,000 ppm to as much as 260,000 ppm and temperatures of 200°C to as high as 388°C. The predominant anion of the dissolved solids is chloride along with lesser amounts of sulfate and carbonate. Sodium and potassium are the main cations with a smaller amount of calcium and sometimes magnesium. Up to 800 ppm of silica may be present which, along with several ppm of fluoride and several tens of ppm of boron, are troublesome in the disposal of these high enthalpy fluids.

Wells drilled into this type of reservoir produce a mixture of water and steam; the steam may be separated at a suitable pressure to operate a turbine. Noncondensable gas in the separated steam is usually below 1%.

The best developed high-enthalpy liquid-dominated reservoir is located at Wairakei, New Zealand, where wells are drilled into a permeable pumiceous volcanic rock capped by an impermeable sedimentary formation. Temperature of the fluid is about 260°C and about 20%

Fig. 13. An industrial plant processing seaweed, located in northern Iceland. (*Photo by Mats Wibe Lund.*)

is flashed to steam for power production. Another such system still being developed is the Cerro Prieto reservoir in the Mexicali Valley of Mexico north of the Gulf of California. Electric power is now being produced at Cerro Prieto from a fluid having a temperature of 300°C or more and salinities of 15,000 to 25,000 ppm, in a reservoir of permeable sedimentary rock. To the north in the Imperial Valley of California, wells have been drilled to 2,500 meters into reservoirs with similar characteristics except the Salton Sea reservoir which contains a concentrated brine having as much as 260,000 ppm total solids.

Low-enthalpy liquid-dominated systems have properties more variable than those known for the high enthalpy systems. In some the sulfate anion may be dominant and in others carbonate-bicarbonate. The salinities tend to be lower and some could be considered potable. Dissolved silica content, which is a function of temperature, is less and the toxic elements fluorine and boron also are generally diminished. Temperatures in low-enthalpy systems range from about 10°C above average annual temperature to the previously mentioned arbitrary division at 200°C.

Included in this category are low-enthalpy waters found in some deep sedimentary basins where the overlying rocks have a low conductivity. Temperatures may range from 50 to 60°C to 120°C, but the reservoirs are very large. The Hungarian basin and several in the U.S.S.R. are examples of this type. Along the Gulf Coast of the United States similar reservoirs exist within sands in undercompacted sediments. Temperatures above 200°C have been reported and pressures much above hydrostatic. Deep wells, 2,000 meters or more in depth, are required to tap the thermal waters of these basins. Since no connection exists with young volcanism in these basins, heat is thought to be supplied by a slightly above normal terrestrial heat flow coupled with the insulating effect of the overlying sediments.

Iceland is perhaps best known for the many low-enthalpy reservoirs which have been discovered and are being utilized. Numerous other reservoirs are known throughout the world, among which several are being utilized in the United States, notably in Oregon, Idaho, and California. In general, the close association of these reservoirs with young volcanism suggests magmatic heat as the source.

Exploration for Geothermal Energy Sources. The known higher-enthalpy geothermal systems or resources of the world are located where faulting has created uplift and subsidence of the crust with attendant mass transfer of heat from depth by magmas and geothermal convection systems. These activities are closely associated with geologically recent movement of the lithospheric plates.

The United States has a broad region covering the western conterminous states which has been distributed by recent interaction of the plates and changes in direction of their motions. Investigations over the last few years show large areas of this region to have above normal heat flow with numerous hot springs and wells.

Surface displays of heat offer the simplest and easiest means of exploring for geothermal resources. Yet hot springs or geysers may be some distance from a reservoir and drilling a deep test well at a spring may prove nonproductive. Also, some reservoirs may have little or no surface display. Therefore, geologic and geophysical methods must be used to enhance the chances of discovery.

Geologic studies can help to show the structure and stratigraphy which may outline domed areas, grabens, and calderas prospective for geothermal resources. Aerial and satellite photography and imagery are very important in the geologic investigations of such things as fault patterns and recent volcanism.

Of the many geophysical methods, measurement of the geothermal gradient and determination of heat flow from shallow drill holes is most valuable. Care must be used in extrapolating the data for greater depths, and groundwater migration can introduce serious discrepancies, but the method is direct in outlining thermal anomalies.

Gravimetric studies can indicate the presence of intrusive rock, which may be a heat source, or contrasting densities which may define a caldera or graben. Small gravity anomalies are associated with several known thermal anomalies in the Imperial Valley, California.

Rocks which are hot and also saturated with saline waters have very low electrical resistivities. Low resistivities are characteristic of high-temperature, liquid-dominated systems and electrical and electromagnetic methods are useful in the search for and delineation of their size. Practical results have been obtained on newly discovered reservoirs in the Imperial Valley and in New Zealand.

Passive seismic surveys, including ground noise, have been performed over a number of known geothermal reservoirs. One method involves the recording of microearthquakes which many geothermal systems seem to generate. The activity probably arises from the highly faulted nature of the reservoirs and their association with regions of young tectonism. On the other hand, ground noise or geothermal noise surveys record the acoustic signals within a narrow range of amplitude and frequency. Results so far suggest that individual geothermal systems produce characteristic signals and are related to reservoir depth and temperature gradients. If the reliability of this method can be demonstrated, it could become very important in geothermal exploration because of its simplicity and economy.

Active seismic surveys, generating and recording seismic waves produced by explosions or shock, are useful in determining subsurface structure and faults. Either the reflection or refraction method can be employed depending upon which is most suitable for a particular location and problem. Some recent work indicates that attenuation of seismic waves may occur in geothermal systems. Further development of this procedure may increase the usefulness of active seismic investigations for geothermal resources.

Magnetic surveys involve measuring the magnetic properties of the underlying rocks. Positive magnetic anomalies often are associated with intrusive rocks and negative anomalies occur over rocks in which the magnetic minerals have been altered by geothermal fluids. A magnetic survey would thus seem to be useful for seeking geothermal reservoirs but so many complicating factors arise that results are generally very difficult to interpret.

Many geochemical and isotopic investigations have been performed on samples of spring waters and geothermal fluids throughout the world. As a result, certain constituents or ratios of these constituents may be used to indicate probable reservoir temperatures of liquid-dominated systems. Silica content and the sodium-potassium-calcium rates are the best indicators. High chloride content (above 50 ppm) in springs suggests that the system is liquid-dominated. Springs associated with vapor-dominated systems are said to contain less than 20 ppm chloride.

Isotopic analyses of hydrogen and oxygen in geothermal waters provide a means of determining the origin of the waters. It is now known that geothermal fluids are meteoric in origin and any volcanic or magmatic addition is minor. By this method the hydrology of an area can be appraised concerning the recharge of water to a geothermal reservoir.

None of these exploration methods can prove the existence and size of a geothermal reservoir. Only the drilling of deep wells and testing of the product found will determine if successful development and utilization can be obtained.

Hot Dry Rock Geothermal Technology

Of a much longer-range potential is the possible exploitation of heat energy contained in hot dry rocks (HDR) of the earth's crust. These HDR regions are not directly related to the belts of geothermal energy activity previously described, but are reasonably well distributed under the land areas of continents, rather than concentrated in earthquake and volcano belts. While much effort remains in the development of exploitive technology, the gross estimates of HDR energy resources tend to be almost unbelievably high. Exploitation of such resources, of course, is essentially a matter of the geothermal gradient. This is the factor which determines the depth of drilling required to reach a specified temperature. The HDR bases generally has been defined to include crustal rock that is hotter than 150°C and at depths of less than 10 kilometers, which is essentially at the edge of present commercially feasible drilling and recovery technology. Although a temperature of only 100°C would be attractive for space-heating needs, a minimum temperature of 200–250°C is desirable for using such energy in the generation of electricity. Scientists at the Los Alamos Scientific Laboratory (L.A.S.L.) have developed a chart which indicates the best sector of useful heat, with depth plotted against temperature.

From temperature data obtained from deep gas wells, the geothermal gradient has been determined or estimated for several regions of the

United States. Such regions have been found or are postulated for central and eastern Oregon, southern California, southwestern and central Arizona, western South Dakota and eastern Montana, Nebraska, much of Colorado, some pockets in Indiana and Illinois, additional pockets in southeastern Texas along the Gulf Coast, various pockets in eastern Pennsylvania and New England, among others. High geothermal gradients occur on the Atlantic Coast from the Delmarva Peninsula southward to Georgia. These are regions where it is believed that the geothermal temperature gradient is 36.5°C or higher per kilometer of depth. Immediately adjacent to these regions and frequently of much larger area, the gradient lies between 29.2 and 36.5°C. Some scientists at the U.S. Geological Survey (U.S.G.S.) have observed that rock underlying about 5% of the total United States land area may have geothermal gradients of 40°C or greater. They also believe that it can be assumed conservatively that over a third of the land area in the United States has above-average heat flow with thermal gradients ranging from 30 to 36°C per kilometer. It has been pointed out that igneous rock systems to depths of 10 kilometers under the United States (excluding Alaska and Hawaii) contain some 105×11^{21} joules (J), which is equivalent to 105,000 quads. (A quad equals 1 quadrillion Btus.)

If one uses an average geothermal temperature gradient of 22°C for the entire United States, it is further estimated that the energy, if available, would amount to 13×10^{24} J. This is equivalent to 13,000,000 quads. By comparison, the current annual energy consumption of the United States approximates 80 quads. Scientists postulate that if only 0.2% of such energy were made available, it would be comparable to all of the coal remaining in the United States.

Basically, two techniques have been suggested for mining HDR geothermal heat: one method for rock formations with low permeability, and one for highly permeable rocks. As pointed out by Cummings et al. (1979), "If the permeability of the formation is low, an artificial circulation system can be created by fracturing the rock in the reservoir to provide many flow passages with a large heat-transfer surface area. A fluid—for example, water—is then circulated through the fractured reservoir to recover the energy. Most of the injected fluid is recovered in a second production wellbore simply because of the low natural permeability of the formation. Large fracture surface areas are required because rock conducts heat rather poorly, and it quickly controls the rate of heat transfer to the fluid contained in the fracture zone. Such an HDR reservoir will most likely be formed by injecting fluid through a wellbore at pressures sufficient to fracture the rock. Under ideal conditions, the fracture would be vertically oriented, circular in shape, with a maximum radius of typically 100 meters or more, and a width or opening of only a few millimeters."

Some of the facets related to the technical feasibility of HDR systems have been demonstrated in experiments conducted by L.A.S.L. at the Fenton Hill site in the Jemez Mountains of New Mexico. A hydraulically fractured reservoir in low-permeability crystalline basement rock at about 185°C was created and flow tested for 75 days at an energy extraction rate of about 5 thermal megawatts (MWt). Additional tests are underway on an expanded scale. For greater detail on this program, see the Cummings et al. reference.

Obviously, considerable further research, particularly of an engineering nature, is required to fully demonstrate the potential of HDR as a future major energy source. Environmental impact studies also will be required.

Acknowledgments. This technical summary on geothermal energy was made possible by information and portions of the text furnished by: R. G. Bowen, consulting geologist, Portland, Oregon and E. A. Groh, private geologist, Portland, Oregon (geological aspects and exploration); R. S. Bolton, chief geothermal engineer, Ministry of Works New Zealand, Wellington North, New Zealand; Pacific Gas and Electric Company, San Francisco, California (The Geysers); and Baldur Líndal, director, Virkir Consulting Group Limited, Reykjavík, Iceland (geothermal energy in Iceland—and space and process heating).

References

Armstead, H. C. H., Editor: "Geothermal Energy: Review of Research and Development," UNESCO Earth Sciences Publication **12**, Paris, 1973.

Bloomster, C. H.: "Economic Analysis of Geothermal Energy," *Chem. Eng. Progress*, **72**, 7, 89–95 (1976).

Blair, A. G., et al.: "L.A.S.L. Hot Dry Rock Geothermal Energy Development Project," Los Alamos Scientific Laboratory Report, LA-6525-PR, Los Alamos, New Mexico, 1978.

Considine, D. M. (editor): "Energy Technology Handbook," McGraw-Hill, New York, 1976.

Coury, G. E.: "Two-Phase Flow in Geothermal Brine Wells," *Chem. Eng. Progress*, **73**, 7, 87–88 (1977).

Coury, G. E., and M. Vorum: "Removing H_2S from Geothermal Steam," *Chem. Eng. Progress*, **73**, 9, 93–98 (1977).

Cummings, R. G., et al.: "Mining Earth's Heat: Hot Dry Rock Geothermal Energy," *Technology Review* (*MIT*), **81**, 4, 58–78 (1979).

Einarson, S.: "Geothermal District Heating," in "Geothermal Energy," UNESCO, New York, 1973.

Ellis, A. J., and W. A. J. Mahon: Chemistry and Geothermal Systems," Academic, New York, 1977.

Hornburg, C. D.: "Geothermal Development of the Salton Sea," *Chem. Eng. Progress*, **73**, 7, 89–94 (1977).

Iyer, H. M., Oppenheimer, D. H., and T. Hitchcock: "Abnormal *P*-Wave Delays in The Geysers–Clear Lake Geothermal Area, California," *Science*, **204**, 495–497 (1979).

Kruger, P., and C. Otte: "Geothermal Energy," Stanford University Press, Palo Alto, California, 1980.

Nemat-Nasser, S.: "Thermal Cracks and Their Effects on the Heat Extraction Process," The Technological Institute, Northwestern University, Evanston, Illinois, 1980.

Otte, C.: "The Potential for Geothermal Steam," *Chem. Eng. Progress*, **72**, 7, 80–82 (1976).

Ramachandrian, G., et al.: "Economic Analysis of Geothermal Energy Development in California," Vol. 1, Stanford Research Institute Report ECU 5013, Menlo Park, California, 1977.

Sherwood, T. K.: "Process Design for Thermal Power," *Chem. Eng. Progress*, **72**, 7, 83–88 (1976).

Sigvaldason, G. E.: "Geochemical Methods in Geothermal Exploration," in Geothermal Energy," UNESCO, New York, 1973.

Spera, F.: "Thermal Evolution of Plutons: A Parameterized Approach," *Science*, **207**, 299–301 (1980).

Staff: "Assessment of Geothermal Resources of the United States," U.S. Geological Survey Circular 726, U.S. Geological Survey, Reston, Virginia, 1975.

Staff: "Heat Extraction from Fractured Geothermal Reservoirs," Los Alamos Scientific Laboratory Report LA-UR-78-917, Los Alamos, New Mexico, 1978.

Staff: "Hydro and Geo," *Sci. Amer.*, **242**, 4, 64–66 (1980).

Swearingen, J. S.: "Power from Hot Geothermal Brines," *Chem. Eng. Progress*, **73**, 7, 83–86 (1977).

Tester, J. W. (editor): "Phase I—Energy Extraction Field Test Results from the Fenton Hill Hot Dry Rock Geothermal System–Segments 1–2," Los Alamos Scientific Laboratory, Los Alamos, New Mexico, 1980.

Tilman, J. E.: "Eastern Geothermal Resources: Should We Pursue Them?" *Science*, **210**, 595–600 (1980).

Wilson, J. S.: "A Geothermal Energy Plant," *Chem. Eng. Progress*, **73**, 11, 95–98 (1977).

Wunder, R., and H. D. Murphy: "Thermal Drawdown and Recovery of Singly and Multiply Fractured Hot Dry Rock Reservoirs," Los Alamos Scientific Laboratory Report LA-7219-MS, Los Alamos, New Mexico, 1978.

GEOTHERMAL GRADIENT. **Thermal Gradient.**

GEOTROPISM. The response of living things to the effects of gravity. The term is used especially with reference to the response of roots and stems of plants. Most roots are said to be positively geotropic,

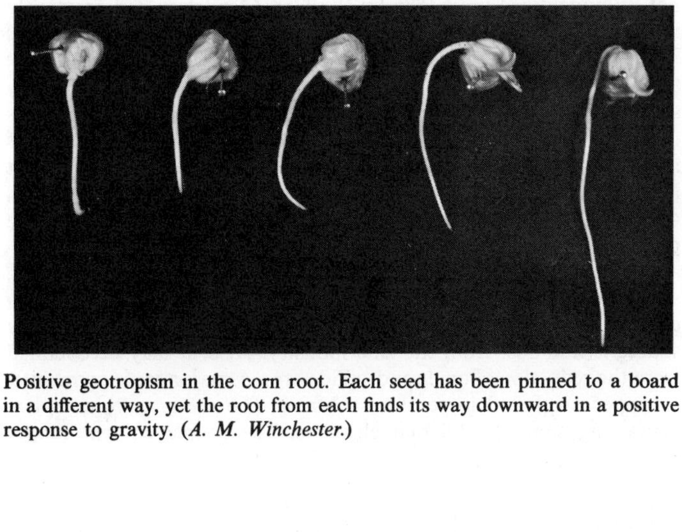

Positive geotropism in the corn root. Each seed has been pinned to a board in a different way, yet the root from each finds its way downward in a positive response to gravity. (*A. M. Winchester.*)

that is, they grow in the direction of the pull of gravity. See accompanying diagram. Most stems are negatively geotropic, that is, they grow away from the pull of gravity. Such a combination of responses causes the roots to grow down and the stems to grow in an upward direction.

These responses are explained on the basis of auxins. These are produced in the tips of the stems, in young leaves, and, in lesser quantities, in the tips of the roots. When a plant is turned on its side, gravity causes auxin to accumulate on the lower side of the stem. This extra auxin acts as a stimulant to the cells in the region of elongation and the stem turns upward. The roots are much more sensitive to auxin than stems, and the same concentration that stimulates stems acts to inhibit the growth of the cells of the root in the region of elongation. Thus the growth is greater on the upper surface and the root turns down. The possible relationship between geotropism and ethylene is described by Wheeler and Salisbury (*Science*, **209**, 1126–1127, 1980).

See also **Plant Growth Modification and Regulation.**

GEOTROPISM (Plant Hormones). Plant Growth Modification and Regulation.

GEPHYREA. Large marine worms. As adults they are not segmented but since the young show evidence of metameric segmentation they have been placed with the segmented worms in the phylum *Annelida*, but the three groups are now classified as separate phyla. They have a large body cavity, nephridia, and in some species a few setae. The internal organs are not metamerically arranged.

The group is divided into:

Phylum *Echinoidea.* With a pair of setae near the anterior end. Body cylindrical with a slender anterior protuberance, the prostomium.

Phylum *Sipunculoidea.* No setae. Body slender, with a protrusible proboscis and a group of tentacles near the mouth.

Phylum *Priapuloidea.* No setae or tentacles.

GERBIL. Rodentia.

GERENUK. Antelope.

GERM. 1. A microorganism (microbe), commonly used to refer to bacteria and their relations. 2. The reproductive material; for instance, the germ plasm is the plasm of reproductive material that links the generations. 3. The embryo of seeds; for instance, wheat germ is made from the region of the wheat kernel that contains the embryo.

GERMANIUM. Chemical element symbol Ge, at. no. 32, at. wt. 72.59, periodic table group 4a, mp 937°C, bp 2830°C, density 5.36 g/cm^3 (20°C). Elemental germanium has a diamond cubic crystal structure. Germanium is a silver-white, lustrous, hard, brittle metal. When heated in oxygen to 730°C, the metal is partially oxidized to dioxide. The element is unaffected by solutions of acids and bases, but is soluble in fused NaOH. In the form of powder (dull gray), combines readily with chlorine to form the volatile tetrachloride. Although predicted by Mendeleev as early as 1871, the element was not fully identified until 1886 by Winkler. Mendeleev had previously termed the missing element *eka-silicon.* There are five natural isotopes ^{70}Ge, ^{72}Ge through ^{74}Ge, and ^{76}Ge. Seven radioactive isotopes include ^{67}Ge through ^{69}Ge, ^{71}Ge, ^{75}Ge, ^{77}Ge, and ^{78}Ge. All have a relatively short half-life, the longest, ^{68}Ge with a half-life of 275 days. In terms of abundance, germanium ranks 32nd among the element and thus is about as abundant as gallium, selenium, arsenic, and bromine. First ionization potential 8.13 eV; second, 15.86 eV; third, 31.97 eV; fourth, 45.5 eV. Other important physical characteristics of germanium are given under **Chemical Elements.**

Germanium occurs in very small amounts in many sulfide ores, such as American zinc ores (0.25% GeO_2), and the rare mineral argyrodite (silver germanium sulfide) of Saxony and Bolivia. The primary source is flue dust from the zinc industry. Also, it may be obtained from the reduction of oxide and sulfide ores. A major ore is germanite, a copper ore found in southwest Africa. The ore is quite complex, containing some 20 different elements. The copper content ranges as high as 45%, sulfur up to 30%, whereas the germanium content is from 6 to 9%. The ore also contains up to 1% gallium. A major sulfide ore is renierite which contains up to about 8% germanium. Small quantities of germanium are found in lepidolite, sphalerite, and spodumene. Some English coals contain as much as 1.6% germanium oxide. The germanium metal of 99.99 + % purity is obtained by zone melting. In this system, electric heating coils are moved slowly along the length of an ingot. Impurities in the metal tend to raise or lower the freezing point of the molten alloy. By progressively melting the metal along the length of the ingot, the impurities which tend to lower the melting point will be swept to the last portion of the ingot to freeze, whereas the impurities which tend to raise the melting point will concentrate in the first region to freeze.

Uses: The principal uses of germanium have been in solid-state electronic devices, notably transistors, which can be used as amplifiers and oscillators. The electrical properties of germanium metal which have brought about its wide use in semiconductors are its high specific resistance at ordinary temperatures and the narrow gap between its filled energy band and its conduction band. Thus, germanium is an intrinsic semiconductor, wherein an increase of temperature or the addition of very small amounts of group 3 or group 5 elements can cause electrons to move readily to the conduction band to form "holes," thus making the material conductive. A key to the manufacture of semiconductor devices is making materials of high purity, great uniformity, and in sufficient quantity. See also **Semiconductor.**

The addition of as little as 0.35% germanium to tin doubles the hardness of tin. Similarly, germanium improves the strength and hardness of aluminum and magnesium alloys. These applications are limited, however, because of the current high costs of germanium. Germanium-silicon alloys are under intensive study for use in thermoelectric generators. Advantages claimed for these metals include better thermoelectric qualities above 600°C, an improved efficiency per unit weight factor, and virtually no corrosion or decomposition.

Chemistry and Compounds: Germanium forms compounds in which the oxidation states are (II) and (IV). The divalent ones are unstable. Thus the monoxide is readily oxidized by air when hydrated. However, when completely dehydrated it resists the action of H_2SO_4 and potassium hydroxide, and reacts only slowly with fuming HNO_3. On heating in an inert atmosphere it disproportionates to the elements and germanium dioxide, GeO_2. The latter resembles silicon dioxide in existing in more than one form, with a difference in chemical properties. The stable form at room temperature has the rutile structure, but just below the melting point the stable form has the cristobalite structure. Germanium(IV) oxide, GeO_2, prepared by hydrolysis of germanium-(IV) chloride, $GeCl_4$, is somewhat soluble in water, acids, and alkalis, but GeO_2 from heating of germanic acid is insoluble. Like silicon dioxide, GeO_2 forms gels readily.

Germanium(II) hydroxide, $Ge(OH)_2$, is obtained by action of alkali hydroxides upon germanium(II) chloride, $GeCl_2$, solutions; it is amphiprotic, dissolving in excess of the alkali. Moreover, the acid form, sometimes called germanous acid, is obtained upon heating the hydroxide: $Ge(OH)_2 \rightarrow HGe(O)H$. GeO_2 is slightly acid in solution and when freshly precipitated ($pK_A = 9.4$). There is no experimental evidence for the existence of a definite hydrate, although melting point diagrams of germanate salts have indicated the existence of ortho($\equiv GeO_4$), meta($=GeO_3$), and tetra($=Ge_4O_9$) compounds.

Germanium forms dihalides and tetrahalides with all four of the common halogens. In general, the dihalides readily react with halogens or other oxidizing agents to form tetravalent germanium compounds, and some, e.g., the iodide, disproportionate to the metal and tetravalent compound.

Suggestive of carbon and silicon is the existence of hydrides of germanium, though they are much fewer in number. The compound GeH_4 is called germane (mp −165°C, bp −90°C). Compounds having the general formula Ge_nH_{2n+2} ($n = 2, 3$, etc.) are called digermane, trigermane, etc., according to the number of germanium atoms present. The first three compounds in this series have been obtained by treatment of magnesium germanide with ammonium bromide in liquid ammonia. Compounds such as $GeHCl_3$ and alkylgermanes are also known. Germane and the alkyl- and aryl-substituted germanes retaining at least one hydrogen atom are somewhat more acidic than the corresponding silanes in nonaqueous media, easily forming alkali salts,

R_3GeM and even dialkali salts R_2GeM_2 under some circumstances. Germane, GeH_4, appears to be thermodynamically stable, although no quantitative data are available on its heat of formation. It decomposes at about 285°C.

Germanium resembles the other main group IV elements in forming organometallic compounds. Over two hundred have been reported, from chloromethyl trichlorogermane, $ClCH_2GeCl_3$ to cyclotetrakis (diphenyl germanoxane), $[(C_6H_5)_2GeO]_4$.

See list of references at end of entry on **Chemical Elements.**

GERMANIUM TRANSISTOR. Transistor.

GERMAN SILVER. Copper.

GERMICIDE. Any substance or agent, physical or chemical, which is destructive to germs (bacteria). See also **Antiseptic.**

GERMINATION (Seed). Seed.

GERM LAYER. The three tissues resulting from the first differentiation in the embryonic development of multicellular animals. They are formed from the presumptive areas of the blastula during gastrulation by a redistribution which results in three layers. The three are an outer ectoderm, an inner endoderm, and between the two the mesoderm.

Animals of the phyla *Porifera*, *Coelenterata*, and according to one interpretation the *Ctenophora*, develop only the first two germ layers and are said to be diploblastic. Multicellular forms of all other phyla have all three and are therefore triploblastic.

The chief parts of the body formed from the various germ layers are as follows (in these lists the terms are chosen to embrace both vertebrates and invertebrates and so do not all apply to the same animal):

Ectoderm: Outer cellular layers of the integument, their glandular derivatives, and the cuticula. Exoskeleton and exoskeletal structures such as setae, scales, feathers, hair, claws, hoofs and nails. Parts of sensory organs including the cornea and lenses of all types of eyes, external and internal ears of vertebrates. Lining of oral cavities and salivary glands, and in vertebrates the enamel of the teeth. Lining of the posterior part of the alimentary tract. The entire nervous system of most animals, including the nervous structures in the sense organs. A limited amount of muscular tissue. Organs of reproduction of some animals. Lining or covering of organs of respiration of many invertebrates.

Mesoderm: Lining of body cavity, circulatory system, water vascular system of echinoderms, and parts of excretory system. Muscular tissue. Bone. Teeth, except the enamel. The mesenchymal tissues such as cartilage, connective tissue, adipose tissue, and tendon. Blood. A limited part of the nervous system of starfishes. Reproductive organs.

Endoderm: Lining of the enteric cavity, including most of the alimentary system of vertebrates and the limited midintestine of arthropods. Respiratory epithelium of vertebrates. Lining of parts of vertebrate excretory system. Reproductive organs and cells.

GERM PLASM. The essential reproductive tissue and the germ cells that it produces.

The concept of the germ plasm has been emphasized chiefly in the field of organic development. Since the germ cells of one generation produce both the body (soma, somatoplasm) and the germ plasm of the next, the continuity of this material is evident. It has been interpreted as the perpetual living substance, whereas the material of the body appears as an offshoot in each generation. In the one-celled animals, however, there is no differentiation.

In the one-celled animals, however, there is no differentiation. Obtaining and retaining germ plasm from very old plants is essential to the development of new species and strains of crop plants. See also **Plant Breeding and Genetics.**

GERM PLASM BANKS. Plant Breeding.

GERM PLASM (Plant). Plant Breeding.

GERONTOLOGY. The scientific study of aging. Aging represents the progressive changes which take place in a cell, tissue, organ or organism with the passage of time. The changes which occur after attainment of maturity are of primary interest in gerontology. Geriatrics is the branch of medical science concerned with the prevention and treatment of the diseases of older people and is part of the broader field of gerontology. For the most part, age changes represent a gradual loss in functional capacity which ultimately results in the death of the cell or organism. In fact, for humans and many other animals such as rats, mice, dogs, cats, and even insects, the probability of death increases logarithmically with age.

With reference to gerontological research, Ludwig (1980) stated, in part, "Contemporary medicine cures or prevents damage wrought by the environment. It achieves this by neutralizing pathogens or by compensating for the lack of something the environment normally supplies. Even genetic disease is dealt with in this fashion, be it by intercepting some environmental trigger or by prosthetic means. Medicine's thrust is ecological. Man himself remains beyond its reach. But with increasing age, the causation of disease shifts away from the environment to originate more and more in the organism itself. At the same time, man's capability to counter this intrinsic pathogenesis by ecological means, which has been so effective up to now, is approaching its limits, in spite of further sophisticated (and socially inconsequential) advances. Medical care, one might say, remains in its infancy as long as it cannot forestall intrinsic pathogenesis as effectively as that originating in the environment. To overcome this limitation is the true aim of gerontological research. In initiating the revolutionary step from an environmentally oriented health care to one centered on man himself, it becomes the very foundation of future scientific medicine."

Although the average expectation of life has increased, the life span of persons who have enjoyed life-long health and who have been reasonably well endowed genetically has not increased perceptibly. There always have been examples of persons who exceed the Biblical "threescore years and ten" by a significant number of years. Throughout history, there always have been a few centenarians. As observed by Hayflick (1979), "Even if all the major causes of death today were eliminated, the length of human life would still be fixed at 90 or 100 years."

Although there are many aspects of gerontology, biochemistry must ultimately play a key role in understanding the mechanisms of impairments in cellular function and death which occur even in metazoans such as man. Although biochemists have greatly expanded our knowledge of the mechanisms of energy transformation and protein synthesis, they have given little attention to the effects of age of the cell or tissue on these processes. The determination of such information is of great importance for the science of gerontology and the understanding of aging.

Currently, there are several biochemical and gentic hypotheses, some of which incorporate similar elements.

Exhaustion or Accumulation Hypotheses. These concepts assume that aging results from the exhaustion of some essential material or the accumulation of toxic or deleterious materials in cells. In view of the turnover rates which have been shown for most cellular constituents, it is doubtful whether the exhaustion of any specific material can be the cause of aging. Of course, specific cells in a metazoan may die when they are deprived of their normal source of nutrients by interference with the blood supply. This, however, is usually based on pathological processes, such as arteriosclerosis, and is not a basic mechanism of aging. If there is an impairment in the synthetic mechanisms required for the turnover and replacement of key molecules, exhaustion may occur. However, exhaustion is not the primary factor. The breakdown in the replacement mechanisms is the primary process, and it will be considered later in connection with "error" theories.

The presumption that the accumulation of deleterious substances contributes to aging receives support in the studies of Carrel who found that tissue cultures of chicken heart could not be maintained if serum from old chickens was used in preparing the culture medium. It still remains for biochemists to isolate such a substance from the blood of senescent animals. More recently, it has been found that highly insoluble granules accumulate with advancing age in cells from certain tissues, such as the heart and nervous system. These granules, called "age pigments," are composed of varying proportions of lipid and protein and may occupy a substantial part of the cell at advanced ages. The accumulation of peroxides and free radicals as well as S—S groups in the tissues of old animals has also been proposed as a cause of aging. However, although a slight increase in life span of rats fed various antioxidants (to remove peroxides and free radicals) has been reported, no direct evidence is available on the effect of age on the concentration of peroxides or free radicals in tissues.

The fact that alterations in environmental temperatures can significantly alter longevity in poikilothermic animals has also been interpreted as evidence for the exhaustion or accumulation theory of aging, since it is assumed that changes in temperature will influence the rate of chemical reactions in the animal and hence the rate of utilization of essential materials or the rate of formation and accumulation of deleterious substances. The life spans of *Drosophila* and *Daphnia* are significantly shorter in animals reared at 27°C than in those reared at 15°C. Exposure to high temperature for a short period of time does not influence the mortality curve of the surviving *Drosophila* so that the life shortening effect of the high temperature cannot be attributed to denaturation of essential proteins, but must be related to the rates of chemical reactions in the animal.

In the rat, as well as some other species such as the rotifer, restriction in the amount of food given increases life span. The dietary restriction must begin early in life (at weaning in the rat) and continue throughout life in order to be effective. This treatment prolongs the period of growth, but no firm biochemical explanation for increased longevity can be given although some investigators have assumed that starvation, by prolonging the growth period, delays the accumulation of unknown deleterious products.

Error Hypothesis. This concept was formulated by Medvedev (Medical Research Council, London) and further developed by Orgel (Stalk Institute). This concept explains a number of the facts of aging which are known at present, although it is far from proved. The error theory is based on the hypothesis that information with regard to the synthesis of cellular proteins resides in the DNA molecule within the nucleus of the cell. This information is transmitted by messenger RNA from the nucleus to the sites of protein formation in the ribosomes of cells.

It is assumed that with increasing age, slightly atypical molecules of messenger RNA are formed so that errors occur in the formation of protein molecules. If the protein molecules which contain errors are enzymes, either they may be completely incapable of participating in the essential chemical reactions in cells, or, they may do so at slower rates. The result of either condition would be an accumulation of the substrates on which the enzymes act. Because of the feedback mechanisms which operate in cells, the accumulation of substrates may stimulate an increased production of messenger RNA and enzymes. When this increase is insufficient to produce adequate amounts of functional enzymes the cell dies.

Evidence for the error theory stems from a number of sources. Some studies have indicated the predicted increase in messenger RNA in cells of some tissues as well as decreases in the activities of some enzymes with advancing age in the rat. For example, the succinoxidase activity of cardiac muscle and renal cells of the rat declines by 15–20% between adult (age 10–12 months) and senescent (age 24 months) animals. However, other enzymes such as D-amino acid oxidase and pyrophosphatase remain relatively stable throughout the life span. More detailed studies on isolated mitochondria have failed to show a significant age decrement in specific enzyme activity. It seems probable that any age decrement in mitochondrial enzymatic activity at the cellular level is due primarily to a reduction in the number of mitochondria present in cells.

The error theory also receives support from the observed life-shortening effects of exposure to non-lethal amounts of radiation. Radiation shows its primary biological effects in inducing alterations in DNA which produce mutations in cell lines. These mutations are essentially errors which reduce the viability of the daughter cells. Recently it has been shown that the incidence of atypical chromosomes in liver cells in the rat is increased both by aging and by exposure to radiation.

The important role of DNA in aging is also reflected in genetic differences in life span between different species of animals as well as differences within the same species. The May fly is reported to have a life span of one day in contrast to the potential human life span of 100 years. The range of life spans among different species is much greater than the range of individual life spans within a species and can be regarded as a reflection of differences in DNA. However, selective inbreeding within a species will separate strains with significantly different life spans. In genreal, inbred strains have shorter life spans than their heterozygous ancestors.

Gene Exhaustion. Scientists have estimated that only about 0.4% of the information in the DNA of the cell nucleus is utilized by a given cell in its lifetime. Because the genes along the DNA molecule are repeated in identical sequences, the genetic message is highly redundant. But, perhaps after a very long period, these messages are exhausted, permitting errors to accumulate at an accelerated rate and ultimately leading to death. This concept ties in well with the previous observation that species with short lives would have DNA much less redundant than that in species with long lives.

Age-Programming Genes. Perhaps the entire aging process is preprogrammed so that at various periods during life, processes which we recognize as endogenous aging phenomena are, in effect, working precisely according to plan. Examples along these lines include graying of hair and menopause, among others, which are not regarded as diseases, but rather as expected changes with age.

Eversion Hypotheses. These concepts are based upon the changes in structure and configuration of molecules that take place with the passage of time after they have been formed. Evidence for this theory is based primarily on the changes which take place in connective tissue with advancing age. Collagen from old animals is less readily solubilized than that from young. Its thermal contractility is reduced and in general it attains a more rigid physical and chemical structure. These changes in properties have been attributed to the formation of cross linkages in the collagen molecule which are similar to those induced in the tanning of leather. Some investigators have ascribed these changes to the presence of aldehydes in the body which serve as effective cross-linking agents. Molecular changes also take place in elastin which result in decreased elasticity in many tissues, such as skin, blood vessels, etc., with advancing age. The formation of cross-links in the elastin molecule is regarded as the basic mechanism of these age changes. Thus there is good evidence that molecular changes occur with aging in extra-cellular proteins such as collagen and elastin. Similar changes may also occur in intracellular proteins. Preliminary experiments indicated a significant age difference in the melting temperature of DNA isolated from thymus glands of old and young cattle, which led to the presumption that structural changes in the DNA molecule had taken place with aging. However, subsequent experiments showed that the differences in melting temperatures were due to differences in the histones associated with the DNA rather than to changes in the DNA molecule itself. There is thus some evidence for alterations in an intracellular structural protein or in the DNA-protein complex. Small changes in molecular structure of other intracellular proteins, such as enzymes, might well interfere with their participation in essential biochemical reactions in cells with advancing age.

Progeria. This is a very poorly understood condition in some people in whom there is premature development of the characteristics usually associated with old age. At an early age, affected children show evidence of the process and, at a time prior to the period of puberty of normal children, these individuals literally resemble little old men and women, and their life span is short. The condition is rare, but research with these individuals may enlighten our understanding of the normal aging process.

References

Bell, E., et al.: "Loss of Division Potential in Vitro: Aging or Differentiation," *Science*, **202**, 1158–1163 (1978).

Bowden, D. M.: "Aging in Nonhuman Primates," Van Nostrand Reinhold, New York, 1979.

Comfort, A.: "Biological Theories of Aging," *Human Develop.*, **13**, 217–239 (1970).

Finch, C. E., and L. Hayflick: "Handbook of the Biology of Aging," Van Nostrand Reinhold, New York, 1977.

Hayflick, L.: "The Strategy of Senescence," *Gerontologist*, **14**, 1, 37–45 (1974).

Hayflick, L.: "Current Theories of Biological Aging," *Federation Proceedings*, **34**, 1, 9–13 (1975).

Hayflick, L.: "The Cell Biology of Human Aging," *N. Engl. J. Med.*, **295**, 23, 1302–1308 (1976).
Hayflick, L.: "The Cell Biology of Human Aging," *Sci. Amer.*, **242**, 1, 58–65 (1979).
Holden, C.: "National Institute of Aging," *Science*, **192**, 1081–1084 (1976).
Holliday, R., et al.: "Testing the Commitment Theory of Cellular Aging," *Science*, **198**, 366–372 (1977).
Kohn, R. R.: "Principles of Mammalian Aging," Prentice-Hall, Englewood Cliffs, New Jersey, 1971.
Ludwig, F. C.: "What to Expect from Gerontological Research?" *Science*, **209**, 1071 (1980).
Marx, J. L.: "Hormones and Their Effects in the Aging Body," *Science*, **206**, 805–806 (1979).
Strehler, B. L: "Time, Cells and Aging," Academic, New York, 1977.

GERSDORFFITE. A mineral related to cobaltite and ullmannite in the cobaltite group. A sulfide-arsenide of nickel, NiAsS. Crystallizes in the isometric system. Hardness, 5.5; specific gravity, 5.9; color, white to gray with metallic luster; opaque. Named after von Gersdorf (1842) at Schladming.

GESTATION. The period of intrauterine fetal development. See also **Pregnancy.** Pregnancy in humans is usually about 280 days. The period varies considerably over the spectrum of mammals—from a few weeks to well over a year.

GETTER. Barium.

GETTERING. The absorption of gas by a getter film. When this process occurs during the dispersal of the getter through an evacuated system (such as an electron tube), it is called dispersal gettering; when by action of the already dispersed film, it is called contact gettering. In electric-discharge gettering, the process is accelerated by passing an ionizing electron discharge through the gas. The gas is ionized, and the ions are neutralized when they impinge on an electrode, so that the final product is neutral gas atoms. These are then easily absorbed by the getter.

A getter film is a metallic deposit in a vacuum system with the function of absorbing residual gas. Electropositive metals, such as sodium, potassium, magnesium, calcium, strontium, and barium have been used as getters. The process of depositing a getter film upon a surface may be done in various ways. In the distillation method, the metal to be deposited is volatilized into the vacuum system from a side tube provided with constructions for sealing-off when the process is completed. The electrolytic method is applicable where the metal to be deposited is sodium, and where the system is made of soda-lime glass. It is well known that sodium may be electrolyzed through lime-soda glass. If, therefore, a thermionic source of electrons is provided inside an evacuated sealed-off vessel, part of which is dipped into a suitable liquid kept at a high potential relative to the source of electrons, a current will pass, carried by electrons between the thermionic cathode and the inner surface of the glass, and by ions within the glass. The only ions in the glass that are mobile are sodium ions, and thus pure sodium is released at the inner surface of the envelope.

Other modern getter materials include cesium-rubidium alloys, tantalum, titanium, zirconium, and several of the rare-earth elements, such as hafnium.

GETTER-ION VACUUM PUMP. Vacuum Pumps.

GEYSER. Derived from the Icelandic word *geysa*, meaning gush and descriptive of hot springs which at regular, or irregular, intervals throw a column of steam and hot water into the air. Geyser waters usually build up tubes or conduits of siliceous sinter. Geyser waters have been proved to be mainly vadose with approximately 10% of juvenile or magmatic water. Geyser action is the result of vadose water coming in contact with steam arising from the solidifying magma, and periodically returning to the surface through the geyser tube, for the same reason that water is suddenly expelled from a test tube when heated too rapidly. The mechanics of geyser action are illustrated in Fig. 1.

The principal geyser fields are in the western United States, notably

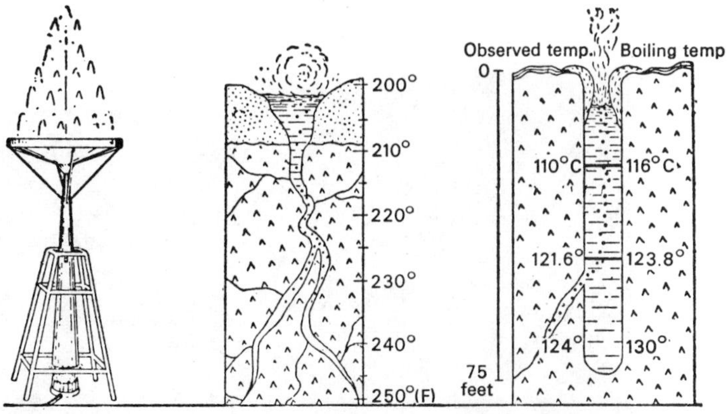

Fig. 1. The mechanics of geyser action, as illustrated by laboratory experiment, and the hypothetical cross sections of natural geysers. (*Field, "Outline of Geology," Barnes & Noble.*)

Wyoming (Yellowstone National Park), California, and in New Zealand, and Iceland. Geysers and other sources of geothermal energy are receiving increasing attention as the result of diminishing supplies of fossil fuels. Such exploitation of geothermal energy, of course, is not recent, but extends back for many years in Iceland, New Zealand, and Italy. A portion of the geothermal steam field at *The Geysers* (California) is shown in Fig. 2 on next page. More details will be found under **Geothermal Energy.**

Yellowstone Park claims the world's largest geyser area with approximately 3,000 geysers and hot springs.

GEYSERITE. A loose or compact, sometimes concretionary, siliceous deposit, formed by geysers and hot springs from the material held in solution by the thermal waters.

GEYSERS (The) GEOTHERMAL FIELD. Geothermal Energy.

GHATTI GUM. Gums and Mucilages.

GHOST IMAGE. Two of the uses in science of this term are: 1. In spectroscopy, false images of a spectral line produced by irregularities in the ruling of diffraction gratings. Rowland ghosts are false images grouped symmetrically on both sides of the true line. Lyman ghosts are false orders of spectra for which the order is not an integer. 2. In television, a second image appearing on the receiver screen, superimposed on the desired signal. These images are caused by reflected rays arriving at the receiving antenna some small interval after the desired wave. A single, reflected ray from a stationary object will produce a single, clear ghost, while a number of reflected rays arriving at assorted times creates an effect known as "smearing" or "smear ghost." Ghosts may also be produced with intensity reversal (white becomes black and vice-versa) due to a suitable phase of the secondary signal with respect to the primary signal, occurring on a suitable amplitude range of the received primary signal. This ghost is customarily called a negative ghost.

GIANT AND DWARF STARS. During the first two decades of this century, it was found, on the basis of parallax and photometric studies, that stars of similar spectral characteristics and temperatures diversified into two essentially distinct classes. This separation being on the basis of absolute magnitude, it was surmised by E. Hertzprung and independently by H. N. Russell that the difference must be due to a larger radius for the brighter stars at the same color, or effective temperature. The terms *giant* and *dwarf* were applied to the two groups. Intermediate groupings are now also recognized, which are *supergiants* and *subgiants* for the most luminous stars, and *white dwarfs* and *subdwarfs* for those of lower luminosity.

Largely due to the work of Adams at Mount Wilson and Morgan at Yerkes, it was recognized by the 1930s that the spectral characteristics of the giants also differ from dwarfs, in that the giants always show narrower lines and often, at the same effective temperature, appear to have an earlier spectral type. In addition, there is a steady

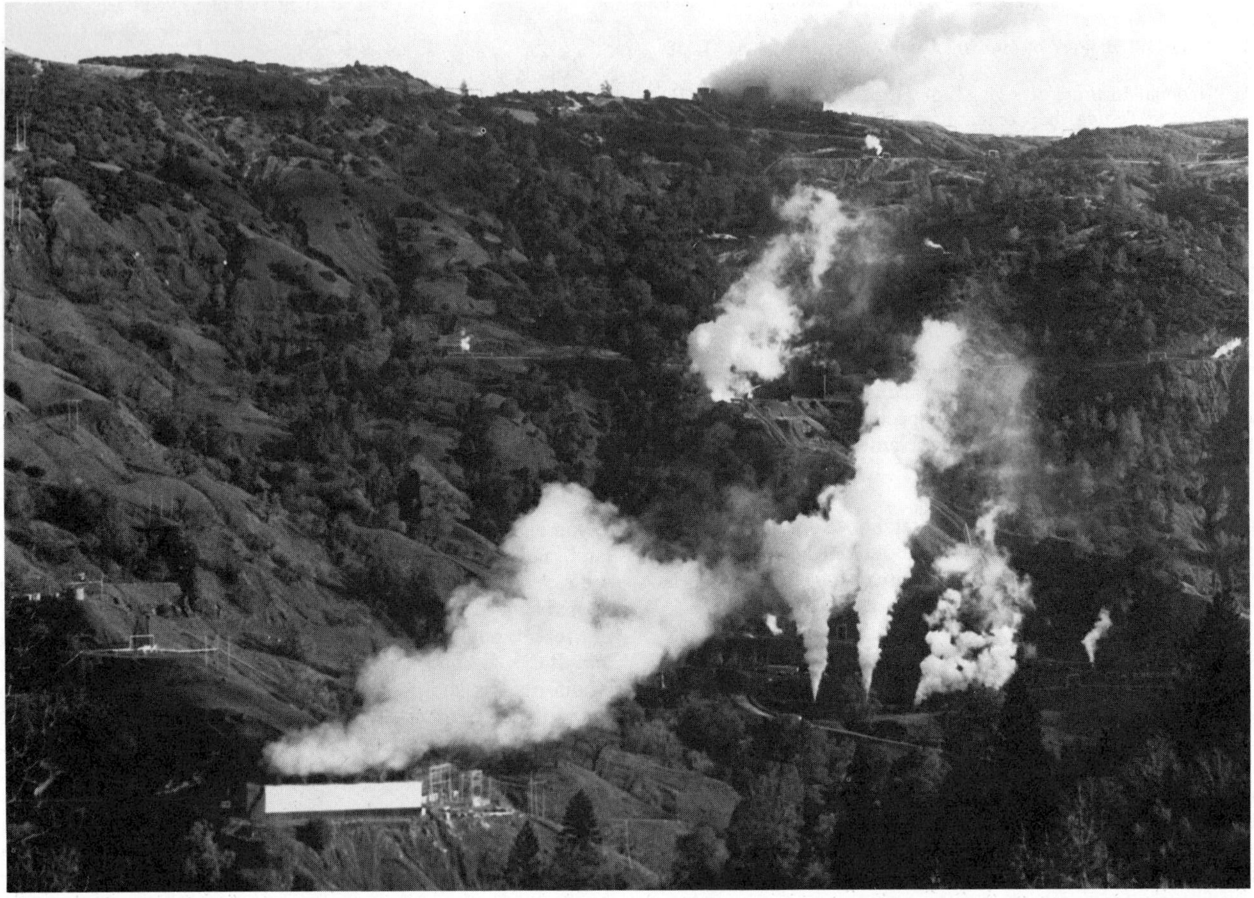

Fig. 2. Section of geothermal steam field, located in northeastern Sonoma County, California (about 90 miles; 145 kilometers north of San Francisco), which generates in excess of 400,000 kilowatts of electricity. Two of the generating stations are shown on the ridge of the hill. Below and in the foreground are additional steam-electric plants. (*Pacific Gas and Electric Co.*)

progression in the strength of certain lines on the basis of increasing or decreasing strength with increasing luminosity.

This is the basis of the second dimension of spectral classification in the MK (Morgan-Keenan) system, which adds a "luminosity class" to the temperature class of the Harvard (HD) system. In the MK system, luminosity classes run from I (supergiants) through V (dwarfs), and have temperature classes (in order of decreasing temperature) of O, B, A, F, G, K, M with additional classes R and S being reserved for the carbon stars. The sun, with an absolute magnitude of +4.6 and a surface temperature of about 5800 K is a G2V star, that is, a G2 dwarf, while δ Cygni is of a similar temperature, but has an absolute magnitude of −4.7 and is an F8Ib supergiant. The standard star for photometry, Vega (α Lyrae) is defined to be an AOV star having an absolute magnitude of +0.5. The MK system of classification proceeds by comparison of a given unknown stellar spectrum with agreed-upon standards and so is internally consistent. This behavior can be explained on the basis of a difference in surface gravity, and consequently pressure in the atmospheres of these stars. The lower pressure of the giant envelope produces less line broadening due to fewer perturbing collisions between radiating atoms, while the lower electron density causes an increase in the ionization at the same temperature.

The dwarf stars correspond to members of the main sequence, which is the hydrogen core burning stage of stellar evolution. It should be noted that the number of stars in any region of the Hertzprung-Russell (H-R) diagrams is approximately proportional to the period of a star's life during which it resides at that temperature and luminosity. The main sequence can thus be shown to be the longest lived stage of a star's life. The subdwarf population corresponds to the older, more metal poor, main sequence of the halo and old disk and is similar in characteristics to that observed in the globular clusters like 47 Tuc and ω Centauri. The subgiants, which are the first "post-main sequence" phase, are hydrogen core exhaustion and shell burning stars, and represent a transition between the main sequence and the giants. The brightness of giants is not as regularly correlated with mass as in the main sequence, for which a mass-luminosity relation exists (the more massive main sequence stars are brighter). The giants and supergiants are helium core and shell burning stars (and possibly double-shell sources), having ignited the spent helium core relic from the main sequence stage. These stars will eventually (depending upon

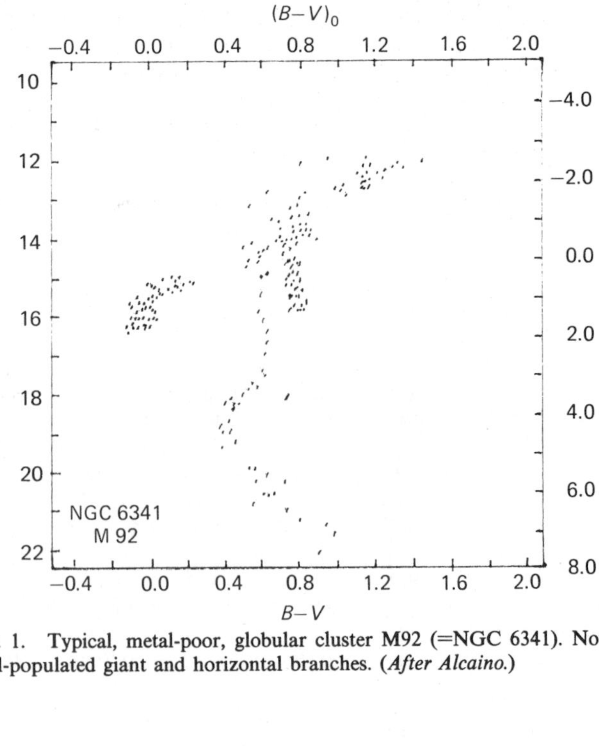

Fig. 1. Typical, metal-poor, globular cluster M92 (=NGC 6341). Note the well-populated giant and horizontal branches. (*After Alcaino.*)

mass) evolve into planetary nebulae and white dwarfs, supernovae, or if massive enough perhaps into black holes.

The giant and dwarf stars differ also in other important characteristics. While only the most massive main sequence stars show any evidence of stellar winds of any appreciable strength (greater than 10^{-9} solar masses/year), many red and blue supergiants show evidence of substantial mass loss. Blue supergiants like P Cygni, and red supergiants like α Ori and α Her show considerable envelopes, with characteristic velocities of hundreds of kilometers per second. Some also display radio continuum emission, another indication of mass loss. Only the δ Sct and β Cep stars are on or near the main sequence, while the giants and supergiants show most of the other classes of variable stars. See also **Variable Star.**

While the majority of dwarf stars in the galactic disk show abundances similar to the Sun (to within a factor of 2), giants show a wide range, indicative of considerable mixing of interior material which has undergone nuclear processing. At least one giant, FG Sge, has shown atmospheric abundance changes with time, an increase of heavy metals (rare earths) which are produced by neutron irradiation with subsequent mixing. The giant stars in globular clusters also show evidence for some time-dependent mixing processes.

The giants are best studied in globular clusters, where they form the horizontal branch population. Brighter stars, observed in several of the oldest of the clusters, lie on the asymptotic branch, parallel to the giant branch, but slightly bluer and brighter. The main sequence stars in these clusters are often too faint for careful study. The main sequence is best observed in galactic or open clusters, like η and χ Per, Coma, the Pleiades, and the Hyades. The H-R diagrams of a typical globular cluster is shown in Fig. 1, while a diagram for a typical galactic cluster is shown in Fig. 2.

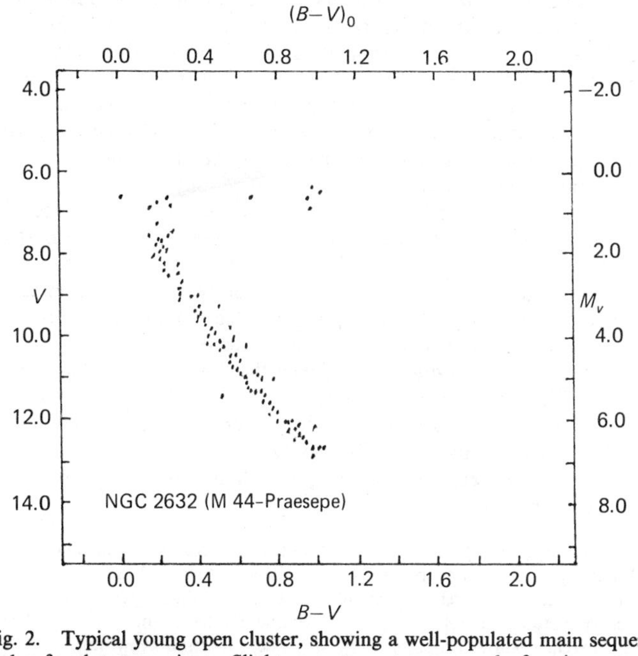

Fig. 2. Typical young open cluster, showing a well-populated main sequence and a few late-type giants. Slight curvature at upper end of main sequence indicates these stars have begun to exhaust their hydrogen cores and evolve away from the main sequence stage. (*After Hagen.*)

References

Allen, C. W.: "Astrophysical Quantities," 3rd edition, Athlone, London, 1973.

Clayton, D. D.; "Stellar Evolution and Nucleosynthesis," McGraw-Hill, New York, 1968.

Cox, J. P., and R. T. Giuli: "Principles of Stellar Structure," Gordon and Breach, New York, 1968.

Iben, L.: "Globular Cluster Stars," *Sci. Amer.*, **227**, 7, 26 (1970).

Keenan, P. C.: "Classification of Stellar Spectra," in "Basic Astronomical Data," (K. Aa. Strand, editor), Univ. of Chicago Press, Chicago, Illinois, 1963.

Meadows, A. J.: "Stellar Evolution," Pergamon Press, Oxford, 1978.

Schwarzschild, M.: "The Structure and Evolution of the Stars," Dover, New York, 1958.

Taylor, R. J.: "The Stars: Their Structure and Evolution," Sykeham, London, 1970.

Steven N. Shore, Assistant Professor of Astronomy, Warner and Swasey Observatory, Case Western Reserve University, Cleveland, Ohio.

GIANT ANTEATER. Edentata.

GIANT CELL ARTERITIS. Cerebrovascular Diseases.

GIANT CELL TUMOR. Bone.

GIANTISM. Hormones; Pituitary Gland.

GIANT PANDA. Raccoons and Pandas.

GIANT SEQUOIA. Of the family *Taxodiaceae* (swamp cypress family), genus *Sequoiadendron*, the Giant Sequoia (*S. giganteum*) is the only species of this genus. The champion tree, as selected by The American Forestry Association, is the "General Sherman," located in Sequoia National Park, California. See Fig. 1. This specimen has a circumference of 83 feet (25.6 meters) 11 inches at $4\frac{1}{2}$ feet (1.4 meters) above ground level, a height of 272 feet (82.9 meters) and a spread of 90 feet (27.4 meters)—as measured in 1972. In dimensions, the Giant Sequoia is rivaled only by the coast redwoods. See **Redwood (Coast).**

The Giant Sequoias are found on the western slopes of the Sierra Nevada Mountains of California at an altitude of from 4,500 to 8,000 feet (1,372 to 2,438 meters). There are over 25 isolated groves in which the trees occur, the taller and more dense trees being found on the northwestern slopes. The first grove was found in 1852 by a miner, A. T. Dowd. Now known as the Calaveras North Grove, it consists of about 50 acres (20 hectares) of these trees.

The bark is from 1 to 2 feet (0.3 to 0.6 meters) thick with furrows 4 to 5 inches (10 to 12.5 centimeters) wide. The bark is a red-brown color. The outer scales are fibrous and grayish-purple in color; the inner scales are a cinnamon red. The bark provides outstanding protection against the hazards of fire. The cones are deeply pitted and are of a red-brown color. The twig also is a cinnamon color and is scaly. The flower is green-gold, with pollen raining down profusely when in bloom. However, most new trees rise from shoots from stumps or roots. The leaf is from $\frac{1}{8}$ to $\frac{1}{4}$ inch (3 to 6 millimeters) long, overlapping the twig. The leaf is sharply pointed, dark green, and glossy. The wood is light in weight and not considered prime timber because it is soft, brittle but spongy, weak, and coarse-grained. At one time, the trees were cut for timber, but are now protected. Timbering operations are now concentrated on the coastal redwoods where extensive reforestation programs have been in effect for a number of years.

The Giant Sequoias also are referred to as the "Big Trees." It is important that a distinction be drawn between the coastal redwoods and the Giant Sequoias because the nomenclature can be quite confusing. Collectively, both genera are frequently referred to as redwoods. The physical differences between the two genera, however, are clearly obvious from Fig. 2.

GIANT SEQUOIA. Genus: *Sequoiadendron*; Species: *giganteum*
—also called "Big Tree," or Sierra Redwood
—Grows inland on the slopes of the Sierra Nevada Mountains

COAST REDWOOD. Genus: *Sequoia*; Species: *sempervirens*
—Sometimes also called California Redwood. The timber usually is simply referred to as redwood.
—Grows along a comparatively narrow coastal fog strip

Although the coast redwoods are taller, the extremely large girth of the Giant Sequoia qualifies it as the largest, most massive of living things. It is estimated that the tree lives for 3,000 to 4,000 years or more and thus is second only to the bristlecone pines as among the oldest living species. See **Pine Trees.**

The "Big Tree" is highly regarded in Europe, where it was introduced shortly after its discovery in California. It is known in Europe as the "Wellingtonia." Weather and soil conditions in Britain in particular appear to be well suited to the growth of the tree. As of the mid-1970s, the tallest of these introduced trees had attained a height of over 165 feet (50.3 meters). It is located in Devonshire.

Fig. 1. The "General Sherman" tree, revered specimen of the Giant Sequoia, located in Sequoia National Park, California. (*National Park Service.*)

In 1864, President Abraham Lincoln authorized a federal grant transferring the area known as the Yosemite Valley and the Mariposa Grove of Redwoods to California. This act marked the beginning of the state park concept, not just for California, but the entire nation. These properties were subsequently returned to the federal government to become part of Yosemite National Park. The first of California's present-day parks, the California Redwood Park at Big Basin, Santa Cruz County, was created in 1902, following public pressure to preserve the redwoods. This was followed by state acquisition of other notable redwood groves.

See also **Conifers;** and **Redwood (Coast).** For references, see **Tree.**

GIARDIASIS. Intestinal Protozoa.

Fig. 2. *Sequoia sempervirens* (coast redwood) at left; *Sequoiadendron giganteum* (the Giant Sequoia) at right.

GIBBERELLIC ACID AND GIBBERELLIN PLANT GROWTH HORMONES. These organic chemical compounds, first isolated from the parasitic fungus *Gibberella fujikuori* in Japan in the late 1930s, produce unusual results when applied to plants, including various food crops. The results can be advantageous or disadvantageous. The phenomena of the gibberellins were uncovered as the result of studying the excessive leaf elongation in rice plants. This fungus disease of rice is sometimes referred to as the "foolish seedling" disease in rice. When infected with this fungus, the rice plants grow ridiculously tall and the stems break before the plants can flower and produce seed. When experimentally applied to higher plants, the gibberellins have varied effects. The most common reaction is the rapid lengthening of the stems. The stems of citrus trees, for example have been stimulated to grow at a rate six times greater than normal. When applied to the young fruit of seedless grapes, the gibberellins cause the fruit to grow much larger and to stay on the vine longer. Although some results can be predicted from experience with other species, generally results must be observed through long trial-and-error experimentation with many plants and many different concentrations and forms of the chemical growth hormones. The gibberellins are but one category of several kinds of plant hormones which affect food crop production. See also **Plant Growth Modification and Regulation.**

Since the 1960s, commercial gibberellin formulations have been available. These take several forms, ranging from liquid concentrates through tablets and powders. In some countries, registration is required of these compounds. The following practical results, among others, have been achieved when gibberellins are used properly on certain food plants:

Artichoke: prolongs picking period
Barley: enzyme content increased
Bean: more rapid emergence of plant
Blueberry: better fruit set
Celery: extends winter crop
Cherry (sour): combats cherry yellow virus
Cucumber: produces staminate flowers
Grape: loosens and elongates clusters; increases grape size
Hops: increases yields; aids harvesting
Lemon: delays yellow color development
Oats: promotes more rapid emergence of plant
Orange (navel): retards aging of rind
Lettuce: increases seed production; effects uniform bolting
Potato: stimulates sprouting
Prune (Italian): increases yield; reduces internal browning
Rhubarb: for forced crops, increases yield
Rye: promotes more rapid emergence of plant
Soybean: Promotes more rapid emergence of plant
Sugarcane: increases sucrose yield
Tangerine: increases yield and fruit set
Wheat: promotes more rapid emergence of plant

The gibberellins are actually a family of closely related substances. To date, structures have been determined for well over a dozen of these and a number have been isolated from higher plants. See accompanying diagrams. The structure of three fused saturated or nearly saturated rings, with two additional rings perpendicular to them, suggests relationship to the diterpens for which there is strong isotopic evidence. For example, C^{14}-kaurene is readily converted to gibberellic acid (GA_3) by *Gibberella* cultures. The biosynthesis is apparently inhibited by chlorocholine, which is suspected as the basis for the dwarf-

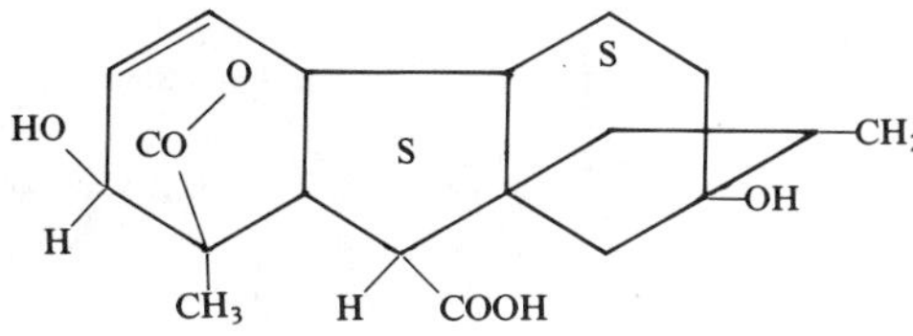

Gibberellic acid (GA_3)

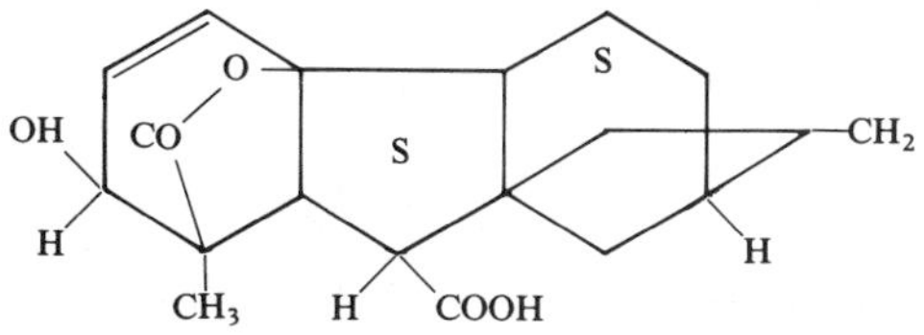

Gibberellic acid (GA_7)

ing action of this compound. GA_7 to date has had the highest activity in most tests.

Gibberellins cause rapid elongation of shoots; many of the dwarf forms of maize (corn), bean, pea, and morning glory (closely allied to sweet potato) are caused to grow into tall forms indistinguishable from their tall genetic relatives. Many long-day plants are brought into flower in short days by gibberellin, and some biennials, including *Hyoscyamus* (henbane), are made to flower in one year. This process depends on the activation of cell divisions in the shoot apex. Like auxins (other plant hormones), gibberellins produce parthenocarpic fruits, especially on tomato, but unlike auxins, they do not inhibit lateral bud development, but they inhibit rooting of cuttings and promote the germination of many seeds. Their transport shows no polarity. They are active at concentrations comparable to those of the auxins. There is good evidence that the gibberellins act only when auxin is present.

In their biological function, it is believed that the gibberellins destroy or bypass naturally occurring inhibitors which normally prevent premature germination. However, high concentrations of the gibberellins and like substances actually prevent germination in certain varieties of seed.

GIBBON. Anthropoids.

GIBBS DIVISION SURFACE. Consider a system consisting of two homogeneous bulk phases α and β separated by a surface phase. The concentrations vary continuously through the surface phase from those of the interior of one phase to those of the interior of the other. In order to give a well-defined meaning to the thermodynamic functions of the surface phase, independently of the exact position of the boundaries of the surface layer, it is useful, following Gibbs, to replace the real surface phase by a geometrical surface. The bulk phases are considered to be homogeneous up to this geometrical surface, which is called the Gibbs division surface. See figure.

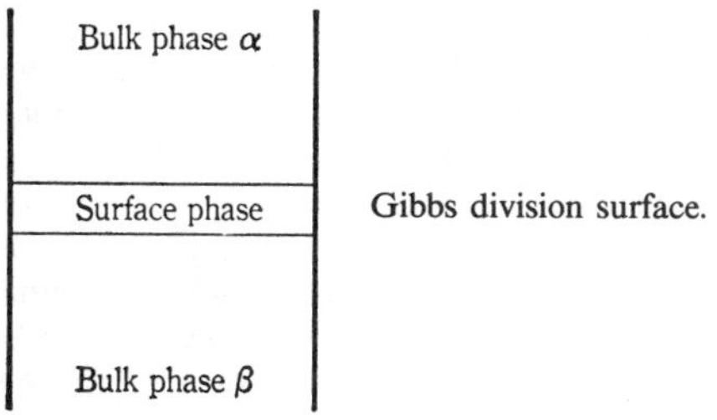

GIBBS-DUHEM EQUATION. In a system of two or more components at constant temperature and pressure, the sum of the changes for the various components, of any partial molar quantity, each multiplied by the number of moles of the component present, is zero. The special case of two components is the basis of the Gibbs-Duhen equation of the form:

$$n_1\, d\overline{X}_1 = -n_2\, d\overline{X}_2$$

in which n_1 and n_2 are the number of moles of the respective components and $\overline{X}_1$ and $\overline{X}_2$ are the partial molar values of any extensive property of the components.

GIBBS FORMULA (Surface Tension). Surface Tension (Gibbs Formula).

GIBBS FUNCTION. Thermodynamics.

GIBBS-HELMHOLTZ EQUATION. A thermodynamic relationship useful in calculating changes in the energy or enthalpy (heat content) of a system, from certain other data. Two useful general forms of this equation are:

$$\Delta A - \Delta U = T\left(\frac{\partial(\Delta A)}{\partial T}\right)_V$$

$$\Delta G - \Delta H = T\left(\frac{\partial(\Delta G)}{\partial T}\right)_P$$

in which A is the Helmholtz free energy, defined in this book under Free Energy (2), U is the internal energy of the system, T is the absolute temperature, V is the volume, P is the pressure, G is the Gibbs free energy (see **Free Energy**), and H is the heat content of the system.

For a reversible cell, if the heat of the chemical reaction taking place in the cell is ΔH, F is the Faraday constant and the reaction takes place by the migration of an ion bearing a charge j, then

$$\Delta H = jF\left(\epsilon - T\frac{d\epsilon}{dT}\right)$$

where ϵ is the emf of the cell.

GIBBS-KONOVALOV THEOREMS. Consider a binary system containing two phases (i.e., liquid and vapor). Both components can pass from one phase to another. The Gibbs-Konovalov theorems refer to the properties of the phase diagrams of such systems (see **Azeotropic System**). The first theorem is: *At constant pressure, the temperature of coexistence passes through an extreme value (maximum, minimum or inflexion with a horizontal value) if the composition of the two phases is the same, and conversely at a point at which the temperature passes through an extreme value, the phases have the same composition.* The second theorem is similar. It refers to the coexistence pressure at constant temperature.

GIBBS PARADOX. When two samples of the same gas at a given temperature and pressure are allowed to mingle by the removal of a separating partition, the entropy of the resulting system is equal to the sum of the entropies of the two original parts, and there is no extra term which arises when the two original systems are composed of different gases. This paradoxical absence is called the Gibbs paradox; it can be explained by using the theory of grand canonical ensembles.

GIBBS PHASE RULE. Phase Rule.

GIESELER PLASTICITY. Coal.

GIGA PREFIX. Units and Standards.

GILA MONSTER (*Reptilia, Sauria*). A poisonous lizard, *Heloderma suspectum*, of the southwestern deserts. It attains a length of 18 inches

Gila monster. (*A. M. Winchester.*)

(45.7 centimeters), is thick-bodied, with a stubby tail. The skin bears rounded tubercles instead of flat scales and is black or blackish with pink to yellow markings.

GILBERT. Units and Standards.

GILBERT'S DISEASE. Bile.

GILL. A respiratory organ for the extraction of oxygen from the water and for the liberation of carbon dioxide.

Many small aquatic animals absorb oxygen through the surface of the body generally but the more complex forms have localized respiratory organs formed to present an adequate surface. They are usually thin plates of tissue or slender tufted processes and, with the exception of some aquatic insects, they contain blood or coelomic fluid which absorbs oxygen through their thin walls. In the insects a unique type of respiratory organ is the tracheal gill which contains air tubes. The oxygen of these tubes is renewed in the gills.

Gills are developed in starfishes and sea urchins (see **Echinoidea**) as thin protuberances on the surface of the body containing diverticula of the water vascular system. In the crustaceans, mollusks, and some insects they are tufted or plate-like structures at the surface of the body in which blood circulates. The gills of other insects are of the tracheal type and also include both thin plates and tufted structures, and in the larval dragonfly the wall of the caudal end of the alimentary tract (rectum) is richly supplied with tracheae as a rectal gill. Water pumped into and out of the rectum supplies oxygen to the closed tracheae.

Gills of vertebrates are developed in the walls of the pharynx along a series of gill slits opening to the exterior. Water taken into the mouth passes out of the slits, bathing the gills as it passes. Some fishes utilize the gills for the excretion of electrolytes. In some of the amphibians the gills occupy a similar position on the body but protrude as external tufts.

Gill Chamber. A partially enclosed space containing gills. In many invertebrates external gills project from the surface of the body. Such structures are very delicate and in many species are protected by folds of the body wall. The crayfish offers a good example, with the carapace extended down on each side of the body to form the outer wall of a chamber in which the gills lie.

Gill Filament. A thread-like component of a gill. Also the ciliated ridges of the gills of bivalve mollusks.

Gill Plate. The respiratory organ of some bivalve mollusks. It is formed of two thin plates or lamellae, each made up of united ctenidial filaments (ctenidium), and contains passages communicating with the mantle cavity and with the chamber above the gills. Water passes into these passages from the mantle cavity.

Gill Raker. A comb-like structure along the inner margin of the gill arches of fishes. These combs prevent the passage of food into the gill slits and direct it toward the esophagus.

Gill Slit. A perforation of the body wall of vertebrates opening into the pharynx. In the fishes and amphibians the slits are associated with the gills, but in terrestrial vertebrates they occur only in the embryo, and in mammals they usually fail to open. The gill slits are paired, opening as a series on each side of the body. In the lampreys and most cartilaginous fishes (sharks, etc.), the openings are externally separate. In the bony fishes, those of each side are covered by an operculum.

See also **Fishes.**

GILSONITE (or Uintaite). The mineral Gilsonite, named for S. H. Gilson of Salt Lake City, is a variety of asphaltum that occurs in Uinta County, Utah. It is found in black lustrous masses which ignite easily. A less frequently used name for it is uintaite.

GILSONITE INSULATION. Insulation (Thermal).

GIMBAL. 1. A device with two mutually perpendicular and intersecting axes of rotation, thus giving free angular movement in two directions, on which an engine or other object may be mounted. 2. In a gyroscope, a support which provides the spin axis with a degree of freedom. The outer and inner gimbals of a pendulous two-axis gyro are shown in the accompanying diagram. See also **Gyroscope.**

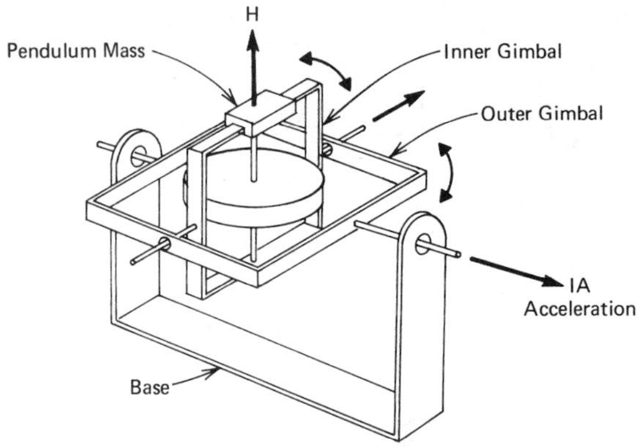

Gimbal arrangement in a pendulous two-axis gyro.

GIN (Cotton). Cotton.

GINGER. The dried rhizome (rootlike stem) of a perennial monocotyledonous plant, probably native to tropical Asia. Ginger is used mainly as a condiment and as an aromatic stimulant. Several volatile oils are responsible for the characteristic odor, while a ketone (gingerol or zingerone) produces the hot, biting taste. Ginger is used in the preparation of ginger ale and a variety of food products. Ginger also appears on the market as preserved ginger, a Chinese product made from uncured rhizomes.

The species of ginger plant used is *Zingiber officinale,* a member of the family *Zingiberaceae* (ginger family). The plant has a fleshy, irregularly branched rhizome from which arise erect leafy stems from 2 to 3 feet (0.6 to 0.9 meter) in height. The leaves are grasslike. The flowers, borne on a separate stem, are yellow and of a distinctive shape resembling that of orchids. The inside tissues of the rhizome are white and richly spotted with resin dots. The plant is propagated by means of rhizome-cuttings, each cutting having an eye or bud which produces an erect stem.

When the leaves begin to turn yellow the plant is ready to harvest. The rhizomes are dug up and cleaned, then immersed in boiling water to kill the buds or eyes, and also to loosen the periderm or outer portion. The rhizomes are then peeled and dried.

GINGIVITIS. Periodontitis.

GINGKO TREE. Maidenhair Tree.

GINI MEAN DIFFERENCE. A measure of dispersion, defined as the average absolute difference between all possible pairs of observations in a sample. As an estimate of the standard deviation of a normal distribution, it is slightly more efficient than the mean deviation but much more difficult to compute.

GINSENG. Of the family *Araliaceae* (ginseng family), this is a relatively small group of herbs and a few small shrubs (or trees), probably best known because of the curative powers which over many centuries the Chinese have attributed to the roots of notably two species, *Panax schinseng* and *P. quinquefolium.* Scientifically, these powers have not been dramatically proved or disproved. The plants are low-growing perennial herbs having compound leaves and compound umbels of small white flowers. They grow best in rich shady woods of hardwood trees. When mature, the thick fleshy roots are removed from the ground, very carefully to avoid any damage. They are then dried and marketed for use in making various brews.

Other members of the ginseng family growing in North America include the *Aralia spinosa,* the Angelica tree or Hercules' club. This plant can grow to a height of nearly 40 feet (12 meters) and has very large doubly compounded leaves, ranging from 2 to 4 feet (0.6 to 1.2 meters) in length. The flowers also are large, white, in clusters approaching 20 inches (51 centimeters) in length. The tree bears a very small black berry which occurs in clusters. The tree is sometimes planted in gardens and for landscaping effects. It occurs naturally from New York south to Florida and Texas and is found in the midwestern states. The devil's club, *Fatsia horrida*, is a rather high shrub, ranging up to about 15 feet (4.5 meters) in height. The leaves are large, the flowers occur in terminal clusters and are of a greenish-white coloration. The shrub prefers rocky soils and ranges widely from the Great Lakes region westward into California, Oregon, and southern Alaska. The devil's club is also found in Japan.

GIRAFFE AND OKAPI (*Mammalia, Artiodactyla*). The group of *Giraffines* is one of the smaller in the order of *Artiodactyla* (even-toed hoofed animals). There are two types of giraffines remaining today: (1) Giraffes (*Giraffinae*) and (2) Okapis (*Palaeotraginae*). Because of their extremely long necks, they represent unusual natural solutions to anatomical and physiological problems.

The giraffe is the tallest of all mammals, the head rising about $18\frac{1}{2}$ feet (5.5 meters) above the ground. The head is long with a wide range of movements. The tail is long, slender, and tufted. There are seven cervical vertebra, each of extra length, giving the animal its greatly elongated neck. The tongue is up to 18 inches (46 centimeters) in length and quite elastic; it can be shaped to a point to reach tiny branches. The animal prefers the leaves of the mimosa and acacia trees. Because of its great height and long legs, the giraffe must stand with its legs far apart when grazing or drinking. The animal has an ambling walk, with the legs on the same side moving together. When galloping the giraffe can attain a speed of some 30 miles (48 kilometers) per hour. For protection against certain types of predators, the giraffe can kick hard and fast with its front legs. Most giraffes are of a white-to-sandy color with darker maplike patterning. In both sexes, there are two protuberances between the ears that appear much like horns, but are more like raised lumps with skin and tufts of hair on them. However, in some species, a third protuberance or horn is present, making a total of three "bumps" in all. These horns are used only in sparring when rival males engage in what might be termed "necking" combat.

Giraffes. (*A. M. Winchester.*)

Giraffes are found over most of tropical Africa. They do not frequent the closed-canopy forest, but prefer to remain on the drier savannas. They live in communities. Considered to be of a mild disposition, giraffes rely essentially on their keen vision and speed to avoid and escape danger.

There is a misconception that giraffes are voiceless. They can make whimpering and whistling sounds used when calling their young. It is interesting to note that giraffes can go for extended periods without water, essentially rivaling the camel in this respect. These animals cannot swim and are not known to wade even the shallowest of streams or ponds.

The gestation period of the giraffe is 15 months. Well developed before birth, the young giraffe can stand within a few minutes and run within two days. A baby giraffe weighs about 85 pounds (38.5 kilograms). Multiple births are rare.

The Okapi is quite different from the giraffe. Existence of this animal was not learned until early in this century. A skin of one of the animals was returned to England in 1901 by Sir Harry Johnson. Several years followed before complete specimens were located. The okapi is about the size of an ox, standing some 5 feet at the shoulders. It is of a purple coloration that blends in extremely well with the dense forests of Africa. There are wide horizontal stripes on the hind quarters. Small horns with polished tips are present only in the males. They are browsers, preferring the leaves of small shrubs and trees. While the okapi has an elongated neck, it is quite ungiraffe-like in appearance. Proportionately, the head is larger than the giraffe, coloration and markings are entirely different, legs are much shorter, and the body is heavier.

For references, see **Mammalia.**

GIRDER. A girder is a large heavy beam capable of carrying both concentrated and uniformly distributed loads. Large rolled steel beams are frequently called girders although the name is generally applied to large beams which are made up of rolled steel sections connected by rivets or welding. In concrete construction the large beams which are used to support smaller beams are called girders. A girder, like a beam, resists transverse bending, and is loaded, ordinarily, by gravity load which is transferred by the girder to its supports. The common plate girder is a compound steel structure composed of plates and angles, bound together in one structure by the use of rivets or welding. Plate girders are used where strength requirements cannot be met by the largest available rolled steel sections. Due to their adaptability, plate girders are to be found in almost every form of construction embodying steel. Bridges, cranes, and buildings, show many examples of the plate girder.

The built-up plate girder roughly resembles an I-beam in shape. Its area may be thought of as subdivided into area of flanges and area of web. The flange sections are most useful in withstanding the bending, and the web resists most of the shear to which a girder is subjected. The arrangement of plates and angles in a plate girder is shown in the accompanying figure. The girder is built up of a web

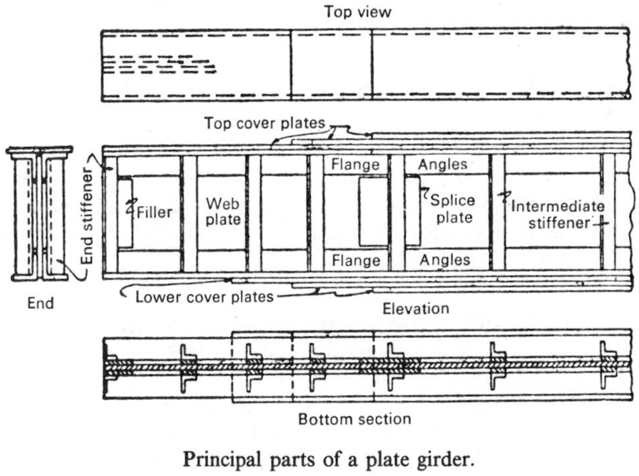

Principal parts of a plate girder.

plate whose depth is nearly equal to the full depth of the girder, flange angles which are riveted near the top and bottom of the web plate, and cover plates that are riveted to the flange angles. Since the flange chiefly resists bending, and bending moment is greatest at the center of a girder (for ordinary load conditions), the cover plate could be of a thickness increasing from minimum at the abutment to maximum at midspan. It is not practicable to specify a tapered plate, but the same effect is achieved by subdividing the total maximum required cover plate area into a number of plates in laminar arrangement, and achieving the taper effect by cutting off the plates where reduction of bending stress permits. Localized buckling of the web must be resisted in order to permit the girder to develop its full strength. For this purpose, stiffeners, consisting of angles arranged vertically, and riveted to the web and to the flange angles, are spaced periodically along the length of the girder. These are called stiffener angles, and may be smaller than the flange angles.

As the girder carries load by beam action, the flexure theory applies. The problem of design of plate girders begins with the computation of bending moment and shear. Generally, bending moment governs the design. A cross-section of the girder is then assumed and the moment of inertia of the same computed. The value of the moment of inertia must be such that the unit stress on the extreme fiber, as computed by the common flexure formula, is not greater than the allowable.

Most authorities require that the design of an important girder be carried through with an exact computation of the moment of inertia of some assumed section. If a determination of an economic section is made by trial and error, this moment of inertia method of design may become quite tedious. The number of trials can be greatly shortened if some approximation, which would guide the designer towards a correct selection of the proper structural shapes, could be employed. Such a method is outlined below. It is based on the assumption that a girder is made up of a simple rectangular web connecting rectangular flanges. Let the area of the web be A_w and the area of each flange A_F, while the distance between the centers of gravity of the area of the flanges is h. The moment of inertia of this assumed area about the neutral axis which is taken to be on the axis of symmetry is

$$I = \frac{h^2}{2}(A_F + A_w/6)$$

If this expression be substituted in the flexure formula the flange area is found to be given by the following equation:

$$A_F = \frac{M}{fh} - \frac{A_w}{6}$$

in which f represents the allowable stress.

As ordinarily given in structural texts, this formula represents the net flange area (area with rivet holes deducted). Consequently it has A_w divided by 8 instead of 6, the difference being accounted for by deduction of a certain amount of web area to account for rivet holes. If the approximate flange area is obtained by some rapid estimating system, such as this flange area method, an arrangement of commercially procurable steel shapes can be set up, and the exact moment of inertia accurately established by the principles of mechanics.

The complete design of a steel plate girder includes also such problems as determining the riveting pitch in the flanges, the design of splices in the web plate, the spacing and riveting of stiffeners, and the strengthening of the ends by end stiffeners where the girder bears on its supports.

GIRDLE. 1. The part of the mantle of a chiton which surrounds the shell plates and contains the spicules characteristic of the group. 2. The skeletal structures of vertebrates by which the appendages are associated with the trunk.

GIZZARD. In some animals, a region of the alimentary tract with thick muscular walls and some adaptation for grinding food. The gizzards of birds are the best known examples. They have a tough lining and their grinding action depends on the movements of hard particles such as gravel contained in them. One of the fishes, the gizzard shad, has a stomach of similar nature. Many insects also have a gizzard but in this organ the supposed grinding structures are chitinous folds and teeth projecting into the cavity. The grinding action of the organ has been questioned by some observers.

GIZZARD SHAD (*Osteichthyes*). A widely distributed North American fish whose stomach is developed like the gizzard of a bird. It occurs in both fresh and salt water. This fish is of the order *Isospondyli*, family *Dorosomidae.* Maximum length is usually about 20 inches (51 centimeters). They are deep-bodied and appear something like a herring. The Atlantic gizzard shad (*Dorosoma cepedianum*) has been successfully introduced as a forage fish in several areas of the United States, notably in the central and eastern regions. The small gizzard shad (*Dorosoma nasus*) is mainly a saltwater species. Found in Australian waters, they are from 6 to 15 inches (15 to 38 centimeters) in length.

GLACIAL DEPOSITS (or Drift). The general term for glacial deposits, or sands, gravels, boulders, etc., which are the result of mountain or continental glaciation. Drift is classified as either stratified drift, the result of deposition by waters from the melting glacier, or, till (unstratified drift) which is apt to be coarsely graded sediments composed of clay, sand, gravel and boulders. Till may grade, in places, into stratified drift, but is principally transported and deposited by the ice. Both stratified drift and till also form distinctive topographic features, to such an extent that both mountain ranges and even broad continental areas which have been subjected to glaciation cannot be described as having been subjected to the normal cycle of erosion.

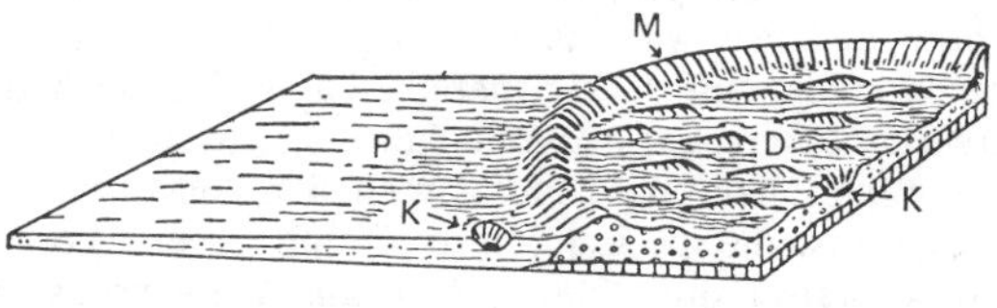

Block diagram showing: (M) a terminal moraine; (P) an outwash plain; (D) drumlins; and (K) kettle holes. (*After A. Penck.*)

When a glacier advances over old drift it may form cigar-shaped hills, called drumlins, whose longer axes are relatively parallel with the movement of the ice. Till, which is built up into long mounds and ridges at the frontal margin of the ice sheet, forms significant topographic features called moraines. The waters coming off from the front of a melting ice-sheet deposit great sheets of stratified gravels, sands and clays. If ice-blocks have been covered by the outwash, when these ice-blocks finally melt they leave depressions in the outwash plain which fill with ground water to form ponds and lakes. These depressions are called kettle holes.

GLACIATION. **Clouds and Cloud Formation; Glacier, Polar Research.**

GLACIER. Wherever upon the earth's surface the temperature is sufficiently low and there is sufficient precipitation to produce a permanent snow field, glaciers may be found. Other things being equal, perpetual snow is more likely to be found in high latitudes and high altitudes; as examples we have the extensive snow and ice field on Greenland and the Antarctic continent as well as valley glaciers of the Alps, of Alaska, the Rocky Mountains, the Andes, the Himalayas and elsewhere. Repeated thawing and freezing of the snow in perpetual snow fields permit the formation of coarse granular ice called nevé which passes into ice of the usual sort. On slopes, the accumulated ice will eventually begin to move, and as it fills a mountain valley, becoming literally a river of ice, it may be called a valley glacier. Even in the absence of great slopes ice will only accumulate to a limited thickness before it commences to spread out in all directions from its place of accumulation. Such a mass of ice is called a continental ice sheet or continental glacier; Greenland is an example of such a sheet of continental ice.

Among well-known glaciers are the Zermatt, Stechelberg, Grindelwald, Trient, Les Diablerets, and Rhone in Switzerland; the Nigards, Gaupne, Fanarak, Lom, and Bøver in Norway; the Wright, Taylor, and Wilson Piedmont glaciers in Antarctica; the Bossons Glacier in

France; and, in the United States, the Emmons and Nisqually glaciers on Mt. Rainier, Washington; Grinnell glacier in Glacier National Park, Montana, the Dinwoody glacier in the Wind River Mountains, Wyoming, the Teton glacier in Teton National Park, Wyoming. And, of course, there are numerous glaciers in the Canadian Rockies.

See also **Polar Research;** and **Ocean.**

References

England, J., and R. S. Bradley: "Past Glacial Activity in the Canadian High Arctic," *Science*, **200**, 265–270 (1978).

Evans, D. L., and H. J. Freeland: "Variations in the Earth's Orbit: Pacemaker of the Ice Ages?" *Science*, **198**, 528–529 (1977).

Hartline, B. K.: "Icebergs and Oil Tankers Soon to Mix," *Science*, **209**, 381 (1980).

Hays, J. D., Imbrie, J., and N. J. Shackleton: "Variations in the Earth's Orbit: Pacemaker of the Ice Ages," *Science*, **194**, 1121–1132 (1976).

Ruddiman, W. F., and A. McIntyre: "Warmth of the Subpolar North Atlantic Ocean during Northern Hemisphere Ice-Sheet Growth," *Science*, **204**, 173–175 (1979).

Staff: "The Surface of the Ice-Age Earth," *Science*, **191**, 1131–1137 (1976).

Stuiver, M., Heusser, C. J., and I. C. Yang: "The North American Glacial History Extended to 75,000 Years Ago," *Science*, **200**, 16–21 (1978).

Sugden, D. W., and B. S. John: "Glaciers and Landscape. A Geomorphological Approach," Arnold, London, 1976.

Whillans, I. M.: "Inland Ice Sheet Thinning Due to Holocene Warmth," *Science*, **201**, 1014–1016 (1978).

GLAN AIR-SPACED PRISM. Prism (Optics).

GLANCING ANGLE. Two common uses of the term glancing angle are: 1. The angle between a ray and the tangent plane to a surface. The complement of the angle of incidence. 2. The term is often used as a modifier, to indicate the incidence of a beam at a very small angle with the surface.

GLAND. 1. In valve and piping terminology, a gland is a movable part that compresses the packing on a stuffing box. 2. In biology and medicine, a gland is an organ of epithelial structure which produces secretions necessary to the body, or which excretes waste materials from the system. Glands vary greatly in form and complexity and in the nature of their products.

The simplest glands are unicellular. In the glandular lining of the intestine, for example, are isolated cells which secrete mucus. They are known as goblet cells because the mucus accumulates in a clear ovoid mass above the constricted base of the cell, approximating the form of a goblet.

Multicellular glands develop from the epithelial layers by local increase of cells and consequent expansion of the layer into adjacent spaces or tissues. They include tubular, acinous, and alveolar structures. Tubular glands are slender tubes lined with glandular epithelium; acini are rounded groups of cells with a small central cavity; and alveoli are larger rounded chambers lined with glandular cells. Many of the larger glands of the body, including the pancreas and salivary glands, are made of great numbers of acini borne by complex branching ducts. These glands are said to be compound. In the most complex forms, the secretion may leave the cells by minute canals, or similar canals between the cells may conduct it to the cavity of the acinus. This cavity empties into a short secretory duct lined with gland cells, and this in turn into the excretory duct. These smaller ducts join to form larger and larger passages, ultimately reaching the main duct which delivers the secretion of the entire gland to its destination. See accompanying diagram.

Classification of Glands. Glands may be divided into three major types: (1) Glands of external secretion whose products are discharged through ducts—also identified as *exocrine glands* and include such glands as sweat, stomach, and salivary glands; (2) glands of internal secretion, the *endocrine* or *ductless glands* (see **Endocrine System; Hormones**); and (3) glands that have both external and internal secretion.

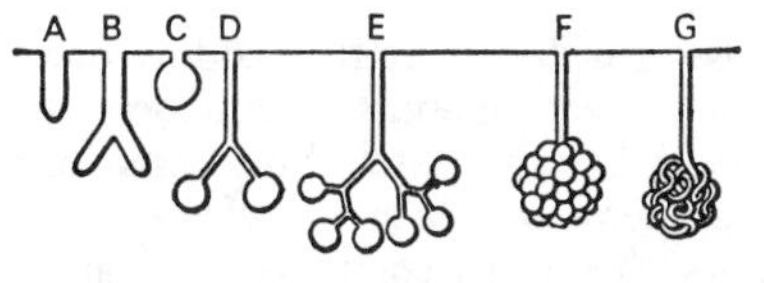

Types of glands: (A) Simple tubular; (B) branched tubular; (C) simple acinous; (D and E) branched acinous; (F) compound acinous; (G) composed tubular. (*Kimber and Gray, "Textbook of Anatomy and Physiology," The Macmillan Co.*)

Glands which produce cells are known as cytogenic glands. They include the reproductive glands which produce germ cells and the spleen, lymph glands, and red bone marrow, in which blood cells develop. See also **Blood;** and **Gonads.**

Special glands are derived from all germ layers and are associated with all organic systems. They serve for hormone production, for lubrication, to prevent drying, for defense, in reproduction, and in numerous other biochemical ways in practically all forms of life.

GLAND (Packing). Fluid Seal.

GLANDS (Endocrine). Endocrine System; Hormones.

GLANDULAR FEVER. Infectious Mononucleosis.

GLAN THOMPSON PRISM. Prism (Optics).

GLASS. Glass is an inorganic product of fusion which has cooled to a rigid solid without undergoing crystallization. It is a solid. It may be transparent, translucent, or opaque, and it may be colored. The chemical composition and corresponding properties may vary over a wide range. Glass will support a load, and may be shaped, broken, or cut. It is much like other solid materials, and yet it is unique.

Its uniqueness becomes obvious when it is examined on a submicroscopic level. Most solids have regular, orderly patterns for the arrangement of atoms, molecules, and ions, but glassy materials are highly disordered. There is some short-range order in glass, but beyond one or two atoms or ions the ordering may be described as random. Thus, on a submicroscopic level, glassy solids look more like liquids than solids.

Since glasses do not have ordered structures with correspondingly specific bonding energies between rows, stacks, planes, or discrete ions, they do not have definite melting points. When a glassy material is heated, it softens slowly and transforms to the liquid state. Crystalline solids generally transform from a solid to a liquid at a single specific temperature, the melting point. On cooling, a material that has a tendency to crystallize to solid will do so at the same temperature at which it transformed to a liquid. When a glass is cooled from a high temperature, it becomes increasingly viscous in a manner which is related to the inverse of the temperature until it becomes a rigid solid again. Thus, a specific temperature where melting or freezing takes place cannot be found for glass; i.e., glass does not have a melting point.

Most glasses can be made to crystallize if they are subjected to the right conditions of temperature and rate of cooling, which suggests that the glassy state is like a supercooled liquid. This is not borne out by measurements of density and other volume properties, which do not decrease in a linear manner as glass is cooled below its crystallization temperature.

Why is it that some melts when cooled through a crystallization temperature form glasses while others do not? It is simply a question of whether the melt can be cooled through the temperature range of maximum crystal growth rate faster than the crystals can grow. Thus table salt cannot be formed as a glass, but sand, or SiO_2, can be. The maximum crystal growth rate is normally just below the melting point of the material, but materials that tend to form glasses easily are much more viscous at these temperatures. For example, in the extreme cases of salt and sand, the differences in viscosities at their respective melting points is about eight orders!

The two-dimensional drawing in Fig. 1 shows SiO_2 in the ordered, or crystalline, and in the random, or glassy, state to illustrate the difference on a submicroscopic scale. Figure 2 shows how the volume properties of a material would respond to temperature if they could be prepared as a glass, a supercooled liquid, or crystalline material.

Most glasses are composed of inorganic oxides, and most commercial glasses contain SiO_2 as their major constituent, but there are

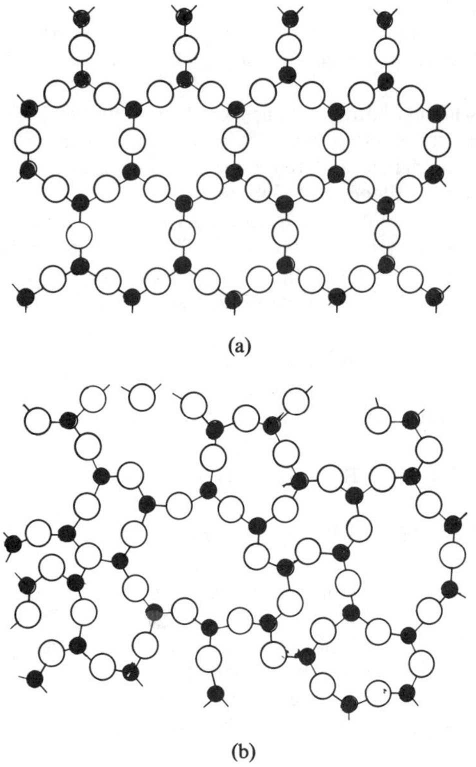

Fig. 1. Silicon dioxide (SiO_2): (a) crystalline; and (b) glassy state. ● = silicon; ○ = oxygen.

organic glasses and elemental metallic glasses. Glass is typically hard and brittle, and exhibits a conchoidal fracture. Most commercial glasses are transparent or translucent in the visible portion of the spectrum.

Types of Glasses

A wide range of glass products exists, each type having special properties. The properties of glass are determined primarily by chemical composition, and since the composition may be varied almost infinitely, there are many thousands of different glasses. However, they may be generally classified into soda-lime-silica glasses; lead glasses; borosilicate glasses; and a number of special glasses, including solder glasses, laser glasses, silica glass, glass-ceramics, and colored glasses. These types essentially bracket the commercial glasses.

Soda-Lime-Silica Glasses. This is the most important group in terms of tonnage melted and variety of use. The combination of silica sand, soda ash, and limestone produces a glass that is easily melted and shaped and has good chemical durability. The raw materials are indigenous to most areas of the world and inexpensive. Soda-lime glasses are particularly suited to automatic-machine-forming methods and are the basis for most of the bottle-, sheet-, and window-glass industry.

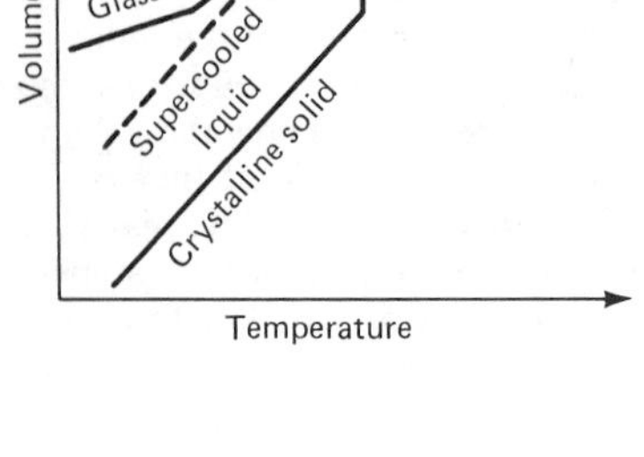

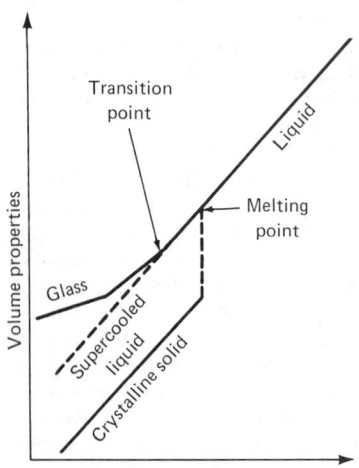

Fig. 2. Volume properties of glass in contract with crystalline solids as a function of temperature.

Very small amounts (often less than 1% of the total batch) of alumina, magnesia, boric oxide, and other chemicals are added to act as stabilizers and to increase durability.

Lead Glasses. The glasses of this group, composed basically of silica sand and lead oxide, have a high refractive index and high electrical resistivity. Potash is present as a significant constituent in most of these glasses. The slow rate of increase in viscosity with decrease in temperature makes lead glass particularly suitable to hand fabrication. The amount of lead may vary considerably, even up to 92% lead oxide; it is a more expensive glass, as the raw materials are relatively expensive and special care is needed in melting to avoid bubbles and seeds. Glasses of this type are used in high-quality art and tableware and for special electrical applications.

Borosilicate Glasses. This group of glasses is basically a combination of silica sand with boric oxide and soda ash. The glasses have excellent chemical durability and electrical properties, and their low thermal expansion yeilds a glass with a high resistance to thermal shock. High durability makes them ideal for demanding industrial and domestic use, such as chemical laboratory ware, cook ware, and pharmaceutical ware. These glasses were developed in the early part of this century to cope with the problem of cold rain on hot railway-signal lights.

Special Glasses

Solder Glasses. These glasses have low softening and annealing temperatures together with expansion characteristics which permit them to be used as intermediate glasses in making seals between two glass surfaces, between a glass and a metal, or between two ceramic surfaces. In fact, solder glass might be described as a high-grade glass glue. Normally, sealing temperatures are well below the annealing temperature of the glass being sealed, and there is little permanent effect on the glass parts being joined. The major constituents of these glasses include lead oxide, boric oxide, and zinc oxide.

Laser Glasses. Glass has various characteristics which make it an ideal laser host material. Its random structure permits broad emission and absorption bands, which provide higher efficiency, more energy storage, and greater energy per pulse than any other material. In addition, most lasing ions are easily soluble in the glass, and rods, fibers, or disks of any size and of high optical quality are easily fabricated. Of the several rare-earth ions which have been made to lase in a glass host, only neodymium has received commercial application. When a neodymium glass lases, it emits light at a rather fixed wavelength of 1.06 nm. Neodymium-doped silica and phosphate glasses have been used to provide the energy source for laser fusion research throughout the world.

Silica Glass. A glass composed of silicon dioxide as the only constituent has a very high softening temperature and a very low thermal expansion. It is costly to make and fabricate because temperature in excess of 1800°C is required to manufacture it. However, its refractory character coupled with its very high resistance to thermal shock makes it ideal for special laboratory equipment, windows in high-temperature environments, and instruments.

Glass-Ceramics. These materials are formed in the same manner as conventional glasses and then subjected to heat treatments which caused controlled nucleation and crystallization. Although nearly completely crystalline, their properties can range from transparent to opaque; electrically insulating to weakly conducting; hard to machineable; and with zero or negative thermal expansions depending upon the composition and heat treatment. This family of materials is based on glasses whose major constituents are magnesium oxide, lithium oxide, aluminum oxide, and silicon dioxide. The crystalline phase or phases and their morphology control the properties of the materials, but the starting chemical composition and the heat treatment determine which crystalline phases will result. Glass-ceramics, which are the result of recent research efforts, have found applications as household cooking ware, reflective optics substrates, chemical processing components, and cooking-stove tops.

Colored Glasses. Nearly all glasses can be colored by adding one or more colorants to the batch in correct amounts. Production of some colors requires, or is enhanced by, the state of oxidation of the coloring agents and the atmospheres in which the glasses are melted. Table 1 indicates the colors obtainable, colorants used, and chemical states required or utilized.

TABLE 1. COMMONLY USED INGREDIENTS FOR COLORING GLASS

Glass Color	Coloring Agent	State
Red	Cadmium sulfide, cadmium selenide	Reduced
	Cuprous oxide	Reduced
	Gold (metal)	
Yellow	Cerium oxide with titanium oxide	
Yellow-green ..	Chromic oxide	Oxidized
Blue-green	Iron chromite	Reduced
Blue	Cobalt oxide	
Purple	Neodymium oxide	
Gray	Nickel oxide with titanium oxide	
Black	Copper, cobalt, nickel, and iron oxides in combinations of two or more	
Amber	Iron sulfide	Reduced
Flint (or colorless)	Selenium and cobalt oxide*	Oxidized

* Selenium and cobalt are used in flint glass to add red and blue hues in amounts only sufficient to balance the green hue resulting from iron oxide present as impurity in most naturally occurring raw materials. The intended result is an even light transmission over the whole visible spectrum.

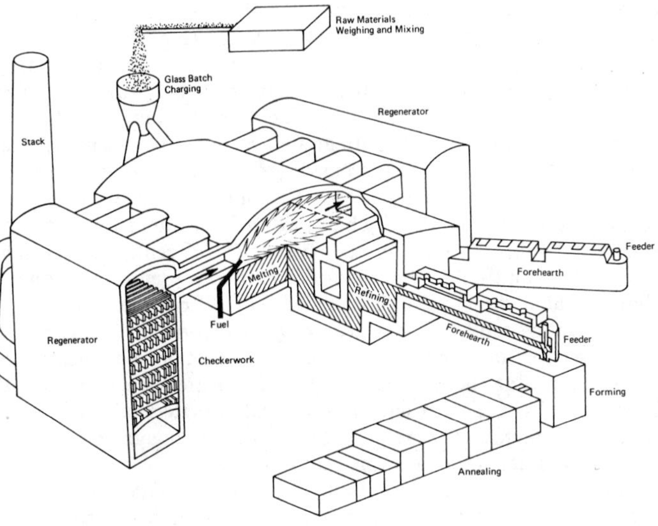

Fig. 3. Representative glass-producing facility.

While the preceding paragraphs describe several classes of glass, within each class there can be infinite composition variations to fit the exact requirements of the user. Table 2 shows typical composition ranges for commercial glasses.

Manufacturing Processes

Glass products are many and varied, and glass compositions range rather widely, depending on the desired products. Figure 3 shows a typical cross section of a glass manufacturing facility. Raw-materials weighing, mixing, charging, and melting are common requirements regardless of the forming operation that is to follow. Most melting furnaces have a primary melting area, followed by a refining or homogenizing section, which is connected to the forming operation by channels called feeders. Although fiber glass is not passed through an annealing furnace after it is formed, most other glass products are annealed to relieve stresses caused by uneven cooling during and immediately after forming.

It is apparent that although there are several similar steps in all glass-manufacturing processes, the forming operations are the most diverse.

Batch Preparation. This begins with the selection, procurement, and storage of an adequate quantity of the raw materials. Selection is made on the basis of the oxides which each material contains and will provide to a glass and on the basis of purity and grain size. Naturally occurring raw materials are used wherever possible for economy, e.g., silica sand, limestones, feldspars, borates, soda ash, boric acid, potash, and barium carbonate. The prescribed quantities of these raw materials, depending on their chemical composition, are measured carefully and mixed together to provide a homogeneous batch. Such mixing is done on an intermittent or a continuous basis, depending on the volume of batch needed to charge the furnaces. The batch is conveyed by a variety of means to the furnaces but always in such a way that segregation is avoided. The importance of grain size of the various raw materials becomes evident in preventing dusting and/or segregation.

Furnaces. A variety of furnaces are used in the industry to melt the batch to produce glass. They must all accomplish the two purposes of confining the heat to the necessary area and containing the melted glass within the furnace. Crucibles or pots are sometimes used to

TABLE 2. COMPOSITION OF COMMERCIAL GLASSES (Weight Percent)

	Soda-Lime-Silica Glass								
	Containers	Plate and Window Glass	Tableware	Fiber Glass Fabrics and Insulation	Borosilicate Glass	Laser Glass	Solder Glass	Lead Glass	Glass-Ceramics
SiO_2	70–74	71–74	71–74	65–74	70–82	61–69	0.5–16	35–70	62–70
Al_2O_3	1.5–2.5	1–2	0.5–2	2–4.5	2–7.5	0–5	0.1–4	0.5–2.0	17–22
B_2O_3				3–5.5	9–14		7–20		
Li_2O									3–5
Na_2O				8–16		12–24		4–8	
	13–16	12–15	13–15		3–8				
K_2O				0–1				5–10	
CaO			5.5–7.5	5–16					0–5
	10–14	8–12			0.1–1.2	3–10			
MgO			4.0–6.5	3–5.5					0–7
BaO					0–2.5		0–4		
ZnO							7–62		
PbO							4–77	12–60	
CuO							0–10		
Nd_2O_3						1–6			
CeO_2						0.1–1			
F_2							0–2		
ZrO_2 and TiO_2									3–10

contain the batch and the melted glass, in which cases the furnace merely retains heat; however, tank furnaces (Fig. 3) are far more common. They are so constructed that the lower portion contains the glass and the superstructure retains the heat and provides combustion space for the fuels used. "Day" tanks are used in some instances where the operation is intermittent and the quantity of glass is small. The great majority of glass produced is melted in continuous furnaces, which are charged initially with batch and cullet (broken-up pieces of previously melted glass) which are melted, filling the tank to a specified depth, sometimes up to 66 inches (~168 centimeters). Thereafter, batch and cullet are charged continuously at a rate equal to that at which the molten glass is withdrawn from the working end.

Continuous tank furnaces are designed to provide for a separate melter section and a refiner or conditioning section. The melting end is maintained at the necessary high temperatures to acoomplish the melting and chemical reactions of the batch materials. The refining, or conditioning, section retains the glass long enough for it to cool to the necessary lower working temperatures.

Glass-melting furnaces are built of refractory materials of various types which will withstand the severe conditions to which they are exposed. The lower portion of the melter section, for instance, must be of the highest quality to withstand the corrosive action of the glass as well as the high temperatures used. Some sections may use lower-quality refractories because the temperature or corrosion conditions are not as severe.

Fuels used in today's furnaces in the United States are natural gas or oil. The fuel is fed to burners that project flames over the surface of the glass. Nearly all continuous furnaces utilize regenerators, which reclaim a portion of the heat from the exhausting combustion gases. Although some glass is melted entirely by the use of electric power, it is generally too expensive to use as a source of energy. When electric power is used to augment the fossil fuels, it is called electric boosting.

For the areas that do have sufficiently low-cost electric power, the furnaces are constructed with conventional bottoms but with superstructure only adequate for initial heat-up. They depend on a blanket of batch floating on the surface of the glass to retain the heat within the tank that is provided by the submerged electrodes. Fresh batch is added to the blanket at a rate equal to the rate of melted glass withdrawn.

Melting. This provides the mutual solution of the oxide materials at high temperatures to yield a homogeneous liquid. Temperatures may range from 1427°C to over 1593°C, depending on the glass composition. Water vapor, entrapped air, and CO_2 are given off, some of which become entrapped in the glass, resulting, initially, in a foamy mass. As the melt moves to the higher-temperature regions, the viscosity is lowered and the gases escape. Deliberate hot spots enhance the natural convection currents, promoting homogeneity. More modern furnaces utilize bubblers, which introduce controlled pulses of air through the furnace bottom, further enhancing convection. This is particularly valuable for increasing temperatures near the tank bottom in melting those glasses which are more opaque to infrared radiation.

The glass is essentially free from bubbles (or seeds) when it reaches the end of the melting chamber. It then passes under floaters in some furnaces, or through submerged throats in most, to the so-called refining section (more properly, the conditioning section). Here the refining or conditioning consists of allowing the glass to increase to a more usable viscosity level by uniformly lowering the temperature, which also allows the remaining tiny seeds or gaseous inclusions to dissolve.

Furnaces supply glass to up to eight forming machines. Forehearths or alcoves serve to channel the glass to the individual machines or machine locations and to further change the temperature and viscosity.

Forming Operations. These are many and varied, involving two, three, or four major steps. The first is a further temperature conditioning to place the glass in the exact viscosity range, sometimes wide but often quite narrow, suitable for the selected primary forming operation. The second step is the primary forming itself, followed usually, but not always, by an annealing step. Single or multiple secondary operations may ensue. Only the major forming processes of drawing, pressing, blowing, and casting will be discussed.

Drawing is one of the simpler forming methods by which thousands of tons of window glass and millions of feet of rod and tubing are produced annually. Drawing window glass frequently utilizes a rectangular refractory frame, called a debiteuse, placed on the surface of the conditioned glass. It has a slot roughly 4–8 in. (10–20 cm) wide and 8 ft (2.4 m) or more long through which the glass is pulled vertically. The width and length of the slot in the debiteuse, together with the drawing speed, aid materially in controlling the width and thickness of the sheet. The upward draw may continue until the sheet is nearly cold, when it can be stored and cracked off in suitable lengths, or it may be bent over a large roller at nearly the last moment it will withstand bending and conveyed horizontally into the annealing lehr.

Glass tubing may be drawn vertically in a manner similar to that for window glass. Another common method is the Danner process, in which a suitable stream of glass is flowed onto a conical rotating mandrel supported with its small end downward and its axis at a suitable angle to the horizontal. The tubing is drawn from the small end, through which sufficient air is blown to retain the desired cross section of the tubing. Drawing continues horizontally over rollers until the tubing can be cracked off in lengths at the cold end.

Plate glass may be formed by flowing the molten glass over the lip of the discharge end of the furnace between a set of large water-cooled rollers and then pulling it away by means of driven rollers. The resulting sheet is up to 1 in. (2.5 cm) or more thick and 10–12 ft. (3–3.7 m) wide. However, most flat glass made throughout the world today is made by the recently developed *float-glass* process. In this process the molten glass is formed into a sheet by floating it on a bath of molten metal such as tin. The glass flowing onto the bath of tin is pulled across the surface and cooled to the temperature at which it is rigid while still on the molten metal. The outstanding advantage of this process is that it produces a plate of glass both surfaces of which require no further polishing.

Modern methods of pressing, blowing, and casting usually involve an intermediate step, the formation of a suitable charge of glass, or gob, for the ensuing operation. The most common method involves a gob feeder located at the end of the forehearth. This consists of a bowl, or spout, kept full of glass by flow from the forehearth and having an orifice in its bottom and a refractory tube suspended in the bowl over the spout. The tube may be lowered to shut off the flow of glass or raised to permit flow at a selected rate. A refractory plunger operates vertically inside the tube. It provides a pumping action on its upstroke, momentarily restraining the flow of the glass. Its downstroke forces the accumulated glass out of the orifice, where it is sheared off. The result is a charge of glass, called a gob, of controlled size which is delivered to the forming machine by gravity.

Pressing, or press-forming, operations normally are used for relatively shallow, heavy-walled products. Pressing is accomplished by means of a metal mold (usually iron or steel), a ring which is centered on top of the mold, and a plunger which is forced into the mold through the ring. The mold shapes the exterior of the product, the ring the top, and the plunger the interior. A pressing machine may have many molds mounted on its circular rotating table, a ring for each mold or, more commonly, a single ring mounted on the same mechanism as the plunger, and a single plunger. After a gob is charged into the mold, the machine indexes one station under the plunger and the plunger moves down into the mold, dwells momentarily, then retracts. It is noteworthy that the plunger action flows the glass into the mold cavity rather than stamping out the product by a quick movement. Since considerable heat is removed from the glass by the plunger, it is cooled with water internally. The product remains in the mold for about half the revolution of the press table before removal to allow it to cool below its deformation temperature. The molds may be cooled by forced air.

Blowing methods work best for deep products and frequently must be used for thin-walled items. A common procedure, called the blow and blow, involves two steps, of which the first is shaping the glass charge into a form called a blank or parison. Gob-fed machines receive the gob in the parison mold, where it is shaped into a cylinder about two-thirds the height of the bottle. The finish, or top, of the bottle is formed in the same operation at the bottom of the mold by action of a small plunger entering the mold from below and delivering a puff of air. A transfer mechanism holding the parison by the completed

Fig. 4. A high-productivity IS machine manufacturing three bottles on each section at the same time. (*Owens-Illinois, Inc.*)

finish then swings and inverts it into a second mold for the second step, blowing the glass into its final shape. A cross section of the molds shows this process in Fig. 4. The most modern machinery for rapidly forming containers and bottles commercially are individual section (IS) machines. Each section is capable of forming up to three gobs at the same time and there are as many as ten sections per machine. The individual sections can be sequenced electronically to produce 300 or more bottles per minute on a 10-section machine. See also Fig. 5.

The *Owens process* employs vacuum to charge the glass into the blank or parison mold. Here, a blank mold dips into a shallow pot of molten glass, a vacuum is applied, and a charge of viscous glass is pulled into the blank mold. The finish is formed simultaneously at the top of the blank. This blank or parison is subsequently transferred into the blow mold, where the bottle is blown into its final form. See Fig. 6.

In another modern machine, the glass flows downward from an orifice in a continuous stream which passes between rollers that flatten it into a ribbon with alternate thick and thin spots. The ribbon is

Fig. 5. Three white-hot bottles immediately after being formed on a section of an IS machine. The bottles will be immediately transferred to an annealing lehr for cooling and annealing. (*Owens-Illinois, Inc.*)

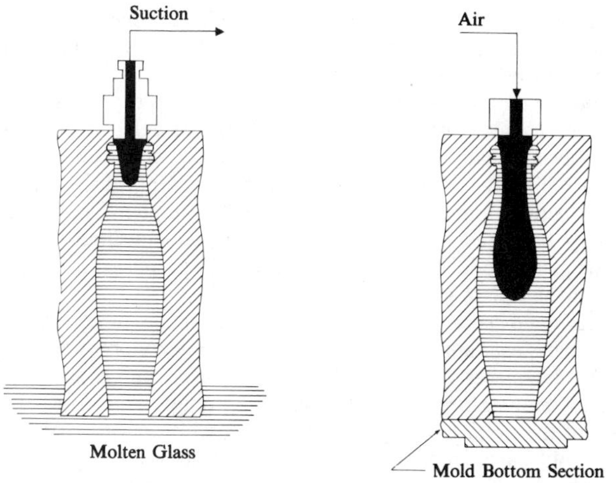

Fig. 6. Owens process. *Left*: Blank mold is dipped into the surface of molten glass, where it is filled by a vacuum suction. As the mold is lifted from the glass, a knife cuts off the glass and closes the mold. *Right*: The blank mold opens, and a puff of air is introduced to shape the parison before transferring it to the blow mold, where it is blown to its final shape.

picked up by a horizontally moving support in which voids coincide with the thick portions of the ribbon. Blow heads on an endless belt operating from above the ribbon provide puffs of air to aid in producing a bulbous sagging in the thick portion of the ribbon. After sufficient sagging, molds on an endless belt close around the sagging glass from below, and air from the blow heads blows the glass into the shape of the mold. After the molds open, the product, frequently light bulbs or Christmas ornaments, can be cracked off the ribbon.

Casting is usually restricted to two types of operations. The first involves the simple pouring of molten glass into molds. Examples include such massive shapes as the borosilicate mirror blank for the Mt. Palomar telescope and the large glass-ceramic mirror blanks for observatories in Australia and South America. The molds are specially constructed for refractory materials.

The second type of casting is spin casting, in which a gob from a gob feeder is fed into the bottom of a metal mold supported so that it can be rotated rapidly or spun on its vertical axis. The centrifugal force thus generated causes the glass to flow up the inclined sides of the mold, producing a conical shape. The initial movement of the glass is aided by insertion of a conical plunger into the glass at the bottom of the mold when spinning is begun. Mold speeds of up to 1,600 rpm are attained within one second. The funnel portion of television tubes are sometimes produced by this method.

Annealing. As with most substances on cooling, the temperature differential between the surface and interior layers of a piece of glass establishes temporary stresses, and the higher this differential the greater the stresses. Fracturing can occur when the stresses exceed the tensile strength of the glass. Permanent stresses can be avoided by carefully controlled cooling from a little below the annealing point to the strain point. This is the annealing range. Thereafter, the rate of cooling need only be such that the temporary stresses do not exceed the tensil strength of the glass. Glass manufacturers have learned to take advantage of these phenomena.

Annealing immediately follows glass-forming operations. In continuous processes, the ware is placed on an endless belt, which carries it through the lehr, a tunnel in which the temperature is carefully controlled. Temperature of the ware is raised initially to near the softening point, then lowered slowly through the annealing range and thereafter at a more rapid rate to the point where it can be packed or stored. The process is designed to result in the degree of permanent stresses desired. Optical glass must be annealed very thoroughly to produce an essentially distortion- and strain-free lens; however, some stresses can be tolerated or become beneficial to most other products. Small rods and tubing, for instance, are strong enough because of their regular cross section to require no annealing, while tempered glass has uniformly controlled stresses to increase its mechanical performance.

Secondary Operations. Lampworking is one of the many and varied operations utilized to produce glassware following the initial forming. The materials used are rod and tubing, which are softened in the flame of burners and shaped or blown as desired.

Grinding and polishing are important steps in many glass-manufacturing processes. Use of a sequence of increasingly finer gradations of abrasives, usually ending with jewelers' rouge or cerium oxide powder for polishing, products the desired results. Optical lenses, prisms, and reflective optics parts are prominent examples. The plate-glass industry has used long lines of grinding and polishing equipment, but the glass produced by the float process has replaced much ground and polished plate glass.

Bending procedures are utilized to produce shapes otherwise difficult to fabricate, e.g., automotive windshields. They are produced by placing the flat pieces of proper shape and size on molds and exposing them to temperatures above the softening point. The glass takes the shape of the mold by sagging or slumping with or without assistance from mold parts contacting the glass from above. Temperatures are maintained sufficiently low and the mold material is such that the surface of the glass is unaffected.

Laminating to produce safety-glass parts, as for automotive windows, is a common practice. A sheet of resin such as polyvinyl butyral is placed between properly sized sheets of glass and the whole exposed to slightly elevated temperatures and pressures to bond the glass tightly to the resin.

Coating of glass products such as containers is quite common, the objective being to protect the container from abuse to which it is subjected in handling during filling and shipping. A coating which is not visible, can be labeled, protects the surface, and provides lubricity is required and usually calls for a two-layer coating such as tin or titanium oxide, followed by a lubricious coating such as polyethylene. The oxide coatings are obtained by subjecting the hot container to a vapor of chloride which oxidizes to the oxide. Thick opaque or translucent oxide and metallic coatings are sometimes used to provide attractive color effects or light protection. Many precision optical lenses are coated with thin, vapor-deposited layers which reduce the light losses by reflection from the surface, and some architectural glass is coated to provide attractive colors and reflect undesirable infrared radiation.

Decorating glass or glassware is an old art that takes many and varied forms. Cutting, grinding, and mechanical or chemical polishing or etching are well known. Opaque, translucent, and transparent enamels can be applied by silk screens or other means in multiple colors and in almost any pattern. Low-melting vitreous enamels have been used for many years, and when properly fired, they provide good durability. More recently, organic polymers have been substituted for the vitreous enamel. They are not quite as durable as vitreous enamels, but they do not require high curing temperatures.

Tempering is the direct reverse of annealing; i.e., high permanent stress is induced in the glass. Rapid cooling or quenching is applied to the glass surfaces at a temperature slightly below the softening point, placing the surfaces in a high degree of compression while the balancing tensile forces are confined to the interior. Since glass always breaks in tension, very considerable strength is incorporated. Typical products are glass doors, windows, goggles, spectacles, and table ware. Tempering must be the final step in the production line. Other products can be strengthened by judicious control of the degree of annealing if their shapes permit it.

Sealing glasses to each other or to other materials must take into account the thermal expansion-and-contraction characteristics. Many glasses have thermal-expansion properties which allow them to be sealed to metals, but each metal usually requires a different glass composition. Solder glasses are used to seal two pieces of glass to each other, two pieces of metal, or a piece of metal and a piece of glass. The glass seals on light bulbs and vacuum tubes are examples of commercial glass-metal seals, while color TV tubes are sealed together with solder glass at a temperature at which the phosphors are not degraded.

See also **Ceramics.**

References

Babcock, C. L.: "Silicate Glass Technology Methods," Wiley, New York, 1977.
Chaudhari, P., and D. Turnbull: "Structure and Properties of Metallic Glasses," *Science*, **199**, 11–21 (1978).
McMillan, P. W.: "Glass Ceramics," 2nd edition, Academic, New York, 1979.
Tooley, F. V. (editor): "The Handbook of Glass Manufacturing," Volumes 1 and 2, Books for Industry, Inc. and *The Glass Industry Magazine*, New York, 1974.
Uhlman, D. R., and N. J. Kreidl (editors): "Glass: Science and Technology," Vol. 5: "Elasticity and Strength in Glasses," Academic, New York, 1980.

Earl D. Dietz, Owens-Illinois, Inc., Toledo, Ohio.

GLASS (Age Determination). **Radioactivity and Other Dating Techniques.**

GLASS-CERAMICS. **Glass.**

GLASS ELECTRODE. **pH (Hydrogen Ion Concentration).**

GLASS FIBERS. **Fiber Glass.**

GLASS FIBERS (Optical). **Optical Fibers; Telephony.**

GLASS INSULATION. **Insulation (Thermal).**

GLASS (Obsidian Volcanic). **Obsidian Volcanic Glass.**

GLASS (Optical). **Optical Glass.**

GLASS SNAKE (*Reptilia, Sauria*). A legless lizard, *Ophisaurus ventralis*, whose tail is exceptionally brittle. Although snake-like, it may be recognized as a lizard by its small ventral scales and its eyelids. Its habitat is chiefly the central and southern part of the United States.

European glass snake. (*New York Zoological Society.*)

GLASS SPONGES. **Hexactinellida.**

GLASS THERMOMETER. **Liquid-in-Glass Thermometer.**

GLASS (Vitreous). **Vitreous State.**

GLAUBERITE. This anhydrous sulfate of sodium and calcium mineral, $Na_2Ca(SO_4)_2$, crystallizes in the monoclinic system. Hardness of 2.5–3, specific gravity of 2.8, with vitreous luster and pale yellow to gray in color. Grades from transparent to translucent. Perfect basal pinacoidal cleavage with conchoidal fracture. Glauberite is a product of salt lake evaporation. World occurrences include the Stassfurt, Germany saline deposits, and Borax Lake in San Bernardino County, California.

GLAUBER'S SALT. **Sodium.**

GLAUCOMA. A disease of the eye characterized by increase in pressure within the eye which causes atrophy of the optic nerve with resultant gradual loss of vision and blindness. It is caused by an inability to eliminate, at an adequate rate, fluid produced by the ciliary body of the eye. As the pressure within the eye rises, the blood supply to the optic nerve is hampered and vision is reduced.

In the United States, it is estimated that 2% of persons over 40 years of age have chronic glaucoma. Because glaucoma is not always discovered and treated promptly, some 4000 persons per year become fully or partially blind. However, where glaucoma is discovered and treated early, the prognosis for useful vision over the life span is excellent. The disease is rare among young people, but the incidence increases with age. Diabetics run a twofold greater risk of having glaucoma than nondiabetics. Infrequently, the appearance of glaucoma is related to glucocorticoid therapy.

In the eye, there is a constant flow of fluid (*aqueous humor*) into

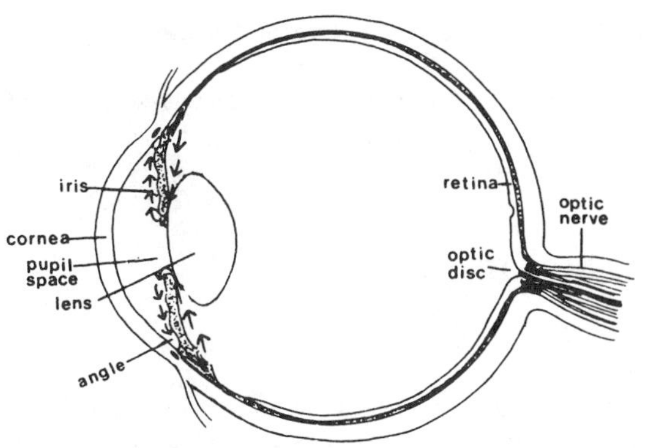

Constant flow of aqueous fluid is indicated by arrows. (*Wills Eye Hospital, Philadelphia, Pennsylvania.*)

and out of the eye. This fluid keeps the eye firm and clear so that the eyeball functions well visually. There is also a constant flow of blood into and out of the eye. The relative state of inflow and outflow of blood and of aqueous humor largely determines how firm the eye is. If the outflow of aqueous humor is blocked, the pressure inside the eye increases. The constant flow of aqueous humor is indicated by the accompanying diagram. This flow may be blocked at any point. Nerve damage occurs first at the optic disc. Elevated intraocular pressure can directly damage the nerves that transmit the electrical impulses from the light-sensitive element of the eye (*retina*) to the brain, where the electrical impulses are processed into images. This pressure also can squeeze out of the eye the blood required to keep the nerves healthy and can, in this fashion, damage the nerves.

The outflow of aqueous humor can be impeded in several ways: (1) the hole (*pupil*), through which the aqueous humor flows as it passes from the back to the front of the iris (colored part of eye) can be blocked by adhesions, or by a cataract. (2) The sieve out of which the aqueous humor exists can become blocked by debris caused by inflammation, or by deposits which are due to aging, or by abnormal material which is the result of certain drugs, or by the iris itself. (3) The veins into which the aqueous humor flows when it leaves the eye can be partially blocked by heart disease, or by pressure on the large veins in the orbit.

Some authorities place glaucoma into four categories: (1) *Chronic open-angle glaucoma*; (2) *acute angle-closure glaucoma*; (3) *congenital glaucoma*; and (4) *secondary glaucoma.* Obviously, the treatment of these various types of glaucoma will be different. Some types require surgery; some need medication; some require attention to other organs of the body; some require that certain medications be halted.

Chronic open-angle glaucoma represents nearly 90% of all cases of the disease and is slow and insidious in its onset. The condition can destroy vision without causing any symptoms of blurring or discomfort. The eye pressure usually rises gradually over a long time span (months or years) due to increased resistance to the outflow of aqueous humor. Loss of vision commences in the periphery or edge of the field of vision and is often not noticed until it has nearly reached the center. The optic nerve may permanently lose most of its function without any discomfort, blurring, or other symptoms. There are, however, subtle symptoms, such as a vague aching around the eyes, haloes, watery eyes, and frequent need to change glasses. When presented with such complaints, the physician will inquire about past incidences of glaucoma among family members. Firm diagnosis will require notation of changes in the optic nerve inside the eye as seen with an ophthalmoscope, changes in the field of vision, usually shrinkage of side vision, and elevated intraocular pressure. The latter measurement is made with a *tonometer,* a simple device with a footplate which rests gently on the cornea (after administration of a local anesthetic). This instrument accurately gages the pressure within the eyeball. Medication is directed toward (1) improving the drainage of fluid from the eye and (2) slowing down the production of fluid. When a patient does not respond to medication, a surgical procedure (*iridectomy*) usually will be performed to relieve the pressure.

Occurring less frequently, *acute angle-closure glaucoma* constitutes a true medical emergency. In this form of the disease, the pressure rises abruptly in one or both eyes from a normal to a very high level. Ocular pain, blurring of vision, haloes around lights, and vomiting are usually, but not always, present in varying degrees. Unless the pressure is relieved in a matter of hours, irretrievable visual loss may occur. Even one day of delay may have a disastrous effect upon a given eye. Treatment is directed first toward reducing the pressure by medical means. As soon as it has been brought to a safe level, a simple but delicate surgical operation is usually performed. If the patient receives expert treatment within a few hours after onset of attack, this operation is likely to result in a permanent cure and the eye may remain glaucoma-free from that time forward. In some cases, however, chronic glaucoma may persist for months or years.

Very rarely, *congenital glaucoma* is present in newborn infants, or appears shortly after birth. Infants thus afflicted are frequently, but not always, born with enlarged eyes. Tearing and unusual sensitivity to light are important signs of infantile glaucoma.

Secondary glaucoma occurs in connection with certain ocular inflammations, tumors, injuries, and hypermature cataracts, among others.

GLAUCONITE. Glauconite is a hydrous silicate of potassium, iron, and aluminum with considerable ionic substitution, crystallizing in the monoclinic system. A general formula is (K,Na) $(Al,Fe^{3+},Mg)_2(Al,Si)_4O_{10}(OH)_2$. It possesses perfect basal cleavage; hardness, 2; specific gravity, 2.4–2.95; color dull green to blue-green; and is often a constituent of marine deposits, forming "green sands." It is believed to have been produced through the alteration of iron-bearing silicates, chiefly biotite and possibly augite and hornblende. It occurs along the Atlantic Coastal Plain of the United States. Frequently found filling the interiors of the shells of *Globigerina,* a common genus of the foraminifera (*Protozoa*). Since *Globigerina* occurs as a deep-sea deposit, some European geologists have claimed that glauconite is only found in deep water. On the other hand, typical green sands occur associated with sand and clays which are certainly of shallow marine origin. Glauconite derives its name from the Greek word meaning *bluish-green.*

Glauconite, being of sedimentary origin, can be used to determine the age of those sediments by evaluating its $^{40}K/^{40}A$ ratio (potassium-argon isotope ratio). See also **Ocean Resources (Mineral).**

GLAUCOPHANE. Glaucophane, essentially a complex silicate of sodium, and iron or aluminum, $Na_2(Mg, Fe)_3Al_2Si_8O_{22}(OH)_2$, is a rather rare mineral although it has been noted from widely separated occurrences. It is monoclinic and ordinarily is fibrous or granular. It is brittle; hardness, 6; specific gravity, 3–3.1; color, azure blue, blackish-blue or gray; luster, vitreous to pearly; translucent to opaque. Glaucophane is found only in the metamorphic rocks sometimes forming glaucophane schists. It is found in Switzerland, Italy, Siberia, Japan and in the United States chiefly in the rocks of the Coast Ranges in California and Oregon. The name glaucophane is derived from the Greek words meaning *bluish-green,* and *appear.*

GLAZES. Ceramics.

GLIAL CELLS. Brain and Nervous System.

GLIDE PATH. Air Traffic Control.

GLIDE PLANE. In solid state physics, this term denotes: 1. a symmetry element of a space lattice, such that the lattice remains unchanged after a reflection in the plane, followed by a translation parallel to the same plane; 2. a slip plane as defined in the theory of dislocations.

GLIDER. Sinking Speed (Aircraft).

GLIDING MAMMALS. Dermaptera.

GLISSON (Capsule of). Liver.

GLOBAR (or Globar Lamp). A ceramic rod consisting largely of silicon carbide (carborundum) which has some electrical conductivity at room temperature and which can be heated to an almost white heat in air without rapid deterioration. It radiates almost like a black body. Globars are used as a radiation source like the Nernst glower in infrared spectrometers.

They have the advantage over Nernst glowers of not requiring a secondary heat source for starting and in being more rugged; however they cannot be made as small as Nernst glowers and, in general, some sort of cooling device, such as water jacket, is necessary.

GLOBULINS. Proteins that are insoluble in water, but that dissolve readily in aqueous salt solutions. The term globulins is applied to certain subgroups of the plasma proteins. See also **Antibody;** and **Blood.**

GLOCHIDIUM. A larval form of certain clams. The eggs hatch into the glochidia larvae which must find a fish to continue their existence. The larvae clamp onto the skin or gills of the fish and live there as parasites until they develop into small clams. This serves as a means of distribution as well as protection and nourishment. The fish will carry these parasites for great distances before they drop off.

GLOGER'S RULE. Adaptation (Ecology).

GLO-LITE TETRA. Characids.

GLOMAR CHALLENGER. Ocean; Ocean Research Vessels.

GLOMERATE. The textural term, proposed by R. M. Field, for a sedimentary rock with a coarse and poorly graded texture, when the origin of the shape of the larger constituents has either been undetermined or is indeterminable.

GLOMERULAR FILTRATION. Urine.

GLOMERULAR FILTRATION RATE (GFR). Kidney and Urinary Tract.

GLOMERULONEPHRITIS. Kindey and Urinary Tract.

GLOMERULUS. Kindey and Urinary Tract.

GLORY RING. Atmospheric Optical Phenomena.

GLOSSITIS. An inflammation of the tongue resulting from nutritional deficiencies or bacterial infections. Taste buds disappear and the tongue becomes smooth and shiny. The condition may indicate pernicious anemia and vitamin B deficiencies.

GLOSSMETER. An instrument for measuring the ratio of the light regularly or specularly reflected from a surface, to the total light reflected.

GLOSSOPTERIS. A genus of fossil plants belonging to the Cycadofilicales.

GLOW DISCHARGE PROCESS. Semiconductor.

GLOW WORM (*Insecta, Coleoptera*). Wingless females of certain beetles. They resemble larvae throughout life and are luminous.

Glow worm also refers to the larvae of the firefly.

GLUCAGON. Diabetes Mellitus.

GLUCOCORTICOIDS. Adrenal Glands; Corticosteroids; Rheumatic Arthritis.

GLUCOCORTICOIDS (In Osteoporosis). Bone.

GLUCOSE. Carbohydrates; Starches; Sweeteners.

GLUCOSE (Brain Utilization of). Brain and Nervous System.

GLUCOSE (Hyperglycemia). Diabetes Mellitus.

GLUCOSE (Hypoglycemia). Adrenal Glands; Hormones; Insulin.

GLUCOSE TESTS. Benedict Solution; Diabetes Mellites; Fehling's Solution.

GLUCOSIDE PIGMENTS. Pigmentation (Planets).

GLUCOSINOLATES. Protein.

GLUCURONIC ACID. Bile.

GLUE. Glue is the product of hydrolysis of proteins, usually of animal origin. Waste products of slaughterhouses (i.e., hide clippings, hoofs, etc.) are the source of most commercial glue. See **Adhesives.**

GLUTAMIC ACID. Amino Acids; Brain and Nervous System; Vitamins.

GLUTAMINE. Amino Acids.

GLUTEN. Starch.

GLUTTON. Mustelines.

GLYCEROL. Glycerol, propanetriol, glycyl alcohol, "glycerine" $CH_2OH \cdot CHOH \cdot CH_2OH$ is a colorless, viscous liquid, of sweetish taste, odorless, boiling point 290°C, or at 12 mm pressure 170°C, gradually solidifies at 0°C to solid, melting point 18°C, miscible in all proportions with water or alcohol, insoluble in ether or chloroform, absorbs water on exposure to the atmosphere. Glycerol reacts (1) with phosphorus pentachloride to form glyceryl trichloride $CH_2Cl \cdot CHCl \cdot CH_2Cl$, (2) with acids to form esters, e.g., glycerol monoacetate $CH_2OH \cdot CHOH \cdot CH_2OOCCH_3$, glycerol diacetate $C_3H_5(OH)(OCOCH_3)_2$, glycerol triacetate, triacetin $CH_2OOCCH_3 \cdot CHOOCCH_3 \cdot CH_2OOCCH_3$, glycerol mononitrates (alpha, $CH_2OH \cdot CHOH \cdot CH_2ONO_2$; beta, $CH_2OH \cdot CHONO_2 \cdot CH_2OH$), glycerol dinitrates (1,2,$CH_2OH \cdot CHONO_2 \cdot CH_2ONO_2$; 1,3, $CH_2ONO_2 \cdot CHOH \cdot CH_2ONO_2$), glyceryl trinitrate, "nitroglycerine" $CH_2ONO_2 \cdot CHONO_2 \cdot CH_2ONO_2$, glyceryl tristearate, tristearin $CH_2OOCC_{17}H_{35} \cdot CHOOCC_{17}H_{35} \cdot CH_2OOCC_{17}H_{35}$, indirectly, glycerol monophosphates (alpha, $CH_2OH \cdot CHOH \cdot CH_2OPO(OH)_2$, beta, $CH_2OH \cdot CHOPO(OH)_2 \cdot CH_2OH$, (3) with oxidizing agents, e.g., dilute nitric acid, to form glyceric acid $CH_2OH \cdot CHOH \cdot COOH$, tartronic acid $COOH \cdot CHOH \cdot COOH$, mesoxalic acid $COOH \cdot CO \cdot COOH$, (4) with phosphorus plus iodine, to form allyl iodide $CH_2{:}CHCH_2I$, which with hydrogen iodide yields propylene $CH_2{:}CHCH_3$ and then isopropyl iodide CH_3CHICH_3, (5) with sodium or sodium hydroxide to form alcoholates, (6) with sodium hydrogen sulfate or phosphorous pentoxide heated, to form acrolein $CH_2{:}CHCHO$. Glycide alcohol

is obtained by treatment of glycerol alphamonochlorohydrin $CH_2OH \cdot CHOH \cdot CH_2Cl$, which is made by reaction of hypochlorous acid and allyl alcohol with barium hydroxide. With hydrogen chloride, glycide alcohol yields epichlorohydrin

Glycerol is obtained (1) from vegetable and animal oils and fats, most of which are mainly glycerol esters of stearic, palmitic and oleic acids by treatment with alkali (sodium hydroxide commonly used), acid (sulfobenzone—or naphthalene—stearic acid, "Twitchell's reagent"), superheated steam, or an enzyme (lipase of castor beans). Glycerol is recovered from the water solution by evaporation under diminished pressure, and purified by treatment with decolorizing carbon followed by filtration, (2) by fermentation of glucose in the presence of yeast and sodium sulfite.

Glycerol may be detected by the characteristic odor of acrolein, found on heating with potassium bisulfate.

Glycerol is used (1) in the manufacture of high explosives, e.g., glyceryl trinitrate ("nitroglycerin"), which is the main component of dynamite, (2) in antifreeze solutions, especially for automobile radiators, (3) to maintain a moist condition in fruits and tobacco, (4) in cosmetics and skin preparations, (5) to prepare glycerol phosphoric acid, used in medicine, and "boroglyceride" used as a preservative. See accompanying table.

CHARACTERISTICS OF WATER SOLUTIONS OF GLYCEROL

% Glycerol by Weight	Specific Gravity (15.6°C/60°F)	Freezing Point, °C
20	1.049	−5.0
40	1.103	−15.6
60	1.158	−34.0

GLYCINE. Amino Acids; Brain and Nervous System.

GLYCOGEN. Carbohydrates.

GLYCOL. A dihydric alcohol (i.e., a compound containing two alcoholic hydroxyl groups). The chemical properties are represented by those of the simplest members of the class, ethylene glycol, 1,2-ethanediol, $CH_2OH \cdot CH_2OH$, which is a colorless, viscous liquid, of sweetish taste, odorless, boiling point 197°C, miscible in all proportions with water or alcohol, slightly soluble in ether. Like ethyl alcohol, ethylene glycol is often called by the class name.

Glycol reacts (1) with sodium to form sodium glycol $CH_2OH \cdot CH_2ONa$ and disodium glycol $CH_2ONa \cdot CH_2ONa$; (2) with phosphorus pentachloride to form ethylene dichloride $CH_2Cl \cdot CH_2Cl$; (3) with carboxy acids to form mono- and disubstituted esters, e.g., glycol monoacetate $CH_2OH \cdot CH_2OOCCH_3$, glycol diacetate $CH_3COOCH_2 \cdot CH_2OOCCH_3$; (4) with nitric acid (with sulfuric acid), glycol mononitrate $CH_2OH \cdot CH_2ONO_2$, glycol dinitrate $CH_2ONO_2 \cdot CH_2ONO_2$; (5) with hydrogen chloride, heated, to form glycol chlorohydrin (ethylene chlorohydrin, $CH_2OH \cdot CHCl$); (6) upon regulated oxidation to form glycollic aldehyde $CH_2OH \cdot CHO$, glyoxal $CHO \cdot CHO$, glycollic acid $CH_2OH \cdot COOH$, glyoxalic acid $CHO \cdot COOH$, oxalic acid $COOH \cdot COOH$.

In the preparation of glycol derivatives, important substances are: (a) the sodium glycols; (b) ethylene dichloride 1,2-dichloroethane, best prepared by reaction of ethylene and chlorine; (c) ethylene chlorohydrin 1-hydroxy-2-chloroethane, best prepared by reaction of ethylene and hypochlorous acid. From these are readily made, respectively: (a) ethers, e.g., glycol monoethyl ether $CH_2OH \cdot CH_2OC_2H_5$, glycol diethyl ether $CH_2OC_2H_5 \cdot CH_2OC_2H_5$; (b) ethylene diamine $CH_2NH_2 \cdot CH_2NH_2$, ethylene dicyanide $CH_2CN \cdot CH_2CN$, glycol itself; (c) hydroxyethylamine $CH_2OH \cdot CH_2NH_2$, ethylene cyanhydrin $CH_2OH \cdot CH_2CN$, ethylene oxide by sodium hydroxide

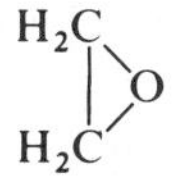

Glycol is made by reaction of ethylene and chlorine or hypochlorous acid to form ethylene dichloride or ethylene chlorohydrin, respectively, followed by treatment of either of these with sodium carbonate solution heated under pressure. Glycol is also formed when ethylene is treated with potassium permanganate.

CHARACTERISTICS OF WATER SOLUTIONS OF GLYCOL

% Glycol by Volume	Specific Gravity (15.6°C/60°F)	Freezing Point, °C
17	1.026	−6.7
32.5	1.048	−17.8
44	1.063	−28.9

Glycol is used (1) in antifreeze solutions, especially for automobile radiators; (2) in the preparation of ethers and esters, especially nitrate for explosive; (3) as a solvent substitute for glycerol.

See accompanying table.

GLYCOLYSIS. A series of about 10 enzyme-stimulated reactions in which glucose is broken down into pyruvic acid in cell respiration. No oxygen is needed for glycolysis and it is used as the sole energy source for anaerobic organisms. In aerobic metabolism, however, the pyruvic acid is then taken through the tricarboxylic acid (TCA) cycle, and the balance of the energy is extracted. It appears that glycolysis can take place free in the cytoplasm, but that the tricarboxylic acid cycle must take place within the mitochondria of the cell.

Glycolysis was defined in the late 1920s by Otto Warburg as "the splitting of carbohydrate into lactic acid." This type of lactic acid fermentation was well known to Berzelius, Liebig, Pasteur, and Claude Bernard in the mid-1800s, as was also alcoholic fermentation. Various kinds of carbohydrates may serve as substrates for glycolysis. It is remarkable that although glycolysis is the sum of a very large number of consecutive intermediate compounds, enzymes, and coenzymes, knowledge of these components and their sequences was acquired many years ago. For most animal cells studied, the biochemical sequence from glucose may be summarized as shown in the accompanying table.

EMBDEN-MEYERHOF PATHWAY

Step	Product	By way of
	Glucose (start)	
1	D-Glucose	Glucokinase, ATP, Mg^{2+}, insulin: anti-insulin regulators
2	D-Glucose-6-phosphate	Phosphoglucoisomerase
3	D-Fructose-6-phosphate	Phosphofructokinase, ATP, Mg^{2+}
4	D-Fructose-1,6-diphosphate	Fructaldose
5	D-Glyceraldehyde-3-phosphate	Glyceraldehyde-3-phosphate dehydrogenase, DPN, $HOPO_3^{2-}$
6	1,3-Diphospho-D-glycerate	3-Phosphoglycerate kinase, ADP, Mg^{2+}
7	3-Phospho-D-glycerate	Phosphoglycerate mutase, Mg^{2+}
8	2-Phospho-D-glycerate	Enolase, Mg^{2+}
9	Phosphoenolpyruvate	Pyruvate kinase, ADP, Mg^{2+}
10	Pyruvate	Pyruvate reductase = lactate dehydrogenase, $DPNH_2$
11	L-Lactate	

The splitting of sugar to lactic acid is thus, briefly, the shifting of hydrogen be means of the nicotinamide moiety of diphosphopyridine nucleotide (also termed nicotinamide adenine dinucleotide). Nicotinamide in DPN takes away two atoms of hydrogen from phosphorylated carbohydrate, and after dephosphorylation gives back two hydrogens (in $DPNH_2$) to pyruvic acid.

The biochemical importance of the foregoing sequence in glycolysis is at least twofold: (1) Each one of the intermediate compounds formed leads to one or more important possible side reactions also, and these, in turn, lead to innumerable reactions that are indispensable to life processes, including respiration; and (2) in the entire sequence, and also in some of its parts, comparatively large amounts of free energy are made available—up to a maximum of 28,000 cal/mole lactate formed under common *in vivo* conditions from one-half mole of glucose. This free energy available is considerably larger than the approximately 9,000 calories free energy available from hydrolysis of the high-energy ATP to ADP and inorganic phosphate, although much smaller than the free energy of combustion of a mole of lactate to carbon dioxide and water, some 332,000 calories. Whereas the free and heat energies of combustion of lactate are nearly equal, lactic acid fermentation from glucose represents an instance of the relatively rare situation in which the free energy liberated is considerably greater (about 50%) than the heat energy liberated, owing to the large entropy change involved in the formation of the additional carbonyl (=C=O) bond in two lactates derived from one glucose molecule.

The foregoing reaction sequence, commonly called the Embden-Meyerhof pathway after its initial investigators, was in due course worked out in greater detail by Warburg. This pathway is also common to ethyl alcohol fermentation down to the pyruvate stage, which then branches off (via carboxylase) to form acetaldehyde and finally (via alcohol dehydrogenase, $DPNH_2$) to ethanol. Alcoholic fermentation is sometimes erroneously referred to as glycolysis. Ordinary respiration, by this same reasoning, could be called glycolysis, since it too shares the common pathway down to pyruvate. Just as lactate fermentation is the most common fermentation met with in animal cells, so alcoholic fermentation is the most common fermentation met with in plant cells, a distinction most easily observed under anaerobic conditions.

See also **Carbohydrates.**

GLYCOSIDES. Substances that by reaction with water, either in the presence of certain enzymes or of dilute acids or alkalis, yield a sugar (see **Carbohydrates**) as one of the products, plus a *principle* (see accompanying table) characteristic of the individual glycoside. When the sugar is glucose, the parent compound is called a glucoside, and further called an alpha- or beta-glucoside according to the type of glucose produced. Analogous terms are the alpha and beta glycosides, applied generally to this class of compounds yielding sugars on hydrolysis. Most glycosides are soluble in cold or hot water, and in alcohol (95% C_2H_5OH), and insoluble or slightly soluble in ether (used to separate from alcohol solution). Most optically-active glycosides are levorotatory. The di- and polysaccharides are to be considered as glycosides. Glycosides occur in plants, especially in leaves, buds, young shoots where metabolism is active, and in the bark and seeds. Anthocyanins, the plant colors of flowers, are glycosides, as are also some tannins.

GLYCOSIDES (Food Poisons). Foodborne Diseases.

GLYCOSURIA. Diabetes Mellitus; Kidney and Urinary Tract.

GLYCOSYLATION. Diabetes Mellitus.

GLYCYRRHIZINS. Sweeteners.

GLYOXYLATE SHUNT PATHWAY. Carbohydrates.

GNAT (*Insecta, Diptera*). A term applied to many small 2-winged flies. In such names as buffalo gnat, gall gnat, and fungus gnat it applies to specific groups.

SELECTED REPRESENTATIVE GLYCOSIDES

Glycoside	Formula	Melting Point, °C	Hydrolysis	
			Sugar	Principle
1. Aesculin in horsechestnut bark	$C_{15}H_{16}O_9 \cdot 1\frac{1}{2}H_2O$	205	glucose	aesculetin
2. Amygdalin in peach kernels, cherry laurel leaves, bitter almonds	$C_{12}H_{16}O_7 \cdot 3H_2O$	200 (anhyd.)	glucose glucose	mandelocyanides benzaldehyde + hydrocyanic acid
3. Arbutin in bearberry leaves	$C_{12}H_{16}O_7 \cdot \frac{1}{2}H_2O$	165	glucose	hydroquinone
4. Coniferin in sap of coniferous trees	$C_{16}H_{22}O_8$	185	glucose	coniferyl alcohol
5. Dhurrin in sorghum seedlings, millet	$C_{14}H_{17}O_7N$	—	glucose	para-hydroxy-benzaldehyde + hydrocyanic acid
6. Digitalin in digitalis	$C_{35}H_{56}O_{14}$	217	glucose	digitaligenin, digitalose
7. Digitonin in digitalis	$C_{55}H_{90}O_{29}$	235 approx. decom.	glucose galactose	digitogenin
8. Digitoxin in digitalis	$C_{34}H_{54}O_{11}$	240 (anhyd.)	digotoxose	digitoxigenin
9. Helleborein	$C_{37}H_{56}O_{18}$	200–230 decom.	glucose	helleboretin
10. Hesperidin in unripe oranges	$C_{50}H_{60}O_{27}$	251	glucose rhamnose	hesperetin
11. Indican in natural indigo	$C_{14}H_{17}O_6N \cdot 3H_2O$	100 (anhyd.)	glucose	indigo
12. Phloridzin in bark of fruit trees	$C_{21}H_{24}O_{10} \cdot 2H_2O$	108 Remelts 170 decom.	glucose	phloretin
13. Quercitrin	$C_{21}H_{22}O_{12} \cdot 2H_2O$	168 decom. (anhyd.)	glucose rhamnose	quercitin
14. Salicin in willow trees; used in medicine	$C_{13}H_{18}O_7$	201 Remelts 235 approx.	glucose	saligenin
15. Saponin in soapwort root, forms foam with water, toxic to cold blooded animals	$C_{32}H_{52}O_{17}$	195 decom.	sugar	sapogenin
Tannins in nut galls	—	—	glucose	gallic acid
Anthocyanins Red (with acids), violet (free), blue (with alkalis) pigments of flowers	—	—	—	anthocyanidins
Cyanin	$C_{15}H_{10}O_6$	—	glucose	cyanidin
Idaein	—	—	galactose	cyanidin
Pelargonin	$C_{15}H_{10}O_5$	—	—	pelargonidin
Delphinin	$C_{15}H_{10}O_7$	—	glucose	delphinidin + para-hydroxy-benzoic acid

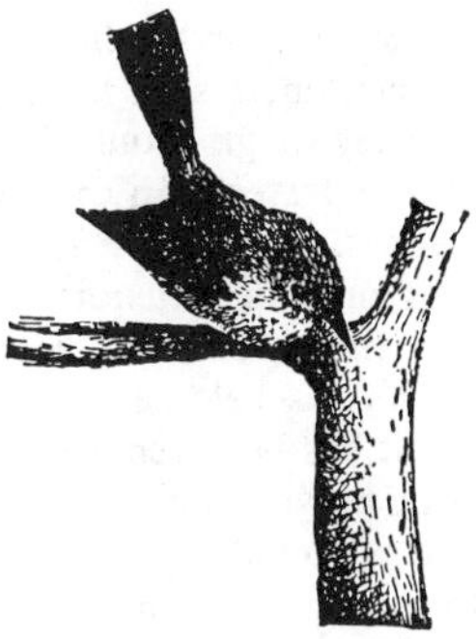

Gnatcatcher (*Polioptila caerulea*).

GNATCATCHER (*Aves, Passeriformes*). Small birds related to the kinglets. One, the blue-gray gnatcatcher, *Polioptila caerulea*, ranges over North America east of the Rockies. It is 4½ inches (11 centimeters) long, bluish gray above, grayish white beneath, with white outer and black inner tailfeathers, and a narrow black border on the front and sides of the head. Two other species occur in the southwestern states.

GNATHOCHILARIUM. The posterior element of the mouth parts of millipedes. It is formed of a pair of appendages and is similar to the labium of insects.

GNATHOSTOMATA. A term applied collectively to the members of the subphylum *Vertebrata* which have hinged jaws, in contrast with the funnel-shaped mouth of the class *Cyclostomata*. It embraces the fishes, amphibians, reptiles, birds and mammals.

GNEISS. The gneisses are common and widely distributed rocks which have been derived by metamorphic processes from pre-existing formations that were originally either igneous or sedimentary rocks. Gneissic rocks are coarsely laminated and largely recrystallized but do not carry excessive quantities of the micas, chlorite or other platy minerals. Gneisses that are metamorphosed igneous rocks or their equivalent are termed granite gneisses, diorite gneisses, etc.; however depending upon their mineralogical composition, they may be called garnet gneiss, biotite gneiss, albite gneiss and so on. Orthogneiss designates a gneiss derived from an igneous rock; paragneiss, one from a sedimentary rock. The word gneiss is from an old Saxon mining term which seems to have meant decayed or rotten, or possibly worthless material.

GNOMONIC PROJECTION. A type of projection used in producing, for navigation, especially, what are frequently referred to as great-circle charts, so called because of the fact that great circles (geodesic lines) on the surface of the earth are projected as straight lines. In the gnomonic projection, the chart is constructed by placing a plane tangent to the surface of the earth at some selected point and then projecting the surface features by extending radii from the center of the earth until they meet the plane.

In the gnomonic projection, the distortion of both shape and size is very severe except for a very limited area immediately about the point of tangency with the earth. The great value of the charts lies in the fact that the shortest distance, even between very widely separated points, will be projected as a straight line. A series of charts are available on this type of projection for all the principal cruising areas of the world, and they are of immense value to navigators for determining at a glance whether or not the following of the shortest course between two points (great-circle course) is practicable. See also **Great-Circle Course.**

GNU. Antelope.

GOA (Cedar of). Cedar Trees.

GOAT (Digestive System). Digestive System (Ruminants).

GOATS AND SHEEP (*Mammalia, Artiodactyla*). The goats and sheep (*Caprines*) comprise a significant group in the order *Artiodactyla* (even-toed hoofed mammals). Because of the general familiarity with

Fig. 1. Common American goat. (*USDA photo.*)

the domesticated goat, this description will start with that animal. As is true of the dog, sheep (discussed later in this description), and several other domesticated animals, the ancestry of the "farm yard" goat is not entirely clear. See Fig. 1. No creatures exactly like or even very closely resembling the domesticted goat exist in the wild today. All known wild species may be described as being exaggerated forms and these appear among the Tahrs, Markhors, Ibexes, and Turs. Some authorities believe that the domesticated goat is a descendant of *Capra aegagrus* and that these animals were Persian in origin. Variations in the domesticated goat now are generally identified in terms of the country from which they originally came—thus, Swiss goats; Nubians (from Egypt and north Africa); Indian goats; and Israeli and Syrian goats, etc.

Goats are, of course, very important commercially. They can produce very large quantities of milk. A Great Britain saanen (Swiss) goat is on record of having produced 6,400 pounds (2,903 kilograms) of milk in 365 days of lactation. In the United States, a saanen produced 4,900 pounds (2,223 kilograms) of milk, representing 150 pounds (68 kilograms) of butterfat, in 305 days of lactation. See Fig. 2. In California, a Nubian goat produced just under 4,250 pounds of milk, representing 185 pounds (84 kilograms) of butterfat, during a similar period. Where there are extremes of temperature (tropical or arctic), the milk from goats is considered superior to that from cows. The milk, pure white in color, is easily digested and is used for some infants and invalids, as well as by people who are allergic

Fig. 2. Purebred Saanen buck goat. (*USDA photo.*)

Fig. 3. French Alpine buck goat. (*USDA photo.*)

to cow's milk. The curds are smaller, more soluble, and the fat globules are finer and more easily assimilated making homogenization usually unnecessary. Of course, cheese from goat milk is made on a high-tonnage basis, particularly in Europe.

The flesh of the goat is edible and, in particular, that of the young kids. The hair is used (mohair) and the skin is used for leather.

Goats produce a litter of two, although triplets are fairly common. The female is sometimes referred to as the "nanny" or doe and comes in heat once every three weeks. The gestation period is from 21 to 22 weeks. The life span of the goat ranges from 8 to 12 years.

Some authorities believe that the best breeding goats are Swiss. A majority of the French and German goats stem from Swiss stock, as do the goats in Scandanavia and the Netherlands where the goat is held in high esteem. See Fig. 3. The Maltese goat is considered to have blood strains of eastern goats.

Nubians are large goats with short legs, lop ears, "Roman" noses, and are short-haired. They are partially colored or spotted. Syrian goats have long hair and large lop ears, colored black with or without patches of white. Most goats found in Great Britain are fairly small with short legs, long hair, and gray in color. The breeding of fine goats, with the importation of excellent Swiss specimens, commenced in earnest in the United States in about 1910.

Organization of the caprines is shown in the accompanying table. The following paragraphs of brief description follow the order of that table.

Gazelle-goats are not to be confused with the goat-gazelles which are described under **Antelope.** At one time, gazelle-goats were placed formally with the gazelles, but now are considered to be typical members of the caprines. The Chiru is of moderate size with long horns, ringed in the basal half, and is fawn-gray color with white underneath.

GENERAL ORGANIZATION OF THE GOATS AND SHEEP

CAPRINES

GAZELLE-GOATS (*Saiginae*)
The Chiru (*Panthalops*)
The Saiga (*Saiga*)

ROCK-GOATS (*Rupicaprinae*)
The Goral (*Naemorhedus*)
Serows (*Capricornis*)
Chamois (*Rupicapra*)
Rocky Mountain Goats (*Oreamnos*)

OX-GOATS (*Ovibovinae*)
Takins (*Budorcas*)
Muskox (*Ovibos*)

TRUE GOATS (*Caprinae*)
Domesticates (*C. hircus*)
Markhors (*C. falconeri*)
The Tur (*C. caucasica*)
Ibexes (*C. ibex*)
Tahrs (*Hemitragus*)

SHEEP (*Ovinae*)
The Aoudad (*Ammotragus*)
—Maned or Barbary Sheep
The Bharal (*Pseudois*)
True Sheep (*Ovis*)
—Argalis (*O. ammon*)
—Mouflon (*O. musimon*)

The animal is somewhat sheep-shaped and lives on the high plateaus of Tibet. The male may be 30 inches (76 centimeters) at the shoulders with horns as long as the animal is tall. They weigh about 120 pounds (54.5 kilograms). The speed of the Chiru is faster than that of a dog or wolf, but not as fast as some antelopes. The males may have a harem of from 10 to 20 females at mating time and often fierce battles take place among competing males. Mating occurs in the autumn and the fawns are born in May. The Chiru is held sacred by many Tibetans—they do not eat the flesh, but it has been reported as quite good. The Saiga is a rather ugly-appearing beast and considered somewhat clumsy. The animal is small and is found on the steppes of western Asia and eastern Europe. Its most conspicuous feature is the peculiarly swollen face with nostrils that point straight downward.

The Rock-goats are widely distributed over the northern hemisphere. They like to climb and dwell in rocky country. Their performance in climbing and walking along narrow ledges and precipices has been described by some authorities as unbelievable. The Goral is fairly small, dull olive brown with backwardly curving horns, and prefers mountainsides, ranging from the Himalayas to Amuria and Korea. The Serow is widely distributed in eastern and southeastern Asia, habitating hilly or mountainous country, sometimes at an altitude of 12,000 feet (3,658 meters). Unofficially, they are sometimes called "goat-antelopes." The Chamois is mainly native to the barren mountain ranges of Europe—the Alps, Carpathians, and the Caucasus Mountains, preferring the edges of the tree line where it can scurry for protection when in danger. The animal, originally of Switzerland, is about the size of a male deer. The herds are usually small. The chamois has two small horns between the ears. The horns turn backward and are sharply pointed. The ears are long, alert, and tapered to a point. The tail is about 4 to 5 inches (10 to 12.5 centimeters) in length, the face, back, and tail have black and white markings. The coat is chestnut brown in summer, turning to gray in the winter. The animal is timid and is protected from hunters by law in many localities. At one time, the widely-used chamois skin was derived from this animal, but the product used today, if not synthetic, usually comes from kid, sheep, or buckskin. However, even today, for high performance, the original chamois skin is preferred particularly for drying off expensively decorated surfaces. The Rocky Mountain Goat is well distributed through the Rocky Mountains of the United States—from Alaska southward through Canada and into Montana. Unlike the gorals, these goats do not descend to the tree areas, but prefer remaining in the barren, rocky areas at all times. Their diet is stunted growth, mosses, and lichens.

Of all the caprines, the Ox-goats are the least goat-like in appearance and are considered carry-overs from very early species. The Takin is a moderately large animal of heavy build, with strong curved horns. The animal ranges in mountainous country from the eastern Himalayas through Assam and northern Burma to eastern Tibet, and Szechwan, Kansu, and Shensi provinces in the People's Republic of China. They are not truly mountain animals, but prefer giant bamboo forests and thick woodlands.

The Muskox is about two-thirds as large as the American bison and is clothed with long shaggy hair. It has a thick coating of under-wool. The horns are broad at the base, but become rapidly narrower as they curve downward from the forehead over the sides of the head. The more slender tips turn abruptly upward. The muskox lives on the treeless Arctic tundras and snowfields. The animals are hunted by the Eskimos for their hides and flesh. The animal population has shrunk in recent years and is no longer spread across northern Canada from Labrador to Alaska as it once was, but now is found in the vicinity of Hudson Bay and the Mackenzie River. Another herd is found on the islands to the north—from Banks Island in the west, eastward to Greenland. The muskox has huge feet and widely splayed hoofs; the legs are short and stout.

The muskox travels in herds of from 20 to 50 animals. When attacked by wolves or other predators, the animals form rings around the attacker(s) with their sharp horns pointing toward the center of the ring. This method of protection is considered quite effective. The muskox diets on scrub grasses, stunted growth, lichens, and mosses. The muskox is not to be confused with the musk deer, the latter which is well known for the highly aromatic material which it secretes and which is used as a fixative in perfumes.

Fig. 4. Siberian ibex. (*New York Zoological Society.*)

Members of the true-goats, other than the domesticated varieties previously described, include the Markhors which are rather magnificient animals, standing proudly upright, with a heavy mane and beautifully twisted horns. The horns appear something like a twisted or spiral candle. There are several variations of the Tur. These animals inhabit the Caucasus Mountains. The Tur is of a rich-brown coloration, with short hair and a forwardly-brushed beard and huge horns. There are several variations of the Ibex and they are distributed widely in locations of a mountainous nature. A Siberian ibex is shown in Fig. 4.

In briefly describing the sub-family of sheep, it should be noted that, as with the goat, it is difficult to look into the wildlife known today and to find what would seem to be a "wild" sheep of the type known so well by way of the millions of domesticated animals; or, in fact, to identify what appears to be an ancestor of the domestic sheep. However, zoologically, as indicated by the accompanying table, there are three broad classes of sheep. The Aoudad (also Udad), also known as the maned or Barbary sheep, is the only indigenous sheep of Africa and is found around the Sahara. The animals are powerful, with large and very thick horns, and are avid rock-climbers. See Fig. 5.

The true sheep have horns that resemble those of domestic breeds. The Argalis is found in east central Asia. The Mouflon is a wild sheep of Europe and inhabits the areas around the Mediterranean, including Corsica and Sardinia. They are reddish in coloration and quite distinguished, with large sweeping horns.

A group of wild sheep occurs in North America, with two distinct

Fig. 5. Udad (Barbary Wild Sheep). (*New York Zoological Society.*)

Fig. 6. Registered Rambouillet ram, 15 months old. (*American Rambouillet Sheep Breeders' Association.*)

species. One of these is the Canadian, Rocky Mountain, or Bighorn sheep, which is found from British Columbia southward to Lower California and as far east as the western Mexican mainland. Close relatives of this sheep are found in eastern Siberia. Another species, Dall's sheep, is closely related to the Bighorn.

The Sha is an Asiatic sheep with a very wide range. It is found from Iran into India and northward through Tibet. The name more widely used is *urial*, the term *sha* applying to a large variety of sheep ranging from northern Tibet to Afghanistan. The sheep inhabits areas up to altitudes of 14,000 feet (4,267 meters).

Two of many domesticated breeds of sheep are shown in Figs. 6 and 7.

More detail on goats and sheep can be found in the "Foods and Food Production Encyclopedia," (D. M. Considine, editor), Van Nostrand Reinhold, New York, 1982.

Fig. 7. Columbia yearling ram. (*The Columbia Sheep Breeders' Association of America.*)

GOATSUCKER. Nightjars and Nighthawks.

GOBIES (*Osteichthyes*). Of the suborder *Gobioidea* and family *Gobiidae*, there are numerous species of gobioid fishes. They are characterized by a sucker which is present under the forward part of the body, and by two dorsal fins. Some of the over-400 species are quite small, ranging from ½-inch (13 millimeters) long up to about 4 inches (10 centimeters). Most gobies are quite colorful. *Pandaka pygmaea*, a freshwater fish found in the Philippines, is considered by some authorities to be the smallest vertebrate animal in terms of length. The species *Eviota* found in the Indo-Pacific is also quite small. Even the freshwater gobies spawn in saline waters and advantage of this fact is taken by fisheries in the Philippines to capture extremely large schools of the genus *Paragobiodon*. A fermented paste (bagoong) is made from the tiny fish (ipon). Several species of gobies develop a symbiotic relationship with other creatures, such as crabs, burrowing worms, and notably shrimp, wherein the gobies share shelters with their hosts, but also serving to warn their hosts of impending danger. The *Elecatinus oceanops* is a small 2-inch (5-centimeter) neon goby that is noted for its ability to clean parasites from larger fishes. Several species of gobies are favorites among tropical-fish fanciers. There are about 15 species of eel gobies (*Taenioididae*). They are elongated with a maximum length of just over a foot (0.3 meter). They are found in tropical Indo-Pacific waters. The loach goby (*Rhyacichthys aspro*) is the only member of the family *Rhyacichthyidae*, growing to a length of about 9 inches (23 centimeters) and found in large streams and rivers of the Philippines and Indonesia. Were it not for its spiny first dorsal fin, this species would be difficult to distinguish from the homalopterid loaches. See **Loaches (Osteichthyes).**

GODWIT. **Waders, Shorebirds and Gulls.**

GOETHITE. The mineral goethite is a hydroxide of iron corresponding to the formula FeO(OH) crystallizing in the orthorhombic system. It occurs in prisms, but is often found in foliated or other massive forms. When observable it shows one good cleavage parallel to the prism; fracture, uneven; hardness, 5–5.5; specific gravity, 3.3–4.3; luster, adamantine to dull; color; yellowish, reddish, brownish to nearly black; translucent to opaque. It is found associated with hematite and limonite, being perhaps in part an alteration product of the latter mineral. Goethite is used as an ore of iron. There are many European localities, including Bohemia, Saxony, Westphalia, and Cornwall. In the United States it is found in the hematite mines of the Lake Superior region and in Colorado. This mineral was named in honor of the German poet Johannes Wolfgang von Goethe.

GOITER (Iodine Correlation). **Iodine (In Biological Systems); Thyroid Gland.**

GOLAY PNEUMATIC CELL. A small transparent cell containing gas which is used to detect radiation. A very thin film within the cell absorbs incident radiation, which increases the cell temperature and pressure. Changes in pressure are recorded as indications of the amount of incident radiation.

GOLD. Chemical element symbol Au (from Latin *aurum*), at. no. 79, at. wt. 196.967, periodic table group lb (transition metals), mp 1,064.43°C, bp approximately 2,805–2,809°C, density 19.32 g/cm^3 (20°C). Elemental gold has a face-centered cubic crystal structure.

Gold is a yellow metal, soft, and extremely malleable. The purity of gold (sometimes referred to as "fineness") is expressed in karats. Pure gold is 24 karat. Gold has 19 isotopes, ^{185}Au through ^{203}Au. Only one of these, ^{100}Au, is stable. See also **Radioactivity.** In terms of cosmic abundance, the estimate of Harold C. Urey (1952), using silicon as a base with a figure of 10,000, gold was ranked number 79 among the elements, with an abundance figure of 0.0015. In terms of abundance in seawater, gold is ranked number 59 among the elements, with an estimated content of 38 pounds per cubic mile (4 kilograms per cubic kilometer) of seawater.

Electronic configuration is

$$1s^2 2s^2 2p^6 3s^2 3p^6 3d^{10} 4s^2 4p^6 4d^{10} 4f^{14} 5s^2 5p^6 5d^{10} 6s^1.$$

First ionization potential is 9.223 eV; second 19.95 eV. Oxidation potentials: $Au \rightarrow Au^{1+}$, $E° = -1.68$ V; $Au \rightarrow Au^{3+}$, $E° = -1.50$ V. Other important physical properties of gold are given under **Chemical Elements.**

Gold is one of the most ancient metals. Gold jewelry and ornaments made as early as 3,500 B.C. have been discovered at Ur in Mesopotamia. During the period from 3,000 to 2,000 B.C., lead cupellation was used to purify gold and most modern jewelry techniques were developed during that time.

Occurrence and Processing: Gold is found chiefly as the free metal scattered through gravel (*placer gold*) or disseminated in veins of quartz (*vein gold*). Small quantities also are found in lead and copper sulfide ores. Nuggets of native gold, varying in size from that of a tiny pebble to a mass weighing as much as 248 pounds (112.5 kilograms) have been found. In a combined state, gold occurs in sylvanite, a telluride of gold and silver, $(Au, Ag)Te_2$, a rich ore found in Colorado. The bulk of the gold ores contain very little gold (about 5 to 15 grams/metric ton). Some of the richest ores found in Africa contain from 20 to 30 grams/metric ton. Almost all countries produce some gold. The leader, by far, is the Republic of South Africa, followed by the U.S.S.R. and Canada. Far behind, other producers include the United States, Australia, Ghana, and Zimbabwe. See also **Mineralogy.**

The treatment of gold ores involves: (1) grinding, amalgamation, and/or cyanidation of those ores containing coarse free gold, and (2) the very fine grinding, flotation, roasting, and amalgamation and/or cyanidation of those ores containing gold telluride or sulfide. These processes produce an impure gold metal containing considerable silver and some copper plus other base metals. The impure gold is purified by melting and oxidizing the base metals or by melting and chlorinating (Miller process) which removes the base metals and silver. The silver-containing oxidized gold is purified by the electrolysis of gold chloride solutions containing an HCl solution (Wohlwill process). In the latter process, the anode is the alloy (gold-silver) and the cathode is pure gold. The gold deposits then on the cathode and the silver forms silver chloride and remains as a deposit about the anode.

Uses of Gold: The monetary aspects of gold have long dominated commercial interest in the metal. Gold through history has provided a common base from which the value of materials and services can be measured. Gold probably became a medium of exchange as early as 3,400 B.C.

Jewelry is the largest commercial user of gold, accounting for nearly 65% of the total consumption. Most jewelry is made by the "lost wax process," a casting method that dates to 3,000 B.C. or earlier. Usually these jewelry products employ karat golds which contain 10 and 14 karats, and less commonly 18 karat, of gold (41.7, 58.3 and 75.0 weight percent of gold, respectively). These gold alloys are of two general types. Red, yellow and green golds are basically alloys of gold, copper, and silver. A wide variety of color shades can be produced by varying composition within this ternary alloy system, with reddish hues provided by high copper to silver ratios, and pale green tint when silver is predominant. These alloys almost always contain minor amounts of zinc and deoxidizers or grain refiners to facilitate fabrication. The second widely used class is the white karat golds, which are produced in two basic alloy types. These are the original gold-nickel-zinc-copper (18 karat) and the gold-copper-nickel-zinc (10 and 14 karats) alloys, and the more recent gold-palladium-silver-copper, and gold-copper-nickel-palladium-silver alloys which are usually 14 and 10 karat alloys. The pink golds are derived from the system gold-silver-copper-nickel-zinc. These are essentially red golds, which are "whitened" by the addition of silver, nickel, and zinc.

Considerable brazing is done by jewelry manufacturers and the solders that are used may be of a lower karat content than the alloy being brazed. Usually they contain much more silver and zinc than the alloys themselves.

The use of gold in the electrical, electronic, and other industrial fields has grown considerably in recent years, estimated at about 25%. The electrical and thermal conductivity, resistance to oxidation, and ease of being electroplated make gold an excellent coating for electrical contacts. This has been particularly true in metallized ceramics for use in microelectronics and other electronic components. Here gold does not migrate into the ceramic as does silver. Gold is widely used as a conductor in thin and thick film circuitry. It is also useful as bonding wire for integrated circuit electrical connections and mechanical packaging of semiconductor chips (die bonding).

Gold is used extensively in many industrial solders and brazing alloys. These range from the low-melting eutectics of gold with germanium, silicon, and tin to gold-copper, gold-nickel, and gold-palladium-nickel alloys. The latter brazing materials have the ability to withstand long use at high temperatures and are particularly applicable to jet engine fabrication.

Gold is also used in dentistry. This application has declined in recent years; however, it still accounts for about 7% of gold consumption. Gold alloys, such as gold-silver-copper with varying amounts of platinum and palladium, are used for restorations and for bridges, inlays, and partial dentures. These are cast with much more precision than jewelry, and have in fact, replaced wrought gold wire in many of these dental appliances. Gold wire is now used principally in orthodontic and prosthetic appliances. These are complex alloys containing gold, platinum, palladium, silver, copper, nickel, and zinc.

Some of the minor commercial uses of gold are among the most interesting. Gold is used to produce a very beautiful ruby glass. When an oxidizing glass is melted with a gold salt, the gold dissolves forming colorless ions. If reducing agents like Sn, Sb, Bi, Pb, Se, or Te are present, the glass will become red after heating at temperatures between 600–700°C, as a result of the precipitation of minute particles of gold. Gold films deposited on glass by evaporation are superior to other metals for reflectivity in the infrared. Mirrors thus coated have application in spectroscopy and space science. Thin films applied to plate glass give adequate transmission of light combined with good infrared reflectivity, reducing the overheating of office windows during hot weather. Gold is extremely malleable. It can be rolled and beaten into foil less than 5 millionths of an inch (0.00013 millimeter) thick. Such foil has been used for indoor and outdoor decoration for centuries. One of the most conspicuous examples is the gold leaf dome, an architectural highlight in many important structures.

Chemistry of Gold: Gold has a $5d^{10}6s^1$ electron configuration, like the similar ones at lower levels of copper and silver, and thus the *d* electrons can take part in bonding. However, for gold the +3 oxidation state is the most stable, and the +1 state next to it in stability, so that Au^{3+} as well as Au^+ are found both in simple compounds and in complexes. As with copper and silver, the bonds in most gold compounds, including the oxides, are largely covalent. In most of its compounds gold is univalent or trivalent. While a few compounds are known in which it is divalent, some of these are considered to consist of Au(I) and Au(III), rather than Au(II). Thus, the compound with cesium and chlorine, $CsAuCl_3$, is black and diamagnetic, and so contains both Au(I) and Au(III). A similar compound with cesium, silver, and chlorine, $Cs_2AuAgCl_6$, yields $[AuCl_4]$ and $[AgCl_2]^-$ ions on hydrolysis. However, the sulfide, AuS, probably contains divalent gold.

Gold does not combine directly with oxygen. Gold(I) oxide, Au_2O, formed by heating AuOH to 200°C, is very easily reduced to gold. It is essentially covalent. Gold(I) hydroxide, AuOH, is prepared from a gold(I) solution by the addition of potassium hydroxide solution in theoretical amounts. It forms a deep blue "solution" believed to be a colloidal sol. It dissolves in excess alkali to form aurates(I), such as $KAu(OH)_2$. Gold(III) oxide, Au_2O_3, is formed by heating $Au(OH)_3$ at 100°C in the presence of a dehydrating agent. Like Au_2O, it is easily reduced to gold. It dissolves in hydrochloric, hydrobromic, and hydriodic acids, forming the haloauric acids, $HAuX_4$. It also dissolves in excess of alkali hydroxide, forming an aurate, containing the ion $[Au(OH)_4]^-$. Gold(III) hydroxide, $Au(OH)_3$, is precipitated by the addition of potassium hydroxide solution in equivalent amount, to a solution of chloroauric acid (obtained by dissolution of gold in aqua regia). It is insoluble in H_2O, gives many of the reactions of Au_2O_3, and may be a hydrous form of that compound. Gold(II) oxide, AuO, formed by the action of potassium bicarbonate upon solutions of chloroauric acid, is believed, as stated above, to consist of gold(I) and gold(III), based on properties of other divalent gold compounds.

Gold does not react directly with fluorine, but dissolves in bromine trifluoride BrF_3, to form BrF_2AuF_4, which loses BrF_3 at a 120°C to give gold(III) trifluoride, AuF_3, which decomposes into the elements at about 500°C. Water decomposes AuF_3 into hydrogen fluoride and $Au(OH)_3$. The chlorides, on the other hand, are the most important of the gold salts. Gold(I) chloride, AuCl, may be produced by heating gold(III) chloride, $AuCl_3$, in air at 170°C; it is hydrolyzed by H_2O to $AuCl_3$ and gold. Gold(III) chloride, $AuCl_3$, is formed directly from the elements at 200°C; unlike AuCl, it is soluble in H_2O, forming initially $H[AuCl_3(OH)]$, which then undergoes further hydrolysis. With hydrochloric acid, $AuCl_3$ forms tetrachloroauric(III) acid, $H[AuCl_4]$, of which many salts are known. Gold(I) bromide, AuBr, is formed by continued heating of bromoauric(III) acid above 100°C. Like the AuCl, it readily undergoes hydrolysis. Gold(III) bromide is formed by the action of bromine water upon gold. The equivalence of its three Au–Br bonds have been proved by a tracer technique with radioactive bromine. With hydrobromic acid it forms $H[AuBr_4]$. Gold(I) iodide is prepared from the elements at 50°C, or by the slow decomposition of AuI_3 at room temperature. It decomposes on heating above 120°C. It dissolves in potassium iodide, KI, solution, forming $KAuI_2$, which then decomposes to gold and $KAuI_4$. Gold(III) iodide, obtained by evaporation of an 1:1 hydriodic acid solution of $AuCl_3$ is unstable, decomposing when dry or when heated with H_2O, into the elements. It dissolves in hydriodic acid as $H[AuI_4]$. The gold(I) halides are the least soluble of the univalent halides except for silver iodide. The solubility product constants are AuI, 1.6×10^{-23}; AuBr, 5.0×10^{-17}; AgI, 8.30×10^{-17}; AuCl, 2.0×10^{-13}; and AgBr, 4.27×10^{-13}.

There are many gold complexes. The gold(I) and gold(III) halocomplexes, involving the groups $[AuX_2]^-$ and $[AuX_4]^-$ have already been discussed. Apparently, there are no other gold(I) halocomplexes than the chloro-compound. There are also fluorocomplexes of the form $M[AuF_4]$ formed by fluorination of $M[AuCl_4]$ where M is an alkali metal or ammonium. Due to the polar character of the AuF bond, they are readily hydrolyzed. Many hexachloroaurates, such as $Cs_2M[AuCl_6]$ are known.

Gold forms complexes with ammonia much less readily than do copper and silver. A few ammonia complexes of gold(III), such as $KAuCl_4 \cdot 3NH_3$ have been prepared. Gold(I) halides react more readily. AuCl forms $[Au(NH_3)_2]Cl$, while AuBr and AuI react, but only with anhydrous ammonia, to form $[Au(NH_3)_2]Br$ and $[Au(NH_3)_6]$. Gold(I) cyanide dissolves in excess cyanide to form the very stable ion $[Au(CN)_2]^- K_{inst} = 10^{-38.3}$. (This complex is so stable that gold metal dissolves in potassium cyanide solution in the presence of air. This is of importance in the separation of gold from its ores); while $Au(CN)_3$ reacts to form $[Au(CN)_4]^- K_{inst} = 10^{-56}$. Treatment of salts of this ion with sulfites, gold(III) forms such complexes as $K_5[Au(SO_3)_4] \cdot 5H_2O$ and $Na_5[Au(SO_3)_4] \cdot 14H_2O$. In these complexes, the sulfito group is monodentate and is attached to the gold atom through the sulfur atom (really an aurisulfonate ion); however, a bidentate compound is also known.

Gold(III) chloride or tetrachloroaurates(III) also form thiosulfate complexes, especially in the presence of NaI, of the form $Na_3[Au(S_2O_3)_2]$, in which the gold is monovalent.

Gold forms thiocyanate complexes $M[Au(SCN)_2]$ and $M[Au(SCN)_4]$.

A striking difference between gold and copper or silver, is the fact that its oxyacid compounds do not exist in stable form, and few have been isolated. Among the few that are known are the gold(III) orthoarsenite, $AuAsO_3 \cdot H_2O$, the gold(III) selenate, $Au_2(SeO_4)_3$, and the gold(III) iodate, $Au(IO_3)_3$. Nevertheless a number of complexes of oxyacids are known, including $M[Au(NO_3)_4] \cdot 2H_2O$ ($M = H_3O^+$, NH_4^+, K^+, Rb^+), $Mg[Au(CH_3CO_2)_4]$.

Gold is unique among the coinage metals in forming true (i.e., sigma-bonded) stable organometallics. The action of methyllithium on $AuBr_3$ in ether at −65°C, produces a solution of $(CH_3)_3Au$, which begins to decompose at −35°C into gold, ethane, and methane. The presence of benzylamine or ethylenediamine, however, stabilizes the solution up to room temperature. Triethylgold is less stable than trimethylgold. The action of a hydrogen halide on a trialkylgold or the action of an alkyl Grignard reagent in pyridine on gold(III) halides produces dialkylgold halides, which are much more stable. Appropriate metathetical reactions of these produce the corresponding cyanides, sulfates, etc. These are all covalent compounds, as attested by the solubility of the sulfates, $(R_2Au)_2SO_4$, in benzene and chloroform. The melting points of a few dialkylgold compounds are: $(CH_3)_2AuBr$, 68°C; $(C_2H_5)_2AuCl$, 48°C; $(C_2H_5)_2AuBr$, 58°C; $(C_2H_5)_2AuCN$, 103–105°C; $(n\text{-}C_3H_7)_2AuCN$, 94–95°C; $(i\text{-}C_3H_7)_2$ AuCN, 88–90°C; $(i\text{-}C_5H_{11})_2AuCN$, 70°C; $(C_6H_5CH_2)_2AuCl$, 100°C decomposes; $(C_6H_5CH_2CH_2)_2AuBr$, 112.5°C. The *n*-propyl chloride and bromide, and the *n*-butyl, *i*-butyl, and *i*-amyl bromides are liquid at room temperature.

The dialkylgold halides are dimeric, having the planar structure

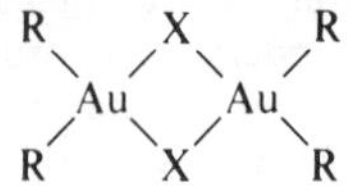

The cyanides, on the other hand are tetrameric, having the structure shown below.

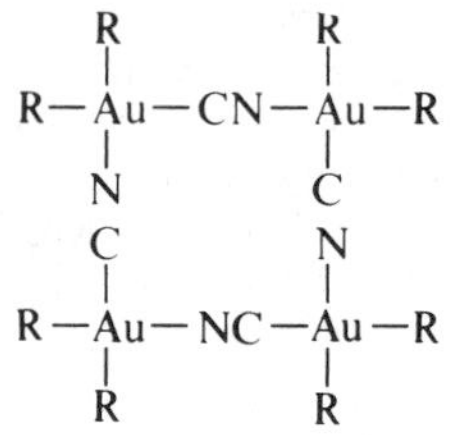

References

Chynoweth, A. G.: "Electronic Materials: Functional Substitutions," *Science*, **191**, 725–732 (1976).
Pudde-Phatt, R. J.: "The Chemistry of Gold," Elsevier, New York, 1978.
Smithells, C. J.: "Metal Reference Book," Vol. 3, Plenum, New York, 1967.
Wise, E. M. (editor): "Gold: Recovery, Properties and Applications," Van Nostrand Reinhold, New York, 1964.

Donald A. Corrigan, Handy & Harman, Fairfield, Connecticut.

GOLDEN BERYL. **Chrysoberyl; Yellow Beryl.**

GOLDEN-EYE. 1. *Insecta, Neuroptera.* The lace-wing, adult of the aphis-lion. These insects are small and delicate, with large many-veined wings of yellowish or green color and shining eyes. They have a disagreeable odor. 2. *Aves, Anseriformes.* A North American and European duck, *Bucephala.* See **Eagle.**

GOLDEN RATIO AND GOLDEN RECTANGLE. **Fibonacci Numbers.**

GOLDFISH. **Carp.**

GOLDLEAF ELECTROSCOPE. **Electrical Instruments; Electrometer.**

GOLD NUMBER. When certain colloids (hydrophilic), such as gelatine, are added to a gold sol, the gold sol is strongly protected against the flocculating action of electrolytes. This protective action on red gold sols may be measured by utilizing the color change red to blue which indicates the first stage of coagulation. The "gold number" as defined by Zsigmondy is the weight in milligrams of protective colloid which is just sufficient to prevent the change from red to blue in 10 cm^3 of a standard gold sol (0.0053 to 0.0058 percent Au) after the addition of 1 cm^3 of a 10 percent sodium chloride solution.

GOLDSCHMIDT DETINNING PROCESS. A process for the recovery of tin, based on the action of dry chlorine on scrap tinplate. The tin reacts readily to form stannic chloride, and the iron reacts only slightly. By fractional distillation the stannic chloride is separated from the small amount of ferric chloride that is formed.

GOLDSCHMIDT LAW. The structure of a crystal is determined by the ratio of the numbers, the ratio of the sizes, and the properties of polarization of its structural units, i.e., atoms or ions.

GOLDSCHMIDT REDUCTION PROCESS. 1. Reaction of oxides of various metals with aluminum to yield aluminum oxide and the free metal. This reaction has been used to produce certain metals, e.g., chromium and zirconium, from oxide ores; and it is also used in welding (iron oxide plus aluminum giving metallic iron and aluminum oxide, plus considerable heat.) (Thermite process.) 2. A method of producing formates by heating sodium hydroxide with carbon monoxide under pressure. 3. A process for recovery of tin, by treatment of scrap tinplate with dry chlorine, better known as the Goldschmidt detinning process.

GOLD TELLURIDE. **Calaverite; Sylvanite.**

GOLGI APPARATUS. **Liver.**

GOLGI BODY. **Cell (Biology).**

GOLGI STAINING TECHNIQUE. **Brain and Nervous System.**

GONADOTROPHIC HORMONE. **Endocrine System; Hormones; Pituitary Gland.**

GONADS. Both the female sex gland (*ovary*) and the male sex gland (*testis*) are referred to by the general word, *gonads.* Not only are the gonads the fundamental organs of reproduction, but they also produce several hormones. The two testes are made up of tissues that specialize in producing the male germ cells and tissues that manufacture the male hormone. The two ovaries provide the egg (*ovum*) and several hormones that are involved in the regulation of sexual function. Because the ovaries and testes produce hormones, they are considered endocrine glands. See also **Endocrine System.** Collectively, these male and female hormones are called gonadal hormones.

The hormones produced by the ovaries are called female hormones. The name female or male hormone does not imply that these substances are produced exclusively by either sex, but that they are produced predominantly by one sex. Thus, certain structures in males, especially the adrenals, can and do produce female hormones that are excreted in the urine. Women also produce male hormones and in some instances where the balance is disturbed by disease the effects of overproduction of male hormone become evident. In such instances, women develop signs of masculinity.

Sex hormones act primarily upon the reproductive system, which in both men and women is made up of the gonads and the accessory or secondary sex organs. The proper development and functioning of the accessory sex organs are dependent upon the production of sex hormones. In women, the accessory sex organs are the breasts, the womb (uterus), Fallopian tubes, vagina, vulva, and clitoris; each serves a particular function in the complex process of reproduction. In men, the secondary or accessory sex organs are represented by a series of tubes or ducts that convey the germ cells from the testes through the penis to the outside of the body, plus several glands located at different points. These glands are the prostate glands, the seminal vesicles, and Cowper's glands. Again, each performs a particular function, and each is dependent on the male hormone for its proper functioning. Castration results in a decrease in size of all these structures, and eventually they cease to function. The effect is the result of removing the source of male hormone.

Secondary Sex Characteristics. Male and female hormones are poured into the blood like all other hormones. They exert different actions on different parts of the body, imparting qualities that are typical of each sex. Thus, the distribution of hair on the body, particularly pubic hair, varies greatly. In women pubic hair is limited above by a horizontal line, and the hair may grow in a triangular zone. In men, the growth of pubic hair may extend from the navel to the anus. The hair on other parts of the body is more abundant in men. The female voice is high pitched, and the larynx is less developed than in the male. Other qualities, such as breast development, shape of pelvis, and distribution of fat are also different in the sexes as a result of different sex hormone production.

Both male and female sex hormones belong to the group of substances called *steroids* to which also belong the hormones produced by the adrenal cortex. See also **Steroid.** Pregnant women excrete large quantities of certain sex hormones in their urine. In the past, urine from pregnant women was used as a source of a female sex hormone (*estrone*). Pregnant mares also eliminate large amounts of estrone in their urine. Sex hormones produced by animals are identical with those produced by humans. Urine obtained from postmenopausal women has been utilized on an industrial scale to obtain a hormone that stimulates the gonads. It is termed *human menopausal gonadotrophin.*

The Testes

Production of sperm cells takes place in the testes. The testes are two oval-shaped organs located outside the abdominal cavity below the penis, and held by a pouch called the *scrotum.* In addition to the reproduction function, the testes produce male sex hormones which are secreted into the bloodstream. Rarely is an individual born having both testes and ovaries. When such occurs this is *true hermaphroditism.*

Before a boy is born, the testes are present within the abdominal cavity where they have been formed and descend gradually until, by the time of birth, they make their exit through a passage called the *inguinal canal* and have become localized in the scrotum.

For the testes to function effectively, they must be at a lower temperature than that of the abdomen. When the temperature increases, the testes do not produce mature spermatozoa. Because they are located within the scrotum outside the abdominal cavity, the testes are kept at a temperature a few degrees lower than that of the body. When the outside temperature is lowered, the spermatic cord that is attached to the testes and the scrotum draws upward, keeping the testes close to the body and allowing them to be warmed by the body's heat. The reverse occurs when the outside temperature is raised.

The surface of the testes is covered by a layer of fibrous tissue called the *tunica vaginalis.* The internal structure of the testes is divided into sections separated by thin membranes. Within each section are long, thin, tube-like strands, called the *seminiferous tubules.* It is within these tubules that the spermatozoa are produced. In the spaces or interstices that exist between the tubules are the interstitial cells which produce the male hormone. If a section of the testes is observed with a powerful microscope, a number of circular structures representing cross sections of the tubules can be seen. Within the circular structures are seen the spermatozoa at different stages of development. Toward the center of the tubules are seen the mature spermatozoa with complete heads and tails.

During the maturing process, the spermatozoa pass into multiple small tubes (*vasa efferentia*) which lead to the *epididymis.* The epididymis is a long, thin duct (*ductus* or *vas deferens*). Upward in its course toward the abdomen, the vas deferens is joined by the resticular arteries, veins, lymphatics, and nerves to form a thick tube, the *spermatic cord.* The spermatic cord, containing the vas deferens and other vessels, passes into the abdomen through the inguinal canal, and descends by the side of the urinary bladder to the prostate, through which it passes to reach the urethra. It is there joined by the small duct of the *seminal vesicles.* For each testis, there is one spermatic cord, one vas deferens, and one seminal vesicle.

The seminal vesicles are two pouches located between the bladder and the rectum, although not connected to either. The lower ends of the two seminal vesicles unite to form two short ducts that serve to carry the spermatic fluid to the large duct in the penis (urethra) and outside the body. These are the *ejaculatory ducts,* which are two small ducts that penetrate the prostate. From this point, both the semen and the urine share the same passage, the remaining portion of the urethra.

The prostate is an organ located at the base of the bladder; it completely surrounds the portion of the urethra that leads from the bladder. The prostate is an accessory organ of reproduction, containing numerous glands that produce the *prostatic fluid,* an important component of the *semen.* The secretion is produced at a low, but constant rate, and is poured into the urethra in small amounts; small quantities escape into the urine. Sexual stimulation accelerates production of prostatic fluid. During ejaculation, the prostatic fluid is delivered in larger quantities and is mixed with the seminal plasma to form the semen. In addition to serving as a housing and transporting vehicle for the sperm, the prostatic fluid appears to be necessary to maintain viable spermatozoa in the vagina, possibly by protecting the sperm from the acid condition of the vagina.

A single ejaculation may contain over a quarter of a billion spermatozoa. If fertilization does not occur, all of these cells die; if fertilization does occur, only one spermatozoon will survive; it will fertilize the egg. Occasionally, two ova may be produced within a short period of time and two spermatozoa will fertilize them, producing *fraternal twins. Identical twins* develop from a single ovum. Fraternal twins may be of different sexes, but identical twins are of the same sex and look alike. The sperm cells which swim in the semen are microscopic. Their propulsion is brought about by movements of their tails. When sperm are deposited in the vagina during sexual intercourse, they move gradually upward toward the womb. The fatality rate of the sperm is high, but the chances of one arriving alive in the womb are usually good. The life span of a sperm cell is not precisely known, but it is believed that the sperm has the ability to penetrate and fertilize an ovum for only about 48 hours. The energy necessary for maintenance and propulsion of spermatozoa is derived mostly from the various types of nourishment present in the seminal plasma.

The Penis. In sexual intercourse, the penis serves to convey the semen into the vagina of the female. The shape of the penis varies greatly depending on whether it is flaccid or erect. In the flaccid state, the penis is cylindrical, but when erect, it assumes a triangular shape in cross section. The organ consists of three cylindrical masses of erectile tissue held together by fibrous tissue and covered by skin. Two of the cylindrical bodies lie side by side, and the third, which holds the urethra, is located underneath the other two. The lower cylinder ends in a cone-shaped body (the *glans*), which constitutes the free end of the penis; in the center of the glans is the opening of the urethra. The skin that covers the penis is thin and has no hairs except near the root of the organ, but possesses numerous glands that produce secretion.

The glans of the penis is covered by a circular fold of skin called the *prepuce.* In many instances, the prepuce, or foreskin, may cover the entire glans, obstructing the passage of urine. Under these conditions, the secretion of the skin glands accumulates, creating a constant source of irritation and infection. Therefore, surgical removal of the foreskin (*circumcision*) may be desirable as a prophylactic measure, and is usually performed shortly after birth. The operation was performed in ancient Egypt before it was introduced among the Hebrews. Today, it is practiced among the Jews and Mohammedans as a religious rite. However, it is practiced widely as a hygienic measure by peoples of all continents.

Erection is necessary for normal transmission of semen into the body of the female. Sexual stimulus, either mental or physical, sets off a series of reactions that culminate in erection. The sexual stimulus received by the nervous system causes a flow of blood from the arteries that lead to the penis and within the penis, to the many vessels and cavities of the erectile tissue to occur at a faster rate than the blood flows from the penis via the veins. The penis becomes engorged with blood, thus becoming firm and erect. The organ returns to its original flaccid state when the process is reversed after erection.

Male Sex Hormones. The male sex hormone is produced after complete development of the testes. At puberty, the secondary sexual characteristics make their appearance rapidly. In normal boys, signs of puberty may appear at any age between 10 and 17 years. The average onset is 12 to 13 years. A related problem in the development of sexual characteristics in boys is *cryptorchidism,* or undescended testes.

When the output of male hormone is less than normal, a condition known as *hypogonadism* develops. A patient who is of adolescent age or younger may develop symptoms characterized by effeminate traits and retarded development of the sexual organs. In men who have attained maturity, the signs of *androgen* (male sex hormone) deficiency are less conspicuous. The most common events are reduction in prostatic size, diminished growth of the beard and body hair, the appearance of fine wrinkles around the eyes, and a pasty, sallow complexion. Also, semen volume is reduced.

Klinefelter's syndrome is a common form of hypogonadism. Feminine characteristics and infertility may exist. Patients with this condition are often tall with disproportionately long lower extremities. Mental retardation and psychopathic behavior are not uncommon, and men with this syndrome are often poorly adapted socially. In 1956, it was discovered that Klinefelter's syndrome is the result of a genetically determined defect. Treatment for patients with this condition must be closely supervised by a physician, as the use of hormones is usually involved.

In the male, with age, sexual activity declines gradually. The climacteric (*change of life*) is not as conspicuous as it is in women, and the age at which it occurs varies over a wider range. At the time of the male climacteric, sexual activity declines to a lower level.

Tumors of the testes are uncommon. The greatest incidence occurs in men in their twenties and thirties. The most common testicular tumor is called seminoma. Generally, this tumor is relatively slow growing and responds well to radiotherapy.

See also **Pituitary Gland.**

The Ovaries

Located on each side of the womb, the ovaries are two almond-shaped organs. Each is about the size of a walnut. The ovaries, unlike the testes, produce several hormones. Although different, they are grouped under the term *female sex hormones.* These substances regulate various functions of the body, but their major duty is regulation of the female reproductive system. Two chemically determined types of ovarian hormones are (1) the estrogenic steroids or *estrogens* (*estradiol, estrone,* etc.), and (2) the *progestagens* (*progesterone,* etc.). Within recent years, it has been possible to produce these hormones synthetically.

The control that ovarian hormones exert upon the reproductive system is not limited to the accessory or secondary sex organs, i.e., the womb, Fallopian tubes, vagina, vulva, and clitoris. In an indirect sense, the ovaries themselves are affected by their own secretions, since a reciprocal ovary-pituitary relationship is of importance in the regulation of the ovaries. The maturation of the eggs, ovulation, and other changes that occur in the ovaries are dependent, then, to some degree, on the hormones from the ovaries. See also **Pituitary Gland.**

The sexual cycle in women is well-regulated as long as the production and secretion of both the gonadotrophic hormones of the pituitary gland and the sex hormones from the ovaries are normal. This occurs most of the time, but occasionally the pituitary gland, the ovaries, or both may vary in their production of hormones. When the pituitary gland becomes underactive as a result of disease, the production of all the pituitary hormones is affected.

The two ovaries establish contact with the uterus by means of the two Fallopian tubes which convey the egg cells from the ovaries to the womb. The womb (uterus) is a muscular organ with great capacity for expansion. The inside of the womb is hollow and the walls are covered by a mucous membrane known as the *endometrium.* Here, the fertilized ovum develops into a baby.

The hollow portion of the female reproductive system constitutes a continuous structure, so that the ovaries, tubes, and womb may be regarded as a unit. The uterus forms the center of this unit, and is located in the pelvic cavity between the urinary bladder and the rectum, and the tubes form a passageway to the ovaries which are located on each side of the uterus.

The female reproductive system does not produce a fluid corresponding to the male seminal fluid. Under the influence of sexual stimulation, however, the walls of the vagina secrete fluids which serve as lubricants that facilitate intercourse.

The egg cells or ova are periodically produced in the ovaries at intervals of approximately 4 weeks. At the end of each 4-week period, one egg reaches maturity and passes into one of the Fallopian tubes. The egg descends gradually and remains viable for a short while. Following intercourse, the sperm cells swim toward the tubes, in one of which fertilization may take place. Since neither the male nor the female reproductive cells live long, successful fertilization can occur only during a short period of time each month. This period of maximum fertility in women can be ascertained by various means, including temperature measurements.

If the egg is fertilized by the sperm, the fertilized ovum enters the uterus and becomes attached to the uterine wall where the child develops. Ordinarily, only one egg is produced each month, although more than one egg may be produced and, in some cases, may lead to multiple birth. If pregnancy occurs, usually no eggs are produced until after the child is born, or pregnancy is interrupted.

The maturing of the egg is a continuous process regulated by the endocrine system. Within the ovary, there is a layer of cells called the *germinal epithelium.* Here, the potential egg begins its existence and continues to develop until a *primary follicle* is formed around it, which is a clump of cells isolated from the main layer. The central cell of the clump is the egg, the remaining cells forming a ring around the egg. During a lifetime, each ovary forms between 200,000 and 400,000 follicles. Of all these potential eggs, only a few develop into mature eggs; most of them degenerate at the follicle stage. Those follicles that do not degenerate increase in size; meanwhile the egg cell itself enlarges until the original size is doubled. The one-ring layer of cells around the egg then multiplies and forms several layers. Fluid begins to accumulate in little pools which merge and form larger ones until one large pool is formed with the egg inside of it.

Other changes occur in the areas adjacent to the follicle. As the follicle matures, it moves toward the surface of the ovary; when the maturation process is complete, the follicle protrudes from the surface of the ovary. At this time ovulation occurs. The follicle bursts and the egg, with its fluid, is expelled from the surface of the ovary, leaving a cavity. Consequently, the adult woman who has ovulated many times possesses ovaries that have a pitted appearance. See also **Gamete.**

The Uterus. Commonly known as the womb, this is a pear-shaped organ the size of a small fist and is located in the pelvic cavity of the female. The uterus is the organ that receives the fertilized egg from the Fallopian tube and provides the necessary nourishment and protection of the fetus during the various stages of pregnancy, and expels the developed child by the action of its muscular walls. The walls of the uterus are elastic, allowing for distention during pregnancy and return to the original thickness after childbirth.

The cavity of the womb is lined with the endometrium, a mucous membrane. The endometrium is not of the same thickness and consistency all the time, but varies considerably during the menstrual cycle. During menstruation, the endometrium disintegrates and is expelled with the menstrual blood, but a new endometrial lining begins to form immediately following each menstruation. The womb possesses two parts called the "body" (*fundus*) and the "neck" (*cervix*). The cervix is below the fundus and connects with the vagina at a right angle. The position of the womb is not always the same. In general, the long axis of the womb extends from front to back and slightly downward. The neck of the womb is then pointed toward the rectum and meets the vagina at a right angle. The urinary bladder lies in front and the rectum in the back of the womb.

The cervix, or neck of the womb, is an important organ that has numerous functions in the reproductive system. During pregnancy, the cervix protects the fetus, and during childbirth it distends to permit passage of the child. The cervix may be the origin of a variety of disorders and the site of numerous infections.

The Vagina. During sexual intercourse, the vagina receives the male sperm cells. The organ is made up of muscular tissue which possesses a considerable degree of elasticity. This permits distention without tearing when the child passes from the womb to the exterior of the body. The vagina is located between the urinary bladder and the rectum, although it is not directly connected to either. The vagina serves as a passageway between the opening of the vulva and the opening of the cervix.

In the adult woman, the size of the vagina varies but the average length is approximately 3 inches (7.5 centimeters). When the woman is in a standing position, the direction of the vagina is backward and upward, forming almost a right angle with the long axis of the uterus. The outer opening of the vagina is surrounded by a mucous membrane called the *hymen.* In the virgin woman, the hymen covers a considerable area of the vaginal opening; in rare instances, it may cover it entirely (*imperforate hymen*) causing retention of the menstrual flow. The hymen varies considerably in shape, but in general is semicircular. If the hymen is intact at the incident of first intercourse, it is usually ruptured at that time, although not always; sometimes it does not tear, but merely stretches. Consequently, absence of a hymen or a ruptured hymen should not be construed to mean that a woman is not a virgin.

The lining of the vagina secretes a fluid that is acid in nature and serves as a cleanser and lubricant. In an acid environment only certain types of bacteria can live, most of which are harmless and even helpful. The vaginal lining is smooth only in women that have borne children or after the menopause in childless women. In the young woman, the lining forms a series of folds.

The Vulva. Vulva is a collective name applied to the external female organs of reproduction and includes the mons pubis, labia majora, labia minora, clitoris, vestibular bulbs, vestibule, Bartholin's glands,

Skene's glands, and hymen. The urethra, which is part of the urinary system, is often regarded as a structure of the vulva.

The *mons pubis* is located on top of the pubic bone just above the genital organs. This is a pad of fatty tissue covering the underlying bone. It forms an inverted triangular area which is covered with hair in the adult woman. The sides of the triangular area are delimited by the groins. From the top of the triangle, the mons pubis bends gradually downward and backward, dividing in the center to form two distinct sides that eventually, toward the perineum, become indistinguishable from the labia majora. The mons pubis contains many erogeneous nerve endings which, when stimulated, add to the female's excitement.

Labia majora means "major lips," and as the name indicates they are two large folds of tissue located around the vaginal opening. When the woman is in the erect position, the labia majora conceal most of the other external organs of reproduction. Extending downward they gradually decrease in thickness until they disappear into the region of the perineum. The perineum is the area between the vulva and the anus. When the labia majora are pulled aside, the remainder of the female organs of reproduction become visible.

Within the labia majora lie the *labia minora*, which means minor lips. These are folds of skin which form an angle. The area bounded by this angle is called the vestibule, and within this area is located the opening of the vagina. The labia minora have an abundance of erogeneous nerve endings. When stimulated during sexual excitement, the labia minora thicken two to three times their normal size.

The *clitoris*, which is located at the apex of the triangular area delimited by the labia minora, is a relatively small organ made up of erectile tissue. Erectile tissue becomes firm and engorged with blood in response to stimulation. The clitoris in the female and the penis in the male are somewhat similar in structure and response. The clitoris is covered by a fold of skin, which is known as the prepuce; the tip of the clitoris is called the glans.

The opening of the urethra and the vagina are located in the vestibule. The urethral opening and openings of the Skene's glands lie just below the clitoris. Below these lies the opening of the vagina. Skene's glands secrete an alkaline substance which reduces the acidity of the vagina. The Bartholin's glands are located in the lower portion of the vestibule and are not normally conspicuous, but become prominent when inflamed and infected. Bartholin's glands produce a drop or so of mucous secretion which at one time was thought to serve as a lubricant during sexual intercourse. However, this secretion is insufficient for that purpose.

Ovarian Tumors. The diseases not related to the endocrine system that affect the ovaries comprise a large number, of which tumor formation is the most important. Tumor does not necessarily imply cancer and actually most ovarian tumors are not cancers.

Most ovarian tumors develop without presenting symptoms, except those that produce hormones. Eventually, pain is caused by the tumor pressing against neighboring organs, tension of the tumor mass, rupture, or infection. When a positive diagnosis of tumor has been made, surgical exploration becomes necessary in almost every case. Abdominal exploration is necessary to secure a complete diagnosis and to remove the tumor. All ovarian tumors may be dangerous if not removed, because it is almost impossible to determine which will or will not develop into a cancer. The extensive growth of tumors of the ovary can be prevented only by early discovery and removal. Therefore, periodic pelvic examinations are extremely important in the early detection of cancer. Also, a pelvic examination is of great importance for early detection of cancer of the cervix. An examination of the cervix by a physician is a simple procedure. A procedure known as the "Pap" test is a cytologic examination and was developed chiefly by the late Dr. George N. Papanicolaou. The test involves the microscopic examination of cells collected from the vagina. These are cells shed from the uterus into the vagina as a part of the normal life process. If microscopic examination of the smear reveals any abnormal cells, bits of tissue are taken from the cervix for further microscopic study.

Infection and Tumors of the Fallopian Tubes. Infections of the Fallopian tubes frequently cause permanent sterility. The Fallopian tubes are attacked most often by the organisms causing gonorrhea, infections produced during childbirth, tuberculosis, and a variety of systemic infections. These infections may be acute or chronic. In some instances, they may involve the entire reproductive system. Tumors may develop in the Fallopian tubes, usually as a secondary growth which originated in some other organ of the body. Tumors of the Fallopian tubes are relatively rare.

Tubal Pregnancy. The Fallopian tube is at times the site of an abnormal type of pregnancy, called tubal pregnancy. In these cases, the embryo fails to descend into the womb and develops instead in the Fallopian tube. As the fertilized egg grows within the tube, the tension increases, and the tube may rupture, causing death of the fetus. Once the existence of tubal pregnancy has been established, surgical intervention to remove the tube and the embryo is usually required. Often, there may be no symptoms of tubal pregnancy prior to rupture. This condition endangers the patient because hemorrhage is imminent in nearly every case. Tubal pregnancy is not the only form of abnormal pregnancy that takes place outside the womb, but it is perhaps the most common abnormal type. Other types include abdominal and ovarian pregnancies.

Retrodisplacement of the Uterus. The uterus is held in place by the floor of the pelvis and a series of tough bands of tissue (ligaments). Thus, the womb is not rigidly fixed in one position, but is movable. Abnormal displacements may occur when the position of the womb changes beyond certain limits. The uterus can turn backward, causing retrodisplacement. The most common cause is childbirth. During labor there is often considerable stretching of the supports that keep the womb in place. To avoid displacement, the physician instructs the mother to lie on her abdomen or side during convalescence. Once the condition has been discovered, the physician institutes treatment. This generally consists of bringing the uterus to a normal position by manual manipulation and maintaining it in a normal position by some mechanical support. Such supports vary in design and shape and are called *pessaries*, usually consisting of a flexible ring made of rubber or plastic.

Prolapse. At childbirth, the stretching of the uterine supports may cause both retrodisplacement and *prolapse* of the uterus. In the latter condition, the womb falls from the normal position and the cervix pushes far into the vagina. Severe prolapse can cause the womb to push the cervix through the vagina. Complications ensue, usually associated with ulcerations of the cervix as a result of irritation produced by continuous contact with the clothing of the patient. The pressure exerted by the prolapsed womb upon the urinary bladder causes an inability to retain urine. Frequently, incontinence is the complaint that induces the patient to consult the physician. Prolapse is corrected with pessaries and by surgical means. The restoration of the normal position of the womb does not necessarily involve loss of reproductive function.

Endometriosis. The lining (endometrium) of the womb sometimes behaves abnormally and grows not only on the walls of the womb, but within the walls, or on adjacent pelvic organs, causing a condition known as *endometriosis*. The patient with this condition may suffer irregularities in the menstrual cycle. Menstruation is often painful and copious. The manner in which bits of lining are transported from the womb and lodge in other parts of the body is not fully understood. Apparently, they can be transported by way of the Fallopian tubes, the blood, and the lymph. External endometriosis may necessitate surgical treatment. The results are satisfactory in most cases.

Uterine Tumors. The uterus is one of the most frequent sites of tumor formation, being second only to the breast. Tumors develop in nearly any part of the organ. Tumors of the fundus are of many types, but most common are fibroids (*leiomyomata*) of the uterus which develop from muscle tissue. The patient may have a group of small fibroids for many years and suffer no ill effects. However, the size of the tumors varies, sometimes reaching large proportions. Treatment of patients who have fibroids varies according to type and size of the tumors. If small and cause no symptoms, no treatment may be deemed necessary. Others that may endanger health usually are removed surgically. There are many other forms of tumors that can grow in the fundus and that can arise from any of its component tissues. In their early stages, many of these growths can be treated successfully either surgically or radiologically. Some, such as choriocarcinoma, respond to chemotherapy.

Cervical Cancer. When cancer develops in the cervix, it is at first confined to this organ, but, depending on the type of growth, spreads at different rates to the adjacent organs. In the eartly stages of the disease, there are no specific symptoms except perhaps irregular bleeding and discharge. The patient may delay examination until she is sure that the bleeding will not disappear. After such delay, the cancer may have advanced beyond hope of cure. Any unusual bleeding or discharge, other irregularities in the menstrual cycle, periods in which there is profuse bleeding, and the recurrence of a period after several months without periods should be recognized as danger signals. Cervical cancer rarely appears in women under age 20; sometimes before age 30; but most commonly in women around 45 years of age.

Trichomonas vaginalis, a parasitic protozoan, may infect the vagina, producing an irritative discharge. *Monilasis*, a fungus infection caused by *Candida albicans*, may affect the vaginal wall causing a white discharge and white patches. Nonspecific infections, caused by a number of bacteria, may be present in the vagina. Bacterial infections can usually be controlled by administration of one of the antibiotic drugs. Vaginal tumors are relatively uncommon. The most common type is called "inclusion cyst," which in most instances is not serious.

Vulvitis. Inflammation of the vulva may be caused by a number of factors. Since the external portion of the vulva is covered by skin, many skin conditions, such as eczema, ringworm, crysipelas, contact dermatitis, etc., may occur. Acute vulvitis occurs in children and obese women because of constant irritation. Vulvitis occurring in diabetic patients is caused by increased sugar content of the urine which produces irritation and provides a favorable environment for the growth of yeasts and fungi.

The reproductive organs of both male and female, of course, are affected by venereal diseases, described elsewhere in this volume.

Menopause. Cessation of menstruation marks the commencement of the *menopause.* This is a period when there are numerous biochemical and hormonal changes in the body, the symptoms of which vary widely from one woman to the next. In addition to physical changes, there are frequently accompanying psychological features in many cases. Some women experience no symptoms, whereas others require varying degrees of medical assistance in making the adjustments. Physiologically, as the result of ovarian failure, the amount of estrogens produced declines. The average age of ovarian failure is 48 years (statistic for the United States). Some women become amenorrheic in their earlier forties; others continue to menstruate and ovulate regularly into their fifties. The functions of estrogen still are not fully understood, but a number of the signs of menopause are associated with estrogen deficiency. These include vascular symptoms (among these are "hot flashes") in the shorter term and, extended over a period of time, consequences of estrogen deficiency may include an acceleration of atherogenesis (see **Arterial and Venous Disorders**); osteoporosis (see **Bone**); and urethral and vaginal atrophy.

Because the wide range of symptoms and their degree of severity, there is no universal approach to treating them. Several years ago, estrogen replacement therapy was welcomed by patients and physicians alike as an excellent pathway to alleviating many of the problems of menopause. Several studies were published in the mid- and late-1970s linking estrogens to increased incidence of endometrial (lining of the uterus) carcinoma. Studies have convinced many physicians that there are some risks in estrogen therapy—risks that must be weighed against the specific symptoms and needs of the patient. Thus, the present situation is one of using estrogens with the utmost of discretion. However, in the case of younger women who have lost their ovaries surgically, some physicians suggest that they should have full estrogen replacement until the time of a natural menopause, at which time the continuation of the therapy must be reevaluated.

References

Carmichael, D. E.: "The Pap Smear," Charles C. Thomas, Springfield, Illinois, 1973.

Chakravarti, S., et al.: "Hormonal Profiles after the Menopause," *Br. Med. J.*, **2**, 784 (1976).

Federman, D. D.: "Abnormal Sexual Development," W. B. Saunders, Philadelphia, 1967.

Federman, D. D.: "The Testis," in "The Year in Endocrinology (1977)," Plenum, New York, 1978.

Gray, L. A., Christopherson, W. M., and R. N. Hoover: "Estrogens and Endometrial Carcinoma," *Obstet. Gynecol.*, **49**, 385 (1977).

Gray, L. A.: "Endometrial Carcinoma and Its Treatment," Charles C. Thomas, Springfield, Illinois, 1977.

Greenwald, P., Caputo, T. A., and P. E. Wolfgang: "Endometrial Cancer after Menopausal Use of Estrogens," *Obstet. Gynecol.*, **50**, 239 (1977).

Klopper, A.: "Endocrine and Metabolic Diseases: Treatment of Infertility and Menopausal Symptoms," *Br. Med. J.*, **2**, 414 (1976).

Lipsett, M. B.: "Estrogen Use and Cancer Risk." *JAMA*, **237**, 1112 (1977).

McCann, S. M.: "Luteinizing-Hormone-Releasing Hormone," *N. Engl. J. Med.*, **296**, 797 (1977).

Odell, W. D., and R. S. Swerdloff: "Male Hypogonadism," *West. J. Med.*, **124**, 446 (1976).

Paulsen, C. A.: "The Testes," in "Textbook of Endocrinology," 5th edition (R. H. Williams, editor), Saunders, Philadelphia, 1974.

Robertson, J. R.: "Genitourinary Problems in Women," Charles C. Thomas, Springfield, Illinois, 1978.

Ross, G. T.: "The Ovaries," in "Textbook of Endocrinology," 5th edition (R. H. Williams, editor), Saunders, Philadelphia, 1974.

Stearns, E. L., et al.: "Declining Testicular Function with Age," *Amer. J. Med.*, **57**, 761 (1974).

GONAPOPHYSIS. Appendages of the insect abdomen which serve as accessory organs of reproduction. Copulatory organs of the male and egg-laying organs of the female. They are probably derived from jointed appendages.

GONDWANALAND. Earthquakes, Seismology, and Plate Tectonics; Ocean.

GONIOMETER. 1. An instrument for measuring the angles between the reflecting surfaces of a crystal or a prism. Parallel rays from a collimator, impinging upon the polished surfaces, are reflected in different directions. Two methods may be used. In one, the crystal or prism is held stationary and the angle between the reflected beams from the two faces, received in succession by a telescope moving around a graduated circle, is measured on the circle; the angle between the two faces is then $\frac{1}{2}$ of this (see Fig. 1). In the other method, the telescope is clamped in some convenient position and the crystal or prism is rotated so that first one and then the other face reflects light into it; the angle between the faces is the supplement of the angle through which the prism mounting is turned. An ordinary spectrometer may be used for the purpose. An instrument similar in geometrical principle, but employing x-rays instead of light and an ionization chamber instead of a telescope, is used for measuring angles between the atomic planes within crystals.

2. For approximate measurement of interfacial angles on larger crystals, a simpler instrument, known as a *contact goniometer*, can be used. This instrument consists of a protractor with a movable arm attached to its base at a point exactly perpendicular to its 90° reading. The crystal is held between the base of the protractor and the movable arm with the interfacial plane surfaces making parallel contact with the protractor base and the arm. The corresponding external angle is then read directly from the protractor. In using this instrument, it is required that it be held perpendicular to the interfacial crystal planes being measured. The angle desired will be the *internal* angle, which as in the example of the preceding paragraph, will be the supplement of the measured *external* angle.

3. Sometimes in the use of a loop antenna for directional purposes it is convenient or impossible to rotate the loop. This is especially true for transmitting loops where the size becomes appreciable in order to improve the radiation efficiency. To overcome this difficulty the goniometer is used. As shown in (Fig. 2), the instrument consists

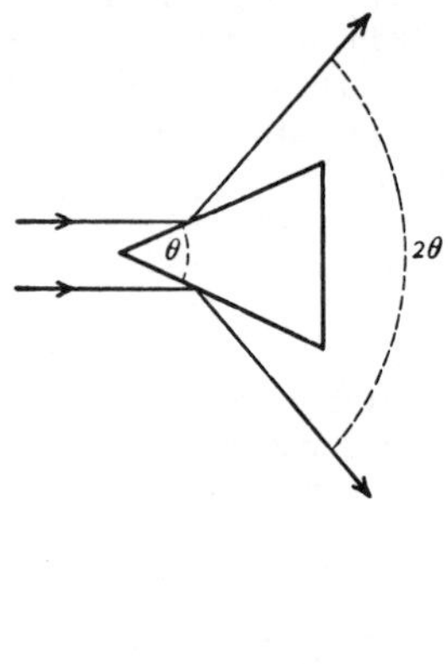

Fig. 1. Angle between reflected rays is twice the angle between prism faces.

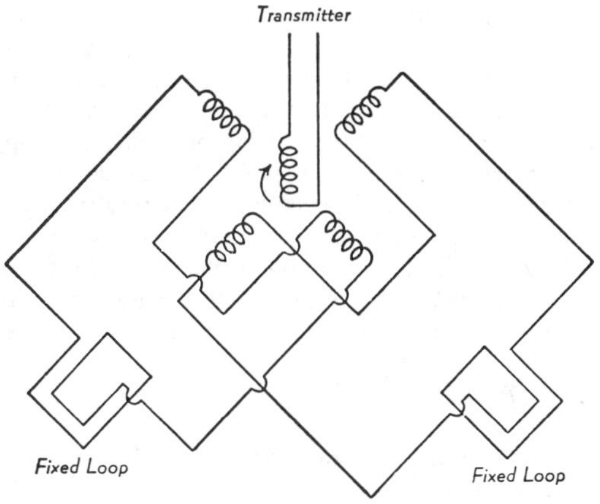

Fig. 2. Goniometer (radio).

of crossed stationary coils feeding fixed, crossed (90°) loops with the coupling to the moving coil proportional to the cosine of the angle of rotation. The movable coil is fed from the transmitter. The amount of energy transferred from the rotating coil to the fixed coils and hence to their antennas is determined by the position of the moving coil with respect to the others. The effect as far as the resultant field pattern of the antennas is concerned is exactly the same as if a single-loop antenna had been physically rotated. An extension of this principle is used with two sets of crossed loops in many radio range systems. A further advantage of the goniometer is that the antennas may be connected through a transmission line and thus it is not necessary that the operating position be near the antenna. It may be used equally well for reception.

GO/NO-GO DETECTOR. An instrument which has only two stable states of indication, and which therefore will give full response to any stimulus capable of actuating it. For example, a common fuse is a go/no-go detector, since either it is intact, or it is burned out. An ammeter, however, can respond continuously to the same current.

Go/no-go detectors are widely used in automated sorting and inspecting machines.

GONORRHEA AND GONOCOCCEMIA. Caused by the diplococcus *Neisseria gonorrhoeae*, gonorrhea is one of the most common and widespread of genital tract infections (venereal diseases) and is worldwide in occurrence.

Direct contact between persons, usually of a sexual nature, is required for the transmission of gonorrhea. The disease is pandemic in some countries, including the United States, where between 2.5 and 3 million or more cases per year are reported. Most cases (90%) of gonorrhea occur in persons under 30 years of age and, of these, one-fourth are in the teenage bracket. The disease, however, is not age-group, racially, or economically limited among sexually active individuals. Persons who engage in fleeting sexual relationships with numerous partners over relatively short time spans run the greatest risk of contracting the disease.

At one time, it was reasonably well accepted that many females essentially were reservoirs of the disease without being aware of having the disease—because of lack of symptoms, i.e., the disease was spread mainly by asymptomatic females to males who almost always became symptomatic. It was believed that one female could infect several males within a short or long period prior to her awareness of disease in her body. Because symptom awareness in males is more vivid than in females, there is some strength to this earlier observation. However, it is now also realized that, although most males who develop urethral gonococcal infection recognize it and seek medical attention shortly after symptoms develop, there are also some males who can be infected for extended periods and thus available to infect females over extended periods without experiencing the usual vivid symptoms. Or, the symptoms may be so mild that medical attention is not sought. Thus, some members of both sexes from a practical standpoint can serve as carriers of the disease, spreading infection to dozens or even scores of sexual partners if they are promiscuous and very active sexually. A few thousand individuals in these categories can create a pandemic or epidemic situation.

Symptoms. The symptoms of gonorrhea in heterosexual males are anterior urethritis with a purulent urethral exudate and dysuria (difficult and/or painful urination). Incubation requires 3 to 4 days, but can be as many as 14 days or more. Without treatment, the course of gonorrhea and its complications can include epididymitis (testicular infection), prostatitis (prostate gland infection), infection of the paraurethral glands, and sometimes urethral stricture. There is also always the possibility of the development of disseminated gonococcal disease, described a bit later. In homosexual males, there is gonococcal infection of the urethra, as well as infection of the anal canal (30–55% of cases) and infection of the pharynx (throat) in some 21% of the cases. Anal infections are frequently asymptomatic, but may include pain (upon defecation), a feeling of rectal fullness, and rectal discharge. Pharyngeal infection can lead to acute exudative pharyngitis.

The symptoms of gonorrhea in heterosexual females are quite versatile. As previously mentioned, they may go unnoticed (asymptomatic) for a long period. The infection may be associated with vaginal discharge, discomfort in the area of the lower abdomen, as well as abnormal uterine bleeding. The anal canal may be infected by vaginal secretion. Other symptoms may include dysuria, pyuria (pus in urine), and less frequently, hematuria (blood in urine). Untreated, the course of gonorrhea and its complications may include abscess of Bartholin's glands (vestibular glands in vagina), acute pelvic inflammatory disease, conjunctivitis, and disseminated gonococcal disease.

Gonococci may infect an infant during birth, notably the eyes. Gonococcal ophthalmia neonatorum is prevented by the administration of silver nitrate solution to the infant's eyes, but this procedure alone does not guarantee full and permanent protection of the infant. Some mothers may have an asymptomatic gonococcal infection during pregnancy. Thus, hematogenous gonococcal arthritis may occur in the neonate by infection of the anogenital, oropharyngeal, or umbilical area during birth. Thus the need to screen pregnant women, particularly in instances where multiple sexual relationships may be suspected, for possible gonococcal infection during the prenatal period.

Gonorrhea is also known, infrequently, among prepubertal children, sometimes the result of sexual molestation or precocious childhood sexual activity. The characteristics of the disease are consistent with that of adults.

Diagnosis. The definitive diagnosis of gonorrhea is contingent on the recovery of gonococci from the patient. Gram's-stained smears from the urethra or freshly cleansed cervix can be used for tentative diagnosis. When typical Gram-negative diplococci are seen within three or more polymorphonuclear leukocytes, the degree of certainty is 90% in males and females, although sensitivity is only about 65%. Definitive confirmation requires selective culture media. These include Thayer-Martin and transglow media. Cultures are obtained from all clinically infected sites, whether or not local symptoms are present. In women, a culture of the endocervix is an effective screening test. These tests are positive in from 80–90% of infected females. In instances where fellatio is practiced, pharyngeal cultures should also be obtained. Cultures of material from the female urethra are usually omitted. In men, urethral cultures are paramount. In the case of homosexual males, rectal and pharyngeal cultures are also made.

Treatment. Persons to be treated for gonorrhea should be screened for evidence of syphilis because this will alter the course of treatment. Antibiotic therapy of uncomplicated gonorrhea is effective in over 90% of cases treated. Several regimens are available to the treating physician. Where penicillin G can be used, it is generally the drug of choice. In penicillin-allergic patients, tetracycline (not for pregnant women or small children) or spectinomycin can be used. Ampicillin and amoxicillin are also used. Gonococci, as the result of chromosomal mutations, have built up a resistance to some of the antibiotics, but to date this has not been so extreme as to reduce the use of traditional antibiotics. However, this situation is of great concern, as one day the conventional antibiotics may largely become ineffective in treating gonorrhea. An interesting development known as *bacterial pili* may, at some time, be effective against *N. gonorrhoeae* and a number of other pathogens. This is described a bit later.

The special forms of penicillin, such as the long-acting benzathine penicillin, phenoxymethyl penicillin (V), and the semisynthesized penicillinase-resistant penicillin, are not as effective against gonorrhea as penicillin G. See also **Antibiotic.** For pharyngeal gonococcal infection, aqueous procaine penicillin G or tetracycline are commonly used. Some authorities do not consider tetracycline effective for anorectal gonorrhea in men.

Individuals with a recent known exposure to gonorrhea should receive the same treatment used for the established disease.

Disseminated Gonococcal Disease. Also called the arthritis-dermatitis syndrome, disseminated gonococcal infection (gonococcemia) is the most common cause of infectious arthritis in young adults. This condition develops as the result of gonococci invading the bloodstream. Many more cases are seen in women than in men. Symptoms include tender, pustular skin lesions (5 to 25 per patient) of a distinctive and repelling appearance, ranging from 5 to 15 millimeters ($\frac{1}{5}$–$\frac{3}{5}$-inch) in diameter. They often have a necrotic center. In the later phase of gonococcemia (a week or more after onset), purulent arthritis involving one or two joints will appear, with gonococci present in the synovial fluid in over 50% of the cases. The pattern of disease development varies from one patient to the next. Sexually active persons who display skin rashes and acute arthritis at the same time should be considered arthritis-dermatitis syndrome suspects because relatively few other diseases mimic this condition. High-risk diseases, such as meningitis and endocarditis, although infrequent, may develop as a consequence of gonococcemia. Gonococcal meningitis and endocarditis require prolonged intravenous penicillin therapy with accompanying supportive measures. For treatment of endocarditis in patients allergic to penicillin G, a procedure to desensitize the patient to the antibiotic may be required. Chloramphenicol can be effective against gonococcal meningitis.

Bacterial Pili. The human body has many protective mechanisms for blocking off invading bacteria. Secretions, for example, protect the urogenital tract, as well as the respiratory and digestive tracts. And yet pathogens such as *N. Gonorrhoeae* do successfully colonize parts of the body that are usually not invaded by many other microorganisms. Scientists at the University of Michigan have recently suggested that the special ability of some bacteria to colonize can be attributed to a property for adhering to cell surfaces. Once the bacteria adhere, they can multiply. Discovered by researchers at the University of Pittsburgh some 25 years ago, bacterial structures called *pili* may be responsible for this adherance phenomenon. This has opened up the possibility for developing vaccines that will elicit antibodies against the pili. Currently, some vaccines are being tested in human volunteers for immunization against gonorrhea as well as bacterial diarrhea. More detail on this development can be found in the Marx (1980) reference listed.

References

Berg, W. S., et al.: "Cefoxitin as a Single-Dose Treatment for Urehtritis Caused by Penicillinase-producing *Neisseria gonorrhoeae*," *New Engl. J. Med.*, **301**, 509 (1979).

Handsfield, H. H., et al.: "Asymptomatic Gonorrhea in Men," *New Engl. J. Med.*, **290**, 117 (1974).

Klein, E. J., et al.: "Anorectal Gonococcal Infection," *Ann. Intern. Med.*, **86**, 340 (1977).

Marx, J. L.: "Vaccinating with Bacterial Pili," *Science*, **209**, 1103–1105 (1980).

McCormack, W. M., et al.: "Clinical Spectrum of Gonococcal Infection in Women," *Lancet*, **1**, 1182 (1977).

McCormack, W. M.: "Treatment of Gonorrhea," *Ann. Intern. Med.*, **90**, 845 (1979).

Nelson, J. D., et al.: "Gonorrhea in Preschool- and School-aged Children," *J. Amer. Med. Assn.*, **236**, 1359 (1976).

Phillips, I.: "Beta-Lactamase-producing Penicillin-resistant Gonococcus," *Lancet*, **2**, 656 (1976).

GOOSANDER. Waterfowl.

GOOSE. Poultry; Waterfowl.

GOOSEFISHES. Anglerfishes.

GOPHER. Squirrels and Other Sciuromorphs.

GORAL. Goats and Sheep.

GORILLA. Anthropoids.

GOSSAN. This term is applied to the decomposed upper parts of mineral veins and ore deposits. It usually consists chiefly of hydrated iron oxide resulting from the weathering of pyrite, chalcopyrite, etc. Gossans have been important sources for the release of the relatively insoluble precious metals and gems which are washed away to form placer deposits. Many valuable gold ore bodies have been traced to their source by means of their derived placers. Also, secondary enriched sulfide ores of copper have been discovered beneath gossans which were originally prospected for the more precious metals.

GOSSYPOL. Protein.

GOUGE. A term used to designate soft or clay-like material between the sides of a mineral vein or ore deposit and the wall rock; also (structural geology), a layer of finely comminuted material between the walls of a fault.

GOURAMI. Labyrinth Fishes.

GOURDS. Cucurbitaceae.

GOUT. A syndrome made up of a number of physical and chemical factors. These include abnormally high levels of uric acid (hyperuricemia) in the blood, usually symptomatic of gout, but which can occur from a few other causes; attacks of acute arthritis, with the presence of deposits of uric acid salts in and within the region of joints and tendons as well as in the kidney parenchyma; and formation of uric acid stones in the urinary tract and renal collecting system. The latter condition may lead to occasional kidney failure. The patient with acute gout is usually debilitated for a period, the length of time depending upon promptness of treatment and response to therapy. In Europe and America, the incidence of gout is about 3 cases per 1000 population. The disease occurs in males about ten times more often than in females. In the latter, the disease rarely occurs before menopause. The Maoris of New Zealand are particularly prone to gout, with the disease found in about one out of every ten males. It is estimated that many gouty people go undiagnosed unless the complications of serious joint and renal changes occur. Although the genetics of the disease are not accurately known, experience has shown familial connections.

Gout is a manifestation of faulty purine metabolism. Uric acid is a product of the purines. These occur in all tissues and are characteristic constituents of the nucleoproteins. Nucleoproteins are found in the nuclei and cytoplasm of all living tissues, plant and animal. In the breakdown of nucleoproteins, nucleic acids are released, and purines are located in these portions of the nucleoproteins. The purines have as their end product, in humans, an oxidized purine, namely, uric acid, In addition, there is a pathway for the formation of uric acid which does not involve purines, but which does involve glycine and other simple products.

Normally, the excretion of uric acid by the kidney keeps pace with its formation from purines of the food, purine metabolism of the tissues, and synthesis of uric acid. An elevation of serum uric acid may occur if the kidney cannot eliminate it at a normal rate, or if the rate of tissue breakdown is accelerated. This is usually accompanied by rises in other nitrogenous constituents of the blood, e.g., urea, and may not always be associated with gout. In gout, the only nitrogenous constituent of serum which rises characteristically is uric acid. As a result of this increased amount in blood, uric acid precipitates out in various locations of the body. The onset of the first attack is usually a very severe pain in the joint of a finger or toe. The joint becomes red, swollen, and extremely tender. Other joints are sometimes affected and frequently more than one finger or toe is involved. In severe cases, knoblike deformities around the affected joints appear, due to the deposition of uric acid to form "tophi."

Other conditions which may cause hyperuricemia sometimes interfere with an accurate diagnosis, particularly in the milder cases. These include individuals who have drastically lowered their carbohydrate intake in connection with dieting, rheumatic fever, rheumatoid arthritis, septic arthritis, cellulitis, and bursitis. Pseudogout sometimes is seen in older people and is easily distinguished by x-ray examination, which shows calcification of tissues, often involving the knee joint.

Gout therapy frequently involves the use of colchicine, which is reasonably specific to acute gouty arthritis. See also **Alkaloids.** Oral doses (sometimes intravenous) are given frequently for several hours until vomiting or diarrhea is induced. Within 12 to 24 hours, considerable improvement usually will be noted. Colchicine is then resumed in small dosages at less frequent intervals. The physician is aware of possible gastrointertinal side effects of this drug. Other drugs used include indomethacin and phenylbutazone. When the rare patient does not respond to any of these drugs, parenteral glucocorticoids may be used.

To inhibit *interval gout,* the plasma urate concentration will be maintained at proper levels. This is usually done with colchicine therapy. In the treatment of *chronic gout,* several objectives must be met: (1) further precipitation of monosodium urate crystals in tissues must be prevented; (2) dissolution of crystalline deposits already formed must occur; and (3) the function of affected joints must be restored. The drug probenecid acts effectively in most patients as a uricosuric agent. Other drugs of this type are sulfinpyrazone and allopurinol. Sometimes probenecid and colchicine are administered in combination. This type of therapy, although sometimes prolonged, usually is successful. It is uncommon to have to surgically remove the tophaceous deposits. Dietary procedures appear to be of little avail, probably because uric acid can be synthesized from generously available very small molecules.

References

Seegmiller, J. E.: "Disorders of Purine and Pyrimidine Metabolism," in "The Year in Metabolism." (N. Freinkel, editor), Plenum, New York, 1976.

Stanbury, J. B., Wyngaarden, J. B., and D. B. Fredrickson (editors): "The Metabolic Basis of Inherited Disease," 4th edition, McGraw-Hill, New York, 1977.

Wyngaarden, J. B., and W. N. Kelley: "Gout and Hyperuricemia," Grune & Stratton, New York, 1976.

Yu, T., and A. B. Gutman: "Uric Acid Nephrolithiasis in Gout: Predisposing Factors," *Ann. Intern. Med.,* **67,** 1133 (1967).

GOUT (Kidney Involvement). Kidney and Urinary Tract.

GOVERNOR. An automatic controller for maintaining the rotative speed of a machine. The governor senses the speed, compares the measured value with the desired value, and acts to correct any error between these two values—most often by adjusting the flow of energy to the machine. The two major types of governors are: (1) Designs wherein the speed-sensing element operates an energy-metering device directly; and (2) a design which employs one or more stages of power amplification between the speed-sensing element and the energy-control device. The first type usually gives stable control on an engine or other prime mover. The second type requires some stabilizing factor to prevent continual oscillation of the speed (*hunting*).

A direct-acting type centrifugal governor is shown in Fig. 1. The desired speed is set by the operator by varying the compression of the reference spring. If the engine exceeds this speed, the centrifugal force of the flyweights overcomes the spring force, thus lifting the fuel-metering rod to decrease the flow of fuel to the engine. If the speed drops below the desired value, the reverse action occurs, the spring force overcoming the flyweight force and thus the flow of fuel to the engine is increased. This design is reminiscent of the early whirling flyballs used on the steam engine of James Watt. However, even though the principle dates back 200 years or more, the principle is still used. The friction governor in a telephone dial system, for example, uses this principle. Other examples of use for direct-acting governors are vane governors used on small engines, matching the force of the cooling-fan blast against a spring, "velocity" governors on automotive engines, weighing the impact of the induction-air stream against a spring-loaded plate. Governors of this type require a change in engine speed in order to obtain the force needed to overcome friction in the fuel-metering system. This results in hysteresis and inaccuracy. These designs also exhibit *droop,* i.e., the governed speed of the engine droops off as the load is increased. Further, such designs do not have much actuating power.

The advent of hydraulic turbines required extensive redesign of

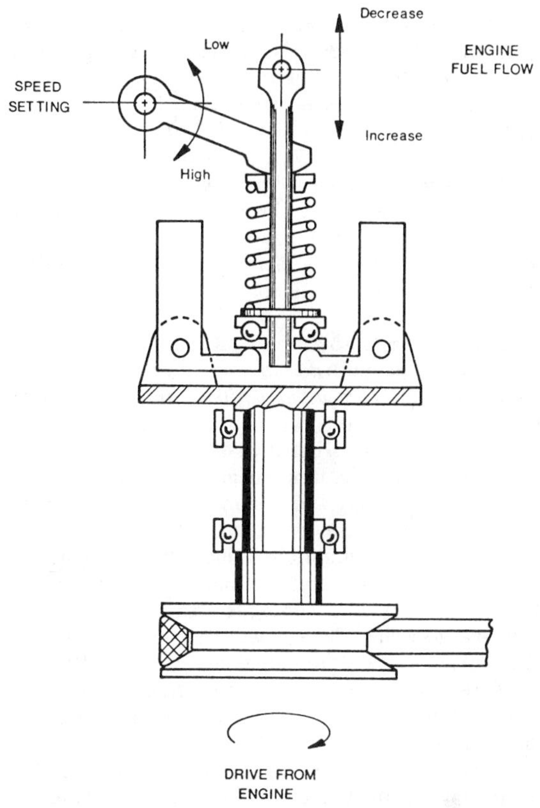

Fig. 1. Direct-acting type centrifugal governor.

governing mechanisms. For example, the governor had to develop sufficient power to operate the wicket gates against the force of the water and the high-friction forces. Early power-amplifier governors operated mechanically wherein the flyweights controlled the power through clutches or differential gearing. The power was derived from the turbine wheel and numerous ingenious designs were developed. Hydraulic power and amplification is used in contemporary governors.

Figure 2 shows a single-stage power-amplified governor. The speed-sensing centrifugal element operates a small pilot valve which controls the flow of pressurized fluid to or from the power piston. Because of the force available, the accuracy of this design is not materially

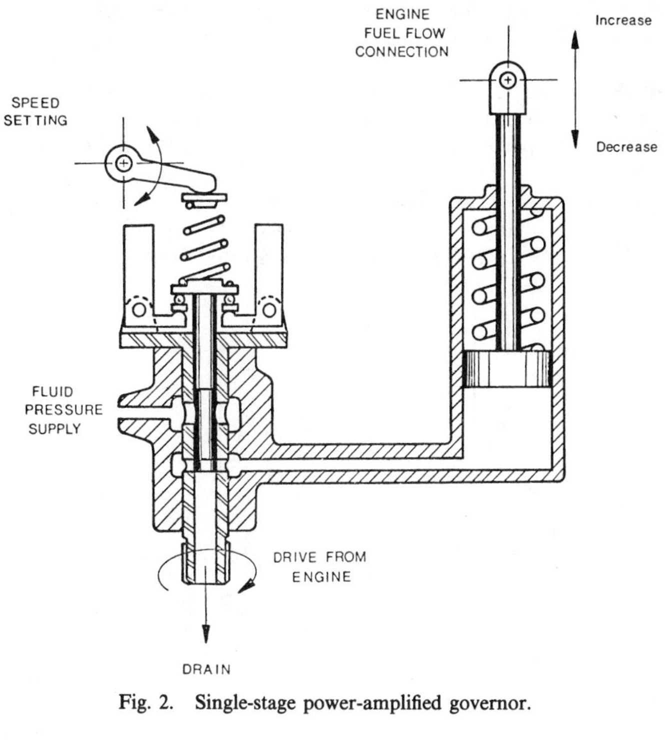

Fig. 2. Single-stage power-amplified governor.

affected by friction or other forces on the device controlling the flow of energy to the prime mover. Also, there is no droop. The governor will hold the speed for which it is set, namely, the speed at which the pilot valve is centered regardless of the load on the prime mover and the corresponding position demanded of the power piston. This type of governor has been used to control the speed of propeller-type aircraft power plants. These loads are inherently stable because increased torque is required by them when the speed increases and no stabilizing device is required in the governor. In most other applications, however, such stabilizing factors are not present in the engine or its load and the design will not control the system in a stable fashion.

Feedback is required to overcome oscillation for the reasons that (1) the response of the governor to a speed error is a *velocity of the power piston toward the required position*; and (2) the response of the engine to a new throttle position is an *acceleration toward the required speed.* It should be noted that position is the time integral of velocity; and speed is the time integral of acceleration. Thus once such a system is disturbed and commences to hunt, an oscillation will continue with the system continuing to hunt. The speed describes a sine curve about the equilibrium value, while the governor power piston will follow a cosine curve, the integral of the sine-wave speed curve, and lagging it by 90°. Adding feedback from the throttle position to the speed-sensing element will short-cut the engine lag and reduce the system phase shift to something less than 180°, and stability will result. However, it should be noted that the drooping characteristic of the direct-acting governor has been reintroduced. Hence the only gain is in increased power and governing accuracy.

Inasmuch as the droop is needed only to stabilize the system during an upset, it may be allowed to dissipate at a rate which will not again disturb the system. In another design, the pressure oil flowing to or from the power piston displaces the buffer piston against its springs. The resulting pressure differential appears across the compensating piston on the pilot valve, adding to, or subtracting from, the reference-spring force to cause speed droop. Once the system is stabilized, the compensating pressure equalizes again by leakage through the reset needle valve, and the governor holds a constant speed regardless of load on the engine. Inasmuch as such a governor, when used on an engine-generator set, will maintain constant frequency and clock time, it is called isochronous.

In some cases, speed may be sensed electrically and sometimes more simply and better than mechanically. Electronic circuits also facilitate combining speed, load, and other input signals in an optimum manner, and feedback signals, derivative, proportional, and integral, in any combination, are readily obtained electrically. However, hydraulic power amplification for the final stages is superior in rigidity, smoothness, and energy-storage capability.

GRAB BUCKET. A grab bucket is an apparatus which is able to pick up a load of bulk material by "biting" into the surface of the material. The particular usefulness of the grab bucket is that it may be lowered from the end of a boom onto the surface of the material to be moved, where it is operated to bite into this material, picking up a load, which can then be raised and deposited where wanted. Figure 1 shows a grab bucket in open and closed positions. The procedure by which this bucket is caused to close upon a load of material is based on the differential action of a two-step drum. A rope, to which power may be applied by winding it around a drum, passes to the bucket and has its end wrapped around the larger drum of the bucket. After power is applied to this rope, it unwraps from the drum on the bucket, turning that drum and wrapping the chain on the smaller diameter portion. The added leverage thus attained is sufficient to enable the drum to wind itself up on the chain, thus closing the bucket. The axis of the drum is pivoted to the center arms of the bucket. The digging power of a grab bucket of this kind is a function of the weight of the bucket, the sharpness of its cutting edges, the power applied to the operating rope, and the resistance of the material which it digs. Buckets of this type are not suitable for hard-packed material such as earth in original embankment, but are suited to handling grain, coal, ore, etc. They range in capacities from $\frac{1}{2}$ to 5 cubic yards (≈0.4 to 4.0 cubic meters). See Fig. 2.

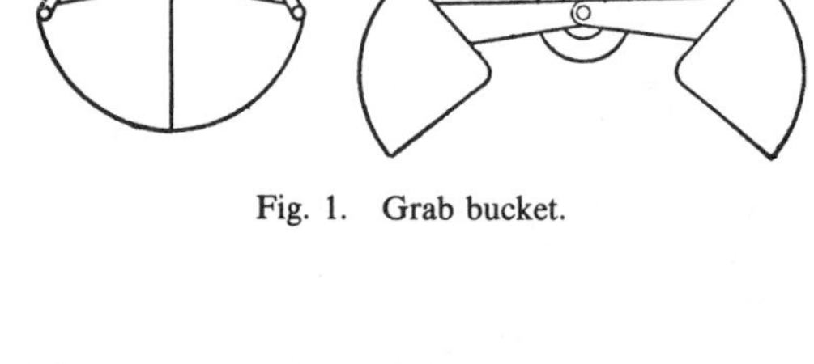

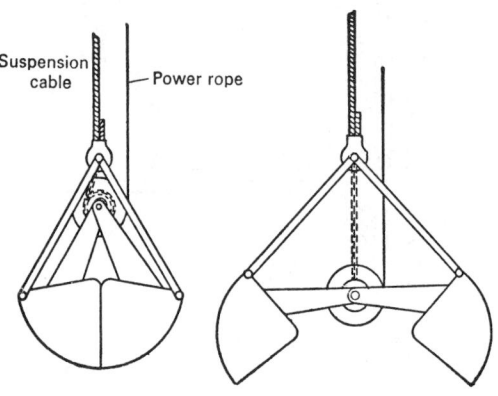

Fig. 1. Grab bucket.

Fig. 2. Use of a very large grab bucket in connection with an ore bridge at the steel plant of Ford Motor Company, Ltd works at Dagenham, England.

GRABEN. **Fault.**

GRACKLE (*Aves, Passeriformes*). In North America, several species of birds with black plumage and iridescent metallic luster, related to the orioles and blackbirds. The great-tailed grackle, *Cassidix mexicanus*, which ranges from Texas into South America, is also called the jackdaw. It should not be confused with the European jackdaw. In India the hill mynas and related species are called grackles.

GRADED BEDDING. A geological term denoting a type of bedding or stratification characterized by a cyclic or rhythmic deposition of coarse to fine sediments. A helpful criterion for determining the original position of the strata after they have been deformed. Graded bedding is generally supposed to be characteristic of offshore rather than inshore deposition.

GRADED-INDEX FIBER. **Optical Fibers; Telephony.**

GRADE (Engineering). In highway, railway, or municipal engineering, the slope of a line is called the grade. Grades are usually expressed as percentages preceded by a plus or minus sign. As an example, a +2% grade indicates a rise of 2 feet in every 100 feet (2 meters in every 100 meters) measured horizontally in the direction of travel; a −2% grade indicates a drop of 2 feet in every 100 feet (2 meters in every 100 meters). A curve known as a vertical curve is used to make the transition at a point of change in the grade of a highway or railroad. A second-degree parabola is used because it is the only curve in which the rate of change of slope is constant. The length is a function of the difference of the connected grades and the allowable rate of change of slope of the parabola per hundred feet measured horizontally. In the case of highways, the length of a vertical curve, at a point where the grade changes from plus to minus (at the crest of a hill), is governed by the safe sight distance.

GRADIENT CURRENT. In oceanography, a current associated with horizontal pressure gradients in the ocean and determined by the condition that the pressure force due to the distribution of mass balances the Coriolis force due to the earth's rotation. The gradient current corresponds to the geostrophic wind in meteorology.

GRADIENT FLOW. Horizontal frictionless flow in which isobars and streamlines coincide; or equivalently, in which the tangential acceleration is everywhere zero. Important special cases of gradient flow, in which two of the normal forces predominate over the third, are: (1) *Cyclostrophic flow,* in which the centripetal acceleration exactly balances the horizontal pressure force; (2) *Geostrophic flow,* where the Coriolis force exactly balances the horizontal pressure force; (3) *Inertial flow,* which is flow in the absence of external forces; in meterology, frictionless flow in a geopotential surface in which there is no pressure gradient, so that centripetal and Coriolis accelerations must be equal and opposite.

GRADIENT FLOW (Meteorology). Atmosphere (Earth).

GRADIENT (Geology). The term is applied to streams to refer to the slope of their beds, as steep, gentle, or in terms of so many feet per mile or meters per kilometer. The term is synonymous with *grade* as used in engineering. A stream valley is said to have become graded when its longitudinal profile is a smooth curve without waterfalls or rapids. The term grade is also used by students of sedimentary rocks, in a textural sense, to designate those grains of any sediment or sedimentary rock which are of the same size. The classification of grade-sizes is as follows:

Name of Grade		Range of Diameters
Pebbles		Greater than 10 mm
Gravel		10 mm to 2 mm
Sand	Very Coarse	2 mm to 1 mm
	Coarse	1 mm to 0.5 mm
	Medium	0.5 mm to 0.25 mm
	Fine	0.25 mm to 0.1 mm
Silt		0.1 mm to 0.01 mm
Clay		Less than 0.01 mm

GRADIENT (Mathematics). A vector obtained by the application of the vector differential operator del (∇) to a scalar point function. In rectangular coordinates, it is

$$\text{grad}\ \phi = \nabla\phi = \mathbf{i}\frac{\partial\phi}{\partial x} + \mathbf{j}\frac{\partial\phi}{\partial y} + \mathbf{k}\frac{\partial\phi}{\partial z}$$

where **i, j, k** are unit vectors. It expresses, both in magnitude and direction, the greatest space rate of change of the scalar ϕ. At any point, P, it is normal to the surface $\phi(x, y, z)$ = constant, which passes through P.

GRADIENT WIND. Winds and Air Movement.

GRAEFFE METHOD. A procedure for obtaining approximate values of the roots of a polynomial. Known also by the names of Dandelin and Lobachevsky, its advantages over other methods are that no first order approximation to the root is needed and that all roots, both real and complex, are obtained in one operation.

The given polynomial, conveniently written in the form $x^n + a_1x^{n-1} + \cdots + a_n = 0$, is repeatedly squared and the coefficients of equal powers of x are collected. After this has been done s times, the coefficients are the numbers 1, $m_1, m_2, \ldots, m_n$. If $r_1, r_2, \ldots, r_n$ are the real roots of the polynomial, their absolute magnitudes are $|r_1|^p = m_1$; $|r_2|^p = m_2/m_1$; $\ldots$; $|r_n|^p = m_n/m_{n-1}$, where $p = 2^s$. The signs of the roots must be found in some other way. Appropriate modification of the method also furnishes the complex roots, which occur in pairs.

See also **Polynomial;** and terms listed under **Mathematics.**

GRAFTING AND BUDDING. Grafting is the process of inserting a part of one plant into another in such manner that the two unite and the inserted piece continues to grow. The part which is inserted is called the scion, the plant into which it is inserted is the stock. Budding is a similar process in which the part inserted consists of a bud with some of the bark adjoining it.

This process is possible because of the cambium cells. The successful union of the two pieces is caused by the formation of callus tissue by the cambium cells. Callus tissue is composed of a mass of parenchyma cells which fill in or grow over wounds, thus repairing the injury. In graft unions, the cells of the callus tissue soon begin maturing into cells of various types, as xylem and phloem cells, while others become typical cambium cells joining the cambium layer of stock and scion. In grafting, the cambium layers of the two parts are to be brought as closely together as is possible.

There are several methods of grafting. A very common method is known as cleft grafting. In this method a small twig having several buds is removed from the plant which is selected as desirable. The lower end of this twig is cut wedge-shaped. A branch of the plant used as stock is cut off, and a vertical cut made in the end. Into this cut the prepared scion is inserted in such position that its cambium layer and that of the stock come together. To prevent drying of the tissues the entire cut surface is covered with a prepared wax. Usually, several scions are inserted in a branch of the stock. When union has taken place and the scion started to grow, all but one may be cut off.

Another method is whip grafting, which is used when the stock is too small for successful cleft grafting. In whip grafting, both stock and scion are cut in a long oblique cut. In the cut surface of each a vertical cut is made. They are then fitted together so that the parts of one slide into and against those of the other, with the cambium of one in contact with that of the other. The two parts are then bound firmly together and the whole covered with wax.

In budding, a small bit of bark bearing a bud is removed from the selected plant. Usually, little wood is taken with this. In the stem of the stock, a T-shaped cut is made in the bark and the flaps so formed loosened. The prepared bud is inserted under the flaps, which are then pressed down over it and bound tightly in place to insure contact between the two cambium layers. Wax is used here also to prevent loss of water.

In modern horticulture, grafting is a very important practice. Many plants, for instance, do not come true when grown from seed. It becomes necessary, therefore, to propagate such desirable plants vegetatively. This may be done in two ways. One is by means of cuttings, pieces of the plant which are rooted and grown into new plants. The other method is grafting, which is now done on an immense scale. Vegetative propagation must be used also in those plants which do not bear seed, as seedless oranges and seedless grapes.

Commonly, the stock used in such cases is not a mature plant but a seedling. This is often chosen for its hardness or its resistance to diseases and pests. The seedlings are allowed to grow until their roots are well established. The graft is then inserted at the base of the stem. As soon as union has taken place and the scion started to grow, the shoot of the stock is cut off, so that all substances absorbed by the root are sent into the scion. Grafting of this sort is used in producing nursery stock for rubber plantations, as well as nearly all common fruit trees.

Successful grafting can only take place between plants of the same kind or closely related. Others fail entirely to develop any union between the two parts. In nearly all cases, the nature of the scion is constant after grafting, so that one can be sure of the product which will result. Because of this, it is possible to graft several different scions on a single stock. Now infrequently one sees an apple tree bearing many different kinds of apples maturing at different times of the year. Dwarf apple and pear trees are produced by budding, using quince as stock. Grafting also hastens the time of fruiting, grafted plants coming into bearing earlier than those growing from seed.

Bridge grafting is done for a very different reason. Often, trees are completely girdled at the surface of the ground by rodents, especially during the winter months. Damage of this sort is fatal to the trees unless quickly corrected. Correction is done by bridge grafting. This is done by trimming the edges of the girdled region and inserting small twigs across the gap in the bark in such a way that the cambium region of the strips is in contact with that of the tree in which it is inserted. Long sloping ends greatly increase the probability of such contact. These "bridges" unite with the damaged tissues and allow movement of materials to occur. Gradually the damaged and new tissues fill in the gap, and the damage is repaired.

See also **Budding.**

GRAHAM LAW. The rates of diffusion of two gases are inversely proportional to the square roots of their densities.

GRAIN (As Energy Source). Biomass and Wanes as Energy Sources.

GRAIN BOUNDARY. The surface separating two regions of a solid in which the crystal axes are differently oriented. It has been shown that such a boundary may be thought of as built up of an array, or network of dislocations, whose spacing depends on the tilt θ of the axes across the surface. The energy (per unit area) of a grain boundary is given by

$$E/E_m = (\theta/\theta_m)\{1 - \ln(\theta/\theta_m)\}$$

where E_m and θ_m are parameters depending on the material.

Grain boundary relaxation is a source of internal friction in solids due to the motion of grain boundaries under stress.

GRAIN REFINER. An additive agent used to obtain finer grains in a casting added to a molten metal before casting.

GRAINS. Grasses.

GRAIN SIZE. In metallurgy, it is common practice to call the crystals of a polycrystalline metal its grains. The grain or crystal size of metals is determined by microscopic examination of a suitably prepared section. There are two principal standards of grain size in use in the United States. Both are standards of the American Society for Testing and Materials.

For most non-ferrous alloys, particularly brass and bronze and other alloys having homogeneous grain structures with twin bands, a set of ten photomicrographs having average grain diameters ranging from 0.010 to 0.200 millimeter are used for direct comparison with microstructures at a magnification of 75 times.

The A.S.T.M. standard grain size chart for steels covers about the same range of average grain diameters but the comparison is made at 100 times magnification and the grain size is expressed by numbers from 1 to 8. The following single equation relates the grain size number to the grain sizes:

$$n = 2^{N-1}$$

where N is the grain size number and n the number of grains per square inch. In general, grain sizes 1 to 3 are considered coarse, 4 to 6 intermediate, and 7 to 8 fine. The grain size of steel can also be judged from a clean fracture if the steel can be fractured without appreciable plastic deformation because the fracture surface mirrors the grain structure. This is possible with most heat-treated machine steels and tool steels, but low-carbon steels are often too tough to break with a crystalline fracture. A series of standard fractures is available for direct visual comparison, and the numbering system for these standards coincides with that of the charts used for microscopic determination of grain size.

The grain size of metals is related to many important properties. In general, fine grain size is an indication of relatively high strength, hardness, and toughness while coarse grain indicates softness and plasticity. However, the hardenability of steels by heat treatment is highest for coarse grain steel. Coarse grain size is usually desirable for creep strength at elevated temperatures.

In the case of sheet and strip for drawing or stamping, coarse grain may give a rough surface. On the other hand, metal with too fine a grain size may lack plasticity and crack in the dies; therefore, a compromise must be reached.

The grain size of castings is generally much coarser than that of wrought products such as rod or sheet. In the case of steel castings the original coarse structure may be refined by heat treatment. This is not possible in the case of most non-ferrous alloys because they do not undergo a change in type of crystal structure on heating or cooling.

In the case of hot-rolled or forged metals, the finishing temperature has an important influence on grain size. A high finish-forging temperature, for example, will permit grain growth after recrystallization. In the case of metals finished by cold-working processes, the final annealing temperature establishes the grain size. A high annealing temperature results in coarse grain size.

GRAIN-STORAGE INSECTS. Attack on stored grain varies from region to region. This damage in the United States is divided into four regions as shown by Fig. 1. Damage is heaviest in the southern region, where long summers and high temperatures permit development of many insect generations during the year. In colder climates, stored-grain insects are generally fewer and less troublesome. However, a large infestation can heat up the grain and cause it to remain active, even in cold weather.

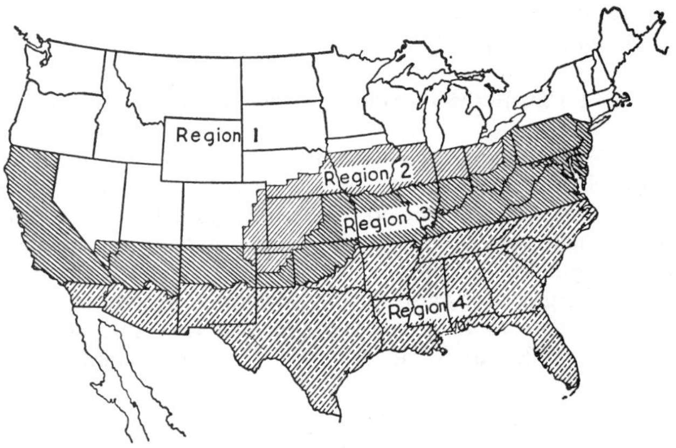

Fig. 1. The map shows, by regions, the degrees to which farmstored grain in the United States is subject to insect attack: *Region 1*: Little if any damage occurs to grain on the farm during the first season's storage. *Region 2*: Insects may be troublesome during the first season. *Region 3*: Insects are troublesome every year. *Region 4*: Insects are a serious problem through the storage period.

Treatment

In the United States, the rice weevil, the red flour beetle, the lesser grain borer, the saw-toothed grain beetle, and the granary weevil are among the most destructive insect pests of stored grains. In many other areas of the world, the khapra beetle is also very destructive. See **Khapra Beetle.**

Precautionary measures for reducing insect populations in storage areas are simple and straightforward, but must be observed if infestations are to be avoided. Fundamental rules include: (1) Clean the storage bin and area around it; (2) spray bins before storing grain in them; and (3) treat and inspect the grain regularly. When a spray mixture is used to treat bins, only one day's supply should be prepared at one time. To make sufficient spray for treating a given storage area, any of the mixtures listed in Table 1 can be used.

TABLE 1. SPRAY REQUIREMENTS FOR STORAGE BINS

Treatment for Area of: 100 Square Meters	1000 Square Feet
Mixture No. 1	
$\frac{1}{4}$ liter of 57% premium-grade malathion emulsifiable concentrate and 8.2 liters of water.	$\frac{1}{2}$ pint of 57% premium-grade malathion emulsifiable concentrate and 2 gallons of water.
Mixture No. 2	
367 grams of 50% methoxychlor wettable powder and 8.2 liters of water.	12 ounces of 50% methoxychlor wettable powder and 2 gallons of water.
Mixture No. 3	
1 liter of 25% methoxychlor emulsifiable concentrate and 8.2 liters of water.	1 quart of 25% methoxychlor emulsifiable concentrate and 2 gallons of water.
Mixture No. 4	
$\frac{2}{3}$ liter of 6% pyrethrins combined with 60% piperonyl butoxide and 8.2 liters of water.	$1\frac{1}{3}$ pints of 6% pyrethrins combined with 60% piperonyl butoxide and 2 gallons of water.

After the bins and storage area have been thoroughly cleaned and sprayed, further steps can be taken to protect against infestation, including: (1) Applying insecticide to the grain as it goes into the bin; (2) applying a surface dressing to the grain after it is in the bin; or (3) fumigating the grain to disinfect it. Such protection will last for about one season. Because of the warmer temperatures in region 4 (see Fig. 1), sprays or dusts are less effective than in the other regions. Table 2 lists effective grain sprays and dusts.

To get rid of an infestation that is already established, fumigation is almost always necessary. Fumigants are sold under various trade names with ingredients listed on the label. Representative fumigant requirements are given in Table 3.

Fumigants should be applied only by a trained operator wearing a gas mask and equipped with a fresh canister. Before fumigating, grain surface should be level to ensure even distribution of the fumigant. Any crust on the surface of the grain should be broken up. An assistant always should be present during the fumigating procedure. Fumigation should occur within 2 weeks after binning the grain if installation is in region 4 (Fig. 1); with 6 weeks for regions 2 and 3; and only when required by inspection in region 1. Samples of grain from the center of the bin should be taken once per month for insect inspection. The samples should be sifted through a screen with mesh large enough to let most kinds of insects fall through, but small enough to hold back the grain. Most stored-grain insects are smaller than the grain. Fumigate at once if even only one granary weevil, rice weevil, or lesser grain borer is present. Methyl bromide is commonly used to fumigate farm-type bins of wheat and corn (maize).[1]

TABLE 2. SURFACE DRESSINGS FOR STORED GRAIN

Treatment for Quantity of:* 25,000 Kilograms	1000 Bushels
Sprays	
Mixture No. 1	
0.4 liter of 57% premium-grade malathion emulsifiable concentrate and 10.5 to 17.4 liters of water.	1 pint of 57% premium-grade malathion emulsifiable concentrate and 3 to 5 gallons of water.
Mixture No. 2	
0.8 liter of 6% pyrethrins, combined with 60% piperonyl butoxide emulsifiable concentrate and 10.5 17.4 liters of water.	1 quart of 6% pyrethrins combined with 60% piperonyl butoxide emulsifiable concentrate and 3 to 5 gallons of water.
Dusts	
Compound No. 1	
25.1 kilograms of 1% premium-grade malathion dust on wheat base.	60 pounds of 1% premium-grade malathion dust on wheat base.
Compound No. 2	
31.3 kilograms of 0.08% pyrethrins combined with 1.1% piperonyl butoxide dust on wheat base.	75 pounds of 0.08% pyrethrins combined with 1.1% piperonyl butoxide dust on wheat base.

* Assuming wheat = 60 pounds/bushel.

Important Pests

Confused flour beetle (*Tribolium confusum*, Duval). A very common insect of grain storage areas. This insect is also a pest in grocery stores and warehouses and can be a serious pest in flour mills. The beetles are small, about $\frac{1}{7}$-inch (3–3.5 millimeters) long, with elongated bodies and a reddish-brown coloration. Large numbers of them will appear whenever the stored product is slightly disturbed. Mixed with the adult beetles may be found the brownish-white, flat, 6-legged larvae

[1] This application is well described in Report 929, U.S. Department of Agriculture, Washington, D.C. (revised periodically).

TABLE 3. FUMIGANT REQUIREMENTS PER QUANTITY OF STORED GRAIN

Fumigant Mixture	Metal Bin: Wheat, rye	Metal Bin: Shelled corn, oats barley	Metal Bin: Grain sorghum	Wood Bin: Wheat, rye	Wood Bin: Shelled corn, oats barley	Wood Bin: Grain sorghum
80% carbon tetrachloride and 20% carbon bisulfide	3 gal 11 lit	3.5 gal 12.2 lit	5 gal 17.4 lit	6 gal 20.8 lit	7 gal 24.4 lit	10 gal 34.8 lit
75% ethylene dichloride and 25% carbon tetrachloride	3.5 gal 12.2 lit	5 gal 17.4 lit	— —	7 gal 24.4 lit	10 gal 34.8 lit	— —
5% ethylene dibromide, 35% ethylene dichloride, 60% carbon tetrachloride	3 gal 11 lit	3.5 gal 12.2 lit	5 gal 17.4 lit	6 gal 20.8 lit	7 gal 24.4 lit	10 gal 34.8 lit

NOTE: Gallonage (gal) figures are for 1000 bushels of grain; liter (lit) figures are for 25,000 kilograms of grain.

SPECIAL NOTES: Carbon bisulfide is explosive when used without a fire suppressant, such as carbon tetrachloride. Apply mixture only when grain temperature is above 60°F (15.6°C).

The ethylene dichloride–carbon tetrachloride mixture should be applied when grain temperature is above 70°F (21.1°C).

Grain treated with fumigants containing ethylene dibromide should be thoroughly aerated before feeding to laying hens.

that feed on the inside of the grain kernels. The larvae sometimes are called "barn bugs." In addition to all kinds of grain products, these insects like anything of a starchy nature, including beans, baking powder, peas, dried plant roots, dried fruits, nuts, chocolate, certain drugs, snuff, cayenne pepper, and many other substances. It is interesting to note that they are pests on insect collections. These insects occur worldwide and were first noted in the United States in 1893. The beetle is found more frequently in the northern United States than in the southern states.

Red flour beetle (*Tribolium castaneum*, Herbst). See Fig. 2. This beetle is closely allied with the confused flour beetle, but occurs more in the southern climes than in the north. It is seldom encountered north of latitude 41° N. Under the best circumstances, from four to five generations of both of these beetles will take place per year.

Saw-toothed grain beetle (*Oryzaephilus surinamensis*, Linne). See Fig. 3. The feeding habits are much the same as those for the confused flour and red beetle. It can penetrate packages that would seem to be tightly sealed. Often these beetles will follow the damage of other insects because it cannot successfully devour sound seeds. Distribution is worldwide. Only the adult stage overwinters in unheated structures. Normally there are from four to six generations per year. Under the best of conditions, the entire life cycle of this insect could occur in less than one month.

Granary weevil (*Sitophilus granarius*, Linne). The granary and the rice weevil are considered by some experts as the most destructive of all grain insects. They are true weevils. If undisturbed these weevils can cause almost complete destruction of grain stored in elevators, ships, and on the farm. A telltale of infested grain is a rise of the surface temperature as well as wetness. Sometimes, sprouting of seed will be noted. The beetles have prominent snouts, which are an advantage in feeding upon grain. The larvae prefer the interior of the kernels. Substances particularly attractive to the granary weevil include buckwheat, barley, maize (corn), macaroni, oats, kaffir seed, and wheat. The weevil is distributed widely throughout the world, but less abundant in tropical and semitropical areas.

The weevil overwinters as adult or larva. The adult can withstand subzero temperatures for many hours. The adult weevil is dark brown or nearly black and has ridged wing-covers and long snout extending downward from the front of the head. Length is about $\frac{1}{16}$-inch (1.5 millimeters). The female weevil deposits her eggs (from 300 to 400 small and white) in small cavities found in the grain kernels. Legless, soft, fleshy, white grubs are hatched within a few days and immediately

Fig. 2. Red flour beetle. (*USDA photo.*)

Fig. 3. Saw-toothed grain beetle. (*USDA photo.*)

commence feeding on the interior of the kernel. When fully grown, the larvae are about $\frac{1}{8}$-inch (3 millimeters) long. The full life cycle ranges from 4 to 7 weeks. The adults can go for long periods without food, if necessary, and it has been estimated that they can live over 2 years on a starvation diet. There are four to five generations of granary weevils per year.

Rice weevil (*Sitophilus oryza*, Linne). See Fig. 4. This beetle is very similar in construction and habits to the granary beetle. A major difference is that the rice weevil has well developed wings and can fly, and frequently does so, particularly under warm conditions. The granary weevil's wing covers are grown together, keeping it from flying.

Mealyworms (*Tenebrio molitor*, Linne, yellow; and *T. obscurus*, Fabricius, dark colored). These worms have shiny bodies, yellow-to-brown in color, with smooth coats. They have some resemblance to wireworms of black beetles. They are relatively large, about 1 inch (2.5 centimeters) in length. They are found in dark, damp locations where grain or other attractive substances have been stored for a long period. In addition to grain and grain products, mealyworms enjoy feathers, dead insects, and scraps from meat-packing operations. Native to Europe, mealyworms are distributed worldwide. As adults the two species are much alike and often are difficult to distinguish. From about 1 to 3 weeks are requried for eggs to hatch. The eggs are placed in stored food substances. One female may lay as many as 250–1000 eggs.

Cadelle (*Tenebroides mauritanicus*, Linne). If not present in overabundance, the cadelle beetle can help in controlling the population of other damaging beetles, because the adults often kill and feed upon other insects. However, in this regard, they are not considered to be predaceous insects. On balance, the cadelle is a serious pest of grain bins and like storage places. This insect is notably damaging in flour mills, not only consuming grain, but also destroying flour sacs, cloth used in machinery, cardboard containers, etc. The insect may overwinter as an adult or larva, but not as a pupa. The black adult beetle ranges in length from $\frac{1}{3}$ to $\frac{1}{2}$ inch (8 to 12 millimeters). A single female can lay up to 1300 eggs in cracks and crevices. Hatching occurs within 1 to 2 weeks. The resulting larvae are an off-white color, have prominent black heads, some black spots, and two hooks at the rear of the body (a bit like an earwig). When fully developed the larvae are about $\frac{2}{3}$-inch (17 millimeters) in length. These larvae have the additional bad habit of boring into wood as may be found in grain bins, ships holds, etc. Tunnels in wood make excellent hiding places, where the pupal stage is passed. The full development period of the cadelle is considerably longer than most of the other grain pests, ranging from 7 to 14 months, although the average time span is 2 to 3

Fig. 4. Rice weevil. (*USDA photo.*)

months. The cadelle adult has been known to live as long as 3.5 years.

Lesser grain borer (*Rhyzopertha dominica*, Fabricius). Also known as the *Australian wheat weevil*, the insect is widely distributed throughout the southern and midwestern United States. It is rarely found in the northern states. Adult beetles are brown-to-black, cylindrical in shape, about $\frac{1}{8}$-inch (3 millimeters) long by $\frac{1}{4}$-inch (6 millimeters) wide. The larvae appear as grubs and assume a curved posture. They are about 1/10 inch (2.5 millimeters) long. The habits of this insect are similar to the other insects described, but they have a wider range of attractive feeding substances. In addition to grain, they like seeds, certain drugs, dry roots, cork, wood, and paper boxes. But, it thrives in wheat and is one of the most common of the wheat pests. The *larger grain borer* (*Dinoderus truncatus*, Horn) is similar to the lesser grain borer in most respects, with exception that it is a bit larger, and prefers corn to wheat. It does not occur widely in the United States, with exception of a few locations in the southern states.

Angoumois grain moth (*Sitotroga cerealella*, Olivier). A buff-colored and delicate adult moth having a wingspread of from $\frac{1}{2}$ to $\frac{2}{3}$ inch (12 to 18 millimeters) was first found to be a damaging insect on wheat, corn (maize), and other grains in France (Province of Angoumois) in about 1736. It occurs in many parts of the world, including all of the United States. The other stages of the insect are seldom seen because the larvae and pupae habitate the internals of seeds and the eggs are extremely tiny. The fully grown larva is about 1/5 inch (5 millimeters) in length. The eggs are deposited by the female moths in the hundreds on grain in the shock or on the heads in the field. Only 1 to 4 weeks is required for hatching, at which time the larvae burrow into the kernel. The full life cycle is about 5 weeks. The larvae may overwinter in the grain. Thus, the insect goes with the grain into storage and adults thus may appear from time to time while the grain is in storage. Reproduction can continue during storage. In mild latitudes, there are about two generations per year, whereas in southern climates, there may be as many as six generations per year. The insect not only destroys corn in the crib, but also damages ripening grain in the field before storage.

Mediterranean flour moth (*Ephestia* or *Anagasta künniella*, Zeller). At one time this was a very serious pest in flour milling operations. Fumigating procedures have largely brought the insect under control. Conveyors and chutes that carry flour may be webbed over when the small caterpillars are present. A telltale is a number of small gray moths that will be present in infested structures. Although the insect prefers flour, it will also feed upon breakfast cereals, maize, bran, and whole grain wheat. It will also feed on pollen in beehives. Although widely distributed in the United States and Canada (first reported in 1889), the insect is also found in many other regions of the world. The life cycle requires from 9 to 10 weeks. The eggs are laid in crevices, cracks, undisturbed accumulations of flour, etc. The eggs hatch within less than a week, after which the caterpillars spin silken threads to form small tubes in which they live and feed. The web-spinning and the clogging of machinery that results is the principal damaged caused by the insect.

Indian meal moth (*Plodia interpunctella*, Hübner). In addition to feeding on grain, this insect (native to Europe) feeds on breakfast cereals, soybean, nuts, seeds, dried roots, dead insects, powdered milk, beehive pollen, and soybean. The insect is also a pest in museums where it attacks specimens. The Indian meal moth is also a serious pest in confection factories. Distribution is throughout the United States and many other regions of the world. All phases of the life cycle (4 to 6 weeks) can be present at the same time, with exception of unheated structures during winter, under which conditions the insect winters over as a larva. As with the Mediterranean flour moth, a principal damaging aspect of this moth is its web-spinning and its binding together dirt with larvae excreta in the nearness of processed foods and processing machinery which is subject to clogging and jamming by the webs.

The *flour mite* is described under **Mite.**

See also **Khapra beetle.**

GRAIN (Unit). Units and Standards.

GRAM. Units and Standards.

GRAMA GRASS. Grasses.

GRAM-ATOM. That quantity of an element having a mass in grams numerically equal to the atomic weight. One gram atom contains the Avogadro number of atoms.

GRAM-ATOMIC WEIGHT. The weight of a gram-atom.

GRAM-CHARLIER SERIES. This series attempts to represent frequency functions in statistics by an expansion, resembling a Taylor series, in terms of derivatives of the normal (Gaussian) distribution.

$$F(x) = \sum_{k=0}^{\infty} c_k e^{-x^2/2} H_k(x)$$

where the constants c_k depend on the frequency function represented over the interval $[-\infty, \infty]$ and the $H_k(x)$ are the Hermite polynomials. The Gram-Charlier series is similar to the Edgeworth series, and indeed the two are identical for infinite series; their difference arises in regard to the stoppage point when a finite number of terms only is taken as an approximation, in which case Edgeworth's form is probably preferable. See also **Edgeworth Series.**

GRAM DETERMINANT. A means of testing functions for linear dependence or independence. Let vectors $\mathbf{u}_1, \mathbf{u}_2, \ldots, \mathbf{u}_n$ be given in an n-dimensional space. Then the Gram determinant has as elements $\mathbf{u}_i \cdot \mathbf{u}_k$, the scalar product or the Hermitian scalar product if the vector space is complex. If the determinant vanishes, this is a necessary and sufficient condition that the vectors be linearly dependent; if the determinant does not vanish, the vectors are independent. A similar procedure may be used for n functions $\phi_1, \phi_2, \ldots, \phi_n$ and the elements of the determinant become

$$\int \phi_i \phi_k^* \, dt$$

The determinant is never negative. It is a generalization of the Schwartz inequality, which applies in the case of two vectors.

See also **Schwartz Inequality;** and **Wronskian.**

GRAM-EQUIVALENT. The gram-atomic weight of an element (or formula weight of a radical) divided by its valence. In the case of multivalent substances there will be more than one value for the gram-equivalent, viz., Fe(II) = 27.92 grams, Fe(III) = 18.61 grams, and the proper value for the particular reaction must be chosen.

GRAM-EQUIVALENT WEIGHT. Normal Concentration.

GRAMINÈAE. Wheat.

GRAM-MOLE. Mole; Mole Volume.

GRAM-MOLECULAR WEIGHT. That amount of a pure substance having a weight in grams numerically equal to the molecular weight. One gram-molecular weight contains the Avogadro number of molecules. It is also designated as the mole or mol.

GRAM-NEGATIVE. Gram Stain.

GRAM-POSITIVE. Gram Stain.

GRAMPUS. Whales, Dolphins, and Porpoises.

GRAM STAIN. A method of staining microorganisms which enables such organisms to be classified into two main groups, those which retain the stain being described as Gram-positive, and those from which the stain is decolorized being described as Gram-negative. The organisms are first stained with either gentian violet, or its analogue, crystal violet, and then treated with a solution of iodine. An organic solvent, usually alcohol, is then applied, which washes out the stain from Gram-negative organisms, leaving Gram-positive organisms with the violet stain unaffected. A counterstain of some contrasting color is then applied to demonstrate the Gram-negative organisms. Gram-positive organisms include staphylococci, streptococci, pneumococci; among the Gram-negative are gonococci, meningococci, *Bacillus coli*, and the salmonella.

GRAND CONJUNCTION. Conjunction (Astronomy).

GRAND FIR. Fir Trees.

GRANDLURE. Boll Weevil.

GRAND MAL. Seizure (Neurological).

GRANATIC SHELL (Earth). Earth.

GRANITE. This name is applied to a common and widely occurring group of deep-seated igneous rocks consisting of orthoclase, plagioclase, quartz, hornblende, biotite, muscovite and minor accessories such as magnetite, garnet, zircon and apatite. Rarely, a pyroxene is present. Ordinary granite always carries a small amount of plagioclase, but when this is absent the rock is then referred to as an alkali-granite. An increasing proportion of plagioclase feldspar causes granite to pass into granodiorite. A rock consisting of equal proportions of orthoclase and plagioclase plus quartz may be considered a quartz monzonite. A granite containing both muscovite and biotite micas is called a binary granite.

The word granite comes from the Latin *granum*, a grain, in reference to the grained structure of such a crystalline rock.

Granite occurs as stock-like masses and as batholiths often associated with mountain ranges and frequently of great extent. Granite has been intruded into the crust of the earth during all geologic periods, except perhaps the most recent; much of it is of pre-Cambrian age. Granite is widely distributed throughout the earth.

Graphic granite is a coarsely crystalline variety of granite or pegmatite composed almost entirely of quartz and feldspar which have intergrown in such a manner as to simulate Semitic or cuneiform characters.

GRANITOID. A textural term derived from granite and signifying the relatively uniform and coarse-grain of batholithic rocks, such as granite, syenite, anorthosite, etc. In a typical granitoid rock, each species of mineral occurs as a single generation; the silicates crystallize first, and any surplus of free silica crystallizes last in the form of quartz, or is finally driven off with the surplus water to form quartz veins.

GRANULITE (also Leptite). This is a general term for a group of rocks that vary considerably in composition but for the most part seem to be derived by metamorphic processes from quartz-feldspar rocks. The classic locality for granulite is in Saxony, where there occurs a granular gneiss of quartz and feldspar plus such accessory minerals as pyroxene and garnet, with occasionally small quantities of kyanite, spinel and similar minerals. The Saxon granulites have a decided banded structure and seem to resemble injection gneisses. It appears reasonable to suppose that these and other granulites may have been derived from sedimentary formations severely altered by igneous processes. Leptite is a term used in the Scandinavian countries for fine-grained granulites that originally were rhyolitic tuffs and lavas.

Other than in Saxony and Scandinavia, these rocks are found in the northern highlands of Scotland, India, West Africa, and Canada.

GRANULOCYTES. Blood.

GRANULOCYTIC LEUKEMIA. Leukemias.

GRANULOMATOUS ILEITIS. Colitis and Other Inflammatory Bowel Diseases.

GRAPE. Of the family *Vitaceae* (grape family), grapes are climbing plants of numerous species which have been cultivated for centuries for their fruits and the various products obtainable from them.

Climbing in grapes is made possible by tendrils, modified stems which coil tightly around any suitable support. These tendrils are usually interpreted as terminal portions of the stem which have been pushed to one side by the more rapid growth of an axillary bud. The leaves of grapes are simple, palmately lobed and alternate, with small stipules. The stems elongate rapidly and are of a coarse porous nature; the internodes of young stems are frequently hollow, the nodes solid. The flowers are borne in compact panicles. Each flower is small and inconspicuous. The calyx is a mere rim around the tip of the pedicel; the corolla five-parted and greenish. When the flower opens, the petals, united at their tips but free at the base, are forced away

Grapes ready to harvest. (*USDA photo.*)

from the base of the flower and drop off. There are five stamens and a single pistil. The fruit is a 2-celled berry. See accompanying illustration.

Commercial grapes are largely derived from three species, *Vitis vinifera*, the wine grape of Europe, a native of Asia, *Vitis labrusca*, the northern fox grape of eastern North America, and *Vitis rotundifolia*, the southern fox grape. Many varieties and hybrids of these exist, as well as hybrids with other wild species. In commercial vineyards, grapevines are variously pruned to increase yield and improve quality. Pruning cuts are made through the nodes, to prevent the leaving of hollow internodes in which disease might gain entrance to the plant. Propagation of the grape is mainly by means of stem cuttings, a method which has been used in Europe for centuries.

Grapes are used as a table fruit, as raisins when dried, and for making wine. Fewer than a dozen important varieties of grapes are grown for table grapes. Most of the sweet juice produced in North America is from the *Concord*. Only a few varieties are used for canning. *Concord* grapes are used extensively for juice and also for jams, jellies, puddings, and pies. Table grapes, such as *Emperor, Thompson Seedless, Tokay, Cardinal, Ribier*, and others are mostly eaten out of hand, but are also used in salads, fruit cups, pies, puddings, cakes, stewed fruit, and as meat accompaniments. Dried or raisin grapes are mainly *Thompson Seedless* (also known as *Sultanina*), *Black Corinth*, and *Muscat of Alexandria*. A variety closely related to *Thompson Seedless* is important and dominates the raisin vineyards of Greece, Iran, and Turkey. Remarkably few grapes are well suited for wine as well as fresh (table) use or raisin production. Worldwide, the *Muscat* grape is considered a triple-purpose grape. There are numerous subvarieties of the *Muscat*, but all possess the characteristic *Muscat* odor and flavor. For wines, the *Muscat* is used principally in making sweet, fortified wines. In California, the *Thompson Seedless* grape plays the three roles and some production is used in making wines. Wines from this grape, however, tend to be rather neutral and bland and thus are mainly used for blending purposes. Among the better known wine grapes are *Cabernet-Sauvignon, Chardonnay, Chenin Blanc, Gamay, Grenache, Grignolino, Gutadel, Müller-Thurgau, Pinot Noir, Riesling, Sauvignon Blanc, Sémillon, Silvaner, Trollinger*, and *Zinfandel*. As indicated by their names, most of the famous wine-variety grapes were originated in Europe, notably in France, Germany, Italy, Spain, and Austria.

Raisins are either sun-dried or artificially dried. Because of the risk of rainfall occurring during the drying season, artificial drying has become increasingly popular among growers. The principal problem is the requirement for additional energy. When this process is used, the fresh grapes go through a hot caustic solution which removes the waxy coating (bloom) and makes tiny cracks in the skins. The grapes are then spread onto long, shallow wooden trays. In the case of golden seedless raisins, grapes for processing are transferred to a chamber where they are exposed to sulfur dioxide for about five hours. This treatment prevents darkening during drying. The grapes are then transferred to dehydrating tunnels where they are exposed to warm, dry air for about 18 hours. Raisins require a residual moisture content because most consumers do not like a thoroughly dry or crispy raisin. Residual moisture encourages mold and yeast growth. Protection can be obtained by dipping the fruit in weak solutions of potassium sorbate, thus leaving a fine coating of the antimicrobial agent on the fruit pieces.

Depending upon the variety of grape, the water content of a ripe berry will range between 70 and 80%. Most of this water is contained in the *pulp* of the berry, that is, the fleshy and juicy part. But, there are also liquid and some semiliquid components in the *skins* (peels, husks, or hulls) and in the *stems*; these liquids are freed when the total mass of berries is subjected to considerable pressure (squeezing force). Further, in any crushing, macerating, or pressing operation applied to a mass of berries, there is an inevitable mixing of both solid and liquid components—so that a reasonably complete separation of liquid (juice) components from the grape requires more than one crushing or squeezing operation. The purest juice (from the pulp) is obtained from the first squeezing action. This is known as *free-run juice*. Many of the traditional hydraulically operated basket presses have been replaced by roller-type crushers, Garolla blade-type crushers, or disintegrators.

The grape juice and/or the mass of crushed grapes on the way to wine production is referred to as *must*. The grape pressings (skins, seeds, etc.) after the juice has been fully extracted is known as *marc* or *pomace*. The antiseptic and antioxidant properties of sulfur dioxide are used effectively in the treatment of musts prior to fermentation and later in the winemaking process. Many winemakers prefer compressed SO_2 gas, but sulfurous acid or sodium or potassium metabisulfite may be used. These essentially sterilize the must, which can be later reinoculated with a specially selected yeast culture.

One of the continuing and fundamental problems of winemaking is the lack of uniformity of the grapes used, from one season to the next—a lack of consistency which in some years is responsible for truly exceptional and great wines and, in other years, wines that are only passable to good. Two important factors are sugar content and acidity. When these factors are purposely altered after the grapes are picked by way of adding sugar, water, or acid, the process is referred to as *amelioration*. While practiced in some wine-producing regions, the practice is frowned upon and is outlawed in some regions. Where permitted, amelioration is strictly regulated. When water is the only additive, the term *gallisation* is used. The term *chaptalisation* (French) refers to the addition of sugar.

It has been observed since ancient times that grape juice at ordinary temperatures does not retain its freshness, but instead commences to turn to wine, that is, the juice (or must) exhibits the aromatic characteristics of alcohol, but when retained for a still longer period it turns into vinegar. Centuries ago, the Latin word *fermentum*, from *fervere* (to boil) was first used to describe the bubbling nature of the process, and thus the word *fermentation* became a part of the language. A few centuries of research have gone into the process, particularly as applied in winemaking. Considering the total time span of winemaking, the addition of yeast cultures purposely by the winemaker to the fermentation vats is a relatively recent action. From whence did the yeast organisms come that made wine possible during all of those earlier centuries? Yeast occurs naturally on grapes, but there are numerous species—some desirable and many more undesirable for making the best wines. The species most favorable to the winemaker is *Saccharomyces cervisae* var. *ellipsoideus*. Numerous molds are found on green grapes and, as grapes ripen, so-called wild yeasts appear. These yeasts can cause many problems, including off-flavors, off-colors, spoilage, and a host of problems that sometimes are difficult to trace. Poor or unacceptable wines are sometimes referred to as greasy or ropy, wines that become cloudy and pour like oil. Other wines may be flat or bitter. Although Pasteur offered a depth of understanding to the entire winemaking process, his recommendation for overcoming the problems associated with undesired microorganisms was a shortcut solution, namely, *pasteurization*. Over the years, of course, winemakers have regarded pasteurizing with mixed feelings. While pasteurization kills many undesirable microorganisms, rendering wine stable and suitable for long-term storage, it also eliminates or interferes with the possibilities of improving the product during normal aging. Pasteurization is widely used for common table and dessert wines produced in high volume for early consumption. In contrast, winemakers who target to superior and excellent wines regard pasteurization with much caution. The longer, more painstaking procedure for overcoming contamination by wild yeasts involves sulfiting, the addition of specially selected yeast cultures to the sterilized must, and the careful manipulation of all process variables which favor the type and degree of fermentation desired for any given kind of wine.

Must fermentation occurs in three stages: (1) an initial slow stage during which the yeast cells are multiplying; (2) a very vigorous stage, accompanied by bubbling and a marked rise in temperature; and (3) quiet fermentation that can proceed for quite a long time at a lower and lower rate. The main fermentation stages (1 and 2) take place in a variety of vessels, ranging from concrete vats (not often glass-lined) or in wooden tanks (oak, redwood), and ranging from 10,000 to 60,000 gallons (380–2,280 hectoliters) and more. While some continuous fermenting systems have been built, by and large fermenting remains a batch operation. Fermenting may range from 2 to 20 days, depending upon numerous variables. With alcohol-tolerant yeasts, fermentation proceeds rapidly to completion, producing from 10 to 12.5% alcohol by volume. When the sugar content exceeds 23%, this may inhibit fermentation rate as well as full completion of fermentation.

At total acidities of less than 1% (pH greater than 3), alcohol fermentation is not inhibited. Yeasts require a number of amino acids, but fortunately these are present in most grapes in ample amounts. Some winemakers will sometimes add nitrogen-bearing substances in small quantities as yeast food.

Temperature is quite critical to the fermenting process. Each winemaker may have opinions as to which temperature is best for any given type of wine. For white wines, the optimum temperature ranges between 50 and 60°F (10 and 15.6°C); for sherry, the optimum is about 80°F (26.7°C); for red wines, about 85°F (29.4°C); for wines from Pinot noir grapes, 70–80°F (21.1–26.7°C); and for Cabernet-Sauvignon grapes, 70°F (21.1°C). For some wines, retardation of fermentation commences at about 85°F (29.4°C), and for all must fermentations the action is greatly weakened with a temperature rise to 95°F (35°C); above 100–105°F (37.8–40.5°C), fermentation essentially ceases. At temperatures above 90°F (32.2°C), it is likely that wine flavor and bouquet will be injured. The end of fermentation is signaled by a clearing of the liquid, by a vinous taste and aroma, and by a drop in temperature, and can be confirmed by checking sugar residual. It is interesting to note that fermentation can be halted as the result of a temperature too high or too low. In this case, the condition is referred to as a "stuck wine." If a batch is stuck at a low temperature, warming will usually cause fermentation to resume. In the case of a batch stuck because of high temperature, cooling alone may not suffice. The addition of small quantities of ammonium phosphate will usually help to restart fermentation.

To date, no substitute for time has been found in the transformation of the green wine (after drawn off the fermenters) into an acceptable product. Considerable settling of finely divided solid particles and colloidal materials is required, the subtle and slow chemical reactions involving aldehydes, esters, etc. that enter into the ultimate bouquet of a wine—all are time-related events, much more critical with some wines than others. There is a requirement for all wines for a minimum of clarification, stabilizing, and settling that occurs at the winery prior to containerizing for the market; there is the additional aging that goes on once a wine has reached the market. Popular writers interested in viniculture tend to overemphasize the aging aspects of wine, considering that 90% or more of the wine produced is for the mass market and relatively early consumption. For example, most of the common table wines in Spain are consumed when less than two years old. It has been shown that common California red wines can be adequately matured in a year or less. In contrast, a fine Cabernet requires up to a minimum of three years in wood. Such wines should be further aged a year in the bottle at the winery prior to labeling and releasing for sale. Such wines usually will continue to improve for a period of from 5 to 15 years in the bottle.

One authority estimates that about 75% of all wine produced is as good when about two years old as it is likely to be and deterioration is likely to commence after three years. Wine recommended for consumption within three to five years include: Vin Rosé (California, France, etc.); most California white wines, with the exception of a select few prepared from Chardonnay, Chenin Blanc, Pinot Blanc, Sauvignon Blanc, and Johannesberg Riesling; most white Burgundies, with the exception of those from the excellent vineyards in good vintage years; and nearly all Italian wines, except a few select red wines.

Fining agents that bring about clarification of wine include gelatin, casein, tannin, and bentonite. Fining is most efficiently accomplished in relatively small vessels, including barrels. Because of so many variables involved, a careful laboratory examination of the wine is made prior to selection and determination of the amount of fining agent to be used.

The French use the term *maderisé* to describe a wine that is overage (past its prime condition), which has become partially oxidized, and which has frequently acquired a brownish tinge, and an aroma and flavor remindful of Madeira (not desirable except in a Madeira wine). The term is more commonly applied to white and rosé wines.

Other terms used in winemaking include: *Racking*—the drawing off of the clear portion of a young wine from one vessel and transferring it to another vessel. In this process, the lees and sediment formed during the prior storage period are separated. To hasten the total process, more rackings are required. Winemakers have found that refrigeration helps to hurry the aging process. *Binning* involves laying away bottled wine for aging. Always with table and sparkling wines, the bottles should be stored on their side so that the wine is in constant contact with the cork. The wine should be stored at a cool temperature. There is a relationship between the size of the container and the time of aging. A half-bottle will be ready earlier than a full bottle. *Blending* is widely used in connection with high-volume wines and where year-to-year quality is important to consumer acceptance. *Filtering* is commonly practiced in connection with high-volume wines and with most other wines with relatively few exceptions. Many winemakers prefer a lighter filtration so that the wine will not take on what is known as a character of *numbness*, that is, removal of some constituents that help prior to their ultimately becoming sediment upon aging in the bottle. It is a well accepted fact that discriminating wine consumers do not look upon sediment, particularly in certain wines, such as old red wines, as a defect, but rather as a natural result of proper aging. Sediment in white and rosé wines is in the form of colorless crystals of cream of tartar, which is tasteless and harmless and often disappears when the wine is slightly warmed. The sediment in red wines is of larger amount and complexity, made up of pigments, small quantities of mineral salts and tannins, all of which can be removed by careful decanting.

Fortification signifies a wine that contains more alcohol than is obtainable through natural fermentation. Fortified wine is not grape juice to which alcohol has been added (known as *mistelle*). Port is a fortified wine, to which about half-way through the fermentation, juice is drawn off and put into vessels that contain high-proof grape brandy of a predetermined volume. Sherry is also fortified with high-proof brandy. Not regarded as fortification, but effective in adding a few percentage points of alcohol to wine, some winemakers use alcohol-tolerant yeasts. *Brandy* is made by distilling wine.

GRAPEFRUIT TREE. **Citrus Trees.**

GRAPE-LEAF FOLDER (*Insecta*, Lepidoptera). A moth, *Desmia funeralis*, whose larva eats the leaves of grape vines and lives in a fold fastened with silk. It is not an important pest. Arsenical sprays used for other insects destroy it.

GRAPE-LEAF SKELETONIZER (*Insecta*, Lepidoptera). A moth, *Harrisina americana*, whose larvae, working in groups, destroy the soft tissues of the grape leaf, leaving the network of veins. It is rarely an important pest.

GRAPE PHYLLOXERA (*Insecta*, *Homoptera*). A sucking insect related to the plant lice aphids and scale insects. The many species make up a subfamily which, with the adelgids, constitutes the family *Phylloxeridae*. They differ from the aphids in that all females lay eggs and form the scales in their more complex structure, including the four wings of the winged stages.

The most important phylloxerid is a species (*Phylloxera vitifoliae*, Fitch) which attacks grapevines, working on the leaves and roots. It once threatened to ruin the vineyards of France and has destroyed millions of acres of vines. The use of roots of certain American grapes which are not seriously harmed by the pest has greatly lessened the danger from its attack. Tender varieties are grafted onto the resistant roots.

GRAPE SUGAR. **Carbohydrates.**

GRAPH COMPONENT. A component of a graph G is a nonseparable maximal connected subgraph. The decomposition of a graph into components is unique.

GRAPHICAL STATICS. **Statics (Graphical).**

GRAPHITE. An allotropic form of carbon, graphite occurs in nature and also is produced artificially. Graphite crystallizes in the hexagonal system, often in the form of scales or plates, or in large foliated masses. Graphite has a perfect basal cleavage, is soft (hardness between 0.5–1 on the Mohs scale—similar to talc), and feels greasy to the touch. Specific gravity 2–2.2, black to steel gray, lusterous metallic appearance, very opaque. Graphite finds many uses: (1) in the manufacture of "lead" pencils, graphite (the marking medium) is mixed with clay as a bonder, the amount of clay used determining the hardness of the pencil lead; (2) in the manufacture of self-lubricative metals in which graphite is mixed with copper, lead, and tin, after which the mix is sintered and subjected to powder metallurgy techniques to form alloys which will hold relatively large volumes of lubricating oil over long periods of use; (3) in the construction of heat-resistance structures, such as rocket casings and chemical process equipment, allowing operating temperatures up to 3,000°C and greater; (4) in the manufacture of corrosion-resistant apparatus for chemical processing; (5) in the manufacture of packings where the lubricative and corrosion-resistant characteristics of graphite are advantageous; (6) in the production of electrodes for electric furnaces and electrolysis equipment; and (7) a special pyrolytic graphite, with excellent electrical and thermal conductivity properties, good tensile strength at temperatures up to about 2,800°C, and impervious to gases and liquids, finds use in various electrical apparatus and, when mixed with boron, makes an effective nuclear radiation shield. Graphite slows the flow of neutrons without capturing them.

Graphite is formed during the metallurgical operations of producing pig iron, cast iron, malleable cast iron, and some special die steels and has a marked effect upon the characteristics of these materials. See also **Iron Metals, Alloys, and Steels.** The effects may be positive or negative. When present in cast iron in excessive amounts, or in the form of large interlocking flakes or films, graphite reduces the tensile strength.

Graphite is a rather widely distributed mineral and is found in a variety of rocks. It occurs in marbles, gneisses or schists; granites and other igneous rocks often carry graphite. It has been noted in pegmatites. It is likely that graphite has been formed by different processes, by magmatic separation of the graphite as an original constituent or as the result of assimilation of carbonacous rocks, by pneumatolytic action, or by the metamorphism of sedimentary rocks that contained original carbonacous matter. Well-known localities are in Siberia, on the Island of Ceylon, which is the chief producing district at present; England, Malagasy Republic, Mexico, and Canada. In the United States it is found in the Adirondack region of New York State, in Massachusetts, Rhode Island, Pennsylvania, Alabama, New Mexico, and Montana. Natural graphite sometimes is referred to as plumbago, black lead, and Flanders stone.

Graphite is made artificially by heating coke to a very high temperature, usually in an electric furnace. To prevent oxidation, the coke is covered with a layer of sand.

The German mineralogist, A. G. Werner, devised the name graphite from the Greek meaning *to write,* with reference to its use in pencils.

For a comparison of the characteristics and crystalline structure of graphite and diamond, see **Carbon;** and **Diamond.**

GRAPHITE MODERATOR. Nuclear Reactor.

GRAPH (Mathematics). Generally, a curve or surface on which the locus of a function is shown on a series of coordinates which are set at right angles to each other.

Graph (*Complete*). A complete graph G is a linear graph in which every two distinct vertices are endpoints of an edge in G. Figure 1 is a complete graph with four vertices. The total number N of distinct labeled trees in a complete graph containing v vertices is $N = v^{v-2}$, a result due to Caylet. Thus, this example has 16 trees.

Graph (*Connected*). A graph is connected if there exists a path between any two vertices. Stated in another way, any two distinct vertices β_1 and β_2 are the terminal vertices of some path.

Graph (*Directed*). See **Digraph.**

Graph (*Dual*). The linear graph G_2 is the dual of the linear graph G_1 if the conditions enumerated below are satisfied:

1. The edges of G_1 and G_2 are in one-to-one correspondence.

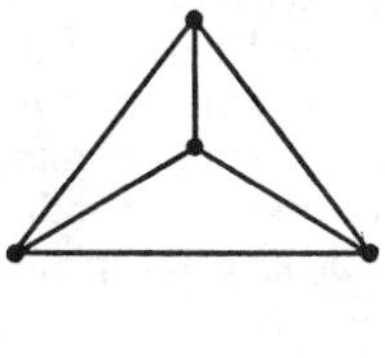

Fig. 1. Complete graph with four vertices.

2. If H_1 is any subgraph of G_1 and H_2 is the complement of the corresponding subgraph in G_2.

$$r_2 = R_2 - n_1$$

where r_2 is the graph rank of H_2, R_2 is the rank of G_2 and n_1 is the nullity of H_1.

It follows easily from this definition that rank G_1 = nullity G_2 and rank G_2 = nullity G_1. Furthermore if G_2 is the dual of G_1, G_1 is the dual of G_2.

Two extremely useful and significant results are that the dual of a nonseparable graph is nonseparable and that a linear graph is planar if and only if it possesses a dual.

The usual geometric procedure for finding the dual of a planar graph G involves three steps:

1. Choose a set of fundamental circuits. See **Circuits, Fundamental (Mathematics).**
2. Put a node in each such circuit and a node outside the graph.
3. Connect any two nodes which are on opposite sides of a branch by a line segment.

The resulting graph is the dual of G. These rules are illustrated in Fig. 2, in which the dual appears dotted.

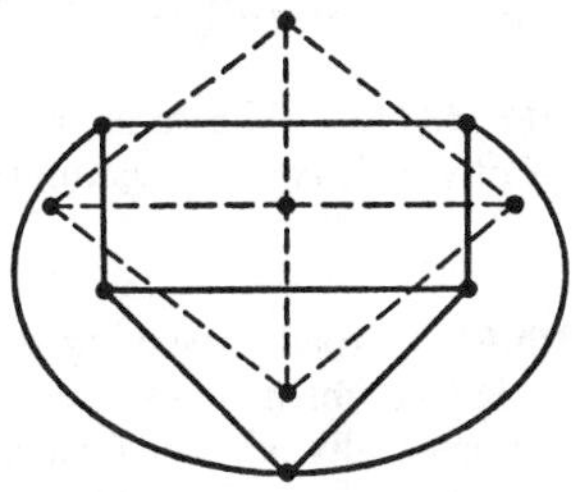

Fig. 2. Dual graph (dotted).

Graph (*Finite*). A finite graph contains only a finite number of line segments and vertices.

Graph (*Homeomorphic*). Two graphs G and G' are homeomorphic if there exists a one-to-one bicontinuous mapping between the two point-sets defined by G and G'. Refer to description of planar graph.

Graph (*Infinite*). Graph containing an infinite number of line segments and vertices. Such graphs have many interesting mathematical properties.

Graph (*Isomorphic*). Two graphs G and G' are said to be isomorphic if there exists a one-to-one transformation which maps the vertices of G onto the vertices of G' and the edges of G onto the edges of G' in such a way as to preserve incidence relationships. Thus, if vertex B and edge ϵ are incident in G, the respective images β' and ϵ' are incident in G'. The one-to-one transformation is an isomorphism of G with G'.

Graph (*Linear*). A collection of edges no two of which have a point in common that is not a vertex. The words linear-complex and 1-complex are frequently used alternatives. As defined here, a graph is an abstract graph devoid of any geometric significance. It is true, however, that a graph can be interpreted as a configuration in three-dimensional Euclidean space.

Graph (*Nonoriented*). A linear graph in which the elements have not been assigned an orientation is said to be nonoriented. A graph of this type also is called *ordinary.*

Graph (*Nonseparable*). A graph of which every subgraph has at least two vertices in common with its complement.

Graph (*Nullity*). The nullity μ of a graph G possessing v vertices, e edges and P maximal connected subgraphs is

$$\mu = e - v + P \geq 0$$

Graph (*Oriented*). A linear graph is oriented when an orientation has been assigned to each of its elements. By long-standing convention the phrase "oriented graph" is applied only to graphs which possess at most one directed segment between any two vertices. (For the more general case in which parallel edges are permitted see **Digraph.**)

Graph (*Planar*). A linear graph G can be viewed from either a geometric or a topological standpoint. In the first, it is considered a collection of edges, no two of which have a point in common that is not a vertex. In the latter, it is thought of as defining a set of

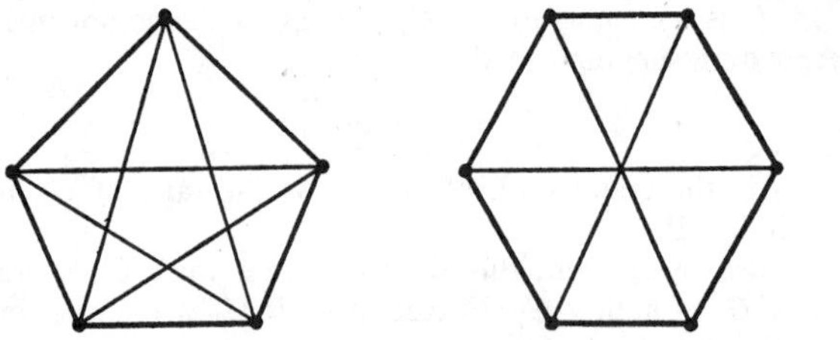

Fig. 3. Kuratowski graphs.

points in three dimensions, whose members are the points which make up the edges of the graph. This point set is the topological graph G^* corresponding to the linear graph G. G is said to be planar if G^* can be mapped on a plane by a one-to-one continuous transformation in such a way that no two image edges have a point in common that is not the image of a vertex in G.

It has been shown by Kuratowski that a linear graph is planar if and only if it does not contain either of the two graphs shown in Fig. 3 as subgraphs.

Graph (*Separable*). A connected graph is separable if it contains at least one subgraph which has only one vertex in common with its complement. Otherwise the graph is nonseparable.

GRAPH RANK. The rank of a graph G is $v - P$ where v is the number of vertices and P the number of maximal connected subgraphs of G.

GRAPNEL. Broadly speaking, a grapnel is any device used to grapple with an object which is obscured to view, such as a submarine object. Grapnels generally take the form of grapnel hooks, which have several flukes, so that they will be certain to hook into any object with which they may come in contact.

GRAPTOLITES (Fossils). Invertebrate Paleontology.

GRAS. In the United States, the acronym for "generally recognized as safe," used for designating foods and materials used in food products with regard to their impact upon human health. During recent years, there has been a gradual erosion of the list of GRAS substances as the result of research efforts on the part of various government regulatory bodies, in Canada, France, Germany, the United Kingdom, etc., as well as in the United States, and also on the parts of various industry self-regulating bodies. Research activities have been directed essentially in terms of determining and confirming possible carcinogenic qualities of such substances. Some GRAS substances have been eliminated and there is a trend toward lowering the levels of usage generally recognized as safe. The parts per million (ppm) levels range considerably from one type of food substance to the next. For a number of years, at periodic intervals, the Institute of Food Technologists (U.S.) has reported summaries of current progress in the consideration of flavoring ingredients under the Food Additives Amendment (U.S.). These summaries appear in *Food Technology* magazine. Lists of GRAS substances are also obtainable from the U.S. Food and Drug Administration, Washington, D.C., and from its counterparts of other governments in many major countries.

GRASHOF NUMBER. A nondimensional parameter appearing in the theory of flows caused by free convection. It is

$$G = \frac{\alpha \theta g d^3}{\nu^2}$$

where θ is the temperature difference producing the convection, α is the coefficient of thermal expansion of the fluid, d is the length scale of the system, and ν is the kinematic viscosity. Flows without large density changes caused by the temperature differences are dynamically similar if the Grashof and Prandtl numbers are equal. Similar nondimensional numbers include the Froude number, the Mach number, and the Reynolds number.

GRASS-CAT. Cats.

GRASS (CRT). Cathode-Ray Tube.

GRASSES. Of all plant families, the grass family (*Gramineae*) is one of the most important economically. With the many thousands of species of grasses, this is one of the largest families in the plant kingdom. Members of the grass family were probably among the first plants to be cultivated by humans. Grasses are found just about everywhere plants can grow, ranging from the polar regions to the tropics and to the upper limits of vegetation on mountains.

Most grasses are herbaceous plants of low stature. A few, notably the Bamboos, become woody plants of great height, and a small number are of clambering or trailing habit. See also **Bamboo.** The cereals, and many other grasses, are annuals, completing their growth in a single growing season; others are perennial plants. Some of the former are winter annuals, plants which start growth in one season, remain dormant over winter, and complete growth and fruit in the following season. Winter wheat is an example.

Among the earliest records of the grasses are those of the Old Testament, all of which emphasize the importance of the grasses to populations thousands of years ago. *Genesis* 1:12: "And the earth brought forth grass . . . whose seed was in itself, after its kind; and God saw that it was good." *Deuteronomy* 11:15: "And I will seed grass in thy fields for the cattle, that thou mayest eat and be full." *Proverbs* 19:12: "The king's wrath is as the roaring of a lion; but his favor is as dew upon the grass." *Isaiah* 15:6: "For the waters of Nimrim shall be desolate; for the hay is withered away, the grass faileth, there is no green thing." And, in the New Testament, *Revelation* 9:4: "And it was commanded that they should not hurt the grass of the earth, neither any green thing. . . ."

Grasses are important to food production in several ways: (1) The cereal grasses, such as barley, corn (maize), grain sorghum, some millets, oats, rice, rye, and wheat, furnish the cereal grains, are basic foodstuffs and which frequently are the sources of important food byproducts, such as edible oils. Cereals also become part of feedstuffs for livestock. (2) The *forage grasses*, such as the bluegrasses, the bromegrasses, the fescues, the ryegrasses, timothy, and wheatgrasses, among many others, along with a number of legumes, comprise pasturage, fodder, green feed, hay, and silage for consumption by livestock, and are the basic ingredients for processed feedstuffs consumed by livestock of many kinds, including beef and dairy cattle, sheep, and poultry. (3) The grasses also aid in the production of field food crops by playing an important role in soil conservation. Grasses are highly effective in reducing erosion and runoff. It is generally agreed among experts that a mat of grass and grass roots has no equal in holding soil. The establishment of grass waterways is an accepted procedure in many areas for routing excessive rainfall.

Botany of the Grasses

The characteristic growth of the principal elements of a representative grass plant are shown in Fig. 1.

The *root system* of a grass plant is made up entirely of fine fibrous roots, which enlarge but little, remaining about the same diameter throughout their length. These roots are mainly adventitious, arising from the lowermost nodes of the stem. The roots of many grasses penetrate deeply into the ground, thus reaching supplies of moisture which enable the plant to live in dry regions where surface moisture may be rare.

The *stems* of grasses, frequently called *culms*, are cylindrical and in most genera hollow except in the region of the nodes, where solid plugs occur. When young, the stem is solid, but as growth continues the central portion fails to keep pace with the outer and gradually becomes hollow. Maize (corn) is an exception, the stems being permanently solid in the plant. In most grasses, the stem grows erect, but frequently falls over during the growing season, because of climatic disturbances or to lack of suitable nutrient sources to give it strength. Such fallen stems do not remain flat, but gradually become erect through renewed growth in the nodal regions. The cause of such a growth is not definitely known. The upward bend, negative geotropism, may be produced by auxin which accumulates in the lower half of the node and stimulates overgrowth in that region. In many species of grass the lowermost nodes normally give rise to a number of buds which develop into lateral branches which give the plant a tufted

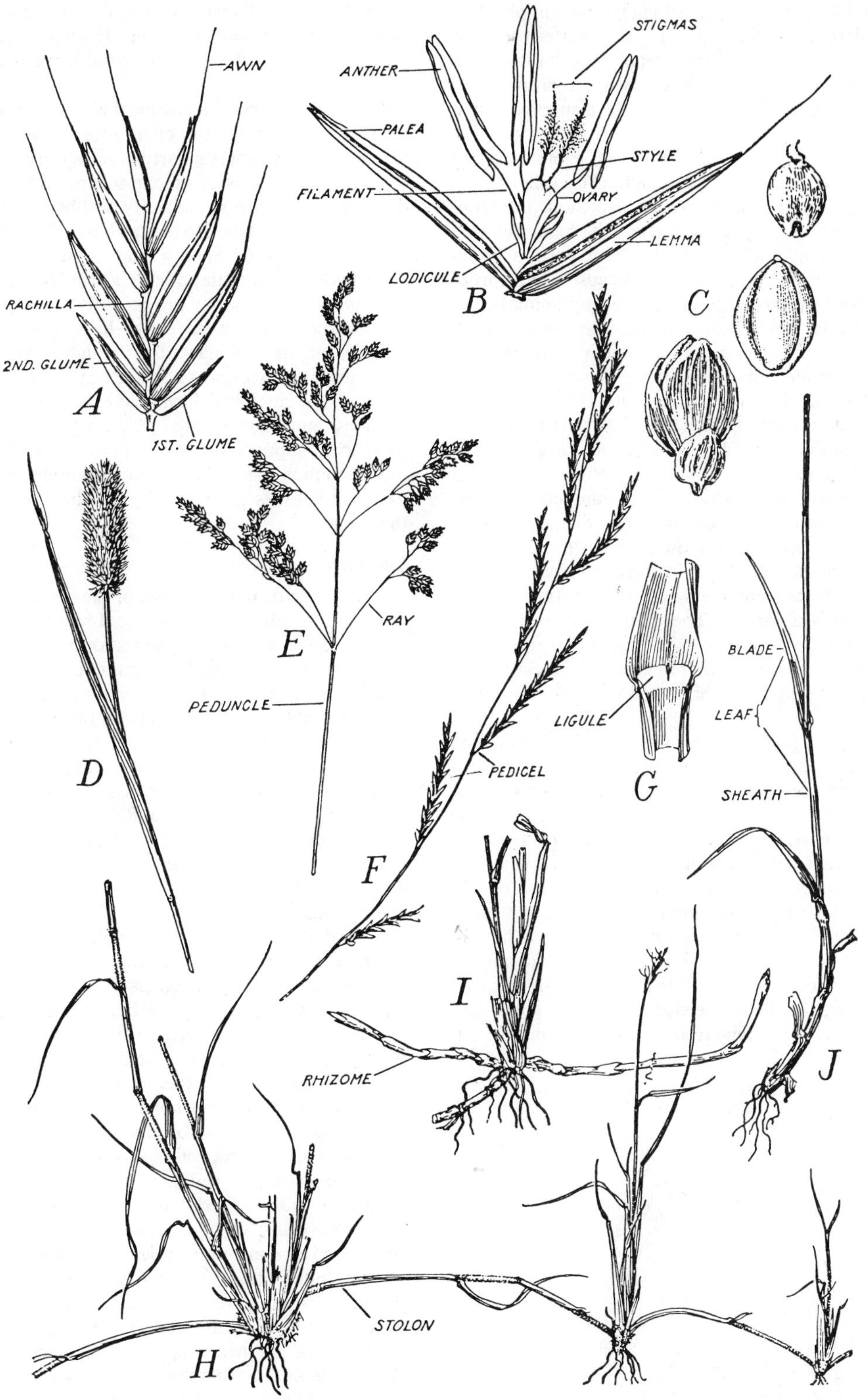

Fig. 1. Characteristic growth of the parts of a representative grass plant: (A) Flowers in a spikelet arranged on a central axis enclosed in two empty glumes or bracts; (B) the different parts of a grass flower; (C) the developed fruit or seed (a caryopsia). This is shown successively enclosed in the outer glumes, with the lemma and pales both closely adhering and free; (D) spikelets arranged in a terminal spike; (E) spikelets arranged in a panicle; (F) spikelets in a raceme; (G) a ligule, at the junction of the leaf blade and leaf sheath; (H, I, J) means of propagating or spreading—stolon, rhizome, and bulb, respectively. (*USDA diagram.*)

appearance. Such basal branches are known as tillers, stools, and the habit of forming them as tillering or stooling. It is a valuable property of many cereals, and undesirable in others, for example, corn, where it causes a considerable reduction in yield. In a few grasses, the basal portion of the stem becomes enlarged by an accumulation of reserve food material, the plant being known as a bulbous grass. Many grasses develop underground stems known as rhizomes, from the nodes of which erect branch stems may develop, as well as numerous adventitious roots. These rhizomes may be short and the erect branches numerous, producing a tufted grass, or they may be long and wide spreading, as in the case of witch grass, *Agropyron repens*, also called quack grass. Due to the readiness with which the joints of the rhizomes of the latter grass strike root and develop to erect stems, it becomes a pestiferous weed. Eradication by chopping up the rhizome with a hoe only serves to increase its numbers, each joint or node producing a new plant. Only by preventing the green tops from forming can the plant be controlled and eliminated, or of course by complete removal of the entire underground rhizome. In some grasses the stem grows out over the surface of the ground, being then known as a stolon. Rhizomes and stolons form an effective way of propagating the plant, and in many species insure considerable dispersal over a limited area.

The leaves of grasses are composed of two parts, a basal sheath which enwraps the stem and a flat elongate blade. The veins of the leaf are all parallel to one another, with few inconspicuous interconnecting veinlets. The blades of grasses grow from the bases, so that the apical portion is older and the cells of the basal portion retain for some time the ability to divide and increase. Because of this property grasses can be mowed by machines or cropped by animals, the upper portions of the blades being removed and the basal portion growing to renew the blade. Each node bears a single leaf, which is often reduced to a small scale, especially in the lowermost nodes, and in modified stems, such as rhizomes. At the junction of the sheath with the blade there occurs in many grasses a distinct structure called the ligule. This appears on the stem side of the leaf, and is a membranous or cartilaginous fringe or ring.

The inflorescence, in grasses, is composed of large numbers of groups of flowers, called spikelets, attached to the main stem or rachis. These spikelets are variously arranged. If they grow directly from the main stem and the latter is unbranched, the inflorescence is said to be a spike. If the main stem produces many branches, which in turn branch, the resulting inflorescence is a panicle. The nature of the branches, whether long or short, spreading or appressed, determines the nature of the panicle. In other grasses the inflorescence is a raceme, the spikelets being borne on short unbranched lateral branches. See Fig. 2.

The individual spikelet of a grass is composed of a short axis called a rachilla from which arise a series of opposite overlapping bracts. The two lowermost bracts are called glumes; these are empty, that is, have no flowers formed in their axils. The next bract above the glumes is the lemma, in the axil of which is borne a flower. In many grasses, each spikelet contains several lemmas, each with its associated flower. Opposite the lemma is the palea, which is not borne on the rachilla, but on a short pedicel, or flowerstalk. Opposite the palea and at the base of the ovary appear two minute scales, the lodicules. Three stamens, each with a long slender filament and a large anther, come next, while a single pistil grows at the apex of the pedicel. The pistil is composed of a 1-celled, 1-seeded ovary, two styles and two feathery stigmas. Many variations from the typical spikelet described occur in different species, the number of parts being increased, or parts being completely absent. In many species of grass, conspicuous prolongations on the glumes or the lemmas are noted—these are the awns.

Fig. 2. A grass (red top, *Agrostis alba*): (1) Panicle of flowers; (2) single flower, consisting of three stamens and one pistil with two branching feathery styles all enclosed by scales.

Pollination in grasses is almost entirely by wind, the light dry pollen being scattered from the open anthers, often in conspicuous clouds. Grass pollen is a particularly common cause of hay fever.

The fruit of grasses is one-seeded, dry and indehiscent, that is, does not split open at maturity to liberate the seed. The ovary wall, or pericarp, is attached to the seedcoat. Within the latter is an abundant starchy endosperm. Such a fruit is known as a grain or a karyopsis.

Considerable speculation has been advanced as to the probable origin of grasses, whether they are primitive monocotyledonous plants from which others such as lilies may have developed, or whether they are reduced plants. To many the available evidence indicates reduction from lily-like ancestors, a reduction in which two of the three pistil lobes of the ancestral form have been lost, also an entire whorl of stamens, and many of the perianth parts. The anatomy of the floral parts lends support to this conception; the vascular bundles suggesting that reduction has occurred. For example, in the pistil there are three vascular bundles, two passing to the styles, and the third bearing the ovule.

The Forage Grasses

The forage or pasture types of grasses can be classified in a number of ways—as annual warm or cool season grasses; as perennial warm or cool season grasses; as grasses for humid regions or dry-land conditions; etc. It is extremely difficult to classify the grasses in terms of relative importance (quantity grown, etc.) because of the wide range of adaptabilities and preferences throughout the world. The scope of this book does not permit detailed descriptions of all major forage grasses. Among other references, these grasses are described in some detail in "Foods and Food Production Encyclopedia," (D. M. Considine, editor), Van Nostrand Reinhold, New York, 1981. Following are brief descriptions of representative forage grasses.

Bahiagrass (genus *Paspalum*). A deep-rooted perennial that forms dense beds even on sandy soils. The rhizomes are short, stout, and woody and reach out horizontally. Once a good sod of Bahiagrass is formed, it is difficult for other plants to encroach. See Fig. 3. Bahiagrass ranks between carpetgrass and Bermudagrass in productivity and nutritive value. Bahiagrass may become a pest in certain pastures because of its aggressive growth habits and prolific seeding. It is important to note that seeds germinate even after passage through the digestive system of cattle. In some regions, this has caused Bahiagrass to ultimately crowd out other desirable grasses. Bahiagrass is suitable to range conditions, but not fully drought resistant.

Bermudagrass (*Cynodon dactylon*). This grass is commonly found in tropical and subtropical regions of the world. Because Bermudagrass grows so widely in India, it was believed for a long time that the species originated there. However, recent research indicates a much greater diversity of types found in Africa, and thus Africa is now considered by many authorities as the original source of Bermudagrass. In the United States, Bermudagrass is found mainly in the southern portions, ranging from southern California eastward to the North Carolina coast. The Midland variety ranges a bit further north, particularly east of the Mississippi River, where it is found in Kentucky, West Virginia, and northward to southern New England.

Bermudagrass has been an important pasture cover since the early 1800s. It is believed that it was first introduced to Savannah, Georgia as early as 1751. Common Bermudagrass is a fast-spreading grass that can be used effectively to prevent soil erosion. Common Bermudagrass is established from either seed or vegetative sprigs. When grazed closely, common Bermudagrass will grow in association with lespedeza, improved white clovers, vetches, crimson clover, and arrowleaf clover.

A number of hybrid Bermudagrasses have been developed, including Coastal Bermuda, which is superior to common bermuda and is adapted for moderately well-drained soils. Coastcross Bermuda is another hybrid. Suwanee and Midland bermuda are also hybrids, developed for particular conditions.

Bluegrasses (genus *Poa*). The bluegrasses are found widely distributed throughout the world in temperate and cooler regions. There

Fig. 3. Argentine Bahiagrass (*Paspalum notatum*) growing on a well-fertilized hill soil in Florida. Bahia grass has endured heavy grazing and pasture mismanagement better than most other grasses. (*USDA photo.*)

are some 200 species of *Poa*, of which about one-third are native to North America. See Fig. 4. Although the word "blue" has been used to describe these grasses for at least a couple of centuries, the exact reason is unknown. Some authorities believe the association arose from the fact that some of these grasses take on a somewhat bluish appearance when in bloom. Other attribute this to the vaguely blue color of the leaf of *Canada* bluegrass.

Kentucky bluegrass (*Poa pratensis*), also known as *June* grass, is one of the most widely grown grasses in parts of North America. The grass is found throughout the United States and ranks as one of the important forage plants. It is most commonly found in the northeastern quadrant of the United States, ranging eastward from the eastern Dakotas to the Atlantic seaboard and as far south as Kentucky, Tennessee, and western North Carolina. The grass was first reported at Grassy Lick, Kentucky in 1775 and referred to as abundant at that time. Some authorities believe that grazing animals, such as the elk and buffalo, which were commonly found east of the Mississippi River at that time, helped to spread the grass westward. Kentucky bluegrass is also commonly found in the meadows of eastern Europe and western Asia. Where the soil pH is 5 or higher and of high fertility, Kentucky blue grass will dominate other plants. The grass can survive severe droughts. In recent years, some authorities have grown less enthusiastic about Kentucky bluegrass because of its low midseason yield, aggressiveness, and high fertility requirements. These objections have been partially met through the development of several new varieties—*Delta, Merion, Newport, Park*, etc. Kentucky bluegrass, because it is quite variable in chromosome numbers and other genetic factors, is prone to produce aberrant plants.

Canada bluegrass (*Poa compressa*) is native to eastern Europe and western Asia. It was first reported in North America about 1792 and generally followed the same pattern of spread across the continent as in the case of Kentucky bluegrass. The grass generally ranges from northern Michigan and Ontario westward to the Rocky Mountains. It is an erect growing perennial bunchgrass.

Bluestems. These are among the truly native forage grasses of the United States that have been cultivated since the 1930s. Prior to that time, the only native grass of any significance was slender wheatgrass. Use of native grasses commenced as the result of the dust bowl conditions of the 1930s. It was found that soil erosion in very low-rainfall areas could be controlled by the use of native grasses. There are several

Fig. 4. Specimen of Canbyi bluegrass (*Poa canbyi*) in the soft dough stage. (*USDA photo.*)

bluestems, including *Big* bluestem, *Little* bluestem, *Sand* bluestem, *Yellow* bluestem, and *Caucasian* bluestem.

Bromegrass (genus *Bromous*). Smooth brome, also known as *Austrian* brome, *Hungarian* brome, and *Russian* brome, has been grown in the United States since about 1880. It is very tolerant of heat and drought and consequently is used widely in many of the dry regions west of the Mississippi River, but usually north of a latitude of about 36° N. Records indicate that the grass was first cultivated in the west and widely in California, but that persistent periods of drought in the midwestern United States progressively brought attention to the desirable properties of this grass. A common procedure is to plant smooth brome with a legume for hay, followed by use as a pasture. This is an excellent combination because nitrogen available from the legume provides a nitrogen supply for the grass for several years.

Brome grass may be described as an extremely hardy perennial that grows to a height of 3 to 4 feet (0.9 to 1.2 meters). The root system is highly branched and sometimes reaches a depth of 6 to 8 feet (1.8 to 2.4 meters).

Varieties of brome, in addition to smooth brome, include field bromegrass, cheat bromegrass, nodding brome, and fescuegrass. See Figs. 5, 6, and 7.

Buffalograss. This is highly regarded as a range pasture plant. The grass has numerous qualities that are attractive to stockmen—very palatable and nutritious when green in summer, but also retains a good feeding value when dried and cured for winter feeding. It tolerates heavy grazing.

Carpetgrass (*Axonopus affinis*). This is a low-growing, creeping perennial that makes a dense sod. Native to Central America and the West Indies, the grass was introduced into the United States in the early 1830s and first reported in the New Orleans area. The grass is well suited to sandy or sandy loam soils. It is a prolific seeder and

Fig. 5. Soft chess bromegrass (*Bromus mollis*), a naturalized Mediterranean annual now widely distributed in California. It is a valuable species in the annual forage range and can be used for cover croppings. The grass is self-seeding and favored in some areas for soil conservation. (*USDA photo.*)

Fig. 6. Fescuegrass or prairie brome (*Bromus catharticus*), a valued grass for grazing and soil conservation. (*USDA photo.*)

does not require high fertility. The grass does not do well in swampy areas. In the United States, carpetgrass is found mainly in the southeastern coastal area. The grass will tolerate close grazing, but is not as nutritious and productive as many other pasture plants.

Dallisgrass (genus *Paspalum*). Also known as *watergrass*, this is a fast-growing, rather stout perennial primarily utilized for pasture in the southeastern United States, ranging as far west as Texas. Dallisgrass is native to South America, ranging from Brazil to Argentina. It is believed that the grass was accidentally introduced into the United States in the mid-1800s. It is not suitable for hay production.

Fescues (genus *Festuca*). Made up of both annuals and perennials, there are some one-hundred or more species of fescue. The growth habit may be creeping or erect. Of the species, tall fescue (*Festuca arundinacea*) is one of the more important forage grasses in the western, northwestern, and southeastern United States. See Fig. 8. It is also widely used in other grassland regions throughout the world. Tall fescue is a deep-rooted, strongly tufted, winter-hardy perennial with broad basal leaves that are dark green, coarse, and flat. It will tolerate a high water table and may be used in areas too low and wet for other pasture plants. Also see Fig. 9.

A few years after tall fescue was introduced to New Zealand from Europe (early 1800s), livestock that grazed on the grass for extensive periods were noted to develop a lameness, a condition called *fescue foot*. The situation became widespread and serious and a program was undertaken to eradicate the grass from that country. Although exacting conclusions may not have been drawn, some authorities believe that it was a peculiar grouping of circumstances rather than the qualities of the grass. For example, where fescue foot was observed the areas usually were wet, low, and swampy with extensive deficiencies of minerals.

Other varieties of fescue include *Meadow* fescue, *Sheep* fescue, *Red* fescue, *Chewings* fescue, and *Idaho* fescue, among others.

Foxtail. This grass is commonly used in Europe for pasture and hay and is well adapted to wet lands. Records indicate that it was first used in the mid-1700s. In the United States, it performs well in the northwestern states, including Alaska. It prefers a cool, moist climate and does not resist high-temperature and drought conditions. There is a superficial resemblance of foxtail with timothy. The palatability of the grass, both as pasture and hay, is very good. Varieties of foxtail include meadow, creeping, and reed foxtail.

Grama Grasses. Two species of the grama grasses are of significance in the Great Plains regions of the United States—a bunch form and perennial known as *sideoats grama* and a more drought-resistant form known as *blue grama*. Grama grasses are palatable and retain their flavor and nutrition well into the winter months. However, grama grasses are not suitable for hay. There are a number of important native varieties of grama which are cultivated for forage locally.

Johnsongrass (*Sorghum halapense*). Not commonly considered a cultivated grass, but more often as a weed by some food crop growers, nevertheless Johnson grass is an important hay grass in the southeastern United States. See Fig. 10. This grass also can be used as an effective soil-conserving crop. It requries relatively fertile and loose soil and does not endure close grazing.

Lovegrasses. The principal attractions of the lovegrasses are their toleration of low fertility and sandy soils. These grasses produce abundant quantities of seed which germinate readily. In the United States, one native and three introduced species occur. Native to the central southern Great Plains is *sand lovegrass*. The value of the grass was not formally recognized until the late 1930s, after the dust bowl period.

Millet Grasses. These grasses, of several species, offer the advantage of only requiring 60 to 70 days from seeding to maturity. Some authorities have found that the foxtail millets (not to be confused with foxtail grass) exceed all other crops in their efficient use of water.

Napiergrass (*Pennisetum purpureum*). A grass native to equatorial Africa and introduced into the United States in 1913. This grass is

Fig. 7. Bromer mountain bromegrass (*Bromus marginatus*), a development of the Washington Agriculture Experiment Station. (*USDA and Soil Conservation Service photo.*)

Fig. 8. Specimen of tall fescue (*Festuca arundinacea*). (*USDA and Soil Conservation Service photo.*)

adapted to the Gulf coastal region from Texas to and including all of Florida. It also does well in southern California. The grass will grow on almost any soil that will support ordinary food crops. The useful area of the grass can be extended northward if planted on rather fertile soils.

Natalgrass (*Tricholaena rosea* Nees). A grass native to South Africa and introduced into the United States in the late 1860s. It is also known as *Hawaiian redtop* and *Australian redtop*. First attention was brought to the grass because of its ornamental potential. The grass is suited to well-drained, poor, sandy soils. An outstanding advantage of Natalgrass is its resistance to attack by nematodes. The grass often succeeds as a forage crop in areas where no other forage grass can grow. It can be cut for hay.

Oatgrass. Tall oatgrass, at one time, was very important as a forage grass in Europe. It was introduced into the United States in the early 1800s. Of secondary importance, the grass is found mainly in the northwestern United States. It is not drought or heat resistant.

Orchardgrass (*Cactylis glomerata* L.) This grass is native to western and central Europe, but has been cultivation in the United States since 1760. It is a cool season perennial that grows in clumps producing an open stand. It makes excellent hay. It is tolerant of partial shade and grows well in mixtures with white clover. However, the grass is highly susceptible to a number of diseases. The flowering culms of the plant reach a height of 2 to 4 feet (0.6 to 1.2 meters). See Fig. 11. The importance of this grass in North America has increased manyfold since the early 1930s. In some states, such as Virginia, Kentucky, and Tennessee, this is the major forage grass. It is frequently part of a mixture, particularly with red clover or alfalfa for hay. Throughout the United States, in terms of quantity, orchardgrass probably is exceeded only by smooth bromegrass, timothy, and Kentucky bluegrass, although reliable figures are difficult to obtain. Persistance of the grass under continuous grazing is limited. Rotational grazing is the best practice for orchardgrass.

The use of orchardgrass in the British Isles has increased considerably during the last couple of decades. Orchardgrass possesses much versatility, being adapted for harvest for hay or silage as well as for grazing. Much orchardgrass seed is produced in Oregon, Washington, and California. High applications of nitrogen can increase seed production by a factor of 100%. Shattering is a problem in seed processing. There are numerous varieties of orchardgrass. The *Akaroa* variety was released for use in western Washington in 1951 and for use in California in 1952. This grass has long been popular in New Zealand. It is well adapted to all of the Pacific coastal states, but must be irrigated in California. See Fig. 11.

Redtop (*Agrostis alba* L.). Of the same genus as the bentgrasses, redtop at one time (until 1940s) was second only to Kentucky bluegrass as an important forage and pasture grass in North America. Since that time, redtop has been significantly displaced by a number of other grasses and grass–legume mixtures. In addition to forage uses, redtop finds application for lawns, recreational areas, highway plantings, etc. Redtop is most common in the northeastern quadant of the United States. Most frequently, redtop is sown with legumes and other grasses.

Reed Canarygrass (*Phalaris arundinacea*). This is an important grass, not only as a hay and silage crop, but also for use in soil conservation programs. The grass will frequently produce good yeilds of forage from soils that are too wet or poorly drained for other grasses and legumes. Variations of reed canarygrass include ribbongrass and Hardinggrass.

Fig. 9. Sheep fescue (*Festuca ovina*) of a form grown in Kenya and Turkey. The grass is densely-tufted, fine-leaved, and somewhat dwarfed. The grass develops slowly, but is drought-resistant. (*USDA and Soil Conservation Service photo.*)

Fig. 10. Johnson grass (*Sorghum halapense*), an important forage grass for the southeastern United States. A difficult grass to control where not desired. (*USDA and Soil Conservation Service photo.*)

Fig. 11. Akaroa orchardgrass (*Dactylis glomerata*), a variety developed for the western United States. (*USDA and Soil Conservation Service photo.*)

Ryegrasses (genus *Lolium*). These are hardy winter annual bunch grasses with glossy, dark-green foliage. Ryegrass furnishes grazing in the late fall, winter, and spring. The grass is most often used in mixtures of small grain and annual clover. Ryegrass adds to nutritive value when grown with wheat for silage. It will extend the grazing period in the late spring. Ryegrass does best when heavily fertilized, especially with nitrogen. The greatest concentrations of ryegrass are found in the Gulf coast states, as well as Georgia, South Carolina, and parts of North Carolina. Ryegrass is not extensively used in Florida. See Fig. 12.

Saint Augustinegrass (*Stenotaphrum secundatum*). This grass is native to the West Indies, and possibly to Australia and southern Mexico. The grass is also found in South Africa. It was introduced into France and Italy from Africa and probably introduced into the United States from Cuba. This grass is also called saltgrass, sheepgrass, and jointgrass. The grass does well in most kinds of soil, but requires a lot of moisture. It is notably well adapted to mucky soils and partially shaded areas.

Sudangrass (genus *Sorgos*). The grass sorghums include a number of varieties, one of the most important being Sudangrass. This is an excellent annual grass and used extensively in the United States, with exception of the far north and southeastern states. In these areas, because of frost or disease problems, Sudangrass is essentially replaced by pearl millet. For clarity, it should be pointed out that there are many kinds of sorghum—grain sorghum, forage sorghum, sirup sorghum, grass sorghum, and broomcorn.

Timothy (*Phelum pratense*). At one time, timothy was the most important and widely used of the many forage grasses. The existence of timothy dates back to antiquity. The grass is native to most of Europe, eastward through Siberia and north to a latitude of 70° N.

Fig. 12. Wimmara ryegrass (*Lolium subulatum*). One of the best self-seeding annual grasses for wet-winter and dry-summer climates of the Pacific coast region, notably parts of California. The plant is easy to establish, grows rapidly with autumn rains, matures and shatters seed early. This grass, long used in Australia in short-lay pastures, was introduced into the United States in the early 1950s. In addition to agricultural uses, the grass has been successfully used for reseeding burned-over brush lands. (*USDA and Soil Conservation Service photo.*)

Fig. 13. Drummond timothy (*Phelum pratense*) in the early heading stage. (*USDA and Soil Conservation Service photo.*)

The grass also occurs naturally in the Caucasus region and in Algeria. In the New England states, timothy is sometimes called herdgrass. Timothy grows best in a cool and humid climate. Although it may survive in some hot humid or hot dry climates, it does not yield well. Best results are achieved when the plant is grown on clay or silt loam soils that are fairly well drained. Timothy roots are shallow and fibrous. Timothy is a bunch grass with erect culms, ranging from 20 to 40 inches (51 to 102 centimeters) in height. It produces a dense, cylindrical, spikelike inflorescence (the head). See Figs. 13 and 14.

Although it is grown alone, more often timothy is sown in mixtures with legumes, such as medium-read or Alsike clover. Principal regions for plantings in the United States are in the northeastern quadrant. Timothy is grown mainly for hay. Improved varieties of timothy have been developed in recent years; *Essex* (Cornell University); *Clair* (Kentucky Agricultural Experiment Station); *Climax* (Canadian Department of Agriculture); *Marietta, Lorain,* and *Hopkins* (Ohio Agricultural Experiment Station); *Drummond* and *Milton* (Macdonald College, Quebec, Canada); *Medon* and *Patton* (Ontario Agricultural College); *Dural* (University of Manitoba); and *Swallow* (University of Alberta).

Wheatgrasses (genus *Agropyron,* tribe *Hordeae*). At one time, the wheatgrasses were considered to be in the same genus as wheat. The common name stems from the fact that the seed heads resemble those of wheat. The wheatgrasses are widely distributed through the temperature regions of the world. Of the 150 known species, about 30 are native to North America. Most species originated in eastern Europe and western Asia in desert or steppe soils and in climates ranging from semihumid to arid. A few species are confined to South America. In North America, most of the wheatgrasses are found in the northwestern quadrant, including British Columbia. They range eastward as far as Minnesota and Ontario and as far south as northern Texas. These are cool-season grasses and are highly valued where suited as important sources of very nutritious early-season forage. They also are highly regarded for control of wind and water erosion. Some au-

Fig. 14. Close-up of timothy in heading stage. (*USDA and Soil Conservation Service photo.*)

Fig. 15. Crested wheatgrass (*Agropyron desertorum*). (*USDA and Soil Conservation Service photo.*)

Fig. 17. Volga wildrye (*Elymus gigantus*) introduced into the United States from the U.S.S.R. in 1952. (*USDA and Soil Conservation Service photo.*)

Fig. 16. Species of wheatgrass (*Agropyron michei*) introduced into the United States from Manchuria for application in the Great Plains region. (*USDA photo.*)

thorities have estimated that natural wheatgrasses in the United States are found on 300 million or more acres (120 million hectares). Species introduced into the United States include Crested wheatgrass (*Agropyron desertorum*), a hardy, drought-resistant bunchgrass native to eastern Russia, western Siberia, and central Asia. See Fig. 15. *Fairway* was introduced from Canada. *Siberian* wheatgrass, a drought-resistant bunchgrass, was introduced from the U.S.S.R. in 1934. There were many other introductions. See Fig. 16.

Wildrye Grasses. These grasses are closely related to the wheatgrasses, differing mainly by the fact that the wildryes have two spikelets at each rachis node. Among the wildrye grasses in North America, some are native, but several have been introduced and are now used in grassland agriculture in the United States and Canada. *Russian* wildrye is particularly well adapted to the northern Great Plains region. See Fig. 17. This grass does well in several of the Canadian provinces. This variety is drought-resistant. Other varieties include *Canada* wildrye, *Virginia* wildrye, *Basin* wildrye, and *Beardless* wildrye, among others.

The *cereal grasses* are described in separate alphabetical entries on oat, rye, wheat, etc.

A relatively recent publication, "Grasses: An Identification Guide," Houghton Mifflin, 1979, may be of interest to some readers.

GRASSHOPPER (*Insecta, Orthoptera*). Also known as locusts and of many species of the family *Locustidae*, grasshoppers have been known since ancient times and associated with devastating crop losses and resulting famines. See Fig. 1. Practically no plant, cultivated or wild, is immune from attack by one or several species of grasshopper. These insects occur worldwide. In the United States, serious outbreaks of grasshopper seldom develop east of the Mississippi River, but they are not uncommon in the western two-thirds of the country. Grasshoppers often severely damage range grasses. Their feeding is one of the

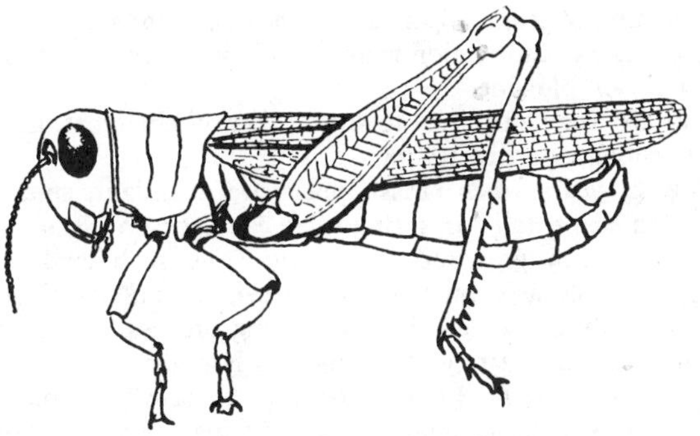

Fig. 1. Grasshopper (also known as locust). (*USDA diagram.*)

main reasons for loss of productive grasslands in many of the western states.

When range grass is scarce and outbreaks are severe, grasshoppers often migrate into and severely damage the foliage of alfalfa, clover, corn (maize), small grains, potato, and fruit trees. In fruit orchards, grasshoppers sometimes fully strip the leaves and may kill young trees. Both insecticides and cultural practices can be effective and must be used for effective grasshopper control.

Many species of grasshopper winter in the egg stage. The eggs are laid in masses that are found from $\frac{1}{2}$ to 3 inches (about 2 to 8 centimeters) below the soil surface. Each mass will have from 20 to 120 elongated eggs, held together securely by cement. One female may deposit from 8 to 25 egg masses. The eggs usually are deposited in uncultivated ground, often in alfalfa, clover, and stubble fields. The egg-laying procedure varies from one species of grasshopper to the next. The pellucid grasshopper prefers sod land and heavy soil. The migratory grasshopper prefers crop land. Other species prefer uncultivated ground, as previously mentioned.

The red-legged grasshopper (*Melanoplus femur-rubrum*, De Geer) is a small species, ranging up to 1 inch (2.5 centimeters) in length when fully developed. The insect is severely destructive of legumes, notably soybean. The color is a brown-red. The hind tibiae are a pinkish-red with black spines.

The migratory grasshopper (*Melanoplus bilituralus*, Walker) is the most destructive and widespread of all species. The insect has a great ability to survive in dry and waste lands. This insect is also about 1 inch (2.5 centimeters) long when fully grown. The term migratory is used in describing the species because the partially developed nymphs normally travel or migrate from their breeding ground to find more attractive vegetation. See Fig. 2. The adults also may fly for many miles in search of more attractive feeding areas.

The clear-winged grasshopper (*Camnula pellucida*, Scudder). In terms of damage, this insect is only second to the migratory grasshopper. It occurs throughout the United States, but is most common in the west and it seems to prefer relatively high elevations. It is well adapted to survive heat and drought. The hind wings are nearly transparent.

The differential grasshopper (*Melanoplus differentiallis*, Thomas). This insect prefers cultivated areas and does not survive long dry periods. In such times, they will be found only near ditches and irrigated areas. The insect ranges from $1\frac{1}{2}$ to $1\frac{3}{4}$ inch (about 3.5 to 4.5 centimeters) in length and of a brown-green color with yellow underparts. The differential grasshopper is a severe destroyer of corn (maize).

The two striped grasshopper (*Melanoplus bivittalus*, Say). This is a strong species and is of an olive-green color with yellow stripes on each side. The species is frequently found in clover fields.

The Carolina grasshopper (*Dissosteira carolina*, Linne). One of the largest of the grasshoppers, attaining a length of about 2 inches (5 centimeters). One of the most commonly observed species, although somewhat less destructive of crops.

Closely allied to the grasshopper are the **Cicada; Katydid;** and **Locust;** see separate entries under these headings.

Chemical Insecticides

The best chemical control practice for grasshopper varies from one type of crop to the next.

Alfalfa. Commonly used insecticides are carbaryl, diazinon, malathion, and mevinphos. When an entire field is severely infested, it

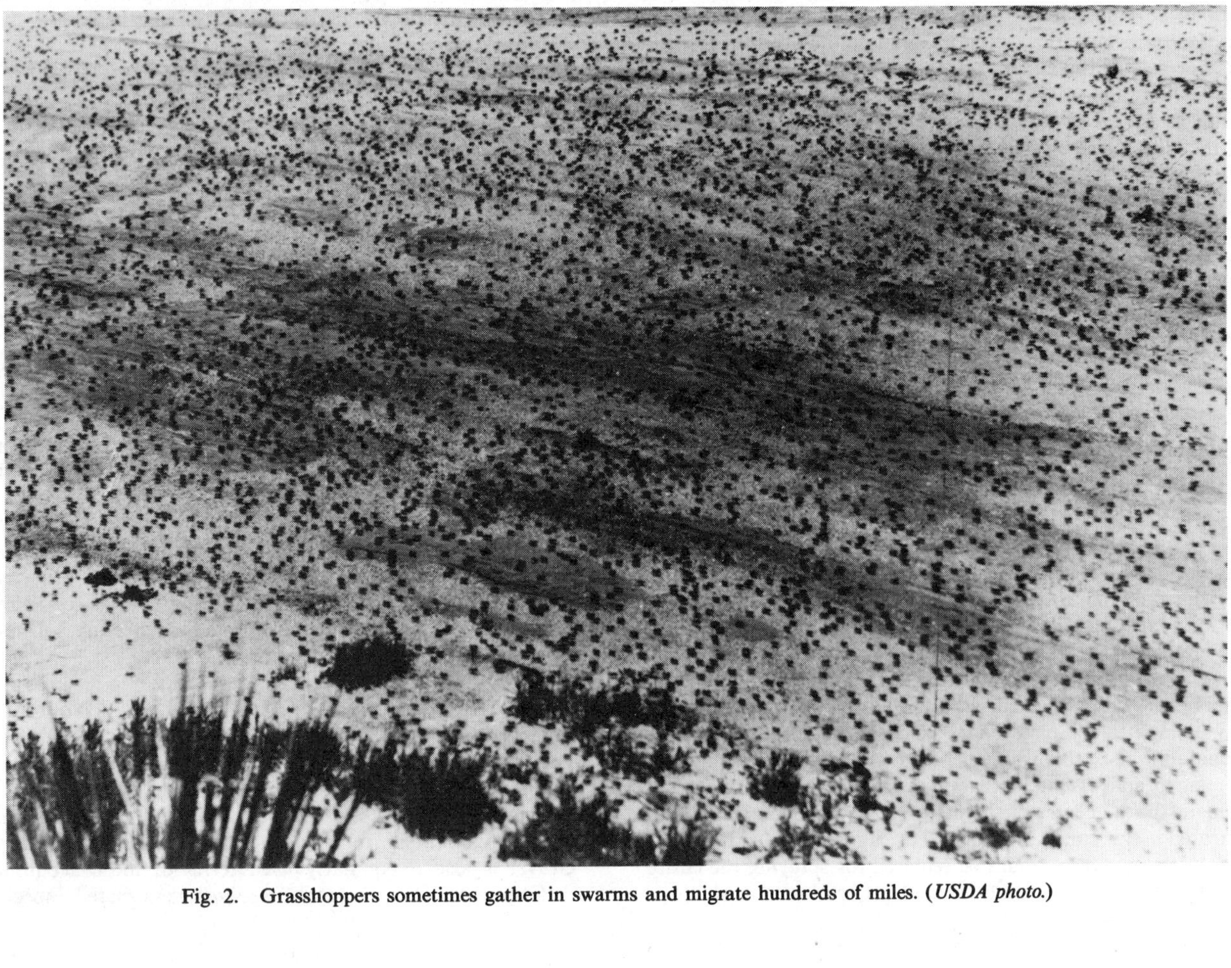

Fig. 2. Grasshoppers sometimes gather in swarms and migrate hundreds of miles. (*USDA photo.*)

usually is most economical to cut the alfalfa and apply the insecticide to protect the next cutting. Field margins, ditchbanks, patches of weeds, and uncut strips of alfalfa, where grasshoppers congregate, must be thoroughly treated. Grasshoppers often hatch in large numbers after the first crop has been harvested. To control these insects, the next crop should be sprayed before the new growth is 6 inches (15 centimeters) high.

Clover. The same chemicals as for alfalfa are used. Fields should be treated when the grasshoppers are young. Frequently, a second application of control chemicals will be required.

Corn (*maize*). Control chemicals frequently used include carbaryl, malathion, and toxaphene. To prevent damage to corn, margins of the field and other adjacent infested hatching areas should be sprayed prior to the time when the grasshoppers are ready to migrate into the field.

Grassland, range, and pasture. Carbaryl or malathion are most effective. Treatment should be commenced when young grasshoppers first appear and before eggs are laid.

Lettuce. Carbaryl as a bait or malathion as a foliage spray should be applied immediately upon detecting first damage.

Potato. Malathion should be applied as a foliage spray immediately upon noticing first damage.

Barley, Oats, Rye, and Wheat. Malathion or toxaphene are effective. To protect small grains, hatching areas should be sprayed when the nymphs first appear. To prevent damage to fall-seeded grain, field margins should be sprayed as far into the field as the grasshoppers have moved. To protect the field margins, phorate granules should be applied at fall seeding of wheat to the outer two or three drills widths of the field.

Sugarbeets. Carbaryl or malathion are effective. Hatching areas, infested fence rows, idle land, and roadsides adjoining the field should be treated. Carbaryl can be used for broadcast application as a bait. Malathion can be used as a spray.

The usual precautions taken for applying insecticides should be followed in grasshopper control.

Cultural Practices

Grasshoppers, particularly those that lay their eggs in fields planted to crops, may be controlled to some extent by tillage and seeding operations. Cultural operations do not eliminate the need for insecticides, but they reduce the amount of chemicals needed.

Tillage. Working the soil kills grasshoppers in several ways. It can bury their eggs so deep that young grasshoppers do not hatch. It can bring the eggs to the surface where they are destroyed by drying of sun and wind. Tillage also discourages egg-laying, preventing dispersal of the pests and forcing grasshoppers scattered over a field to concentrate in a smaller area. Proper tillage before eggs have hatched often gives excellent control of threatening grain-stubble infestations. Fall tillage is preferable, but spring tillage can be effective. Tillage immediately after harvest will make the soil less attractive to egg-laying and will assist in destroying eggs already laid.

Shallow cultivation is less effective than moldboard plowing, but it will destroy many of the eggs by exposing them to sun and wind. The one-way disk is the best implement for this operation. The duck-foot cultivator, the single or double harrow, and the one-way harrow, also are satisfactory. Blade tillers used in stubble-mulch farming are less effective than the others. Shallow cultivation is most effective during dry weather.

Grasshopper-infested grain stubble that is to be summer-fallowed should be worked before the eggs hatch. If tillage is delayed until after the young grasshoppers appear, it still may be useful in preventing the insect from moving to nearby crops. This tillage can be accomplished by cultivating a guard strip 3 rods (about 5 meters) wide around the entire field. If the strip is kept cleanly fallowed, the young grasshoppers can usually be held within the field for a week or two. There may be time to complete tillage operations before they escape. Tillage done after the establishment of the guard strip should start next to the strip and extend until only a small block of unworked stubble remains in the center of the field. The grasshoppers will then be concentrated in this small area. Here they can be killed with insecticide at much less cost than would be required for spraying the entire field. Large tracts of sod or idle land should not be plowed or shallow-tilled for control of grasshoppers unless the land is intended for seeding or summer-fallow. Cultivation ruins such land for pasture and makes it subject to soil blowing.

Aircraft are frequently used to spread control chemicals in connection with grasshopper infestations.

Seeding. In years when grasshoppers are abundant, small grains may be planted on fall- or spring-tilled land, or on clean summer-fallowed land. Few grasshoppers emerge from such land. A grain drill should not be used on heavily infested, unworked stubble. This will destroy only a few eggs by the seeding process. When the eggs hatch, the field will swarm with young grasshoppers. Then, immediate spraying of the entire field will be required to save the crop.

Early spring seeding is important in reducing grasshopper damage. These crops make considerable growth before grasshoppers hatch. Thus, they withstand a longer period of feeding than late-seeded crops and also provide a better opportunity to kill the grasshoppers with chemicals.

When small grains are ripening, flying grasshoppers frequently congregate in late-seeded crops that are still green and succulent. Such crops are often severely damaged before the grasshoppers are noticed. Well advanced crops are much less attractive to the pests. Barley, oats, and wheat that have headed can withstand considerable defoliation without serious reduction in yield of grain.

Regrassing Field Margins. Weedy field margins, including roadsides and fence rows, contain more grasshopper eggs than other habitats. Replacing broad-leaved weeds with perennial grasses greatly reduces the number of grasshoppers in such locations. Crested wheatgrass can be used for this purpose. It is easily and quickly established and is less attractive for egg-laying than native grasses. Elimination of weeds and prevention of soil erosion are additional benefits of grassed field margins.

Immune Crops. Some of the sorghums, such as sorgo and kafir, after reaching a height of 8 to 10 inches (20 to 25 centimeters), are practically immune to grasshopper attack. They can be planted rather late in the season to provide valuable feed for livestock.

Irrigation. When alfalfa and other legumes are irrigated, large numbers of grasshoppers are sometimes driven to ditchbanks and other dry places. Here, they can be killed with sprays at very low cost. Flooding hay meadows where grasshopper eggs have recently hatched will destroy many young grasshoppers.

GRASSLAND BIOME. **Biome.**

GRASS (Revegetation). **Revegetation.**

GRASS STAGGERS or TETANY. **Magnesium (In Biological Systems).**

GRATE STOKER. **Burner.**

GRATICULE. 1. A graticule is a reticle composed of lines ruled on a transparent plate, instead of the usual fine threads or wires. 2. By extension, the pattern of lines representing parallels of latitude and meridians of longitude on a map or chart is known as the graticule of the chart. A person familiar with the various types of map projection can usually tell by examination of the graticule the type of projection that was used in constructing the sheet.

GRATING. Any framework or latticework, consisting of a regular arrangement of bars, rods, or other long, narrow objects with interstices between them. A diffraction grating consists of rulings upon the surface of a light-transmitting or light-reflecting substance; it is used for the production of spectra.

GRAVEL. An unconsolidated, natural accumulation of rounded rock fragments resulting from erosion, consisting predominantly of particles larger than sand (diameter greater than 2 millimeters; $\frac{1}{12}$ inch), such as boulders, cobbles, pebbles, granules, or any combination of these fragments; the unconsolidated equivalent of conglomerate. In the United Kingdom, the range of 2–10 millimeters has been specified.

Gravel is also a popularly used term for loose accumulation of rock fragments, such as detrital sediment associated especially **with** streams or beaches, composed predominantly of more or less rounded pebbles and small stones, and mixed with sand that may compose 50–70% of the total mass.

Gravel is also a term for rock or mineral particles having a diameter in the range of 2–50 millimeters. In the United States, the term is used for rounded rock or mineral soil particles having a diameter in the range of 2–75 millimeters $\frac{1}{6}$ to 3 inches); formerly the term applied to fragments having diameters ranging from 1–2 millimeters.

See also **Ocean Resources (Mineral).**

GRAVE'S DISEASE. Thyroid Gland.

GRAVIMETRIC ANALYSIS. Analysis (Chemical); Weighing.

GRAVIMETRIC ANALYSIS (Thermal). Thermogravimetric Analysis.

GRAVIMETRY. Earth.

GRAVIRECEPTORS. Highly specialized nerve endings and receptor organs located in skeletal muscles, tendons, joints, and in the inner ear which furnish information to the brain with respect to body position, equilibrium, and the direction of gravitational forces.

GRAVITATION. A phenomenon characterized by the mutual attraction of any two physical bodies.[1] This universal character of the gravitational force was first recognized by Sir Isaac Newton who also gave its quantitative expression. For point masses or spherical bodies, a simple expression results:

$$F=\frac{GM_1M_2}{R^2} \tag{1}$$

In addition to the masses M_1, M_2 of the two bodies and their distance apart R, the force depends only on a constant $G = 6.670 \times 10^{-8}$ dyne cm^2 gm^{-2} which is independent of all properties of the particular bodies involved. The same force law describes the motion of the planets around the sun, of the moon around the earth, as well as the falling of an apple to the earth. A body moving under an inverse square law as given in Equation (1) satisfies the three laws established by Kepler for the motion of the planets around the sun:

1. The planets move in elliptical orbits with the sun at one focus (the general orbit is a conic section) (Fig. 1).
2. The radius vector sweeps out equal areas in equal times.
3. The square of the period of revolution is proportional to the cube of the semi-major axis: $a^3 = (2\pi)^{-2}GM_{\odot}T^2$. Here $M_{\odot}$ is the mass of the sun and T is the period of the planet.

These results together with a detailed analysis of anomalies in the motion of the moon established the correctness of the Newtonian theory of gravitation.

The *weight* of a body of mass M on the earth is the force with which it is attracted to the center of the earth. On the surface of the earth the weight is given by

$$W=Mg$$

where the *acceleration due to gravity* is obtained from Equation (1):

$$g=\frac{GM_E}{R_E{}^2}=980.665 \text{ cm/sec}^2$$

$$=32.174 \text{ ft/sec}^2$$

All freely falling bodies near the surface of the earth are accelerated at the same rate g. It is for this reason that Galileo found that both light and heavy objects take the same time to reach the ground when dropped from the Leaning Tower of Pisa.

An astronaut is said to be in a state of *weightlessness* when in orbit. Strictly speaking, the body still has weight for the earth's gravity still acts on it. Otherwise the astronaut would fly off into outer space.

[1] Einstein's general relativity theory is essentially the modern statement of gravity and reference to the entry on **Relativity and Relativity Theory** is also suggested, where the topic is approached from a somewhat different direction and viewpoint.

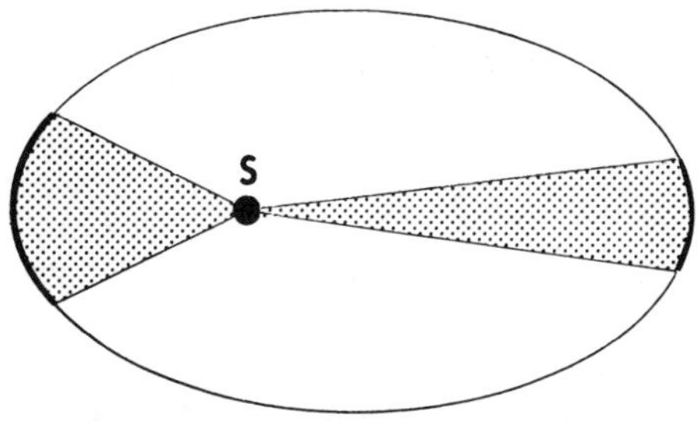

Fig. 1. An elliptical orbit for a planet around the sun. The shaded areas indicate equal areas swept out in equal times at different parts of the orbit. Clearly, the speed of the planet varies with its position in its orbit.

However, when in free fall, the local effects of the gravitational field are eliminated for the astronaut. Objects which are released fall together with him and hence remain in his vicinity unlike the situation on the ground. Therefore, the organs of the body respond as though the gravitational field were absent and this gives the sensation of weightlessness.

Gravitational Field. According to Newtonian theory, the sun exerts the gravitational force directly on the earth without an intervening medium for transmitting that force. The behavior of such forces is called "action at a distance." To overcome the conceptual difficulty of a force acting directly over large distances, one assumes that a *gravitational field* fills all space. The force acting on any mass is determined by the gravitational field in its neighborhood. Thus, at the point P a distance R from the center of the earth, the gravitational field has the magnitude

$$\mathscr{G}=\frac{GM_E}{R^2}$$

and magnitude of the force on a mass M at P is simply $F = M\mathscr{G}$. Note that the field is to exist at P even in the absence of the mass M.

It is sometimes convenient to introduce the gravitational potential which determines the field through its gradient. For a spherical earth, it is defined as

$$\phi=-\frac{GM_E}{R}, \qquad \mathscr{G}=-\text{grad } \phi$$

In general ϕ will satisfy Poisson's equation

$$\frac{\partial^2\phi}{\partial x^2}+\frac{\partial^2\phi}{\partial y^2}+\frac{\partial^2\phi}{\partial z^2}=4\pi\rho \tag{2}$$

ρ is the density of matter. The potential energy of a mass M, in the field is simply expressed in terms of ϕ,

$$V=M\phi$$

Although one can introduce the gravitational field, it is an auxiliary concept in Newtonian theory for the field has no independent dynamical behavior as is true of the electromagnetic field (e.g., electromagnetic waves). At any time, the Newtonian gravitational field is determined by the configuration of masses at that instant and does not depend on previous history or state of motion. Thus if the sun were to vanish, the gravitational force on the earth would immediately be removed. This property may be thought of in terms of an infinite velocity of propagation for the gravitational field. Letting the velocity of light become infinite in Maxwell's equations eliminates all independent dynamical behavior for the electromagnetic field. In that case there could be no radio or television. The special theory of relativity which is based on the velocity of light in vacuum being the maximum velocity for the transmission of energy, implies that Newton's theory requires modification.

Principle of Equivalence. The mass of a body may be measured either by weighing $W = Mg$ (*gravitational mass*) or by observing its motion under a known applied force using Newton's second law of motion $F = MA$ (*inertial mass*). The equality of these two differently defined masses has been measured by R. H. Dicke to an accuracy of 1×10^{-11} improving an earlier measurement by Eötvös. It is this equality which distinguishes the gravitational force from all other forces in giving all bodies the same acceleration. The discussion of

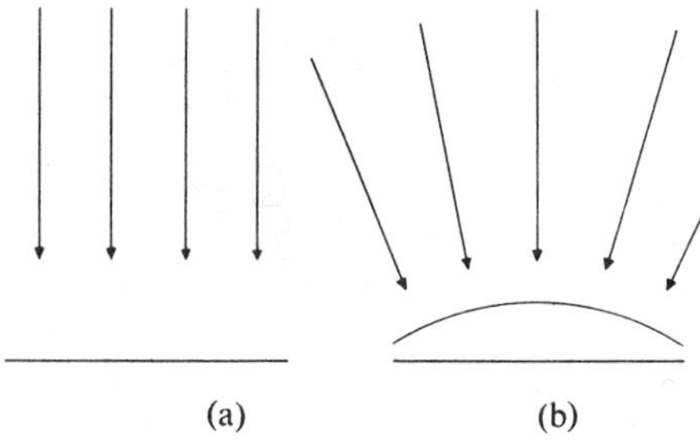

Fig. 2. (a) The paths of particles released in an acceleration field (the acceleration is up, the apparent force is down); (b) the paths of particles released in a gravitational field showing convergence toward the source.

weightlessness pointed out that local effects of the gravitational field are eliminated for an observer in free fall precisely because all bodies fall at the same rate. It follows that the gravitational field *measured* by an observer will depend on his state of motion. In a sense *there is an equivalence between a gravitational field down and an acceleration up for the observer.* However, the equivalence is not complete, for real gravitational fields converge on their sources so that two particles released at the same time will drift closer together as they fall. On the other hand, acceleration fields have no effect on the separation of particles moving on parallel paths (Fig. 2). In a curved space, initially parallel geodesics—the "straight lines"—do not maintain a constant separation (e.g., great circles on a sphere). Thus, the gravitational field may have its explanation in the geometry of a curved space-time.

Red Shift. According to the quantum theory, a photon of frequency ν has an energy $h\nu$ (h is Planck's constant), and by the relation $E = mc^2$, this quantum has a mass $m = h\nu/c^2$. To lift a mass m a height H requires expenditure of the energy mgH. Therefore, a photon emitted at the surface of the earth arrives at the height H with the energy

$$h\nu - (h\nu/c^2)gH = h\nu\left(1 - \frac{gH}{c^2}\right) = h\nu'$$

At the surface of the earth, the frequency shift amounts to

$$\frac{\Delta\nu}{\nu} = 1.1 \times 10^{-16} H \text{ (}H\text{ in meters)}$$

This shift was measured by Pound and Rebka using the Mössbauer effect in good agreement with the prediction. As time standards are determined by frequency, it follows that if the same photon were emitted at the height H, it would be measured to have the frequency ν, not ν'. Therefore, an observer at H must conclude that his clock is running faster than the same clock would run on the surface of the earth in the ratio $\Delta T/T = -\Delta\nu/\nu$ (Fig. 3).

Einstein's theory of gravitation. Albert Einstein assumed that gravitation is a physical effect produced by the curvature of a four-dimensional space-time. The generalization of Newton's gravitational potential is the metric tensor $g_{\mu\nu}$ in terms of which the four-dimensional distance, and hence the geometry of space-time, is determined:

$$ds^2 = \sum_{\mu,\nu=1}^{4} g_{\mu\nu}\, dx^\mu\, dx^\nu$$

The curvature of space-time is defined in terms of a four index tensor $R^\mu_{\nu\rho\sigma}$, the curvature tensor. The vanishing of the curvature tensor means that no real gravitational field is present. The field equations are ten linear combinations of the curvature components which are of the second order in the derivatives of the metric tensor and are a generalization of Poisson's equation [Equation (2)]. Symbolically these equations are written

$$G^{\mu\nu} = 8\pi\kappa T^{\mu\nu}$$

where $T^{\mu\nu}$ is a symmetric tensor which describes the distribution of matter and energy throughout space-time and $\kappa = G/c^2$. In a weak field static approximation, these equations contain Newton's theory of gravitation with the Newtonian gravitational potential. Given by $2\phi = 1 - g_{44}$.

The metric tensor outside a static spherically symmetric mass distribution is given by the Schwarzschild solution:

$$ds^2 = \left(1 - \frac{2\kappa m}{r}\right) dt^2 - \left(1 - \frac{2\kappa m}{r}\right)^{-1} dr^2 - r^2\, d\theta^2 - r^2 \sin^2\theta\, d\phi^2$$

This geometry exhibits the red shift described above and in addition shows three other effects:

1. The bending of a ray of light passing near the sun's edge by

$$\delta\theta = 1.75''$$

2. The precession of the perihelion of Mercury by

$$\delta\phi = 43''.03/\text{century}$$

3. The retardation of signals passing near the sun; for a radar pulse reflected from Mercury, this amounts to a maximum time delay

$$\Delta t = 1.6 \times 10^{-4} \text{ sec}$$

Observations and experiments to check these predictions are still in progress.

Since one can see stars near the sun's edge only during an eclipse, the optical data on the bending of light have been slow and difficult to obtain and such measurements have poor reliability—about 10–25%. A group under H. Hill set up equipment using photomultiplier tubes sensitive to a narrow spectral range so that the solar background can be filtered out. As a result, measurements at a fixed site can be made continuously as the sun moves into and out of a selected field of stars. Therefore, much improved accuracy is possible. Using radio frequency measurements, Shapiro observed the angular position of two sources, 3C279 and 3C273, which have an angular separation of about 10°. The latter source acts as the reference, as 3C279 is occulted by the sun each year on October 8. Results gave agreement with predicted value within 20%.

Shapiro also reevaluated the optical data with regard to the solar system and established new data, using radar ranging. In both cases, he found agreement with the predicted value for the perihelion precession of Mercury within 3%. By combining the data, the error can be reduced to 1%. As another test of Einstein's theory of general relativity, Shapiro suggested measuring the retardation of radar echo signals from Mercury when the planet moves into a position of superior conjunction. The gravitational field of the sun, as represented by the Schwartzschild solution, not only produces a bending of the ray, but also affects the time of flight of the signal. Therefore, the time delay between the transmission of a radar pulse to Mercury and the reception of the reflected signal will depend not only on the relative positions of the earth and Mercury in their respective orbits, but also on whether the radar signals pass near the sun. See Fig. 4. Measurements have given agreement within 5%.

Gravitational Collapse. The gravitational force between any two masses is attractive. Therefore, given a quantity of matter, under action of gravity alone it will become as compact as possible. In the planets,

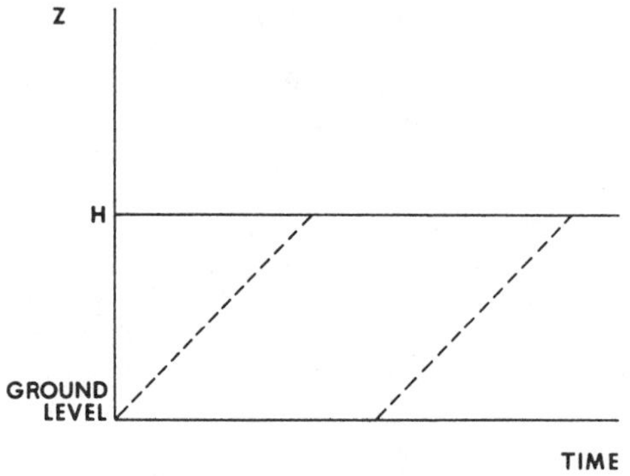

Fig. 3. Photons are emitted on the ground and are received at the height H. Between the two dotted lines representing the beginning and end of a pulse, the same number of oscillations, n, are received at H as are emitted at the ground level. Because of the red shift, the interval t' between oscillations at H is greater than the interval t between oscillations on the ground. Therefore, the time measured at H for the reception of the n oscillations is greater than the time required for their emission on the ground: $nt' > nt$. This result implies that clocks run faster at H than on the ground.

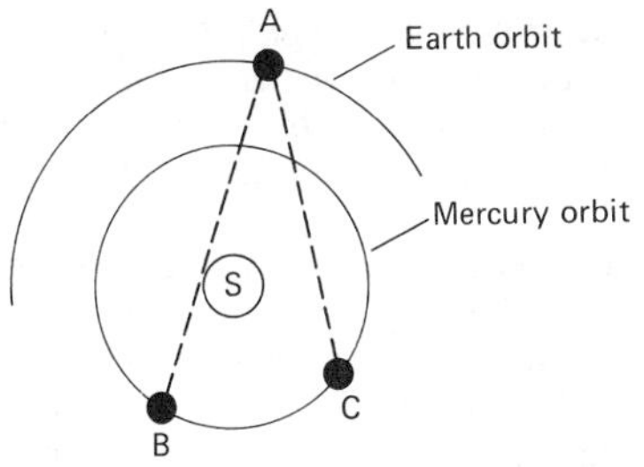

Fig. 4. Conditions for testing Einstein's theory of general relativity. S = sun.

the compaction process is stopped by the electrical forces which act between atoms and molecules in close range. The pressure in the sun, however, is much too great to be supported by such solid body forces. The tremendous pressure is balanced primarily by the counterpressure of electromagnetic radiation which is produced by the nuclear processes at the sun's center. Stars in which the nuclear processes have ended undergo a further contraction which is stopped by the pressure of free electrons at the densities associated with white dwarfs. This pressure, which occurs because electrons obey the Pauli exclusion principle, is capable of supporting up to 1.4 solar masses within a volume of 10^{-4} to 10^{-8} of the solar volume. Objects which are more massive continue the crush. Neutrons become the most stable particles in the interior and the contraction is stopped by repulsive nuclear forces when a neutron occupies only about 10^{-39} cubic centimeter, the nuclear volume. If the resulting neutron star is one solar mass, its radius is just 10 kilometers and its volume 10^{-15} the sun's volume. Objects with more than about 1.2 solar masses cannot be stable as neutron stars. They continue to contract. Beyond this point, the situation is confused by the abundance of exotic elementary particles, but there is no theoretical evidence that the contraction can be stopped.

One might have hoped that Einstein's theory of gravitation would contain a short-range repulsion which would stop this endless contraction. However, the opposite is the case. First of all, all forms of energy contribute to the attractive mass in general relativity, and secondly, the fact that matter determines the geometry indicates that there should be peculiarities in the space when the body is highly collapsed. There are several general theorems, particularly by Penrose and Hawkins, whose general conclusion seems to be that a long as the energy density remains everywhere positive, collapse is inevitable. This does not mean that collapse actually occurs in nature. As a very massive star proceeds through the various stages indicated in the foregoing paragraph, it may become unstable and throw off enough mass through an explosive process, such as a supernova, that it may settle down at a planetary size, or as a white dwarf, or as a neutron star. There is evidence for the existence of these objects. A pulsar is considered to be a rapidly rotating neutron star. And, thus it is unlikely that everything continues to collapse. But there are many very massive stars and, in the absence of more information, it is not unreasonable to rule out the possibility that some indeed go through an indefinite collapse or that some may have already done so.

What physical effects result from the collapse? It was pointed out (Eq. 4) that at the Schwarzschild radius, the escape velocity from a point mass is the velocity of light. Thus, no signal can escape from a body which has collapsed below R_s. This result can be deduced from the Schwarzschild solution of the Einstein equations. As a result, knowledge of events is limited at the Schwarzschild radius; the surface $r = R_s$ is an *absolute event horizon.* Because no light or other signal can be received from a source which has collapsed below its Schwarzschild radius, it has been called a *black hole.*

A neutron star of one solar mass has a radius of 10 kilometers, while $R_s = 3$ kilometers; a neutron star of 10 solar masses will have a radius of 30 kilometers. Thus, there is observational evidence for the existence of objects which are very nearly black holes. See also **Black Hole;** and **Cosmology.**

Gravitational Waves. Einstein's field equations require that the gravitational field have a finite velocity of propagation—the same as that for light. Therefore, the gravitational field has independent dynamical degrees of freedom which permit gravitational waves to exist in two states of polarization. These states are wholly transverse, i.e., the waves act on matter only in planes which are orthogonal to the direction of propagation. In passing through matter, one state produces oscillations such that there is a compression followed by elongation along one axis and a corresponding elongation followed by compression along the perpendicular axis. See Fig. 5. For a periodic wave this process repeats at the frequency of the wave. The other state of polarization has the same effect along axes rotated by 45°. This character for the modes is caused by the tensor nature of the potentials g_{uv} which limits the lowest order of gravitational waves to quadrupole radiation. A crude estimate of the energy radiated by the earth–sun system per year amounts to 10^{16} ergs (about 10^6 k Wh). Radiating at this rate, the earth has lost about 10^{-15} of its available mechanical energy since its formation possibly some 5×10^9 years ago. Presumably there are stronger sources of gravitational waves available in the universe.

Experiments to detect gravitation radiation were begun in 1958 by Weber. For a detector, Weber used an aluminum cylinder which is suspended in the earth's gravitational field. An incident gravitational wave sets up transverse oscillations in the cylinder. These oscillations are transformed into electrical signals by piezoelectric crystals which are bonded to the surface of the cylinder. The apparatus is acoustically insulated from outside interferences.

The initial detection program used principally two identical cylinders, 153 centimeters long and 66 centimeters in diameter. These were located at the University of Maryland and the Argonne National Laboratory, respectively, some 1000 kilometers apart. The electronic recording system was narrowly tuned to 1660 Hz, which has an acoustic half-wavelength of 153 centimeters in aluminum. Thermal oscillations are randomly generated and one would not expect correlation between the outputs of two detectors 100 kilometers apart. Therefore, Weber looked for coincidences in the output signals of the two detectors. The observation technique was to record each signal separately at its own location and, at the same time, to transmit the Argonne signal to Maryland where it could be compared directly with the Maryland signal. Coincidences of a certain pulse height were then marked. The coincidence rate due to random fluctuations was correlated with the observed rate by careful statistical analysis. Weber concluded that there is a "significant coincidence rate of about one every two days."

Both cylinders were lined up in an east-west direction. Therefore, some directional information was available by studying the change in coincidence rate as the earth rotated on its axis. The information was not very precise, as the two-cylinder array was a broad-beam detector and there was twelve-hour symmetry in orientation because the earth does not absorb much energy. Nonetheless, there was a definite indication that a source of radiation lies in the direction of the center of the galaxy.

Gravitational Wave Antennas. As described further in the entry on **Quantum Mechanics,** the principles of quantum mechanics were introduced in the 1920s and, among other guidelines, states that when the property of an electron or other microparticle is measured, the state of that particle will inevitably be disturbed—and disturbed in some unpredictable fashion. It follows that the more accurate the measurement, the greater and more unpredictable will be the disturbance (Heisenberg uncertainty principle). These ground rules contribute to the complexity of designing antennas and detectors used in gravity-wave research.

Typically, gravity-wave detectors are made of aluminum, sapphire, or silicon bars that weigh as little as 10 kilograms and up to several hundred kilograms. With an instrumental ability of measuring end-to-end vibrations with the accuracy required (10^{-19} centimeter), the

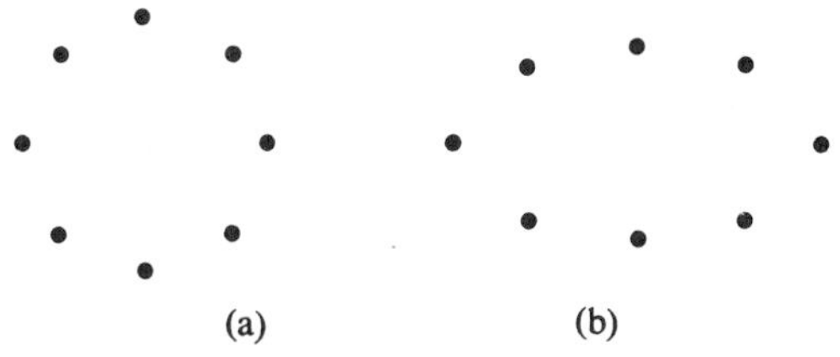

Fig. 5. (a) A circular arrangement of dust particles before a gravitational wave arrives; (b) the same particles after a passage of a wave consisting of one mode. The second mode would produce the same effect, rotated at 45°.

device will behave quantum mechanically. Scientists in the U.S.S.R. and California have proposed a quantum nondemolition (QND) method to circumvent the effects of the Heisenberg uncertainty principle. It has been proposed that instead of measuring the position of a 10-ton bar (visualized for future experiments), the momentum of the bar would be measured. The bar would purposely be set in motion so that the effects of a passing gravity wave on the bar's momentum could be detected. As pointed out by Braginsky (1980), actual laboratory work on QND measurement schemes is only now beginning to get under way, and the levels of sensitivity are so great that we cannot hope for any laboratory results until several years from now. Nevertheless, it is reasonable to expect QND measurements to be a routine part of gravity-wave technology in the late 1980s.

Neutron Interferometer. Prior to the mid-1970s, little tangible experimentation occurred that would permit the establishment of a good relationship between quantum mechanics and the general theory of relativity (the modern theory of gravitation). For one thing, there is a vast gap of scale between quantum theory and the general theory of relativity, with quantum mechanics concerned with particles at the atomic scale of 10^{-8} centimeter, whereas the effects of gravity appear significant only in terms of a stellar or cosmic scale. Among ways to narrow this gap and to learn more about gravity is the neutron interferometer. As early as 1964, Bonse and Hart (Cornell University) constructed an x-ray interferometer, but it was not felt at that time that an instrument of this type would work in the case of neutron beams. Thus, the first neutron interferometer was not constructed until 1974 (Bonse, Rauch, Triemer—Austrian Nuclear Institute). The instrument was constructed essentially from a single, perfect crystal of silicon. The crystal about 10 centimeters long, was free of dislocations and other defects in its atomic structure. Since then, other similar instruments have been built, as by Shull (Massachusetts Institute of Technology). See Fig. 6. This one-piece instrument is cut from a cylindrical crystal approximately 8 centimeters long and features three ears that are about 0.5 centimeter thick and somewhat less than 3 centimeters apart. Because of the perfection of the crystal, the atoms of the three ears all line up exactly. Thus, the coherence of the neutron beam entering the instrument is not disturbed.

Scattering of the neutron beams does not occur from the surface of the ears; rather, they are scattered by the planes of atoms in the crystal. The behavior of neutron beams in the interferometer is somewhat complex and is well explained by Greenberger/Overhauser (1980).

In assessing the use of the neutron interferometer as a means of detecting gravitational effects, it is important to note that although neutron waves have much in common with light waves and water waves (reinforcement and cancellation when exactly in or out of phase, respectively), there are some basic differences. The neutron possesses both mass and a magnetic moment; the photon does not. Thus, a neutron is affected by a magnetic field and can be caused to rotate, whereas such a field has no effect on a photon. It follows that the characteristic of the neutron wave is much that it will be affected much more strongly by gravity than will a light wave, the measurable gravity–photon interactions of which can be observed only on a cosmic scale. The shorter neutron wave length (10^{-8} centimeter) compared with the longer light wave (10^{-5} centimeter) permits resolution of effects on a smaller scale.

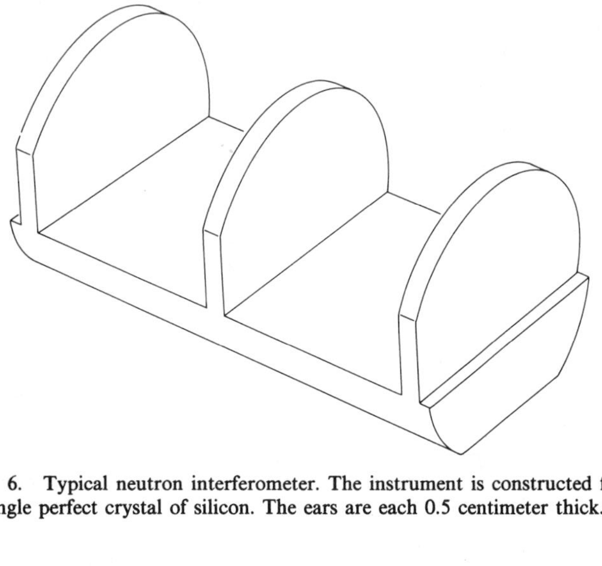

Fig. 6. Typical neutron interferometer. The instrument is constructed from a single perfect crystal of silicon. The ears are each 0.5 centimeter thick.

In 1975, Coella, Overhauser, and Werner conducted an experiment (termed COW for the initials of the investigators) to measure the effect of the earth's gravity on the phase of the neutron wave. As pointed out by Greenberger/Overhauser (1980), "it was already known experimentally that the neutron falls in the earth's gravitation field as any other massive particle does. That fall, however, is strictly Galilean, or classical. The question is whether one can observe an effect of gravity on the wave nature of the neutron. The way to do this is through an interference effect, for which the neutron interferometer is ideally suited (provided the effect is large enough to detect)." It is interesting to note that it has been estimated that the force of gravity at the earth's surface is derived from some 10^{52} protons and neutrons of which the earth is comprised. Also, it has been established that the electric repulsion between two protons is 10^{36} times greater than their gravitational attraction. And, two protons at an atomic distance of 10^{-8} from each other have an electric force on each other that is some 10^{16} times greater than the gravitational force exerted on them by the entire earth. Thus, the investigators had the task of proving that such a weak gravitational force could produce measurable effects in the neutron interferometer.

The neutron wave, as previously mentioned, maintains its coherency over the full 10-centimeter length of the instrument crystal. During this distance, the wave oscillates 10^9 times. It was possible with the instrument to observe 100 additional oscillations—these extra oscillations attributed to gravity effects. As the scientists pointed out, "As weak as gravity is, it has a measurable effect on the wave function because the neutron wave is coherent on a macroscopic scale."

The experimental data obtained agreed precisely with the amount predicted by the Schrödinger equation. In their explanation of the experiment, the scientists describe why it is believed that the measurement is due to gravitational force and is not a manifestation of the time difference or red shift effect described by Einstein in 1916, i.e., in the case of this experiment, the difference between the time on a clock moving along with one beam and the time on a clock moving along with the other beam. Since the COW experiment, a number of other sophisticated experiments have been conducted with the neutron interferometer. Their complexity is beyond the scope of this encyclopedia, but details can be found in some of the references listed.

Gravity Lens. It is currently believed that the comparatively weak forces of gravity waves require a cosmic scale to observe their effects. What was believed to be twin quasars were photographed in the early 1950s, using the 1.2-meter Schmidt telescope on Palomar Mountain (California). In these early views, the image of the bodies appeared fused because of the motion of the earth's atmosphere. Scientists have observed that had the telescope been above the earth's atmosphere, it could have resolved objects 60 times closer together than the twins. But, until March of 1979, these bodies were considered twins. Subsequent research involving the 2.1-meter telescope at the Kitt Peak National Observatory and the 2.3 meter telescope of the University of Arizona yielded spectral information that was strikingly similar for both bodies. A red shift 1.4 was measured for each body and this, coupled with the similarity of spectral data puzzled the astronomers. The spectral and velocity measurements were further confirmed, using a multiple-mirror telescope of the Smithsonian Astrophysical Observatory and the University of Arizona. Later, data were gathered by the National Radio Astronomy Observator's Very Large Array (near Socorro, New Mexico). A computer-generated display on a cathode-ray tube of one image of a quasar whose radiation has been deflected to form two images by a gravitational lens is shown in Fig. 7. It is believed that an elliptical galaxy is acting as a gravitational lens. As pointed out by Chaffee (1980), "Eight months of theoretical work and intensive investigation with the largest optical and radio telescopes has demonstrated that these "twin" quasars are not two distinct objects at all. Rather, they are a single object whose light has been split into two images by the gravitational field of a galaxy between the quasar and our galaxy; a kind of optical illusion on a cosmic scale." Several technical objections were raised concerning

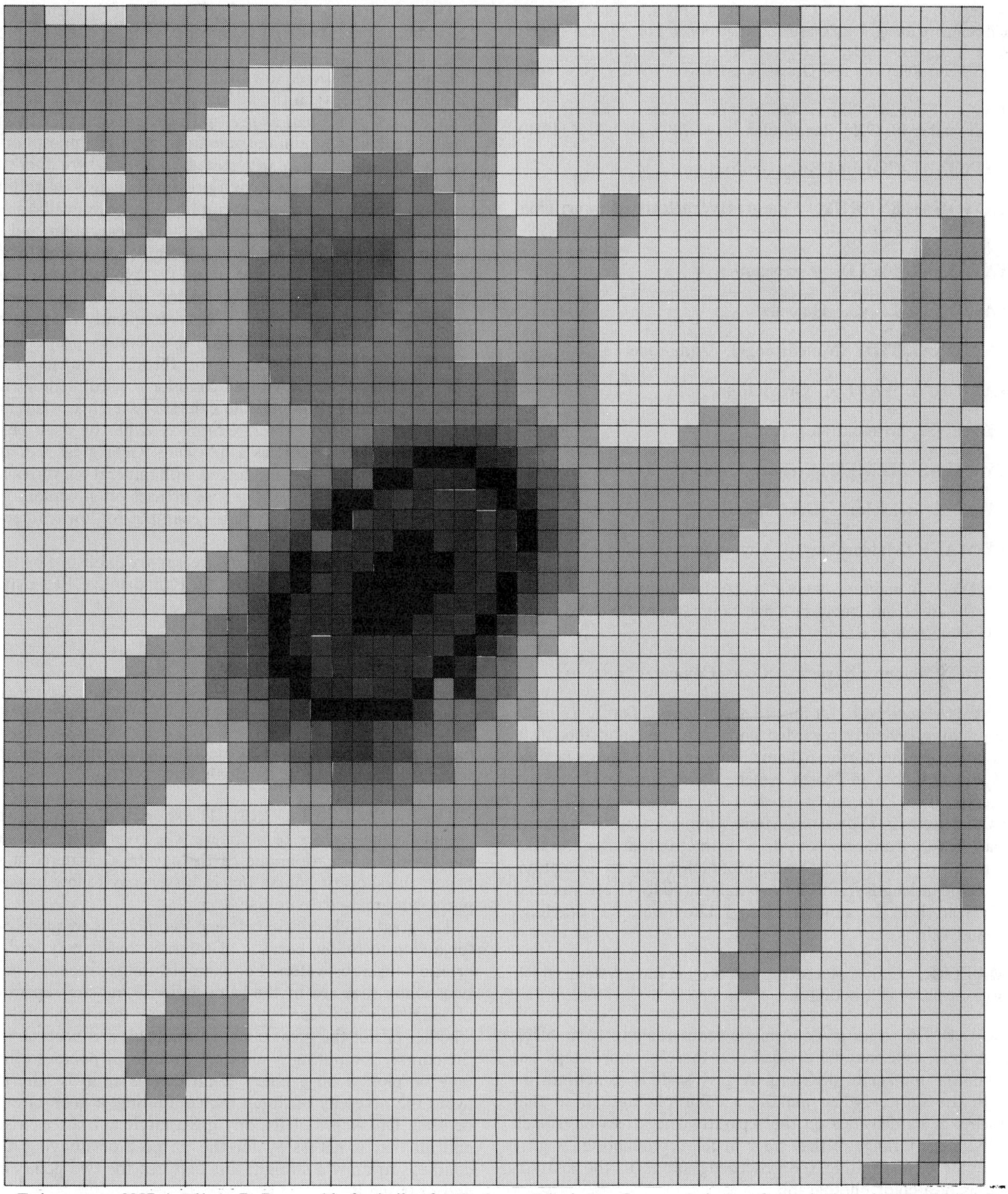

Fig. 7. Twin quasars 0957 + 561 A, B. Reasonable facsimile of cathode-ray tube image. Increased shades of gray indicate increased intensity of radio waves (wavelength = 6 centimeters).

the conclusion that a gravitational lens is involved, most of which have since been resolved.

References

Bergman, P. G.: "The Riddle of Gravitation," Chas. Scribner and Sons, New York, 1968.

Braginsky, V. B., Vorontsov, Y. I., and K. S. Thorne: "Quantum Nondemolition Measurements," *Science*, **209**, 547–557 (1980).

Chaffee, F. H., Jr.: "The Discovery of a Gravitational Lens," *Sci. Amer.*, **243**, 5, 70–78 (1980).

Chagas, C.: "Einstein," Einstein Session of the Pontifical Academy, Vatican City (November 10, 1979). Reprinted in *Science*, **207**, 1159–1161 (1980).

Dirac, P. A. M.: "Einstein," Einstein Session of the Pontifical Academy, Vatican City (November 10, 1979). Reprinted in *Science*, **207**, 1161–1162 (1980).

Frankel, T.: "Gravitational Curvature," Freeman, San Francisco, 1979.

French, A. P.: "Einstein: A Centenary Volume," Harvard Univ. Press, Cambridge, Massachusetts, 1979.

Greenberger, D. M. and A. W. Overhauser: "The Role of Gravity in Quantum Theory," *Sci. Amer.*, **242**, 5, 66–76 (1980).

Hawking, S. W., and W. Israel (editors): "General Relativity," Cambridge Univ. Press, New York, 1979.

Held, A., Editor: "General Relativity and Gravitation," Plenum, New York, 1980.

Lebovitz, N. R., Reid, W. H., and P. O. Vandervoort (editors): "Theoretical Principles in Astrophysics and Relativity," Univ. of Chicago Press, Chicago, Illinois, 1978.

Levy, M., and S. Deser (editors): "Recent Developments in Gravitation," Plenum, New York, 1979.

Pauli, W.: "Theory of Relativity," Pergamon, London (1958).

Pinheiro, R.: "Gravity Wave Astronomy," *Astronomy*, **7**, 6, 6–14 (1979).

Pope John Paul II: "Einstein," Einstein Session of the Pontifical Academy, Vatican City (November 10, 1979). Reprinted in *Science*, **207**, 1165–1167 (1980).

Sachs, R. K.: "General Relativity and Cosmology," Academic, New York, 1971.

Sciama, D.: "Modern Cosmology," Cambridge Univ. Press, Cambridge, 1971.
Staff: "Solid Support for Gravity's Mirage," *Sci. News*, **116**, 19, 324 (1979).
Staff: "Gravity's Repulsive Side," *Sci. News*, **117**, 10, 148 (1980).
Weber, J.: "General Relativity and Gravitational Waves," Wiley, New York, 1961.
Weisskopf, V. F.: "Einstein," Einstein Session of the Pontifical Academy, Vatican City (November 10, 1979). Reprinted in *Science*, **207**, 1163–1164 (1980).

GRAVITATIONAL COLLAPSE. **Gravitation.**

GRAVITATIONAL ENERGY. **Energy; Hydroelectric Power; Tidal Power.**

GRAVITATIONAL FIELD. **Astronautics.**

GRAVITATIONAL LENS. **Quasars.**

GRAVITATIONAL TIDE (Meteorology). **Atmosphere (Earth).**

GRAVITATIONAL WAVES. **Gravitation.**

GRAVITY LENS. **Gravitation.**

GRAVITY MEASUREMENTS (Earth). **Earth; Torsion Balance.**

GRAVITY (Plant Behavior). **Geotropism.**

GRAVITY WAVE (Meteorology). **Atmosphere (Earth).**

GRAY BODY. A radiator whose spectral emissivity is constant throughout the spectrum, being in a constant ratio to that of a black body at the same temperature.

GRAY CODE (Computer System). **Code (Computer System).**

GRAYLING (*Osteichthyes*). Of the order *Isospondyli*, family *Thymallidae*, the grayling is highly regarded both as a sport and food fish. Graylings occur in the northern hemisphere and are found in cold lakes and streams, both in North America and Eurasia. There are several species. The European species is *Thymallus thymallus* and the American species is *T. arcticus*. At one time, graylings were found in Michigan, but they are now considered extinct in that area. Availability is limited in the United States, Montana being an exception. Graylings range in length from 12 to 16 inches (30 to 41 centimeters) and weigh from 1 to 2 pounds (0.5 to 1 kilogram). All graylings are freshwater fish.

GRAY MATTER. Gray colored part of the nervous system which is composed of nerve cell bodies.

GRAY SCALE. A series of achromatic tones ranging from black to white. A gray scale may be divided into three or more steps but 10 is a common number of divisions. A gray scale is sometimes included with the subject when making a color photograph so that measurements of its densities on the separation negatives or tripack will give the density range of that stage in the reproduction. A gray scale is helpful in controlling the processing stages in the analysis and synthesis of a color photograph.

GRAYWACKE (or Grauwacke). This term is of British origin and is not used extensively outside of western Europe. As originally defined graywacke designates hard, dark-colored, coarse sandstones and grits having an argillaceous matrix or cement and occurring among the lower Paleozoic formations of Wales, England. Many typical graywackes are similar to basic arkoses, the dark color being due to a preponderance of the femic minerals and plagioclase feldspar.

GREASE. A lubricating agent of higher viscosity than oils, consisting originally of a calcium or sodium soap jelly emulsified with mineral oil. Greases are employed where heavy pressures exist, where oil drip from the bearings is undesirable, and where the motion of the contacting surfaces is discontinuous so that it is difficult to maintain a separating film in the bearing. Grease-lubricated bearings have greater frictional characteristics at the beginning of operation, causing a temperature rise which tends to melt the grease and give the effect of an oil-lubricated bearing.

The principal categories of greases are: (1) calcium soap greases; (2) sodium soap greases; (3) complex soap greases—combinations of soaps and fatty acids used to impart high-temperature properties and moisture resistance. A low-molecular-weight soap can be used as a binding agent between the oil and soap in place of water. (4) Lithium soap greases—excellent as multipurpose greases; (5) extreme-pressure greases, usually containing some form of sulfur, phosphorus, or other reactive agent—particularly suited to uses where there are sudden shock loads or continuous high pressures, as in steel rollingmill bearings; (6) nonsoap greases—exemplified by organically modified clays which hold the lubricating oil both by absorption and adsorption. Such greases are often used in high-temperature applications because they actually have no melting point; (7) Asphalt-base greases—blends of asphaltic materials with lubricating oil, enabling a wide range of consistencies; (8) filler-type greases—frequently calcium-base greases that contain solid materials having unctuous properties. The filler essentially serves as a cushion for absorbing impacts. Calcium and sodium base greases are most commonly used; sodium base greases have higher melting point than calcium base greases but are not resistant to the action of water. Graphite, either by itself or mixed with grease, is also employed as a lubricant. Gear greases consist of rosin oil, thickened with lime and mixed with mineral oil, with some percentage of water. The special-purpose greases often contain glycerol and sorbitan esters. They are used, for example, for low temperature conditions. See also **Lubricant.**

Standard methods for testing greases are published by the American Society for Testing and Materials, Philadelphia, Pennsylvania.

GREAT BLUE SHARK. **Sharks.**

GREAT CATS. **Cats.**

GREAT-CIRCLE CHART. Among navigators, the gnomonic projection is commonly known as a great-circle chart because of the fact that, on this type of projection, great circles are projected as straight lines. See also **Gnomonic Projection.**

GREAT-CIRCLE COURSE. The shortest distance between any two points on the surface of a sphere is a great circle. For all practical purposes of navigation, the earth may be considered a sphere, and hence, the shortest course that a vessel may follow between any two ports is a great-circle course.

The great-circle course between two ports is frequently impractical for a ship to follow because of the fact that it may lead across land or into dangerous waters. For example, the great-circle course between two points that are in the same latitude but are separated by 180° of longitude will lead across a pole of the earth. Before deciding whether or not the great circle is practical, it is necessary to compute the course, computing a sufficient number of points so that the track may be plotted on a chart. Such computation is laborious, and to avoid the necessity of doing the computing, a great-circle chart may be used. On such a chart, any great circle appears as a straight line, and all that is necessary for the purpose of studying a great-circle course is to draw a straight line between the two points on the chart and examine it.

Even when the great-circle course does not lead the ship into danger, it is a very difficult course to follow because it makes a different angle with each successive meridian and requires the helmsman to continually change his course. To avoid this difficulty, as well as to avoid dangers, and yet to still approximate as closely as practical the shortest distance between the ports, the composite course is the type almost universally followed by vessels and aircraft on long-distance flights.

See also **Course; Gnomonic Projection;** and **Navigation.**

GREAT NEBULA IN ORION. **Nebula.**

GREAT RED SPOT. **Jupiter.**

GREAT WHITE SHARK. **Sharks.**

GREBE (*Aves, Podicepediformes, Podicipedidae*). This order of birds has a long geological history. They evolved in the Northern Hemisphere but now inhabit all continents except the Antarctic. They are from thrush to duck size; the length is 20–78 centimeters (8–31 inches), and the weight is 120–1500 grams (4–53 ounces). There are 17 to 21 cervical vertebrae. Some thoracic vertebrae are fused. The legs are positioned far back on the trunk. The tarsus is laterally compressed with a sharp front edge; on the back a double row of horny sawteeth is found, which is not known in any other group of birds. The lobed membranes along one side of the toes is 1 centimeter wide (0.4 inches). The claw of the mid-toe resembles a fingernail and is somewhat comblike at the tip; possibly this is used to clean the plumage. Tail feathers are small and soft (unlike most birds), and so these birds appear tailess. See accompanying illustration.

Grebe: Pied-billed grebe (*Podilymbus podiceps*).

There are four genera with nine species: (a) Grebes (*Podiceps*) with six species; (b) Pied-Billed Grebes (*Podilymbus*) with two species; (c) the Running Grebes have only one species, the Western Grebe (*Aechmophorus occidentalis*); (d) Titicaca Grebes have only one species (*Centropelma micropterum*).

Grebes move on land only when they have no other choice, such as when building nests, in order to incubate, or to get from one open water hole to another in severe frost. Although they generally escape pursuers by diving, they are not bad fliers. During migration the species which is capable of flight can cover great distances.

They are all excellent divers, although they dive neither as deeply nor for as long as the loons, generally for less than half a minute and less then 7 meters (23 feet) deep. They live in still, fresh water and are seen at sea only outside the breeding season. The feltlike, thick, silky-soft contour feathers protect the underside against the water.

The nests are built of rotting plants, they float, and they are anchored to reeds of branches. A clutch consists of at least three eggs; and incubating birds always cover the eggs when leaving the nest. The eggs are at first snow-white and covered with chalky calcium carbonate, but soon they become chocolate brown on the wet plants on which they lie. The downy young are generally colorfully marked and striped, often producing a clownlike effect; right after hatching they move under the wings into the furlike back plumage of whichever parent happens to be on the nest. Thus protected, they swim and dive with their parents weeks before they can dive themselves. In the breeding season of the second year of their lives, they usually resemble their parents. See also **Podicipediformes.**

GREEN FLASH (or Blue Flash; Blue-green Flame; Green Segment). A brilliant green coloration of the upper limb of the sun, occasionally observed just as the sun's apparent disk is about to sink below a distant horizon.

GREEN FORMULA. Green Function.

GREEN FUNCTION. The name of George Green (1793–1841), an English mathematician, is attached to several different mathematical results and not always consistently by different writers. The relation called Green's theorem, for example, may be called Green's equation by another. It has seemed useful to collect all of these results in one item. The names given are chosen in accordance with what seems to be the most prevalent usage, but they are uniquely determined only by the accompanying equations.

1. Green function. A symmetric kernel $G(x, z)$ used to convert a Sturm-Liouville equation and its boundary conditions into an integral equation. It is defined to have the properties: (a) continuity over the range $a < x < b$ and with continuous derivatives of orders up to $(n-2)$, where n is the order of the differential equation; (b) its derivative of order $(n-1)$ is discontinuous at a point z within the range (a, b); (c) it satisfies the differential equation everywhere except at $x = z$.

2. Green formula. In the general theory of the nth order linear differential operator, the linear differential operator, L and its adjoint, $\bar{L}$ are of interest. See **Differential Equation (Integral Solution of).** Then the homogeneous equation $L(u) = 0$ is adjoint to $\bar{L}(v) = 0$ and Green's formula is

$$\int_a^b [vL(u) - uL(v)]\, dx = [P(u, v)]_a^b$$

where the left-hand side is the Lagrange identity. The right-hand side is a bilinear form in the $2n$ quantities $u(a), u'(a), \ldots, u^{(n-1)}(a); u(b), u'(b), \ldots, u^{(n-1)}(b); v(a), \ldots, v^{(n-1)}(a); v(b), \ldots, v^{(n-1)}(b)$. Its determinant does not vanish and $P(u, v)$ is called the bilinear concomitant.

3. Green theorem. In vector analysis, there are several relations between single and multiple integrals. If u, v are scalar functions, S indicates a double and τ a triple integral, the Gauss theorem in vector form is

$$\int_\phi \nabla u \cdot \nabla v\, d\tau + \int_\phi u \nabla^2 v\, d\tau = \int_S u \nabla v \cdot d\mathbf{S}$$

On exchanging u and v and subtracting the result from this equation, the Green theorem results

$$\int_\phi (u \nabla^2 v - v \nabla^2 u)\, d\tau = \int_S (u \nabla v - v \nabla u) \cdot d\mathbf{S}$$

These relations, which correspond to integration by parts in scalar calculus, are also known as Gauss theorems for the divergence theorem.

See also **Divergence (Mathematics);** and terms listed under **Mathematics.**

"GREENHOUSE" EFFECT. Climate; Venus.

GREENHOUSE HEATING (Geothermal). Geothermal Energy.

GREENLAND HALIBUT. Flatfishes.

GREEN MANURE. Fertilizer.

GREENOCKITE. The mineral greenockite is cadmium CdS, sulfide and is used as an ore of that metal. It is found rarely in hexagonal crystals, sometimes as earthy coatings on other minerals. Its hardness is 3–3.5; specific gravity, 4.9–5.0; luster, adamantine to earthy; color, yellow to yellowish-orange; subtransparent. It is found in Scotland, Bohemia, and France; also, in the United States, at Franklin Furnace. New Jersey; and Marion County, Arkansas, where it occurs as a yellow coloring matter in smithsonite; and in Mono County, California. It was named for Lord Greenock.

GREEN PIGMENTS. Chlorophylls.

GREEN REVOLUTION. A popular term used mainly in the 1965–1975 period to describe the results of technology transfer to the growing of certain crops in some of the developing countries, such as India, Mexico, Pakistan, and the Phillipines, this new technology increasing yields beyond the expectations of many experts. However, enthusiasm for the green revolution has been tempered somewhat in recent years. Generally credited with these productivity improvements is the work done by Borlaug and his associates on wheat genetics at the International Maize and Wheat Improvement Center (CIMMYT) in Mexico. Originally sponsored by the Rockefeller Foundation, the Center developed HYVs (high-yielding varieties) of wheat. Some of the current semidwarf HYVs, of course, are the offspring of varieties developed from similar ancestors in other breeding programs. The relatively short and stiff stalk of the semidwarfs means that they respond to improved cultural practiaces through increased yields rather than through increased plant growth, which would also result in lodging (falling over of the plant). The semidwarf varieties in use as of the early 1980s, while considered by some to be revolutionary in their impact, are the product of a long developmental process. Semidwarf wheats were noticed in Japan in the 1800s.

In 1946, S. C. Salmon, a U.S. Department of Agriculture scientist acting as agricultural advisor to the occupation army in Japan, noticed *Norin 10* growing at the Morioka Branch Research Station in northern Honshu. The stems were short, but produced many full-sized heads. Salmon brought 16 varieties of this plant back to the United States. They were grown in a detention nursery for a year and then made available to breeders in seven locations. Although *Norin 10* was not satisfactory for direct use in the United States, it was useful for breeding. O. A. Vogel, a U.S. Department of Agriculture scientist stationed at Washington State University, was the first to recognize its worth and to use it in a breeding program as early as 1949.

In the interim, word about the short-strawed germ plasm had reached Borlaug in Mexico. His breeding efforts had run into a yield plateau because of lodging under high levels of nitrogen fertilization. Introduction of the *Norin 10* genes led to the development of a number of Mexican dwarf and semidwarf bread varieties of wheat. International diffusion of these varieties began very quickly at the experimental level and India and Pakistan were the first countries to be substantially involved.

The first Mexican wheats arrived in India in 1962 by way of the international nursery system. They became of immediate interest to M. S. Swaminathan of the Indian Agricultural Research Institute (IARI) in the spring of 1963. Borlaug, at the request of IARI, toured wheat areas in India and, upon his return to Mexico, he sent 100 kilograms of each of four varieties and small samples of over 600 other selections. The material was grown and studied at seven locations during the 1963–1964 season, as a part of the All-India Coordinated Wheat Trials. In 1965, two varieties, *Lerma Rojo* and *Sonora 64,* were released for general cultivation.

In another undertaking, in the spring of 1962, Borlaug gave some of the improved seeds to two trainees from Pakistan. The seeds were subsequently planted at the Agricultural Research Institute near Lyallpur. Borlaug visited Lyallpur in the spring of 1963 and later sent 203 kilograms of experimental Mexican seed to Pakistan. In the spring of 1964, Borlaug again visited Pakistan and soon secured government and foundation support for the varieties. Pakistan purchased several hundred tons of Mexican seed for planting during the 1965–1966 and 1967–1968 seasons.

The Mexican varieties proved remarkably adapted to India and Pakistan—for several reasons: (1) They had been bred in Mexico with alternate generations in different climatic and daylength regimes, primarily in order to get two generations each year. A valuable side-effect of this system was to establish a good degree of insensitivity to photoperiod. (2) Selection for disease resistance had also been practiced and the stocks introduced were found to show a remarkable level of resistance under the conditions in India and Pakistan. (3) The original stocks incorporated diversity. They had not been bred to pure line standards and there remained in them a reservoir of genetic potential that Indian wheat breeders were quick to exploit.

By the mid-to-latter 1970s, the process of varietal change had gone through four stages in India. A large percentage of plantings in India, Pakistan, Afghanistan, and Nepal, among other less-developed countries, is planted to varieties of Mexican origin. Exceptional increases in yield were obtained.

A number of improved varieties of corn (maize) also came out of the outstanding research done in Mexico. Dr. Borlaug, one of the leading researchers in an international crop improvement program, received the Nobel Peace Prize in 1970 in recognition for his efforts.

Research into the genetics of rice with an objective of improving yields also occurred during the green revolution period. The activities of the International Rice Research Institute (IRRI), established in 1962 in Los Banos, Phillippines, and of the Indian Council of Agricultural Research, are particularly well known. Since the inception of those programs, numerous new varieties have been introduced. The rice situation was well summarized by K. L. Bachman of the Food and Agricutlure Organization (United Nations): "The most important factor influencing the adoption of the new strains was their potential to give much higher yields than traditional and improved local varieties. With the new varieties, it now paid to apply more fertilizers and pesticides and to devote more time and money to improved cultural practices; with the older varieties, it was risky to use even modest amounts of fertilizers owing to the danger of lodging, particularly in the wet season."

As evidence of the successes achieved during the green revolution, Pakistan's 1971 wheat production which was up 76% from its 1961–1965 average; Latin American corn (maize) production was up more than 50%; the Indian wheat crop of 1971 was almost double that of six years earlier; and Pakistan's 1974 rice crop set an all-time record.

Progress from improved varieties has, in some instances, reduced the nutritional level of people in farming areas because more emphasis was placed on wheat, rice, and corn (maize), and the production of food legumes was lowered. Recent years have indicated that much more than improved crop varieties is required to improve the food status of many of the underdeveloped countries. Better means of storage are needed as well; it has been estimated that 15% of all the rice and other cereal crops raised in the Orient is destroyed by rats, either in the field or in storage. Better means of distribution and processing are also required.

An excellent summary of the green revolution era of agriculture is contained in Report No. 95, "Development and Spread of High-yielding Varieties of Wheat and Rice in the Less Developed Countries," by D. G. Dalrymple, U.S. Department of Agriculture, Washington, D.C., 1976.

See also **Plant Breeding.**

GREENSHANK. Shorebirds and Gulls.

GREENSTICK FRACTURE. Bone.

GREENSTONE. Greenstone is an old field term for more or less altered basalts and dolerites, which, because of the development of chlorite, or perhaps hornblende or epidote, develop a characteristic green color. Many diabases and epidiorites have been called greenstones.

GREEN THEOREM. Gauss Theorem; Green Function.

GREENWICH MEAN TIME (GMT). Time.

GREGARINIDEA. An order of one-celled animals, parasitic in various invertebrates. See **Sporozoa.**

GREGARIOUSNESS. An association of animals of the same species which may be of benefit to the individual but is not essential. The incidental grouping of animals, as in the swarms of maggots in a dead body, is not an association of this type, but the grouping of caterpillars of certain moths, even though the group originates in a like manner by the deposition of eggs in a mass, must be regarded as a gregarious association because the maintenance of the group is due to the behavior of the individuals. They are free to scatter but do not.

Herds of grazing animals cooperate for the common defense and such animals as the killer whale and the wolves are able to attack large animals by hunting in groups, but in all such cases the individual is able to subsist without the assistance of his fellows.

GREGORIAN CALENDAR. Calendar; Time.

GREGORIAN TELESCOPE. A reflecting telescope with a concave secondary mirror, located extrafocally, that reflects the light through an opening in the primary mirror and forms a real image behind the primary mirror. See also **Telescope.**

GREGORY FORMULA. A formula for the numerical evaluation of an integral. It is obtained from the Newton formula for interpolation and may be written

$$\int_a^b f(x)\,dx = h\left[\frac{y_0}{2} + y_1 + y_2 + \cdots + y_{n-1} + \frac{y_n}{2}\right]$$
$$- \frac{h}{12}(\Delta y_{n-1} - \Delta y_0) - \frac{h}{24}(\Delta^2 y_{n-2} + \Delta^2 y_0)$$
$$- \frac{19h}{720}(\Delta^3 y_{n-3} - \Delta^3 y_0) - \frac{3h}{160}(\Delta^4 y_{n-4} + \Delta^4 y_0) - \cdots$$

where h is the interval between equally-spaced values of the independent variable x and the quantities $\Delta^m y_k$ are finite differences. Gregory's formula is equivalent to the trapezoidal rule, with correction terms in these differences.

GREISEN. An old German petrological term originally proposed by Werner for an igneous rock of granitic or aplitic texture composed principally of quartz, alkali feldspar, the fluorine-rich micas, and sometimes containing topaz. Greisens are pneumatolytically altered granites which are closely associated with the development of the tin ore mineral, cassiterite.

GREVY'S ZEBRA. Horses, Asses, and Zebras.

GRIBBLE (*Crustacea, Isopoda*). A small marine crustacean, *Limnoria lignorum*, which bores into submerged timbers. A source of serious damage to docks and piling.

GRIEBE AND SCHIEBE METHOD. A method for observing the piezoelectric characteristics of small crystals. In this procedure, a number of grains of the substance are inserted between two electrodes which are placed across the resonant circuit of an oscillator. The frequency of the oscillator is changed by changing the resonant circuit, and, if a resonance of one of the piezoelectric crystals occurs near the oscillator frequency, the frequency of the oscillator will briefly be controlled by the crystal resonance. As the resonant circuit is tuned further, the natural frequency of the oscillator becomes far enough away from the crystal resonance so that it cannot control the oscillator frequency, and a jump occurs from the crystal frequency to a different frequency controlled by the oscillator constants. This jump in frequency is accompanied by a change in the plate current, so that if a pair of headphones or a loudspeaker is attached to the plate-circuit of the oscillator, a click is heard.

GRIEF. Depression (Psychiatric).

GRIFFITH CRACK THEORY. A theory relating to the brittle fracture of solids. The observed strength of ordinary window glass is less than one-hundredth of its theoretical strength. This discrepancy led Griffith to postulate that the low observed strength was due to the presence of small cracks or flaws in the glass. Because the ends of cracks have the ability to act as stress raisers, Griffith assumed that the theoretical strength was obtained at the ends of a crack, even though the average stress was still far below the theoretical strength. Fracture, according to this concept, occurs when the stress at the ends of the cracks exceeds the theoretical stress. When this occurs, the crack expands catastrophically. With the aid of the additional assumption that the strain energy released by the spreading of a crack is converted into the energy of the surfaces created by the fracture, it is possible to derive the following equation

$$S_n = \left(\frac{\sigma E}{2c}\right)^{1/2}$$

where S_n is the average applied stress necessary to make a crack spread, σ is the specific surface energy, $2c$ is the crack length, and E is Young's modulus.

GRIFULVIN. Antibiotic.

GRIGNARD REACTIONS. Very important to the synthesis of numerous organic compounds, both in the laboratory and on a large scale in industry, is a two-step reaction involving the use of organomagnesium halides. These reactions were studied intensively by Victor Grignard during the early 1900s and for this work he was awarded the Nobel Prize in Chemistry in 1912. The reactions are referred to universally as Grignard reactions and the many magnesium compounds required by the reactions are known as Grignard reagents. Grignard's work steemed from a discovery by Barbier in 1899 that dimethylheptenol could be prepared by reacting methyl iodide, dimethylheptenone, and magnesium in ethyl ether. In studying the mechanics of Barbier's reaction, Grignard found that the reaction proceeds in two steps: (1) the reaction of magnesium and an alkyl halide to form the corresponding alkyl magnesium halids; and (2) the reaction of the alkyl magnesium halide with a compound containing a carbonyl group to form a new carbon-carbon bond. Through subsequent years of experience, researchers have learned that nearly all alkyl and aryl halides react with magnesium to form Grignard reagents. However, the aryl and vinyl derivatives are more difficultly achieved. In the mid-1950s, Normant and Ramsden showed that some of the less reactive halides, such as vinyl chloride and chlorobenzene will form a Grignard reagent with comparative ease if tetrahydrofuran is used as the solvent. See accompanying table.

REACTIONS OF GRIGNARD REAGENTS

Grignard Reagents React with	To Yield
H_2O, alcohols, primary or secondary amines	Hydrocarbons
Oxygen	Alcohols and phenols
CO_2	Carboxylic acids
Nitriles	Ketones
Metal halides	Organometallic compounds
NH_3	Hydrocarbons
γ-Lactones	Glycols
Acid esters	Tertiary alcohols (except formic acid which yields secondary alcohols or aldehydes)
Aldehydes	Secondary alcohols (except formaldehyde which yields primary alcohols)
Carboxylic acids	Tertiary alcohols
Acid halides	Tertiary alcohols or ketones
Ketones	Tertiary alcohols
Hydrogen halides	Hydrocarbons
Sulfur	Mercaptans

Because of the importance of the Grignard reaction techniques, they have received much study and numerous proposals have been made concerning the detailed mechanics involved. Originally, Grignard represented a Grignard reagent by RMgX, where R is the alkyl or aryl radical and X is the halide. Thus, magnesium ethyl bromide, a Grignard reagent, would appear in Grignard's symbolism as C_2H_5MgBr. Two of the main factors which make Grignard reagents so important are: (1) the many kinds of reagents that can be formulated, considering the substitution possibilities of the R and the X in the formula; and (2) the variety of reactions in which the Grignard reagents participate to yield numerous kinds of compounds. This versatility is demonstrated partially by the accompanying table.

In addition to the mono-Grignard reagent RMgX, di-Grignard reagents have proved valuable in organic synthesis. These may be symbolized by XMgRMgX. Most important of these for the synthesis of heterocyclic compounds have been $BrMg(CH_2)_4MgBr$ and $BrMg(CH_2)_5MgBr$. The di-Grignard reagents of *o*-bromiodobenzene also have been used in the synthesis of *o*-phenylene tertiary diphosphines.

Among industrial and commercial products that involve Grignard reactions in their synthesis are certain vitamins, pharmaceuticals, hormones, motor fuel additives, insecticides, organometallic compounds, and synthetic perfumes.

GRILLAGE. A grillage is a system of timber or steel beams which is used under columns to spread the loads over a comparatively large area. Timber grillages, consisting of layers of wooden beams, laid at right angles to each other, are generally used for temporary construction, although there are instances in which they have been enclosed in concrete for permanent construction. If this grillage is used for permanent foundation it should be either entirely submerged or creosoted to withstand deterioration.

The steel grillage consists of one or more layers or tiers of beams which are encased in concrete. If there are two or more tiers the beams in one tier are laid at right angles to those in the next tier. The individual beams in each tier are held in place by rods and pipe separators, cast iron separators or steel diaphragms. Since the concrete-encased steel grillage has more resistance to bending than the ordinary reinforced concrete spread footing it can be used to distribute heavy column loads over large areas.

GRISON. Mustelines.

GRIT. An old term for coarse-grained sandstones whose components are angular or "gritty." There is a tendency to use it for any coarse-grained sandstone without regard to the angularity of the fragments.

GRIZZLY BEAR. Bears.

GROCERY CHECKOUT SYSTEM. Input/Output Devices (Computing System).

GROSSULARITE. Garnet.

GROTTHUSS-DRAPER LAW. Only those radiations which are absorbed by the reacting system are effective in inducing chemical change.

GROUND CUSHION. Helicopters and V/STOL Craft.

GROUND-EFFECT MACHINE. Sometimes also referred to as air-cushion vehicle or hover craft, the ground-effect machine essentially "traps" a volume of air between itself and the ground or water beneath it. Depending upon the design, the vehicle can be lifted from a fraction of an inch up to several feet above the underlying surface, with sustaining pressures, or the equivalent in lifting force, of some 36 psi (2.4 atmospheres) or more. Normally, operational economy requires that the machine be kept as close to the surface over which it is to travel as may be possible.

Machines of this type created much interest in the late 1950s and early 1960s and great expectations were expounded. As of the mid-1970s, however, ground-effect machines enjoy only limited and very specialized applications. The ground-effect principle has been employed in vehicles for traveling over water and land, for industrial conveyors, and for industrial towing vehicles, but numerous practical problems remain to be worked out prior to a widespread application of this principle.

GROUND (Electrical). A ground is a conductor connected to earth, or a large conductor whose potential is taken as zero (e.g., the steel frame of a car). A ground may be an undesirable, inadvertent, or accidental path taken by an electrical current in its effort to reach ground potential; or it may be the deliberate provision of conductors well connected to the ground by means of plates buried therein, or similar device.

There is always the possibility that, during the life of an insulated conductor, the insulation may be punctured or broken down and a ground occurs. Usually, a ground develops rapidly into a low-resistance path through which currents of damaging magnitude may flow. Insulation may be damaged in many ways—by the effect of moisture, or chemical vapors, by age, heat, abrasion, breaking, or crushing. Two-wire dc systems are permanently grounded on one side of the line, three-wire dc systems permanently grounded on the neutral wire. The same applies to two- or three-wire single-phase ac systems. The common grounding point of station three-phase lines is the generator neutral.

The grounding system of the ac generating station fulfills two distinct functions. The first is the grounding of noncurrent-carrying parts, the second is the furnishing of a ground connection for generator or transformer neutral to provide for the operation of a ground protection system. A common ground bus is employed, to which are connected the frames of all electric machines, the cases of instruments, transformers, circuit breakers, the secondaries of current and potential transformers, the switchboard ground bus, conduits, insulator bases, building structural steel, etc. Thus, if the grounding system is effective, a zero, or earth, potential will be established on all metal parts which might otherwise be dangerous in case a ground developed. To the common ground bus is also connected the fault bus, when used.

Grounds should be detected as soon as possible after they occur and the defective section immediately taken out of service. If the ground persists, in a short time an otherwise small repair job may become a large one. Lamp type ground detectors are used to a considerable extent on low-voltage circuits because they are reliable and cheap. It is important that the low-voltage control circuits be kept as free of grounds as the main circuits. They may be applied to high-voltage lines through the interposition, between lamps and line, of potential transformers. Also the vacuum tube type and the electrostatic type of ground detector are available for direct connection to the high-voltage lines.

In the terminology of building construction, a ground is a strip of wood about 2 inches (5 centimeters) wide and as thick as the plaster, which is applied to the framing when nailing room is needed. Grounds are used in such places as around windows and doors, at baseboards, and at cornices.

Connection to or insulation from grounds are also very important to the successful operation of various instrumentation, data processing, and telemetry systems. See also **Common-Mode Rejection Ratio; Common-Mode Voltage.**

GROUND FOG. Fog and Fog Clearing.

GROUND LOOP (Aircraft). An uncontrollable violent turn of an airplane usually with one wing tip dragging on the ground during the landing or take-off run or while taxing.

GROUND MORAINE. When a valley glacier melts completely away the debris carried on or within it is dropped upon the valley floor, forming a deposit called ground moraine. The ground moraine from the melting of the great Pleistocene ice sheets is usually spoken of as till.

GROUNDNUT OIL. Vegetable Oils (Edible).

GROUND PEARL (*Insecta, Homoptera*). The iridescent covering secreted by some of the scale insects which live on the roots of plants. Used as ornaments.

GROUND PINES AND HEMLOCKS. Lycopsida.

GROUND POSITION (GP). Navigation.

GROUND-SUPPORT EQUIPMENT (or GSE). That equipment on the ground, including all implements, tools, and devices (mobile or fixed), required to inspect, text, adjust, calibrate, appraise, gauge, measure, repair, overhaul, assemble, disassemble, transport, safeguard, record, store, or otherwise function in support of a rocket, space vehicle, or the like, either in the research and development phase or in an operational phase, or in support of the guidance system used with the missile, vehicle, or the like. The GSE is not considered to include land or buildings; nor does it include the guidance-station equipment itself, but it does include the test and check out equipment required for operation of the guidance-station equipment.

GROUNDWATER. At varying depths below the surface of the earth, depending upon wet or dry seasons, underground structures, and other natural and unnatural factors, is a zone which is saturated with water most of which comes from rain which has penetrated the ground. The upper surface of this saturated zone is called the water table, and the water itself, the groundwater or the sub-surface water. The region above the upper surface of the water table is called the zone of aeration or vadose zone.

There is a lower limit to the saturated zone as well as an upper limit. Little groundwater exists at depths below 2,000–3,000 feet (610–914 meters). Deep down in the earth's crust the pressure must be

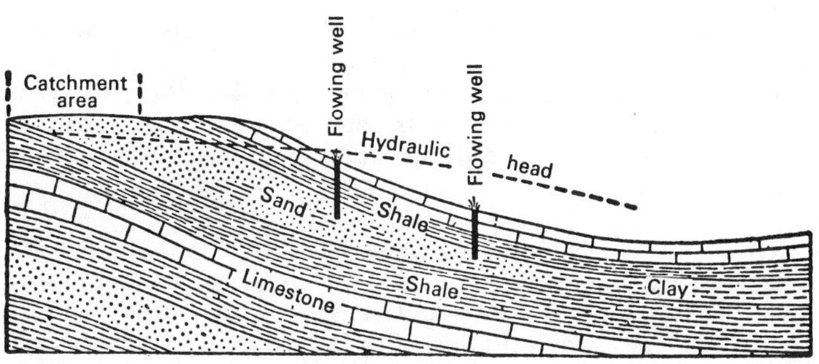

Structure section illustrating flowing artesian wells in a monocline. (*U.S. Geological Survey.*)

so great that all pores in the rocks are completely closed; thus at depths of several miles below the surface there could exist no zone of saturation.

The groundwater moves through the rocks and unconsolidated materials of the earth near the surface, constantly seeping into streams and lakes to maintain these bodies of water between rains. If this seepage is sufficiently strong on hillsides or elsewhere springs may result. A well is simply an opening dug deep enough to encounter the zone of saturation.

In certain cases, the groundwater will flow through porous tilted beds called aquifers from higher to lower localities, establishing a "head" which is sometimes sufficiently great to cause the water to flow out under pressure and rise above the surface of the ground, when the aquifer is penetrated by a drill. Such a source of water is called an artesian well, from Artois, France, a classic locality for such waters. Artesian conditions exist along much of the Atlantic Coastal Plain of the United States and in North and South Dakota, Nebraska, Kansas, Illinois, Indiana, Missouri, and Arkansas. Since the supply of underground water is largely dependent upon structure, the geology of water supply is one of the most important economic phases of the earth sciences. From the point of view of their origin, groundwaters are classified as juvenile, connate, and meteoric. Juvenile waters are of volcanic or magmatic origin, hence original. Connate waters are those in which the sediments were originally deposited. Meteoric waters are those of atmospheric origin.

All pure water, and most of all of the underground waters are of meteoric or surface-water origin. See also **Hydrology.**

GROUNDWATER (Hydrology). Hydrology.

GROUND WAVE. The energy which reaches the radio receiving antenna from the transmitter by travel along the surface of the earth rather than by reflection from the ionosphere. The ground wave is unaffected by seasonal or diurnal variations and is consequently very reliable for communication. However, it is attenuated by absorption of the earth and gradually becomes too weak to furnish a reliable signal. This attenuation depends in a complicated way upon the frequency, the soil conductivity and dielectric constant, but increases markedly with frequency. Thus, while it is suitable for communication over several thousand miles at the lower radio frequencies, over a hundred or two in the broadcast band, it becomes almost useless at the high frequencies. See **Fading** for its effect on the total received signal.

GROUP. A set of elements, finite or infinite in number, satisfying the following conditions: (1) There is a defined operation by which to each ordered pair of elements A and B in the group **G** there is associated an element C of **G**, denoted by $C = AB$, and called the product of A and B. (2) For this operation the associative law holds: $(AB)C = A(BC) = ABC$ for any three elements A, B, C of **G**. There exists: (3) a unit element E in **G** such that $EA = A$ for every element A of **G**, and (4) to each element A of **G** a reciprocal (or inverse) element A^{-1} of **G** such that $A^{-1}A = E$.

It must be understood that product, as defined in (1), is a convenient word to use for the result of combining two or more elements in a group but the law of combination is not confined to multiplication. For example, let the group elements be the integers $0, \pm 1, \pm 2, \ldots$ and let the combination law be addition, then the product of any two elements is their algebraic sum. These integers, regarded as elements of a group, will be seen to satisfy the requirements (1)–(4).

Infinite groups are discrete if the elements are denumerable; continuous, if they contain a non-denumerable infinity of elements. A finite group containing n elements is of order n. If $m < n$ elements satisfy the requirements of (1)–(4), they form a subgroup. Every group contains at least two subgroups: the unit element and the group itself.

The elements of a group may be symbols only, with no meaning attached to them and one then speaks of an abstract group. However, the elements may be numbers, matrices, geometrical operations, etc., and these are special groups (see **Representation of Groups**). For definitions and properties of some special groups, see **Group (Full Linear).**

If X is an element of a group **G** not contained in one of its subgroups **H**, then the set of elements $\mathbf{H}X$ is called a right coset and $X\mathbf{H}$ is a left coset. Cosets are not groups because they do not contain E, the unit element. Nevertheless, they are called "Nebengruppen" in German. If A, B, X are three elements of a group, then $B = X^{-1}AX$ is the transform of A by X and A, B are conjugate to each other. The complete set of group elements conjugate among themselves is a class of the group.

If **H** is a subgroup of the group **G** and X is an element of **G**, but not necessarily contained in **H**, then $X^{-1}\mathbf{H}X$ is also a subgroup of **G** and a conjugate subgroup to **H**. If **H** and $\mathbf{H}' = X^{-1}\mathbf{H}X$ are conjugate then these two subgroups are invariant if $\mathbf{H} = \mathbf{H}'$. It is also called a normal subgroup or a normal divisor.

Suppose **H** is an invariant subgroup of a group **G** and that $\mathbf{H}X$, $\mathbf{H}Y, \ldots$ are its cosets. The elements of **H** can be considered collectively as the unit element of another group and the various cosets as the remaining elements. It is called the quotient or factor group and is often designated by **G/H**. The multiplication properties of this group are similar to those of **G**.

Given a group **G**′ of order m with elements $A_1, A_2, \ldots, A_m$ and a second group **G**″ of order n with elements $B_1, B_2, \ldots, B_n$ such that every element of **G**′ commutes with every element of **G**″, then the mn element A_iB_j for a group $\mathbf{G} = \mathbf{G}' \times \mathbf{G}''$ is of order mn and is called the direct product of **G**′ and **G**″. (See **Lie Group.**)

Many other types of groups have been studied. They are of interest in geometry, differential equations, topology, and other branches of mathematics. In physics and chemistry, groups are used in the study of quantum mechanics; molecular, crystal, and nuclear structure; electrical circuits, etc.

See also classified index under **Mathematics.**

GROUPERS. Bass.

GROUP (Full Linear). The collection of all nonsingular matrices of order n, with matrix multiplication as the law of combination. Its order is infinite since its elements are the infinite number of linear transformations of one n-dimensional vector into another.

It contains several subgroups, also of finite order: (1) The unitary group with as elements all n-dimensional unitary matrices. (2) The orthogonal group. Its elements are all real n-dimensional square unitary matrices. The determinants of these matrices can be only ± 1; if $+1$, they are proper orthogonal matrices; if -1, improper orthogonal matrices; hence, one speaks of proper or improper groups. The subgroup of the orthogonal group containing only proper orthogonal matrices is the rotation group of order n. If $n = 3$, the proper orthogonal matrices correspond to rotations about three mutually perpendicular coordinate axes, while the improper matrices correspond to such rotations, followed by reflections in a plane perpendicular to the axis of rotation. These conceptions may be generalized for n-dimensions and specialized for two dimensions.

Examples of finite subgroups, which find application to physical or chemical problems, are: (1) The Abelian group. It is a commutative group, thus $AB = BA$, where A, B are any two elements contained in it. It is formed from a single element A and its integral powers: $A, A^2, A^3, \ldots, A^n = E$. (2) The permutation group. See **Permutation.** The space groups. Consideration of the ways in which identical atoms, ions, or molecules can be placed in a crystal lattice shows that the number is finite, actually 230. These are the space groups of interest in the study of crystal structure. Subgroups, 32 in number, are point groups. These describe the possible types of symmetry shown by atomic or molecular species composing the crystal lattice. They also describe the symmetry of free molecules but more than 32 point groups can then occur.

See **Lie Group;** and terms listed under **Mathematics.**

GROUP MODULATION. In some telephone carrier circuits, and in radio links of telephone circuits, several carrier channels are treated as a single group and the whole modulated upon a new carrier. This is known as group modulation. In the reception of such a system, of course, the received signal must first be demodulated into the various channels, and then, in another step, each channel must be demodulated to obtain the original voice currents.

GROUP (Permutation). Permutation Group.

GROUP (Rotation). The group of all real unitary matrices with determinant equal to plus one. (See **Lie Group.**)

GROUP (Symmetry). A group whose elements are those operations which transform regular bodies into themselves, i.e., the tetrahedral group.

GROUP (Translation). The totality of operations by which a crystal lattice may be transformed into itself by bodily displacements without rotation.

GROUP VELOCITY (Wave Train). The velocity of propagation of an interference pattern between two or more wave trains traveling in the same direction with different speeds. It may be quite different from the velocity of any one of the component wave trains. If there are more than two components, the character (waveform) of the resultant wave changes as the "group" progresses, so that the group velocity becomes ambiguous. For two components, the analysis is fairly simple.

To illustrate, first suppose for the moment that the wave train A of shorter wavelength λ is standing still, and the other, B of wavelength $\lambda + \Delta\lambda$ is moving past it in the positive direction (see figure). For example, let $\lambda = 1$ cm and $\lambda + \Delta\lambda = 1.1$ cm, and let the velocity Δv of the train B relative to the (stationary) train A be +3 cm per sec. As often as B moves forward 0.1 cm, the coincidence or beat maximum X moves backward 1 cm; consequently, X moves with respect to A with the velocity −30 cm per sec, which is −10 times, or, in general $\lambda/\Delta\lambda$ times, the velocity Δv with which B moves. (The analogy to a vernier should be quite apparent.) Now suppose that an additional velocity v is imposed upon both wave trains, so that now A moves with velocity v and B with velocity $v + \Delta v$. If $v = +100$ cm per sec, A moves with this velocity, B moves 103 cm per sec, but X moves only 100 − 30 = 70 cm per sec. That is, the velocity of the interference maximum X is $u = v - \lambda \cdot \Delta v/\Delta\lambda$. This is the group velocity, usually written

$$u = v - \lambda \frac{dv}{d\lambda}$$

In the case of media in which there is dispersion, v is a function of λ; where there is no dispersion, $u = v$, since $(dv/d\lambda)\ dv$ is then zero.

Take the case of sodium light traveling through carbon bisulfide. This light has two close components with respective wavelengths 5,890Å and 5,896Å (in air). The refractive index for the 5.890Å component being about 1.64, the velocity v of this component in CS_2 is about 1.83×10^{10} cm per sec. Now the dispersion of CS_2 in this part of the spectrum is such that $dv/d\lambda$ is readily computed to be 3.81×10^{13} cm per sec per cm, while the wavelength λ in CS_2 is 3,590 Å or 3.59×10^{-5} cm. Hence the group velocity u is 1.83×10^{10} cm per sec -3.59×10^{-5} cm $\times 3.81 \times 10^{13}$ cm per see per cm $= 1.69 \times 10^{10}$ cm per see.

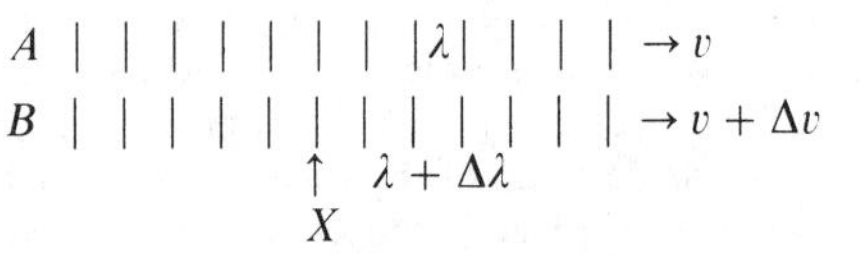

Two sets of waves traveling at different velocities. Resultant maximum is at X.

Michelson, using the same revolving-mirror method as in measuring the speed of light in vacuo, actually obtained this velocity in carbon bisulfide, showing that it is the group velocity which this method really measures.

GROUSE (*Aves*, Galliformes). Game birds with compact rounded bodies and legs feathered to the feet. The closely related ptarmigans have both legs and feet feathered. Grouse are birds of the northern hemisphere. The ptarmigans, including the red grouse of the British Isles and the willow grouse, are found at high altitudes and in the north. Most of these birds have white plumage in the winter. Grouse vary in habits, some frequenting woodlands and others open ground.

The blackcock is the same as the heathcock. It is a large grouse (*Tetrao tetrix*) of Europe, named for its glossy black feathers. It is sometimes called black grouse. The hen is gray with mixed darker colors. She is called gray hen or heath hen.

The sage grouse (*Centrocercus urophasianus*) is the largest grouse in North America. It measures about 2 feet (0.6 meter) in length, largely comprised of tail. The male weighs from 6 to 8 pounds (3 to 3½ kilograms). All of the male grouse have air sacs at the neck, some as large as golf balls and brightly colored.

The ruffled grouse (*Bonasa umbellus*) has plumage of a rich-brown coloration. The birds nest on the ground with 11 to 12 eggs at incubation time. Hatching requires 21 days.

The prairie chicken (*Tympanuchus cupido*) is of a pale-brown color and is found from Canada to Texas. The eastern heath hen is extinct in the United States.

Grouse are well known for their courtship dance. During this dance, the colored air sacs are inflated and feathers stand straight up to encase most of the fowl's body. The dance occurs just before daylight when the males of the field gather to be chosen for mates. As the males go into the dance, they are about 6 feet (1.8 meters) apart and start shuffling their feet, dancing back and forth, making loud, deep, pumping-like noises all during the dance. The females, attracted by these maneuvers, gather around to ultimately select their choice of the brightest, strongest male for a mate. Once the selection has been made, the female immediately starts to build a nest. The female incubates the eggs. The young remain in the nest about one week after hatching, after which time the young poults follow the female in a covey.

The capercaillie is a large woodland grouse of Scandinavian stock and is found in northern and central Europe and Asia. The male measures about 3 feet (0.9 meter) in length, averaging about 1 foot (0.3 meter) longer than the females. The species (*Tetrao urogallus*) is also known as the capercally, capercailizie, wood-grouse, and cock-of-the-walk. These birds are very shy and are clever in avoiding hunters. However, the birds tend to enter a hypnotic state during the courtship dance and are comparatively easy to capture during such display maneuvers.

The characteristics and habits of most all grouse are much alike. Different coloring and slight variations are visible, but mainly all are about the same. See also **Galliformes;** and **Ptarmigans.**

GROUT. Grout is a mixture of cement and water, or of cement, sand and water, of such consistency that it will flow, or may be forced by pressure, into small confined spaces.

It is widely used at present to seal geological faults, cracks, crevices, or other cavities in the rock foundations of dams. Grout is also excellent for use in connection with column footings and machine foundations which require level bearing surfaces. Since a column footing can never be poured to an exact elevation, it is usually built up to within about an inch (~3 centimeters) of its final elevation. Steel shims (fillers) are placed on the top of the footing so as to provide the correct elevation for the bottom of the base plate of the column. The column base which is an integral part of the column is then set on the shims. After the anchor bolts, which are used to fasten the column to the footing, have been tightened, the space between the top of the footing, and the bottom of the base plate is filled with grout.

The meaning of the term *grout* has expanded within recent years to include various polymeric and silicone substances used for household and other purposes.

GROWING PROCESS (Human). Endocrine System; Hormones; Pituitary Gland.

GROWTH. Increase in size and complexity. Growth of living structures depends upon increase in the number of cells or in the bulk of cells and intercellular material. It is based on the process of intussusception through which materials received as food become an integral part of the structures already present. Accretional growth is of very limited occurrence in living things and is not independent of intussusception.

Most animals exhibit determinate growth; that is, they increase in size until they approximate a limit characteristic of their kind. A few mature within rather wide limits according to the amount of food available. In the adult body the capacity of various tissues to continue their growth varies, but in all cases tissues which are worn away in the course of normal life have the power of renewal and some, such as the bone-producing cells of vertebrates, are capable of becoming active for the restoration of damaged structures. These aspects of growth are closely associated with regeneration.

The rate of growth in different parts of the body also varies, as also does the rate of total growth at different periods of life. Most mammals increase in size rapidly during early life and gradually slow down as maturity is approached; whereas man grows rapidly during infancy, slowly during childhood, rapidly again during youth, and more slowly toward the completion of his size. In the human body, the nervous system most rapidly approaches its maximum size, and the reproductive system lags until the onset of maturity. Some of the glandular tissues increase rapidly before maturity and then decrease in bulk. The balance of all these processes when normal food is available results in the gradual process of general growth, and the attainment of stability in adult life is a result of their correlation with external factors. Although no one factor is wholly responsible for growth, hormones of the pituitary and thyroid glands are of great importance in its regulation in vertebrates. Deficiency of either gland may result in dwarfing, and pituitary excess sometimes causes human beings to attain unusual height. Heights of more than 7 feet (2.1 meters) are probably due in all cases to such abnormality.

Plant growth is indeterminate. In the higher plants, primary growth is confined to the tips of stems and roots, secondary growth to cambium layers which produce wood and bark. The cambiums and the undifferentiated tissues at the tips of stems and roots are called meristems. Meristem cells divide rapidly and some of them finally become the mature cells of the plant. Each cell starts to grow, like an animal cell, by adding more protoplasm but finally increases tremendously in size by taking up a quantity of water to form a large central vacuole. Tissues are differentiated by the accumulation of excess food (cellulose, lignin, suberin) on the outside of each cell in the form of a cell wall. Certain columns of cells thicken their side walls, digest their end walls, and then die, leaving long tubes (vessels) which conduct water. Other cells die from an excess accumulation of impervious wall material and become fibers or cork cells. Others remain alive for a season or two and manufacture, transport, or store food, much more food than the plant can ever use. Some few cells become concerned with the isolation of meristems in reproductive organs (ovules, seeds). These isolated meristems produce the cells of new plants. The life of a plant need never terminate. There is no adult stage as in animals. Propagation may serve to keep a single set of meristems in action continuously.

See also **Tree.**

Reviewed and updated for the 6th Edition by Ann C. Vickery, Ph.D., University of South Florida, College of Medicine, Tampa, Florida.

GROWTH CURVE. 1. An activity curve in which the activity increases with time, or that portion of an activity curve showing such an increase. 2. A theoretical or experimental curve showing, as a function of time, the number of atoms, or the mass, or the activity of a nuclide being produced in a radioactive transformation or in an induced nuclear reaction. See also **Logistic Curve.**

GROWTH (Gall). Gall (Botany).

GROWTH HORMONE. Endocrine System; Hormones; Pituitary Gland.

GROWTH REGULATOR (Plant). Plant Growth Modification and Regulation.

GROWTH (Root). Root (Plant).

GRUB. The larva of certain insects, usually of beetles and flies. The term *worm* is sometimes applied to a grub. The grub is frequently the most damaging stage in the life cycle of an insect. See also **White Grubs.**

GRUIFORMES (*Aves*). The cranes and their relatives form this order of wading and swimming birds. Hardly any other order among birds has so little uniformity. Cranes cover a wide variety of forms, such as the common moorhens and coots, the long-legged cranes, the heavy bustards, and the peculiar seriemas. Even in appearance the various familes do not resemble each other very much.

All cranes are covered with down and able to run about when newly born. The length is 10–150 centimeters (4–59 inches), and the weight is 5 grams to 16 kilograms (2 ounces to 35 pounds). The cranes are characterized by the absence of horny ridges in the beak, of ramicorn over the nostril sheath, of elongated patellae, of a crop, and of fully developed toe-webbings.

There are eleven families: The Rails (*Rallidae*); The Stilt Rails (*Mesitornithidae*); Sun Bitterns (*Eurypygidae*); Finfoots (*Heliornithidae*); Kagus (*Rhynochetidae*); Cranes (*Gruidae*); Limpkins (*Aramidae*); Trumpeters (*Psophiidae*); Bustards (*Otididae*); Seriemas (*Cariamidae*); and Buttonquails (*Turnicidae*).

Rails, cranes, bustards, and buttonquails all inhabit northern parts of both the Old and New Worlds. The rest of the families are confined to warm regions: stilt rails in Madagascar, limpkins in Central and South America, trumpeters, sunbittern, and seriemas in South America and Central and South Africa, and the kagus in New Caledonia. See also **Rails, Coots, and Cranes.**

GRUNERITE. Amphibole.

GRUNION. Silversides.

GRUNTS (*Osteichthyes*). Of the order *Percomorphi*, suborder *Percoidea*, family *Pomadasyidae*, grunts are named after sounds which they produce, much the same way that croakers, somewhat related, received their name for acoustic reasons. In the grunt, the noise stems from sharp pharyngeal teeth which when ground together and assisted by a nearby air bladder acting as a resonator create deep vibrations. The sounds can be picked up underwater by a hydrophone and can also be heard when the fish is taken out of water. In appearance, the grunts look quite a lot like snappers. They favor tropical marine waters.

The grunts include white grunts (*Haemulon plumieri*) and French grunts (*H. flavolineatum*) both of which inhabit American Atlantic waters. The latter is considered a beautiful fish. The porkfish (*Anisotremus* virginicus) is also quite a spectacular fish in this group. The *Anisotremus davidsoni* is the only western species and may be described as a dull silver fish that attains a length of about 20 inches (51 centimeters). Gruntlike fishes in Indo-Australian waters are tropical marine varieties, sometimes called sweetlips.

See also **Fishes.**

GRUS (the crane). A southern constellation located between Tucana and Piscis Australis.

GUANACO. Camels and Llamas.

GUANIDINE. Guanidine, or carbamidine or iminourea $(NH_2)_2C{=}NH$, is formed (1) by heating ammonium thiocyanate to 180°C, (2) by ammonolysis of orthocarbonates, $C(OC_2H_5)_4 + 3NH_3 \rightarrow (NH_2)_2C{=}NH + 4C_2H_5OH$, (3) by ammonolysis of chloropicrin, $Cl_3CNO_2 + 7NH_3 \rightarrow (NH_2)_2C{=}NH + 3NH_4Cl + N_2 + 3H_2O$. (4) by ammonolysis of cyanogen chloride, $ClCN + NH_3 \rightarrow ClC(NH_2){=}NH \rightarrow HN{=}C{=}NH \rightarrow (NH_2)_2C{=}NH$.

Guanidine forms salts with acids, e.g., guanidine nitrate, $HNC(NH_2)_2 \cdot HNO_3$. By heating at 120°C for several hours, a mixture of ammonium thiocyanate and dicyanodiamide, guanidine thiocyanate solution is obtained by extracting with water. Treating guanidine with a mixture of nitric and sulfuric acids forms nitroguanidine

$$\left(HN{:}C\begin{matrix} NH\cdot NO_2 \\ NH_2 \end{matrix}\right)$$

which is reduced by zinc and acetic acid to aminoguanidine

$$\left(HN{:}C\begin{matrix} NH\cdot NH_2 \\ NH_2 \end{matrix}\right)$$

By treating aminoguanidine (1) with dilute acid or alkali, there is obtained first, semicarbazide, finally hydrazine; (2) with nitrous acid, diazoguanidine

$$\left(HN{:}C\begin{matrix} NHN{:}NOH \\ NH_2 \end{matrix}\right),$$

which is decomposed by alkali into alkali azide (e.g., NaN_3) plus cyanamide ($H_2N \cdot CN$) plus water.

In the Pauling theory of its structure, guanidine is a resonance compound of the molecular structure cited [$(NH_2)_2C{=}NH$] and two ionic structures in which the nitrogen of the imino group gains an electron lost by one of the amino groups.

The monoalkyl- and N,N-dialkyl-guanidines are somewhat weaker bases than guanidine, because resonance of the double bond to the substituted $-NH_2$ group is restricted by the fact that carbon is more electronegative than hydrogen, and renders more difficult the acquisition of a positive charge by an adjacent nitrogen atom. This effect is still more marked with the N,N′-dialkyl guanidines, while in contrast, the N,N′,N″-trialkyl guanidines are essentially as strong bases as guanidine.

The accompanying table lists seven representative substituted guanidines.

GUANO. **Anchovy and Anchoveta.**

GUANOSINE TRIPHOSPHATE (GTP). **Carbohydrates.**

GUAR GUM. **Gums and Mucilages.**

GUAVA TREES. Of the family *Myrtaceae* (myrtle family), there are some 150 species of guava trees and shrubs, including *Psidium guajava* and *P. cattleyanum.* These plants are indigeous to tropical America. It is recorded that the guava was one of the favorite foods of the Aztecan and Incan Indians. In South American countries, the fruit is called the *guayaba.* These plants have oblong, short-petioled leaves, and white flowers. The fruits are aromatic and slightly acid. The seedy pulp is used for making guava jelly as well as a blending agent by ice cream manufacturers. Significant quantities of the fruits are also consumed fresh. The fruit is very rich in ascorbic acid (vitamin C), having about ten times the quantity contained in an average orange. The guava is also an excellent source of vitamin B_1. The tree has been widely introduced into tropical countries throughout the world, including Florida, Hawaii, and southern California.

GUAYULÈ. **Rubber (Natural).**

GUEMALS. **Deer.**

GUIANA CURRENT. An ocean current flowing northwestward along the northern coast of South America (the Guianas).

The Guiana current is an extension of the south equatorial current (flowing west across the ocean between the equator and 20°S), which crosses the equator and approaches the coast of South America. Eventually, it is joined by part of the north equatorial current and becomes, successively, the Caribbean current and the Florida current.

GUIDANCE AND CONTROL (Space Vehicle). **Space Vehicle Guidance and Control.**

GUIDANCE (Bird). **Birds.**

GUIDANCE (Map-Matching). **Map-Matching Guidance.**

GUIDANCE SYSTEM ACCELEROMETER. **Acceleration Measurement.**

GUIDED MISSILE. 1. Broadly, any missile that is subject to, or capable of, some degree of guidance or direction after having been launched, fired, or otherwise set in motion. 2. Specifically, an unmanned, self-propelled flying vehicle (such as a pilotless aircraft or rocket) carrying a destructive load and capable of being directed or of directing itself after launching or take-off, responding either to external direction or to direction originating from devices within the missile itself. 3. Loosely by extension, any steerable projectile.

GUILLEMIN EFFECT. **Magnetostriction.**

GUANIDINES

Guanidine	Formula	Melting Point °C.	Boiling Point °C.
1. Guanidine	$HN{:}C\begin{matrix} NH_2 \\ NH_2 \end{matrix}$		
2. 1,3-diphenylguanidine	$HN{:}C\begin{matrix} NHC_6H_5 \\ NHC_6H_5 \end{matrix}$	147	
3. 1,1,3,3-tetraphenylguanidine	$HN{:}C\begin{matrix} N(C_6H_5)_2 \\ N(C_6H_5)_2 \end{matrix}$	130	
4. 1,2,3-triphenylguanidine	$C_6H_5N{:}C\begin{matrix} NHC_6H_5 \\ NHC_6H_5 \end{matrix}$	144	
5. 1,1,3-triphenylguanidine	$HN{:}C\begin{matrix} NHC_6H_5 \\ N(C_6H_5)_2 \end{matrix}$	131	
6. Guanylurea	$HN{:}C\begin{matrix} NH_2 \\ NHCONH_2 \end{matrix}$	105	160 decomposes
7. Aminoguanidine	$HN{:}C\begin{matrix} NHNH_2 \\ NH_2 \end{matrix}$	decomposes	

GUILLEMOT. **Shorebirds and Gulls.**

GUINEA FOWL. **Pheasant.**

GUINEA PIG. **Rodentia.**

GUINEA WORM (*Nemathelminthes, Nematoda*). A large roundworm, *Dracunculus* (*Filaria*) *medinensis*, parasitic in man. It sometimes reaches a length of more than a yard. The worm lives in the superficial tissues, especially of the legs, forming an abscess open to the surface, and can be removed by gradual traction on the end of the body exposed in this opening.

GUINIER-PRESTON ZONE. Guinier-Preston zones occur as the first step of precipitation in some precipitation hardening alloys. They are regions of unusually high concentration of solute atoms, but are not characterized by a definite crystal structure of their own.

GUITARFISHES. **Skates and Rays.**

GULF STREAM. As the North Equatorial Current in the Atlantic Ocean moves westward, it is deflected, first, by the continental land mass and, second, by the Coriolis effect. This intensification, turning clockwise in the Northern Hemisphere, results in a warm, powerful current known as the *Gulf Stream.* Originating in the Gulf of Mexico, the stream passes through the Straits of Florida, and flows northeast parallel to the U.S. coastline. Finally, it slows down and spreads out to become the North Atlantic Drift, an eastward movement of warm water that is responsible for the warmth of Western Europe commonly attributed to the Gulf Stream. Presently, the ocean thermal differences existing along the Gulf Stream and similar ocean currents are being considered as energy sources for solar sea power stations. See also **Irminger Current; Ocean;** and **Ocean Resources (Energy).**

GULF STREAM COUNTERCURRENT. A density ocean current flowing southwestward in the vicinity of Cape Hatteras and skirting the Bahamas. It flows at a depth of approximately 6,000–9,000 feet (1,830–2,745 meters) and at a rate of about 8 miles (12.8 kilometers) a day.

GULL. **Petrels and Albatrosses; Shorebirds and Gulls.**

GUM ARABIC. **Acacia Trees; Gums and Mucilages.**

GUMBO. **Malvaceae.**

GUM GUIAC. **Gums and Mucilages.**

GUMMA. A soft spongy inflammatory tumor which occurs in late untreated syphilis. The common sites for gummata are the skin, liver, brain, and lung. Treatment with antisyphilitic drugs causes them to disappear.

GUM RESINS. **Resins (Natural).**

GUMS AND MUCILAGES. Natural gums and mucilages are carbohydrate polymers of high molecular weight obtained from plants. They can be dispersed in cold water to give viscous or mucilaginous solutions which normally do not gel. They are composed of acidic and/or neutral monosaccharide building units joined by glycosidic bonds. The acid groups (—CO_2H, —SO_3H) are usually present as salts of calcium, magnesium, sodium, and potassium; in certain cases substituents such as acetyl (karaya gum) and methyl groups (mesquite gum) may be present as well. Pyruvic acid residues, linked as ketals, are present in several cases (such as agar). The properties of several gums are described in the accompanying table.

Gums are of particular importance in the food processing field where they perform at least three functions—emulsifying, stabilizing, and thickening. A few also function as gelling agents, bodying agents, foam enhancers, and suspension agents. Gum guiac also serves as an antioxidant and preservative.

Sources of Gums. Gums and mucilages may be found either in the *intracellular parts* of plants or as *extracellular exudates.* Those found within plant cells represent storage material in seeds and roots. They also serve as a water reservoir and as protection for germinating seed. The polysaccharides found as extracellular exudates of higher plants appear to be produced as a result of injury caused by mechanical means or by insects. It has not been well established whether the exudates are formed at the site of the injury, or whether they are generated elsewhere and then transported to the injured area.

The true exudates, such as gum arabic and the East African and Indian gums are picked by hand. Seldom are commercial samples pure. This is a serious disadvantage in product control. They are classified according to grade, which, in turn, depends upon color and contamination with foreign bodies, such as wood and bark. The exudates are processed simply by grinding, their only prior treatment being sorting and sometimes bleaching under the sun. In some cases, they are purified by extraction with water and precipitated by alcohol.

Gums and mucilages present in roots, tubers and seaweeds are usually extracted with hot water, dried, and marketed as a powder. Those gums found on the inner side of the seed coat as vitreous layers (e.g., locust bean, guar bean, etc.) are best obtained by a suitable milling process which first removes the seed coat and then makes use of the fact that the gum layer is very hard and tough as compared with the seed endosperm. The intracellular gums and mucilages can be purified by precipitation with alcohol from aqueous solution as in the case of the plant gum exudates, or by a process such as acetylation. In a similar way, the bacterial polysaccharides can be precipiated from the cell-free culture fluid with alcohol, or as the salt of a quaternary ammonium compound where acidic groups are present.

Characteristics of Gums. The extracellular plant gums and mucilages (gum arabic, karaya gum, and tragacanth, for example) generally have a more complex structure than the intracellular types. They are made up of a number of different sugar-building units linked together by a variety of glycosidic bonds. They possess a central core or nucleus composed mainly of D-galactose and D-glucuronic acid units joined by glycosidic bonds which are relatively stable to hydrolysis by acids. To this central nucleus are attached as side chains those sugar units which are removed by mild acid hydrolysis. Thus, in the case of gum arabic, the acid-resistant portion of the molecule is composed of D-glucuronic acid and D-galactose and to this nucleus are attached units of L-arabinose, L-rhamnose, and D-galactopyranosyl (1 $\longrightarrow$ 3) L-arabinose.

The neutral mucilages and gums, such as mannans, glactomannans, and glucomannans extracted from seed and roots, have a relatively simple structure. The kinds of building units are fewer and the molecules are much less branched. The galactomannans are usually composed of a backbone of linear chains of D-mannose units jointed by 1,6-glycosidic bonds, to which are attached at regular intervals side chains of D-galactose residues. The glucomannans are essentially linear polymers united by 1,4-linkages.

The algal polysaccharides resembled the relatively simplified structures of the neutral mucilages, as in the case of carrageenan. A wider spectrum of structures is found in the bacterial gums, which are generally of the highly branched type exuded by higher plants.

Food processing and other industrial applications of gums and mucilages take advantage of their physical properties, especially the viscosity and colloidal nature. They are substances of high molecular weight. For example, gum arabic has a molecular weight of 250,000 to 300,000. The gums and mucilages which possess relatively linear molecules, such as gum tragacanth, form more viscous solutions than the more spherically shaped gums, such as gum arabic, when at the same concentration. Consequently, for some applications, the gums with linear molecules are more economic to use. Due also to the elongated molecular shape of the seed gums and mucilages, the viscosity of their aqueous solutions varies widely with concentration. They exhibit structure viscosity. In contrast, the gums and mucilages of more spherical shape, i.e., the exudates, give solutions whose viscosities do not depend so much upon concentration.

Gums and mucilages influence each other. Mixing of two gums of the same viscosity may result in a mixture with a different viscosity. The viscosity of solutions of gums and the mucilages is dependent upon the pH, especially for those containing acid groups. In certain cases, the viscosity decreases upon standing as the result of enzymatic

GUMS AND MUCILAGES—PROPERTIES AND APPLICATIONS

ACACIA GUM (arabic gum)
The dried water-soluble exudate from stems of *Acacia senegal* or related species. Thin flakes, powder, granules, or angular fragments; color white to yellowish white; almost odorless; mucilaginous taste. Completely soluble in hot and cold water, yielding a viscous solution of mucilage; insoluble in alcohol. Aqueous solution is acid to litmus. Produced in the Sudan, Nigeria, and other parts of west Africa. Used in adhesives, inks, textile printing, cosmetics; as a thickening agent and colloidal stabilizer in confectionery and other food products.

ALGINIC ACID $(C_6H_8O_6)_n$
White to yellow powder, possessing marked hydrophilic colloidal properties for suspending, thickening, emulsifying, and stabilizing. Insoluble in organic solvents; slowly soluble in alkaline solutions. Used in food industry as thickener and emulsifier; as a protective colloid; in tooth paste, cosmetics, pharmaceuticals, textile sizing, coatings; as a waterproofing agent for concrete; in boiler water treatment; in oil-well drilling muds; in storage of gasoline as a solid.

AGAR
Thin, translucent, membranous pieces or pale bluff powder. Strongly hydrophilic—absorbs 20 times its weight of cold water with swelling; forms strong gels at about 40°C. Agar (sometimes called agar-agar) is a phycocolloid derived from red algae, such as *Gelidium* and *Gracilaria.* It is a polysaccharide mixture of agarose and agaropectin. Agar is used as a culture medium in microbiology and bacteriology; as an antistaling agent in bakery products; in confectionery; in meats and poultry; as a gelation agent; in desserts and beverages; as a protective colloid in ice cream; in pet foods, health foods; as a laxative; in pharmaceuticals; for making dental impressions; as a laboratory reagent; in photographic emulsions.

CALCIUM ALGINATE
White or cream-colored powder, or filaments, grains, or granules. Slight odor and taste. Insoluble in water; insoluble in acids, but soluble in alkaline solutions. It is used in pharmaceutical products; as a food additive; as a thickening agent and stabilizer in ice cream, cheese products, canned fruits, and sausage casings also used in synthetic fibers.

CARRAGEENAN
A yellowish to colorless, coarse to fine powder, practically odorless, but with a mucilaginous taste. Moderately soluble (1 gram in 100 milliliters of water at 27°C), forming a viscous, clear, or slightly opalescent solution which flows readily. Carrageenan disperses in water more readily if first moistened with alcohol, glycerin, or a saturated solution of sucrose in water. Carrageenan is a hydrocolloid consisting mainly of a sulfated polysaccharide, the dominant hexose units of which are galactose and anhydrogalactose. It is a two-component, polyanionic colloid. The *kappa* and *lambda* components occur in varying proportions and degrees of polymerization and are associated with ammonium, calcium, potassium, or sodium ions, or with a combination of these four. Varying proportions alter the physical qualities of the substance. Carrageenan is obtained by extraction with water of members of the *Gigartinaceae* and *Solieriaceae* families of the class *Rhodophyceae* (red seaweed). The seaweed is also called Irish Moss and is prevalent off the coasts of Canada, New England, and New Jersey, but is found in other parts of the world. Carageenan is used as an emulsifier in food products, especially chocolate milk; in toothpastes, cosmetics, pharmaceuticals; as a protective colloid; and as a stabilizing aid in ice cream (0.02%).

GUAR GUM
Yellowish-white powder. Dispersible in hot or cold water. It possesses 5–8 times the thickening power of starch. Reduces friction drag of water on metals. Guar gum is obtained from the ground endosperms of *Cyanopsis tetragonoloba,* which is cultivated in Pakistan and used there as a livestock feed. The water-soluble portion of the flour (85%) is called *guaran* and consists of 35% galactose, 63% mannose, probably combined in a polysaccharide, and 5–7⁄8% protein. Guar gum is used in paper manufacture; cosmetics; pharmaceuticals; as an interior coating of fire-hose nozzles; as a fracturing aid in oil wells; in textiles, printing, polishing; as a thickener and emulsifier in food products.

GUIAC GUM
Moderate yellow-brown powder, becoming olive brown upon exposure to air. Odor is balsamic. Taste is slightly acrid. Dissolves incompletely but readily in alcohol, ether, chloroform, and in solutions of alkalies. Slightly soluble in carbon disulfide and benzene. Occurs as irregular masses enclosing fragments of vegetable tissues, or in large, nearly homogenous masses. Source is resin of the wood of *Guajacum officinale,* principally found in Central America.

KARAYA GUM
A pale yellow to pinkish brown, translucent, and horny gum with a slightly acetous odor and a mucilaginous and slightly acetous taste. In powdered form it is light gray to pinkish gray. Karaya gum is insoluble in alcohol, but swells in water to form a gel. Karaya gum is obtained as a dried gummy exudate from *Sterculia urens* and other species of *Sterculiaceae* family, or from *Cochlospermum gossypium.* It occurs in tears of variable size or in broken irregular pieces having a somewhat crystalline appearance. The properties depend upon freshness and time of storage. Viscosity greatly decreases over a 6-month period. The gum is used in pharmaceuticals, textile coatings, ice cream and other food products, adhesives; as a protective colloid, stabilizer, thickener, and emulsifier.

LOCUST BEAN GUM (carob-bean gum)
White to yellowish-white, nearly odorless powder. It is dispersible in either hot or cold water, forming a sol, having a pH between 5.4 and 7.0, which may be converted to a gel by the addition of small amounts of sodium borate. It has a molecular weight of about 310,000. The gum swells in water, but viscosity increases when heated. Insoluble in organic solvents. The gum is extracted from the ground endosperms of *Ceratonia siliqua* of the *Leguminosae* family. The gum is used in foods as a stabilizer, thickener, and emulsifier; in packaging material, cosmetics, sizing and finishes for textiles, pharmaceuticals, paints.

POTASSIUM ALGINATE
Occurs in filamentous, grainy, granular, and powdered forms. It is colorless or slightly yellow and may have a slight characteristic odor and taste. Slowly soluble in water, forming a viscous solution; insoluble in alcohol. The gum is used as a thickening agent and stabilizer in dairy products, canned fruits, and sausage casings. It is variously used as an emulsifier.

SODIUM ALGINATE
A colorless or slightly yellow solid occurring in filamentous, granular, and powdered form. Forms a viscous colloidal solution with water; insoluble in alcohol, ether, and chloroform. It is extracted from brown seaweeds. The gum is used as a thickener, stabilizer, and emulsifier in foods, especially ice cream. Also used in boiler compounds, pharamaceuticals, textile printing, cement compositions, paper coatings, and in some water-base paints.

TRAGACANTH GUM
Dull white, translucent plates or yellowish powder. Soluble in alkaline solutions, aqueous hydrogen peroxide solution; strongly hydrophilic; insoluble in alcohol. One gram in 50 milliliters of water swells to form a smooth, stiff, opalsescent mucilage free from cellular fragments. It is obtained as a dried gummy exudate from *Astragalus gummifer,* or other Asiatic species of *Astragalus* (*Leguminosae* family). The gum is used in pharmaceutical emulsions, adhesives, leather dressings, textile printing and sizing, dyes, food products (notably ice cream and desserts), toothpastes; for coating soap chips and powders; and in hair wave preparations.

XANTHAN GUM
See separate entry on **Xanthan Gum.**

breakdown of the molecules. The molecules can undergo large changes in shape and size under the osmotic influence of opposing ions. Some of them, such as carrageenan from Irish Moss, can be fractionated by dilute salt solutions (potassium chloride) and the poly-β-glucosan from barley grain may be precipitated with ammonium sulfate. Gum arabic shows the phenomenon of coacervation when mixed with gelatin. See **Coacervation.**

The specific uses of gums are wide and diverse. By way of a few examples, seaweed gums (e.g., carrageenan) and seed mucilages (guar gum) are used as stabilizers in dairy products, such as ice cream and certain cheeses. They are used in confectionery, in making jams, jellies, and in stabilizing citrus oil emulsions and salad dressings. They have been used as fixatives for 2,3-butanedione in the baking industry. Outside the food field gums and mucilages find scores of applications.

In 1974, the Northern Regional Research Center (Peoria, Illinois) of the U.S. Department of Agriculture and the Kelco Company were joint recipients of the Institute of Food Technologists award for the development and commercialization of xanthan gum. See also **Xanthan Gum.** This gum differs by virtue of its production by pure-culture fermentation of a carbohydrate as contrasted with refining a naturally occurring substance.

See list of references under **Colloidal Systems.** A particularly good reference covering the physical properties and procedures for testing various gums and mucilages is 'Food Chemicals Codex,' published by the National Academy of Sciences, Washington, D.C. (revised periodically).

GUM TREES. Eucalyptus Trees.

GUN (Cathode-Ray Tube). Cathode-Ray Tube.

GUNN DIODE. Semiconductor.

GUPPY. Viviparous Topminnows.

GURNARDS (*Osteichthyes*). Of the suborder *Dactylopteroidea*, family *Dactylopteridae*, these are tropical marine fishes frequently called flying gurnards because of their apparent ability to propel themselves out of water. There is little documentation available to indicate their abilities at flight as in the instances of the true flying fishes. See also **Hatchet Fishes.**

The *flying gurnards* are characterized by greatly developed pectoral fins, the rear portion of which has become a large, winglike structure. The front of the pectoral fins is short. There are two long individual spines in front of the first dorsal fin. The body is elongate with firmly attached scales. The pre-opercle gill cover has strong spines; the opercle has no spine. The jaws bear small teeth. Flying gurnards are very similar to sea robbins. See also **Sea Robbins.** However, they differ from them in the arrangement of the skull bones. The snout is short and very steep. The top of the skull is flat. The gill openings are very small. One species is found in the Atlantic Ocean and Mediterranean Sea, while three species are found in the Indo-Pacific region. Flying gurnards prefer warm-to-subtropical seas.

Juvenile flying gurnards, with small pectoral fins, and adults, with winglike pectoral fins, are so different that the young were once classed in another genus. In older studies, the flying gurnards were often confused with the flying fishes. See entry on **Flying Fishes.** The chief enemies of flying gurnards are sea breams and mackerel; but while in the air they are fed upon by frigate birds, gulls, white-tailed sea eagles, procellariids, and tropical birds.

Many travelers have reported flying gurnard schools some 13 to 16 feet (4 to 5 meters) in the air for flights extending up to 300 feet (90 meters). This spectacle is repeated continuously. One group flies out of the water, leans forward, and then disappears again into the sea, while a second group has already shot into the air; then comes a third, and so forth. When flying gurnards leap out of the water at night, they glow in a phosphorescent light. When the sea is calm, the rushing sound of their beating pectoral fins can be heard, as well as the whistling sound of the air shooting through the gill openings.

Can flying gurnards actually fly? Most marines researchers say not. They claim that these fishes could never lift themselves out of the sea and fly over it because of the armored, spiny skull, the heavy body with its thick scales, and the caudal fin with its two small tips. Further detailed research is required to form a definite conclusion in the opinion of other authorities.

GUSHER. Natural Gas.

GUSSET PLATE. A gusset plate is a flat plate connecting two or more structural members where they meet at a joint. Stress is transferred between the members through the gusset plate by riveted, bolted, or welded connections. A gusset plate should be of a shape giving a minimum waste of material, and which can be fabricated in the shop with minimum amount of labor. For this reason it should be cut

Gusset plate at joint.

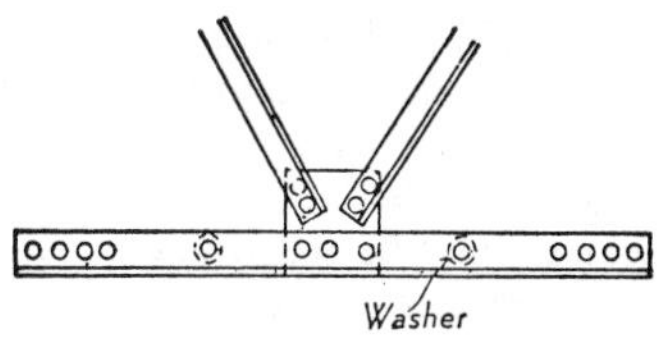

with straight edges. The thickness of a gusset plate should be sufficient to give *bearing* value, so that the material or the rivet will not be crushed. Minimum thicknesses of gusset plates are usually $\frac{1}{4}$-inch (6 millimeters) for inside protected structures and $\frac{3}{8}$-inch (9 millimeters) for outside exposed structures. The area between rivet holes should be great enough to transmit the stress from one member to another. Examples of gusset plates are to be found in all types of welded and riveted steel structures, and in gussets which strengthen and make the joints in the rib structure of an airplane wing.

GUST. Winds and Air Movements.

GUST FRONT. Thunderstorms and some showers are accompanied by small-area but frequently intense rain and sometimes hail. The precipitation originates well up in the cumulonimbus clouds and cascades earthward, accompanied by a downdraft of cold air which arrives at the earth's surface significantly colder than environing air. The temperature difference may be as much as 27°F (15°C). See accompanying illustration.

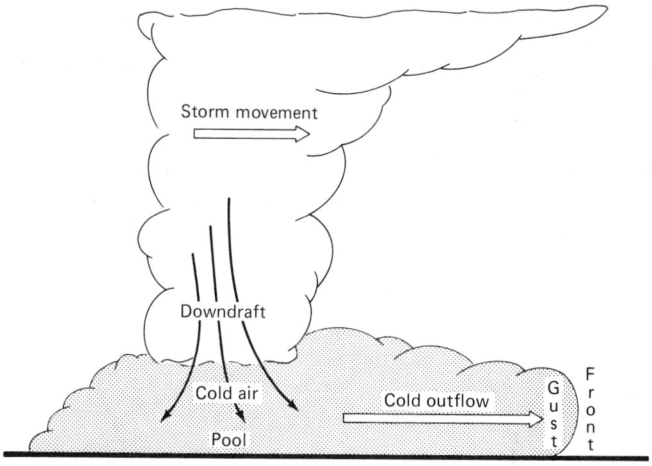

Schematic cross section showing the mechanics of a gust front. Vertical and lateral dimensions are not to scale.

The cold air accumulates under the downdraft and forms a pool of air which is heavier (more dense) than the environing air by reason of its lower temperature. Very quickly the cold air begins to flow away from the area of accumulation under the influence of gravity, that is, a gravity-induced flow of a heavier fluid into a region of lighter and less dense fluid. The leading edge of the outflowing cold air becomes a *gust front,* along which there is a wind shift, often vigorous, and a temperature drop.

Gust fronts tend to be most vigorous near the cold air source region and diminish as they move outward and away. The most intense wind shift is usually on the side toward which the storm is moving. Gust fronts have been observed as many as 15 miles (24 kilometers) from the parent storm. Gust fronts associated with a line of thunderstorms tend to form a common front and move as a *squall line,* triggering new thunderstorms as the front moves.

References

FAA: "Gust Front Analytical Study," Final Report No. FAA-RD-77-119, Federal Aviation Administration, Washington, D.C., December 1977.

FAA: "Thunderstorm Gust Fronts—Observing and Modeling," Final Report No. FAA-RD-78-145, Federal Aviation Administration, Washington, D.C., December 1978).

See also references listed at ends of entries on **Climate;** and **Meteorology.**

Peter E. Kraght, Certified Consulting Metereologist, Mabank, Texas.

GUTTA PERCHA (*Palaquium Gutta*, and related species; *Sapotaceae*). Gutta percha is prepared from the latex found in the stem and leaves of certain trees native in Malaysia and various South Sea Islands. To obtain the latex, which does not flow readily from living trees, the tree may be felled and a series of rings cut in the bark. From these the latex oozes and may be gathered. Such a method is naturally very destructive to continued production. A more desirable method is practiced in plantations of today. Fresh leaves are gathered and chopped up and crushed. The crushed mass is then boiled in water and the gun removed and pressed into blocks.

In South America a related tree, *Mimusops Balata* (*Sapotaceae*) yields a similar gum of somewhat inferior quality. This tree is usually tapped by cutting a row of zigzag gashes which connect one with another. Down these the latex flows, to be gathered in a cup at the bottom, and later coagulated in trays.

Gutta percha is a yellowish or brownish somewhat leathery solid containing up to 90% of a hydrocarbon gutta. On heating, it becomes plastic and is very resistant to water.

GUTTATION. The loss of liquid water from intact plants is called guttation. This process should not be confused with transpiration which is the loss of water vapor. Guttation occurs most commonly from the leaves, the exuded drops of water appearing at the tips or margins of the leaves. The water is not pure but contains traces of sugars and other solutes. Guttation occurs through distinctive structures, called hydathodes or water stomates. In external structure a hydathode resembles an enlarged stomate. In temperate regions, guttation can most often be observed on cool, late spring mornings following a warm day. Exuded drops of water can be observed at the margins or tips of many, but by no means all, kinds of herbaceous plants at this season. The exudation of water is believed to result from a root pressure (see **Ascent of Sap**) which is imposed on the sap in the xylem ducts. The drops of water exuded in this process are often erroneously considered to be dew. The quantities of water lost by most species of plants in guttation are negligible compared with the quantities lost in transpiration.

GUYOTS. **Seamount.**

GYMNOPHIONA. A class of amphibians made up of legless burrowing species which resemble worms or small snakes. They live in the tropics of South America and the Old World. About 40 species are known.

GYMNOSPERMS. The characteristic feature of the gymnosperms is the occurrence of the ovule on the surface of the scale which bears it, and not surrounded by an ovary wall. In most gymnosperms, the reproductive bodies are borne in cones. The gymnosperms are the most primitive of seed plants. Arising early in geological time, these plants became abundant and widespread in the Carboniferous period. From that period to the present, gymnosperms have decreased in numbers, many groups becoming entirely extinct. There remain some 500 species, occurring in nearly all parts of the world, but attaining their greatest development in the temperate zones. They often form a dominant forest tree.

The gymnosperms are woody plants. The majority of them are trees, often attaining immense size, as exemplified by the Giant Sequoias of California. See also **Giant Sequoia.** However, a few are low shrubby plants, and a very small number of vine-like species still exist. Nearly all gymnosperms are plants of xerophytic habit, that is, fitted to survive in regions in which water is not abundant. Some, like the Welwitschia of the arid deserts of southwestern Africa, live in regions where the annual rainfall is less than 0.5 inch (12 millimeters).

GYMNOSPORE. Asexual reproductive cells which are naked and capable of active locomotion by amoeboid movement or by cilia or flagella.

GYMNOTID EELS (*Osteichthyes*). These eels, along with knifefishes and the electric eel, are members of the order *Ostariophysi* (which includes characins, minnows, and catfishes), and the family *Gymnotidae*. They are not true eels. Characteristics of the gymnotids include: (1) dimunitive beady eyes; (2) no true dorsal fin with fin rays; (3) presence of a long, undulating anal fin, extending the greater length of the fish; (4) thin cylindrical body sometimes resembling a ribbon; and (5) a thin, often pointed tail. The long tail, of course, accounts for the extreme ability of the gymnotids to move in all directions speedily and easily. Gymnotids essentially are habitants of Central and South American waters, southward at least to Paraguay. There are probably less than 50 species of gymnotids, of which there are four convenient groups: (1) the *Rhamphichthys rostratus*, a food fish that may attain a length up to about $4\frac{1}{2}$ feet (1.4 meters); (2) the knifefishes (*stenarchids*), some of which are sought by tropical fish hobbyists; (3) other knifefishes, including the banded knifefish (*Gymnotus carapo*); and (4) *Electrophorus electricus*, the well known electric "eel."

The electric organs of the electric eel are so powerful that it appears to have no enemies other than people. As an air breather, the fish must surface about every 15 minutes. Rather than lungs or truly functioning gills, this fish has a unique tissue lining in its mouth which permits obtaining oxygen directly from air. Thus the fish can be left out of water for many hours as long as moisture is provided to keep the special tissue moist. Advantage has been taken of this fact by experimental biologists.

Electrically, the fish is positive toward the head; negative toward the tail—just the opposite of the electrical profile of the electric catfish. Authorities have recorded outputs as high as 650 volts, but the average is about 350 volts for a 3-foot long eel. The ability to generate voltage levels off with age, but amperage increases slightly. The electric eel possesses a combination of battery power. The principal battery occupies most of the body of the fish and creates the highest voltage. Discharge of this battery takes the form of a train of waves of about 0.002 second duration each. The train may consist of six or more waves, each varying some in time interval and voltage. The fish employs its main battery for immobilizing food sources and for defense. Another battery, identified as Hunters organ, works in concert with the main battery. Still another battery is used in some fashion by the fish as a detector.

Because the amperage is low (0.5 to 0.75 amperes), a shock from an electric eel is not necessarily lethal, depending of course upon the size and physical characteristics of the victim.

Although electric eels have been kept in captivity, they have not been bred. Apparently because of a protective antibiotic exuded by the electric eel, they survive best in water that is not frequently changed. The electric eel has effective eyes when young, but these tend to become cloudy with age and it is theorized that this may be due to the effects of electrical discharges by other eels. Thus, the older electric eels must use their electrical form of detection to find potential sources of nourishment. The electric eel is found in the Amazon River and tributaries.

GYNANDROMORPH. An abnormal individual whose boby shows the characteristics of the two sexes in different parts. Not synonymous with hermaphrodite although this term is sometimes applied to these abnormalities. It is due to abnormalities in the distribution of the chromosomes, especially the sex chromosome, in cell division during development.

Gynandromorphs are fairly common among the insects, where they are often of the bilateral type. Such individuals have one side of the body male and the other female, with a sharp boundary in the median line. Mosaic gynandromorphs present an irregular distribution of the sexual characters.

GYNECOLOGIC CANCERS. **Cancer and Oncology.**

GYNECOLOGY. The study, diagnosis and treatment of diseases and disorders of the female genital organs.

GYPSUM. The mineral gypsum is hydrous calcium sulfate, $CaSO_4 \cdot 2H_2O$. It occurs as flattened monoclinic crystals, often twinned, transparent cleavable masses, called selenite, or silky and fibrous, called satin spar; it may also be granular or quite compact. It is a soft mineral, hardness 2; has two good cleavages which yield rhombic plates whose angles are 66° and 114°. Its specific gravity is 2.31–2.33; luster, vitreous to silky or pearly; color, colorless to white and gray, may be tinted red, yellow, blue, brown, etc., by impurities; transparent to opaque. A very fine-grained white or lightly tinted variety of gypsum is called alabaster, and prized for ornamental work of various sorts.

Gypsum is a very common mineral, thick and extensive beds of which are associated with sedimentary rocks. The largest deposits known occur in strata of Permian age. Besides being a result of deposition in sea and lake waters, gypsum has been deposited by hot springs, from volcanic vapors, and by sulfate solutions in veins. Notable localities for gypsum are in Greece, Czechoslovakia, Austria, Saxony, Bavaria, Italy, France, Spain, England and Mexico. In the United States, well-known localities are at Lockport, New York; the Mammoth Cave, Kentucky; Ellsworth, Ohio; Grand Rapids, Michigan; Hermosa, South Dakota; Wayne County, Utah; and San Bernardino County, California. In Canada, the Provinces of New Brunswick and Nova Scotia have large gypsum deposits. Because the gypsum from the quarries of the Montmartre district of Paris has long furnished burnt gypsum used for various purposes, this material has been called plaster of Paris. See also classified index under **Mineralogy.**

Often, there is confusion between the mineral gypsum, $CaSO_4 \cdot 2H_2O$, and the useful product of partial dehydration, $CaSO_4 \cdot 1/2H_2O$. See accompanying table. There are numerous commercial products based upon gypsum. *Plaster*, made from gypsum, is widely used for the economical fabrication of building products. Importantly, the setting time of gypsum plaster can be carefully controlled through the addition of fractional percentages of *accelerators* (typically water-soluble salts, such as K_2SO_4, or finely-ground gypsum) and *retarders*, which frequently are modified organic substances, such as glue, casein, blood, hair, and hoof meal; or citric, boric, and phosphoric acids and their salts. Accelerators are believed to function by providing additional nuclei for crystallization, whereas retarders are believed to provide protective colloids or insoluble salts which block water access to the plaster particle. A controlled rate of reaction can be obtained by incorporating a combination of retarders and accelerators in the gypsum plaster mix.

Wallboard is the largest single user of gypsum. The product usually consists of a core of gypsum sandwiched between two layers of paper. Characteristics of the product include fire resistance, dimensional stability, low cost, and easy workability. Wallboard conventionally measures $\frac{1}{2}$ inch (1.3 centimeters) thick, 48 inches (1.2 meters) wide, and 8 to 20 feet (2.4 to 6 meters) in length. The average weight is 1.8 pounds per square feet. In manufacture, foamed plaster slurry is mixed and discharged on a moving web of paper. The edges of the bottom paper are scored and folded so that the slurry is completely contained between that sheet and the top paper, which is laid on the slurry. The paper surfaces not only provide strength and paintability to the finished board, but also form a continuous mold within which the gypsum is cast. The board machine operates continuously. Within five minutes after forming, the gypsum is sufficiently hard to be cut, after which the sheets are dried further before storage and shipment. Fibers may be added to provide crack resistance and additional fire resistance. Water-repellent chemicals may be added to the board core or to the paper surface. Also, decorative and functional finishes may be factory-applied.

Industrial plasters of a gypsum base include dental plasters, used in making tooth impressions, orthopedic plasters for immobilizing broken bones, pottery plasters, oil-well cements, permeable plasters for casting nonferrous metals, art and statuary casting, lamp bases, patching and grouting compounds, insulating-brick production, and pattern and model making for the aircraft and automotive industries. Water-reducing additives and reinforcing resins and cements may be added to achieve a compressive strength of over 15,000 pounds per square inch (1021 atmospheres).

Portland cement also consumes large quantities of gypsum. About 5% of gypsum is added to the cement clinker before grinding. Addition of gypsum aids in increasing the early strength of the cement and prevents undesirable false set.

Agriculturally, gypsum serves as a soil conditioner, providing a source of available calcium and sulfate, assisting the retention of organic nitrogen, without the addition of acidity or alkalinity to the soil. Gypsum is widely used in areas where the soils are deficient in sulfur. Gypsum also has been used in mixed fertilizers and animal feeds.

Terra alba or dead-burned, fine white gypsum is used as a paper filler, in plastics, and as an extender for titanium dioxide. Pharmaceutically-pure gypsum can be added to bread and other bakery products, finds use in beer production, and as a pharmaceutical-tablet diluent. In Japan, calcium sulfate is used in making *tofu*, a soybean curd.

Gypsum may be a potential source of sulfur and sulfuric acid. Some European plants make portland cement and sulfuric acid from gypsum or anhydrite. In the Muller-Kuhne process, gypsum is mixed with clay and silica in quantities necessary to make cement, along with coke to reduce $CaSO_4$ to CaO. In equipment similar to that for portland-cement manufacture, the SO_2 is driven off and converted to sulfuric acid by the contact process.

References

Schroeder, H. J.: "Mineral Facts and Problems (Gypsum)," *Bur. Mines Bull.* (U.S.), **650**, 1970.

Lane, M. K.: Disintegration of Plaster Particles in Water, *Rock Prod.*, **71** (3), 60 and (4), 73, 1968.

TERMINOLOGY AND PROPERTIES OF CALCIUM SULFATE-WATER COMPOUNDS

Chemical Formula	Designations Commonly Used	Properties
$CaSO_4 \cdot 2H_2O$	Calcium sulfate dihydrate; rock gypsum; chemical gypsum; alabaster (white fine-grained); selenite (translucent, platey); satin spar (fibrous); land plaster (pulverized gypsum).	All forms (natural, synthetic, and recrystallized) are thermodynamically and crystallographically equivalent. Habit may be needles, plates, or prisms.
$CaSO_4 \cdot 1/2H_2O$	Calcium sulfate hemihydrate; calcined gypsum; stucco; plaster of Paris; molding plaster; gypsum plaster; chemical hemihydrate.	Alpha and beta types exist, depending upon conditions of calcination. Alpha type is more stable, crystalline, of lower energy. Beta type is less stable, disordered, of higher energy.
$CaSO_4$	Anhydrite	
I	Anhydrite I; high-temperature anhydrite.	Produced by high-temperature (> 1,000°C) calcining. Contains free CaO.
II	Anhydrite II; insoluble anhydrite; inactive anhydrite; dead-burned gypsum; chemical anhydrite; mineral anhydrite.	Produced by calcining at 250–1,000°C. Relatively inert. Reactivity depends upon calcining-time-temperature relationship and particle size.
III	Anhydrite III; soluble anhydrite; active anhydrite; dehydrated hemihydrate.	Produced by low-temperature (175–250°C) dehydration of hemihydrate. Reacts vigorously with water and moist air to form hemihydrate.

SOURCE: United States Gypsum Company, Des Plaines, Illinois.

Edinger, S. E.: "The Chemistry of Gypsum and Its Dehydration Products," *U.S. Dept. Commerce Bull.* NTIS, PB-203, September 1971.
Kelly, K. K., Southard, J. C., and C. T. Anderson: Thermodynamic Properties of Gypsum and Its Dehydration Products, *Bur. Mines Tech. Paper* (U.S.), **625**, 1941.
Hansen, W. C. and J. S. Offutt: "Gypsum and Anhydrite in Portland Cement," 2nd edition, U.S. Gypsum Co., Chicago, Illinois, 1969.

GYPSY MOTH (*Insecta, Lepidoptera*). A moth, *Lymantria* (*Porthetria*) *dispar*, introduced from Europe and now a serious pest in the northeastern United States. The caterpillars are able to defoliate shade and forest trees and also attack apple and sometimes the conifers. The damage and control are the same as in the case of the browntail moth.

The female moth does not fly. It measures about 2 inches (5 centimeters) from wing tip to wing tip, has black markings on the wings, and is creamy white in color. Usually from 300 to 500 eggs are deposited on the underside of a branch, in the bark of a tree, or along tree roots where they are hidden from view. The larvae feed on leaves and can cause serious damage. After the caterpillars transform to pupae, they soon emerge as adult insects, requiring a period of about ten days. Several insects help to control the population of the gypsy moth, but nevertheless effective means of eradicating the insect is under intense investigation. One approach under study is that of destroying the reproductivity of the insect.

GYRE. **Ocean Resources (Energy).**

GYROMAGNETIC RATIO. Two important uses of this term are: 1. The ratio of the magnetic moment of a system to its angular momentum. 2. The ratio of moment of momentum to magnetic moment. An electron traveling around a circular orbit f times per second generates a magnetic moment equal to the product of the orbit area and the equivalent current:

$$\mu_0 = ef\pi r^2/c$$

Since the charge is negative, the mechanical angular momentum is in the opposite direction and has the magnitude

$$L_0 = 0\pi fmr^2$$

yielding the gyromagnetic ratio, for orbital motion

$$G_0 = \frac{\mu_0}{L_0} = \frac{e}{2mc}$$

The factor c disappears throughout when mksa units are used. For an electron spinning about its own center, the quantum-theory values of magnetic moment and mechanical angular momentum yield

$$G_s = 2G_0 = e/mc$$

twice that for orbital motion, leading to a g factor that has a magnitude of 2. Similarly, nuclear gyromagnetic ratios are ratios of magnetic moment and angular momentum for atomic nuclei.

GYROSCOPE. A heavy symmetrical disk free to rotate about an axis which itself is confined within a framework that is free to rotate about one axis or two. The two qualities of a gyroscope which account for its usefulness are: the axis of a free gyroscope will remain fixed with respect to space, provided no external forces act upon it; and a gyroscope can be made to deliver a torque (or a signal) which is proportional to the angular velocity about a perpendicular axis. Both qualities stem from the principle of conservation of angular momentum, which may be stated as follows: in any system of particles, the total angular momentum of the system relative to any point fixed in space remains constant, provided no external forces act on the system.

Gyroscopes are frequently spoken of as having one or two degrees of freedom, or as being *free gyroscopes.* This terminology is confusing because it res.ılts from the conventional use of the number of degrees of freedom of the vector of angular momentum rather than from the actual degrees of rotational freedom. Figure 1a shows diagrammatically the mounting of what is commonly called a *single-degree-of-freedom*, or "rate," gyroscope. Although there are obviously two rotational axes involved, in its use it is a single-degree-of-freedom system.

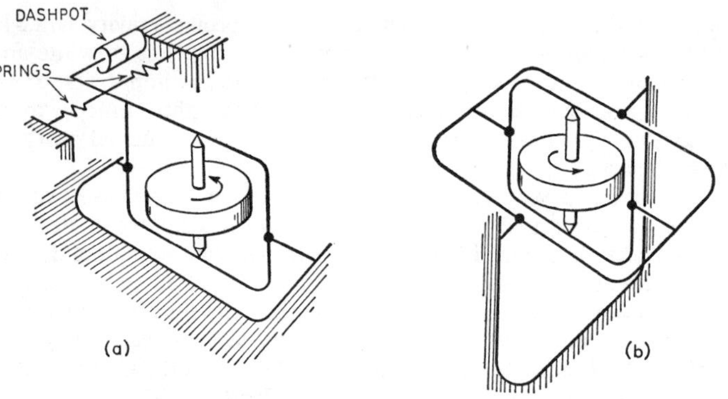

Fig. 1. Gyroscope gimbal mounting systems: (a) single-degree-of-freedom gyro; (b) two-degrees-of-freedom gyro.

Figure 1b illustrates the gimballing arrangement for what is sometimes called a *two-degree-of-freedom* gyroscope. As can be seen, a gyro wheel so mounted has three degrees of rotational freedom, except when all three axes are in the same plane. When the measurements of motion are made only from two coordinate axes, or when the outer axes lie in the same plane, this arrangement is frequently called a two-degree-of-freedom gyroscope. A free gyroscope is defined as one wherein the wheel has three degrees of rotational freedom and is unconstrained with respect to rotation. Although the wheel illustrated in Fig. 1b fulfills this definition as long as the axes are not aligned, a wheel so mounted as to be capable of rotation about five intersecting axes has three degrees of rotational freedom, whatever the direction of the axes.

Precession. The phenomenon of gyroscopic precession is explained readily by Newton's law of motion for rotation, which may be stated: The time rate of change of angular momentum about any given axis is equal to the torque applied about the given axis. When a torque is applied about the input axis of the gyroscope illustrated in Fig. 2 and the speed of the wheel is held constant, the angular momentum of the rotor may be changed only by rotating the projection of the spin axis with respect to the input axis, i.e., the rate of rotation of the spin axis about the output axis is proportional to the applied torque. This may be stated in equation as

$$T = I\omega_r\Omega$$

where

T = torque
I = inertia of the gyroscope rotor about the spin axis
ω_r = rotor speed
Ω = angular velocity about the output axis

The rule for determination of the direction of precession about the output axis is: Precession is always in such direction as to align the direction of rotation of the rotor with the direction of rotation of the applied torque. This is illustrated in Fig. 2, which indicates the direction of precession about the output axis as a result of the

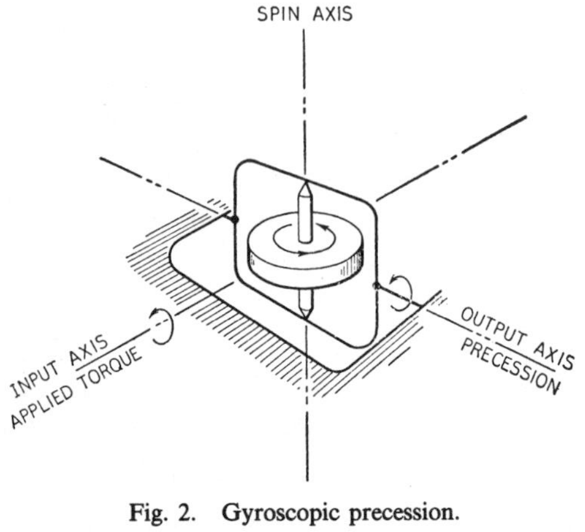

Fig. 2. Gyroscopic precession.

to temperate marine waters from shallow levels down to about 3,000 feet (900 meters). They cannot tolerate fresh or brackish waters.

See also **Cyclostomata;** and **Fishes.**

HAIDINGER FRINGES. Optical interference fringes seen with thick, flat plates near normal incidence. The fringes of the Fabry-Perot interferometer are of this type. They are also known as constant angle or constant deviation fringes.

HAIL. Precipitation and Hydrometeors.

HAIR. There are several kinds of hair on the human body. The appearance depends on age and body location. The so-called *lanugo* is that hair which develops on the unborn child. Usually, it is shed before birth, or within the first few months after birth. The lanugo is immediately replaced by secondary hair which is fine and soft and is often called "baby hair." The coarser hair of later life is called *tertiary hair.* Hairs are continually lost from all parts of the body throughout life, and up to a certain age, those which replace them often are coarser than their predecessors.

There are about 125,000 hairs on the scalp of the average person. Darker persons usually have fewer scalp hairs than blonds. Scalp hair usually grows from 3 to 5 inches (7.5 to 12.5 centimeters) per year and, if permitted, can become as long as 2 to 3 feet (0.6 to 0.9 meter), or even longer.

The hairs of the body originate from hair follicles embedded in the skin. The lower part of the follicle extends into the dermis where it is supplied with blood vessels. Generally, only one hair grows from a single follicle. That part of the hair beneath the surface of the skin is termed the *root,* while that part extending outward from the skin is called the *shaft.* The sebaceous glands of the skin have their openings in the hair follicles. These glands secrete a substance (sebum) which is responsible for the oily appearance of the skin or scalp. Persons with oily skin possess overactive sebaceous glands. When the hair follicle becomes plugged, the sebum collects within it, turns dark at the surface, and becomes a "blackhead."

Minute muscles (*erectors pilorum*) are connected to the hair follicle. When these muscles contract, they temporarily displace the entire follicle, causing the hair to "stand on end." The skin surrounding the hair is also elevated by the contraction of these muscles, giving the skin a prickled appearance, sometimes called "goose pimples." Contraction of the muscles also exerts pressure on the sebaceous glands, causing the emission of extra amounts of sebum. Thus, this set of reactions aids in protecting the body from sudden cold, the hairs forming better insulation when standing erect, and the sebum coats the skin with a further barrier against the cold.

The partial or complete absence of hair from the body is called *alopecia.*

In recent years, the value of hair as a diagnostic tool to complement blood serum and urine has gained recognition. Trace elements, in particular, are accumulated in hair at concentrations that are generally about ten times higher than those present in blood serum or urine and can provide a historical record of nutritional status and exposure to heavy metal pollutants. Some organic chemicals also are prone to collect in the hair. The best results have been obtained with heavy metal pollutants, such as lead, arsenic, cadmium, and mercury. Studies to date have shown excellent correlation between hair samples and the exposure of individuals to heavy metals. One scientist has suggested that possibly animal hair could be used to monitor pollution over substantial periods of time. Other relationships have been uncovered. Researchers have found that persons who suffer with celiac disease have less sodium and potassium in their hair than normal levels. Zinc deficiencies can be diagnosed by determining zinc concentrations of hair. Other researchers have demonstrated below-normal concentrations of chromium in the hair of persons with juvenile-onset diabetes. This finding is supported by biochemical evidence that demonstrates low concentrations of chromium in the blood of some patients with juvenile-onset diabetes. On the other hand, chromium concentrations are normal in the hair of maturity-onset diabetics from the Pima Indian nation, which has a very high incidence of diabetes—this further demonstrating major differences between juvenile-onset and maturity-onset diabetes. See also **Diabetes Mellitus.** Some researchers are attempting to adapt highly sensitive chromatography-mass spectrometry techniques to the study of hair and thus broaden the use of hair as a diagnostic tool.

The use of hair analysis and examination in the forensic sciences has been known for a number of years.

HAIR (Abnormal Growth). Hirsutism.

HAIR HYGROMETER. Hygrometer.

HAIR (Identification). Bertillon System.

HAIRSTREAK (*Insecta, Lepidoptera*). Small butterflies, those of the temperate zone dull-colored and those of the tropics often brilliant. The hind wings of most species bear hairlike tails. With the coppers and blues they make up the family *Lycaenidae.*

HAIRTAIL. Cutlassfishes.

HAIRWORM (*Nematomorpha* or *Gordiacea*; formerly placed in the phylum Nemathelminthes with the *Nematoda*). Long slender roundworms of small size, which live as parasites in the bodies of invertebrates, chiefly insects.

HAKE (*Osteichthyes*). Of the family *Merluccidae,* the hakes are closely related to the codfishes, but have a special systematic position due to their unusual distribution. The family has just one genus, *Merluccius.* The slender body, skull structure, and the large-toothed mouth give this carnivorous fish a garlike appearance. There are two dorsal fins and one long anal fin, which is almost the mirror image of the second dorsal fin in shape, size, and position.

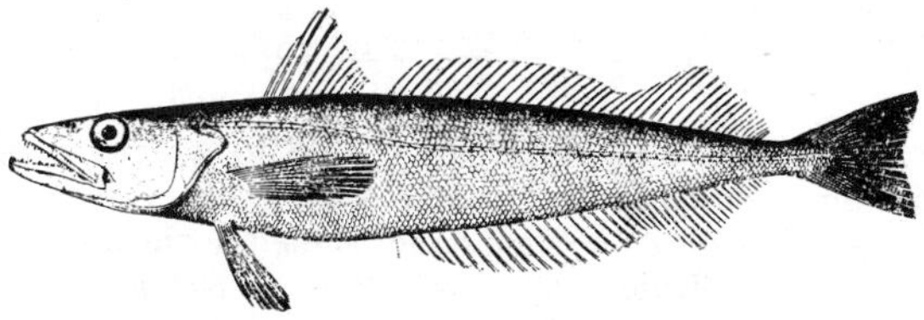

Silver hake (*Merluccius bilnearis*).

The hake (*Merluccius merluccius*) is found in the northeastern Atlantic Ocean off the western and southwestern coasts of Europe, along the continental shelf. The northern border of the distribution is formed where branches of the Gulf Stream meet masses of polar waters. This is also the northern limit of the *American hake* or *silve hake* (*Merluccius bilnearis*). See accompanying figure. Living in deep water has enabled the hakes to penetrate the tropical Atlantic Ocean and inhabit oceanic regions in the southern hemisphere with temperate to subtropical conditions. This accounts for the large South Atlantic populations of *stockfish* (*Merluccius capensis*) off southwestern Africa; and *Merluccius hubbsi* from the coasts of southern Brazil and Argentina. There are also Pacific Ocean species: *Merluccius gayi* and *M. productus,* off the western coasts of North and South America. Their presence has been explained by a presumed migration around Cape Horn. The New Zealand species, *Merluccius australis,* may also have come by this route.

Hakes can be over 3 feet (1 meter) in length, but there are small- and medium-sized species as well. They are predators, feeding chiefly on herring and other schooling fishes. The European hake seeks its prey at night in the upper water levels. During the day, it is less active and stays near the floor, at which time it can be caught easily, even with a dragnet. This species spawns in spring, apparently without preferred spawning sites. The floating eggs then drift within the hake distribution region. The commercial importance of hake has increased since the early 1960s.

The *Cape hake* in South African waters is a whitefish (*M. capensis*) and it appears that there are two distinct populations on the trawling grounds. About half of the population attains sexual maturity at an age of 3 to 4 years. Peak spawning occurs during spring and early summer. Preliminary studies indicate that in the case of the Cape hake, diurnal vertical migration is much less pronounced than in the

case of the European cod. The diet of the adult hake is comprised of rattails, maasbanker, and squid, and cannibalism is quite common.
See also **Fishes.**

HALF-ADDER (Computer System). A circuit having two output points, *S* and *C*, representing sum without carry and carry, and two input points, *A* and *B*, representing addend and augend, such that the output is related to the input according to the following table:

INPUT		OUTPUT	
A	*B*	*S*	*C*
0	0	0	0
0	1	1	0
1	0	1	0
1	1	0	1

Two half-adders and an Inclusive-OR circuit, properly connected, can provide a Full-Adder having two inputs (augend and addend) and a carry input which produces a sum output (without carry) and a carry output. See **Full-Adder (Computer System).**
See also terms listed under **Data Processing.**

HALF-CELL. An electrochemical system consisting of a single electrode and an electrolytic solution, with usually a (reversible) ionization process in progress between electrode and electrolyte. See also **Galvanic Cell.**

HALF-LIFE (Biological). The time of survival of half the individual members of an unstable system. The half-life $t_{1/2}$ of the system is related to the decay constant λ and the mean life τ by the relation:

$$t_{1/2} = \frac{\ln 2}{\lambda} = \frac{0.693}{\lambda} = 0.693\tau$$

The term half-life is most commonly applied to systems of radionuclides but may also be applied to other systems that decay.

The biological half-life of a substance is the time in which a living tissue, organ or individual eliminates, through biological processes, one-half of a given amount of a substance which has been introduced into it. The effective half-life is a term usually applied to a radioactive substance in a biological organism. It is defined in terms of the half-life of the radioactive substance itself, and its biological half-life in the organism, by the following expression:

$$\text{effective half-life} = \frac{\text{radioactive half-life} \times \text{biological half-life}}{\text{radioactive half-life} + \text{biological half-life}}$$

HALF-SHADE PLATE. A semicircular, half-wave plate of quartz set between the polarizer and analyzer and close to the former. Useful in making precision settings with a polariscope.

HALF-LIFE (Elements). Chemical Elements.

HALF-SILVERED SURFACE. A surface coated with a metallic film of such thickness that it transmits approximately half of the light falling on it at normal incidence and reflects approximately half.

HALF-THICKNESS (Absorber). The thickness of a particular absorber that will reduce the intensity of a beam of radiation to one-half its initial value. If the absorption is exponential, the half-thickness is related to the linear or mass absorption coefficient and the mean free path as follows:

$$d_{1/2} = \frac{\ln 2}{\mu} = \frac{0.693}{\mu} = 0.693l$$

where $d_{1/2}$ is the half-thickness, μ is the absorption coefficient and l is the mean free path.

HALFWIDTH OF A SPECTRAL LINE. The intensity within a spectral line may be expressed as $I(x)$, where x is a measure of wavelength, frequency or wave number, and where $I(x)\,dx$ is a measure of the contribution to the intensity between x and $x + dx$. The halfwidth of the line is the halfwidth of the function $I(x)$.

HALIBUT. Flatfishes.

HALIDES. A compound made up of a halogen (astatine, bromine, chlorine, fluorine, or iodine) and another element or radical may be termed a *halide.* Fundamentally, there are three classes: (1) the *ionic* (saline) halides, (2) the *covalent* (acid) halides, and (3) the *complex* halides. The ionic halides are most sharply characterized by the halides of the alkali and alkaline earth metals, plus those of certain Lanthanide and Actinide metals. They form ionic or semi-ionic crystals in the solid state, have high boiling points and melting points, and are soluble in polar solvents. Their bonding is electrovalent, varying in degree with the difference between the electronegativities of the halogen and the metal. Potassium iodide and silver fluoride are ionic, but silver iodide is essentially covalent. The fluorides exhibit a primarily ionic character for most of the metals, but the other halogens form fewer ionic compounds. The degree of ionicity varies down as well as across the periodic table.

The covalent (acid) halides have low boiling and melting points, are soluble in nonpolar solvents and insoluble in polar solvents, although they often react with the latter. The degree of covalence generally is greatest for the nonmetals. For a given nonmetal, the boiling point depends upon both the number of atoms of the halogen with which it is combined and the symmetry of the molecule. For example, the boiling points of bromine(I) fluoride, bromine(III) trifluoride, and bromine(V) pentafluoride, BrF, BrF_3 and BrF_5, are 20, 135, and 40.5°C, respectively.

The complex halides are very numerous, because of the readiness with which halide ions form coordination compounds with metals. In general, stability of these complexes depends upon the size and electronic structure of the metal ion—the smaller cations form their more stable compounds with the smaller halide ions, notably with fluoride, while with larger cations the order of stability is that of polarizability of the halide, i.e., decreasing from iodide to fluoride. The more electronegative transition elements form especially stable complexes; e.g., those of palladium, platinum, etc., $PdCl_4^{2-}$, PtF_6^{2-}, etc. The most common halo complexes have four or six halogen ions coordinated with the cation, although such complexes as those of copper, gold and mercury, e.g., CuI_2^-, $AuCl_2^-$, $HgCl_3^-$, etc., are notable exceptions.

See also **Bromine; Carbon; Chlorine; Chlorinated Organics; Fluorine;** and **Iodine.**

HALITE (Rock Salt). The mineral halite (rock salt) is naturally occurring sodium chloride, NaCl, common salt. It is isometric with cubic habit and cleavage. It is brittle; hardness, 2.5; specific gravity, 2.168; luster, vitreous; colorless when pure, but usually white, yellow, red, or blue. It is soluble in water. Halite occurs interbedded with sedimentary rocks in all parts of the world and in all but the very oldest rocks. It frequently occurs in association with anhydrite and gypsum. In the United States this type of "salt beds" has been exploited in Michigan, New York, Ohio, and Pennsylvania. Louisiana produces salt from great subsurface dome-shaped masses, often 2,000–4,000 feet thick. The salt domes of the Gulf Coastal Plain are particularly important as subsurface structures, on the flanks of which are apt to occur large and important pools of petroleum. Poland, Saxony, Austria, and France possess well-known deposits of salt, as well as the U.S.S.R., England, Algeria, India, and China. Salt is chiefly used in cooking and as a preservative; in the manufacture of soda ash for the glass industry; and as a source of many sodium compounds. It derives its name from the halogen group of elements to which chlorine belongs.

See also **Sodium Chloride.**

HALL EFFECT. In 1879, Hall, at Johns Hopkins University, discovered that if a strip of gold leaf, carrying an electric current longitudinally, was placed in a magnetic field with the plane of the strip perpendicular to the direction of the field, the points directly opposite each other on the edges of the strip acquired different electric potentials; and that if such points were joined through a sensitive galvanometer, a feeble current would be indicated. In other words, the equipotential lines, ordinarily running across at right angles to the edges, were skewed into an oblique position, and the electric lines of flow in the plane of the strip were deflected to one side.

If one looks along the strip in the direction of the current, with the magnetic field directed downward, then, with strips of antimony, cobalt, zinc, or iron, the electric potential drop is toward the right and the effect is said to be positive; while with gold, silver, platinum, nickel, bismuth, copper, and aluminum, it is toward the left, and the effect is called negative. The transverse electric potential gradient per unit magnetic field intensity per unit current density is called the "Hall coefficient" for the metal in question. Thus, the Hall coefficient R_H is defined as

$$R_H = \frac{E_y}{j_x H_z}$$

where E_y is the electric field developed in the y direction when a current of current density j_x flows in the x direction through a magnetic field H_z in the z direction. According to the free electron theory of metals, the Hall coefficient should be given by

$$R_H = \frac{1}{Nce}$$

where N is the number of free electrons per unit volume, of charge e (in esu), and c is the velocity of light. The observed result that for some metals the carriers would seem to have positive charges is explained by the band theory of solids. In a nearly filled band, the wave functions of the electrons near the top of the band are so modified that it is the holes in the band that behave like particles. Since a hole represents the absence of negative charge, it behaves as if positively charged. The Hall angle is the ratio of E_y (defined above) to the field E_x, generating the current in the magnetic field H_z. The Hall mobility is the mobility of the electrons or holes in a semiconductor as measured by the Hall effect.

A number of transducers utilize the Hall effect. Shown in the accompanying diagram is a direct-current oscilloscope probe based on the effect. A steady direct current I_c is applied to one axis of the Hall generator and a magnetic field B, proportional to the current through the conductor, is applied to a second axis. An output voltage V_c is taken across the third axis of the Hall generator. The output voltage can be calculated from:

$$V_c = \frac{10^{-5} R_H}{t} I_c B$$

where

V_c = Hall voltage, volts
R_H = Hall coefficient, cm^3/coulomb
t = thickness, cm
B = magnetic field density, kilogauss

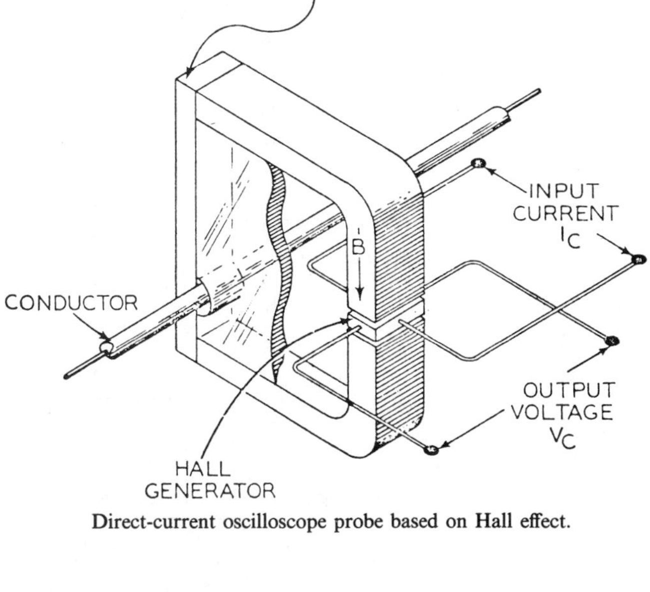

Direct-current oscilloscope probe based on Hall effect.

HÄLLEFLINTA. A Swedish term for hard, dense, metamorphic rocks composed chiefly of microscopic crystals of quartz and feldspar with occasional phenocrysts. Accessory minerals may be hornblende, chlorite, hematite or magnetite. The texture and composition of hälleflinta suggests that it is the metamorphosed equivalent of acid lava flows or tuffs.

HALLEY'S COMET. This is probably the most famous of all the comets. It is the brightest periodic comet, and so was the first to have its return predicted. In 1705, Edmund Halley (whose name rhymes with *tally*) computed the orbit of the great comet that he and others observed in 1682, and found the elements to be almost identical with those that he derived for the prominent comets observed by Kepler and Longomontanus in 1607 and by Peter Apian in 1531. He noted that the intervals between the three dates were not quite identical, and correctly attributed the difference to gravitational perturbations of the comet's motion by the planets. Celestial mechanics had not yet advanced to the stage at which planetary perturbations could be readily evaluated, either to prove the aforementioned conclusion about the slight inequality in the intervals of the comet's return, or to enable a completely accurate prediction of the next return after 1682. Nonetheless Halley's predicted time for the next return, late 1758 or early 1759, did partially allow for the effect of Jupiter, the most important of the perturbing planets. The comet was actually first seen again on Christmas night of 1758. By that time, with the aid of improved mathematical methods, perihelion passage had been computed beforehand by Clairaut, Lalande, and Madame Lepaute, and predicted to be within a month of mid-April, 1759. The comet actually passed perihelion on March 13 of that year. With each successive return since 1682, Halley's comet has had its position determined with methods of increasing precision. This, coupled with increasingly sophisticated computational procedures, has led to increasingly accurate predictions for subsequent returns: in 1835 it passed perihelion within a few days of the predicted time, and for the return of 1910 the agreement was better still. However, the possibility of nongravitational forces (see entry on **Comet**) remains a small element of uncertainty, increasing over long periods of time.

Following the apparition of 1910, it was possible to compute the dates of perihelion passage backward for many centuries. Examination of ancient records—which, prior to the fourteenth century, are mainly Chinese—has enabled observations of Halley's comet to be identified with confidence as far back as 87 B.C. This is no small achievement, in view of the planetary perturbation problem on the one hand, and, on the other, of the fact that many comets as bright as or brighter than Halley's have been recorded.

At its return in 1910, Halley's comet was first picked up by Wolf at Heidelberg, on Sept. 11, 1909, at 5×10^8 km from the sun, intermediate between the distances of Mars and Jupiter. It approached the sun in the evening sky as a telescopic object, passing within the earth's orbit on the far side from the sun, and passed perihelion, at 0.59 au, on April 20. It emerged into the morning sky in the first weeks of May as a beautiful naked-eye object for those possessing a dark sky. See Figs. 1 and 2. It then turned eastward and passed between the earth and sun; according to computation the head of the comet actually transited the solar disk on May 19 (universal time), although it was undetectable in doing so. Around this time, experienced observers with good skies traced the tail to a distance of 120° from the below-horizon head. The earth grazed the comet's tail and probably passed through it, and patent-medicine vendors advertised concoctions for warding off the effects of the comet's tail, which had, however, no detectable terrestrial effects. A few days later the head of the comet was visible to the naked eye in the evening sky, despite interference by bright moonlight (except for a total lunar eclipse on May 23). It remained a naked-eye object, for experienced observers with good skies, throughout most of June. It was followed photographically, at that epoch already a more sensitive method than visual detection, until July 1, 1911, when it was 8.3×10^8 km from the sun or slightly more distant than Jupiter. It should have reached aphelion, more distant than the planet Neptune, in 1949.

According to a careful computation done in 1971, the next perihelion passage will occur on Feb. 9, 1986. At this time the comet, as seen from Earth, will be almost directly behind the sun. This circumstance

Fig. 1. Visual aspect of Halley's comet, morning of May 12, 1910. The Square of Pegasus is on the left, and the planet Venus on the right. The comet tail is about 32 degrees long. The observer was in Mexico.

Fig. 2. Lowell Observatory photograph of Halley's comet, May 13, 1910. Venus, its image greatly enlarged by overexposure, is on the right.

should cause the 1986 apparition to provide the least favorable naked-eye views of the comet of any return since well before Halley's time. Nevertheless it will be subjected to intensive scientific study, and, given the increase in size and sensitivity of astronomical telescopes during the 76 years since the last return, the object as it returns will undoubtedly be recovered at a considerably greater distance than was possible in 1909.

The future brightness of a comet is notoriously unpredictable. One thing that may safely be said is that any attempt to view the comet, either by naked eye or with optical aid, against the brightly lit sky of a major population center will be vastly less satisfactory than the views obtainable remote from such a center. As a rough guide, an effective viewing location will be one which can provide a fine view of the Milky Way, at such times as the Milky Way is well up in the sky.

The accompanying table lists some representative predicted positions of the comet, in the horizon system of coordinates, during the expected period of maximum brightness at the 1985–86 return. During the listed 1985 dates the brightness should be increasing, while it should be on the wane during the other listed dates. The table shows the comet approaching the sun from the evening sky and leaving it from the morning sky, ultimately passing again into the late evening sky. If it is a naked-eye object during the approach phase, it is likely to be only feebly so. The table also shows the comet poorly placed well after perihelion, and actually lost for a time; this is solely the result of its motion carrying it far to the south. It should be a very fine object soon after perihelion, for southern hemisphere observers, but will have faded greatly once it is again well placed for mid-northern latitudes.

Many plans have been made for the study of Halley's comet at its next return. The National Aeronautics and Space Administration had hoped to conduct an unmanned rendezvous mission for the close study of the comet. This mission, which would have involved matching the space probe's velocity to that of the comet, depended upon early

PREDICTED ALTITUDES AND AZIMUTHS OF HALLEY'S COMET AT 40° NORTH LATITUDE

Date	Hours after Local Sunset	Altitude (Degrees)	Azimuth (North through West) (Degrees)
1985, Nov. 26	3	54	241
Dec. 1	3	61	209
6	3	60	175
11	2	56	176
16	2	51	157
21	1	50	164
26	1	45	150
31	1	40	140
1986, Jan. 5	1	35	132
10	1	29	125
15	1	22	118
Too near the sun			
1986, Mar. 11	1	12	130
16	1	13	136
21	1	14	144
26	1	13	153
31	1	11	166
Low or below horizon			
1986, Apr. 20	7	17	182
25	7	23	195
30	7	26	206
Becoming faint			

funding for expanded development of the ion-propulsion engine. The failure of this funding to materialize left open the possibility of a flyby (quick-pass) mission, but this has since been cancelled. Observations from earth orbit will provide access to the ultraviolet—where, for example, the bulk of the radiation from atomic hydrogen occurs. There will also be observations by radio and, of course, optical telescopes from the ground, and there will undoubtedly be attempts to detect the comet by radar. Questions being investigated include the size of the cometary nucleus and, above all, its composition; the manner in which comets originated; the details of the nongravitational forces; the interaction between comets and solar radiation of all kinds; and the nature of the reservoir from which these infrequent visitors are drawn.

The next perihelion passage after 1986 has been calculated as July 29, 2061.

References

Halley's summary of his comet work was published in *Synopsis Astronomiae Cometicae*, Oxford 1705, and *Tabulae Astronomicae*, London 1749. See also Shapley and Howarth, *A Source Book in Astronomy*. Contemporary English-language accounts of the 1910 apparition of Halley's comet are given in the journals *The Observatory*, Vol. 33, and *Popular Astronomy*, Vol. 18. All the securely identified perihelion passages of the comet are listed in Marsden's *Catalogue of Cometary Orbits* (Smithsonian Astrophysical Observatory, 3d ed., 1979). The 1971 calculation of the 1985–86 ephemeris is by Brady and Carpenter, published in *The Astronomical Journal*, Vol. 76, p. 728 ff.; our table of azimuths and altitudes, calculated by the undersigned, is based on this ephemeris. Calculations by Yeomans, leading to the date of the 2061 return, are discussed in *The Astronomical Journal*, Vol. 82, p. 435 ff.; this paper also lists in detail the early basic papers. See also "Giotto's Portrait of Halley's Comet," by R. J. M. Olson, *Sci. Amer.*, **240**, 5, 160–170 (1979).

C. Bruce Stephenson, Case Western Reserve University, Cleveland, Ohio.

HALLUCINOGENS. There are many substances which will, if taken in appropriate quantities, produce distortion of perception, vivid images, or hallucinations. Most of these substances will produce powerful peripheral as well as the central effects. Some few agents are characterized by the predominance of their actions on mental and psychic functions. This group of drugs has been called hallucinogens, psychotomimetics, psycholytics, and psychodelics, among several ambiguous terms. None of these names is adequately descriptive of these compounds.

Hallucinogens may be classified into five groups of chemically distinct compounds: (1) lysergic acid derivatives of which lysergic acid diethylamide (LSD-25) is the prototype; (2) phenylethylamines, such as mescaline; (3) indolealkylamines, which include psilocybin, psilocin, and bufotenin; (4) piperidyl benzilate esters, typified by ditran (a 70 : 30 mixture of N-ethyl-2-pyrrolidymethyl phenylcyclopentylglcolate and N-ethyl-3-piperidyl phenylcyclopentylglycolate), and (5) phenylcyclohexyl piperidines (sernyl). The chemical structures of these compounds is shown in the accompanying figure.

Drugs from the first three groups have been isolated from naturally-occurring sources. LSD-25 is a molecular component of ergot, a fungus which infects cereal grains. Mescaline, historically the oldest hallucinogen, was isolated from a Mexican peyote cactus. Psilocybin and psilocin were isolated from the Mexican mushroom, *Psilocybe mexicana*. Bufotenin is found in some varieties of toadstools. The indole derivatives are chemically closely related to serotonin (5-hydroxytryptamine), a compound which plays an important, yet unknown role in the central nervous system.

The piperidyl benzilate esters and phenylcyclohexyl piperidines are synthetic compounds, and have not been shown to occur naturally. Some authorities do not consider them to be hallucinogens, but active researchers in the field include them among the most active psychotomimetics.

Clinical syndromes from LSD-25, mescaline, and the indoleamines are similar. Somatic symptoms are nausea, dizziness, loss of appetite, blurred vision, paresthesia, weakness, drowsiness, and trembling. These result frequently and are usually associated with sympathomimetic effects, such as increased pulse rate and slight temperature elevation. Perceptual and psychic changes are marked. Visual illusions and vivid hallucinations, decreased concentration, slow thinking, depersonalization, dreamy states, changes in mood, and often anxiety are commonly found.

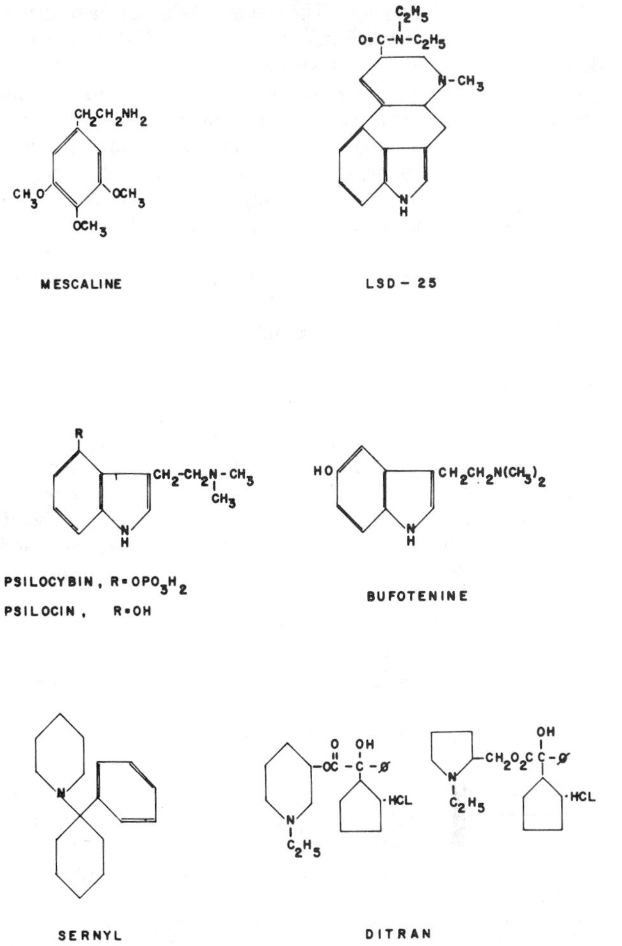

Structures of some hallucinogenic drugs.

The clinical syndromes from ditran are different from those produced by the aforementioned drugs in some respects. Disorganization of thought, disorientation, confusion, mood changes, and visual and auditory hallucinations are observed. The piperidyl benzilate esters are central anticholinergics, and mental states produced by them are reminiscent of those from other anticholinergics, such as scopolamine.

The effects of phenylcyclohexyl derivatives are also distinctive. Comparatively minor somatic symptoms are evoked. Psychic effects predominate, being typically characterized by feelings of unreality, depression, anxiety, and delusional or illusional experiences. The effects of these drugs are said to be more analogous to natural psychoses than those of the other drugs; however, the same claim has been made for ditran.

When LSD-25 was discovered, it was believed that the drug would provide an extremely useful tool in the investigation of psychoses and mental illness. However, therapeutically, the hallucinogens, including LSD-25, have been of little value to psychiatrists.

References

Lincoff, G., and D. H. Mitchel: "Toxic and Hallucinogenic Mushroom Poisoning," Van Nostrand Reinhold, New York, 1977.

Seiden, L. S., and L. A. Dykstra: "Psychopharmacology," Van Nostrand Reinhold, New York, 1977.

HALO. **Atmospheric Optical Phenomena.**

HALOCARBONS (Ozone Depletion). **Oxygen.**

HALOGENATED COMPOUNDS. **Chlorinated Organics; Organic Chemistry.**

HALOGEN GROUP (The). The elements of group 7a of the periodic classification sometimes are referred to as the Halogen Group. The individual elements commonly are called *halogens.* In order of increasing atomic number, they are fluorine, chlorine, bromine, iodine, and astatine. The elements of this group are characterized by the presence of seven electrons in an outer shell, and hence have the ability to gain an electron to form negative ions with a completed octet of valence electrons. The halogens present striking similarities of chemical behavior, all being very reactive and, in particular, readily form substitution compounds with numerous organic compounds. Although these elements also have other valences all have a -1 valence in common.

HALOGENS (Budde Effect). Budde Effect.

HALOTHANE. Anesthesia.

HALTERE. The vestigial hind wing of a 2-winged fly. In this order the 2-winged condition is due to the loss of the hind wings as organs of flight. They persist, however, as small knobbed organs, in some species large enough to be seen with the naked eye just behind the bases of the functional wings. They contain sense organs, and both static and chordotonal functions have been attributed to them. The evidence seems inconclusive.

HAMILTONIAN (or Hamiltonian Function of a System). Generally denoted by the symbol H, the Hamiltonian is defined by the equation

$$H(q_k, p_k, t) = -L(q_k, p_k, t) + \sum_{l=1}^{3n} p_l \dot{q}_l(q_k, p_k, t)$$

L is the Lagrangian function of the system, expressed as a function of the coordinates, momenta and time. $\dot{q}_l$ stands for the generalized velocities, also expressed as functions of the coordinates, momenta and time, where q are the coordinates of position, p, those of momentum, and the dot means the derivative with respect to time. n is the number of particles of the system. If the time does not occur explicitly, the system is called conservative, and H is identical with the total energy of the system. See also terms listed under **Mathematics.**

HAMMER FORGING. Forging; Iron Metals, Alloys, and Steels.

HAMMERHEAD SHARKS. Sharks.

HAMMETT ACIDITY FUNCTION. The function

$$H_0 = pK_{BH^+} - \log(C_{BH^+}/C_B)$$

where C_B is the concentration of an indicator C_{BH^+} is the concentration of its protonated form, of which K_{BH^+} is the ionization constant.

HAMSTER. Rodentia.

HAND. The terminal portion of the pectoral appendage of mammals, developed for grasping and in some species largely freed from locomotor uses. True hands appear only in the primates.

The skeletal structure of the hand includes the series of five bones, the metacarpals, which attach it to the wrist, and the five divergent series of phalanges located in the digits. Of the digits, one, the thumb, is placed and articulated so that it can be opposed to the other four, which are fingers. As a result, the appendage can be used for grasping like a forceps, and also as a prehensile organ by folding the fingers back against the palm. In some of the monkeys, the prehensile method of grasping is more important in moving through the trees, and the thumb has shifted and become smaller so that it can no longer be opposed.

The human hand is the most versatile grasping organ in the animal kingdom.

HANDEDNESS (Right- and Left-). Defined in terms of the motion of a screw. A *right-handed* screw, when rotated in the sense of Fig. 1(a) (counterclockwise looking down at the page), will move out of the page; when rotated in the sense of Fig. 1(b), a right-handed screw will move into the page. A *left-handed* screw will move into the page

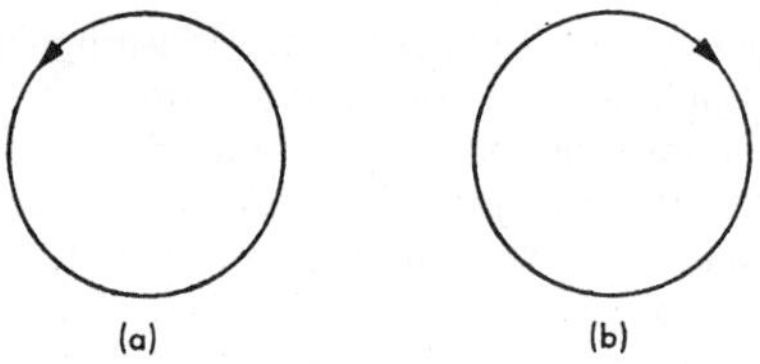

Fig. 1. Screw rotations.

in (a) and out of the page in (b). The mirror image of a right-handed screw is left-handed and vice versa.

The vector product is also defined in terms of a right-handed screw. Thus $\mathbf{A} \times \mathbf{B} = \mathbf{C}$ where the magnitude of $\mathbf{C}$ is $|\mathbf{A}||\mathbf{B}| \sin \theta$, and the direction of $\mathbf{C}$ is given by the direction of progression of a right-handed screw rotating in the sense of rotating $\mathbf{A}$ into $\mathbf{B}$ through the smaller angle (θ). $\mathbf{C}$ is thus a vector pointing into the page in Fig. 2.

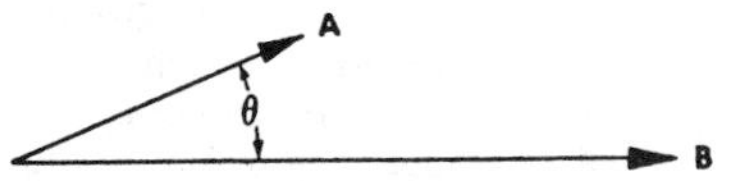

Fig. 2. Vector product.

Coordinate systems are also classed as right or left handed. In Fig. 3, coordinate system (a) is right-handed, since rotation of the unit vector $\mathbf{i}$ into the unit vector $\mathbf{j}$ would make a right-handed screw progress in the direction of $\mathbf{k}$, $\mathbf{i} \times \mathbf{j} = \mathbf{k}$ and also $\mathbf{j} \times \mathbf{k} = \mathbf{i}$, $\mathbf{k} \times \mathbf{i} = \mathbf{j}$. Thus in a right-handed coordinate system, the above cyclic relations among the unit vectors hold. Coordinate system (b), on the other hand, is left-handed, i.e., it would take a left-handed screw to carry the unit vectors into each other in cyclic order of vector multiplication.

Circular polarization of electromagnetic waves is described as right-handed or left-handed depending on whether the direction of rotation of the electric vector and the direction of progression of the electromagnetic wave are related to a right-handed or a left-handed screw.

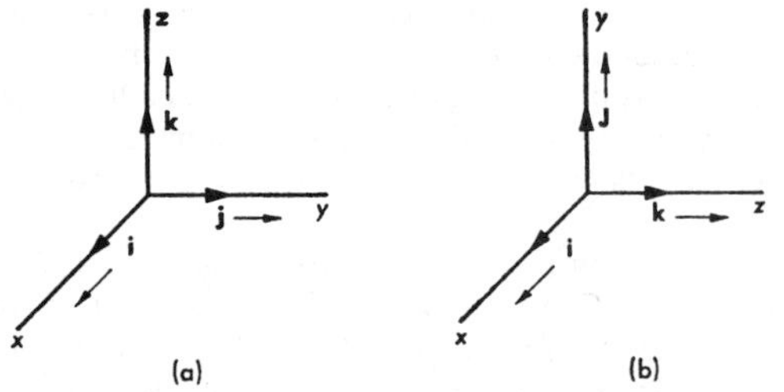

Fig. 3. Coordinate system.

HAND-SCHÜLLER-CHRISTIAN SYNDROME. Lipidoses.

HANDSET. This is the part of the modern telephone which contains the transmitter and receiver, i.e., the part which the user holds when talking. While this placing of the transmitter and receiver on the same handle seems a rather insignificant accomplishment, the necessity of avoiding feedback from the receiver into the transmitter makes very careful acoustical design necessary. In addition the electrical and mechanical construction of the transmitter must be such that it can be operated in almost any position as the user will hold the instrument in innumerable positions.

HANGING VALLEY. Under normal conditions a tributary stream enters the main stream at grade, that is, at the same level. Under certain circumstances the tributary valley may be at a greater elevation than the main valley into which the tributary stream will plunge, forming a waterfall. In such cases the tributary valley is called a hanging valley, and the stream in it is said to be out of adjustment with the main stream.

Hanging valleys originate in the following ways: by glacial action, the main glacier cutting down its valley faster than a tributary glacier; by river action, the main stream eroding its bed faster than the tributary stream; by faulting, the tributary stream flowing off the upthrown block. A fourth type of hanging valley, much less common, may result from a stream plunging over wave-cut cliffs or other escarpments into a lake or ocean basin.

HANKEL FUNCTION. Mathieu Equation.

HANTZSCH-WIDMAN NAME. Organic Chemistry.

HAPLOID. Cell (Biology).

HAPLOMI. Fishes.

HAPTEN. Immune System and Immunology.

HARDENABILITY OF STEEL. The hardenability of steel refers to the ease with which it can be hardened rather than the maximum hardness value attainable. For example, a 1-inch diameter bar of a certain 0.20% carbon alloy steel can be hardened to 50 Rockwell "C" in the center by quenching in oil. A similar bar of plain carbon steel requires a drastic quench in brine to attain the same hardness, and therefore, has a lower hardenability. Neither bar can be quenched to a greater hardness because 50 Rockwell "C" is the maximum attainable for a 0.20% carbon steel. A 0.40% carbon steel can be hardened to a maximum of about 60 Rockwell "C" and the maximum for high-carbon steel is about 65 Rockwell "C."

Of the several methods for determining the relative hardenability of steels, the Jominy test is the most widely used. A cylindrical specimen 1 inch (2.5 centimeters) in diameter and about 3 inches (7.6 centimeters) long is heated to the hardening temperature and quenched in a special fixture which holds the specimen in a vertical position and directs a stream of water on the bottom surface. The stream takes an "umbrella" shape and does not wet the sides. Cooling occurs progressively from the bottom to the top of the cylinder and the cooling rate at any distance from the bottom is known and reproducible from one sample to another. The hardness along the length of a quenched Jominy bar decreases from bottom to top. The distance from the bottom, expressed in sixteenths of an inch, to the point where the hardness is 50 Rockwell "C" is one method of reporting the hardenability.

One of the principal functions of alloying elements in steel, such as manganese, chromium, nickel, molybdenum, etc., is to increase the hardenability. Whereas prodigious amounts of expensive alloys were formerly used to insure full hardening, especially in medium and heavy sections, wartime shortages focused attention on the use of as little alloy as possible within the hardenability requirements. A large number of steels were developed containing relatively small additions of a number of elements, and a number of these steels have continued in use.

HARDENING OF METALS. There are three principal methods of hardening metals and alloys: cold working by (see **Cold-Worked Metal**) plastic deformation, precipitation hardening, and quench hardening as applied to steel. The last two methods involve heating and cooling operations. A pure metal may also be hardened through the addition of alloying elements. When a solid solution is formed it is normally harder than the pure metal. If additional phases are formed by alloying, these may also be harder than the pure metal and contribute to the hardness of the metal.

HARDENING OF THE ARTERIES (Atherosclerosis). Arterial and Venous Disorders.

HARDENING (Precipitation). Precipitation Hardening.

HARD FACING. Deposition of a hard wear-resistant alloy on a metal surface. The material to be deposited is generally in the form of a welding rod and may be applied by gas or arc welding. Such surfaces are usually finished by grinding.

While hard facing or hard surfacing is usually a maintenance operation, it is also used in new production. The surfacing material may be cemented carbides, nonferrous Stellite-type alloys, or iron-base alloys with alloying additions such as chromium, tungsten, manganese, silicon, nickel, and carbon. While hard facing is most often applied to steel, cast iron and some of the nonferrous alloys such as Monel metal can also be coated. Typical applications are metal-working dies, oil well drilling tools, excavating equipment, shafting, and rolling mill rolls.

HARDNESS. The significance of this term as applied to solids has various interpretations. Commonly, it refers to the resistance of the substance to surface abrasion, so that of two solids, the one that will scratch the other, as diamond scratches glass, is the harder. Again, it may denote rigidity, or lack of plasticity, or even strength; in some cases a combination of several such properties. The original Mohs' Scale of Hardness is delineated in Table 1 and further described under **Mineralogy.**

In metallurgy and engineering, hardness is determined by methods based on resistance to penetration by an indenter of greater hardness than the material being tested. Aluminum, copper, lead, magnesium, tin, and their alloys, as well as plastics are generally indented by hardened steel balls ranging in size in the various tests from $\frac{1}{16}$ inch to 10 millimeters in diameter. The same methods may be used for soft steels and irons, but for heat-treated steels and all other alloys which develop high hardness special diamond indenters, or in some cases sintered tungsten carbide balls, are used. In all of the technological tests, the indenters are impressed into the test material under carefully regulated loads; thus, the relative size of the resulting indentation becomes a measure of hardness. (See Table 2.) The operating principles of the instruments most widely used in this country follow:

Brinell. The indenter is a 10-millimeter diameter hardened steel ball. A sintered tungsten-carbide ball is also coming into use, especially for testing hard metals. The load applied is generally 500 kilograms for soft metals and 3,000 kilograms for steels and hard metals. Brinell hardness is equal to the load (kilogram) divided by the surface area (square millimeter) of the impression made in the test material. Tables are available for direct conversion to hardness from the diameter of the indentation as measured with a calibrated magnifier after removal of the piece from the testing machine.

Rockwell. Indenter is $\frac{1}{16}$-, $\frac{1}{8}$-, or $\frac{1}{4}$-inch-diameter (1.6, 3.2, or 6.4 millimeter) steel ball or a conical diamond having an apex angle of 120° and a slightly rounded point. The various scales used are designated by letters. Rockwell "B," for example, indicates a 100-kilogram load on a $\frac{1}{16}$-inch (1.6 millimeter) diameter ball. Rockwell "C" indicates a 150-kilogram load on the diamond indenter. Rockwell "30T" designates a load of 30 kilograms on a $\frac{1}{16}$-inch (1.6 millimeter) diameter ball. (An instrument of higher sensitivity known as the Rockwell Superficial Tester is used for loads of 15, 30, and 45 kilograms.) The size of the indentation is measured by a dial gauge as the final depth minus a small preliminary penetration produced by a minor preload of 10 kilograms. The Rockwell hardness values are arbitrary numbers having an inverse relationship to the depth of the indentation.

Vickers. Also known as Diamond Pyramid Hardness. Indenter is a square-based diamond pyramid with included angle between faces of 136°. Loads may vary from 1 to 120 kilograms with 10, 30, and 50 kilograms in common use. Hardness is equal to load (kilograms) divided by surface area (square millimeter) of the permanent indentation. It is determined directly from optical measurements of the diagonals of the indentation which appears square at the surface of the metal.

Tukon. A highly sensitive instrument for determining hardness under very light loads down to 25 grams. The small indentations are measured at high magnifications up to 1,000 times. The indenter is a diamond pyramid that makes an elongated impression, one diagonal being 7 times the other in length.

Eberbach. Also used for very light loads. Consists of a spring-loaded, Vickers-type diamond pyramid indenter arranged for use on a metallurgical microscope.

Scleroscope. Depends on the height of rebound of a diamond-tipped body falling under the force of gravity from a fixed height. The instrument is relatively small and is portable. One type reads directly on a graduated dial.

While there is overlapping in the field of useful application of the various hardness tests, each has certain special qualifications. The

TABLE 1. HARDNESS SCALES

Mohs' Scale	Ridgway's Extension of Mohs' Scale	Metal Equivalent	Others
1. Talc			
2. Gypsum			
			2.5. Finger Nail
3. Calcite			
4. Fluorite			
5. Apatite			
			5.5. Window Glass
6. Feldspar (Orthoclase)	6. Orthoclase or Periclase		
			6.5. Steel (Knife Blade; File)
7. Quartz	7. Vitreous Pure Silica		
8. Topaz	8. Quartz	8. Stellite	
9. Corundum or Sapphire	9. Garnet		
	10. Topaz		
	11. Fused Zirconia	11. Tantalum Carbide	
	12. Fused Alumina	12. Tungsten Carbide	
	13. Silicon Carbide		
	14. Boron Carbide		
10. Diamond	15. Diamond		

1. In the above scales each abrasive is capable of scratching all others above it in each scale and may be scratched by all abrasives below it.
2. The gap between 9 and 10 in the original Mohs' scale is much greater than that between 1 and 9 in the same scale.
3. Various additional hardness scales have been devised by different investigators; in general, different materials maintain the same order of hardness in all these scales.

TABLE 2. TYPICAL HARDNESS VALUES

Material	Brinell 500 kg	Brinell 3000 kg	Rockwell	Vickers 50 kg
Aluminum, annealed	23		H 45	25
Magnesium alloy	63		B 21	63
Armco iron	66	73	B 31	71
Yellow brass, annealed	72	82	B 40	77
Copper, cold rolled	99	83	B 55	110
Mild steel, annealed	107	117	B 70	123
Aluminum alloy, 24st	130	144	B 78	146
Stainless steel, annealed	121	145	B 80	153
Yellow brass, cold rolled	174	178	B 91	189
Ni-Moly steel, quenched in water, tempered at 1200°F (649°C)		241	C 23	255
Same, 1000°F (538°C)		293	C 31	310
Same, 800°F (427°C)		363	C 38	380
High-speed tool steel		684	C 62	740

Brinell test makes a large indentation, giving an average hardness value for several grains even in rather coarse-grained metals; however, it cannot be used on small or thin specimens. The various Rockwell tests are widely used, especially for rapid production inspection of parts. The Vickers test, which originated in England, is less rapid than the Rockwell but has the advantage of a single scale covering the hardness of all metals from lead to the hardest tool materials. The Tukon test makes it possible to determine the hardness of very thin sheets and of thin metallic coatings such as chromium plate, or zinc on galvanized steel. The Scleroscope test is used principally on heavy forgings or castings which cannot be placed in an indentation-type instrument, or for field tests where a portable instrument is required.

HARDNESS (Mineral). **Mineralogy.**

HARDNESS (Water). **Feedwater (Boiler).**

HARDGROVE GRINDABILITY (Coal).

HARDPAN. The term which prospectors and miners give to the subsurface or basal layers of placer deposits in which the gold-bearing gravels have been cemented and hardened. The same term is also used to designate till or boulder clay which has been cemented by limonite.

HARDWARE (Computer System). The physical equipment or devices forming a computer and peripheral equipment. This is contrasted with the program, procedures, rules, and associated documentation which, collectively are generally termed *software*. See also **Software (Computer System).**

HARDWOODS. Wood.

HARE. Rabbits and Hares.

HARELIP. A congenital deformity in which there is a failure of fusion of the maxillary and median nasal processes, resulting in a cleft in the upper lip. This is part of the same defect associated with cleft palate. In some cases, the harelip may be double, in which case there is a division on either side of the mid-line of the lip. Correction of the deformity must be by surgery, and this is best accomplished at a very early age. Usually, correction of the lip is done first, thus enabling the infant to suck. Surgery on the palate normally is undertaken just as soon as there is sufficient tissue to cover over the bony palate after repair. This usually occurs at the age of 18 to 24 months, well before abnormal speech habits are formed. In some cases, the operation must be performed in several stages. Cosmetically and functionally, the surgical results usually are excellent.

HARLEQUIN BUG (*Insecta, Hemiptera*). Of the family *Pentatomidae* (stink bug), also known as the "fire bug" and the "calico black" bug, this insect is of major economic importance on cabbage and cruciferous crops in the southern United States. When not controlled, an entire crop can be destroyed. The bug kills plants by sucking sap from the underground portions of the plant. In addition to cabbage, the harlequin is injurious to Brussels sprouts, collard, cauliflower, horseradish, kohlrabi, mustard, radish, and turnip. If these preferred sources of nourishment are not immediately available to the insect, it will attack asparagus, bean, eggplant, okra, potato, and tomato. The insect also can damage numerous other garden crops, certain weeds, small fruit trees, and field crops.

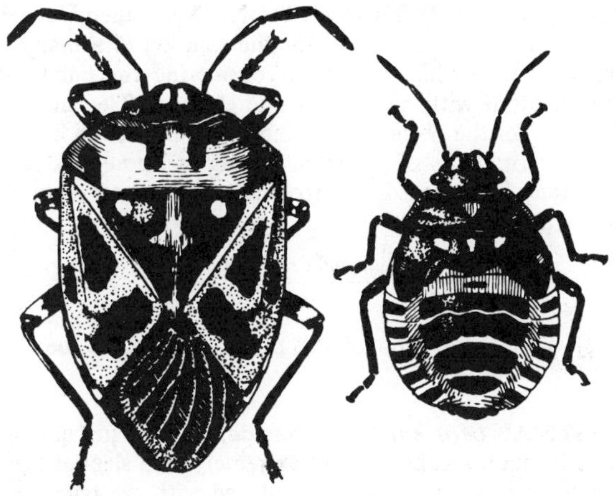

Harlequin bug. Adult at left; nymph at right. (*USDA diagram.*)

The harlequin bug is colorful, with red and black spots, of a shield shape, and about $\frac{3}{8}$ inch (9 to 10 millimeters) long. The nymphs, somewhat smaller, have a similar appearance. All stages of the insect may be found from the early to the late months of the year. In the United States, the pest is found in the southern portion in all areas from the Atlantic to the Pacific coasts. The insect is believed to be native to Mexico. In more northern areas, the bug winters as an adult. During the first warm spell of spring, eggs are deposited on the underside of leaves. They appear something like tiny white beer kegs, standing on end and usually in a double row. There are two black bands around each "keg." Hatching of the eggs occurs within 1 to 4 weeks, depending upon temperature. Immediately, the nymphs commence feeding and destroying target plants. During a period ranging from 4 to 9 weeks, the insect passes through five instars, after which it is ready to mate and lay eggs for the next generation. Normally there are three to four generations per year.

In addition to control chemicals, populations can be controlled by destroying weeds that attract the insects. These include *Amaranthus* and wild mustard. Advantage of the insect's preference for certain crops, such as kale, mustard, radish, and turnip, can be taken by planting a small area to such plants either very early in the season or after harvest. Large concentrations of the insects thus can be killed with relatively small amounts of insecticide. The debris from such decoy crops should be burned.

HARMONIC. A sinusoidal frequency component of a waveform. The harmonic has a frequency that is an integral multiple of the fundamental frequency. The frequency of the second harmonic will be double that of the fundamental frequency (first harmonic).

Harmonic distortion is nonlinear distortion characterized by the appearance in the output of harmonics other than the fundamental component when the input wave is sinusoidal. Harmonic distortion is sometimes called amplitude distortion.

HARMONIC ANALYSIS. Not only is it possible to combine two or more simple harmonic motions of different period, amplitude, and phase to form a complex motion, but there are also means of analyzing the resultant motion, when the latter is given, to find its component harmonics. For example, if the wave form of such a complex tone as that produced by a bell or a saxophone is accurately graphed by means of a phonodeik the equation of the vibratory motion can be deduced in such form as to show the separate components. Fourier showed that the same analysis is possible for any periodic motion, however complicated. The equation, called Fourier's series, may be written

$$y = a \sin 2\pi nt + b \cos 2\pi nt + c \sin 4\pi nt + d \cos 4\pi nt + e \sin 6\pi nt + f \cos 6\pi nt + \cdots$$

in which y is the displacement of the vibrating particle and t is the time. The fundamental frequency n and the constants a, b, c, d, etc., must be calculated from the given wave form or the data from which it is plotted. There is a type of instrument, called a "harmonic analyzer," which automatically computes the coefficients; or it may be done mathematically, though the process is very laborious. The accompanying figure shows the wave form and the twelve components of complex tone, analyzed by Professor D. C. Miller. See also **Musical Sound.**

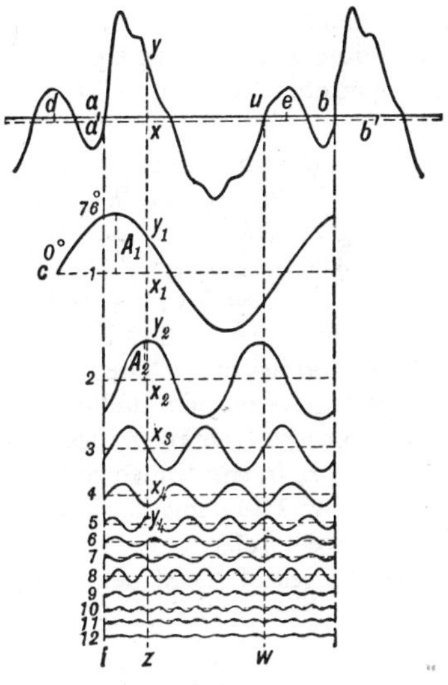

Records of a complex sound and twelve of its components.

HARMONIC MEAN. Average.

HARMONIC MOTION. A distinct type of periodic motion, or vibration, characteristic of elastic bodies; illustrated by a bird-cage bobbing up and down at the end of a spiral spring, or (approximately) by the piston of the steam engine. It may be either simple, with only one frequency and amplitude, or made up of two or more simple components and consequently of more complex character. The essential feature of simple harmonic motion is that, with its range extending to equal distances on both sides of an equilibrium position or origin, the acceleration is always toward the origin and directly proportional to the distance from it. With elastic vibrations this is easily seen to follow from Hooke's law, since the force tending to restore the deformed body to equilibrium is proportional to the deformation. See **Elasticity.** The motion is called "harmonic" undoubtedly because the vibrations of bodies emitting musical sounds are of this character. Any simple harmonic motion may be represented by the equation

$$y = a \cos(2\pi nt + \phi)$$

in which y is the distance at time, t, a is the amplitude, n is the frequency or number of vibrations per unit time, and ϕ is the phase constant, such that when $t = 0$, $y = a \cos \phi$.

It is interesting to note the relationship between harmonic and circular motion. If a peg is inserted in the face of a circular disk or wheel and the latter uniformly rotated, the motion of the peg, as viewed with the wheel seen edgewise, is simple harmonic. In fact, uniform circular motion is made up of two simple harmonic components of the same period and amplitude at right angles, one being a quarter-period ahead of the other in phase. If the two harmonic components have a phase difference other than a quarter-period, the resultant in general is motion in an ellipse; while if they have unequal periods, the path is one of a class of more or less complicated loci called "Lissajous' curves."

HARMONIC OPERATION. *Impeded* harmonic operation is constrained magnetization or forced magnetization. It is the type of operation which takes place in a magnetic amplifier in which the impedance of the control circuit and any circuit closely coupled to it is so great as to substantially prevent the flow of all harmonic currents in such circuits.

Unimpeded harmonic operation is natural magnetization or free magnetization. It is the type of operation that takes place in a magnetic amplifier in which the impedance of the control circuit or any circuit closely coupled to it is so small as to permit substantially unimpeded flow of all harmonic currents in such circuit.

See also **Amplifier.**

HARMONIC PROGRESSION. Progression.

HARMONIC SYNTHESIZER. A machine which combines elementary harmonic constituents into a single periodic function. A machine performing the opposite function is called a harmonic analyzer.

HARMOTOME. The mineral harmotome is a zeolite, composition approximately $(Ba, K)(Al, Si)_2Si_6O_{16} \cdot 6H_2O$; it is monoclinic but often forms double twins giving the effect of a square prism. It is a brittle mineral; hardness, 4.5; specific gravity, 2.41–2.50; luster, vitreous; color, white to gray or perhaps yellow, red or brown; white streak; translucent. Harmotome like other zeolites is found in cavities in basalts and similar rocks, sometimes in trachytes or in gneisses, occasionally as a gangue mineral in veins of metallic minerals. Some well-known localities are in Bavaria; the Harz Mountains; Norway; and Scotland. Harmotome occurs in the United States with stilbite, near Port Arthur, Lake Superior. The name harmotome comes from the Greek meaning joint and to cut, referring to the division of the pyramid formed by the prismatic faces of the mineral when in the twinned position.

HARPY. Eagle.

HARRIER. Eagle.

HARTEBEEST. Antelope.

HARTLEY. In information theory, a unit of logarithmic measures of information equal to the decision content of a set of ten mutually exclusive events expressed by the logarithm with the base ten. For example, the decision content of a character set 8 characters equals $\log_{10} 8$, or 0.903 Hartley. Synonymous with information content decimal unit. (*American National Dictionary for Information Processing*) See also **Shannon.**

HARTLEY OSCILLATOR. Oscillator.

HARTLEY PRINCIPLES (Transmission). The amount of information that can be transmitted is proportional to the width of the frequency range, and the time it is available. Information content is equated to the total number of code elements, multiplied by the logarithm of the number of possible values a code element may assume. Information content is independent of how the code elements are grouped. By quantizing, the continuous magnitude-time function used in ordinary telephony may be transmitted by a succession of code symbols such as are employed in telegraphy. To obtain the maximum rate of transmission of information, the signal elements need to be spaced uniformly.

Time-Frequency Duality. As implied by the Fourier integral, a time function cannot be confined within a small region on the time scale when the steady-state transmission characteristic is confined to a narrow range on the time scale. For example, it is well known that, if a telegraph dot is made narrower and narrower, its corresponding significant-frequency spectrum becomes broader and broader until, in the limit when the dot becomes an impluse, its significant-frequency spectrum is of infinite intent.

HARTMANN TEST. Hartmann devised various optical tests, including the following: (1) Hartmann test for telescope mirrors. For a perfect mirror, light from all points on the mirror should come to the same focus. By covering the mirror with a screen, in which regularly spaced holes have been cut, and then permitting the reflected light to strike a photographic plate placed near the focus, the failure of dots on the plate to be regularly spaced indicates a fault of the mirror. (2) Hartmann test for spectrometers. Light is passed through different parts of the entrance slit. Any change in the spectrum as different parts of the slit are used indicates a fault of the instrument. A "Hartmann diaphragm" is one device for using only one part of the entrance slit at a time.

HARTREE-FOCK APPROXIMATION. Also called Hartree-Fock-Slater approximation. A method for the solution of a many electron problem, e.g., that which arises in considering the band theory of solids or an atom with more than one electron. The antisymmetric wave function for the N-electron system is expanded as a linear combination of determinants of order N, having as elements one electron wave functions. This procedure introduces exchange terms in the Hamiltonian, of the form:

$$e^2 \int \left[\frac{\psi_i(r_1)\psi_j^*(r_2)}{r_{12}} \, d\tau_2 \right] \psi_j(r_1)$$

where r_{12} is the separation of the points defined by the vectors r_1 and r_2.

HARVESTMAN (*Arachnida, Phalangida*). Spider-like animals, most species with small oval bodies and extremely long slender legs. Those with shorter legs are more easily confused with the true spiders but all may be recognized by the segmented abdomen. Daddy longlegs.

HASHIMOTO'S STRUMA. Thyroid Gland.

HASHISH. Drug Addiction; Marijuana.

HASTELLOY. Nickel.

HATCHET FISHES (*Osteichthyes*). Of the order *Isospondyli*, family *Sternoptychidae*, hatchet fishes are small, rarely exceeding 3½ inches (9 centimeters) in length. They are silvery and are so named because of their hatched-head appearance. They possess photophores (light organs) on their sides and undersurfaces. The genus *Argyroplecus* features telescopic eyes which are aimed in an upward direction. They are a food source for tuna, but are not nearly so abundant as their relatives, the bristlemouths. Some species of hatchet fishes have been favorites among tropical-fish fanciers.

The flying hatchet fishes of South America of the order *Ostariophysi*, family *Characidae* are the only fishes credited with performing true flight. See also **Characids (Osteichthyes).**

Hatchet fishes are fully adapted to a life near the water surface. Like speedboats, the front of which rises off the water at high speed, these fishes can also rise off the water surface. They literally fly several yards (meters) through the air, after a starting movement of a few yards (meters). The initiation of the movement has been seen in nature, but the actual flying is difficult to observe. Laboratory investigations, however, indicate that the process is accompanied by a humming sound.

HAUSDORFF SPACE. Topological Space.

HAVERSINE. Trigonometric Function.

HAWAIIAN AGENT (Virus). Norwalk Virus.

HAWK (*Aves, Falconiformes*). Birds of prey with hooked beaks and large curved claws, closely related to the eagles, falcons, harriers, and others and not sharply distinguished as a group. Hawks are found on all continents. North America has many species, including buzzards, harriers, goshawks and other forms. Most of them are beneficial as destroyers of vermin but the sharp-shinned (*Accipiter velox*), and Cooper (*A. cooperi*) hawks destroy too many birds, including poultry, to be regarded as friends.

There are numerous species of hawks, at least 25 of these occurring in North America, particularly north of Mexico. The hawks are swift in flight, seek their prey by day, have remarkable vision, and eat only what they kill. They are very bold, pouncing upon their prey in a rapid swoop, using claws and talons, firmly clinching the victim. The hawk prefers to take its victim to a private location for consumption.

Cooper's hawk. (*American Museum of Natural History.*)

In the African rain forest, three species are known as darters.

The osprey is a large bird of prey of almost worldwide distribution. It is a skillful fisher and is known in North America as the fish hawk, *Pandion haliaëtus.*

Buzzards are birds of prey of several species that belong to the genus *Buteo.* The North American representatives are commonly called hawks, as Swainson's hawk. The same may be said of the nearly related rough-legged buzzards; American representatives of the genus are the rough-legged hawks. The name is incorrectly, although commonly, applied to the turkey buzzard, which is a vulture. See also **Falconiformes.**

HAWK MOTH (*Insecta, Lepidoptera*). Large moths composing the family *Sphingidae,* one of the largest of the order. These moths have a long, rather stout body projecting beyond the narrow wings. The front wings are much longer than the hinder pair, and because of their limited surface they are vibrated rapidly in flight. The moths have long tongues and visit deep-throated flowers. From their habit of hovering as they probe the flower for nectar they are also called hummingbird moths. Another common name is sphinx moth.

HAWTHORN TREES AND SHRUBS. Rose Family.

HAY BRIDGE. Bridge Circuits (Electrical).

HAY FEVER. Allergy.

HAZE. Precipitation and Hydrometeors.

HAZELNUT SHRUBS. Of the family *Corylaceae,* genus *Corylus,* hazelnut shrubs (rarely trees) are deciduous, and characterized by male catkins that hang from the tree during most of the winter months, and by their edible and tasty fruit, an ovoid nut in a toothed container, simply known as the hazelnut of commerce. The term *filbert* is generally reserved for use with reference to the fruits of two European hazelnut plants, *Corylus avellana pontica* and *C. maxima.* The American hazelnut (*C. americana*) is a shrub ranging from 3 to 8 feet (0.9 to 2.4 meters) in height. It is commonly found in thickets and hedgerows. The leaves are narrow, heart-shaped or sometimes ovate, with abrupt points. They are of a lackluster dark green color and from 3 to 5 inches (7.6 to 12.7 centimeters) in length. The stems are short. The staminate catkins are from 3 to 4 inches (7.6 to 10.1 centimeters) in length. This shrub ranges from Maine westward to Alberta and Kansas and southward to Florida. The beaked hazelnut (*C. rostrata*) ranges from 3 to 8 feet (0.9 to 2.4 meters) in height and commonly occurs along the road in thickets, ranging throughout Canada from Quebec westward to the Pacific slopes and south into the United States to Missouri, Michigan, and Ohio, and Delaware in the east. It is found in the mountains as far south as northern Georgia. The fruit is edible and sweet and in the form of an ovoid nut. The nut in enclosed in a bristly cup which has a beak-like termination, hence the name. The California hazelnut (*C. cornuta* or *californica*) is well known for its velvety leaves and makes an attractive garden shrub. For purple coloration in gardens, the purple hazel (*C. maxima purpurea*) is sometimes used. The Turkish hazel (*C. colurna*) can be classified as a tree, in that it can attain a height up to 75 feet (22.5 meters), but in most respects it is similar to the lesser hazelnut shrubs. Another species is the corkscrew hazel or Harry Lauder's walking stick, a shrub which attains a height up to 10 feet (3 meters). It makes an attractive shrub, particularly in winter months when the catkins are on display.

Slected as a champion shrub by The American Forestry Association in this class in 1972 is the *C. cornuta,* located in the Siuslaw National Forest, Oregon. This shrub/tree has a circumference (at $4\frac{1}{2}$ feet) (1.4 meters) of 1 foot, 9 inches (0.53 meter) a height of 22 feet (6.6 meters), and a spread of 20 feet (6 meters).

HEADACHE. Head pain and ache is a symptom and not a disease. Headache is one of the most common symptoms of a disorder, not only of the nervous system, but other parts of the body as well. Consequently, discovery of the primary cause of headache is often difficult. The degree of pain associated with headache does not necessarily correlate with seriousness of a cause, a violent headache sometimes being associated with a relatively minor injury. Diagnosis of headache complaint can be facilitated by providing accurate information to the physician—events occurring before the headache, such as emotional stress, exertion, eating, and so on; the time of day or night when headache usually occurs; and other symptoms that may accompany headache, such as nausea, flashes of light, ringing in the ears, rapid or slow onset of the headache, as well as how the headache usually ceases.

The large veins (venous sinuses) and their tributaries that drain the surface of the brain are sensitive to pain, as are the arteries. The brain substance itself apparently is not sensitive to pain, but the coverings of the brain are. The sinuses, teeth, ears, and muscles in the area of the head may be affected so that pain from them, at first local, later covers a wider area.

At least eight pain mechanisms have been identified as causative factors in headache: (1) dilation of the cranial arteries; (2) pulling or traction upon pain-sensitive intracranial structures; (3) traction on and dilation of intracranial blood vessels; (4) inflammation of structures within the skull; (5) contraction of skeletal muscles over the head and neck; (6) spread of pain from stimulation elsewhere in the head; (7) pain from allergenic reaction; and (8) mentally-produced (*psychogenic*) pain. The majority of headaches for which medical attention is sought arise either from dilation of the cranial arteries or contraction of the muscles of the head and neck, or by combinations of these factors. Fortunately, headaches of this type arise from conditions that usually are easy to correct.

Vascular headache is the term applied to the condition caused by dilation of the cranial arteries. It is associated with general infections, migraine headaches, or those resulting from taking certain drugs; and is largely responsible for so-called hunger and hangover headaches. The headaches of suddenly increased blood pressure are in this group, as well as headaches which follow convulsive seizures or head injury. Headaches of this type usually have a throbbing quality, but this may not be present if the headache is prolonged.

Treatment of vascular headache is generally directed to the underlying cause. The inhalation of high concentrations of oxygen are particularly helpful to persons whose headaches are caused by lack of oxygen. Headaches caused by traction or pressure on intracranial structures are associated with expanding intracranial masses, with brain tumors, abscesses, and hematomas, as examples. Such headaches are aggravated by coughing or straining and are not relieved by drugs which constrict the arteries. Headache associated with brain tumor may be intermittent and mild to moderate in severity and usually does not interfere with sleep.

The headache produced by a hematoma (swelling or tumor filled with blood) is dull, steady, and felt throughout the head. The pain from brain abscess is similar to that of tumor. However, the abscess must be of sufficient size to cause traction before pain is felt.

Headaches caused by traction upon and dilation of the intracranial vessels are typified by the headache which frequently follows lumbar spinal puncture. Despite precautions, at times there may be slow leakage of the spinal fluid through the hole made by the needle. This results in headaches which are ordinarily mild, but can be severe. Once the headache develops, bed rest is about all that is required. The condition heals spontaneously.

Headaches resulting from inflammation of cranial structures are experienced if the patient has any infection within the skull, such as meningitis or encephalitis. Such a headache also occurs as a result of the inflammation that follows brain hemorrhages. These headaches may be intense and require narcotics.

Typical of the headache which may occur secondary to pain arising elsewhere is that associated with sinusitis. Certain diseases of the eye, or eye strain resulting from long use, or excessive attempts at accommodation of eyes that require glasses, also cause this type of headache. Infections which involve muscles, especially rheumatic fever, cause this type of head pain. Arthritis of the upper part of the neck or tumors in this area also produce this type of headache.

Headaches also may be caused by sustained contraction of the muscles of the head and neck. The contraction may result from local muscle or nerve injury. Muscle contraction sometimes is associated with emotional tension.

Headaches resulting from muscle contraction are located frequently over the back and lower part of the head and upper part of the neck. The pain is a steady, deep ache often associated with a feeling of pulling or tightness. The headache may be aggravated by movements of the head. The muscles themselves may be tender and tight. Massaging the muscles and application of heat often relieve the pain.

Headaches caused by allergy often go unrecognized because they are not distinguished from other types by location and duration of pain.

Headache caused by emotional stress is not well understood. This type of headache can be bizarre. The patient may experience a wide variety of unusual sensations. While the headache may be described as being terribly intense, it ordinarily is of moderate severity and usually does not respond to medications which relieve other forms of headache. In many cases, the treatment needed is primarily psychiatric.

Migraine headaches have been reported since ancient times. Migraine has been termed one of the most common complaints of civilized people. The onset of migraine headaches usually occurs between the ages of 12 to 25, but they can begin at any age. Persons who perform mental work are more likely to be affected than blue collar workers. Also, urban dwellers seem to be more affected than people in rural areas. Often, the migraine victim will be an ambitious, hard driving, meticulous, and exceptionally intelligent individual.

An outstanding feature of migraine headache, thus differentiating it from other types, is that it affects one side of the head. Other distinctions are the periodic recurrence. There is some evidence that migraine may be hereditary. In most instances, the headaches occur about once every two weeks. In women, it may be associated with the menstrual period. Attacks in some persons, however, do not show this regularity, with headaches being separated by months or even years. A migraine headache may last from a few hours to more than a week. In any individual, the characteristic pain, accompanying symptoms, and length of time are usually about the same for each attack. Some sufferers can generally predict such experiences.

The typical migraine headache commences in the temple, eyeball, or forehead, and soon spreads to include either the left- or right-half of the head. The pain may involve the face and neck and sometimes the arms. Sometimes the headache is preceded by disturbances of vision (dullness of vision, blinding flashes of light, sensitivity to light or sound, or dizziness). As the attack begins, the patient may notice a blind spot, that is, several words in a printed sentence may not be seen. This spot, in rare instances, increases in size until vision in one field is fully gone. The patient may regain the ability to see in the later stages of the attack, but he may still be troubled with dazzling black and white flashes of light. During the attack, the victim's face usually is pale and sallow and the skin may be sweaty and clammy. The arms and legs may feel cool to the patient, even though there may be fever. Nausea and violent vomiting often mark the climax of the attack. After the attack has run its course, if there has been no vomiting, the patient usually feels relaxed and relieved and may be filled with energy and tend to be overactive, although a dull headache may persist for a day or two.

Many migraine sufferers report that attacks seem to occur in relation to periods of let-down or of exhilaration. Many have noted that their headaches commence on weekends, the first day of a holiday, or on days of planned social engagements or travel. Often, on the eve of onset, the victim may be in high spirits, with an unusually increased appetite. However, on the following morning, the victim may arise with a very depressed or melancholic attitude. The victim may become restless, irritable, and confused, with an inability to concentrate on routine tasks, or to make decisions. There may be a tendency to absent-mindedness.

Many theories have been offered as to the cause of migraine headaches. Most authorities agree that distention of the cranial arteries in the scalp is the immediate cause, but the causes of such distension are not well understood. There appears to be a relation between the personality traits of migraine patients with those having high blood pressure. It has been noted that about 50% of the children of migraine patients also suffer from migraine.

Some investigators believe that migraine headaches occur as the result of allergy, probably to certain protein-containing foods. According to some patients, chocolate has been an offender. In some patients, migraine may be akin to asthma. However, in other patients, the headaches are thought to occur because of eyestrain; and in others, to an imbalance of the endocrine system.

Generally, the patient should be left alone in a quiet, darkened room because most migraine patients are extremely sensitive to light and odors. An ice bag on the head and hot water bottle at the feet may provide some relief from pain. In some patients, sitting in an upright position rather than lying down reduces the intensity of the pain. Ergotamine tartrate, to be prescribed only by a physician, has been found helpful in terminating the headache in many instances, if given at the beginning of an attack. Inhalation of 100% oxygen may alleviate pain. Strong drugs should not be taken for a migraine attack unless prescribed by a physician, for it is too easy for migraine sufferers to develop a drug habit. The victim should make every effort to determine the factors associated with attacks and to avoid wherever possible such factors. Avoidance of fatigue, late hours, strain, and worry tend to reduce frequency or severity of migraine attacks. In most persons, physical or mental tension is often the immediate cause of an attack.

Women who suffer attacks of migraine usually do not have any episodes during pregnancy; and the attacks may disappear entirely after the change of life. The disease may disappear in men and women at all ages, but most frequently attacks cease at around 50 years of age, when the elasticity of the blood vessels has diminished, so that the dilation previously described in the etiology of migraine has decreased.

See also **Brain (Injury).**

A classical review of headache can be found in "Headache and Other Head Pain," 3rd Edition, by H. G. Wolff (D. J. Dalessio, editor), Oxford Univ. Press, New York, 1972.

HEAD (Data Processing). A device that reads, records, or erases information in a storage medium, usually a small electromagnet used to read, write or erase information on a magnetic drum or tape or the set of perforating or reading fingers and block assembly for punching or reading holes in paper tape or cards.

HEADER. Any pipe, conduit, duct, or channel, which acts as a central point of distribution of a fluid flow to several branch lines, is a header.

HEAD FLOWMETER. Flow Measurement.

HEAD (Hydrostatic). Hydrostatic Pressure.

HEADING. The direction of the forward end of the keel of a ship (either airborne or seaborne) is known as the heading of the ship. Unless a qualifying adjective is used with the term heading, it means direction with reference to true north. Compass heading, or magnetic heading, may be converted to heading by applying the compass corrections.

See also **Compass (Navigation); Course;** and **Navigation.**

HEAD WIND. Jet Streams.

HEAD (Zoology). The region of a bilaterally symmetrical animal body lying at the front end in relation to the ordinary direction of locomotion, or, in bipedal vertebrates like humans and some of the birds, at the highest level.

The development of a head is indicated in animals which are without sharply separated body regions, such as the flatworms. This process of cephalization is closely correlated with bilateral symmetry. The portion of a bilateral animal which goes first inevitably is the first to encounter new sources of stimuli, and shows some concentration of sense organs. Usually the chief nerve center, a cerebral ganglion or brain, also develops here. The concentration of sense organs and nervous control in the head remains characteristic of the region throughout the animal kingdom and in most groups is accompanied by the location of the mouth in the head, together with associated structures for securing food.

HEARING AND THE EAR. The role of the sense organ of hearing (the ear) is to code acoustic disturbances into neural signals suitable for transmission to the brain. The study of this process necessarily involves anatomy and physiology of the ear, the nature of auditory pathways and central nervous system activity in hearing, properties of acoustic signals that elicit auditory responses, and observed phenomena of auditory behavior. These aspects serve to define and delineate areas for investigations of hearing.

The truly phenomenal aspects of hearing can be observed in such behavior as localization of sounds, speech perception and particularly the understanding of one voice in the noisy environment of many, and the recognition of acoustic events that only last a few milliseconds. These and other behavioral phenomena remain to be fully accounted for in theories of hearing.

Emphasis should also be given to the fact that the vestibular apparatus of the ear serves a nonhearing function of large importance, namely, for sensing orientation, balance, and posture.

Function and Structure of the Human Ear

The ear is a highly complex, intricate organ and, because of the several phenomena taking place in the chain of actions from the outer ear through the middle ear to the inner ear and hence to the brain, it is difficult and cumbersome to construct one or more analogs of its function so as to relate ear function to known, mechanical, electrical, or hydraulic systems to assist in the understanding of the total hearing system. Further, not all of the phenomena occurring in the ear are yet fully understood in a highly detailed fashion.

Although physiological functions may correspond in a general way to anatomical sequences, several physiological functions may occur in the same anatomical structure, or a single function may require several anatomical units. In a similar manner, psychological functions cannot usually be identified with specific physiological functions, and it is recognized that the central nervous system, as well as the auditory system, is involved in any auditory response. The correlations between and knowledge about structure and functions are best developed for peripheral, rather than central, parts of the auditory system because the ear is more accessible for examination and study than are the more central parts of the auditory system.

Traditional theories of hearing have been largely concerned with pitch perception or loudness of pure tones. However, within the last few decades, there has been an increased awareness that any comprehensive theory of hearing needs to encompass various experimental phenomena of hearing. In this regard, increased attention has been given to auditory processing of complex signals, to the examination of binaural inputs to the system, and to the study of pathological hearing conditions.

Principal Chambers of the Ear. When the structure of the ear is examined, it is convenient to consider the (1) external, (2) middle, and (3) inner ear separately. However, from a functional standpoint, the ear may be divided into an outer and an inner part. Some of the terms used here are indicated by the highly schematic representation of Fig. 1. The term *external* as used in connection with the ear encompasses the auricle (that part of the ear which can be seen) as

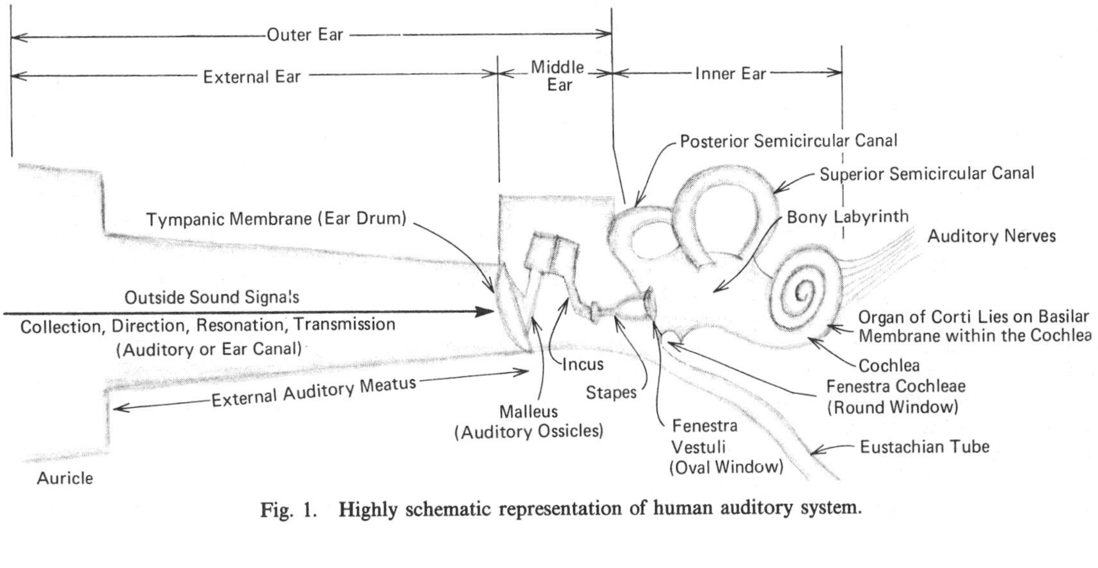

Fig. 1. Highly schematic representation of human auditory system.

well as the auditory or ear canal which extends to the *tympanic membrane* (eardrum). The outer ear is concerned with the transformation of acoustic energy into mechanical energy. The inner ear is concerned with the transduction of mechanical energy into neural impulses.

The auricle and *external auditory meatus* constitute the external ear. The meatus is an irregularly-shaped tube approximately 27 millimeters long with a diameter of about 7 millimeters, terminated by the tympanic membrane. The ear canal is an acoustic resonator, and frequencies in the range of 3,000 to 4,000 Hz are increased in pressure at the eardrum, as compared to the pressure at the entrance to the canal. The eardrum is in a protected position at the end of the canal, and humidity and temperature conditions at the drum are relatively independent of those external to the ear.

The middle ear is an irregular, air-filled space in the pestrous portion of the temporal bone. The three auditory ossicles of the middle ear (a) the *malleus*, (b) the *incus*, and (c) the *stapes* provide mechanical linkage between the tympanic membrane and the *fenestra vestuli*, an opening in the vestibule of the inner ear, commonly referred to as the oval window. The auditory ossicles are shown greatly enlarged in Fig. 2. The handle of the malleus attaches to the tympanic membrane, and the footplate of the stapes attaches to the oval window. Two important functions are provided by the middle ear. The first is to amplify and deliver sound vibrations from the drum to the inner ear, and the second is that of protecting the inner ear from very loud sounds. The amplification of sound waves is accomplished by apparent lever action of the ossicles that produces a greater force at the oval window than the force at the drum, and because of the gain in force that results from the relationship between the larger drum area to the smaller stapedial footplate area. The area of the drum is approximately 25 times that of the oval window. The amplification gain of these two factors is approximately 25 dB. The effectiveness of the middle ear action in increasing hearing sensitivity is evidenced in middle ear pathologies where the ossicular chain is disrupted. A hearing loss of 25 dB or more occurs. The second function of the middle ear, that of protecting the inner ear from loud sounds, is accomplished by reflex action of the middle ear musculature, the tensor tympani, and the stapedius. The action of the muscles is to retract the eardrum, draw the stapes away from the oval window, and change ossicle vibrations in such a way as to decrease the transmitted pressure. Latency of muscle contraction and possible muscle fatigue limit protection of the inner ear by these mechanisms. Middle ear air pressure is equalized by virture of the Eustachian tube which connects the middle ear and the nasopharynx. The pressure equalization is necessary for normal ear drum movement.

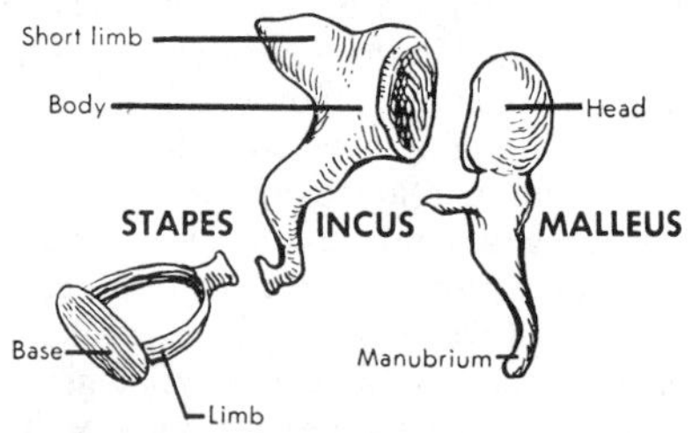

Fig. 2. Greatly enlarged sketches of the malleus, incus, and stapes. ("*Anatomy of the Ear*," *Grace Hewitt.*)

The inner ear is a system of cavities in the dense petrous portion of the temporal bone. One of the cavities is the cochlea, a bony labyrinth that is approximately 35 millimeters in length coiled around a central core for two and three-quarters turns. The spiral-shaped cochlea is divided into three ducts, two bony and one membraneous. (See Figs. 3 and 4.) The upper bony duct, the scala vestibuli and the lower bony duct, the scala tympani, are separated from each other by a membranous labyrinth, the cochlear duct. The cochlear duct is bound on top by Reissner's membrane and is bound below by the basilar membrane. The cochlear duct is filled with a viscous fluid called endolymph, and the duct is surrounded by a fluid called perilymph that has about twice the viscosity of water. The scala vestibuli and the scala tympani join at the apical end of the cochlea at a passage called the helicotrema. The scala tympani terminates at the basal end at the round window, a membrane-covered opening into the middle ear. The scala vestibuli is continuous with the vestibule; the oval window opens into the vestibule. Vibrations at the footplace of the stapes are transmitted into the fluid adjacent to the oval window. Vibration of the stapes and resultant disturbances in cochlear fluids results in movement of the basilar membrane. The cochlear duct contains the sensory receptors, specifically the organ of Corti, which lies upon the basilar membrane. There are about 25,000 hair cells; one end of each rests on the basilar membrane. The other ends of the hair cells are the cilia, very fine hairline processes, which make contact with the tectorial membrane, a membrane that overlaps the organ of Corti and that functionally behaves as if it were hinged at the cochlear wall. There are three rows of outer, and one row of inner, hair cells along most of the length of the basilar membrane. When vibrations are introduced into the inner and cause displacement of the basilar membrane, a shearing of the action of the cilia occurs that results in neural activity. It is assumed that amplification occurs in the inner ear in that small pressures on the basilar membrane result in a shearing force of considerably greater magnitude that distorts the hair cells. The result is increased sensitivity of the hearing system. Physical properties of the cochlea are such that different frequencies tend to localize at different points along the basilar membrane. The basilar membrane is narrowest and stiffest at the basal end, and most lax and widest at the apical end of the cochlea. High-frequency sounds result in

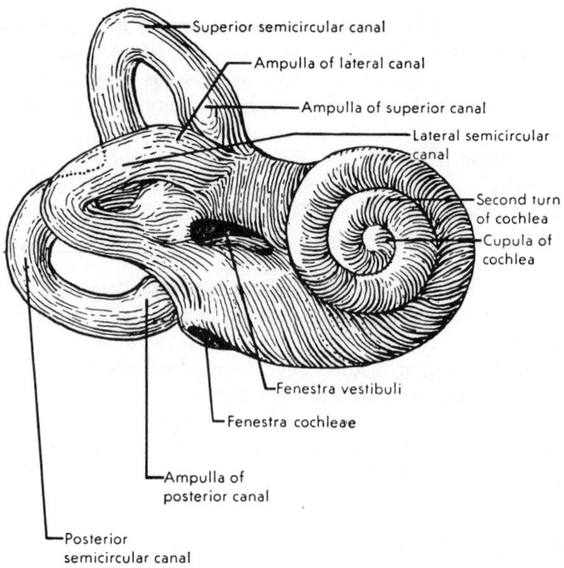

Fig. 3. The bony labyrinth. ("*Anatomy of the Ear*," *Grace Hewitt.*)

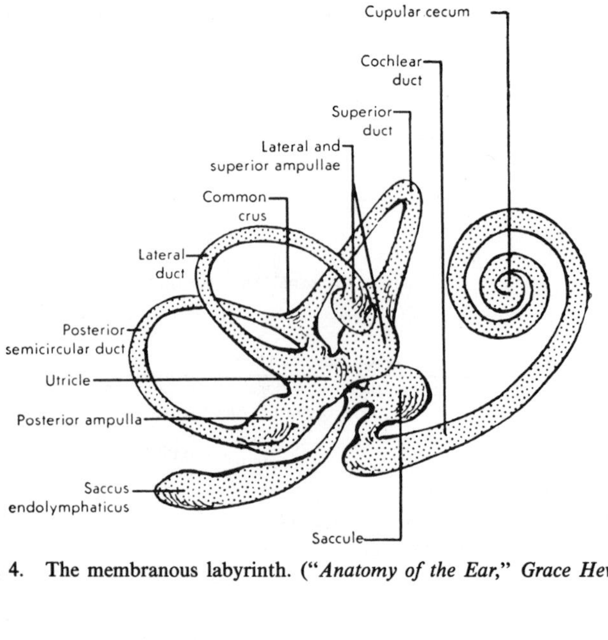

Fig. 4. The membranous labyrinth. ("*Anatomy of the Ear*," *Grace Hewitt.*)

the greatest disturbances near the basal end, and low-frequency sounds tend to localize near the apical end. When the role of the cochlea in pitch and loudness analyses is considered it is now realized that more is involved in pitch perception than the place of localization on the basilar membrane, although the particular neural fibers involved are probably relevant. Loudness is probably related to the total number of neural impulses per unit time.

The auditory pathways provide for the neural impulses from the ear to be trasmitted to the cerebral centers of the auditory cortex. Processing of the neural signals probably occurs at synaptic connections as well as in the cortex. The cell bodies of the receptor neurons are located in the spiral ganglion. Neurons of the auditory nerve make synaptic connections with the hair cells of the cochlea. Nerve fibers typically innervate many hair cells, and more than one nerve fiber may make a connection with the same hair cell. There is recent evidence to indicate that there are also descending neural pathways as well as ascending ones. The central nervous system may thus be involved in auditory processing at the cochlea. Spiral ganglion axons make synaptic connections with cells of the central nervous system at the cochlear nucleus. At this point, there is interconnection between the pathways for the two ears. Other synaptic stations between this point and the auditory cortex include the inferior colliculus and the medial geniculate body. Evidence from pathological auditory systems is of particular interest with respect to the auditory pathways. An impaired cochlea, for example, may result in a better than normal response to small amplitude changes in a sound. A lesion of the VIIIth Nerve is frequently manifested by a rapid decrease in the ability to respond under sustained stimulation. The ability to process speech is markedly affected when there is an involvement of the lower central nervous system. Cortical involvement does not affect usual speech or pure tone inputs.

Sound involves a disturbance in the air that is a forward and backward, rarefaction and compression, movement of air particles. The unit of force usually used in acoustics is the dyne. Sound pressure is frequently expressed in dynes per square centimeter. Intensities of sounds are usually measured on a decibel scale, a logarithmic ratio scale. The tremendous loudness range of the ear is exemplified by the fact that the most intense sound that can be tolerated is a million million times greater in intensity than a sound that is just audible. This is a range of approximately 120 dB. The frequency range of hearing is frequently given as 16 to 20,000 Hz. The ear is most sensitive in the middle frequency range of 1,000 to 6,000 Hz. In terms of discrimination of frequency and intensity, it is possible for about 1,400 pitches and 280 intensity levels to be distinguished.

Sense of Balance and Orientation. The vestibular apparatus of the ear participates in a major way to the sense of balance. The operation is part of a multifaceted system, which also includes the eyes and the somatic sensors in the skin and joints. It is interesting to note that even though the vestibular apparatus is very important to this function in the normal individual, a person can be trained to function without them, as in the case of vestibular damage. Many other vertebrate animals possess a vestibular system along the lines of that found in humans. Although the detailed functioning of the vestibular apparatus remains somewhat obscure, quite a lot of knowledge has been obtained in recent years from scanning electron micrographs of components of the apparatus. Located within the membranous labyrinth of the inner ear, the vestibular apparatus incorporates a series of fluid-filled sacs and tubes. Although extremely miniature, the means for generating signals are essentially of a mechanical nature.

Two types of receptors have been described: (1) Organs which respond to linear accelerations and known as *otoliths* or *statoliths*; and (2) semicircular canals which respond to angular accelerations. Small crystals of calcium carbonate which have a density of just 3 times greater than water act as an inertial mass to resist the application of external forces. These are located on the hair cells of the utricle, giving the hairs weight and thus causing them to be sensitive to change in position. Whenever an otolith pulls or pushes on a hair, a signal is generated. As shown in Fig. 3, the three semicircular canals are positioned such that each is lying in a different plane, approximately at right angles to the other two. The plane of rotation of the head determines whether one, two, or all three canals will be affected by head movement. An endolymph fluid provides inertial mass in the canals much as the crystals provide mass in the utricle. However, fine hairs also detect any movement, generating a signal. Several pages would be required to describe what is known or postulated concerning these unique and complex mechanisms. Even more complex and less understood is the manner in which nerve signals from three groups of receptors (vestibular, visual, and somatic) are processed by the brain to provide an integrated output reading in terms of one's balance and orientation. The Parker (1980) reference explores the topic in considerably more detail.

Diseases of the Ear

Common earache may arise from many causes and occurs in numerous forms. The most frequent cause of pain, aside from mechanical injuries, arises from some kind of bacterial infection. A physician should be consulted when an earache persists over several hours.

Otitis Media—Acute. An infection of the middle ear. Normally, the middle ear is sterile. The problem occurs rather commonly in early childhood, but the incidence decreases with increasing age. The disease take a number of forms. When bacteria ascend from the nose and throat to the middle ear, the condition is referred to as *purulent otitis media.* The predominant symptoms are pain, fever, and often diminished hearing. Perforation of the tympanic membrane (See Fig. 1) and otorrhea (discharge) may occur. One key to diagnosis is a bulging tympanic membrane with accompanying obscuration of the bony landmarks. The most frequent cause of purulent otitis media is pneumococcus, followed in order by *Haemophilus influenzae.* Anaerobic bacteria, although prominent in the normal flora of the upper respiratory tract, rarely cause acute otitis. Staphylococci are rarely involved. *H. influenzae* is a common cause in young children and is seen in about one-third of the patients between 5 and 9 years of age. A number of other microorganisms can be involved (streptococci, *Neisseria catarrhalis,* and *S. epidermis,* among others).

In past years, acute suppurative mastoiditis and, less frequently, meningitis sometimes followed acute otitis media. With current antibiotic therapy, these are rare occurrences. Therapy includes pain relievers (analgesics), decongestants, and antibiotics. Ampicillin is frequently the drug of choice for children, and penicillin for adults. Where ampicillin-resistant strains of *H. influenzae* are encountered and where there is allergic response to penicillin, combinations of erythromycin and sulfisoxazole or trimethoprim and sulfamethoxazole are used. The latter drug alone is sometimes used for the chemoprophylaxis of recurrent otitis in children.

Otitis Media—Chronic. This condition usually results from neglected or recurrent acute otitis media and is seen in all age groups. Pain and fever may be absent, with hearing loss and foul discharge being the major symptoms. The tympanic membrane will be perforated. A number of microorganisms (staphylococci, streptococci, *Pseudomonas aeruginosa,* and enteric gram-negative bacilli, among others) may be cultured from the discharge. Antibiotics generally are ineffective. Surgery may be required in advanced cases. In some tropical and developing countries, where clostridia may be introduced with dirty cloths used for removing ear drainage, otogenous tetanus may develop. Otitis media can be a predisposing cause of bacterial meningitis and also may follow a measles infection.

Serous Otitis Media. Also called *secretory otitis,* this condition is characterized by the collection of fluid in the middle ear. This fluid may be either clear (serous) or gluelike (mucous). The predominant symptom is impaired hearing, which varies from a slight to almost total loss. Children who have serous otitis media may be subject to frequent upper respiratory infections and often have enlarged lymphoid tissue in the nasopharynx. If there is an underlying allergy or infection, appropriate antihistamines, antibiotics, or sulfonamides may be administered. Draining the fluid through an incision in the eardrum may relieve the condition. When there are repeated attacks, tiny plastic tubes can be inserted into the middle ear to provide adequate aeration, a procedure that requires a hospital environment. These tubes may be left in place for 3 to 4 months. Many cases of severely impaired hearing in adults can be attributed to middle ear infections in childhood. In infants and children, the Eustachian tube is shorter and more nearly horizontal than in adults, thus making the tube more likely to be an avenue of infection.

Otitis Externa. This disorder originates from the same causes as all middle ear infections, but it differs in the type of inflammation and the changes that occur in the tissues. A head cold may precede the infection. The attack of inflammation is sudden and causes congestion in the linings of the ear spaces, Eustachian tube, and mastoid cells. The ear itself fills with fluid, which gradually becomes puslike. Pain is the main symptom and can be severe, radiating, and throbbing. In children, early symptoms may include refusal to eat, nausea and vomiting, rolling the head, or tugging at the ear. Temperature generally runs high. A ringing sensation and dizziness may be present. Hearing is impaired as long as pus remains in the middle ear. If the condition is left untreated, after several days the eardrum ruptures spontaneously. For as long as three weeks, fluid seeps through the canal and then subsides. The parts of the middle ear are so intricate and delicate that infection spreads easily. Pain resulting from movement of the external ear assists in distinguishing otitis externa from otitis media. Topical therapy includes polymyxin B and neomycin, usually with excellent results. Where true cellulitis of the external ear develops (infrequently), systemic antibiotics and possibly debridement of infected cartilage may be indicated. Very rare neurologic complications of this condition can be life-threatening and require parenteral therapy with tobramycin and carbenicillin, as well as surgical debridement.

Aero-Otitis Media. In this disorder, the structures of the middle ear are affected by changes of pressure which occur during airplane flights. In milder cases, there is a sensation of stuffiness in the ears, with a slight inflammation of the eardrum, and perhaps some minor hearing impairment. Excruciating pain and hemorrhages in the tympanic membrane may occur in more severe cases. Although the condition still may occur among sensitive individuals, pressurized aircraft cabins have greatly alleviated the problem. If one senses this developing, chewing gum or moving the lower jaw with the mouth open will usually prevent it by opening the Eustachian tube which will equalize the pressure. The problem is more common with persons who have upper respiratory infection or severe nasal allergy.

Mastoiditis. The middle ear is generally involved when there is an infection of the mastoid process of the temporal bone. The acute form of this disease (*acute mastoiditis*) has been practically eliminated since antibiotic drugs became available to combat middle ear infections.

The inflammation in mastoiditis involves the lining of the mastoid cells. The infection may enter the bone, which becomes soft and decayed. The causes of mastoiditis include respiratory infection, abnormal anatomy of the ear in infants and children, improper channels for ear drainage, and lowered resistance to infection. Mastoiditis may occur as a secondary infection to various diseases. The predominating symptom is pain, which may be either continuous or intermittent. If the patient is not treated, the intense pain could persist for 6 or more days, which may not be true for middle ear infection. Also unlike middle ear infection, mastoiditis is characterized by a definite, localized tenderness over the mastoid process.

In *chronic mastoiditis*, which now occurs more often than the acute type, drainage from the ear (*otorrhea*) is the principal symptom. Fever may or may not be present. If acute mastoiditis should occur, the physician may perform a *mastoidectomy*. In this operation, the infected mastoid cells are removed through an incision in the area behind the ear, or in the external auditory meatus.

Punctured Eardrum. The most common cause of a punctured eardrum is the insertion of a sharp object into the ear. Violent explosions near the ear may cause the drum to tear or rupture. Decreased air pressure during or after descent from high altitudes, severe sneezing, diving, and increased pressure frequently are responsible for damaged membranes. Sometimes, diagnosis is difficult. The pain accompanying a puncture is sharp and intermittent. Blood may ooze from the injury, but this is not positive proof of a drum tear, because the same symptom may be present in a skull fracture. Dizziness, ringing sounds, and headaches also are significant symptoms. A tear in the eardrum may heal without treatment within a period of a few weeks, but there may be aftereffects which may not be noticed, even for as long as a year. A grafting operation known as *tympanoplasty* can be employed in cases in which the tear does not close.

Growth on the Eardrum. Following rupture or perforation of the eardrum, small chalky (lime) deposits may form at the site of healing as a result of repeated attacks of middle ear infection. If they form from a healed perforation, they mark the path of least resistance for a future rupture. It is the general opinion of physicians that such deposits do not affect normal hearing. There is no successful way of removing the chalk deposits without injuring the eardrum seriously or depressing the hearing. Hence, it is rarely attempted.

Boils or Furnucles. When present in the external ear, these often produce severe pain because the skin in this region normally adheres closely to the underlying cartilage and bone. If infection is allowed to persist, perforations of the eardrum may occur. Through them, infection may spread to the middle ear, the inner ear, or the mastoid area. An x-ray will assist in determining the nature of any secondary complication.

Fungus Infection. Otomycosis is a fungus infection of the outer ear and canal. The inside of the ear appears dirty and crusty, and fluid seeps out continually. When the crusts and scales are removed, the skin beneath is raw and bleeds easily. Itching causes much discomfort. Pain is usually present because of the swelling of the canal; hearing may be impaired. Treatment is by specific solutions and ointments. Home remedies are not recommended.

Tinnitus. Most persons, at one time or another, experience this disorder, a sensation of ear noise which is more noticeable in a quiet environment. Such sounds may seem to be in the head rather than the ear, and may affect one or both ears. The symptom is associated with many conditions, including middle ear infection. Ménière's syndrome, exposure to intense noise, circulatory diseases, otosclerosis, and neutritis of the auditory nerve. The symptom also may be caused by excessive amounts of coffee, tobacco, or alcohol. Quinine, certain antibiotics, or large doses of aspirin also may produce tinnitus. Such sounds occur most often in persons between ages 50 and 70. The reason for the sensation has not been established. Inasmuch as the symptom could be an early warning of hearing damage, it should be investigated.

Cauliflower Ear. Known as *hematoma of the auricle*, this disorder has long been recognized as the badge of the prizefighter. It is caused by injury to the external ear. A hard blow may cause bleeding below the skin. If this accumulation of blood remains for sometime, it become fibrous tissue and eventually will be converted into a bone-like or cartilaginous substance. Thus, the ear will be deformed by this irregular mass of extra tissue. For prevention, the blood should be removed before it clots. Plastic surgery also is used for restoration of affected ears.

Congenital Malformations. These occur rather frequently, but generally they are not gross enough to impair hearing. They may be unsightly. Absence of the lobe or the outer rim of the ear (*helix*), large protruding ears, and irregular shapes are among the more common malformations. Plastic surgery can restore most of these conditions to normal appearance. Occasionally, a congenital defect, such as an obstruction in the canal, may have to be removed before hearing improves. In rare instances, the ears may be displaced on the head, and in some extreme cases when the lower jaw is grossly misshapen, they may even be fused together (*synotia* or *otocephaly*).

Disturbances of Equilibrium

As previously mentioned, the semicircular canals of the inner ear are partially responsible for adjusting the body to changes in motion. The rate of these changes normally allows sufficient time for the canals to maintain bodily equilibrium. When rapid, irregular, and continuous waves of motion persist, the canals are not able to perform properly. The result is called *motion sickness.* Seasickness, airsickness, car sickness, and elevator sickness are forms of motion sickness. The usual symptoms are dizziness, nausea, vomiting, and thirst. Sometimes profuse sweating occurs. Despite the extreme unpleasantness experienced, recovery from motion sickness is rapid once the cause is removed. For many years, dimenhydrinate (Dramamine®) has been available as a remedy for motion sickness. Dimenhydrinate is the chlorotheophylline salt of the antihistaminic agent diphenylhydramine and, while the precise mode of action of the drug is not understood, it has a depressant action on hyperstimulated labyrinthine function and thus is effective for motion sickness. Dramamine should not be taken in conjunction with an antibiotic except under advice from a physician. A normal dosage will provide relief for about 4 hours. Drowsiness is a side effect, and thus it should not be taken by persons operating automobiles or other vehicles and dangerous machinery.

Ménière's Syndrome. Prosper Ménière described this malady in 1861 and correctly attributed its origin to the inner ear. Its characteristic symptoms are sudden severe episodes of *vertigo* (dizziness), tinnitus, and fluctuating hearing loss. The term syndrome continues to be used because the exact causes of the disorder have not been fully established. Persons in the middle age group are more commonly affected by the syndrome. The vertigo associated with an attack may be so severe that the simplest activities become impossible. Usually, the patient has a sensation that objects are whirling about. The same type of dizziness occurs with certain cardiovascular disorders and middle ear infections. Attacks may last for minutes or weeks. The tinnitus, usually a roaring noise, sometimes persists between attacks. Nausea and vomiting are also usual symptoms.

The course of the syndrome is unpredictable. Remissions of up to several years often occur. About two-thirds of the patients improve or recover regardless of treatment. No single form of therapy has been fully successful. Certain drugs, such as Dramamine®, often help control the vertigo. Sedatives or tranquilizers are occasionally helpful. If the condition is disabling and unilateral, the diseased parts of the labyrinth may be surgically removed. The procedure stops the vertigo, but balance is impaired and hearing loss in the affected ear is total. Ultrasonic radiation has been used to irradiate the labyrinth with the objective of destroying the diseased portions. For relief of severe vertigo, some surgeons recommend the Tack operation to drain the saccule, which contains endolymph. A tack, a small pointed piece of metal, is placed through the footplate into the sac, thus allowing drainage. According to one theory, this syndrome is related to an imbalance of pressure between the perilymph and the endolymph. Another innovation has been the use of surgical instruments which are maintained at temperatures as low as −140°C. With these instruments a surgical procedure should be less likely to damage the cochlea.

Vestibular Neuronitis. This is a comparatively common syndrome, the manifestations of which are vertigo, vomiting, and imbalance. Some authorities believe the disorder results from irritation of the vestibular portion of the eighth cranial nerve. Although vestibular neuronitis resembles Ménière's syndrome in many respects, there are no audiologic symptoms, and in particular no hearing loss. The disease is benign and there is no specific treatment.

Deafness

Deafness means nearly complete or total loss of hearing. There are two types: (1) congenital, and (2) acquired. In the congenital type, the person is born deaf or later becomes deaf because of an inborn defect. Hard of hearing is a term that applies to those who lose some of the ability to hear later in life, but who have learned how to speak before the loss occurred.

Causes of deafness are many. Some conditions which may cause deafness or milder hearing difficulties include (1) temporary or chronic infections in one or both ears; (2) secondary complications of disease elsewhere in the body; (3) direct damage or defect in some part of the hearing system; (4) aging; (5) occlusion of the auditory canal; (6) aero-otitis media; (7) Ménières syndrome; (8) ostosclerosis; (9) noise; and (10) certain toxic drugs. Side effects of the loop diuretics, ethacrynic acid and furosemide, include transient hearing impairment. Complete deafness has been reported after intravenous administration of ethacrynic acid and a permanent hearing deficit after chronic use.

Conductive deafness results when sound waves are not transmitted properly through the outer and the middle ear. If the damage is to the inner ear or the nerve pathway to the brain, a *sensorineural* (also called *nerve* or *perceptive*) *deafness* occurs. The latter type is generally a greater handicap and usually cannot be reversed. In *mixed hearing loss,* there are elements of both conductive and sensorineural types of loss. Some deafness is caused by a disorder in the central nervous system.

Over the years, rubella (German measles), if occurring in the first trimester of pregnancy, can cause partial hearing defects in the newborn. This congenital deafness may be masked by even more severe birth defects. A number of children suffered hearing loss as the result of outbreaks of rubella in 1964 and 1969. Since that time, there has been a greater awareness of the situation and rubella vacination is widely practiced. Rubella vaccine has been licensed in the United States since 1969.

Otosclerosis. Usually first detected during early adulthood, *otosclerosis* can cause a conductive type of hearing loss. Bony growths form just inside the inner ear where the middle ear's stirrup (*stapes*) enters it. Eventually, the footplate of the stapes becomes anchored and no longer conducts sound waves to the inner ear. About 10% of the population is affected to some extent in this way, although they may have no hearing loss for many years. Experience indicates that the disorder may become arrested at any stage. Heredity appears to be an important factor. Middle ear infections are not a cause. The disorder occurs about twice as often in females as in males.

Since 1952, an operation in the tiny stapes itself has been used for many patients with advanced otosclerosis. In this *stapedectomy,* all or part of the fixed stapes is removed and replaced with an artificial device. The operation has since been modified in which only part of the stapes is removed.

Noise. Particularly since the late 1950s, there has been a growing public and professional awareness of the damage wrought by noise on the hearing system. Hearing loss among war veterans had been known for many decades, but the effects of noise in the working place, of vehicles, and even of unduly loud music in the theater and residence, have only been well studied and understood in recent years. For example, it was only a few years ago that an occupational noise exposure standard was established in the United States. See accompanying table. See also the entry on **Acoustics.**

GENERAL INDUSTRY STANDARD FOR OCCUPATIONAL NOISE EXPOSURE

DURATION PER DAY (Hours)	SOUND LEVEL (dBA Slow Response)
8	90
6	92
4	95
3	97
2	100
1.5	102
1	105
0.5	110
0.25 (or less)	115

NOTE: When the daily noise exposure is composed of two or more periods of noise exposure of different levels, their combined effect should be considered, rather than the individual effect of each. If the sum of the following fractions: $(C_1/T_1) + (C_2/T_2) + \cdots + (C_n/T_n)$ exceeds unity, then the mixed exposure should be considered to exceed the limit value. C_n = total time of exposure at a specified noise level, and T_n = total time of exposure allowed at that level. Exposure to impulsive or impact noise should not exceed 140 dB peak sound pressure level.

SOURCE: Miner and Peters (1977).

Instrumental Measurements of the Ear

The most common measurement of hearing function is the pure-tone audiogram in which a frequency from 125 to 8,000 Hz is plotted against hearing loss in decibels. The audiogram displays the ability of the ear to hear a pure sine-wave tone at a given frequency compared with a "normal" ear. The unit of loudness is the decibel, defined as $10 \times \log_{10}(P_1/P_2)$, where P_1 is the power of the sound being applied and P_2 is the just-audible power required at the given frequency for the "normal" ear to hear. The standard audiometer contains a frequency-selection knob, an attenuator calibrated in 5-dB increments, and a key which connects the output of the instrument to the earphones placed on the subject's head. The procedure is to increase the amplitude slowly while depressing the key in short pulses until the subject reports that the sound can just be detected.

In addition to pure tones, speech sounds are also used as test signals. Using +9 dB (referred to 0.0002 dyne/square centimeter) as a 0-dB threshold level, it is possible to determine the extent of the hearing loss for speech using specially selected two-syllable words having approximately equal stress on each syllable (called "spondaic" words). The equipment used for this measurement consists of a microphone, audio amplifier, and a pair of headsets, the system having a float frequency response between 125 Hz and 8 kHz. Sensitivity, or gain, of the amplifier is controlled by a step attenuator calibrated in 1-dB steps, and the output is arranged to go into either ear separately, or both ears simultaneously.

In the von Békésy pure-tone audiometer, the amplitude control is run up and down by a motor while the subject operates a key. The amplitude is slowly increased until the subject hears the sound, which reverses the motor. The frequency is similarly increased slowly and automatically. The resulting curve is somewhat sawtooth in form and more accurately brackets the threshold values.

In designing and using audiometers, great care must be given to the elimination of background noise and hum. If more than one tone is presented at a time, "masking effects" may occur, giving different results than would be obtained with each sound separately.

Hearing losses are generally greatest at the higher frequencies (above 2 kHz). Thus, the sibilants and percussives in speech are the most difficult to perceive. In addition, there usually is a loss in dynamic range so that not only must sound be louder to be perceived, but the range between threshold of perception and threshold of pain is much narrower. Thus, the ear cannot accommodate in many cases to an undue amount of amplification. This condition is known as "recruitment." Any hearing aid should take these two factors into account and hence should not be a simple audio amplifier if best results are to be achieved. Furthermore, stereophonic sound location is lost with a single-channel hearing aid. Thus, some patients find that binaural hearing aids are of much value. In some cases, the user can discriminate a desired conversation from background noise in a dramatic fashion.

As a further aid in helping the profoundly deaf, whose hearing may be cut off as low as 1,000 Hz, a device known as a transposer may be used. This translates sounds at the upper frequencies made by sibilants and percussives into the lower-frequency region where they can be amplified sufficiently to be audible. There are several variations on the basic scheme, but in principle the devices contain filters which detect only the upper frequencies and then trigger sound generators below 1,000 Hz which indicate to the user the presence of the high frequencies.

References

Barr, D. F.: "Auditory Perceptual Disorders," Charles C. Thomas, Springfield, Illinois, 1973.

Clark, J. G.: "Audiology for the School Speech-Language Clinician," Charles C. Thomas, Springfield, Illinois, 1980.

Etter, L. E.: "Roentgenography and Roentgenology of the Temporal Bone, Middle Ear, and Mastoid Process," 2nd edition, Charles C. Thomas, Springfield, Illinois, 1972.

Evans, E. F., and J. P. Wilson (editors): "Psychophysics and Physiology of Hearing," Academic, New York, 1977.

Graham, M. D.: "Cleft Palate: Middle Ear disease and Hearing Loss," Charles C. Thomas, Springfield, Illinois, 1978.

Haug, O., and S. Haug: "Help for the Hard-of-Hearing," Charles C. Thomas, Springfield, Illinois, 1977.

Howard, I. P., and W. B. Templeton: "Human Spatial Orientation," Wiley, New York, 1966.

Miller, A. L.: "Hearing Loss, Hearing Aids, and Your Child," Charles C. Thomas, Springfield, Illinois, 1980.

Miner, L. E., and K. R. Peters: "A Model Hearing Conservation Program," *Job Safety and Health*, 15–20 (February 1977).

Parker, D. E.: "The Vestibular Apparatus," *Sci. Amer.*, **243**, 5, 118–135 (1980).

Robinson, D. W.: "Occupational Hearing Loss," Academic, New York, 1971.

Schneider, B., Trehub, S. E., and D. Bull: "High-Frequency Sensitivity in Infants," *Science*, **207**, 1003–1004 (1980).

Smith, C., and A. and J. Vernon: "Handbook of Auditory and Vestibular Research Methods," Charles C. Thomas, Springfield, Illinois, 1976.

Wilson, V. J.: "The Labyrinth, the Brain, and Posture," *American Scientist*, **63**, 3, 325–332 (1975).

Wilson, V. J., and G. M. Jones: "Mammalian Vestibular Physiology," Plenum, New York, 1979.

Young, L. R.: "Role of the Vestibular System in Posture and Movement," in *Medical Physiology* (V. B. Mountcastle, editor), C. V. Mosby Co., 1974.

HEARING (Fishes). Fishes.

HEARING ORGANS. Sensory Organs.

HEART AND CIRCULATORY SYSTEM (Human). The circulatory or cardiovascular system of the body is comprised of the heart and the blood vessels (arteries, veins, and capillaries). These organs are highly interdependent. The study, diagnosis, and treatment of diseases and disorders of this system fall under the general classification of *cardiovascular medicine.*

Heart

The heart is the muscular organ that pumps blood through various conduits to and from all parts of the body. Depending upon the size of the adult individual, the human heart weighs somewhat less than three-quarters of a pound (about 340 grams). The organ essentially is a hollow muscle capable of contraction like other muscles. A contraction of the heart is referred to in general terms as a *heartbeat.* The rate of the heartbeats can be changed by two different sets of nerves: (1) The accelerating nerves are connected to the spinal cord and are a part of the sympathetic nervous system; (2) the *vagus nerve* depresses the rate and is connected to the brain stem. The beating of the heart commences long before birth and continues as long as life continues. Beats occur at the rate of 70–80 times per minute in adults, but may increase to 100 beats per minute during exertion, or in the presence of emotional disturbance. During a 70-year life span, it is estimated that the heart beats some 3 billion times, an average of about 42 million beats per year. Each contraction of the heart moves slightly more than 2 fluid ounces (~59 cubic centimeters) out into the arteries, providing a change of blood over the body about once every minute. During a lifetime of 70 years, a total of 250 million quarts (~236.5 million liters) of blood are moved, almost enough to fill a large football stadium. There are only a little over 6 quarts (~5.7 liters) of blood in the average human body, so that this blood requires not only rapid circulation, but also a fine adjustment of controls to assure the proper and effective distribution required by the body.

The highly schematic diagram of the heart given in Fig. 1 indicates

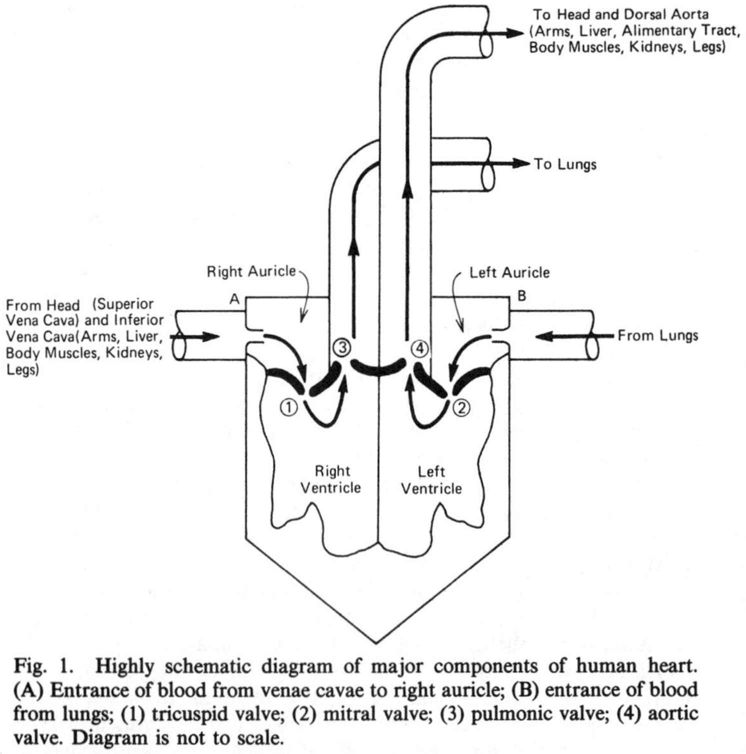

Fig. 1. Highly schematic diagram of major components of human heart. (A) Entrance of blood from venae cavae to right auricle; (B) entrance of blood from lungs; (1) tricuspid valve; (2) mitral valve; (3) pulmonic valve; (4) aortic valve. Diagram is not to scale.

the principal components of the heart structure. The heart is divided into four chambers—two auricles, referred to as the right and the left auricle; and two ventricles, referred to as the right and the left ventricle. The flow of blood through these chambers is controlled by four valves, as numbered in the diagram: (1) the tricuspid valve; (2) the mitral valve; (3) the pulmonic valve; and (4) the aortic valve.

Blood coming from over the body through the large veins (venae cavae) enters the right auricle at A. This blood has been partially depleted of its oxygen. As the lower, thick-muscled ventricles expand, this blood enters the right ventricle through the tricuspid valve. Then, the ventricle contracts and forces the blood into the pulmonary artery toward the capillaries of the lungs and is prevented from running back into the heart by the closure of the pulmonic valve. In the meantime, the purified blood in the left auricle has just arrived from the lungs through the pulmonary veins, at B. From here it passes into the thick-walled left ventricle through the mitral valve. When the right ventricle forces blood out into the pulmonary artery, the left ventricle at the same time contracts and sends blood out into the arteries of the body, passing through the aortic valve into the aorta. The auricles thus act as collecting chambers, while the ventricles serve as pumps. The right side of the heart collects the blood and forces it through the lungs; while the left side collects it from the lungs and forces it through the body as a whole. The four valves between the various chambers of the heart prevent the blood from flowing backward and maintain the pressure between heartbeats because of the closed system that results.

In order that blood can be moved forward in an orderly manner, it is important that the heart muscles expand and contract at just the right time and that all the valves open and close completely at the proper time during the cycle. This control is accomplished by a special structure known as the *sino-auricular node.* This is the pacemaker of the heart. It is not entirely dependent upon the general nervous system, and it has been known to function for some time after breathing has ceased. Sudden changes in temperature, unusual nervous stimuli, fright, a sense of impending danger, or a happy thought can affect this heart center and, thereby cause speeding or slowing of the heart action. All warm-blooded animals have such a fine adjustment that acceleration or retardation may occur within 1/100th second.

The sino-auricular node lies in the wall of the right auricle, embedded within the muscular tissue. A heavy partition extends between the left and right side of the heart, so that there is no direct connection between them except for a group of structures consisting of the auriculo-ventricular node, the *common bundle* and its left and right branches. The auriculo-ventricular node transmits impulses from the common bundle, also known as the *buncle of His,* thence to the two branches, and from there to a network of muscle fibers which covers the inside of each ventricle. The network extends to the outer covering of the heart and is called the *Purkinje system.* This system assures an almost instantaneous response of the muscles of the ventricles once the impulse has passed into it. Although the heartbeat is not entirely independent of the general nervous system, it may carry on for some time without the ordinary nerve impulses. This is illustrated by the fact that the heart of a rabbit, for example, may continue to beat long after the animal has died. This automaticity of the heartbeat allows for cardiac transplantation.

The normal beating of the heart is associated with the production of bioelectric currents in the organ. Although these currents are not strong, they are carried to the surface of the body where they may be measured by a sensitive instrument, the electrocardiograph. The beat of a normal heart shows a characteristic pattern of electrical responses.

There are thousands of small muscle fibers interwoven to make up the walls of the heart. The organ also has its own circulatory system to provide the muscle with nourishment. The whole structure is sheathed with a tough sac, the *pericardium,* containing a small amount of fluid. This provides for lubrication of the rapidly moving heart.

The 70 or 80 normal heartbeats per minute do not allow much time between the expansion and contraction of the four heart chambers. The period of relaxation of the muscles, during which the heart fills, is about equal to that of contraction, when it empties. This period of relaxation permits the heart to recover fully from its work period. The contraction of the heart is called *systole*; the relaxation is called *diastole.*

To make the blood move in only one direction, there not only are the valves inside the heart, but also valves in the veins. In addition, in the small veins there is a constricting type of valve which helps adjust the rate of blood flow and the distribution of blood between the several organs in accordance with need. The capillaries act as the final speed control, by being so small that only one or two rows of blood cells may pass through at a time. Here the speed of the flow is so reduced that time is allowed for rebalancing the mineral content of the area, the exchange of oxygen for carbon dioxide, and soluble food for waste materials.

Blood-carrying Vessels

The aggregate length of conduits required to transport blood throughout the human body would be measured in terms of miles or kilometers. The beat of the heart forces a temporarily increased amount of blood into the arteries. The arterial walls are elastic and expand to accommodate this larger volume of blood. Between beats, the walls gradually contract, forcing the blood through the capillaries at an approximately constant rate. In this manner, the arteries act as a reservoir which prevents the blood from flowing through the tissues in gushes.

Blood passing from the heart through the lungs has only about one-sixth of the pressure of the blood as it is forced out over the body through the *aorta.* See Fig. 2. The pressure is still sufficient,

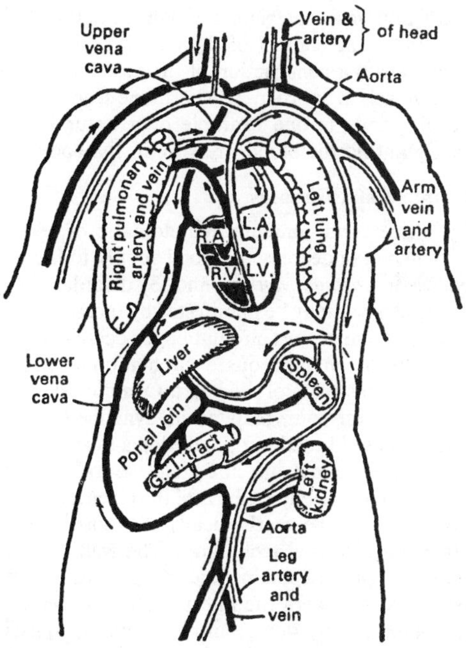

Fig. 2. Highly schematic representation of circulatory system of human body. R.A. = right auricle; L.A. = left auricle; R.V. = right ventricle; L.V. = left ventricle; G.I. = gastrointestinal tract. (*Carlson and Johnson, "The Machinery of the Body," Univ. Chicago Press.*)

however, to cause flow through the multitude of capillaries in the walls of the lungs. The lungs are composed of innumerable small sacs which have a supply of changing air. In the lung or pulmonary capillaries, the blood releases carbon dioxide and takes up oxygen.

The blood continues to flow back through the pulmonary veins and into the left auricle for distribution over the body. The loss of carbon dioxide and the assimilation of oxygen is accompanied by a change of color in the blood, from a dark to a bright red.

Although the liver does not have a special connection with the heart, it acts as a storage organ for blood. Blood is carried to the liver from the stomach and intestinal tract by the portal vein and from the rest of the body by the hepatic artery. It has been estimated that the liver and portal vein drainage system may hold as much as one-third of all the blood in the body. When the body is inactive and requires a smaller amount of blood, the liver and portal vein system relieves the remainder of the system by holding a large part of the excess. Some impurities are removed in the liver and excreted into the digestive tract. The hepatic vein returns the blood from the liver to the larger *vena cava* and heart for distribution over the circulatory system.

The blood supply of the heart itself is by way of special *coronary* arteries. These are necessary to supply the thick heart muscles with the large amounts of food and oxygen necessary for their continuous activity. The walls of the blood vessels themselves contain small canals through which blood is transported to nourish the cells of these tissues.

In addition to its function in the transportation of materials throughout the body, the circulatory system is important in temperature regulation. This arises by virtue of the ability of the muscular walls of the blood vessels to expand or contract, thereby changing the diameter of the vessels. When the capillaries in the skin are expanded or dilated, a larger amount of blood flows through them. If the temperature outside of the body is below body temperature, the blood in these capillaries is cooled. This cooled blood is then transported to the interior of the body where it is able to counterbalance any tendency toward a rise in temperature. On a cold day, these surface capillaries will be constricted so that the blood will not lose undue amounts of heat to the atmosphere.

The size of the various blood vessels thus varies automatically with the particular needs of the body. Drugs which cause a constriction of the blood vessels (*vasoconstrictors*) bring about a rise in blood pressure even though blood content remains fixed. By contrast, *vasodilators* generally bring about a reduction in blood pressure. Physiological changes in the sizes of the blood vessels are in part under the control of vasodilators and vasoconstrictors produced naturally in the body, and partially under the control of the nervous system. Sometimes, a substance that causes a constriction of the blood vessels in one tissue may dilate the vessels in another. The hormone secreted by the medulla of the adrenal glands is one example of a natural vasoconstrictor that aids in regulating the blood pressure in the body.

Cardiac Disorders and Diseases

Generally, three conditions are symptomatic of heart disease: (1) *Myocardial ischemia* (a decrease in blood supply to the heart muscle); (2) disturbances in *cardiac rhythm*; and (3) disorders in the *pumping efficiency* of the heart as may be manifested by increased filling pressure which causes upstream venous circulation, or decreased systolic pumping which results in an inadequate circulation of blood to organs that are located downstream of the heart. Common symptoms of heart problems include chest pain, palpitation, syncope (loss of consciousness), dyspnea (labored breathing), and edema (accumulation of fluid). These conditions, of course, are not exclusive to heart conditions.

Except in emergencies, when time is of the essence, milder symptoms of heart problems will be methodically diagnosed through the use of a number of instrumental techniques. The well-established cornerstone of heart diagnosis remains the electrocardiogram, preferably using twelve leads. Although the interpretation of electrocardiograms has been computerized to a degree and has been found useful in studies of mass populations, the input of an experienced cardiologist is considered mandatory in the analysis of specific patients with possible heart problems. During recent years, the two-step exercise procedure has largely been replaced by treadmill exercise. Ambulatory electrocardiographic measurements also have been emphasized in recent years. See **Electrocardiography.**

The use of ultrasound in a technique known as echocardiography has been available since the late 1960s and is growing in acceptance, being of particular value in the diagnosis of such conditions as pericardial effusion, mitral valve prolapse, and left atrial tumors, among others. See **Echocardiography.**

Another relatively new diagnostic tool is *isotope imaging.* Examples include radionuclide angiocardiography, using radioactive technetium, myocardial scanning with techetium pyrophosphate, and myocardial perfusion scanning with radioactive thallium. See **Radioisotopes in Medicine.**

Invasive procedures are still required in the diagnosis of many cardiac problems. These include cardiac catheterization, angiocardiography, coronary arteriography, intracardiac electrophysiological studies, and myocardial biopsy. These methods generally are limited to situations of an advanced, more serious nature and where other diagnostic procedures do not suffice. Principal limiting factors in their use are the risks generally attendant to invasive procedures, patient discomfort, and cost.

The major cardiac disorders and diseases are described in separate entries in this encyclopedia. See **Arrythmias (Cardiac); Arterial and Venous Disorders; Congestive Heart Failure; Endocarditis; Heart Failure; Hypertension (High Blood Pressure); Ischemic Heart Disease** (angina pectoris and acute myocardial infarction); and **Shock Syndrome.** See also the list of entries and cross references at the end of this entry.

Valvular Heart Disease. The function of the valves of the heart has previously been described in this entry. See Fig. 1. At one time, most heart valvular damage was ascribed to rheumatic fever. See **Rheumatic Fever.** It has since been established that there are over twenty forms of nonrheumatic valvular diseases.

In *rheumatic heart disease,* there is fibrotic scarring of the valvular tissue which ultimately produces *stenosis* or *regurgitation.* In nearly all cases, some stenosis is present. With exception of rare congenital causes, *stenosis of the mitral valve* is usually considered of rheumatic origin. Stenosis is defined as the narrowing or contraction of a passage or opening. Regurgitation is the abnormal backward progression of fluids; in the case of the heart, the backward return of blood through the valves of the heart. Stenosis adds an extra load on the heart because of increased pressure required to overcome resistance to flow; regurgitation reduces the efficiency of the heart as a pump.

In *nonrheumatic mitral regurgitation,* there is the *floppy valve syndrome,* a dysfunction related to coronary artery disease as well. In floppy valve syndrome, in what is described as an idiopathic pathologic process, there is a loss of fibrous and elastic tissue; this is sometimes called *myxomatous degeneration.* Mitral regurgitation also may result from rupture of the papillary muscle (*papillae* are conical projections from the walls of the cardiac ventricles attached to the cusps of the atrioventricular valves by the *chordae tendineae*). Rupture may occur as the result of infarction or ischemia and thus contributes to mitral regurgitation.

Aortic valve stenosis in adults (particularly the elderly) is considered of nonrheumatic origin and results from a gradual but progressive degenerative thickening and calcification of the leaflets in the valves. The disease process is considered to be somewhat like that occurring in atherogenesis. See **Arterial and Venous Disorders.**

Aortic regurgitation is also generally considered a nonrheumatic disorder and frequently occurs as a secondary manifestation of other diseases (syphilis, ankylosing spondylitis, aortic dissection, aortic aneurysm, and inherited diseases that affect connective tissue).

Prevention and therapy in valvular heart disease include the long-term administration of prophylactic antibiotics to decrease the possibilities of a return of rheumatic fever for persons who previously have had acute rheumatic fever. The length of time during which such prophylaxis should be given is debatable among authorities. In persons with rheumatic heart disease featuring aortic regurgitation or a bicuspid aortal valve, most specialists suggest the administration of antibiotics to prevent the development of bacterial endocarditis after dental and surgical procedures. In valvular disease, the physician will be aware of the risk of systemic embolism which sometimes develops in connection with rheumatic heart disease. Long-term administration of anticoagulants may be indicated in such cases. Cardiac arrhythmias arising from valvular disease will be handled as described in the entry on **Arrhythmias (Cardiac).**

Surgery is frequently indicated in valvular heart disease. This may range from repair of malfunctioning parts to valve replacement. Over the years, over three dozen designs of *artificial valves* have been used. Designs of preference in recent years have included the Starr-Edwards, the Smeloff-Cutter, and the Björk-Shiley valves. These valves are considered to have ample durability. The principal problems sometimes involved include thrombus formation and embolism, and thus long-term anticoagulant therapy is usually indicated.

In the United States, the natural tissue valve preferred is the porcine aortic valve. In Europe, some valves are configured from dura mater (outermost membrane of the brain and spinal cord), pericardium (membrane enclosing the heart), and fascia lata (wide, dense sheath of the thigh muscles). As of the early 1980s, it is estimated that about 20,000 persons receive heart valve transplants and about 50,000 people worldwide undergo this operation in a year. The valves from pig hearts make excellent replacements for human heart valves. They are durable, resistant to infection, and not readily rejected by the human body. There has been a shortage of valves of the proper size from this source. The valves are taken from pigs of various sizes, with most of the animals weighing less than 80 pounds (36 kilograms). Since most pigs in the United States are slaughtered at around 200 pounds (90.7

kilograms), the supply of hearts from small pigs is limited. Also, only about one of every ten valves is suitable for placement in the human heart. The cost of raising pigs strictly for their heart valves has proved prohibitive. A number of countries slaughter pigs weighing less than 80 pounds (36 kilograms) and the hearts from these pigs can be obtained rather inexpensively at slaughterhouses. However, they have the potential of introducing exotic diseases of swine into the United States. Such diseases as African swine fever, hog cholera, foot and mouth disease, and swine vesicular disease could devastate the pork industry in the United States. To prevent the introduction of such diseases, scientists at the U.S. Department of Agriculture (Plum Island, New York) have developed a method for inactivating these viruses. They have found that glutaraldehyde, a substance used to stabilize pig heart valves prior to their transplantation in humans, will kill the viruses associated with these diseases. Nevertheless, great care must be exercised in making certain that all porcine valves are fully free of such viruses prior to surgery.

Cardiomyopathies. Dysfunctions of the heart muscle (*myocardium*) that are *not* related to coronary atherosclerosis, hypertension, or valvular problems, fall into four categories which when considered as a group are called *cardiomyopathies.* From the standpoint of hemodynamics (study of movements of the blood), these categories (Goodwin, 1970) are: (1) *congestive*; (2) *hypertrophic*; (3) *restrictive*; and (4) *obliterative.*

In *congestive cardiomyopathy,* the contractility of the heart muscle is subnormal. Common symptoms include dyspnea (labored or difficult breathing) and fatigue. Often pulmonary congestion accompanies the disorder. There is often mild elevation of blood pressure. Cardiac enlargement is common. This condition must be differentiated from acute myocarditis. The usual course of congestive cardiomyopathy is to congestive heart failure, ultimately the cause of death of persons with the condition. The prognosis is variable. Therapy includes salt reduction, digitalis glycosides, and diuretics.

For many years, it has been observed that congestive cardiomyopathy is frequently seen in alcoholics. The term *alcoholic cardiomyopathy* now frequently appears in the literature. Alcohol has not been definitely identified as the cause; possibly the malnutrition usually associated with alcoholism may be the major contributor. In the midwestern United States and Canada in the 1960s, there was an epidemic of cardiomyopathy, but this was ultimately traced to cobalt toxicity derived from an additive used in making the beer consumed in the region.

For reasons still not understood, a *peripartum cardiomyopathy* may appear in a woman during the third or fourth month after completion of pregnancy (Demakis, et al., 1971).

In recent years, there has been considerable rethinking as regards the possible connection between cardiomyopathy and coronary artery disease; in the past, the presence of cardiomyopathy by definition ruled out coronary artery disease.

A common cause of congestive cardiomyopathy in certain regions of South America is **Chaga's Disease,** which see.

Hypertrophic cardiomyopathy has been known for many years, but possibly well defined for the first time by Teare (1958), who termed the disorder "asymmetrical hypertrophy of the heart." Hypertrophy is an increase in the volume of a tissue or organ caused entirely by enlargement of existing cells. Asymmetry refers to the disproportionate hypertrophy of the left ventricle which effectively reduces the size of the left ventricular chamber. The result is obstruction to left ventricular outflow. In recent years, new names have been given to the disease—*muscular subaortic stenosis*; and *idiopathic hypertrophic subaortic stenosis.* Symptoms include angina, syncope, palpitations, and congestive heart failure. See **Congestive Heart Failure.** Although the symptoms of the disease worsen with time, the process may be slow—a span of years. In some cases, however, sudden death may occur, particularly in children and men with a family history of this condition. About 15% of cases are treatable by surgery. Drug therapy is not universal, but is directed toward the profile of symptoms presented.

In *restrictive cardiomyopathy,* the myocardium loses its resilience and becomes rigid—conditions which offer resistance to ventricular filling and elevate cardiac filling pressures. The condition tends to mimic constrictive pericarditis. Symptoms are those of congestive heart failure. There are no fixed therapies for this disease that have proven effective. Some authorities believe that removal of excess iron in the body by phlebotomy (incision of a vein) may provide some relief.

In *obliterative cardiomyopathy,* there is a massive fibrosis (formation of fibrous tissue) of the endocardium. This reduces the size of the ventricular cavities. Although the disease, of unknown etiology, is frequently seen in eastern Africa, it is seldom encountered in Europe and the Western world.

Pericarditis. Inflammation of the membrane enclosing the heart (*pericardium*) may take three fundamental forms, all of which are generally termed *pericarditis. Acute pericarditis* is usually associated with a viral infection. There is chest pain which increases with inspiration (contrast with myocardial infarction), a low-grade fever, and sometimes tachycardia. The physician will listen for the sounds of a characteristic pericardial friction rub. Where a bacterial infection is diagnosed, antibiotics will be used; for neoplasms, radiation or chemotherapy may be indicated. In *pericardial effusion,* fluids accumulate in the pericardial cavity. Echocardiography is commonly used in diagnosis. In acute forms of cardiac tamponade (compression of heart due to collection of fluid in pericardium), as may arise from an injury, an aortic dissection, or rupture of an aortic aneurysm, prompt surgery may be indicated. In a less severe situation, pericardiocentesis (puncture and aspiration) may be used. In *constrictive pericarditis,* diastolic filling of the heart is impeded, the results of which are an increase in venous pressure and reduced cardiac output. At one time, this condition was almost exclusively attributed to a tuberculous lesion. A majority of cases are classified as idiopathic, but some are related to radiation exposure, to rheumatoid arthritis, or uremia. Surgical removal of the pericardium is sometime indicated.

Congenital Disorders and Anomalies

Most congenital disorders of the circulatory system appear in the embryo as the result of some defect in development, usually between the fifth and eight week of pregnancy. An infection in the mother during pregnancy, or rubella (German measles), may be responsible for the abnormality. In some cases, the heart may be located in the right side of the body, although this seldom causes any difficulty and may not be noticed immediately. More serious defects are those which involve the size and development of the chambers of the heart, its valves, and connecting vessels. In some patients, such congenital defects may manifest themselves only after many years, and cause nothing more than a slight discomfort in breathing. In other instances, the defects may be such as to inhibit seriously the flow of blood through the heart and lungs.

In one of the malformations (*patent ductus arteriosus*), a small duct connecting the aorta and the pulmonary artery fails to close at birth. Since the pressure is higher in the aorta, blood will flow from this vessel to the pulmonary artery and back to the lungs, from which it had just come. This means that even when the lungs are working at full capacity, all of the oxygenated blood is not being circulated to the body. Difficulty in breathing and palpitation are outstanding symptoms. Once it is discovered, this defect can be repaired surgically by tying or dividing and sewing the open ends of the duct.

If defects exist which allow a mixing of arterial and venous blood, the patient frequently has a bluish or *cyanotic* appearance. This condition, if not corrected, may limit the life of the patient to a relatively few years. Best known of the cyanotic congenital heart defects are those that are found in "blue babies." One of the most common conditions causing blue babies is really a combination of four malformations (*tetralogy of Fallot*). In this disorder, the prenatal partition (*septum*) between the two pumping chambers (*ventricles*) of the heart has failed to close at birth. In addition, the major artery (*aorta*) leading from the heart is slightly out of place, and the artery leading from the heart to the lungs is constricted. The right ventricle, therefore, not only must pump blood through the lungs, but also must work directly against pressure from the left, so that the ventricle becomes enlarged because of the extra work. Blood which has been through the lungs becomes mixed with that which has not. An increase in the number of red blood cells may occur to compensate for the circulatory insufficiency. The child's fingers may be club-shaped and there may be a failure on the part of the child to develop physically in a normal manner. Breathlessness is common.

At one time, the treatment of blue babies was limited and consisted mainly in preventing infection and overactivity of the patient. The span of life was short. Now, in a special surgical procedure, one of the arteries—the *aorta, common carotid, subclavian,* or *innominate*—is connected to the pulmonary artery. There is then an increase of the blood flow to the lungs sufficient to permit the patient maximum activity without placing undue strain on the heart. This operation, when needed, is performed during the very early years of childhood. At a later date, the individual can be fully corrected with a second operation, utilizing the heart-lung machine.

Congenital Anomalies. Each of the four valves of the heart may have congenital anomalies. The *tricuspid valve* may have a deformity of the leaflets, known as *Ebstein's malformation of the tricuspid valve.* Or there may be *tricuspid atresia,* in which the valve never forms, preventing the normal flow of blood from the right auricle into the right ventricle. Instead, it flows from the right auricle into the left auricle through a hole in the wall between the two upper chambers of the heart. The *pulmonary valve* cusps are partially fused in some individuals and prevent the proper flow of blood, *pulmonary stenosis.* This condition can be caused by narrowing of the orifice leading to the valve or fusion of the leaves of the valve itself. In the normal heart, the systolic pressure is the same on both sides of the valve. If the pressure is found to be lower in the pulmonary artery than in the right ventricle, the physician knows that *pulmonary stenosis* exists. The mitral valve may have *atresia, incompetence,* or *stenosis,* although isolated cases of these conditions are rare. The aortic valve in the heart may have a congenital narrowing of the orifice or fusion of the cusps, known as *aortic stenosis.* Most of these abnormalities of heart valves can be corrected surgically.

The most common congenital malformation occurring as a single lesion is *ventricular septal defect,* in which there is a hole in the wall between the left and right ventricles. Following diagnosis, this abnormality can be corrected surgically by sewing a patch composed of a tough, resilient plastic material over the opening. A hole between the two auricles, *atrial septal defect,* allows blood to flow from the left side of the heart as the result of pressure differences. This defect can be corrected by directly suturing the edges of the defect.

A more complicated group of defects occurs when there is a hole between both the upper chambers (*atria*) and lower chambers (*ventricula*) with malformed intervening tissue and one or both valves between the atria and ventricula. These most difficult lesions can be corrected with the use of the heart-lung machine and require the use of a patch and sometimes a prosthetic valve.

In some cases, the oxygenated blood from the lungs returns partially or totally to the right side of the heart instead of draining into the left auricle. This type of malformation, *anomalous drainage of pulmonary veins,* is characterized by an abnormal condition—the same amount of oxygen being present in all the chambers of the heart, the pulmonary artery, and the aorta. This condition can be corrected by various surgical procedures in which the anomalous drainage is redirected into the correct left auricle.

In *coarctation of the aorta,* another rather common genital heart defect, the main artery leaving the heart is constricted to such an extent that the flow of blood to all parts of the body is restricted. When the diagnosis of this condition has been confirmed, the constriction can be removed surgically and the ends of the aorta reunited or the defect bridged with a synthetic vessel, thus allowing the blood to flow freely.

Heart Transplants

As of the early 1980s, socio-economic considerations rather than scientific factors have become the key to the future of heart transplantation technology. The first human heart transplantation was accomplished by Christiaan Barnard, a South African surgeon, in December 1967. The early success rates of this procedure were dismal. For example, heart transplantations were suspended in the United Kingdom for several years and it was not until the early 1980s that surgeons at Cambridge decided to resume work in the field. Scientists, notably at Stanford University Medical Center, gradually improved the success rate of transplants by vigorous patient selection, better diagnosis of early signs of rejection, and extremely careful and thorough attention to immunosuppression. As of the early 1980s, over twenty patients per year are receiving transplanted hearts at Stanford with about a 65% chance of living at least one year after the operation—versus a 100% chance of dying within about six months without the operation. Knox (1980) reports that one Stanford patient is doing well eleven years after the operation. A few survivors have had two and three hearts. Actuarial projections put the overall 5-year survival at somewhat better than 50%.

A number of factors have slowed heart transplant technology:

(1) *Economics*—the governments of a number of countries including the United States and the United Kingdom have been studying heart transplants in terms of cost-effectiveness and cost-benefit ratios, particularly as heart transplant costs are supported by public funds. Many penetrating questions have been raised in recent years, such as "Does transplantation imply diverting funds from other useful programs, such as immunization of the elderly against pneumonia?

(2) *Removing technical problems*—which obviously are numerous, for if they were not, the survival rate would be many times better than has been achieved to date. Some scientists argue that many of the technical improvements and refinements can only come from additional experience with the procedure. The question is then posed, "Should not most of the current costs of transplantation be charged against research rather than accounted for mainly in terms of present socio-economic benefits?"

(3) *Availability of donor hearts is limited*—a fact that will impede heart transplantation from becoming a "runaway" technology. Some authorities estimate that, at best, only about a thousand donor hearts can be made available in the United States each year. Thus, rather than widely spread the technology, involving costly equipment and personnel, over a large number of hospitals, it may be optimal to create a relatively few heart transplant centers—at least for the intermediate future.

(4) *Competition with other new medical technologies* is another factor that seriously affects the availability of facilities and personnel. With budgets that are always limited, universitites and hospitals have other new proven or promising technologies that compete for money and/or people. These would include kidney dialysis, coronary artery surgery, and computerized tomography, not to mention transplantations of other important organs, such as the liver. There also are new technical developments, such as combined heart-lung transplants and the ultimate refinement of "artificial hearts" that may render heart transplantation technology obsolete as presently conceived. It is interesting to note that in February 1980, the trustees of the Massachusetts General Hospital voted not to permit heart transplants at that institution at the present time. The trustees stated, "in an age where technology so pervades the medical community, there is a clear responsibility to evaluate new procedures in terms of the greatest good for the greatest number." Not all authorities, of course, have agreed with the decision.

Cardiopulmonary bypass technology has been a key not only to heart transplantations, but to all procedures that involve "open heart" surgery. This technology (the so-called heart-lung machine) permits surgeons to operate on the heart for long periods of time in a dry, bloodless field, under direct vision. A pump draws blood from the vena cava, through tubes which are connected to these veins before they enter the heart. The blood is pumped under controlled pressure and flows to an "artificial lung" usually a plastic, membranous structure, where it is allowed to contact a steady stream of oxygen. The oxygenated blood is then pumped through another tube into the arterial system. The oxygen content, temperature, degree of alkalinity or acidity, rate of flow, and pressure, among other instrumental variables, must be carefully regulated throughout the entire surgical procedure. Checks on the circulation in the extremities are made continuously during the bypass procedure to prevent death of any tissues because of inadequate blood supply.

Artificial Heart. The serious quest for an artificial heart dates back over a half century. Actually, the concept goes back as early as 1812, when Julien-Jean César La Gallois observed, "if one could substitute for the heart a kind of injection (of arterial blood), one would succeed easily in maintaining alive indefinitely any part of the body." Mechanical perfusion experiments with heart and lung organs date back a century ago (1880), exemplified by the work of Henry Martin. Martin's work prepared the foundation for modern cardiopulmonary bypass technology. Jarvik (1981), an outstanding researcher in the field, observes that by 1951 over thirty artificial heart-lung designs had been

TOPICS RELATED TO HEART AND CIRCULATORY SYSTEM

Topic	Title of Entry in this Encyclopedia
Acute myocardial infarction	**Ischemic Heart Disease**
Aneurysm	**Aneurysm**
Angina pectoris	**Ischemic Heart Disease**
Angiography	**Angiography**
Angiotensin	**Hypertension (High Blood Pressure)**
Antiarrhythmic agents	**Arrhythmias (Cardiac); Ischemic Heart Disease**
Antihypertensive drugs	**Hypertension (High Blood Pressure)**
Aorta	**Aorta; Heart and Circulatory System (Human)**
Aortal aneurysm	**Aneurysm**
Aortic valve stenosis and regurgitation	**Heart and Circulatory System (Human)**
Arterial and venous disorders	**Arterial and Venous Disorders**
Arteriosclerosis	**Arterial and Venous Disorders**
Artery	**Artery; Heart and Circulatory System (Human); Arterial and Venous Disorders**
Arrhythmias (Cardiac)	**Arrhythmias (Cardiac)**
Artificial heart	**Heart and Circulatory System (Human)**
Artificial heart valves	**Heart and Circulatory System (Human)**
Ascites	**Ascites**
Atherogenesis	**Arterial and Venous Disorders**
Atherosclerosis	**Arterial and Venous Disorders**
Blood clot	**Anticoagulants; Arterial and Venous Disorders**
Blood pressure	**Blood Pressure; Hypertension (High Blood Pressure); Hypotension**
Bradycardia	**Arrhythmias (Cardiac); Ischemic Heart Disease**
Buerger's disease	**Arterial and Venous Disorders**
Bundle branch heart block	**Arrhythmias (Cardiac)**
Cardiac arrhythmias	**Arrhythmias (Cardiac); Ischemic Heart Disease**
Cardiac massage	**Cardiac Massage**
Cardiomyopathies	**Heart and Circulatory System (Human)**
Cardiopulmonary bypass	**Heart and Circulatory System (Human)**
Cardiopulminary resuscitation (CPR)	**Arrhythmias (Cardiac)**
Cardioversion	**Arrhythmias (Cardiac)**
Cerebral aneurysm	**Aneurysm; Cerebrovascular Diseases**
Cerebral embolism	**Cerebrovascular Diseases**
Cerebral hemorrhage	**Cerebrovascular Diseases**
Cerebral thrombosis	**Cerebrovascular Diseases**
Cerebral transient ischemic attacks	**Anticoagulants; Cerebrovascular Diseases**
Cerebrovascular diseases	**Cerebrovascular Diseases**
Cholesterol (Arteriosclerosis)	**Arterial and Venous Disorders**
Circulatory system (Animals)	**Circulatory System**
Coarctation (Aorta)	**Heart and Circulatory System (Human)**
Collateral circulation	**Collateral Circulation**
Congenital disorders (heart)	**Heart and Circulatory System (Human)**
Congestive cardiomyopathy	**Heart and Circulatory System (Human)**
Congestive heart failure	**Congestive Heart Failure**
Contractility (heart)	**Congestive Heart Failure**
Coronary angiography	**Angiography**
Coronary artery bypass	**Ischemic Heart Disease**
Coronary artery spasms	**Ischemic Heart Disease**
Coronary care unit (CCU)	**Ischemic Heart Disease**
Coronary thrombosis	**Ischemic Heart Disease**
Deep venous thrombi	**Arterial and Venous Disorders**
Diastole	**Diastole; Heart and Circulatory System (Human); Hypertension (High Blood Pressure)**
Digitalis	**Arrhythmias (Cardiac); Congestive Heart Failure**
Diuretics	**Diuretics; Hypertension (High Blood Pressure)**
Donor (heart)	**Heart and Circulatory System (Human)**
Dyspnea	**Congestive Heart Failure**
Ebstein's anomaly	**Arrhythmias (Cardiac)**
Echocardiography	**Echocardiography**
Electrocardiography	**Electrocardiography**
Embolism	**Arterial and Venous Disorders; Cerebrovascular Diseases; Ischemic Heart Disease**
Endocarditis	**Endocarditis**
Endothelium	**Endothelium**
Epicardium	**Epicardium**
Epidemiology (heart attack)	**Ischemic Heart Disease**
Essential hypertension	**Hypertension (High Blood Pressure)**
First aid (cardiac)	**Cardiac Massage**
Floppy valve syndrome	**Heart and Circulatory System (Human)**
Hardening of the arteries	**Arterial and Venous Disorders**
Heart failure	**Heart Failure**
Heart failure (congestive)	**Congestive Heart Failure**
Heart-lung machine	**Heart and Circulatory System (Human)**
Heart massage	**Cardiac Massage**
Heart transplant	**Heart and Circulatory System (Human)**
Hemorrhage	**Cerebrovascular Diseases; Hemorrhage**
Hemorrhoids	**Arterial and Venous Disorders**
High blood pressure	**Hypertension (High Blood Pressure)**
Hypercholesteremia	**Arterial and Venous Disorders**
Hyperlipidemia	**Arterial and Venous Disorders**
Hypertension (High Blood Pressure)	**Hypertension (High Blood Pressure)**
Hypertrophic cardiomyopathy	**Heart and Circulatory System (Human)**
Hyperchloremic alkalosis	**Diuretics**
Hypokalemia	**Diuretics**
Hypoprothrombinemia	**Hypoprothrombinemia**
Hypotension	**Hypotension**
Ischemic heart disease	**Ischemic Heart Disease**
Infarction	**Arterial and Venous Disorders; Ischemic Heart Disease**
Intra-aortic balloon counterpulsation	**Intra-Aortic Balloon Counterpulsation**
Loop diuretics	**Diuretics**
Lipoproteins	**Arterial and Venous Disorders**
Magnesium deficiency (heart attack)	**Ischemic Heart Disease**
Massive pulmonary embolism	**Arterial and Venous Disorders**
Mitral regurgitation	**Heart and Circulatory System (Human)**
Mortality (heart attack)	**Ischemic Heart Disease**
Multiple-risk hypothesis	**Arterial and Venous Disorders**
Myocardial anoxia and infarction	**Ischemic Heart Disease**
Norepinephrine	**Hypertension (High Blood Pressure)**
Open-heart surgery	**Heart and Circulatory System (Human)**
Orthopnea	**Congestive Heart Failure**
Pacemaker (heart)	**Arrhythmias (Cardiac)**
Percutaneous transluminal coronary angioplasty (PTCA)	**Arterial and Venous Disorders**
Pericarditis	**Heart and Circulatory System (Human)**
Phlebitis	**Arterial and Venous Disorders**
Phlebotomy	**Phlebotomy**
Plaque (arterial)	**Arterial and Venous Disorders**
Platelets (heart attack)	**Ischemic Heart Disease**
Potassium depletion	**Diuretics**
Potassium-sparing agents	**Diuretics**
Preexcitation syndromes	**Arrhythmias (Cardiac)**

TOPICS RELATED TO HEART AND CIRCULATORY SYSTEM
(*continued*)

Topic	Title of Entry in this Encyclopedia
Prinzmetal's angina	**Ischemic Heart Disease**
Prosthetic valve endocarditis	**Anticoagulants; Endocarditis**
PTCA	**Arterial and Venous Disorders**
Pulmonary embolism	**Arterial and Venous Disorders**
Pulse	**Pulse**
Radioisotopes in medicine	**Radioisotopes in Medicine**
Raynaud's disease	**Arterial and Venous Disorders**
Regurgitation (heart)	**Heart and Circulatory System (Human)**
Renin-angiotensin-aldosterone system	**Hypertension (High Blood Pressure)**
Rheumatic heart disease	**Heart and Circulatory System (Human)**
Risk factor hypothesis	**Arterial and Venous Disorders**
Sick sinus syndrome	**Arrhythmias (Cardiac)**
Spasms (coronary artery)	**Ischemic Heart Disease**
Sphygmomanometer	**Sphygmomanometer**
Stenosis	**Heart and Circulatory System (Human)**
Stone heart	**Ischemic Heart Disease**
Streptokinase	**Ischemic Heart Disease**
Stroke	**Cerebrovascular Diseases**
Sympatholytic agents	**Hypertension (High Blood Pressure)**
Syncope	**Syncope**
Systole	**Hypertension (High Blood Pressure)**
Tachycardia	**Arrhythmias (Cardiac); Ischemic Heart Disease**
Thiazides	**Diuretics**
Thrombocytopenia	**Anticoagulants**
Thrombophlebitis	**Arterial and Venous Disorders**
Thrombosis	**Anticoagulants; Arterial and Venous Disorders; Cerebrovascular Diseases**
Thrombosis (deep vein)	**Anticoagulants; Arterial and Venous Disorders**
Thrombus	**Arterial and Venous Disorders; Cerebrovascular Diseases; Ischemic Heart Disease**
Transplant (heart)	**Heart and Circulatory System (Human)**
Truncus arteriosus	**Truncus Arteriosus**
Valve (artificial heart)	**Heart and Circulatory System (Human)**
Valvular heart disease	**Heart and Circulatory System (Human)**
Variococele	**Arterial and Venous Disorders**
Varicose veins	**Arterial and Venous Disorders**
Vascular tissue	**Vascular Tissue**
Vasoconstrictor	**Hypertension (High Blood Pressure)**
Vasomotor nerves	**Arterial and Venous Disorders**
Veins (disorders)	**Arterial and Venous Disorders**
Venous thrombosis	**Arterial and Venous Disorders**
Ventricular fibrillation	**Arrhythmias (Cardiac); Ischemic Heart Disease**
Wolff-Parkinson-White syndrome	**Arrhythmias (Cardiac)**

proposed, some researching the experimental stage. Well known is the work of Lindbergh and Carrel in the 1930s in connection with their perfusion pump, reported by the news media at that time as a "robot heart."

Prominent in the search for an artifical heart has been the concept of a device that will be implanted in the human body and take over all heart functions. Lindbergh and Carrel added an interesting new dimension to this objective, as suggested by the following quotation from one of their publications: "We can perhaps dream of removing diseased organs from the body and placing them in the Lindbergh pump as the patients are placed in a hospital. There [the organs] could be treated far more energetically than within the organism and, if cured, replanted in the patient."

Jarvik (1981) lists at least six criteria for what may be termed "the total artificial heart": (1) *Small size,* to fit into the existing human cardiac cavity; (2) *work output* ample to provide all needs supplied by a natural heart; (3) a *variable output* in accordance with the changing rate of body requirements (range from rest to vigorous exercise); (4) *gentle handling of blood* to avoid hemolysis (disintegration of the elements of the blood); (5) *ease of sterilization*; and (6) *durability.* There are, of course, numerous other criteria, certainly one of which is economics.

From a bioengineering standpoint, it is interesting to note that the total power output of the human heart is about 2.5 watts, of which 80% is required by the left ventricle (the output side of the heart which pumps blood into the arteries and ultimately to the capillaries). The pressure parameters are well established. See the entry on **Hypertension (High Blood Pressure),** which gives diastolic and systolic pressures.

Working essentially with these criteria in mind, a number of teams have been researching and experimenting with artificial hearts. These include work at the Cleveland Clinic, dating back to the 1950s. In 1957, these researchers were able to keep dogs alive for about 1.5 hours with a plastic polyvinyl chloride heart energized by compressed air. It should be noted that these experiments were conducted at a time before attempts at human heart transplantation had been made, and when open-heart surgery was in the early pioneering stage. Research on artificial heart valves had just commenced. Progressively, these and other workers refined their designs and selection of materials of construction as well as various energy supplies, including electrically driven apparatus. Nuclear power as a source was considered. By the mid-1960s, researchers at the Cleveland Clinic were able to keep calves alive for 1.5 days with an artificial heart.

It should be mentioned that in England in 1928, Dale and Schuster built a pump with the objective of temporarily bypassing the heart during heart surgery. Dodrill (General Motors Corporation), in 1952, developed a mechanical heart which was used for nearly an hour during human heart surgery. Jarvik stresses, however, that open-heart surgery as it is known today requires the heart-lung machine, not simply a pump to replace the heart.

In 1969, for the first time, an artificial heart was installed in a human being. This artifact was designed by Liotta and Hall (Texas Heart Institute) and sustained life for about 64 hours, during which time a natural heart was being sought for transplantation.

Progress in artificial hearts has continued, but delineation of these details is beyond the scope of this encyclopedia. An excellent summary is given by Jarvik (1981). It should be mentioned that a unit known as the Jarvik-7 was designed to fit the human anatomy, although experiments conducted thus far have involved calves and sheep. As described by Jarvik, the two ventricles are made of polyurethane supported on aluminum bases. Rings of polycarbonate support tilting-disk valves. In one test, this artificial heart sustained the growth of a calf from 200 pounds (91 kilograms) at surgery to over 350 pounds (159 kilograms) over a 7-month period. In the experiment, it was found that the calf could walk on a treadmill for an hour even after it had grown to twice the size of the originally intended recipient, a 150-pound (68 kilogram) human.

References

General

Akera, T.: "Membrane Adenosinetriphosphatase: A Digitalis Receptor?" *Science*, **198**, 569–575 (1977).

Baltaxe, H. A., Amplatz, K., and D. C. Levin: "Coronary Angiography," Charles C. Thomas, Springfield, Illinois, 1976.

Fletcher, G. F., and J. D. Cantwell: "Exercise and Coronary Heart Disease," Charles C. Thomas, Springfield, Illinois, 1979.

Gresham, G. A.: "Reversing Atherosclerosis," Charles C. Thomas, Springfield, Illinois, 1980.

Gutstein, W. H., et al.: "Neutral Factors Contribute to Atherogenesis," *Science*, **199**, 449–451 (1978).

Harvey, W.: "De Motu Cordis: Anatomical Studies on the Motion of the Heart and Blood," 5th edition (translation by C. D. Leake), Charles C. Thomas, Springfield, Illinois, 1978.

Kolata, G. B., and J. L. Marx: "Epidemiology of Heart Disease: Searches for Causes," *Science*, **194**, 509–512 (1976).

Kolata, G. B.: "Detection of Heart Disease: Promising New Methods," *Science*, **194**, 1029–1031 (1976).
Kolata, G. B.: "The Aging Heart," *Science*, **195**, 166–167 (1977).
Kolata, G. B.: "New Treatment for Coronary Artery Disease," *Science*, **206**, 917 (1979).
Kolata, G. B.: "Treatment Reduces Deaths from Hypertension," *Science*, **206**, 1386–1387 (1979).
Linask, J., Votta, J., and M. Willis: "Perfusion Preservation of Hearts for 6 to 9 Days at Room Temperature," *Science*, **199**, 299–301 (1978).
Marx, J. L.: "Hypertension: A Complex Disease with Complex Causes," *Science*, **194**, 821–825 (1976).
Marx, J. L.: "After the Heart Attack: Limiting the Damage," *Science*, **194**, 1147–1150 (1976).
Marx, J. L.: "Sudden Death: Strategies for Prevention," *Science*, **195**, 39–41 (1977).
Marx, J. L.: "Coronary Artery Spasms and Heart Disease," *Science*, **208**, 1127–1130 (1980).
Marx, J. L., and G. B. Kolata: "Combating the #1 Killer," American Association for the Advancement of Science, Washington, D.C., 1980.
McDonald, I. G.: "Introduction ot Echocardiography," Charles C. Thomas, Springfield, Illinois, 1976.
McGill, H. C., Jr., et al.: "The Heart is a Target Organ for Androgen," *Science*, **207**, 775–777 (1980).
Morad, M. (editor): "Biophysical Aspects of Cardiac Muscle," Academic, New York, 1979.
Pollack, G. H.: "Cardiac Pacemaking: An Obligatory Role of Catecholamines?" *Science*, **196**, 731–738 (1977).
Selye, H.: "Selye's Guide to Stress Research," Van Nostrand Reinhold, New York, 1979.
Staff: "The American Heart Association Heartbook: A Guide to Prevention and Treatment of Cardiovascular Diseases," Dutton, New York, 1980.
Stallones, R. A.: "The Rise and Fall of Ischemic Heart Disease," *Sci. Amer.*, **243**, 5, 53–59 (1980).
Stumpf, W. E., Sar, M., and G. Aumüller: "The Heart: A Target Organ for Estradiol," *Science*, **196**, 319–321 (1977).
Turlapaty, P. D. M. V., and B. M. Altura: "Magnesium Deficiency Produces Spasms of Coronary Arteries: Relationship to Etiology of Sudden Death Ischemic Heart Disease," *Science*, **208**, 198–200 (1980).
Voors, A. W., et al.: "Racial Differences in Blood Pressure Control," *Science*, **204**, 1091–1095 (1979).
Zoneraich, S.: "Diabetes and the Heart," Charles C. Thomas, Springfield, Illinois, 1978.

ARTIFICIAL HEART AND TRANSPLANTS
Curran, W. J.: "The First Mechanical Heart Transplant: Informed Consent and Experimentation," *N. Engl. J. Med.*, **291**, 1015–1016 (1974).
Jarvik, R. K., et al.: "Criteria for Human Total Artificial Heart Implantation Based on Steady State Animal Data," *Trans. Amer. Soc. Artif. Int. Organs*, **23**, 535–542 (1977).
Jarvik, R. K.: "The Total Artificial Heart," *Sci. Amer.*, **244**, 1, 74–80 (1981).
Knox, R. A.: "Heart Transplants: To Pay or Not to Pay," *Science*, **209**, 570–575 (1980).
Knox, R. A.: "Massachusetts General: No Heart Transplants Here," *Science*, **209**, 574 (1980).
Kolff, W. J.: "An Artificial Heart Inside the Body," *Sci. Amer.*, **213**, 5, 38–46 (1965).
Murakami, T. et al.: "Transient and Permanent Problems Associated with the Total Artificial Heart Implantation," *Trans. Amer. Soc. Artif. Int. Organs*, **25**, 239–248 (1979).
Pierce, W. S., et al.: "Calcification Inside Artificial Hearts: Inhibition by Warfarin-Sodium," *Science*, **208**, 601–602 (1980).
Unger, F.: "Assisted Circulation," Springer-Verlag, New York, 1979.

CARDIOMYOPATHIES
Demakis, J. G., et al.: "Natural Course of Peripartum Cardiomyopathy," *Circulation*, **44**, 1053 (1971).
Fowler, N. O. (editor): "Myocardial Disease," Grune and Stratton, Inc., New York, 1973.
Goodwin, J. F.: "Congestive and Hypertrophic Cardiomyopathies," *Lancet*, **1**, 731 (1970).
Morrow, A. G., et al.: "Operative Treatment in Hypertrophic Subaortic Stenosis," *Circulation*, **52**, 88 (1975).
Pohost, G., et al.: "Hypertrophic Cardiomyopathy," *Circulation*, **55**, 92 (1977).
Teare, D.: "Asymmetrical Hypertrophy of the Heart in Young Adults," *Br. Heart J.*, **20**, 1 (1958).

CONGENITAL HEART DISORDERS
Bayer, L., and M. P. Honzik: "Children with Congenital Intracardiac Defects," Charles C. Thomas, Springfield, Illinois, 1976.
Elliott, L. P., and G. L. Schiebler: "The X-ray Diagnosis of Congenital Heart Diseases in Infants, Children, and Adults," Charles C. Thomas, Springfield, Illinois, 1979.
Engle, M. A.: "Cyanotic Congenital Heart Disease," *Am. J. Cardiol.*, **37**, 283 (1976).
Gyepes, M. T., and W. R. Vincent: Cardiac Catherization and Angiocardiography in Severe Neonatal Heart Disease," Charles C. Thomas, Springfield, Illinois, 1974.
Kidd, B. S., Keith, L., and J. D. Keith: "The Natural History and Progress in Treatment of Congenital Heart Defects," Charles C. Thomas, Springfield, Illinois, 1971.
Nadas, A. S., and D. C. Fyler: "Pediatric Cardiology," 3rd edition, Saunders, Philadelphia, 1972.
Nora, J., and A. H. Nora: "Genetics and Counseling in Cardiovascular Diseases," Charles C. Thomas, Springfield, Illinois, 1978.
Perloff, J. K.: "Pediatric Congenital Cardiac Becomes a Postoperative Adult," *Circulation*, **46**, 606 (1973).
Rowe, R. D.: "Evaluation of Late Results of Surgical Treatment of Congenital Heart Disease," *Canad. Med. Assoc. J.*, **113**, 853 (1975).

PERICARDITIS
Cortes, F. M. (editor): "The Pericardium and Its Disorders," Charles C. Thomas, Springfield, Illinois, 1971.
Spodick, D. H.: "Pericardial Diseases," F. A. Davis Co., Philadelphia, Pennsylvania, 1976.
Krikorian, J. G., and E. W. Hancock: "Pericardiocentesis," *Am. J. Med.*, **65**, 808 (1978).

VALVULAR HEART DISEASE
Bristow, J. D., and E. L. Kremkau: "Hemodynamic Changes after Valve Replacement with Starr-Edwards Prosthesis," *Am. J. Cardiol.*, **35**, 716 (1975).
CDC: "Isolation of Mycobacteria Species from Porcine Heart Valve Prostheses," *Morbidity Mortality Weekly Report*, **26**, 42–43, Center for Disease Control, Atlanta, Georgia (February 11, 1977).
Hancock, E. W.: "Aortic Stenosis, Angina Pectoris, and Coronary Artery Disease," *Am. Heart J.*, **93**, 382 (1977).
Laskowski, L. F., et al.; "Fastidious Mycobacteria Grown from Porcine Prosthetic-Heart-Valve Cultures," *N. Engl. J. Med.*, **297**, 101–102 (1977).
Mehlman, D. J., and L. Resnekov: "A Guide to the Radiographic Identification of Prosthetic Heart Valves," *Circulation*, **57**, 613 (1978).
Moreno-Cabral, R. J., et al.: "Acute Thrombotic Obstruction with Björk-Shiley Valves," *J. Thorac. Cardiovasc. Surg.*, **75**, 321 (1978).
Pierce, W. S., et al.: "Calcification Inside Artificial Hearts: Inhibition by Warfarin-Sodium," *Science*, **208**, 601–603 (1980).
Petersdorf, R. G.: "Antimicrobial Prophylaxis of Bacterial Endocarditis," *Am. J. Med.*, **65**, 220 (1978).
Roberts, W. C., Dangel, J. C., and B. H. Bulkley: "Nonrheumatic Valvular Cardiac Disease," *Cardiovasc. Clin.*, **5**, 334 (1973).
Selzer, A., and K. E. Cohn: "Natural History of Mitral Stenosis," *Circulation*, **45**, 878 (1972).
Stinson, E. B., et al.: "Long-Term Experience with Porcine Aortic Valve Xenografts," *J. Thorac. Cardiovasc. Surg.*, **73**, 54 (1977).
Stollerman, G. H.: "Rheumatic Fever and Streptococcal Infection," Grune and Stratton, Inc., New York, 1973.
Ward, C.: "Virus Valvular Heart Disease," *Am. Heart J.*, **90**, 804 (1975).

HEARTBURN. **Esophagus.**

HEART FAILURE. This term does not normally signify complete failure or stoppage of heart action, but rather it describes the inability (failure) of the heart to pump the amount of blood required for proper circulation. In essence, the heart or some of its chambers fail to discharge their contents properly. The exact mechanism of the disease is not fully understood, but may result from a variety of diseases in which there is heart muscle involvement. When the left ventricle fails, the pressure rises in the left auricle and in the pulmonary veins. Heart failure is not necessarily fatal, and many persons who at one time have suffered from it may live for many years.

Symptoms of heart failure include difficulty in breathing and generalized enlargement of the veins caused by increased pressure in the right auricle and the veins. The liver becomes enlarged and fluids accumulate in the tissues with marked swelling (*edema*, commonly referred to as *dropsy*) of areas, such as the feet and ankles. The dropsy can usually be corrected by a low-salt diet and diuretics, a group of drugs which stimulate the kidneys to excrete water and sodium.

Although there usually is no unconsciousness, heart failure is a condition which frequently requires first aid. The symptoms vary, depending upon the cause. There are three main categories of symptoms: (1) *Heart failure resembling fainting*—the patient may be conscious, the face pale, and the pulse weak. If no pain is present about the heart, the condition is distinguished from simple fainting by the failure of the person to recover fairly quickly after lying down. (2) *Heart failure characterized by pain*—the patient may have an agonizing pain in the region of the heart, usually behind the breastbone rather than on the left side. The pain may also go down the left arm. The patient is usually conscious and very apprehensive. (3) *Heart failure characterized by shortness of breath*—the patient cannot lie down and often there is congestion of the face. The patient usually insists on sitting up, leaning forward in order to breathe. In the first two instances, the patient should be kept quite in a lying position; in the latter type, the patient should be propped up to permit better breathing. A physician should be called immediately and if police, fire, or paramedic personnel can be alerted by telephone, this should be done immediately so that full attention can be provided to the patient prior to arrival of a physician or attendance at a hospital.

Should the heart collapse completely, the patient loses respiration, the pulse stops in all major blood vessels, and the pupils become dilated. In case of collapse, mouth-to-mouth artificial respiration should be given along with external cardiac massage. See **Cardiac Massage.**

See also list of entries in entry on **Heart and Circulatory System (Human).**

HEART FAILURE (Congestive). Congestive Heart Failure.

HEART-LUNG MACHINE. Heart and Circulatory System (Human).

HEART MASSAGE. Cardiac Massage.

HEART (Systemic). Systemic Heart.

HEART TRANSPLANT. Heart and Circulatory System (Human).

HEARTWORM DISEASE (Dirofilariasis). This is a serious and potentially fatal disease in dogs. It is caused by a worm (*Dirofilaria immitis*) which is found in the animal's heart and large adjacent vessels. The female worm is 6–14 inches (15.2–35.6 centimeters) long and about $\frac{1}{8}$ inch (3 millimeters) wide. The male is smaller. One dog may have as many as 300 worms. Adult heartworms live in the animal up to 5 years and during that period the female produces millions of young *microfilariae.* These microfilariae live in the bloodstream, mainly in the small blood vessels. They cannot grow to adults without passing through an intermediate host (a mosquito). As many as 30 species of mosquito can serve as host. The microfilariae develop for 10–30 days in the mosquito and then enter the saliva of the insect. At this point, the organisms are *infective larvae* because at this stage of development they will grow to adults when they enter a dog. The mosquito bites the dog, mostly on the abdomen where the haircoat is thinnest.

Adult worms cause disease by clogging the heart and major blood vessels leading from the heart. They interfere with the valve action. By clogging the main blood vessels, the blood supply to other organs of the body is reduced, particularly the lungs, liver, and kidneys. Most dogs infected with heartworms do not show external signs of the disease. When symptoms develop after some period of infection, these will include soft dry chronic cough, shortness of breath, weakness, nervousness, listlessness, and loss of stamina. These features are noticed particularly after exercise. The microfilariae circulate throughout the body, but remain mainly in the small blood vessels, which they tend to clog. Ultimately there is destruction of lung and kidney tissue.

An arsenical drug is used in treatment and usually requires a hospital environment. The treatment requires injections of the drug over a period of 2–3 days. About 6 weeks after the adult worms have been eradicated, further injections of other drugs are required to eradicate the microfilariae. Some veterinarians prefer to eradicate the microfilariae first. To prevent heartworms, many veterinarians recommend the use of diethylcarbamazine citrate (e.g., Filarbits®) in the pet's diet during the mosquito season.

HEAT. The agency whose addition to or removal from a physical system is the cause of thermal changes of various types. These include rise and fall of temperature, changes in length and volume, changes of physical states, such as melting, evaporations, etc.

During the eighteenth century heat was assumed to be a subtle fluid called *caloric,* filling the interstices between the ultimate particles of matter and, under conditions of isolation from the surroundings, known to satisfy a conservation law. The production of heat by friction as well as its disappearance during the performance of external mechanical work established its essential physical nature as another form of *energy* and led to the overthrow of the caloric theory. Nevertheless, we still speak of the *flow* of heat as though it were a fluid and have retained the methods of measuring the *quantity of heat* originally devised by the upholders of the caloric view.

Our direct knowledge of heat is provided by the sensation of hotness and coldness when we come in contact with various physical bodies. It is possible to arrange a set of bodies in a sequence such that A feels hotter than B, B hotter than C, etc. We say that A has a higher *temperature* than B, B a higher one than C, and so on. Of course our sensations are qualitative and are considerably influenced by the thermal conductivity of the body we touch. Thus, on a frosty morning, the head of an ax being metal feels considerably colder than the wooden handle though the two are presumably at the same temperature. To obtain a continuous and reproducible physical scale of temperature, various types of thermometers have been devised of which the mercury-in-glass or colored-alcohol-in-glass are familiar examples. The two temperature scales in common use are the Fahrenheit scale and the Celsius scale. The first assigns values of 32° and 212° to the normal freezing and boiling points of pure water, respectively, and divides this interval into 180 equal sub-intervals or degrees. The Celsius, formerly called the Centigrade scale assigns the respective values of 0° and 100° to the above fixed points; the standard interval is then divided into 100 equal degrees.

Temperature changes are produced by the addition or subtraction of heat from a body. Thus, temperature may be regarded as a measure of the concentration or *intensity* of heat. In general, the more heat we add to a given body the more its temperature rises.

Measurement of Heat. Since heat is imponderable and not directly observable, it is necessary to measure the size of a given quantity of heat by its effect on another body. If this effect is the production of a rise in temperature from some initial temperature, t_1, to a final temperature, t, then the rise $(t - t_1)$ is found to vary inversely with the mass of the test body. It is thus natural therefore, following the calorists, to regard the quantity of heat, say Q, as determined by the product of m and $(t - t_1)$. Thus we say

$$Q \text{ is proportional to } m \times (t - t_1)$$

To make this statement into an equation we write

$$Q = \text{constant} \times m \times (t - t_1) \quad (1)$$

where the constant of proportionality depends on the substance, being large for some materials and small for others. This constant for water, for example, is about 33 times as great as for lead; water is said therefore to have a greater *heat capacity* than lead. Notice that the constant in Equation (1) actually gives the numerical value of Q which is required to warm a unit mass of the substance through a temperature interval of exactly 1°. This constant is accordingly called the *specific heat capacity* (usually abbreviated to *specific heat*) and is indicated by c. Since it is found that the value of the specific heat, particularly for gases, but in principle for all materials, depends on the conditions under which the heat is absorbed, this must be indicated. We thus have c_p and c_v, for example, for the two important cases of absorption at constant pressure and constant volume, respectively. Since the former characterizes the common laboratory case of working under atmospheric pressure, we accordingly rewrite Equation (1) as

$$Q_p = c_p m (t - t_1) \quad (2)$$

Q_p now measures the heat absorbed under constant pressure, and c_p is the constant pressure specific heat. Since the right side of Equation (2) contains *three* quantities, a mere choice of a mass unit and a degree unit is insufficient to establish a unit of heat. It is necessary to select some substance as a standard reference body and assign an arbitrary value of, say c_p equal to unity for it. Water is the universal choice for this standard body due not only to its cheapness and ease of purification but also to its large heat capacity.

With the selection of water as the standard with $c_p = 1$, the left side of Equation (2) clearly becomes of unit value when m and $(t - t_1)$ are each of unit value. In the English system we accordingly have the *British thermal-unit* (or Btu) as the heat required to warm 1 pound of pure water through an interval of 1°F. In the metric system, the corresponding unit is the *calorie*, the heat required to warm 1 gram of water 1°C. A large or *kilocalorie* corresponding to 1,000 ordinary calories is also frequently used in scientific work.

Specific Heats. Use of Equation (2) reveals that the values of c_p obtained experimentally depend on the temperature interval used, indicating a dependence of c_p on temperature. Thus, if c_p for water were actually unity throughout the 0 to 100°C range, a mass of water at 100°C mixed with an equal mass at 10°C would give a final mixture at exactly 50°C. The actual value is near 50.05°; this difference although small, indicates the need to specify the calorie at some particular temperature. For this purpose, we suppose a system of mass m is warmed from t to $t + \Delta t$ by the addition at constant pressure of an increment of heat ΔQ_p. Then Equation (2) becomes

$$\Delta Q_p = m\bar{c}_p \, \Delta t \tag{3}$$

where now $\bar{c}_p$ is an average value of c_p over this interval. Then we define the *instantaneous* heat capacity, c_p at t by the following relation

$$c_p = \frac{1}{m} \lim_{\Delta t \to 0} \frac{\Delta Q_p}{\Delta t} = \frac{1}{m} \frac{dQ_p}{dt}$$

i.e., the heat absorbed per unit mass per degree as the interval becomes smaller and smaller without limit. This leads to the differential form of Equation (3)

$$dQ_p = mc_p \, dt \tag{4}$$

where dQ_p is the differential heat absorption which produces a differential temperature rise dt in a body of mass m and specific heat c_p.

The standard or 15° calorie is now defined as the rate of absorption of heat per gram per degree at 15°C and in practice is essentially the same as the average calorie over the 1° interval from 14.5 to 15.5°C.

If a mass m of water is warmed from t_1 to t, the integral of Equation (4) gives for the total heat absorbed in 15° calories

$$Q_p = \int_{t_1}^{t} dQ_p = m \int_{t_1}^{t} c_p \, dt = m\left[\int_{0^\circ}^{t} c_p \, dt - \int_{0^\circ}^{t_1} c_p \, dt\right] \tag{5}$$

where the integral of c_p over the range t_1 to t has been written as the difference of two integrals from a common lower limit of 0°C. If, therefore, we evaluate an integral of the type $\int_{0^\circ}^{t} c_p \, dt$ with t varying in 1° steps and arrange these in a table, the right side of Equation (5) may be evaluated by merely subtracting appropriate entries.

In the accompanying figure, the value of c_p in 15° calories per gram per degree is plotted graphically from 0 to 100°C, and the integrals on the right of Equation (5) are represented by appropriate areas under the c_p curve. Thus the integral from 0° to t is hatched with lines sloping up to the right, while that from 0° to t_1 has the lines sloping up to the left. The value of Q_p is then the singly hatched area.

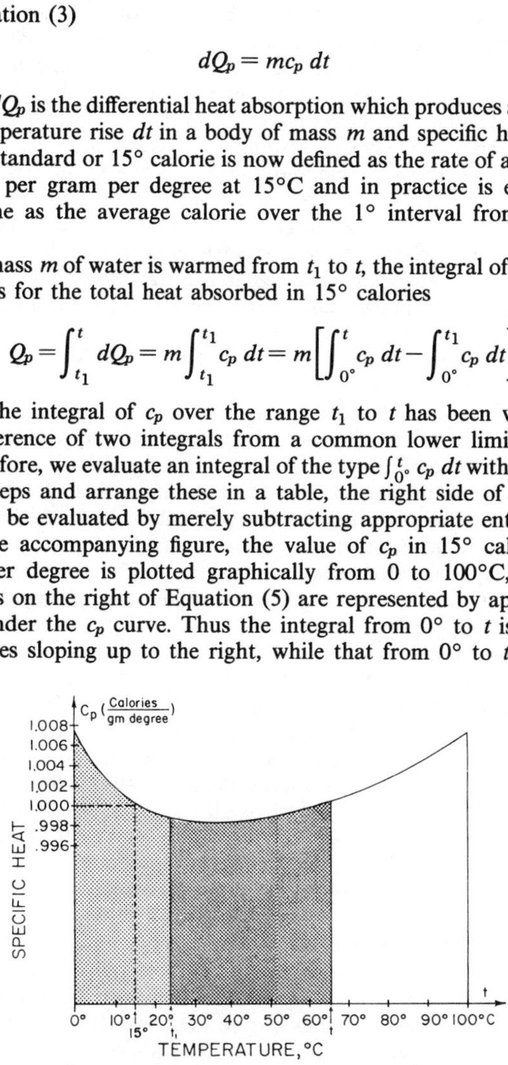

Specific heat of water versus temperature.

With heat quantities measured in 15° calories, from the observed rise or fall of temperature in known masses of water, the specific heats of various substances, the heats absorbed on melting solids to liquids (heats of fusion), the heats absorbed on passage from the liquid to the vapor state (heats of vaporization), the heats evolved on combination of various substances, and the heats absorbed or evolved in chemical changes are at once determinable (see **Calorimetry**). For the present purpose, the accompanying table gives the values of the

APPROXIMATE CONSTANT-PRESSURE SPECIFIC HEAT OF SELECTED MATERIALS

Substance	State	c_p(cal/g deg)
Water	Vapor	0.48
Water	Liquid	1.00
Water	Solid	0.50
Ethyl alcohol	Liquid	.54
Hydrogen	Gas	3.44
Air	Gas	.24
Aluminum	Solid	.22
Iron	Solid	.11
Lead	Solid	.03

constant pressure heat capacities of a few typical substances, variations with temperature being disregarded. Notice that c_p although expressed in terms of calories per gram per degree is in fact independent of the system of units since water is the reference body in all systems. Thus the specific heat of water in the English system would be 1 Btu per pound per degree Fahrenheit.

The Mechanical Nature of Heat. The conservation of heat *per se* is observed only for systems involving the performance of no mechanical or electrical work. (Count Rumford (ca. 1800) was the first to establish this fact in his famous cannon-boring experiments carried out in the arsenal of the Dutchy of Bavaria in Munich. He observed that when his drills became dull, heat was produced in great quantities limited only by the amount of work done against friction. He concluded that the large scale mechanical energy used in overcoming friction could only be converted into the motions of the ultimate particles of matter, a motion not directly observable but detected by our senses as heat. His results were confirmed and extended by the later work of Joule and Helmholtz, in particular, and also provided a more reliable value for the so-called *mechanical equivalent of heat.* This is taken as the amount of mechanical (or electrical) energy which when converted into heat is equivalent to exactly 1 calorie. The presently accepted value for this important constant is 4.185 joules per 15° calorie. Here the joule is the work performed when power is expended at the rate of 1 watt for 1 second. Thus an ordinary 100-watt lamp bulb converts 100 joules of electrical energy to thermal each second; this amounts to 100/4.185 or about 24 calories.

As a result of experiments such as these and a host of others, we are forced to recognize that heat is merely another form of the universal quantity *energy.* Its transformation always occurs at the rate of 4.185 joules per calorie whether heat goes into external work or work is dissipated through friction into heat.

See also entries which follow; and **Thermodynamics.**

References

NOTE: See also list of references at end of entries on **Heat Transfer; Thermodynamics.**

Holman, J. P.: "Thermodynamics," McGraw-Hill, New York, 1969.

Siegel, R.: "Thermal Radiation Heat Transfer," McGraw-Hill, New York, 1971.

Welty, J. R., Wicks, C. E., and R. E. Wilson: "Fundamentals of Momentum, Heat, and Mass Transfer," Wiley, New York, 1969.

Zemansky, M. W.: "Heat and Thermodynamics," 5th edition, McGraw-Hill, New York, 1968.

HEAT (Atomic). Atomic Heat.

HEAT BALANCE (Distillation). Distillation.

HEAT BALANCE (Planet). The equilibrium which exists on the average between the radiation received by a planet and its atmosphere from the sun and that emitted by the planet and atmosphere. That the equilibrium does exist in the mean is demonstrated by the observed long-term constancy of the earth's surface temperature. On the average, regions of the earth nearer the equator than about 35° latitude receive more energy from the sun than they are able to radiate, whereas latitudes higher than 35° received less. The excess of heat is carried from low latitudes to higher latitudes by atmospheric and oceanic circulations and is reradiated there.

HEAT BALANCE (Process). A heat balance is a method of accounting for all heat units in a process or change during which heat is transferred. Examples of cases where heat balances might be undertaken are: (1) Determining the nature and the magnitude of the various losses which occur when fuel is burned in a steam boiler furnace. (2) Accounting for all heat units during the operation of a prime mover, such as a Diesel engine or a steam turbine. (3) Determining the distribution of heat in a static heating device, such as a water heater supplied with steam.

Heat balance work is based upon the first law of thermodynamics, a statement of which is: Energy may not be created or destroyed, but may be converted from one form to another. The significance of this law applied to the heat balance is that the total energy may be accounted for by straight addition, hence striking a heat balance resembles bookkeeping, with heat supplied on the credit side of the ledger, and various heats usefully employed on the debit side. One way of showing a heat balance is a tabular form; another shows the heat as a stream, properly branched and subdivided to indicate the distribution of heat. Briefly, a heat balance might be said to be the bookkeeping by which heat supplied is shown to be equal to the sum of heat utilized and lost.

(It should be added that the above statement of the First Law, while adequate for many engineering calculations, is subject to modification in accordance with the principle of mass-energy equivalence.)

HEAT CAPACITY. The amount of heat necessary to raise the temperature of a system, entity, or substance by one degree of temperature. It is most frequently expressed in calories per degree centigrade, or Btu per degree Fahrenheit. If the mass of a substance is specified, then certain derived values of the heat capacity can be obtained, such as the atomic heat, molar heat, or specific heat.

HEAT CAPACITY EQUATION (Einstein). A quantum relationship for the heat capacity at constant volume of an element of the form:

$$C_v = 3R\left(\frac{h\nu}{kT}\right)^2\left(\frac{e^{h\nu/kT}}{(e^{h\nu/kT}-1)^2}\right)$$

in which C_v is the heat capacity at constant volume for one gram-atom of an element, R is the gas constant, h is Planck's constant, k is the Boltzmann constant, ν is the characteristic frequency of oscillation of the atoms of the element, T is the absolute temperature, and e is the natural logarithmic base.

The Einstein equation was the first approximation to a quantum theoretical explanation of the variation of specific heat with temperature. It was later replaced by the Debye theory of specific heat and its modifications.

HEAT CAPACITY MAPPING MISSION (HCMM). Earth Resources Satellites and Geologic Remote Sensors.

HEAT CAPACITY (Quantum Theory). Quantum Theory of Heat Capacity.

HEAT (Chemical Reaction). Gibbs-Helmholtz Equation.

HEAT CONSERVATION. Insulation (Thermal).

HEAT CONTENT. Enthalpy.

HEAT (Earth's Interior). Geothermal Energy.

HEAT ENGINE. As used in thermodynamics the term denotes a thermodynamic system, e.g., a sample of gas, carried through a cyclic process in such a way that a closed path is traced out on a pressure-volume (P-V) diagram, and positive work is done *by* the system. If Q_1 is the positive amount of heat energy absorbed by the system, Q_2 the positive amount of heat energy rejected by the system and W the net amount of work done by the system, then the first law of thermodynamics (conservation of energy) gives $W = Q_1 - Q_2$. The efficiency of the engine is defined as

$$\eta = \frac{W}{Q_1} = 1 - \frac{Q_2}{Q_1}$$

For an engine following a reversible Carnot cycle (Carnot engine), the efficiency is given by $\eta = (T_1 - T_2)/T_1$, where T_1 is the Kelvin temperature of the reservoir at which Q_1 is absorbed, and T_2 is the Kelvin temperature at which Q_2 is rejected. The second law of thermodynamics states that no engine working between these same two temperatures can have a greater efficiency than that of the Carnot engine.

A thermodynamic engine run backwards becomes a *refrigerator.* Thus a positive amount of heat Q_2 is absorbed at a low temperature, work W is done, and positive heat Q_1 is rejected at a higher temperature. The first law now gives $Q_1 = W + Q_2$. The ratio Q_2/W is known as the *coefficient of performance* of the refrigerator.

HEATER (Hysteresis). Hysteresis Heater.

HEAT EXCHANGERS. Heat Transfer.

HEAT EXHAUSTION AND HEAT STROKE. Weakness, mental fogginess, incapacity for work, and irritability are characteristics of heat exhaustion. These symptoms also appear in dehydration, alcoholism, and periods of insufficient rest and sleep. Heat exhaustion is produced in some persons when they are confined to an uncomfortably warm environment for an extensive period, during which time body fluids and salt may be depleted. The usual immediate symptoms are subnormal body temperature, clammy skin, gastic muscular spasms, and, less frequently, vomiting and diarrhea. Some authorities suggest that this syndrome may be the result of a sharp curtailment of heat production within the body, with a corresponding suppression of other functions of the adrenal cortex. Incidences of heat exhaustion can be prevented in many instances by curtailing vigorous physical exercise (as in the case of military training exercises in hot, dry or hot, humid areas) when the temperature exceeds 100°F (38°C). Another preventive measure is the scheduling of frequent rest periods in cooler locations where this is practical.

Heatstroke is a much more serious manifestation of similar factors and usually occurs when the body is subjected to very high temperatures and high humidities over relatively long periods. Epidemics of heatstroke occur in metropolitan areas during a heat wave, causing the deaths of hundreds of persons, particularly the elderly. Many such deaths go unreported and are not always identified with the cause. Poorly ventilated areas, such as barracks, sauna baths, and crowded facilities, as found in old nursing homes for the aged, aggravate the underlying conditions.

Heatstroke frequently may be manifested quite precipitously, with delirium, impaired senses, seizures, and coma. Heatstroke victims may be found with body temperatures as high as 104°F (40°C). Sweating is not always present. There may be extreme tachycardia, circulation may fail, and pulmonary edema and shock may result. Dehydration is often present. Most patients have vomiting and diarrhea.

Many deaths result from heatstroke because persons are not found and advised in time to initiate treatment. The initial step in treatment is removal of the person from the causative hot, humid environment. This should be followed by immersion in ice baths, application of ice packs, or sponging with alcohol in a relatively cool environment where there is good movement of air. Preferably with a rectal temperature sensor, the body temperature should be carefully monitored so that the patient will not be overcooled or allowed to reaccumulate risky heat loads. Phenothiazines may be administered to prevent excessive shivering or seizures, both of which conditions tend to increase body temperature. Water and electrolyte deficiencies should be corrected. The patient should be checked for possible renal failure and disseminated intravascular coagulation.

Certain drugs predispose heatstroke. These include anticholinergic agents, tricyclic antidepressants, monoamine oxidase inhibitors, inhalation anesthetics, and succinylcholine. Heatstroke occurring during anesthesia (malignant hyperthermia) is described in the entry on **Myopathy.**

References

Gary, J.: "Prevention of Heat Injuries during Distance Running," *J. Sports Med.*, **3**, 194 (1975).

Hamilton, D.: "The Immediate Treatment of Heatstroke," *Anaesthesia*, **31**, 270 (1976).

O'Donnell, T. F., Jr.: "Acute Heat Stroke," *JAMA*, **234**, 824 (1975).

Shibolet, S. M., Lancaster, M. C., and Y. Danon: "Heat Stroke," *Aviat. Space Environ. Med.*, **47**, 280 (1976).

HEAT-FLOW RESISTANCE. **Insulation (Thermal).**

HEATHER SHRUBS AND TREES. The heather or heath family (*Ericaceae*) is comprised of a number of genera, many species, hybrids, clones, and cultivars. Three of the main genera are: *Arbutus*, small to large evergreen trees, of which the strawberry tree (not to be confused with the fruit-bearing plant of the rose family) and the madrona tree are examples; *Clethra*, a small genus of deciduous or evergreen trees or shrubs, of which the Lily-of-the-valley clethra is representative; and *Rhododendron*, evergreen or deciduous shrubs (usually) or trees, of which azaleas and rhododenrons are members.

The strawberry tree (*Arbutus unedo*) is characterized by small whitish flowers occurring in clusters, a small fruit, about $\frac{1}{2}$ inch (1.2 centimeters) in diameter, which appears something like a strawberry, and narrow, oval leaves of medium length. This tree, which can reach a height of 40 feet (12 meters), does well in southwestern Ireland and the Mediterranean region. However, the tree can withstand somewhat colder climes and is found in parts of Britain and North America. Some authorities describe the fruit more as a roughened cherry than as a strawberry. Flowering occurs in late autumn, a definite attraction to the gardener. The *A. andrachne* is the strawberry tree of the eastern Mediterranean region. It is a slightly smaller tree, generally attaining a height of about 30 to 35 feet (9 to 10.5 meters). The flowers are an off-white and occur in broad clusters. Flowering occurs in the spring. The fruit is similar to the *A. unedo.*

The strawberry tree of western and southwestern North America is the *A. menziestii*, or, as commonly termed, the *madrona* or *madrone* tree. As shown by the accompanying table, there are closely related, localized species, such as the *A. arizonica* and *A. texana.* As noted, these trees are capable of achieving excellent heights under favorable conditions. The leaves are of medium length, oval, with dark green coloration above and a bluish-white color underneath. The tree usually flowers late in the spring. The fruit is a pea-sized berry. The tree is also found in Europe, where it may achieve a height of 50 to 55 feet (15 to 16.5 meters).

Not previously mentioned, the genus *Oxydendrum* claims the sorrel tree (*O. arboreum*), which occurs in the eastern United States and is related to the strawberry tree. The sorrel is a relatively small tree, ranging from 20 to 55 feet (6 to 16.5 meters) in height, with a trunk diameter up to about 20 inches (50.8 centimeters). The bark is gray-brown, somewhat furrowed. The branches are pendulous. Leaves are elliptically shaped, pointed, dark green, and finely-toothed. The flowers are white and occur in drooping clusters. The tree occurs from Pennsylvania westward to Indiana and southward along the Alleghany Mountains into Louisiana and western Florida. Some of the tallest specimens are found on the eastern slopes of the Blue Ridge Mountains.

Of the genus *Clethra*, the Lily-of-the-valley clethra attains a height of about 30 feet (9 meters) and can be classified as a small tree. The tree is found on the Island of Madeira, but has been introduced elsewhere. The leaves are dark green and alternate; the flowers are white and fragrant. The tree is sensitive to climate and soil, requiring a rich and acid mix.

It is interesting to note that prior to 1820, rhododendrons other than the European and American varieties were unknown. In particular, the *R. maximum* of the eastern United States was best known then. In that year, the *Rhododendron arboreum* was brought out of the Himalayas. During the intervening years, numerous species from Asia have been found and propagated.

The *R. maximum*, also sometimes referred to as the Great Laurel or rose bay, is a shrub/tree that can attain a height of 40 feet (12 meters) under favorable conditions, with a trunk diameter approaching a foot (0.3 meter). The bark is gray-brown, smooth, with minor scaling. The leaves are evergreen and lustrous, quite large—from 4 to 8 inches (10.1 to 20.3 centimeters) in length. The flowers occur in large clusters and may be described as pale pink with spots of coloration in the upper part of the throat. In nature, the plant is found, usually in damp woodsy areas or along streams, from Nova Scotia westward through Quebec and Ontario to Ohio and Lake Erie, and as they range southward, they become more numerous, notably through the Alleghany Mountains and in the southeastern states as far as Georgia.

Other species of rhododendrons occurring in the mountains and woods of the United States, notably east of the Rocky Mountains include: *R. viscosum*, also known as clammy azalea or white swamp honeysuckle; *R. nudiflorum*, also known as Pinxter flower; *R. arborescens* or the smooth azalea; *R. canescens* or mountain azalea; *R. calendulaceum* or flame azalea; *R. canadense* or rhodora; *R. catawbiense* or rose bay; *R. lapponicum* or Lapland rose bay; and *R. hispida* or rose acacia of the southeastern United States. Rhododendrons from Asia include the previously mentioned *R. arboreum.* This plant is found in the Himalayas in the Khasia Hills, Ceylon and from Kashmir to Bhutan. It displays bell-shaped flowers of dark red color. The height ranges from 30 to 40 feet (9 to 12 meters). The *R. barbatum* is found in Bhutan, Nepal, and Sikkim and is capable of growing to a height of 40 feet. The flowering is similar to that of *R. arboreum.* The *R. calophytum*, with similar flowers but ranging from white to rose pink and with a characteristic maroon blotch, is found in western China.

RECORD MADRONE TREES IN THE UNITED STATES[1]
(Heather or Heath Family)

Specimen	Circumference[2]		Height		Spread		Location
	(Inches)	(Centimeters)	(Feet)	(Meters)	(Feet)	(Meters)	
Arizona madrone (1970) (*Arbutus arizonica*)	126	320	45	13.5	42	12.6	Arizona
Pacific madrone (1974) (*Arbutus menziesii*)	384 (at 36 inches)	975 (at 91 centimeters)	79	23.7	123	36.9	California
Texas madrone (1972) (*Arbutus texana*) (Co-Champion)	60	152	34	10.2	40	12	Texas
Texas madrone (1971) (*Arbutus texana*) (Co-Champion)	60	152	36	10.8	29	8.7	Texas

[1] From the "Social Register of Big Trees," The American Forestry Association (by permission).
[2] At 4.5 feet (1.4 meters).

It can attain a height of about 35 feet (10.5 meters). Also from the Bhutan, Nepal, and Sikkim regions is the *R. falconeri*, with purple-blotched white flowers, and reaching a height of about 30 feet (9 meters). The *R. giganteum* of Yunnan is well named because it ranges in height between 40 and 80 feet (12 and 24 meters). The flowers are of a deep rose-crimson color. Also of Yunnan and of southeastern Tibet and upper Burma is the *R. sinogrande* which can rise to a height of about 45 feet (13.5 meters), displaying white/yellow flowers.

In a brief description of the heather family, certainly the common heather shrub (*Culluna vulgaris*) should not be omitted. This is a small, straggly shrub ranging from about 6 to 16 inches (15.2 to 40.6 centimeters) in height. It was introduced into America from Europe. The leaves are small, gray-green, perhaps $\frac{1}{8}$ inch (0.3 centimeters) in length and overlap the branches. Tiny bell-shaped flowers, white or of a deep pink color, occur as spikes. The Scottish heather counterpart is the *Erica cineria*. Another heather, the *Erica Tetralix*, was introduced along with the aforementioned two species into Nantucket Island, at the same time the Scots pine was introduced, during the years 1875–1877.

HEAT INDEX (Stellar). Stellar Heat Index.

HEATING (Aerodynamic). Supersonic Aerodynamics.

HEATING DEGREE DAY. Climate.

HEATING (Geothermal). Geothermal Energy.

HEATING OILS. Petroleum.

HEATING (Solar). Solar Energy.

HEATING VALUE (Coal). Coal.

HEATING VALUE (Fuels). Fuel.

HEATING VALUE (Natural Gas). Natural Gas.

HEATING VALUE (SNG). Substitute Natural Gas (SNG).

HEAT (Mechanical Equivalent of). Mechanical Equivalent of Heat.

HEAT (Molecular). Molar Heat.

HEAT (Nernst Theorem). Nernst Heat Theorem.

HEAT OF COMBUSTION. Calorimetry; Combustion.

HEAT OF FUSION. Compton Rule; Fusion (Heat of).

HEAT OF REACTION. Titration (Thermometric).

HEAT OF SUBLIMATION. Sublimation.

HEAT OF VAPORIZATION. Vaporization (Heat of).

HEAT PUMP. A system involving a compressor, heat exchangers, a refrigerant, and a flow restriction that can be used to supply or remove heat. In its cooling cycle, the heat pump operates very much like a conventional air conditioner. Although the principle of the heat pump has been known for decades, it has received renewed interest in recent years in connection with the search for more energy efficient systems. The fundamentals of the heat pump for both its heating and cooling cycles are described briefly in connection with the accompanying diagrams.

The heat pump has been well established as an efficient user of electrical power. However, heat pumps are not universally applicable to year-round heating/cooling applications if they are to operate at maximum efficiency and thus outperform separate heating and cooling equipment. Comparisons can be made by determining the coefficient of performance (COP) of a given heat pump for a given application. The COP is a ratio, namely, of the amount of heat required by the condenser (exchanger *A* in diagram *b*), or Q_c, divided by the quantity of electrical energy consumed to power the pump, or *W*. That is, $COP = Q_c/W$.

Assuming that the average efficiency of an electric generating and distributing utility is 30%, the use of a heat pump can bring the overall heating efficiency up to 75%. It is interesting to note that a heat pump can yield more thermal energy than the electrical energy which it consumes when operating under certain conditions. This in no way defies the law of energy conservation because the heat pump picks up increments of thermal energy from the evaporator when used in the heating cycle. As pointed out in diagrams *a* and *b*, when on the heating cycle, a heat pump system, by virtue of absorbing heat from already cold ambient air, dumps air that is below ambient temperature to the atmosphere. And, when on the cooling cycle, a heat pump system, by viture of absoring heat from already warm room air, dumps air that is above ambient temperature to the atmosphere.

For a given capacity heat pump, the volume flow rate of refrigerant vapor through the compressor is approximately constant. It will be noted from diagrams *a* and *b* that in either the heating cycle or cooling cycle the medium taken into the compressor is in the vapor phase at a comparatively low temperature and pressure; and that the medium exiting the compressor is in the vapor phase at a comparatively high temperature and pressure. It is evident from diagram *a* that the temperature of the cold ambient air in its effect on heat exchanger *A* determines the temperature of the medium exiting the exchanger and thus entering the compressor. Thus, the colder the ambient air, the lower will be the vapor pressure and the density of the medium. These conditions reduce the effective mass of the medium moving through the compressor. This decreased mass flow rate lowers the thermal capacity of the medium, thus reducing the quantity of heat energy which it can transfer from the compressor to heat exchanger *B* and thence to heat the conditioned space. Depending upon the base capacity of the unit, the point may be reached where the ambient temperature is just too low to permit the heat pump to heat the conditioned space adequately. In the present stae of the art, this leaves the designer two options—either use a unit with greater capacity (larger initial investment, etc.) or arrange to furnish for auxiliary heating during abnormally cold periods when such conditions may persist. Depending upon the form of the auxiliary energy and the efficiency of the auxiliary system, some or all of the advantages of the heat pump (cost, normal operating efficiency, etc.) may be negated. In general, it has been a common practice to size the heat pump to the summer's cooling requirements and permit the winter heating to fall where it may, with dependence upon auxiliary heating. In southern climes, this may be acceptable because relatively little if any heating may be required of the unit during the winter season. In norther climes, designing to the summer cooling load will normally lead to the requirement for auxiliary heating.

As pointed out by Glicksman (1978), a "wide disparity between ideal heat pump performance and the actual performance of today's equipment leaves much room for improvement. Some of the present performance shortfall originates in the emphasis on minimizing manufacturing costs and consumer prices. For example, heat exchangers—condensers and evaporators—are made relatively small to keep costs down. But small heat exchangers have limited capacity to transfer heat between refrigerant and air. To achieve a high rate of heat transfer with a small heat exchanger requires a rather large temperature difference between the refrigerant and the air. This means that, in heating mode, the refrigerant in the condenser must be much warmer than the indoor air and the refrigerant in the evaporator must be much colder than outside ambient air. To maintain these exaggerated temperature differences, the compressor must literally work overtime. The expected result is that the COP is lower than it would be with larger—more expensive—heat exchangers."

To improve heat performance, several concepts have been proposed. These are discussed in some detail in the Glicksman reference: (1) Use of a volume of water to store and provide low-temperature heat—with some ice forming in the evaporator and deliberately used. The concept has been called the "annual cycle energy system." (2) Use of a heat pump to supplement a solar collector system. (3) Combined use of a thermal storage system and heat pump to partially solve

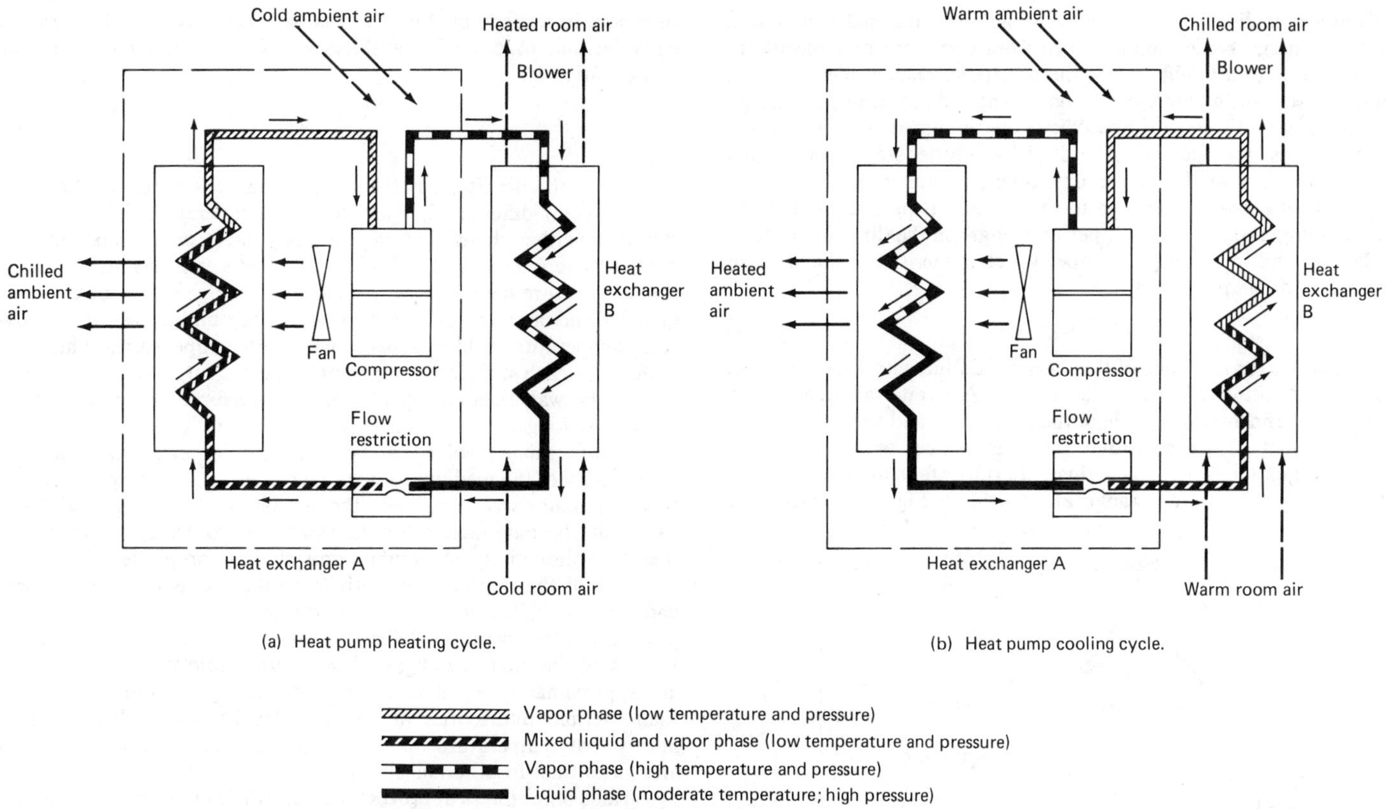

(a) Heat pump heating cycle.

(b) Heat pump cooling cycle.

Operating cycles of heat pump. Although the heat pump, which can be used for both heating and cooling, is a relatively simple concept, its operation is sometimes not fully understood from the first reading of a description. Rather than begin with the usual comparison of a heat pump with an air conditioning unit and making reference to the terms condenser and evaporator, the present description commences with the heating cycle and refers to heat exchangers.

(a) There are four principal elements of equipment—a compressor, a flow restriction, and two heat exchangers, *A* and *B*. These are represented very schematically in the diagram. The heat-exchange medium (a refrigerant liquid, such as Freon®) exits the compressor in the vapor phase at a high temperature and pressure. It passes to heat exchanger *B*, where it is gradually cooled by the cold room air, which it in turn warms. The medium exits heat exchanger *B* in the liquid phase at moderate temperature, but still under high pressure, then proceeds to a flow restriction which effects a pressure drop. The medium exits the restriction as mixed liquid and vapor phase at a much lower temperature and pressure. (The lower temperature is the result of cooling caused by expansion.) This liquid/vapor mixture enters heat exchanger *A*, where it absorbs some heat from cold ambient air, making the ambient air just a bit colder in the vicinity of the unit. The medium exits the exchanger *A* as a vapor at low temperature and pressure, and returns to the compressor, where the cycle begins anew.

(b) By using a simple arrangement of valves, the flows can be reversed to make the heat pump a means for chilling room air rather than warming it. The heat-exchange medium exits the compressor in the vapor phase at high temperature and pressure. It passes to heat exchanger *A*, from which it exits in the liquid phase at moderate temperature and high pressure, then passes through the flow restriction, from which it exits as a mixed liquid and vapor at a low temperature and pressure. Again, the cooling is the result of expansion of the vapor. This mixed liquid and vapor phase enters heat exchanger *B*, where it is gradually warmed by warm room air, which it in turn cools. The medium exits heat exchanger *B* in the vapor phase at low temperature and pressure, and thence returns to the compressor, where the cooling cycle begins anew.

It will be noted that during the heating cycle the medium actually extracts heat from already cold ambient air as it passes through heat exchanger *A*. In contrast, during the cooling cycle, the medium actually adds heat to already warm ambient air as it passes through heat exchanger *A*.

It will be evident that, when operating in the cooling cycle, the heat pump operates like a conventional air conditioner. Heat exchanger *A* is the evaporator and heat exchanger *B* is the condenser. These units reverse their roles for the heating cycle.

the problems of oversize heat pumps for cooling, particularly in northern climates. (4) A system for varying the capacity of the system by throttling down large heat pumps when their full capacity is too great for either heating or cooling requirements. (5) Use of high-efficiency natural gas-fired heat pumps as the energy source.

As heat pump efficiencies improve, the costs probably will become comparable with the costs of conventional fossil-fuel heating and cooling systems. Any marked increase, relative to other fuels, of natural gas, for example, could make the heat pump more competitive on a capital and operating costs basis. Of course, when the air conditioning or cooling cycle is not included in the plans for a heat pump system, the system becomes less competitive with conventional approaches.

See also **Solar Energy.**

References

Glicksman, L. R.: "Heat Pumps," *Technology Review* (*MIT*), **80**, 7, 64–70 (1978).

Hiller, C. C., and L. R. Glicksman: "Improving Heat Pump Performance via Compressor Capacity Control," Energy Laboratory Report MIT El-76-001 (NTIS-TB-250592/AS), January 1976.

Kirschbaum, H. S., and S. E. Veyo: "An Investigation of Methods to Improve Heat Pump Performance and Reliability in a Northern Climate," EPRI EM-319, Electric Power Research Institute, Palo Alto, California, January 1977.

Sporn, P., Ambrose, E. R., and Baumeister: "Heat Pumps," Wiley, New York, 1947.

Staff: "Handbook of Fundamentals," Amer. Soc. Heating, Refrigeration and Air Conditioning Engineers, New York, 1972.

HEAT-RESISTING STEELS. **Iron Metals, Alloys, and Steels.**

HEAT STORAGE (Solar). **Solar Energy.**

HEAT TRANSFER. Although there are three generally accepted methods for transferring heat from one medium to another, or from one locale to another within a given medium, it is uncommon for one method to act unilaterally. Particularly where convection may predominate, some conduction of heat will be involved. In conduction, heat must diffuse through material substances; in convection, heat is essentially carried from one locale to another by actual movement of the transport medium; in radiation, heat transfer involves radiant wave energy.

Conduction. From a microscopic standpoint, thermal conduction refers to energy being handed down from one atom or molecule to the next one. In a liquid or gas, these particles change their position continuously even without visible movement and they transport energy also in this way. From a macroscopic or continuum viewpoint, thermal conduction is quantitatively described by Fourier's equation, which states that the heat flux q per unit time and unit area through an area element arbitrarily located in the medium is proportional to the drop in temperature, $-\text{grad } T$, per unit length in the direction normal to the area and to a transport property k characteristic of the medium and called *thermal conductivity*:

$$q = -k \text{ grad } T \tag{1}$$

Predictions for the value of the thermal conductivity k can be made from considerations of the atomic structure. Accurate values, however, require experimentation in which the heat flux q and the temperature gradient, grad T, are measured and these values are inserted into Fourier's equation. Thermal conductivity values for a number of media over a large temperature range are shown in Fig. 1. Metals have

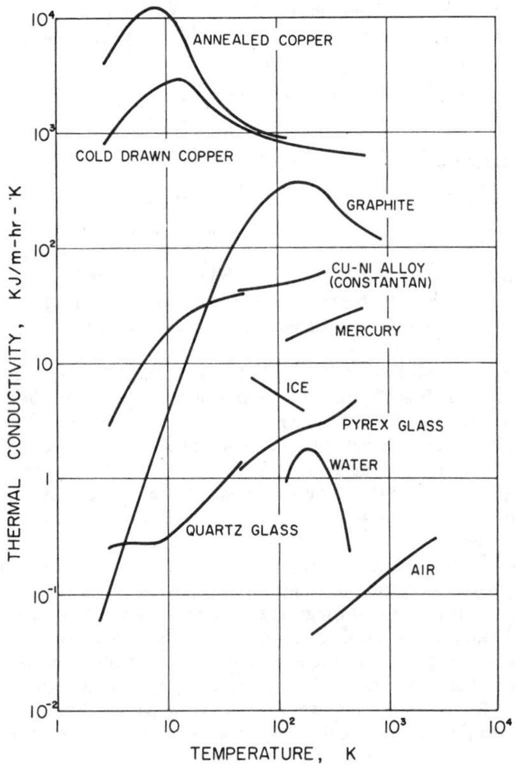

Fig. 1. Thermal conductivity values for a wide range of substances and over a temperature range of 1 to 10^4 K.

the largest conductivities and, among these, pure metals have larger values than alloys. Gases, in contrast, have very low heat conductivity values. Electrically nonconducting solids and liquids are arranged in between. The low thermal conductivity of air is utilized in the development of thermally insulating materials. Such materials, like cork or glass fiber, consist of a solid substance with a very large number of small spaces filled by air. The thermal transport occurs then essentially through the air spaces, and the solid structure only supplies the framework which prevents convective currents. It will be noted that the thermal conductivities indicated in Fig. 1 (at ambient temperature) extends through five powers of 10. This range is still small when compared with the range for the electric conductivity of various substances, where electric conductors have values which are larger by 25 powers of 10 than electric insulators. As a consequence, it is much easier to channel electricity along a desired path than to do so with heat, a fact which accounts for the difficulty in accurate experimentation in the field of heat transfer.

Fourier's equation can be used together with a statement on energy conservation to derive a differential equation describing the temperature field in a medium. Fourier was the first person to develop this equation and to device means for its solution. In vector notation, this equation is:

$$\rho c = \frac{\partial T}{\partial t} = \nabla(k \nabla T) \tag{2}$$

where ρ is the density, c is the specific heat, t is time, and ∇ is the Nabla (vector differential) operator. The temperature field in a substance can either change in time (unsteady state), or it can be independent of time (steady state, $\partial T/\partial t = 0$). For a steady-state situation, the temperature field depends primarily on the geometry of the body involved and on the boundary conditions. The simplest case of a steady-state temperature field is a plane wall with temperatures which are uniform on each surface, but different at the two surfaces. The temperature in the wall then changes linearly in the direction of the surface normal as long as the variation of the thermal conductivity in the temperature range involved can be neglected. For an unsteady process, the capacity of the medium to store energy enters the energy conservation equation; correspondingly, the specific heat of the material and its density become factors for the conduction process, as well as the thermal conductivity. A combination of these properties, defined as the ratio of the thermal conductivity to the product of specific heat and density, called *thermal diffusivity* ($k/\rho c$), then determines how fast existing temperature differences in a medium equalizes in time. It is found that metals and gases have thermal diffusivity values which are approximately equal in magnitude and are considerably higher than thermal diffusivities of liquid and solid nonconductors. This means that temperature differences equalize much faster in metals and gases than in other substances.

Various other physical processes lead in their mathematical description to equations of the same form as Eq. (2), especially in its steady-state form. Such processes include the conduction of electricity in a conductor, or the shape of a thin membrane stretched over a curved boundary. This situation has led to the development of analogies (electric analogy, soap film analogy) to heat conduction processes which are useful because they often offer the advantages of simpler experimentation.

Convection. When energy is transported by convection in fluids, conduction usually takes care of the transport of heat from one stream tube to another and is the dominating mode of transfer near solid walls. Convection transports heat along the stream lines and is dominating in the main body of the fluid where the velocities are large. In many situations, the flow is turbulent; this means that unsteady mixing motions are superimposed on the mean flow. These mixing motions contribute also to a transport of heat between stream tubes, a process which can be described by an "effective" conductivity which often has values by several powers of ten larger than the actual conductivity of the fluid.

Movement of the fluid may be generated by means external to the heat transfer process, as by fans, blowers, or pumps. It may also be created by density differences connected with the heat transfer process itself. The first mode is called *forced convection*; the second one *natural* or *free convection*. Convection heat transfer may also be classifed as heat transfer in *duct flow*, or in *internal flow* (over cylinders, spheres, air foils, and similar objects). In the case of external flow, the heat transfer process is essentially concentrated in a thin fluid layer surrounding the object (boundary layer).

Of special interest in such heat transfer processes is the knowledge of the heat flux from the surface of a solid object exposed to the flow. This heat flux q_w per unit area and time is conventionally described by Newton's equation:

$$q_w = h(T_w - T_f) \tag{3}$$

where T_w is the surface temperature and T_f is a characteristic temperature in the fluid. This equation defining the heat transfer coefficient h is convenient because in many siutations the heat flux is at least approximately proportional to the temperature difference $T_w - T_f$. Information on the heat transfer coefficients can be obtained by a solution of the Navier-Stokes equation describing the flow of a viscous fluid and the related energy equation, or they are found by experimentation. Computers enhance the ability to study heat transfer analytically at least for laminar flow, whereas in turbulent flow the bulk of the information is determined experimentally.

Experimentation is difficult because of the large number of parameters involved. Dimensional analysis has been applied to reduce the number of influencing parameters, and relations for convective heat transfer are correspondingly presented in many handbooks as relations between dimensionless parameters. Such an analysis demonstrates that heat transfer in forced flow can be described by a relation of the form:

$$Nu = f(Re, Pr) \quad (4)$$

in which the Nusselt number Nu is a dimensionless parameter hL/k, containing the heat transfer coefficient h, the Reynolds number $Re = \rho(VL/\mu)$ describes essentially the nature of the flow, and the Prandtl number $Pr = c_p\mu/k$ can be considered a dimensionless transport property characterizing the fluid involved. L and V are an arbitrarily selected characteristic length and velocity, respectively; ρ denotes the density, μ the viscosity, and c_p the specific heat of the fluid at constant pressure. See also **Reynolds Number.**

Convection is frequently thought of in terms of space heating and industrial heat-exchange processes. It should be pointed out that convection plays a cosmic role (in the sun's photosphere, for example), and a very large role in connection with the atmosphere of the earth and some other planetary bodies. For example, when normal convective transport is inadequate, temperature inversions occur and create smog hazards over large cities. See **Atmosphere (Earth).**

Attempts to develop a theory for convection date back at least to the 1790s when Thompson (Count Rumford) introduced the concept of heat convection. Very little theoretical work was undertaken, however, until the early 1900s, when Bénard (France) undertook experimental investigations. Modern convection physics stems from the work of Lord Rayleigh, who first published on the subject in 1916. In current times, advanced convection research studies have been undertaken by Velarde and Normand (1980), among others. See reference listed. See also **Boiler;** and **Heat.**

Radiation. In the transfer of energy from one location to another in the form of photons (electromagnetic waves), usually a multiplicity of wavelengths is involved. In vacuum, all waves regardless of their wavelength move with the same speed (2.9977×10^8 meters per second). In various substances, the wave velocity c changes somewhat with wavelength, and the ratio of the wave velocity in vacuum to the velocity in a substance is equal to the optical refraction index. Air and generally all gases have refractive indices which differ from one only in the fourth decimal. Their wave velocity is therefore practically equal to that in vacuum. See also **Waves and Wave Mechanics.**

Prévost's principle states that the amount of energy emitted by a volume element within a radiating substance is completely independent of its surroundings. Whether the volume element increases or decreases its temperature by the process of radiation depends upon whether it absorbs more foreign radiation than it emits or vice versa. One refers to thermal radiation when the emission of photons is thermally excited, i.e., when the substance within the volume element is nearly in thermodynamic equilibrium. For such radiation, Kirchhoff was able to derive a number of relations by consideration of a system of media in thermodynamic equilibrium. If $j\nu$ indicates the coefficient of emission, i.e., the radiative flux at the frequency ν* emitted per unit volume into a unit solid angle, and η is the coefficient of absorption at the same frequency, i.e., the fraction of the intensity of a radiant beam which is absorbed per unit path length, then one of these relations states:

$$c^2 \frac{j\nu}{\eta_\nu} = f(T, \nu) \quad (5)$$

with c dennoting the wave velocity. According to this relation, the combination of parameters on the left-hand side of Eq. (5) is a function of temperature T and frequency ν of the radiation only, but does not depend upon the substance under consideration. Kirchhoff's law can also be expressed in parameters which refer to the interface of two media (1 and 2). It then takes the form:

$$c^2 \frac{i_\nu}{\alpha_\nu} = f(T, \nu) \quad (6)$$

in which i_ν is the monochromatic intensity of the radiative flux at frequency ν originating in medium 2 and traveling through the interface into medium 1 per unit solid angle and area normal to the direction of the radiant beam. α_ν is the monochromatic absorptance or absorptivity, i.e., that fraction of a radiant beam approaching the interface in the medium 1 in the opposite direction that is absorbed in medium 2. The wave velocity in medium 1 is c. Kirchhoff's law states that the combination of the parameters on the left-hand side of Eq. (6) is again a function of temperature and frequency only, but does not depend upon the nature of the medium. A medium which absorbs all the radiation traveling into it through an interface ($\alpha_\nu = 1$) is called a *blackbody.* The intensity of radiation emitted by an arbitrary medium is, according to Eq. (6), in the following way related to the intensity of radiation $i_{b\nu}$ emitted by a black body at the same temperature and frequency:

$$\frac{i_\nu}{\nu} = i_{b\nu} \quad (7)$$

See also **Planck's Radiation Formula.**

The amount of heat transferred by radiation can be determined by use of the *Stefan-Boltzmann law*:

$$Q = bA(T_1^4 - T_2^4) \quad (8)$$

where Q is the amount of heat transferred per unit time, b is a constant, A is the area of the radiating surface, T_1 is the absolute temperature of the radiating body and T_2 is the absolute temperature of the receiving body. Various correction factors are introduced into the formula to account for the shape of the bodies, their thermal radiation characteristics and the properties of the media through which the radiant rays must pass while traveling from radiator to absorber. The thermal radiation characteristics are its emissivity, a measure of its ability to radiate at a given temperature, its absorptivity, a measure of its ability to absorb heat and its reflectivity, which measures its ability to reflect without absorbing.

Radiant energy travels in a straight line. Therefore to transmit it to an object out of sight of the radiator requries a reflector, such as a furnace wall, to deflect the rays to their objective.

It is possible to set up controlled laboratory radiation between simple plane surfaces and determine therefrom accurate coefficients to incorporate into radiation equations. However, the radiation of heat from furnace gases, consisting of non-luminous gases, luminous carbon particles in flame, ash globules, etc., to the walls and tubes of a steam generator in commercial operation at variable load, is another matter. Here, empirical data which are gathered and interpreted from field tests on similar equipment, must still be resorted to however great the designer's urge to go back to basic laws of heat transfer.

Radiant heat transfer in furnaces is roughly proportioned to the difference in the fourth power of the absolute temperatures of the radiating and receiving surfaces. The water wall surface is approximately at boiler saturation temperature, while the superheater surface varies from this to somewhat above the temperature of the steam at the superheater outlet. However, the mean radiating temperature of the furnace gases is usually over 1204°C. The fourth power of the receiving surface temperature is thus seen to be small compared to the fourth power of the transmitting surface temperature; consequently the latter controls the transmittance, and boiler tube temperature does not need to be considered a variable to be accounted for.

Figure 2 shows some of the arrangments in which radiant heat-absorbing surface is disposed. It may be used to illustrate another of the difficulties which beset the designer in following a rational or semi-rational form of radiation analysis. Projected radiant surface is one thing; actual radiant energy receiving surface may be quite a different area. For example, suppose the tubes of case (a) to be sepa-

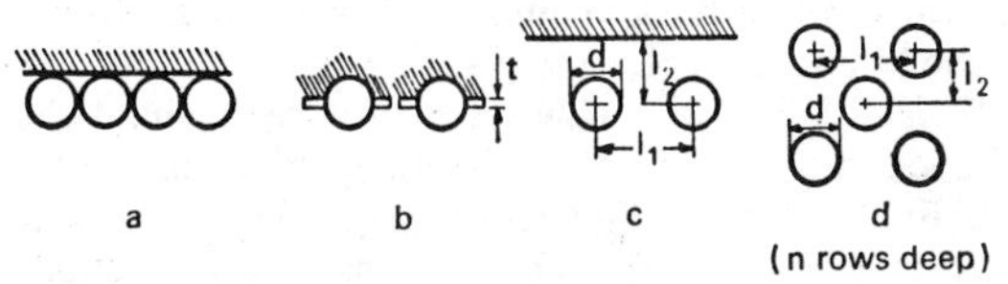

Fig. 2. Arrangements of radiant heat-absorbing surface.

rated and spaced l_1 inches on centers. The *projected* areas of cases (a) and (c) would then be the same, but it seem obvious that re-radiation from the wall causes more of a (c) tube to receive radiant energy than is the case with an (a) tube. Also, if δ is a factor correcting projected area to *equivalent* absorbing surface, what value should be assigned to it in the case of a bank of tubes which may receive by re-radiation some radiant energy deep in the tube bank? Here δ has a minimum value of 1, but some investigators have derived expressions which indicate that δ may have a magnitude of 3 or more.

Industrial Heat Transfer. Plate-type heat exchangers are widely used in the process industries. For example, in the dairy industry alone, they serve such purposes as heating milk with hot water or steam (especially for high temperature-short time pasteurization), heat recovery by milk to milk heat transfer and milk cooling with water or brine. In the modern plate heat exchanger, the plates are so shaped that at frequent intervals the passage of milk is retarded. This produces a pulsating flow, which gives high heat transfer with low pumping power.

Some of the more common cases of industrial heat transfer are:

1. Radiation from fuel beds and luminous gases to absorptive surfaces such as boilers, cylinder walls, etc.
2. Radiation from heat generators such as drying lamps.
3. Convection of heat out of combustion regions.
4. Convection of heat from hot surfaces under either free or forced convection.
5. Conduction of heat through the tubes of boilers, heaters, heat exchangers, condensers, etc.
6. Conduction in walls, pipe covering, and other so-called "heat insulators."
7. Conduction of heat through the plates of plate-type heat exchangers and regenerators.

Types of Heat Exchangers. Heat exchangers perform many functions within a manufacturing plant. See Fig. 3. Often they are given special

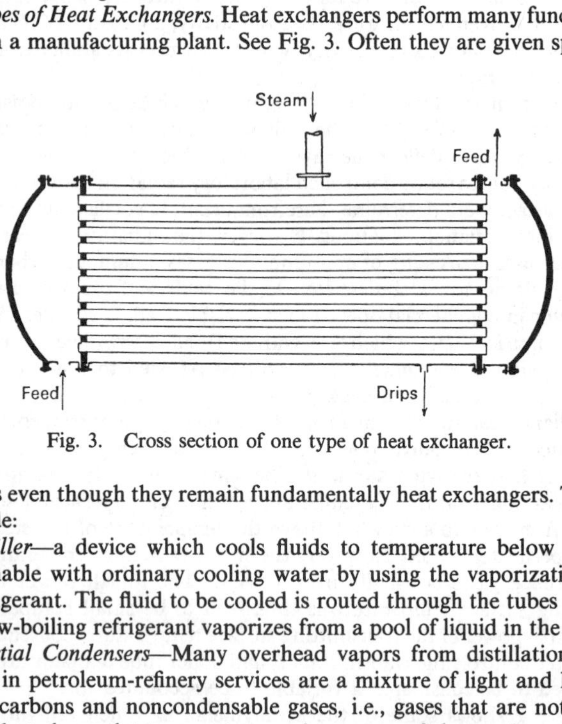

Fig. 3. Cross section of one type of heat exchanger.

names even though they remain fundamentally heat exchangers. These include:

Chiller—a device which cools fluids to temperature below those obtainable with ordinary cooling water by using the vaporization of a refrigerant. The fluid to be cooled is routed through the tubes while the low-boiling refrigerant vaporizes from a pool of liquid in the shell.

Partial Condensers—Many overhead vapors from distillation columns in petroleum-refinery services are a mixture of light and heavy hydrocarbons and noncondensable gases, i.e., gases that are not condensed at the outlet temperature and pressure of the condenser (air, hydrogen sulfide, methane, and other light ends). These vapors are routed through the shell side while water is used as the cooling medium on the tube side of the unit. Condensation on the shell side begins at the saturation temperature of the heavy components and continues over a decreasing temperature range until part of the lighter components are condensed. Part of the existing liquid is sent back to the tower as reflux, while the remainder is further refined or passes to the trim cooler and storage.

Trim Cooler—This unit condenses the last remaining light-end vapors and cools the liquid to the ultimate storage temperature (often about 100°F: 38°C) by using cooling water. This cooling usually is not conducted in the main condenser because it would reduce column pressure.

Thermosiphon Reboiler—Flow of the vaporizing fluid depends upon the difference in static head between the column of liquid flowing from the tower to the reboiler and the partially vaporized column of liquid returning from the exchanger to the tower.

Reboilers—These exchangers operate in conjunction with a distillation tower to vaporize enough liquid to assure vaporization of the overhead product. A hot process stream of steam may be used as the heating medium. Most reboilers are shell-and-tube exchangers located at the base of the tower. The vaporizing fluid is routed through the shell side of the exchanger.

Forced-circulation Reboiler—A pump is used to provide more positive circulation than available with the thermosiphon effect, e.g., in the vaporization of viscous fluids.

Vapor Heat Exchanger—Units of this type preheat a cool stream of process fluid by using heat from partially condensing vapor. The objective is to conserve heat and eliminate the requirement for a separate preheater.

Air-cooled Exchanger—As used in the petroleum industry, air-cooled exchangers normally comprise two headers joined by a horizontal bank of finned tubes. Usually two motor-driven fans located above (induced draft) or below (forced draft) the tubes are used to circulate the air over the finned surface.

Superheater—A unit of this type heats vapor above the saturation temperature.

Waste-heat Boiler—A unit of this type generates steam and is similar to a regular steam generator except that hot gas or liquid produced by a chemical reaction (often combustion) is the heating medium.

Heat Storage

It is often necessary to store heat in rather large quantities in specially designed apparatus. Hot water, of course, is one of the easiest forms in which to store thermal energy that is immediately available. As contrasted with hot water, electric energy and steam have to be generated on an as-needed basis. The blast furnace poses a difficult heat storage problem which obviously cannot be handled by storing heat in water. Great amounts of hot gas are required on a cyclic basis. To heat such quantities of air on a continuous, as-required, basis would be quite impractical with the present stage of the art. The solution used involves several stoves which are quite large, often

Fig. 4. A 30-foot (9-meter) diameter blast furnace is shown at right. The blast furnace stoves, lined with checkerwork and three in number, are shown at left. The size of these stoves can be appreciated by noting the two persons on the catwalk above the stove at the right.

over 100 feet (30 meters) in height and about 25 feet (7.5 meters) in diameter. The blast temperature of approximately 1,000°F (538°C) is accomplished by preheating the stove checkerwork to a much higher temperature. Checkerwork is comprised of refractory material forms constructed in high walls in checkerboard fashion to permit free passage of air through the interstices when under pressure. The gas passing through the stove exhausts initially at 2000°F (1093°C). Mixing this with unheated air produces the required blast temperature for the blast furnace. The stoves usually are heated for a period of three hours and exhaust (termed "on wind") for a period of about one hour. See Fig. 4. A similar system of checkerwork regenerators is used in connection with glass-tank heat-storage systems.

Flowing streams of pebbles also have been used in the chemical industry for removing heat from gases. Pebbles and stones are also used in some solar energy storage systems. See **Solar Energy.**

References

Chandrasekhar, S.: Hydrodynamic and Hydromagnetic Stability," Oxford Univ. Press, New York, 1961.

Kittel, C.: "Thermal Physics," 2nd edition, Freeman, San Francisco, 1980.

Normand, C., Pomeau, Y., and M. G. Velarde: "Convective Instability: A Physicist's Approach," *Rev. Mod. Phys.*, **49**, 3, 581–624 (1977).

Prigogine, I.: "Time, Structure, and Fluctuations," *Science*, **201**, 777–785 (1978).

Staff: "Heat Transfer," Special issues of *Chem. Eng. Progress*: **74**, 7, 41–46 (1978); **75**, 7, 41–91 (1979).

Staff: "Practical Aspects of Heat Transfer," Amer. Inst. of Chem. Engineers, New York, 1978.

Turner, J. S.: "Buoyancy Effects in Fluids," Cambridge Univ. Press, Cambridge, England, 1973.

Velarde, M. G., and C. Normand: "Convection," *Sci. Amer.*, **243**, 1, 92–108 (1980).

HEAT TRANSFER (Nusselt Number). Nusselt Number.

HEAT TRANSFER (Solar). Solar Energy.

HEAT TREATING. Heating and cooling of metals to effect changes in properties. Annealing and normalizing are generally for the purpose of softening or improving the grain structure. Patenting is also a softening process in which cold drawn carbon-steel wire is heated above its critical temperature range followed by cooling to below this range in a molten lead or molten salt bath, with subsequent cooling to room temperature.

While heat treating includes the softening treatments, it most often implies hardening and strengthening. In the case of steels this requires heating to above the critical temperature range followed by rapid cooling (quenching) in oil, water, or brine, except in the case of special grades which harden on cooling in air. This is followed by tempering, a low-temperature reheating treatment which reduces the internal stresses caused by the hardening treatment. Tempering may be carried to a high enough temperature to reduce somewhat the extreme hardness of the as-quenched steel and increase the toughness and ductility, depending on the requirements of the part. See **Iron Metals, Alloys, and Steels.**

Another important form of heat treatment for hardening is precipitation hardening. See also **Annealing; Carbonitriding; Carburizing; Case Hardening;** and **Nitriding.**

HEAT UNITS. Units and Standards.

HEAVISIDE LAYER. Ionosphere.

HEAVISIDE UNIT RAMP INPUT. Laplace Transform.

HEAVY HYDROGEN. Deuteron.

HEAVY WATER REACTOR. Nuclear Reactor.

HEBE. Asteroid.

HEBERDEN'S NODES. Osteoarthritis.

HECTOCOTYLUS ARM. One of the arms or tentacles of the male of certain species of cephalopod mollusks, specialized for the insemination of the female. It may serve as an intromittent organ, introducing spermatophores into the mantle cavity of the female, and in the paper nautilus, *Argonauta*, it breaks away from the body and enters the mantle cavity of the female, remaining inside for some time.

HECTO PREFIX. Units and Standards.

HEDGEHOG. Moles and Shrews; Rodentia.

HEEL. The prominence at the posterior end of the foot. It is based on the projection of one bone, the calcaneum, behing the articulation of the bones of the lower leg. In the long-footed mammals, both the hoofed species and the clawed forms which walk on the toes, the heel is well above the ground at the apex of the angular joint known as the hock or hough. In plentigrade species it rests on the ground.

HEIGHT CHECKING. Thickness Measurement and Gaging Systems.

HEILGENSCHEIN. Precipitation and Hydrometeors.

HEINE FORMULA. Legendre Differential Equation.

HEISENBERG FORCE. The Heisenberg force is a phenomenologically postulated force between two nucleons derivable from a potential in which there appears an operator which exchanges the spins and positions of the two particles.

HEISENBERG REPRESENTATION. Representation of the equations of motion in quantum mechanics and in quantized field theory where the vector describing the state is treated as constant and the time dependence is transferred to the operators which operate on this state vector. This may be represented in Hilbert space by keeping the state vector constant and allowing the axes to rotate with time as the motion of the system develops. Matrices representing operators referred to these axes are thus time dependent and obey the Heisenberg equation of motion. The theory developed in this representation is therefore called matrix mechanics.

HEISENBERG THEORY OF FERROMAGNETISM. Ferromagnetism (Heisenberg Theory).

HEISENBERG UNCERTAINTY PRINCIPLE. Uncertainty Principle.

HELICAL GEARING. For high pitch-line velocities and heavy loads, some form of "twisted tooth" gear is generally used. Two important types are helical gears and double-helical or herringbone gears. Both helical and herringbone gears are essentially spur gears with teeth twisted across the face in the form of a helix about the axis of rotation.

When spur gear teeth engage, the contact extends across the entire tooth on a line parallel to the axis of rotation, and may result in noise and shock at high speeds. In helical gear engagement, contact begins at one end of the entering tooth and gradually extends along a diagonal line across the tooth face as the gears rotate. The nature of the contact is such that with sufficient face width, two or more teeth are in contact and are carrying the load at all times. Helical gears are therefore used for transmission ratios as high as 10:1, and at pitchline velocities up to 2,000 feet (610 meters) per minute for commercially-cut units. Herringbone gear sets of special design have been successfully operated at pitch-line speeds of 12,000 feet (3,660 meters) per minute or above.

Tooth elements of helical gears are similar to those of spur gears. (See accompanying figure). The *helix angle H* of the tooth is measured between the line tangent to the tooth helix at the pitch circle and the shaft axis. In any pair, the gears have teeth with mating right-hand and left-hand helices. The usual method of tooth measurement is by diametral pitch P_d, which corresponds to circular pitch P_c in the diametral plane, perpendicular to the axis of rotation. By using

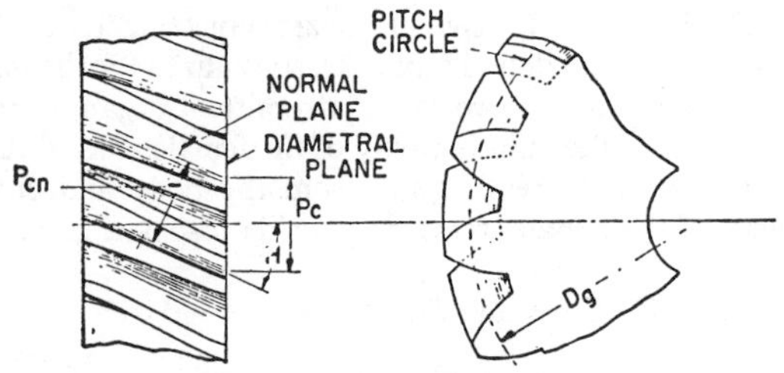

Tooth elements of helical gear.

standard pitches, the pitch diameters (and therefore the center distance of helical gear sets) can be given in commonly used fractions or integers; consequently, a spur gear set of a certain size can be replaced directly by a similar helical gear set. Actual tooth thickness depends upon the pitch and the size of the helix angle; if the circular pitch P_c be held constant, the actual tooth thickness measured perpendicular to its elements will decrease as the helix angle H is increased. A different cutter is required for every change in helix angle, although the pitch may remain constant. To eliminate an extensive variety of cutters, commercially available helical gears are made in several standard helix angles, among which are 7° 30′, 15°, and 23°.

By using a standard pitch in a plane normal to the tooth helix, the pitch diameter of a helical gear can be varied to suit a particular center distance by changing the helix angle. In this method of tooth measurement, the normal diameter pitch P_n corresponds to a normal circular pitch P_{cn} in the normal plane. Helical gear teeth designed with normal diametral pitches may be cut with standard spur gear cutters or hobs.

The pitch diameter D_g of a helical gear, based upon normal pitch P_n, is given by

$$D_g = N_g/P_n \cos H$$

where N_g is the number of teeth in the gear. The power transmitting capacity of helical and herringbone gears may be found by methods analogous to those used for spur gearing.

End thrust inherent in single helical gears can be eliminated by the use of herringbone gears that consist virtually of two integral single helical gears of opposite hand, which absorb the axial thrust within the gear. Herringbone gears are used for hoisting and mining machinery, rolling mills, sugar mill and lumber machinery, turbine and compressor drives.

HELIARC WELDING. **Welding.**

HELIARC BOURDON. **Bourdon Tube.**

HELICOPTERS AND V/STOL CRAFT. Although both fixed- and rotary-winged aircraft use airfoils to produce lift, in fixed-wing craft the wings can move no faster than the fuselage to produce lift and in order to fly; the whole aircraft must maintain considerable forward speed at all times. In the helicopter the wings (called rotor blades) are rotated at high speed, and there is no relationship between blade speed and fuselage speed. The helicopter, employing one or more horizontal rotors to give both lift and translation, can rise and descend vertically from the ground, hover over a spot on the ground, and fly backward and sideward as well as forward. With these flight characteristics, the helicopter does not require a prepared runway or landing area. A clearing about the size of a tennis court is adequate for landing, even though the surface be rough or uneven.

V/STOL aircraft are of more conventional lines. V/STOL is an abbreviation for a vertical or short take-off and landing.

Rotary-Wing Aerodynamics

The aerodynamics of rotary-wing and fixed-wing aircraft are basically the same. See also **Aerodynamics.** Both types of aircraft employ airfoils to produce lift; and both are subjected to identical fundamental forces of lift, drag, thrust, and gravity. It is true, however, that the flight characteristics of the helicopter differ widely from those of the fixed-wing craft.

Lift. Weight and lift are closely associated inasmuch as weight tends to pull the helicopter down and lift acts to hold it up. The similarity between the fixed-wing airplane and the helicopter is apparent; both are heavier than air, and both are sustained in flight by reaction of airflow over airfoils. The helicopter's airfoils are rotor blades which are turned at high speed. In a fixed-wing aircraft, if the angle of attack is increased, lift is increased until the stalling angle is reached; and for a given angle of attack, the greater the speed, the greater the lift. The helicopter rotates the rotor blades at high speed in rpm, to establish high-speed airflow over the airfoils. It is normal for the tip speed of the rotor blades to be as much as 350 miles (563 kilometers) per hour when the speed of the fuselage is zero. This explains why the helicopter does not require forward speed to produce lift, and why it can hover or fly backward, sideward, or forward.

Airflow. During normal operation conditions, the direction of airflow is from the top down through the main rotor system. As the blades are rotated with a positive angle of attack, they, in effect, screw upward into the air; thus a downwash of air (Fig. 1) is established through the rotor system. Notice that the leading edge of each blade bites into air throughout the complete cycle of rotation, forcing the air downward.

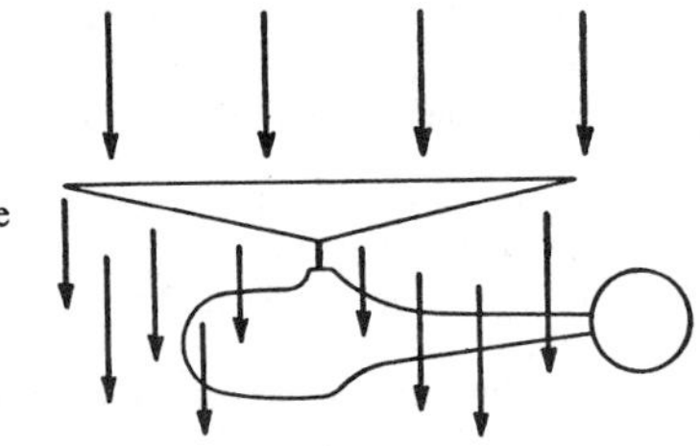

Fig. 1. Downwash through the rotor system.

At the root of the blade, airflow is slightly more than zero, but the velocity progressively increases throughout the length of the blade and at the tip may be 350 miles (563 kilometers) per hour or higher. It is the blade velocity that determines the resultant strength and direction of the relative wind at a positive angle of attack. The helicopter changes the angle of attack by varying the pitch of the main rotor blades. In a helicopter, the relative wind is developed throughout the complete cycle of 360 degrees by rotation of the rotor system, and it usually varies considerably. This variation is dependent upon flight conditions. See Figs. 2 through 6.

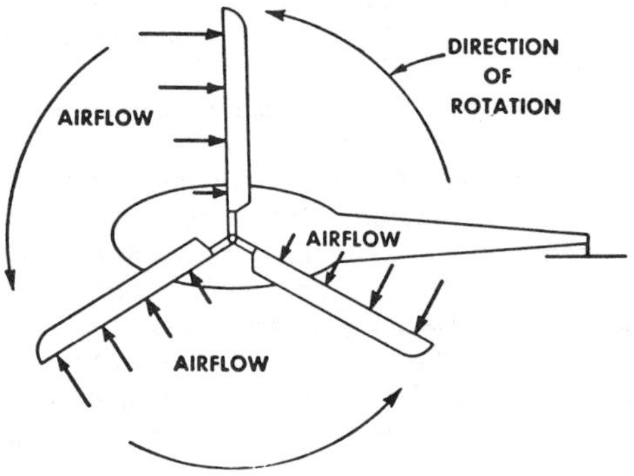

Fig. 2. Airflow in the rotor system.

Angle of Incidence. This is the angle formed by the chord of the airfoil and the longitudinal axis of the aircraft. The conventional airplane's angle of incidence is built into the aircraft by the designer and in most aircraft cannot be changed. In the case of the helicopter, however, the pilot continually changes the angle of incidence during flight by increasing or decreasing the pitch of the main rotor blades. See Fig. 7.

Airfoil Section. The type of wing used on conventional airplanes varies considerably; the airfoils may be symmetrical or unsymmetrical, usually dependent upon some specific requirements. The unsymmetrical airfoil may be efficient for an airplane wing, but it has one disadvantage that makes it unsatisfactory for use as a rotor blade. It is normal for the center of pressure to "walk" forward and rearward as the angle of attack is changed. The center of pressure is the imaginary point on the airfoil where all the aerodynamic forces are considered to be concentrated. See Fig. 8.

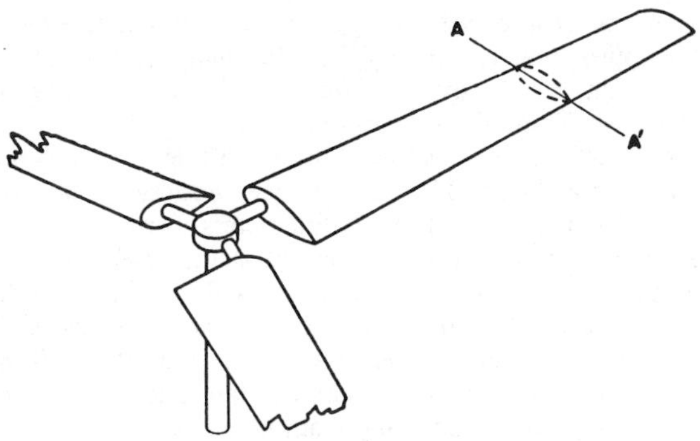

Fig. 3. High-speed blade section.

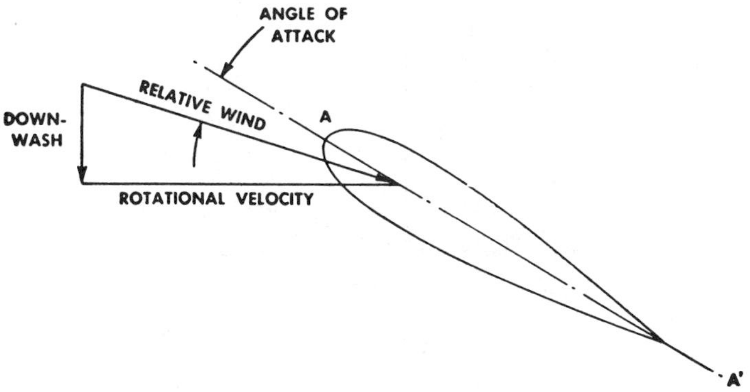

Fig. 4. Relative wind components.

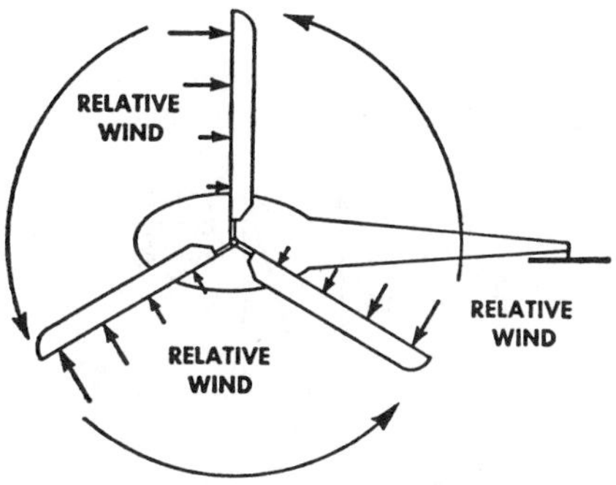

Fig. 5. Direction of relative wind.

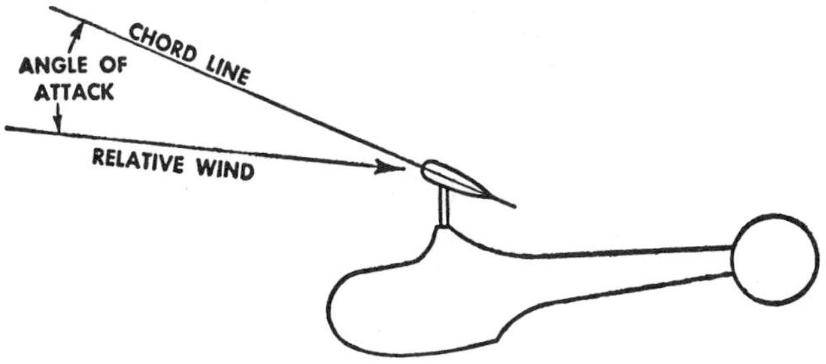

Fig. 6. Rotor blade angle of attack.

The airfoil section used for rotor blades is symmetrical, having equal camber above and below the chord line. Normally the greatest thickness of the blade is at a point about one-fourth of the way back from the leading edge. It is at this point that the center of pressure is located. There are several reasons for using the symmetrical airfoil for rotor blades: (1) There is a restricted migration of the center of pressure on a symmetrical airfoil when the angle of attack is changed; (2) the lift-drag ratio is very good even though the velocity of the blade varies from root to tip; and (3) the symmetrical airfoils permit ease of construction. If the center of pressure were permitted to travel

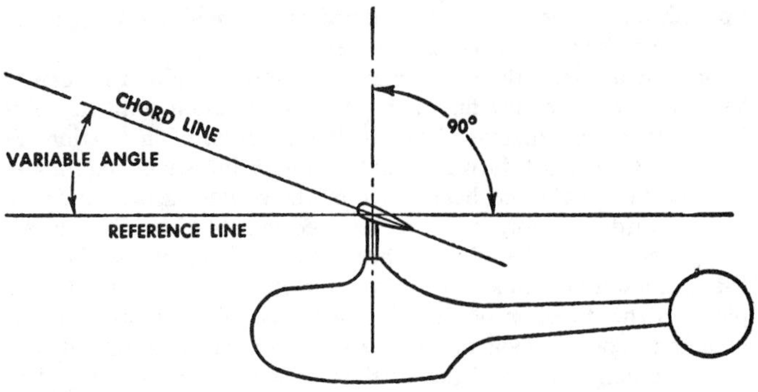

Fig. 7. Rotor-blade angle of incidence.

during angle of attack variations, pitching moments would be introduced into the rotor system; this condition would set up violent vibrations. Good lift-drag ratio throughout a wide range of velocities is important because it is necessary to have the lift forces spread over a wide area in order to equalize stresses. Usually a slight twist is built into the blade to help equalize these forces.

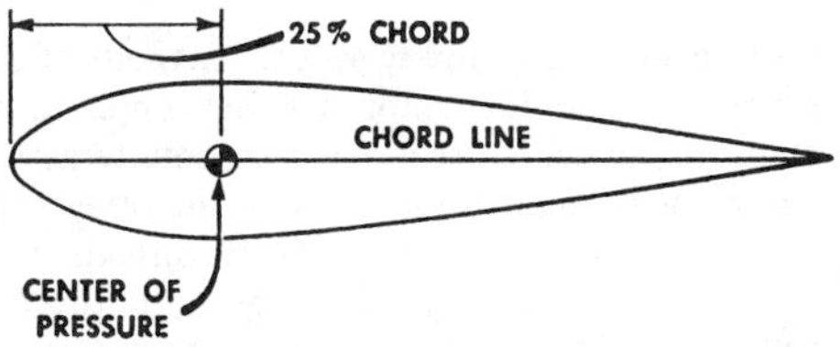

Fig. 8. Airfoils.

Thrust and Drag. As weight and lift are closely associated, so are thrust and drag. Thrust moves the helicopter in a designated direction and drag tends to hold it back. The helicopter develops both lift and thrust in the main rotor system. In vertical ascent, thrust acts upward in a vertical direction; drag, the opposing force, acts vertically downward. Lift sustains the weight of the helicopter; and excess thrust is available to give translation or vertical acceleration. During vertical ascent, drag is considerably increased by the downwash of the main rotor system striking the fuselage. Thrust must be sufficient to overcome both drag and downwash. The force representing the total reaction of the airfoils with the air is divided into two components: (1) Lift, and (2) thrust. However, drag is a separate force from weight, as shown in Fig. 9.

At all times, the lift forces of the rotor system are perpendicular to the tip-path plane. The tip-path plane is the imaginary plane described by the tips of the blades in making a cycle of rotation. The lift on the individual blade is perpendicular to the airfoil, but the resultant lift developed by the several blades is perpendicular to the

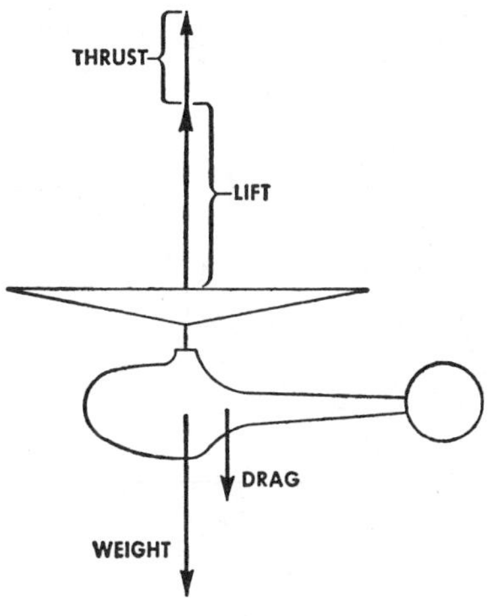

Fig. 9. Forces in vertical ascent.

tip-path plane. See Fig. 10. Lift increases in magnitude from root to tip of blade because of increase in velocity.

In vertical flight, the tip-path plane is horizontal; and in forward, backward, or sideward flight, the plane of rotation is tilted off the horizontal, thus inducing thrust in the direction of inclination. For example, to establish forward flight, resultant lift is inclined forward. See Fig. 11. Total force, being tilted off the vertical, acts both upward and forward; therefore it can be resolved into two components. One component is lift; the other is thrust. Likewise, flight may be established sideward, or in any horizontal direction, by tilting the tip-path plane in the direction of desired flight. Also, the rate of movement or speed depends upon the degree of tilt of the resultant lift force. Note the magnitude of thrust at the two speeds shown in Fig. 12.

Torque. Torque effect is displayed in a helicopter by the turning of the fuselage in the opposite direction to the rotation of the main rotor system. This reaction is in accord with Newton's third law of motion (to every action there is an equal and opposite reaction). The engine is the initiating force that drives the rotor system in a counterclockwise direction, and the reaction to this driving force would cause the fuselage of the helicopter to rotate with an equal force in a clockwise direction. See Fig. 13. Torque is of real concern to both the pilot and designer. Adequate means must be provided not only to counteract torque, but also for positive control over its effect during flight.

The designers of helicopters employ several methods of compensating for torque reaction. The dual-rotor type helicopter turns the two main rotor systems in opposite directions, thus counteracting the torque effect of one rotor by the torque effect of the other. The coaxial configuration likewise turns its rotors in opposite directions to equalize the torque effect. In the case of jet helicopters, if the engines are mounted on the tips of the rotor blades, no torque reaction is transmitted to the fuselage because the reaction is directly between the blade and the air. In the single main rotor helicopter, torque is usually counterbalanced by a vertically-mounted tail rotor which is located on the outboard end of the tail-boom extension. See Fig. 14. The tail rotor develops horizontal thrust that opposes the torque reaction. The pilot can vary the amount of horizontal thrust by activating foot pedals which are linked by cables to a pitch changing mechanism in the tail rotor system.

Ground Cushion. Also called ground effect, this is a volume of packed air built up between the rotor blades and the ground when the helicopter hovers near the ground. The downward flow of air strikes the ground and is partially trapped under the main rotor system. The air packs because it cannot escape as rapidly as the downward flow; therefore a cushion of slightly compressed air is established. The packed air is denser, thus increasing the efficiency of both the engine and the rotor system. The ground cushion is effective to a height of approximately one-half the rotor diameter; above this height, the air cannot be effectively trapped: Also, the ground-cushioning effect is lost at airspeeds in excess of ten miles per hour.

Translational Lift. This is the additional lift developed by a helicopter in horizontal flight. This lift becomes noticeably effective at an airspeed of 10 to 15 miles (16 to 24 kilometers) per hour and it continues to increase in magnitude as speed is increased. As horizontal flight is progressively induced, a higher inflow of air is established through the rotor disk, and greater lift is produced because of increased rotor efficiency. However, when a speed of from 45 to 50 miles (72 to 80 kilometers) per hour is reached, translational lift is canceled by fuselage drag. When hovering from 6 to 8 feet (1.8 to 2.4 meters) above the ground, the helicopter is aided by the ground-cushion effect.

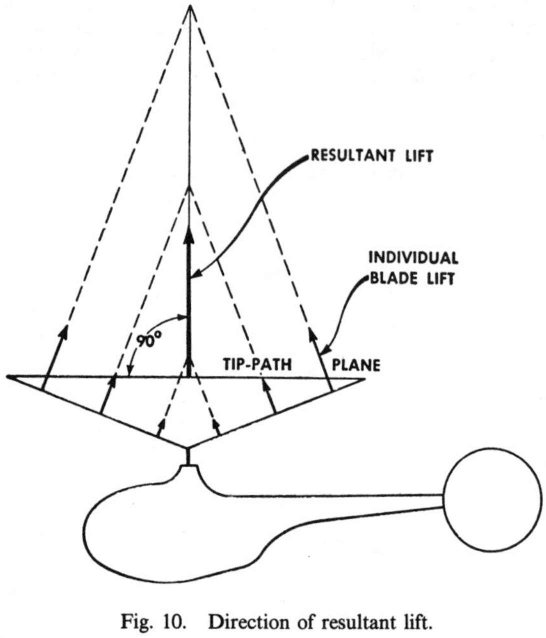

Fig. 10. Direction of resultant lift.

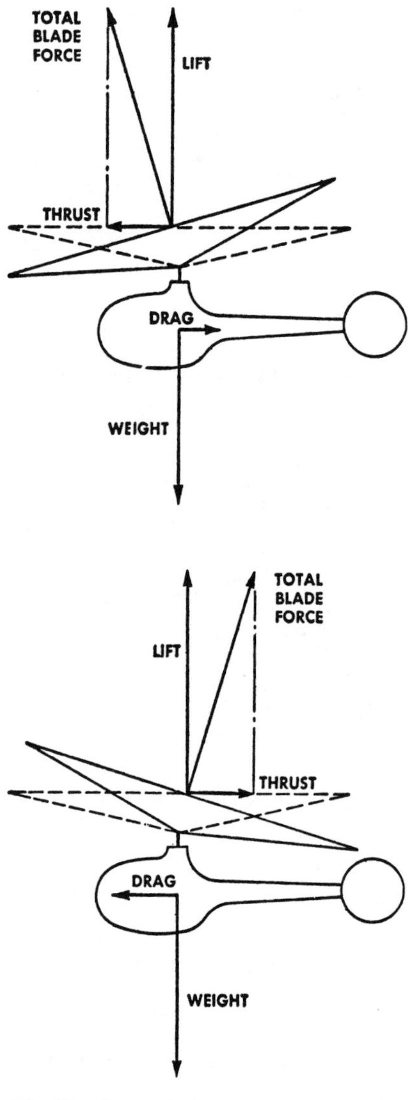

Fig. 11. Forces in forward and rear flight.

Dissymmetry of Lift. This is the unequal lift that develops between the advancing half of the disk area and the retreating half of the disk area during horizontal flight. The tip-speed rotational velocity is usually constant when the helicopter is hovering in a no-wind condition. Lift is equal on the advancing and retreating halves of the disk area when the craft is hovering because the angle of attack is constant and velocity airflow over the rotor blades is the same. When the helicopter enters forward flight, however, there will be a difference in airspeed between the advancing half and the retreating half of the disk area. To the rotational velocity on the advancing side of the disk is added the forward speed; the latter is subtracted on the retreating side. Forward flight of 50 miles (80 kilometers) per hour, therefore,

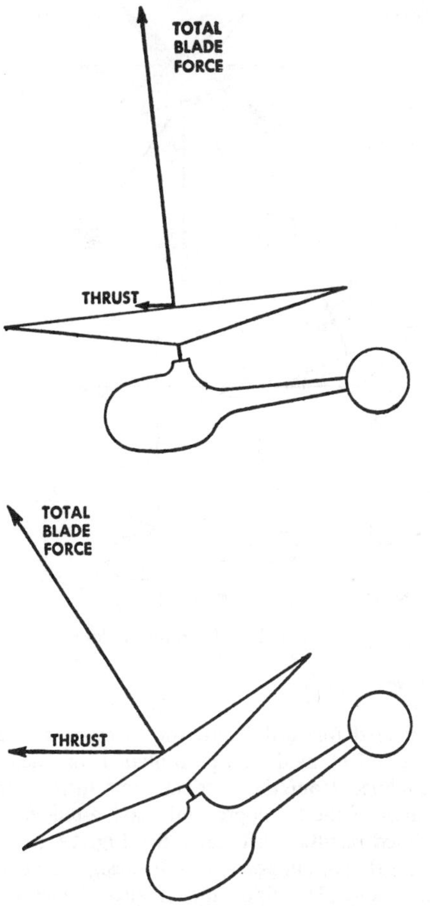

Fig. 12. Effects of slow and high speed.

would establish a differential of 100 miles (161 kilometers) per hour, a condition, if uncorrected, would develop unequal lift and the helicopter would turn over.

It is normal in rotor-head design to incorporate a flapping hinge, a device which permits the rotor blade to flap upward. Under normal operational conditions, the high-speed rotation of the rotor system develops a centrifugal force of approximately 20,000 pounds (88,930 N) on each blade. Centrifugal force holds the blades in a horizontal plane, but lift will cause the blade to rise vertically. The rotor blade will take the resultant position between centrifugal force and lift. It is normal for the rotor blades to take this coned-up attitude. See Fig. 15. The centrifugal force will be constant throughout the complete cycle of 360 degrees. Lift will vary between the advancing and retreating portion of the disk area in forward flight because of difference in air flow velocity. During forward flight, the advancing blade will flap higher because it has greater lift, and the retreating blade will flap to a lower angle because it has less lift. As the rotor blade flaps up, the effective lift area is lessened; and vice versa. On the retreating half of the disk area, however, reduced airspeed developed less lift. Therefore the retreating blade will assume a more horizontal attitude. Experience has proved that blades free to flap will assume a position which develops symmetry of lift between the advancing and retreating portions of the disk area.

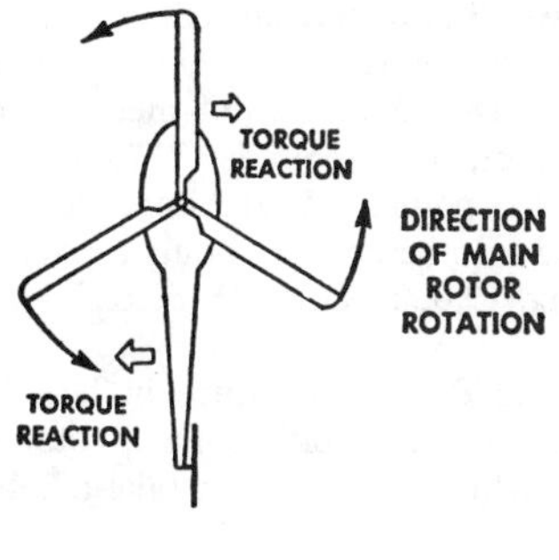

Fig. 13. Torque reaction.

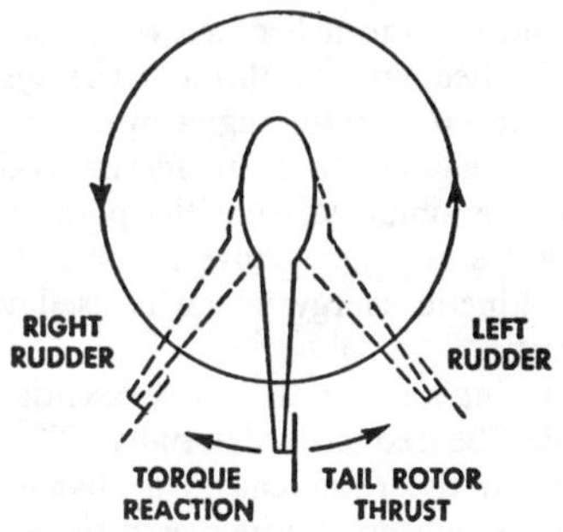

Fig. 14. Torque correction.

Gyroscopic Precession. This is the innate quality of all rotating bodies by which application of a force perpendicular to the plane of rotation will produce a maximum displacement of the plane approximately 90 degrees later in the direction of rotation. See Fig. 16. Thus, if a downward force is applied to the right side of a rotating disk, gyroscopic precession will cause the disk plane to tilt to the front, provided the disk is turning from right to left. Maximum resulting displacement occurs approximately 90 degrees further in the direction of turning, but speed of rotation, weight, and diameter of the disk, and friction are factors which determine the actual displacement of the system. The main rotor system of a helicopter displays the phenomenon of gyroscopic precession. The applied force is introduced by pitch change on the main rotor system. As the pilot moves the cyclic stick forward,

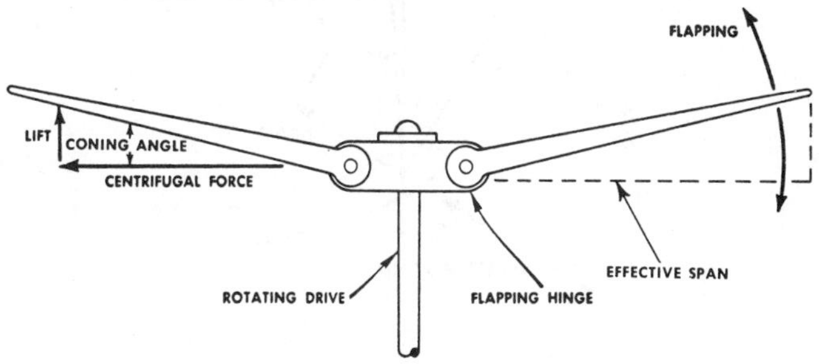

Fig. 15. Flapping hinge device.

it causes the control plane to tilt forward, thus introducing an equal but opposite pitch change at points 180 degrees apart in the cycle of rotation. Thus, if a linkage were not provided to take care of precession, the helicopter would fly 90 degrees out of phase. Forward stick movement would cause the craft to fly to the left. Thus, it is common practice to set the cyclic pitch change back approximately 90 degrees in the cycle of rotation. Various types of linkages are used.

Autorotation. This is the process of producing lift with rotor blades that freely rotate because of the developed aerodynamic forces resulting from the flow of air up through the rotor system. Under power-off

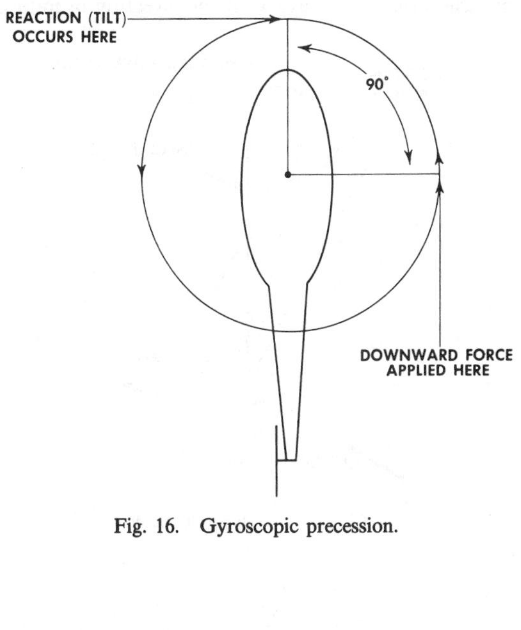

Fig. 16. Gyroscopic precession.

conditions, the helicopter will descend; thus the flow of air will be established upward through the system. The rotor is automatically disengaged from the engine by a free-wheeling device and the necessary power required to overcome parasitic and induced drag of the rotor blades is obtained from the potential energy due to the helicopter's weight and height above ground. This potential energy is converted into kinetic energy which is used to drive the rotor system during descent.

During autorotation, it is essential that the pitch angle of the rotor blades be reduced materially. The change in direction of airflow through the rotor causes a change in the direction of the relative wind which greatly increases the angle of attack at which the rotor blades are operating. If the pitch were not reduced, the blade would stall for much the same reason that a conventional airplane's wing stalls when the nose of the aircraft is pulled up to high. When the pitch angle of the blade is low and the angle of attack is large, the resultant lift force lies ahead of the axis of rotation of the blades, tending to keep the blades turning in their normal direction. See Figs. 17 and 18. If, on the other hand, the pitch angle remains high, drag

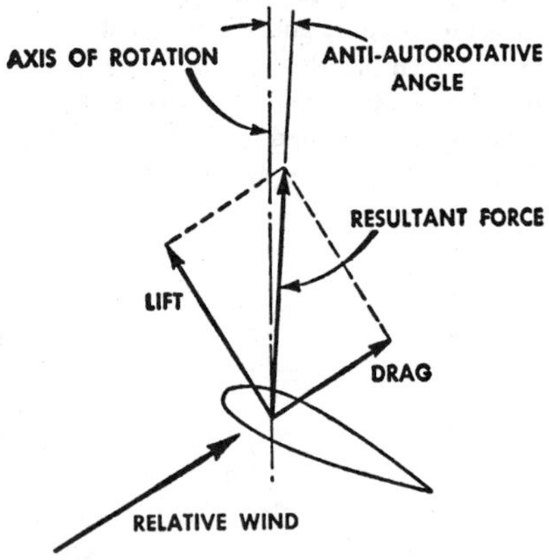

Fig. 17. Low-pitch angle.

is increased and the resultant lift force lies behind the axis of rotation, tending to slow and stop the rotor. Autorotation is an emergency procedure that permits the helicopter to make a safe landing in case of engine failure. It is necessary to maintain the speed of the rotor at sufficient rpm to provide not only adequate airflow over the rotor blades, but also the required centrifugal force to hold the blades in an extended attitude; otherwise the blades would fold up and the helicopter would tumble out of control.

Pendular Action. The fuselage of the helicopter is suspended from the drive shaft that mounts the main rotor head. Because the fuselage is bulky and suspended from a single point of attachment, it is free to oscillate laterally and longitudinally much in the same fashion as a freely-swinging pendulum. As the rotor system introduces horizontal translation, the fuselage is dragged in the direction of induced flight.

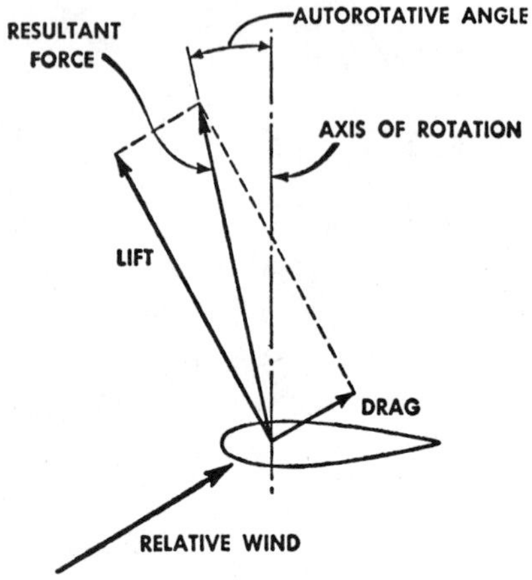

Fig. 18. High-pitch angle.

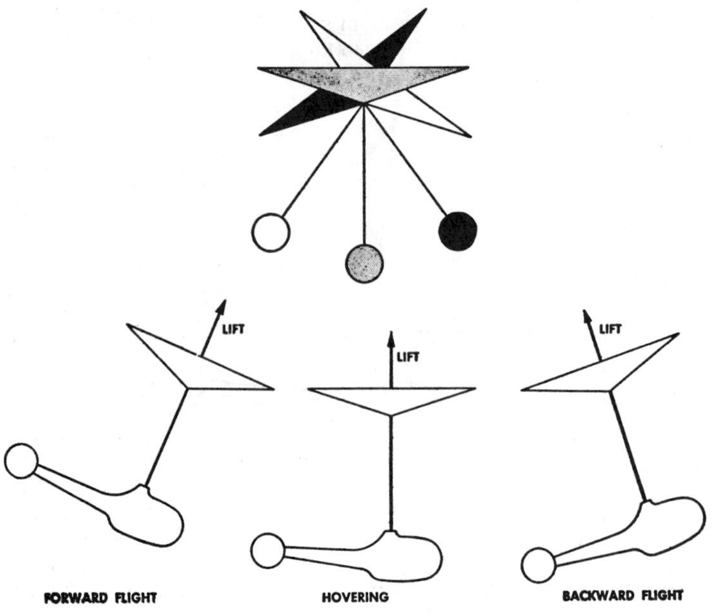

Fig. 19. Pendular action.

During established forward flight, the fuselage will assume a nose-low attitude. In effect, as the tip-path plane is inclined forward, resultant lift is inclined from the vertical, thus introducing thrust. The main drive shaft of the helicopter will have a tendency to align itself with the inclined resultant lift force. See Fig. 19.

Other factors that are important to the design of a helicopter include (1) Resonance; and (2) weight and balance. Generally, sympathetic resonance has been well overcome by controlling design features of gear boxes and other mechanisms. Ground resonance always has been a knotty problem. This is a self-excited vibration which develops when the landing gear repeatedly strikes the ground, thus unseating the center of mass of the main rotor system. The pounding effect of the landing gear is prone to occur during take-off and landing when the helicopter is from 87 to 93% airborne. The aircraft, being light on the landing gear, bounces from one wheel to another in rapid succession, setting up a pendular oscillation of the fuselage. The succession of shocks is transmitted to the main rotor system, and the main rotor blades straddling the pounding wheel are forced to change their angular relationship. This condition unbalances the main rotor system, which in turn transmits the shock back to the landing gear. To control this potentially damaging condition, various dampening devices are used to control the unbalancing of the rotor system, and helicopter pilots are trained to avoid critical maneuvers conducive to agitating ground resonance.

Representative Helicopters. An abridged list of important helicopters designed and operated over the past several years would include:

(*Bell UH-1N*)—United States, operational in 1970; multipurpose craft; one 1800-horsepower turbine engine; rotor blade diameter, 42.8 feet (14.7 meters); length, 42.1 feet (12.8 meters); height, 14.4 feet (4.4 meters); empty weight, 6120 pounds (2776 kilograms); maximum weight on take-off, 20,000 pounds (4536 kilograms); maximum speed, 121 miles (195 kilometers) per hour; range, about 300 miles (483 kilometers); ceiling, 11,500 feet (3505 meters); requires a crew of one; passengers, 14.

(*Bell OH-58A Kiowau*)—United States, introduced in late 1960s; multipurpose craft; one 317-horsepower turbine engine; rotor blade diameter, 35.4 feet (10.8 meters); length, 32.3 feet (9.8 meters); height, 9.6 feet (2.9 meters); empty weight, 1580 pounds (717 kilograms); maximum weight on take-off, 3000 pounds (1361 kilograms); maximum speed about 140 miles (225 kilometers) per hour; range, about 350 miles (563 kilometers); ceiling, 19,000 feet (5791 meters); requires a crew of 2; passengers, 2.

(*Bell AH-1 J Seacobra*)—United States, introduce in late 1960s; combat helicopter; one 1800-horsepower turbine engine; rotor blade diameter, 44 feet (13.4 meters); maximum weight on take-off, 10,000 pounds (4536 kilograms); maximum speed, nearly 210 miles (338 kilometers) per hour; range, about 360 miles (579 kilometers); ceiling, 10,550 feet (3216 meters); combat crew, 2.

(*Dornier Do 132*)—Germany, introduced in 1971; passenger helicopter, one 720-horsepower engine; rotor blade diameter, 35.1 feet (10.7) meters); length, nearly 25 feet (7.6 meters); height, just over 9 feet (2.7 meters); empty weight, about 1500 pounds (680 kilograms); maximum weight on take-off, about 3635 pounds (1649 kilograms); maximum speed, just over 140 miles (225 kilometers) per hour; range, 275 miles (442 kilometers); crew of one; passengers, 4.

(*Agusta A-106*)—Italy, introduced in late 1960s, antisubmarine helicopter; one 350-horsepower turbine engine; rotor blade diameter, just over 31 feet (9.4 meters); height, just over 8 feet (2.4 meters); length, nearly 29 feet. (8.8 meters); empty weight, about 1520 pounds (689 kilograms); maximum weight on take-off, about 3085 pounds (1399 kilograms); maximum speed, 110 miles (177 kilometers) per hour; range, 460 miles (740 kilometers); crew of 1 (total).

(*Aérospatiale-Westland SA-300 Puma*)—France, operational in 1970; transport helicopter; two 1320-horsepower turbine engines; rotor blade diameter, just over 49 feet (14.9 meters); length, just over 46 feet (14 meters); height, nearly 14 feet (4.3 meters); empty weight, about 7560 pounds (3429 kilograms); maximum weight on take-off, 14,110 pounds (6400 kilograms); maximum speed, about 175 miles (282 kilometers) per hour; range, nearly 400 miles (644 kilometers); ceiling, 15,750 feet (4801 meters); crew of two; passengers, 16.

(*Westland WG 13N Lynx*)—Great Britain, introduced in early 1970s; passenger helicopter; two 900-horsepower turbine engines; rotor blade diameter, 42 feet (12.8 meters); length, just over 38 feet (11.6 meters); height, just over 11 feet (3.4 meters); empty weight, nearly 7500 pounds (3402 kilograms); maximum weight on take-off, about 8780 pounds (3983 kilograms); maximum speed, nearly 185 miles (298 kilometers) per hour; range, 150 miles (241 kilometers); crew of two; plus passengers.

V/STOL Craft

The concept of taking off and landing vertically inspired early efforts on helicopters just described. Attempts during the early 1900s included: (1) The Gyroplane, designed and built in Europe in 1907 by Bréguet and Richet, managed to lift only a few feet off the ground. The engines were inadequate. Following these experiments, Bréguet decided to concentrate, and successfully, on the design of more conventional airplanes. (2) Frenchman Paul Cornu, while aboard his craft, succeeded in lifting a fragile design off the ground about one foot for a period of about 20 seconds in 1907. (3) Igor Sikorsky, in 1909, constructed a helicopter prototype using an Anzani engine, but the latter proved inadequate to lift the craft off the ground. For several years thereafter, Sikorsky concentrated successfully on the design of seaplanes. (4) In Denmark, Ellehammer, a designer of prior fixed-winged airplanes, managed to lift his helicopter design prototype a few inches above the ground in 1916. This craft was equipped with two coaxial rotors. (5) Frenchman Etienne Oemichen succeeded with his design to take off vertically for a flight of a few hundred yards in mid-1924; (6) The Marquis Raul Patteras Pescara (Spain) in the same year managed a flight of 2,415 feet (736 meters). Pescara continued with other models, but ceased activity upon the successful demonstration of Juan de la Cierva's autogiro which, at that time, appeared to provide the answers being sought from helicopters; (7) In 1930, Italian D'Ascanio's design, piloted by Marinello Nelli, broke the prior distance record, achieving a flight of some 3,535 feet (1077 meters), and at a record height of 59 feet (16 meters) above ground. But probably success for the first practical helicopter designs should go to the Germans with the development of the Focke-Wulf FW-61, which appeared in 1936; and one year later when Igor Sikorsky built and flew the VS-300.

Autogiro. This craft was developed in Spain and flown in 1923. The autogiro consists of a wingless fuselage mounting a pylon which contains a rotating head to which are affixed three or four balanced blades of airfoil section resembling a propeller configuration. The blades are rotated by the action of the relative wind and thus are self-rotating, hence the name of the craft. The autogiro is equipped with a regular power plant and propeller that produce the necessary forward thrust to get the craft in motion and keep it in motion when in flight. The angle of the rotor to the fuselage is controlled by the pilot, and this takes the place of the normal control surfaces of the conventional airplane. The blades rotate at speeds that give an average air velocity over them considerably in excess of the autogiro's airspeed, and so the autogiro may be flown at speeds lower than the stalling speed of the airfoil section. This is an advantage, in that it permits the autogiro to land in small fields, with short landing runs, to descend almost vertically, and to approach "hovering" flight. Although autogiro development essentially has been abandoned, primarily because of the successes of later helicopter designs, much of the information gained in autogiro experiments has proven of much value to helicopter pioneers and later VTOL designers.

During the past few decades, a number of V/STOL have been proposed and built, either experimentally or in production. Principal configurations include: (1) *Compound or convertible aircraft*—essentially helicopters with propellers for horizontal flight. At take-off, the engines power the rotors; at altitude, the power is transferred to the propellers. A serious disadvantage of this concept is the substantial drag presented by the rotors during horizontal flight. (2) *Vertiplanes*—large, propeller-equipped airplanes, with a jet engine for effecting vertical take-off. In appearance, past designs of these craft look like an essentially conventional fixed-wing airplane. (3) *Tilt-Prop or Tilt-Wing Airplanes*—essentially airplanes with wings and jets or propellers, either of which can be tilted by 90 degrees during take-off and landing. (4) *Bidirectional Jet Aircraft*—jet-propelled aircraft in which the jet engine(s) can be directed downward for vertical take-off and turned toward the rear for horizontal flight when at altitude.

The AV-8B V/STOL. Developed for the United States Marine Corps, this aircraft is a single-seat, single-turbofan-powered craft for close air support and interdiction missions. The craft is powered by a Rolls Royce Pegasus 11 vectored thrust turbofan, with 21,500 pounds of thrust (95,600 N) without afterburning. The length is 46.3 feet (14.1 meters); height is 11.6 feet (3.5 meters); the wingspan is 30.3 feet (9.2 meters); and the wing area is 230 square feet (21.4 square meters). Take-off distance is 0–1100 feet (0–305 meters). The first flight of the YAV-8B prototype occurred in November 1978. Combat range is 600+ nautical miles (1112 kilometers); ferry range is 2460 nautical miles (4558 kilometers). See Fig. 20. The aircraft can carry

Fig. 20. United States Marine Corps' AV-8B V/STOL aircraft for close air support and interdiction missions. (*McDonnell Aircraft Company; McDonnell Douglas Corporation.*)

armament up to 9200 pounds (4175 kilograms) on several stations: two 30 millimeter cannons; laser or TV-guided weapons; as well as up to four 300-gallon (1136-liter) fuel tanks. The Marine Corps V/STOL close air support uses three types of land sites as well as a variety of ships. Closest to the battle area would be a number of V/STOL forward sites, accommodating two to four AV-8Bs. These sites would normally have fuel and ordnance for turnaround operations only. If a STO (short take-off) strip is not available, the aircraft would operate in the VTOL (vertical take-off and landing) mode. Located about 50 miles (80 kilometers) from the battlefield would be a V/STOL facility with at least a 600-foot (182-meter) strip, providing six to ten aircraft turnaround, support, and maintenance. The V/STOL main base would have at least a 1500-foot (457-meter) strip and all the logistics and support assets for prolonged AV-8B operations.

The AV-8B's vertical takeoff and landing ability is derived from four exhaust nozzles—two on either side of the aircraft—positioned

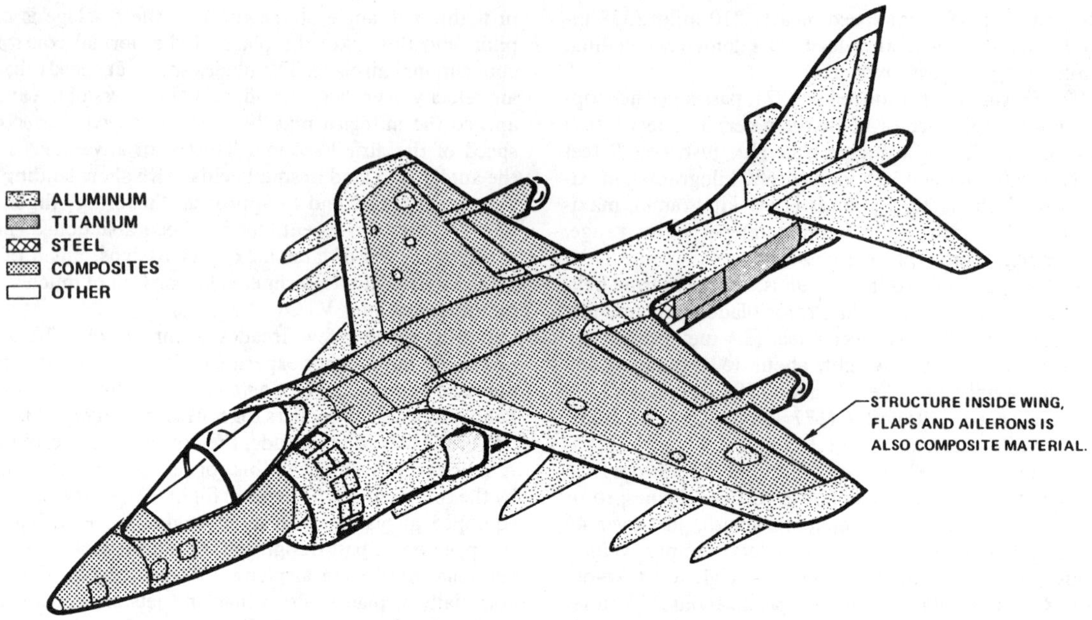

Fig. 21. Types of materials used in the AV-8B V/STOL aircraft. As a percent of structural weight: Aluminum, 48.4%; titanium, 8.5%; steel, 14.5%; composites, 23.3%; other materials, 5.3%. Pylons not included in these figures. Approximately 1186 pounds (540 kilograms) of graphite epoxy are used in the structure. (*McDonnell Aircraft Company*; *McDonnell Douglas Corporation.*)

around the plane's center of gravity. These nozzles can be rotated from the full-aft position, for forward flight, to a full-down position for vertical operations. Within the engine, some rotating parts turn clockwise while others turn counterclockwise. This is necessary to prevent gyroscopic effects which, if all parts rotate in the same direction, can make the aircraft difficult or impossible to control during hover, and during transition from hover to forward flight or from forward flight to hover.

All that is required for take-off or landing is an amphibious assault ship, or a clearing large enough for a 72-foot (22-meter) square aluminum mat, a section of two-lane road, or even a damaged airfield. As with the trend in the design and construction of modern aircraft, much stress is being given to composites and graphite epoxy materials. See Fig. 21.

References

NOTE: See also references at ends of entries on **Aerodynamics**; and **Airplane.**
Angelucci, E.: "Air Planes: From the Dawn of Flight to the Present Day," McGraw-Hill, New York, 1973.
Lambermont, P., and A. Pirie: "Helicopters and Autogyros of the World," Cassel, London, 1970.
Staff: "Jane's All the World's Aircraft," Sampsom Low, Merston, London (Published annually).
Van Sickle, N. D.: "Modern Airmanship," 3rd edition, Van Nostrand Reinhold, New York, 1966.

HELIDES. Rocket Propellants.

HELIOPAUSE. That distant location in outer space at which the sun's influence ends and is replaced by the interstellar medium. Latest estimates indicate that this is about 15 billion kilometers (9.3 billion miles) out from the sun.

HELIOSTAT. An arrangement of mirrors, driven by clockwork, used to reflect a beam of sunlight in a fixed direction as the sun moves across the sky. The heliostat is used in the control of some astronomical instruments as well as in some solar energy systems for tracking the sun. See also **Solar Energy.**

HELIOSTAT (Solar Telescope). Sun (The).

HELIOTROPE. Bloodstone.

HELIOTROPIC WIND. Winds and Air Movement.

HELIUM. Chemical element symbol He, at. no. 2, at. wt. 4.0026, periodic table group 0 (inert or noble gases), mp −272.2°C (20 atmospheres), bp −268.93°C (4.2144K), specific gravity 0.124 at 4.2144K. The element has no triple point and can be solidified only by applying high pressure to the liquid phase. Described later, liquid helium undergoes a change in its physical properties at 2.178K, known as the *lambda point.* Solid helium has a close-packed hexagonal crystal structure (subject to further study and confirmation). At standard conditions, helium is a colorless, tasteless, odorless gas. There are two natural isotopes ^3He and ^4He, with ^4He being slightly less than 100% abundant. The boiling point is 3.2K for ^3He. Radioactive ^5He and ^6He have extremely short half-lives. See also **Radioactivity.**The first ionization potential for helium is 24.58 eV; second, 54.14 eV. Other physical properties of helium are described under **Chemical Elements.**

Like the other rare gases, helium exhibits negative chemical properties with ordinary materials under normal conditions. Under the influence of electric glow discharge or electron bombardment, helium forms compounds with tungsten and other metals, as well as with iodine, sulfur, and phosphorus. In a vacuum electric discharge tube shows green to canary-yellow glow. Discovered first in the vapors surrounding the sun by Lockyer in 1868, through the yellow spectral line near the two yellow lines of sodium, then by Ramsay in 1895 in the mineral clevite.

Helium occurs (1) in minerals of uranium and thorium, such as clevites, pitchblende, carnotite, monazite, and also in beryl, (2) in mineral waters (1 part He per thousand of water, in some Iceland waters), (3) in volcanic gases, (4) especially in certain natural gases of the United States. The first discovery of this kind was made in Kansas. The richest helium wells are in Utah. In northeastern Texas, four wells have produced 55 million ft^3 of helium. The fields in which are located the wells having the greatest percentage (1.3–8.0%) of helium are now held as government reserves. (5) in ordinary air, about 1 part in 200,000.

Uses: Industrially, helium is used to provide an inert gaseous shield for arc welding, for growing transistor crystals, in the production of titanium and zirconium, to fill the space between optical lenses in instruments, as the carrier gas in some chromatographic apparatus, as a liquid bath for masers and cryotrons, as a refrigerant for furnishing the low temperature required for superconducting electrical equipment, in lasers, as a diluent gas in deep-sea diving applications, as a heat-transfer medium in gas-cooled nuclear reactors, and as a leak-detecting medium for testing pressure and vacuum equipment. Now, to a rather limited extent, helium is used as a lifting gas for airships and for balloons used in meteorological investigations. Helium is used in aerospace programs in several ways, including its use in propellant tanks as a compressed gas which expands and takes the place of fuel as the fuel is consumed, in ground-support equipment, and in communication satellites for providing the low temperature required for sensitive electronic systems. In medicine, helium sometimes is mixed with oxygen for patients with certain respiratory ailments and also it is mixed with certain anesthetics to reduce the hazards of forming an explosive mixture with air.

The liquefaction of helium was accomplished by Onnes in 1908 in Leiden, and Keosom in 1926 succeeded in solidifying helium in the same laboratory. Relatively recently, helium has been solidified at room temperature. The melting pressure at 24°C is 115 kilobars, in complete agreement with the Simon equation. Besson and Pinceaux (1979) developed an original apparatus for the experiment, which allowed loading of the cell at room temperature. Diamond anvil cells were used in the procedure.

Liquid Helium II: Upon cooling, ^4He liquefies at atmospheric pressure at 4.216K to form an essentially normal liquid, liquid helium I. On further cooling to the lambda-point, 2.178K at one atmosphere, a change occurs to liquid helium II. The latter has a very low viscosity (hence the name "superfluid") and a very high thermal conductivity, which produce such phenomena as the creeping of a film over the edge of the container, and the fountain effect, in which the liquid sprays out of a capillary. Superfluidity is commonly explained in terms of a two-fluid theory. Thus, London and Tirza attribute the properties of helium II to a mathematical peculiarity in the distribution function of Bose-Einstein statistics, whereby below the λ-point, a finite fraction of the atoms fall into a ground state of zero thermal energy. In this state they would have the properties of a superfluid. However, this theory has not yielded good quantitative predictions of the properties of the aggregare liquid helium. ^3He, which follows Fermi-Dirac statistics, does not have a superfluid state.

Landau treats liquid helium by an approach similar to that of the Debye theory of solids. The longitudinal and transverse sound waves, which are the elementary excitations of that theory of solids, correspond in the case of liquid helium to phonons and rotons. The *phonons* are the longitudinal sound waves, while the *rotons* are another type of elementary excitation postulated by Landau to represent the rotational motion of the liquid, because a liquid cannot support transverse waves. The specific heat can be expressed as the sum of contributions from phonons and rotons. Landau derived expressions for these which fit the data and experiments quite closely up to 1.6K.

Feynman developed wave functions to provide an atomistic interpretation of Landau's spectrum of elementary excitations.

The complexity of the helium II problem is apparent at once when one attempts to extend the equations of classical hydrodynamics to this two-component system, in which each component has its own density and velocity. Khalatnikov derived such equations by ignoring terms of second order.

Still another area of investigation has been that of the properties of ^3He-^4He mixtures. As stated above, ^3He exhibits no λ-transition and no superfluidity. It has a critical temperature of 3.35K and a boiling point of 3.2K, against values of 5.2K and 4.216K for ^4He.

The most abundant helium atoms, ^4He, are bosons, but the ^3He atoms are fermions. This has as a consequence that liquid ^3He does not show superfluidity—a property very probably connected with the Bose-Einstein statistics obeyed by the ^4He atoms.

Chemistry: The most striking properties of helium are its emission as the positively charged (+2) alpha particles in radioactive changes, its formation in radioactive change by uranium-radium and thorium-containing substances, emitting alpha particles, later losing the charge to become helium, and its production artificially by bombardment of lithium or boron with high-velocity protons or alpha rays.

Unlike the other inert gases, helium gives little evidence of compound formation with organic substances. Like neon, but unlike the others, it forms no hydrate. However, it forms compounds much more readily under excitation, due apparently to unpairing of its 1*s* electrons and promoting of one of them to the 2*s* state. The 460 kcal/g-atom of energy is readily obtained by electric discharge or electron bombardment. Under such conditions the helium molecule-ion, He_2^+, with a pair of bonding electrons (1*s*) and a single antibonding electron (1*s*), is formed, as are combinations of the type of HeH^+ and HeH_2^+. In a mercury discharge tube, the compound $HgHe_{10}$ has been found, and with various metallic electrodes corresponding helides, such as the compounds of tungsten, platinum, iron, palladium, bismuth, etc., e.g., WHe_2, Pt_3He, FeHe, PdHe, BiHe, etc., have been formed.

Helium in Resources Planning. The Tip Top natural gas field in Sublette County, Wyoming contains the highest known helium content in a natural gas, namely, 0.8%. There are a few helium-rich fields in Alaska and Canada. The percentage of helium in natural gas drops off fast for most fields. The very large gas fields of the world have a helium content of below 0.10%. Most gas fields with an average content as high as 0.3% are found in Kansas, Colorado, Texas, and Oklahoma. Natural gas fields in coastal zones contain as little as 0.007% helium. Outside North America, helium is found in West Germany and South Africa, Algeria, Poland, and the British sector of the North Sea, the content ranging between 0.10 and 0.3%.

As pointed out by Hurley (1954), helium has a geologic occurrence and distribution unique among the elements. It is a product of radioactive disintegration of uranium and thorium within the earth's mantle and crust, but flows to the surface at a rate less than that of its generation, because most of it is driven into crystal structures of rock minerals until released by alpha radiation damage near radioactive concentrations. Mobile helium rising through the crust may then be trapped, along with other gases, beneath relatively impermeable barriers. Nitrogen is almost always associated with helium in natural gases, although this has not been fully explained. Also, carbon dioxide is abundant in some helium-rich gas mixtures.

As the helium-rich gas is burned toward exhaustion, most of its contained helium is dissipated into the atmosphere. It has been estimated that the recovery of helium from the atmosphere would cost 800 times (with 1980 technology) that of separating it from natural gas, prior to releasing the natural gas to pipelines. But, as pointed out by Cook (1979), controversy over the need for a government-directed helium-conservation program reflects fundamental differences in viewpoints on the economic future of industrial society, on the limits of substitution and labor and capital for a depleting resource, and on intergenerational equity and risk-bearing.

Cook (1979) stresses that future demand for helium on a much larger scale than the demands of the past will depend upon development and deployment of technologies either in their infancy, or not yet even conceived. Among these are magnetic containment systems for fusion reactors, breeder and high-temperature gas reactors, high-temperature gas turbines, laser-based missile-defense systems, magnetic propulsion units for new transport systems, helium refrigeration systems for military aircraft, advanced energy conversion cycles (mainly magnetohydrodynamic systems), and low-temperature energy transmission, distribution, and storage.

As summarized by Cook, helium conservation is a national issue in which thermodynamic certainty collides with economic uncertainty. Low-entropy helium is wasting into the atmosphere at least in part because there is no way to prove that future generations will be better off it is saved for them. A helium conservation program was started, then aborted. A decision to resume storing helium will be a political decision based more upon prevailing ideas of fairness in intergenerational risk-bearing and equality, and on current view of the qualitative impact on future society of materials scarcities, than on any quantitative forecasts of future needs and costs.

References

Besson, J. M., and J. P. Pinceaux: "Melting of Helium at Room Temperature and High Pressure," *Science*, **206**, 1073–1075 (1979).

Cohen, E. G. D.: "Quantum Statistics and Liquid Helium-3–Helium-4 Mixtures," *Science*, **197**, 11–16 (1977).

Cook, E.: "The Helium Question," *Science*, **206**, 1141–1148 (1979).

Ebisch, R.: "Helium," *Science News*, **116**, 3, 50 (1979).

Staff: "Superfluid Helium," *Science News*, **116**, 8, 135 (1979).

HELIUM COOLANT. **Nuclear Reactor.**

HELIUM DILUTION REFRIGERATION SYSTEM. **Refrigeration.**

HELIUM LEAK DETECTION. **Mass Spectrometry.**

HELIUM (Para-State). Para-State.

HELIUM PRECURSOR (Earthquake). Earthquakes, Seismology, and Plate Tectonics.

HELIUM (Solar). Sun (The).

HELIUM (Superfluid). Cryogenics; Superfluidity.

HELIX. A space curve traced on a cylinder or conical surface in such a way that all elements of the surface are cut at a constant angle. A circular helix lies on a right-circular cylindrical surface. In parametric form, its equation is $x = a \cos \theta$, $y = a \sin \theta$, $z = b\theta$ where a, b are constants and θ is the parameter. The thread of a screw is often a circular helix.

See **Conical Surface** and terms listed under **Mathematics.**

HELIX FEEDER. Feeder (Volumetric).

HELLBENDER (*Amphibia, Urodela*). A large aquatic salamander, *Cryptobranchus allegheniensis,* of the Mississippi river system. It reaches a length of 18 inches and has a flattened head and body, short legs, and a compressed tail. The gills are concealed, but otherwise it resembles the mudpuppy.

HELLGRAMMITE (*Insecta, Neuroptera*). The large aquatic larva of the dobson fly, *Corydalus.* It lives in running water and is an excellent bait for bass.

HELLIGE TURBIDIMETER. Turbidimetry.

HELMHOLTZ EQUATION. An equation of the form

$$n_1 y_1 \tan \theta_1 = n_2 y_2 \tan \theta_2$$

expressing the relation between the linear and the angular magnification at a spherical refracting interface. y_1, y_2 are linear dimensions of object and image, θ_1, θ_2 the angles made by focal rays and axis at object and image points and n_1, n_2 are refractive indices of object and image space. Also called Lagrange-Helmholtz equation. (See, however, the **Abbe Sine Condition.**) A spherical surface cannot satisfy both these equations for finite angles. Hence a spherical surface can never make a perfect image.

HELMHOLTZ FUNCTION. Thermodynamics.

HELMHOLTZ INSTABILITY. Atmosphere (Earth).

HELMHOLTZ LAW. Thomson (Law of).

HELMHOLTZ RESONATOR. An enclosure communicating with the external medium through an opening of small cross-sectional area. Such a device resonates at a single frequency dependent on the geometry of the resonator.

HELMHOLTZ SOUND SYNTHESIS. Musical Sound.

HELMHOLTZ THEOREM. The statement that if **F** is a vector field satisfying certain quite general mathematical conditions, then **F** is the sum of two vectors, one which is irrotational (has no vorticity), the other solenoidal (has no divergence).

HELMINTHOLOGY. A biological science dealing with the worms, more particularly parasitic flatworms and roundworms. Since many worms are parasitic, the term parasitology is more commonly used. The study of roundworms is important in agriculture and has resulted in the science of nematology (see **Nematoda**) which is properly a subsidiary of helminthology.

HEMATITE. The mineral hematite, ferric oxide, Fe_2O_3, occurs as thick or thin tabular rhombohedral forms, sometimes in pyramids but rarely in hexagonal prisms. It also assumes botryoidal, columnar and lamellar shapes, and may be granular or compact. Its hardness is 5.6; specific gravity, 5.26; luster, metallic to earthy or dull; color, dark gray to black; earthy forms may be different shades of red; streak, red to red-brown; translucent (in very thin flakes) to opaque. Hematite with a metallic luster is called specular iron.

It is a widely distributed and common mineral, found in igneous, sedimentary and metamorphic rocks as beds and veins, having probably been formed in many different ways under very different conditions. Beautifully crystalized hematite has been found in the Urals of the U.S.S.R.; Rumania; Switzerland; the Island of Elba; Alsace, France; Cumberland, England. Extremely rich, large hematite ore bodies have been found and are being worked in Minas Gerais, Brazil; Cerro de Mercado, Durango, Mexico; Quebec and Labrador in Canada. The hematite ore deposits which lay along the southern and northwestern sides of Lake Superior in Michigan, Wisconsin and Minnesota have been worked to near depletion. Extensive beds of hematite are found throughout the Appalachian region from New York to Alabama, being mined near Birmingham in the latter state. Hematite occurs in quantity in Nova Scotia and Newfoundland. It is the most important ore of iron, and has other industrial uses in paint manufacture and polishing compounds. The name hematite is derived from the Greek word meaning blood.

See also terms listed under **Mineralogy.**

HEMAOTOCRIT. Blood.

HEMATOGENOUS OSTEOMYELITIS. Bone.

HEMATOLOGY. That branch of medicine having to do with the study of the blood, the blood-forming tissues and the diseases of the blood.

HEMATOMA. An accumulation of free blood in the body tissues forming a localized mass. This usually follows an injury in which rupture of blood vessels takes place. See also **Brain (Injury);** and **Cerebrovascular Diseases.**

HEMATOPOIESIS. Blood.

HEMATURIA. The presence of blood in the urine. This condition is found in certain forms of nephritis and with injury, tumors, stones, or calculi in the urinary tract. It is also seen in scurvy and in some cases of severe sepsis.

HEME. Cytochromes.

HEMI- AND HOLOCELLULOSE. Pulp (Wood) Production and Processing.

HEMICHORDATA. A subphylum of the phylum *Chordata* containing only a few primitive marine animals without common names. The genus *Balanoglossus* has lent its name to the forms most commonly seen, although some belong to other genera. They are worm-like animals which live in mud and sand at the bottom of the ocean. The central nervous system is dorsal in this group but it remains partly or wholly at the surface. The notochord is limited to the anterior part of the body and is sometimes connected with the alimentary tract. Gill slits vary from one to many pairs. The group is also commonly named Enteropneusta and rarely Adelochorda.

There are two orders:

Order *Balanoglossida.* Worm-like animals with many gill slits and with a fleshy proboscis before the mouth. *Balanoglossus* and related forms.

Order *Pterobranchia* (*Cephalodisca*). Sessile animals, some solitary and some colonial. One pair of gill slits. A proboscis and branching tentacles lie before the mouth and the intestine is U-shaped. *Cephalodiscus* and *Rhabdopleura.*

HEMICOLLOID. A colloid composed of particles of small size, i.e., ranging from 0.005 to 0.0025 micrometer in length.

HEMIGALES. Viverrines.

HEMIHEDRITY. A term describing crystal symmetry operations, to indicate that only half of a symmetrical structure undergoes modification. For example, if, in truncating a cube, the process is carried out symmetrically on four out of the eight solid angles, the resulting structure exhibits hemihedral symmetry.

HEMIMETABOLA. A division of the insects characterized by incomplete metamorphosis. The immature insect differs conspicuously from the adult in form and is adapted to an entirely different mode of life; in this the group resembles the *Holometabola.* The young have compound eyes, however, and the wings develop externally as in the *Paurometabola.* The group includes the three orders, *Plecoptera, Ephemerida,* and *Odonata,* all with aquatic larvae which are called naiads.

HEMIMORPHITE. This mineral is zinc silicate, $Zn_4Si_2O_7(OH)_2 \cdot 2H_2O$, occurring in tabular and prismatic orthorhombic crystals, although often in massive and fibrous forms. There is a perfect cleavage parallel to the prism; it is brittle with a subconchoidal fracture; hardness, 4.5–5; specific gravity, 3.40–3.50; luster, vitreous; color, white, tending to translucent. Hemimorphite differs from willemite, also a zinc silicate, in that the former contains considerable water which may be driven off when heated to a high temperature.

There are many localities for hemimorphite in Europe, fine specimens having come from Saxony, Sardinia; Cumberland, Alston Moor and Derbyshire, England. It is found in Siberia, Slgeria, and Mexico. In the United States, hemimorphite has been found at Sterling Hill, New Jersey; in Lehigh County, Pennsylvania, and in Virginia, Missouri, Montana, Colorado, Utah, New Mexico, and Nevada.

The mineral is so named because of the tendency to form doubly terminated crystals showing a different grouping of faces at either end. The name is derived from the Greek meaning half and form.

HEMIPLEGIA. Loss of voluntary movement on one side of the body, commonly resulting from damage to the cerebral cortex on the opposite side of the body, or to the nervous pathways leading from it. Transient hemiplegias occur in epilepsy and hysteria but the majority are persistent and are due to hemorrhage or sometimes tumor compressing or destroying the cerebral cortex and associated tracts of nerve fibers.

HEMIPODE. Rails, Coots, and Cranes.

HEMIPTERA. The true bugs, an order of insects containing about 21,000 species, many of economic importance. They have piercing and sucking mouth parts and live on the blood or juices of animals or the sap of plants. The wings, when present, are usually distinctive. The basal half is thicker than the terminal, and the tips overlap partially so that the margins of the wings form an X on the back. Metamorphosis usually gradual. The chinch bug and bedbug are species of economic importance.

Bugs of several families are aquatic and some forms live on the surface of the water, supported by the surface film. The swimming forms are the water boatmen, back swimmers, and giant water bugs and the water striders skate on the surface. One of the last, *Halobates,* is the only marine insect known. Shore forms include the toad bugs. On dry land the order is represented in almost every possible habitat. The main families of *Hemiptera* include:

Belostomatidae	Giant water bugs
Cimicidae	Bed bugs
Coreidae	Squash bug
Corixidae	Water boatmen
Gerridae (also called *Hydrobatidae*)	Water striders
Lygaeidae	Chinch bugs
Miridae (also called *Capsidae*)	Leaf bugs
Nabidae	Damsel bugs
Nepidae	Water scorpions
Notonectidae	Back swimmers
Pentatomidae	Stink bugs
Phymatidae	Ambush bugs
Reduviidae	Assassin or kissing bugs
Tingidae	Lace bugs

HEMITROPIC. A term used by mineralogists for a crystal that appears to be composed of two halves of the same crystal turned partly around.

HEMLOCK TREES. Members of the family *Pinaceae* (pine family), these trees are of the genus *Tsuga.* The trees are sometimes referred to as hemlock spruces or hemlock firs. The hemlocks are evergreen trees, broadly conical. They are well known for their immunity to disease, with the exception of normal decay with age. The trees are quite tolerant of shade. The principal species include:

Black hemlock	*Tsuga martensiana*
Canadian hemlock	*T. canadensis*
Low weeper form	*T.c.* '*Pendula*'
Carolina hemlock	*T. caroliniana*
Eastern hemlock	See Canadian hemlock
Formosan hemlock	*T. formosana*
Himalayan hemlock	*T. dumosa*
Mountain hemlock	See Black hemlock
Northern Japanese hemlock	*T. diversifolia*
Southern Japanese hemlock	*T. sieboldii*
Western hemlock	*T. heterophylla*

The champion hemlock trees in the United States, as selected by The American Forestry Association are listed in the accompanying table.

The Canadian or Eastern hemlock normally attains a height between 50 and 80 feet (15 to 24 meters), but under favorable conditions can approach 100 feet. This species sometimes has several stems which form a spreading tree. The foliage may be described as feathery or plumelike. The bark is a dull brown. The tree is commonly found in swamps, ravines, rocky woods, and the mountain slopes of cold areas. It is found in some of the eastern mountains up to an altitude of about 2,000 feet (600 meters). The natural range of the tree is from Labrador, Newfoundland, and Nova Scotia westward to Michigan and Minnesota and southward to Delaware and Maryland and on the mountain slopes as far south as Georgia and Alabama. It is

RECORD HEMLOCKS IN THE UNITED STATES[1]

Specimen	Circumference (Inches)	Circumference (Centimeters)	Height (Feet)	Height (Meters)	Spread (Feet)	Spread (Meters)	Location
Carolina hemlock (1972) (*Tsuga caroliniana*)	139	353	88	26.4	54	16.2	North Carolina
Eastern hemlock (1972) (*Tsuga canadensis*)	239	607	100	30	70	21	Tennessee
Mountain hemlock (1972) (*Tsuga mertensiana*)	277	704	113	33.9	44	13.2	California
Western hemlock (1972) (*Tsuga heterophylla*)	328	833	164	49.2	62	18.6	

[1] From the "Social Register of Big Trees," The American Forestry Association (by permission).

[2] At 4.5 feet (1.4 meters).

found throughout most of New England and is particularly common in the central portions of Maine. Commercially the tree is often called hemlock spruce in the northern states; and spruce pine in the southern states. Timber from the tree is used, but is not considered a high-grade wood. It is of uneven texture, tending to splinter easily. Major uses are for pulp wood, and boxes and crating. In the green condition, Eastern hemlock wood has a moisture content of 111% and weighs 50 pounds per cubic foot (801 kilograms per cubic meter). When air-dried to 12% moisture content, the weight is 28 pounds per cubic foot (448.5 kilograms per cubic meter) or 1,000 board-feet (2.36 cubic meter) weigh 2,330 pounds (1057 kilograms). The compressive or crushing strength of the dry wood, parallel to the grain, is 5,410 pounds per square inch (37.3 MPa); the tensile strength perpendicular to the grain in the green wood is 230 pounds per square inch (1.6 MPa).

The tree makes an excellent hedge and is often used for this purpose in landscaping.

The Carolina hemlock is essentially exclusive to the Alleghany Mountains. It is a smaller tree, but has many of the characteristics of the Eastern hemlock. Normal height is about 50–60 feet (15 to 18 meters), but can grow higher under favorable conditions. The tree prefers dry, rocky mountain soil as found in Virginia, North and South Carolina, Tennessee, and Georgia.

The Western hemlock is considered the master tree of the genus and can attain a height well in excess of 150 feet (45 meters). The branches are slender and pendulous. The crown is narrow and pyramidal. The needles are dark green. This tree ranges widely from central California northward to Oregon, Washington, and British Columbia on into southern Alaska and eastward to the Rocky Mountains, mainly in Idaho and Montana. Along with the mountain pine, Douglas fir, white fir, and Engelmann's spruce, the Western hemlock makes up a significant portion of the forests of western United States and Canada. The tree is a major source of timber and commercially may be called West Coast hemlock, hemlock spruce, Prince Albert fir, gray fir, Western hemlock fir, or Alaskan pine. The wood has a slight pinkish tinge, is moderately soft, straight-grained, and nonresinous. Select grades are free of knots and suitable for preferred construction uses. However, although the wood is easy to work, it does not plane smoothly. Unfortunately, the wood has frequent dark streaks from heart rot, particularly common in the older trees. In the green condition, Western hemlock wood has a moisture content of 74% and weighs 41 pounds per cubic foot (657 kilograms per cubic meter). When air-dried to 12% moisture content, the weight is 29 pounds per cubic foot (465 kilograms per cubic meter); or 2,420 pounds (1098 kilograms) for 1,000 board-feet (2.36 cubic meter). The compression or crushing strength parallel to the gram is 2,990 pounds per square inch (20.6 MPa) for the green wood; 6,210 pounds per square inch (42.8 MPa) for the dried wood. The tensile strength perpendicular to the grain for the green wood is 310 pounds per square foot (1513 kilograms per square meter) and about the same for the dried wood. Commercially, hemlock timber is commonly mixed with Douglas fir. The bark of the Western hemlock contains 22% tannin. However, most hemlock-bark extract is obtained from the Eastern hemlock.

As will be noted by the names given in the prior list of hemlock species, the hemlocks also occur in Asia at similar latitudes. See also **Conifers.**

For references, see **Tree.**

HEMOCHROMATOSIS. Anemias; Liver.

HEMOCOELE. A cavity in which blood or hemolymph circulates. Well developed in the arthropods and mollusks where it superficially resembles a true body cavity. It is associated with some tubular blood vessels whose contractions propel the blood in it, and these movements are supplemented by the shifting of its contents as a result of body movements.

HEMOCYANIN. An oxygen-absorbing substance in the plasma of the blood of the crayfish and many other arthropods. It is a clear material in the blood, but turns blue when removed and allowed to stand for a time. It serves a purpose similar to hemoglobin such as is found in higher forms of animal life.

HEMODIALYSIS. Kidney and Urinary Tract.

HEMOGLOBIN. The main function of the hemoglobin molecule is oxygen transport. The hemoglobin molecules from each species of organism which has been examined differ in the sequence of amino acids in their polypeptide chains unless they are very closely related. Chimpanzee and human hemoglobins are apparently identical. Sometimes two or more different kinds of hemoglobin are found simultaneously in the same organism. These structural variations may give rise to differences in the physiological properties which help to determine the efficiency of oxygen transport by the blood from lungs or gills to the tissues. Hemoglobin also plays an important role in carbon dioxide transport. See also **Blood.**

Vertebrate hemoglobins are usually composed of four polypeptide chains of two types, called α and β. The molecules can, therefore, be described as $\alpha_2\beta_2$. An iron porphyrin moiety, *heme*, is associated with each chain. Evidence indicates that combination of the heme with oxygen results in structural changes in the protein to which it is bound. Studies of single crystals of horse and human hemoglobins by x-ray diffraction show that removal of oxygen from the iron atoms of the four hemes results in a separation of the β-chains from one another; the relative positions of the α-chains do not appear to change. Although the molecular basis is not fully understood, the consequences are important. It is certain that any change in the mutual relationships of the polypeptide chains will alter the environment of many amino acid residues. These environmental changes are probably responsible for the degree of oxygenation to the oxygen pressure; and the dependence of the oxygenation upon pH and upon carbon dioxide concentration.

Mutations which alter the amino acid sequence can occur in either the α- or the β-chain of the adult. However, most mutations are deleterious and changes in the α-chain would be more severely selected against in a process of natural selection because any change in the α-chain would affect the sensitive fetus, whereas changes in the β-chain would affect only the adult. This means that evolution tends to favor changes in the β-chain over changes in the α-chain. These considerations indicate that molecular adaptation of hemoglobin, at least in mammals, may involve changes more in the β-chain than in the α-chain.

Hemoglobins can be dissociated into their α- and β-subunits. Not only are hemoglobins capable of dissociating into their polypeptide subunits, but certain hemoglobins are also capable of polymerization. Many reptiles and amphibians and certain mice possess hemoglobins which polymerize to form double molecules $(\alpha_2\beta_2)_2$ and sometimes triple or quadruple molecules. Many hemoglobins from invertebrate animals have very large molecular weights and are composed of a large number of subunits—as many as 180 in some species. The nature of the forces holding these large aggregates together is under study.

The amino acid sequences of hemoglobins have been extensively altered by mutation during evolution. Data on the amino acid sequences of the chains from a variety of mammalian and other vertebrate hemoglobins show that the sequence can be varied extensively without drastic change in function. There appears to exist a hierarchy in the functional importance of different parts of a protein. Substitutions in different segments of a polypeptide chain may, according to the type and position of the substitution, exhibit a spectrum of effects, ranging from detectable to catastrophic. For example, the single substitution of valine for glutamic acid in the 6th position of the β-chain in human sickle cell hemoglobin results in a large decrease in the solubility of deoxygenated hemoglobin within the red cells. The hemoglobin, by forming a gel, distorts the red cell shape ("sickle") in such a way that flow through the capillaries is retarded. Such drastic consequences do not result if the substitution is lysine rather than glutamic acid (hemoglobin C). Histidine in position 63 of the human β-chain has an essential role stabilizing the ferrous state of the heme iron. Substitution by tyrosine (in hemoglobins "M") results in the loss of this stability because the ferric iron can form a strong linkage with the —OH group of tyrosine. Such a substitution results in a complete loss of capacity to combine reversibly with oxygen.

The foregoing are radical substitutions. Most effective substitutions appear to be relatively conservative and do not drastically affect the oxygen transport function. Therefore, the number of differences between homologous chains appears to be related not to functional differences, but to the time which has elapsed since the chains diverged from a hypothetical polypeptide ancestor. The mean number of differences between the hemoglobin chains of man, horse, pig, rabbit, and cattle is approximately 11. The common ancestor of these mammals may have existed some 80 million years ago. Thus, approximately 11 effective mutations per chain occurred in 80 million years, or 1 substitution per chain in 7 million years. Zuckerkandl and Pauling, using standard probability theory, have used this figure to estimate the time at which the different human hemoglobin chains (α, β, γ, and δ) are believed to have arisen by gene duplication. These estimates are shown in the accompanying table.

DIVERGENCE OF HEMOGLOBIN CHAINS WITH TIME

Type of Chain Divergence	Number of Differences	Estimated Time since Divergence
β-δ	10	35 million years
β-γ	37	150 million years
β-α	76	380 million years
(α-β)-myoglobin	~ 135	650 million years

Estimates like these indicate that hemoglobins are very old and that it may be possible to find relatives of vertebrate hemoglobins in invertebrate animals. They also suggest that the gene duplication believed to be responsible for the divergence of the α- and β-chains took place in the Devonian period at the time of the appearance of early amphibians and the dominance of fish.

The suggested relationship between numbers of differences and evolutionary time is not wholly secure. It assumes uniformity in the rate of effective amino acid substitution, but this rate may be neither uniform with time, nor uniform in different parts of the polypeptide chain. Differences in the rate of effective substitution along the polypeptide chain may be due not only to restrictions imposed by the required tertiary structure, but also to differences in the rate at which various parts of the DNA or the gene mutate. The evolution of hemoglobin may be contrasted with that of cytochrome *c* in which approximately 50% of the molecule appears to have remained invariant during the time yeast and man have evolved.

See the list of references at the end of the entry on **Blood.**

HEMOGLOBINOMETER. Photometers.

HEMOGLOBINURIA. The presence of hemoglobin in the urine. This occurs when red cells of the blood are destroyed at such a rate that the hemoglobin set free cannot be disposed of by the normal processes, but appears unchanged in the urine. Myohemoglobin, the pigment of muscle cells, may similarly appear in urine, especially after extensive crush injury, from mis-matched blood transfusion, from allergy to the bean, *vicia faba* and in certain rare conditions, e.g., after exposure to cold in certain persons whose blood contains a hemolytic agent active only when the blood is cooled (Donath-Landsteiner reaction), in certain otherwise normal persons after exercise (march hemoglobinuria) and in a rare type of hemolytic anemia (Marchiafava-Micheli anemia) in which the hemoglobinuria occurs only in sleep. See also **Kidney and Urinary Tract.**

HEMOLYMPH. The blood of higher inveertebrates, consisting of a clear plasma and white cells but without red cells. Respiratory pigments are dissolved in the plasma. It contains a lower percentage of water than the blood of more primitive forms.

HEMOLYSIS. Blood.

HEMOLYTIC ANEMIAS. Anemias.

HEMOPHILIA. A hereditary blood condition in which the blood fails to coagulate; an abnormal tendency to bleed. Transmitted by females, but occurs in severe form only in males. This familial blood disease has been recognized for hundreds of years. Because of the frequency of the disease in many of the royal families of Europe, particularly those of Spain and Russia, the incidence of hemophilia has changed world history. Hemophilia in women is practically unknown. A woman may carry the genetic factor producing hemophilia without having any of the symptoms; she may display the symptoms if each of her parents carried this factor for the disease. It behaves as a sex-linked Mendelian recessive. See **Heredity.** A female capable of transmitting the disease does so to about two-thirds of her male children, while two-thirds of her female offspring are conductors of the disease. See also **Sex-Linked Inheritance.**

Hemophilia is almost always apparent in the first year of life, and is generally recognized without difficulty because of its prior occurrence in the family. On rare occasions, it occurs in families which have no history of the condition. Therefore, unusually severe bleeding from a seemingly minor injury should be checked out by a physician. The hereditary nature of hemophilia should serve as a warning to members of the families in which the disease has occurred.

Hemophiliacs do not usually die from the first severe bleeding because of their reserve stores of blood cells. Subsequent hemorrhage may prove fatal. Patients who receive no medical treatment in cases of bleeding seldom live beyond their 20th year, while those who obtain proper care have an excellent chance for a long life.

The immediate treatment of patients with hemophilia is frequently self-administered. The victim should administer clot-stimulating materials, as previously prescribed by a physician, applying them directly to any cut or scratch. The usual methods for stopping blood flow have little or no effect.

If bleeding cannot be stopped by applying such substances, then an injection of antihemophilic factor VIII, the special clot-forming protein that is missing from the blood of hemophiliacs, is required. Potent doses of this protein can be prepared by freezing, thawing, and then centrifuging fresh plasma. This procedure is to be contrasted with massive plasma transfusions which may have to be repeated and carry the risk of hepatitis. VIII concentrate can be administered quickly by syringe and is especially valuable for the hemophiliac who may need an emergency operation.

Fortunately, for the hemophilic patient, there may be remissions in the disease, during which time nearly normal clotting activity will occur for weeks, or even years. A life of moderate activity with some precautions, prompt attention to bleeding, and VIII injections when necessary are the measures that will increase life span.

HEMOPHILIA (Inheritance Link). Sex-Linked Inheritance.

HEMOPTYSIS. The bringing up of blood from the larynx, trachea, bronchi or lungs. The commonest cause is pulmonary tuberculosis; carcinoma of the bronchus is a frequent cause; it may also occur in any chronic bronchial or pulmonary disease and certain varieties of heart disease, especially mitral stenosis, in which the pulmonary blood pressure is persistently raised; and in aneurysm of the aorta. Other diseases which may predispose hemoptysis include polyarteritis nodosa (a subacute or chronic, remittent, disseminated vascular disease characterized by focal necrotizing inflammation of the walls of medium- and small-sized arteries and arterioles); Weil's disease (a severe form of leptospirosis); and wool sorter's disease. See also **Leptospirosis.**

HEMORRHAGE. Bleeding. Escape of blood from the vessels. Anemias caused by sudden blood loss as in traumatic injury are generally normocytic, that is, the cells are of normal size, but reduced in number. When the blood is lost over a longer period of time, as from bleeding hemorrhoids, peptic ulcer, in hookworm disease, and in excessive menstrual bleeding (menorrhagia), a microcytic anemia may result. Following hemorrhage, body fluids seep into the blood which restore it to its former volume; consequently, dilution of the blood occurs, and anemia may result. It may require some time for the body to manufacture the necessary red cells and other substances necessary to return the blood to normal. The symptoms of such a blood-loss anemia include a general weakness, dizziness, and faintness. In more severe cases, there may be vomiting and a great thirst, the heart rate may be rapid, and the breathing weak and shallow.

The first step in the treatment of persons with a posthermorrhagic anemia is to stop the loss of blood. Blood transfusions may be given to return the blood to its proper volume before excessive dilution occurs. In milder hemorrhages, however, the body may be able to restore the lost blood without transfusion. This is often accomplished by ample rest and a good diet, including adequate amounts of iron and protein necessary for red cell building.

Hemophilia, a rather rare hemorrhagic disease, is described under **Hemophilia.** Other conditions exist in which unusually large amounts of blood may be lost. In many such cases, bleeding may take place into the skin, as in a bruise. This symptom is referred to as *purpura.*

Essential thrombocytopenic purpura is a disease characterized by hemorrhage, and caused by a deficiency in the number of blood platelets. The spleen may be responsible for this disease by destroying the blood platelets. Corticosteroid therapy helps control the bleeding, and in most patients, is regarded as a desirable precaution prior to removal of the spleen.

Purpura and excessive bleeding may occur in persons suffering from deficiency of vitamins C and K. Some newborn infants contract a hemorrhagic disease which once was frequently fatal; the victims now recover rapidly when treated with vitamin K. Purpura may occur in persons receiving antitoxin treatments, or as a symptom of snakebite poisoning, or with some types of food poisoning. The taking of certain drugs may bring about abnormal bleeding. Purpura is occasionally a symptom of such varied conditions as meningitis, scarlet fever, severe measles, chronic kidney disease, endocrine disorders, liver disease, macrocytic anemias, allergies, typhus fever, and a specific bacterial heart disease. The symptom disappears in each case when the primary cause is removed.

See also **Arterial and Venous Disorders; Cerebrovascular Diseases;** and **Ischemic Heart Disease.**

HEMORRHOIDS. Arterial and Venous Disorders.

HEMOSIDERINURIA. Anemias.

HEMOTOXIN. Snake.

HEMP. The fibers of the hemp plant, *Cannabis sativa* of the family *Cannabinaceae* (hemp family) are coarse and rather harsh and much less pliable than flax fibers. They are dark colored and not easily bleached without injury. The fibers are used mainly for making rope and coarse twine, warp of carpet, belt and upholstery webbing, and wherever strength and durability without appearance are of importance. Short fibers of hemp, called tow, are used in packing joints in pipes, for pump packing, and for stuffing upholstery. The woody waste from hemp fiber is sometimes used in the manufacture of certain papers. See also **Rope.**

Hemp is obtained from the stem pericycle of a tall hollow-stemmed annual which is a native of central and western Asia. In cultivation, the slight branching, which characterizes the plant, is considerably reduced by planting thickly. The plants grow from 5 to 16 feet high. They have digitately compound dark green leaves and small inconspicuous flowers, which are of two kinds, occurring on different plants. The staminate flowers appear in small axillary clusters on male plants, and the pistillate flowers are borne in leafy spikes on female plants. The fruit, an achene, is a hard ovoid structure, often called hemp seed. Hemp grows best in regions having a warm humid growing season of about 5 months. The plants grow rapidly soon shading the ground so effectively as to suppress other plants, and thus plantings of hemp have been used as a means to eradicate weeds. When the staminate flowers are mature, the plants are ready for harvest. To delay after that is not desirable, since the male plants die soon after flowering. After flowering, the fibers become coarser. Harvesting and the treatment of the plants after harvesting are similar to the procedures used with flax plants. See **Flax.** The hemp plants are cut off or pulled up, denuded of leaves, roots and tops, and tied in bunches and left to dry for about 2 weeks. They are then immersed in water to ret. In retting, the intercellular substance of the stems is acted upon by bacteria and softened so that the fibers are readily cleaned of surrounding tissues. Scutching removes the woody tissue, after which the rough hemp fibers are hackled, or drawn over coarse combs which pull out the fibers.

In recent years, the cultivation of hemp has been subject to controls because marijuana is prepared from the dried leaves and flowers of the plant, which then are smoked in the form of cigarettes as a narcotic. The species *Cannabis indica* is usually used in this connection. See **Marijuana.**

HENNA SHRUB. Of the family *Lythraceae,* the *Lawsonia inermis* is a small shrub native to Africa and Asia and cultivated in tropical countries. Well known since the time of the early Egyptians as a red dye for hair, nails, hoofs of animals, etc., the leaves of this plant are powdered and made into a paste which is then used as a dyeing medium. The small flowers of the plant are inconspicuous, but fragrant.

HENRY. Units and Standards.

HEPARIN. A complex organic acid (mucopolysaccharide) present in mammalian tissue; a strong inhibitor of blood coagulation. Precise chemical formula has not been fully established, but the formula $(C_{12}H_{16}NS_2Na_3)_{20}$, with a molecular weight of 12,000, has been suggested for sodium heparinate. The drug is derived from animal livers or lungs. Heparin is used in deep venous thrombosis therapy. It is also used in rodenticides which cause internal hemorrhaging. Pets exposed to such poisons must receive immediate treatment with the administration of vitamin K, also sometimes called the antihemorrhagic vitamin. See **Anticoagulants; Vitamin K.**

HEPATIC SYSTEM. Bile; Gallbladder and Biliary Tract Diseases; Jaundice; Liver; Wilson's Disease.

HEPATITIC VIRUSES. Virus.

HEPATITIS. Liver.

HERBICIDE. A substance that kills or interferes markedly in the life cycle of certain plants and is used with other control chemicals, such as fungicides and insecticides, to increase the yield and quality of crops. In addition to eliminating or greatly stunting the growth of those plants (weeds) that compete with crops for water and soil nutrients, herbicides achieve a number of objectives. Usually lumped under the phrase *weed control,* the advantages of herbicides include: (1) Eliminate weeds that serve as harboring places for insects which attack crop plants. See accompanying illustration. (2) Eliminate perennial plants that may serve as hosts for survival and build-up of virus diseases. An example is the corn stunt virus, which overwinters in johnson-grass rhizomes. Insects able to carry virus diseases may feed on weeds and move to crop plants, causing infection by damaging virus diseases. (3) Eliminate weeds that serve as traps for moisture. Easy availability of moisture encourages fungus diseases that can be spread easily from weeds to crop plants by wind movement of fungus spores. (4) Eliminate honeysuckle, kudzu, and other plants that grow on fences and that are severely damaging to fencing and other minor structures because of the sheer weight of their foliage. Damaged fences adversely affect livestock production. (5) In so-called "no-till" planting, herbicides are used exclusively, eliminating the mechanical removal of weeds by cultivating equipment. The advantages of herbicides in this regard are time and labor savings.

Classification of Herbicides

There are several ways in which herbicides can be grouped.

Target plant selectivity is a measure of the effectiveness of a herbicide against a range of plants to be destroyed. *Nonselective* herbicides are not difficult to create. There are hundreds of chemicals that will kill just about any living plant within range. These are extremely *wide-spectrum* substances, not only destroying or stunting both broadleaf and grasslike weeds, but woody plants as well. Of course, some control over these very powerful chemical substances can be exerted by regulating concentration. Dilute applications may result in desired defoliating, for example, without fully destroying a stand of plants, such as trees. *Broad-spectrum* and nonselective herbicides are sometimes regarded as the same, but more generally, broad-spectrum refers to a compound that does not differentiate between broad- and narrowleaf plants. A *selective* or *narrow-spectrum* herbicide is customized to make this selection and, in fact, to differentiate even more closely. In operating with crop-rotation programs, it is usually advantageous to select herbicides with relatively narrow spectrums of effectiveness so that later crops may not be adversely affected by any residues from a prior crop. Herbicide manufacturers continue in their research toward the devel-

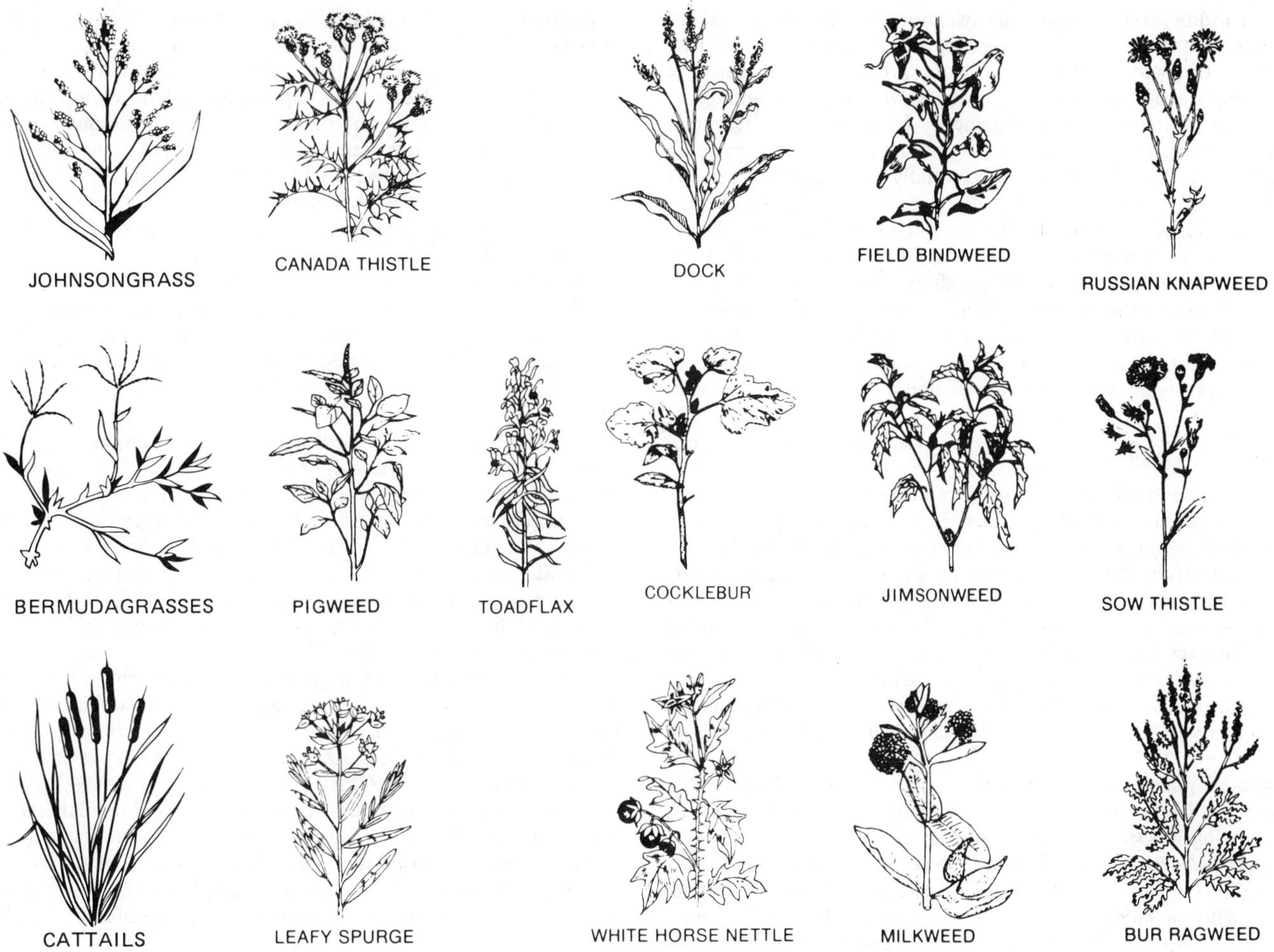

Examples of important target grasses and weeds that can be controlled by herbicides. (*The Dow Chemical Company.*)

opment of crop-specific control chemicals. Just one example—a herbicide to control wild oats in connection with wheat production.

Timing. Herbicides are usually designated for a *pre-plant* or *pre-emergence* use or for *post-emergence* application. The terms tend to be self-descriptive—pre-emergence signifying the use of the herbicide on the land prior to the cracking stage or emergence of weeds or desired crop above the soil line. The herbicide, possibly in granule or liquid form, may be incorporated into the soil a number of weeks before planting, in which case, the term pre-plant is used. For effective control over the growing season, some land areas or crops may have to be treated a number of times between emergence and harvest. Thus, the term post-emergence. Several days must elapse between application and harvesting to avoid contamination of the crop when gathered.

Stability. A number of factors determine the stability of a herbicide. For example, some of the control chemicals decompose (and thus become ineffective) when applied at temperatures in excess of about 90°F (32.3°C), or during long periods of intense sunlight. Most herbicides are more efficient when applied on cool, partly cloudy days. In the case of many other herbicides, the presence of moisture (after an irrigation, rain, or during a generally wet period) greatly reduces or destroys their effectiveness. The presence of certain chemicals also affects application success. Some control chemicals are adversely affected by any mixture with acidic materials; others by the presence of alkaline materials, or certain metals, such as iron or copper. Essentially, these materials interact with the original chemical composition and alter it so that, as a result, instead of applying an effective herbicide, for this purpose the material may be essentially inert. Careful preparation, particularly of emulsions for spraying, cannot be overemphasized. If the stability and effectiveness of a herbicide is long term, it can be designated as a *persistent herbicide,* meaning effectiveness over a period of several months. Those substances that break down within several days to a few weeks (biodegradeable in a sense) are *nonpersistent herbicides.* This is an important factor in selecting a herbicide. Some control chemicals can essentially sterilize a plot of land for a period of years, and should one's objectives change after such an application, neutralization or removal of the substance from the affected area can be costly and quite difficult.

All control chemicals can be categorized as either *contact-type* or *systemic* substances. Although these designations are more commonly applied to insecticides, they also are operable in terms of herbicides. In a contact-type substance, the killing action is largely limited to the area of actual contact between chemical and plant (for example, a defoliant that damages a bush or plant without completely destroying the whole plant). In a systemic substance, contact of the substance with part of the plant is progressively spread throughout the plant as, for example, by the plant's vascular system.

Physical Form. A large number of herbicides are available in several forms, including granules, powders and dusts, wettable powders, emulsifiable concentrates, slurries, etc. Some of these are factory-prepared; others can be prepared locally by the user. Sprays and dusts are widely used for foliar applications, whereas dusts and granules may be preferred for soil applications.

Chemical Structure. As with fungicides, insecticides, and other pesticide chemicals, herbicides are usually complex, often synthetic organic chemicals and of a widely varying composition, ranging from carbamates, to anilides, to organic acids, salts, etc. Chemical make-up is discussed further a bit later in this entry.

Nomenclature of Herbicides. There are well over 100,000 pesticides and agricultural control chemical formulations; perhaps 25% of these fall into the sphere of herbicides. As with insecticides, although the basic chemicals used in the formulation of herbicides may number in the several hundreds, many thousands of possible formulations arise from the various physical formats offered, as well as minor differences provided by many manufacturers in brand name products. Each manufacturer markets products under tradenames—names that are essentially coined for their marketing charisma and infrequently connoting much about the content or purpose of the product. Thus, there are scores of equivalent (or essentially equivalent) products, adding to the difficulty of selecting these chemicals. Unfortunately, from this standpoint, the generic chemical names of the majority of herbicide chemicals are long and complex and essentially meaningless to persons who are not well versed in organic and biochemistry. Helpful listings of this type can be found in the "Foods and Food Production Encyclopedia" (D. M. Considine, editor), Van Nostrand Reinhold, New York, 1982. There are also a number of frequently revised directories of control chemicals and considerable information available from various government agencies and universities. See list of references at end of this entry. This situation of nomenclature is quite similar to that which applies to generic and tradename drugs and pharmaceuticals.

Chemistry of Herbicides

Aromatic Carboxylic Acids. Considerable research has gone into investigations of the physiological activity of these acids on plants, including benzoic, phenylacetic, and naphthoic acids. Among the benzoic acid derivatives, the greatest activity is shown by those compounds containing substituents in the 2, 3, and 6 positions; and only to a slightly lesser degree, by those substituents in other positions. Included among commercial herbicides in this category are: 2,3,6-trichlorobenzoic acid; 2-methoxy-3,6-dichlorobenzoic acid; 2,5-dichloro-3-nitrobenzoic acid; and 2,5-dichloro-3-aminobenzoic acids. Slightly less active are: 2-bromo-3,5-dichlorobenzoic acid; and 2,3,5-triiodobenzoic acid.

Substituted phenylacetic acids have a high activity. Considerable activity is shown by monohalogen-substituted acids. Introduction of a second halogen does not markedly affect the degree of activity.

Aryloxyalkylcarboxylic Acids and Derivatives. Research has shown that the physiological activity of phenoxyacetic acid toward plants increases when a halogen atom is incorporated into the molecule. The strongest effects are displayed by fluorine and chlorine. The position of the substituent also affects physiological power, with the 4-halophenoxyacetic acids displaying the greatest activity. It is interesting to note that the activity of this compound is about ten times greater than the case of the 2-isomer. And, the activity is further reduced in the 3-chlorophenoxyacetic acid. There are numerous herbicides in this chemical structural category, including 2,4-D, 3,4-D, and MCPA (4-chloro-2-methylphenoxyacetic acid). It is also interesting to note that while MCPA is very effective as an agricultural control chemical, the very closely related compound, 4-chloro-2-chloromethylphenoxyacetic acid is not of great value.

Derivatives of Carbamic Acid. Whereas the aryl esters of *N*-methylcarbamic acid find wide application as insecticides, the alkyl esters are strong herbicides against monocotyledonous weeds. Their actions against dicotyledonous plants is much weaker. Because of these differences, these herbicides are effective controlling monocotyledonous weeds in such crops as carrot, cotton, and sugar beet. Research has shown that the esters of naphthylcarbamic, diphenylcarbamic, and other polycyclocarbamic acids are not effective herbicides. It has been found that the arylcarbamic acid ester derivatives of unsaturated alcohols are stronger herbicides than the corresponding esters of saturated alcohols. The carbamates have an ability to form hydrogen bonds with the chlorophyll molecule or proteins of plants, accounting for their effective herbicidal activity.

Derivatives of Thio- and Dithiocarbamic Acids. Research has indicated that the derivatives of the thiocarbamic acids are good penetrants of plants, moving easily through the xylem. Among this structural class of herbicides, the *S*-alkyl-*N*-dialkylthiocarbamates are the most effective. Most of these compounds are selective herbicides against annual grasses and a few dicotyledons. They have been applied successfully in connection with such crops as bean, beet, other vegetables, and sugarcane. The thiocarbamates are usually mixed with the soil as pre-emergence herbicides. In terms of effectiveness, this usually decreases as the number of carbon atoms in the ester radical increases, particularly in excess of five carbon atoms. The activity also decreases when the total carbon atoms in the alkyl radicals on the nitrogen atom is greater than six.

Derivatives of Urea and Thiourea. A great deal of investigation has gone into the effectiveness of these compounds and, as the result, a number of urea derivatives have found use as effective herbicides as well as growth regulators. This is particularly true among the trialkylureas that contain simple and complex hydrocarbon radicals. Examples include: 3-(3,4-dichlorophenyl)-1,1-dimethylurea (*Diuron*); 3-(3,4-dichlorophenyl)-1-methoxy-1-methylurea (*Linuron*); 1,3-dimethyl-3-3(2 benzothiazoyl) urea (*Methabenzthiazuron*); 3-(4-bromophenyl)-1-methoxy-1-methylurea (*Metobromuron*); and *N*-benzyl-*N*-(dichloro-3,4-phenyl)-*N,N*-dimethylurea (*Phenobenzuron*).

The salts of aryldialkylureas tend to be more active than the ureas.

Thiocyanates and Isothiocyanates. At one time, ammonium thiocyanate was a commonly applied, nonselective, contact-type herbicide and desiccant. The compound is less important now because of the development of other organic herbicides that do not decompose so readily.

Sulfuric and Sulfurous Acid Derivatives. Earlier, sulfuric acid was applied as a herbicide and still finds some use as a desiccant for potato plant tops prior to mechanical harvesting. The primary drawback of sulfuric acid, not experienced with more recently developed herbicides, is the large amount of acidity which it adds to the soil. Ammonium sulfamate continues to find use as an effective herbicide in some areas, both for the elimination of weeds and as a sterilant for soil. This compound hydrolyzes in soil to form ammonium sulfate, a source of ammonia and nitrogen.

Several other classes of organic chemicals, including heterocyclic compounds, are represented among the scores of herbicides commercially available. As of the late 1970s, in terms of tonnage usage, the Food and Agriculture Organization (United Nations) listed the following major categories: MCPA, 2,4-D, 2,4,5-T, triazines, carbamates, and urea derivatives. It should be stressed here that regulations pertaining to the use of herbicides vary widely from one country to another—and from one time period to another. The proliferation of new products tends to offset those prior compounds that have been banned in some countries, or where usage has been severely curtailed.

References

Andrilenas, P. A.: "Farmers' Use of Pesticides," U.S. Department of Agriculture, Washington, D.C. (Revised periodically).

Crafts, A. S.: "Modern Weed Control," Univ. California Press, Berkeley, California, 1977.

Considine, D. M. (editor): "Foods and Food Production Encyclopedia," Van Nostrand Reinhold, New York, 1981.

Melnikov, N. N. (F. A. Gunther and J. D. Gunther, editors): "Chemistry of Pesticides," Springer-Verlag, New York, 1971.

Staff: "2,4-D Increases Insect and Pathogen Pests on Corn," *Science*, **192**, 239–240 (1976).

Thomson, W. T.: "Agricultural Chemicals—Herbicides," Thomson Publications, Fresno, California, 1980.

HERBICIDES (Pyridine Derivatives). Pyridine and Derivatives.

HERBS. ***Compositae* or Composite Family.**

HERCULES. A large constellation lying between Lyra and Corona Borealis. Hercules contains no strikingly bright stars, and hence is somewhat difficult to locate. Once found, however, it is a fertile field for a small telescope. In 1934, this constellation received considerable notice because of the brilliant nova that appeared in it just before Christmas. Perhaps the most interesting object within it is a remarkable star cluster, which was first noted by Halley, in 1714. Although this cluster can be distinguished as such in a telescope of only 2-inch aperture, it requires a telescope larger than 6 inches to appreciate the magnificence of the object. (See map accompanying entry on **Constellations.**)

HERCULES X-1. Neutron Stars.

HEREDITARY HEMORRHAGIC TOLANGIECTASIA. Purpura.

HEREDITARY MATERIAL. Cell (Biology).

HEREDITARY MECHANICS. The field of mechanics involving boundary conditions extending over continuous intervals of space and time and demanding integrals for their representation. For example, in the application of stress to a deformable elastic medium, the final strain at any instant depends not only on the stress at that instant but on the whole previous stress to which the medium has been exposed. Analytically,

$$\delta(t) = kX(t) + \int_{t_0}^{t} \theta(t, \tau)X(\tau)\, d\tau$$

where δ is the final strain at time t, $X(t)$ is the instantaneous stress at time t and the integral represents the effect of the stress heredity of the system. The quantity $\theta(t, \tau)$ is called the coefficient of heredity. The above equation may be considered an integral equation for the evaluation of X when δ is known.

HEREDITY. The transmission of developmental potentialities from one generation of living things to the next and following generations through the natural process of reproduction. The materials of the parent bodies from which a new individual develops are its actual heritage. During its own embryonic development, the potentialities of this heritage are expressed in the structural characteristics of the new body, normally like those of the parents or those of a more remote generation of ancestors. This fact leads to the statement that the organism inherits certain characters; although this may not be precisely true, the interpretation is permissible for ordinary purposes of description.

Genetics is that branch of biology which deals with the phenomena of heredity and the variations between parents and offspring. See also **Genes; Genetics.**

Work of Mendel. The first steps in genetics were taken by plant hybridizers of the eighteenth and nineteenth centuries, chiefly in Europe, and culminated in the experiments of Gregor Johann Mendel, a monk at Brŋo, Czechoslovakia, then Brünn in Austria. Mendel's results were published in 1866 and lay almost unnoticed until 1900, when they were corroborated by three scientists in the birth of modern genetics. The published report of Mendel's work repeated the significant observations of his predecessors and added a simple mathematical analysis that had not been previously expressed. As a result of the importance of this work, the term Mendelian heredeity is applied to the established fundamentals with which subsequent discoveries have been correlated.

Mendelian heredity depends on three fundamental concepts: (1) The organism is a mosaic of unit characters capable of separate hereditary transmission. (2) A unit character may mask a related unit character completely when the potentialities for the development of both are present in the same individual. This principle is called dominance, and the masked character is said to be recessive. (3) Unit characters may be segregated during reproduction, regardless of the combinations in which they have been associated.

To these concepts, later scientists in the field added that the association of different related unit characters in one individual may result in the development of both in different parts of the body, in a mosaic inheritance, or in an intermediate condition.

Some characters, particularly of a quantitative nature, are due to multiple genes. Such characters must be studied by statistical methods. They were the foundation of another attempt to formulate laws of inheritance made by Sir Francis Galton, from which we retain the law of ancestral inheritance and the law of filial regression. The former indicates that each parent contributes one-quarter of the total heritage of the individual, each grand-parent one-sixteenth, and so on in a rapidly diminishing percentage. The law indicates the great reduction of the possibility of a hereditary character reappearing after a lapse of generations. Filial regression is the tendency of extreme parents to produce offspring less extreme than themselves. Thus tall parents beget tall children, but usually shorter than themselves. Galton studied human inheritance and in addition to his mathematical analyses, so necessary in this field, took the initial steps in proposing deliberate control, which led to the science of eugenics.

Modern science has also added to early discoveries the definite recognition that hereditary potentialities are resident in the chromosomes of body cells and that definitely located genes within these chromosomes are the determiners through which specific unit characters are brought to expression. The behavior of chromosomes is strictly in harmony with the transmission of characters by Mendelian heredity. Since nothing was known of chromosomes during Mendel's life, this correlation had to await further advances in cytology.

Mendel's chief contributions were derived from the study of garden peas, in which he observed seven pairs of unit characters, all similar in behavior. He noted, for example, that seed colors included two unit characters, yellow and green. When he crossed parent plants of the two strains the resulting hybrid seeds were entirely yellow, indicating the dominance of this color over green. He then inbred the hybrids, and in their offspring both yellow and green seeds appeared in the ratio of three yellow to one green. Related unit characters of this kind are said to be alleles or allelomorphs. It is now known that their genes occupy the same position in the paired chromosomes of the cells, while only one can be represented in the single chromosome of a germ cell. Since each parent contributes one chromosome to each pair in its offspring, it may also contribute one gene of an allelic pair. The one parent plant contributed a gene for yellow, the other for green, and through dominance the offspring were yellow. Segregation, however, enabled these hybrids to transmit either yellow or green during their reproduction, and through random fertilization all possible combinations of these determiners were established. The characters are commonly represented by symbols, using a capital letter for the dominant and a small letter for the related recessive, as Y and *y* for yellow and green, respectively. For the pair of characters mentioned, the following diagram is representative:

Parental generation (P):	YY	yy
Germ cells:	Y	y
Hybrids of first filial generation (F_1):	Yy	
Gametes of F_1 generation	Y	y

and their combinations in the F_2 generation, in a Punnett square:

	Y	y
Y	YY	Yy
y	Yy	yy

The YY and *yy* individuals in this diagram are homozygous, and the Y*y* individuals are heterozygous. Since all YY and Y*y* individuals look alike, due to the dominance of Y, they belong to the same phenotype, but since their hereditary potentialities are different they belong to different genotypes. The *yy* individuals from hybrid parents are known as extracted recessives. There are twice as many heterozygotes as homozygotes of either kind in this 3 : 1 ratio because similar individuals in this category result from reciprocal combinations of genes, half of the individuals receiving the dominant from one parent and half from the other. Examples of this kind, involving only one pair of allelic characters, are known as monohybrids.

Additional complexity arises in dihybrids, trihybrids, and polyhybrids of still more characters through the free reassortment of the unrelated pairs of alleles. Thus peas from smooth yellow seeds crossed with others from wrinkled green seeds, a dihybrid combination, produce only yellow smooth seeds in the F_1 generation, but when inbred these plants give rise in the F_2 generation to the four possible combinations: smooth yellow, smooth green, wrinkled yellow, and wrinkled green, in the ratio 9:3:3:1. The reason is evident in the following diagram:

	SY	Sy	sY	sy
SY	SY SY	Sy SY	sY SY	sy SY
Sy	SY Sy	Sy Sy	sY Sy	sy Sy
sY	SY sY	Sy sY	sY sY	sy sY
sy	SY sy	Sy sy	sY sy	sy sy

In this diagram, each pair of symbols above and at the left side represents the contribution of one parent in one of its germ cells, and in the small squares the possible combinations from the two parents are shown. Dominance prevails as in the monohybrid.

In a trihybrid, free reassortment results in an F_2 ratio of 27:9:9:9:3:3:3:1. The number of phenotypes is always a power of two indicated by the number of pairs of alleles under consideration. See also **Cell (Biology).**

Studies of Fruit Fly. The study of heredity in animals has shown that these principles are applicable in that kingdom as well as in plants, but relatively few animals are sufficiently prolific to demonstrate complex ratios. The fruit fly, *Drosophila melanogaster,* has been the most productive of all genetic subjects, whereas man and the domestic animals yield very limited Mendelian data.

Modern genetics, largely from studies of the fruit fly, has disclosed many principles as corollaries of simple Mendelian heredity. The more important are as follows:

Multiple alleles: More than two unit characters may be related to each other as alleles. In such cases only two of the series may be present in any one individual, and dominance is in a graded series, as may be determined by experimental results.

Multiple genes: More than one gene may be necessary for the production of a single unit character. If two genes are essential for its appearance and either alone is incapable of expression, they are said to be complementary. If one expresses itself alone, a gene that modifies this expression is supplementary. If two are capable of producing the same effect whether present singly or in combination, so that the resulting character is absent only from homozygous recessives, they are said to be duplicate genes. In all cases, recombination of the genes during reproduction follows the same course as in simple Mendelian heredity, but the resulting phenotypic ratios differ because fewer unit characters are involved.

Lethal genes: Some genes completely inhibit development or modify it in such a way that the individual dies. They also modify the usual ratios of associated characters.

Linkage: Some characters, although not allelic, are inherited in definite groups; they are said to be linked. Modern genetics shows that linkage is due to the presence of genes for the linked characters in the same chromosomes.

Crossing over: Linkage relations are sometimes interrupted in a limited number of individuals, permitting some reassortment of normally grouped characters. This change is due to the breaking of paired chromosomes in synapsis and the reunion of their fragments in new combinations to form similar chromosomes, sometimes with new combinations of genes.

Translocation: This change is a shifting of the relations of genes in the chromosomes, due to looping, fusion, and rupture, or to the attachment of fragments to other chromosomes. It may result in the duplication of genes within a chromosome or in a change in the serial arrangement of the included genes.

The inheritance of sex has also been shown in many cases to depend on a simple chromosomal mechanism. Males of many species have an X chromosome without a synaptic mate or with a Y chromosome mate that is evidently abortive. The females of such species have two X chromosomes. In the formation of germ cells all eggs receive an X chromosome while half of the sperm cells receive an X chromosome and half a Y or none. Random combination of these cells restores the XX combination in one-half and X or XY in the other, thus producing half females and half males. Other investigations have shown that the quantitative balance between the sex and other chromosomes is the active factor in conditioning the differentiation of the sexes.

This disclosure also explains the phenomenon of sex linkage. Genes lying in the sex chromosomes, mostly in the X chromosomes but a few in the Y, are inevitably transmitted and expressed in some definite relation with sex; hence they are said to be sex linked. Such characters need have no active sexual role.

Plant and Animal Breeding. The findings of genetics have been of value in plant and animal breeding. Although the improvement of cultivated plants and domestic animals by selection preceded by many years the formulation of scientific principles of heredity, the discovery of these principles has made possible much more precise and efficient procedure in the establishment of useful strains. Hybridization and selection together are the chief means of improvement. Applied by scientists they have brought about many modifications of living things and have disclosed many facts concerning heredity. Corn has been studied in detail and subjected to many experiments, both practical and purely scientific. Tomatoes, radishes, various cereals, and flowers of many species have also commanded attention. More has been done with plants than with animals because the domestic animals are less amenable to experiment. From the practical point of view plants are more satisfactory subjects because desirable hybrid strains may often be propagated by cuttings, grafting, and other asexual methods which avoid the segregation that is inevitable in sexual processes. Only rigid selection can establish desired hybrid combinations in plants or animals that must be produced sexually.

The study of human heredity depends entirely on observation of the family, since controlled mating is impossible. Genealogical records have furnished a large amount of valuable material and the records of public institutions such as prisons and asylums have been equally useful to the geneticist. Such records are not to be compared with scientifically assembled experimental data, but they leave no doubt that the principles of heredity worked out in the study of other organisms are also applicable to man. In a few cases they have also disclosed adequate evidence of a specific type of Mendelian inheritance of human characters.

The clearest evidences of human heredity are found in the behavior of simple structural defects, such as the appearance of extra digits (polydactylism), the fusion of bones in the digits (symphalangism), and shortness of the fingers (brachydactylism). These defects are transmitted as Mendelian unit characters allelic to normal structure. Red-green color-blindness (vision) is one of the most striking examples of inheritance in man. It is a sex-linked recessive allele of normal vision. Both X chromosomes of the female must carry the gene for the defect if she is to be color-blind, whereas the male may be color-blind if he receives such a gene in his one X chromosome. Females may be heterozygous carriers of the defect, with normal vision; males are either strictly normal or defective. In this type of inheritance the male always receives the genes for his characters from his mother; therefore a carrier mother may have some color-blind sons. A color-blind man and a genotypically normal woman cannot produce color-blind children, but all of their daughters are carriers. On the other hand, a color-blind woman and a normal man will produce carrier daughters and color-blind sons. Hemophilia is inherited in a like manner, except that the recessive genes for hemophilia are lethal in the homozygous condition in the female.

Pigmentation of the skin is controlled by multiple-factor inheritance. Since variations of skin color within a race are not always readily identified and may be partly environmental, knowledge of skin color inheritance has had to come mainly from study of black-white marriages. When a black person without white ancestry marries a white person without black ancestry, their children are typically intermediate in color, or mulattoes. Children from the marriage of a typical mulatto to another typical mulatto may vary in skin color from the black of the black grandparent to the light color of the white grandparent. It has been estimated that the color differences in blacks and whites are controlled by from two to four pairs of alleles. It is possible for a white-skinned person of black-white ancestry to have all the genes of the white genotype. Children from such a person married to a white or similar near-white should be all-white. Children from the marriage between two near-whites are seldom much darker than their parents, and some would have light skin color. If a near-white marries a white, their children are usually no darker than their near-white parents; there is no well-established evidence that a very dark or black child could be born to them.

Albinism is a rare inherited condition in which the skin, hair, and eyes lack the melanin pigment normally present. It results from a biochemical deficiency in which specialized skin cells called melanocytes are unable to synthesize melanin from the amino acid tyrosine. See also **Albinism.**

There is evidence that some allergies may be inherited. Inherited weaknesses in the tissues may make it easier for some antigens to enter the body of certain persons. See also **Allergy.**

Some geneticists believe that diabetes is heritable, but others disagree because an acceptable pattern for the inheritance of the disease has not been presented.

Muscular dystrophy is a group of hereditary diseases in which muscle fibers undergo progressive damage and eventual destruction. In the early stages of aggressive muscular dystrophy, the symptoms of which usually become evident within the first five years of life, high levels of certain enzymes are almost always found in the blood. Creatine phosphokinase (CPK) is the enzyme which indicates both the active disease and carriers of the disease. Sisters of patients with known Duchenne muscular dystrophy or mothers of one child with the disease often also have high levels of CPK in their blood. Such women, with no clinical evidence of the disease, are carriers of this dystrophy. Half of the carrier's sons will have the disease and half of her daughters will also be carriers. Known carriers, at least two-thirds of whom can be detected by measuring the level of CPK in the blood, should be advised not to have children. The incidence of this disease could be decreased if such genetic counseling is followed. Unfortunately, no specific treatment is available for muscular dystrophy patients.

Nearsightedness has long been known to occur with relatively great frequency among members of certain families; consequently, heredity is recognized as one of the possible causes of the disorder. The exact method of inheritance, however, is not known.

Sickle-cell anemia is a hereditary condition in which the red blood cells are easily destroyed in the body. Sickle-cell anemia is often fatal and usually at an early age; few victims live beyond the third decade. A single pair of abnormal genes, one from each parent, is responsible for the disease. Should an individual inherit only one abnormal gene, the sickling trait will be present, but without the anemia the chances are good for a reasonably healthy life. See also **Anemia.**

Degenerative diseases of the nervous system include many disorders whose cause is unknown. Hereditary factors are obvious in certain diseases of the nervous system; in these disorders, the specific disease may not be inherited, but rather a general inherent weakness of the nervous system. Important examples of hereditary diseases of the nervous system include amaurotic idiocy, Huntington's chorea, familial periodic paralysis, hereditary spastic paralysis, Friedreich's and Marie's ataxia, hereditary tremor, progressive lenticular degeneration, feeble mindedness, and certain psychoses. See also **Hip.**

Extranuclear Inheritance. The existence of cytoplasmic genes was suggested as long ago as 1909 when the first examples of non-Mendelian inheritance were described by Correns and Baur. However, the demonstration that chloroplasts, mitochondria, and the kinetoplasts of trypanosomes contain specific DNA of their own came as a surprise to most biologists. It is now recognized that organelle DNAs are present in the cell in small amounts, perhaps 1–10% of the total cellular DNA. Organelle DNAs are also distinct entities, as indicated by average nucleotide compositions different from nuclear DNA. All organelle DNAs examined thus far consist of covalently closed circles and exhibit autonomous replication. Although the functions of such organelle DNAs remain largely unknown, it appears that ribosomal RNAs and most if not all tRNAs of chloroplasts and mitochondria are transcribed from the corresponding DNAs. Specific proteins either coded by organelle genes or synthesized with the organelle have been more difficult to identify. The importance to the cell of organelle DNA and the resultant extranuclear inheritance of genes present in this DNA is illustrated by the petite mutants of yeasts. These mutants contain an altered mitochondrial DNA which results in lack of mitochondrial respiratory function. Thus, to survive, petite mutants must utilize an alternative source of energy such as anaerobic fermentation of carbohydrates. Genetic analysis has established that inheritance patterns of the defect are consistent with cytoplasmic inheritance.

For references, see list at end of entry on **Biology.**

Reviewed and updated for the 6th Edition by Ann C. Vickery, Ph.D., University of South Florida, College of Medicine, Tampa, Florida.

HEREDITY (Diabetes). Diabetes Mellitus.

HERMANN-MAUGUIN SYMBOLS. A notation sometimes used to describe the symmetry classes of crystals. Two-, three-, four-, and six-fold rotation axes are represented by the numbers 2, 3, 4 and 6. Three-, four- and six-fold inversion axes have symbols 3, 4, 6. Asymmetry has the symbol 1. A center of symmetry has the symbol 1. A plane of symmetry is represented by *m* (mirror). The first number denotes the principal axis. If a plane of symmetry is perpendicular to an axis, this is represented by *n*/*m* (e.g., 2/*m*, 4/*m*, 6/*m*). Then follow the symbols for the secondary axes, if any, and then any other symmetry planes.

HERMAPHRODITE. An animal with functional reproductive organs of both sexes. The condition is common among the flatworms and segmented worms and occurs in a few species of echinoderms and mollusks (*Mollusca*). Among the vertebrates the occurrence of both sexes in one individual is rare and the sexes appear at different periods. Animals in which such a transition is possible are sometimes influenced by external conditions and during the transformation may be functionally hermaphrodite. This is true of some fishes and amphibians.

HERMAPHRODITISM. A condition characterized by the presence of both ovarian and testicular tissue. Because of overactivity of the adrenal glands, excessive hormones can be produced. In some patients with overactive adrenals, there is an excessive development of fat, accompanied by sexual disturbances. The symptoms vary according to age and sex. If the disease develops during fetal life and the child is a female, a form of hermaphroditism, or dual sexuality, may result, in which the clitoris is enlarged and resembles the penis. Other signs of masculinization accompany this condition. Sometimes a true hermaphrodite may appear to be a normal female, but who is found at surgery to possess testes in the groin region. Only about a dozen cases of true hermaphroditism in the human race have been reported. This term signifies the presence of all of the functioning genital organs of both sexes in one individual. The reported cases were claimed to have both testicles and ovaries present. However, the ability to impregnate as well as to conceive has never been reported in one individual.

Many cases of pseudo-hermaphroditism have been seen. In this condition the genital organs, internal or external, do not conform either totally or in part with the sexual glands (testicles or ovaries) present. In the male hermaphrodite, testicles are present but may be abdominal in position. The penis is small and more nearly resembles a large clitoris; the scrotum is divided by a cleft resembling the female labia with a small short vagina. Uterus and tubes are not present.

The female hermaphrodite has a large clitoris more like a small penis, rudimentary vagina, a uterus and ovaries. Various inbetween stages may be present, given a very bizarre picture where the sex can only be determined by microscopic study of sex characteristics shown by the nuclei of the tissue cells. Such cells may be examined by biopsy of the skin, and a definite decision as to sex given with accuracy of a high degree; where biopsy is not desired or facilities are not available, the nuclei of epithelial cells scraped from the inside of the mouth, or even of polymorphonuclear leucocytes in the blood will furnish a slightly less reliable answer. Such sexing should be done as soon as possible after birth in any infant in whom the identity of the sex organs appears dubious; by this means mistakes in naming and upbringing can be avoided. Where such a decision as to sex is not made in very early life, it is probably wise to bring up the child according to the sex which seems most apparent and defer final decision until the onset of puberty, when the development of sex consciousness may reveal psychological orientation to one sex or the other.

Dewald et al. (Mayo Clinic and Mayo Foundation), using chromosome heteromorphisms and blood cell types as genetic markers, demonstrated chimerism in a chi46, XX/46,XY true hermaphrodite. The pattern of inheritance of the chromosome heteromorphisms indicated that this individual was probably conceived by the fertilization, by two different spermatozoa, of an ovum and the second meiotic division polar body derived from the ovum and subsequent fusion of the two zygotes. A chimera may be defined as an individual with two or more genetic cell types resulting from the fusion of different zygotes. As described by Dewald et al., "Chimeras can be readily classified as wholebody or partial chimeras according to their mode of origin. Partial chimeras can arise by placental cross-fertilization between dizygotic twins, maternal-fetal transplacental exchange, transfusions, or grafting. Because of lack of suitable studies, the origin of wholebody chimeras is less clear. Theoretically, they can arise by (1) early fusion of different embryos, (2) fertilization of an ovum and any polar body by two different sperm and subsequent fusion of the zygotes, (3) fertilization of a haploid ovum or polar body and subsequent fusion with a diploid polar body or ovum, or (4) fusion of a diploid sperm with an embryo." Most reported chimeras have sexual abnormalities, such as clitoral hypertrophy or true hermaphroditism. More detail will be found in Dewald, G., et al.: "Origin of chi46, XX/46,XY Chimerism in a Human True Hermaphrodite," *Science*, **207**, 321–323 (1980).

HERMITE EQUATION. A second-order differential equation

$$y'' - 2xy' + 2ny = 0$$

where n is a constant. The Hermite polynomials (see **Generating function**) are solutions. The equation occurs in the quantum mechanical problem of the harmonic oscillator. (See also **Weber Equation,** from which the Hermite equation can be obtained by a change of variable; **Whittaker Differential Equation,** another related form; **Gauss Hypergeometric Equation,** from which it can be derived by confluence of singular points.)

See also terms listed under **Mathematics.**

HERNIA. At one time commonly called rupture, a hernia is an abnormal protrusion of a part or organ through the containing wall of its cavity. In common usage, the term hernia usually applies to the abdominal cavity and implies a covering or sac over the protrusion. There are two main classes of hernias: (1) *congenital* hernia in which the sac was present before birth; and (2) *acquired* hernia in which the sac is formed after birth and pushes through an opening in the muscle wall which failed to close at birth, or that was formed following an incision. A large percentage of acquired hernias result from injury or strain, such as those hernias which occur when a person lifts a heavy object. Hernias may occur in the groin, the navel, the membrane separating the abdominal and chest cavities (diaphragm), in surgical incisions, and elsewhere. All herniation takes place through a normal opening, or through an opening that should have been eliminated at some period of development, or through an opening which had closed and then reopened in later life.

The hernial sac has a mouth, a neck, and a body. The mouth connects with the abdominal cavity and is called the hernial ring; the body is the pouch or sac that projects outside the abdominal wall; and the neck connects the mouth and body of the sac.

The contents of the sac might be any of the abdominal organs, in whole or part; loops of the intestine are commonly found in hernias. The sac and its contents are subject to injury which can lead to serious complications. The skin surface is vulnerable to blows, falls, pressure, irritation from binders or trusses, or may become inflamed, infected, or abscessed. From within, the contents of the sac are prone to strangulation when the blood supply is cut off by a narrow or constricted hernial ring; gangrene may set in if treatment is not sought promptly.

Hernias are considered *reducible* or *irreducible.* Reduction may be spontaneous; for example, sac contents may return unaided to the abdominal cavity when the patient lies flat. If the patient remains untreated, however, a reducible hernia may become irreducible, that is, the contents of the sac can no longer be returned to the abdominal cavity. Irreducibility may be caused by increased size of the hernia, formation of adhesions, or development of a small or constricted hernial ring. Hernias of enormous size, hanging down to the knees, have been reported. An irreducible hernia is a constant source of danger.

Hernias occurring in the groin are either *inguinal* hernias or *femoral* hernias. Inguinal hernias account for about 92% of all hernias. Superficially, inguinal and femoral hernias look alike because the bulge is in the groin. However, they differ anatomically. Inguinal hernias slip through the normal openings for the passage of nerves or organs of the reproductive system. Femoral hernias occur through the passageway for nerves and vessels to the thigh.

Normally, the deep and shallow layers of muscles and ligaments on the abdominal wall protect these normal openings against herniation. With rise of intra-abdominal tension, as by straining, coughing, or lifting, the muscles contract and flatten like a shutter in a normal situation. But if the muscles and/or other protective structures of these openings are weak, the shutter action fails and an increase of intra-abdominal tension may push part of the abdominal organs through the opening into the preformed sac, and thus a hernia is begun. Successive incidents of tension increase the size of the sac by forcing additional intra-abdominal tissue into it.

Hernias of the navel are called *umbilical* hernias. The navel is an opening that should close in the process of development. After birth, it is a scar formed of interlaced muscle fibers of the contracted umbilical ring. Sometimes a defect occurring before birth prevents its closing, and the baby is born with a hernia, or may soon acquire one. In adults between 25 and 40 years of age, obesity and pregnancy are the most common predisposing causes of this form of hernia.

Hiatus hernia is described under **Esophagus.**

Obese women are the most frequent subjects of hernia in the site of a surgical incision. Some incisional hernias are caused by failure of the layers of deep muscle and fascia to knit firmly after surgery. Blood clot, infection, exudate, and swelling in the line of incision, as well as increased intra-abdominal tension, also are factors favoring herniation. The neck of the incisional hernia is a firm ring of scar tissue. Because of the large hernial ring, these hernias are difficult to control by a truss. Large incisional hernias may cause invalidism unless surgical relief is obtained.

The treatment of a hernia patient can be accomplished by a mechanical device (truss), or surgery. Most authorities agree that a truss is a makeshift which is acceptable only when surgery would be hazardous. Improved techniques have made possible the surgical repair of hernias which not many decades ago would have been irreparable. It is often necessary to close the opening with a fascial graft, or an inert foreign material, such as polypropylene mesh.

HEROIN. Alkaloids.

HERON (*Aves, Ciconiiformes*). Long-legged wading birds (*Aves*) with a sharp slender beak and when adult with plumes or a crest. They live chiefly on fish.

Herons are found throughout the world. The most widely known North American species are the great blue heron, *Ardea herodias,* the green heron, *Butorides virescens,* and the egret, *Egretta.* The last is a white bird which bears beautiful plumes known as aigrettes during the breeding season. It was once threatened with extinction through the use of these plumes as ornaments for hats, but the remaining birds are adequately protected. See illustration.

Egret. (*Grant M. Haist, National Audubon Society.*)

These birds are remarkable for the down which they produce. It is exceptionally light and fluffy and grows all over the breast, rump, and flanks. The birds roost in tall trees during the day and feed mostly at night. They have long legs and long beaks. Sometimes they will reach a height of about 20 inches (51 centimeters). Some species are gray with black on head and neck. However, there is a wide variation both in size and coloration. See also **Ciconiiformes.**

HERPES SIMPLEX VIRUS DISEASES. Four major herpesviruses cause infections in humans: (1) Herpes simplex; (2) varicella-zoster; (3) cytomegalovirus; and (4) Epstein-Barr virus. These are among the most widespread of all human pathogens and, characteristically, tend to follow cycles of dormancy and activity within an individual, such cycles often extending over long periods. A long span of dormancy may be interrupted by a flareup resulting from unusual physical or psychological stress. There is no known effective treatment for achieving their full eradication. Incidence of herpesvirus infections tend to occur more frequently in immunosuppressed patients. See **Immune System and Immunology.** These viruses are described further in the entry on **Virus.** See also **Cancer Research.**

There are two types of herpes simplex virus (HSV), with multiple strains of each type. Type 1 infects mucous membranes of the oral cavity, perioral skin, eyes, and skin above the waist. Type 2 usually causes a genital infection, an infection which is the second most common venereal disease found in the United States and a number of other countries. It has been estimated that between 30 and 90% of young adults carry antibody to one or both types of herpes simplex virus. Most infections from Type 1 HSV occur during childhood, but may occur any time during life. Type 2 usually does not appear before puberty and the commencement of sexual activity. The incidence of Type 2 antibody peaks by age 35 years. Type 2 antibody can be found in 20–35% of the general population. Certain occupational groups, such as health professionals and prostitutes, are at greater risk for HSV infection—simply because of the greater number of possible contacts with the virus. However, the infection is found in all segments of society.

Both types of herpes simplex virus appear to be spread by close contact between infected and susceptible individuals. Incubation period ranges from 2 to 20 days. Even in persons with antibody to the virus, second infections are observed. Virus may be shed from the oral or genital mucous membranes. Where there is adequate immunity, the shed virus may produce either subclinical infection or observable clinical disease. In persons who are highly immunosuppressed (as in cases of persons who have received organ transplants), the infection may be widespread and not heal for many months.

The most serious infection of Type 1 HSV is herpes keratitis, which can lead to destruction of the cornea. Other primary infections of Type 1 include stomatitis, pharyngitis, tracheobronchitis, and dermatitis. Type 2 usually involves vulvovaginitis or balanitis. Some clinical studies have shown that, in addition to causing oral, ocular, and genital lesions, HSV infections may involve visceral sites, such as the throat, lungs, esophagus, brain, meninges, liver, spleen, and pancreas. It has been observed that organ transplantation and the widespread use of cancer chemotherapy have increased the frequency of herpes simplex visceral infection.

The neonate is seriously threatened in cases of maternal genital infection. Such infections can be fatal in about half of the cases of the newborn. Studies have shown that even when the mother is asymptomatic, there is a risk to the infant. Some authorities have suggested that the incidence of neonatal Type 2 disease could be lowered by performing a cesarean section in symptomatic women, shortly after the onset of labor, when the membranes are still intact. This, however, is essentially an unproven procedure as of the early 1980s.

Treatment for herpes simplex infections has been most successful in cases of Type 1 herpes keratitis, where adenine arabinoside and idoxuridine may be administered. Because of asymptomatic shedding of the virus, there are presently no known methods of certain prevention. Considerable research is being directed toward finding more effective chemical agents or vaccines to prevent replication and shedding.

References

Kaplan, A. S.: "The Herpesviruses," Academic, New York, 1973.

Nahmias, A. J., and W. E. Josey: "Epidemiology of Herpes Simplex Viruses 1 and 2" in Viral Infections of Humans," (A. S. Evans, editor), Plenum, New York, 1976.

Nicholas, L.: "Sexually Transmitted Diseases," Charles C. Thomas, Springfield, Illinois, 1973.

HERPETOLOGY. The study of amphibians and reptiles, often mistakenly believed to be a study of reptiles only and snakes in particular. See also **Snakes.**

HERPOLHODE. The curve along which the cone traced out by the angular velocity vector intersects the invariable plane tangent to the momental ellipsoid and perpendicular to the angular momentum vector, in the case of a rotating rigid body not subject to any external torque. The concept is useful in studying the dynamics of a rigid body.

HERRING (*Osteichthyes*). Of the order *Clupeiformes*, herring are a characteristic fish group of the oceans, but also include a number of species that inhabit tropical fresh water. They form schools and are found near shores as well as in the open sea. Many of them are migratory. Herrings can be distinguished from other species by several characteristics—there are no rayed canals on the gill cover bones; lateral line pores are absent; there are keel scales along the medial line of the belly. Noteworthy skull characteristics include a suprabranchial organ with unknown function which joins the fourth and fifth gill arches; there is little dentition in the mouth, since most species feed on plankton; and there are no teeth on the parasphenoid (a bone at the base of the skull).

Commercially important herring are of the suborder Clupeoidei and most statistics report the catch of clupeoids without distinguishing specific subtypes. About 25 herring genera, with some 100 species, live in the sea. Until recent times, herring was considered the most important commercial fish catch.

Atlantic Herring. This fish (*Clupea harengus*), shown in Fig. 1, is

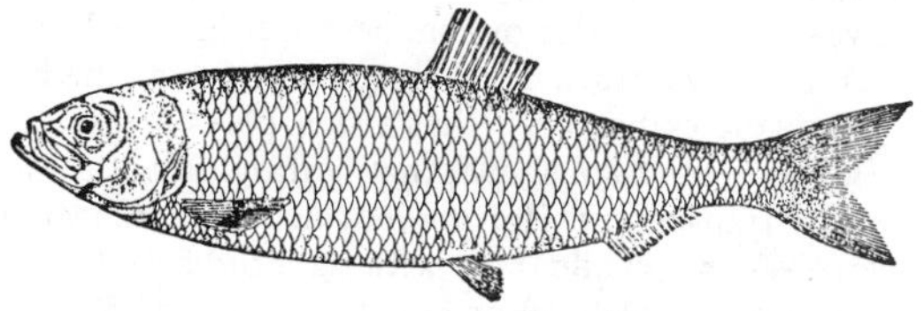

Fig. 1. Atlantic herring (*Clupea harengus*).

one of the most important of commercial fishes in the northeastern Atlantic Ocean. Originally, it was presumed that the entire herring population in the northeastern Atlantic Ocean was a unified group, inhabiting the region from the Arctic Ocean to the English Channel. From here, this group presumably migrated extensively to the north and south (and back) during the course of a year. But in recent years structural differences in herring caught in different regions have been noted. Based upon these and other studies, herring researchers met in Copenhagen in 1956 and proposed the following classification:

Class A Herring—Known as the *Atlanto-scandian herring*, which inhabit the open Atlantic Ocean and which spawn on the Atlantic coasts of northern Europe in mid-winter, spring and possibly in early summer. These fishes are of appreciable size and are characterized by an intermediate number of vertebrae (57 or more).

Class B Herring—Known as *shelf herring*, which inhabit the North Sea on the shelf west of the British Isles and in the transition zone between the North and Baltic Seas. These fishes spawn between August and January along the coasts. They reach a smaller size and have an intermediate number of vertebrae.

Class C Herring—Distributed inside coastal waters of the North Sea, in the traditional area between the North and Baltic Seas, and in the Baltic Sea. They spawn in shallow water during winter and spring. Body size and vertebral number are smaller than in the Class B herring.

Class D Herring—Found in the most northeastern part of the Atlantic Ocean and which, at one time, were classified with the Pacific herring.

The eggs of the Atlantic herring have a diameter of about 1 to 2 millimeters. In general, the winter-spring spawners have a relatively lower fertility and larger eggs, while the opposite is true of the summer-fall spawners. Winter-spring spawners lay from 22,000 to 40,000 eggs, whereas the summer-fall spawners lay from 48,000 up to 70,000 eggs. Freshly hatched larvae are found in tremendous masses on the spawning grounds and vicinity. They are transparent and very slender. Even when the fish is only about 0.8 inch (2 centimeters) in length, the distinguishing characteristics of the herring can be observed.

The availability of suitable plankton is the most important determinant for the development and ultimate survival of the larvae. The prey must be as close as 0.2 inch (0.5 centimeter) from the herring larvae in order to be perceived and eaten. The larvae feed only on moving organisms.

Summer and fall spawners of the North Sea reach sexual maturity in the third or fourth year, at which time they have a length of about 9.5 inches (24 centimeters). Life expectancy of the summer-fall spawners is from 12 to 16 years, while late-winter spawners of the Norwegian coast may live from 23 to 25 years.

Herrings have been found in schools ranging from hundreds to thousands of individuals, in all sizes from young to sexually mature adults. The individuals within a particular school are generally of equal size and age. It is not known how long they remain together. Studies have indicated that herring in the North Sea spend the day at the floor. With the beginning of dusk, they ascend to depths of some 100 to 165 feet (30 to 50 meters) into a warmer-temperature zone. Light plays an important role in this movement. During darkness, they seem to remain scattered at the higher level, but with the onset of dawn, they collect and return to the floor. In recent years, herring behavior at darkness has been studied closely. Soviet researchers have observed daily activities from a submarine and report that the Atlanto-scandian herring spends the night motionless at the surface of the water in an oblique position, as if sleeping. They become active shortly before dawn and begin their move to greater depths. Other researchers believe that schooling behavior ceases at night.

In recent years, the behavior of commercial fishes has been studied intensively with a view toward developing better fishing methods. One finding of these studies has been that vision plays a significant role in herring. Experiments have shown that herring do not avoid plastic sheets if they are transparent. Herring in which the eyes were covered could not perceive a net, while those with sight did detect the obstruction. During the day, a wall of air bubbles can act as an obstacle to herring, but they will swim right through it at night. Herring can also detect noises and vibrations, to which they respond with fright behavior. This has been confirmed with echolocation tracking. When a ship moves toward a herring school, the school sinks to a depth of many meters. Fright can also be induced in an aquarium by tapping on the wall.

During a few months of the year, herring can be found in certain regions in tremendous quantities, while at other times these same areas are completely devoid of the fishes. On the other hand, herring can be caught in some places throughout the year, but the catch varies from year to year. Marking studies have shown that Atlanto-scandian herring migrate between feeding grounds off Iceland and spawning grounds on the Norwegian coast. It has been established that this major herring population has 3 major growth and development areas, i.e., in the Norwegian fjords, the Barents Sea, and in the southern and eastern parts of the ocean off northern Europe. Young herring which have developed in the fjords migrate to the sea at an age of 2 to 3 years, where they meet those herring which have been developing there.

Herring feed on plankton, which is not simply filtered, but selected—a phenomenon found by stomach content investigations and aquarium studies. Some researchers indicate that herring select food visually and then again test it in the mouth. Materials that are useless or of bad taste are immediately rejected.

Three stocks of herring are taken in Icelandic waters. Two of them are of Icelandic origin and the third of Norwegian origin. The Newfoundland herring fishery is entirely coastal, particularly concentrated in the west and south coast regions. The fish are caught during winter and spring with gillnets and purse seines. The industry in Newfoundland has increased by 200% to 300% during the past 20 years.

Pacific Herring. This fish (*Clupea pallastii*) inhabits the coasts of the northern Pacific Ocean from the Bering Strait to Korea and in the Arctic Sea to the mouth of the Lena River. On the North American coast, its distribution extends from California to Nome, Alaska. Herring in the White Sea and from Cape Kanin to the Kara Sea are very similar to the Pacific herring. This species differs from the Atlantic herring by the smaller number of vertebrae, among other features. Generally, the eggs are laid in brackish water on plants. Pacific herring form spawning groups whose distribution is limited to very specific narrow zones. They apparently do not migrate to a great extent. On the Asiatic coast, 10 spawning groups are known in the region from Korea to the Sea of Okhotsk. Of these, the *Hokkaido-Sakhalin herring* is commercially the most important. The principal herring fisheries on the Pacific coast of the United States are in the bays and channels of southeastern and central Alaska. A general downward trend of the herring catch in Alaska, as contrasted with British Columbia, commenced in the early 1950s.

Sprat. This fish (genus *Sprattus*) is closely related to the herring. Six well-defined species have been identified. The majority are found in the southern hemisphere. Length ranges up to about 8 inches (20 centimeters). Coloration of the best known species (*Sprattus sprattus*) resembles herring and is irridescent. These fishes are found in the northern hemisphere on the European coast from Tromsö to the Baltic Sea and the Bay of Biscay, as well as in the Mediterranean and in the bordering waters of the Black Sea. Sprats do not undertake long migrations like herring. They generally stay near the coast and in river mouths. In the Baltic Sea, the sprat is found in water with low salt content. The sprat apparently avoids areas far from the coast. It reaches sexual maturity at an age of 2 to 3 years. Spawning takes place some distance from the coast and occurs in the North Sea from April to July; in the Kattegat and Skagerrak from May to June; in the Baltic Sea from May to August. Sprats, like herring, are commercially important fishes.

Sardines. Also related to sprats and herrings and of large commercial importance are the *Sardinops* sardines, of which there are at least 5 significant species: (1) *Pacific sardine* (*Sardinops caerulea*); (2) *South American sardine* (*S. sagax*); (3) *Japanese sardine* (*S. melanosticta*); (4) *Australian sardine* (*S. neo-pilchardus*); and (5) *South African sardine* (*S. ocellata*). Commercially, the latter species is of the least commercial importance.

The *Pacific sardine* is of large commercial importance for the United States and is found on the east coast of the Pacific Ocean from Baja California to British Columbia. The species lives in schools near the surface of the water and spawns between January and June, chiefly in March and April. Spawning takes place on the high seas off Baja California and southern California, as much as 300 nautical miles (556 kilometers) from shore. The larvae hatch 3 to 4 days after spawning and they migrate to the coast at a length of 3 to 5 inches (7.5 to 12.5 centimeters). They are caught in great masses and used as bait for tuna. At a length of about 6.5+ inches (17 centimeters), they leave the feeding grounds off the coast and meet the adults swimming on the open sea. Sexual maturity is attained at a length between 6.5 and 9.8 inches (17 and 25 centimeters), which occurs at an age of 2 to 3 years. The species can reach an age of 13 years. In California waters, the sardine catch has decreased dramatically since the late 1930s.

Sardine Ecology. The California sardine fishery has become the classic example of an ecologically complex community modified by an intensive fishery. A simple matter of overfishing might, on theoretical grounds, be overcome by abstention from fishing for an appropriate period, but apparently an ecologically related and evidently competitive species, the anchovy, has occupied the gap left by the exploitation of the sardines (the gap was evidently increased by harvesting during a period of conditions unfavorable for reproductive success). In the absence of a similar market for anchovies, a reduced technology for processing sardines, and legislative restrictions on harvesting anchovies, the situation reached the stage where a sardine fishery of nearly any magnitude further decreased the stock. Possibly this imbalance could be redressed by an unpredictable alteration in natural conditions in favor of the sardine, but this does not appear likely. It is interesting to note that the Pacific sardine supported the largest fishery in the Western Hemisphere in the early 1930s (exceeding over 1 billion pounds; 0.45 billion kilograms) taken from California waters, as compared with lower catches of just a few million pounds annually in recent years. From an economic standpoint, much of the loss of production of California and Oregon sardines has been compensated by large increases in menhaden catches in the south Atlantic Ocean and off the Gulf states.

The *South American sardine* and the *South African sardine* have increased in production since World War II. The South African sardine, sometimes referred to as the *pilchard* (*Sardinops ocellata*), has a wide geographical distribution and is known from St. Lucia Bay (north of Durban) to Bahia dos Tigres on the Angolan coast. The main commercial concentrations are limited to the Walvis Bay region, the waters off St. Helena Bay, and the area between Cape Point and Cape Agulhas. The species is normally found within 25 miles (40 kilometers) of the coastline, but occasionally schools have been reported up to 80 miles (129 kilometers) offshore.

The South African pilchard is a fast-growing fish and reaches sexual maturity at the age of about 2.5 years, by which time it attains a length of some 8.25 inches (21 centimeters). The main spawning seasons are spring and early summer. Spawning occurs offshore and three main grounds have been identified—those off Walvis Bay; near St. Helena Bay; and east of Cape Point. The pilchard is a filter-feeder, its diet consisting of both phytoplankton and zooplankton. Tagging experiments have established that there is periodically an influx of pilchards from the Walvis region into Cape waters.

The *Japanese sardine* is a warm-water species which attains a length of about 11.5 inches (29 centimeters). Distribution is chiefly in a temperature of from 59 to 79°F (15 to 26°C). Spawning grounds are off the south coast of Japan and Korea at some distance from the coast. Japanese sardines spawn from December to May on the high seas at a water temperature of 55 to 68°F (13 to 20°C). The number of eggs varies between 27,000 and 84,000. After spawning, a migration to the north takes place on the far eastern coast, where a large fishing industry has developed. These sardines feed chiefly on plankton. The annual catch (mainly by Japanese, Russian, and Korean fishers) varies considerably from year to year.

True Sardine. Only one species belongs to the genus of the true sardines, the *Sardina pilchardus*, also called *pilchard*, and not to be confused with the South African fish previously described. See Fig. 2. The true sardine reaches a length of about 11.8 inches (30 centimeters), but is generally from 9 to 9.8 inches (23 to 25 centimeters) long. Commonly, the larger sizes are called pilchards, while the smaller fishes are called sardines, the latter ranging between 5 and 6 inches (13 and 16 centimeters) in length. Distribution is on the coasts of west and southwest Europe and north Africa, from southern Ireland, the southern part of the North Sea, and the Kattegat in the north to Madeira and the Canary islands in the south. Distribution also includes the northern parts of the Mediterranean and bordering waters. There are two subspecies. The spawning period of the pilchard is rather extended. Off the Iberian peninsula, spawning takes place from February to March; in the North Sea, from July to August; off the coast of west Britanny, November to June; in the Mediterranean, September to May; and in the Black Sea, July and August.

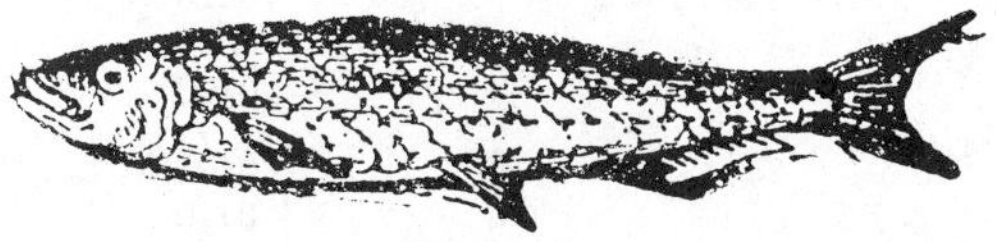

Fig. 2. True sardine (*Sardina pilchardus*).

The distribution of the true sardine is approximately limited by the 68°F (20°C) isotherm. Until 1930, the Strait of Dover was apparently the northern limit for pilchards. Those found in Norwegian waters and in the Kattegat only occurred in small numbers. Since that time, the northern population has increased significantly. In the late 1930s, large quantities of sardine eggs were reported off the East Frisian islands. In late 1940s, the first large quantities were found in the area near Amrum island (in the North Frisians). Climatic changes are probably responsible for the extension of the northern limit.

Adult pilchards feed primarily on zooplankton. The catch of pilchards has been increasing progressively over a number of years and is very important to Portugal, Morocco, Spain, France, and Yugoslavia, most of the catch being used in production of canned sardine oil.

Shad. This fish of the genus *Alosa* is also related to the herring. Shads have a compressed upper body and the keel scales form a sharp keel. The teeth in the jaw are either small or absent altogether, and in adults there are no vomerine teeth. There are four species in the north Atlantic Ocean, Mediterranean, and in the northern Pacific Ocean. These species migrate into fresh water. The best known species in Europe is the shad (*Alosa alosa*). The length of the shad exceeds 27.5 inches (70 centimeters) and the jaw protrudes forward. The scales are not as lightly attached as in the herring. Distribution is on the European coast from Norway to the Iberian peninsula, and along the north African coast to Morocco, as well as the western part of the Baltic Sea and the Mediterranean. In March, shad migrate from the sea to spawn in the rivers well upstream. Earlier, they were found in the Neckar River, Germany. During recent years, shad have disappeared from much of their original habitat, largely as the result of pollution.

The *American shad* (*A. sapidissima*) is found mainly in the Atlantic Ocean, from the Gulf of St. Lawrence to Florida. In 1871, the species was introduced to the Sacramento and Columbia Rivers in the western United States and, by 1876, shad were caught off Vancouver island. Since that time, the species has spread along the entire coast of the Pacific Ocean from southern California to Alaska and Kamchatka. It is a prevalent fish in the California rivers. Since the earliest settlements in North America, the American shad has been an important and valuable commercial fish, while the Alabama species has enjoyed much regional acclaim. Over the years, however, the shad population has decreased.

See also **Fishes.**

HERRINGBONE GEAR. Helical Gearing.

HERTZ. Frequency; Units and Standards.

HERTZSPRUNG-RUSSELL SYSTEM. Giant and Dwarf Stars.

HESSIAN. A functional determinant, related to the Jacobian and defined for six variables by the equation

$$H(F) = \frac{\partial(u, v, w)}{\partial(x, y, z)} = \begin{vmatrix} F_{xx} & F_{xy} & F_{xz} \\ F_{xy} & F_{yy} & F_{yz} \\ F_{xz} & F_{yz} & F_{zz} \end{vmatrix}$$

where u, v, w are differential coefficients of another function, $F(x, y, z)$ and $u = \partial F/\partial x = F_x$, $v = F_y$, $w = F_z$; $\partial^2 F/\partial x^2 = F_{xx}$, etc.

It can be generalized for any number of variables. The Hessian of two binary quantics is a covariant; hence, it is useful in studying the invariants of algebraic functions. As an example, the Hessian of a quadratic is its discriminant.

See also **Determinant;** and terms listed under **Mathematics.**

HESSIAN FLY (*Insecta, Diptera*). One of the worst pests of wheat. It is a small two-winged fly *Mayetiola* (*Phytophaga*) *destructor*, a member of the gall-gnat family, which was introduced into the United States in the Revolutionary period. The larva lives between the base of a leaf and the stem of the wheat plant and either kills or weakens the plant so that no grain develops. Other cereals are attacked to some extent.

Fall plowing and burning stubble aid in destroying many insects. The most effective means of avoiding damage to winter wheat is to sow late enough to avoid the attack of most of the adults. They live no more than ten days and the date of emergence is known for various regions; hence, late planting subjects the crop only to the light infestation due to the eggs deposited by the relatively few flies which emerge late. Phorate is an effective chemical control.

HESSITE. A mineral telluride of silver, Ag_2Te, with some gold, crystallizing in the monoclinic system at normal temperatures; isometric system above 149.5°F (65.3°C). Crystalline form not obvious at normal temperatures. Hardness, 2–3; specific gravity, 8.24–8.45; color, gray with metallic luster; opaque. Named after G. H. Hess (1802–1850).

HETEROCYCLIC COMPOUNDS. Compound (Chemical); Organic Chemistry.

HETERODYNE. This term is used in communications terminology as an adjective or a verb, but in either case it concerns the beating together in an electrical circuit of two frequencies to produce new frequencies which are the sum or difference of the original ones. When two voltages of different frequencies are applied simultaneously to a circuit containing a non-linear impedance, for example, one in which the signal current varies as the square or higher power of the input signal voltage, the output of the circuit will contain new frequencies, among them one equal to the sum and another equal to the difference of the applied frequencies. Either one or both of these may be selected by properly tuning or filtering the output.

HETERODYNE FREQUENCY METER. **Frequency Measurement.**

HETERODYNE WHISTLE. The steady tone heard in the output of an amplitude-modulation receiver due to the beating of two carriers having a small frequency difference.

HETEROGAMY. The occurrence or union of male and female gametes of different size and structure; anisogamy. The alternation of two sexual generations, one true sexual, the other parthenogenetic.

HETEROMI. **Eels.**

HETEROMORPHOSIS. Deviation from normal form. Malformation or deformity and also less extreme departures incidental to slightly different conditions in the animal or its environment.

HETEROPOLYACIDS. Acids derived from two or more other acids, under such conditions that the negative radicals of the individual acids retain their structural identity within the complex radical or molecule formed. The term heteropolyacids is usually restricted to complex acids in which both radicals are derived from oxides, such as phosphomolybdic acid.

HETEROSIS. **Bovines.**

HETEROSOMATA. **Flatfishes.**

HETEROSPORY. The production of two distinct types of spores by a plant, in contrast to homospory, which is the production of only one type of spore. The two kinds of spores produced in heterospory are known as microspores and megaspores. The microspores are very small and grow into the male gametophyte. The megapores are much larger and form the female gametophyte. All of the seed plants have heterospory and a few of the minor subphyla of vascular plants do also. The *Lycopsida* is one of these subphyla. See also **Lycopsida.**

HETEROSTYLOUS FLOWERS. **Pollination.**

HETEROZYGOUS. Bearing two allelic genes of a different nature. The opposite of homozygous, which means to bear allelic genes of the same kind. For instance, if a person is homozygous for the recessive gene for albinism, the person will bear two such genes, represented as *aa*. The person will be an albino. A person who is heterozygous, however, will bear one gene for normal pigmentation and one gene for albinism, represented as *Aa*. A person can also be homozygous for the dominant gene, *AA*. Both heterozygous and homozygous persons will have normal pigmentation.

HEULANDITE. The mineral heulandite is a monoclinic zeolite whose crystals are often quite suggestive of orthorhombic forms. Its chemical composition is probably $(Na, Ca)_{4-6}Al_6(Al, Si)_4Si_{26}O_{72} \cdot 24H_2O$; strontium may be present. Heulandite has one good cleavage; is brittle with a conchoidal fracture; hardness, 3.4–4; specific gravity, 2.18–2.22; luster, vitreous to pearly; color, white to gray, red or brown; streak, white; transparent to translucent. Occurs chiefly in cavities in basaltic rocks with other zeolites, but may be found in granites, pegmatites, gneisses, and schists. Famous localities are in Iceland, India, the Harz Mountains, Italy, Switzerland, Scotland, Nova Scotia; and in the United States at Bergen Hill and West Paterson, New Jersey. This mineral was named for the English mineralogist Heuland.

HEVEA BRASILIENSIS. **Euphorbiaceae; Latex.**

HEVELIAN HALO. **Atmospheric Optical Phenomena.**

HEXACTINELLIDA. The glass sponges, constituting a class of the phylum *Porifera.* The spicules of the skeleton are silicious and of six-rayed form. Many of the species have a large central cavity, resulting in a tubular or vase-like form, and when freed of organic matter appear to be made of spun glass. These sponges are found in deep water in the ocean. Venus' flower basket, *Euplectella,* and the glass-rope sponge. *Hyalonema,* are the most common examples.

HEXADECIMAL NUMBER. In computer design, the hexadecimal (radix 16) numbering system is used as a convenient method for representing large binary numbers, which often consist of long strings of zeros and ones. The latter are difficult to handle in a computer. Each hexadecimal digit stands for four binary digits.

Hexadecimal notation calls for the use of 16 symbols to represent 16 number values. Inasmuch as the decimal system provides only 10 number symbols (0 to 9), six additional marks thus are needed to represent the remaining values. The letters A, B, C, D, E, and F are used for this purpose. As shown in the accompanying table, the

COMPARISON OF DECIMAL, HEXADECIMAL, AND BINARY NOTATION

Decimal	Hexadecimal	Binary
0	0	0000
1	1	0001
2	2	0010
3	3	0011
4	4	0100
5	5	0101
6	6	0110
7	7	0111
8	8	1000
9	9	1001
10	A	1010
11	B	1011
12	C	1100
13	D	1101
14	E	1110
15	F	1111
16	10	10000
17	11	10001
18	12	10010
19	13	10011
20	14	10100
21	15	10101
22	16	10110
23	17	10111
24	18	11000
25	19	11001
26	1A	11010
27	1B	11011
28	1C	11100
29	1D	11101
30	1E	11110
31	1F	11111

list of hexadecimal symbols is comprised of 0, 1, 2, 3, 4, 5, 6, 7, 8, 9, A, B, C, D, E, and F, in ascending sequence. From the table, note that upon reaching decimal 16, the hexadecimal symbols are used up and hence a "1 carry" must be placed in front of each hexadecimal symbol during its second cycle, i.e., from decimal 16 to decimal 31.

Binary numbers are converted to hexadecimal notation simply by dividing the number into groups of four binary digits, commencing from the right, and replacing each group by the corresponding hexadecimal symbol. Where the left-hand group is incomplete, zeros are filled in as required. This is illustrated by the following example.

$$
\begin{aligned}
1111100110110100 11 &= 0011/1110/0110/1101/0011\\
&= 3\quad E\quad 6\quad D\quad 3\\
&= (3E6D3)_{16}
\end{aligned}
$$

Hexadecimal numbers are best understood in terms of expansion in powers of 16. In the case of hexadecimal number 2CA.B6, for example, when decimals are substituted for hexadecimal symbols, it is evaluated as

$$
\begin{aligned}
&2 \times 16^2 + 12 \times 16^1 + 10 \times 16^0 + 11 \times 16^{-1} + 6 \times 16^{-2}\\
&= 2 \times 256 + 12 \times 16 + 10 \times 1 + 11/16 + 6/256\\
&= 512 + 192 + 10 + 0.6875 + 0.0234375\\
&= 714 + 0.7109375\\
&= (714.7109375)_{10}
\end{aligned}
$$

HEXAGONAL CRYSTAL. Crystal.

HEXAHEDRON. Polyhedron.

HEXAMETHYLENETETRAMINE. Hexamine.

HEXAMINE. $(CH_2)_6N_4$, formula weight 140.19, white crystalline solid, mp 280°C, decomposes at higher temperatures. Also known as hexamethylenetetramine, methenamine, and urotropine, the compound is soluble in H_2O and only very slightly soluble in alcohol or ether. Although used to some extent in medicine as an internal antiseptic, the primary use of hexamine is in the manufacture of synthetic resins where the compound is a substitute for formalin (aqueous solution of paraformaldehyde) and its NaOH catalyst. Hexamine also is used as an accelerator for rubber.

On a commercial scale, hexamine is manufactured from anhydrous NH_3 and a 45% solution of methanol-free formaldehyde. These raw materials, plus recycle mother liquor, are charged continuously at carefully controlled rates to a high-velocity reactor. The reaction is exothermic. The reactor effluent is discharged into a vacuum evaporator which also serves as a crystallizer. The hexamine crystals then are washed, dried, and screened. Average yield of the process is about 96% conversion of ingredients to produce hexamine.

HEXAPODA. Synonymous with Insecta.

HEXOSE MONOPHOSPHATE OXIDATIVE PATHWAY. Carbohydrates.

HEXOSES. Carbohydrates.

HEYDWEILER BRIDGE. Bridge Circuits (Electrical).

HEYN STRESSES. Microscopic stresses in a metal.

HIATUS HERNIA. Esophagus.

HIBERNATION (Anorexia). Anorexia.

HICKORY AND WINGNUT TREES. Of the family *Juglandaceae* (walnut family), hickory trees are of the genus *Carya* and are one of the most, if not the most distinctly North American tree. They are relatively unknown on the other continents. One authority aptly describes the hickories as walnuts with greater height and grace. The principal species are indicated in the accompanying table. It should be appreciated that the tree dimensions given in the table represent record specimens and that the average tree, most likely growing under somewhat more adverse conditions, will not attain such dimensions.

The bitternut or swamp hickory (*C. cordiformis*) has a light-brown or gray-brown thin bark. The fissures are shallow. The leaves are compound. The leaflets are of a deep yellow-green color, somewhat lighter underneath. The ovoid fruit is about an inch long and is contained in a thin husk. The tree prefers a rich woodsy environment, but will tolerate a variety of soils. The tree ranges from southern Maine and western Quebec westward to the Great Lakes and Minnesota and south through Nebraska, Kansas, Oklahoma, and Texas. It ranges eastward and south to Florida. The tree is found commonly only in southern New England, with only occasional representation in Vermont and New Hampshire. The tree attains its greatest height in the mountains of the Carolinas. As compared with other hickories, the wood is considered inferior.

The nutmeg hickory (*C. myristicaeformis*) prefers alluvial soil. It is found in the southeastern states and westward through Arkansas. The fruit is a little over an inch long and contained within a thin husk. The shell is very hard; the kernel is not edible. The wood is strong, hard, and is of a light-brown color.

RECORD HICKORY TREES IN THE UNITED STATES[1]

Specimen	Circumference[2]		Height		Spread		Location
	(Inches)	(Centimeters)	(Feet)	(Meters)	(Feet)	(Meters)	
Bitternut hickory (1975) (*Carya cordiformis*)	172	437	120	36	81	24.3	Virginia
Black hickory (1961) (*Carya texana*)	112	284	90	27	86	25.8	Texas
Coast Pignut hickory (1972) (*Carya glabra var. megacarpa*)	173	439	126	37.8	91	27.3	Louisiana
Mockernut hickory (1971) (*Carya tomentosa*)	161	409	110	33	96	28.8	Florida
Nutmeg hickory (1971) (*Carya myristiciformis*)	87	221	105	31.5	60	18	South Carolina
Pignut hickory (1972) (*Carya glabra var. glabra*)	183	464	125	37.5	87	26.1	Georgia
Sand hickory (1969) (*Carya pallida*)	112	284	100	30	75	22.5	South Carolina
Scarit hickory (1972) (*Carya × coliina*)	73	185	63	18.9	52	15.6	Missouri
Shagbark hickory (1977) (*Carya ovata*)	137	348	100	30	113	33.9	Maryland
Shellbark hickory (1974) (*Carya laciniosa*)	131	333	134	40.2	52	15.6	Indiana
Water hickory (1967) (*Carya aquatica*)	266	676	150	45	87	26.1	Florida

[1] From the "Social Register of Big Trees," The American Forestry Association (by permission).
[2] At 4.5 feet (1.4 meters).

The mockernut or bigbud hickory (*C. tomentosa*) occurs in southeastern Canada and the eastern United States. It tends to be a very tall tree with a round head. The dark-green leaves are long, with from five to nine toothed and pointed leaflets. The male catkins are from 3 to 5 inches (7.6 to 12.7 centimeters) long.

The water hickory or bitter pecan (*C. aquatica*) is generally a tree of the coastal plain, ranging from Virginia southward to Florida and then westward into Texas. The tree is capable of attaining great heights and is generally quite slender. The bark is an ashen gray, thin, and often quite shaggy on older trees. The leaves are compound. Leaflets have sharp points, of a deep yellow-green color, with slightly lighter coloration underneath. The wood is considered inferior as compared with other hickory species.

The shagbark or shellbark hickory (*C. ovata*) is a tree of stature and beauty and of great utility. It is capable of attaining great height, as evident from the accompanying table. It is valued for its wood and nuts. The bark is a pale-brown/gray and very shredded and shaggy—hence the name. Often, the bark will hang loosely in strips of a foot or more in length. The branches are pendulous and the foliage is a deep green. The leaves are large, from 4 to 6 inches (10.1 to 15.2 centimeters) in length. The staminate catkins occur in clusters of three and are green. The fruit may be described as globular in shape, with a very thick husk. The nut is white, thin-shelled and the kernel is sweet. This is the most important of the hickory nuts marketed, not of course including the pecans to be described shortly.

The shagbark hickory ranges from the Saint Lawrence River valley southward into Maine and generally following the Appalachian Mountains into the southeastern states. The tree ranges westward through Michigan and Minnesota and southward into Kansas, Oklahoma, and Texas. The tree does very well in certain parts of New England and particularly well in the Piedmont region of North Carolina. The wood is well known for its hardness, density, toughness, and close-grain. It is of a pale-brown color and remains the preferred wood for quality tool and implement handles and other heavy-duty applications. In the green state, the wood has a moisture content of 57% and a weight of 63 pounds per cubic foot (1009 kilograms per cubic meter). After air-drying to 12% moisture content, the weight per cubic foot is 51 pounds (817 kilograms per cubic meter) and 1,000 board-feet (2.36 cubic meter) of nominal sizes weigh 4,250 pounds (1927 kilograms). Crushing strength of the green wood when compression is applied parallel to the grain is 4,570 pounds per square inch (31.5 MPa); of the dry wood, 8,970 psi (61.9 MPa). The wood has 30% greater strength than white oak and double the shock resistance.

The species *C. illinoensis* is well known for its production of pecans. The tree ranges through much of the eastern United States. In particular, these trees are extensively cultivated in the southern states for their nut crop. The trees have huge branches and are capable of attaining a height of 100 feet (30 meters) or more. The head is rounded. There are 11 to 17 toothed, pointed leaflets. The tree was introduced to Europe many years ago and does well in the central and southern parts of France. For top production of pecans, the tree requires hot summers.

Wingnut trees are of the genus *Pterocarya*, deciduous, with large alternate, pinnate leaves. The leaflets are toothed. There are unisexual flowers in separate catkins appearing on the same tree. The trees bear small, winged nuts which occur on long hanging spikes. The trees are fast-growing and not too particular about soil. The Caucasian wingnut tree (*Pterocarya*) occurs in the Caucasus. It can attain a height of 100 feet (30 meters), is broad and spreading, with large oblong leaves from 8 to 14 inches (20.3 to 35.5 centimeters) in length. The *P. stenoptera* occurs in China and also can attain a height of 100 feet (30 meters). The *P.* × *rehdrena* is a hybrid of the two aforementioned species and is a shorter (up to 40 feet (12 meters) in height) broad-domed tree.

HIDDENITE. **Spodumene.**

HIDING POWER. **Paint.**

HIGH. **Atmosphere (Earth).**

HIGH-ALTITUDE WIND MEASUREMENTS. **Wind and Air Velocity Measurements.**

HIGH BLOOD PRESSURE. **Hypertension (High Blood Pressure).**

HIGH ENERGY ASTRONOMICAL OBSERVATORIES. **Gamma-Ray Astronomy; Neutron Stars; X-Ray Astronomy.**

HIGH FIDELITY. The quality of a sound reproducing system such that the acoustical characteristics of the reproduced sounds (usually musical) match as closely as possible the characteristics of the original sounds when made under their normal conditions. Thus a high fidelity reproduction of a symphonic work should sound the same to the listener as if he were present in a concert auditorium, listening to the orchestra directly, even though the sounds used in the recording were actually transcribed in a recording studio with extremely artificial acoustical characteristics.

See also **Musical Sound.**

HIGH-G ACCELEROMETER. **Acceleration Measurement.**

HIGH HEATING VALUE. **Coal; Combustion.**

HIGH-INTENSITY DISCHARGE LAMPS. **Illumination.**

HIGH-LIFT DEVICES. **Aerodynamics.**

HIGH LIMITING CONTROL. **Control Action.**

HIGH-PASS FILTER. **Filter (Communications System).**

HIGH-PRESSURE, HIGH-TEMPERATURE BOILER. **Boiler; Feedwater (Boiler).**

HIGH-PRESSURE TECHNOLOGY. **Pressure.**

HIGH-TEMPERATURE FUEL CELL. **Fuel Cells.**

HIGH-TEMPERATURE GAS-COOLED REACTOR. **Nuclear Reactor.**

HIGH-TEMPERATURE RESEARCH. **Solar Energy.**

HIGH VACUUM. **Vacuum.**

HIGHWAY BANKING. **Superelevation.**

HILDEBRAND RULE. The entropy of vaporization, i.e., the ratio of the heat of vaporization to the temperature at which it occurs, is a constant for many substances if it is determined at the same molal concentration of vapor for each substance.

HILL'S DETERMINANT. **Mathieu Equation.**

HINGE LIGAMENT. A tough elastic connection between the dorsal margins of the two valves of the shell of bivalve mollusks.

HINNY. **Horses, Asses, and Zebras.**

HIP. The joint at the attachment of the human thigh to the body. Also, the adjacent portion of the thigh where it merges with the buttocks and less commonly the corresponding part of the leg in various animals.

Congenital dislocation of the hip, caused by improper development during the fetal life, is thought to be a heritable condition. Females are much more likely to be afflicted than males, and the dislocation may be of one or both hips. The condition is often difficult to diagnose before the child begins to walk, although it is during infancy that treatment is most useful. The first symptoms may be a more pronounced rotation of the femur than in normal infants. When only one hip is affected, the creases in the infant's thighs may not be symmetrical. Upon starting to walk, the individual may develop a limp and marked lordosis (forward curvature of the spine). Later, there is a shortening of the thigh and a wide space between the thighs when the child stands with feet together. Usually there is no pain associated with the dislocation until adulthood and that is usually a low-back pain resulting from the lordosis. Treatment involves a long tedious procedure employing casts and weights to gradually correct the dislocation. Surgical treatment may be required if soft tissues have developed in the space to which the head of the femur is to be restored.

Accidental dislocation of the hip, as with other bones, causes pain and limitation of movement. Nerves also may be severely injured. In hip dislocations, the major motor nerve may be paralyzed; and if nutrient-supplying blood vessels are torn, the head of the thighbone may become necrotic, soft, and die, or osteoarthritis may develop.

Degeneration of the hip and other joints is described in the entry on **Bone.** See also **Osteoarthritis;** and **Rheumatoid Arthritis.**

HIPPOCAMPUS. Brain and Nervous System; Memory.

HIPPOCASTANACEAE. Horse Chestnut and Buckeye Trees.

HIPPOPOTAMUS (*Mammalia, Artiodactyla*). The group of *Hippotamines* is one of the smaller in the order *Artiodactyla* (even-toed hoofed animals). There are two extant species: (1) the greater Hippopotamus (*Hippopotamus*); and (2) the Pigmy Hippopotamus (*Choeropsis*).

The greater hippopotamus is a large, rather commonly occurring animal of the rivers of tropical Africa. The body is large, barrel-shaped, bulky, with short strong legs, and a very broad muzzle. The beast may attain a length of about 14 feet (4.2 meters) and a height of about 4 feet (1.2 meters). Four tons is usually the figure quoted for the larger specimens. They are almost hairless, and of a gray-black coloration with white underneath. Hippopotamuses are largely aquatic in habits and live entirely on vegetation, including both water plants and terrestrial species. They have been known to do extensive damage to crops in their roving. The animals feed mostly at night. The greater hippopotamus is the largest of the living nonruminating even-toed mammals. They are fast swimmers and can run about as fast as humans on land. They can issue loud grunts and bellows. Multiple births are rather uncommon. The gestation period is about eight months. The baby animal is known as a calf. See accompanying photo.

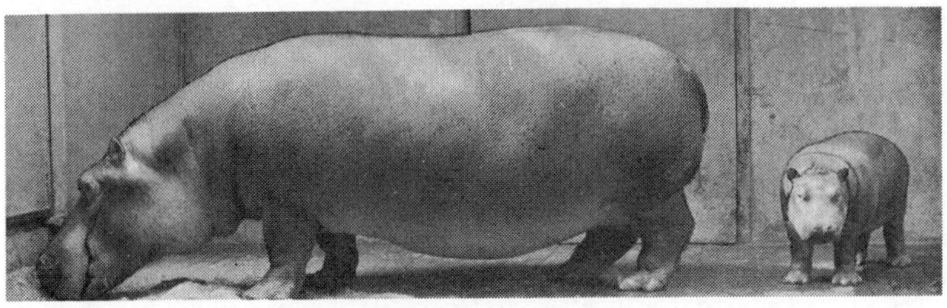

Female hippopotamus and baby. (*A. M. Winchester.*)

In some regions, the hippopotamus has been a staple food among tribesmen who harpoon the animal from small canoes. The manner in which the animal is prepared for consumption by some tribesmen is rather offensive to most people. After the body of the animal is dragged to the shore of the river, it is allowed to "ripen" under the hot tropical sun for a few days. The animal is then ripped open and the natives tear apart the softened carcass, gorging themselves on the rotten flesh.

Normally, the hippopotamus has a reasonably good disposition, but is proprietary concerning staked out stretches of the river. Usually, the animal will move out of the way of boats, floating just beneath the surface, but watching the passerby with use of the periscopic eyes which are just above the water line. However, the animal has been known to attack boats for unknown reasons, stomping and chewing the boat and occupants. Considering the size of the creature's mouth, the bite of a hippopotamus is no less than ghastly and usually terminal. The animal also is unpredictable when encountered on land, particularly at night.

The pigmy hippopotamus attains the size of a large pig. It has a large mouth equipped with fang-like teeth. For habitat, the pigmy hippopotamus prefers small lakes and rivers, ponds and stagnant pools and seldom wanders far from its aquatic habitat. The animal cannot afford to stay out of water very long because the skin is equipped with very large pores and, unless kept moist, the skin cracks easily. Apparently, the animals use these pores for absorbing water into their system, rather than taking all of their liquid input by mouth.

Naturalists always have suspected a relationship between the *Hippopotamines* and the *Suines*, but fossil remains have failed to yield evidence for this connection. And thus they remain in a separate classification. It is interesting to note that the Romans regarded them as pigs of a very special nature.

For references, see **Mammalia.**

HIRAYAMA FAMILIES. Asteroid.

HIRSUTISM. Abnormal growth of hair, particularly on the face of women. Although not well understood, causative factors appear to include a predisposition to the condition by inheritance, variations in endocrine activity, and imbalance of the metabolic processes. The condition usually does not appear until middle age. Treatment essentially is of a cosmetic nature.

Hirsutism is one of the principal features of polycistic ovary syndrome. Control of this hirsutism is extremely difficult. Best results have been obtained with therapy directed to suppressing adrenal and ovarian functions. Hirsutism is also seen in Cushing's syndrome (hyperfunction of adrenal cortex). See also **Androgens.**

HIRUDINEA. The leeches, a class of segmented worms (phylum *Annelida*), well known for their habit of sucking blood. Marine and freshwater species are known, and in the moist tropical forests terrestrial species occur. They often attach themselves to bathers.

The members of this class are distinguished from other annelids by the following characteristics: (1) The body is relatively short, usually with 32 segments. (2) The external segments are annuli, numbering from 2 to 14 to each metamere. (3) Each end of the body bears a sucker. (4) The mouth is usually provided with three toothed plates or jaws. (5) The alimentary tract is provided with an enormous pouched crop in which blood is stored prior to digestion. (6) The anus opens dorsally to the posterior sucker. (7) The coelom is partially obliterated by a peculiar mesenchymal tissue. (8) At the anterior end of the ventral nerve cord, several ganglia (ganglion) are fused to form a large mass.

Leeches were once extensively used in medicine for letting blood and are still of minor importance for this purpose. Otherwise they are of no importance to man save as an occasional annoyance. They eat small aquatic animals as well as the blood of vertebrates, and some species are entirely predacious.

Two orders are recognized:

Order *Rhynchobdellida.* With a protrusible proboscis, colorless blood, and no jaws. Marine and freshwater.

Order *Gnathobdellida.* With jaws and red blood. No proboscis. Freshwater and terrestrial. The medicinal leech belongs to this order. It is native to Europe but is naturalized in ponds and streams of the eastern United States.

HISTAMINE. A powerful vasodilator which is released in anaphylactic (hypersensitivity to protein) shock and occurs in blood and tissues in minute amounts.

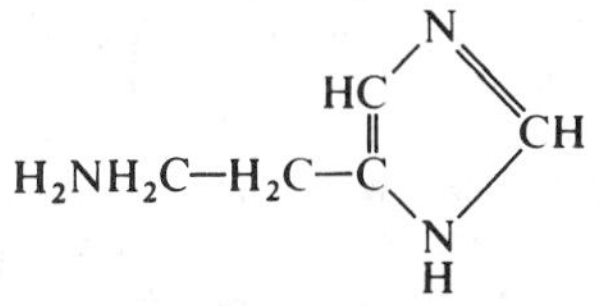

The injection of 1 microgram intravenously in humans is said to bring about a sharp drop in blood pressure. Its close relationship to histidine is emphasized by the fact that the amino acid can be decarboxylated by certain intestinal bacteria to produce it.

Histamine is a product of the degradation of histidine and is liberated by injury to the tissue, or whenever a protein is decomposed by putrefactive bacteria. Histamine's biological role is both positive and negative. The production of excessive histamine gave rise to the formulation of drugs for countering such excesses. See also **Antihistamine.** Histamine can cause pulmonary edema of noncardiac etiology. Histamine causes both constriction of bronchial smooth muscle and edema of bronchial mucosa by increasing the permeability of small bronchial veins. Histamine is one of several humoral mediators that affect bronchial tissue. Most cells contain and release histamine. Histamine stimulates connective tissue regeneration by producing edema.

HISTAMINE ANALOGUES. **Ulcer.**

HISTIDINE. **Amino Acids.**

HISTOGRAM. A histogram is a graphical representation of a grouped frequency distribution. Rectangles are formed by using the class interval as the base and the frequency of the class as the height. Equal areas represent equal frequencies. See accompanying figure.

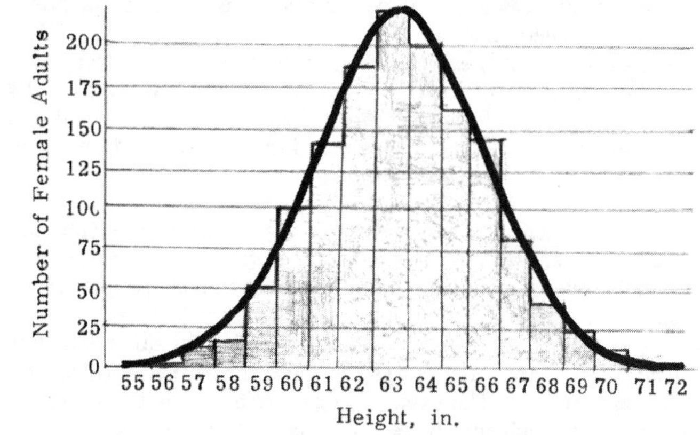

Histogram showing heights of female adults prepared from a survey of approximately 1400 women. Normal distribution curve is fitted to the histogram. This familiar bell-shaped curve typifies numerous empirical distributions found in biology.

HISTOLOGY. The science that deals with the minute structure of living things. Microscopic morphology. The study of the structure and functions of cells is the special province of cytology, leaving the study of special forms of cells and their association in tissues and organs as the field of histology, but histology necessarily includes much cytological matter.

The science is made up of two subordinate fields, general and special histology. In the former are considered the specialization of cells in the multicellular body and the characteristics and classification of the tissues in which they are grouped. The details of minute structure of the organs and organ systems are the materials of the latter. This field of histology is necessarily extensive and detailed, even in the study of a single species.

Histology recognizes five principal kinds of animal tissues, epithelium, nervous tissue, mesenchymal (connective and supporting) tissues, muscular tissue, and vascular tissue. All organs are made up of these components.

The tissues of plants are not so easily separable, but plant histologists recognize epithelial tissue, vascular tissue, supporting tissue, and parenchymous tissue. Plant tissue is studied more by its location than by the particular kinds of tissue.

HISTONES. Basic proteins which occur in the nuclei of both plant and animal cells. They are less basic than the protamines, having isolectric points at about PH 11. Some investigators restrict the term *histone* to only those basic proteins anatomically and chemically associated with DNA (deoxyribonucleic acids). The close associations of histones with DNA led to the hypothesis that histones might play a role in the control of genetic expression at the cellular level. Advances in molecular biology have permitted more detailed mechanisms for such control to be proposed. Histones, by blocking some areas of the DNA molecule, may permit only part of the DNA base sequences to act as templates for the formation of messenger RNA. Thus histones, by controlling messenger RNA formation, may ultimately control protein biosynthesis within the cell. Or, the primary role of histone may be structural, histone being essential for stabilizing the DNA helix, for the integration of DNA strands into more complex chromosomal structures, and for fixing and maintaining during cell division chromosomal changes occurring during differentiation and development. The foregoing two concepts are not mutually exclusive, i.e., histone may fix chromosomal structure in a specific configuration in which the position of the histone molecules also limit RNA formation. The possibility that histones play a role in genetic mechanisms suggests the possibility that histone changes may initiate or accompany early cellular changes, leading to the formation of tumors. Further investigation is needed to ascertain whether or not tumor histones differ from those of corresponding normal tissues. See also **Cell (Biology).**

HISTOPLASMOSIS. An infection of humans or animals by the fungus *Histoplasma capsulatum*, resulting from breathing in the spores which occur in soil. Although of worldwide occurrence, cases are reported most frequently in the United States, particularly in the Mississippi and Ohio River valleys. In humans, two forms appear: (1) the disseminated form with ulcerative lesions in skin and mucous membranes, and (2) the pulmonary form which simulates tuberculosis. The condition often is difficult to differentiate from tuberculosis. Infants are highly susceptible and the disease can be fatal in the young. In the limited form, the lesions appear most commonly around the mouth, on buccal mucous membranes, and on the penis. Drugs which have been effective in some cases include 2-hydroxystilbamidine and amphotericin B, the latter administered intravenously. In the late 1970s, increasing interest was shown in the use of immunotherapy in selected patients with progressive disease who are receiving conventional therapy and have a demonstrable immune defect. Initial experience with transfer factor therapy has shown some promise and research continues.

HISTOSOLS. **Soil.**

HITTORF PRINCIPLE. An application of the Paschen law. The Hittorf principle states that discharge between electrodes in gas at a given pressure will not always occur between the closest points of the electrodes if the distance between these points corresponds to a point to the left of the minimum of the ignition potential curve.

H LINES. A contour along which the electromagnetic field strength is constant with respect to some reference plane.

HIVES. **Urticaria.**

HMX ROCKET PROPELLANTS. **Rocket Propellants.**

HOARFROST. **Precipitation and Hydrometeors.**

HOATZINS (*Aves, Galliformes*). A very strange bird, which at first sight look like small curassows, belonging to the suborder *Opisthocomi.* There is only one species, the Hoatzin (*Opisthocomus hoazin*). The size is approximately that of a crow, the length is 60 centimeters ($23\frac{1}{2}$ inches) and the weight is about 800 grams (28 ounces).

The hoatzins lay claim to a special position among all birds: first, as a result of their specialized diet, and second, because of the ability of their young, while they are still very undeveloped, to climb about a network of branches on all fours with the aid of the primary wing feathers which have talons. Young hoatzins have particularly long and movable first and second digits; each has a strong claw which retrogresses later. The oldest bird so far known, *Archaeopteryx* from the Jurassic period, also had such flexible fingers with claws. We assume that it, too used them to climb around in trees. Therefore, young hoatzins look primitive when they move around in the branches like reptiles. They not only climb, but also swim and dive with all fours when danger impels them to drop into the water.

Old hoatzins, in contrast, avoid the water and almost never touch the ground. Yet they, too, give the impression of being "primitive" when they flit about in tree branches or awkwardly fly short distances. But that has to do with their diet, in which their crop plays a peculiar part. Hoatzins primarily eat leaves of various arum types which they pick or from which they tear off large pieces with their beaks. They form the pieces into a ball in their mouths and swallow these large

Hoatzin (*Opisthocomus hoazin*).

chunks. The leaves are ground into a fine mash in their huge crops, which are extremely muscular, with horny ridges, and are divided into several sections. The mash them passes through the small gizzard and the short intestine. The crop is fifty times as large as the gizzard and represents 13% of the entire weight of the bird. In no other bird is the crop comparatively as large. See also **Galliformes.**

HOB. A milling cutter with form-type teeth of helicoidal shape and with profiles such that conjugate surfaces on cylindrical parts may be machined by rotating the work and the hob at a constant velocity ratio. Hobs are extensively used for cutting spur gears, and hobbing is the only really precise method of cutting heavy-duty worm wheels. Two types of gear hobs are commonly used; the radial or infeed type, and the tapered or tangential feed hob. The latter is superior, particularly for hobbing worm gears with high helix angles and high pressure angle. Hobbing processes are also used for spline cutting, and for generating ratchet teeth.

The term hobbing is also used to designate a method of die sinking, in which a hardened master punch, a duplicate of the part to be formed, is pressed into an unheated die blank so that the shape of the hob is reproduced in the die impression. This method of producing die cavities is simpler than die sinking by cutting away the material, since it is considerably easier to machine the surface of the hob than to machine the die cavity. It is also advantageous in the production of multiple die cavities, since a single hob can be used for a series of duplicate dies. The process is also referred to as "hubbing."

HOBBING. Worm Gearing.

HOCK (or Hough). The joint at the attachment of the foot and the leg in animals which walk on the toes (digitigrade or unguligrade), commonly applied to domestic animals. It corresponds to the ankle joint of other species. Also the back of the human knee.

HODGKIN'S DISEASE. A malady characterized by a painless localized enlargement of lymph nodes, usually beginning in one side of the neck. The patient may develop fever, and generalized itching, or an eruption on the skin. Anemia is uncommon except in advanced cases and tests of peripheral blood appear to have no diagnostic value. Diagnosis is made by biopsy. Surgical excision, irradiation, and chemotherapy all have a place in the treatment of patients with Hodgkin's disease. Surgical excision can be used when the condition is localized, followed sometimes by local irradiation and/or chemotherapy. X-irradiation alone is valuable for localized disease. Massive doses in the early stages can produce dramatic results, including the rapid disappearance of masses and long remissions. Nitrogen mustard has been beneficial in patients with disseminated disease. Other drugs which have been used include chlorambucil, cyclophosphamide, and vinblastine sulfate.

Hodgkin's disease has a bimodal, age-specific incidence rate in the United States and northern Europe. There is a high rate between the ages of 15 and 34 and after the age of 50 years. The first age mode appears to be absent in Japan. Hodgkin's disease in children under 10 years of age is seen much more frequently in some of the developing countries of Latin America and the Middle East than in the United States and is found in boys from 8 to 10 times more frequently than in girls.

The clinical course of the disease can be extremely variable. In addition, almost all patients receive treatment that may profoundly affect the course of the disease. Sometimes the treatment results in apparent cure, and sometimes it produces complications that become difficult to separate from the disease itself. However, in time, nearly all patients with untreated or uncontrollable Hodgkin's disease develop increasingly severe systemic symptoms. High continuous fever, drenching night sweats, malaise, fatigue, anorexia, and weight loss characterize the terminal picture.

Hodgkin's disease no longer can be considered inevitably fatal. No matter what the stage of disease, patients now have the potential for cure, although the probability of cure ranges between 25 and 90%.

References

NOTE: See also the list of references at the end of the entry on **Cancer.**

Coleman, C. N., et al.: "Hematologic Neoplasia in Patients Treated for Hodgkin's Disease," *N. Engl. J. Med.*, **297**, 1249 (1977).

MacMahon, B.: "Epidemiology of Hodgkin's Disease," *Cancer Res.*, **26**, 1189 (1966).

Rosenberg, S. A., and H. S. Kaplan: "Hodgkin's Disease and Other Malignant Lymphomas," *Calif. Med.*, **113**, 23 (1970).

Siber, G. R., et al.: "Impaired Antibody Response to Pneumococcal Vaccine after Treatment for Hodgkin's Disease," *N. Engl. J. Med.*, **299**, 442 (1978).

HODOGRAPH. In general (mathematics), the locus of one end of a variable vector as the other end remains fixed. A common hodograph in meteorology represents the vertical distribution of the horizontal wind.

HOFFMAN REARRANGEMENT. Rearrangement (Organic Chemistry).

HOFMEISTER SERIES. Coagulation (Hofmeister Series).

HOGBACK. Ridge-like topographic features, the result of the differential erosion of highly tilted hard and soft strata. The steeper, or dip-slope, side is developed on the harder or less soluble formation, while the gentler slope is developed on the opposite side, on the softer rocks.

HOIST. Any device for lifting materials, weights, articles, etc., may be called a hoist. Hoists often compose a part of other apparatus whose purpose may extend to movement of material other than vertically. For example, the bridge crane incorporates within it a hoist for vertical lift. The energy required for lifting is derived ultimately from a number of various sources. For example, in the hoisting field one finds such varied power sources as compressed air, internal combustion engines, hydraulic power, steam and electric power. The pneumatic drives may be either a direct lift supplied by air acting on a piston connected directly to the load, or it may be employed in compressed air engines, whose crankshaft is geared to the hoisting apparatus. In the internal combustion engine type hoist, the gasoline engine is generally used for the light-capacity hoist, and the Diesel engine for heavier hoists. It has the advantage over other drives for portable service, such as locomotive cranes, and power shovels.

The essential parts of a hoist are a rope or chain which is wrapped around a drum or drive sheave. A hook, grapnel magnet, or other device for handling the load is attached to the free end. The rotation of the drum winds up the rope, thus shortening the distance between the drum and the load. If the drum is fixed in position over the load, naturally the load must be hoisted. To drive the drum, one of the power supplies just mentioned is connected with the drum through a suitable speed-reducing, torque-increasing mechanism. A gear train is often used. These component parts when supplied with a brake controlling the speed during lowering of weights, are the essential elements of all hoists except the direct-acting.

HOLDING BEAM (CRT). Cathode-Ray Tube.

HOLE THEORY (Liquids). Liquids differ from solids in having a sufficient number of unoccupied positions, "holes," in the lattice that comparatively free movement of molecules is possible by movement into unoccupied sites. The volume of the holes is the free volume.

HOLLERITH. Pertaining to a widely used system of encoding alphanumeric information onto cards (described by American National Standard ANSI X3.26–1970). The term Hollerith cards is synonymous with punch cards. Such cards were first used in 1890 for the United States Census and were named after Herman Hollerith, their originator. See also **Input/Output Devices.**

HOLLOW-TOOTHED DEER. Deer.

HOLLYHOCK. Malvaceae.

HOLLY TREES AND SHRUBS. Of the family *Aquifoliaceae* (holly family), genus *Ilex*, there are numerous species of hollies and many hybrids and cultivars, making both nomenclature and generalization difficult. The plants may be deciduous or evergreen. They often are spiny with leathery leaves. The flowers frequently are white, usually polygamous. They bear small fruit and can withstand full sun or partial shade. The hollies tend to me more resistant to pests than most plants. Some of the important varieties include:

American holly	*Ilex opaca*
Azorean holly	*I. perado*
Chinese holly	*I. pernyi*
Dahoon holly	*I. cassine*
English holly	*I. aquifolium*
Highclere hybrid holly	*I. × altacierensis*
Longstalk holly	*I. pedunculosa*
Posshmhaw holly	*I. decidua*
Tarajo holly	*I. latifolia*

Depending upon height, the American holly may be considered a shrub or a tree. The plant can range from about 15 to 30 feet (4.5 to 9 meters) in height, although as shown by the accompanying table, under favorable conditions, the plant can develop into a very sizeable tree. The foliage may be described as being of a bronze-green or olive-green color. The leaves are glossy and quite spiny, but less so than the English species. The leaves are from 2 to 3 inches (5 to 7.6 centimeters) in length. The fruit is a scarlet red, sometimes (in the 'Xanthocarpa') a bright yellow, and a little over $\frac{1}{4}$-inch (0.6 centimeters) in diameter. It is berrylike on short stems and often clings to the plant throughout most of the winter months. With proper care, there are numerous areas in the United States where the plant does quite well. The natural occurrence generally follows the coastal regions. Sheltered locations are preferred. The plant ranges from Massachusetts southward into Florida and westward to the Mississippi Valley south of lower Indiana and Illinois.

As the name suggests, Azorean holly is found on the Azores and also on the Canary Islands. This species is a small evergreen tree with dark green foliage and deep red berries. *I. pernyi* is found in central and westward China and is a narrow tree of pyramidal form that can rise to a height of about 30 feet (9 meters). The leaves are small, leathery, and of a lustrous dark green color. The tree bears clusters of small red berries.

The Dahoon holly is found in the southeastern United States and is characterized by a somewhat heavier trunk than found on most hollies. The leaves are evergreen and narrow, about 2 to 4 inches (5 to 10 centimeters) in length. Their color is dark green. The flowers and fruit are similar to the American holly. The plant ranges from the southern part of Virginia to Florida and along the Gulf coast west to Louisiana.

English holly occurs naturally in western Asia, northern Africa, and southern Europe, as well as the British Isles from which it derives its name. However, several forms of this holly are not so hardy in England as they may be on the continent. Because of the numerous hybrids, a great variety of leaf and fruit colorations, as well as other characteristics of the shrubs and trees, is obtainable. Some of the more important varieties include: Perry's weeping silver holly ('Argenteomarginata Pendula,' which has silver foliage and lots of berries; the silver milk-boy ('Argenteo-Medio Picta'), which grows to a height of about 30 feet (9 meters) and has dark green, spiny leaves with cream-colored spots in their central portion; the golden queen ('Aurea Regina'), with yellow-edged dark green leaves; and the silver hedgehog holly ('Ferrox Argentea'), which has leaves featuring white spines and margins, and ranging up to 15 feet (4.5 meters). Other varieties include the 'Bacciflavia,' the 'Crispa,' the 'Elegantissima,' the 'Ferox,' and the 'Hastata.'

The highclere hybrid hollies are known for their vigor. They have evergreen leaves, quite large. They are known for their toleration of industrial and seaside environments. They range in height from small bushes to trees of 50 feet (15 meters). These hybrids were obtained by crossing the English holly with the Azorean holly. Some of the more important varieties include: 'Camellifolia,' a tree of conical contour, characterized by a purple bark, almost spineless evergreen leaves that are purple when young, later turning a dark green; the 'Golden

RECORD HOLLIES IN THE UNITED STATES[1]

Specimen	Circumference[2]		Height		Spread		Location
	(Inches)	(Centimeters)	(Feet)	(Meters)	(Feet)	(Meters)	
American holly (1972) (*Ilex opaca*)	169	429	53	15.9	61	18.3	Texas
Carolina holly (1971) (*Ilex ambiqua*)	108	274	17	5.1	12	3.6	Florida
Gallberry holly (1973) (*Ilex coriacea*)	60	152	17	5.1	12	3.6	Texas
Silver Varigated holly (1977) (*Ilex aquifolium*)	75	191	40	12	22	6.6	Oregon
Tawnyberry holly (1973) (*Ilex krugiana*)	34	86	55	16.5	22	6.6	Florida
Mountain winterberry holly (1972) (*Ilex montana*)	25	63	40	12	80	24	New York
Winterberry holly (1971) (*Ilex verticillata*)	28	71	40	12	16	4.8	Florida
Yaupon holly (1972) (*Ilex vomitoria*)	49	124	45	13.5	40	12	Texas

[1] From the "Social Register of Big Trees," The American Forestry Association (by permission).

[2] At 4.5 feet (1.4 meters).

King,' which has green leaves with yellow edges, nearly spineless; the 'J. C. van Tol,' almost spineless leaves of dark-green color and produces large quantities of berries; the 'Lawsoniana,' which has large leaves with yellow borders and marbleized centers; the 'Purple Shaft,' which is known for its vigor and large quantities of berries; and the 'Silver Sentinel,' which has mottled leaves that are flat and almost spineless.

The longstalk holly is found in Japan. It ranges from a shrub to a small tree of about 30 feet (9 meters) in height. The plant has evergreen leaves and small red fruits.

The Posshmhaw holly is found in the southeastern United States. It is also sometimes referred to as the swamp holly. Normally, it is a small shrub or tree, but as shown by the accompanying table, the plant can attain very respectable dimensions under favorable conditions. The leaves are a lustrous deep green, deciduous, and from 1½ to 3 inches (7.6 centimeters) in length. The flowers are similar to those of the American holly. Its natural range is between the Atlantic coast and the Appalachian Mountains south of Virginia and into western Florida and westward to Arkansas, Missouri, and Texas.

The Tarajo holly is found in Japan. This species can attain a height of 60 feet (18 meters) or more and features the largest leaves of any holly. The leaves are evergreen of a dark green color, yellow underneath. The fruit occurs in large numbers of orange-red clusters.

HOLMIUM. Chemical element symbol Ho, at. no. 67, at. wt. 164.93, tenth in the Lanthanide Series in the periodic table, mp 1,472°C, bp 2700°C, density 8.795 g/cm^3 (20°C). Elemental holmium has a close-packed hexagonal crystal structure at 25°C. The pure holmium is silver-gray in color, slow to tarnish or oxidize at room temperature in normal atmospheres. Even at relatively high temperatures, the metal is slow to oxidize. Under a vacuum of about 10 torr, holmium will react when hot with water vapor, CO_2, NH_3, and hydrocarbons. Holmium is soft and can be worked by conventional equipment. There is one natural isotope of holmium, ^{165}Ho and 18 artificial isotopes have been produced. The natural isotope is not radioactive. In terms of abundance, holmium is present on the average of 1.2 ppm in the earth's crust, ranking ahead of bismuth, antimony, cadmium, and mercury in potential availability. The element was first identified by P. T. Cleve and J. L. Soret in 1879. The metal has a low acute-toxicity rating. Electronic configuration $1s^22s^22p^63s^23p^63d^{10}4s^24p^64d^{10}4f^{10}5s^25p^65d^16s^2$. Ionic radius Ho^{3+} 0.894 Å. Metallic radius 1.766 Å. Other important physical properties of holmium are given under **Rare-Earth Elements and Metals.**

Holmium occurs in apatite, xenotime, and yttrium-and heavy rare-earth minerals. The element of a purity of 99.9% can be obtained through organic ion-exchange techniques. Supplies of holmium are available commercially as the result of yttrium production. To date, the applications for holmium have been very limited. When added to orthoferrites, it has shown promise for use in electronic circuits. Uses in semiconductors, lasers, thermoelectric devices, phosphors, and ferrite bubble devices currently are being studied.

See references listed at ends of entries on **Chemical Elements;** and **Rare-Earth Elements and Metals.**

NOTE: This 6th Edition entry was revised and updated by K. A. Gschneidner, Jr., Director, and B. Evans, Assistant Chemist, Rare-Earth Information Center, Energy and Mineral Resources Research Institute, Iowa State University, Ames, Iowa. Original 5th Edition entry was prepared by J. G. Cannon, Molycorp, Inc.

HOLOCRYSTALLINE. The term applied by petrologists to igneous rocks composed entirely of crystals; in contradistinction to igneous rocks which are partly or entirely composed of natural glass, such as obsidian.

HOLOENZYME. Coenzymes.

HOLOGRAPHY. The technique of holography is similar to photography in many respects, yet it is fundamentally different. With photography, one generally records, by means of lens and film, the two-dimensional irradiance distribution in the image of an object. With holography, one records not the optically formed image of an object, but the object wave itself. This wave is recorded (frequently on photographic film) in such a way that a subsequent illumination of this record, called a *hologram,* reconstructs the original object wave. A visual observation of this reconstructed wavefront then yields a view of the object which is practically indiscernible from the original, including three-dimensional parallax effects. The process was discovered by Gabor[1] (England) in 1948. It was then identified as a two-step method of optical imagery. During the past couple of decades, holography has become widely known and a limited number of practical uses for it have been developed. This later progress is attributed to the general availability of the laser, with the outstanding temporal and spatial coherence of its light. Much of the work in adapting the laser to holography was carried out by Upatnieks and Leith (University of Michigan) during the early 1960s.

With reference to Fig. 1(a), one starts with a single, monochromatic beam of light that has originated from a very small source. This single beam is split into two components, one of which is directed toward the object and the other to a suitable recording medium, most commonly a photographic emulsion. The component that is incident on the object is scattered by it, and this scattered radiation, now called the object wave, impinges on the recording medium. The wave that proceeds directly to the recording medium is called the *reference wave.* Since the object and reference waves originate from the same source, they are mutually coherent and form a stable interference pattern when they meet at the recording medium. The detailed record of this interference pattern constitutes the hologram.

When the hologram is illuminated with a beam similar to the original reference wave, it modulates the phase and/or amplitude of the illuminating wave in such a way that the transmitted wave divided into three separate components, one of which exactly duplicates the original object wave.

If the two interfering beams are traveling in substantially the same direction, the recording of the interference pattern is said to be a *Gabor hologram* or *in-line hologram.* If the two interfering beams arrive at the recording medium from substantially different directions, the recording is a *Leith-Upatnieks* or *off-axis hologram.* If the two interfering beams are traveling in essentially opposite directions, the recorded hologram is said to be a *Lippmann* or *reflection* hologram, first invented by Denisyuk.

Electromagnetic radiation is most commonly used, although acoustic radiation can be used. The most common electromagnetic radiation employed is light, but holograms have also been recorded successfully with electron beams, x-radiation, and microwaves.

Holograms can be classified by the way they diffract light. In an *amplitude hologram,* the varying irradiance distribution of the interference pattern is recorded as a density variation of the recording medium. In this type of hologram, the illuminating wave is always partially absorbed, i.e., the illuminating wave is *amplitude-modulated.* In the *phase hologram,* a *phase modulation* is imposed on the illuminating beam which, in turn, results in diffraction of the light. Phase modulation occurs when the optical path (thickness × index) varies with position. A phase hologram results from either relief-image or index variation, or both.

Either phase or amplitude holograms can be classified further as *Fresnel holograms* or an *Fraunhofer holograms.* Generally speaking, if the object is reasonably close to the recording medium, say just a few hologram or object diameters distant, the field at the hologram plane is the Fresnel diffraction pattern of the object. A hologram recorded in this manner is termed a *Fresnel hologram.*

If the object and hologram are separated by many object or hologram diameters, the field at the hologram due to the object alone is the Fraunhofer diffraction pattern of the object. A hologram recorded in this manner is termed a *Fraunhofer hologram.*

Any of these holograms types may be recorded as either a *thick* or a *thin* hologram. A thin hologram is one for which the thickness of the recording medium is thin compared to the space between the recorded interference fringes. A thick or volume hologram is one in which the thickness of the recording medium is of the order of or greater than the spacing of the recorded fringes.

[1] For which he received the Nobel Prize in physics (1971).

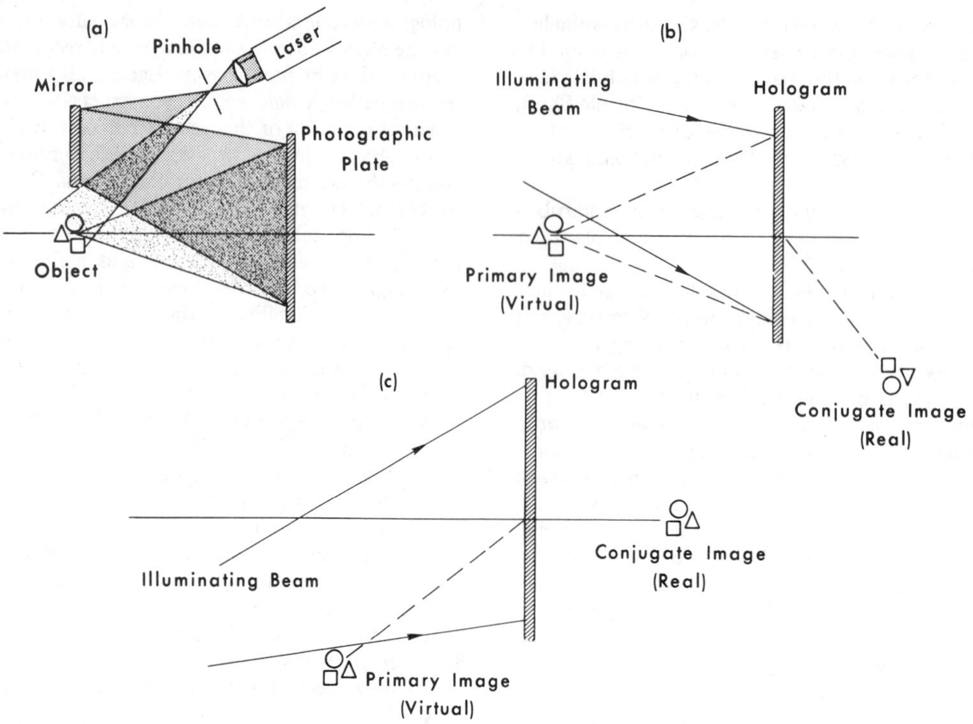

Fig. 1. A typical holographic arrangement: (a) Recording the hologram; (b) reconstructing the primary object wave; (c) reconstructing an undistorted conjugate wave.

Conceptually, the simplest form of an off-axis hologram is one for which the object is just a single, infinitely distant point so that the object wave at the recording medium is a plane wave. If the reference wave is also plane, and incident on the recording medium at an angle to the object wave, the hologram will consist of a series of Young's interference fringes. These recorded fringes are equally spaced straight lines running perpendicular to the plane of incidence. Since the hologram consists of a series of alternating clear and opaque strips, it is in the form of a diffraction grating. When the hologram is illuminated with a plane wave, the transmitted light consists of a zero-order wave traveling in the direction of the illuminating wave, plus two first-order waves. The higher diffracted orders are generally missing or very weak, inasmuch as the irradiance distribution of a two-beam interference pattern is sinusoidal. As long as the recording is essentially linear (irradiance proportional to final amplitude transmittance), the hologram will be a diffraction grating varying sinusoidally in amplitude transmittance, and only the first diffracted orders will be observed. One of these first-order waves will be traveling in the same direction as the object wave. This is the reconstructed wave.

The recording of a hologram and the subsequent reconstruction is shown in Fig. 1. In Fig. 1(a), the laser beam is first expanded and then divided by a mirror, which directs part of the beam directly onto the photographic plate; the rest of the light is reflected from the object. After processing, the hologram plate may be replaced in its original position (Fig. 1(b)), and the object removed. The light diffracted by the hologram forms, in part, the same wavefront that was originally scattered by the object. A viewer looking through the hologram will see an undistorted view of the object, just as if it were still present.

In addition to the *virtual* or *primary image*, a real, or *conjugate image* will be formed on the observer's side of the hologram. This image will appear unsharp and highly distorted, and it will also be inverted in depth, i.e., reversed front to back, as shown in Fig. 1(b). However, a distortion-free real image can be formed by changing the position of the illuminating beam so that all of the rays of the reference beam are reversed in direction. In this way, an undistorted, real, three-dimensional image of the object scene appears in front of the hologram, as shown in Fig. 1(c).

Holograms may be recorded with diverging, parallel, or converging reference beams. If care is taken to maintain the recording geometry during reconstruction, it is possible to form holograms with an arbitrary reference beam, the only requirement being that it be coherent with the object beam.

Color holograms can be produced by recording three separate holograms on a single photographic plate, each in a different color. Subsequent illumination with a three-color beam yields three separate wavefronts, one in each of the three colors representing the portion of the object corresponding to that color.

Holograms also can be made that can be viewed in reflection. This is done by allowing the reference and object beams to enter the recording medium from opposite sides. The fringes formed are planes lying approximately parallel to the plane of the hologram. When such a hologram is illuminated by a beam similar to the reference wave, a reflected wave is formed which exactly duplicates the object wave. The image is viewed in reflected light. This type of hologram can be illuminated with white light. The interference planes filter the light by acting as a $\lambda/2$ multilayer interference filter, in the same way as in Lippmann color photography.

One of the most striking aspects of the modern hologram is the three-dimensional image that it is capable of producing. The three-dimensional image indicates that there is a large amount of information contained in a single hologram—much more than is contained in a conventional photograph of the same size. Because of the many perspectives available, the hologram is well suited to display purposes. With a hologram, one can present all of the observable characteristics of a three-dimensional object clearly and concisely. Complex molecular or anatomical structure can be simply presented with a single holographic image, with little chance of error or misinterpretation on the part of the viewer. Thus holograms may reduce the number of conventional drawings or photographs to illustrate a single object. It has been proposed that the use of holograms in textbooks would be an aid to readers, particularly in fields where three dimensions are important. Holograms can be made to be viewed with a small penlight and a colored filter.

Holographs have been applied to microscopy and have been used in interferometry and vibration analysis. A number of artists have investigated the hologram as a new art form (*holoart*).

During the last few years, considerable interest has been shown in employing holography as a means for filing and storing information. The most common way to store data, as of the early 1980s, is through the use of magnetic media (discs, spools, reels, etc.). Although far superior to most other means, the magnetic methods also have their disadvantages. Holograms provide additional security—because if half of a hologram (for example) is destroyed, half of the information is not lost as in the case of a photograph. The image from half a hologram can still be reproduced, with a slight decrease in the signal- (information) -to-noise ratio. Thus, binary data may be stored in a hologram without the problems associated, for example with conventional microfiche. Because of the natural redundancy of the hologram, a hologram on a microfiche can be considerably smaller than a microphotographed page. It has been estimated (Maugh, 1978) that a 4 × 6-inch (10 × 15-centimeter) holofiche will store up to 200 megabits of information, as contained in 20,000 individuals holograms, and that a file of holofiche may contain up to 208,000 times the data as an equal volume of punched cards.

Figure 2 illustrates a page of text illuminated by coherent light

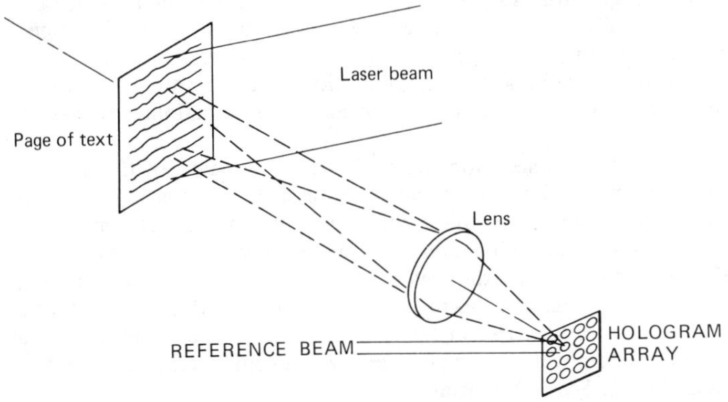

Fig. 2. Holographic storage and retrieval of documents. (*Harris Corporation.*)

derived from a laser. Light reflected from the page of print, along with a reference beam derived from the same laser, exposes a small area of a sheet of film. The interference between the reference beam and the signal beam captures the phase and amplitude information from the page of text so that, after illumination by the reference beam of the developed film, the page of text can be reconstructed as in conventional holograph.[2]

The surface roughness of the paper unfortunately causes the reflected light to scatter over wide angles so that a rather large hologram is required to capture all the information contained in the text. In principle, all parts of the signal contribute to all parts of the hologram; it should be possible therefore to reconstruct the page of text from any small portion of the hologram. In practice, however, it is found that the reconstructed image has a coarse speckled pattern superimposed on the text, making it less legible than desired.

A solution to the speckle problem is to first microfilm the page at a demagnification of 10×, which requires significantly less stringent local tolerances than would a very high demagnification. A hologram is then made from the microfilm transparency and the speckle pattern does not appear in the reconstructed image because the transparency does not have surface roughness. Furthermore, because the text originally black on white, the microfilm can be made so that the text is white on black, the net effect being that all the light contributing to the hologram is generated by the characters and no light is contributed by the background.

With this process, an additional 20× reduction in size is feasible, for an overall reduction of 200×. By using a step-and-repeat exposure sequence, it is feasible to record as many as 14,000 pages of text on a standard 4 × 6-inch microfiche. Such a reduction factor is extremely difficult to achieve with conventional microfilm techniques. Generally, demagnifications of over 48× using conventional microfilming techniques are also done in two steps, with the initial reduction being 10× to 20×. The chief advantage of holography is that depth of focus is not a factor in the second reduction process, so that if the hologram is not exactly in the correct position, the resolution of the reconstructed image is not thereby affected. Reconstruction can be done in a one-step process; it is not necessary to magnify the image in two steps as when the hologram was generated. Finally, a characteristic of a certain type of hologram known as a *Fourier transform hologram* is that, upon construction, a lateral shift of the hologram does not cause a magnified lateral shift in the reconstructed image. Both focal and positional tolerances of the hologram upon retrieval are thus significantly eased.

As the resolution of the input material increases so too does the necessary hologram size. At some point, the gain in demagnification that holography provides is offset by the need for better resolution. For ordinary textual material, engineering drawings, and low-resolution photography, the advantage of holography is appreciable, but for very high resolution images these advantages largely disappear. Still, there are no attractive alternatives to the high density storage and retrieval of such information except to leave them in their original state or to convert them to a digital data base.

Holography can also be used to store digital data which consists of a set of zeros and ones that are generally organized into *N*-bit words. Figure 3 shows the basic holographic recording and reconstruc-

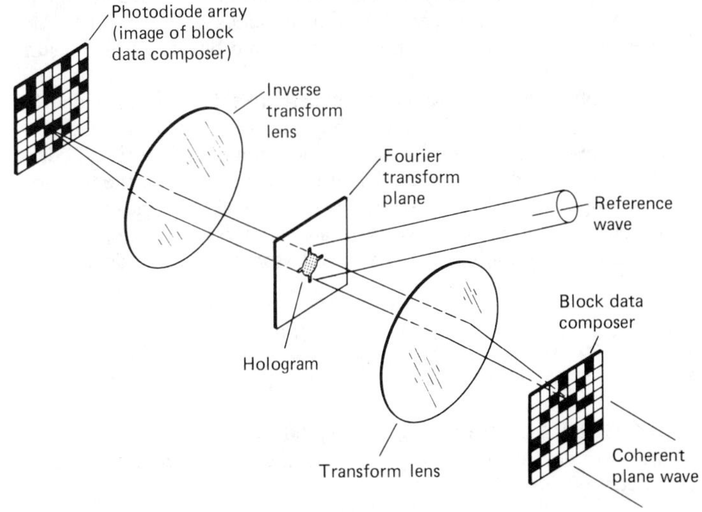

Fig. 3. Basic elements of a block-oriented memory. (*Harris Corporation.*)

tion process for digital data. The electrical impulses representing the ones in the data set activate a device called a *block data composer* so that it is transparent where the data set has ones and opaque where the data set has zeros. As shown in the figure, the data may be reorganized by the block data composer so that each row contains an *N*-bit word; if there are *M* rows, the number of bits temporarily stored in the block data composer is *NM*.

Coherent light from a laser is spatially modulated by the block data composer and a lens produces the Fourier transform of the data pattern at the plane of the hologram. After the reference beam is added to the signal beam an exposure is made to record the hologram. The process is then repeated with a new block of *NM* bits and another hologram is recorded at a new location either by deflecting both the signal and reference beam to the new location or by moving the recording medium to the new location.

The retrieval process consists of addressing the proper hologram location with the same (or similar) reference beam, which reconstructs the original data pattern onto a photodiode array which converts the optical pattern into electrical impulses denoting ones and zeros. The data can be retrieved in the same format as it was recorded or reformated, depending on the characteristics of the photodiode array.

For those applications where it is desirable to record one-dimensional holograms, both the block data composer and the photodiode array consists of just one row. The holograms are also, of course, one-dimensional and the recording medium usually is moved with a uniform linear velocity during both the recording and reconstruction process.

A feature of holographic storage mentioned before is that light from each bit in the block data composer contributes to the entire area of the hologram. Dust particles or small imperfections in the

[2] Several following paragraphs contributed by A. VanderLugt, Senior Scientist, Advanced Technology Department, Harris Corporation, Melbourne, Florida.

recording media, therefore, do not destroy any single bit; rather, all bits are reconstructed with slightly reduced contrast, which is not evident after the output of the photodetector array is thresholded. Another feature is the relaxed tolerances on focal and lateral positioning of the hologram when reconstructed.

To see why the redundancy and relaxed positioned tolerances characteristics of holography arise, consider a data set consisting of only two bits which are equivalent to point sources in the block data composer plane. At the plane of the hologram, each point source produces a plane wave that illuminates the entire hologram aperture. Since the plane waves travel in slightly different directions, an interference fringe pattern is produced over the hologram aperture. Since the fringes are not localized at any particular plane, the depth of focus is relaxed. Since the fringes exist over the entire hologram, the hologram is said to have redundancy and a small imperfection removes only a small amount of energy in the reconstruction process. Finally, since the fringe pattern has no apparent central point, a lateral error in the position of the hologram upon reconstruction does not change the position or intensity of the reconstructed bits. The only effect of a positional error is a phase change at the photodiode array plane which does not affect the *intensity* of the reconstructed bits. This argument can be valid for any number of bits in the block data composer; the interference pattern becomes more complicated but it is still the sum of elementary fringe patterns.

In the early 1970s the reduced geometrical tolerances provided by the holographic process, along with high storage density, sparked considerable research and development of holographic memories. Today, the trend is toward direct optical spot recording based on the technology developed for video disc recording (q.v.) in which some of geometrical tolerance problems have been solved. Both technologies offer an attractive alternative to the archival storage of large data bases.

References

Holden, C.: "Holoart," *Science*, **204**, 40–41 (1980).
Maugh, T. H., II: "Holographic Filing," *Science*, **201**, 431–432 (1978).
Walker, J.: "Holograms," *Sci. Amer.*, **242**, 158–170 (February 1980).

HOLOHEDRAL CRYSTAL. A crystal in which the full number of faces are developed, corresponding to the maximum and complete symmetry of the system. See **Mineralogy.**

HOLOTHUROIDEA. The sea cucumbers, a class of the phylum *Echinodermata.*

These animals differ from other echinoderms in several particulars: (1) The principal axis is elongated and the animal rests on its side. (2) The body wall is soft because of the reduction of the calcareous ossicles. (3) A branching respiratory tree extends from the alimentary tract into the body cavity.

Sea cucumbers are used as food in the Oriental region. They are dried for the market and in this form are called trepang or bêche-de-mer.

The class includes five orders:

Order *Aspidochirota.* Tropical species with shield-shaped tentacles. In shallow water.
Order *Elasipoda.* Benthonic species of deep water.
Order *Dendrochirota.* Shallow water species with branching tentacles.
Order *Molpadonia.* Burrowing species. Tentacles unbranched or slightly branched.
Order *Apoda.* (*Synaptida, Paractinopoda.*) Burrowing species without respiratory trees.

HOLOTYPE. A term used by biologists and paleontologists to mean the specimen to which all others should ultimately refer to determine the species. The holotype does not necessarily have to be the originally described species (type) and frequently is not.

HOME HEATING (Hydrogen). Hydrogen (Fuel).

HOME HEATING OIL. Petroleum.

HOMEOMORPHIC GRAPH. Graph (Mathematics).

HOMEOSTASIS. Maintenance of the steady state. As applied to living organisms, this refers to the many adjustments which are constantly being made to keep the organism in a rather constant environment internally in spite of the fact that there may be many variations in external environment. Life within individual cells can continue only within a rather narrow range of conditions, and each form of life possesses many self-regulating systems whereby it can maintain a favorable internal environment in spite of the great variations in its surroundings. When a person goes from bright sunlight into a dark room the eyes undergo certain changes as a result of automatic internal adjustments which permit the eyes to function in spite of the greatly reduced light intensity. A person living in arctic regions of the north and persons on a tropical beach have an internal temperature that does not vary more than a fraction of a degree. Internal thermostatic adjustments regulate the body temperature to keep it at such a constant level. The human brain must have a blood supply at a constant pressure; a slight drop in pressure brings a "blackout" and too great a pressure will cause the bursting of capillaries and a "stroke." Homeostatic mechanisms change the beat of the heart and the force of the blood to the head and thus regulate the blood pressure at a constant level. We sometimes say that these mechanisms maintain the steady state.

Frequently, homeostatic regulations are by means of the negative feedback mechanism. As an example the male hormones, androgens, of vertebrate animals inhibit the production of gonadotropin from the pituitary gland. Gonadotropin, on the other hand, stimulates androgen production by the testes. In this way, the level of androgens in the blood is kept within rather close tolerances. A castrated animal will show a sudden rise in the gonadotropin in the blood and urine due to the removal of the androgens which usually serve as a control. (See Fig. 1). See also **Hormones.**

A good example of homeostasis in plants concerns the maintenance of the water balance in the leaves. The leaf must maintain a steady state of water concentration in spite of great variations in the amount of water available in the soil and variations in the humidity and temperature of the air, which affect the loss of water from the leaves through transpiration. The leaves have tiny stomata which admit air to the leaf. This air is needed to supply the carbon dioxide for photosynthesis, but it can also carry moisture from the leaf. The guard cells surrounding the stomata minimize the loss of water by opening and closing the stomata in accordance with the amount of water in the leaf. The guard cells have a tough inner portion that, as the cells swell with turgor pressure, becomes more convex and opens the stoma in between. When the guard cells lose turgor pressure, the inner portions become less convex and the stoma is closed, thus preventing the loss of more water when the water level drops in the cells. The guard cells also function according to the usage of carbon dioxide during photosynthesis. As the carbon dioxide level drops, some of the starch is converted to sugar. This increase of solutes within the

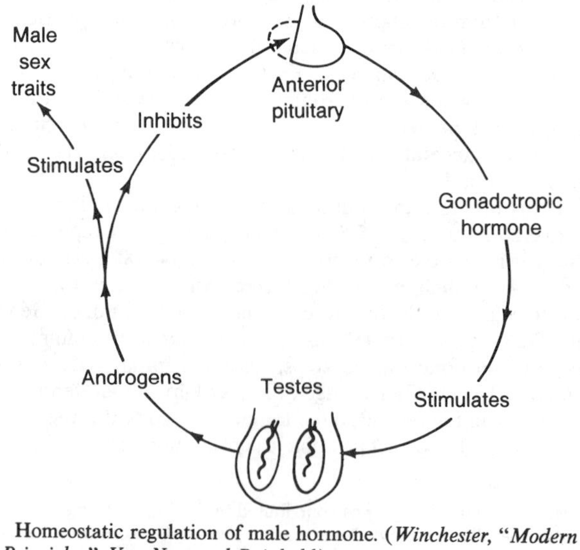

Fig. 1. Homeostatic regulation of male hormone. (*Winchester, "Modern Biological Principles," Van Nostrand Reinhold.*)

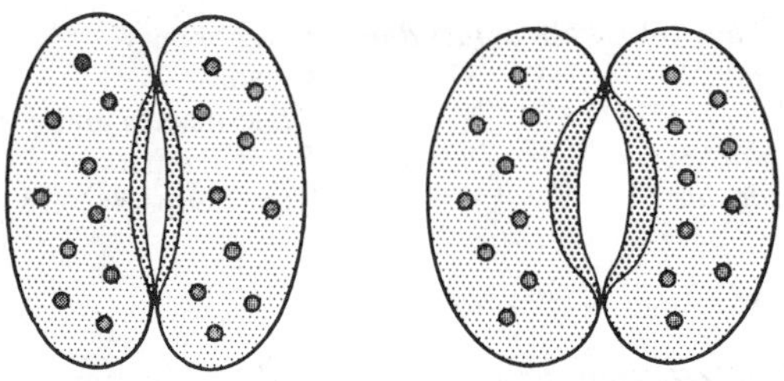

Fig. 2. Guard cell closing. (*Winchester, "Modern Biological Principles," Van Nostrand Reinhold.*)

cell causes the cell to absorb more water from the surrounding cells and the stoma is opened. At night when no carbon dioxide is being used, the sugar concentration is low and the stoma is closed. The concentration of carbonic acid, from the carbon dioxide in the cells, is the factor which activates the starch-splitting enzymes. With low carbonic acid, there is an inactivity of the enzymes, with high acid, the enzymes are most active. (See Fig. 2.)

There can also be homeostasis of a genetic nature. The balance of genes within a gene pool is homeostatically regulated. Suppose a certain harmful gene is continually being added to a population through mutation. As the gene is expressed it will reduce the reproductive potential of the individual and there will be a gradual elimination of the gene from the population. Soon the input through mutation is exactly balanced by the outgo through genetic death and the gene remains at a stable level in the population. If we increase the mutation rate, say, by radiation, then the concentration in the gene pool will increase. If we reduce the rate of elimination by medical means, we can also increase the gene pool.

Ecological homeostasis concerns the balance of nature. When certain plants which serve as food for herbivores increase, the number of herbivores increases. This, in turn, results in an increase of the predators. A balance is established that will vary according to variations in environmental factors which cause an increase or decrease of any one of the organisms in the complicated food web.

HOMEOTYPE. A term used by biologists and paleontologists for a specimen which has been identified by an authority by comparing it with the type.

HOME TERMINAL. Television.

HOMING SYSTEMS. Navigation.

HOMOCENTRIC RAYS. Rays having the same focal point. (It may be at infinity; in other words, the rays may be parallel.)

HOMOCLINE. Group of strata which dip in one and the same direction. Never a complete structure and usually representing the limb of an anticline or syncline.

HOMODYNE RECEPTION. In this system, used in connection with radio reception for suppressed-carrier systems of radiotelephony, the receiver generates a voltage which has the original carrier frequency. This is combined with the incoming signal. The term zero-beat reception is also used.

HOMOGENEOUS ATMOSPHERE. Atmosphere (Earth).

HOMOGENEOUS CIRCUIT (Law of). Thermocouple.

HOMOGENEOUS (Mathematics). The term has several meanings. Inhomogeneous is the opposite of homogeneous.

A function $f(x_1, x_2, \ldots, x_n)$ is homogeneous in all of its variables if, for any parameter t, $f(tx_1, tx_2, \ldots, tx_n) = t^n f(x_1, x_2, \ldots, x_n)$. The exponent n is the degree or order of the function. The behavior of such a function is known as the Euler theorem on homogeneous functions.

The term is used with two meanings for a differential equation:

1. A first order equation, $y' = M(x, y)/N(x, y)$ is a homogeneous equation if M, N are homogeneous functions of the same degree.
2. The general equation, $f(x, y, y', y'', \ldots) = 0$ is homogeneous and linear if f is a homogeneous function of y and all its derivatives. If the right-hand side equals a function of x, the independent variable, it is still linear but now inhomogeneous.

An integral equation, a boundary condition, or a system of simultaneous linear algebraic equations can also be homogeneous or inhomogeneous in a similar way.

See also terms listed under **Mathematics.**

HOMOGENIZING. A process for reducing the size of particles in a liquid and useful in the preparation of numerous food substances, including milk, ice cream, salad dressings, various fruit juices, flavor concentrates, infant foods, among others.

A reduction of particle or globule size in a mixture of two immiscible liquids makes an emulsion possible. If an emulsifying agent is present, a more stable emulsion can be produced and coalescence of the dispersed phase is prevented. The homogenizer is also used to produce dispersions by reducing the particle size in solid-in-liquid mixtures. As in the preparation of an emulsion, a dispersing agent is needed to maintain a homogeneous mixture.

Typically, a homogenizer consists of a high-pressure, positive-displacement pump and an adjustable orifice. The pump is a piston or plunger type, usually consisting of three plungers, although some homogenizers are made with five or even seven plungers. The cylinder for each plunger has an inlet and discharge valve. The plunger pump must push the product through the homogenizing valve (adjustable orifice). For two-stage homogenization, two valves are arranged in series.

A typical homogenizing valve consists of a seat and plug of very hard abrasion-resistant materials (alloys such as Stellite are used). The seating surfaces must be lapped smooth and be parallel. In operation, the plug is spring-loaded against the seat. Spring compression is adjusted so that when the product flows, energy in the form of pressure is required to lift the plug. Although many products can be homogenized at pressures below 3000 pounds per square inch (204 atmospheres), machines are made to develop pressures in excess of 8000 pounds per square inch (544 atmospheres). In another design, a valve uses a compressed cone of stainless-steel wire inserted into a socket, the product being homogenized by flowing between the wires.

A number of theories have been proposed as to what actually breaks up the particles in the homogenizer: (1) As the product enters the area between the lapped surfaces, it is suddenly accelerated to velocities as high as 30,000 feet per minute (9,144 meters per minute) at a pressure of 5,000 pounds per square inch (340 atmospheres). When acceleration is this sudden, the particle (especially the liquid particle) is stretched or elongated to the point of breaking. (2) At this high velocity, there are shear forces between layers of liquids under flow that break up particles. (3) Cavitation may be the major cause of homogenization. When the pressure energy is converted into velocity energy, the vapor pressure of the product exceeds product pressure, resulting in the formation of vapor cavities which collapse upon leaving the valve at higher pressures. This collapsing, or implosion, of cavitation exerts tremendous force, breaking up the particles. Most homogenizers are designed to incorporate one or more of the foregoing principles.

References

Becher, P.: "Emulsions: Theory and Practice," Van Nostrand Reinhold, New York, 1963.
Farral, A. W.: "Food Engineering Systems," AVI, Westport, Connecticut, 1976.
Harper, W. J., and C. W. Hall: "Dairy Technology and Engineering," AVI, Westport, Connecticut, 1976.
Loo, C. C., Slatter, W. L., and R. W. Powell: "Study of the Cavitation Effect in the Homogenization of Dairy Products," *J. Dairy Sci.*, **33**, 672 (1950).
Selitzer, R. (editor): "The Dairy Industry in America," Dairy and Ice Cream Field, New York, 1977.

HOMOGRAPHIC TRANSFORMATION. Transformation (Mathematics).

HOMOIOTHERMY. Warm-bloodedness. The maintenance of a body temperature above that of the environment is common among animals; hence the usual terms warm-blooded and cold-blooded are inaccurate. Cold-blooded forms are those whose body temperature fluctuates with that of the surrounding air or water, so that the animal's activity is directly conditioned by external temperatures. They are more accurately described as poikilothermal. In contrast, homoiothermal animals tend to maintain a constant body temperature in spite of external fluctuations. Fluctuations are normal, although the human body usually maintains a constant temperature.

Only birds and mammals are homoiothermal. Both regulate the body temperature by producing excess heat and by regulating its radiation from the surface. Regulation is accomplished by nervous control of the blood vessels near the surface, by insulating vestiture, and by the evaporation of water from the body. When the surrounding air is warm, the blood flows more freely near the surface of the body and more heat is radiated, but when the air is cold, less blood reaches the surface and the heat is conserved. In air too warm to permit adequate radiation, the animal reduces its activity, exposes as much surface as possible, and either sweats or pants. The evaporation of water either from the mouth or from the sweat glands absorbs heat from the underlying tissues. Vestiture plays a passive role as an insulating coat, but it is capable of some regulation, especially in the birds. The erection of the feathers provides a thicker and looser covering of high insulating value, and their depression results in less interference with radiation.

Homoiothermy is one of the highest adaptations of living things, since it provides for the maintenance of optimum conditions for the vital processes of the body. Through it the animal becomes virtually independent of one of the most important of the fluctuating environmental conditions.

HOMOLOGOUS SERIES. Two organic compounds are said to be homologous if their molecular formulas differ by CH_2, or a multiple of CH_2. For example, the alkane series has the general formula, C_nH_{2n+2}, its first three members being methane, CH_4, ethane, C_2H_6, and propane, C_3H_8.

HOMOLOGY. Fundamental structural relationship, based on similarity of embryological development and evolutionary history. The antithesis of analogy, which is superficial likeness based on adaptation for similar uses.

The anterior appendages of terrestrial vertebrates, for example, are regarded as fundamentally similar structures, derived from the pentadactyl appendage; yet they include the wings of birds, flippers of aquatic mammals, and a great variety of less extreme adaptations, including the legs of animals and the arms of man. In contrast, the wings of birds and of insects are broad thin structures used for flight, but in structure and origin they show no resemblance beyond this point and so are analogous.

HOMOLYTIC REACTION. Organic Chemistry.

HOMOMETRIC PAIRS. Crystal (Homometric Pairs).

HOMOPOLAR BOND. A covalent bond which has no resultant dipole moment. See also **Chemical Elements.**

HOMOPTERA. The cicadas, leaf hoppers, plant lice, scale insects, and numerous other forms, constituting a large order of insects. They have sucking mouths which differ from those of most bugs in that the slender proboscis arises from the hind margin of the head and extends back between the legs. The wings, when present, are membraneous. The order includes about 16,000 species.

Many members of this order, particularly the plant lice, scale insects, and phylloxerans, are economically important.

The main families of Homoptera include:

Aleyrodidae	White flies
Aphidae	Aphids or plant lice
Cercopidae	Spittle bugs
Chermidae (also called *Psyllidae*)	Jumping plant lice
Cicadellidae (also called *Jassidae*)	Leafhoppers
Cicadidae	Cicadas
Coccidae	Scale insects and mealybugs
Fulgoridae	Plant hoppers
Membracidae	Tree hoppers

HOMOZYGOUS. Heterozygous.

HONEY. Raw, unprocessed honey is a thick, viscous, high-density, very sweet, hygroscopic liquid that is formed by honeybees from the nectar of flowers and, to a limited extent, from the juices of fruits and honeydew. Honey is available commercially as a liquid, as crystallized honey, as comb honey, as chunk honey, and as powdered honey. Honey contains a large percentage of simple sugars, as well as essential oils of the flowers from which it is derived, plus about 20% water. The flavor of honey depends upon the flowers from which the nectar is derived, upon manufacturing conditions if it is processed, upon the season and climate during which it is gathered and stored by the honeybees, and upon its age. Under appropriate conditions, honey is one of the most storable of foods and can be kept for many years, particularly in a frozen state.

In food processing, honey is frequently used because of its qualities as a humectant and a source of reducing sugars. Bakers and candy makers prefer honey for these reasons plus the fact that it promotes caramelization and aids in obtaining uniform browning of baked goods, as well as providing clarity to glazes. In addition to bakery products and confections, honey is used in the manufacture of breakfast foods, snacks, sauces, and syrups, as well as a sweetener and bodying agent in some canned fruits, jams, jellies, and spreads. Honey is a common ingredient of graham crackers, where it blends with the dark whole-wheat flours, as it also does with whole-wheat breads.

Honeybees

There was little scientific knowledge of honeybees until the 1850s. Even though European scientists of the 1700s and early 1800s developed an understanding of the biological aspects of honey and wax production by bees, this knowledge did not contribute in a major way to the practical aspects of bee-keeping. It was not until 1852 that an American minister (L. L. Langstroth) discovered what became known as "bee space." This concept led to development of the first practical, movable-frame hives. This breakthrough, coupled with other important equipment developments, such as the centrifugal honey extractor (commercialized in 1870), the bee smoker, bee escape, and queen excluder, transformed beekeeping during the latter half of the 1800s from a minor activity on the part of many farmers to a serious business capable of centralization and full-time management. Prior to that time, there was little if any organization in terms of producing and marketing honey.

As of the early 1980s, it is estimated that 60–70% of the honey produced in the United States comes from 1200 to 1500 fulltime beekeepers who operate 35–40% of the nation's nearly 5 million bee colonies. Production of honey in the world is estimated at 683,000 metric tons, of which the United States accounts for about 114,000 metric tons per year. Other large honey producing countries include the U.S.S.R., China, Mexico, Uruguay, Canada, Turkey, and Ethiopia. The principal honey producing states in the United States are Florida, North Dakota, Minnesota, California, and Wisconsin. However, honey production is found in practically all of the states to some degree.

A fulltime beekeeper maintains a minimum of 400 hives, but the average is about 1,200 hives. Large operators will maintain 20,000 or more hives. The optimal hive density for good honey yield usually is insufficient for efficient crop pollination. Thus crop growers compensate beekeepers for reduced honey crops resulting from maintaining relatively high hive density and also for the costs of moving colonies from one farm to the next during pollinating season. One of the specialty sectors of beekeeping is that of raising bees and queens for sale to beekeepers and growers of crops. This activity is concentrated in the Gulf states and southeastern states.

Strains and Hybrids. The most common strain of bees kept in North America are of Italian origin. These bees, yellow to brown in color, are industrious and relatively unexcitable. The Caucasian strain, even more gentle than the Italian, is also kept. This honeybee is gray-to-black in coloration. An import from Japan in the late 1970s, the species *Osmia cornifrons*, is under intensive study by the U.S. Department of Agriculture. It is believed that the species may be of particular value for solving pollination problems for small-farm fruit tree growers. The usual honeybee suffers from diseases and predator pests; they swarm; they sometimes abscond; they require considerable attention; and they sting. The Japanese honeybee, about two-thirds the size of the Italian or Caucasian bees, has few of the foregoing disadvantages. They produce no honey or beeswax, they are more active at cooler temperatures, they rarely range more than 300 feet (90 meters) from their nests, they are docile, and their sting is about like that of a mosquito bite. Although they produce no honey, they appear to be quite ideal for pollinating.

Biological Aspects. Bees are of the subgroup *Aculeata* within the order of *Hymenoptera* (*Insecta*). *Aculeata* also includes wasps and ants, these insects all incorporating the sting, which was developed in the course of evolution from the ovipositor apparatus. This modification involved a change of function, for it no longer serves for egg-laying; rather it is employed as an effective weapon of defense or as an injection cannula for the paralysis of prey. In view of the origin of the device, it is obvious that only female *Hymenoptera* can sting. The sting consists of several reciprocally movable chitinous elements. Into it there open the ducts of two glands, one of which produces poison. In the stinging *Hymenoptera* the eggs are ejected from the opening of the genital chamber at the base of the ovipositor.

Bees are members of the superfamily *Apoidea* and include, in addition to the universally known honeybee, more than 20,000 other species. They are distributed almost worldwide, and in the north their range extends well beyond the Arctic Circle. The smallest bees measure barely 2 millimeters, while the largest approach 4 centimeters in body length. They are among the most economically important animals, particularly since they play a critical role in the reproduction of numerous cultivated plants. Moreover, the study of their life history has charms all its own, for they have evolved fascinating forms of social interaction.

Bees are classified into six families, of which two, the *Halictidae* and the *Apidae*, have evolved social species. In addition to the honeybee, there are plasterer bees, mining bees or burrowing bees, mason bees, leaf-cutter bees, carpenter bees, and bumblebees.

At present, four species of *true honeybees* (tribe *Apini*) are known, all of them native to tropical southeastern Asia. But the so-called domesticated honeybee (*Apis mellifera*) has been distributed by humans all over the world. Before cane and beet sugar came into use, honey was the primary sweetener for foods. For that reason, the bees were most highly prized for their honey and wax. But, in more recent years, they have become even more prized for their beneficial activity in pollinating flowering crops.

All honeybees (see Fig. 1) build their combs of pure wax, which can be produced only by the workers. The combs are hung up vertically either in the open or in enclosed areas, and on both sides they have geometrically perfect hexagonal cells, a shape which minimizes the construction material required. The cells for the worker brood and those for the storage of honey and pollen are similar, while those for the males (drones) are larger. The queens are raised in special vertically hanging chambers, as shown by Fig. 2. Only the workers have the organs and instincts necessary for the activities of construction and foraging. The queen, somewhat larger than the workers, has as her only task the laying of eggs. Fertilized eggs give rise to workers or queens, depending upon the food given to the larvae, while the drones come from unfertilized eggs. The time at which reproductive

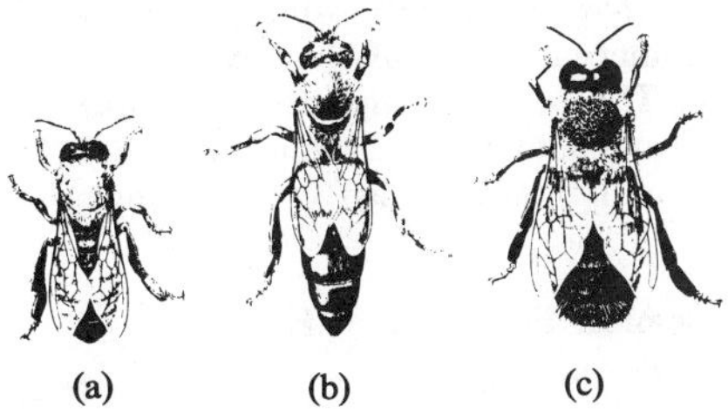

Fig. 1. Relative sizes of honeybees: (a) worker, (b) queen, (c) drone. (*U.S. Department of Agriculture diagram.*)

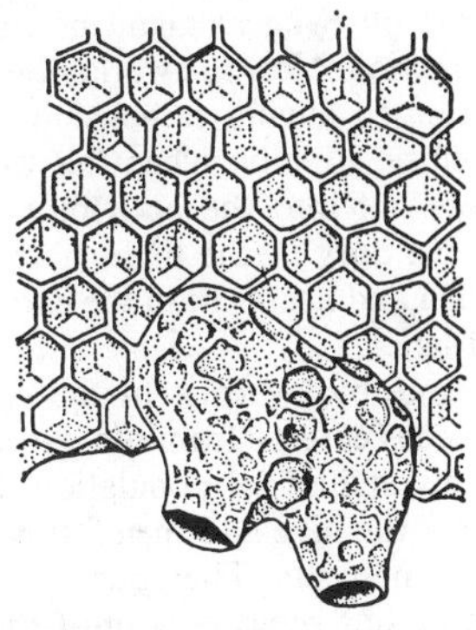

Fig. 2. Section of a honeybee comb. Drone cells are above and two queen cells are shown at the lower edge.

individuals are to be produced is determined by the workers, who then prepare the appropriate cells and food. All the larvae are fed and cared for continually during their development. A queen can lay as many as 2000 eggs in a single day. In her 4 to 5 years of life, she produces about 2 million eggs. More, than 80,000 bees can live in a colony.

The tasks of the workers are manifold. They supply the colony with food, guard the nest, and build the combs. They also keep the combs clean, for these are used several times for the brood. By fanning with their wings, the workers cool the nest, and by their muscular activity they warm it. Thus, they ensure that the temperature in the brood area stays close to 35°C (95°F). Although every worker bee is, should special circumstances demand it, capable of performing any of these tasks, ordinarily there is an orderly division of labor, corresponding to the age of the workers. In their first days as workers, they act as janitors, keeping the combs clean. Next, after the pharyngeal gland in the head (See Fig. 3) has matured, they devote themselves to the larvae, feeding them first with the secretion of this gland and later with pollen and honey as well.

The food given by worker honeybees to the young larvae during the first 3 days of their existence and to the larvae of queens until they are fully developed is known as *royal jelly*. This is a thick, white liquid formed in the stomach of the worker by partial digestion of honey and pollen and is apparently a highly-concentrated food. Queen cells are supplied with the material in excess of the needs of the larvae. If conditions within the colony deprive queen larvae of this abundance, they fail to become large and, in some cases, they may revert to development as intermediate forms between queen and worker. Such individuals may, however, have the instincts of queens and so may mate and lay fertile eggs. The change from royal jelly to a less concentrated food in the case of worker larvae apparently is responsible for the development of worker bees, since both queens and workers may develop from identical eggs.

At about the tenth day, the workers fly out briefly for the first time and become acquainted with the surroundings of the hive. In

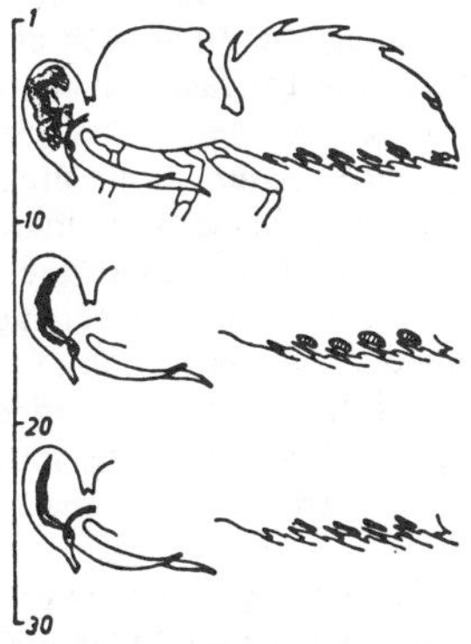

Fig. 3. Physiological changes having behavioral effects are correlated here with the age of the honeybee worker, the numbers at the left indicating the age of the worker. To the right of the ordinate is shown the stage of development of both the pharyngeal gland in the head and the wax glands in the abdomen.

the following days, the pharyngeal gland becomes reduced and the wax glands begin to function. Now the worker bee becomes a construction worker, in addition to having responsibility for the food stores and carrying away refuse. Finally, when the worker approaches the age of 20 days, it goes out into the open more and more often. At first, it takes over guard duty at the flight hole; then it forages with great industry for pollen and nectar until the end of its life, in summer, after 4 to 5 weeks of labor.

Bee colonies propagate by swarms. In the early summer, shortly before one or more queens emerge from the queen cells, the old queen leaves the hive with about half its population. She first gathers her court into a cluster near the hive and then follows the advance guard, which has found a new nest site. The young queens which have matured in the old hive fly out repeatedly until they have mated, each with several drones. The supply of sperm thus accumulated must last the queen a lifetime. They mate in flight, at assembly places where the drones often congregate in great numbers. There the drones fling themselves on every female which flies past and has the appropriate scent signal. The inseminated queen returns to the nest and stings to death any rivals which may still be present. The drones too meet their fate in late summer. The workers drive them out of the nest with bites and stings, after which they are left to starve.

Chemical signals are involved not only in mating. They also play a role in many aspects of communications by bees. Each queen secretes a substance by which the workers are continually reassured that the colony is not without a mother. Scent signals spread the alarm when the bees are in danger and also serve to identify food sources and the entrance to the hive.

Bees store up honey collected from various sources. Blossom honey is the thickened nectar of millions of flowers, which has passed through the stomachs of many bees and has been altered by glandular secretions. Bees also collect the sweet secretions of aphids which stick to leaves. With this, they form the leaf honey or pine honey, which is considered a delicacy. Several hundred kinds of plants produce nectar, which the bees also use for honey, but only a few kinds are common enough, or produce enough nectar, to be considered as major sources. The best sources of nectar for producing surplus honey vary from place to place. Some plants that are major nectar sources in the United States include: Alfalfa, aster, buckwheat, catclaw, citrus fruit, clover, cotton, fireweed, goldenrod, holly, horsemint, locust, mesquite, palmetto, tulip tree, tupelo, sage, sourwood, star thistle, sweet clover, sumac, and willow. The varying qualities of these sources is reflected in the color and flavor of the raw honey taken from the hive.

Since a good colony can store as much as 1 kilogram (2.2 pounds) of honey per day, the number of foraging flights undertaken in a day is astronomical. This efficiency depends upon two special achievements: (1) The bees can orient themselves; and (2) they can communicate with one another. Considering orientation, each bee must cover the distance between hive and collecting place as quickly and accurately as possible. This phenomenon was studied by Karl von Frisch and his coworkers for over 50 years. It was found that the bee uses all its senses—color leads it to the bright flowers; sense of smell enables it to distinguish different species of flowers. On the flight out and back, it not only observes conspicuous landmarks, but also uses the sun as a compass by keeping the flight path at the proper angles to the direction of the sun. Even the fact that the sun seemingly moves across the sky does not confuse the bee. Its sense of time permits it to take the time of day into account in its flights, correctly altering the setting of the sun compass. This sense of time also makes it possible for the bee to go at any time of the day to the sort of flower that is producing nectar then. Even if the sun is covered by clouds, the forager is not helpless. The direction of polarization of the light waves which penetrate from a patch of blue sky is dependent upon the position of the sun. Since the eyes of the honeybee, unlike human eyes, can measure this direction of polarization, the bee can orient to that just as well as to the sun itself.

If a source of food is to be really efficiently exploited, its discoverer must communicate its location to the other foragers. The discovery of the way the honeybees do this is among the most exciting revelations of modern biology. The 1973 Nobel Prize winner, Karl von Frisch, who developed these findings, described the various kinds of aerial dances which the bees used to communicate with each other. For example, there are the waggle dances and the round dances, as indicated schematically in Fig. 4. Details are beyond the scope of this volume, but reference to Grzimek (listed at end of entry) is suggested. Although these performances of the honeybee appear to border on the miraculous, they have been thoroughly observed and documented by Dr. von Frisch and other researchers.

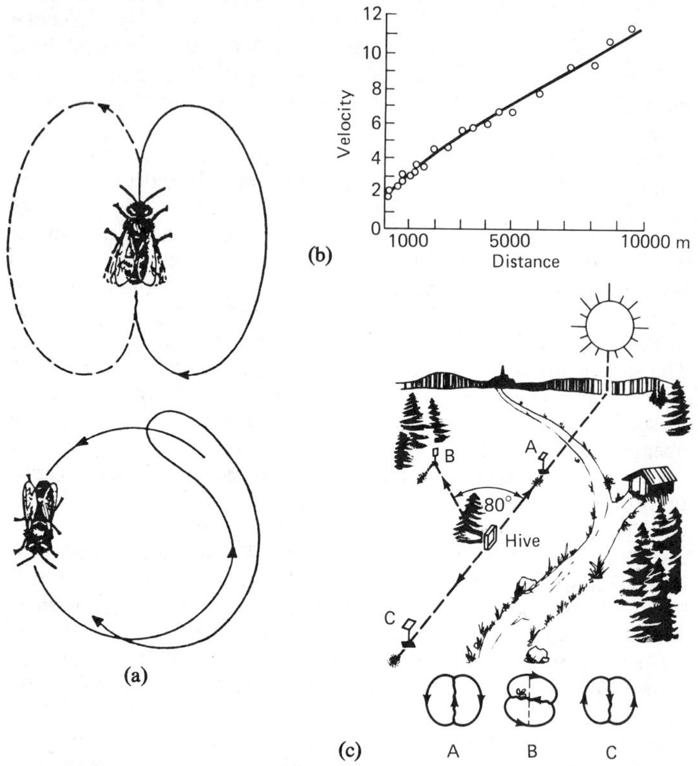

Fig. 4 (a). Dances with which honeybees communicate. The "waggle" dance (top) is performed by long-distance foragers to communicate location of food sources. Both direction and distance are communicated. Direction is illustrated in (c). Distance is related to the speed or frequency with which individual dances are repeated. The round dance, indicated by the lower diagram, announces to the bees in the hive that a rich harvest has been located. Foragers repeat this dance each time they return to the hive until a more important new source is located. It is important to realize that the dance is normally done in the pitch darkness of a closed beehive, so that the dancer is invisible to its comrades, who depend upon their senses of touch (vibrations) and smell to follow the tortuous course of the dance.

(b) Velocity of the waggle dance (ordinate) as a function of the distance (abscissa) between feeding place and hive.

(c) Relationship between flight angle with respect to sun and dance angle in the hive with respect to gravity. The latter is diagrammed for the three feeding sites shown (A, B, and C).

These abilities of the honeybee were determined by observing the domestic honeybee (*Apis mellifera*), whose many subspecies have been distributed by human intervention throughout the world. In structure and habits, the Indian subspecies (*Apis mellifera derana*) is quite similar. The bees *A. mellifera dorsata* and *A. mellifera florea*, both of which also live in southeastern Asia, represent a somewhat lower level of performance. Each of these species builds only one comb and hangs it in the open on a branch. These subspecies are not important commercially in terms of honey production, nor is the African honeybee (*A. mellifera adansonii*), which is dreaded by natives. So severe are the numerous stings of this bee that it may cause death among livestock. The bee was imported into Brazil in 1956 and is now spreading throughout northern South America. Since that time, over 300 deaths of humans caused by the stings of this bee have been reported.

Physical and Chemical Properties of Honey

Although a seemingly simple substance, honey is relatively complex and requires some rather exacting conditions when it is processed. Important factors include: (1) moisture content which, if excessive, causes the honey to ferment over a period of time; (2) the tendency of the glucose to crystallize out of the liquid phase, a process known in the trade as *granulation*; and (3) the presence of nitrogeneous substances, which even in very small amounts cause the honey to darken with age. Because water content is important to ultimate quality of the honey, including the possibility of fermentation, the water content is strictly regulated by most countries. In the United States, U.S. Grade A (Fancy) and Grade B (Choice) cannot contain over 18.6% water. Grade C (Standard) for reprocessing may contain up to 20% water. Honey with greater amounts of water is Grade D (Substandard).

Beeswax. This is a commercial byproduct of honey production. The wax represents approximately 1.9% of the weight of honey produced. In the United States, beeswax production ranges between 3 and 5 million pounds (1.4 and 2.3 million kilograms) per year. Freshly made wax is of a light yellow color, but becomes brown with age. However, the wax may be bleached by sunlight or with acids. Beeswax is made up mainly of a complex long-chain ester, myricil palmitate, $C_{15}H_{31}COOC_{30}H_{61}$, and cerotic acid, $C_{25}H_{51}COOH$; specific gravity, 0.965–0.969; mp 63°C. The wax is easily colored with dyes and finds numerous uses, as in polishes, candles, leather dressings, adhesives, cosmetics, and molded articles.

Mead (Honey Wine). This is one of the most ancient of fermented products and wines. Although regarded by some people as a curiosity in modern times, there is some demand for mead in various parts of the world. It is usually prepared and consumed on a regional basis.

Honey and Infant Botulism. Honey has been implicated in about one-third of the cases of infant botulism. Consequently, some physicians recommend that honey not be fed to infants under one year of age.

References

Detroy, B. F.: "Shade and Water for the Honey Bee Colony," Leaflet 530, U.S. Department of Agriculture, Washington, D.C., 1977.

Grzimek, B. (editor): "Grzimek's Animal Life Encyclopedia," Vol. 2, Chapter 18, Van Nostrand Reinhold, New York, 1975.

Smith, L. W.: "New Bee . . . for Small-scale Farmers," *Agricultural Research*, **27**, 12, 3–4 (1979).

Staff: "Agricultural Statistics," U.S. Department of Agriculture, Washington, D.C. (Issued annually)

Watt, B. K., and A. L. Merrill: "Composition of Foods," Agriculture Handbook 8, U.S. Department of Agriculture, Washington, D.C., 1975.

HONEYBEE. Honey.

HONEY BUZZARD. Eagle.

HONEYCOMB COMPOSITES. Fiber-Reinforced Composites.

HONEYDEW. A sweet secretion produced by aphids, which, when abundant on trees, sometimes cause spotting on the leaves and anything below the tree like a heavy dew. Honeydew is freely sought by ants and is sometimes gathered by bees, but it makes a very inferior honey.

HONEY GUIDE. Woodpeckers and Toucans.

HOODOO. In geology, a columnar or pillar-like erosional remnant which has been carved and sculptured from relatively horizontal formations by differential erosion. The form and subsidiary features of Hoodoos may be partly governed by joint planes and the differential hardness of the stratified sediments. The term applies particularly to eccentric and peculiar forms which are especially noticeable because of their fancied resemblance to animals and artifacts.

HOOKE'S LAW. Elasticity.

HOOKWORM. Dermatitis and Dermatosis.

HOOPOE. Kingfishers and Other Coraciiformes.

HOPEITE. Conversion Coatings.

HOPLOPHONEUS. Fossil Reptilian Mammals.

HOPHORNBEAM TREES. Hornbeam Trees.

HOPS. Mulberry Family.

HORDEOLUM. Eyelid.

HORIZON (Artificial). Artificial Horizon.

HORIZON (Astronomical). Also called sensible horizon; real horizon. The plane that passes through the observer's eye and is perpendicular to the zenith at that point; or, the intersection of that plane with the celestial sphere (i.e., a great circle on the celestial sphere equidistant from the observer's zenith and nadir). It is the projection of a horizontal plane in every direction from the point of orientation. See also **Celestial Sphere.**

HORIZON (Celestial). Also called rational horizon; geometrical horizon; true horizon. The plane, through the center of the earth, perpendicular to a radius of the earth that passes through the point of observation on the earth's surface; or, the intersection of that plane with the celestial sphere.

In astronomy, the term horizon is used to describe the great circle cut out on the celestial sphere by a plane perpendicular to the direction of gravity. If this plane is tangent to the surface of the earth, the horizon so described is the *apparent horizon*; if the plane passes through the center of the earth, we have the *geocentric horizon.*

The difference in direction between the visible horizon (as the line where earth and sky meet) and the apparent astronomical horizon is known as the dip of the horizon. In the figure, *O′ H′* represents the direction of the visible horizon from an observer at a station *O′* elevated above the surface of the earth by an amount *h*. *OH* represents the direction of the horizon for the observer on the surface of the earth at *O*. The angle HAH′ is the dip of the horizon, and may be given, very approximately, by this relation: the dip of the horizon (expressed in minutes of arc) is equal to the square root of the height of the observer above the surface of the earth (expressed in feet). The distance *O′T* from the observer to the visible horizon is approximately as follows: the distance of the visible horizon (expressed in miles) is given by the square root of $\frac{3}{2}$ the height of the observer above the surface of the earth (expressed in feet).

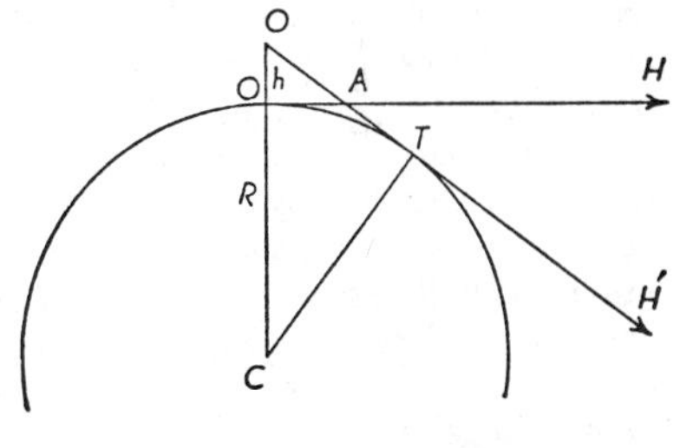

Visible and astronomical horizons.

This distance is frequently very much increased by an effect known as looming of the horizon, produced by refraction of light in heated (or cooled) layers of the air near the surface. Both the expressions for the dip and distance of the horizon are applicable only when the point of observation of the visible horizon (the point *T*) is actually on the surface of the earth. See also **Celestial Sphere.**

HORIZON (Geographic). Also called apparent horizon, local horizon, and visible horizon, the distant line along which earth and sky appear to meet. In both popular usage and weather observing, this is the usual conception of horizon. Nearby prominences are said to obscure the horizon and are not considered to be a part of it. For observational reference, the minimum desirable horizon distance should be of the order of three miles.

Sea-level horizon, also called ideal horizon, sensible horizon, sea horizon, visible horizon, and apparent horizon, is the apparent junction of the sky and the sea-level surface of the earth; the horizon that is actually observed at sea. This type of horizon is used as the reference for establishing times of sunrise and sunset.

In these definitions of horizon, the zenith is considered to be at right angles to the horizon.

HORIZON (Soil). Soil.

HORIZONTAL COORDINATE SYSTEM (Astronomy). A system of spherical coordinates on the celestial sphere, which uses the horizon as a fundamental plane. Planes perpendicular to the horizon cut out great circles on the celestial sphere known as vertical circles. The fundamental direction selected in the fundamental plane is true south. The azimuth of a point on the celestial sphere is the angular distance, measured in the plane of the horizon, from the true south direction to the point of intersection of the vertical circle through the object with the horizon. There are several different methods for expressing azimuth, but the astronomical method is to measure azimuth from the south through the west through 360°. The altitude of a point on the celestial sphere is the angular distance, measured along the vertical circle through the point, from the plane of the horizon to the point.

The horizontal system of spherical coordinates is frequently referred to as the altazimuth system.

See also **Celestial Sphere.**

HORMONES. In animals, hormones are organic compounds, usually of considerable complexity and even after years of research not fully understood, that are secreted by endocrine (ductless) glands, such as the adrenal gland, the thyroid and parathyroid glands, the pituitary gland, and the gonads, among others. See **Endocrine System.** Hormones are sometimes commonly called by the names of the gland which secrete them. Thus, there are adrenal cortical hormones, thyroid and parathyroid hormones, etc. Hormones are regulators of physiological processes within the body, exerting control over such processes as metabolism, growth, reproduction, molting, pigmentation, and electrolytic and osmotic balance, among other processes. Apparently, hormones achieve these objectives chemically and electrically, although the mechanisms are not fully understood and, in fact, the mechanisms may vary from one situation to the next. At one time, hormones were loosely called "chemical messengers" because they are transported from point to point within the organism and thus effect actions at distances from the region where they are made. If one visualizes secreting glands as sensors of a type detecting need for correction of some physiological process, then the hormones might be visualized as both the transmitters or carriers of this information and the initiators of actions as well. The conventional concept is that cells have receptors on their surface which sense the presence of specific hormones. At one time, it was firmly believed that hormones, particularly polypeptide hormones, such as insulin, prolactin, and growth hormone, all of which are large charged molecules, could not penetrate through the cell's membrane and actually enter the cell. This belief has since been altered because researchers have shown that insulin, for example, can enter into the cell. Referring to this process as "internalization," one investigator in 1978 suggested that the internalization of polypeptide hormones will be one of the most active topics in cell biology for a number of years.

Research during the late 1970s and continuing into the 1980s has shown that hormones and/or their receptors may be degraded. As gross examples of this type of situation, it is known that many obese people with high concentrations of insulin in their blood also have normal concentrations of blood sugar. Why doesn't the insulin decrease the blood sugar concentration in these cases? It is well known that pregnant women produce much angiotensin II, which normally increases blood pressure, but these women usually do not have hypertension. What alterations in the hormone-cell mechanism provide this result? There are also instances where males have tumors which secrete large quantities of a hormone that stimulates the production of testosterone, and yet there is no evidence of abnormal amounts of testosterone. At least two questions can be posed: Do certain hormones lose their effectiveness with time? Or, are there changes in target receptor cells? Research has indicated that there may be a relationship between concentration of hormones and the surface receptors which bind them, such receptors being inactive for a time or possibly disappearing from the cell surface altogether. A number of investigators have observed that a better understanding of the manner in which hormones affect their own and other receptors possibly may result in new ways to treat certain diseases, including insulin-resistant diabetes.

In recent years, it has been shown that a wide variety of receptors are regulated by hormones. Some receptors are sensitive to only one hormone; this appears to be the case with insulin. Others appear to be regulated not only by one hormone, but others as well. For example, it has been shown that the receptors for TRH (thyrotropin-releasing hormone) are not exclusively regulated by TRH, but also by other hormones. Receptors for gonadotropins (pituitary hormones that act on the gonads) appear to be regulated by hormones in addition to gonadotropin. There are numerous other instances of this kind.

Marx (1979) reports on the exploratory consideration being given to the overall process of aging in terms of hormones and their receptors by endocrinologists. Further details are given by Korenman (1979).

Hormones display not only great variations in function, but also in their chemical nature, of which there is a great diversity. Some are steroids, such as estrogen, progesterone, cortisone, etc., while others are amino acids (thyroxine), polypeptides (vasopressin), low-molecular-weight proteins, and conjugated proteins. Amino acid and steroid hormones have been isolated and many, including insulin, have been synthesized. Other types are prepared directly from the endocrine organs of animals.

Hormones produced by one species usually show similar activity in other species. The hormones showing greatest species specificity are proteins or conjugated proteins.

Hormones are markedly affected by deficiencies or excesses of the various vitamins and other dietary essentials.

Because of the great complexity of a number of the natural hormones, conventional approaches of organic synthesis which have been used so successfully over the years in connection with many drugs have not proved viable to date with some of the hormones. Insulin is an example. Presently, millions of diabetics still depend upon animal insulin as extracted from the pancreatic glands of slaughtered pigs. If diabetes mellitus continues to become more prevalent, as it has over the past several years, natural sources may not be sufficient. Further, means to lower the cost of hormones such as insulin would be beneficial. In the late 1970s and continuing into the 1980s, researchers at several institutions have been probing the possibilities of applying genetic engineering to this problem. One group has been successful in inducing the bacterium *Escherichia coli* to manufacture and secrete rat proinsulin, an immediate precursor of rat insulin that incorporates insulin itself. Research like this is an important step toward the objective of developing bacterium-based industrial systems that can replace animal and human tissues as the source of medically useful proteins, such as insulin, growth hormone, and clotting factor.

Classes of Hormones

Hormones may be grouped into two distinct types: (1) *direct-acting*; and (2) *stimulating*-substances that stimulate other organs to produce their own characteristic hormones. The latter group is sometimes called the *tropic hormones.* See accompanying tabular summary of hormones.

Thyroid Hormones. These are compounds of the amino acid *thyronine.* They are present in the free form only to a slight extent, existing chiefly as constituents of the protein thyroglobulin. The most important of these acids in terms of hormone action are the 3,5,3′-tri, and the 3,5,3′,5′-tetraiodocompounds, *triiodothyronine* and *thyroxin,* the structures of which are given in the accompanying table. See also **Thyroid Gland.** The action of thyroid hormones is to accelerate cellular reactions and to increase the metabolic rate and oxygen consumption of tissues. They effect this action by stimulating many of the enzyme systems, not only the glucose oxidation system and the cytochrome chain for dehydrogenating the coenzyme NADPH, but other processes, such as the synthesis of proteins from amino acids. Their effects are clearly apparent in the pathological changes in the organism caused by their excess or deficiency. The thyrotrophic hormone and other biochemical interactions with the thyroid gland are discussed later in this entry.

Parathyroid Hormones. The influence of the parathyroid glands on the regulation of calcium concentrations in the blood of mammals was first recognized by MacCullum and Voegtlin in 1909.

More recently, several groups of investigators have succeeded in purifying and partially identifying the structure of the hormone, variously called *parathormone* and *parathyroid hormone.* This is a single-

Hormone Common Names, (Synonyms), Structure and Production Site	Principal Physiologic Functions	Interrelationships with Vitamins
Adrenocorticotropic Hormone (ACTH) (Adrenocorticotrophin; corticotrophic hormone) Straight-chain, simple polypeptide, 39 amino acids, no S—S bridges. (See text, Fig. 1.) Molecular Weight ~4500 Production Site: Anterior pituitary	Maintenance of adrenal cortex Promotes secretion of steroids, oxidative phosphorylation in adrenal cortex Mobilizes and increases oxidation of free fatty acids in adipose tissue Increases gluconeogenesis in liver; increases cyclic adenosine monophosphate (AMP) in adrenal cortex Decreases urea formation in liver	Ascorbic acid: depleted in adrenal cortex on stimulation by ACTH Biotin and vitamin A: adrenocortical insufficiency noted in biotin and vitamin A deficiency Niacin: production of reduced nicotinamide adenine dinucleotide (phosphate) (NADPH) by ACTH via cyclic adenosine monophosphate (AMP) Niacin and pantothenic acid: synergistic with ACTH in steroid hormone synthesis Vitamin D: antagonized directly by ACTH via cortisol action
Aldosterone (Aldocortin; electrocortin; mineralocorticoid; 18-oxo-corticosterone) 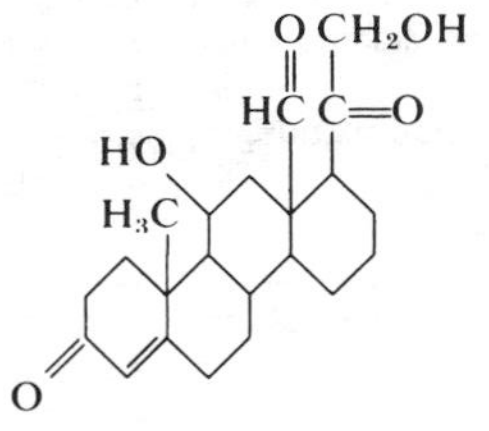 Molecular Weight 360.4 Production Site: Adrenal cortex	Maintenance of normal electrolyte blood balances Prolongs survival of adrenalectomized animals Accelerates gluconeogenesis Regulates kidney function	Ascorbic acid: adrenal cortex depleted of ascorbic acid on production of aldosterone Biotin: prolongs life in adrenalectomized rats Niacin: nicotinamide adenine dinucleotide (phosphate) (NADPH) involved in synthesis of aldosterone
Cortisol (Hydrocortisone, 17-hydroxycorticosterone) 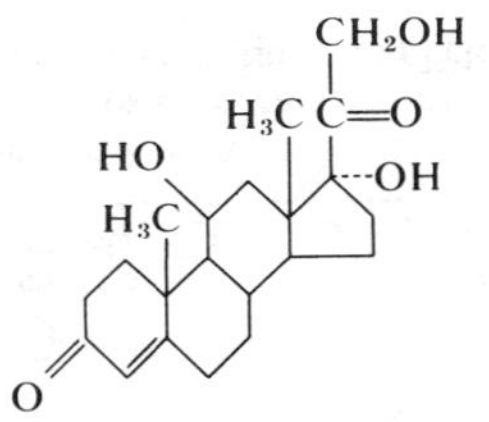 Molecular Weight 362.5 Production Site: Adrenal cortex	Increases (1) protein catabolism (excepting liver) gluconeogenesis; (2) carbohydrate anabolism (liver); (3) blood sugar; (4) glucose absorption; (5) brain excitation; (6) spread of infections; (7) urinary glucose and nitrogen; (8) stress tolerance; (9) lactation; (10) water diuresis Decreases (1) fat anabolism; (2) growth rate; (3) inflammation; (4) eosinophils; (5) lymphocytes; (6) antigen sensitivity; (7) respiratory quotient; (8) ketosis; (9) wound healing; (10) skin pigmentation; (11) RBC hemolysis. Regulates general adaptation syndrome, water balance, blood pressure, and hormone release.	Ascorbic acid: may be required for steroid hormone biosynthesis; depleted from adrenal cortex on cortical secretion Biotin: adrenocortical insufficiency noted in biotin deficiency Folic acid and pantothenic acids maintain secretions of steroids by adrenal cortex Niacin: nicotinamide adenine dinucleotide (phosphate) (NADPH) required for steroid hormone biosynthesis Vitamin A: deficiency causes cortical necrosis Vitamin D: action antagonized by cortisol by reducing calcium absorption in intestine
Epinephrine (Adrenaline, adrenin, suprarenin, vasotonin, vasoconstrictine, adrenamine, levorenine) 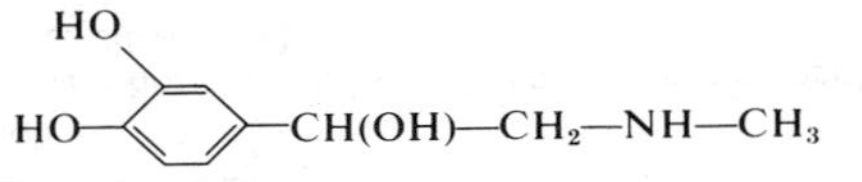 Molecular Weight 183.2 Production Site: Adrenal medulla and chromaffin cells in gut	Blood circulation: increases blood pressure; peripheral vasodilator; increases heart output and rate; flow increased in brain, liver, and skeletal muscle Central nervous system: causes restlessness, anxiety Kidney: reduces glomerular filtration rate Lung, intestine, genital system: inhibited motility Metabolic effects: increases oxygen consumption, temperature, basal metabolic rate, gluconeogenesis Pituitary effects: stimulates production and release of ACTH and corticoids	Ascorbic acid: maintains reduced state of epinephrine Ascorbic acid, folic acid, and vitamins B_6, and B_{12} are cofactors in synthesis of epinephrine from phenylalanine

Estradiol (Female hormone; dihydrotheelin; dihydrofollicular hormone; dihydrofolliculin) 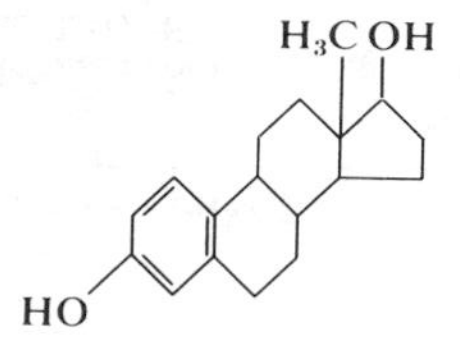 Molecular Weight 272.4 Production Sites: Ovarian follicles; testes; corpus luteum; adrenal cortex; placenta	Regulates menstrual cycle, female sex behavior Maintains secondary sex characteristics Affects antibody properties Induces estrus, uterine hypertrophy, vaginal cornification; potentiates and stimulates calcitonin secretion	Folic acid: involved in mitotic effect of estradiol Niacin, diphosphopyridine nucleotide (DPN), triphosphopyridine nucleotide (TPN): involved in increased respiration and in cholesterol precursor synthesis Pyridoxine: competes as cofactor with estrogen sulfate in kynurenine aminotransferase activity Vitamin D: synergistic in calcium metabolism with estradiol Vitamin E: involved in follotropin production or release
Follicle-Stimulating Hormone (FSH) (Follotropin, luteoantine, thylakentrin, Prolan A, gonadotropin 1, gametogenic hormone, follicle ripening hormone, gametokinetic hormone) Structure: Not fully definitized. Production Site: Anterior pituitary.	Female: stimulates ovarian follicles to grow and to develop, forming multiple layers and antra Male: stimulates seminferous tubules; stimulates spermatogenesis	Ascorbic acid: depletion in ovary due to follicle-stimulating hormone and luteinizing hormone action Vitamin E: required to maintenance of membranes in sex organs
Glucagon (HGF) (Hyperglycemic-glycogenolytic factor; glukagon; HG-factor) Structure: Polypeptide, 29 amino acids (structure determined). No S—S bridges. Molecular Weight ~3500 Production Site: Alpha cells in pancreas.	Increases: blood sugar; blood K^+, oxygen consumption, liver glycogenolysis, gluconeogenesis, nitrogen and salt excretion Decreases: liver glycogen, protein formation, gastric juice, fatty acid synthesis	Ascorbic acid: depletion of adrenal ascorbic acid by glucagon
Insulin (no synonyms) Structure: 51 amino acids. Known and synthesized. 3 S—S bridges. (See text, Fig. 4) Molecular Weight 5,734 (monomer); 12,000–48,000 (polymer), depending upon pH. Production Site: Beta cells of islets of pancreas.	Regulates carbohydrate and fat metabolism, especially glucose and fat oxidations Stimulates amino acid and glucose transport into cells and protein synthesis	Ascorbic acid: acts similarly to alloxan (i.e., antangonist)
Luteinizing Hormone (LH) (Luteotrophin, ISCH) Structure: Globular glycoprotein with S—S bridges. Molecular Weight 26,000 Production Site: Anterior pituitary.	Female: promotes estrogen and progesterone secretion, ovulation; maintains ovarian tissues Male: stimulates Leydig cells to secrete testosterone; gametogenic with follotropin (FSH)	Ascrobic acid: ovarian depletion on LH stimulation Vitamin E: involved in spermatogenesis
Melanocyte-stimulating Hormone (MSH) (Melanotrophin, chromatophorotropic hormone; pigmentation hormone) Structure: Polypeptide; purified, synthesized; alpha and beta forms; straight chains. Molecular Weight: 1500 (alpha) 2100–2600 (beta) Production Site: Intermediate lobe of pituitary.	Mammals: exerts small effect on skin pigmentation (protection from sunlight not fully proved) Expands or contracts pigments in various chromatophores Expands melanophore pigments with color changes in amphibia (adaptation to environment) Lower vertebrates: increases sensitivity to light; decreases dark adaptation time	Ascorbic acid: adrenal cortex depleted on ACTH and MSH activity Vitamin A: MSH decreases dark adaptation time.

Norepinephrine (Arterenol; noradrenaline; levarterenol) 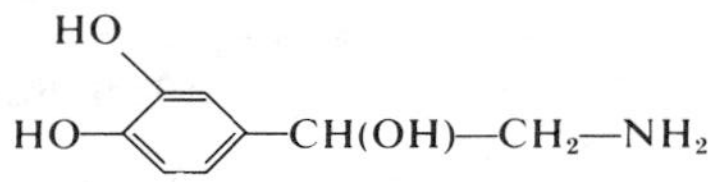 Molecular Weight 169.2 Production Site: Adrenal medulla; adrenergic nerve endings; chromaffin cells.	Blood circulation: increases blood pressure; peripheral vasoconstrictor without change or slight decrease in output and heart rate. No flow increase in brain, liver, or muscle Central nervous system effects: adrenergic transmitter agent at synapses; no brain excitation Kidney: decreases glomerular filtration rate Lung, intestine, genital system: inhibited Metabolic effects: weak epinephrine effect	Ascorbic acid: protects against oxidation of norepinephrine Ascorbic acid, folic acid, and vitamin B_6 are cofactors in synthesis of norepinephrine from phenylalanine
Oxytocin (Oxytocic hormone; pitocin; uteracon; α-hypophamine) Cys—Tyr—Ile—Gln—Asn—Cys—Pro—Leu—Gly NH_2 (Cys to Cys bridged) Molecular Weight 1007 Production Site: Hypothalamus.	Uterine contraction, milk ejection, facilitates sperm ascent in female tract Decreases membrane potential of myometrium, basic metabolic rate, and liver glycogen Stimulates oviposition in hen, releases luteinizing hormone (LH) Increases blood sugar and urinary sodium and potassium	Findings on interrelationships with vitamins are not extensive
Parathyroid Hormone (PTH) (Parathormone) Structure: Simple polypeptide (83 amino acids), sequence determined; straight chain; No S—S bridges. Production Site: Parathyroid glands.	Increases blood calcium, kidney calcium reabsorption, phosphate excretion, and blood citrate level Mobilizes calcium and phosphate from bone Activates calcium and phosphate absorption from the gastrointestinal tract (for which vitamin D is required) Increases osteoclast formation	Vitamin D: synergistic with PTH in maintenance of serum calcium
Progesterone (Progestin, luteosterone) 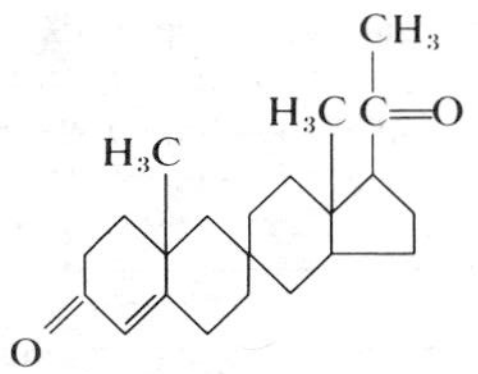 Molecular Weight 314.5 Production Sites: Ovary (follicles, corpus luteum); testicles; adrenal cortex; placenta	In low concentrations: prepares uterus for blastocyst implantation; promotes ovulation and mammary gland development; regulates female sex accessory organs; weak corticosteroid properties; precursor to sex hormones In high concentrations: maintains pregnancy; represses ovulation and sex activity; inhibits vaginal cornification and parturition; decreases myometrial excitation	Ascorbic acid: depleted from adrenal cortex or ovary on progesterone formation Niacin: diphosphopyridine nucleotide (DPN) involved in progesterone synthesis
Prolactin LTH (Lactogenic hormone; lactogen; galactin; mammotropin) Structure: Single-chain protein, 205 amino acids Molecular Weight 23,000–25,000 Production Site: Anterior pituitary	Initiates lactation Develops mammary glands in female Increases weight and growth (similar to somatotrophin in some species) Participates in nidation of zygote Protein anabolism (some species) Growth and secretion of crop gland (birds) Luteotropic (only in mouse, rat) Promotes maternal behavior	Not fully determined. Generally participates with other substances having growth action
Relaxin (Releasin, cervilaxin) Structure: Polypeptide (4 peptides with activity have been isolated); about 30–40 amino acids in each peptide Molecular Weight 4000–5000 Production Site: Corpus luteum in pregnancy	Enlarges birth canal in preparation for parturition Separation of symphysis pubis, loss of rigidity in pelvic bones Decreases uterine motility Maintains pregnancy Increases sensitivity to oxytocin; releases oxytocin Stimulates mammary gland Stimulates imbibition of water in uterus Inhibits uterine contraction	Ascorbic acid: maintains mucoprotein ground substance in connective tissue, affected by relaxin

REPRESENTATIVE HUMAN HORMONES (*continued*)

Somatotrophin (STH) (Growth hormone, GH; somatotrophic hormone; hypophyseal growth hormone) Structure: Known and synthesized; coiled, unbranched protein; 188 amino acid residues; 2 S—S bridges Molecular Weight 21,500 Production Site: Anterior pituitary	Promotes general growth of organism Promotes skeletal growth, protein anabolism, fat metabolism, carbohydrate metabolism, water, and salt metabolism	Relates with all vitamins in connection with growth actions
Testosterone (17 beta-hydroxy-4-androsten-3-one) 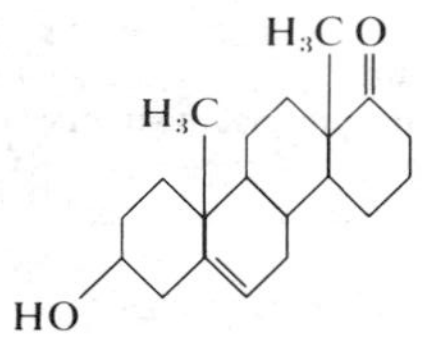 Molecular Weight 288.4 Production Sites: Interstitial cells of ovary and testis; adrenal cortex; embryonic placenta	Controls secondary male sex characteristics Maintains functional competence of male reproductive ducts and glands Increases protein anabolism; maintains spermatogenesis; inhibits follotropin Increases male sex behavior; increases closure of epiphyseal plates	Ascorbic acid, folic acid, vitamins A and E are synergists with testosterone for maturation of germ cells and increased anabolic activity
Thyroid-stimulating Hormone (TSH) (Thyrotrophic hormone, thyrotrophin) Structure: Glycoprotein (300 amino acids) Molecular Weight 26,000–30,000 Production Site: S^2 type cell, anterior pituitary	Regulates body temperature via thyroxine Maintains thyroid gland and its secretory activity (colloid discharge) Maintains iodine uptake by thyroid gland Promotes differentiation in embryo during development via thyroxine Stimulates coupling of diodotyrosine to form thyroxine	Ascorbic acid, thiamine, riboflavin, and vitamin B_{12}: requirements increase in hyperthyroidism; tissue concentrations reduced Vitamin A: massive doses of vitamin A inhibit secretion of TSH; thyroid hormones required for carotene and retinene conversions Vitamins A, D, E, and K: requirements increased in hyperthyroidism; tissue concentrations reduced in hyperthyroidism Vitamin B_6, niacin: conversion to phosphorylated reactive forms impaired in hyperthyroidism
Thyroxine (T_4) (3,5,3′,5′ tetraiodothyronine) 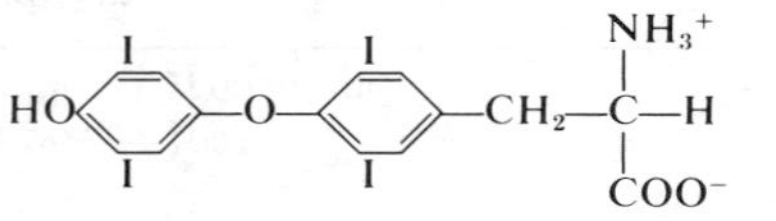 Molecular Weight 776.9 Production Site: Thyroid gland	Regulates growth, differentiation, oxidative metabolism, electrolytic balance Increases carbohydrate metabolism, calorigenesis, protein anabolism, basal metabolic rate, oxygen consumption, fat catabolism, fertility Sensitizes nervous system	Ascorbic acid: synergist in cold survival Niacin: synergist in mitochondrial metabolism Vitamin A: T_4 is required for vitamin A synthesis in liver Vitamin B_{12}: T_4 aids in B_{12} absorption B complex vitamins: deficiencies develop in hyperthyroidism
Vasopressin (Arginine vasopressin; antidiuretic hormone; ADH; pitressin; tonephin; vasophysin) Cys—Tyr—Phe—Glu NH_2—Asp NH_2—Cys—Pro—Leu—Gly NH_2 Vasopressin Molecular Weight 1084 (arginine-vasopressin) Production Site: Hypothalamus	Elevates blood pressure (mammals) (reverse effect in birds) Decreases kidney blood flow Antiduretic, releases ACTH Increases sodium chloride and urea excretion Regulates water balance Stimulates contraction of smooth muscles Increases renal tubular water reabsorption Releases anterior pituitary hormones	Not fully determined

SOURCE: Adapted from R. J. Kutsky (see reference list)

chain peptide hormone with a molecular weight of about 8,000. A second parathyroid hormone, *calcitonin*, was postulated by Copp (1961). Subsequent research has indicated that this hormone is actually the hormone which is now known to be produced by the thyroid gland. However, a parathyroid calcitonin may exist in certain species.

The more classical function of parathyroid hormone is concerned with its control of the maintenance of constant circulating calcium levels. Its action is on (1) the kidney, where it increases the phosphate in the urine, (2) the skeletal system, where it causes calcium resorption from bone, and (3) the digestive system, where it accelerates (stimulates) calcium absorption into the blood. The hormone and gland exhibit characteristics of feedback control; when the concentration of calcium ions in the blood falls, the secretion of the hormone increases, and when their concentration rises, the secretion of hormone decreases. See **Parathyroid Glands.**

Adrenal Cortical Hormones. The adrenal gland is made up of two parts, the medulla and the cortex, each of which secretes characteristic hormones. The hormones of the adrenal medulla are the catecholamines, epinephrine (adrenalin) and norepinephrine (noradrenalin), which are closely related chemically, differing only in that epinephrine has an added methyl group. See accompanying table. In fact, animal experiments have established a metabolic pathway for the biosynthesis of both compounds from the amino acid phenylalanine, which involves enzymatic oxidation and decarboxylation reactions. It is also to be noted that the isomeric form of norepinephrine is most important; the natural D-form (which incidentally, is levorotatory) has many times the activity of the synthetic isomer. Epinephrine has a pronounced action upon the circulatory system, increasing both blood pressure and pulse rate, and hence the cardiac output by its direct action upon the heart muscle, and especially because it causes constriction of the arterioles. However, its effects upon smooth muscles vary; it relaxes the muscles of the digestive system, but contracts the pyloric sphincter.

Norepinephrine does not affect the cardiac output, although it does raise the blood pressure by constricting the arterioles. Its muscular effects are less pronounced. Both epinephrine and norepinephrine release free fatty acids from adipose tissue, so raising its level in the blood. This effect is due to the action of the hormones in accelerating enzymatic reactions whereby the esters of the fatty acids are hydrolyzed. The third type of action of epinephrine is its effect upon the carbohydrate metabolism, notably the acceleration of the hydrolysis of glycogen in muscular tissue and the liver, and so raising the glucose level in the blood, and the rate of glucose oxidation, with resulting increase in oxygen utilization, carbon dioxide production, and body temperature. See **Adrenal Glands.**

The hormones of the adrenal cortex are steroids. See also **Steroid.** Among them there are a number of hormones with androgenic activity, such as adrenosterone and 17α-hydroxyprogesterone, which are discussed under the sex hormones later in this entry. In all, over ten steroids have been identified in the adrenal cortex, including seven of characteristic cortical activity. These are corticosterone, from which the others are named, 17α-hydroxyl-11-dehydrocorticosterone (cortisone), 17α-hydroxycorticosterone (cortisol or hydrocortisone, and 18-oxocorticosterone (aldosterone). Only two hormones, cortisol and corticosterone, are normally released in fairly large quantities, and another, aldosterone, deserves mention because of its somewhat different effects, even though it is released to a far lesser extent.

All of these hormones are synthesized from cholesterol in the adrenal cortex, by an extended series of reactions which include many related compounds. Although these hormones have widespread effects throughtout the organism, their primary mechanism is not known, so that many of the effects may be indirect. Much of the knowledge of their action arises from studies of insufficiency or hyperactivity of the adrenal cortex, which produces a wide variety of pathological conditions. See accompanying table.

It is generally considered that aldosterone, and to some extent the other hormones, have a regulatory effect upon the metabolism of electrolytes and water, particularly upon the concentration of the ions of the alkali metals in intracellular fluids. Administration of steroids also increases the concentration of calcium ions in those fluids. However, all three of these hormones have a number of other effects, roughly in the order of potency—cortisol, corticosterone, aldosterone. They produce changes in the metabolism of carbohydrates, proteins, and fats.

For the carbohydrates alone, three major effects are evident—increase in the rate of formation of glucose, increase in the rate of release of glucose from the liver, and increase in the rate of utilization of glucose. These hormones affect the digestive system, increasing the secretion of hydrochloric acid, pepsinogen, and trypsinogen. They prevent inflammatory responses to bacterial or even chemical stimuli; they counteract anaphylactic shock, and other effects of hypersensitivity. Obviously, these properties have led to their widespread therapeutic use.

There are relationships between the adrenal cortical hormones and the thyroid and pituitary glands. Depression of the function of the adrenals produces thyroid deficiency, whereas administration of thyroxine stimulates the ACTH-adrenal cortical mechanism.

Pituitary Hormones. The hormones of the hypophysis (pituitary gland) are quite numerous, being secreted variously in three parts of the gland—the neurohypophysis (posterior lobe), the adenohypophysis (anterior lobe), and the *pars intermedia*, which connects the other two.

The chief hormones of the neurohypophysis are the polypeptides oxytocin and vasopressin. The hormone characteristic of the *pars intermedia* is the melanocyte-stimulating hormone. It is usually spoken of in the plural, since in most mammals both alpha and beta forms are known. The structures of the first two are shown in the accompanying table. See also **Diabetes Insipidus.**

The most prominent effect of oxytocin is the contraction of smooth muscle, especially of the uterus. It also has a major effect upon the muscles about the breast, and so stimulates the ejection of milk in lactating animals. It has a definite stimulating effect upon the muscles of the ureter, urinary bladder, intestine, and gall bladder.

The most prominent effect of vasopressin is upon the kidneys, where it stimulates the resorption of water in the tubules (which by repeated release and absorption concentrate the urine). It also constricts the coronary arteries, raises the blood pressure, and exhibits the effect of oxytocin upon smooth muscles, but generally to a lesser degree.

The action of the melanocyte-stimulating hormones has been established by studies of animals, in which they cause dispersal of certain black pigments from the cells that contain them, with resulting darkening of the skin.

The adenohypophysis is the part of the gland in which the tropic hormones are secreted. They include the adrenocorticotropic hormone (ACTH), the thyrotropic hormone (TSH), and somatotropin, as well as three hormones with pronounced effects upon the gonads: the hormone prolactin, the follicle-stimulating hormone (FSH) and the luteinizing or interstitial cell stimulating hormone (LH or ISCH).

ACTH. Adrenocorticotropin (ACTH) in humans is a polypeptide containing a sequence of 39 amino acids, although work with animal forms of it and with degradation products of the human form have shown that not all of them are essential to the activity of the hormone. This sequence for the human ACTH is shown in Fig. 1.

The primary function of ACTH is the stimulation of the adrenal cortex to produce its hormones, which have already been discussed. This is evident from the therapeutic effect of administration of ACTH, which is closely similar to that of these hormones, so that if the action of only one of them is sought, its administration is preferable. Moreover, ACTH stimulates secretion of the androgenic substances mentioned as produced by the adrenal cortex.

Thyrotrophic Hormone. This hormone (TSH) stimulates the development of the thyroid and controls its secretion. Although purified prepa-

Ser—Tyr—Ser—Met—Glu—His—Phe—Arg—Tyr—Gly—Lys—Pro—

Val—Gly—Lys—Lys—Arg—Arg—Pro—Val—Lys—Val—Tyr—Pro—

NH_2 (on Glu)

Asp—Ala—Gly—Glu—Asp—Glu—Ser—Ala—Glu—Ala—Phe—Pro—

Leu—Glu—Phe

Fig. 1. Amino acid sequence of human adrenocorticotrophin (ACTH).

rations of it have been obtained, they consist of a mixture of proteins of high mean molecular weight (about 30,000). Some of their amino acids have been determined, as well as their carbohydrates, but the structures have not been elucidated.

Growth Hormone. Somatotropin is the growth hormone. Purified preparations of extracts of it from the human adenohypophysis have been crystalized. They are known to be proteins, of mean molecular weight 21,000, and containing a single polypeptide chain. This hormone differs from the others of its group in not acting primarily upon the other endocrine glands, but in controlling the gain in body weight and the rate of skeletal growth. The growth abnormalities, such as dwarfism and giantism, have been shown to result from its hypo- and hypersecretion. In addition to its effect upon growth and anabolism generally, it has been found to affect the kidneys and pancreas, and to influence glucose, galactose, and lipid metabolism. See also **Pituitary Gland.**

Gonadotrophic Hormones. These include follicle stimulating hormones (FSH), luteinizing or interstitial cell stimulating hormone (LH or ISCH), and prolactin. Their structures are not known; the molecular weight of human LH is about 26,000, that of human FSH is about 30,000, and that of human prolactin is uncertain. They are proteins, with variable amounts of carbohydrates. FSH induces the growth of Graafian follicles in the ovary and the production of spermatozoa in the testis. LH stimulates the final development of the ovarian follicles, the appearance of estrus, and the change of the follicles to corpora lutea. In the male, it stimulates the secretion of testosterone. Since these effects are due to the effect of this hormone upon interstitial cells, it is also called ISCH. Prolactin stimulates lactation after birth, acts with estrogen to promote the growth of the mammary gland, and influences the activity of the corpora lutea.

Male Hormones. The androgenic hormones produced in the testes (and adrenal gland) have a widespread effect upon the development of secondary sexual characteristics (musculature, facial hair, larynx, etc.), as well as upon the sexual organs and responses themselves. They also promote anabolism to a marked degree by their effect upon nitrogen and calcium metabolism. The structure of testosterone is shown in the accompanying table. See **Gonads.**

Female Hormones. Closely related to the male androgenic hormones, and probably synthesized from them in the female organism, are the estrogenic hormones which are produced principally in the ovary. Although β-estradiol is the normally secreted ovarian hormone, a number of other estrogenic substances have been isolated from urine and from animal studies. They include α-estradiol, estriol, and estrone. The structures of these hormones are given in Fig. 2.

These hormones are important in both the menstrual cycle and the reproductive cycle, and of course play an important role in oral contraceptives (the "pill"). They induce growth of the vaginal epithelium, secretion of mucus by the glands of the cervix, and initiate the growth of the endometrium, which is taken over by progesterone (from the corpus luteum) later in the cycle. They activate the proliferation of the mammary gland during pregnancy. As the androgens do

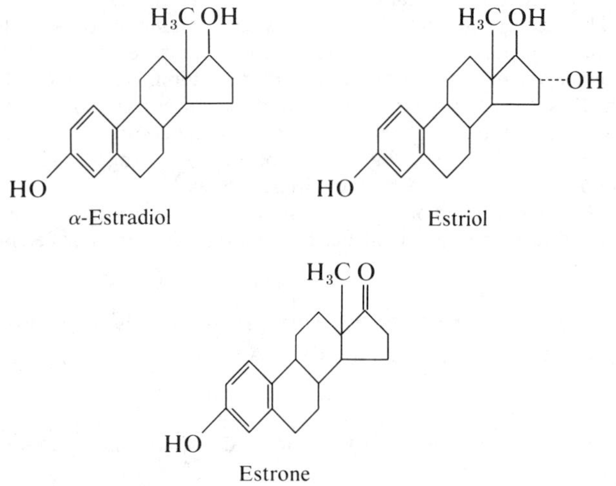

Fig. 2. Major ovarian hormones.

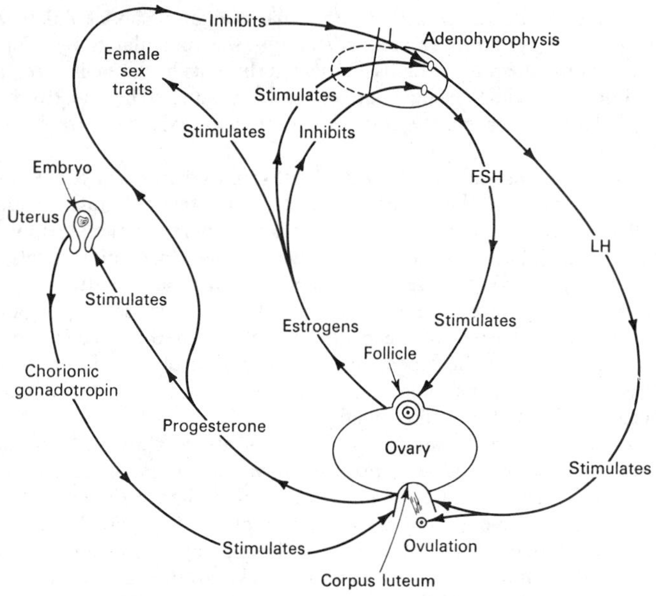

Fig. 3. Cycle of hormone adjustment in human female.

for the male, the estrogens bring about the secondary sexual characteristics of the female. They have a number of effects upon metabolism, notably that of calcium and phosphorus, and of lipids and proteins. A number of other estrogens, some made synthetically and others obtained from animals, are known.

The corpus luteum produces two hormones, progesterone and relaxin. The structures of these hormones are shown in the accompanying table.

Progesterone acts to complete the proliferation of the endometrium, which was initiated by the estrogenic hormones, and to prepare it for the ovum. In pregnancy the continued action of progesterone is necessary. It aids the growth of the breasts and has a definite effect against ovulation. It is also the biosynthetic precursor of some of the estrogenic hormones. Relaxin has been shown to have a relaxing effect on the cartilaginous junction of the pubic bones in preparation for parturition. See also **Embryo.**

Feedback in Hormone Control Systems. Not only do the hormones initiate or stimulate biological processes, both directly and by bringing about production of other hormones in other glands, but they also act to maintain the organism in a steady state, or *homeostasis.* Thus the gonadotropic hormones from the hypophysis stimulate the testes, but the resulting production there of androgens like testosterone, inhibits the action of the hypophysis in producing the gonadotropic hormones. The complicated cycle of adjustment in the human female is shown in the cycle illustrated in Fig. 3.

As shown in the figure, the regulation of the ovarian hormones in the human female involves both positive and negative feedback. The follicle-stimulating hormone (FSH) from the adenohypophysis stimulates the Graafian follicles, which thus produce estrogens. These not only inhibit FSH production through negative feedback, but also stimulate the adenohypophysis to increase its production of luteinizing hormone (LH) through positive feedback. This hormone in turn brings about ovulation from the Graafian follicle. After the ova are discharged, the LH stimulates the empty follicle, now the corpus luteum, to produce progesterone.

This hormone brings about the changes in the reproductive organs required for the development of the embryo. Then the progesterone partly inhibits the adenohypophysis from producing further LH, an example of negative feedback; as a result, there is no further ovulation. The progesterone also acts as a positive feedback and stimulates the production of FSH.

When pregnancy intervenes, a new feedback mechanism must be introduced, or the embryo would be expelled by the shedding of the lining of the uterus in menstruation. Here the placenta (chorion) of the embryo itself produces hormones, as already noted. Its LH stimulates continuing production of progesterone from the corpus luteum,

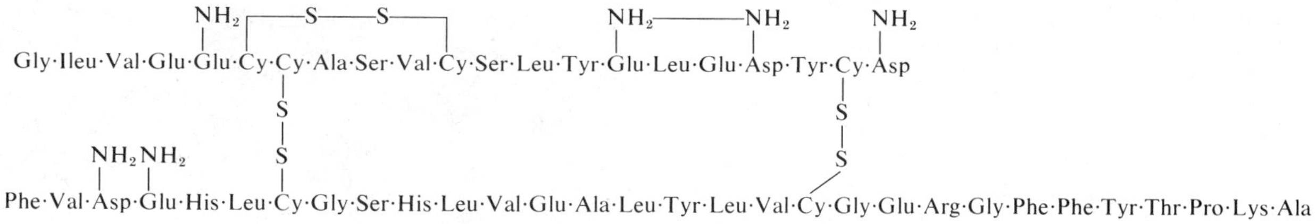

Fig. 4. Primary structure of bovine insulin. (Abbreviations of amino acids will be found under **Amino Acids.**)

Fig. 5. Glucagon. (Abbreviations of amino acids will be found under **Amino Acids.**)

thus preventing menstruation and stimulating the continuing development of the uterus as needed by the growing embryo. The extra progesterone also inhibits further ovulation in spite of the presence of the gonadotropin from the placenta (chorion).

Pancreas and Nonendocrine Hormone Sources. In addition to producing hormones, the pancreas also generates digestive fluids (*pancreatic juice*). It is the hormone function which makes the pancreas a part of the endocrine system. See **Endocrine System; Pancreas.** The pancreas secretes *insulin* and *glucagon,* both hormones. The primary structure of bovine insulin is shown in Fig. 4. Abbreviations of the amino acids will be found under **Amino Acids.** The upper chain is termed the A chain; the lower chain, the B chain. In the spatial configuration of the molecule, these chains occur as coiled helices. The structure of glucagon is considerably simpler inasmuch as it consists of a single chain of amino acids. See Fig. 5.

These two hormones have two opposing effects. That of insulin is *hypoglycemic,* i.e., it increases the rate of utilization of glucose, the probable process being an effect of insulin to increase the penetration of glucose through the cell walls as well as increased phosphorylation. The overall result of action of insulin in its relation to glucose is to increase the rate of the reactions by which glucose is oxidized, but also its transformation to glycogen. The enzyme glucagon raises blood glucose levels by increasing the rate of hydrolysis of glycogen (in the liver) to increase the formation ultimately of glucose. Insulin increases the rate of entry of amino acids into cells and their rate of protein biosynthesis. Insulin also accelerates the formation of lipids from carbohydrates, whereas glucagon stimulates the formulation of keto compounds from lipids, inhibits the synthesis of fatty acids, and accelerates the breakdown of various phosphorus and nitrogen compounds. The primary result of insulin deficiency is diabetes mellitus. See **Diabetes Mellitus.**

Hormones may be produced by organs other than the endocrine glands. Conspicuous among such organs is the placenta, the organ on the wall of the uterus to which the umbilical cord is attached. It has been found to produce the same estrogenic hormones as the ovary, the same hormones (progesterone and relaxin) as does the corpus luteum, and gonadotropic hormones (and luteinizing hormones) similar to, but not identical with, those produced by the adenohypophysis.

Other hormones which do not originate in endocrine glands are the cholecystokinin of the intestine, and the enterogastrone and gastrin of the stomach. The first is produced by the upper intestinal mucosa and causes the gall bladder to contract; the enterogastrone is produced in the same tissue and inhibits gastric motility and secretion; it also excites secretion of digestive fluids, principally hydrochloric acid.

In addition to the entries covering specific endocrine glands, see also **Brain and the Nervous System; Endocrine System;** and **Steroids.**

In plants, a *plant hormone* or "phytohormone" is an organic compound produced by the plant, controlling growth and other functions at sites remote from where the hormone is produced. Plant hormones also act in very minute amounts. Plant hormones include the auxins, gibberellins, and kinetins. These are described in the entries on **Gibberellic Acid and Gibberellin Plant Growth Hormones;** and **Plant Growth Modification and Regulation.** Plant hormones are also mentioned in a number of specific plant-related entries.

The use of hormones in cancer chemotherapy is described in entry on **Cancer and Oncology.**

References

NOTE: In addition to the references listed here, see the reference lists at the ends of the other entries mentioned in this description.

Barrington, E. J. W. (editor): "Trends in Comparative Endocrinology," American Society of Zoologists, Thousand Oaks, California, 1975.

Brownstein, M. J., Russell, J. T., and H. Gainer: "Synthesis, Transport, and Release of Posterior Pituitary Hormones," *Science,* **207**, 373–378 (1980).

Crews, D.: "The Hormonal Control of Behavior in a Lizard," *Sci. Amer.,* **242**, 2, 180–187 (1979).

Kolata, G. B.: "Hormone Receptors," *Science,* **196**, 747–748 (1977).

Kolata, G. B.: "Polypeptide Hormones," *Science,* **201**, 895–897 (1978).

Kolata, G. B.: "Sex Hormones and Brain Development," *Science,* **205**, 985–987 (1979).

Korenman, S.: "Conference on the Endocrine Aspects of Aging," National Institute of Aging, the Endocrine Society, and the Veterans Administration, Bethesda, Maryland (October 18–20, 1979). Veterans Administration Hospital, Sepulveda, California, 1979.

Krieger, D. T., and A. S. Liotta: "Pituitary Hormones in Brain," *Science,* **205**, 366–372 (1979).

Kutsky, J. F.: "Handbook of Vitamins and Hormones," Van Nostrand Reinhold, New York, 1973.

Martial, J. A., et al.: "Human Growth Hormone," *Science,* **205**, 602–607 (1979).

Marx, J. L.: "Hormones and Their Effects in the Aging Body," *Science,* **206**, 805–806 (1979).

Mettes, J.: "The 1977 Nobel Prize (Hormone Research)," *Science,* **198**, 594–596 (1977).

Meites, J., Donovan, B. T., and S. M. McCann (editors): "Pioneers in Neuroendrocrinology," Plenum, New York, 1978.

Oppenheimer, J. H.: "Thyroid Hormone Action at the Cellular Level," *Science,* **203**, 971–979 (1979).

Staff: "Insulin by Bacterium," *Sci. Amer.,* **239**, 4, 85–90 (1978).

Turakulov, Y. K. (editor): "Thyroid Hormones," Plenum, New York, 1975.

White, B. A., and C. S. Nicoll: "Prolactin Receptors in *Rana catesbeiana* during Development and Metamorphosis," *Science,* **204**, 851–853 (1979).

HORMONES (Insecticides). Insecticide and Pesticide Technology.

HORMONES (Plant). Plant Growth Modification and Regulation.

HORNBEAM TREES. Of the family *Carpinaceae* (hornbeam family), these trees are of the genus *Carpinus,* except the hophornbeams, which are of the genus *Ostrya.* These trees once were classified with the birches (family *Betulaceae*). There are both shrubs and trees within *Carpinus,* all deciduous and hardy. The common hornbeam of Europe is the *C. betulus,* which can attain a height up to 75 feet (22.5 meters). The tree is generally of a pyramidal contour. The flowers are unisexual. In terms of landscaping, one authority attributes much to what may be termed the interesting texture of the trunk and base, where there are innumerable muscle and tendon-like configurations. It is also of interest to note that the common European hornbeam has been used in formal gardens as a form of "hedge-on-stilts." There are several variations of the common European hornbeam, including the 'Columnaris,' the 'Fastigiata,' the 'Incisa,' and the 'Intertexta,' thus providing a range of size, coloration, and density.

The common hornbeam native to and common in North America is the *C. caroliniana,* also sometimes called the American hornbeam, the blue beech, or the water beech. This plant may be described as a tall shrub or small tree and ranges up to 30 or 40 feet (9 to 12 meters) in height. One excellent specimen selected by The American Forestry Association in 1975 as a champion for its records is located in Minton, New York. The tree has a circumference at 4½ feet (1.4 meters) of 86 inches (218 centimeters), a height of 65 feet (19.5 meters), and a spread of 66 feet (19.8 meters). The branches of the American hornbeam are slender and extend nearly horizontally from the trunk. The leaf is narrow, ovate, sharp-pointed and a dull light green, lighter color underneath. The staminate catkins are about 1½ inches (3.8 centimeters) long. The bark is scaly and gray-brown. This tree ranges from Nova Scotia and Quebec west to the Great Lakes and south through Nebraska, Kansas, and Oklahoma, to Texas in the southwest and Florida in the southeast. The tree is quite common in New England.

The Japanese hornbeam (*C. japonica*) reaches a height of about 50 feet (15 meters) and is wide-spreading. It has male catkins from 1 to 2 inches (2.5 to 5 centimeters) in length. A more ornamental Japanese hornbeam is the *C. laxiflora*. Other species include the *C. orientalis*, a bushy shrub of southeastern Europe and Asia Minor; and the *C. turczaninowil*, a thin, spindly hornbeam.

Closely related to the hornbeams are the hophornbeams. See accompanying photo. Generally, these trees are medium-to-large in size and are deciduous. One excellent specimen selected by The American Forestry Association is the Eastern Hophornbeam (*Ostrya virginiana*) located in Monroe Center, Michigan 1976. The tree has a circumference at 4½ feet (1.4 meters) of 112 inches (284 centimeters), a height of 73 feet (21.9 meters), and a spread of 88 feet (26.4 meters). They are also characterized by drooping male catkins and upright female catkins. Their fruit is in the form of a nutlet contained in a husk. The hophornbeam of southern Europe and Asia Minor is the *Ostrya carpinifolia*, which can attain a height up to 65 feet (19.5 meters). The common hophornbeam in America is the American hornbeam (*O. virginiana*), sometimes also called the ironwood or leverwood. Generally, this is a fairly small tree, ranging from 30 to 45 feet (9 to 13.5 meters) in height, but under favorable conditions the tree can do much better. The bark is gray-brown, scaly, and has perpendicular scoring. The leaves are from 3 to 4 inches (7.6 to 10 centimeters) in length, narrow, ovate, double-toothed, sharp-pointed, and of a dull light-green color. The staminate flowers normally occur in three drooping catkins. The tree prefers a dry soil and open woods. It ranges from Nova Scotia and New Brunswick southward along the Saint Lawrence and Lower Ottawa Rivers, westward to Lake Huron, and northwest to Minnesota and the Dakotas, thence southward as far as Kansas and Nebraska. In the east, it is found in the Alleghany Mountains and south into Florida. In the southwest, the tree also is found in eastern Texas. Thus, the tree has a broad climatic range.

The *Ostrya knowltoni* is found on the southern slopes of the Colorado River canyon in Arizona and northward to Flagstaff. It is abundant to the 6,000 to 7,000-foot (1800 to 2100-meter) level. The height of the tree ranges from 25 to 40 feet (7.5 to 12 meters). The trunk is about 15 inches (38 centimeters) in diameter. The bark is light brown and scaly, with bright orange underneath. The leaf is small, 1½ to 2 inches (3.8 to 5 centimeters) long, ovate, soft, hairy above and smooth underneath.

The Catalina ironwood (*Lyonothamnus Gray*) or Lyon tree is of interest because it is so rare and because of its unusual location. The tree grows on the canyon slopes of the steep shores of Santa Catalina Island, just off the coast of southern California. The tree was discovered by William Lyon, a young forester, in 1884 and thus so named. It is postulated that the tree was on the island at the time when the sea level was many hundreds of feet lower than it is today. There are a number of species of the tree, particularly on the eastern side of the island. They are also found on Santa Rosa and Santa Cruz islands nearby. At one time, geologists postulate that these islands were a connected land mass. However, the tree leaf differs from one location to the next. The tree has compound foliage on Santa Rosa and Santa Cruz, whereas this is not the case on Santa Catalina.

If the trees were numerous, the wood would be of considerable economic value because it is very strong and tough, weighing about 50 pounds per cubic foot (801 kilograms per cubic meter). It is believed that the Canalino Indians once used the wood for handles and shaft wood. The trees grow straight and tall, varying greatly in size, some towering up to 60 feet (18 meters) in height, with trunks of about

The former champion eastern hophornbeam tree (*Ostrya virginiana*).

1½ feet (0.5 meter) in diameter. They are found at an elevation between 500 and 2,000 feet (150 and 600 meters). The flowers are small and lacy, in groups on branches that are somewhat flat. They bloom in June and July.

The leaf has a simple-bladed fern-like foliage, projecting an aura of antiquity. It is blade-shaped with teeth coarsely cut. The seed is oblong and light-brown. The bark is dark brown, but appears to weather to a lighter color. It is often tattered, with the underbark showing through. The tree is difficult to propagate from cuttings or seeds. Root sprouts thus far have proved most effective.

HORNBILL. Kingfishes and Other Coraciiformes.

HORNBLENDE. The mineral hornblende is a complex silicate which is probably an isomorphous mixture of three molecules, a calcium-iron-magnesium silicate, an aluminum-iron-magnesium silicate and an iron-magnesium silicate. A general formula is

$$(Ca,Na,K)_{2-3}(Mg,Fe^{2+},Fe^{3+},Al)_5(Al,Si)_8O_{22}(OH)_2.$$

Manganese and alkalis are sometimes present as is also titanium. It is monoclinic, with prismatic crystals, often pseudo-hexagonal. Bladed, fibrous, columnar, granular and compact massive varieties also are common. It has a perfect prismatic cleavage; hardness, 5–6; specific gravity, 3.02–3.27; color, green, greenish-brown, brown and black; luster, vitreous to silky; transparent to opaque.

Hornblende is a common constituent of many of the igneous rocks such as granite, syenite, diorite, or gabbro, of gneisses and schists and is the principal mineral of the amphibolites. Hornblende alters easily to chlorite and epidote. A variety of hornblende that contains little (less than 5%) of iron oxides is gray to white in color and named edenite, from its locality in Edenville, New York. Very dark brown to black hornblendes, which contain titanium, ordinarily are called basaltic hornblende from the fact that they are usually a constituent of basalts and similar rocks.

Well-known localities for hornblende are in Czechoslovakia, Mount Vesuvius, Italy, Norway, Sweden, and, in the United States, in Massachusetts, New Hampshire, and New York. Black hornblende is found in Renfrew County, Canada. The word hornblende is derived from the German *horn*, and *blende*, to blind or dazzle. The term blende was often used to refer to a brilliant nonmetallic luster, i.e., zinc blende.

See also terms listed under **Mineralogy.**

Elmer B. Rowley, Union College, Schenectady, New York.

HORNBLENDITE. A coarse-grained rock related to gabbro which consists almost wholly of hornblende. Olivine being present, this rock may grade into a hornblende-peridotite (cortlandtite). Hornblendite is a rare rock type and of relatively little importance.

HORNED TOAD (*Reptilia, Sauria; Phrynosoma*). Small spiny lizards of the southwestern states and Mexico. They have short broad bodies and short tails, hence the confusion of terms in the common name. Horned lizard is a better term.

Horned toad. (*A. M. Winchester.*)

Horned toads are desert animals and are capable of living for incredibly long periods without food or water. They cannot, however, survive for the long periods of years as has sometimes been claimed.

HORN (Electromagnetic). Horn radiators are used to obtain directional radiation characteristics which cannot be obtained as conveniently with simple antennae. As such directors, they are used both with conventional antennae and with wave guides, but in either case they serve to direct the radiation in a pattern from the open end of the horn in a manner determined by the dimensions of the horn. The important dimensions are the horn opening (in terms of wavelength of the radiation) and the flare angle. While theoretically an infinitely long horn will give a radiation pattern whose angle conforms to that of the horn, those of practical length do not confine the beam to quite this degree. For example, a horn with an angle of 15° may give a radiation pattern which spreads 23°. The common horns may be divided into three classes, sectoral, pyramidal and biconical. The sectoral horn has two sides which are parallel and the other two flared. The pyramidal horn has all sides flared. The conical horn is really a pyramidal horn with a circular cross section. The biconical horn consists of two cones with their vertices coinciding or adjacent to one another. The first two types are used where a singly directed beam of radiation is desired, the exact pattern in both vertical and horizontal planes being determined by the dimensions. The biconical horn gives a uniform pattern in a plane perpendicular to the axis and highly directional in any plane containing the axis.

Sectoral and pyramidal horns may be excited by more or less conventional antennae or by waveguides. In the former case, a short section of waveguide is attached to the end of the horn and this is excited by the antennae. In the latter case, the horn is really a flared extension

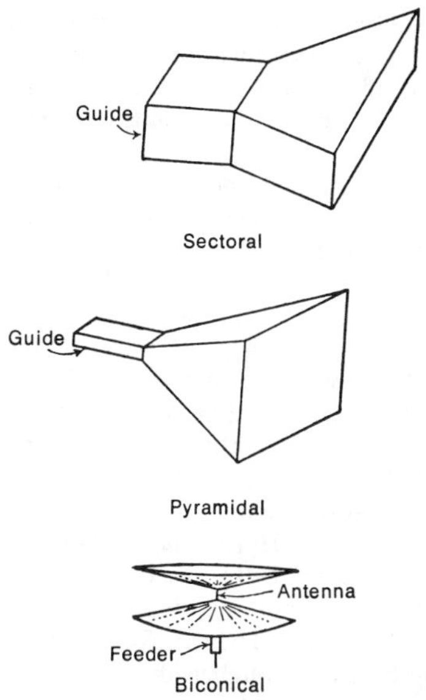

Types of electromagnetic horns.

of the guide and may be looked upon as an impedance-transforming section for matching the impedance of the guide to that of free space. Biconical horns are excited by a variety of antenna arrangements in the space between vertices of the horns. Because of space considerations, horns are not feasible except at ultra-high frequencies, but in the microwave region they are widely used as radiators.

HORNET (*Insecta, Hymenoptera*). A name loosely applied to many of the larger wasps, particularly to the species which build paper nests and have a severe sting.

The true hornet is a European wasp (*Vespa crabro*). In America, the term may refer to any form of large stinging wasp that makes paper nests. Hornets are all social. In the southern United States, a smaller species (*V. carolina*) goes by the name of hornet. The white-faced hornet is the common American hornet (*V. maculata*).

The hornet is usually yellow and black in coloration and has a pugnacious spirit. Its sting is severe. The nest is usually pear-shaped, but sometimes round, and suspends from a branch of a tree or roof of a building. The nest consists of horizontal cones all facing downward. A small hole is left in the side of the nest as an entrance. The nest may accommodate from a few hundred to over 5,000 hornets.

These insects live much as honeybees, with similar work habits. Their food is the nectar of flowers. Adult hornets prefer carbohydrate food sources; the young prefer protein foods from caterpillars. Hornets are susceptible to bacterial and fungus disease and have numerous insect enemies, factors which keep their numbers under control. Among their worst enemies are other hornets which pillage the food they store.

HORNFELS. A more or less general term applied to fine-grained, massive, and frequently speckled rock, the result of contact metamorphism developed in slates by granitic intrusions.

HORN FLY (*Insecta, Diptera*). This insect is most irritating and injurious to cattle, but also will attack goat, horse, and sheep. On occasions, the fly will also be a pest on dog. The species *Haemotobia* or *Siphona irritans* (Linne) pierces the animal's skin and sucks blood. The associated pain and irritation trouble the animals during resting and feeding periods and, as a result, the cattle lose weight, milk cows have a lower milk yield, and the general health of the animals deteriorates, particularly when the irritation continues over a long period. The horn fly is about half the size of the common house fly, but appears very much like it. Although not fully proven, some scientists believe that the horn fly may carry anthrax disease.

The insect was first noted in the United States in Philadelphia in 1887, but has since spread throughout the continental United States and also to Hawaii.

Control is usually by spraying the animals along their back and flanks with one of several formulations, including methoxychlor, ronnel, or toxaphene. Treatment of dairy cows should be confined to handrubbing a suitable formulation around the neck area, thus avoiding possible contamination of milk. Pyrethrins and allethrin also have proved effective. Where practical, housing the animals in darkened structures equipped with entrance curtains that help to remove the flies can be helpful.

The fly maggots depend largely on animal dung for their food. Thus, cleanliness about shelters frequented by the animals cannot be overstressed.

HORN SILVER. Chlorargyrite.

HORN (Substance). A hard translucent material formed by the development of epidermal cells containing a substance known as keratin. The outer layers of the skin are keratinized and the nails, claws and hoofs of mammals are formed of similar material. Horn is also developed in large amounts in the appendages of the head which go by the same name. Horns may be bony cores sheathed in horn or solid bony growths. The former occur in cattle and the latter in deer. The median horn or horns borne on the head of rhinoceroses are quite unlike true horns. They are formed of aggregated hair-like components firmly based on roughened areas of the underlying bones.

HORN-TAIL (*Insecta, Hymenoptera*). Large sawflies (woodwasps) whose larvae bore in the trunks of trees. The adults have a cylindrical body and in the female sex a short strong ovipositor which is the source of the name horn-tail. With this organ holes are drilled into the wood of the tree for the deposition of the eggs.

HORNWORM (*Insecta, Lepidoptera*). Closely related species of this worm are known as tobacco worms and tomato hornworms. The adult moths do not injure plants, but their larvae are quite damaging.

Southern or tobacco hornworm (*Protoparce sexta*, Johanssen). See Fig. 1. This insect is found in most of the United States and ranges southward into South America.

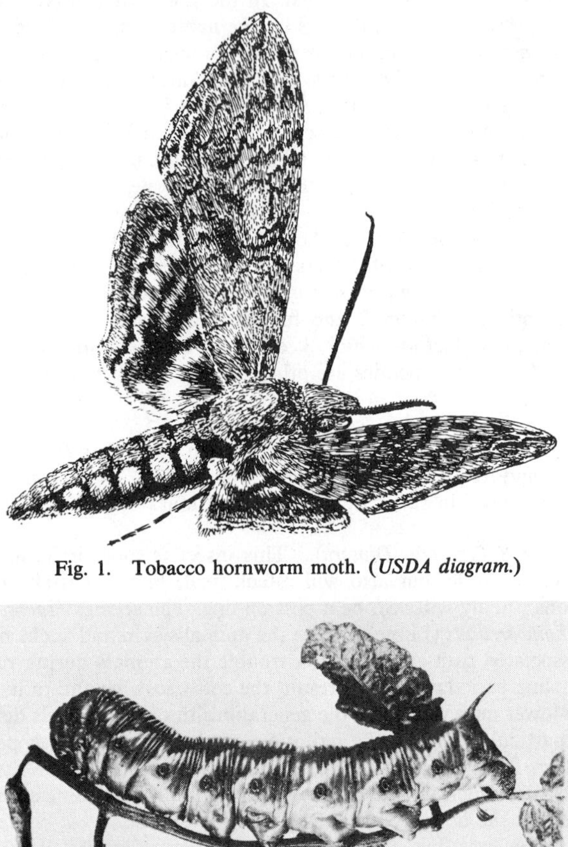

Fig. 1. Tobacco hornworm moth. (*USDA diagram.*)

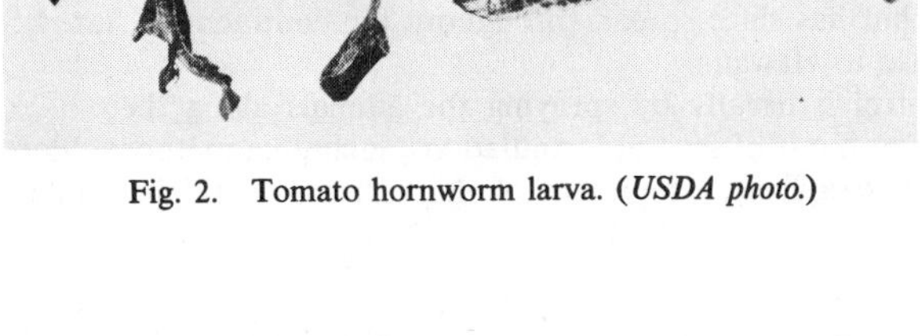

Fig. 2. Tomato hornworm larva. (*USDA photo.*)

Northern or tomato hornworm (*Protoparce quinquemaculata*, Haworth). See Fig. 2. This insect also is found throughout the United States and ranges northward into Canada.

In habit and damage, the two species are strikingly similar and, in fact, both species may be found attacking the same crop. In addition to tobacco and tomato, the insect is injurious to eggplant, pepper, and tomato. The worms also feed on a number of weeds often associated with these crops.

The worms are green with diagonal lines on their sides. They have a prominent horn (red horn on tobacco hornworm; black horn on tomato hornworm) on the rear end. They range up to 4 inches (10 centimeters) in length. Damage is caused by their eating of foliage and fruit. While widely distributed, infestations are usually localized. Because the worms are so large, handpicking is comparatively easy. Control chemicals used are carbaryl, endosulfan, and toxaphene. A natural enemy is the braconid wasp *Apanteles congregatus* (Say).

HOROLOGIUM. A southern constellation situated near Eridanus.

HORSE-ANTELOPES. Antelope.

HORSE CHESTNUT AND BUCKEYE TREES. Members of the family *Hippocastanaceae* (horse chestnut family), these trees are of the genus *Aesculus* and are of several species. These trees are deciduous shrubs or trees with compound leaves. Important species include:

Baumann's horse chestnut	*Aesculus hippocastanum* "*Baumannii*"
California buckeye	*A. californica*
Common horse chestnut	*A. hippocastanum*
Damask horse chestnut	*A. carnea* "*Plantierensis*"
Dwarf buckeye	*A. parviflora*
Indian horse chestnut	*A. indica*
Japanese horse chestnut	*A. turbinata*
Ohio buckeye	*A. glabra*
Painted buckeye	*A. sylvatica*
Red buckeye	*A. pavia*
Red horse chestnut	*A. carnea*
Sweet buckeye	*A. flava*
Yellow buckeye	*A. octandra*

The preferred geographic regions for some of the foregoing species are obvious from their common names. The common horse chestnut is found in Greece and Albania and grows to a height of about 120 feet (36 meters). The Damask horse chestnut is a backcross between *A. hippocastanum* (three-fourths) and *A. carnea* (one-fourth). The dwarf buckeye is found in the southeastern United States and grows to a height of about 15 feet (4.5 meters) and is shade tolerant. The Indian horse chestnut is found in the northwestern Himalayas and grows to a height of about 100 feet (30 meters). The Ohio buckeye (Ohio is called the buckeye state) is found in the central and southeastern portions of the United States. The tree grows to a height of about 30 feet (9 meters). The red buckeye grows in the southern regions of the United States and reaches a height of about 20 feet (6 meters) and thus can be called a shrub or small tree. The red horse chestnut reaches a height of about 70 feet (21 meters). The sweet buckeye is found in the southeastern United States and reaches a height of about 90 feet (27 meters). Champion buckeyes as reported by The American Forestry Association are listed in the accompanying table.

The wood of the American horse chestnut (Ohio buckeye) is dense, whitish in color, and is used for making furniture. It is particularly suited for the construction of artificial limbs. The kernel of this chestnut is poisonous.

Budding of the buckeye is described under **Bud.**

HORSEFLESH ORE. Bornite.

HORSE FLY (*Insecta, Diptera*). These insects are of special species (*Tabanua* and *Chrysops* spp.) and are irritating and injurious to horses, mules, cattle, hogs, deer, and other wild animals. On occasion, they are pests to humans. The flies look something like bees and are of a tan to brown coloration. The wings are faintly spotted. During spring and summer, the adult flies severely irritate the aforementioned animals and hover closely about the head, neck, and forequarters

RECORD BUCKEYE TREES IN THE UNITED STATES[1]

Specimen	Circumference[2] (Inches)	Circumference[2] (Centimeters)	Height (Feet)	Height (Meters)	Spread (Feet)	Spread (Meters)	Location
California buckeye (1972) (*Aesculus californica*)	174 (at 36 inches)	442 (at 91 centimeters)	48	14.4	78	23.4	California
Ohio Buckeye (1973) (*Aesculus glabra*)	143	363	146	43.8	54	16.2	Kentucky
Painted buckeye (1972) (*Aesculus sylvatica*)	159	404	144	43.2	61	18.3	Georgia
Red buckeye (1972) (*Aesculus pavia*)	45	114	35	10.5	33	9.9	South Carolina
Yellow buckeye (1973) (*Aesculus octandra*)	154	391	140	42	54	16.2	Kentucky

[1] From the "Social Register of Big Trees," The American Forestry Association (by permission).
[2] At 4.5 feet (1.4 meters).

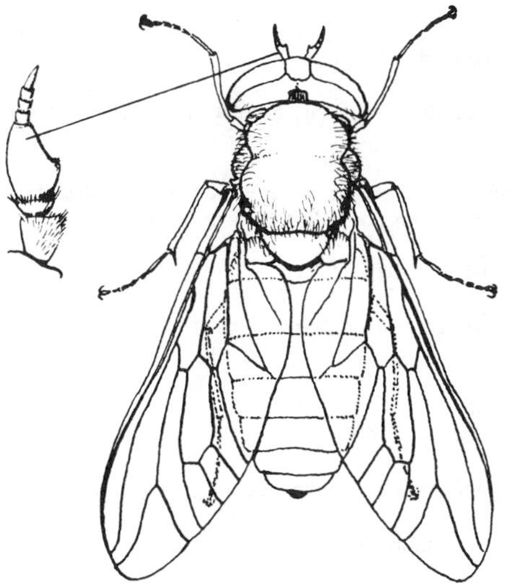

Horse fly showing detail of biting structure. (*USDA diagram.*)

of the host, waiting for an opportunity to strike and lay their eggs. The bite is painful because the mouth parts of the insect are very sharp. The fly will suck blood from the animal's neck or back for several minutes. The animal may twitch and run about in an unmanageable fashion. Some scientists have estimated that in regions where the horse fly is abundant, an animal in the field may lose as much as 3 ounces (about 90 grams) of blood per day to horse fly inflictions. Some species of horse fly carry such diseases as tularaemia, Calabar swellings (filariasis), and el debab, a disease of camels and horses that occurs in Algeria. Evidence also indicates that these flies are carriers of swamp fever, surra, and anthrax.

Control measures usually are directed against the maggots. The insect usually winters as a fully grown larva in some wet area near a stream. The maggot is pointed at both ends and is about 2 inches (5 centimeters) in length. They are fully grown by late spring, after which they pupate until early summer in dried mud. Adult flies commence to appear in early summer. Depending upon species, there are one or two generations per year.

Preventive measures, such as draining wet places and taking precautions similar to those for mosquito control, are effective. Spraying of the animals with allethrin, incorporating piperonyl butoxide, can be effective. Broadcasting a granular insecticide over areas frequented by animals also is effective.

HORSEPOWER. **Units and Standards.**

HORSEPOWER (Brake). **Dynamometer.**

HORSERADISH. **Flavorings.**

HORSES, ASSES, AND ZEBRAS (*Mammalia, Perissodactyla*). With exception of Grevy's Zebra, all horses, asses, and zebras are placed in the genus *Equus* (Equines).

The horse is a hoofed animal with a single toe on each foot, encased in a massive hoof. The teeth are very high-crowned grinding structures. Two wild species of central Asia, the Tarpan and Przewalskii's horse (*Equus przewalskii*), are most closely related to the domestic horse and probably represent the original stock from which the domestic horse (*Equus caballus*) was derived. Although much of the developmental history of the horses is known from North American fossils, there is no evidence to indicate that horses were on the continent when it was explored for the first time by Europeans. The wild horses of the American West are feral descendents of the domesticated horses imported originally from Spain. See accompanying illustration.

Under domestication, many varieties of horses have been developed for riding, driving, draft animals, and other uses. They also have been crossed with the domestic ass to produce mules for various uses, and have been hybridized experimentally with other species. Some authorities believe that the horse has no equal in its capacity and adaptability to withstand extreme climatic conditions and the various uses to which it has been placed by humans.

There are several ways to classify horses, as, for example, the draft horse (for working and pulling heavy loads); the light horse (light loading and riding); and the pony (a small horse, normally not over 14 hands high). (A hand is considered four inches (10 centimeters) in breadth; thus $4 \times 14 = 56 = 4$ feet, 8 inches, or (1.42 meters). Horses also are commonly designated by color. Bay is considered the more or less standard color and this is reddish-brown. Other colors of horses include brown, gray, chestnut (a particular reddish-brown), and roan (various—reddish-brown with a significant sprinkling of gray or white). Pinto signifies marked with spots of two or more colors; Palomino designates a golden or brownish-gray horse with an ivory or silvery-white mane and tail.

There are numerous breeds of horses, often classified generally as light, coach, and heavy breeds. The coach breeds originated in England and include the Cleveland Bay, the Yorkshire Coach, and the Hackney. The latter is possibly the most popular and is $14\frac{1}{2}$ to $15\frac{1}{2}$ hands high, quite strong, with high carriage of the tail, and usually of a dark color. Heavy breeds include the Clydesdale, developed in Scotland and named for the Clyde River water. The height is 16 to $16\frac{1}{2}$ hands high; color is dark brown or bay; noted for its heavy fetlock (tuft of hair on back of leg just above hoof), and high action. The Belgian is another heavy breed, originated in Belgium, with a height of from 16 to 17 hands, weight from 1,800 to 2,200 pounds (816 to 997 kilograms); color chestnut or sorrel (particular reddish-brown shade); flaxen mane and tail. The Percheron is a famous draft breed that originated in Normandy. The head is small; the contour is Arabian; height is from 16 to 17 hands; weight from 1,900 to 2,100 pounds (861 to 953 kilograms); heavy, but active and supple; excellent for agricultural jobs.

Thoroughbred race horses were originally developed by crossing English with Turkish and Arabic horses. Many special racing and

Przewalskii's horse. (*A. M. Winchester.*)

riding breeds have been developed, much too numerous for description here. Commonly known breeds include the Tennessee walking horse which has a distinctive gait—the running walk; the American saddle horse; the American quarter horse, particularly adapted for running short distances; and numerous others.

Przewalskii's horse is believed to be the only truly ancestral horse of the domestic horses that can be found in the wild today. This is a small, quite heavy-set animal, reddish-brown in coloration. Many years ago, it commonly roamed the plains of Eurasia. It is found today only in Siberia (Altai Mountains and environs) and in western Mongolia. It probably is destined to extinction as a pure species because of continuous interbreeding with feral domestic horses. The tarpan is a wild horse of the steppes of central Asia. This species has been regarded as feral rather than a natural species, but this interpretation has been disputed. It is in any case closely related to the domestic horse and may be an ancestral form. The western American mustang, a spirited, agile horse, is exemplary of interbreeding in the wild and is believed to be descended from stock introduced by Spanish explorers. The mustang also is known as the Indian pony or broncho. A worldwide classification of common horse and pony breeds is given in the accompanying table.

One breed of wild asses occurs in Asia and two breeds occur in Africa. However, all are believed to have been derived from the Nubian Wild Ass. Possibly the best known is the Onager (*Equus onager*), of a rust color, and very horse-like in appearance. The animal is found in the more arid regions of India and throughout Iran and Afghanistan. Traveling in small herds, the animals appear to prefer a semi-desert habitat. Possibly, the Kiang (*E. hemionus*) accounts for the greatest population of wild asses today. This animal is found in Tibet and Mongolia and is considerably larger than the Onager. In winter, the animal has a coat of dark shaggy hair and a white and red coloration in summer. Another breed or two, small in stature and in number, can be found in the semiarid parts of Mongolia. The asses found in Africa are usually of a gray tone, with white muzzles and white under-

COMMON HORSE AND PONY BREEDS WORLDWIDE

Horse Breeds

AFRICA
Egyptian Arabian, Libyan Berber, Berber, Fulani horse, Nigerian horse, Basuto pony
ARGENTINA
Criollo
AUSTRALIA AND NEW ZEALAND
Brumby, Wales horse, New Zealand pony
AUSTRIA
Lipizzan
BELGIUM
Belgian warm-blooded horse, Brabant, Ardenner
BRAZIL
Crioulo, campolino, mangalarga
CANADA
Royal Canadian mounted police horse, Stable Island pony, Canadian cutting horse
CHINA
China pony
CZECHOSLOVAKIA
Kladrub
DENMARK
Fjord horse, Frederiksborger, Knabstniper, Jutländer
FINLAND
Finnish universal, Finnish draft horse
FRANCE
French thoroughbred, half-bred trotting horse, Norman trotter, Norman horse, Anglo-Arabian Camargue horse, Ardenne horse, trait du nord, Bretonne, Percheron, Boullonnais, Poitevine
GERMANY
East Prussian horse, Hannoveraner, Oldenburger, Holsteiner, Dülmener, Württemberger, Rhinish, Schleswiger, Pinzgauer
GREAT BRITAIN
Exmoor pony, new forest pony, Dartmoor pony, fell pony, dales pony, Welsh mountain pony, Welsh pony, highland pony, Shetland pony, Welsh cob, English thoroughbred, Shire, Clydesdale, Suffolk punch
GREECE
Peneia pony, Pindos pony, Skyros pony
HAITI
Haiti pony
HUNGARY
Nonius, furioso, lipizzan, Arabian (shagya)
ICELAND
Icelandic pony
INDIA
Kathiawar pony, Marwar pony
INDONESIA
Sumba pony, sandalwood pony, Sumbawa pony, Java pony, Timor pony, Batak pony, Bali pony
IRAN
Persian Arabian, darashoori (schiras horse), jaf, tchenarani, Turkomane, Polo ponies, Turkmen pony, British pony
IRELAND
Irish Clydesdale
ITALY
Salerner, Kalabrier, Aveligneser
MEXICO
Mexican horse
NETHERLANDS
Frisian horse, gelderse, Groninger
NORWAY
Fjord horse, Döle horse, Gudbrandsdaler, Döle trotting horse
PERU
Criollo (costeno), morochuco
POLAND
Huzuler, komik, Sokólsker, Mazure, Poznan horse Arabian, Anglo-Arabian
PORTUGAL
Lusitanian
SOVIET UNION
Viatka, zemaitnika (petschora), tori, Bukhyonii horse, Kirghiz horse, karabagh, lokai, yomud, akhal-tekkiner, Arabian, Métis, Orlov trotting horse, Russian-American trotting horse, Vladimir
SPAIN
Sorreia, Andalusian, Arabian
SWEDEN
Gotlander (Skogruss), Swedish warm-blooded horse
SWITZERLAND
Freiburger, Einsiedler horse, Anglo-Norman, Holsteiner
TIBET
Tibetan horse
TURKEY
Anatolian pony, Karakabeyer
UNITED STATES
American thoroughbred, quarter horse, Kentucky saddle horse, Tennessee walking horse, Missouri fox-trotting horse, Morgan, standardbred trotting horse, Spanish mustang (the wild horse of Wyoming), galiceno, Chincoteague pony, Assateague pony, American Cleveland-bay, American hackney, Welsh and Shetland ponies, American pony (P.O.A. = Pony of America), pinto, appaloosa, palomino, albino, American percheron, American Belgian
VENEZUELA
Llanero (prairie) horse

Donkey Breeds

Poitou ass
Puli ass
Spanish giant ass
Gascogne ass
Savoy ass
Sicilian ass
Macedonian ass
Maskat ass

sides. The Nubian is characterized by distinct black shoulder stripes. The African asses prefer mountain and desert terrain. A domesticated ass is referred to as a donkey.

The Quagga (couagga) is now an extinct species. It was a South African animal (*E. quagga*) related to the zebras and asses. It was reddish-brown above, blending to white on the legs, and marked with dark brown stripes on the head, neck, and fore part of the body. The last known specimen died in 1875 at the Berlin Zoo, although hopefully some wild specimens still may be roaming in the secluded areas of South Africa. The animal was overharvested for use as native labor and food.

The Zebra (*E. burchelli*, . . .) is related to the asses and the quagga and is distinguished by the complete or nearly complete transverse striping of the body and legs. The stripes vary in the several species from white to yellow brown, alternating with dark brown to black. They range throughout much of Africa south of the Sahara Desert, but are concentrated in the south. Sub-species include: The Common Zebra (characterized by a V-shaped junction-pattern occurring in the middle of the sides); the Damaraland Zebra (Zaire, Zambia, Botswana regions); the East African Zebra (Rhodesia, Abyssinia, the Sudan); and Selou's Zebra (central and southeast Africa). General characteristics of zebras include: Grazing like horses; traveling in big herds; mixing with other animals; a main dietary item for lions; shy and nervous; can be quite pugnacious in self-defense, kicking vigorously with either front or hind feet and inflict severe bites; coloration and striping puts them at disadvantage when viewed against a green backdrop, but is an advantage in tall grass and open plains; frequently harbor intestinal parasites which are believed to aid in their digestion.

Grevy's Zebra (*Dolichohippus*) is not classified with *Equus*. However, it appears much as a horse, with large head and big ears. The animal prefers desert scrub foliage and requires a minimum of water. It is found in Somalia. Certain anatomical distinctions set it in its own genus.

A mule is a hybrid between the domestic horse and ass, produced by mating a mare and a jack. The reciprocal cross of stallion and she ass is called a hinny. Mules have large ears, small hoofs, and tufted tail characteristic of the ass and the stature of the horse. They are strong and hardy, resistant to disease and generally adverse conditions. Since they are infertile, they are always bred by crossing the two species. Some authorities claim that hinnys are smaller and lacking in the qualities desired in mules; while others claim that both forms fall within the range of variation to be expected in the hybrid.

Other members of the order of *Perissodactyla* are listed under **Perissodactyla.** For references, see **Mammalia.**

HORSE SERUM. Antitoxin.

HORST. Fault.

HORTICULTURE (Grafting and Budding). Grafting and Budding.

HORTVET CRYOSCOPE. Cryoscope.

HOST. An animal which is used as a source of food by a parasite. The parasite may live on the surface of the body or within it and may be harmless or harmful, but in all cases the host is the source of its food.

HOT LIME ZEOLITE SOFTENING. Feedwater (Boiler).

HOT DRY ROCK (HDR) GEOTHERMAL ENERGY TECHNOLOGY. Geothermal Energy.

HOT SHORTNESS. Brittleness of metals in the hot working temperature range.

HOT-SPOT VOLCANO. Earthquakes, Seismology, and Plate Tectonics; Volcano.

HOT SPRING (Submarine). Earthquakes, Seismology, and Plate Tectonics; Ocean; Ocean Resources (Energy); Ocean Resources (Mineral).

HOT STANDBY. Gas and Expansion Turbines.

HOTWELL. A hotwell is a tank or container in which heated liquid collects. An example is the hotwell attached to and made part of a steam condenser of the surface type. As the steam is condensed, the condensate drops to the bottom of the condenser shell and flows into the hotwell, from which it is pumped.

"HOT WIRE" ANALYZERS. Gas Analyzers (Combustion-Type); Gas Analyzers (Thermal-Conductivity-Type).

HOT WORKING. Plastic deformation of metals at temperatures sufficiently elevated so that the effects of the working are nullified by concurrent softening processes. Thus, when steel is hot worked by rolling at a white heat, the metal recrystallizes and softens almost immediately after it is deformed. Similarly, the deformation of lead at room temperature is also hot working and accounts for the fact that it is not possible to work harden this material at this temperature. An empirical rule states that the lower limit of the hot-working temperature range is the recrystallization temperature.

Forging, rolling, pressing, extruding, swaging, drawing, or forming of metals at temperatures above their recrystallization temperatures are examples of hot working.

HOUR. Units and Standards.

HOUR ANGLE. In reference to a celestial object, the spherical coordinate, in the equatorial system of coordinates, measured in the plane of the celestial equator from the local meridian, in the direction of apparent rotation of the celestial sphere, to the intersection of the hour circle through the object with the equator. Since time and hour angle are practically synonymous (e.g., the hour angle of the mean sun is local mean time), the determination of hour angle is vitally necessary for the determination of local time and, hence, longitude.

At sea, hour angle is determined by measuring the altitude of the object by means of the sextant, reducing the observed altitude to true geocentric, and solving the astronomical triangle. For the solution of the triangle, both the declination of the object and the latitude of the observer must be known. The declination may be immediately obtained from the tabulated coordinates of the object, but the latitude can be obtained only by some previous observation. If the ship is in motion, the latitude must be obtained by dead reckoning from the previously determined position.

See also **Celestial Sphere.**

HOUR CIRCLE. Celestial Sphere.

HOUSEFLY (*Insecta, Diptera*). A true fly, *Musca domestica*, well known for its habit of frequenting houses and alighting on all kinds of food. Since it also visits filth of any kind, it is an important carrier of disease, especially typhoid fever, and has been the object of public health crusades for many years. With the improvement of sanitation, the danger has been lessened, although it has not been entirely eliminated.

The housefly breeds in horse manure and in various kinds of decaying organic matter. Proper disposal of such wastes is an important measure in the control of the insect.

The housefly is found worldwide except at high altitudes. Hair covers most of the head of the housefly. Its jaws work horizontally and the stubby snout has a piercing stylet. It does not bite, but pierces and sucks its food. The eyes are large, covering about three-fourths of the facial area. The eyes of a fly are made up of approximately 4000 six-sided facets. All facets together frame the total object of view. However, the fly does not focus for a sharp image, nor does it have the ability to close its eyes. The vision is believed to be reasonably sharp for distances of 2 to 3 feet (up to 1 meter) or less.

Several generations of flies may be born in one season. They live for only a few weeks, but their eggs live on in fertile debris until spring.

The housefly is a principal agent for transmitting many diseases in areas where it is not carefully controlled. The legs are hairy and well designed to carry filth. Infection may be transferred by the piercing proboscis of the fly, or the fly may vomit its own food, leaving a trail of infection. As many as several million microorganisms may be found in the intestines of a housefly. The flying speed of the housefly is approximately 5 miles (8 kilometers) per hour.

HOUSEKEEPING OPERATION (Computer System). A general term for those computer operations which must be performed for a machine run but do not contribute directly to the solution of a problem, contributing rather to the operation of the computer. Examples of housekeeping operations include: establishing controlling marks; setting up auxiliary storage units; reading in the first record for processing; initializing; set-up verification operations; and file identification.

HOVER CRAFT. **Ground-Effect Machine; Hydrofoil Ship.**

H-THEOREM. **Kinetic Theory.**

HUBBLE'S LAW. **Cosmology.**

HUBNERITE. **Wolframite.**

HUDSON-JARDI RAIN GAGE. **Precipitation and Hydrometeors.**

HUE. The attribute of color perception that determines whether it is red, yellow, green, blue, purple, or the like. White, black, and gray are not considered hues.

HUE (Abney Effect). **Abney Effect.**

HUMAN FACTORS ENGINEERING (Personal Subsystem Development).* Acceptable human performance in complex systems depends upon precise human-machine interaction. Such interaction is the focus of attention of design engineers and computer programmers, for example, on the one hand, and of human performance psychologists on the other. The meaning and extent of that interaction has evolved and expanded over the years. We now commonly find computers of various sizes on the machine side of the human-machine interface, and their presence has changed human performance considerations markedly. On the human side we have seen an accelerating emphasis on the person as an information processor, thus adding many considerations to the older, but persistent, anthropomorphic concerns.

In this entry, the "human factors engineering" field is reviewed briefly, followed by a discussion in some detail of the requirements that computers have put on people and human performance technology in computer-based systems. For examples, and in the citation of solutions, the authors draw heavily upon their experience in the Bell system.

There have been human-machine interaction concerns of a sort ever since primitive people first extended their own abilities with simple weapons and tools. In more recent history, the industrial revolution accelerated greatly the transfer of work functions from people to machines and complicated human-machine interface problems considerably.

Human Factors Engineering

It is generally agreed that World War II marked the beginning of a professional approach to what came to be called human factors engineering, that is, a systematic approach to studying problems of human-machine interaction and to arriving at practical solutions on a scientific basis. Before the war, going back into the late 19th century, systematic work had been done by psychologists, but it tended to focus upon selecting or training people to interact with machines. But the tremendous industrial and military expansion brought on by World War II, and the greatly increased complexity of the weapons systems being produced, complicated human-machine interaction considerably, so that the selection and training approach no longer was sufficient. For example, it was a simple matter to get a relatively small number of men to fly the slow uncomplicated fighter aircraft of World War I as compared to getting thousands of men to perform satisfactorily in the high performance P-38s and P-51s of World War II. Thus, a recognized professional specialty, usually known as human factors psychology or human engineering, was spawned.

Since human factors psychology came into being during World War II, it is to be expected that it would continue to thrive in a military environment after the end of that war. The Army, Navy, and Air Force all established substantial centers for the study and application of this discipline, and many of them exist today. Human factors practitioners spread into the industries that supplied military equipment and systems. There was a particularly large concentration in the aerospace industry. The movement also took hold in many nonmilitary fields such as transportation, telephony, and occupational safety.

In 1953, Paul Fitts and others typified the work being done in the 1940s and 1950s as follows: "In the design of equipment, human engineering places major emphasis upon efficiency as measured by speed and accuracy of human performance in the use of the equipment. Allied with efficiency are the safety and comfort of the operator. The successful design of equipment for human use requires consideration of the man's basic characteristics, among them his sensory capacities, his muscular strength and coordination, his body dimensions, his perception and judgment, his native skills, his capacity for learning new skills, his optimum work load, and his basic requirements for comfort, safety, and freedom from environment stress" (Lindgren, 1966).

Thus traditional human factors engineering concerns itself with data gathering and experimentation meant to yield precise information about human capabilities. With such information, machines can be built to fit humans. For example, studies revealed design requirements for visual displays of information so that correct decisions or control actions can be made without delay. Comfortable physical fit between person and machine can be established with the use of such information as the average human's physical size, strength, and reach. Such information has been stored in handbooks for the use of equipment designers and for human factors personnel who work with equipment designers.

Personnel Subsystem Development

By 1960 there was a growing recognition on the part of behavioral scientists that profound changes needed to be made in the organization and execution of complex system designs. Traditionally, human factors engineers had focused attention on the design and/or adaptation of machines for human use, and on performing studies which would increase knowledge of human characteristics and requirements. But attention began to turn to having the human factors engineers concentrate their efforts on working with other categories of system designers to produce an optimal system. Van Cott said, "This approach changes the typical role of the human engineer from that of a design critic to that of a component specialist and a member of the system design team" (Van Cott, 1960).

By 1960, too, not only was there a healthy appreciation of complex weapons systems and the important role of human factors engineering in their designs, but the importance of the computer was being realized and there was a rising concern about the complexity of the interaction between humans and computers. In that year, an article by Licklider appeared in the first issue of the *IRE Transactions* on "Human Factors in Electronics," in which he raised three issues of "man-computer symbiosis": the language mismatch between computers and people, the physical interface (computer console and terminals), and the speed-cost mismatch between human and computer (Licklider, 1960).

Today, we no longer hear much said about Licklider's three problem areas, at least not expressed in the same terms. We have been able to compensate for the speed-cost mismatch between man and computer by such approaches as time-sharing. In information systems, our concern often is the opposite: the response time of a computer to an operator inquiry sometimes is not fast enough for optimum system performance.

With regard to the physical interface, the variety of available well-designed terminals is much better than it was a few years ago. This improvement may be due at least partly to the negative reactions to many of the early terminals, which were so poorly designed.

The language mismatch between people and computers—at least one aspect of it—still is a major problem. The matter of preparing

* Adapted from the article, "Human Performance Considerations in Complex Systems," *Science*, **195**, 1205–1209 (1977), with permission.

and debugging programs, especially for large software projects, still is, as one authority puts it, "in the category of a cottage craft" (Vyssotsky, private communication).

But today there is another major problem of "human-computer symbiosis." It concerns the fit between the computer and its software (computer subsystem) and the people (personnel subsystem) in the same system. The term "personnel subsystem" refers to the "people part" or manual part of the total system. The personnel subsystem is an integral part of the complete operational system and is composed of personnel who interact with the computer subsystem and/or who use the system documentation and performance aids.

We will now consider the issues raised by Van Cott and Licklider, and give illustrations from our own experiences of how these issues have been addressed in the development of computer-based systems, particularly in the past 12 years. To repeat the issues: Van Cott pointed out that the human factors engineer should be concerned with producing an optimal system, and that ". . . This approach changes the . . . role of the human engineer from that of a design critic to that of a component specialist and a member of the system design team" (Van Cott, 1960). Licklider raised the issue of human-computer symbiosis, which we interpret to include the difficult task of coordinating human information processing and computer data processing into an integrated system. It is very important for computer-based system designers to appreciate that there are two interactive logic streams in the operating system: the fixed ones represented in the computer programs, and the flexible ones existing in the brains of human system components. When these two streams fail to mesh smoothly, as, for example, when an operator cannot understand an error message, system operation may halt—or, at least, be impaired.

System developers have addressed the issue of human-computer symbiosis in varying ways. One approach has been to pay only casual attention to human factors and personnel subsystem considerations. A second approach has been to devote extensive resources to the development of training courses for the personnel. A third approach has been to concentrate on engineering the human-computer interface for human use. The fourth approach has been to design the complete personnel subsystem concurrently with the computer subsystem. We will discuss each of these approaches separately.

Develop Hardware and Software, but Very Little Personnel Subsystem

Many new systems involve the application of a new technology to existing business or scientific functions. Making full use of a new technology—for example, computer technology—is a challenging undertaking. Development organizations sometimes become preoccupied with applying the new technology; all their attention and resources are devoted to it, and little or no attention is given to the personnel subsystem.

The worst problems caused by inattention to the personnel subsystem tend to be caused by input errors. When system software detects errors entering the system, error messages are then fed back to personnel in the error-correction feedback loop. There is a snowballing effect: "temporary" personnel are added to correct errors; system performance keeps dropping; the rate of system throughput drops because most of the data processing effort is expended in the error-correction process.

Abnormally high error rates have three consequences. First, if outputs are delayed, customer schedules (and customers) are upset. Second, the cost of extra processing required to correct errors can be high; for example, a study of errors on customer service orders showed the total cost to correct each error to be $7.25. Such costs can wipe out the savings which the new system was to effect. Third, high error rates demoralize system users. We once asked a clerk whose sole job all day long was to prepare a complex input form, how many forms were returned for correction of errors. The answer: "Almost all of them, and some more than once."

Another consequence of the single-technology approach is that system users often are unable or unwilling to use system features which have been devised at great expense. Licklider, describing the SAGE (semi-automatic ground environment) system, reported: "According to credible reports, there were makeshift plastic overlays on the cathode-ray displays, and the scope watchers were bypassing the elaborate electronics—operating more or less in the same 'manual mode' used in World War II" (Licklider, 1969).

Produce Extensive Training Course Materials

Most designers realize the value of training people "to use the system." Typically, though, they commit two important errors. First, what they refer to as "the system" really is the computer (or machine) subsystem. Their design has omitted the personnel subsystem. Second, they appear to believe that training (usually *lots* of training) can compensate for their lack of attention to the human element in the design of the system. Actually, human performance in systems is the complex interaction of many variables, only one of which is training. Fox, citing experience with developmental testing, described eight major variables affecting human performance in systems; only one was training (Fox, 1974). The other variables are design, human-machine interface, information transfer, environment, personal factors, supervision, and documentation.

Second, simplistic as it may sound, system personnel *should* be trained to do the manual procedures of the system. However, the most carefully developed training courses cannot compensate for a poorly designed or suboptimized personnel subsystem. The procedures themselves may be incorrect; for example, some information should be stored in performance aids instead of memorized. No matter how much training they receive, system personnel probably will be unable to overcome many of the error-producing procedural and interface problems inadvertently incorporated in the system design.

During the 1970s, instructional technology has become widely used. This method of training course development features thorough analysis of tasks to be taught. We have found that a task analysis, done for training purposes, often reveals system design deficiencies. It also often reveals the need for new performance aids and procedural documents. The net effect of such an analysis usually is reduced training time.

For example, we were asked to help with training course development for a new system which provides standardized circuit designs. A task analysis revealed the need to add performance aids and restructure user documentation. The result was a 30 percent reduction in training costs, improved human performance, and a reduction in the volume of system documentation that was required.

Design the Human-Machine Interface

Systems designers are often familiar with some of the terminology and concepts of human factors engineering; therefore, the "human factoring" of interfaces has traditionally received support in both large and smaller systems.

One such project was MECHSIM (mechanical simulation of a computer-based directory assistance system). This project had as its ultimate goal the reduction of the amount of time an operator spends responding to a request for directory assistance. The MECHSIM study followed some 10 years of low budget research and experimentation directed to reducing operator work time.

Traditionally directory assistance operators have used essentially the same data base as telephone customers have available to them, that is, the listings found in telephone directories. But MECHSIM assumed that the listings would be in a computer data base and dealt primarily with the problems operators would have in interfacing with such a data base, e.g., problems of keyboards and displays.

In a 1966 article describing MECHSIM, Lindgren wrote: "The Bell project also presents an interesting sidelight. Human factors engineers always stress the importance of being in on design studies as early as possible so that their investigations can properly influence the final system design. In the Bell case, the engineering department is not even going to begin to implement the computer look-up system until they have seen the complete results of the human factors study . . . which amounts to the realization of a human factors dream" (Lindgren, 1966).

The dream, if that is what it is, has continued. The mechanical simulation of computer-assisted directory assistance, when it was completed in 1966, indicated that there probably would be sufficient savings in operator look-up time to justify the development of an actual computer-based system. As a result, a computer-based human factors experiment was begun in late 1970. It was referred to as the "live traffic experiment" because it involved ten operators who worked at specially designed CRT terminals and who received actual "live" customer requests for directory assistance. They used the terminals to locate customer telephone numbers which were stored in a computer memory. (Fig. 1)

Operators talked directly with customers and, using keys on the CRT console, keyed into the computer the necessary trigrams indicating name and address information. The computer then activated a display on the CRT from which the operator selected the desired telephone number and relayed it to the customer.

The study was in two parts. The first was an experiment or, more correctly, the last stage of an experiment that had several preceding steps in other parts of the laboratories going back over some 12 years. Operators systematically tried eight search (keying) strategies to retrieve telephone numbers from computer storage. The best strategy was then selected for the second part of the experiment, a pilot study of the computer-based directory assistance operation. Its purpose was to see if sufficient savings in work time could be realized to justify the use of the computer subsystem for directory assistance.

Fig. 1. Directory assistance operator working at specially designed terminal of experimental computer-based system. (*Bell Laboratories.*)

The outcome of the pilot study indicated that the computer-based directory assistance would indeed save sufficient operator work time to justify a change to that type of system when development resources became available.

More important, perhaps, is the fact that the live traffic experiment represents in several ways the coming of age of human factors psychology or "human performance technology." For example, the human factors experiment and pilot study were undertaken before huge sums were committed to system development. Next, there was the specific commitment of designer resources to the attainment of an integrated operator-computer interaction. This required an explicit understanding that software designers were dependent upon human performance requirements developed by human performance psychologists.

Finally, the live traffic experiment served to call management attention to the principle that the personnel subsystem of a computer-based system requires careful design attention just as does the computer subsystem.

In information systems, the human-computer design is important for error prevention. Human-caused errors are anathema to system designers. A large system that we studied illustrates the effect of human error on system performance; it also illustrates why, when systems have human-error problems, designers tend to focus their attention on interface and data display designs.

Almost immediately after the change to the new system, human errors snowballed. Computer-processed transactions bogged down to the point where the computer was bypassed entirely; later, the errors were located and corrected, and the computer data base and manual protective records updated. The system's error-correction process quickly became overburdened. System performance then fell below acceptable or even tolerable levels.

Most of the human errors were detected by computer edit of manual inputs. System designers proposed a two-part solution: Replace the input clerks with more skilled people; and substitute CRT terminals for input forms. Before this solution was implemented, a special task force worked with the system personnel. As a result, CRTs were not installed, nor were more skilled people hired. Instead, many of the paper forms were redesigned, performance aids developed, and system personnel trained. Error rates immediately dropped, and system performance was soon above acceptable levels, where it remained.

A key point, however, is this: The most visible changes made were to the forms used by humans to transmit information to the computer; however, task force members believed the other personnel subsystem work (procedures, performance aids, and training courses) to have been much more important.

This final example points out a critical issue. On many systems, the design of the man-computer interface itself is not enough. After all, many large computer-based systems use off-the-shelf terminal equipment which satisfactorily meets human capabilities and stereotypes. As DeGreene says, in regard to design of data displays: "Many of the baffling human factors problems, then, relate not so much to vision as to cognition. The display can be considered a direct extension not only of the inner structure and workings on the computer, but also of what is going on within the user's (operator's) problem-solving head" (DeGreene, 1970).

For most system designers, human-machine interface design should focus on integrated, complementary human-machine procedures. Human procedures and machine procedures come together at the interface, whether console, CRT display, or paper form. The interface design should take into account these interactions, and should be consistent with both human and hardware capabilities.

Design the Subsystems Concurrently

Human performance technology integrates elements of human engineering, engineering psychology, instructional technology, and industrial and organizational psychology. The application of human performance technology is called personnel subsystem development (PSD), which means the integrated design of human procedures, human-machine interfaces, training, performance aids, and documentation as part of the total system. It also includes the rigorous testing of the personnel subsystem together with the computer and hardware subsystem counterparts. The products of PSD are the controlled utilization of human resources, and the means (procedures, training, performance aids, interfaces, for example) for obtaining required human performance in systems.

A recent Bell Laboratories example of designing the personnel subsystem simultaneously with the computer subsystem was the enhancement of a computer-based message switching system. During design there was continual interaction between personnel subsystem and software designers. The personnel system and software were fully tested before installation. During a 3-month period following installation of the enhanced systems, only 6% of the problems encountered required design changes, and none of these involved human procedures or a human-machine interface. Previous experience suggested that, if integrated, concurrent design of the computer and personnel subsystem had not been done, 50% of the encountered problems would have required changes to the personnel subsystem (Soth, 1976).

One final example deals with the conversion of a system which maintains an inventory of available facilities, and assigns these facilities to customer orders for telephone service. The conversion requires the transcription of the records of 120 million telephones from paper to computer data base, at an estimated cost of $305 million.

The computer and personnel subsystems of the conversion system were designed and implemented concurrently and systematically. Training and documentation were developed, and the personnel subsystem was tested as part of the design process. When the testing program was completed, the transcription error rate had been reduced from a projected 10% to an overall 5%. Thus, when the necessary records have been converted to the computer data base, there will have been an estimated cost avoidance of $46 million.

Signs of Progress

As of this writing, based upon our observations, we judge that many, if not most, computer-based and hardware-based systems being developed still do not give sufficient attention to the human element. Probably a major reason for this is that hardware/software engineers dominate the design/development process, whereas major impetus for personnel subsystem development tends to come from human performance persons who represent a minority group which hasn't much clout. Thus, most personnel subsystem developments which came into being tend to be retrofit designs, i.e., they are begun after the hardware/software design is fairly well along. See Fig. 2.

Fig. 2. Customer records being checked for accuracy during conversion from existing manual system to new computer-based system. (*Bell Laboratories.*)

There is strong evidence of change, however. A recent book entitled "Humanized Input: Techniques for Reliable Keyed Input" (Gilb/Weinberg, 1977) addresses itself to design problems and solutions of that part of the personnel subsystem dealing with human input to computer-based systems. Since, as we pointed out earlier, the most serious problems with large computer-based systems are with poorly designed input procedures and display design, the appearance of such a book is of considerable importance. The fact that it is authored by computer software experts, rather than human performance experts, is highly significant. That is, when a sufficient number of hardware/software developers become convinced that design for the human element of systems is important, it is much more likely that it will be done.

Other evidence of progress in this regard comes from a development organization where a large amount of development effort goes into computer-based systems for deployment in various associated companies. System developers there now are working under guidelines which define the personnel subsystem, and list a number of documents which should be delivered to system recipients to enable them to activate a viable personnel subsystem as part of the operating system. As a matter of fact, parts of the development organization have been developing and documenting the personnel subsystem for some twelve years, but now that is to be done for all operations systems.

Computer based instruction seems finally to be coming of age as an integral part of computer-based systems. While its potential has been apparent since the late 1950s, it is generally recognized that computer-based instruction (computer assisted instruction) is far short of meeting that potential. Hardware costs, inadequate software for authors, and teacher resistance have been important causes.

But for a long time it has seemed eminently reasonable that computer-based instruction should be used in the orientation and training of the human components of computer-based systems. By definition, computers and terminals are parts of such systems, so additional hardware costs are minimal. Recently, some satisfactory software for authors has become available, as well as computer graphics, thus making the development of training materials easier. Since computer-based instruction is individualized instruction, it has all the advantages of that mode: no waiting for classes to be formed, uniform learning, timeliness, etc. Such a use of computer-based instruction logically means that attention to the personnel subsystem is improved. (Fig. 3)

Fig. 3. Computer-based instructional packages for a large computer-based system provide efficient, economical, hands-on training at individual terminals without interfering with the system's normal operations. (*Bell Laboratories.*)

In summary, the years since World War II have been increasing complexity of human-machine interaction. Machines have become much more complicated and the responses required of humans may be said to have moved from "brawn to brains," that is, from the physical interaction between person and machine to an interaction which depends more on the information processing abilities of human beings.

In our experience with computer-based systems, we observe an increasing emphasis upon personnel subsystem development. This is largely because today's complicated systems, particularly those involving computers, put increasing demands upon people in the system for fast and nearly error-free performance. Careful design attention must be given to the human element if such performance is to be attained.

See also **Nuclear Reactor.**

References

DeGreene, K. B.: "Systems Psychology," p. 319, McGraw-Hill, New York, 1970.

Gilb, T., and G. M. Weinberg: "Humanized Input: Techniques for Reliable Keyed Input," Winthrop, Cambridge, Massachusetts, 1977.

Licklider, J. C. R.: "Human Factors Electronics," *IRE Transaction HFE-1, PA-11* (March 1960).

Licklider, J. C. R.: *Comput. Autom.*, **18**, 48 (August 1969).

Lindgren, N.: "Human Engineering in the National Defense," *Panel on Human Engineering and Psychophysiology, Research and Development Board, HPS 205/1* (June 29, 1953), cited by Lindgren in *IEEE Spectrum*, **3**, 3, 133 (1966).

Lindgren, N.: *IEEE Spectrum*, **3**, 4, 70 (April 1966).

Soth, M. W.: "The Human Work Module: A Structural Entity for Personnel Subsystem Design," Bell Telephone Laboratories, Murray Hill, New Jersey, 1976.

Van Cott, H. P.: "Human Factors Methods for System Design," foreword, American Institutes for Research, Washington, D.C., 1960.

H. O. Holt and F. L. Stevenson, Bell Laboratories, Piscataway, New Jersey.

HUMAN-MACHINE COMMUNICATION. Telephony; Voice Recognition and Synthesis.

HUMBOLDT CURRENT (also called Peru Current). The cold ocean current flowing north along the coasts of Chile and Peru. It is one of the swiftest of ocean currents. The Peru current originates where part of the water that flows toward the east across the subantarctic Pacific Ocean is deflected toward the north as it approaches South America. The northern limit of the current can be placed a little south of the equator, where the flow turns toward the west, joining the south equatorial current.

The southern portion of the Humboldt current is sometimes called the *Chile current.*

HUMECTANTS AND MOISTURE-RETAINING AGENTS. Substances that have affinity for water, with stabilizing action on the water content of a material, are called *humectants* or moisture-retaining agents. Ideally, a humectant maintains within a rather narrow range the moisture content caused by humidity fluctuations. These materials are widely used in certain food products, as well as tobacco, and in recent years have taken on increasing importance in the case of intermediate-moisture foods. Traditionally, hummectants have been used to retain moisture in foods like coconut and marshmallows which otherwise would quickly dry and become tasteless. For example, flaked coconut is kept moist in the container by adding glycerine and glyceryll monostearate.

Among the most commonly used hemectants are glycerine, potassium polymetaphosphate, propylene glycol, sodium chloride, sorbitol, sucrose, and triacetin. Also, phosphates are added to the pickling solutions used to treat cured meats, such as ham, bacon, corned beef, etc., by soaking or injection. Their principal purpose is for moisture binding to reduce the loss of fluids during curing and cooking.

During the last few years, important research has gone into the addition of multiple humectants and water to food systems. Studies have shown that a hysteresis effect may occur with certain humectants, i.e., a different rate of moisture absorption than the rate for moisture desorption. Multiple humectants tend to compensate these hysteresis effects, giving uniform rates in both directions.

References

Gee, M., Farkas, D., and A. R. Rahman: "Some Concepts for the Development of Intermediate-Moisture Foods," *Food Technology,* **31**, 58–64 (1977).
Labuza, T. P., et al.: "Water Activity Determination," *J. Food Sci.,* **41**, 910 (1976).
Staff: "Food Chemicals Codex," National Academy of Sciences, Washington, D.C. (Issued periodically).

HUME-ROTHERY RULES. When alloy systems form distinct phases, it is found that the ratio of the number of valence electrons to the number of atoms is characteristic of the phase (e.g., β-, γ-, ϵ-) whatever the actual elements making up the alloy. Thus, both $Na_{31}Pb_8$ and Ni_5Zn_{21} are γ-structures, with the electron-atom ratio 21:13. The rules are explained by the tendency to form a structure in which all the Brillouin zones are nearly full, or else entirely empty.

HUMIDITY. Generally, some measure of water-vapor content of air. *Absolute humidity* is the ratio of the mass of water vapor present to the volume occupied by the mixture; that is, the density of the water vapor component. The percentage of water vapor in the total composition of the air may be determined by passing a measured quantity of air through a tube containing an absorbing substance that removes all the vapor, and which can be weighed before and after the absorption.

Absolute humidity is usually expressed in grams of water vapor per cubic meter or, in engineering practice, in grains per cubic foot. Because this measure of atmospheric humidity is not conservative with respect to adiabatic expansion or compression, it is not commonly used by meteorologists. As occasionally used in air-conditioning practice, absolute humidity refers to the number of grains of water vapor per pound of moist air, which is dimensionally identical with the specific humidity (defined below).

Critical humidity is the point at which the partial pressure of water vapor in the atmosphere is equal to the saturation vapor pressure. Condensation on suitable nuclei will occur when the humidity reaches or exceeds this value.

Relative humidity is the ratio of the actual vapor pressure of the air, at any temperature, to the maximum of saturation vapor pressure at the same temperature. It expresses the vapor content as a fraction or percentage of the concentration necessary to render the vapor saturated at the given temperature. At the dew point, the relative humidity is 100%. A rise of temperature without the addition of more vapor reduces the relative humidity (but not the absolute humidity), while a fall of temperature increases it and may bring about saturation. Relative humidity is measured by the hygrometer.

Specific humidity is the (dimensionless) ratio of the mass of water vapor to the total mass of the system. It may be approximated by the mixing ratio for many purposes:

$$q=\frac{w}{1+w}$$

where q is the specific humidity and w the mixing ratio.

See also **Psychrometric Chart.**

HUMIDITY (Atmospheric). Precipitation and Hydrometeors.

HUMITE. Chondrodite.

HUMMINGBIRD. Swifts and Hummingbirds.

HUMORAL IMMUNITY. Immune System and Immunology.

HUMORAL TRANSMISSION. Brain and Nervous System.

"HUMP-BACK." Bone.

HUMBACK SALMON. Salmon.

HUMPHRIES EQUATION. Specific Heat (Humphries Equation).

HUMUS. Soil.

HUND'S CASE. Energy Level.

HUNTING. The tendency of a rotating mechanism which normally should operate at constant speed to pulsate in speed above and below the normal point, is known as hunting. It may occur in prime movers controlled by governors which are too isosynchronous, or in electric apparatus where rotating and stationary parts are electrically coupled. The nature of such coupling is essentially elastic, and may, under certain circumstances, lead to hunting action on the part of the rotor. Governors which hunt must be corrected by the use of dash pots or other damping devices, and the introduction to the governor characteristic of a slight amount of speed regulation.

See also **Governor.**

HUNTINGTON'S CHOREA. Chorea (Hungington's).

HURRICANE. Fronts and Storms; Hurricane Prediction.

HURRICANE PREDICTION. Since weather satellites were first launched, an average of about 100 small disturbances per year in the atmosphere over the tropical Atlantic Ocean have been observed. These range from dust clouds off the African coast to rain squalls near the equator. In their evolution or dissipation, about 10% of these disturbances develop into tropical storms. Of these, five or six become storms hundreds of kilometers across and with peak winds ranging from 33 to 90 meters per second (75–200 mph). Although, as described in other entries in this volume, much has been learned in recent years pertaining to the physics of these phenomena and in locating and monitoring them, less-than-encouraging progress has been made toward predicting the ultimate path of many hurricanes. Essentially utilizing conventional weather forecasting techniques, it has been claimed that in about 70% of the cases, the movement of a hurricane can be predicted by systematic extrapolation of observed movements over the prior 24 to 36 hours. Unfortunately, the remaining 30% can represent several killer hurricanes in a year's time and it is to these that much scientific effort has been and will continue to be addressed.

Considering the tremendous loss of life (mostly by drowning) and of property resulting from the impact of a hurricane on the coastline, this is indeed an important scientific challenge. The National Hurricane Center (NHC in Miami, Florida) of the National Weather Service makes hurrican predictions largely based upon statistical methods, climatological inference, analogs with the tracks of earlier hurricanes, tracking extrapolation from immediately prior periods as previously mentioned, and calculations using a simplified dynamic model of the storm. Accuracies as of the mid-1970s leave much to be desired. More precise forecasts could not only reduce loss of life and property damage, but also lessen the scope of evacuation areas and procedures which, in themselves, are costly. Obviously what is required is a forecast of the height and extent of the storm surge as it hits land—and, most importantly, a reasonably accurate prediction of the landfall. It is interesting to note that the accuracy of 24-hour forecasts has improved by only 30% since the 1950s.

Forecasts have been limited by the time needed to compute a model. General circulation models have not sufficed and more specific models of storms have been confined to idealized two-dimensional calculations which have not included interactions between the storm and its environment. It is hoped that availability of larger computers to this field and work along the lines already commenced at the NHC, the National Center for Atmospheric Research (Boulder, Colorado), and the National Meteorological Center (Washington, D.C.) will yield positive results within a few years. Most extensive testing of the hurricane models in three dimensions has taken place at the NHC.

Even less encouraging and more controversial than hurricane forecasting are proposed scientific methods for hurricane control. Storm seeding with dry ice or silver iodide is much in doubt. It is interesting to note that when the first attempt was made to seed a hurricane in 1947, the storm had passed Florida and was headed east into the Atlantic, but after seeding the storm abruptly turned around and struck the coastline with renewed fury. This action most likely was coincidental in light of meteorological evidence, but the incident did have the effect of tarnishing the reputation of the seeding technique. Much debate has arisen pertaining to what part of the storm should be seeded. An earlier *Stormfury* concept (involving seeding the eyewall clouds and thus changing the distribution of atmospheric pressure within the storm and, in turn, reducing maximum winds by 10–15%) has been proved scientifically groundless. Some contemporary thinking, although considered questionable by some experts, is concerned with striking a balance between (1) seeding far enough from the center of the storm to disrupt the eyewall and (2) the logistical difficulties of covering larger and larger areas and of finding enough seedable clouds at these distances. Current thinking is to seed just outside the eyewall with expectation of a sufficiently large result.

Despite discouragement, a group of scientists remains dedicated to at least obtaining partial solutions to the hurricane prediction problem—and in taking advantage of larger computers in improving forecasting accuracy. Perhaps, as with so many other scientific problems, the ultimate solutions may stem from very fundamental research as, in this case, from continuing investigations into cloud microphysics.

As early as 1977, high-frequency skywave radar (WARF = Wide Aperture Research Facility) was used to monitor hurricane activity from a remote location. Scientists at the Remote Measurements Laboratory of SRI (Stanford Research Institute) International in Menlo Park, California monitored hurricane Anita in the Gulf of Mexico. The hurricane was tracked over a 5-day period with WARF. Sea backscatter at distances exceeding 300 kilometers (186 miles) was recorded, using single F-layer ionosphere reflection. Real-time maps of surface wind direction within a radial distance of 200 kilometers (124 miles) of the storm center were compiled. In comparing WARF measurements with those of data obtained from the ocean-moored buoy EB 70 (26.0°N, 93.5°W), operated by the National Data Buoy Office of the National Oceanic and Atmospheric Administration, agreement of measurements was within 10% for significant wave height and surface wind speed and direction of all quadrants of the storm. The WARF estimates of longshore coastal surface currents correlated well with those of a moored current meter. The sea backscatter received at the WARF is coherently processed in range and Doppler shift to produce a sea-echo Doppler spectrum. Considerably more detail on the system is given in papers presented at the meeting of the Advisory Group for Aerospace Research and Development (AGARD), Lisbon, Portugal (May 28–June 1, 1979) and the 16th Conference on Coastal Engineering, Hamburg, Germany (August 27–September 3, 1978). A summary is given by Maresca and Carlson in *Science*, **209**, 1189–1196 (1980).

As observed by Maresca and Carlson, "The supportive surface data supplied by WARF radar would prove particularly useful for tracking during early formative stages of hurricanes when multiple centers may be observed or when cirrus shielding may obscure visual location by satellite cloud photography. The high-resolution, large-coverage area, real-time sterring, and continuous monitoring capabilities are unique to skywave radar. The hurricane data obtained from skywave radar complements data obtained from satellites, aircraft, and buoys."

See also **Atmosphere (Earth); Atmosphere-Ocean Interface; Clouds and Cloud Formation; Fronts and Storms; Precipitation and Hydrometeors; Tsunami; Weather Observations and Forecasting;** and **Winds and Air Movement.**

HURWITZ CRITERION. Stability (System).

HUTIA. Rodentia.

HUYGEN'S PENDULUM. Pendulum Clock; Tantochrone.

HUYGEN'S PRINCIPLE. Interference (Wave); Wave Propagation. (Huygen's Principle).

H WAVE. A transverse electric wave.

HYACINTH (Water). Water Hyacinth.

HYADES. An open, V-shaped, moving cluster of stars in the constellation of Taurus. References to the Hyades are to be found in all the ancient literatures, Virgil referring to them as the "rainy Hyades." The group is exceedingly rich in double stars, which, even with a small telescope and low magnifying power, present a beautiful appearance.

The Hyades form one of the best known of the so-called moving star clusters. The brightest star of the Hyades, Aldebaran, is not a member of the cluster, but has an independent motion through space and just happens to be in its present position at this time.

HYALITE. Opal.

HYBRID. An organism produced by parents belonging to different species or to different strains of the same species. A hybrid combines characteristics derived from the two parent stocks and in some cases is more desirable than either. Beauty of flowers, productivity of various plants, and appearance and hardiness of animals have been enhanced by controlled hybridization.

When a hybrid is once secured its propagation is hampered by the fact that the diverse hereditary characters are reassorted in hereditary transmission by sexual reproduction. Hybrids are often infertile but even when they are capable of producing offspring they rarely breed true. The mule is the only animal hybrid of great value, and it is produced always by parents of the two species, horse and ass. Plant hybrids are not subject to this limitation, for they can usually be propagated by bulbs, cuttings, or grafts. Plants produced in this fashion are sometimes referred to as cultivars. See also **Plant Breeding.**

HYBRIDIZATION. Heredity.

HYBRIDIZATION (Plant). Plant Breeding.

HYBRIDOMA. Cancer Research.

HYBRID TRANSFORMER. Also called hybrid coil or hybrid repeater, a bridging transformer used in coupling a two-way telephone circuit to the repeater station or for coupling two one-way circuits to a two-way circuit. The coil is so sound that when the line is properly balanced by a balancing network there is no reaction between the output and input connections of the transformer.

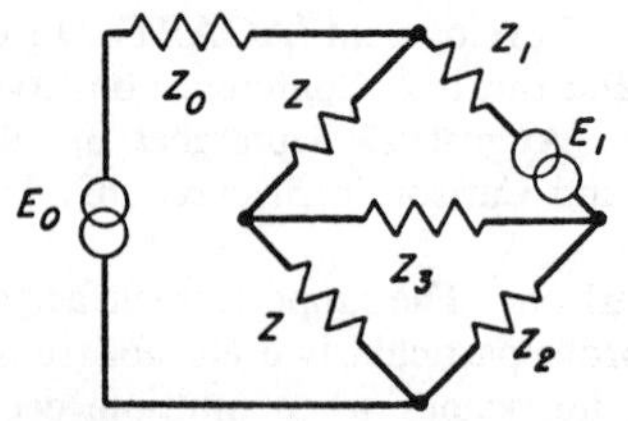

Fig. 1. Wheatstone bridge circuit.

Consider the Wheatstone bridge circuit shown in Fig. 1. If the bridge is balanced ($Z_1 = Z_2$) then no current from the generator E_0 will flow in Z_3 although the generator E_1 will produce a current in Z_3 and the generator E_0 will produce a current in both Z_1 and Z_2. The bridge circuit accordingly provides a means of transmitting from E_0 to Z_1 and Z_2 and from E_1 to Z_3 while at the same time preventing transmission from E_0 to Z_3. This is exactly the problem presented in making connections to a repeater amplifier used on telephone lines. In such a situation Z_1 might represent a line from the East, Z_2 a line from the West, Z_3 the input to the repeater, and E_0 the output of the repeater. Were the output signal of the repeater to be transmitted back to its input, undesired oscillations would result. The condition of balance, however, eliminates this possibility.

The hybrid coil provides a means of achieving the balanced bridge condition. In its simplest form it consists of a two winding transformer with a center tapped secondary as shown in Fig. 2. Z_1 and Z_2 may

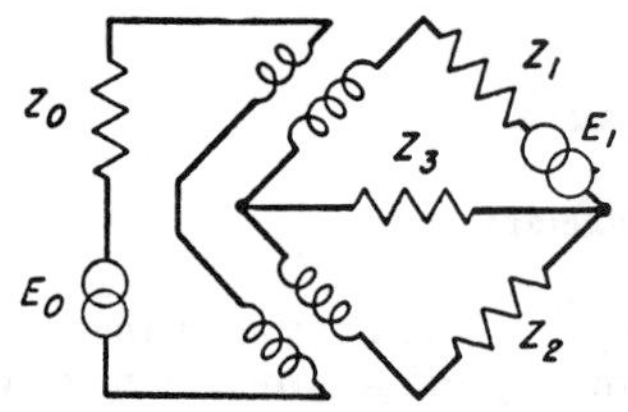

Fig. 2. Simple form of hybrid coil.

both represent telephone lines or alternatively one may be a line and the other a network designed to have an impedance equal to that of the line over the desired frequency range. To provide for telephone applications where a circuit balanced to ground is needed, the hybrid transformer is provided with three windings connected as indicated in Fig. 3 to achieve the results just discussed.

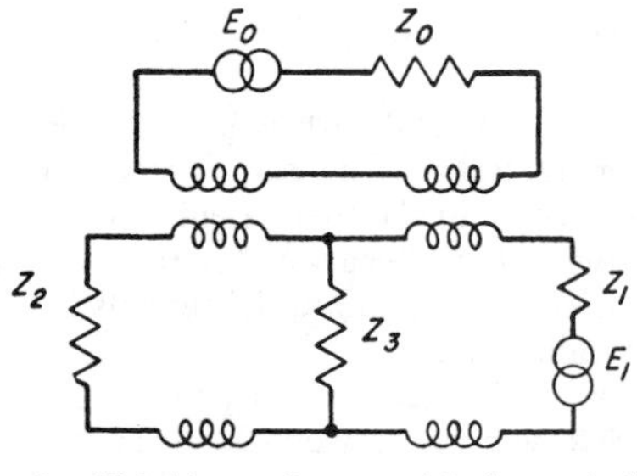

Fig. 3. Hybrid transformer with three windings.

HYDANTOIN PROCESS. Amino Acids.

HYDATID. The bladderworm or cysticercus of the cestodes *Echinococcus granulosus* and *E. multioccularis.* See **Hydatid Disease;** and **Tapeworm.**

HYDATID DISEASE. Also referred to as *echinococcosis,* this is an infection with *Echinococcus granulosus* or *E. multiocularis,* which are cestodes. These worms live in the intestines of dogs and wolves, whose feces include infective eggs. Such material may inadvertently find its way to a substance that is ingested by sheep, cattle, or humans. Infection with *E. granulosus* is most commonly found in regions where sheep and cattle are produced as, for example, in the western United States, parts of Canada, and Alaska. Upon ingestion of eggs, oncospheres are carried by the bloodstream to the liver, lungs, and other organs. These cause the development of cysts, often with neurologic symptoms. The cysts may grow to a diameter of 6 inches (15 centimeters), and contain many worms.

Mice and small animals are intermediate hosts to *E. multiocularis* which resides in the intestines of foxes and dogs. These cestodes are found mainly in the Northern Hemisphere—Europe, Canada, Alaska, and north-central United States. They produce extensive alveolar hydatid cysts, frequently resulting in jaundice.

Frequently, the therapy for hydatid disease involves surgical excision of cysts. Cryosurgery is frequently used. Some success has been reported with the drug mebendazole in the treatment of the disease. See also **Tapeworm.**

HYDATOGENESIS. A term used by petrologists to designate the process by which rocks are formed from highly aqueous solutions. Some petrologists limit the use of the term to rocks which have been deposited from water-rich magmatic solutions.

HYDRA. Asexual Reproduction.

HYDRA (the serpent). A southern constellation that forms the outline of a serpent.

HYDRANTH. A form of individual which receives and digests food in colonies of hydrozoan coelenterates. It is a polyp attached to the remainder of the colony at its base and with a mouth surrounded by a circlet of tentacles at the free end.

HYDRATE. Excluding the loose usages in which the term hydrate indicates merely the presence of water or of its elements in 2 : 1 ratio, as in carbohydrate, the term hydrate denotes the appearance of water in compounds. There are a number of ways in which water may appear in stoichiometric proportions in compounds. Moreover, these ways may be described from more than one point of view. A somewhat systematic approach is to view these compounds from the point of view of the extent of integration of the water, or its elements, into the compound.

The term "water of constitution" is a somewhat old usage, applied to compounds in which no H_2O groupings appear in the structure of the compound, but the compound may undergo reaction, usually reversible, in which water is one of the products. Magnesium hydroxide and sulfuric acid could thus be said to have "water of constitution," even though it appears in their structure as hydroxyl groups, or hydroxyl groups and hydrogen atoms (protons).

The term "cationic water" may be used to describe the situation in which water appears in coordination compounds apparently joined to cations by covalent bonds. However, the fact that a number of such compounds exhibit "hydrate isomerism" is evidence for cationic bonding, as well as it is for the existence of other forms of these compounds in which the presence of water is due to electrostatic attractions or crystal stability requirements.

The term "anionic water" describes the situation in which water is joined to anions through covalent bonds, or more frequently, through hydrogen bonds. The type case is copper(II) sulfate pentahydrate, where the cation has a coordination number of four and presumably the fifth molecule of H_2O is bound to the sulfate ion (as well as to other H_2O molecules) by hydrogen bonds.

The term "lattice water" is commonly applied to cases in which the water molecules are occupying definite positions in the crystal lattice but are apparently not coordinated with either cations or anions. Again, clear-cut cases are those in which the compound is so highly hydrated that both lattice water and "ion water" are present.

The water in crystals may, however, be present in other than definite lattice positions. For example, the water molecules may be found in holes in the lattices, or they may occupy random positions in the lattices. The latter situation is often found in ion exchange resins where loss of water, up to a certain point, does not materially change the lattice structure.

Finally, in essentially noncrystalline materials, such as hydrous precipitates and colloidal gels, the water present is at the limiting case of being a hydrate, in which virtually no bonding, in the chemical sense, exists.

HYDRATUBA. The attached form, resembling a polyp, which develops from the first larval stage of some jellyfishes. It may bud off other similar individuals and at certain seasons is subdivided by constrictions to form ephyrae which become jellyfishes.

HYDRAULIC ACCUMULATOR. **Accumulator (Hydraulic).**

HYDRAULIC CONTROL (Governor). **Governor.**

HYDRAULIC CONTROLLER. A device that uses a liquid control medium to provide an output signal which is a function of an input error signal. Aside from the use of a liquid controlling medium, hydraulic controllers are similar in operating principle to electric, electronic, and pneumatic controllers. In fact, there are striking similarities between hydraulic control and pneumatic control. Because a liquid control medium is essentially incompressible, there is an excellent speed of response between controller and final actuating element. Hydraulic control systems also are characterized by high power gain inasmuch as liquids can be converted readily to high pressures or flows through the use of various types of pumps. The final actuators are comparatively simple; most outputs are two hydraulic lines that can be tied directly to a straight-type cylinder to provide a linear mechanical output. Inasmuch as the parts of a hydraulic system are essentially self-lubricating, they have a long life when properly designed.

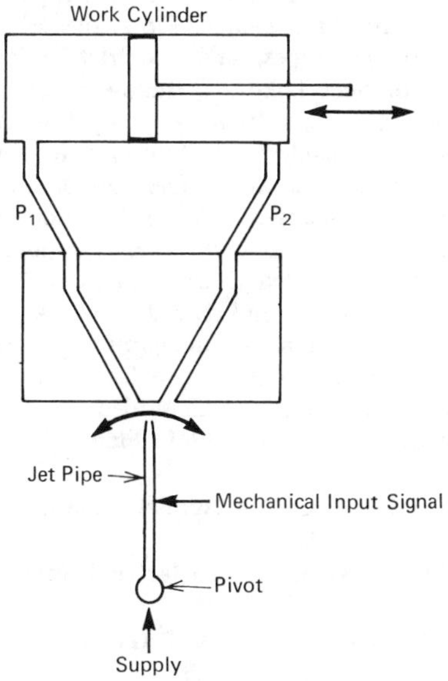

Fig. 1. Jet-pipe valve used in hydraulic control systems.

Limitations of hydraulic control systems include special maintenance problems in connection with hydraulic fluids—fire hazard and leakage, and somewhat higher cost, dependent upon the size of the equipment.

Hydraulic controllers are extensively used as liquid pipeline-pressure controllers where a pipeline control valve can be operated against sudden pressure surges. Edge-guiding control systems are also common. For example, a hydraulic system can control the edge of a moving steel strip (typical strip velocity of 1,000 feet; 300 meters per minute) to plus or minus $\frac{1}{64}$-inch (0.4 millimeter) and accomplish this by shifting a coil of steel weighing up to 50,000 pounds (22,680 kilograms).

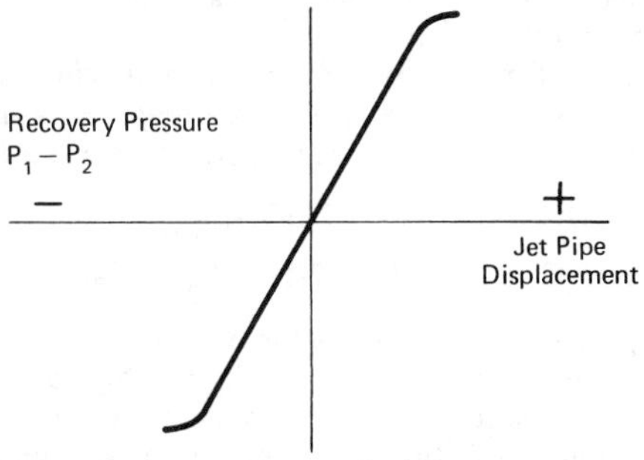

Fig. 2. Jet-pipe motion and recovery pressure relationship.

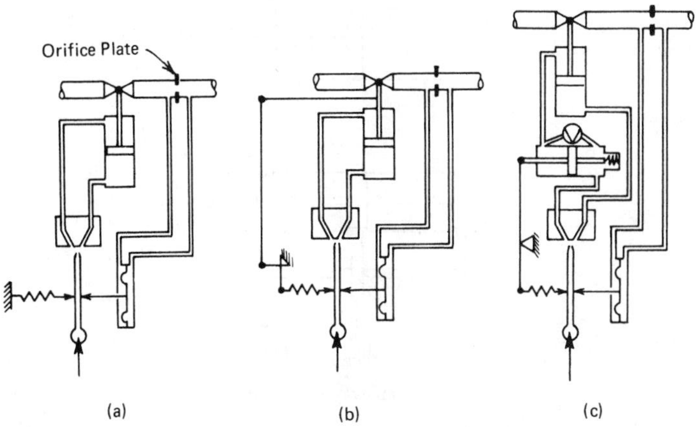

Fig. 3. Hydraulic controllers: (a) Proportional speed floating control; (b) proportional position control; (c) proportional plus reset control.

The hydraulic relay is the heart of a hydraulic control system. Commonly, a jet-pipe valve is used—as shown in Fig. 1. By pivoting a jet pipe, a fluid jet can be directed from one recovery port to another. The fluid energy is converted entirely into a velocity head as it leaves the jet-pipe tip and then is reconverted into a pressure head as it is recovered by the recovery ports. The relationship between jet-pipe motion and recovery pressure is shown in Fig. 2. Although the jet pipe can be used at higher pressures, most applications are less than 800 psi (~54 atmospheres). The proportional operation of the jet pipe makes it useful in proportional-speed floating systems (integral control) as indicated in Fig. 3(a). See also **Integral (Reset) Control.** Position feedback can be provided by rebalancing the jet pipe from the work cylinder as shown in Fig. 3(b). A proportional-plus-reset arrangement is shown in Fig. 3(c). In this last instance, the proportional feedback is reduced to zero as the oil bleeds through the needle valve. The hydraulic flow obtainable from a jet pipe is a function of the pressure drop across the jet pipe.

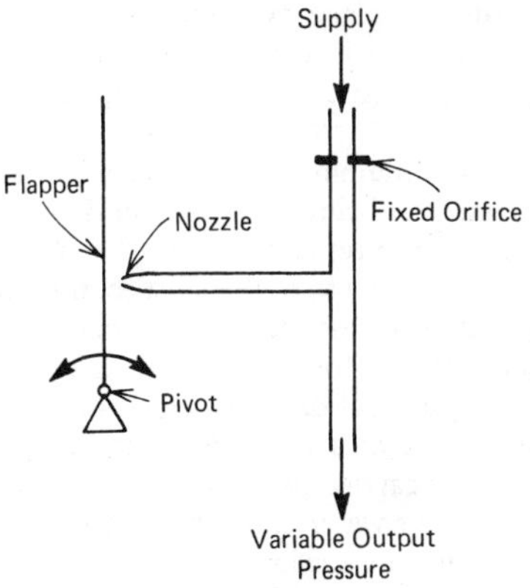

Fig. 4. Single flapper valve.

Flapper valves of the type shown in Fig. 4 also are used. The spool valve, shown in Fig. 5, when used as a hydraulic relay usually is

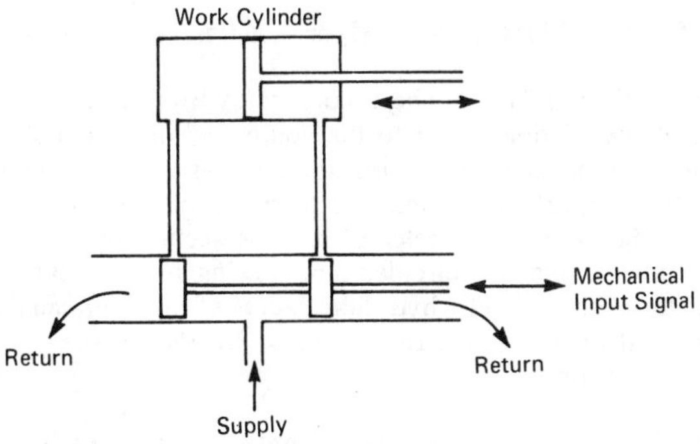

Fig. 5. Spool valve.

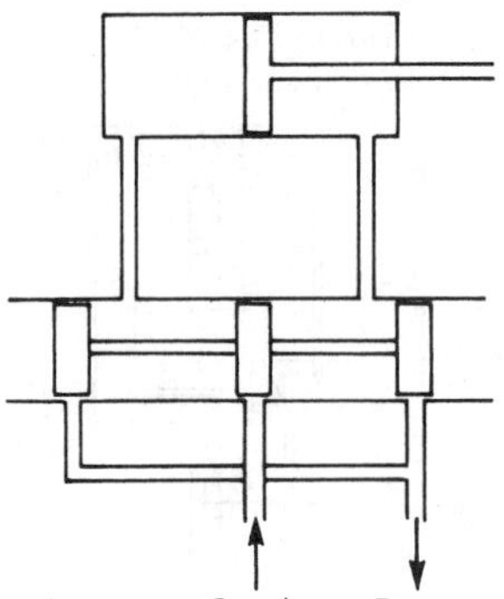

Fig. 6. Four-way spool valve.

constructed in either a three-way or a four-way valve-porting arrangement. See Fig. 6. The mechanical displacement of the spools allows the hydraulic-pressure supply to be ported in a fashion that will displace the work cylinder in either direction, depending upon the spool displacement.

References

Henke, R.: "Speed and Displacement: Keys to Comparing LSHT Hydraulic Motors," *Control Engineering,* **24**, 2, 37–39 (1977).

Henke, R.: "Energy Costs Forcing Fluid Power Trends," *Control Engineering,* **25**, 8, 37–38 (1978).

Heubner, J. R.: "Hydraulic Controllers," in "Process Instruments and Controls Handbook," 2nd edition (D. M. Considine, editor), McGraw-Hill, New York, 1974.

Holzbock, W. G.: "Comparing Electromagnetic Servovalve Actuators for Hydraulic Power Control," *Control Engineering,* **23**, 5, 41–44 (1976).

Pluhar, K.: "Computer Control in Fluid Power," *Control Engineering,* **26**, 10, 63–67 (1979).

HYDRAULIC FILL. An embankment or other fill in which the materials are deposited in place by a flowing stream of water, with the deposition being selective, is termed hydraulic fill. Gravity, coupled with velocity control, is used to effect the selected deposition of the material.

Where borrow pits containing suitable material are accessible at an elevation such that, after being washed from the bank by a powerful stream from a large, high-pressure nozzle, the earth can be sluiced to the fill, hydraulic fill is likely to be the most economical construction. Even where the elevation is not realized, the material can be washed into pools, then elevated to the sluice with a dredge pump. In the construction of a hydraulic fill dam, the edges of the dam are defined by low embankments or dykes which are carried upwards as the fill proceeds. The sluices are carried parallel to and just inside these dykes. The sluices discharge their water-earth mixture at intervals, the water then fanning out and flowing towards the central pool which is maintained at the desired level by discharge control. While flowing from the sluices, the coarse material is first deposited, then, as the central pool is reached, the water velocity is diminished and the fine materials are deposited to form an impervious central section. The water flow must be well controlled at all times, otherwise the central section may be bridged by tongues of coarse material which would facilitate seepage through the dam of the water later impounded behind the dam.

HYDRAULIC GRADE LINE. **Hydrokinetics.**

HYDRAULIC RADIUS. The theory of hydraulics indicates that the ratio of the frictional area to the volume of the liquid stream is an important dimension governing the friction loss. The hydraulic radius, which expresses this fact, is the cross-sectional area of flow divided by the wetted perimeter of a cross section of the conduit. The hydraulic radius of a circular pipe flowing full of water is one-fourth of the diameter. The hydraulic radius of an open canal is the cross-sectional area of the stream divided by the wetted perimeter of the cross section.

HYDRAULICS. Hydraulics is the dynamics of liquids (hydrodynamics), especially applied to the practical problems of engineering. Although this general definition is entirely correct, in common usage hydraulics is the study of water at rest or in motion. This conception of hydraulics is used in this article. The mechanics of fluids (liquids and gases) in general is termed fluid mechanics. A basic proposition of hydraulics is that water is incompressible. While this condition is not completely met in fact, the compressibility of water is so small as to be negligible for practically all propositions of hydraulics. The viscosity of water varies with the temperature and is one reason for change of conditions of water flow in pipes with changing temperature. The unit weight of fresh water is usually taken as 62.4 pounds per cubic foot.

The science of hydraulics is divisible into hydrostatics and hydrokinetics. Hydrostatics is the hydrodynamics of liquids considered apart from their motion: hydrokinetics is the hydrodynamics of moving, especially flowing liquids. Among the subjects included in any study of hydrostatics are the following: (1) the pressure on a submerged area of any shape or inclination, (2) the measurement of pressure on water at rest by manometers or pressure gauges, (3) buoyancy and flotation. Practical application of (1) is to be found in problems associated with water gates, large valves, pressure against dams, tanks, hydraulic presses, etc.

Hydrokinetics includes a great many different phases of hydraulics. Most of these will be found treated in specialized articles, references to which are given below. The flow of fluids supplies many cases of the application of hydraulic science. Flows of steady, uniform, unsteady and non-uniform types, and the friction losses occasioned thereby, in closed or open conduits; the measurement of flows and the discharges under given conditions, are part of this phase of hydraulics; also, there is to be considered the flow of water through openings, such as orifices, nozzles, and weirs. The flow of water in pipe lines offers a great many problems in addition to friction: the discharge through different sections of branching and looping pipes, siphons, fittings, valves, etc., is included. Measurement of discharge of large amounts of water, as in stream and river flow, offers problems different from those met in closed conduits. Furthermore, the forces occasioned by deviated flows of water, as met in hydraulic turbines, the pump, and other hydraulic machinery, are fit subjects to be included in any study of hydromechanics. See also **Fluid Flow.**

HYDRAULIC TORQUE CONVERTER. **Torque Converter.**

HYDRAULIC TURBINES. **Hydroelectric Power.**

HYDRAULIC UPLIFT (Dam). **Uplift (Hydraulic).**

HYDRAZINE. $H_2N \cdot NH_2$, formula weight 32.04, colorless, fuming liquid, mp 1°C, bp 113°C, sp gr 1.011, decomposes when heated above 350°C at atmospheric pressure into N_2 and NH_2, also decomposes in presence of a catalyst (e.g., platinum) into N_2 and NH_3. Hydrazine burns when ignited in air with a violet-colored flame. The compound is soluble in all proportions with H_2O and is soluble in alcohol. Hydrazine forms a hydrate with one molecule of H_2O. Upon moderate heating or in a vacuum, the hydrate yields hydrazine and H_2O. Hydrazine is a base slightly weaker than NH_4OH.

Hydrazine is a tonnage chemical with numerous uses, including that of a propellant for rockets, yielding exhaust products at a high temperature and of a low-molecular weight; use as a strong reducing agent in the manufacture of various chemicals; and as a blowing agent for foamed rubber. The compound reacts with citric acid to form *Continazin*, an antituberculan drug.

Although the earlier processes for the commercial production of hydrazine used urea as a raw material, modern processes employ direct ammonia oxidation. In one such process, reactions occur in two steps:

$$(1)\ NH_3 + NaOCl \rightarrow NH_2Cl + NaOH,$$

$$(2)\ NH_3 + NH_2Cl + NaOH \rightarrow H_2N \cdot NH_2 + NaCl + H_2O.$$

Highgrade hypochlorite is required for Step 1. Special agents, such as gelatin, ethylenediamine tetracetic acid, glue, high alcohols, or formaldehyde are required to inhibit undesirable side reactions that would reduce the hydrazine yield through formation of ammonium chloride and N_2. In another hydrazine process, chlorine, NH_3, and H_2SO_4, along with methylethyl ketone, are used as the charge. The products of this process include hydrazine hydrate, hydrazine sulfate, ketazine, and dialkyldiazacyclopropane. Hydrazine also is used as a start-up ingredient in the preparation of cooling water for nuclear reactors where it is desired to keep the oxygen content of the water to an absolute minimum and thus decrease corrosion. Oxygen reacts with hydrazine. $H_2N \cdot NH_2 + O_2 \rightarrow N_2 + 2H_2O$. When no oxygen is present in the water, the hydrazine acts as a sink for dissolved oxygen that may enter later, by maintaining metal oxides at their lower oxidation states.

Hydrazine forms two series of salts (1) hydrazinium (1+) chloride, $H_2NNH_3^+Cl^-$, nitrate $H_2NNH_3^+NO_3^-$, hemisulfate $(H_2NNH_3^+)_2SO_4^{2-}$, (2) hydrazinium (2+) chloride $H_3NNH_3^{2+}(Cl^-)_2$, dinitrate $H_3NNH_3^{2+}(NO_3^-)_2$, hydrogen sulfate $H_3NNH_3^{2+}(HSO_4^-)_2$, all soluble in H_2O. This last is produced when hydrogen azide reacts with concentrated H_2SO_4. It is very hygroscopic and decomposes in aqueous solution to give the slightly soluble monosulfate and H_2SO_4. The monosulfate and difluoride, which have been thought to have the structures $N_2H_5^+HSO_4^-$ and $N_2H_5^+ HF_2^-$ in the solids, have been shown in fact to be $N_2H_6^{2+}SO_4^{2-}$ and $N_2H_6^{2+}(F^-)_2$. Hydrazinium azide $N_2H_5^+N_3^-$ is a soluble solid.

In the laboratory, hydrazine can be prepared by converting one-half of a given amount of NH_3 into chloramine NH_2Cl by sodium hypochlorite solution in the presence of a colloid and heating. The remaining one-half of the NH_3 reacts with chloramine to form hydrazine. The product is then cooled to 0°C and H_2SO_4 added in amount to react with the hydrazine to form hydrazine sulfate $N_2H_6SO_4$, insoluble solid. Hydrazine hemisulfate $(N_2H_5)_2SO_4$ is soluble in H_2O. It can also be made by the reaction of NH_3 and hydroxylamine-*O*-sulfonic acid.

Phenylhydrazine is a colorless liquid, slightly soluble in H_2O, miscible in all proportions with alcohol or ether, forms salts with acids, e.g., phenylhydrazine hydrochloride or phenylhydrazinium chloride $C_6H_5NHNH_3Cl$, is a powerful reducing agent, with alkaline copper(II) salt solution (Fehling's solution) yields copper(I) oxide precipitate, reacts with carbonyl group of aldehydes or ketones yielding phenylhydrazones, white solids, of definite melting point and utilized in identification of aldehydes and ketones, e.g., acetaldehyde phenylhydrazone $CH_3CH{:}NNHC_6H_5$.

Phenylhydrazine, as hydrochloride solution plus sodium acetate react with polyhydroxy aldehydes or ketones yielding *osazones* or diphenylhydrazones, yellow solids, of definite melting point and utilized in identification of sugars, e.g., phenyl-d-glucosazone $CH_2OH(CHOH)_3C{:}(NNHC_6H_5)CH{:}(NNHC_6H_5)$ plus aniline $C_6H_5NH_2$ plus NH_3.

Attention should be given to the difference between osazones and osones. An *osone* is formed by reaction of an osazone with HCl, e.g., glucosone $CH_2OH(CHOH)_3CO \cdot CHO$.

1,1-Diphenylhydrazine is made by reduction of diphenylnitrosamine $(C_6H_5)_2N \cdot NO$ by zinc plus acetic acid, the nitrosamine being formed by reaction of diphenylamine $(C_6H_5)_2NH$ and nitrous acid.

Tetraphenylhydrazine is a white solid, soluble in chloroform, acetone, benzene, or toluene, and upon standing is changed into triphenylamine plus azobenzene. In solution, tetraphenylhydrazine dissociates into nitrogen diphenyl $(C_6H_5)_2N\cdot$, free radical, which in toluene at 90°C reacts with nitric oxide NO. Tetraphenylhydrazine is formed by oxidation of diphenylamine $(C_6H_5)_2NH$ by lead dioxide.

Hydrazine reacts with ketones to form *azines.*

HYDRAZINE (Boiler Water). Feedwater (Boiler).

HYDRAZINE (Fuel Cell). Fuel Cells.

HYDRAZINE (Rocket Fuel). Rocket Propellants.

HYDRAZOIC ACID. HN_3, formula weight 43.03, colorless, odorous, poisonous liquid, mp −80°C, bp 37°C, explodes with marked violence. Also known as azoimide and hydronitric acid, the compound is miscible in all proportions with H_2O, alcohol, and ether. Hydrazoic acid reacts (1) with metals, e.g., magnesium, aluminum, zinc, iron, to form azides or hydrazoates (or trinitrides), (2) with heavy metal salt solutions to form insoluble azides, e.g., silver azide AgN_3, mercury(I) azide HgN_3, lead azide PbN_6. Silver, mercury(I), and copper(I), azides decompose in the light to form nitrogen plus the metal. (3) It reacts with NH_4OH to form ammonium azide $NH_4 \cdot N_3$, (4) with hydrazine to form hydrazine azide $N_2H_4 \cdot HN_3$, (5) with sodium hypochlorite plus acetic acid to form chlorazide ClN_3, explosive, (6) with sodium amalgam to form NH_3 with some hydrazine, (7) with potassium permanganate to form nitrogen and H_2O.

Hydrazoic acid is formed (1) by reaction of sodium nitrate with molten sodamide, (2) by reaction of nitrous oxide with molten sodamide, (3) by reaction of nitrous acid and hydrazinium ion $(N_2H_5^+)$, (4) by oxidation of hydrazinium salts, (5) by reaction of ethyl nitrite with NaOH solution and acidifying. See also **Azides.**

HYDRAZONES. The products of the reaction between an aldehyde or a ketone with phenylhydrazine are termed *hydrazones.* Sometimes the compounds are referred to as phenylhydrazones.

$$\underset{\text{(acetaldehyde)}}{CH_3 \cdot CHO} + \underset{\text{(phenylhydrazine)}}{C_6H_5 \cdot NH \cdot NH_2} \rightarrow \underset{\text{(acetaldehyde hydrazone)}}{CH_3 \cdot CH{:}N \cdot NH \cdot C_6H_5} + H_2O$$

$$\underset{\text{(benzaldehyde)}}{C_6H_5CHO} + C_6H_5 \cdot NH \cdot NH_2 \rightarrow \underset{\text{(benzylidenehydrazone)}}{C_6H_5 \cdot CH{:}N \cdot NH \cdot C_6H_5} + H_2O$$

$$\underset{\text{(acetone)}}{(CH_3)_2CO} + C_6H_5 \cdot NH \cdot NH_2 \rightarrow \underset{\text{(acetone hydrazone)}}{(CH_3)_2C{:}N \cdot NH \cdot C_6H_5} + H_2O$$

$$\underset{\text{(acetophenone)}}{C_6H_5 \cdot CO \cdot CH_3} + C_6H_5 \cdot NH \cdot NH_2 \rightarrow \underset{\text{(acetophenonehydrazone)}}{(C_6H_5)(CH_3)C{:}N \cdot NH \cdot C_6H_5} + H_2O$$

Several of the hydrazones may be decomposed by strong acids whereupon the original aldehyde or ketone is regenerated, along with the formation of a phenylhydrazine salt. When reduced, hydrazones yield primary amines.

HYDRIDE. A binary compound of hydrogen. Hydrides traditionally have been classified into three groups. In modern terminology, these are conveniently designated as covalent, electrovalent and metallic, although reference to the entries for hydrogen and the various hydrogen halides shows that a number of binary hydrogen compounds are partly ionic and partly covalent. See also **Hydrogen.**

Covalent hydrides are formed by the non-metals. In general, the elements of main groups III to VII form single compounds consisting of a single atom of the element combined with a number of hydrogen atoms equal to the number of electrons which the element needs to complete its octet. Exceptions are beryllium, aluminum, and indium, which have polymeric hydrides, and boron and gallium, which have dimeric hydrides. Then also the elements of lower atomic number in main groups IV, V and VI (carbon, silicon, germanium, nitrogen, phosphorus, oxygen, and sulfur) and boron form more than one hydride. The covalent hydrides are volatile with low melting points and low boiling points (except as those properties are modified, as in the case of hydrogen fluoride, water and ammonia, by hydrogen bonding). They are nonconductors of electricity in the liquid state or when dissolved in nonpolar solvents.

Complex hydrides are formed by some elements (particularly in main group III) having too few electrons to attain an octet in the neutral hydrides. These are structurally similar to the corresponding complex chlorides and are all excellent reducing agents. The most important are the tetrahydroborate, BH_4^-, -aluminate (frequently called alanate in the European literature), AlH_4^-, -gallate, GaH_4^-, and -indate, InH_4^-. There is evidence for polymeric ions, such as $B_2H_7^-$. Anions derived from higher hydrides are also known, e.g., $B_4H_{11}^-$.

Only the strongly electropositive elements, the alkali metals, the alkaline earth metals, and certain lanthanide and actinide metals, form electrovalent hydrides. The compounds are definitely crystalline, the alkali hydrides being cubic, but the structure increasing in complexity in going from main group 1 to main group 2 and to the lanthanides and actinides. In fact, hydrides of the last two groups, while approaching the alkali and alkaline earth hydrides in electropositive character, and while also giving evidence of the presence of H^- ions in their structures, are usually non-stoichiometric, compositions such as $CeH_{2.70}$, $PrH_{2.85}$ and $ThH_{3.07}$ being found. In this respect, those compounds approach in character the metallic hydrides.

This gradation in properties extends to the metallic hydrides themselves, some of which, such as copper hydride, approaches closely, but never quite reaches, a 1:1 atomic ratio of hydrogen to copper. In the case of palladium, the pressure-composition graph at temperatures below 200°C indicates a wide range of composition at little or no increase in pressure. At higher temperatures the flat portion shortens, and two breaks develop before and after it, indicating solid solutions, one of which approaches a 2:1 atomic ratio of H to Pd.

A group of complex metal hydrides have been used successfully for the preparation on an industrial scale of many organic and metallorganic compounds. Among these complex hydrides are highly reactive lithium aluminum hydride and the related sodium aluminum hydride and magnesium aluminum hydride. A more selectively reactive group of complex hydrides are the lithium, sodium, and potassium borohydrides. These compounds also have properties which make them useful as high energy fuels and rocket propellants.

Lithium aluminum hydride, $LiAlH_4$, also known as lithium aluminohydride and often abbreviated as LAH is prepared by the reaction of lithium hydride and aluminum chloride in ether solution

$$4LiH + AlCl_3 \rightarrow LiAlH_4 + 3LiCl$$

with some prior prepared complex hydride used as a seeding material to control the reaction rate. Lithium aluminum hydride forms a microcrystalline powder which is stable in dry air but decomposes above 125°C. It is soluble in many organic compounds like ether, dimethyl Cellosolve, tetrahydrofuran but is only slightly soluble in dioxane. It reacts vigorously with water, yielding hydrogen in a manner similar to the reaction of the simple hydrides:

$$LiAlH_4 + 4H_2O \rightarrow 4H_2 + LiOH + Al(OH)_3$$

It reacts with carbon dioxide to form methyl alcohol, or formaldehyde, or formic acid. It is a powerful reducing agent and reduces aldehydes, ketones, quinones, acids, esters, anhydrides, lactones, epoxides, and acid chlorides to the corresponding alcohols; amides, lactams, imides, nitriles, isocyanides, oximes, hydroxylamines, and related compounds to amines; dithiols, disulfides, polysulfides, sulfoxides, sulfones, and related compounds to the mercaptan or sulfide (see **Sulfur**); and aromatic nitro compounds to azo compounds. Olefinic bonds are not attacked unless conjugated with a nitrile, phenyl, or carbonyl group.

Sodium aluminum hydride can be prepared like the lithium analogue but tetrahydrofuran is used as the solvent because the sodium complex is insoluble in ether. The sodium compound produces virtually the same reductions as the lithium compound.

Magnesium aluminum hydride may be prepared by treating an etherate of magnesium bromide with an ether solution of lithium aluminum hydride

$$2LiAlH_4 + MgBr_2 \rightarrow Mg(AlH_4)_2 + 2LiBr$$

or by use of an excess of magnesium hydride in ether solution with an ether solution of aluminum chloride. While the reducing activity of magnesium aluminum hydride is similar to that of the lithium complex in that polar double and triple bonds such as carbonyl and nitrile groups are reduced whereas nonpolar groups are not attacked, the magnesium complex, however, does not reduce the triple bond of propargyl aldehyde nor the double bond of cinnamic acid in contrast to the lithium complex.

Lithium borohydride, $LiBH_4$, may be prepared by the reaction of aluminum borohydride on ethyllithium or by the action of diborane on ethyllithium. It forms orthorhombic crystals which decompose at 250 to 272°C and while it is stable under usual conditions it is decomposed by humid air. It reacts readily with water and is a strong reducing agent.

Aluminum borohydride, AlB_3H_{12}, is a liquid which boils at about 44.5°C. It ignites in air. It can be prepared by the reaction of diborane with trimethylaluminum. It reacts readily with hydrogen chloride and water to yield hydrogen. It can be used in organic syntheses.

Metals like palladium and platinum absorb hydrogen forming mixtures which may be considered as alloys. Such mixtures may be placed into two groups: those in which the absorption takes place with decrease in temperature like those mixtures of hydrogen with palladium and tantalum; and those which absorb hydrogen with an increase in temperature like those with calcium, iron, nickel, and platinum.

HYDROCARBONS. Organic Chemistry.

HYDROCARBONS (Carcinogens). Carcinogens.

HYDROCARBONS (Pollutants). Pollution (Air).

HYDROCEPHALUS. A condition characterized by abnormally large amounts of cerebrospinal fluid around or within the brain, usually associated with enlargement of the cerebral ventricles. See **Meningitis.**

HYDROCHLORIC ACID. HCl (hydrogen chloride gas) in aqueous solution, colorless when pure. Commercial grades of HCl (also known as muriatic acid) generally are marketed in three concentrations: (1) 18° Bé (sp gr 1.1417 at 15.6°C, 27.92% HCl); (2) 20° Bé (sp gr 1.160, 31.45% HCl); and (3) 22° Bé (sp gr 1.1789, 35.21% HCl). Frequently the commercial grades are slightly yellow because of impurities, notably dissolved iron. Fuming hydrochloric acid contains about 37% HCl, with a sp gr 1.194. Reagent grade hydrochloric acid usually is of this latter high strength, is perfectly clear and colorless. The maximum limits set on impurities commonly are: NH_4 0.003%; arsenic 0.000001%; free chlorine 0.0001%; heavy metals, such as lead 0.001%; iron 0.00002%; sulfates 0.0001%; sulfites 0.0001%; and residue after ignition 0.0005%. A mixture of three parts HCl and one part HNO_3 is known as *aqua regia*, a powerful solvent and oxidizing agent which will dissolve materials that may be unaffected by either acid alone. Gold and platinum are soluble in aqua regia.

Hydrochloric acid is a very high-tonnage chemical, finding major uses in (1) the cleaning and preparation of metals prior to application of coatings, (2) the recovery of zinc from galvanized iron scrap, (3) the production of numerous chlorides, and (4) production of chlorine. At one time, HCl was extensively used as a source of both hydrogen and chlorine by way of electrolysis. This process was made obsolete many years ago when the chlor-alkali process (electrolysis of sodium chloride brines) was introduced for the production of chlorine. In recent years, however, the production of by-product HCl, resulting from chlorination of numerous organic compounds, has increased. In some of these instances, the installation of a HCl electrolysis plant may be economically feasible. For industrial consumption anhydrous HCl gas also is available in steel cylinders under a pressure of 1,000 psi (68 atmospheres). Hydrochloric acid forms a constant-boiling solution with H_2O (20.22% HCl) which has a bp 108.58°C (760 mm Hg).

Dilute HCl reacts (1) with many hydroxides, e.g., NaOH, to yield the corresponding chloride, e.g., sodium chloride, solution, (2) with many ordinary oxides, e.g., magnesium oxide, to yield the corresponding chloride, e.g., magnesium chloride, solution, (3) with many carbonates, e.g., calcium carbonate, to yield the corresponding chloride, e.g., calcium chloride solution plus CO_2, (4) with many sulfides, e.g., ferrous sulfide, to yield the corresponding chloride, e.g., ferrous chloride, solution plus H_2S, (5) with many metals, e.g., zinc (but not copper) to yield the corresponding chloride, e.g., zinc chloride, solution plus hydrogen gas, (6) with some special oxides, e.g., lead or manganese dioxide, to yield lead or manganese chloride plus chlorine gas, (7) with solution of some salts, e.g., silver nitrate, to yield the corresponding chloride, silver chloride, precipitate. Higher strengths of hydrochloric acid usually react similarly to the dilute. Hydrochloric acid sometimes reacts as a reducing acid, e.g., (6) above.

All metallic chlorides, except silver chloride and mercurous chloride, are soluble in H_2O, but lead chloride, cuprous chloride and thallium chloride are only slightly soluble. Metallic chlorides when heated melt, and volatilize or decompose, e.g., sodium chloride, mp 804°C; calcium, strontium, barium chloride volatilize at red heat; magnesium chloride crystals yield magnesium oxide residue and hydrogen chloride; cupric chloride yields cuprous chloride and chlorine. See also **Chlorine; Chlorinated Organics; Halides; Hypochlorites;** and **Sodium Chloride.**

Hydrogen Chloride: This is a colorless gas, heavier than air, density 1.639 g/l at standard conditions. The gas is poisonous and quickly causes suffocation. Formula weight 36.47, mp −111°C, bp −85°C, critical pressure 83 atm, critical temperature 51.3°C. The gas is very soluble in H_2O, accounting for the high concentrations of hydrochloric acid obtainable. Although hydrogen chloride gas may be used directly in some industrial operations, normally it is generated for the purpose of dissolving in H_2O to form hydrochloric acid. The most common route to HCl is by reacting sodium chloride with H_2SO_4. This is a two-step, exothermic reaction: (1) $NaCl + H_2SO_4 \rightarrow NaHSO_4 +$

HCl, and (2) $NaCl + NaHSO_4 \rightarrow Na_2SO_4 + HCl$. Preparation of hydrochloric acid from the gas involves an absorption tower where the gas meets a fine spray of H_2O. Ratio controllers are used to assure maximum yield of the acid of desired concentration. These controls are easily adjusted for obtaining different concentrations. In most chlorinations of organic compounds, only half of the chlorine is used to substitute for hydrogen atoms, the remaining chlorine forming HCl. Frequently, this by-product HCl is recycled or recovered.

HYDROCORTISONE. **Adrenal Glands; Pituitary Gland; Steroids.**

HYDROCRACKING (SNG). **Substitute Natural Gas (SNG).**

HYDROCYCLONE. **Coal.**

HYDRODESULFURIZATION. **Substitute Natural Gas (SNG).**

HYDRODYNAMIC CAVITATION. **Cavitation.**

HYDROELECTRIC POWER. Electric power derived from generators driven by hydraulic turbines or water wheels. The principal elements of a hydroelectric plant are a site to provide flow and head of water, a turbine-generator machine, and a discharge channel. Gravitational energy available for the generation of electricity from water flowing from a higher level to a lower level is manifested on the earth in essentially two forms: (1) descending natural watercourses, created by precipitation of rain and snow, which flow from mountains, hills, and plateaus to sea level; and (2) the changes in levels of estuaries and other ocean-associated bodies of water which occur as the result of actions of the tides. In terms of total hydrogravitational energy available, the successful exploitation of tidal energy as of the early 1980s, is quite small by comparison with what can be achieved through a considerable expenditure of capital and engineering effort. The availability of tidal energy requires special topography at the ocean-land interface and, unfortunately, such sites are quite limited in North America. However, they are particularly attractive in some regions, such as in the vicinity of the English Channel. This topic is discussed further in the entry on **Tidal Power.**

The utilization of energy available from watercourses, on the other hand, has historically contributed quite significantly to the total amount of electricity produced worldwide.

The major hydroelectric plants worldwide are listed in Table 1. This listing includes facilities with generating capacity, present or ultimate, that exceeds 1000 megawatts. Medium-capacity plants in the United States are listed in Table 2. The exploitation of hydroelectric potential varies considerably from one country to the next and some of the countries that do not have a few large facilities, but rather several medium-size and smaller facilities, as in the cases of Spain, Italy, and France, generate an impressive percentage of their electricity requirements from hydropower. In terms of installed capacity as a percentage of the country's electric power needs, Canada leads with an estimated 65.6%; followed by Spain (58.5%); Italy (42.2%); France (37.3%); the U.S.S.R. (20–22%); the United States (11–13%); West Germany (9%); and the United Kingdom (3–4%). Percentage of total power generated by hydro facilities varies from one year to the next, depending upon precipitation. This percentage for the United States has been decreasing since it peaked in the 1930s—a result of diminishing hydroelectric power plant construction and the overall growth of electric power needs that have largely been met in recent years by the advent of nuclear power and increasing dependence upon fossil-fueled power plants. See Fig. 1. Considerable potential remains for new siting of dams and reservoirs, but in recent years, much greater

TABLE 1. MAJOR HYDROELECTRIC PLANTS WORLDWIDE

Country and Name of Dam or Power Plant	Year of Dedication	Present Capacity (Megawatts)	Ultimate Capacity (Megawatts)
UNITED STATES			
Grand Coulee	1941	4163	10,080
John Day	1968	2160	2700
Chief Joseph	1956	1024	2069
Robert Moses–Niagara	1961	1950	1950
Ludington	1973	1872	1872
Saint Lawrence Power Dam (with Canada)	1958	1824	1824
The Dalles	1957	1807	1807
Blue Ridge	NC		1600
Raccoon Mountain	1975	1530	1530
McNary	1953	980	1406
Hoover	1936	1345	1345
Wanapum	1963	831	1330
Priest Rapids	1959	789	1262
Castaic	1974	1060	1250
Rocky Reach	1961	712	1215
Dworshak	NC		1060
Northfield Mountain	NC		1000
CANADA			
La Grande No. 2	NC		5328
Churchill Falls	NC		5225
W. A. C. Bennett	1969	1816	2270
Mica	NC		1740
Kemano	1954	813	1670
Beauharnois	1950	1021	1574
Sir Adam Beck No. 2	1954	900	1370
Daniel Johnson	1970	650	1292
Kettle Rapids	NC	714	1224
Manicouagan No. 3	NC		1176
Bersimis No. 1	1956	1050	1050
Manicouagan No. 2	1965	1016	1016
U.S.S.R.			
Sayanskaya	NC		6400
Krasnoyarsk	1968	6096	6096
Bratsk	1964	4100	4600

TABLE 1. MAJOR HYDROELECTRIC PLANTS WORLDWIDE (*continued*)

Country and Name of Dam or Power Plant	Year of Dedication	Present Capacity (Megawatts)	Ultimate Capacity (Megawatts)
Sukhovo	NC		4500
Ust-Illimsk	1975	720	4320
Rogunsky	NC		3600
Volgograd	NC		2700
Volga-22nd Congress	1958	2560	2560
Volga-V.I.Lenin	1955	2300	2300
Cheboksary	1972	1404	1404
Saratov	1967	1360	1360
Inguri	NC		1300
Zeya	1975	300	1290
Nizhne-Kamskaya	1973	624	1248
Toktogul	NC		1200
Votkinsk	1961		1000
Chirkey	NC		1000
BRAZIL			
Itaipu (with Paraguay)	NC		12,870
Tucari	NC		3960
Paulo Afonso	1955	1524	3409
Solteira	1973	3200	3200
Foz Do Areia	NC		2250
Itumbiara	NC		2080
Salto Santiago	NC		1998
Itaparica	NC		1700
Marimbondo	1975	1400	1400
Jupia	1961	1400	1400
Agua Vemelha	NC		1380
Furnas	1963	1200	1200
MEXICO			
Chicoasen	NC		2400
Malpaso	1968	830	1245
TURKEY			
Kara Kaya	NC		1800
Keban	1974	620	1240
PAKISTAN			
Tarbella	1977	700	2100
Mangla	NC	300	1000
CHINA			
Liukiahsia	1963	1225	1225
Sanmen Hsia	NC		1100
AUSTRALIA			
Tumut-3	1972	1500	1500
Talbingo	NC		1500
ARGENTINA			
Salto Grande (with Urguay)	NC		1890
El Chocan	NC		1200
COLOMBIA			
San Carlos	NC		1550
Chivor	NC	500	1000
OTHER COUNTRIES			
Guri (Venezuela)	1967	6500	6500
Cabora Basa (Mozambique)	1975	2000	4000
Inga I (Zaire)	1974	300	2820
Iron Gate (Rumania/Yugoslavia)	1970	2300	2300
Saad El-Aali–High Aswan (Egypt)	1967	2100	2100
Dinorwick (United Kingdom)	NC		1880
Grand Maison (France)	NC	1200	1800
Kariba (Zimbabwe/Zambia)	1959	1266	1566
Takase (Japan)	NC		1280
Kaniji (Nigeria)	NC		1000

NOTE: There are many more hydroelectric facilities of <1000 megawatts located in the foregoing and other countries.

NC = not completed.

TABLE 2. MEDIUM-CAPACITY HYDROELECTRIC PLANTS IN THE UNITED STATES
(<1000, >150 megawatts)

Name of Dam or Power Plant and State	Present Capacity (Megawatts)
Wells (Washington)	774
Boundary (Washington)	551
Conowingo (Maryland)	474
Hells Canyon (Oregon)	392
Brownlee (Idaho)	360
Edward G. Hyatt (California)	351
Cowans Ford (North Carolina)	350
Smith Mountain (Virginia)	300
Mossyrock (Washington)	300
New Colgate (California)	284
Noxon Rapids (Montana)	283
Round Butte (Oregon)	247
Safe Harbor (Pennsylvania)	227
Rock Island (Washington)	212
Swift No. 1 (Washington)	204
Cabinet Gorge (Idaho)	200
Saluda (South Carolina)	198
Oxbow (Oregon)	190
White Rock (California)	190
Caribou No. 1 and No. 2 (California)	185
Gaston (North Carolina)	178
Lay Dam (Alabama)	177
Osage (Missouri)	172
Kent (Montana)	168
Lewis Smith (Alabama)	158
Keowee (South Carolina)	158
Martin Dam (Alabama)	155

attention has been given to their overall environmental impact. Hydropower facilities, once looked upon very positively by the lay public because of their concurrent creation of boating and recreational facilities, no longer enjoy universal favor. The complexities of obtaining approval for facilities now extend over months and years of environmental studies and litigation in the courts, making it easier and faster to turn to fossil-fuel energy sources. See entry on **Perches and Darters.**

As of the early 1980s, approximately 45% of the hydroelectric facilities are located in the Pacific Northwest and California: Washington (21%), California (14%), and Oregon (10%). The Mountain Region accounts for nearly 12% of the nation's total hydropower facilities, with Arizona, Montana, and Idaho in the lead. The Middle Atlantic States, mainly New York and Pennsylvania, account for about 10.5% of total hydropower production. These are followed by South Atlantic states (9.7%); East South Central states (9.3%); West North Central states (5.5%); West South Central states (4.0%); New England states (2.6%); and East North Central states (1.5%). Because of the large investment required as well as legal problems, sometimes involving

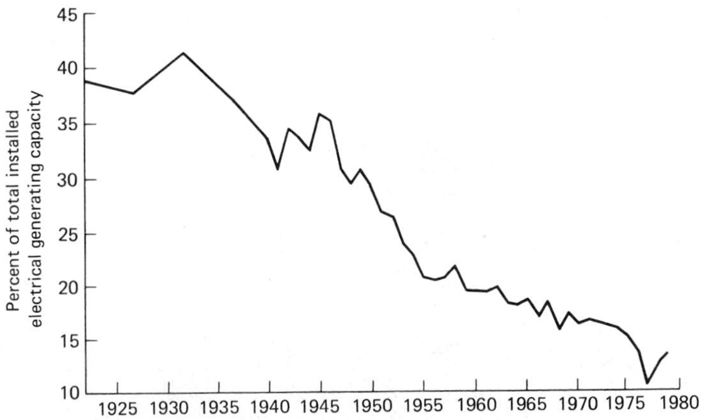

Fig. 1. Trend of hydroelectric power as a percentage of the total installed electric generating capacity in the United States.

TABLE 3. RELATIVE ADVANTAGES AND LIMITATIONS OF HYDROELECTRIC POWER INSTALLATIONS

ADVANTAGES

- Continuous low-cost power production except when droughts occur.
- Low maintenance costs.
- No consumption of irreplaceable fossil fuel.
- No air pollution.
- Reservoir lakes can be used for recreation in majority, but not in all cases.
- Reservoirs can provide considerable, but not complete flood protection to downstream areas.
- Reservoirs are capable of storing large quantities of water for long periods of time, but not indefinitely.
- Downstream flow can be managed to aid in water-quality control and to level out the extremes of winter versus summer stream conditions.
- Ground-water reserves are increased by recharging from the reservoir.

LIMITATIONS

- High initial cost of construction.
- Recreational facilities can be adversely affected in reservoirs where drawdown in the dry season lowers the water level.
- Flood protection can best be provided by an *empty* reservoir, while power production is best from a *full* reservoir. A full reservoir cannot retain a major flood; an empty reservoir generates no power. The compromise then, is to retain enough water in a reservoir to insure continuous power generation, but leave a margin of free board to take the major surges out of a sudden torrential rain storm.
- Loss of land suitable for agricuiture.
- Power production may be curtailed or even discontinued in time of drought.
- Original stream valley is inundated.
- Some water is lost by evaporation from the reservoir surface.
- In coastal areas, such as Oregon and Washington, the construction of dams prohibits the upstream migration of anadromous fish, such as the Pacific Salmon, unless some arrangement, such as a "fish ladder" is provided.

SOURCE: Battelle Memorial Institute, Columbus, Ohio.

the rights and interests of more than one region or state, the majority of hydroelectric generation has been essentially government-sponsored. Traditionally, investor-owned facilities have represented only about one-third of the total.

Relative Advantages and Limitations of Hydroelectric Power. These are listed in Table 3.* Some of the obstacles which large hydroelectric projects usually encounter include: (1) high initial costs of a major dam(s), including usually one main structure, but often requiring smaller dams to complete the reservoir complex; (2) at least to some extent, creation of a large reservoir requires relocation of highways, railroads, and power lines and, in some instances, of small communities; (3) unless located in wilderness areas, numerous home and farm sites must be moved; (4) very long delivery time for major equipment, such as turbines; (5) frequently, the abandonment of large areas of farmland; and (6) increasing pressures by special interest groups.

There are two forms of underdeveloped hydropotential, not only in the United States, but throughout the world: (1) areas of large hydropotential where no dams have yet been built; and (2) where dams have been constructed, but where full generating potential has not been fully exploited. This latter factor is clearly reflected in Table 1. Even though there are obstacles, the further development of hydroelectric power represents a sound approach whereby, within a period of 20 years or less, the hydroelectric generating capacity of the United States could be doubled—and essentially with no new technology required, as is true of most other proposals for additional energy.

It is interesting to note that although 45% of the present hydroelectric generating capacity in the United States is located in the Pacific region, there are other important hydro reservoirs in the eastern por-

* Information in this and the four following paragraphs is based essentially upon observations made by Battelle Memorial Institute, Columbus, Ohio.

tion of the nation. The Allegheny River watershed is an example. This important stream drains much of southwestern New York and northwestern Pennsylvania and, with the Monongahela River, it forms the Ohio River. The region drained by the Allegheny lies in the snowbelt at the eastern end of Lake Erie. It is in this watershed that the snow melt-water, often augmented by early spring rains, produces the surges of high water which numerous times have flooded the cities of Pittsburgh, Wheeling, and Cincinnati. Part of the Allegheny watershed is already protected by dams, but part of it, a major tributary, the Clarion River, has a substantial, unused hydroelectric potential.

Also, the upper reaches of the Susquehanna have not been dammed. Some of the disastrous flooding following as aftereffects of east coast hurricanes, might have been avoided had flood-control projects been in existence upstream from Harrisburg and Wilkes-Barre.

A discussion of hydroelectric projects should not overlook the water-supply situation. It is well established that water consumption is increasing and, as in the case of southern California, when consumption exceeds local supply, water must be imported. Projects comparable to delivery of water to southern California from the north will inevitably be constructed in other parts of the nation and world. Great tunnel systems will be cut through mountain ranges and streams will be diverted, in some cases to flow in the opposite direction.

In British Columbia, for example, the Nechako River, draining the high snow fields of the Coast Range and originally flowing east, has been dammed. The water now flows west through a tunnel, then drops several hundred feet to sea level, generating thousands of kilowatts, before flowing into the Pacific Ocean. The concept of drainage reversal to gain a hydroelectric advantage has been used successfully elsewhere, but not in the United States. A possible site for such a plant could be considered for western New York, the "panhandle" of Pennsylvania, and northeastern Ohio. Through a series of interconnecting reservoirs, some already in existence, water which would normally flow into the Allegheny and Ohio Rivers could be rerouted northwestward over a divide and down into Lake Erie, thereby creating a hydraulic "head" of some 700 feet (213 meters). Much-needed water-storage facilities could result from such a project, plus added flood control on both the Allegheny and Ohio Rivers.

Classification of Hydroelectric Plants

Hydroelectric plants can be classified by:

1. *Extent of impounded volume*
   (a) Storage plants
   (b) Run-of-river plants
2. *Status of the facility in the total power system*
   (a) Peak load plant
   (b) Base load plant
   (c) Isolated plant

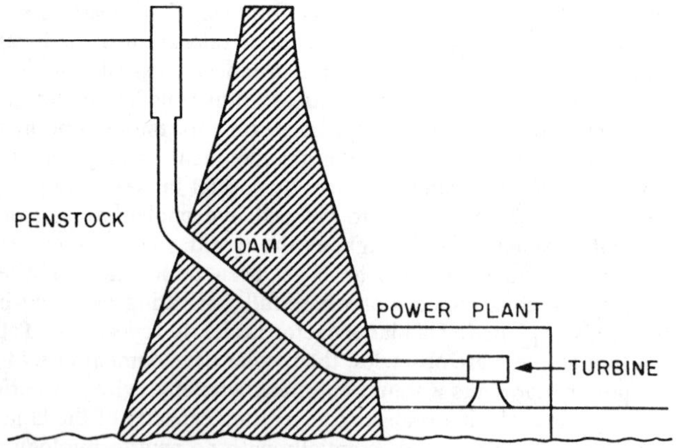

Fig. 2. River plant (high head). Typified by the Hoover and Niagara installations.

3. *Head*
   (a) High-head development (Fig. 2)
   (b) Medium-head development
   (c) Low-head development (Fig. 3)

See also Fig. 4.

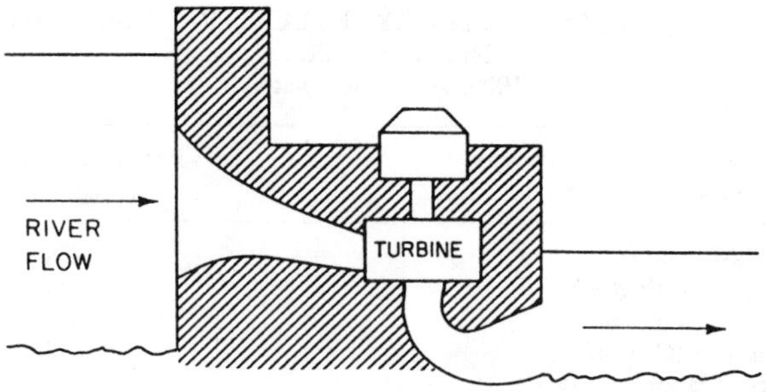

Fig. 3. River plant (low head). Typified by the Bonneville and St. Lawrence projects.

The low-head plant has a characteristic design differing in all essentials from the high-head plant. The medium-head plant may partake of the characteristics of either the high- or low-head plants as its working head approaches either the high- or low-head range. There is no definite line of demarcation between high, medium, and low heads; however, a head of more than 500 feet (152 meters) can be considered a high-head development, and one lower than 50 feet (15 meters) a low-head development. Briefly, the characteristics of the low-head plant are: vertical, reaction type, runners using large volumes of water and requiring large water passages. Substructure is both extensive and expensive, and intake works are large and complicated. Large diameter generators are made necessary by the low rotational speeds. Characteristics of the high-head plant are: horizontal impulse turbines, small volumes of water at high pressures, plant at some distance from the dam. The advantage of smaller and simpler substructure is offset by the presence of a long water conduit, or penstock, between dam and plant. The turbines are high-speed and allow smaller generator diameter. The high-speed is accounted for by the high heads used. Inherently, the impulse turbine has a low characteristic speed.

The possible hydroelectric development sites along the flow of a stream are of two types, namely, those suitable for run-of-the-river plants and those offering natural impounding basins for storage plants. In general, the run-of-the-river plant is cheaper than the storage plant of equal capacity, but it suffers seasonal variation of output more or less proportional to the variation of stream flow.

Storage plants give a greater proportion of firm power which can be delivered day by day on a regular schedule. This firm power is in more or less direct ratio to the degree of regulation of the flow of the stream and this in turn is a function of the impounded volume. Complete regulation of stream flow is rarely possible or practical, although 80–90% regulation is not infrequent.

In any storage plant, the theoretical energy or power available over and beyond the firm power developed is known as flash power or flood peak power. Firm power commands, commercially, a considerably higher rate than flash power.

When integrated with steam generating plants, hydroplants are frequently used to give peak power outputs to take care of peak load conditions and thus avoid the expensive standby service of additional steam generating equipment. Such service, of course, may still permit the delivery of a certain amount of firm power.

If all the run-of-river plants were located upstream from the storage plants they would be operated continuously on a base load plan, be-

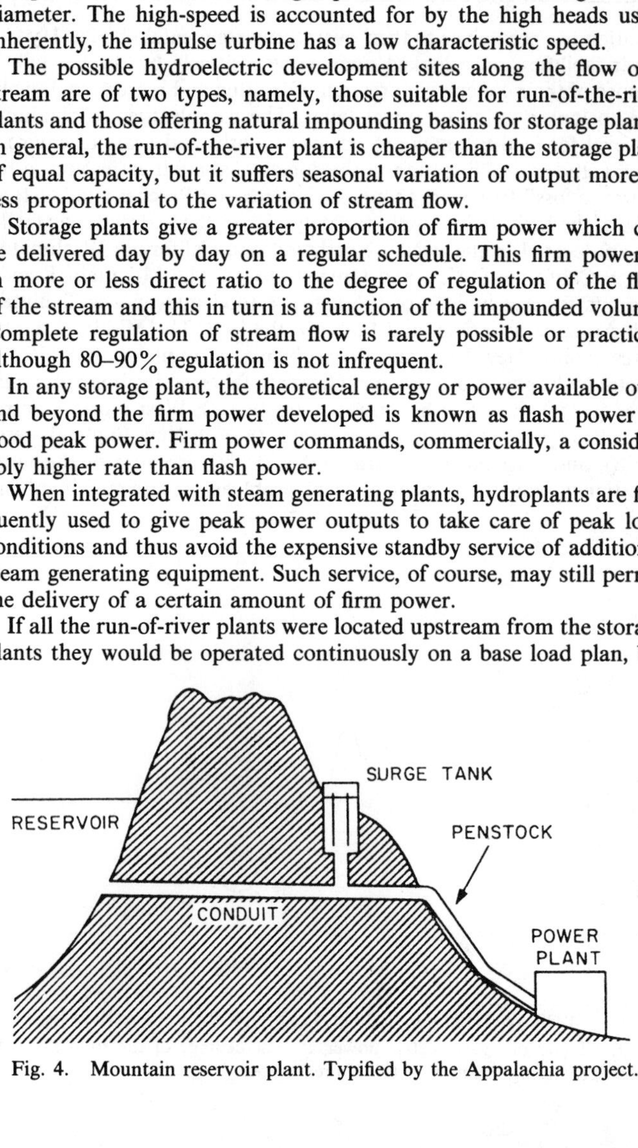

Fig. 4. Mountain reservoir plant. Typified by the Appalachia project.

cause, were they idle, their small reservoirs would quickly overflow and water would be wasted over the crest gates. If, however, they are located between storage plants, the run of the river, as far as they are concerned, is just what the storage plants are passing on to them. Thus, located downstream from a storage plant, a run-of-river plant will produce an increase in output when the storage plant increases its output.

In the hydroelectric plant the turbines and generators are the main items of equipment. The hydroelectric superstructure, as usually laid out, has one large building housing the main units and an electrical bay, or wing, of one or more stories in which are located the switching equipment, offices, storerooms, and most of the auxiliary equipment.

Hydro sites that are developed to use but part of the normal stream flow are exceptions to the general rule. Only rarely is a development made where conservation of the water and its use in the most efficient manner are not paramount features of operation. Failure to give due cognizance to this feature may wipe out the net operating profit; hence a continuous, watchful scrutiny of all natural factors which can affect the station operation is a duty of the operating personnel.

A hydraulic turbine suffers loss of efficiency at heads above or below the designed value because of shock losses. At the correct head there will be one point of best efficiency, somewhere between 80–95% of full load. When a number of units are installed in a plant, and when steam reserve is available, it is generally possible to operate the units near the point of best efficiency. There are four faults of operation and maintenance which can reduce the maximum energy production of a plant:

1. Waste of water over spillways.
2. Improper distribution of the load between the station units.
3. Water leakage through valves, gates, dam or flow line.
4. Wear on moving parts, especially corrosion or erosion of the runner.

The relative simplicity of hydroelectric equipment makes hydraulic efficiency of the turbine the principal consideration.

Hydraulic Turbines

The fundamentals of the turbine were incorporated into the wheels built before the turn of the nineteenth century, but its principal development has occurred since that time. Beginning with Fourenyon and his outward flow turbine, Jonval, Boyden, Swain, and Francis rapidly brought the reaction turbine to an advanced stage of development. By 1875 the inward flow turbine, as perfected by Francis, and which now bears his name, had established itself in the lead, a position which it maintained until about 1900, when the impulse, or Pelton, type of wheel had progressed to the point of dominating the high head field.

The inherent slow speed of the Francis-type runner on low heads was a fault that the propeller-type runner was designed to cure. During the decade 1910–1920 progress was made with this type of wheel, and by 1920 the propeller-type runner, often called the Nagler runner, was definitely established in the hydroelectric field. Later it was arranged so that the blades could be adjusted and set at different angles to accommodate changes in elevation of the forebay level without undue loss of efficiency. The success of the propeller-type turbine encouraged American adoption of the Kaplan turbine, on which the blade adjustment is performed automatically, being under the same control as the turbine gates.

As between impulse and reaction types, the action in the impulse turbine is easiest to understand. There is no difficulty in visualizing the transformation of pressure head into velocity head at the nozzle, nor of understanding the push, or impulse, that is given to the buckets by the stream of water. The jet is directed upon the rotor tangentially, and hence this type is also called the tangential turbine. The velocity of the jet of water is only slightly less than the free spouting velocity under the effective head *h*. Impulse buckets are divided into two halves by a "splitter" and the axial thrusts which would otherwise have to be borne by special bearings are equalized.

The essential difference between the impulse and reaction types is that in the former the entire energy received by the wheel is in the velocity form, while in the latter it may be partially in the velocity form, but is also, in a large measure, still in the pressure form. The reaction of conversion of residual pressure into velocity in the runner is the source of much of the torque delivered to the reaction turbine. If the turbine were blocked stationary and had its gates opened, the water would issue from the turbine as from a nozzle. Now, by removing the blocking, let these nozzles begin to rotate, and the absolute velocity of water leaving them is found to be diminishing, the energy having been absorbed by the runner. At the best speed the final velocity will be just sufficient to enable the water to clear the runner. At this time, the wheel may be absorbing from 90–95% of the energy that the water had in the pressure form just before reaching the turbine gates.

In a Francis turbine, the water flows inward, then downward and into the draft tube. See Fig. 5.

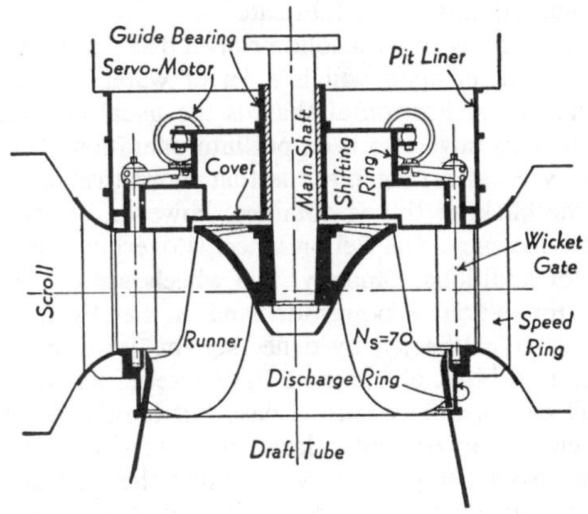

Fig. 5. Cross section showing component parts of a Francis turbine.

A convenient classification of hydraulic turbines is:

1. *Reaction Turbine* (Water under pressure is only partially converted into velocity before it enters the turbine runner.)
   (a) Francis Turbine
   (b) Propeller Turbine
   —1 Fixed-blade
   —2 Adjustable-blade (*Example*: Kaplan turbine)
   —3 Axial-flow (*Example*: Dariaz turbine)
   —4 Diagonal-flow
2. *Impulse Turbine* (*Example*: Pelton Wheel) (Water under pressure is entirely converted into velocity before it enters the turbine runner.)

The Francis turbine is rarely a horizontal shaft machine, except in small sizes and where it is desired to avoid the expense of excavation for a vertical setting. The standard runner consists of two crowns between which the buckets or blades are placed. It is best adapted to vertical setting. In order to pass the large discharges possible in a high specific speed wheel, the buckets are curved downward. Some axial flow action is present in runners of high specific speed. Water is admitted to the runner through guide vanes and gates.

Loss of efficiency at part load is sometimes a serious fault as, for instance, where only one or two units are installed in an isolated plant. The feature of the Kaplan turbine is that the blade angles and gates are adjusted simultaneously by the governor mechanism so that the blades are always in the position best suited for full utilization of the flow, through the reduction of eddying and shock losses. The result is that the efficiency at part load holds up remarkably well.

Conveying the water from the penstock and directing the proper amount of it correctly against the runner requires first, a scroll case; second, a speed ring; and third, turbine gates.

The scroll case for medium and high head development is circular in form. In plan, it leads from the penstock and wraps, in spiral form, around the speed ring. The cross section of the spiral at any point should be such that the water flows with uniform velocity. This leads to the spiral form, because the water is being delivered to the turbine uniformly around the entire circumference.

The speed ring is that part of the turbine which joins the discharge ring with the turbine cover and pit liner. The ribs between the top and bottom portions must be strong enough to support the dead weight above the casing, consisting of concrete, generator, and turbine rotative parts; hence the speed ring is a very important part of the turbine.

Inside the speed ring, and rigidly bolted to it, is the inlet gate mechanism. The mechanism is operated by the governor which, by opening or closing the gates, can maintain a control of speed under variable load. Gates are of the guide vane type and, while various types of gates have been used, the wicket gate is in general use at the present time. Its principal advantage is its efficiency. Shock losses at part gate opening are reduced to a minimum in the wicket gate. It is not particularly tight and has many wearing parts, most of which are bronze bushed and grease lubricated.

The Pelton wheel is either a solid or open disk, to the rim of which are attached buckets upon which a jet of water is played from a stationary nozzle. A horizontal shaft is the usual arrangement, but vertical shaft units have also been put into operation. The advantage of using the vertical arrangement is that more than one jet can be played on the buckets; this is obtained, however, at the expense of some loss of efficiency. The Pelton wheel is overhung on the bearing and often, for additional capacity, two wheels are overhung on the same generator. Variable power demand is met by decreasing the amount of water in the jet, by deflecting the jet from the buckets, or both. Some turbines of this type have a relief jet which opens as the main jet closes. Afterwards, a dash pot slowly closes the relief jet, slowly enough to prevent a large pressure rise in the penstock. The same is also accomplished by deflecting the jet from the wheel upon loss of load, then slowly closing the valve controlling the jet.

Draft Tube. Hydraulic turbines frequently discharge the water with considerably more velocity than would be economical from the efficiency viewpoint, were it not possible to recover a great deal of that energy by the proper use of a diffusing chamber at the outlet. The diffusing chamber or tube is known as the draft tube, and there are a variety of types. However, the main objective is to convert the velocity head residing in the water leaving the turbine into pressure head. If this can be done efficiently, the turbine can be set somewhat below normal tailwater level.

The greater the specific speed of a turbine runner the higher will be the velocity of the water discharged into the draft tube, and the more important the recovery of this velocity by draft tube design.

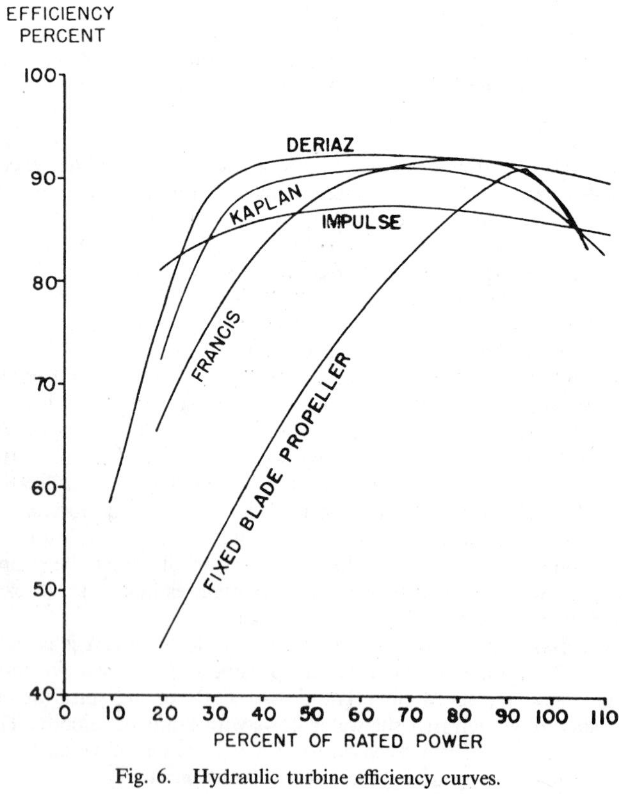

Fig. 6. Hydraulic turbine efficiency curves.

The draft tube is to take the water from the turbine at a point where the pressure is considerably less than atmospheric, and, by efficiently reducing the velocity, convert it into pressure head so that it can emerge smoothly into the tailrace at atmospheric pressure. By "efficiently" is meant without shock or whirl loss. Not all the velocity head can be recovered, for the water must be given to the tailrace at normal tailrace velocity to prevent its backing up into the turbine. Also, whatever friction loss occurs in the draft tube adds to this reduction of useful head.

Typical turbine efficiency curves are shown in Fig. 6. A schematic diagram of a typical hydro governor system is shown in Fig. 7.

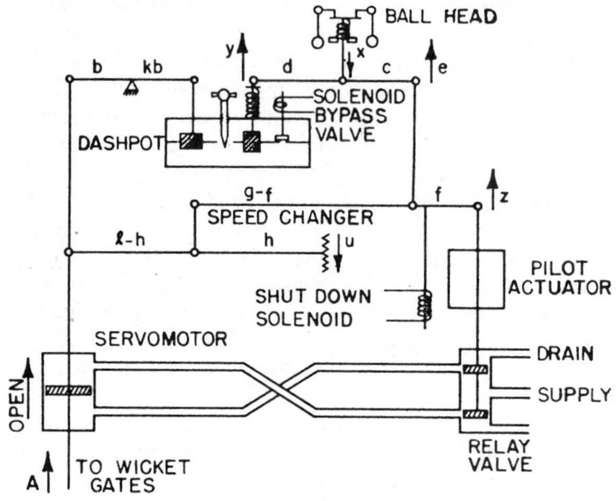

Fig. 7. Typical hydro governor system.

Pumped-Storage Plants

Growing emphasis over the past couple of decades has been placed upon the use of special hydro plants as a means of storing energy in the form of a head of water—pumped into an upper reservoir during offpeak hours. The history of pumped-storage plants dates back to the late 1920s when several plants were first installed in Europe. The Rocky River plant of Connecticut Light and Power Company was the first to be built in the United States during that early period. From the viewpoint of plant location, there are three categories of pumped-storage installations:

1. *Combined with conventional hydro plant.* Plants of this type are used in locations suitable for conventional hydro plants, but where rainfall or water availability and system demand are out of phase. For example, in Switzerland, demand is highest in winter when water is scarce and lowest in summer where there is an abundance of water. The available energy in summer can be used to fill the reservoirs and store the available water for later use in meeting the winter demand. A situation similar to this exists at Niagara where the cycle time is one day rather than one year.

2. *Pure pumped storage.* The advantages of this type of plant are its flexibility of location, in that the upper reservoir need have no source of water other than what is pumped into it, and the possibility of developing large plants with a small reservoir and high head. This type of plant is commonly used in steam-based systems which lack the many advantages of available hydro generation. As well as providing fast and reliable peaking power, pumped-storage units have the added advantage of smoothing the weekly load curve and enabling more of the efficient base-loaded steam plants to be operated continuously.

3. *Pumped storage with diversion.* This situation arises when available water must be shared between power generation and irrigation use. Water which must be pumped to a higher reservoir to feed an irrigation canal can be used as a source of peaking power if allowed to run back down through a pump-turbine.

Design Configurations. The three fundamental configurations include: (1) separate pump and turbine on the same shaft; (2) reversible pump-turbines; and (3) axial-flow units.

The turbines used in the first class are Francis type for heads in the range 100 to 1,000 feet (30.5 to 305 meters) and Pelton wheels for heads up to 3,000 feet (914 meters). These units are usually mounted on a horizontal shaft with a clutch (hydraulic or friction) between the motor-generator and the pump. The turbine is usually rigidly connected to the shaft and is dewatered, using compressed air during pumping. Sometimes small impulse turbines are installed on the shaft for starting and braking.

Reversible pump-turbines are of radial or mixed flow type—Francis or Deriaz—and have been designed to operate at a wide range of heads.

Axial-flow units are designed to operate at low heads (around 20 feet (6 meters) and have adjustable blades similar to those of a Kaplan turbine. The bulb type is used in Europe and the tube type has been designed and built in the United States. Both can operate as turbines or as pumps in both directions of flow by reversing the pitch of the blades. A 9-megawatt bulb unit was installed at Saint Malo, France, and in the Rance tidal project near Saint Malo, there are twenty-four 10-megawatt bulb units. These are described in entry on **Tidal Power.**

References

In addition to specific references indicated below, statistics on hydroelectric power are continuously updated and available from such organizations as: U.S. Federal Power Commission, U.S. Army Corps of Engineers, Bureau of Reclamation, U.S. Department of the Interior—all in Washington, D.C. Also, Edison Electric Institute, New York; and Electric Power Research Institute, Palo Alto, California.

Baumeister, T. (editor): "Marks' Standard Handbook for Mechanical Engineers," 8th edition, McGraw-Hill, New York, 1978.

Considine, D. M. (editor): "Energy Technology Handbook," McGraw-Hill, New York, 1977.

Fink, D. G., and J. M. Carroll (editors): "Standard Handbook for Electrical Engineers," 11th edition, McGraw-Hill, New York, 1979.

Loftness, R. L.: "Energy Handbook," Van Nostrand Reinhold, New York, 1978.

Parker, A.: "Planning and Estimating Dam Construction," McGraw-Hill, New York, 1971.

HYDROFLUORIC ACID. HF (hydrogen fluoride gas) in aqueous solution, colorless when pure, fuming (dependent on concentration), highly corrosive, extremely reactive, available commercially in 30, 52, 60, and 80% HF concentrations. There is a maximum constant boiling point 111°C (750 torr) at 43% HF (distillate) for mixtures of HF and water. Because HF attacks glass and many other container materials, the laboratory HF reagent is packaged in polyethylene bottles or carboys. Larger containers for industrial use usually are steel drums or tanks with a polyethylene lining. Anhydrous HF is available in tank cars of 22- or 42-ton (20- or 38-metric ton) capacity, as well as in steel cylinders of 100- or 200-pound (45- or 90-kilogram) capacity.

The formula weight of HF is 20.01 (calculated). However, its apparent molecular weight ranges widely with temperature and pressure. The molecular weight of saturated HF vapor at 19.51°C is 78.24; at 100°C, the value is 49.08. Because of strong hydrogen bonding between molecules, significant polymerization occurs, thus resulting in marked departures from ideal behavior, both in the gaseous and liquid phases. The polymerization mechanism has not been fully determined, but both ring- and chain-type structures have been suggested. Of interest is the comparison of the high boiling point of HF (+19.5°C) with the boiling points of other acids in the halogen series: HCl, −85°C; HBr, −65°C; HI, −36°C. Because of this polymerization, the formula H_2F_2 often has been used for hydrogen fluoride, although the polymers in the gas appear to be chiefly $(HF)_6$. Evidence of the stability of these hydrogen bonds is furnished by the existence of the hydrogen fluoride ion (HF_2^-) in ionic crystals and acid fluoride solutions.

Liquid hydrogen fluoride is one of the three binary hydrides (the others are H_2O and NH_3) which are self-ionized and highly associated (Trouton constant 26.6, bp 19.54°C, mp −83.7°C). It has a dielectric constant of 83.6 at 0°C and is an excellent ionizing solvent. Because of its very high acidity, most oxygen-containing substances are protonated in solution in HF, forming substituted oxonium ions or oxonium ion itself by solvolysis. It has a very low viscosity—0.256 centipoises at 0°C. Its surface tension is also exceptionally low. Its density at the boiling point is 0.991 g/ml.

The effects of hydrogen fluoride on glass were observed by A. S. Marggraf in 1764. In 1771, Scheele established that a new acid had been discovered. In 1814, Davy showed that the acid contained a newly found element, fluorine. Fluorine was not isolated until 1886 by Moissan. The first anhydrous HF was not prepared until 1856 by Frémy. The first commercial shipment of HF (anhydrous acid) was not made until 1936.

Because of the strong affinity of HF for H_2O, there is no known chemical substance that can be used for drying it. HF immediately reacts or complexes with the drying agents. Thus, the compound can be dehydrated only by electrolysis. Even though HF is highly reactive, it has the characteristics of a weak acid, due to the extensive polymerization of the HF.

Anhydrous hydrogen fluoride is prepared commercially by reacting calcium fluoride (acid-grade fluorspar) with concentrated H_2SO_4 in a heated reactor. The presence of silica in the calcium fluoride is highly objectionable inasmuch as each pound of silica present will consume 2.6 lb of CaF_2 to form silicon tetrafluoride. When the latter compound is absorbed with HF in H_2O, fluosilicic acid is formed, representing a further loss of net HF produced: (1) $CaF_2 + H_2SO_4 \rightarrow CaSO_4 + 2HF$; (2) $4HF + SiO_2 \rightarrow SiF_4 + 2H_2O$; (3) $2HF + SiF_4 \rightarrow H_2SiF_6$. After reacting at a temperature of 200–250°C, the HF is treated to remove dust and H_2SO_4 fumes and then condensed as 99% HF.

Uses: Principal uses for HF include: (1) the production of aluminum fluoride and synthetic cryolite required for aluminum production, (2) the production of fluorinated organics of several types and for several applications, including aerosol propellants, special-purpose solvents, refrigerants, and plastics (polytetrafluoroethylene, polyvinylidene fluoride, polychlorotrifluoroethylene), (3) in the formulation of atomic-energy feed materials, (4) as an alkylation catalyst in petroleum processing, (5) as a pickling acid in stainless-steel and nonferrous metals manufacture, (6) as an agent for etching and polishing glass, (7) as a reactant in several organic syntheses, (8) in the manufacture of elemental fluorine, and (9) as a starting material for the preparation of fluorides and fluoborates.

Toxicity and Hazards: All forms of HF, liquid and gas, are very corrosive to mucous membranes, lungs, skin, eyes. Exposure results in very painful burns and often deep-seated ulceration. All manner of protective clothing and equipment must be used. All users of HF should study the first or equivalent references given below.

References

Hydrofluoric Acid (Anhydrous and Aqueous), Chemical Safety Data Sheet SD-25, Manufacturing Chemists Association, Inc., Washington, D.C., updated periodically.

Kautsch, Oscar: "Flussaure, Kieselflussaure und deren Metallsalze," *Enke's Bibliothek für Chemie und Technik*, vol. 24, Enke, Stuttgart, 1936.

Methods of Analysis for Anhydrous Hydrofluoric Acid, Procedures Recommended by Manufacturing Chemists Association, *Ind. Eng. Chem., Anal. Ed.*, **16**, 483–486, 1944.

Sax, N. I.: "Dangerous Properties of Industrial Materials," Van Nostrand Reinhold, New York, 1979.

Simons, J. H.: "Fluorine Chemistry," vol. 1, pp. 225–259, Academic Press, New York, 1950.

HYDROFOIL SHIP. Seagoing vessels equipped with wing-like transverse surfaces suspended below the hull. As the craft moves forward, hydrodynamic forces are produced, just as aerodynamic forces are produced in air, to give lift. At the start, the vessel operates as a normal displacement vessel, but as the speed increases the lift on the hydrofoils increases to raise the craft out of the water, thereby decreasing the water resistance to that of the hydrofoils alone. Higher speeds can be obtained, since the water resistance for the required lift is considerably less for the gear still submerged in the water. To be most effective, propulsive units, such as engine-propeller combinations, or jet engines operating well above the water, are necessary.

Hydrofoil craft are of a general class of designs collectively referred to as *interface vehicles.*

HYDROFORMULATION. Organic Chemistry.

HYDROGASIFICATION (Coal). Coal.

HYDROGASIFICATION (SNG). Substitute Natural Gas (SNG).

HYDROGEN. Chemical element symbol H, at. no. 1, at. wt. 1.0080, periodic table group la, mp −259.14°C, bp −252.87°C, density 0.089 (solid at 4.2K), 0.071 (liquid at 20.4K), sp gr 0.0696 (air = 1.0000). Solid hydrogen has a hexagonal crystal structure. Hydrogen at standard conditions is a colorless, odorless, tasteless gas, suffocating, but not toxic. Hydrogen occurs chiefly combined with oxygen in H_2O, with carbon in hydrocarbons, with carbon and oxygen, and with carbon and several other elements, including oxygen, nitrogen, sulfur, phosphorus, and most metals in a vast variety of hundreds of thousands of organic compounds. See also **Organic Chemistry.** Hydrogen is considered by some scientists as the primordial substance from which all other elements in the universe were developed. In terms of cosmic abundance, with a rating of silicon = 10,000, it has been estimated that the figure for hydrogen is about 3.5×10^8, this figure compared with that of carbon = 80,000, nitrogen = 160,000, and oxygen = 220,000. For further comparison, the figure for gold is 0.0015 and for uranium it is 0.0002. In terms of abundance of the chemical elements in seawater, hydrogen ranks second (behind oxygen) with an estimated 510 million tons per cubic mile (~109 million metric tons per cubic kilometer). Hydrogen ranks eleventh in terms of content in igneous rocks in the earth's crust, the estimate of average content being 0.13%. Although free hydrogen escaped from the earth's lower atmosphere, some of the planets appear to have significant amounts, including the atmospheres of Jupiter, Saturn, and Uranus. At an altitude of 1,000 miles (1609 kilometers) above the surface of the earth, there is a greater abundance of hydrogen atoms than of nitrogen or oxygen atoms.

Hydrogen was first identified by Cavendish in 1766. The element was named by Lavoisier in 1783. However, it was not until 1931 that a second isotope of hydrogen (deuterium) with a mass number 2 was discovered by Urey. In 1934, Rutherford, Oliphant, and Harteck prepared a third isotope (tritium) with a mass number 3. Normal hydrogen (protium) and deuterium are stable, whereas tritium is radioactive, with a half-life of 12.26 years. Tritium emits a negative electron to form ^3He. It is estimated that the isotopic abundance of ^1H (protium) in natural occurring hydrogen is 99.9851% and on the basis of carbon = 12 (atomic weight scale), protium has a mass of 1.007825 amu. The isotopic abundance of ^2H (deuterium) is estimated at 0.0149% with a mass of 2.014101 amu. The artificially-prepared ^3H (tritium), $^9Be + {}^2H \rightarrow 2\ {}^4He + {}^3H$, has a mass of 3.01605 amu. Heavy water is deuterium oxide, ^2H_2O, usually written D_2O. Deuterium and deuterium oxide gained prominence largely because of their excellent properties as moderators in nuclear reactors. The ionization potential of hydrogen is 13.59765 ± 0.00022 eV. Other physical properties of hydrogen are given under **Chemical Elements.** See also **Deuteron; Deuterium.**

When ignited, hydrogen burns in air with a pale blue to colorless, nonluminous flame, yielding H_2O. When mixed with air, the flammability limit is 4–74% hydrogen. When mixed with oxygen, the flammability limit is 4–94% hydrogen. Care always must be exercised where there may be hydrogen mixtures with air or oxygen because violent explosions may occur. In sunlight or magnesium light, hydrogen combines with chlorine with violent release of energy, forming hydrogen chloride HCl. When hydrogen is heated with sodium, calcium, and several other metals, the corresponding hydride is formed. In the presence of a catalyst, hydrogen reacts with nitrogen to form ammonia NH_3. Upon heating sulfur in the presence of hydrogen, hydrogen sulfide, H_2S, is formed. At elevated temperatures, hydrogen will reduce many of the metal oxides to the metal, notably copper, iron, nickel, tin, and lead. The oxides of zinc, aluminum, and magnesium are not so reduced. Hydrogen reacts with unsaturated organic compounds in most cases to form saturated compounds. For example, in the presence of a catalyst, hydrogen will add to oleic acid $C_{17}H_{33}COOH$ to form stearic acid $C_{17}H_{35}COOH$. See also **Hydrogenation.**

Production of Hydrogen: For chemical and petroleum processes, hydrogen is an extremely high-tonnage and one of the most fundamental raw materials. Sources of hydrogen and processes for producing it are described in entry on **Hydrogen (Fuel).**

Uses: In terms of consumption, NH_3 is by far the largest user of hydrogen. Petroleum refining processes and methanol synthesis are the next largest consumers. Hydrogen needs for these uses are almost always fulfilled by hydrogen-generation capacity on the premises. What might be termed commodity hydrogen is shipped from hydrogen plants to various users. Some of the more important uses include the hydrogenation of numerous organic compounds, such as vegetable and animal oils, the oxyhydrogen and atomic-hydrogen welding applications, the reduction of several metallic oxides, such as iron, copper, nickel, cobalt, tungsten, and molybdenum, and the use of liquid hydrogen as a rocket fuel. See also **Ammonia; Hydrogenation; Methyl Alcohol; Petrochemicals;** and **Synthesis Gas.** For the potential role as a fuel, see **Hydrogen (Fuel).**

Ortho- and Para-Hydrogen: On the basis of nuclear spin, two forms of hydrogen are known: *ortho-hydrogen,* in which the two nuclei in the H_2 molecule have parallel spins, and *para-hydrogen,* in which the nuclear spins are anti-parallel. At ordinary temperatures (and above) ortho-hydrogen is present to the extent of about 75%; at lower temperatures, the ortho changes to para-hydrogen, until at very low temperatures, as that of liquid hydrogen, the para form is present to the extent of 99.7%. There is some difference in properties between the two, notably in thermal conductivity.

The transition from ortho- to para-hydrogen releases heat in amount of 168 cal/g. The heat of vaporization of liquid hydrogen is 107 cal/g. Thus, more than ample heat is released to revaporize liquid hydrogen. Knowledge of the existence of the ortho-para transition and the development of catalysts to equilibriate the liquid during liquefaction essentially have made possible the very large-scale manufacture, use, and storage of liquid hydrogen.

Below −220°C the specific heat of hydrogen is that of a monatomic gas like helium (He). Practically pure para-hydrogen may be obtained by adsorption of ordinary hydrogen, which is three-fourths ortho and one-fourth para, on charcoal at about −225°C. The mp of para-hydrogen is 0.13°C lower (ortho-hydrogen 0.04°C higher) than ordinary hydrogen, and the bp at 60 mm pressure is 0.13°C lower (ortho-hydrogen 0.04°C higher) than ordinary hydrogen. Para-hydrogen reverts slowly to ordinary hydrogen, but immediately in the presence of platinized asbestos.

Atomic Hydrogen: At high temperatures, the loss of heat from a glowing wire in hydrogen is larger than expected on regular assumptions. This is believed to be due to dissociation of ordinary hydrogen into atomic hydrogen (H). See accompanying table.

DISSOCIATION OF HYDROGEN

Temperature, °C	Pressure At 760 mm	At 1 mm
1730	0.33%	8.7%
2230	3.1	57.5
2730	34	99.3

When hydrogen is passed through an electric arc between tungsten poles, a considerable transformation into atomic hydrogen occurs, and when a stream of this gas strikes a surface a large evolution of heat takes place through recombination to ordinary hydrogen. This atomic hydrogen flame is of temperature sufficiently high to melt tungsten (mp 3,370°C). The half-life of the hydrogen atom is one-third second at 0.5 mm pressure. This reaction is endothermic, values of 98–105 kcal per mole having been reported for it. It is an active reducing agent, reducing many metallic oxides and halides to the free metals, and forming hydrides with many nonmetals. The energy of its exothermic recombination is utilized, in combination with the energy released by the oxidation of the H_2 formed, by atmospheric oxygen, in the oxyhydrogen welding process.

Ionization: The ionization potential of hydrogen is 13.59765 ± .00022 eV, and the ionization process (in the case of protium) yields an electron and a free proton. The electric field of the proton is strong, due to its small radius, so that it readily combines with polarizable atoms. Thus, in aqueous solution, it shares an unshared pair of electrons of the oxygen atom of H_2O to form H_3O^+, the hydronium ion; with NH_3 it forms NH_4^+, the ammonium ion; with phosphine it forms the phosphonium ion, PH_4^+, etc. The hydrogen atom can also add an electron, to form the hydride anion, H^-, this potential (electron affinity) being only about 0.7 eV. Hydride ions have been shown (by electrolysis, crystal structure, etc.) to exist in the hydrides which hydrogen forms with the alkali metals and some of the other metals on the left side of the periodic table. While most other hydrogen compounds are essentially covalent, the binary compounds with the halogens and some of the other elements on the right side of the periodic table exhibit a considerable degree of ionicity, varying considerably in the same group.

The hydrogen atoms in many compounds tend to be shared between the electronegative atom or group to which they are attached and similar groups on other molecules. These hydrogen bonds increase the intermolecular forces and boiling points of hydrogen fluoride, water, organic acids and alcohols, etc. A descriptive explanation of the process is the positive polarity of the H atom that is attached to the electronegative atom or group, which gives it an effective coordination number of 2, so that it can attract an unshared electron pair of a fluorine, oxygen, nitrogen, atom of another molecule. The atom having the unshared pair must be negatively polarized or easily polarizable. For example, tertiary arsines form stronger hydrogen bonds with phenols than do tertiary phosphines.

A number of hydrogen compounds ionize to yield solvated protons, i.e., $2H_2O \leftrightarrows OH_3^+ + OH^-$, and $2NH_3$ (liq.) $\rightleftarrows NH_4^+ + NH_2^-$. Moreover, many hydrogen compounds, when dissolved in such solvents, ionize more or less completely to give solvated protons and anions. In the case of polybasic acids, ionization constants are reported for each step in this dissociation.

Hydrides: See section on hydrides in entry on **Hydrogen (Fuel);** and separate entry on **Hydride.**

Water and Acids: The properties of the most prevailing hydrogen-bearing compound, water, are given under **Water.** The characteristics of acids are attributed essentially to the presence of hydrogen ions. These topics are treated under **Acids and Bases;** and **pH (Hydrogen Ion Concentration).**

For references see list at end of entry on **Hydrogen (Fuels).**

HYDROGEN (Ammonia). Ammonia.

HYDROGENATION. In its simplest interpretation, to hydrogenate is to add hydrogen. There are scores of examples where hydrogenation is used as a unit process throughout the chemical and process industries. Generally, the process is associated with relatively high pressure, elevated temperature, and the presence of a catalyst.

Nickel, prepared in finely divided form by reduction of nickel oxide in a stream of hydrogen gas at about 300°C, was introduced by Sabatier (1897) as a catalyst for the reaction of hydrogen with unsaturated organic substances to be conducted at about 175°C. Nickel proved to be one of the most successful catalysts for such reactions. The unsaturated organic substances that are hydrogenated are usually those containing a double bond, but those containing a triple bond also may be hydrogenated. Platinum black, palladium black, copper metal, copper oxide (Adkin catalyst), nickel oxide, aluminum, and other materials have subsequently been developed as hydrogenation catalysts. Temperatures and pressures have been increased in many instances to improve yields of desired product. The hydrogenation of methyl ester to fatty alcohol and methanol, for example, occurs at about 3,000 psig (204 atmospheres) and 290–315°C. In the hydrotreating of liquid hydrocarbon fuels to improve quality, the reaction may take place in fixed-bed reactors at pressures ranging from 100 to 3,000 (7 to 204 atmospheres) psig. Many hydrogenation processes are of a proprietary nature, with numerous combinations of catalysts, temperature, and pressure possible.

Among the better known products of hydrogenation are hydrogenated vegetable and fish oils which may be hardened or solidified by catalytic hydrogenation. Some of these oils can be partially hydrogenated to clarify and deodorize them. Fatty oils, such as oleic acid, may be converted into stearic acid by hydrogenation. Through hydrogenation, peanut oil, cottonseed oil, and coconut oil can be converted to materials that taste, appear, and smell like lard; or by varying the process, they can be made to resemble tallow. Most synthetic shortenings are comprised of hydrogenated oils. Usually, hydrogenated oils will have higher melting points and lower iodine values than the natural untreated oils.

Hydrogenation of Coal and Crudes. The interest in hydrogenation has been greatly intensified since the early- and mid-1970s in connection with the synthesis of new types of fuels to augment the world energy supplies. Basically, however, the hydrogenation of coal is not a new concept, but dates back at least a half-century to the time when manufactured gas (artificial, illuminating, producer, water gas, etc.) was used prior to the more general availability of low-cost, cleaner natural gas. In 1927, a White Paper was published discussing the processes then available for production of oil from coal. One of the first large-scale applications of the Fischer-Tropsch process for the production of oil from coal was that of the South African Coal, Oil and Gas Corporation's plant in Sasolburg. Republic of South Africa, constructed in the mid-1950s and expanded and improved several times during the interim.

Similarly, sour crudes, heavy residuums, and other petroleum-base starting materials can be hydrogenated, sometimes coupled with other processes, to sweeten, reduce viscosity, and otherwise improve the materials for better use as fuels. See also **Coal; Hydrotreating;** and **Petroleum.**

HYDROGENATION (Coal). Coal.

HYDROGENATION (Vegetable Oils). Vegetable Oils (Edible).

HYDROGEN ATOM (Energy Levels). Energy Level.

HYDROGEN BOND. Chemical Elements.

HYDROGEN BROMIDE. Bromine.

HYDROGEN CHLORIDE (Photochemistry). Photochemistry and Photolysis.

HYDROGEN ELECTRODE. pH (Hydrogen Ion Concentration).

HYDROGEN FLUORIDE. Fluorine.

HYDROGEN CYANIDE. HCN, formula weight 27.03, colorless gas with characteristic odor, very poisonous, mp −14°C, bp 26°C, critical temperature 183.5°C, critical pressure 50 atmospheres, density 0.20 g/cm³, sp gr 0.697 (18°C). There are two isomeric forms: (1) HCN which forms cyanides, (2) HNC (inferred from its derivatives) which forms isocyanides. Hydrogen cyanide is soluble in H_2O, or alcohol, or ether in all proportions. The compound usually is marketed as an aqueous solution containing 2–10% (weight) HCN. For many process uses, it is frequently more convenient to generate HCN as needed and thus avoid storage and handling problems. HCN burns with a red-blue flame, yielding CO_2, nitrogen, and H_2O. Aqueous solutions of HCN decompose slowly, yielding ammonium formate: $HCN + 2H_2O \rightarrow HCOONH_4$. Decomposition is slowed by storage in dark locations. Peaches, apricots, bitter almonds, cherries, and plums contain some HCN derivatives in their kernels, frequently in combination with glucose and benzaldehyde as a glucoside (amylgdalin). The bitter almond fragrance of HCN and its derivatives sometimes can be detected in such kernels.

Production: Hydrogen cyanide can be prepared from a mixture of NH_3, methane, and air by partial combustion in the presence of a platinum catalyst:

$$HN_3 + CH_4 + 1.5\,O_2 + 6N_2 \rightarrow HCN + 3H_2O + 6N_2.$$

The process is carried out at about 900–1,000°C; yield ranges from 55–60%. IN another process, methane (contained in natural gas) is reacted with NH_3 over a platinum catalyst at from 1,200–1,300°C, the reaction requiring considerable heat input. In still another process, a mixture of methane and propane is reacted with NH_3: $C_3H_8 + 3NH_3 \rightarrow 3HCN + 7H_2$; or $CH_4 + NH_3 \rightarrow HCN + 3H_2$. An electrically-heated fluidized bed reactor is used. Reaction temperature is approximately 1,510°C.

The high-tonnage uses of HCN are in the preparation of numerous chemical products and intermediates for organic syntheses. As a gas, HCN sometimes is applied as a disinfectant; or cellulosic disks impregnated with HCN may be used. In ore processing and metal treating, cyanides are widely used.

Hydrogen cyanide reacts with hydrogen at 140°C in the presence of a catalyst, e.g., platinum black, to form methyl amine CH_3NH_2; when burned in air, produces a pale violet flame; when heated with dilute sulfuric acid forms formamide $HCONH_2$ and ammonium formate $HCOONH_4$; when exposed to sunlight with chlorine forms cyanogen chloride CNCl, plus hydrogen chloride. An important reaction of hydrogen cyanide is that with aldehydes or ketones, whereby cyanhydrins are formed, e.g., acetaldehyde cyanhydrin $CH_3CHOH \cdot CH$, and the resulting cyanhydrins are readily converted into alpha-hydroxy acids, e.g., alpha-hydroxypropionic acid $CH_3 \cdot CHOH \cdot COOH$.

Metallic cyanides are (1) soluble, e.g., sodium cyanide NaCN, potassium cyanide KCN, calcium cyanide $Ca(CN)_2$, mercuric cyanide $Hg(CN)_2$, aurous cyanide AuCN, (2) insoluble, e.g., silver cyanide AgCN, cuprous cyanide CuCN, (3) complex, (a) decomposed by dilute H_2SO_4 and not affected by dilute NaOH, e.g., sodium silver cyanide $NaAg(CN)_2$ solution, sodium cuprous cyanide $NaCu(CN)_2$ colorless solution, (b) changed only to acid by dilute H_2SO_4 and reactive with dilute NaOH, e.g., potassium hexacyanoferrate(II) $K_4Fe(CN)_6$ yields, with dilute H_2SO_4, hexacyanoferric(II) acid, cupric hexacyanoferrate(II) $Cu_2Fe(CN)_6$ yields, with dilute NaOH, cupric hydroxide.

Sodium cyanide solution dissolves certain metals (1) with absorption of oxygen, e.g., gold, silver, mercury, lead, (2) with evolution of hydrogen, e.g., copper, nickel, iron, zinc, aluminum, magnesium; and solid sodium cyanide, when heated with certain oxides, e.g., lead monoxide PbO, stannic oxide SnO_2, yields the metal of the oxide, e.g., lead, tin, respectively, and sodium cyanate NaCNO. Two classes of esters are known, cyanides or nitriles, and isocyanides, isonitriles or carbylamines, the latter being very poisonous and of marked nauseating odor.

Methyl cyanide CH_3CN, bp 82°C, formed by reaction of (1) methyl iodide and potassium cyanide, (2) acetamide and phosphorus pentoxide. Methyl isocyanide CH_3NC, bp 60°C, formed by reaction (1) of methyl iodide and silver cyanide, (2) of methylamine, chloroform and NaOH solution warmed. Ethyl isocyanide C_2H_5NC, bp 78°C. Phenyl isocyanide C_6H_5NC, bp 78°C at 40 torr pressure.

Oxidation of cyanide ion (e.g., by copper(II) gives cyanogen or oxalonitrile NCCN, poisonous colorless gas, bp −21°C. This reacts with organic compounds and bases like a halogen, for example, disproportionating in aqueous alkali to cyanide and cyanate. In aqueous acid, hydrolysis to oxalamide and ultimately oxalic acid takes place. Oxidation of cyanides by oxygen donors (e.g., lead monoxide or dioxide, manganese dioxide or dichromate) a little below red heat produces cyanates.

HYDROGEN (Fuel). Because of the wide use of hydrogen in the processing industries and for the hydrogenation of various oils and fats in the food and related industries, hydrogen has become much better understood during the past several decades. For many years, hydrogen has served as a specialized fuel for certain applications, such as oxy-hydrogen cutting and welding torches. But generally, until the late 1960s, the possible role of hydrogen as a major energy source fuel was rarely discussed. The word *hydrogen* took on a negative connotation with the development of the hydrogen bomb, as it also did some years ago when the hydrogen-filled dirigible Hindenburg exploded as it moved towards its mooring mast in Lakewood, New Jersey in 1937.

The probable future of hydrogen in the world's energy system was the subject of prophecy over one-hundred years ago. In 1874, Jules Verne wrote: "I believe that water will one day be employed as a fuel; that hydrogen or oxygen, which constitute it, used singly or together, will furnish an inexhaustible source of heat and light." And, in the early 1900s, Britain's Lord Haldane said: "It is axiomatic that the exhaustion of our coal and oil fields is a matter of centuries only. . . . As it has often been assumed that their exhaustion would lead to the collapse of industrial civilization, I may perhaps be pardoned if I give some of the reasons which led me to doubt this proposition." Haldane envisioned networks of windmills generating the electricity needed to separate hydrogen from water. The hydrogen would then be liquefied and stored underground.

Some readily apparent advantages of hydrogen, both as a direct and an indirect fuel (discussed later) have been extrapolated into terms of a future hydrogen economy. As the result of continuing and concentrated research and development in the energy field, many experts see a hydrogen energy economy gradually emerging.

Like other energy proposals, and there have been many in the past decade or so, three factors will likely determine the pace of hydrogen energy technology: (1) the manner in which, step-by-step, hydrogen-oriented systems and subsystems will compete economically and environmentally with other energy source, conservation, and utilization proposals; (2) the pace of technological advancement in related fields, such as nuclear engineering, upon which hydrogen systems may depend; and (3) the pace of unilateral efforts on behalf of hydrogen-oriented systems, including the refinement of current planning-purpose data and opinions into actual operating information relating to hydrogen generation, transportation, conversion and/or end-utilization, and safety. Without the funding of a series of "crash programs," unilateral developments probably will be relatively slow. Most likely, the information bank for hydrogen systems will stem from an increasing awareness of the energy characteristics of hydrogen and the progressive use of hydrogen subsystems in situations where they are eminently superior.

The present concept of a hydrogen fuel economy includes a primary energy source, such as a nuclear fission or fusion reactor, a geothermal source, or a solar-powered source, with hydrogen being produced as the portable energy carrier. See accompanying figure. Thermal energy

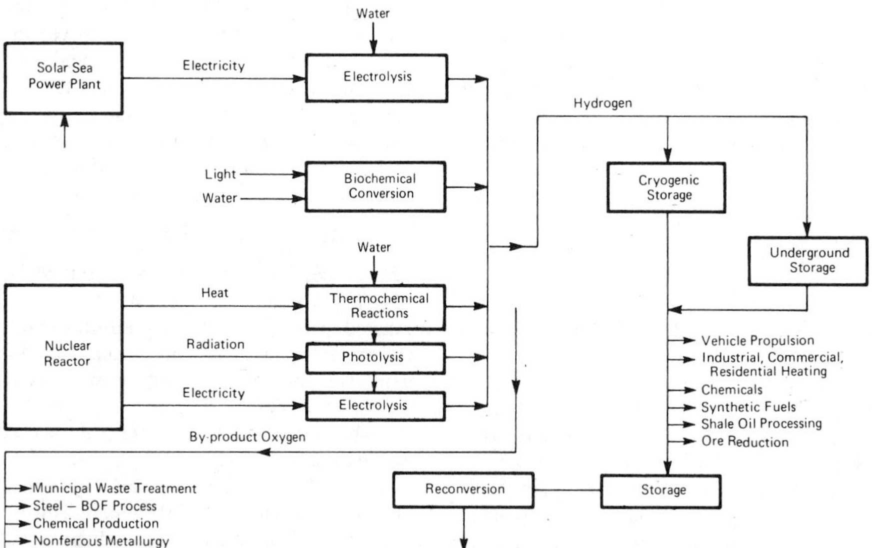

Major elements of hydrogen fuel economy.

from nuclear sources would be used to generate electricity that would then be used to electrolyze water for the production of hydrogen and oxygen. The hydrogen would be distributed by pipeline to distant points of use, with storage provided by underground gas storage, or by liquefaction and refrigerated storage.

Fuel-related Background of Hydrogen. Although the abundant hydrogen isotope *protium* is the simplest known atom, it forms two diatomic molecules, namely, *ortho-hydrogen,* in which the two atomic nuclei spin in the same direction; and *para-hydrogen,* in which the nuclei spin in opposite directions. While the equilibrium composition of hydrogen gas is 75% ortho at ambient temperature, it changes to 99.8% para in the liquid state. The transition from ortho- to para-hydrogen is exothermic (168 cal/gram), so that the heat released is more than enough to revaporize liquid hydrogen (heat of vaporization 107 cal/gram). Recognition of the existence of the ortho-para transition and the development of catalysts to equilibrate the liquid during liquefaction have made possible the large-scale production, use, and storage of liquid hydrogen.

Hydrogen molecules dissociate to atoms endothermally at high temperatures (heat of dissociation about 103 cal/gram mole), in an electric arc, or by irradiation. This property is used to effect atomic-hydrogen arc welding, in which hydrogen gas is dissociated by an ac electric arc between two tungsten electrodes, the hydrogen atoms recombining at the metal surface to provide the heat required for welding.

Pertinent properties of hydrogen are given in Table 1.

Actual and potential uses for hydrogen can be predicted by inspection of its properties. Its low density, 7% that of air, plus its high thermal conductivity, 6.7 times that of air, have led to its use as a coolant in large rotating electrical equipment. The low density reduces windage friction losses to less than 10% those with air, while its high thermal conductivity and heat capacity permit more efficient heat transfer, the result being an overall increase in generator efficiency of as much as 1%.

TABLE 1. FUEL PROPERTIES OF HYDROGEN

Melting point, K	13.96
Heat of fusion at 14.0 K, calories/gram	14.0
Boiling point at 1 atmosphere, K	20.39
Heat of vaporization at 20.4K, calories/gram	107
Density, grams/cubic centimeter	
Solid at 4.2K	0.089
Liquid at 20.4K	0.071
Critical temperature K	33.3
Critical pressure, atmospheres abolute	12.8
Critical volume, cubic centimeters/mole	65.0
Critical density, grams/cubic centimeter	0.031
Heat of transition, ortho to para at 20.4K calories/gram	168
Specific heat (At constant pressure C_p, calories/gram)	
Liquid at 17.2K	1.93
Solid at 13.4K	0.63
0–200°C	3.44
Specific heat (At constant volume C_v (0–200°C) calories/gram	2.46
Specific heat: Ratio C_p/C_v (0–200°C)	1.40
Gas density, 0°C and 1 atmosphere, grams/liter	0.0899
Gas specific gravity (Air = 1.0)	0.0695
Gas thermal conductivity, 25°C (cal)(cm)/(s)(cm^2)(°C)	0.00044
Gas viscosity, 25°C and 1 atmosphere, centipoise	0.0089
Coefficient of thermal expansion per °C	0.00356
Heat of combustion at 25°C, kcal/gram mole	
Gross	63.3174
Net	57.7976
Energy release upon combustion, calories/gram	29,000
calories/cubic centimeter	2,050
joule/gram	1.21×10^5
Flame temperature, K	2,483
Autoignition temperature, K	858
Heat of formation of HF at 25°C, kcal/gram mole ΔH	−64.2
Flammability limit, percent	
In oxygen	4 to 94
In air	4 to 74

The high heats of reaction of hydrogen with oxygen or fluorine, plus the low molecular weights of the product gases, have made hydrogen a prime fuel for rocket propulsion, since rocket thrust increases directly with the temperature and inversely with the molecular weight of the exhaust gases. Liquid hydrogen and oxygen were used in the second- and third-stage Saturn engines in the Apollo moon flights. The low atomic weight of hydrogen has made it the preferred propellant for nuclear rockets, in which nuclear emission provides heat for exhausting hydrogen gas at high temperatures.

Some studies have indicated that the cost of transporting and distributing hydrogen by pipeline may be less than the cost of transporting and distributing electric power. Presumably existing natural gas pipelines and distribution systems can be adapted to the use of hydrogen. Although hydrogen has a net heating value of only 275 Btus per cubic foot (2448 Calories per cubic meter) as compared with 913 Btus per cubic foot (8126 Calories per cubic meter) for methane, the lower density and viscosity of hydrogen make it possible for a pipeline to deliver about the same amount of thermal energy as with methane, at a somewhat greater compression cost. The thermal energy in hydrogen can be utilized more efficiently in home heating than natural gas, because hydrogen can be burned in nonvented heaters, with no loss of heat, since its only primary combustion product is water. By using flameless catalytic heaters, nitrogen oxide formation can be eliminated. However, oxygen depletion of closed spaces will still present a hazard.

One advantage of hydrogen as a source of thermal energy, as compared with electricity, is that it can be stored for later use—it is a commodity with weight and volume. Electricity, although it can be converted into chemical energy in batteries, essentially is a form of energy that must be used as it is generated. Hydrogen, like natural gas or substitute natural gases, may be stored and transported as a refrigerated liquid, or stored as a gas under pressure in underground systems. Hydrogen also may be stored as a metallic hydride.

Categories of Energy-related Hydrogen Uses. The probable functions of hydrogen in future energy technology may be put into two major categories: (1) *direct functions* in which hydrogen serves as a fuel, that is, as the source of heat, power, and light without prior conversion to some other energy form; and (2) *indirect functions* in which hydrogen is an important component of the total energy system, but before the end-use of that energy, the hydrogen is involved in some conversion, possibly chemically, used in the creation of a synthetic fuel, such as substitute natural gas, or possibly converted into electrical energy which becomes the final end-energy used. One of the major indirect or secondary roles proposed for hydrogen is that of an energy transporter, wherein in one scheme, other forms of energy would be consumed to generate hydrogen which then would be pipelined and stored at distant points available for another conversion step—for example, converted into electrical energy as needed.

Hydrogen as Energy Source for Motive Power

Aside from their relatively low costs until the mid-1970s and continuing into the 1980s, the hydrocarbon fuels, notably gasoline and kerosine, have offered convenience in handling and transportability for use in connection with powered vehicles. And, during the past decade, the political factors that arise from the striking geographic imbalance between petroleum resources and petroleum consumption in most regions of the world have provided ample incentives to strike out for alternative sources of vehicular power.

Hydrogen, when cost competitive, can provide many of the advantages of petroleum liquids and offer the additional attraction of decreasing air pollution.

Because of its low density, the net storage volume required would be at least as much as for gasoline. The storage tank must be maintained at a temperature of −423°F (−253°C), which is the boiling point of hydrogen at atmospheric pressure. This would require insulation that would increase the overall size of the storage container. Vaporization losses from the storage tank, amounting to perhaps 2% or more per day, must be vented so that no ignition of the vented hydrogen gas can occur, and no accumulation of explosive hydrogen-air mixtures are possible. The lower explosive limit of hydrogen in air is 4%, so adequate ventilation must be provided. Fortunately, hydrogen gas, being the lightest gas with a specific gravity of 0.07 referred to air, will rise and diffuse rapidly and thus can be easily dispersed. Service stations for dispensing liquid hydrogen will require more expensive storage and pumping facilities than required for gasoline.

TABLE 2. SOME HYDROGEN STORAGE OPTIONS FOR VEHICLES

Storage System	Relative System Weight[a]	Relative Contained Volume[a]
Gaseous phase, 2,000 psi (136 atmospheres)	~30.0	~24.0
Solid (as magnesium hydride with 40% porosity)	4.6	4.0
Liquid phase at 37°R	2.4	3.8

[a] Relative to gasoline, as unity for same energy content.

The estimated weights and volumes expressed in Table 2 are relative to the same energy content of gasoline. Relative weight includes that of containers. The data indicate that magnesium hydride would be at a 4.6 weight disadvantage and thus require four times the tankage in comparison with the use of gasoline in a conventional automobile. New hydrides, as described later, may change this. This would be equivalent to 450 pounds (204 kilograms) of added vehicle weight and 60 more gallons (227 liters) (2 × 2 × 2 feet storage; 0.2 cubic meter) over that required for a vehicle with a 20-gallon (76 liters) gasoline tank. Burst upon collision for liquid storage can be overcome by using containers capable of withstanding 30 Gs, which are presently available.

If a designer were to elect the option of using hydrogen in the gaseous phase at 2000 psi (136 atmospheres), this would require a metal container weighing some 30 times and requiring a volume of some 24 times that required for an energy equivalent volume of a hydrocarbon fuel. Also important in the total energy equation is the additional energy required to compress hydrogen (gaseous phase) or to liquefy it.

It is most likely that the first major use of liquid hydrogen as an energy source for motive power will be jet aircraft, largely because of the excellent weight advantage and the less serious nature of the boil-off loss and distribution problems as compared with other forms of transportation. City buses and long-haul motor trucks, already equipped mainly with hydride hydrogen power, have been tested in the United States and West Germany, among other countries. These may follow as candidates wherein refueling may be effected through replacement of entire storage tanks (dewars). Because the private motorcar presents the most crucial logistics problems, including the small-capacity fuel system, concern with safety, boil-off loss of fuel even when vehicle is not in use, and the education and acceptance involving millions of users, it probably will follow rather than lead the use of hydrogen in other modes of transportation. However, during the last few years, a few firms have offered hydrogen-powered private motor vehicles, set up to switch from hydrocarbon fuel to hydrogen and vice versa, but at a cost that is not competitive with mass-marketed vehicles.

Metal Hydrides. For a number of years many scientists and advanced planners have considered the possible use of metal hydrides to store hydrogen at atmospheric or reasonable pressures and at relatively low temperatures (comparable to current metal temperatures in some conventional engines). The fact that hydrogen will form hydrides with most metals has been known for many years, during which time, a number of these hydrides have been formed and tested. Until comparatively recently, magnesium hydride and a hydride of a rare-earth metal plus nickel, such as $LaNi_5$, appeared to be best suited as hydrogen storage media.

The hydride-forming reaction is exothermic and reversible: Metal + Hydrogen $\rightleftharpoons$ Metal Hydride + Heat. Thus, when it is desired to call for the separation of hydrogen from the hydride, heat (of decomposition) is required. As may be expected, the heat of decomposition is roughly proportional to the stability of the particular hydride. It is thus evident that for a metal hydride to serve as an efficient and viable means for storing hydrogen, it should be capable of decomposition at a relatively low temperature—say 300°C of lower. At the same time, the hydride must be reasonably stable and, of course, not require a high hydrogen pressure to manufacture it. It is further evident that the metal portion of the hydride must be comparatively inexpensive and thus common and readily available in the quantities that may be required. Metal hydride storage, operating as it does in terms of hydrogen as a battery operates in terms of electricity, must be capable of easy and efficient replenishing or "recharging" cycles. In every respect, the hydride storage element must be as safe as current vehicular fuel systems.

Researchers have investigated a large number of known binary hydrides, i.e., compounds which contain one metal and hydrogen. Investigators now regard magnesium hydride (MgH_2) as a borderline possibility. This binary hydride evolves hydrogen at a pressure of one atmosphere and requires a decomposition temperature of 289°C.

In comparatively recent research, much has been learned concerning the manner in which hydride compounds hold hydrogen. It has been known for a long time, of course, that the metal portion of the hydride should be comprised of tiny particles so that there is a large surface area available for reaction. In searching for reasons why hydrides permit such a high density of hydrogen, Reilly/Sandrock (1980) have observed that it is possible to pack more hydrogen into a metal hydride than into the same volume of liquid hydrogen. When the subject metal is first exposed to diatomic hydrogen (H_2), the hydrogen atoms are adsorbed onto the surface of the metal. Immediately, some of the hydrogen is dissociated into monoatomic hydrogen (H). This permits the monoatomic hydrogen to penetrate deeply into the crystal lattice of the metal and to occupy what are known as *interstitial sites.* Investigators have found that these sites must have a critical minimum volume if they are to easily receive the hydrogen atom. Upon increasing the pressure of the hydrogen applied, the metal reaches a saturated phase—the metal hydride phase. It has been found that under certain conditions and with certain metals, the number of hydrogen atoms contained in the crystal will range from 2 to 3 times the number of metal atoms.

The most recent experimentation with metal hydrides has involved multiple-metal hydrides. It has been known for some time that hydrogen reacts with alloy metal combinations. Considerable research has gone forth in connection with ternary hydrides (2 metals + hydrogen), and one of the most promising of these compounds as of the early 1980s is iron-titanium hydride ($FeTiH_x$), where x may range from 1 to 2. Reilly/Sandrock report that the hydrogen storage capacity by weight percent of this ternary hydride is 1.75 and by volume (grams per milliliter) is 0.096. The energy density by weight is 593 calories per gram; and the energy density by volume is 3254 calories per milliliter. Thus, this ternary hydride has a higher hydrogen-storage capacity than an equal volume of liquid or gaseous hydrogen (at 100 atmospheres). Another promising intermetallic hydride is lanthanum-pentanickel hydride ($LaNi_5H_x$), although it is more costly to produce. In this hydride, x may range from 1 to 6. Both the iron-titanium hydride and the lanthanum-pentanickel hydride have low temperatures of formation and decomposition, contributing to easy charging and discharging at ambient temperature.

In connection with hydrogen engine design, it is assumed that the heat of decomposition required by the hydride can be furnished from the inevitable waste heat generated by any engine. See also **Hydride.**

Hydrogen as a Heating Fuel

The routine use of hydrogen as a heating fuel for industry and commercial-residential installations entails even greater complications and would appear to be much more dependent upon the overall economic and technical aspects of a so-called hydrogen fuel economy. From many standpoints, assuming availability, hydrogen can be an excellent fuel for almost any heating application. Hydrogen can be used in the home for cooking and heating (and even lighting) and likewise in commerce and industry. Compared with natural gas, hydrogen burns with a faster, hotter flame. Hydrogen-air mixtures are flammable over wider limits of mixtures. Hydrogen burns without producing noxious exhaust products, allowing unvented appliances except where water vapor and resulting increased humidity may be objectionable. In winter, the additional humidity can, in fact, be highly desirable.

But, in humid locations in summer, the water vapor produced could be objectionable. Adequate ventilation must be provided to prevent depletion of oxygen in closed spaces.

But, generally because of the absence of hazards from carbon monoxide and other fumes, large savings could be achieved from the elimination or at least simplification of flues. Some experts suggest that not only construction costs could be lowered as the result of clean burning, but that an increase of some 30% in the efficiency of a gas-fired home heating system could be achieved. The concept of peripherally placed unflued devices, particularly through the use of catalytic "flameless" heaters, could ultimately lead to a serious revision of the widely accepted central heating concept. By maintaining the temperature of a catalytic bed as low as 100°C, the production of nitrogen oxides would be virtually eliminated.

Because hydrogen burns with a hotter flame, some design features of heating apparatus would require change. The energy content per unit mass of liquid hydrogen is about 2.75 times greater than that of hydrocarbon fuels. On the other hand, there are only 325 Btus per standard cubic foot (2893 Calories per cubic meter) of hydrogen as compared with about 1,000 Btus per standard cubic foot (8900 Calories per cubic meter) of natural gas, thus dictating further design changes. The ignition energy of hydrogen is about 0.02 millijoules, which is less than 7% that of natural gas, a major factor in making low-temperature catalytic burners possible; also a major factor in designing for safe operation.

Despite the numerous advantages of hydrogen as a direct heating fuel, particularly in the home, the application of hydrogen must be viewed in terms of the total energy concept of an exclusively hydrogen-supplied (all-hydrogen home) installation. Where the direct use of hydrogen for heating is large, the economy will be most favorable. If a substantial amount of the hydrogen must be converted into electrical energy, as by a fuel cell, then economic justification becomes more difficult.

Lighting in the all-hydrogen home may be accomplished by condoluminescence, a cold process. A phosphor is spread on the inside of a tube similar to the conventional fluorescent lamp. Upon coming in contact with the phosphor, small amounts of hydrogen combine with the oxygen in the air to excite bright luminescence in the phosphor.

Conversion of burners and other design aspects of heating systems and appliances to pure hydrogen, or to a hydrogen-enriched natural or substitute natural gas supply, while costly and inconvenient, is certainly not in the economically insurmountable category. Similar alterations over the years were made in the United States when communities switched from manufactured gas (about 50% hydrogen) to natural gas. Such switchovers are even more recent in European communities.

As more hydrogen becomes available for transportation use and as more hydrogen is pipelined regionally or transcontinentally, depending largely on the demand placed upon the supply of hydrogen for industrial uses, it may be that the hydrogen content of community gas supplies will be progressively enriched with hydrogen (in a periodic, stepwise manner because of switchover problems) and thus contribute in a gradual manner to less pollution and to the conservation of natural gas.

Hydrogen as an Energy Transporter

With the possible use of hydrogen as a source of motive power in the transportation field, the *non*chemical interest in hydrogen in the total energy picture is directed to the use of hydrogen as a means or mode of storing and transporting energy. It is in this area that hydrogen directly confronts the past ever-increasing trend toward a fully-electrical energy economy. Undeniably, hydrogen energy has a major starting advantage over electrical energy, namely, hydrogen is a storable energy form. Investigations are showing that hydrogen in pipelines may cost less to transport than electricity flowing over long power lines. Thus, hydrogen may play an important future role simply as a mode of storage and transport, even though source and terminal energy conversions may be required.

Electrical power plants are most efficient when operated at constant output at full-rate load. Because of wide fluctuations in consumer load (daily and seasonally), generating rates require constant adjustment. Communication systems and some emergency systems employ batteries for interim storage of electrical energy, but these applications are minuscule when compared with the total electrical generating and distribution system. The principal means of large-scale storage is the use of pumped storage, i.e., in essence a reversible hydroelectric station wherein electrical energy is temporarily converted to a hydraulic head by pumping water to an elevated reservoir. Unfortunately, the topography has to be suitable for such an installation and thus this approach is limited to comparatively few power-generating sites. See also **Hydroelectric Power.**

The high-voltage cables required to transmit electricity from generating stations to load centers are costly. The cost of going to underground cables for transmitting bulk current ranges from 9 to 20 times that of overhead configurations. The effective use of cryogenic superconducting cables may lower underground costs considerably, but much research remains to be completed before this is possible.

Because of the tremendous volumes of fuel that can be moved in pipelines, the construction, maintenance, and operating costs of a buried pipeline are much less in terms of a percentage of the total product moved. Pipeline operations have been profitable even at the relatively low price ranges for liquid and gaseous fuels prevailing prior to the price rises of the 1970s and 1980s. Pipeline technology, of course, is well established—with several hundred thousand miles of trunklines installed and operating in the United States. These lines transport nearly 23 trillion cubic feet (0.65 trillion cubic meters) of gas. Typical pipelines range from 600 to 1,000 miles (965–1609 kilometers) in length and are up to 48 inches (1.2 meters) in diameter. Line pressures may range from 600 to 800 psi (41–54 atmospheres) but go up to 1,000 psi (68 atmospheres). A representative 36-inch (0.9 meter) pipeline will carry a gaseous fuel with the equivalent of 37,500 billion Btus (9450 billion Calories) per hour. The electrical energy equivalent would be 11,000 megawatts. By comparison, this is ten times the energy-carrying capacity of a single-circuit 500-kilovolt overhead transmission line.

The figures for pipeline transportation of pure hydrogen are not quite so attractive, but nevertheless the comparison with electric transmission costs remains highly significant.

One study shows that the pipeline transmission costs for hydrogen will range from 30% to 50% more than for natural gas. Conversion of an existing natural gas line to hydrogen service is estimated to require a rise of compressor capacity by a factor of 3.8 and compressor horsepower by 5.5.

Obviously, hydrogen transmission costs represent but one part of a total system. Should the costs of generating hydrogen in the first place, and the subsequent conversion of hydrogen into electricity at the terminal end of the system remain excessively high, then the savings in energy transportation costs, of course, become academic.

Sources of Hydrogen

The major source of chemical hydrogen over the past several decades has been natural gas. In strictly terms of chemical needs, where economic factors are favorable, natural gas has served this need well. Obviously, in terms of total energy conservation, where hydrogen is looked to as a means of conserving fossil-fuel sources, a much less costly and much more abundant hydrogen-containing raw material must be sought. The logical candidate is water. Particularly in areas of the world where hydrocarbons are not readily available, reasonably large water electrolysis installations have been made, notably in locations with low electricity costs.

In addition to electrolysis, the principal means under consideration for deriving hydrogen from water is that of thermochemical splitting. The waste heat and high temperature available from certain types of nuclear reactors would effect a series of chemical reactions, still much in the research phase, to free hydrogen and oxygen from water. Additional proposals have included the use of ultraviolet radiation from the plasma of a fusion reactor for the direct photolysis of water vapor (Department of Energy) and the use of some forms of algae, under the stimulation of light, to convert hydrogen ions to hydrogen gas by a complex chain of biochemical reactions (L. O. Krampitz, Case Western Reserve University).

Electrolysis. Because of years of operating experience, electrolysis is possibly an order of magnitude ahead of other proposals from a technological standpoint. Although simple in concept, electrolysis is costly—hence the research efforts to find other ways of splitting water carry a high incentive. Nevertheless, this side of one or more breakthroughs in other areas, most likely electrolysis operations will continue to serve as the basis for costs in extending the use of hydrogen in the relatively near term.

As of the early 1980s, industrial electrolyzers ranged in size from 500 standard cubic feet (14.2 cubic meters) of hydrogen production per day, consuming 3 kilowatts of electricity, to more than 40 million standard cubic feet (~1.1 million cubic meters) of hydrogen per day, consuming 240,000 kilowatts. Most common installations are from 10,000 to 500,000 standard cubic feet (283–14,160 cubic meters) of hydrogen per day. Two factors generally characterize an electrolyzer installation: (1) access to comparatively low-cost electricity, as found in some areas served by hydroelectric installations; and (2) need for the oxygen which accompanies the production of the hydrogen. Industrial electrolyzers usually operate at efficiencies of about 60% to 70%. Some high-pressure prototype models have reached 85%. It has been pointed out (D. P. Gregory, Institute of Gas Technology) that, in theory, electrolyzers can approach a maximum electrical efficiency of nearly 120% as the result of the ideal unit absorbing ambient heat and also converting this energy into hydrogen. A reasonable, practical target for an improved electrolyzer appears to be around 100%. Thus, the production of electrolytic hydrogen would be limited only by the efficiency of electric current generation, namely, between 35% and 45%. An estimate has been made (E. C. Tanner, Princeton University; R. Huse, Public Service Electric & Gas Co.) that the overall conversion efficiency of electricity-to-hydrogen-to-electricity will approximate 38%. The theoretical power required to produce hydrogen from water is 79 kilowatts per 1,000 cubic feet (~28 cubic meters) of hydrogen gas. One of the largest electrolyzers operating commercially is that of Cominco, Limited (British Columbia). This is a 90-megawatt installation that produces approximately 36 tons (32.4 metric tons) of hydrogen gas per day for use in ammonia synthesis. Other large plants are located in Norway and Egypt.

Two main types of electrolyzers are in commercial use: (1) Tank cells with monopolar electrodes. Porous diaphragms separate the alternate cathodes and anodes to prevent gas mixing. The anodes and cathodes are connected in parallel to keep the required voltage at approximately 2 volts and to permit high current densities. This arrangement requires a large floor area; (2) bipolar electrodes, connected in series and suitably insulated. The electrodes are cathodic on one side; anodic on the other side. This arrangement requires less floor space, is more complex, and requires high voltages.

High pressure can increase efficiency and this concept has been under development for many years. A commercial electrolyzer (Lurgi) is available which operates at a pressure of 30 atmospheres and 90°C, requiring 300 amperes of electric current at 217 volts. In the mid-1960s, bipolar cells of porous nickel electrodes were developed which operate at current densities of 800 and 1600 amperes per square foot (0.09 square meter).

In the mid-1960s, electric-high-temperature, vapor-phase electrolysis (General Electric Co.) was developed. In this process, the electrolyte is solid, porous zirconia which contains dopants. Operating temperature ranges from 500° to 800°C. A modification of the process is under development which will produce only hydrogen by consuming by-product oxygen.

Among electrolyzer design improvements that may occur are better electrodes which may result as a spinoff from fuel-cell work. There are indications that electrode improvement could cut the costs of electrolytic hydrogen by about 20% to 25%. Electrolysis looms high in consideration of utilization of ocean thermal gradients and thus these two technologies are closely interacting.

Thermochemical Splitting. The major objective is to find one or more series of chemical reactions that will result in the satisfactory separation of hydrogen (and oxygen) from water. Considerable work has been going forth at the Nuclear Research Center, Julich, Federal Republic of Germany, where much attention has been given to sulfur- and chlorine-base thermochemical cycles. Other researchers (Institute of Gas Technology; General Electric Co.; European Atomic Energy Community) have been probing various combinations of at least 56 chemical elements, including over 700 different compounds, that may show promise in various schemes for a closed-water-splitting cycle. It is understood that approximately 20 promising schemes have emerged, mainly centered in chlorine compounds. The most frequent flaw encountered among prospective reactions is the large amount of free energy required to force one or possibly two of the series of reactions; and the appearance of reactions that produce stable compounds incapable of regeneration.

Some of these reactions would rely upon a nuclear reactor as a heat source and would not have to await the emergence of a practical, operating fusion reactor. One sequence of reactions, in particular, is of interest:

$$CaBr_2 + 2H_2O \rightarrow Ca(OH)_2 + 2HBr$$
$$Hg + 2HBr \rightarrow HgBr_2 + H_2$$
$$HgBr_2 + Ca(OH)_2 \rightarrow CaBr_2 + HgO + H_2O$$
$$HgO \rightarrow Hg + \tfrac{1}{2}O_2$$

A drawback of this sequence is its use of highly corrosive hydrogen bromide. The scheme also requires a large inventory of mercury.

Of major concern to investigators in the thermochemical splitting schemes is the availability of appropriate materials of construction. Heat exchangers between the nuclear side and the chemical side must withstand both corrosion and radioactive contamination. The conventional nickel-chromium alloys are capable up to about 1050K; exotic, but available alloys, up to about 1400K. Above these temperatures, ceramics and new alloys may have to be used. Considerable materials research along these lines is going forth at the Los Alamos Scientific Laboratory.

Conventional Hydrogen Uses. Even before its serious consideration in the fuel economy, the demand for hydrogen grew at a rate of about 15% annually since World War II. About 3 trillion standard cubic feet (~85 million cubic meters) of hydrogen (8 million tons; 7.2 million metric tons) were produced in the United States in 1970. Not including energy applications, the chemical requirements for hydrogen are expected to increase by about 7% per year through the year 2000. Among demands for hydrogen include petroleum refining, plastics, elastomers, increased desulfurization of fuel oils, increase use in iron ore reduction, aerospace uses, and hydrogen/air fuel cells. About 42% of the hydrogen produced now is consumed in ammonia production; about 38% is used in petroleum refining. The other large consumers are metallurgical and food processing.

In terms of presently nonconventional fuels that will require increasing quantities of hydrogen as new processes develop, it is estimated that (1) synthetic crude oil from coal will require 6,500 standard cubic feet (184 cubic meters) of hydrogen per barrel of oil; (2) 1,300 standard cubic feet (37 cubic meters) of hydrogen will be required per barrel of oil from shale; and (3) 1,500 standard cubic feet (42 cubic meters) of hydrogen will be required for every 1,000 standard cubic feet (~28 cubic meters) of synthetic pipeline gas produced from the gasification of coal. Petroleum refining use of hydrogen is expected to increase to 610 standard cubic feet (~17.3 cubic meters) per barrel of crude refined. Direct iron ore reduction use of hydrogen is expected to increase to 20,000 standard cubic feet (566 cubic meters) per ton (0.9 metric ton) of iron. If there were not other hydrogen sources available, the hydrogen needs could be met by using approximately 10% of the natural gas production.

References

Bassett, L. C., and R. S. Natarajan: "Hydrogen—Buy It or Make It?" *Chem. Eng. Progress*, **76**, 3, 93–98 (1980).

Considine, D. M. (editor): "Energy Technology Handbook," McGraw-Hill, New York, 1977.

Corniel, H. G., Heinzelmann, F. J., and E. W. S. Nicholson: Production Economics for Hydrogen, Ammonia, and Methanol During the 1980–2000 Period," Exxon Research and Engineering, New York, 1977.

Cox, K. E., and K. D. Williamson: "Hydrogen: Its Technology and Implications, Vol. 2: Transmission and Storage of Hydrogen," CRC Press, Boca Raton, Florida, 1977.

Gregory, D. P.: "The Hydrogen Economy," *Sci. Amer.*, **228**, 1, 13–21 (1973).

Hänsch, T. W., Schawlow, A. L., and G. W. Series: "The Spectrum of Atomic Hydrogen," *Sci. Amer.*, **240**, 3, 94–110 (1979).

Mao, H. K., and P. M. Bell: "Observations of Hydrogen at Room Temperature (25°C) and High Pressure (to 500 kilobars)," *Science*, **203**, 1004–1006 (1979).

Reilly, J. J., and G. D. Sandrock: "Hydrogen Storage in Metal Hydrides," *Sci. Amer.*, **242**, 2, 118–129 (1980).

Ross, M., and C. Shishkevish: "Molecular and Metallic Hydrogen," Paper R-2056-ARPA, Rand Corporation, Santa Monica, California, 1977.

Sandrock, G. D., Reilly, J. J., and J. R. Johnson: "Metallurgical Considerations in the Production and Use of FeTi Alloys for Hydrogen Storage," *11th Intersociety Energy Conversion Engineering Conference Proceedings*, Vol. 1, pages 965–971 (September 1976).

HYDROGEN IODIDE. Iodine.

HYDROGEN ION CONCENTRATION. pH (Hydrogen Ion Concentration).

HYDROGEN (Metallic). Jupiter, Saturn.

HYDROGENOLYSIS. Organic Chemistry.

HYDROGEN-OXYGEN FUEL CELL. Fuel Cells.

HYDROGEN PEROXIDE. H_2O_2, formula weight 34.02, in pure, anhydrous form is a viscous, colorless liquid, sp gr 1.44, mp −0.89°C, bp 151.4°C. Hydrogen peroxide is soluble in H_2O in all proportions, soluble in alcohol, or ether, but not in hydrocarbons. Reagent, chemically-pure (CP) grade H_2O_2 is a solution of 90% H_2O_2 and 10% H_2O, sp gr 1.39. This concentration contains 42% active oxygen by weight. One volume yields 410 volumes of oxygen. Hydrogen peroxide solutions are high-tonnage chemicals and are supplied commercially in several strengths, ranging from 3–35% H_2O_2 by weight. Commercial grades for oxidation and bleaching normally contain 27.5–35% H_2O_2.

To reduce the tendency of H_2O_2 solutions to decompose, storage must be at comparatively low temperatures and in light-tight containers. Often, an organic material, such as acetanalide, will retard degradation. H_2O_2 has been used as an oxidizer in liquid bipropellant systems, or as a monopropellant through controlled catalytic decomposition, in supplying oxygen to various fuel mixtures for rockets and torpedoes. Low-concentration (normally 3% H_2O_2) solutions have been used for many years as antiseptics in medical applications. Bleaching is a primary outlet for H_2O_2, particularly in connection with cotton, wool, groundwood pulp—as well as hair-bleaching formulations. The compound is used as a source of gas in foaming rubber plastics. The highly reactive H_2O_2 molecule readily participates in oxidation, epoxidation, and hydroxylation reactions and is frequently used in an intermediate capacity in chemical syntheses. In restoring old paintings, H_2O_2 has been used to convert black PbS tarnish into the original white lead sulfate.

Industrial Production of Hydrogen Peroxide: The traditional process for manufacturing H_2O_2 has been the electrolysis of aqueous solutions of $KHSO_4$, H_2SO_4, or NH_4HSO_4. In recent years, chemical autoxidation processes have grown in favor, largely because of energy costs. In these processes, the feedstock may be an alkylated quinone, alkylated anthraquinone, and hydroquinone solvents, together with hydrogen, air or oxygen, H_2O, and a nickel, palladium, or platinum catalyst. The process yields a 15–75% solution of H_2O_2 in H_2O, depending upon adjustment of process concentrations and conditions to provide desired concentration. The yield for this type of process is about 90% of theoretical. The process proceeds essentially in two steps. In the first step, anthraquinone contained in a solvent is hydrogenated at a temperature of about 40°C and a pressure of 1–3 atmospheres. The anthraquinone is reduced to hydroquinone (*p*-dihydroxybenzene):

$$C_6H_4{:}(CO)_2{:}C_6H_3 + H_2 \rightarrow C_6H_4{:}(COH)_2{:}C_6H_3R.$$

R is a radical such as ethyl or tertiary butyl. In the second step, the hydroquinone solution is oxidized with air or oxygen: $C_6H_4{:}(COH)_2{:}C_6H_3R + O_2 \rightarrow C_6H_4{:}(CO)_2{:}C_6H_3R + H_2O_2$. In theory, the process consumes only hydrogen, atmospheric oxygen, and H_2O. A solvent must be used that will minimize side reactions during hydrogenation while also dissolving both the hydrogenated and oxidized forms of the organic compound. Solvents referred to in this connection are benzene-methyl-cyclohexanol mixtures and primary and secondary nonyl alcohols. Very tight purity precautions are required because any impurities in the H_2O_2 cause spontaneous catalytic decomposition of the product. As the result of these necessary precautions, the resulting H_2O_2 is one of the purest of commercial chemicals.

The process is highly corrosive. At one time, enameled steel vessels were standard for H_2O_2 processing. Aluminum, once properly passified through pickling and treatment after fabrication, has been found satisfactory.

Hydrogen peroxide reacts (1) with alkalis to form peroxides, (2) with potassium iodide solution, in presence of ferrous sulfate, to liberate iodine. This reaction serves to indicate the presence of as small an amount as 1 part by weight of hydrogen peroxide in 25,000,000 parts of H_2O, (3) with lead sulfide PbS, brown solid, to form lead sulfate $PbSO_4$, white solid, and sometimes used to brighten the lead pigment of darkened oil paintings, (4) with lead dioxide to form lead oxide, (5) with sulfites, especially in alkaline solution, to form sulfates, (6) with nitrites to form nitrates, (7) with arsenites for form arsenates, (8) with ferrous compounds to form ferric, (9) with chromic compounds to form chromates (see **Chromium**), (10) with permanganates in acid solution to form manganous compounds plus oxygen of twice the volume available from the hydrogen peroxide, (11) with dichromates in acid solution cold to form perchromic acid, blue solution, more soluble in ether than in acid, (12) with titanic salt solutions to form pertitanic acid, yellow solution, (13) with colored organic materials, e.g., litmus, indigo, to destroy the color, and thus used for bleaching hair, silk, feathers, straw, ivory, teeth, bones, gelatin, flour. When hydrogen peroxide solution is treated with finely divided platinum or other substances, or comes in contact with rough surfaces, e.g., ground glass, oxygen is evolved (water also formed).

In the laboratory, hydrogen peroxide is prepared from barium peroxide by treatment with ice-cold dilute acid; when H_2SO_4 is used barium sulfate insoluble may be separated by filtration. Other peroxides, e.g., sodium peroxide, react similarly with acids to form hydrogen peroxide plus the salt corresponding to the peroxide and acid used. Hydrogen peroxide is formed when ether is exposed to sunlight, when a hydrogen-oxygen flame impinges on ice, and when H_2O in a quartz vessel is exposed to ultraviolet light.

HYDROGEN PEROXIDE (Foods). Antimicrobial Agents (Foods).

HYDROGEN PEROXIDE (Fuel Cell). Fuel Cells.

HYDROGEN PEROXIDE (Rocket Fuel Oxidizer). Rocket Propellants.

HYDROGEN (Rocket Fuel). Rocket Propellants.

HYDROGEN SCALE. 1. A thermometric scale. (See **Temperature.**) 2. Since there is no reliable method for determining the absolute potential of a single electrode, electrode potentials are measured against a reference electrode whose potential is arbitrarily taken as zero. The arbitrary zero in general use is the potential of a reversible hydrogen electrode, with gas at 1 atmosphere pressure, in a solution of hydrogen ions of unit activity, or other electrodes calibrated against the hydrogen electrode.

HYDROGEN SELENIDE. Selenium.

HYDROGEN (SNG). Substitute Natural Gas (SNG).

HYDROGEN (Solar). Sun (The).

HYDROGEN SULFIDE. H_2S, formula weight 34.08, colorless, odorous gas, mp −82.9°C, bp −59.6°C, sp gr 1.1895 (air = 1). The gas must be handled carefully because of (1) its toxic properties (particularly dangerous because it may paralyze the olfactory nerves), and (2) its explosive tendencies (low ignition temperature of 260°C and wide flammability range from 4.3 to 44% by volume in air). Hydrogen sulfide liberates considerable heat upon burning (6,230 calories/liter at 15.6°C). The gas is produced by acid hydrolysis of many sulfides and by water hydrolysis of those elements higher in the hydrogen scale.

An aqueous solution of hydrogen sulfide is termed hydrosulfuric acid which undergoes slow atmospheric oxidation to sulfur. The acid is a strong reducing agent, usually with the separation of sulfur, e.g., with nitric acid (nitric oxide formed), with concentrated H_2SO_4(SO_2 is formed), with permanganate (manganous ion formed in the presence of acid), dichromate (chromic ion formed in the presence of acid).

Fluorine, chlorine, bromine, and iodine react with H_2S to form the corresponding halogen acid. Metal sulfides are formed when H_2S is passed into solutions of the heavy metals, such as Ag, Pb, Cu, and Mn. This reaction is responsible for the tarnishing of Ag and is the basis for the separation of these metals in classical wet qualitative analytical methods. Hydrogen sulfide reacts with many organic compounds.

The gas results from the decomposition of metal sulfides and albuminous matter and is found in the areas of mineral springs, sewers, and in some mines where it is referred to as "stink damp." H_2S also is a by-product of several industrial processes, including synthetic rubber, viscose rayon, petroleum refining, dyeing, and leather-treating operations. In the laboratory, H_2S usually is prepared by treating a sulfide with an acid, such as iron pyrites and HCl, or by heating thioacetamide $CH_3C(:S)NH_2$. Three processes are used industrially to produce H_2S in large quantities: (1) treating a sulfide with an acid, $2NaHS + H_2SO_4 \longrightarrow 2H_2S + Na_2SO_4$, (2) reacting sulfur with an alkali, $4S + 2NaOH + 2H_2O \longrightarrow 2H_2S + Na_2S_2O_3$, and (3) directly reacting sulfur with hydrogen, $S + H_2 \longrightarrow H_2S$. Large quantities of by-product H_2S usually are converted into elemental sulfur or H_2SO_4.

Industrial uses for H_2S include (1) the preparation of sulfides, such as sodium sulfide and sodium hydrosulfide, (2) the production of sulfur-bearing organic compounds, such as thiophenes, mercaptans, and organic sulfides, (3) the removal of Cu, Cd, and Ti from spent catalysts where the gas acts as a precipitant, (4) the formulation of extreme-pressure lubricants, and (5) the preparation of rare-earth phosphors used in color TV tubes.

See also **Coal;** and Substitute Natural Gas (SNG).

HYDROGEN THIOCYANATE. Thiocyanic Acid.

HYDROGRAPH. By graphing the discharge of a stream as ordinate against time sequence as the abscissa, a hydrograph of the stream flow is obtained. The hydrograph proves to be an important source of information in the design of sewerage and water supply systems and in the design of hydroelectric power projects. The reliability of the information it contains increases as the period of time over which the hydrograph extends is lengthened. Hydrographs extending over periods of less than 10 years are liable to be deceptive in the information they convey regarding maximum and minimum flows. The United States Geological Survey water supply papers form a valuable and important reference source for data upon which hydrographs are constructed.

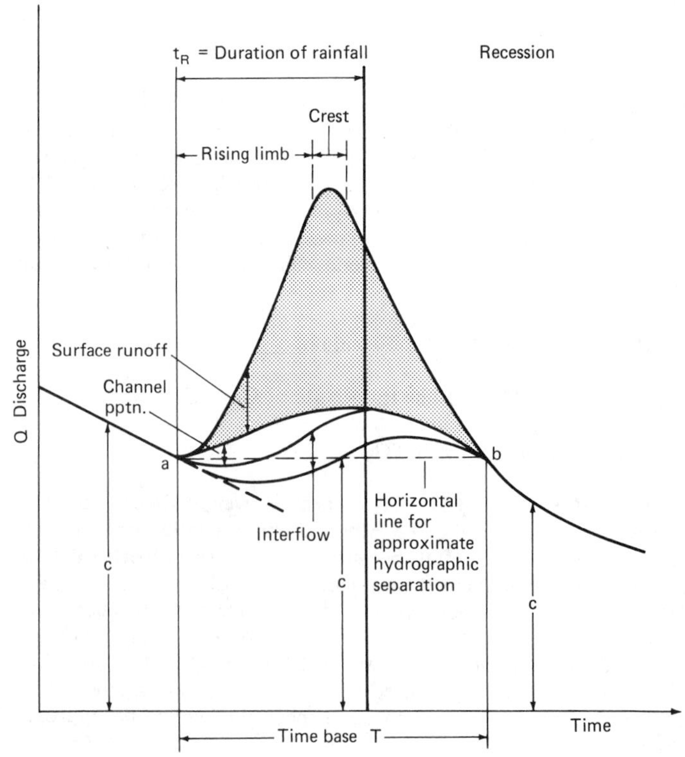

Hydrograph parts and flow contributions. (a) and (b) are reference points; (c) is groundwater flow.

Runoff and *stream flow* are synonymous. Surface runoff, interflow, and groundwater flow in varying proportions make up the total runoff in stream channels.

The *direct runoff* is that runoff which enters the stream promptly after rainfall or melting of snow. It is equal to the *surface runoff*, the *prompt subsurface runoff*, plus the *channel precipitation* which falls directly on the water surfaces of lakes and streams. Surface runoff is commonly represented in the form of a hydrograph similar to that shown by the accompanying figure.

Interflow is that part of the precipitation which infiltrates the surface soil, moves laterally through the upper soil horizons as ephemeral, shallow, perched groundwater above the main groundwater level, and reaches the stream before it reaches the water table. This lateral movement results from the presence of relatively impervious horizons near the surface.

See also **Drainage Systems;** and **Hydrology.**

HYDROID. One of the two forms of individuals in the coelenterates. The polyp. This form is a tubular or sac-like individual whose body wall is composed of two cellular layers separated by a thin mesogloea. The latter contains some cells derived from the other layers but is not developed as a third cellular layer. The hydroid is usually attached to the stalk of a colony or directly to a supporting surface. Its cavity opens at the free end of the body, and the mouth is surrounded by a circlet of slender tentacles except in specialized individuals found in some colonial species. The other form of coelenterate individual is the medusa.

HYDROKINETICS. The flowing of liquids is due to three principal causes: pressure difference, gravity, and inertia. Bernoulli's law expresses an ideal condition fulfilled by the three components of "head" corresponding to these three causes. The value of this head (whether constant or not) is, at a given point (x, y, z) of the liquid,

$$e + \frac{p}{\rho g} + \frac{v^2}{2g} = F(x, y, z) \quad (1)$$

The terms of this expression represents lengths, usually given in centimeters or feet. The assumption of constant density requires that the product of the speed of flow by the cross section of any conserved portion of the stream shall be constant and that the streamlines (paths of the moving particles) therefore converge as the speed increases. If one could assume that the function F is really constant, or if it were possible to obtain F as a known function of the coordinates of the moving particle, then all hydrokinetic problems could be solved by applying suitable mathematics to the equation which would thus develop from (1).

Various attempts have been made to do this. Useful formulae result from assuming F constant (Bernoulli's law) and applying the equation to special cases. But when such formulas are tested, the calculated results are found to be in error, in every case indicating that appreciable energy has been lost in friction. While some improvement is obtained by introducing a friction factor, it has on the whole been found more satisfactory to employ empirical formulas adapted to each type of problem. Thus, we have the Darcy formula,

$$v = D\sqrt{\frac{d(F_1 - F_2)}{l}} \quad (2)$$

for the speed of flow in a pipe of length l and diameter d, running full, and with a difference of total head $F_1 - F_2$ at the two ends; D being a constant to be determined by experiment. Also, the Chézy formula,

$$v = C\sqrt{\frac{as}{u}} \quad (3)$$

giving the speed of flow in an inclined channel, like a ditch or a sewer; a being the cross section of the flow, u the length of channel perimeter covered by the liquid, s the fall per unit length, and C an experimental constant.

The "hydraulic grade line" is a convenient concept in connection with flow through pipes. This is an imaginary line so drawn that each point of it lies vertically above (or below) the pipe at a distance equal to the pressure head $p/\rho g$ at the corresponding point of the pipe. In the case of a siphon, part, at least, of the conduit rises above this line, which means that the pressure in this portion is less than atmospheric.

Among the more difficult problems are those of vortex motion (like a whirlpool) and turbulent flow; and the general treatment of flow through a cavity of given shape under given boundary conditions, which presents some analogies to the electric current and the conduction of heat.

HYDROLASES. **Enzyme.**

HYDROLIQUEFACTION. **Coal Conversion Processes.**

HYDROLOGIC DATA. **Earth Resources Satellites and Geologic Remote Sensors.**

HYDROLOGY. The science, or study, of water, especially in relation to its occurrence in streams, lakes, underground structures, and as snow. The study of glaciers, their origin and geological effects is usually included under the heading of Glaciology. The term hydrology is derived from the Greek meaning water, and reason, hence the science of water, including its discovery, uses, control and conservation. Since water ranks first of all the natural resources, the science of hydrology is of great practical importance. The basis of hydrology is the hydrologic cycle. All terrestrial (fresh) waters are derived from the great oceanic reservoirs through evaporation and precipitation.

The Hydrologic Cycle. Also known as the water cycle, this is the never-ending circulation of water and water vapor over the entire earth. This circulation penetrates the three parts of the total earth system: the atmosphere (gaseous envelope above the hydrosphere), the hydrosphere (water covering the surface of the earth), and the lithosphere (solid rock beneath the hydrosphere). Solar energy and gravity provide the energy for the circulation.

Water is evaporated from the oceans and the land, with the former providing the largest amounts. The evaporated water is carried into the atmosphere, usually drifting tens to hundreds of miles before being returned to the earth as rain, snow, hail, or sleet. This precipitated water may be intercepted by plants, may run over the ground surface and into streams, may infiltrate into the ground, or fall back into the oceans. A considerable part of the water intercepted and transpired by plants and the surface runoff returns to the air by evaporation. The infiltrated water may seep down to deeper zones of the earth, forming groundwater storage which may later flow out to streams as runoff and finally evaporate into the atmosphere to complete the hydrologic cycle. Thus, the main processes involved in the hydrologic cycle are evaporation, precipitation, interception, transpiration, infiltration, seepage, storage, and runoff.

The quantity of water going through the hydrologic cycle during a given period for an area can be evaluated by the hydrologic or continuity equation:

$$I - O = \Delta S$$

where

I = total inflow of surface runoff, groundwater and total precipitation

O = total outflow, which includes evapotranspiration and subsurface and surface runoff from the area

ΔS = the change in storage in the various forms of retention and interception

A qualitative representation of the hydrologic cycle is given in Fig. 1, as first depicted by R. E. Horton (1931). The cycle is also shown diagrammatically in Fig. 2.

Magnitude of the Hydrologic Cycle. Each year, approximately 96,000 cubic miles (4×10^5 cubic kilometers or 4×10^{20} grams) of water is evaporated from the earth's surface. Of this amount, the oceans account for 84.4%, and inland water bodies and wet soils providing the remaining 15.6%. Most of the inland evaporation occurs into relatively dry air masses. Much of the water evaporated from the oceans is transported by maritime air masses (which can hold considerably more water vapor than continental air masses) to the continents, where total precipitation amounts to 24,000 cubic miles/year (100,000 cubic kilometers/year). This amount of water would cover the entire state of Texas (267,339 square miles; 692,408 square kilometers) to a depth of 475 feet (144.8 meters). Of the 24,000 cubic miles of water precipitated, 9,000 cubic miles (37.5%) returns to the sea as runoff to balance the excess precipitation over evaporation inland.

E. Reichel calculated (1952) that the mean annual precipitation for the entire world is 34 inches (86.4 centimeters), which is balanced by a comparable amount of evaporation. It is estimated that 97% of

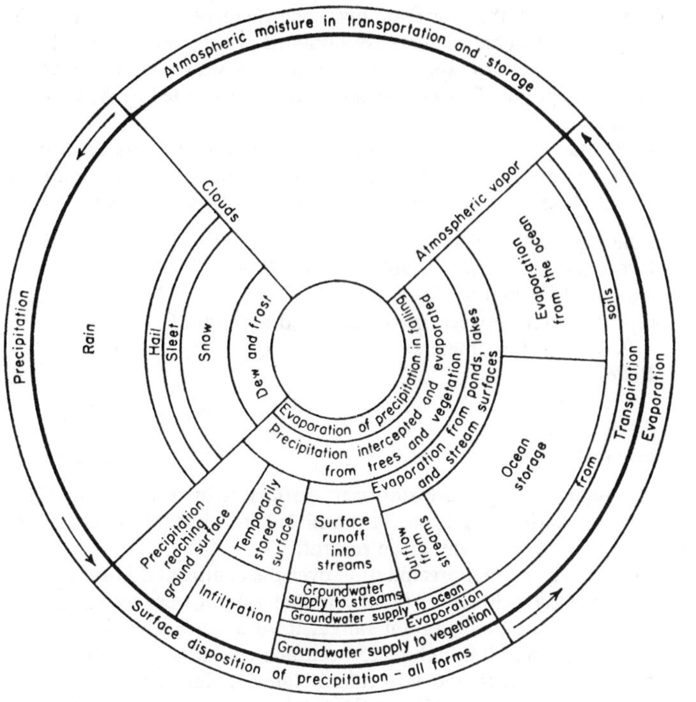

Fig. 1. The hydrologic cycle as conceived by Horton in 1931.

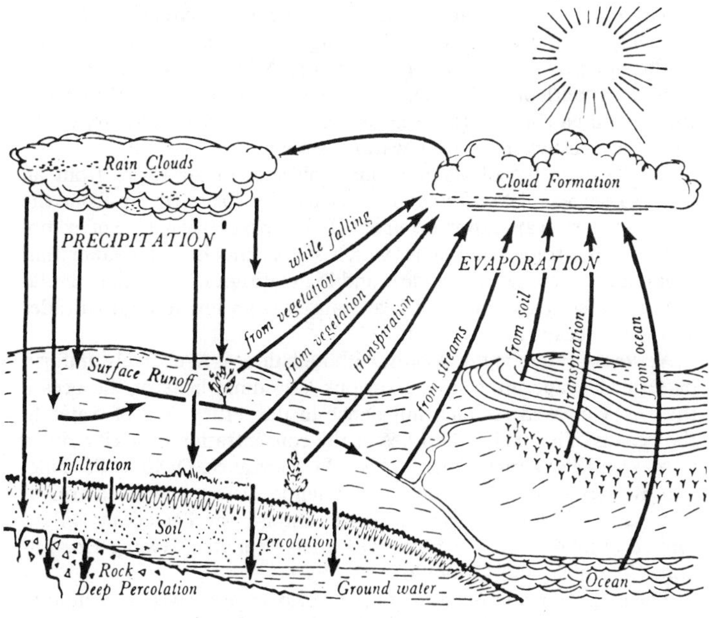

Fig. 2. The hydrologic cycle as presented by Ackermann, Colman, and Ogrosky in 1955.

all the water in the world or over one quadrillion (10^{15}) acre-feet (1,234 × 10^{15} cubic meters) is contained within the oceans. If the earth were a uniform sphere, this volume of water would cover the earth to a depth of 800 feet (243.8 meters), as estimated by A. Wolman (1962).

The total volume of fresh water on the earth is estimated at 33 trillion acre-feet (4.1 × 10^{15} cubic meters; or 4.1 × 10^{21} gallons) distributed, as estimated by V. T. Chow (1964), as follows:

Polar ice and glaciers	75%
Groundwater between 2,500 and 12,500 feet (762 and 3,810 meters)	14
Groundwater between the surface and a depth of 2,500 feet (762 meters)	11
Lakes	0.3
Soil moisture	0.06
Atmosphere	0.035
Streams	0.03

The foregoing figures are stationary estimates of distribution. Huge amounts of water pass through the atmosphere while the water content is relatively small at any given instant.

The average annual precipitation over the continental United States would amount to 30 inches (76.2 centimeters) if it were spread evenly. In actuality, the precipitation ranges from a few inches in the arid southwest to over 100 inches (254 centimeters) in parts of the Pacific northwest. Although the 17 western states contain 60% of the land area, they receive only 25% of the total precipitation. The 30 inches of water for the United States represents 4,800,000,000 acre-feet/year or 4,300 billion gallons/day. Of this amount, 21.5 inches (54.6 centimeters), or 71.7% is returned to the atmosphere by processes of evapotranspiration. The remaining 8.5 inches (21.6 centimeters) (28.3%) becomes surface and groundwater runoff into the oceans. The foregoing estimates were made by C. J. Robinove (1963).

Mount Waialeale, Hawaii (Kauai) receives the most rain of any location in the world, averaging 460 inches (1,168 centimeters). There is also a wide range of precipitation over Canada, from about 10 inches (25.4 centimeters) in parts of Yukon to nearly 70 inches (178 centimeters) in Halifax, Nova Scotia. St. John's, Newfoundland also receives over 60 inches (152 centimeters) of precipitation.

Function of the Hydrologic Cycle. If the atmosphere and the earth were considered as separate entities, radiation and conduction fail to provide balanced heat budgets, because the earth's surface has a net gain and the free atmosphere a net loss. The link between the gain and loss is the hydrologic cycle.

Some of the heat absorbed by the earth's surface is expended in evaporation, and therefore is transferred to latent heat, which is later realized as sensible heat and released to the atmosphere when the vapor condenses to clouds. Evaporation is high where relatively cool air sweeps over warmer oceans. The highest evaporation values found in the northern hemisphere occur in the Atlantic and Pacific trade wind belts south of 30°N. High values also occur over the northwestern Pacific and North Atlantic oceans during winter when cold, dry continental air masses move over warmer waters.

The average life of water vapor molecules in air varies from an hour to several days. Latent heat is usually liberated far from the regions where evaporation occurred. This is particularly true of evaporation in the trade wind belts, which supply much of the vapor that eventually precipitates in middle and high latitudes. Thus, the circulation of water is a key part of heat transfer from low to high latitudes and from oceans to continents.

Return of Water to the Oceans. Although there is a relatively uniform pattern of evaporation in the various latitudinal belts of the ocean, there is a marked regional imbalance in the return flow of water to the oceans. The explanation lies in the concentration of major rivers (Amazon, Mississippi, Congo, Niger, St. Lawrence, Danube, Po, Nile, and Rhine) which drain into the Atlantic Ocean and its marginal seas (Gulf of Mexico, Black Sea). In contrast, the Pacific has only a limited number of major discharge outlets (Yangtze, Hwang-Ho, Yukon, Columbia, and Colorado).

The mean annual discharges of the world's major rivers are summarized by Livingstone (U.S. Geological Survey Professional Paper 440-G) in Table 1. The information in Table 2 also provides further evidence that the Atlantic not only drains the largest portion of the earth's land surface, but has the highest proportion of land area drained to ocean area.

TABLE 1. ESTIMATED RUNOFF OF MAJOR RIVERS OF THE WORLD

River	Cubic Feet/ Second (Thousands)	Cubic Meters/ Second (Hundreds)
Rivers discharging into Atlantic Ocean		
Eastern North America		
Mississippi	620	175.5
St. Lawrence	500	141.5
South Atlantic slope	325	92.0
North Atlantic slope	210	59.4
	1655	468.4
Europe		
Danube	225	63.7
Rhine	76	21.5
Rhone	59	16.7
Dnieper	59	16.7
Elbe	24	6.8
Garonne	24	6.8
Don	24	6.8
	491	139.0
South America		
Amazon	3600	1018.8
Orinoco	600	169.8
Parana	526	148.9
Uruguay	136	38.5
	4862	1376.0
Africa		
Congo	1600	452.8
Niger	326	92.3
Orange and Zambezi	352	99.6
Nile	100	28.3
	2378	673.0
TOTAL for Atlantic Ocean	9386	2656.4
Rivers discharging into Pacific Ocean		
Columbia	345	97.6
Colorado	23	6.5
Yukon	180	50.9
Australia	354	100.2
Japan and Korea	225	63.7
Middle latitude Asian rivers	2250	636.8
TOTAL for Pacific Ocean	3377	955.7
GRAND TOTAL		
Atlantic and Pacific Oceans	12763	3612.1

Basic Principles of Hydrology. Since the eighteenth century, the development of hydrologic principles has been aimed at refinement of the understanding of each distinct phase of the hydrologic cycle and of the relationship between the phases. A few of the more important principles are listed as follows, not necessarily in order of importance or discovery:

1. The recognition that groundwater moves from points of high pressure to points of low pressure (down gradient) and that gradients are often, but not exclusively, related to rock type and structure.
2. The fact that the velocity of flowing water, on the surface or underground, is governed by the differences in pressure head, or slope, and the resistance of the confining channel or of the aquifer.
3. The knowledge that water is capable of dissolving and carrying large amounts of mineral matter that changes composition as the water comes in contact with various types of potential solutes.
4. The geologic recognition that water transports and deposits vast

TABLE 2. OCEANIC AND LAND-DRAINAGE AREAS
(In Millions of Square Miles and Millions of Square Kilometers)

	Area		Land Area Drained			
Ocean	Square Miles	Square Kilometers	Square Miles	Square Kilometers	Percent of Total Land Area	Percent of Area Drained to Ocean Area
Atlantic	37.8	98	25.9	67	45.3	68.5
Indian	25.3	65.5	6.6	17	11.5	26.1
Antarctic	12.4	32	5.4	14	9.4	43.5
Pacific	63.7	165	6.9	18	12.1	10.8
Interior Drainage	—	—	12.4	32	21.7	—
Total	139.2	360.5	57.2	148	100.0	

quantities of solid rock waste and is a major agent in the modification of land forms and in chemical alterations underground.

5. The fact that natural (underground) or artificial (surface) storage of water modifies the regimen of water in an area by changing the time of flow.

With these basic principles in mind, scientists, geographers, and engineers, who practice in the field of hydrology, are constantly attempting to refine two areas of knowledge: (1) an inventory or description of the water resources of the world—the amount of water in storage, rates and volumes of precipitation, recharge and discharge, the quantitative availability and suitability of water for use and the effects of water in terms of floods and droughts; and (2) a full understanding of water in all of its properties and cycles.

Hydrology of Coastal Terrain. In coastal districts, the fresh water in the water table migrates slowly downhill to the sea. Because of their different densities, the fresh water and salt water do not generally mix, except in the ocean where the tides, waves, and currents do the mixing. In the aquifers in coastal districts, the less dense fresh water tends to float on the more dense fresh water sometimes like an iceberg. The shape of the fresh water lens on a sandy island, assuming that fresh water is being replenished by rainfall, is shown in Fig. 3. The relationship between the thickness of the fresh water body (a) and the depth of the lowest part of the fresh water body below sea level (b) is:

$$b/a = \frac{\text{Specific gravity of fresh water}}{\text{Specific gravity of seawater}} = 40/41$$

Thus, for every foot the fresh water stands above sea level, the surface of the salt water lies some forty times as many feet below sea level. These figures, of course, only approximate the condition and depend upon the salinity of the seawater and the purity of the fresh water. The flow lines, i.e., the paths of water movement, for the fresh water contained within the lens are indicated in Fig. 4. Both the lens and the underlying salt water will rise and fall with the tide unless there is a barrier between the underground water and the sea. The time of the peaks and troughs of the fluctuations becomes later as traced inland, just as the time of high and low tide becomes progressively later as it is traced up the tidal portion of a river. The time between the peaks and troughs will remain the same, while the time lag will be constant for a given well.

Some mixing of the fresh and salt water does occur at the interface. Usually this is negligible, but it can be appreciable when favored by certain conditions. This produces a brackish water zone which may be quite thick. This zone occurs where there are considerable fluctuations in the level of the interface due to tidal action or irregular heavy rains. Thus, a strong development of a brackish zone is found in the basalt aquifers along the coast of Oahu in Hawaii. It is also increased by pumping the wells in these regions. Like the fresh water, the brackish water lens moves slowly downslope.

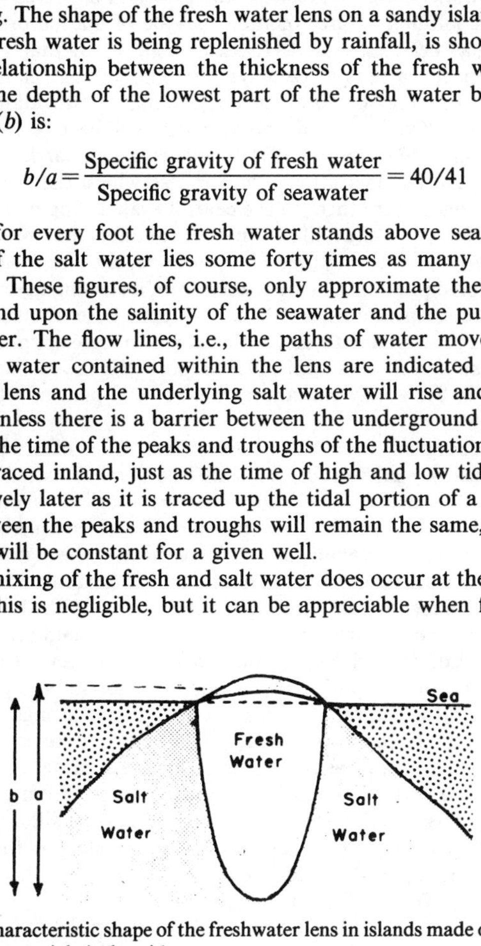

Fig. 3. Characteristic shape of the freshwater lens in islands made of uniformly permeable materials in humid areas.

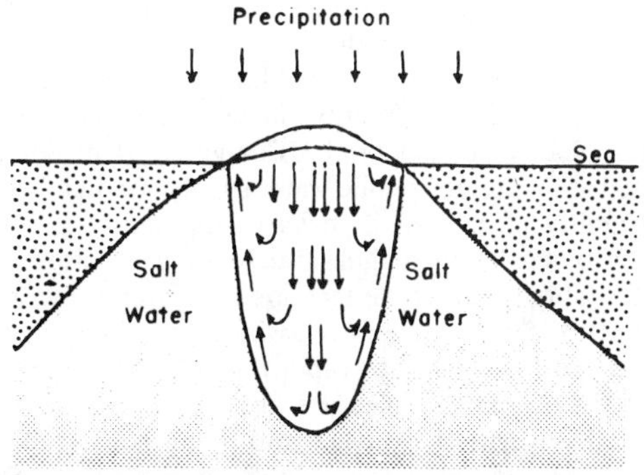

Fig. 4. Flow lines within the freshwater lens on the island shown in Fig. 3. The lens is recharged by infiltration of rain and snow into the ground, but loses water by diffusion and by flow into the sea.

The worldwide rise of sea level as the glaciers melt has an important effect on the water tables in coastal areas. As the sea level rises, so does the water table and the saline water. This increases the tendency of tides to sweep saline water into rivers and it tends to push the fresh water shoreward. In the case of the Atlantic seaboard of the United States, the sea is rising at a rate equivalent to about 2 feet (0.6 meter) per century. It has been estimated that this small rise will cause the fresh water in the artesian aquifer of New Jersey to recede inland at the rate of 1 to 4 miles (1.6 to 6.4 kilometers) per century, depending on the dip of the aquifer.

The thin layer of fresh water underlain by salt water means that great care must be taken in exploiting the fresh water. The usual method of exploitation is by well from which the water flows or is pumped. When the water is taken from a well, the surface of the water table is lowered close to the well by an amount depending on the output of the well and the porosity of the aquifer, as well as other factors. As mentioned before, for every foot of lowering of the surface of the water table, the fresh-saltwater boundary moves 40 feet (12 meters) nearer the surface. Thus, it does not take a great output of water to cause the bottom of the fresh water layer to rise to the bottom of the well. Thereafter, the water produced by the well will be saline. By limiting the production, contamination can be prevented.

Particular care must be taken in the case of artesian basins in coastal regions. The quantity of fresh water that is stored is finite, and the amount of recharge is limited. Overpumping will cause influx of saline water, as has occurred in the Savannah area of Georgia and South Carolina. Only restricted pumping or artificial recharge will prevent the eventual salinization of such a productive fresh water source.

Hydrology of Limestone Terrain. The primary factor in the hydrology of limestone terrains is the solubility of carbonate rocks in aqueous solutions, which leads to underground networks of pipes and channels known as *karst*. Limestone terrains are defined as regions where carbonate rock formations extend from near the surface to below the water table. The soil, and the material directly beneath, strongly influence the hydrology of these terrains. An impermeable cover overlying a thick limestone sequence can cause most of the precipitation to pass out of the area as surface runoff and the effect of the limestone is minimized. A pervious soil horizon, or erosion of the impermeable cover, will facilitate infiltration to the limestone substrate.

The hydrology of a carbonate region can range from a karst area with many interconnecting passages and caves, readily absorbing surface waters and transmitting these waters fairly rapidly from source area to discharge or storage areas, to a situation where a limestone formation of low permeability acts as an aquiclude and the terrain has high surface runoff and little available groundwater. The variability of the hydrology is as significant as the unique hydrology of a fully developed karst terrain and makes the concept of an average hydrology rather hypothetical. This variability can be demonstrated by comparing an area in Texas where the Edwards limestone absorbs an estimated 156,000 acre-feet (192.4×10^6 cubic meters)/day from surface stream flow, with a chalky portion of the Cooper Marl in South Carolina which is used as an unlined public water supply conduit for Charleston. The permeability of carbonate rocks varies greatly as a function of its purity, ratio of calcium/magnesium, texture, structure, and history.

The water table in limestone terrains which have undergone extensive karstification fluctuates greatly, rising in rapid response to precipitation input and falling due to rapid flow of water through solution passages to the discharge zones. Because there may be perched flow or storage on impermeable layers and along some bedding planes, some hydrologists question the validity of the water table concept in limestone terrains.

Hydrology of Semiarid Regions. The semiarid regions of the earth's surface occur as transition zones between the arid deserts and the subhumid belts. Water movement will shape the landscape, according to its geology and past topography, and will work in conjunction with wind erosion, solar insolation, temperature changes, and soils (stable or in movement), as well as with the vegetation and the animals which live thereon, to produce an ecological balance of all factors, either in a temporary or a permanent sense.

In defining arid and semiarid areas, Peveril Meigs (1952) used only three factors: humidity, season of precipitation, and temperature. An extremely arid region is defined as an area with at least one entirely rainless month per year; arid regions are defined as regions where precipitation is less than potential evaporation. A typical semiarid region is defined as one in which precipitation occurs in a cold winter, with the coldest month in the 0–10°C average range and the hottest month in the 20–30°C range. These conditions are typical of Mediterranean semiarid climate, occurring in Morocco, Algeria, Lebanon, northern Iran and also on the western coast of the United States around latitude 35°N.

Arid and semiarid lands account for over one-third of the land surface of the earth, while cultivated lands account for but one-tenth of the whole. The greatest belt of arid and semiarid regions extends across North Africa as the Sahara, through the Arabian Peninsula with the "Empty Quarter" of extreme aridity, into the Salt Desert of Iran, and the Takla Makan of Central Asia. In North America, the Great American Desert falls generally within this classification.

Precipitation in the semiarid regions is restricted and kept low by the inability of moisture-bearing winds to penetrate into, and cool down within, such regions. Zones of high pressure may prevent the entry of winds, and the great desert areas are mainly associated with this meteorological phenomenon. Such winds as do enter arid and semiarid regions may have had no opportunity of acquiring moisture by passage over oceans or sea, or they may have been forced to lose their moisture in passing over high mountains, as in the "rainshadow" deserts of Imperial Valley, California and the Jordan-Syrian steppe. Again, lack of orogenetic effects within the regions, combined with high heat reflected from the ground, may prevent cooling of the incoming winds so that no moisture condenses to form clouds of precipitation, as in the coastal deserts of Chile, southern California, Morocco, and western Australia.

Almost all precipitation in the semiarid and arid regions occurs as rain, except in higher altitudes. Dew and even hoarfrost are also of importance, and are due to the great differences between day and night temperatures. Where infiltration conditions are good, as over coastal sand dunes, or suitable vegetation exists, such dew may make a permanent addition to the useful water resources of the area because of this type of moisture intake by certain vegetation.

Wind-wells have been reported in use in some areas of the Crimea and the south of France, wherein airborne moisture condenses on rapidly cooling stones. It has been reported that the Byzantines had irrigated vines by planting them at the base of an octahedron of open stones, the upper pyramid above ground surface to condense moisture and the lower inverted pyramid leading the condensate down to the vine root.

High evaporation and high transpiration in vegetated areas is the dominant hydrological characteristic of the semiarid and arid zones. Both transpiration and evaporation are high because abundant heat energy is supplied to change the limited amounts of liquid water into water vapor, either directly or through the biological processes. In this way, the heat balance of the area is maintained.

The inability of surface waters to maintain themselves against evaporation has a longterm effect in that it permits the formation of basins of inland drainage. Any basin of closed drainage will cease to exist if the average annual storage of surface water exceeds evaporation from its central lake system, for then the lake will rise and spread each year till it overtops the lowest point of the encircling water divide over which it will discharge to the ocean level and also cut its way down so as to reduce the size of the lake. Basins of closed drainage are characteristic of the semiarid lands of the world. The ability of the Nile, the Euphrates-Tigris, the Indus, the Colorado, and similar rivers, to keep open their basins is due to the fact that the amount of incoming surface waters (originating in nonarid regions) exceeds the evaporation losses.

Precipitation is never pure water, but contains salts and gases in solution. The salts are dissociated into cations, mainly calcium, magnesium, sodium, and potassium, while the anions are bicarbonate, chloride, and sulfate. Carbon dioxide is the principal dissolved gas. These elements in solution in precipitation may be of marine or terrestrial origin. It has been estimated (E. Ericksson, 1958) that the annual precipitation of sea salts to be about three kilograms per hectare for the drier steppe regions south of the Sahara; as two kilograms per hectare in the Kalahari; and one kilogram per hectare for parts of Iraq and Iran. Full evaporation of the water which carries these salts will result in their deposition more or less where they fell; surface runoff will concentrate them in the central evaporating pans in basins of closed drainage, while infiltration to the aquifers may be with water which already is far from pure. Thus, in Syria, the chemical composition of the precipitation may be altered from a starting figure of about 20 ppm to concentrations of 100 to 200 ppm by evapotranspiration and leaching of precipitates in the zones of precipitation. Thus, D. J. Burdon and S. Mazloum (1958) report that the recharge waters to aquifers in Syria may contain from 50 to 200 ppm of total soluble salts.

In some aquifers, such as the Fars Formation of Iraq-Syria (of lagoonal facies), such soluble salts will be very abundant, while in other aquifers, such as the continental arkositic sandstones of the Sahara and Arabia, soluble minerals are almost completely absent. When the amount of groundwater flowing through the aquifer is large, such soluble salts tend to be removed and the aquifer flushed and cleaned out; likewise, fast-moving groundwater will flush an aquifer quicker than slow-moving water. Since the amount and often the rate of movement of groundwater in the semiarid zone tends to be small, mineralization by dissolution of the aquifer tends to be high.

At the point of natural discharge from aquifers, springs or marshy ground occurs. If the spring is large, a perennial river carries off the discharge, and an oasis is formed, or else a large city, such as Damascus (fed by the Barada River flowing mainly from Ain Figeh) comes into existence. If the discharge is small or diffuse, a saline marsh tends to form, of which one of the greatest is the Qatarra Depression in Egypt, the probable discharge zone for the sandstone aquifer of the Western Desert of Egypt.

As pointed out by D. J. Burdon ("The Encyclopedia of Geochemistry and Environmental Sciences," Van Nostrand Reinhold, New York,

1972), serious studies and efforts are underway to improve the overall hydrological conditions of semiarid zones, including: (1) Surface management—directed to making use of the water before it is lost by evaporation—increasing transpiration through useful vegetation; (2) control of storage of surface runoff—often intensive and short-lived—possibly spreading the water over large areas by diverting it from the wadi or stream bed and controlling it behind earth banks in such a way that its flow velocity is never sufficient to erode the retaining structures; (3) control of aquifers—use of proper techniques in connection with extraction from galleries, wells, and bore holes; (4) underground storage of groundwater—borrowing some of the successful techniques used in storing surplus natural gas and oil underground; (5) possibly using weather modification techniques, such as cloud seeding, and through the construction of dams and other holding means to allow large surfaces of water to form upwind of the semiarid area under consideration (Would introduction of the Mediterranean Sea to the Qatarra depression increase precipitation along the Alexandrian coast?); and (6) desalting of brackish waters.

References

Alessi, J. J.: "Mountain Gullies," *Soil Conservation*, **44**, 3, 18–19 (1978).

Burdon, D. J.: "Hydrology, Semiarid Regions," in "The Encyclopedia of Geochemistry and Environmental Sciences," Vol. IVA (R. W. Fairbridge, editor), Van Nostrand Reinhold, New York, 1972.

Grizzell, R. A.: "Flood Effects on Stream Ecosystems," *J. Soil Water Cons.*, **31**, 6, 283–285 (1976).

Holtan, H. N., Stiltner, G. J., Henson, W. H., and N. C. Lopez: "Revised Model of Watershed Hydrology," Tech. Bul. 1518, U.S. Department of Agriculture, Washington, D.C., 1975.

Schuhart, A.: "Wetlands Inventory," *Soil Conservation*, **44**, 4, 17–19 (1978).

Staff: "Soil and Water Conservation Research," U.S. Department of Agriculture, Washington, D.C. (Revised periodically).

HYDROLYMPH. The watery body fluid or blood of lower invertebrates. It carries nutriment to organs and tissues and removes waste; it has no respiratory function generally, though may contain proteins able to function as oxygen carriers.

HYDROLYSIS. A chemical reaction in which water reacts with another substance to form two or more substances. This involves ionization of the water molecules as well as splitting of the compound hydrolyzed, e.g., $CH_3COOC_2H_5 + H \cdot OH \rightarrow CH_3COOH + C_2H_5OH$. Examples are conversion of starch to glucose by water in the presence of suitable catalysts; or the conversion of sucrose (cane sugar) to glucose and fructose by reaction with water in the presence of an enzyme or acid catalyst; or conversion of natural fats into fatty acids and glycerin by reaction with water, as occurs in one stage of soap manufacturing; or the reaction of the ions of a dissolved salt to form various products, such as acids, complex ions, etc. See also **Cellulose Ester Plastics (Organic); Organic Chemistry;** and **Starch.**

HYDROMAGNETIC EQUATIONS. The time dependent equations which describe the behavior of a plasma in a magnetic field, assuming that the plasma is a compressible fluid and the plasma pressure P is a scalar. These equations are:

$$\rho \frac{d\mathbf{V}}{dt} = \mathbf{j} \times \mathbf{B} + q\mathbf{E} - \nabla P + \rho \mathbf{g} \tag{1}$$

$$\nabla \cdot (\rho \mathbf{V}) = -\frac{\partial \rho}{\partial t} \tag{2}$$

$$\mathbf{E} + \frac{\mathbf{V}}{c} \times \mathbf{B} = \frac{1}{\sigma}(c\mathbf{j} - q\mathbf{V}) \tag{3}$$

$$\frac{1}{P}\frac{dP}{dt} = \frac{\gamma}{\rho}\frac{d\rho}{dt} \tag{4}$$

$$\nabla \times \mathbf{B} = 4\pi \mathbf{j} + \frac{1}{c}\frac{\partial \mathbf{E}}{\partial t} \tag{5}$$

$$\nabla \cdot \mathbf{B} = 0 \tag{6}$$

$$\nabla \times \mathbf{E} = -\frac{1}{c}\frac{\partial \mathbf{B}}{\partial t} \tag{7}$$

$$\nabla \cdot \mathbf{E} = 4\pi q \tag{8}$$

The first equation is a force equation including gravitational forces. The second equation is a statement of mass conservation, while the third is analogous to Ohm's law. Number four is a statement of the adiabatic condition of the motion where γ is the ratio of specific heats of the plasma. The next four equations are the familiar Maxwell equations with no distinction made for **B** and **H** and **D** and **E** because all currents and charges are treated explicitly. The electromagnetic quantities are given in mixed Gaussian units and the conductivity σ in esu.

HYDROMETEOROLOGY. Meteorology.

HYDROMETEORS. Clouds and Cloud Formation; Precipitation and Hydrometeors.

HYDROMETERS. Specific Gravity.

HYDRONIUM ION. An ion found in water and all its solutions, which has the formula H_3O^+ and which consists of a proton combined with a water molecule. It has been established that hydrogen ions do not exist free in aqueous solution, but are present as hydronium ions. Formation of such ions is statistically rare, resulting from the interaction of water molecules in a ratio of 1 to 556 million.

HYDROPHILE-LIPOPHILE BALANCE. Paint.

HYDROPHILIC. Having a strong tendency to bind or absorb water, which results in swelling and formation of reversible gels. This property is characteristic of carbohydrates, such as algin, vegetable gums, pectins, starches, and of complex proteins, such as gelatin and collagen. See also **Colloidal Systems.**

HYDROPHOBE. Colloidal System; Detergents.

HYDROPHOBIC. Antagonistic to water; incapable of dissolving in water. This property is characteristic of oils, fats, waxes, and many resins, as well as of finely divided powders, such as carbon black and magnesium carbonate. Some interesting concepts are explored in "The Hydrophobic Effect and the Organization of Living Matter," by C. Tanford, *Science*, **200**, 1012–1018 (1978). See also **Colloidal Systems.**

HYDROPHONE. A transducer which responds to water-borne sound waves and, if of electroacoustic design, produces equivalent electric waves as output. Types of hydrophones include:

Line Hydrophone. A directional hydrophone consisting of a single straight line element, or an array of contiguous or spaced electroacoustic transducing elements disposed on a straight line, or the acoustic equivalent of such an array.

Split Hydrophone. A directional hydrophone in which electroacoustic transducing elements are so divided and arranged that each division may induce a separate electromotive force between its own electric terminals.

Directional Hydrophone. A hydrophone the response of which varies significantly with the direction of incoming sound.

HYDROPHYTES. Sometimes called water plants, these plants can grow only where there is an abundance of water, essentially growing in water or saturated soil. Those hydrophytes which are flowering plants are probably those which have reverted to an aquatic habitat. The reverting land plants may first have become marsh plants and then gradually developed into definite hydrophytes.

An aqueous environment presents conditions far more constant than an aerial one does. In the tropics, such conditions permit the plants to grow throughout the year. In colder regions there is a definite winter period when growth must cease. Many hydrophytes of temperate regions merely sink to the bottom and remain dormant during the winter. Others accumulate food reserves in rhizomes, which remain rooted in the bottom and renew growth in the spring. Still others form winter buds, consisting of large apical buds surrounded by many closely packed leaves containing much reserve food material. A few hydrophytes form small tubers.

The stems of hydrophytes contain a very small amount of vascular tissue, since support is largely afforded by the water, and conduction is not a great problem. In many of these plants the stem is very porous so that the plant floats in the water. The leaves of hydrophytes are of two types. Submerged leaves are thin and of various shapes; some, like eel grass leaves, are long and ribbonlike; others, like bladderworts, are finely dissected; while others are reduced to awl-shaped structures of small size. Floating leaves are usually large, undivided, and with stomata on the upper surface.

Reproduction in hydrophytes occurs both asexually and sexually. The flowers of nearly all hydrophytes are wind and insect pollinated, apparently a hangover from the time when they lived on land. A few have become modified to such an extent that pollination takes place on the surface of the water, the pollen floating about thereon and eventually reaching the stigma. A small number of hydrophytes are pollinated under water.

Nearly all algae and many fungi are hydrophytes; so also are some of the higher plants. In the flowering plants there are many water plants, such as the water lilies, bladderwort, eel grass and pondweeds. Many are very interesting plants; several are aquarium plants, serving to oxygenate the water; few are of any economic value. See **Algae;** and **Fungus.**

HYDROPONICS. The soilless culture of plants. In this technique, plants are grown with their roots immersed in a solution containing the necessary mineral salts or rooted in a sand medium which is kept moistened with such a solution. The soilless culture of plants is similar in principle but larger in scale. In one version of the method, the plants are supported in a matrix of peat, excelsior or some similar material on a wire screen with their roots dipping into the solution below. Aeration of the solution must also be provided if the best results are to be obtained. In another method, the plants are rooted in a medium of sand, gravel, or some similar material contained in a shallow tank into which the solution is automatically pumped at suitable intervals. Between pumpings, the solution gradually drains back into a reservoir tank.

The elements known to be necessary in chemically detectable amounts for the development of plants are carbon, oxygen, hydrogen, nitrogen, phosphorus, sulfur, potassium, magnesium, calcium, iron, manganese, boron, copper, zinc, and perhaps molybdenum. The first three of these elements are obtained by the plant from atmospheric gases or from water absorbed from the soil. The others are all absorbed in the form of mineral salts from the soil. Of the elements absorbed as salts the iron, manganese, boron, copper, zinc, and molybdenum are required in relatively minute quantities and are often called micronutrient elements. The principal elements which must be provided in the form of dissolved salts in hydroponic techniques, therefore, are nitrogen, phosphorus, sulfur, potassium, calcium, and magnesium.

Numerous solutions have been devised for use in the solution or sand culture of plants on both large and small scales. One solution which has been widely and successfully used for such purposes is made as follows: To each liter of water (preferably distilled or rain) add 1 M solution of the following salts as indicated: 1 cubic centimeter KH_2PO_4, 5 cubic centimeters KNO_3, 5 cubic centimeters $Ca(NO_3)_2$, and 2 cubic centimeters $MgSO_4$. To this solution then add 1 cubic centimeter per liter of a solution of micronutrients made as follows: 2.5 grams H_3BO_3, 1.8 grams $MnCl_2 \cdot 4H_2O$, 0.1 gram $ZnCl_2$, 0.05 gram $CuCl_2 \cdot 2H_2O$, and 0.075 gram MoO_3 per liter of distilled water. Also add to each liter of the solution made as described first, 1 cubic centimeter of a 0.5% solution of iron tartrate. The solution must be replaced with a fresh one at suitable intervals, and it is often necessary to add more of the iron solution between replacements.

Crop yields of at least some kinds of plants fully equal to those obtained on fertile soils can be obtained by hydroponic methods. The raising of crops by this method, however, probably will prove to be economically sound only for certain intensive types of agriculture or under certain special conditions. Some greenhouse floricultural and horticultural crops are now being grown successfully by this method. In regions where there is no soil, or where the soil is extremely infertile, but in which the climate is suitable to the development of plants, it seems likely that hydroponic techniques may prove useful. They have been used with some success, for example, on some of the coral islands of the Pacific Ocean.

See also **Aquaculture.**

HYDRORHIZAE. The root-like processes by which colonies of hydrozoan coelenterates are attached to a supporting surface. This term is in the plural. The singular is *hydrorhiza.*

HYDROSPHERE. The discontinuous envelope of water, both fresh and salt, which covers a major portion of the lithosphere. The bulk of the hydrosphere is contained within the deeper depressions of the surface of the earth. These depressions are termed ocean basins, and the water within them, oceans. Since the ocean basins are not large enough to hold the entire hydrosphere, seas are formed by the overflow of the oceanic waters on the continents. Geologists classify seas as epicontinental (epeiric) or relict. Technically, lakes, rivers and underground waters are also part of the hydrosphere. In general, therefore, the term hydrosphere is used mainly to distinguish the watery covering of the earth from the lithosphere on which, and in which, in part, it rests. See also **Earth; Hydrology; Ocean;** and **Polar Research.**

HYDROSTATIC EQUATION (Atmosphere). Atmosphere (Earth); Atmospheric Pressure.

HYDROSTATIC PRESSURE. The pressure created by a superimposed layer of a liquid is hydrostatic pressure. The intensity of hydrostatic pressure is commonly expressed as pounds per square inch. A head of 2.31 feet of fresh water creates a hydrostatic pressure of 1 pound per square inch. At a given depth of immersion in water, the pressure acts with equal intensity in all directions, that is, hydrostatic pressure is not directional in effect. Hydrostatic pressures are measured by means of pressure gauges of the Bourden tube type, or manometees of the U-tube type. Hydrostatic pressure sometimes is referred to as hydrostatic head.

In compressible flow, the hydrostatic pressure must be defined more carefully. A suitable definition is in terms of the Helmholtz free energy, $A = U - TS$,

$$p = -\left(\frac{\partial A}{\partial(1/\rho)}\right)_T$$

HYDROSTATICS. This branch of mechanics has to do with the equilibrium of liquids and the laws relating to liquid pressure. A study of these laws makes it clear that the components of pressure in a liquid fall naturally into two classes, according to the way in which they are produced; namely, (1) pressures due to forces applied externally, as by the atmosphere or by the piston of a pump, and (2) those due to causes operating throughout the body of liquid, such as gravity or inertia.

Pascal's law applied only to the first class, and states that any pressure in an enclosed liquid, originating in forces applied at its boundary, is communicated with unaltered intensity to all parts of the liquid. A familiar illustration of this fundamental law is the hydraulic press, which consists of two communicating cylinders, usually of different diameter, fitted with pistons, the force acting upon one piston and the force exerted by the other being in proportion to their areas.

The pressure in an enclosed liquid due to its own weight, on the other hand, increases uniformly with the depth below its highest point, and is equal to the product of the depth by the weight per unit volume. For fresh water, the pressure at depth h feet is 62.4 h pounds per square foot. The total pressure of water against a submerged plane are area is equal to the intensity of pressure on the center of gravity times the area.

Problems of flotation, draft, and buoyant stability always involve the density of the liquid and the volume and shape of the floating object. A floating body of mass m in a liquid of density ρ, will float with a volume v submerged, v determined by the relationship $v = (m/\rho)$. That is, a floating body displaces a volume of liquid having the same weight as the body.

The buoyancy may be said to be the force which is equivalent to the weight of the liquid displaced by the submerged portion of the floating object. Buoyancy and weight do not, in general, act in the same vertical line. The weight acts at the center of gravity of the floating object, the buoyancy at the center of gravity of the displaced liquid, called the center of buoyancy. The relative positions of the buoyancy and the weight when a floating object is disturbed from an upright floating position, determines whether it is stable or unstable flotation. If the vertical drawn through the center of buoyancy passes above the center of gravity of the body, there is a righting moment, and the body is stable, whereas if it passes below the center of gravity, it is unstable in that the buoyancy tends to tip the object still further. The intersection of the line of buoyancy with the axis of symmetry of the floating body is the metacenter, and the distance from the metacenter to the center of gravity is the metacentric height. The latter is used to measure the stability of a hull. Another case of flotation is illustrated by the balance of the hydrometer. It is apparent from the above that with a given weight, the volume of immersion varies inversely with the density of the liquid. In other words, a floating body rides higher in a denser liquid. This fact is put to use in the hydrometer, which has a given weight and which is immersed in fluids to measure their density. The hydrometer is calibrated to read the volume submerged directly in terms of density of the liquid.

An important general principle of hydrostatics is that which determines the free liquid surface in equilibrium. The direction of the surface at any point is perpendicular to the resultant of all forces acting upon a particle at that point. Thus, if only gravity is acting, the surface is horizontal or "level"; but if there are capillary forces, or if the external pressure is not uniform, the surface is inclined. An interesting case is that of a liquid rotating uniformly in a cylindrical tub; the surface then assumes the form of a paraboloid of revolution, symmetrical about the vertical axis.

Hydrostatic Equation. The form assumed by the vertical component of the vector equation of fluid motion when Coriolis, earth curvature, frictional, and vertical acceleration terms are considered negligible compared with those involving the vertical pressure force and the force of gravity. Thus,

$$\frac{\partial p}{\partial z} = -\rho g$$

where p is the pressure, ρ the density, g the acceleration of gravity, and z the geometric height.

In compressible flow, the hydrostatic pressure must be defined more carefully. A suitable definition is in terms of the Helmholtz free energy, $A = U - TS$,

$$p = -\left(\frac{\partial A}{\partial(1/\rho)}\right)_T$$

Hydrostatic Equilibrium. The state of a fluid whose surfaces of constant pressure and constant mass (or density) coincide and are horizontal throughout. Complete balance exists between the force of gravity and the pressure force. The relation between the pressure and the geometric height is given by the hydrostatic equation. The analysis of atmospheric stability has been developed most completely for an atmosphere in hydrostatic equilibrium.

Hydrostatic Pressure. The pressure created by a superimposed layer of liquid is hydrostatic pressure. See **Hydrostatic Pressure.**

HYDROTHERMAL DEPOSITS (Submarine). Ocean; Ocean Resources (Mineral).

HYDROTREATING. A specialized kind of hydrogenation in which the quality of liquid hydrocarbon streams is improved by subjecting them to mild or severe conditions of hydrogen pressure in the presence of a catalyst. The objective is to convert undesirable material in the feedstock to either desired materials or easily disposed byproducts, on a highly selective basis. As of the early 1980s about 45% of the crude oil refined in the United States is hydrotreated. Some applications of hydrotreating include: (1) improvement of the burning quality of jet fuels, kerosines, and diesel fuels; (2) purification of light aromatic by-products from pyrolysis operations; (3) pretreatment of naphtha feeds for catalytic reforming units; (4) reduction in sulfur content of residual fuel oils; (5) pretreatment of catalytic cracking feeds and cycle oils by removal of metals, sulfur, nitrogen, and reduction of polycyclic aromatics; (6) desulfurization of distillate fuels; (7) upgrading of lubricating oil quality; and (8) improvement of color, odor, and storage stability of various fuels.

Some of the specific reactions involved include: (1) hydrogenation of monoaromatics to naphthenes to improve burning quality of certain fuels; (2) removal of nitrogen as ammonia from its organic combinations; (3) removal of oxygen from its organic combinations as water; (4) hydrogenation of polycyclic aromatics so that only one aromatic ring remains in the molecule; (5) hydrogenation of diolefins and olefins to paraffins or naphthenes; (6) removal of sulfur from its organic combinations in various types of sulfur compounds by hydrodesulfurization to form hydrogen sulfide; and (7) decomposition and removal of organometals, such as arsenic compounds in naphthas, by retention of these metals on the catalyst. Vanadium and nickel also can be removed.

In the hydrotreating process shown by the accompanying diagram,

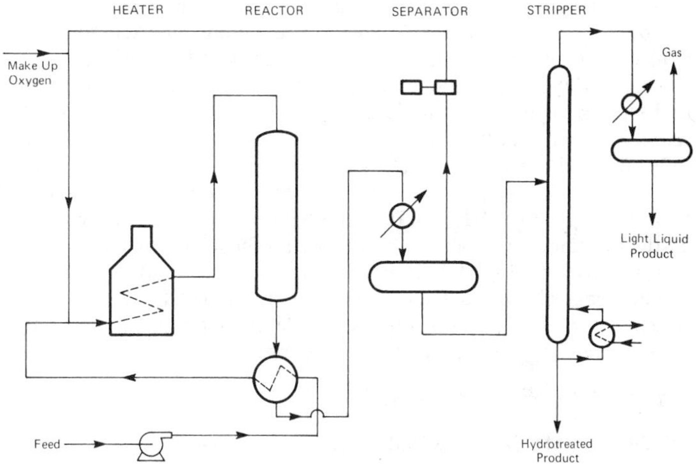

Representative hydrotreating unit. (*UOP Inc.*)

the liquid feed is preheated by exchange with the reactor effluent. It is then heated to the desired reactor-inlet temperature in a fired heater. At this point, recycle hydrogen joins the feedstock. An excess of hydrogen is used to suppress accumulation of deactivating carbonaceous deposits on the catalyst. Fresh makeup hydrogen enters the process to maintain a sufficient supply and also pressure on the system. Cooled effluent from the reactor goes to a separator vessel at which point the recycle or net hydrogen is removed. The liquid then goes to a stripper or stabilizer where hydrogen, hydrogen sulfide, ammonia, water, and light hydrocarbons dissolved in the separator liquid are removed. The stabilized hydrotreated liquid, free of dissolved, unwanted contaminants, is routed to subsequent processing or to product fuel blending.

It is interesting to note that there are over 25 proprietary versions of this basic process. Numerous modifications are required, depending upon the nature of the feedstock and desired end-products.

Technical Staff, UOP Inc., Des Plaines, Illinois.

HYDROXYCARBOXYLIC ACIDS. Carboxylic Acids.

HYDROXYLAMINE. H_2NOH, formula weight 33.02, white, odorless solid, mp 33°C, bp 56°C (22 mm pressure), explosive, soluble in all proportions in H_2O or alcohol. Hydroxylamine is: (1) A weak base forming with acids soluble salts that decompose more or less violently when heated, e.g., hydroxylamine hydrochloride (hydroxylammonium chloride, $H_2NOH \cdot HCl$), mp 151°C, nitrate $H_2NOH \cdot HNO_3$, hemisulfate $H_2NOH \cdot \frac{1}{2}H_2SO_4$. Dihydroxylamine oxalate and trihydroxylamine phosphate are insoluble in H_2O. Hydroxylamine hydrochloride is soluble in alcohol. (2) A weak acid forming with bases soluble salts, e.g., sodium hydroxylamite H_2NONa. Hydroxylamine salt solution is a powerful reducing agent, more especially in alkaline than in acid solution, for example, cupric salt solutions changed to cuprous oxide, silver salt solutions to silver, mercuric chloride solution to mercurous chloride, ferric salt solutions (in acid) to ferrous. Ferrous hydroxide in sodium hydroxide is, however, oxidized by hydroxylamine to ferric hydroxide plus NH_3.

Hydroxylamine reacts with carbonyl group =CO of aldehydes, ketones or quinones, yielding *oximes*, white solids, of definite melting point and used in identification of aldehydes and ketones, e.g., acetaldehyde oxime $CH_3CH{:}NOH$:

Beta-phenylhydroxylamine, N-phenylhydroxylamine, is a white solid, slightly soluble in water, very soluble in alcohol or ether, forms salts with acids, e.g., beta-phenylhydroxylamine hydrochloride $C_6H_5NHOH \cdot HCl$, upon exposure to air the water solution forms azobenzene $C_6H_5N{:}NC_6H_5$. Beta-phenylhydroxylamine reacts (1) with oxidizing agents, such as chromic acid or ferric chloride, to form nitrosobenzene C_6H_5NO, (2) with reducing agents, such as tin plus hydrochloric acid, to form aniline $C_6H_5NH_2$, (3) with alkaline cupric salt solution (Fehling's solution) at room temperature to form cuprous oxide, (4) with ammonio-silver salt solution (Tollen's solution) at room temperature to form silver, (5) in the presence of hydrochloric acid to form paraminophenol $HO \cdot C_6H_4 \cdot NH_2(1,4)$.

Beta-phenylhydroxylamine is formed by reduction of nitrobenzene (1) by zinc and calcium chloride or ammonium chloride solution, (2) by electrolysis in acetic acid plus sodium acetate solution.

Diphenylhydroxylamine is prepared by reaction of nitrosobenzene and phenylmagnesium bromide in anhydrous ether, followed by treatment with H_2O (magnesium hydroxybromide also formed).

When hydroxylamine reacts with aldehydes, the resulting compounds are termed *aldoximes* as, for example, acetaldoxime. $CH_3 \cdot CHO + H_2NOH \rightarrow CH_3 \cdot CH{:}N \cdot OH$ (acetaldoxime) $+ H_2O$. Hydroxylamine reactions with ketones produce *ketoximes*. $(CH_3)_2CO + H_2NOH \rightarrow (CH_3)_2C{:}N \cdot OH$ (dimethylketoxime) $+ H_2O$.

The lower aldoximes are essentially odorless, volatile liquids, and miscible with H_2O in all proportions. The higher members are only slightly soluble. Ketoximes have similar properties.

HYDROXYLASES. Enzyme.

HYDROXYMETHYLPROGESTERONE. Steroids.

HYDROXYPROLINE. Amino Acids.

HYDROZOA. A class of the phylum *Coelenterata* composed chiefly of small animals without common names. Many species are colonial and the colonies of a few, such as the Portuguese man-of-war, are quite large.

The class differs from the other coelenterates in the occurrence of both hydroid polyps and medusae in the same species, usually in alternating generations. In many colonies, additional specialization occurs among the polyps for the performance of different functions; the gonozooids, gastrozooids, and dactylozooids of the siphonophores are such individuals. Hydrozoan medusae differ from jellyfishes and are called medusoids. In some cases, they are specialized forms which remain attached to the colony and show no resemblance to medusae, but the free-swimming forms differ from medusae only in details of structure. The medusoids are sexual reproductive individuals.

Relatively few species of hydrozoans live in fresh water. *Hydra*, the most widely known genus, includes a number of species without a medusa stage. The polyps are solitary and carry on both asexual and sexual reproduction. Several freshwater medusae for which no polyp stage has been discovered are also known from lakes in Europe, Africa, and the Americas. The marine species are numerous.

The class can be conveniently classified as follows:

Order 1. *Hydroidea.* Fixed zoophyte stage.
  Suborder a. *Anthomedusae* (*Athecata*). Polyps and reproductive zooids not protected.
  Suborder b. *Leptomedusae* (*Thecata*). Polyps protected by hydrothecae and reproductive zooids by gonothecase.
Order 2. *Hydrocorallina.* Massive skeleton of calcium carbonate secreted from coenosarc—Hydroid corals.
Order 3. *Trachylinae.* No fixed zoophyte stage, all members being locomotive medusae.
  Suborder a. *Trachymedusae.* Tentacles from margin of umbrella and gonads develop in connection in the radial canals.
  Suborder b. *Narcomedusae.* Tentacles from ex-umbrella away from margin and gonads develop in connection with the manabrium.
Order 4. *Siphonophora.* Pelagic forms; colony usually exhibits polymorphism of its zooids.

HYENA (*Mammalia, Carnivora*). A large animal slightly resembling a wolf, but more closely related to the civets. The Aard-Wolf (*Protelinae*) and the Hyena (*Hyaeninae*) make up a small, special grouping in the order of *Carnivora.* The aard-wolf is believed by some authorities to bridge the gap between the Hyaenines and the Viverrines. See also **Viverrines.** Some years ago, investigators believed the aard-wolf was a type of civet and belonged with the Viverrines. This animal is difficult to describe, appearing something like a clumsy dog, with a rather fox-like face, with wooly hair along the back in a form of a permanently erected crest, giving it something of a skunk-like appearance. The aard-wolf differs from the hyena, in that it has five toes on the forefeet; four on the hind feet. Because the animal's teeth are widely set and reduced in number and size, the principal diet is comprised of insects or very decomposed meat or newly born animals. The aard-wolf is found uncommonly in eastern Africa (north of the Kalahari Desert) and on the west coast of Africa as far north as Angola.

At one time, hyenas ranged over much of Europe in climes south of Scotland and the Scandinavian countries and reached eastward through eastern Europe and central Asia. They are found today in most parts of Africa south of the great deserts.

The Striped Hyena (*Hyaena*) is found in Africa as described, as well as in fairly large numbers in northern India. It is also characterized by a crest along its back and by striping on the flanks, with cross-stripes on the legs. The animal is sturdily built, with massive head and somewhat disproportionately long front legs.

The Spotted Hyena (*Crocuta*) is larger than the striped hyena and its limbs are better proportioned. The animal is extremely powerful and can crack large bones, including those of the elephant and hippopotamus. It can put up a good fight with the Big Cats. The striped hyena is considered of very poor disposition, sometimes described as sneaky and generally unpleasant. However, except when in danger, it is described as cowardly. In nature, the animal is extremely dirty, carrying around a most offensive odor. Surprisingly, however, the animal is reported to make an excellent, docile, and trustworthy pet if taken young and trained.

The hyenas are known for their bloodcurdling howling and noise like an insane laugh which usually occurs at the end of a barking streak. On flat ground, their speed exceeds that of a horse.

HYGEIA. Asteroid.

HYGROMETER. Any apparatus for measuring atmospheric humidity, either absolute or relative. There are six basically different means of transduction used in measuring the water vapor content of the atmosphere, and an equal number of types of hydrometers: (1) The thermodynamic method of transduction is represented by the *psychrometer.* See Fig. 1. This instrument consists of two thermometers, one of which (the dry bulb) is an ordinary glass thermometer, while the other (the wet bulb) has its bulb covered with a jacket of clean muslin, which is saturated with distilled water prior to an observation.

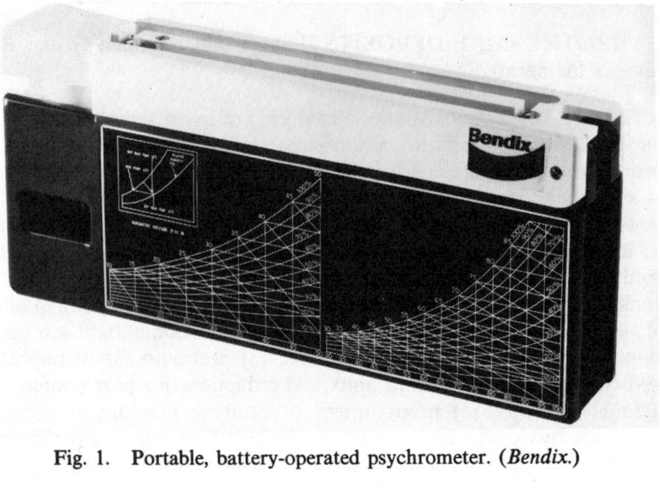

Fig. 1. Portable, battery-operated psychrometer. (*Bendix.*)

When the bulbs are suitably ventilated, they indicate the thermodynamic wet- and dry-bulb temperatures of the atmosphere. A *sling psychrometer* has the wet- and dry-bulb thermometers mounted upon a frame connected to a handle at one end by means of a bearing or a length of chain. Thus, it may be whirled by hand in order to provide the necessary ventilation. Or, a hand-aspirated instrument of the type shown in Fig. 2 may be used.

Fig. 2. Hand-operated psychrometer. (*Bendix.*)

(2) There are several variations of hygrometers that depend upon a change of physical dimensions of a sensitive element due to the absorption of moisture: (a) The *hair hygrometer* is one in which the strands of human hair, mounted under tension, expand with increasing relative humidity and contract with decreasing relative humidity. One end of a set of strands is fixed, and the other end operates a set of levers whose initial movement is magnified mechanically to cause a pointer or other indicating device to ride over a calibrated scale. The length variation of a properly treated hair is approximately logarithmic between relative humidity limits of 20 to 100%. The lag time of the response of the hair increases with decreasing temperatures and becomes virtually infinite at temperatures below −40°C. (b) In the *torsion hygrometer,* the rotation of the hygrometric element is a function of the humidity. Such hygrometers are constructed by taking a substance whose length is a function of the humidity and twisting or spiraling it under tension in such a manner that a change in length will cause a further rotation of the element. (c) The *goldbeater's-skin hygrometer* relies upon variations of the physical dimensions of goldbeater's skin, which is the prepared outside membrane of the large intestine of an ox, used in goldbeating to separate the leaves of the metal.

(3) Some hygrometers depend upon the condensation of moisture: (a) The *dew-point hygrometer* is used to determine the dew point; that is, a mean of the temperature at time of formation and evaporation of dew. A parcel of air is cooled at constant pressure, usually by contact with a refrigerated polished metal surface. Condensation appears upon the metal surface at a temperature slightly below that of the thermodynamic dew point of the air. The observed dew point will differ from the thermodynamic dew point, depending upon the nature of the condensing surface, the condensation nuclei, and the sensitivity of the condensate-detecting apparatus. Some instruments require visual observation of dew drops on the instrument and the disappearance of the same drops but others use photoelectric cells to determine the time of dew formation and evaporation. (b) The *frost-point hygrometer* is used to determine the frost-point, i.e., the highest temperature at which atmospheric moisture will sublimate in the form of hoarfrost on a cooled polished surface. Air under test is passed continuously across a polished surface, whose temperature is adjusted so that a thin deposit of frost is formed, in equilibrium with the air. The surface temperature is varied by directing a jet of cooled liquid against the under surface or by passing an electrical current through a heating coil surrounding the surface. The deposit is observed either visually or by means of photoelectric cells.

(4) Other hygrometers depend upon the change of chemical or electrical properties due to the absorption of moisture: (a) The *absorption hygrometer* utilizes a hygroscopic chemical to absorb water vapor in the atmosphere. The amount of vapor absorbed may be determined in an absolute manner by weighing the hygroscopic material, or in a non-absolute manner by measuring a physical property of the substance that varies with the amount of water vapor absorbed. (b) The *electrical hygrometer* uses a transducing element whose electrical properties are a function of atmospheric water vapor content. For example, the carbon-film hygrometer element is a plastic strip coated with a film of carbon black dispersed in a hygroscopic binder. Variations in atmospheric moisture content vary the volume of the binder and thus change the resistance of the carbon coating. Similarly, the hygroscopic coating of the flat plastic humidity strip can be measured for electrical resistance as a function of the amount of moisture absorbed and the temperature of the strip.

(5) The *diffusion hygrometer* depends upon the diffusion of water vapor through a porous membrane. In its simplest form, this instrument consists of a closed chamber having porous walls and containing a hygroscopic compound. The absorption of water vapor by the compound causes a pressure drop within the chamber, which is measured by a manometer.

(6) The *spectral hygrometer* depends upon measurements of the absorption spectra of water vapor. The amount of precipitable moisture in a given region of the atmosphere is determined by measuring the attenuation of radiant energy caused by the absorption bands of water vapor.

See also **Precipitation and Hydrometeors.** For references see entries on **Climate;** and **Meteorology.**

HYGROSCOPIC. 1. Pertaining to a marked ability to accelerate the condensation of water vapor. In meteorology, this term is applied principally to those condensation nuclei composed of salts that yield aqueous solutions of a very low equilibrium vapor pressure compared with that of pure water at the same temperature. Condensation on hygroscopic nuclei may begin at a relative humidity much lower than 100% (about 75% for sodium chloride); while on so-called non-hygroscopic nuclei, which merely furnish sufficiently large (by molecular standards) wettable surfaces, relative humidities of nearly 100% are required.

2. Descriptive of a substance, the physical characteristics of which are appreciably altered by effects of water vapor. The hygroscopicity of certain materials has been advantageously utilized in humidity measurement and control devices; for example, the hair element of a hair hygrometer.

HYGROTHERMOGRAPH. Precipitation and Hydrometeors.

HyL STEELMAKING PROCESS. Iron Metals, Alloys, and Steels.

HYMEN. Gonads.

HYMENOPTERA. One of the large orders of insects, including ants, bees, wasps, sawflies, and many species without common names. The mouth is formed for biting or for biting and sucking and the wings, when present, are four in number and membranous. Metamorphosis is complete. The order includes plant-eating, parasitic, and predacious species, and in the ants and bees displays some of the finest examples of social organization. The order includes about 70,000 species.

Owing to its extent and diversity this division of the insects includes many species of economic importance. Some of the sawflies and gall wasps are harmful to plants and on the other hand the fig insects are beneficial and the galls produced by some gall wasps are of commercial value. Many parasitic species are of undoubted value in holding in check important insect pests. Ants are sometimes very troublesome and the large carpenter bee sometimes damages wood in construction. The most important single species is the honeybee, which is of great

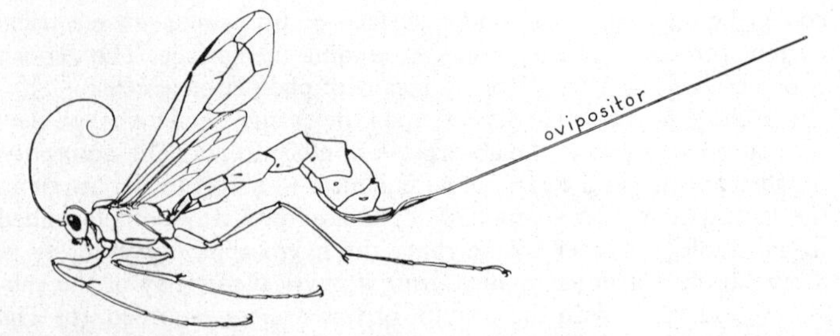

The degrees of specialization represented by members of *Hymenoptera* are exemplified by this ichneumon wasp, which incorporates a greatly extended proboscis for placing eggs in the bodies of other insects, notably caterpillars. (*USDA diagram.*)

value as a producer of honey and wax and in the cross pollination of fruit trees.

The main families of Hymenoptera include:

Andrenidae (also called *Halictidae*)	Mining bees, sweat bees
Apidae	Honeybees
Anthophoridae	Anthophorid bees
Bombidae (also called *Bremidae*)	Bumblebees
Braconidae	Braconid wasps
Cephidae	Stem sawflies
Ceratinidae	Small carpenter bees
Chalcididae	Chalcid wasps
Chrysididae	Cuckoo wasps
Colletidae (also called *Hylaeidae*)	Bifid-tongued bees or plasterer bees
Cynipidae	Gall wasps
Dryinidae	Dryinid wasps
Eumenidae	Mud or potter wasps
Evaniidae	Ensign wasps
Formicidae	Ants
Ichneumonidae	Ichneumon wasps
Megachilidae	Leaf-cutting bees and mason bees
Mutillidae	Velvet ants
Nomadidae	Cuckoo bees
Pompilidae (also called *Psammocharidae*)	Spider asps
Proctotrupidae (also called *Serphiodea*)	Egg-parasite wasps
Prosopidae (also called *Hylaeidae*)	Obtuse-tongued bees or wasplike bees
Scoliidae	Vespoid digger wasps
Siricidae	Horn-tails
Sphecidae	Digger wasps, mud-daubers, thread-waisted wasps
Tenthredinidae	Sawflies
Vespidae	Social wasps, paper-nest wasps, hornets, yellow jackets
Xylocopidae	Large carpenter bees

HYOIDS. Cats.

HYPABYSSAL. A general term sometimes used by structural geologists and petrologists to designate those igneous rocks such as sills and dikes which have congealed under less pressure than the plutonic or deep-seated rocks, but under greater pressure than the effusive rocks (lavas).

HYPERBARIC OXYGEN THERAPY. Gangrene.

HYPERBOLA. A conic section obtained by a plane cutting both nappes of a right-circular conical surface. It is the locus of a point which moves so that the difference of its distances from two foci is a constant. Its eccentricity is greater than unity.

The standard equation may be taken as $x^2/a^2 - y^2/b^2 = 1$. The curve is a central conic for it is symmetric about both the *X*- and *Y*-axes when placed in this standard position and the coordinate origin is its center. The transverse axis, coincident with the *X*-axis, is of length $2a$; the conjugate axis, along the *Y*-axis, has length $2b$ ($b < a$). The distance from the center of the hyperbola to either focus is $\sqrt{a^2 + b^2}$; the eccentricity, $e = \sqrt{a^2 + b^2}\ a$; the length of the latus rectum is $2b^2/a$; the equations for the directrices are $x = \pm a/e$, the same as for the ellipse. The distance from any point on the hyperbola to a focus is a focal radius and the differences between any two focal radii equals $2a$. The lines $y = \pm bx/a$ are asymptotes to the hyperbola. If the length of the transverse axis becomes equal to that of the conjugate axis ($a = b$), the curve is an equilateral or rectangular hyperbola. In this case, the asymptotes are perpendicular to each other. If the coordinate axes are rotated so that they coincide with the asymptotes, the equation for the rectangular hyperbola becomes $xy = a^2/2$, a form which is familiar to students of physical chemistry as Boyle's law.

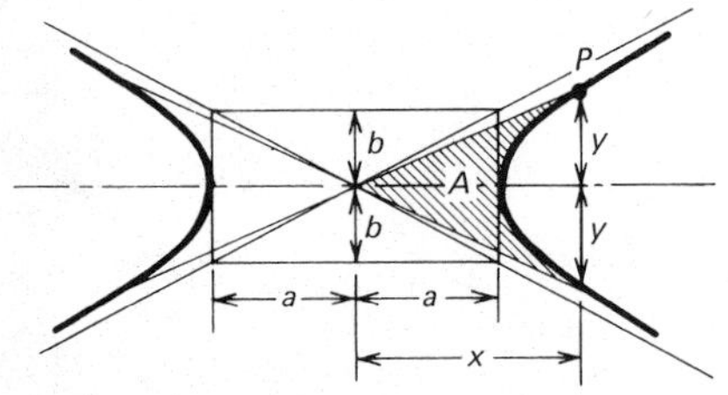

Major parameters of hyperbola.

The polar equation of the hyperbola is $r = a(e^2 - 1)/(e \cos\theta - 1)$ and its parametric equations are $x = a \cosh u$, $y = b \sinh u$ or $x = a \sec\phi$, $y = b \tan\phi$. Its evolute is similar to that of the ellipse $X^{2/3} - Y^{2/3} = 1$, where $X = ax/e^2$, $Y = by/e^2$.

With reference to the accompanying diagram, the shaded area $= ab \log_e(x/a + y/b)$. In an equilateral hyperbola, $a = b$, in which case the shaded area $= a^2 \log((x + y)/a) = a^2 \log(a/(x - y)) = a^2 \sin h^{-1}(y/a)$, or $a^2 \cos^{-1}(x/a)$.

See also **Conic Section;** and terms listed under **Mathematics.**

HYPERBOLIC FUNCTION. Combinations of $e^{\pm z}$ with properties similar to those of the trigonometric functions. They are defined by:

$$\sinh z = (e^z - e^{-z})/2 = z + \frac{z^3}{3!} + \frac{z^5}{5!} + \cdots$$

$$\cosh z = (e^z + e^{-z})/2 = 1 + \frac{z^2}{2!} + \frac{z^4}{4!} + \cdots$$

$$\tanh z = \sinh z/\cosh z; \quad \coth z = 1/\tanh z$$

$$\operatorname{sech} z = 1/\cosh z; \quad \operatorname{csch} z = 1/\sinh z$$

If n is a positive integer, $i = \sqrt{-1}$, $u = n\pi i$; $\sinh u = \tanh u = 0$; $\cosh u = (-1)^n$; $\sinh(z + u) = (-1)^n \sinh z$;

$$\cosh(z + u) = (-1)^n \cosh u.$$

For real z, the hyperbolic functions are related to the hyperbola in the same way that the trigonometric functions are related to a circle. If $x^2 \pm y^2 = a^2$ is the equation for a circle of radius a, or with the minus sign, a rectangular hyperbola, the parametric equations are $x = a \cos\phi$, $y = a \sin\phi$ and $x = a \cosh z$, $y = a \sinh z$, respectively. The equations $x = a \sec\phi$, $y = a \tan\phi$ also apply to the hyperbola, and ϕ is the same angle as that for the circle. Comparison of these results shows that $\cosh z = \sec\phi$, $\sinh z = \tan\phi$, with $-\pi/2 < \phi < \pi/2$. Further relations are obtained from the equations defining the hyperbolic functions: $\operatorname{sech} z = \cos\phi$, $\operatorname{csch} z = \cot\phi$, $\tanh z = \sinh\phi$, $\coth z = \csc\phi$. The equation $\sinh z = \tan\phi$ determines ϕ as a function of z. It is called the gudermannian of z, and thus $\phi = \tan^{-1} \sinh z = \operatorname{gd} z$.

Again, with real z, hyperbolic and circular (trigonometric) functions are related as follows: $\sinh iz = i \sin z$; $\cosh iz = i \cos z$; $\tanh iz = i \tan z$. Additional formulas, similar to those familiar from trigonometry, are: $\cosh^2 z - \sinh^2 z = 1$; $1 - \tanh^2 z = \operatorname{sech}^2 z$; $\cosh^2 z + \sinh^2 z = \cosh 2z$; $2 \sinh z \cosh z = \sinh 2z$.

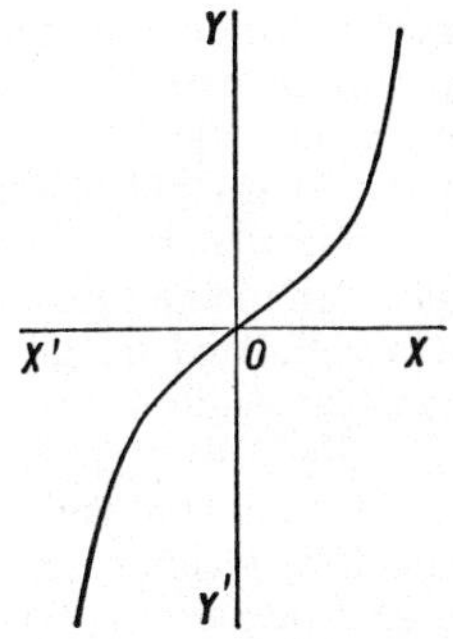

$y = \sinh\ x.$

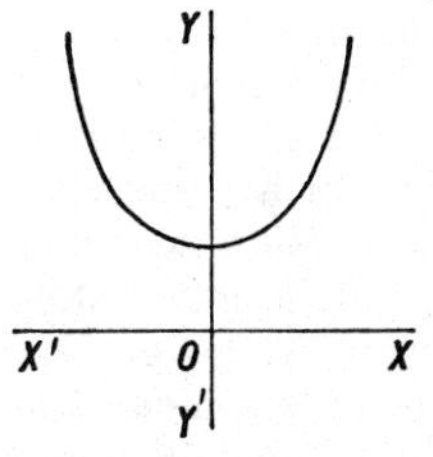

$y = \cosh\ x.$

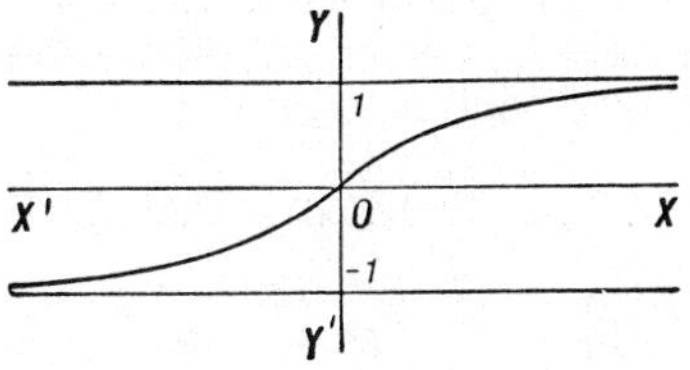

$y = \tanh\ x.$

Major hyperbolic functions.

The inverse hyperbolic functions are also denoted in a manner similar to that for the inverse trigonometric functions. Thus, if $y = \sinh z$, the inverse function is the angle whose hyperbolic sine is y, or $z = \sinh^{-1} y = \text{arc sinh } y$. The following relations may be obtained from the definitions of the various functions:

$$\sinh^{-1} z = \ln(z + \sqrt{z^2 + 1})$$

$$\cosh^{-1} z = \ln(z \pm \sqrt{z^2 - 1}); \quad z \geq 1$$

$$\tanh^{-1} z = \frac{1}{2} \ln \frac{1+z}{1-z}; \quad z^2 < 1$$

$$\coth^{-1} z = \frac{1}{2} \ln \frac{z+1}{z-1}; \quad z^2 > 1$$

$$\operatorname{sech}^{-1} z = \ln \frac{1 \pm \sqrt{1 - z^2}}{z}; \quad 0 < z \leq 1$$

$$\operatorname{csch}^{-1} z = \ln \frac{1 + \sqrt{1 + z^2}}{z}$$

See also terms listed under **Mathematics.**

HYPERBOLIC GEOMETRY. Geometry.

HYPERBOLIC NAVIGATION. Navigation.

HYPERBOLIC SPIRAL. A transcendental plane curve, also known as a reciprocal spiral, with polar equation $r\theta = a$ and thus inverse to Archimedes' spiral. It begins at an infinite point from the pole, but as it winds around it never reaches the pole. It has an asymptote $y = a$. Its equation can also be taken as $xt = a \cos t$, $yt = a \sin t$, where t is a parameter.

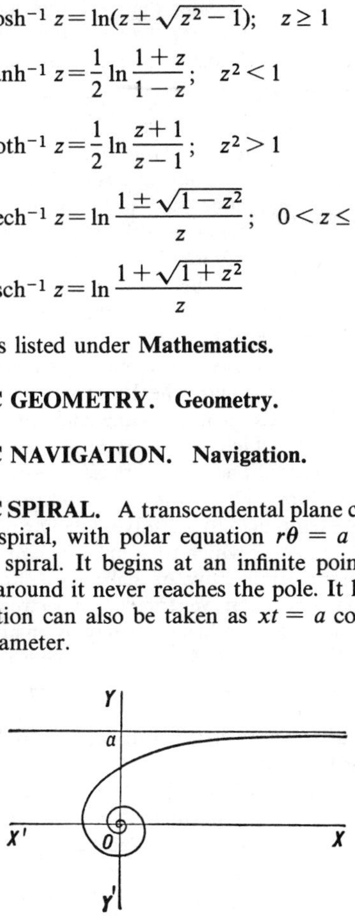

Hyperbolic spiral.

HYPERBOLOID. A central quadric surface with one or two negative terms in its equation. If there is only one, so that $x^2/a^2 + y^2/b^2 - z^2/c^2 = 1$, the surface is a hyperboloid on one sheet. It is given this name because any point on the surface may be reached from any other point on the surface. A plane parallel to the XY-plane cuts out an ellipse but if the sections are parallel to the XZ- or YZ-planes the results are hyperbolas. When $a = b$, the sections by planes $z =$ constant are circles and the surfaces can be generated by revolving the hyperbola, $x^2/a^2 - z^2/c^2 = 1$ about its conjugate axis, the Z-axis.

If there are two negative terms in the equation, $x^2/a^2 - y^2/b^2 - z^2/c^2 = 1$, the surface is a hyperboloid of two sheets, separated into two parts symmetrically located above and below the planes $x =$ constant. Traces parallel to the XY- and XZ-planes are hyperbolas and traces parallel to the YZ-planes are ellipses, provided $x > a$. When $b = c$, the sections by planes $x =$ constant are circles and a surface of revolution results when the hyperbola $x^2/a^2 - y^2/c^2 = 1$ is rotated about its X- or transverse axis.

See also **Quadric Surface;** and terms listed **Mathematics.**

HYPERCALCEMIA. Parathyroid Glands.

HYPERCALCIURIA. Kidney and Urinary Tract.

HYPERCHOLESTEREMIA. Arterial and Venous Disorders.

HYPERCOMPLEX NUMBER. Quaternion.

HYPERCONJUGATION. The description of the properties of a molecule in terms of resonance structures in which an atom or group is not joined by any sort of bond to the atom to which it is ordinarily considered linked. Also called no-bond resonance. The hypothesis of hyperconjugation has been advanced to interpret some properties of substances containing but 1 double bond by analogy with those of substances containing conjugated double bonds. Consider a substance with a terminal structure $H_3C{-}CH{=}CH{-}\ldots$. One of the possible resonating structures of this group is

$$H_3{\equiv}C{-}CH{=}CH{-} \leftrightarrow H_3{=}C{=}CH{-}CH{-}$$

the dotted line indicating two unpaired electrons with opposite spins.

HYPEREUTECTIC ALLOY. An alloy with a composition falling on the right of the eutectic point of a binary phase diagram that freezes with a structure containing some eutectic.

HYPERFINE STRUCTURE. In general, a set of very closely spaced lines in atomic spectra or other kinds of spectra. There may be many causes of hyperfine structure: (1) for a single atomic species or nuclide, the occurrence of spectral lines as doublets, triplets, etc., due to the interaction, or coupling, of the total angular momentum of the orbital electrons with the nuclear spin and associated magnetic moment; (2) for an element consisting of several isotopes, the occurrence of components for each spectral line that is observable under high resolution, each isotope contributing one or more components. This type of hyperfine structure is often called isotope structure to differentiate it from the first type of hyperfine structure discussed above. See also **Atomic Spectrum.**

HYPERGEOMETRIC DISTRIBUTION. A distribution of a discrete random variable generally associated with sampling from a finite population without replacement. The frequency of r "successes" and $n - r$ "failures" in a sample of n so drawn from a population of N in which there are N_p "successes" and N_q "failures" $(p + q = 1)$ is

$$\frac{1}{N^n} \binom{n}{r} (N_p)^{[r]} (N_q)^{[n-r]}$$

where $N^{[r]} = N(N - 1) \cdots (N - r + 1)$. As N tends to infinity the distribution tends to the ordinary binomial form. The distribution derives its name from the fact that the probability generating function may be put in the form of a hypergeometric series.

HYPERGLYCEMIA. Diabetes Mellitus.

HYPERINSULISM. **Diabetes Mellitus.**

HYPERKALEMIA. **Kidney and Urinary Tract.**

HYPERKINESIA. **Kinesiology.**

HYPERLIPIDEMIA. **Arterial and Venous Disorders; Lipids.**

HYPERONS. These are subatomic particles that are more massive than nucleons (protons and neutrons). *Strangeness* is a property of elementary particles found useful in classifying hyperons. Each particle is assigned a strangeness quantum number S which is related to the electric charge Q, the isospin number T, and the baryon number B by the formula $Q = T + (S + B)/2$. ($T = \frac{1}{2}$ for a proton and $-\frac{1}{2}$ for a neutron; other particles may have $T = 0$ or $T = 1$, depending on the type.) Strangeness is conserved in reactions involving the strong interaction. The selection rules resulting from strangeness conservation are important in understanding why some reactions take place much more slowly than others. See also **Particles (Subatomic).**

HYPEROPIA. **Vision and the Eye.**

HYPERPARATHYROIDISM. **Parathyroid Glands.**

HYPERPROLACTINEMIA. **Infertility.**

HYPERSENSITIVITY. **Allergy; Immune System and Immunology.**

HYPERSONIC FLOW. In aerodynamics, flow of a fluid over a body at speeds much greater than the speed of sound and in which the shock waves start at a finite distance from the surface of the body.

HYPERSTHENE. The mineral hypersthene is an orthorhombic pyroxene, chemically a ferro-magnesian silicate, differing from enstatite in that the iron content is considerable (FeO being greater than 15%). A general formula is $(Mg, Fe)SiO_3$. It is usually found as a massive mineral, whose crystals tend to be prismatic or tabular in habit. It has a distinct prismatic cleavage; fracture, uneven; brittle; hardness, 5–6; specific gravity, 3.42–3.84; luster, pearly to somewhat metallic; color, brownish-green, brown, greenish-black to grayish-black; streak, grayish-brown; translucent to opaque. Hypersthene is often associated with labradorite in gabbro and norite and in extrusive rocks like andesite. It is occasionally encountered in meteorites. Hypersthene is associated with pyrrhotite in Bavaria, with labradorite on the Isle St. Paul, Labrador. It is also found in Montmorency County, Quebec; and in the United States in the rocks of the Cortlandt series in the Hudson River Valley, and the andesites of Colorado and northern California. Superb crystals of exceptional size and quality have been found growing into and within the almandine-pyrope garnets at Gore Mountain, North River, New York. The rarity of hypersthene in crystal form makes this occurrence noteworthy. The word hypersthene comes from the Greek words meaning *strong* or *tough.*

See also **Pyroxene.**

HYPERTENSION (High Blood Pressure). Commonly regarded and treated as a disorder in itself, high blood pressure may be more accurately described as a major symptom of a complex of disorders. Not all of these disorders are present in one person. There is a variety of patterns of these underlying disorders which readily explains what was a puzzle for many years—namely, the manner in which different people with the symptom of hypertension react differently to various drug therapies. Considering the numerous body systems—circulatory, nervous, endocrine, excretory, among others—that interact in different ways to produce a universal symptom (hypertension), it would indeed be surprising if all persons with high blood pressure did react precisely in the same manner to therapy.

Considering the heart as a pump, (1) the *systole* is the period of the heart's contraction, or the contraction itself—the systolic pressure represents the highest arterial blood pressure; (2) the *diastole* is the period of the heart's dilation—the diastolic pressure represents the lowest arterial blood pressure that occurs between the pulse waves. The arterial blood pressure in humans is usually measured in the arm at the brachial artery, preferably with the patient seated or lying down with the arm slightly flexed and at heart level. In a thorough examination for hypertension, multiple readings will be made—both arms and legs. With the aid of a stethoscope, the examiner will determine both systolic and diastolic pressure with the *sphygmomanometer,* an instrument whose pressure scale is calibrated in millimeters of mercury. See **Manometer;** and **Sphygmomanometer.**

Statistically, over decades, expected ranges for these pressures in healthy persons have been established. These ranges have become established standards against which individual readings are compared and from which a diagnosis of high blood pressure (*hypertension*) or low blood pressure (*hypotension*) is made. The pressures in the arteries and the veins were first measured as early as 1733 by Stephen Hales, who used a rather crude measurement technique in making determinations of these pressures in a mare. By 1828, Hales had developed a method using a U-tube manometer, the prototype of current instruments.

The statistical averages for blood pressure of healthy adults are:

Systolic	110–120 mm mercury
Diastolic	65–80 mm mercury

Some observers have reported that the systolic pressure is higher in men than in women. The normal upper limits of systolic pressure are

140 mm mercury (men)
130 mm mercury (women)

In a conservative approach to hypertension, treatment is indicated as follows:

Age	*Systolic Pressure*
Under 35	Greater than 140 mm mercury
35–59	Greater than 150 mm mercury
60+	Greater than 160 mm mercury

According to Koch-Wester (1973), *hypertension* exists when the systolic pressure exceeds 150 mm mercury and the diastolic pressure is greater than 90 mm mercury. *Borderline* hypertension has been defined as the intermittent elevation of systolic or diastolic pressure above the accepted normal value for a person's age and sex.

Hypertension has been variously estimated to affect between 20 and 35 million persons in the United States alone and thus it is considered the most common of the chronic disorders. Of the millions of people affected, it is estimated that, as of the early 1980s, only about 50% of cases have been diagnosed and are known to the individuals. Of the remaining 50% of cases, only about half are being treated. Hypertension is considered a major health problem because the disorder predisposes individuals to debilitating and often fatal diseases, the major categories of which are heart attack and heart diseases, stroke, and kidney failure. The risk of these consequences is *greatly reduced* with proper therapy for lowering blood pressure.

The foregoing is exemplified by a study made in Framingham, Massachusetts several years ago, which took into account five risk factors: (1) Glucose intolerance (indicative of diabetes), (2) cholesterol level, (3) cigarette smoking, (4) left ventricular hypertrophy (increase in volume of a tissue or organ produced entirely by enlargement of existing cells), and (5) hypertension. See accompanying table.

Primary and Secondary Hypertension. Traditionally, authorities in the field have made a distinction between two forms of hypertension. *Primary* or *essential* hypertension is a disorder of unknown etiology. This form accounts for approximately 90% of all cases of hypertension. In *secondary* hypertension, a cause for the disorder can be identified. The causes of secondary hypertension are many and include: various drugs, such as amphetamines, oral contraceptives, estrogens, steroids, and thyroid hormones; increased intracranial pressure; certain tumors, such as pheochromocytoma; primary aldosteronism (abnormal aldosterone secretion by adrenal cortex, causing excessive loads of potassium and muscular weakness); several renal diseases, such as chronic pyelonephritis, diabetic nephropathy, glomerulonephritis, gout, polycysystic disease, vasculitis, and renovascular hypertension, among others. Hypertension is also associated with toxemia of pregnancy, acute

EFFECT OF VARIOUS RISK FACTORS (Including Hypertension) ON OCCURRENCE OF CARDIOVASCULAR DISEASE
(Within eight years in a 45-year-old male)

Risk Factor Effects	Glucose Intolerance	Cholesterol Level[1]	Smoke Cigarettes	Left Ventricular Hypertrophy	Probable Cases per Thousand at a Systolic Blood Pressure of (Millimeters Mercury): 105	135	165	195
ONE RISK FACTOR PRESENT								
None present	No	Low	No	No	22	35	54	84
Glucose intolerance (GI)	Yes	Low	No	No	39	61	95	143
High cholesterol (HC)	No	High	No	No	44	68	105	158
Smoking (SM)	No	Low	Yes	No	38	59	91	138
Left ventricular hypertrophy (LVH)	No	Low	No	Yes	60	93	141	208
COMBINED MULTIPLE RISK FACTORS PRESENT								
GI + HC	Yes	High	No	No	145	214	304	411
GI + HC + SM	Yes	High	Yes	No	229	323	433	550
GI + HC + SM + LVH	Yes	High	Yes	Yes	460	577	686	778

NOTES: The data are based upon a follow-up of patients in the Framingham Study. Framingham males in the study had the following characteristics at the age of 45 years:

Average systolic blood pressure, 131 millimeters mercury; average serum cholesterol level, 234 milligrams/100 milliliters; 0.7% have definite left ventricular hypertrophy as shown by an electrocardiogram; 3.9% have glucose intolerance. Considering these average values, the probability of having cardiovascular disease within 8 years is 75/1000.

[1] Low cholesterol level is considered 185 milligrams/100 milliliters; high cholesterol level, 335 milligrams/100 milliliters.

pulmonary edema, acute myocardial infarction, dissecting aortic aneurysm, and cerebral hemorrhage.

Primary (Essential) Hypertension

Although considerable progress has been made during the past few years in understanding the complex root causes which contribute to primary hypertension, much of this information has stemmed from observing the actions of various drugs used in the therapy of hypertension.

The Renin-Angiotensin-Aldosterone System. In the early 1900s, Tigerstedt and Bergman suggested that *renin*,[1] a proteolytic enzyme elicited by ischemia of the kidneys or by diminished pulse pressure, played an important role in blood pressure homeostasis and in the pathogenesis of hypertension. In a revival of interest in the role of renin, which commenced during the 1960s, considerable new information has been gained, with an increasing implication of the kidneys in hypertension. Additional reninlike enzymes have been identified and their possible functions are being researched. In a rather complex pathway, renin cleaves renin substrate to yield *angiotensin I*, a decapeptide and apparently quite inactive physiologically. Through further enzymatic action during its passage through the pulmonary circulation, angiotensin I is cleaved to produce *angiotensin II*. This substance is the most potent vasoconstrictor known, causing constriction of the arterioles (small arteries that branch to form the capillaries—the smallest of blood vessels at the sites where the blood exchanges nutrients and waste products with the tissues).

Angiotensin also stimulates adrenocortical production of aldosterone, which promotes the reabsorption of sodium and water by the renal tubules. The resulting augmentation of the fluid content of the circulatory system elevates the blood pressure. The action of the angiotensins also involves the nervous system, releasing the neurotransmitter norepinephrine by the nerve terminals of the sympathetic nervous system and by the adrenal medulla (inner portion of adrenal gland). This action potentiates the action of the norepinephrine, which causes increased blood pressure as the result of constriction of the arterioles. This complex is known as the *renin-angiotensin-aldosterone system.* When this system is functioning normally, a decrease of pressure of the blood flowing through the kidneys will stimulate renin release, while an increase of that pressure "signals" the kidney to halt the release of renin as well as angiotensin, an action accomplished by a feedback mechanism. Thus, any increase in blood pressure will be transient. But, when the feedback apparatus dysfunctions, chronic hypertension may result.

[1] Not to be confused with rennin, an enzyme secreted by the glands of the stomach which causes curdling of milk.

Statistics show that about 90% of persons with primary hypertension have elevated levels of renin in the blood, although they usually do not exhibit symptoms of kidney damage. One researcher has found that persons with high plasma renin activity run the highest risk of heart attack, stroke, and kidney failure. The reverse situation holds for those persons with low plasma renin activity.

Some researchers have learned that angiotensin II must combine with specific receptors on target organs before its effects can be produced. Most peptides are angiotensin antagonists and thus, by binding with the receptors, prevent angiotensin from binding. These antagonist substances have been synthesized by several investigators (Cleveland Clinic Foundation; Washington University Medical School). One of these antagonists is the octapeptide called Saralasin® (first synthesized by Norwich Pharmacal Company). It has been found that intravenous injection of this drug will lower the blood pressure to near-normal levels if the cause of hypertension is renin. This kind of information has proved helpful in determining the most effective drug therapy for primary hypertension.

Other investigators have found that the destruction of a portion of the mid-brain (*subnucleus medialis*) will negate the blood pressure response, probably because the area is involved in the control of peripheral resistance to blood flow. Apparently angiotensin II has the ability to cross the blood-brain barrier. More recently, some researchers have suggested that angiotensin is synthesized in the brain. Much of the experimentation to date has been carried on with laboratory dogs.

Involvement of the nervous system in hypertension has attracted much interest in recent years and ultimately may prove or disprove the association of stress with hypertension. Other investigators have found that prostaglandin E_2 tends to decrease blood pressure by countering the angiotensin-induced constriction of the blood vessels.

Treatment of Primary Hypertension. With increasing knowledge, new classes of drugs may be introduced within a few years. As of the early 1980s, drug therapy consists mainly of diuretics, sympatholytics, vasodilators, and renin-blocking agents. Of these antihypertensive drugs, diuretics are most commonly used. Most often, a treatment regimen will commence with diuretics. The function of diuretics is described in the entry on **Diuretics.**

When diuretic therapy is not successful, the next step taken is usually the administration of *sympatholytic agents.* Several of these agents, with varying side effects, are available and include methyldopa (Aldomet®), clonidine (Catapres®), propanolol (Inderal®), metoprolol (Lopressor®), guanethidine (Ismelin®), respirine (Raudixin®); Serpasil®; Sandril®), and prazosin (Minipress®). Of these, propranolol and methyldopa are most frequently used. Some physicians initiate antihypertensive therapy with propanolol rather than diuretics, particularly in younger patients. A sympatholytic agent blocks transmission of impulses from the postganglionic fibers to effector organs or tissues, inhibiting smooth muscle contraction and glandular secretion. In some cases, a vasodilator, such as hydralazine (Apresoline®) will be used in combination with propranolol and a diuretic. The resin-blocking agents are still in an experimental phase and, in the United States, have yet to be approved by the Food and Drug Administration.

In accelerated hypertension and hypertensive emergencies, as may occur in malignant hypertension, hypertensive encephalopathy, acute left ventricular failure, or cardiac ischemia, more potent drugs may be required. These include diazoxide (Hyperstat®), sodium nitroprusside (Nipride®), and the previously mentioned hydralazine—all vasodilators. Trimethaphan (Arfonad®) is sometimes the drug of choice when nitroprusside cannot be used.

Diagnosis. In the diagnosis of both primary and secondary hypertension, a rather large number of determinations and procedures may be involved. These include: complete blood count; urinalysis; serum potassium; blood urea nitrogen; fasting blood sugar; serum cholesterol; serum uric acid; serum sodium, chloride, carbon dioxide, creatinine, and triglycerides; urine potassium, creatinine, metanephrine, and vanillylmandelic acid; plasma renin activity; urine aldosterone, plasma aldosterone, plasma cortisol—as well as electrocardiography, chest x-ray, intravenous pyelogram, radioisotope renal scan, renal arteriography, and selective renal or adrenal vein catheterization, depending upon the complexity of a particular case.

Racial Differences in Blood Pressure Control. As reported by Voors, et al. (1979), a sample of children from an entire biracial geographical population was examined for blood pressure. In the sample, stratified by diastolic blood pressure and reexamined 1 to 2 years later, it was found that dopamine beta-hydroxylase, renin activity, and resting heart rate differed in black and white children. In the group with high blood pressure, whites had higher heart rates and greater renin activity than blacks. Dopamine beta-hydroxylase concentrations in blacks were lower than in whites over the entire spectrum of blood pressure levels. Thus it appears that high blood pressure may have a different metabolic background in the two races. This may influence the early natural history of essential hypertension. It is suggested by the investigators that the rationale of prevention and possibly treatment of early hypertension in blacks and whites may differ.

References

Araoye, M. A., et al.: "Furosemide Compared with Hydrochlorothiazide: Long-term Treatment for Hypertension," *JAMA*, **240**, 1863 (1978).

Earley, L. E., and C. W. Gottschalk (editors): "Strauss and Welt's Diseases of the Kidney," Little, Brown & Co., Boston, Massachusetts, 1979.

Garraway, W. M. et al.: "The Declining Incidence of Stroke," *N. Engl. J. Med.*, **300**, 449 (1979).

Goldman, L., and D. L. Caldera: "Risks of General Anesthesia and Elective Surgery in the Hypertensive Patient," *Anesthesiology*, **50**, 285 (1979).

Haber, E.: "The Renin-Angiotensin System and Hypertension," *Kidney Int.*, **15**, 427 (1979).

Julius, S., and M. D. Esler: "The Nervous System in Arterial Hypertension," Charles C. Thomas, Springfield, Illinois, 1976.

Junco, L. I.: "Physiological Basis of Diuretic Therapy," Charles C. Thomas, Springfield, Illinois, 1979.

Kassirer, J. P., and J. T. Harrington: "Diuretics and Potassium Metabolism," *Kidney Int.*, **11**, 505 (1977).

Koch-Weser, J.: "Correlation of Pathophysiology and Pharmacotherapy in Primary Hypertension," *Am. J. Cardiol.*, **32**, 499 (1973).

Kolata, G. B.: "Is Labile Hypertension a Myth?" *Science*, **204**, 489 (1979).

Kolata, G. B.: "Treatment Reduces Deaths from Hypertension," *Science*, **206**, 1386–1387 (1979).

Marx, J. L.: "Hypertension: A Complex Disease with Complex Causes," *Science*, **194**, 821–825 (1976).

Moser, M., et al.: "Report of the Joint National Committee on Detection, Evaluation, and Treatment of High Blood Pressure," *JAMA*, **237**, 255 (1977).

Reisin, E., et al.: "Effect of Weight loss without Salt Restriction on the Reduction of Blood Pressure in Overweight Hypertensive Patients," *N. Engl. J. Med.*, **298**, 1 (1978).

Slater, E. E., and R. W. DeSanctis: "Dissection of the Aorta," Symposium on Cardiac Emergencies, *Med. Clin. North Am.*, **63**, 141 (1979).

Stehbens, W. E.: "Hemodynamics and the Blood Vessel Wall," Charles C. Thomas, Springfield, Illinois, 1979.

Stokes, G. S., and H. F. Oates: "Prazosin: New Alpha-Adrenergic Blocking Agent in Treatment of Hypertension," *Cardiovascular Medicine*, **3**, 41 (1978).

Voors, A. W., et al.: "Racial Differences in Blood Pressure Control," *Science*, **204**, 1091–1094 (1979).

Waldo, R.: "Prazosin Relieves Raynaud's Vasopasm," *JAMA*, **241**, 1037 (1979).

HYPERTENSIVE ENCEPHALOPATHY. Cerebrovascular Diseases.

HYPERTHECOSIS. Androgens.

HYPERTHYROIDISM. Thyroid Gland.

HYPERTHYROIDISM (In Osteoporosis). Bone.

HYPERTROPHIC CARDIOMYOPATHY. Heart and Circulatory System (Human).

HYPERTROPHY. Breast.

HYPERURICEMIA. Gout.

HYPERVITAMINOSIS. Vitamins.

HYPNOSIS. A trancelike state resembling sleep, induced by means of verbal suggestion or intense concentration on some object. Hypnosis is characterized by the subject's extreme responsiveness to suggestions made by the hypnotist. A role of hypnosis is in assisting to remove underlying neuroses which are reflected in psychosomatic illness, with such symptoms as bronchitis, asthma, neuritis, headaches, and some allergies. Often, the psychiatrist or physician can convince the patient while under hypnosis that his disease has no organic basis. When the patient gains insight into emotional stress, the symptoms often begin to regress.

The results of using hypnosis in the treatment of obesity may be described as transient in most cases. Hypnosis has been used in the therapy of spasmodic torticollis (involuntary movement of head and other parts under control of nuchal muscles) but has not proved particularly successful.

HYPO. Sodium Thiosulfate.

HYPOBARIC (Controlled-Atmosphere) SYSTEMS. Sensitive materials, notably fresh foods, normally cannot withstand long periods of transportation and storage prior to consumption. Over the years, much of the effort extended toward offering produce in marketplaces far distant from the source was concentrated on reducing the time for delivery. Thus, the extensive use of air express and air freight. Conventional refrigeration systems for trucks and railway cars also were especially adapted for use during transport. But even with all of these improvements in technology, certain transporting feats (as, for example, shipping midwestern pork to the California and even Hawaii markets) were difficult to achieve. The controlled-atmosphere hypobaric concept has greatly extended the potential for distant shipping of delicate, perishable materials (not necessarily limited to foodstuffs).

In 1964, the Institute of Food Technologists annual award was given in recognition of the development of a controlled-atmosphere storage system. Essentially, the process was designed to reduce the rate of deterioration of certain fruits and vegetables in refrigerated storage by reducing the oxygen level, increasing the carbon dioxide level, and maintaining the relative humidity close to 100%. In an initial design, the conditions were created by using a home-furnace size catalytic generator which burned natural gas or propane gas to create the atmosphere that essentially halts the natural respiration of the stored products. See also **Heat of Respiration.** Later, the gas generator was replaced by cryogenic liquefied gases, allowing additional flexibility and the creation of any desired gas mixture. Atmospheres can be tailored to particular perishables. For example, an atmosphere of 15–20% carbon dioxide and 80–85% nitrogen is optimal for strawberries. For iceberg lettuce, an atmosphere of 8–10% oxygen, less than 10% carbon dioxide, with the remainder nitrogen is used. As of the early 1980s, the system has been installed on over ten thousand rail cars and 7000 sea vans.

In 1979, another IFT award was given in recognition of a hypobaric transport and storage system for fresh meats and meat products. Hypobarics is defined as a precisely controlled combination of low pressure, low temperature, high humidity, and ventilation which, when properly applied, extends up to six times the length of time a perishable commodity remains fresh. This makes possible the shipment of perishable items by way of relatively low-cost surface transportation to distant points. In developing the concept, it was observed that refrigerated storage of fruits in closed containers will result in accumulation of gases generated by the fruit, i.e., ethylene and carbon dioxide, an atmosphere which hastens ripening and spoilage. Although ventilation of fruit containers can prevent accumulation of the gases, the gases are not removed from within the product itself—with no prevention of accumulation of gases within the cells of the fruit. The researchers made the supposition that by drawing a partial vacuum on a closed vessel containing the fruit, the low pressure would increase the diffusivity of the gases, thus promoting release and removal of the gases. At the same time, a reduction of pressure would reduce the oxygen concentration, thus retarding respiration and attendant spoilage. Combined with refrigeration, this would decelerate the metabolic processes, not only of the fruit, but also of any bacteria present. Humidification of the chamber would prevent any drying of the fruit. After testing the concept on bananas and other perishables, the system was patented.

The capability and flexibility of the system has been illustrated a number of times. Early in 1976, a shipment of 14,000 pounds (6350 kilograms) of fresh chicken was shipped in a hypobaric container from Arkansas to Arizona, a trip requiring 48 hours. The shelf life of the chicken was extended by 10 to 14 days over that attainable by conventional refrigeration. In the fall of 1977, 21,000 pounds (9,525 kilograms) of fresh pork cuts were shipped by truck from South Dakota to California, and then by ship to Hawaii, requiring a total of 7 days. Official inspectors judged the meat to be in excellent condition and superior to that shipped by conventional means. In the spring of 1978, 14,000 pounds (6350 kilograms) of fresh lamb carcasses, 1400 pounds (635 kilograms) of fresh beef, and 900 pounds (408 kilograms) of fresh veal were shipped by truck from Texas to Maryland and then by ship to Iran, requiring a total of 42 days. The load was judged by Iranian officials to be in excellent condition upon receipt.

In addition to the foregoing examples, the hypobaric concept has been shown to be practical for various seafoods, avocados, cherries, limes, mangos, papayas, mushrooms, tomatoes, and peppers. Generally the storage temperature for meats is about −1°C, and up to 10 or 12°C for various fruits and vegetables. In all cases, the relative humidity is controlled at about 95%. Pressure ranges between 10 and 80 millimeters of mercury. Lower pressures are maintained for meats and seafoods; somewhat higher pressures for fruits and vegetables.

HYPOBROMOUS ACID. Bromine.

HYPOCHLOREMIC ALKALOSIS. Diuretics.

HYPOCHLORINATION. Organic Chemistry.

HYPOCHLORITES. When chlorine is reacted with an alkali, a hypochlorite is formed. These compounds are very high-tonnage chemicals for sanitizing and bleaching purposes. Commercial sodium hypochlorite $NaClO$ usually is available in two strengths (1) the familiar household liquid bleach which contains about 5.25% (weight) $NaClO$, and (2) commercial bleach which contains about 13% (weight) $NaClO$. The latter compound sometimes is referred to as 15% bleach because the chlorine content is approximately 150 grams/liter of available chlorine. The term "liquid chlorine" usually refers to a solution of $NaClO$ (up to 10%) used in the swimming-pool trade. "Dry chlorine" is part of the registered trade-mark of a proprietary calcium hypochlorite product containing 70% available chlorine. See also **Bleaching Agents.**

Sodium hypochlorite normally is manufactured in batches by diluting caustic soda to the proper starting concentration. This is approximately 6.8% $NaOH$ for the 5.25% bleach; and about 18.5% $NaOH$ for the 15% bleach. After cooling the caustic soda solution, chlorine gas is added through a sparger pipe until the desired concentration is reached. This usually is determined by making a series of titration analyses. Bleaching powder $CaOCl_2$ is made by passing chlorine gas over slaked lime. This was the first type of chlorine bleaching agent made and dates back to 1799. The product usually contains about 30% available chlorine. Over the years, it was used extensively in the bleaching of textiles and for sanitizing even though the compound is unstable and difficult to use. The original bleaching powder largely has been replaced by an improved calcium hypochlorite product which contains about 70% available chlorine. The compound essentially is a calcium hypochlorite dihydrate and, in one process, is made by chlorinating a slurry of lime and caustic soda. The crystals which precipitate out are mixed with calcium chloride and chlorinated lime. When warmed, the calcium hypochlorite dihydrate precipitates, with sodium chloride remaining in solution. After filtering, the cake is dried, granulated, sized, and packaged. In addition to use in swimming pools, products of this type are used widely for water purification, algae control, and sanitation. On a very high-tonnage basis, calcium hypochlorite $Ca(ClO)_2 \cdot 4H_2O$ is used for pulp bleaching in the paper industry. Bleach liquor containing from 20–40% available chlorine may be produced in batches or continuously. In a continuous system, the flow of chlorine is controlled by making frequent (or continuous) measurements of oxidation-reduction potential.

A common means of detecting hypochlorites is the production of a blue color (caused by free iodine) with starch iodide paper by hypochlorites in weakly alkaline solution. Silver nitrate also precipitates part of the hypochlorite in solutions as white silver chloride.

Hypochlorous Acid: This compound, $HOCl$, is prepared by the reaction of (1) chlorine monoxide Cl_2O with H_2O, (2) sodium hypochlorite and an acid, excess acid yielding chlorine and oxygen, and (3) chlorine with mercuric oxide suspended in water, mercuric chloride being formed simultaneously. Hypochlorous acid is a yellow solution of characteristic odor. It decomposes upon standing, the rate depending upon (1) concentration, (2) exposure to light, (3) presence of a catalyst (cobaltous hydroxide, for example, promotes the evolution of oxygen), and (4) acidity or alkalinity. Hypochlorous acid is a powerful oxidizing agent and sometimes used as a bleaching agent for organic colors.

Perchloric Acid: This compound, $HClO_4$, is a colorless, fuming, oily liquid, miscible with H_2O, volatile under diminished pressure. A maximum constant-boiling solution (203°C, 760 millimeters Hg) results when the concentration of $HClO_4$ reaches 73% in H_2O. Cold dilute perchloric acid reacts with such metals as zinc and iron, yielding hydrogen gas and the corresponding perchlorate in solution; is stable from the point of view of oxidation and reduction (except that iodine is oxidized to periodic acid, with liberation of chlorine, ferrous salt solutions to ferric, titanous salt solutions to titanic). Concentrated hot perchloric acid, on the other hand, is a powerful oxidizing agent, exploding violently in contact with charcoal, paper, alcohol; causes serious wounds in contact with the skin.

Prepared by distilling ammonium perchlorate with HNO_3 and HCl.

Metallic perchlorates are soluble in water, except that potassium perchlorate is slightly soluble. Potassium perchlorate is, however, insoluble in alcohol containing perchloric acid, a property made use of in the qualitative recognition and quantitative estimation of potassium in salt solutions. Perchlorates, when heated, evolve oxygen and leave the chloride as a residue. Potassium perchlorate decomposes at 400°C.

HYPOCHROMIC ANEMIA. Anemias.

HYPOCYCLOID. A special case of a cyclic curve, thus a higher plane curve and, in particular, the case where a circle of radius r rolls around inside a fixed circle of radius R. Its parametric equations are

$$x = (R - r)\cos\phi + r\cos\frac{(R - r)\phi}{r}$$

$$y = (R - r)\sin\phi - r\sin\frac{(R - r)\phi}{r}$$

Reference to the corresponding equations for the epicycloid, where the circle rolls around the outside of the fixed circle, will show that the hypocycloid (R, r) is identical with the hypocycloid $(R, R - r)$ or the epicycloid $(R, r - R)$.

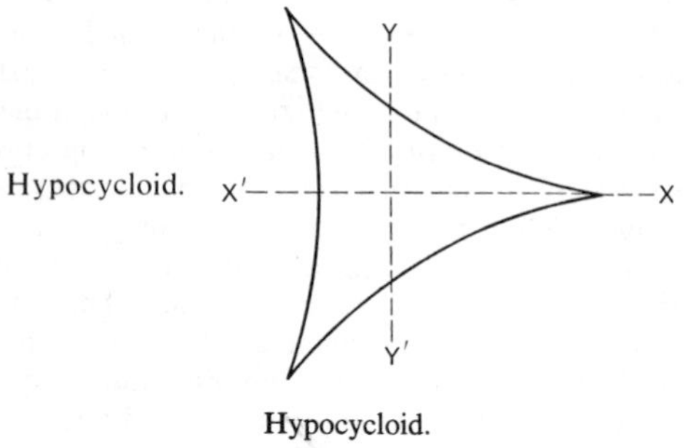

Hypocycloid.

Considerations similar to those used for the epicycloid will also show that the curve may or may not repeat itself and that it will produce cusps when its generating point touches the fixed circle. The special case is that in which $R = 4r$ has four cusps, and is called the asteroid.

See also **Asteroid (Mathematics); Curve (Higher Plane);** and terms listed under **Mathematics.**

HYPODERMIS. The cellular layer of the integument (integumentary system) in the invertebrates, which secretes the outer cuticula.

HYPOEUTECTIC ALLOY. An alloy to the left of the eutectic point in a binary phase diagram that freezes with a structure containing some eutectic.

HYPOFLUORITE. Any compound containing the group —OF. The simple anion FO^- is unknown. A number of covalent hypofluorites are known, including such compounds with carbon, oxygen, nitrogen, sulfur, chlorine and arsenic (uncertain), CF_3OF, CF_3COOF, C_2F_5COOF, NO_2OF, OF_2, O_2F_2, O_3F_2, SF_5OF, FSO_2OF, ClO_3OF and possibly AsF_4OF. These are all powerful fluorinating agents. They react violently with water yielding OF_2 as one product. The oxygen fluorides O_3F_2 and O_2F_2 decompose about −158°C and −100°C, respectively, the former into the latter and the latter into the elements. Nitryl and perchloryl hypofluorites (fluorine nitrate and fluorine perchlorate) easily detonate. The perfluoracyl hypofluorites are much more stable but may also decompose violently. The others appear to be stable.

HYPOGENE. Originated by the geologist Charles Lyell for all igneous rocks which assumed their form, fabric and texture at great depths beneath the surface of the lithosphere.

HYPOGLYCEMIA. Adrenal Glands; Hormones; Insulin.

HYPOGLYCEMIC AGENT. Diabetes Mellitus.

HYPOGONADISM. Gonads.

HYPOIODOUS ACID AND HYPOIODITES. Hypoiodous acid (HOI) is a greenish-yellow solution, of characteristic odor. It is unstable, and cannot be distilled unchanged.

Prepared by reaction (1) of iodine and mercuric oxide (see **Mercury**) suspension in water, mercuric iodide being simultaneously formed, (2) of sodium hypoiodite and an acid, excess acid yielding iodine.

Sodium hydroxide solution reacts with iodine to form iodide and hypoiodite, the latter decomposing in a few hours at ordinary temperatures to form iodide and iodate.

HYPOKALEMIA. Diuretics; Kidney and Urinary Tract.

HYPONATREMIA. Water.

HYPONITROUS ACID AND HYPONITRITES. Hyponitrous acid $H_2N_2O_2$ is a white solid, explosive even at as low a temperature as 0°C, soluble in water, more soluble in ether, can thus be extracted from water solution by ether and the latter evaporated, water solution decomposes quickly into nitrous oxide plus water. Hyponitrous acid is nonreactive with hydriodic acid (a strong reducing agent), but reactive with permanganic acid (a strong oxidizing agent) to form nitrous or nitric acid.

Prepared (1) by reaction of silver hyponitrite $Ag_2N_2O_2$ and hydrogen chloride in anhydrous ether, an evaporation of the resulting solution, (2) by reaction of hydroxylamine H_2NOH plus nitrous acid HONO.

Sodium hyponitrite $Na_2N_2O_2$ is formed (1) by reaction of sodium nitrate or nitrite solution with sodium amalgam (sodium dissolved in mercury), after which acetic acid is added to neutralize the alkali. Sodium stannite ferrous hydroxide, or electrolytic reduction with mercury cathode may also be utilized, (2) by reaction of hydroxylamine sulfonic acid and sodium hydroxide. Silver hyponitrite is formed by reaction of silver nitrate solution and sodium hyponitrite.

HYPOPARATHYROIDISM. Parathyroid Glands.

HYPOPHOSPHORIC ACID AND HYPOPHOSPHATES. Hypophosphoric acid (H_2PO_3 or $H_4P_2O_6$) is a solid, melting point 55°C, decomposing in solution to form phosphorous plus phosphoric acids. Hypophosphoric acid is used in solution and is a reducing agent, but only with strong oxidizing agents, such as potassium permanganate; and the acid is unaffected by zinc and dilute sulfuric acid (distinction from phosphorous acid). Dehydration of hypophosphoric acid does not yield phosphorus tetroxide; hydration of phosphorus tetroxide does not yield hypophosphoric acid but phosphorous plus phosphoric acids.

Hypophosphoric acid is formed by reaction (1) of yellow phosphorous and potassium permanganate in sodium hydroxide medium, (2) of red phosphorus and calcium hypochlorite solution, (3) also one of the products of slow oxidation at ordinary temperatures of phosphorus in moist air.

There are recorded the following sodium hypophosphates: Na_2PO_3 (or $Na_4P_2O_6$), $NaHPO_3$ (or $Na_2H_2P_2O_6$), $Na_3H(PO_3)_2$ (or $Na_3HP_2O_6$), and $(NaH_3PO_3)_2$ (or $NaH_3P_2O_6$). There is evidence in support of each of the formulas H_2PO_3, $H_4P_2O_6$ for hypophosphoric acid.

Ester: Dimethyl hypophosphate $(CH_3)_2PO_3$ or $(CH_3O)_2PO$. See also **Phosphorus.**

HYPOPHOSPHOROUS ACID AND HYPOPHOSPHITES. Hypophosphorous acid (H_3PO_2, or $H \cdot PO_2H_2$) is a colorless liquid, melting point 26.5°C, density 1.493.

Hypophosphorous acid is miscible with water in all proportions and a commercial strength is 30% H_3PO_2. Hypophosphites are used in medicine.

Hypophosphorous acid is a powerful reducing agent, e.g., with copper sulfate forms cuprous hydride Cu_2H_2, brown precipitate, which evolves hydrogen gas and leaves copper on warming; with silver nitrate yields finely divided silver; with sulfurous acid yields sulfur and some hydrogen sulfide; with sulfuric acid yields sulfurous acid, which reacts as above; forms manganous immediately with permanganate.

Hypophosphorous acid is formed by reaction of barium hypophosphite and sulfuric acid, and filtering off barium sulfate. By evaporation of the solution in vacuum at 80°C, and then cooling to 0°C, hypophosphorous acid crystallizes.

Sodium hypophosphite $NaPO_2H_2$, the only sodium hypophosphite, is formed (1) by reaction of yellow phosphorus and sodium hydroxide solution (phosphine simultaneously formed), (2) by reaction of hypophosphorous acid and sodium hydroxide, and evaporating. Sodium hypophosphite, upon heating, yields sodium phosphate and sodium phosphide. Common tests for the hypophosphites are as follows:

1) Zinc reduces dilute sulfuric acid solution of hypophosphites to phosphine recognizable by odor (difference from phosphates).

2) Barium chloride produces no precipitate (difference from phosphites). See also **Phosphorus.**

HYPOPHYSIS. Pituitary Gland.

HYPOPLASIA. Defective or insufficient development of any tissue. Thymic hypoplasia, also known as DiGeorge's syndrome, results from embryopathy of third and fourth pharyngeal pouch area. There are deficiencies of cell-mediated immunity (CMI) and impaired antibodies. Attendant features of the condition are hypoparathyroidism, abnormal feces, and cardiovascular abnormalities. See also **Immunology and Immunization.**

HYPOPROTHROMBINEMIA. Lack of adequate amounts of prothrombin in the blood resulting in tendency to hemorrhage from impairment of the clotting mechanism.

HYPOPUS. A larval form of certain mites. It has eight legs but no mouth. Ventral suckers enable it to attach itself to another animal for transportation.

HYPOPYGIUM. The protruding male genital organs of some flies. In some species they form a conspicuous appendage at the tip of the abdomen, much like an additional segment.

HYPOSULFUROUS ACID AND HYPOSULFITES. Hyposulfurous acid $H_2S_2O_4$ is a yellow solution rapidly oxidized in air to sulfurous acid and then to sulfuric acid. Commercially known as hydrosulfurous acid and its salts as hydrosulfites (but not to be confused with "hypo" which is sodium thiosulfate).

Hyposulfurous acid is a powerful reducing agent, e.g., with copper sulfate forms cuprous hydride Cu_2H_2, brown precipitate, which evolves hydrogen gas and leaves copper on warming, with silver nitrate yields finely divided silver, with permanganate yields manganous compounds. Hyposulfurous acid is formed by reaction of sodium hyposulfite and an acid.

Sodium hyposulfite, sodium hydrosulfite $Na_2S_2O_4 \cdot 2H_2O$ is formed (1) by reaction of zinc and sulfurous acid (or sodium hydrogen sulfite), yielding zinc hyposulfite and then converted by sodium chloride into sodium hyposulfite, (2) by electrolysis of sodium hydrogen sulfite and then addition of sodium chloride.

Sodium hyposulfite is used to bleach sugar, indigo, wood pulp. With moist hydrogen sulfide, sulfur is precipitated and sodium thiosulfate simultaneously formed.

HYPOTENSION. When the systolic arterial pressure is consistently below 100 millimeters of mercury, low blood pressure (hypotension) is said to exist. Many healthy individuals have a blood pressure that is somewhat below average. A moderately low value is usually considered conducive to longer life. When no cause for the low pressure can be found, the condition is referred to as *essential hypotension.* There often are no significant symptoms.

In *orthostatic* or *postural hypotension*, the regulatory mechanism does not function properly so that a person with this condition may suffer unconsciousness simply in changing from a reclining or sitting position to a standing position—as the result of the action causing an abnormal drop in blood pressure. Some normal individuals may from time to time experience a slight giddiness when standing up quickly, but the severe changes in postural hypotension are such that they should be called to the attention of a physician.

Frequently, unrelated diseases, largely degenerative in nature, may cause hypotension as a secondary symptom. Such conditions include acute fevers. Addison's disease, heart failure, hypothyroidism, malnutrition, hyperinsulinism, and anemia. Sometimes associated with transient hypotension are internal hemorrhage, shock, fainting, and anesthesia. In most situations of this type, the blood pressure returns to normal upon removal of the original causative condition. In most instances, hypotension is of major significance only when the blood pressure falls below that required to produce adequate filtration through the kidneys.

See also **Heart and Circulatory System (Human); Hypertension (High Blood Pressure);** and **Shock.**

HYPOTENUSE. Trigonometric Function.

HYPOTHALAMUS. Anorexia; Brain and Nervous System; Hormones.

HYPOTHYROIDISM. Thyroid Gland.

HYPOVOLEMIC SHOCK. Shock Syndrome.

HYRAXES (*Hyracoidea*). A very small group of *Mammalia*, hyraxes are small animals and are of two genera: Dassies (*Procavia*) and Tree-Hyraxes (*Dendrohyrax*). These rabbit-shaped animals are popularly termed Coneys, a term used in the Bible. Classification of these animals has been a problem for zoologists over the years, finally solved by creating a separate small group. At one time, they were considered to be rodents closely associated with guinea-pigs. At another time, they were classified with the Pachyderms, at a time when elephants, rhinoceroses, and hippopotamuses were all grouped together. In the mean time, all of the aforementioned mammals have been reclassified, Pachyderm being an obsolete term.

Hyraxes are nocturnal in nature and are considered highly aggressive and essentially mean, attempting to bite anything that gets close to them. These animals also are quite noisy and can issue a number of different sounds, including whistles and screams. The fur contrasts in color on their mid-backs. The fur is thick and coarse. They are good jumpers. Vacuum cups in their padded feet enable them to cling to vertical surfaces. They have daggerlike teeth. The Dassies are found in Africa south of the Sahara Desert. The Coney mentioned in the Bible is found in the Sinai Peninsula, Palestine, and Syria. They are rock dwellers and live in fur-lined nests. The Tree-Hyraxes prefer a mountain habitat and frequently are found at relatively high altitudes—7,000 to 10,000 feet (2100 to 3000 meters). They are omnivorous. They prefer closed-canopy forests in the central and west-central regions of Africa. See accompanying photo.

Adult cape hyrax with young. (*New York Zoological Society photo.*)

HYSTERECTOMY. Total or partial removal of the uterus.

HYSTERESIS. In general, the phenomenon exhibited by a system whose state depends on its previous history. This term usually refers to magnetic hysteresis, of importance in alternating-current machinery. When a ferromagnetic material such as iron is placed in a magnetic field, a certain amount of energy is involved in bringing about its magnetization. If the field is a rapidly alternating one, the material may become noticeably warm. It appears that the repeated changes of orientation in whatever it is within the substance that responds to the reversals of field are opposed by something like viscous friction.

A quantitative study of the process indicates that, as the field intensity H increases, the magnetic induction B also increases in a manner characteristic of the substance. This is conveniently represented by a graph, which is called the magnetization curve (see figure). Its initial slope is the initial permeability (μ_0). If H is carried to some maximum value H_m and then reduced (to $-H_m$), B follows the dotted hysteresis curve. B does not fall off as it was built up (solid line); the residual induction B_r is the induction remaining when H has been reduced to zero; the reverse H needed to reduce B to zero is called the coercive force (H_c). From this point the cycle proceeds to describe the closed curve shown by the dotted lines, which is called the hysteresis loop. The initial portion (solid line) is not retraced. The amount of energy converted into heat is proportional to the area of the cycle.

Electric hysteresis is a somewhat analogous phenomenon exhibited

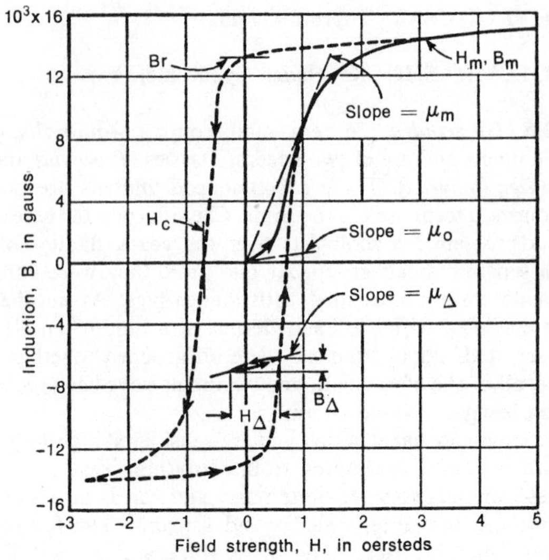

Hysteresis loop (dotted). Some important magnetic quantities are shown.

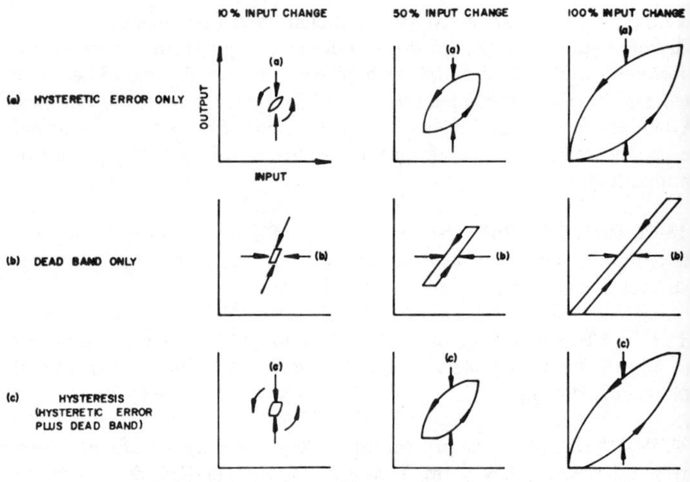

Hysteretic error, dead band, and hysteresis.

by dielectrics in the electric field and gives rise to heating in capacitors.

Some solids exhibit what is called elastic hysteresis, in which the variables corresponding to H and B in the magnetic case are the stress and the strain or deformation. Elastic bodies such as metals operating at stresses below the proportional limit also undergo hysteresis.

Hysteresis energy is that energy used per cycle of operation to overcome the effect of hysteresis.

HYSTERESIS DISTORTION. The distortion of voltage and/or current waveforms in circuits containing magnetic components, which is caused by the non-linear hysteresis effect.

HYSTERESIS HEATER. An induction device in which a charge or a muffle about the charge is heated principally by hysteresis losses due to a magnetic flux which is produced in it. A distinction should be made between hysteresis heating and the enhanced induction heating in a magnetic charge.

HYSTERESIS (Instrument). With reference to industrial and scientific instruments, the Scientific Apparatus Makers Association defines hysteresis as:

1. When used as a performance specification, the maximum difference for the same input between the upscale and downscale output values during a full range traverse in each direction. See (c) of accompanying diagram. This is a common usage definition which includes hysteretic error and dead band. That portion of the difference which is dependent on the history of prior excursion is hysteretic error, while that portion due to dead band may be determined by a conventional dead band test.

2. When describing a physical property, that property of an element evidenced by the dependence of the value of the output, for a given excursion of the input, upon the history of prior excursions and the direction of the current traverse. Some reversal of the output will occur on any small reversal of the input if a device exhibits hysteretic error without dead band.

Hysteretic Error. That portion of hysteresis due to energy absorption in the elements of a measuring instrument. It is obtained by subtracting the value of dead band from the corresponding value of hysteresis for a given input. See (a) of accompanying diagram. The energy absorbed is conceived as produced by molecular friction and appears as heat in dynamic cycling when cyclic mechanical force is applied to a spring or cyclic magnetizing force to a magnetic material.

See also **Backlash;** and **Core Loss.**

HYSTERIA. A conversion reaction characterized by loss of normal control of bodily or emotional function without structural disease of the nervous system. Caused by unconscious emotional conflict. Hysteria may precipitate noncardiovascular syncope (sudden temporary loss of consciousness). Trismus (lockjaw) may be an infrequent manifestation of hysteria and thus mimic the trismus associated with tetanus.

HYSTRICOMORPHS. Rodentia.